# Machinery's Handbook

## 32nd Edition

## Volume

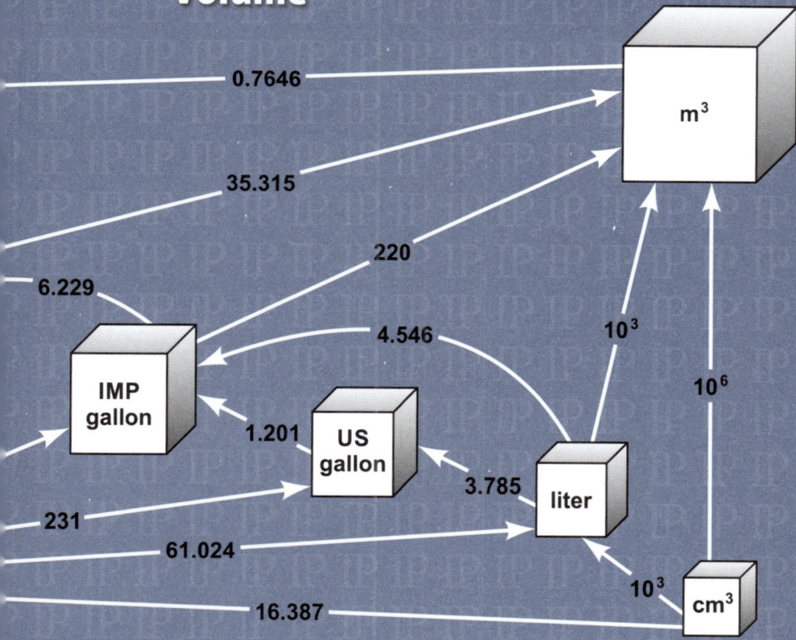


kW = kilowatt
lb = pound
M = mega
m = meter
mm = millimeter
MPa = megapascal
N = newton
oz = ounce
Pa = pascal
PS = pferdestarke (metric hp)
s = second
US = United States (gallon)
W = watt
yd = yard

**Example:** Convert 10 $m^2$ to $yd^2$

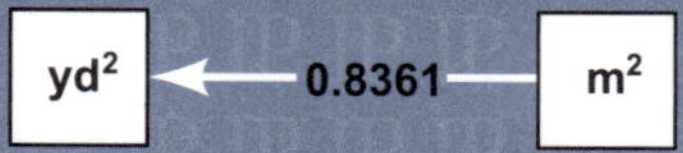


10 divided by 0.8361 = 11.96 $yd^2$

**Example:** Convert 10 $yd^2$ to $m^2$

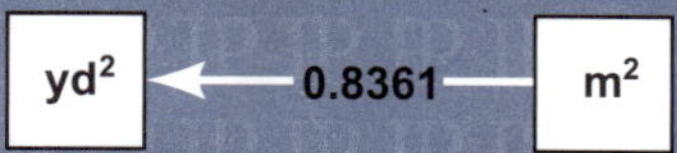


10 multiplied by 0.8361 = 8.36 $m^2$

*Continued on the inside back cover...*

A REFERENCE BOOK
FOR THE MANUFACTURING AND MECHANICAL ENGINEER, DESIGNER, DRAFTER, METALWORKER, TOOLMAKER, MACHINIST, EDUCATOR, STUDENT, AND HOBBYIST

# Machinery's Handbook

## 32nd Edition

BY ERIK OBERG, FRANKLIN D. JONES, HOLBROOK L. HORTON, HENRY H. RYFFEL, AND CHRISTOPHER J. MCCAULEY

LAURA BRENGELMAN, EDITOR

2024

INDUSTRIAL PRESS, INC.

**INDUSTRIAL PRESS, INC.**
1 Chestnut Street
South Norwalk, Connecticut 06854 U.S.A.
Phone: 203-956-5593
Toll-Free: 888-528-7852
Email: info@industrialpress.com

Title: *Machinery's Handbook, 32nd Edition*
Authors: Erik Oberg, Franklin D. Jones, Holbrook L. Horton, Henry H. Ryffel, and Christopher J. McCauley
Library of Congress Control Number: Toolbox: 2023949276; Large Print: 2023948948



ISBN 978-0-8311-3732-8 (Toolbox, Thumb-Indexed, 4.6 × 7 in., 11.7 × 17.8 cm)
ISBN 978-0-8311-3832-5 (Large Print, Thumb-Indexed, 7 × 10 in., 17.8 × 25.4 cm)
ISBN 978-0-8311-3932-2 (Digital Edition, Physical Package, with Access Code)
ISBN 978-0-8311-4032-8 (Digital Edition Upgrade, Physical Package, with Access Code)
ISBN 978-0-8311-9732-2 (Digital Edition, Online, at IP eBookStore)
ISBN 978-0-8311-9832-9 (Digital Edition Upgrade, Online, at IP eBookStore)
ISBN 978-0-8311-4132-5 (Toolbox and Digital Edition Combo, 4.6 × 7 in., 11.7 × 17.8 cm)
ISBN 978-0-8311-4232-2 (Large Print and Digital Edition Combo, 7 × 10 in., 17.8 × 25.4 cm)

For ISBNs of other packages including this title, see books.industrialpress.com.



Printed and bound by Thomson Press.

**MACHINERY'S HANDBOOK**
32ND EDITION
Second Printing

books.industrialpress.com
ebooks.industrialpress.com

# TABLE OF CONTENTS



Each section has a detailed Table of Contents located on the page indicated.

## TABLE OF CONTENTS



Each section has a detailed Table of Contents located on the page indicated.

# PREFACE

Industrial Press is celebrating 140 years of excellence and 110 years of continuous publication of the *Machinery's Handbook*. Launched in 1914, this technical masterpiece has since served as the principal reference work in the manufacturing and mechanical industries. Found worldwide in countless engineering and design departments, manufacturing and industrial facilities, machine, metalworking, and other workshops of all sizes, and in general, trade, technical, and engineering schools, colleges, and universities, it is the bestselling and most referenced engineering resource of all time.

Over the *Handbook's* many editions, the editors have strived to carefully craft and perpetually improve a comprehensive, practical resource, covering the most basic yet critical aspects of crucial manufacturing processes. As such, the *Handbook* serves as an invaluable tool, to be ulilized in much the same way as other tools: to design, make, maintain, repair, and otherwise optimize the manufacturing, assembly, and use of parts and machinery of the highest quality, at the lowest cost, as efficiently as possible.

Accordingly, the original edition was designed to fit inside a traditional toolbox. To this day, the compact "toolbox edition" remains the most popular version of the *Handbook*. For those who appreciate the toolbox edition's portability but find the smaller fonts challenging, we now offer it in a bundle with a convenient, flat magnifier.

In 1997, at the request of users, the "desktop" or "large print edition" of the *Handbook* was introduced. The large print version is 40 percent larger than the original toolbox edition, but other than size, the two versions are identical. (The primary typeface in the large print edition is standard reference size, not a large font for visually impaired readers.)

Longtime readers of the *Handbook* will note many changes in recent editions. Yet an enduring goal of the editors is to ensure this encyclopedic reference is as accessible and easy to use as possible. Thus, the *Handbook* continues to incorporate the popular, time-saving thumb tabs. In addition to the table of contents in the front matter, detailed sectional contents begin each major section, and a lengthy index further guides readers.

In 1998, Christopher McCauley launched the *Machinery's Handbook CD-ROM,* containing the contents of the printed book, plus additional indexes and material from earlier editions. Now the *Machinery's Handbook 32 Digital Edition,* this versatile format offers view options, rapid searching, and quick navigation aids in the form of clickable links and cross references to related pages and topics elsewhere in the *Handbook* and in the included *Guide to the Machinery's Handbook*. The growing family of *Machinery's Handbook* products includes printed and digital versions of the *Guide* and the concise *Pocket Companion,* as well as an attractive array of useful combination packages.

As an authoritative reference, the *Machinery's Handbook, 32nd Edition*, may be used to gain vital knowledge of materials, processes, and interrelated considerations; to understand and explore machining and manufacturing options; to reference essential industry data and calculations for part and procedural design and implementation; and to place users on the path to discovering and effectively applying additional data from trusted industry sources. Ultimately, in this age of proliferation of online information, the continued mission of the *Handbook* is to be a carefully curated collection of technical data and information that serves the needs of the manufacturing and mechanical engineering community. To achieve that mission, it must include material of proven and everlasting worth, across the vast spectrum of included topics.

At the same time, the *Machinery's Handbook* (MH) Team editors and world-class experts are faced with the ongoing challenge of selecting suitable material from the enormous body of available data related to traditional yet rapidly evolving fields. The MH Team relies to a great extent on conversations and written communications with users of the *Handbook,* as well as research and comments from subject matter experts. This all-important guidance helps us make difficult decisions about topics and specifics to be introduced, revised, expanded, shortened, or omitted in each new edition.

# PREFACE

Thanks to such input and extensive work by the MH Team, this new edition incorporates thousands of individual changes and more than 100 new and revised tables and figures. While we have added a number of pages of expanded and new material to key sections, we also have tightened spacing in many places to prevent the book from growing beyond the physical limits of one volume. In addition, some material excised from the printed books has been moved to the archival collection of legacy material in the *Machinery's Handbook 32 Digital Edition*.

The final result of our concerted efforts, this updated, edited, and reset *Machinery's Handbook, 32nd Edition*, has expanded to 3,008 pages. Among major revisions of existing content and additions of new material are the following:

First and foremost, hundreds of references, explanations, and specifications based on the most current ABMA, ANSI, ASME, ASTM, BS, ISO, SAE, and other industry Standards are updated throughout this edition. Some content changes involved simply verifying and adjusting dates for the latest Standards versions issued by these authoritative institutions. Other changes, such as those derived from the significantly revised ANSI/ASME B18.8.2-2020 Standard, required important data adjustments and procedural updates. These and other vital changes make the *32nd Edition* a must-have resource.

In *PROPERTIES, TREATMENT, AND TESTING OF MATERIALS*, the legacy elements table on page 362 is revised with modern scientific data and is followed by a new table giving an overview of the most commonly used materials. Expanding on the subject of material properties and selection is the timely addition of *Sustainability Considerations for Materials and Processes*, addressing lifecycle, energy, and supply chain; calculating production costs and carbon footprint; and best practices for material recycling and reclamation.

Following industry advances, the *PLASTICS* section, beginning on page 553, also encompasses new and updated information on materials and manufacturing, including the *Rotomolding* production process on page 583.

A new *Polymer Composites* section, on page 585, examines polymer matrices engineered for superior strength and modulus. An associated addition, *CNC Machining of Carbon-Fiber Reinforced Polymers*, on page 1385, covers cutting, drilling, and milling multi-layered matrix materials.

In *DIMENSIONING, GAGING, AND MEASURING*, the instructions and equations for *Tolerance Analysis and Assignment*, beginning on page 685, have been refined. In *MEASURING, INSTRUMENTS, AND INSPECTION METHODS*, the discussion of *Reading Verniers and Micrometers*, beginning on page 691, has been rewritten to facilitate understanding and usage. *MICROMETERS AND CALIPERS*, starting on page 766, also has been reworked with an emphasis on features and use of inch and metric manual and digital devices.

In *MACHINING OPERATIONS*, expert updates can be found throughout the *CNC PROGRAMMING AND CAD/CAM* section, starting on page 1338. Among the expanded and new material is *Shop-Floor Programming*, on page 1341, *Simulation and Verification of CNC Programs*, on page 1384, and *Feature-Based Machining*, on page 1385.

*MANUFACTURING PROCESSES* also has benefited from additions and refinements. *METALWORKING DIES*, on page 1445, includes new information and figures on classification and types of dies, operations and caveats, and sheet metalworking applications. In *WELDING*, see page 1612 for useful modifications related to positioning of welded joint components.

In *METAL CASTING, MOLDING, AND EXTRUSION*, see *Using Computer Modeling to Optimize Casting Processes*, on page 1518, for more on utilizing this time-saving method for efficient design and production of cast components.

*Metal Additive Manufacturing*, starting on page 1555, offers an extended list of international *Standards for Additive Manufacturing* that provide guidance for this burgeoning manufacturing segment.

## PREFACE

Throughout the *FASTENERS* and *THREADS AND THREADING* sections, specifications have been carefully reviewed and revised, per the current Standards. Among the many updates are specifications for ANSI/ASME, BS, and ISO metric fasteners, starting on page 1713. ISO tables also have been added, including the *Hexalobular Internal Driving Feature* (commonly known as Torx®) table on page 1732.

In *MACHINE ELEMENTS*, coverage of *BALL, ROLLER, AND NEEDLE BEARINGS*, beginning on page 2446, also has been revisited, with new tables, figures, and text on applicable Standards, common configurations, and capabilities of antifriction bearings.

Another section that has received significant revisions is *LUBRICATION*, beginning on page 2502. See page 2526 for *Food-Grade Lubricants* used for machinery and equipment in food and beverage production and processing plants.

In *COUPLINGS, CLUTCHES, BRAKES*, beginning on page 2527, added text and a new table illustrate common types of rigid and flexible couplings.

*SHAFT ALIGNMENT*, on page 2726, has been revised to include a critical Standard, plus added advice on correcting misalignment. Among other changes, see the *Reverse Dial Alignment Method*, on page 2755; *Laser Alignment* on page 2757; and concluding comments on pros and cons of various alignment methods.

Finally, as mentioned above, the addition of new and revised *Handbook* topics often requires removal of older topics to gain space. Absent in the printed *32nd Edition* are mathematical tables for higher value prime numbers, as well as information on the manufacture of measuring instruments. This extracted material now appears in the *Machinery's Handbook 32 Digital Edition*. Users requiring this information, or wishing to comment on materials removed from this or previous print editions, may reference the *Digital Edition* or contact the editors.

The editors are greatly indebted to all who contact us about possible errors and defects in the current edition of the *Handbook*, offer suggestions for including new or revised material, or pose questions about specific content in the context of modern-day manufacturing problems. Queries involving *Handbook* material usually entail an in-depth review, help us identify subject matter that may require clarification or expansion, and frequently result in improved or new material.

As we constantly seek to improve the *Handbook*, we also welcome new contributors to the team working on each edition, joining the long line of erudite experts who have worked on this master reference. We invite topical experts in industry and academia to reach out to the editors for a discussion of potential future contributions.

Our perpetual goal is to increase the usefulness of the *Handbook* as much as possible. We welcome input and encourage you to send us your thoughts and feedback. The best way to reach us is by emailing MHTeam@industrialpress.com.

## ACKNOWLEDGMENTS AND SPECIAL THANKS

The *Machinery's Handbook* is indebted to the whole mechanical field for the data contained in this master reference work. On behalf of the *Handbook* editors past and present, we wish to express our appreciation to all who have assisted us by furnishing data and contributed ideas, corrections, and other commentary.

Most importantly, we thank the thousands of readers who have contacted our team over the years with constructive criticism and suggestions regarding *Handbook* topics and presentation. Your detailed comments, on this edition and about past and future ones, are invaluable.

# ACKNOWLEDGMENTS AND SPECIAL THANKS

Many of the American National Standards Institute (ANSI) Standards that deal with mechanical engineering, extracts from which are included in the *Handbook,* are published by the American Society of Mechanical Engineers (ASME). The editors would like to particularly thank those at ASME for their exceptional collaboration in helping us identify and bring essential data up to date, according to the latest, definitive, industry Standards.

While information referencing and discussing various Standards data and nomenclature may be found throughout the *Handbook,* it should be noted that all official Standards and related publications are copyrighted by the issuing industry organizations. For instance, the ASME Standards and all portions and derivative works thereof referenced are the copyrighted property of ASME used under license. Readers should contact ASME directly for further information regarding current editions of the ASME Standards. In fact, the *Handbook* editors always encourage readers to contact the official Standards bodies and to refer to the most up-to-the-minute, complete versions available of all published Standards and related documentation.

On the following pages are brief biographies for the *Machinery's Handbook, 32nd Edition*, Team—an impressive roster of editorial advisory board members and contributors. These esteemed colleagues have played a crucial role in guiding content decisions and advising on specific engineering questions and content challenges. Their lifelong educational and industry experience, impressive technical knowledge and expertise, meticulous research, and exemplary professional collaboration, all have immeasurably enhanced the content of this edition.

We also wish to thank those behind the scenes, our tremendous editorial and production team, without whom this edition would not be possible: the incomparable page layout and production mastermind Jason Hughes; the eagle-eyed trio Stella Bonifazi, Anna Fuller, and Kristen Joseph; and the rest of Scribe's extended editorial and design team; our dedicated team members Susan Gilpin, Gerald Murray, and Rayne Stone; talented cover designer Jeff Weeks; and, as always, our masterful printing and binding partners at Thomson Press.

The *Handbook* is indebted to so many others in industry organizations and associations, educational institutions, commercial enterprises, and private practice. In this context, we cannot thank everyone who has written in, helped resolve a question, or has otherwise spent their valuable time and effort assisting us in improving past and present editions. Therefore, the "Machinery's Handbook Hall of Fame" page—proudly posted in the *Machinery's Handbook Resources* section of our main website, books.industrialpress.com—honors and acclaims these most important participants in the community surrounding this legendary product.

Once again, we look forward to hearing from you and hope that you will share with us how the *Machinery's Handbook* product family informs, supports, and enhances your involvement in this endlessly fascinating field.

**Laura Brengelman**
**Editor**

# MACHINERY'S HANDBOOK, 32ND EDITION, TEAM

## Editorial Advisory Board

**Steve Heather,** an acclaimed mechanical engineer, worked in the aircraft, automobile, defense, and lighting industries for more than 30 years. More recently, he taught machine shop engineering and computer-aided design (CAD) at the college level to engineering and architectural students. A multi-disciplinary expert, his specialties include AutoCAD®, CAD/CAM, and CNC programming, multistage press and other tool design, and precision machining. He has authored a number of leading technical titles, including *Engineers Precision Data Pocket Reference, AutoCAD® 3D Modeling,* and the *AutoCAD® 2D Resource Kit,* and is coauthor of the multi-book series *Beginning AutoCAD® Exercise Workbook, Advanced AutoCAD® Exercise Workbook,* and *AutoCAD® Pocket Reference,* all published by Industrial Press. An invaluable engineering advisor, exceptional technical consultant, and top contributor of essential material throughout the *Machinery's Handbook,* as well as a highly skilled illustrator, his work enhances numerous pages of the current edition.

**David O. Kazmer** is Professor and former chair of the Plastics Engineering at the University of Massachusetts Lowell. His teaching and research are focused on product, process, and machine design, as supported by mechanistic and statistical modeling, design for manufacturing, simulation and optimization, and process control techniques. His ongoing work involves developing new manufacturing processes with improved efficiency and performance. A fellow of the American Society of Mechanical Engineers, and a former chair of its Design for Manufacturing and the Life Cycle Technical Committee, he is a fellow of the Society of Plastics Engineers, having served as chair of the Special Interest Group on Process Monitoring and Control. He is the recipient of over twenty recognition awards, an inventor with more than two dozen patents, and author of more than 300 technical publications. His far-reaching knowledge and expert contributions significantly enhance and improve this edition of the *Machinery's Handbook.*

**Howard Kuhn** most recently served as adjunct professor at the University of Pittsburgh, Swanson School of Engineering, where he taught courses in traditional and additive manufacturing, product realization, and engineering entrepreneurship and performed research on additive manufacturing for tissue engineering. He also has served as a technical advisor to America Makes (National Additive Manufacturing Innovation Institute), where he previously was acting deputy director. Specializing in advanced technology implementation, he has engaged in the design and application of multiple additive manufacturing technologies for government and private industry. He also has developed undergraduate and graduate courses in engineering design, failure analysis, deformation processing, and powder and mechanical metallurgy; written about and conducted tailored training courses on additive manufacturing and more for government agencies, trade organizations, and major companies; and has contributed expert consultation and authoritative revisions to this edition.

## Contributors

**Vukota Boljanovic** received his B.S., M.S., and Ph.D. in mechanical engineering and has nearly 50 years of experience in applied engineering. His impressive career in the aircraft and automotive industries included serving as vice president for research and development with a major aircraft company. He has taught aerospace engineering, among other subjects, and has performed extensive research in development and manufacturing engineering, including aircraft design, assembly, and inspection, the impact of design modification on tooling and production practices, and die design and related material and process selection. He is widely recognized by both academia and industry both for his analytical brilliance and outstanding contributions to manufacturing processes and procedures. In addition to consulting and writing for multiple editions of the *Handbook,* he has authored numerous technical papers and titles, including the definitive Industrial Press references *Applied Mathematical and Physical Formulas Pocket Reference; Die Design Fundamentals; Metal Shaping Processes: Casting and Molding, Particulate Processing, Deformation Processes, and Metal Removal; Sheet Metal Forming Processes and Die Design;* and *Sheet Metal Stamping Die Designs: Die Design and Die-Making Practices.*

**Ron Brook's** career in vibration analysis of rotating machinery started in 1975. His in-depth industry experience includes working with such firms as Nicolet Scientific, Reliance Electric, REM Technologies, Rockwell Automation, and Zonic. He currently serves as Supervisor of Engineering Services, Emeritus, at Integrated Power Services. His years of experience with problem solving and maintenance and reliability challenges have played a part in his acquiring broad technical knowledge and keen subject matter understanding, as well as developing a master set of professional analysis and troubleshooting tools and technical processes for use in the field. This has led to him holding five U.S. patents for homopolar inductive high-speed motors and cracked shaft detection in nuclear reactor coolant pump shafts, as well as his authoring the deep-dive case study, diagnosis, and solutions manual *Rotating Machinery Reliability for Technicians,* published by Industrial Press.

**Arief Era** received his B.S. and M.S. degrees in mechanical engineering at Columbia University in New York. As a structural analysis engineer, he worked on the designs of commercial aircraft for Boeing. In the field of energy distribution, he has participated in several key projects to improve the aging gas delivery infrastructure serving the greater New York City area. More recently, his efforts in the Energy Efficiency and Demand Management Department of Consolidated Edison of New York have involved implementation of clean energy initiatives, ranging from smart building control systems to heat pump installations. As an exacting, multidisciplinarian, keen-eyed researcher, technical editor, and wordsmith, his countless contributions to this edition—resolving queries from industry experts and hands-on users of the *Machinery's Handbook,* developing, refining, and updating vital information related to industry standards, working on other critical improvements in legacy sections, and more—have been invaluable.

**Ken Evans** has had an illustrious career working in aerospace and manufacturing as a CNC programmer, innovation center director, prototype lab manager, quality control inspector, technical instructor, sales engineer, machinist, and tool, die, and mold maker. He has more than twenty-five years of experience using and teaching about CAD/CAM programing and processes. In addition to widespread, expert industry consulting, he has served as advisor of multiple medalists in the Precision Machining Technology area at SKILLS USA. He has published several top-selling titles with Industrial Press, including the *CNC Machining Certification Exam Guide: Setup, Operation, Programming; Programming of CNC Machines* and its accompanying *Student Workbook;* and *Interpretation of Geometric Dimensioning and Tolerancing.*

**Charles Gillis** received his B.S. in mechanical engineering from Worcester Polytechnic Institute and his M.S. in mechanical engineering from Northeastern University, and has nearly 30 years of machine design experience. He previously served as a mechanical engineer for the Gillette Company, designing automated machinery for manufacturing blade and razor products. Through his ongoing work with Dynamic Design Consulting, he helps clients with product development and documentation, design for assemblability and manufacturability, and tolerance stack-up analysis. In addition to being a licensed P.E., he holds a Geometric Dimensioning and Tolerancing Professional Certificate–Senior Level (GDTP-S) from ASME. He has been training engineers in geometric dimensioning and tolerancing, print reading, and related mechanical design and documentation topics for the last decade and a half. Author of the bestselling *Hammer's Blueprint Reading Basics* and *Blueprint Reading Basics Instructor's Resource Kit,* he also has contributed to other Industrial Press titles, including *Cam Design and Manufacturing Handbook* and *Machine Designers Reference.*

## MACHINERY'S HANDBOOK, 32ND EDITION, TEAM

**Michael D. Holloway** earned a B.S. in chemistry and a B.A. in philosophy from Salve Regina University, and an M.S. in polymer engineering from the University of Massachusetts. He has held positions as a research engineering and product development chemist, applications engineering specialist, aerospace materials and polymer development engineer, and director of technical development and reliability with such firms as ALS Tribology, General Electric Plastics, NCH Corporation, Olin Chemical, Rohm & Haas Electronic Chemicals, and W.R. Grace. Recently, he joined SGS (Société Générale de Surveillance) of Geneva, Switzerland, as global technical manager. Among his specialties, he has focused on lubricants used in heavy industrial equipment, mining, and construction. Known for conducting on-site seminars, professional training, and speaking at industry events, he has served as North Texas section chair for the Society of Tribologists and Lubrication Engineers, national section chair for the International Council for Machinery Lubrication, and adjunct professor at the University of North Texas, and holds multiple professional certifications. He has authored more than a dozen books, including the *Machinery Lubrication Technician (MLT) I and II Certification Exam Guide*, published by Industrial Press.

**Krishna Murty Kommajosyula** draws on more than 50 years of successful industry experience. Upon retiring from Coromandel International, after 38 years in maintenance management, he was asked to return and served as head of technical training for another 9 years. As a leading maintenance and reliability management expert and root cause failure analyst, he has provided detailed assessments and process improvement recommendations for power, metal, cement, sugar, and petrochemical industries throughout India and has worked with Dr. Reddy's Laboratories as an engineering consultant. With his extensive experience and broad expertise, he is a highly sought-after technical trainer, renowned for his popular science books, electronics and science magazine articles, and radio broadcasts. He also has authored a number of bestselling titles, including the Industrial Press classic *All-in-One Manual of Piping Practice: On-The-Job Solutions, Tips and Insights* and his more recent practical guide *Why Industrial Bearings Fail: Analysis, Maintenance, and Prevention*.

**David R. Quinonez** has over 25 years of experience in welding, welding inspection, and nondestructive testing (NDT). His impressive experience in NDT began with nuclear submarines and aircraft carriers. Subsequent positions included NDT, welding inspection, and other work with Rolls-Royce gas turbine engines, Lockheed Martin F-22 stealth fighter airframes, and a wide range of missile defense, defense/commercial rocket, pipeline, and structural steel projects for the private sector and public works. A Certified Welding Inspector, he regularly performs visual welding inspection, detailed dimensional verification, and NDT procedures. His Level II certifications include MT, PT, and UT; past certifications have included Level II RT, ET, and ASNT Level III MT and PT. As an expert in welding processes, he has authored the highly effective teaching and learning text *1,001 Questions & Answers for the CWI Exam: Welding Metallurgy and Visual Inspection Study Guide*, published by Industrial Press.

**Richard Rex** has worked on lathes and milling machines for decades, starting in his teen years in a home workshop and later working with various production machines. He still regularly uses his home machine shop. For 10 years, he worked on product development with Hewlett-Packard and Brown Boveri in the United Kingdom. As chief executive officer of several manufacturing companies in the United States, he established a number of engineering model labs with the usual complement of lathes, milling machines, and metrology equipment. In recent years, he has provided expert consultation and written numerous manuals and technical bulletins for machine tool developers and suppliers. Drawing on his wealth of experience in the metalworking industry, he recently authored two excellent instructional guides, including useful information for metal shops and operators at every level, *Choosing & Using the Right Metal Shop Lathe* and *Choosing & Using the Right Milling Machine*, both published by Industrial Press.

# TABLE OF CONTENTS
## MATHEMATICS

### REAL NUMBERS AND THEIR OPERATIONS



### ALGEBRA



### ALGEBRA
*(Continued)*



### GEOMETRY

## TABLE OF CONTENTS

## MATHEMATICS

### TRIGONOMETRY: SOLUTION OF TRIANGLES



### MATRICES



### CALCULUS



### STATISTICAL ANALYSIS OF MANUFACTURING DATA



### ENGINEERING ECONOMICS

## REAL NUMBERS AND THEIR OPERATIONS

### Real Numbers

Most mathematical computation is performed in the *real number system*. The universal set of the "reals" includes the subsets: *naturals, whole numbers, integers, rationals,* and *irrationals*. The naturals (also called *counting numbers*): {1, 2, 3, . . .} are included in the whole numbers: {0, 1, 2, 3, . . .}, which are included in the integers (or signed whole numbers): {. . . ,–2,–1,0, 1,2,. . .}. And all of these subsets are included in the rationals.

Rational numbers, including integers, can be written in fraction form. Since all fractions can be divided numerator by denominator, their decimal form either terminates or repeats. Examples of rational numbers: –4/1, 3/5 = 0.6, 1/3 = 0.333. . . .

The only set in the real numbers larger than the naturals that does not contain any of the other sets is the irrationals. These are *not* expressible as ratios. An irrational number's decimal representation does not terminate and it has no pattern of repetition. Examples of irrational numbers are roots that cannot be simplified, such as $\sqrt{6}$ and $\sqrt[3]{70}$, as well as quantities like π and the natural log base *e*. The entire real number set is the union of the rationals and the irrationals.

**Properties of Real Numbers.**—Though often obvious and followed almost automatically, the properties of real numbers are critical to mathematical reasoning. These properties justify various steps in solving algebraic problems, such as those in this *Handbook*. Equivalence properties (symmetry, reflexivity, transitivity) and operational properties of numbers are summarized here.

*Equivalence Properties:* The properties of equivalence relations are the basis of equation solving.

Reflexive: $a = a$.
Symmetric: If $a = b$, then $b = a$.
Transitive: If $a = b$ and $b = c$, then $a = c$.
Substitution: If $a = b$, then *a* may be replaced by *b* in any equation or expression.

*Operational Properties:* These concern addition, subtraction, multiplication, and division, as summarized in the table below.

| *Property* | *Addition* | *Multiplication* |
|---|---|---|
| Commutative: | $a + b = b + a$ | $a \times b = b \times a$ |
| Associative: | $(a + b) + c = a + (b + c)$ | $(a \times b) \times c = a \times (b \times c)$ |
| Identity: | $a + 0 = 0 + a = a$ | $1 \times a = a \times 1 = a$ |
| Inverse: | $a + (-a) = 0$ | $a \times 1/a = 1$ |

*Other Properties:*

| | | |
|---|---|---|
| Distributive of multiplication over addition: | $a \times (b + c) = (a \times b) + (a \times c)$ | $(a + b) \times c = (a \times c) + (b \times c)$ |
| Zero property of multiplication: | If $a \times b = 0$, then either $a = 0$ or $b = 0$ | |
| Zero property of division: | If $a/b = 0$, then $a = 0$ $(b \neq 0)$ | |

**Integers (Signed Numbers).**—Positive whole numbers extend to the right of zero on the number line. Negative whole numbers extend to the left of zero. Together with zero, these make up the *integers* (sometimes called *signed numbers*): {. . . ,–2,–1,0, 1,2,. . .}.

The sciences (as well as economics and other fields) deal with negative as well as non-negative quantities. Temperature is an obvious example; so is land altitude, which can be above, at, or below sea level. Angles can be negative, too, as explained in

*TRIGONOMETRY: SOLUTION OF TRIANGLES.* Calculators facilitate computation that involves integers (signed numbers). Knowing the rules of integer operations prevents errors that might occur when a calculator is used.

*Absolute Value:* A number's absolute value, sometimes called its *magnitude*, is the number's distance from zero on the number line. Whether a number is positive or negative, its absolute value is positive. For example, the absolute value of both 5 and –5 is 5. The absolute value of $n$ is notated $|n|$; thus, $|5| = 5$ and $|-5| = 5$. Absolute value helps explain the rules of signed number addition and subtraction.

*Real Number Line:* The real number line is generally shown with only the integers marked off (though all numbers are included). A number line is useful for conveying how signed numbers are added or subtracted.

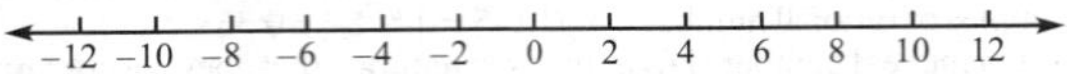


*Operations on Signed Numbers:* The following rules of operations apply to rational and irrational numbers as well. For simplicity, only integers are given as examples.

*Addition and Subtraction:* Adding a negative number is equivalent to subtracting its absolute value. When a larger number is subtracted from a smaller number, the result is negative. The rules for adding and subtracting integers are illustrated with an example using four values: 7, 11, –7, and –11. The following examples illustrate the rules:

| *Examples, Addition* | *Examples, Subtraction* |
|---|---|
| $7 + 11 = 18$ | $7 - 11 = -4$ |
| $7 + (-11) = 7 - 11 = -4$ | $7 - (-11) = 7 + 11 = 18$ |
| $(-7) + 11 = 11 + (-7) = 11 - 7 = 4$ | $(-7) - (-11) = (-7) + 11 = 11 + (-7) = 11 - 7 = 4$ |
| $(-7) + (-11) = -18$ | $-7 - 11 = -18$ |

*Multiplication and Division:* Multiplication or division of numbers with the same sign results in a positive answer. Opposite signed numbers result in negative answers when multiplied or divided. The following examples illustrate the rules:

| *Examples, Multiplication* | *Examples, Division* |
|---|---|
| $5 \times 2 = 10$ | $12 \div 3 = 4$ |
| $5 \times (-2) = -10$ | $(-12) \div 3 = -4$ |
| $(-5) \times 2 = -10$ | $(12) \div (-3) = -4$ |
| $(-5) \times (-2) = 10$ | $(-12) \div (-3) = 4$ |

**Order of Operations.**—Mathematical operations are performed on numbers in a particular order, commonly referred to as PEMDAS, which stands for "**P**arentheses, **E**xponents, **M**ultiplication, **D**ivision, **A**ddition, **S**ubtraction." First, when there are no parentheses or other grouping symbols, multiplication and division are done before addition and subtraction. Then, proceeding from left to right, the addition and subtraction are done in the order they appear. For example:

$$100 - 26 + 7 \times 2 - 100 \div 4 = 100 - 26 + 14 - 25 = 74 + 14 - 25 = 88 - 25 = 63$$

Parentheses ( ) and brackets [ ]—called *grouping symbols*—indicate if addition and subtraction are to occur before multiplication and division. The operations are performed from the innermost to the outermost grouping symbols. For example:

$$[6 \times (15 - 7)] \div 2 = [6 \times 8] \div 2 = 48 \div 2 = 24$$

Exponents are a multiplication operation, but unless parentheses or brackets are present, exponents are applied before multiplication. For example:

$$4 \times 9^2 = 4 \times 81 = 324$$

Also, when parentheses are present next to a multiplication, the × can be omitted:

$$5(8 - 3) = 5(5) = 25$$

As explained in *Fractions*, the horizontal line in a fraction implies division. The top number (called the *numerator*) is divided by the bottom number (called the *denominator*). For example,

$$\frac{50}{10} = 50 \div 10 = 5$$

In formulas, the multiplication sign (×) may be omitted (when letters—called variables—are multiplied) or replaced by parentheses, which serve the same purpose.

$$A \times B = AB, \qquad 6 \times 4 = (6)(4), \qquad 8 \times a = 8a$$

A multiplication dot (·) is also sometimes used.

**Fractions**

Rational numbers can be written as *common fractions* or as *decimal fractions*. Common fractions are written as $\frac{a}{b}$ or $a/b$, where $a$ (the numerator) and $b$ (the denominator) are integers (but $b$ cannot be 0, since division by zero is not defined). The denominator represents the number of equal parts that a whole quantity is broken into. The numerator is the number of these parts under consideration. For example, $\frac{2}{5}$ indicates the whole of something is broken into 5 equal parts, and 2 of these parts are being considered. Any integer is a fraction with a denominator of 1. For example, $\frac{6}{1} = 6$.

The implied operation in a fraction is division. Thus, $\frac{a}{b}$ means $a \div b$.

*Multiple:* A multiple of a number $n$ is the result of multiplying $n$ by positive integer 1, 2, 3, . . . Thus, the multiples of 3 are 3, 6, 9, 12, . . . The *least common multiple* (LCM) of two or more numbers is the smallest multiple the numbers have in common. In the example below, the first few multiples of 6 and 20 are shown, with the LCM indicated in bold:

6: 6, 12, 18, 24, 30, 36, 42, 48, 54, **60**, 66, . . .

20: 20, 40, **60**, 80, . . .

Thus, 60 is the LCM of 6 and 20.

*Factor:* An integer $a$ is a factor of $n$ if there is no remainder when $n$ is divided by $a$. That is, if the result of $n \div a$ is an integer. For example, 3 is a factor of 12 because $^{12}/_{3} = 4$. The greatest common factor (GCF) of two or more numbers is the largest of their common factors. Thus, the common factors of 12 and 18 are 2, 3, and 6; 6 is the GCF.

*Unit Fraction:* A fraction having the same numerator and denominator is the unit fraction, 1 (or "one whole"). For example, $^2/_2$, $^4/_4$, $^8/_8$, $^{16}/_{16}$, $^{32}/_{32}$, and $^{64}/_{64}$ all equal 1.

*Proper Fraction:* A fraction whose numerator is less than its denominator. $^1/_4$, $^1/_2$, and $^{47}/_{64}$ are examples of proper fractions. The value of any proper fraction is less than 1.

*Improper Fraction:* A fraction whose numerator is greater than its denominator. $^3/_2$, $^5/_4$, and $^{-17}/_8$ are examples of improper fractions. The absolute value of any improper fraction is greater than 1.

*Reducible Fraction:* A reducible fraction is a common fraction in which numerator and denominator have a common factor and so can be reduced to lowest terms by dividing both numerator and denominator by this common factor. For example, in the fraction $^{12}/_{18}$, the numerator and denominator have a GCF of 6. Thus, $^{12}/_{18}$ reduces to $^2/_3$ by dividing each part of the fraction by 6. A fraction such as $^{16}/_{21}$ cannot be reduced, since 16 and 21 do not have a common factor.

*Mixed Number:* A mixed number is a combination of a whole number and a proper fraction. The implied operation between them is addition. For example, $4\frac{2}{9}$ means $4 + \frac{2}{9}$. A mixed number is converted to an improper fraction by multiplying the whole number part with the denominator and adding the numerator to obtain the numerator of the final fraction; the denominator remains the same.

*Examples:*

$$5\frac{2}{3} = 5 + \frac{2}{3} = \frac{15}{3} + \frac{2}{3} = \frac{17}{3} \qquad 9\frac{1}{2} = 9 + \frac{1}{2} = \frac{18}{2} + \frac{1}{2} = \frac{19}{2}$$

To convert mixed numbers to improper fractions, multiply the whole number by the denominator and add the numerator to obtain the new numerator. The denominator remains the same. For example,

$$2\frac{1}{2} = \frac{2 \times 2 + 1}{2} = \frac{5}{2}$$

$$3\frac{7}{16} = \frac{3 \times 16 + 7}{16} = \frac{55}{16}$$

An improper fraction is converted to its mixed number form by dividing the numerator by denominator and placing the remainder over the denominator. Sometimes the fraction part can be reduced, as the second example shows.

$$\frac{17}{8} = 17 \div 8 = 2\frac{1}{8}$$

$$\frac{26}{16} = 26 \div 16 = 1\frac{10}{16} = 1\frac{5}{8}$$

*Equivalent Fractions:* A fraction raised to its equivalent form ("higher terms") by multiplying numerator and denominator by the same number (that is, by multiplying by a form of 1). For example, $^1/_4 \times {^4/_4} = {^4/_{16}}$ and $^3/_8 \times {^4/_4} = {^{12}/_{32}}$.

Any integer $n$ can be expressed as a fraction with a chosen denominator value of $m$ by simply writing $n$ as $n/1$ and multiplying by $m/m$.

*Example:* To express 4 as an equivalent fraction with a denominator of 16, write $^4/_1 \times {^{16}/_{16}} = {^{64}/_{16}}$

*Reciprocal:* The *reciprocal* of any number $a$ other than 0 is $1/a$. (0 has no reciprocal, since 1/0 is undefined.) The reciprocal also is called the *multiplicative inverse*, since $a \times 1/a = 1$. For example, the reciprocal of 8 is $^1/_8$; the reciprocal of $^4/_7$ is $^7/_4$.

*Least Common Denominator:* Fractions cannot be added or subtracted without a common denominator. For example, $\frac{2}{5} + \frac{1}{5} = \frac{2+1}{5} = \frac{3}{5}$, a simple computation, since the denominator in the answer is the same denominator seen in the fractions. In general, $\frac{a}{c} + \frac{b}{c} = \frac{a+b}{c}$. But fractions with different denominators cannot be added or subtracted until they are converted to *equivalent forms* that have common denominators. This is done by raising the fractions to higher terms (as explained previously). While any common multiple serves as a common denominator, it is preferable to use the *least common multiple* (LCM) of the denominator, referred to as the *least common denominator* (LCD). For example, 18 is the LCD of $\frac{2}{9}$ and $\frac{5}{6}$, since the LCM of 9 and 6 is 18. Raising each fraction to its equivalent form having a denominator of 18 is shown:

$$\frac{2}{9} \times \frac{2}{2} = \frac{4}{18} \quad \text{and} \quad \frac{5}{6} \times \frac{3}{3} = \frac{15}{18}$$

*Example:* In the case of $\frac{9}{11}$ and $\frac{7}{10}$ the LCD is the product of the denominators, 11 × 10 = 110. Raising each fraction to its equivalent form is shown:

$$\frac{9}{11} \times \frac{10}{10} = \frac{90}{110} \quad \text{and} \quad \frac{7}{10} \times \frac{11}{11} = \frac{77}{110}$$

**Adding and Subtracting Fractions and Mixed Numbers**

*To Add or Subtract Common Fractions:* 1) Convert each fraction to terms of the least common denominator; 2) add or subtract numerators; 3) if answer is an improper fraction, change it to a mixed number; and 4) reduce fraction part if necessary.

*Example, Addition of Common Fractions*

$$\begin{array}{ccrcr} \frac{1}{4} & \underrightarrow{\text{LCD} = 16} & \frac{1}{4}\times\frac{4}{4} & = & \frac{4}{16} \\ \frac{3}{16} & & \frac{3}{16}\times\frac{1}{1} & = & \frac{3}{16} \\ \underline{+\frac{7}{8}} & & \underline{+\frac{7}{8}\times\frac{2}{2}} & = & \underline{+\frac{14}{16}} \\ & & & & \frac{21}{16} \end{array}$$

*Example, Subtraction of Common Fractions*

$$\begin{array}{ccrcr} \frac{15}{16} & \underrightarrow{\text{LCD} = 48} & \frac{15}{16}\times\frac{3}{3} & = & \frac{45}{48} \\ \underline{-\frac{7}{12}} & & \underline{-\frac{7}{12}\times\frac{4}{4}} & = & \underline{-\frac{28}{48}} \\ & & & & \frac{17}{48} \end{array}$$

*To Add Mixed Numbers:* Two methods for adding mixed numbers are shown below the explanations.

First method: 1) Raise fraction parts to the higher terms of the LCD; 2) add whole number parts and fraction parts separately; 3) if result has an improper fraction, convert it to a mixed number and add the whole number parts.

Second method: 1) Convert mixed numbers to improper fractions; 2) raise resulting fractions to the higher terms of the LCD; 3) add fractions as usual and convert back to a mixed number; reduce, if needed.

*Examples, Addition of Mixed Numbers*

*Method 1*

$$\begin{array}{rcrcr} 2\frac{1}{2} & & 2\frac{1}{2}\times\frac{16}{16} & = & 2\frac{16}{32} \\ 4\frac{1}{4} & \rightarrow & 4\frac{1}{4}\times\frac{8}{8} & = & 4\frac{8}{32} \\ \underline{+1\frac{15}{32}} & & \underline{+1\frac{15}{32}\times\frac{1}{1}} & = & \underline{+1\frac{15}{32}} \end{array}$$

$$7\tfrac{39}{32} = 7 + 1\tfrac{7}{32} = 8\tfrac{7}{32}$$

*Method 2*

$$\begin{array}{rcrcr} 2\frac{1}{2} & & \frac{5}{2} = \frac{5}{2}\times\frac{16}{16} & = & \frac{80}{32} \\ 4\frac{1}{4} & \rightarrow & \frac{17}{4} = \frac{17}{4}\times\frac{8}{8} & = & \frac{136}{32} \\ \underline{+1\frac{15}{32}} & & \underline{+\frac{47}{32} = \frac{47}{32}\times\frac{1}{1}} & = & \underline{+\frac{47}{32}} \end{array}$$

$$\tfrac{263}{32} = 8\tfrac{7}{32}$$

*To Subtract Mixed Numbers:* The methods are similar to those for adding, except the fraction part may need to "borrow" from the whole number. The examples show the details.

1) Convert fraction parts to equivalent fractions with LCD; 2) subtract whole number and fraction parts separately, *unless* the first fraction's numerator is *smaller* than the second. In that case, proceed as shown in the second and third examples below, borrowing 1 in the form of a fraction and then subtracting.

*Examples, Subtraction of Mixed Numbers*

*Example 1*

$$\begin{array}{r} 12\frac{4}{5} \\ \underline{-4\frac{1}{5}} \\ 8\frac{3}{5} \end{array}$$

*Example 2*

$$\begin{array}{rcr} 43\frac{14}{15} & = & 43\frac{14}{15} \\ \underline{-19\frac{3}{5}} & = & \underline{-19\frac{9}{15}} \\ & & 24\frac{5}{15} = 24\frac{1}{3} \end{array}$$

*Example 3*

$$\begin{array}{rcrcr} 20 & = & 19+1 & = & 19\frac{2}{2} \\ \underline{-7\frac{1}{2}} & = & \underline{-7\frac{1}{2}} & = & \underline{-7\frac{1}{2}} \\ & & & & 12\frac{1}{2} \end{array}$$

*Example 4*

$$\begin{array}{rcrcr} 8\frac{2}{9} & = & 7+1\frac{2}{9} & = & 7\frac{11}{9} \\ \underline{-1\frac{4}{9}} & = & \underline{-1\frac{4}{9}} & = & \underline{-1\frac{4}{9}} \\ & & & & 6\frac{7}{9} \end{array}$$

**Multiplying Fractions and Mixed Numbers**

*To Multiply Common Fractions:* 1) Multiply numerators; 2) multiply denominators; and 3) convert improper fractions to mixed numbers, if necessary.

*To Multiply Mixed Numbers:* 1) Convert mixed numbers to improper fractions; 2) multiply numerators; 3) multiply denominators; and 4) convert improper fractions to mixed numbers, if necessary.

*Examples, Multiplication of Fractions*

$$\frac{2}{3}\times\frac{8}{15} = \frac{2\times 8}{3\times 15} = \frac{16}{45} \qquad 4\frac{2}{5}\times 1\frac{1}{3} = \frac{22}{5}\times\frac{4}{3} = \frac{22\times 4}{5\times 3} = \frac{88}{15}$$

**Dividing Fractions and Mixed Numbers**

*To Divide Common Fractions:* 1) Take the reciprocal of the dividing fraction; 2) multiply the numerators and denominators; and 3) convert improper fractions to mixed numbers, if necessary.

*To Divide Mixed Numbers:* 1) Convert the mixed numbers to improper fractions; 2) take the reciprocal of the dividing fraction; 3) multiply numerators and denominators; and 4) convert improper fractions to mixed numbers, if necessary.

*Examples, Division of Fractions*

$$\frac{2}{7} \div \frac{5}{21} = \frac{2}{7} \times \frac{21}{5} = \frac{2 \times 21}{7 \times 5} = \frac{42}{35} = \frac{6}{5} \qquad 3\frac{1}{3} \div 2\frac{4}{5} = \frac{10}{3} \div \frac{14}{5} = \frac{10}{3} \times \frac{5}{14} = \frac{50}{42} = \frac{25}{21}$$

**Decimal Numbers.—**Decimal fractions are fractional parts of a whole whose implied denominators are multiples of 10. A decimal fraction of 0.1 has a value of 1/10, 0.01 has a value of 1/100, 0.001 has a value of 1/1000, and so on. Thus, the value of the digit in the first place right of the decimal point is expressed in tenths, a digit two places to the right is expressed in hundredths, a digit three places to the right is expressed in thousandths, and so on. Because the denominator is implied, the number to the right of the decimal point indicates the numerator of the decimal fraction. For example, 0.125 is equivalent to 125/1000.

In industry, most decimal fractions are expressed in terms of thousandths rather than tenths or hundredths. For example, a decimal fraction of 0.2 is written as 0.200 and read as "200 thousandths" rather than "2 tenths"; a value of 0.75 is written as 0.750, and read as "750 thousandths" rather than "75 hundredths." In the case of four place decimals, the values are expressed in terms of ten-thousandths. So a value of 0.1875 is read as "1875 ten-thousandths."

Just as a mixed number is the sum of a whole number and a fraction, a decimal number greater than 1 has a whole and a decimal part. For example, $10.125 = 10^{125}/_{1000}$, which is read as "10 and 125 thousandths."

*Adding or Subtracting Decimal Numbers:* To add or subtract decimal numbers, align the decimal points and add or subtract the digits as usual. The decimal point in the answer is aligned with the decimal points in the numbers added or subtracted.

*Examples, Adding Decimal Fractions*

| | |
|---|---|
| 0.125 | 1.750 |
| 1.0625 | 0.875 |
| 2.50 | 0.125 |
| +0.1875 | +2.0005 |
| 3.8750 | 4.7505 |

*Examples, Subtracting Decimal Fractions*

| | |
|---|---|
| 1.750 | 2.625 |
| −0.250 | −1.125 |
| 1.500 | 1.500 |

*Multiplying Decimal Numbers:* In setting up decimal multiplication, the decimal points do not have to be aligned. Long multiplication is done as usual, but the decimal point in the answer is placed so that the number of digits on its right is the same as the total number of digits on the right of the numbers multiplied.

*Examples, Decimal Number Multiplication*

| | |
|---|---|
| 24.035 | three decimal places |
| × 0.08 | two decimal places |
| 1.92280 | five decimal places |

| | |
|---|---|
| 6.002 | three decimal places |
| × 41.3 | one decimal place |
| 18006 | |
| 60020 | |
| + 2400800 | |
| 247.8826 | four decimal places |

*Dividing Decimal Numbers:* There are several types of decimal division problems: (1) a whole number divided by a decimal number; (2) a decimal number divided by a whole number; and (3) a decimal number divided by a decimal number. For all situations, if the divisor is a decimal, its decimal point must first be moved right to make it a whole number, and the dividend's decimal likewise moved, before the operation is performed. Examples of each type are: $18 \div 0.3 = 180 \div 3 = 60$; $1.8 \div 3 = 0.6$; $1.8 \div 0.003 = 1800 \div 3 = 600$.

**Ratio and Proportion.**—A *ratio* of quantities *a* to *b* is written *a*:*b* or as a fraction *a*/*b*. For example, the ratio of 12 to 3 is written 12:3 or 12/3. Ratios, like fractions, can be reduced: 12:3 is 4:1. The *inverse* (or reciprocal) ratio of *a*:*b* is *b*:*a*. Thus, the inverse ratio of 12:3 is 3:12.

When two or more ratios are multiplied, the ratio obtained is a called a *compound ratio.* The compound ratio of *a*:*b*, *c*:*d*, and *e*:*f* is the ratio *ace*:*bdf*. For example, the compound ratio of 8:2, 9:3, and 10:5 is $8 \times 9 \times 10 : 2 \times 3 \times 5$, or 720:30.

An equality of ratios, $a/b = c/d$, is called a *proportion*, which can be written as *a*:*b*::*c*:*d*, read as "*a* is to *b* as *c* is to *d*." Thus, 6:3::10:5 because 6/3 and 10/5 both reduce to 2/1 or 2. In a proportion *a*:*b*::*c*:*d*, the first and last terms (which can be variables or numbers) are called the *extremes*, and the second and third are the *means*.

Note that if $a/b = c/d$, then the rules of algebra show that $ad = bc$. Thus, the proportion *a*:*b*::*c*:*d* is equivalent to $a \times d = b \times c$. So the proportion 6:3::10:5 is equivalent to $6 \times 5 = 3 \times 10$.

Often, some part of a proportion is an unknown. For example, in the proportion 2:3::*n*:4 (2 is to 3 as *n* is to 4), *n* is found by setting up a proportion. According to the basic rules of algebra, 2:3::*n*:4 means $(2)(4) = 3n$, and hence, $8 = 3n$, so $n = 8/3$. A full discussion of the rules for solving equations can be found in *ALGEBRA*.

If the second and third terms in a proportion are the same, that term is the *mean proportional* of the other two. Thus, in the proportion 8:4::4:2, 4 is the mean proportional of 8 and 2. The mean proportional of any two numbers may be found by multiplying them and extracting the square root of the product. Thus, the mean proportional of 3 and 12 is 6, because $3 \times 12 = 36$, which is $6^2$.

*Example 1, Involving Proportion:* If it takes 18 days to assemble 4 lathes, how many days would it take to assemble 14 lathes?

*Solution:* Let *x* be the number of days to be found. The proportion is written 4:18 :: 14:*x*, where *x* is the number of days to be found. Setting this up as an equation and solving:

$$\frac{4}{18} = \frac{14}{x}$$

$$x = \frac{18 \times 14}{4} = 63 \text{ days}$$

*Example 2, Involving Direct (Simple) Proportion:* 10 linear meters (32.81 feet) of bar stock are required as blanks for 100 clamping bolts. What total length *x* of stock, in meters and feet, is required for 912 bolts?

*Solution:* The setup to solve the proportional meters-to-bolts problem comes from the way this proportion is read: "10 meters is to 100 bolts as how many meters is to 912 bolts." It is solved accordingly:

$$10 : 100 :: x : 912, \text{ that is, } \frac{10}{100} = \frac{x}{912} \qquad \text{Solving for } x\text{: } x = \frac{10 \times 912}{100} = \frac{9120}{100} = 91.2 \text{ meters}$$

Likewise, the setup to solve the feet-to-bolts problem comes from reading it as: "32.81 feet is to 100 bolts as how many feet is to 912 bolts." Thus:

$$32.81 : 100 :: x : 912, \text{ that is, } \frac{32.81}{100} = \frac{x}{912} \qquad \text{Solving for } x\text{: } \quad x = \frac{32.81 \times 912}{100} = \frac{29{,}922.72}{100} = 299.2 \text{ feet}$$

*Inverse Proportion:* Quantities with an inversely proportional relationship behave in such a way that as one increases the other decreases. For example, a factory employing 270 workers completes a given number of automotive components weekly, the number of working hours being 44 per week. If the hours are reduced, then more workers will be required to do the same amount of work. How many employees would be required for the same production if the working hours were reduced to 40 per week?

The hours per week is inversely proportional to the number of workers; fewer hours per worker means more workers are required. Letting $x$ be the number of workers needed when time is reduced, the inverse proportion is written:

$$270 : x :: 40 : 44$$

Thus

$$\frac{270}{x} = \frac{40}{44} \qquad \text{and} \qquad x = \frac{270 \times 44}{40} = 297 \text{ workers}$$

*Problems Involving Both Direct and Inverse Proportions:* If two groups of data are related by both direct (simple) and inverse proportions among the various quantities, a simple mathematical relation may be used to solve the problem as follows:

$$\frac{\text{Product of all directly proportional items in first group}}{\text{Product of all inversely proportional items in first group}}$$

$$= \frac{\text{Product of all directly proportional items in second group}}{\text{Product of all inversely proportional items in second group}}$$

*Example:* If a worker capable of turning 65 studs in a 10-hour day is paid $13.50 per hour, how much per hour should a worker be paid who turns 72 studs in a 9-hour day if compensated in the same proportion as the first worker?

*Solution:* The first group of data in this problem consists of the number of hours worked, the hourly wage of the first worker, and the number of studs produced per day; the second group contains similar data for the second worker, except the hourly wage is unknown, so it is indicated by $x$.

The labor cost per stud, as may be seen, is directly proportional to the number of hours worked and the hourly wage. These quantities, therefore, are used in the numerators of the fractions in the formula. The labor cost per stud is inversely proportional to the number of studs produced per day. (The greater the number of studs produced in a given time the less the cost per stud.) The numbers of studs per day, therefore, are placed in the denominators of the fractions in the formula. Thus,

$$\frac{(10)(13.50)}{65} = \frac{9x}{72}$$

$$x = \frac{(10)(13.50)(72)}{(65)(9)} = \$16.62 \text{ per hour}$$

**Percentage.—**A percentage is a ratio expressed as a part of 100. For example, if out of 100 manufactured parts, 12 do not pass inspection, then 12 percent (12 of the 100) are rejected. The symbol % indicates percentage.

The percent of gain (or loss) with respect to a base (original) quantity is found by dividing the amount of gain (or loss) by the base quantity and multiplying the quotient by 100. For example, if a quantity of steel is bought for $2000 and sold for $2500, the profit is $500/2000 × 100, or 25 percent of the invested amount.

*Example:* Out of a total output of 280 castings a day, 30 castings are, on average, rejected. What is the percentage of bad castings?

$$\frac{30}{280} \times 100 = 10.71 \text{ percent}$$

*Percent Change:* Any increase or decrease in some measured quantity can be expressed as a *percent change* using the formula: $\frac{\text{final} - \text{original amount}}{\text{original}} \times 100 = \text{percent change}$. If in the previous example, production increased from 280 to 300, then the percent change would be:

$$\frac{\text{final} - \text{original amount}}{\text{original}} \times 100 = \frac{300 - 280}{280} \times 100 = \frac{20}{280} \times 100 = 7.14\%.$$

The denominator is always the original amount. Percent change also can be negative. If production decreased from 280 to 245, the percent change would be:

$$\frac{245 - 280}{280} \times 100 = \frac{-35}{280} \times 100 = -12.5\%.$$

## Powers and Roots

*Powers:* The *square* or *second power* of a number (or quantity) is the product of that number multiplied by itself. Thus, the square of 9 is $9 \times 9$. The square of a number is indicated by the *exponent*$^2$, thus: $9^2 = 9 \times 9 = 81$.

The *cube* or *third power* of a number $n$ is the product $n \times n \times n$, or $n^3$. Thus, the cube of 4 is $4 \times 4 \times 4 = 64$, and is written $4^3$.

In general, the $n$th power of $a$ is written $a^n$, where $a$ is the *base* and $n$ is the *exponent.*

*Roots:* The *square root* of a given number is the positive number which, when multiplied by itself, will produce the given number. The square root of 16 (written $\sqrt{16}$ ) is 4 because $4 \times 4 = 16$. The other root of 16 is −4, but the use of the square root symbol indicates the positive (principal) square root only.

Similarly, the *cube root* of a given number is the number which, when used as a factor three times, will produce the given number. Thus, the cube root of 64 (written $\sqrt[3]{64}$ ) is 4 because $4 \times 4 \times 4 = 64$.

In general, the $n$th root of $a$ is written as $\sqrt[n]{a}$ or $a^{1/n}$.

### Properties of Exponents

$$a^n a^m = a^{n+m} \qquad \frac{a^n}{a^m} = a^{(n-m)} \qquad (a^m)^n = a^{mn} \qquad (ab)^m = a^m b^m$$

$$\left(\frac{a}{b}\right)^n = \frac{a^n}{b^n} \qquad a^{m/n} = (a^{1/n})^m \text{ or } (a^m)^{1/n} \qquad a^{-n} = \frac{1}{a^n} \qquad \frac{1}{a^{-n}} = a^n$$

$$a^0 = 1\ (a \neq 0) \qquad a^{1/n} = \sqrt[n]{a} \qquad a^{m/n} = (\sqrt[n]{a})^m = \left(\sqrt[n]{a^m}\right)$$ unless $a < 0$, and $m$ and $n$ are both even

*Examples:*

$$3^1 3^2 = 3^{1+2} = 3^3 = 27 \qquad (x)(x^3) = x^{(1+3)} = x^4$$

$$\frac{5^4}{5^2} = 5^{4-2} = 5^2 = 25 \qquad \frac{x^9}{x^6} = x^{(9-6)} = x^3$$

$$(2^4)^2 = 2^{(4)(2)} = 2^8 = 256 \qquad (x^3)^3 = x^{(3)(3)} = x^9$$

$$(9x)^2 = 9^2 x^2 = 81x^2 \qquad (ab^4)^2 = a^2 b^8$$

$$32^{3/5} = (32^{1/5})^3 = (\sqrt[5]{32})^3 = 2^3 = 8 \qquad 4^{-3} = \frac{1}{4^3} = \frac{1}{64}$$

$$\frac{1}{2^{-5}} = 2^5 = 32 \qquad 9x^0 = 9(1) = 9$$

Using logarithms can greatly facilitate the process of raising a number to a power or extracting its root. As shown in *Logarithms* on page 31, this is especially true if the power is not an integer. For example, the square root of 137.1 can only be found with a degree of accuracy through logarithms, a scientific calculator, or Taylor series polynomials.

**Scientific Notation.**—Calculations involving both large and small magnitude numbers are facilitated by *scientific notation*. In this system, a number is expressed by two factors: (1) an integer from 1 to 9, possibly followed by a decimal, and (2) a power of 10. Large numbers in standard form are converted to scientific notation as shown in the following examples:

$$50{,}000 = 5 \times 10^4 \qquad 273.15 = 2.7315 \times 10^2$$

In the example, 50,000 becomes $5 \times 10^4$ because the positive exponent on 10 is the number of places to the right that the decimal point moves so that the first factor falls between 1 and 10. Numbers less than 1 are converted to scientific notation as shown in the following examples:

$$0.840 = 8.40 \times 10^{-1} \qquad 0.0000001 = 1 \times 10^{-7}$$

The negative exponent shows the number of places to the left that the decimal point moves, so that the first factor falls between 1 and 10.

Science and engineering quantities—which are often quite large or small—lend themselves to representation in scientific notation. For instance, *Avogadro's number*, which is the number of particles in one mole of a substance, is $6.024 \times 10^{23}$. The metric (SI) pressure unit of 1 pascal (Pa) is equivalent to 0.00000986923 atmosphere (atm) or 0.0001450377 pound/square inch (psi). In scientific notation, these figures are $9.86923 \times 10^{-6}$ atm and $1.450377 \times 10^{-4}$ psi, respectively.

*Engineering notation* is a version of scientific notation in which the exponent of 10 is always a multiple of 3. (See *MEASURING UNITS* on page 2836 for a table of this system.)

*Multiplication in Scientific Notation:* The procedure is as follows:

1) Multiply the first factors of the numbers to obtain the first factor of the product.

2) Add the exponents of the factors of 10 to obtain the product's factor of 10. Thus:

$$(4.31 \times 10^{-2}) \times (9.01 \times 10) = (4.31 \times 9.01) \times 10^{-2+1} = 38.8331 \times 10^{-1}$$

$$(5.98 \times 10^4) \times (4.37 \times 10^3) = (5.98 \times 4.37) \times 10^{4+3} = 26.1326 \times 10^7$$

3) Write the final in conventional scientific notation, as explained in the previous section. So, for the two examples:

$38.8331 \times 10^{-1} = 3.88331 \times 10^0 = 3.88331$, because $10^0 = 1$, and $26.1326 \times 10^7 = 2.61326 \times 10^8$.

When multiplying several numbers written in this notation, the procedure is the same. Thus, $(4.02 \times 10^{-3}) \times (3.987 \times 10) \times (4.863 \times 10^5) = (4.02 \times 3.987 \times 4.863) \times 10^{(-3+1+5)} = 77.94 \times 10^3 = 7.79 \times 10^4$, rounding off the first factor to two decimal places.

*Division in Scientific Notation:* The procedure is as follows:

1) Divide the first factor of the dividend (the first number) by the first factor of the divisor (the second number) to get the first factor of the quotient.

2) Subtract the exponents of the factors of 10 to obtain the product's factor of 10:

$$(4.31 \times 10^{-2}) \div (9.0125 \times 10) =$$

$$(4.31 \div 9.0125) \times (10^{-2-1}) = 0.4782 \times 10^{-3} = 4.782 \times 10^{-4}$$

It can be seen that this system of notation is helpful where several numbers of different magnitudes are to be multiplied and divided.

*Example:* Find the solution of $\dfrac{250 \times 4698 \times 0.00039}{43678 \times 0.002 \times 0.0147}$

*Solution:* Changing all these numbers to powers of 10 notation and performing the operations indicated:

$$\frac{(2.5 \times 10^2) \times (4.698 \times 10^3) \times (3.9 \times 10^{-4})}{(4.3678 \times 10^4) \times (2 \times 10^{-3}) \times (1.47 \times 10^{-2})}$$

$$= \frac{(2.5 \times 4.698 \times 3.9)(10^{2+3-4})}{(4.3678 \times 2 \times 1.47)(10^{4-3-2})} = \frac{45.8055 \times 10}{12.8413 \times 10^{-1}}$$

$$= 3.5670 \times 10^{1-(-1)} = 3.5670 \times 10^2 = 356.70 \text{ (rounded)}$$

**Factorial Notation.**—A factorial is a mathematical shortcut denoted by the symbol **!** following a number (for example, 3! is "three factorial"). $n!$ is found by multiplying together all the positive integers less than or equal to the factorial number $n$. Zero factorial (0!) is defined as 1. For example: $3! = 1 \times 2 \times 3 = 6$; $4! = 1 \times 2 \times 3 \times 4 = 24$; $7! = 1 \times 2 \times 3 \times 4 \times 5 \times 6 \times 7 = 5040$; etc. Factorial notation is used in certain areas, including probability and analysis. The following two topics (permutations and combinations) relate to probability and statistics.

**Permutation.**—A permutation is an arrangement of objects of a set into a sequence or order. In mathematics, the number of arrangements of $n$ objects is given by $n!$. For example, 4 objects can be arranged 4! ways, that is, $4 \times 3 \times 2 \times 1 = 24$ ways. The number of ways $r$ objects can be arranged (that is, ordered) from a set of $n$ is given by the *permutation* formula $_nP_r = \dfrac{n!}{(n-r)!}$

*Example:* How many ways can the letters X, Y, and Z be arranged?

*Solution:* Three objects ($r = 3$) out of a set of 3 ($n = 3$) are being arranged. The numbers of possible arrangements for the three letters are $3!/(3-3)! = (3 \times 2 \times 1)/1 = 6$. Listing them is not difficult, since there are so few: XYZ, XZY, YXZ, YZX, ZXY, ZYX.

*Example:* There are 10 people participating in a foot race. How many arrangements of first, second, and third place winners are there?

*Solution:* Here $r$ is 3 and $n$ is 10. The number of possible arrangements of winners are:

$$_{10}P_3 = \frac{10!}{(10-3)!} = \frac{10!}{7!} = 10 \times 9 \times 8 = 720$$

**Combination.**—This is the number of ways $r$ objects can be chosen from $n$ in a way that order does not matter. It is expressed as "$n$ choose $r$." There are fewer combinations than permutations of $r$ objects out of $n$, since it does not matter in what order the three objects are chosen. So in a combination, choosing ABC is the same as choosing ACB or BAC and so on. The formula is $_nC_r = \dfrac{n!}{(n-r)!r!}$

*Example:* How many possible sets of 6 numbers can be picked with no regard for order from the numbers 1 to 52?

*Solution:* Here $r$ is 6 and $n$ is 52. So the possible number of combinations is:

$$_{52}C_6 = \frac{52!}{(52-6)!6!} = \frac{52!}{46!6!} = \frac{52 \times 51 \times 50 \times 49 \times 48 \times 47}{1 \times 2 \times 3 \times 4 \times 5 \times 6} = 20{,}358{,}520$$

**Prime Factorization of Numbers.**—Tables of prime numbers and factors of numbers are particularly useful for calculations involving change-gear ratios for compound gearing, dividing heads, gear-generating machines, and mechanical designs having gear trains.

*Definition:* $p$ is a *factor* of a number $n$ if the division $n/p$ leaves no remainder. Thus, any number $n$ has factors of itself and 1, because $n/n = 1$ and $n/1 = n$. Other factors of a number are found as follows:

2 is a factor of any even number. Thus, $28 = 2 \times 14$, and $210 = 2 \times 105$.

3 is a factor of any number where the sum of its digits is divisible by 3. Thus, 3 is a factor of 2397, because $2 + 3 + 9 + 7 = 21$, and $21 \div 3 = 7$.

4 is a factor of any number in which the last two digits are a number divisible by 4. Thus, 1844 has a factor 4, because $44 \div 4 = 11$. 761 does not have a factor of 4, since 61 is not divisible by 4.

5 is a factor of any number that has a ones digit that is either 0 or 5.

A *prime number* is one that has no factors except itself and 1. Thus, 2, 3, 5, 7, 11, etc., are prime numbers. 2 is the only even prime number. A factor which itself is a prime number is called a *prime factor*. All numbers can be expressed as a product of their prime factors.

It can be determined if 7 is a factor of a number according to this process: Remove the last digit from the number, double it, and subtract it from the remaining number. If the result is divisible by 7 (e.g., 14, 7, 0, −7, etc.), then 7 is a prime factor of the number.

The *prime factorization* of a number is done by expressing the number as a product of its primes. For example, the prime factors of 20 are 2 and 5. The prime factorization is $2 \times 2 \times 5 = 20$.

*Table 1* gives the smallest prime factor of all odd numbers from 1 to 2399. This table can be used for finding all the factors for numbers up to this odd number. Where no factor is given for a number in the table, the letter **P** indicates that the number is a prime number. *Table 2* lists prime numbers from 2 through 15277 and can be used to identify unfactorable numbers in that range. (More prime number tables are included in the *ADDITIONAL MATERIAL* in the *Machinery's Handbook 32 Digital Edition*.)

*Example 1:* Find the factors of 833. Use the table on page 15 as illustrated below.

*Solution:* The table on page 15 indicates that 7 is the smallest prime factor of 833, shown at the row-column intersection for 833. This leaves another factor, because $833 \div 7 = 119$.

| From | 0 | 100 | 200 | 300 | 400 | 500 | 600 | 700 | 800 | 900 | 1000 | 1100 |
|---|---|---|---|---|---|---|---|---|---|---|---|---|
| To | 100 | 200 | 300 | 400 | 500 | 600 | 700 | 800 | 900 | 1000 | 1100 | 1200 |
| 33 | 3 | 7 | P | 3 | P | 13 | 3 | P | 7 | 3 | P | 11 |

Looking up 119, it also shows that 7 is a prime factor of 119, leaving a factor $119 \div 7 = 17$.

| From | 0 | 100 | 200 | 300 | 400 | 500 | 600 | 700 | 800 | 900 | 1000 | 1100 |
|---|---|---|---|---|---|---|---|---|---|---|---|---|
| To | 100 | 200 | 300 | 400 | 500 | 600 | 700 | 800 | 900 | 1000 | 1100 | 1200 |
| 19 | P | 7 | 3 | 11 | P | 3 | P | P | 3 | P | P | 3 |

Looking up 17, **P** indicates that 17 is a prime number and no other prime factors of 833 exist.

| From | 0 | 100 | 200 | 300 | 400 | 500 | 600 | 700 | 800 | 900 | 1000 | 1100 |
|---|---|---|---|---|---|---|---|---|---|---|---|---|
| To | 100 | 200 | 300 | 400 | 500 | 600 | 700 | 800 | 900 | 1000 | 1100 | 1200 |
| 17 | P | 3 | 7 | P | 3 | 11 | P | 3 | 19 | 7 | 3 | P |

Hence, the prime factorization of 833 is $7 \times 7 \times 17$.

*Example 2:* A set of four gears is required in a mechanical design to provide an overall gear ratio of $3332 \div 1200$. Furthermore, no gear in the set is to have more than 120 teeth or less than 24 teeth. Determine the tooth numbers.

*Solution:* The prime factorization of 3332 is determined to be $2 \times 2 \times 7 \times 7 \times 17 = 3332$. The prime factorization of 1200 is $2 \times 2 \times 2 \times 2 \times 5 \times 5 \times 3 = 1200$. Therefore,

$$\frac{3332}{1200} = \frac{2 \times 2 \times 7 \times 7 \times 17}{2 \times 2 \times 2 \times 2 \times 5 \times 5 \times 3} = \left(\frac{68}{48}\right)\left(\frac{49}{25}\right)$$

Each resulting factor represents the number of teeth that fulfill the requirement. Note that the first gear set having 68 and 48 teeth can be reduced in size to gears having 34 and 24 teeth, or even 17 and 12 teeth, while preserving the same gear ratio, since both 3332 and 1200 are divisible by 4.

*Example 3:* Factor 21,194 into two numbers, neither of which is larger than 200.

*Solution:* The smallest factor of 21,194 is obviously 2, leaving $21{,}194 \div 2 = 10{,}597$ to be factored further. However, in Table 2a, starting on page 17, which includes the prime numbers from 2 to 15,277, it can be seen that 10,597 is a prime number; therefore, no other factors exist. So the factorization named is not possible.

An approximate factorization with two numbers can be quickly found by taking the square root of 21,194, which equals 145.582. The two factors 145 and 146 provide a product of 21,170, which has an error of 0.113 percent, relative to the desired number 21,194.

**Table 1a. Prime Number and Factor Table for 1 to 1199**

| From<br>To | 0<br>100 | 100<br>200 | 200<br>300 | 300<br>400 | 400<br>500 | 500<br>600 | 600<br>700 | 700<br>800 | 800<br>900 | 900<br>1000 | 1000<br>1100 | 1100<br>1200 |
|---|---|---|---|---|---|---|---|---|---|---|---|---|
| 1 | **P** | **P** | 3 | 7 | **P** | 3 | **P** | **P** | 3 | 17 | 7 | 3 |
| 2 | **P** | 2 | 2 | 2 | 2 | 2 | 2 | 2 | 2 | 2 | 2 | 2 |
| 3 | **P** | **P** | 7 | 3 | 13 | **P** | 3 | 19 | 11 | 3 | 17 | **P** |
| 5 | **P** | 3 | 5 | 5 | 3 | 5 | 5 | 3 | 5 | 5 | 3 | 5 |
| 7 | **P** | **P** | 3 | **P** | 11 | 3 | **P** | 7 | 3 | **P** | 19 | 3 |
| 9 | 3 | **P** | 11 | 3 | **P** | **P** | 3 | **P** | **P** | 3 | **P** | **P** |
| 11 | **P** | 3 | **P** | **P** | 3 | 7 | 13 | 3 | **P** | **P** | 3 | 11 |
| 13 | **P** | **P** | 3 | **P** | 7 | 3 | **P** | 23 | 3 | 11 | **P** | 3 |
| 15 | 3 | 5 | 5 | 3 | 5 | 5 | 3 | 5 | 5 | 3 | 5 | 5 |
| 17 | **P** | 3 | 7 | **P** | 3 | 11 | **P** | 3 | 19 | 7 | 3 | **P** |
| 19 | **P** | 7 | 3 | 11 | **P** | 3 | **P** | **P** | 3 | **P** | **P** | 3 |
| 21 | 3 | 11 | 13 | 3 | **P** | **P** | 3 | 7 | **P** | 3 | **P** | 19 |
| 23 | **P** | 3 | **P** | 17 | 3 | **P** | 7 | 3 | **P** | 13 | 3 | **P** |
| 25 | 5 | 5 | 3 | 5 | 5 | 3 | 5 | 5 | 3 | 5 | 5 | 3 |
| 27 | 3 | **P** | **P** | 3 | 7 | 17 | 3 | **P** | **P** | 3 | 13 | 7 |
| 29 | **P** | 3 | **P** | 7 | 3 | 23 | 17 | 3 | **P** | **P** | 3 | **P** |
| 31 | **P** | **P** | 3 | **P** | **P** | 3 | **P** | 17 | 3 | 7 | **P** | 3 |
| 33 | 3 | 7 | **P** | 3 | **P** | 13 | 3 | **P** | 7 | 3 | **P** | 11 |
| 35 | 5 | 3 | 5 | 5 | 3 | 5 | 5 | 3 | 5 | 5 | 3 | 5 |
| 37 | **P** | **P** | 3 | **P** | 19 | 3 | 7 | 11 | 3 | **P** | 17 | 3 |
| 39 | 3 | **P** | **P** | 3 | **P** | 7 | 3 | **P** | **P** | 3 | **P** | 17 |
| 41 | **P** | 3 | **P** | 11 | 3 | **P** | **P** | 3 | 29 | **P** | 3 | 7 |
| 43 | **P** | 11 | 3 | 7 | **P** | 3 | **P** | **P** | 3 | 23 | 7 | 3 |
| 45 | 3 | 5 | 5 | 3 | 5 | 5 | 3 | 5 | 5 | 3 | 5 | 5 |
| 47 | **P** | 3 | 13 | **P** | 3 | **P** | **P** | 3 | 7 | **P** | 3 | 31 |
| 49 | 7 | **P** | 3 | **P** | **P** | 3 | 11 | 7 | 3 | 13 | **P** | 3 |
| 51 | 3 | **P** | **P** | 3 | 11 | 19 | 3 | **P** | 23 | 3 | **P** | **P** |
| 53 | **P** | 3 | 11 | **P** | 3 | 7 | **P** | 3 | **P** | **P** | 3 | **P** |
| 55 | 5 | 5 | 3 | 5 | 5 | 3 | 5 | 5 | 3 | 5 | 5 | 3 |
| 57 | 3 | **P** | **P** | 3 | **P** | **P** | 3 | **P** | **P** | 3 | 7 | 13 |
| 59 | **P** | 3 | 7 | **P** | 3 | 13 | **P** | 3 | **P** | 7 | 3 | 19 |
| 61 | **P** | 7 | 3 | 19 | **P** | 3 | **P** | **P** | 3 | 31 | **P** | 3 |
| 63 | 3 | **P** | **P** | 3 | **P** | **P** | 3 | 7 | **P** | 3 | **P** | **P** |
| 65 | 5 | 3 | 5 | 5 | 3 | 5 | 5 | 3 | 5 | 5 | 3 | 5 |
| 67 | **P** | **P** | 3 | **P** | **P** | 3 | 23 | 13 | 3 | **P** | 11 | 3 |
| 69 | 3 | 13 | **P** | 3 | 7 | **P** | 3 | **P** | 11 | 3 | **P** | 7 |
| 71 | **P** | 3 | **P** | 7 | 3 | **P** | 11 | 3 | 13 | **P** | 3 | **P** |
| 73 | **P** | **P** | 3 | **P** | 11 | 3 | **P** | **P** | 3 | 7 | 29 | 3 |
| 75 | 3 | 5 | 5 | 3 | 5 | 5 | 3 | 5 | 5 | 3 | 5 | 5 |
| 77 | 7 | 3 | **P** | 13 | 3 | **P** | **P** | 3 | **P** | **P** | 3 | 11 |
| 79 | **P** | **P** | 3 | **P** | **P** | 3 | 7 | 19 | 3 | 11 | 13 | 3 |
| 81 | 3 | **P** | **P** | 3 | 13 | 7 | 3 | 11 | **P** | 3 | 23 | **P** |
| 83 | **P** | 3 | **P** | **P** | 3 | 11 | **P** | 3 | **P** | **P** | 3 | 7 |
| 85 | 5 | 5 | 3 | 5 | 5 | 3 | 5 | 5 | 3 | 5 | 5 | 3 |
| 87 | 3 | 11 | 7 | 3 | **P** | **P** | 3 | **P** | **P** | 3 | **P** | **P** |
| 89 | **P** | 3 | 17 | **P** | 3 | 19 | 13 | 3 | 7 | 23 | 3 | 29 |
| 91 | 7 | **P** | 3 | 17 | **P** | 3 | **P** | 7 | 3 | **P** | **P** | 3 |
| 93 | 3 | **P** | **P** | 3 | 17 | **P** | 3 | 13 | 19 | 3 | **P** | **P** |
| 95 | 5 | 3 | 5 | 5 | 3 | 5 | 5 | 3 | 5 | 5 | 3 | 5 |
| 97 | **P** | **P** | 3 | **P** | 7 | 3 | 17 | **P** | 3 | **P** | **P** | 3 |
| 99 | 3 | **P** | 13 | 3 | **P** | **P** | 3 | 17 | 29 | 3 | 7 | 11 |

**Table 1b. Prime Number and Factor Table for 1201 to 2399**

| From<br>To | 1200<br>1300 | 1300<br>1400 | 1400<br>1500 | 1500<br>1600 | 1600<br>1700 | 1700<br>1800 | 1800<br>1900 | 1900<br>2000 | 2000<br>2100 | 2100<br>2200 | 2200<br>2300 | 2300<br>2400 |
|---|---|---|---|---|---|---|---|---|---|---|---|---|
| 1 | **P** | **P** | 3 | 19 | **P** | 3 | **P** | **P** | 3 | 11 | 31 | 3 |
| 3 | 3 | **P** | 23 | 3 | 7 | 13 | 3 | 11 | **P** | 3 | **P** | 7 |
| 5 | 5 | 3 | 5 | 5 | 3 | 5 | 5 | 3 | 5 | 5 | 3 | 5 |
| 7 | 17 | **P** | 3 | 11 | **P** | 3 | 13 | **P** | 3 | 7 | **P** | 3 |
| 9 | 3 | 7 | **P** | 3 | **P** | **P** | 3 | 23 | 7 | 3 | 47 | **P** |
| 11 | 7 | 3 | 17 | **P** | 3 | 29 | **P** | 3 | **P** | **P** | 3 | **P** |
| 13 | **P** | 13 | 3 | 17 | **P** | 3 | 7 | **P** | 3 | **P** | **P** | 3 |
| 15 | 3 | 5 | 5 | 3 | 5 | 5 | 3 | 5 | 5 | 3 | 5 | 5 |
| 17 | **P** | 3 | 13 | 37 | 3 | 17 | 23 | 3 | **P** | 29 | 3 | 7 |
| 19 | 23 | **P** | 3 | 7 | **P** | 3 | 17 | 19 | 3 | 13 | 7 | 3 |
| 21 | 3 | **P** | 7 | 3 | **P** | **P** | 3 | 17 | 43 | 3 | **P** | 11 |
| 23 | **P** | 3 | **P** | **P** | 3 | **P** | **P** | 3 | 7 | 11 | 3 | 23 |
| 25 | 5 | 5 | 3 | 5 | 5 | 3 | 5 | 5 | 3 | 5 | 5 | 3 |
| 27 | 3 | **P** | **P** | 3 | **P** | 11 | 3 | 41 | **P** | 3 | 17 | 13 |
| 29 | **P** | 3 | **P** | 11 | 3 | 7 | 31 | 3 | **P** | **P** | 3 | 17 |
| 31 | **P** | 11 | 3 | **P** | 7 | 3 | **P** | **P** | 3 | **P** | 23 | 3 |
| 33 | 3 | 31 | **P** | 3 | 23 | **P** | 3 | **P** | 19 | 3 | 7 | **P** |
| 35 | 5 | 3 | 5 | 5 | 3 | 5 | 5 | 3 | 5 | 5 | 3 | 5 |
| 37 | **P** | 7 | 3 | 29 | **P** | 3 | 11 | 13 | 3 | **P** | **P** | 3 |
| 39 | 3 | 13 | **P** | 3 | 11 | 37 | 3 | 7 | **P** | 3 | **P** | **P** |
| 41 | 17 | 3 | 11 | 23 | 3 | **P** | 7 | 3 | 13 | **P** | 3 | **P** |
| 43 | 11 | 17 | 3 | **P** | 31 | 3 | 19 | 29 | 3 | **P** | **P** | 3 |
| 45 | 3 | 5 | 5 | 3 | 5 | 5 | 3 | 5 | 5 | 3 | 5 | 5 |
| 47 | 29 | 3 | **P** | 7 | 3 | **P** | **P** | 3 | 23 | 19 | 3 | **P** |
| 49 | **P** | 19 | 3 | **P** | 17 | 3 | 43 | **P** | 3 | 7 | 13 | 3 |
| 51 | 3 | 7 | **P** | 3 | 13 | 17 | 3 | **P** | 7 | 3 | **P** | **P** |
| 53 | 7 | 3 | **P** | **P** | 3 | **P** | 17 | 3 | **P** | **P** | 3 | 13 |
| 55 | 5 | 5 | 3 | 5 | 5 | 3 | 5 | 5 | 3 | 5 | 5 | 3 |
| 57 | 3 | 23 | 31 | 3 | **P** | 7 | 3 | 19 | 11 | 3 | 37 | **P** |
| 59 | **P** | 3 | **P** | **P** | 3 | **P** | 11 | 3 | 29 | 17 | 3 | 7 |
| 61 | 13 | **P** | 3 | 7 | 11 | 3 | **P** | 37 | 3 | **P** | 7 | 3 |
| 63 | 3 | 29 | 7 | 3 | **P** | 41 | 3 | 13 | **P** | 3 | 31 | 17 |
| 65 | 5 | 3 | 5 | 5 | 3 | 5 | 5 | 3 | 5 | 5 | 3 | 5 |
| 67 | 7 | **P** | 3 | **P** | **P** | 3 | **P** | 7 | 3 | 11 | **P** | 3 |
| 69 | 3 | 37 | 13 | 3 | **P** | 29 | 3 | 11 | **P** | 3 | **P** | 23 |
| 71 | 31 | 3 | **P** | **P** | 3 | 7 | **P** | 3 | 19 | 13 | 3 | **P** |
| 73 | 19 | **P** | 3 | 11 | 7 | 3 | **P** | **P** | 3 | 41 | **P** | 3 |
| 75 | 3 | 5 | 5 | 3 | 5 | 5 | 3 | 5 | 5 | 3 | 5 | 5 |
| 77 | **P** | 3 | 7 | 19 | 3 | **P** | **P** | 3 | 31 | 7 | 3 | **P** |
| 79 | **P** | 7 | 3 | **P** | 23 | 3 | **P** | **P** | 3 | **P** | 43 | 3 |
| 81 | 3 | **P** | **P** | 3 | 41 | 13 | 3 | 7 | **P** | 3 | **P** | **P** |
| 83 | **P** | 3 | **P** | **P** | 3 | **P** | 7 | 3 | **P** | 37 | 3 | **P** |
| 85 | 5 | 5 | 3 | 5 | 5 | 3 | 5 | 5 | 3 | 5 | 5 | 3 |
| 87 | 3 | 19 | **P** | 3 | 7 | **P** | 3 | **P** | **P** | 3 | **P** | 7 |
| 89 | **P** | 3 | **P** | 7 | 3 | **P** | **P** | 3 | **P** | 11 | 3 | **P** |
| 91 | **P** | 13 | 3 | 37 | 19 | 3 | 31 | 11 | 3 | 7 | 29 | 3 |
| 93 | 3 | 7 | **P** | 3 | **P** | 11 | 3 | **P** | 7 | 3 | **P** | **P** |
| 95 | 5 | 3 | 5 | 5 | 3 | 5 | 5 | 3 | 5 | 5 | 3 | 5 |
| 97 | **P** | 11 | 3 | **P** | **P** | 3 | 7 | **P** | 3 | 13 | **P** | 3 |
| 99 | 3 | **P** | **P** | 3 | **P** | 7 | 3 | **P** | **P** | 3 | 11 | **P** |

**Table 2a. Prime Numbers from 2 to 6869**

| | | | | | | | | | | | | |
|---|---|---|---|---|---|---|---|---|---|---|---|---|
| 2 | 347 | 773 | 1259 | 1753 | 2293 | 2803 | 3389 | 3929 | 4517 | 5099 | 5689 | 6277 |
| 3 | 349 | 787 | 1277 | 1759 | 2297 | 2819 | 3391 | 3931 | 4519 | 5101 | 5693 | 6287 |
| 5 | 353 | 797 | 1279 | 1777 | 2309 | 2833 | 3407 | 3943 | 4523 | 5107 | 5701 | 6299 |
| 7 | 359 | 809 | 1283 | 1783 | 2311 | 2837 | 3413 | 3947 | 4547 | 5113 | 5711 | 6301 |
| 11 | 367 | 811 | 1289 | 1787 | 2333 | 2843 | 3433 | 3967 | 4549 | 5119 | 5717 | 6311 |
| 13 | 373 | 821 | 1291 | 1789 | 2339 | 2851 | 3449 | 3989 | 4561 | 5147 | 5737 | 6317 |
| 17 | 379 | 823 | 1297 | 1801 | 2341 | 2857 | 3457 | 4001 | 4567 | 5153 | 5741 | 6323 |
| 19 | 383 | 827 | 1301 | 1811 | 2347 | 2861 | 3461 | 4003 | 4583 | 5167 | 5743 | 6329 |
| 23 | 389 | 829 | 1303 | 1823 | 2351 | 2879 | 3463 | 4007 | 4591 | 5171 | 5749 | 6337 |
| 29 | 397 | 839 | 1307 | 1831 | 2357 | 2887 | 3467 | 4013 | 4597 | 5179 | 5779 | 6343 |
| 31 | 401 | 853 | 1319 | 1847 | 2371 | 2897 | 3469 | 4019 | 4603 | 5189 | 5783 | 6353 |
| 37 | 409 | 857 | 1321 | 1861 | 2377 | 2903 | 3491 | 4021 | 4621 | 5197 | 5791 | 6359 |
| 41 | 419 | 859 | 1327 | 1867 | 2381 | 2909 | 3499 | 4027 | 4637 | 5209 | 5801 | 6361 |
| 43 | 421 | 863 | 1361 | 1871 | 2383 | 2917 | 3511 | 4049 | 4639 | 5227 | 5807 | 6367 |
| 47 | 431 | 877 | 1367 | 1873 | 2389 | 2927 | 3517 | 4051 | 4643 | 5231 | 5813 | 6373 |
| 53 | 433 | 881 | 1373 | 1877 | 2393 | 2939 | 3527 | 4057 | 4649 | 5233 | 5821 | 6379 |
| 59 | 439 | 883 | 1381 | 1879 | 2399 | 2953 | 3529 | 4073 | 4651 | 5237 | 5827 | 6389 |
| 61 | 443 | 887 | 1399 | 1889 | 2411 | 2957 | 3533 | 4079 | 4657 | 5261 | 5839 | 6397 |
| 67 | 449 | 907 | 1409 | 1901 | 2417 | 2963 | 3539 | 4091 | 4663 | 5273 | 5843 | 6421 |
| 71 | 457 | 911 | 1423 | 1907 | 2423 | 2969 | 3541 | 4093 | 4673 | 5279 | 5849 | 6427 |
| 73 | 461 | 919 | 1427 | 1913 | 2437 | 2971 | 3547 | 4099 | 4679 | 5281 | 5851 | 6449 |
| 79 | 463 | 929 | 1429 | 1931 | 2441 | 2999 | 3557 | 4111 | 4691 | 5297 | 5857 | 6451 |
| 83 | 467 | 937 | 1433 | 1933 | 2447 | 3001 | 3559 | 4127 | 4703 | 5303 | 5861 | 6469 |
| 89 | 479 | 941 | 1439 | 1949 | 2459 | 3011 | 3571 | 4129 | 4721 | 5309 | 5867 | 6473 |
| 97 | 487 | 947 | 1447 | 1951 | 2467 | 3019 | 3581 | 4133 | 4723 | 5323 | 5869 | 6481 |
| 101 | 491 | 953 | 1451 | 1973 | 2473 | 3023 | 3583 | 4139 | 4729 | 5333 | 5879 | 6491 |
| 103 | 499 | 967 | 1453 | 1979 | 2477 | 3037 | 3593 | 4153 | 4733 | 5347 | 5881 | 6521 |
| 107 | 503 | 971 | 1459 | 1987 | 2503 | 3041 | 3607 | 4157 | 4751 | 5351 | 5897 | 6529 |
| 109 | 509 | 977 | 1471 | 1993 | 2521 | 3049 | 3613 | 4159 | 4759 | 5381 | 5903 | 6547 |
| 113 | 521 | 983 | 1481 | 1997 | 2531 | 3061 | 3617 | 4177 | 4783 | 5387 | 5923 | 6551 |
| 127 | 523 | 991 | 1483 | 1999 | 2539 | 3067 | 3623 | 4201 | 4787 | 5393 | 5927 | 6553 |
| 131 | 541 | 997 | 1487 | 2003 | 2543 | 3079 | 3631 | 4211 | 4789 | 5399 | 5939 | 6563 |
| 137 | 547 | 1009 | 1489 | 2011 | 2549 | 3083 | 3637 | 4217 | 4793 | 5407 | 5953 | 6569 |
| 139 | 557 | 1013 | 1493 | 2017 | 2551 | 3089 | 3643 | 4219 | 4799 | 5413 | 5981 | 6571 |
| 149 | 563 | 1019 | 1499 | 2027 | 2557 | 3109 | 3659 | 4229 | 4801 | 5417 | 5987 | 6577 |
| 151 | 569 | 1021 | 1511 | 2029 | 2579 | 3119 | 3671 | 4231 | 4813 | 5419 | 6007 | 6581 |
| 157 | 571 | 1031 | 1523 | 2039 | 2591 | 3121 | 3673 | 4241 | 4817 | 5431 | 6011 | 6599 |
| 163 | 577 | 1033 | 1531 | 2053 | 2593 | 3137 | 3677 | 4243 | 4831 | 5437 | 6029 | 6607 |
| 167 | 587 | 1039 | 1543 | 2063 | 2609 | 3163 | 3691 | 4253 | 4861 | 5441 | 6037 | 6619 |
| 173 | 593 | 1049 | 1549 | 2069 | 2617 | 3167 | 3697 | 4259 | 4871 | 5443 | 6043 | 6637 |
| 179 | 599 | 1051 | 1553 | 2081 | 2621 | 3169 | 3701 | 4261 | 4877 | 5449 | 6047 | 6653 |
| 181 | 601 | 1061 | 1559 | 2083 | 2633 | 3181 | 3709 | 4271 | 4889 | 5471 | 6053 | 6659 |
| 191 | 607 | 1063 | 1567 | 2087 | 2647 | 3187 | 3719 | 4273 | 4903 | 5477 | 6067 | 6661 |
| 193 | 613 | 1069 | 1571 | 2089 | 2657 | 3191 | 3727 | 4283 | 4909 | 5479 | 6073 | 6673 |
| 197 | 617 | 1087 | 1579 | 2099 | 2659 | 3203 | 3733 | 4289 | 4919 | 5483 | 6079 | 6679 |
| 199 | 619 | 1091 | 1583 | 2111 | 2663 | 3209 | 3739 | 4297 | 4931 | 5501 | 6089 | 6689 |
| 211 | 631 | 1093 | 1597 | 2113 | 2671 | 3217 | 3761 | 4327 | 4933 | 5503 | 6091 | 6691 |
| 223 | 641 | 1097 | 1601 | 2129 | 2677 | 3221 | 3767 | 4337 | 4937 | 5507 | 6101 | 6701 |
| 227 | 643 | 1103 | 1607 | 2131 | 2683 | 3229 | 3769 | 4339 | 4943 | 5519 | 6113 | 6703 |
| 229 | 647 | 1109 | 1609 | 2137 | 2687 | 3251 | 3779 | 4349 | 4951 | 5521 | 6121 | 6709 |
| 233 | 653 | 1117 | 1613 | 2141 | 2689 | 3253 | 3793 | 4357 | 4957 | 5527 | 6131 | 6719 |
| 239 | 659 | 1123 | 1619 | 2143 | 2693 | 3257 | 3797 | 4363 | 4967 | 5531 | 6133 | 6733 |
| 241 | 661 | 1129 | 1621 | 2153 | 2699 | 3259 | 3803 | 4373 | 4969 | 5557 | 6143 | 6737 |
| 251 | 673 | 1151 | 1627 | 2161 | 2707 | 3271 | 3821 | 4391 | 4973 | 5563 | 6151 | 6761 |
| 257 | 677 | 1153 | 1637 | 2179 | 2711 | 3299 | 3823 | 4397 | 4987 | 5569 | 6163 | 6763 |
| 263 | 683 | 1163 | 1657 | 2203 | 2713 | 3301 | 3833 | 4409 | 4993 | 5573 | 6173 | 6779 |
| 269 | 691 | 1171 | 1663 | 2207 | 2719 | 3307 | 3847 | 4421 | 4999 | 5581 | 6197 | 6781 |
| 271 | 701 | 1181 | 1667 | 2213 | 2729 | 3313 | 3851 | 4423 | 5003 | 5591 | 6199 | 6791 |
| 277 | 709 | 1187 | 1669 | 2221 | 2731 | 3319 | 3853 | 4441 | 5009 | 5623 | 6203 | 6793 |
| 281 | 719 | 1193 | 1693 | 2237 | 2741 | 3323 | 3863 | 4447 | 5011 | 5639 | 6211 | 6803 |
| 283 | 727 | 1201 | 1697 | 2239 | 2749 | 3329 | 3877 | 4451 | 5021 | 5641 | 6217 | 6823 |
| 293 | 733 | 1213 | 1699 | 2243 | 2753 | 3331 | 3881 | 4457 | 5023 | 5647 | 6221 | 6827 |
| 307 | 739 | 1217 | 1709 | 2251 | 2767 | 3343 | 3889 | 4463 | 5039 | 5651 | 6229 | 6829 |
| 311 | 743 | 1223 | 1721 | 2267 | 2777 | 3347 | 3907 | 4481 | 5051 | 5653 | 6247 | 6833 |
| 313 | 751 | 1229 | 1723 | 2269 | 2789 | 3359 | 3911 | 4483 | 5059 | 5657 | 6257 | 6841 |
| 317 | 757 | 1231 | 1733 | 2273 | 2791 | 3361 | 3917 | 4493 | 5077 | 5659 | 6263 | 6857 |
| 331 | 761 | 1237 | 1741 | 2281 | 2797 | 3371 | 3919 | 4507 | 5081 | 5669 | 6269 | 6863 |
| 337 | 769 | 1249 | 1747 | 2287 | 2801 | 3373 | 3923 | 4513 | 5087 | 5683 | 6271 | 6869 |

## Table 2b. Prime Numbers from 6871 to 15277

| | | | | | | | | | | | | |
|---|---|---|---|---|---|---|---|---|---|---|---|---|
| 6871 | 7523 | 8117 | 8737 | 9343 | 9931 | 10601 | 11257 | 11927 | 12539 | 13159 | 13807 | 14533 |
| 6883 | 7529 | 8123 | 8741 | 9349 | 9941 | 10607 | 11261 | 11933 | 12541 | 13163 | 13829 | 14537 |
| 6899 | 7537 | 8147 | 8747 | 9371 | 9949 | 10613 | 11273 | 11939 | 12547 | 13171 | 13831 | 14543 |
| 6907 | 7541 | 8161 | 8753 | 9377 | 9967 | 10627 | 11279 | 11941 | 12553 | 13177 | 13841 | 14549 |
| 6911 | 7547 | 8167 | 8761 | 9391 | 9973 | 10631 | 11287 | 11953 | 12569 | 13183 | 13859 | 14551 |
| 6917 | 7549 | 8171 | 8779 | 9397 | 10007 | 10639 | 11299 | 11959 | 12577 | 13187 | 13873 | 14557 |
| 6947 | 7559 | 8179 | 8783 | 9403 | 10009 | 10651 | 11311 | 11969 | 12583 | 13217 | 13877 | 14561 |
| 6949 | 7561 | 8191 | 8803 | 9413 | 10037 | 10657 | 11317 | 11971 | 12589 | 13219 | 13879 | 14563 |
| 6959 | 7573 | 8209 | 8807 | 9419 | 10039 | 10663 | 11321 | 11981 | 12601 | 13229 | 13883 | 14591 |
| 6961 | 7577 | 8219 | 8819 | 9421 | 10061 | 10667 | 11329 | 11987 | 12611 | 13241 | 13901 | 14593 |
| 6967 | 7583 | 8221 | 8821 | 9431 | 10067 | 10687 | 11351 | 12007 | 12613 | 13249 | 13903 | 14621 |
| 6971 | 7589 | 8231 | 8831 | 9433 | 10069 | 10691 | 11353 | 12011 | 12619 | 13259 | 13907 | 14627 |
| 6977 | 7591 | 8233 | 8837 | 9437 | 10079 | 10709 | 11369 | 12037 | 12637 | 13267 | 13913 | 14629 |
| 6983 | 7603 | 8237 | 8839 | 9439 | 10091 | 10711 | 11383 | 12041 | 12641 | 13291 | 13921 | 14633 |
| 6991 | 7607 | 8243 | 8849 | 9461 | 10093 | 10723 | 11393 | 12043 | 12647 | 13297 | 13931 | 14639 |
| 6997 | 7621 | 8263 | 8861 | 9463 | 10099 | 10729 | 11399 | 12049 | 12653 | 13309 | 13933 | 14653 |
| 7001 | 7639 | 8269 | 8863 | 9467 | 10103 | 10733 | 11411 | 12071 | 12659 | 13313 | 13963 | 14657 |
| 7013 | 7643 | 8273 | 8867 | 9473 | 10111 | 10739 | 11423 | 12073 | 12671 | 13327 | 13967 | 14669 |
| 7019 | 7649 | 8287 | 8887 | 9479 | 10133 | 10753 | 11437 | 12097 | 12689 | 13331 | 13997 | 14683 |
| 7027 | 7669 | 8291 | 8893 | 9491 | 10139 | 10771 | 11443 | 12101 | 12697 | 13337 | 13999 | 14699 |
| 7039 | 7673 | 8293 | 8923 | 9497 | 10141 | 10781 | 11447 | 12107 | 12703 | 13339 | 14009 | 14713 |
| 7043 | 7681 | 8297 | 8929 | 9511 | 10151 | 10789 | 11467 | 12109 | 12713 | 13367 | 14011 | 14717 |
| 7057 | 7687 | 8311 | 8933 | 9521 | 10159 | 10799 | 11471 | 12113 | 12721 | 13381 | 14029 | 14723 |
| 7069 | 7691 | 8317 | 8941 | 9533 | 10163 | 10831 | 11483 | 12119 | 12739 | 13397 | 14033 | 14731 |
| 7079 | 7699 | 8329 | 8951 | 9539 | 10169 | 10837 | 11489 | 12143 | 12743 | 13399 | 14051 | 14737 |
| 7103 | 7703 | 8353 | 8963 | 9547 | 10177 | 10847 | 11491 | 12149 | 12757 | 13411 | 14057 | 14741 |
| 7109 | 7717 | 8363 | 8969 | 9551 | 10181 | 10853 | 11497 | 12157 | 12763 | 13417 | 14071 | 14747 |
| 7121 | 7723 | 8369 | 8971 | 9587 | 10193 | 10859 | 11503 | 12161 | 12781 | 13421 | 14081 | 14753 |
| 7127 | 7727 | 8377 | 8999 | 9601 | 10211 | 10861 | 11519 | 12163 | 12791 | 13441 | 14083 | 14759 |
| 7129 | 7741 | 8387 | 9001 | 9613 | 10223 | 10867 | 11527 | 12197 | 12799 | 13451 | 14087 | 14767 |
| 7151 | 7753 | 8389 | 9007 | 9619 | 10243 | 10883 | 11549 | 12203 | 12809 | 13457 | 14107 | 14771 |
| 7159 | 7757 | 8419 | 9011 | 9623 | 10247 | 10889 | 11551 | 12211 | 12821 | 13463 | 14143 | 14779 |
| 7177 | 7759 | 8423 | 9013 | 9629 | 10253 | 10891 | 11579 | 12227 | 12823 | 13469 | 14149 | 14783 |
| 7187 | 7789 | 8429 | 9029 | 9631 | 10259 | 10903 | 11587 | 12239 | 12829 | 13477 | 14153 | 14797 |
| 7193 | 7793 | 8431 | 9041 | 9643 | 10267 | 10909 | 11593 | 12241 | 12841 | 13487 | 14159 | 14813 |
| 7207 | 7817 | 8443 | 9043 | 9649 | 10271 | 10937 | 11597 | 12251 | 12853 | 13499 | 14173 | 14821 |
| 7211 | 7823 | 8447 | 9049 | 9661 | 10273 | 10939 | 11617 | 12253 | 12889 | 13513 | 14177 | 14827 |
| 7213 | 7829 | 8461 | 9059 | 9677 | 10289 | 10949 | 11621 | 12263 | 12893 | 13523 | 14197 | 14831 |
| 7219 | 7841 | 8467 | 9067 | 9679 | 10301 | 10957 | 11633 | 12269 | 12899 | 13537 | 14207 | 14843 |
| 7229 | 7853 | 8501 | 9091 | 9689 | 10303 | 10973 | 11657 | 12277 | 12907 | 13553 | 14221 | 14851 |
| 7237 | 7867 | 8513 | 9103 | 9697 | 10313 | 10979 | 11677 | 12281 | 12911 | 13567 | 14243 | 14867 |
| 7243 | 7873 | 8521 | 9109 | 9719 | 10321 | 10987 | 11681 | 12289 | 12917 | 13577 | 14249 | 14869 |
| 7247 | 7877 | 8527 | 9127 | 9721 | 10331 | 10993 | 11689 | 12301 | 12919 | 13591 | 14251 | 14879 |
| 7253 | 7879 | 8537 | 9133 | 9733 | 10333 | 11003 | 11699 | 12323 | 12923 | 13597 | 14281 | 14887 |
| 7283 | 7883 | 8539 | 9137 | 9739 | 10337 | 11027 | 11701 | 12329 | 12941 | 13613 | 14293 | 14891 |
| 7297 | 7901 | 8543 | 9151 | 9743 | 10343 | 11047 | 11717 | 12343 | 12953 | 13619 | 14303 | 14897 |
| 7307 | 7907 | 8563 | 9157 | 9749 | 10357 | 11057 | 11719 | 12347 | 12959 | 13627 | 14321 | 14923 |
| 7309 | 7919 | 8573 | 9161 | 9767 | 10369 | 11059 | 11731 | 12373 | 12967 | 13633 | 14323 | 14929 |
| 7321 | 7927 | 8581 | 9173 | 9769 | 10391 | 11069 | 11743 | 12377 | 12973 | 13649 | 14327 | 14939 |
| 7331 | 7933 | 8597 | 9181 | 9781 | 10399 | 11071 | 11777 | 12379 | 12979 | 13669 | 14341 | 15121 |
| 7333 | 7937 | 8599 | 9187 | 9787 | 10427 | 11083 | 11779 | 12391 | 12983 | 13679 | 14347 | 15131 |
| 7349 | 7949 | 8609 | 9199 | 9791 | 10429 | 11087 | 11783 | 12401 | 13001 | 13681 | 14369 | 15137 |
| 7351 | 7951 | 8623 | 9203 | 9803 | 10433 | 11093 | 11789 | 12409 | 13003 | 13687 | 14387 | 15139 |
| 7369 | 7963 | 8627 | 9209 | 9811 | 10453 | 11113 | 11801 | 12413 | 13007 | 13691 | 14389 | 15149 |
| 7393 | 7993 | 8629 | 9221 | 9817 | 10457 | 11117 | 11807 | 12421 | 13009 | 13693 | 14401 | 15161 |
| 7411 | 8009 | 8641 | 9227 | 9829 | 10459 | 11119 | 11813 | 12433 | 13033 | 13697 | 14407 | 15173 |
| 7417 | 8011 | 8647 | 9239 | 9833 | 10463 | 11131 | 11821 | 12437 | 13037 | 13709 | 14411 | 15187 |
| 7433 | 8017 | 8663 | 9241 | 9839 | 10477 | 11149 | 11827 | 12451 | 13043 | 13711 | 14419 | 15193 |
| 7451 | 8039 | 8669 | 9257 | 9851 | 10487 | 11159 | 11831 | 12457 | 13049 | 13721 | 14423 | 15199 |
| 7457 | 8053 | 8677 | 9277 | 9857 | 10499 | 11161 | 11833 | 12473 | 13063 | 13723 | 14431 | 15217 |
| 7459 | 8059 | 8681 | 9281 | 9859 | 10501 | 11171 | 11839 | 12479 | 13093 | 13729 | 14437 | 15227 |
| 7477 | 8069 | 8689 | 9283 | 9871 | 10513 | 11173 | 11863 | 12487 | 13099 | 13751 | 14447 | 15233 |
| 7481 | 8081 | 8693 | 9293 | 9883 | 10529 | 11177 | 11867 | 12491 | 13103 | 13757 | 14449 | 15241 |
| 7487 | 8087 | 8699 | 9311 | 9887 | 10531 | 11197 | 11887 | 12497 | 13109 | 13759 | 14461 | 15259 |
| 7489 | 8089 | 8707 | 9319 | 9901 | 10559 | 11213 | 11897 | 12503 | 13121 | 13763 | 14479 | 15263 |
| 7499 | 8093 | 8713 | 9323 | 9907 | 10567 | 11239 | 11903 | 12511 | 13127 | 13781 | 14489 | 15269 |
| 7507 | 8101 | 8719 | 9337 | 9923 | 10589 | 11243 | 11909 | 12517 | 13147 | 13789 | 14503 | 15271 |
| 7517 | 8111 | 8731 | 9341 | 9929 | 10597 | 11251 | 11923 | 12527 | 13151 | 13799 | 14519 | 15277 |

## ALGEBRA

In engineering, manufacturing, and industrial applications, physical laws govern the behavior of all quantities. Algebraic formulas (equations) are the models for these laws. They usually consist of algebraic expressions, the most common being polynomials, rational expressions, and radicals. Most of the formulas used in this *Handbook* contain one or more of these. This section gives a foundation for understanding the algebra indispensable to solving equations.

**Definitions.**—The vocabulary of algebra extends to all mathematics. The essential definitions are given here.

*Operation:* Addition, subtraction, multiplication, division, root-taking, raising to a power, taking a logarithm.

*Inverse Operation:* An operation that reverses another operation. Addition and subtraction are inverse operations, as are multiplication and division. Taking the *n*th root is the inverse of raising a number to a power. Finding an antilogarithm is the inverse of finding a logarithm.

*Constant:* A known quantity, either a number standing alone or a letter that is assumed to be given or known in an application. In $5x + 14$, 14 is the constant. Usually, the letters *a*, *b*, and *c* are used to represent constants, as in the linear equation $ax + by = c$. Note: *e* and π are commonly seen constants.

*Variable:* An unknown quantity, represented by a letter such as $n, x, y, x, t$. Note: *e* and π are not variables.

*Exponent:* The power to which a variable or number is raised.

*Monomial:* A single variable or number or a product of numbers and variables. Examples: $5, x, -4y^2, 12xy^2z^3$. Exponents in monomials are whole numbers, 0, 1, 2, 3, . . . , so $x^{-1} = 1/x$ and $x^{1/2} = \sqrt{x}$ are not monomials.

*Coefficient:* The numerical factor in a term. Examples (in bold): **5**$x$, **16**$n$, **−2**$r$. The coefficient of a variable standing alone is understood to be 1, for example, $x = 1x$; the coefficient of $-x$ is $-1$.

*Term:* Monomials are terms, but so are expressions that are not monomials: $1/x$, $\sqrt{x}$, $8x^{1/3}$, $\log x$, and so on.

*Like Terms:* This usually refers to monomials with the same variable and exponent, and having any real number coefficients, such as $x$ and $7x$; $2n^2$ and $n^2/4$; $2rst/5$ and $14rst$, and so on. Any constant *a* can be written as $ax^0$, so all constants are like terms. But $x^{1/2}$ and $4x^{1/2}$ also are like terms.

*Expression:* Numbers and variables with *operators* (addition: +, multiplication: × or ·, √ etc.).

*Equation:* Two expressions set equal to one another with the equal sign = .

*Examples:* $5x = x^2 - 6$; $\sqrt{3x} = 14$. Solving equations for the unknown is the foundation of algebra.

*Inequality:* Two expressions set against one another by $>, <, \geq, \leq$, or $\neq$.

**Evaluating Algebraic Expressions.**—An expression is *evaluated* by substituting given values of the variable. For example, $x^2 - 2x + 7$ evaluated at $x = -3$ is $(-3)^2 - 2(-3) + 7 = 9 + 6 + 7 = 22$.

Another example, $\sqrt{1 - x^2}$ evaluated at $x = \frac{1}{3}$ is $\sqrt{1 - (\frac{1}{3})^2} = \sqrt{1 - \frac{1}{9}} = \sqrt{\frac{8}{9}}$.

**Combining Like Terms.**—Like terms are added and subtracted by combining their coefficients and leaving the rest of the term as is. For example:

$$3x + x - 7x = -3x \qquad 2n^2 + \frac{n^2}{4} - n^2 = \frac{5n^2}{4} \qquad 2rst - 14rst + rst = -11rst$$

**Solving an Equation for an Unknown.—**Solving an equation for an unknown (say, $x$) requires isolating $x$ from the other terms. This is accomplished by applying techniques of *inverse operations*, *combining like terms*, and *factoring*, as explained throughout *ALGEBRA, GEOMETRY*, and *TRIGONOMETRY: SOLUTION OF TRIANGLES*.

The simplest equations contain only one or two variables, such as linear equations and other polynomials in two dimensions. Others may contain many variables, such as formulas in physics and engineering. A *formula* is an equation that determines a physical quantity based on other known quantities. An example is area of a rectangle, $A = lw$, where $l$ is length and $w$ is width in like units. Another is horsepower transmitted by belting, $P = SVW/33{,}000$, where $S$, $V$, and $W$ are, respectively, working stress of the belt, belt velocity, and belt width, all in appropriate units.

The following examples give an overview of equation solving with general terms $A$, $B$, $C$, and $D$. The same processes apply for solving polynomial and other algebraic equations for an unknown.

*Solving by Adding or Subtracting*:

Given: $B - C = A - D$

To solve for $A$: $B - C + D = A$, that is, $A = B - C + D$

To solve for $B$: $B = A - D + C$

To solve for $C$: $B = A - D + C \rightarrow B - A + D = C$, that is, $C = B - A + D$

To solve for $D$: $B - C + D = A \rightarrow D = A - B + C$

In the last two, $C$ and $D$ are moved first so they do not have a minus sign before them in the answer. But *keeping in mind that subtracting a term is the same as adding its negative* (see *Integers (Signed Numbers)* on page 3), another way to solve for $C$ or $D$ is given below. The subtraction "minus $C$" is treated as "negative $C$." Then, multiplying through by –1, all signs change but equality is maintained, leaving $C$:

Solve for $C$: $B - C = A - D \rightarrow -C = A - D - B \rightarrow (-1)(-C) = (-1)(A - D - B) \rightarrow C = -A + D + B$

Solve for $D$: $B - C = A - D \rightarrow B - C - A = -D \rightarrow (-1)(B - C - A) = (-1)(-D) \rightarrow -B + C + A = D$, or $D = A - B + C$

Changing the sign of each term by multiplying through by –1 is a common technique.

*Solving by Multiplying or Dividing:* A variable that is a factor or a divisor within a term is isolated using division (*multiplying by the reciprocal*) or multiplication (*cross-multiplying*).

*Example*: In the equation $\frac{AB}{C} = D$, each variable is isolated using multiplication.

To isolate $A$, multiply each side by the reciprocal of $\frac{B}{C}$:

$$\frac{C}{B} \times \frac{AB}{C} = \frac{C}{B} \times D \rightarrow A = \frac{C}{B} \times \frac{D}{1} = \frac{CD}{B}, \text{ so } A = \frac{CD}{B}$$

To isolate $B$, multiply each side by the reciprocal of $\frac{A}{C}$:

$$\frac{C}{A} \times \frac{AB}{C} = \frac{C}{A} \times D \rightarrow B = \frac{C}{A} \times \frac{D}{1} = \frac{CD}{A}, \text{ so } B = \frac{CD}{A}$$

To isolate $C$, multiply each side by $C$: $C \times \frac{AB}{C} = C \times D \rightarrow AB = CD$

Then divide by $D$ (multiply by the reciprocal of $D$):

$$\frac{AB}{1} \times \frac{1}{D} = \frac{CD}{1} \times \frac{1}{D} \rightarrow \frac{AB}{D} = C, \quad \text{so } C = \frac{AB}{D}$$

Equations with more complexity require more steps to solve for an unknown.

*Examples:*

Solve $\frac{A+B}{C} = D$ for $A$:

| | | |
|---|---|---|
| Multiply both sides by $C$ | $\frac{A+B}{C} \cdot C = D \cdot C \rightarrow A+B = DC$ | |
| Subtract $B$ from both sides | $A+B-B = DC-B \rightarrow A = DC-B$ or $CD-B$ | |

Solve $\frac{A+B}{C} = B-D$ for $D$:

| | | |
|---|---|---|
| Add $D$ to both sides | $\frac{A+B}{C} + D = B$ | |
| Subtract $\frac{A+B}{C}$ from both sides | $D = B - \frac{A+B}{C}$ | |

Solve $\frac{A+B}{C} = B-A$ for $A$:

| | | |
|---|---|---|
| Add $A$ to both sides | $\frac{A+B}{C} + A = B$ | |
| Multiply both sides by $C$ | $C\left(\frac{A+B}{C} + A\right) = CB$ | |
| Distribute $C$ | $A+B+CA = CB$ | Like terms $A$ and $CA$ are isolated from $B$ |
| Subtract $B$ from both sides | $A+CA = CB-B$ | Since $A$ and $CA$ cannot be combined, $A$ must be factored out |
| Factor out $A$ | $A(1+C) = CB-B$ | |
| Divide both sides by $1+C$ | $A = \frac{CB-B}{1+C}$ | |

Solving an equation for an unknown is the basic technique for working with formulas.

## Rearrangement and Transposition of Terms in Formulas

A *formula* is a rule for a calculation expressed by using letters and signs, instead of writing out the rule in words. By this means, it is possible to condense, in a small space, the essentials of long and cumbersome rules.

As an example, the formula for horsepower transmitted by belting may be written

$$P = SVW/33000$$

where $P$ = horsepower transmitted; $S$ = working stress of belt per inch of width in pounds; $V$ = velocity of belt in feet per minute; $W$ = width of belt in inches; and, 33,000 = a constant that is part of the formula for horsepower with units of hp/(lb-in$^2$/min).

If the working stress $S$, velocity $V$, and width $W$ are known, horsepower can be found directly from this formula by inserting the given values. For example, if $S = 33$, $V = 600$, and $W = 5$. Then,

$$P = 33 \times 600 \times 5/33000 = 3 \text{ hp}$$

Assume that horsepower $P$, stress $S$, and velocity $V$ are known, and that the width of belt $W$ is to be found. The formula must then be rearranged so that the symbol $W$ will be alone on one side of the equation. This is accomplished by isolating $W$ by moving the other variables and the number to the other side of the equation:

From $P = \dfrac{SVW}{33,000}$, multiply both sides by 33,000: $33,000P = SVW$

Divide both sides by $SV$: $\dfrac{33,000P}{SV} = W$, or $W = \dfrac{33,000P}{SV}$

## Algebraic Operations

Algebraic operations rely on the symbols and properties of real numbers, reviewed below from *ARITHMETIC* and used throughout this section. Key skills include exponent use (for both powers and roots), polynomial operations, and solving equations for a variable (unknown). The same skills carry over to areas of geometry, trigonometry, and calculus, where a strong foundation in algebra is essential.

### Properties of Monomials and Exponents

$$a + \cdots + a = na \quad (n \text{ terms}) \qquad a + 0 = a \qquad a + (-a) = 0 \qquad a \times b = ab$$

$$a \times a \times \cdots \times a = a^n \quad (n \text{ factors}) \qquad a^n a^m = a^{(n+m)} \qquad \frac{a^n}{a^m} = a^{(n-m)} \qquad (a^m)^n = a^{mn}$$

$$a^{m/n} = (a^{1/n})^m \qquad a^{-n} = \frac{1}{a^n} \qquad \frac{1}{a^{-n}} = a^n \qquad a^0 = 1 \ (\text{for } a \neq 0)$$

*Examples:*

$$a + a = 2a \qquad -3a - a = -4a \qquad 4a \times 3a = 12aa = 12a^2$$

$$a^2 a^3 = a^{2+3} = a^5 \qquad (a^2)^3 = a^{2 \times 3} = a^6$$

$$a^{-3} = \frac{1}{a^3} = \left(\frac{1}{a}\right)^3 \qquad (ab)^2 = a^2 b^2 \qquad \left(\frac{a}{b}\right)^3 = \frac{a^3}{b^3} \qquad \frac{a^4}{a^3} = a^{4-3} = a$$

### Properties of Radicals

$$\sqrt{a} \times \sqrt{a} = a \qquad \sqrt{a} \times \sqrt{b} = \sqrt{ab} \qquad \frac{\sqrt{a}}{\sqrt{b}} = \sqrt{\frac{a}{b}}$$

$$(\sqrt[n]{a})^n = a \qquad \sqrt[n]{a^m} = (\sqrt[n]{a})^m = a^{m/n}, \text{ unless } a < 0,\ m \text{ and } n \text{ are both even}$$

$$\sqrt[n]{\frac{1}{a}} = \frac{1}{\sqrt[n]{a}} = a^{-1/n} \qquad (\sqrt{a} + \sqrt{b})^2 = (\sqrt{a} + \sqrt{b})(\sqrt{a} + \sqrt{b}) = a + 2\sqrt{ab} + b$$

*Examples:*

$$\sqrt[3]{a} \times \sqrt[3]{a} \times \sqrt[3]{a} = (\sqrt[3]{a})^3 = a \qquad \sqrt[3]{ab} = \sqrt[3]{a} \times \sqrt[3]{b} \qquad \sqrt[3]{\frac{a}{b}} = \frac{\sqrt[3]{a}}{\sqrt[3]{b}}$$

$$\sqrt[3]{a^2} = (\sqrt[3]{a})^2 = a^{2/3} \qquad \sqrt[3]{\frac{1}{a}} = \frac{1}{\sqrt[3]{a}} = \frac{1}{a^{1/3}} = a^{-1/3} \qquad \sqrt[4]{\sqrt[3]{a}} = \sqrt[4 \times 3]{a} = \sqrt[3]{\sqrt[4]{a}} \qquad (a^{1/4})^{1/3}$$

$$\sqrt{x} + 9\sqrt{x} - 4\sqrt{x} = (1 + 9 - 4)\sqrt{x} = 6\sqrt{x} \qquad \sqrt{x^2} = x \qquad \sqrt{x^3} = \sqrt{x^2 x} = \sqrt{x^2}\sqrt{x} = x\sqrt{x}$$

$$\sqrt{100x} = \sqrt{100}\sqrt{x} = 10\sqrt{x} \qquad \sqrt[5]{x^3} = \left(\sqrt[5]{x}\right)^3 = x^{5/3} \qquad \sqrt{20x^3} = \sqrt{4.5.x^2.x} = \sqrt{4x^2}\sqrt{5x} = 2x\sqrt{5x}$$

$$4\sqrt{5x}\left(\sqrt{x} + 3x\right) = \sqrt{5x \cdot x} + 12x\sqrt{5x} = 4\sqrt{5x^2} + 12x\sqrt{5x} = 4x\sqrt{5} + 12x\sqrt{5x}$$

$$\sqrt{\frac{36}{x^4}} = \frac{\sqrt{36}}{\sqrt{x^4}} = \frac{6}{x^2} \qquad \sqrt{x+1} \text{ cannot be simplified}$$

## Polynomials

Polynomials in a single variable are expressions of the form $a_n x^n + a_{n-1} x^{n-1} + a_{n-2} x^{n-2} + \cdots + a_1 x^1 + a_0$, where *n is a non-negative integer*, and the coefficients $a_n, a_{n-1}, \ldots, a_0$ are real numbers (the subscripts are simply labels that correspond to the variables ($x^n$, $x^{n-1}$, . . .). $n$ is the *degree* (or *order*) of the polynomial; $a_n$ is the *leading coefficient*; $a_0$ is the *constant coefficient*. A *first-degree polynomial* has $n = 1$; a *second-degree* polynomial has $n = 2$, and so on.

*Example*: $\frac{x}{3} + \frac{1}{6}$ is degree 1, $a_1 = \frac{1}{3}$ is the leading coefficient, $a_0 = \frac{1}{6}$ is the constant coefficient.

*Example*: $4x^2 - 7$ is degree 2, $a_2 = 4$ is the leading coefficient, $a_0 = -7$ is the constant coefficient.

*Example*: $x^3 + 5x^2 - x + \frac{1}{3}$ is degree 3, and the coefficients are $a_3 = 1, a_2 = 5, a_1 = -1$, and $a_0 = \frac{1}{3}$.

A *monomial* is a single-term polynomial. For example, $8x^5$ is degree 5 with coefficient 8. A constant is a polynomial, by the definition. Its degree is 0. For example, 19 is a 0-degree polynomial, since it can be written as $19x^0$. A *binomial* is a polynomial with two terms, such as $x + 9$ and $5 - x^2$. A *trinomial* has three terms, such as, $x^2 + x - 6$.

Finally, second-degree polynomials are called *quadratic* polynomials. They are important because they model so many processes in engineering and other technical and scientific fields.

**Operations on Polynomials.**—Polynomials can be added, subtracted, or multiplied ("expanded"), with the result being another polynomial. If polynomials are divided, the result is a *rational expression*. Polynomials also may be *factored*. That is, they may be written as a product of lower-degree polynomials.

*Combining (Adding and Subtracting) Polynomials:* Two or more polynomials are added by combining like terms. For example:

$$\left(x^4 + 6x^3 - 3x^2 - 5x - 11\right) + \left(x^3 - 12x^2 + x + 28\right) = x^4 + 7x^3 - 15x^2 - 4x + 17$$

*Multiplying (Expanding) Polynomials:* Taking the product of two or more polynomials relies on the distributive property of multiplication over addition (or subtraction): $a(b + c) = ab + ac$, where the rules of exponents are followed (see *Properties of Monomials and Exponents* on page 22).

*Examples of simple distributive case*:

$$2x\left(8x^4 + 3x^2 - 6\right) = 2 \cdot 8x^{1+4} + 2 \cdot 3x^{1+2} - 2 \cdot 6x = 16x^5 + 6x^3 - 12x$$

*Note*: This technique also is used for multiplying other algebraic expressions, such as radical (root) and rational expressions. Such terms are not polynomials, since their exponents are other than non-negative integers.

*Examples of Distributive Property for Root and Rational Expressions*:

$$2x^{1/2}\left(7x + 3\right) = 14x^{1/2+1} + 6x^{1/2} = 14x^{3/2} + 6x^{1/2}$$

$$\frac{1}{x}\left(x^3 - 7x^2 + 3x\right) = x^{-1+3} - 7x^{-1+2} + 3x^{-1+1} = x^2 - 7x^1 + 3x^0 = x^2 - 7x + 3$$

"FOIL" is a version of the distributive property in which two binomials are multiplied. FOIL stands for **F**irst **O**uter **I**nner **L**ast, the order in which terms are multiplied. The first is $ac$; outer is $ad$, inner is $bc$, and last is $bd$.

*FOIL multiplication:* $(a + b)(c + d) = ac + ad + bc + bd$

*Examples*:

$(x-6)(x+1)=x\cdot x+x\cdot 1-6\cdot x-6\cdot 1=x^2+1x-6x-6=x^2-5x-6$

$(3x^2+2x)(4x^3-5)=3x^2\cdot 4x^3-5\cdot 3x^2+2x\cdot 4x^3+2x\cdot(-5)=12x^5-15x^2+8x^4-10x$

$(x-1)(x+1)=x\cdot x+x\cdot 1-1\cdot x-1\cdot 1=x^2+1x-1x-1=x^2-1$ (the middle terms drop out)

$(x+1)^2=(x+1)(x+1)=x^2+x+x+1=x^2+2x+1$

In the first and third examples, the products of the outer and inner terms are like terms, so they are combined. The third example is a *special product* (in this case, the difference of squares). The fourth is an example of raising a polynomial to an exponent.

*Raising a Polynomial to an Exponent:* Consider first $(a+b)^2$, which is $(a+b)(a+b)$, and this expands by FOIL to $a^2+ab+ab+b^2=a^2+2ab+b^2$. In general, an expansion like $(a+b)^2, (a-b)^2, (a+b)^3, (a-b)^3$ is called a *binomial expansion*, $(a+b)^n$. For example:

$$(a+b)^n=\underbrace{(a+b)(a+b)(a+b)\cdots(a+b)}_{n\text{ factors}}$$

*Caution*: A common but *serious* mistake is distributing the exponent to each term in the parentheses:

$$(a+b)^2=(a+b)(a+b)=a^2+2ab+b^2 \quad NOT \quad a^2+b^2$$

**Factoring Polynomials.—**The result of two or more polynomials being multiplied is a higher-degree polynomial. *Factoring a polynomial* breaks it into its lesser-degree factors. This section explains how to factor a polynomial, a skill needed for solving equations and graphing functions (see *Graphs of Functions* on page 30). The categories are: dividing by a *common factor* (i.e., distributive property in reverse); factoring by *reverse FOIL* (with leading coefficient of 1 and otherwise); factoring *special products*.

*Common Factors:* Each term is divided by the *greatest common factor* (GCF) and written as shown: $ab+ac=a(b+c)$

*Examples of GCF factoring*:

$6x^3+2x^2-10x=2x(3x^2+x-5)$ GCF

$mn+m^2n-mn^2=mn(1+m-n)$ GCF

$-7abc-21bc^2+4bc=-bc(7a+21c-4)$ GCF

*Note*: The variable in the GCF is the one with the lowest exponent in common to all terms. If two terms have a GCF, but the others do not, then that factor cannot be pulled out. And, if the leading coefficient is negative (third example), it is customary to factor out –1 in the GCF redundant. This is so the expression in parentheses has a positive leading coefficient, which makes it easier to factor further.

*Reverse FOIL of Form $x^2+bx+c$*: The basic technique is demonstrated for a second-degree trinomial with a leading coefficient of 1. The task is to factor $x^2+bx+c$ as $(x+\square)(x+\square)$, using only integers $\{\ldots,-2,-1,0,1,2,\ldots\}$ in the boxes.

*Example*: Consider $x^2+4x+3$. To factor it as $(x+\square)(x+\square)$, the integers must have a product of +3 (the last term of $x^2+4x+3$) and a sum of +4 (the middle term). 1 and 3 are correct, as verified by FOIL:

$$x^2+4x+3=(x+1)(x+3).$$

As long as the leading coefficient is 1, reverse FOIL works. If one or both operations in the trinomial are negative, the process is the same, but some trial and error may be needed, as shown in the examples below.

*Example*: Three similar trinomials, only two are factorable:

$x^2 + 2x - 3 = (x + 3)(x - 1)$, since $(3)(-1) = -3$ and $3 + -1 = 2$

$x^2 - 2x - 3 = (x - 3)(x + 1)$, since $(-3)(1) = -3$ and $-3 + 1 = -2$

$x^2 - 2x + 3$ Does not factor over the real numbers, since no two integers have a product of $+3$ and a sum of $-2$

*Examples*: Factoring by reverse FOIL, along with the explanation:

$x^2 - 8x - 20 = (x - 10)(x + 2)$ The two integers whose product is –20 and whose sum is –8 are –10 and 2.

$x^2 - 21x + 20 = (x - 20)(x - 1)$ The two integers whose product is 20 and whose sum is –21 are –20 and –1.

Reverse *FOIL for polynomials of Form* $ax^2 + bx + c$: If the leading coefficient is other than 1, reverse FOIL works, but a different tactic is needed. Sometimes called "magic factoring," the procedure is shown in this example:

$$3x^2 + 13x + 12, \quad a = 3, b = 13, c = 12$$

| | |
|---|---|
| *Step 1*: $ac = (3)(12) = 36$ | |
| *Step 2*: Write all factor pairs of $ac = 36$: | 1,36; 2,18; 3,12; 4,9; 6,6 |
| *Step 3*: Choose the factor pair whose sum is $b = 13$: | 4,9 |
| *Step 4*: Rewrite polynomial with middle term $13x$ as $4x + 9x$: | $3x^2 + 4x + 9x + 12$ |
| *Step 5*: Group first two terms and second two terms in parentheses: | $(3x^2 + 4x) + (9x + 12)$ |
| *Step 6*: Take the common factor out of each group: | $x(3x + 4) + 3(3x + 4)$ |
| *Step 7*: Take the common *binomial* factor out of each large term: | $(3x + 4)(x + 3)$ Correct factorization |
| *Step 8*: Check the factorization found: | $3x^2 + 9x + 4x + 12 = 3x^2 + 13x + 12$ |

*Special Products*: Certain binomials are factored according to formulas, which can be checked by multiplying. The *difference of squares* is perhaps the most important because it comes up so often in applications, as do many quadratic (second-degree) polynomials. It comes from multiplying the *conjugate pair* of binomials, $(a + b)$ with $(a - b)$ to get $a^2 - b^2$.

| | |
|---|---|
| Difference of squares: | $a^2 - b^2 = (a + b)(a - b)$ |
| Difference of cubes: | $a^3 - b^3 = (a - b)(a^2 + ab + b^2)$ |
| Sum of cubes: | $a^3 + b^3 = (a + b)(a^2 - ab + b^2)$ |
| Square of a sum: | $(a + b)^2 = (a + b)(a + b) = a^2 + 2ab + b^2$ |
| Difference of a sum: | $(a - b)^2 = (a - b)(a - b) = a^2 - 2ab + b^2$ |

*Note*: *The sum of squares* $a^2 + b^2$ *is not a factorable binomial over the set of real numbers*. However, it is factorable over the set of complex numbers (see *Complex Numbers* on page 54).

*Examples*: Several special product factorizations are shown:

| Difference of squares: $a^2 - b^2$ | Differerence of cubes: $a^3 - b^3$ | Sum of cubes: $a^3 + b^3$ |
|---|---|---|
| $x^2 - 81 = (x + 9)(x - 9)$ | $x^3 - 1 = (x - 1)(x^2 + x + 1)$ | $x^3 + 1 = (x + 1)(x^2 - x + 1)$ |
| $100 - 49x^2 = (10 + 7x)(10 - 7x)$ | $8 - x^3 = (2 - x)(4 + 2x + x^2)$ | $27 + x^3 = (3 + x)(9 - 3x + x^2)$ |

Factorization of all other cases is done using the *quadratic formula*, as shown in the next section.

## Equation Solving

An equation is a statement of equality between two expressions, such as one monomial set equal to another, like $5x = 105$. The unknown, or variable, is frequently designated by $x$. Other unknowns (if any) are designated by letters also usually selected from the end of the alphabet: $y, z, u, t$, etc.

Equations, like expressions, have a degree. A *first-degree* equation is one in which the variable is raised to the first power, as in $3x = 9$. A *second-degree* equation, also called a *quadratic* equation, is one in which the highest power of the variable is two; for example, $x^2 + 3x = 10$.

Solving a first-degree equation requires isolating the unknown. In the example below, $x$ is the unknown variable. To get $x$ alone (to isolate $x$), constants are combined on one side of the equation and variable terms are combined on the other. The steps are:

| | |
|---|---|
| Given: | $10x - 14 = 8 - 2x$ |
| Add $2x$ to both sides: | $10x - 14 + 2x = 8 - 2x + 2x$ |
| | $12x - 14 = 8$ |
| Add 14 to both sides: | $12x - 14 + 14 = 8 + 14$ |
| | $12x = 22$ |
| Divide by 12 (multiply by 1/12): | $12x/12 = 22/12$ |
| Simplify: | $x = 22/12 = 11/6$ |

Any answer can be checked by substituting it into the original equation to see that it satisfies it.

**Solving a System of Linear Equations.**—More involved than solving a single-variable equation is the process of solving a *system* of linear equations. A simple linear system represents two lines in the plane that behave in one of three ways: they *intersect at one point,* in which case they have a *unique solution*, $(x, y)$; they intersect everywhere—that is, they are *collinear*—so all points $(x, y)$ of one line satisfy the other; or they are *parallel* and thus intersect nowhere, hence, there is no solution.

The methods for solving a system of linear equations are *substitution* and *elimination*, as shown next.

*Substitution:* In this method, one of the variables is expressed in terms of the other variable by isolating it. This expression is then substituted into the second equation, converting it to a single-variable equation. It is solved for this variable, and the solution is substituted back into the either of the original two equations to find the value of the other variable.

*Example (Unique Solution):* Find the ordered pair $(x, y)$ that satisfies the system of equations:

$$2x + y = 7$$
$$x - 2y = -4$$

First, solve either equation for one variable in terms of the other. Say, solve the second equation for $x$:

$$x = 2y - 4$$

Then, substitute this expression for $x$ into the first equation and solve it for $y$:

$$2(2y - 4) + y = 7 \quad \rightarrow \quad 4y - 8 + y = 7 \quad \rightarrow \quad 5y = 15 \quad \rightarrow \quad y = 3$$

Finally, substitute $y = 3$ into the second equation: $x = 2(3) - 4 = 6 - 4 = 2$. The solution (that is, the point at which the lines intersect) is $(2, 3)$.

*Example (Infinite Solutions):* Find the ordered pair $(x, y)$ that satisfies the system of equations:

$$7x - y = 3$$
$$14x - 2y = 6$$

The first equation is rearranged to $y = 7x - 3$. Substituting into the second equation yields: $14x - 2(7x - 3) = 6 \rightarrow 14x - 14x + 6 = 6 \rightarrow 6 = 6$.

If a system results in $a = a$, then the lines are the same, or collinear. All of the points on either line are solutions to (points on) the other line. The indication that lines are collinear is that one is a multiple of the other, term by term.

*Example (No Solution):* Parallel lines have the same slope, and so they have no point of intersection. This is seen in the substitution of one equation's variable into the other equation, giving a false statement, such as $1 = 3$. An example is the system $y = -x + 4$ and $y = -x - 1$. Substitution of the first into the second gives $-x + 4 = -x - 1$, or $4 = -1$. This clearly is not true, hence the system has no solution.

A final way to find the solution to a linear system is to put the coefficients of the equations into the following formulas and to solve for $x$ and $y$, as follows:

$$a_1x + b_1y = c_1$$
$$a_2x + b_2y = c_2$$

Then,

$$x = \frac{c_1b_2 - c_2b_1}{a_1b_2 - a_2b_1} \qquad y = \frac{a_1c_2 - a_2c_1}{a_1b_2 - a_2b_1}$$

*Example:*

$$3x + 4y = 17$$
$$5x - 2y = 11$$

$$x = \frac{(17)(-2) - (11)(4)}{(3)(-2) - (5)(4)} = \frac{-34 - 44}{-6 - 20} = \frac{-78}{-26} = 3$$

The value of $y$ can now be most easily found by inserting the value of $x$ in one of the equations:

$$(5)(3) - 2y = 11, \qquad 2y = 15 - 11 = 4, \qquad y = 2$$

Checking the solution by putting these values into the original system shows that (3, 2) is the solution of this linear system.

**Solving a Second-Degree (Quadratic) Equation.**—A second-degree equation is also called a *quadratic equation*. To *solve* a second-degree equation is to find the $x$ values (or points) at which the parabola intersects the $x$-axis. These are the *roots* or *zeros* of the parabola.

There are several ways to solve a quadratic equation for its roots. If an equation can be factored, then each factor is set equal to zero and the solution is thus found. This is according to the *zero property of multiplication*, which states that if $AB = 0$, then either $A = 0$ or $B = 0$.

If the equation is not readily or obviously factored (or even if it is), the *quadratic formula* can be used to find the roots, as explained in the next section, *Using the Quadratic Formula*.

The simplest quadratic equation has the form $x^2 = c$, where $c$ is a constant. Solving this entails simply taking the square root of both sides:

$x^2 = c$, therefore, $x = \sqrt{c}$ and $-\sqrt{c}$. For example, $x^2 = 36$ has two solutions, $x = 6$ and $x = -6$. If the quadratic has the form, like $ax^2 = c$, the solution is also straightforward, $x^2 = a/c$, and so $x = \pm\sqrt{a/c}$.

*Note:* The square root of a number as it stands alone is understood to be its positive root; that is, the square root of 9 is 3. But *in an equation* the solution includes both positive and negative roots. This makes sense because the equation represents a parabola, which can intersect the $x$-axis in two places, its two roots. The other two possibilities occur when the parabola (a quadratic function) intersects the $x$-axis once (touches it) or does not intersect it. All three possibilities are revealed in solutions to the quadratic equation.

Designating the roots of a quadratic equation (or any polynomial) by $r$, $x^2 = c$ can be regarded as $x^2 = r^2$, so $x = r$ and $-r$, the roots of the parabola. Another way to rewrite $x^2 = r^2$

is $x^2 - r^2 = 0$, and then factor as the difference of squares (see *Polynomials* on page 23): $(x+r)(x-r) = 0$, hence, $x = -r$ and $r$.

Verifying this to be the factorization of the difference of squares is simply a matter of applying the distributive property of multiplication over addition to each term:

$$(x+r)(x-r) = xx - xr + xr - rr = x^2 - r^2$$

As previously explained, this process is called expanding by FOIL, for the **F**irst, **O**uter, **I**nner, and **L**ast terms of each binomial, which are multiplied in that order to get the product. From the zero property of multiplication,

$$(x+r)(x-r) = 0 \quad \text{implies} \quad x+r=0 \quad \text{or} \quad x-r=0, \text{thus } x=-r \quad \text{or} \quad x=r.$$

If a quadratic cannot be factored as a difference of squares, then factorization is approached as shown in *Factoring Polynomials* on page 24. Given a quadratic equation in the form $ax^2 + bx + c = 0$, first obtain the product $ac$ from the coefficients $a$ and $c$; then determine two numbers whose product is $ac$ and whose sum is $b$.

*Example:* Find the solution to $x^2 - 5x + 6 = 0$ by factoring.

In this example, $a = 1$, $b = -5$, $c = 6$ and $ac = (1)(6) = 6$. The factors of 6 whose sum is $-5$ are $-2$ and $-3$. The equation is factored as $x^2 - 5x + 6 = (x-2)(x-3) = 0$. Then by the zero property of multiplication, the roots of the equation are $x = 2$ and $x = 3$. The parabola intersects the $x$-axis at these two values of $x$.

A more difficult example has a leading coefficient of $x$ other than 1.

*Example:* Factor $8x^2 + 22x + 5 = 0$ and find the values of $x$ that satisfy the equation.

*Solution:* Here, $a = 8$, $b = 22$, and $c = 5$. Therefore, $ac = 8 \times 5 = 40$, and $ac$ is positive, so we are looking for two factors of $ac$, namely $f_1$ and $f_2$, such that $f_1 + f_2 = 22$.

The possible combinations of numbers with product of 40 are 20 and 2, 8 and 5, 4 and 10, and 40 and 1. The requirements that satisfy a sum of 22 are 20 and 2, since $20 \times 2 = 40$ and $20 + 2 = 22$. Hence:

$$8x^2 + 22x + 5 = 0$$

$$8x^2 + 20x + 2x + 5 = 0$$

$$4x(2x+5) + 1(2x+5) = 0$$

$$(2x+5)(4x+1) = 0$$

On the second line, the common factor of $4x$ in the first two terms is factored out, so the common binomial factor of $2x + 5$ is then apparent, to be factored out of the larger terms. Checking the answer is a matter of simply remultiplying the factors to produce the original expression.

Because the product of the two factors equals zero, each of the factors also equals zero. Thus, $2x + 5 = 0$ and $4x + 1 = 0$. Rearranging and solving, $x = -5/2$ or $x = -1/4$.

*Example:* Factor $8x^2 + 3x - 5 = 0$ and find the solutions of the equation.

*Solution:* Here $a = 8$, $b = 3$, $c = -5$, and $ac = 8 \times (-5) = -40$. The required numbers must have a product of $-40$ and a sum of 3.

As in the previous example, the possible combinations are 20 and $-2$, $-20$ and 2, $-8$ and 5, 8 and $-5$, 40 and $-1$, and $-40$ and 1. Only 8 and $-5$ satisfy the requirements because $8 \times (-5) = -40$, and $8 + (-5) = 3$. Notice that $3x$ in the first line is thus written as $8x - 5x$ in the second line, making it possible to rearrange and simplify the expression.

$$8x^2 + 3x - 5 = 0$$

$$8x^2 + 8x - 5x - 5 = 0$$

$$8x(x+1) - 5(x+1) = 0$$

$$(x+1)(8x-5) = 0$$

Solving, for $x + 1 = 0$, $x = -1$; and, for $8x - 5 = 0$, $x = 5/8$.

**Solving by Completing the Square.—**An equation of the form $x^2 + bx + c = 0$ can be turned into the square of a sum. The steps are:

- Move constant to the right side of the equal sign: $x^2 + bx = -c$
- Add $(b/2)^2$ to both sides: $x^2 + bx + (b/2)^2 = -c + (b/2)^2$
- Note the left side is the square of a sum, that is: $(x + b/2)^2$
- The right is a new constant, call it $d$: $-c + (b/2)^2 = d$
- The equation is thus converted to: $(x + b/2)^2 = d$
- Take the positive and negative square root of both sides: $x + \dfrac{b}{2} = \pm\sqrt{d}$
- Solve for $x$: $x = -\dfrac{b}{2} \pm \sqrt{d}$

*Example:* Complete the square to solve $x^2 + 4x - 3 = 0$.

- $x^2 + 4x = 3$
- $x^2 + 4x + (4/2)^2 = 3 + (4/2)^2$, that is, $x^2 + 4x + 4 = 3 + 4$
- Left side: $x^2 + 4x + 4 = (x + 2)^2$
- Right side: $3 + 4 = 7$
- So, $(x + 2)^2 = 7$
- $x + 2 = \pm\sqrt{7}$
- $x = -2 \pm\sqrt{7}$

Completing the square is more involved if the leading coefficient $a$ is not 1. Then it is preferable to use the *quadratic formula*, which can be used for any value of $a$ in a quadratic equation.

**Using the Quadratic Formula.—**The method of completing the square leads to the quadratic formula for finding the roots of equations with the form $ax^2 + bx + c = 0$:

$$x = \frac{-b \pm \sqrt{b^2 - 4ac}}{2a}$$

*Example:* In the equation, $x^2 + 6x + 5 = 0$, $a = 1$, $b = 6$, and $c = 5$.

$$x = \frac{-6 \pm \sqrt{6^2 - (4)(1)(5)}}{(2)(1)} = \frac{(-6) + 4}{2} = -1 \quad \text{or} \quad \frac{(-6) - 4}{2} = -5$$

*Example:* A right triangle has a hypotenuse of 5 cm and one leg that is 1 cm longer than the other; find the lengths of the two legs.

Let $x$ be the length of one leg and $x + 1$ be the length of the other; then by Pythagorean theorem (see page 57), $x^2 + (x + 1)^2 = 5^2$. Expanding this and setting all terms equal to zero gives $x^2 + x - 12 = 0$. Now referring to the basic formula, $ax^2 + bx + c = 0$, here, $a = 1$, $b = 1$, and $c = -12$. Substituting these values into the quadratic formula:

$$x = \frac{-1 \pm \sqrt{1 - (4)(1)(-12)}}{(2)(1)} = \frac{(-1) + 7}{2} = 3 \quad \text{or} \quad x = \frac{(-1) - 7}{2} = -4$$

Since only the positive value, 3, makes sense in this case, the lengths of the two sides are $x = 3$ cm and $x + 1 = 4$ cm.

**Solving a Cubic Equation.—**Just as quadratic equations may be simple to solve if they are of the form $x^2 = c$, a cubic equation of the form $x^3 = c$ also is simple to solve. But, the given equation has the form: $x^3 + ax + b = 0$ then one of the real roots is:

$$x = \left(-\frac{b}{2} + \sqrt{\frac{a^3}{27} + \frac{b^2}{4}}\right)^{1/3} + \left(-\frac{b}{2} - \sqrt{\frac{a^3}{27} + \frac{b^2}{4}}\right)^{1/3}$$

The equation $x^3 + px^2 + qx + r = 0$ may be rewritten in the form $x_1{}^3 + ax_1 + b = 0$ by substituting $x_1 - \dfrac{p}{3}$ for $x$ in the given equation.

### Functions

Functions are understood through both equations and graphs. Graphs are drawn on the $(x, y)$-coordinate system (or rectangular coordinate system), which is described fully in *Analytic Geometry* section of *GEOMETRY* starting on page 37.

A *function* consists of two sets (generally called $X$ and $Y$) of numbers and a *rule* that assigns (or sends) each element (usually, a real number) $x$ in $X$ to a *unique* element (another real number) $y$ in $Y$. "$y$ is a function of $x$" is commonly expressed as $y = f(x)$. Countless physical processes are represented by functions. For example, displacement, $s$, is a function of time, $t$, so it is represented as $s(t)$. Velocity also is a function of time, $v(t)$.

The set $X$ is called the *domain* of a function, which is the set of all real number values for which a function is *defined*. When the function acts on these numbers, the result is another real number $y$.

As an example, $y = f(x) = 2x + 5$ is a linear function whose domain is the set of all real numbers, since when 5 is added to the product of 2 and any real number $x$, the result is always a real number $y$. Lines are first-degree polynomials; in fact, the domain of *any* polynomial is the set of reals (see *Polynomials* on page 23).

Two more examples: $f(x) = 1/x$ is not defined at $x = 0$, since $1/0$ is not a number, so its domain is the set of all real numbers except 0. Since negative numbers do not have square roots in the reals, the domain of $f(x) = \sqrt{x}$ is the set of non-negative real numbers.

In $y = f(x)$, $x$ is called the *input* to the function. The $y$ value resulting from $x$ is the *output*, or *function value*. Since $y$ depends on $x$, it is the *dependent variable*; $x$ is the *independent variable*.

*Interval Notation:* Domain is often expressed in *interval notation*, indicating the portion of the number line that contains a function's valid input values. The interval notation for set of all reals is $(-\infty, \infty)$; for non-negative numbers, it is $[0, \infty)$; for positive numbers, it is $(0, \infty)$; and for all real numbers except 0, it is $(-\infty, 0) \cup (0, \infty)$. The domain of a polynomial is therefore $(-\infty, \infty)$; the domain of $f(x) = 1/x$ is $(-\infty, 0) \cup (0, \infty)$; and the domain of $f(x) = \sqrt{x}$ is $[0, \infty)$.

**Graphs of Functions.**—The *graph* of a function is drawn through the points $(x, y)$ on the rectangular coordinate system (see page 37) so that the curve satisfies the equation $y = f(x)$. A graph depicts the relationship between $x$ and $y$. The graph of a function will always pass the *vertical line test*, by which any vertical line intersects the graph at most once. Thus, any line other than a vertical line is expressible in function form, $f(x) = mx + b$, where the slope is $m$ and the $y$-intercept is $b$. (The several *forms* of linear equations are explained in *Equation Forms of a Line* on page 39.) A parabola is represented by a polynomial function of the form $f(x) = ax^2 + bx + c$. All polynomials pass the vertical line test (Fig. 1a).

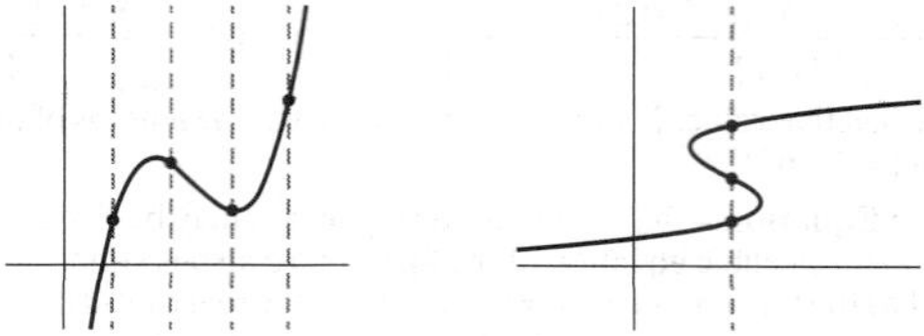

Fig. 1. (a) Passes Vertical Line Test—Represents a Function;
(b) Fails the Test—Does Not Represent a Function.

*Sketches of Basic Functions:* Sketches of basic functions and domain interval notation are shown in Fig. 2. It is obvious that each function passes the vertical line test.

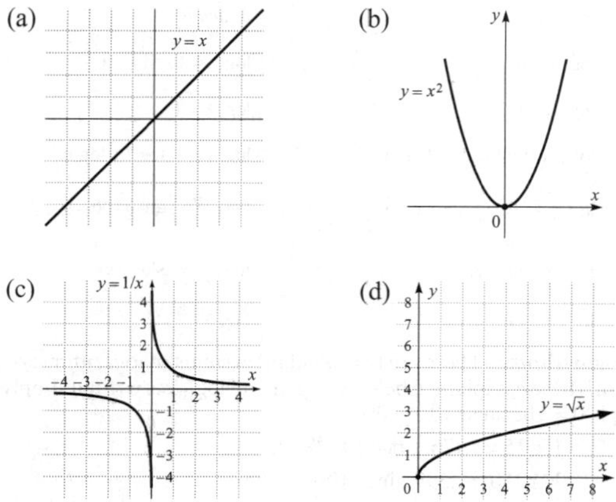


Fig. 2. (a) $f(x) = x, D_f = (-\infty, \infty)$; (b) $f(x) = x^2, D_f = (-\infty, \infty)$; (c) $f(x) = \sqrt{x}, D_f = [0, \infty)$; (d) $f(x) = 1/x, D_f = (-\infty, 0) \cup (0, \infty)$.

## Logarithms

Logarithms are of value in many engineering and shop calculations because they make it possible to solve cumbersome and difficult problems that otherwise would require more complex mathematical methods. Because of this, "logs" (for short) have long been used to facilitate and shorten calculations involving multiplication, division, the extraction of roots, and obtaining powers of numbers. Since the advent of handheld calculators, however, logarithms are rarely used to do these basic operations. The *Guide* and *Logarithms* in the *ADDITIONAL MATERIAL* in the *Machinery's Handbook 32 Digital Edition* include explanations and examples of how the log tables are used for computation.

Log properties and principles are still necessary in many areas. Logarithmic growth and its inverse, exponential growth (and decay), are essential to investigating processes in technical fields and science, in general. The main principles and properties of logarithms are covered here, along with representative examples. In most cases, a calculator is used to arrive at the answers.

**Meaning of Logarithm.**—The logarithm of a given number is the exponent to which a stated base must be raised to produce the given number. A formulaic definition of logarithm is:

$$\log_b x = y \quad \text{means} \quad b^y = x$$

which is read, "The logarithm of *x* in base *b* is *y*; that is, *b* raised to the *y* power equals *x*." *y* is the logarithm and *x* is the *antilogarithm* ("antilog"). Base *b* is always greater than 1. The antilog must be positive, since a positive number *b* raised to any power cannot give zero or a negative number. Some examples:

$\log_2 8 = y$ means $2^y = 8$, so $y = 3$

$\log_b 100 = 2$ means $b^2 = 100$, so $b = 10$

$\log_3 x = -1$ means $3^{-1} = x$, so $x = 1/3$

**Properties of Logarithms.**—The definition along with exponent rules covered previously lead to the *properties of logarithms*, given here with examples. When no base is shown, base 10 is implied. (See *Common Logarithms* on page 32.)

| *Property* | *Example* |
|---|---|
| $\log_b 1 = 0$ | $\log_3 1 = 0$ |
| $\log_b b = 1$ | $\log_8 8 = 1$ |
| $\log_b (xy) = \log_b x + \log_b y$ | $\log(2y) = \log 2 + \log y$ |
| $\log_b \left(\frac{x}{y}\right) = \log_b x - \log_b y$ | $\log\left(\frac{100}{3}\right) = \log 100 - \log 3$ |
| $\log_b x^r = r \log_b x$ | $\log_2 x^3 = 3 \log_2 x$ |
| $\log_b b^r = r$ | $\log_5 5^{11} = 11$ |

**Common Logarithms.—**There are two standard systems of logarithms: *common* (base 10) and *natural* (base $e$, explained below). In general, $\log_{10} x$ is written simply as $\log x$. For example, $\log 100 = 2$ because $10^2 = 100$.

*Examples*: $\log 15 = 1.176$, since $10^{1.176} \approx 15$
$\log 250 = 2.397$, since $10^{2.397} \approx 250$
$\log 4000 = 3.602$, since $10^{3.602} \approx 4000$

The log values seen here can be found by using either log tables or a scientific calculator. Because most logarithms are irrational numbers, the values given in any example and in the tables are approximations, rounded to several decimal places. When a calculator is used, the answer should be rounded to four or five decimal places.

The whole number part of a logarithm is called the *characteristic*; the decimal portion is called the *mantissa*. In the examples above, the characteristics are 1, 2, and 3, respectively, which correspond to the power of 10 of the antilog when it is written in scientific notation:

$$\log 15 = \log(1.5 \times 10^1) = \log 1.5 + \log 10^1 = 0.176 + \mathbf{1} \approx 1.176$$
$$\log 250 = \log(2.5 \times 10^2) = \log 2.5 + \log 10^2 = 0.397 + \mathbf{2} \approx 2.397$$
$$\log 4000 = \log(4.0 \times 10^3) = \log 4.0 + \log 10^3 = 0.602 + \mathbf{3} \approx 3.602$$

The property $\log(ab) = \log a + \log b$ has been applied; the "$\log b$" portion (characteristic) is quickly determined, as it is simply the power of 10 (bold). The "$\log a$" portion (mantissa) is read from the log table, which gives logs of numbers from 1 to 10 up to a certain number of decimal places. If the log of a number less than 1 is to be found, again, the antilog is represented in scientific notation to get to the answer, only now the characteristic is negative, so a subtraction is involved. For example:

$$\log 0.63 = \log(6.3 \times 10^{-1}) = (\log 6.3) + \log(\mathbf{-1}) = 0.799 + (\mathbf{-1}) \approx -0.201$$

**Natural Logarithms.—**In certain formulas and in some branches of mathematical analysis, use is made of the *natural* logarithm. The base of this system is given as $e$, which is the symbol for the irrational number that is approximately equal to 2.7182818284. (Recall that an irrational number cannot be represented by a repeating or terminating decimal.) $e$ is the base of exponential growth phenomena such as populations and compound interest, among others. Though $e$ was first conceptualized by John Napier (who developed logarithmic calculation) and developed further by Jacob Bernoulli, the use of $e$ credits eighteenth-century Swiss mathematician Leonhard Euler (pronounced "oiler"), who developed mathematical analysis, in part, through his discovery of the so-called Euler identity, $e^{i\pi} = -1$.

It is conventional to write $\log_e x$ as "$\ln x$"; hence:

$$\ln x = y \text{ means } \log_e x = y, \text{ that is, } x = e^y$$

So, for example, $\ln e = 1$, since $e^1 = e$; $\ln 1 = 0$, since $e^0 = 1$; $\ln e^3 = 3 \ln e = 3 \cdot 1 = 3$. And, the process of finding logarithm values in base $e$ is the same as in base 10.

*Example*: To represent ln 0.239 as a sum of natural logs: $\ln(2.39 \times 10^{-1}) = \ln 2.39 + \ln 10^{-1} = \ln 2.39 + (-1)\ln 10$.

Logarithms in any base can be converted to another base by the formula: $\log_b x = \dfrac{\log_a x}{\log_a b}$

Logarithms of bases other than 10 and *e* were often converted to these bases, since the values were easy to look up on the common or natural log tables. The formula also can convert between the two most-often used bases:

Base *e* to base 10: $\ln x = \dfrac{\log x}{\log e} \approx \dfrac{\log x}{0.4343}$ so $\log x \approx 0.4343 \ln x$

Base 10 to base *e*: $\log x = \dfrac{\ln x}{\ln 10} \approx \dfrac{\ln x}{2.3026}$ so $\ln x \approx 2.3026 \log x$

*Example*: Convert ln 4 to log 4 using the conversion, $\log x \approx 0.4343 \ln x$:

$$\log 4 \approx 0.4343 \ln 4 \approx (0.4343)(1.3863) \approx 0.60207$$

*Example*: Express $\log_2 9$ in terms of the natural logarithm:

$$\log_2 9 = \frac{\log_e 9}{\log_e 2} = \frac{\ln 9}{\ln 2} \approx 3.170$$

**Using Calculators to Solve Logarithms.**—To find a common logarithm on a scientific calculator (handheld or online), the **log** key is used. To find a natural logarithm the **ln** key is used. Depending on the calculator used, either the number is entered before the log key is pressed, or log is entered before the number. The correct sequence can be seen if an error message results. For example, log 6 is found by this sequence on the typical scientific calculator:

*Example*: To find log 6, press in sequence, [6] [log] to get display

[0.77815125038 . . .]

To find the common antilog of a given number, the **10^x** key is used. The find the natural antilog, the **e^x** key is used. This kind of problem is often asked as $\ln x = 4$, meaning $e^4 = x$, so one must know the definition of $\log_b x = y$ to get the answer by calculator.

*Example*: To find *x* in $\ln x = 4$, press in sequence, [4] [e^x] get the display

[54.5980015003 . . .]

On calculators without the **10^x** and **e^x** keys, the **x^y** key enables, substituting 10 or *e* (2.718281 . . .) for *x* and the logarithm of the number sought for *y*. On some calculators, while the **log** and **ln** keys are used to find common and natural logarithms, the same keys in combination with the **INV**, or inverse, key are used to find the number corresponding to a given logarithm.

**Solving an Equation Using Logarithms.**—Solving exponential and logarithmic equations is possible because of the following properties of logs and exponents, which are true for *any* base:

If $x = y$, and $x, y > 0$, then $\log x = \log y$. If $x = y$, then $a^x = a^y$ for $a > 0$, $a \neq 1$.

Both statements are true in the other direction, too:

If $\log x = \log y$, then $x = y$. If $a^x = a^y$ then $x = y$.

*Example 1:* Find the square root of 754.

*Solution:* Let $x = \sqrt{754}$. Then $\log x = \log\sqrt{754} = \log 754^{1/2} = \dfrac{1}{2}\log 754 \approx \dfrac{2.8774}{2} \approx 1.4387$.

So, $\log x \approx 1.4387$, hence, $x = 10^{1.4387} \approx 27.460$. That is, $\sqrt{754} \approx 27.460$.

*Example 2:* Solve $4^x = 7^{x-3}$ for *x*.

*Solution:* $4^x = 7^{x-3} \underset{\text{apply property}}{\longrightarrow} \log 4^x = \log 7^{x-3} \longrightarrow x\log 4 = (x-3)(\log 7)$

$\underset{\text{distribute on the right}}{\longrightarrow} x\log 4 = x\log 7 - 3\log 7 \underset{\text{by calculator}}{\longrightarrow} 0.6021x = 0.8451x - 3(0.8451)$

$\underset{\text{proceed with algebra}}{\longrightarrow} 3(0.8451) = 0.8451x - 0.6021x \longrightarrow 2.5353 = 0.243x$, so $x \approx 10.433$

The technique of taking the log (either common, natural, or other) of both sides of an equation is used often to solve for unknown exponents, as happens with compounding of interest (see *ENGINEERING ECONOMICS* on page 138).

### Arithmetic Sequence

An arithmetic sequence (also called an arithmetic progression) is a sequence of numbers in which each term differs from the preceding one by a fixed amount, called the *common difference*, $d$. Thus, 1, 3, 5, 7, etc. is an arithmetic sequence where the difference $d$ is 2. Here, the consecutive terms of the sequence are increasing by 2. In the sequence 13, 10, 7, 4, etc., the difference is –3, and the sequence is decreasing. In any arithmetic progression (or portion of one):

$a$ = first term of the sequence, also called the $a_1$ term
$l$ = last term considered, also called $a_n$ for the $n$th term
$n$ = number of terms
$d$ = common difference
$S_n$ = sum of $n$ terms

The formula for the last term is $l = a + (n-1)d$, or $a_n = a_1 + (n-1)d$. The sum of an arithmetic sequence with $n$ terms is $S_n = \frac{n(a+l)}{2}$ or $\frac{n(a_1 + a_n)}{2}$.

In these formulas, $d$ is positive when the progression is increasing and negative when it is decreasing. When any three of the preceding five quantities are given, the other two can be found by the formulas in the table *Arithmetic Sequence Formulas* on page 35. Often, however, the desired quantity can be determined by working with the information given.

*Example 1:* In a given arithmetic progression, the first term is 5 and the last term 40, and the difference between terms is 7. To find the sum of the progression, first the number of terms has to be found. This is done by considering the difference between the first and last: 40 – 5 = 35. Dividing this by the difference between terms gives the number of intervals between the terms: 35 ÷ 7 = 5. Finally, adding 1 gives the number of terms in the sequence: $n = 5 + 1 = 6$. The sum of the sequence is:

$$S = \frac{n}{2}(a + l) = \frac{6}{2}(5 + 40) = 3(45) = 135$$

### Geometric Sequence

A geometric sequence or progression is a sequence of numbers in which each term is derived by multiplying the preceding term by a constant multiplier called the *ratio*. When this ratio is greater than 1, the progression is increasing; when less than 1, it is decreasing. Thus, the sequence 2, 6, 18, 54, etc. is an increasing geometric sequence with a ratio of 3, and the sequence 24, 12, 6, etc. is a decreasing sequence with a ratio of 1/2.

In any geometric progression (or part of progression):

$a$ = first term of the sequence
$l$ = last (or $n$th) term of the sequence
$n$ = number of terms
$r$ = ratio of the progression
$S_n$ = sum of $n$ terms

The general formulas for the $n$th term: $l = ar^{n-1}$ and $S = \frac{rl - a}{r - 1}$

When any three of the preceding five quantities are given, the other two can be found by the formulas in the table *Geometric Sequence Formulas* on page 36. Geometric progressions are used for finding the successive speeds in machine tool drives, and in interest calculations.

*Example 2:* The lowest speed of a lathe is 20 rpm. The highest speed is 225 rpm. There are 18 speeds. Find the ratio between successive speeds.

$$\text{Ratio } r = \sqrt[n-1]{\frac{l}{a}} = \sqrt[17]{\frac{225}{20}} = \sqrt[17]{11.25} = 1.153$$

**Arithmetic Sequence Formulas**

| To Find | Given | | | Use Formula |
|---|---|---|---|---|
| $a$ | $d$ | $l$ | $n$ | $a = l-(n-1)d$ |
| | $d$ | $n$ | $S$ | $a = \frac{S}{n} - \frac{n-1}{2} \times d$ |
| | $d$ | $l$ | $S$ | $a = \frac{d}{2} \pm \frac{1}{2}\sqrt{(2l+d)^2 - 8dS}$ |
| | $l$ | $n$ | $S$ | $a = \frac{2S}{n} - l$ |
| $d$ | $a$ | $l$ | $n$ | $d = \frac{l-a}{n-1}$ |
| | $a$ | $n$ | $S$ | $d = \frac{2S-2an}{n(n-1)}$ |
| | $a$ | $l$ | $S$ | $d = \frac{l^2-a^2}{2S-l-a}$ |
| | $l$ | $n$ | $S$ | $d = \frac{2nl-2S}{n(n-1)}$ |
| $l$ | $a$ | $d$ | $n$ | $l = a+(n-1)d$ |
| | $a$ | $d$ | $S$ | $l = -\frac{d}{2} \pm \frac{1}{2}\sqrt{8dS+(2a-d)^2}$ |
| | $a$ | $n$ | $S$ | $l = \frac{2S}{n} - a$ |
| | $d$ | $n$ | $S$ | $l = \frac{S}{n} + \frac{n-1}{2} \times d$ |
| $n$ | $a$ | $d$ | $l$ | $n = 1 + \frac{l-a}{d}$ |
| | $a$ | $d$ | $S$ | $n = \frac{d-2a}{2d} \pm \frac{1}{2d}\sqrt{8dS+(2a-d)^2}$ |
| | $a$ | $l$ | $S$ | $n = \frac{2S}{a+l}$ |
| | $d$ | $l$ | $S$ | $n = \frac{2l+d}{2d} \pm \frac{1}{2d}\sqrt{(2l+d)^2 - 8dS}$ |
| $S$ | $a$ | $d$ | $n$ | $S = \frac{n}{2}[2a+(n-1)d]$ |
| | $a$ | $d$ | $l$ | $S = \frac{a+l}{2} + \frac{l^2-a^2}{2d} = \frac{a+l}{2d}(l+d-a)$ |
| | $a$ | $l$ | $n$ | $S_n = \frac{n(a+l)}{2}$ |
| | $d$ | $l$ | $n$ | $S = \frac{n}{2}[2l-(n-1)d]$ |

**Geometric Sequence Formulas**

| To Find | Given | | | Use Formula |
|---|---|---|---|---|
| $a$ | $l$ | $n$ | $r$ | $a = \dfrac{l}{r^{n-1}}$ |
| | $n$ | $r$ | $S$ | $a = \dfrac{(r-1)S}{r^n - 1}$ |
| | $l$ | $r$ | $S$ | $a = lr - (r-1)S$ |
| | $l$ | $n$ | $S$ | $a(S-a)^{n-1} = l(S-l)^{n-1}$ |
| $l$ | $a$ | $n$ | $r$ | $l = ar^{n-1}$ |
| | $a$ | $r$ | $S$ | $l = \dfrac{1}{r}[a + (r-1)S]$ |
| | $a$ | $n$ | $S$ | $l(S-l)^{n-1} = a(S-a)^{n-1}$ |
| | $n$ | $r$ | $S$ | $l = \dfrac{S(r-1)r^{n-1}}{r^n - 1}$ |
| $n$ | $a$ | $l$ | $r$ | $n = \dfrac{\log l - \log a}{\log r} + 1$ |
| | $a$ | $r$ | $S$ | $n = \dfrac{\log[a + (r-1)S] - \log a}{\log r}$ |
| | $a$ | $l$ | $S$ | $n = \dfrac{\log l - \log a}{\log(S-a) - \log(S-l)} + 1$ |
| | $l$ | $r$ | $S$ | $n = \dfrac{\log l - \log[lr - (r-1)S]}{\log r} + 1$ |
| $r$ | $a$ | $l$ | $n$ | $r = \sqrt[n-1]{\dfrac{l}{n}}$ |
| | $a$ | $n$ | $S$ | $r^n = \dfrac{Sr}{a} + \dfrac{a-S}{a}$ |
| | $a$ | $l$ | $S$ | $r = \dfrac{S-a}{S-l}$ |
| | $l$ | $n$ | $S$ | $r^n = \dfrac{Sr^{n-1}}{S-l} - \dfrac{l}{S-l}$ |
| $S$ | $a$ | $n$ | $r$ | $S = \dfrac{a(r^n - 1)}{r-1}$ |
| | $a$ | $l$ | $r$ | $S = \dfrac{lr - a}{r-1}$ |
| | $a$ | $l$ | $n$ | $S = \dfrac{\sqrt[n-1]{l^n} - \sqrt[n-1]{a^n}}{\sqrt[n-1]{l} - \sqrt[n-1]{a}}$ |
| | $l$ | $n$ | $r$ | $S = \dfrac{l(r^n - 1)}{(r-1)r^{n-1}}$ |

## GEOMETRY

Geometry is the branch of mathematics that studies the features of two- and three-dimensional figures. (The word "geometry" means "measure of the earth.") This branch can be separated into *pure geometry* and *analytic geometry*. Pure geometry is concerned with the propositions of shape, size, and relative position of figures, as well as their constructions. Analytic geometry studies geometry using the coordinate system, relying heavily on algebraic principles. The results of geometry apply to many areas of industry.

The first part of this section, addressing analytic geometry, may be considered a continuation of the material in *ALGEBRA*, which begins on page 19. The rest of the section focuses on pure geometry, particularly the formulas for figure dimensions and figure construction; these are based on the concepts introduced by the Greek mathematician Euclid, in the fourth century BCE. Included are examples showing how measures of diameter, perimeter, area, surface area, volume, angle, and more are determined. In the case of angle measure, use of trigonometry is often necessary. Explanations of trigonometric relations are found in the next section, *TRIGONOMETRY: SOLUTION OF TRIANGLES*, beginning on page 89.

### Analytic Geometry

Analytic geometry uses algebra to model geometric objects, such as points, lines, and circles, on the rectangular coordinate system.

**Rectangular Coordinate System.**—The *rectangular coordinate system* (also called the *xy-plane* or the *Cartesian plane*) is a grid formed by intersecting two real number lines at right angles (Fig. 1a). The horizontal $x$-axis (labeled $X$) intersects the vertical $y$-axis (labeled $Y$) at the point $(0, 0)$, the *origin*. Any point $P$ on the plane can be so identified by its $x$-coordinate and its $y$-coordinate in the *ordered pair* $(x, y)$.

The four quadrants formed by the $x$- and $y$-axes are numbered counterclockwise (see Fig. 1a). In Quadrant 1, both $x$ and $y$ coordinates are positive; in Quadrant 2, the $x$ is negative and $y$ is positive; in Quadrant 3, both coordinates are negative; and in Quadrant 4, the $x$-coordinate is positive and the $y$ negative. Several representative points are pictured in Fig. 1b.

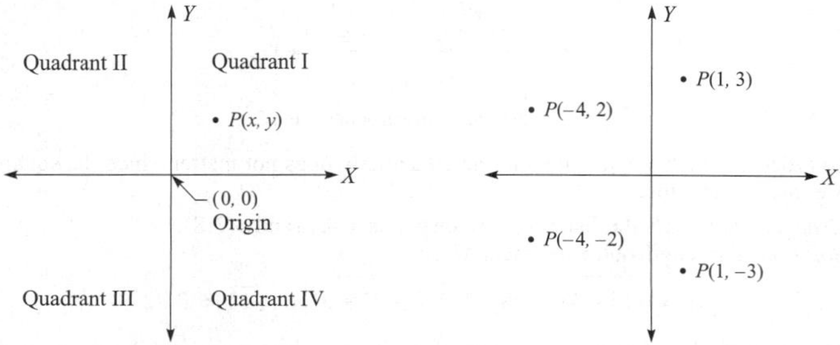


Fig. 1. (a) Rectangular Coordinate System; (b) Examples of Points in Each Quadrant.

The rectangular coordinate system is used to illustrate ideas in algebra, analytic geometry, and trigonometry.

**Slope of a Line.**—The *slope* of the line passing through any two points $(x_1, y_1)$ and $(x_2, y_2)$ in the plane is given by $m = \frac{\Delta y}{\Delta x} = \frac{y_2 - y_1}{x_2 - x_1}$, where $\Delta$ represents difference. Fig. 2a shows a line with a positive slope, Fig. 2b a line with a negative slope. A horizontal line (Fig. 2c) has a no slope, since $\Delta y = 0$. Hence, $m = 0/\Delta x = 0$. Between any two points on a vertical line, $\Delta x = 0$ (Fig. 2d). So, $m = \Delta y/0$, which is undefined.

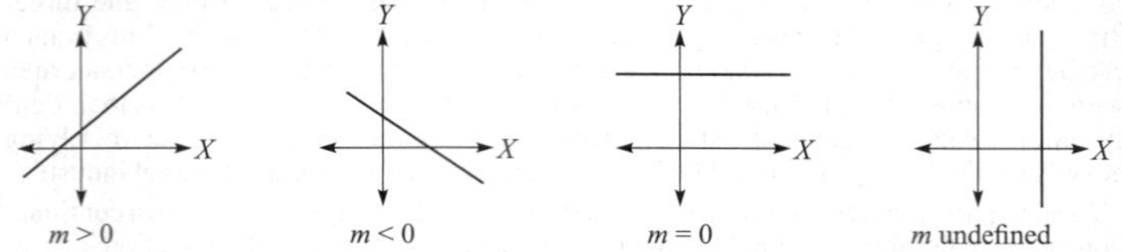


Fig. 2. Lines with (a) Positive Slope, $m > 0$; (b) Negative Slope, $m < 0$; (c) No Slope, $m = 0$; (d) Undefined Slope.

**Lines and Line Segments.**—A *line* in the plane is the shortest path between two known points, extending indefinitely in both directions. "Line *AB*" is notated $\overleftrightarrow{AB}$. A *line segment* is the portion of the line between *A* and *B*. The line segment *AB* is notated $\overline{AB}$; its *length* is indicated without the bar, as *AB*. The distinction between the actual line and its length is helpful to keep in mind when referring to the formulas for distance, midpoint, and the other concepts in this section.

*Distance Between Two Points:* The distance *d* between two points *A* and *B* is the length of the line segment connecting them. The formula comes from the Pythagorean theorem, which says that the sum of the squares of the leg measures is the square of the hypotenuse. As labeled in Fig. 3, the legs are lengths $x_2 - x_1$ and $y_2 - y_1$, and the hypotenuse is the distance *d* between these points. Its formula is:

$$d(A, B) = AB = \sqrt{(x_2 - x_1)^2 + (y_2 - y_1)^2}$$

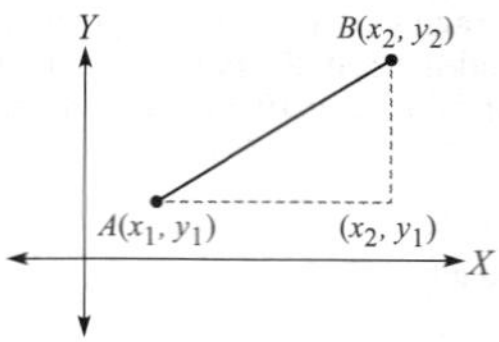


Fig. 3. Distance between points *A* and *B*.

The order in which *x* and *y* are subtracted actually does not matter, since the squared difference is the same.

*Example 1:* Find *AB*, the distance between points *A*(4, 5) and *B*(7, 8).
*Solution:* The length of line segment *AB* is:

$$d = \sqrt{(7-4)^2 + (8-5)^2} = \sqrt{3^2 + 3^2} = \sqrt{18} = 3\sqrt{2}$$

*Midpoint of a Line Segment:* The midpoint *M(x, y)* of line segment *AB* is found by the coordinate formulas (Fig. 4a):

$$x = \frac{x_1 + x_2}{2} \quad \text{and} \quad y = \frac{y_1 + y_2}{2}$$

*Internal Division of a Line Segment:* Point *P* divides line segment *AB* (Fig. 4b) in the ratio *m:n*. That is, *P* is such a point that *AP:PB* as *m:n*. Then, the coordinates of *P* are given by:

$$x = \frac{mx_1 + nx_2}{m + n} \qquad \text{and} \qquad y = \frac{my_1 + ny_2}{m + n}$$

If the desired point is the *midpoint* of $AB$, then $m = n = 1$, and the coordinates of $P$ are:

$$x = \frac{x_1 + x_2}{2} \qquad \text{and} \qquad y = \frac{y_1 + y_2}{2}$$

*Example 2:* Find the coordinates of a point $P$ that divides the line segment defined by $A(0,0)$ and $B(8,6)$ at a ratio of 5:3.

*Solution:* $x = \dfrac{5 \times 0 + 3 \times 8}{5 + 3} = \dfrac{24}{8} = 3 \qquad y = \dfrac{(5)(0) + (3)(6)}{5 + 3} = \dfrac{18}{8} = 2.25$

*External Division of a Line Segment:* If the line segment $AB$ is extended to point $Q$ (Fig. 4c), and $Q$ is such a point that $AQ{:}QB$ as $m{:}n$, then the coordinates of $Q$ are given by:

$$x = \frac{mx_1 - nx_2}{m - n} \qquad \text{and} \qquad y = \frac{my_1 - ny_2}{m - n}$$

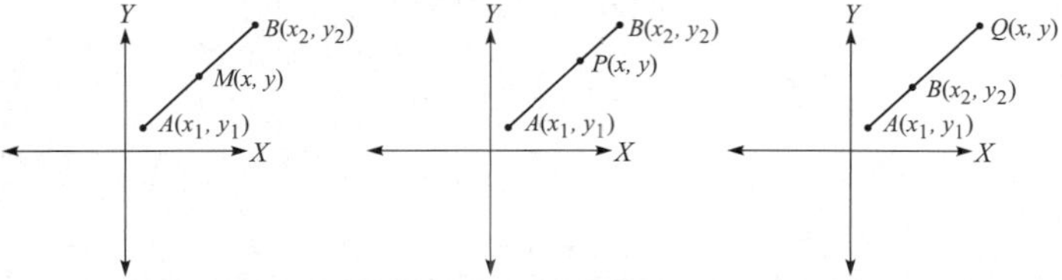


Fig. 4. Divisions of Line Segment $AB$:
(a) Midpoint $M$; (b) Internal Division Point $P$; (c) External Division Point $Q$.

**Equation Forms of a Line.—**Given any two known points, a linear equation can be expressed in either *point-slope* or *slope-intercept* form.

*Point-Slope Form:* Consider that $m = \dfrac{y_2 - y_1}{x_2 - x_1}$ for any two points $(x, y)$ on a line. If the known point $(x_2, y_2)$ is replaced with an arbitrary (not specified) point $(x, y)$, the slope becomes $m = \dfrac{y - y_1}{x - x_1}$, which rearranges to $y - y_1 = m(x - x_1)$. This is the *point-slope form* of the line. If $x_1 = y_1 = 0$ (that is, the line passes through the origin (0,0)), then the equation becomes $y - 0 = m(x - 0)$, or $y = mx$. For any line, the $y$-intercept is the point of intersection of the line with the $y$-axis, and the $x$-intercept is the point of intersection of the line with the $x$-axis. Thus, the points $(0, y_1)$ and $(x_1, 0)$ are the $y$- and $x$-intercepts, respectively.

Suppose $\overleftrightarrow{AB}$ intersects the $x$-axis at point $A(a, 0)$ and the $y$-axis at point $B(0, b)$; then $m = \dfrac{0 - b}{a - 0} = \dfrac{-b}{a}$. Substitution of either $(a,0)$ or $(0,b)$ into the formula and rearranging terms gives the equation for $\overleftrightarrow{AB}$: $y - 0 = -\frac{b}{a}(x - a) \rightarrow ay = -bx + ab \rightarrow ay + bx = ab \rightarrow \frac{y}{b} + \frac{x}{a} = 1$.

Generally, two points are known, and from these, $m$ is determined. Then, either of the points are substituted into the point-slope form along with the slope to get the equation of the line.

*Example 3:* Find the equation of the line that passes through (4, 1) and (–2, 5). So, $m = \frac{1-5}{4-(-2)} = \frac{-4}{6} = \frac{-2}{3}$.

Substituting this value for $m$ and the point (4, 1) for $(x_1, y_1)$ into the point-slope form gives $y - 1 = -\frac{2}{3}(x - 4)$.

*Point-Point Form:* Substituting $m = \frac{y_2 - y_1}{x_2 - x_1}$ into the point-slope form and rearranging terms gives $y - y_1 = \frac{y_1 - y_2}{x_1 - x_2}(x - x_1)$, the *point-point form*.

*Slope-Intercept Form:* Another often-used form is $y = mx + b$, where $m$ is the slope and $b$ is the $y$-intercept. (That is, when $x = 0$, then $y = m(0) + b = b$.) If the previous example is rearranged and solved for $y$, then the *slope-intercept form* is arrived at: $y = -\frac{2}{3}(x-4) + 1 = -\frac{2}{3}x + \frac{8}{3} + 1 = -\frac{2}{3}x + \frac{11}{3}$.

*Standard Form of a Line:* In vector representations (see *Complex Numbers* on page 54), lines usually take the standard form, $Ax + By = C$, where $A, B, C$. For example, $7x - 3y = 4$ is a line in standard form, with $A = 7, B = -3$, and $C = 4$.

*Example 4:* What is the standard form equation of a line $AB$ between points $A(4, 5)$ and $B(7,8)$?

*Solution:* Using the point-point form of the line, where (4,5) is $(x_1, y_1)$ and (7,8) is $(x_2, y_2)$,

$$y - y_1 = \frac{y_1 - y_2}{x_1 - x_2}(x - x_1)$$

$$y - 5 = \frac{5-8}{4-7}(x - 4)$$

$$y - 5 = x - 4$$

$$x - y = -1$$

*Example 5:* Find the slope-intercept equation of a line passing through the points (3, 2) and (5,6). The $y$-intercept is the intersection point of the line with the $y$-axis.

*Solution:* First, find the slope:

$$m = \frac{\Delta y}{\Delta x} = \frac{6-2}{5-3} = \frac{4}{2} = 2$$

The slope-intercept form of the line is $y = 2x + b$, and the value of the constant $b$ can be determined by substituting the coordinates of a point on the line into the general form. Using the coordinates of the point (3,2) gives $2 = (2)(3) + b$ and rearranging, $b = 2 - 6 = -4$. As a check, using another point on the line, (5,6), yields the same result, $y = 6 = (2)(5) + b$ and $b = 6 - 10 = -4$.

The equation of the line, therefore, is $y = 2x - 4$, indicating that line $y = 2x - 4$ intersects the $y$-axis at point (0,–4), the $y$-intercept.

*Example 6:* Use the point-slope form to find the equation of the line passing through the point (3,2) and having a slope of 2.

$$\begin{aligned} y - 2 &= 2(x-3) \\ y &= 2x - 6 + 2 \\ y &= 2x - 4 \end{aligned}$$

The slope of this line is positive and crosses the $y$-axis at the $y$-intercept, point (0,–4).

*Parallel Lines:* The two lines, $l_1$ and $l_2$, are parallel if they have the same slope, that is, if $m_1 = m_2$.

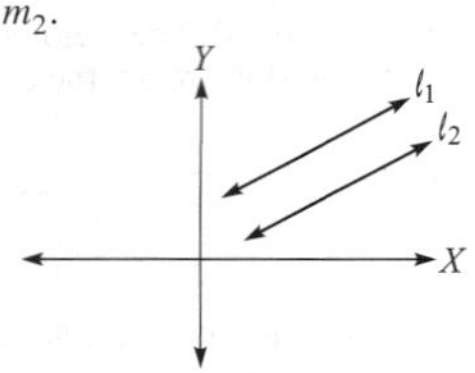


Fig. 5. (a) Parallel lines, $l_1$ and $l_2$.

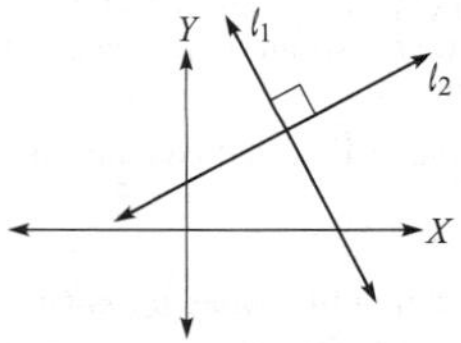


(b) Perpendicular lines, $l_1$ and $l_2$.

*Perpendicular Lines:* The two lines, $l_1$ and $l_2$, are perpendicular if the product of their slopes is −1, that is, if $m_1 m_2 = -1$, since $m_1 = -1/m_2$. (The slopes are negative reciprocals of one another.)

*Example 7:* (a) Find the equation of the line that passes through the point (3, 4) and is parallel to line $2x - 3y = 16$. (b) Find the equation of the line perpendicular to the given line and through the same point.

*Solution (a):* Line $2x - 3y = 16$ in slope-intercept form is $y = {}^2\!/_3 x - {}^{16}\!/_3$, so the equation of a line passing through (3,4) is $y - 4 = m(x - 3)$.

Parallel lines have equal slope. Thus, from the point-slope form, $y - 4 = {}^2\!/_3(x - 3)$ is parallel to line $2x - 3y = -6$ and passes through point (3,4).

*Solution (b):* As illustrated in part (*a*), line $2x - 3y = -6$ has a slope of ${}^2\!/_3$. The product of the slopes of perpendicular lines is −1, thus the slope $m$ of a line passing through point (3,4) and perpendicular to $2x - 3y = -6$ must satisfy the following:

$$m = \frac{-1}{m_1} = \frac{-1}{{}^2\!/_3} = -\frac{3}{2}$$

The equation of a line passing through point (3,4) and perpendicular to the line $2x - 3y = 16$ is $y - 4 = {}^{-3}\!/_2(x - 3)$, which rewritten is $3x + 2y = 17$.

*Angle Between Two Lines:* For two non-perpendicular lines with slopes $m_1$ and $m_2$, the angle θ between the two lines is found by first applying trigonometric equation:

$$\tan\theta = \left|\frac{m_1 - m_2}{1 + m_1 m_2}\right|$$

The discussion of how the angle is determined by this relation is found in *TRIGONOMETRY: SOLUTION OF TRIANGLES*, which begins on page 89.

*Example 8:* Find the angle between the lines: $2x - y = 4$ and $3x + 4y = 12$.

*Solution:* Rearranging each to be in the slope-intercept form shows the slopes are 2 and $-{}^3\!/_4$, respectively. The angle between two lines is given by

$$\tan\theta = \left|\frac{m_1 - m_2}{1 + m_1 m_2}\right| = \left|\frac{2 - \left(-\frac{3}{4}\right)}{1 + 2\left(-\frac{3}{4}\right)}\right| = \left|\frac{2 + \frac{3}{4}}{1 - \frac{6}{4}}\right| = \left|\frac{\frac{8+3}{4}}{\frac{4-6}{4}}\right| = \left|\frac{11}{-2}\right| = \frac{11}{2}$$

$$\theta = \tan^{-1}\left(\frac{11}{2}\right) = 79.70°, \text{ by trigonometry.}$$

*Distance Between a Point and a Line:* The distance between a point $(x_1, y_1)$ and a line given in the standard form $Ax + By + C = 0$ is

$$d = \frac{|Ax_1 + By_1 + C|}{\sqrt{A^2 + B^2}}$$

*Example 9:* Find the distance between the point (4,6) and the line $2x + 3y - 9 = 0$.

*Solution:* Using the formula:

$$d = \frac{|Ax_1 + By_1 + C|}{\sqrt{A^2 + B^2}} = \frac{|2 \times 4 + 3 \times 6 - 9|}{\sqrt{2^2 + 3^2}} = \frac{|8 + 18 - 9|}{\sqrt{4 + 9}} = \frac{17}{\sqrt{13}}$$

**Changing Coordinate Systems.—**For simplicity it may be assumed that the origin in the Cartesian coordinate system coincides with the pole on a polar coordinate system and its *x*-axis with the polar horizontal axis. Then, if point *P* has polar coordinates of (*r*, θ) and Cartesian coordinates of (*x*, *y*), by trigonometry $x = r\cos\theta$ and $y = r\sin\theta$.

Furthermore, by the Pythagorean theorem and trigonometry, $r = \sqrt{x^2 + y^2}$ and $\theta = \tan^{-1}\left(\frac{y}{x}\right)$. See *TRIGONOMETRY: SOLUTION OF TRIANGLES* on page 89 for a discussion of the related principles of trigonometry.

*Example 1:* Convert the Cartesian coordinate (3, 2) into polar coordinates.

$$r = \sqrt{3^2 + 2^2} = \sqrt{9 + 4} = \sqrt{13} \approx 3.6 \qquad \theta = \tan^{-1}\left(\frac{2}{3}\right) = 33.69°$$

Therefore the point is located approximately 3.6 units from the origin at an angle of about 33.69°. Thus, (3.6, 33.69°) is the polar form of the Cartesian point (3, 2).

Graphically, the polar and Cartesian coordinates are related in the following figure:

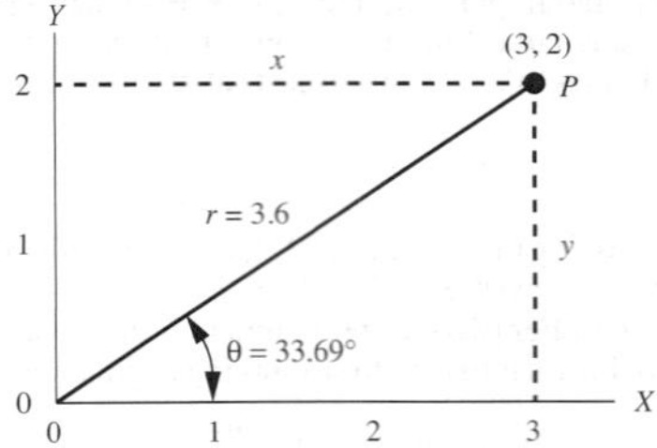


*Example 2:* Convert the polar form (5, 608°) to Cartesian coordinates. First note that this point lies 5 units from the origin at an angle of 608°. As explained on page 100, in *Trigonometric Functions*, this locates the point in Quadrant IV. By trigonometry, $x = r\cos\theta$ and $y = r\sin\theta$. Then, $x = 5\cos(608°) = -1.873$ and $y = 5\sin(608°) = -4.636$. Therefore, the Cartesian point equivalent is (−1.873, −4.636). This point lies in the fourth quadrant, where both coordinates are negative.

**Spherical Coordinates.—**It is convenient in certain problems, for example, those concerned with spherical surfaces and therefore three-dimensional, to introduce spherical coordinates. In three-dimensional space, as the figure on the right shows, the $x,y$-plane is like a floor in a room, and the third dimension is given by the $z$-axis, which is where the walls of the room meet. An arbitrary point $P$ in this space is described by three rectangular coordinates $(x, y, z)$, converted to the following *spherical coordinates*: the distance $r$ between point $P$ and the origin $O$, the angle $\phi$ that $OP'$ makes with the $x, y$-plane, and the angle $\lambda$ that the projection $OP'$ (the "shadow" of the segment $OP$ on the $x, y$-plane) makes with the positive $x$-axis.

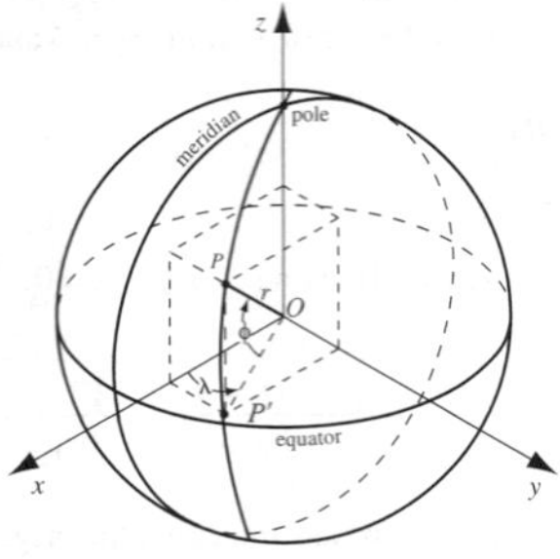


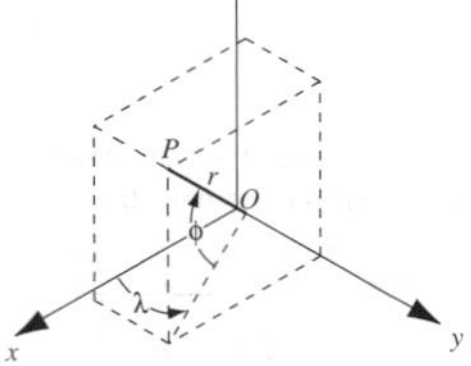


The rectangular coordinates of a point in space can therefore be calculated from the spherical coordinates, and vice versa, by use of the formulas in the following table.

**Relationship Between Spherical and Rectangular Coordinates**

| Spherical to Rectangular | Rectangular to Spherical |
|---|---|
| $x = r\cos\phi\cos\lambda$<br>$y = r\cos\phi\sin\lambda$<br>$z = r\sin\phi$ | $r = \sqrt{x^2+y^2+z^2}$<br>$\phi = \tan^{-1}\frac{z}{\sqrt{x^2+y^2}}$ for $x^2+y^2 \neq 0$<br>$\lambda = \tan^{-1}\left(\frac{y}{x}\right)$ for $x>0, y>0$<br>$\lambda = \pi + \tan^{-1}\left(\frac{y}{x}\right)$ for $x<0$<br>$\lambda = 2\pi + \tan^{-1}\left(\frac{y}{x}\right)$ for $x>0, y<0$ |

*Example 3:* Find the spherical coordinates of the point $P(3,-4,-12)$.

$$r = \sqrt{3^2+(-4)^2+(-12)^2} = 13$$

$$\phi = \tan^{-1}\frac{-12}{\sqrt{3^2+(-4)^2}} = \tan^{-1}\left(-\frac{12}{5}\right) = -67.38°$$

$$\lambda = 360° + \tan^{-1}\left(-\frac{4}{3}\right) = 360° - 53.13° = 306.87°$$

The spherical coordinates of $P$ are therefore $r=13, \phi=-67.38°$, and $\lambda=306.87°$.

*Cylindrical Coordinates:* For problems in which points lie on the surface of a cylinder it is convenient to use cylindrical coordinates. The cylindrical coordinates $r$, $\theta$, $z$ of $P$ coincide with the polar coordinates of the point $P'$ in the $x$, $y$-plane and the rectangular $z$-coordinate of $P$. Formulas for $\theta$ hold only if $x^2+y^2 \neq 0$; $\theta$ is undetermined if $x=y=0$.

| Cylindrical to Rectangular | Rectangular to Cylindrical |
|---|---|
| $x = r\cos\theta$<br>$y = r\sin\theta$<br>$z = z$ | $r = \frac{\sqrt{x^2+y^2}}{1}$<br>$\cos\theta = \frac{x}{\sqrt{x^2+y^2}}$<br>$\sin\theta = \frac{y}{\sqrt{x^2+y^2}}$<br>$z = z$ |

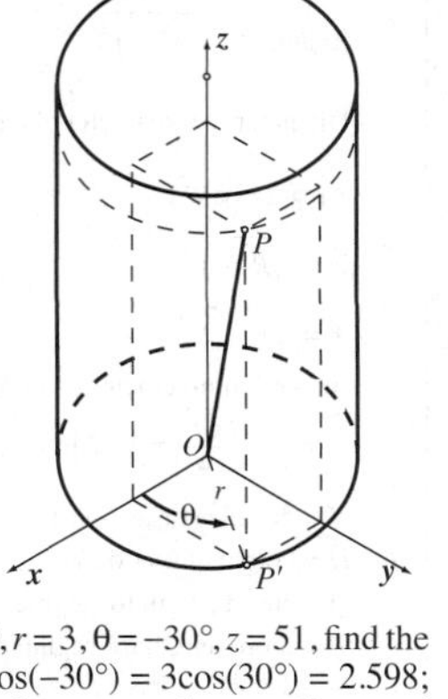


*Example 4:* Given the cylindrical coordinates of a point $P$, $r=3, \theta=-30°, z=51$, find the rectangular coordinates. Using the above formulas $x = 3\cos(-30°) = 3\cos(30°) = 2.598$; $y = 3\sin(-30°) = -3\sin(30°) = -1.5$; and $z = 51$. Therefore, the rectangular coordinates of point $P$ are $x=2.598, y=-1.5$, and $z=51$.

**Circle.—**A circle is the set of points equidistant from a given point in the plane. Another name for this set of points is *locus*, which is a curve formed by all the points satisfying a particular equation. The general form for the equation of a circle is $x^2 + y^2 + 2gx + 2fy + c = 0$, where $-g$ and $-f$ are the coordinates of the center and the radius is $r = \sqrt{g^2 + f^2 - c}$.

The standard form of a circle (center-radius form) is

$$(x-h)^2 + (y-k)^2 = r^2$$

where $r$ = length of the radius and point $(h, k)$ is the center.

When the center of circle is at point (0, 0), the equation reduces to $x^2 + y^2 = r^2$ or $r = \sqrt{x^2 + y^2}$

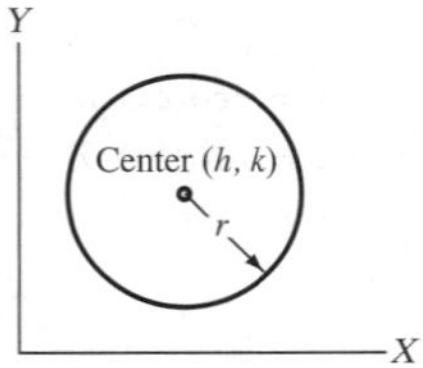


*Example 1:* Point (4, 6) lies on a circle whose center $(h, k)$ is at point (–2, 3). Find the circle's equation.

*Solution:* The radius is the distance from the center point (–2, 3) to point (4, 6), found using the method of Example 1 on page 38.

$$r = \sqrt{[4-(-2)]^2 + (6-3)^2} = \sqrt{6^2 + 3^2} = \sqrt{45}$$

Using the form $(x-h)^2 + (y-k)^2 = r^2$ and substituting $h = -2$, $k = 3$, and $r^2 = 45$:

$$(x+2)^2 + (y-3)^2 = x^2 + 4x + 4 + y^2 - 6y + 9 = 45$$

$$x^2 + y^2 + 4x - 6y - 32 = 0$$

*Additional Formulas:* Listed below are additional formulas for determining the geometry of plane circles and arcs. Although trigonometry and circular measure are related, they deal with angles in entirely different ways. In each of these formulas, the entered measure of the angles are in degrees, and the formulas convert them to radian measure (see page 89, in *TRIGONOMETRY: SOLUTION OF TRIANGLES*).

$C = \pi D = 2\pi R$

Radius $R = \sqrt{X^2 + Y^2}$

Diameter $D$ = diameter of circle = $2R = \frac{C}{\pi}$

Area $A = \pi R^2$

$X = \sqrt{R^2 - Y^2}$

$Y = \sqrt{R^2 - X^2}$

Area of complement sector $M$

$= R^2 - \frac{\pi R^2}{4} = 0.2146R^2$

$I$ = distance to section $T$

$H$ = height of section $T$

$Q$ = chord length for segment $S$

$P$ = chord length for segment section $T$

$T + S$ = area of segment = $R^2 \sin^{-1}\left(\frac{P}{2R}\right) - \frac{IP}{2}$

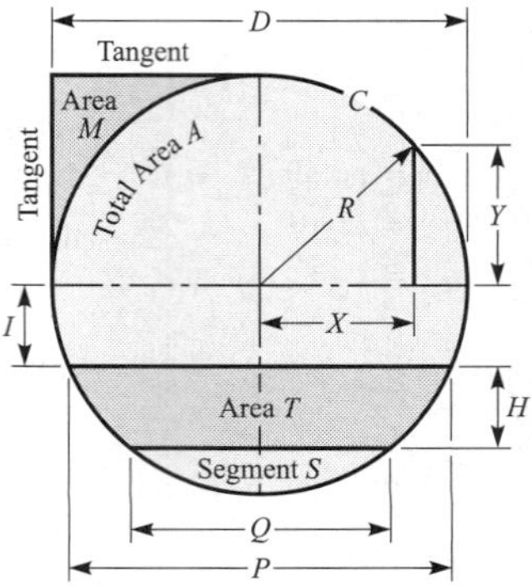


Fig. 6a.

Central angle $\phi$, in degrees $= \dfrac{L}{R} \cdot \dfrac{\pi}{180}$

Arc length $L = R\phi \cdot \dfrac{\pi}{180}$

Radius $R = \dfrac{L}{\phi} \cdot \dfrac{\pi}{180} = \dfrac{2K}{L} = \dfrac{E^2 + 4F^2}{8F}$

Area of sector $K = \phi R^2 \cdot \dfrac{\pi}{360} = \dfrac{RL}{2}$

Area of segment $S = \dfrac{RL}{2} - \dfrac{E(R-F)}{2}$

Chord length $E = 2\sqrt{F(2R-F)} = D \sin\left(\dfrac{\phi}{2}\right)$

Depth $F = R - \dfrac{\sqrt{4R^2 - E^2}}{2} = R\left(1 - \cos\left(\dfrac{\phi}{2}\right)\right)$

$\tan\left(\dfrac{\phi}{2}\right) = \dfrac{E/2}{R-F}$

$\sin\left(\dfrac{\phi}{2}\right) = \dfrac{E}{2R}$

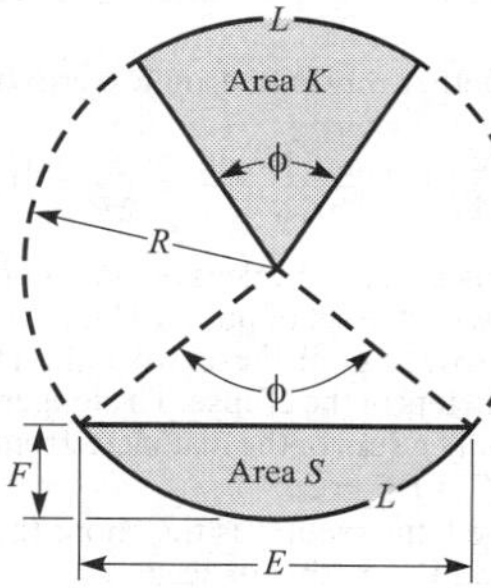


Fig. 6b.

*Annulus*

$R_1$ = radius of outer circle

$R_2$ = radius of inner circle

Area of annulus $W = \pi\left(R_1^2 - R_2^2\right)$

Area of annulus segment $U = \dfrac{\phi}{360}\pi\left(R_1^2 - R_2^2\right)$

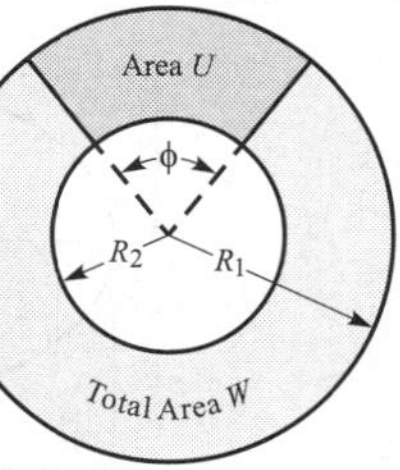


Fig. 6c.

*Example 2:* Find the area of a sector of a circle having a central angle of 30° and a radius of 7 cm.

*Solution:* Referring to Fig. 6b, $K = \left(\dfrac{\phi°}{360}\right)\pi \times R^2 = \left(\dfrac{30}{360}\right)\pi \times 7^2 = 12.83\ \text{cm}^2$

*Example 3:* Find the chord length $E$ of a circular segment (Fig. 6b) with a depth of 2 cm at the center that is formed in a circle whose radius 12 cm.

*Solution:* The chord length is

$$E = 2\sqrt{F(2R-F)} = 2\sqrt{2[(2)(12)-2]} = 2\sqrt{44}$$
$$= 4\sqrt{11} = 13.27\ \text{cm}$$

*Example 4:* Find the area $S$ of the segment in Example 3.

*Solution:* First determine angle $\phi$, then find arc length $L$ of the segment, and then solve for area $S$, as follows:

$$\tan\left(\frac{\phi}{2}\right) = \frac{E/2}{R-F} = \frac{13.27/2}{12-2} = 0.6635, \quad \frac{\phi}{2} = \tan^{-1}0.6635 = 33.56°, \quad \phi = 67.13°$$

$$L = R\phi \times \frac{\pi}{180} = 12 \times 67.13° \times \frac{\pi}{180} = 14.06 \text{ cm}$$

$$\text{Area } S = \frac{RL}{2} - \frac{E(R-F)}{2} = \frac{12(14.06)}{2} - \frac{13.27(10)}{2} = 84.36 - 66.35 = 18.01 \text{ cm}^2$$

Another way to find angle $\phi$ is divide the chord length by twice the radius to obtain

$$\sin\left(\frac{\phi}{2}\right) = \frac{\text{chord length}}{2R} = \frac{E}{2R} = 0.5529, \quad \frac{\phi}{2} = \sin^{-1} 0.5529 = 33.5662°, \quad \phi = 67.13°$$

**Ellipse.**—As a circle is the locus of points equidistant from a single point in the plane, an *ellipse* is the set of points whose location is established by *two* points in the plane. Referring to the figure, these two points are the *foci*, $F_1$ and $F_2$, which lie on the longer of the two diameters of the ellipse. The longer diameter $a$ is the major axis; the shorter $b$ is the minor axis. The sum of the distances from the foci to any point $P$ on the ellipse is constant. That is $PF_1 + PF_2 = 2a$.

The latus rectum is the chord through the focus and perpendicular to the major axis. $V_1$ and $V_2$, are the vertices.

Like the circle, there is a general form and a standard form of the equation of an ellipse. The general form is:

$$Ax^2 + Cy^2 + Dx + Ey + F = 0\Big|_{AC>0} \text{ and } A \neq C$$

The constant $F$ in this equation is not related to the foci, which are not numbers, but labels for points.

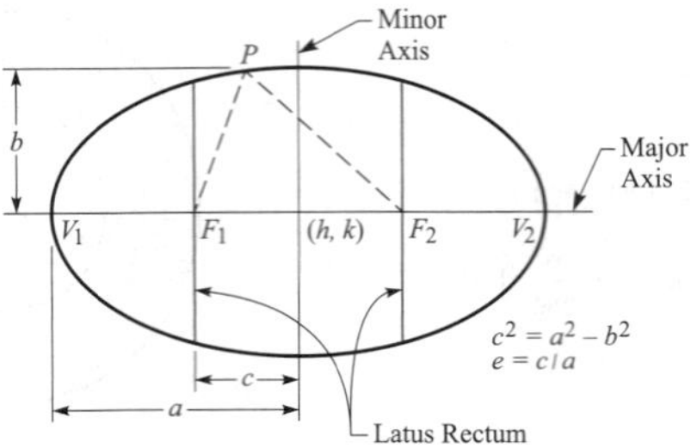


Ellipse

If $(h, k)$ is the center, the standard equation of an ellipse is $\dfrac{(x-h)^2}{a^2} + \dfrac{(y-k)^2}{b^2} = 1.$

The eccentricity $e$ of the ellipse is given by $e = c/a$, where $c^2 = a^2 - b^2$. This is *not* the same $e$ as the exponential base, which is a constant. Rather, eccentricity varies with the figure. $e = \dfrac{\sqrt{a^2 - b^2}}{a} = \dfrac{c}{a}$ is a measure of the elongation of the ellipse and is always less than 1. The distance between the two foci is $2c = 2\sqrt{a^2 - b^2}$. The aspect ratio of the ellipse is $a/b$. The equation of an ellipse centered at $(h, k) = (0, 0)$ with foci at $(\pm c, 0)$ is $\dfrac{x^2}{a^2} + \dfrac{y^2}{b^2} = 1$, and the ellipse is symmetric about both coordinate axes. Its $x$-intercepts are $(\pm a, 0)$ and $y$-intercepts are $(0, \pm b)$. The line segment joining $(0, b)$ and $(0, -b)$ is called the minor axis. The major vertices of the ellipse are $(\pm a, 0)$, and the line segment joining vertices $V_1$ and $V_2$ is the major axis of the ellipse.

*Example 1:* Determine the values of $h, k, a, b, c$, and $e$ of the following ellipse:

$$3x^2 + 5y^2 - 12x + 30y + 42 = 0$$

*Solution:* Rearrange the ellipse equation into the standard form as follows, using the method of completing the square (see page 29):

$$3x^2 + 5y^2 - 12x + 30y + 42 = 3x^2 - 12x + 5y^2 + 30y + 42 = 0$$

$$3(x^2 - 4x + 2^2) + 5(y^2 + 6y + 3^2) = 15$$

$$\frac{3(x-2)^2}{15} + \frac{5(y+3)^2}{15} = \frac{(x-2)^2}{(\sqrt{5})^2} + \frac{(y+3)^2}{(\sqrt{3})^2} = 1$$

Comparing to the form $\frac{(x-h)^2}{a^2} + \frac{(y-k)^2}{b^2} = 1$ and solving for $c$ and $e$ gives:

$$h = 2, \quad k = -3, \quad a = \sqrt{5}, \quad b = \sqrt{3}, \quad c = \sqrt{2}, \quad e = \sqrt{\frac{2}{5}}$$

*Additional Formulas:* An ellipse can be represented "parametrically" by the equations $x = a\cos\theta$ and $y = b\sin\theta$, where $x$ and $y$ are the rectangular coordinates of any point on the ellipse, and the parameter $\theta$ is the angle at the center measured from the $x$-axis counter-clockwise. The following figures correspond to the formulas below them for other measurements of the ellipse.

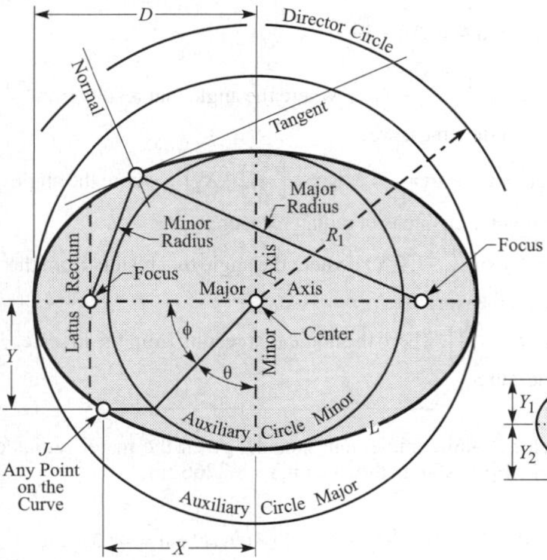


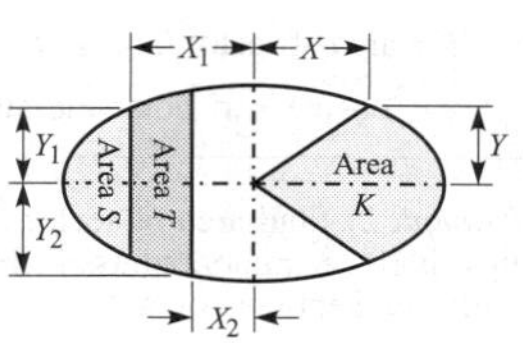


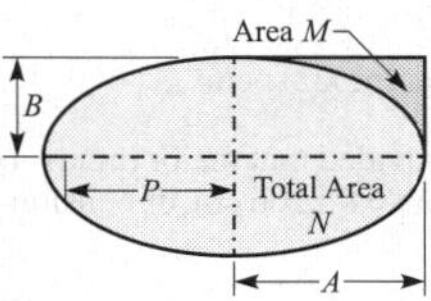


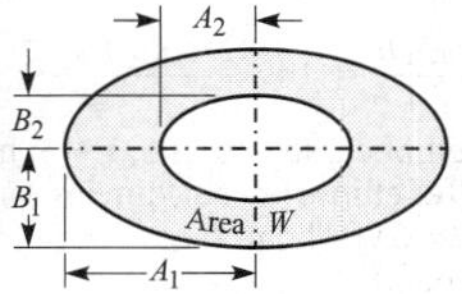


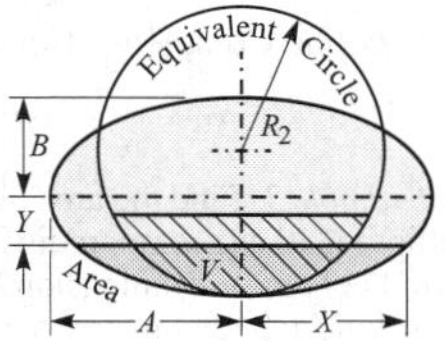

*Length, Point, and Angle Calculations*

$R_1$ = radius of director circle = $\sqrt{A^2 + B^2}$ , $A$ = major radius = $\sqrt{B^2 + P^2}$

$R_2$ = radius of equivalent circle = $\sqrt{AB}$ , $B$ = minor radius = $\sqrt{A^2 - P^2}$

$P$ = center to focus distance = $\sqrt{A^2 - B^2}$ , $\dfrac{2B^2}{A}$ = distance, origin to latus rectum

$J$ = any point $(X, Y)$ on curve where $X = A\sin\theta = A\cos\phi$, and $Y = B\cos\theta = B\sin\phi$

$\phi$ = angle with major axis = $\sin^{-1}\left(\dfrac{Y}{B}\right) = \cos^{-1}\left(\dfrac{X}{A}\right)$, $\theta$ = angle with minor axis = $90° - \phi$

$L$ = total perimeter (approximate) = $A\left[1.2\left(\dfrac{B}{A}\right)^2 + 1.1\left(\dfrac{B}{A}\right) + 4\right]$

$L$ = perimeter (sections) = $\left(\dfrac{\pi}{180}\right)2\phi\sqrt{AB}$

*Area Calculations*

$N$ = total surface area of ellipse = $\pi AB$

$W$ = area between outer and inner ellipse = $\pi(A_1B_1 - A_2B_2)$

$M$ = area of complement section $M = AB - \dfrac{\pi AB}{4}$

$S$ = area of segment $S = AB\cos^{-1}\left(\dfrac{X_1}{A}\right) - X_1Y_1$, where the angle that results from the inverse cosine is in radian measure

$T+S$ = combined area of segment $S$ + area $T = AB\cos^{-1}\left(\dfrac{X_2}{A}\right) - X_2Y_2$, where the angle that results from the inverse cosine is in radian measure

$V$ = area of segment $V = (R_2)^2\sin^{-1}\left(\dfrac{X}{A}\right) - XY$, where the angle that results from the inverse sine is in radian measure

$K$ = area of sector $K = AB\cos^{-1}\left(\dfrac{X}{A}\right)$, where the angle that results from the inverse cosine is in radian measure

*Example 2:* Find area of sector $K$ and complement area $M$, given the major radius of ellipse is 10 cm, minor radius of ellipse is 7 cm, dimension $X = 8.2266$ cm.

*Solution:* Sectional area $K$:

$$\text{Area } K = AB\cos^{-1}\left(\frac{X_1}{A}\right) = 10 \times 7 \times \cos^{-1}\left(\frac{8.2266}{10}\right) = 70 \times 0.6047 \text{ rad} = 42.33 \text{ cm}^2$$

*Solution:* Complement area $M$:

$$\text{Area } M = AB - \frac{\pi AB}{4} = 10 \times 7 - \frac{\pi \times 10 \times 7}{4} = 15.0221 \text{ cm}^2$$

*Example 3:* Find the area of elliptical segments $S$, $T + S$, provided that major radius $A$ of ellipse is 10 cm, minor radius $B$ of ellipse is 7 cm, dimension $X_1 = 8.2266$ cm, dimension $Y_1 = 4.4717$ cm, and dimension $X_2 = 6.0041$ cm.

*Solution:* Segment area $S$ is found:

$$S = AB\cos^{-1}\left(\frac{X_1}{A}\right) - X_1Y_1 = 10 \times 7 \times \cos^{-1}\left(\frac{8.2266}{10}\right) - 8.2266 \times 4.4717 = 5.5437 \text{ cm}^2$$

*Solution:* Segment area $T + S$:

$$\phi = \cos^{-1}\left(\frac{X_2}{A}\right) = 1.51072 \text{ rad}, \quad Y_2 = B\sin\phi = 7\sin(1.51072 \text{ rad}) = 10.57504$$

$$T + S = AB\cos^{-1}\left(\frac{X_2}{A}\right) - X_2Y_2 = 10 \times 7 \times 0.9268 - (6.0041 \times 5.5978)$$

$$= 64.876 - 33.6097 = 31.266 \text{ cm}^2$$

*Example 4:* Find the area of elliptical segment $V$ if major radius of ellipse is 4 inches, minor radius is 3 inches, dimension $X = 2.3688$ inches, dimension $Y = 2.4231$ inches.

*Solution:* Segment area $V$:

$$R_2 = \sqrt{AB}, \quad (R_2)^2 = AB = 3 \times 4 = 12$$

$$V = (R_2)^2\sin^{-1}\left(\frac{X}{A}\right) - XY = 12\sin^{-1}\left(\frac{2.3688}{4}\right) - (2.3688 \times 2.4231)$$

$$= 7.6054 - 5.7398 = 1.8656 \text{ in}^2$$

**Four-Arc Oval Approximating an Ellipse*.—**The method of constructing an approximate ellipse by circular arcs, described on page 64, fails when the ratio of the major to minor diameter is 4 or greater. Additionally, it is reported that the method always produces a somewhat larger minor axis than intended. The method described below presents an alternative.

An oval that approximates an ellipse, illustrated in Fig. 7, can be constructed from the following equations:

$$r = \frac{B^2}{2A}\left(\frac{A}{B}\right)^{0.38} \quad (1)$$

where $A$ and $B$ are dimensions of the major and minor axis, respectively, and $r$ is the radius of the curve at the long ends.

The radius $R$ and its location are found from Equations (2) and (3):

$$X = \frac{\frac{A^2}{4} - Ar + Br - \frac{B^2}{4}}{B - 2r} \quad (2) \qquad R = \frac{B}{2} + X \quad (3)$$

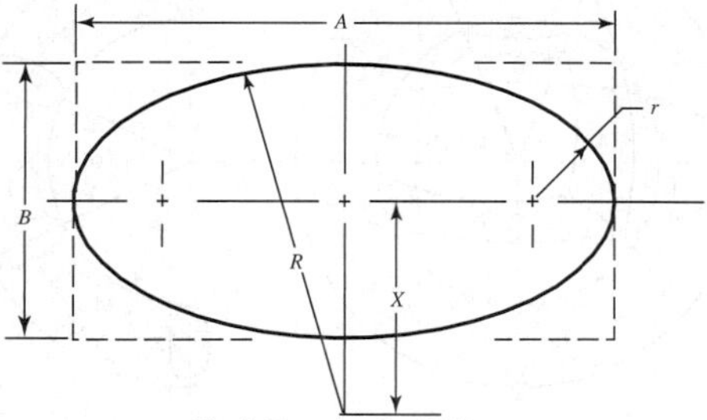


Fig. 7. Four-Arc Oval Ellipse

To make an oval thinner or fatter than what is given, select a smaller or larger radius $r$ is chosen than that calculated by Equation (1), and then $X$ and $R$ are found using Equations (2) and (3).

* *Four-Arc Oval* material contributed by Manfred K. Brueckner

**Sphere.**—A sphere is the locus of points equidistant from a given point (the sphere's center) in three-dimensional space. Similar to the circle, the standard form for the equation of a sphere of radius $R$ and centered at point $(h, k, l)$ is:

$$(x-h)^2 + (y-k)^2 + (z-l)^2 = R^2$$

The general form for the equation of a sphere can be written as follows, where $A \neq 0$:

$$Ax^2 + Ay^2 + Az^2 + Bx + Cy + Dz + E = 0$$

The general and standard forms of the sphere equations are related as follows:

$$h = \frac{-B}{2A} \qquad k = \frac{-C}{2A} \qquad l = \frac{-D}{2A} \qquad R = \sqrt{\frac{B^2 + C^2 + D^2}{4A^2} - \frac{E}{A}}$$

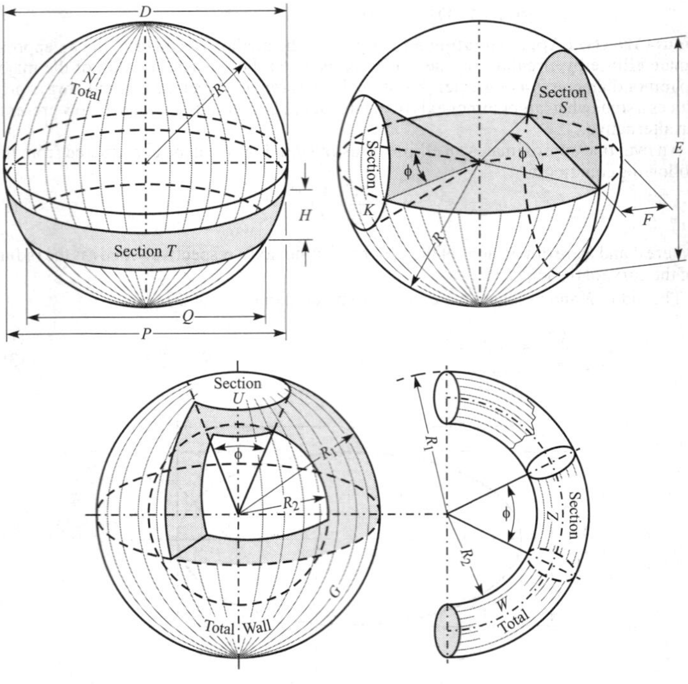


$R$ = radius of sphere
$D$ = diameter of sphere
$N_s$ = total surface area of sphere
$N_v$ = total volume of sphere

$R_1$ = radius of outer sphere
$R_2$ = radius of inner sphere
$G_a, K_a, S_a, T_a, U_a, W_a, Z_a$ = sectional surface areas
$G_v, K_v, S_v, T_v, U_v, W_v, Z_v$ = sectional volumes

**Formulas for Spherical Radius**

| To Find | Formula | To Find | Formula |
|---|---|---|---|
| Radius of sphere from volume $N_v$ | $R_N = \sqrt[3]{\frac{3N_v}{4\pi}}$ | Radius of section $T$ | $R_T = \sqrt{\left(\frac{P^2 - Q^2 - 4H^2}{8H}\right)^2 + \frac{P^2}{4}}$ |

**Formulas for Spherical Areas and Volumes**

| Section | Area Formula | Volume Formula |
|---|---|---|
| Entire sphere | $N_a = 4\pi R^2$ | $N_v = \left(\frac{\pi}{6}\right)D^3 = \frac{4\pi}{3}R^3$ |
| Section $G$ | $G_a = 4\pi R_1^2 + 4\pi R_2^2$ | $G_v = \frac{4\pi}{3}(R_1^3 - R_2^3)$ |
| Section $K$ | $K_a = 2\pi R^2\left(1 - \cos\left(\frac{\phi}{2}\right)\right)$ | $K_v = \frac{2\pi R^2 F}{3}$ |
| Section $S$ | $S_a = \pi\left(F^2 + \frac{E^2}{4}\right)$ | $S_v = \pi F\left(\frac{E^2}{8} + \frac{F^2}{6}\right)$ |
| Section $T$ | $T_a = 2\pi RH$ | $T_v = H\left(\frac{\pi}{6}\right)\left(H^2 + \frac{3Q^2}{4} + \frac{3P^2}{4}\right)$ |
| Section $U$ | $U_a = 2\pi(R_1^2 + R_2^2)\left(1 - \cos\left(\frac{\phi}{2}\right)\right)$ | $U_v = 2\pi(R_1^3 - R_2^3)\left(1 - \cos\left(\frac{\phi}{2}\right)\right)$ |
| Section $W$ | $W_a = 4\pi^2 R_1 R_2$ | $W_v = 2\pi^2 R_1 R_2^2$ |
| Section $Z$ | $Z_a = (4\pi^2 R_1 R_2)\frac{\phi}{360}$ | $Z_v = (2\pi^2 R_1 R_2^2)\left(\frac{\phi}{360}\right)$ |

*Example 1:* Find the inside and outside surface area $G_a$ and volume $G_v$ of wall $G$, provided that $R_1$ is 12.5 cm and $R_2$ is 10.0 cm.

*Solution:* Sectional area $G_a$ and sectional volume $G_v$:

$$G_a = 4\pi R_1^2 + 4\pi R_2^2 = 4\pi(12.5)^2 + 4\pi(10)^2 = 3220.13 \text{ cm}^2$$

$$G_v = \frac{4\pi}{3}(R_1^3 - R_2^3) = \frac{4\pi}{3}(12.5^3 - 10^3) = 3992.44 \text{ cm}^3$$

*Example 2:* Find the surface area $K_a$ and volume $K_v$ of section $K$ of a sphere of radius 15.0 cm, if included angle $\phi = 90°$ and depth $F = 5.0$ cm.

*Solution:* Sectional area $K_a$ and sectional volume $K_v$:

$$K_a = 2\pi R^2\left(1 - \cos\left(\frac{\phi}{2}\right)\right) = 2\pi(15^2)\left(1 - \cos\left(\frac{90°}{2}\right)\right) = 414.07 \text{ cm}^2$$

$$K_v = \frac{2\pi R^2 F}{3} = \frac{2\pi(15)^2(5)}{3} = 2356.19 \text{ cm}^3$$

*Example 3:* Find the outside surface area $S_a$ and sectional volume $S_v$ of section $S$ of a sphere if $E = 20.0$ cm and $F = 5.0$ cm.

*Solution:* Sectional area $S_a$ and sectional volume $S_v$:

$$S_a = \pi\left(F^2 + \frac{E^2}{4}\right) = \pi\left(5^2 + \frac{20^2}{4}\right) = 392.70 \text{ cm}^2$$

$$S_v = \pi F\left(\frac{E^2}{8} + \frac{F^2}{6}\right) = \pi \times 5\left(\frac{20^2}{8} + \frac{5^2}{6}\right) = 850.85 \text{ cm}^3$$

*Example 4:* Find the outside and inside surface area $U_a$ and volume $U_v$ of section $U$ of a sphere if $R_1 = 5.0$ inches, $R_2 = 4.0$ inches, and included angle $\phi = 30°$.

*Solution:* Sectional area $U_a$ and sectional volume $U_v$:

$$U_a = 2\pi(R_1^2 + R_2^2)\left(1 - \cos\left(\frac{\phi}{2}\right)\right) = 2\pi(5^2 + 4^2)\left(1 - \cos\left(\frac{30°}{2}\right)\right) = 8.78 \text{ in}^2$$

$$U_v = 2\pi(R_1^3 - R_2^3)\left(1 - \cos\left(\frac{\phi}{2}\right)\right) = 2\pi(5^3 - 4^3)\left(1 - \cos\left(\frac{30°}{2}\right)\right) = 13.06 \text{ in}^3$$

*Example 5:* Find the total surface area $W_a$ and volume $W_v$ of ring $W$, if $R_1 = 5.0$ inches and $R_2 = 4.0$ inches.

*Solution:* Sectional area $W_a$ and sectional volume $W_v$:

$$W_a = 4\pi^2 R_1 R_2 = 4\pi^2 \times 5 \times 4 = 789.57 \text{ in}^2$$

$$W_v = 2\pi^2 R_1 R_2^2 = 2\pi^2 \times 5 \times 4^2 = 1579.14 \text{ in}^3$$

**Parabola.**—A parabola is the set of all points $P$ in the plane that are equidistant from a focus $F$ and a line called the directrix. The parts of the parabola are labeled in the figure below.

The general equation of a parabola horizontal axis is given by $(y-k)^2 = 4p(x-h)$, where the vertex is located at point $(h, k)$, the focus $F$ at point $(h+p, k)$, and the directrix is the line $x$-axis at $x = h - p$. The latus rectum is the vertical line segment through the focus; its endpoints are on the parabola. Hence, it lies on the line $x = h + p$. The length of the latus rectum is four times the absolute value of the $x$-coordinate of the focus, hence $|4(h+p)|$.

*Example:* Determine the focus, directrix, parabolic axis, vertex, and length of the latus rectum of the parabola

$$4y^2 - 8x - 12y + 1 = 0$$

*Solution:* Rewrite the equation in the general form of a parabolic equation (see *Solving by Completing the Square* on page 29).

$$4y^2 - 8x - 12y + 1 = 0$$

$$4y^2 - 12y = 8x - 1$$

$$y^2 - 3y = 2x - \frac{1}{4}$$

$$y^2 - 3y\frac{3}{2} + \left(\frac{3}{2}\right)^2 = 2x - \frac{1}{4} + \frac{9}{4}$$

$$\left(y - \frac{3}{2}\right)^2 = 2(x+1)$$

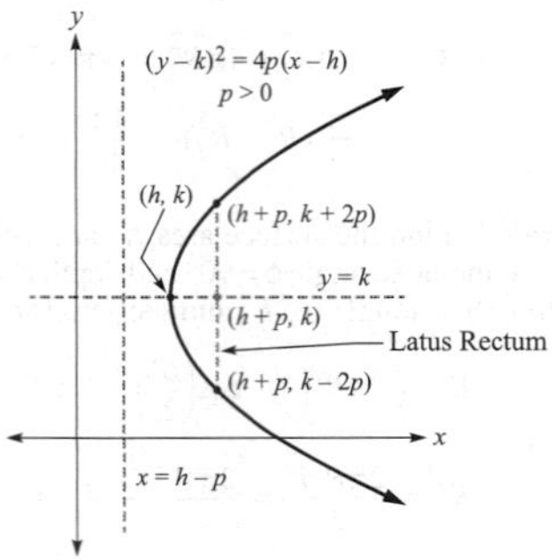


Parabola

Thus, $k = 3/2$, $h = -1$ and $p = 1/2$. Focus $F$ is located at point $(h + p, k) = (1/2, 3/2)$; the directrix is located at $x = h - p = -1 - 1/2 = -3/2$; the parabolic axis is the horizontal line $y = 3/2$; the vertex $V(h, k)$ is located at point $(-1, 3/2)$; and the latus rectum lies on the line $x = h + p = -1/2$. Its length is $4|-1/2| = 4(1/2) = 2$.

**Hyperbola.**—Referring to the figure on the left, below, a hyperbola is the set of all points such that $|d_1 - d_2|$ is constant. That is, the difference between the distances from any point $(x, y)$ to the *foci*, marked $F_1(-c,0)$ and $F_2(c,0)$ does not change. The distance between the *vertices*, $V_1(-a, 0)$ and $V_2(a,0)$ (the turning points of the hyperbola) is $2a$. Therefore, $|d_1 - d_2| = 2a$ for any two points on the hyperbola.

The figure on the right shows more detail. The slopes of the asymptotes (lines of approach) relate to the transverse and conjugate axis lengths, $2a$ and $2b$. The center of a hyperbola is the point of intersection of the asymptotes. In the figure, the center is shown as the origin, (0, 0). The standard form of the hyperbola, as derived from the foci with center at the origin, is $\frac{x^2}{a^2} - \frac{y^2}{b^2} = 1$. For any center $(h, k)$, the equation is $\frac{(x-h)^2}{a^2} - \frac{(y-k)^2}{b^2} = 1$.

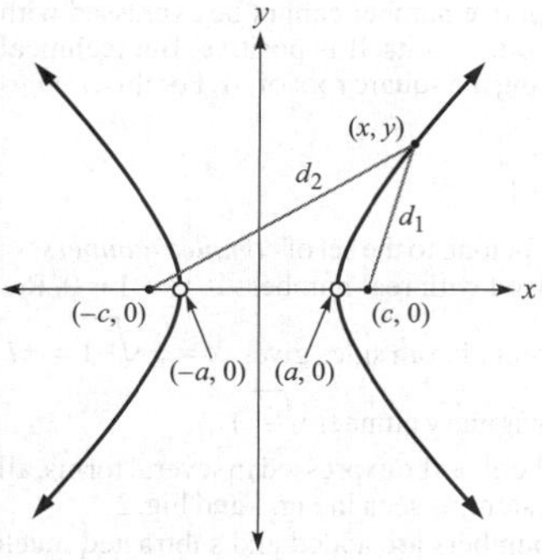


Hyperbola 1

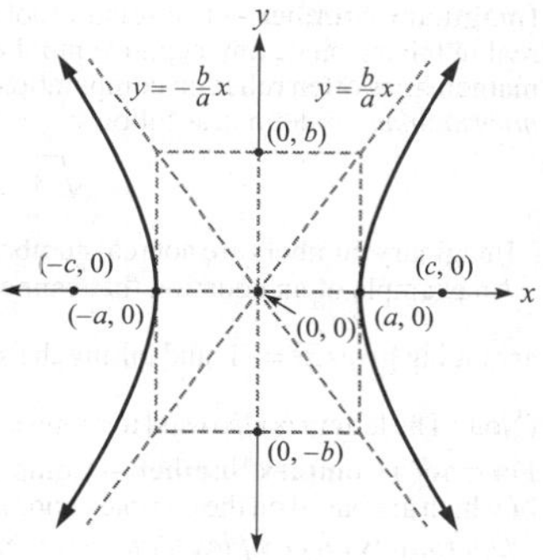


Hyperbola 2

The general form of the hyperbola is given by $Ax^2 + By^2 + Cx + Dy + E = 0$, where $AB < 0$ and $AB \neq 0$.

Also, the eccentricity of a hyperbola, $e = \frac{\sqrt{a^2 + b^2}}{a}$, is always less than 1.

The distance $2c$ between the two foci is given by the relation: $2c = 2\sqrt{a^2 + b^2}$.

*Example:* Determine the values of $h, k, a, b, c$, and $e$ of the hyperbola general form:

$$9x^2 - 4y^2 - 36x + 8y - 4 = 0$$

*Solution:* Convert the hyperbola equation into the standard form (see *Solving by Completing the Square* on page 29):

$$9x^2 - 4y^2 - 36x + 8y - 4 = (9x^2 - 36x) - (4y^2 - 8y) = 4$$

$$9(x^2 - 4x + 4) - 4(y^2 - 2y + 1) = 36$$

$$\frac{9(x-2)^2}{36} - \frac{4(y-1)^2}{36} = \frac{(x-2)^2}{2^2} - \frac{(y-1)^2}{3^2} = 1$$

Comparing the results above with the form $\dfrac{(x-h)^2}{a^2} - \dfrac{(y-k)^2}{b^2} = 1$ and calculating

eccentricity from $e = \dfrac{\sqrt{a^2 + b^2}}{a}$ and distance $c$ from $c = \sqrt{a^2 + b^2}$ gives

$$h = 2, \qquad k = 1, \qquad a = 2, \qquad b = 3, \qquad c = \sqrt{13}, \qquad e = \frac{\sqrt{13}}{2}$$

## Complex Numbers

**Imaginary Number.**—The square root of a negative number cannot be expressed with real numbers, since any negative number multiplied by itself is positive. But technical mathematics often relies on computation involving the square root of –1. For this, *imaginary number i* is defined as follows:

$$i = \sqrt{-1}, \text{ so } \left(\sqrt{-1}\right)^2 = -1$$

Imaginary numbers are not real numbers; they belong to the set of *complex numbers*.

An example of an equation that cannot be solved with real numbers is $x^2 + 1 = 0$. Rearranging gives $x^2 = -1$, and taking the square root of both sides gives $x = \pm\sqrt{-1} = \pm i$.

(Note: The letter $j$ is also used to represent the imaginary number $\sqrt{-1}$.)

**Forms of a Complex Number.**—Complex numbers can be expressed in several forms, all of which are based on the complex coordinate system, as seen in Fig. 1 and Fig. 2.

*Operations on complex numbers:* Complex numbers are added and subtracted much like real numbers, but with real parts added to real parts and imaginary to imaginary:

$$(a+bi)+(c+di)=(a+c)+(bi+di)=(a+c)+(b+d)i$$

$$(a+bi)-(c+di)=(a-c)+(bi-di)=(a-c)+(b-d)i$$

where coefficients $a$, $b$, $c$, and $d$ are real numbers.

*Example 1:* $(3+4i)+(2-i)=(3+2)+(4i-i)=5+3i$

Complex numbers are multiplied as binomials are, by FOIL:

$$(a+bi)(c+di)=ac+adi+bci+bdi^2=ac+(ad+bc)i+bd(-1)=ac+(ad+bc)i-bd$$

*Example 2:* $(1+2i)(5-7i)=5-7i+10i-14i^2=5+3i-(14)(-1)=5+3i+14=19+3i$

*Standard (Rectangular) Form of a Complex Number:* A complex number $z$ has a real part and an imaginary part. Its standard form is $z = a + bi$, where $a$ is the real part and $bi$ the imaginary part. Fig. 1 shows how the complex plane is similar to the real plane (see page 37 in *Analytic Geometry*), except here only the horizontal axis, $x$, is real, whereas the vertical axis, $yi$, is imaginary. Fig. 2 shows examples of complex numbers in rectangular form.

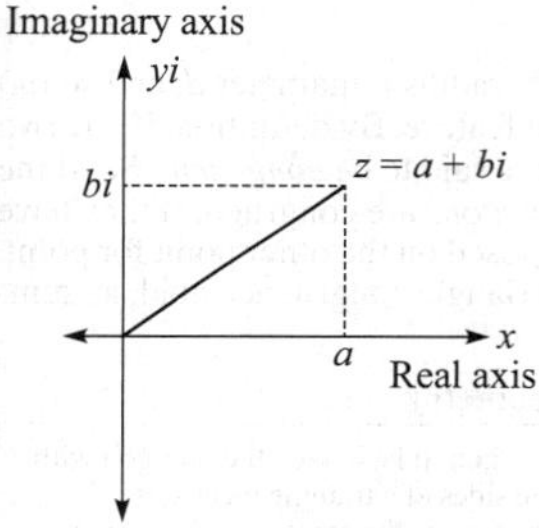


Fig. 1. Complex coordinate system

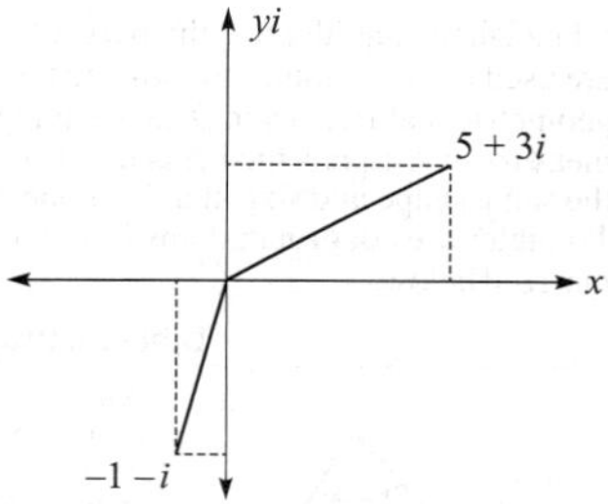


Fig. 2. Examples of standard (rectangular) form, $z = a + bi$

$z = 5 + 3i$ lies in the first quadrant, with $a$ and $b$ both positive; $z = -1 - i$ is in the third quadrant. Complex numbers can be converted from the standard form to any of three *vector forms*: polar, trigonometric, and exponential.

*Polar Form of a Complex Number:* Vectors are objects that have both magnitude ($r$) and direction ($\theta$). Vectors are essential in electrical engineering and other fields for representing many processes such as alternating current and voltage. They are represented graphically by an arrow, as seen in Fig. 3.

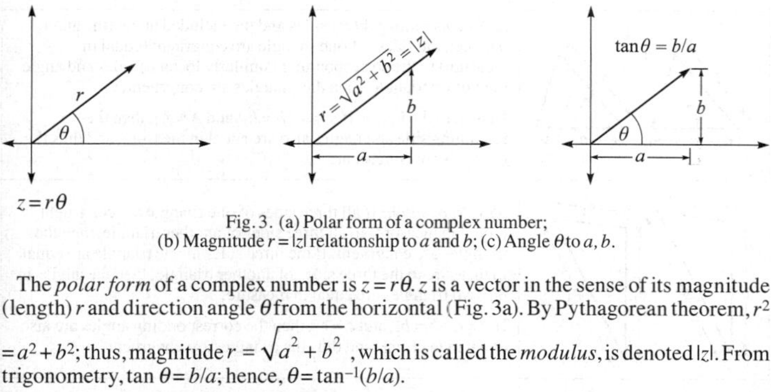


Fig. 3. (a) Polar form of a complex number;
(b) Magnitude $r = |z|$ relationship to $a$ and $b$; (c) Angle $\theta$ to $a$, $b$.

The *polar form* of a complex number is $z = r\theta$. $z$ is a vector in the sense of its magnitude (length) $r$ and direction angle $\theta$ from the horizontal (Fig. 3a). By Pythagorean theorem, $r^2 = a^2 + b^2$; thus, magnitude $r = \sqrt{a^2 + b^2}$, which is called the *modulus*, is denoted $|z|$. From trigonometry, $\tan\theta = b/a$; hence, $\theta = \tan^{-1}(b/a)$.

*Trigonometric Form of a Complex Number:* Another form that shows the directional nature of complex numbers is the trigonometric form of $z$. From trigonometry, $\cos\theta = a/r$ and $\sin\theta = b/r$; hence,

$$a = r\cos\theta, \quad b = r\sin\theta, \quad \text{so}, a + bi = r\cos\theta + ir\sin\theta = r(a\cos\theta + i\sin\theta)$$

*Exponential Form of a Complex Number:* Recall that the number $e$ is the base of the natural logarithm (see page 32). A complex number is represented in exponential form through Euler's formulas, which are used widely in electrical engineering applications:

$$e^{i\theta} = \cos\theta + i\sin\theta, \qquad \cos\theta = \frac{e^{i\theta} + e^{-i\theta}}{2}, \qquad \sin\theta = \frac{e^{i\theta} - e^{-i\theta}}{2i}$$

## Pure Geometry

The labels that identify the parts of a figure (angle $A$, radius $r$, diameter $d$, and so on) are used in the formulas to indicate the *measure* of that feature. By definition, if any two geometric features $A$ and $B$ have equal measure, they are said to be *congruent*. So, if the measure of $A$ equals the measure of $B$, then $A \cong cB$. Polygons are congruent if they have the same shape and size, that is, if one can be superimposed on the other point for point. Triangles are congruent if any of the propositions for triangle congruence hold, as summarized below.

**Table 1a. Propositions of Geometry**

| | |
|---|---|
| 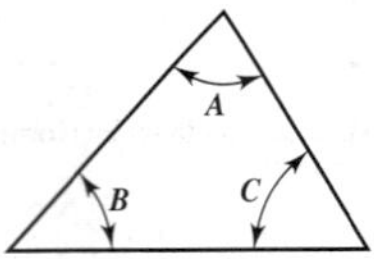 | A *triangle* is a three-sided polygon. It is, in fact, the polygon with the least number of sides. The sides of a triangle meet at its *vertices* (singular *vertex*). The sum of the measures of all three angles of a triangle is 180 degrees. Hence, if the measures of any two angles are known, the third angle measure can always be found.<br>$A + B + C = 180°$  $A = 180° - (B + C)$<br>$B = 180° - (A + C)$  $C = 180° - (A + B)$ |
| 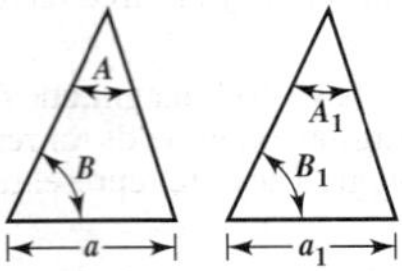 | *AAS Proposition:* If two angles and the non-included side of one triangle are congruent to the corresponding (similarly located) angles and sides of another triangle, the triangles are congruent.<br>Hence, if $a = a_1$, $A = A_1$, and $B = B_1$, the other corresponding side and angle are equal in measure, and thus the triangles are congruent. |
| 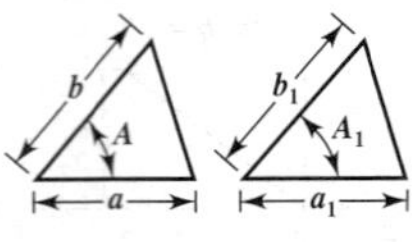 | *SAS Proposition:* If two sides and the included angle (the angle between the sides) of one triangle are congruent (equal in measure) to the corresponding (similarly located) sides and angle of another triangle, then the triangles are congruent.<br>Hence, in the figure, if $a = a_1$, $b = b_1$, and $A = A_1$, then the remaining side and angles also are equal in measure, and thus the triangles are congruent. |
| 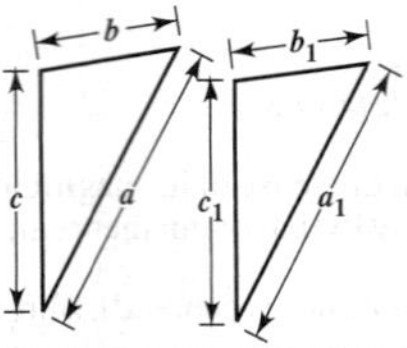 | *SSS Proposition:* If all three sides of one triangle are congruent (equal in measure) to all three sides of another triangle, then the triangles are congruent. If the three sides in one triangle are equal in measure to the three sides of another triangle, then the angles in the two triangles are equal in measure.<br>If $a = a_1$, $b = b_1$, and $c = c_1$, then the corresponding angles are also equal in measure, and thus the triangles are congruent. |
| 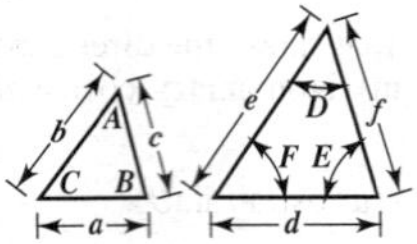 | If the three sides of a triangle are proportional to corresponding sides of another triangle, then the triangles are *similar*, and the angles in the one are congruent (equal in measure) to the angles in the other.<br>Hence, if $a/d = b/e = c/f$ then $A = D$, $B = E$, $C = F$ |
| 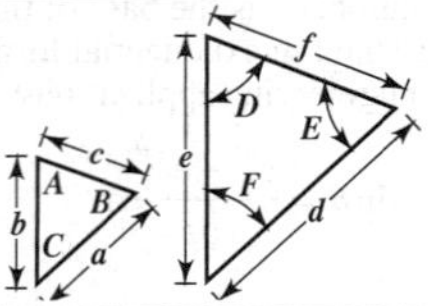 | *Similar triangles* are ones whose corresponding angles are congruent. If this is true then the corresponding sides are proportional. If the angles of one triangle are congruent (equal in measure) to the angles of another triangle, then the triangles are similar and their corresponding sides are proportional.<br>Hence, if $A = D$, $B = E$, and $C = F$ then $a/d = b/e = c/f$ |

**Table 1b. Propositions of Geometry**

| | |
|---|---|
| 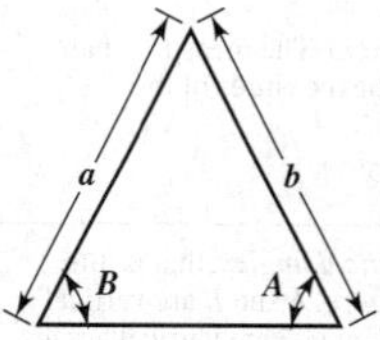  | In an *isosceles triangle*, two sides *a*, *b* are congruent, hence the two angles opposite them (the base angles) are congruent. An equilateral (equiangular) triangle is also an isosceles triangle. |
| 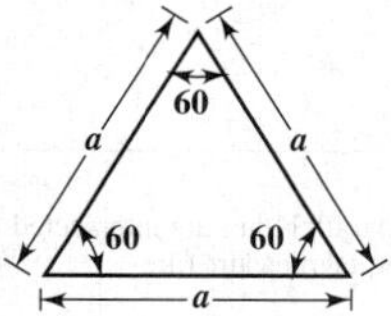  | In an *equilateral (equiangular) triangle*, all three sides (angles) are congruent. Since the sum of angle measures in any triangle is 180 degrees, then each angle in an equilateral (equiangular) triangle measures 60 degrees. |
| 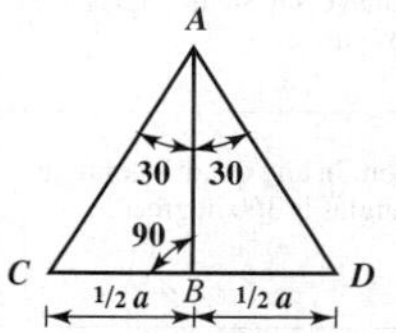  | A line in an equilateral triangle that bisects any of the angles (that is, divides it into two 30-degree angles), also bisects the side opposite the bisected angle and is perpendicular (at right angles) to it.<br><br>Thus, if line *AB* bisects angle *CAD*, it also bisects line *CD* into two equal parts and is perpendicular to it. |
| 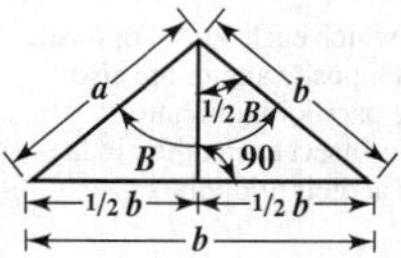  | If a line in an isosceles triangle drawn from the vertex where the two congruent sides meet in such a way that is bisects the third side (or base), then it also bisects the angle at the vertex from which it is drawn. |
| 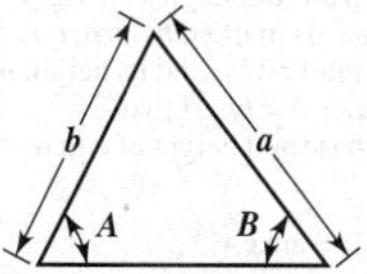  | In every triangle, the greatest angle is opposite the longest side. And, the longest side is opposite the greatest angle.<br><br>Thus, if *a* is longer than *b*, then the measure of angle *A* is greater than that of angle *B*. And, if angle *A* measure is greater than *B*, then side *a* is longer than side *b*. |
| 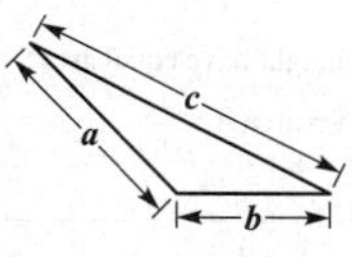  | According to the *triangle inequality*, for any triangle, the sum of the lengths of any two sides must be greater than or equal to the length of the remaining side.<br><br>Thus, $a + b \geq c$. |
| 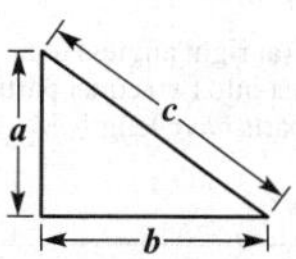  | The *Pythagorean theorem* states that in a right-angle triangle, the square of the *hypotenuse*, that is, the side opposite the right angle, is equal to the sum of the squares on the two sides that form the right angle. These sides are the *legs* of the right triangle.<br><br>Thus, $a^2 + b^2 = c^2$ |

**Table 1c. Propositions of Geometry**

| | |
|---|---|
| 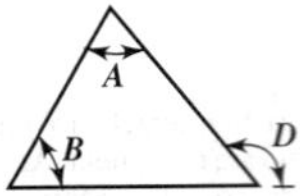  | In the figure, angle $D$ is an *exterior angle*. The measure of an exterior angle is equal to the sum of the measures of the non-adjacent interior angles. Thus,<br>$A + B = D$ |
| 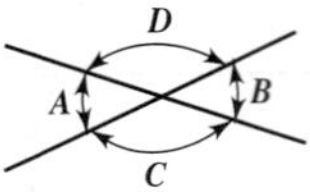  | Intersecting lines form congruent *vertical angles*, that is, the angles opposite one another. In the figure, $A$ and $B$ are vertical angles and therefore congruent, and $C$ and $D$ are vertical angles and therefore congruent. That is,<br>$A \cong B$<br>$C \cong D$ |
| 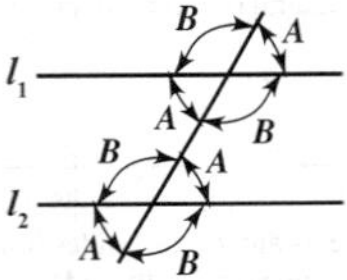  | Corresponding angles formed when parallel lines are intersected by another line (a *transversal*) have equal measure (are congruent).<br>Lines $l_1$ and $l_2$ are intersected by *s* transversal, so all angles $A$ are congruent, and all angles $B$ are congruent. |
| 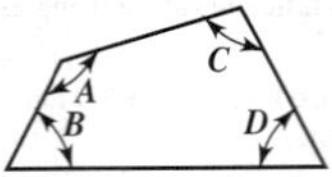  | A quadrilateral is a four-sided polygon. In any quadrilateral, the sum of the measures of the interior angles is 360 degrees.<br>Thus, $A + B + C + D = 360$ degrees |
| 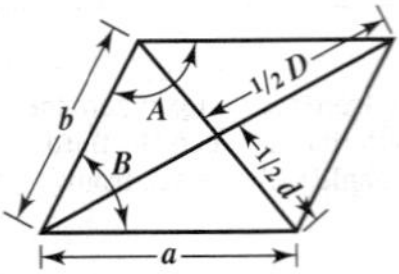  | A parallelogram is a quadrilateral in which each pair of opposite sides is parallel and thus congruent. Opposite angles are also congruent. Each diagonal divides the parallelogram into congruent triangles, and the diagonals bisect each other. In the figure, diagonal $D$ bisects diagonal $d$ at their midpoints. |
| 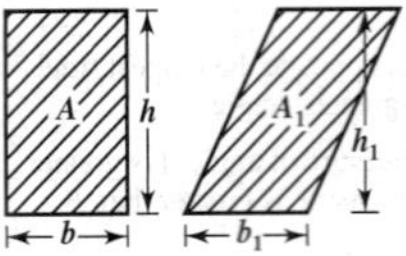  | A rectangle is a parallelogram in which all four angles are right angles. In the figure, the rectangle's *base* is marked $b$; its *height* is marked $h$; the parallelogram's base is marked $b_1$, and its height is marked $h_1$. The area of a parallelogram is $A = bh$. If two parallelograms have a corresponding base and height of equal measure, their areas are equal.<br>Thus, if $b = b_1$ and $h$ and $h_1$, then area $A$ = area $A_1$. |
| 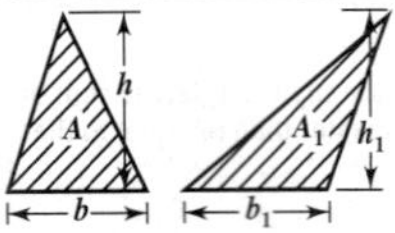  | Triangles having equal base and equal height have equal area.<br>Thus, if $b = b_1$ and $h$ and $h_1$, then area $A$ = area $A_1$. |
| 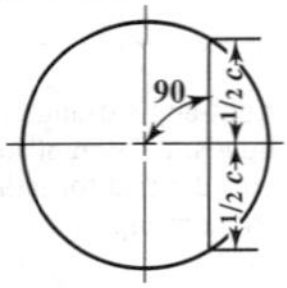  | If a diameter of a circle is perpendicular (at right angles) to a chord, then it bisects the chord, dividing it into two equal parts. In the figure, the chord has length $c$, so its parts have length $c/2$. |

**Table 1d. Propositions of Geometry**

| | |
|---|---|
| 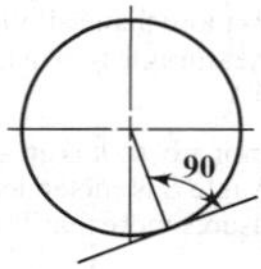  | A line tangent to a circle lies perpendicular (at right angles) to a radius drawn to meet it at the point of tangency. |
| 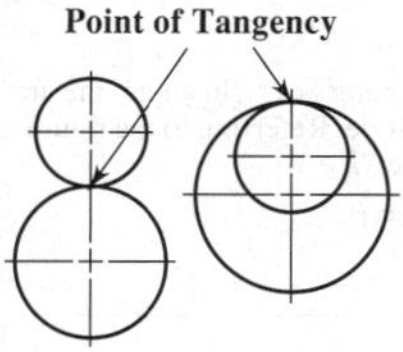  | The figure shows two ways that circles can be tangent. A line drawn through the center of each circle (the diameter) will pass through the point of tangency. |
| 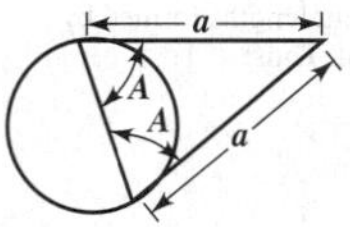  | Two tangents drawn from a single point outside a circle will be equal in length ($a$), and the angles they make with the chord that connects the points of tangency ($A$) will be equal in measure. The figure shows this congruency. |
| 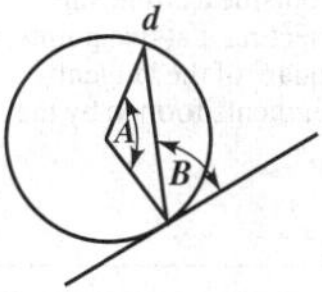  | The angle formed by a tangent and a chord drawn from the point of tangency measures one-half the central angle subtended by that chord.<br>That is, $B = A/2$ |
| 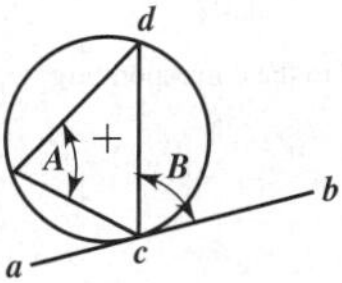  | The angle formed by a tangent and a chord drawn from the point of tangency has measure equal to the inscribed angle subtended by the chord.<br>That is, $B = A$. |
| 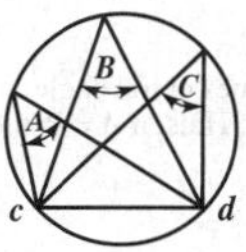  | All inscribed angles subtended by the same chord in a circle are congruent (equal in measure).<br>In the figure, $A$, $B$, and $C$ are subtended by chord $cd$, and so are equal in measure. |
|   | Referring to the figure, inscribed angle $A$ and central angle $B$ are subtended by the same arc. The measure of inscribed $A$ is half the measure of central angle $B$. In the figure,<br>$A = B/2$ |

**Table 1e. Propositions of Geometry**

| | |
|---|---|
| 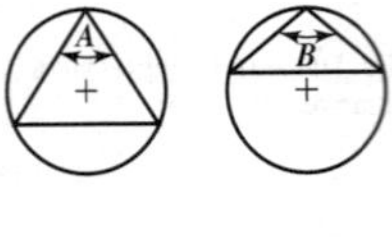  | A *major arc* of a circle is one that measures more than half the circumference of the circle. A *minor arc* measures less than half the circumference of the circle.<br>In the figure, angle *A* is subtended by a minor arc, so it is an acute angle (it measures less than 90 degrees). Angle *B* is subtended by a major arc, so it is an obtuse angle (it measures more than 90 degrees). |
| 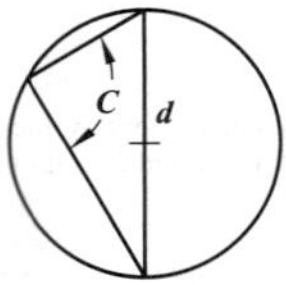  | An angle subtended by a circle's diameter is a right angle; the arc described by the diameter is a semicircle. Referring to the figure, right angle *A* is subtended by diameter *d*.<br>Angle $C = 90°$ |
| 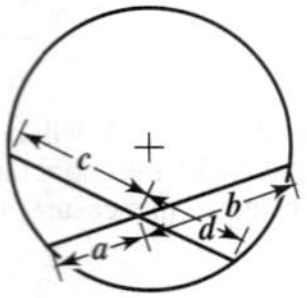  | In the figure, the product of line segment lengths formed by intersecting chords in a circle are equal. Thus:<br>$ab = cd$ |
| 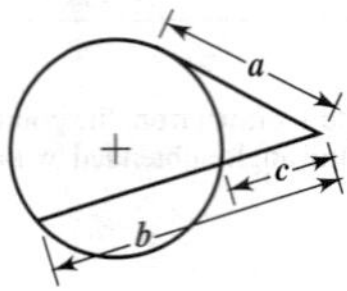  | When two lines are drawn from a point outside a circle, one tangent and one through the circle intersecting it at two points, the line segment lengths are such that the square of the tangent segment is equal to the product of the segments formed by the other line. Thus:<br>$a^2 = bc$ |
| 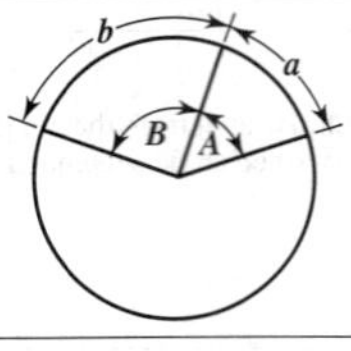  | Arc lengths of a circle are proportional to the corresponding central angle measures. Thus:<br>$A{:}B = a{:}b$ |
| 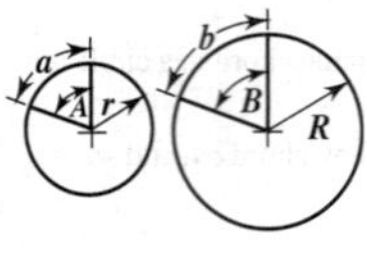  | The lengths of circular arcs having the same central angle are proportional to the lengths of the radii. Thus, if $A = B$, then $a/b = r/R$ |
| 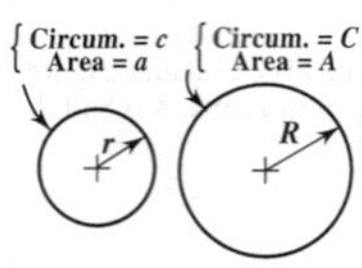  | The ratio of the circumferences of two circles is proportional to the ratio of their radii.<br>$c{:}C :: r{:}R$ so $c/C = r/R$<br>The ratio of the areas of two circles is proportional to the ratio of the squares of their radii.<br>$a{:}A :: r^2{:}R^2$ so $a/A = r^2/R^2$ |

The geometric constructions below are produced with a compass, protractor, and straight edge. The compass point marks the "centers" in the constructions that follow. The techniques described also apply for electronic drafting programs. As the section progresses, some constructions rely on previous ones.

**Table 2a. Geometric Constructions**

| | |
|---|---|
| 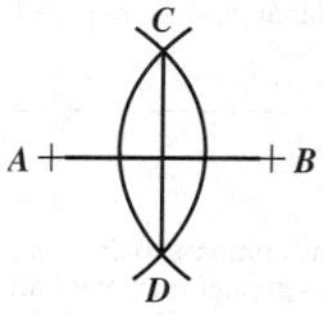 | To bisect a line segment (divide it into two equal parts):<br>With the ends *A* and *B* as centers and a radius greater than one-half the line segment, use the compass to draw circular arcs. Through intersections *C* and *D*, use the straight edge draw line *CD*. This line bisects *AB* and is also perpendicular to *AB*. |
| 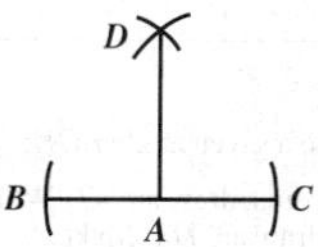 | To draw a perpendicular to a line from a point *A* on that line:<br>With *A* as the center and with any radius, use the compass to draw circular arcs intersecting the given line at *B* and *C*. Then, with *B* and *C* as centers and a radius longer than *AB*, draw circular arcs intersecting at *D*. With the straight edge, draw line *DA*. This line is perpendicular to *BC*. |
| 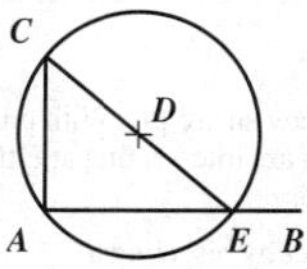 | To draw a perpendicular line from a point *A* to *C* at the end of line *AB*:<br>Draw line *AB*, and then with any point *D* not on *AB* as center, use the compass to draw a circle intersecting *AB* at *E*. Draw a line through *E* and *D* intersecting the circle at *C*; then join *AC*. This line is the required perpendicular. |
| 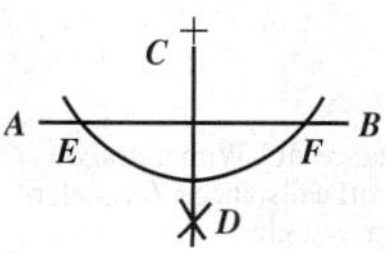 | To draw a perpendicular to a line *AB* from a point *C* at a distance from it:<br>With *C* as center, use the compass to draw an arc intersecting the given line at *E* and *F*. With *E* and *F* as centers, draw circular arcs with a radius longer than one-half the distance between *E* and *F*. These arcs intersect at *D*. Line *CD*, drawn with the straight edge, is the required perpendicular. |
| 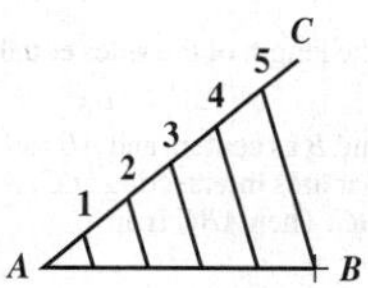 | To divide a line segment *AB* into a number of equal parts:<br>Assume line segment *AB* is to be divided into five smaller segments of equal measure. Draw line *AC* at an acute angle with *AB*. Mark off on *AC* five parts of equal measure, any convenient length. With the straight edge, draw the segment *B*-5 and then draw lines parallel to *B*-5 through the other division points on *AC*. The points where these lines intersect *AB* are the required division points. |

**Table 2b. Geometric Constructions**

| | |
|---|---|
| 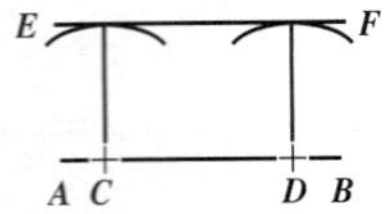  | To draw a line parallel to a given line *AB*, at a given distance from it:<br><br>Using the compass, with any points *C* and *D* on *AB* as centers, draw with the compass circular arcs with the given distance as radius. Line *EF*, drawn with the straight edge to be tangent to (that is, touch without intersecting) the circular arcs, is the required parallel line. |
| 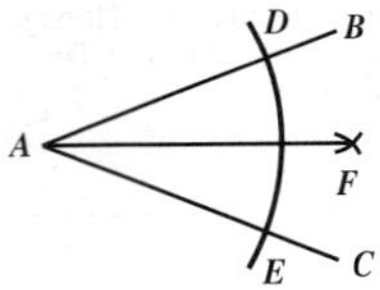  | To bisect an angle *BAC*:<br><br>With *A* as a center and any radius, use the compass to draw arc *DE*. With *D* and *E* as centers and a radius greater than one-half *DE*, use the compass to draw circular arcs intersecting at *F*. Line *AF*, drawn with the straight edge, divides the angle into two equal parts. |
| 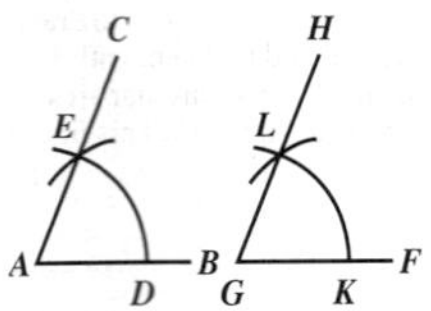  | To draw an angle upon a line *AB* equal to a given angle *FGH*:<br><br>With point *G* as a center and with any radius, draw arc *KL*. With *A* as a center and with the same radius, draw arc *DE*. Make arc *DE* equal in length to *KL* and draw *AC* through *E*. Angle *BAC* then is equal in length to angle *FGH*. |
| 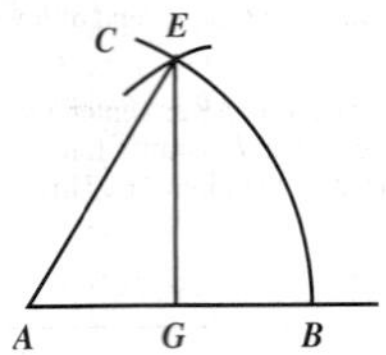  | To lay out a 60-degree angle:<br><br>With *A* as a center and any radius *AB*, draw an arc *BC*. With point *B* as a center and *AB* as a radius, draw an arc intersecting at *E* the arc just drawn. *EAB* is a 60-degree angle.<br><br>A 30-degree angle may be obtained either by bisecting a 60-degree angle or by drawing a line *EG* perpendicular to *AB*. Angle *AEG* is then 30 degrees. |
| 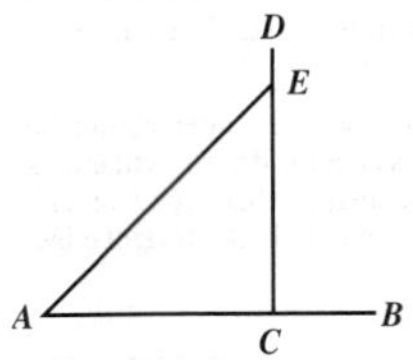  | To draw a 45-degree angle:<br><br>From point *A* on line *AB*, set off a distance *AC*. With a straight edge, draw perpendicular *DC* and set off a distance *CE* equal to *AC*. Draw *AE*. Angle *EAC* is a 45-degree angle. |
| 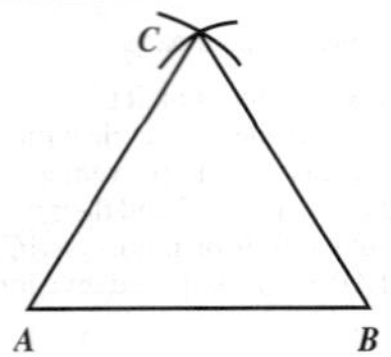  | To draw an equilateral triangle, with the length of the sides equal to *AB*:<br><br>Draw *AB* with straight edge. With *A* and *B* as centers and *AB* as radius, use the compass to draw circular arcs intersecting at *C*. With the straight edge, draw *AC* and *BC*. Then *ABC* is an equilateral triangle. |

**Table 2c. Geometric Constructions**

| | |
|---|---|
| 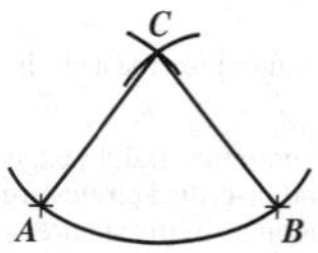 | To draw a circular arc with a given radius through two given points *A* and *B*:<br>With *A* and *B* as centers, with the compass draw two circular arcs with the given radius intersecting at *C*. With *C* as center and the same radius, draw a circular arc through *A* and *B*. |
| 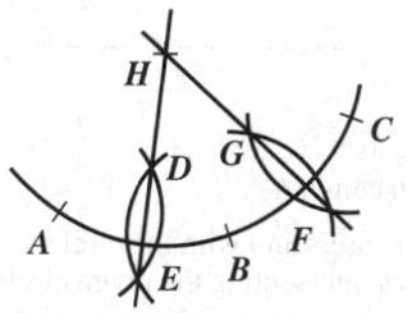 | To locate the center of an arc of a circle:<br>Select three points on the periphery of the circle, as *A*, *B*, and *C*. With each of these points as a center and setting the same radius with the compass, draw arcs intersecting each other. Through the points of intersection, draw lines *DE* and *GF*. Point *H*, where these lines intersect, is the center of the circle. |
| 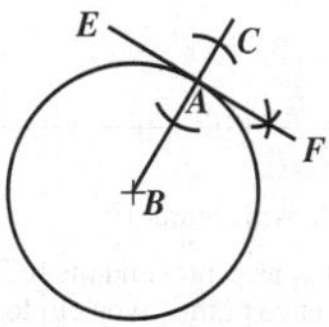 | To draw a tangent to a circle through a given point on the circumference:<br>Through a chosen point of tangency *A* on a circle with center *B*, draw radius *BC*. At point *A*, draw a line *EF* at right angles to *BC*. This line is the required tangent. |
| 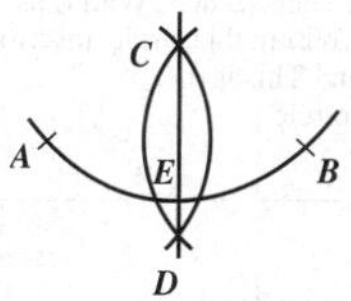 | To divide a circular arc *AB* into two equal parts:<br>With *A* and *B* as centers, and a radius larger than half the distance between *A* and *B*, with the compass draw circular arcs intersecting at *C* and *D*. Line *CD* divides arc *AB* into two equal parts at *E*. |
| 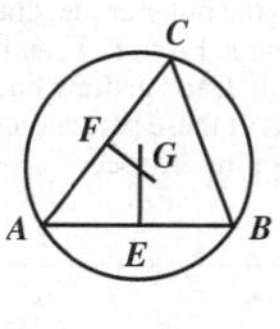 | To circumscribe a circle about a triangle *ABC*:<br>Bisect the sides *AB* and *AC*, and from the midpoints *E* and *F* draw lines at right angles to the sides. These lines intersect at *G*. With *G* as a center and *GA* as a radius, draw circle *ABC*. |
| 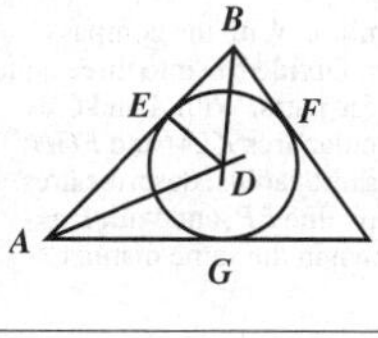 | To inscribe a circle in a triangle *ABC*:<br>Bisect two of the angles, *A* and *B*, by lines intersecting at *D*. From *D*, draw a line *DE* perpendicular to one of the sides, and with *DE* as a radius, use the compass to draw circle *EFG*. |

**Table 2d. Geometric Constructions**

| | |
|---|---|
| 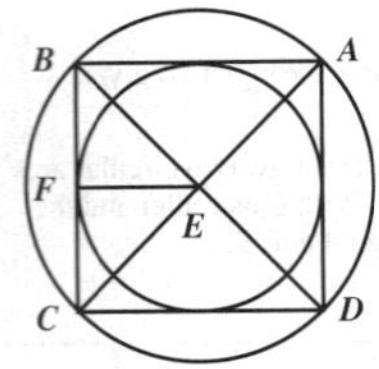 | To circumscribe a circle about a square and to inscribe a circle in a square *ABCD*:<br><br>Draw the square's diagonals *AC* and *BD* with the straight edge. The centers of both the circumscribed and inscribed circles are located at the point *E*, where the two diagonals of the square intersect. The radius of the circumscribed circle is *AE*, and of the inscribed circle, *EF*. |
| 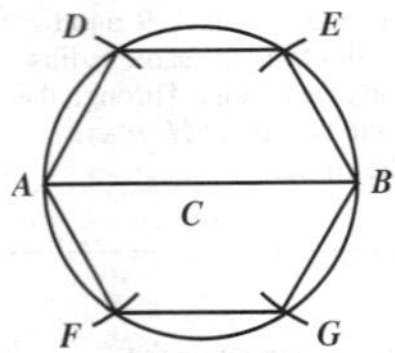 | To inscribe a hexagon in a circle with center *C*:<br><br>Draw diameter *AB*. With *A* and *B* as centers and with the circle's radius as radius, describe circular arcs intersecting the given circle at *D*, *E*, *F*, and *G*. Draw chords *AD*, *DE*, etc., forming the required hexagon *ABCDEFG*. |
| 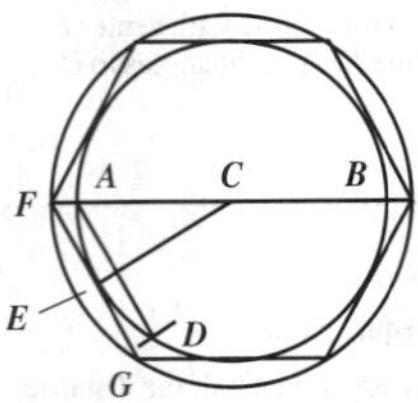 | To circumscribe a hexagon about a circle with center *C*:<br><br>Draw the circle's diameter *AB*, and with *A* as center and the radius of the circle as radius, cut the circumference of the given circle at *D*. Draw chord *AD* and bisect it with radius *CE*. Through *E*, draw *FG* parallel to *AD* and intersecting diameter *AB* at *F*. With *C* as center and *CF* as radius, draw a circle. Within this circle, inscribe the hexagon as in the preceding problem. This is the circumscribed hexagon about the first circle. |
| 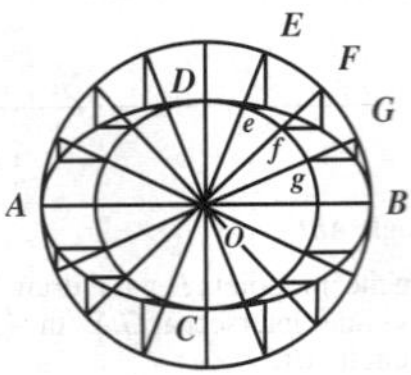 | To describe an ellipse with the given axes *AB* and *CD*:<br><br>Describe circles with *O* as center and *AB* and *CD* as diameters. From a number of points, *E*, *F*, *G*, etc., on the outer circle, draw radii intersecting the inner circle at *e*, *f*, and *g*. From *E*, *F*, and *G*, draw lines perpendicular to *AB*, and from *e*, *f*, and *g*, draw line segments parallel to *AB*. The intersections of these perpendicular and parallel lines are points on the curve of the ellipse. |
| 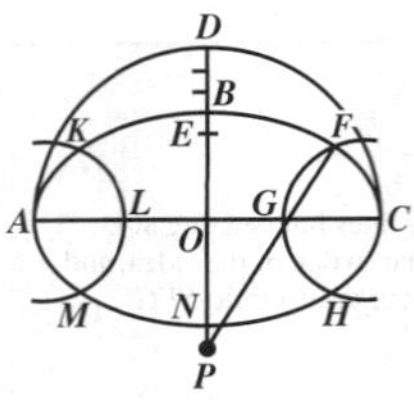 | To construct an approximate ellipse by circular arcs:<br><br>Let *AC* be the major axis and *BN* the minor. With the compass, draw semicircle *ADC* with *O* as center. Divide *BD* into three equal parts and set off *BE* equal to one of these parts. With *A* and *C* as centers and *OE* as radius, describe circular arcs *KLM* and *FGH*; with *G* and *L* as centers, and with the same radius, describe arcs *FCH* and *KAM*. Through *F* and *G*, draw line *FP*, and with *P* as center draw arc *FBK*. Arc *HNM* is drawn in the same manner. |

**Table 2e. Geometric Constructions**

<table>
<tr><td>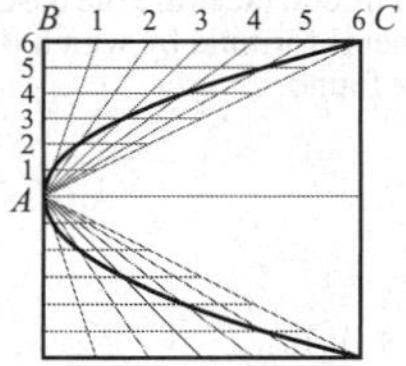
</td><td>To construct a parabola:<br><br>Divide line segment AB into a number of equal parts and divide BC into the same number of parts. From the division points on AB, draw horizontal lines. From the division points on BC, draw lines to point A. The points of intersection of the lines drawn from points numbered alike are points on the parabola.</td></tr>
<tr><td>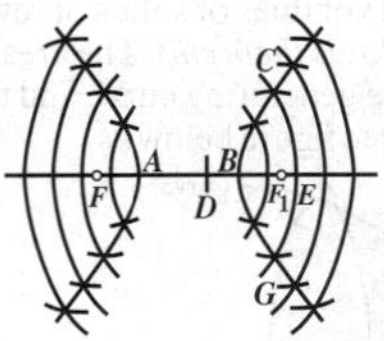
</td><td>To construct a hyperbola:<br><br>From focus F, lay off a distance FD to be the transverse axis, or the distance AB between the two branches of the curve. With F as center and any distance FE greater than FB as a radius, describe a circular arc. Then with $F_1$ as center and DE as radius, describe arcs intersecting at C and G the arc just described. C and G are points on the hyperbola. Any number of points can be found in a similar manner; when a sufficient number of points are found, draw a smooth curve through them.</td></tr>
<tr><td>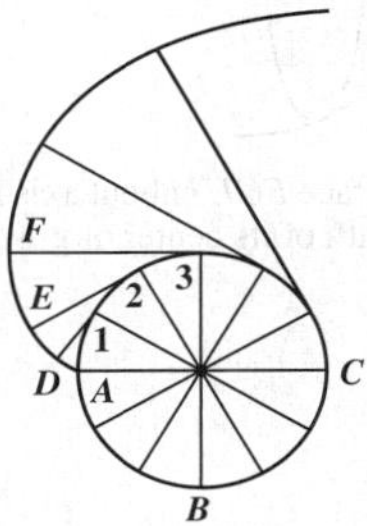
</td><td>To construct an involute:<br>Divide the circumference of the base circle ABC into a number of equal parts. Through the division points 1, 2, 3, etc., draw tangents to the circle and make the lengths D-1, E-2, F-3, etc., of these tangents equal to the actual length of the arcs A-1, A-2, A-3, etc. Connect the ends of these tangents with a smooth curve.</td></tr>
<tr><td>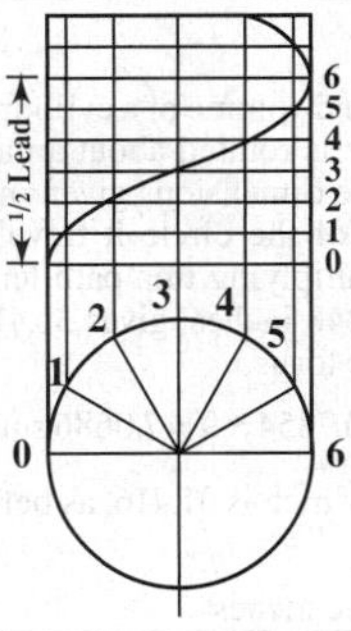
</td><td>To construct a helix:<br><br>Divide half the circumference of the cylinder on the surface of which the helix is to be described into a number of equal parts. Divide half the lead of the helix into the same number of equal parts. From the division points on the circle representing the cylinder, draw vertical lines using the straight edge, and from the division points on the lead, draw horizontal lines as shown. The intersections between lines numbered alike are points on the helix. Connect these points with a smooth curve.</td></tr>
</table>

## Area and Volume

**The Prismoidal Formula.—**A *right prism* is a three-dimensional figure composed of (two-dimensional) polygons that form the faces of the prism. The end faces are the bases, the other faces are the sides. The prismoidal formula is a general formula by which the volume of any prism, pyramid, or frustum of a pyramid may be found.

$A_1, A_2 =$ end areas of the body

$A_m =$ area of faces between the two end surfaces

$h =$ height of body

Then, volume of the body is calculated as $V = \frac{h}{6}(A_1 + 4A_m + A_2)$ .

**Pappus-Guldinus Rules.—**A *surface of revolution* is generated when a curve is revolved about an external axis. (The curve must lie wholly on one side of the axis of revolution and in the same plane.) The mathematics for finding the curve's length and the length of the path of the centroid involves calculus. Some surface areas and volumes of solids of revolution can then be determined by the rules of the *Pappus-Guldinus theorems*. The area of the resulting surface is equal to the product of the length of the generating curve and the distance traveled by the curve's center of gravity, or *centroid* (see figure below).

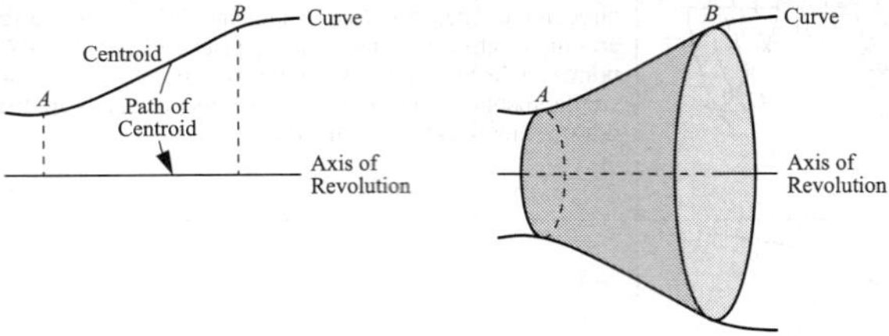


The volume of a solid body formed by the revolution of a surface *FGHJ* about axis *KL* equals the product of the surface area and the length of the path of its center of gravity about axis *KL*.

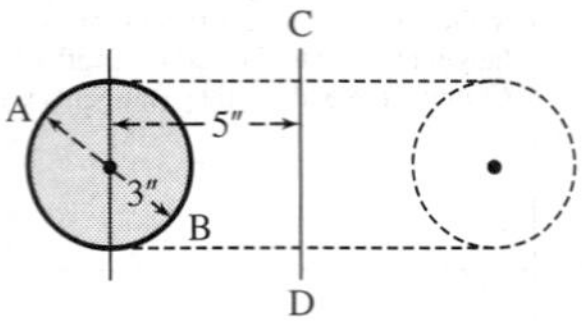


*Example:* By means of the Pappus-Guldinus rules, the area and volume of a cylindrical ring, or torus, may be found. A torus is formed when a circle is rotated about an axis. The center of gravity of the circle is its center. Hence, with the dimensions given in the illustration, the length of the path of the center of gravity of the circle it travels is the circumference $2\pi r = 2 \times 3.1416 \times 5 = 31.416$ inches. Multiplying this path length by the circumference of the circle, which is $3.1416 \times 3 = 9.4248$ inches, gives $31.416 \times 9.4248 = 296.089$ square inches. This is the surface area of the torus.

The volume of the torus equals the area of the circle, $\frac{\pi d^2}{4} = 0.7854 \times 9 = 7.0686$ square inches, multiplied by the path length of the center of gravity, which is 31.416, as before; hence,

$$\text{Volume} = 7.0686 \times 31.416 = 222.067 \text{ cubic inches}$$

**Approximate Method for Finding the Area of a Surface of Revolution.—**The illustration below is an example of the approximate method based on Guldinus rule for finding the surface area of a symmetrical body. In the illustration, the dimensions in common fractions are the known dimensions; those in decimals are found by actual measurements on a figure drawn to scale.

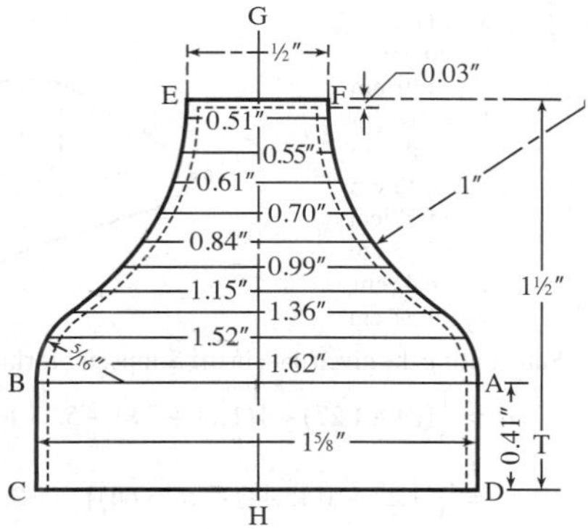


The surface area is found as follows: First, the entire form is separated into such areas as are cylindrical, conical, or spherical, since their surface areas can be found by exact formulas. In the illustration, the three-dimensional portion marked in the plane by *ABCD* is a cylinder, the area of the surface of which can be easily found. The top area *EF* is simply a circular area and can thus be computed separately. The remainder of the surface generated by rotating line *AF* about the axis *GH* is found by the approximate method. From point *A*, equal distances are set off on line *AF*. In the illustration, each division indicated is ⅛ inch long. From the central or middle point of each of these parts a line is drawn at right angles to the axis of rotation *GH*, the length of these lines or diameters (the length of each is given in decimals) is measured, all these lengths are added together and the sum is multiplied by the length of one division set off on line *AF* (in this case, ⅛ inch), and this product is multiplied by $\pi$ to give the approximate area of the surface of revolution.

In setting off divisions ⅛ inch long along line *AF*, the last division does not reach all the way to point *F*, but only to a point 0.03 inch below it. The part 0.03 inch high at the top of the cup can be considered as a cylinder measuring ½ inch in diameter and 0.03 inch in height, with the area of the cylindrical surface being easily computed. By adding the various surfaces together, the total surface of the cup is found as follows:

| | |
|---|---|
| Cylinder, 1⅝ in. diameter, 0.41 in. height | 2.093 in$^2$ |
| Circle, ½ in. diameter | 0.196 in$^2$ |
| Cylinder, ½ in. diameter, 0.03 in. height | 0.047 in$^2$ |
| Irregular surface | 3.868 in$^2$ |
| Total | 6.204 in$^2$ |

**Area of Irregular Plane Figure.—**One of the most useful and accurate methods for determining the approximate area of a plane figure or irregular outline is known as *Simpson's rule*. In applying Simpson's rule to find an area, the work is done in four steps:

1) The area is divided into an *even* number $N$ of parallel strips of equal width $W$; for example, in the accompanying diagram, the area has been divided into 8 strips of equal width.

2) The sides of the strips are labeled $V_0, V_1, V_2$, etc., up to $V_N$.

3) These heights $V_0, V_1, V_2, \ldots, V_N$ are measured.

4) The values $V_0, V_1$, etc. are substituted in the following formula to find the area $A$ of the figure:

$$A = \frac{W}{3}[(V_0 + V_N) + 4(V_1 + V_3 + \cdots + V_{N-1}) + 2(V_2 + V_4 + \cdots + V_{N-2})]$$

*Example:* The area of the accompanying figure was divided into 8 strips on a full-size drawing and the following data obtained. Calculate the area using Simpson's rule.

$W = 1$ cm
$V_0 = 0$ cm
$V_1 = 1.91$ cm
$V_2 = 3.18$ cm
$V_3 = 3.81$ cm
$V_4 = 4.13$ cm
$V_5 = 5.27$ cm
$V_6 = 6.35$ cm
$V_7 = 4.45$ cm
$V_8 = 1.27$ cm

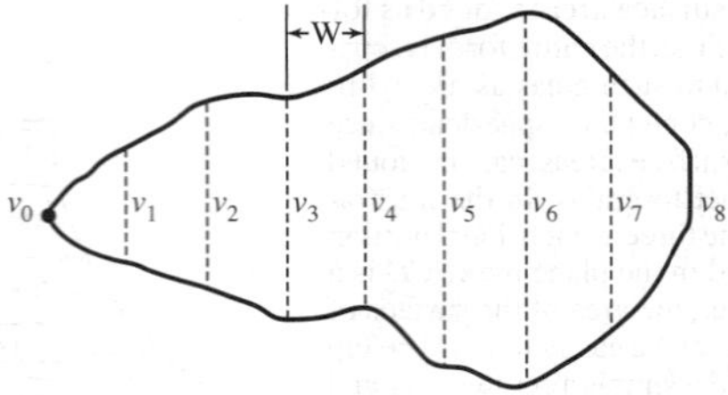


Substituting the given data in the Simpson's formula,

$$A = \frac{1}{3}[(0 + 1.27) + 4(1.91 + 3.81 + 5.27 + 4.45) + 2(3.18 + 4.13 + 6.35)]$$

$$= \frac{1}{3}[1.27 + 4(15.44) + 2(13.66)] = 30.12 \text{ cm}^2$$

In applying Simpson's rule, it should be noted that the larger the number of strips into which the area is divided the more accurate the results obtained.

**Areas Enclosed by Cycloidal Curves.**—The area between a cycloid and the line upon which the generating circle rolls equals three times the area of the generating circle (see diagram, page 74). The areas between epicycloidal and hypocycloidal curves and the "fixed circle" upon which the generating circle is rolled may be determined by the following formulas, in which $a$ = radius of the fixed circle upon which the generating circle rolls, $b$ = radius of the generating circle, $A$ = the area for the epicycloidal curve, and $A_1$ = the area for the hypocycloidal curve.

$$A = \frac{\pi b^2(3a + 2b)}{a} \qquad A_1 = \frac{\pi b^2(3a - 2b)}{a}$$

**Contents of Cylindrical Tanks at Different Levels.**—In conjunction with the table *Segments of Circles for Radius = 1* starting on page 79, the following relations can give a close approximation of the liquid contents, at any level, in a cylindrical tank.

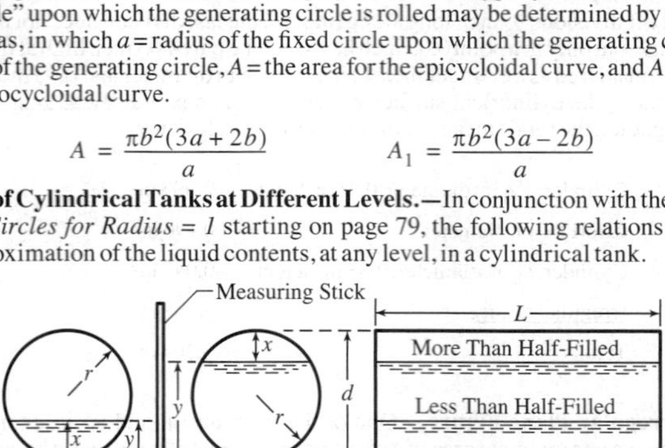


A long measuring rule calibrated in length units or a plain stick can be used for measuring contents at a particular level. In turn, the rule or stick can be graduated to serve as a volume gauge for the tank in question. The requirements are that the tank must have a circular cross section; the dimensions of the tank must be known; the gauge rod has to be inserted vertically through the top center of the tank so that it rests precisely in the center at the bottom of the tank; and the calculations must be done using consistent metric *or* US customary (also called *English*) units. The formulas and parameters are:

$K = Cr^2L$ = tank constant (the same for any given tank) (1)

$V_T = \pi K$, for a tank that is completely full (2)

$V_s = KA$ (3)

$V = V_s$, for a tank that is less than half full (4)

$V = V_T - V_s = V_T - KA$, for a tank that is more than half full (5)

where: $C$ = liquid volume conversion factor; the exact value depends on the length and liquid volume units used during measurement: 0.00433 US gal/in$^3$; 7.48 US gal/ft$^3$; 0.00360 UK gal/in$^3$; 6.23 UK gal/ft$^3$; 0.001 liter/cm$^3$; or 1000 liters/m$^3$

$V_T$ = total volume of liquid tank can hold

$V_s$ = volume formed by segment of circle having depth $x$ in the given tank (see diagram)

$V$ = volume of liquid at particular level in tank

$L$ = length of tank; $r$ = radius of tank ( = ½ diameter)

$A$ = segment area of a corresponding unit circle taken from the table starting on page 79.

$y$ = actual depth of contents in tank as shown on a gauge rod or stick

$x$ = depth of the segment of a circle to be considered in given tank. As can be seen in the diagram, both $x$ and $y$ are the actual depth of contents when the tank is less than half full, but $x$ is the depth of the void $(d - y)$ above the contents when the tank is more than half full. In the discussion of the unit circle, page 79, $r = 1$, and so the height of a segment of a corresponding unit circle is $x/r$.

*Example:* A tank is 20 feet long and 6 feet in diameter. Convert a long stick graduated in inches into a gauge graduated at two points, 1000 and 3000 US gallons.

*Solution:* $L = 20 \times 12 = 240$ in. $r = {}^6\!/_2 \times 12 = 36$ in.

From Formula (1): $K = 0.00433 \times 36^2 \times 240 = 1346.80$

From Formula (2): $V_T = 3.1416 \times 1347 = 4231.7$ US gal.

The 72-inch mark from the bottom on the inch-stick can be graduated for the rounded full volume of 4230 and the halfway point 36 in. for 4230/2 or 2115. It can be seen that the 1000-gal mark would be below the halfway mark. From Formulas (3) and (4):

$$A_{1000} = \frac{1000}{1347} = 0.7424$$ ; from the table starting on page 79, $h$ can be interpolated as 0.5724; and $x = y = 36 \times 0.5724 = 20.61$. If the desired level of accuracy permits, interpolation can be omitted by choosing $h$ directly from the table on page 79 for the value of $A$ nearest that calculated above.

Therefore, the 1000-gal mark is graduated 20⅝ in. from bottom of rod.

It can be seen that the 3000 mark would be above the halfway mark. Therefore, the circular segment considered is the cross section of the void space at the top of the tank. From Formulas (3) and (5):

$$A_{3000} = \frac{4230 - 3000}{1347} = 0.9131, \quad h = 0.6648, x = 36 \times 0.6648 = 23.93 \text{ in.}$$

Therefore, the 3000-gal mark is $72.00 - 23.93 = 48.07$, or at the 48$\frac{1}{16}$-in. mark from the bottom.

**Dimensions of Plane Figures**

The following pages contain diagrams of plane figures, along with formulas for finding their dimensions. Illustrations are labeled with the variables given in the formulas. Formulas for one dimension are derived from another by rearranging terms or substituting the formula for one dimension with another, where appropriate.

Formulas that include decimals are derived either from the square root of 2 or $\frac{1}{2}$. As such, they are approximations of the dimension being solved for. Formulas that include $\pi$ are approximated as well.

The dimensions generally are as follows, with distinctions shown as needed for each figure: $A$ = area; $s$ = side; $d$ = diagonal or diameter; $h, H$ = height; $a, b, c$ = sides or other segments as shown; $r, R$ = radius; $l$ = arc length or other.

*Square:*

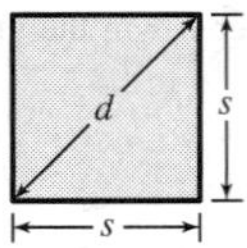


$$A = s^2 = \tfrac{1}{2}d^2$$
$$s = 0.7071d = \sqrt{A}$$
$$d = 1.414s = 1.414\sqrt{A}$$

*Example:* Side $s$ of a square is 15 in. Find the area of the square and the length of its diagonal.

$$A = s^2 = 15^2 = 225 \text{ in}^2$$
$$d = 1.414s = 1.414 \times 15 = 21.21 \text{ in}$$

*Example:* The area of a square is 625 $\text{cm}^2$. Find the length of side $s$ and diagonal $d$.

$$s = \sqrt{A} = \sqrt{625} = 25 \text{ cm}$$
$$d = 1.414\sqrt{A} = 1.414 \times 25 = 35.35 \text{ cm}$$

*Rectangle:*

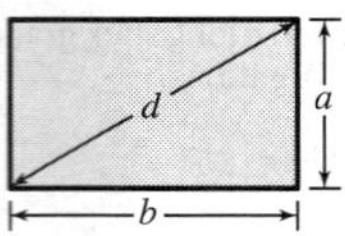


$$A = ab = a\sqrt{d^2 - a^2} = b\sqrt{d^2 - b^2}$$
$$d = \sqrt{a^2 + b^2}$$
$$a = \sqrt{d^2 - b^2} = A/b$$
$$b = \sqrt{d^2 - a^2} = A/a$$

*Example:* Side $a$ of a rectangle is 12 cm, and the area is 70.5 $\text{cm}^2$. Find the length of side $b$ and diagonal $d$.

$$b = A/a = 70.5/12 = 5.875 \text{ cm}$$
$$d = \sqrt{a^2 + b^2} = \sqrt{12^2 + 5.875^2} = \sqrt{178.516} = 13.361 \text{ cm}$$

*Example:* The sides of a rectangle are 30.5 and 11 cm. Find the area.

$$A = ab = 30.5 \times 11 = 335.5 \text{ cm}^2$$

*Parallelogram:*

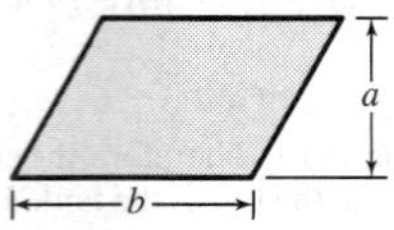


$$A = ab$$
$$a = A/b$$
$$b = A/a$$

*Note:* The dimension $a$ is the length of the vertical drawn at a right angle to side $b$. Dimension $a$ is also considered the height of the parallelogram.

*Example:* Base $b$ of a parallelogram is 16 ft. Height $a$ is 5.5 ft. Find the area.

$$A = ab = 5.5 \times 16 = 88 \text{ ft}^2$$

*Example:* The area of a parallelogram is 12 $\text{in}^2$. The height is 1.5 in. Find the length of the base $b$.

$$b = A/a = 12/1.5 = 8 \text{ in.}$$

*Right Triangle (one angle is a 90-degree angle):*

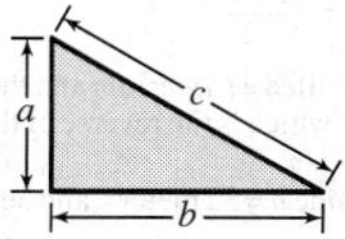


From the Pythagorean theorem, $a^2 + b^2 = c^2$, thus $A = \dfrac{ab}{2}$

$c = \sqrt{a^2 + b^2} \qquad a = \sqrt{c^2 - b^2} \qquad b = \sqrt{c^2 - a^2}$

*Example:* Side $a$ is 6 in. and side $b$ is 8 in. Find side $c$ and area $A$:

$$c = \sqrt{a^2 + b^2} = \sqrt{6^2 + 8^2} = \sqrt{36 + 64} = \sqrt{100} = 10 \text{ in.}$$

$$A = \frac{ab}{2} = \frac{6 \times 8}{2} = \frac{48}{2} = 24 \text{ in}^2$$

*Example:* Side $c = 10$ and side $a = 6$. Find side $b$:

$$b \text{ in.} = \sqrt{c^2 - a^2} = \sqrt{10^2 - 6^2} = \sqrt{100 - 36} = \sqrt{64} = 8 \text{ in.}$$

*Acute Triangle (all three angles measure less than 90 degrees):*

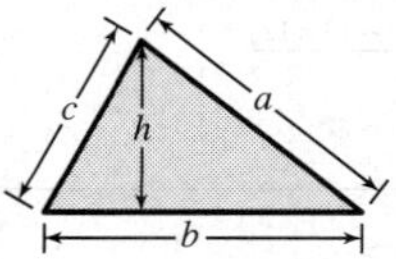


$$A = \frac{bh}{2}, \quad h = \sqrt{a^2 - \left(\frac{a^2 + b^2 - c^2}{2b}\right)^2}, \quad \text{so } A = \frac{b}{2}\sqrt{a^2 - \left(\frac{a^2 + b^2 - c^2}{2b}\right)^2}$$

$$A = \sqrt{S(S-a)(S-b)(S-c)}, \quad \text{where } S = \frac{a+b+c}{2}$$

*Example:* Side $b = 7$ inches, $h = 4$ inches, so $A = bh/2 = (7 \text{ in.} \times 4 \text{ in.})/2 = 28 \text{ in}^2/2 = 14 \text{ in}^2$

*Example:* Side $a = 10$ cm, $b = 9$ cm, and $c = 8$ cm. Find the area.

$$A = \frac{b}{2}\sqrt{a^2 - \left(\frac{a^2 + b^2 - c^2}{2b}\right)^2} = \frac{9}{2}\sqrt{10^2 - \left(\frac{10^2 + 9^2 - 8^2}{2 \times 9}\right)^2} = 4.5\sqrt{100 - \left(\frac{117}{18}\right)^2}$$

$$= 4.5\sqrt{100 - 42.25} = 4.5\sqrt{57.75} = 4.5 \times 7.60 = 34.20 \text{ cm}^2$$

*Obtuse Triangle (one angle measures greater than 90 degrees):*

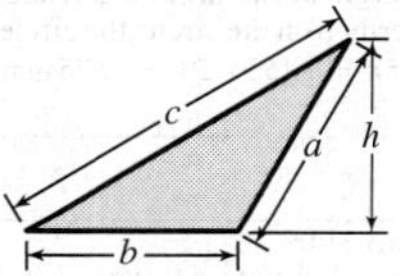


$$A = \frac{bh}{2}, \quad h = \sqrt{a^2 - \left(\frac{a^2 + b^2 - c^2}{2b}\right)^2}, \quad \text{so } A = \frac{b}{2}\sqrt{a^2 - \left(\frac{a^2 + b^2 - c^2}{2b}\right)^2}$$

$$A = \sqrt{S(S-a)(S-b)(S-c)}, \quad \text{where } S = \frac{a+b+c}{2}$$

*Example:* If $b = 5$ cm and $h = 3$ cm, then $A = bh/2 = (5 \text{ cm} \times 3 \text{ cm})/2 = 15 \text{ cm}^2/2 = 7.5 \text{ cm}^2$

*Example:* Side $a = 5$ in., side $b = 4$ in., and side $c = 8$ in. Find the area.

$$S = (a + b + c)/2 = (5 + 4 + 8)/2 = 17/2 = 8.5$$

$$A = \sqrt{S(S-a)(S-b)(S-c)} = \sqrt{8.5(8.5-5)(8.5-4)(8.5-8)}$$

$$= \sqrt{8.5 \times 3.5 \times 4.5 \times 0.5} = \sqrt{66.937} = 8.18 \text{ in}^2$$

*Trapezoid:*

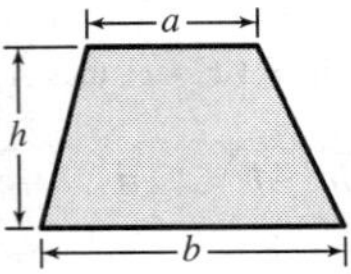


$$\text{Area} = A = \frac{(a+b)h}{2}$$

*Note:* In Britain, this figure is called a *trapezium* and the figure below it is known as a *trapezoid*, which is the reverse of the US terms.

*Example:* Side $a = 23$ meters, side $b = 32$ meters, and height $h = 12$ meters. Find the area.

$$A = \frac{(a+b)h}{2} = \frac{(23+32)\times 12}{2} = \frac{55\times 12}{2} = 330 \text{ m}^2$$

*Trapezium:*

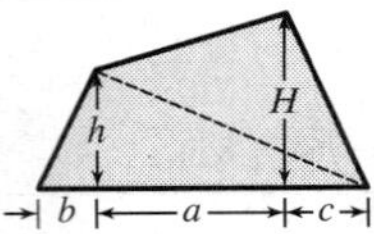


$$\text{Area} = A = \frac{(H+h)a + bh + cH}{2}$$

The area of a trapezium also can be found by dividing it into two triangles, as indicated by the dashed line. Each area is added to give the total area of the trapezium.

*Example:* Let $a = 10$ in., $b = 2$, $c = 3$ in., $h = 8$ in., and $H = 12$ in. Find the area.

$$A = \frac{(H+h)a + bh + cH}{2} = \frac{(12+8)\times 10 + (2\times 8) + (3\times 12)}{2}$$

$$= \frac{(20\times 10) + 16 + 36}{2} = \frac{252}{2} = 126 \text{ in}^2$$

*Regular Hexagon:*

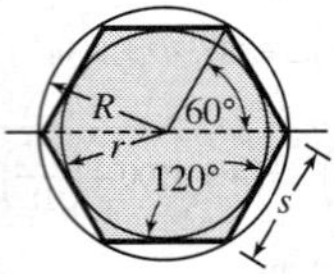


$A = 2.598s^2 = 2.598R^2 = 3.464r^2$

$R = s =$ radius of circumscribed circle $= 1.155r$

$r =$ radius of inscribed circle $= 0.866s = 0.866R$

$s = R = 1.155r$

*Example:* The side $s$ of a regular hexagon is 40 millimeters. Find the area and the radius $r$ of the inscribed (drawn inside) circle.

$$A = 2.598s^2 = 2.598\times 40^2 = 2.598\times 1600 = 4156.8 \text{ mm}^2$$

$$r = 0.866s = 0.866\times 40 = 34.64 \text{ mm}$$

*Example:* What is the length of the side of a hexagon circumscribed on (drawn around) a circle of 50 millimeters radius? In this case, because the hexagon is circumscribed on the circle, the circle is *inscribed* (drawn within) the hexagon. Hence, $r = 50$ mm and $s = 1.155r = 1.155\times 50 = 57.75$ mm

*Regular Octagon:*

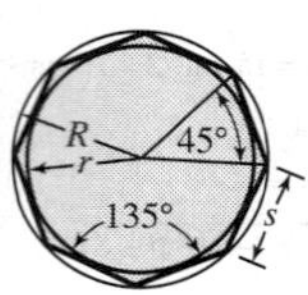


$A =$ area $= 4.828s^2 = 2.828R^2 = 3.314r^2$

$R =$ radius of circumscribed circle $= 1.307s = 1.082r$

$r =$ radius of inscribed circle $= 1.207s = 0.924R$

$s = 0.765R = 0.828r$

*Example:* Find the area and the length of the side of an octagon inscribed (drawn inside) in a circle of 12 inches diameter.

Diameter of circumscribed (drawn around) circle = 12 inches; hence, $R = 6$ in.

$$A = 2.828R^2 = 2.828\times 6^2 = 2.828\times 36 = 101.81 \text{ in}^2$$

$$s = 0.765R = 0.765\times 6 = 4.590 \text{ in.}$$

*Regular Polygon:*

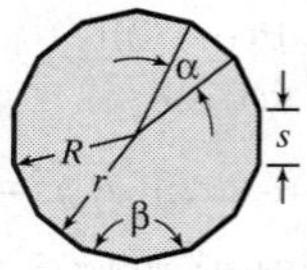


$A$ = area $\quad n$ = number of sides

$\alpha = 360° \div n \qquad \beta = 180° - \alpha$

$$A = \frac{nsr}{2} = \frac{ns}{2}\sqrt{R^2 - \frac{s^2}{4}}$$

$$R = \sqrt{r^2 + \frac{s^2}{4}} \qquad r = \sqrt{R^2 - \frac{s^2}{4}} \qquad s = 2\sqrt{R^2 - r^2}$$

*Example:* Find the area of a polygon having 12 sides, inscribed in a circle with radius of 8 centimeters. The length of the side $s$ is 4.141 centimeters.

$$A = \frac{ns}{2}\sqrt{R^2 - \frac{s^2}{4}} = \frac{12 \times 4.141}{2}\sqrt{8^2 - \frac{4.141^2}{4}} = 24.846\sqrt{59.713}$$

$$= 24.846 \times 7.727 = 191.98 \text{ cm}^2$$

*Circle:*

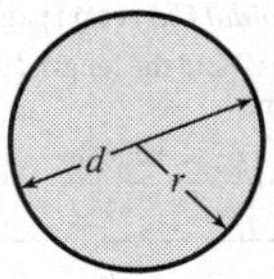


$$\text{Area} = A = \pi r^2 = 3.1416r^2 = 0.7854d^2$$

$$\text{Circumference} = C = 2\pi r = 6.2832r = 3.1416d$$

$$r = C \div 6.2832 = \sqrt{A \div 3.1416} = 0.564\sqrt{A}$$

$$d = C \div 3.1416 = \sqrt{A \div 0.7854} = 1.128\sqrt{A}$$

Length of arc for center angle of 1° = $0.008727d$

Length of arc for center angle of $n°$ = $0.008727nd$

*Example:* Find area $A$ and circumference $C$ of a circle with a diameter of $2\frac{3}{4}$ inches.

$$A = 0.7854d^2 = 0.7854 \times 2.75^2 = 0.7854 \times 2.75 \times 2.75 = 5.9396 \text{ in}^2$$

$$C = 3.1416d = 3.1416 \times 2.75 = 8.6394 \text{ in}$$

*Example:* The area of a circle is 16.8 in$^2$. Find its diameter.

$$d = 1.128\sqrt{A} = 1.128\sqrt{16.8} = 1.128 \times 4.099 = 4.624 \text{ in.}$$

*Sector of a Circle:*

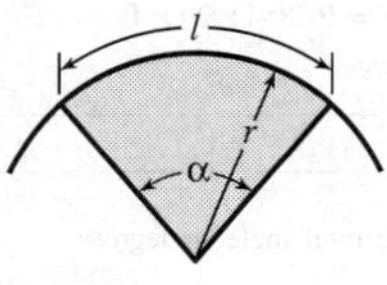


$$\text{Length of arc} = l = \frac{3.1416\, r\, \alpha}{180} = 0.01745r\alpha = \frac{2A}{r}$$

$$\text{Area} = A = \tfrac{1}{2}rl = 0.008727\alpha r^2$$

$$\text{Central angle, in degrees} = \alpha = \frac{57.296\, l}{r}, \qquad r = \frac{2A}{l} = \frac{57.296\, l}{\alpha}$$

*Example:* The radius of a circle is 35 millimeters, and angle $\alpha$ of a sector of the circle is 60 degrees. Find the area of the sector and the length of arc $l$.

$$A = 0.008727\alpha r^2 = 0.008727 \times 60 \times 35^2 = 641.41 \text{mm}^2 = 6.41 \text{cm}^2$$

$$l = 0.01745r\alpha = 0.01745 \times 35 \times 60 = 36.645 \text{ mm}$$

*Segment of a Circle:*

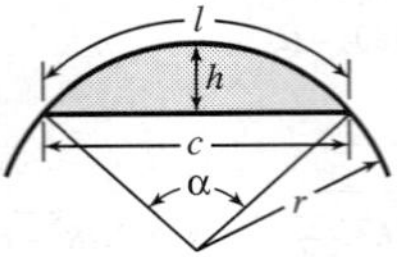


$A$ = area $l$ = length of arc $\alpha$ = angle, in degrees

$c = 2\sqrt{h(2r-h)}$ $A = \frac{1}{2}[rl - c(r-h)]$

$r = \dfrac{c^2 + 4h^2}{8h}$ $l = 0.01745r\alpha$

$h = r - \frac{1}{2}\sqrt{4r^2 - c^2} = r[1 - \cos(\alpha/2)]$ $\alpha = \dfrac{57.296\ l}{r}$

See also *Segments of a Circle* starting on page 78.

*Example:* The radius $r$ is 60 inches and the height $h$ is 8 inches. Find the length of the chord $c$.

$$c = 2\sqrt{h(2r-h)} = 2\sqrt{8 \times (2 \times 60 - 8)} = 2\sqrt{896} = 2 \times 29.93 = 59.86 \text{ in.}$$

*Example:* If $c = 16$, and $h = 6$ inches, what is the radius of the circle of which the segment is a part?

$$r = \frac{c^2 + 4h^2}{8h} = \frac{16^2 + 4 \times 6^2}{8 \times 6} = \frac{256 + 144}{48} = \frac{400}{48} = 8\tfrac{1}{3} \text{ in.}$$

*Cycloid:*

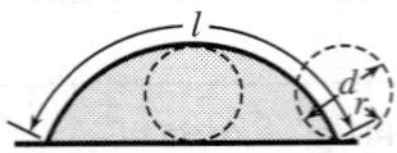


Area $= A = 3\pi r^2 = 9.4248r^2 = 2.3562d^2$

$= 3 \times$ area of generating circle

Length of cycloid $= l = 8r = 4d$

See also *Areas Enclosed by Cycloidal Curves* on page 68.

*Example:* The diameter of the generating circle of a cycloid is 6 inches. Find the length $l$ of the cycloidal curve and the area enclosed between the curve and the base line.

$l = 4d = 4 \times 6 = 24$ in $A = 2.3562d^2 = 2.3562 \times 6^2 = 84.82$ in²

*Circular Ring (Annulus):*

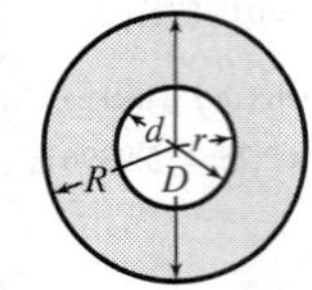


$$\text{Area} = A = \pi(R^2 - r^2) = 3.1416(R^2 - r^2)$$
$$= 3.1416(R + r)(R - r)$$
$$= 0.7854(D^2 - d^2) = 0.7854(D + d)(D - d)$$

*Example:* Let the outside diameter $D = 12$ centimeters and the inside diameter $d = 8$ centimeters. Find the area of the ring.

$$A = 0.7854(D^2 - d^2) = 0.7854(12^2 - 8^2) = 0.7854(144 - 64) = 0.7854 \times 80$$
$$= 62.83 \text{ cm}^2$$

By the alternative formula:

$$A = 0.7854(D + d)(D - d) = 0.7854(12 + 8)(12 - 8) = 0.7854 \times 20 \times 4$$
$$= 62.83 \text{ cm}^2$$

*Sector of Circular Ring:*

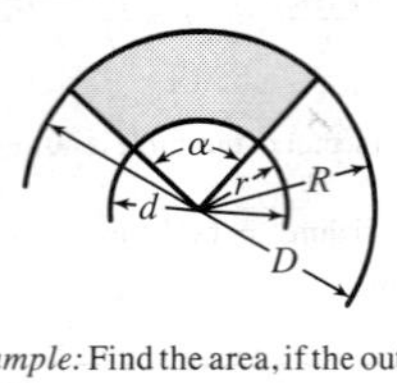


$A$ = area, $\alpha$ = central angle, in degrees

$$A = \frac{\alpha\pi}{360}(R^2 - r^2) = 0.00873\alpha(R^2 - r^2)$$
$$= \frac{\alpha\pi}{4 \times 360}(D^2 - d^2) = 0.00218\alpha(D^2 - d^2)$$

*Example:* Find the area, if the outside radius $R = 5$ inches, the inside radius $r = 2$ inches, and $\alpha = 72$ degrees.

$$A = 0.00873\alpha(R^2 - r^2) = 0.00873 \times 72(5^2 - 2^2)$$
$$= 0.6286(25 - 4) = 0.6286 \times 21 = 13.2 \text{ in}^2$$

*Spandrel or Fillet:*

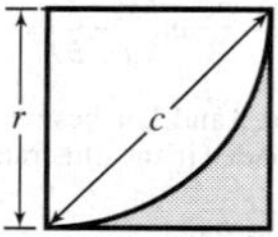


The shaded region is the spandrel (fillet).

$$\text{Area} = A = r^2 - \frac{\pi r^2}{4} = 0.215r^2 = 0.1075c^2$$

*Example:* Find the area of a spandrel, the radius of which is 0.7 inch.

$$A = 0.215r^2 = 0.215 \times 0.7^2 = 0.105 \text{ in}^2$$

*Example:* If chord $c$ were given as 2.2 inches, what would be the area?

$$A = 0.1075c^2 = 0.1075 \times 2.2^2 = 0.520 \text{ in}^2$$

*Parabola:*

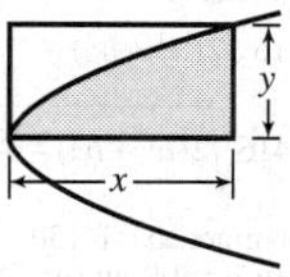


$$\text{Area} = A = \tfrac{2}{3}xy$$

The area of the shaded portion is equal to two-thirds of a rectangle which has $x$ for its base and $y$ for its height.

*Example:* Let $x$ in the illustration be 15 centimeters, and $y$ be 9 centimeters. Find the area of the shaded portion of the parabola.

$$A = \tfrac{2}{3}xy = \tfrac{2}{3} \times 15 \times 9 = 10 \times 9 = 90 \text{ cm}^2$$

*Parabola:*

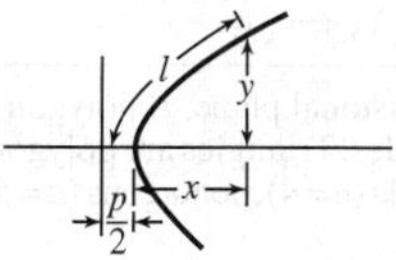


$$l = \text{length of arc} = \frac{p}{2}\left[\sqrt{\frac{2x}{p}\left(1 + \frac{2x}{p}\right)} + \ln\left(\sqrt{\frac{2x}{p}} + \sqrt{1 + \frac{2x}{p}}\right)\right]$$

When $x$ is small in proportion to $y$, the following is a close approximation:

$$l = y\left[1 + \frac{2}{3}\left(\frac{x}{y}\right)^2 - \frac{2}{5}\left(\frac{x}{y}\right)^4\right] \quad \text{or} \quad l = \sqrt{y^2 + \frac{4}{3}x^2}$$

*Example:* If $x = 2$ feet and $y = 24$ feet, what is the approximate length $l$ of the parabolic curve?

$$l = y\left[1 + \frac{2}{3}\left(\frac{x}{y}\right)^2 - \frac{2}{5}\left(\frac{x}{y}\right)^4\right] = 24\left[1 + \frac{2}{3}\left(\frac{2}{24}\right)^2 - \frac{2}{5}\left(\frac{2}{24}\right)^4\right]$$

$$= 24 \times 1.0046 = 24.04 \text{ ft}$$

*Segment of Parabola:*

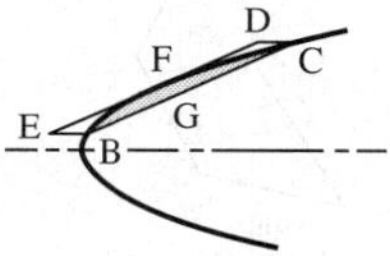


$$\text{Area } BFC = A = \tfrac{2}{3} \times \text{Area of parallelogram BCDE}$$

If $FG$ is the height of the segment, measured at right angles to $BC$, then:

$$\text{Area of segment } BFC = \tfrac{2}{3}(BC)(FG)$$

*Example:* Suppose the length of the chord $BC = 19.5$ inches, and the distance between lines $BC$ and $DE$, measured at right angles to $BC$, is 2.25 inches. Find the area of the segment.

$$\text{Area} = A = \tfrac{2}{3}(BC)(FG) = \tfrac{2}{3} \times 19.5 \times 2.25 = 29.25 \text{ in}^2$$

*Hyperbola:*

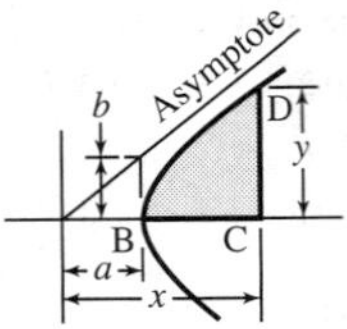


$$\text{Area } BCD = A = \frac{xy}{2} - \frac{ab}{2}\ln\left(\frac{x}{a} + \frac{y}{b}\right)$$

*Example:* The half-axes $a$ and $b$ are 3 and 2 inches, respectively. Find the area shown shaded in the illustration for $x = 8$ inches and $y = 5$ inches.

Inserting the known values in the formula:

$$\text{Area} = A = \frac{8 \times 5}{2} - \frac{3 \times 2}{2} \times \ln\left(\frac{8}{3} + \frac{5}{2}\right) = 20 - 3 \times \ln(5.167)$$

$$= 20 - 3 \times 1.6423 = 20 - 4.927 = 15.073 \text{ in}^2$$

*Ellipse:*

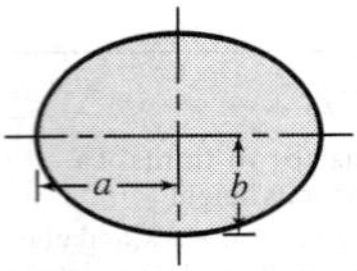


$$\text{Area} = A = \pi ab = 3.1416ab$$

An approximate formula for the perimeter is

$$\text{Perimeter} = P = 3.1416\sqrt{2(a^2 + b^2)}$$

A closer approximation is $P = 3.1416\sqrt{2(a^2 + b^2) - \frac{(a-b)^2}{2.2}}$

*Example:* The larger, or major, axis is 200 millimeters. The smaller, or minor, axis is 150 millimeters. Find the area and the approximate circumference. Here, then, $a = 100$, and $b = 75$.

$$A = 3.1416ab = 3.1416 \times 100 \times 75 = 23{,}562 \text{ mm}^2 = 235.62 \text{ cm}^2$$

$$P = 3.1416\sqrt{2(a^2 + b^2)} = 3.1416\sqrt{2(100^2 + 75^2)} = 3.1416\sqrt{2 \times 15{,}625}$$

$$= 3.1416\sqrt{31{,}250} = 3.1416 \times 176.78 = 555.37 \text{ mm} = 55.537 \text{ cm}$$

**Polygons.**—A *polygon* is a many-sided figure in a two-dimensional plane. A polygon is sometimes referred to as an $n$-gon, where $n$ is the number of sides. Triangles are polygons with the least number of sides ($n = 3$), followed by quadrilaterals ($n = 4$), pentagons ($n = 5$), hexagons ($n = 6$), heptagons ($n = 7$), octagons ($n = 8$), and so on.

A *regular* polygon has congruent sides (all sides are of equal measure) and, hence, its interior angles are congruent. In Fig. 4a, β is the measure of each interior angle, 180 – β is the exterior angle measure at each vertex. α is a measure of the central angle. *Irregular polygons* (Fig. 4b) are polygons whose sides are not all congruent.

Both Fig. 4a and Fig. 4b show convex polygons, in which all exterior angles measure less than 180 degrees. A concave polygon has some interior angles that measure greater than 180 degrees (see Fig. 4c). All the formulas in this section concern convex regular polygons.

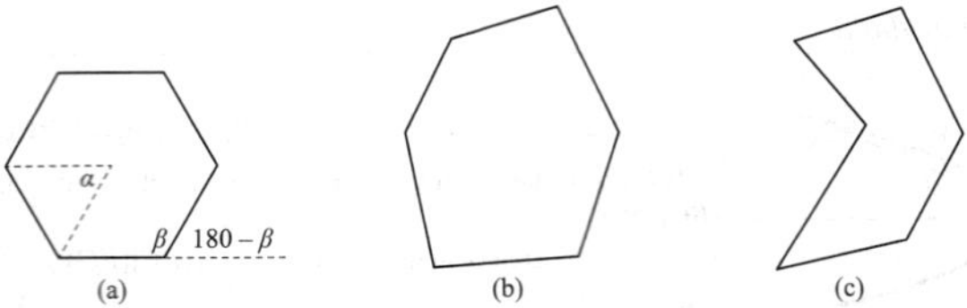


Fig. 4. Polygons: (a) Convex Regular; (b) Convex Irregular; (c) Concave Irregular.

Fig. 5 shows how a regular polygon is either *inscribed* (drawn inside) or *circumscribed* (drawn around) a circle. A polygon inscribed within a circle, as shown in Fig. 5a, is drawn so that all its vertices touch the circle. Its radius is marked $r$. A polygon that circumscribes a circle, as shown in Fig. 5b, is drawn so that the circle touches each of the sides of the polygon. Its radius is marked $R$.

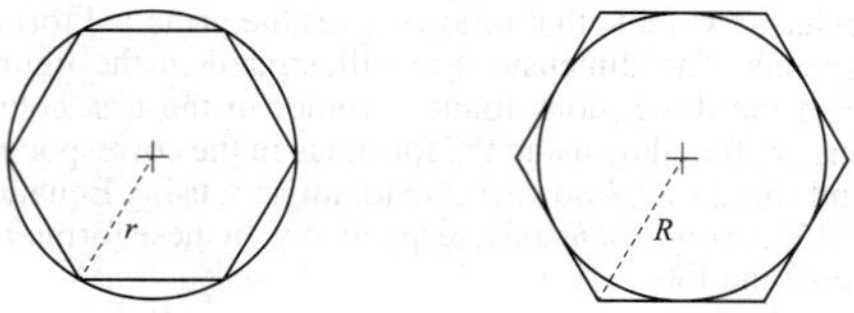

(a) (b)

Fig. 5. Hexagons: (a) Inscribed (Drawn Inside) a Circle; (b) Circumscribed (Drawn Around) a Circle.

*Formulas and Table for Regular Polygons:* The following formulas and table can be used to calculate area, side length, and radii of inscribed and circumscribed polygons on a circle.

$$A = \frac{nS^2 \cot\alpha}{4} = nR^2 \sin\alpha \cos\alpha = nR^2 \tan\alpha, \qquad r = R\cos\alpha = \frac{S\cot\alpha}{2} = \sqrt{\frac{A\cot\alpha}{n}}$$

$$R = \frac{S}{2\sin\alpha} = \frac{r}{\cos\alpha} = \sqrt{\frac{A}{N\sin\alpha\cos\alpha}}, \qquad S = 2R\sin\alpha = 2r\tan\alpha = 2\sqrt{\frac{A\tan\alpha}{N}}$$

**Area, Length of Side, and Inscribed and Circumscribed Radii of Regular Polygons**

| No. of Sides, $n$ | $\frac{A}{S^2}$ | $\frac{A}{R^2}$ | $\frac{A}{r^2}$ | $\frac{R}{S}$ | $\frac{R}{r}$ | $\frac{S}{R}$ | $\frac{S}{r}$ | $\frac{r}{R}$ | $\frac{r}{S}$ |
|---|---|---|---|---|---|---|---|---|---|
| 3 | 0.4330 | 1.2990 | 5.1962 | 0.5774 | 2.0000 | 1.7321 | 3.4641 | 0.5000 | 0.2887 |
| 4 | 1.0000 | 2.0000 | 4.0000 | 0.7071 | 1.4142 | 1.4142 | 2.0000 | 0.7071 | 0.5000 |
| 5 | 1.7205 | 2.3776 | 3.6327 | 0.8507 | 1.2361 | 1.1756 | 1.4531 | 0.8090 | 0.6882 |
| 6 | 2.5981 | 2.5981 | 3.4641 | 1.0000 | 1.1547 | 1.0000 | 1.1547 | 0.8660 | 0.8660 |
| 7 | 3.6339 | 2.7364 | 3.3710 | 1.1524 | 1.1099 | 0.8678 | 0.9631 | 0.9010 | 1.0383 |
| 8 | 4.8284 | 2.8284 | 3.3137 | 1.3066 | 1.0824 | 0.7654 | 0.8284 | 0.9239 | 1.2071 |
| 9 | 6.1818 | 2.8925 | 3.2757 | 1.4619 | 1.0642 | 0.6840 | 0.7279 | 0.9397 | 1.3737 |
| 10 | 7.6942 | 2.9389 | 3.2492 | 1.6180 | 1.0515 | 0.6180 | 0.6498 | 0.9511 | 1.5388 |
| 12 | 11.196 | 3.0000 | 3.2154 | 1.9319 | 1.0353 | 0.5176 | 0.5359 | 0.9659 | 1.8660 |
| 16 | 20.109 | 3.0615 | 3.1826 | 2.5629 | 1.0196 | 0.3902 | 0.3978 | 0.9808 | 2.5137 |
| 20 | 31.569 | 3.0902 | 3.1677 | 3.1962 | 1.0125 | 0.3129 | 0.3168 | 0.9877 | 3.1569 |
| 24 | 45.575 | 3.1058 | 3.1597 | 3.8306 | 1.0086 | 0.2611 | 0.2633 | 0.9914 | 3.7979 |
| 32 | 81.225 | 3.1214 | 3.1517 | 5.1011 | 1.0048 | 0.1960 | 0.1970 | 0.9952 | 5.0766 |
| 48 | 183.08 | 3.1326 | 3.1461 | 7.6449 | 1.0021 | 0.1308 | 0.1311 | 0.9979 | 7.6285 |
| 64 | 325.69 | 3.1365 | 3.1441 | 10.190 | 1.0012 | 0.0981 | 0.0983 | 0.9988 | 10.178 |

*Example 1:* A regular hexagon is inscribed in a circle of 6 in. diameter. Find the area and the radius of an inscribed circle. Here, $R = 3$ in. and $n = 6$. From the table, area $A = 2.5981R^2 = 2.5981 \times 9 = 23.3829$ in$^2$. Radius of inscribed circle, $r = 0.866R = 0.866 \times 3 = 2.598$ in.

*Example 2:* An octagon is inscribed in a circle of 100 mm diameter. Thus, $R = 50$ mm and $n = 8$. Find the area and radius of an inscribed circle. $A = 2.8284R^2 = 2.8284 \times 2500 = 7071$ mm$^2$ = 70.7 cm$^2$. Radius of inscribed circle, $r = 0.9239R = 09239 \times 50 = 46.195$ mm.

*Example 3:* Thirty-two bolts are to be equally spaced on the periphery of a 16-in. diameter bolt-circle. Find the chordal distance between the bolts. Chordal distance equals the side length $S$ of a polygon with $n = 32$ sides and $R = 8$. Hence, $S = 0.196R = 0.196 \times 8 = 1.568$ in.

*Example 4:* Sixteen bolts are to be equally spaced on the periphery of a bolt-circle, 250 mm diameter. Find the chordal distance between the bolts. Chordal distance equals the side length $S$ of a polygon with 16 sides. $R = 125$. Thus, $S = 0.3902R = 0.3902 \times 125 = 48.775$ mm.

**Segments of a Circle.**—The table that follows gives the principal formulas for dimensions of circle segments. The dimensions are illustrated in the figures on pages 74 and 79. When two of the dimensions found together in the first column are known, the other dimensions are found by using the formulas in the corresponding row. For example, if radius $r$ and chord $c$ are known, solve for angle $\alpha$ using Equation (13), then use Equations (14) and (15) to solve for $h$ and $l$, respectively. In these formulas, the value of $\alpha$ is in degrees between 0 and 180°.

**Formulas for Segments of a Circle**

| Given | Formulas | | |
|---|---|---|---|
| $\alpha, r$ | $c = 2r\sin\left(\frac{\alpha}{2}\right)$ (1) | $h = r\left(1 - \cos\left(\frac{\alpha}{2}\right)\right)$ (2) | $l = \frac{\pi r\alpha}{180}$ (3) |
| $\alpha, c$ | $r = \frac{c}{2\sin\left(\frac{\alpha}{2}\right)}$ (4) | $h = \frac{c}{2}\tan\left(\frac{\alpha}{4}\right)$ (5) | $l = \frac{\pi c\alpha}{360\sin\left(\frac{\alpha}{2}\right)}$ (6) |
| $\alpha, h$ | $r = \frac{h}{1 - \cos\left(\frac{\alpha}{2}\right)}$ (7) | $c = \frac{2h}{\tan\left(\frac{\alpha}{4}\right)}$ (8) | $l = \frac{\pi h\alpha}{180\left(1 - \cos\left(\frac{\alpha}{2}\right)\right)}$ (9) |
| $\alpha, l$ | $r = \frac{180}{\pi}\frac{l}{\alpha}$ (10) | $c = \frac{360 l\sin\left(\frac{\alpha}{2}\right)}{\pi\alpha}$ (11) | $h = \frac{180 l\left(1 - \cos\left(\frac{\alpha}{2}\right)\right)}{\pi\alpha}$ (12) |
| $r, c$ | $\alpha = \cos^{-1}\left(1 - \frac{c^2}{2r^2}\right)$ (13) | $h = r - \frac{\sqrt{4r^2 - c^2}}{2}$ (14) | $l = \frac{\pi}{90}r\sin^{-1}\left(\frac{c}{2r}\right)$ (15) |
| $r, h$ | $\alpha = 2\cos^{-1}\left(1 - \frac{h}{r}\right)$ (16) | $c = 2\sqrt{h(2r - h)}$ (17) | $l = \frac{\pi}{90}r\cos^{-1}\left(1 - \frac{h}{r}\right)$ (18) |
| $r, l$ | $\alpha = \frac{180}{\pi}\frac{l}{r}$ (19) | $c = 2r\sin\left(\frac{90 l}{\pi R}\right)$ (20) | $h = r\left(1 - \cos\left(\frac{90 l}{\pi r}\right)\right)$ (21) |
| $c, h$ | $\alpha = 4\tan^{-1}\left(\frac{2h}{c}\right)$ (22) | $r = \frac{c^2 + 4h^2}{8h}$ (23) | $l = \pi\left(\frac{c^2 + 4h^2}{360h}\right)\tan^{-1}\left(\frac{2h}{c}\right)$ (24) |

| Given | Formula To Find | Given | Formula To Find |
|---|---|---|---|
| $c, l$ | $\frac{360}{\pi}\frac{l}{c} = \frac{\alpha}{\sin\left(\frac{\alpha}{2}\right)}$ (25)<br>Solve Equation (25) for $\alpha$ by iteration[a], then<br>$r$ = Equation (10)<br>$h$ = Equation (5) | $h, l$ | $\frac{180}{\pi}\frac{l}{h} = \frac{\alpha}{1 - \cos\left(\frac{\alpha}{2}\right)}$ (26)<br>Solve Equation (26) for $\alpha$ by iteration[a], then<br>$r$ = Equation (10)<br>$c$ = Equation (11) |

[a] Equations (25) and (26) cannot be easily solved by ordinary means. To solve these equations, test various values of $\alpha$ until the left side of the equation equals the right side. For example, if given $c = 4$ and $l = 5$, the left side of Equation (25) equals 143.24, and testing various values of $\alpha$ reveals the right side equals 143.24 when $\alpha = 129.62°$.

Angle $\alpha$ is in degrees, $0 < \alpha < 180$

*Formulas for Circular Segments* contributed by Manfred Brueckner.

**Segments of Circles for Radius = 1.**—Formulas for segments of circles are given on pages 74 and 78. When central angle α and radius *r* are known, the following table can be used to find the length of arc *l*, height of segment *h* (*x* in the discussion on page 74), chord length *c*, and segment area *A*. Column *A*/π is the ratio of segment area *A* to the area of a circle with radius *r* = 1, in percent.

When angle α and radius *r* are not known, but segment height *h* and chord length *c* are known, ratio *h*/*c* can be used to find α, *l*, and *A* by linear interpolation. Radius *r* is found by the formula on page 74 or 78. The value of *l* is then multiplied by the radius *r* and the area *A* by $r^2$.

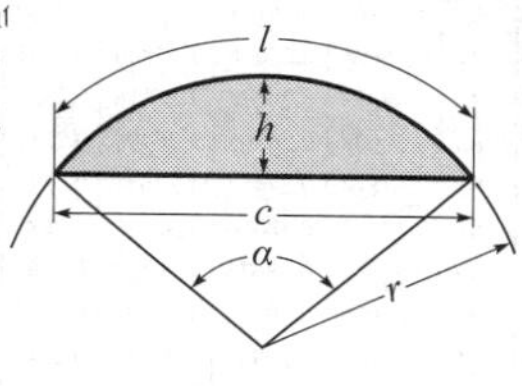


Angle α can be found thus with an accuracy of about 0.001 degree; arc length *l* with an error of about 0.02 percent; and area *A* with an error ranging from about 0.02 percent for the highest entry value of *h*/*c* to about 1 percent for values of *h*/*c* of about 0.050. For lower values of *h*/*c*, and where greater accuracy is required, area *A* should be found by the formula on page 74.

*Example:* A 3-foot diameter cylindrical tank, mounted horizontally, contains fuel. What is the fuel depth, in inches, when the tank is 20% full? *Solution:* Locate 20% in table column *A*/π%. The depth equals *h* multiplied by the radius: $hr = 0.50758 \times 1.5 \times 12 = 9.14$ inches

**Segments of Circles for Radius = 1** (US Customary or Metric Units)

| α, Deg. | *l* | *h* | *c* | Area *A* | *A*/π % | *h*/*c* | α, Deg. | *l* | *h* | *c* | Area *A* | *A*/π % | *h*/*c* |
|---|---|---|---|---|---|---|---|---|---|---|---|---|---|
| 1 | 0.01745 | 0.00004 | 0.01745 | 0.0000 | 0.0 | 0.00218 | 41 | 0.71558 | 0.06333 | 0.70041 | 0.0298 | 0.9 | 0.09041 |
| 2 | 0.03491 | 0.00015 | 0.03490 | 0.0000 | 0.0 | 0.00436 | 42 | 0.73304 | 0.06642 | 0.71674 | 0.0320 | 1.0 | 0.09267 |
| 3 | 0.05236 | 0.00034 | 0.05235 | 0.0000 | 0.0 | 0.00655 | 43 | 0.75049 | 0.06958 | 0.73300 | 0.0342 | 1.1 | 0.09493 |
| 4 | 0.06981 | 0.00061 | 0.06980 | 0.0000 | 0.0 | 0.00873 | 44 | 0.76794 | 0.07282 | 0.74921 | 0.0366 | 1.2 | 0.09719 |
| 5 | 0.08727 | 0.00095 | 0.08724 | 0.0001 | 0.0 | 0.01091 | 45 | 0.78540 | 0.07612 | 0.76537 | 0.0391 | 1.2 | 0.09946 |
| 6 | 0.10472 | 0.00137 | 0.10467 | 0.0001 | 0.0 | 0.01309 | 46 | 0.80285 | 0.07950 | 0.78146 | 0.0418 | 1.3 | 0.10173 |
| 7 | 0.12217 | 0.00187 | 0.12210 | 0.0002 | 0.0 | 0.01528 | 47 | 0.82030 | 0.08294 | 0.79750 | 0.0445 | 1.4 | 0.10400 |
| 8 | 0.13963 | 0.00244 | 0.13951 | 0.0002 | 0.0 | 0.01746 | 48 | 0.83776 | 0.08645 | 0.81347 | 0.0473 | 1.5 | 0.10628 |
| 9 | 0.15708 | 0.00308 | 0.15692 | 0.0003 | 0.0 | 0.01965 | 49 | 0.85521 | 0.09004 | 0.82939 | 0.0503 | 1.6 | 0.10856 |
| 10 | 0.17453 | 0.00381 | 0.17431 | 0.0004 | 0.0 | 0.02183 | 50 | 0.87266 | 0.09369 | 0.84524 | 0.0533 | 1.7 | 0.11085 |
| 11 | 0.19199 | 0.00460 | 0.19169 | 0.0006 | 0.0 | 0.02402 | 51 | 0.89012 | 0.09741 | 0.86102 | 0.0565 | 1.8 | 0.11314 |
| 12 | 0.20944 | 0.00548 | 0.20906 | 0.0008 | 0.0 | 0.02620 | 52 | 0.90757 | 0.10121 | 0.87674 | 0.0598 | 1.9 | 0.11543 |
| 13 | 0.22689 | 0.00643 | 0.22641 | 0.0010 | 0.0 | 0.02839 | 53 | 0.92502 | 0.10507 | 0.89240 | 0.0632 | 2.0 | 0.11773 |
| 14 | 0.24435 | 0.00745 | 0.24374 | 0.0012 | 0.0 | 0.03058 | 54 | 0.94248 | 0.10899 | 0.90798 | 0.0667 | 2.1 | 0.12004 |
| 15 | 0.26180 | 0.00856 | 0.26105 | 0.0015 | 0.0 | 0.03277 | 55 | 0.95993 | 0.11299 | 0.92350 | 0.0704 | 2.2 | 0.12235 |
| 16 | 0.27925 | 0.00973 | 0.27835 | 0.0018 | 0.1 | 0.03496 | 56 | 0.97738 | 0.11705 | 0.93894 | 0.0742 | 2.4 | 0.12466 |
| 17 | 0.29671 | 0.01098 | 0.29562 | 0.0022 | 0.1 | 0.03716 | 57 | 0.99484 | 0.12118 | 0.95432 | 0.0781 | 2.5 | 0.12698 |
| 18 | 0.31416 | 0.01231 | 0.31287 | 0.0026 | 0.1 | 0.03935 | 58 | 1.01229 | 0.12538 | 0.96962 | 0.0821 | 2.6 | 0.12931 |
| 19 | 0.33161 | 0.01371 | 0.33010 | 0.0030 | 0.1 | 0.04155 | 59 | 1.02974 | 0.12964 | 0.98485 | 0.0863 | 2.7 | 0.13164 |
| 20 | 0.34907 | 0.01519 | 0.34730 | 0.0035 | 0.1 | 0.04374 | 60 | 1.04720 | 0.13397 | 1.00000 | 0.0906 | 2.9 | 0.13397 |
| 21 | 0.36652 | 0.01675 | 0.36447 | 0.0041 | 0.1 | 0.04594 | 61 | 1.06465 | 0.13837 | 1.01508 | 0.0950 | 3.0 | 0.13632 |
| 22 | 0.38397 | 0.01837 | 0.38162 | 0.0047 | 0.1 | 0.04814 | 62 | 1.08210 | 0.14283 | 1.03008 | 0.0996 | 3.2 | 0.13866 |
| 23 | 0.40143 | 0.02008 | 0.39874 | 0.0053 | 0.2 | 0.05035 | 63 | 1.09956 | 0.14736 | 1.04500 | 0.1043 | 3.3 | 0.14101 |
| 24 | 0.41888 | 0.02185 | 0.41582 | 0.0061 | 0.2 | 0.05255 | 64 | 1.11701 | 0.15195 | 1.05984 | 0.1091 | 3.5 | 0.14337 |
| 25 | 0.43633 | 0.02370 | 0.43288 | 0.0069 | 0.2 | 0.05476 | 65 | 1.13446 | 0.15661 | 1.07460 | 0.1141 | 3.6 | 0.14574 |
| 26 | 0.45379 | 0.02563 | 0.44990 | 0.0077 | 0.2 | 0.05697 | 66 | 1.15192 | 0.16133 | 1.08928 | 0.1192 | 3.8 | 0.14811 |
| 27 | 0.47124 | 0.02763 | 0.46689 | 0.0086 | 0.3 | 0.05918 | 67 | 1.16937 | 0.16611 | 1.10387 | 0.1244 | 4.0 | 0.15048 |
| 28 | 0.48869 | 0.02970 | 0.48384 | 0.0096 | 0.3 | 0.06139 | 68 | 1.18682 | 0.17096 | 1.11839 | 0.1298 | 4.1 | 0.15287 |
| 29 | 0.50615 | 0.03185 | 0.50076 | 0.0107 | 0.3 | 0.06361 | 69 | 1.20428 | 0.17587 | 1.13281 | 0.1353 | 4.3 | 0.15525 |
| 30 | 0.52360 | 0.03407 | 0.51764 | 0.0118 | 0.4 | 0.06583 | 70 | 1.22173 | 0.18085 | 1.14715 | 0.1410 | 4.5 | 0.15765 |
| 31 | 0.54105 | 0.03637 | 0.53448 | 0.0130 | 0.4 | 0.06805 | 71 | 1.23918 | 0.18588 | 1.16141 | 0.1468 | 4.7 | 0.16005 |
| 32 | 0.55851 | 0.03874 | 0.55127 | 0.0143 | 0.5 | 0.07027 | 72 | 1.25664 | 0.19098 | 1.17557 | 0.1528 | 4.9 | 0.16246 |
| 33 | 0.57596 | 0.04118 | 0.56803 | 0.0157 | 0.5 | 0.07250 | 73 | 1.27409 | 0.19614 | 1.18965 | 0.1589 | 5.1 | 0.16488 |
| 34 | 0.59341 | 0.04370 | 0.58474 | 0.0171 | 0.5 | 0.07473 | 74 | 1.29154 | 0.20136 | 1.20363 | 0.1651 | 5.3 | 0.16730 |
| 35 | 0.61087 | 0.04628 | 0.60141 | 0.0186 | 0.6 | 0.07696 | 75 | 1.30900 | 0.20665 | 1.21752 | 0.1715 | 5.5 | 0.16973 |
| 36 | 0.62832 | 0.04894 | 0.61803 | 0.0203 | 0.6 | 0.07919 | 76 | 1.32645 | 0.21199 | 1.23132 | 0.1781 | 5.7 | 0.17216 |
| 37 | 0.64577 | 0.05168 | 0.63461 | 0.0220 | 0.7 | 0.08143 | 77 | 1.34390 | 0.21739 | 1.24503 | 0.1848 | 5.9 | 0.17461 |
| 38 | 0.66323 | 0.05448 | 0.65114 | 0.0238 | 0.8 | 0.08367 | 78 | 1.36136 | 0.22285 | 1.25864 | 0.1916 | 6.1 | 0.17706 |
| 39 | 0.68068 | 0.05736 | 0.66761 | 0.0257 | 0.8 | 0.08592 | 79 | 1.37881 | 0.22838 | 1.27216 | 0.1986 | 6.3 | 0.17952 |
| 40 | 0.69813 | 0.06031 | 0.68404 | 0.0277 | 0.9 | 0.08816 | 80 | 1.39626 | 0.23396 | 1.28558 | 0.2057 | 6.5 | 0.18199 |

**Segments of Circles for Radius = 1** (US Customary or Metric Units) *(Continued)*

| α, Deg. | *l* | *h* | *c* | *Area A* | *A/π* % | *h/c* | α, Deg. | *l* | *h* | *c* | *Area A* | *A/π* % | *h/c* |
|---|---|---|---|---|---|---|---|---|---|---|---|---|---|
| 81 | 1.41372 | 0.23959 | 1.29890 | 0.2130 | 6.8 | 0.18446 | 131 | 2.28638 | 0.58531 | 1.81992 | 0.7658 | 24.4 | 0.32161 |
| 82 | 1.43117 | 0.24529 | 1.31212 | 0.2205 | 7.0 | 0.18694 | 132 | 2.30383 | 0.59326 | 1.82709 | 0.7803 | 24.8 | 0.32470 |
| 83 | 1.44862 | 0.25104 | 1.32524 | 0.2280 | 7.3 | 0.18943 | 133 | 2.32129 | 0.60125 | 1.83412 | 0.7950 | 25.3 | 0.32781 |
| 84 | 1.46608 | 0.25686 | 1.33826 | 0.2358 | 7.5 | 0.19193 | 134 | 2.33874 | 0.60927 | 1.84101 | 0.8097 | 25.8 | 0.33094 |
| 85 | 1.48353 | 0.26272 | 1.35118 | 0.2437 | 7.8 | 0.19444 | 135 | 2.35619 | 0.61732 | 1.84776 | 0.8245 | 26.2 | 0.33409 |
| 86 | 1.50098 | 0.26865 | 1.36400 | 0.2517 | 8.0 | 0.19696 | 136 | 2.37365 | 0.62539 | 1.85437 | 0.8395 | 26.7 | 0.33725 |
| 87 | 1.51844 | 0.27463 | 1.37671 | 0.2599 | 8.3 | 0.19948 | 137 | 2.39110 | 0.63350 | 1.86084 | 0.8546 | 27.2 | 0.34044 |
| 88 | 1.53589 | 0.28066 | 1.38932 | 0.2682 | 8.5 | 0.20201 | 138 | 2.40855 | 0.64163 | 1.86716 | 0.8697 | 27.7 | 0.34364 |
| 89 | 1.55334 | 0.28675 | 1.40182 | 0.2767 | 8.8 | 0.20456 | 139 | 2.42601 | 0.64979 | 1.87334 | 0.8850 | 28.2 | 0.34686 |
| 90 | 1.57080 | 0.29289 | 1.41421 | 0.2854 | 9.1 | 0.20711 | 140 | 2.44346 | 0.65798 | 1.87939 | 0.9003 | 28.7 | 0.35010 |
| 91 | 1.58825 | 0.29909 | 1.42650 | 0.2942 | 9.4 | 0.20967 | 141 | 2.46091 | 0.66619 | 1.88528 | 0.9158 | 29.2 | 0.35337 |
| 92 | 1.60570 | 0.30534 | 1.43868 | 0.3032 | 9.7 | 0.21224 | 142 | 2.47837 | 0.67443 | 1.89104 | 0.9314 | 29.6 | 0.35665 |
| 93 | 1.62316 | 0.31165 | 1.45075 | 0.3123 | 9.9 | 0.21482 | 143 | 2.49582 | 0.68270 | 1.89665 | 0.9470 | 30.1 | 0.35995 |
| 94 | 1.64061 | 0.31800 | 1.46271 | 0.3215 | 10.2 | 0.21741 | 144 | 2.51327 | 0.69098 | 1.90211 | 0.9627 | 30.6 | 0.36327 |
| 95 | 1.65806 | 0.32441 | 1.47455 | 0.3309 | 10.5 | 0.22001 | 145 | 2.53073 | 0.69929 | 1.90743 | 0.9786 | 31.1 | 0.36662 |
| 96 | 1.67552 | 0.33087 | 1.48629 | 0.3405 | 10.8 | 0.22261 | 146 | 2.54818 | 0.70763 | 1.91261 | 0.9945 | 31.7 | 0.36998 |
| 97 | 1.69297 | 0.33738 | 1.49791 | 0.3502 | 11.1 | 0.22523 | 147 | 2.56563 | 0.71598 | 1.91764 | 1.0105 | 32.2 | 0.37337 |
| 98 | 1.71042 | 0.34394 | 1.50942 | 0.3601 | 11.5 | 0.22786 | 148 | 2.58309 | 0.72436 | 1.92252 | 1.0266 | 32.7 | 0.37678 |
| 99 | 1.72788 | 0.35055 | 1.52081 | 0.3701 | 11.8 | 0.23050 | 149 | 2.60054 | 0.73276 | 1.92726 | 1.0428 | 33.2 | 0.38021 |
| 100 | 1.74533 | 0.35721 | 1.53209 | 0.3803 | 12.1 | 0.23315 | 150 | 2.61799 | 0.74118 | 1.93185 | 1.0590 | 33.7 | 0.38366 |
| 101 | 1.76278 | 0.36392 | 1.54325 | 0.3906 | 12.4 | 0.23582 | 151 | 2.63545 | 0.74962 | 1.93630 | 1.0753 | 34.2 | 0.38714 |
| 102 | 1.78024 | 0.37068 | 1.55429 | 0.4010 | 12.8 | 0.23849 | 152 | 2.65290 | 0.75808 | 1.94059 | 1.0917 | 34.7 | 0.39064 |
| 103 | 1.79769 | 0.37749 | 1.56522 | 0.4117 | 13.1 | 0.24117 | 153 | 2.67035 | 0.76655 | 1.94474 | 1.1082 | 35.3 | 0.39417 |
| 104 | 1.81514 | 0.38434 | 1.57602 | 0.4224 | 13.4 | 0.24387 | 154 | 2.68781 | 0.77505 | 1.94874 | 1.1247 | 35.8 | 0.39772 |
| 105 | 1.83260 | 0.39124 | 1.58671 | 0.4333 | 13.8 | 0.24657 | 155 | 2.70526 | 0.78356 | 1.95259 | 1.1413 | 36.3 | 0.40129 |
| 106 | 1.85005 | 0.39818 | 1.59727 | 0.4444 | 14.1 | 0.24929 | 156 | 2.72271 | 0.79209 | 1.95630 | 1.1580 | 36.9 | 0.40489 |
| 107 | 1.86750 | 0.40518 | 1.60771 | 0.4556 | 14.5 | 0.25202 | 157 | 2.74017 | 0.80063 | 1.95985 | 1.1747 | 37.4 | 0.40852 |
| 108 | 1.88496 | 0.41221 | 1.61803 | 0.4669 | 14.9 | 0.25476 | 158 | 2.75762 | 0.80919 | 1.96325 | 1.1915 | 37.9 | 0.41217 |
| 109 | 1.90241 | 0.41930 | 1.62823 | 0.4784 | 15.2 | 0.25752 | 159 | 2.77507 | 0.81776 | 1.96651 | 1.2084 | 38.5 | 0.41585 |
| 110 | 1.91986 | 0.42642 | 1.63830 | 0.4901 | 15.6 | 0.26028 | 160 | 2.79253 | 0.82635 | 1.96962 | 1.2253 | 39.0 | 0.41955 |
| 111 | 1.93732 | 0.43359 | 1.64825 | 0.5019 | 16.0 | 0.26306 | 161 | 2.80998 | 0.83495 | 1.97257 | 1.2422 | 39.5 | 0.42328 |
| 112 | 1.95477 | 0.44081 | 1.65808 | 0.5138 | 16.4 | 0.26585 | 162 | 2.82743 | 0.84357 | 1.97538 | 1.2592 | 40.1 | 0.42704 |
| 113 | 1.97222 | 0.44806 | 1.66777 | 0.5259 | 16.7 | 0.26866 | 163 | 2.84489 | 0.85219 | 1.97803 | 1.2763 | 40.6 | 0.43083 |
| 114 | 1.98968 | 0.45536 | 1.67734 | 0.5381 | 17.1 | 0.27148 | 164 | 2.86234 | 0.86083 | 1.98054 | 1.2934 | 41.2 | 0.43464 |
| 115 | 2.00713 | 0.46270 | 1.68678 | 0.5504 | 17.5 | 0.27431 | 165 | 2.87979 | 0.86947 | 1.98289 | 1.3105 | 41.7 | 0.43849 |
| 116 | 2.02458 | 0.47008 | 1.69610 | 0.5629 | 17.9 | 0.27715 | 166 | 2.89725 | 0.87813 | 1.98509 | 1.3277 | 42.3 | 0.44236 |
| 117 | 2.04204 | 0.47750 | 1.70528 | 0.5755 | 18.3 | 0.28001 | 167 | 2.91470 | 0.88680 | 1.98714 | 1.3449 | 42.8 | 0.44627 |
| 118 | 2.05949 | 0.48496 | 1.71433 | 0.5883 | 18.7 | 0.28289 | 168 | 2.93215 | 0.89547 | 1.98904 | 1.3621 | 43.4 | 0.45020 |
| 119 | 2.07694 | 0.49246 | 1.72326 | 0.6012 | 19.1 | 0.28577 | 169 | 2.94961 | 0.90415 | 1.99079 | 1.3794 | 43.9 | 0.45417 |
| 120 | 2.09440 | 0.50000 | 1.73205 | 0.6142 | 19.6 | 0.28868 | 170 | 2.96706 | 0.91284 | 1.99239 | 1.3967 | 44.5 | 0.45817 |
| 121 | 2.11185 | 0.50758 | 1.74071 | 0.6273 | 20.0 | 0.29159 | 171 | 2.98451 | 0.92154 | 1.99383 | 1.4140 | 45.0 | 0.46220 |
| 122 | 2.12930 | 0.51519 | 1.74924 | 0.6406 | 20.4 | 0.29452 | 172 | 3.00197 | 0.93024 | 1.99513 | 1.4314 | 45.6 | 0.46626 |
| 123 | 2.14675 | 0.52284 | 1.75763 | 0.6540 | 20.8 | 0.29747 | 173 | 3.01942 | 0.93895 | 1.99627 | 1.4488 | 46.1 | 0.47035 |
| 124 | 2.16421 | 0.53053 | 1.76590 | 0.6676 | 21.3 | 0.30043 | 174 | 3.03687 | 0.94766 | 1.99726 | 1.4662 | 46.7 | 0.47448 |
| 125 | 2.18166 | 0.53825 | 1.77402 | 0.6813 | 21.7 | 0.30341 | 175 | 3.05433 | 0.95638 | 1.99810 | 1.4836 | 47.2 | 0.47865 |
| 126 | 2.19911 | 0.54601 | 1.78201 | 0.6950 | 22.1 | 0.30640 | 176 | 3.07178 | 0.96510 | 1.99878 | 1.5010 | 47.8 | 0.48284 |
| 127 | 2.21657 | 0.55380 | 1.78987 | 0.7090 | 22.6 | 0.30941 | 177 | 3.08923 | 0.97382 | 1.99931 | 1.5184 | 48.3 | 0.48708 |
| 128 | 2.23402 | 0.56163 | 1.79759 | 0.7230 | 23.0 | 0.31243 | 178 | 3.10669 | 0.98255 | 1.99970 | 1.5359 | 48.9 | 0.49135 |
| 129 | 2.25147 | 0.56949 | 1.80517 | 0.7372 | 23.5 | 0.31548 | 179 | 3.12414 | 0.99127 | 1.99992 | 1.5533 | 49.4 | 0.49566 |
| 130 | 2.26893 | 0.57738 | 1.81262 | 0.7514 | 23.9 | 0.31854 | 180 | 3.14159 | 1.00000 | 2.00000 | 1.5708 | 50.0 | 0.50000 |

## Diameters of Circles and Sides of Squares of Equal Area
(US Customary or Metric Units)

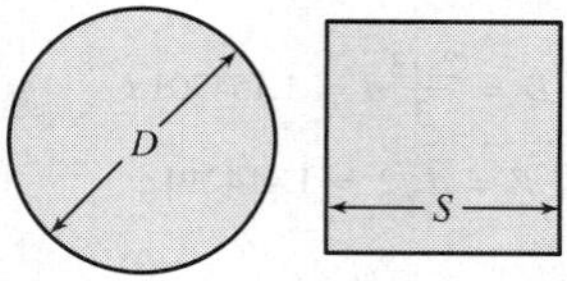


The table below will be found useful for determining the diameter of a circle of an area equal to that of a square, the side of which is known, or for determining the side of a square which has an area equal to that of a circle, the area or diameter of which is known. For example, if the diameter of a circle is 17½ inches, it is found from the table by reading across from the first column that the side of a square of the same area is 15.51 inches. And both have area 240.53 $in^2$.

| Dia. of Circle, *D* | Side of Square, *S* | Area of Circle or Square | Dia. of Circle, *D* | Side of Square, *S* | Area of Circle or Square | Dia. of Circle, *D* | Side of Square, *S* | Area of Circle or Square |
|---|---|---|---|---|---|---|---|---|
| ½ | 0.44 | 0.196 | 20½ | 18.17 | 330.06 | 40½ | 35.89 | 1288.25 |
| 1 | 0.89 | 0.785 | 21 | 18.61 | 346.36 | 41 | 36.34 | 1320.25 |
| 1½ | 1.33 | 1.767 | 21½ | 19.05 | 363.05 | 41½ | 36.78 | 1352.65 |
| 2 | 1.77 | 3.142 | 22 | 19.50 | 380.13 | 42 | 37.22 | 1385.44 |
| 2½ | 2.22 | 4.909 | 22½ | 19.94 | 397.61 | 42½ | 37.66 | 1418.63 |
| 3 | 2.66 | 7.069 | 23 | 20.38 | 415.48 | 43 | 38.11 | 1452.20 |
| 3½ | 3.10 | 9.621 | 23½ | 20.83 | 433.74 | 43½ | 38.55 | 1486.17 |
| 4 | 3.54 | 12.566 | 24 | 21.27 | 452.39 | 44 | 38.99 | 1520.53 |
| 4½ | 3.99 | 15.904 | 24½ | 21.71 | 471.44 | 44½ | 39.44 | 1555.28 |
| 5 | 4.43 | 19.635 | 25 | 22.16 | 490.87 | 45 | 39.88 | 1590.43 |
| 5½ | 4.87 | 23.758 | 25½ | 22.60 | 510.71 | 45½ | 40.32 | 1625.97 |
| 6 | 5.32 | 28.274 | 26 | 23.04 | 530.93 | 46 | 40.77 | 1661.90 |
| 6½ | 5.76 | 33.183 | 26½ | 23.49 | 551.55 | 46½ | 41.21 | 1698.23 |
| 7 | 6.20 | 38.485 | 27 | 23.93 | 572.56 | 47 | 41.65 | 1734.94 |
| 7½ | 6.65 | 44.179 | 27½ | 24.37 | 593.96 | 47½ | 42.10 | 1772.05 |
| 8 | 7.09 | 50.265 | 28 | 24.81 | 615.75 | 48 | 42.54 | 1809.56 |
| 8½ | 7.53 | 56.745 | 28½ | 25.26 | 637.94 | 48½ | 42.98 | 1847.45 |
| 9 | 7.98 | 63.617 | 29 | 25.70 | 660.52 | 49 | 43.43 | 1885.74 |
| 9½ | 8.42 | 70.882 | 29½ | 26.14 | 683.49 | 49½ | 43.87 | 1924.42 |
| 10 | 8.86 | 78.540 | 30 | 26.59 | 706.86 | 50 | 44.31 | 1963.50 |
| 10½ | 9.31 | 86.590 | 30½ | 27.03 | 730.62 | 50½ | 44.75 | 2002.96 |
| 11 | 9.75 | 95.033 | 31 | 27.47 | 754.77 | 51 | 45.20 | 2042.82 |
| 11½ | 10.19 | 103.87 | 31½ | 27.92 | 779.31 | 51½ | 45.64 | 2083.07 |
| 12 | 10.63 | 113.10 | 32 | 28.36 | 804.25 | 52 | 46.08 | 2123.72 |
| 12½ | 11.08 | 122.72 | 32½ | 28.80 | 829.58 | 52½ | 46.53 | 2164.75 |
| 13 | 11.52 | 132.73 | 33 | 29.25 | 855.30 | 53 | 46.97 | 2206.18 |
| 13½ | 11.96 | 143.14 | 33½ | 29.69 | 881.41 | 53½ | 47.41 | 2248.01 |
| 14 | 12.41 | 153.94 | 34 | 30.13 | 907.92 | 54 | 47.86 | 2290.22 |
| 14½ | 12.85 | 165.13 | 34½ | 30.57 | 934.82 | 54½ | 48.30 | 2332.83 |
| 15 | 13.29 | 176.71 | 35 | 31.02 | 962.11 | 55 | 48.74 | 2375.83 |
| 15½ | 13.74 | 188.69 | 35½ | 31.46 | 989.80 | 55½ | 49.19 | 2419.22 |
| 16 | 14.18 | 201.06 | 36 | 31.90 | 1017.88 | 56 | 49.63 | 2463.01 |
| 16½ | 14.62 | 213.82 | 36½ | 32.35 | 1046.35 | 56½ | 50.07 | 2507.19 |
| 17 | 15.07 | 226.98 | 37 | 32.79 | 1075.21 | 57 | 50.51 | 2551.76 |
| 17½ | 15.51 | 240.53 | 37½ | 33.23 | 1104.47 | 57½ | 50.96 | 2596.72 |
| 18 | 15.95 | 254.47 | 38 | 33.68 | 1134.11 | 58 | 51.40 | 2642.08 |
| 18½ | 16.40 | 268.80 | 38½ | 34.12 | 1164.16 | 58½ | 51.84 | 2687.83 |
| 19 | 16.84 | 283.53 | 39 | 34.56 | 1194.59 | 59 | 52.29 | 2733.97 |
| 19½ | 17.28 | 298.65 | 39½ | 35.01 | 1225.42 | 59½ | 52.73 | 2780.51 |
| 20 | 17.72 | 314.16 | 40 | 35.45 | 1256.64 | 60 | 53.17 | 2827.43 |

**Diagonals of Squares and Hexagons.**—The table below gives values of dimensions *D* and *E* described in the figures and equations that follow.

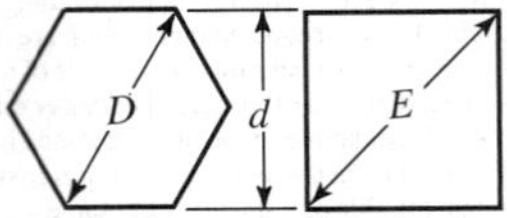


$$D = \frac{2\sqrt{3}}{3}d = 1.154701d$$

$$E = d\sqrt{2} = 1.414214d$$

A value not given in the table can be obtained from the equations above, or by the simple addition of two or more values taken directly from the table. Further values can be obtained by shifting the decimal point.

*Example 1:* Find *D* when $d = 2\,{}^5/_{16}$ inches. From the table, for $d = 2$, $D = 2.3094$, and for $d = {}^5/_{16}$, $D = 0.3608$. Therefore, $D = 2.3094 + 0.3608 = 2.6702$ inches.

*Example 2:* Find *E* when $d = 20.25$ millimeters. From the table, for $d = 20$, $E = 28.2843$; for $d = 0.25$, that is, ${}^1/_4$, $E = 0.3536$. Thus, $E = 28.2843 + 0.3536 = 28.6379$ millimeters.

**Diagonals of Squares and Hexagons** (US Customary or Metric Units)

| *d* | *D* | *E* | *d* | *D* | *E* | *d* | *D* | *E* | *d* | *D* | *E* |
|---|---|---|---|---|---|---|---|---|---|---|---|
| 1/32 | 0.0361 | 0.0442 | 0.9 | 1.0392 | 1.2728 | 32 | 36.9504 | 45.2548 | 67 | 77.3650 | 94.7523 |
| 1/16 | 0.0722 | 0.0884 | 29/32 | 1.0464 | 1.2816 | 33 | 38.1051 | 46.6691 | 68 | 78.5197 | 96.1666 |
| 3/32 | 0.1083 | 0.1326 | 15/16 | 1.0825 | 1.3258 | 34 | 39.2598 | 48.0833 | 69 | 79.6744 | 97.5808 |
| 0.1 | 0.1155 | 0.1414 | 31/32 | 1.1186 | 1.3700 | 35 | 40.4145 | 49.4975 | 70 | 80.8291 | 98.9950 |
| 1/8 | 0.1443 | 0.1768 | 1.0 | 1.1547 | 1.4142 | 36 | 41.5692 | 50.9117 | 71 | 81.9838 | 100.409 |
| 5/32 | 0.1804 | 0.2210 | 2.0 | 2.3094 | 2.8284 | 37 | 42.7239 | 52.3259 | 72 | 83.1385 | 101.823 |
| 3/16 | 0.2165 | 0.2652 | 3.0 | 3.4641 | 4.2426 | 38 | 43.8786 | 53.7401 | 73 | 84.2932 | 103.238 |
| 0.2 | 0.2309 | 0.2828 | 4.0 | 4.6188 | 5.6569 | 39 | 45.0333 | 55.1543 | 74 | 85.4479 | 104.652 |
| 7/32 | 0.2526 | 0.3094 | 5.0 | 5.7735 | 7.0711 | 40 | 46.1880 | 56.5686 | 75 | 86.6026 | 106.066 |
| 1/4 | 0.2887 | 0.3536 | 6.0 | 6.9282 | 8.4853 | 41 | 47.3427 | 57.9828 | 76 | 87.7573 | 107.480 |
| 9/32 | 0.3248 | 0.3977 | 7.0 | 8.0829 | 9.8995 | 42 | 48.4974 | 59.3970 | 77 | 88.9120 | 108.894 |
| 0.3 | 0.3464 | 0.4243 | 8.0 | 9.2376 | 11.3137 | 43 | 49.6521 | 60.8112 | 78 | 90.0667 | 110.309 |
| 5/16 | 0.3608 | 0.4419 | 9.0 | 10.3923 | 12.7279 | 44 | 50.8068 | 62.2254 | 79 | 91.2214 | 111.723 |
| 11/32 | 0.3969 | 0.4861 | 10 | 11.5470 | 14.1421 | 45 | 51.9615 | 63.6396 | 80 | 92.3761 | 113.137 |
| 3/8 | 0.4330 | 0.5303 | 11 | 12.7017 | 15.5564 | 46 | 53.1162 | 65.0538 | 81 | 93.5308 | 114.551 |
| 0.4 | 0.4619 | 0.5657 | 12 | 13.8564 | 16.9706 | 47 | 54.2709 | 66.4681 | 82 | 94.6855 | 115.966 |
| 13/32 | 0.4691 | 0.5745 | 13 | 15.0111 | 18.3848 | 48 | 55.4256 | 67.8823 | 83 | 95.8402 | 117.380 |
| 7/16 | 0.5052 | 0.6187 | 14 | 16.1658 | 19.7990 | 49 | 56.5803 | 69.2965 | 84 | 96.9949 | 118.794 |
| 15/32 | 0.5413 | 0.6629 | 15 | 17.3205 | 21.2132 | 50 | 57.7351 | 70.7107 | 85 | 98.1496 | 120.208 |
| 0.5 | 0.5774 | 0.7071 | 16 | 18.4752 | 22.6274 | 51 | 58.8898 | 72.1249 | 86 | 99.3043 | 121.622 |
| 17/32 | 0.6134 | 0.7513 | 17 | 19.6299 | 24.0416 | 52 | 60.0445 | 73.5391 | 87 | 100.459 | 123.037 |
| 9/16 | 0.6495 | 0.7955 | 18 | 20.7846 | 25.4559 | 53 | 61.1992 | 74.9533 | 88 | 101.614 | 124.451 |
| 19/32 | 0.6856 | 0.8397 | 19 | 21.9393 | 26.8701 | 54 | 62.3539 | 76.3676 | 89 | 102.768 | 125.865 |
| 0.6 | 0.6928 | 0.8485 | 20 | 23.0940 | 28.2843 | 55 | 63.5086 | 77.7818 | 90 | 103.923 | 127.279 |
| 5/8 | 0.7217 | 0.8839 | 21 | 24.2487 | 29.6985 | 56 | 64.6633 | 79.1960 | 91 | 105.078 | 128.693 |
| 21/32 | 0.7578 | 0.9281 | 22 | 25.4034 | 31.1127 | 57 | 65.8180 | 80.6102 | 92 | 106.232 | 130.108 |
| 11/16 | 0.7939 | 0.9723 | 23 | 26.5581 | 32.5269 | 58 | 66.9727 | 82.0244 | 93 | 107.387 | 131.522 |
| 0.7 | 0.8083 | 0.9899 | 24 | 27.7128 | 33.9411 | 59 | 68.1274 | 83.4386 | 94 | 108.542 | 132.936 |
| 23/32 | 0.8299 | 1.0165 | 25 | 28.8675 | 35.3554 | 60 | 69.2821 | 84.8528 | 95 | 109.697 | 134.350 |
| 3/4 | 0.8660 | 1.0607 | 26 | 30.0222 | 36.7696 | 61 | 70.4368 | 86.2671 | 96 | 110.851 | 135.765 |
| 25/32 | 0.9021 | 1.1049 | 27 | 31.1769 | 38.1838 | 62 | 71.5915 | 87.6813 | 97 | 112.006 | 137.179 |
| 0.8 | 0.9238 | 1.1314 | 28 | 32.3316 | 39.5980 | 63 | 72.7462 | 89.0955 | 98 | 113.161 | 138.593 |
| 13/16 | 0.9382 | 1.1490 | 29 | 33.4863 | 41.0122 | 64 | 73.9009 | 90.5097 | 99 | 114.315 | 140.007 |
| 27/32 | 0.9743 | 1.1932 | 30 | 34.6410 | 42.4264 | 65 | 75.0556 | 91.9239 | 100 | 115.470 | 141.421 |
| 7/8 | 1.0104 | 1.2374 | 31 | 35.7957 | 43.8406 | 66 | 76.2103 | 93.3381 | … | … | … |

## Volumes of Solids

*Cube:*

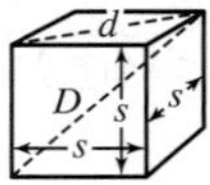


Diagonal of cube face $= d = s\sqrt{2}$

Diagonal of cube $= D = \sqrt{\dfrac{3d^2}{2}} = s\sqrt{3} = 1.732s$

Volume $= V = s^3$

Side $s = \sqrt[3]{V}$

*Example:* The side of a cube *s* measures 9.5 centimeters. Find its volume.

$$\text{Volume} = V = s^3 = 9.5^3 = 9.5 \times 9.5 \times 9.5 = 857.375\ \text{cm}^3$$

*Example:* The volume of cube is 231 cubic centimeters. What is the length of the side?

$$s = \sqrt[3]{V} = \sqrt[3]{231} = 6.136\ \text{cm}$$

*Rectangular Prism:*

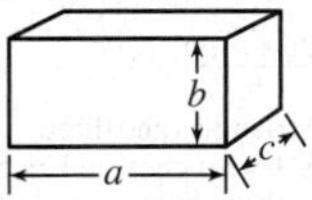


Volume $= V = abc$

$$a = \frac{V}{bc} \qquad b = \frac{V}{ac} \qquad c = \frac{V}{ab}$$

*Example:* In a rectangular prism, $a = 6$ in., $b = 5$ in., $c = 4$ in. Find the volume.

$$V = abc = 6 \times 5 \times 4 = 120\ \text{in}^3$$

*Example:* What should the height of a box be if it is to contain 25 cubic feet and if it is 4 feet long and $2\frac{1}{2}$ feet wide? Here, $a = 4$, $c = 2.5$, and $V = 25$. Then,

$$b = \text{height} = \frac{V}{ac} = \frac{25}{4 \times 2.5} = \frac{25}{10} = 2.5\ \text{ft}$$

*General Right Prism:*

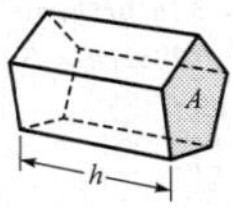


$h$ = edge length

$A$ = area of end surface

$V = Ah$

The area $A$ of the end surface is found by the formulas for areas of plane figures on the preceding pages.

*Example:* A right prism having for its base a regular hexagon with a side *s* of 7.5 centimeters is 25 centimeters high. Find the volume.

$$\text{Area of hexagon} = A = 2.598s^2 = 2.598 \times 56.25 = 146.14\ \text{cm}^2$$

$$\text{Volume of prism} = Ah = 25 \times 146.14 = 3653.5\ \text{cm}^3$$

*Right Pyramid:*

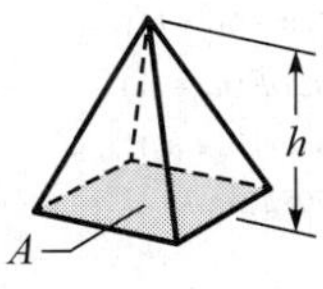


$$\text{Volume} = V = \left(\frac{1}{3}\right)(\text{Base area} \times h)$$

A square pyramid is pictured. In general, any right pyramid of height $h$, whose base is a regular, $n$-sided polygon of side length $s$ has volume:

$$V = \frac{nsrh}{6} = \frac{nsh}{6}\sqrt{R^2 - \frac{s^2}{4}}$$

where $r$ is the radius of the circle inscribed in the base, and $R$ is the radius of the circle circumscribed on the base.

*Example:* A pyramid having a height of 5 feet has a base formed by a square, the sides of which are 3 feet. Find the volume.

$$\text{Area of base} = 3 \times 3 = 9\ \text{square feet}, \qquad h = 5\ \text{feet}$$

$$\text{Volume} = V = (\tfrac{1}{3})(\text{Base area} \times \text{h}) = \tfrac{1}{3} \times 9 \times 5$$

*Frustum of Pyramid:*

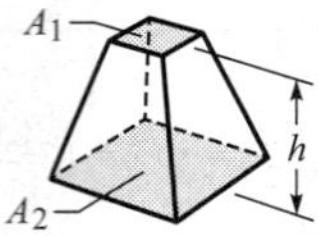


$$\text{Volume} = V = \frac{h}{3}\left(A_1 + A_2 + \sqrt{A_1A_2}\right)$$

*Example:* The pyramid in the previous example is cut off 4.5 feet from the base, and the upper part removed. The sides of the rectangle forming the top surface of the frustum are, then, 1 and 1.5 feet long, respectively. Find the volume of the frustum.

$$\text{Area of top} = A_1 = 1 \times 1.5 = 1.5 \text{ ft}^2 \qquad \text{Area of base} = A_2 = 2 \times 3 = 6 \text{ ft}^2$$

$$V = \frac{4 \times 5}{3}(1.5 + 6 + \sqrt{1.5 \times 6}) = 1.5(7.5 + \sqrt{9}) = 1.5 \times 10.5 = 15.75 \text{ ft}^3$$

*Wedge:*

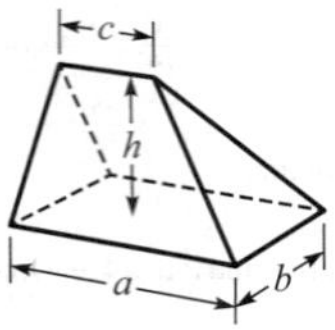


$$\text{Volume} = V = \frac{(2a + c)bh}{6}$$

A wedge has five faces, two of them triangles and three trapezoids. *Example:* Let $a = 4$ inches, $b = 3$ inches, and $c = 5$ inches. The height $h = 4.5$ inches. Find the volume in $\text{in}^3$ and $\text{cm}^3$.

$$V = \frac{(2a + c)bh}{6} = \frac{(2 \times 4 + 5) \times 3 \times 4.5}{6} = \frac{175.5}{6} = 29.25 \text{ in}^3$$

$$V = 29.25 \text{ in}^3 \times 16.387064 \frac{\text{cm}^3}{\text{in}^3} = 479.32162 \text{ cm}^3$$

*Cylinder:*

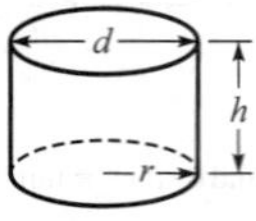


$$\text{Volume} = V = 3.1416r^2h = 0.7854d^2h$$

$$\text{Surface area of open-ended cylinder} = S = 6.2832rh = 3.1416dh$$

Total surface area $A$ of closed cylinder:

$$A = 6.2832r(r + h)$$

*Example:* Diameter $d$ of a cylinder is 2.5 inches. Height $h$ is 20 inches. Find the volume and the area of the open-ended cylindrical surface $S$.

$$V = 0.7854d^2h = 0.7854 \times 2.5^2 \times 20 = 0.7854 \times 6.25 \times 20 = 98.17 \text{ in}^3$$

$$S = 3.1416dh = 3.1416 \times 2.5 \times 20 = 157.08 \text{ in}^2$$

*Cylinder Portion:*

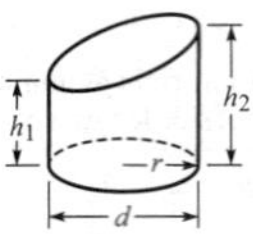


$$\text{Volume} = V = 1.5708r^2(h_1 + h_2) = 0.3927d^2(h_1 + h_2)$$

$$\text{Surface area} = S = 3.1416r(h_1 + h_2) = 1.5708d(h_1 + h_2)$$

*Example:* A cylinder 125 millimeters in diameter is cut off at an angle, as shown in the illustration. Dimension $h_1 = 100$ mm, and $h_2 = 150$ mm. Find the volume and the surface area $S$ of the cylinder.

$$V = 0.3927d^2(h_1 + h_2) = 0.3927 \times 125^2 \times (100 + 150)$$

$$= 0.3927 \times 15{,}625 \times 250 = 1{,}533{,}984 \text{ mm}^3 = 1534 \text{ cm}^3$$

$$S = 1.5708d(h_1 + h_2) = 1.5708 \times 125 \times 250$$

$$= 49{,}087.5 \text{ mm}^2 = 490.9 \text{ cm}^2$$

*Segment of a Cylinder:*

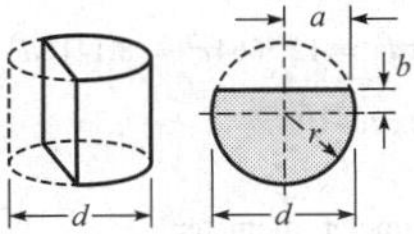


In the formulas below, segment area $A$ and arc length $l$ refer to the shaded segment of the circle. The formulas of these dimensions, found on page 70, assume the central angle measure is known.

$$\text{Volume} = V = (\tfrac{2}{3}a^3 \pm b \times \text{segment area})\frac{L}{r \pm b}$$

$$\text{Surface area} = S = (ad \pm b \times \text{arc length area})\frac{L}{r \pm b}$$

Use + when shaded segment is greater than or equal to one-half the base circle area, use – when it is less than the base circle area.

*Example:* Find the volume of a cylinder segment of a 2-inch high cylinder with a diameter of 5 inches when it is cut so that line $AC$ passes through the center of the base circle—that is, the base area is a half-circle.

In this case, $b = 0$, and $a = r$; that is, $a = 2.5$ inches. Thus, $ABC$ is a semicircle, and area $ABC = 3.1416 \times r^2 \times \tfrac{1}{2} = 3.1416 \times (2.5)^2 \times \tfrac{1}{2} = 9.82$ in².

$$V = \left(\frac{2}{3} \times 2.5^3 + 0 \times 9.82\right)\frac{2}{2.5 + 0} = \frac{2}{3} \times 15.625 \times 0.8 = 8.33 \text{ in}^3$$

*Hollow Cylinder:*

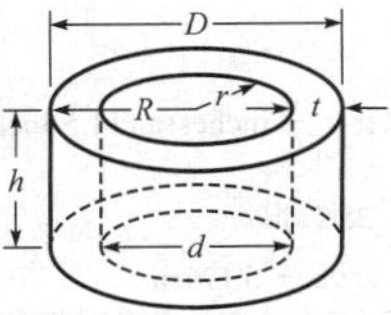


$$\begin{aligned} \text{Volume} = V &= 3.1416h(R^2 - r^2) = 0.7854h(D^2 - d^2) \\ &= 3.1416ht(2R - t) = 3.1416ht(D - t) \\ &= 3.1416ht(2r + t) = 3.1416ht(d + t) \\ &= 3.1416ht(R + r) = 1.5708ht(D + d) \end{aligned}$$

*Example:* A cylindrical shell, 28 cm high, is 36 cm in outside diameter, and 4 cm thick. Find its volume.

$$\begin{aligned} V = 3.1416ht(D - t) &= 3.1416 \times 28 \times 4(36 - 4) = 3.1416 \times 28 \times 4 \times 32 \\ &= 11{,}259.5 \text{ cm}^3 \end{aligned}$$

*Cone:*

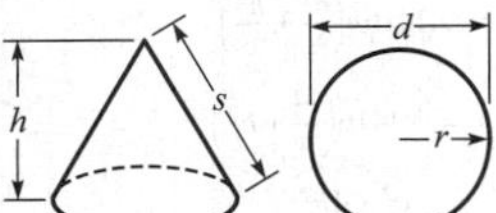


$$\text{Volume} = V = \frac{3.1416r^2h}{3} = 1.0472r^2h = 0.2618d^2h$$

$$\begin{aligned} \text{Conical surface area} = A &= 3.1416r\sqrt{r^2 + h^2} = 3.1416rs \\ &= 1.5708ds \end{aligned}$$

$$s = \sqrt{r^2 + h^2} = \sqrt{\frac{d^2}{4} + h^2}$$

*Example:* Find the volume and surface area of a cone, the base of which is a circle of 6 inches diameter and the height of which is 4 inches.

$$V = 0.2618d^2h = 0.2618 \times 6^2 \times 4 = 0.2618 \times 36 \times 4 = 37.7 \text{ in}^3$$

$$\begin{aligned} A &= 3.1416r\sqrt{r^2 + h^2} = 3.1416 \times 3 \times \sqrt{3^2 + 4^2} = 9.4248 \times \sqrt{25} \\ &= 47.124 \text{ in}^2 \end{aligned}$$

*Frustum of Cone:*

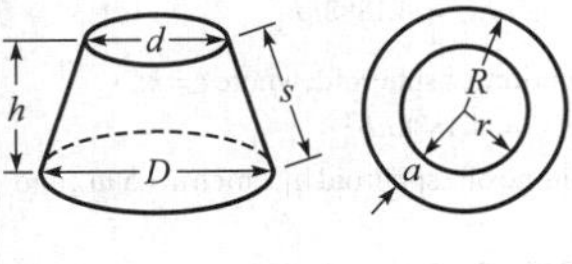


$V$ = volume  $A$ = area of conical surface

$$V = 1.0472h(R^2 + Rr + r^2) = 0.2618h(D^2 + Dd + d^2)$$

$$A = 3.1416s(R + r) = 1.5708s(D + d)$$

$$a = R - r \qquad s = \sqrt{a^2 + h^2} = \sqrt{(R - r)^2 + h^2}$$

*Example:* Find the volume of a frustum of a cone of the following dimensions: $D = 8$ centimeters; $d = 4$ centimeters; $h = 5$ centimeters.

$$\begin{aligned} V &= 0.2618 \times 5(8^2 + 8 \times 4 + 4^2) = 0.2618 \times 5(64 + 32 + 16) \\ &= 0.2618 \times 5 \times 112 = 146.61 \text{ cm}^3 \end{aligned}$$

*Sphere:*

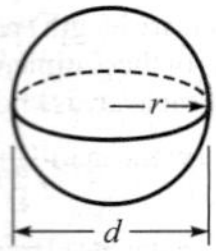


$$\text{Volume} = V = \frac{4\pi r^3}{3} = \frac{\pi d^3}{6} = 4.1888r^3 = 0.5236d^3$$

$$\text{Surface area} = A = 4\pi r^2 = \pi d^2 = 12.5664r^2 = 3.1416d^2$$

$$r = \sqrt[3]{\frac{3V}{4\pi}} = 0.6204\sqrt[3]{V}$$

*Example:* Find the volume and the surface area of a sphere 6.5 centimeters diameter.

$$V = 0.5236d^3 = 0.5236 \times 6.5^3 = 0.5236 \times 6.5 \times 6.5 \times 6.5 = 143.79 \text{ cm}^3$$

$$A = 3.1416d^2 = 3.1416 \times 6.5^2 = 3.1416 \times 6.5 \times 6.5 = 132.73 \text{ cm}^2$$

*Example:* The volume of a sphere is 64 cubic centimeters. Find its radius.

$$r = 0.6204\sqrt[3]{64} = 0.6204 \times 4 = 2.4816 \text{ cm}$$

*Spherical Sector:*

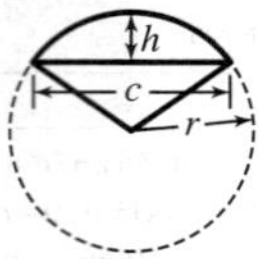


$$V = \frac{2\pi r^2 h}{3} = 2.0944r^2h$$

= total area of conical and spherical surface

$$A = 3.1416r(2h + c/2)$$

$$c = 2\sqrt{h(2r - h)}$$

*Example:* Find the volume of a sector of a sphere with a 6-inch diameter ($r = 3$ inches) and 1.5-inch height $h$. Also find the length of chord $c$.

$$V = 2.0944r^2h = 2.0944 \times 3^2 \times 1.5 = 2.0944 \times 9 \times 1.5 = 28.27 \text{ in}^3$$

$$c = 2\sqrt{h(2r - h)} = 2\sqrt{1.5(2 \times 3 - 1.5)} = 2\sqrt{6.75} = 2 \times 2.598 = 5.196 \text{ in}$$

*Spherical Segment:*

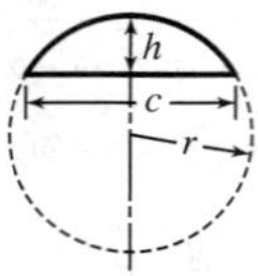


$V$ = volume $\qquad A$ = area of spherical surface

$$V = 3.1416h^2\left(r - \frac{h}{3}\right) = 3.1416h\left(\frac{c^2}{8} + \frac{h^2}{6}\right)$$

$$A = 2\pi rh = 6.2832rh = 3.1416\left(\frac{c^2}{4} + h^2\right)$$

$$c = 2\sqrt{h(2r - h)} \qquad r = \frac{c^2 + 4h^2}{8h}$$

*Example:* A segment of a sphere has the following dimensions: $h = 50$ millimeters; $c = 125$ millimeters. Find the volume $V$ and the radius of the sphere of which the segment is a part.

$$V = 3.1416 \times 50 \times \left(\frac{125^2}{8} + \frac{50^2}{6}\right) = 157.08 \times \left(\frac{15{,}625}{8} + \frac{2500}{6}\right) = 372{,}247 \text{ mm}^3 = 372 \text{ cm}^3$$

$$r = \frac{125^2 + 4 \times 50^2}{8 \times 50} = \frac{15{,}625 + 10{,}000}{400} = \frac{25{,}625}{400} = 64 \text{ mm}$$

*Ellipsoid:*

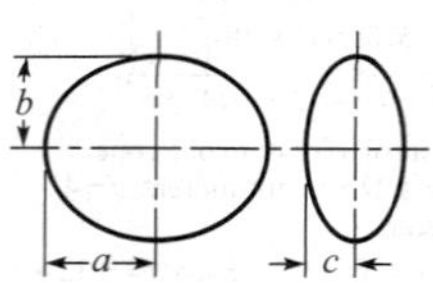


$$V = \frac{4\pi}{3}abc = 4.1888abc$$

In an ellipsoid of revolution, or spheroid, where $c = b$:

$$V = 4.1888ab^2$$

*Example:* Find the volume of a spheroid in which $a = 5$ in., and $b = c = 1.5$ in.

$$V = 4.1888 \times 5 \times 1.5^2 = 47.124 \text{ in}^3$$

*Spherical Zone:*

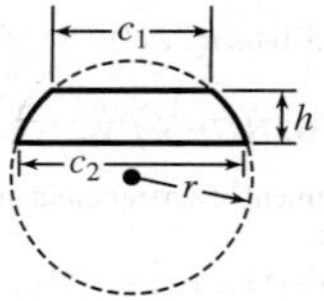


$$V = 0.5236h\left(\frac{3c_1^2}{4} + \frac{3c_2^2}{4} + h^2\right)$$

$$A = 2\pi rh = 6.2832rh = \text{area of spherical surface}$$

$$r = \sqrt{\frac{c_2^2}{4} + \left(\frac{c_2^2 - c_1^2 - 4h^2}{8h}\right)^2}$$

*Example:* In a spherical zone, let $c_1 = 3$; $c_2 = 4$; and $h = 1.5$ in. Find the volume.

$$V = 0.5236 \times 1.5 \times \left(\frac{3 \times 3^2}{4} + \frac{3 \times 4^2}{4} + 1.5^2\right) = 0.5236 \times 1.5 \times \left(\frac{27}{4} + \frac{48}{4} + 2.25\right) = 16.493 \text{ in}^3$$

*Spherical Wedge:*

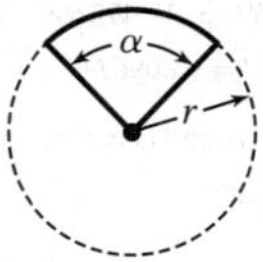


$V$ = volume $A$ = area of spherical surface

$\alpha$ = center angle in degrees

$$V = \frac{\alpha}{360} \times \frac{4\pi r^3}{3} = 0.0116\alpha r^3$$

$$A = \frac{\alpha}{360} \times 4\pi r^2 = 0.0349\alpha r^2$$

*Example:* Find the area of the spherical surface and the volume of a wedge of a sphere. The diameter of the sphere is 100 mm, and the center angle $\alpha$ is 45 degrees.

$$V = 0.0116 \times 45 \times 50^3 = 0.0116 \times 45 \times 125{,}000 = 65{,}250 \text{ mm}^3 = 65.25 \text{ cm}^3$$

$$A = 0.0349 \times 45 \times 50^2 = 3926.25 \text{ mm}^2 = 39.26 \text{ cm}^2$$

*Hollow Sphere:*

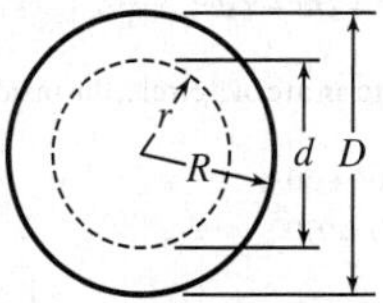


$V$ = volume of material used to make hollow sphere

$$V = \frac{4\pi}{3}(R^3 - r^3) = 4.1888(R^3 - r^3)$$

$$= \frac{\pi}{6}(D^3 - d^3) = 0.5236(D^3 - d^3)$$

*Example:* Find the volume of a hollow sphere, 8 in. in outside diameter, with a thickness of material of 1.5 in.

Here $R = 4$; $r = 4 - 1.5 = 2.5$.

$$V = 4.1888(4^3 - 2.5^3) = 4.1888(64 - 15.625) = 4.1888 \times 48.375 = 202.63 \text{ in}^3$$

*Paraboloid:*

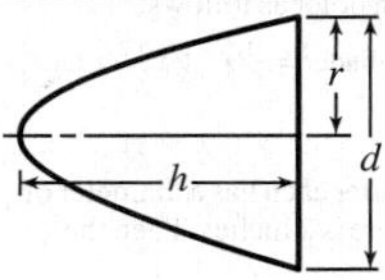


$$\text{Volume} = V = \tfrac{1}{2}\pi r^2 h = 0.3927d^2h$$

$$\text{Area} = A = \frac{2\pi}{3p}\left[\sqrt{\left(\frac{d^2}{4} + p^2\right)^3} - p^3\right]$$

$$\text{where } p = \frac{d^2}{8h}$$

*Example:* Find the volume of a paraboloid in which $h = 300$ millimeters and $d = 125$ millimeters.

$$V = 0.3927d^2h = 0.3927 \times 125^2 \times 300 = 1{,}840{,}781 \text{ mm}^3 = 1{,}840.8 \text{ cm}^3$$

*Paraboloid Segment:*

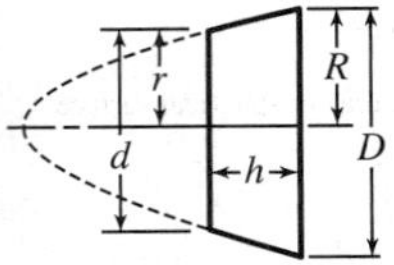


$$V = \frac{\pi}{2}h(R^2 + r^2) = 1.5708h(R^2 + r^2)$$

$$= \frac{\pi}{8}h(D^2 + d^2) = 0.3927h(D^2 + d^2)$$

*Example:* Find the volume of a segment of a paraboloid in which $D = 5$ in., $d = 3$ in., and $h = 6$ in.

$$V = 0.3927h(D^2 + d^2) = 0.3927 \times 6 \times (5^2 + 3^2)$$

$$= 0.3927 \times 6 \times 34 = 80.11 \text{ in}^3$$

*Torus:*

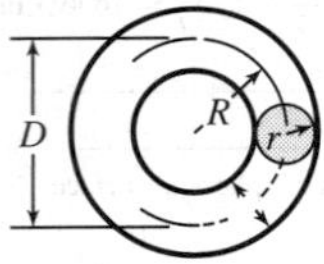


$$V = 2\pi^2 Rr^2 = 19.739Rr^2$$

$$= \frac{\pi^2}{4}Dd^2 = 2.4674Dd^2$$

$$\text{Area of surface} = A = 4\pi^2 Rr = 39.478Rr$$

$$= \pi^2 Dd = 9.8696Dd$$

*Example:* Find the volume and area of surface of a torus in which $d = 1.5$ in. and $D = 5$ in.

$$V = 2.4674 \times 5 \times 1.5^2 = 2.4674 \times 5 \times 2.25 = 27.76 \text{ in}^3$$

$$A = 9.8696 \times 5 \times 1.5 = 74.022 \text{ in}^2$$

*Barrel:*

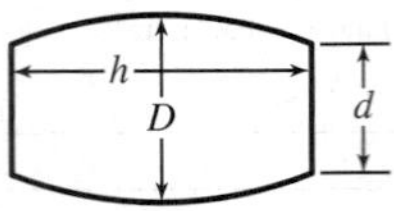


$V$ = approximate volume

If the sides are bent to the arc of a circle:

$$V = \frac{1}{12}\pi h(2D^2 + d^2) = 0.262h(2D^2 + d^2)$$

If the sides are bent to the arc of a parabola:

$$V = 0.209h(2D^2 + Dd + \tfrac{3}{4}d^2)$$

*Example:* Find the approximate contents of a barrel with sides bent to an arc of a circle, the inside dimensions of which are $D = 60$ cm, $d = 50$ cm, $h = 120$ cm.

$$V = 0.262h(2D^2 + d^2) = 0.262 \times 120 \times (2 \times 60^2 + 50^2)$$

$$= 0.262 \times 120 \times (7200 + 2500) = 0.262 \times 120 \times 9700$$

$$= 304{,}968 \text{ cm}^3 = 0.305 \text{ m}^3$$

*Ratio of Volumes:*

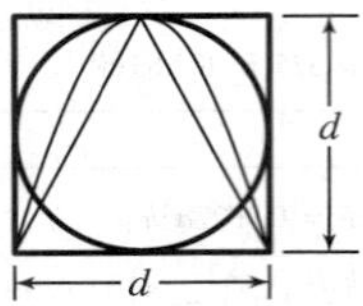


If $d$ = base diameter and height of a cone, a paraboloid and a cylinder, and the diameter of a sphere, then the proportions of the volumes of these solids are to each other as follows:

$$\text{Cone} : \text{Paraboloid} : \text{Sphere} : \text{Cylinder} = \tfrac{1}{3} : \tfrac{1}{2} : \tfrac{2}{3} : 1$$

*Example:* Assume, as an example, that a cone, paraboloid, and cylinder each has a diameter of 2 inches, and that the height is 2 inches, and that the diameter of a sphere is 2 inches. Then the volumes, written in formula form, are as follows:

$$\underset{\text{Cone}}{\frac{3.1416 \times 2^2 \times 2}{12}} : \underset{\text{Paraboloid}}{\frac{3.1416 \times (2p)^2 \times 2}{8}} : \underset{\text{Sphere}}{\frac{3.1416 \times 2^3}{6}} : \underset{\text{Cylinder}}{\frac{3.1416 \times 2^2 \times 2}{4}} = \tfrac{1}{3} : \tfrac{1}{2} : \tfrac{2}{3} : 1$$

## TRIGONOMETRY: SOLUTION OF TRIANGLES

**Terminology.**—A triangle is a polygon with three sides. The sum of the angle measures in any triangle in a plane is 180 degrees. The *triangle inequality* states that the sum of any two side lengths of a triangle is always greater than the length of the third side. It is not possible to construct a triangle that violates the *triangle inequality*

Triangles are either *right* or *oblique*. A right triangle has a right angle, which measures 90 degrees. Oblique triangles do not contain a right angle.

As with any polygon (see page 55), parts with equal measure are called *congruent*. Thus, a triangle with congruent sides is one whose sides have the same measure. A triangle with two congruent sides is called an *isosceles* triangle; a triangle with all three sides congruent is *equilateral* and hence *equiangular*. Each angle measures 180 degrees/3 = 60 degrees.

Two positive angles whose measures total 90 degrees are called *complementary* angles. The two acute angles in any right triangle are *complements* of each other. Two positive angles whose sum is 180 degrees are called *supplementary angles*.

An *isosceles* triangle has at least two congruent sides and angles (an equilateral triangle is also isosceles). Angles opposite the congruent sides are congruent angles. An *obtuse* triangle has one angle measuring greater than 90 degrees. An *acute* triangle has all three angles measuring less than 90 degrees; hence, an equilateral (equiangular) triangle is also acute.

**Degree and Radian Angle Measure.**—Two modes of measuring angles are *degree* measure and *radian* measure. 1 radian is the measure of a circle's central angle whose arc is the same length as the radius of the circle. For any size circle, 1 radian is approximately 57.3 degrees. Conversion between degree and radian measure is based on the relation $360° = 2\pi$ radians, or $180° = \pi$ radians. ($\pi$ is the ratio of circumference to diameter, $C/d$, and is approximately 22/7 or 3.1415926. The actual value of $\pi$ is an irrational number.) Degree measure is converted to its equivalent radian measure by the formula:

$$\text{Degree measure} \times \pi/180 = \text{Radian measure}$$

For example, $45° \times \pi/180 = \pi/4$.

Radian measure is converted to degree by the formula:

$$\text{Radian measure} \times 180/\pi = \text{Degree measure}$$

For example, $\pi/3 \times 180° = 60°$. Radian measure is actually unitless, but it is customary to write "rad" to indicate when radian measure is used. Conversions for the essential degree measures of the circle are shown in the chart on page 90.

**Trigonometric Ratios of Essential Angles.**—An acute angle can be any degree measure between 0° and 90°, but the trigonometric values for base angles (designated as $\theta$ in the table below the diagrams) 30°, 45°, and 60° are usually memorized. They are derived from right triangles constructed so that the shortest side has length = 1. The other dimensions follow from the geometric construction of the angles (see *Table 2a. Geometric Constructions* starting on page 61) and the Pythagorean theorem. The triangles and trigonometric values of these angle measures, as well as those with base angles 0° and 90° (envisioned by a horizontal and a vertical line segment, respectively) are given in the diagrams and table below. These five are the essential angles.

In the development of the essential angles, the length of the shortest side in each of the triangles is designated as 1, so by construction the other lengths follow. Derivations of the main three functions for the five angle measures is shown. Decimal values of square roots are rounded to three decimal places. In the complete trigonometry tables (Table 2a, Table 2b, and Table 2c), irrational values are carried out to six decimal places, for greater accuracy. For quick estimates, it is useful to memorize the truncated values in the table below.

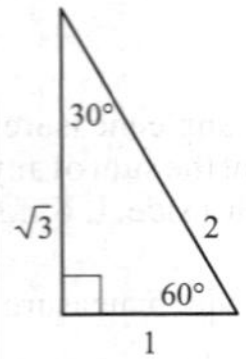


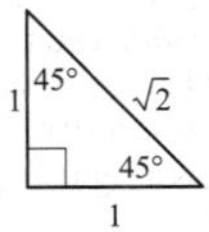


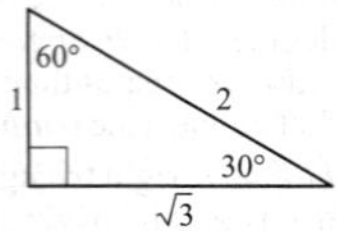


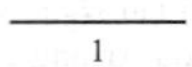


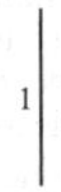


| 60°-30°-90° | 45°-45°-90° | 30°-60°-90° | 0°-90°-90° | 90°-0°-90° |
|---|---|---|---|---|
| $\sqrt{3}^2 + 1^2 = 2^2$ | $1^2 + 1^2 = \sqrt{2}^2$ | $1^2 + \sqrt{3}^2 = 2^2$ | $1^2 + 0^2 = 1^2$ | $0^2 + 1^2 = 1^2$ |
| Base angle = 60° | Base angle = 45° | Base angle = 30° | Base angle = 0° | Base angle = 90° |
| $\sin 60 = \frac{\sqrt{3}}{2} \approx 0.866$ | $\sin 45 = \frac{1}{\sqrt{2}} \approx 0.707$ | $\sin 30 = \frac{1}{2} = 0.5$ | $\sin 0 = \frac{0}{1} = 0$ | $\sin 90 = \frac{1}{1} = 1$ |
| $\cos 60 = \frac{1}{2} = 0.5$ | $\cos 45 = \frac{1}{\sqrt{2}} \approx 0.707$ | $\cos 30 = \frac{\sqrt{3}}{2} \approx 0.866$ | $\cos 0 = \frac{1}{1} = 1$ | $\cos 90 = \frac{0}{1} = 0$ |
| $\tan 60 = \frac{\sqrt{3}}{1} \approx 1.732$ | $\tan 45 = \frac{1}{1} = 1$ | $\tan 30 = \frac{1}{2} = 0.5$ | $\tan 0 = \frac{0}{1} = 0$ | $\tan 90 = \frac{1}{0}$ UND |

*Note:* Base angles are designated as θ in the table below. The values are approximated to three decimal places. If more accuracy is needed, then calculator values carried out further can be used.

| Angle θ, degrees | Angle θ, radians | sin θ | cos θ | tan θ | csc θ | sec θ | cot θ |
|---|---|---|---|---|---|---|---|
| 0 | 0 | 0 | 1 | 0 | UND | 1 | UND |
| 30 | $\pi/6$ | $1/2 = 0.5$ | $\sqrt{3}/2 \approx 0.866$ | $1/\sqrt{3} \approx 0.577$ | 2 | $2/\sqrt{3} \approx 1.155$ | $\sqrt{3}/1 \approx 1.732$ |
| 45 | $\pi/4$ | $1/\sqrt{2} \approx 0.707$ | $1/\sqrt{2} \approx 0.707$ | 1 | $\sqrt{2} \approx 1.414$ | $\sqrt{2} \approx 1.414$ | 1 |
| 60 | $\pi/3$ | $\sqrt{3}/2 \approx 0.866$ | $1/2 = 0.5$ | $\sqrt{3}/1 \approx 1.732$ | $2/\sqrt{3} \approx 1.155$ | 2 | $1/\sqrt{3} \approx 0.577$ |
| 90 | $\pi/2$ | 1 | 0 | UND | 1 | UND | 0 |

Note: UND = undefined

## Functions of Angles

**Right Triangle Ratios.**—Trigonometry is the branch of mathematics that addresses the relations of the sides and angles of triangles and the resulting functions of angles. The properties of trigonometry are derived, however, from the relationships of angles and sides in the *right* triangle, as pictured here.

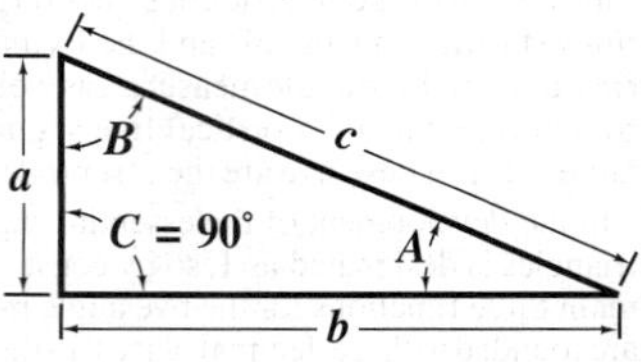


In a right triangle, the 90° angle, $C$, is opposite the longest side, $c$, the *hypotenuse*. Since the sum of angle measures in any triangle is 180°, the sum of the two acute angles is 90°. The shorter sides, $a$ and $b$, are called the *legs*. They are opposite to angles $A$ and $B$, respectively. By the Pythagorean theorem, the sum of the squares of the legs is equal to the square of the hypotenuse; that is,

$$a^2 + b^2 = c^2$$

Referring to the figure at the bottom of page 90, the ratios of the sides of a right triangle with respect to angle *A* are named *sine A*, *cosine A*, and *tangent A* (abbreviated sin *A*, sin *A*, and tan *A*). Ratios and reciprocal ratios, named *cosecant A*, *secant A*, and *cotangent A* (abbreviated csc*A*, sec*A*, and cot*A*) are defined as:

$$\sin A = \frac{\text{opposite}}{\text{hypotenuse}} = \frac{a}{c} \qquad \csc A = \frac{\text{hypotenuse}}{\text{opposite}} = \frac{c}{a} = \frac{1}{\sin A}$$

$$\cos A = \frac{\text{adjacent}}{\text{hypotenuse}} = \frac{b}{c} \qquad \sec A = \frac{\text{hypotenuse}}{\text{adjacent}} = \frac{c}{b} = \frac{1}{\cos A}$$

$$\tan A = \frac{\text{opposite}}{\text{adjacent}} = \frac{a}{b} \qquad \cot A = \frac{\text{adjacent}}{\text{opposite}} = \frac{b}{a} = \frac{1}{\tan A}$$

Similar ratios are defined for angle *B*:

$$\sin B = \frac{\text{opposite}}{\text{hypotenuse}} = \frac{b}{c} \qquad \csc B = \frac{\text{hypotenuse}}{\text{opposite}} = \frac{c}{b} = \frac{1}{\sin B}$$

$$\cos B = \frac{\text{adjacent}}{\text{hypotenuse}} = \frac{a}{c} \qquad \sec B = \frac{\text{hypotenuse}}{\text{adjacent}} = \frac{c}{a} = \frac{1}{\cos B}$$

$$\tan B = \frac{\text{opposite}}{\text{adjacent}} = \frac{b}{a} \qquad \cot B = \frac{\text{adjacent}}{\text{opposite}} = \frac{a}{b} = \frac{1}{\tan B}$$

Thus, in a given right triangle, sin *A* = cos *B*, cos *A* = sin *B*, tan *A* = cot *B*. Similarly, csc *A* = sec *B*, sec *A* = csc *B*, cot *a* = tan *B*.

**Law of Sines.**—In any triangle, if *a*, *b*, and *c* are the sides, and *A*, *B*, and *C* their opposite angles, respectively, then:

$$\frac{a}{\sin A} = \frac{b}{\sin B} = \frac{c}{\sin C} \quad \text{so that:}$$

$$a = \frac{b \sin A}{\sin B} \quad \text{or} \quad a = \frac{c \sin A}{\sin C}$$

$$b = \frac{a \sin B}{\sin A} \quad \text{or} \quad b = \frac{c \sin B}{\sin C}$$

$$c = \frac{a \sin C}{\sin A} \quad \text{or} \quad c = \frac{b \sin C}{\sin B}$$

**Law of Cosines.**—In any triangle, if *a*, *b* and *c* are the sides and *A*, *B*, and *C* are the opposite angles, respectively, then:

$$a^2 = b^2 + c^2 - 2bc\cos A$$

$$b^2 = a^2 + c^2 - 2ac\cos B$$

$$c^2 = a^2 + b^2 - 2ab\cos C$$

The sine and cosine laws together with the proposition that the sum of the measures of the three angles is 180 degrees are the basis of all formulas relating to the solution of triangles.

Formulas and examples for the solution of right-angle and oblique-angle triangles, arranged in tabular form, are given on the following pages.

**Trigonometric Identities.**—It is possible to express trigonometric ratios in terms of other ratios by way of trigonometric identities. For example, sin(*A* + *B*) = sin *A* cos *B* + cos *A* sin *B*. It may be helpful to use an identity to quickly evaluate a trigonometric function, as shown by the examples given below the trigonometric identities and formulas derived from them.

*Reciprocal Identities:*

$$\tan A = \frac{\sin A}{\cos A} = \frac{1}{\cot A} \qquad \sec A = \frac{1}{\cos A} \qquad \csc A = \frac{1}{\sin A}$$

*Negative-Angle Identities:*

$$\sin(-A) = -\sin A \qquad \cos(-A) = \cos A \qquad \tan(-A) = -\tan A$$

*Cofunction Identities:*

$$\sin(90 - A) = \cos A \qquad \tan(90 - A) = \cot A \qquad \sec(90 - A) = \csc A$$

$$\cos(90 - A) = \sin A \qquad \cot(90 - A) = \tan A \qquad \csc(90 - A) = \sec A$$

*Pythagorean Identities:*

$$\sin^2 A + \cos^2 A = 1 \qquad 1 + \tan^2 A = \sec^2 A \qquad 1 + \cot^2 A = \csc^2 A$$

*Sum- and Difference-of-Angles Formulas:*

$$\sin(A + B) = \sin A \cos B + \cos A \sin B \qquad \sin(A - B) = \sin A \cos B - \cos A \sin B$$

$$\cos(A + B) = \cos A \cos B - \sin A \sin B \qquad \cos(A - B) = \cos A \cos B + \sin A \sin B$$

$$\tan(A + B) = \frac{\tan A + \tan B}{1 - \tan A \tan B} \qquad \tan(A - B) = \frac{\tan A - \tan B}{1 + \tan A \tan B}$$

$$\cot(A + B) = \frac{\cot A \cot B - 1}{\cot B + \cot A} \qquad \cot(A - B) = \frac{\cot A \cot B + 1}{\cot B - \cot A}$$

*Double-Angle Formulas:*

$$\cos(2A) = \cos^2 A - \sin^2 A = 2\cos^2 A - 1 = 1 - 2\sin^2 A$$

$$\sin(2A) = 2\sin A \cos A \qquad \tan(2A) = \frac{2\tan A}{1 - \tan^2 A} = \frac{2}{\cot A - \tan A}$$

*Half-Angle Formulas:*

$$\sin(A/2) = \pm\sqrt{(1 - \cos A)/2} \qquad \cos(A/2) = \pm\sqrt{(1 + \cos A)/2}$$

$$\tan(A/2) = \pm\sqrt{\frac{1 - \cos A}{1 + \cos A}} = \frac{1 - \cos A}{\sin A} = \frac{\sin A}{1 + \cos A}$$

*Product-to-Sum Formulas:*

$$\sin A \cos B = \tfrac{1}{2}[\sin(A + B) + \sin(A - B)]$$

$$\cos A \cos B = \tfrac{1}{2}[\cos(A + B) + \cos(A - B)]$$

$$\sin A \sin B = \tfrac{1}{2}[\cos(A - B) - \cos(A + B)]$$

$$\tan A \tan B = \frac{\tan A + \tan B}{\cot A + \cot B}$$

*Sum and Difference of Functions Formulas:*

$$\sin A + \sin B = 2\sin\left(\frac{A + B}{2}\right)\cos\left(\frac{A - B}{2}\right) \qquad \tan A + \tan B = \frac{\sin(A + B)}{\cos A \cos B}$$

$$\sin A - \sin B = 2\sin\left(\frac{A - B}{2}\right)\cos\left(\frac{A + B}{2}\right) \qquad \tan A - \tan B = \frac{\sin(A - B)}{\cos A \cos B}$$

$$\cos A + \cos B = 2\cos\left(\frac{A + B}{2}\right)\cos\frac{(A - B)}{2} \qquad \cot A + \cot B = \frac{\sin(B + A)}{\sin A \sin B}$$

$$\cos A - \cos B = 2\sin\left(\frac{A + B}{2}\right)\sin\left(\frac{A - B}{2}\right) \qquad \cot A - \cot B = \frac{\sin(B - A)}{\sin A \sin B}$$

**Solution of Right Triangles**

Right triangles (one angle measures 90°) are the easiest to "solve" (that is, to find the missing sides or angles), thanks to the Pythagorean theorem and right triangle ratios.

| | | | |
|---|---|---|---|
| 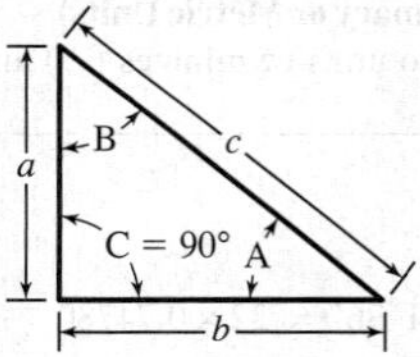 | The figure to the left shows right triangle *ABC*. Sides opposite corresponding angles are labeled *a*, *b*, *c*. The formulas in the table are for finding an unknown side or angle from given information as shown. There are several ways to solve for the missing dimension through one of the three basic trigonometric functions.<br><br>Right angle *C* is always known (90°), thus, $A = 90 - B$ and $B = 90 - A$. | | |
| Sides and Angle Known | Formulas for Sides, Angles to be Found | | |
| Sides *a* and *b* | $c = \sqrt{a^2 + b^2}$ | $A = \tan^{-1}(a/b)$ | $B = 90° - A$ |
| Side *a*, hypotenuse *c* | $b = \sqrt{c^2 - a^2}$ | $A = \sin^{-1}(a/c)$ or $A = \cos^{-1}(b/c)$ | $B = 90° - A$ |
| Side *b*, hypotenuse *c* | $a = \sqrt{c^2 - b^2}$ | $B = \sin^{-1}(b/c)$ or $B = \cos^{-1}(a/c)$ | $A = 90° - B$ |
| Hypotenuse *c*, angle *B* | $b = c \sin B$ or $a = c \cos B$ | $a = c \cos B$ or $b = c \sin B$ | $A = 90° - B$ |
| Hypotenuse *c*, angle *A* | $b = c \cos A$ or $a = c \sin A$ | $a = c \sin A$ or $b = c \cos A$ | $B = 90° - A$ |
| Side *b*, angle *B* | $c = \frac{b}{\sin B}$ or $c = \frac{a}{\cos B}$ | $a = \frac{b}{\tan B}$ | $A = 90° - B$ |
| Side *b*, angle *A* | $c = \frac{b}{\cos A}$ or $c = \frac{a}{\sin A}$ | $a = b \tan A$ | $B = 90° - A$ |
| Side *a*, angle *B* | $c = \frac{a}{\cos B}$ | $b = a \tan B$ | $A = 90° - B$ |
| Side *a*, angle *A* | $c = \frac{a}{\sin A}$ | $b = \frac{a}{\tan A}$ | $B = 90° - A$ |

The solutions of triangles—both right and oblique—rely not only on the Pythagorean theorem and the laws of cosines and of sines, but also on correct use of the trigonometry tables, either when given an angle and seeking a trigonometric ratio, or "in reverse," when given a ratio (side lengths) and seeking an angle. A full explanation of trigonometric functions and the use of their tables begins on page 100.

**Solution and Examples of Right Triangles (US Customary or Metric Units)**

Examples below show angle measure further divided into units of minutes ( ′ ) and seconds ( ″ ). 1° = 60′; 1′ = 60″, rather than decimal parts.

<table>
<tr>
<td>
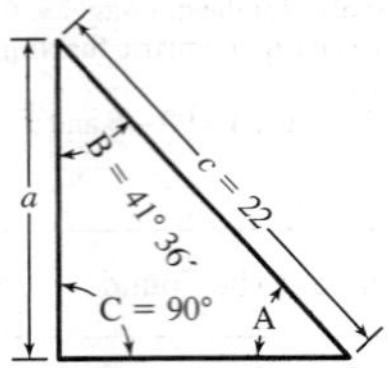

Hypotenuse and One Acute Angle Known
</td>
<td>

$c = 22$ inches; $B = 41° 36'$

$$a = c\cos B = 22 \times \cos 41°36' = 22 \times 0.74780 = 16.4516 \text{ in.}$$

$$b = c\sin B = 22 \times \sin 41°36' = 22 \times 0.66393 = 14.6065 \text{ in.}$$

$$A = 90° - B = 90° - 41°36' = 48°24'$$

</td>
</tr>
<tr>
<td>
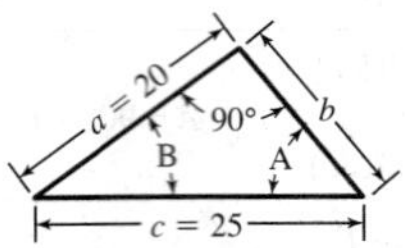

Hypotenuse and One Side Known
</td>
<td>

$c = 25$ centimeters; $a = 20$ centimeters

$$b = \sqrt{c^2 - a^2} = \sqrt{25^2 - 20^2} = \sqrt{625 - 400} = \sqrt{225} = 15 \text{ cm}$$

$$\sin A = \frac{a}{c} = \frac{20}{25} = 0.8$$

$$\text{Hence, } A = \sin^{-1}(0.8) = 53°8'$$

$$B = 90° - A = 90° - 53°8' = 36°52'$$

</td>
</tr>
<tr>
<td>
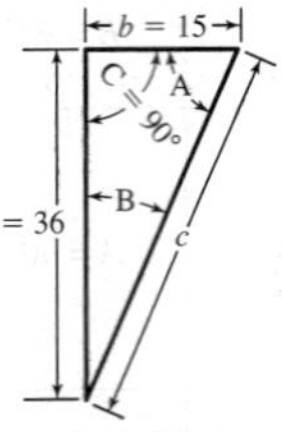

Two Sides Known
</td>
<td>

$a = 36$ mm; $b = 15$ mm

$$c = \sqrt{a^2 + b^2} = \sqrt{36^2 + 15^2} = \sqrt{1296 + 225} = \sqrt{1521} = 39 \text{ mm}$$

$$\tan A = \frac{a}{b} = \frac{36}{15} = 2.4$$

$$\text{Hence, } A = \tan^{-1}(2.4) = 67°23'$$

$$B = 90° - A = 90° - 67°23' = 22°37'$$

</td>
</tr>
<tr>
<td>
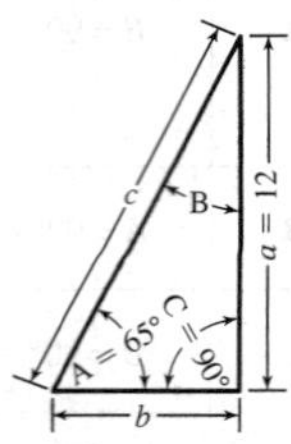

One Side and One Angle Known
</td>
<td>

$a = 12$ meters; $A = 65°$

$$c = \frac{a}{\sin A} = \frac{12}{\sin 65°} = \frac{12}{0.90631} = 13.2405 \text{ m}$$

$$b = \frac{a}{\tan A} = 12 \times \cot 65° = 12 \times 0.46631 = 5.5957 \text{ m}$$

$$B = 90° - A = 90° - 65° = 25°$$

</td>
</tr>
</table>

**Solution and Examples of Oblique Triangles (US Customary or Metric Units)**

*One Side and Two Excluded (Not Between) Angles Known (Law of Sines):*

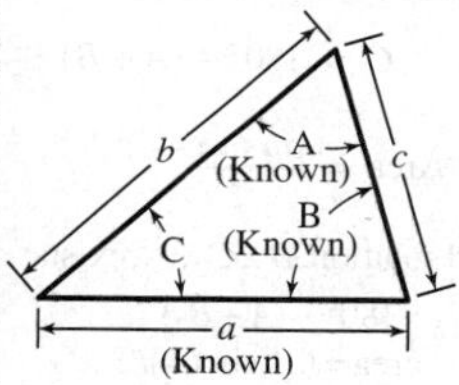


One Side and Two Excluded Angles Known

If side $a$, angle $A$ opposite it, and angle $B$, are known:

$$C = 180° - (A + B)$$

$$b = \frac{a \sin B}{\sin A} \qquad c = \frac{a \sin C}{\sin A}$$

$$\text{Area} = \frac{ab \sin C}{2}$$

If angles $B$ and $C$ are known, but not $A$, then $A = 180 - (B + C)$.

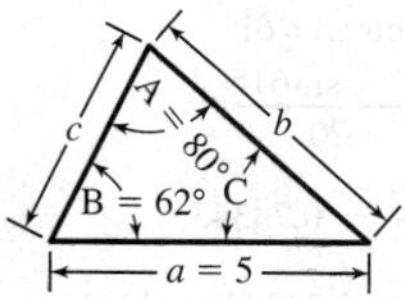


Side and Angles Known

*Example:* $a = 5$ cm; $A = 80°$; $B = 62°$

$$C = 180° - (80° + 62°) = 180° - 142° = 38°$$

$$b = \frac{a \sin B}{\sin A} = \frac{5 \times \sin 62°}{\sin 80°} = \frac{5 \times 0.88295}{0.98481}$$

$$= 4.483 \text{ cm}$$

$$c = \frac{a \sin C}{\sin A} = \frac{5 \times \sin 38°}{\sin 80°} = \frac{5 \times 0.61566}{0.98481}$$

$$= 3.126 \text{ cm}$$

*Two Sides and Included Angle Known:*

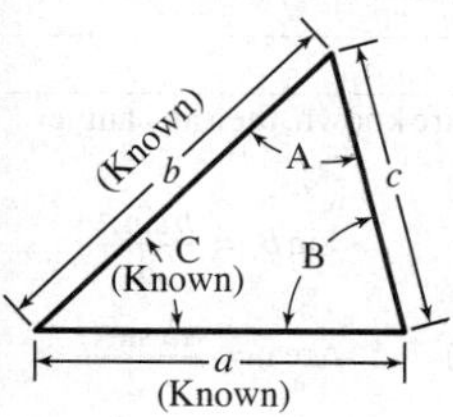


Two Sides and One Included Angle Known

If sides $a$ and $b$, and angle $C$ between them are known:

$$\tan A = \frac{a \sin C}{b - (a \cos C)}, \text{ so } A = \tan^{-1} \frac{a \sin C}{b - (a \cos C)}$$

$$B = 180° - (A + C) \qquad c = \frac{a \sin C}{\sin A}$$

Side $c$ may also be found directly as below:

$$c = \sqrt{a^2 + b^2 - (2ab \cos C)}$$

$$\text{Area} = \frac{ab \sin C}{2}$$

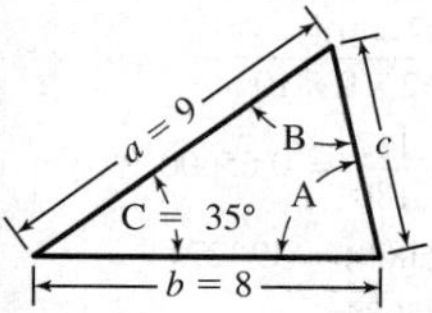


Sides and Angle Known

*Example:* $a = 9$ inches; $b = 8$ inches; $C = 35°$

$$\tan A = \frac{a \sin C}{b - (a \cos C)} = \frac{9 \times \sin 35°}{8 - (9 \times \cos 35°)}$$

$$= \frac{9 \times 0.57358}{8 - (9 \times 0.81915)} = \frac{5.16222}{0.62765} = 8.22468$$

Hence, $A = \tan^{-1}(8.22468) = 83°4'$

$$B = 180° - (A + C) = 180° - 118°4' = 61°56'$$

$$c = \frac{a \sin C}{\sin A} = \frac{9 \times 0.57358}{0.99269} = 5.2 \text{ inches}$$

*Two Sides and the Angle Opposite One of the Sides Known:*

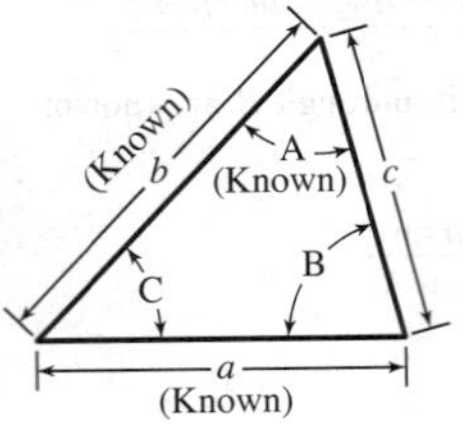


Two Sides and Angle Opposite One Side Known

If angle $A$, opposite side $a$, and other side $b$ are known:

$$\sin B = \frac{b \sin A}{a} \qquad C = 180° - (A + B)$$

$$c = \frac{a \sin C}{\sin A} \qquad \text{Area} = \frac{ab \sin C}{2}$$

If $B > A$ but $< 90°$, a second solution, $B_2, C_2, c_2$, exists:

$$B_2 = 180° - B, \quad C_2 = 180° - (A + B_2)$$

$$c_2 = (a \sin C_2)/\sin A, \quad \text{area} = (ab \sin C_2)/2$$

If $a \geq b \sin A$, then only the first solution exists.
If $a < b \sin A$, then no solution exists.

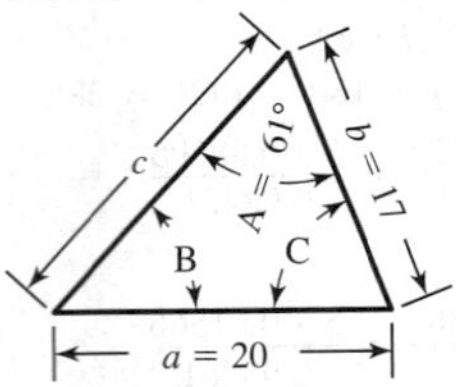


Sides and Angle Known

*Example:* $a = 20$ cm; $b = 17$ cm; $A = 61°$

$$\sin B = \frac{b \sin A}{a} = \frac{17 \times \sin 61°}{20}$$

$$= \frac{17 \times 0.87462}{20} = 0.74343$$

Hence, $B = \sin^{-1}(0.74343) = 48°1'$

$$C = 180° - (A + B) = 180° - 109°1' = 70°59'$$

$$c = \frac{a \sin C}{\sin A} = \frac{20 \times \sin 70°59'}{\sin 61°} = \frac{20 \times 0.94542}{0.87462}$$

$$= 21.62 \text{ cm}$$

*All Three Sides Known:*

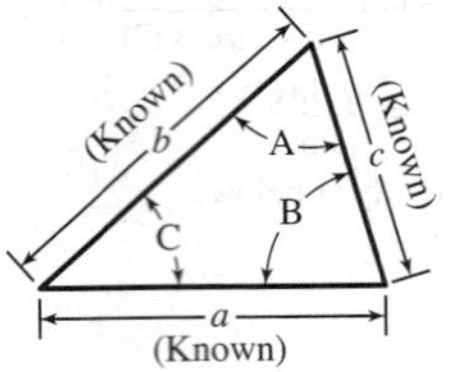


All Three Sides Known

If all three sides $a, b$, and $c$ are known, then any angle can be found:

$$\cos A = \frac{b^2 + c^2 - a^2}{2bc} \qquad \sin B = \frac{b \sin A}{a}$$

$$C = 180° - (A + B) \qquad \text{Area} = \frac{ab \sin C}{2}$$

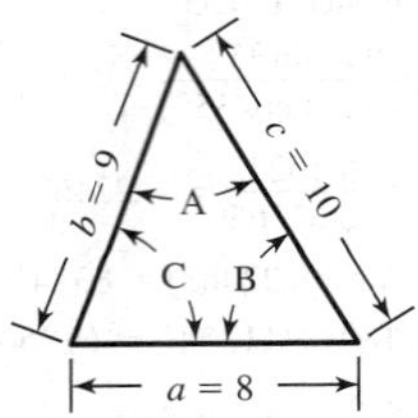


Sides Known

*Example:* $a = 8$ in.; $b = 9$ in.; $c = 10$ in.

$$\cos A = \frac{b^2 + c^2 - a^2}{2bc} = \frac{9^2 + 10^2 - 8^2}{2 \times 9 \times 10}$$

$$= \frac{81 + 100 - 64}{180} = \frac{117}{180} = 0.65000$$

Hence, $A = \cos^{-1}(0.65000) = 49°27'$

$$\sin B = \frac{b \sin A}{a} = \frac{9 \times 0.75984}{8} = 0.85482$$

Hence, $B = \sin^{-1}(0.85482) = 58°44'$

$$C = 180° - (A + B) = 180° - 108°11' = 71°49'$$

## Rapid Solution of Right and Oblique Triangles

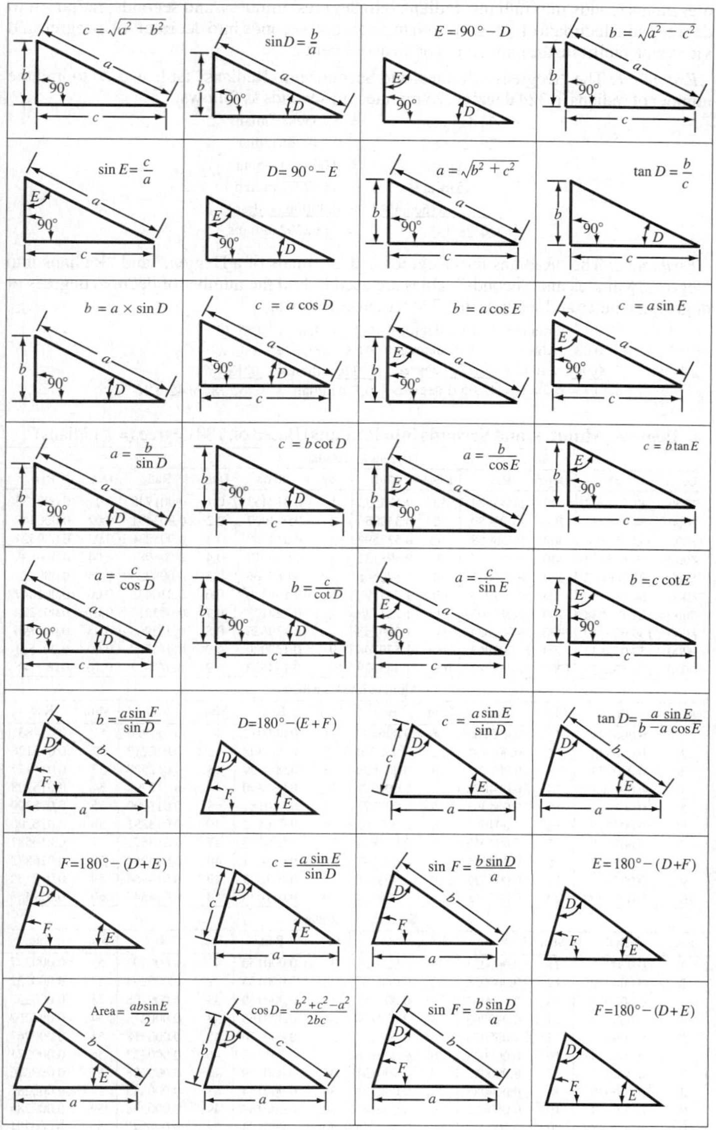

**Conversion Tables of Angular Measure.—**The accompanying tables of degrees, minutes, and seconds into radians; radians into degrees, minutes, and seconds; radians into degrees and decimals of a degree; and minutes and seconds into decimals of a degree and vice versa facilitate the conversion of measurements.

*Example 1:* The "Degrees, Minutes, and Seconds into Radians" table is used to find the number of radians in 324 degrees, 25 minutes, 13 seconds as follows:

| | |
|---|---|
| 300 degrees | = 5.235988 radians |
| 20 degrees | = 0.349066 radian |
| 4 degrees | = 0.069813 radian |
| 25 minutes | = 0.007272 radian |
| 13 seconds | = 0.000063 radian |
| 324°25′13″ | = 5.662202 radians |

*Example 2:* The "Radians into Degrees and Decimals of a Degree," and "Radians into Degrees, Minutes and Seconds" tables are used to find the number of decimal degrees or degrees, minutes and seconds in 0.734 radian as follows:

| | |
|---|---|
| 0.7 radian = 40.1070 degrees | 0.7 radian = 40°6′25″ |
| 0.03 radian = 1.7189 degrees | 0.03 radian = 1°43′8″ |
| 0.004 radian = 0.2292 degree | 0.004 radian = 0°13′45″ |
| 0.734 radian = 42.0551 degrees | 0.734 radian = 41°62′78″ or 42°3′18″ |

**Degrees, Minutes, and Seconds into Radians** (Based on 180 degrees = π radians)

Degrees into Radians

| Deg. | Rad. | Deg. | Rad. | Deg. | Rad. | Deg. | Rad. | Deg. | Rad. | Deg. | Rad. |
|---|---|---|---|---|---|---|---|---|---|---|---|
| 1000 | 17.453293 | 100 | 1.745329 | 10 | 0.174533 | 1 | 0.017453 | 0.1 | 0.001745 | 0.01 | 0.000175 |
| 2000 | 34.906585 | 200 | 3.490659 | 20 | 0.349066 | 2 | 0.034907 | 0.2 | 0.003491 | 0.02 | 0.000349 |
| 3000 | 52.359878 | 300 | 5.235988 | 30 | 0.523599 | 3 | 0.052360 | 0.3 | 0.005236 | 0.03 | 0.000524 |
| 4000 | 69.813170 | 400 | 6.981317 | 40 | 0.698132 | 4 | 0.069813 | 0.4 | 0.006981 | 0.04 | 0.000698 |
| 5000 | 87.266463 | 500 | 8.726646 | 50 | 0.872665 | 5 | 0.087266 | 0.5 | 0.008727 | 0.05 | 0.000873 |
| 6000 | 104.719755 | 600 | 10.471976 | 60 | 1.047198 | 6 | 0.104720 | 0.6 | 0.010472 | 0.06 | 0.001047 |
| 7000 | 122.173048 | 700 | 12.217305 | 70 | 1.221730 | 7 | 0.122173 | 0.7 | 0.012217 | 0.07 | 0.001222 |
| 8000 | 139.626340 | 800 | 13.962634 | 80 | 1.396263 | 8 | 0.139626 | 0.8 | 0.013963 | 0.08 | 0.001396 |
| 9000 | 157.079633 | 900 | 15.707963 | 90 | 1.570796 | 9 | 0.157080 | 0.9 | 0.015708 | 0.09 | 0.001571 |
| 10000 | 174.532925 | 1000 | 17.453293 | 100 | 1.745329 | 10 | 0.174533 | 1.0 | 0.017453 | 0.10 | 0.001745 |

Minutes into Radians

| Min. | Rad. | Min. | Rad. | Min. | Rad. | Min. | Rad. | Min. | Rad. | Min. | Rad. |
|---|---|---|---|---|---|---|---|---|---|---|---|
| 1 | 0.000291 | 11 | 0.003200 | 21 | 0.006109 | 31 | 0.009018 | 41 | 0.011926 | 51 | 0.014835 |
| 2 | 0.000582 | 12 | 0.003491 | 22 | 0.006400 | 32 | 0.009308 | 42 | 0.012217 | 52 | 0.015126 |
| 3 | 0.000873 | 13 | 0.003782 | 23 | 0.006690 | 33 | 0.009599 | 43 | 0.012508 | 53 | 0.015417 |
| 4 | 0.001164 | 14 | 0.004072 | 24 | 0.006981 | 34 | 0.009890 | 44 | 0.012799 | 54 | 0.015708 |
| 5 | 0.001454 | 15 | 0.004363 | 25 | 0.007272 | 35 | 0.010181 | 45 | 0.013090 | 55 | 0.015999 |
| 6 | 0.001745 | 16 | 0.004654 | 26 | 0.007563 | 36 | 0.010472 | 46 | 0.013381 | 56 | 0.016290 |
| 7 | 0.002036 | 17 | 0.004945 | 27 | 0.007854 | 37 | 0.010763 | 47 | 0.013672 | 57 | 0.016581 |
| 8 | 0.002327 | 18 | 0.005236 | 28 | 0.008145 | 38 | 0.011054 | 48 | 0.013963 | 58 | 0.016872 |
| 9 | 0.002618 | 19 | 0.005527 | 29 | 0.008436 | 39 | 0.011345 | 49 | 0.014254 | 59 | 0.017162 |
| 10 | 0.002909 | 20 | 0.005818 | 30 | 0.008727 | 40 | 0.011636 | 50 | 0.014544 | 60 | 0.017453 |

Seconds into Radians

| Sec. | Rad. | Sec. | Rad. | Sec. | Rad. | Sec. | Rad. | Sec. | Rad. | Sec. | Rad. |
|---|---|---|---|---|---|---|---|---|---|---|---|
| 1 | 0.000005 | 11 | 0.000053 | 21 | 0.000102 | 31 | 0.000150 | 41 | 0.000199 | 51 | 0.000247 |
| 2 | 0.000010 | 12 | 0.000058 | 22 | 0.000107 | 32 | 0.000155 | 42 | 0.000204 | 52 | 0.000252 |
| 3 | 0.000015 | 13 | 0.000063 | 23 | 0.000112 | 33 | 0.000160 | 43 | 0.000208 | 53 | 0.000257 |
| 4 | 0.000019 | 14 | 0.000068 | 24 | 0.000116 | 34 | 0.000165 | 44 | 0.000213 | 54 | 0.000262 |
| 5 | 0.000024 | 15 | 0.000073 | 25 | 0.000121 | 35 | 0.000170 | 45 | 0.000218 | 55 | 0.000267 |
| 6 | 0.000029 | 16 | 0.000078 | 26 | 0.000126 | 36 | 0.000175 | 46 | 0.000223 | 56 | 0.000271 |
| 7 | 0.000034 | 17 | 0.000082 | 27 | 0.000131 | 37 | 0.000179 | 47 | 0.000228 | 57 | 0.000276 |
| 8 | 0.000039 | 18 | 0.000087 | 28 | 0.000136 | 38 | 0.000184 | 48 | 0.000233 | 58 | 0.000281 |
| 9 | 0.000044 | 19 | 0.000092 | 29 | 0.000141 | 39 | 0.000189 | 49 | 0.000238 | 59 | 0.000286 |
| 10 | 0.000048 | 20 | 0.000097 | 30 | 0.000145 | 40 | 0.000194 | 50 | 0.000242 | 60 | 0.000291 |

## Radians into Degrees and Decimals of a Degree

(Based on π radians = 180 degrees)

| Rad. | Deg. | Rad. | Deg. | Rad. | Deg. | Rad. | Deg. | Rad. | Deg. | Rad. | Deg. |
|---|---|---|---|---|---|---|---|---|---|---|---|
| 10 | 572.9578 | 1 | 57.2958 | 0.1 | 5.7296 | 0.01 | 0.5730 | 0.001 | 0.0573 | 0.0001 | 0.0057 |
| 20 | 1145.9156 | 2 | 114.5916 | 0.2 | 11.4592 | 0.02 | 1.1459 | 0.002 | 0.1146 | 0.0002 | 0.0115 |
| 30 | 1718.8734 | 3 | 171.8873 | 0.3 | 17.1887 | 0.03 | 1.7189 | 0.003 | 0.1719 | 0.0003 | 0.0172 |
| 40 | 2291.8312 | 4 | 229.1831 | 0.4 | 22.9183 | 0.04 | 2.2918 | 0.004 | 0.2292 | 0.0004 | 0.0229 |
| 50 | 2864.7890 | 5 | 286.4789 | 0.5 | 28.6479 | 0.05 | 2.8648 | 0.005 | 0.2865 | 0.0005 | 0.0286 |
| 60 | 3437.7468 | 6 | 343.7747 | 0.6 | 34.3775 | 0.06 | 3.4377 | 0.006 | 0.3438 | 0.0006 | 0.0344 |
| 70 | 4010.7046 | 7 | 401.0705 | 0.7 | 40.1070 | 0.07 | 4.0107 | 0.007 | 0.4011 | 0.0007 | 0.0401 |
| 80 | 4583.6624 | 8 | 458.3662 | 0.8 | 45.8366 | 0.08 | 4.5837 | 0.008 | 0.4584 | 0.0008 | 0.0458 |
| 90 | 5156.6202 | 9 | 515.6620 | 0.9 | 51.5662 | 0.09 | 5.1566 | 0.009 | 0.5157 | 0.0009 | 0.0516 |
| 100 | 5729.5780 | 10 | 572.9578 | 1.0 | 57.2958 | 0.10 | 5.7296 | 0.010 | 0.5730 | 0.0010 | 0.0573 |

## Radians into Degrees, Minutes, and Seconds

(Based on π radians = 180 degrees)

| Rad. | Angle | Rad. | Angle | Rad. | Angle | Rad. | Angle | Rad. | Angle | Rad. | Angle |
|---|---|---|---|---|---|---|---|---|---|---|---|
| 10 | 572°57′28″ | 1 | 57°17′45″ | 0.1 | 5°43′46″ | 0.01 | 0°34′23″ | 0.001 | 0°3′26″ | 0.0001 | 0°0′21″ |
| 20 | 1145°54′56″ | 2 | 114°35′30″ | 0.2 | 11°27′33″ | 0.02 | 1°8′45″ | 0.002 | 0°6′53″ | 0.0002 | 0°0′41″ |
| 30 | 1718°52′24″ | 3 | 171°53′14″ | 0.3 | 17°11′19″ | 0.03 | 1°43′8″ | 0.003 | 0°10′19″ | 0.0003 | 0°1′2″ |
| 40 | 2291°49′52″ | 4 | 229°10′59″ | 0.4 | 22°55′6″ | 0.04 | 2°17′31″ | 0.004 | 0°13′45″ | 0.0004 | 0°1′23″ |
| 50 | 2864°47′20″ | 5 | 286°28′44″ | 0.5 | 28°38′52″ | 0.05 | 2°51′53″ | 0.005 | 0°17′11″ | 0.0005 | 0°1′43″ |
| 60 | 3437°44′48″ | 6 | 343°46′29″ | 0.6 | 34°22′39″ | 0.06 | 3°26′16″ | 0.006 | 0°20′38″ | 0.0006 | 0°2′4″ |
| 70 | 4010°42′16″ | 7 | 401°4′14″ | 0.7 | 40°6′25″ | 0.07 | 4°0′39″ | 0.007 | 0°24′4″ | 0.0007 | 0°2′24″ |
| 80 | 4583°39′44″ | 8 | 458°21′58″ | 0.8 | 45°50′12″ | 0.08 | 4°35′1″ | 0.008 | 0°27′30″ | 0.0008 | 0°2′45″ |
| 90 | 5156°37′13″ | 9 | 515°39′43″ | 0.9 | 51°33′58″ | 0.09 | 5°9′24″ | 0.009 | 0°30′56″ | 0.0009 | 0°3′6″ |
| 100 | 5729°34′41″ | 10 | 572°57′28″ | 1.0 | 57°17′45″ | 0.10 | 5°43′46″ | 0.010 | 0°34′23″ | 0.0010 | 0°3′26″ |

## Minutes and Seconds into Decimal of a Degree and Vice Versa

(Based on 1 second = 0.00027778 degree)

| Minutes into Decimals of a Degree | | | | | | Seconds into Decimals of a Degree | | | | | |
|---|---|---|---|---|---|---|---|---|---|---|---|
| Min. | Deg. | Min. | Deg. | Min. | Deg. | Sec. | Deg. | Sec. | Deg. | Sec. | Deg. |
| 1 | 0.0167 | 21 | 0.3500 | 41 | 0.6833 | 1 | 0.0003 | 21 | 0.0058 | 41 | 0.0114 |
| 2 | 0.0333 | 22 | 0.3667 | 42 | 0.7000 | 2 | 0.0006 | 22 | 0.0061 | 42 | 0.0117 |
| 3 | 0.0500 | 23 | 0.3833 | 43 | 0.7167 | 3 | 0.0008 | 23 | 0.0064 | 43 | 0.0119 |
| 4 | 0.0667 | 24 | 0.4000 | 44 | 0.7333 | 4 | 0.0011 | 24 | 0.0067 | 44 | 0.0122 |
| 5 | 0.0833 | 25 | 0.4167 | 45 | 0.7500 | 5 | 0.0014 | 25 | 0.0069 | 45 | 0.0125 |
| 6 | 0.1000 | 26 | 0.4333 | 46 | 0.7667 | 6 | 0.0017 | 26 | 0.0072 | 46 | 0.0128 |
| 7 | 0.1167 | 27 | 0.4500 | 47 | 0.7833 | 7 | 0.0019 | 27 | 0.0075 | 47 | 0.0131 |
| 8 | 0.1333 | 28 | 0.4667 | 48 | 0.8000 | 8 | 0.0022 | 28 | 0.0078 | 48 | 0.0133 |
| 9 | 0.1500 | 29 | 0.4833 | 49 | 0.8167 | 9 | 0.0025 | 29 | 0.0081 | 49 | 0.0136 |
| 10 | 0.1667 | 30 | 0.5000 | 50 | 0.8333 | 10 | 0.0028 | 30 | 0.0083 | 50 | 0.0139 |
| 11 | 0.1833 | 31 | 0.5167 | 51 | 0.8500 | 11 | 0.0031 | 31 | 0.0086 | 51 | 0.0142 |
| 12 | 0.2000 | 32 | 0.5333 | 52 | 0.8667 | 12 | 0.0033 | 32 | 0.0089 | 52 | 0.0144 |
| 13 | 0.2167 | 33 | 0.5500 | 53 | 0.8833 | 13 | 0.0036 | 33 | 0.0092 | 53 | 0.0147 |
| 14 | 0.2333 | 34 | 0.5667 | 54 | 0.9000 | 14 | 0.0039 | 34 | 0.0094 | 54 | 0.0150 |
| 15 | 0.2500 | 35 | 0.5833 | 55 | 0.9167 | 15 | 0.0042 | 35 | 0.0097 | 55 | 0.0153 |
| 16 | 0.2667 | 36 | 0.6000 | 56 | 0.9333 | 16 | 0.0044 | 36 | 0.0100 | 56 | 0.0156 |
| 17 | 0.2833 | 37 | 0.6167 | 57 | 0.9500 | 17 | 0.0047 | 37 | 0.0103 | 57 | 0.0158 |
| 18 | 0.3000 | 38 | 0.6333 | 58 | 0.9667 | 18 | 0.0050 | 38 | 0.0106 | 58 | 0.0161 |
| 19 | 0.3167 | 39 | 0.6500 | 59 | 0.9833 | 19 | 0.0053 | 39 | 0.0108 | 59 | 0.0164 |
| 20 | 0.3333 | 40 | 0.6667 | 60 | 1.0000 | 20 | 0.0056 | 40 | 0.0111 | 60 | 0.0167 |

*Example 3:* Convert 11′37″ to decimals of a degree. From the "Min. into Dec. Deg." table, 11′ = 0.1833 degree. From the "Sec. into Dec. Deg." table, 37″ = 0.0103 degree. Adding, 11′37″ = 0.1833 + 0.0103 = 0.1936 degree.

*Example 4:* Convert 0.1234 degree to minutes and seconds. From the "Min. into Dec. Deg." table, 0.1167 degree = 7′. Subtracting 0.1167 from 0.1234 gives 0.0067. From the "Sec. into Dec. Deg." table, 0.0067 = 24″ so that 0.1234 = 7′24″.

**Trigonometric Functions.**—Like algebraic functions, trigonometric functions define a relationship between input values (angle measure, expressed in radians) and output values (the trigonometric ratios associated with the angles). Recalling function notation, $y = f(x)$ (see *ALGEBRA* on page 19), the main trigonometric functions are: $f(x) = \sin x$, $f(x) = \cos x$, and $f(x) = \tan x$. The trigonometric graphs are derived from the points on the unit circle (circle of radius = 1) "translated" onto the $(x, y)$-coordinate system. Because the radius is 1, $\cos\theta = x/1 = x$, and $\sin\theta = y/1 = y$. Thus, the rectangular coordinate ordered pair $(x, y)$ on the unit circle corresponds to the trigonometric coordinates $(\cos\theta, \sin\theta)$.

Possible confusion arises from the different roles of $x$ and $y$ in the two graphs. On the unit circle (Fig. 1), $x$ and $y$ are the coordinates of the points that correspond to the adjacent (cosine) and opposite (sine) sides of the right triangle with radius 1. But on the function graphs (Fig. 2a), $x$ is the *angle measure*, the same as $\theta$, and $y$ is the *value* of the function being plotted. So, "sin $x$" is the same as "sin $\theta$." Angle measure is marked in radians (*not degrees*). The graphs of $y = \sin x$ and $y = \cos x$ are seen in the figure on the same set of axes (Fig. 2a). $y = \tan x$ is graphed on its own set of axes (Fig. 2b).

The *domain* of a trigonometric function is the set of angle measures for which the function is defined. The domain of both $y = \sin x$ and $y = \cos x$ is the set of all real numbers, since it makes sense to substitute any angle—positive or negative—into these functions. The answer is always a real number. Negative angles indicate an angle measured clockwise from the horizontal, whereas a positive angle (the usual situation) is measured counterclockwise. The domain of $y = \tan x$, however, does not include the odd multiples of $\pi/2$, since $\tan x = \sin x/\cos x$, and so, when $\cos x = 0$, $\tan x$ is undefined. This happens at $\pm\pi/2$, $\pm 3\pi/2$, $\pm 5\pi/2$, . . . As Fig. 2b shows, the tangent function approaches the dotted lines increasingly closer, never meeting them.

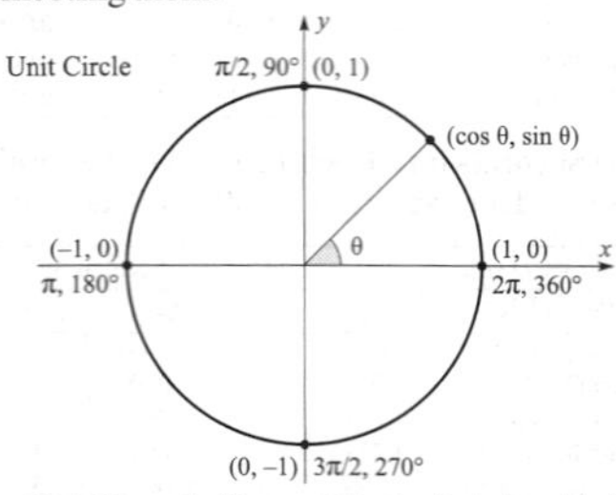


Fig. 1. The Unit Circle, which Gives the Sine and Cosine Relationship for All Angle Measures.

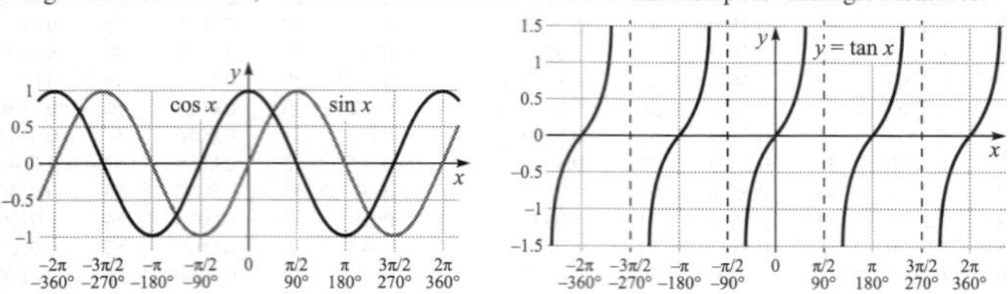


Fig. 2. (a) Graph of $y = \sin x$ and $y = \cos x$; (b) Graph of $y = \tan x$.
*Note:* For any real number $x$, $\sin x$ (for example) is defined to be sin($x$ radians). Equivalent degree measures are shown here for reference.

The signs of the three main trigonometric functions (positive or negative) are shown in the diagram at the top of page 101. The names of the positive-signed functions of angles that lie in a particular quadrant are shown. All the functions are positive in the first, only $\sin x$ in the second, only $\tan x$ in the third, and only $\cos x$ is positive in the fourth. The mnemonic device "ASTC" for "All Students Take Calculus" is a simple way to remember the signs. $\csc x$, $\sec x$, and $\cot x$ have the same signs as their respective reciprocal functions.

| Sin | All |
|---|---|
| Tan | Cos |

Fig. 3. Signs of the Three Main Trigonometric Functions

Fig. 3 shows the values of angles and ratios on the unit circle in great detail, as well as the signs of the functions in each of the four quadrants. Radian and degree measure are also marked off.

The figure also shows the sign (+ or –) and the values of trigonometric functions for angles in each of the four quadrants, 0 to 90, 90 to 180, 180 to 270, and 270 to 360 degrees. A $\pi$ radian is approximately 3.14, and $2\pi$ is approximately 6.3 radian measure, as the completed circle indicates. The corresponding degree measures are marked as well, as are the pure radian equivalents. The chart indicates, for example, that all the functions are positive for angles between 0 and 90 degrees. This is because $x$ and $y$ are both positive in the first quadrant, and all the trigonometric ratios are therefore also positive. In the same way, the cotangent of an angle between 180 and 270 degrees is positive and has a value between infinity and 0; in other words, the cotangent for 180 degrees is infinitely large and then the cotangent gradually decreases for increasing angles, so that the cotangent for 270 degrees equals 0. The cosine, tangent, and cotangent for angles between 90 and 180 degrees are negative, although they have the same absolute values as their respective angles from 0 to 90 degrees. Negative trigonometric values are preceded by a minus sign; thus, tan 123° 20′ = –1.5204.

*Inverse Trigonometric Functions:* If the value of the sides of a triangle are known but an angle is unknown, the trigonometric function inverse is used to work backwards to find the angle measure. Inverse trigonometric functions are known by either the prefix "arc" before the name, or by the superscript –1 after the name. The notation is not to be confused with the meaning of a negative exponent.

Trigonometric functions and their inverses are shown below. The roles of $x$ and $y$ are reversed in the inverse functions: $x$ is the ratio given and $y$ is the angle measure that is sought:

| *Function* | *Inverse Function* |
|---|---|
| $y = \sin x$ | $y = \arcsin x$, or $y = \sin^{-1} x$ |
| $y = \cos x$ | $y = \arccos x$, or $y = \cos^{-1} x$ |
| $y = \tan x$ | $y = \arctan x$, or $y = \tan^{-1} x$ |

The value of the angle that corresponds to the ratio is intended. Any angle can be found using the inverse trig function and relying on trig tables or a calculator. The examples give a few of the essential, familiar ratios for which the angles are readily known.

*Examples:* Given $\sin x = 1$, arcsin(1), or $\sin^{-1}(1) = 90°$, or $\pi/2$.
Given $\cos x = 0.866$, arccos(0.866), or $\cos^{-1}(0.866) = 60°$, or $\pi/3$.
Given $\tan x = 1$, arctan(1), or $\tan^{-1}(1) = 45°$, or $\pi/4$.

**Trigonometry Tables.**—Table 2a, Table 2b, and Table 2c, starting on page 103, contain the values of the sine, cosine, tangent, and cotangent functions of angles from 0 to 90 degrees. Commonly referred to as "trig tables," these values also are accessible on standard scientific calculators. Function values for all other angles can be obtained from the trig tables by applying the rules for signs of trigonometric functions and the useful relationships among angles given in the following. Secant and cosecant functions can be found from $\sec A = 1/\cos A$ and $\csc A = 1/\sin A$.

The trig tables are divided by a double line. The body of each half table consists of four labeled columns of data between the columns that contain the angles. The angles left of the data increase moving down the table, and angles right of the data increase moving up the table. Column labels above the data identify the trig functions for angles listed in the left column of each half table. Columns labels below the data are for angles listed in the right column of each half table. To find the value of a function for a particular angle, first the angle is located in the table, then the appropriate function label across the top or bottom row of the table is located. At the intersection of the angle row and label column is the function value. Angles on opposite sides of each table are complementary angles (i.e., their sum is 90°) and are related (see *Cofunction Identities:*, page 92). For example, sin 10° = cos 80° and cos 10° = sin 80°. Expanded trig tables are also available in the *ADDITIONAL MATERIAL* in the *Machinery's Handbook 32 Digital Edition*.

Angle measures greater than 90 degrees are converted to their *reference angle* (that is, its angle equivalent) before a trigonometric value can be found. If the angle θ is between 90 and 180, its reference angle is 180 – θ; if θ is between 180 and 270, then θ – 180 is its reference angle; and if it is between 270 and 360 degrees, 360 – θ is its acute angle equivalent. To determine trigonometric values of functions of angles greater than 90°, subtract 90, 180, 270, or 360 from the angle to get the reference angle less than 90° and use Table 1, *Useful Relationships Among Angles*, to find the equivalent first-quadrant function and angle to look up in the trig tables.

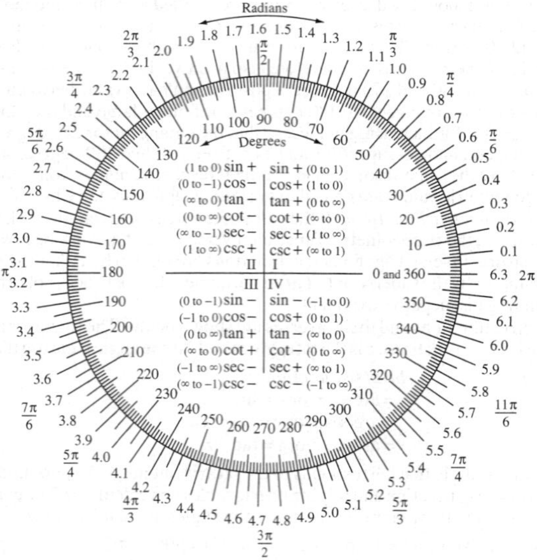


Fig. 4. Signs of Trigonometric Functions, Fractions of π, and Degree-Radian Conversion

**Table 1. Useful Relationships Among Angles**

| Angle Function | θ | –θ | 90° ± θ | 180° ± θ | 270° ± θ | 360° ± θ |
|---|---|---|---|---|---|---|
| sine (sin) | sin θ | –sin θ | +cos θ | ∓sin θ | –cos θ | ±sin θ |
| cosine (cos) | cos θ | +cos θ | ∓sin θ | –cos θ | ±sin θ | +cos θ |
| tangent (tan) | tan θ | –tan θ | ∓cot θ | ±tan θ | ∓cot θ | ±tan θ |
| cotangent (cot) | cot θ | –cot θ | ∓tan θ | ±cot θ | ∓tan θ | ±cot θ |
| secant (sec) | sec θ | +sec θ | ∓csc θ | –sec θ | ±csc θ | +sec θ |
| cosecant (csc) | csc θ | –csc θ | +sec θ | ∓csc θ | –sec θ | ±csc θ |

*Examples:* cos (270° – θ) = –sin θ; tan (90° + θ) = –cot θ.

*Example:* Find the cosine of 336°40′. Fig. 4 shows that the cosine of every angle in Quadrant IV (270° to 360°) is positive. To find the angle and trig function to use when entering the trig table, subtract 270 from 336 to get cos 336°40′ = cos (270° + 66°40′) and then find the intersection of the "cos row" and the 270 ± θ column in Table 1. Because cos (270 ± θ) in the fourth quadrant is equal to ± sin θ in the first quadrant, find sin 66°40′ in the trig table. Therefore, cos 336°40′ = sin 66°40′ = 0.918216.

## Table 2a. Trigonometric Values of Angles from 0° to 15° and 75° to 90°

| Angle | sin | cos | tan | cot | | Angle | sin | cos | tan | cot | |
|---|---|---|---|---|---|---|---|---|---|---|---|
| **0° 0′** | 0.000000 | 1.000000 | 0.000000 | — | **90° 0′** | **7° 30′** | 0.130526 | 0.991445 | 0.131652 | 7.595754 | **82° 30′** |
| **10** | 0.002909 | 0.999996 | 0.002909 | 343.7737 | **50** | **40** | 0.133410 | 0.991061 | 0.134613 | 7.428706 | **20** |
| **20** | 0.005818 | 0.999983 | 0.005818 | 171.8854 | **40** | **50** | 0.136292 | 0.990669 | 0.137576 | 7.268725 | **10** |
| **30** | 0.008727 | 0.999962 | 0.008727 | 114.5887 | **30** | **8° 0′** | 0.139173 | 0.990268 | 0.140541 | 7.115370 | **82° 0′** |
| **40** | 0.011635 | 0.999932 | 0.011636 | 85.93979 | **20** | **10** | 0.142053 | 0.989859 | 0.143508 | 6.968234 | **50** |
| **50** | 0.014544 | 0.999894 | 0.014545 | 68.75009 | **10** | **20** | 0.144932 | 0.989442 | 0.146478 | 6.826944 | **40** |
| **1° 0′** | 0.017452 | 0.999848 | 0.017455 | 57.28996 | **89° 0′** | **30** | 0.147809 | 0.989016 | 0.149451 | 6.691156 | **30** |
| **10** | 0.020361 | 0.999793 | 0.020365 | 49.10388 | **50** | **40** | 0.150686 | 0.988582 | 0.152426 | 6.560554 | **20** |
| **20** | 0.023269 | 0.999729 | 0.023275 | 42.96408 | **40** | **50** | 0.153561 | 0.988139 | 0.155404 | 6.434843 | **10** |
| **30** | 0.026177 | 0.999657 | 0.026186 | 38.18846 | **30** | **9° 0′** | 0.156434 | 0.987688 | 0.158384 | 6.313752 | **81° 0′** |
| **40** | 0.029085 | 0.999577 | 0.029097 | 34.36777 | **20** | **10** | 0.159307 | 0.987229 | 0.161368 | 6.197028 | **50** |
| **50** | 0.031992 | 0.999488 | 0.032009 | 31.24158 | **10** | **20** | 0.162178 | 0.986762 | 0.164354 | 6.084438 | **40** |
| **2° 0′** | 0.034899 | 0.999391 | 0.034921 | 28.63625 | **88° 0′** | **30** | 0.165048 | 0.986286 | 0.167343 | 5.975764 | **30** |
| **10** | 0.037806 | 0.999285 | 0.037834 | 26.43160 | **50** | **40** | 0.167916 | 0.985801 | 0.170334 | 5.870804 | **20** |
| **20** | 0.040713 | 0.999171 | 0.040747 | 24.54176 | **40** | **50** | 0.170783 | 0.985309 | 0.173329 | 5.769369 | **10** |
| **30** | 0.043619 | 0.999048 | 0.043661 | 22.90377 | **30** | **10° 0′** | 0.173648 | 0.984808 | 0.176327 | 5.671282 | **80° 0′** |
| **40** | 0.046525 | 0.998917 | 0.046576 | 21.47040 | **20** | **10** | 0.176512 | 0.984298 | 0.179328 | 5.576379 | **50** |
| **50** | 0.049431 | 0.998778 | 0.049491 | 20.20555 | **10** | **20** | 0.179375 | 0.983781 | 0.182332 | 5.484505 | **40** |
| **3° 0′** | 0.052336 | 0.998630 | 0.052408 | 19.08114 | **87° 0′** | **30** | 0.182236 | 0.983255 | 0.185339 | 5.395517 | **30** |
| **10** | 0.055241 | 0.998473 | 0.055325 | 18.07498 | **50** | **40** | 0.185095 | 0.982721 | 0.188349 | 5.309279 | **20** |
| **20** | 0.058145 | 0.998308 | 0.058243 | 17.16934 | **40** | **50** | 0.187953 | 0.982178 | 0.191363 | 5.225665 | **10** |
| **30** | 0.061049 | 0.998135 | 0.061163 | 16.34986 | **30** | **11° 0′** | 0.190809 | 0.981627 | 0.194380 | 5.144554 | **79° 0′** |
| **40** | 0.063952 | 0.997953 | 0.064083 | 15.60478 | **20** | **10** | 0.193664 | 0.981068 | 0.197401 | 5.065835 | **50** |
| **50** | 0.066854 | 0.997763 | 0.067004 | 14.92442 | **10** | **20** | 0.196517 | 0.980500 | 0.200425 | 4.989403 | **40** |
| **4° 0′** | 0.069756 | 0.997564 | 0.069927 | 14.30067 | **86° 0′** | **30** | 0.199368 | 0.979925 | 0.203452 | 4.915157 | **30** |
| **10** | 0.072658 | 0.997357 | 0.072851 | 13.72674 | **50** | **40** | 0.202218 | 0.979341 | 0.206483 | 4.843005 | **20** |
| **20** | 0.075559 | 0.997141 | 0.075775 | 13.19688 | **40** | **50** | 0.205065 | 0.978748 | 0.209518 | 4.772857 | **10** |
| **30** | 0.078459 | 0.996917 | 0.078702 | 12.70621 | **30** | **12° 0′** | 0.207912 | 0.978148 | 0.212557 | 4.704630 | **78° 0′** |
| **40** | 0.081359 | 0.996685 | 0.081629 | 12.25051 | **20** | **10** | 0.210756 | 0.977539 | 0.215599 | 4.638246 | **50** |
| **50** | 0.084258 | 0.996444 | 0.084558 | 11.82617 | **10** | **20** | 0.213599 | 0.976921 | 0.218645 | 4.573629 | **40** |
| **5° 0′** | 0.087156 | 0.996195 | 0.087489 | 11.43005 | **85° 0′** | **30** | 0.216440 | 0.976296 | 0.221695 | 4.510709 | **30** |
| **10** | 0.090053 | 0.995937 | 0.090421 | 11.05943 | **50** | **40** | 0.219279 | 0.975662 | 0.224748 | 4.449418 | **20** |
| **20** | 0.092950 | 0.995671 | 0.093354 | 10.71191 | **40** | **50** | 0.222116 | 0.975020 | 0.227806 | 4.389694 | **10** |
| **30** | 0.095846 | 0.995396 | 0.096289 | 10.38540 | **30** | **13° 0′** | 0.224951 | 0.974370 | 0.230868 | 4.331476 | **77° 0′** |
| **40** | 0.098741 | 0.995113 | 0.099226 | 10.07803 | **20** | **10** | 0.227784 | 0.973712 | 0.233934 | 4.274707 | **50** |
| **50** | 0.101635 | 0.994822 | 0.102164 | 9.788173 | **10** | **20** | 0.230616 | 0.973045 | 0.237004 | 4.219332 | **40** |
| **6° 0′** | 0.104528 | 0.994522 | 0.105104 | 9.514364 | **84° 0′** | **30** | 0.233445 | 0.972370 | 0.240079 | 4.165300 | **30** |
| **10** | 0.107421 | 0.994214 | 0.108046 | 9.255304 | **50** | **40** | 0.236273 | 0.971687 | 0.243157 | 4.112561 | **20** |
| **20** | 0.110313 | 0.993897 | 0.110990 | 9.009826 | **40** | **50** | 0.239098 | 0.970995 | 0.246241 | 4.061070 | **10** |
| **30** | 0.113203 | 0.993572 | 0.113936 | 8.776887 | **30** | **14° 0′** | 0.241922 | 0.970296 | 0.249328 | 4.010781 | **76° 0′** |
| **40** | 0.116093 | 0.993238 | 0.116883 | 8.555547 | **20** | **10** | 0.244743 | 0.969588 | 0.252420 | 3.961652 | **50** |
| **50** | 0.118982 | 0.992896 | 0.119833 | 8.344956 | **10** | **20** | 0.247563 | 0.968872 | 0.255516 | 3.913642 | **40** |
| **7° 0′** | 0.121869 | 0.992546 | 0.122785 | 8.144346 | **83° 0′** | **30** | 0.250380 | 0.968148 | 0.258618 | 3.866713 | **30** |
| **10** | 0.124756 | 0.992187 | 0.125738 | 7.953022 | **50** | **40** | 0.253195 | 0.967415 | 0.261723 | 3.820828 | **20** |
| **20** | 0.127642 | 0.991820 | 0.128694 | 7.770351 | **40** | **50** | 0.256008 | 0.966675 | 0.264834 | 3.775952 | **10** |
| **7° 30′** | 0.130526 | 0.991445 | 0.131652 | 7.595754 | **82° 30** | **15° 0′** | 0.258819 | 0.965926 | 0.267949 | 3.732051 | **75° 0′** |
| | cos | sin | cot | tan | Angle | | cos | sin | cot | tan | Angle |

For angles 0° to 15° 0′ (angles found in a column to the left of the data), use the column labels at the top of the table; for angles 75° to 90° 0′ (angles found in a column to the right of the data), use the column labels at the bottom of the table.

## Table 2b. Trigonometric Values of Angles from 15° to 30° and 60° to 75°

| Angle | sin | cos | tan | cot | | Angle | sin | cos | tan | cot | |
|---|---|---|---|---|---|---|---|---|---|---|---|
| **15° 0′** | 0.258819 | 0.965926 | 0.267949 | 3.732051 | **75° 0′** | **22° 30′** | 0.382683 | 0.923880 | 0.414214 | 2.414214 | **67° 30** |
| **10** | 0.261628 | 0.965169 | 0.271069 | 3.689093 | **50** | **40** | 0.385369 | 0.922762 | 0.417626 | 2.394489 | **20** |
| **20** | 0.264434 | 0.964404 | 0.274194 | 3.647047 | **40** | **50** | 0.388052 | 0.921638 | 0.421046 | 2.375037 | **10** |
| **30** | 0.267238 | 0.963630 | 0.277325 | 3.605884 | **30** | **23° 0′** | 0.390731 | 0.920505 | 0.424475 | 2.355852 | **67° 0′** |
| **40** | 0.270040 | 0.962849 | 0.280460 | 3.565575 | **20** | **10** | 0.393407 | 0.919364 | 0.427912 | 2.336929 | **50** |
| **50** | 0.272840 | 0.962059 | 0.283600 | 3.526094 | **10** | **20** | 0.396080 | 0.918216 | 0.431358 | 2.318261 | **40** |
| **16° 0′** | 0.275637 | 0.961262 | 0.286745 | 3.487414 | **74° 0′** | **30** | 0.398749 | 0.917060 | 0.434812 | 2.299843 | **30** |
| **10** | 0.278432 | 0.960456 | 0.289896 | 3.449512 | **50** | **40** | 0.401415 | 0.915896 | 0.438276 | 2.281669 | **20** |
| **20** | 0.281225 | 0.959642 | 0.293052 | 3.412363 | **40** | **50** | 0.404078 | 0.914725 | 0.441748 | 2.263736 | **10** |
| **30** | 0.284015 | 0.958820 | 0.296213 | 3.375943 | **30** | **24° 0′** | 0.406737 | 0.913545 | 0.445229 | 2.246037 | **66° 0′** |
| **40** | 0.286803 | 0.957990 | 0.299380 | 3.340233 | **20** | **10** | 0.409392 | 0.912358 | 0.448719 | 2.228568 | **50** |
| **50** | 0.289589 | 0.957151 | 0.302553 | 3.305209 | **10** | **20** | 0.412045 | 0.911164 | 0.452218 | 2.211323 | **40** |
| **17° 0′** | 0.292372 | 0.956305 | 0.305731 | 3.270853 | **73° 0′** | **30** | 0.414693 | 0.909961 | 0.455726 | 2.194300 | **30** |
| **10** | 0.295152 | 0.955450 | 0.308914 | 3.237144 | **50** | **40** | 0.417338 | 0.908751 | 0.459244 | 2.177492 | **20** |
| **20** | 0.297930 | 0.954588 | 0.312104 | 3.204064 | **40** | **50** | 0.419980 | 0.907533 | 0.462771 | 2.160896 | **10** |
| **30** | 0.300706 | 0.953717 | 0.315299 | 3.171595 | **30** | **25° 0′** | 0.422618 | 0.906308 | 0.466308 | 2.144507 | **65° 0′** |
| **40** | 0.303479 | 0.952838 | 0.318500 | 3.139719 | **20** | **10** | 0.425253 | 0.905075 | 0.469854 | 2.128321 | **50** |
| **50** | 0.306249 | 0.951951 | 0.321707 | 3.108421 | **10** | **20** | 0.427884 | 0.903834 | 0.473410 | 2.112335 | **40** |
| **18° 0′** | 0.309017 | 0.951057 | 0.324920 | 3.077684 | **72° 0′** | **30** | 0.430511 | 0.902585 | 0.476976 | 2.096544 | **30** |
| **10** | 0.311782 | 0.950154 | 0.328139 | 3.047492 | **50** | **40** | 0.433135 | 0.901329 | 0.480551 | 2.080944 | **20** |
| **20** | 0.314545 | 0.949243 | 0.331364 | 3.017830 | **40** | **50** | 0.435755 | 0.900065 | 0.484137 | 2.065532 | **10** |
| **30** | 0.317305 | 0.948324 | 0.334595 | 2.988685 | **30** | **26° 0′** | 0.438371 | 0.898794 | 0.487733 | 2.050304 | **64° 0′** |
| **40** | 0.320062 | 0.947397 | 0.337833 | 2.960042 | **20** | **10** | 0.440984 | 0.897515 | 0.491339 | 2.035256 | **50** |
| **50** | 0.322816 | 0.946462 | 0.341077 | 2.931888 | **10** | **20** | 0.443593 | 0.896229 | 0.494955 | 2.020386 | **40** |
| **19° 0′** | 0.325568 | 0.945519 | 0.344328 | 2.904211 | **71° 0′** | **30** | 0.446198 | 0.894934 | 0.498582 | 2.005690 | **30** |
| **10** | 0.328317 | 0.944568 | 0.347585 | 2.876997 | **50** | **40** | 0.448799 | 0.893633 | 0.502219 | 1.991164 | **20** |
| **20** | 0.331063 | 0.943609 | 0.350848 | 2.850235 | **40** | **50** | 0.451397 | 0.892323 | 0.505867 | 1.976805 | **10** |
| **30** | 0.333807 | 0.942641 | 0.354119 | 2.823913 | **30** | **27° 0′** | 0.453990 | 0.891007 | 0.509525 | 1.962611 | **63° 0′** |
| **40** | 0.336547 | 0.941666 | 0.357396 | 2.798020 | **20** | **10** | 0.456580 | 0.889682 | 0.513195 | 1.948577 | **50** |
| **50** | 0.339285 | 0.940684 | 0.360679 | 2.772545 | **10** | **20** | 0.459166 | 0.888350 | 0.516875 | 1.934702 | **40** |
| **20° 0′** | 0.342020 | 0.939693 | 0.363970 | 2.747477 | **70° 0′** | **30** | 0.461749 | 0.887011 | 0.520567 | 1.920982 | **30** |
| **10** | 0.344752 | 0.938694 | 0.367268 | 2.722808 | **50** | **40** | 0.464327 | 0.885664 | 0.524270 | 1.907415 | **20** |
| **20** | 0.347481 | 0.937687 | 0.370573 | 2.698525 | **40** | **50** | 0.466901 | 0.884309 | 0.527984 | 1.893997 | **10** |
| **30** | 0.350207 | 0.936672 | 0.373885 | 2.674621 | **30** | **28° 0′** | 0.469472 | 0.882948 | 0.531709 | 1.880726 | **62° 0′** |
| **40** | 0.352931 | 0.935650 | 0.377204 | 2.651087 | **20** | **10** | 0.472038 | 0.881578 | 0.535446 | 1.867600 | **50** |
| **50** | 0.355651 | 0.934619 | 0.380530 | 2.627912 | **10** | **20** | 0.474600 | 0.880201 | 0.539195 | 1.854616 | **40** |
| **21° 0′** | 0.358368 | 0.933580 | 0.383864 | 2.605089 | **69° 0′** | **30** | 0.477159 | 0.878817 | 0.542956 | 1.841771 | **30** |
| **10** | 0.361082 | 0.932534 | 0.387205 | 2.582609 | **50** | **40** | 0.479713 | 0.877425 | 0.546728 | 1.829063 | **20** |
| **20** | 0.363793 | 0.931480 | 0.390554 | 2.560465 | **40** | **50** | 0.482263 | 0.876026 | 0.550513 | 1.816489 | **10** |
| **30** | 0.366501 | 0.930418 | 0.393910 | 2.538648 | **30** | **29° 0′** | 0.484810 | 0.874620 | 0.554309 | 1.804048 | **61° 0′** |
| **40** | 0.369206 | 0.929348 | 0.397275 | 2.517151 | **20** | **10** | 0.487352 | 0.873206 | 0.558118 | 1.791736 | **50** |
| **50** | 0.371908 | 0.928270 | 0.400646 | 2.495966 | **10** | **20** | 0.489890 | 0.871784 | 0.561939 | 1.779552 | **40** |
| **22° 0′** | 0.374607 | 0.927184 | 0.404026 | 2.475087 | **68° 0′** | **30** | 0.492424 | 0.870356 | 0.565773 | 1.767494 | **30** |
| **10** | 0.377302 | 0.926090 | 0.407414 | 2.454506 | **50** | **40** | 0.494953 | 0.868920 | 0.569619 | 1.755559 | **20** |
| **20** | 0.379994 | 0.924989 | 0.410810 | 2.434217 | **40** | **50** | 0.497479 | 0.867476 | 0.573478 | 1.743745 | **10** |
| **22° 30** | 0.382683 | 0.923880 | 0.414214 | 2.414214 | **67° 30** | **30° 0′** | 0.500000 | 0.866025 | 0.577350 | 1.732051 | **60° 0′** |
| | cos | sin | cot | tan | Angle | | cos | sin | cot | tan | Angle |

For angles 15° to 30° 0′ (angles found in a column to the left of the data), use the column labels at the top of the table; for angles 60° to 75° 0′ (angles found in a column to the right of the data), use the column labels at the bottom of the table.

## Table 2c. Trigonometric Values of Angles from 30° to 60°

| Angle | sin | cos | tan | cot | | Angle | sin | cos | tan | cot | |
|---|---|---|---|---|---|---|---|---|---|---|---|
| **30° 0′** | 0.500000 | 0.866025 | 0.577350 | 1.732051 | **60° 0′** | **37° 30′** | 0.608761 | 0.793353 | 0.767327 | 1.303225 | **52° 30′** |
| **10** | 0.502517 | 0.864567 | 0.581235 | 1.720474 | **50** | **40** | 0.611067 | 0.791579 | 0.771959 | 1.295406 | **20** |
| **20** | 0.505030 | 0.863102 | 0.585134 | 1.709012 | **40** | **50** | 0.613367 | 0.789798 | 0.776612 | 1.287645 | **10** |
| **30** | 0.507538 | 0.861629 | 0.589045 | 1.697663 | **30** | **38° 0′** | 0.615661 | 0.788011 | 0.781286 | 1.279942 | **52° 0′** |
| **40** | 0.510043 | 0.860149 | 0.592970 | 1.686426 | **20** | **10** | 0.617951 | 0.786217 | 0.785981 | 1.272296 | **50** |
| **50** | 0.512543 | 0.858662 | 0.596908 | 1.675299 | **10** | **20** | 0.620235 | 0.784416 | 0.790697 | 1.264706 | **40** |
| **31° 0′** | 0.515038 | 0.857167 | 0.600861 | 1.664279 | **59° 0′** | **30** | 0.622515 | 0.782608 | 0.795436 | 1.257172 | **30** |
| **10** | 0.517529 | 0.855665 | 0.604827 | 1.653366 | **50** | **40** | 0.624789 | 0.780794 | 0.800196 | 1.249693 | **20** |
| **20** | 0.520016 | 0.854156 | 0.608807 | 1.642558 | **40** | **50** | 0.627057 | 0.778973 | 0.804979 | 1.242268 | **10** |
| **30** | 0.522499 | 0.852640 | 0.612801 | 1.631852 | **30** | **39° 0′** | 0.629320 | 0.777146 | 0.809784 | 1.234897 | **51° 0′** |
| **40** | 0.524977 | 0.851117 | 0.616809 | 1.621247 | **20** | **10** | 0.631578 | 0.775312 | 0.814612 | 1.227579 | **50** |
| **50** | 0.527450 | 0.849586 | 0.620832 | 1.610742 | **10** | **20** | 0.633831 | 0.773472 | 0.819463 | 1.220312 | **40** |
| **32° 0′** | 0.529919 | 0.848048 | 0.624869 | 1.600335 | **58° 0′** | **30** | 0.636078 | 0.771625 | 0.824336 | 1.213097 | **30** |
| **10** | 0.532384 | 0.846503 | 0.628921 | 1.590024 | **50** | **40** | 0.638320 | 0.769771 | 0.829234 | 1.205933 | **20** |
| **20** | 0.534844 | 0.844951 | 0.632988 | 1.579808 | **40** | **50** | 0.640557 | 0.767911 | 0.834155 | 1.198818 | **10** |
| **30** | 0.537300 | 0.843391 | 0.637070 | 1.569686 | **30** | **40° 0′** | 0.642788 | 0.766044 | 0.839100 | 1.191754 | **50° 0′** |
| **40** | 0.539751 | 0.841825 | 0.641167 | 1.559655 | **20** | **10** | 0.645013 | 0.764171 | 0.844069 | 1.184738 | **50** |
| **50** | 0.542197 | 0.840251 | 0.645280 | 1.549715 | **10** | **20** | 0.647233 | 0.762292 | 0.849062 | 1.177770 | **40** |
| **33° 0′** | 0.544639 | 0.838671 | 0.649408 | 1.539865 | **57° 0′** | **30** | 0.649448 | 0.760406 | 0.854081 | 1.170850 | **30** |
| **10** | 0.547076 | 0.837083 | 0.653551 | 1.530102 | **50** | **40** | 0.651657 | 0.758514 | 0.859124 | 1.163976 | **20** |
| **20** | 0.549509 | 0.835488 | 0.657710 | 1.520426 | **40** | **50** | 0.653861 | 0.756615 | 0.864193 | 1.157149 | **10** |
| **30** | 0.551937 | 0.833886 | 0.661886 | 1.510835 | **30** | **41° 0′** | 0.656059 | 0.754710 | 0.869287 | 1.150368 | **49° 0′** |
| **40** | 0.554360 | 0.832277 | 0.666077 | 1.501328 | **20** | **10** | 0.658252 | 0.752798 | 0.874407 | 1.143633 | **50** |
| **50** | 0.556779 | 0.830661 | 0.670284 | 1.491904 | **10** | **20** | 0.660439 | 0.750880 | 0.879553 | 1.136941 | **40** |
| **34° 0′** | 0.559193 | 0.829038 | 0.674509 | 1.482561 | **56° 0′** | **30** | 0.662620 | 0.748956 | 0.884725 | 1.130294 | **30** |
| **10** | 0.561602 | 0.827407 | 0.678749 | 1.473298 | **50** | **40** | 0.664796 | 0.747025 | 0.889924 | 1.123691 | **20** |
| **20** | 0.564007 | 0.825770 | 0.683007 | 1.464115 | **40** | **50** | 0.666966 | 0.745088 | 0.895151 | 1.117130 | **10** |
| **30** | 0.566406 | 0.824126 | 0.687281 | 1.455009 | **30** | **42° 0′** | 0.669131 | 0.743145 | 0.900404 | 1.110613 | **48° 0′** |
| **40** | 0.568801 | 0.822475 | 0.691572 | 1.445980 | **20** | **10** | 0.671289 | 0.741195 | 0.905685 | 1.104137 | **50** |
| **50** | 0.571191 | 0.820817 | 0.695881 | 1.437027 | **10** | **20** | 0.673443 | 0.739239 | 0.910994 | 1.097702 | **40** |
| **35° 0′** | 0.573576 | 0.819152 | 0.700208 | 1.428148 | **55° 0′** | **30** | 0.675590 | 0.737277 | 0.916331 | 1.091309 | **30** |
| **10** | 0.575957 | 0.817480 | 0.704551 | 1.419343 | **50** | **40** | 0.677732 | 0.735309 | 0.921697 | 1.084955 | **20** |
| **20** | 0.578332 | 0.815801 | 0.708913 | 1.410610 | **40** | **50** | 0.679868 | 0.733334 | 0.927091 | 1.078642 | **10** |
| **30** | 0.580703 | 0.814116 | 0.713293 | 1.401948 | **30** | **43° 0′** | 0.681998 | 0.731354 | 0.932515 | 1.072369 | **47° 0′** |
| **40** | 0.583069 | 0.812423 | 0.717691 | 1.393357 | **20** | **10** | 0.684123 | 0.729367 | 0.937968 | 1.066134 | **50** |
| **50** | 0.585429 | 0.810723 | 0.722108 | 1.384835 | **10** | **20** | 0.686242 | 0.727374 | 0.943451 | 1.059938 | **40** |
| **36° 0′** | 0.587785 | 0.809017 | 0.726543 | 1.376382 | **54° 0′** | **30** | 0.688355 | 0.725374 | 0.948965 | 1.053780 | **30** |
| **10** | 0.590136 | 0.807304 | 0.730996 | 1.367996 | **50** | **40** | 0.690462 | 0.723369 | 0.954508 | 1.047660 | **20** |
| **20** | 0.592482 | 0.805584 | 0.735469 | 1.359676 | **40** | **50** | 0.692563 | 0.721357 | 0.960083 | 1.041577 | **10** |
| **30** | 0.594823 | 0.803857 | 0.739961 | 1.351422 | **30** | **44° 0′** | 0.694658 | 0.719340 | 0.965689 | 1.035530 | **46° 0′** |
| **40** | 0.597159 | 0.802123 | 0.744472 | 1.343233 | **20** | **10** | 0.696748 | 0.717316 | 0.971326 | 1.029520 | **50** |
| **50** | 0.599489 | 0.800383 | 0.749003 | 1.335108 | **10** | **20** | 0.698832 | 0.715286 | 0.976996 | 1.023546 | **40** |
| **37° 0′** | 0.601815 | 0.798636 | 0.753554 | 1.327045 | **53° 0′** | **30** | 0.700909 | 0.713250 | 0.982697 | 1.017607 | **30** |
| **10** | 0.604136 | 0.796882 | 0.758125 | 1.319044 | **50** | **40** | 0.702981 | 0.711209 | 0.988432 | 1.011704 | **20** |
| **20** | 0.606451 | 0.795121 | 0.762716 | 1.311105 | **40** | **50** | 0.705047 | 0.709161 | 0.994199 | 1.005835 | **10** |
| **37° 30** | 0.608761 | 0.793353 | 0.767327 | 1.303225 | **52° 30** | **45° 0′** | 0.707107 | 0.707107 | 1.000000 | 1.000000 | **45° 0′** |
| | cos | sin | cot | tan | Angle | | cos | sin | cot | tan | Angle |

For angles 30° to 45° 0′ (angles found in a column to the left of the data), use the column labels at the top of the table; for angles 45° to 60° 0′ (angles found in a column to the right of the data), use the column labels at the bottom of the table.

**Using a Calculator to Find Trigonometric Function Values.—**Scientific calculators are quicker and more accurate than tables for finding trigonometric ratios or angles than relying on trigonometric tables. Inputting an angle, in either degree (**DEG**) or radian (**RAD**) measure, and pressing **SIN**, **COS**, or **TAN** key will produce the ratio value to a many decimal places, which can be rounded to the desired accuracy. Though reciprocal function keys are not usually included on the calculator, using the main three functions and the **1/x** key will produce these ratios as well, since $\csc x = 1/\sin x$, $\sec x = 1/\cos x$, and $\cot x = 1/\tan x$.

If the triangle's side dimensions are known and the angle measure is sought, then ratio is entered using the keys labeled **sin**$^{-1}$, **cos**$^{-1}$, and **tan**$^{-1}$. If these keys are not present, then **INV** is used with **SIN**, **COS**, or **TAN**. Again, the correct units, whether degree or radian measure, can be chosen.

An advantage of using a calculator instead of a trigonometry table to find function values is that both positive and negative degree measures can be entered into a calculator; also, angles greater than 90 degrees do not need to be converted to their acute angle equivalent. Interpolation for angles whose measures fall between the values available in the tables is also not necessary.

*Example:* Use a calculator to find all six of the trigonometry function values, to four decimal places, of 172°.

*Solution:* The degree measure can be entered as is, without having to first inspect its quadrant and reference (acute) angle. Enter 172 in DEG mode, then, for the three main values, choose SIN, COS, TAN, each time rounding answer to four decimal places.

$$\sin(172°) = 0.1392 \qquad \csc(172°) = 1/\sin(172°) = 7.1853$$
$$\cos(172°) = -0.9903 \qquad \sec(172°) = 1/\cos(172°) = -1.0098$$
$$\tan(172°) = -0.1405 \qquad \cot(172°) = 1/\tan(172°) = -7.1153$$

If a scientific calculator or computer is not available, tables are the easiest way to find trig values. However, trigonometric function values can be calculated very accurately without a scientific calculator by using the infinite series formulas (see *CALCULUS*):

$$\sin A = A - \frac{A^3}{3!} + \frac{A^5}{5!} - \frac{A^7}{7!} \pm \cdots \qquad \cos A = 1 - \frac{A^2}{2!} + \frac{A^4}{4!} - \frac{A^6}{6!} \pm \cdots$$

$$\sin^{-1} A = A + \frac{1}{2} \times \frac{A^3}{3} + \frac{1}{2} \times \frac{3}{4} \times \frac{A^5}{5} + \cdots \qquad \tan^{-1} A = A - \frac{A^3}{3} + \frac{A^5}{5} - \frac{A^7}{7} \pm \cdots$$

where angle $A$ is expressed in radians (multiplying degrees by $\pi/180 = 0.0174533$ gives radian measure). Generally, calculating just three or four terms of the expression is sufficient for accuracy. In these formulas, a number followed by the symbol **!** is called a *factorial* (see *Factorial Notation* on page 13). Except for 0!, which equals 1, $n! = n(n-1)(n-2) \ldots$ down to 1. For example, $4! = 4 \times 3 \times 2 \times 1 = 24$. As an example, $\sin 42° = \sin(42 \times 0.0174533) = \sin(0.733) = 0.733 - (0.733)^3/3! + (0.733)^5/5! - \ldots \approx 0.66912$, which is close to the calculator answer of higher accuracy, 0.669130606.

**Versed Sine and Versed Cosine.—**These functions are sometimes used in formulas for segments of a circle and may be obtained using the relationships:

$$\text{versed } \sin\theta = 1 - \cos\theta; \qquad \text{versed } \cos\theta = 1 - \sin\theta.$$

**Sevolute Functions.—**Sevolute functions are used in calculating the form diameter of involute splines. They are computed by subtracting the involute function of an angle from the secant of the angle ($1/\cos\theta = \sec\theta$). For example, sevolute of 20° = secant of 20° – involute function of 20° = 1.064178 – 0.014904 = 1.049274.

**Involute Functions.—**Involute functions are used in certain formulas relating to the design and measurement of gear teeth as well as measurement of threads over wires. (See, for example, pages 2133 through 2136, 2288, and 2352).

The value of an involute function is calculated from the following formulas:

$$\text{Involute of } \theta = (\tan\theta) - \theta \qquad \text{for } \theta \text{ in radians, and}$$
$$\text{Involute of } \theta = (\tan\theta) - \pi \times \theta/180 \qquad \text{for } \theta \text{ in degrees.}$$

*Example:* For an angle of 14 degrees and 10 minutes (14° 10′), the involute function is found as follows: 10 minutes = 10/60 = 0.166666°, 14 + 0.166666 = 14.166666°, so that the involute of 14.166666° = (tan 14.166666) – $\pi$ × 14.166666/180 = 0.252420 – 0.247255 = 0.005165.

The same result would be obtained by using the conversion tables beginning on page 107 to convert 14° 10′ to radians and then applying the first involute formula for radian measure given above.

## Involute Function Values for Angles from 14 to 23 Degrees

| | Degrees | | | | | | | | |
|---|---|---|---|---|---|---|---|---|---|
| | 14 | 15 | 16 | 17 | 18 | 19 | 20 | 21 | 22 |
| Minutes | Involute Function Values | | | | | | | | |
| 0 | 0.004982 | 0.006150 | 0.007493 | 0.009025 | 0.010760 | 0.012715 | 0.014904 | 0.017345 | 0.020054 |
| 1 | 0.005000 | 0.006171 | 0.007517 | 0.009052 | 0.010791 | 0.012750 | 0.014943 | 0.017388 | 0.020101 |
| 2 | 0.005018 | 0.006192 | 0.007541 | 0.009079 | 0.010822 | 0.012784 | 0.014982 | 0.017431 | 0.020149 |
| 3 | 0.005036 | 0.006213 | 0.007565 | 0.009107 | 0.010853 | 0.012819 | 0.015020 | 0.017474 | 0.020197 |
| 4 | 0.005055 | 0.006234 | 0.007589 | 0.009134 | 0.010884 | 0.012854 | 0.015059 | 0.017517 | 0.020244 |
| 5 | 0.005073 | 0.006255 | 0.007613 | 0.009161 | 0.010915 | 0.012888 | 0.015098 | 0.017560 | 0.020292 |
| 6 | 0.005091 | 0.006276 | 0.007637 | 0.009189 | 0.010946 | 0.012923 | 0.015137 | 0.017603 | 0.020340 |
| 7 | 0.005110 | 0.006297 | 0.007661 | 0.009216 | 0.010977 | 0.012958 | 0.015176 | 0.017647 | 0.020388 |
| 8 | 0.005128 | 0.006318 | 0.007686 | 0.009244 | 0.011008 | 0.012993 | 0.015215 | 0.017690 | 0.020436 |
| 9 | 0.005146 | 0.006340 | 0.007710 | 0.009272 | 0.011039 | 0.013028 | 0.015254 | 0.017734 | 0.020484 |
| 10 | 0.005165 | 0.006361 | 0.007735 | 0.009299 | 0.011071 | 0.013063 | 0.015293 | 0.017777 | 0.020533 |
| 11 | 0.005184 | 0.006382 | 0.007759 | 0.009327 | 0.011102 | 0.013098 | 0.015333 | 0.017821 | 0.020581 |
| 12 | 0.005202 | 0.006404 | 0.007784 | 0.009355 | 0.011133 | 0.013134 | 0.015372 | 0.017865 | 0.020629 |
| 13 | 0.005221 | 0.006425 | 0.007808 | 0.009383 | 0.011165 | 0.013169 | 0.015411 | 0.017908 | 0.020678 |
| 14 | 0.005239 | 0.006447 | 0.007833 | 0.009411 | 0.011196 | 0.013204 | 0.015451 | 0.017952 | 0.020726 |
| 15 | 0.005258 | 0.006469 | 0.007857 | 0.009439 | 0.011228 | 0.013240 | 0.015490 | 0.017996 | 0.020775 |
| 16 | 0.005277 | 0.006490 | 0.007882 | 0.009467 | 0.011260 | 0.013275 | 0.015530 | 0.018040 | 0.020824 |
| 17 | 0.005296 | 0.006512 | 0.007907 | 0.009495 | 0.011291 | 0.013311 | 0.015570 | 0.018084 | 0.020873 |
| 18 | 0.005315 | 0.006534 | 0.007932 | 0.009523 | 0.011323 | 0.013346 | 0.015609 | 0.018129 | 0.020921 |
| 19 | 0.005334 | 0.006555 | 0.007957 | 0.009552 | 0.011355 | 0.013382 | 0.015649 | 0.018173 | 0.020970 |
| 20 | 0.005353 | 0.006577 | 0.007982 | 0.009580 | 0.011387 | 0.013418 | 0.015689 | 0.018217 | 0.021019 |
| 21 | 0.005372 | 0.006599 | 0.008007 | 0.009608 | 0.011419 | 0.013454 | 0.015729 | 0.018262 | 0.021069 |
| 22 | 0.005391 | 0.006621 | 0.008032 | 0.009637 | 0.011451 | 0.013490 | 0.015769 | 0.018306 | 0.021118 |
| 23 | 0.005410 | 0.006643 | 0.008057 | 0.009665 | 0.011483 | 0.013526 | 0.015809 | 0.018351 | 0.021167 |
| 24 | 0.005429 | 0.006665 | 0.008082 | 0.009694 | 0.011515 | 0.013562 | 0.015850 | 0.018395 | 0.021217 |
| 25 | 0.005448 | 0.006687 | 0.008107 | 0.009722 | 0.011547 | 0.013598 | 0.015890 | 0.018440 | 0.021266 |
| 26 | 0.005467 | 0.006709 | 0.008133 | 0.009751 | 0.011580 | 0.013634 | 0.015930 | 0.018485 | 0.021316 |
| 27 | 0.005487 | 0.006732 | 0.008158 | 0.009780 | 0.011612 | 0.013670 | 0.015971 | 0.018530 | 0.021365 |
| 28 | 0.005506 | 0.006754 | 0.008183 | 0.009808 | 0.011644 | 0.013707 | 0.016011 | 0.018575 | 0.021415 |
| 29 | 0.005525 | 0.006776 | 0.008209 | 0.009837 | 0.011677 | 0.013743 | 0.016052 | 0.018620 | 0.021465 |
| 30 | 0.005545 | 0.006799 | 0.008234 | 0.009866 | 0.011709 | 0.013779 | 0.016092 | 0.018665 | 0.021514 |
| 31 | 0.005564 | 0.006821 | 0.008260 | 0.009895 | 0.011742 | 0.013816 | 0.016133 | 0.018710 | 0.021564 |
| 32 | 0.005584 | 0.006843 | 0.008285 | 0.009924 | 0.011775 | 0.013852 | 0.016174 | 0.018755 | 0.021614 |
| 33 | 0.005603 | 0.006866 | 0.008311 | 0.009953 | 0.011807 | 0.013889 | 0.016215 | 0.018800 | 0.021665 |
| 34 | 0.005623 | 0.006888 | 0.008337 | 0.009982 | 0.011840 | 0.013926 | 0.016255 | 0.018846 | 0.021715 |
| 35 | 0.005643 | 0.006911 | 0.008362 | 0.010011 | 0.011873 | 0.013963 | 0.016296 | 0.018891 | 0.021765 |
| 36 | 0.005662 | 0.006934 | 0.008388 | 0.010041 | 0.011906 | 0.013999 | 0.016337 | 0.018937 | 0.021815 |
| 37 | 0.005682 | 0.006956 | 0.008414 | 0.010070 | 0.011939 | 0.014036 | 0.016379 | 0.018983 | 0.021866 |
| 38 | 0.005702 | 0.006979 | 0.008440 | 0.010099 | 0.011972 | 0.014073 | 0.016420 | 0.019028 | 0.021916 |
| 39 | 0.005722 | 0.007002 | 0.008466 | 0.010129 | 0.012005 | 0.014110 | 0.016461 | 0.019074 | 0.021967 |
| 40 | 0.005742 | 0.007025 | 0.008492 | 0.010158 | 0.012038 | 0.014148 | 0.016502 | 0.019120 | 0.022018 |
| 41 | 0.005762 | 0.007048 | 0.008518 | 0.010188 | 0.012071 | 0.014185 | 0.016544 | 0.019166 | 0.022068 |
| 42 | 0.005782 | 0.007071 | 0.008544 | 0.010217 | 0.012105 | 0.014222 | 0.016585 | 0.019212 | 0.022119 |
| 43 | 0.005802 | 0.007094 | 0.008571 | 0.010247 | 0.012138 | 0.014259 | 0.016627 | 0.019258 | 0.022170 |
| 44 | 0.005822 | 0.007117 | 0.008597 | 0.010277 | 0.012172 | 0.014297 | 0.016669 | 0.019304 | 0.022221 |
| 45 | 0.005842 | 0.007140 | 0.008623 | 0.010307 | 0.012205 | 0.014334 | 0.016710 | 0.019350 | 0.022272 |
| 46 | 0.005862 | 0.007163 | 0.008650 | 0.010336 | 0.012239 | 0.014372 | 0.016752 | 0.019397 | 0.022324 |
| 47 | 0.005882 | 0.007186 | 0.008676 | 0.010366 | 0.012272 | 0.014409 | 0.016794 | 0.019443 | 0.022375 |
| 48 | 0.005903 | 0.007209 | 0.008702 | 0.010396 | 0.012306 | 0.014447 | 0.016836 | 0.019490 | 0.022426 |
| 49 | 0.005923 | 0.007233 | 0.008729 | 0.010426 | 0.012340 | 0.014485 | 0.016878 | 0.019536 | 0.022478 |
| 50 | 0.005943 | 0.007256 | 0.008756 | 0.010456 | 0.012373 | 0.014523 | 0.016920 | 0.019583 | 0.022529 |
| 51 | 0.005964 | 0.007280 | 0.008782 | 0.010486 | 0.012407 | 0.014560 | 0.016962 | 0.019630 | 0.022581 |
| 52 | 0.005984 | 0.007303 | 0.008809 | 0.010517 | 0.012441 | 0.014598 | 0.017004 | 0.019676 | 0.022633 |
| 53 | 0.006005 | 0.007327 | 0.008836 | 0.010547 | 0.012475 | 0.014636 | 0.017047 | 0.019723 | 0.022684 |
| 54 | 0.006025 | 0.007350 | 0.008863 | 0.010577 | 0.012509 | 0.014674 | 0.017089 | 0.019770 | 0.022736 |
| 55 | 0.006046 | 0.007374 | 0.008889 | 0.010608 | 0.012543 | 0.014713 | 0.017132 | 0.019817 | 0.022788 |
| 56 | 0.006067 | 0.007397 | 0.008916 | 0.010638 | 0.012578 | 0.014751 | 0.017174 | 0.019864 | 0.022840 |
| 57 | 0.006087 | 0.007421 | 0.008943 | 0.010669 | 0.012612 | 0.014789 | 0.017217 | 0.019912 | 0.022892 |
| 58 | 0.006108 | 0.007445 | 0.008970 | 0.010699 | 0.012646 | 0.014827 | 0.017259 | 0.019959 | 0.022944 |
| 59 | 0.006129 | 0.007469 | 0.008998 | 0.010730 | 0.012681 | 0.014866 | 0.017302 | 0.020006 | 0.022997 |
| 60 | 0.006150 | 0.007493 | 0.009025 | 0.010760 | 0.012715 | 0.014904 | 0.017345 | 0.020054 | 0.023049 |

## Involute Function Values for Angles from 23 to 32 Degrees

| Minutes | Degrees | | | | | | | | |
|---|---|---|---|---|---|---|---|---|---|
| | 23 | 24 | 25 | 26 | 27 | 28 | 29 | 30 | 31 |
| | Involute Function Values | | | | | | | | |
| 0 | 0.023049 | 0.026350 | 0.029975 | 0.033947 | 0.038287 | 0.043017 | 0.048164 | 0.053752 | 0.059809 |
| 1 | 0.023102 | 0.026407 | 0.030039 | 0.034016 | 0.038362 | 0.043100 | 0.048253 | 0.053849 | 0.059914 |
| 2 | 0.023154 | 0.026465 | 0.030102 | 0.034086 | 0.038438 | 0.043182 | 0.048343 | 0.053946 | 0.060019 |
| 3 | 0.023207 | 0.026523 | 0.030166 | 0.034155 | 0.038514 | 0.043264 | 0.048432 | 0.054043 | 0.060124 |
| 4 | 0.023259 | 0.026581 | 0.030229 | 0.034225 | 0.038590 | 0.043347 | 0.048522 | 0.054140 | 0.060230 |
| 5 | 0.023312 | 0.026639 | 0.030293 | 0.034294 | 0.038666 | 0.043430 | 0.048612 | 0.054238 | 0.060335 |
| 6 | 0.023365 | 0.026697 | 0.030357 | 0.034364 | 0.038742 | 0.043513 | 0.048702 | 0.054336 | 0.060441 |
| 7 | 0.023418 | 0.026756 | 0.030420 | 0.034434 | 0.038818 | 0.043596 | 0.048792 | 0.054433 | 0.060547 |
| 8 | 0.023471 | 0.026814 | 0.030484 | 0.034504 | 0.038894 | 0.043679 | 0.048883 | 0.054531 | 0.060653 |
| 9 | 0.023524 | 0.026872 | 0.030549 | 0.034574 | 0.038971 | 0.043762 | 0.048973 | 0.054629 | 0.060759 |
| 10 | 0.023577 | 0.026931 | 0.030613 | 0.034644 | 0.039047 | 0.043845 | 0.049064 | 0.054728 | 0.060866 |
| 11 | 0.023631 | 0.026989 | 0.030677 | 0.034714 | 0.039124 | 0.043929 | 0.049154 | 0.054826 | 0.060972 |
| 12 | 0.023684 | 0.027048 | 0.030741 | 0.034785 | 0.039201 | 0.044012 | 0.049245 | 0.054924 | 0.061079 |
| 13 | 0.023738 | 0.027107 | 0.030806 | 0.034855 | 0.039278 | 0.044096 | 0.049336 | 0.055023 | 0.061186 |
| 14 | 0.023791 | 0.027166 | 0.030870 | 0.034926 | 0.039355 | 0.044180 | 0.049427 | 0.055122 | 0.061292 |
| 15 | 0.023845 | 0.027225 | 0.030935 | 0.034997 | 0.039432 | 0.044264 | 0.049518 | 0.055221 | 0.061400 |
| 16 | 0.023899 | 0.027284 | 0.031000 | 0.035067 | 0.039509 | 0.044348 | 0.049609 | 0.055320 | 0.061507 |
| 17 | 0.023952 | 0.027343 | 0.031065 | 0.035138 | 0.039586 | 0.044432 | 0.049701 | 0.055419 | 0.061614 |
| 18 | 0.024006 | 0.027402 | 0.031130 | 0.035209 | 0.039664 | 0.044516 | 0.049792 | 0.055518 | 0.061721 |
| 19 | 0.024060 | 0.027462 | 0.031195 | 0.035280 | 0.039741 | 0.044601 | 0.049884 | 0.055617 | 0.061829 |
| 20 | 0.024114 | 0.027521 | 0.031260 | 0.035352 | 0.039819 | 0.044685 | 0.049976 | 0.055717 | 0.061937 |
| 21 | 0.024169 | 0.027581 | 0.031325 | 0.035423 | 0.039897 | 0.044770 | 0.050068 | 0.055817 | 0.062045 |
| 22 | 0.024223 | 0.027640 | 0.031390 | 0.035494 | 0.039974 | 0.044855 | 0.050160 | 0.055916 | 0.062153 |
| 23 | 0.024277 | 0.027700 | 0.031456 | 0.035566 | 0.040052 | 0.044940 | 0.050252 | 0.056016 | 0.062261 |
| 24 | 0.024332 | 0.027760 | 0.031521 | 0.035637 | 0.040131 | 0.045024 | 0.050344 | 0.056116 | 0.062369 |
| 25 | 0.024386 | 0.027820 | 0.031587 | 0.035709 | 0.040209 | 0.045110 | 0.050437 | 0.056217 | 0.062478 |
| 26 | 0.024441 | 0.027880 | 0.031653 | 0.035781 | 0.040287 | 0.045195 | 0.050529 | 0.056317 | 0.062586 |
| 27 | 0.024495 | 0.027940 | 0.031718 | 0.035853 | 0.040366 | 0.045280 | 0.050622 | 0.056417 | 0.062695 |
| 28 | 0.024550 | 0.028000 | 0.031784 | 0.035925 | 0.040444 | 0.045366 | 0.050715 | 0.056518 | 0.062804 |
| 29 | 0.024605 | 0.028060 | 0.031850 | 0.035997 | 0.040523 | 0.045451 | 0.050808 | 0.056619 | 0.062913 |
| 30 | 0.024660 | 0.028121 | 0.031917 | 0.036069 | 0.040602 | 0.045537 | 0.050901 | 0.056720 | 0.063022 |
| 31 | 0.024715 | 0.028181 | 0.031983 | 0.036142 | 0.040680 | 0.045623 | 0.050994 | 0.056821 | 0.063131 |
| 32 | 0.024770 | 0.028242 | 0.032049 | 0.036214 | 0.040759 | 0.045709 | 0.051087 | 0.056922 | 0.063241 |
| 33 | 0.024825 | 0.028302 | 0.032116 | 0.036287 | 0.040839 | 0.045795 | 0.051181 | 0.057023 | 0.063350 |
| 34 | 0.024881 | 0.028363 | 0.032182 | 0.036359 | 0.040918 | 0.045881 | 0.051274 | 0.057124 | 0.063460 |
| 35 | 0.024936 | 0.028424 | 0.032249 | 0.036432 | 0.040997 | 0.045967 | 0.051368 | 0.057226 | 0.063570 |
| 36 | 0.024992 | 0.028485 | 0.032315 | 0.036505 | 0.041077 | 0.046054 | 0.051462 | 0.057328 | 0.063680 |
| 37 | 0.025047 | 0.028546 | 0.032382 | 0.036578 | 0.041156 | 0.046140 | 0.051556 | 0.057429 | 0.063790 |
| 38 | 0.025103 | 0.028607 | 0.032449 | 0.036651 | 0.041236 | 0.046227 | 0.051650 | 0.057531 | 0.063901 |
| 39 | 0.025159 | 0.028668 | 0.032516 | 0.036724 | 0.041316 | 0.046313 | 0.051744 | 0.057633 | 0.064011 |
| 40 | 0.025214 | 0.028729 | 0.032583 | 0.036798 | 0.041395 | 0.046400 | 0.051838 | 0.057736 | 0.064122 |
| 41 | 0.025270 | 0.028791 | 0.032651 | 0.036871 | 0.041475 | 0.046487 | 0.051933 | 0.057838 | 0.064232 |
| 42 | 0.025326 | 0.028852 | 0.032718 | 0.036945 | 0.041556 | 0.046575 | 0.052027 | 0.057940 | 0.064343 |
| 43 | 0.025382 | 0.028914 | 0.032785 | 0.037018 | 0.041636 | 0.046662 | 0.052122 | 0.058043 | 0.064454 |
| 44 | 0.025439 | 0.028976 | 0.032853 | 0.037092 | 0.041716 | 0.046749 | 0.052217 | 0.058146 | 0.064565 |
| 45 | 0.025495 | 0.029037 | 0.032920 | 0.037166 | 0.041797 | 0.046837 | 0.052312 | 0.058249 | 0.064677 |
| 46 | 0.025551 | 0.029099 | 0.032988 | 0.037240 | 0.041877 | 0.046924 | 0.052407 | 0.058352 | 0.064788 |
| 47 | 0.025608 | 0.029161 | 0.033056 | 0.037314 | 0.041958 | 0.047012 | 0.052502 | 0.058455 | 0.064900 |
| 48 | 0.025664 | 0.029223 | 0.033124 | 0.037388 | 0.042039 | 0.047100 | 0.052597 | 0.058558 | 0.065012 |
| 49 | 0.025721 | 0.029285 | 0.033192 | 0.037462 | 0.042120 | 0.047188 | 0.052693 | 0.058662 | 0.065123 |
| 50 | 0.025778 | 0.029348 | 0.033260 | 0.037537 | 0.042201 | 0.047276 | 0.052788 | 0.058765 | 0.065236 |
| 51 | 0.025834 | 0.029410 | 0.033328 | 0.037611 | 0.042282 | 0.047364 | 0.052884 | 0.058869 | 0.065348 |
| 52 | 0.025891 | 0.029472 | 0.033397 | 0.037686 | 0.042363 | 0.047452 | 0.052980 | .058973 | 0.065460 |
| 53 | 0.025948 | 0.029535 | 0.033465 | 0.037761 | 0.042444 | 0.047541 | 0.053076 | 0.059077 | 0.065573 |
| 54 | 0.026005 | 0.029598 | 0.033534 | 0.037835 | 0.042526 | 0.047630 | 0.053172 | 0.059181 | 0.065685 |
| 55 | 0.026062 | 0.029660 | 0.033602 | 0.037910 | 0.042608 | 0.047718 | 0.053268 | 0.059285 | 0.065798 |
| 56 | 0.026120 | 0.029723 | 0.033671 | 0.037985 | 0.042689 | 0.047807 | 0.053365 | 0.059390 | 0.065911 |
| 57 | 0.026177 | 0.029786 | 0.033740 | 0.038060 | 0.042771 | 0.047896 | 0.053461 | 0.059494 | 0.066024 |
| 58 | 0.026235 | 0.029849 | 0.033809 | 0.038136 | 0.042853 | 0.047985 | 0.053558 | 0.059599 | 0.066137 |
| 59 | 0.026292 | 0.029912 | 0.033878 | 0.038211 | 0.042935 | 0.048074 | 0.053655 | 0.059704 | 0.066251 |
| 60 | 0.026350 | 0.029975 | 0.033947 | 0.038287 | 0.043017 | 0.048164 | 0.053752 | 0.059809 | 0.066364 |

## Involute Function Values for Angles from 32 to 41 Degrees

| | Degrees | | | | | | | | |
|---|---|---|---|---|---|---|---|---|---|
| | 32 | 33 | 34 | 35 | 36 | 37 | 38 | 39 | 40 |
| Minutes | Involute Function Values | | | | | | | | |
| 0 | 0.066364 | 0.073449 | 0.081097 | 0.089342 | 0.098224 | 0.107782 | 0.118061 | 0.129106 | 0.140968 |
| 1 | 0.066478 | 0.073572 | 0.081229 | 0.089485 | 0.098378 | 0.107948 | 0.118238 | 0.129297 | 0.141173 |
| 2 | 0.066591 | 0.073695 | 0.081362 | 0.089628 | 0.098532 | 0.108113 | 0.118416 | 0.129488 | 0.141378 |
| 3 | 0.066705 | 0.073818 | 0.081494 | 0.089771 | 0.098686 | 0.108279 | 0.118594 | 0.129679 | 0.141584 |
| 4 | 0.066820 | 0.073941 | 0.081627 | 0.089914 | 0.098840 | 0.108445 | 0.118773 | 0.129870 | 0.141789 |
| 5 | 0.066934 | 0.074064 | 0.081760 | 0.090058 | 0.098994 | 0.108611 | 0.118951 | 0.130062 | 0.141995 |
| 6 | 0.067048 | 0.074188 | 0.081894 | 0.090201 | 0.099149 | 0.108777 | 0.119130 | 0.130254 | 0.142201 |
| 7 | 0.067163 | 0.074312 | 0.082027 | 0.090345 | 0.099303 | 0.108943 | 0.119309 | 0.130446 | 0.142408 |
| 8 | 0.067277 | 0.074435 | 0.082161 | 0.090489 | 0.099458 | 0.109110 | 0.119488 | 0.130639 | 0.142614 |
| 9 | 0.067392 | 0.074559 | 0.082294 | 0.090633 | 0.099614 | 0.109277 | 0.119667 | 0.130832 | 0.142821 |
| 10 | 0.067507 | 0.074684 | 0.082428 | 0.090777 | 0.099769 | 0.109444 | 0.119847 | 0.131025 | 0.143028 |
| 11 | 0.067622 | 0.074808 | 0.082562 | 0.090922 | 0.099924 | 0.109611 | 0.120027 | 0.131218 | 0.143236 |
| 12 | 0.067738 | 0.074932 | 0.082697 | 0.091067 | 0.100080 | 0.109779 | 0.120207 | 0.131411 | 0.143443 |
| 13 | 0.067853 | 0.075057 | 0.082831 | 0.091211 | 0.100236 | 0.109947 | 0.120387 | 0.131605 | 0.143651 |
| 14 | 0.067969 | 0.075182 | 0.082966 | 0.091356 | 0.100392 | 0.110114 | 0.120567 | 0.131799 | 0.143859 |
| 15 | 0.068084 | 0.075307 | 0.083101 | 0.091502 | 0.100549 | 0.110283 | 0.120748 | 0.131993 | 0.144068 |
| 16 | 0.068200 | 0.075432 | 0.083235 | 0.091647 | 0.100705 | 0.110451 | 0.120929 | 0.132187 | 0.144276 |
| 17 | 0.068316 | 0.075557 | 0.083371 | 0.091793 | 0.100862 | 0.110619 | 0.121110 | 0.132381 | 0.144485 |
| 18 | 0.068432 | 0.075683 | 0.083506 | 0.091938 | 0.101019 | 0.110788 | 0.121291 | 0.132576 | 0.144694 |
| 19 | 0.068549 | 0.075808 | 0.083641 | 0.092084 | 0.101176 | 0.110957 | 0.121473 | 0.132771 | 0.144903 |
| 20 | 0.068665 | 0.075934 | 0.083777 | 0.092230 | 0.101333 | 0.111126 | 0.121655 | 0.132966 | 0.145113 |
| 21 | 0.068782 | 0.076060 | 0.083913 | 0.092377 | 0.101490 | 0.111295 | 0.121837 | 0.133162 | 0.145323 |
| 22 | 0.068899 | 0.076186 | 0.084049 | 0.092523 | 0.101648 | 0.111465 | 0.122019 | 0.133358 | 0.145533 |
| 23 | 0.069016 | 0.076312 | 0.084185 | 0.092670 | 0.101806 | 0.111635 | 0.122201 | 0.133553 | 0.145743 |
| 24 | 0.069133 | 0.076439 | 0.084321 | 0.092816 | 0.101964 | 0.111805 | 0.122384 | 0.133750 | 0.145954 |
| 25 | 0.069250 | 0.076565 | 0.084458 | 0.092963 | 0.102122 | 0.111975 | 0.122567 | 0.133946 | 0.146165 |
| 26 | 0.069367 | 0.076692 | 0.084594 | 0.093111 | 0.102280 | 0.112145 | 0.122750 | 0.134143 | 0.146376 |
| 27 | 0.069485 | 0.076819 | 0.084731 | 0.093258 | 0.102439 | 0.112316 | 0.122933 | 0.134339 | 0.146587 |
| 28 | 0.069602 | 0.076946 | 0.084868 | 0.093406 | 0.102598 | 0.112486 | 0.123117 | 0.134537 | 0.146799 |
| 29 | 0.069720 | 0.077073 | 0.085005 | 0.093553 | 0.102757 | 0.112657 | 0.123300 | 0.134734 | 0.147010 |
| 30 | 0.069838 | 0.077200 | 0.085142 | 0.093701 | 0.102916 | 0.112829 | 0.123484 | 0.134931 | 0.147222 |
| 31 | 0.069956 | 0.077328 | 0.085280 | 0.093849 | 0.103075 | 0.113000 | 0.123668 | 0.135129 | 0.147435 |
| 32 | 0.070075 | 0.077455 | 0.085418 | 0.093998 | 0.103235 | 0.113172 | 0.123853 | 0.135327 | 0.147647 |
| 33 | 0.070193 | 0.077583 | 0.085555 | 0.094146 | 0.103395 | 0.113343 | 0.124037 | 0.135525 | 0.147860 |
| 34 | 0.070312 | 0.077711 | 0.085693 | 0.094295 | 0.103555 | 0.113515 | 0.124222 | 0.135724 | 0.148073 |
| 35 | 0.070430 | 0.077839 | 0.085832 | 0.094443 | 0.103715 | 0.113688 | 0.124407 | 0.135923 | 0.148286 |
| 36 | 0.070549 | 0.077968 | 0.085970 | 0.094593 | 0.103875 | 0.113860 | 0.124592 | 0.136122 | 0.148500 |
| 37 | 0.070668 | 0.078096 | 0.086108 | 0.094742 | 0.104036 | 0.114033 | 0.124778 | 0.136321 | 0.148714 |
| 38 | 0.070788 | 0.078225 | 0.086247 | 0.094891 | 0.104196 | 0.114205 | 0.124964 | 0.136520 | 0.148928 |
| 39 | 0.070907 | 0.078354 | 0.086386 | 0.095041 | 0.104357 | 0.114378 | 0.125150 | 0.136720 | 0.149142 |
| 40 | 0.071026 | 0.078483 | 0.086525 | 0.095190 | 0.104518 | 0.114552 | 0.125336 | 0.136920 | 0.149357 |
| 41 | 0.071146 | 0.078612 | 0.086664 | 0.095340 | 0.104680 | 0.114725 | 0.125522 | 0.137120 | 0.149572 |
| 42 | 0.071266 | 0.078741 | 0.086804 | 0.095490 | 0.104841 | 0.114899 | 0.125709 | 0.137320 | 0.149787 |
| 43 | 0.071386 | 0.078871 | 0.086943 | 0.095641 | 0.105003 | 0.115073 | 0.125896 | 0.137521 | 0.150002 |
| 44 | 0.071506 | 0.079000 | 0.087083 | 0.095791 | 0.105165 | 0.115247 | 0.126083 | 0.137722 | 0.150218 |
| 45 | 0.071626 | 0.079130 | 0.087223 | 0.095942 | 0.105327 | 0.115421 | 0.126270 | 0.137923 | 0.150434 |
| 46 | 0.071747 | 0.079260 | 0.087363 | 0.096093 | 0.105489 | 0.115595 | 0.126457 | 0.138124 | 0.150650 |
| 47 | 0.071867 | 0.079390 | 0.087503 | 0.096244 | 0.105652 | 0.115770 | 0.126645 | 0.138326 | 0.150866 |
| 48 | 0.071988 | 0.079520 | 0.087644 | 0.096395 | 0.105814 | 0.115945 | 0.126833 | 0.138528 | 0.151083 |
| 49 | 0.072109 | 0.079651 | 0.087784 | 0.096546 | 0.105977 | 0.116120 | 0.127021 | 0.138730 | 0.151299 |
| 50 | 0.072230 | 0.079781 | 0.087925 | 0.096698 | 0.106140 | 0.116296 | 0.127209 | 0.138932 | 0.151517 |
| 51 | 0.072351 | 0.079912 | 0.088066 | 0.096850 | 0.106304 | 0.116471 | 0.127398 | 0.139134 | 0.151734 |
| 52 | 0.072473 | 0.080043 | 0.088207 | 0.097002 | 0.106467 | 0.116647 | 0.127587 | 0.139337 | 0.151952 |
| 53 | 0.072594 | 0.080174 | 0.088348 | 0.097154 | 0.106631 | 0.116823 | 0.127776 | 0.139540 | 0.152169 |
| 54 | 0.072716 | 0.080306 | 0.088490 | 0.097306 | 0.106795 | 0.116999 | 0.127965 | 0.139743 | 0.152388 |
| 55 | 0.072838 | 0.080437 | 0.088631 | 0.097459 | 0.106959 | 0.117175 | 0.128155 | 0.139947 | 0.152606 |
| 56 | 0.072960 | 0.080569 | 0.088773 | 0.097611 | 0.107123 | 0.117352 | 0.128344 | 0.140151 | 0.152825 |
| 57 | 0.073082 | 0.080700 | 0.088915 | 0.097764 | 0.107288 | 0.117529 | 0.128534 | 0.140355 | 0.153044 |
| 58 | 0.073204 | 0.080832 | 0.089057 | 0.097917 | 0.107452 | 0.117706 | 0.128725 | 0.140559 | 0.153263 |
| 59 | 0.073326 | 0.080964 | 0.089200 | 0.098071 | 0.107617 | 0.117883 | 0.128915 | 0.140763 | 0.153482 |
| 60 | 0.073449 | 0.081097 | 0.089342 | 0.098224 | 0.107782 | 0.118061 | 0.129106 | 0.140968 | 0.153702 |

## Involute Function Values for Angles from 41 to 50 Degrees

| Minutes | Degrees | | | | | | | | |
|---|---|---|---|---|---|---|---|---|---|
| | 41 | 42 | 43 | 44 | 45 | 46 | 47 | 48 | 49 |
| | Involute Function Values | | | | | | | | |
| 0 | 0.153702 | 0.167366 | 0.182024 | 0.197744 | 0.214602 | 0.232679 | 0.252064 | 0.272855 | 0.295157 |
| 1 | 0.153922 | 0.167602 | 0.182277 | 0.198015 | 0.214893 | 0.232991 | 0.252399 | 0.273214 | 0.295542 |
| 2 | 0.154142 | 0.167838 | 0.182530 | 0.198287 | 0.215184 | 0.233304 | 0.252734 | 0.273573 | 0.295928 |
| 3 | 0.154362 | 0.168075 | 0.182784 | 0.198559 | 0.215476 | 0.233616 | 0.253069 | 0.273933 | 0.296314 |
| 4 | 0.154583 | 0.168311 | 0.183038 | 0.198832 | 0.215768 | 0.233930 | 0.253405 | 0.274293 | 0.296701 |
| 5 | 0.154804 | 0.168548 | 0.183292 | 0.199104 | 0.216061 | 0.234243 | 0.253742 | 0.274654 | 0.297088 |
| 6 | 0.155025 | 0.168786 | 0.183547 | 0.199377 | 0.216353 | 0.234557 | 0.254078 | 0.275015 | 0.297475 |
| 7 | 0.155247 | 0.169023 | 0.183801 | 0.199651 | 0.216646 | 0.234871 | 0.254415 | 0.275376 | 0.297863 |
| 8 | 0.155469 | 0.169261 | 0.184057 | 0.199924 | 0.216940 | 0.235186 | 0.254753 | 0.275738 | 0.298251 |
| 9 | 0.155691 | 0.169500 | 0.184312 | 0.200198 | 0.217234 | 0.235501 | 0.255091 | 0.276101 | 0.298640 |
| 10 | 0.155913 | 0.169738 | 0.184568 | 0.200473 | 0.217528 | 0.235816 | 0.255429 | 0.276464 | 0.299029 |
| 11 | 0.156135 | 0.169977 | 0.184824 | 0.200747 | 0.217822 | 0.236132 | 0.255767 | 0.276827 | 0.299419 |
| 12 | 0.156358 | 0.170216 | 0.185080 | 0.201022 | 0.218117 | 0.236448 | 0.256106 | 0.277191 | 0.299809 |
| 13 | 0.156581 | 0.170455 | 0.185337 | 0.201297 | 0.218412 | 0.236765 | 0.256446 | 0.277555 | 0.300200 |
| 14 | 0.156805 | 0.170695 | 0.185594 | 0.201573 | 0.218708 | 0.237082 | 0.256786 | 0.277919 | 0.300591 |
| 15 | 0.157028 | 0.170935 | 0.185851 | 0.201849 | 0.219004 | 0.237399 | 0.257126 | 0.278284 | 0.300983 |
| 16 | 0.157252 | 0.171175 | 0.186109 | 0.202125 | 0.219300 | 0.237717 | 0.257467 | 0.278649 | 0.301375 |
| 17 | 0.157476 | 0.171415 | 0.186367 | 0.202401 | 0.219596 | 0.238035 | 0.257808 | 0.279015 | 0.301767 |
| 18 | 0.157701 | 0.171656 | 0.186625 | 0.202678 | 0.219893 | 0.238353 | 0.258149 | 0.279381 | 0.302160 |
| 19 | 0.157925 | 0.171897 | 0.186883 | 0.202956 | 0.220190 | 0.238672 | 0.258491 | 0.279748 | 0.302553 |
| 20 | 0.158150 | 0.172138 | 0.187142 | 0.203233 | 0.220488 | 0.238991 | 0.258833 | 0.280115 | 0.302947 |
| 21 | 0.158375 | 0.172380 | 0.187401 | 0.203511 | 0.220786 | 0.239310 | 0.259176 | 0.280483 | 0.303342 |
| 22 | 0.158601 | 0.172621 | 0.187661 | 0.203789 | 0.221084 | 0.239630 | 0.259519 | 0.280851 | 0.303736 |
| 23 | 0.158826 | 0.172864 | 0.187920 | 0.204067 | 0.221383 | 0.239950 | 0.259862 | 0.281219 | 0.304132 |
| 24 | 0.159052 | 0.173106 | 0.188180 | 0.204346 | 0.221682 | 0.240271 | 0.260206 | 0.281588 | 0.304527 |
| 25 | 0.159279 | 0.173349 | 0.188440 | 0.204625 | 0.221981 | 0.240592 | 0.260550 | 0.281957 | 0.304924 |
| 26 | 0.159505 | 0.173592 | 0.188701 | 0.204905 | 0.222281 | 0.240913 | 0.260895 | 0.282327 | 0.305320 |
| 27 | 0.159732 | 0.173835 | 0.188962 | 0.205185 | 0.222581 | 0.241235 | 0.261240 | 0.282697 | 0.305718 |
| 28 | 0.159959 | 0.174078 | 0.189223 | 0.205465 | 0.222881 | 0.241557 | 0.261585 | 0.283067 | 0.306115 |
| 29 | 0.160186 | 0.174322 | 0.189485 | 0.205745 | 0.223182 | 0.241879 | 0.261931 | 0.283438 | 0.306513 |
| 30 | 0.160414 | 0.174566 | 0.189746 | 0.206026 | 0.223483 | 0.242202 | 0.262277 | 0.283810 | 0.306912 |
| 31 | 0.160642 | 0.174811 | 0.190009 | 0.206307 | 0.223784 | 0.242525 | 0.262624 | 0.284182 | 0.307311 |
| 32 | 0.160870 | 0.175055 | 0.190271 | 0.206588 | 0.224086 | 0.242849 | 0.262971 | 0.284554 | 0.307710 |
| 33 | 0.161098 | 0.175300 | 0.190534 | 0.206870 | 0.224388 | 0.243173 | 0.263318 | 0.284927 | 0.308110 |
| 34 | 0.161327 | 0.175546 | 0.190797 | 0.207152 | 0.224690 | 0.243497 | 0.263666 | 0.285300 | 0.308511 |
| 35 | 0.161555 | 0.175791 | 0.191060 | 0.207434 | 0.224993 | 0.243822 | 0.264014 | 0.285673 | 0.308911 |
| 36 | 0.161785 | 0.176037 | 0.191324 | 0.207717 | 0.225296 | 0.244147 | 0.264363 | 0.286047 | 0.309313 |
| 37 | 0.162014 | 0.176283 | 0.191588 | 0.208000 | 0.225600 | 0.244472 | 0.264712 | 0.286422 | 0.309715 |
| 38 | 0.162244 | 0.176529 | 0.191852 | 0.208284 | 0.225904 | 0.244798 | 0.265062 | 0.286797 | 0.310117 |
| 39 | 0.162474 | 0.176776 | 0.192116 | 0.208567 | 0.226208 | 0.245125 | 0.265412 | 0.287172 | 0.310520 |
| 40 | 0.162704 | 0.177023 | 0.192381 | 0.208851 | 0.226512 | 0.245451 | 0.265762 | 0.287548 | 0.310923 |
| 41 | 0.162934 | 0.177270 | 0.192646 | 0.209136 | 0.226817 | 0.245778 | 0.266113 | 0.287924 | 0.311327 |
| 42 | 0.163165 | 0.177518 | 0.192912 | 0.209420 | 0.227123 | 0.246106 | 0.266464 | 0.288301 | 0.311731 |
| 43 | 0.163396 | 0.177766 | 0.193178 | 0.209705 | 0.227428 | 0.246433 | 0.266815 | 0.288678 | 0.312136 |
| 44 | 0.163628 | 0.178014 | 0.193444 | 0.209991 | 0.227734 | 0.246761 | 0.267167 | 0.289056 | 0.312541 |
| 45 | 0.163859 | 0.178262 | 0.193710 | 0.210276 | 0.228041 | 0.247090 | 0.267520 | 0.289434 | 0.312947 |
| 46 | 0.164091 | 0.178511 | 0.193977 | 0.210562 | 0.228347 | 0.247419 | 0.267872 | 0.289812 | 0.313353 |
| 47 | 0.164323 | 0.178760 | 0.194244 | 0.210849 | 0.228654 | 0.247748 | 0.268225 | 0.290191 | 0.313759 |
| 48 | 0.164556 | 0.179009 | 0.194511 | 0.211136 | 0.228962 | 0.248078 | 0.268579 | 0.290570 | 0.314166 |
| 49 | 0.164788 | 0.179259 | 0.194779 | 0.211423 | 0.229270 | 0.248408 | 0.268933 | 0.290950 | 0.314574 |
| 50 | 0.165021 | 0.179509 | 0.195047 | 0.211710 | 0.229578 | 0.248738 | 0.269287 | 0.291330 | 0.314982 |
| 51 | 0.165254 | 0.179759 | 0.195315 | 0.211998 | 0.229886 | 0.249069 | 0.269642 | 0.291711 | 0.315391 |
| 52 | 0.165488 | 0.180009 | 0.195584 | 0.212286 | 0.230195 | 0.249400 | 0.269998 | 0.292092 | 0.315800 |
| 53 | 0.165722 | 0.180260 | 0.195853 | 0.212574 | 0.230504 | 0.249732 | 0.270353 | 0.292474 | 0.316209 |
| 54 | 0.165956 | 0.180511 | 0.196122 | 0.212863 | 0.230814 | 0.250064 | 0.270709 | 0.292856 | 0.316619 |
| 55 | 0.166190 | 0.180763 | 0.196392 | 0.213152 | 0.231124 | 0.250396 | 0.271066 | 0.293238 | 0.317029 |
| 56 | 0.166425 | 0.181014 | 0.196661 | 0.213441 | 0.231434 | 0.250729 | 0.271423 | 0.293621 | 0.317440 |
| 57 | 0.166660 | 0.181266 | 0.196932 | 0.213731 | 0.231745 | 0.251062 | 0.271780 | 0.294004 | 0.317852 |
| 58 | 0.166895 | 0.181518 | 0.197202 | 0.214021 | 0.232056 | 0.251396 | 0.272138 | 0.294388 | 0.318264 |
| 59 | 0.167130 | 0.181771 | 0.197473 | 0.214311 | 0.232367 | 0.251730 | 0.272496 | 0.294772 | 0.318676 |
| 60 | 0.167366 | 0.182024 | 0.197744 | 0.214602 | 0.232679 | 0.252064 | 0.272855 | 0.295157 | 0.319089 |

## Spherical Trigonometry

Spherical trigonometry deals with the measurement of angles of polygons—triangles in particular—that lie on the surface of spheres. The sides of a spherical triangle conform to the surface of the sphere, and, unlike a plane triangle, the sum of angle measures of a spherical triangle ranges from 180° to 540°.

**Right-Angle Spherical Trigonometry.—**The solid black lines $A$, $B$, and $C$ of Fig. 1 represent the sides of a *right* spherical triangle. The dashed lines $J$ and $K$ are radii of the sphere extending from the center of the sphere to the triangle's vertices. The several plane triangles, indicated by the various broken lines, are formed from the radii and vertices of the spherical triangle. $J$ and $K$ are radii and thus have the same value.

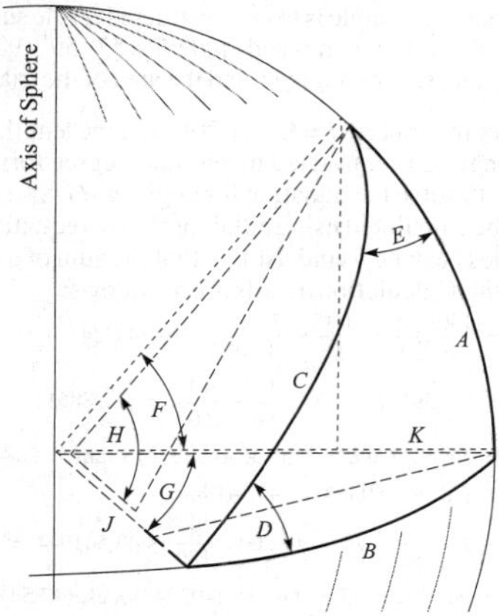


Fig. 1. Right Spherical Triangle

### Formulas for Right Spherical Triangles

| Formulas for Lengths | | | | |
|---|---|---|---|---|
| $A = K \times \frac{\pi}{180} \times F°$ | $B = J \times \frac{\pi}{180} \times G°$ | $C = J \times \frac{\pi}{180} \times H°$ | $J = \frac{180}{\pi} \times \frac{B}{G°}$ | $K = \frac{180}{\pi} \times \frac{A}{F°}$ |

| Formulas for Angles | | |
|---|---|---|
| $F° = \frac{180}{\pi} \times \frac{A}{K}$ | $G° = \frac{180}{\pi} \times \frac{B}{J}$ | $H° = \frac{180}{\pi} \times \frac{C}{J}$ |

| Angle | Angular Relationships | | |
|---|---|---|---|
| $D$ | $\sin D = \sin F \times \csc H$ | $\cos D = \tan G \times \cot H$ | $\tan D = \tan F \times \csc G$ |
| $E$ | | $\cos E = \cos G \times \sin D$ | $\tan E = \tan G \times \csc F$ |
| $F$ | $\sin F = \tan G \times \cot E$ | $\cos F = \sec G \times \cos H$ | $\tan F = \tan D \times \sin G$ |
| $G$ | | $\cos G = \cos H \times \sec F$ | $\tan G = \sin F \times \tan E$ |
| $H$ | $\cos H = \cos G \times \cos F$ | $\cos H = \cot D \times \cot E$ | |

| Area Formula |
|---|
| $\text{Area} = K^2 \times \frac{\pi}{180}(D° + E° + 90° - 180°) = K^2 \times \frac{\pi}{180}(D° + E° - 90°)$ |

The side and angle labels in examples that follow refer to those of the right spherical triangle in Fig. 1.

*Example 1:* Find the length of arc $A$ of a right spherical triangle on the surface of a sphere where radius $K = 30.00$ inches and angle $F = 10°$.

*Solution:* $A = K \times \frac{\pi}{180} \times F = 30 \times \frac{\pi}{180} \times 10 = 5.2359$ in.

*Example 2:* Find length of arc $B$ on a sphere of radius $J = 11.20$ inches if angle $G = 10°$.

*Solution:* $B = J \times \frac{\pi}{180} \times G = 11.20 \times \frac{\pi}{180} \times 10 = 1.9547$ in.

*Example 3:* A right spherical triangle is to be constructed on the surface of a sphere 22.400 inches in diameter. Side $A$ is 7.125 inches and angle $E$ is 57° 59′ 19″. Determine the lengths of sides $B$ and $C$, and the measure of angle $D$, and the area of the triangle.

*Solution:* The radius of the sphere, $J = K = 11.200$, and the length of side $A$ is used to find the value of angle $F$. Angle $E$ is converted to decimal degree format for simplicity; then angles $E$ and $F$ are used to solve the equation for angle $\tan G$. Side $B$ and angle $D$ can then be found. Angle $H$ can be calculated using either of the two equations given for $\cos H$, and finally the length of side $C$ can be found. Notice that the sum of angles $D + E + 90°$ is not equal to 180°, but 194.98°. Calculation details are as follows:

$$F° = \frac{180}{\pi} \times \frac{A}{K} = \frac{180}{\pi} \times \frac{7.125}{11.200} = 36.449324°$$

$$E = 57°59'19'' = 57 + \frac{59}{60} + \frac{19}{3600} = 57.988611°$$

$$\tan G = \sin F \times \tan E = \sin(36.449324°) \times \tan(57.988611°) = 0.950357$$

$$G = \tan^{-1}(0.950357) = 43.541944°$$

$$B = J \times \frac{\pi}{180} \times G° = 11.200 \times \frac{\pi}{180} \times 43.541944 = 8.511443$$

$$\tan D = \tan F \times \csc G = \tan(36.449324°) \times \csc(43.541944°) = 1.0721569$$

$$D = \frac{180}{\pi} \times \tan^{-1}(1.0721569) = 46.994354°$$

$$\cos H = \cos G \times \cos F = \cos(43.541944°) \times \cos(36.449324°) = 0.58307306$$

$$H = \frac{180}{\pi} \times \cos^{-1}(0.58307306) = 54.333023°$$

$$C = J \times \frac{\pi}{180} \times H° = 11.200 \times \frac{\pi}{180} \times 54.333023° = 10.62085$$

$$\text{Angles}(D + E + 90°) = 46.994354° + 57.988611° + 90° = 194.98297°$$

$$\text{Area} = 11.200^2 \times (194.98297 - 180) = 50.142591 \text{ in}^2$$

*Example 4:* A right spherical triangle on a 20-mm diameter sphere has two 90° angles, and the distance $B$ between the 90° angles is ⅓ the circumference of the sphere. Find angle $E$, the area of the triangle, and check using the conventional formula for area of a sphere.

*Solution:* By inspection, angle $G$ is 360°/3 = 120°. Because angles $D$ and $G$ are known, angle $E$ can be calculated using $\cos E = \cos G \times \sin D$. Therefore,

$$\cos E = \cos G \times \sin D = \cos(120°) \times \sin(90°) = -0.5$$

$$E = \cos^{-1}(-0.5) = 120°$$

$$\text{Area} = 10^2 \times \frac{\pi}{180}(120° + 90° + 90° - 180°) = 100 \times 2.0943951 = 209.4 \text{ mm}^2$$

$$\text{Check: Total area of 20-mm diameter sphere/6} = \frac{4\pi R^2}{6} = \frac{4\pi(100)}{6} = 209.4 \text{ mm}^2$$

**Oblique Spherical Trigonometry.**—The heavy solid lines $B$, $C$, and $S$ of Fig. 2 represent the sides of an *oblique* spherical triangle. The dashed lines $J$ and $L$ are radii of the sphere extending from the center of the sphere to the vertices of the triangle. The several plane triangles, indicated by the various broken lines, are formed from the radii and vertices of the spherical triangle. $J$ and $L$ are radii and thus have the same value.

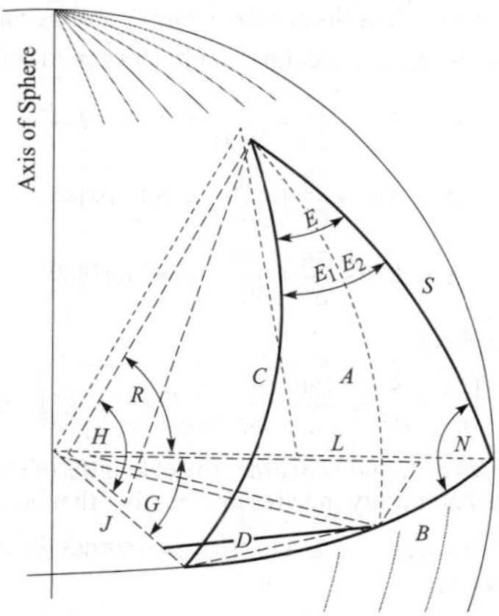


Fig. 2. Oblique Spherical Triangle

## Formulas for Oblique Spherical Triangles

| Formulas for Lengths | | | | |
|---|---|---|---|---|
| $B = J \times \frac{\pi}{180} \times G°$ | $C = J \times \frac{\pi}{180} \times H°$ | $S = L \times \frac{\pi}{180} \times R°$ | $J = \frac{180}{\pi} \times \frac{B}{G°}$ | $L = \frac{180}{\pi} \times \frac{S}{R°}$ |

| Formulas for Angles | | |
|---|---|---|
| $G° = \frac{180}{\pi} \times \frac{B}{J}$ | $H° = \frac{180}{\pi} \times \frac{C}{J}$ | $R° = \frac{180}{\pi} \times \frac{S}{L}$ |

| Angular Relationships | | | |
|---|---|---|---|
| Angle | Relationships | Angle | Relationships |
| $D$ | $\sin D = \sin R \times \sin E \times \csc G$ | $E$ | $\sin E = \sin D \times \sin G \times \csc R$ |
| $G$ | $\sin G = \sin R \times \sin E \times \csc D$ | $E_1$ | $\cot E_1 = \tan D \times \cos H$ |
| $N$ | $\cos N = \cos D \times \csc E_1 \times \sin E_2$ | $E_2$ | $\cot E_2 = \tan N \times \cos R$ |
| $N$ | $\cot\left(\frac{N}{2}\right) = \frac{\sin\left(\frac{R+G}{2}\right)}{\sin\left(\frac{R-G}{2}\right)} \times \tan\left(\frac{D-E}{2}\right)$ | $H$ | $\tan\left(\frac{H}{2}\right) = \frac{\sin\left(\frac{D+E}{2}\right)}{\sin\left(\frac{D-E}{2}\right)} \times \tan\left(\frac{R-G}{2}\right)$ |
| $R$ | $\sin R = \sin D \times \sin G \times \csc E$ | | |

| Area Formula |
|---|
| $\text{Area} = L^2 \times \frac{\pi}{180}(D + E + N - 180°)$ |

The side and angle labels in the examples that follow refer to those of the oblique spherical triangle in Fig. 2.

*Example:* An oblique spherical triangle is to be constructed on the surface of a sphere of unknown size. The length of side *S* will be 5.470 inches; the spherical angle of arc *S* must be 51° 17′ 31″ (angle *R* in Fig. 2). Angle *D* must be 59° 55′ 10″, and angle *E* must be 85° 36′ 32″. Find the size of the sphere, lengths of sides *B* and *C*, and the value of angle *N*.

*Solution:* Convert known angles to decimal degree format to simplify calculations:

$$R = 51° + \frac{17}{60} + \frac{31}{3600} = 51.291944°$$

$$D = 59° + \frac{55}{60} + \frac{10}{3600} = 59.919444°$$

$$E = 85° + \frac{36}{60} + \frac{32}{3600} = 85.608889°$$

Find the radius of the sphere:

$$L = \frac{180}{\pi} \times \frac{S}{R°} = \frac{180}{\pi} \times \frac{5.470}{51.291944°} = 6.11 \text{ inches}$$

Find the values of angles of *G* and *H* in order to get lengths of sides *B* and *C*. Then solve for the value of angle *N*, and finally the area. Remember that both *J* and *L* are radii, thus $J = L$.

$$\sin G = \sin R \times \sin E \times \csc D = 0.780342 \times 0.997065 \times 1.15564$$

$$= 0.899148$$

$$G = \sin^{-1}(0.899148) = 64.046301°$$

$$B = J \times \frac{\pi}{180} \times G° = 6.11 \times \frac{\pi}{180} \times 64.046301° = 6.829873 \text{ inches}$$

$$\tan\left(\frac{H}{2}\right) = \frac{\sin\left(\frac{D+E}{2}\right)}{\sin\left(\frac{D-E}{2}\right)} \times \tan\left(\frac{R-G}{2}\right) = \frac{\sin(72.76417)}{\sin(-12.844723)} \times \tan(-6.377185)$$

$$= \frac{0.955093}{-0.222310}(-0.111765) = 0.480167$$

$$\frac{H}{2} = \tan^{-1}(0.480167) = 25.648772°, \qquad H = 51.297543°$$

$$C = J \times \frac{\pi}{180} \times H° = 6.11 \times \frac{\pi}{180} \times 51.297543° = 5.470350 \text{ inches}$$

$$\cot\left(\frac{N}{2}\right) = \frac{\sin\left(\frac{R+G}{2}\right)}{\sin\left(\frac{R-G}{2}\right)} \times \tan\left(\frac{D-E}{2}\right) = \frac{\sin(57.669123)}{\sin(-6.377185)} \times \tan(-12.844723)$$

$$= \frac{0.844974}{-0.111073}(-0.228015) = 1.7345957$$

$$\frac{N}{2} = \cot^{-1}(1.7345957) = 29.963587°, \qquad N = 59.927175°$$

$$\text{Area} = L^2 \times \frac{\pi}{180}(D + E + N - 180°) = 16.585 \text{ in}^2$$

The triangle is an isosceles spherical triangle with legs *B* and *C* each being 5.470 inches.

If angle $E_1$ or $E_2$ is known, then any problem involving oblique spherical triangles can be solved as two right spherical triangles; in that case, the equations for right spherical triangles are used.

## Compound Angles

Referring to the *Formulas for Compound Angles* on page 116, in Fig. 1 is shown what might be considered as a thread-cutting tool without front clearance. $A$ is a known angle in plane $y$-$y$ of the top surface. $C$ is the corresponding angle in plane $x$-$x$ that is at some given angle $B$ with plane $y$-$y$. Thus, angles $A$ and $B$ are components of the compound angle $C$.

*Example Referring to* Fig. 1: Angle $2A$ in plane $y$-$y$ is known, as is angle $B$ between planes $x$-$x$ and $y$-$y$. It is required to find compound angle $2C$ in plane $x$-$x$.

*Solution:* Let $2A = 60$ and $B = 15$; then

$$\tan C = \tan A \times \cos B = \tan 30° \times \cos 15°$$

$$\tan C = 0.57735 \times 0.96592 = 0.55767$$

$$C = 29° \ 8.8' \qquad 2C = 58° \ 17.6'$$

Fig. 2 shows a thread-cutting tool with front clearance angle $B$. Angle $A$ is one-half the angle between the cutting edges in plane $y$-$y$ of the top surface, and compound angle $C$ is one-half the angle between the cutting edges in a plane $x$-$x$ at right angles to the inclined front edge of the tool. The angle between planes $y$-$y$ and $x$-$x$ is, therefore, equal to clearance angle $B$.

*Example Referring to* Fig. 2: Find the angle $2C$ between the front faces of a thread-cutting tool having a known clearance angle $B$ that will permit the grinding of these faces so their top edges form the desired angle $2A$ for cutting the thread.

*Solution:* Let $2A = 60$ and $B = 15$; then

$$\tan C = \frac{\tan A}{\cos B} = \frac{\tan 30°}{\cos 15°} = \frac{0.57735}{0.96592}$$

$$\tan C = 0.59772$$

$$C = 30° \ 52' \qquad 2C = 61° \ 44'$$

In Fig. 3 is shown a form-cutting tool in which the angle $A$ is one-half the angle between the cutting edges in plane $y$-$y$ of the top surface; $B$ is the front clearance angle; and $C$ is one-half the angle between the cutting edges in plane $x$-$x$ at right angles to the front edges of the tool. The formula for finding angle $C$ when angles $A$ and $B$ are known is the same as that for Fig. 2.

*Example Referring to* Fig. 3: Find the angle $2C$ between the front faces of a form-cutting tool having a known clearance angle $B$ that will permit the grinding of these faces so their top edges form the desired angle $2A$ for form cutting.

*Solution:* Let $2A = 46$ and $B = 12$; then

$$\tan C = \frac{\tan A}{\cos B} = \frac{\tan 23°}{\cos 12°} = \frac{0.42447}{0.97815}$$

$$\tan C = 0.43395$$

$$C = 23° \ 27.5' \qquad 2C = 46° \ 55'$$

In Fig. 4 is shown a wedge-shaped block, the top surface of which is inclined at compound angle $C$ with the base in a plane at right angles with the base and at angle $R$ with the front edge. Angle $A$, in the vertical plane of the front of the plate, and angle $B$, in the vertical plane of one side that is at right angles to the front, are components of angle $C$.

## Formulas for Compound Angles

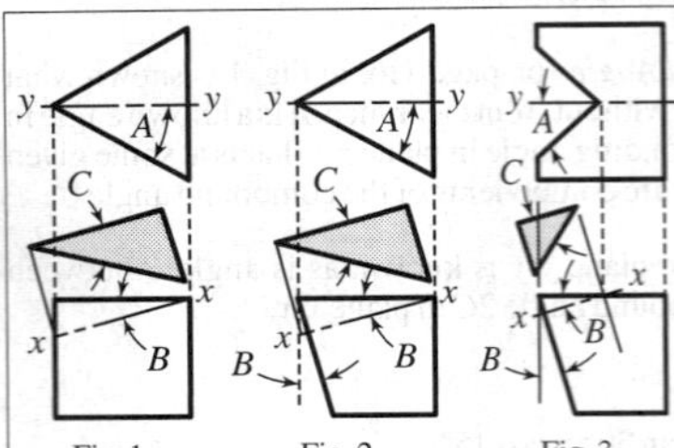


Fig. 1. Fig. 2. Fig. 3.

For given angles $A$ and $B$, find the resultant angle $C$ in plane $x$-$x$. Angle $B$ is measured in vertical plane $y$-$y$ of midsection.

Fig. 1 $\tan C = \tan A \times \cos B$

Fig. 2 $\tan C = \dfrac{\tan A}{\cos B}$

Fig. 3 (Same formula as for Fig. 2)

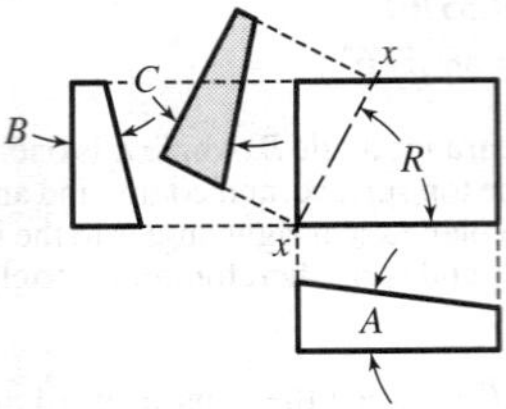


Fig. 4.

Fig. 4. In machining a plate to angles $A$ and $B$, the plate is held at angle $C$ in plane $x$-$x$. Angle of rotation $R$ in plane parallel to base (or complement of $R$) is for locating plate so that plane $x$-$x$ is perpendicular to axis of pivot on angle-plate or work-holding vise.

$$\tan R = \frac{\tan B}{\tan A}, \quad \tan C = \frac{\tan A}{\cos R}$$

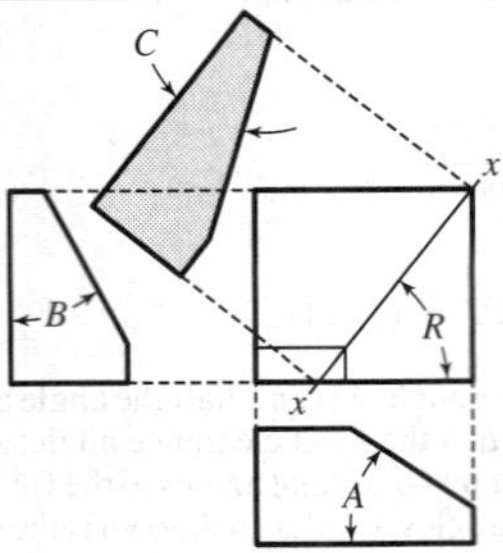


Fig. 5.

Fig. 5. Angle $R$ in horizontal plane parallel to base is angle from plane $x$-$x$ to side having angle $A$.

$$\tan R = \frac{\tan A}{\tan B}$$

$$\tan C = \tan A \times \cos R = \tan B \times \sin R$$

Compound angle $C$ is angle in plane $x$-$x$ from base to corner formed by intersection of planes inclined to angles $A$ and $B$. This formula for $C$ may be used to find cotangent of complement of $C_1$, Fig. 6.

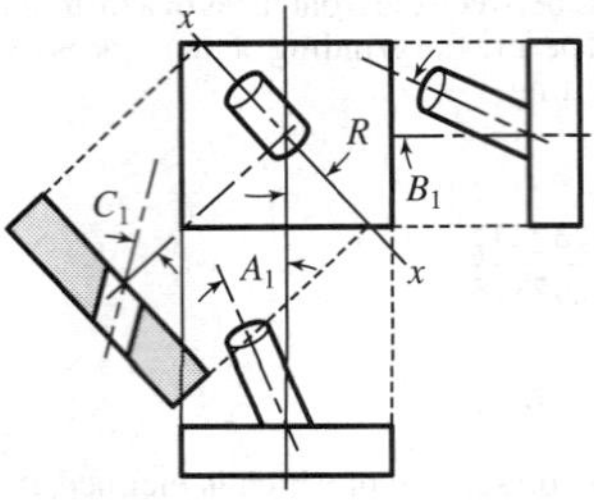


Fig. 6.

Fig. 6. Angles $A_1$ and $B_1$ are measured in vertical planes of front and side elevations. Plane $x$-$x$ is located by angle $R$ from center-line or from plane of angle $B_1$.

$$\tan R = \frac{\tan A_1}{\tan B_1}$$

$$\tan C_1 = \frac{\tan A_1}{\sin R} = \frac{\tan B_1}{\cos R}$$

The resultant angle $C_1$ would be required in drilling hole for pin.

*Example Referring to* Fig. 4*:* Find the compound angle $C$ of a wedge-shaped block having known component angles $A$ and $B$ in sides at right angles to each other.

$C$ = compound angle in plane $x$-$x$, which is the resultant of angles $A$ and $B$

*Solution:* Let $A = 47°14'$ and $B = 38°10'$; then

$$\tan R = \frac{\tan B}{\tan A} = \frac{\tan 38°10'}{\tan 47°14'} = \frac{0.78598}{1.0812} = 0.72695 \qquad R = 36°0.9'$$

$$\tan C = \frac{\tan A}{\cos R} = \frac{\tan 47°14'}{\cos 36°0.9'} = \frac{1.0812}{0.80887} = 1.3367 \qquad C = 53°12'$$

In Fig. 5 is shown a four-sided block, two sides of which are at right angles to each other and to the base of the block. The other two sides are inclined at an oblique angle with the base. Angle $C$ is a compound angle formed by the intersection of these two inclined sides and the intersection of a vertical plane passing through $x$-$x$, and the base of the block. The components of angle $C$ are angles $A$ and $B$, and angle $R$ is the angle in the base plane of the block between the plane of angle $C$ and the plane of angle $A$.

*Example Referring to* Fig. 5*:* Find the angles $C$ and $R$ in the block shown in Fig. 5 when angles $A$ and $B$ are known.

*Solution:* Let angle $A = 27°$ and $B = 36°$; then

$$\cot C = \sqrt{\cot^2 A + \cot^2 B} = \sqrt{1.9626^2 + 1.3764^2} = \sqrt{5.74627572} = 2.3971$$

$$C = 22°38.6'$$

$$\tan R = \frac{\cot B}{\cot A} = \frac{\cot 36°}{\cot 27°} = \frac{1.3764}{1.9626} = 0.70131 \qquad R = 35°2.5'$$

*Example Referring to* Fig. 6*:* A rod or pipe is inserted into a rectangular block at an angle. Angle $C_1$ is the compound angle of inclination (measured from the vertical) in a plane passing through the center line of the rod or pipe and at right angles to the top surface of the block. Angles $A_1$ and $B_1$ are the angles of inclination of the rod or pipe when viewed respectively in the front and side planes of the block. Angle $R$ is the angle between the plane of angle $C_1$ and the plane of angle $B_1$. Find angles $C_1$ and $R$ when a rod or pipe is inclined at known angles $A_1$ and $B_1$.

*Solution:* Let $A_1 = 39°$ and $B_1 = 34°$; then

$$\tan C_1 = \sqrt{\tan^2 A_1 + \tan^2 B_1} = \sqrt{0.80978^2 + 0.67451^2} = 1.0539$$

$$C_1 = 46°30.2'$$

$$\tan R = \frac{\tan A_1}{\tan B_1} = \frac{0.80978}{0.67451} = 1.2005 \qquad R = 50°12.4'$$

**Interpolation.**—In mathematics, interpolation is the process of finding a value in a table or in a mathematical expression which falls between two given tabulated or known values. In engineering handbooks, the values of trigonometric functions are usually given only in degrees and minutes; hence, if the angle is given in degrees, minutes and seconds, the value of the function is determined from the nearest given values by interpolation.

*Interpolation to Find Functions of an Angle:* Assume that the sine of 14°22′26″ is to be determined. It is evident that this value lies between the sin 14° 22′ and the sin 14° 23′. sin 14° 23′ = 0.24841 and sin 14° 22′ = 0.24813. The difference is 0.24841 − 0.24813 = 0.00028. Consider this difference as a whole number (28) and multiply it by a fraction having as its numerator the number of seconds (26) in the given angle, and as its denominator 60 (number of seconds in one minute). Thus $^{26}/_{60} \times 28 = 12$ (nearly); hence, by adding 0.00012 to sin 14° 22′ we find that sin 14°22′26″ ≈ 0.24813 + 0.00012 = 0.24825. The correction value (represented in this example by 0.00012) is *added* to the function of the *smaller* angle nearest the given angle in dealing with *sines* or *tangents,* but this correction value is *subtracted* in dealing with cosines or cotangents.

## MATRICES

Multiple variables are often present in technology and engineering scenarios; some examples include electrical circuitry, in which there is a long series of resistors, and cost analysis (labor, materials, capital) in industrial economics.

A system of linear equations can be solved by several methods, such as substitution and elimination (see pages 27 to 28, *Solving a System of Linear Equations* in *ALGEBRA*). Another way to solve a system, useful especially when more than two variables are involved, is to set up a *matrix* of the equations' coefficients.

A matrix consists of real numbers arranged in horizontal rows and vertical columns to form a rectangular array. An array of $m$ rows and $n$ columns is an $m \times n$ matrix (read as "$m$ by $n$") and is written as

$$\begin{pmatrix} a_{11} & \cdots & 0 \\ \vdots & \ddots & \vdots \\ 0 & \cdots & a_{mn} \end{pmatrix}$$

The $a_{ij}$ terms are called the entries or elements of the matrix. The subscript $i$ identifies the row position of an entry, and the subscript $j$ identifies its column position in the matrix. For example, in the matrix below, $a_{11} = 3, a_{12} = 4, a_{21} = -1$, and $a_{22} = 2$:

$$\begin{bmatrix} 3 & 4 \\ -1 & 2 \end{bmatrix}$$

Special matrices used in matrix operations include:

$$\begin{pmatrix} a_{11} \\ \vdots \\ a_{m1} \end{pmatrix} \qquad (a_{11} \ \cdots \ a_{1n}) \qquad \begin{pmatrix} a_1 & \cdots & a_{1n} \\ \vdots & \ddots & \vdots \\ a_{n1} & \cdots & a_{nn} \end{pmatrix} \qquad \begin{pmatrix} 0 & \cdots & 0 \\ \vdots & \ddots & \vdots \\ 0 & \cdots & 0 \end{pmatrix}$$

*Column Matrix* ($m \times 1$) | *Row Matrix* ($1 \times n$) | *Square Matrix* ($n \times n$) | *Zero, or Null, Matrix* ($n \times n$)

The two types of special square matrices include:

$$\begin{pmatrix} a_1 & \cdots & 0 \\ \vdots & \ddots & \vdots \\ 0 & \cdots & a_{nn} \end{pmatrix} \qquad \begin{pmatrix} 1 & \cdots & a_{1n} \\ \vdots & \ddots & \vdots \\ a_{n1} & \cdots & 1 \end{pmatrix}$$

*Diagonal Matrix* ($n \times n$)
All entries are 0, except possibly those on the diagonal.

*Identity Matrix* ($n \times n$)
All diagonal entries are 1, all others are 0.

## Matrix Operations

**Matrix Addition and Subtraction.**—The sum or difference of matrices is determined simply by adding or subtracting the corresponding elements of each matrix.

So, matrix $C$ is the result of adding or subtracting matrices $A$ and $B$; the entries are combined as follows:

$$a_{ij}+b_{ij}=c_{ij} \quad \text{or} \quad a_{ij}-b_{ij}=c_{ij}$$

Matrices must be the same size, $m \times n$, to be combined this way. That is, $A_{mn}+B_{mn}=C_{mn}$; $A_{mn}-B_{mn}=C_{mn}$.

An efficient way to indicate both + and – is with the symbol ±. Thus, in the matrix display below, $a_{ij} \pm b_{ij}$ covers both matrix operations.

*Example 1:*

$$\begin{bmatrix} 4 & 6 & -5 \\ 5 & -7 & 8 \\ -8 & 6 & -7 \end{bmatrix} + \begin{bmatrix} 8 & -2 & 6 \\ -6 & 9 & 5 \\ 9 & -2 & 2 \end{bmatrix} = \begin{bmatrix} (4+8) & (6-2) & (-5+6) \\ (5-6) & (-7+9) & (8+5) \\ (-8+9) & (6-2) & (-7+2) \end{bmatrix} = \begin{bmatrix} 12 & 4 & 1 \\ -1 & 2 & 13 \\ 1 & 4 & -5 \end{bmatrix}$$

**Matrix Multiplication.**—Two matrices can be multiplied *only* if the number of columns in the first matrix equals the number of rows of the second matrix. For example, a $1 \times 3$ matrix can multiplied by a $3 \times 2$ matrix, but not the other way around. Or a $2 \times 4$ by a $4 \times 3$, in that order. In general, an $m \times n$ and an $n \times p$ can be multiplied; the product matrix is an $m \times p$ matrix. Matrix multiplication is not commutative, that is, $A \times B$ is not necessarily equal to $B \times A$.

The steps in matrix multiplication are shown in the instructive example below for a general $3 \times 2$ matrix multiplied by a $2 \times 1$ matrix. The result is a $3 \times 1$ matrix.

$$\begin{pmatrix} 1 & 4 \\ -3 & 1 \\ 2 & -1 \end{pmatrix} \times \begin{pmatrix} 0 \\ -2 \end{pmatrix} = \begin{pmatrix} 1 \cdot 0 + 4 \cdot -2 \\ -3 \cdot 0 + 1 \cdot -2 \\ 2 \cdot 0 + -1 \cdot -2 \end{pmatrix} = \begin{pmatrix} -8 \\ -2 \\ 2 \end{pmatrix}$$

In general, *each entry* $(AB)_{ij}$ in the product matrix (where $i$ is the row number and $j$ is the column number) is equal to $a_{i1}b_{1j}+a_{i2}b_{2j}+\ldots+a_{in}b_{nj}$. For example:

$$\underset{\substack{3 \times 2 \\ \text{matrix}}}{\begin{pmatrix} a_{11} & a_{12} \\ a_{21} & a_{22} \\ a_{31} & a_{32} \end{pmatrix}} \times \underset{\substack{2 \times 2 \\ \text{matrix}}}{\begin{pmatrix} b_{11} & b_{12} \\ b_{21} & b_{22} \end{pmatrix}} = \underset{\substack{3 \times 2 \\ \text{matrix}}}{\begin{pmatrix} a_{11}b_{11}+a_{12}b_{21} & a_{11}b_{12}+a_{12}b_{22} \\ a_{21}b_{11}+a_{22}b_{21} & a_{21}b_{12}+a_{22}b_{22} \\ a_{31}b_{11}+a_{32}b_{21} & a_{31}b_{12}+a_{32}b_{22} \end{pmatrix}}$$

*Example 2:*

$$\begin{bmatrix} 1 & 2 & 3 \\ 4 & 5 & 6 \\ 3 & 2 & 1 \end{bmatrix} \times \begin{bmatrix} 7 & 8 & 9 \\ 1 & 2 & 3 \\ 4 & 5 & 7 \end{bmatrix} = \begin{bmatrix} (1 \cdot 7 + 2 \cdot 1 + 3 \cdot 4) & (1 \cdot 8 + 2 \cdot 2 + 3 \cdot 5) & (1 \cdot 9 + 2 \cdot 3 + 3 \cdot 7) \\ (4 \cdot 7 + 5 \cdot 1 + 6 \cdot 4) & (4 \cdot 8 + 5 \cdot 2 + 6 \cdot 5) & (4 \cdot 9 + 5 \cdot 3 + 6 \cdot 7) \\ (3 \cdot 7 + 2 \cdot 1 + 1 \cdot 4) & (3 \cdot 8 + 2 \cdot 2 + 1 \cdot 5) & (3 \cdot 9 + 2 \cdot 3 + 1 \cdot 7) \end{bmatrix}$$

$$= \begin{bmatrix} (7+2+12) & (8+4+15) & (9+6+21) \\ (28+5+24) & (32+10+30) & (36+15+42) \\ (21+2+4) & (24+4+5) & (27+6+7) \end{bmatrix} = \begin{bmatrix} 21 & 27 & 36 \\ 57 & 72 & 93 \\ 27 & 33 & 40 \end{bmatrix}$$

**Transpose of a Matrix.**—If the rows of a matrix $A_{mn}$ are interchanged with its columns, the new matrix is called the *transpose of matrix* $A$, or $A^T$. The first row of the matrix becomes the first column in the transposed matrix, the second row of the matrix becomes the second column, and the third row of the matrix becomes the third column.

*Example 3:*

$$A = \begin{bmatrix} 1 & 2 & 3 \\ 4 & 5 & 6 \\ 7 & 8 & 9 \end{bmatrix} \qquad A^T = \begin{bmatrix} 1 & 4 & 7 \\ 2 & 5 & 8 \\ 3 & 6 & 9 \end{bmatrix}$$

**Determinant of a Square Matrix.**—Every square matrix $A$ is associated with a real number, its determinant, which may be written $\det A$ or $|A|$.

$$\text{For } A = \begin{bmatrix} a & b \\ c & d \end{bmatrix}, \det A = |A| = ad - bc.$$

*Example 4:*

$$A = \begin{bmatrix} 2 & -1 \\ 1 & -3 \end{bmatrix} \qquad \det A = \begin{vmatrix} 2 & -1 \\ 1 & -3 \end{vmatrix} = (2)(-3) - (1)(-1) = -5$$

The process for taking the determinant of a 3 × 3 matrix is shown next. It entails multiplying the first entry of each column by the determinant of the remaining 2 × 2 matrix and alternately adding or subtracting the product.

$$B = \begin{pmatrix} a & b & c \\ d & e & f \\ g & h & i \end{pmatrix}, \qquad \det B = \begin{vmatrix} a & b & c \\ d & e & f \\ g & h & i \end{vmatrix} = a\begin{vmatrix} e & f \\ h & i \end{vmatrix} - b\begin{vmatrix} d & f \\ g & i \end{vmatrix} + c\begin{vmatrix} d & e \\ g & h \end{vmatrix}$$

*Example 5:* Find the determinant of the following matrix.

$$A = \begin{bmatrix} 5 & 6 & 7 \\ 1 & 2 & 3 \\ 4 & 5 & 6 \end{bmatrix}$$

*Solution:*

$$\det A = 5(12 - 15) - 6(6 - 12) + 7(5 - 8)$$
$$= 5(-3) - 6(-6) + 7(-3) = -15 + 36 - 21 = 0$$

**Minors and Cofactors.**—The *minor* $M_{ij}$ of a matrix $A$ is the determinant of a submatrix resulting from the elimination of row $i$ and column $j$. If $A$ is a square matrix, the minor $M_{ij}$ of the entry $a_{ij}$ is the determinant of the matrix obtained by deleting the $i$th row and $j$th column of $A$.

The *cofactor* $C_{ij}$ of the entry $a_{ij}$ is given by $C_{ij} = (-1)^{(i+j)} M_{ij}$. Thus, the sign of cofactor $a_{ij}$ alternates across the row it lies in. The matrix formed by its cofactors is called the *cofactor matrix*.

*Example 6:* Find the minors and cofactors of $A = \begin{bmatrix} 1 & 2 & 3 \\ 4 & 5 & 6 \\ 3 & 2 & 1 \end{bmatrix}$

*Solution:* To determine the minor $M_{11}$, delete the first row and first column of $A$ and find the determinant of the resulting matrix.

$$M_{11} = \begin{vmatrix} 5 & 6 \\ 2 & 1 \end{vmatrix} = (5 \times 1) - (6 \times 2) = 5 - 12 = -7$$

Similarly to find $M_{12}$, delete the first row and second column of $A$ and find the determinant of the resulting matrix.

$$M_{12} = \begin{vmatrix} 4 & 6 \\ 3 & 1 \end{vmatrix} = (4 \times 1) - (6 \times 3) = 4 - 18 = -14$$

Continuing this way, we obtain the following minors:

$$\begin{array}{lll} M_{11} = -7 & M_{12} = -14 & M_{13} = -7 \\ M_{21} = -4 & M_{22} = -8 & M_{23} = -4 \\ M_{31} = -3 & M_{32} = -6 & M_{33} = -3 \end{array}$$

To find the cofactor, calculate $C_{ij} = (-1)^{(i+j)} \times M_{ij}$, thus $C_{11} = (-1)^{(1+1)} \times M_{11} = 1 \times (-7) = -7$.
Similarly $C_{12} = (-1)^{(1+2)} \times M_{12} = (-1)(-14) = 14$, and continuing this way we obtain the following cofactors

$$\begin{array}{lll} C_{11} = -7 & C_{12} = 14 & C_{13} = -7 \\ C_{21} = 4 & C_{22} = -8 & C_{23} = 4 \\ C_{31} = -3 & C_{32} = 6 & C_{33} = -3 \end{array}$$

Thus, the cofactor matrix is $\begin{pmatrix} -7 & 14 & -7 \\ 4 & -8 & 4 \\ -3 & 6 & -3 \end{pmatrix}$

**Adjoint of a Matrix.—**The transpose of cofactor matrix is called the *adjoint* matrix. To obtain the adjoint matrix, the cofactor matrix is determined and then transposed.

*Example 7:* Find the adjoint matrix of $A$:

$$A = \begin{bmatrix} 1 & 2 & 3 \\ 4 & 5 & 6 \\ 3 & 2 & 1 \end{bmatrix}$$

*Solution:* The cofactor matrix from the above example is shown below at the left, and the adjoint matrix is shown on the right.

$$\text{cofactor}(A) = \begin{bmatrix} -7 & 14 & -7 \\ 4 & -8 & 4 \\ -3 & 6 & -3 \end{bmatrix} \qquad \text{adj}(A) = \begin{bmatrix} -7 & 14 & -7 \\ 4 & -8 & 4 \\ -3 & 6 & -3 \end{bmatrix}^T = \begin{bmatrix} -7 & 4 & -3 \\ 14 & -8 & 6 \\ -7 & 4 & -3 \end{bmatrix}$$

**Singularity and Rank of a Matrix.—**A *singular* matrix is one whose determinant is zero. The *rank* of a matrix is the maximum number of linearly independent row or column vectors it contains. In *ALGEBRA, Solving a System of Linear Equations*, it is explained that a system with a unique solution is a linearly independent system. Such systems are vital to depicting many real-life processes. Dependent systems are those with infinite solutions (the lines are collinear), and their determinant is zero. Linearly independent vectors have a non-zero determinant.

**Inverse of a Matrix.—**A square non-singular matrix $A$ has an inverse $A^{-1}$ such that the product of matrix $A$ and inverse matrix $A^{-1}$ is the identity matrix $I$. The operation is commutative as well. Thus, $AA^{-1} = A^{-1}A = I$. The inverse is the ratio of adjoint of the matrix and the determinant of that matrix.

$$A^{-1} = \frac{\text{adj}(A)}{|A|}$$

*Example 8:* What is the inverse of the following matrix?

$$A = \begin{bmatrix} 2 & 3 & 5 \\ 4 & 1 & 6 \\ 1 & 4 & 0 \end{bmatrix}$$

*Solution:* The basic formula of an inverse of a matrix is

$$A^{-1} = \frac{\text{adj}(A)}{|A|}$$

The determinant of $A$ is

$$
\begin{aligned}
|A| &= 2(1\times 0-4\times 6)-3(4\times 0-1\times 6)+5(4\times 4-1\times 1)\\
&= 2(0-24)-3(0-6)+5(16-1)\\
&= -48+18+75 = 45
\end{aligned}
$$

The cofactors are

$$
a_{11} = (-1)^{1+1}\begin{vmatrix}1 & 6\\ 4 & 0\end{vmatrix} = -24 \qquad a_{12} = (-1)^{1+2}\begin{vmatrix}4 & 6\\ 1 & 0\end{vmatrix} = 6 \qquad a_{13} = (-1)^{1+3}\begin{vmatrix}4 & 1\\ 1 & 4\end{vmatrix} = 15
$$

$$
a_{21} = (-1)^{2+1}\begin{vmatrix}3 & 5\\ 4 & 0\end{vmatrix} = 20 \qquad a_{22} = (-1)^{2+2}\begin{vmatrix}2 & 5\\ 1 & 0\end{vmatrix} = -5 \qquad a_{23} = (-1)^{2+3}\begin{vmatrix}2 & 3\\ 1 & 4\end{vmatrix} = -5
$$

$$
a_{31} = (-1)^{3+1}\begin{vmatrix}3 & 5\\ 1 & 6\end{vmatrix} = 13 \qquad a_{32} = (-1)^{3+2}\begin{vmatrix}2 & 5\\ 4 & 6\end{vmatrix} = 8 \qquad a_{33} = (-1)^{3+3}\begin{vmatrix}2 & 3\\ 4 & 1\end{vmatrix} = -10
$$

The matrix of cofactors is $\begin{bmatrix}-24 & 6 & 15\\ 20 & -5 & -5\\ 13 & 8 & -10\end{bmatrix}$ and the adjoint matrix is $\begin{bmatrix}-24 & 20 & 13\\ 6 & -5 & 8\\ 15 & -5 & -10\end{bmatrix}$

Then the inverse of matrix $A$ is

$$
A^{-1} = \frac{\text{adj}(A)}{|A|} = \frac{1}{45}\begin{bmatrix}-24 & 20 & 13\\ 6 & -5 & 8\\ 15 & -5 & -10\end{bmatrix}
$$

Multiplying $A^{-1}$ by $A$ results in the identity matrix:

$$
\frac{1}{45}\begin{pmatrix}-24 & 20 & 13\\ 6 & -5 & 8\\ 15 & -5 & -10\end{pmatrix}\begin{pmatrix}2 & 3 & 5\\ 4 & 1 & 6\\ 1 & 4 & 0\end{pmatrix} = \frac{1}{45}\begin{pmatrix}45 & 0 & 0\\ 0 & 45 & 0\\ 0 & 0 & 45\end{pmatrix} = \begin{pmatrix}1 & 0 & 0\\ 0 & 1 & 0\\ 0 & 0 & 1\end{pmatrix}
$$

**Solving a System of Equations.—**Matrices can be used to solve systems of simultaneous equations with a large number of unknowns. Geometrically, the solution of a system is the point in the plane or in space where the lines intersect. Variables may represent the unknowns in industrial and other engineering applications: a series of resistances in a circuit, for example, or, perhaps, factors in a manufacturing process, such as labor costs, materials, and equipment. Generally, this method is less cumbersome than using substitution methods (see *Solving a System of Linear Equations* in *ALGEBRA*). The coefficients of the equations are placed in matrix form. The matrix is then transformed by row and column operations into the identity matrix to yield a solution. The process, as described in the example, is called *matrix reduction to row echelon form*. It is done using any or all of three valid vector operations: 1) multiplying each entry in a row by a constant; 2) adding or subtracting two rows; 3) changing the order of rows.

*Example 9:* Solve the system of linear equations in three dimensions using matrix operations.

$$
\begin{aligned}
-4x_1+8x_2+12x_3 &= 16\\
3x_1-x_2+2x_3 &= 5\\
x_1+7x_2+6x_3 &= 10
\end{aligned}
$$

*Solution:* First, the equation coefficients and constants are placed into what is called an *augmented matrix*. The object is to transform the matrix of the original coefficients into the following form, thereby obtaining a solution $(x_1, x_2, x_3)$ to the system of equations.

$$\left[\begin{array}{ccc|c} -4 & 8 & 12 & 16 \\ 3 & -1 & 2 & 5 \\ 1 & 7 & 6 & 10 \end{array}\right] \Leftrightarrow \left[\begin{array}{ccc|c} 1 & 0 & 0 & x_1 \\ 0 & 1 & 0 & x_2 \\ 0 & 0 & 1 & x_3 \end{array}\right]$$

The coefficient matrix is transformed so that element $c_{11}$ is 1 and all other elements in the first column are 0, as follows: a) divide row 1 ($R_1$) by –4; b) multiply new $R_1$ by –3, then add to $R_2$; and c) multiply $R_1$ by –1, then add to $R_3$.

$$\left[\begin{array}{ccc|c} -\frac{4}{-4} & \frac{8}{-4} & \frac{12}{-4} & \frac{16}{-4} \\ 3 & -1 & 2 & 5 \\ 1 & 7 & 6 & 10 \end{array}\right] \Rightarrow \left[\begin{array}{ccc|c} 1 & -2 & -3 & -4 \\ (3-3) & (-1+6) & (2+9) & (5+12) \\ (1-1) & (7+2) & (6+3) & (10+4) \end{array}\right] \Rightarrow \left[\begin{array}{ccc|c} 1 & -2 & -3 & -4 \\ 0 & 5 & 11 & 17 \\ 0 & 9 & 9 & 14 \end{array}\right]$$

The resulting matrix is transformed so that element $c_{22}$ is 1 and all other elements in the second column are 0, as follows: a) divide $R_3$ by 9; b) multiply new $R_3$ by –5, then add to $R_2$; c) multiply $R_3$ by 2, then add to $R_1$; and d) swap $R_2$ and $R_3$.

$$\left[\begin{array}{ccc|c} 1 & -2 & -3 & -4 \\ 0 & 5 & 11 & 17 \\ \frac{0}{9} & \frac{9}{9} & \frac{9}{9} & \frac{14}{9} \end{array}\right] \Rightarrow \left[\begin{array}{ccc|c} 1 & (-2+2) & (-3+2) & \left(-4+\frac{28}{9}\right) \\ 0 & (5-5) & (11-5) & \left(17-\frac{70}{9}\right) \\ 0 & 1 & 1 & \frac{14}{9} \end{array}\right] \Rightarrow \left[\begin{array}{ccc|c} 1 & 0 & -1 & -\frac{8}{9} \\ 0 & 0 & 6 & \frac{83}{9} \\ 0 & 1 & 1 & \frac{14}{9} \end{array}\right] \Rightarrow \left[\begin{array}{ccc|c} 1 & 0 & -1 & -\frac{8}{9} \\ 0 & 1 & 1 & \frac{14}{9} \\ 0 & 0 & 6 & \frac{83}{9} \end{array}\right]$$

The resulting matrix is finally reduced so that element $c_{33}$ is 1 and all other elements in the third column are 0, as follows: a) divide $R_3$ by 6; b) multiply new $R_3$ by –1, then add to $R_2$; and c) add $R_3$ to $R_1$.

$$\left[\begin{array}{ccc|c} 1 & 0 & -1 & -\frac{8}{9} \\ 0 & 1 & 1 & \frac{14}{9} \\ 0 & 0 & \frac{6}{6} & \frac{83}{9(6)} \end{array}\right] \Rightarrow \left[\begin{array}{ccc|c} 1 & 0 & (-1+1) & \left(-\frac{8}{9}+\frac{83}{54}\right) \\ 0 & 1 & (1-1) & \left(\frac{14}{9}-\frac{83}{54}\right) \\ 0 & 0 & 1 & \frac{83}{54} \end{array}\right] \Rightarrow \left[\begin{array}{ccc|c} 1 & 0 & 0 & \frac{35}{54} \\ 0 & 1 & 0 & \frac{1}{54} \\ 0 & 0 & 1 & \frac{83}{54} \end{array}\right]$$

When the identity matrix has been formed, the last column contains the values of $x_1$, $x_2$, and $x_3$ that satisfy the original equations.

$$x_1 = \frac{35}{54} \qquad x_2 = \frac{1}{54} \qquad x_3 = \frac{83}{54}$$

Checking that the solutions satisfy the original system:

$$-4\left(\frac{35}{54}\right) + 8\left(\frac{1}{54}\right) + 12\left(\frac{83}{54}\right) = \frac{864}{54} = 16$$

$$3\left(\frac{35}{54}\right) - 1\left(\frac{1}{54}\right) + 2\left(\frac{83}{54}\right) = \frac{270}{54} = 5$$

$$1\left(\frac{35}{54}\right) + 7\left(\frac{1}{54}\right) + 6\left(\frac{83}{54}\right) = \frac{540}{54} = 10$$

*Example 10:* Use matrix operations to find the amperages of the currents $(I_1, I_2, I_3)$ in the following electrical network.

By Kirchoff's first law, concerning the sum of currents: $I_1 + I_2 = I_3$

that is, $I_1 + I_2 - I_3 = 0$

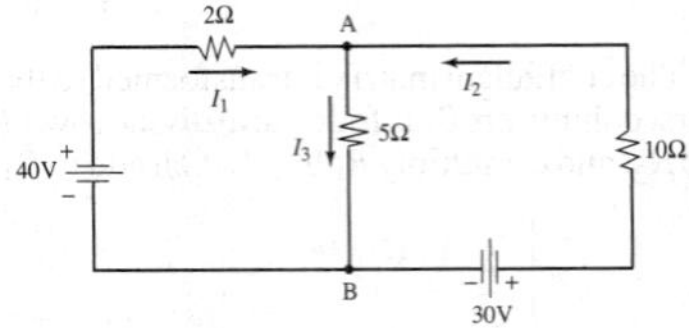


By Kirchoff's second law, concerning the sum of voltages, and Ohm's law, that voltage is the product of current and resistance:

$$2I_1 + 5I_3 - 40 = 0$$
$$10I_2 + 5I_3 - 30 = 0$$

By combining the three equations, a linear system of independent equations is formed. A system is linearly independent if no one equation is a constant multiple of any other. That is, the equations represent distinct lines, but they are not parallel (hence, they have a solution, that is, a point of intersection). Solve the system for the currents $I_1, I_2$, and $I_3$:

$$\begin{aligned} I_1 + I_2 - I_3 &= 0 \\ 2I_1 + 5I_3 &= 40 \\ 10I_2 + 5I_3 &= 30 \end{aligned}$$

*Solution:* If $A$ is the matrix of coefficients of the currents, $B$ is the matrix of the currents themselves (the variables), and $C$ is the matrix of constants from the right side of the equations (the voltages), then the problem can be written in the following form: $AB = C$, or equivalently, $B = A^{-1}C$, where $A^{-1}$ is the inverse of matrix $A$.

Thus,

$$A = \begin{bmatrix} 1 & 1 & -1 \\ 2 & 0 & 5 \\ 0 & 10 & 5 \end{bmatrix} \quad B = \begin{bmatrix} I_1 \\ I_2 \\ I_3 \end{bmatrix} \quad C = \begin{bmatrix} 0 \\ 40 \\ 30 \end{bmatrix} \quad \text{and} \quad \begin{bmatrix} I_1 \\ I_2 \\ I_3 \end{bmatrix} = \begin{bmatrix} 1 & 1 & -1 \\ 2 & 0 & 5 \\ 0 & 10 & 5 \end{bmatrix}^{-1} \begin{bmatrix} 0 \\ 40 \\ 30 \end{bmatrix}$$

Using the method of Example 8, the inverse of matrix $A$ is

$$A^{-1} = \begin{bmatrix} 1 & 1 & -1 \\ 2 & 0 & 5 \\ 0 & 10 & 5 \end{bmatrix}^{-1} = -\frac{1}{80}\begin{bmatrix} 50 & 15 & -5 \\ 10 & -5 & 7 \\ -20 & 10 & 2 \end{bmatrix} = \begin{bmatrix} \frac{5}{8} & \frac{3}{16} & -\frac{1}{16} \\ \frac{1}{8} & -\frac{1}{16} & \frac{7}{80} \\ -\frac{1}{4} & \frac{1}{8} & \frac{1}{40} \end{bmatrix}$$

and finally, matrix $B$ can be found by matrix multiplication:

$$B = A^{-1}C = \begin{bmatrix} \frac{5}{8} & \frac{3}{16} & -\frac{1}{16} \\ \frac{1}{8} & -\frac{1}{16} & \frac{7}{80} \\ -\frac{1}{4} & \frac{1}{8} & \frac{1}{40} \end{bmatrix} \begin{bmatrix} 0 \\ 40 \\ 30 \end{bmatrix} = \begin{bmatrix} 5.625 \\ 0.125 \\ 5.75 \end{bmatrix}$$

Thus, $I_1 = 5.625$ amps, $I_2 = 0.125$ amp, and $I_3 = 5.75$ amps.

## CALCULUS

Problems in engineering and other sciences are often modeled by functions other than simple first- and second-degree polynomials. The features of higher order polynomials, as well as trigonometric and other non-algebraic functions, are found using techniques of calculus. The essential operations of calculus are *differentiation* and *antidifferentiation* (or *integration*). Brief explanations of these processes, as well as detailed formulas, are given here.

### Derivatives

Between any two points of a linear function $y = mx + b$, the change in $y$ with respect to $x$ is the constant $m$. Most functions of interest, however, are nonlinear. For example, the velocity of a projectile is not modeled by a line, since objects are subject to gravity, among other outside forces, and therefore accelerate as they fall. The *instantaneous rate of change* of a function $y = f(x)$ at a single point of a curve is a critical feature for any model. This quantity is called the *derivative of f with respect to x* (or other independent variable).

A derivative is notated in one of three ways: $y'$, $f'(x)$, or $dy/dx$. As an example, the instantaneous rate of change of *displacement s* of an object at any instant in time $t$ is its derivative function, *velocity*; that is, $s'(t)$ or $ds/dt = v(t)$.

Graphically, the derivative function gives the *slope* of the line tangent to a point of the graph of $f$. That is, $y' = m_{tan}$. Any constant function $y = c$ has a slope of zero, since its graph is a horizontal line. Hence, $y' = 0$ at every point of a line. Fig. 1a shows a portion of a nonlinear function. The tangent line drawn to the point of the curve at $x$ has a slope equal to the derivative $f'(x)$.

Any group of curves that represent a *family* of functions, such as the parabolas $f_1, f_2$, and $f_3$ in Fig. 1b, have the same derivative function, since the slope of the tangent lines at any given value of $x$ is the same for each curve. The slope of the tangent line drawn to each of the curves at $x = 0$ is $f'(0) = 2(0) + 2 = 2$.

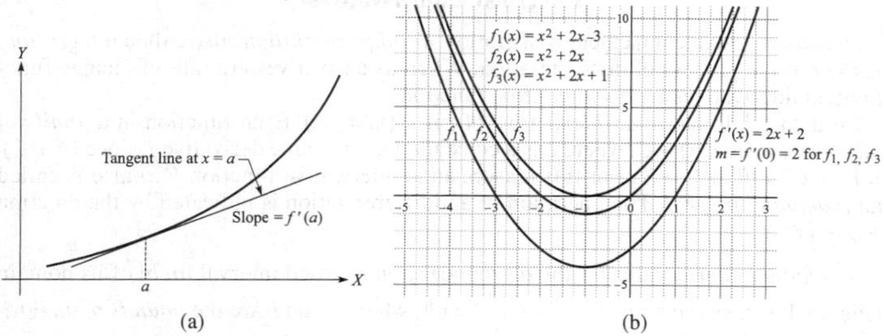


Fig. 1. (a) $f'(x)$ gives the slope of the tangent line to a curve $f(x)$ for any $x$ in the domain; (b) A family of parabola functions, all with the same derivative function, $f'(x) = 2x + 2$.

Any continuous, smooth function is *differentiable* (that is, it has a derivative) at each point on its domain. (Roughly speaking, a "continuous" function has no breaks, and a "smooth" function has no sharp corners.) Polynomial, trigonometric, exponential, and logarithmic functions are differentiable everywhere on their domains. To "differentiate" a function means to find (or "take") its derivative.

**Derivative Formulas.—**The formulas used most often in derivative applications are:

*Constant:* If $y = c$, then $y' = 0$.

*Coefficient:* If $y = cx$, then $y' = c$.

*Power:* If $y = x^n$, then $y' = nx^{n-1}$, for any $n$ in the set of real numbers.

*Note:* The power rules when $n = -1$ and when $n = ½$ are often stated as their own rules:

$$y' = x^{-1}, \quad y' = -x^{-2} = \frac{-1}{x^2} \qquad \text{and} \qquad y = \sqrt{x} = x^{\frac{1}{2}}, \quad y' = \tfrac{1}{2}x^{-\frac{1}{2}} = \frac{1}{2x^{\frac{1}{2}}} = \frac{1}{2\sqrt{x}}$$

*Logarithmic:*

For any base $a$, if $y = \log_a x$, then $y' = \frac{1}{x\ln a}$; for natural base $e$, $y = \ln x$, then $y' = \frac{1}{x}$.

*Exponential:* If $y = a^x$, then $y' = a^x \ln a$. If $y = e^x$, then $y' = e^x$.

*Trigonometric:*

If $y = \sin x$, then $y' = \cos x$. If $y = \cos x$, then $y' = -\sin x$. If $y = \tan x$, then $y' = \sec^2 x$.

For a complete list of the differentiation formulas, see *Table of Derivatives and Integrals.*

**Derivative Rules.**—Just like other functions, derivatives have certain properties:

| | *Rule* | *Example* |
|---|---|---|
| Sum or difference | $(f \pm g)'(x) = f'(x) \pm g'(x)$ | $y = x^5 + \frac{1}{x} - 9, \quad y' = 5x^4 - \frac{1}{x^2}$ |
| Product | $(fg)'(x) = f'(x)g(x) + g'(x)f(x)$ | $y = 2x\ln x, \quad y' = 2\ln x + (2x)\left(\frac{1}{x}\right) = 2\ln x + 2$ |
| Quotient | $\left(\frac{f(x)}{g(x)}\right)' = \frac{f'(x)g(x) - g'(x)f(x)}{[g(x)]^2}$ | $y = \frac{4x-9}{x^2}, \quad y' = \frac{4x^2 - (4x-9)(2x)}{(x^2)^2} = \frac{4x^2 - 8x^2 + 18x}{x^4}$ |
| Chain rule | $[f(u(x))]' = f'(u(x))u'(x)$ | $y = \sqrt{x^2 + 5x - 1} = (x^2 + 5x - 1)^{1/2}$ |
| | | $y' = \frac{1}{2}(x^2 + 5x - 1)^{-1/2}(2x + 5) = \frac{2x+5}{2\,(x^2 + 5x - 1)^{1/2}}$ |

### Integrals (Antiderivatives)

The other fundamental calculus operation, *antidifferentiation*, also called *integration*, is the reverse process of differentiation. Whereas derivatives are rate-of-change functions, antiderivatives are *accumulation* functions.

We define $F$ to be *an antiderivative* of $f$ if $F'(x) = f(x)$. Each function in a *family of antiderivatives* $F(x) + C$, where $c$ is a constant, has the same derivative $f$, since $(F + C)' = F' + C' = F' + 0 = F'$. For this reason, an antiderivative function $F(x)$ also is called an *indefinite integral*. The operation of antidifferentiation is indicated by the notation $F(x) = \int f(x)dx$.

A *definite integral* indicates the *area* under $f$ on a closed interval $[a, b]$ of its domain (Fig. 2). It is denoted by $F(b) - F(a) = \int_a^b f(x)dx$, where $a$ and $b$ are the *bounds of integration*, and $F(a)$ and $F(b)$ are the antiderivative values of $f$ at these bounds.

Just as velocity is the derivative of displacement, so displacement is the antiderivative of velocity. On a time interval $[t_1, t_2]$, displacement is found by $s(t_2) - s(t_1) = \int_{t_1}^{t_2} v(t)\,dt$.

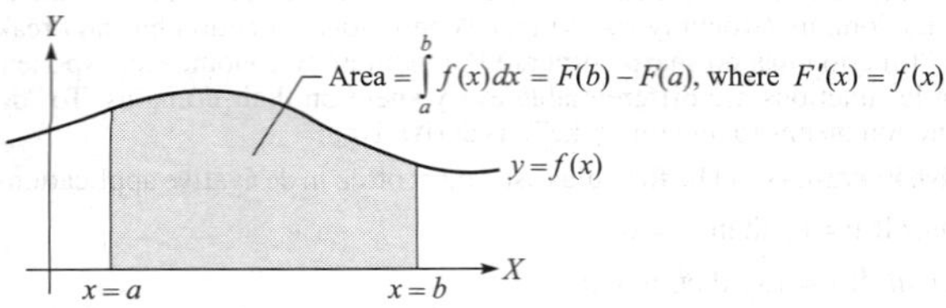


Fig. 2. The definite integral on $[a, b]$ is the area under the curve from $x = a$ to $x = b$.

**Integral Rules.**—Integration rules are somewhat similar to differentiation rules, since they undo what the derivative does. The analogous processes for the chain, product, and quotient rules of derivatives are covered by either $u$-substitution or integration by parts for integrals. *Table of Derivatives and Integrals* contains integration formulas.

**Newton's Method for Solving Equations.**—Algebraic (polynomial, rational, and root) and transcendental (trigonometric, exponential and logarithmic) equations often can be quickly solved directly, using the processes described in *ALGEBRA* and *TRIGONOMETRY: SOLUTION OF TRIANGLES* (for example, $\cos x = 1$ or $\log x = 4$). But, long before there were calculators that could solve less straightforward equations to a high degree of accuracy, *approximation methods* were developed. One such method is *Newton's* (or the *Newton-Raphson*) method, which produces excellent approximations of the solution of more difficult equations, with the help of differentiation. Some equations that can be solved by Newton's method are:

$$x^2 = 101 \qquad x^3 - 2x^2 = 5 \qquad \cos x = x$$

Rewriting any of these equations as a function $f(x) = 0$ converts the problem into one of finding the roots of $f(x)$—that is, those values where the function crosses the $x$-axis. For example, $x^2 = 101$ is rewritten as $f(x) = x^2 - 101 = 0$. This function has two real roots (see *ALGEBRA*). A good first estimate of the positive root is 10. Rewriting the other two equations, $x^3 - 2x^2 = 5$ and $\cos x = x$ in the $f(x) = 0$ form does not give as obvious an estimate of the root(s), but by inspection (trial and error) or a rough graph, an estimate can be made. In each case, this first estimate is called $r_1$. From these, successive estimates $r_2, r_3$, . . . are made, each progressively closer to the exact value of the root.

After estimating $r_1$, the first derivative of the function, $f'(x)$, is found. This is the equation for the function's instantaneous rate of change at any value of $x$. $f'(x)$ is the equation that gives the slope of the line tangent to the function's curve at a given $x$.

In the above examples, $f'(x)$ is, respectively, $2x$, $3x^2 - 4x$, and $-\sin x + 1$. These were found by the methods described in *Table of Derivatives and Integrals* on page 128.

Starting with the first estimate, the steps of Newton's method are as follows:

$r_1$ is the first estimate of the value of the root of $f(x) = 0$.

Find $f(r_1)$, the value of $f(x)$ at $x = r_1$.

Find $f'(x)$, the first derivative of $f(x)$.

Find $f'(r_1)$, the value of $f'(x)$ at $x = r_1$.

Get the second approximation of the root of $f(x) = 0$, $r_2$, by calculating

$$r_2 = r_1 - f(r_1)/f'(r_1)$$

and, further approximations,

$$r_n = r_{n-1} - f(r_{n-1})/f'(r_{n-1})$$

*Example:* Find the square root of 101 using the Newton-Raphson method.

*Solution:* The problem is restated as the algebraic equation $x^2 = 101$, rewritten as $x^2 - 101 = 0$, and solved for the positive root. $r_1 = 10$ is a good first estimate. Then, apply the steps of the algorithm:

$$f(r_1) = f(10) = 10^2 - 101 = -1$$

Step 2. The first derivative, $f'(x)$, of $x^2 - 101$ is $2x$, as stated previously, so that $f'(10) = 2(10) = 20$. Then,

$r_2 = r_1 - f(r_1)/f'(r_1) = 10 - (-1)/20 = 10 + 0.05 = 10.05$

*Check:* $10.05^2 = 101.0025$; a calculator determination of $\sqrt{101}$ gives 10.0498756; the error using Newton's method is 0.0001244.

Step 3. The next, better approximation is

$$r_3 = r_2 - f(r_2)/f'(r_2) = 10.05 - f(10.05)/f'(10.05)$$

$$= 10.05 - (10.05^2 - 101)/2(10.05) = 10.049875$$

*Check:* $10.049875^2 = 100.9999875$; error $= 0.0000125$

Closer approximations result from subsequent applications of the algorithm.

**Formulas for Differential and Integral Calculus.—**The following are formulas for obtaining the derivatives and integrals of basic mathematical functions. In these formulas, the letters $a$, $b$, and $c$ denote constants; the letter $x$ denotes a variable; and the letters $u$ and $v$ denote functions of the variable $x$. The expression $d/dx$ means the derivative with respect to $x$, and as such applies to whatever expression in parentheses follows it. Thus, $d(cx)/dx$ means the derivative with respect to the variable $x$ of the product $cx$ where $c$ is a constant.

**Table of Derivatives and Integrals**

| Derivatives | Integrals |
|---|---|
| $\frac{d}{dx}(c) = 0$ | $\int c\,dx = cx + \text{constant}$ |
| $\frac{d}{dx}(x) = 1$ | $\int 1\,dx = x + C$ |
| $\frac{d}{dx}(x^n) = nx^{n-1}$, for all real numbers $n$ | $\int x^n dx = \frac{x^{n+1}}{n+1} + C$, for all real numbers $n \neq -1$ |
| $\frac{d}{dx}\left(\frac{1}{x}\right) = -\frac{1}{x^2} + C$ | $\int \frac{dx}{ax+b} = \frac{1}{a}\ln\lvert ax+b\rvert + C$ |
| $\frac{d}{dx}(\log_a x) = \frac{1}{x \ln a}$ | $\int \frac{1}{x} dx = \ln\lvert x\rvert + C$ |
| $\frac{d}{dx}(e^x) = e^x$ | $\int e^x dx = e^x + C$ |
| $\frac{d}{dx}(e^{ax}) = ae^{ax}$ | $e^{ax} dx = \frac{e^{ax}}{a} + C$ |
| $\frac{d}{dx}(a^x) = a^x \ln a$ | $\int a^x dx = \frac{a^x}{\ln a} + C$ |
| $\frac{d}{dx}(\sqrt{x}) = \frac{1}{2\sqrt{x}}$ | $\int \frac{dx}{\sqrt{x}} = 2\sqrt{x} + C$ |
| $\frac{d}{dx}(\ln x) = \frac{1}{x}$ | $\int \ln x\,dx = x \ln x - x + C$ |
| $\frac{d}{dx}(\sin x) = \cos x$ | $\int \cos x\,dx = \sin x + C$ |

**Table of Derivatives and Integrals** *(Continued)*

| Derivatives | Integrals |
|---|---|
| $\frac{d}{dx}(\cos x) = -\sin x$ | $\int \sin x dx = -\cos x + C$ |
| $\frac{d}{dx}(\tan x) = \sec^2 x$ | $\int \tan x dx = -\ln\|\cos x\| + C$ |
| $\frac{d}{dx}(\cot x) = -\csc^2 x$ | $\int \cot x dx = \ln\|\sin x\| + C$ |
| $\frac{d}{dx}(\sec x) = \sec x \tan x$ | $\int \sin^2 x dx = -\frac{1}{4}\sin(2x) + \frac{1}{2}x + C$ |
| $\frac{d}{dx}(\csc x) = -\csc x \cot x$ | $\int \cos^2 x dx = \frac{1}{4}\sin(2x) + \frac{1}{2}x + C$ |
| $\frac{d}{dx}(\sin^{-1} x) = \frac{1}{\sqrt{1-x^2}}\ x \neq 1, -1$ | $\int \frac{dx}{\sqrt{b^2-x^2}} = \sin^{-1}\left(\frac{x}{b}\right) + C$ |
| $\frac{d}{dx}(\cos^{-1} x) = \frac{-1}{\sqrt{1-x^2}}\ x \neq 1, -1$ | $\int \frac{dx}{\sqrt{x^2-b^2}} = \ln\left(\frac{x}{b} + \frac{\sqrt{x^2-b^2}}{b}\right) + C$ |
| $\frac{d}{dx}(\tan^{-1} x) = \frac{1}{1+x^2}$ | $\int \frac{dx}{b^2+x^2} = \frac{1}{b}\tan^{-1}\left(\frac{x}{b}\right) + C$ |
| $\frac{d}{dx}(\cot^{-1} x) = \frac{-1}{1+x^2}$ | $\int \frac{dx}{b^2-x^2} = \frac{1}{b}\tan^{-1}\left(\frac{x}{b}\right) = \frac{-1}{2b}\log\left[\frac{\|x-b\|}{\|x+b\|}\right] + C$ |
| $\frac{d}{dx}(\sec^{-1} x) = \frac{1}{x\sqrt{x^2-1}}\ x \neq 1, -1$ | $\int \frac{dx}{x^2-b^2} = -\frac{1}{b}\cos^{-1}\left(\frac{x}{b}\right) = \frac{1}{2b}\log\left[\frac{\|x-b\|}{\|x+b\|}\right] + C$ |
| $\frac{d}{dx}(\csc^{-1} x) = \frac{-1}{x\sqrt{x^2-1}}\ x \neq 1, -1$ | $\int \frac{dx}{ax^2+bx+c} = \frac{2}{\sqrt{4ac-b^2}}\tan^{-1}\left[\frac{2ax+b}{\sqrt{4ac-b^2}}\right] + C$ |
| $\frac{d}{dx}(\ln(\sin x)) = \cot x$ | $\int e^{ax}\sin(bx)dx = \frac{a\sin(bx) - b\cos(bx)}{a^2+b^2}e^{ax} + C$ |
| $\frac{d}{dx}(\ln(\cos x)) = -\tan x$ | $\int e^{ax}\cos(bx)dx = \frac{a\cos(bx) + b\sin(bx)}{a^2+b^2}e^{ax} + C$ |
| $\frac{d}{dx}(\ln(\tan x)) = \frac{2}{\sin(2x)}$ | $\int \frac{1}{\sin x}dx = \ln\|\csc x - \cot x\| + C$ |
| $\frac{d}{dx}(\ln(\cot x)) = \frac{-2}{\sin(2x)}$ | $\int \frac{1}{\cos x}dx = \ln\|\sec x + \tan x\| + C$ |
| $\frac{d}{dx}(\ln(ax+b)) = \frac{a}{ax+b}$ | $\int \frac{1}{1+\cos x}dx = \csc x - \cot x + C$ |

## Series Representation of a Function

Some hand calculations, as well as computer programs of certain types of mathematical problems, may be facilitated by the use of an appropriate series. For example, in some gear problems, the angle corresponding to a given or calculated involute function (see *TRIGONOMETRY: SOLUTION OF TRIANGLES* on page 89) is found by using a series together with an iterative procedure such as the Newton-Raphson method described in the previous section. The following are those series most commonly used for such purposes. In the series for trigonometric functions, the angles $x$ are in radians (1 radian = 180/π degrees, or about 57.3 degrees). The expression $\exp(-x^2)$ means that the base $e$ of the natural logarithm system is raised to the $-x^2$ power, where $e = 2.7182818$.

A sum of terms in a sequence of terms is called a *series*. In its simplest notation, the infinite sum of the sequence $a_0, a_1, a_2, \ldots$ is the series $\sum_{n=0}^{\infty} a_n = a_0 + a_1 + a_2 + \ldots$. For example, for the sequence of $x, x^2, x^3, \ldots$, the series is given as $\sum_{n=1}^{\infty} x^n = x^1 + x^2 + \ldots = x^1 + x^2 + \ldots$ .

In calculus, it is helpful to represent certain functions by a special series called the Taylor series. In a manner similar to the Newton-Raphson method, the terms of a Taylor series include the derivatives of the function being approximated. In the table below, common functions and their corresponding infinite series are shown. For any $x$ in the domain (shown to the right of each series), the value of the function can be found.

**Common Series**

| | |
|---|---|
| $\sin x = x - \frac{x^3}{3!} + \frac{x^5}{5!} - \frac{x^7}{7!} + \cdots$ | for all real $x$ |
| $\cos x = 1 - \frac{x^2}{2!} + \frac{x^4}{4!} - \frac{x^6}{6!} + \cdots$ | for all real $x$ |
| $\tan x = x + \frac{x^3}{3} + \frac{2x^5}{15} - \frac{17x^7}{315} + \cdots$ | for $\lvert x \rvert < \frac{\pi}{2}$ |
| $\sin^{-1} x = x + \frac{x^3}{6} + \frac{3x^5}{40} + \frac{15x^5}{336} + \cdots$ | for $\lvert x \rvert \le 1$ |
| $\cos^{-1} x = \frac{\pi}{2} - (x + \frac{x^3}{6} + \frac{3x^5}{40} + \frac{15x^5}{336} + \cdots) = \frac{\pi}{2} - \sin^{-1} x$ | for $\lvert x \rvert \le 1$ |
| $\tan^{-1} x = x - \frac{x^3}{3} + \frac{x^5}{5} - \frac{x^7}{7} + \cdots$ | for $\lvert x \rvert \le 1$ |
| $\frac{\pi}{4} = 1 - \frac{1}{3} + \frac{1}{5} - \frac{1}{7} + \cdots$ | |
| $e = 1 + \frac{1}{1!} + \frac{1}{2!} + \frac{1}{3!} + \cdots$ | |
| $e^x = 1 + \frac{x}{1!} + \frac{x}{2!} + \frac{x}{3!} + \cdots$ | for all real $x$ |
| $a^x = 1 + \frac{x \ln a}{1!} + \frac{(x \ln a)^2}{2!} + \frac{(x \ln a)^3}{3!} + \cdots$ | for all real $x$ |
| $\ln(1+x) = x - \frac{x^2}{2} + \frac{x^3}{3} - \frac{x^4}{4} + \frac{x^5}{5} - \cdots$ | for $-1 < x \le 1$ |
| $\frac{1}{1+x} = 1 - x + x^2 - x^3 + x^4 - \cdots$ | for $\lvert x \rvert \le 1$ |
| $\frac{1}{1-x} = 1 + x + x^2 + x^3 + x^4 + \cdots$ | for $\lvert x \rvert \le 1$ |
| $\frac{1}{(1+x)^2} = 1 - 2x + 3x^2 - 4x^3 + 5x^4 - \cdots$ | for $\lvert x \rvert \le 1$ |
| $\frac{1}{(1-x)^2} = 1 + 2x + 3x^2 + 4x^3 + 5x^4 + \cdots$ | for $\lvert x \rvert \le 1$ |
| $\frac{1}{\sqrt{1+x}} = 1 - \frac{x}{2} + \frac{1 \cdot 3x^2}{2 \cdot 4} - \frac{1 \cdot 3 \cdot 5x^3}{2 \cdot 4 \cdot 6} - \ldots$ | for $\lvert x \rvert \le 1$ |
| $(a+x)^n = a^n + \frac{na^{(n-1)}x}{1!} + \frac{(n-1)na^{n-2}x^2}{2!} + \frac{(n-2)(n-1)a^{n-3}x^3}{3!} + \ldots$ | for all real $x$ |

## STATISTICAL ANALYSIS OF MANUFACTURING DATA

### Statistics Theory in Brief

High-volume manufacturing production, unlike prototype design work, typically involves repeating machining operations and processes hundreds, thousands, or even millions of times during a given product or product family's production run. Understanding the failure mechanisms in a product's tooling and improving the efficiency of these operations by adjusting manufacturing parameters brings many essential benefits to the manufacturing process. Seeing where failure and inefficiency occur can save on tool wear of indexable inserts, milling cutters, reamers, twist drills, etc.; improve speed, feeds, and power consumption profiles; reduce machine tool accuracy drift; and reduce lubrication and other maintenance-related failures. Improving these and other related process, even by a tiny amount, can result in huge cost savings in large production run environments.

To begin this process, measurements and other production process values must be collected so that patterns can be found. This is the collection of *raw* (or *source*) *data*. The information is put into tables (tabular form), to be processed (usually) by a computer program that analyzes and interprets it by rigorous *statistical analysis*. Without statistics theory, it would be impossible to know whether or not the testing was comprehensive enough to offer valid experimental conclusions that can then be used to make manufacturing process changes.

**Probability.**—In commercial enterprises, predictions about a particular outcome or event are made regularly to determine best practices and to minimize risk. Predictions are formed by observing events and collecting data, which is then interpreted using statistical analysis, based on concepts of *probability theory*. Probability is the likelihood of an event happening randomly among all possible events. Simple experiments, such as observing the number of times heads or tails turn up in a coin toss, and complex ones, like locating the position of an electron in an atom, are studied using probability. Data analysis enables development of a detailed statistical picture of the event being studied, and probability quantifies how likely the particular event is.

In industry, statistical analysis is used to investigate data to evaluate the success and/or efficiency of a manufacturing process. Statistical methods are described in this section using these terms:

*Experiment:* A well-defined action that is repeatable.

*Data:* Information collected during an experiment.

*Trial:* A particular performance of an experiment.

*Set:* A collection of events (or elements), such as may occur in an experiment.

*Sample space:* The set of all possible outcomes.

*Event:* Any subset of the sample space; one outcome or a collection of outcomes.

*Probability of an event:* Ratio of number of outcomes of a particular event to total number of outcomes possible in the sample space of events.

The probability of an event can be expressed as a *decimal* between 0 and 1, as a *fraction*, or as a *percent* between 0 and 100. The sum of all probabilities for outcomes of a particular experiment is 1; that is, it is 100% likely that one of the outcomes will occur.

In the coin toss example, each toss is a trial, and there are two possible outcomes: heads or tails (as long as the coin is fair, that is, both sides have an equal chance of turning up). So the probability of tossing a head is $1/2$ or 0.5 (50%). As another example with two outcomes, consider a water treatment plant with 34 female workers (set A) and 36 male workers (set B); the universal set is $34 + 36 = 70$ workers. The probability that a female worker is chosen at random from the universal set is 34/70, or approximately 0.486 (48.6%).

In manufacturing parts, the probability of a defective part occurring in a production run would be the number of parts found to be defective (the "event") in relation to the total number of parts in the run (the "trials"); defective parts ought to be far fewer than acceptable ones. So, to say there is a 3% probability of a defect means that for every 100 parts that are produced, on average, 3 will be defective and the other 97 will be acceptable. In other words, the predicted defect rate is 3% and the yield of acceptable parts is 97%.

Such simple examples are not that useful in explaining *manufacturing statistics*, where any event can have many values—for instance, in the hardness of an alloy or duration of tool life. However, more complex statistical analysis starts with the same concepts: whether a company is inspecting for defects or for a range of values of physical characteristics, the analysis involves how data from an experiment is regarded and how it can be used to determine the probability of various outcomes.

Statistical analysis depends on the type of statistical distributions that apply to various properties of the data being examined.

Some common statistical distributions are: 1) normal; 2) log-normal; 3) exponential; 4) binomial; 5) Weibull; and 6) Poisson. Normal distribution is described here.

**Normal Distribution Analysis.**—The *normal* distribution is the most widely used statistical distribution for modeling mechanical, physical, electrical, and chemical properties that scatter randomly about a well-defined mean value without either positive or negative bias. This curve is frequently called a bell curve. In the following discussion, the characteristics of the normal distribution curve are assumed.

Statistical analysis of raw data is a crucially important scientific and engineering tool that summarizes the characteristics of samples (number of observations, trials, data points, etc.). If a sample of data is randomly selected from the population, its statistical characteristics converge towards the statistical characteristics of the population as the sample size increases. Because economic constraints, such as testing time and cost, prevent a large number of repeat tests, it is important to understand how a sample of data represents an approximation of the real population of data. The following parameters must be calculated to evaluate the sample of data with respect to the population of data:

$\bar{X}$ = sample mean  $S$ = sample standard deviation  $V$ = coefficient of variation

$A_x$ = absolute error of the sample mean

$R_x$ = relative error of the sample mean

$t$ = critical value of $t$-distribution (or Student's $t$-distribution)

$\mu$ = population mean  $\sigma$ = population standard deviation

$\bar{X} \pm t A_x$ = confidence interval for the population mean

*Sample Mean* ($\bar{X}$): There are several types of *average* measures, the most common being the *arithmetic mean*, or *sample mean*. It is the value about which all data are "centered." The sample mean $\bar{X}$ is calculated as:

$$\bar{X} = \frac{1}{n}\sum_{i=1}^{n} x_i \tag{1}$$

where $x_i$ = individual data point values, and $n$ = number of data points. According to the *central limit theorem*, when a large number of samples is taken independently from a population, the sample *means* follow a normal distribution, regardless of how data in each sample are distributed.

*Sample Standard Deviation (S):* A measure of the dispersion of data about its standard mean $\bar{X}$. The sample standard deviation is calculated by the formula:

$$S = \sqrt{\frac{\sum_{i=1}^{n}(x_i - \bar{X})^2}{n-1}} \tag{2}$$

where $n - 1$ = the number of *degrees of freedom* (*d.f.*)

*Degrees of Freedom (d.f.):* The number of observations made in excess of the minimum needed to estimate a statistical parameter or quantity. For example, only one measurement is required to identify the width of an indexable insert's flank wear that occurs while machining a workpiece. If the measurements are repeated seven times, then the sample variance of flank wear measurement has six degrees of freedom.

*Coefficient of Variation (V):* Used to evaluate or control the variability in data points. The coefficient of variation is calculated by dividing the sample standard deviation $S$ by the sample mean $\overline{X}$ and expressing the result as a percent:

$$V = \frac{S}{\overline{X}} 100\% \tag{3}$$

*Absolute Error of the Sample Mean ($A_x$):* Calculated by dividing the sample standard deviation by the square root of the number of data points. The result is expressed in the same unit of measure as the sample standard deviation and the sample mean:

$$A_x = \frac{S}{\sqrt{n}} \tag{4}$$

*Relative Error of the Sample Mean ($R_x$):* Calculated by dividing the absolute error of the sample mean by the sample mean and expressing the result as a percent:

$$R_x = \frac{A_x}{\overline{X}}(100) \tag{5}$$

*Critical Value of "t-Distribution" ("Student's t-distribution"):* The *t-distribution* was discovered in 1908, by W. S. Gosset, who wrote under the name "Student" (the brewery he worked for did not allow him to publish under his own name). The critical value of $t$ depends on the number of degrees of freedom and the probability of error. If a 95% two-sided confidence is used for statistical analysis, then the probability of error is 5%, or 2.5% per side. A 5% probability of error provides practical accuracy, which is commonly acceptable in various engineering calculations. The $t$-distribution is broader than the normal distribution for small sample sizes, but approaches the shape of a normal distribution as sample size increases, with the difference negligible even for moderately large sample sizes ($n > 30$).

For a 5% probability of error, the critical value of $t$-distribution can be determined from Table 1, page 135, at the intersection of the column under the heading $t_{0.025}$ and the row corresponding to the number of degrees of freedom shown in the column heading *d.f.*

*Population Mean (μ):* The normal distribution has two parameters: the population mean μ and the population standard deviation $S$. The sample mean $\overline{X}$ is an estimate of the population mean ($\overline{X} \approx \mu$), and the sample standard deviation is an estimate of the population standard deviation ($\sigma \approx S$). A graph of the normal distribution is symmetric about its mean μ. Virtually all of the area (99.74%) under the graph is contained within the interval:

$$(\mu - 3\sigma, \mu + 3\sigma)$$

The population mean lies at the center of the normal distribution curve. Thus, almost all of the probability associated with a normal distribution falls within ± three standard deviations of the population mean μ. Also, 95.44% of the area falls within ±two standard deviations of μ, and 68.26% within ± one standard deviation.

*Confidence Interval for the Population Mean:* The interval of the normal curve associated with the probability of a parameter occurring is the *confidence interval*. The probability itself is the *confidence level*. The ends of this interval are the *confidence limits*. Larger samples tend to give better estimates of a population parameter. A higher confidence level tends to produce a wider confidence interval. Confidence levels of 90%, 95%, and 99% are commonly used. For example, a 95% confidence level implies that the true population parameter (say, a measurement being looked at) has a 95% probability of falling within the presented interval.

Equations (1) through (5) describe a sample mean that is only an estimate of the true (population) mean. Therefore, it is important to define a confidence interval that determines a range within which the population mean lies. Such an interval depends on the sample mean, $\bar{X}$, absolute error of the sample mean, $A_x$, and $t$-distribution (Student's) value. A confidence interval for the population mean satisfies the inequality:

$$\bar{X} - A_x t \le \mu \le \bar{X} + A_x t \qquad (6)$$

## Applying Statistics

**Minimum Number of Test or Data Points.—**The minimum number of data points needed to represent a sample size can be determined through the formulas for the coefficient of variation $V$, Equation (3), the absolute error of the sample mean $A_x$, Equation (4), and the relative error of the sample mean $R_x$, Equation (5).

According to Equation (4), the absolute error of the sample mean is:

$$A_x = \frac{S}{\sqrt{n}}$$

The other expression for the absolute error of the sample mean from Equation (5) is:

$$A_x = \bar{X}R_x \qquad (7)$$

Because the values to the left of the equal sign in Equations (4) and (7) are equal, the values on the right are also equal and, therefore:

$$\frac{S}{\sqrt{n}} = \bar{X}R_x \qquad (8)$$

Solving for $\sqrt{n}$ in Equation (8) produces:

$$\sqrt{n} = \frac{S}{\bar{X}R_x} \qquad (9)$$

Because $S/\bar{X}$ is the coefficient of variation $V$, see Equation (3), then:

$$\sqrt{n} = \frac{V}{R_x} \quad \text{and} \quad n = \frac{V^2}{(R_x)^2} \qquad (10)$$

The coefficient of variation of the sample mean must be known or selected according to previously collected data of a similar kind, or, if necessary, preliminary tests should be conducted to estimate its value. Based on numerous studies of cutting tool performance and publications on mechanical properties of cutting tool materials, values of the coefficient of variation within 25 to 45% are considered typical. A relative error of the sample mean between 6 and 12% is also considered typical. The coefficient of variation and the relative error are used to estimate how many tests are required. For example, if $V = 30\%$ and $R_x = 8\%$, then the number of tests required is $n = 30^2/8^2 = 14$.

**Comparing Products with Respect to Average Performance.—**Lab and field tests are usually conducted to compare the average performance of two or more products. The term "average performance" is a quantitative value, which can be any mechanical, physical, or chemical characteristic of a product. For example, the average tool life of drills and indexable cutting inserts or the average hardness of cemented carbide grades, etc. (*Tool life* usually refers to the usable life of the tool based on wear, a judgment that is made by the manufacturer for a specific product based on various factors.) The products may differ in manufacturing procedure (CVD or PVD coatings), in chemical composition (alloying elements and their amount), and in other parameters. Data collected during experiments must be statistically treated to determine whether products have the same performance characteristics or not.

Statistical treatment of data obtained from experiments with two products includes the following steps:

a) Calculation of the sample means $\overline{X}_1$ and $\overline{X}_2$ using Equation (1)

b) Calculation of the sample standard deviations $S_1$ and $S_2$ using Equation (2)

c) Calculation of a weighted, or pooled standard deviation using the following formula:

$$S_p = \sqrt{\frac{(n_1-1)S_1^2 + (n_2-1)S_2^2}{(n_1-1)+(n_2-1)}} \tag{11}$$

where $n_1$ and $n_2$ represent the number of data points for products 1 and 2 respectively.

d) Selection of a confidence level. A 95% two-sided confidence level is recommended. At this confidence level, the probability of error is ±2.5% per side. The values of $t$-distribution versus *degrees of freedom* ($d.f.$) are provided in Table 1, and for a 95% confidence level are located in the column under the heading "$t_{0.025}$" with respect to given degrees of freedom ($d.f. = n_1 + n_2 - 2$).

e) Calculation of *decision criterion* ($d.c.$) using the following formula:

$$d.c. = t_{0.025} \times S_p \sqrt{\frac{n_1+n_2}{n_1 n_2}} \tag{12}$$

f) Comparison of the value of the decision criterion with the difference of the sample means: $\overline{X}_1 - \overline{X}_2$ if $\overline{X}_1 > \overline{X}_2$, or $\overline{X}_2 - \overline{X}_1$ if $\overline{X}_2 > \overline{X}_1$.

A product's average performance is statistically significant if the difference in the two sample means is greater than the decision criterion $d.c.$

$$\overline{X}_1 - \overline{X}_2 > d.c. \quad \text{or} \quad \overline{X}_2 - \overline{X}_1 > d.c.$$

**Table 1. Critical Values of *t*-Distribution**

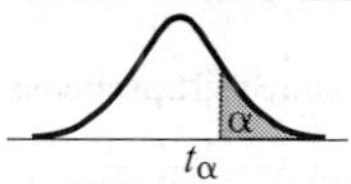


| $d.f.$ | $t_{0.100}$ | $t_{0.050}$ | $t_{0.025}$ | $t_{0.010}$ | $t_{0.005}$ | $d.f.$ | $d.f.$ | $t_{0.100}$ | $t_{0.050}$ | $t_{0.025}$ | $t_{0.010}$ | $t_{0.005}$ | $d.f.$ |
|---|---|---|---|---|---|---|---|---|---|---|---|---|---|
| 1 | 3.078 | 6.314 | 12.706 | 31.821 | 63.657 | 1 | 16 | 1.337 | 1.746 | 2.120 | 2.583 | 2.921 | 16 |
| 2 | 1.886 | 2.920 | 4.303 | 6.965 | 9.925 | 2 | 17 | 1.333 | 1.740 | 2.110 | 2.567 | 2.898 | 17 |
| 3 | 1.638 | 2.353 | 3.182 | 4.541 | 5.841 | 3 | 18 | 1.330 | 1.734 | 2.101 | 2.552 | 2.878 | 18 |
| 4 | 1.533 | 2.132 | 2.776 | 3.747 | 4.604 | 4 | 19 | 1.328 | 1.729 | 2.093 | 2.539 | 2.861 | 19 |
| 5 | 1.476 | 2.015 | 2.571 | 3.365 | 4.032 | 5 | 20 | 1.325 | 1.725 | 2.086 | 2.528 | 2.845 | 20 |
| 6 | 1.440 | 1.943 | 2.447 | 3.143 | 3.707 | 6 | 21 | 1.323 | 1.721 | 2.080 | 2.518 | 2.831 | 21 |
| 7 | 1.415 | 1.895 | 2.365 | 2.998 | 3.499 | 7 | 22 | 1.321 | 1.717 | 2.074 | 2.508 | 2.819 | 22 |
| 8 | 1.397 | 1.860 | 2.306 | 2.896 | 3.355 | 8 | 23 | 1.319 | 1.714 | 2.069 | 2.500 | 2.807 | 23 |
| 9 | 1.383 | 1.833 | 2.262 | 2.821 | 3.250 | 9 | 24 | 1.318 | 1.711 | 2.064 | 2.492 | 2.797 | 24 |
| 10 | 1.372 | 1.812 | 2.228 | 2.764 | 3.169 | 10 | 25 | 1.316 | 1.708 | 2.060 | 2.485 | 2.787 | 25 |
| 11 | 1.363 | 1.796 | 2.201 | 2.718 | 3.106 | 11 | 26 | 1.315 | 1.706 | 2.056 | 2.479 | 2.779 | 26 |
| 12 | 1.356 | 1.782 | 2.179 | 2.681 | 3.055 | 12 | 27 | 1.314 | 1.703 | 2.052 | 2.473 | 2.771 | 27 |
| 13 | 1.350 | 1.771 | 2.160 | 2.650 | 3.012 | 13 | 28 | 1.313 | 1.701 | 2.048 | 2.467 | 2.763 | 28 |
| 14 | 1.345 | 1.761 | 2.145 | 2.624 | 2.977 | 14 | 29 | 1.311 | 1.699 | 2.045 | 2.462 | 2.756 | 29 |
| 15 | 1.341 | 1.753 | 2.131 | 2.602 | 2.947 | 15 | Inf. | 1.282 | 1.645 | 1.960 | 2.326 | 2.576 | Inf. |

*Example 1:*

## Tool Life Tests of CNMG-432 Indexable Inserts Made of Two Different Carbide Grades (A and B)

**Carbide Grade A**

| Test Number $n_i$ | Data, $x_i$ minutes | Treatment of Data $x_i - \overline{X}$ | $(x_i - \overline{X})^2$ |
|---|---|---|---|
| 1 | 15.0 | –2.1 | 4.41 |
| 2 | 19.0 | 1.9 | 3.61 |
| 3 | 16.9 | –0.2 | 0.04 |
| 4 | 16.6 | –0.5 | 0.25 |
| 5 | 16.6 | –0.5 | 0.25 |
| 6 | 16.1 | –1.0 | 1.00 |
| 7 | 19.4 | 2.3 | 5.29 |
| $n = 7$ | | $\overline{X} = 17.1$ | |
| $\Sigma x_i = 119.6$ | | $\Sigma(x_i - \overline{X})^2 = 14.85$ | |

**Carbide Grade B**

| Test Number $n_i$ | Data, $x_i$ minutes | Treatment of Data $x_i - \overline{X}$ | $(x_i - \overline{X})^2$ |
|---|---|---|---|
| 1 | 14.6 | –0.2 | 0.04 |
| 2 | 13.5 | –1.3 | 1.69 |
| 3 | 15.6 | 0.8 | 0.64 |
| 4 | 12.4 | –2.4 | 5.76 |
| 5 | 14.6 | –0.2 | 0.04 |
| 6 | 13.8 | –1.0 | 1.00 |
| 7 | 19.1 | 4.3 | 18.49 |
| $n = 7$ | | $\overline{X} = 14.8$ | |
| $\Sigma x_i = 103.6$ | | $\Sigma(x_i - \overline{X})^2 = 27.66$ | |

| Characteristics of Normal Distribution | Carbide Grades A | B |
|---|---|---|
| Number of data points, $n$ | 7 | 7 |
| Number of degrees of freedom, $n - 1$ | 6 | 6 |
| Sample mean (s.m.), Equation (1) | 17.1 | 14.8 |
| Sample standard deviation, Equation (2) | 1.6 | 2.1 |
| Coefficient of variation, Equation (3) | 9.4% | 14.2% |
| Absolute error of the s.m., Equation (4) | 0.6 | 0.8 |
| Relative error of the s.m., Equation (5) | 3.5% | 5.4% |
| $t$-value at 95% confidence level at given degrees of freedom, d.f. = 6, Table 1 | 2.447 | 2.447 |
| Population mean is greater than: | 15.6 | 12.8 |
| Population mean is less than: | 18.6 | 16.8 |

Comparison of Grades A and B

| Comparison of Grades A and B | |
|---|---|
| Pooled standard deviation, Equation (11) | 1.9 |
| $t$-value at 95% confidence level at given degrees of freedom, d.f. = 6 + 6 = 12, Table 1 | 2.179 |
| Decision criterion, Equation (12) | 2.2 |
| Difference between the two sample means | 2.3 |

**Conclusion:**

Sample means of the tool life of carbide grades A and B are statistically *significant* at the 95% confidence level, since the difference of the sample means (17.1 – 14.8 = 2.3 min) is **greater** than the *decision criterion* (2.2 min).

Note: $n_i$ = $i$th test or data point, $x_i$ = $i$th value of the data point

*Example 2:*

## Tensile Strength of Carbon Steel Specimens versus Heat Treatment

**Carbon Steel Sample A**

| Test Number $n_i$ | Data, $x_i$ MPa | Treatment of Data $x_i - \overline{X}$ | $(x_i - \overline{X})^2$ |
|---|---|---|---|
| 1 | 522.0 | 6.9 | 47.61 |
| 2 | 511.0 | –4.1 | 16.81 |
| 3 | 488.9 | –26.2 | 686.44 |
| 4 | 553.7 | 38.6 | 1490.00 |
| 5 | 499.9 | –15.1 | 228.01 |
| $n = 5$ | | $\overline{X} = 515.1$ | |
| $\Sigma x_i = 2575.5$ | | $\Sigma(x_i - \overline{X})^2 = 2468.9$ | |

**Carbon Steel Sample B**

| Test Number $n_i$ | Data, $x_i$ MPa | Treatment of Data $x_i - \overline{X}$ | $(x_i - \overline{X})^2$ |
|---|---|---|---|
| 1 | 517.1 | 12.4 | 153.76 |
| 2 | 490.2 | –14.5 | 210.25 |
| 3 | 499.1 | –5.6 | 31.36 |
| 4 | 514.4 | 9.7 | 94.09 |
| 5 | 502.6 | –2.1 | 4.41 |
| $n = 5$ | | $\overline{X} = 504.7$ | |
| $\Sigma x_i = 2523.4$ | | $\Sigma(x_i - \overline{X})^2 = 493.87$ | |

Note: $n_i$ = $i$th test or data point, $x_i$ = $i$th value of the data point

| Characteristics of Normal Distribution | Samples A | B |
|---|---|---|
| Number of data points, $n$ | 5 | 5 |
| Number of degrees of freedom, $n - 1$ | 4 | 4 |
| Sample mean (s.m.), Equation (1) | 515.1 | 504.7 |
| Sample standard deviation, Equation (2) | 24.8 | 11.1 |
| Coefficient of variation, Equation (3) | 4.8% | 2.2% |
| Absolute error of the s.m., Equation (4) | 11.1 | 5.0 |
| Relative error of the s.m., Equation (5) | 2.2% | 1.0% |
| $t$-value at 95% confidence level at given degrees of freedom, d.f. = 4, Table 1 | 2.776 | 2.776 |
| Population mean is greater than: | 484.3 | 490.8 |
| Population mean is less than: | 545.9 | 518.6 |

| Comparison of Samples A and B | |
|---|---|
| Pooled standard deviation, Equation (11) | 19.2 |
| $t$-value at 95% confidence level at given degrees of freedom, d.f. = 4 + 4 = 8, Table 1 | 2.306 |
| Decision criterion, Equation (12) | 28.0 |
| Difference between the two sample means | 10.4 |

**Conclusion:**

Sample means of the tensile strength of samples A and B are statistically *insignificant* at the 95% confidence level, since the difference of the sample means (515.1 – 504.7 = 10.4 MPa) is **less** than the *decision criterion* (28.0 MPa).

*Example 3:*

## Tool Life Tests of 6.0-mm Diameter Drills with Different Web Thickness

**Drills with 2.0 mm Web, Group A**

| Test Number $n_i$ | Data, $x_i$ minutes | Treatment of Data $x_i - \overline{X}$ | $(x_i - \overline{X})^2$ |
|---|---|---|---|
| 1 | 15.68 | −11.06 | 122.3236 |
| 2 | 18.88 | −7.86 | 61.7796 |
| 3 | 19.20 | −7.54 | 56.8516 |
| 4 | 22.56 | −4.18 | 17.4724 |
| 5 | 23.20 | −3.54 | 12.5316 |
| 6 | 24.40 | −2.34 | 5.4756 |
| 7 | 24.64 | −2.10 | 4.4100 |
| 8 | 26.56 | 0.18 | 0.0324 |
| 9 | 27.20 | 0.46 | 0.2116 |
| 10 | 30.24 | 3.50 | 12.2500 |
| 11 | 32.16 | 5.42 | 29.3764 |
| 12 | 33.60 | 6.86 | 47.0596 |
| 13 | 36.80 | 10.06 | 101.2036 |
| 14 | 39.20 | 12.46 | 155.2516 |
| $n = 14$ | | $\overline{X} = 26.74$ | |
| $\Sigma x_i = 374.32$ | | $\Sigma(x_i - \overline{X})^2 = 626.2296$ | |

| Characteristics of Normal Distribution | Drill Groups A | B |
|---|---|---|
| Number of data points, $n$ | 14 | 16 |
| Number of degrees of freedom, $n - 1$ | 13 | 15 |
| Sample mean (s.m.), Equation (1) | 26.74 | 15.01 |
| Sample standard deviation, Equation (2) | 6.94 | 7.30 |
| Coefficient of variation, Equation (3) | 26.0% | 48.6% |
| Absolute error of the s.m., Equation (4) | 1.85 | 1.83 |
| Relative error of the s.m., Equation (5) | 6.9% | 12.2% |
| $t$-value at 95% confidence level at given degrees of freedom, Table 1 | 2.160 d.f.=13 | 2.131 d.f.=15 |
| Population mean is greater than: | 22.74 | 11.11 |
| Population mean is less than: | 30.74 | 18.91 |

**Comparison of Grades A and B**

| | |
|---|---|
| Pooled standard deviation, Equation (11) | 7.14 |
| $t$-value at 95% confidence level at given degrees of freedom, d.f. = 13 + 15 = 28, Table 1 | 2.048 |
| Decision criterion, Equation (12) | 5.35 |
| Difference between the two sample means | 11.73 |

**Drills with 0.9 mm Web, Group B**

| Test Number $n_i$ | Data, $x_i$ minutes | Treatment of Data $x_i - \overline{X}$ | $(x_i - \overline{X})^2$ |
|---|---|---|---|
| 1 | 5.04 | −9.97 | 99.4009 |
| 2 | 6.48 | −8.53 | 72.7609 |
| 3 | 7.12 | −7.89 | 62.2521 |
| 4 | 7.20 | −7.81 | 60.9961 |
| 5 | 9.44 | −5.57 | 31.0249 |
| 6 | 11.36 | −3.65 | 13.3225 |
| 7 | 12.16 | −2.85 | 8.1225 |
| 8 | 14.24 | −0.77 | 0.5929 |
| 9 | 15.68 | 0.67 | 0.4489 |
| 10 | 16.32 | 1.31 | 1.7161 |
| 11 | 17.84 | 2.83 | 8.0089 |
| 12 | 18.00 | 2.99 | 8.9401 |
| 13 | 21.28 | 6.27 | 39.3129 |
| 14 | 23.04 | 8.03 | 64.4809 |
| 15 | 24.60 | 9.59 | 91.9681 |
| 16 | 30.40 | 15.39 | 236.8521 |
| $n = 16$ | | $\overline{X} = 15.01$ | |
| $x_i = 240.20$ | | $(x_i - \overline{X})^2 = 800.2008$ | |

**Conclusion:**

Sample means of the tool life of the drills in Group A and B are statistically *significant* at the 95% confidence level, since the difference of the sample means (26.74 − 15.01 = 11.73 min) is **greater** than the decision criterion (5.35 min Difference). Note: $n_i$ = $i$th test or data point, $x_i$ = $i$th value of the data point

*Regarding Machinability and Hardness:* In cutting steels, the allowable cutting speed for a given tool life between grindings is, as a general rule, inversely proportional to the hardness of a given steel. To illustrate, tests in turning an alloy steel with a high-speed steel tool showed a cutting speed of 70 feet per minute (21.3 meters per minute) when the hardness of the steel was 180 BHN (Brinell Hardness Number); the cutting speed had to be reduced to about 35 feet per minute (10.7 meters per minute) when the hardness was increased to 360 BHN, the life between tool grindings for these tests being 20 minutes in each case. The machinability of other steels of the same hardness might vary. For example, the tests just referred to showed more or less variation in the cutting speeds for steels of the same hardness, but having different compositions or properties. Thus, while there is a constant relationship between the hardness of a steel and its tensile strength, there is not the same constant relationship between steel hardness and machinability as applied to different steels.

## ENGINEERING ECONOMICS

Engineers, managers, purchasing agents, and others are often required to plan and evaluate project alternatives and make economic decisions that may greatly affect the success or failure of a project. In industry, equipment has to be justified before it can be put into use; multiple projects demand money. The topic of this section is how decisions are made on which projects to prefer over others, based on financial considerations.

The goals of a project, such as reducing manufacturing cost or increasing production, selection of machine tool alternatives, or reduction of tooling, labor and other costs, determine which of the available alternatives may bring the most attractive economic return.

Various cost analysis techniques that may be used to obtain the desired outcome are discussed in the material that follows.

### Interest

*Interest* is money paid out or earned for the use of money that has been borrowed or lent (the *principal*). *Simple interest* is interest paid only on the principal; *compound interest* is interest paid on the principal and any previous interest earned. Principal multiplied by annual interest rate, expressed as a decimal, gives *accrued interest*. *Annual percentage rate* (APR) is percent of principal charged or earned in a year, also referred to as *nominal interest rate per year*. An APR is stated when no intermediate compounding occurs within a year. *Effective rate per year* is calculated by considering multiple compoundings occurring in periods throughout a year, using the nominal interest rate. Thus, effective rate is higher than nominal rate, and the more compounding periods per year, the higher the interest paid over the duration of the loan. Most commercial financial dealings these days involve compound interest.

For example, a nominal interest rate of 10% would be 10/100 = 0.1 as a decimal. Adding the principal to the interest that results from this calculation gives the total value of the account after a year. This is the basis of the formulas for both simple and compound interest.

An example illustrates the difference between simple and compound calculations: A person borrows $10,000 from a bank for 2 years at an annual rate of 10%. At the end of the first year, the net amount owed, using both the simple interest and the compound interest calculation, is $10,000 + ($10,000 × 0.1) = $11,000. Hence, the interest paid for the first year is $1,000, for both ways of calculating. In the second year, if simple interest is applied, the interest owed is still calculated on just the amount borrowed, $10,000. So, at the end of the second year, the interest paid will be the principal of $10,000 plus $2,000 in interest, or a total of $12,000. But, if compound interest is applied, for the second year, the borrower pays an interest of $11,000 × 0.1 = $1,100. Hence, the total paid to pay the bank is $10,000, plus this first-year interest of $1,000, plus the second-year accrued interest of $1,100, for a total of $12,100.

**Variables.**—The symbols used in the formulas to calculate various types of interest are:

$P$ = principal amount of money lent, invested, or borrowed; sometimes also present value ($PV$)

$I$ = nominal annual interest rate stated as a percentage, e.g., 10 percent per annum

$EIR$ = effective annual interest rate when interest is compounded more often than once a year

$i$ = nominal annual interest rate percent expressed as a decimal, e.g., if $I$ = 12 percent, then $i = 12/100 = 0.12$

$n$ = number of annual interest periods (that is, number of years)

$m$ = number of interest compounding periods in one year

$F$ = future sum of money at the end of $n$ interest periods from the present date that is equivalent to $P$ with added interest $i$

$A$ = the payment at the end of each period in a uniform series of payments continuing for $n$ periods, the entire series equivalent to $P$ at interest rate $i$

**Simple Interest.—**Over $n$ years, the simple interest formulas are:

Value of interest ($I$): $I = Pni$

Future account value ($F$): $F = P + I = P + Pni = P(1 + ni)$

*Example:* \$250 is lent for three years at 6 percent simple interest. $P = 250$, $i = 6/100 = 0.06$, $n = 3$. Then:

$$F = 250 + (250 \times 3 \times 0.06) = 250[1 + (0.06 \times 3)] = 250[1.18] = \$295$$

**Compound Interest.—**If compound interest is calculated multiple times in a year, say $m$ times a year, then the future value $F$ of principal $P$ over $n$ years at annual rate $i$ is found by the formula:

$$F = P\left(1 + \frac{i}{m}\right)^{mn}$$

*Example:* Suppose \$10,000 is borrowed from a bank as discussed above for 2 years at 10% APR or nominal interest rate; there are three ways the bank can calculate interest:

Simple interest over 2 years: The borrower will pay $F = P + Pni = 10{,}000 + (10{,}000 \times 2 \times 0.1) = 10{,}000 + 2{,}000 = \$12{,}000$.

Compound interest over 2 years with a nominal interest rate of 10% compounded only at the end of the year: $F = P\,[1 + (i/m)]^{mn} = 10{,}000\,[1 + (0.10/1)]^{1\times2} = 10{,}000 \times 1.1^2 = 10{,}000 \times 1.21 = \$12{,}100$. This is the same the calculation at the beginning of the discussion.

Compound interest over 2 years with a nominal interest rate of 10% compounded four times a year, or $m = 4$: $F = P\,[1 + (i/m)]^{mn} = 10{,}000\,[1 + (0.10/4)]^{4\times2} = 10{,}000 \times 1.025^{4\times2} = 10{,}000 \times 1.2184 = \$12{,}184$. Hence, the interest compounded multiple times ($m = 4$) in a year gives more interest than if compounded only once a year. The effect of the four compounds is that the interest is higher by \$84.

Clearly, compounding yields more interest than simple interest. This example shows the effect of once a year compounding over a number of years, compared to a single calculation at the end of the same number years.

**Determining Principal, Rate, or Time.—**What principal should be invested to reach a known or desired future value, based on a given interest rate and time? At what annual rate should an investment be made when the other parameters are known? How much time will a principal take to reach a future value given an annual rate of interest? Solutions to these questions are found by transposing the main formula to isolate the unknown in question.

*Note:* In the formulas below, the units for time are years. If time is given in months or any other unit, it must be converted to years before substituting it into the formula.

*Principal Value of an Investment or Loan (P):* These days all commercial banks use compound interest only, so only the formulas and steps for calculating compound interest are shown:

Simple interest:

$$F = P(1 + ni), \text{ so } P = \frac{F}{1 + ni} = F(1 + ni)^{-1}$$

Compound interest, $m = 1$:

$$F = P(1 + i)^n, \text{ so } P = \frac{F}{(1+i)^n}, \text{ or } P = F(1 + i)^{-n}$$

Compound interest, $m > 1$:

$$F = P\left(1 + \frac{i}{m}\right)^{mn}, \text{ so } P = \frac{F}{\left(1 + \frac{i}{m}\right)^{mn}} \text{ or } P = F\left(1 + \frac{i}{m}\right)^{-mn}$$

*Example:* Determine the initial deposit (principal) required for an account to reach \$8,000 ($F$), in 10 years ($n$), at an annual interest rate of 4% ($i = 0.04$), compounded monthly ($m = 12$).

*Solution:* Substituting known values into the formula:

$$P = F\left(1 + \frac{r}{n}\right)^{-nt} = 8000\left(1 + \frac{0.04}{12}\right)^{-12\times10} = 8000(1.003333)^{-120} = \$5{,}366$$

*Example:* Determine the principal needed for an account to have a future value of \$25,000 ($F$) in 10 years ($n$), given that interest is compounded *quarterly* (four times a year; so $m = 4$) at an annual rate of 2.5% ($i = 0.025$).

*Solution:* $P$ is sought, so the formula is:

$$P = F\left(1+\frac{i}{m}\right)^{-mn}; \quad P = 25000\left(1+\frac{0.025}{4}\right)^{-4\times10} = 25000(0.7794) = \$19{,}485.17$$

To find $i$ when the other parameters of the loan or investment are known, use the formulas:

At compound interest: $i = \sqrt[mn]{\frac{F}{P}} - 1$

*Solving for Time (n):* Solving the future value formulas for $n$ requires logarithms. The process for compound interest is shown (for properties of logarithms, see page 31):

$$F = P\left(1+\frac{i}{m}\right)^{mn} \rightarrow \frac{F}{P} = \left(1+\frac{i}{m}\right)^{mn} \rightarrow \log\left(\frac{F}{P}\right) = \log\left(1+\frac{i}{m}\right)^{mn}$$

$$\rightarrow \log\left(\frac{F}{P}\right) = mn\log\left(1+\frac{i}{m}\right) \rightarrow n = \frac{\log\left(\frac{F}{P}\right)}{n\log\left(1+\frac{i}{m}\right)}$$

*NOTE:* From logarithm properties, $\log(a/b) = \log a - \log b$.

*Time to Increase Investment (n):* A related approach is taken when calculating time to reach a specific total value:

Compound interest: $n = \dfrac{\log\left(\frac{F}{P}\right)}{n\log\left(1+\frac{i}{m}\right)}$

*Example:* Determine the time required to double \$500, when interest is compounded monthly at a rate of 5%.

*Solution:* Here, $i = 0.05$, $m = 12$, $P = \$500$, and $F = 2P = \$1000$.

$$n = \frac{\log\left(\frac{1000}{500}\right)}{12\log\left(1+\frac{0.05}{12}\right)} = \frac{\log 2}{12\log(1.0042)} = 13.9 \text{ years}$$

*Doubling Time for an Investment:* The time needed to double an investment is the same for any principal, since the variables $F$ and $P$ drop out.

Compound interest: $n_{\text{double}} = \dfrac{\log 2}{n\log\left(1+\frac{i}{n}\right)}$

**Nominal versus Effective Interest Rates.**—Deposits in savings banks, automobile loans, interest on bonds, and many other transactions of this type involve computation of interest due and payable more often than once a year. For such instances, there is a difference between the *nominal* annual interest rate (the stated rate) for the cost of borrowed money and the *effective annual interest rate* (the amount that is actually accrued).

The formula for calculating the effective interest rate (EIR) is:

$$EIR = (1 + i/m)^m - 1$$

The nominal interest rate is also called the "APR," or the annual percentage rate. APR is quoted without compounding frequency per year. If compounding is done only at the end of the year, the APR is the same as the effective rate. If compounding happens more than once in a year, then the effective rate will be greater than the APR. In this case, it is essential the frequency of compounding be stated when APR is quoted. For example, simply saying APR is 10% is not sufficient. Rather, it must be stated that the APR is 10% compounded yearly, for example, or that it is compounded monthly.

*Example:* For a nominal per annum rate of 12%, with monthly compounding, the effective per annum rate is:

$$EIR = (1 + 0.12/12)^{12} - 1 = 0.1268 = 12.7\% \text{ effective annual interest rate}$$

*Example:* Same as before but with quarterly compounding:

$$EIR = (1 + 0.12/4)^{4} - 1 = 0.1255 = 12.6\% \text{ effective annual interest rate}$$

## Cash Flow and Equivalence

*Cash flow* refers to the sum of money received or disbursed, as shown in a project's financial report. Due to the time value of money, the timing of cash flows over a project's life span plays a vital role in project success. Engineering economics problems involve four patterns of cash flow, both separately and in combination. Two cash flow patterns are *equivalent* if they have the same value at a particular time.

**Present Value and Discount.**—The *present value* or present worth $P$ of a given future or final sum $F$ is the amount $P$ that, when placed at interest $i$ for a given time $n$, will produce the given amount $F$.

Simple interest: $P = \frac{F}{1+ni} = F(1+ni)^{-1}$ Compound interest: $P = \frac{F}{(1+i)^n} = F(1+i)^{-n}$

The *true discount* $D$ is the difference between $F$ and $P$:

$$D = F - P.$$

*Example:* Find the present value and discount of \$500 due in six months at 6% simple interest. Here, $F = 500$; $n = 6/12 = 0.5$ year; $i = 0.06$. Then, $P = 500/(1 + 0.5 \times 0.06) =$ \$485.44. The discount is $D = 500 - 485.44 = \$14.56$.

*Example:* Find the sum that, placed at 5% compound interest, will in three years produce \$5,000. Here, $F = 5000$, $i = 0.05$, $n = 3$. Then,

$$P = \frac{5000}{(1 + 0.05)^3} = \$4{,}319.19$$

**Annuities.**—An *annuity* is a fixed sum paid at regular (uniform) intervals. In the formulas that follow, yearly payments are assumed. When a situation entails a monthly payment, such as rent, the interest rate and duration are modified as needed. It is customary to calculate annuities on the basis of compound interest. If an annuity $A$ is to be paid out for $n$ consecutive years, the interest rate being $i$, then the present value $P$ of the annuity is

$$P = A\frac{(1+i)^n - 1}{i(1+i)^n}$$

If at the *beginning* of each year a sum $A$ is set aside at an interest rate $i$, the total value $F$ of the sum set aside, with interest, at the end of $n$ years, will be

$$F = A\frac{(1+i)[(1+i)^n - 1]}{i}$$

If at the *end* of each year a sum $A$ is set aside at an interest rate $i$, then the total value $F$ of the principal, with interest, at the end of $n$ years will be

$$F = A\frac{(1+i)^n - 1}{i}$$

If a principal $P$ is increased or decreased by a sum $A$ at the end of each year, then the value of the principal after $n$ years will be

$$F = P(1+i)^n \pm A\frac{(1+i)^n - 1}{i}$$

If the sum $A$ by which the principal $P$ is decreased each year is greater than the total yearly interest on the principal, then the principal, with the accumulated interest, will be entirely used up in $n$ years:

$$n = \frac{\log\left(\frac{A}{A-iP}\right)}{\log(i+1)}$$

*Example:* If an annuity of $200 is to be paid for 10 years, what is the present amount of money that needs to be deposited if the interest is 5%? Here, $A = 200$, $i = 0.05$, $n = 10$:

$$P = 200\frac{(1+0.05)^{10}-1}{0.05(1+0.05)^{10}} = \$1{,}544.35$$

The annuity a principal $P$ drawing interest at the rate $i$ will give for a period of $n$ years is

$$A = P\frac{i(1+i)^n}{(1+i)^n-1}$$

*Example:* A sum of $10,000 is placed at 4%. What is the amount of the annuity payable for 20 years out of this sum: Here, $P = 10000$, $i = 0.04$, $n = 20$:

$$A = 10000\frac{0.04(1+0.04)^{20}}{(1+0.04)^{20}-1} = \$735.82$$

**Sinking Funds.—**Amortization is the extinction of debt, usually by means of a *sinking fund*. The sinking fund is created by a fixed investment $A$ placed each year at compound interest for a term of years $n$, and is therefore an annuity of sufficient size to produce at the end of the term of years the amount $F$ necessary for the repayment of the principal of the debt, or to provide a definite sum for other purposes. Then,

$$F = A\frac{(1+i)^n-1}{i} \quad \text{and} \quad A = F\frac{i}{(1+i)^n-1}$$

*Example:* If $2,000 is invested annually for 10 years at 4% compound interest, as a sinking fund, what would be the total amount of the fund at the expiration of the term? Here, $A = 2000$, $n = 10$, $i = 0.04$:

$$F = 2000\frac{(1+0.04)^{10}-1}{0.04} = \$24{,}012.21$$

**Cash Flow Diagrams.—**The following conventions are used to standardize cash flow diagrams. The horizontal (time) axis (as seen in several patterns in Table 1 and Table 2) is marked off in equal increments, representing periods, up to the duration of the project. Receipts are represented by arrows directed upwards and disbursements are represented by arrows directed downwards. Arrow length is proportional to magnitude of cash flow. In the following, $A$ = uniform annuity, $i$ = interest rate, and $n$ = number of payments or periods.

**Table 1. Cash Flow Patterns**

| | | |
|---|---|---|
| *P*-pattern<br>$P$ = present value | A single amount $P$ occurring at the beginning of $n$ years. $P$ represents *Present* amount. | 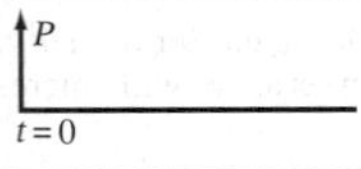  |
| *F*-pattern<br>$F$ = future value | A single amount $F$ occurring at the end of $n$ years. $F$ represents *Future* amount. | 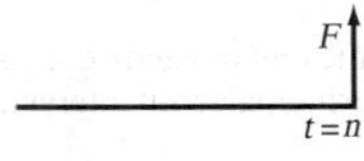  |
| *A*-pattern<br>$A$ = annual value | Equal amounts $A$ occurring at the end of each of $n$ years. $A$ represents *Annual* amount. | 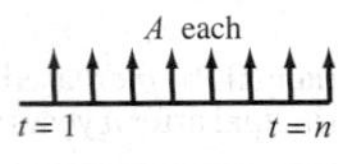  |
| *G*-pattern<br>$G$ = uniform gradient of expense | $G$ is increasing by an equal amount for each time increment (e.g., month) over the period of life $n$. $G$ represents *Gradient* amount. | 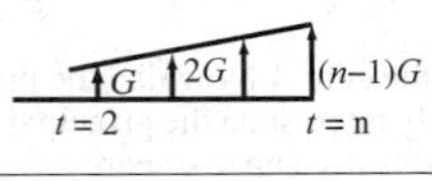  |

**Table 2. Standard Cash Flow Factors**

| Symbol | To Find | Formula | Symbol |
|---|---|---|---|
| 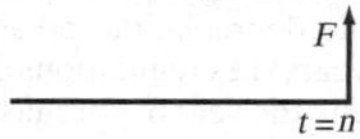  | $F_{\text{given } P}$ | $F = P(1+i)^n$ | 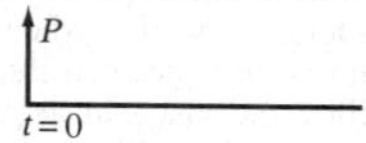  |
|   | $P_{\text{given } F}$ | $P = \dfrac{F}{(1+i)^n}$ | 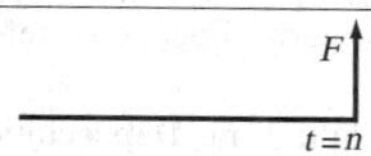  |
| 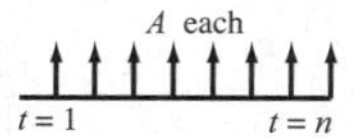  | $A_{\text{given } P}$ | $A = P\dfrac{i(1+i)^n}{(1+i)^n - 1}$ |   |
| 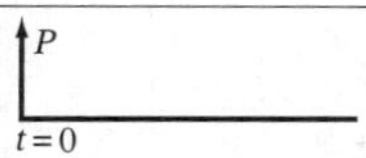  | $P_{\text{given } A}$ | $P = A\dfrac{(1+i)^n - 1}{i(1+i)^n}$ | 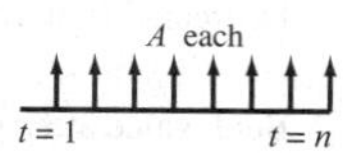  |
| 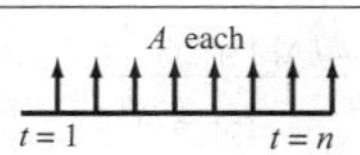  | $A_{\text{given } F}$ | $A = F\dfrac{i}{(1+i)^n - 1}$ |   |
| 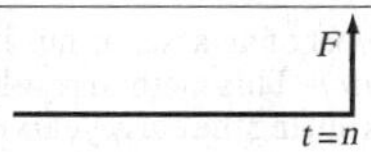  | $F_{\text{given } A}$ | $F = A\dfrac{(1+i)^n - 1}{i}$ | 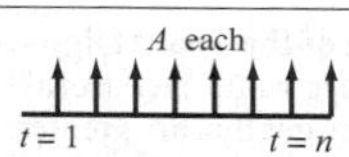  |
| 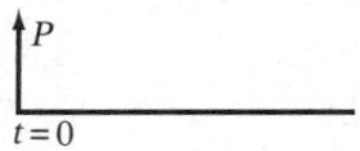  | $P_{\text{given } G}$ | $P = G\dfrac{1}{i}\left[\dfrac{(1+i)^n - 1}{i(1+i)^n} - \dfrac{n}{(1+i)^n}\right]$ | 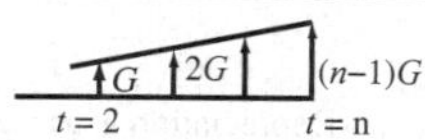  |
| 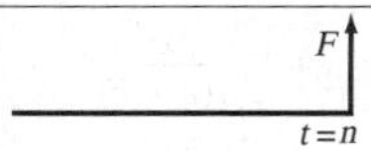  | $F_{\text{given } G}$ | $F = G\dfrac{1}{i}\left[\dfrac{(1+i)^n - 1}{i} - n\right]$ | 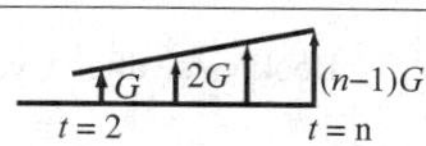  |
| 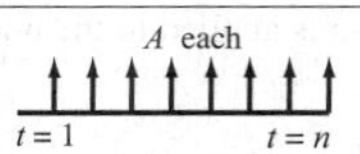  | $A_{\text{given } G}$ | $A = G\left[\dfrac{1}{i} - \dfrac{n}{(1+i)^n - 1}\right]$ | 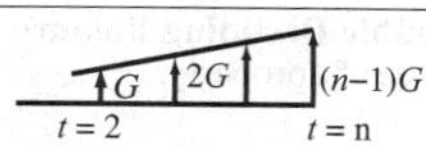  |

*Example:* A rental property pays \$2,000/month with a \$10 per month increase starting the second year. Based on a 10-year period and 8% annual interest, compute the uniform average annuity $A$ from the rental of this property, considering the gradient.

*Solution:* The gradient $G$ = \$10 per month; the interest each month is 1/12 of the rate $i$ = 0.08; the duration $n$ =120 months (using the basis of the annuity, which is monthly rent).

$$A = G\left[\frac{1}{i} - \frac{n}{(1+i)^n - 1}\right] + A_{\text{initial}}$$

$$= 10\left[\frac{1}{(.08/12)} - \frac{120}{(1+(.08/12))^{120} - 1}\right] + 2000$$

$$= 10(150 - 98.39) = 516 + 2000 = \$2{,}516$$

## Depreciation

Depreciation is an accounting method that allocates the cost of an asset over its useful life. A machine may decline in value because it is wearing out and no longer performing its function as well as when it was new. As an economics technique, depreciation spreads the purchase price of an asset or other property over a number of years. Tax regulations do not allow the cost of an asset to be treated as a deductible expense in the year of purchase; rather, a portion of the expense is allocated to each year of the asset's depreciation period. This yearly allocation is referred to as *depreciation*.

**Straight Line Depreciation.**—Straight line depreciation is a constant depreciation charge over the period of life. If $P$ is principal value, $L$ is salvage value and $n$ is period of life, then depreciation is calculated:

Depreciated value of an asset at year $n$: $A = P(1 - x) + L$

Depreciation at $x$th year: $$D_x = \frac{P-L}{n}$$

Book value after $x$ years: $$BV_x = \frac{(P-L)(n-x)}{n} + L$$

After tax depreciation recovery: $$ATDR = TR\left(\frac{P-L}{n}\right)\frac{(1+i)^n - 1}{i(1+i)^n}$$

**Sum of the Years Digits.**—Another method for allocating the cost of an asset minus its salvage value over useful life is *sum of the years digits depreciation*. This method results in charges that are greater than straight line depreciation charges during the early years of an asset and smaller charges near the end period.

Depreciation at $x$th year: $$D_x = \frac{2(P-L)(n-x+1)}{n(n+1)}$$

Book value after $x$ years: $$BV_x = P - (P-L)(2n-x+1)\frac{x}{n(n+1)}$$

**Double Declining Balance Method.**—A constant depreciation is applied to the book value of a property.

Depreciation at $x$th year: $$D_x = 2\left(\frac{P}{n}\right)\left(\frac{n-2}{n}\right)^{(x-1)}$$

Book value after $x$ years: $$BV_x = P\left(\frac{n-2}{n}\right)^x$$

**Statutory Depreciation System.**—A depreciation method used for income tax purposes is *accelerated cost recovery system* (ACRS) depreciation. The first step in ACRS is to determine the property class of the asset being depreciated. All personal property falls into one of six classes. Reading across Table 3 after identifying the correct property class, principal $P$ for year $x$ is substituted into the following formula:

Depreciation at $x$th year: $$D_x = P \times \textit{Depreciation Factor}$$

**Table 3. Property Class and Factor**

| ACRS Classes of Depreciable Property | | Year | Depreciation Factor for Recovery Period ($n$) | | | |
|---|---|---|---|---|---|---|
| Property Class | Personal Property | ($x$) | 3 Years | 5 Years | 7 Years | 10 Years |
| 3 | Handling device for food and beverage manufacture, plastic products, fabricated metal products | 1 | 33.33% | 20.00% | 14.29% | 10.00% |
| | | 2 | 44.45% | 32.00% | 24.49% | 18.00% |
| | | 3 | 14.81% | 19.20% | 17.49% | 14.40% |
| 5 | Automobiles, trucks, computer, aircraft, petroleum drilling equipment, research and experimentation equipment | 4 | 7.41% | 11.52% | 12.49% | 11.52% |
| | | 5 | | 11.52% | 8.93% | 9.22% |
| 7 | Office furniture, fixtures, and equipment | 6 | | 5.76% | 8.92% | 7.37% |
| 10 | Railroad cars, manufacture of tobacco products | 7 | | | 8.93% | 6.55% |
| | | 8 | | | 4.46% | 6.55% |
| 15 | Telephone distribution line, municipal sewers plant | 9 | | | | 6.56% |
| | | 10 | | | | 6.55% |
| 20 | Municipal sewers | 11 | | | | 3.28% |

## Evaluating Alternative Investments

When two or more mutually exclusive investments compete for limited funds, a number of ways are available to select the best investment options from a group of proposals. This section discusses such strategies for analyzing alternatives and making decisions to maximize net value.

**Net Present Value.—**One of the easiest ways to compare mutually exclusive investment alternatives is to calculate and compare the net present value of each. *Net present value* (NPV) is most frequently used to determine the present value of future money receipts and disbursements. Three economic criteria for present worth analysis are described: (1) If investment cost is the same, only the output money is considered. (2) If the output result is known, then the investment cost is minimized. (3) If neither input nor output is fixed, then the output minus the input is maximized. The NPV method is widely applied when the life spans of the alternative proposals are the same.

In selecting the project with the highest NPV from several projects that all achieve the same outcome, a common rate of interest called the minimum attractive rate of return (MARR) is used. The MARR is a rate that companies strive to earn on all their investments. This rate usually is a rate above the interest rate that a company's idle cash will be earning in a risk-free environment. Small-scale companies also may seek investments based on the MARR to achieve returns slightly above the average rate the same money would earn in the stock market on a long-term basis.

The symbols used for the following calculations are defined:

$P$ = present value
$AE$ = annual expense
$i$ = interest rate (as a decimal)
$NPV$ = net present value
$G_A$ = uniform gradient of expense given annual amount $A$
$n$ = number of payments or periods
$AR$ = annual revenue
$TR$ = tax rate as percentage
$L$ = salvage value

With uniform annual expense before tax

$$NPV = -P + (AR - AE)\left[\frac{(1+i)^n - 1}{i(1+i)^n}\right] + \frac{L}{(1+i)^n}$$

With uniform gradient on annual expense before tax

$$NPV = -P + \left(AR - AE - A_G\right)\left[\frac{(1+i)^n - 1}{i(1+i)^n}\right] + \frac{L}{(1+i)^n}$$

With uniform annual expense after tax

$$NPV = -P + (AR - AE)(1 - TR)\left[\frac{(1+i)^n - 1}{i(1+i)^n}\right] + \frac{L}{(1+i)^n}$$

With uniform gradient on annual expense after tax

$$NPV = -P + \left(AR - AE - A_G\right)(1 - TR)\left[\frac{(1+i)^n - 1}{i(1+i)^n}\right] + \frac{L}{(1+i)^n}$$

To include depreciation, the after tax depreciation recovery (ATDR) must be added to get the net present value.

*Example:* A pharmaceutical company makes a product from a composition of chemicals. Two mixing processes, batch and continuous, are available.

| Process | Continuous | Batch |
|---|---|---|
| Initial cost (*P*) | \$75,000 | \$35,000 |
| Lifetime (years) (*n*) | 10 | 10 |
| Maintenance (per year) | \$5,000 | \$8,000 |
| Capacity (units/year) | 25,000 | 20,000 |

The company uses straight line depreciation, pays 40% of its net income as income tax, and has an after tax minimum attractive rate of return of 15%. The company can sell the product at \$1.00 per unit. Which manufacturing process should the company invest in?

*Solution:* Because the lifetimes of the batch and continuous processes are equal at 10 years each, we can make a comparison using the present worth method by applying the formulas for *NPV* and also for *ATDR*.

$$NPV_{\text{Continuous}} = -P + (AR - AE)(1 - TR)\left(\frac{(1+i)^n - 1}{i(1+i)^n}\right) + TR\left(\frac{P-L}{n}\right)\left(\frac{(1+i)^n - 1}{i(1+i)^n}\right)$$

$$= -75000 + ([25000 \times 1] - 5000)(1 - 0.40)\left(\frac{\left(1+0.15\right)^{10} - 1}{0.15\left(1+0.15\right)^{10}}\right) + 0.40\left(\frac{75000}{10}\right)\left(\frac{\left(1+0.15\right)^{10} - 1}{0.15\left(1+0.15\right)^{10}}\right)$$

$$= -14775 + 15056 = 281$$

$$NPV_{\text{Batch}} = -P + (AR - AE)(1 - TR)\left(\frac{(1+i)^n - 1}{i(1+i)^n}\right) + TR\left(\frac{P-L}{n}\right)\left(\frac{(1+i)^n - 1}{i(1+i)^n}\right)$$

$$= -35000 + [(20000 \times 1) - 8000]\,(1 - 0.40)\left(\frac{\left(1+0.15\right)^{10} - 1}{0.15\left(1+0.15\right)^{10}}\right) + 0.40\left(\frac{35000}{10}\right)\left(\frac{\left(1+0.15\right)^{10} - 1}{0.15\left(1+0.15\right)^{10}}\right)$$

$$= 1135 + 7026 = 8161$$

Based on above calculations, the batch production process is selected because it gives a greater net present value (*NPV*) than the continuous process.

**Capitalized Cost.**—In governmental analyses, there are some circumstances where a service is required for an infinite period of time, such as with roads, dams, and pipelines. Present worth of a project with an infinite life is known as *capitalized cost* (CC). Capitalized cost is the amount of money at $n = 0$ needed to perpetually support projected expenses with earned interest only. Capitalized cost is the present sum of money that would need to be set aside now, at some interest rate, to yield the funds required to provide the service.

$$CC = P + (A)(P_{\text{given }A}) - (L)(P_{\text{given }F}) + (G)(P_{\text{given }G})$$

Without periodical replacement: $$CC = P + \frac{A}{i}$$

With 100% periodical replacement: $$CC = P + \frac{P - L}{(1+i)^n - 1} + \frac{A}{i}$$

With periodical renovation cost: $$CC = P + \frac{RC}{(1+i)^n - 1} + \frac{A}{i}$$

where $CC$ = capitalized cost; $P$ = initial cost; $L$ = salvage value; $A$ = annual cost; $RC$ = renovation cost; $i$ = interest rate; and, $n$ = effective period of time.

**Equivalent Uniform Annual Cost.**—This method is applied when the alternatives have unequal periods of life. To avoid unequal periods of time, the present value and future value is converted to an annual value. The alternatives must be mutually exclusive and repeatedly renewed up to the duration of the longest lived alternative.

$$A = (P)(A_{\text{given } P}) - (L)(A_{\text{given } F}) + (G)(A_{\text{given } G}) + AE$$

With sinking fund depreciation:

$$A = (P - L)\frac{i(1+i)^n}{(1+i)^n - 1} + Li + AE$$

With sinking fund depreciation and uniform gradient $G$:

$$A = (P - L)\frac{i(1+i)^n}{(1+i)^n - 1} + Li + AE + G\left[\frac{1}{i} - \frac{n}{(1+i)^n - 1}\right]$$

Straight line depreciation:

$$A = \frac{P - L}{n} + Li + AE + \frac{(P - L)(n+1)i}{2n}$$

*Example:* An investment of $15,000 is being considered to reduce labor and labor-associated costs in a materials handling operation from $8,200 a year to $3,300. This operation is expected to be used for 10 years before being changed or discontinued entirely. In addition to the initial investment of $15,000 and the annual cost of $3,300 for labor, there are additional annual combined costs for power, maintenance, insurance, and property taxes of $1,800 associated with the revised operation. Based on comparisons of annual costs, should the $15,000 investment be made or the present operation continued?

*Solution:* The present annual cost of the operation is $8,200 for labor and labor-associated costs. The proposed operation has an annual cost of $3,300 for labor and labor extras plus $1,800 for additional power, maintenance, insurance, and taxes, plus the annual cost of recovering the initial investment of $15,000 at some interest rate (minimum acceptable rate of return).

Assuming that 10 percent would be an acceptable rate of return on this investment over a period of 10 years, the annual amount to be recovered on the initial investment would be $15,000 multiplied by the capital recovery factor.

Putting this value into $(A_{\text{given } P})$ yields:

$$A = \frac{i(1+i)^n}{(1+i)^n - 1}P + AE = \frac{(0.10)(1+0.10)^{10}}{(1+0.10)^{10} - 1}15000 + 5100 = 7541.18$$

This amount is less than the present annual cost of $8,200. Thus, the investment is justified unless there are other considerations, such as the effects of income taxes, salvage values, expected life, uncertainty about the required rate of return, changes in the cost of borrowed funds, and others.

A tabulation of annual costs of alternative plans A, B, C, etc. is a good way to compare costs item by item. For this example:

| | Item | Plan A | Plan B |
|---|---|---|---|
| 1 | Labor and labor extras | $8,200 | $3,300 |
| 2 | Annual cost of $15,000 investment | | 2,442 |
| 3 | Power | | 400 |
| 4 | Maintenance | | 1,100 |
| 5 | Property taxes and insurance | | 300 |
| | Total annual cost | $8,200 | $7,542 |

*Example, Annual Cost Considering Salvage Value:* If in the previous example the salvage value of the equipment installed was $5,000 at the end of 10 years, what is the effect on the annual cost of the proposed investment of $15,000?

*Solution:* The only item in the annual cost of the previous example that will be affected is the capital recovery amount of $2,442. The following formula gives the amount of annual capital recovery when salvage value is considered:

$$A = (P-L)\frac{i(1+i)^n}{(1+i)^n-1} + Li + AE$$

$$= (15000-5000)\frac{(0.10)(1+0.10)^{10}}{(1+0.10)^{10}-1} + 5000(0.10) + 5100 = 7227.45$$

This amount is $315 less than the previous annual cost of $7,542 for the proposed investment.

**Rate of Return.—**This is the estimated interest rate produced by an investment. *Rate of return (ROR)* is the interest rate at which benefits are equivalent to costs. It is defined as the interest rate paid on the unpaid balance of a loan in such a way that the payment schedule makes the unpaid loan balance equal zero when the final payment is made. It may be computed by finding the interest rate in such a way that the estimated expenditures are equal to the capital gain. *Net Present Worth* = 0, or equivalently, *PW of benefits* – *PW of costs* = 0

$$\frac{(1+ROR)^n-1}{ROR(1+ROR)^n}(AR-AE) + \frac{L}{(1+ROR)^n} = P$$

The rate of return can only be calculated through trial and error. To find out the present worth, a reasonable interest rate is selected and the present worth is calculated. Then another rate is chosen and the present worth is calculated. The value of the *ROR* is interpolated or extrapolated to find the zero value of present worth.

**Benefit-Cost Ratio.—**This is the ratio of present worth of benefit to present worth of cost. This method is applied to municipal project evaluations where benefits (*B*) and costs (*C*) accrue to different segments of the community. A project is considered acceptable if the ratio equals or exceeds 1. For fixed input, $B/C \geq 1$ is maximized, and for fixed output, $B/C \geq 1$ is maximized, and if neither input nor output is fixed, to compute the incremental benefit-cost ratio ($\Delta B/\Delta C$), choose $\Delta B/\Delta C \geq 1$.

*Example:* To build a bridge over a river costs $1,200,000, with benefits of $2,000,000, and disbenefits of $500,000. (a) What is the benefit-cost ratio? (b) What is the excess of benefits over costs?

*Solution:* (a) The benefit-cost ratio *B*/*C* is regarded as (*B* – *D*)/*C*. Here *B* = benefits, *D* = disbenefits, and *C* = cost.

Benefit-cost ratio: (2,000,000 – 500,000)/1,200,000 = 1.25

As the ratio is greater than 1, the project is worth undertaking.

(b) The excess of benefits over cost = 2,000,000 – 1,200,000 – 500,000 = 300,000.

**Payback Period.—**This is the period required to obtain benefits or cost savings equal to the initial investment made. For example, if a piece of equipment costs $10,000 and the company can obtain cost savings of $2,500 per year by procuring that equipment, the payback period can be said to be $10,000 divided by $2,500, which is 4 years. Payback period is a quick way of evaluating whether an investment is worth making. For example, a plant manager who has a payback period of 5 years in her mind might think the $10,000 investment is worth making.

Payback period ignores the rate of interest or assumes that rate to be 0%. Also, payback period ignores benefits coming in a non-linear manner over the life of the equipment.

## TABLE OF CONTENTS

## MECHANICS AND STRENGTH OF MATERIALS

### MECHANICS



### VELOCITY, ACCELERATION, WORK, AND ENERGY



### VELOCITY, ACCELERATION, WORK, AND ENERGY

*(Continued)*



### STRENGTH OF MATERIALS



### RIGID BODY PARAMETERS



MATH & MECHANICS

## TABLE OF CONTENTS

## MECHANICS AND STRENGTH OF MATERIALS

### BEAMS



### COLUMNS



### PLATES, SHELLS, AND CYLINDERS



### SHAFTS



### SHAFTS

*(Continued)*



### SPRINGS



### DISC SPRINGS

## MECHANICS

**Throughout this section of the *Handbook*, both English and metric SI data and formulas are given to cover the requirements of working in either system of measurement. Except for the passage entitled** *Use of Metric SI System in Mechanics Calculations*, **formulas and text relating exclusively to SI are given in bold face type below.**

### Terms and Definitions

**Definitions.—**The science of mechanics is concerned with the effects of forces in causing or preventing motion. *Statics* is the branch of mechanics that deals with bodies in equilibrium, i.e., the forces acting on bodies cause them to remain at rest or to move with uniform velocity. *Dynamics* deals with bodies not in equilibrium, i.e., the forces acting on bodies cause them to move with non-uniform velocity. *Kinetics* is the branch of dynamics that concerns both the forces acting on bodies and the motions that they cause. *Kinematics* is the branch of dynamics that deals only with the motions of bodies without reference to the forces that cause them.

Definitions of certain terms and quantities as used in mechanics follow:

*Force* may be defined simply as a push or a pull, which may result from the force of contact between bodies or from a force, such as magnetism or gravitation, in which no direct contact takes place.

*Matter* is any substance that occupies space; gases, liquids, solids, electrons, atoms, molecules, etc., are all classified as matter.

*Inertia* is that property by which matter tends to resist any change in its motion or state of rest.

*Mass* is a measure of the inertia of a body.

*Work*, in mechanics, is the product of force and distance and is expressed by a combination of units of force and distance, expressed as foot-pound, inch-pound, meter-kilogram, etc. **The metric SI unit of work is the joule (J), which is the work done when the point of application of a force of one newton is displaced through a distance of one meter in the direction of the force.**

*Power*, in mechanics, is the rate of change of work per unit time; it is the product of force and distance divided by time; it measures the amount of work performed in a given time period. It is the rate of doing work and is expressed in foot-pound per minute, foot-pound per second, kilogram-meter per second, etc. **The metric SI unit is the watt (W), which is equivalent to one joule per second.**

*Horsepower* is the unit of power that has been adopted for engineering work. One horsepower (hp) is equal to 33,000 foot-pounds per minute or 550 foot-pounds per second. The *kilowatt* (kW), used in electrical work, equals 1.34 horsepower; or 1 horsepower equals 0.746 kilowatt. **However, in the metric SI, the term "horsepower" is not used, and the basic unit of power is the watt. This unit, and the derived units milliwatt and kilowatt, for example, are the same as those used in electrical work.**

*Torque or moment* of a force is a measure of the tendency of a force to rotate a body upon which it acts about an axis. The magnitude of the moment due to a force acting in a plane perpendicular to some axis is obtained by multiplying the force by the perpendicular distance from the axis to the line of action of the force. (If the axis of rotation is not perpendicular to the plane of the force, then the components of the force in a plane perpendicular to the axis of rotation are used to find the resultant moment of the force by finding the moment of each component and adding the component moments algebraically.) Moment or torque is commonly expressed in pound-feet, pound-inches, kilogram-meters, etc. **The metric SI unit is the newton-meter (N·m).**

*Velocity* is the rate of change of distance; it is expressed as distance divided by time, that is, feet per second, miles per hour, centimeters per second, meters per second, etc.

*Acceleration* is the time-rate of change of velocity and is expressed as velocity divided by time or as distance divided by time squared, that is, in feet per second per second or feet per second squared ($ft/sec^2$); inches per second per second or inches per second squared ($in/sec^2$); centimeters per second per second or centimeters per second squared ($cm/sec^2$); etc. **The metric SI unit is the meter per second squared ($m/sec^2$).**

**Unit Abbreviations.—**Standard abbreviations for the units of physical quantities are used throughout the *Handbook*. Comprehensive tables of unit abbreviations are found starting on page 2836 for US units, and on page 2841 for metric units.

**Unit Systems.—**In mechanics calculations, both *absolute* and *gravitational* systems of units are employed. The fundamental quantities in absolute systems are *length*, *time*, and *mass*, and from these the dimension of *force* is derived. Two absolute systems that have been in use for many years are the CGS (centimeter-gram-second) and the MKS (meter-kilogram-second) systems. They are named for the fundamental units of length, mass and time, respectively. Another system known as MKSA (meter-kilogram-second-ampere) links the MKS system of units of mechanics with electromagnetic units.

**The General Conference of Weights and Measures (CGPM), which is the body responsible for all international matters concerning the metric system, adopted in 1954 a rationalized and coherent system of units based on the four MKSA units, including the kelvin as the unit of temperature and the candela as the unit of luminous intensity. In 1960, the CGPM formally named this system the "Système International d'Unités," for which the abbreviation is SI in all languages. In 1971, the 14th CGPM adopted a seventh base unit, the *mole*, which is the unit of quantity ("amount of substance"). Further details of the SI are given in the section** *MEASURING UNITS* starting on page 2840**, and its application in mechanics calculations, contrasted with the use of the English system, is considered below.**

The fundamental quantities in gravitational systems are *length*, *time*, and *force*, and from these units, the dimension of *mass* is derived. In the gravitational system most widely used in English measure countries, the units of length, time, and force are, respectively, the foot (ft), the second (s or sec), and the pound (lb). The corresponding unit of mass, commonly called the *slug*, is equal to 1 lb-s$^2$/ft and is derived from the formula, $M = W/g$ in which $M$ = mass in slugs, $W$ = weight in pounds, and $g$ = acceleration due to gravity, commonly taken as 32.16 ft/s$^2$. A body that weighs 32.16 lbs on the surface of the earth has, therefore, a mass of 1 slug.

Many engineering calculations utilize a system of units consisting of the inch, the second, and the pound. The corresponding units of mass are pounds second squared per inch (lb-s$^2$/in) and the value of $g$ is taken as 386 in/s$^2$.

In a gravitational system that has been widely used in metric countries, the units of length, time, and force are, respectively, the meter, the second, and the kilogram-force (1 kgf = 9.80665 N). The corresponding units of mass are kgf-s$^2$/m and the value of $g$ is taken as 9.81 m/s$^2$.

**Acceleration of Gravity *g* Used in Mechanics Formulas.—**The acceleration of a freely falling body varies according to location on the earth's surface as well as the height from which the body falls. Its value measured at sea level at the equator is 32.09 ft/sec$^2$ while at the poles is 32.26 ft/sec$^2$. In the United States it is customary to regard 32.16 as satisfactory for most practical purposes in engineering calculations.

*Standard Pound Force:* For use in defining the magnitude of a standard unit of force, known as the *pound force*, a fixed value of 32.1740 ft/sec$^2$, designated by the symbol $g_0$, has been adopted by international agreement. As a result of this agreement, whenever the term mass, $M$, appears in a mechanics formula and the substitution $M = W/g$ is made, use of the standard value $g_0$ = 32.1740 ft/sec$^2$ is implied, although as stated previously, it is customary to use approximate values for $g$ except where extreme accuracy is required.

**Use of Metric SI System in Mechanics Calculations.—**The SI system is a development of the traditional metric system based on decimal arithmetic; fractions are avoided. For each physical quantity, units of different sizes are formed by multiplying or dividing a single base value by powers of 10. Conversions can be made simply by adding zeros or shifting decimal points. For example, the meter is the basic unit of length; *kilo*meter is a multiple (1000 meters); and *milli*meter is a sub-multiple (1/1000 of a meter).

In the older metric system, the simplicity of a series of units linked by powers of 10 is an advantage for plain quantities such as length, but this simplicity is lost as soon as more complex units are encountered. For example, in different branches of science and engineering, energy may appear as the erg, calorie, kilogram-meter, liter-atmosphere, or horsepower-hour. In contrast, the metric SI provides only one basic unit for each physical quantity, and universality is thus achieved.

There are seven *base units,* and in mechanics calculations three are used, which are for the basic quantities of length, mass, and time, expressed as meter (m), kilogram (kg), and second (s). The other four base-units are ampere (A) for electric current, kelvin (K) for thermodynamic temperature, candela (cd) for luminous intensity, and mole (mol) for amount of substance.

The SI is a coherent system. A system of units is said to be coherent if the product or quotient of any two unit quantities in the system is the unit of the resultant quantity. For example, in a coherent system in which the foot is a unit of length, the square foot is the unit of area, whereas the acre is not. Further details of the SI, and definitions of the units are given in the section *MEASURING UNITS* starting on page 2840, near the end of the *Handbook*.

Additional units of physical quantities are derived from the base-units. For example, the unit of velocity is meter per second (m/s), which is a combination of the base-units of length and time. The unit of acceleration is meter per second squared ($m/s^2$). By applying Newton's second law of motion, *force is proportional to mass multiplied by acceleration:*

The unit of force is obtained, which is $kg{\cdot}m/s^2$. This unit is the *newton,* or N.

*Work,* the product of force and distance, has units of $kg{\cdot}m^2/s^2$, which is the *joule,* or J. (1 J = 1 N·m) and energy is also expressed in these terms.

*Power,* or work per unit time, has units of $kg{\cdot}m^2/s^3$, which is the *watt,* or W. (1 W = 1 J/s = 1 N·m/s).

Information on Newton's laws may be found in *Newton's Laws of Motion* on page 176.

The coherence of SI units has two important advantages. The first, that of uniqueness and therefore universality, has been explained. The second is that it greatly simplifies technical calculations. Equations representing physical principles can be applied without introducing such numbers as 550 in power calculations, which, in the English system of measurement have to be used to convert units. Thus, conversion factors largely disappear from calculations carried out in SI units, with a great saving in time and labor.

*Mass, Weight, Force, Load:* SI is an absolute system (see *Unit Systems* on page 152), and consequently it is necessary to make a clear distinction between mass and weight. The *mass* of a body is a measure of its inertia, which is unaffected by gravity whereas the *weight* of a body is the *force* exerted on it by gravity. In a fixed gravitational field, weight is directly proportional to mass, and the distinction between mass and weight can be easily overlooked. However, if a body is moved to a different gravitational field, for example, that of the moon, its weight alters, but its mass remains unchanged. Because the gravitational field on earth varies from place to place by only a small amount, and because weight is proportional to mass, it is practical to use the weight of unit mass as a unit of force, and this procedure is adopted in both the English and older metric systems of measurement. In common usage, they are given the same names, and we say that a mass of 1 pound has a weight of 1 pound. In the former case the pound is being used as a unit of mass, and in the latter case, as a unit of force. This procedure is convenient in some branches of engineering, but leads to confusion in others.

As mentioned earlier, Newton's second law of motion states that force is proportional to mass times acceleration. Because an unsupported body on the earth's surface falls with acceleration $g$ (32 $ft/s^2$ approximately), the pound force (lbf) is that force which will impart an acceleration of $g$ $ft/s^2$ to a 1 pound mass (lbm). Similarly, the kilogram force (kgf) is that force which will impart an acceleration of $g$ (9.8 $m/s^2$ approximately), to a mass of 1 kg. In the SI, the *newton* is that force which will impart *unit acceleration* (1 $m/s^2$) to a mass of one kilogram. The newton is therefore smaller than the kilogram (force) in the ratio 1:$g$ (about 1:9.8). This fact has important consequences in engineering calculations. The factor $g$ now disappears from a wide range of formulas in dynamics, but appears in many formulas in statics where it was formerly absent. It is however not quite the same $g$, for reasons which will now be explained.

In the article on page 180, the mass of a body is referred to as $M$, but it is immediately replaced in subsequent formulas by $W/g$, where $W$ is the weight in pounds (force), which leads to familiar expressions such as $WV^2/2g$ for kinetic energy. In this treatment, the $M$ is really expressed in terms of the slug (page 152), a unit normally used only in aeronautical engineering. In everyday engineers' language, weight and mass are regarded as synonymous and expressions such as $WV^2 / 2g$ are used without pondering the distinction. Nevertheless, on reflection it seems odd that $g$ should appear in a formula which has nothing to do with gravity at all. In fact the $g$ used here is not the true, local value of the acceleration due to gravity, but an arbitrary standard value chosen as part of the definition of the pound force and is more properly designated $g_0$ (page 152). Its function is not to indicate the strength of the local gravitational field but to convert from one unit to another.

In SI the unit of mass is the *kilogram*, and the unit of force (and weight) is the *newton*.

The following are typical statements in dynamics expressed in SI units:

A force of $R$ newtons acting on a mass of $M$ kilograms produces an acceleration of $R/M$ m/s². The kinetic energy of a mass of $M$ kg moving with velocity $V$ m/s is $\frac{1}{2} MV^2$ kg (m/s)² or $\frac{1}{2} MV^2$ joules. The work done by a force of $R$ newtons moving a distance $L$ meters is $RL$ N·m, or $RL$ joules. If this work were converted entirely into kinetic energy we could write $RL = \frac{1}{2} MV^2$ and it is instructive to consider the units. Remembering that the N is the same as the kg·m/s², we have (kg·m/s²) m = kg (m/s)², which is clearly correct. It will be noted that $g$ does not appear anywhere in these statements.

In contrast, in many branches of engineering where the weight of a body is important, rather than its mass, using SI units, $g$ does appear where formerly it was absent. Thus, if a rope hangs vertically supporting a mass of $M$ kilograms the tension in the rope is $Mg$ N. Here $g$ is acceleration due to gravity, and its units are m/s². The ordinary numerical value of 9.81 will be sufficiently accurate for most purposes on earth. The expression is still valid elsewhere, for example, on the moon, provided the proper value of $g$ is used. The maximum tension the rope can safely withstand (and other similar properties) will also be specified in terms of the newton, so that direct comparison may be made with the tension predicted.

Words like load and weight have to be used with greater care. In everyday language we might say "a lift carries a load of five people of average weight 70 kg," but in precise technical language we say that if the average mass is 70 kg, then the average weight is $70g$ N, and the total load (that is force) on the lift is $350g$ N.

If the lift starts to rise with acceleration $a$ m/s², the load becomes 350 $(g + a)$ N; both $g$ and $a$ have units of m/s², the mass is in kg, so the load is in terms of kg·m/s², which is the same as the newton.

*Pressure and stress:* These quantities are expressed in terms of force per unit area. In SI the unit is the *pascal* (Pa), which expressed in terms of SI derived and base units is the newton per meter squared (N/m²). The pascal is very small—equivalent to $0.15 \times 10^{-3}$ lb/in² — hence the kilopascal (kPa = 1000 pascals) and the megapascal (MPa = $10^6$ pascals) may be more convenient multiples in practice. Thus, note: 1 N/mm² = 1 MN/m² = 1 MPa.

In addition to the pascal, the *bar,* a non-SI unit, is in use in the field of pressure measurement in some countries, including England. Thus, in view of existing practice, the International Committee of Weights and Measures (CIPM) decided in 1969 to retain this unit for a limited time for use with those of SI. 1 bar = $10^5$ pascals and 1 hectobar = $10^7$ pascals.

## Force Systems

**Scalar and Vector Quantities.—**The quantities dealt with in mechanics are of two kinds according to whether magnitude alone or direction as well as magnitude must be known in order to completely specify them. Quantities such as time, volume and density are completely specified when their magnitude is known. Such quantities are called *scalar* quantities. Quantities such as force, velocity, acceleration, moment, and displacement must, in order to be specified completely, include direction as well as magnitude, are called *vector* quantities.

**Graphical Representation of Forces.—**A force has three characteristics which, when known, determine it. They are *direction, point of application*, and *magnitude*. The direction of a force is the direction it tends to move the body upon which it acts. The point of application is the place on the line of action where the force is applied. Forces may conveniently be represented by straight lines and arrow heads. The arrow head indicates the direction of the force, and the length of the line, its magnitude to any suitable scale. The point of application may be at any point on the line, but it is generally convenient to assume it to be at one end. In Fig. 1, a force is supposed to act along line $AB$ in a direction from left to right. The length of line $AB$ shows the magnitude of the force. If point $A$ is the point of application, the force is exerted as a pull, but if point *B* be assumed to be the point of application, it would indicate that the force is exerted as a push.

Fig. 1. Vector

Velocities, moments, displacements, etc. may similarly be represented and manipulated graphically because they are all of the same class of quantities, *vectors*. (See *Scalar and Vector Quantities* above.)

*Addition and Subtraction of Forces:* The resultant of two forces applied at the same point and acting in the same direction, as in Fig. 2, is equal to the sum of the forces. For example, if forces *AB* and *AC*, one equal to 2 lbs the other equal to 3 lbs, are applied at point *A*, then their resultant *AD* equals the sum of these forces, or 5 lbs.

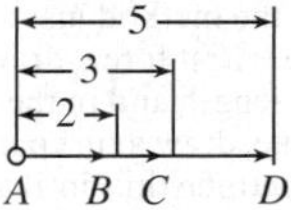


Fig. 2.

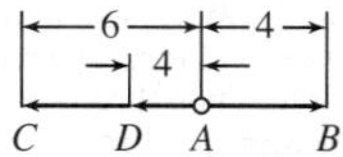


Fig. 3.

If two forces act in opposite directions, as in Fig. 3, then their resultant is equal to their difference, and the direction of the resultant is the same as the direction of the greater of the two forces. For example, *AB* and *AC* are both applied at point *A*; then, if *AB* equals 4 N and *AC* equals 6 N, the resultant force *AD* equals 2 N and acts in the direction of *AC*.

*Parallelogram of Forces:* If two forces applied at a point are represented in magnitude and direction by the adjacent sides of a parallelogram (*AB* and *AC* in Fig. 4), their resultant will be represented in magnitude and direction by the diagonal *AR* drawn from the intersection of the two component forces.

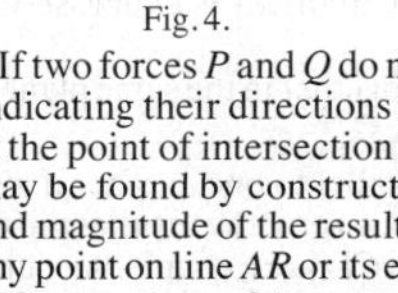


Fig. 4.

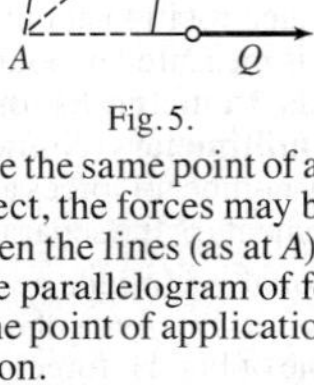


Fig. 5.

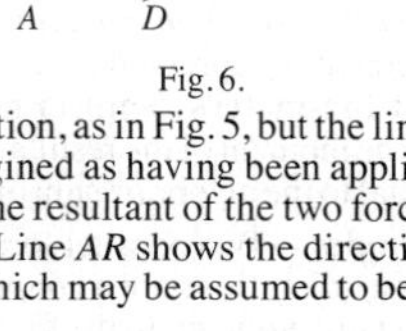


Fig. 6.

If two forces *P* and *Q* do not have the same point of application, as in Fig. 5, but the lines indicating their directions intersect, the forces may be imagined as having been applied at the point of intersection between the lines (as at *A*), and the resultant of the two forces may be found by constructing the parallelogram of forces. Line *AR* shows the direction and magnitude of the resultant, the point of application of which may be assumed to be at any point on line *AR* or its extension.

If the resultant of three or more forces having the same point of application is to be found, as in Fig. 6, first the resultant of any two of the forces (*AB* and *AC*) is found. Then, the resultant of the resultant just found ($AR_1$) and the third force (*AD*) is found. If there are more than three forces, continue in this manner until the resultant of all remaining forces has been found.

*Parallel Forces:* If two forces are parallel and act in the same direction, as in Fig. 7, then their resultant is parallel to both lines, is located between them, and is equal to the sum of the two components. The point of application of the resultant divides the line joining the points of application of the components inversely as the magnitude of the components. Thus,

$$AB{:}CE = CD{:}AD$$

The resultant of two parallel and unequal forces acting in opposite directions, Fig. 8, is parallel to both lines, is located outside of them on the side of the greater of the components, has the same direction as the greater component, and is equal in magnitude to the difference between the two components. The point of application on the line *AC* produced is found from the proportion:

$$AB:CD = CE:AE$$

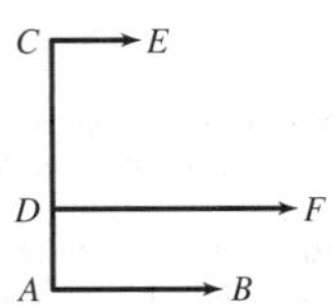


Fig. 7.

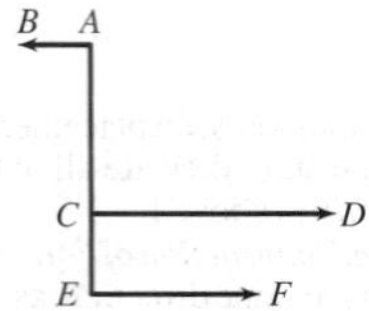


Fig. 8.

*Polygon of Forces:* When several forces are applied at a point and act in a single plane, Fig. 9, their resultant may be found more simply than by the method just described, as follows: From the extreme end of the line representing the first force, draw a line representing the second force, parallel to it and of the same length and in the direction of the second force. Then through the extreme end of this line, draw a line parallel to, and of the same length and direction as the third force, and continue this until all the forces have been thus represented. Then draw a line from the point of application of the forces (as *A*) to the extreme point (as $5_1$) of the line last drawn. This line ($A5_1$) is the resultant of the forces.

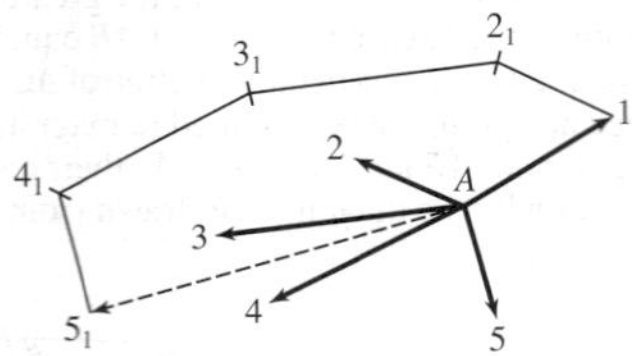


Fig. 9.

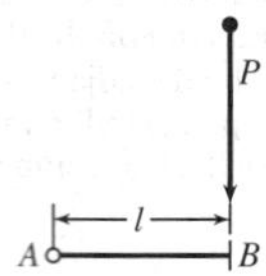


Fig. 10.

*Moment of a Force:* The moment of a force with respect to a point is the magnitude of the force multiplied by the perpendicular distance from the given point to the direction of the force. In Fig. 10, the moment of the force *P* with relation to point *A* is $P \times AB$. The perpendicular distance *AB* is called the *lever-arm* of the force. The moment is the measure of the tendency of the force to produce rotation about the given point, which is termed the *center-of-moments*. If the force is measured in pounds and the distance in inches, the moment is expressed in inch-pounds. **In metric SI units, the moment is expressed in newton-meters (N·m), or newton-millimeters (N·mm).**

The moment of the resultant of any number of forces acting together in the same plane is equal to the algebraic sum of the moments of the separate forces.

**Couples.**—Two parallel forces of equal magnitude acting parallel to one another in opposite directions are a *couple*. The *resultant force* of a couple is zero; the resultant itself is a *pure moment*. In the first example of Fig. 11, forces *AB* and *CD* are a couple. A couple tends to produce *rotation*. The measure of this tendency is called the moment of the couple; it is the product of one of the forces multiplied by the distance between the two.

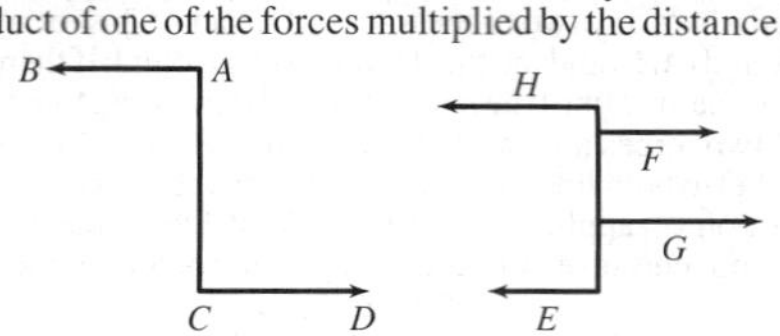


Fig. 11. Two Examples of Couples

As a couple has zero resultant force, no single force can balance or counteract the tendency of the couple to produce rotation. To prevent the rotation of a body acted upon by a couple, two other forces are required, forming a second couple. In the second illustration of Fig. 11, *E* and *F* form one couple and *G* and *H* are the balancing couple. The body on which they act is in equilibrium if the moments of the two couples are equal and tend to rotate the body in opposite directions. A couple may also be represented by a vector in the direction of the axis about which the couple acts. The length of the vector, to some scale, represents the magnitude of the couple, and the direction of the vector is that in which a right-hand screw would advance if it were to be rotated by the couple.

**Composition of a Single Force and Couple.—**A single force and a couple in the same plane or in parallel planes may be replaced by another single force equal in magnitude and parallel to the first force, at a distance from it equal to the moment of the couple divided by the magnitude of the force. The new single force is located so that the moment of the resultant about the point of application of the original force is of the same sign as the moment of the couple.

Fig. 12 illustrates this relationship. Forces *N* and –*N* are a couple. The moment of this couple is $N(ac + bc)$. The resultant of *N*, –*N*, and a single force *P* is a force *O*, equal in magnitude to *P* and acting on an axis through a point *c* such that the moment of *P* and –*N* are equal to the moment of *N*. That is: $(P - N) \times ac = N \times bc$.

Thus,

$$ac = \frac{N(ac + bc)}{P} = \frac{\text{Moment of Couple}}{P}$$

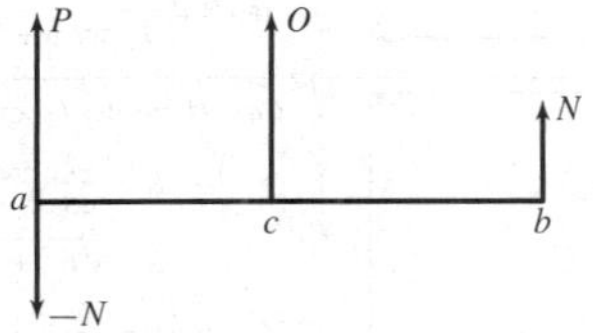


Fig. 12. Single Force and Couple Composition

**Algebraic Composition and Resolution of Force Systems.—**The graphical methods given beginning on page 154 are convenient for solving problems involving force systems in which all of the forces lie in the same plane and only a few forces are involved. If many forces are involved, however, or the forces do not lie in the same plane, it is better to use algebraic methods to avoid complicated space diagrams. Systematic procedures for solving force problems by algebraic methods are outlined beginning on page 157. In connection with the use of these procedures, several terms applicable to force systems in general must be defined.

As has been illustrated, single force that produces the same effect upon a body as two or more forces acting together is called their *resultant*. The separate forces which can be so combined are called the *components*. Finding the resultant of two or more forces is called the *composition* of forces, and finding two or more components of a given force, the *resolution* of forces. Forces are said to be *concurrent* when their lines of action can be extended to meet at a common point; forces that are *parallel* are, of course, *nonconcurrent*. Two forces having the same line of action are said to be *collinear*. Two forces equal in magnitude, parallel, and in opposite directions constitute a *couple*. Forces all in the same plane are said to be *coplanar;* if not in the same plane, they are called *noncoplanar* forces.

The resultant of a system of forces is the simplest equivalent system that can be determined. It may be a single force, a couple, or a noncoplanar force and a couple. This last type of resultant, a noncoplanar force and a couple, may be replaced, if desired, by two *skewed* forces (forces that are nonconcurrent, nonparallel, and noncoplanar). When the resultant of a system of forces is zero, the system is in equilibrium, that is, the body on which the force system acts remains at rest or continues to move with uniform velocity.

## Algebraic Solution of Force Systems — All Forces in the Same Plane

*Finding Two Concurrent Components of a Single Force:*

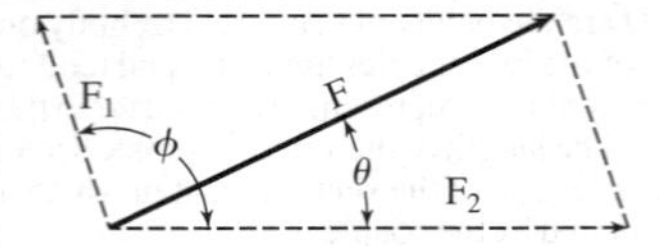


*Case I:* To find two components $F_1$ and $F_2$ at angles $\theta$ and $\phi$, $\phi$ not being 90°.

$$F_1 = \frac{F \sin\theta}{\sin\phi}$$

$$F_2 = \frac{F \sin(\phi - \theta)}{\sin\phi}$$

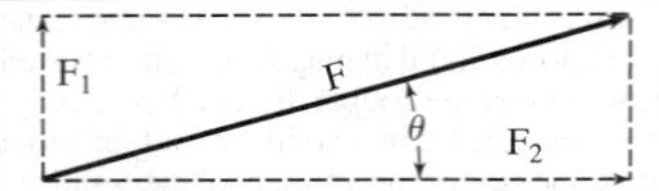


*Case II:* Components $F_1$ and $F_2$ form 90° angle.

$$F_1 = F \sin\theta$$

$$F_2 = F \cos\theta$$

*Finding the Resultant of Two Concurrent Forces:*

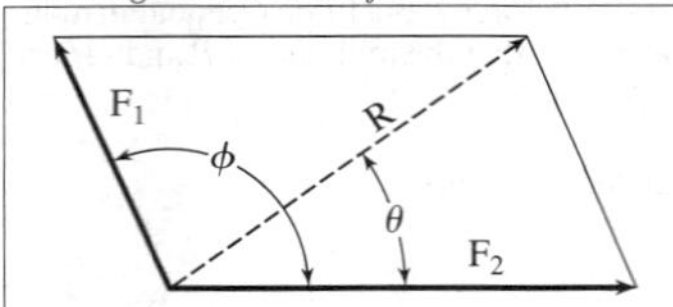


*Case I:* Forces $F_1$ and $F_2$ do not form 90° angle.

$$R = \frac{F_1 \sin\phi}{\sin\theta} \quad \text{or} \quad R = \frac{F_2 \sin\phi}{\sin(\phi - \theta)} \quad \text{or}$$

$$R = \sqrt{F_1^2 + F_2^2 + 2F_1 F_2 \cos\phi}$$

$$\tan\theta = \frac{F_1 \sin\phi}{F_1 \cos\phi + F_2}$$

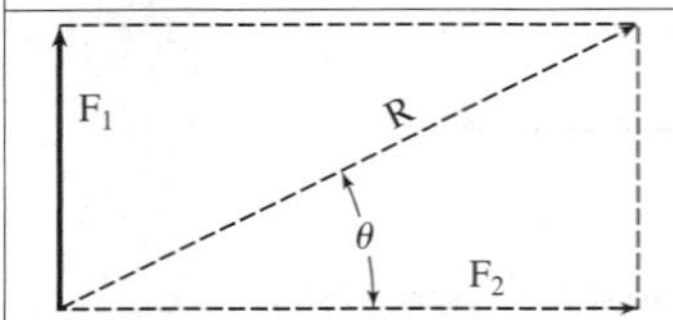


*Case II:* Forces $F_1$ and $F_2$ form 90° angle.

$$R = \frac{F_2}{\cos\theta} \quad \text{or} \quad R = \frac{F_1}{\sin\theta} \quad \text{or}$$

$$R = \sqrt{F_1^2 + F_2^2}$$

$$\tan\theta = \frac{F_1}{F_2}$$

*Finding the Resultant of Three or More Concurrent Forces:*

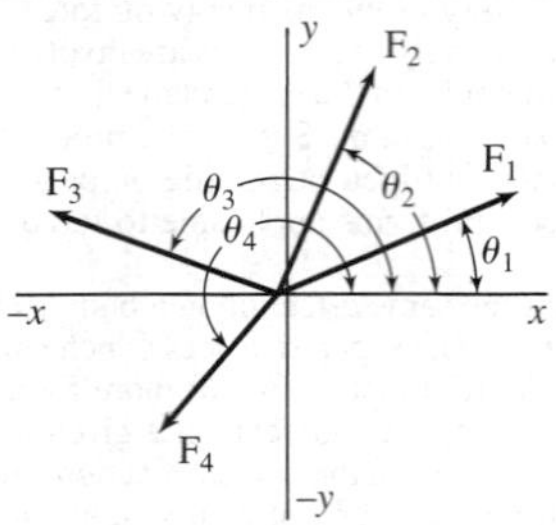


To determine resultant of forces $F_1, F_2, F_3$, etc. making angles, respectively, of $\theta_1, \theta_2, \theta_3$, etc. with the $x$-axis, find the $x$ and $y$ components $F_x$ and $F_y$ of each force and arrange in a table similar to that shown below for a system of three forces. Find the algebraic sum of the $F_x$ and $F_y$ components ($\Sigma F_x$ and $\Sigma F_y$) and use these to determine resultant $R$.

| Force | $F_x$ | $F_y$ |
|---|---|---|
| $F_1$ | $F_1 \cos\theta_1$ | $F_1 \sin\theta_1$ |
| $F_2$ | $F_2 \cos\theta_2$ | $F_2 \sin\theta_2$ |
| $F_3$ | $F_3 \cos\theta_3$ | $F_3 \sin\theta_3$ |
| | $\Sigma F_x$ | $\Sigma F_y$ |

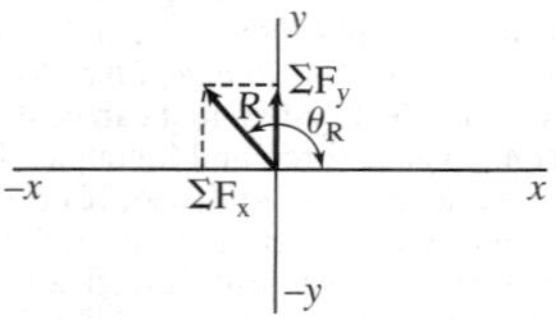


$$R = \sqrt{(\Sigma F_x)^2 + (\Sigma F_y)^2}$$

$$\cos\theta_R = \frac{\Sigma F_x}{R}$$

$$\text{or} \quad \tan\theta_R = \frac{\Sigma F_y}{\Sigma F_x}$$

*Finding a Force and a Couple which Together Are Equivalent to a Single Force:*

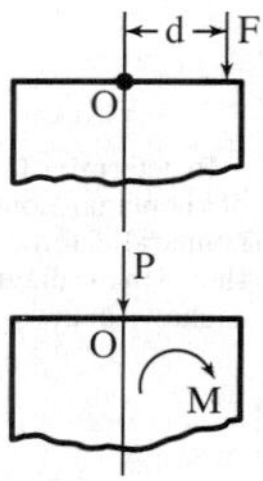


To resolve a single force $F$ into a couple of moment $M$ and a force $P$ passing through any chosen point $O$ at a distance $d$ from the original force $F$, use the relations

$$P = F$$

$$M = F \times d$$

The moment $M$ must, of course, tend to produce rotation about $O$ in the same direction as the original force. Thus, as seen in the diagram, $F$ tends to produce clockwise rotation; hence $M$ is shown clockwise.

*Finding the Resultant of a Single Force and a Couple:*

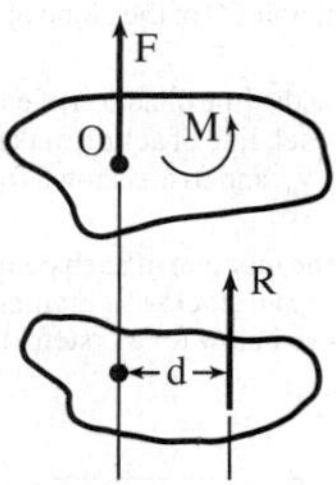


The resultant of a single force $F$ and a couple $M$ is a single force $R$ equal in magnitude and direction to $F$ and parallel to it at a distance $d$ to the left or right of $F$.

$$R = F$$

$$d = M \div R$$

Resultant $R$ is placed to the left or right of point of application $O$ of the original force $F$ depending on which position will give $R$ the same direction of moment about $O$ as the original couple $M$.

*Finding the Resultant of a System of Parallel Forces:*

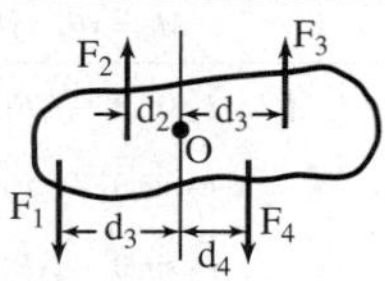


To find the resultant of a system of coplanar parallel forces, proceed as indicated below.

1) Select any convenient point $O$ from which perpendicular distances $d_1, d_2, d_3$, etc. to parallel forces $F_1, F_2, F_3$, etc. can be specified or calculated.

2) Find the algebraic sum of all the forces; this will give the magnitude of the resultant of the system.

$$R = \Sigma F = F_1 + F_2 + F_3 + \ldots$$

3) Find the algebraic sum of the moments of the forces about $O$; clockwise moments may be taken as negative and counterclockwise moments as positive:

$$\Sigma M_O = F_1 d_1 + F_2 d_2 + \ldots$$

4) Calculate the distance $d$ from $O$ to the line of action of resultant R:

$$d = \frac{\Sigma M_O}{R}$$

This distance is measured to the left or right from $O$ depending on which position will give the moment of $R$ the same direction of rotation about $O$ as the couple $\Sigma M_O$, that is, if $\Sigma M_O$ is negative, then $d$ is left or right of $O$ depending on which direction will make $R \times d$ negative.

*Note Concerning Interpretation of Results:* If $R = 0$, then the resultant of the system is a couple $\Sigma M_O$; if $\Sigma M_O = 0$ then the resultant is a single force $R$; if both $R$ and $\Sigma M_O = 0$, then the system is in equilibrium.

*Finding the Resultant of Forces Not Intersecting at a Common Point:*

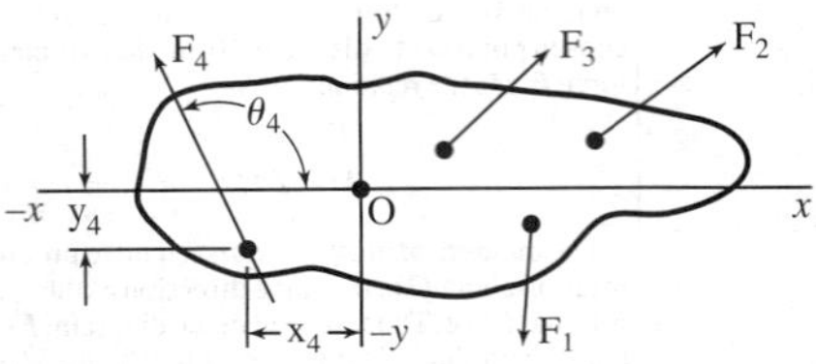


To determine the resultant of a coplanar, nonconcurrent, nonparallel force system as shown in the diagram, proceed as shown below.

1) Draw a set of $x$ and $y$ coordinate axes through any convenient point $O$ in the plane of the forces as shown in the diagram.
2) Determine the $x$ and $y$ coordinates of any convenient point on the line of action of each force and the angle $\theta$, measured in a counterclockwise direction, that each line of action makes with the positive $x$-axis. For example, in the diagram, coordinates $x_4$, $y_4$, and $\theta_4$ are shown for $F_4$. Similar data should be known for each of the forces of the system.
3) Calculate the $x$ and $y$ components ($F_x$, $F_y$) of each force and the moment of each component about $O$. Counterclockwise moments are considered positive and clockwise moments are negative. Tabulate all results in a manner similar to that shown below for a system of three forces and find $\Sigma F_x$, $\Sigma F_y$, $\Sigma M_O$ by algebraic addition.

| Force | Coordinates of $F$ | | | Components of $F$ | | Moment of $F$ about $O$ |
|---|---|---|---|---|---|---|
| $F$ | $x$ | $y$ | $\theta$ | $F_x$ | $F_y$ | $M_O = xF_y - yF_x$ |
| $F_1$ | $x_1$ | $y_1$ | $\theta_1$ | $F_1 \cos \theta_1$ | $F_1 \sin \theta_1$ | $x_1F_1 \sin \theta_1 - y_1F_1 \cos \theta_1$ |
| $F_2$ | $x_2$ | $y_2$ | $\theta_2$ | $F_2 \cos \theta_2$ | $F_2 \sin \theta_2$ | $x_2F_2 \sin \theta_2 - y_2F_2 \cos \theta_2$ |
| $F_3$ | $x_3$ | $y_3$ | $\theta_3$ | $F_3 \cos \theta_3$ | $F_3 \sin \theta_3$ | $x_3F_3 \sin \theta_3 - y_3F_3 \cos \theta_3$ |
| | | | | $\Sigma F_x$ | $\Sigma F_y$ | $\Sigma M_O$ |

4) Compute the resultant of the system and the angle $\theta_R$ it makes with the $x$-axis by using the formulas:

$$R = \sqrt{(\Sigma F_x)^2 + (\Sigma F_y)^2}$$

$$\cos \theta_R = \Sigma F_x \div R \quad \text{or} \quad \tan \theta_R = \Sigma F_y \div \Sigma F_x$$

5) Calculate the distance $d$ from $O$ to the line of action of the resultant $R$:

$$d = \Sigma M_O \div R$$

Distance $d$ is in such direction from $O$ as will make the moment of $R$ about $O$ have the same sign as $\Sigma M_O$.

*Note Concerning Interpretation of Results:* If $R = 0$, then the resultant is a couple $\Sigma M_O$; if $\Sigma M_O = 0$, then $R$ passes through $O$; if both $R = 0$ and $\Sigma M_O = 0$, then the system is in equilibrium.

*Example:* Find the resultant of three coplanar nonconcurrent forces for which the following data are given.

$$F_1 = 10 \text{ lbs}; \; x_1 = 5 \text{ in.}; \; y_1 = -1 \text{ in.}; \; \theta_1 = 270°$$
$$F_2 = 20 \text{ lbs}; \; x_2 = 4 \text{ in.}; \; y_2 = 1.5 \text{ in.}; \; \theta_2 = 50°$$
$$F_3 = 30 \text{ lbs}; \; x_3 = 2 \text{ in.}; \; y_3 = 2 \text{ in.}; \; \theta_3 = 60°$$

$$F_{x_1} = 10\cos 270° = 10 \times 0 = 0 \text{ lbs.}$$
$$F_{x_2} = 20\cos 50° = 20 \times 0.64279 = 12.86 \text{ lbs.}$$
$$F_{x_3} = 30\cos 60° = 30 \times 0.5000 = 15.00 \text{ lbs.}$$

$$F_{y_1} = 10 \times \sin 270° = 10 \times (-1) = -10.00 \text{ lbs.}$$
$$F_{y_2} = 20 \times \sin 50° = 20 \times 0.76604 = 15.32 \text{ lbs.}$$
$$F_{y_3} = 30 \times \sin 60° = 30 \times 0.86603 = 25.98 \text{ lbs.}$$

$$M_{O_1} = 5 \times (-10) - (-1) \times 0 = -50 \text{ in. lbs.}$$
$$M_{O_2} = 4 \times 15.32 - 1.5 \times 12.86 = 41.99 \text{ in. lbs.}$$
$$M_{O_3} = 2 \times 25.98 - 2 \times 15 = 21.96 \text{ in. lbs.}$$

***Note:* When working in metric SI units, pounds are replaced by newtons (N); inches by meters or millimeters, and inch-pounds by newton-meters (N · m) or newton-millimeters (N · mm).**

| Force $F$ | Coordinates of $F$ $x$ | $y$ | $\theta$ | Components of $F$ $F_x$ | $F_y$ | Moment of $F$ about $O$ |
|---|---|---|---|---|---|---|
| $F_1 = 10$ | 5 | –1 | 270° | 0 | –10.00 | –50.00 |
| $F_2 = 20$ | 4 | 1.5 | 50° | 12.86 | 15.32 | 41.99 |
| $F_3 = 30$ | 2 | 2 | 60° | 15.00 | 25.98 | 21.96 |
| | | | | $\Sigma F_x = 27.86$ | $\Sigma F_y = 31.30$ | $\Sigma M_O = 13.95$ |

$$R = \sqrt{(27.86)^2 + (31.30)^2}$$
$$= 41.90 \text{ lbs}$$
$$\tan\theta_R = \frac{31.30}{27.86} = 1.1235$$
$$\theta_R = 48°20'$$
$$d = \frac{13.95}{41.90} = 0.33 \text{ inch}$$

measured as shown on the diagram.

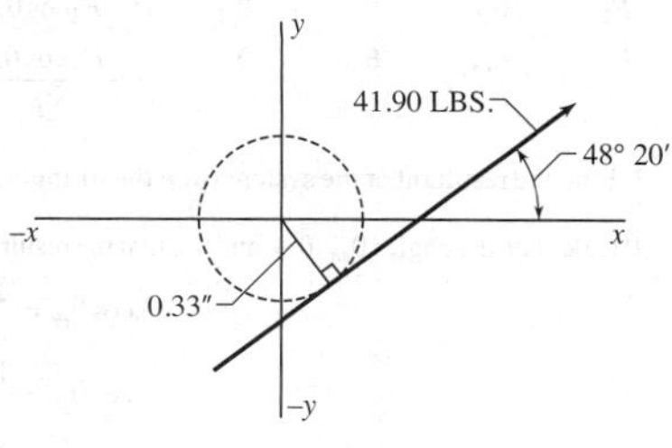

## Algebraic Solution of Force Systems — Forces Not in Same Plane

*Resolving a Single Force Into Its Three Rectangular Components:*

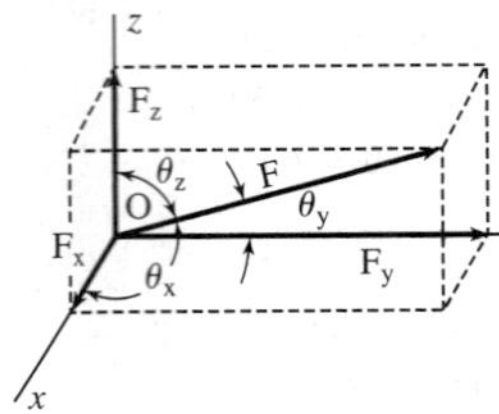


$$F_x = F\cos\theta_x$$
$$F_y = F\cos\theta_y$$
$$F_z = F\cos\theta_z$$
$$F = \sqrt{F_x^2 + F_y^2 + F_z^2}$$

The diagram shows how a force $F$ may be resolved at any point $O$ on its line of action into three concurrent components each of which is perpendicular to the other two.

The $x, y, z$ components $F_x, F_y, F_z$ of force $F$ are determined from the accompanying relations in which $\theta_x, \theta_y, \theta_z$ are the angles that the force $F$ makes with the $x, y, z$ axes.

*Finding the Resultant of Any Number of Concurrent Forces:*

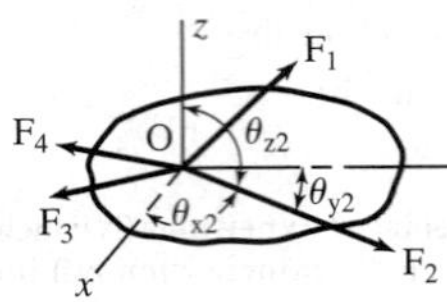


To find the resultant of any number of noncoplanar concurrent forces $F_1, F_2, F_3$, etc., use the procedure outlined below.

1) Draw a set of $x, y, z$ axes at $O$, the point of concurrency of the forces. The angles each force makes measured counterclockwise from the positive $x$, $y$, and $z$ coordinate axes must be known in addition to the magnitudes of the forces. For force $F_2$, for example, the angles are $\theta_{x2}, \theta_{y2}, \theta_{z2}$ as indicated on the diagram.
2) Apply the first three formulas given under the heading "Resolving a Single Force Into Its Three Rectangular Components" to each force to find its $x$, $y$, and $z$ components. Tabulate these calculations as shown below for a system of three forces. Algebraically add the calculated components to find $\Sigma F_x, \Sigma F_y$, and $\Sigma F_z$ which are the components of the resultant.

| Force | Angles | | | Components of Forces | | |
|---|---|---|---|---|---|---|
| $F$ | $\theta_x$ | $\theta_y$ | $\theta_z$ | $F_x$ | $F_y$ | $F_z$ |
| $F_1$ | $\theta_{x1}$ | $\theta_{y1}$ | $\theta_{z1}$ | $F_1\cos\theta_{x1}$ | $F_1\cos\theta_{y1}$ | $F_1\cos\theta_{z1}$ |
| $F_2$ | $\theta_{x2}$ | $\theta_{y2}$ | $\theta_{z2}$ | $F_2\cos\theta_{x2}$ | $F_2\cos\theta_{y2}$ | $F_2\cos\theta_{z2}$ |
| $F_3$ | $\theta_{x3}$ | $\theta_{y3}$ | $\theta_{z3}$ | $F_3\cos\theta_{x3}$ | $F_3\cos\theta_{y3}$ | $F_3\cos\theta_{z3}$ |
| | | | | $\Sigma F_x$ | $\Sigma F_y$ | $\Sigma F_z$ |

3) Find the resultant of the system from the formula $R = \sqrt{(\Sigma F_x)^2 + (\Sigma F_y)^2 + (\Sigma F_z)^2}$

4) Calculate the angles $\theta_{xR}, \theta_{yR}$, and $\theta_{zR}$ that the resultant $R$ makes with the respective coordinate axes:

$$\cos\theta_{xR} = \frac{\Sigma F_x}{R}$$

$$\cos\theta_{yR} = \frac{\Sigma F_y}{R}$$

$$\cos\theta_{zR} = \frac{\Sigma F_z}{R}$$

*Finding the Resultant of Parallel Forces Not in the Same Plane:*

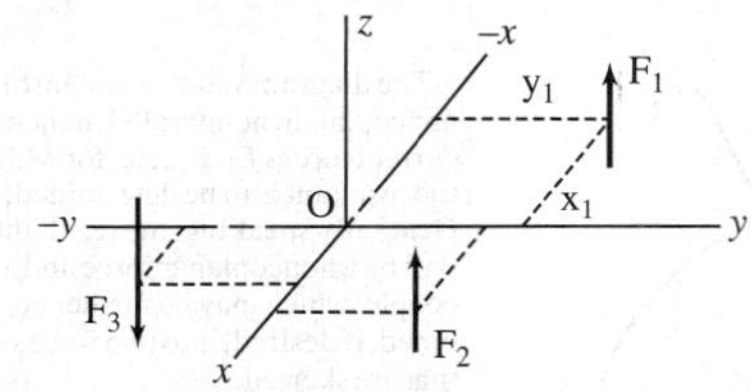


In the diagram, forces $F_1, F_2$, etc. represent a system of noncoplanar parallel forces. To find the resultant of such systems, use the procedure shown below.

1) Draw a set of $x$, $y$, and $z$ coordinate axes through any point $O$ in such a way that one of these axes, say the $z$-axis, is parallel to the lines of action of the forces. The $x$- and $y$-axes then will be perpendicular to the forces.
2) Set the distances of each force from the $x$- and $y$-axes in a table as shown below. For example, $x_1$ and $y_1$ are the $x$ and $y$ distances for $F_1$ shown in the diagram.
3) Calculate the moment of each force about the $x$- and $y$-axes and set the results in the table as shown for a system consisting of three forces. The algebraic sums of the moments $\Sigma M_x$ and $\Sigma M_y$ are then obtained. (In taking moments about the $x$- and $y$-axes, assign counterclockwise moments a plus (+) sign and clockwise moments a minus (−) sign. In deciding whether a moment is counterclockwise or clockwise, look from the positive side of the axis in question toward the negative side.)

| Force | Coordinates of Force $F$ | | Moments $M_x$ and $M_y$ due to $F$ | |
|---|---|---|---|---|
| $F$ | $x$ | $y$ | $M_x$ | $M_y$ |
| $F_1$ | $x_1$ | $y_1$ | $F_1y_1$ | $F_1x_1$ |
| $F_2$ | $x_2$ | $y_2$ | $F_2y_2$ | $F_2x_2$ |
| $F_3$ | $x_3$ | $y_3$ | $F_3y_3$ | $F_3x_3$ |
| $\Sigma F$ | | | $\Sigma M_x$ | $\Sigma M_y$ |

4) Find the algebraic sum $\Sigma F$ of all the forces; this will be the resultant $R$ of the system.

$$R = \Sigma F = F_1 + F_2 + \ldots$$

5) Calculate $x_R$ and $y_R$, the moment arms of the resultant:

$$x_R = \Sigma M_y \div R$$

$$y_R = \Sigma M_x \div R$$

These moment arms are measured in a direction along the $x$- and $y$-axes as will give the resultant a moment of the same direction of rotation as $\Sigma M_x$ and $\Sigma M_y$.

*Note Concerning Interpretation of Results:* If $\Sigma M_x$ and $\Sigma M_y$ are both 0, then the resultant is a single force $R$ along the $z$-axis; if $R$ is also 0, then the system is in equilibrium. If $R$ is 0 but $\Sigma M_x$ and $\Sigma M_y$ are not both 0, then the resultant is a couple

$$M_R = \sqrt{(\Sigma M_x)^2 + (\Sigma M_y)^2}$$

that lies in a plane parallel to the $z$-axis and making an angle $\theta_R$ measured in a counterclockwise direction from the positive $x$-axis and calculated from the following formula:

$$\sin \theta_R = \frac{\Sigma M_x}{M_R}$$

*Finding the Resultant of Nonparallel Forces Not Meeting at a Common Point:*

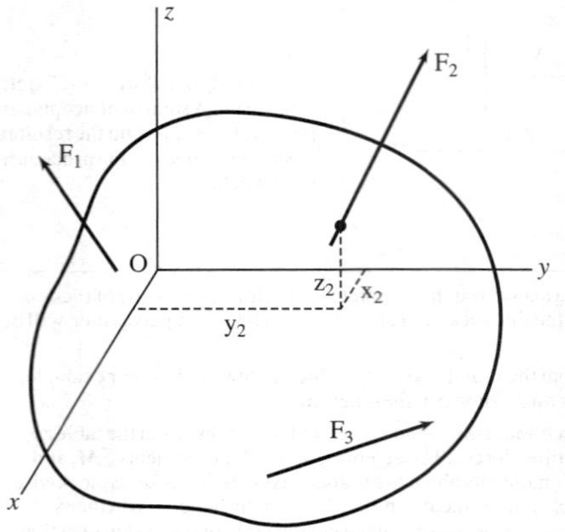


The diagram shows a system of noncoplanar, nonparallel, noncon-current forces $F_1, F_2$, etc. for which the resultant is to be determined. Generally speaking, the resultant will be a noncoplanar force and a couple, which may be further com-bined, if desired, into two forces that are skewed.

This is the most general force system that can be devised, so each of the other systems so far described represents a special, simpler case of this general force system. The method of solution described below for a system of three forces applies for any number of forces.

1) Select a set of coordinate $x$-, $y$-, and $z$-axes at any desired point $O$ in the body as shown in the diagram.
2) Determine the $x, y$, and $z$ coordinates of any convenient point on the line of action of each force as shown for $F_2$. Also determine the angles, $\theta_x, \theta_y, \theta_z$ that each force makes with each coordinate axis. These angles are measured counterclockwise from the positive direction of the $x, y$, and $z$ axes. The data is tabulated, as shown in the table accompanying Step 3, for convenient use in subsequent calculations.
3) Calculate the $x, y$, and $z$ components of each force using the formulas given in the accompanying table. Add these components algebraically to get $\Sigma F_x, \Sigma F_y$ and $\Sigma F_z$, which are the components of the resultant $R$ given by the formula,

$$R = \sqrt{(\Sigma F_x)^2 + (\Sigma F_y)^2 + (\Sigma F_z)^2}$$

| Force | Coordinates of Force $F$ | | | | | | Components of $F$ | | |
|---|---|---|---|---|---|---|---|---|---|
| $F$ | $x$ | $y$ | $z$ | $\theta_x$ | $\theta_y$ | $\theta_z$ | $F_x$ | $F_y$ | $F_z$ |
| $F_1$ | $x_1$ | $y_1$ | $z_1$ | $\theta_{x1}$ | $\theta_{y1}$ | $\theta_{z1}$ | $F_1 \cos \theta_{x1}$ | $F_1 \cos \theta_{y1}$ | $F_1 \cos \theta_{z1}$ |
| $F_2$ | $x_2$ | $y_2$ | $z_2$ | $\theta_{x2}$ | $\theta_{y2}$ | $\theta_{z2}$ | $F_2 \cos \theta_{x2}$ | $F_2 \cos \theta_{y2}$ | $F_2 \cos \theta_{z2}$ |
| $F_3$ | $x_3$ | $y_3$ | $z_3$ | $\theta_{x3}$ | $\theta_{y3}$ | $\theta_{z3}$ | $F_3 \cos \theta_{x3}$ | $F_3 \cos \theta_{y3}$ | $F_3 \cos \theta_{z3}$ |
| | | | | | | | $\Sigma F_x$ | $\Sigma F_y$ | $\Sigma F_z$ |

The resultant force $R$ makes angles of $\theta_{xR}, \theta_{yR}$, and $\theta_{zR}$ with the $x, y$, and $z$ axes, respectively, and passes through the selected point $O$. These angles are determined from the formulas,

$$\cos \theta_{xR} = \Sigma F_x \div R$$
$$\cos \theta_{yR} = \Sigma F_y \div R$$
$$\cos \theta_{zR} = \Sigma F_z \div R$$

4) Calculate the moments $M_x, M_y, M_z$ about $x$-, $y$-, and $z$-axes, respectively, due to the $F_x, F_y$, and $F_z$ components of each force and set them in tabular form. The formulas to use are given in the accompanying table.

In interpreting moments about the $x$-, $y$-, and $z$-axes, consider counterclockwise moments a plus (+) sign and clockwise moments a minus (−) sign. In deciding whether a moment is counterclockwise or clockwise, look from the positive side of the axis in question toward the negative side.

| Force | Moments of Components of $F$ $(F_x, F_y, F_z)$ about $x$-, $y$-, and $z$-axes | | |
|---|---|---|---|
| $F$ | $M_x = yF_z - zF_y$ | $M_y = zF_x - xF_z$ | $M_z = xF_y - yF_x$ |
| $F_1$ | $M_{x1} = y_1F_{z1} - z_1F_{y1}$ | $M_{y1} = z_1F_{x1} - x_1F_{z1}$ | $M_{z1} = x_1F_{y1} - y_1F_{x1}$ |
| $F_2$ | $M_{x2} = y_2F_{z2} - z_2F_{y2}$ | $M_{y2} = z_2F_{x2} - x_2F_{z2}$ | $M_{z2} = x_2F_{y2} - y_2F_{x2}$ |
| $F_3$ | $M_{x3} = y_3F_{z3} - z_3F_{y3}$ | $M_{y3} = z_3F_{x3} - x_3F_{z3}$ | $M_{z3} = x_3F_{y3} - y_3F_{x3}$ |
| | $\Sigma M_x$ | $\Sigma M_y$ | $\Sigma M_z$ |

5) Add the component moments algebraically to get $\Sigma M_x$, $\Sigma M_y$ and $\Sigma M_z$, which are the components of the resultant couple, $M$, given by the formula,

$$M = \sqrt{(\Sigma M_x)^2 + (\Sigma M_y)^2 + (\Sigma M_z)^2}$$

The resultant couple $M$ will tend to produce rotation about an axis making angles of $\beta_x$, $\beta_y$, and $\beta_z$ with the $x$-, $y$-, and $z$-axes, respectively. These angles are determined from the formulas,

$$\cos\beta_x = \frac{\Sigma M_x}{M} \quad \cos\beta_y = \frac{\Sigma M_y}{M} \quad \cos\beta_z = \frac{\Sigma M_z}{M}$$

*General Method of Locating Resultant When Its Components Are Known:* To determine the position of the resultant force of a system of forces, proceed as follows:

From the origin, point $O$, of a set of coordinate axes $x, y, z$, lay off on the $x$-axis a length $A$ representing the algebraic sum $\Sigma F_x$ of the $x$ components of all the forces. From the end of line $A$ lay off a line $B$ representing $\Sigma F_y$, the algebraic sum of the $y$ components; this line $B$ is drawn in a direction parallel to the $y$-axis. From the end of line $B$ lay off a line $C$ representing $\Sigma F_z$. Finally, draw a line $R$ from $O$ to the end of $C$; $R$ will be the resultant of the system. Notice that $R$ is in the three-dimensional space bounded by the $xy$, $xz$, and $yz$ planes.

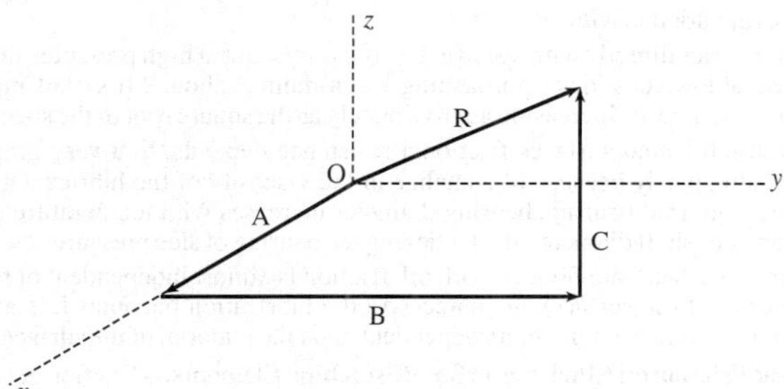

## Friction

**Properties of Friction.—**Friction is the resistance to motion that takes place when one body is moved upon another, and is generally defined as "that force which acts between two bodies at their surface of contact, so as to resist their sliding on each other." According to the conditions under which sliding occurs, the force of friction, $F$, bears a certain relation to the force between the two bodies called the normal force $N$. The relation between force of friction and normal force is given by the *coefficient of friction*, generally denoted by the Greek letter $\mu$. Thus:

$$F = \mu \times N \quad \text{and} \quad \mu = \frac{F}{N}$$

*Example:* A body weighing 28 pounds rests on a horizontal surface. The force required to keep it in motion along the surface is 7 pounds. Find the coefficient of friction.

$$\mu = \frac{F}{N} = \frac{7}{28} = 0.25$$

If a body is placed on an inclined plane, the friction between the body and the plane will prevent it from sliding down the inclined surface, provided the angle of the plane with the horizontal is not too great. There is a certain angle, however, at which the body will just barely be able to remain stationary, the frictional resistance being very nearly overcome by the tendency of the body to slide down. This angle is termed the *angle of repose,* frequently denoted by the Greek letter $\theta$, and the tangent of this angle is the coefficient of friction. Thus, $\mu = \tan\theta$.

A greater force is required to start a body moving from a state of rest than to merely keep it in motion, because *static (resting) friction* is greater than *sliding (motion) friction.*

**Laws of Friction.—***Unlubricated or Dry Surfaces:*

1) At low pressures (low normal force per unit area) friction is directly proportional to the normal force between the two surfaces. As the pressure increases, the friction does not rise proportionally; but when pressure becomes abnormally high, friction increases at a rapid rate until seizing takes place.

2) Friction, both in its total amount and its coefficient, is independent of the area in contact, so long as the normal force remains the same. This is true for moderate pressures only. For high pressures, this law is modified in the same way as in the first case.

3) At very low velocities friction is independent of the velocity of rubbing. As velocity increases, friction decreases.

*Lubricated Surfaces:* For well-lubricated surfaces, the laws of friction are considerably different from those governing dry or poorly lubricated surfaces.

1) Frictional resistance is almost independent of pressure (normal force per unit area) if the surfaces are flooded with oil.

2) Friction varies directly with speed at low pressures; but at high pressures the friction is very great at low velocities, approaching a minimum at about 2 ft/s (0.61 m/s), linear velocity, and afterwards increasing approximately as the square root of the speed.

3) For well-lubricated surfaces frictional resistance depends, to a very great extent, on temperature, partly because of a change in the viscosity of the lubricant and partly because, for a journal bearing, bearing diameter increases with temperature rise more rapidly than does shaft diameter, thus relieving the bearing of side pressure.

4) If bearing surfaces are flooded with oil, friction is almost independent of the nature of the material of the surfaces in contact. As the lubrication becomes less ample, the coefficient of friction becomes more dependent upon the material of the surfaces.

**Influence of Friction on Efficiency of Small Machine Elements.—**Friction between machine parts lowers the efficiency of a machine. Average values of efficiency, in percent, of the most common machine elements when carefully made are: ordinary bearings, 95 to 98; roller bearings, 98; ball bearings, 99; spur gears with cut teeth, including bearings, 99; bevel gears with cut teeth, including bearings, 98; belting, 96 to 98; high-class silent power transmission chain, 97 to 99; roller chains, 95 to 97.

**Coefficients of Friction.**—Table 1 and Table 2 provide representative values of static friction for various combinations of materials with dry (clean, unlubricated) and lubricated surfaces. The values for static, or breakaway, friction shown in these tables will generally be higher than the subsequent sliding friction. Typically, the steel-on-steel static coefficient of 0.8 unlubricated will drop to 0.4 when sliding has been initiated; with oil lubrication, the value will drop from 0.16 to 0.03.

Many factors affect friction, and even slight deviations from normal or test conditions can produce wide variations. Accordingly, when friction coefficients are used in design calculations, due allowance or factors of safety should be considered, and in critical applications, specific tests should be conducted to provide specific coefficients for material, geometry, and/or lubricant combinations.

**Table 1. Coefficients of Static Friction for Steel on Various Materials**

| Material | Coefficient of Friction, μ | | Material | Coefficient of Friction, μ | |
|---|---|---|---|---|---|
| | Clean | Lubricated | | Clean | Lubricated |
| Steel | 0.8 | 0.16 | Hard carbon | 0.14 | 0.11–0.14 |
| Copper-lead alloy | 0.22 | | Graphite | 0.1 | 0.1 |
| Phosphor-bronze | 0.35 | | Tungsten carbide | 0.4–0.6 | 0.1–0.2 |
| Aluminum-bronze | 0.45 | | Plexiglas | 0.4–0.5 | 0.4–0.5 |
| Brass | 0.35 | 0.19 | Polystyrene | 0.3–0.35 | 0.3–0.35 |
| Cast iron | 0.4 | 0.21 | Polythene | 0.2 | 0.2 |
| Bronze | | 0.16 | Teflon | 0.04 | 0.04 |
| Sintered bronze | | 0.13 | | | |

Table 1 and Table 2 used with permission from *The Friction and Lubrication of Solids*, Vol. 1, by Bowden and Tabor, Clarendon Press, Oxford, 1950.

**Table 2. Coefficients of Static Friction for Various Materials Combinations**

| Material Combination | Coefficient of Friction, μ | | Material Combination | Coefficient of Friction, μ | |
|---|---|---|---|---|---|
| | Clean | Lubricated | | Clean | Lubricated |
| Aluminum-aluminum | 1.35 | 0.30 | Tungsten carbide-tungsten carbide | 0.2–0.25 | 0.12 |
| Cadmium-cadmium | 0.5 | 0.05 | Plexiglas-Plexiglas | 0.8 | 0.8 |
| Chromium-chromium | 0.41 | 0.34 | Polystyrene-polystyrene | 0.5 | 0.5 |
| Copper-copper | 1.0 | 0.08 | Teflon-Teflon | 0.04 | 0.04 |
| Iron-iron | 1.0 | 0.15–0.20 | Nylon-nylon | 0.15–0.25 | |
| Magnesium-magnesium | 0.6 | 0.08 | Solids on rubber | 1–4 | |
| Nickel-nickel | 0.7 | 0.28 | Wood on wood (clean) | 0.25–0.5 | |
| Platinum-platinum | 1.2 | 0.25 | Wood on wood (wet) | 0.2 | |
| Silver-silver | 1.4 | 0.55 | Wood on metals (clean) | 0.2–0.6 | |
| Zinc-zinc | 0.6 | 0.04 | Wood on metals (wet) | 0.2 | |
| Glass-glass | 0.9–1.0 | 0.1–0.6 | Brick on wood | 0.6 | |
| Glass-metal | 0.5–0.7 | 0.2–0.3 | Leather on wood | 0.3–0.4 | |
| Diamond-diamond | 0.1 | 0.05–0.1 | Leather on metal (clean) | 0.6 | |
| Diamond-metal | 0.1–0.15 | 0.1 | Leather on metal (wet) | 0.4 | |
| Sapphire-sapphire | 0.2 | 0.2 | Leather on metal (greasy) | 0.2 | |
| Hard carbon on carbon | 0.16 | 0.12–0.14 | Brake material on cast iron | 0.4 | |
| Graphite-graphite (in vacuum) | 0.5–0.8 | | Brake material on cast iron (wet) | 0.2 | |
| Graphite-graphite | 0.1 | 0.1 | | | |

**Rolling Friction.**—When a body rolls on a surface, the force resisting the motion is termed *rolling friction* or *rolling resistance*. Let $W$ = total weight of rolling body or load on wheel, in pounds (newtons); $r$ = radius of wheel, in inches (centimeters); $f$ = coefficient of rolling resistance, in inches (centimeters). Then, resistance to rolling, in pounds (newtons) is given as:

$$F_r = Wf/r$$

Coefficient of rolling resistance varies with conditions. For wood on wood, $f$ = 0.06 inch (0.152 cm) may be used; for iron on iron, 0.02 inch (0.051 cm); iron on granite, 0.085 inch (0.216 cm); iron on asphalt, 0.15 inch (0.381 cm); and iron on wood, 0.22 inch (0.559 cm).

The coefficient of rolling resistance $f$ is given in inches (or centimeters) and is not the same as the sliding or static coefficient of friction given in Table 1 and Table 2, in which μ is a dimensionless ratio between frictional resistance and normal load. Various investigators are not in close agreement on the true values for these coefficients and the foregoing values should only be used for the approximate calculation of rolling resistance.

## Simple Mechanisms

### Levers

| Types of Levers | Examples |
|---|---|
| 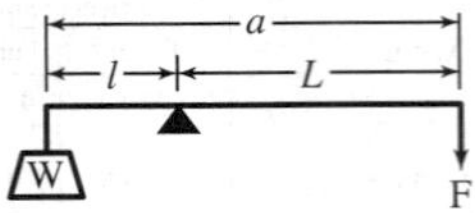<br>$F:W = l:L$ $\quad F \times L = W \times l$<br>$F = \frac{W \times l}{L}$ $\quad W = \frac{F \times L}{l}$<br>$L = \frac{W \times a}{W + F} = \frac{W \times l}{F}$ $\quad l = \frac{F \times a}{W + F} = \frac{F \times L}{W}$ | A pull of 80 pounds is exerted at the end of the lever, at $W$; $l$ = 12 inches and $L$ = 32 inches.<br>Find the value of force $F$ required to balance the lever.<br>$F = \frac{80 \times 12}{32} = \frac{960}{32} = 30$ pounds<br>If $F$ = 20; $W$ = 180; and $l$ = 3; how long must $L$ be made to secure equilibrium?<br>$L = \frac{180 \times 3}{20} = 27$ |
| 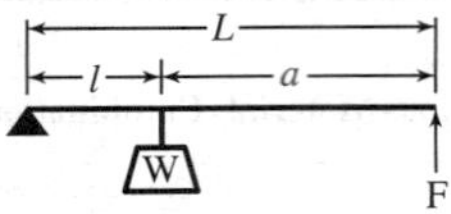<br>$F:W = l:L$ $\quad F \times L = W \times l$<br>$F = \frac{W \times l}{L}$ $\quad W = \frac{F \times L}{l}$<br>$L = \frac{W \times a}{W - F} = \frac{W \times l}{F}$ $\quad l = \frac{F \times a}{W - F} = \frac{F \times L}{W}$ | Total length $L$ of a lever is 25 cm. A weight of 400 N, supported at $W$; $l$ is 10 cm. Find the value of $F$.<br>$F = \frac{(400)(10)}{25} = 160$ N<br>If $F$ = 400 N, $W$ = 900 lbs, and $a$ = 1.5 m, what should $L$ equal to secure equilibrium?<br>$L = \frac{(900)(1.5)}{900 - 400} = 2.7$ m |
| 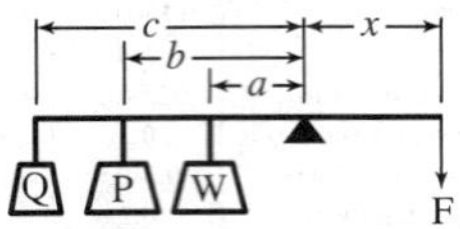<br>When three or more forces act on lever:<br>$F \times x = W \times a + P \times b + Q \times c$<br>$x = \frac{W \times a + P \times b + Q \times c}{F}$<br>$F = \frac{W \times a + P \times b + Q \times c}{x}$ | Let $W$ = 20, $P$ = 30, and $Q$ = 15 lbs; $a$ = 4, $b$ = 7, and $c$ = 10 in.<br>If $x$ = 6 in., find $F$.<br>$F = \frac{20 \times 4 + 30 \times 7 + 15 \times 10}{6} = 73\frac{1}{3}$ lbs<br>Assuming $F$ = 20 lbs in the example above, how long must lever arm $x$ be made?<br>$x = \frac{20 \times 4 + 30 \times 7 + 15 \times 10}{20} = 22$ inches |

**The above formulas are valid using metric SI units, with forces expressed in newtons, and lengths in meters. However, it should be noted that the weight of a mass W kilograms**

**is equal to a force of *Wg* newtons, where *g* is approximately 9.81 m/s². Thus, supposing that in the first example $l = 0.4$ m, $L = 1.2$ m, and $W = 30$ kg, then the weight of *W* is 30*g* newtons, so that the force *F* required to balance the lever is** $F = \frac{30g \times 0.4}{1.2} = 10g = 98.1\,\text{N}$**. This force could be produced by suspending a mass of 10 kg at location *F*.**

## Table of Forces on Inclined Planes

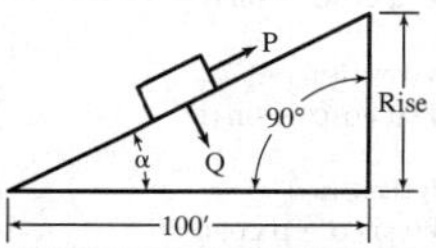


The table below is used to find the force required to move a load on an inclined plane. Friction on the plane is not taken into account. The column headed "Tension Force *P* per Ton" gives the pull in pounds required for moving one ton (2000 lbs) along the inclined surface. The fourth column gives the perpendicular or normal force of the load. If the coefficient of friction μ is known, the added pull to overcome friction is:

Additional pull per ton = $Q \times \mu$

*Example:* A 200 lb. body pulled up a 14% grade when friction (μ) is 0.3 requires a force $P = (0.1)(277.3) + (0.1)(1980.68)(0.3) = 87.15$ lbs.

| Percent of Grade, Rise in ft per 100 ft | Angle α | Tension Force *P* per Ton | Perpendicular Pressure or Normal Force *Q* on Plane per Ton | Percent of Grade, Rise in ft per 100 ft | Angle α | Tension Force *P* in Cable per Ton | Perpendicular Pressure or Normal Force *Q* on Plane per Ton |
|---|---|---|---|---|---|---|---|
| 1 | 0.57 | 20.00 | 1999.90 | 51 | 27.02 | 908.65 | 1781.67 |
| 2 | 1.15 | 39.99 | 1999.60 | 52 | 27.47 | 922.71 | 1774.43 |
| 3 | 1.72 | 59.97 | 1999.10 | 53 | 27.92 | 936.59 | 1767.15 |
| 4 | 2.29 | 79.94 | 1998.40 | 54 | 28.37 | 950.30 | 1759.81 |
| 5 | 2.86 | 99.88 | 1997.50 | 55 | 28.81 | 963.84 | 1752.43 |
| 6 | 3.43 | 119.78 | 1996.41 | 56 | 29.25 | 977.21 | 1745.01 |
| 7 | 4.00 | 139.66 | 1995.12 | 57 | 29.68 | 990.41 | 1737.55 |
| 8 | 4.57 | 159.49 | 1993.63 | 58 | 30.11 | 1003.44 | 1730.06 |
| 9 | 5.14 | 179.28 | 1991.95 | 59 | 30.54 | 1016.30 | 1722.54 |
| 10 | 5.71 | 199.01 | 1990.07 | 60 | 30.96 | 1028.99 | 1714.99 |
| 11 | 6.28 | 218.68 | 1988.01 | 61 | 31.38 | 1041.52 | 1707.41 |
| 12 | 6.84 | 238.29 | 1985.75 | 62 | 31.80 | 1053.88 | 1699.81 |
| 13 | 7.41 | 257.83 | 1983.31 | 63 | 32.21 | 1066.08 | 1692.18 |
| 14 | 7.97 | 277.30 | 1980.68 | 64 | 32.62 | 1078.11 | 1684.54 |
| 15 | 8.53 | 296.68 | 1977.87 | 65 | 33.02 | 1089.98 | 1676.89 |
| 16 | 9.09 | 315.98 | 1974.88 | 66 | 33.42 | 1101.68 | 1669.22 |
| 17 | 9.65 | 335.19 | 1971.71 | 67 | 33.82 | 1113.23 | 1661.54 |
| 18 | 10.20 | 354.31 | 1968.37 | 68 | 34.22 | 1124.62 | 1653.85 |
| 19 | 10.76 | 373.32 | 1964.85 | 69 | 34.61 | 1135.85 | 1646.16 |
| 20 | 11.31 | 392.23 | 1961.16 | 70 | 34.99 | 1146.92 | 1638.46 |
| 21 | 11.86 | 411.03 | 1957.31 | 71 | 35.37 | 1157.84 | 1630.77 |
| 22 | 12.41 | 429.72 | 1953.29 | 72 | 35.75 | 1168.61 | 1623.07 |
| 23 | 12.95 | 448.30 | 1949.11 | 73 | 36.13 | 1179.22 | 1615.37 |
| 24 | 13.50 | 466.75 | 1944.77 | 74 | 36.50 | 1189.69 | 1607.68 |
| 25 | 14.04 | 485.07 | 1940.29 | 75 | 36.87 | 1200.00 | 1600.00 |
| 26 | 14.57 | 503.27 | 1935.65 | 76 | 37.23 | 1210.17 | 1592.32 |
| 27 | 15.11 | 521.33 | 1930.86 | 77 | 37.60 | 1220.19 | 1584.66 |
| 28 | 15.64 | 539.26 | 1925.93 | 78 | 37.95 | 1230.06 | 1577.00 |
| 29 | 16.17 | 557.05 | 1920.86 | 79 | 38.31 | 1239.80 | 1569.36 |
| 30 | 16.70 | 574.70 | 1915.65 | 80 | 38.66 | 1249.39 | 1561.74 |
| 31 | 17.22 | 592.20 | 1910.31 | 81 | 39.01 | 1258.84 | 1554.13 |
| 32 | 17.74 | 609.55 | 1904.85 | 82 | 39.35 | 1268.16 | 1546.54 |
| 33 | 18.26 | 626.76 | 1899.26 | 83 | 39.69 | 1277.34 | 1538.96 |
| 34 | 18.78 | 643.81 | 1893.55 | 84 | 40.03 | 1286.38 | 1531.41 |
| 35 | 19.29 | 660.70 | 1887.72 | 85 | 40.36 | 1295.30 | 1523.88 |
| 36 | 19.80 | 677.44 | 1881.77 | 86 | 40.70 | 1304.08 | 1516.37 |
| 37 | 20.30 | 694.02 | 1875.72 | 87 | 41.02 | 1312.73 | 1508.89 |
| 38 | 20.81 | 710.44 | 1869.57 | 88 | 41.35 | 1321.26 | 1501.43 |
| 39 | 21.31 | 726.69 | 1863.31 | 89 | 41.67 | 1329.65 | 1493.99 |
| 40 | 21.80 | 742.78 | 1856.95 | 90 | 41.99 | 1337.93 | 1486.59 |
| 41 | 22.29 | 758.71 | 1850.50 | 91 | 42.30 | 1346.08 | 1479.21 |
| 42 | 22.78 | 774.47 | 1843.96 | 92 | 42.61 | 1354.11 | 1471.86 |
| 43 | 23.27 | 790.06 | 1837.34 | 93 | 42.92 | 1362.03 | 1464.54 |
| 44 | 23.75 | 805.48 | 1830.63 | 94 | 43.23 | 1369.82 | 1457.26 |
| 45 | 24.23 | 820.73 | 1823.84 | 95 | 43.53 | 1377.50 | 1450.00 |
| 46 | 24.70 | 835.81 | 1816.98 | 96 | 43.83 | 1385.06 | 1442.77 |
| 47 | 25.17 | 850.72 | 1810.05 | 97 | 44.13 | 1392.52 | 1435.58 |
| 48 | 25.64 | 865.46 | 1803.05 | 98 | 44.42 | 1399.86 | 1428.43 |
| 49 | 26.10 | 880.03 | 1795.98 | 99 | 44.71 | 1407.09 | 1421.30 |
| 50 | 26.57 | 894.43 | 1788.85 | 100 | 45.00 | 1414.21 | 1414.21 |

Tension and Normal Forces in pounds per 1 ton (2000 lbs) load weight.

## Inclined Plane—Wedge

*W* = weight of body

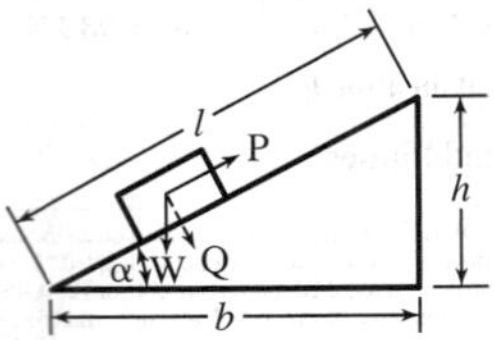


If friction is taken into account, then

Force *P* to pull body up is:

$$P = W(\mu \cos\alpha + \sin\alpha)$$

Force $P_1$ to pull body down is:

$$P_1 = W(\mu \cos\alpha - \sin\alpha)$$

Force $P_2$ to hold body stationary:

$$P_2 = W(\sin\alpha - \mu\cos\alpha)$$

in which μ is the coefficient of friction.

Neglecting friction:

$$P = W \times \frac{h}{l} = W \times \sin\alpha$$

$$W = P \times \frac{l}{h} = \frac{P}{\sin\alpha} = P \times \operatorname{cosec}\alpha$$

$$Q = W \times \frac{b}{l} = W \times \cos\alpha$$

*W* = weight of body

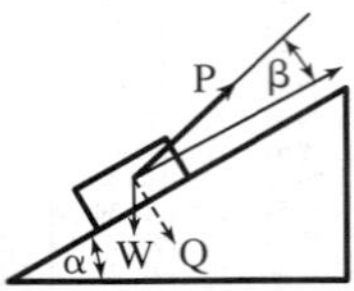


Neglecting friction:

$$P = W \times \frac{\sin\alpha}{\cos\beta}$$

$$W = P \times \frac{\cos\beta}{\sin\alpha}$$

$$Q = W \times \frac{\cos(\alpha + \beta)}{\cos\beta}$$

With friction:

Coefficient of friction

$$\mu = \tan\phi$$

$$P = W \times \frac{\sin(\alpha + \phi)}{\cos(\beta - \phi)}$$

*W* = weight of body

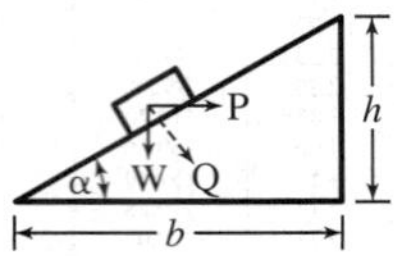


Neglecting friction:

$$P = W \times \frac{h}{b} = W \times \tan\alpha$$

$$W = P \times \frac{b}{h} = P \times \cot\alpha$$

$$Q = \frac{W}{\cos\alpha} = W \times \sec\alpha$$

With friction:

Coefficient of friction

$$\mu = \tan\phi$$

$$P = W\tan(\alpha + \phi)$$

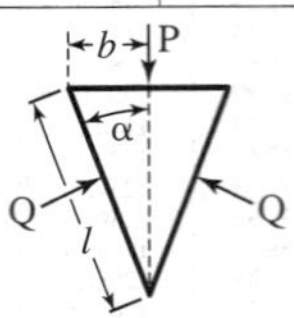


Neglecting friction:

$$P = 2Q \times \frac{b}{l} = 2Q \times \sin\alpha$$

$$Q = P \times \frac{l}{2b} = \frac{1}{2}P \times \csc\alpha$$

With friction: Coefficient of friction = μ.

$$P = 2Q(\mu\cos\alpha + \sin\alpha)$$

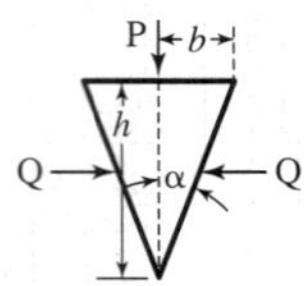


Neglecting friction:

$$P = 2Q \times \frac{b}{h} = 2Q \times \tan\alpha$$

$$Q = P \times \frac{h}{2b} = \frac{1}{2}P \times \cot\alpha$$

With friction: Coefficient of friction = μ = tan ϕ.

$$P = 2Q\tan(\alpha + \phi)$$

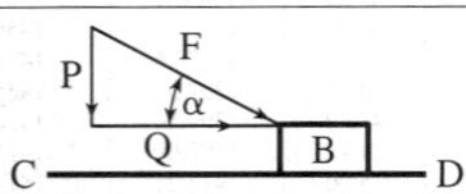


**Force Moving Body on Horizontal Plane.**—*F* tends to move *B* along line *CD*; *Q* is the component which actually moves *B*; *P* is the pressure, due to *F*, of the body on *CD*.

$$Q = F \times \cos\alpha \quad P = \sqrt{F^2 - Q^2}$$

## Wheels and Pulleys

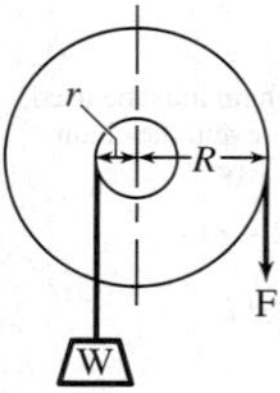


$$F{:}W = r{:}R$$
$$F \times R = W \times r$$
$$F = \frac{W \times r}{R}$$
$$W = \frac{F \times R}{r}$$
$$R = \frac{W \times r}{F}$$
$$r = \frac{F \times R}{W}$$

The radius of a drum on which is wound the lifting rope of a windlass is 2 in. What force will be exerted at the periphery of a gear of 24 in. diameter, mounted on the same shaft as the drum and transmitting power to it, if one ton (2000 lbs) is to be lifted? Here $W = 2000$; $R = 12$; $r = 2$.

$$F = \frac{2000 \times 2}{12} = 333 \text{ lbs}$$

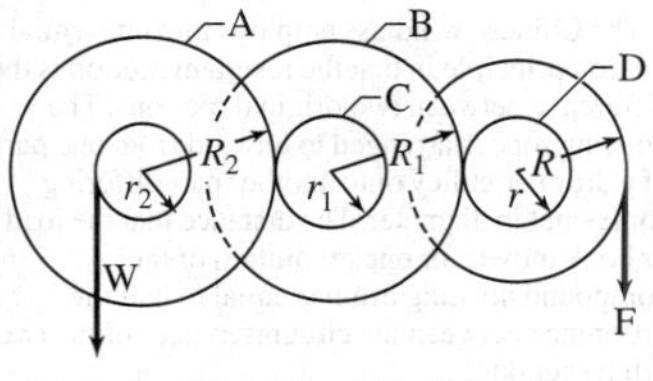


$A, B, C$ and $D$ are the pitch circles of gears.

$$F = \frac{W \times r \times r_1 \times r_2}{R \times R_1 \times R_2}$$

$$W = \frac{F \times R \times R_1 \times R_2}{r \times r_1 \times r_2}$$

Let the pitch diameters of gears $A, B, C$ and $D$ be 30, 28, 12 and 10 in., respectively. Then $R_2 = 15$; $R_1 = 14$; $r_1 = 6$; and $r = 5$. Let $R = 12$, and $r_2 = 4$. The force $F$ required to lift a weight $W$ of 2000 lbs, friction being neglected, is:

$$F = \frac{2000 \times 5 \times 6 \times 4}{12 \times 14 \times 15} = 95 \text{ pounds}$$

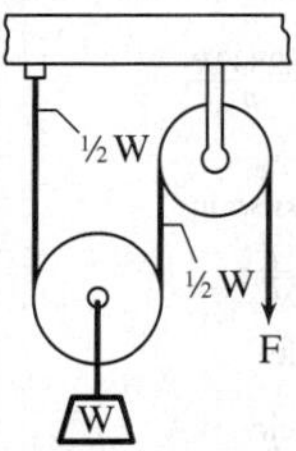


$$F = \frac{1}{2} W$$

The velocity with which weight $W$ will be raised is one-half the velocity of the force applied at $F$.

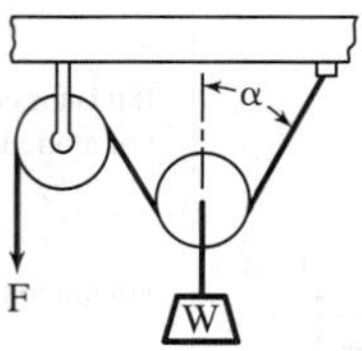


$$F{:}W = \sec\alpha{:}2$$
$$F = \frac{W \times \sec\alpha}{2}$$
$$W = 2F \times \cos\alpha$$

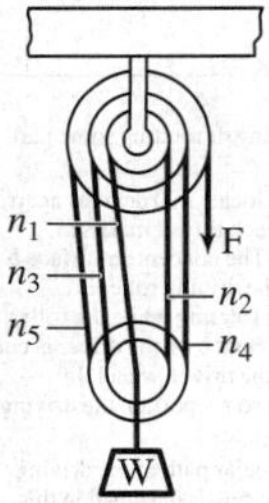


$n$ = number of strands or parts of rope ($n_1, n_2$, etc.).

$$F = \frac{1}{n} \times W$$

The velocity with which $W$ will be raised equals $\frac{1}{n}$ of the velocity of the force applied at $F$.

In the illustration is shown a combination of a double and triple block. The pulleys each turn freely on a pin as axis, and are drawn with different diameters, to show the parts of the rope more clearly. There are 5 parts of rope. Therefore, if 200 lbs is to be lifted, the force $F$ required at the end of the rope is:

$$F = \frac{1}{5} \times 200 = 40 \text{ lbs}$$

***Note:* The above formulas are valid using metric SI units, with forces expressed in newtons, and lengths in meters or millimeters. (See note on** page 168 **concerning weight and mass.)**

## Differential Pulley

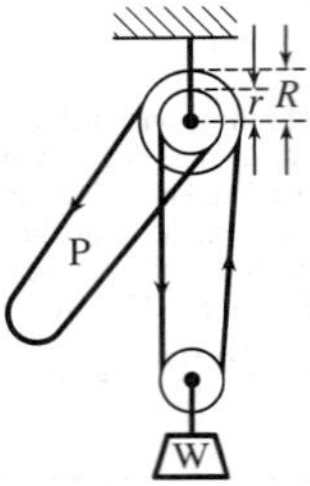


In the differential pulley a chain must be used, engaging sprockets, so as to prevent the chain from slipping over the pulley faces.

$$PR = \frac{1}{2}W(R - r)$$

$$P = \frac{W(R - r)}{2R}$$

$$W = \frac{2PR}{R - r}$$

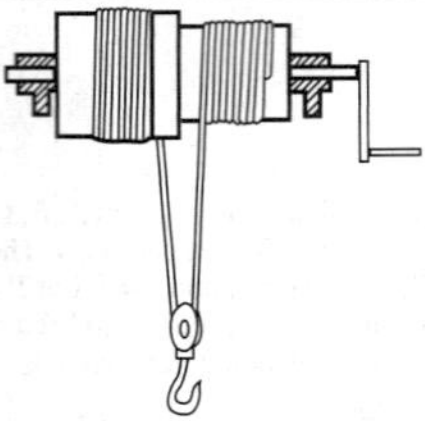

Chinese Windlass

The Chinese windlass employs the differential motion principle in that the resultant motion is the difference between two original motions. The hoisting rope is arranged to unwind from one part of a drum or pulley onto another part differing somewhat in diameter. The distance that the load or hook moves for one revolution of the compound hoisting drum is equal to half the difference between the circumferences of the two drum sections.

## Screw

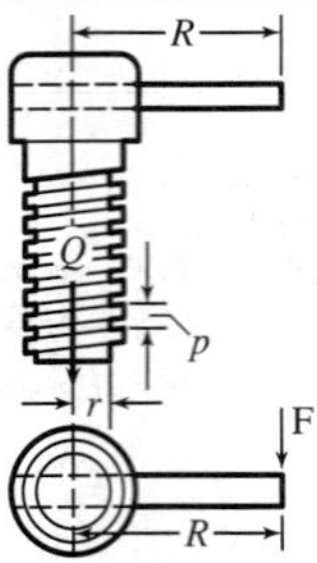


$F$ = force at end of handle or wrench; $R$ = lever-arm of $F$; $r$ = pitch radius of screw; $p$ = lead of thread; $Q$ = load. Then, neglecting friction:

$$F = Q \times \frac{p}{6.2832R} \qquad Q = F \times \frac{6.2832R}{p}$$

If $\mu$ is the coefficient of friction, then:
For motion in direction of load $Q$ which *assists* it:

$$F = Q \times \frac{6.2832\mu r - p}{6.2832r + \mu p} \times \frac{r}{R}$$

For motion opposite load $Q$ which *resists* it:

$$F = Q \times \frac{p + 6.2832\mu r}{6.2832r - \mu p} \times \frac{r}{R}$$

## Geneva Wheel

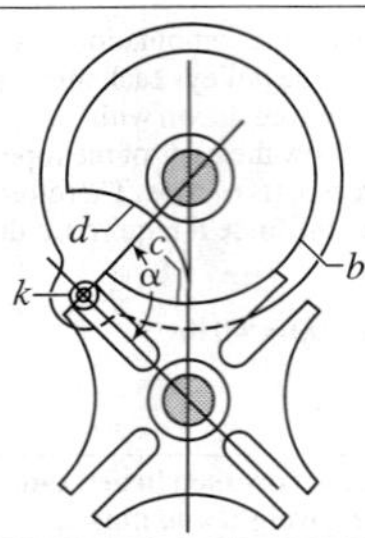


Geneva wheels are frequently used on machine tools for indexing or rotating some part of the machine through a fractional part of a revolution.

The driven wheel shown in the illustration has four radial slots located 90 degrees apart, and the driver carries a roller *k* which engages one of these slots each time it makes a revolution, thus turning the driven wheel one-quarter revolution. The concentric surface *b* engages the concave surface *c* between each pair of slots before the driving roller is disengaged from the driven wheel, which prevents the latter from rotating while the roller is moving around to engage the next successive slot. The circular boss *b* on the driver is cut away at *d* to provide a clearance space for the projecting arms of the driven wheel. In designing gearing of the general type illustrated, it is advisable to so proportion the driving and driven members that the angle $\alpha$ will be approximately 90 degrees.

The radial slots in the driven part will then be tangent to the circular path of the driving roller at the time the roller enters and leaves the slot. When the gearing is designed in this way, the driven wheel is started gradually from a state of rest and the motion is also gradually checked.

## Toggle Joint

A link mechanism commonly known as a toggle joint is applied to machines of different types, such as drawing and embossing presses, stone crushers, etc., for securing great pressure. The principle of the toggle joint is shown by Fig. 1. There are two links, *b* and *c*, which are connected at the center. Link *b* is free to swivel about a fixed pin or bearing at *d*, and link *c* is connected to a sliding member *e*. Rod *f* joins links *b* and *c* at the central connection. When force is applied to rod *f* in a direction at right angles to centerline *xx*, along which the driven member *e* moves, this force is greatly multiplied at *e*, because a movement at the joint *g* produces a relatively slight movement at *e*. As the angle α becomes less, motion at *e* decreases and the force increases until the links are in line. If $R$ = the resistance at *e*, $P$ = the applied power or force, and α = the angle between each link, and a line *x-x* passing through the axes of the pins, then:

$$2R \sin \alpha = P \cos \alpha$$

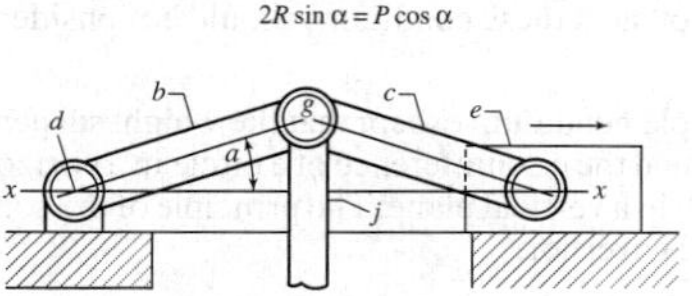


Fig. 1. Toggle Joint Principle

If arms *ED* and *EH* are of unequal length then

$$P = \frac{Fa}{b}$$

The relation between $P$ and $F$ changes constantly as $F$ moves downward.

If arms *ED* and *EH* are equal, then

$$P = \frac{Fa}{2h}$$

A double toggle-joint does not increase the pressure exerted so long as the relative distances moved by $F$ and $P$ remain the same.

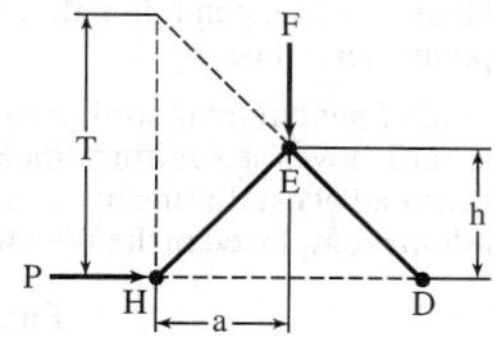


## Toggle-joints with Equal Arms

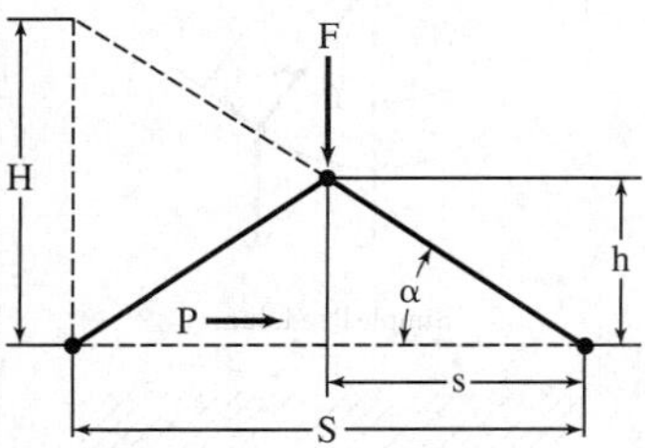


$$2P \sin \alpha = F \cos \alpha$$

$$\frac{P}{F} = \frac{\cos \alpha}{2 \sin \alpha} = \text{coefficient}$$

$$P = F \times \text{coefficient}$$

where $F$ = force applied; $P$ = resistance; and, α = given angle.

Equivalent expressions (see diagram):

$$P = \frac{FS}{4h} \quad P = \frac{Fs}{H}$$

To use the table, measure angle α, and find the coefficient in the table that corresponds to it. The coefficient is the ratio of the resistance to the force applied, and multiplying the force applied by the coefficient gives the resistance, neglecting friction.

| Angle ° | Coefficient | Angle ° | Coefficient | Angle ° | Coefficient | Angle ° | Coefficient |
|---|---|---|---|---|---|---|---|
| 0.01 | 2864.79 | 1.00 | 28.64 | 5.25 | 5.44 | 23 | 1.18 |
| 0.02 | 1432.39 | 1.10 | 26.04 | 5.50 | 5.19 | 24 | 1.12 |
| 0.03 | 954.93 | 1.20 | 23.87 | 5.75 | 4.97 | 25 | 1.07 |
| 0.04 | 716.20 | 1.30 | 22.03 | 6.00 | 4.76 | 26 | 1.03 |
| 0.05 | 572.96 | 1.40 | 20.46 | 6.50 | 4.39 | 27 | 0.98 |
| 0.10 | 286.48 | 1.50 | 19.09 | 7.00 | 4.07 | 28 | 0.94 |
| 0.15 | 190.99 | 1.60 | 17.90 | 7.50 | 3.80 | 29 | 0.90 |
| 0.20 | 143.24 | 1.70 | 16.85 | 8.00 | 3.56 | 30 | 0.87 |
| 0.25 | 114.59 | 1.80 | 15.91 | 8.50 | 3.35 | 31 | 0.83 |
| 0.30 | 95.49 | 1.90 | 15.07 | 9.00 | 3.16 | 32 | 0.80 |
| 0.35 | 81.85 | 2.00 | 14.32 | 10.00 | 2.84 | 33 | 0.77 |
| 0.40 | 71.62 | 2.25 | 12.73 | 11.00 | 2.57 | 34 | 0.74 |
| 0.45 | 63.66 | 2.50 | 11.45 | 12.00 | 2.35 | 35 | 0.71 |
| 0.50 | 57.29 | 2.75 | 10.41 | 13.00 | 2.17 | 36 | 0.69 |
| 0.55 | 52.09 | 3.00 | 9.54 | 14.00 | 2.01 | 37 | 0.66 |
| 0.60 | 47.74 | 3.25 | 8.81 | 15.00 | 1.87 | 38 | 0.64 |
| 0.65 | 44.07 | 3.50 | 8.17 | 16.00 | 1.74 | 39 | 0.62 |
| 0.70 | 40.92 | 3.75 | 7.63 | 17.00 | 1.64 | 40 | 0.60 |
| 0.75 | 38.20 | 4.00 | 7.15 | 18.00 | 1.54 | 41 | 0.58 |
| 0.80 | 35.81 | 4.25 | 6.73 | 19.00 | 1.45 | 42 | 0.56 |
| 0.85 | 33.70 | 4.50 | 6.35 | 20.00 | 1.37 | 43 | 0.54 |
| 0.90 | 31.83 | 4.75 | 6.02 | 21.00 | 1.30 | 44 | 0.52 |
| 0.95 | 30.15 | 5.00 | 5.72 | 22.00 | 1.24 | 45 | 0.50 |

## Pendulums

A *compound* or *physical* pendulum consists of any rigid body suspended from a fixed horizontal axis about which the body may oscillate in a vertical plane due to the action of gravity.

A *simple* or *mathematical* pendulum is similar to a compound pendulum except that the mass of the body is concentrated at a single point which is suspended from a fixed horizontal axis by a weightless cord. Actually, a simple pendulum cannot be constructed since it is impossible to have either a weightless cord or a body whose mass is entirely concentrated at one point. A good approximation, however, consists of a small, heavy bob suspended by a light, fine wire. A pendulum that does not meet these conditions should be considered a compound pendulum.

A *conical* pendulum is similar to a simple pendulum except that the weight suspended by the cord moves at a uniform speed around the circumference of a circle in a horizontal plane instead of oscillating back and forth in a vertical plane. The principle of the conical pendulum is employed in the Watt fly-ball governor.

### Four Types of Pendulum

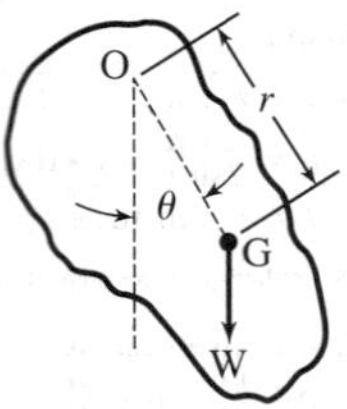

Physical Pendulum

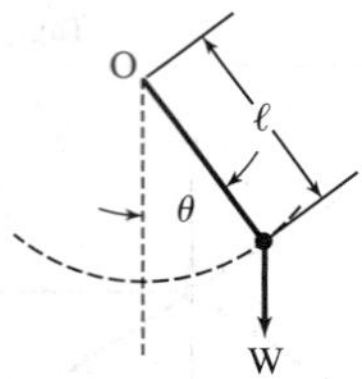

Simple Pendulum

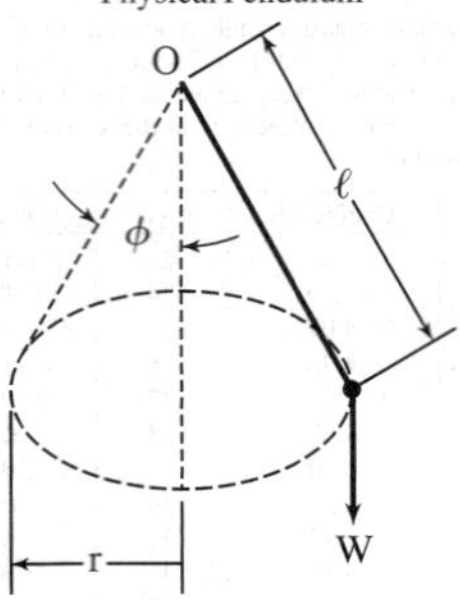

Conical Pendulum

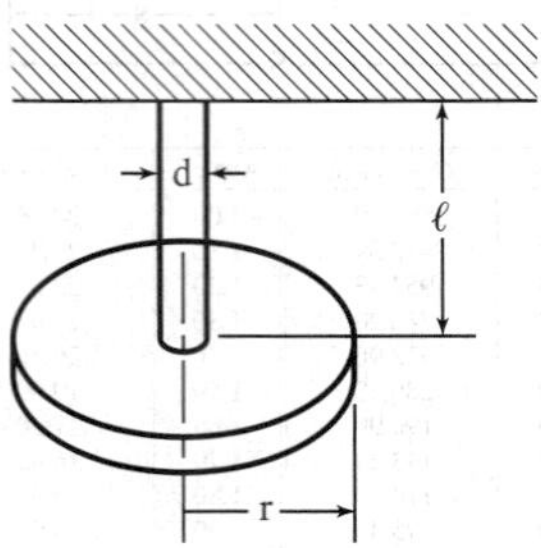

Torsional Pendulum

**W** = Weight of Disk

A *torsional* pendulum in its simplest form consists of a disk fixed to a slender rod, the other end of which is fastened to a fixed frame. When the disc is twisted through some angle and released, it will oscillate back and forth about the axis of the rod because of the torque exerted by the rod.

**Pendulum Formulas.—**From the formulas that follow, the period of vibration or time required for one complete cycle back and forth may be determined for the types of pendulums shown in the accompanying diagram.

For a *simple* pendulum,

$$T = 2\pi\sqrt{\frac{l}{g}} \tag{1}$$

where $T$ = period in seconds for one complete cycle; $g$ = acceleration due to gravity = 32.17 ft/sec$^2$ (approximately); and $l$ is the length of the pendulum in feet as shown on the accompanying diagram.

For a *physical* or *compound* pendulum,

$$T = 2\pi\sqrt{\frac{k_o^2}{gr}} \tag{2}$$

where $k_o$ = radius of gyration of the pendulum about the axis of rotation in feet, and $r$ is the distance from the axis of rotation to the center of gravity in feet.

**The metric SI units that can be used in the two above formulas are $T$ = time in seconds; $g$ = approximately 9.81 m/s$^2$, which is the value for acceleration due to gravity; $l$ = the length of the pendulum in meters; $k_o$ = the radius of gyration in meters, and $r$ = the distance from the axis of rotation to the center of gravity in meters.**

Formulas (1) and (2) are accurate when the angle of oscillation $\theta$ shown in the diagram is very small. For $\theta$ equal to 22 degrees, these formulas give results that are too small by 1 percent; for $\theta$ equal to 32 degrees, by 2 percent.

For a *conical* pendulum, the time in seconds for one revolution is:

$$T = 2\pi\sqrt{\frac{l\cos\phi}{g}} \tag{3a}$$

or

$$T = 2\pi\sqrt{\frac{r\cot\phi}{g}} \tag{3b}$$

For a *torsional* pendulum consisting of a thin rod and a disk as shown in the figure

$$T = \frac{2}{3}\sqrt{\frac{\pi W r^2 l}{g d^4 G}} \tag{4}$$

where $W$ = weight of disk in pounds; $r$ = radius of disk in feet; $l$ = length of rod in feet; $d$ = diameter of rod in feet; and $G$ = modulus of elasticity in shear of the rod material in pounds per square inch (psi or lb/in$^2$).

**The formula using metric SI units is:**

$$T = 8\sqrt{\frac{\pi M r^2 l}{d^4 G}}$$

**where $T$ = time in seconds for one complete oscillation; $M$ = mass in kilograms; $r$ = radius in meters; $l$ = length of rod in meters; $d$ = diameter of rod in meters; $G$ = modulus of elasticity in shear of the rod material in pascals (newtons per meter squared). The same formula can be applied using millimeters, provided dimensions are expressed in millimeters throughout, and the modulus of elasticity in megapascals (newtons per millimeter squared).**

**Harmonic.**—A harmonic is any component of a periodic quantity which is an integral multiple of the fundamental frequency. For example, a component the frequency of which is twice the fundamental frequency is called the second harmonic.

A harmonic, in electricity, is an alternating-current electromotive force wave of higher frequency than the fundamental, and superimposed on the same so as to distort it from a true sine-wave shape. It is caused by the slots, the shape of the pole pieces, and the pulsation of the armature reaction. The third and the fifth harmonics, i.e., with a frequency three and five times the fundamental, are generally the predominating ones in three-phase machines.

## VELOCITY, ACCELERATION, WORK, AND ENERGY

### Velocity and Acceleration

Motion is a progressive change of position of a body. Velocity is the rate of motion, that is, the rate of change of a body's position. It is expressed in units of distance over time, such as feet per second, miles per hour, and kilometers per second. *Uniform motion* indicates that the velocity of a body is the same at every moment during which the motion takes place. When velocity is variable and constantly increasing, the rate at which it changes is called *acceleration*. Acceleration is the rate at which the velocity of a body changes in a unit of time, as the change of feet or meters per second in one second. When the motion is decreasing instead of increasing, it is called *retarded* motion, and the rate at which the motion is retarded is frequently called *deceleration*. If the acceleration is uniform, the motion is called *uniformly accelerated* motion. An example of such motion is that of falling bodies.

**Newton's Laws of Motion.**—The first clear statement of the fundamental relations existing between force and motion was made in the seventeenth century by Sir Isaac Newton, the English mathematician and physicist. It was put in the form of three laws, which are given as originally stated by Newton:

1) Every body continues in its state of rest, or uniform motion in a straight line, except in so far as it may be compelled by force to change that state.

2) Change of motion is proportional to the force applied and takes place in the direction in which that force acts.

3) To every action there is always an equal reaction; or, the mutual actions of two bodies are always equal and oppositely directed.

**Motion with Constant Velocity.**—In the formulas that follow, $S$ = distance moved; $V$ = velocity; $t$ = time of motion, $\theta$ = angle of rotation, and $\omega$ = angular velocity; the usual units for these quantities in the US customary system are, respectively, feet, feet per second, seconds, radians, and radians per second. The usual metric units are meters, meters per second, seconds, radians, and radians per second. Any consistent set of units may be employed.

*Constant Linear Velocity:* $S = Vt$  $V = S/t$  $t = S/V$

*Constant Angular Velocity:* $\theta = \omega t$  $\omega = \theta/t$  $t = \theta/\omega$

*Relation between Angular Motion and Linear Motion:* The relation between the angular velocity of a rotating body and the linear velocity of a point at a distance $r$ from the center of rotation is:

$$V(\text{ft/s}) = r(\text{ft}) \times \omega(\text{radians/s}) \qquad V(\text{m/s}) = r(\text{m}) \times \omega(\text{radians/s})$$

Similarly, the distance moved by the point during rotation through angle $\theta$ is:

$$S(\text{ft}) = r(\text{ft}) \times \theta(\text{radians}) \qquad S(\text{m}) = r(\text{m}) \times \theta(\text{radians})$$

**Linear Motion with Constant Acceleration.**—The relations between distance, velocity, and time for linear motion with constant or uniform acceleration are given by the formulas in the accompanying Table 1. In these formulas, the acceleration is assumed to be in the same direction as the initial velocity; hence, if the acceleration in a particular problem should happen to be in a direction opposite that of the initial velocity, then $a$ should be replaced by $-a$. Thus, for example, the formula $V_f = V_o + at$ becomes $V_f = V_o - at$ when $a$ and $V_o$ are opposite in direction.

*Example:* A car is moving at 100 km/h when the brakes are suddenly locked and the car begins to skid. If it takes 2 seconds to slow the car to 50 km/h, at what rate is it being decelerated, how long is it before the car comes to a halt, and how far will it have traveled?

*Solution:* The initial velocity $V_o$ of the car is 100 km/h or 27.78 m/s and the acceleration $a$ due to braking is opposite in direction to $V_o$, since the car is slowed to 50 km/h or 13.89 m/s.

**Table 1. Linear Motion with Constant Acceleration**

| To Find | Known | Formula | To Find | Known | Formula |
|---|---|---|---|---|---|
| Motion Uniformly Accelerated from Rest ($V_o = 0$) | | | | | |
| $S$ | $a, t$ | $S = \frac{1}{2}at^2$ | $t$ | $S, V_f$ | $t = 2S \div V_f$ |
| | $V_f, t$ | $S = \frac{1}{2}V_f t$ | | $S, a$ | $t = \sqrt{2S \div a}$ |
| | $V_f, a$ | $S = V_f^2 \div 2a$ | | $a, V_f$ | $t = V_f \div a$ |
| $V_f$ | $a, t$ | $V_f = at$ | $a$ | $S, t$ | $a = 2S \div t^2$ |
| | $S, t$ | $V_f = 2S \div t$ | | $S, V$ | $a = V_f^2 \div 2S$ |
| | $a, S$ | $V_f = \sqrt{2aS}$ | | $V_f, t$ | $a = V_f \div t$ |
| Motion Uniformly Accelerated from Initial Velocity $V_o$ | | | | | |
| $S$ | $a, t, V_o$ | $S = V_o t + \frac{1}{2}at^2$ | $t$ | $V_o, V_f, a$ | $t = (V_f - V_o) \div a$ |
| | $V_o, V_f, t$ | $S = (V_f + V_o)t \div 2$ | | $V_o, V_f, S$ | $t = 2S \div (V_f + V_o)$ |
| | $V_o, V_f, a$ | $S = (V_f^2 - V_o^2) \div 2a$ | $a$ | $V_o, V_f, S$ | $a = (V_f^2 - V_o^2) \div 2S$ |
| | $V_f, a, t$ | $S = V_f t - \frac{1}{2}at^2$ | | $V_o, V_f, t$ | $a = (V_f - V_o) \div t$ |
| $V_f$ | $V_o, a, t$ | $V_f = V_o + at$ | | $V_o, S, t$ | $a = 2(S - V_o t) \div t^2$ |
| | $V_o, S, t$ | $V_f = (2S \div t) - V_o$ | | $V_f, S, t$ | $a = 2(V_f t - S) \div t^2$ |
| | $V_o, a, S$ | $V_f = \sqrt{V_o^2 + 2aS}$ | *Meanings of Symbols* | | |
| | $S, a, t$ | $V_f = (S \div t) + \frac{1}{2}at$ | | | |
| $V_o$ | $V_f, a, S$ | $V_o = \sqrt{V_f^2 - 2aS}$ | $S$ = distance moved in ft or m | | |
| | $V_f, S, t$ | $V_o = (2S \div t) - V_f$ | $V_f$ = final velocity, ft/s or m/s | | |
| | $V_f, a, t$ | $V_o = V_f - at$ | $V_o$ = initial velocity, ft/s or m/s | | |
| | $S, a, t$ | $V_o = (S \div t) - \frac{1}{2}at$ | $a$ = acceleration, ft/s² or m/s² | | |
| | | | $t$ = time of acceleration, sec or s | | |

Since $V_o$, $V_f$, and $t$ are known, $a$ can be determined from the formula

$$a = (V_f - V_o) \div t = (13.89 - 27.78) \div 2 = -6.95 \text{m/s}^2$$

The time required to stop the car can be determined from the formula

$$t = (V_f - V_o) \div a = (0 - 27.78) \div (-6.95) = 4 \text{ seconds}$$

The distance traveled by the car is obtained from the formula

$$S = V_o t + \frac{1}{2}at^2 = (27.78)(4) + \frac{1}{2}(-6.95)4^2 = (111.12 - 55.6) = 55.52 \text{ m}$$

**Angular Velocity of a Rotating Body.**—The angular velocity of a rotating body is the angle through which the body turns in a unit of time. Angular velocity is commonly expressed in terms of revolutions per minute rpm, but in certain engineering applications it is necessary to express it as radians per second (rad/s). By definition there are $2\pi$ radians in 360 degrees, or one revolution, so that one radian = $360 / 2\pi \approx 57.3$ degrees. To convert angular velocity $n$ in rpm to angular velocity $\omega$ in radians per second, and vice versa, use Equation (1):

$$\omega = \frac{\pi n}{30} \text{ rad/s} \qquad n = \frac{30\omega}{\pi} \text{ rpm} \qquad (1)$$

Table 2 can be used to obtain angular velocity in radians per second for all whole numbers of revolutions per minute from 1 to 239.

**Table 2. Angular Velocity in Revolutions per Minute Converted to Radians per Second**

| RPM | Angular Velocity in Radians per Second (rad/s) | | | | | | | | | |
|---|---|---|---|---|---|---|---|---|---|---|
| | 0 | 1 | 2 | 3 | 4 | 5 | 6 | 7 | 8 | 9 |
| 0 | 0.00 | 0.10 | 0.21 | 0.31 | 0.42 | 0.52 | 0.63 | 0.73 | 0.84 | 0.94 |
| 10 | 1.05 | 1.15 | 1.26 | 1.36 | 1.47 | 1.57 | 1.67 | 1.78 | 1.88 | 1.99 |
| 20 | 2.09 | 2.20 | 2.30 | 2.41 | 2.51 | 2.62 | 2.72 | 2.83 | 2.93 | 3.04 |
| 30 | 3.14 | 3.25 | 3.35 | 3.46 | 3.56 | 3.66 | 3.77 | 3.87 | 3.98 | 4.08 |
| 40 | 4.19 | 4.29 | 4.40 | 4.50 | 4.61 | 4.71 | 4.82 | 4.92 | 5.03 | 5.13 |
| 50 | 5.24 | 5.34 | 5.44 | 5.55 | 5.65 | 5.76 | 5.86 | 5.97 | 6.07 | 6.18 |
| 60 | 6.28 | 6.39 | 6.49 | 6.60 | 6.70 | 6.81 | 6.91 | 7.02 | 7.12 | 7.23 |
| 70 | 7.33 | 7.43 | 7.54 | 7.64 | 7.75 | 7.85 | 7.96 | 8.06 | 8.17 | 8.27 |
| 80 | 8.38 | 8.48 | 8.59 | 8.69 | 8.80 | 8.90 | 9.01 | 9.11 | 9.21 | 9.32 |
| 90 | 9.42 | 9.53 | 9.63 | 9.74 | 9.84 | 9.95 | 10.05 | 10.16 | 10.26 | 10.37 |
| 100 | 10.47 | 10.58 | 10.68 | 10.79 | 10.89 | 11.00 | 11.10 | 11.20 | 11.31 | 11.41 |
| 110 | 11.52 | 11.62 | 11.73 | 11.83 | 11.94 | 12.04 | 12.15 | 12.25 | 12.36 | 12.46 |
| 120 | 12.57 | 12.67 | 12.78 | 12.88 | 12.98 | 13.09 | 13.19 | 13.30 | 13.40 | 13.51 |
| 130 | 13.61 | 13.72 | 13.82 | 13.93 | 14.03 | 14.14 | 14.24 | 14.35 | 14.45 | 14.56 |
| 140 | 14.66 | 14.76 | 14.87 | 14.97 | 15.08 | 15.18 | 15.29 | 15.39 | 15.50 | 15.60 |
| 150 | 15.71 | 15.81 | 15.92 | 16.02 | 16.13 | 16.23 | 16.34 | 16.44 | 16.55 | 16.65 |
| 160 | 16.75 | 16.86 | 16.96 | 17.07 | 17.17 | 17.28 | 17.38 | 17.49 | 17.59 | 17.70 |
| 170 | 17.80 | 17.91 | 18.01 | 18.12 | 18.22 | 18.33 | 18.43 | 18.53 | 18.64 | 18.74 |
| 180 | 18.85 | 18.95 | 19.06 | 19.16 | 19.27 | 19.37 | 19.48 | 19.58 | 19.69 | 19.79 |
| 190 | 19.90 | 20.00 | 20.11 | 20.21 | 20.32 | 20.42 | 20.52 | 20.63 | 20.73 | 20.84 |
| 200 | 20.94 | 21.05 | 21.15 | 21.26 | 21.36 | 21.47 | 21.57 | 21.68 | 21.78 | 21.89 |
| 210 | 21.99 | 22.10 | 22.20 | 22.30 | 22.41 | 22.51 | 22.62 | 22.72 | 22.83 | 22.93 |
| 220 | 23.04 | 23.14 | 23.25 | 23.35 | 23.46 | 23.56 | 23.67 | 23.77 | 23.88 | 23.98 |
| 230 | 24.09 | 24.19 | 24.29 | 24.40 | 24.50 | 24.61 | 24.71 | 24.82 | 24.92 | 25.03 |

*Example:* To find the angular velocity in radians per second of a flywheel making 97 revolutions per minute, locate 90 in the left-hand column and 7 at the top of the columns; at the intersection of the two lines, the angular velocity is read off as equal to 10.16 radians per second.

**Linear Velocity of Points on a Rotating Body.**—The linear velocity, $v$, of any point on a rotating body expressed in feet per second may be found by multiplying the angular velocity of the body in radians per second, $\omega$, by the radius, $r$, in feet from the center of rotation to the point:

$$v = \omega r \tag{2}$$

**The metric SI units are $v$ = meters per second; $\omega$ = radians per second, $r$ = meters.**

**Rotary Motion with Constant Acceleration.**—The relations among angle of rotation, angular velocity, and time for rotation with constant or uniform acceleration are given in the accompanying Table 3.

In these formulas, the acceleration is assumed to be in the same direction as the initial angular velocity; hence, if the acceleration in a particular problem should happen to be in a direction opposite that of the initial angular velocity, then $\alpha$ should be replaced by $-\alpha$. Thus, for example, the formula $\omega_f = \omega_o + \alpha t$ becomes $\omega_f = \omega_o - \alpha t$ when $\alpha$ and $\omega_o$ are opposite in direction.

*Linear Acceleration of a Point on a Rotating Body:* A point, $P$, on a body rotating about a fixed axis has a linear acceleration $a$ that is the resultant of two component accelerations. The first component is the centripetal, or normal, acceleration, which is directed from the point $P$ toward the axis of rotation; its magnitude is $r\omega^2$, where $r$ is the radius from the axis to the point $P$ and $\omega$ is the angular velocity of the body at the time acceleration $a$ is to be determined. The second component of $a$ is the tangential acceleration, which is equal to $r\alpha$, where $\alpha$ is the angular acceleration of the body.

**Table 3. Rotary Motion with Constant Acceleration**

| To Find | Known | Formula | To Find | Known | Formula |
|---|---|---|---|---|---|
| Motion Uniformly Accelerated from Rest ($\omega_o = 0$) | | | | | |
| $\theta$ | $\alpha, t$ | $\theta = \frac{1}{2}\alpha t^2$ | $t$ | $\theta, \omega_f$ | $t = 2\theta \div \omega_f$ |
| | $\omega_f, t$ | $\theta = \frac{1}{2}\omega_f t$ | | $\theta, \alpha$ | $t = \sqrt{2\theta \div \alpha}$ |
| | $\omega_f, \alpha$ | $\theta = \omega_f^2 \div 2\alpha$ | | $\alpha, \omega_f$ | $t = \omega_f \div \alpha$ |
| $\omega_f$ | $\alpha, t$ | $\omega_f = \alpha t$ | $\alpha$ | $\theta, t$ | $\alpha = 2\theta \div t^2$ |
| | $\theta, t$ | $\omega_f = 2\theta \div t$ | | $\theta, \omega_f$ | $\alpha = \omega_f^2 \div 2\theta$ |
| | $\alpha, \theta$ | $\omega_f = \sqrt{2\alpha\theta}$ | | $\omega_f, t$ | $\alpha = \omega_f \div t$ |
| Motion Uniformly Accelerated from Initial Velocity $\omega_o$ | | | | | |
| $\theta$ | $\alpha, t, \omega_o$ | $\theta = \omega_o t + \frac{1}{2}\alpha t^2$ | $\alpha$ | $\omega_o, \omega_f, \theta$ | $\alpha = (\omega_f^2 - \omega_o^2) \div 2\theta$ |
| | $\omega_o, \omega_f, t$ | $\theta = (\omega_f + \omega_o)t \div 2$ | | $\omega_o, \omega_f, t$ | $\alpha = (\omega_f - \omega_o) \div t$ |
| | $\omega_o, \omega_f, \alpha$ | $\theta = (\omega_f^2 - \omega_o^2) \div 2\alpha$ | | $\omega_o, \theta, t$ | $\alpha = 2(\theta - \omega_o t) \div t^2$ |
| | $\omega_f, \alpha, t$ | $\theta = \omega_f t - \frac{1}{2}\alpha t^2$ | | $\omega_f, \theta, t$ | $\alpha = 2(\omega_f t - \theta) \div t^2$ |
| $\omega_f$ | $\omega_o, \alpha, t$ | $\omega_f = \omega_o + \alpha t$ | *Meanings of Symbols* | | |
| | $\omega_o, \theta, t$ | $\omega_f = (2\theta \div t) - \omega_o$ | $\theta$ = angle of rotation, radians | | |
| | $\omega_o, \alpha, \theta$ | $\omega_f = \sqrt{\omega_o^2 + 2\alpha\theta}$ | $\omega_f$ = final angular velocity, rad/s | | |
| | $\theta, \alpha, t$ | $\omega_f = (\theta \div t) + \frac{1}{2}\alpha t$ | $\omega_o$ = initial angular velocity, rad/s | | |
| $\omega_o$ | $\omega_f, \alpha, \theta$ | $\omega_o = \sqrt{\omega_f^2 - 2\alpha\theta}$ | $\alpha$ = angular acceleration, rad/s$^2$ | | |
| | $\omega_f, \theta, t$ | $\omega_o = (2\theta \div t) - \omega_f$ | $t$ = time in seconds | | |
| | $\omega_f, \alpha, t$ | $\omega_o = \omega_f - \alpha t$ | | | |
| | $\theta, \alpha, t$ | $\omega_o = (\theta \div t) - \frac{1}{2}\alpha t$ | | | |
| $t$ | $\omega_o, \omega_f, \alpha$ | $t = (\omega_f - \omega_o) \div \alpha$ | 1 degree = 0.01745 radian | | |
| | $\omega_o, \omega_f, \theta$ | $t = 2\theta \div (\omega_f + \omega_o)$ | (See degree-radian conversion table on page 98) | | |

The acceleration of point $P$ is the resultant of $r\omega^2$ and $r\alpha$ and is given by the formula

$$a = \sqrt{(r\omega^2)^2 + (r\alpha)^2}$$

When $\alpha = 0$, this formula reduces to: $a = r\omega^2$

*Example:* A flywheel on a press rotating at 120 rpm is slowed to 102 rpm during a punching operation that requires ¾ second for the punching portion of the cycle. What angular deceleration does the flywheel experience?

*Solution:* From the table on page 178, angular velocities corresponding to 120 and 102 rpm, respectively, are 12.57 and 10.68 radians per second. Therefore, using the formula:

$$\begin{aligned}\alpha &= (\omega_f - \omega_o) \div t \\ &= (10.68 - 12.57) \div \tfrac{3}{4} = -1.89 \div \tfrac{3}{4} \\ &= -2.52 \text{ rad/s}^2\end{aligned}$$

which is, from the table on page 178, −24 rpm per second. The minus sign in the answer indicates that acceleration $\alpha$ acts to slow the flywheel, that is, the flywheel is decelerating.

## Force, Work, Energy, and Momentum

**Accelerations Resulting from Unbalanced Forces.—**In the section describing the resolution and composition of forces, page 157, it was stated that when the resultant of a system of forces is zero, the system is in equilibrium, that is, the body on which the force system acts remains at rest or continues to move with uniform velocity. If, however, the resultant of a system of forces is not zero, the body on which the forces act will be accelerated in the direction of the unbalanced force. To determine the relation between the unbalanced force and the resulting acceleration, Newton's laws of motion must be applied. These laws may be stated as follows:

*First Law:* Every body continues in a state of rest or in uniform motion in a straight line until it is compelled by a force to change its state of rest or motion.

*Second Law:* Change of motion is proportional to the force applied and takes place along the straight line in which the force acts. The "force applied" represents the resultant of *all* the forces acting on the body. This law is sometimes worded: An unbalanced force acting on a body causes an acceleration of the body in the direction of the force and of magnitude proportional to the force and inversely proportional to the mass of the body. Stated as a formula, $R = Ma$ where $R$ is the resultant of *all* the forces acting on the body, $M$ is the mass of the body (mass = weight $W$ divided by acceleration due to gravity $g$), and $a$ is the acceleration of the body resulting from application of force $R$.

*Third Law:* To every action there is always an equal reaction; or, if a force acts to change the state of motion of a body, the body offers a resistance equal and directly opposite to the force.

Newton's second law may be used to calculate linear and angular accelerations of a body produced by unbalanced forces and torques acting on the body; however, it is necessary first to use the methods described under *Algebraic Composition and Resolution of Force Systems* starting on page 157 to determine the magnitude and direction of the resultant of *all* forces acting on the body. Then, for a body moving with pure translation,

$$R = Ma = \frac{W}{g}a$$

where $R$ is the resultant force in pounds acting on a body weighing $W$ pounds; $g$ is the gravitational constant, usually taken as 32.16 ft/sec$^2$, approximately; and $a$ is the resulting acceleration in ft/sec$^2$ of the body due to $R$ and in the same direction as $R$.

**Using metric SI units, the formula is $R = Ma$, where $R$ = force in newtons (N), $M$ = mass in kilograms (kg), and $a$ = acceleration in meters/second squared (m/s$^2$). It should be noted that the weight of a body of mass $M$ kg is $Mg$ newtons, where $g$ is approximately 9.81 m/s$^2$.**

*Free Body Diagram:* In order to correctly determine the effect of forces on the motion of a body it is necessary to construct a *free body diagram*. This diagram shows 1) the body removed or isolated from contact with all other bodies that exert force on it; and 2) *all* the forces acting on the body.

For example, in Fig. 1a the block being pulled up the plane is acted upon by certain forces; the free body diagram of this block is shown at Fig. 1b. Note that all forces acting on the block are indicated. These forces include: 1) the force of gravity (weight); 2) the pull of the cable, $P$; 3) the normal component, $W \cos\phi$, of the force exerted on the block by the plane; and 4) the friction force, $\mu W \cos\phi$, of the plane on the block.

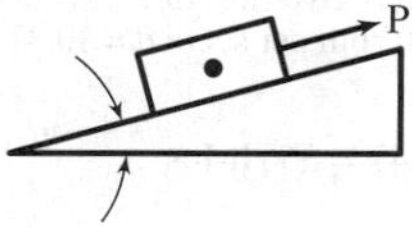


Fig. 1a.

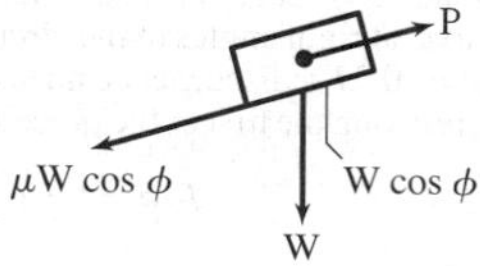


Fig. 1b.

In preparing a free body diagram, it is important to understand that only those forces exerted *on* the body being considered are shown; forces exerted by the body on other bodies are disregarded. This feature makes the free body diagram an invaluable aid in the solution of problems in mechanics.

*Example:* A 100-pound body is being hoisted by a winch, the tension in the hoisting cable being kept constant at 110 pounds. At what rate is the body accelerated?

*Solution:* Two forces are acting on the body, its weight, 100 pounds downward, and the pull of the cable, 110 pounds upward. The resultant force *R*, from a free body diagram, is therefore 110 – 100. Thus, applying Newton's second law,

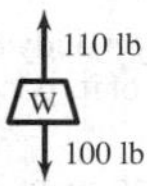


$$110 - 100 = \frac{100}{32.16}a$$

$$a = \frac{32.16 \times 10}{100} = 3.216 \text{ ft/sec}^2 \text{ upward}$$

It should be noted that since in this problem the resultant force *R* was positive (110 – 100 = +10), the acceleration *a* is also positive, that is, *a* is in the same direction as *R*, which is in accord with Newton's second law.

***Example using SI metric units:* A body of mass 50 kilograms is being hoisted by a winch, and the tension in the cable is 600 newtons. What is the acceleration? The weight of the 50 kg body is 50*g* newtons, where *g* = approximately 9.81 m/s² (see *Note* on page 188). Applying the formula *R* = *Ma*, the calculation is: (600 – 50*g*) = 50*a*. Thus,**

$$a = \frac{600 - 50g}{50} = \frac{600 - (50 \times 9.81)}{50} = 2.19 \text{ m/s}^2$$

*Formulas Relating Torque and Angular Acceleration:* For a body rotating about a fixed axis the relation between the unbalanced torque acting to produce rotation and the resulting angular acceleration may be determined from any one of the following formulas, each based on Newton's second law:

$$T_o = J_M \alpha$$

$$T_o = M k_o^2 \alpha$$

$$T_o = \frac{W k_o^2 \alpha}{g} = \frac{W k_o^2 \alpha}{32.16}$$

where $T_o$ is the unbalanced torque in pounds-feet; $J_M$ in ft-lbs-sec² is the moment of inertia of the body about the axis of rotation; $k_o$ in feet is the radius of gyration of the body with respect to the axis of rotation, and α rad/sec², is the angular acceleration of the body.

*Example:* A flywheel has a diameter of 3 feet and weighs 1000 pounds. What torque must be applied, neglecting bearing friction, to accelerate the flywheel at the rate of 100 revolutions per minute per second (rpm/s)?

*Solution:* From page 247 the moment of inertia of a solid cylinder with respect to a gravity axis at right angles to the circular cross section is given as $\frac{1}{2}Mr^2$. From page 178, 100 rpm = 10.47 rad/sec, hence an acceleration of 100 rpm per second = 10.47 rad/sec$^2$. Therefore, using the first of the preceding formulas,

$$T_o = J_M \alpha = \left(\frac{1}{2}\right)\frac{1000}{32.16}\left(\frac{3}{2}\right)^2 \times 10.47 = 366 \text{ ft-lbs}$$

**Using metric SI units, the formulas are: $T_o = J_M\alpha = Mk_o^2\alpha$, where $T_o$ = torque in newton-meters; $J_M$ = the moment of inertia in kg·m$^2$, and $\alpha$ = the angular acceleration in rad/s$^2$.**

***Example:* A flywheel has a diameter of 1.5 m and a mass of 800 kg. What torque is needed to produce an angular acceleration of 100 rpm/s? As in the preceding example, $\alpha$ = 10.47 rad/s$^2$. Thus:**

$$J_M = \frac{1}{2}Mr^2 = \frac{1}{2} \times 800 \times 0.75^2 = 225 \text{ kg·m}^2$$

**Therefore: $T_o = J_M\alpha = 225 \times 10.47 = 2356$ N·m.**

**Energy.**—A body is said to possess energy when it is capable of doing work or overcoming resistance. The energy may be either mechanical or non-mechanical, the latter including chemical, electrical, thermal, and atomic energy.

Mechanical energy includes *kinetic energy* (energy possessed by a body because of its motion) and *potential energy* (energy possessed by a body because of its position in a field of force and/or its elastic deformation).

*Kinetic Energy:* The motion of a body may be one of pure translation, pure rotation, or a combination of rotation and translation. By translation is meant motion in which every line in the body remains parallel to its original position throughout the motion; that is, no rotation is associated with the motion of the body.

The kinetic energy of a translating body is given by the formula

$$\text{Kinetic Energy in ft-lbs due to translation} = E_{KT} = \frac{1}{2}MV^2 = \frac{WV^2}{2g} \quad (3a)$$

where $M$ = mass of body ($M = W/g$); $V$ = velocity of the center of gravity of the body in ft/sec; $W$ = weight of body in lbs; and $g$ = acceleration due to gravity = 32.16 ft/sec$^2$.

The kinetic energy of a body rotating about a fixed axis $O$ is expressed by the formula:

$$\text{Kinetic Energy in ft-lbs due to rotation} = E_{KR} = \frac{1}{2}J_{MO}\omega^2 \quad (3b)$$

where $J_{MO}$ is the moment of inertia of the body about the fixed axis $O$ in lbs-ft-sec$^2$, and $\omega$ = angular velocity in rad/sec.

For a body that is moving with both translation and rotation, the total kinetic energy is given by the following formula as the sum of the kinetic energy due to translation of the center of gravity and the kinetic energy due to rotation about the center of gravity:

$$\text{Total Kinetic Energy in ft-lbs} = E_T = \frac{1}{2}MV^2 + \frac{1}{2}J_{MG}\omega^2 \quad (3c)$$

$$= \frac{WV^2}{2g} + \frac{1}{2}J_{MG}\omega^2 = \frac{WV^2}{2g} + \frac{1}{2}\frac{Wk^2\omega^2}{g} = \frac{W}{2g}(V^2 + k^2\omega^2)$$

where $J_{MG}$ is the moment of inertia of the body about its gravity axis in lbs-ft-sec$^2$, $k$ is the radius of gyration in feet with respect to an axis through the center of gravity, and the other quantities are as previously defined.

**In the metric SI system, energy is expressed as the joule (J). One joule = 1 newton-meter. The kinetic energy of a translating body is given by the formula $E_{KT} = \frac{1}{2}MV^2$,**

**where $M$ = mass in kilograms, and $V$ = velocity in meters per second. Kinetic energy due to rotation is expressed by the formula $E_{KR} = \frac{1}{2} J_{MO}\omega^2$, where $J_{MO}$ = moment of inertia in kg·m², and $\omega$ = the angular velocity in radians per second. Total kinetic energy $ET = \frac{1}{2}MV^2 + \frac{1}{2} J_{MO}\omega^2$ joules = $\frac{1}{2}M(V^2 + k^2\omega^2)$ joules, where $k$ = radius of gyration in meters.**

*Potential Energy:* A common example of a body having potential energy because of its position in a field of force is that of a body elevated to some height above the earth. The field of force is the gravitational field of the earth. The potential energy of a body weighing $W$ pounds elevated to height $S$ feet above the earth's surface is $E_{PF} = WS$ lb. If the body is permitted to drop from this height its potential energy will be converted to kinetic energy. Thus, after falling through height $S$ the kinetic energy of the body will be $WS$ ft-lbs.

**In metric SI units, the potential energy $E_{PF}$ of a body of mass $M$ kilograms elevated to a height of $S$ meters, is $MgS$ joules. After it has fallen a distance $S$, the kinetic energy gained will thus be $MgS$ joules.**

Another type of potential energy is *elastic potential energy*, such as possessed by a spring that has been compressed or extended. The amount of work in ft-lbs done in compressing the spring $S$ feet is equal to $KS^2/2$, where $K$ is the spring constant in pounds per foot. Thus, when the spring is released to act against some resistance, it can perform $KS^2/2$ ft-lbs of work, which is the amount of elastic potential energy $E_{PE}$ stored in the spring.

**Using metric SI units, the amount of work done in compressing the spring a distance $S$ meters is $KS^2/2$ joules, where $K$ is the spring constant in newtons per meter.**

**Work Performed by Forces and Couples.**—The work $U$ done by a force $F$ in moving an object along some path is the product of the distance $S$ the body is moved and the component $F \cos \alpha$ of the force $F$ in the direction of $S$.

$$U = FS\cos\alpha$$

where $U$ = work in ft-lbs; $S$ = distance moved in feet; $F$ = force in lbs; and $\alpha$ = angle between line of action of force and the path of $S$.

If the force is in the same direction as the motion, then $\cos \alpha = \cos 0 = 1$, and this formula reduces to:

$$U = FS$$

Similarly, the work done by a couple $T$ turning an object through an angle $\theta$ is:

$$U = T\theta$$

where $T$ = torque of couple in pounds-feet and $\theta$ = the angular rotation in radians.

**The above formulas can be used with metric SI units: $U$ in joules; $S$ in meters; $F$ in newtons, and $T$ in newton-meters.**

**Relation Between Work and Energy.**—Theoretically, when work is performed on a body and there are no energy losses (such as those due to friction, air resistance, etc.), the energy acquired by the body is equal to the work performed on the body; this energy may be either potential, kinetic, or a combination of both.

In actual situations, however, there may be energy losses that must be taken into account. Thus, the relation between work done on a body, energy losses, and the energy acquired by the body can be stated as:

$$\text{Work Performed} - \text{Energy Losses} = \text{Energy Acquired}$$

$$U - \text{Losses} = E_T$$

*Example 1:* A 12-inch cube of steel weighing 490 pounds is being moved on a horizontal conveyor belt at a speed of 6 miles per hour (8.8 feet per second). What is the kinetic energy of the cube?

*Solution:* Since the block is not rotating, Formula (3a) for the kinetic energy of a body moving with pure translation applies:

$$\text{Kinetic Energy} = \frac{WV^2}{2g} = \frac{490 \times 8.8^2}{2 \times 32.16} = 590 \text{ ft-lbs}$$

**A similar example using metric SI units is as follows: If a cube of mass 200 kg is being moved on a conveyor belt at a speed of 3 meters per second, what is the kinetic energy of the cube? It is:**

$$\textbf{Kinetic Energy} = \frac{1}{2}MV^2 = \frac{1}{2} \times 200 \times 3^2 = 900 \text{ joules}$$

*Example 2:* If the conveyor in Example 1 is brought to an abrupt stop, how long would it take for the steel block to come to a stop and how far along the belt would it slide before stopping if the coefficient of friction μ between the block and the conveyor belt is 0.2 and the block slides without tipping over?

*Solution:* The only force acting to slow the motion of the block is the friction force between the block and the belt. This force $F$ is equal to the weight of the block $W$ multiplied by the coefficient of friction; $F = \mu W = 0.2 \times 490 = 98$ lbs.

The time required to bring the block to a stop can be determined from the impulse-momentum Formula (4c) on page 185.

$$R \times t = \frac{W}{g}(V_f - V_o) = (-98)t = \frac{490}{32.16}(0 - 8.8)$$

$$t = \frac{490 \times 8.8}{98 \times 32.16} = 1.37 \text{ seconds}$$

The distance the block slides before stopping can be determined by equating the kinetic energy of the block and the work done by friction in stopping it:

$$\text{Kinetic energy of block}\left(\frac{WV^2}{2g}\right) = \text{Work done by friction } (F \times S)$$

$$590 = 98 \times S$$

$$S = \frac{590}{98} = 6.0 \text{ feet}$$

**If metric SI units are used, the calculation is as follows (for the cube of 200 kg mass): The friction force = μ multiplied by the weight $Mg$ where $g$ = approximately 9.81 m/s². Thus, $\mu Mg = 0.2 \times 200g = 392.4$ newtons. The time $t$ required to bring the block to a stop is $(-392.4)t = 200(0 - 3)$. Therefore,**

$$t = \frac{200 \times 3}{392.4} = 1.53 \text{ seconds}$$

**The kinetic energy of the block is equal to the work done by friction, that is 392.4 × $S$ = 900 joules. Thus, the distance S which the block moves before stopping is**

$$S = \frac{900}{392.4} = 2.29 \text{ meters}$$

**Force of a Blow.**—A body that weighs $W$ pounds and falls $S$ feet from an initial position of rest is capable of doing $WS$ foot-pounds of work. The work performed during its fall may, for example, be that necessary to drive a pile a distance $d$ into the ground. Neglecting losses in the form of dissipated heat and strain energy, the work done in driving the pile is equal to the product of the impact force acting on the pile and the distance $d$ which the pile is driven. Since the impact force is not accurately known, an average value, called the "average force of the blow," may be assumed. Equating the work done on the pile and the work done by the falling body, which in this case is a pile driver:

$$\text{Average force of blow} \times d = WS$$

or,

$$\text{Average force of blow} = \frac{WS}{d}$$

where $S$ = total height in feet through which the driver falls, including the distance $d$ that the pile is driven
$W$ = weight of driver in pounds
$d$ = distance in feet which pile is driven

**When using metric SI units, it should be noted that a body of mass $M$ kilograms has a weight of $Mg$ newtons, where $g$ = approximately 9.81 m/s$^2$. If the body falls a distance $S$ meters, it can do work equal to $MgS$ joules. The average force of the blow is $MgS/d$ newtons, where $d$ is the distance in meters that the pile is driven.**

*Example:* A pile driver weighing 200 pounds strikes the top of the pile after having fallen from a height of 20 feet. It forces the pile into the ground a distance of ½ foot. Before the ram is brought to rest, it will do 200 × (20 + ½) = 4100 foot-pounds of work, and as this energy is expended in a distance of ½ foot, the average force of the blow equals 4100 ÷ ½ = 8200 pounds.

**A similar example using metric SI units is as follows: A pile driver of mass 100 kilograms falls 10 meters and moves the pile a distance of 0.3 meter. The work done = $100g(10 + 0.3)$ joules, and it is expended in 0.3 meter. Thus, the average force is**

$$\frac{100g \times 10.3}{0.3} = 33680 \text{ newtons or } 33.68 \text{ kN}$$

**Impulse and Momentum.**—The *linear momentum* of a body is defined as the product of the mass $M$ of the body and the velocity $V$ of the center of gravity of the body:

$$\text{Linear momentum} = MV \text{ or since } M = \frac{W}{g}$$

$$\text{Linear momentum} = \frac{WV}{g} \quad (4a)$$

It should be noted that linear momentum is a vector quantity, the momentum being in the same direction as $V$.

*Linear impulse* is defined as the product of the resultant $R$ of *all* the forces acting on a body and the time $t$ that the resultant acts:

$$\text{Linear Impulse} = Rt \quad (4b)$$

The change in the linear momentum of a body is numerically equal to the linear impulse that causes the change in momentum:

$$\text{Linear Impulse} = \text{change in Linear Momentum}$$

$$Rt = \frac{W}{g}V_f - \frac{W}{g}V_o = \frac{W}{g}(V_f - V_o) \quad (4c)$$

where $V_f$, the final velocity of the body after time $t$, and $V_o$, the initial velocity of the body, are both in the same direction as the applied force $R$. If $V_o$ and $V_f$ are in opposite directions, then the minus sign in the formula becomes a plus sign.

**In metric SI units, the formulas are: Linear Momentum = $MV$ kg·m/s, where $M$ = mass in kg, and $V$ = velocity in meters per second; and Linear Impulse = $Rt$ newton-seconds (N·s), where $R$ = force in newtons, and $t$ = time in seconds. In Formula (4c) above, $W/g$ is replaced by $M$ when SI units are used.**

*Example:* A 1000-pound block is pulled up a 2-degree incline by a cable exerting a constant force $F$ of 600 lbs. If the coefficient of friction $\mu$ between the block and the plane is 0.5, how fast will the block be moving up the plane 10 seconds after the pull is applied?

*Solution:* The resultant force $R$ causing the body to be accelerated up the plane is the difference between $F$, the force acting up the plane, and $P$, the force acting to resist motion up the plane. This latter force for a body on a plane is given by the formula at the top of page 170 as $P = W(\mu \cos \alpha + \sin \alpha)$ where $\alpha$ is the angle of the incline.

Thus,
$$\begin{aligned} R &= F - P = F - W(\mu \cos \alpha + \sin \alpha) \\ &= 600 - 1000(0.5 \cos 2° + \sin 2°) = 600 - 1000(0.5 \times 0.99939 + 0.03490) \\ R &= 600 - 535 = 65 \text{ pounds.} \end{aligned}$$

Formula (4c) can now be applied to determine the speed at which the body will be moving up the plane after 10 seconds:

$$\begin{aligned} Rt &= \frac{W}{g}V_f - \frac{W}{g}V_o \\ 65 \times 10 &= \frac{1000}{32.2}V_f - \frac{1000}{32.2} \times 0 \\ V_f &= \frac{65 \times 10 \times 32.2}{1000} = 20.9 \text{ ft/sec} = 14.3 \text{ mph} \end{aligned}$$

**A similar example using metric SI units is as follows: A 500 kg block is pulled up a 2 degree incline by a constant force *F* of 4 kN. The coefficient of friction $\mu$ between the block and the plane is 0.5. How fast will the block be moving 10 seconds after the pull is applied?**

**The resultant force *R* is:**

$$\begin{aligned} R &= F - Mg(\mu \cos\alpha + \sin \alpha) \\ &= 4000 - 500 \times 9.81(0.5 \times 0.99939 + 0.03490) = 1378\text{N or } 1.378 \text{ kN} \end{aligned}$$

**Formula (4c) can now be applied to determine the speed at which the body will be moving up the plane after 10 seconds. Replacing *W/g* by *M* in the formula, the calculation is:**

$$\begin{aligned} Rt &= MV_f - MV_o \\ 1378 \times 10 &= 500(V_f - 0) \\ V_f &= \frac{1378 \times 10}{500} = 27.6 \text{ m/s} \end{aligned}$$

*Angular Impulse and Momentum:* In a manner similar to that for linear impulse and moment, the formulas for angular impulse and momentum for a body rotating about a fixed axis are:

$$\text{Angular momentum} = J_M \omega \qquad (5a)$$

$$\text{Angular impulse} = T_o t \qquad (5b)$$

where $J_M$ is the moment of inertia of the body about the axis of rotation in lb-ft-sec$^2$, $\omega$ is the angular velocity in rad/sec, $T_o$ is the torque in lb-ft about the axis of rotation, and $t$ is the time in seconds that $T_o$ acts.

The change in angular momentum of a body is numerically equal to the angular impulse that causes the change in angular momentum:

$$\text{Angular Impulse} = \text{Change in Angular Momentum}$$

$$T_o t = J_M \omega_f - J_M \omega_o = J_M(\omega_f - \omega_o) \qquad (5c)$$

where $\omega_f$ and $\omega_o$ are the final and initial angular velocities, respectively.

*Example:* A flywheel having a moment of inertia of 25 lbs-ft-sec$^2$ is revolving with an angular velocity of 10 rad/sec, when a constant torque of 20 lbs-ft is applied to reverse its direction of rotation. For what length of time must this constant torque act to stop the flywheel and bring it up to a reverse speed of 5 rad/sec?

*Solution:* Applying Formula (5c),

$$T_o t = J_M(\omega_f - \omega_o)$$
$$20t = 25(10 - [-5]) = 250 + 125$$
$$t = 375 \div 20 = 18.8 \text{ seconds}$$

**A similar example using metric SI units is as follows: A flywheel with a moment of inertia of 20 kg-m$^2$ is revolving with an angular velocity of 10 rad/s when a constant torque of 30 N-m is applied to reverse its direction of rotation. For what length of time must this constant torque act to stop the flywheel and bring it up to a reverse speed of 5 rad/s? Applying Formula (5c), the calculation is:**

$$T_o t = J_M(\omega_f - \omega_o)$$
$$30t = 20(10 - -5)$$
$$\text{Thus, } t = \frac{20 \times 15}{30} = 10\text{s}$$

**Formulas for Work and Power.**—The formulas in the accompanying Table 4 may be used to determine work and power in terms of the applied force and the velocity at the point of application of the force.

**Table 4. Formulas[a] for Work and Power**

| To Find | Known | Formula |
|---|---|---|
| | $P, t, F$ | $S = P \times t \div F$ |
| $S$ | $K, F$ | $S = K \div F$ |
| | $t, F, hp$ | $S = 550 \times t \times hp \div F$ |
| | $P, F$ | $V = P \div F$ |
| $V$ | $K, F, t$ | $V = K \div (F \times t)$ |
| | $F, hp$ | $V = 550 \times hp \div F$ |
| | $F, S, P$ | $t = F \times S \div P$ |
| $t$ | $K, F, V$ | $t = K \div (F \times V)$ |
| | $F, S, hp$ | $t = F \times S \div (550 \times hp)$ |
| | $P, V$ | $F = P \div V$ |
| $F$ | $K, S$ | $F = K \div S$ |
| | $K, V, t$ | $F = K \div (V \times t)$ |
| | $V, hp$ | $F = 550 \times hp \div V$ |
| | $F, V$ | $P = F \times V$ |
| | $F, S, t$ | $P = F \times S \div t$ |
| $P$ | $K, t$ | $P = K \div t$ |
| | $hp$ | $P = 550 \times hp$ |

| To Find | Known | Formula |
|---|---|---|
| | $F, S$ | $K = F \times S$ |
| $K$ | $P, t$ | $K = P \times t$ |
| | $F, V, t$ | $K = F \times V \times t$ |
| | $t, hp$ | $K = 550 \times t \times hp$ |
| | $F, S, t$ | $hp = F \times S \div (550 \times t)$ |
| $hp$ | $P$ | $hp = P \div 550$ |
| | $F, V$ | $hp = F \times V \div 550$ |
| | $K, t$ | $hp = K \div (550 \times t)$ |

*Meanings of Symbols:* (metric units see note [a])

$S$ = distance in feet
$V$ = constant or average velocity in ft/sec
$t$ = time in seconds
$F$ = constant or average force in pounds
$P$ = power in foot-pounds per second
$hp$ = horsepower
$K$ = work in foot-pounds

[a] ***Note:* The metric SI unit of work is the joule (one joule = 1 newton-meter), and the unit of power is the watt (one watt = 1 joule per second = 1 N·m/s). The term horsepower is not used. Thus, those formulas above that involve horsepower and the factor 550 are not applicable when working in SI units. The remaining formulas can be used, and the units are: $S$ = distance in meters; $V$ = constant or average velocity in meters per second; $t$ = time in seconds; $F$ = force in newtons; $P$ = power in watts; $K$ = work in joules.**

*Example:* A casting weighing 300 pounds is to be lifted by means of an overhead crane. The casting is lifted 10 feet in 12 seconds. What is the horsepower developed?

*Solution:* Here $F = 300$ lbs, $S = 10$ ft, and $t = 12$ sec.

$$\text{hp} = \frac{F \times S}{550t} = \frac{300 \times 10}{550 \times 12} = 0.45$$

**A similar example using metric SI units is as follows: A casting of mass 150 kg is lifted 4 meters in 15 seconds by means of a crane. What is the power? Here $F = 150g$ N, $S = 4$ m, and $t = 15$ s. Thus:**

$$\textbf{Power} = \frac{FS}{t} = \frac{150g \times 4}{15} = \frac{150 \times 9.81 \times 4}{15} = \textbf{392 watts or 0.392 kW}$$

## Centrifugal Force

When a body rotates about an axis other than the one at its center of mass, the inertia of its rotating body exerts an apparent outward force, drawing it away from that center. This is called *centrifugal force.*

$F$ = centrifugal force in pounds

$W$ = weight of revolving body in pounds

$v$ = velocity at radius $R$ on body in feet/second

$n$ = number of revolutions per minute

$g$ = acceleration due to gravity = 32.16 feet/second$^2$

$R$ = perpendicular distance in feet from axis of rotation to center of mass, or for practical use, to center of gravity of revolving body

*Note:* If a body rotates about its own center of mass, $R$ equals zero and $v$ equals zero. This means that the *resultant* of the centrifugal forces of all the elements of the body is equal to zero or, in other words, no centrifugal force is exerted on the axis of rotation. The centrifugal force of any part or element of such a body is found by the equations given below, where $R$ is the radius to the center of gravity of the part or element. In a flywheel rim, $R$ is the mean radius of the rim because it is the radius to the center of gravity of a thin radial section.

$$F = \frac{Wv^2}{gR} = \frac{Wv^2}{32.16R} = \frac{4WR\pi^2 n^2}{60 \times 60g} = \frac{WRn^2}{2933} = 0.000341WRn^2$$

$$W = \frac{FRg}{v^2} = \frac{2933F}{Rn^2} \qquad v = \sqrt{\frac{FRg}{W}}$$

$$R = \frac{Wv^2}{Fg} = \frac{2933F}{Wn^2} \qquad n = \sqrt{\frac{2933F}{WR}}$$

(If $n$ is the number of revolutions per second instead of per minute, then $F = 1227WRn^2$.)

**If metric SI units are used in the foregoing formulas, $W/g$ is replaced by $M$, which is the mass in kilograms; $F$ = centrifugal force in newtons; $v$ = velocity in meters per second; $n$ = number of revolutions per minute; and $R$ = the radius in meters. Thus:**

$$F = \frac{Mv^2}{R} = \frac{Mn^2(2\pi R^2)}{60^2 R} = 0.01097\ MRn^2$$

**If the rate of rotation is expressed as $n_1$ = revolutions per second, then $F = 39.48\,MRn_1^{\,2}$; if it is expressed as ω radians per second, then $F = MR\omega^2$.**

**Calculating Centrifugal Force.**—In the ordinary formula for centrifugal force, $F = 0.000341\ WRn^2$; the mean radius $R$ of the flywheel or pulley rim is given in feet. For small dimensions, it is more convenient to have the formula in the form:

$$F = 0.2842 \times 10^{-4} Wrn^2$$

in which $F$ = centrifugal force, in pounds; $W$ = weight of rim, in pounds; $r$ = mean radius of rim, in inches; $n$ = number of revolutions per minute.

In this formula let $C = 0.000028416n^2$. This, then, is the centrifugal force of 1 pound, 1 inch from the axis. The formula can now be written in the form,

$$F = WrC$$

$C$ is calculated for various values of the revolutions per minute $n$, and the calculated values of $C$ are given in Table 5. To find the centrifugal force in any given case, simply find the value of $C$ in the table and multiply it by the product of $W$ and $r$; the four multiplications in the original formula given are thus reduced to two.

*Example:* A cast-iron flywheel with a mean rim radius of 9 inches is rotated at a speed of 800 revolutions per minute. If the weight of the rim is 20 pounds, what is the centrifugal force?

*Solution:* From Table 5, for $n$ = 800 revolutions per minute, the value of $C$ is 18.1862.

Thus,

$$F = WrC = 20 \times 9 \times 18.1862 = 3273.52 \text{ pounds}$$

**Using metric SI units, $0.01097n^2$ is the centrifugal force acting on a body of 1 kilogram mass rotating at $n$ revolutions per minute at a distance of 1 meter from the axis. If this value is designated $C_1$, then the centrifugal force of mass $M$ kilograms rotating at this speed at a distance from the axis of $R$ meters is $C_1MR$ newtons. To simplify calculations, values for $C_1$ are given in Table 6. If it is required to work in terms of millimeters, the force is $0.001\,C_1MR_1$ newtons, where $R_1$ is the radius in millimeters.**

***Example:* A steel pulley with a mean rim radius of 120 mm is rotated at a speed of 1100 rpm. If the mass of the rim is 5 kg, what is the centrifugal force?**

**From Table 6, for $n$ = 1100 rpm, the value of $C_1$ is 13,269.1.**

**Thus,**

$$F = 0.001\,C_1MR_1 = 0.001 \times 13{,}269.1 \times 5 \times 120 = 7961.50 \text{ N}$$

**Centrifugal Casting.**—The centrifugal casting of metals is an old art. This process has become important in such work as the manufacture of paper-mill rolls, railroad car wheels, and cast-iron pipe. The centrifugal casting process has been successfully applied in the production of nonmetallic tubes, such as concrete pipe, in the production of solid castings by locating the molds around the rim of a spinning wheel, and to a limited extent in the production of solid ingots by a largely similar process. Hollow objects such as cast-iron pipe are cast by introducing molten metal into a spinning mold. If the chilling of the metal is extremely rapid, for example in casting cast-iron pipe against a water-cooled chilled mold, it is imperative that a movable spout be used. The particular feature that determines the field of application of hot-mold centrifugal casting is the ability to produce long cast shapes of comparatively thin metal.

Table 5. **Factors *C* for Calculating Centrifugal Force (English units)**

| *n* | *C* | *n* | *C* | *n* | *C* | *n* | *C* |
|---|---|---|---|---|---|---|---|
| 50 | 0.07104 | 100 | 0.28416 | 470 | 6.2770 | 5200 | 768.369 |
| 51 | 0.07391 | 101 | 0.28987 | 480 | 6.5470 | 5300 | 798.205 |
| 52 | 0.07684 | 102 | 0.29564 | 490 | 6.8227 | 5400 | 828.611 |
| 53 | 0.07982 | 103 | 0.30147 | 500 | 7.1040 | 5500 | 859.584 |
| 54 | 0.08286 | 104 | 0.30735 | 600 | 10.2298 | 5600 | 891.126 |
| 55 | 0.08596 | 105 | 0.31328 | 700 | 13.9238 | 5700 | 923.236 |
| 56 | 0.08911 | 106 | 0.31928 | 800 | 18.1862 | 5800 | 955.914 |
| 57 | 0.09232 | 107 | 0.32533 | 900 | 23.0170 | 5900 | 989.161 |
| 58 | 0.09559 | 108 | 0.33144 | 1000 | 28.4160 | 6000 | 1022.980 |
| 59 | 0.09892 | 109 | 0.33761 | 1100 | 34.3834 | 6100 | 1057.360 |
| 60 | 0.10230 | 110 | 0.34383 | 1200 | 40.9190 | 6200 | 1092.310 |
| 61 | 0.10573 | 115 | 0.37580 | 1300 | 48.0230 | 6300 | 1127.830 |
| 62 | 0.10923 | 120 | 0.40921 | 1400 | 55.6954 | 6400 | 1163.920 |
| 63 | 0.11278 | 125 | 0.44400 | 1500 | 63.9360 | 6500 | 1200.580 |
| 64 | 0.11639 | 130 | 0.48023 | 1600 | 72.7450 | 6600 | 1237.800 |
| 65 | 0.12006 | 135 | 0.51788 | 1700 | 82.1222 | 6700 | 1275.590 |
| 66 | 0.12378 | 140 | 0.55695 | 1800 | 92.0678 | 6800 | 1313.960 |
| 67 | 0.12756 | 145 | 0.59744 | 1900 | 102.5820 | 6900 | 1352.890 |
| 68 | 0.13140 | 150 | 0.63936 | 2000 | 113.6640 | 7000 | 1392.380 |
| 69 | 0.13529 | 160 | 0.72745 | 2100 | 125.3150 | 7100 | 1432.450 |
| 70 | 0.13924 | 170 | 0.82122 | 2200 | 137.5330 | 7200 | 1473.090 |
| 71 | 0.14325 | 180 | 0.92067 | 2300 | 150.3210 | 7300 | 1514.290 |
| 72 | 0.14731 | 190 | 1.02590 | 2400 | 163.6760 | 7400 | 1556.060 |
| 73 | 0.15143 | 200 | 1.1367 | 2500 | 177.6000 | 7500 | 1598.400 |
| 74 | 0.15561 | 210 | 1.2531 | 2600 | 192.0920 | 7600 | 1641.310 |
| 75 | 0.15984 | 220 | 1.3753 | 2700 | 207.1530 | 7700 | 1684.780 |
| 76 | 0.16413 | 230 | 1.5032 | 2800 | 222.7810 | 7800 | 1728.830 |
| 77 | 0.16848 | 240 | 1.6358 | 2900 | 238.9790 | 7900 | 1773.440 |
| 78 | 0.17288 | 250 | 1.7760 | 3000 | 255.7400 | 8000 | 1818.620 |
| 79 | 0.17734 | 260 | 1.9209 | 3100 | 273.0780 | 8100 | 1864.370 |
| 80 | 0.18186 | 270 | 2.0715 | 3200 | 290.9800 | 8200 | 1910.690 |
| 81 | 0.18644 | 280 | 2.2278 | 3300 | 309.4500 | 8300 | 1957.580 |
| 82 | 0.19107 | 290 | 2.3898 | 3400 | 328.4890 | 8400 | 2005.030 |
| 83 | 0.19576 | 300 | 2.5574 | 3500 | 348.0960 | 8500 | 2053.060 |
| 84 | 0.20050 | 310 | 2.7308 | 3600 | 368.2710 | 8600 | 2101.650 |
| 85 | 0.20530 | 320 | 2.9098 | 3700 | 389.0150 | 8700 | 2150.810 |
| 86 | 0.21016 | 330 | 3.0945 | 3800 | 410.3270 | 8800 | 2200.540 |
| 87 | 0.21508 | 340 | 3.2849 | 3900 | 432.2070 | 8900 | 2250.830 |
| 88 | 0.22005 | 350 | 3.4809 | 4000 | 454.6560 | 9000 | 2301.700 |
| 89 | 0.22508 | 360 | 3.6823 | 4100 | 477.6730 | 9100 | 2353.130 |
| 90 | 0.23017 | 370 | 3.8901 | 4200 | 501.2580 | 9200 | 2405.130 |
| 91 | 0.23531 | 380 | 4.1032 | 4300 | 525.4120 | 9300 | 2457.700 |
| 92 | 0.24051 | 390 | 4.3220 | 4400 | 550.1340 | 9400 | 2510.840 |
| 93 | 0.24577 | 400 | 4.5466 | 4500 | 575.4240 | 9500 | 2564.540 |
| 94 | 0.25108 | 410 | 4.7767 | 4600 | 601.2830 | 9600 | 2618.820 |
| 95 | 0.25645 | 420 | 5.0126 | 4700 | 627.7090 | 9700 | 2673.660 |
| 96 | 0.26188 | 430 | 5.2541 | 4800 | 654.7050 | 9800 | 2729.070 |
| 97 | 0.26737 | 440 | 5.5013 | 4900 | 682.2680 | 9900 | 2785.050 |
| 98 | 0.27291 | 450 | 5.7542 | 5000 | 710.4000 | 10000 | 2841.600 |
| 99 | 0.27851 | 460 | 6.0128 | 5100 | 739.1000 | | |

**Table 6. Factors $C_1$ for Calculating Centrifugal Force (Metric SI units)**

| *n* | $C_1$ | *n* | $C_1$ | *n* | $C_1$ | *n* | $C_1$ |
|---|---|---|---|---|---|---|---|
| 50 | 27.4156 | 100 | 109.662 | 470 | 2,422.44 | 5200 | 296,527 |
| 51 | 28.5232 | 101 | 111.867 | 480 | 2,526.62 | 5300 | 308,041 |
| 52 | 29.6527 | 102 | 114.093 | 490 | 2,632.99 | 5400 | 319,775 |
| 53 | 30.8041 | 103 | 116.341 | 500 | 2,741.56 | 5500 | 331,728 |
| 54 | 31.9775 | 104 | 118.611 | 600 | 3,947.84 | 5600 | 343,901 |
| 55 | 33.1728 | 105 | 120.903 | 700 | 5,373.45 | 5700 | 356,293 |
| 56 | 34.3901 | 106 | 123.217 | 800 | 7,018.39 | 5800 | 368,904 |
| 57 | 35.6293 | 107 | 125.552 | 900 | 8,882.64 | 5900 | 381,734 |
| 58 | 36.8904 | 108 | 127.910 | 1000 | 10,966.2 | 6000 | 394,784 |
| 59 | 38.1734 | 109 | 130.290 | 1100 | 13,269.1 | 6100 | 408,053 |
| 60 | 39.4784 | 110 | 132.691 | 1200 | 15,791.4 | 6200 | 421,542 |
| 61 | 40.8053 | 115 | 145.028 | 1300 | 18,532.9 | 6300 | 435,250 |
| 62 | 42.1542 | 120 | 157.914 | 1400 | 21,493.8 | 6400 | 449,177 |
| 63 | 43.5250 | 125 | 171.347 | 1500 | 24,674.0 | 6500 | 463,323 |
| 64 | 44.9177 | 130 | 185.329 | 1600 | 28,073.5 | 6600 | 477,689 |
| 65 | 46.3323 | 135 | 199.860 | 1700 | 31,692.4 | 6700 | 492,274 |
| 66 | 47.7689 | 140 | 214.938 | 1800 | 35,530.6 | 6800 | 507,078 |
| 67 | 49.2274 | 145 | 230.565 | 1900 | 39,588.1 | 6900 | 522,102 |
| 68 | 50.7078 | 150 | 246.740 | 2000 | 43,864.9 | 7000 | 537,345 |
| 69 | 52.2102 | 160 | 280.735 | 2100 | 48,361.1 | 7100 | 552,808 |
| 70 | 53.7345 | 170 | 316.924 | 2200 | 53,076.5 | 7200 | 568,489 |
| 71 | 55.2808 | 180 | 355.306 | 2300 | 58,011.3 | 7300 | 584,390 |
| 72 | 56.8489 | 190 | 395.881 | 2400 | 63,165.5 | 7400 | 600,511 |
| 73 | 58.4390 | 200 | 438.649 | 2500 | 68,538.9 | 7500 | 616,850 |
| 74 | 60.0511 | 210 | 483.611 | 2600 | 74,131.7 | 7600 | 633,409 |
| 75 | 61.6850 | 220 | 530.765 | 2700 | 79,943.8 | 7700 | 650,188 |
| 76 | 63.3409 | 230 | 580.113 | 2800 | 85,975.2 | 7800 | 667,185 |
| 77 | 65.0188 | 240 | 631.655 | 2900 | 92,226.0 | 7900 | 684,402 |
| 78 | 66.7185 | 250 | 685.389 | 3000 | 98,696.0 | 8000 | 701,839 |
| 79 | 68.4402 | 260 | 741.317 | 3100 | 105,385 | 8100 | 719,494 |
| 80 | 70.1839 | 270 | 799.438 | 3200 | 112,294 | 8200 | 737,369 |
| 81 | 71.9494 | 280 | 859.752 | 3300 | 119,422 | 8300 | 755,463 |
| 82 | 73.7369 | 290 | 922.260 | 3400 | 126,770 | 8400 | 773,777 |
| 83 | 75.5463 | 300 | 986.960 | 3500 | 134,336 | 8500 | 792,310 |
| 84 | 77.3777 | 310 | 1,053.85 | 3600 | 142,122 | 8600 | 811,062 |
| 85 | 79.2310 | 320 | 1,122.94 | 3700 | 150,128 | 8700 | 830,034 |
| 86 | 81.1062 | 330 | 1,194.22 | 3800 | 158,352 | 8800 | 849,225 |
| 87 | 83.0034 | 340 | 1,267.70 | 3900 | 166,796 | 8900 | 868,635 |
| 88 | 84.9225 | 350 | 1,343.36 | 4000 | 175,460 | 9000 | 888,264 |
| 89 | 86.8635 | 360 | 1,421.22 | 4100 | 184,342 | 9100 | 908,113 |
| 90 | 88.8264 | 370 | 1,501.28 | 4200 | 193,444 | 9200 | 928,182 |
| 91 | 90.8113 | 380 | 1,583.52 | 4300 | 202,766 | 9300 | 948,469 |
| 92 | 92.8182 | 390 | 1,667.96 | 4400 | 212,306 | 9400 | 968,976 |
| 93 | 94.8469 | 400 | 1,754.60 | 4500 | 222,066 | 9500 | 989,702 |
| 94 | 96.8976 | 410 | 1,843.42 | 4600 | 232,045 | 9600 | 1,010,650 |
| 95 | 98.9702 | 420 | 1,934.44 | 4700 | 242,244 | 9700 | 1,031,810 |
| 96 | 101.065 | 430 | 2,027.66 | 4800 | 252,662 | 9800 | 1,053,200 |
| 97 | 103.181 | 440 | 2,123.06 | 4900 | 263,299 | 9900 | 1,074,800 |
| 98 | 105.320 | 450 | 2,220.66 | 5000 | 274,156 | 10000 | 1,096,620 |
| 99 | 107.480 | 460 | 2,320.45 | 5100 | 285,232 | | |

## Balancing Rotating Parts

**Static Balancing.—**There are several methods of testing the standing or static balance of a rotating part. A simple method that is sometimes used for flywheels, etc., is illustrated by the diagram, Fig. 1. An accurate shaft is inserted through the bore of the finished wheel, which is then mounted on carefully leveled "parallels" A. If the wheel is in an unbalanced state, it will turn until the heavy side is downward. When it will stand in any position as the result of counterbalancing and reducing the heavy portions, it is said to be in standing or static balance. Another test which is used for disk-shaped parts is shown in Fig. 2. The disk D is mounted on a vertical arbor attached to an adjustable cross-slide B. The latter is carried by a table C, which is supported by a knife-edged bearing. A pendulum having an adjustable screw-weight W at the lower end is suspended from cross-slide B. To test the static balance of disk D, slide B is adjusted until pointer E of the pendulum coincides with the center of a stationary scale F. Disk D is then turned halfway around without moving the slide, and if the indicator remains stationary, it shows that the disk is in balance for this particular position. The test is then repeated for ten or twelve other positions, and the heavy sides are reduced, usually by drilling out the required amount of metal. Several other devices for testing static balance are designed on this same principle.

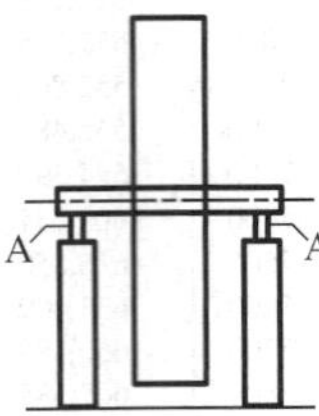


Fig. 1.

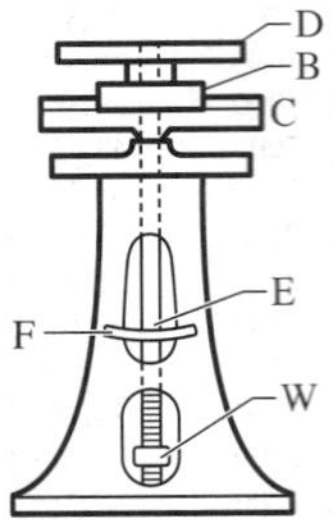


Fig. 2.

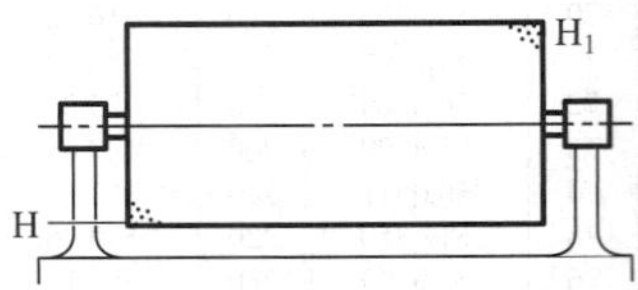


Fig. 3.

**Running or Dynamic Balance.—**A cylindrical body may be in perfect static balance yet not be in a balanced state when rotating at high speed. If the part is in the form of a thin disk, static balancing, if carefully done, may be accurate enough for high speeds, but if the rotating part is long in proportion to its diameter, and the unbalanced portions are at opposite ends or in different planes, the balancing must be done so as to counteract the centrifugal force of these heavy parts when they are rotating rapidly. This process is known as a *running balance* or *dynamic balancing*. To illustrate, if a heavy section is located at H (Fig. 3), and another correspondingly heavy section at $H_1$, one may exactly counterbalance the other when the cylinder is stationary, and this static balance may be sufficient for a part rigidly mounted and rotating at a comparatively slow speed; but when the speed is very high, as in turbine rotors, etc., the heavy masses H and $H_1$, being in different planes, are in an unbalanced state owing to the effect of centrifugal force, which results in excessive strains and injurious vibrations. Theoretically, to obtain a perfect running balance, the exact positions of the heavy sections should be located and the balancing effected either by reducing their weight or by adding counterweights opposite each section and in the same plane at the proper radius; but if the rotating part is rigidly mounted on a stiff shaft, a running balance that is sufficiently accurate for practical purposes can be obtained by means of comparatively few counterbalancing weights located with reference to the unbalanced parts.

**Balancing Calculations.—**As indicated previously, centrifugal forces caused by an unbalanced mass or masses in a rotating machine member cause additional loads on the bearings, which are transmitted to the housing or frame and to other machine members.

Such dynamically unbalanced conditions can occur even though static balance (balance at zero speed) exists. Dynamic balance can be achieved by the addition of one or two masses rotating about the same axis and at the same speed as the unbalanced masses. A single unbalanced mass can be balanced by one counterbalancing mass located 180 degrees opposite and in the same plane of rotation as the unbalanced mass if the product of their respective radii and masses are equal; i.e., $M_1r_1 = M_2r_2$. Two or more unbalanced masses rotating in the same plane can be balanced by a single mass rotating in the same plane, or by two masses rotating about the same axis in two separate planes. Likewise, two or more unbalanced masses rotating in different planes about a common axis can be balanced by two masses rotating about the same axis in separate planes. When the unbalanced masses are in separate planes they may be in static balance but not in dynamic balance; i.e., they may be balanced when not rotating but unbalanced when rotating. If a system is in dynamic balance, it will remain in balance at all speeds, although this is not strictly true at the critical speed of the system. (See *Critical Speeds* on page 198.)

In all the equations that follow, the symbol $M$ denotes either mass in kilograms or in slugs, or weight in pounds. Either mass or weight units may be used; the equations may be used with metric or with customary English units without change; however, in a given problem the units must be all metric or all customary English.

**Counterbalancing Several Masses Located in a Single Plane.**—In all balancing problems, the product of the counterbalancing mass (or weight) and its radius are calculated; it is thus necessary to select either the mass or the radius and then calculate the other value from the product of the two quantities. Design considerations usually make this decision self-evident. The angular position of the counterbalancing mass must also be calculated. Referring to Fig. 4:

$$M_B r_B = \sqrt{(\Sigma Mr\cos\theta)^2 + (\Sigma Mr\sin\theta)^2} \quad (1)$$

$$\tan\theta_B = \frac{-(\Sigma Mr\sin\theta)}{-(\Sigma Mr\cos\theta)} = \frac{y}{x} \quad (2)$$

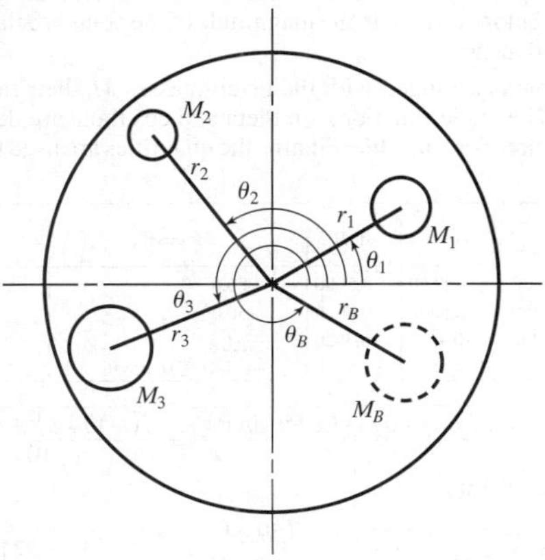


Fig. 4.

where:

$M_1, M_2, M_3, \ldots, M_n$ = any unbalanced mass or weight, kg or lb

$M_B$ = counterbalancing mass or weight, kg or lb

$r$ = radius to center of gravity of any unbalanced mass or weight, mm or inch

$r_B$ = radius to center of gravity of counterbalancing mass or weight, mm or inch

$\theta$ = angular position of $r$ of any unbalanced mass or weight, degrees

$\theta_B$ = angular position of $r_B$ of counterbalancing mass or weight, degrees

$x$ and $y$ = position with sign according to Table 1

Table 1 is helpful in finding the angular position of the counterbalancing mass or weight. It indicates the range of the angles within which this angular position occurs by noting the plus and minus signs of the numerator and the denominator of the terms in Equation (2). In a like manner, Table 1 is useful in determining the *sign* of the sine or cosine functions for angles ranging from 0 to 360 degrees. Balancing problems are usually solved most conveniently by arranging the arithmetical calculations in a tabular form.

**Table 1. Relationship of Angle Function Signs to Quadrant in Which They Occur**

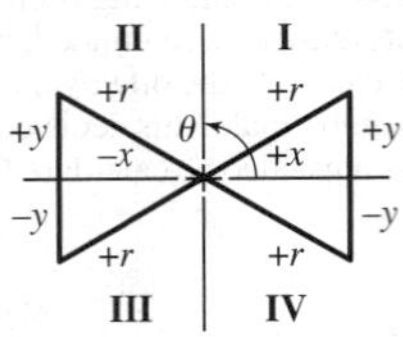


| | Angle θ | | | |
|---|---|---|---|---|
| | 0° to 90° | 90° to 180° | 180° to 270° | 270° to 360° |
| | Signs of the Functions | | | |
| tan | $\frac{+y}{+x}$ | $\frac{+y}{-x}$ | $\frac{-y}{-x}$ | $\frac{-y}{+x}$ |
| sine | $\frac{+y}{+r}$ | $\frac{+y}{+r}$ | $\frac{-y}{+r}$ | $\frac{-y}{+r}$ |
| cosine | $\frac{+x}{+r}$ | $\frac{-x}{+r}$ | $\frac{-x}{+r}$ | $\frac{+x}{+r}$ |

*Example:* Referring to Fig. 4, particular values of unbalanced weights $M_1, M_2, M_3$ are entered in the table below. Calculate the magnitude of the counterbalancing weight $M_B$ if its radius $r$ is to be 10 inches.

*Solution:* The table is arranged with the given masses $M$, their radii $r$, and their angular positions $\theta$. The angle functions sin theta and cos theta are determined from the angles using a calculator or trig table. Finally, the quantities are used to fill in the last two columns.

| M | | r | θ | cos θ | sin θ | Mr cos θ | Mr sin θ |
|---|---|---|---|---|---|---|---|
| No. | lb. | in. | deg. | | | | |
| 1 | 10 | 10 | 30 | 0.8660 | 0.5000 | 86.6 | 50.0 |
| 2 | 5 | 20 | 120 | −0.5000 | 0.8660 | −50.0 | 86.6 |
| 3 | 15 | 15 | 200 | −0.9397 | −0.3420 | −211.4 | −77.0 |
| | | | | | | −174.8 = ΣMr cos θ | 59.6 = ΣMr sin θ |

$$M_B = \frac{\sqrt{(\Sigma Mr\cos\theta)^2 + (\Sigma Mr\sin\theta)^2}}{r_B} = \frac{\sqrt{(-174.8)^2 + (59.6)^2}}{10}$$

$$M_B = 18.5 \text{ lbs}$$

$$\tan\theta_B = \frac{-(\Sigma Mr\sin\theta)}{-(\Sigma Mr\cos\theta)} = \frac{-(59.6)}{-(-174.8)} = \frac{-y}{+x};\quad \theta_B = 341°10'$$

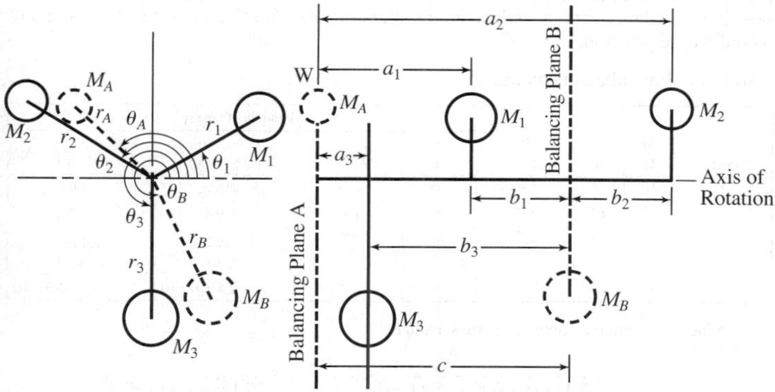


Fig. 5.

**Counterbalancing Masses Located in Two or More Planes.**—Unbalanced masses or weights rotating about a common axis in two separate planes of rotation form a couple, which must be counterbalanced by masses or weights, also located in two separate planes, say, planes *A* and *B*, and rotating about a common axis (see *Couples* on page 156). In addition, they must be balanced in the direction perpendicular to the axis, as before. Since two counterbalancing masses are required, two separate equations are required to calculate the product of each mass or weight and its radius, and two additional equations are required to calculate the angular positions. The planes *A* and *B* selected as balancing planes may be any two planes separated by any convenient distance *c*, along the axis of rotation. In Fig. 5:

For balancing plane *A*:

$$M_A r_A = \frac{\sqrt{(\Sigma Mrb\cos\theta)^2 + (\Sigma Mrb\sin\theta)^2}}{c} \quad (3)$$

$$\tan\theta_A = \frac{-(\Sigma Mrb\sin\theta)}{-(\Sigma Mrb\cos\theta)} = \frac{y}{x} \quad (4)$$

For balancing plane *B*:

$$M_B r_B = \frac{\sqrt{(\Sigma Mra\cos\theta)^2 + (\Sigma Mra\sin\theta)^2}}{c} \quad (5)$$

$$\tan\theta_B = \frac{-(\Sigma Mra\sin\theta)}{-(\Sigma Mra\cos\theta)} = \frac{y}{x} \quad (6)$$

where: $M_A$ and $M_B$ are the mass or weight of the counterbalancing masses in the balancing planes *A* and *B*, respectively; $r_A$ and $r_B$ are the radii; and $\theta_A$ and $\theta_B$ are the angular positions of the balancing masses in these planes. *M*, *r*, and $\theta$ are the mass or weight, radius, and angular positions of the unbalanced masses, with the subscripts defining the particular mass to which the values are assigned. The distance, *c*, between the balancing planes is always a positive value. The axial dimensions, *a* and *b*, may be either positive or negative, depending upon their position relative to the balancing plane; for example, in Fig. 5, the dimension $b_2$ would be negative.

*Example:* Referring to Fig. 5, a set of values for the masses and dimensions has been selected and put into convenient table form below. The separation of balancing planes, *c*, is chosen as 15 inches. If in balancing plane *A*, the radius of the counterbalancing weight is selected to be 10 inches; calculate the magnitude of the counterbalancing mass and its position. In balancing plane *B* the counterbalancing mass is selected to be 10 lb; calculate its radius and position.

*Solution:* For balancing plane *A*:

| | | | | Balancing Plane *A* | | | |
|---|---|---|---|---|---|---|---|
| Plane | *M* lb | *r* in. | θ deg. | *b* in. | *Mrb* | *Mrb* cos θ | *Mrb* sin θ |
| 1 | 10 | 8 | 30 | 6 | 480 | 415.7 | 240.0 |
| 2 | 8 | 10 | 135 | –6 | –480 | 339.4 | –339.4 |
| 3 | 12 | 9 | 270 | 12 | 1296 | 0.0 | –1296.0 |
| *A* | ? | 10 | ? | 15[a] | | 755.1 | –1395.4 |
| *B* | 10 | ? | ? | 0 | | = Σ*Mrb* cos θ | = Σ*Mrb* sin θ |

[a] 15 inches = distance *c* between planes *A* and *B*.

$$M_A = \frac{\sqrt{(\Sigma Mrb\cos\theta)^2 + (\Sigma Mrb\sin\theta)^2}}{r_A c} = \frac{\sqrt{(755.1)^2 + (-1395.4)^2}}{10(15)}$$

$$M_A = 10.6 \text{ lb}$$

$$\tan\theta_A = \frac{-(\Sigma Mrb\sin\theta)}{-(\Sigma Mrb\cos\theta)} = \frac{-(-1395.4)}{-(755.1)} = \frac{+y}{-x}$$

$$\theta_A = 118° \; 25'$$

For balancing plane *B*:

| | | | | Balancing Plane *B* | | | |
|---|---|---|---|---|---|---|---|
| Plane | *M* lb | *r* in. | θ deg. | *a* in. | *Mra* | *Mra* cos θ | *Mra* sin θ |
| 1 | 10 | 8 | 30 | 9 | 720 | 623.5 | 360.0 |
| 2 | 8 | 10 | 135 | 21 | 1680 | –1187.9 | 1187.9 |
| 3 | 12 | 9 | 270 | 3 | 324 | 0.0 | –324.0 |
| *A* | ? | 10 | ? | 0 | | –564.4 | 1223.9 |
| *B* | 10 | ? | ? | 15[a] | | = Σ*Mra* cos θ | = Σ*Mra* sin θ |

[a] 15 inches = distance *c* between planes *A* and *B*.

$$r_B = \frac{\sqrt{(\Sigma Mra\cos\theta)^2 + (\Sigma Mra\sin\theta)^2}}{M_B c} = \frac{\sqrt{(-564.4)^2 + (1223.9)^2}}{10(15)}$$

$$= 8.985 \text{ in.}$$

$$\tan\theta_B = \frac{-(\Sigma Mra\sin\theta)}{-(\Sigma Mra\cos\theta)} = \frac{-(1223.9)}{-(-564.4)} = \frac{-y}{+x}$$

$$\theta_B = 294° \; 45'$$

**Balancing Lathe Fixtures.**—Lathe fixtures rotating at a high speed require balancing. Often it is assumed that the center of gravity of the workpiece and fixture and of the counterbalancing masses are in the same plane; however, this is not usually the case. Counterbalancing masses are required in two separate planes to prevent excessive vibration or bearing loads at high speeds.

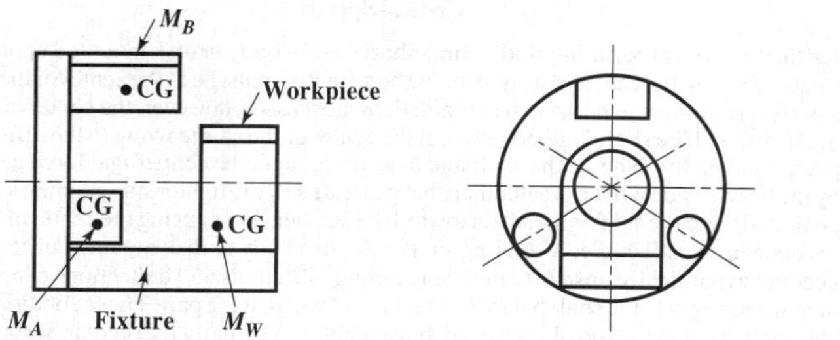


**Lathe Fixture**

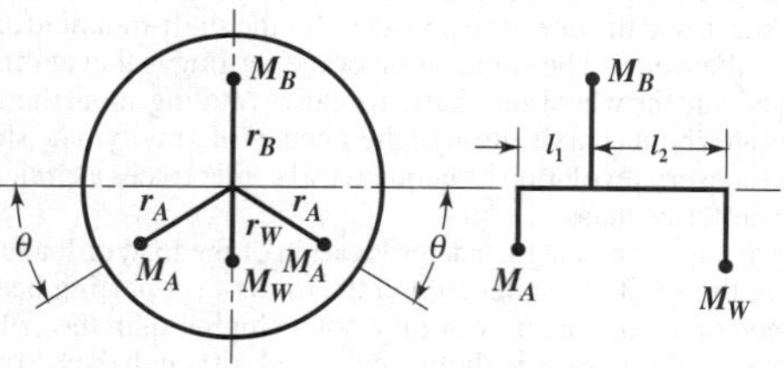


**Schematic View**

Fig. 6.

Usually a single counterbalancing mass is placed in one plane selected to be 180 degrees directly opposite the combined center of gravity of the workpiece and the fixture. Two equal counterbalancing masses are then placed in the second counterbalancing plane, equally spaced on each side of the fixture. Referring to Fig. 6, the two counterbalancing masses $M_A$ and the two angles $\theta$ are equal. For the design in this illustration, the following formulas can be used to calculate the magnitude of the counterbalancing masses. Since their angular positions are fixed by the design, they are not calculated.

$$M_B = \frac{M_W r_W (l_1 + l_2)}{r_B l_1} \tag{7}$$

$$M_A = \frac{M_B r_B - M_W r_W}{2 r_A \sin\theta} \tag{8}$$

In these formulas $M_W$ and $r_W$ denote the mass or weight and the radius of the combined center of gravity of the workpiece and the fixture.

*Example:* In Fig. 6 the combined weight of the workpiece and the fixture is 18.5 lb. The following dimensions were determined from the layout of the fixture and by calculating the centers of gravity: $r_W = 2$ in.; $r_A = 6.25$ in.; $r_B = 6$ in.; $l_1 = 3$ in.; $l_2 = 5$ in.; and $\theta = 30°$. Calculate the weights of the counterbalancing masses.

$$M_B = \frac{M_w r_w (l_1 + l_2)}{r_B l_1} = \frac{18.5 \times 2 \times 8}{6 \times 3} = 16.44 \text{ lb}$$

$$M_A = \frac{M_B r_B - M_w r_w}{2 r_A \sin\theta} = \frac{(16.44 \times 6) - (18.5 \times 2)}{(2 \times 6.25) \sin 30°} = 9.86 \text{ lb (each weight)}$$

## Critical Speeds

**Critical Speeds of Rotating Bodies and Shafts.—**If a body or disk mounted upon a shaft rotates about it, the center of gravity of the body or disk must be at the center of the shaft if a perfect running balance is to be obtained. In most cases, however, the center of gravity of the disk will be slightly removed from the center of the shaft, owing to the difficulty of perfect balancing. Now, if the shaft and disk are rotated, the centrifugal force generated by the heavier side will be greater than that generated by the lighter side geometrically opposite to it, and the shaft will deflect toward the heavier side, causing the center of the disk to rotate in a small circle. A rotating shaft without a body or disk mounted on it can also become dynamically unstable, and the resulting vibrations and deflections can result in damage not only to the shaft but to the machine of which it is a part. These conditions hold true up to a comparatively high speed, but a point is eventually reached (at several thousand revolutions per minute) when momentarily there will be excessive vibration, and then the parts will run quietly again. The speed at which this occurs is called the *critical speed* of the wheel or shaft, and the phenomenon itself for the shaft-mounted disk or body is called the *settling* of the wheel. The explanation of the settling is that at this speed the axis of rotation changes, and the wheel and shaft, instead of rotating about their geometrical center, begin to rotate about an axis through their center of gravity. The shaft itself is then deflected so that for every revolution its geometrical center traces a circle around the center of gravity of the rotating mass.

Critical speeds depend upon the magnitude or location of the load or loads carried by the shaft, the length of the shaft, its diameter and the kind of supporting bearings. The normal operating speed of a machine may or may not be higher than the critical speed. For instance, some steam turbines exceed the critical speed, although they do not run long enough at the critical speed for the vibrations to build up to an excessive amplitude. The practice of the General Electric Co. at Schenectady is to keep below the critical speeds. It is assumed that the maximum speed of a machine may be within 20 percent above or below the critical speed without vibration troubles. Thus, in a design of steam turbine sets, critical speed is a factor that determines the size of the shafts for both the generators and turbines. Although a machine may run very close to the critical speed, the alignment and play of the bearings, the balance and construction generally will require extra care, resulting in a more expensive machine; moreover, while such a machine may run smoothly for a considerable time, any looseness or play that may develop later, causing a slight imbalance, will immediately set up excessive vibrations.

The formulas commonly used to determine critical speeds are sufficiently accurate for general purposes. There are cases, however, where the torque applied to a shaft has an important effect on its critical speed. Investigations have shown that the critical speeds of a uniform shaft are decreased as the applied torque is increased and that there exist critical torques which will reduce the corresponding critical speed of the shaft to zero. A detailed analysis of the effects of applied torques on critical speeds may be found in a paper, "Critical Speeds of Uniform Shafts under Axial Torque," by Golumb and Rosenberg, presented at the First US National Congress of Applied Mechanics in 1951.

**Formulas for Critical Speeds.—**The critical speed formulas given in the accompanying table (from the paper on Critical Speed Calculation presented before the ASME by S. H. Weaver) apply to (1) shafts with single concentrated loads and (2) shafts carrying uniformly distributed loads. These formulas also cover different conditions as regards bearings. If the bearings are self-aligning or very short, the shaft is considered to be supported at the ends; whereas, if the bearings are long and rigid, the shaft is considered fixed. These formulas, for both concentrated and distributed loads, apply to vertical shafts as well as horizontal shafts, the critical speeds having the same value in both cases. The data required for the solution of critical speed problems are the same as for shaft deflection. As the shaft is usually of variable diameter and its stiffness is increased by a long hub, an ideal shaft of uniform diameter and equal stiffness must be assumed.

## Critical Speed Formulas

| Formulas for Single Concentrated Load | | |
|---|---|---|
| 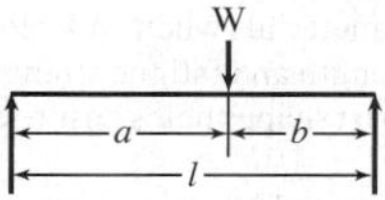<br>$N = 387{,}000\frac{d^2}{ab}\sqrt{\frac{l}{W}}$<br>Bearings supported | 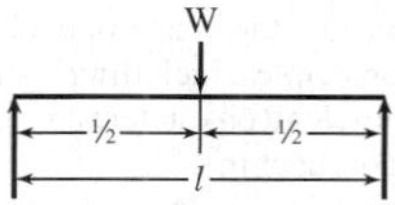<br>$N = 1{,}550{,}500\frac{d^2}{l\sqrt{Wl}}$<br>Bearings supported | 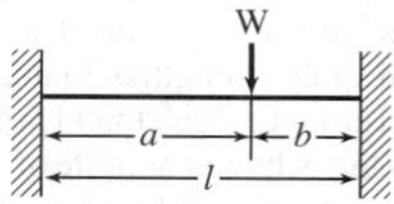<br>$N = 387{,}000\frac{d^2 l}{ab}\sqrt{\frac{l}{Wab}}$<br>Bearings fixed |
| 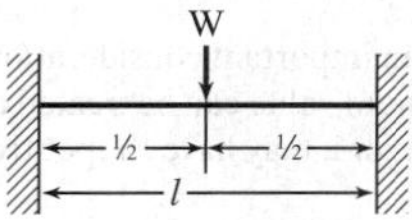<br>$N = 3{,}100{,}850\frac{d^2}{l\sqrt{Wl}}$<br>Bearings fixed | 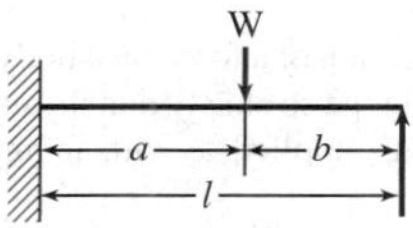<br>$N = 775{,}200\frac{d^2 l}{ab}\sqrt{\frac{l}{Wa\,(3l+b)}}$<br>One-fixed — One supported | 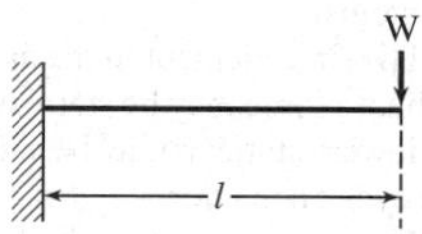<br>$N = 387{,}000\frac{d^2}{l\sqrt{Wl}}$<br>One fixed — One free end |
| **Formulas for Distributed Loads—First Critical Speed** | | |
| 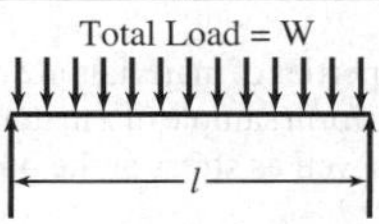<br>$N = 2{,}232{,}500\frac{d^2}{l\sqrt{Wl}}$<br>$N_1 = 4{,}760{,}000\frac{d}{l^2}$<br>Bearings supported | 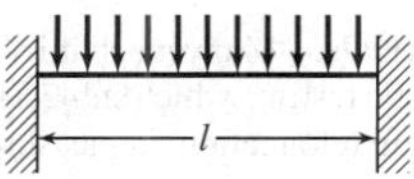<br>$N = 4{,}979{,}250\frac{d^2}{l\sqrt{Wl}}$<br>$N_1 = 10{,}616{,}740\frac{d}{l^2}$<br>Bearings fixed | 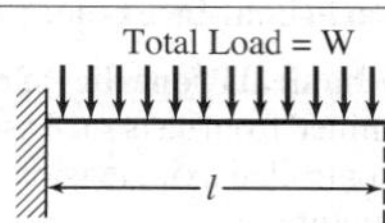<br>$N = 795{,}200\frac{d^2}{l\sqrt{Wl}}$<br>$N_1 = 1{,}695{,}500\frac{d}{l^2}$<br>One fixed—One free end |

$N$ = critical speed, RPM
$N_1$ = critical speed of shaft alone
$d$ = diameter of shaft, in inches
$W$ = load applied to shaft, in pounds
$l$ = distance between centers of bearings, in inches
$a$ and $b$ = distances from bearings to load

In calculating critical speeds, the weight of the shaft is either neglected or, say, ½ to ⅔ of the weight is added to the concentrated load. The formulas apply to steel shafts having a modulus of elasticity $E = 29{,}000{,}000$. Although a shaft carrying a number of loads or a distributed load may have an infinite number of critical speeds, ordinarily it is the first critical speed that is of importance in engineering work. The first critical speed is obtained by the formulas given in the distributed loads portion of the table *Critical Speed Formulas*.

## STRENGTH OF MATERIALS

Strength of materials deals with the relations between elastic bodies, external forces applied, and resulting deformations and stresses. In the design of structures and machines, mechanical engineering must consider the behavior of chosen materials when under load and apply strength of materials principles, including static strength and fatigue strength. Optimizing mechanical properties, load characteristics, and part shapes helps ensure expected behavior and life of a part or machine.

Forces are produced by the action of gravity, by accelerations and impacts of moving parts, by gasses and fluids under pressure, and by the transmission of mechanical power, among others. In order to analyze the stresses and deflections of a body, the magnitudes, directions and points of application of forces acting on the body must be known. Information given previously in the Mechanics section provides the basis for evaluating force systems.

The time element in the application of a force on a body is an important consideration. Thus, a force may be static or change so slowly that its maximum value can be treated as if it were static; it may be suddenly applied, as with an impact; or it may have a repetitive or cyclic behavior.

The environment in which forces act on a machine or part is also important. Such factors as high and low temperatures; the presence of corrosive gases, vapors and liquids; radiation, etc. may have a marked effect on how well parts are able to resist stresses.

**Throughout this and other sections of the *Handbook*, both English and metric SI data and formulas are given to cover the requirements of working in either system of measurement. In many cases, formulas and text relating exclusively to SI units are given in bold-face type.**

**Mechanical Properties of Materials.—**Many mechanical properties of materials are determined from tests such as tensile testing, which subjects a uniform sample of a material to controlled axial tension. Strain (elongation) is measured as well as stress at the point of rupture.

*Engineering stress* is the applied load on a tensile specimen divided by its original cross-sectional area. *True stress* uses the actual cross-sectional area. The maximum stress a material can withstand before breaking may be called *ultimate tensile stress* or *breaking stress*. *Engineering strain* is the amount by which a tensile specimen of a given material changes when the body is subjected to a load, divided by the original value of the dimension. *True strain* is the natural log of the ratio of current length over original length. Instead of *unit strain*, the simpler term *strain* often is used.

The shape of an engineering stress-strain curve (see Fig. 1) reveals how material will behave under tensile loading, assuming the tensile specimen has a constant cross-sectional area as loading is applied. (True stress and strain both reference the instantaneous cross-sectional area, which can be more accurate, but in practice, engineering stress-strain curves are used almost exclusively.) Values beyond the elastic limit must be used with caution, as the cross-sectional area begins to change and introduce error.

Fig. 1 (1) shows the curve shape for a ductile ferrous material similar to low-carbon steel. The (upper or first) yield point is well defined and followed by some elongation with little additional stress. Then the curve turns upward again, toward an inflection point; strain hardening accounts for the increasing stress in this segment. When the curve again turns downward at the ultimate strength point, necking has begun. Fig. 1 (2) shows a curve typical of a ductile nonferrous material such as annealed copper. A well-defined yield point is absent, so an offset yield point is constructed. Fig. 1 (3) illustrates an offset yield point, as discussed below.

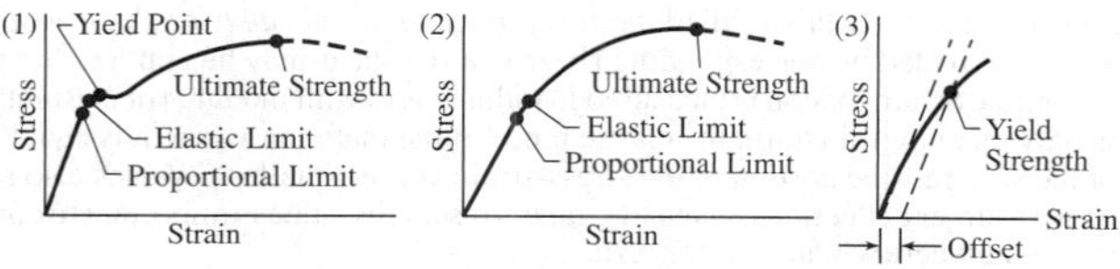


Fig. 1. Stress-Strain Curves

**In the SI, the unit of stress is the pascal (Pa), the newton per meter squared (N/m²). The megapascal (newtons per millimeter squared) is often an appropriate sub-multiple for use in practice.**

*Proportional limit* is the point on a stress-strain curve at which it begins to deviate from the straight-line relationship between stress and strain. Prior to this point, the curve has a constant slope equal to the modulus of elasticity for that material.

*Elastic limit* is the maximum tensile stress to which a test specimen may be subjected and still return to its original length upon release of the load. A material is said to be stressed within the *elastic region* when the working stress does not exceed the elastic limit, and to be stressed in the *plastic region* when the working stress does exceed the elastic limit. The elastic limit for steel is for all practical purposes the same as its proportional limit.

*Yield point* is a point on the stress-strain curve at which there is a sudden increase in strain without a corresponding increase in stress. While not all materials have a yield point, for those that do, this point tends to occur at or near the onset of plastic or permanent deformation. Modern standards may refer to this as *upper yield strength*.

*Yield strength, $S_y$,* is the stress associated with a specified small amount of permanent strain. Ideally, this is the maximum stress that can be applied with no permanent deformation of the material; in practice, it is not easily measurable. When a material does not have a clearly defined upper yield strength (yield point), an offset yield strength may be defined instead, as illustrated in Fig. 1 (3). In such cases, yield strength is the stress value on the stress-strain curve corresponding to a defined amount of permanent set or strain, usually 0.1%, 0.2%, or 0.5% of the original dimension. A line drawn parallel to the elastic range curve (with slope equal to the modulus of elasticity) intersects the $x$-axis at the target strain. The intersection of the stress-strain curve with that offset line is the offset yield point (again 0.2%). Yield strength data for various materials are given in tables starting on pages 386, 388, 416, 417, 419, 421, 425, 508, 510, 514, 523, 524, 529, 530, 534, 542, 544, 545, and elsewhere.

*Ultimate tensile strength, $S_u$* (also called *ultimate strength*), is the maximum stress value obtained in a tensile test for a given material. This value has significance in designs with brittle materials; design calculations for ductile materials use yield strength instead.

*Modulus of elasticity, E* (also called *Young's modulus*), is a measurement of the elasticity of a given material. Calculated according to Hooke's law by dividing stress by the corresponding strain within a material's elastic range, this value is equal to the slope of the linear portion of the stress-strain curve. Smaller $E$ values indicate larger deflections under a given stress. Young's modulus has significant implications for parts in use, as well as those being machined or processed. Some representative values of Young's modulus in both US customary and metric units are as follows:

| Material | Young's Modulus | | Material | Young's Modulus | |
|---|---|---|---|---|---|
| | $10^6$ psi | $10^9$ Pa | | $10^6$ psi | $10^9$ Pa |
| Aluminum, cast, pure | 9 | 62.1 | Magnesium, AZ80A-T5 | 6.5 | 44.8 |
| Aluminum, wrought, 2014-T6 | 10.6 | 73.1 | Titanium, pure | 15.5 | 106.9 |
| Beryllium copper | 19 | 131 | Titanium, alloy, 5 Al, 2.5 Sn | 17 | 117.2 |
| Brass, naval | 15 | 103.4 | Steel for bridges and buildings, ASTM A7-61T, all shapes | 29 | 199.9 |
| Bronze, phosphor, ASTM B159 | 15 | 103.4 | | | |
| Cast iron, malleable | 26 | 179.3 | Steel, castings, high strength, for structural purposes, ASTM A148-60 | 29 | 199.9 |
| Cast iron, nodular | 23.5 | 162 | | | |

*Modulus of rigidity, G* (also called *modulus of elasticity in shear* or *shear modulus*), is shear stress divided by corresponding shear strain. (Shear may be applied directly or through torsion. Strain torsion is measured in radians.) As with modulus of elasticity, this value is only meaningful within the elastic range of the material, where it is equal to the slope of the straight-line portion of the stress-strain curve. Smaller $G$ values also are associated with larger deflections, given the same stress. This value is important to consider in shaft designs where windup is a concern.

*Poisson's ratio, μ,* is the value of the strain transverse to the load direction, divided by the strain in the load direction. This ratio is determined when uniaxial force is applied, with tensile force having a negative sign and compressive force a positive sign. Values of Poisson's ratio for common materials are as follows:

| | | | |
|---|---|---|---|
| Aluminum | 0.334 | Nickel silver | 0.322 |
| Beryllium copper | 0.285 | Phosphor bronze | 0.349 |
| Brass | 0.340 | Rubber | 0.500 |
| Cast iron, gray | 0.211 | Steel, cast | 0.265 |
| Copper | 0.340 | high carbon | 0.295 |
| Inconel | 0.290 | mild | 0.303 |
| Lead | 0.431 | nickel | 0.291 |
| Magnesium | 0.350 | Wrought iron | 0.278 |
| Monel metal | 0.320 | Zinc | 0.331 |

**Compressive and Shear Properties.—***Compressive yield strength,* $S_{cy}$, and *compressive ultimate strength,* $S_{cu}$, are determined from compression tests. It is common to assume that ultimate compressive strength is equal to ultimate tensile strength. But ductile materials in compression usually fail by excessive yielding well before they fracture; as they are weaker in shear than in tension/compression, shear failure is more likely to occur. For brittle materials, which are stronger in shear, tensile and compressive stresses are of particular concern.

The properties of *shear yield strength,* $S_{sy}$, and *shear ultimate strength,* $S_{su}$, are determined by direct shear and torsional tests. The shear yield strength of ductile materials is approximated as 0.577 times the tensile yield strength.

**Creep.—**Continuing changes in dimensions of a stressed material over time is called creep; it varies with different materials and periods under stress, and with temperature. Creep tests may take some time, as it is necessary to apply a constant tensile load to a specimen under a selected temperature. Measurements are taken to record the resulting elongation at time periods sufficiently long for a relationship to be established. The data are then plotted as elongation against time. The load is applied to the specimen only after it has reached the testing temperature, and causes an initial elastic elongation that includes some plastic deformation if the load is above the proportional limit for the material.

Some combinations of stress and temperature may cause failure of the specimen. Others show initial high rates of deformation, followed by decreasing, then constant, rates over long periods. Generally, testing times to arrive at the constant rate of deformation are over 1000 hours.

**Creep Rupture.—**Tests for creep rupture are similar to creep tests but are prolonged until the specimen fails. Further data to be obtained from these tests include time to rupture, amount of elongation, and reduction of area. Stress-rupture tests are performed without measuring the elongation, so that no strain data are recorded, time to failure, elongation and reduction of area being sufficient. Sometimes, a V-notch is cut in the specimen to allow measurement of notch sensitivity under the testing conditions.

**Stress and Strain Analysis.—**Stresses, deflections (strains), and loads may be characterized by applying strain gages or indicating coatings to the surface of a part and then subjecting it to loads simulating those encountered in service. Strain gages, available in various configurations, are attached to the part's surface and often calibrated before use. Such gages convert tension, pressure, force, and so on into electrical resistance, which can be measured to indicate strain. Indicating coatings include brittle lacquers that crack in patterns indicative of strain direction and magnitude, as well as color-changing coatings that indicate the amount of strain at each point, identifying areas requiring component modification. Alternatively, modern materials testing systems offer automated, multi-axial, and high-precision capabilities. Testing may involve use of specific wavelengths of light or be performed at extreme temperatures.

Most of these tests have been standardized by the American Society for Testing and Materials (ASTM) and are published in their *Book of Standards* in separate sections for metals, plastics, rubber, and wood. Many of the test methods are also adopted by the American National Standards Institute (ANSI).

**Fatigue Properties.—**When a material is subjected to many cycles of stress reversal or fluctuation (variation in magnitude without reversal), failure may occur, even though the maximum stress at any cycle is considerably less than the value at which failure would occur if the stress were constant. Fatigue properties are determined by subjecting test specimens to stress cycles and counting the number of cycles to failure. From a series of such tests in which maximum stress values are progressively reduced, diagrams of stress versus number of cycles until failure (*S-N* diagrams) can be plotted as illustrated by the accompanying figures. The *S-N* diagram Fig. 2a shows the behavior of a material for which there is an *endurance limit* $S_{en}$. Endurance limit is the stress value at which the number of cycles to failure is infinite. Steels have endurance limits that vary according to hardness, composition, and quality, but many nonferrous metals do not. The *S-N* diagram Fig. 2b does not have an endurance limit. For a metal that does not have an endurance limit, it is standard practice to specify fatigue strength as the stress value corresponding to a specific number of stress reversals, usually 100,000,000 or 500,000,000.

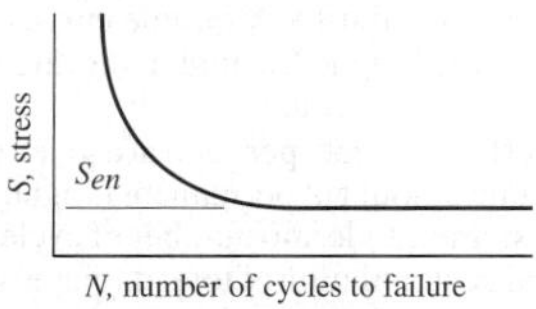


Fig. 2a. *S-N* Endurance Limit

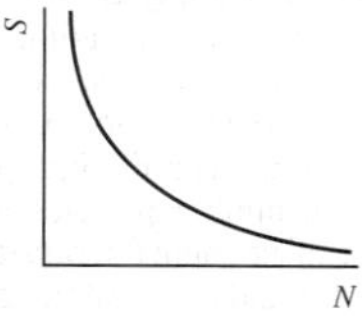


Fig. 2b. *S-N* No Endurance Limit

**Influence of Mean Stress on Fatigue.—**Most published data on the fatigue properties of metals are for completely reversed alternating stresses, that is, the mean stress of the cycle is equal to zero. However, if a structure is subjected to stresses that fluctuate between different values of tension and compression, then the mean stress is not zero.

When fatigue data for a specified mean stress and design life are not available for a material, the influence of nonzero mean stress can be estimated from empirical relationships that relate failure at a given life, under zero mean stress, to failure at the same life under zero mean cyclic stress. One widely used formula is Goodman's linear relationship,

$$S_a = S\left(1 - S_m/S_u\right)$$

where $S_a$ is the alternating stress associated with some nonzero mean stress $S_m$. $S$ is the alternating fatigue strength at zero mean stress. $S_u$ is the ultimate tensile strength.

Goodman's linear relationship is usually represented graphically on a so-called *Goodman Diagram*, shown in Fig. 3a. The alternating fatigue strength or the alternating stress for a given number of endurance cycles is plotted on the ordinate (*y*-axis) and the static tensile strength is plotted on the abscissa (*x*-axis). The straight line joining the alternating fatigue strength $S$ and the tensile strength $S_u$ is the Goodman line.

The value of an alternating stress $S_{ax}$ at a known value of mean stress $S_{mx}$ is determined as shown by the dashed lines on the diagram.

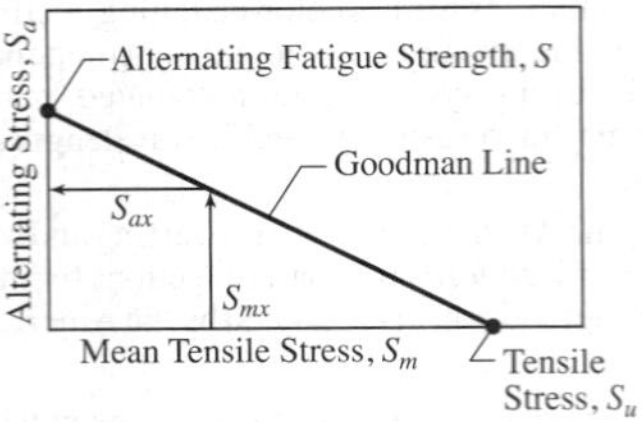


Fig. 3a. Goodman Diagram

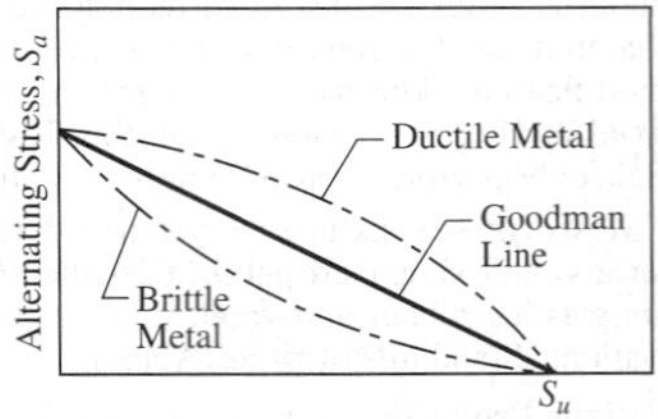


Fig. 3b. Mean Tensile Stress

For ductile materials, the Goodman law is usually conservative, since approximately 90 percent of actual test data for most ferrous and nonferrous alloys fall above the Goodman line, even at low endurance values where the yield strength is exceeded. For many brittle materials, however, actual test values can fall below the Goodman line, as illustrated in Fig. 3b.

As a rule of thumb, materials having an elongation of less than 5 percent in a tensile test may be regarded as brittle. Those having an elongation of 5 percent or more may be regarded as ductile.

**Cumulative Fatigue Damage.—**Most data are determined from tests at a constant stress amplitude. This is easy to do experimentally, and the data can be presented in a straightforward manner. In actual engineering applications, however, the alternating stress amplitude usually changes in some way during service operation. Such changes, referred to as "spectrum loading," make direct use of standard *S-N* fatigue curves inappropriate. A problem exists, therefore, in predicting the fatigue life under varying stress amplitude from conventional, constant-amplitude *S-N* fatigue data.

The assumption in predicting spectrum loading effects is that operation at a given stress amplitude and number of cycles will produce a certain amount of permanent fatigue damage and that subsequent operation at different stress amplitude and number of cycles will produce additional fatigue damage and a sequential accumulation of total damage, which at a critical value will cause fatigue failure. Although the assumption appears simple, the amount of damage incurred at any stress amplitude and number of cycles has proven difficult to determine, and several "cumulative damage" theories have been advanced.

One of the first and simplest methods for evaluating cumulative damage is known as *Miner's law* or the *linear damage rule*, where it is assumed that $n_1$ cycles at a stress of $S_1$, for which the average number of cycles to failure is $N_1$, causes an amount of damage $n_1/N_1$. Failure is predicted to occur when

$$\sum_{1}^{N} \frac{n_i}{N_i} = 1$$

The term $n/N$ is known as the "cycle ratio" or the damage fraction.

The greatest advantages of the Miner rule are its simplicity and prediction reliability, which approximates that of more complex theories. For these reasons the rule is widely used. It should be noted, however, that it does not account for all influences, and errors are to be expected in failure prediction ability.

**Modes of Fatigue Failure.—**Several modes of fatigue failure are:

*Low/High-Cycle Fatigue:* This fatigue process covers cyclic loading in two significantly different domains, with different physical mechanisms of failure. One domain is characterized by relatively low cyclic loads, strain cycles confined largely to the elastic range, and long lives, that is, a high number of cycles to failure; traditionally, this has been called "high-cycle fatigue." The other domain has cyclic loads that are relatively high, significant amounts of plastic strain induced during each cycle, and short lives, that is, a low number of cycles to failure. This domain has commonly been called "low-cycle fatigue" or cyclic strain-controlled fatigue.

The transition from low- to high-cycle fatigue behavior occurs in the range from approximately 10,000 to 100,000 cycles. Many define low-cycle fatigue as failure that occurs in 50,000 cycles or less.

*Thermal Fatigue:* Cyclic temperature changes in a machine part will produce cyclic stresses and strains if natural thermal expansions and contractions are either wholly or partially constrained. These cyclic strains produce fatigue failure just as if they were produced by external mechanical loading. When strain cycling is produced by a fluctuating temperature field, the failure process is termed "thermal fatigue."

While thermal fatigue and mechanical fatigue phenomena are very similar, and can be mathematically expressed by the same types of equations, the use of mechanical fatigue results to predict thermal fatigue performance must be done with care. For equal values of plastic strain range, the number of cycles to failure is usually up to 2.5 times lower for thermally cycled than for mechanically cycled samples.

*Corrosion Fatigue:* Corrosion fatigue is a failure mode where cyclic stresses and a corrosion-producing environment combine to initiate and propagate cracks in fewer stress cycles and at lower stress amplitudes than would be required in a more inert environment. The corrosion process forms pits and surface discontinuities that act as stress raisers to accelerate fatigue cracking. Cyclic loads may also cause cracking and flaking of the corrosion layer, baring fresh metal to the corrosive environment. Each process accelerates the other, making the cumulative result more serious.

*Surface or Contact Fatigue:* Surface fatigue failure is usually associated with rolling surfaces in contact, and results in pitting, cracking, and spalling of the contacting surfaces from cyclic Hertz contact stresses that cause the maximum values of cyclic shear stresses to be slightly below the surface. The cyclic subsurface shear stresses generate cracks that propagate to the contacting surface, dislodging particles in the process.

*Combined Creep and Fatigue:* In this failure mode, all of the conditions for both creep failure and fatigue failure exist simultaneously. Each process influences the other in producing failure, but this interaction is not well understood.

**Factors of Safety.—**There is always a risk that the working stress to which a member is subjected will exceed the strength of its material. The purpose of a factor of safety is to minimize this risk.

Factors of safety can be incorporated into design calculations in many ways. For most calculations the following equation is used:

$$s_w = \frac{S_m}{f_s} \qquad (1)$$

where $f_s$ is the factor of safety, $S_m$ is the strength of the material in pounds per square inch, and $s_w$ is the allowable working stress, also in pounds per square inch. Since the factor of safety is greater than 1, the allowable working stress will be less than the strength of the material.

In general, $S_m$ is based on yield strength for ductile materials, ultimate strength for brittle materials, and fatigue strength for parts subjected to cyclic stressing. Most strength values are obtained by testing standard specimens at 68° F in normal atmospheres. If, however, the character of the stress or environment differs significantly from that used in obtaining standard strength data, then special data must be obtained. If special data are not available, standard data must be suitably modified.

General recommendations for values of factors of safety $f_s$ are given in the following list.

| $f_s$ | Application |
|---|---|
| 1.3–1.5 | For use with highly reliable materials where loading and environmental conditions are not severe, and where weight is an important consideration. |
| 1.5–2 | For applications using reliable materials where loading and environmental conditions are not severe. |

2–2.5 For use with ordinary materials where loading and environmental conditions are not severe.

2.5–3 For less tried and for brittle materials where loading and environmental conditions are not severe.

3–4 For applications in which material properties are not reliable and where loading and environmental conditions are not severe, or where reliable materials are to be used under difficult loading and environmental conditions.

**Stress and Strain.—**Stress is force per unit area of a material associated with a given strain (stress = force/area), normally expressed in pounds per square inch (psi) or newtons per square meter ($N/m^2$, or pascals, Pa). Strain is the deformation or extension of a material or part under load (strain = extension/length). Stress and strain are linearly related in some materials over all or part of their elastic ranges. This linear relationship is expressed as Hooke's law, where stress/strain = $E$, with $E$ being the elastic modulus of the material. Stresses can be axial (tensile or compressive), shear (simple or torsional), isotropic (uniform loading on all surfaces), or combined (any combination of types). Tensile and compressive stress act at right angles to (normal to) the stressed area; shear stress acts along the plane of the affected area (at right angles to compressive or tensile stresses).

**Working Stress.—**Calculated working stresses are the products of calculated nominal stress values and stress concentration factors. Calculated nominal stress values are based on the assumption of idealized stress distributions. Such nominal stresses may be simple stresses, combined stresses, or cyclic stresses. Depending on the nature of the nominal stress, one of the following equations applies:

$$s_w = K\sigma \quad (2) \qquad s_w = K\sigma' \quad (4) \qquad s_w = K\sigma_{cy} \quad (6)$$

$$s_w = K\tau \quad (3) \qquad s_w = K\tau' \quad (5) \qquad s_w = K\tau_{cy} \quad (7)$$

where $K$ is a stress concentration factor; $\sigma$ and $\tau$ are, respectively, simple normal (tensile or compressive) and shear stresses; $\sigma'$ and $\tau'$ are combined normal and shear stresses; $\sigma_{cy}$ and $\tau_{cy}$ are cyclic normal and shear stresses.

Where uneven stress distribution occurs, as illustrated in the table (on page 210) of simple stresses for Cases 3, 4 and 6, the maximum stress is the one to which the stress concentration factor is applied in computing working stresses. The location of the maximum stress in each case is discussed under the section *Simple Stresses* and the formulas for these maximum stresses are given in the *Table of Simple Stresses* on page 210.

**Stress Concentration Factors.—**Stress concentration is related to type of material, the nature of the stress, environmental conditions, and the geometry of parts. When stress concentration factors that specifically match all of the foregoing conditions are not available, the following equation may be used:

$$K = 1 + q(K_t - 1) \quad (8)$$

$K_t$ is a theoretical stress concentration factor that is a function only of the geometry of a part and the nature of the stress; $q$ is the *index of sensitivity* of the material. If the geometry is such as to provide no theoretical stress concentration, $K_t = 1$.

Curves for evaluating $K_t$ are on pages 207 through 210. For constant stresses in cast iron and in ductile materials, $q = 0$ (hence, $K = 1$). For constant stresses in brittle materials such as hardened steel, $q$ may be taken as 0.15; for very brittle materials such as steels that have been quenched but not drawn, $q$ may be taken as 0.25. When stresses are suddenly applied (impact stresses) $q$ ranges from 0.4 to 0.6 for ductile materials; for cast iron it is taken as 0.5; and, for brittle materials, 1.

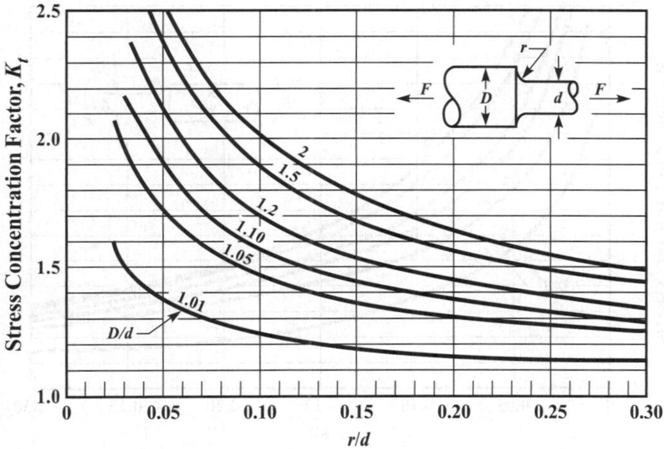


Fig. 4. Stress Concentration Factor $K_t$ for a Filleted Shaft in Tension

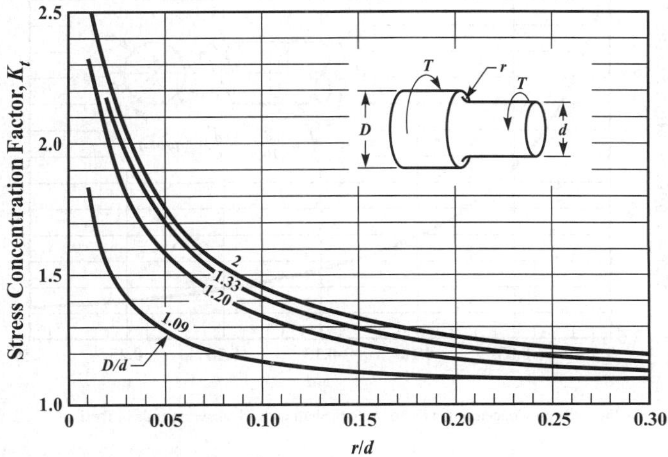


Fig. 5. Stress Concentration Factor $K_t$ for a Filleted Shaft in Torsion[a]

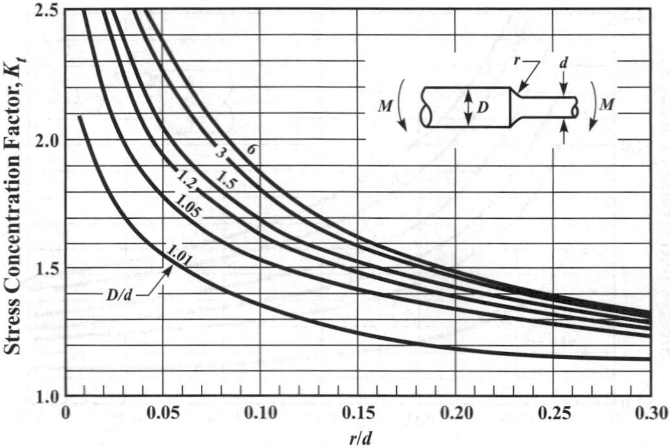


Fig. 6. Stress Concentration Factor $K_t$ for a Shaft with Shoulder Fillet in Bending[a]

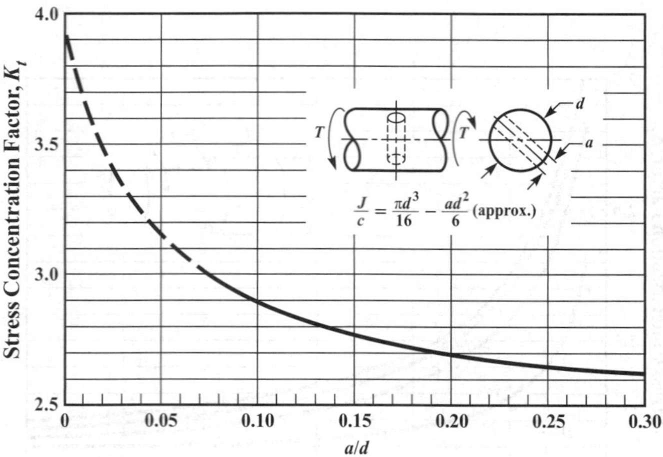


Fig. 7. Stress Concentration Factor $K_t$ for a Shaft with a Transverse Hole in Torsion[a]

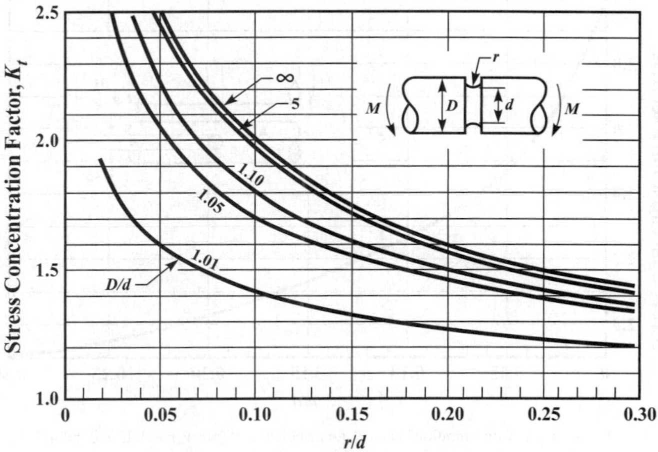


Fig. 8. Stress Concentration Factor $K_t$ for a Grooved Shaft in Bending[a]

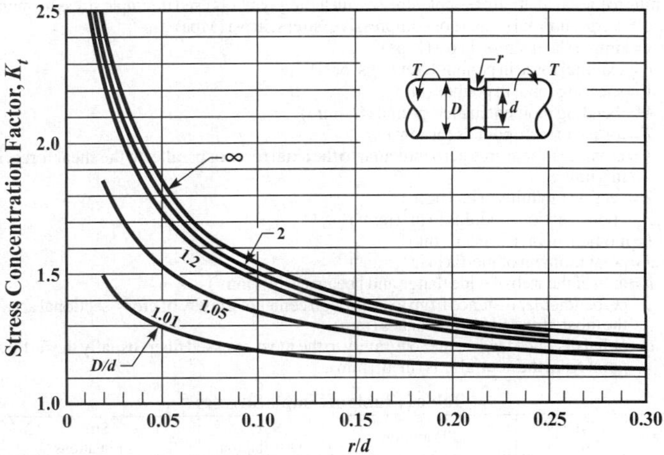


Fig. 9. Stress Concentration Factor $K_t$ for a Grooved Shaft in Torsion[a]

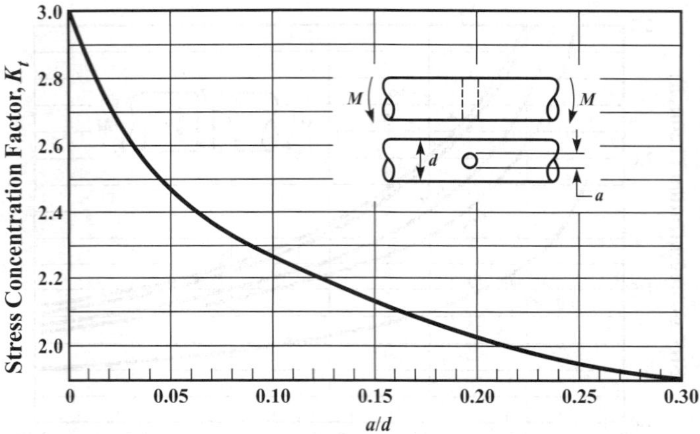


Fig. 10. Stress Concentration Factor $K_t$ for a Shaft with a Transverse Hole in Bending[a]

[a] Source: R. E. Peterson, "Design Factors for Stress Concentration," *Machine Design*, vol. 23, 1951. For other stress concentration charts, see Lipson and Juvinall, *The Handbook of Stress and Strength*, The Macmillan Co., 1963.

**Simple Stresses.**—Simple stresses are produced by constant conditions of loading on elements that can be represented as beams, rods, or bars. Table 2 summarizes information pertaining to the calculation of simple stresses, including symbols used in simple stress formulae:

$\sigma$ = simple normal (tensile or compressive) stress, in psi (Mpa)
$\tau$ = simple shear stress, in psi (Mpa)
$F$ = external force in pounds (Newtons, or N)
$V$ = shearing force in pounds (N)
$M$ = bending moment in inch-pounds (N-mm)
$T$ = torque in inch-pounds (N-mm)
$A$ = cross-sectional area perpendicular to the axial force or parallel to the shear force, in $in^2$ ($mm^2$)
$Z$ = section modulus in $in^3$ ($mm^3$)
$Z_p$ = polar section modulus in $in^3$ ($mm^3$)
$I$ = moment of inertia in $in^4$ ($mm^4$)
$J$ = polar moment of inertia in $in^4$ ($mm^4$)
$a$ = area of the web of wide flange and I beams in $in^2$ ($mm^2$)
$y$ = perpendicular distance from axis through center of gravity of cross-sectional area to the most-stressed fiber in inches (mm)
$c$ = radial distance from center of gravity to the most-stressed fiber (usually the farthest point from the neutral axis) in in. (mm).

**Table 2. Table of Simple Stresses**

| Case | Type of Loading | Illustration | Stress Distribution | Stress Equations |
|---|---|---|---|---|
| 1 | Direct tension | 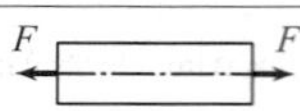 | Uniform | $\sigma = \frac{F}{A}$ (9) |
| 2 | Direct compression | 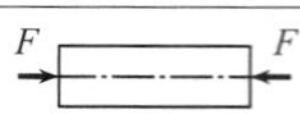 | Uniform | $\sigma = -\frac{F}{A}$ (10) |

**Table 2.** *(Continued)* **Table of Simple Stresses**

| Case | Type of Loading | Illustration | Stress Distribution | Stress Equations |
|---|---|---|---|---|
| 3 | Bending | 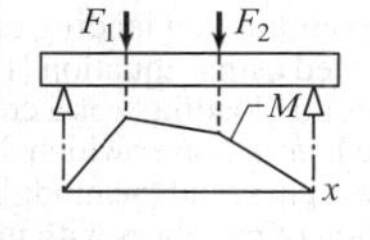 Bending moment diagram | 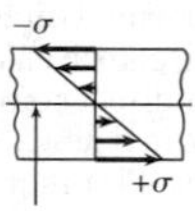 Neutral plane | $\sigma = \pm \frac{M}{Z} = \pm \frac{My}{I}$ (11) |
| 4 | Shear | 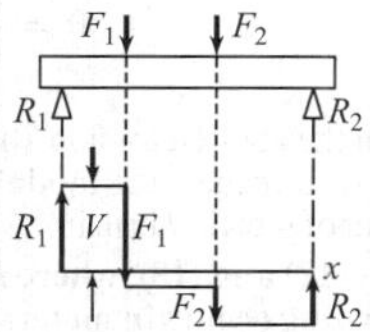 Shearing force diagram | 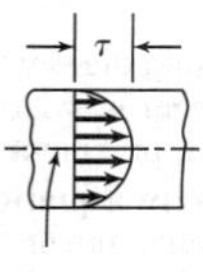 Neutral plane | For beams of rectangular cross section: $\tau = \frac{3V}{2A}$ (12)<br>For beams of solid circular cross section: $\tau = \frac{4V}{3A}$ (13)<br>For wide flange and I beams (approximately): $\tau = \frac{V}{a}$ (14) |
| 5 | Direct shear | 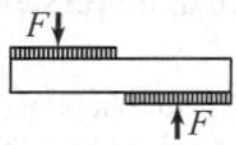 | Uniform | $\tau = \frac{F}{A}$ (15) |
| 6 | Torsional Shear | 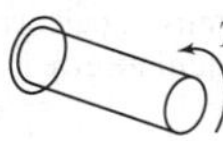 | 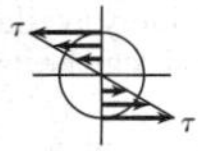 | $\tau = \frac{T}{Zp} = \frac{Tc}{J}$ (16) |

For direct tension and direct compression loading, Cases 1 and 2 in the table on page 210, the force $F$ must act along a line through the center of gravity of the section at which the stress is calculated. The equation for direct compression loading applies only to members for which the ratio of length to least radius of gyration is relatively small, approximately 20; otherwise the member must be treated as a column.

The table *Stresses and Deflections in Beams* starting on page 251 give equations for calculating stresses due to bending for common types of beams and conditions of loading. Where these tables are not applicable, stress may be calculated using Equation (11) in the table on page 210. In using this equation it is necessary to determine the value of the bending moment at the point where the stress is to be calculated. For beams of constant cross-section, stress is ordinarily calculated at the point coinciding with the maximum value of bending moment. Bending loading results in the characteristic stress distribution shown in the table for Case 3. It will be noted that the maximum stress values are at the surfaces farthest from the neutral plane. One of the surfaces is stressed in tension and the other in compression. It is for this reason that the ± sign is used in Equation (11). Numerous tables for evaluating section moduli are given in the section starting on page 233.

Shear stresses caused by bending have maximum values at neutral planes and zero values at the surfaces farthest from the neutral axis, as indicated by the stress distribution diagram shown for Case 4 in the *Table of Simple Stresses*. Values for $V$ in Equations (12), (13) and (14) can be determined from shearing force diagrams. The shearing force diagram shown in Case 4 corresponds to the bending moment diagram for Case 3. As shown in this diagram, the value taken for $V$ is represented by the greatest vertical distance from the

$x$-axis. The shear stress caused by direct shear loading, Case 5, has a uniform distribution. However, the shear stress caused by torsion loading, Case 6, has a zero value at the axis and a maximum value at the surface farthest from the axis.

**Deflections.—**For direct tension and direct compression loading on members with uniform cross sections, deflection can be calculated using Equation (17). For direct tension loading, $e$ is an elongation; for direct compression loading, $e$ is a contraction. Deflection is in inches when the load $F$ is in pounds, the length $L$ over which deflection occurs is in inches, the cross-sectional area $A$ is in square inches, and the modulus of elasticity $E$ is in pounds per square inch. The angular deflection of members with uniform circular cross sections subject to torsion loading can be calculated with Equation (18).

$$e = \frac{FL}{AE} \qquad (17) \qquad\qquad \theta = \frac{TL}{GJ} \qquad (18)$$

The angular deflection θ is in radians when the applied torsion $T$ is in inch-pounds, the length $L$ over which the member is twisted is in inches, the modulus of rigidity $G$ is in pounds per square inch, and the polar moment of inertia $J$ is in $in.^4$

**Metric SI units can be used in Equations (17) and (18), where $F$ = force in newtons (N); $L$ = length over which deflection or twisting occurs in meters; $A$ = cross-sectional area in $meters^2$; $E$ = the modulus of elasticity in $N/m^2$; θ = radians; $T$ = the applied torsion in newton-meters (N·m); $G$ = modulus of rigidity, in pascals (Pa); and $J$ = the polar moment of inertia in $meters^4$. If the load ($F$) is applied as a weight, it should be noted that the weight of a mass $M$ kilograms is $Mg$ newtons, where $g = 9.81\ m/s^2$. Millimeters can be used in the calculations in place of meters, provided the treatment is consistent throughout.**

**Combined Stresses.—**A member may be loaded in such a way that a combination of simple stresses acts at a point. Three general cases are shown in the accompanying illustration, Fig. 11. Strength data is widely available for tensile performance, and the *von Mises stress* method and other methods can be used to calculate effective tensile stress at a point under combined loading.

*von Mises Effective Tensile Stress:* This method of calculating an equivalent tensile stress for combined stress scenarios is applicable to ductile material. The von Mises effective tensile stress, σ', can be directly compared to the tensile yield strength of the material to predict performance for two-dimensional loading (Equation 19) and three-dimensional loading (Equation 20):

$$\sigma' = \sqrt{\sigma_x^2 + \sigma_y^2 - \sigma_x\sigma_y + 3\tau_{xy}^2} \qquad (19)$$

$$\sigma' = \sqrt{\frac{(\sigma_x - \sigma_y)^2 + (\sigma_y - \sigma_z)^2 + (\sigma_z - \sigma_x)^2 + 6(\tau_{xy}^2 + \tau_{yz}^2 + \sigma_{zx}^2)}{2}} \qquad (20)$$

*Superposition of Stresses:* This method may be used in cases where the deformations are small and within the elastic range. Fig. 11 at (1) illustrates a common situation that results in simple stresses combining by superposition at points **a** and **b**. The equal and opposite forces $F_1$ will cause a compressive stress $\sigma_1 = -F_1/A$. Force $F_2$ will cause a bending moment $M$ to exist in the plane of points **a** and **b**. The resulting stress $\sigma_2 = \pm M/Z$. The combined stress at point **a**,

$$\sigma_a' = -\frac{F_1}{A} - \frac{M}{Z} \qquad (21) \qquad \text{and at } \mathbf{b}, \qquad \sigma_b' = -\frac{F_1}{A} + \frac{M}{Z} \qquad (22)$$

where the minus sign indicates a compressive stress and the plus sign a tensile stress. Thus, the stress at **a** will be compressive and at **b** either tensile or compressive, depending on which term in the equation for $\sigma_b'$ has the greatest value.

*Normal Stresses at Right Angles:* This is shown in Fig. 11 at (2). This combination of stresses occurs, for example, in tanks subjected to internal or external pressure. According to the Maximum Shear Stress theory, failure will occur in ductile materials due to shear, so this stress is critical to evaluate. The principal normal stresses are $\sigma_x = F_1/A_1$, $\sigma_y = F_2/A_2$, and $\sigma_z = 0$ in this plane stress problem. Determine the values of these three stresses with their signs, order them algebraically, and then calculate the maximum shear stress:

$$\tau = \frac{\sigma_{largest} - \sigma_{smallest}}{2} \quad (23)$$

*Normal and Shear Stresses:* The example in Fig. 11 at (3) shows a member subjected to a torsional shear stress, $\tau = T/Z_p$, and a direct compressive stress, $\sigma = -F/A$. At some point **a** on the member the principal normal stresses are calculated using the equation,

$$\sigma' = \frac{\sigma}{2} \pm \sqrt{\left(\frac{\sigma}{2}\right)^2 + \tau^2} \quad (24)$$

The maximum shear stress is calculated by using the equation,

$$\tau' = \sqrt{\left(\frac{\sigma}{2}\right)^2 + \tau^2} \quad (25)$$

The point **a** should ordinarily be selected where stress is a maximum value. For the example shown in Fig. 11 at (3), the point **a** can be anywhere on the cylindrical surface because the combined stress has the same value anywhere on that surface.

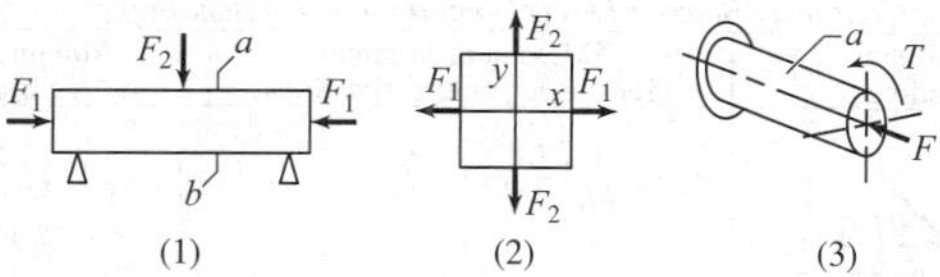


Fig. 11. Types of Combined Loading

**Tables of Combined Stresses.—**Beginning on page 214, these tables list equations for maximum nominal tensile or compressive (normal) stresses, and maximum nominal shear stresses for common machine elements. These equations were derived using general Equations (21), (22), (24), and (25). The equations apply to the critical points indicated on the figures. Cases 1, 2, 3, and 4 are cantilever beams. These may be loaded with a combination of a vertical and horizontal force, or by a single oblique force. If the single oblique force $F$ and the angle $\theta$ are given, then horizontal and vertical forces can be calculated using the equations $F_x = F\cos\theta$ and $F_y = F\sin\theta$. In cases 9 and 10 of the table, the equations for $\sigma_a'$ can give a tensile and a compressive stress because of the ± sign in front of the radical. Equations involving direct compression are valid only if machine elements have relatively short lengths with respect to their sections; otherwise column equations apply.

*Calculation of Worst Stress Condition:* Stress failure can occur at any critical point if the tensile, compressive, or shear stress properties of the material are exceeded by the corresponding working stress. It is necessary to evaluate the safety factor for each possible failure condition. If working with von Mises equivalent stress, only tensile yield need be considered.

The following rules apply to calculations using equations in the *Table of Simple Stresses* on page 210 and to calculations based on Equations (21) and (22). *Rule 1:* For every calculated normal stress there is a corresponding induced shear stress; the value of the shear stress is equal to half that of the normal stress. *Rule 2:* For every calculated shear stress there is a corresponding induced normal stress; the value of the normal stress is equal to that of the shear stress. The tables of combined stress formulas below include equations for calculating both maximum nominal tensile or compressive stresses, and maximum nominal shear stresses.

## Formulas for Combined Stresses

(1) *Circular Cantilever Beam in Direct Compression and Bending:*

| Type of Beam and Loading | Maximum Nominal Tensile or Compressive Stress | Maximum Nominal Shear Stress |
|---|---|---|
| 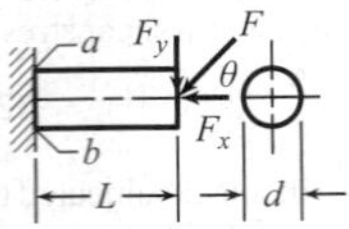 | $\sigma_a' = \frac{1.273}{d^2}\left(\frac{8LF_y}{d} - F_x\right)$ | $\tau_a' = 0.5\sigma_a'$ |
| | $\sigma_b' = -\frac{1.273}{d^2}\left(\frac{8LF_y}{d} + F_x\right)$ | $\tau_b' = 0.5\sigma_b'$ |

(2) *Circular Cantilever Beam in Direct Tension and Bending:*

| Type of Beam and Loading | Maximum Nominal Tensile or Compressive Stress | Maximum Nominal Shear Stress |
|---|---|---|
| 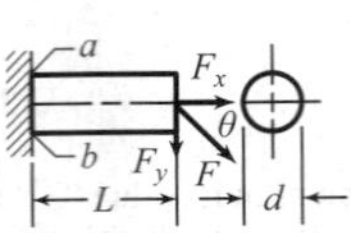 | $\sigma_a' = \frac{1.273}{d^2}\left(F_x + \frac{8LF_y}{d}\right)$ | $\tau_a' = 0.5\sigma_a'$ |
| | $\sigma_b' = \frac{1.273}{d^2}\left(F_x - \frac{8LF_y}{d}\right)$ | $\tau_b' = 0.5\sigma_b'$ |

(3) *Rectangular Cantilever Beam in Direct Compression and Bending:*

| Type of Beam and Loading | Maximum Nominal Tensile or Compressive Stress | Maximum Nominal Shear Stress |
|---|---|---|
| 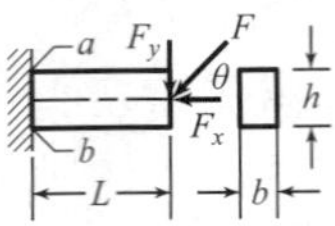 | $\sigma_a' = \frac{1}{bh}\left(\frac{6LF_y}{h} - F_x\right)$ | $\tau_a' = 0.5\sigma_a'$ |
| | $\sigma_b' = -\frac{1}{bh}\left(\frac{6LF_y}{h} + F_x\right)$ | $\tau_b' = 0.5\sigma_b'$ |

(4) *Rectangular Cantilever Beam in Direct Tension and Bending:*

| Type of Beam and Loading | Maximum Nominal Tensile or Compressive Stress | Maximum Nominal Shear Stress |
|---|---|---|
| 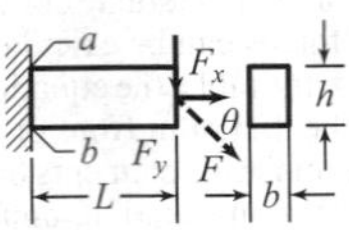 | $\sigma_a' = \frac{1}{bh}\left(F_x + \frac{6LF_y}{h}\right)$ | $\tau_a' = 0.5\sigma_a'$ |
| | $\sigma_b' = \frac{1}{bh}\left(F_x - \frac{6LF_y}{h}\right)$ | $\tau_b' = 0.5\sigma_b'$ |

(5) *Circular Beam or Shaft in Direct Compression and Bending:*

| Type of Beam and Loading | Maximum Nominal Tensile or Compressive Stress | Maximum Nominal Shear Stress |
|---|---|---|
| 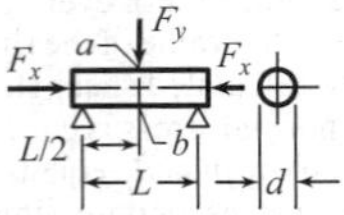 | $\sigma_a' = -\frac{1.273}{d^2}\left(\frac{2LF_y}{d} + F_x\right)$ | $\tau_a' = 0.5\sigma_a'$ |
| | $\sigma_b' = \frac{1.273}{d^2}\left(\frac{2LF_y}{d} - F_x\right)$ | $\tau_b' = 0.5\sigma_b'$ |

(6) *Circular Beam or Shaft in Direct Tension and Bending:*

| Type of Beam and Loading | Maximum Nominal Tensile or Compressive Stress | Maximum Nominal Shear Stress |
|---|---|---|
| 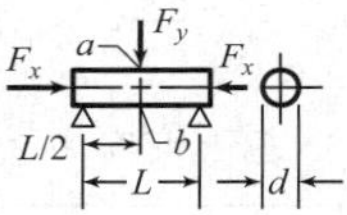  | $\sigma_a' = \frac{1.273}{d^2}\left(F_x - \frac{2LF_y}{d}\right)$<br>$\sigma_b' = \frac{1.273}{d^2}\left(F_x + \frac{2LF_y}{d}\right)$ | $\tau_a' = 0.5\sigma_a'$<br>$\tau_b' = 0.5\sigma_b'$ |

(7) *Rectangular Beam or Shaft in Direct Compression and Bending:*

| Type of Beam and Loading | Maximum Nominal Tensile or Compressive Stress | Maximum Nominal Shear Stress |
|---|---|---|
| 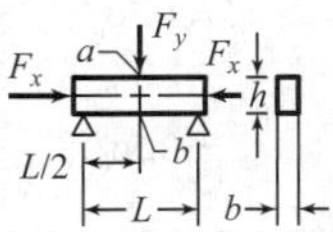  | $\sigma_a' = -\frac{1}{bh}\left(\frac{3LF_y}{2h} + F_x\right)$<br>$\sigma_b' = \frac{1}{bh}\left(-\frac{3LF_y}{2h} - F_x\right)$ | $\tau_a' = 0.5\sigma_a'$<br>$\tau_b' = 0.5\sigma_b'$ |

(8) *Rectangular Beam or Shaft in Direct Tension and Bending:*

| Type of Beam and Loading | Maximum Nominal Tensile or Compressive Stress | Maximum Nominal Shear Stress |
|---|---|---|
| 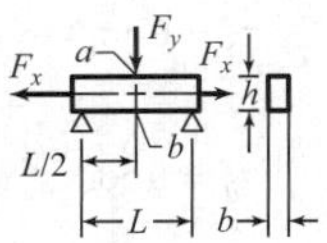  | $\sigma_a' = \frac{1}{bh}\left(F_x - \frac{3LF_y}{2h}\right)$<br>$\sigma_b' = \frac{1}{bh}\left(F_x + \frac{3LF_y}{2h}\right)$ | $\tau_a' = 0.5\sigma_a'$<br>$\tau_b' = 0.5\sigma_b'$ |

(9) *Circular Shaft in Direct Compression and Torsion:*

| Type of Beam and Loading | Maximum Nominal Tensile or Compressive Stress | Maximum Nominal Shear Stress |
|---|---|---|
| 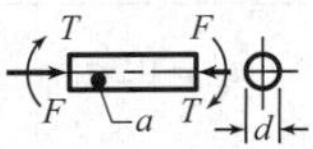  **a** anywhere on surface | $\sigma_a' = -\frac{0.637}{d^2}\left[F \pm \sqrt{F^2 + \left(\frac{8T}{d}\right)^2}\right]$ | $\tau_a' = -\frac{0.637}{d^2}\sqrt{F^2 + \left(\frac{8T}{d}\right)^2}$ |

(10) *Circular Shaft in Direct Tension and Torsion:*

| Type of Beam and Loading | Maximum Nominal Tensile or Compressive Stress | Maximum Nominal Shear Stress |
|---|---|---|
| 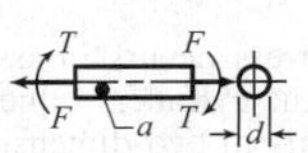  **a** anywhere on surface | $\sigma_a' =$<br>$\frac{0.637}{d^2}\left[F \pm \sqrt{F^2 + \left(\frac{8T}{d}\right)^2}\right]$ | $\tau_a' =$<br>$\frac{0.637}{d^2}\sqrt{F^2 + \left(\frac{8T}{d}\right)^2}$ |

(11) *Offset Link, Circular Cross Section, in Direct Tension:*

| Type of Beam and Loading | Maximum Nominal Tensile or Compressive Stress | Maximum Nominal Shear Stress |
|---|---|---|
| 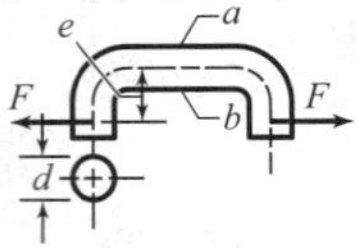 | $\sigma_a' = \frac{1.273F}{d^2}\left(1 - \frac{8e}{d}\right)$<br>$\sigma_b' = \frac{1.273F}{d^2}\left(1 + \frac{8e}{d}\right)$ | $\tau_a' = 0.5\sigma_a'$<br>$\tau_b' = 0.5\sigma_b'$ |

(12) *Offset Link, Circular Cross Section, in Direct Compression:*

| Type of Beam and Loading | Maximum Nominal Tensile or Compressive Stress | Maximum Nominal Shear Stress |
|---|---|---|
| 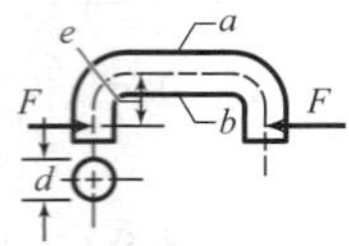 | $\sigma_a' = \frac{1.273F}{d^2}\left(\frac{8e}{d} - 1\right)$<br>$\sigma_b' = -\frac{1.273F}{d^2}\left(\frac{8e}{d} + 1\right)$ | $\tau_a' = 0.5\sigma_a'$<br>$\tau_b' = 0.5\sigma_b'$ |

(13) *Offset Link, Rectangular Section, in Direct Tension:*

| Type of Beam and Loading | Maximum Nominal Tensile or Compressive Stress | Maximum Nominal Shear Stress |
|---|---|---|
| 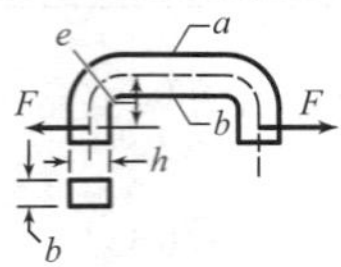 | $\sigma_a' = \frac{F}{bh}\left(1 - \frac{6e}{h}\right)$<br>$\sigma_b' = \frac{F}{bh}\left(1 + \frac{6e}{h}\right)$ | $\tau_a' = 0.5\sigma_a'$<br>$\tau_b' = 0.5\sigma_b'$ |

(14) *Offset Link, Rectangular Section, in Direct Compression:*

| Type of Beam and Loading | Maximum Nominal Tensile or Compressive Stress | Maximum Nominal Shear Stress |
|---|---|---|
| 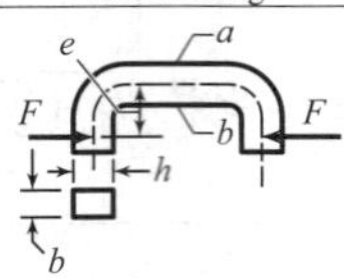 | $\sigma_a' = \frac{F}{bh}\left(\frac{6e}{h} - 1\right)$<br>$\sigma_b' = -\frac{F}{bh}\left(\frac{6e}{h} + 1\right)$ | $\tau_a' = 0.5\sigma_a'$<br>$\tau_b' = 0.5\sigma_b'$ |

**Formulas from the simple and combined stress tables, as well as tension and shear factors, can be applied without change in calculations using metric SI units. Stresses are given in $N/m^2$ or in $N/mm^2$.**

**Three-Dimensional Stress.—**Three-dimensional or triaxial stress occurs in assemblies such as a shaft press-fitted into a gear bore, or in pipes and cylinders subjected to internal or external fluid pressure. Triaxial stress also occurs in two-dimensional stress problems if the loads produce normal stresses that are either both tensile or both compressive. In either case the calculated maximum shear stress, based on the corresponding two-dimensional theory, will be less than the true maximum value because of three-dimensional effects.

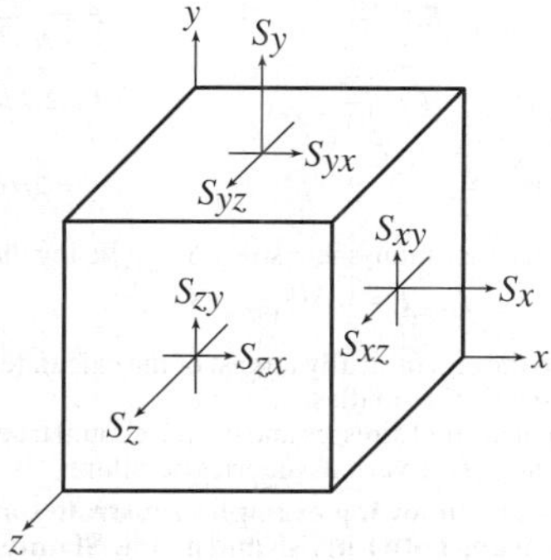


Fig. 12. *x, y, z*-Coordinate System Showing Positive Directions of Stresses

The von Mises effective tensile stress method can be used to evaluate three-dimensional stress failure of ductile materials. If stress analysis is based on the maximum-shear-stress theory of failure, the triaxial stress cubic equation is used first to calculate the three principal stresses; true maximum shear stress can then be determined. The following procedure provides the principal maximum normal tensile and compressive stresses and the true maximum shear stress at any point on a body subjected to any combination of loads. The basis for the procedure is the stress cubic equation

$$S^3 - AS^2 + BS - C = 0$$

in which:

$$A = S_x + S_y + S_z$$
$$B = S_x S_y + S_y S_z + S_z S_x - S_{xy}^{\ 2} - S_{yz}^{\ 2} - S_{zx}^{\ 2}$$
$$C = S_x S_y S_z + 2S_{xy} S_{yz} S_{zx} - S_x S_{yz}^{\ 2} - S_y S_{zx}^{\ 2} - S_z S_{xy}^{\ 2}$$

and $S_x$, $S_y$, etc. are as shown in Fig. 12.

The *x, y, z* coordinate system in Fig. 12 shows the positive directions of the normal and shear stress components on an elementary cube of material. Only six of the nine components shown are needed for the calculations: the normal stresses $S_x$, $S_y$, and $S_z$ on three of the faces of the cube; and the three shear stresses $S_{xy}$, $S_{yz}$, and $S_{zx}$. The remaining three shear stresses are known because $S_{yx} = S_{xy}$, $S_{zy} = S_{yz}$, and $S_{xz} = S_{zx}$. The normal stresses $S_x$, $S_y$, and $S_z$ are shown as positive (tensile) stresses; the opposite direction is negative (compressive). The first subscript of each shear stress identifies the coordinate axis perpendicular to the plane of the shear stress; the second subscript identifies the axis to which the stress is parallel. Thus, $S_{xy}$ is the shear stress in the *yz*-plane to which the *x*-axis is perpendicular, and the stress is parallel to the *y*-axis.

Step 1. Draw a diagram of the hardware to be analyzed, and show the applied loads *P*, *T*, and any others.

Step 2. For any point at which the stresses are to be analyzed, draw a coordinate diagram similar to Fig. 12 and show the magnitudes of the stresses resulting from the applied loads (these stresses may be calculated by using standard basic equations from strength of materials, and should include any stress concentration factors).

Step 3. Substitute the values of the six stresses $S_x$, $S_y$, $S_z$, $S_{xy}$, $S_{yz}$, and $S_{zx}$, including zero values, into the formulas for the quantities *A* through *K*. The quantities *I*, *J*, and *K* represent the principal normal stresses at the point analyzed. As a check, if the algebraic sum $I + J + K$ equals *A*, within rounding errors, then the calculations up to this point should be correct.

$$D = \frac{A^2}{3} - B \qquad E = \frac{AB}{3} - C - \frac{2A^3}{27} \qquad F = \sqrt{\frac{D^3}{27}}$$

$$G = \arccos\left(\frac{-E}{2F}\right) \qquad H = \sqrt{\frac{D}{3}} \qquad I = 2H\cos\left(\frac{G}{3}\right) + \frac{A}{3}$$

$$J = 2H\cos\left(\frac{G}{3} + 120°\right) + \frac{A}{3} \qquad K = 2H\cos\left(\frac{G}{3} + 240°\right) + \frac{A}{3}$$

Step 4. Calculate the true maximum shear stress $S_{s(\max)}$ using the formula

$$S_{s(\max)} = 0.5(S_{\text{large}} - S_{\text{small}})$$

in which $S_{\text{large}}$ is equal to the algebraically largest of the calculated principal stresses *I*, *J*, or *K* and $S_{\text{small}}$ is algebraically the smallest.

The maximum principal normal stresses and the maximum true shear stress calculated above may be used with any of the various theories of failure.

**Sample Calculations.—The following examples illustrate some typical strength of materials calculations, using both English and metric SI units of measurement.**

*Example 1(a):* A round bar made from SAE 1025 low carbon steel is to support a direct tension load of 50,000 lbs. Using a factor of safety of 4, and assuming that the stress concentration factor $K = 1$, a suitable standard diameter is to be determined. Calculations are to be based on a yield strength of 40,000 psi.

*Solution:* Because the factor of safety and strength of the material are known, the allowable working stress $s_w$ may be calculated using Equation (1): 40,000/4 = 10,000 psi. The relationship between working stress $s_w$ and nominal stress $\sigma$ is given by Equation (2). Since $K = 1$, $\sigma = 10{,}000$ psi. Applying Equation (9) in the *Table of Simple Stresses* on page 210, the area of the bar can be solved for: $A = 50{,}000/10{,}000$ or 5 square inches. The next largest standard diameter corresponding to this area is $2\frac{9}{16}$ inches.

***Example 1(b):* A similar example to that given in Example 1(a), using metric SI units, is as follows. A round steel bar of 300 meganewtons/meter $^2$ yield strength is to withstand a direct tension of 200 kilonewtons. Using a safety factor of 4, and assuming that the stress concentration factor $K = 1$, a suitable diameter is to be determined.**

**Because the factor of safety and the strength of the material are known, the allowable working stress $s_w$ may be calculated using Equation (1): 300/4 = 75 meganewtons/meter$^2$. The relationship between working stress and nominal stress $\sigma$ is given by Equation (2). Since $K = 1$, $\sigma = 75$ MN/m$^2$. Applying Equation (9) in the *Table of Simple Stresses* on page 210, the area of the bar can be determined from:**

$$A = \frac{200 \text{ kN}}{75 \text{ MN/m}^2} = \frac{200{,}000\, N}{75{,}000{,}000 \text{ N/m}^2} = 0.00267\text{m}^2$$

**The diameter corresponding to this area is 0.058 meters, or approximately 0.06 m. Millimeters can be employed in the calculations in place of meters, provided the treatment is consistent throughout. In this instance the diameter would be 60 mm.**

***Note:* If the tension in the bar is produced by hanging a mass of *M* kilograms from its end, the value is *Mg* newtons, where *g* = approximately 9.81 meters per second$^2$.**

*Example 2(a) :* What would the total elongation of the bar in Example 1(a) be if its length were 60 inches? Applying Equation (18) on page 212,

$$e = \frac{50{,}000 \times 60}{5.157 \times 30{,}000{,}000} = 0.019 \text{ inch}$$

***Example 2(b):* What would be the total elongation of the bar in Example 1(b) if its length were 1.5 meters? The problem is solved by applying Equation (17) in which $F = 200$ kilonewtons; $L = 1.5$ meters; $A = \pi 0.06^2/4 = 0.00283$ m$^2$. Assuming a modulus of elasticity *E* of 200 giganewtons/meter$^2$, then the calculation is:**

$$e = \frac{200{,}000 \times 1.5}{0.00283 \times 200{,}000{,}000{,}000} = 0.000530 \text{ m}$$

**The calculation is less unwieldy if carried out using millimeters in place of meters; then $F = 200$ kN; $L = 1500$ mm; $A = 2830$ mm$^2$, and $E = 200{,}000$ N/mm$^2$. Thus:**

$$e = \frac{200{,}000 \times 1500}{2830 \times 200{,}000} = 0.530 \text{ mm}$$

*Example 3(a):* Determine the size for the section of a square bar which is to be held firmly at one end and is to support a load of 3000 pounds at the other end. The bar is to be 30 inches long and is to be made from SAE 1045 medium carbon steel with a yield point of 60,000 psi. A factor of safety of 3 and a stress concentration factor of 1.3 are to be used.

*Solution:* From Equation (1) the allowable working stress $s_w = 60{,}000/3 = 20{,}000$ psi. The applicable equation relating working stress and nominal stress is Equation (2); hence, $\sigma = 20{,}000/1.3 = 15{,}400$ psi. The member must be treated as a cantilever beam subject to a bending moment of 30 × 3000 or 90,000 inch-pounds. Solving Equation (11) in the *Table of Simple Stresses* for section modulus: $Z = 90{,}000/15{,}400 = 5.85$ inch$^3$. The section modulus for a square section with neutral axis equidistant from either side is $a^3/6$, where $a$ is the dimension of the square, so $a = \sqrt[3]{35.1} = 3.27$ inches. The bar size can be $3\frac{5}{16}$ inches.

***Example 3(b):* A similar example to that given in Example 3(a), using metric SI units is as follows. Determine the size for the section of a square bar which is to be held firmly at one end and is to support a load of 1600 kilograms at the other end. The bar is to be 1 meter long and is to be made from steel with a yield strength of 500 newtons/mm$^2$. A factor of safety of 3 and a stress concentration factor of 1.3 are to be used. The calculation can be performed using millimeters throughout.**

***Solution:* From Equation (1) the allowable working stress $s_w$ = 500 N/mm$^2$/3 = 167 N/mm$^2$. The formula relating working stress and nominal stress is Equation (2); hence $\sigma = 167/1.3 = 128$ N/mm$^2$. Since a mass of 1600 kg equals a weight of 1600$g$ newtons, where $g = 9.81$ meters/second$^2$, the force acting on the bar is 15,700 newtons. The bending moment on the bar, which must be treated as a cantilever beam, is thus 1000 mm × 15,700 N = 15,700,000 N·mm. Solving Equation (11) in the** *Table of Simple Stresses* **for section modulus: $Z = M/\sigma = 15{,}700{,}000/128 = 123{,}000$ mm$^3$. Since the section modulus for a square section with neutral axis equidistant from either side is $a^3/6$, where $a$ is the dimension of the square,**

$$a = \sqrt[3]{6 \times 123{,}000} = 90.4 \text{ mm}$$

*Example 4(a):* Find the working stress in a 2-inch diameter shaft through which a transverse hole $\frac{1}{4}$ inch in diameter has been drilled. The shaft is subject to a torsional moment of 80,000 inch-pounds and is made from hardened steel so that the index of sensitivity $q = 0.2$.

*Solution:* The polar section modulus is calculated using the equation shown in the stress concentration curve for a Round Shaft in Torsion with Transverse Hole, Fig. 7, page 208.

$$\frac{J}{c} = Z_p = \frac{\pi \times 2^3}{16} - \frac{2^2}{4 \times 6} = 1.4 \text{ inches}^3$$

The nominal shear stress due to torsion loading is computed using Equation (16) in the *Table of Simple Stresses*:

$$\tau = \frac{80{,}000}{1.4} = 57{,}200 \text{ psi}$$

Referring to the previously mentioned stress concentration curve on page 208, $K_t$ is 2.82 since $d/D$ is 0.125. The stress concentration factor may now be calculated by means of Equation (8): $K = 1 + 0.2(2.82 - 1) = 1.36$. Working stress calculated with Equation (3) is $s_w = 1.36 \times 57{,}200 = 77{,}800$ psi.

***Example 4(b):* A similar example to that given in** Example 4(a), **using metric SI units, is as follows. Find the working stress in a 50 mm diameter shaft through which a transverse hole 6 mm in diameter has been drilled. The shaft is subject to a torsional moment of 8000 N·m, and has an index of sensitivity of $q = 0.2$. If the calculation is made in millimeters, the torsional moment is 8,000,000 N·mm.**

***Solution:*** **The polar section modulus is calculated using the equation shown in the stress concentration curve for a Round Shaft in Torsion with Transverse Hole, Fig. 7, page 208:**

$$\frac{J}{c} = Z_p = \frac{\pi \times 50^3}{16} - \frac{6 \times 50^2}{6} = 24{,}544 - 2500 = 22{,}044 \text{ mm}^3$$

**The nominal shear stress due to torsion loading is computed using Equation (16) in the** ***Table of Simple Stresses*****:**

$$\tau = \frac{8{,}000{,}000}{22{,}000} = 363\frac{\text{N}}{\text{mm}^2} = 363 \text{ megapascals}$$

**Referring to the previously mentioned stress concentration curve on page 208, $K_t$ is 2.85, since $a/d = 6/50 = 0.12$. The stress concentration factor may now be calculated by means of Equation (8): $K = 1 + 0.2(2.85 - 1) = 1.37$. From Equation (3), working stress $s_w = 1.37 \times 363 = 497$ N/mm² = 497 MPa.**

*Example 5(a)* : For Case 3 in the *Tables of Combined Stresses*, calculate the least factor of safety for a 5052-H32 aluminum beam that is 10 inches long, 1 inch wide, and 2 inches high. Yield strengths are 23,000 psi tension; 21,000 psi compression; 13,000 psi shear. The stress concentration factor is 1.5; $F_y$ is 600 lbs; $F_x$ 500 lbs.

*Solution:* From *Tables of Combined Stresses*, Case 3:

$$\sigma_b{}' = -\frac{1}{1 \times 2}\left(\frac{6 \times 10 \times 600}{2} + 500\right) = -9250 \text{ psi (in compression)}$$

The other formulas for Case 3 give $\sigma_a{}' = 8750$ psi (in tension); $\tau_a{}' = 4375$ psi, and $\tau_b{}' = 4625$ psi. Using Equation (4) for the nominal compressive stress of 9250 psi: $S_w = 1.5 \times 9250 = 13{,}900$ psi. From Equation (1) $f_s = 21{,}000/13{,}900 = 1.51$. Applying Equations (1), (4) and (5) in appropriate fashion to the other calculated nominal stress values for tension and shear will show that the factor of safety of 1.51, governed by the compressive stress at *b* on the beam, is minimum.

***Example 5(b):*** **What maximum $F$ can be applied in Case 3 if the aluminum beam is 200 mm long; 20 mm wide; 40 mm high; $\theta = 30°$; $f_s = 2$, governing for compression, $K = 1.5$, and $S_m = 144$ N/mm² for compression.**

***Solution:*** **From Equation (1) $S_w = -144$ N/mm². Therefore, from Equation (4), $\sigma_b{}' = -72/1.5 = -48$ N/mm². Since $F_x = F\cos 30° = 0.866F$, and $F_y = F\sin 30° = 0.5F$:**

$$-48 = -\frac{1}{20 \times 40}\left(0.866F + \frac{6 \times 200 \times 0.5F}{40}\right) \qquad F = 2420 \text{ N}$$

**Stresses and Deflections in a Loaded Ring.**—For *thin* rings, that is, rings in which the dimension *d* shown in the accompanying diagram is small compared with *D*, the maximum stress in the ring is due primarily to bending moments produced by the forces *P*. The maximum stress due to bending is:

$$S = \frac{PDd}{4\pi I} \qquad (26)$$

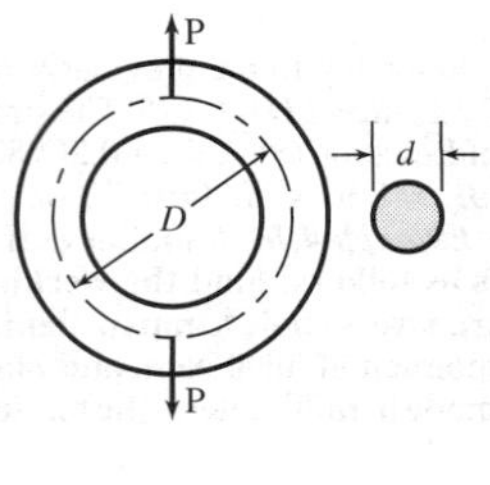


For a ring of circular cross section where *d* is the diameter of the bar from which the ring is made,

$$S = \frac{1.621PD}{d^3} \quad \text{or} \quad P = \frac{0.617Sd^3}{D} \qquad (27)$$

The increase in the vertical diameter of the ring due to load *P* is:

$$\text{Increase in vertical diameter} = \frac{0.0186PD^3}{EI} \text{ inches} \qquad (28)$$

The *decrease* in the horizontal diameter will be about 92 percent of the increase in the vertical diameter given by Formula (28). In the above formulas, $P$ = load on ring in pounds; $D$ = mean diameter of ring in inches; $S$ = tensile stress in psi, $I$ = moment of inertia of section in inches$^4$; and $E$ = modulus of elasticity of material in psi.

**Strength of Taper Pins.—**The mean diameter of taper pin required to safely transmit a known torque may be found from the formulas:

$$d = 1.13\sqrt{\frac{T}{DS}} \quad (29) \quad \text{and} \quad d = 283\sqrt{\frac{HP}{NDS}} \quad (30a)$$

where $T$ = torque in inch-pounds; $S$ = safe unit stress in psi; HP = horsepower transmitted; $N$ = number of revolutions per minute; and $d$ and $D$ denote dimensions shown in the figure.

**Formula (29) can be used with metric SI units where $d$ and $D$ denote dimensions shown in the figure in millimeters; $T$ = torque in newton-millimeters (N·mm); and $S$ = safe unit stress in newtons per millimeter$^2$ (N/mm$^2$). Formula (30a) is replaced by:**

$$d = 110.3\sqrt{\frac{Power}{NDS}} \qquad (30b)$$

**where $d$ and $D$ denote dimensions shown in the figure in millimeters; $S$ = safe unit stress in N/mm$^2$; $N$ = number of revolutions per minute, and Power = power transmitted in watts.**

*Example 6(a):* A lever secured to a 2-inch round shaft by a steel tapered pin (dimension $d = 3/8$ inch) has a pull of 50 pounds at a 30-inch radius from shaft center. Find $S$, the unit working stress on the pin. By rearranging Formula (29):

$$S = \frac{1.27T}{Dd^2} = \frac{1.27 \times 50 \times 30}{2 \times \left(\frac{3}{8}\right)^2} \cong 6770 \text{ psi}$$

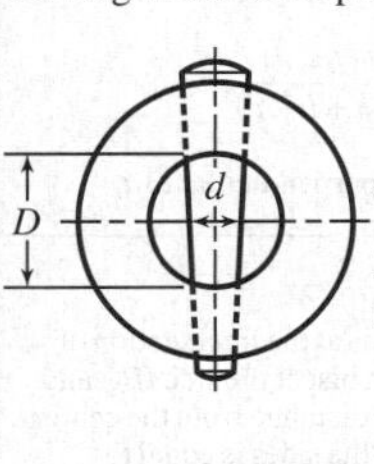


6770 psi is a safe unit working stress for machine steel in shear.

Let $P$ = 50 pounds, $R$ = 30 inches, $D$ = 2 inches, and $S$ = 6000 pounds unit working stress. Using Formula (29) to find $d$:

$$d = 1.13\sqrt{\frac{T}{DS}} = 1.13\sqrt{\frac{50 \times 30}{2 \times 6000}} = 1.13\sqrt{\frac{1}{8}} = 0.4 \text{ inch}$$

*Example 6(b):* **A similar example using SI units is as follows: A lever secured to a 50 mm round shaft by a steel tapered pin ($d$ = 10 mm) has a pull of 200 newtons at a radius of 800 mm. Find $S$, the working stress on the pin. By rearranging Formula (29):**

$$S = \frac{1.27T}{Dd^2} = \frac{1.27 \times 200 \times 800}{50 \times 10^2} = 40.6 \text{ N / mm}^2 = 40.6 \text{ megapascals}$$

**If a shaft of 50 mm diameter is to transmit power of 12 kilowatts at a speed of 500 rpm, find the mean diameter of the pin for a material having a safe unit stress of 40 N/mm$^2$. Using Equation (30b):**

$$d = 110.3\sqrt{\frac{Power}{NDS}} \quad \text{then} \quad d = 110.3\sqrt{\frac{12{,}000}{500 \times 50 \times 40}}$$

$$= 110.3 \times 0.1096 = 12.09 \text{ mm}$$

## RIGID BODY PARAMETERS

### Center of Gravity

The center of gravity of a body, volume, area, or line is that point at which if the body, volume, area, or line were suspended it would be perfectly balanced in all positions. For symmetrical bodies of uniform material it is at the geometric center. The center of gravity of a uniform round rod, for example, is at the center of its diameter halfway along its length; the center of gravity of a sphere is at the center of the sphere. For solids, areas, and arcs that are not symmetrical, the determination of the center of gravity may be made experimentally or may be calculated by the use of formulas.

The tables that follow give such formulas for some of the more important shapes. For more complicated and unsymmetrical shapes the methods outlined on page 228 may be used.

*Example:* A piece of wire is bent into the form of a semi-circular arc of 10-inch (25.4-cm) radius. How far from the center of the arc is the center of gravity located?

*Solution:* Accompanying the *Circular Arc* diagram on page 223 is a formula for the distance from the center of gravity of an arc to the center of the arc: $a = 2r \div \pi$. Therefore,

$$a = \frac{2 \times 10}{3.1416} = 6.366 \text{ inches} \qquad a = \frac{2 \times 25.4}{3.1416} = 16.17 \text{ cm}$$

### Formulas for Center of Gravity

*Triangle:*

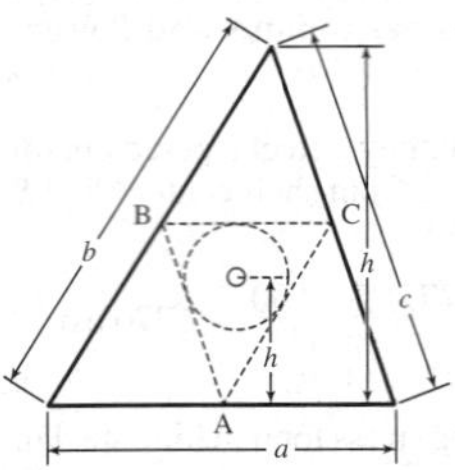


*Perimeter*

If $A$, $B$ and $C$ are the midpoints of the sides of the triangle, then the center of gravity is at the center of the circle that can be inscribed in triangle $ABC$. The distance $d$ of the center of gravity from side $a$ is:

$$d = \frac{h(b + c)}{2(a + b + c)}$$

where $h$ is the height perpendicular to $a$.

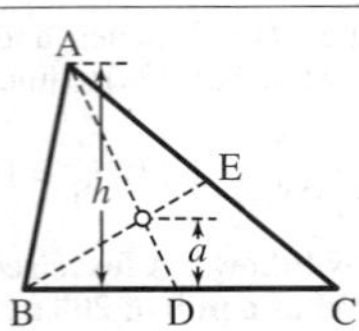


*Area*

The center of gravity is at the intersection of lines $AD$ and $BE$, which bisect the sides $BC$ and $AC$. The perpendicular distance from the center of gravity to any one of the sides is equal to one-third the height perpendicular to that side. Hence, $a = h \div 3$.

*Perimeter or Area of a Parallelogram:*

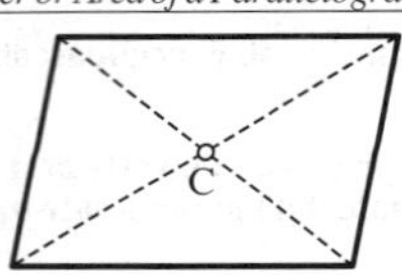


The center of gravity is at the intersection of the diagonals.

*Area of Trapezoid:*

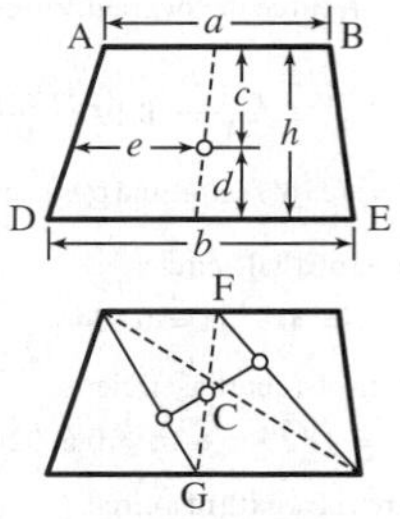


The center of gravity is on the line joining the midpoints of parallel lines *AB* and *DE*.

$$c = \frac{h(a+2b)}{3(a+b)} \quad d = \frac{h(2a+b)}{3(a+b)}$$

$$e = \frac{a^2 + ab + b^2}{3(a+b)}$$

The trapezoid can also be divided into two triangles. The center of gravity is at the intersection of the line joining the centers of gravity of the triangles and the middle line *FG*.

*Any Four-Sided Figure:*

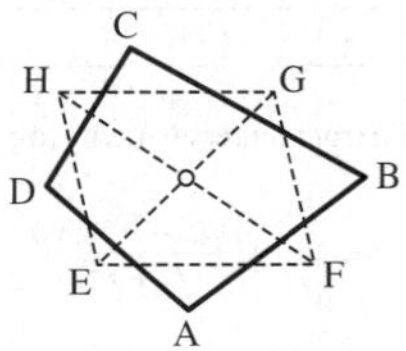


Two cases are possible, as shown in the illustration. To find the center of gravity of the four-sided figure *ABCD*, each of the sides is divided into three equal parts. A line is then drawn through each pair of division points next to the points of intersection *A*, *B*, *C*, and *D* of the sides of the figure. These lines form a parallelogram *EFGH*; the intersection of the diagonals *EG* and *FH* locates center of gravity.

*Circular Arc:*

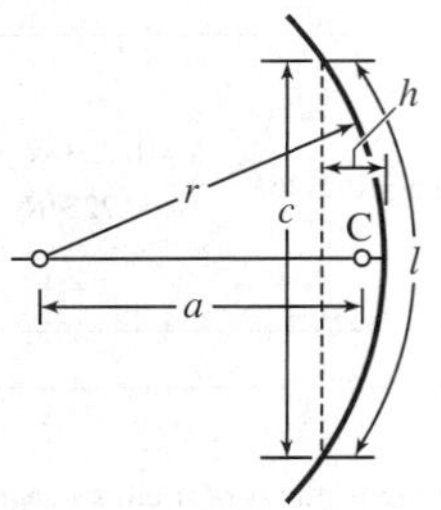


The center of gravity is on the line that bisects the arc, at a distance $a = \frac{r \times c}{l} = \frac{c(c^2 + 4h^2)}{8lh}$ from the center of the circle.

For an arc equal to ½ the periphery:

$$a = 2r \div \pi = 0.6366r$$

For an arc equal to ¼ of the periphery:

$$a = 2r\sqrt{2} \div \pi = 0.9003r$$

For an arc equal to ⅙ of the periphery:

$$a = 3r \div \pi = 0.9549r$$

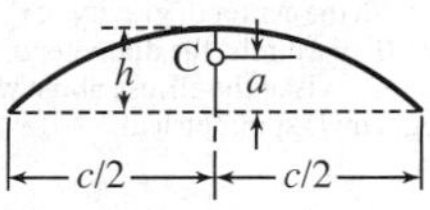


An approximate formula is very nearly exact for all arcs less than ¼ the periphery is:

$$a = \frac{2}{3}h$$

The error is only about 1 percent for a quarter circle, and it decreases for smaller arcs.

*Circle Segment:*

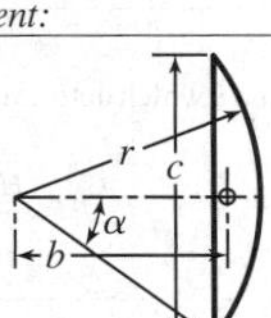


The distance of the center of gravity from the center of the circle is:

$$b = \frac{c^3}{12A} = \frac{2}{3} \times \frac{r^3 \sin \alpha^3}{A}$$

in which $A$ = area of segment.

*Circle Sector:*

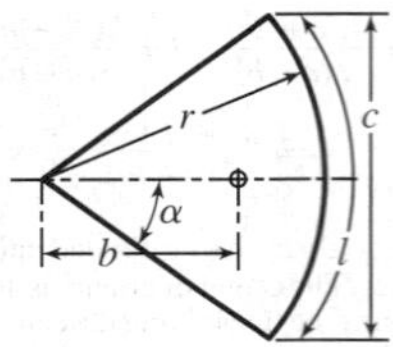


Distance $b$ from center of gravity to center of circle is:

$$b = \frac{2rc}{3l} = \frac{r^2 c}{3A} = 38.197\frac{r \sin\alpha}{\alpha}$$

in which $A$ = area of sector, and $\alpha$ is expressed in degrees.

For the area of a half-circle:

$$b = 4r \div 3\pi = 0.4244r$$

For the area of a quarter-circle:

$$b = 4\sqrt{2} \times r \div 3\pi = 0.6002r$$

For the area of a sixth of a circle:

$$b = 2r \div \pi = 0.6366r$$

*Part of Circle Ring:*

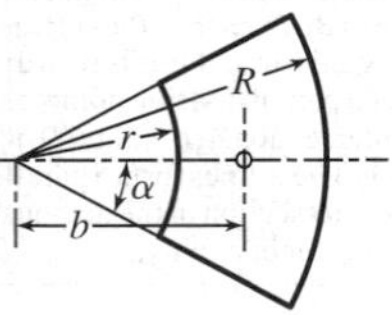


Distance $b$ from center of gravity to center of circle is:

$$b = 38.197\frac{(R^3 - r^3)\sin\alpha}{(R^2 - r^2)\alpha}$$

Angle $\alpha$ is expressed in degrees.

*Spandrel or Fillet:*

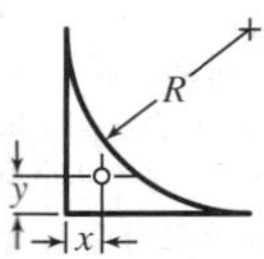


$$\text{Area} = 0.2146R^2 \qquad x = 0.2234R \qquad y = 0.2234R$$

*Segment of an Ellipse:*

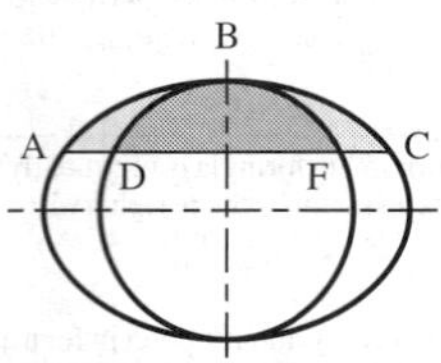


The center of gravity of an elliptic segment $ABC$, symmetrical about one of the axes, coincides with the center of gravity of the segment $DBF$ of a circle, the diameter of which is equal to that axis of the ellipse about which the elliptic segment is symmetrical.

*Spherical Surface of Segments and Zones of Spheres:*

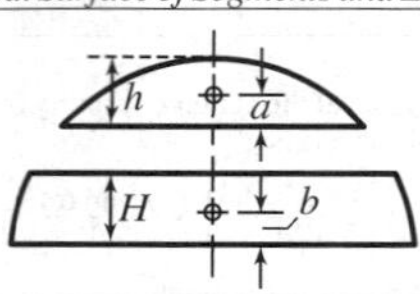


Distances $a$ and $b$ which determine the center of gravity are:

$$a = \frac{h}{2} \qquad b = \frac{H}{2}$$

*Area of a Parabola:*

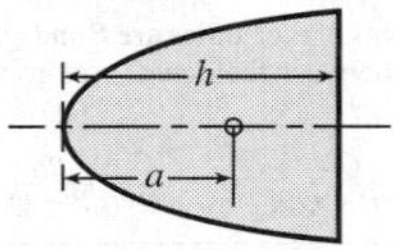


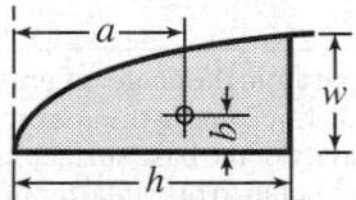


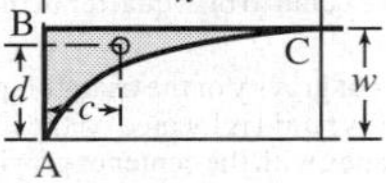


For the complete parabolic area, the center of gravity is on the center line or axis, and

$$a = \frac{3h}{5}$$

For one-half of the parabola:

$$a = \frac{3h}{5} \quad \text{and} \quad b = \frac{3w}{8}$$

For the complement area $ABC$:

$$c = 0.3h \quad \text{and} \quad d = 0.75w$$

*Cylinder:*

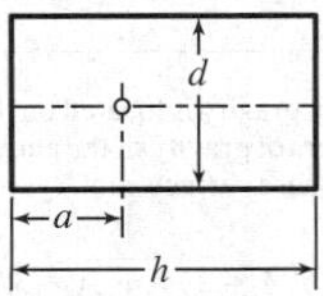


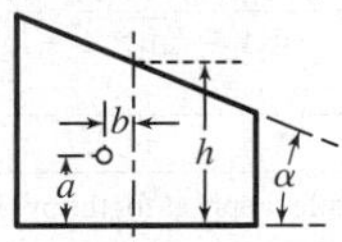


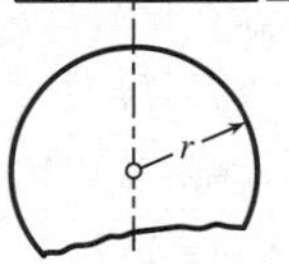


The center of gravity of a solid cylinder (or prism) with parallel end surfaces is located at the middle of the line that joins the centers of gravity of the end surfaces.

The center of gravity of a cylindrical surface or shell with the base or end surface in one end is found from:

$$a = \frac{2h^2}{4h + d}$$

The center of gravity of a cylinder cut off by an inclined plane is located by:

$$a = \frac{h}{2} + \frac{r^2 \tan^2\alpha}{8h} \qquad b = \frac{r^2 \tan\alpha}{4h}$$

where $\alpha$ is the angle between the obliquely cut off surface and the base surface.

*Portion of Cylinder:*

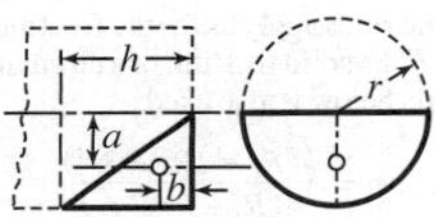


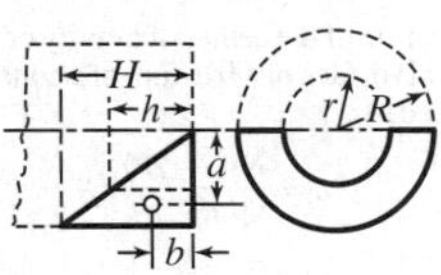


For a solid portion of a cylinder, as shown, the center of gravity is determined by:

$$a = \frac{3}{16} \times 3.1416r \quad b = \frac{3}{32} \times 3.1416h$$

For the cylindrical surface only:

$$a = \frac{1}{4} \times 3.1416r \quad b = \frac{1}{8} \times 3.1416h$$

If the cylinder is hollow, the center of gravity of the solid shell is found by:

$$a = \frac{3}{16} \times 3.1416\frac{R^4 - r^4}{R^3 - r^3}$$

$$b = \frac{3}{32} \times 3.1416\frac{H^4 - h^4}{H^3 - h^3}$$

*Center of Gravity of Two Bodies:*

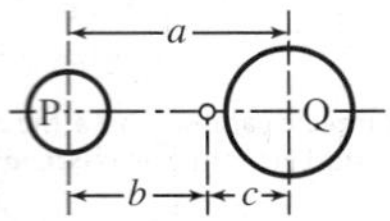


If the weights of the bodies are $P$ and $Q$, and the distance between their centers of gravity is $a$, then:

$$b = \frac{Qa}{P+Q} \qquad c = \frac{Pa}{P+Q}$$

*Pyramid:*

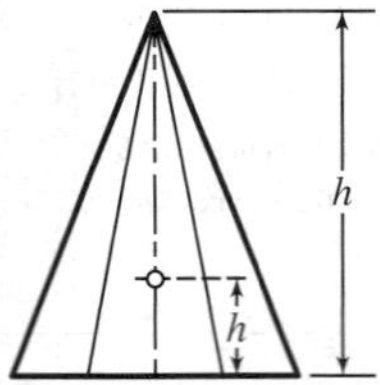


In a solid pyramid the center of gravity is located on the line joining the apex with the center of gravity of the base surface, at a distance from the base equal to one-quarter of the height; or $a = \frac{1}{4} h$.

The center of gravity of the triangular surfaces forming the pyramid is located on the line joining the apex with the center of gravity of the base surface at a distance from the base equal to one-third the height: $a = \frac{1}{3} h$.

*Frustum of Pyramid:*

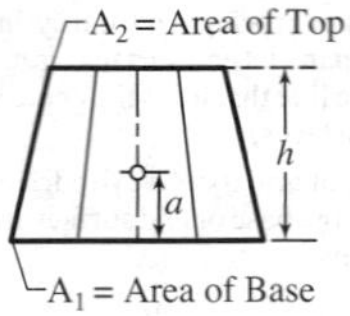


The center of gravity is located on the line that joins the centers of gravity of the end surfaces. If $A_1$ = area of base surface, and $A_2$ area of top surface,

$$a = \frac{h\left(A_1 + 2\sqrt{A_1 \times A_2} + 3A_2\right)}{4\left(A_1 + \sqrt{A_1 \times A_2} + A_2\right)}$$

*Cone:*

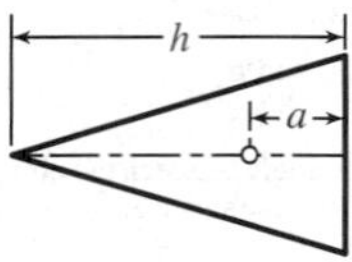


The same rules apply as for the pyramid.

For the solid cone:

$$a = \frac{1}{4}h$$

For the conical surface:

$$a = \frac{1}{3}h$$

*Frustum of Cone:*

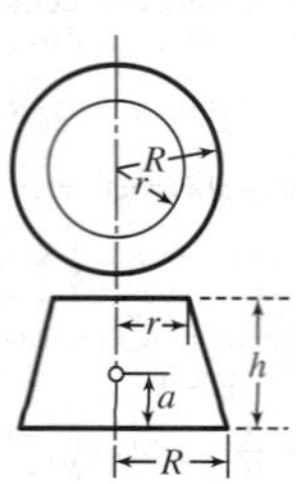


The same rules apply as for the frustum of a pyramid. For a solid frustum of a circular cone the formula below is also used:

$$a = \frac{h(R^2 + 2Rr + 3r^2)}{4(R^2 + Rr + r^2)}$$

The location of the center of gravity of the conical surface of a frustum of a cone is determined by:

$$a = \frac{h(R + 2r)}{3(R + r)}$$

*Wedge:*

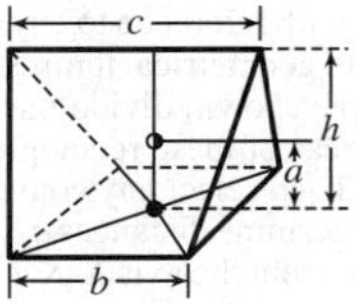


The center of gravity is on the line joining the center of gravity of the base with the middle point of the edge, and is located at:

$$a = \frac{h(b+c)}{2(2b+c)}$$

*Paraboloid:*

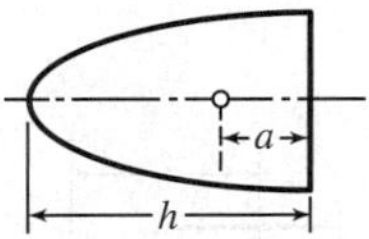


The center of gravity of a solid paraboloid of rotation is at:

$$a = \frac{1}{3}h$$

*Half of a Hollow Sphere:*

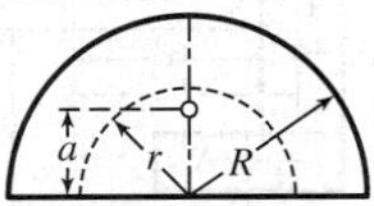


The center of gravity is located at:

$$a = \frac{3(R^4 - r^4)}{8(R^3 - r^3)}$$

*Spherical Segment:*

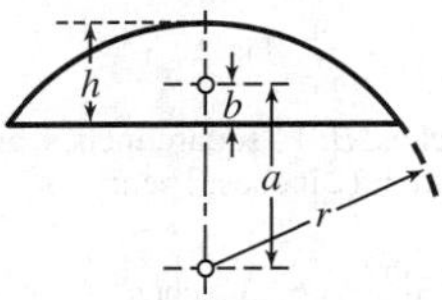


The center of gravity of a solid segment is determined by:

$$a = \frac{3(2r-h)^2}{4(3r-h)}$$

$$b = \frac{h(4r-h)}{4(3r-h)}$$

For a half-sphere, $a = b = \frac{3}{8}r$

*Spherical Sector:*

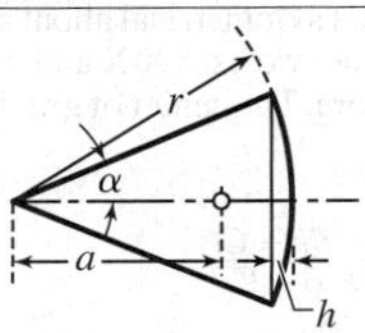


The center of gravity of a solid sector is at:

$$a = \frac{3}{8}(1+\cos\alpha)r = \frac{3}{8}(2r-h)$$

*Segment of Ellipsoid or Spheroid:*

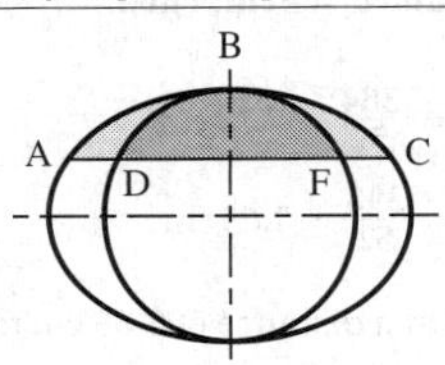


The center of gravity of a solid segment *ABC*, symmetrical about the axis of rotation, coincides with the center of gravity of the segment *DBF* of a sphere, the diameter of which is equal to the axis of rotation of the spheroid.

**Center of Gravity of Figures of Any Outline.**—If the figure is symmetrical about a center line, as in Fig. 1, the center of gravity will be located on that line. To find the exact location on that line, the simplest method is by taking moments with reference to any convenient axis at right angles to this center line. Divide the area into geometrical figures the centers of gravity of which can be easily found. In the example shown, divide the figure into three rectangles KLMN, EFGH and OPRS. Call the areas of these rectangles *A*, *B* and *C*, respectively, and find the center of gravity of each. Then select any convenient axis, such as X-X, at right angles to the center line Y-Y, and determine distances *a*, *b* and *c*. The distance *y* of the center of gravity of the complete figure from the axis X-X is then found from the equation:

$$y = \frac{Aa + Bb + Cc}{A + B + C}$$

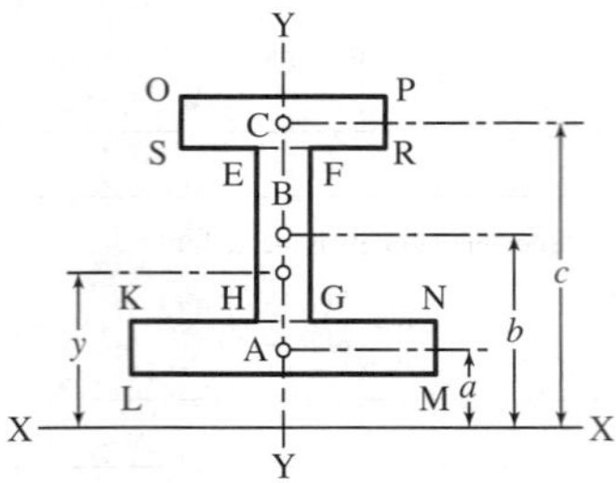


Fig. 1.

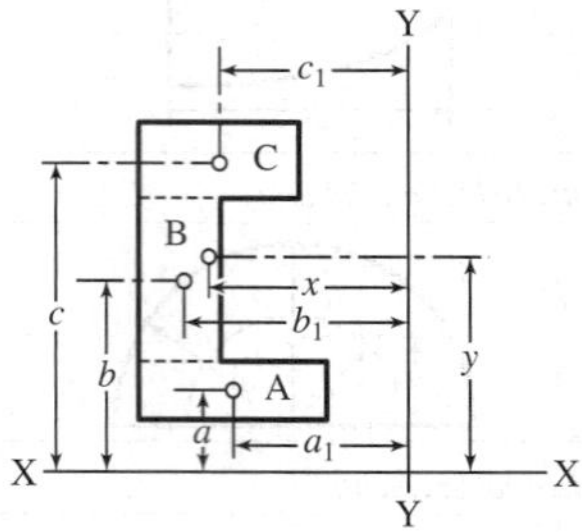


Fig. 2.

*Example 1:* Assume that the area *A* is 24 square inches, *B*, 14 square inches, and *C*, 16 square inches, and that $a = 3$ inches, $b = 7.5$ inches, and $c = 12$ inches. Then:

$$y = \frac{24 \times 3 + 14 \times 7.5 + 16 \times 12}{24 + 14 + 16} = \frac{369}{54} = 6.83 \text{ inches}$$

If the figure whose center of gravity is to be found is not symmetrical about any axis, as in Fig. 2, then moments must be taken with relation to the two axes X-X and Y-Y, centers of gravity of which can be easily found, the same as before. The center of gravity is determined by the equations:

$$x = \frac{Aa_1 + Bb_1 + Cc_1}{A + B + C} \qquad y = \frac{Aa + Bb + Cc}{A + B + C}$$

*Example 2:* **In Fig. 2, let $A = 14\text{ cm}^2$, $B = 18\text{ cm}^2$, and $C = 20\text{ cm}^2$. Let $a = 3$ cm, $b = 7$ cm, and $c = 11.5$ cm. Let $a_1 = 6.5$ cm, $b_1 = 8.5$ cm, and $c_1 = 7$ cm. Then:**

$$x = \frac{14 \times 6.5 + 18 \times 8.5 + 20 \times 7}{14 + 18 + 20} = \frac{384}{52} = 7.38 \text{ cm}$$

$$y = \frac{14 \times 3 + 18 \times 7 + 20 \times 11.5}{14 + 18 + 20} = \frac{398}{52} = 7.65 \text{ cm}$$

**In other words, the center of gravity is located at a distance of 7.65 cm from the axis X-X and 7.38 cm from the axis Y-Y.**

## Radius of Gyration

The radius of gyration with reference to an axis is that distance from the axis at which the entire mass of a body may be considered as concentrated, the moment of inertia, meanwhile, remaining unchanged. If $W$ is the weight of a body; $J_M$ its moment of inertia with respect to some axis; and $k_o$ the radius of gyration with respect to the same axis, then:

$$k_o = \sqrt{\frac{J_M g}{W}} \quad \text{and} \quad J_M = \frac{Wk_o^2}{g}$$

**When using metric SI units, the formulas are:**

$$\boldsymbol{k_o = \sqrt{\frac{J_M}{M}} \quad \text{and} \quad J_M = Mk_o^2}$$

**where $k_o$ = the radius of gyration in meters, $J_M$ = moment of inertia in kilogram-meter$^2$ (kg·m$^2$), and $M$ = mass in kilograms.**

To find the radius of gyration of an area, such as for the cross section of a beam, divide the moment of inertia of the area by the area and extract the square root.

When the axis, the reference to which the radius of gyration is taken, passes through the center of gravity, the radius of gyration is the least possible and is called the *principal* radius of gyration. If $k$ is the radius of gyration with respect to such an axis passing through the center of gravity of a body, then the radius of gyration $k_o$ with respect to a parallel axis at a distance $d$ from the gravity axis is given by: $k_o = \sqrt{k^2 + d^2}$

Tables of radii of gyration for various bodies and axes follows.

### Formulas for Radius of Gyration

*Bar of Small Diameter:*

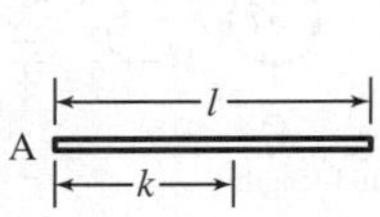


$k = 0.5773l$

$k^2 = \frac{1}{3}l^2$

Axis at end

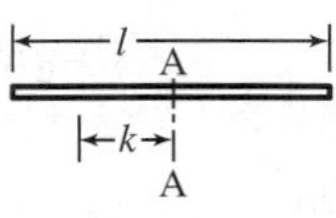


$k = 0.2886l$

$k^2 = \frac{1}{12}l^2$

Axis at center

*Bar of Small Diameter Bent to Circular Shape:*

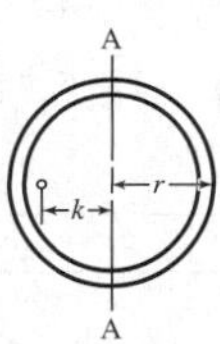


$k = 0.7071r$

$k^2 = \frac{1}{2}r^2$

Axis, a diameter of the ring

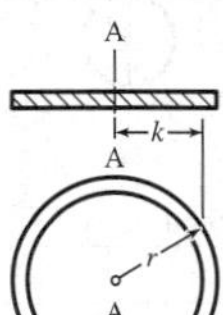


$k = r$

$k^2 = r^2$

Axis through center of ring

*Parallelogram (Thin Flat Plate):*

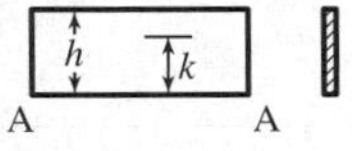


$k = 0.5773h$

$k^2 = \frac{1}{3}h^2$

Axis at base

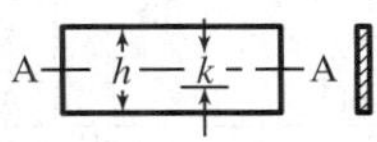


$k = 0.2886h$

$k^2 = \frac{1}{12}h^2$

Axis at mid-height

*Thin Circular Disk:*

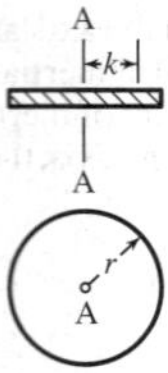


$$k = 0.7071r$$

$$k^2 = \frac{1}{2}r^2$$

Axis through center

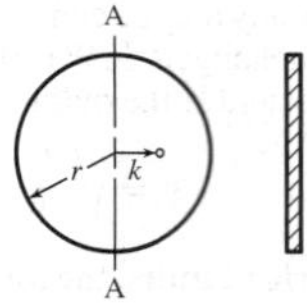


$$k = \frac{1}{2}r$$

$$k^2 = \frac{1}{4}r^2$$

Axis its diameter

*Thin, Flat, Circular Ring:*

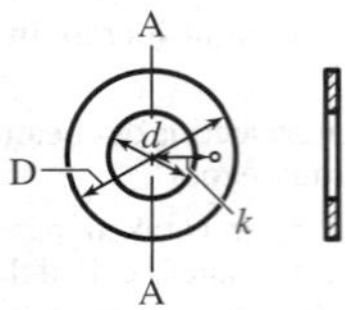


$$k = \frac{1}{4}\sqrt{D^2 + d^2}$$

$$k^2 = \frac{D^2 + d^2}{16}$$

Axis its diameter

*Cylinder:*

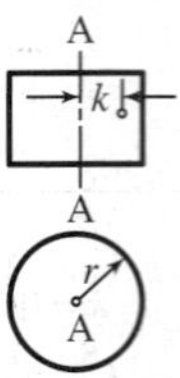


$$k = \frac{r}{\sqrt{2}}$$

$$k^2 = \frac{1}{2}r^2$$

Axis through center

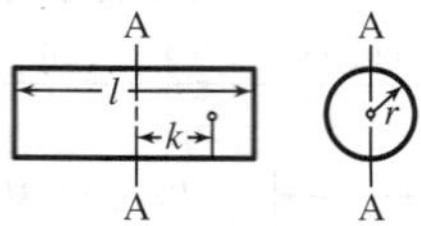


$$k = \frac{\sqrt{l^2 + 3r^2}}{\sqrt{12}}$$

$$k^2 = \frac{l^2}{12} + \frac{r^2}{4}$$

Axis, diameter at mid-length

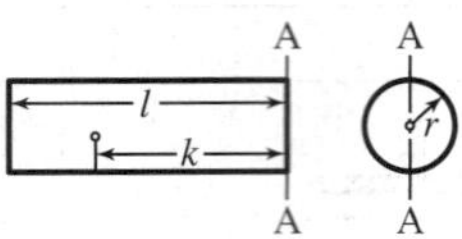


$$k = \frac{\sqrt{4l^2 + 3r^2}}{\sqrt{12}}$$

$$k^2 = \frac{l^2}{3} + \frac{r^2}{4}$$

Axis, diameter at end

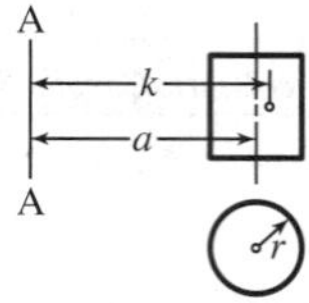


$$k = \sqrt{a^2 + \frac{1}{2}r^2}$$

$$k^2 = a^2 + \frac{1}{2}r^2$$

Axis at a distance

*Parallelepiped:*

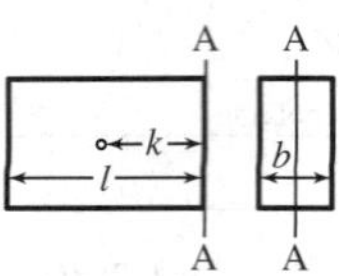


$$k = \frac{\sqrt{4l^2 + b^2}}{\sqrt{12}}$$

$$k^2 = \frac{4l^2 + b^2}{12}$$

Axis at one end, central

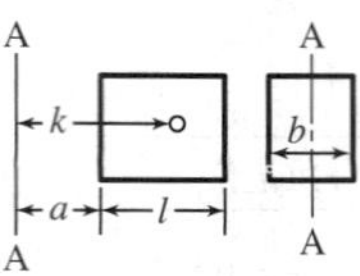


$$k = \sqrt{\frac{4l^2 + b^2}{12} + a^2 + al}$$

Axis at distance from end

*Rectangular Prism:*

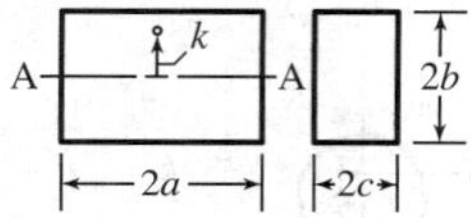


$$k = 0.577\sqrt{b^2 + c^2}$$

$$k^2 = \frac{1}{3}(b^2 + c^2)$$

Axis through center

*Thin Hollow Cylinder:*

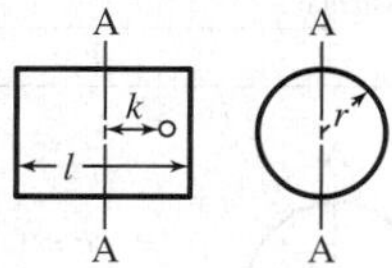


$$k = 0.289\sqrt{l^2 + 6r^2}$$

$$k^2 = \frac{l^2}{12} + \frac{r^2}{2}$$

Axis, diameter at mid-length

*Hollow Cylinder:*

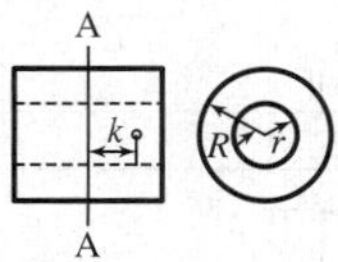


$$k = \frac{\sqrt{l^2 + 3(R^2 + r^2)}}{\sqrt{12}}$$

$$k^2 = \frac{l^2}{12} + \frac{R^2 + r^2}{4}$$

Axis, diameter at mid-length

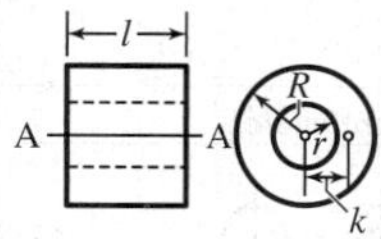


$$k = \frac{\sqrt{R^2 + r^2}}{\sqrt{2}}$$

$$k^2 = \frac{1}{2}(R^2 + r^2)$$

Longitudinal axis

*Cone:*

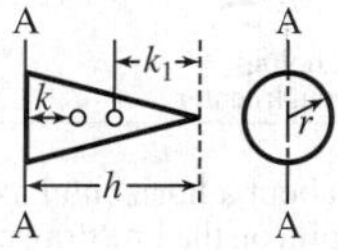


Axis at base $\quad k = \sqrt{\dfrac{2h^2 + 3r^2}{20}}$

Axis at apex $\quad k_1 = \sqrt{\dfrac{12h^2 + 3r^2}{20}}$

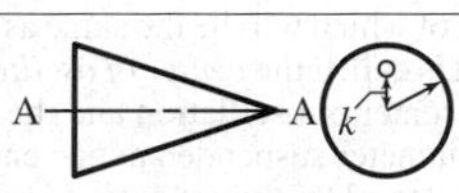


$$k = 0.5477r$$

$$k^2 = 0.3r^2$$

Axis through its center line

*Frustum of Cone:*

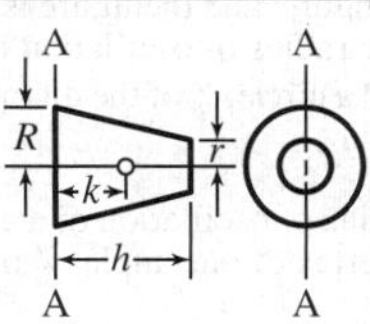


$$k = \sqrt{\frac{h^2}{10}\left(\frac{R^2 + 3Rr + 6r^2}{R^2 + Rr + r^2}\right) + \frac{3}{20}\left(\frac{R^5 - r^5}{R^3 - r^3}\right)}$$

Axis at large end

*Sphere:*

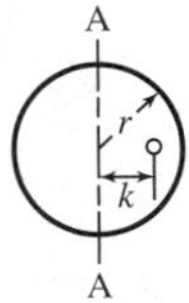


$$k = 0.6325r$$

$$k^2 = \frac{2}{5}r^2$$

Axis its diameter

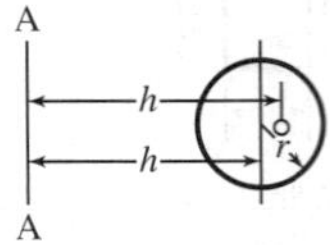


$$k = \sqrt{a^2 + \frac{2}{5}r^2}$$

$$k^2 = a^2 + \frac{2}{5}r^2$$

Axis at a distance

*Hollow Sphere and Thin Spherical Shell:*

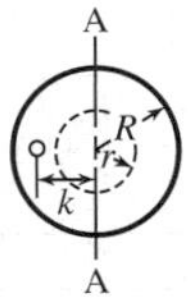


$$k = 0.6325\sqrt{\frac{R^5 - r^5}{R^3 - r^3}}$$

$$k^2 = \frac{2(R^5 - r^5)}{5(R^3 - r^3)}$$

Hollow Sphere
Axis its diameter

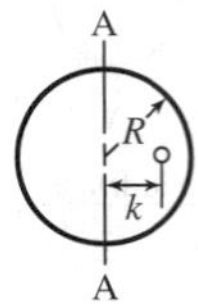


$$k = 0.8165r$$

$$k^2 = \frac{2}{3}r^2$$

Thin Spherical Shell

*Ellipsoid and Paraboloid:*

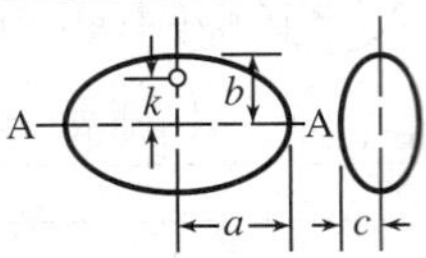


$$k = \frac{\sqrt{b^2 + c^2}}{\sqrt{5}}$$

$$k^2 = \frac{1}{5}(b^2 + c^2)$$

Ellipsoid
Axis through center

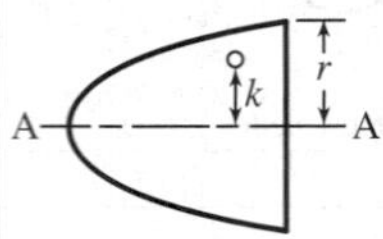


$$k = 0.5773r$$

$$k^2 = \frac{1}{3}r^2$$

Paraboloid
Axis through center

**Center and Radius of Oscillation.—**If a body oscillates about a horizontal axis which does not pass through its center of gravity, there will be a point on the line drawn from the center of gravity, perpendicular to the axis, the motion of which will be the same as if the whole mass were concentrated at that point. This point is called the *center of oscillation*. The *radius of oscillation* is the distance between the center of oscillation and the point of suspension. In a straight line, or in a bar of small diameter suspended at one end and oscillating about it, the center of oscillation is at ⅔ the length of the rod from the end by which it is suspended.

When the vibrations are perpendicular to the plane of the figure, and the figure is suspended by the vertex of an angle or its uppermost point, the radius of oscillation of an isosceles triangle is equal to ¾ of the height of the triangle; of a circle, ⅝ of the diameter; of a parabola, 5⁄7 of the height.

If the vibrations are in the plane of the figure, then the radius of oscillation of a circle equals ¾ of the diameter; of a rectangle suspended at the vertex of one angle, ⅔ of the diagonal.

**Center of Percussion.—**For a body that moves without rotation, the resultant of all the forces acting on the body passes through the center of gravity. On the other hand, for a body that rotates about some *fixed axis*, the resultant of all the forces acting on it does not pass through the center of gravity of the body but through a point called the *center of* *percussion*. The center of percussion is useful in determining the position of the resultant in mechanics problems involving angular acceleration of bodies about a fixed axis.

*Finding the Center of Percussion when Radius of Gyration and Location of Center of Gravity are Known:* The center of percussion lies on a line drawn through the center of rotation and the center of gravity. The distance from the axis of rotation to the center of percussion may be calculated from the following formula

$$q = k_o^2 \div r$$

in which $q$ = distance from the axis of rotation to the center of percussion; $k_o$ = radius of gyration of the body with respect to the axis of rotation; and $r$ = distance from the axis of rotation to the center of gravity of the body.

## Moment of Inertia

An important property of areas and solid bodies is moment of inertia. Standard formulas are derived by multiplying elementary particles of area or mass by the squares of their distances from reference axes. Moments of inertia, therefore, depend on the location of reference axes. Values are minimum when these axes pass through the centers of gravity.

Three kinds of moments of inertia occur in engineering formulas:

1) *Moments of inertia of plane area*, $I$, in which the axis is in the plane of the area are found in formulas for calculating deflections and stresses in beams. When dimensions are given in inches, the units of $I$ are $\text{inches}^4$. A table of formulas for calculating the $I$ of common areas can be found beginning on page 235.

2) *Polar moments of inertia of plane areas*, $J$, in which the axis is at right angles to the plane of the area, occur in formulas for the torsional strength of shafting. When dimensions are given in inches, the units of $J$ are $\text{inches}^4$. If moments of inertia, $I$, are known for a plane area with respect to both $x$ and $y$ axes, then the polar moment for the $z$-axis may be calculated using the equation, $J_z = I_x + I_y$

A table of formulas for calculating $J$ for common areas can be found on page 246 in this section.

**When metric SI units are used, the formulas referred to in (1) and (2) above are valid if the dimensions are given consistently in meters or millimeters. If meters are used, the units of $I$ and $J$ are in $\text{meters}^4$; if millimeters are used, these units are in $\text{millimeters}^4$.**

3) *Polar moments of inertia of masses*, $J_M$*, appear in dynamics equations involving rotational motion. $J_M$ bears the same relationship to angular acceleration as mass does to linear acceleration. If units are in the foot-pound-second system, the units of $J_M$ are ft-lbs-$\text{sec}^2$ or slug-$\text{ft}^2$ (1 slug = 1 lb-$\text{sec}^2$/ft). If units are in the inch-pound-second system, the units of $J_M$ are inch-lbs-$\text{sec}^2$.

**If metric SI values are used, the units of $J_M$ are kilogram-meter squared (kg-$\text{m}^2$).** Formulas for calculating $J_M$ for various bodies are given beginning on page 247. If the polar moment of inertia $J$ is known for the area of a body of constant cross section, $J_M$ may be calculated using the equation,

$$J_M = \frac{\rho L}{g} J$$

where ρ is the density of the material, $L$ the length of the part, and $g$ the gravitational constant. If dimensions are in the foot-pound-second system, ρ is in lbs/$\text{ft}^3$, $L$ is in ft, $g$ is

* In some books the symbol $I$ denotes the polar moment of inertia of masses; $J_M$ is used in this *Handbook* to avoid confusion with moments of inertia of plane areas.

32.16 ft/sec$^2$, and $J$ is in ft$^4$. If dimensions are in the inch-pound-second system, ρ is in lbs/in$^3$, $L$ is in inches, $g$ is 386 in/sec$^2$, and $J$ is in inches$^4$.

**Using metric SI units, the above formula becomes $J_M$ = ρ$LJ$, where ρ = density in kilograms/meter$^3$, $L$ = length in meters, and $J$ = polar moment of inertia in meters$^4$. The units of $J_M$ are kg·m$^2$.**

**Moment of Inertia of Built-Up Sections.—**The usual method of calculating the moment of inertia of a built-up section involves the calculations of the moment of inertia for each element of the section about its own neutral axis and the transferring of this moment of inertia to the previously found neutral axis of the whole built-up section. A much simpler method, called the tabular method, can be used in the case of any section that can be divided into rectangular elements bounded by lines parallel and perpendicular to the neutral axis. It is based upon the formula: $I = b(h_1{}^3 - h^3)/3$ in which $I$ = the moment of inertia about axis $D$-$E$, Fig. 1, and $b$, $h$ and $h_1$ are dimensions as given in the same illustration.

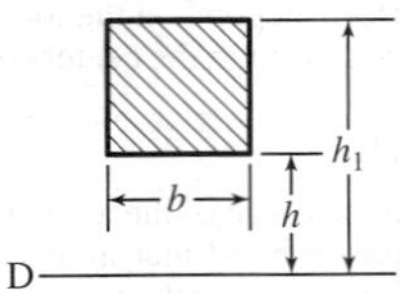


Fig. 1.

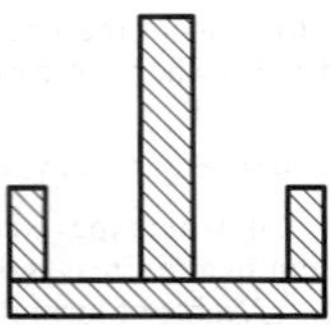

Fig. 2.

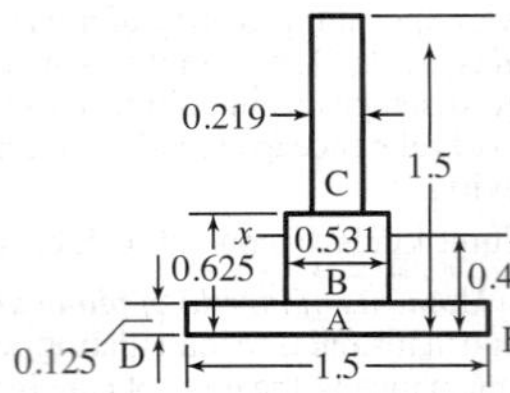


Fig. 3.

The method may be illustrated by applying it to the section shown in Fig. 2, and for simplicity of calculation shown "massed" in Fig. 3. The calculation may then be tabulated as shown in the accompanying table. The distance from the axis $D$-$E$ to the neutral axis $x$-$x$ (which will be designated as $d$) is found by dividing the sum of the geometrical moments by the area. The moment of inertia about the neutral axis is then found in the usual way by subtracting the area multiplied by $d^2$ from the moment of inertia about the axis $D$-$E$.

**Tabulated Calculation of Moment of Inertia**

| Section | Breadth $b$ | Height $h_1$ | Area $b(h_1 - h)$ | $h_1{}^2$ | Moment $\frac{b(h_1^2 - h^2)}{2}$ | $h_1{}^3$ | $I$ About Axis $D$-$E$ $\frac{b(h_1^3 - h^3)}{3}$ |
|---|---|---|---|---|---|---|---|
| $A$ | 1.500 | 0.125 | 0.187 | 0.016 | 0.012 | 0.002 | 0.001 |
| $B$ | 0.531 | 0.625 | 0.266 | 0.391 | 0.100 | 0.244 | 0.043 |
| $C$ | 0.219 | 1.500 | 0.191 | 2.250 | 0.203 | 3.375 | 0.228 |
| | | | $\Sigma A = 0.644$ | | $\Sigma M = 0.315$ | | $\Sigma I_{DE} = 0.272$ |

The distance $d$ from $D$-$E$, the axis at the base of the configuration, to the neutral axis $x$-$x$ is:

$$d = \frac{M}{A} = \frac{0.315}{0.644} = 0.49$$

The moment of inertia of the entire section with reference to the neutral axis $x$-$x$ is:

$$\begin{aligned} I_N &= I_{DE} - Ad^2 \\ &= 0.272 - 0.644 \times 0.49^2 \\ &= 0.117 \end{aligned}$$

**Formulas for Moments of Inertia, Section Moduli, etc.—**On the following pages are given formulas for moments of inertia and other properties of forty-two different cross sections. The formulas give the area of the section $A$, and the distance $y$ from the neutral axis to the extreme fiber, for each example. Where the formulas for the section modulus and radius of gyration are very lengthy, the formula for the section modulus, for example, has been simply given as $I \div y$. The radius of gyration is sometimes given as $\sqrt{I \div A}$ to save space.

**Moments of Inertia, Section Moduli, and Radii of Gyration**

| Section<br>$A$ = area<br>$y$ = distance from axis to extreme fiber | Moment of Inertia<br>$I$ | Section Modulus<br>$Z = \frac{I}{y}$ | Radius of Gyration<br>$k = \sqrt{\frac{I}{A}}$ |
|---|---|---|---|
| Square and Rectangular Sections | | | |
| 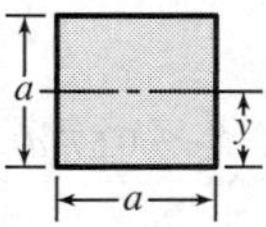<br>$A = a^2 \quad y = a/2$ | $\frac{a^4}{12}$ | $\frac{a^3}{6}$ | $\frac{a}{\sqrt{12}} = 0.289a$ |
| 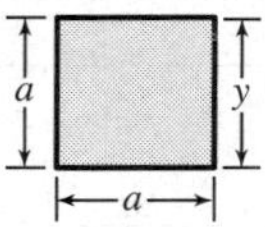<br>$A = a^2 \quad y = a$ | $\frac{a^4}{3}$ | $\frac{a^3}{3}$ | $\frac{a}{\sqrt{3}} = 0.577a$ |
| 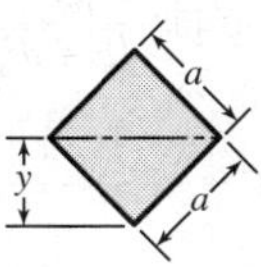<br>$A = a^2$<br>$y = \frac{a}{\sqrt{2}} = 0.707a$ | $\frac{a^4}{12}$ | $\frac{a^3}{6\sqrt{2}} = 0.118a^3$ | $\frac{a}{\sqrt{12}} = 0.289a$ |
| 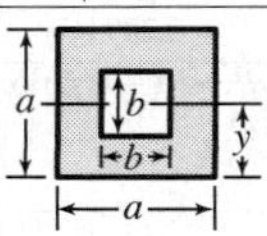<br>$A = a^2 - b^2 \quad y = a/2$ | $\frac{a^4 - b^4}{12}$ | $\frac{a^4 - b^4}{6a}$ | $\sqrt{\frac{a^2 + b^2}{12}}$<br>$= 0.289\sqrt{a^2 + b^2}$ |
| 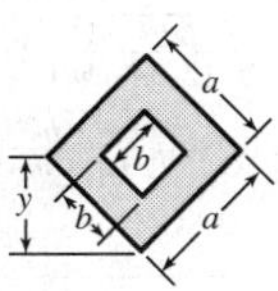<br>$A = a^2 - b^2 \qquad y = \frac{a}{\sqrt{2}}$ | $\frac{a^4 - b^4}{12}$ | $\frac{\sqrt{2}(a^4 - b^4)}{12a}$<br>$= 0.118\frac{a^4 - b^4}{a}$ | $\sqrt{\frac{a^2 + b^2}{12}}$<br>$= 0.289\sqrt{a^2 + b^2}$ |

**Moments of Inertia, Section Moduli, and Radii of Gyration** *(Continued)*

| Section<br>$A$ = area<br>$y$ = distance from axis to extreme fiber | Moment of Inertia<br>$I$ | Section Modulus<br>$Z = \frac{I}{y}$ | Radius of Gyration<br>$k = \sqrt{\frac{I}{A}}$ |
|---|---|---|---|
| Square and Rectangular Sections | | | |
| 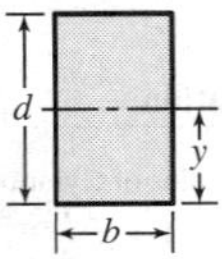<br>$A = bd \quad y = d/2$ | $\frac{bd^3}{12}$ | $\frac{bd^2}{6}$ | $\frac{d}{\sqrt{12}} = 0.289d$ |
| 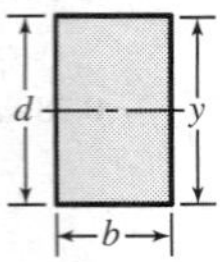<br>$A = bd \quad y = d$ | $\frac{bd^3}{3}$ | $\frac{bd^2}{3}$ | $\frac{d}{\sqrt{3}} = 0.577d$ |
| 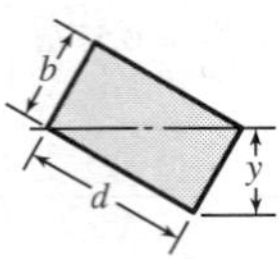<br>$A = bd$<br>$y = \frac{bd}{\sqrt{b^2 + d^2}}$ | $\frac{b^3 d^3}{6(b^2 + d^2)}$ | $\frac{b^2 d^2}{6\sqrt{b^2 + d^2}}$ | $\frac{bd}{\sqrt{6(b^2 + d^2)}}$<br>$= 0.408\frac{bd}{\sqrt{b^2 + d^2}}$ |
| 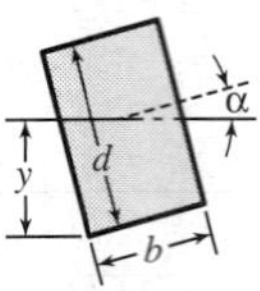<br>$A = bd$<br>$y = \frac{1}{2}(d\cos\alpha + b\sin\alpha)$ | $\frac{bd}{12}(d^2\cos^2\alpha$<br>$+ b^2\sin^2\alpha)$ | $\frac{bd}{6} \times$<br>$\left(\frac{d^2\cos^2\alpha + b^2\sin^2\alpha}{d\cos\alpha + b\sin\alpha}\right)$ | $\sqrt{\frac{d^2\cos^2\alpha + b^2\sin^2\alpha}{12}}$<br>$= 0.289 \times$<br>$\sqrt{d^2\cos^2\alpha + b^2\sin^2\alpha}$ |
| 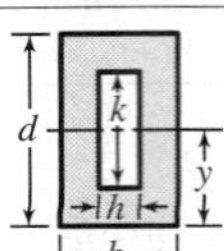<br>$A = bd - hk$<br>$y = d/2$ | $\frac{bd^3 - hk^3}{12}$ | $\frac{bd^3 - hk^3}{6d}$ | $\sqrt{\frac{bd^3 - hk^3}{12(bd - hk)}}$<br>$= 0.289\sqrt{\frac{bd^3 - hk^3}{bd - hk}}$ |

## Moments of Inertia, Section Moduli, and Radii of Gyration *(Continued)*

| Section | Area of Section, $A$ | Distance from Neutral Axis to Extreme Fiber, $y$ | Moment of Inertia, $I$ | Section Modulus, $Z = \frac{I}{y}$ | Radius of Gyration, $k = \sqrt{\frac{I}{A}}$ |
|---|---|---|---|---|---|
| **Triangular Sections** | | | | | |
| 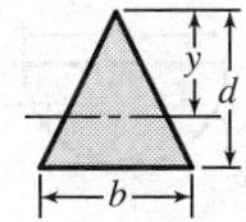 | $\frac{1}{2}bd$ | $\frac{2}{3}d$ | $\frac{bd^3}{36}$ | $\frac{bd^2}{24}$ | $\frac{d}{\sqrt{18}} = 0.236d$ |
| 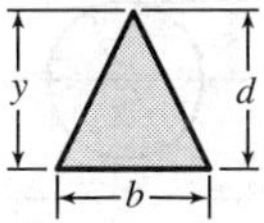 | $\frac{1}{2}bd$ | $d$ | $\frac{bd^3}{12}$ | $\frac{bd^2}{12}$ | $\frac{d}{\sqrt{6}} = 0.408d$ |
| **Polygon Sections** | | | | | |
| 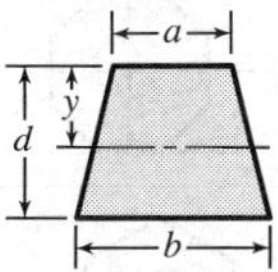 | $\frac{d(a+b)}{2}$ | $\frac{d(a+2b)}{3(a+b)}$ | $\frac{d^3(a^2+4ab+b^2)}{36(a+b)}$ | $\frac{d^2(a^2+4ab+b^2)}{12(a+2b)}$ | $\sqrt{\frac{d^2(a^2+4ab+b^2)}{18(a+b)^2}}$ |
| 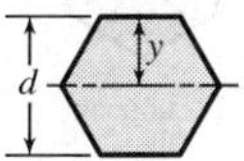 | $\frac{3d^2\tan 30°}{2} = 0.866d^2$ | $\frac{d}{2}$ | $\frac{A}{12}\left[\frac{d^2(1+2\cos^2 30°)}{4\cos^2 30°}\right] = 0.06d^4$ | $\frac{A}{6}\left[\frac{d(1+2\cos^2 30°)}{4\cos^2 30°}\right] = 0.12d^3$ | $\sqrt{\frac{d^2(1+2\cos^2 30°)}{48\cos^2 30°}} = 0.264d$ |

**Moments of Inertia, Section Moduli, and Radii of Gyration** *(Continued)*

| Section | Area of Section, $A$ | Distance from Neutral Axis to Extreme Fiber, $y$ | Moment of Inertia, $I$ | Section Modulus, $Z = \frac{I}{y}$ | Radius of Gyration, $k = \sqrt{\frac{I}{A}}$ |
|---|---|---|---|---|---|
| 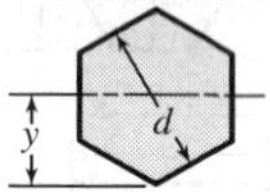 | $\frac{3d^2 \tan 30°}{2}$ $= 0.866d^2$ | $\frac{d}{2 \cos 30°} = 0.577d$ | $\frac{A}{12}\left[\frac{d^2(1 + 2\cos^2 30°)}{4\cos^2 30°}\right]$ $= 0.06d^4$ | $\frac{A}{6.9}\left[\frac{d(1 + 2\cos^2 30°)}{4\cos^2 30°}\right]$ $= 0.104d^3$ | $\sqrt{\frac{d^2(1 + 2\cos^2 30°)}{48\cos^2 30°}}$ $= 0.264d$ |
| 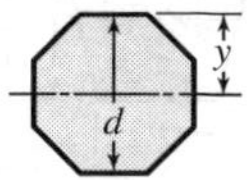 | $2d^2 \tan 22\frac{1}{2}° = 0.828d^2$ | $\frac{d}{2}$ | $\frac{A}{12}\left[\frac{d^2(1 + 2\cos^2 22\frac{1}{2}°)}{4\cos^2 22\frac{1}{2}°}\right]$ $= 0.055d^4$ | $\frac{A}{6}\left[\frac{d(1 + 2\cos^2 22\frac{1}{2}°)}{4\cos^2 22\frac{1}{2}°}\right]$ $= 0.109d^3$ | $\sqrt{\frac{d^2(1 + 2\cos^2 22\frac{1}{2}°)}{48\cos^2 22\frac{1}{2}°}}$ $= 0.257d$ |
| Circular, Elliptical, and Circular Arc Sections | | | | | |
| 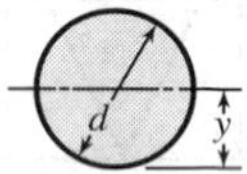 | $\frac{\pi d^2}{4} = 0.7854d^2$ | $\frac{d}{2}$ | $\frac{\pi d^4}{64} = 0.049d^4$ | $\frac{\pi d^3}{32} = 0.098d^3$ | $\frac{d}{4}$ |
| 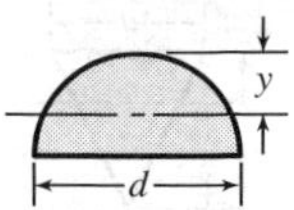 | $\frac{\pi d^2}{8} = 0.393d^2$ | $\frac{(3\pi - 4)d}{6\pi}$ $= 0.288d$ | $\frac{(9\pi^2 - 64)d^4}{1152\pi}$ $= 0.007d^4$ | $\frac{(9\pi^2 - 64)d^3}{192(3\pi - 4)}$ $= 0.024d^3$ | $\frac{\sqrt{(9\pi^2 - 64)d^2}}{12\pi}$ $= 0.132d$ |
| 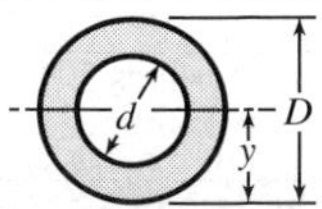 | $\frac{\pi(D^2 - d^2)}{4}$ $= 0.7854(D^2 - d^2)$ | $\frac{D}{2}$ | $\frac{\pi(D^4 - d^4)}{64}$ $= 0.049(D^4 - d^4)$ | $\frac{\pi(D^4 - d^4)}{32D}$ $= 0.098\frac{D^4 - d^4}{D}$ | $\frac{\sqrt{D^2 + d^2}}{4}$ |

**Moments of Inertia, Section Moduli, and Radii of Gyration** *(Continued)*

| Section | Area of Section, $A$ | Distance from Neutral Axis to Extreme Fiber, $y$ | Moment of Inertia, $I$ | Section Modulus, $Z = \frac{I}{y}$ | Radius of Gyration, $k = \sqrt{\frac{I}{A}}$ |
|---|---|---|---|---|---|
| 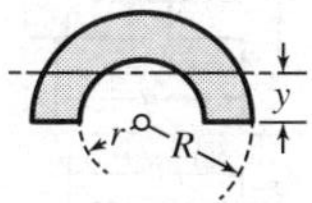  | $\frac{\pi(R^2 - r^2)}{2}$ $= 1.5708(R^2 - r^2)$ | $\frac{4(R^3 - r^3)}{3\pi(R^2 - r^2)}$ $= 0.424\frac{R^3 - r^3}{R^2 - r^2}$ | $0.1098(R^4 - r^4)$ $- \frac{0.283R^2r^2(R - r)}{R + r}$ | $\frac{I}{y}$ | $\sqrt{\frac{I}{A}}$ |
| 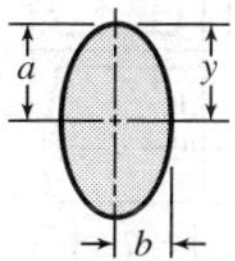  | $\pi ab = 3.1416ab$ | $a$ | $\frac{\pi a^3 b}{4} = 0.7854a^3b$ | $\frac{\pi a^2 b}{4} = 0.7854a^2b$ | $\frac{a}{2}$ |
| 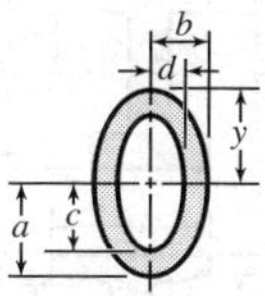  | $\pi(ab - cd)$ $= 3.1416(ab - cd)$ | $a$ | $\frac{\pi}{4}(a^3b - c^3d)$ $= 0.7854(a^3b - c^3d)$ | $\frac{\pi(a^3b - c^3d)}{4a}$ $= 0.7854\frac{a^3b - c^3d}{a}$ | $\frac{1}{2}\sqrt{\frac{a^3b - c^3d}{ab - cd}}$ |
| I-Sections | | | | | |
| 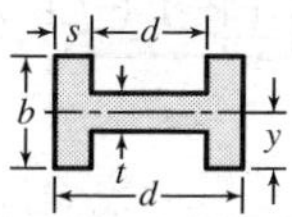  | $bd - h(b - t)$ | $\frac{b}{2}$ | $\frac{2sb^3 + ht^3}{12}$ | $\frac{2sb^3 + ht^3}{6b}$ | $\sqrt{\frac{2sb^3 + ht^3}{12[bd - h(b - t)]}}$ |

**Moments of Inertia, Section Moduli, and Radii of Gyration** *(Continued)*

| Section | Area of Section, $A$ | Distance from Neutral Axis to Extreme Fiber, $y$ | Moment of Inertia, $I$ | Section Modulus, $Z = \frac{I}{y}$ | Radius of Gyration, $k = \sqrt{\frac{I}{A}}$ |
|---|---|---|---|---|---|
| 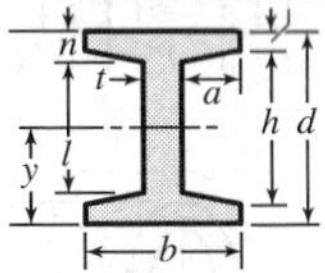 | $dt + 2a(s + n)$ | $\frac{d}{2}$ | $\frac{1}{12}\left[bd^3 - \frac{1}{4g}(h^4 - l^4)\right]$ in which $g$ = slope of flange = $(h - l)/(b - t)$ = $^{100}/_{6}$ for standard I-beams | $\frac{1}{6d}\left[bd^3 - \frac{1}{4g}(h^4 - l^4)\right]$ | $\sqrt{\frac{\frac{1}{12}\left[bd^3 - \frac{1}{4g}(h^4 - l^4)\right]}{dt + 2a(s + n)}}$ |
| 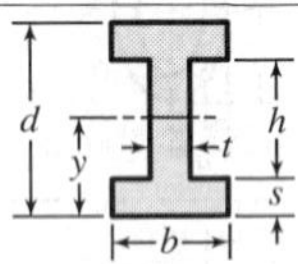 | $bd - h(b - t)$ | $\frac{d}{2}$ | $\frac{bd^3 - h^3(b - t)}{12}$ | $\frac{bd^3 - h^3(b - t)}{6d}$ | $\sqrt{\frac{bd^3 - h^3(b - t)}{12[bd - h(b - t)]}}$ |
| 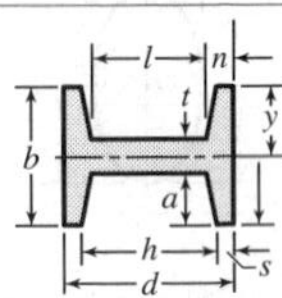 | $dt + 2a(s + n)$ | $\frac{b}{2}$ | $\frac{1}{12}[b^3(d - h) + lt^3 + \frac{g}{4}(b^4 - t^4)]$ in which $g$ = slope of flange = $(h - l)/(b - t)$ = $\frac{1}{6}$ for standard I-beams | $\frac{1}{6b}[b^3(d - h) + lt^3 + \frac{g}{4}(b^4 - t^4)]$ | $\sqrt{\frac{I}{A}}$ |
| 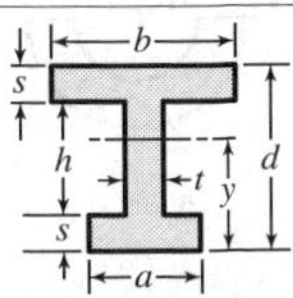 | $bs + ht + as$ | $d - [td^2 + s^2(b - t) + s(a - t)(2d - s)] \div 2A$ | $\frac{1}{3}[b(d - y)^3 + ay^3 - (b - t)(d - y - s)^3 - (a - t)(y - s)^3]$ | $\frac{I}{y}$ | $\sqrt{\frac{I}{A}}$ |

**Moments of Inertia, Section Moduli, and Radii of Gyration** *(Continued)*

| Section | Area of Section, $A$ | Distance from Neutral Axis to Extreme Fiber, $y$ | Moment of Inertia, $I$ | Section Modulus, $Z = \frac{I}{y}$ | Radius of Gyration, $k = \sqrt{\frac{I}{A}}$ |
|---|---|---|---|---|---|
| C-Sections | | | | | |
| 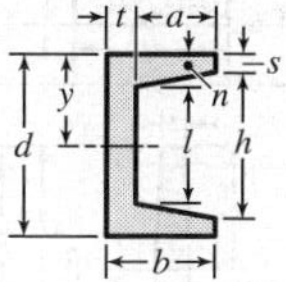 | $dt + a(s+n)$ | $\frac{d}{2}$ | $\frac{1}{12}\left[bd^3 - \frac{1}{8g}(h^4 - l^4)\right]$ <br> $g$ = slope of flange <br> $= \frac{h-l}{2(b-t)} = \frac{1}{6}$ <br> for standard channels | $\frac{1}{6d}\left[bd^3 - \frac{1}{8g}(h^4 - l^4)\right]$ | $\sqrt{\frac{\frac{1}{12}\left[bd^3 - \frac{1}{8g}(h^4 - l^4)\right]}{dt + a(s+n)}}$ |
| 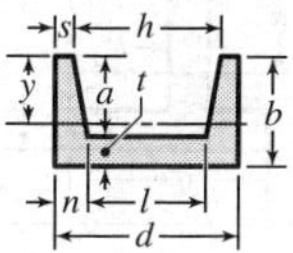 | $dt + a(s+n)$ | $b - [b^2 s + \frac{ht^2}{2}$ <br> $+ \frac{g}{3}(b-t)^2$ <br> $\times (b+2t)] \div A$ <br> $g$ = slope of flange <br> $= \frac{h-l}{2(b-t)}$ | $\frac{1}{3}\left[2sb^3 + lt^3 + \frac{g}{2}(b^4 - t^4)\right]$ <br> $-A(b-y)^2$ <br> $g$ = slope of flange <br> $= \frac{h-l}{2(b-t)} = \frac{1}{6}$ <br> for standard channels | $\frac{I}{y}$ | $\sqrt{\frac{I}{A}}$ |
| 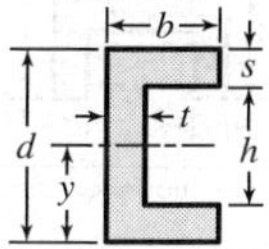 | $bd - h(b-t)$ | $\frac{d}{2}$ | $\frac{bd^3 - h^3(b-t)}{12}$ | $\frac{bd^3 - h^3(b-t)}{6d}$ | $\sqrt{\frac{bd^3 - h^3(b-t)}{12[bd - h(b-t)]}}$ |

**Moments of Inertia, Section Moduli, and Radii of Gyration** *(Continued)*

| Section | Area of Section, $A$ | Distance from Neutral Axis to Extreme Fiber, $y$ | Moment of Inertia, $I$ | Section Modulus, $Z = \frac{I}{y}$ | Radius of Gyration, $k = \sqrt{\frac{I}{A}}$ |
|---|---|---|---|---|---|
| 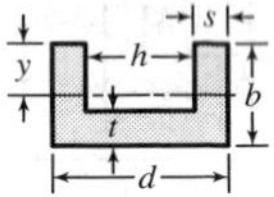 | $bd - h(b - t)$ | $b - \frac{2b^2s + ht^2}{2bd - 2h(b - t)}$ | $\frac{2sb^3 + ht^3}{3} - A(b - y)^2$ | $\frac{I}{y}$ | $\sqrt{\frac{I}{A}}$ |
| | | T-Sections | | | |
| 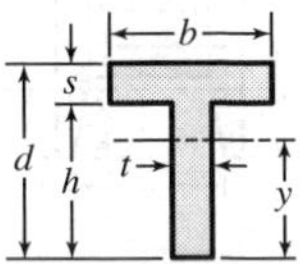 | $bs + ht$ | $d - \frac{d^2t + s^2(b - t)}{2(bs + ht)}$ | $\frac{1}{3}[ty^3 + b(d - y)^3 - (b - t)(d - y - s)^3]$ | $\frac{I}{y}$ | $\sqrt{\frac{1}{3(bs + ht)}[ty^3 + b(d - y)^3 - (b - t)(d - y - s)^3]}$ |
| 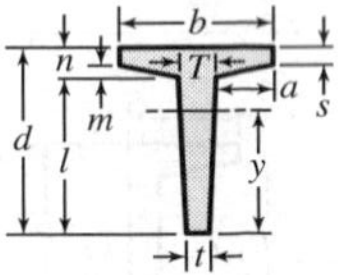 | $\frac{l(T + t)}{2} + Tn + a(s + n)$ | $d - [3s^2(b - T) + 2am(m + 3s) + 3Td^2 - l(T - t)(3d - l)] \div 6A$ | $\frac{1}{12}[l^3(T + 3t) + 4bn^3 - 2am^3] - A(d - y - n)^2$ | $\frac{I}{y}$ | $\sqrt{\frac{I}{A}}$ |
| 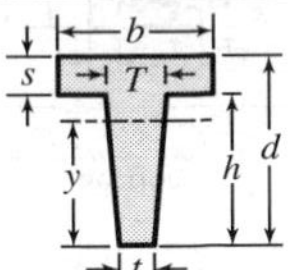 | $bs + \frac{h(T + t)}{2}$ | $d - [3bs^2 + 3ht(d + s) + h(T - t)(h + 3s)] \div 6A$ | $\frac{1}{12}[4bs^3 + h^3(3t + T)] - A(d - y - s)^2$ | $\frac{I}{y}$ | $\sqrt{\frac{I}{A}}$ |

**Moments of Inertia, Section Moduli, and Radii of Gyration** *(Continued)*

| Section | Area of Section, $A$ | Distance from Neutral Axis to Extreme Fiber, $y$ | Moment of Inertia, $I$ | Section Modulus, $Z = \frac{I}{y}$ | Radius of Gyration, $k = \sqrt{\frac{I}{A}}$ |
|---|---|---|---|---|---|
| 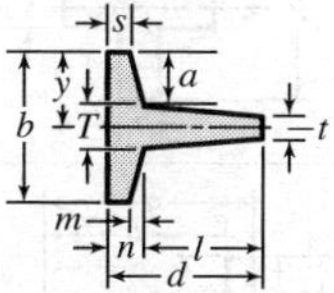  | $\frac{l(T+t)}{2} + Tn + a(s+n)$ | $\frac{b}{2}$ | $\frac{sb^3 + mT^3 + lt^3}{12} + \frac{am[2a^2 + (2a+3T)^2]}{36} + \frac{l(T-t)[(T-t)^2 + 2(T+2t)^2]}{144}$ | $\frac{I}{y}$ | $\sqrt{\frac{I}{A}}$ |
| L-, Z-, and X-Sections | | | | | |
| 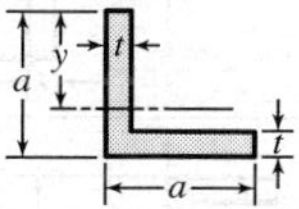  | $t(2a-t)$ | $a - \frac{a^2 + at - t^2}{2(2a-t)}$ | $\frac{1}{3}[ty^3 + a(a-y)^3 - (a-t)(a-y-t)^3]$ | $\frac{I}{y}$ | $\sqrt{\frac{I}{A}}$ |
| 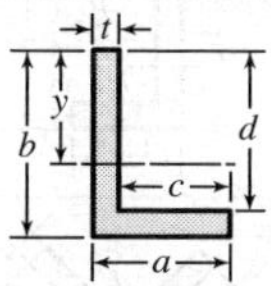  | $t(a+b-t)$ | $b - \frac{t(2d+a)+d^2}{2(d+a)}$ | $\frac{1}{3}[ty^3 + a(b-y)^3 - (a-t)(b-y-t)^3]$ | $\frac{I}{y}$ | $\sqrt{\frac{1}{3t(a+b-t)}[ty^3 + a(b-y)^3 - (a-t)(b-y-t)^3]}$ |
| 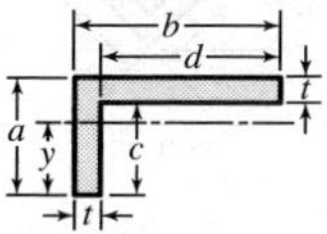  | $t(a+b-t)$ | $a - \frac{t(2c+b)+c^2}{2(c+b)}$ | $\frac{1}{3}[ty^3 + b(a-y)^3 - (b-t)(a-y-t)^3]$ | $\frac{I}{y}$ | $\sqrt{\frac{1}{3t(a+b-t)}[ty^3 + b(a-y)^3 - (b-t)(a-y-t)^3]}$ |

**Moments of Inertia, Section Moduli, and Radii of Gyration** *(Continued)*

| Section | Area of Section, $A$ | Distance from Neutral Axis to Extreme Fiber, $y$ | Moment of Inertia, $I$ | Section Modulus, $Z = \frac{I}{y}$ | Radius of Gyration, $k = \sqrt{\frac{I}{A}}$ |
|---|---|---|---|---|---|
| 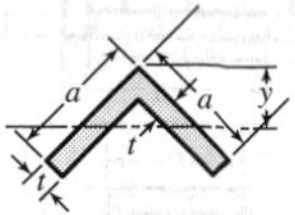 | $t(2a - t)$ | $\frac{a^2 + at - t^2}{2(2a - t)\cos 45°}$ | $\frac{A}{12}[7(a^2 + b^2) - 12y^2]$ $- 2ab^2(a - b)$ in which $b = (a - t)$ | $\frac{I}{y}$ | $\sqrt{\frac{I}{A}}$ |
| 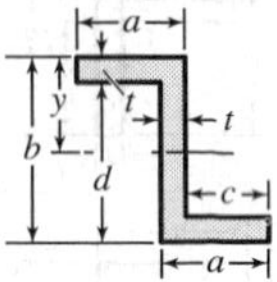 | $t[b + 2(a - t)]$ | $\frac{b}{2}$ | $\frac{ab^3 - c(b - 2t)^3}{12}$ | $\frac{ab^3 - c(b - 2t)^3}{6b}$ | $\sqrt{\frac{ab^3 - c(b - 2t)^3}{12t[b + 2(a - t)]}}$ |
| 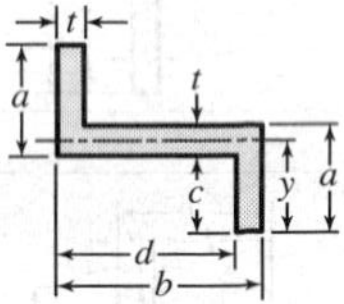 | $t[b + 2(a - t)]$ | $\frac{2a - t}{2}$ | $\frac{b(a + c)^3 - 2c^3d - 6a^2cd}{12}$ | $\frac{b(a + c)^3 - 2c^3d - 6a^2cd}{6(2a - t)}$ | $\sqrt{\frac{b(a + c)^3 - 2c^3d - 6a^2cd}{12t[b + 2(a - t)]}}$ |
| 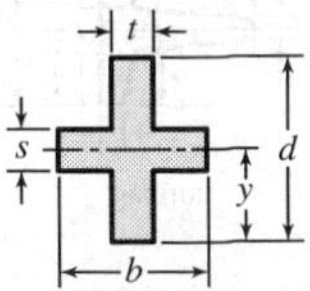 | $dt + s(b - t)$ | $\frac{d}{2}$ | $\frac{td^3 + s^3(b - t)}{12}$ | $\frac{td^3 - s^3(b - t)}{6d}$ | $\sqrt{\frac{td^3 + s^3(b - t)}{12[td + s(b - t)]}}$ |

**Polar Area Moment of Inertia and Section Modulus.—**The *polar moment of inertia J* of a cross section with respect to a polar axis, that is, an axis at right angles to the plane of the cross section, is defined as the moment of inertia of the cross section with respect to the point of intersection of the axis and the plane. The polar moment of inertia may be found by taking the sum of the moments of inertia about two perpendicular axes lying in the plane of the cross section and passing through this point. Thus, for example, the polar moment of inertia of a circular or a square area with respect to a polar axis through the center of gravity is equal to twice the moment of inertia with respect to an axis lying in the plane of the cross section and passing through the center of gravity.

The polar moment of inertia with respect to a polar axis through the center of gravity is required for problems involving the torsional strength of shafts since this axis is usually the axis about which twisting of the shaft takes place.

The *polar section modulus* (also called section modulus of torsion) $Z_p$ for *circular* sections may be found by dividing the polar moment of inertia $J$ by the distance $c$ from the center of gravity to the most remote fiber. This method may be used to find the *approximate* value of the polar section modulus of sections that are *nearly* round. For other than circular cross sections, however, the polar section modulus *does not* equal the polar moment of inertia divided by the distance $c$.

The accompanying table *Polar Moment of Inertia and Polar Section Modulus* on page 246 gives formulas for the polar section modulus for several different cross sections. The polar section modulus multiplied by the allowable torsional shearing stress gives the allowable twisting moment to which a shaft may be subjected; see Formula (7) on page 290.

**Polar Mass Moment of Inertia.—**Starting on page 247, formulas for mass moment of inertia* $J_M$ of various solids are given in a series of tables. The example that follows illustrates the derivaion of $J_M$ for one of the bodies given on page 247.

*Example, Polar Mass Moment of Inertia of a Hollow Circular Section:* Referring to the figure *Hollow Cylinder* on page 247, consider a strip of width $dr$ on a hollow circular section whose inner radius is $r$ and outer radius is $R$.

The mass of the strip = $2\pi r dr \rho$, where $\rho$ is the density of material. In order to get the mass of an individual section, integrate the mass of the strip from $r$ to $R$.

$$M = \int_r^R 2\pi\rho r dr = 2\pi\rho\int_r^R r dr = 2\pi\rho\left[\frac{r^2}{2}\right]_r^R$$

$$= 2\pi\rho\left(\frac{R^2}{2} - \frac{r^2}{2}\right) = \pi\rho(R^2 - r^2)$$

The second moment of the strip about the axis AA = $2\pi r dr \rho r^2$. To find the polar moment of inertia about the axis AA, integrate the 2nd moment from $r$ to $R$.

$$J_M = \int_r^R 2\pi r(dr)\rho r^2 = 2\pi\rho\int_r^R r^3(dr) = 2\pi\rho\left[\frac{r^4}{4}\right]_r^R$$

$$= 2\pi\rho\left(\frac{R^4}{4} - \frac{r^4}{4}\right) = \frac{\pi\rho}{2}(R^2 - r^2)(R^2 + r^2)$$

$$= \pi\rho(R^2 - r^2)\frac{(R^2 + r^2)}{2} = \frac{M(R^2 + r^2)}{2}$$

*In some books the symbol $I$ denotes the polar moment of inertia of masses; $J_M$ is used in this *Handbook* to avoid confusion with moments of inertia of plane areas.

**Polar Moment of Inertia and Polar Section Modulus**

| Section | Polar Moment of Inertia, $J$ | Polar Section Modulus, $Z_p$ |
|---|---|---|
| 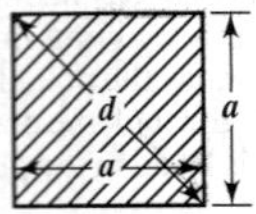 | $\frac{a^4}{6} = 0.1667a^4$ | $0.208a^3 = 0.074d^3$ |
| 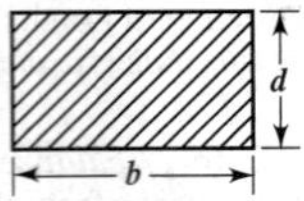 | $\frac{bd(b^2+d^2)}{12}$ | $\frac{bd^2}{3+1.8\frac{d}{b}}$ (*d* is the shorter side) |
| 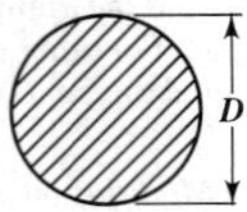 | $\frac{\pi D^4}{32} = 0.098D^4$ | $\frac{\pi D^3}{16} = 0.196D^3$ |
| 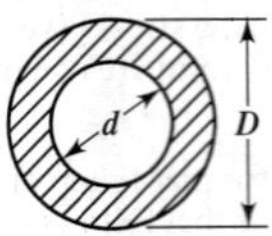 | $\frac{\pi}{32}(D^4 - d^4)$ $= 0.098(D^4 - d^4)$ | $\frac{\pi}{16}\left(\frac{D^4 - d^4}{D}\right)$ $= 0.196\left(\frac{D^4 - d^4}{D}\right)$ |
| 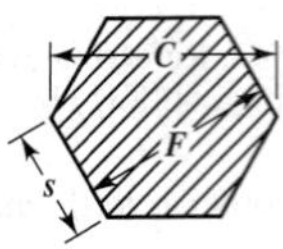 | $\frac{5\sqrt{3}}{8}s^4 = 1.0825s^4$ $= 0.12F^4$ | $0.20F^3$ |
| 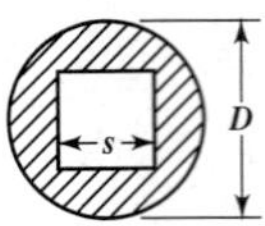 | $\frac{\pi D^4}{32} - \frac{s^4}{6}$ $= 0.098D^4 - 0.167s^4$ | $\frac{\pi D^3}{16} - \frac{s^4}{3D}$ $= 0.196D^3 - 0.333\frac{s^4}{D}$ |
| 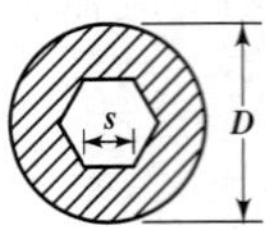 | $\frac{\pi D^4}{32} - \frac{5\sqrt{3}}{8}s^4$ $= 0.098D^4 - 1.0825s^4$ | $\frac{\pi D^3}{16} - \frac{5\sqrt{3}}{4D}s^4$ $= 0.196D^3 - 2.165\frac{s^4}{D}$ |
| 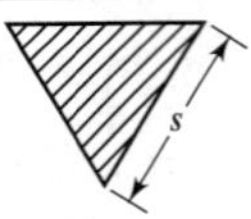 | $\frac{\sqrt{3}}{48}s^4 = 0.036s^4$ | $\frac{s^3}{20} = 0.05s^3$ |

*Polar Section Modulus* for a shaft of given diameter can be obtained by multiplying its section modulus by 2. *Polar Moment of Inertia* can be obtained by multiplying its moment of inertia by 2.

## Formulas for Polar Moment of Inertia of Masses, $J_M$

*Prism:*

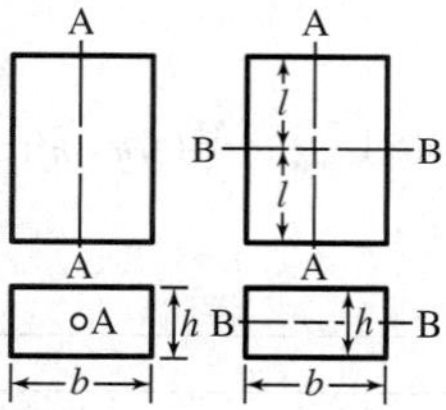


With reference to axis AA: $J_M = \frac{M}{12}(h^2 + b^2)$

With reference to axis BB: $J_M = M\left(\frac{l^2}{3} + \frac{h^2}{12}\right)$

*Cylinder:*

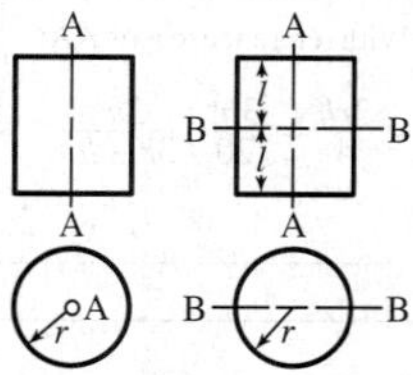


With reference to axis AA: $J_M = \frac{1}{2}Mr^2$

With reference to axis BB: $J_M = M\left(\frac{l^2}{3} + \frac{r^2}{4}\right)$

*Hollow Cylinder:*

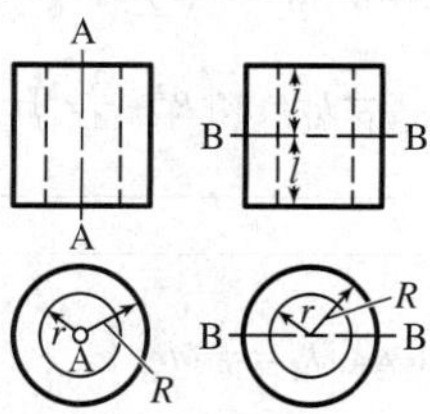


With reference to axis AA: $J_M = \frac{1}{2}M(R^2 + r^2)$

With reference to axis BB:

$$J_M = M\left(\frac{l^2}{3} + \frac{R^2 + r^2}{4}\right)$$

*Pyramid, Rectangular Base:*

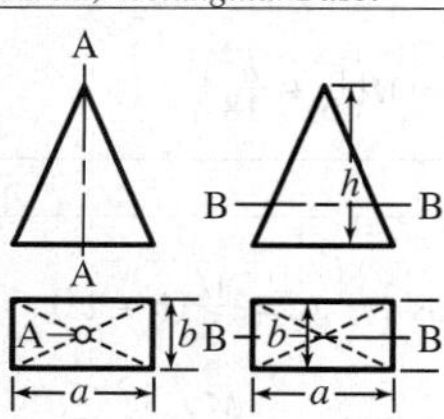


With reference to axis AA: $J_M = \frac{M}{20}(a^2 + b^2)$

With reference to axis BB (through the center of gravity):

$$J_M = M\left(\frac{3}{80}h^2 + \frac{b^2}{20}\right)$$

*Sphere:*

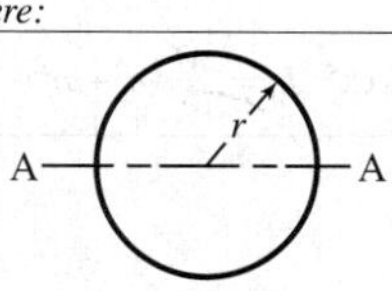


With reference to any axis through the center:

$$J_M = \frac{2}{5}Mr^2$$

*Spherical Sector:*

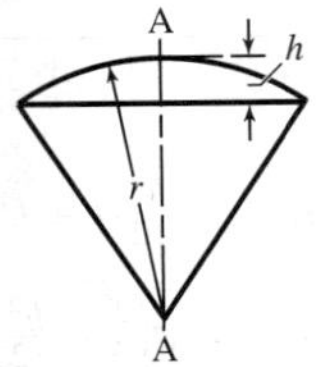


With reference to axis AA: $J_M = \frac{M}{5}(3rh - h^2)$

*Spherical Segment:*

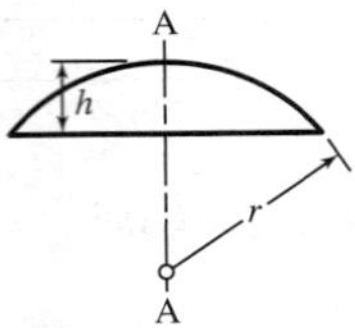


*Spherical Segment*: With reference to axis AA:

$$J_M = M\left(r^2 - \frac{3rh}{4} + \frac{3h^2}{20}\right)\frac{2h}{3r - h}$$

*Torus:*

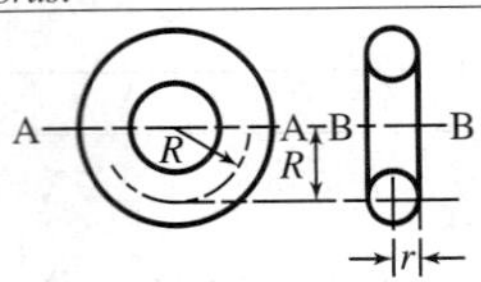


With reference to axis AA: $J_M = M\left(\frac{R^2}{2} + \frac{5r^2}{8}\right)$

With reference to axis BB: $J_M = M\left(R^2 + \frac{3}{4}r^2\right)$

*Paraboloid:*

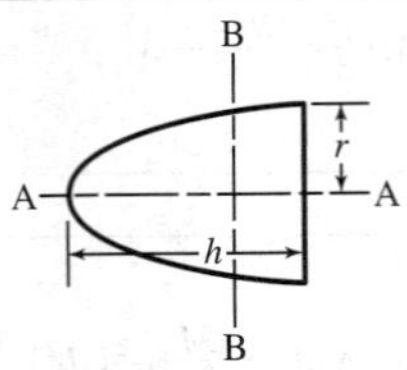


With reference to axis AA: $J_M = \frac{1}{3}Mr^2$

With reference to axis BB (through the center of gravity):

$$J_M = M\left(\frac{r^2}{6} + \frac{h^2}{18}\right)$$

*Ellipsoid:*

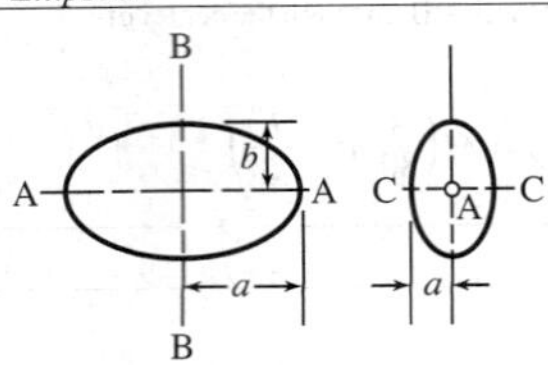


With reference to axis AA: $J_M = \frac{M}{5}(b^2 + c^2)$

With reference to axis BB: $J_M = \frac{M}{5}(a^2 + c^2)$

With reference to axis CC: $J_M = \frac{M}{5}(a^2 + b^2)$

*Cone:*

| | |
|---|---|
| 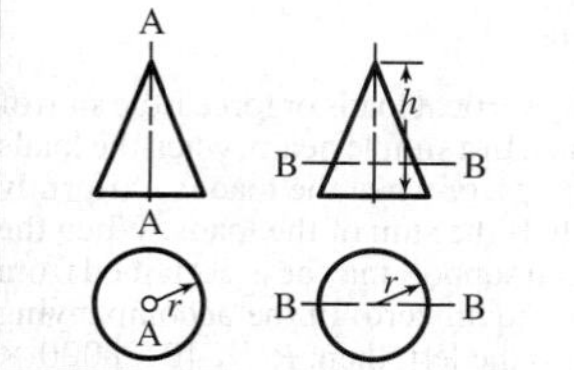 | With reference to axis AA: $J_M = \frac{3M}{10}r^2$<br>With reference to axis BB (through the center of gravity):<br>$J_M = \frac{3M}{20}\left(r^2 + \frac{h^2}{4}\right)$ |

*Frustum of Cone:*

| | |
|---|---|
| 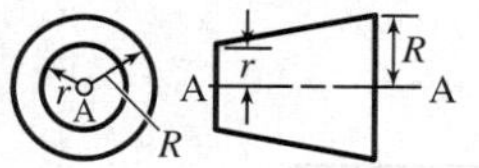 | With reference to axis AA: $J_M = \frac{3M(R^5 - r^5)}{10(R^3 - r^3)}$ |

*Moments of Inertia of Complex Areas and Masses* may be evaluated by the addition and subtraction of elementary areas and masses. For example, the accompanying figure shows a complex mass at (1); its mass polar moment of inertia can be determined by adding the moments of inertia of the bodies shown at (2) and (3), and subtracting that at (4). Thus, $J_{M1} = J_{M2} + J_{M3} - J_{M4}$. All of these moments of inertia are with respect to the axis of rotation $z-z$. Formulas for $J_{M2}$ and $J_{M3}$ can be obtained from the tables beginning on page 247. The moment of inertia for the body at (4) can be evaluated by using the following transfer-axis equation: $J_{M4} = J_{M4}' + d^2M$. The term $J_{M4}'$ is the moment of inertia with respect to axis $z'-z'$; it may be evaluated using the same equation that applies to $J_{M2}$ where $d$ is the distance between the $z-z$ and the $z'-z'$ axes, and $M$ is the mass of the body (= weight in lbs ÷ $g$).

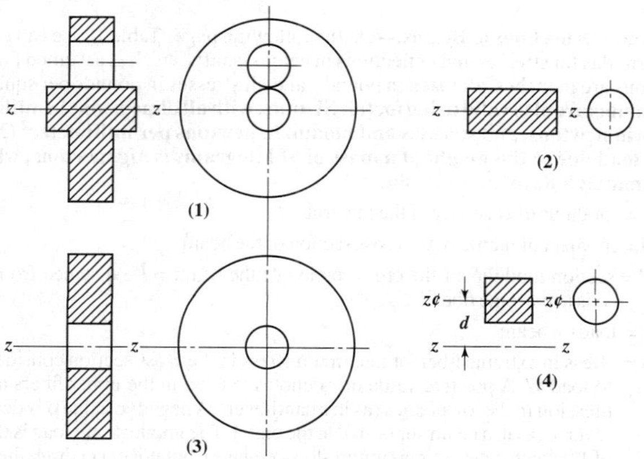


Moments of Inertia of Complex Masses

Similar calculations can be made when calculating $I$ and $J$ for complex areas using the appropriate transfer-axis equations $I = I' + d^2A$ and $J = J' + d^2A$. The primed term, $I'$ or $J'$, is with respect to the center of gravity of the corresponding area $A$; $d$ is the distance between the axis through the center of gravity and the axis to which $I$ or $J$ is referred.

# BEAMS

## Beam Calculations

**Reaction at the Supports.—**When a beam is loaded by vertical loads or forces, the sum of the reactions at the supports equals the sum of the loads. In a simple beam, when the loads are symmetrically placed with reference to the supports, or when the load is uniformly distributed, the reaction at each end will equal one-half the sum of the loads. When the loads are not symmetrically placed, the reaction at each support may be ascertained from the fact that the algebraic sum of the moments must equal zero. In the accompanying illustration, if moments are taken about the support to the left, then: $R_2 \times 40 - 8000 \times 10 - 10{,}000 \times 16 - 20{,}000 \times 20 = 0$; $R_2 = 16{,}000$ pounds. In the same way, moments taken about the support at the right give $R_1 = 22{,}000$ pounds.

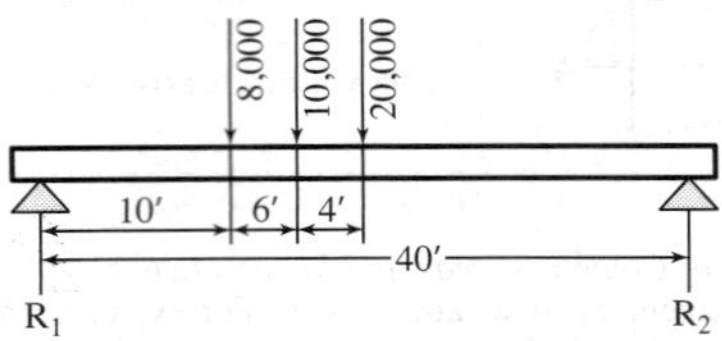


The sum of the reactions equals 38,000 pounds, which is also the sum of the loads. If part of the load is uniformly distributed over the beam, this part is first equally divided between the two supports, or the uniform load may be considered as concentrated at its center of gravity.

**If metric SI units are used for the calculations, distances may be expressed in meters or millimeters, provided the treatment is consistent, and loads in newtons.** ***Note:*** **If the load is given in kilograms, the value referred to is the mass. A mass of *M* kilograms has a weight (applies a force) of *Mg* newtons, where $g$ = approximately 9.81 m/s$^2$.**

**Stresses and Deflections in Beams.—**On the following pages Table 1 gives an extensive list of formulas for stresses and deflections in beams, shafts, etc. It is assumed that all the dimensions are in inches, all loads in pounds, and all stresses in pounds per square inch. **The formulas are also valid using metric SI units, with all dimensions in millimeters, all loads in newtons, and stresses and moduli in newtons per millimeter$^2$ (N/mm$^2$).** ***Note:*** **A load due to the weight of a mass of *M* kilograms is *Mg* newtons, where $g$ = approximately 9.81 m/s$^2$.** In the tables:

$E$ = modulus of elasticity of the material

$I$ = moment of inertia of the cross section of the beam

$Z$ = section modulus of the cross section of the beam = $I$ ÷ distance from neutral axis to extreme fiber

$W$ = load on beam

$s$ = stress in extreme fiber, or maximum stress in the cross section considered, due to load $W$. A positive value of $s$ denotes tension in the upper fibers and compression in the lower ones (as in a cantilever). A negative value of $s$ denotes the reverse (as in a beam supported at the ends). The greatest safe load is that value of $W$ which causes a maximum stress equal to, but not exceeding, the greatest safe value of $s$

$y$ = deflection measured from the position occupied if the load causing the deflection were removed. A positive value of $y$ denotes deflection below this position; a negative value, deflection upward

$u, v, w, x$ = variable distances along the beam from a given support to any point

## Table 1. Stresses and Deflections in Beams

| Type of Beam | Stresses: General Formula for Stress at any Point | Stresses: Stresses at Critical Points | Deflections: General Formula for Deflection at any Point[a] | Deflections: Deflections at Critical Points[a] |
|---|---|---|---|---|
| Case 1. — Supported at Both Ends, Uniform Load | | | | |
| 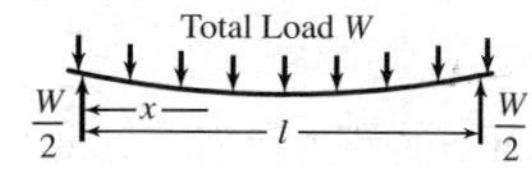 $W = wl$ (where $w$ = load per unit length) | $s = -\frac{W}{2Zl}x(l-x)$ | Stress at center, $-\frac{Wl}{8Z}$ If cross section is constant, this is the maximum stress. | $y = \frac{Wx(l-x)}{24EIl}[l^2 + x(l-x)]$ | Maximum deflection, at center, $\frac{5}{384}\frac{Wl^3}{EI}$ |
| Case 2. — Supported at Both Ends, Load at Center | | | | |
| 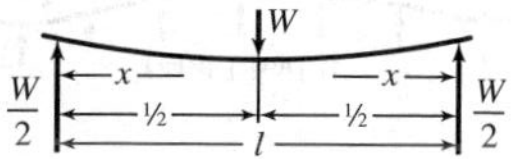 | Between each support and load, $s = -\frac{Wx}{2Z}$ | Stress at center, $-\frac{Wl}{4Z}$ If cross section is constant, this is the maximum stress. | Between each support and load, $y = \frac{Wx}{48EI}(3l^2 - 4x^2)$ | Maximum deflection, at load, $\frac{Wl^3}{48EI}$ |
| Case 3. — Supported at Both Ends, Load at any Point | | | | |
| 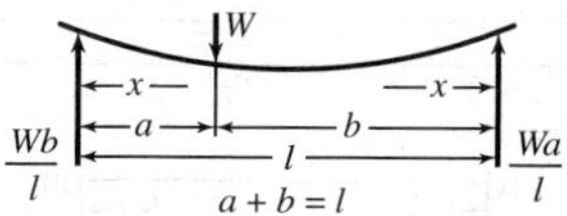 | For segment of length $a$, $s = -\frac{Wbx}{Zl}$ For segment of length $b$, $s = -\frac{Wav}{Zl}$ | Stress at load, $-\frac{Wab}{Zl}$ If cross section is constant, this is the maximum stress. | For segment of length $a$, $y = \frac{Wbx}{6EIl}(l^2 - x^2 - b^2)$ For segment of length $b$, $y = \frac{Wav}{6EIl}(l^2 - v^2 - a^2)$ | Deflection at load, $\frac{Wa^2b^2}{3EIl}$ Let $a$ be the length of the shorter segment and $b$ of the longer one. The maximum deflection $\frac{Wav_1^3}{3EIl}$ is in the longer segment, at $v = b\sqrt{\frac{1}{3} + \frac{2a}{3b}} = v_1$ |

## Table 1. *(Continued)* Stresses and Deflections in Beams

| Type of Beam | Stresses: General Formula for Stress at any Point | Stresses: Stresses at Critical Points | Deflections: General Formula for Deflection at any Point[a] | Deflections: Deflections at Critical Points[a] |
|---|---|---|---|---|
| Case 4. — Supported at Both Ends, Two Symmetrical Loads | | | | |
| 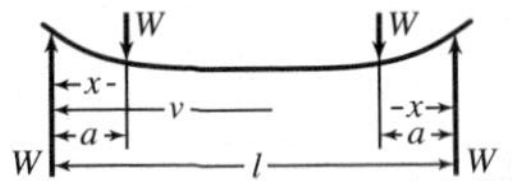 | Between each support and adjacent load,<br>$s = -\frac{Wx}{Z}$<br>Between loads,<br>$s = -\frac{Wa}{Z}$ | Stress at each load, and at all points between, $-\frac{Wa}{Z}$ | Between each support and adjacent load,<br>$y = \frac{Wx}{6EI}[3a(l-a) - x^2]$<br>Between loads,<br>$y = \frac{Wa}{6EI}[3v(l-v) - a^2]$ | Maximum deflection at center,<br>$\frac{Wa}{24EI}(3l^2 - 4a^2)$<br>Deflection at loads<br>$\frac{Wa^2}{6EI}(3l - 4a)$ |
| Case 5. — Both Ends Overhanging Supports Symmetrically, Uniform Load | | | | |
| 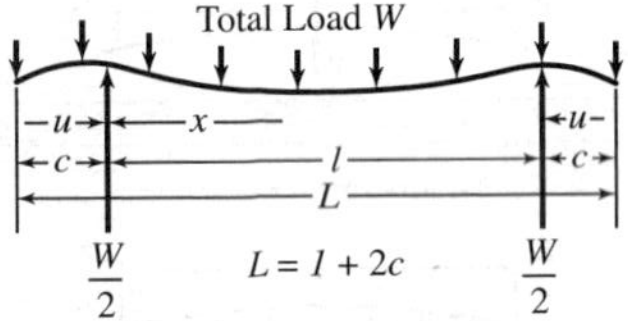 | Between each support and adjacent end,<br>$s = \frac{W}{2Zl}(c-u)^2$<br>Between supports,<br>$s = \frac{W}{2ZL}[c^2 - x(l-x)]$ | Stress at each support,<br>$\frac{Wc^2}{2ZL}$<br>Stress at center,<br>$\frac{W}{2ZL}\left(c^2 - \frac{1}{4}l^2\right)$<br>If cross section is constant, the greater of these is the maximum stress.<br>If $l$ is greater than $2c$, the stress is zero at points<br>$\sqrt{\frac{1}{4}l^2 - c^2}$ on both sides of the center.<br>If cross section is constant and if $l = 2.828c$, the stresses at supports and center are equal and opposite, and are<br>$\pm\frac{WL}{46.62Z}$ | Between each support and adjacent end,<br>$y = \frac{Wu}{24EIL}[6c^2(l+u) - u^2(4c-u) - l^3]$<br>Between supports,<br>$y = \frac{Wx(l-x)}{24EIL}[x(l-x) + l^2 - 6c^2]$ | Deflection at ends,<br>$\frac{Wc}{24EIL}[3c^2(c+2l) - l^3]$<br>Deflection at center,<br>$\frac{Wl^2}{384EIL}(5l^2 - 24c^2)$<br>If $l$ is between $2c$ and $2.449c$, there are maximum upward deflections at points<br>$\sqrt{3\left(\frac{1}{4}l^2 - c^2\right)}$ on both sides of the center, which are,<br>$-\frac{W}{96EIL}(6c^2 - l^2)^2$ |

**Table 1.** *(Continued)* **Stresses and Deflections in Beams**

| Type of Beam | Stresses: General Formula for Stress at any Point | Stresses: Stresses at Critical Points | Deflections: General Formula for Deflection at any Point[a] | Deflections: Deflections at Critical Points[a] |
|---|---|---|---|---|
| Case 6. — Both Ends Overhanging Supports Unsymmetrically, Uniform Load | | | | |
| 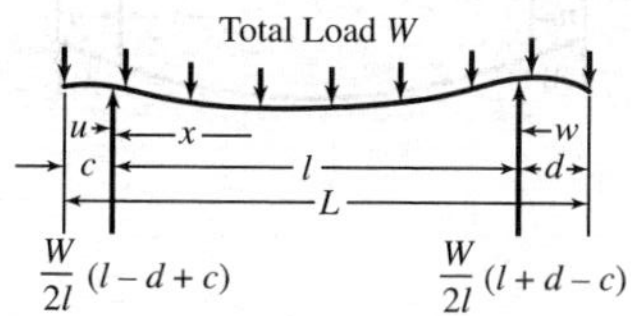<br> | For overhanging end of length $c$,<br>$s = \frac{W}{2ZL}(c-u)^2$<br>Between supports,<br>$s = \frac{W}{2ZL}\left\{c^2\left(\frac{l-x}{l}\right) + d^2\frac{x}{l} - x(l-x)\right\}$<br>For overhanging end of length $d$,<br>$s = \frac{W}{2ZL}(d-w)^2$ | Stress at support next to end of length $c$, $\frac{Wc^2}{2ZL}$<br>Critical stress between supports is at<br>$x = \frac{l^2 + c^2 - d^2}{2l} = x_1$<br>and is $\frac{W}{2ZL}(c^2 - x_1^2)$<br>Stress at support next to end of length $d$, $\frac{Wd^2}{2ZL}$<br>If cross section is constant, the greatest of these three is the maximum stress.<br>If $x_1 > c$, the stress is zero at points $\sqrt{x_1^2 - c^2}$ on both sides of $x = x_1$. | For overhanging end of length $c$,<br>$y = \frac{Wu}{24EIL}[2l(d^2 + 2c^2) + 6c^2u - u^2(4c-u) - l^3]$<br>Between supports,<br>$y = \frac{Wx(l-x)}{24EIL}\{x(l-x) + l^2 - 2(d^2+c^2) - \frac{2}{l}[d^2x + c^2(l-x)]\}$<br>For overhanging end of length $d$,<br>$y = \frac{Ww}{24EIL}[2l(c^2+2d^2) + 6d^2w - w^2(4d-w) - l^3]$ | Deflection at end $c$,<br>$\frac{Wc}{24EIL}[2l(d^2+2c^2) + 3c^3 - l^3]$<br>Deflection at end $d$,<br>$\frac{Wd}{24EIL}[2l(c^2+2d^2) + 3d^3 - l^3]$<br>This case is so complicated that convenient general expressions for the critical deflections between supports cannot be obtained. |
| Case 7. — Both Ends Overhanging Supports, Load at any Point Between | | | | |
| 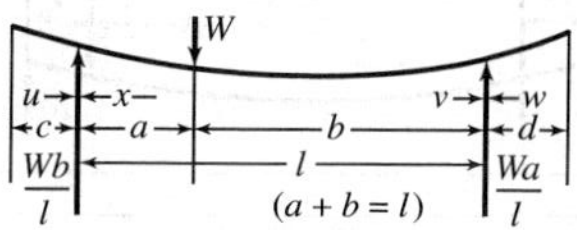<br> | Between supports:<br>For segment of length $a$,<br>$s = -\frac{Wbx}{Zl}$<br>For segment of length $b$,<br>$s = -\frac{Wav}{Zl}$<br>Beyond supports $s = 0$. | Stress at load,<br>$-\frac{Wab}{Zl}$<br>If cross section is constant, this is the maximum stress. | Between supports, same as Case 3.<br>For overhanging end of length $c$,<br>$y = -\frac{Wabu}{6EIl}(l+b)$<br>For overhanging end of length $d$,<br>$y = -\frac{Wabw}{6EIl}(l+a)$ | Between supports, same as Case 3.<br>Deflection at end $c$,<br>$-\frac{Wabc}{6EIl}(l+b)$<br>Deflection at end $d$,<br>$-\frac{Wabd}{6EIl}(l+a)$ |

## Table 1. *(Continued)* Stresses and Deflections in Beams

| Type of Beam | Stresses: General Formula for Stress at any Point | Stresses: Stresses at Critical Points | Deflections: General Formula for Deflection at any Point[a] | Deflections: Deflections at Critical Points[a] |
|---|---|---|---|---|
| **Case 8. — Both Ends Overhanging Supports, Single Overhanging Load** | | | | |
| 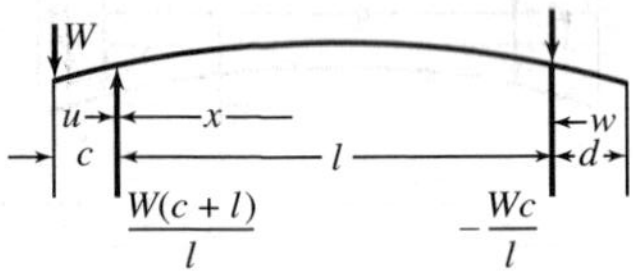 | Between load and adjacent support, $s = \frac{W}{Z}(c-u)$<br>Between supports, $s = \frac{Wc}{Zl}(l-x)$<br>Between unloaded end and adjacent supports, $s = 0$. | Stress at support adjacent to load, $\frac{Wc}{Z}$<br>If cross section is constant, this is the maximum stress.<br>Stress is zero at other support. | Between load and adjacent support, $y = \frac{Wu}{6EI}(3cu - u^2 + 2cl)$<br>Between supports, $y = -\frac{Wcx}{6EIl}(l-x)(2l-x)$<br>Between unloaded end and adjacent support, $y = \frac{Wclw}{6EI}$ | Deflection at load, $\frac{Wc^2}{3EI}(c+l)$<br>Maximum upward deflection is at $x = .42265l$, and is $-\frac{Wcl^2}{15.55EI}$<br>Deflection at unloaded end, $\frac{Wcld}{6EI}$ |
| **Case 9. — Both Ends Overhanging Supports, Symmetrical Overhanging Loads** | | | | |
| 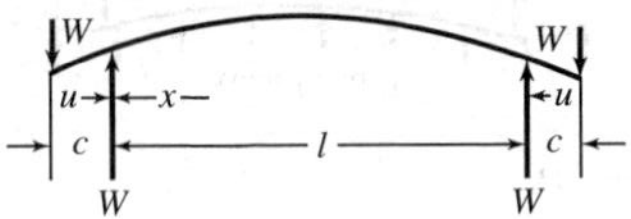 | Between each load and adjacent support, $s = \frac{W}{Z}(c-u)$<br>Between supports, $s = \frac{Wc}{Z}$ | Stress at supports and at all points between, $\frac{Wc}{Z}$<br>If cross section is constant, this is the maximum stress. | Between each load and adjacent support, $y = \frac{Wu}{6EI}[3c(l+u) - u^2]$<br>Between supports, $y = -\frac{Wcx}{2EI}(l-x)$ | Deflections at loads, $\frac{Wc^2}{6EI}(2c+3l)$<br>Deflection at center, $-\frac{Wcl^2}{8EI}$ |

The above expressions involve the usual approximations of the theory of flexure, and hold only for small deflections. Exact expressions for deflections of any magnitude are as follows:

Between supports the curve is a circle of radius $r = \frac{EI}{Wc}$

Deflection at any point $x$ between supports

$$y = \sqrt{r^2 - \frac{1}{4}l^2} - \sqrt{r^2 - \left(\frac{1}{2}l - x\right)^2}$$

Deflection at center, $\sqrt{r^2 - \frac{1}{4}l^2} - r$

**Table 1.** *(Continued)* **Stresses and Deflections in Beams**

| Type of Beam | Stresses: General Formula for Stress at any Point | Stresses: Stresses at Critical Points | Deflections: General Formula for Deflection at any Point[a] | Deflections: Deflections at Critical Points[a] |
|---|---|---|---|---|
| Case 10. — Fixed at One End, Uniform Load | | | | |
| 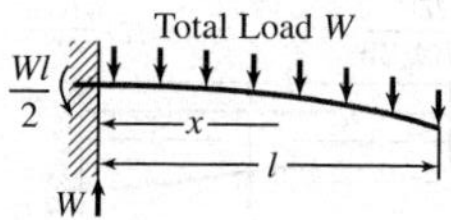 | $s = \frac{W}{2Zl}(l - x)^2$ | Stress at support, $\frac{Wl}{2Z}$ If cross section is constant, this is the maximum stress. | $y = \frac{Wx^2}{24EIl}[2l^2 + (2l - x)^2]$ | Maximum deflection, at end, $\frac{Wl^3}{8EI}$ |
| Case 11. — Fixed at One End, Load at Other | | | | |
| 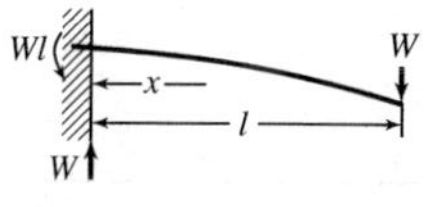 | $s = \frac{W}{Z}(l - x)$ | Stress at support, $\frac{Wl}{Z}$ If cross section is constant, this is the maximum stress. | $y = \frac{Wx^2}{6EI}(3l - x)$ | Maximum deflection, at end, $\frac{Wl^3}{3EI}$ |
| Case 12. — Fixed at One End, Intermediate Load | | | | |
| 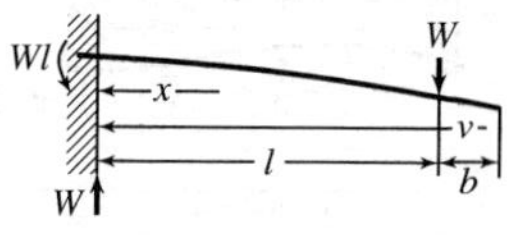 | Between support and load, $s = \frac{W}{Z}(l - x)$ Beyond load, $s = 0$. | Stress at support, $\frac{Wl}{Z}$ If cross section is constant, this is the maximum stress. | Between support and load, $y = \frac{Wx^2}{6EI}(3l - x)$ Beyond load, $y = \frac{Wl^2}{6EI}(3v - l)$ | Deflection at load, $\frac{Wl^3}{3EI}$ Maximum deflection, at end, $\frac{Wl^2}{6EI}(2l + 3b)$ |

**Table 1.** *(Continued)* **Stresses and Deflections in Beams**

| Type of Beam | Stresses: General Formula for Stress at any Point | Stresses: Stresses at Critical Points | Deflections: General Formula for Deflection at any Point[a] | Deflections: Deflections at Critical Points[a] |
|---|---|---|---|---|
| **Case 13. — Fixed at One End, Supported at the Other, Load at Center** | | | | |
| 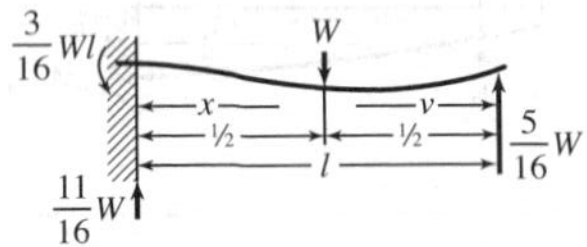 | Between point of fixture and load,<br>$s = \frac{W}{16Z}(3l - 11x)$<br>Between support and load,<br>$s = -\frac{5}{16}\frac{Wv}{Z}$ | Maximum stress at point of fixture, $\frac{3}{16}\frac{Wl}{Z}$<br>Stress is zero at $x = \frac{3}{11}l$<br>Greatest negative stress at center, $-\frac{5}{32}\frac{Wl}{Z}$ | Between point of fixture and load,<br>$y = \frac{Wx^2}{96EI}(9l - 11x)$<br>Between support and load,<br>$y = \frac{Wv}{96EI}(3l^2 - 5v^2)$ | Maximum deflection is at $v = 0.4472l$, and is $\frac{Wl^3}{107.33EI}$<br>Deflection at load,<br>$\frac{7}{768}\frac{Wl^3}{EI}$ |
| **Case 14. — Fixed at One End, Supported at the Other, Load at any Point** | | | | |
| $m = (l + a)(l + b) + al$<br>$n = al(l + b)$<br>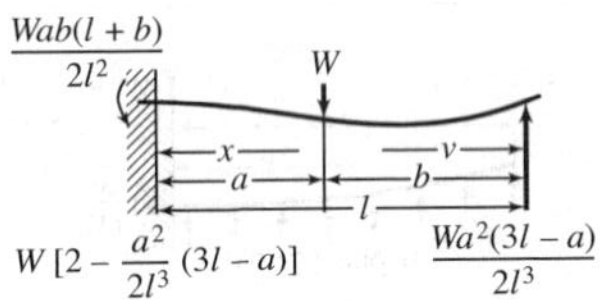 | Between point of fixture and load,<br>$s = \frac{Wb}{2Zl^3}(n - mx)$<br>Between support and load,<br>$s = -\frac{Wa^2v}{2Zl^3}(3l - a)$ | Greatest positive stress, at point of fixture,<br>$\frac{Wab}{2Zl^2}(l + b)$<br>Greatest negative stress, at load,<br>$-\frac{Wa^2b}{2Zl^3}(3l - a)$<br>If $a < 0.5858l$, the first is the maximum stress. If $a = 0.5858l$, the two are equal and are $\pm\frac{Wl}{5.83Z}$<br>If $a > 0.5858l$, the second is the maximum stress.<br>Stress is zero at $x = \frac{n}{m}$ | Between point of fixture and load,<br>$y = \frac{Wx^2b}{12EIl^3}(3n - mx)$<br>Between support and load,<br>$y = \frac{Wa^2v}{12EIl^3}[3l^2b - v^2(3l - a)]$ | Deflection at load,<br>$\frac{Wa^3b^2}{12EIl^3}(3l + b)$<br>If $a < 0.5858l$, maximum deflection is<br>$\frac{Wa^2b}{6EI}\sqrt{\frac{b}{2l + b}}$<br>and located between load and support, at $v = l\sqrt{\frac{b}{2l + b}}$<br>If $a = 0.5858l$, maximum deflection is at load and is $\frac{Wl^3}{101.9EI}$<br>If $a > 0.5858l$, maximum deflection is $\frac{Wbn^3}{3EIm^2l^3}$ and located between load and point of fixture, at $x = \frac{2n}{m}$ |

**Table 1.** *(Continued)* **Stresses and Deflections in Beams**

| Type of Beam | Stresses: General Formula for Stress at any Point | Stresses: Stresses at Critical Points | Deflections: General Formula for Deflection at any Point[a] | Deflections: Deflections at Critical Points[a] |
|---|---|---|---|---|
| Case 15. — Fixed at One End, Supported at the Other, Uniform Load | | | | |
| 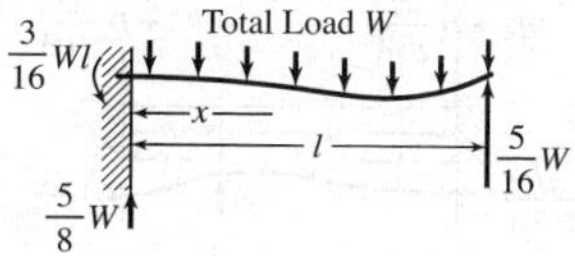 | $s = \frac{W(l-x)}{2Zl}\left(\frac{1}{4}l - x\right)$ | Maximum stress at point of fixture, $\frac{Wl}{8Z}$<br>Stress is zero at $x = \frac{1}{4}l$.<br>Greatest negative stress is at $x = \frac{5}{8}l$ and is $-\frac{9}{128}\frac{Wl}{Z}$ | $y = \frac{Wx^2(l-x)}{48EIl}(3l-2x)$ | Maximum deflection is at $x = 0.5785l$, and is $\frac{Wl^3}{185EI}$<br>Deflection at center, $\frac{Wl^3}{192EI}$<br>Deflection at point of greatest negative stress, at $x = \frac{5}{8}l$ is $\frac{Wl^3}{187EI}$ |
| Case 16. — Fixed at One End, Free but Guided at the Other, Uniform Load | | | | |
| 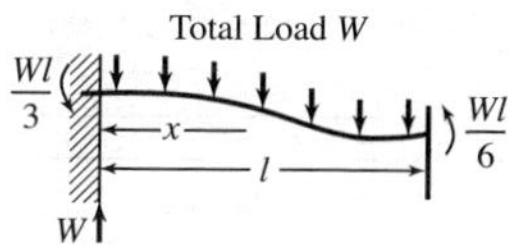 | $s = \frac{Wl}{Z}\left\{\frac{1}{3} - \frac{x}{l} + \frac{1}{2}\left(\frac{x}{l}\right)^2\right\}$ | Maximum stress, at support, $\frac{Wl}{3Z}$<br>Stress is zero at $x = 0.4227l$<br>Greatest negative stress, at free end, $-\frac{Wl}{6Z}$ | $y = \frac{Wx^2}{24EIl}(2l - x)^2$ | Maximum deflection, at free end, $\frac{Wl^3}{24EI}$ |
| Case 17. — Fixed at One End, Free but Guided at the Other, with Load | | | | |
| 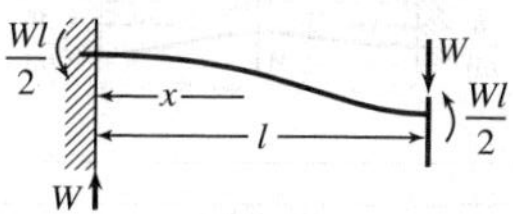 | $s = \frac{W}{Z}\left(\frac{1}{2}l - x\right)$ | Stress at support, $\frac{Wl}{2Z}$<br>Stress at free end, $-\frac{Wl}{2Z}$<br>These are the maximum stresses and are equal and opposite.<br>Stress is zero at center. | $y = \frac{Wx^2}{12EI}(3l - 2x)$ | Maximum deflection, at free end, $\frac{Wl^3}{12EI}$ |

**Table 1.** *(Continued)* **Stresses and Deflections in Beams**

| Type of Beam | Stresses: General Formula for Stress at any Point | Stresses: Stresses at Critical Points | Deflections: General Formula for Deflection at any Point[a] | Deflections: Deflections at Critical Points[a] |
|---|---|---|---|---|
| Case 18. — Fixed at Both Ends, Load at Center | | | | |
| 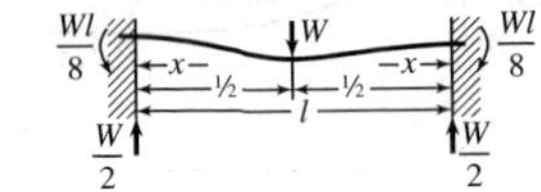 | Between each end and load, $s = \frac{W}{2Z}\left(\frac{1}{4}l - x\right)$ | Stress at ends, $\frac{Wl}{8Z}$ <br> Stress at load, $-\frac{Wl}{8Z}$ <br> These are the maximum stresses and are equal and opposite. <br> Stress is zero at $x = \frac{1}{4}l$ | $y = \frac{Wx^2}{48EI}(3l - 4x)$ | Maximum deflection, at load, $\frac{Wl^3}{192EI}$ |
| Case 19. — Fixed at Both Ends, Load at any Point | | | | |
| 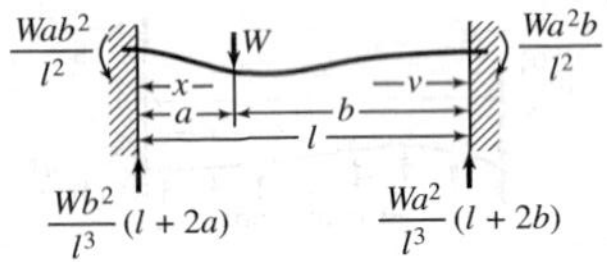 | For segment of length $a$, $s = \frac{Wb^2}{Zl^3}[al - x(l + 2a)]$ <br> For segment of length $b$, $s = \frac{Wa^2}{Zl^3}[bl - v(l + 2b)]$ | Stress at end next to segment of length $a$, $\frac{Wab^2}{Zl^2}$ <br> Stress at end next to segment of length $b$, $\frac{Wa^2b}{Zl^2}$ <br> Maximum stress is at end next to shorter segment. <br> Stress is zero at $x = \frac{al}{l + 2a}$ and $v = \frac{bl}{l + 2b}$ <br> Greatest negative stress, at load, $-\frac{2Wa^2b^2}{Zl^3}$ | For segment of length $a$, $y = \frac{Wx^2b^2}{6EIl^3}[2a(l - x) + l(a - x)]$ <br> For segment of length $b$, $y = \frac{Wv^2a^2}{6EIl^3}[2b(l - v) + l(b - v)]$ | Deflection at load, $\frac{Wa^3b^3}{3EIl^3}$ <br> Let $b$ be the length of the longer segment and $a$ of the shorter one. <br> The maximum deflection is in the longer segment, at $v = \frac{2bl}{l + 2b}$ and is $\frac{2Wa^2b^3}{3EI(l + 2b)^2}$ |

## Table 1. *(Continued)* Stresses and Deflections in Beams

| Type of Beam | Stresses: General Formula for Stress at any Point | Stresses: Stresses at Critical Points | Deflections: General Formula for Deflection at any Point[a] | Deflections: Deflections at Critical Points[a] |
|---|---|---|---|---|
| Case 20. — Fixed at Both Ends, Uniform Load | | | | |
| 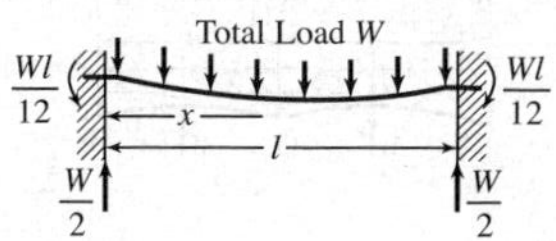 | $s = \frac{Wl}{2Z}\left\{\frac{1}{6} - \frac{x}{l} + \left(\frac{x}{l}\right)^2\right\}$ | Maximum stress, at ends, $\frac{Wl}{12Z}$<br>Stress is zero at $x = 0.7887l$ and at $x = 0.2113l$<br>Greatest negative stress, at center, $-\frac{Wl}{24Z}$ | $y = \frac{Wx^2}{24EIl}(l - x)^2$ | Maximum deflection, at center, $\frac{Wl^3}{384EI}$ |
| Case 21. — Continuous Beam, with Two Unequal Spans, Unequal, Uniform Loads | | | | |
| 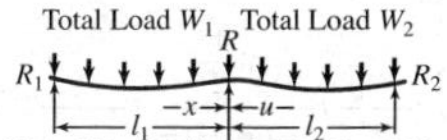$\frac{l_1 W_1(3l_1 + 4l_2) - W_2 l_2^2}{8l_1(l_1 + l_2)}$ $\frac{l_2 W_2(3l_2 + 4l_1) - W_1 l_1^2}{8l_2(l_1 + l_2)}$<br>$\left(\frac{W_1 + W_2}{2}\right) + \frac{1}{8}\left(\frac{W_1 l_1}{l_2} + \frac{W_2 l_2}{l_1}\right)$ | Between $R_1$ and $R$,<br>$s = \frac{l_1 - x}{Z}\left\{\frac{(l_1 - x)W_1}{2l_1} - R_1\right\}$<br>Between $R_2$ and $R$,<br>$s = \frac{l_2 - u}{Z}\left\{\frac{(l_2 - u)W_2}{2l_2} - R_2\right\}$ | Stress at support $R$,<br>$\frac{W_1 l_1^2 + W_2 l_2^2}{8Z(l_1 + l_2)}$<br>Greatest stress in the first span is at<br>$x = \frac{l_1}{W_1}(W_1 - R_1)$<br>and is $-\frac{R_1^2 l_1}{2ZW_1}$<br>Greatest stress in the second span is at<br>$u = \frac{l_2}{W_2}(W_2 - R_2)$<br>and is, $-\frac{R_2^2 l_2}{2ZW_2}$ | Between $R_1$ and $R$,<br>$y = \frac{x(l_1 - x)}{24EI}\left\{(2l_1 - x)(4R_1 - W_1) - \frac{W_1(l_1 - x)^2}{l_1}\right\}$<br>Between $R_2$ and $R$,<br>$y = \frac{u(l_2 - u)}{24EI}\left\{(2l_2 - u)(4R_2 - W_2) - \frac{W_2(l_2 - u)^2}{l_2}\right\}$ | This case is so complicated that convenient general expressions for the critical deflections cannot be obtained. |

**Table 1.** *(Continued)* **Stresses and Deflections in Beams**

| Type of Beam | Stresses: General Formula for Stress at any Point | Stresses: Stresses at Critical Points | Deflections: General Formula for Deflection at any Point[a] | Deflections: Deflections at Critical Points[a] |
|---|---|---|---|---|
| Case 22. — Continuous Beam, with Two Equal Spans, Uniform Load | | | | |
| 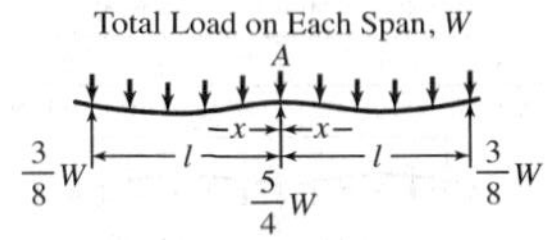 | $s = \frac{W(l-x)}{2Zl}\left(\frac{1}{4}l - x\right)$ | Maximum stress at point $A$, $\frac{Wl}{8Z}$<br>Stress is zero at $x = ¼l$<br>Greatest negative stress is at $x = ⅝l$ and is, $-\frac{9}{128}\frac{Wl}{Z}$ | $y = \frac{Wx^2(l-x)}{48EIl}(3l - 2x)$ | Maximum deflection is at $x = 0.5785l$, and is $\frac{Wl^3}{185EI}$<br>Deflection at center of span, $\frac{Wl^3}{192EI}$<br>Deflection at point of greatest negative stress, at $x = ⅝l$ is $\frac{Wl^3}{187EI}$ |
| Case 23. — Continuous Beam, with Two Equal Spans, Equal Loads at Center of Each | | | | |
| 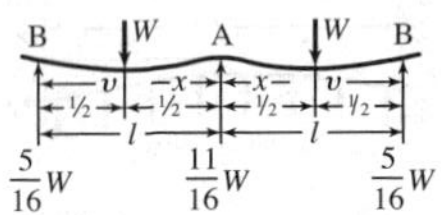 | Between point $A$ and load, $s = \frac{W}{16Z}(3l - 11x)$<br>Between point $B$ and load, $s = -\frac{5}{16}\frac{Wv}{Z}$ | Maximum stress at point $A$, $\frac{3}{16}\frac{Wl}{Z}$<br>Stress is zero at $x = \frac{3}{11}l$<br>Greatest negative stress at center of span, $-\frac{5}{32}\frac{Wl}{Z}$ | Between point $A$ and load, $y = \frac{Wx^2}{96EI}(9l - 11x)$<br>Between point $B$ and load, $y = \frac{Wv}{96EI}(3l^2 - 5v^2)$ | Maximum deflection is at $v = 0.4472l$, and is $\frac{Wl^3}{107.33EI}$<br>Deflection at load, $\frac{7}{768}\frac{Wl^3}{EI}$ |

**Table 1.** *(Continued)* **Stresses and Deflections in Beams**

| Type of Beam | Stresses: General Formula for Stress at any Point | Stresses: Stresses at Critical Points | Deflections: General Formula for Deflection at any Point[a] | Deflections: Deflections at Critical Points[a] |
|---|---|---|---|---|
| Case 24. — Continuous Beam, with Two Unequal Spans, Unequal Loads at any Point of Each | | | | |
| $m = \frac{1}{2(l_1+l_2)}\left(\frac{W_1a_1b_1}{l_1}(l_1+a_1)+\frac{W_2a_2b_2}{l_2}(l_2+a_2)\right)$ 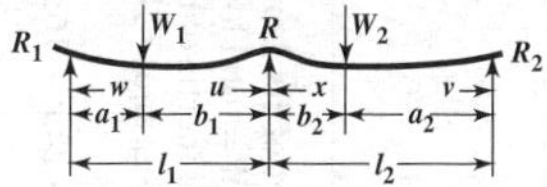 $\frac{W_1b_1-m}{l_1} = r_1$; $\left(\frac{W_1a_1+m}{l_1}+\frac{W_2a_2+m}{l_2}\right) = r$; $\frac{W_2b_2-m}{l_2} = r_2$ | Between $R_1$ and $W_1$, $s = -\frac{wr_1}{Z}$ Between $R$ and $W_1$, $s = \frac{1}{l_1Z}[m(l_1-u)-W_1a_1u]$ Between $R$ and $W_2$, $s = \frac{1}{l_2Z}[m(l_2-x)-W_2a_2x]$ Between $R_2$ and $W_2$, $s = -\frac{vr_2}{Z}$ | Stress at load $W_1$, $-\frac{a_1r_1}{Z}$ Stress at support $R$, $\frac{m}{Z}$ Stress at load $W_2$, $-\frac{a_2r_2}{Z}$ The greatest of these is the maximum stress. | Between $R_1$ and $W_1$, $y = \frac{w}{6EI}\left\{(l_1-w)(l_1+w)r_1-\frac{W_1b_1^3}{l_1}\right\}$ Between $R$ and $W_1$, $y = \frac{u}{6EIl_1}[W_1a_1b_1(l_1+a_1)-W_1a_1u^2-m(2l_1-u)(l_1-u)]$ Between $R$ and $W_2$ $y = \frac{x}{6EIl_2}[W_2a_2b_2(l_2+a_2)-W_2a_2x^2-m(2l_2-x)(l_2-x)]$ Between $R_2$ and $W_2$, $y = \frac{v}{6EI}\left\{(l_2-v)(l_2+v)r_2-\frac{W_2b_2^3}{l_2}\right\}$ | Deflection at load $W_1$, $\frac{a_1b_1}{6EIl_1}[2a_1b_1W_1-m(l_1+a_1)]$ Deflection at load $W_2$, $\frac{a_2b_2}{6EIl_2}[2a_2b_2W_2-m(l_2+a_2)]$ This case is so complicated that convenient general expressions for the maximum deflections cannot be obtained. |

[a] The deflections apply only to cases where the cross section of the beam is constant for its entire length.

In the diagrammatical illustrations of the beams and their loading, the values indicated near, but below, the supports are the "reactions" or upward forces at the supports. For Cases 1 to 12, inclusive, the reactions, as well as the formulas for the stresses, are the same whether the beam is of constant or variable cross section. For the other cases, the reactions and the stresses given are for constant cross section beams only.

The bending moment at any point in inch-pounds (newton-meters if metric units are used) is $s \times Z$ and can be found by omitting the divisor $Z$ in the formula for the stress given in the tables. A positive value of the bending moment denotes tension in the upper fibers and compression in the lower ones. A negative value denotes the reverse, The value of $W$ corresponding to a given stress is found by transposition of the formula. For example, in Case 1, the stress at the critical point is $s = -Wl \div 8Z$. From this formula we find $W = -8Zs \div l$. Of course, the negative sign of $W$ may be ignored.

In Table 1, if there are several kinds of loads, as, for instance, a uniform load and a load at any point, or separate loads at different points, the total stress and the total deflection at any point is found by adding together the various stresses or deflections at the point considered due to each load acting by itself. If the stress or deflection due to any one of the loads is negative, it must be subtracted instead of added.

Table 2a and Table 2b give expressions for determining dimensions of rectangular and round beams in terms of beam stresses and load.

**Table 2a. Rectangular Solid Beams**

| Style of Loading and Support | Breadth of Beam, $b$ inch (mm) | Beam Height, $h$ inch (mm) | Stress in Extreme Fibers, $f$ lb/in² (N/mm²) | Beam Length, $l$ inch (mm) | Total Load, $W$ lb (N) |
|---|---|---|---|---|---|
| | Beam fixed at one end, loaded at the other | | | | |
| 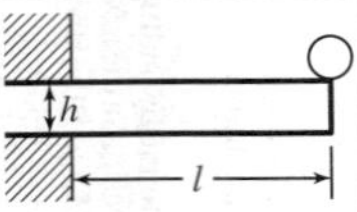  | $\frac{6lW}{fh^2} = b$ | $\sqrt{\frac{6lW}{bf}} = h$ | $\frac{6lW}{bh^2} = f$ | $\frac{bfh^2}{6W} = l$ | $\frac{bfh^2}{6l} = W$ |
| | Beam fixed at one end, uniformly loaded | | | | |
| 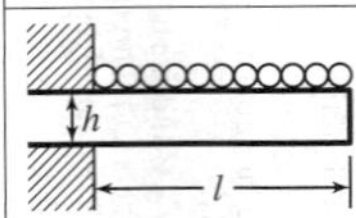  | $\frac{3lW}{fh^2} = b$ | $\sqrt{\frac{3lW}{bf}} = h$ | $\frac{3lW}{bh^2} = f$ | $\frac{bfh^2}{3W} = l$ | $\frac{bfh^2}{3l} = W$ |
| | Beam supported at both ends, single load in middle | | | | |
| 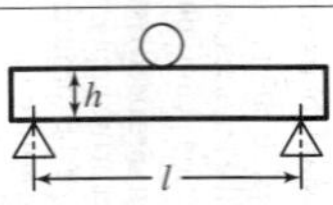  | $\frac{3lW}{2fh^2} = b$ | $\sqrt{\frac{3lW}{2bf}} = h$ | $\frac{3lW}{2bh^2} = f$ | $\frac{2bfh^2}{3W} = l$ | $\frac{2bfh^2}{3l} = W$ |
| | Beam supported at both ends, uniformly loaded | | | | |
| 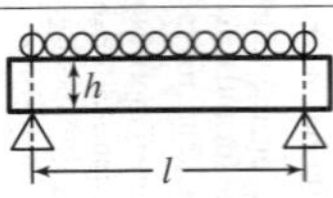  | $\frac{3lW}{4fh^2} = b$ | $\sqrt{\frac{3lW}{4bf}} = h$ | $\frac{3lW}{4bh^2} = f$ | $\frac{4bfh^2}{3W} = l$ | $\frac{4bfh^2}{3l} = W$ |
| | Beam supported at both ends, single unsymmetrical load | | | | |
| 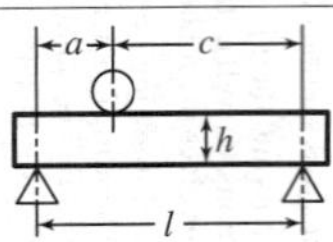  | $\frac{6Wac}{fh^2 l} = b$ | $\sqrt{\frac{6Wac}{bfl}} = h$ | $\frac{6Wac}{bh^2 l} = f$ | $a + c = l$ | $\frac{bh^2 fl}{6ac} = W$ |
| | Beam supported at both ends, two symmetrical loads | | | | |
| 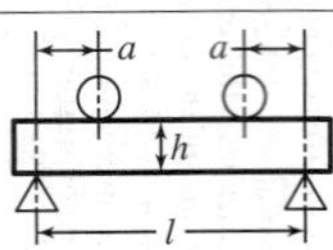  | $\frac{3Wa}{fh^2} = b$ | $\sqrt{\frac{3Wa}{bf}} = h$ | $\frac{3Wa}{bh^2} = f$ | $l$, any length $\frac{bh^2 f}{3W} = a$ | $\frac{bh^2 f}{3a} = W$ |

**Deflection of Beam Uniformly Loaded for Part of Its Length.**—In the following formulas, lengths are in inches, weights in pounds. $W$ = total load; $L$ = total length between supports; $E$ = modulus of elasticity; $I$ = moment of inertia of beam section; $a$ = fraction of length of beam at each end, that is not loaded = $b \div L$; and $f$ = deflection.

$$f = \frac{WL^3}{384EI(1-2a)}(5 - 24a^2 + 16a^4)$$

The expression for maximum bending moment is: $M_{max} = \frac{1}{8}WL(1 + 2a)$.

## Table 2b. Round Solid Beams

| Style of Loading and Support | Diameter of Beam, $d$ inch (mm) | Stress in Extreme Fibers, $f$ lb/in² (N/mm²) | Beam Length, $l$ inch (mm) | Total Load, $W$ lb (N) |
|---|---|---|---|---|
| 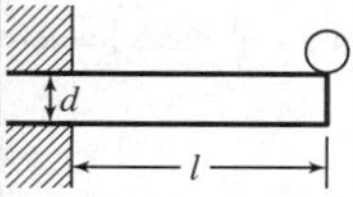 | Beam fixed at one end, loaded at the other | | | |
| | $\sqrt[3]{\frac{10.18lW}{f}} = d$ | $\frac{10.18lW}{d^3} = f$ | $\frac{d^3f}{10.18W} = l$ | $\frac{d^3f}{10.18l} = W$ |
| 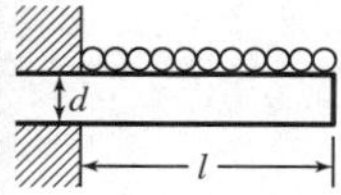 | Beam fixed at one end, uniformly loaded | | | |
| | $\sqrt[3]{\frac{5.092Wl}{f}} = d$ | $\frac{5.092Wl}{d^3} = f$ | $\frac{d^3f}{5.092W} = l$ | $\frac{d^3f}{5.092l} = W$ |
| 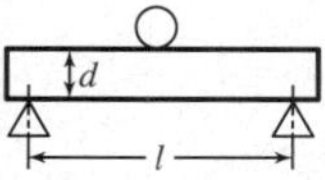 | Beam supported at both ends, single load in middle | | | |
| | $\sqrt[3]{\frac{2.546Wl}{f}} = d$ | $\frac{2.546Wl}{d^3} = f$ | $\frac{d^3f}{2.546W} = l$ | $\frac{d^3f}{2.546l} = W$ |
| 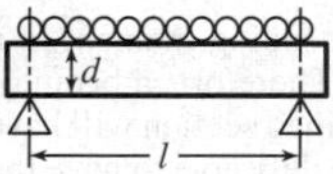 | Beam supported at both ends, uniformly loaded | | | |
| | $\sqrt[3]{\frac{1.273Wl}{f}} = d$ | $\frac{1.273Wl}{d^3} = f$ | $\frac{d^3f}{1.273W} = l$ | $\frac{d^3f}{1.273l} = W$ |
| 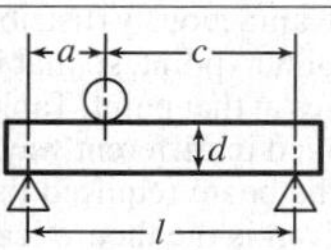 | Beam supported at both ends, single unsymmetrical load | | | |
| | $\sqrt[3]{\frac{10.18Wac}{fl}} = d$ | $\frac{10.18Wac}{d^3l} = f$ | $a + c = l$ | $\frac{d^3fl}{10.18ac} = W$ |
| 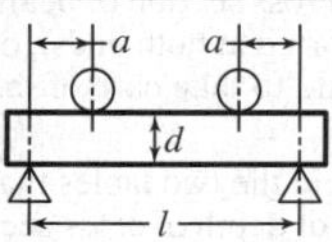 | Beam supported at both ends, two symmetrical loads | | | |
| | $\sqrt[3]{\frac{5.092Wa}{f}} = d$ | $\frac{5.092Wa}{d^3} = f$ | $l$, any length $\frac{d^3f}{5.092W} = a$ | $\frac{d^3f}{5.092a} = W$ |

These formulas apply to simple beams resting on supports at the ends.

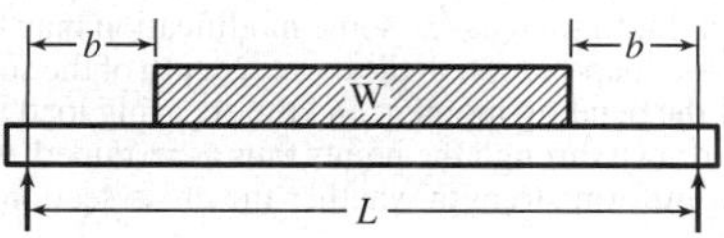


**If the formulas are used with metric SI units, $W$ = total load in newtons (N); $L$ = total length between supports in millimeters; $E$ = modulus of elasticity in newtons per millimeter² (N/mm²); $I$ = moment of inertia of beam section in mm⁴; $a$ = fraction of length of beam at each end, that is not loaded = $b \div L$; and $f$ = deflection in mm. The bending moment $M_{max}$ is in newton-millimeters (N·mm).**

***Note:* A load due to the weight of a mass of $M$ kilograms is $Mg$ newtons, where $g$ = approximately 9.81 m/s².**

**Bending Stress Due to an Oblique Transverse Force.—**The following illustration shows a beam and a channel being subjected to a transverse force acting at an angle $\phi$ to the center of gravity. To find the bending stress, the moments of inertia $I$ around axes 3-3 and 4-4 are computed from the following equations: $I_3 = I_x \sin^2\phi + I_y \cos^2\phi$, and $I_4 = I_x \cos^2\phi + I_y \sin^2\phi$.

The computed bending stress $f_b$ is then found from $f_b = M\left(\frac{y}{I_x}\sin\phi + \frac{x}{I_y}\cos\phi\right)$ where $M$ is the bending moment due to force $F$.

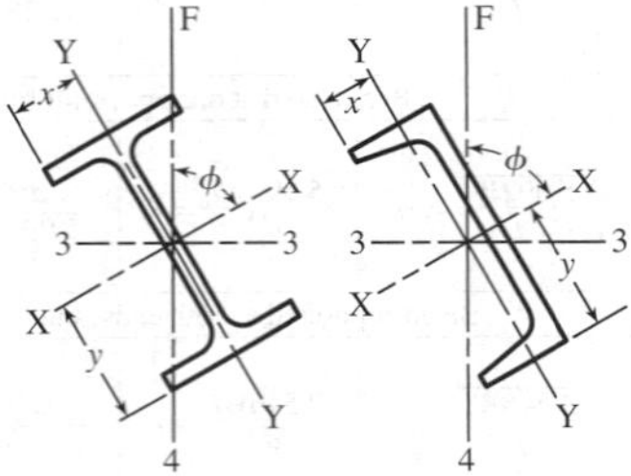


**Beams of Uniform Strength Throughout Their Length.—**The bending moment in a beam is generally not uniform throughout its length, but varies. Therefore, a beam of uniform cross section which is made strong enough at its most strained section will have an excess of material at every other section. Sometimes it may be desirable to have the cross section uniform, but at other times the metal can be more advantageously distributed if the beam is so designed that its cross section varies from point to point, so that it is at every point just great enough to take care of the bending stresses at that point. Table 3a and Table 3b are given showing beams in which the load is applied in different ways and which are supported by different methods, and the shape of the beam required for uniform strength is indicated. It should be noted that the shape given is the theoretical shape required to resist bending only. It is apparent that sufficient cross section of beam must also be added either at the points of support (in beams supported at both ends), or at the point of application of the load (in beams loaded at one end), to take care of the vertical shear.

It should be noted that the theoretical shapes of the beams given in the two tables that follow are based on the stated assumptions of uniformity of width or depth of cross section, and unless these are observed in the design the theoretical outlines do not apply without modifications. For example, in a cantilever with the load at one end, the outline is a parabola only when the width of the beam is uniform. It is not correct to use a strictly parabolic shape when the thickness is not uniform, as, for instance, when the beam is made of an I- or T-section. In such cases, some modification may be necessary; but it is evident that whatever the shape adopted, the correct depth of the section can be obtained by an investigation of the bending moment and the shearing load at a number of points, and then a line can be drawn through the points thus ascertained, which will provide for a beam of practically uniform strength whether the cross section be of uniform width or not.

**Table 3a. Beams of Uniform Strength Throughout Their Length**

| Type of Beam | Description | Formula[a] |
|---|---|---|
| 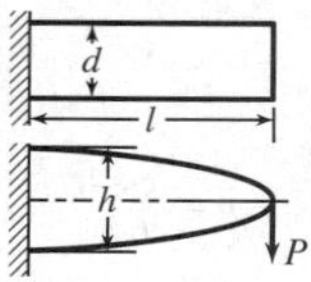  | Load at one end. Width of beam uniform. Depth of beam decreasing towards loaded end. Outline of beam-shape, parabola with vertex at loaded end. | $P = \frac{Sbh^2}{6l}$ |
| 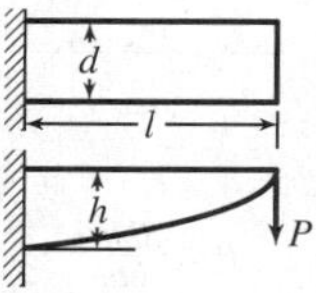  | Load at one end. Width of beam uniform. Depth of beam decreasing towards loaded end. Outline of beam, one-half of a parabola with vertex at loaded end. Beam may be reversed so that upper edge is parabolic. | $P = \frac{Sbh^2}{6l}$ |
| 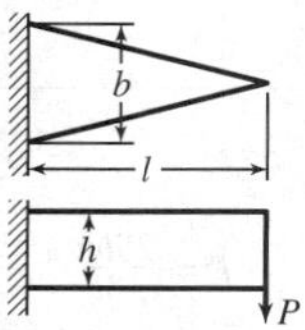  | Load at one end. Depth of beam uniform. Width of beam decreasing towards loaded end. Outline of beam triangular, with apex at loaded end. | $P = \frac{Sbh^2}{6l}$ |
| 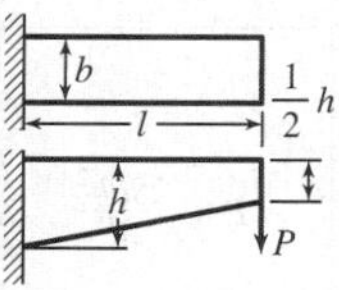  | Beam of *approximately* uniform strength. Load at one end. Width of beam uniform. Depth of beam decreasing towards loaded end, but not tapering to a sharp point. | $P = \frac{Sbh^2}{6l}$ |
| 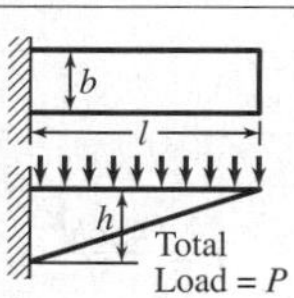  | Uniformly distributed load. Width of beam uniform. Depth of beam decreasing towards outer end. Outline of beam, right-angle triangle. | $P = \frac{Sbh^2}{3l}$ |
| 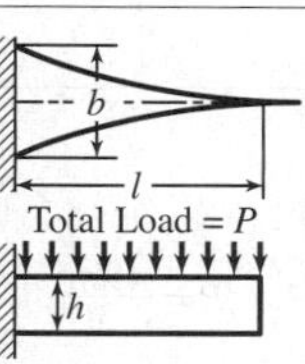  | Uniformly distributed load. Depth of beam uniform. Width of beam gradually decreasing towards outer end. Outline of beam is formed by two parabolas which are tangent to each other at their vertexes at the outer end of the beam. | $P = \frac{Sbh^2}{3l}$ |

[a] In the formulas, $P$ = load in pounds; $S$ = safe stress in lb/in$^2$; and $a, b, c, h,$ and $l$ are in inches. **If metric SI units are used, $P$ is in newtons (N); $S$ = safe stress in N/mm$^2$; and $a, b, c, h,$ and $l$ are in millimeters.**

**Table 3b. Beams of Uniform Strength Throughout Their Length**

| Type of Beam | Description | Formula[a] |
|---|---|---|
| 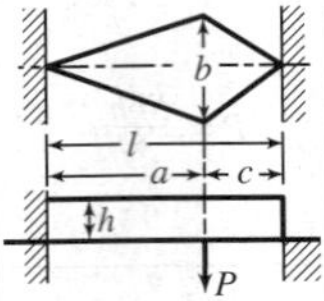 | Beam supported at both ends. Load concentrated at any point. Depth of beam uniform. Width of beam maximum at point of loading. Outline of beam, two triangles with apexes at points of support. | $P = \frac{Sbh^2 l}{6ac}$ |
| 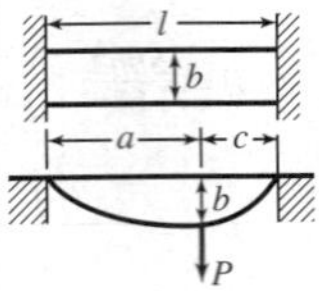 | Beam supported at both ends. Load concentrated at any point. Width of beam uniform. Depth of beam maximum at point of loading. Outline of beam is formed by two parabolas with their vertexes at points of support. | $P = \frac{Sbh^2 l}{6ac}$ |
| 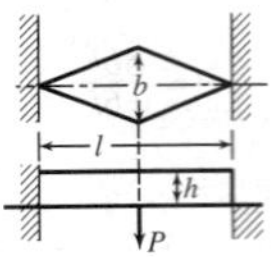 | Beam supported at both ends. Load concentrated in the middle. Depth of beam uniform. Width of beam maximum at point of loading. Outline of beam, two triangles with apexes at points of support. | $P = \frac{2Sbh^2}{3l}$ |
| 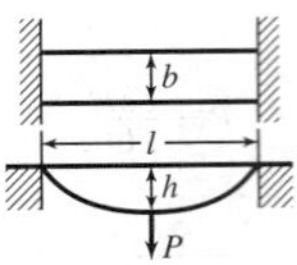 | Beam supported at both ends. Load concentrated at center. Width of beam uniform. Depth of beam maximum at point of loading. Outline of beam, two parabolas with vertices at points of support. | $P = \frac{2Sbh^2}{3l}$ |
| 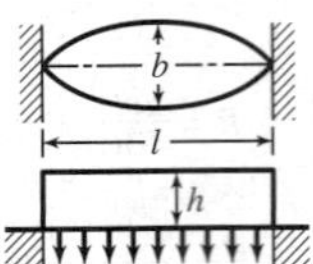 Total Load = $P$ | Beam supported at both ends. Load uniformly distributed. Depth of beam uniform. Width of beam maximum at center. Outline of beam, two parabolas with vertexes at middle of beam. | $P = \frac{4Sbh^2}{3l}$ |
| 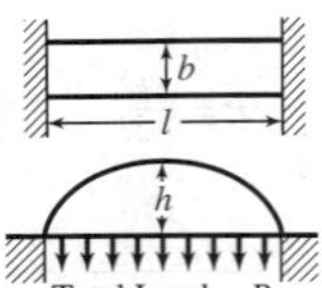 Total Load = $P$ | Beam supported at both ends. Load uniformly distributed. Width of beam uniform. Depth of beam maximum at center. Outline of beam one-half of an ellipse. | $P = \frac{4Sbh^2}{3l}$ |

[a] For details of English and metric SI units used in the formulas, see footnote on page 265.

**Deflection as a Limiting Factor in Beam Design.**—For some applications, a beam must be stronger than required by the maximum load it is to support in order to prevent excessive deflection. Maximum allowable deflections vary widely for different classes of service, so a general formula for determining them cannot be given. When exceptionally stiff girders are required, one rule is to limit the deflection to 1 inch per 100 feet of span; hence, if $l$ = length of span in inches, deflection = $l \div 1200$. According to another formula, deflection limit = $l \div 360$ where beams are adjacent to materials like plaster, which would be broken by excessive beam deflection. Some machine parts of the beam type must be very rigid to maintain alignment under load. For example, the deflection of a punch press column may be limited to 0.010 inch or less. These examples merely illustrate variations in practice. It is impracticable to give general formulas for determining the allowable deflection in any specific application because the allowable amount depends on the conditions governing each class of work.

*Procedure in Designing for Deflection:* Assume that a deflection equal to $l \div 1200$ is to be the limiting factor in selecting a wide-flange (W-shape) beam having a span length of 144 inches. Supports are at both ends and load at center is 15,000 pounds. Deflection $y$ is to be limited to $144 \div 1200 = 0.12$ inch. According to the formula on page 251 (Case 2), in which $W$ = load on beam in pounds, $l$ = length of span in inches, $E$ = modulus of elasticity of material in psi, and $I$ = moment of inertia of cross section in inches$^4$:

$$\text{Deflection } y = \frac{Wl^3}{48EI} \text{ hence, } I = \frac{Wl^3}{48yE} = \frac{15{,}000 \times 144^3}{48 \times 0.12 \times 29{,}000{,}000} = 268.1$$

A structural wide-flange beam, see *Steel Wide-Flange Sections* on page 2700, having a depth of 12 inches and weighing 35 pounds per foot has a moment of inertia $I$ of 285 and a section modulus ($Z$ or $S$) of 45.6. Checking now for maximum stress $s$ (Case 2, page 251):

$$s = \frac{Wl}{4Z} = \frac{15{,}000 \times 144}{4 \times 46.0} = 11{,}842 \text{ lbs/in}^2$$

Although deflection is the limiting factor in this case, the maximum stress is checked to make sure that it is within the allowable limit. As the limiting deflection is decreased, for a given load and length of span, the beam strength and rigidity must be increased, and, consequently, the maximum stress is decreased. Thus, in the preceding example, if the maximum deflection is 0.08 inch instead of 0.12 inch, then the calculated value for the moment of inertia $I$ will be 402; hence a W 12 × 53 beam having an $I$ value of 426 could be used (nearest value above 402). The maximum stress then would be reduced to 7640 lb/in$^2$ and the calculated deflection is 0.076 inch.

**A similar example using metric SI units is as follows. Assume that a deflection equal to $l \div 1000$ mm is to be the limiting factor in selecting a W-beam having a span length of 5 meters. Supports are at both ends and the load at the center is 30 kN. Deflection $y$ is to be limited to $5000 \div 1000 = 5$ mm. The formula on page 251 (Case 2) is applied, and $W$ = load on beam in N; $l$ = length of span in mm; $E$ = modulus of elasticity (assume 200,000 N/mm$^2$ in this example); and $I$ = moment of inertia of cross section in mm$^4$. Thus,**

$$\textbf{Deflection } y = \frac{Wl^3}{48EI}$$

**hence**

$$I = \frac{Wl^3}{48yE} = \frac{30{,}000 \times 5000^3}{48 \times 5 \times 200{,}000} = 78{,}125{,}000 \text{ mm}^4$$

**Although deflection is the limiting factor in this case, the maximum stress is checked to make sure that it is within the allowable limit using the formula from page 251 (Case 2):**

$$s = \frac{Wl}{4Z}$$

**The units of $s$ are newtons per square millimeter; $W$ is the load in newtons; $l$ is the length in mm; and $Z$ = section modulus of the cross section of the beam = $I \div$ distance in mm from neutral axis to extreme fiber.**

**Curved Beams.—**The formula $S = Mc/I$ used to compute stresses due to bending of beams is based on the assumption that the beams are straight before any loads are applied. In beams having initial curvature, however, the stresses may be considerably higher than predicted by the ordinary straight-beam formula because the effect of initial curvature is to shift the neutral axis of a curved member in from the gravity axis toward the center of curvature (the concave side of the beam). This shift in the position of the neutral axis causes an increase in the stress on the concave side of the beam and decreases the stress at the outside fibers.

Hooks, press frames, and other machine members which as a rule have a rather pronounced initial curvature may have a maximum stress at the inside fibers of up to about 3½ times that predicted by the ordinary straight-beam formula.

*Stress Correction Factors for Curved Beams:* A simple method for determining the maximum fiber stress due to bending of curved members consists of 1) calculating the maximum stress using the straight-beam formula $S = Mc/I$; and; and 2) multiplying the calculated stress by a stress correction factor. Table 4 on page 269 gives stress correction factors for some of the common cross sections and proportions used in the design of curved members.

An example in the application of the method using English units of measurement is given at the bottom of the table. **A similar example using metric SI units is as follows: The fiber stresses of a curved rectangular beam are calculated as 40 newtons per millimeter$^2$, using the straight beam formula, $S = Mc/I$. If the beam is 150 mm deep and its radius of curvature is 300 mm, what are the true stresses? $R/c = 300/75 = 4$. From Table 4 on page 269, the $K$ factors corresponding to $R/c = 4$ are 1.20 and 0.85. Thus, the inside fiber stress is $40 \times 1.20 = 48$ N/mm$^2$ = 48 megapascals (MPa); and the outside fiber stress is $40 \times 0.85 = 34$ N/mm$^2$ = 34 MPa.**

*Approximate Formula for Stress Correction Factor:* The stress correction factors given in Table 4 on page 269 were determined by Wilson and Quereau and published in the University of Illinois Engineering Experiment Station Circular No. 16, "A Simple Method of Determining Stress in Curved Flexural Members." In this same publication the authors indicate that the following empirical formula may be used to calculate the value of the stress correction factor for the *inside* fibers of sections not covered by the tabular data to within 5 percent accuracy, except in triangular sections where up to 10 percent deviation may be expected. However, for most engineering calculations, this formula should prove satisfactory for general use in determining the factor for the inside fibers.

$$K = 1.00 + 0.5\frac{I}{bc^2}\left[\frac{1}{R-c} + \frac{1}{R}\right]$$

(Use 1.05 instead of 0.5 in this formula for circular and elliptical sections.)

$I$ = moment of inertia of section about centroidal axis

$b$ = maximum width of section

$c$ = distance from centroidal axis to inside fiber, i.e., to the extreme fiber nearest the center of curvature

$R$ = radius of curvature of centroidal axis of beam

## Table 4. Values of Stress Correction Factor *K* for Various Curved Beam Sections

| Section | $R/c$ | Factor *K* Inside Fiber | Factor *K* Outside Fiber | $y_0$[a] | Section | $R/c$ | Factor *K* Inside Fiber | Factor *K* Outside Fiber | $y_0$[a] |
|---|---|---|---|---|---|---|---|---|---|
| 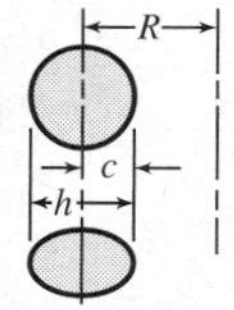 | 1.2 | 3.41 | .54 | .224R | 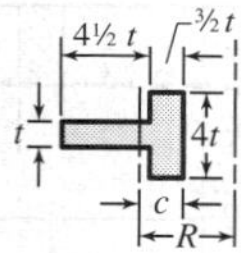 | 1.2 | 3.63 | .58 | .418R |
| | 1.4 | 2.40 | .60 | .151R | | 1.4 | 2.54 | .63 | .299R |
| | 1.6 | 1.96 | .65 | .108R | | 1.6 | 2.14 | .67 | .229R |
| | 1.8 | 1.75 | .68 | .084R | | 1.8 | 1.89 | .70 | .183R |
| | 2.0 | 1.62 | .71 | .069R | | 2.0 | 1.73 | .72 | .149R |
| | 3.0 | 1.33 | .79 | .030R | | 3.0 | 1.41 | .79 | .069R |
| | 4.0 | 1.23 | .84 | .016R | | 4.0 | 1.29 | .83 | .040R |
| | 6.0 | 1.14 | .89 | .0070R | | 6.0 | 1.18 | .88 | .018R |
| | 8.0 | 1.10 | .91 | .0039R | | 8.0 | 1.13 | .91 | .010R |
| | 10.0 | 1.08 | .93 | .0025R | | 10.0 | 1.10 | .92 | .0065R |
| 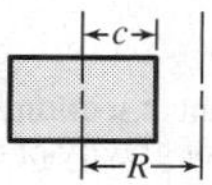 | 1.2 | 2.89 | .57 | .305R | 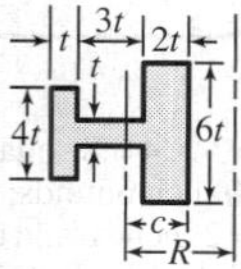 | 1.2 | 3.55 | .67 | .409R |
| | 1.4 | 2.13 | .63 | .204R | | 1.4 | 2.48 | .72 | .292R |
| | 1.6 | 1.79 | .67 | .149R | | 1.6 | 2.07 | .76 | .224R |
| | 1.8 | 1.63 | .70 | .112R | | 1.8 | 1.83 | .78 | .178R |
| | 2.0 | 1.52 | .73 | .090R | | 2.0 | 1.69 | .80 | .144R |
| | 3.0 | 1.30 | .81 | .041R | | 3.0 | 1.38 | .86 | .067R |
| | 4.0 | 1.20 | .85 | .021R | | 4.0 | 1.26 | .89 | .038R |
| | 6.0 | 1.12 | .90 | .0093R | | 6.0 | 1.15 | .92 | .018R |
| | 8.0 | 1.09 | .92 | .0052R | | 8.0 | 1.10 | .94 | .010R |
| | 10.0 | 1.07 | .94 | .0033R | | 10.0 | 1.08 | .95 | .0065R |
| 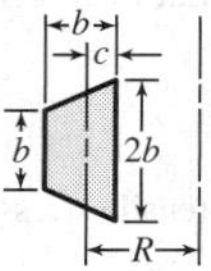 | 1.2 | 3.01 | .54 | .336R | 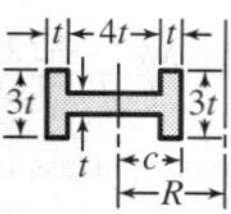 | 1.2 | 2.52 | .67 | .408R |
| | 1.4 | 2.18 | .60 | .229R | | 1.4 | 1.90 | .71 | .285R |
| | 1.6 | 1.87 | .65 | .168R | | 1.6 | 1.63 | .75 | .208R |
| | 1.8 | 1.69 | .68 | .128R | | 1.8 | 1.50 | .77 | .160R |
| | 2.0 | 1.58 | .71 | .102R | | 2.0 | 1.41 | .79 | .127R |
| | 3.0 | 1.33 | .80 | .046R | | 3.0 | 1.23 | .86 | .058R |
| | 4.0 | 1.23 | .84 | .024R | | 4.0 | 1.16 | .89 | .030R |
| | 6.0 | 1.13 | .88 | .011R | | 6.0 | 1.10 | .92 | .013R |
| | 8.0 | 1.10 | .91 | .0060R | | 8.0 | 1.07 | .94 | .0076R |
| | 10.0 | 1.08 | .93 | .0039R | | 10.0 | 1.05 | .95 | .0048R |
| 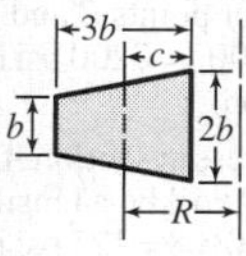 | 1.2 | 3.09 | .56 | .336R | 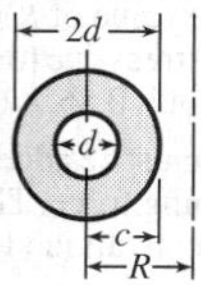 | 1.2 | 3.28 | .58 | .269R |
| | 1.4 | 2.25 | .62 | .229R | | 1.4 | 2.31 | .64 | .182R |
| | 1.6 | 1.91 | .66 | .168R | | 1.6 | 1.89 | .68 | .134R |
| | 1.8 | 1.73 | .70 | .128R | | 1.8 | 1.70 | .71 | .104R |
| | 2.0 | 1.61 | .73 | .102R | | 2.0 | 1.57 | .73 | .083R |
| | 3.0 | 1.37 | .81 | .046R | | 3.0 | 1.31 | .81 | .038R |
| | 4.0 | 1.26 | .86 | .024R | | 4.0 | 1.21 | .85 | .020R |
| | 6.0 | 1.17 | .91 | .011R | | 6.0 | 1.13 | .90 | .0087R |
| | 8.0 | 1.13 | .94 | .0060R | | 8.0 | 1.10 | .92 | .0049R |
| | 10.0 | 1.11 | .95 | .0039R | | 10.0 | 1.07 | .93 | .0031R |
| 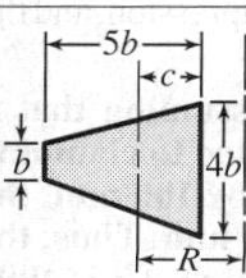 | 1.2 | 3.14 | .52 | .352R | 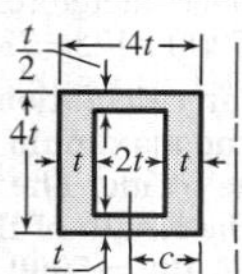 | 1.2 | 2.63 | .68 | .399R |
| | 1.4 | 2.29 | .54 | .243R | | 1.4 | 1.97 | .73 | .280R |
| | 1.6 | 1.93 | .62 | .179R | | 1.6 | 1.66 | .76 | .205R |
| | 1.8 | 1.74 | .65 | .138R | | 1.8 | 1.51 | .78 | .159R |
| | 2.0 | 1.61 | .68 | .110R | | 2.0 | 1.43 | .80 | .127R |
| | 3.0 | 1.34 | .76 | .050R | | 3.0 | 1.23 | .86 | .058R |
| | 4.0 | 1.24 | .82 | .028R | | 4.0 | 1.15 | .89 | .031R |
| | 6.0 | 1.15 | .87 | .012R | | 6.0 | 1.09 | .92 | .014R |
| | 8.0 | 1.12 | .91 | .0060R | | 8.0 | 1.07 | .94 | .0076R |
| | 10.0 | 1.10 | .93 | .0039R | | 10.0 | 1.06 | .95 | .0048R |
| 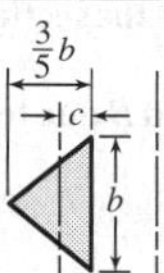 | 1.2 | 3.26 | .44 | .361R | | | | | |
| | 1.4 | 2.39 | .50 | .251R | | | | | |
| | 1.6 | 1.99 | .54 | .186R | | | | | |
| | 1.8 | 1.78 | .57 | .144R | | | | | |
| | 2.0 | 1.66 | .60 | .116R | | | | | |
| | 3.0 | 1.37 | .70 | .052R | | | | | |
| | 4.0 | 1.27 | .75 | .029R | | | | | |
| | 6.0 | 1.16 | .82 | .013R | | | | | |
| | 8.0 | 1.12 | .86 | .0060R | | | | | |
| | 10.0 | 1.09 | .88 | .0039R | | | | | |

*Example:* The fiber stresses of a curved rectangular beam are calculated as 5000 psi using the straight beam formula, $S = Mc/I$. If the beam is 8 inches deep and its radius of curvature is 12 inches, what are the true stresses?

*Solution:* $R/c = 12/4 = 3$. The factors in the table corresponding to $R/c = 3$ are 0.81 and 1.30. Outside fiber stress = 5000 × 0.81 = 4050 psi; inside fiber stress = 5000 × 1.30 = 6500 psi.

[a] $y_0$ is the distance from the centroidal axis to the neutral axis of curved beams subjected to pure bending and is measured from the centroidal axis toward the center of curvature.

The accompanying diagram shows the dimensions of a clamp frame of rectangular cross section. Determine the maximum stress at points *A* and *B* due to a clamping force of 1000 pounds.

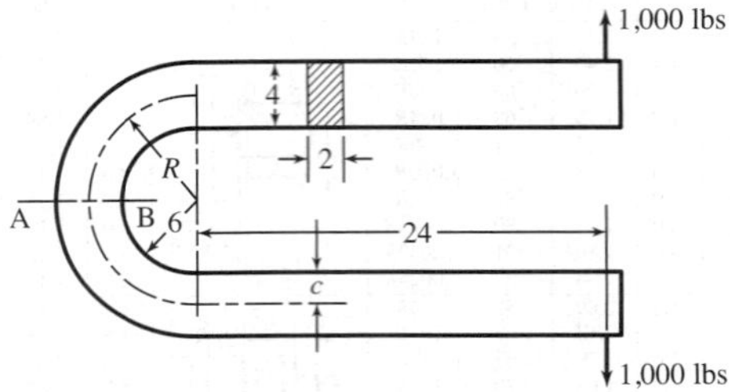


The cross-sectional area = 2 × 4 = 8 square inches; the bending moment at section *AB* is 1000 (24 + 6 + 2) = 32,000 inch pounds; the distance from the center of gravity of the section at *AB* to point *B* is *c* = 2 inches; and using the formula on page 236, the moment of inertia of the section is $2 \times (4)^3 \div 12 = 10.667$ inches$^4$.

Using the straight-beam formula, page 268, the stress at points *A* and *B* due to the bending moment is:

$$S = \frac{Mc}{I} = \frac{32{,}000 \times 2}{10.667} = 6000 \text{ psi}$$

The stress at *A* is a compressive stress of 6000 psi and that at *B* is a tensile stress of 6000 psi.

These values must be corrected to account for the curvature effect. In Table 4 on page 269 for $R/c = (6 + 2)/2 = 4$, the value of *K* is found to be 1.20 and 0.85 for points *B* and *A* respectively. Thus, the actual stress due to bending at point *B* is 1.20 × 6000 = 7200 psi in tension, and the stress at point *A* is 0.85 × 6000 = 5100 psi in compression.

To these stresses at *A* and *B* must be added, algebraically, the direct stress at section *AB* due to the 1000-pound clamping force. The direct stress on section *AB* will be a tensile stress equal to the clamping force divided by the section area. Thus 1000 ÷ 8 = 125 psi in tension.

The maximum unit stress at *A* is, therefore, 5100 − 125 = 4975 psi in compression, and the maximum unit stress at *B* is 7200 + 125 = 7325 psi in tension.

**The following is a similar calculation using metric SI units, assuming that it is required to determine the maximum stress at points *A* and *B* due to clamping force of 4 kN acting on the frame. The frame cross section is 50 by 100 mm, the radius *R* = 200 mm, and the length of the straight portions is 600 mm. Thus, the cross-sectional area = 50 × 100 = 5000 mm$^2$; the bending moment at *AB* is 4000 (600 + 200) = 3,200,000 newton-millimeters; the distance from the center of gravity of the section at *AB* to point *B* is *c* = 50 mm; and the moment of inertia of the section is (from the formula on page 236), $50(100^3)/12 = 4{,}170{,}000$ mm$^4$.**

**Using the straight-beam formula, page 268, the stress at points *A* and *B* due to the bending moment is:**

$$s = \frac{Mc}{I} = \frac{3{,}200{,}000 \times 50}{4{,}170{,}000}$$
$$= 38.4 \text{ newtons per millimeter}^2 = 38.4 \text{ megapascals}$$

**The stress at *A* is a compressive stress of 38.4 N/mm$^2$, while that at *B* is a tensile stress of 38.4 N/mm$^2$. These values must be corrected to account for the curvature**

**effect. From the table on page 269, the $K$ factors are 1.20 and 0.85 for points $A$ and $B$, respectively, derived from $R/c = 200/50 = 4$. Thus, the actual stress due to bending at point $B$ is $1.20 \times 38.4 = 46.1$ N/mm² (46.1 MPa) in tension; and the stress at point $A$ is $0.85 \times 38.4 = 32.6$ N/mm² (32.6 MPa) in compression.**

**To these stresses at $A$ and $B$ must be added, algebraically, the direct stress at section $AB$ due to the 4 kN clamping force. The direct stress on section $AB$ will be a tensile stress equal to the clamping force divided by the section area. Thus, 4000/5000 = 0.8 N/mm². The maximum unit stress at $A$ is, therefore, 32.61 − 0.8 = 31.8 N/mm² (31.8 MPa) in compression, and the maximum unit stress at $B$ is 46.1 + 0.8 = 46.9 N/mm² (46.9 MPa) in tension.**

**Size of Rail Necessary to Carry a Given Load.**—The following formulas may be employed for determining the size of rail and wheel suitable for carrying a given load. Let $A$ = the width of the head of the rail in inches; $B$ = width of the tread of the rail in inches; $C$ = the wheel-load in pounds; $D$ = the diameter of the wheel in inches.

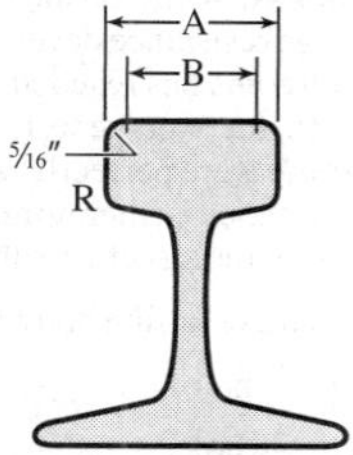


Then the width of the tread of the rail in inches is found from the formula:

$$B = \frac{C}{1250D} \tag{1}$$

The width $A$ of the head equals $B + \frac{5}{8}$ inch. The diameter $D$ of the smallest track wheel that will safely carry the load is found from the formula:

$$D = \frac{C}{A \times K} \tag{2}$$

in which $K = 600$ to 800 for steel castings; $K = 300$ to 400 for cast iron.

As an example, assume that the wheel-load is 10,000 pounds; the diameter of the wheel is 20 inches; and the material is cast steel. Determine the size of rail necessary to carry this load. From Formula (1):

$$B = \frac{10{,}000}{1250 \times 20} = 0.4 \text{ inch}$$

The width of the rail required equals $0.4 + \frac{5}{8}$ inch = 1.025 inch. Determine also whether a wheel 20 inches in diameter is large enough to safely carry the load. From Formula (2):

$$D = \frac{10{,}000}{1.025 \times 600} = 16\frac{1}{4} \text{ inches}$$

This is the smallest diameter of track wheel that will safely carry the load; hence a 20-inch wheel is ample.

**American Railway Engineering Association Formulas.**—The American Railway Engineering Association recommends for safe operation of steel cylinders rolling on steel plates that the allowable load $p$ in pounds per inch of length of the cylinder should not exceed the value calculated from the formula

$$p = \frac{y.s. - 13{,}000}{20{,}000} 600d \text{ for diameter } d \text{ less than 25 inches}$$

This formula is based on steel having a yield strength, y.s., of 32,000 pounds per square inch. For roller or wheel diameters of up to 25 inches, the Hertz stress (contact stress) resulting from the calculated load *p* will be approximately 76,000 pounds per square inch.

For a 10-inch diameter roller the safe load per inch of roller length is

$$p = \frac{32{,}000 - 13{,}000}{20{,}000} 600 \times 10 = 5700 \text{ lbs per inch of length}$$

Therefore, to support a 10,000 pound load the roller or wheel would need to be 10,000/5700 = 1.75 inches wide.

## Stresses Produced by Shocks

**Stresses in Beams Produced by Shocks.**—Any elastic structure subjected to a shock will deflect until the product of the average resistance developed by the deflection and the distance through which it has been overcome has reached a value equal to the energy of the shock. It follows that for a given shock, the average resisting stresses are inversely proportional to the deflection. If the structure were perfectly rigid, the deflection would be zero and the stress infinite. The effect of a shock is, therefore, to a great extent dependent upon the elastic property (the springiness) of the structure subjected to the impact.

The energy of a body in motion, such as a falling body, may be spent in each of four ways:

1) In deforming the body struck as a whole.

2) In deforming the falling body as a whole.

3) In partial deformation of both bodies on the surface of contact (most of this energy will be transformed into heat).

4) Part of the energy will be taken up by the supports, if these are not perfectly rigid and inelastic.

How much energy is spent in the last three ways is usually difficult to determine, and for this reason it is safest to figure as if the whole amount were spent as in Case 1. If a reliable judgment is possible as to what percentage of the energy is spent in other ways than the first, a corresponding fraction of the total energy can be assumed as developing stresses in the body subjected to shocks.

One investigation into the stresses produced by shocks led to the following conclusions: 1) A suddenly applied load will produce the same deflection, and, therefore, the same stress as a static load twice as great; and 2) the unit stress *p* (see formulas in Table 1, *Stresses Produced in Beams by Shocks*) for a given load producing a shock varies directly as the square root of the modulus of elasticity *E* and inversely as the square root of the length *L* of the beam and the area of the section.

Thus, for instance, if the sectional area of a beam is increased by four times, the unit stress will diminish only by half. This result is entirely different from those produced by static loads where the stress would vary inversely with the area, and within certain limits be practically independent of the modulus of elasticity.

In Table 1, the expression for the approximate value of *p*, which is applicable whenever the deflection of the beam is small as compared with the total height *h* through which the body producing the shock is dropped, is always the same for beams supported at both ends and subjected to shock at *any* point between the supports. In the formulas all dimensions are in inches and weights in pounds.

**Table 1. Stresses Produced in Beams by Shocks**

| Method of Support and Point Struck by Falling Body | Fiber (Unit) Stress $p$ Produced by Weight $Q$ Dropped Through a Distance $h$ | Approximate Value of $p$ |
|---|---|---|
| Supported at both ends; struck in center. | $p = \frac{QaL}{4I}\left(1 + \sqrt{1 + \frac{96hEI}{QL^3}}\right)$ | $p = a\sqrt{\frac{6QhE}{LI}}$ |
| Fixed at one end; struck at the other. | $p = \frac{QaL}{I}\left(1 + \sqrt{1 + \frac{6hEI}{QL^3}}\right)$ | $p = a\sqrt{\frac{6QhE}{LI}}$ |
| Fixed at both ends; struck in center. | $p = \frac{QaL}{8I}\left(1 + \sqrt{1 + \frac{384hEI}{QL^3}}\right)$ | $p = a\sqrt{\frac{6QhE}{LI}}$ |

$I$ = moment of inertia of section; $a$ = distance of extreme fiber from neutral axis; $L$ = length of beam; $E$ = modulus of elasticity.

**If metric SI units are used, $p$ is in newtons per square millimeter; $Q$ is in newtons; $E$ = modulus of elasticity in $N/mm^2$; $I$ = moment of inertia of section in $mm^4$; and $h$, $a$, and $L$ in mm. *Note:* If $Q$ is given in kilograms, the value referred to is mass. The weight $Q$ of a mass $M$ kilograms is $Mg$ newtons, where $g$ = approximately 9.81 $m/s^2$.**

*Examples of How Formulas for Stresses Produced by Shocks are Derived:* The general formula from which specific formulas for shock stresses in beams, springs, and other machine and structural members are derived is:

$$p = p_s\left(1 + \sqrt{1 + \frac{2h}{y}}\right) \tag{1}$$

In this formula, $p$ = stress in psi due to shock caused by impact of a moving load; $p_s$ = stress in psi resulting when moving load is applied statically; $h$ = distance in inches that load falls before striking beam, spring, or other member; $y$ = deflection in inches resulting from static load.

As an example of how Formula (1) may be used to obtain a formula for a specific application, suppose that the load $W$ shown applied to the beam in Case 2 on page 251 were dropped on the beam from a height of $h$ inches instead of being gradually applied (static loading). The maximum stress $p_s$ due to load $W$ for Case 2 is given as $Wl \div 4Z$ and the maximum deflection $y$ is given as $Wl^3 \div 48\ EI$. Substituting these values in Formula (1),

$$p = \frac{Wl}{4Z}\left(1 + \sqrt{1 + \frac{2h}{Wl^3 \div 48EI}}\right) = \frac{Wl}{4Z}\left(1 + \sqrt{1 + \frac{96hEI}{Wl^3}}\right) \tag{2}$$

If in Formula (2) the letter $Q$ is used in place of $W$ and if $Z$, the section modulus, is replaced by its equivalent, $I \div$ distance $a$ from neutral axis to extreme fiber of beam, then Formula (2) becomes the first formula given in the accompanying Table 1, *Stresses Produced in Beams by Shocks.*

**Stresses in Helical Springs Produced by Shocks.—**A load suddenly applied on a spring will produce the same deflection, and, therefore, also the same unit stress, as a static load twice as great. When the load drops from a height $h$, the stresses are as given in the accompanying Table 2. The approximate values are applicable when the deflection is small as compared with the height $h$. The formulas show that the fiber stress for a given shock will be greater in a spring made from a square bar than in one made from a round bar, if the diameter of coil is the same and the side of the square bar equals the diameter of the round bar. It is, therefore, more economical to use round stock for springs which must withstand shocks, due to the fact that the deflection for the same fiber stress for a square bar spring is smaller than that for a round bar spring, the ratio being as 4 to 5. The round bar spring is therefore capable of storing more energy than a square bar spring for the same stress.

**Table 2. Stresses Produced in Springs by Shocks**

| Form of Bar from Which Spring is Made | Fiber (Unit) Stress $f$ Produced by Weight $Q$ Dropped a Height $h$ on a Helical Spring | Approximate Value of $f$ |
|---|---|---|
| Round | $f = \frac{8QD}{\pi d^3}\left(1 + \sqrt{1 + \frac{Ghd^4}{4QD^3 n}}\right)$ | $f = 1.27\sqrt{\frac{QhG}{Dd^2 n}}$ |
| Square | $f = \frac{9QD}{4d^3}\left(1 + \sqrt{1 + \frac{Ghd^4}{0.9\pi QD^3 n}}\right)$ | $f = 1.34\sqrt{\frac{QhG}{Dd^2 n}}$ |

$G$ = modulus of elasticity for torsion; $d$ = diameter or side of bar; $D$ = mean diameter of spring; $n$ = number of coils in spring.

**Shocks from Bodies in Motion.**—The formulas given can be applied, in general, to shocks from bodies in motion. A body of weight $W$ moving horizontally with the velocity of $v$ feet per second has a stored-up energy:

$$E_K = \frac{1}{2} \times \frac{Wv^2}{g} \text{ foot-pounds} \quad \text{or} \quad \frac{6Wv^2}{g} \text{ inch-pounds}$$

This expression may be substituted for $Qh$ in the tables in the equations for unit stresses containing this quantity, and the stresses produced by the energy of the moving body thereby determined.

The formulas in the tables give the maximum value of the stresses, providing the designer with some definitive guidance even where there may be justification for assuming that only a part of the energy of the shock is taken up by the member under stress.

**The formulas can also be applied using metric SI units. The stored-up energy of a body of mass $M$ kilograms moving horizontally with the velocity of $v$ meters per second is:**

$$E_K = \frac{1}{2} Mv^2 \text{ newton-meters}$$

**This expression may be substituted for $Qh$ in the appropriate equations in the tables. For calculation in millimeters, $Qh = 1000\,E_K$ newton-millimeters.**

**Fatigue Stresses.**—So-called "fatigue ruptures" occur in parts that are subjected to continually repeated shocks or stresses of small magnitude. Machine parts that are subjected to continual stresses in varying directions, or to repeated shocks, even if of comparatively small magnitude, may fail ultimately if designed from a mere knowledge of the behavior of the material under a steady stress, such as is imposed upon it by ordinary tensile stress testing machines. Examinations of numerous cases of machine parts, broken under actual working conditions, indicate that at least 80 percent of these ruptures are caused by fatigue stresses. Most fatigue ruptures are caused by bending stresses, and frequently by a revolving bending stress. Hence, to test materials for this class of stress the tests should be made to stress the material in a manner similar to that in which it will be stressed under actual working conditions. See *Fatigue Properties* on page 203 for more on this topic.

## COLUMNS

### Strength of Columns or Struts

Structural members which are subject to compression may be so long in proportion to their diameter or lateral dimensions that failure may be the result 1) of both compression and bending; and 2) of bending or buckling to such a degree that compression stress may be ignored.

In such cases, the *slenderness ratio* is important. This ratio equals the length $l$ of the column in inches or millimeters, according to the unit system in use, divided by the least radius of gyration $r$ of the cross section. Various formulas have been used for designing columns which are too slender to be designed for compression only.

**Rankine-Gordon Formula.—**This formula is generally applied when slenderness ratios range between 20 and 100, and sometimes for ratios up to 120. The notation, in English and metric SI units of measurement, is given on page 277.

$$p_{cr} = \frac{S}{1 + K\left(\frac{l}{r}\right)^2} = \text{Critical stress, lb/in}^2 \text{ or N/mm}^2$$

Factor $K$ may be established by tests with a given material and end condition, and for the probable range of $l/r$. If determined by calculation, $K = S/C\pi^2E$. Factor $C$ equals 1 for either rounded or pivoted column ends, 4 for fixed ends, and 1 to 4 for square flat ends. The factors 25,000, 12,500, etc., in the Rankine formulas, arranged as on page 277, equal $1/K$, and have been used extensively.

**Straight-Line Formula.—**This general type of formula is often used in designing compression members for buildings, bridges, or similar structural work. It is convenient especially in designing a number of columns that are made of the same material but vary in size, assuming that factor $B$ is known. This factor is determined by tests.

$$p_{cr} = S_y - B\left(\frac{l}{r}\right) = \text{Critical stress, lb/in}^2$$

$S_y$ equals yield point, lb/in$^2$, and factor $B$ ranges from 50 to 100. Allowable stress $= p_{cr}$ / factor of safety.

**Formulas of American Railway Engineering Association.—**The formulas that follow apply to structural steel having an ultimate strength of 60,000 to 72,000 lb/in$^2$.

For building columns having $l/r$ ratios not greater than 120:

$$\text{Allowable stress} = 17{,}000 - 0.485\, l^2/r^2$$

For columns having $l/r$ ratios greater than 120:

$$\text{Allowable stress} = \frac{18{,}000}{1 + l^2/18{,}000r^2}$$

For bridge compression members centrally loaded and values of $l/r$ not greater than 140:

$$\text{Allowable stress, riveted ends} = 15{,}000 - \frac{1}{4}\frac{l^2}{r^2}$$

$$\text{Allowable stress, pin ends} = 15{,}000 - \frac{1}{3}\frac{l^2}{r^2}$$

**Euler Formula.—**This formula is for columns that are so slender that bending or buckling action predominates and compressive stresses are not taken into account.

$$P_{cr} = \frac{C\pi^2 IE}{l^2} = \text{Critical load, in pounds or newtons}$$

The notation, in English and metric SI units of measurement, is given in the table *Rankine and Euler Formulas for Columns* on page 277. Factors *C* for different end conditions are included in the Euler formulas at the bottom of the table. According to a series of experiments, Euler formulas should be used if the values of $l/r$ exceed the following ratios: Structural steel and flat ends, 195; hinged ends, 155; round ends, 120; cast iron with flat ends, 120; hinged ends, 100; round ends, 75; oak with flat ends, 130. The *critical slenderness ratio*, which marks the dividing line between the shorter columns and those slender enough to warrant using the Euler formula, depends upon the column material and its end conditions. If the Euler formula is applied when the slenderness ratio is too small, the *calculated* ultimate strength will exceed the yield point of the material and, obviously, will be incorrect.

**Eccentrically Loaded Columns.—**In the application of the column formulas previously referred to, it is assumed that the action of the load coincides with the axis of the column. If the load is offset relative to the column axis, the column is said to be eccentrically loaded, and its strength is then calculated by using a modification of the Rankine formula, the quantity $cz/r^2$ being added to the denominator, as shown in the table on the next page. This modified formula is applicable to columns having a slenderness ratio varying from 20 or 30 to about 100.

**Machine Elements Subjected to Compressive Loads.—**As in structural compression members, an unbraced machine member that is relatively slender (i.e., its length is more than, say, six times the least dimension perpendicular to its longitudinal axis) is usually designed as a column because failure due to overloading (assuming a compressive load centrally applied in an axial direction) may occur by buckling or a combination of buckling and compression rather than by direct compression alone. In the design of unbraced steel machine "columns" which are to carry compressive loads applied along their longitudinal axes, two formulas are in general use:

(Euler) $$P_{cr} = \frac{S_y A r^2}{Q} \quad (1)$$

(J. B. Johnson) $$P_{cr} = AS_y\left(1 - \frac{Q}{4r^2}\right) \quad (2) \qquad \text{where} \qquad Q = \frac{S_y l^2}{n\pi^2 E} \quad (3)$$

In these formulas, $P_{cr}$ = critical load in pounds that would result in failure of the column; $A$ = cross sectional area, inches$^2$; $S_y$ = yield point of material, psi; $r$ = least radius of gyration of cross section, inches; $E$ = modulus of elasticity, psi; $l$ = column length, inches; and $n$ = coefficient for end conditions. For both ends fixed, $n = 4$; for one end fixed, one end free, $n = 0.25$; for one end fixed and the other end free but guided, $n = 2$; for round or pinned ends, free but guided, $n = 1$; and for flat ends, $n = 1$ to 4. It should be noted that these values of $n$ represent ideal conditions that are seldom attained in practice; for example, for both ends fixed, a value of $n = 3$ to 3.5 may be more realistic than $n = 4$.

**If metric SI units are used in these formulas, $P_{cr}$ = critical load in newtons that would result in failure of the column; $A$ = cross-sectional area, mm$^2$; $S_y$ = yield point of the material, N/mm$^2$; $r$ = least radius of gyration of cross section, mm; $E$ = modulus of elasticity, N/mm$^2$; $l$ = column length, mm; and $n$ = a coefficient for end conditions. The coefficients given are valid for calculations in metric units.**

## Rankine and Euler Formulas for Columns

| Symbol | Quantity | English Unit | Metric SI Units |
|---|---|---|---|
| $p_{cr}$ | Critical stress | Lbs/in$^2$ | Newtons/mm$^2$ |
| $P_{cr}$ | Critical load | Pounds | Newtons |
| $S$ | Ultimate compressive strength of material | Lbs/in$^2$ | Newtons/mm$^2$ |
| $l$ | Length of column or strut | Inches | Millimeters |
| $r$ | Least radius of gyration | Inches | Millimeters |
| $I$ | Least moment of inertia | Inches$^4$ | Millimeters$^4$ |
| $r^2$ | Moment of inertia/area of section | Inches$^2$ | Millimeters$^2$ |
| $E$ | Modulus of elasticity of material | Lbs/in$^2$ | Newtons/mm$^2$ |
| $c$ | Distance from neutral axis of cross section to side under compression | Inches | Millimeters |
| $z$ | Distance from axis of load to axis coinciding with center of gravity of cross section | Inches | Millimeters |

| Specific Rankine-Gordon Formulas (where $l/r$ is between 20 and 100) | | | |
|---|---|---|---|
| Material | Both Ends of Column Fixed | One End Fixed and One End Rounded | Both Ends Rounded |
| Steel | $p_{cr} = \dfrac{S}{1 + \dfrac{l^2}{25{,}000r^2}}$ | $p_{cr} = \dfrac{S}{1 + \dfrac{l^2}{12{,}500r^2}}$ | $p_{cr} = \dfrac{S}{1 + \dfrac{l^2}{6250r^2}}$ |
| Cast Iron | $p_{cr} = \dfrac{S}{1 + \dfrac{l^2}{5000r^2}}$ | $p_{cr} = \dfrac{S}{1 + \dfrac{l^2}{2500r^2}}$ | $p_{cr} = \dfrac{S}{1 + \dfrac{l^2}{1250r^2}}$ |
| Wrought Iron | $p_{cr} = \dfrac{S}{1 + \dfrac{l^2}{35{,}000r^2}}$ | $p_{cr} = \dfrac{S}{1 + \dfrac{l^2}{17{,}500r^2}}$ | $p_{cr} = \dfrac{S}{1 + \dfrac{l^2}{8750r^2}}$ |
| Timber | $p_{cr} = \dfrac{S}{1 + \dfrac{l^2}{3000r^2}}$ | $p_{cr} = \dfrac{S}{1 + \dfrac{l^2}{1500r^2}}$ | $p_{cr} = \dfrac{S}{1 + \dfrac{l^2}{750r^2}}$ |

| Formulas Modified for Eccentrically Loaded Columns | | | |
|---|---|---|---|
| Material | Both Ends of Column Fixed | One End Fixed and One End Rounded | Both Ends Rounded |
| Steel | $p_{cr} = \dfrac{S}{1 + \dfrac{l^2}{25{,}000r^2} + \dfrac{cz}{r^2}}$ | $p_{cr} = \dfrac{S}{1 + \dfrac{l^2}{12{,}500r^2} + \dfrac{cz}{r^2}}$ | $p_{cr} = \dfrac{S}{1 + \dfrac{l^2}{6250r^2} + \dfrac{cz}{r^2}}$ |
| For materials other than steel, such as cast iron, use the Rankine formulas given in the upper table and add to the denominator the quantity $cz/r^2$ | | | |

| Specific Euler Formulas for Slender Columns (where $l/r$ is greater than 100) | | | |
|---|---|---|---|
| Both Ends of Column Fixed | One End Fixed and One End Rounded | Both Ends Rounded | One End Fixed and One End Free |
| $P_{cr} = \dfrac{4\pi^2 IE}{l^2}$ | $P_{cr} = \dfrac{2\pi^2 IE}{l^2}$ | $P_{cr} = \dfrac{\pi^2 IE}{l^2}$ | $P_{cr} = \dfrac{\pi^2 IE}{4l^2}$ |

*Allowable Loads for Columns:* To find the total allowable working load for a given section, divide the critical load, $P_{cr}$ (or $p_{cr} \times$ area), as found by the appropriate formula above, by a suitable factor of safety.

*Factor of Safety for Machine Columns:* When the conditions of loading and the physical qualities of the material used are accurately known, a factor of safety as low as 1.25 is sometimes used when minimum weight is important. Usually, however, a factor of safety of 2 to 2.5 is applied for steady loads. The factor of safety represents the ratio of the critical load $P_{cr}$ to the working load.

*Application of Euler and Johnson Formulas:* To determine whether the Euler or Johnson formula is applicable in any particular case, it is necessary to determine the value of the quantity $Q \div r^2$. If $Q \div r^2$ is greater than 2, then the Euler Formula (1) should be used; if $Q \div r^2$ is less than 2, then the J. B. Johnson formula is applicable. Most compression members in machine design are in the range of proportions covered by the Johnson formula. For this reason a good procedure is to design machine elements on the basis of the Johnson formula and then as a check calculate $Q \div r^2$ to determine whether the Johnson formula applies or the Euler formula should have been used.

*Example 1, Compression Member Design:* A rectangular machine member 24 inches long and ½ × 1 inch in cross section is to carry a compressive load of 4000 pounds along its axis. What is the factor of safety for this load if the material is machinery steel having a yield point of 40,000 psi, the load is steady, and each end of the rod has a ball connection so that $n = 1$?

*Solution:* From Formula (3)

$$Q = \frac{40{,}000 \times 24 \times 24}{1 \times 3.1416 \times 3.1416 \times 30{,}000{,}000} = 0.0778$$

(The values 40,000 and 30,000,000 were obtained from the table *Strength Data for Iron and Steel* on page 427.)

The radius of gyration $r$ for a rectangular section (page 236) is 0.289 × the dimension in the direction of bending. In columns, bending is most apt to occur in the direction in which the section is the weakest, the ½-inch dimension in this example. Hence, least radius of gyration $r = 0.289 \times ½ = 0.145$ inch.

$$\frac{Q}{r^2} = \frac{0.0778}{(0.145)^2} = 3.70$$

which is more than 2, so the Euler formula will be used.

$$P_{cr} = \frac{s_y A r^2}{Q} = \frac{40{,}000 \times \frac{1}{2} \times 1}{3.70}$$

$$= 5400 \text{ pounds so that the factor of safety is } 5400 \div 4000 = 1.35$$

*Example 2, Compression Member Design:* In the preceding example, the column formulas were used to check the adequacy of a column of known dimensions. The more usual problem involves determining what the dimensions should be to resist a specified load. For example:

A 24-inch long bar of rectangular cross section with width $w$ twice its depth $d$ is to carry a load of 4000 pounds. What must the width and depth be if a factor of safety of 1.35 is to be used?

*Solution:* First determine the critical load $P_{cr}$:

$$P_{cr} = \text{working load} \times \text{factor of safety}$$

$$= 4000 \times 1.35 = 5400 \text{ pounds}$$

Next determine $Q$, which is found as in Example 1, and will again be 0.0778.

Assume Formula (2) applies:

$$P_{cr} = As_y\left(1 - \frac{Q}{4r^2}\right)$$

$$5400 = w \times d \times 40{,}000\left(1 - \frac{0.0778}{4r^2}\right)$$

$$= 2d^2 \times 40{,}000\left(1 - \frac{0.01945}{r^2}\right)$$

$$\frac{5400}{40{,}000 \times 2} = d^2\left(1 - \frac{0.01945}{r^2}\right)$$

As mentioned in Example 1 the least radius of gyration $r$ of a rectangle is equal to 0.289 times the least dimension, $d$, in this case. Therefore, substituting for $d$ the value $r \div 0.289$,

$$\frac{5400}{40{,}000 \times 2} = \left(\frac{r}{0.289}\right)^2\left(1 - \frac{0.01945}{r^2}\right)$$

$$\frac{5400 \times 0.289 \times 0.289}{40{,}000 \times 2} = r^2 - 0.01945$$

$$0.005638 = r^2 - 0.01945$$

$$r^2 = 0.0251$$

Checking to determine if $Q \div r^2$ is greater or less than 2,

$$\frac{Q}{r^2} = \frac{0.0778}{0.0251} = 3.1$$

Therefore, Formula (1) should have been used to determine $r$ and dimensions $w$ and $d$. Using Formula (1),

$$5400 = \frac{40{,}000 \times 2d^2 \times r^2}{Q} = \frac{40{,}000 \times 2 \times \left(\frac{r}{0.289}\right)^2 r^2}{0.0778}$$

$$r^4 = \frac{5400 \times 0.0778 \times 0.289 \times 0.289}{40{,}000 \times 2} = 0.0004386$$

$$d = \frac{0.145}{0.289} = 0.50 \text{ inch}$$

and $w = 2d = 1$ inch as in the previous example.

**American Institute of Steel Construction.—**For main or secondary compression members with $l/r$ ratios up to 120, safe unit stress = $17{,}000 - 0.485l^2/r^2$. For columns and bracing or other secondary members with $l/r$ ratios above 120,

$$\text{For bracing and secondary members, allowable stress, psi} = \frac{18{,}000}{1 + l^2/18{,}000r^2}$$

$$\text{For main members, allowable stress, psi} = \frac{18{,}000}{1 + l^2/18{,}000r^2} \times \left(1.6 - \frac{l/r}{200}\right)$$

*Pipe Columns:* Allowable concentric loads for steel pipe columns based on the above formulas are given in the table on page 280.

## Allowable Concentric Loads for Steel Pipe Columns

| STANDARD STEEL PIPE | | | | | | | | |
|---|---|---|---|---|---|---|---|---|
| Nominal Diameter, Inches | 12 | 10 | 8 | 6 | 5 | 4 | 3½ | 3 |
| Wall Thickness, Inch | 0.375 | 0.365 | 0.322 | 0.280 | 0.258 | 0.237 | 0.226 | 0.216 |
| Weight per Foot, Pounds | 49.56 | 40.48 | 28.55 | 18.97 | 14.62 | 10.79 | 9.11 | 7.58 |
| Effective Length (*KL*), Feet[a] | Allowable Concentric Loads in Thousands of Pounds | | | | | | | |
| 6 | 303 | 246 | 171 | 110 | 83 | 59 | 48 | 38 |
| 7 | 301 | 243 | 168 | 108 | 81 | 57 | 46 | 36 |
| 8 | 299 | 241 | 166 | 106 | 78 | 54 | 44 | 34 |
| 9 | 296 | 238 | 163 | 103 | 76 | 52 | 41 | 31 |
| 10 | 293 | 235 | 161 | 101 | 73 | 49 | 38 | 28 |
| 11 | 291 | 232 | 158 | 98 | 71 | 46 | 35 | 25 |
| 12 | 288 | 229 | 155 | 95 | 68 | 43 | 32 | 22 |
| 13 | 285 | 226 | 152 | 92 | 65 | 40 | 29 | 19 |
| 14 | 282 | 223 | 149 | 89 | 61 | 36 | 25 | 16 |
| 15 | 278 | 220 | 145 | 86 | 58 | 33 | 22 | 14 |
| 16 | 275 | 216 | 142 | 82 | 55 | 29 | 19 | 12 |
| 17 | 272 | 213 | 138 | 79 | 51 | 26 | 17 | 11 |
| 18 | 268 | 209 | 135 | 75 | 47 | 23 | 15 | 10 |
| 19 | 265 | 205 | 131 | 71 | 43 | 21 | 14 | 9 |
| 20 | 261 | 201 | 127 | 67 | 39 | 19 | 12 | |
| 22 | 254 | 193 | 119 | 59 | 32 | 15 | 10 | |
| 24 | 246 | 185 | 111 | 51 | 27 | 13 | | |
| 25 | 242 | 180 | 106 | 47 | 25 | 12 | | |
| 26 | 238 | 176 | 102 | 43 | 23 | | | |
| EXTRA STRONG STEEL PIPE | | | | | | | | |
| Nominal Diameter, Inches | 12 | 10 | 8 | 6 | 5 | 4 | 3½ | 3 |
| Wall Thickness, Inch | 0.500 | 0.500 | 0.500 | 0.432 | 0.375 | 0.337 | 0.318 | 0.300 |
| Weight per Foot, Pounds | 65.42 | 54.74 | 43.39 | 28.57 | 20.78 | 14.98 | 12.50 | 10.25 |
| Effective Length (*KL*), Feet[a] | Allowable Concentric Loads in Thousands of Pounds | | | | | | | |
| 6 | 400 | 332 | 259 | 166 | 118 | 81 | 66 | 52 |
| 7 | 397 | 328 | 255 | 162 | 114 | 78 | 63 | 48 |
| 8 | 394 | 325 | 251 | 159 | 111 | 75 | 59 | 45 |
| 9 | 390 | 321 | 247 | 155 | 107 | 71 | 55 | 41 |
| 10 | 387 | 318 | 243 | 151 | 103 | 67 | 51 | 37 |
| 11 | 383 | 314 | 239 | 146 | 99 | 63 | 47 | 33 |
| 12 | 379 | 309 | 234 | 142 | 95 | 59 | 43 | 28 |
| 13 | 375 | 305 | 229 | 137 | 91 | 54 | 38 | 24 |
| 14 | 371 | 301 | 224 | 132 | 86 | 49 | 33 | 21 |
| 15 | 367 | 296 | 219 | 127 | 81 | 44 | 29 | 18 |
| 16 | 363 | 291 | 214 | 122 | 76 | 39 | 25 | 16 |
| 18 | 353 | 281 | 203 | 111 | 65 | 31 | 20 | 12 |
| 19 | 349 | 276 | 197 | 105 | 59 | 28 | 18 | 11 |
| 20 | 344 | 271 | 191 | 99 | 54 | 25 | 16 | |
| 21 | 337 | 265 | 185 | 92 | 48 | 22 | 14 | |
| 22 | 334 | 260 | 179 | 86 | 44 | 21 | | |
| 24 | 323 | 248 | 166 | 73 | 37 | 17 | | |
| 26 | 312 | 236 | 152 | 62 | 32 | | | |
| 28 | 301 | 224 | 137 | 54 | 27 | | | |

[a] With respect to radius of gyration. The effective length (*KL*) is the actual unbraced length, *L*, in feet, multiplied by the effective length factor (*K*), which is dependent upon the restraint at the ends of the unbraced length and the means available to resist lateral movements. *K* may be determined by referring to the last portion of this table.

## Allowable Concentric Loads for Steel Pipe Columns *(Continued)*

| DOUBLE-EXTRA STRONG STEEL PIPE | | | | | |
|---|---|---|---|---|---|
| Nominal Diameter, Inches | 8 | 6 | 5 | 4 | 3 |
| Wall Thickness, Inch | 0.875 | 0.864 | 0.750 | 0.674 | 0.600 |
| Weight per Foot, Pounds | 72.42 | 53.16 | 38.55 | 27.54 | 18.58 |
| Effective Length (*KL*), Feet[a] | Allowable Concentric Loads in Thousands of Pounds | | | | |
| 6 | 431 | 306 | 216 | 147 | 91 |
| 7 | 424 | 299 | 209 | 140 | 84 |
| 8 | 417 | 292 | 202 | 133 | 77 |
| 9 | 410 | 284 | 195 | 126 | 69 |
| 10 | 403 | 275 | 187 | 118 | 60 |
| 11 | 395 | 266 | 178 | 109 | 51 |
| 12 | 387 | 257 | 170 | 100 | 43 |
| 13 | 378 | 247 | 160 | 91 | 37 |
| 14 | 369 | 237 | 151 | 81 | 32 |
| 15 | 360 | 227 | 141 | 70 | 28 |
| 16 | 351 | 216 | 130 | 62 | 24 |
| 17 | 341 | 205 | 119 | 55 | 22 |
| 18 | 331 | 193 | 108 | 49 | |
| 19 | 321 | 181 | 97 | 44 | |
| 20 | 310 | 168 | 87 | 40 | |
| 22 | 288 | 142 | 72 | 33 | |
| 24 | 264 | 119 | 61 | | |
| 26 | 240 | 102 | 52 | | |
| 28 | 213 | 88 | 44 | | |

| EFFECTIVE LENGTH FACTORS (*K*) FOR VARIOUS COLUMN CONFIGURATIONS | | | | | | |
|---|---|---|---|---|---|---|
| | (a) | (b) | (c) | (d) | (e) | (f) |
| Buckled shape of column is shown by dashed line | 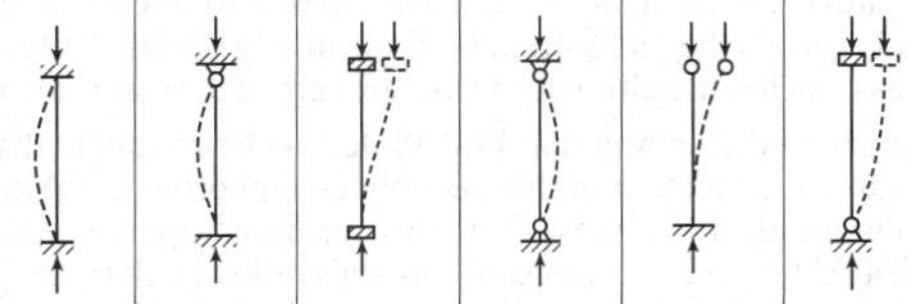 | | | | | |
| Theoretical *K* value | 0.5 | 0.7 | 1.0 | 1.0 | 2.0 | 2.0 |
| Recommended design value when ideal conditions are approximated | 0.65 | 0.80 | 1.2 | 1.0 | 2.10 | 2.0 |
| End condition code |  | Rotation fixed and translation fixed<br>Rotation free and translation fixed<br>Rotation fixed and translation free<br>Rotation free and translation free | | | | |

Load tables are given for 36 ksi yield stress steel. No load values are given below the heavy horizontal lines, because the *Kl/r* ratios (where *l* is the actual unbraced length in inches and *r* is the governing radius of gyration in inches) would exceed 200.

Data from "Manual of Steel Construction," 8th ed., 1980, with permission of the American Institute of Steel Construction (AISC).

## PLATES, SHELLS, AND CYLINDERS

**Flat Stayed Surfaces.**—Large flat areas are often held against pressure by stays distributed at regular intervals over the surface. In boiler work, these stays are usually screwed into the plate and the projecting end riveted over to ensure steam tightness. The US Board of Supervising Inspectors and the American Boiler Manufacturers Association (ABMA) rules give the following formula for flat stayed surfaces:

$$P = \frac{C \times t^2}{S^2}$$

in which $P$ = pressure in pounds per square inch

$C$ = a constant, which equals

112 for plates $^7\!/_{16}$ inch and under

120, for plates over $^7\!/_{16}$ inch thick

140, for plates with stays having a nut and bolt on the inside and outside

160, for plates with stays having washers of at least one-half the plate thickness, and with a diameter at least one-half of the greatest pitch

$t$ = thickness of plate in 16ths of an inch (thickness = $^7\!/_{16}$, $t = 7$)

$S$ = greatest pitch of stays in inches

**Strength and Deflection of Flat Plates.**—Generally, the formulas used to determine stresses and deflections in flat plates are based on certain assumptions that can be closely approximated in practice. These assumptions are:

1) the thickness of the plate is not greater than one-quarter the least width of the plate;

2) the greatest deflection when the plate is loaded is less than one-half the plate thickness;

3) the maximum tensile stress resulting from the load does not exceed the elastic limit of the material; and

4) all loads are perpendicular to the plane of the plate.

Plates of ductile materials fail when the maximum stress resulting from deflection under load exceeds the yield strength; for brittle materials, failure occurs when the maximum stress reaches the ultimate tensile strength of the material involved.

**Square and Rectangular Flat Plates.**—The formulas that follow give the maximum stress and deflection of flat steel plates supported in various ways and subjected to the loading indicated. These formulas are based upon a modulus of elasticity for steel of 30,000,000 psi and a value of Poisson's ratio of 0.3. If the formulas for maximum stress, $S$, are applied without modification to other materials, such as cast iron, aluminum, and brass, for which the range of Poisson's ratio is about 0.26 to 0.34, the maximum stress calculations will be in error by not more than about 3 percent. The deflection formulas may also be applied to materials other than steel by substituting in these formulas the appropriate value for $E$, the modulus of elasticity of the material (see pages 427 and 508). The deflections thus obtained will not be in error by more than about 3 percent.

In the stress and deflection formulas that follow,

$p$ = uniformly distributed load acting on plate, psi

$W$ = total load on plate, pounds; $W = p \times$ area of plate

$L$ = distance between supports (length of plate), inches. For rectangular plates, $L$ = long side, $l$ = short side

$t$ = thickness of plate, inches

$S$ = maximum tensile stress in plate, psi

$d$ = maximum deflection of plate, in.

$E$ = modulus of elasticity in tension. $E$ = 30,000,000 psi for steel

**If metric SI units are used in the formulas, then,**

**$W$ = total load on plate, newtons**

**$L$ = distance between supports (length of plate), mm. For rectangular plates, $L$ = long side, $l$ = short side**

**$t$ = thickness of plate, mm**

**$S$ = maximum tensile stress in plate, N/mm²**

**$d$ = maximum deflection of plate, mm**

**$E$ = modulus of elasticity, N/mm²**

a) Square flat plate supported at top and bottom of all four edges and a uniformly distributed load over the surface of the plate.

$$S = \frac{0.29W}{t^2} \quad (1) \qquad d = \frac{0.0443WL^2}{Et^3} \quad (2)$$

b) Square flat plate supported at the bottom only of all four edges and a uniformly distributed load over the surface of the plate.

$$S = \frac{0.28W}{t^2} \quad (3) \qquad d = \frac{0.0443WL^2}{Et^3} \quad (4)$$

c) Square flat plate with all edges firmly fixed and a uniformly distributed load over the surface of the plate.

$$S = \frac{0.31W}{t^2} \quad (5) \qquad d = \frac{0.0138WL^2}{Et^3} \quad (6)$$

d) Square flat plate with all edges firmly fixed and a uniform load over small circular area at the center. In Equations (7) and (9), $r_0$ = radius of area to which load is applied. If $r_0 < 1.7t$, use $r_s$ where $r_s = \sqrt{1.6r_0^2 + t^2} - 0.675t$.

$$S = \frac{0.62W}{t^2}\log_e\left(\frac{L}{2r_0}\right) \quad (7) \qquad d = \frac{0.0568WL^2}{Et^3} \quad (8)$$

e) Square flat plate with all edges supported above and below, or below only, and a concentrated load at the center. See Item d), above, for definition of $r_0$.

$$S = \frac{0.62W}{t_2}\left[\log_e\left(\frac{L}{2r_0}\right) + 0.577\right] \quad (9) \qquad d = \frac{0.1266WL^2}{Et^3} \quad (10)$$

f) Rectangular plate with all edges supported at top and bottom and a uniformly distributed load over the surface of the plate.

$$S = \frac{0.75W}{t^2\left(\frac{L}{l} + 1.61\frac{l^2}{L^2}\right)} \quad (11) \qquad d = \frac{0.1422W}{Et^3\left(\frac{L}{l^3} + \frac{2.21}{L^2}\right)} \quad (12)$$

g) Rectangular plate with all edges fixed and a uniformly distributed load over the surface of the plate.

$$S = \frac{0.5W}{t^2\left(\frac{L}{l} + \frac{0.623l^5}{L^5}\right)} \quad (13) \qquad d = \frac{0.0284W}{Et^3\left(\frac{L}{l^3} + \frac{1.056l^2}{L^4}\right)} \quad (14)$$

**Circular Flat Plates.**—In the following formulas, $R$ = radius of plate to supporting edge in inches; $W$ = total load in pounds; and other symbols are the same as used for square and rectangular plates.

**If metric SI units are used, $R$ = radius of plate to supporting edge in mm, and the values of other symbols are the same as those used for square and rectangular plates.**

a) Edge supported around the circumference and a uniformly distributed load over the surface of the plate.

$$S = \frac{0.39W}{t^2} \quad (15) \qquad d = \frac{0.221WR^2}{Et^3} \quad (16)$$

b) Edge fixed around circumference and a uniformly distributed load over the surface of the plate.

$$S = \frac{0.24W}{t^2} \quad (17) \qquad d = \frac{0.0543WR^2}{Et^3} \quad (18)$$

c) Edge supported around the circumference and a concentrated load at the center.

$$S = \frac{0.48W}{t^2}\left[1 + 1.3\log_e\frac{R}{0.325t} - 0.0185\frac{t^2}{R^2}\right] \quad (19) \qquad d = \frac{0.55WR^2}{Et^3} \quad (20)$$

d) Edge fixed around circumference and a concentrated load at the center.

$$S = \frac{0.62W}{t^2}\left[\log_e\frac{R}{0.325t} + 0.0264\frac{t^2}{R^2}\right] \quad (21) \qquad d = \frac{0.22WR^2}{Et^3} \quad (22)$$

**Strength of Cylinders Subjected to Internal Pressure.**—In designing a cylinder to withstand internal pressure, the choice of formula to be used depends on 1) the kind of material of which the cylinder is made (whether brittle or ductile); 2) the construction of the cylinder ends (whether open or closed); and 3) whether the cylinder is classed as a thin- or a thick-walled cylinder.

A cylinder is considered to be thin-walled when the ratio of wall thickness to inside diameter is 0.1 or less and thick-walled when this ratio is greater than 0.1. Materials such as cast iron, hard steel, and cast aluminum are considered to be brittle materials; low-carbon steel, brass, bronze, etc. are considered to be ductile.

In the formulas that follow, $p$ = internal pressure, psi; $D$ = inside diameter of cylinder, inches; $t$ = wall thickness of cylinder, inches; $\mu$ = Poisson's ratio, = 0.3 for steel, 0.26 for cast iron, 0.34 for aluminum and brass; and $S$ = allowable tensile stress, psi.

**Metric SI units can be used in Formulas (23), (25), (26), and (27), where $p$ = internal pressure in $N/mm^2$; $D$ = inside diameter of cylinder, mm; $t$ = wall thickness, mm; $\mu$ = Poisson's ratio, = 0.3 for steel, 0.26 for cast iron, and 0.34 for aluminum and brass; and $S$ = allowable tensile stress, $N/mm^2$. For the use of metric SI units in Formula (24), see below.**

*Thin-walled Cylinders:*

$$t = \frac{Dp}{2S} \quad (23)$$

For low-pressure cylinders of cast iron such as are used for certain engine and press applications, a formula in common use is

$$t = \frac{Dp}{2500} + 0.3 \quad (24)$$

This formula is based on allowable stress of 1250 pounds per square inch (psi) and will give a wall thickness 0.3 inch greater than Formula (23) to allow for variations in metal thickness that may result from the casting process.

**If metric SI units are used in Formula (24), $t$ = cylinder wall thickness in mm; $D$ = inside diameter of cylinder, mm; and the allowable stress is in N/mm$^2$. The value of 0.3 inches additional wall thickness is 7.62 mm, and the next highest number in preferred metric basic sizes is 8 mm.**

*Thick-walled Cylinders of Brittle Material, Ends Open or Closed:* Lamé's equation is used when cylinders of this type are subjected to internal pressure.

$$t = \frac{D}{2}\left(\sqrt{\frac{S+p}{S-p}} - 1\right) \tag{25}$$

The table *Ratio of Outside Radius to Inside Radius, Thick Cylinders* on page 286 is for convenience in calculating the dimensions of cylinders under high internal pressure without the use of Formula (25).

*Example, Use of the Table:* Assume that a cylinder of 10 inches inside diameter is to withstand a pressure of 2500 psi; the material is cast iron and the allowable stress is 6000 psi. To solve the problem, locate the allowable stress per square inch in the left-hand column of the table and the working pressure at the top of the columns. Then find the ratio between the outside and inside radii in the body of the table. In this example, the ratio is 1.558, and hence the outside diameter of the cylinder should be 10 × 1.558, or about $15\frac{5}{8}$ inches. The thickness of the cylinder wall will therefore be (15.58 − 10)/2 = 2.79 inches.

Unless very high-grade material is used and sound castings assured, cast iron should not be used for pressures exceeding 2000 pounds psi (13.75 N/mm$^2$). It is well to leave more metal in the bottom of a hydraulic cylinder than is indicated by the results of calculations because a hole of some size must be cored in the bottom to permit the entrance of a boring bar when finishing the cylinder, and when this hole is subsequently tapped and plugged it often gives trouble if there is too little thickness.

For steady or gradually applied stresses, the maximum allowable fiber stress $S$ may be assumed to be from 3500 to 4000 psi (24–27 N/mm$^2$) for cast iron; from 6000 to 7000 psi (41–48 N/mm$^2$) for brass; and 12,000 psi (82 N/mm$^2$) for steel castings. For intermittent stresses, such as in cylinders for steam and hydraulic work, 3000 psi (20 N/mm$^2$) for cast iron; 5000 psi (34 N/mm$^2$) for brass; and 10,000 psi (69 N/mm$^2$) for steel castings is ordinarily used. These values give ample factors of safety.

***Note:* In metric SI units, 1000 pounds per square inch equals 6.895 newtons per square millimeter (1000 lb/in$^2$ = 6.895 N/mm$^2$). Also, one newtons per square millimeter equals one megapascal (1 N/mm$^2$ = 1 MPa).**

*Thick-walled Cylinders of Ductile Material, Closed Ends:* Clavarino's equation is used:

$$t = \frac{D}{2}\left[\sqrt{\frac{S+(1-2\mu)p}{S-(1+\mu)p}} - 1\right] \tag{26}$$

*Thick-walled Cylinders of Ductile Material, Open Ends:* Birnie's equation is used:

$$t = \frac{D}{2}\left[\sqrt{\frac{S+(1-\mu)p}{S-(1+\mu)p}} - 1\right] \tag{27}$$

**Spherical Shells Subjected to Internal Pressure.**—Let:

$D$ = internal diameter of shell in inches

$p$ = internal pressure in psi

$S$ = safe tensile stress per square inch

$t$ = thickness of metal in the shell, in inches. Then, $t = pD \div 4S$ (28)

## Ratio of Outside Radius to Inside Radius, Thick Cylinders

| Allowable Stress per Sq. In. of Section | Working Pressure in Cylinder, Pounds per Square Inch | | | | | | | | | | | | |
|---|---|---|---|---|---|---|---|---|---|---|---|---|---|
| | 1000 | 1500 | 2000 | 2500 | 3000 | 3500 | 4000 | 4500 | 5000 | 5500 | 6000 | 6500 | 7000 |
| 2000 | 1.732 | … | … | … | … | … | … | … | … | … | … | … | … |
| 2500 | 1.528 | 2.000 | … | … | … | … | … | … | … | … | … | … | … |
| 3000 | 1.414 | 1.732 | 2.236 | … | … | … | … | … | … | … | … | … | … |
| 3500 | 1.342 | 1.581 | 1.915 | 2.449 | … | … | … | … | … | … | … | … | … |
| 4000 | 1.291 | 1.483 | 1.732 | 2.082 | 2.646 | … | … | … | … | … | … | … | … |
| 4500 | 1.254 | 1.414 | 1.612 | 1.871 | 2.236 | 2.828 | … | … | … | … | … | … | … |
| 5000 | 1.225 | 1.363 | 1.528 | 1.732 | 2.000 | 2.380 | 3.000 | … | … | … | … | … | … |
| 5500 | 1.202 | 1.323 | 1.464 | 1.633 | 1.844 | 2.121 | 2.517 | 3.162 | … | … | … | … | … |
| 6000 | 1.183 | 1.291 | 1.414 | 1.558 | 1.732 | 1.949 | 2.236 | 2.646 | 3.317 | … | … | … | … |
| 6500 | … | 1.265 | 1.374 | 1.500 | 1.648 | 1.826 | 2.049 | 2.345 | 2.769 | 3.464 | … | … | … |
| 7000 | … | 1.243 | 1.342 | 1.453 | 1.581 | 1.732 | 1.915 | 2.145 | 2.449 | 2.887 | 3.606 | … | … |
| 7500 | … | 1.225 | 1.314 | 1.414 | 1.528 | 1.658 | 1.813 | 2.000 | 2.236 | 2.550 | 3.000 | 3.742 | … |
| 8000 | … | 1.209 | 1.291 | 1.382 | 1.483 | 1.599 | 1.732 | 1.890 | 2.082 | 2.324 | 2.646 | 3.109 | 3.873 |
| 8500 | … | 1.195 | 1.271 | 1.354 | 1.446 | 1.549 | 1.667 | 1.803 | 1.964 | 2.160 | 2.408 | 2.739 | 3.215 |
| 9000 | … | 1.183 | 1.254 | 1.330 | 1.414 | 1.508 | 1.612 | 1.732 | 1.871 | 2.035 | 2.236 | 2.490 | 2.828 |
| 9500 | … | … | 1.238 | 1.309 | 1.387 | 1.472 | 1.567 | 1.673 | 1.795 | 1.936 | 2.104 | 2.309 | 2.569 |
| 10,000 | … | … | 1.225 | 1.291 | 1.363 | 1.441 | 1.528 | 1.624 | 1.732 | 1.856 | 2.000 | 2.171 | 2.380 |
| 10,500 | … | … | 1.213 | 1.275 | 1.342 | 1.414 | 1.494 | 1.581 | 1.679 | 1.789 | 1.915 | 2.062 | 2.236 |
| 11,000 | … | … | 1.202 | 1.260 | 1.323 | 1.390 | 1.464 | 1.544 | 1.633 | 1.732 | 1.844 | 1.972 | 2.121 |
| 11,500 | … | … | 1.192 | 1.247 | 1.306 | 1.369 | 1.438 | 1.512 | 1.593 | 1.683 | 1.784 | 1.897 | 2.028 |
| 12,000 | … | … | 1.183 | 1.235 | 1.291 | 1.350 | 1.414 | 1.483 | 1.558 | 1.641 | 1.732 | 1.834 | 1.949 |
| 12,500 | … | … | … | 1.225 | 1.277 | 1.333 | 1.393 | 1.458 | 1.528 | 1.604 | 1.687 | 1.780 | 1.883 |
| 13,000 | … | … | … | 1.215 | 1.265 | 1.318 | 1.374 | 1.435 | 1.500 | 1.571 | 1.648 | 1.732 | 1.826 |
| 13,500 | … | … | … | 1.206 | 1.254 | 1.304 | 1.357 | 1.414 | 1.475 | 1.541 | 1.612 | 1.690 | 1.776 |
| 14,000 | … | … | … | 1.198 | 1.243 | 1.291 | 1.342 | 1.395 | 1.453 | 1.515 | 1.581 | 1.653 | 1.732 |
| 14,500 | … | … | … | 1.190 | 1.234 | 1.279 | 1.327 | 1.378 | 1.433 | 1.491 | 1.553 | 1.620 | 1.693 |
| 15,000 | … | … | … | 1.183 | 1.225 | 1.268 | 1.314 | 1.363 | 1.414 | 1.469 | 1.528 | 1.590 | 1.658 |
| 16,000 | … | … | … | 1.171 | 1.209 | 1.249 | 1.291 | 1.335 | 1.382 | 1.431 | 1.483 | 1.539 | 1.599 |

Formula (28) also applies to hemi-spherical shells, such as the hemi-spherical head of a cylindrical container subjected to internal pressure, etc.

**If metric SI units are used, then:**

**$D$ = *internal* diameter of shell in millimeters**

**$p$ = internal pressure in newtons per square millimeter**

**$S$ = safe tensile stress in newtons per square millimeter**

**$t$ = thickness of metal in the shell, in millimeters. Use Formula (28).**

**Meters can be used in the formula in place of millimeters, provided the treatment is consistent throughout.**

*Example:* Find the thickness of metal required in the hemi-spherical end of a cylindrical vessel, 2 feet in diameter, subjected to an internal pressure of 500 pounds per square inch.

*Solution:* The material is mild steel and a tensile stress of 10,000 psi is allowable.

$$t = \frac{500 \times 2 \times 12}{4 \times 10{,}000} = 0.3 \text{ inch}$$

***Example:* A similar example using metric SI units is as follows: find the thickness of metal required in the hemi-spherical end of a cylindrical vessel, 750 mm in diameter, subjected to an internal pressure of 3 newtons/mm². The material is mild steel and a tensile stress of 70 newtons/mm² is allowable.**

$$t = \frac{3 \times 750}{4 \times 70} = 8.04 \text{ mm}$$

If the radius of curvature of the domed head of a boiler or container subjected to internal pressure is made equal to the diameter of the boiler, the thickness of the cylindrical shell and of the spherical head should be made the same. For example, if a boiler is 3 feet in diameter, the radius of curvature of its head should also be 3 feet, if material of the same thickness is to be used and the stresses are to be equal in both the head and cylindrical portion.

**Collapsing Pressure of Cylinders and Tubes Subjected to External Pressures.**—The following formulas may be used for finding the collapsing pressures of lap-welded Bessemer steel tubes:

$$P = 86{,}670\frac{t}{D} - 1386 \quad (29)$$

$$P = 50{,}210{,}000\left(\frac{t}{D}\right)^3 \quad (30)$$

in which $P$ = collapsing pressure in psi; $D$ = outside diameter of tube or cylinder in inches; $t$ = thickness of wall in inches.

Formula (29) is for values of $P$ greater than 580 pounds per square inch, and Formula (30) is for values of $P$ less than 580 pounds per square inch. These formulas are substantially correct for all lengths of pipe greater than six diameters between transverse joints that tend to hold the pipe to a circular form. The pressure $P$ found is the actual collapsing pressure, and a suitable factor of safety must be used. Ordinarily, a factor of safety of 5 is sufficient. In cases where there are repeated fluctuations of the pressure, vibration, shocks and other stresses, a factor of safety of from 6 to 12 should be used.

**If metric SI units are used the formulas are:**

$$P = 597.6\frac{t}{D} - 9.556 \quad (31)$$

$$P = 346{,}200\left(\frac{t}{D}\right)^3 \quad (32)$$

**where $P$ = collapsing pressure in newtons per square millimeter; $D$ = outside diameter of tube or cylinder in millimeters; and $t$ = thickness of wall in millimeters. Formula (31) is for values of $P$ greater than 4 N/mm², and Formula (32) is for values of $P$ less than 4 N/mm².**

The table *Tubes Subjected to External Pressure* is based upon the requirements of the Steam Boat Inspection Service of the Department of Commerce and Labor and gives the permissible working pressures and corresponding minimum wall thickness for long, plain, lap-welded and seamless steel flues subjected to external pressure only. The table thicknesses have been calculated from the formula:

$$t = \frac{[(F \times p) + 1386]D}{86,670}$$

in which $D$ = outside diameter of flue or tube in inches; $t$ = thickness of wall in inches; $p$ = working pressure in pounds per square inch; $F$ = factor of safety. The formula is applicable to working pressures greater than 100 pounds per square inch, to outside diameters from 7 to 18 inches, and to temperatures less than 650° F.

The preceding Formulas (29) and (30) were determined by Prof. R. T. Stewart, Dean of the Mechanical Engineering Department of the University of Pittsburgh, in a series of experiments carried out at the plant of the National Tube Co., McKeesport, Pennsylvania.

The apparent fiber stress under which the different tubes failed varied from about 7000 pounds per square inch for the relatively thinnest to 35,000 pounds per square inch for the relatively thickest walls. The average yield point of the material tested was 37,000 pounds and the tensile strength 58,000 pounds per square inch, so it is evident that the strength of a tube subjected to external fluid collapsing pressure is not dependent alone upon the elastic limit or ultimate strength of the material from which it is made.

**Tubes Subjected to External Pressure**

| Outside Diameter of Tube, Inches | Working Pressure in Pounds per Square Inch | | | | | | |
|---|---|---|---|---|---|---|---|
| | 100 | 120 | 140 | 160 | 180 | 200 | 220 |
| | Thickness of Tube in Inches. Safety Factor, 5 | | | | | | |
| 7 | 0.152 | 0.160 | 0.168 | 0.177 | 0.185 | 0.193 | 0.201 |
| 8 | 0.174 | 0.183 | 0.193 | 0.202 | 0.211 | 0.220 | 0.229 |
| 9 | 0.196 | 0.206 | 0.217 | 0.227 | 0.237 | 0.248 | 0.258 |
| 10 | 0.218 | 0.229 | 0.241 | 0.252 | 0.264 | 0.275 | 0.287 |
| 11 | 0.239 | 0.252 | 0.265 | 0.277 | 0.290 | 0.303 | 0.316 |
| 12 | 0.261 | 0.275 | 0.289 | 0.303 | 0.317 | 0.330 | 0.344 |
| 13 | 0.283 | 0.298 | 0.313 | 0.328 | 0.343 | 0.358 | 0.373 |
| 14 | 0.301 | 0.320 | 0.337 | 0.353 | 0.369 | 0.385 | 0.402 |
| 15 | 0.323 | 0.343 | 0.361 | 0.378 | 0.396 | 0.413 | 0.430 |
| 16 | 0.344 | 0.366 | 0.385 | 0.404 | 0.422 | 0.440 | 0.459 |
| 16 | 0.366 | 0.389 | 0.409 | 0.429 | 0.448 | 0.468 | 0.488 |
| 18 | 0.387 | 0.412 | 0.433 | 0.454 | 0.475 | 0.496 | 0.516 |

**Dimensions and Maximum Allowable Pressure of Tubes Subjected to External Pressure**

| Outside Dia., Inches | Thickness of Material, Inches | Max. Pressure Allowed, psi | Outside Dia., Inches | Thickness of Material, Inches | Max. Pressure Allowed, psi | Outside Dia., Inches | Thickness of Material, Inches | Max. Pressure Allowed, psi |
|---|---|---|---|---|---|---|---|---|
| 2 | 0.095 | 427 | 3 | 0.109 | 327 | 4 | 0.134 | 303 |
| 2¼ | 0.095 | 380 | 3¼ | 0.120 | 332 | 4½ | 0.134 | 238 |
| 2½ | 0.109 | 392 | 3½ | 0.120 | 308 | 5 | 0.148 | 235 |
| 2¾ | 0.109 | 356 | 3¾ | 0.120 | 282 | 6 | 0.165 | 199 |

# SHAFTS

## Shaft Calculations

**Torsional Strength of Shafting.**—In the formulas that follow,

$\alpha$ = angular deflection of shaft in degrees

$c$ = distance from center of gravity to extreme fiber

$D$ = diameter of shaft in inches

$G$ = torsional modulus of elasticity = 11,500,000 pounds per square inch for steel

$J$ = polar moment of inertia of shaft cross section (see table)

$l$ = length of shaft in inches

$N$ = angular velocity of shaft in revolutions per minute

$P$ = power transmitted in horsepower

$S_s$ = allowable torsional shearing stress in pounds per square inch

$T$ = torsional or twisting moment in inch-pounds

$Z_p$ = polar section modulus (see table page 246)

The allowable twisting moment for a shaft of any cross section such as circular, square, etc., is:

$$T = S_s \times Z_p \quad (1)$$

For a shaft delivering $P$ horsepower at $N$ revolutions per minute the twisting moment $T$ being transmitted is:

$$T = \frac{63{,}000P}{N} \quad (2)$$

The twisting moment $T$ as determined by Formula (2) should be less than the value determined by using Formula (1) if the maximum allowable stress $S_s$ is not to be exceeded.

The diameter of a solid circular shaft required to transmit a given torque $T$ is:

$$D = \sqrt[3]{\frac{5.1T}{S_s}} \quad (3a) \qquad \text{or} \qquad D = \sqrt[3]{\frac{321{,}000P}{NS_s}} \quad (3b)$$

The allowable stresses that are generally used in practice are: 4000 pounds per square inch for main power-transmitting shafts; 6000 pounds per square inch for lineshafts carrying pulleys; and 8500 pounds per square inch for small, short shafts, countershafts, etc. Using these allowable stresses, the horsepower $P$ transmitted by a shaft of diameter $D$, or the diameter $D$ of a shaft to transmit a given horsepower $P$ may be determined from the following formulas:

For main power-transmitting shafts:

$$P = \frac{D^3N}{80} \quad (4a) \qquad \text{or} \qquad D = \sqrt[3]{\frac{80P}{N}} \quad (4b)$$

For lineshafts carrying pulleys:

$$P = \frac{D^3N}{53.5} \quad (5a) \qquad \text{or} \qquad D = \sqrt[3]{\frac{53.5P}{N}} \quad (5b)$$

For small, short shafts:

$$P = \frac{D^3 N}{38} \quad (6a) \qquad \text{or} \qquad D = \sqrt[3]{\frac{38P}{N}} \quad (6b)$$

Shafts that are subjected to shocks, such as sudden starting and stopping, should be given a greater factor of safety resulting in the use of lower allowable stresses than those just mentioned.

*Example:* What should be the diameter of a lineshaft to transmit 10 horsepower if the shaft is to make 150 revolutions per minute? Using Formula (5b),

$$D = \sqrt[3]{\frac{53.5 \times 10}{150}} = 1.53 \text{ or, say, } 1\frac{9}{16} \text{ inches}$$

*Example:* What horsepower would be transmitted by a short shaft, 2 inches in diameter, carrying two pulleys close to the bearings, if the shaft makes 300 revolutions per minute? Using Formula (6a),

$$P = \frac{2^3 \times 300}{38} = 63 \text{ horsepower}$$

**Torsional Strength of Shafting, Calculations in Metric SI Units.—The allowable twisting moment for a shaft of any cross section such as circular, square, etc. can be calculated from:**

$$T = S_s \times Z_p \quad (7)$$

**where $T$ = torsional or twisting moment in N-mm; $S_s$ = allowable torsional shearing stress in N/mm$^2$; and $Z_p$ = polar section modulus in mm$^3$.**

**For a shaft delivering power of $P$ kilowatts at $N$ revolutions per minute, the twisting moment $T$ being transmitted is:**

$$T = \frac{9.55 \times 10^6 P}{N} \quad (8a) \qquad \text{or} \qquad T = \frac{10^6 P}{\omega} \quad (8b)$$

**where $T$ is in newton-millimeters, and $\omega$ = angular velocity in radians per second.**

**The diameter $D$ of a solid circular shaft required to transmit a given torque $T$ is:**

$$D = \sqrt[3]{\frac{5.1T}{S_s}} \quad (9a) \qquad \text{or} \qquad D = \sqrt[3]{\frac{48.7 \times 10^6 P}{N S_s}} \quad (9b)$$

$$\text{or} \qquad D = \sqrt[3]{\frac{5.1 \times 10^6 P}{\omega S_s}} \quad (9c)$$

**where $D$ is in millimeters; $T$ is in newton-millimeters; $P$ is power in kilowatts; $N$ = revolutions per minute; $S_s$ = allowable torsional shearing stress in newtons per square millimeter, and $\omega$ = angular velocity in radians per second.**

**If 28 newtons/mm$^2$ and 59 newtons/mm$^2$ are taken as the generally allowed stresses for main power-transmitting shafts and small short shafts, respectively, then using these allowable stresses, the power $P$ transmitted by a shaft of diameter $D$, or the diameter $D$ of a shaft to transmit a given power $P$ may be determined from the following formulas:**

**For main power-transmitting shafts:**

$$P = \frac{D^3N}{1.77 \times 10^6} \quad (10a) \qquad \text{or} \qquad D = \sqrt[3]{\frac{1.77 \times 10^6 P}{N}} \quad (10b)$$

**For small, short shafts:**

$$P = \frac{D^3N}{0.83 \times 10^6} \quad (11a) \qquad \text{or} \qquad D = \sqrt[3]{\frac{0.83 \times 10^6 P}{N}} \quad (11b)$$

**where $P$ is in kilowatts, $D$ is in millimeters, and $N$ = revolutions per minute.**

***Example:* What should be the diameter of a power-transmitting shaft to transmit 150 kW at 500 rpm?**

$$D = \sqrt[3]{\frac{1.77 \times 10^6 \times 150}{500}} = 81 \text{ millimeters}$$

***Example:* What power would a short shaft, 50 millimeters in diameter, transmit at 400 rpm?**

$$P = \frac{50^3 \times 400}{0.83 \times 10^6} = 60 \text{ kilowatts}$$

**Torsional Deflection of Circular Shafts.**—Shafting must often be proportioned not only to provide the strength required to transmit a given torque, but also to prevent torsional deflection (twisting) through a greater angle than has been found satisfactory for a given type of service.

For a solid circular shaft the torsional deflection in degrees is given by:

$$\alpha = \frac{584Tl}{D^4G} \quad (12)$$

*Example:* Find the torsional deflection for a solid steel shaft 4 inches in diameter and 48 inches long, subjected to a twisting moment of 24,000 inch-pounds. By Formula (12),

$$\alpha = \frac{584 \times 24{,}000 \times 48}{4^4 \times 11{,}500{,}000} = 0.23 \text{ degree}$$

**Formula (12) can be used with metric SI units, where $\alpha$ = angular deflection of shaft in degrees; $T$ = torsional moment in newton-millimeters; $l$ = length of shaft in millimeters; $D$ = diameter of shaft in millimeters; and $G$ = torsional modulus of elasticity in newtons per square millimeter.**

***Example:* Find the torsional deflection of a solid steel shaft, 100 mm in diameter and 1300 mm long, subjected to a twisting moment of $3 \times 10^6$ newton-millimeters. The torsional modulus of elasticity is 80,000 newtons/mm². By Formula (12)**

$$\alpha = \frac{584 \times 3 \times 10^6 \times 1300}{100^4 \times 80{,}000} = 0.285 \text{ degree}$$

The diameter of a shaft that is to have a maximum torsional deflection $\alpha$ is given by:

$$D = 4.9 \times \sqrt[4]{\frac{Tl}{G\alpha}} \quad (13)$$

**Formula (13) can be used with metric SI units, where $D$ = diameter of shaft in millimeters; $T$ = torsional moment in newton-millimeters; $l$ = length of shaft in**

**millimeters; $G$ = torsional modulus of elasticity in newtons per square millimeter; and $\alpha$ = angular deflection of shaft in degrees.**

According to some authorities, the allowable twist in steel transmission shafting should not exceed 0.08 degree per foot length of the shaft. The diameter $D$ of a shaft that will permit a maximum angular deflection of 0.08 degree per foot of length for a given torque $T$ or for a given horsepower $P$ can be determined from the formulas:

$$D = 0.29\sqrt[4]{T} \quad (14a) \qquad \text{or} \qquad D = 4.6 \times \sqrt[4]{\frac{P}{N}} \quad (14b)$$

**Using metric SI units and assuming an allowable twist in steel transmission shafting of 0.26 degree per meter length, Formulas (14a) and (14b) become:**

$$D = 2.26\sqrt[4]{T} \qquad \text{or} \qquad D = 125.7 \times \sqrt[4]{\frac{P}{N}}$$

**where $D$ = diameter of shaft in millimeters; $T$ = torsional moment in newton-millimeters; $P$ = power in kilowatts; and $N$ = revolutions per minute.**

Another rule that has been generally used in mill practice limits the deflection to 1 degree in a length equal to 20 times the shaft diameter. For a given torque or horsepower, the diameter of a shaft having this maximum deflection is given by:

$$D = 0.1\sqrt[3]{T} \quad (15a) \qquad \text{or} \qquad D = 4.0 \times \sqrt[3]{\frac{P}{N}} \quad (15b)$$

*Example:* Find the diameter of a steel lineshaft to transmit 10 horsepower at 150 revolutions per minute with a torsional deflection not exceeding 0.08 degree per foot of length.

*Solution:* By Formula (14b),

$$D = 4.6 \times \sqrt[4]{\frac{10}{150}} = 2.35 \text{ inches}$$

This diameter is larger than that obtained for the same horsepower and rpm in the example given for Formula (5b) in which the diameter was calculated for strength considerations only. The usual procedure in the design of shafting which is to have a specified maximum angular deflection is to compute the diameter first by means of Formulas (13), (14a), (14b), (15a), or (15b) and then by means of Formulas (3a), (3b), (4b), (5b), or (6b), using the larger of the two diameters thus found.

**Linear Deflection of Shafting.**—For steel line shafting, it is considered good practice to limit the linear deflection to a maximum of 0.010 inch per foot of length. The maximum distance in feet between bearings for average conditions, in order to avoid excessive linear deflection, is determined by the formulas:

$L = 8.95\sqrt[3]{D^2}$ for shafting subject to no bending action except its own weight

$L = 5.2\sqrt[3]{D^2}$ for shafting subject to bending action of pulleys, etc.

in which $D$ = diameter of shaft in inches and $L$ = maximum distance between bearings in feet. Pulleys should be placed as close to the bearings as possible.

*In general, shafting up to 3 inches in diameter* is almost always made from cold-rolled steel. This shafting is true and straight and needs no turning, but if keyways are cut in the shaft, it must usually be straightened afterwards, as the cutting of the keyways relieves the tension on the surface of the shaft produced by the cold-rolling process. Sizes of shafting from 3 to 5 inches in diameter may be either cold-rolled or turned, more frequently the latter, and all larger sizes of shafting must be turned because cold-rolled shafting is not available in diameters larger than 5 inches.

**Diameters of Finished Shafting** (former American Standard *ASA B17.1*)

| Diameters, Inches | | Minus Tolerances, Inches[a] | Diameters, Inches | | Minus Tolerances Inches[a] | Diameters, Inches | | Minus Tolerances, Inches[a] |
|---|---|---|---|---|---|---|---|---|
| Transmission Shafting | Machinery Shafting | | Transmission Shafting | Machinery Shafting | | Transmission Shafting | Machinery Shafting | |
| | 1/2 | 0.002 | | 1 13/16 | 0.003 | | 3 3/4 | 0.004 |
| | 9/16 | 0.002 | | 1 7/8 | 0.003 | | 3 7/8 | 0.004 |
| | 5/8 | 0.002 | 1 15/16 | 1 15/16 | 0.003 | 3 15/16 | 4 | 0.004 |
| | 11/16 | 0.002 | | 2 | 0.003 | | 4 1/4 | 0.005 |
| | 3/4 | 0.002 | | 2 1/16 | 0.004 | 4 7/16 | 4 1/2 | 0.005 |
| | 13/16 | 0.002 | | 2 1/8 | 0.004 | | 4 3/4 | 0.005 |
| | 7/8 | 0.002 | 2 3/16 | 2 3/16 | 0.004 | 4 15/16 | 5 | 0.005 |
| 15/16 | 15/16 | 0.002 | | 2 1/4 | 0.004 | | 5 1/4 | 0.005 |
| | 1 | 0.002 | | 2 5/16 | 0.004 | 5 7/16 | 5 1/2 | 0.005 |
| | 1 1/16 | 0.003 | | 2 3/8 | 0.004 | | 5 3/4 | 0.005 |
| | 1 1/8 | 0.003 | 2 7/16 | 2 7/16 | 0.004 | 5 15/16 | 6 | 0.005 |
| 1 3/16 | 1 3/16 | 0.003 | | 2 1/2 | 0.004 | | 6 1/4 | 0.006 |
| | 1 1/4 | 0.003 | | 2 5/8 | 0.004 | 6 1/2 | 6 1/2 | 0.006 |
| | 1 5/16 | 0.003 | | 2 3/4 | 0.004 | | 6 3/4 | 0.006 |
| | 1 3/8 | 0.003 | 2 15/16 | 2 7/8 | 0.004 | 7 | 7 | 0.006 |
| 1 7/16 | 1 7/16 | 0.003 | | 3 | 0.004 | | 7 1/4 | 0.006 |
| | 1 1/2 | 0.003 | | 3 1/8 | 0.004 | 7 1/2 | 7 1/2 | 0.006 |
| | 1 9/16 | 0.003 | | 3 1/4 | 0.004 | | 7 3/4 | 0.006 |
| | 1 5/8 | 0.003 | | 3 3/8 | 0.004 | 8 | 8 | 0.006 |
| 1 11/16 | 1 11/16 | 0.003 | 3 7/16 | 3 1/2 | 0.004 | … | … | … |
| | 1 3/4 | 0.003 | | 3 5/8 | 0.004 | … | … | … |

[a] *Note:*—These tolerances are *negative* or minus and represent the maximum allowable variation *below* the exact nominal size. For instance the maximum diameter of the $1^{15}/_{16}$ inch shaft is 1.938 inch and its minimum allowable diameter is 1.935 inch. Stock lengths of finished transmission shafting shall be: 16, 20 and 24 feet.

**Design of Transmission Shafting.—** The following guidelines for the design of shafting for transmitting a given amount of power under various conditions of loading are based upon formulas given in the former American Standard ASA B17c Code for the Design of Transmission Shafting. These formulas are based on the *maximum-shear theory* of failure which assumes that the elastic limit of a *ductile* ferrous material in shear is practically one-half its elastic limit in tension. This theory agrees, very nearly, with the results of tests on ductile materials and has gained wide acceptance in practice.

The formulas given apply in all shaft designs including shafts for special machinery. The limitation of these formulas is that they provide only for the strength of shafting and are not concerned with the torsional or lineal deformations which may, in shafts used in machine design, be the controlling factor (see *Torsional Deflection of Circular Shafts* on page 291 and *Linear Deflection of Shafting* on page 292 for deflection considerations). In the formulas that follow,

$B = \sqrt[3]{1 \div (1 - K^4)}$ (see Table 3)

$D$ = outside diameter of shaft in inches

$D_1$ = inside diameter of a hollow shaft in inches

$K_m$ = shock and fatigue factor to be applied in every case to the computed bending moment (see Table 1)

$K_t$ = combined shock and fatigue factor to be applied in every case to the computed torsional moment (see Table 1)

$M$ = maximum bending moment in inch-pounds

$N$ = revolutions per minute

$P$ = maximum power to be transmitted by the shaft in horsepower

$p_t$ = maximum allowable shearing stress under combined loading conditions in pounds per square inch (see Table 2)

$S$ = maximum allowable flexural (bending) stress, in either tension or compression in pounds per square inch (see Table 2)

$S_s$ = maximum allowable torsional shearing stress in pounds per square inch (see Table 2)

$T$ = maximum torsional moment in inch-pounds

$V$ = maximum transverse shearing load in pounds

For shafts subjected to pure torsional loads only,

$$D = B\sqrt[3]{\frac{5.1K_tT}{S_s}} \quad (16a) \qquad \text{or} \qquad D = B\sqrt[3]{\frac{321{,}000K_tP}{S_sN}} \quad (16b)$$

For stationary shafts subjected to bending only,

$$D = B\sqrt[3]{\frac{10.2K_mM}{S}} \quad (17)$$

For shafts subjected to combined torsion and bending,

$$D = B\sqrt[3]{\frac{5.1}{p_t}\sqrt{(K_mM)^2 + (K_tT)^2}} \quad (18a)$$

or

$$D = B \times \sqrt[3]{\frac{5.1}{p_t}\sqrt{(K_mM)^2 + \left(\frac{63{,}000K_tP}{N}\right)^2}} \quad (18b)$$

Formulas (16a) to (18b) may be used for solid shafts or for hollow shafts. For solid shafts the factor $B$ is equal to 1, whereas for hollow shafts the value of $B$ depends on the value of $K$ which, in turn, depends on the ratio of the inside diameter of the shaft to the outside diameter ($D_1 \div D = K$). Table 3 gives values of $B$ corresponding to various values of $K$.

For short solid shafts subjected only to heavy transverse shear, the diameter of shaft required is:

$$D = \sqrt{\frac{1.7V}{S_s}} \quad (19)$$

**Formulas (16a), (17), (18a) and (19), can be used unchanged with metric SI units. Formula (16b) becomes:**

$$D = B\sqrt[3]{\frac{48.7K_tP}{S_sN}}$$ **and Formula (18b) becomes:**

$$D = B\sqrt[3]{\frac{5.1}{p_t}\sqrt{(K_mM)^2 + \left(\frac{9.55K_tP}{N}\right)^2}}$$

**Throughout the formulas, $D$ = outside diameter of shaft in millimeters; $T$ = maximum torsional moment in newton-millimeters; $S_s$ = maximum allowable torsional shearing stress in newtons per millimeter squared (see Table 2); $P$ = maximum power to be transmitted in milliwatts; $N$ = revolutions per minute; $M$ = maximum bending moment in newton-millimeters; $S$ = maximum allowable flexural (bending) stress, either in tension or compression in newtons per millimeter squared (see Table 2); $p_t$ = maximum allowable shearing stress under combined loading conditions in newtons per millimeter squared; and $V$ = maximum transverse shearing load in**

**kilograms. The factors $K_m$, $K_t$, and $B$ are unchanged, and $D_1$ = the inside diameter of a hollow shaft in millimeters.**

**Table 1. Recommended Values of the Combined Shock and Fatigue Factors for Various Types of Load**

| Type of Load | Stationary Shafts $K_m$ | Stationary Shafts $K_t$ | Rotating Shafts $K_m$ | Rotating Shafts $K_t$ |
|---|---|---|---|---|
| Gradually applied and steady | 1.0 | 1.0 | 1.5 | 1.0 |
| Suddenly applied, minor shocks only | 1.5–2.0 | 1.5–2.0 | 1.5–2.0 | 1.0–1.5 |
| Suddenly applied, heavy shocks | … | … | 2.0–3.0 | 1.5–3.0 |

**Table 2. Recommended Maximum Allowable Working Stresses for Shafts Under Various Types of Load**

| Material | Type of Load: Simple Bending | Pure Torsion | Combined Stress |
|---|---|---|---|
| "Commercial Steel" shafting without keyways | $S = 16{,}000$ | $S_s = 8000$ | $p_t = 8000$ |
| "Commercial Steel" shafting with keyways | $S = 12{,}000$ | $S_s = 6000$ | $p_t = 6000$ |
| Steel purchased under definite physical specs. | (See note [a]) | (See note [b]) | (See note [b]) |

[a] $S$ = 60 percent of the elastic limit in tension but not more than 36 percent of the ultimate tensile strength.

[b] $S_s$ and $p_t$ = 30 percent of the elastic limit in tension but not more than 18 percent of the ultimate tensile strength.

**If the values in the Table are converted to metric SI units, note that 1000 pounds per square inch = 6.895 newtons per square millimeter.**

**Table 3. Values of the Factor $B$ Corresponding to Various Values of $K$ for Hollow Shafts**

| $K = \frac{D_1}{D}$ | 0.95 | 0.90 | 0.85 | 0.80 | 0.75 | 0.70 | 0.65 | 0.60 | 0.55 | 0.50 |
|---|---|---|---|---|---|---|---|---|---|---|
| $B = \sqrt[3]{1 \div (1 - K^4)}$ | 1.75 | 1.43 | 1.28 | 1.19 | 1.14 | 1.10 | 1.07 | 1.05 | 1.03 | 1.02 |

For solid shafts, $B = 1$ because $K = 0$, as follows: $B = \sqrt[3]{1 \div (1 - K^4)} = \sqrt[3]{1 \div (1 - 0)} = 1$

**Effect of Keyways on Shaft Strength.**—Keyways cut into a shaft reduce its load-carrying ability, particularly when impact loads or stress reversals are involved. To ensure an adequate factor of safety in the design of a shaft with standard keyway (width, one-quarter, and depth, one-eighth of shaft diameter), the former Code for Transmission Shafting tentatively recommended that shafts with keyways be designed on the basis of a solid circular shaft using not more than 75 percent of the working stress recommended for the solid shaft. See also page 2545.

**Formula for Shafts of Brittle Materials.**—The preceding formulas are applicable to ductile materials and are based on the *maximum-shear theory* of failure which assumes that the elastic limit of a *ductile* material in shear is one-half its elastic limit in tension.

Brittle materials are generally stronger in shear than in tension; therefore, the maximum-shear theory is not applicable. The *maximum-normal-stress theory* of failure is now generally accepted for the design of shafts made from brittle materials. A material may be considered brittle if its elongation in a 2-inch gage length is less than 5 percent. Materials such as cast iron, hardened tool steel, hard bronze, etc. conform to this rule. The diameter of a shaft made of a brittle material may be determined from the following formula which is based on the maximum-normal-stress theory of failure:

$$D = B \sqrt[3]{\frac{5.1}{S_t}[(K_m M) + \sqrt{(K_m M)^2 + (K_t T)^2}]}$$

where $S_t$ is the maximum allowable tensile stress in pounds per square inch and the other quantities are as previously defined.

**The formula can be used unchanged with metric SI units, where $D$ = outside diameter of shaft in millimeters; $S_t$ = the maximum allowable tensile stress in newtons per millimeter squared; $M$ = maximum bending moment in newton-millimeters; and $T$ = maximum torsional moment in newton-millimeters. The factors $K_m$, $K_t$, and $B$ are unchanged.**

**Critical Speed of Rotating Shafts.—**At certain speeds, a rotating shaft will become dynamically unstable and the resulting vibrations and deflections can result in damage not only to the shaft but to the machine of which it is a part. The speeds at which such dynamic instability occurs are called the *critical speeds* of the shaft. On page 199 are given formulas for the critical speeds of shafts subject to various conditions of loading and support. A shaft may be safely operated either above or below its critical speed, good practice indicating that the operating speed be at least 20 percent above or below the critical.

The formulas commonly used to determine critical speeds are sufficiently accurate for general purposes. However, the torque applied to a shaft has an important effect on its critical speed. Investigations have shown that the critical speeds of a uniform shaft are decreased as the applied torque is increased, and that there exist critical torques which will reduce the corresponding critical speed of the shaft to zero. A detailed analysis of the effects of applied torques on critical speeds may be found in a paper, "Critical Speeds of Uniform Shafts under Axial Torque," by Golomb and Rosenberg, presented at the First US National Congress of Applied Mechanics in 1951.

**Shaft Couplings.—**A shaft coupling is a device for fastening together the ends of two shafts, so that the rotary motion of one causes rotary motion of the other. One of the most simple and common forms of coupling is the flange coupling Fig. 1a and Fig. 1b. It consists of two flanged sleeves or hubs, each of which is keyed to the end of one of the two shafts to be connected. The sleeves are held together and prevented from rotating relative to each other by bolts through the flanges as indicated.

**Flange Coupling**

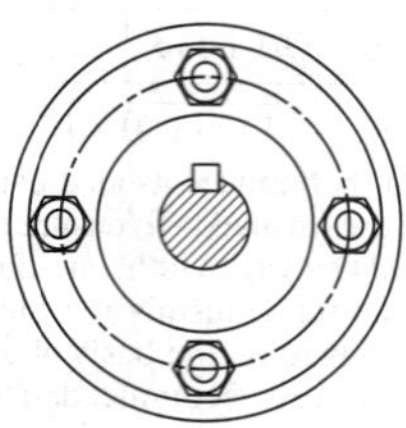

Fig. 1a.

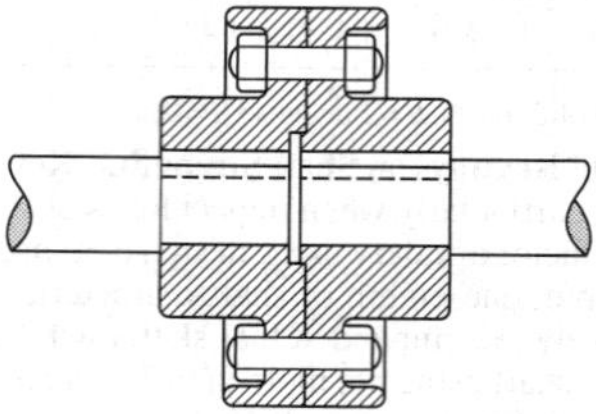

Fig. 1b.

*Flexible Couplings:* Flexible couplings are the most common mechanical means of compensating for unavoidable errors in alignment of shafts and shafting. When correctly applied, they are highly efficient for joining lengths of shafting without causing loss of power from bearing friction due to misalignment, and for use in direct motor drives for all kinds of machinery. Flexible couplings are not intended to be used for connecting a driven shaft and a driving shaft that are purposely placed in different planes or at an angle but are intended simply to overcome slight unavoidable errors in alignment that develop in service. There is a wide variety of flexible coupling designs; most of them consist essentially of two flanged members or hubs, fastened to the shafts and connected by some yielding arrangement. Balance is an important factor in coupling selection or design; it is not sufficient that the coupling be perfectly balanced when installed, but it must remain in balance after wear has taken place.

**Comparison of Hollow and Solid Shafting with Same Outside Diameter.—**Table 4 that follows gives the percent decrease in strength and weight of a hollow shaft relative to the strength and weight of a solid shaft of the same diameter. The upper figures in each line give the percent decrease in strength and the lower figures give the percent decrease in weight.

*Example:* A 4-inch shaft, with a 2-inch hole through it, has a weight 25 percent less than a solid 4-inch shaft, but its strength is decreased only 6.25 percent.

**Table 4. Comparative Torsional Strengths and Weights of Hollow and Solid Shafting with Same Outside Diameter**

| Dia. of Solid and Hollow Shaft, Inches | Diameter of Axial Hole in Hollow Shaft, Inches | | | | | | | | | |
|---|---|---|---|---|---|---|---|---|---|---|
| | 1 | 1¼ | 1½ | 1¾ | 2 | 2½ | 3 | 3½ | 4 | 4½ |
| 1½ | 19.76 | 48.23 | … | … | … | … | … | … | … | … |
| | 44.44 | 69.44 | … | … | … | … | … | … | … | … |
| 1¾ | 10.67 | 26.04 | 53.98 | … | … | … | … | … | … | … |
| | 32.66 | 51.02 | 73.49 | … | … | … | … | … | … | … |
| 2 | 6.25 | 15.26 | 31.65 | 58.62 | … | … | … | … | … | … |
| | 25.00 | 39.07 | 56.25 | 76.54 | … | … | … | … | … | … |
| 2¼ | 3.91 | 9.53 | 19.76 | 36.60 | 62.43 | … | … | … | … | … |
| | 19.75 | 30.87 | 44.44 | 60.49 | 79.00 | … | … | … | … | … |
| 2½ | 2.56 | 6.25 | 12.96 | 24.01 | 40.96 | … | … | … | … | … |
| | 16.00 | 25.00 | 36.00 | 49.00 | 64.00 | … | … | … | … | … |
| 2¾ | 1.75 | 4.28 | 8.86 | 16.40 | 27.98 | 68.30 | … | … | … | … |
| | 13.22 | 20.66 | 29.74 | 40.48 | 52.89 | 82.63 | … | … | … | … |
| 3 | 1.24 | 3.01 | 6.25 | 11.58 | 19.76 | 48.23 | … | … | … | … |
| | 11.11 | 17.36 | 25.00 | 34.01 | 44.44 | 69.44 | … | … | … | … |
| 3¼ | 0.87 | 2.19 | 4.54 | 8.41 | 14.35 | 35.02 | 72.61 | … | … | … |
| | 9.46 | 14.80 | 21.30 | 29.00 | 37.87 | 59.17 | 85.22 | … | … | … |
| 3½ | 0.67 | 1.63 | 3.38 | 6.25 | 10.67 | 26.04 | 53.98 | … | … | … |
| | 8.16 | 12.76 | 18.36 | 25.00 | 32.66 | 51.02 | 73.49 | … | … | … |
| 3¾ | 0.51 | 1.24 | 2.56 | 4.75 | 8.09 | 19.76 | 40.96 | 75.89 | … | … |
| | 7.11 | 11.11 | 16.00 | 21.77 | 28.45 | 44.44 | 64.00 | 87.10 | … | … |
| 4 | 0.40 | 0.96 | 1.98 | 3.68 | 6.25 | 15.26 | 31.65 | 58.62 | … | … |
| | 6.25 | 9.77 | 14.06 | 19.14 | 25.00 | 39.07 | 56.25 | 76.56 | … | … |
| 4¼ | 0.31 | 0.74 | 1.56 | 2.89 | 4.91 | 11.99 | 24.83 | 46.00 | 78.47 | … |
| | 5.54 | 8.65 | 12.45 | 16.95 | 22.15 | 34.61 | 49.85 | 67.83 | 88.59 | … |
| 4½ | 0.25 | 0.70 | 1.24 | 2.29 | 3.91 | 9.53 | 19.76 | 36.60 | 62.43 | … |
| | 4.94 | 7.72 | 11.11 | 15.12 | 19.75 | 30.87 | 44.44 | 60.49 | 79.00 | … |
| 4¾ | 0.20 | 0.50 | 1.00 | 1.85 | 3.15 | 7.68 | 15.92 | 29.48 | 50.29 | 80.56 |
| | 4.43 | 6.93 | 9.97 | 13.57 | 17.73 | 27.70 | 39.90 | 54.29 | 70.91 | 89.75 |
| 5 | 0.16 | 0.40 | 0.81 | 1.51 | 2.56 | 6.25 | 12.96 | 24.01 | 40.96 | 65.61 |
| | 4.00 | 6.25 | 8.10 | 12.25 | 16.00 | 25.00 | 36.00 | 49.00 | 64.00 | 81.00 |
| 5½ | 0.11 | 0.27 | 0.55 | 1.03 | 1.75 | 4.27 | 8.86 | 16.40 | 27.98 | 44.82 |
| | 3.30 | 5.17 | 7.43 | 10.12 | 13.22 | 20.66 | 29.76 | 40.48 | 52.89 | 66.94 |
| 6 | 0.09 | 0.19 | 0.40 | 0.73 | 1.24 | 3.02 | 6.25 | 11.58 | 19.76 | 31.65 |
| | 2.77 | 4.34 | 6.25 | 8.50 | 11.11 | 17.36 | 25.00 | 34.02 | 44.44 | 56.25 |
| 6½ | 0.06 | 0.14 | 0.29 | 0.59 | 0.90 | 2.19 | 4.54 | 8.41 | 14.35 | 23.98 |
| | 2.36 | 3.70 | 5.32 | 7.24 | 9.47 | 14.79 | 21.30 | 28.99 | 37.87 | 47.93 |
| 7 | 0.05 | 0.11 | 0.22 | 0.40 | 0.67 | 1.63 | 3.38 | 6.25 | 10.67 | 17.08 |
| | 2.04 | 3.19 | 4.59 | 6.25 | 8.16 | 12.76 | 18.36 | 25.00 | 32.66 | 41.33 |
| 7½ | 0.04 | 0.08 | 0.16 | 0.30 | 0.51 | 1.24 | 2.56 | 4.75 | 8.09 | 12.96 |
| | 1.77 | 2.77 | 4.00 | 5.44 | 7.11 | 11.11 | 16.00 | 21.77 | 28.45 | 36.00 |
| 8 | 0.03 | 0.06 | 0.13 | 0.23 | 0.40 | 0.96 | 1.98 | 3.68 | 6.25 | 10.02 |
| | 1.56 | 2.44 | 3.51 | 4.78 | 6.25 | 9.77 | 14.06 | 19.14 | 25.00 | 31.64 |

The upper figures in each line give number of percent decrease in strength; the lower figures give percent decrease in weight.

# SPRINGS

## Introduction to Spring Design

Many advances have been made in the spring industry in recent years. For example: developments in materials permit longer fatigue life at higher stresses; simplified design procedures reduce the complexities of design, and improved methods of manufacture help to speed up some of the complicated fabricating procedures and increase production. New types of testing instruments and revised tolerances also permit higher standards of accuracy. Designers should also consider the possibility of using standard springs now available from stock. They can be obtained from spring manufacturing companies located in different areas, and small shipments usually can be made quickly.

Designers of springs require information in the following order of precedence to simplify design procedures.

1) Spring materials and their applications

2) Allowable spring stresses

3) Spring design data with tables of spring characteristics, tables of formulas, and tolerances.

Only the more commonly used types of springs are covered in detail here. Special types and designs rarely used such as torsion bars, volute springs, Belleville washers, constant force, ring and spiral springs and those made from rectangular wire are only described briefly. Belleville and disc springs are discussed in the section *DISC SPRINGS* starting on page 343.

**Notation.—**The following symbols are used in spring equations:

$AC$ = Active coils
$b$ = Widest width of rectangular wire, inches
$CL$ = Compressed length, inches
$D$ = Mean coil diameter, inches = $OD - d$
$d$ = Diameter of wire or side of square, inches
$E$ = Modulus of elasticity in tension, pounds per square inch
$F$ = Deflection, for $N$ coils, inches
$F°$ = Deflection, for $N$ coils, rotary, degrees
$f$ = Deflection, for one active coil
$FL$ = Free length, unloaded spring, inches
$G$ = Modulus of elasticity in torsion, pounds per square inch
$ID$ = Inside diameter
$IT$ = Initial tension, pounds
$K$ = Curvature stress correction factor
$L$ = Active length subject to deflection, inches
$N$ = Number of active coils, total
$OD$ = Outside diameter
$P$ = Load, pounds
$p$ = Pitch, inches
$R$ = Distance from load to central axis, inches
$S$ or $S_t$ = Stress, torsional, pounds per square inch
$S_b$ = Stress, bending, pounds per square inch
$SH$ = Solid height
$S_{it}$ = Stress, torsional, due to initial tension, pounds per square inch
$T$ = Torque = $P \times R$, pound-inches
$TC$ = Total coils
$t$ = Thickness, inches
$U$ = Number of revolutions = $F°/360°$

## Spring Materials

The spring materials most commonly used include high-carbon spring steels, alloy spring steels, stainless spring steels, copper-base spring alloys, and nickel-base spring alloys.

**High-Carbon Spring Steels in Wire Form.—**These spring steels are the most commonly used of all spring materials because they are the least expensive, are easily worked, and are readily available. However, they are not satisfactory for springs operating at high or low temperatures or for shock or impact loading. The following wire forms are available:

*Music Wire, ASTM A228:* (0.80–0.95 percent carbon) This is the most widely used of all spring materials for small springs operating at temperatures up to about 250° F. It is tough, has a high tensile strength, and can withstand high stresses under repeated loading. The material is readily available in round form in diameters ranging from 0.005 to 0.125 inch and in some larger sizes up to 3⁄16 inch. It is not available with high tensile strengths in square or rectangular sections. Music wire can be plated easily and is obtainable pretinned or preplated with cadmium, but plating after spring manufacture is usually preferred for maximum corrosion resistance.

*Oil-Tempered MB Grade, ASTM A229:* (0.60–0.70 percent carbon) This general-purpose spring steel is commonly used for many types of coil springs where the cost of music wire is prohibitive and in sizes larger than are available in music wire. It is readily available in diameters ranging from 0.125 to 0.500 inch, but both smaller and larger sizes may be obtained. The material should not be used under shock and impact loading conditions, at temperatures above 350° F, or at temperatures in the sub-zero range. Square and rectangular sections of wire are obtainable in fractional sizes. Annealed stock also can be obtained for hardening and tempering after coiling. This material has a heat-treating scale that must be removed before plating.

*Oil-Tempered HB Grade, SAE 1080:* (0.75–0.85 percent carbon) This material is similar to the MB Grade except that it has a higher carbon content and a higher tensile strength. It is obtainable in the same sizes and is used for more accurate requirements than the MB Grade, but is not so readily available. In lieu of using this material it may be better to use an alloy spring steel, particularly if a long fatigue life or high endurance properties are needed. Round and square sections are obtainable in the oil-tempered or annealed conditions.

*Hard-Drawn MB Grade, ASTM A227:* (0.60–0.70 percent carbon) This grade is used for general-purpose springs where cost is the most important factor. Although increased use in recent years has resulted in improved quality, it is best not to use this grade where long life and accuracy of loads and deflections are important. It is available in diameters ranging from 0.031 to 0.500 inch and in some smaller and larger sizes also. The material is available in square sections but at reduced tensile strengths. It is readily plated. Applications should be limited to those in the temperature range of 0 to 250° F.

**High-Carbon Spring Steels in Flat Strip Form.—**Two types of thin, flat, high-carbon spring steel strip are most widely used although several other types are obtainable for specific applications in watches, clocks, and certain instruments. These two compositions are used for over 95 percent of all such applications. Thin sections of these materials under 0.015 inch having a carbon content of over 0.85 percent and a hardness of over 47 RC (Rockwell C scale) are susceptible to hydrogen-embrittlement, even though special plating and heating operations are employed. The two types are described as follows:

*Cold-Rolled Spring Steel, Blue-Tempered or Annealed, SAE 1074, also 1064, and 1070:* (0.60 to 0.80 percent carbon) This very popular spring steel is available in thicknesses ranging from 0.005 to 0.062 inch and in some thinner and thicker sections. The material is available in the annealed condition for forming in 4-slide machines and in presses, and can readily be hardened and tempered after forming. It is also available in the heat-treated or blue-tempered condition. The steel is obtainable in several finishes, such as straw color, blue color, black, or plain. Hardnesses ranging from 42 to 46 RC (Rockwell C scale) are recommended for spring applications. Uses include spring clips, flat springs, clock springs, and motor, power, and spiral springs.

*Cold-Rolled Spring Steel, Blue-Tempered Clock Steel, SAE 1095:* (0.90 to 1.05 percent carbon) This popular type should be used principally in the blue-tempered condition. Although obtainable in the annealed condition, it does not always harden properly during heat treatment as it is a "shallow" hardening type. It is used principally in clocks and motor springs. End sections of springs made from this steel are annealed for bending or piercing operations. Hardnesses usually range from 47 to 51 RC.

Other materials available in strip form and used for flat springs are brass, phosphor-bronze, beryllium-copper, stainless steels, and nickel alloys.

**Alloy Spring Steels.**—These spring steels are used for conditions of high stress, and shock or impact loadings. They can withstand both higher and lower temperatures than the high-carbon steels and are obtainable in either the annealed or pretempered conditions.

*Chromium Vanadium, ASTM A231:* This very popular spring steel is used under conditions involving higher stresses than those for which the high-carbon spring steels are recommended and is also used where good fatigue strength and endurance are needed. It behaves well under shock and impact loading. The material is available in diameters ranging from 0.031 to 0.500 inch and in some larger sizes also. In square sections it is available in fractional sizes. Both the annealed and pretempered types are available in round, square, and rectangular sections. It is used extensively in aircraft-engine valve springs and for springs operating at temperatures up to 425° F.

*Silicon Manganese:* This alloy steel is quite popular in Great Britain. It is less expensive than chromium-vanadium steel and is available in round, square, and rectangular sections in both annealed and pretempered conditions in sizes ranging from 0.031 to 0.500 inch. It was formerly used for knee-action springs in automobiles. It is used in flat leaf springs for trucks and as a substitute for more expensive spring steels.

*Chromium Silicon, ASTM A401:* This alloy is used for highly stressed springs that require long life and are subjected to shock loading. It can be heat treated to higher hardnesses than other spring steels so that high tensile strengths are obtainable. The most popular sizes range from 0.031 to 0.500 inch in diameter. Very rarely are square, flat, or rectangular sections used. Hardnesses ranging from 50 to 53 RC are quite common and the alloy may be used at temperatures up to 475° F. This material is usually ordered specially for each job.

**Stainless Spring Steels.**—The use of stainless spring steels has increased and several compositions are available all of which may be used for temperatures up to 550° F. They are all corrosion resistant. Only the stainless 18-8 compositions should be used at sub-zero temperatures.

*Stainless Type 302, ASTM A313:* (18 percent chromium, 8 percent nickel) This stainless spring steel is very popular because it has the highest tensile strength and quite uniform properties. It is cold-drawn to obtain its mechanical properties and cannot be hardened by heat treatment. This material is nonmagnetic only when fully annealed and becomes slightly magnetic due to the cold-working performed to produce spring properties. It is suitable for use at temperatures up to 550° F. and for sub-zero temperatures. It is very corrosion resistant. The material best exhibits its desirable mechanical properties in diameters ranging from 0.005 to 0.1875 inch although some larger diameters are available. It is also available as hard-rolled flat strip. Square and rectangular sections are available but are infrequently used.

*Stainless Type 304, ASTM A313:* (18 percent chromium, 8 percent nickel) This material is quite similar to Type 302, but has better bending properties and about 5 percent lower tensile strength. It is a little easier to draw, due to the slightly lower carbon content.

*Stainless Type 316, ASTM A313:* (18 percent chromium, 12 percent nickel, 2 percent molybdenum) This material is quite similar to Type 302 but is slightly more corrosion resistant because of its higher nickel content. Its tensile strength is 10 to 15 percent lower than Type 302. It is used for aircraft springs.

*Stainless Type 17-7 PH ASTM A313:* (17 percent chromium, 7 percent nickel) This alloy, which also contains small amounts of aluminum and titanium, is formed in a moderately hard state and then precipitation hardened at relatively low temperatures for several hours to produce tensile strengths nearly comparable to music wire. This material is not readily available in all sizes, and has limited applications due to its high manufacturing cost.

*Stainless Type 414, SAE 51414:* (12 percent chromium, 2 percent nickel) This alloy has tensile strengths about 15 percent lower than Type 302 and can be hardened by heat-treatment. For best corrosion resistance it should be highly polished or kept clean. It can be obtained hard drawn in diameters up to 0.1875 inch and is commonly used in flat cold-rolled strip for stampings. The material is not satisfactory for use at low temperatures.

*Stainless Type 420, SAE 51420:* (13 percent chromium) This is the best stainless steel for use in large diameters above 0.1875 inch and is frequently used in smaller sizes. It is formed in the annealed condition and then hardened and tempered. It does not exhibit its stainless properties until after it is hardened. Clean bright surfaces provide the best corrosion resistance, therefore the heat-treating scale must be removed. Bright hardening methods are preferred.

*Stainless Type 431, SAE 51431:* (16 percent chromium, 2 percent nickel) This spring alloy acquires high tensile properties (nearly the same as music wire) by a combination of heat-treatment to harden the wire plus cold-drawing after heat treatment. Its corrosion resistance is not equal to Type 302.

**Copper-Base Spring Alloys.**—Copper-base alloys are important spring materials because of their good electrical properties combined with their good resistance to corrosion. Although these materials are more expensive than the high-carbon and the alloy steels, they nevertheless are frequently used in electrical components and in sub-zero temperatures.

*Spring Brass, ASTM B134:* (70 percent copper, 30 percent zinc) This material is the least expensive and has the highest electrical conductivity of the copper-base alloys. It has a low tensile strength and poor spring qualities, but is extensively used in flat stampings and where sharp bends are needed. It cannot be hardened by heat treatment and should not be used at temperatures above 150° F, but is especially good at sub-zero temperatures. Available in round sections and flat strips, this hard-drawn material is usually used in the "spring hard" temper.

*Phosphor Bronze, ASTM B159:* (95 percent copper, 5 percent tin) This alloy is the most popular of this group because it combines the best qualities of tensile strength, hardness, electrical conductivity, and corrosion resistance with the least cost. It is more expensive than brass, but can withstand stresses 50 percent higher. The material cannot be hardened by heat treatment. It can be used at temperatures up to 212° F and at sub-zero temperatures. It is available in round sections and flat strip, usually in the "extra-hard" or "spring hard" tempers. It is frequently used for contact fingers in switches because of its low arcing properties. An 8 percent tin composition is used for flat springs and a superfine grain composition called "Duraflex" has good endurance properties.

*Beryllium Copper, ASTM B197:* (98 percent copper, 2 percent beryllium) This alloy can be formed in the annealed condition and then precipitation hardened after forming at temperatures around 600° F, for 2 to 3 hours. This treatment produces a high hardness combined with a high tensile strength. After hardening, the material becomes quite brittle and can withstand very little or no forming. It is the most expensive alloy in the group and heat treating is expensive due to the need for holding the parts in fixtures to prevent distortion. The principal use of this alloy is for carrying electric current in switches and in electrical components. Flat strip is frequently used for contact fingers.

**Nickel-Base Spring Alloys.—**Nickel-base alloys are corrosion resistant, withstand both elevated and sub-zero temperatures, and their non-magnetic characteristic makes them useful for such applications as gyroscopes, chronoscopes, and indicating instruments. These materials have a high electrical resistance and therefore should not be used for conductors of electrical current.

*Monel**: (67 percent nickel, 30 percent copper) This material is the least expensive of the nickel-base alloys. It also has the lowest tensile strength but is useful due to its resistance to the corrosive effects of sea water and because it is nearly non-magnetic. The alloy can be subjected to stresses slightly higher than phosphor bronze and nearly as high as beryllium copper. Its high tensile strength and hardness are obtained as a result of cold-drawing and cold-rolling only, since it can not be hardened by heat treatment. It can be used at temperatures ranging from −100 to +425° F at normal operating stresses and is available in round wires up to $\frac{3}{16}$ inch in diameter with quite high tensile strengths. Larger diameters and flat strip are available with lower tensile strengths.

*"K" Monel**: (66 percent nickel, 29 percent copper, 3 percent aluminum) This material is quite similar to Monel except that the addition of the aluminum makes it a precipitation-hardening alloy. It may be formed in the soft or fairly hard condition and then hardened by a long-time age-hardening heat treatment to obtain a tensile strength and hardness above Monel and nearly as high as stainless steel. It is used in sizes larger than those usually used with Monel, is non-magnetic and can be used in temperatures ranging from −100 to +450° F at normal working stresses under 45,000 pounds per square inch.

*Inconel**: (78 percent nickel, 14 percent chromium, 7 percent iron) This is one of the most popular of the non-magnetic nickel-base alloys because of its corrosion resistance and because it can be used at temperatures up to 700° F. It is more expensive than stainless steel but less expensive than beryllium copper. Its hardness and tensile strength is higher than that of "K" Monel and is obtained as a result of cold-drawing and cold-rolling only. It cannot be hardened by heat treatment. Wire diameters up to $\frac{1}{4}$ inch have the best tensile properties. It is often used in steam valves, regulating valves, and for springs in boilers, compressors, turbines, and jet engines.

*Inconel "X"**: (70 percent nickel, 16 percent chromium, 7 percent iron) This material is quite similar to Inconel but the small amounts of titanium, columbium and aluminum in its composition make it a precipitation-hardening alloy. It can be formed in the soft or partially hard condition and then hardened by holding it at 1200° F for 4 hours. It is non-magnetic and is used in larger sections than Inconel. This alloy is used at temperatures up to 850° F and at stresses up to 55,000 pounds per square inch.

*Duranickel**: ("Z" Nickel) (98 percent nickel) This alloy is non-magnetic, corrosion resistant, has a high tensile strength and is hardenable by precipitation hardening at 900° F for 6 hours. It may be used at the same stresses as Inconel but should not be used at temperatures above 500° F.

**Nickel-Base Spring Alloys with Constant Moduli of Elasticity.—**Some special nickel alloys have a constant modulus of elasticity over a wide temperature range. These materials are especially useful where springs undergo temperature changes and must exhibit uniform spring characteristics. These materials have a low or zero thermo-elastic coefficient

*Trade name of the International Nickel Company.

and therefore do not undergo variations in spring stiffness because of modulus changes due to temperature differentials. They also have low hysteresis and creep values which makes them preferred for use in food-weighing scales, precision instruments, gyroscopes, measuring devices, recording instruments and computing scales where the temperature ranges from −50 to +150° F. These materials are expensive, none being regularly stocked in a wide variety of sizes. They should not be specified without prior discussion with spring manufacturers because some suppliers may not fabricate springs from these alloys due to the special manufacturing processes required. All of these alloys are used in small wire diameters and in thin strip only and are covered by US patents. They are more specifically described as follows:

*Elinvar**: (nickel, iron, chromium) This alloy, the first constant-modulus alloy used for hairsprings in watches, is an austenitic alloy hardened only by cold-drawing and cold-rolling. Additions of titanium, tungsten, molybdenum and other alloying elements have brought about improved characteristics and precipitation-hardening abilities. These improved alloys are known by the following trade names: Elinvar Extra, Durinval, Modulvar and Nivarox.

*Ni-Span C**: (nickel, iron, chromium, titanium) This very popular constant-modulus alloy is usually formed in the 50 percent cold-worked condition and precipitation-hardened at 900° F for 8 hours, although heating up to 1250° F for 3 hours produces hardnesses on the Rockwell C scale of 40 to 44 RC, permitting safe torsional stresses of 60,000 to 80,000 pounds per square inch. This material is ferromagnetic up to 400° F; above that temperature it becomes non-magnetic.

*Iso-Elastic*†: (nickel, iron, chromium, molybdenum) This popular alloy is relatively easy to fabricate and is used at safe torsional stresses of 40,000 to 60,000 pounds per square inch and hardnesses of 30 to 36 RC. It is used principally in dynamometers, instruments, and food-weighing scales.

*Elgiloy*‡: (nickel, iron, chromium, cobalt) This alloy, also known by the trade names 8J Alloy, Durapower, and Cobenium, is a non-magnetic alloy suitable for sub-zero temperatures and temperatures up to about 1000° F, provided that torsional stresses are kept under 75,000 pounds per square inch. It is precipitation-hardened at 900° F for 8 hours to produce hardnesses of 48 to 50 RC. The alloy is used in watch and instrument springs.

*Dynavar*§: (nickel, iron, chromium, cobalt) This alloy is a non-magnetic, corrosion-resistant material suitable for sub-zero temperatures and temperatures up to about 750° F, provided that torsional stresses are kept below 75,000 pounds per square inch. It is precipitation-hardened to produce hardnesses of 48 to 50 RC and is used in watch and instrument springs.

## Spring Stresses

**Allowable Working Stresses for Springs.**—The safe working stress for any particular spring depends to a large extent on the following items:

1) Type of spring — whether compression, extension, torsion, etc.

2) Size of spring — small or large, long or short

3) Spring material

4) Size of spring material

5) Type of service — light, average, or severe

6) Stress range — low, average, or high

*Trade name of Soc. Anon. de Commentry Fourchambault et Decazeville, Paris, France.
†Trade name of John Chatillon & Sons.
‡Trade name of Elgin National Watch Company.
§Trade name of Hamilton Watch Company.

7) Loading — static, dynamic, or shock

8) Operating temperature

9) Design of spring — spring index, sharp bends, hooks.

Consideration should also be given to other factors that affect spring life: corrosion, buckling, friction, and hydrogen embrittlement decrease spring life; manufacturing operations such as high-heat stress-equalizing, presetting, and shot-peening increase spring life.

Item 5, the type of service to which a spring is subjected, is a major factor in determining a safe working stress once consideration has been given to type of spring, kind and size of material, temperature, type of loading, and so on. The types of service are:

*Light Service:* This includes springs subjected to static loads or small deflections and seldom-used springs, such as those in bomb fuses, projectiles, and safety devices. This service is for 1000 to 10,000 deflections.

*Average Service:* This includes springs in general use in machine tools, mechanical products, and electrical components. Normal frequency of deflections not exceeding 18,000 per hour permit such springs to withstand 100,000 to 1,000,000 deflections.

*Severe Service:* This includes springs subjected to rapid deflections over long periods of time and to shock loading, such as in pneumatic hammers, hydraulic controls and valves. This service is for 1,000,000 deflections, and above. Lowering the values 10 percent permits 10,000,000 deflections.

Fig. 1 through Fig. 6 show curves that relate the three types of service conditions to allowable working stresses and wire sizes for compression and extension springs, and safe values are provided. Fig. 7 through Fig. 10 provide similar information for helical torsion springs. In each chart, the values obtained from the curves may be increased by 20 percent (but not beyond the top curves on the charts if permanent set is to be avoided) for springs that are baked, and shot-peened, and compression springs that are pressed. Springs stressed slightly above the Light Service curves will take a permanent set.

A curvature correction factor is included in all curves, and is used in spring design calculations (see examples beginning page 311). The curves may be used for materials other than those designated in Fig. 1 through Fig. 10, by applying multiplication factors as given in Table 1.

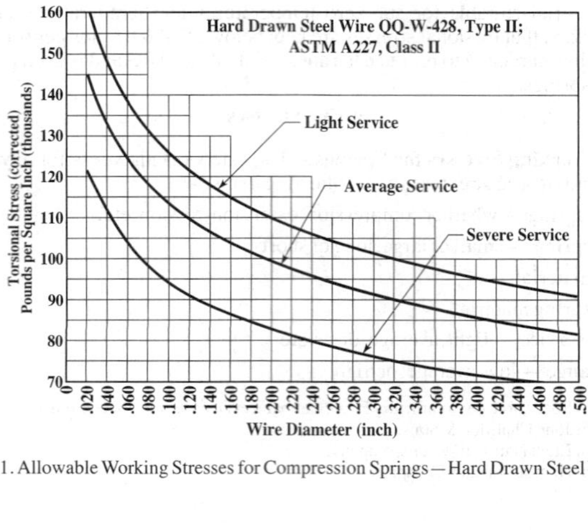


Fig. 1. Allowable Working Stresses for Compression Springs — Hard Drawn Steel Wire[a]

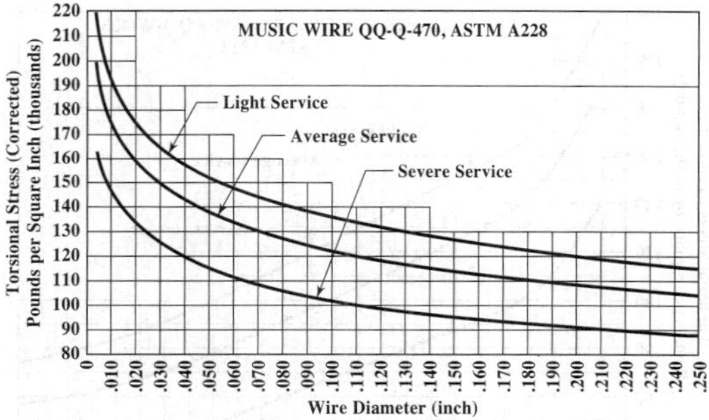


Fig. 2. Allowable Working Stresses for Compression Springs—Music Wire[a]

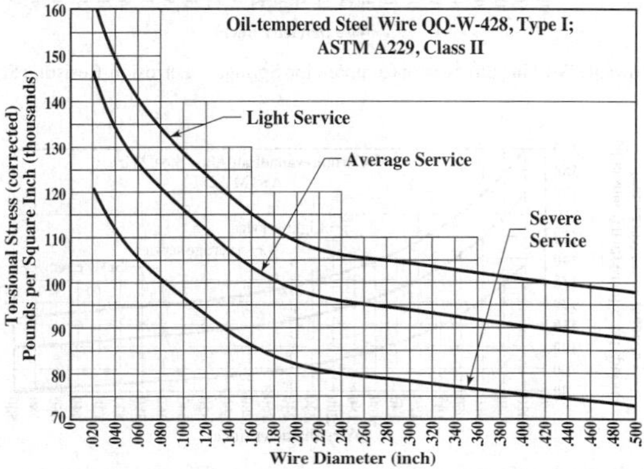


Fig. 3. Allowable Working Stresses for Compression Springs—Oil-Tempered[a]

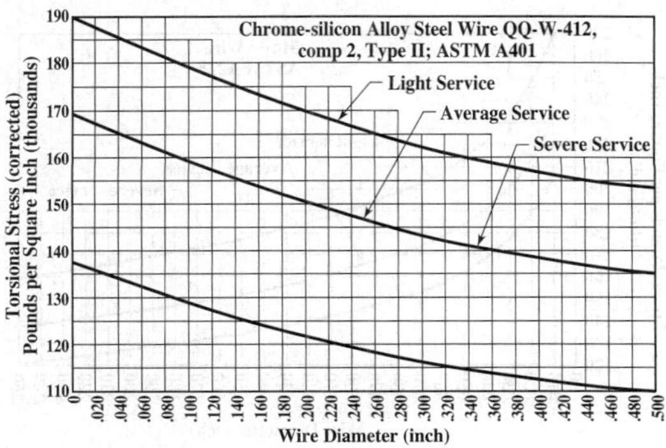


Fig. 4. Allowable Working Stresses for Compression Springs—Chrome-Silicon Alloy Steel Wire[a]

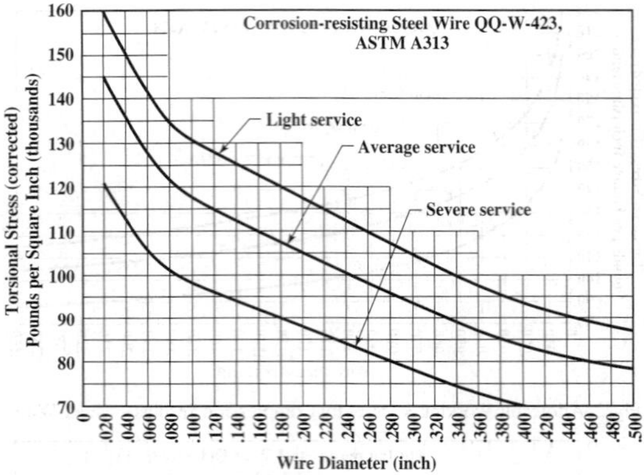


Fig. 5. Allowable Working Stresses for Compression Springs—Corrosion-Resisting Steel Wire[a]

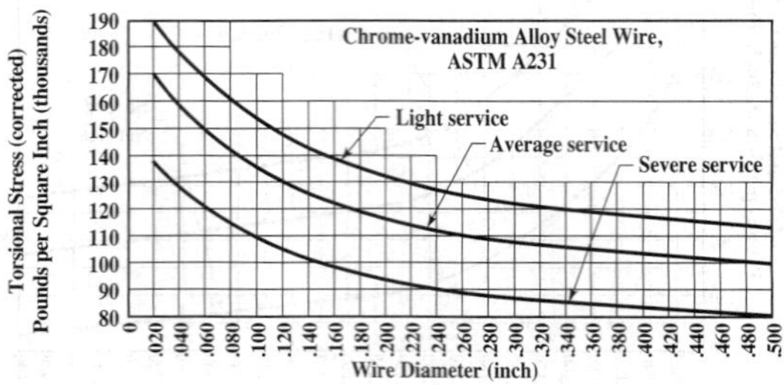


Fig. 6. Allowable Working Stresses for Compression Springs—Chrome-Vanadium Alloy Steel Wire[a]

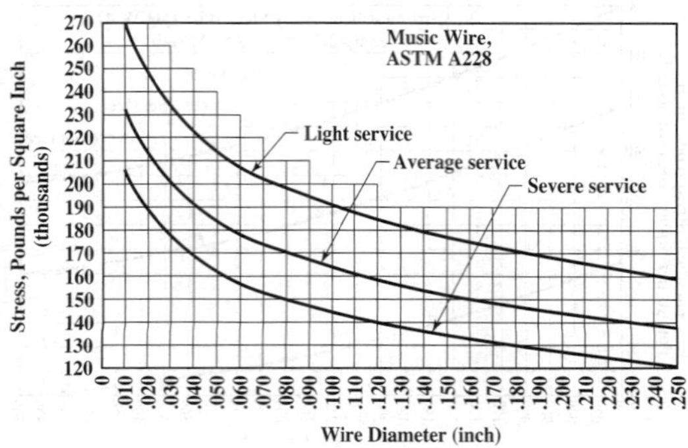


Fig. 7. Recommended Design Stresses in Bending for Helical Torsion Springs—Round Music Wire

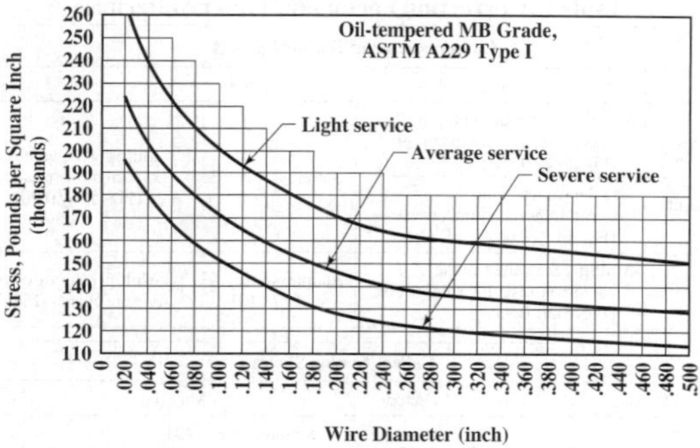


Fig. 8. Recommended Design Stresses in Bending for Helical Torsion Springs—Oil-Tempered MB Round Wire

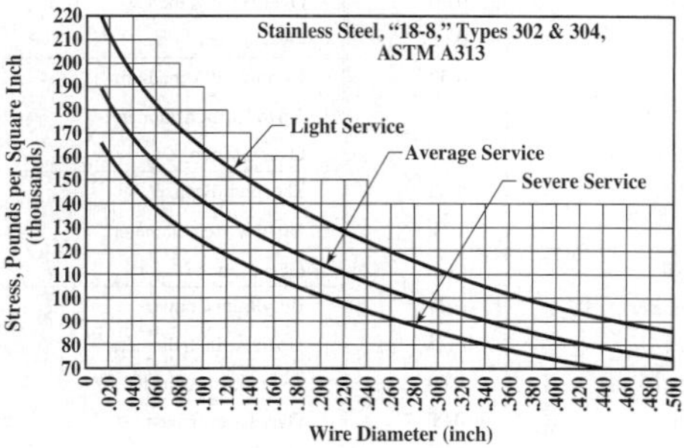


Fig. 9. Recommended Design Stresses in Bending for Helical Torsion Springs—Stainless Steel Round Wire

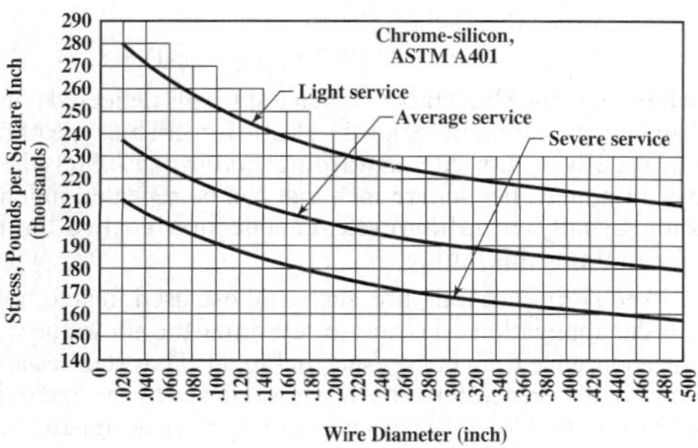


Fig. 10. Recommended Design Stresses in Bending for Helical Torsion Springs—Chrome-Silicon Round Wire

[a]Although Fig. 1 through Fig. 6 are for compression springs, they may also be used for extension springs; for extension springs, *reduce* the values obtained from the curves by 10 to 15 percent.

**Table 1. Correction Factors for Other Materials**

| Compression and Tension Springs | | | |
|---|---|---|---|
| Material | Factor | Material | Factor |
| Silicon-manganese | Multiply the values in the chromium-vanadium curves (Fig. 6) by 0.90 | Stainless Steel, 316 | Multiply the values in the corrosion-resisting steel curves (Fig. 5) by 0.90 |
| Valve-spring quality wire | Use the values in the chromium-vanadium curves (Fig. 6) | | |
| Stainless Steel, 304 and 420 | Multiply the values in the corrosion-resisting steel curves (Fig. 5) by 0.95 | Stainless Steel, 431 and 17-7PH | Multiply the values in the music wire curves (Fig. 2) by 0.90 |

| Helical Torsion Springs | | | |
|---|---|---|---|
| Material | Factor[a] | Material | Factor[a] |
| Hard Drawn MB | 0.70 | Stainless Steel, 431 | |
| Stainless Steel, 316 | | Up to 1⁄32 inch diameter | 0.80 |
| Up to 1⁄32 inch diameter | 0.75 | Over 1⁄32 to 1⁄16 inch | 0.85 |
| Over 1⁄32 to 3⁄16 inch | 0.70 | Over 1⁄16 to 1⁄8 inch | 0.95 |
| Over 3⁄16 to 1⁄4 inch | 0.65 | Over 1⁄8 inch | 1.00 |
| Over 1⁄4 inch | 0.50 | Chromium-Vanadium | |
| Stainless Steel, 17-7 PH | | Up to 1⁄16 inch diameter | 1.05 |
| Up to 1⁄8 inch diameter | 1.00 | Over 1⁄16 inch | 1.10 |
| Over 1⁄8 to 3⁄16 inch | 1.07 | Phosphor Bronze | |
| Over 3⁄16 inch | 1.12 | Up to 1⁄8 inch diameter | 0.45 |
| Stainless Steel, 420 | | Over 1⁄8 inch | 0.55 |
| Up to 1⁄32 inch diameter | 0.70 | Beryllium Copper[b] | |
| Over 1⁄32 to 1⁄16 inch | 0.75 | Up to 1⁄32 inch diameter | 0.55 |
| Over 1⁄16 to 1⁄8 inch | 0.80 | Over 1⁄32 to 1⁄16 inch | 0.60 |
| Over 1⁄8 to 3⁄16 inch | 0.90 | Over 1⁄16 to 1⁄8 inch | 0.70 |
| Over 3⁄16 inch | 1.00 | Over 1⁄8 inch | 0.80 |

[a] Multiply the values in the curves for oil-tempered MB grade ASTM A229 Type 1 steel (Fig. 8) by these factors to obtain required values.

[b] Hard drawn and heat treated after coiling.

For use with design stress curves shown in Fig. 2, Fig. 5, Fig. 6, and Fig. 8.

**Endurance Limit for Spring Materials.—**When a spring is deflected continually it will become "tired" and fail at a stress far below its elastic limit. This type of failure is called *fatigue failure* and usually occurs without warning. *Endurance limit* is the highest stress, or range of stress, in pounds per square inch that can be repeated indefinitely without failure of the spring. Usually ten million cycles of deflection is called "infinite life" and is satisfactory for determining this limit.

For severely worked springs of long life, such as those used in automobile or aircraft engines and in similar applications, it is best to determine the allowable working stresses by referring to the endurance limit curves seen in Fig. 11. These curves are based principally upon the range or difference between the stress caused by the first or initial load and the stress caused by the final load. Experience with springs designed to stresses within the limits of these curves indicates that they should have infinite or unlimited fatigue life. All values include Wahl curvature correction factor. The stress ranges shown may be increased 20 to 30 percent for springs that have been properly heated, pressed to remove set, and then shot peened, provided that the increased values are lower than the torsional elastic limit by at least 10 percent.

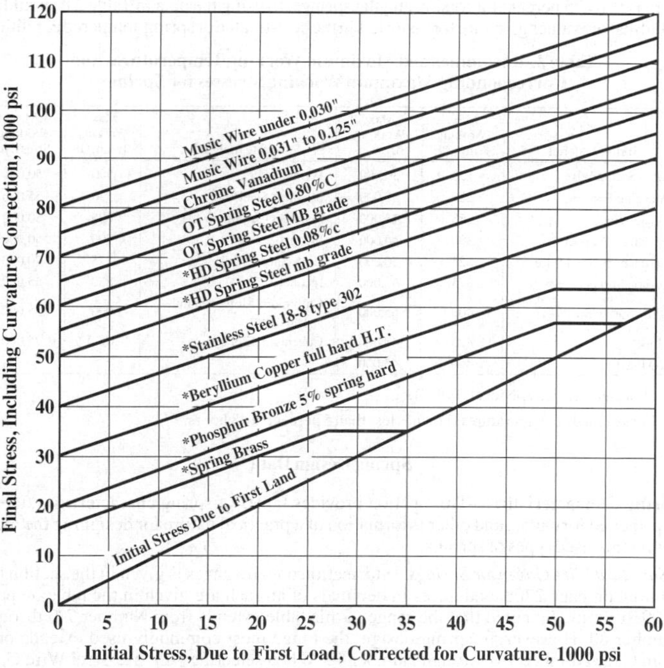


Fig. 11. Endurance Limit Curves for Compression Springs

*Notes:* For commercial spring materials with wire diameters up to ¼ inch except as noted. Stress ranges may be increased by approximately 30 percent for properly heated, preset, shot-peened springs.

Materials preceeded by * are not ordinarily recommended for long continued service under severe operating conditions.

**Working Stresses at Elevated Temperatures.**—Since modulus of elasticity decreases with increase in temperature, springs used at high temperatures exert less load and have larger deflections under load than at room temperature. The torsional modulus of elasticity for steel may be 11,200,000 pounds per square inch at room temperature, but it will drop to 10,600,000 pounds per square inch at 400° F. and will be only 10,000,000 pounds per square inch at 600° F. Also, the elastic limit is reduced, thereby lowering the permissible working stress.

Design stresses should be as low as possible for all springs used at elevated temperatures. In addition, corrosive conditions that usually exist at high temperatures, especially with steam, may require the use of corrosion-resistant material. Table 2 shows the permissible elevated temperatures at which various spring materials may be operated, together with the maximum recommended working stresses at these temperatures. The loss in load at the temperatures shown is less than 5 percent in 48 hours; however, if the temperatures listed are increased by 20 to 40 degrees, the loss of load may be nearer 10 percent. Maximum stresses shown in the table are for compression and extension springs and may be increased by 75 percent for torsion and flat springs. In using the data in Table 2 it should be noted that the values given are for materials in the heat-treated or spring-tempered condition.

**Table 2. Recommended Maximum Working Temperatures and Corresponding Maximum Working Stresses for Springs**

| Spring Material | Max. Working Temp., °F | Max. Working Stress, psi | Spring Material | Max. Working Temp, °F | Max. Working Stress, psi |
|---|---|---|---|---|---|
| Brass Spring Wire | 150 | 30,000 | Permanickel[a] | 500 | 50,000 |
| Phosphor Bronze | 225 | 35,000 | Stainless Steel 18-8 | 550 | 55,000 |
| Music Wire | 250 | 75,000 | Stainless Chromium 431 | 600 | 50,000 |
| Beryllium-Copper | 300 | 40,000 | Inconel | 700 | 50,000 |
| Hard Drawn Steel Wire | 325 | 50,000 | High-Speed Steel | 775 | 70,000 |
| Carbon Spring Steels | 375 | 55,000 | Inconel X | 850 | 55,000 |
| Alloy Spring Steels | 400 | 65,000 | Chromium-Molybdenum-Vanadium | 900 | 55,000 |
| Monel | 425 | 40,000 | Cobenium, Elgiloy | 1000 | 75,000 |
| K-Monel | 450 | 45,000 | | | |

[a] Formerly called Z-Nickel, Type B.

Loss of load at temperatures shown is less than 5 percent in 48 hours.

## Spring Design Data

**Spring Characteristics.—**This section provides tables of spring characteristics, tables of principal formulas, and other information of a practical nature for designing the more commonly used types of springs.

*Standard Wire Gages for Springs:* Information on wire gages is given in the section beginning on page 2708, and gages in decimals of an inch are given in the table on page 2709. It should be noted that the range in this table extends from Number 7/0 through Number 80. However, in spring design, the range most commonly used extends only from Gage Number 4/0 through Number 40. When selecting wire use Steel Wire Gage or Washburn and Moen gage for all carbon steels and alloy steels except music wire; use Brown & Sharpe gage for brass and phosphor bronze wire; use Birmingham gage for flat spring steels, and cold rolled strip; and use piano or music wire gage for music wire.

*Spring Index:* The spring index is the ratio of the mean coil diameter of a spring to the wire diameter ($D/d$). This ratio is one of the most important considerations in spring design because the deflection, stress, number of coils, and selection of either annealed or tempered material depend to a considerable extent on this ratio. The best proportioned springs have an index of 7 through 9. Indexes of 4 through 7, and 9 through 16 are often used. Springs with values larger than 16 require tolerances wider than standard for manufacturing; those with values less than 5 are difficult to coil on automatic coiling machines.

*Direction of Helix:* Unless functional requirements call for a definite hand, the helix of compression and extension springs should be specified as optional. When springs are designed to operate, one inside the other, the helices should be opposite hand to prevent intermeshing. For the same reason, a spring that is to operate freely over a threaded member should have a helix of opposite hand to that of the thread. When a spring is to engage with a screw or bolt, it should, of course, have the same helix as that of the thread.

**Helical Compression Spring Design.—**After selecting a suitable material and a safe stress value for a given spring, designers should next determine the type of end coil formation best suited for the particular application. Springs with unground ends are less expensive but they do not stand perfectly upright; if this requirement has to be met, closed ground ends are used. Helical compression springs with different types of ends are shown in Fig. 12.

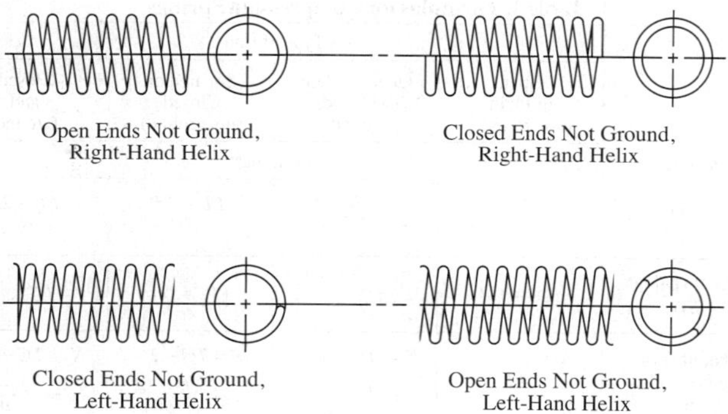


Fig. 12. Types of Helical Compression Spring Ends

*Spring Design Formulas:* Table 3 gives formulas for compression spring dimensional characteristics, and Table 4 gives design formulas for compression and extension springs.

*Curvature Correction:* In addition to the stress obtained from the formulas for load or deflection, there is a direct shearing stress and an increased stress on the inside of the section due to curvature. Therefore, the stress obtained by the usual formulas should be multiplied by a factor $K$ taken from the curve in Fig. 13. The corrected stress thus obtained is used only for comparison with the allowable working stress (fatigue strength) curves to determine if it is a safe stress and should not be used in formulas for deflection. The curvature correction factor $K$ is for compression and extension springs made from round wire. For square wire reduce the $K$ value by approximately 4 percent.

*Design Procedure:* The limiting dimensions of a spring are often determined by the available space in the product or assembly in which it is to be used. The loads and deflections on a spring may also be known or can be estimated, but the wire size and number of coils are usually unknown. Design can be carried out with the aid of the tabular data that appears later in this section (see Table 5, which is a simple method, or by calculation alone using the formulas in Table 3 and Table 4.

*Example:* A compression spring with closed and ground ends is to be made from ASTM A229 high carbon steel wire, as shown in Fig. 14. Determine the wire size and number of coils.

*Method 1, Using Table:* Referring to Table 5, starting on page 315, locate the spring outside diameter ($^{13}/_{16}$ inches, from Fig. 14 on page 313) in the left-hand column. Note from the drawing that the spring load is 36 pounds. Move to the right in the table to the figure nearest this value, which is 41.7 pounds. This is somewhat above the required value but safe. Immediately above the load value, the deflection $f$ is given, which in this instance is 0.1594 inch. This is the deflection of one coil under a load of 41.7 pounds with an uncorrected torsional stress $S$ of 100,000 pounds per square inch for ASTM A229 oil-tempered MB steel. For other spring materials, see the footnotes to Table 5. Moving vertically in Table 5 from the load entry, the wire diameter is found to be 0.0915 inch.

The remaining spring design calculations are completed as follows:

*Step 1:* The stress with a load of 36 pounds is obtained by proportion, as follows: The 36 pound load is 86.3 percent of the 41.7 pound load; therefore, the stress $S$ at 36 pounds = $0.863 \times 100{,}000 = 86{,}300$ pounds per square inch.

**Table 3. Formulas for Compression Springs**

| Feature | Type of End | | | |
|---|---|---|---|---|
| | Open or Plain (not ground) | Open or Plain (with ends ground) | Squared or Closed (not ground) | Closed and Ground |
| | Formula[a] | | | |
| Pitch ($p$) | $\frac{FL-d}{N}$ | $\frac{FL}{TC}$ | $\frac{FL-3d}{N}$ | $\frac{FL-2d}{N}$ |
| Solid Height ($SH$) | $(TC+1)d$ | $TC \times d$ | $(TC+1)d$ | $TC \times d$ |
| Number of Active Coils ($N$) | $N = TC = \frac{FL-d}{p}$ | $N = TC-1 = \frac{FL}{p} - 1$ | $N = TC-2 = \frac{FL-3d}{p}$ | $N = TC-2 = \frac{FL-2d}{p}$ |
| Total Coils ($TC$) | $\frac{FL-d}{p}$ | $\frac{FL}{p}$ | $\frac{FL-3d}{p} + 2$ | $\frac{FL-2d}{p} + 2$ |
| Free Length ($FL$) | $(p \times TC)+d$ | $p \times TC$ | $(p \times N)+3d$ | $(p \times N)+2d$ |

[a] The symbol notation is given on page 298.

**Table 4. Formulas for Compression and Extension Springs**

| Feature | Formula[a, b] | |
|---|---|---|
| | Springs made from round wire | Springs made from square wire |
| Load, $P$ Pounds | $P = \frac{0.393Sd^3}{D} = \frac{Gd^4F}{8ND^3}$ | $P = \frac{0.416Sd^3}{D} = \frac{Gd^4F}{5.58ND^3}$ |
| Stress, Torsional, $S$ Pounds per square inch | $S = \frac{GdF}{\pi ND^2} = \frac{PD}{0.393d^3}$ | $S = \frac{GdF}{2.32ND^2} = P\frac{D}{0.416d^3}$ |
| Deflection, $F$ Inch | $F = \frac{8PND^3}{Gd^4} = \frac{\pi SND^2}{Gd}$ | $F = \frac{5.58PND^3}{Gd^4} = \frac{2.32SND^2}{Gd}$ |
| Number of Active Coils, $N$ | $N = \frac{Gd^4F}{8PD^3} = \frac{GdF}{\pi SD^2}$ | $N = \frac{Gd^4F}{5.58PD^3} = \frac{GdF}{2.32SD^2}$ |
| Wire Diameter, $d$ Inch | $d = \frac{\pi SND^2}{GF} = \sqrt[3]{\frac{2.55PD}{S}}$ | $d = \frac{2.32SND^2}{GF} = \sqrt[3]{\frac{PD}{0.416S}}$ |
| Stress due to Initial Tension, $S_{it}$ | $S_{it} = \frac{S}{P} \times IT$ | $S_{it} = \frac{S}{P} \times IT$ |

[a] The symbol notation is given on page 298.

[b] Two formulas are given for each feature, and designers can use the one found to be appropriate for a given design. The end result from either of any two formulas is the same.

*Step 2:* The 86.3 percent figure is also used to determine the deflection per coil $f$ at 36 pounds load: $0.863 \times 0.1594 = 0.1375$ inch.

*Step 3:* The number of active coils $AC = \frac{F}{f} = \frac{1.25}{0.1375} = 9.1$

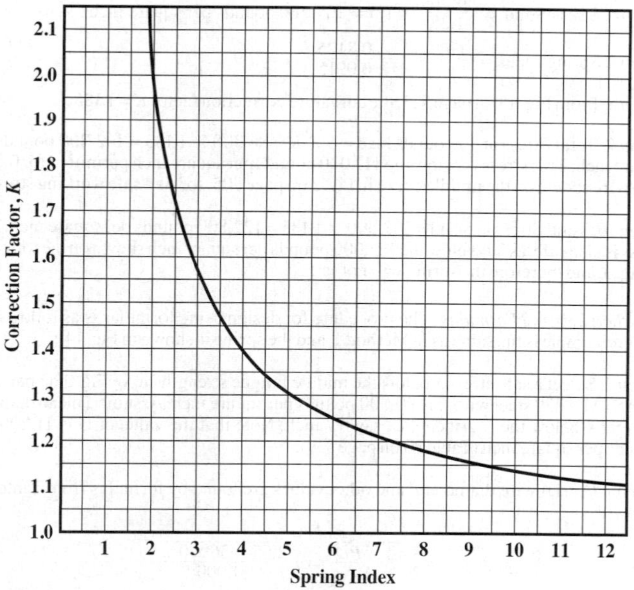


Fig. 13. Compression and Extension Spring-Stress Correction for Curvature[a]

[a]For springs made from round wire. For springs made from square wire, reduce the $K$ factor values by approximately 4 percent.

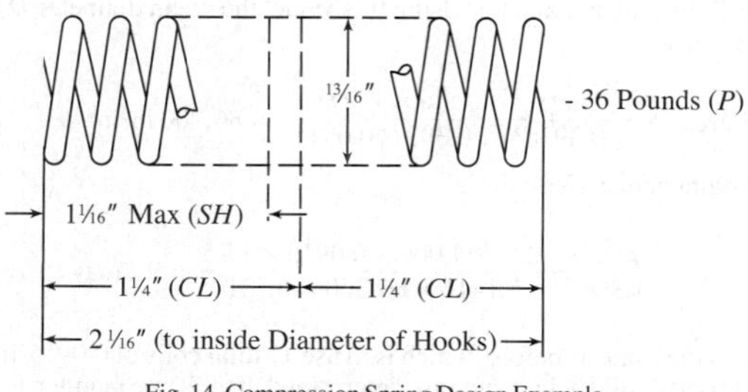


Fig. 14. Compression Spring Design Example

*Step 4:* Total Coils $TC = AC + 2$ (Table 3) $= 9 + 2 = 11$

Therefore, a quick answer is: 11 coils of 0.0915 inch diameter wire. However, the design procedure should be completed by carrying out these remaining steps:

*Step 5:* From Table 3, Solid Height $= SH = TC \times d = 11 \times 0.0915 \cong 1$ inch

Therefore, Total Deflection $= FL - SH = 1.5$ inches

*Step 6:* Stress Solid $= \dfrac{86{,}300}{1.25} \times 1.5 = 103{,}500$ pounds per square inch

*Step 7:* Spring Index $= \dfrac{OD}{d} - 1 = \dfrac{0.8125}{0.0915} - 1 = 7.9$

*Step 8:* From Fig. 13 on page 313, the curvature correction factor $K = 1.185$

*Step 9:* Total Stress at 36 pounds load $= S \times K = 86{,}300 \times 1.185 = 102{,}300$ pounds per square inch. This stress is below the 117,000 pounds per square inch permitted for 0.0915 inch wire shown on the middle curve in Fig. 3 on page 305, so it is a safe working stress.

*Step 10:* Total Stress at Solid $= 103{,}500 \times 1.185 = 122{,}800$ pounds per square inch. This stress is also safe, as it is below the 131,000 pounds per square inch shown on the top curve of Fig. 3, and therefore the spring will not set.

*Method 2, Using Formulas:* The procedure for design using formulas is as follows (the design example is the same as in Method 1, and the spring is shown in Fig. 14):

*Step 1:* Select a safe stress *S* below the middle fatigue strength curve Fig. 3 on page 305 for ASTM A229 steel wire, say 90,000 pounds per square inch. Assume a mean diameter *D* slightly below the $^{13}\!/_{16}$-inch *OD*, say 0.7 inch. Note that the value of *G* is 11,200,000 pounds per square inch (Table 20 on page 339).

*Step 2:* A trial wire diameter *d* and other values are found by formulas from Table 4 as follows:

$$d = \sqrt[3]{\frac{2.55PD}{S}} = \sqrt[3]{\frac{2.55 \times 36 \times 0.7}{90{,}000}}$$
$$= \sqrt[3]{0.000714} = 0.0894 \text{ inch}$$

*Note:* Table 21 on page 340 can be used to avoid solving the cube root.

*Step 3:* From Table 21 (also see the table on page 2709), select the nearest wire gauge size, which is 0.0915 inch diameter. Using this value, the mean diameter $D = {}^{13}\!/_{16}$ inch − 0.0915 = 0.721 inch.

*Step 4:* The stress $S = \dfrac{PD}{0.393d^3} = \dfrac{36 \times 0.721}{0.393 \times 0.0915^3} = 86{,}300 \text{ lb/in}^2$

*Step 5:* The number of active coils is

$$N = \frac{GdF}{\pi SD^2} = \frac{11{,}200{,}000 \times 0.0915 \times 1.25}{3.1416 \times 86{,}300 \times 0.721^2} = 9.1 \text{ (say 9)}$$

The answer is the same as before, which is to use 11 total coils of 0.0915-inch diameter wire. The total coils, solid height, etc., are determined in the same manner as in Method 1.

**Table of Spring Characteristics.—**Table 5 gives characteristics for compression and extension springs made from ASTM A229 oil-tempered MB spring steel having a torsional modulus of elasticity *G* of 11,200,000 pounds per square inch, and an uncorrected torsional stress *S* of 100,000 pounds per square inch. The deflection *f* for one coil under a load *P* is shown in the body of the table. The method of using these data is explained in the problems for compression and extension spring design. The table may be used for other materials by applying factors to *f*. The factors are given in a footnote to the table.

## Table 5. Compression and Extension Spring Deflections[a]

| Spring Outside Dia. | | Wire Size or Washburn and Moen Gauge, and Decimal Equivalent[b] | | | | | | | | | | | | | | | | | | |
|---|---|---|---|---|---|---|---|---|---|---|---|---|---|---|---|---|---|---|---|---|
| | | | | | | | | | | | | | | | | | 19 | 18 | 17 | 16 |
| | | .010 | .012 | .014 | .016 | .018 | .020 | .022 | .024 | .026 | .028 | .030 | .032 | .034 | .036 | .038 | .041 | .0475 | .054 | .0625 |
| Nom. | Dec. | Deflection *f* (inch) per coil, at Load *P* (pounds)[c] | | | | | | | | | | | | | | | | | | |
| 7⁄64 | .1094 | .0277 | .0222 | .01824 | .01529 | .01302 | .01121 | .00974 | .00853 | .00751 | .00664 | .00589 | … | … | … | … | … | … | … | … |
| | | .395 | .697 | 1.130 | 1.722 | 2.51 | 3.52 | 4.79 | 6.36 | 8.28 | 10.59 | 13.35 | … | … | … | … | … | … | … | … |
| 1⁄8 | .125 | .0371 | .0299 | .0247 | .0208 | .01784 | .01548 | .01353 | .01192 | .01058 | .00943 | .00844 | .00758 | .00683 | .00617 | … | … | … | … | … |
| | | .342 | .600 | .971 | 1.475 | 2.14 | 2.99 | 4.06 | 5.37 | 6.97 | 8.89 | 11.16 | 13.83 | 16.95 | 20.6 | … | … | … | … | … |
| 9⁄64 | .1406 | .0478 | .0387 | .0321 | .0272 | .0234 | .0204 | .01794 | .01590 | .01417 | .01271 | .01144 | .01034 | .00937 | .00852 | .00777 | … | … | … | … |
| | | .301 | .528 | .852 | 1.291 | 1.868 | 2.61 | 3.53 | 4.65 | 6.02 | 7.66 | 9.58 | 11.84 | 14.47 | 17.51 | 21.0 | … | … | … | … |
| 5⁄32 | .1563 | .0600 | .0487 | .0406 | .0345 | .0298 | .0261 | .0230 | .0205 | .01832 | 0.1649 | .01491 | .01354 | .01234 | .01128 | .01033 | .00909 | … | … | … |
| | | .268 | .470 | .758 | 1.146 | 1.656 | 2.31 | 3.11 | 4.10 | 5.30 | 6.72 | 8.39 | 10.35 | 12.62 | 15.23 | 18.22 | 23.5 | … | … | … |
| 11⁄64 | .1719 | .0735 | .0598 | .0500 | .0426 | .0369 | .0324 | .0287 | .0256 | .0230 | .0208 | .01883 | .01716 | .01569 | .01439 | .01324 | .01172 | .00914 | … | … |
| | | .243 | .424 | .683 | 1.031 | 1.488 | 2.07 | 2.79 | 3.67 | 4.73 | 5.99 | 7.47 | 9.19 | 11.19 | 13.48 | 16.09 | 21.8 | 33.8 | … | … |
| 3⁄16 | .1875 | .0884 | .0720 | .0603 | .0516 | .0448 | .0394 | .0349 | .0313 | .0281 | .0255 | .0232 | .0212 | .01944 | .01788 | .01650 | .01468 | .01157 | .00926 | … |
| | | .221 | .387 | .621 | .938 | 1.351 | 1.876 | 2.53 | 3.32 | 4.27 | 5.40 | 6.73 | 8.27 | 10.05 | 12.09 | 14.41 | 18.47 | 30.07 | 46.3 | … |
| 13⁄64 | .2031 | .1046 | .0854 | .0717 | .0614 | .0534 | .0470 | .0418 | .0375 | .0338 | .0307 | .0280 | .0257 | .0236 | .0218 | .0201 | .01798 | .01430 | .01155 | … |
| | | .203 | .355 | .570 | .859 | 1.237 | 1.716 | 2.31 | 3.03 | 3.90 | 4.92 | 6.12 | 7.52 | 9.13 | 10.96 | 13.05 | 16.69 | 27.1 | 41.5 | … |
| 7⁄32 | .2188 | … | .1000 | .0841 | .0721 | .0628 | .0555 | .0494 | .0444 | .0401 | .0365 | .0333 | .0306 | .0282 | .0260 | .0241 | .0216 | .01733 | .01411 | .01096 |
| | | … | .328 | .526 | .793 | 1.140 | 1.580 | 2.13 | 2.79 | 3.58 | 4.52 | 5.61 | 6.88 | 8.35 | 10.02 | 11.92 | 15.22 | 24.6 | 37.5 | 61.3 |
| 15⁄64 | .2344 | … | .1156 | .0974 | .0836 | .0730 | .0645 | .0575 | .0518 | .0469 | .0427 | .0391 | .0359 | .0331 | .0307 | .0285 | .0256 | .0206 | .01690 | .01326 |
| | | … | .305 | .489 | .736 | 1.058 | 1.465 | 1.969 | 2.58 | 3.21 | 4.18 | 5.19 | 6.35 | 7.70 | 9.23 | 10.97 | 13.99 | 22.5 | 34.3 | 55.8 |
| 1⁄4 | .250 | … | … | .1116 | .0960 | .0839 | .0742 | .0663 | .0597 | .0541 | .0494 | .0453 | .0417 | .0385 | .0357 | .0332 | .0299 | .0242 | .01996 | .01578 |
| | | … | … | .457 | .687 | .987 | 1.366 | 1.834 | 2.40 | 3.08 | 3.88 | 4.82 | 5.90 | 7.14 | 8.56 | 10.17 | 12.95 | 20.8 | 31.6 | 51.1 |
| 9⁄32 | .2813 | … | … | .1432 | .1234 | .1080 | .0958 | .0857 | .0774 | .0703 | .0643 | .0591 | .0545 | .0505 | .0469 | .0437 | .0395 | .0323 | .0268 | .0215 |
| | | … | … | .403 | .606 | .870 | 1.202 | 1.613 | 2.11 | 2.70 | 3.40 | 4.22 | 5.16 | 6.24 | 7.47 | 8.86 | 11.26 | 18.01 | 27.2 | 43.8 |
| 5⁄16 | .3125 | … | … | … | .1541 | .1351 | .1200 | .1076 | .0973 | .0886 | .0811 | .0746 | .0690 | .0640 | .0596 | .0556 | .0504 | .0415 | .0347 | .0281 |
| | | … | … | … | .542 | .778 | 1.074 | 1.440 | 1.881 | 2.41 | 3.03 | 3.75 | 4.58 | 5.54 | 6.63 | 7.85 | 9.97 | 15.89 | 23.9 | 38.3 |
| 11⁄32 | .3438 | … | … | … | … | .1633 | .1470 | .1321 | .1196 | .1090 | .0999 | .0921 | .0852 | .0792 | .0733 | .0690 | .0627 | .0518 | .0436 | .0355 |
| | | … | … | … | … | .703 | .970 | 1.300 | 1.697 | 2.17 | 2.73 | 3.38 | 4.12 | 4.98 | 5.95 | 7.05 | 8.94 | 14.21 | 21.3 | 34.1 |
| 3⁄8 | .375 | … | … | … | … | … | .1768 | .1589 | .1440 | .1314 | .1206 | .1113 | .1031 | .0960 | .0895 | .0839 | .0764 | .0634 | .0535 | .0438 |
| | | … | … | … | … | … | .885 | 1.185 | 1.546 | 1.978 | 2.48 | 3.07 | 3.75 | 4.53 | 5.40 | 6.40 | 8.10 | 12.85 | 19.27 | 30.7 |

[a] This table is for ASTM A229 oil tempered spring steel with a torsional modulus *G* of 11,200,000 psi, and an uncorrected torsional stress of 100,000 psi. For other materials use the following factors: stainless steel, multiply *f* by 1.067; spring brass, multiply *f* by 2.24; phosphor bronze, multiply *f* by 1.867; Monel metal, multiply *f* by 1.244; beryllium copper, multiply *f* by 1.725; Inconel (non-magnetic), multiply *f* by 1.045.

[b] Round wire. For square wire, multiply *f* by 0.707, and *p*, by 1.2

[c] The upper figure is the deflection and the lower figure the load as read against each spring size. *Note:* Intermediate values can be obtained within reasonable accuracy by interpolation.

## Table 5. *(Continued)* Compression and Extension Spring Deflections[a]

| Spring Outside Dia. | | Wire Size or Washburn and Moen Gauge, and Decimal Equivalent | | | | | | | | | | | | | | | | | |
|---|---|---|---|---|---|---|---|---|---|---|---|---|---|---|---|---|---|---|---|
| | | | | | | | | | 19 | 18 | 17 | 16 | 15 | 14 | 13 | 3/32 | 12 | 11 | 1/8 |
| | | .026 | .028 | .030 | .032 | .034 | .036 | .038 | .041 | .0475 | .054 | .0625 | .072 | .080 | .0915 | .0938 | .1055 | .1205 | .125 |
| Nom. | Dec. | Deflection *f* (inch) per coil, at Load *P* (pounds) | | | | | | | | | | | | | | | | | |
| 13/32 | .4063 | .1560 | .1434 | .1324 | .1228 | .1143 | .1068 | .1001 | .0913 | .0760 | .0645 | .0531 | .0436 | .0373 | .0304 | .0292 | .0241 | … | … |
| | | 1.815 | 2.28 | 2.82 | 3.44 | 4.15 | 4.95 | 5.85 | 7.41 | 11.73 | 17.56 | 27.9 | 43.9 | 61.6 | 95.6 | 103.7 | 153.3 | … | … |
| 7/16 | .4375 | .1827 | .1680 | .1553 | .1441 | .1343 | .1256 | .1178 | .1075 | .0898 | .0764 | .0631 | .0521 | .0448 | .0367 | .0353 | .0293 | .0234 | .0219 |
| | | 1.678 | 2.11 | 2.60 | 3.17 | 3.82 | 4.56 | 5.39 | 6.82 | 10.79 | 16.13 | 25.6 | 40.1 | 56.3 | 86.9 | 94.3 | 138.9 | 217. | 245. |
| 15/32 | .4688 | .212 | .1947 | .1800 | .1673 | .1560 | .1459 | .1370 | .1252 | .1048 | .0894 | .0741 | .0614 | .0530 | .0437 | .0420 | .0351 | .0282 | .0265 |
| | | 1.559 | 1.956 | 2.42 | 2.94 | 3.55 | 4.23 | 5.00 | 6.33 | 9.99 | 14.91 | 23.6 | 37.0 | 51.7 | 79.7 | 86.4 | 126.9 | 197.3 | 223. |
| 1/2 | .500 | .243 | .223 | .207 | .1920 | .1792 | .1678 | .1575 | .1441 | .1209 | .1033 | .0859 | .0714 | .0619 | .0512 | .0494 | .0414 | .0335 | .0316 |
| | | 1.456 | 1.826 | 2.26 | 2.75 | 3.31 | 3.95 | 4.67 | 5.90 | 9.30 | 13.87 | 21.9 | 34.3 | 47.9 | 73.6 | 80.0 | 116.9 | 181.1 | 205. |
| 17/32 | .5313 | .276 | .254 | .235 | .219 | .204 | .1911 | .1796 | .1645 | .1382 | .1183 | .0987 | .0822 | .0714 | .0593 | .0572 | .0482 | .0393 | .0371 |
| | | 1.366 | 1.713 | 2.12 | 2.58 | 3.10 | 3.70 | 4.37 | 5.52 | 8.70 | 12.96 | 20.5 | 31.9 | 44.6 | 68.4 | 74.1 | 108.3 | 167.3 | 188.8 |
| 9/16 | .5625 | … | .286 | .265 | .247 | .230 | .216 | .203 | .1861 | .1566 | .1343 | .1122 | .0937 | .0816 | .0680 | .0657 | .0555 | .0455 | .0430 |
| | | … | 1.613 | 1.991 | 2.42 | 2.92 | 3.48 | 4.11 | 5.19 | 8.18 | 12.16 | 19.17 | 29.9 | 41.7 | 63.9 | 69.1 | 100.9 | 155.5 | 175.3 |
| 19/32 | .5938 | … | … | .297 | .277 | .259 | .242 | .228 | .209 | .1762 | .1514 | .1267 | .1061 | .0926 | .0774 | .0748 | .0634 | .0522 | .0493 |
| | | … | … | 1.880 | 2.29 | 2.76 | 3.28 | 3.88 | 4.90 | 7.71 | 11.46 | 18.04 | 28.1 | 39.1 | 60.0 | 64.8 | 94.4 | 145.2 | 163.6 |
| 5/8 | .625 | … | … | .331 | .308 | .288 | .270 | .254 | .233 | .1969 | .1693 | .1420 | .1191 | .1041 | .0873 | .0844 | .0718 | .0593 | .0561 |
| | | … | … | 1.782 | 2.17 | 2.61 | 3.11 | 3.67 | 4.63 | 7.29 | 10.83 | 17.04 | 26.5 | 36.9 | 56.4 | 61.0 | 88.7 | 136.2 | 153.4 |
| 21/32 | .6563 | … | … | … | .342 | .320 | .300 | .282 | .259 | .219 | .1884 | .1582 | .1330 | .1164 | .0978 | .0946 | .0807 | .0668 | .0634 |
| | | … | … | … | 2.06 | 2.48 | 2.95 | 3.49 | 4.40 | 6.92 | 10.27 | 16.14 | 25.1 | 34.9 | 53.3 | 57.6 | 83.7 | 128.3 | 144.3 |
| 11/16 | .6875 | … | … | … | … | .352 | .331 | .311 | .286 | .242 | .208 | .1753 | .1476 | .1294 | .1089 | .1054 | .0901 | .0748 | .0710 |
| | | … | … | … | … | 2.36 | 2.81 | 3.32 | 4.19 | 6.58 | 9.76 | 15.34 | 23.8 | 33.1 | 50.5 | 54.6 | 79.2 | 121.2 | 136.3 |
| 23/32 | .7188 | … | … | … | … | … | .363 | .342 | .314 | .266 | .230 | .1933 | .1630 | .1431 | .1206 | .1168 | .1000 | .0833 | .0791 |
| | | … | … | … | … | … | 2.68 | 3.17 | 3.99 | 6.27 | 9.31 | 14.61 | 22.7 | 31.5 | 48.0 | 51.9 | 75.2 | 114.9 | 129.2 |
| 3/4 | .750 | … | … | … | … | … | … | .374 | .344 | .291 | .252 | .212 | .1791 | .1574 | .1329 | .1288 | .1105 | .0923 | .0877 |
| | | … | … | … | … | … | … | 3.03 | 3.82 | 5.99 | 8.89 | 13.94 | 21.6 | 30.0 | 45.7 | 49.4 | 71.5 | 109.2 | 122.7 |
| 25/32 | .7813 | … | … | … | … | … | … | … | .375 | .318 | .275 | .232 | .1960 | .1724 | .1459 | .1413 | .1214 | .1017 | .0967 |
| | | … | … | … | … | … | … | … | 3.66 | 5.74 | 8.50 | 13.34 | 20.7 | 28.7 | 43.6 | 47.1 | 68.2 | 104.0 | 116.9 |
| 13/16 | .8125 | … | … | … | … | … | … | … | .407 | .346 | .299 | .253 | .214 | .1881 | .1594 | .1545 | .1329 | .1115 | .1061 |
| | | … | … | … | … | … | … | … | 3.51 | 5.50 | 8.15 | 12.78 | 19.80 | 27.5 | 41.7 | 45.1 | 65.2 | 99.3 | 111.5 |

[a] This table is for ASTM A229 oil tempered spring steel with a torsional modulus *G* of 11,200,000 psi, and an uncorrected torsional stress of 100,000 psi. For other materials, and other important footnotes, see page 315.

**Table 5.** *(Continued)* **Compression and Extension Spring Deflections**[a]

| Spring Outside Dia. | | Wire Size or Washburn and Moen Gauge, and Decimal Equivalent | | | | | | | | | | | | | | | | |
|---|---|---|---|---|---|---|---|---|---|---|---|---|---|---|---|---|---|---|
| | | 15 | 14 | 13 | 3/32 | 12 | 11 | 1/8 | 10 | 9 | 5/32 | 8 | 7 | 3/16 | 6 | 5 | 7/32 | 4 |
| | | .072 | .080 | .0915 | .0938 | .1055 | .1205 | .125 | .135 | .1483 | .1563 | .162 | .177 | .1875 | .192 | .207 | .2188 | .2253 |
| Nom. | Dec. | Deflection $f$ (inch) per coil, at Load $P$ (pounds) | | | | | | | | | | | | | | | | |
| 7/8 | .875 | .251 | .222 | .1882 | .1825 | .1574 | .1325 | .1262 | .1138 | .0999 | .0928 | .0880 | .0772 | .0707 | .0682 | .0605 | .0552 | .0526 |
| | | 18.26 | 25.3 | 39.4 | 41.5 | 59.9 | 91.1 | 102.3 | 130.5 | 176.3 | 209. | 234. | 312. | 377. | 407. | 521. | 626. | 691. |
| 29/32 | .9063 | .271 | .239 | .204 | .1974 | .1705 | .1438 | .1370 | .1236 | .1087 | .1010 | .0959 | .0843 | .0772 | .0746 | .0663 | .0606 | .0577 |
| | | 17.57 | 24.3 | 36.9 | 39.9 | 57.6 | 87.5 | 98.2 | 125.2 | 169.0 | 199.9 | 224. | 299. | 360. | 389. | 498. | 598. | 660. |
| 15/16 | .9375 | .292 | .258 | .219 | .213 | .1841 | .1554 | .1479 | .1338 | .1178 | .1096 | .1041 | .0917 | .0842 | .0812 | .0723 | .0662 | .0632 |
| | | 16.94 | 23.5 | 35.6 | 38.4 | 55.4 | 84.1 | 94.4 | 120.4 | 162.3 | 191.9 | 215. | 286. | 345. | 373. | 477. | 572. | 631. |
| 31/32 | .9688 | .313 | .277 | .236 | .229 | .1982 | .1675 | .1598 | .1445 | .1273 | .1183 | .1127 | .0994 | .0913 | .0882 | .0786 | .0721 | .0688 |
| | | 16.35 | 22.6 | 34.3 | 37.0 | 53.4 | 81.0 | 90.9 | 115.9 | 156.1 | 184.5 | 207. | 275. | 332. | 358. | 457. | 548. | 604. |
| 1 | 1.000 | .336 | .297 | .253 | .246 | .213 | .1801 | .1718 | .1555 | .1372 | .1278 | .1216 | .1074 | .0986 | .0954 | .0852 | .0783 | .0747 |
| | | 15.80 | 21.9 | 33.1 | 35.8 | 51.5 | 78.1 | 87.6 | 111.7 | 150.4 | 177.6 | 198.8 | 264. | 319. | 344. | 439. | 526. | 580. |
| 1 1/32 | 1.031 | .359 | .317 | .271 | .263 | .228 | .1931 | .1843 | .1669 | .1474 | .1374 | .1308 | .1157 | .1065 | .1029 | .0921 | .0845 | .0809 |
| | | 15.28 | 21.1 | 32.0 | 34.6 | 49.8 | 75.5 | 84.6 | 107.8 | 145.1 | 171.3 | 191.6 | 255. | 307. | 331. | 423. | 506. | 557. |
| 1 1/16 | 1.063 | .382 | .338 | .289 | .281 | .244 | .207 | .1972 | .1788 | .1580 | .1474 | .1404 | .1243 | .1145 | .1107 | .0993 | .0913 | .0873 |
| | | 14.80 | 20.5 | 31.0 | 33.5 | 48.2 | 73.0 | 81.8 | 104.2 | 140.1 | 165.4 | 185.0 | 246. | 296. | 319. | 407. | 487. | 537. |
| 1 3/32 | 1.094 | .407 | .360 | .308 | .299 | .260 | .221 | .211 | .1910 | .1691 | .1578 | .1503 | .1332 | .1229 | .1188 | .1066 | .0982 | .0939 |
| | | 14.34 | 19.83 | 30.0 | 32.4 | 46.7 | 70.6 | 79.2 | 100.8 | 135.5 | 159.9 | 178.8 | 238. | 286. | 308. | 393. | 470. | 517. |
| 1 1/8 | 1.125 | .432 | .383 | .328 | .318 | .277 | .235 | .224 | .204 | .1804 | .1685 | .1604 | .1424 | .1315 | .1272 | .1142 | .1053 | .1008 |
| | | 13.92 | 19.24 | 29.1 | 31.4 | 45.2 | 68.4 | 76.7 | 97.6 | 131.2 | 154.7 | 173.0 | 230. | 276. | 298. | 379. | 454. | 499. |
| 1 3/16 | 1.188 | .485 | .431 | .368 | .358 | .311 | .265 | .254 | .231 | .204 | .1908 | .1812 | .1620 | .1496 | .1448 | .1303 | .1203 | .1153 |
| | | 13.14 | 18.15 | 27.5 | 29.6 | 42.6 | 64.4 | 72.1 | 91.7 | 123.3 | 145.4 | 162.4 | 215. | 259. | 279. | 355. | 424. | 467. |
| 1 1/4 | 1.250 | .541 | .480 | .412 | .400 | .349 | .297 | .284 | .258 | .230 | .215 | .205 | .1824 | .1690 | .1635 | .1474 | .1363 | .1308 |
| | | 12.44 | 17.19 | 26.0 | 28.0 | 40.3 | 60.8 | 68.2 | 86.6 | 116.2 | 137.0 | 153.1 | 203. | 244. | 263. | 334. | 399. | 438. |
| 1 5/16 | 1.313 | .600 | .533 | .457 | .444 | .387 | .331 | .317 | .288 | .256 | .240 | .229 | .205 | .1894 | .1836 | .1657 | .1535 | .1472 |
| | | 11.81 | 16.31 | 24.6 | 26.6 | 38.2 | 57.7 | 64.6 | 82.0 | 110.1 | 129.7 | 144.7 | 191.6 | 230. | 248. | 315. | 376. | 413. |
| 1 3/8 | 1.375 | .662 | .588 | .506 | .491 | .429 | .367 | .351 | .320 | .285 | .267 | .255 | .227 | .211 | .204 | .1848 | .1713 | .1650 |
| | | 11.25 | 15.53 | 23.4 | 25.3 | 36.3 | 54.8 | 61.4 | 77.9 | 104.4 | 123.0 | 137.3 | 181.7 | 218. | 235. | 298. | 356. | 391 |
| 1 7/16 | 1.438 | .727 | .647 | .556 | .540 | .472 | .404 | .387 | .353 | .314 | .295 | .282 | .252 | .234 | .227 | .205 | .1905 | .1829 |
| | | 10.73 | 14.81 | 22.3 | 24.1 | 34.6 | 52.2 | 58.4 | 74.1 | 99.4 | 117.0 | 130.6 | 172.6 | 207. | 223. | 283. | 337. | 371. |

[a] This table is for ASTM A229 oil tempered spring steel with a torsional modulus $G$ of 11,200,000 psi, and an uncorrected torsional stress of 100,000 psi. For other materials, and other important footnotes, see page 315.

## Table 5. *(Continued)* Compression and Extension Spring Deflections[a]

| Spring Outside Dia. | | Wire Size or Washburn and Moen Gauge, and Decimal Equivalent | | | | | | | | | | | | | | | | | |
|---|---|---|---|---|---|---|---|---|---|---|---|---|---|---|---|---|---|---|---|
| | | 11 | 1/8 | 10 | 9 | 5/32 | 8 | 7 | 3/16 | 6 | 5 | 7/32 | 4 | 3 | 1/4 | 2 | 9/32 | 0 | 5/16 |
| | | .1205 | .125 | .135 | .1483 | .1563 | .162 | .177 | .1875 | .192 | .207 | .2188 | .2253 | .2437 | .250 | .2625 | .2813 | .3065 | .3125 |
| Nom. | Dec. | Deflection $f$ (inch) per coil, at Load $P$ (pounds) | | | | | | | | | | | | | | | | | |
| 1½ | 1.500 | .443 | .424 | .387 | .350 | .324 | .310 | .277 | .258 | .250 | .227 | .210 | .202 | .1815 | .1754 | .1612 | .1482 | .1305 | .1267 |
| | | 49.8 | 55.8 | 70.8 | 94.8 | 111.5 | 124.5 | 164.6 | 197.1 | 213. | 269. | 321. | 352. | 452. | 499. | 574. | 717. | 947. | 1008. |
| 1⅝ | 1.625 | .527 | .505 | .461 | .413 | .387 | .370 | .332 | .309 | .300 | .273 | .254 | .244 | .220 | .212 | .1986 | .1801 | .1592 | .1547 |
| | | 45.7 | 51.1 | 64.8 | 86.7 | 102.0 | 113.9 | 150.3 | 180.0 | 193.9 | 246. | 292. | 321. | 411. | 446. | 521. | 650. | 858. | 912. |
| 1¾ | 1.750 | .619 | .593 | .542 | .485 | .456 | .437 | .392 | .366 | .355 | .323 | .301 | .290 | .261 | .253 | .237 | .215 | .1908 | .1856 |
| | | 42.2 | 47.2 | 59.8 | 80.0 | 94.0 | 104.9 | 138.5 | 165.6 | 178.4 | 226. | 269. | 295. | 377. | 409. | 477. | 595. | 783. | 833. |
| 1⅞ | 1.875 | .717 | .687 | .629 | .564 | .530 | .508 | .457 | .426 | .414 | .377 | .351 | .339 | .306 | .296 | .278 | .253 | .225 | .219 |
| | | 39.2 | 43.8 | 55.5 | 74.2 | 87.2 | 97.3 | 128.2 | 153.4 | 165.1 | 209. | 248. | 272. | 348. | 378. | 440. | 548. | 721. | 767. |
| 1 15/16 | 1.938 | .769 | .738 | .676 | .605 | .569 | .546 | .492 | .458 | .446 | .405 | .379 | .365 | .331 | .320 | .300 | .273 | .243 | .237 |
| | | 37.8 | 42.3 | 53.6 | 71.6 | 84.2 | 93.8 | 123.6 | 147.9 | 159.2 | 201. | 239. | 262. | 335. | 364. | 425. | 528. | 693. | 737. |
| 2 | 2.000 | .823 | .789 | .723 | .649 | .610 | .585 | .527 | .492 | .478 | .436 | .407 | .392 | .355 | .344 | .323 | .295 | .263 | .256 |
| | | 36.6 | 40.9 | 51.8 | 69.2 | 81.3 | 90.6 | 119.4 | 142.8 | 153.7 | 194.3 | 231. | 253. | 324. | 351. | 409. | 509. | 668. | 710. |
| 2 1/16 | 2.063 | .878 | .843 | .768 | .693 | .652 | .626 | .564 | .526 | .512 | .467 | .436 | .421 | .381 | .369 | .346 | .316 | .282 | .275 |
| | | 35.4 | 39.6 | 50.1 | 66.9 | 78.7 | 87.6 | 115.4 | 138.1 | 148.5 | 187.7 | 223. | 245. | 312. | 339. | 395. | 491. | 644. | 685. |
| 2⅛ | 2.125 | .936 | .898 | .823 | .739 | .696 | .667 | .602 | .562 | .546 | .499 | .466 | .449 | .407 | .395 | .371 | .339 | .303 | .295 |
| | | 34.3 | 38.3 | 48.5 | 64.8 | 76.1 | 84.9 | 111.8 | 133.6 | 143.8 | 181.6 | 216. | 236. | 302. | 327. | 381. | 474. | 622. | 661. |
| 2 3/16 | 2.188 | .995 | .955 | .876 | .786 | .740 | .711 | .641 | .598 | .582 | .532 | .497 | .479 | .435 | .421 | .396 | .362 | .324 | .316 |
| | | 33.3 | 37.2 | 47.1 | 62.8 | 73.8 | 82.2 | 108.3 | 129.5 | 139.2 | 175.8 | 209. | 229. | 292. | 317. | 369. | 459. | 601. | 639. |
| 2¼ | 2.250 | 1.056 | 1.013 | .930 | .835 | .787 | .755 | .681 | .637 | .619 | .566 | .529 | .511 | .463 | .449 | .423 | .387 | .346 | .337 |
| | | 32.3 | 36.1 | 45.7 | 60.9 | 71.6 | 79.8 | 105.7 | 125.5 | 135.0 | 170.5 | 202. | 222. | 283. | 307. | 357. | 444. | 582. | 618. |
| 2 5/16 | 2.313 | 1.119 | 1.074 | .986 | .886 | .834 | .801 | .723 | .676 | .657 | .601 | .562 | .542 | .493 | .478 | .449 | .411 | .368 | .359 |
| | | 31.4 | 35.1 | 44.4 | 59.2 | 69.5 | 77.5 | 101.9 | 121.8 | 131.0 | 165.4 | 196.3 | 215. | 275. | 298. | 347. | 430. | 564. | 599. |
| 2⅜ | 2.375 | 1.184 | 1.136 | 1.043 | .938 | .884 | .848 | .763 | .716 | .696 | .637 | .596 | .576 | .523 | .507 | .477 | .437 | .392 | .382 |
| | | 30.5 | 34.1 | 43.1 | 57.5 | 67.6 | 75.3 | 99.1 | 118.3 | 127.3 | 160.7 | 190.7 | 209. | 267. | 289. | 336. | 417. | 547. | 581. |
| 2 7/16 | 2.438 | … | 1.201 | 1.102 | .991 | .934 | .897 | .810 | .757 | .737 | .674 | .631 | .609 | .554 | .537 | .506 | .464 | .416 | .405 |
| | | … | 33.2 | 42.0 | 56.0 | 65.7 | 73.2 | 96.3 | 115.1 | 123.7 | 156.1 | 185.3 | 203. | 259. | 281. | 327. | 405. | 531. | 564. |
| 2½ | 2.500 | … | 1.266 | 1.162 | 1.046 | .986 | .946 | .855 | .800 | .778 | .713 | .667 | .644 | .586 | .568 | .536 | .491 | .441 | .430 |
| | | … | 32.3 | 40.9 | 54.5 | 64.0 | 71.3 | 93.7 | 111.6 | 120.4 | 151.9 | 180.2 | 197.5 | 252. | 273. | 317. | 394. | 516. | 548. |

[a] This table is for ASTM A229 oil tempered spring steel with a torsional modulus $G$ of 11,200,000 psi, and an uncorrected torsional stress of 100,000 psi. For other materials, and other important footnotes, see page 315.

**Extension Springs.**—About 10 percent of all springs made by many companies are of this type, and they frequently cause trouble because insufficient consideration is given to stress due to initial tension, stress and deflection of hooks, special manufacturing methods, secondary operations and overstretching at assembly. Fig. 15 shows types of ends used on these springs.

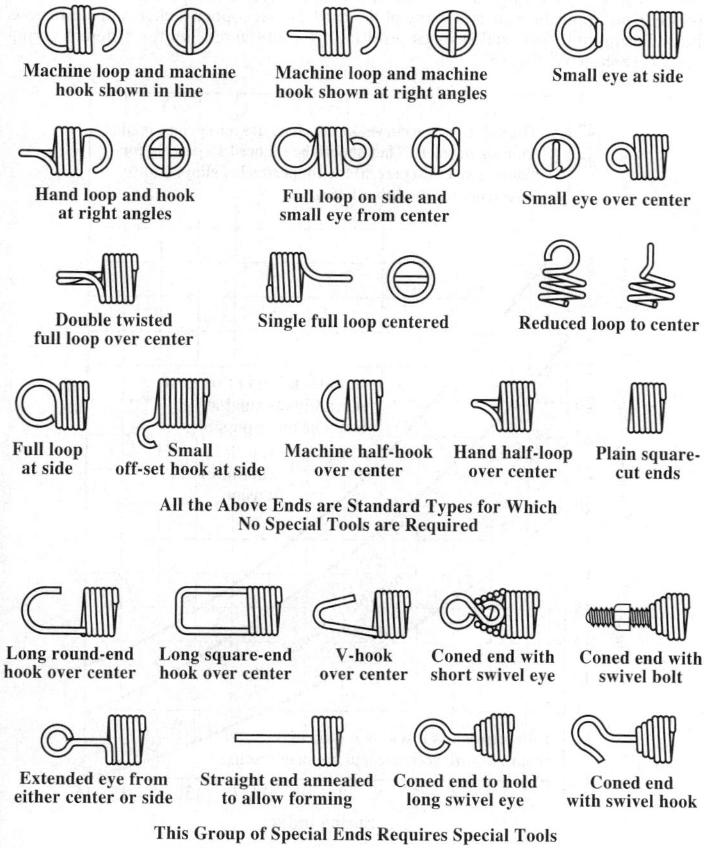


Fig. 15. Types of Helical Extension Spring Ends

*Initial Tension:* In the spring industry, the term "Initial tension" is used to define a force or load, measurable in pounds or ounces, which presses the coils of a close wound extension spring against one another. This force must be overcome before the coils of a spring begin to open up.

Initial tension is wound into extension springs by bending each coil as it is wound away from its normal plane, thereby producing a slight twist in the wire which causes the coil to spring back tightly against the adjacent coil. Initial tension can be wound into cold-coiled extension springs only. Hot-wound springs and springs made from annealed steel are hardened and tempered after coiling, and therefore initial tension cannot be produced. It is possible to make a spring having initial tension only when a high tensile strength, obtained by cold drawing or by heat treatment, is possessed by the material as it is being wound into springs. Materials that possess the required characteristics for the manufacture of such springs include hard-drawn wire, music wire, pre-tempered wire, 18-8 stainless steel, phosphor-bronze, and many of the hard-drawn copper-nickel, and nonferrous alloys. Permissible torsional stresses resulting from initial tension for different spring indexes are shown in Fig. 16.

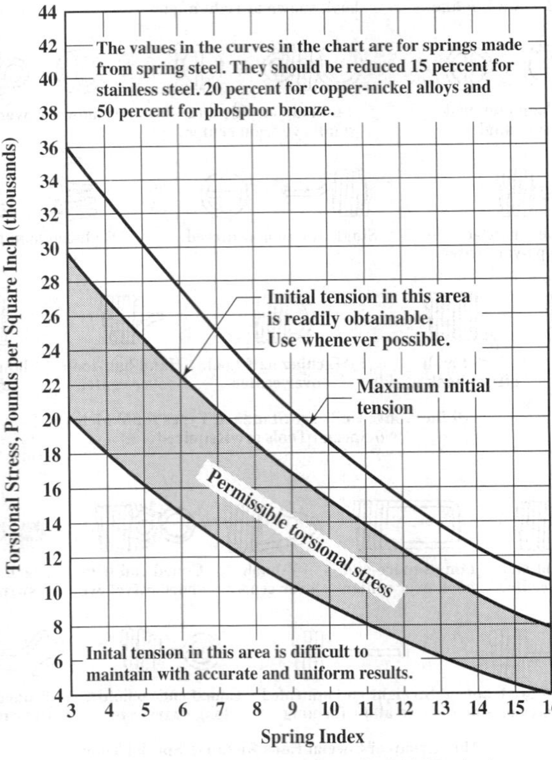


Fig. 16. Permissible Torsional Stress Caused by Initial Tension in Coiled Extension Springs for Different Spring Indexes

*Hook Failure:* The great majority of breakages in extension springs occurs in the hooks. Hooks are subjected to both bending and torsional stresses and have higher stresses than the coils in the spring.

*Stresses in Regular Hooks:* The calculations for the stresses in hooks are quite complicated and lengthy. Also, the radii of the bends are difficult to determine and frequently vary between specifications and actual production samples. However, regular hooks are more highly stressed than the coils in the body and are subjected to a bending stress at section B (see Table 6.) The bending stress $S_b$ at section B should be compared with allowable stresses for torsion springs and with the elastic limit of the material in tension (See Fig. 7 through Fig. 10.)

*Stresses in Cross-Over Hooks:* Results of tests on springs having a normal average index show that the cross-over hooks last longer than regular hooks. These results may not occur on springs of small index or if the cross-over bend is made too sharply.

In as much as both types of hooks have the same bending stress, it would appear that the fatigue life would be the same. However, the large bend radius of the regular hooks causes some torsional stresses to coincide with the bending stresses, thus explaining the earlier breakages. If sharper bends were made on the regular hooks, the life should then be the same as for cross-over hooks.

**Table 6. Formula for Bending Stress at Section B**

| Type of Hook | | Stress in Bending |
|---|---|---|
| 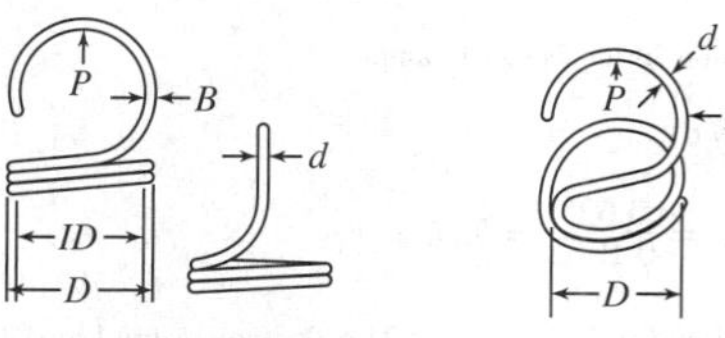 | | $S_b = \dfrac{5PD^2}{IDd^3}$ |
| Regular Hook | Cross-over Hook | |



*Stresses in Half Hooks:* The formulas for regular hooks can also be used for half hooks, because the smaller bend radius allows for the increase in stress. It will therefore be observed that half hooks have the same stress in bending as regular hooks.

Frequently overlooked facts by many designers are that one full hook deflects an amount equal to one half a coil and each half hook deflects an amount equal to one tenth of a coil. Allowances for these deflections should be made when designing springs. Thus, an extension spring, with regular full hooks and having 10 coils, will have a deflection equal to 11 coils, or 10 percent more than the calculated deflection.

**Extension Spring Design.—**The available space in a product or assembly usually determines the limiting dimensions of a spring, but the wire size, number of coils, and initial tension are often unknown.

*Example:* An extension spring is to be made from spring steel ASTM A229, with regular hooks as shown in Fig. 17. Calculate the wire size, number of coils and initial tension.

*Note:* Allow about 20 to 25 percent of the 9 pound load for initial tension, say 2 pounds, and then design for a 7 pound load (not 9 pounds) at ⅝ inch deflection. Also use lower stresses than for a compression spring to allow for overstretching during assembly and to obtain a safe stress on the hooks. Proceed as for compression springs, but locate a load in the tables somewhat higher than the 9 pound load.

*Method 1, Using Table:* From Table 5 locate ¾ inch outside diameter in the left column and move to the right to locate a load $P$ of 13.94 pounds. A deflection $f$ of 0.212 inch appears above this figure. Moving vertically from this position to the top of the column a suitable wire diameter of 0.0625 inch is found.

The remaining design calculations are completed as follows:

*Step 1:* The stress with a load of 7 pounds is obtained as follows:

The 7 pound load is 50.2 percent of the 13.94 pound load. Therefore, the stress $S$ at 7 pounds = 0.502 percent × 100,000 = 50,200 pounds per square inch.

*Step 2:* The 50.2 percent figure is also used to determine the deflection per coil $f$: 0.502 percent × 0.212 = 0.1062 inch.

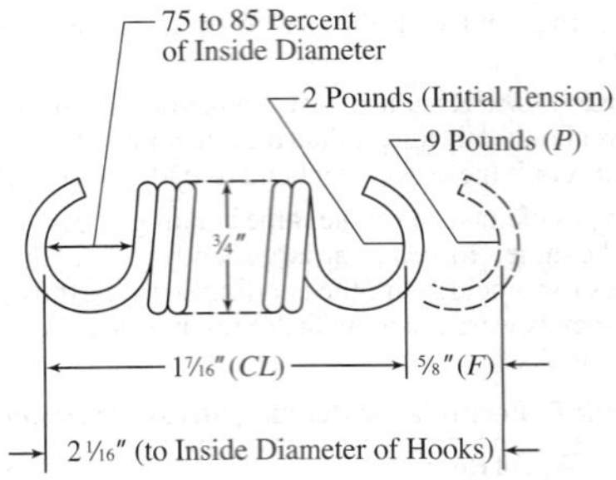


Fig. 17. Extension Spring Design Example

*Step 3:* The number of active coils (say 6)

$$AC = \frac{F}{f} = \frac{0.625}{0.1062} = 5.86$$

This result should be reduced by 1 to allow for deflection of 2 hooks (see notes 1 and 2 that follow these calculations.) Therefore, a quick answer is: 5 coils of 0.0625 inch diameter wire. However, the design procedure should be completed by carrying out the following steps:

*Step 4:* The body length = $(TC + 1) \times d = (5 + 1) \times 0.0625 = \frac{3}{8}$ inch.

*Step 5:* The length from the body to inside hook

$$= \frac{\text{FL} - \text{Body}}{2} = \frac{1.4375 - 0.375}{2} = 0.531 \text{ inch}$$

$$\text{Percentage of ID} = \frac{0.531}{ID} = \frac{0.531}{0.625} = 85 \text{ percent}$$

This length is satisfactory, see Note 3 following this procedure.

*Step 6:*

$$\text{The spring index} = \frac{OD}{d} - 1 = \frac{0.75}{0.0625} - 1 = 11$$

*Step 7:* The initial tension stress is

$$S_{it} = \frac{S \times IT}{P} = \frac{50{,}200 \times 2}{7} = 14{,}340 \text{ pounds per square inch}$$

This stress is satisfactory, as checked against curve in Fig. 16.

*Step 8:* The curvature correction factor $K = 1.12$ (Fig. 13).

*Step 9:* The total stress = $(50{,}200 + 14{,}340) \times 1.12 = 72.285$ pounds per square inch

This result is less than 106,250 pounds per square inch permitted by the middle curve for 0.0625 inch wire in Fig. 3 and therefore is a safe working stress that permits some additional deflection that is usually necessary for assembly purposes.

*Step 10:* The large majority of hook breakage is due to high stress in bending and should be checked as follows:

From Table 6, stress on hook in bending is:

$$S_b = \frac{5PD^2}{IDd^3} = \frac{5 \times 9 \times 0.6875^2}{0.625 \times 0.0625^3} = 139{,}200 \text{ pounds per square inch}$$

This result is less than the top curve value, Fig. 8, for 0.0625 inch diameter wire, and is therefore safe. Also see Note 5 that follows.

*Notes:* The following points should be noted when designing extension springs:

1) All coils are active and thus $AC = TC$.

2) Each full hook deflection is approximately equal to ½ coil. Therefore for 2 hooks, reduce the total coils by 1. (Each half hook deflection is nearly equal to $\frac{1}{10}$ of a coil.)

3) The distance from the body to the inside of a regular full hook equals 75 to 85 percent (90 percent maximum) of the ID. For a cross-over center hook, this distance equals the ID.

4) Some initial tension should usually be used to hold the spring together. Try not to exceed the maximum curve shown on Fig. 16. Without initial tension, a long spring with many coils will have a different length in the horizontal position than it will when hung vertically.

5) The hooks are stressed in bending, therefore their stress should be less than the maximum bending stress as used for torsion springs—use top fatigue strength curves Fig. 7 through Fig. 10.

*Method 2, Using Formulas:* The sequence of steps for designing extension springs by formulas is similar to that for compression springs. The formulas for this method are given in Table 3.

**Tolerances for Compression and Extension Springs.**—Tolerances for coil diameter, free length, squareness, load, and the angle between loop planes for compression and extension springs are given in Table 7 through Table 12. To meet the requirements of load, rate, free length, and solid height, it is necessary to vary the number of coils for compression springs by ± 5 percent. For extension springs, the tolerances on the numbers of coils are: for 3 to 5 coils, ± 20 percent; for 6 to 8 coils, ± 30 percent; for 9 to 12 coils, ± 40 percent. For each additional coil, a further 1½ percent tolerance is added to the extension spring values. Closer tolerances on the number of coils for either type of spring lead to the need for trimming after coiling, and manufacturing time and cost are increased. Fig. 18 shows deviations allowed on the ends of extension springs, and variations in end alignments.

**Table 7. Compression and Extension Spring Coil Diameter Tolerances**

| Wire Diameter, Inch | Spring Index | | | | | | |
|---|---|---|---|---|---|---|---|
| | 4 | 6 | 8 | 10 | 12 | 14 | 16 |
| | Tolerance, ± inch | | | | | | |
| 0.015 | 0.002 | 0.002 | 0.003 | 0.004 | 0.005 | 0.006 | 0.007 |
| 0.023 | 0.002 | 0.003 | 0.004 | 0.006 | 0.007 | 0.008 | 0.010 |
| 0.035 | 0.002 | 0.004 | 0.006 | 0.007 | 0.009 | 0.011 | 0.013 |
| 0.051 | 0.003 | 0.005 | 0.007 | 0.010 | 0.012 | 0.015 | 0.017 |
| 0.076 | 0.004 | 0.007 | 0.010 | 0.013 | 0.016 | 0.019 | 0.022 |
| 0.114 | 0.006 | 0.009 | 0.013 | 0.018 | 0.021 | 0.025 | 0.029 |
| 0.171 | 0.008 | 0.012 | 0.017 | 0.023 | 0.028 | 0.033 | 0.038 |
| 0.250 | 0.011 | 0.015 | 0.021 | 0.028 | 0.035 | 0.042 | 0.049 |
| 0.375 | 0.016 | 0.020 | 0.026 | 0.037 | 0.046 | 0.054 | 0.064 |
| 0.500 | 0.021 | 0.030 | 0.040 | 0.062 | 0.080 | 0.100 | 0.125 |

Courtesy of the Spring Manufacturers Institute

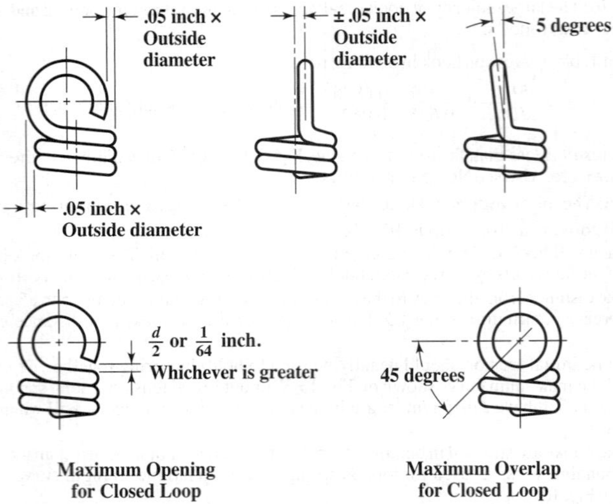


Fig. 18. Maximum Deviations Allowed on Ends and Variation in Alignment of Ends (Loops) for Extension Springs

### Table 8. Compression Spring Normal Free-Length Tolerances, Squared and Ground Ends

| Number of Active Coils per Inch | Spring Index | | | | | | |
|---|---|---|---|---|---|---|---|
| | 4 | 6 | 8 | 10 | 12 | 14 | 16 |
| | Tolerance, ± Inch per Inch of Free Length[a] | | | | | | |
| 0.5 | 0.010 | 0.011 | 0.012 | 0.013 | 0.015 | 0.016 | 0.016 |
| 1 | 0.011 | 0.013 | 0.015 | 0.016 | 0.017 | 0.018 | 0.019 |
| 2 | 0.013 | 0.015 | 0.017 | 0.019 | 0.020 | 0.022 | 0.023 |
| 4 | 0.016 | 0.018 | 0.021 | 0.023 | 0.024 | 0.026 | 0.027 |
| 8 | 0.019 | 0.022 | 0.024 | 0.026 | 0.028 | 0.030 | 0.032 |
| 12 | 0.021 | 0.024 | 0.027 | 0.030 | 0.032 | 0.034 | 0.036 |
| 16 | 0.022 | 0.026 | 0.029 | 0.032 | 0.034 | 0.036 | 0.038 |
| 20 | 0.023 | 0.027 | 0.031 | 0.034 | 0.036 | 0.038 | 0.040 |

[a] For springs less than 0.5 inch long, use the tolerances for 0.5 inch long springs. For springs with unground closed ends, multiply the tolerances by 1.7.

Courtesy of the Spring Manufacturers Institute

### Table 9. Extension Spring Normal Free-Length and End Tolerances

| Free-Length Tolerances | | End Tolerances | | Free-Length Tolerances | | End Tolerances | |
|---|---|---|---|---|---|---|---|
| Spring Free Length (inch) | Tolerance (inch) | Total Number of Coils | Angle Between Loop Planes | Spring Free Length (inch) | Tolerance (inch) | Total Number of Coils | Angle Between Loop Planes |
| Up to 0.5 | ±0.020 | | | Over 4.0 to 8.0 | ±0.093 | 13 to 16 | ±60° |
| Over 0.5 to 1.0 | ±0.030 | 3 to 6 | ±25° | Over 8.0 to 16.0 | ±0.156 | Over 16 | Random |
| Over 1.0 to 2.0 | ±0.040 | 7 to 9 | ±35° | Over 16.0 to 24.0 | ±0.218 | | |
| Over 2.0 to 4.0 | ±0.060 | 10 to 12 | ±45° | | | | |

Courtesy of the Spring Manufacturers Institute

**Table 10. Compression Spring Squareness Tolerances**

| Slenderness Ratio $FL/D$[a] | Spring Index | | | | | | |
|---|---|---|---|---|---|---|---|
| | 4 | 6 | 8 | 10 | 12 | 14 | 16 |
| | Squareness Tolerances (± degrees) | | | | | | |
| 0.5 | 3.0 | 3.0 | 3.5 | 3.5 | 3.5 | 3.5 | 4.0 |
| 1.0 | 2.5 | 3.0 | 3.0 | 3.0 | 3.0 | 3.5 | 3.5 |
| 1.5 | 2.5 | 2.5 | 2.5 | 3.0 | 3.0 | 3.0 | 3.0 |
| 2.0 | 2.5 | 2.5 | 2.5 | 2.5 | 3.0 | 3.0 | 3.0 |
| 3.0 | 2.0 | 2.5 | 2.5 | 2.5 | 2.5 | 2.5 | 3.0 |
| 4.0 | 2.0 | 2.0 | 2.5 | 2.5 | 2.5 | 2.5 | 2.5 |
| 6.0 | 2.0 | 2.0 | 2.0 | 2.5 | 2.5 | 2.5 | 2.5 |
| 8.0 | 2.0 | 2.0 | 2.0 | 2.0 | 2.5 | 2.5 | 2.5 |
| 10.0 | 2.0 | 2.0 | 2.0 | 2.0 | 2.0 | 2.5 | 2.5 |
| 12.0 | 2.0 | 2.0 | 2.0 | 2.0 | 2.0 | 2.0 | 2.5 |

[a] Slenderness Ratio = $FL \div D$

Springs with closed and ground ends, in the free position. Squareness tolerances closer than those shown require special process techniques which increase cost. Springs made from fine wire sizes, and with high spring indices, irregular shapes or long free lengths, require special attention in determining appropriate tolerance and feasibility of grinding ends.

**Table 11. Compression Spring Normal Load Tolerances**

| Length Tolerance, ± inch | Deflection (inch)[a] | | | | | | | | | | | | | | |
|---|---|---|---|---|---|---|---|---|---|---|---|---|---|---|---|
| | 0.05 | 0.10 | 0.15 | 0.20 | 0.25 | 0.30 | 0.40 | 0.50 | 0.75 | 1.00 | 1.50 | 2.00 | 3.00 | 4.00 | 6.00 |
| | Tolerance, ± Percent of Load | | | | | | | | | | | | | | |
| 0.005 | 12 | 7 | 6 | 5 | … | … | … | … | … | … | … | … | … | … | … |
| 0.010 | … | 12 | 8.5 | 7 | 6.5 | 5.5 | 5 | … | … | … | … | … | … | … | … |
| 0.020 | … | 22 | 15.5 | 12 | 10 | 8.5 | 7 | 6 | 5 | … | … | … | … | … | … |
| 0.030 | … | … | 22 | 17 | 14 | 12 | 9.5 | 8 | 6 | 5 | … | … | … | … | … |
| 0.040 | … | … | … | 22 | 18 | 15.5 | 12 | 10 | 7.5 | 6 | 5 | … | … | … | … |
| 0.050 | … | … | … | … | 22 | 19 | 14.5 | 12 | 9 | 7 | 5.5 | … | … | … | … |
| 0.060 | … | … | … | … | 25 | 22 | 17 | 14 | 10 | 8 | 6 | 5 | … | … | … |
| 0.070 | … | … | … | … | … | 25 | 19.5 | 16 | 11 | 9 | 6.5 | 5.5 | … | … | … |
| 0.080 | … | … | … | … | … | … | 22 | 18 | 12.5 | 10 | 7.5 | 6 | 5 | … | … |
| 0.090 | … | … | … | … | … | … | 25 | 20 | 14 | 11 | 8 | 6 | 5 | … | … |
| 0.100 | … | … | … | … | … | … | … | 22 | 15.5 | 12 | 8.5 | 7 | 5.5 | … | … |
| 0.200 | … | … | … | … | … | … | … | … | … | 22 | 15.5 | 12 | 8.5 | 7 | 5.5 |
| 0.300 | … | … | … | … | … | … | … | … | … | … | 22 | 17 | 12 | 9.5 | 7 |
| 0.400 | … | … | … | … | … | … | … | … | … | … | … | 21 | 15 | 12 | 8.5 |
| 0.500 | … | … | … | … | … | … | … | … | … | … | … | 25 | 18.5 | 14.5 | 10.5 |

[a] From free length to loaded position.

**Torsion Spring Design.**—Fig. 19 shows the types of ends most commonly used on torsion springs. To produce them requires only limited tooling. The straight torsion end is the least expensive and should be used whenever possible. After determining the spring load or torque required and selecting the end formations, the designer usually estimates suitable space or size limitations. However, the space should be considered approximate until the wire size and number of coils have been determined. The wire size is dependent principally upon the torque. Design data can be developed with the aid of the tabular data, which is a simple method, or by calculation alone, as shown in the following sections. Many other factors affecting the design and operation of torsion springs are also covered in the section, *Torsion Spring Design Recommendations* on page 330. Design formulas are shown in Table 13.

*Curvature Correction:* In addition to the stress obtained from the formulas for load or deflection, there is a direct shearing stress on the inside of the section due to curvature. Therefore, the stress obtained by the usual formulas should be multiplied by the factor $K$ obtained from the curve in Fig. 20. The corrected stress thus obtained is used only for comparison with the allowable working stress (fatigue strength) curves to determine if it is a safe value, and should not be used in the formulas for deflection.

## Table 12. Extension Spring Normal Load Tolerances

| Spring Index | $\frac{FL}{F}$ | Wire Diameter (inch) | | | | | | | | | | |
|---|---|---|---|---|---|---|---|---|---|---|---|---|
| | | 0.015 | 0.022 | 0.032 | 0.044 | 0.062 | 0.092 | 0.125 | 0.187 | 0.250 | 0.375 | 0.437 |
| | | Tolerance, ± Percent of Load | | | | | | | | | | |
| | 12 | 20.0 | 18.5 | 17.6 | 16.9 | 16.2 | 15.5 | 15.0 | 14.3 | 13.8 | 13.0 | 12.6 |
| | 8 | 18.5 | 17.5 | 16.7 | 15.8 | 15.0 | 14.5 | 14.0 | 13.2 | 12.5 | 11.5 | 11.0 |
| | 6 | 16.8 | 16.1 | 15.5 | 14.7 | 13.8 | 13.2 | 12.7 | 11.8 | 11.2 | 9.9 | 9.4 |
| 4 | 4.5 | 15.0 | 14.7 | 14.1 | 13.5 | 12.6 | 12.0 | 11.5 | 10.3 | 9.7 | 8.4 | 7.9 |
| | 2.5 | 13.1 | 12.4 | 12.1 | 11.8 | 10.6 | 10.0 | 9.1 | 8.5 | 8.0 | 6.8 | 6.2 |
| | 1.5 | 10.2 | 9.9 | 9.3 | 8.9 | 8.0 | 7.5 | 7.0 | 6.5 | 6.1 | 5.3 | 4.8 |
| | 0.5 | 6.2 | 5.4 | 4.8 | 4.6 | 4.3 | 4.1 | 4.0 | 3.8 | 3.6 | 3.3 | 3.2 |
| | 12 | 17.0 | 15.5 | 14.6 | 14.1 | 13.5 | 13.1 | 12.7 | 12.0 | 11.5 | 11.2 | 10.7 |
| | 8 | 16.2 | 14.7 | 13.9 | 13.4 | 12.6 | 12.2 | 11.7 | 11.0 | 10.5 | 10.0 | 9.5 |
| | 6 | 15.2 | 14.0 | 12.9 | 12.3 | 11.6 | 10.9 | 10.7 | 10.0 | 9.4 | 8.8 | 8.3 |
| 6 | 4.5 | 13.7 | 12.4 | 11.5 | 11.0 | 10.5 | 10.0 | 9.6 | 9.0 | 8.3 | 7.6 | 7.1 |
| | 2.5 | 11.9 | 10.8 | 10.2 | 9.8 | 9.4 | 9.0 | 8.5 | 7.9 | 7.2 | 6.2 | 6.0 |
| | 1.5 | 9.9 | 9.0 | 8.3 | 7.7 | 7.3 | 7.0 | 6.7 | 6.4 | 6.0 | 4.9 | 4.7 |
| | 0.5 | 6.3 | 5.5 | 4.9 | 4.7 | 4.5 | 4.3 | 4.1 | 4.0 | 3.7 | 3.5 | 3.4 |
| | 12 | 15.8 | 14.3 | 13.1 | 13.0 | 12.1 | 12.0 | 11.5 | 10.8 | 10.2 | 10.0 | 9.5 |
| | 8 | 15.0 | 13.7 | 12.5 | 12.1 | 11.4 | 11.0 | 10.6 | 10.1 | 9.4 | 9.0 | 8.6 |
| | 6 | 14.2 | 13.0 | 11.7 | 11.2 | 10.6 | 10.0 | 9.7 | 9.3 | 8.6 | 8.1 | 7.6 |
| 8 | 4.5 | 12.8 | 11.7 | 10.7 | 10.1 | 9.7 | 9.0 | 8.7 | 8.3 | 7.8 | 7.2 | 6.6 |
| | 2.5 | 11.2 | 10.2 | 9.5 | 8.8 | 8.3 | 7.9 | 7.7 | 7.4 | 6.9 | 6.1 | 5.6 |
| | 1.5 | 9.5 | 8.6 | 7.8 | 7.1 | 6.9 | 6.7 | 6.5 | 6.2 | 5.8 | 4.9 | 4.5 |
| | 0.5 | 6.3 | 5.6 | 5.0 | 4.8 | 4.5 | 4.4 | 4.2 | 4.1 | 3.9 | 3.6 | 3.5 |
| | 12 | 14.8 | 13.3 | 12.0 | 11.9 | 11.1 | 10.9 | 10.5 | 9.9 | 9.3 | 9.2 | 8.8 |
| | 8 | 14.2 | 12.8 | 11.6 | 11.2 | 10.5 | 10.2 | 9.7 | 9.2 | 8.6 | 8.3 | 8.0 |
| | 6 | 13.4 | 12.1 | 10.8 | 10.5 | 9.8 | 9.3 | 8.9 | 8.6 | 8.0 | 7.6 | 7.2 |
| 10 | 4.5 | 12.3 | 10.8 | 10.0 | 9.5 | 9.0 | 8.5 | 8.1 | 7.8 | 7.3 | 6.8 | 6.4 |
| | 2.5 | 10.8 | 9.6 | 9.0 | 8.4 | 8.0 | 7.7 | 7.3 | 7.0 | 6.5 | 5.9 | 5.5 |
| | 1.5 | 9.2 | 8.3 | 7.5 | 6.9 | 6.7 | 6.5 | 6.3 | 6.0 | 5.6 | 5.0 | 4.6 |
| | 0.5 | 6.4 | 5.7 | 5.1 | 4.9 | 4.7 | 4.5 | 4.3 | 4.2 | 4.0 | 3.8 | 3.7 |
| | 12 | 14.0 | 12.3 | 11.1 | 10.8 | 10.1 | 9.8 | 9.5 | 9.0 | 8.5 | 8.2 | 7.9 |
| | 8 | 13.2 | 11.8 | 10.7 | 10.2 | 9.6 | 9.3 | 8.9 | 8.4 | 7.9 | 7.5 | 7.2 |
| | 6 | 12.6 | 11.2 | 10.2 | 9.7 | 9.0 | 8.5 | 8.2 | 7.9 | 7.4 | 6.9 | 6.4 |
| 12 | 4.5 | 11.7 | 10.2 | 9.4 | 9.0 | 8.4 | 8.0 | 7.6 | 7.2 | 6.8 | 6.3 | 5.8 |
| | 2.5 | 10.5 | 9.2 | 8.5 | 8.0 | 7.8 | 7.4 | 7.0 | 6.6 | 6.1 | 5.6 | 5.2 |
| | 1.5 | 8.9 | 8.0 | 7.2 | 6.8 | 6.5 | 6.3 | 6.1 | 5.7 | 5.4 | 4.8 | 4.5 |
| | 0.5 | 6.5 | 5.8 | 5.3 | 5.1 | 4.9 | 4.7 | 4.5 | 4.3 | 4.2 | 4.0 | 3.3 |
| | 12 | 13.1 | 11.3 | 10.2 | 9.7 | 9.1 | 8.8 | 8.4 | 8.1 | 7.6 | 7.2 | 7.0 |
| | 8 | 12.4 | 10.9 | 9.8 | 9.2 | 8.7 | 8.3 | 8.0 | 7.6 | 7.2 | 6.8 | 6.4 |
| | 6 | 11.8 | 10.4 | 9.3 | 8.8 | 8.3 | 7.7 | 7.5 | 7.2 | 6.8 | 6.3 | 5.9 |
| 14 | 4.5 | 11.1 | 9.7 | 8.7 | 8.2 | 7.8 | 7.2 | 7.0 | 6.7 | 6.3 | 5.8 | 5.4 |
| | 2.5 | 10.1 | 8.8 | 8.1 | 7.6 | 7.1 | 6.7 | 6.5 | 6.2 | 5.7 | 5.2 | 5.0 |
| | 1.5 | 8.6 | 7.7 | 7.0 | 6.7 | 6.3 | 6.0 | 5.8 | 5.5 | 5.2 | 4.7 | 4.5 |
| | 0.5 | 6.6 | 5.9 | 5.4 | 5.2 | 5.0 | 4.8 | 4.6 | 4.4 | 4.3 | 4.2 | 4.0 |
| | 12 | 12.3 | 10.3 | 9.2 | 8.6 | 8.1 | 7.7 | 7.4 | 7.2 | 6.8 | 6.3 | 6.1 |
| | 8 | 11.7 | 10.0 | 8.9 | 8.3 | 7.8 | 7.4 | 7.2 | 6.8 | 6.5 | 6.0 | 5.7 |
| | 6 | 11.0 | 9.6 | 8.5 | 8.0 | 7.5 | 7.1 | 6.9 | 6.5 | 6.2 | 5.7 | 5.4 |
| 16 | 4.5 | 10.5 | 9.1 | 8.1 | 7.5 | 7.2 | 6.8 | 6.5 | 6.2 | 5.8 | 5.3 | 5.1 |
| | 2.5 | 9.7 | 8.4 | 7.6 | 7.0 | 6.7 | 6.3 | 6.1 | 5.7 | 5.4 | 4.9 | 4.7 |
| | 1.5 | 8.3 | 7.4 | 6.6 | 6.2 | 6.0 | 5.8 | 5.6 | 5.3 | 5.1 | 4.6 | 4.4 |
| | 0.5 | 6.7 | 5.9 | 5.5 | 5.3 | 5.1 | 5.0 | 4.8 | 4.6 | 4.5 | 4.3 | 4.1 |

$FL/F$ = the ratio of the spring free length $FL$ to the deflection $F$.

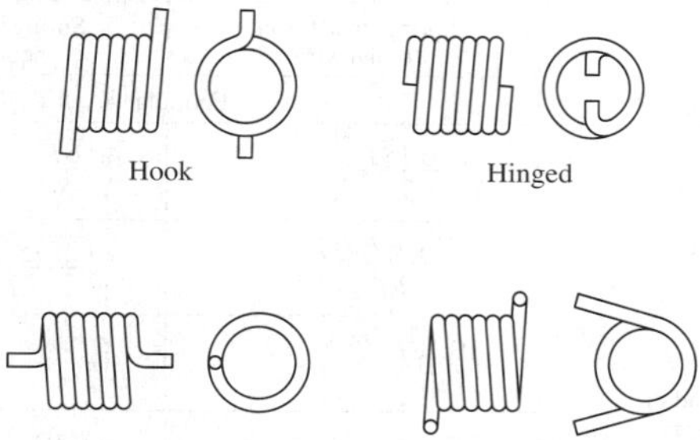


Fig. 19. The Most Commonly Used Types of Ends for Torsion Springs

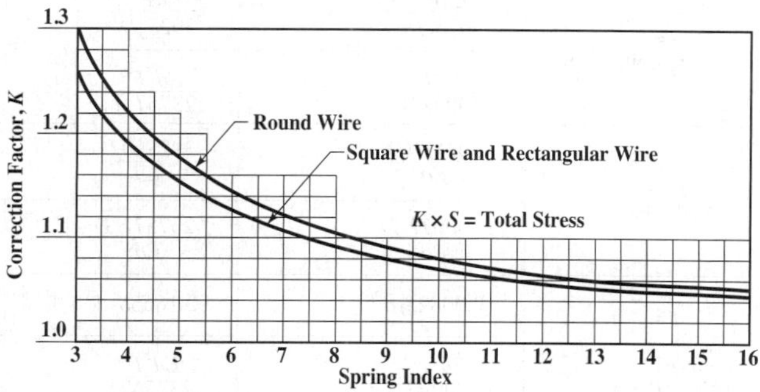


Fig. 20. Torsion Spring Stress Correction for Curvature

*Torque:* Torque is a force applied to a moment arm and tends to produce rotation. Torsion springs exert torque in a circular arc and the arms are rotated about the central axis. It should be noted that the stress produced is in bending, not in torsion. In the spring industry it is customary to specify torque in conjunction with the deflection or with the arms of a spring at a definite position. Formulas for torque are expressed in pound-inches. If ounce-inches are specified, it is necessary to divide this value by 16 in order to use the formulas.

When a load is specified at a distance from a centerline, the torque is, of course, equal to the load multiplied by the distance. The load can be in pounds or ounces with the distances in inches or the load can be in grams or kilograms with the distance in centimeters or millimeters, but to use the design formulas, all values must be converted to pounds and inches. Design formulas for torque are based on the tangent to the arc of rotation and presume that a rod is used to support the spring. The stress in bending caused by the moment $P \times R$ is identical in magnitude to the torque $T$, provided a rod is used.

Theoretically, it makes no difference how or where the load is applied to the arms of torsion springs. Thus, in Fig. 21, the loads shown multiplied by their respective distances produce the same torque; i.e., $20 \times 0.5 = 10$ pound-inches; $10 \times 1 = 10$ pound-inches; and $5 \times 2 = 10$ pound-inches. To further simplify the understanding of torsion spring torque, observe in both Fig. 22 and Fig. 23 that although the turning force is in a circular arc the torque is not equal to $P$ times the radius. The torque in both designs equals $P \times R$ because the spring rests against the support rod at point $a$.

**Table 13. Formulas for Torsion Springs**

| Feature | Springs made from round wire | Springs made from square wire |
|---|---|---|
| | Formula[a, b] | |
| $d =$ Wire diameter, Inches | $\sqrt[3]{\frac{10.18T}{S_b}}$ | $\sqrt[3]{\frac{6T}{S_b}}$ |
| | $\sqrt[4]{\frac{4000TND}{EF^\circ}}$ | $\sqrt[4]{\frac{2375TND}{EF^\circ}}$ |
| $S_b =$ Stress, bending pounds per square inch | $\frac{10.18T}{d^3}$ | $\frac{6T}{d^3}$ |
| | $\frac{EdF^\circ}{392ND}$ | $\frac{EdF^\circ}{392ND}$ |
| $N =$ Active Coils | $\frac{EdF^\circ}{392S_bD}$ | $\frac{EdF^\circ}{392S_bD}$ |
| | $\frac{Ed^4F^\circ}{4000TD}$ | $\frac{Ed^4F^\circ}{2375TD}$ |
| $F^\circ =$ Deflection | $\frac{392S_bND}{Ed}$ | $\frac{392S_bND}{Ed}$ |
| | $\frac{4000TND}{Ed^4}$ | $\frac{2375TND}{Ed^4}$ |
| $T =$ Torque Inch lbs. (Also $= P \times R$) | $0.0982S_bd^3$ | $0.1666S_bd^3$ |
| | $\frac{Ed^4F^\circ}{4000ND}$ | $\frac{Ed^4F^\circ}{2375ND}$ |
| $ID_1 =$ Inside Diameter After Deflection, Inches | $\frac{N(ID\text{ free})}{N + \frac{F^\circ}{360}}$ | $\frac{N(ID\text{ free})}{N + \frac{F^\circ}{360}}$ |

[a] Where two formulas are given for one feature, the designer should use the one found to be appropriate for the given design. The end result from either of any two formulas is the same.
[b] The symbol notation is given on page 298.

*Design Procedure:* Torsion spring designs require more effort than other kinds because consideration has to be given to more details, such as the proper size of a supporting rod, reduction of the inside diameter, increase in length, deflection of arms, allowance for friction, and method of testing.

*Example:* What music wire diameter and how many coils are required for the torsion spring shown in Fig. 24, which is to withstand at least 1000 cycles? Determine the corrected stress and the reduced inside diameter after deflection.

*Method 1, Using Table:* From Table 14, page 332, locate the ½ inch inside diameter for the spring in the left-hand column. Move to the right and then vertically to locate a torque value nearest to the required 10 pound-inches, which is 10.07 pound-inches. At the top of the same column, the music wire diameter is found, which is Number 31 gauge (0.085 inch). At the bottom of the same column the deflection for one coil is found, which

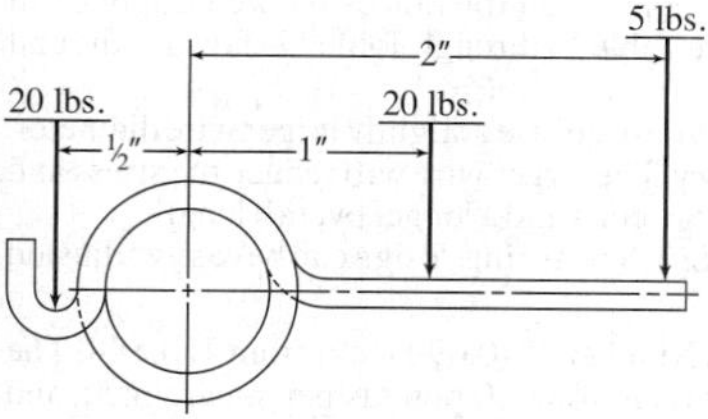


Fig. 21. Right-Hand Torsion Spring

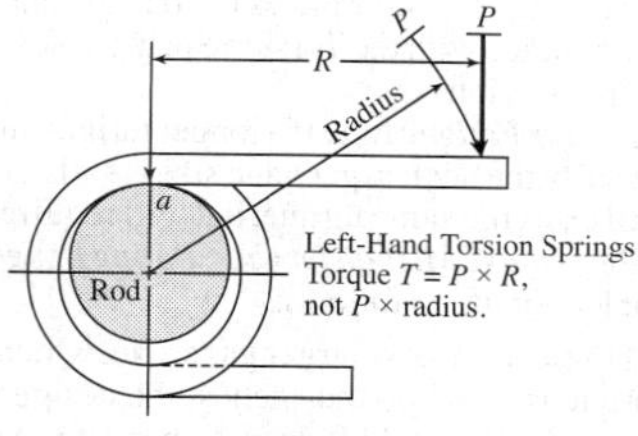


Fig. 22. Left-Hand Torsion Spring

The torque is $T = P \times R$, not $P \times$ radius, because the spring is resting against the support rod at point $a$.

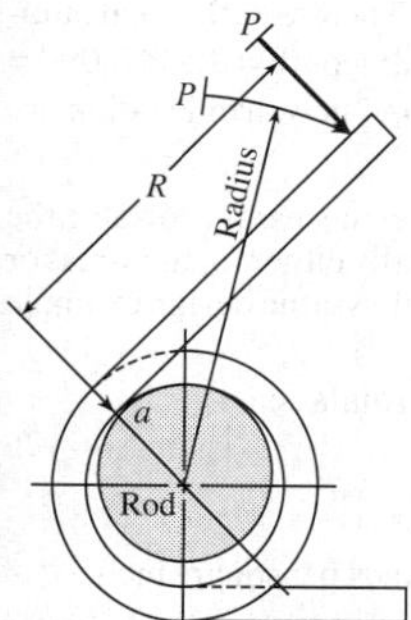


Fig. 23. Left-Hand Torsion Spring

As with the spring in Fig. 22, the torque is $T = P \times R$, not $P \times$ radius, because the support point is at $a$.

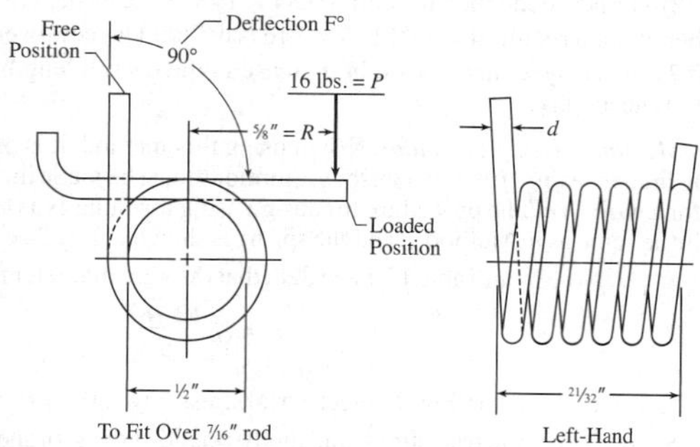


Fig. 24. Torsion Spring Design Example.

The spring is to be assembled on a $\frac{7}{16}$-inch support rod.

is 15.81 degrees. As a 90-degree deflection is required, the number of coils needed is $\frac{90}{15.81} = 5.69$ (say $5\frac{3}{4}$ coils).

The spring index $\frac{D}{d} = \frac{0.500 + 0.085}{0.085} = 6.88$ and thus the curvature correction factor $K$ from Fig. 20 on page 327 = 1.13. Therefore the corrected stress equals $167{,}000 \times 1.13 =$ 188,700 pounds per square inch which is below the Light Service curve (Fig. 7 on page 306) and therefore should provide a fatigue life of over 1,000 cycles. The reduced inside diameter due to deflection is found from the formula in Table 13:

$$ID_1 = \frac{N(ID\text{ free})}{N + \frac{F}{360}} = \frac{5.75 \times 0.500}{5.75 + \frac{90}{360}} = 0.479\text{ in.}$$

This reduced diameter easily clears a suggested $\frac{7}{16}$ inch diameter supporting rod: $0.479 - 0.4375 = 0.041$ inch clearance, and it also allows for the standard tolerance. The overall length of the spring equals the total number of coils plus one, times the wire diameter. Thus, $6\frac{3}{4} \times 0.085 = 0.574$ inch. If a small space of about $\frac{1}{64}$ in. is allowed between the coils to eliminate coil friction, an overall length of $\frac{21}{32}$ inch results.

Although this completes the design calculations, other tolerances should be applied in accordance with the Torsion Spring Tolerance Table 16 through Table 17 shown at the end of this section.

*Longer Fatigue Life:* If a longer fatigue life is desired, use a slightly larger wire diameter. Usually the next larger gage size is satisfactory. The larger wire will reduce the stress and still exert the same torque, but will require more coils and a longer overall length.

*Percentage Method for Calculating Longer Life:* The spring design can be easily adjusted for longer life as follows:

1) Select the next larger gage size, which is Number 32 (0.090 inch) from Table 14. The torque is 11.88 pound-inches, the design stress is 166,000 pounds per square inch, and the deflection is 14.9 degrees per coil. As a percentage the torque is $10/11.88 \times 100 =$ 84 percent.

2) The new stress is $0.84 \times 166{,}000 = 139{,}440$ pounds per square inch. This value is under the bottom or Severe Service curve, Fig. 7, and thus assures longer life.

3) The new deflection per coil is $0.84 \times 14.97 = 12.57$ degrees. Therefore, the total number of coils required $= 90/12.57 = 7.16$ (say $7\frac{1}{8}$). The new overall length $= 8\frac{1}{8} \times 0.090 =$ 0.73 inch (say $\frac{3}{4}$ inch). A slight increase in the overall length and new arm location are thus necessary.

*Method 2, Using Formulas:* When using this method, it is often necessary to solve the formulas several times because assumptions must be made initially either for the stress or for a wire size. The procedure for design using formulas is as follows (the design example is the same as in Method 1, and the spring is shown in Fig. 24):

*Step 1:* Note from Table 13, page 328, that the wire diameter formula is:

$$d = \sqrt[3]{\frac{10.18T}{S_b}}$$

*Step 2:* Referring to Fig. 7, select a trial stress, say 150,000 pounds per square inch.

*Step 3:* Apply the trial stress, and the 10 pound-inches torque value in the wire diameter formula:

$$d = \sqrt[3]{\frac{10.18T}{S_b}} = \sqrt[3]{\frac{10.18 \times 10}{150{,}000}} = \sqrt[3]{0.000679} = 0.0879 \text{ inch}$$

The nearest gauge sizes are 0.085 and 0.090 inch diameter. *Note*: Table 21, page 340, can be used to avoid solving the cube root.

*Step 4:* Select 0.085 inch wire diameter and solve the equation for the actual stress:

$$S_b = \frac{10.18T}{d^3} = \frac{10.18 \times 10}{0.085^3} = 165{,}764 \text{ pounds per square inch}$$

*Step 5:* Calculate the number of coils from the equation, Table 13:

$$N = \frac{EdF^\circ}{392S_bD} = \frac{28{,}500{,}000 \times 0.085 \times 90}{392 \times 165{,}764 \times 0.585} = 5.73 \text{ (say } 5\tfrac{3}{4})$$

*Step 6:* Calculate the total stress. The spring index is 6.88, and the correction factor $K$ is 1.13 therefore total stress $= 165{,}764 \times 1.13 = 187{,}313$ pounds per square inch. *Note*: The corrected stress should not be used in any of the formulas as it does not determine the torque or the deflection.

**Torsion Spring Design Recommendations.—**The following recommendations should be taken into account when designing torsion springs:

*Hand:* The hand or direction of coiling should be specified and the spring designed so deflection causes the spring to wind up and to have more coils. This increase in coils and overall length should be allowed for during design. Deflecting the spring in an unwinding direction produces higher stresses and may cause early failure. When a spring is sighted down the longitudinal axis, it is "right-hand" when the direction of the wire into the spring takes a clockwise direction or if the angle of the coils follows an angle similar to the threads of a standard bolt or screw, otherwise it is "left-hand." A spring must be coiled right-handed to engage the threads of a standard machine screw.

*Rods:* Torsion springs should be supported by a rod running through the center whenever possible. If unsupported, or if held by clamps or lugs, the spring will buckle and the torque will be reduced or unusual stresses may occur.

*Diameter Reduction:* The inside diameter reduces during deflection. This reduction should be computed and proper clearance provided over the supporting rod. Also, allowances should be considered for normal spring diameter tolerances.

*Winding:* The coils of a spring may be closely or loosely wound, but they seldom should be wound with the coils pressed tightly together. Tightly wound springs with initial tension on the coils do not deflect uniformly and are difficult to test accurately. A small space between the coils of about 20 to 25 percent of the wire thickness is desirable. Square and rectangular wire sections should be avoided whenever possible as they are difficult to wind, expensive, and are not always readily available.

*Arm Length:* All the wire in a torsion spring is active between the points where the loads are applied. Deflection of long extended arms can be calculated by allowing one third of the arm length, from the point of load contact to the body of the spring, to be converted into coils. However, if the length of arm is equal to or less than one-half the length of one coil, it can be safely neglected in most applications.

*Total Coils:* Torsion springs having less than three coils frequently buckle and are difficult to test accurately. When thirty or more coils are used, light loads will not deflect all the coils simultaneously due to friction with the supporting rod. To facilitate manufacturing it is usually preferable to specify the total number of coils to the nearest fraction in eighths or quarters, such as $5\frac{1}{8}$, $5\frac{1}{4}$, $5\frac{1}{2}$, etc.

*Double Torsion:* This design consists of one left-hand-wound series of coils and one series of right-hand-wound coils connected at the center. These springs are difficult to manufacture and are expensive, so it often is better to use two separate springs. For torque and stress calculations, each series is calculated separately as individual springs; then the torque values are added together, but the deflections are not added.

*Bends:* Arms should be kept as straight as possible. Bends are difficult to produce and often are made by secondary operations, so they are therefore expensive. Sharp bends raise stresses that cause early failure. Bend radii should be as large as practicable. Hooks tend to open during deflection; their stresses can be calculated by the same procedure as that for tension springs.

*Spring Index:* The spring index must be used with caution. In design formulas it is $D/d$. For shop measurement it is $OD/d$. For arbor design it is $ID/d$. Conversions are easily performed by either adding or subtracting 1 from $D/d$.

*Proportions:* A spring index between 4 and 14 provides the best proportions. Larger ratios may require more than average tolerances. Ratios of 3 or less often cannot be coiled on automatic spring coiling machines because of arbor breakage. Also, springs with smaller or larger spring indexes often do not give the same results as are obtained using the design formulas.

**Table of Torsion Spring Characteristics.—**Table 14 shows design characteristics for the most commonly used torsion springs made from wire of standard gauge sizes. The deflection for one coil at a specified torque and stress is shown in the body of the table. The figures are based on music wire (ASTM A228) and oil-tempered MB grade (ASTM A229), and can be used for several other materials which have similar values for the modulus of elasticity $E$. However, the design stress may be too high or too low, and the design stress, torque, and deflection per coil should each be multiplied by the appropriate correction factor in Table 15 when using any of the materials given in that table.

## Table 14. Torsion Spring Deflections

| AMW Wire Gauge Decimal Equivalent[a] | | 1 .010 | 2 .011 | 3 .012 | 4 .013 | 5 .014 | 6 .016 | 7 .018 | 8 .020 | 9 .022 | 10 .024 | 11 .026 | 12 .029 | 13 .031 | 14 .033 | 15 .035 | 16 .037 |
|---|---|---|---|---|---|---|---|---|---|---|---|---|---|---|---|---|---|
| Design Stress, kpsi | | 232 | 229 | 226 | 224 | 221 | 217 | 214 | 210 | 207 | 205 | 202 | 199 | 197 | 196 | 194 | 192 |
| Torque, pound-inch | | .0228 | .0299 | .0383 | .0483 | .0596 | .0873 | .1226 | .1650 | .2164 | .2783 | .3486 | .4766 | .5763 | .6917 | .8168 | .9550 |
| Inside Diameter, inch | | Deflection, degrees per coil | | | | | | | | | | | | | | | |
| 1/16 | 0.0625 | 22.35 | 20.33 | 18.64 | 17.29 | 16.05 | 14.15 | 18.72 | 11.51 | 10.56 | 9.818 | 9.137 | 8.343 | 7.896 | … | … | … |
| 5/64 | 0.078125 | 27.17 | 24.66 | 22.55 | 20.86 | 19.32 | 16.96 | 15.19 | 13.69 | 12.52 | 11.59 | 10.75 | 9.768 | 9.215 | … | … | … |
| 3/32 | 0.09375 | 31.98 | 28.98 | 26.47 | 24.44 | 22.60 | 19.78 | 17.65 | 15.87 | 14.47 | 13.36 | 12.36 | 11.19 | 10.53 | 10.18 | 9.646 | 9.171 |
| 7/64 | 0.109375 | 36.80 | 33.30 | 30.38 | 28.02 | 25.88 | 22.60 | 20.12 | 18.05 | 16.43 | 15.14 | 13.98 | 12.62 | 11.85 | 11.43 | 10.82 | 10.27 |
| 1/8 | 0.125 | 41.62 | 37.62 | 34.29 | 31.60 | 29.16 | 25.41 | 22.59 | 20.23 | 18.38 | 16.91 | 15.59 | 14.04 | 13.17 | 12.68 | 11.99 | 11.36 |
| 9/64 | 0.140625 | 46.44 | 41.94 | 38.20 | 35.17 | 32.43 | 28.23 | 25.06 | 22.41 | 20.33 | 18.69 | 17.20 | 15.47 | 14.49 | 13.94 | 13.16 | 12.46 |
| 5/32 | 0.15625 | 51.25 | 46.27 | 42.11 | 38.75 | 35.71 | 31.04 | 27.53 | 24.59 | 22.29 | 20.46 | 18.82 | 16.89 | 15.81 | 15.19 | 14.33 | 13.56 |
| 3/16 | 0.1875 | 60.89 | 54.91 | 49.93 | 45.91 | 42.27 | 36.67 | 32.47 | 28.95 | 26.19 | 24.01 | 22.04 | 19.74 | 18.45 | 17.70 | 16.67 | 15.75 |
| 7/32 | 0.21875 | 70.52 | 63.56 | 57.75 | 53.06 | 48.82 | 42.31 | 37.40 | 33.31 | 30.10 | 27.55 | 25.27 | 22.59 | 21.09 | 20.21 | 19.01 | 17.94 |
| 1/4 | 0.250 | 80.15 | 72.20 | 65.57 | 60.22 | 55.38 | 47.94 | 42.34 | 37.67 | 34.01 | 31.10 | 28.49 | 25.44 | 23.73 | 22.72 | 21.35 | 20.13 |

| AMW Wire Gauge Decimal Equivalent[a] | | 17 .039 | 18 .041 | 19 .043 | 20 .045 | 21 .047 | 22 .049 | 23 .051 | 24 .055 | 25 .059 | 26 .063 | 27 .067 | 28 .071 | 29 .075 | 30 .080 | 31 .085 |
|---|---|---|---|---|---|---|---|---|---|---|---|---|---|---|---|---|
| Design Stress, kpsi | | 190 | 188 | 187 | 185 | 184 | 183 | 182 | 180 | 178 | 176 | 174 | 173 | 171 | 169 | 167 |
| Torque, pound-inch | | 1.107 | 1.272 | 1.460 | 1.655 | 1.876 | 2.114 | 2.371 | 2.941 | 3.590 | 4.322 | 5.139 | 6.080 | 7.084 | 8.497 | 10.07 |
| Inside Diameter, inch | | Deflection, degrees per coil | | | | | | | | | | | | | | |
| 1/8 | 0.125 | 10.80 | 10.29 | 9.876 | 9.447 | 9.102 | 8.784 | … | … | … | … | … | … | … | … | … |
| 9/64 | 0.140625 | 11.83 | 11.26 | 10.79 | 10.32 | 9.929 | 9.572 | 9.244 | 8.654 | 8.141 | … | … | … | … | … | … |
| 5/32 | 0.15625 | 12.86 | 12.23 | 11.71 | 11.18 | 10.76 | 10.36 | 9.997 | 9.345 | 8.778 | 8.279 | 7.975 | … | … | … | … |
| 3/16 | 0.1875 | 14.92 | 14.16 | 13.55 | 12.92 | 12.41 | 11.94 | 11.50 | 10.73 | 10.05 | 9.459 | 9.091 | 8.663 | 8.232 | 7.772 | 7.364 |
| 7/32 | 0.21875 | 16.97 | 16.10 | 15.39 | 14.66 | 14.06 | 13.52 | 13.01 | 12.11 | 11.33 | 10.64 | 10.21 | 9.711 | 9.212 | 8.680 | 8.208 |
| 1/4 | 0.250 | 19.03 | 18.04 | 17.22 | 16.39 | 15.72 | 15.09 | 14.52 | 13.49 | 12.60 | 11.82 | 11.32 | 10.76 | 10.19 | 9.588 | 9.053 |

[a] For sizes up to 13 gauge, the table values are for music wire with a modulus $E$ of 29,000,000 psi; and for sizes from 27 to 31 gauge, the values are for oil-tempered MB with a modulus of 28,500,000 psi.

## Table 14. *(Continued)* Torsion Spring Deflections

| AMW Wire Gauge Decimal Equivalent[a] | | 8 .020 | 9 .022 | 10 .024 | 11 .026 | 12 .029 | 13 .031 | 14 .033 | 15 .035 | 16 .037 | 17 .039 | 18 .041 | 19 .043 | 20 .045 | 21 .047 | 22 .049 | 23 .051 |
|---|---|---|---|---|---|---|---|---|---|---|---|---|---|---|---|---|---|
| Design Stress, kpsi | | 210 | 207 | 205 | 202 | 199 | 197 | 196 | 194 | 192 | 190 | 188 | 187 | 185 | 184 | 183 | 182 |
| Torque, pound-inch | | .1650 | .2164 | .2783 | .3486 | .4766 | .5763 | .6917 | .8168 | .9550 | 1.107 | 1.272 | 1.460 | 1.655 | 1.876 | 2.114 | 2.371 |
| Inside Diameter, inch | | Deflection, degrees per coil | | | | | | | | | | | | | | | |
| 9/32 | 0.28125 | 42.03 | 37.92 | 34.65 | 31.72 | 28.29 | 26.37 | 25.23 | 23.69 | 22.32 | 21.09 | 19.97 | 19.06 | 18.13 | 17.37 | 16.67 | 16.03 |
| 5/16 | 0.3125 | 46.39 | 41.82 | 38.19 | 34.95 | 31.14 | 29.01 | 27.74 | 26.04 | 24.51 | 23.15 | 21.91 | 20.90 | 19.87 | 19.02 | 18.25 | 17.53 |
| 11/32 | 0.34375 | 50.75 | 45.73 | 41.74 | 38.17 | 33.99 | 31.65 | 30.25 | 28.38 | 26.71 | 25.21 | 23.85 | 22.73 | 21.60 | 20.68 | 19.83 | 19.04 |
| 3/8 | 0.375 | 55.11 | 49.64 | 45.29 | 41.40 | 36.84 | 34.28 | 32.76 | 30.72 | 28.90 | 27.26 | 25.78 | 24.57 | 23.34 | 22.33 | 21.40 | 20.55 |
| 13/32 | 0.40625 | 59.47 | 53.54 | 48.85 | 44.63 | 39.69 | 36.92 | 35.26 | 33.06 | 31.09 | 29.32 | 27.72 | 26.41 | 25.08 | 23.99 | 22.98 | 22.06 |
| 7/16 | 0.4375 | 63.83 | 57.45 | 52.38 | 47.85 | 42.54 | 39.56 | 37.77 | 35.40 | 33.28 | 31.38 | 29.66 | 28.25 | 26.81 | 25.64 | 24.56 | 23.56 |
| 15/32 | 0.46875 | 68.19 | 61.36 | 55.93 | 51.00 | 45.39 | 42.20 | 40.28 | 37.74 | 35.47 | 33.44 | 31.59 | 30.08 | 28.55 | 27.29 | 26.14 | 25.07 |
| 1/2 | 0.500 | 72.55 | 65.27 | 59.48 | 54.30 | 48.24 | 44.84 | 42.79 | 40.08 | 37.67 | 35.49 | 33.53 | 31.92 | 30.29 | 28.95 | 27.71 | 26.58 |

| AMW Wire Gauge Decimal Equivalent[a] | | 24 .055 | 25 .059 | 26 .063 | 27 .067 | 28 .071 | 29 .075 | 30 .080 | 31 .085 | 32 .090 | 33 .095 | 34 .100 | 35 .106 | 36 .112 | 37 .118 | ⅛ 125 |
|---|---|---|---|---|---|---|---|---|---|---|---|---|---|---|---|---|
| Design Stress, kpsi | | 180 | 178 | 176 | 174 | 173 | 171 | 169 | 167 | 166 | 164 | 163 | 161 | 160 | 158 | 156 |
| Torque, pound-inch | | 2.941 | 3.590 | 4.322 | 5.139 | 6.080 | 7.084 | 8.497 | 10.07 | 11.88 | 13.81 | 16.00 | 18.83 | 22.07 | 25.49 | 29.92 |
| Inside Diameter, inch | | Deflection, degrees per coil | | | | | | | | | | | | | | |
| 9/32 | 0.28125 | 14.88 | 13.88 | 13.00 | 12.44 | 11.81 | 11.17 | 10.50 | 9.897 | 9.418 | 8.934 | 8.547 | 8.090 | 7.727 | 7.353 | 6.973 |
| 5/16 | 0.3125 | 16.26 | 15.15 | 14.18 | 13.56 | 12.85 | 12.15 | 11.40 | 10.74 | 10.21 | 9.676 | 9.248 | 8.743 | 8.341 | 7.929 | 7.510 |
| 11/32 | 0.34375 | 17.64 | 16.42 | 15.36 | 14.67 | 13.90 | 13.13 | 12.31 | 11.59 | 11.00 | 10.42 | 9.948 | 9.396 | 8.955 | 8.504 | 8.046 |
| 3/8 | 0.375 | 19.02 | 17.70 | 16.54 | 15.79 | 14.95 | 14.11 | 13.22 | 12.43 | 11.80 | 11.16 | 10.65 | 10.05 | 9.569 | 9.080 | 8.583 |
| 13/32 | 0.40625 | 20.40 | 18.97 | 17.72 | 16.90 | 15.99 | 15.09 | 14.13 | 13.28 | 12.59 | 11.90 | 11.35 | 10.70 | 10.18 | 9.655 | 9.119 |
| 7/16 | 0.4375 | 21.79 | 20.25 | 18.90 | 18.02 | 17.04 | 16.07 | 15.04 | 14.12 | 13.38 | 12.64 | 12.05 | 11.35 | 10.80 | 10.23 | 9.655 |
| 15/32 | 0.46875 | 23.17 | 21.52 | 20.08 | 19.14 | 18.09 | 17.05 | 15.94 | 14.96 | 14.17 | 13.39 | 12.75 | 12.01 | 11.41 | 10.81 | 10.19 |
| 1/2 | 0.500 | 24.55 | 22.80 | 21.26 | 20.25 | 19.14 | 18.03 | 16.85 | 15.81 | 14.97 | 14.13 | 13.45 | 12.66 | 12.03 | 11.38 | 10.73 |

[a] For sizes up to 13 gauge, the table values are for music wire with a modulus $E$ of 29,000,000 psi; and for sizes from 27 to 31 gauge, the values are for oil-tempered MB with a modulus of 28,500,000 psi.

**Table 14.** *(Continued)* **Torsion Spring Deflections**

| AMW Wire Gauge Decimal Equivalent[a] | | 16 .037 | 17 .039 | 18 .041 | 19 .043 | 20 .045 | 21 .047 | 22 .049 | 23 .051 | 24 .055 | 25 .059 | 26 .063 | 27 .067 | 28 .071 | 29 .075 | 30 .080 |
|---|---|---|---|---|---|---|---|---|---|---|---|---|---|---|---|---|
| Design Stress, kpsi | | 192 | 190 | 188 | 187 | 185 | 184 | 183 | 182 | 180 | 178 | 176 | 174 | 173 | 171 | 169 |
| Torque, pound-inch | | .9550 | 1.107 | 1.272 | 1.460 | 1.655 | 1.876 | 2.114 | 2.371 | 2.941 | 3.590 | 4.322 | 5.139 | 6.080 | 7.084 | 8.497 |
| Inside Diameter, inch | | Deflection, degrees per coil | | | | | | | | | | | | | | |
| 17/32 | 0.53125 | 39.86 | 37.55 | 35.47 | 33.76 | 32.02 | 30.60 | 29.29 | 28.09 | 25.93 | 24.07 | 22.44 | 21.37 | 20.18 | 19.01 | 17.76 |
| 9/16 | 0.5625 | 42.05 | 39.61 | 37.40 | 35.59 | 33.76 | 32.25 | 30.87 | 29.59 | 27.32 | 25.35 | 23.62 | 22.49 | 21.23 | 19.99 | 18.67 |
| 19/32 | 0.59375 | 44.24 | 41.67 | 39.34 | 37.43 | 35.50 | 33.91 | 32.45 | 31.10 | 28.70 | 26.62 | 24.80 | 23.60 | 22.28 | 20.97 | 19.58 |
| 5/8 | 0.625 | 46.43 | 43.73 | 41.28 | 39.27 | 37.23 | 35.56 | 34.02 | 32.61 | 30.08 | 27.89 | 25.98 | 24.72 | 23.33 | 21.95 | 20.48 |
| 21/32 | 0.65625 | 48.63 | 45.78 | 43.22 | 41.10 | 38.97 | 37.22 | 35.60 | 34.12 | 31.46 | 29.17 | 27.16 | 25.83 | 24.37 | 22.93 | 21.39 |
| 11/16 | 0.6875 | 50.82 | 47.84 | 45.15 | 42.94 | 40.71 | 38.87 | 37.18 | 35.62 | 32.85 | 30.44 | 28.34 | 26.95 | 25.42 | 23.91 | 22.30 |
| 23/32 | 0.71875 | 53.01 | 49.90 | 47.09 | 44.78 | 42.44 | 40.52 | 38.76 | 37.13 | 34.23 | 31.72 | 29.52 | 28.07 | 26.47 | 24.89 | 23.21 |
| 3/4 | 0.750 | 55.20 | 51.96 | 49.03 | 46.62 | 44.18 | 42.18 | 40.33 | 38.64 | 35.61 | 32.99 | 30.70 | 29.18 | 27.52 | 25.87 | 24.12 |

| Wire Gauge[a,b] or Size and Decimal Equivalent | | 31 .085 | 32 .090 | 33 .095 | 34 .100 | 35 .106 | 36 .112 | 37 .118 | 1/8 .125 | 10 .135 | 9 .1483 | 5/32 .1563 | 8 .162 | 7 .177 | 3/16 .1875 | 6 .192 | 5 .207 |
|---|---|---|---|---|---|---|---|---|---|---|---|---|---|---|---|---|---|
| Design Stress, kpsi | | 167 | 166 | 164 | 163 | 161 | 160 | 158 | 156 | 161 | 158 | 156 | 154 | 150 | 149 | 146 | 143 |
| Torque, pound-inch | | 10.07 | 11.88 | 13.81 | 16.00 | 18.83 | 22.07 | 25.49 | 29.92 | 38.90 | 50.60 | 58.44 | 64.30 | 81.68 | 96.45 | 101.5 | 124.6 |
| Inside Diameter, inch | | Deflection, degrees per coil | | | | | | | | | | | | | | | |
| 17/32 | 0.53125 | 16.65 | 15.76 | 14.87 | 14.15 | 13.31 | 12.64 | 11.96 | 11.26 | 10.93 | 9.958 | 9.441 | 9.064 | 8.256 | 7.856 | 7.565 | 7.015 |
| 9/16 | 0.5625 | 17.50 | 16.55 | 15.61 | 14.85 | 13.97 | 13.25 | 12.53 | 11.80 | 11.44 | 10.42 | 9.870 | 9.473 | 8.620 | 8.198 | 7.891 | 7.312 |
| 19/32 | 0.59375 | 18.34 | 17.35 | 16.35 | 15.55 | 14.62 | 13.87 | 13.11 | 12.34 | 11.95 | 10.87 | 10.30 | 9.882 | 8.984 | 8.539 | 8.218 | 7.609 |
| 5/8 | 0.625 | 19.19 | 18.14 | 17.10 | 16.25 | 15.27 | 14.48 | 13.68 | 12.87 | 12.47 | 11.33 | 10.73 | 10.29 | 9.348 | 8.881 | 8.545 | 7.906 |
| 21/32 | 0.65625 | 20.03 | 18.93 | 17.84 | 16.95 | 15.92 | 15.10 | 14.26 | 13.41 | 12.98 | 11.79 | 11.16 | 10.70 | 9.713 | 9.222 | 8.872 | 8.202 |
| 11/16 | 0.6875 | 20.88 | 19.72 | 18.58 | 17.65 | 16.58 | 15.71 | 14.83 | 13.95 | 13.49 | 12.25 | 11.59 | 11.11 | 10.08 | 9.564 | 9.199 | 8.499 |
| 23/32 | 0.71875 | 21.72 | 20.52 | 19.32 | 18.36 | 17.23 | 16.32 | 15.41 | 14.48 | 14.00 | 12.71 | 12.02 | 11.52 | 10.44 | 9.905 | 9.526 | 8.796 |
| 3/4 | 0.750 | 22.56 | 21.31 | 20.06 | 19.06 | 17.88 | 16.94 | 15.99 | 15.02 | 14.52 | 13.16 | 12.44 | 11.92 | 10.81 | 10.25 | 9.852 | 9.093 |

[a] For sizes up to 26 gauge, the table values are for music wire with a modulus $E$ of 29,500,000 psi; for sizes from 27 gauge to 1/8 inch diameter the table values are for music wire with a modulus of 28,500,000 psi; for sizes from 10 gauge to 1/8 inch diameter, the values are for oil-tempered MB with a modulus of 28,500,000 psi.

[b] Gauges 31 through 37 are AMW gauges. Gauges 10 through 5 are Washburn and Moen.

## Table 14. *(Continued)* Torsion Spring Deflections

| AMW Wire Gauge Decimal Equivalent[a] | | 24 .055 | 25 .059 | 26 .063 | 27 .067 | 28 .071 | 29 .075 | 30 .080 | 31 .085 | 32 .090 | 33 .095 | 34 .100 | 35 .106 | 36 .112 | 37 .118 | 1/8 .125 |
|---|---|---|---|---|---|---|---|---|---|---|---|---|---|---|---|---|
| Design Stress, kpsi | | 180 | 178 | 176 | 174 | 173 | 171 | 169 | 167 | 166 | 164 | 163 | 161 | 160 | 158 | 156 |
| Torque, pound-inch | | 2.941 | 3.590 | 4.322 | 5.139 | 6.080 | 7.084 | 8.497 | 10.07 | 11.88 | 13.81 | 16.00 | 18.83 | 22.07 | 25.49 | 29.92 |
| Inside Diameter, inch | | Deflection, degrees per coil | | | | | | | | | | | | | | |
| 13/16 | 0.8125 | 38.38 | 35.54 | 33.06 | 31.42 | 29.61 | 27.83 | 25.93 | 24.25 | 22.90 | 21.55 | 20.46 | 19.19 | 18.17 | 17.14 | 16.09 |
| 7/8 | 0.875 | 41.14 | 38.09 | 35.42 | 33.65 | 31.70 | 29.79 | 27.75 | 25.94 | 24.58 | 23.03 | 21.86 | 20.49 | 19.39 | 18.29 | 17.17 |
| 15/16 | 0.9375 | 43.91 | 40.64 | 37.78 | 35.88 | 33.80 | 31.75 | 29.56 | 27.63 | 26.07 | 24.52 | 23.26 | 21.80 | 20.62 | 19.44 | 18.24 |
| 1 | 1.000 | 46.67 | 43.19 | 40.14 | 38.11 | 35.89 | 33.71 | 31.38 | 29.32 | 27.65 | 26.00 | 24.66 | 23.11 | 21.85 | 20.59 | 19.31 |
| 1 1/16 | 1.0625 | 49.44 | 45.74 | 42.50 | 40.35 | 37.99 | 35.67 | 33.20 | 31.01 | 29.24 | 27.48 | 26.06 | 24.41 | 23.08 | 21.74 | 20.38 |
| 1 1/8 | 1.125 | 52.20 | 48.28 | 44.86 | 42.58 | 40.08 | 37.63 | 35.01 | 32.70 | 30.82 | 28.97 | 27.46 | 25.72 | 24.31 | 22.89 | 21.46 |
| 1 3/16 | 1.1875 | 54.97 | 50.83 | 47.22 | 44.81 | 42.18 | 39.59 | 36.83 | 34.39 | 32.41 | 30.45 | 28.86 | 27.02 | 25.53 | 24.04 | 22.53 |
| 1 1/4 | 1.250 | 57.73 | 53.38 | 49.58 | 47.04 | 44.27 | 41.55 | 38.64 | 36.08 | 33.99 | 31.94 | 30.27 | 28.33 | 26.76 | 25.19 | 23.60 |

| Washburn and Moen Gauge or Size and Decimal Equivalent[a] | | 10 .135 | 9 .1483 | 5/32 .1563 | 8 .162 | 7 .177 | 3/16 .1875 | 6 .192 | 5 .207 | 7/32 .2188 | 4 .2253 | 3 .2437 | 1/4 .250 | 9/32 .2813 | 5/16 .3125 | 11/32 .3438 | 3/8 .375 |
|---|---|---|---|---|---|---|---|---|---|---|---|---|---|---|---|---|---|
| Design Stress, kpsi | | 161 | 158 | 156 | 154 | 150 | 149 | 146 | 143 | 142 | 141 | 140 | 139 | 138 | 137 | 136 | 135 |
| Torque, pound-inch | | 38.90 | 50.60 | 58.44 | 64.30 | 81.68 | 96.45 | 101.5 | 124.6 | 146.0 | 158.3 | 199.0 | 213.3 | 301.5 | 410.6 | 542.5 | 700.0 |
| Inside Diameter, inch | | Deflection, degrees per coil | | | | | | | | | | | | | | | |
| 13/16 | 0.8125 | 15.54 | 14.08 | 13.30 | 12.74 | 11.53 | 10.93 | 10.51 | 9.687 | 9.208 | 8.933 | 8.346 | 8.125 | 7.382 | 6.784 | 6.292 | 5.880 |
| 7/8 | 0.875 | 16.57 | 15.00 | 14.16 | 13.56 | 12.26 | 11.61 | 11.16 | 10.28 | 9.766 | 9.471 | 8.840 | 8.603 | 7.803 | 7.161 | 6.632 | 6.189 |
| 15/16 | 0.9375 | 17.59 | 15.91 | 15.02 | 14.38 | 12.99 | 12.30 | 11.81 | 10.87 | 10.32 | 10.01 | 9.333 | 9.081 | 8.225 | 7.537 | 6.972 | 6.499 |
| 1 | 1.000 | 18.62 | 16.83 | 15.88 | 15.19 | 13.72 | 12.98 | 12.47 | 11.47 | 10.88 | 10.55 | 9.827 | 9.559 | 8.647 | 7.914 | 7.312 | 6.808 |
| 1 1/16 | 1.0625 | 19.64 | 17.74 | 16.74 | 16.01 | 14.45 | 13.66 | 13.12 | 12.06 | 11.44 | 11.09 | 10.32 | 10.04 | 9.069 | 8.291 | 7.652 | 7.118 |
| 1 1/8 | 1.125 | 20.67 | 18.66 | 17.59 | 16.83 | 15.18 | 14.35 | 13.77 | 12.66 | 12.00 | 11.62 | 10.81 | 10.52 | 9.491 | 8.668 | 7.993 | 7.427 |
| 1 3/16 | 1.1875 | 21.69 | 19.57 | 18.45 | 17.64 | 15.90 | 15.03 | 14.43 | 13.25 | 12.56 | 12.16 | 11.31 | 10.99 | 9.912 | 9.045 | 8.333 | 7.737 |
| 1 1/4 | 1.250 | 22.72 | 20.49 | 19.31 | 18.46 | 16.63 | 15.71 | 15.08 | 13.84 | 13.11 | 12.70 | 11.80 | 11.47 | 10.33 | 9.422 | 8.673 | 8.046 |

[a] For sizes up to 26 gauge, the table values are for music wire with a modulus $E$ of 29,500,000 psi; for sizes from 27 gauge to 1/8 inch diameter the table values are for music wire with a modulus of 28,500,000 psi; for sizes from 10 gauge to 1/8 inch diameter, the values are for oil-tempered MB with a modulus of 28,500,000 psi.

For an example in the use of the table, see the example starting on page 328. Note: Intermediate values may be interpolated within reasonable accuracy.

**Table 15. Correction Factors for Other Materials**

| Material[a] | Factor | Material[a] | Factor |
|---|---|---|---|
| Hard Drawn MB | 0.75 | Stainless 316 | |
| Chrome-vanadium | 1.10 | Up to ⅛ inch diameter | 0.75 |
| Chrome-silicon | 1.20 | Over ⅛ to ¼ inch diameter | 0.65 |
| Stainless 302 and 304 | | Over ¼ inch diameter | 0.65 |
| Up to ⅛ inch diameter | 0.85 | Stainless 17-7 PH | |
| Over ⅛ to ¼ inch diameter | 0.75 | Up to ⅛ inch diameter | 1.00 |
| Over ¼ inch diameter | 0.65 | Over ⅛ to 3⁄16 inch diameter | 1.07 |
| Stainless 431 | 0.80 | Over 3⁄16 inch diameter | 1.12 |
| Stainless 420 | 0.85 | … | … |

[a] For use with values in Table 14. *Note:* The figures in Table 14 are for music wire (ASTM A228) and oil-tempered MB grade (ASTM A229) and can be used for several other materials that have a similar modulus of elasticity *E*. However, the design stress may be too high or too low, and therefore the design stress, torque, and deflection per coil should each be multiplied by the appropriate correction factor when using any of the materials given in this table (Table 15).

**Torsion Spring Tolerances.—**Torsion springs are coiled in a different manner from other types of coiled springs and therefore different tolerances apply. The commercial tolerance on loads is ±10 percent and is specified with reference to the angular deflection. For example: 100 pound-inches ±10 percent at 45 degrees deflection. One load specified usually suffices. If two loads and two deflections are specified, the manufacturing and testing times are increased. Tolerances smaller than ±10 percent require each spring to be individually tested and adjusted, which adds considerably to manufacturing time and cost. Table 16, Table 17, and Table 18 give, respectively, free angle tolerances, tolerances on the number of coils, and coil diameter tolerances.

**Table 16. Torsion Spring Tolerances for Angular Relationship of Ends**

| Number of Coils (*N*) | Spring Index | | | | | | | | |
|---|---|---|---|---|---|---|---|---|---|
| | 4 | 6 | 8 | 10 | 12 | 14 | 16 | 18 | 20 |
| | Free Angle Tolerance, ± degrees | | | | | | | | |
| 1 | 2 | 3 | 3.5 | 4 | 4.5 | 5 | 5.5 | 5.5 | 6 |
| 2 | 4 | 5 | 6 | 7 | 8 | 8.5 | 9 | 9.5 | 10 |
| 3 | 5.5 | 7 | 8 | 9.5 | 10.5 | 11 | 12 | 13 | 14 |
| 4 | 7 | 9 | 10 | 12 | 14 | 15 | 16 | 16.5 | 17 |
| 5 | 8 | 10 | 12 | 14 | 16 | 18 | 20 | 20.5 | 21 |
| 6 | 9.5 | 12 | 14.5 | 16 | 19 | 20.5 | 21 | 22.5 | 24 |
| 8 | 12 | 15 | 18 | 20.5 | 23 | 25 | 27 | 28 | 29 |
| 10 | 14 | 19 | 21 | 24 | 27 | 29 | 31.5 | 32.5 | 34 |
| 15 | 20 | 25 | 28 | 31 | 34 | 36 | 38 | 40 | 42 |
| 20 | 25 | 30 | 34 | 37 | 41 | 44 | 47 | 49 | 51 |
| 25 | 29 | 35 | 40 | 44 | 48 | 52 | 56 | 60 | 63 |
| 30 | 32 | 38 | 44 | 50 | 55 | 60 | 65 | 68 | 70 |
| 50 | 45 | 55 | 63 | 70 | 77 | 84 | 90 | 95 | 100 |

**Table 17. Torsion Spring Tolerance on Number of Coils**

| Number of Coils | Tolerance | Number of Coils | Tolerance |
|---|---|---|---|
| up to 5 | ±5° | over 10 to 20 | ±15° |
| over 5 to 10 | ±10° | over 20 to 40 | ±30° |

**Table 18. Torsion Spring Coil Diameter Tolerances**

| Wire Diameter, Inch | Spring Index | | | | | | |
|---|---|---|---|---|---|---|---|
| | 4 | 6 | 8 | 10 | 12 | 14 | 16 |
| | Coil Diameter Tolerance, ± inch | | | | | | |
| 0.015 | 0.002 | 0.002 | 0.002 | 0.002 | 0.003 | 0.003 | 0.004 |
| 0.023 | 0.002 | 0.002 | 0.002 | 0.003 | 0.004 | 0.005 | 0.006 |
| 0.035 | 0.002 | 0.002 | 0.003 | 0.004 | 0.006 | 0.007 | 0.009 |
| 0.051 | 0.002 | 0.003 | 0.005 | 0.007 | 0.008 | 0.010 | 0.012 |
| 0.076 | 0.003 | 0.005 | 0.007 | 0.009 | 0.012 | 0.015 | 0.018 |
| 0.114 | 0.004 | 0.007 | 0.010 | 0.013 | 0.018 | 0.022 | 0.028 |
| 0.172 | 0.006 | 0.010 | 0.013 | 0.020 | 0.027 | 0.034 | 0.042 |
| 0.250 | 0.008 | 0.014 | 0.022 | 0.030 | 0.040 | 0.050 | 0.060 |

**Miscellaneous Springs.**—This section provides information on various springs, some in common use, some less commonly used.

*Conical Compression:* These springs taper from top to bottom and are useful where an increasing (instead of a constant) load rate is needed, where solid height must be small, and where vibration must be damped. Conical springs with a uniform pitch are easiest to coil. Load and deflection formulas for compression springs can be used—using the average mean coil diameter, and provided the deflection does not cause the largest active coil to lie against the bottom coil. When this happens, each coil must be calculated separately, using the standard formulas for compression springs.

*Constant Force Springs:* Those springs are made from flat spring steel and are finding more applications each year. Complicated design procedures can be eliminated by selecting a standard design from thousands now available from several spring manufacturers.

*Spiral, Clock, and Motor Springs:* Although often used in wind-up type motors for toys and other products, these springs are difficult to design and results cannot be calculated with precise accuracy. However, many useful designs have been developed and are available from spring manufacturing companies.

*Flat Springs:* These springs are often used to overcome operating space limitations in various products such as electric switches and relays. Table 19 lists formulas for designing flat springs. The formulas are based on standard beam formulas where the deflection is small.

**Table 19. Formulas for Flat Springs**

| Feature | 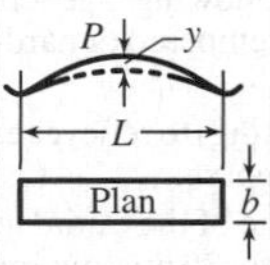 | 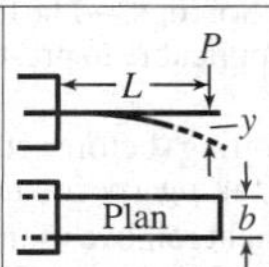 | 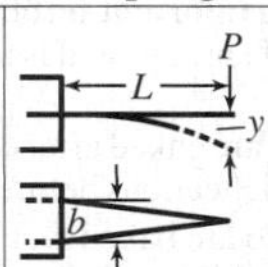 | 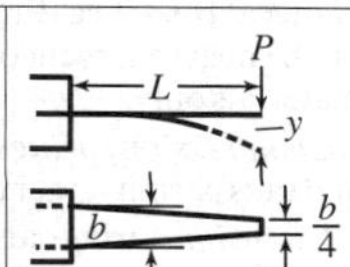 |
|---|---|---|---|---|
| Deflection, *y* Inches | $y = \frac{PL^3}{4Ebt^3}$ <br> $= \frac{S_bL^2}{6Et}$ | $y = \frac{4PL^3}{Ebt^3}$ <br> $= \frac{2S_bL^2}{3Et}$ | $y = \frac{6PL^3}{Ebt^3}$ <br> $= \frac{S_bL^2}{Et}$ | $y = \frac{5.22PL^3}{Ebt^3}$ <br> $= \frac{0.87S_bL^2}{Et}$ |
| Load, *P*, Pounds | $P = \frac{2S_bbt^2}{3L}$ <br> $= \frac{4Ebt^3y}{L^3}$ | $P = \frac{S_bbt^2}{6L}$ <br> $= \frac{Ebt^3y}{4L^3}$ | $P = \frac{S_bbt^2}{6L}$ <br> $= \frac{Ebt^3y}{6L^3}$ | $P = \frac{S_bbt^2}{6L}$ <br> $= \frac{Ebt^3y}{5.22L^3}$ |

**Table 19.** *(Continued)* **Formulas for Flat Springs**

| | | | | |
|---|---|---|---|---|
| Feature | 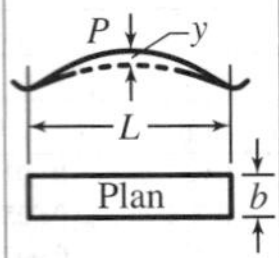 | 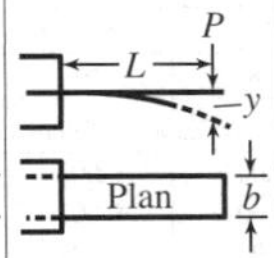 | 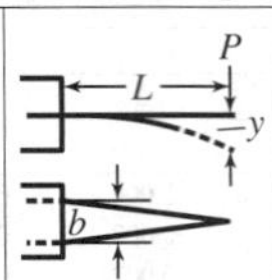 | 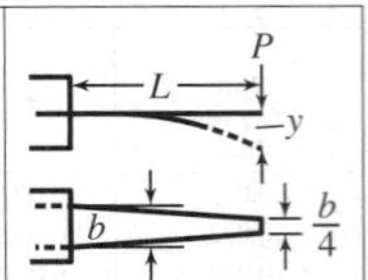 |
| Stress, $S_b$ Bending psi | $S_b = \frac{3PL}{2bt^2} = \frac{6Ety}{L^2}$ | $S_b = \frac{6PL}{bt^2} = \frac{3Ety}{2L^2}$ | $S_b = \frac{6PL}{bt^2} = \frac{Ety}{L^2}$ | $S_b = \frac{6PL}{bt^2} = \frac{Ety}{0.87L^2}$ |
| Thickness, $t$ Inches | $t = \frac{S_b L^2}{6Ey} = \sqrt[3]{\frac{PL^3}{4Eby}}$ | $t = \frac{2S_b L^2}{3Ey} = \sqrt[3]{\frac{4PL^3}{Eby}}$ | $t = \frac{S_b L^2}{Ey} = \sqrt[3]{\frac{6PL^3}{Eby}}$ | $t = \frac{0.87S_b L^2}{Ey} = \sqrt[3]{\frac{5.22PL^3}{Eby}}$ |

Based on standard beam formulas where the deflection is small.
$y$ is deflection, see page 298 for other notation.
*Note:* Where two formulas are given for one feature, the designer should use the one found to be appropriate for the given design. The result from either of any two formulas is the same.

*Belleville Washers or Disc Springs:* These washer type springs can sustain relatively large loads with small deflections, and the loads and deflections can be increased by stacking the springs.

Information on springs of this type is given in the section *DISC SPRINGS* starting on page 343.

*Volute Springs:* These springs are often used on army tanks and heavy field artillery, and seldom find additional uses because of their high cost, long production time, difficulties in manufacture, and unavailability of a wide range of materials and sizes. Small volute springs are often replaced with standard compression springs.

*Torsion Bars:* Although the more simple types are often used on motor cars, the more complicated types with specially forged ends are finding fewer applications.

**Moduli of Elasticity of Spring Materials.**—The modulus of elasticity in tension, denoted by the letter $E$, and the modulus of elasticity in torsion, denoted by the letter $G$, are used in formulas relating to spring design. Values of these moduli for various ferrous and nonferrous spring materials are given in Table 20.

**General Heat-Treating Information for Springs.**—The following is general information on the heat treatment of springs, and is applicable to pre-tempered or hard-drawn spring materials only.

*Compression springs* are baked after coiling (before setting) to relieve residual stresses and thus permit larger deflections before taking a permanent set.

*Extension springs* also are baked, but heat removes some of the initial tension. Allowance should be made for this loss. Baking at 500 degrees F for 30 minutes removes approximately 50 percent of the initial tension. The shrinkage in diameter however, will slightly increase the load and rate.

*Outside diameters shrink* when springs of music wire, pretempered MB, and other carbon or alloy steels are baked. Baking also slightly increases the free length and these changes produce a little stronger load and increase the rate.

*Outside diameters expand* when springs of stainless steel (18-8) are baked. The free length is also reduced slightly and these changes result in a little lighter load and a decrease in the spring rate.

*Inconel, Monel, and nickel alloys* do not change much when baked.

*Beryllium-copper shrinks and deforms* when heated. Such springs usually are baked in fixtures or supported on arbors or rods during heating.

*Brass and phosphor bronze* springs should be given a light heat only. Baking above 450° F will soften the material. Do not heat in salt pots.

*Torsion springs* do not require baking because coiling causes residual stresses in a direction that is helpful, but such springs frequently are baked so that jarring or handling will not cause them to lose the position of their ends.

**Table 20. Moduli of Elasticity in Torsion and Tension of Spring Materials**

| Ferrous Materials | | |
|---|---|---|
| Material (Commercial Name) | Modulus of Elasticity[a], psi | |
| | In Torsion, $G$ | In Tension, $E$ |
| Hard Drawn MB | | |
| Up to 0.032 inch | 11,700,000 | 28,800,000 |
| 0.033 to 0.063 inch | 11,600,000 | 28,700,000 |
| 0.064 to 0.125 inch | 11,500,000 | 28,600,000 |
| 0.126 to 0.625 inch | 11,400,000 | 28,500,000 |
| Music Wire | | |
| Up to 0.032 inch | 12,000,000 | 29,500,000 |
| 0.033 to 0.063 inch | 11,850,000 | 29,000,000 |
| 0.064 to 0.125 inch | 11,750,000 | 28,500,000 |
| 0.126 to 0.250 inch | 11,600,000 | 28,000,000 |
| Oil-Tempered MB | 11,200,000 | 28,500,000 |
| Chrome-Vanadium | 11,200,000 | 28,500,000 |
| Chrome-Silicon | 11,200,000 | 29,500,000 |
| Silicon-Manganese | 10,750,000 | 29,000,000 |
| Stainless Steel | | |
| Types 302, 304, 316 | 10,000,000 | 28,000,000[c] |
| Type 17-7 PH | 10,500,000 | 29,500,000 |
| Type 420 | 11,000,000 | 29,000,000 |
| Type 431 | 11,400,000 | 29,500,000 |

| Nonferrous Materials | | |
|---|---|---|
| Material (Commercial Name) | Modulus of Elasticity[a], psi | |
| | In Torsion, $G$ | In Tension, $E$ |
| Spring Brass | | |
| Type 70-30 | 5,000,000 | 15,000,000 |
| Phosphor Bronze | | |
| 5 percent tin | 6,000,000 | 15,000,000 |
| Beryllium-Copper | | |
| Cold Drawn 4 Nos. | 7,000,000 | 17,000,000 |
| Pretempered, fully hard | 7,250,000 | 19,000,000 |
| Inconel[b] 600 | 10,500,000 | 31,000,000[c] |
| Inconel[b] X 750 | 10,500,000 | 31,000,000[c] |
| Monel[b] 400 | 9,500,000 | 26,000,000 |
| Monel[b] K 500 | 9,500,000 | 26,000,000 |
| Duranickel[b] 300 | 11,000,000 | 30,000,000 |
| Permanickel[b] | 11,000,000 | 30,000,000 |
| Ni Span[b] C 902 | 10,000,000 | 27,500,000 |
| Elgiloy[d] | 12,000,000 | 29,500,000 |
| Iso-Elastic[e] | 9,200,000 | 26,000,000 |

[a] *Note:* Modulus $G$ (shear modulus) is used for compression and extension springs; modulus $E$ (Young's modulus) is used for torsion, flat, and spiral springs.

[b] Trade name of International Nickel Company.

[c] May be 2,000,000 pounds per square inch less if material is not fully hard.

[d] Trade name of Hamilton Watch Company.

[e] Trade name of John Chatillon & Sons.

*Spring brass and phosphor bronze* springs that are not very highly stressed and are not subject to severe operating use may be stress relieved after coiling by immersing them in boiling water for a period of 1 hour.

*Positions of loops* will change with heat. Parallel hooks may change as much as 45 degrees during baking. Torsion spring arms will alter position considerably. These changes should be allowed for during looping or forming.

*Quick heating* after coiling either in a high-temperature salt pot or by passing a spring through a gas flame is not good practice. Samples heated in this way will not conform with production runs that are properly baked. A small, controlled-temperature oven should be used for samples and for small lot orders.

*Plated springs* should always be baked before plating to relieve coiling stresses and again after plating to relieve hydrogen embrittlement.

*Hardness* values fall with high heat—but music wire, hard drawn, and stainless steel will increase 2 to 4 points on the Rockwell C scale.

**Table 21. Squares, Cubes, and Fourth Powers of Wire Diameters**

| Steel Wire Gage (US) | Music or Piano Wire Gage | Diameter, Inch | Section Area | Square | Cube | Fourth Power |
|---|---|---|---|---|---|---|
| 7-0 | … | 0.4900 | 0.1886 | 0.24010 | 0.11765 | 0.05765 |
| 6-0 | … | 0.4615 | 0.1673 | 0.21298 | 0.09829 | 0.04536 |
| 5-0 | … | 0.4305 | 0.1456 | 0.18533 | 0.07978 | 0.03435 |
| 4-0 | … | 0.3938 | 0.1218 | 0.15508 | 0.06107 | 0.02405 |
| 3-0 | … | 0.3625 | 0.1032 | 0.13141 | 0.04763 | 0.01727 |
| 2-0 | … | 0.331 | 0.0860 | 0.10956 | 0.03626 | 0.01200 |
| 1-0 | … | 0.3065 | 0.0738 | 0.09394 | 0.02879 | 0.008825 |
| 1 | … | 0.283 | 0.0629 | 0.08009 | 0.02267 | 0.006414 |
| 2 | … | 0.2625 | 0.0541 | 0.06891 | 0.01809 | 0.004748 |
| 3 | … | 0.2437 | 0.0466 | 0.05939 | 0.01447 | 0.003527 |
| 4 | … | 0.2253 | 0.0399 | 0.05076 | 0.01144 | 0.002577 |
| 5 | … | 0.207 | 0.0337 | 0.04285 | 0.00887 | 0.001836 |
| 6 | … | 0.192 | 0.0290 | 0.03686 | 0.00708 | 0.001359 |
| … | 45 | 0.180 | 0.0254 | 0.03240 | 0.00583 | 0.001050 |
| 7 | … | 0.177 | 0.0246 | 0.03133 | 0.00555 | 0.000982 |
| … | 44 | 0.170 | 0.0227 | 0.02890 | 0.00491 | 0.000835 |
| 8 | 43 | 0.162 | 0.0206 | 0.02624 | 0.00425 | 0.000689 |
| … | 42 | 0.154 | 0.0186 | 0.02372 | 0.00365 | 0.000563 |
| 9 | … | 0.1483 | 0.0173 | 0.02199 | 0.00326 | 0.000484 |
| … | 41 | 0.146 | 0.0167 | 0.02132 | 0.00311 | 0.000455 |
| … | 40 | 0.138 | 0.0150 | 0.01904 | 0.00263 | 0.000363 |
| 10 | … | 0.135 | 0.0143 | 0.01822 | 0.00246 | 0.000332 |
| … | 39 | 0.130 | 0.0133 | 0.01690 | 0.00220 | 0.000286 |
| … | 38 | 0.124 | 0.0121 | 0.01538 | 0.00191 | 0.000237 |
| 11 | … | 0.1205 | 0.0114 | 0.01452 | 0.00175 | 0.000211 |
| … | 37 | 0.118 | 0.0109 | 0.01392 | 0.00164 | 0.000194 |
| … | 36 | 0.112 | 0.0099 | 0.01254 | 0.00140 | 0.000157 |
| … | 35 | 0.106 | 0.0088 | 0.01124 | 0.00119 | 0.000126 |
| 12 | … | 0.1055 | 0.0087 | 0.01113 | 0.001174 | 0.0001239 |
| … | 34 | 0.100 | 0.0078 | 0.0100 | 0.001000 | 0.0001000 |
| … | 33 | 0.095 | 0.0071 | 0.00902 | 0.000857 | 0.0000815 |
| 13 | … | 0.0915 | 0.0066 | 0.00837 | 0.000766 | 0.0000701 |
| … | 32 | 0.090 | 0.0064 | 0.00810 | 0.000729 | 0.0000656 |
| … | 31 | 0.085 | 0.0057 | 0.00722 | 0.000614 | 0.0000522 |
| 14 | 30 | 0.080 | 0.0050 | 0.0064 | 0.000512 | 0.0000410 |
| … | 29 | 0.075 | 0.0044 | 0.00562 | 0.000422 | 0.0000316 |
| 15 | … | 0.072 | 0.0041 | 0.00518 | 0.000373 | 0.0000269 |
| … | 28 | 0.071 | 0.0040 | 0.00504 | 0.000358 | 0.0000254 |
| … | 27 | 0.067 | 0.0035 | 0.00449 | 0.000301 | 0.0000202 |
| … | 26 | 0.063 | 0.0031 | 0.00397 | 0.000250 | 0.0000158 |
| 16 | … | 0.0625 | 0.0031 | 0.00391 | 0.000244 | 0.0000153 |
| … | 25 | 0.059 | 0.0027 | 0.00348 | 0.000205 | 0.0000121 |
| … | 24 | 0.055 | 0.0024 | 0.00302 | 0.000166 | 0.00000915 |
| 17 | … | 0.054 | 0.0023 | 0.00292 | 0.000157 | 0.00000850 |
| … | 23 | 0.051 | 0.0020 | 0.00260 | 0.000133 | 0.00000677 |
| … | 22 | 0.049 | 0.00189 | 0.00240 | 0.000118 | 0.00000576 |
| 18 | … | 0.0475 | 0.00177 | 0.00226 | 0.000107 | 0.00000509 |
| … | 21 | 0.047 | 0.00173 | 0.00221 | 0.000104 | 0.00000488 |
| … | 20 | 0.045 | 0.00159 | 0.00202 | 0.000091 | 0.00000410 |
| … | 19 | 0.043 | 0.00145 | 0.00185 | 0.0000795 | 0.00000342 |
| 19 | 18 | 0.041 | 0.00132 | 0.00168 | 0.0000689 | 0.00000283 |
| … | 17 | 0.039 | 0.00119 | 0.00152 | 0.0000593 | 0.00000231 |
| … | 16 | 0.037 | 0.00108 | 0.00137 | 0.0000507 | 0.00000187 |
| … | 15 | 0.035 | 0.00096 | 0.00122 | 0.0000429 | 0.00000150 |
| 20 | … | 0.0348 | 0.00095 | 0.00121 | 0.0000421 | 0.00000147 |
| … | 14 | 0.033 | 0.00086 | 0.00109 | 0.0000359 | 0.00000119 |
| 21 | … | 0.0317 | 0.00079 | 0.00100 | 0.0000319 | 0.00000101 |
| … | 13 | 0.031 | 0.00075 | 0.00096 | 0.0000298 | 0.000000924 |
| … | 12 | 0.029 | 0.00066 | 0.00084 | 0.0000244 | 0.000000707 |
| 22 | … | 0.0286 | 0.00064 | 0.00082 | 0.0000234 | 0.000000669 |
| … | 11 | 0.026 | 0.00053 | 0.00068 | 0.0000176 | 0.000000457 |
| 23 | … | 0.0258 | 0.00052 | 0.00067 | 0.0000172 | 0.000000443 |
| … | 10 | 0.024 | 0.00045 | 0.00058 | 0.0000138 | 0.000000332 |
| 24 | … | 0.023 | 0.00042 | 0.00053 | 0.0000122 | 0.000000280 |
| … | 9 | 0.022 | 0.00038 | 0.00048 | 0.0000106 | 0.000000234 |

**Spring Failure.—** Spring failure may be breakage, high permanent set, or loss of load. The causes are listed in groups in Table 22. Group 1 covers causes that occur most frequently; Group 2 covers causes that are less frequent; and Group 3 lists causes that occur occasionally.

**Table 22. Causes of Spring Failure**

| | Cause | Comments and Recommendations |
|---|---|---|
| Group 1 | High stress | The majority of spring failures are due to high stresses caused by large deflections and high loads. High stresses should be used only for statically loaded springs. Low stresses lengthen fatigue life. |
| | Hydrogen embrittlement | Improper electroplating methods and acid cleaning of springs, without proper baking treatment, cause spring steels to become brittle, and are a frequent cause of failure. Nonferrous springs are immune. |
| | Sharp bends and holes | Sharp bends on extension, torsion, and flat springs, and holes or notches in flat springs, cause high concentrations of stress, resulting in failure. Bend radii should be as large as possible, and tool marks avoided. |
| | Fatigue | Repeated deflections of springs, especially above 1,000,000 cycles, even with medium stresses, may cause failure. Low stresses should be used if a spring is to be subjected to a very high number of operating cycles. |
| Group 2 | Shock loading | Impact, shock, and rapid loading cause far higher stresses than those computed by the regular spring formulas. High-carbon spring steels do not withstand shock loading as well as do alloy steels. |
| | Corrosion | Slight rusting or pitting caused by acids, alkalis, galvanic corrosion, stress corrosion cracking, or corrosive atmosphere weakens the material and causes higher stresses in the corroded area. |
| | Faulty heat treatment | Keeping spring materials at the hardening temperature for longer periods than necessary causes an undesirable growth in grain structure, resulting in brittleness, even though the hardness may be correct. |
| | Faulty material | Poor material containing inclusions, seams, slivers, and flat material with rough, slit, or torn edges is a cause of early failure. Overdrawn wire, improper hardness, and poor grain structure also cause early failure. |
| Group 3 | High temperature | High operating temperatures reduce spring temper (or hardness) and lower the modulus of elasticity, thereby causing lower loads, reducing the elastic limit, and increasing corrosion. Corrosion-resisting or nickel alloys should be used. |
| | Low temperature | Temperatures below −40 degrees F reduce the ability of carbon steels to withstand shock loads. Carbon steels become brittle at −70 degrees F. Corrosion-resisting, nickel, or nonferrous alloys should be used. |
| | Friction | Close fits on rods or in holes result in a wearing away of material and occasional failure. The outside diameters of compression springs expand during deflection but they become smaller on torsion springs. |
| | Other causes | Enlarged hooks on extension springs increase the stress at the bends. Carrying too much electrical current will cause failure. Welding and soldering frequently destroy the spring temper. Tool marks, nicks, and cuts often raise stresses. Deflecting torsion springs outwardly causes high stresses and winding them tightly causes binding on supporting rods. High speed of deflection, vibration, and surging due to operation near natural periods of vibration or their harmonics cause increased stresses. |

## Table 23. Arbor Diameters for Springs Made from Music Wire

| Wire Dia. (inch) | Spring Outside Diameter (inch) | | | | | | | | | | | | |
|---|---|---|---|---|---|---|---|---|---|---|---|---|---|
| | 1⁄16 | 3⁄32 | 1⁄8 | 5⁄32 | 3⁄16 | 7⁄32 | 1⁄4 | 9⁄32 | 5⁄16 | 11⁄32 | 3⁄8 | 7⁄16 | 1⁄2 |
| | Arbor Diameter (inch) | | | | | | | | | | | | |
| 0.008 | 0.039 | 0.060 | 0.078 | 0.093 | 0.107 | 0.119 | 0.129 | … | … | … | … | … | … |
| 0.010 | 0.037 | 0.060 | 0.080 | 0.099 | 0.115 | 0.129 | 0.142 | 0.154 | 0.164 | … | … | … | … |
| 0.012 | 0.034 | 0.059 | 0.081 | 0.101 | 0.119 | 0.135 | 0.150 | 0.163 | 0.177 | 0.189 | 0.200 | … | … |
| 0.014 | 0.031 | 0.057 | 0.081 | 0.102 | 0.121 | 0.140 | 0.156 | 0.172 | 0.187 | 0.200 | 0.213 | 0.234 | … |
| 0.016 | 0.028 | 0.055 | 0.079 | 0.102 | 0.123 | 0.142 | 0.161 | 0.178 | 0.194 | 0.209 | 0.224 | 0.250 | 0.271 |
| 0.018 | … | 0.053 | 0.077 | 0.101 | 0.124 | 0.144 | 0.161 | 0.182 | 0.200 | 0.215 | 0.231 | 0.259 | 0.284 |
| 0.020 | … | 0.049 | 0.075 | 0.096 | 0.123 | 0.144 | 0.165 | 0.184 | 0.203 | 0.220 | 0.237 | 0.268 | 0.296 |
| 0.022 | … | 0.046 | 0.072 | 0.097 | 0.122 | 0.145 | 0.165 | 0.186 | 0.206 | 0.224 | 0.242 | 0.275 | 0.305 |
| 0.024 | … | 0.043 | 0.070 | 0.095 | 0.120 | 0.144 | 0.166 | 0.187 | 0.207 | 0.226 | 0.245 | 0.280 | 0.312 |
| 0.026 | … | … | 0.067 | 0.093 | 0.118 | 0.143 | 0.166 | 0.187 | 0.208 | 0.228 | 0.248 | 0.285 | 0.318 |
| 0.028 | … | … | 0.064 | 0.091 | 0.115 | 0.141 | 0.165 | 0.187 | 0.208 | 0.229 | 0.250 | 0.288 | 0.323 |
| 0.030 | … | … | 0.061 | 0.088 | 0.113 | 0.138 | 0.163 | 0.187 | 0.209 | 0.229 | 0.251 | 0.291 | 0.328 |
| 0.032 | … | … | 0.057 | 0.085 | 0.111 | 0.136 | 0.161 | 0.185 | 0.209 | 0.229 | 0.251 | 0.292 | 0.331 |
| 0.034 | … | … | … | 0.082 | 0.109 | 0.134 | 0.159 | 0.184 | 0.208 | 0.229 | 0.251 | 0.292 | 0.333 |
| 0.036 | … | … | … | 0.078 | 0.106 | 0.131 | 0.156 | 0.182 | 0.206 | 0.229 | 0.250 | 0.294 | 0.333 |
| 0.038 | … | … | … | 0.075 | 0.103 | 0.129 | 0.154 | 0.179 | 0.205 | 0.227 | 0.251 | 0.293 | 0.335 |
| 0.041 | … | … | … | … | 0.098 | 0.125 | 0.151 | 0.176 | 0.201 | 0.226 | 0.250 | 0.294 | 0.336 |
| 0.0475 | … | … | … | … | 0.087 | 0.115 | 0.142 | 0.168 | 0.194 | 0.220 | 0.244 | 0.293 | 0.337 |
| 0.054 | … | … | … | … | … | 0.103 | 0.132 | 0.160 | 0.187 | 0.212 | 0.245 | 0.287 | 0.336 |
| 0.0625 | … | … | … | … | … | … | 0.108 | 0.146 | 0.169 | 0.201 | 0.228 | 0.280 | 0.330 |
| 0.072 | … | … | … | … | … | … | … | 0.129 | 0.158 | 0.186 | 0.214 | 0.268 | 0.319 |
| 0.080 | … | … | … | … | … | … | … | … | 0.144 | 0.173 | 0.201 | 0.256 | 0.308 |
| 0.0915 | … | … | … | … | … | … | … | … | … | … | 0.181 | 0.238 | 0.293 |
| 0.1055 | … | … | … | … | … | … | … | … | … | … | … | 0.215 | 0.271 |
| 0.1205 | … | … | … | … | … | … | … | … | … | … | … | … | 0.215 |
| 0.125 | … | … | … | … | … | … | … | … | … | … | … | … | 0.239 |

| Wire Dia. (inch) | Spring Outside Diameter (inch) | | | | | | | | | | | | | |
|---|---|---|---|---|---|---|---|---|---|---|---|---|---|---|
| | 9⁄16 | 5⁄8 | 11⁄16 | 3⁄4 | 13⁄16 | 7⁄8 | 15⁄16 | 1 | 1 1⁄8 | 1 1⁄4 | 1 3⁄8 | 1 1⁄2 | 1 3⁄4 | 2 |
| | Arbor Diameter (inches) | | | | | | | | | | | | | |
| 0.022 | 0.332 | 0.357 | 0.380 | … | … | … | … | … | … | … | … | … | … | … |
| 0.024 | 0.341 | 0.367 | 0.393 | 0.415 | … | … | … | … | … | … | … | … | … | … |
| 0.026 | 0.350 | 0.380 | 0.406 | 0.430 | … | … | … | … | … | … | … | … | … | … |
| 0.028 | 0.356 | 0.387 | 0.416 | 0.442 | 0.467 | … | … | … | … | … | … | … | … | … |
| 0.030 | 0.362 | 0.395 | 0.426 | 0.453 | 0.481 | 0.506 | … | … | … | … | … | … | … | … |
| 0.032 | 0.367 | 0.400 | 0.432 | 0.462 | 0.490 | 0.516 | 0.540 | … | … | … | … | … | … | … |
| 0.034 | 0.370 | 0.404 | 0.437 | 0.469 | 0.498 | 0.526 | 0.552 | 0.557 | … | … | … | … | … | … |
| 0.036 | 0.372 | 0.407 | 0.442 | 0.474 | 0.506 | 0.536 | 0.562 | 0.589 | … | … | … | … | … | … |
| 0.038 | 0.375 | 0.412 | 0.448 | 0.481 | 0.512 | 0.543 | 0.572 | 0.600 | 0.650 | … | … | … | … | … |
| 0.041 | 0.378 | 0.416 | 0.456 | 0.489 | 0.522 | 0.554 | 0.586 | 0.615 | 0.670 | 0.718 | … | … | … | … |
| 0.0475 | 0.380 | 0.422 | 0.464 | 0.504 | 0.541 | 0.576 | 0.610 | 0.643 | 0.706 | 0.763 | 0.812 | … | … | … |
| 0.054 | 0.381 | 0.425 | 0.467 | 0.509 | 0.550 | 0.589 | 0.625 | 0.661 | 0.727 | 0.792 | 0.850 | 0.906 | … | … |
| 0.0625 | 0.379 | 0.426 | 0.468 | 0.512 | 0.556 | 0.597 | 0.639 | 0.678 | 0.753 | 0.822 | 0.889 | 0.951 | 1.06 | 1.17 |
| 0.072 | 0.370 | 0.418 | 0.466 | 0.512 | 0.555 | 0.599 | 0.641 | 0.682 | 0.765 | 0.840 | 0.911 | 0.980 | 1.11 | 1.22 |
| 0.080 | 0.360 | 0.411 | 0.461 | 0.509 | 0.554 | 0.599 | 0.641 | 0.685 | 0.772 | 0.851 | 0.930 | 1.00 | 1.13 | 1.26 |
| 0.0915 | 0.347 | 0.398 | 0.448 | 0.500 | 0.547 | 0.597 | 0.640 | 0.685 | 0.776 | 0.860 | 0.942 | 1.02 | 1.16 | 1.30 |
| 0.1055 | 0.327 | 0.381 | 0.433 | 0.485 | 0.535 | 0.586 | 0.630 | 0.683 | 0.775 | 0.865 | 0.952 | 1.04 | 1.20 | 1.35 |
| 0.1205 | 0.303 | 0.358 | 0.414 | 0.468 | 0.520 | 0.571 | 0.622 | 0.673 | 0.772 | 0.864 | 0.955 | 1.04 | 1.22 | 1.38 |
| 0.125 | 0.295 | 0.351 | 0.406 | 0.461 | 0.515 | 0.567 | 0.617 | 0.671 | 0.770 | 0.864 | 0.955 | 1.05 | 1.23 | 1.39 |

# DISC SPRINGS

## Performance of Disc Springs

**Introduction.—**Disc springs, also known as *Belleville springs,* are conically formed from washers and have rectangular cross section. The disc spring concept was invented by a Frenchman Louis Belleville in 1865. His springs were relatively thick and had a small amount of cone height or "dish," which determined axial deflection. At that time, these springs were used in the buffer parts of railway rolling stock, for recoil mechanisms of guns, and some other applications. The use of disc springs will be advantageous when space is limited and high force is required, as these conditions cannot be satisfied by using coil springs. Load-deflection characteristics of disc springs are linear and regressive depending on their dimensions and the type of stacking. A large number of standard sizes are available from disc spring manufacturers and distributors, so that custom sizes may not be required. Therefore, disc springs are widely used today in virtually all branches of engineering with possibilities of new applications.

**Disc Spring Nomenclature.—**Disc spring manufacturers assign their own part number for each disc spring, but the catalog numbers for disc springs are similar, so each item can often be identified regardless of manufacturer. The disc spring identification number is a numerical code that provides basic dimensions in millimeters. Identification numbers represent the primary dimensions of the disc spring and consist of one, two, or three numbers separated from each other by dash marks or spaces. Disc spring manufacturers in the United States also provide dimensions in inches. Dimensions of several typical disc springs are shown in the following table. Basic nomenclature is illustrated in Fig. 1.

| Catalog Number (mm) | Outside Diameter $D$ (mm) | Inside Diameter $d$ (mm) | Thickness $t$ (mm) | Equivalent Catalog Number (inch) |
|---|---|---|---|---|
| 8-4.2-0.4 | 8 | 4.2 | 0.4 | 0.315-0.165-0.0157 |
| 50-25.4-2 | 50 | 25.4 | 2 | 1.97-1.00-0.0787 |
| 200-102-12 | 200 | 102 | 12 | 7.87-4.02-0.472 |

Additional dimensions shown in catalogs are cone (dish) height $h$ at unloaded condition, and overall height $H = h + t$, that combines the cone height and the thickness of a disc spring.

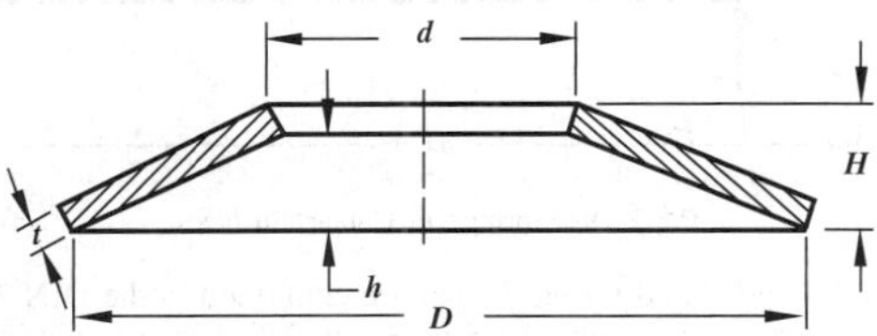


Fig. 1. Disc Spring Nomenclature

**Disc Spring Group Classification.—**Forces and stresses generated by compression depend on disc spring thickness much more than on any other dimensions. Standard DIN 2093 divides all disc springs into three groups in accordance with their thickness:

*Group 1* includes all disc springs with thickness less than 1.25 mm (0.0492 inch).

*Group 2* includes all disc springs with thickness between 1.25 mm and 6.0 mm (0.0492 inch and 0.2362 inch).

*Group 3* includes disc springs with thickness greater than 6.0 mm (0.2362 inch).

There are 87 standard disc spring items, which are manufactured in accordance with DIN 2093 specifications for dimensions and quality requirements. There are 30 standard disc spring items in *Group 1.* The smallest and the largest disc springs in this group are

8-4.2-0.2 and 40-20.4-1, respectively. *Group 2* has 45 standard disc spring items. The smallest and the largest disc springs are 22.5-11.2-1.25 and 200-102-5.5, respectively. *Group 3* includes 12 standard disc spring items. The smallest and the largest disc springs of this group are 125-64-8 and 250-127-14, respectively.

**Summary of Disc Spring Sizes Specified in *DIN 2093***

| Classification | Outside Diameter | | Inside Diameter | | Thickness | |
|---|---|---|---|---|---|---|
| | Min. | Max | Min. | Max | Min. | Max |
| *Group 1* | 6 mm (0.236 in) | 40 mm (1.575 in) | 3.2 mm (0.126 in) | 20.4 mm (0.803 in) | 0.2 mm (0.008 in) | 1.2 mm (0.047 in) |
| *Group 2* | 20 mm (0.787 in) | 225 mm (8.858 in) | 10.2 mm (0.402 in) | 112 mm (4.409 in) | 1.25 mm (0.049 in) | 6 mm (0.236 in) |
| *Group 3* | 125 mm (4.921 in) | 250 mm (9.843 in) | 61 mm (2.402 in) | 127 mm (5.000 in) | 6.5 mm (0.256 in) | 16 mm (0.630 in) |

The number of catalog items by disc spring dimensions depends on the manufacturer. Currently, the smallest disc spring is 6-3.2-0.3 and the largest is 250-127-16. One of the US disc spring manufacturers, Key Bellevilles, Inc. offers 190 catalog items. The greatest number of disc spring items can be found in Christian Bauer GmbH + Co. catalog. There are 291 disc spring catalog items in all three groups.

**Disc Spring Contact Surfaces.—**Disc springs are manufactured with and without contact (also called load-bearing) surfaces. Contact surfaces are small flats at points 1 and 3 in Fig. 2, adjacent to the corner radii of the spring. The width of the contact surfaces $w$ depends on the outside diameter $D$ of the spring, and its value is approximately $w = D/150$.

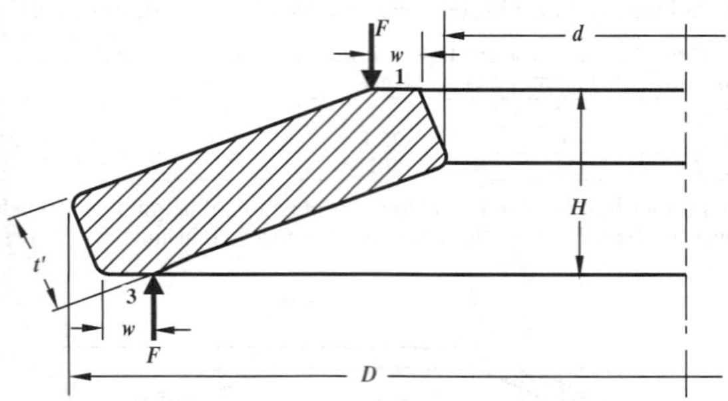


Fig. 2. Disc Spring with Contact Surfaces

Disc springs of *Group 1* and *Group 2*, that are contained in the DIN 2093 Standard, do not have contact surfaces, although some *Group 2* disc springs not included in DIN 2093 are manufactured with contact surfaces. All disc springs of *Group 3* (standard and nonstandard) are manufactured with contact surfaces. Almost all disc springs with contact surfaces are manufactured with reduced thickness.

Disc springs without contact surfaces have a corner radii $r$ whose value depends on the spring thickness $t$. One disc spring manufacturer recommends the following relationship:

$$r = t/6$$

**Disc Spring Materials .—**A wide variety of materials are available for disc springs, but selection of the material depends mainly on application. High-carbon steels are used only for *Group 1* disc springs. AISI 1070 and AISI 1095 carbon steels are used in the United States. Similar high-carbon steels such as DIN 1.1231 and DIN 1.1238 (Germany), and BS 060 A67 and BS 060 A78 (Great Britain) are used in other countries. The most common materials for *Groups 2* and *3* springs operating under normal conditions are chromium-vanadium alloy steels such as AISI 6150 used in the United States. Similar alloys such as DIN 1.8159 and DIN 1.7701 (Germany) and BS 735 A50 (Great Britain) are used in foreign countries. Some disc spring manufacturers in the United States also use chromium alloy steel AISI 5160. The hardness of disc springs in *Groups 2* and *3* should be 42 to 52 RC. The hardness of disc springs in *Group 1* tested by the Vickers method should be 412 to 544 HV.

If disc springs must withstand corrosion and high temperatures, stainless steels and heat-resistant alloys are used. Most commonly used stainless steels in the United States are AISI types 301, 316, and 631, which are similar to foreign material numbers DIN 1.4310, DIN 1.4401, and DIN 1.4568, respectively. The operating temperature range for 631 stainless steel is –330 to 660°F (–200 to 350°C). Among heat-resistant alloys, Inconel 718 and Inconel X750 (similar to DIN 2.4668 and DIN 2.4669, respectively) are the most popular. Operating temperature range for Inconel 718 is –440 to 1290°F (–260 to 700°C).

When disc springs are stacked in large numbers and their total weight becomes a major concern, titanium α-β alloys can be used to reduce weight. In such cases, Ti-6Al-4V alloy is used.

If nonmagnetic and corrosion resistant properties are required and material strength is not an issue, phosphor bronzes and beryllium-coppers are the most popular copper alloys for disc springs. Phosphor bronze C52100, which is similar to DIN material number 2.1030, is used at the ordinary temperature range. Beryllium-coppers C17000 and C17200, similar to material numbers DIN 2.1245 and DIN 2.1247 respectively, works well at very low temperatures.

Strength properties of disc spring materials are characterized by moduli of elasticity and Poisson's ratios. These are summarized in Table 1.

**Table 1. Strength Characteristics of Disc Spring Materials**

| Material | Modulus of Elasticity | | Poisson's Ratio |
|---|---|---|---|
| | $10^6$ psi | N/mm$^2$ | |
| All Steels | 28-31 | 193,000–213,700 | 0.30 |
| Heat-resistant Alloys | | | 0.28–0.29 |
| α-β Titanium Alloys (Ti-6Al-4V) | 17 | 117,200 | 0.32 |
| Phosphor Bronze (C52100) | 16 | 110,300 | 0.35 |
| Beryllium-copper (C17000) | 17 | 117,200 | 0.30 |
| Beryllium-copper (C17200) | 18 | 124,100 | 0.30 |

**Stacking of Disc Springs.—**Individual disc springs can be arranged in series and parallel stacks. Disc springs in series stacking, Fig. 3, provide larger deflection $S_{total}$ under the same load $F$ as a single disc spring would generate. Disc springs in parallel stacking, Fig. 4, generate higher loads $F_{total}$ with the same deflection $s$ than a single disc spring would have.

$n$ = number of disc springs in stack

$s$ = deflection of single spring

$S_{total}$ = total deflection of stack of $n$ springs

$F$ = load generated by a single spring

$F_{total}$ = total load generated by springs in stack

$L_0$ = length of unloaded spring stack

*Series:* For $n$ disc springs arranged in series, Fig. 3, the following equations are applied:

$$F_{total} = F$$

$$S_{total} = s \times n$$

$$L_0 = H \times n = (t + h) \times n \tag{1}$$

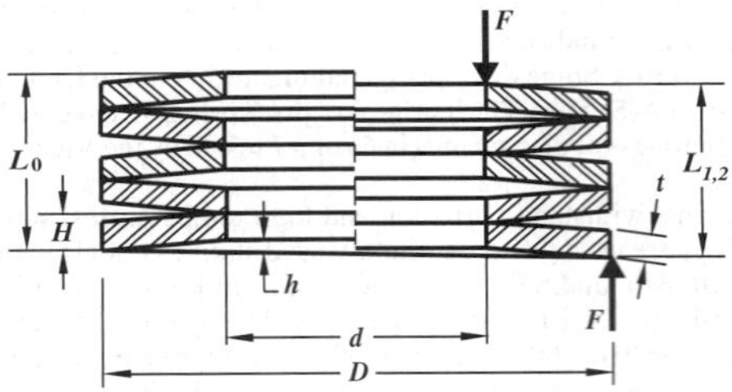


Fig. 3. Disc Springs in Series Stacking
$L_{1,2}$ Indices Indicate Length of Spring Stack under Minimum and Maximum Load.

*Parallel:* Parallel stacking generates a force that is directly proportional to the number of springs arranged in parallel. Two springs in parallel will double the force, three springs in parallel will triple the force, and so on. However, it is a common practice to use two springs in parallel in order to keep the frictional forces between the springs as low as possible. Otherwise, the actual spring force cannot be accurately determined due to deviation from its theoretical value.

For $n$ disc springs arranged in parallel as in Fig. 4, the following equations are applied:

$$F_{total} = F \times n$$

$$S_{total} = s$$

$$L_0 = H + t(n-1) = (h+t) + tn - t = h + tn \tag{2}$$

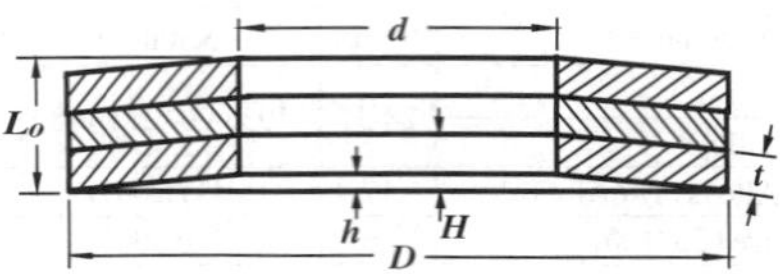


Fig. 4. Disc Springs in Parallel Stacking

*Parallel-Series:* When both higher force and greater deflection are required, disc springs must be arranged in a combined parallel-series stacking as illustrated in Fig. 5.

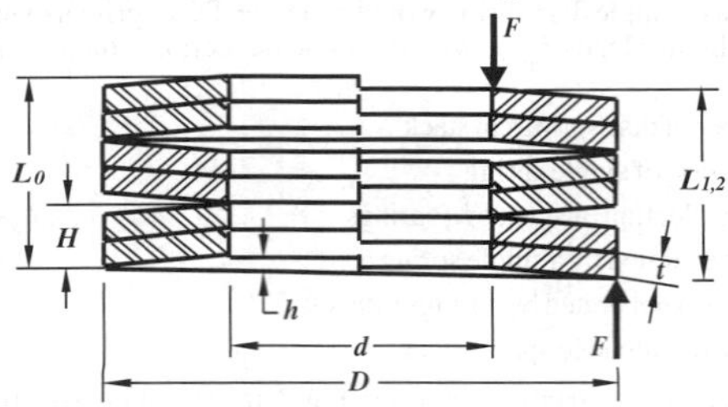


Fig. 5. Disc Springs in Parallel-Series Stacking

Normally, two springs in parallel are nested in series stacking. Two springs in parallel, called a pair, double the force, and the number of pairs, $n_p$, determines the total deflection, $S_{total}$.

For $n_p$ disc spring pairs arranged in series, the following equations are applied:

$$F_{total} = 2 \times F$$
$$S_{total} = s \times n_p$$
$$L_0 = H \times n_p = (2t + h) \times n_p \qquad (3)$$

## Disc Spring Forces and Stresses

Several methods of calculating forces and stresses for given disc spring configurations exist, some very complicated, others of limited accuracy. The theory which is widely used today for force and stress calculations was developed more than 65 years ago by Almen and Laszlo.

The theory is based on the following assumptions: cross sections are rectangular without radii over the entire range of spring deflection; no stresses occur in the radial direction; disc springs are always under elastic deformation during deflection; and due to small cone angles of unloaded disc springs (between 3.5° and 8.6°), mathematical simplifications are applied.

The theory provides accurate results for disc springs with the following ratios: outside-to-inside diameter, $D/d = 1.3$ to 2.5; and cone height-to-thickness, $h/t$ is up to 1.5.

**Force Generated by Disc Springs without Contact Surfaces.—**Disc springs in *Group 1* and most of disc springs in *Group 2* are manufactured without contact (load-bearing) surfaces, but have corner radii.

A single disc spring force applied to points 1 and 3 in Fig. 6 can be found from Equation (4) in which corner radii are not considered:

$$F = \frac{4 \cdot E \cdot s}{(1 - \mu^2) \cdot K_1 \cdot D^2}\left[\left(h - \frac{s}{2}\right) \cdot (h - s) \cdot t + t^3\right] \qquad (4)$$

where $F$ = disc spring force; $E$ = modulus of elasticity of spring material; $\mu$ = Poisson's ratio of spring material; $K_1$ = constant depending on outside-to-inside diameter ratio; $D$ = disc spring nominal outside diameter; $h$ = cone (dish) height; $s$ = disc spring deflection; and, $t$ = disc spring thickness.

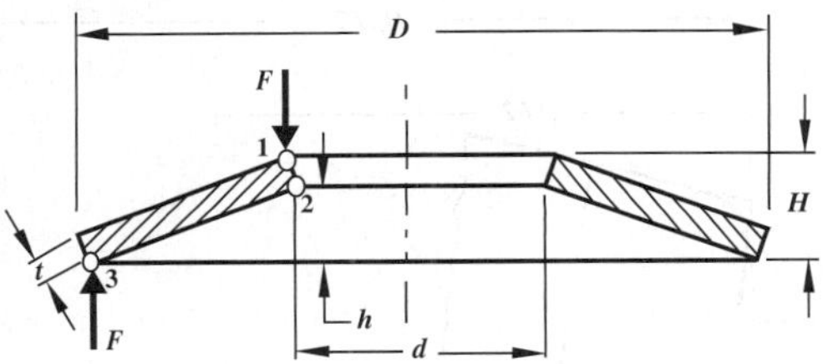


Fig. 6. Schematic of Applied Forces

It has been found that the theoretical forces calculated using Equation (4) are lower than the actual (measured) spring forces, as illustrated in Fig. 7. The difference between theoretical (trace 1) and measured force values (trace 3) was significantly reduced (trace 2) when the actual outside diameter of the spring in loaded condition was used in the calculations.

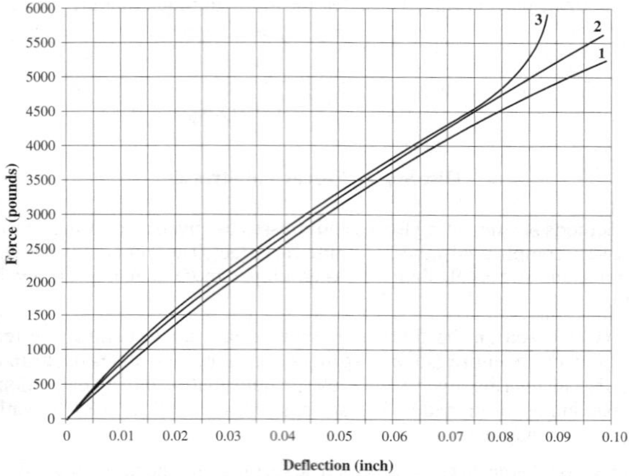


Fig. 7. Force-Deflection Relationships (80-36-3.6 Disc Springs)
1 - Theoretical Force Calculated by Equation (4)
2 - Theoretical Force Calculated by Equation (10)
3 - Measured Force

The actual outside diameter $D_a$ of a disc spring contact circle is smaller than the nominal outside diameter $D$ due to cone angle $\alpha$ and corner radius $r$, as shown in Fig. 8. Diameter $D_a$ cannot be measured, but can be calculated by Equation (9) developed by Dr. Isacov.

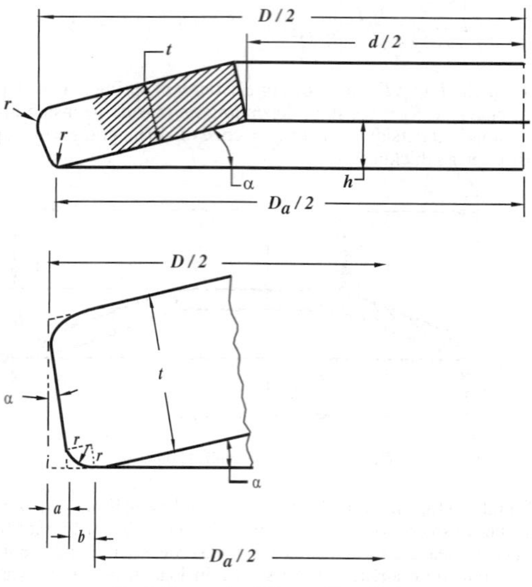


Fig. 8. Conventional Shape of Disc Spring

From Fig. 8,

$$\frac{D_a}{2} = \frac{D}{2} - (a + b) \tag{5}$$

where $a = t\sin\alpha$ and $b = r\cos\alpha$. Substitution of $a$ and $b$ values into Equation (5) gives:

$$\frac{D_a}{2} = \frac{D}{2} - (t\sin\alpha + r\cos\alpha) \tag{6}$$

The cone angle $\alpha$ is found from:

$$\tan\alpha = \frac{h}{\frac{D}{2} - \frac{d}{2}} = \frac{2h}{D - d} \qquad \alpha = \operatorname{atan}\left(\frac{2h}{D - d}\right) \tag{7}$$

Substituting $\alpha$ from Equation (7) and $r = \frac{t}{6}$ into Equation (6) gives:

$$\frac{D_a}{2} = \frac{D}{2} - t\left\{\sin\left[\operatorname{atan}\left(\frac{2h}{D-d}\right)\right] + \frac{1}{6}\cos\left[\operatorname{atan}\left(\frac{2h}{D-d}\right)\right]\right\} \tag{8}$$

Finally,

$$D_a = D - 2t\left\{\sin\left[\operatorname{atan}\left(\frac{2h}{D-d}\right)\right] + \frac{1}{6}\cos\left[\operatorname{atan}\left(\frac{2h}{D-d}\right)\right]\right\} \tag{9}$$

Substituting $D_a$ from Equation (9) for $D$ in Equation (4) yields Equation (10), that provides better accuracy for calculating disc spring forces.

$$F = \frac{4 \cdot E \cdot s}{(1 - \mu^2) K_1 \cdot D_a^2}\left[\left(h - \frac{s}{2}\right) \cdot (h - s) \cdot t + t^3\right] \tag{10}$$

The constant $K_1$ depends on disc spring outside diameter $D$, inside diameter $d$, and their ratio $\delta = D/d$:

$$K_1 = \frac{\left(\frac{\delta - 1}{\delta}\right)^2}{\pi\left(\frac{\delta + 1}{\delta - 1} - \frac{2}{\ln\delta}\right)} \tag{11}$$

Table 2 compares the spring force of a series of disc springs deflected by 75 percent of their cone height, i.e., $s = 0.75h$, as determined from manufacturers' catalogs, calculated forces by use of Equation (10), and measured forces.

**Table 2. Comparison Between Calculated and Measured Disc Spring Forces**

| Disc Spring Catalog Item | *Schnorr* Handbook for Disc Springs | *Christian Bauer* Disc Spring Handbook | *Key Bellevilles* Disc Spring Catalog | Spring Force Calculated by Equation (10) | Measured Disc Spring Force |
|---|---|---|---|---|---|
| 50 – 22.4 – 2.5<br>$S = 1.05$ mm | 8510 N<br>1913 lbf | 8510 N<br>1913 lbf | 8616 N<br>1937 lbf | 9020 N<br>2028 lbf | 9563 N<br>2150 lbf |
| 60 – 30.5 – 2.5<br>$S = 1.35$ mm | 8340 N<br>1875 lbf | 8342 N<br>1875 lbf | 8465 N<br>1903 lbf | 8794 N<br>1977 lbf | 8896 N<br>2000 lbf |
| 60 – 30.5 – 3<br>$S = 1.275$ mm | 13200 N<br>2967 lbf | 13270 N<br>2983 lbf | 13416 N<br>3016 lbf | 14052 N<br>3159 lbf | 13985 N<br>3144 lbf |
| 70 – 35.5 – 3<br>$S = 1.575$ mm | 12300 N<br>2765 lbf | 12320 N<br>2770 lbf | 12397 N<br>2787 lbf | 12971 N<br>2916 lbf | 13287 N<br>2987 lbf |
| 70 – 35.5 – 3.5<br>$S = 1.35$ mm | | 16180 N<br>3637 lbf | | 17170 N<br>3860 lbf | 17304 N<br>3890 lbf |

Comparison made at 75 percent deflection, in newtons (N) and pounds (lbf)

The difference between disc spring forces calculated by Equation (10) and the measured forces varies from −5.7 percent (maximum) to +0.5 percent (minimum). Disc spring forces shown in manufacturers' catalogs tend to be less than measured forces by −11.0 percent (maximum) to −5.6 percent (minimum).

**Force Generated by Disc Spring with Contact Surfaces.—**Some of disc springs in *Group 2* and all disc springs in *Group 3* are manufactured with small contact (load-bearing) surfaces or flats in addition to the corner radii. These flats provide better contact between disc springs, but, at the same time, they reduce the springs outside diameter and generate higher spring force because in Equation (4) force $F$ is inversely proportional to the square of outside diameter $D^2$. To compensate for the undesired force increase, the disc spring thickness is reduced from $t$ to $t'$. Thickness reduction factors $t'/t$ are approximately 0.94 for disc spring series A and B, and approximately 0.96 for series C springs. With such reduction factors, the disc spring force at 75 percent deflection is the same as for equivalent disc spring without contact surfaces. Equation (12), which is similar to Equation (10), has an additional constant $K_4$ that correlates the increase in spring force due to contact surfaces. If disc springs do not have contact surfaces, then $K_4^2 = K_4 = 1$.

$$F = \frac{4 \cdot E \cdot K_4^2 \cdot s}{(1-\mu^2) \cdot K_1 \cdot D_a^2}\left[K_4^2 \cdot \left(h' - \frac{s}{2}\right) \cdot (h' - s) \cdot (t' + t')^3\right] \tag{12}$$

where $t'$ = reduced thickness of a disc spring

$h'$ = cone height adjusted to reduced thickness: $h' = H - t'$ $(h' > h)$

$K_4$ = constant applied to disc springs with contact surfaces.

$K_4^2$ can be calculated as follows:

$$K_4^2 = \frac{-b + \sqrt{b^2 - 4ac}}{2a} \tag{13}$$

where $a = t'(H - 4t' + 3t)(5H - 8t' + 3t)$; $b = 32(t')^3$; and, $c = -t[5(H-t)^2 + 32t^2]$.

**Disc Spring Functional Stresses.—**Disc springs are designed for both static and dynamic load applications. In static load applications, disc springs may be under constant or fluctuating load conditions that change up to 5,000 or 10,000 cycles over long time intervals. Dynamic loads occur when disc springs are under continuously changing deflection between pre-load (approximately 15 to 20 percent of the cone height) and the maximum deflection values over short time intervals. Both static and dynamic loads cause compressive and tensile stresses. The position of critical stress points on a disc spring cross section are shown in Fig. 9.

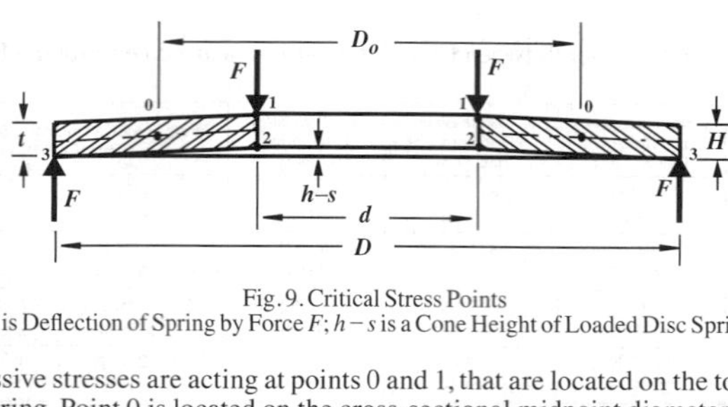


Fig. 9. Critical Stress Points
$s$ is Deflection of Spring by Force $F$; $h - s$ is a Cone Height of Loaded Disc Spring.

Compressive stresses are acting at points 0 and 1, that are located on the top surface of the disc spring. Point 0 is located on the cross-sectional midpoint diameter, and point 1 is located on the top inside diameter. Tensile stresses are acting at points 2 and 3, which are located on the bottom surface of the disc spring. Point 2 is on the bottom inside diameter, and point 3 is on the bottom outside diameter. The following equations are used to calculate stresses. The minus sign "–" indicates that compressive stresses are acting in a direction opposite to the tensile stresses.

Point 0: $$\sigma_0 = -\frac{3}{\pi} \cdot \frac{4E \cdot t \cdot s \cdot K_4}{(1-\mu^2) \cdot K_1 \cdot D_a^2} \qquad (14)$$

Point 1: $$\sigma_1 = \frac{4E \cdot K_4 \cdot s \cdot \left[K_4 \cdot K_2 \cdot \left(h - \frac{s}{2}\right) + K_3 \cdot t\right]}{(1-\mu^2) \cdot K_1 \cdot D_a^2} \qquad (15)$$

Point 2: $$\sigma_2 = \frac{4E \cdot K_4 \cdot s \cdot \left[K_3 \cdot t - K_2 \cdot K_4 \cdot \left(h - \frac{s}{2}\right)\right]}{(1-\mu^2) \cdot K_1 \cdot D_a^2} \qquad (16)$$

Point 3: $$\sigma_3 = \frac{4E \cdot K_4 \cdot s \cdot \left[K_4 \cdot (2K_3 - K_2) \cdot \left(h - \frac{s}{2}\right)\right] + K_3 \cdot t}{(1-\mu^2) \cdot K_1 \cdot D_a^2 \cdot \delta} \qquad (17)$$

$K_2$ and $K_3$ are disc spring dimensional constants, defined as follows:

$$K_2 = \frac{6\left(\frac{\delta - 1}{\ln\delta} - 1\right)}{\pi \cdot \ln\delta} \qquad (18) \qquad\qquad K_3 = \frac{3 \cdot (\delta - 1)}{\pi \cdot \ln\delta} \qquad (19)$$

where $\delta = D/d$ is the outside-to-inside diameter ratio.

In static application, if disc springs are fully flattened (100 percent deflection), compressive stress at point 0 should not exceed the tensile strength of disc spring materials. For most spring steels, the permissible value is $\sigma_0 \leq 1600$ N/mm² or 232,000 psi.

In dynamic applications, certain limitations on tensile stress values are recommended to obtain controlled fatigue life of disc springs utilized in various stacking. Maximum tensile stresses at points 2 and 3 depend on the *Group* number of the disc springs. Stresses $\sigma_2$ and $\sigma_3$ should not exceed the following values:

| | Group 1 | Group 2 | Group 3 |
|---|---|---|---|
| Maximum allowable tensile stresses at points 2 and 3 | 1300 N/mm² (188,000 psi) | 1250 N/mm² (181,000 psi) | 1200 N/mm² (174,000 psi) |

**Fatigue Life of Disc Springs.—**Fatigue life is measured in terms of the maximum number of cycles that dynamically loaded disc springs can sustain prior to failure. Dynamically loaded disc springs are divided into two groups: disc springs with unlimited fatigue life, which exceeds $2 \times 10^6$ cycles without failure, and disc springs with limited fatigue life between $10^4$ cycles and less than $2 \times 10^6$ cycles.

Typically, fatigue life is estimated from three diagrams, each representing one of the three Groups of disc springs (Fig. 10, Fig. 11, and Fig. 12). Fatigue life is found at the intersection of the vertical line representing minimum tensile stress $\sigma_{min}$ with the horizontal line, which represents maximum tensile stress $\sigma_{max}$. The point of intersection of these two lines defines fatigue life expressed in number of cycles $N$ that can be sustained prior to failure.

*Example:* For *Group 2* springs in Fig. 11, the intersection point of the $\sigma_{min} = 500$ N/mm² line with the $\sigma_{max} = 1200$ N/mm² line, is located on the $N = 10^5$ cycles line. The estimated fatigue life is $10^5$ cycles.

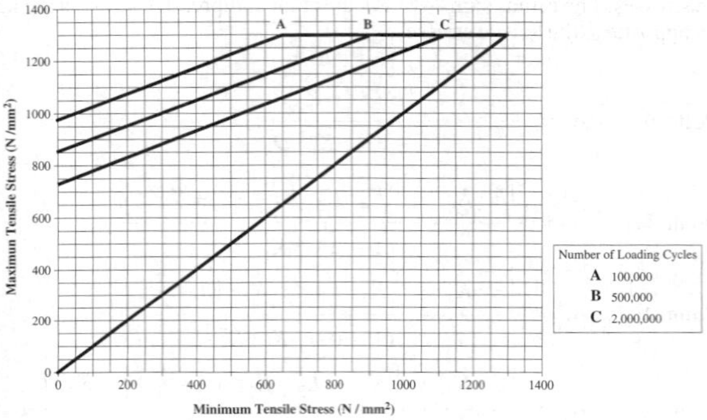


Fig. 10. *Group 1* Diagram for Estimating Fatigue Life of Disc Springs ($0.2 \leq t < 1.25$ mm)

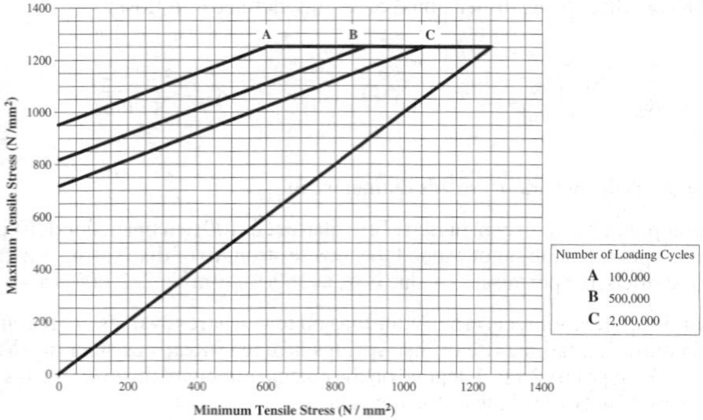


Fig. 11. *Group 2* Diagram for Estimating Fatigue Life of Disc Springs ($1.25 \leq t \leq 6$ mm)

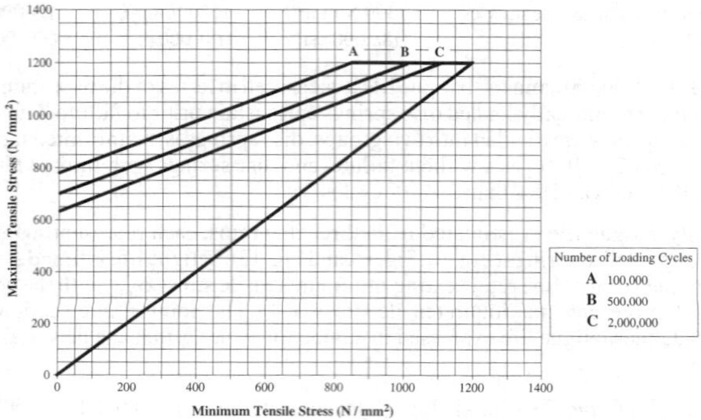


Fig. 12. *Group 3* Diagram for Estimating Fatigue Life of Disc Springs ($6 < t \leq 16$ mm)

When the intersection points of the minimum and maximum stress lines fall inside the areas of each cycle line, only the approximate fatigue life can be estimated by extrapolating the distance from the point of intersection to the nearest cycle line. The extrapolation cannot provide accurate values of fatigue life, because the distance between the cycle lines is expressed in logarithmic scale, and the distance between tensile strength values is expressed in linear scale (Figs. 10, 11, and 12), therefore linear-to-logarithmic scales ratio is not applicable.

When intersection points of minimum and maximum stress lines fall outside the cycle lines area, especially outside the $N = 10^5$ cycles line, the fatigue life cannot be estimated.

Thus, the use of the fatigue life diagrams should be limited to such cases when the minimum and maximum tensile stress lines intersect exactly with each of the cycle lines.

To calculate fatigue life of disc springs without the diagrams, the following equations developed by the author can be used.

Disc Springs in Group 1 $$N = 10^{10.29085532 - 0.00542096\,(\sigma_{max} - 0.5\sigma_{min})} \quad (20)$$

Disc Springs in Group 2 $$N = 10^{10.10734911 - 0.00537616\,(\sigma_{max} - 0.5\sigma_{min})} \quad (21)$$

Disc Springs in Group 3 $$N = 10^{13.23985664 - 0.01084192\,(\sigma_{max} - 0.5\sigma_{min})} \quad (22)$$

As can be seen from Equations (20), (21), and (22), the maximum and minimum tensile stress range affects the fatigue life of disc springs. Since tensile stresses at Points 2 and 3 have different values, see Equations (16) and (17), it is necessary to determine at which critical point the minimum and maximum stresses should be used for calculating fatigue life. The general method is based on the diagram, Fig. 9 from which Point 2 or Point 3 can be found in relationship with disc spring outside-to-inside diameters ratio $D/d$ and disc spring cone height-to-thickness ratio $h/r$. This method requires intermediate calculations of $D/d$ and $h/t$ ratios and is applicable only to disc springs without contact surfaces. The method is not valid for *Group 3* disc springs or for disc springs in *Group 2* that have contact surfaces and reduced thickness.

A simple and accurate method, that is valid for all disc springs, is based on the following statements:

if $(\sigma_{2\,max} - 0.5\,\sigma_{2\,min}) > (\sigma_{3\,max} - 0.5\,\sigma_{3\,min})$, then Point 2 is used, otherwise
if $(\sigma_{3\,max} - 0.5\,\sigma_{3\,min}) > (\sigma_{2\,max} - 0.5\,\sigma_{2\,min})$, then Point 3 is used

The maximum and minimum tensile stress range for disc springs in *Groups 1, 2*, and *3* is found from the following equations.

For disc springs in *Group 1*:

$$\sigma_{max} - 0.5\sigma_{min} = \frac{10.29085532 - \log N}{0.00542096} \quad (23)$$

For disc springs in *Group 2*:

$$\sigma_{max} - 0.5\sigma_{min} = \frac{10.10734911 - \log N}{0.00537616} \quad (24)$$

For disc springs in *Group 3*:

$$\sigma_{max} - 0.5\sigma_{min} = \frac{13.23985664 - \log N}{0.01084192} \quad (25)$$

Thus, Equations (23), (24), and (25) can be used to design any spring stack that provides required fatigue life. The following example illustrates how a maximum-minimum stress range is calculated in relationship with fatigue life of a given disc spring stack.

*Example:* A dynamically loaded stack, which utilizes disc springs in *Group 2*, must have the fatigue life of $5 \times 10^5$ cycles. The maximum allowable tensile stress at Points 2 or 3 is 1250 N/mm². Find the minimum tensile stress value to sustain $N = 5 \times 10^5$ cycles.

*Solution:* Substitution of $\sigma_{max} = 1250$ and $N = 5 \times 10^5$ in Equation (24) gives:

$$1250 - 0.5\sigma_{min} = \frac{10.10734911 - \log(5 \times 10^5)}{0.00537616} = \frac{10.10734911 - 5.69897}{0.00537616} = 820$$

from which $\sigma_{min} = \dfrac{1250 - 820}{0.5} = 860$ N/mm² (124,700 psi)

**Recommended Dimensional Characteristics of Disc Springs.—**Dimensions of disc springs play a very important role in their performance. It is imperative to check selected disc springs for dimensional ratios, that should fall within the following ranges:

1) Diameters ratio, $\delta = D/d = 1.7$ to 2.5.

2) Cone height-to-thickness ratio, $h/t = 0.4$ to 1.3.

3) Outside diameter-to-thickness ratio, $D/t = 18$ to 40.

Small values of $\delta$ correspond with small values of the other two ratios. The $h/t$ ratio determines the shape of force-deflection characteristic graphs, that may be nearly linear or strongly curved. If $h/t = 0.4$ the graph is almost linear during deflection of a disc spring up to its flat position. If $h/t = 1.6$ the graph is strongly curved and its maximum point is at 75 percent deflection. Disc spring deflection from 75 to 100 percent slightly reduces spring force. Within the $h/t = 0.4 - 1.3$ range, disc spring forces increase with the increase in deflection and reach maximum values at 100 percent deflection. In a stack of disc springs with a ratio $h/t > 1.3$ deflection of individual springs may be unequal, and only one disc spring should be used if possible.

## Example Applications of Disc Springs

*Example 1, Disc Springs in Group 2 (no contact surfaces):* A mechanical device that works under dynamic loads must sustain a minimum of 1,000,000 cycles. The applied load varies from its minimum to maximum value every 30 seconds. The maximum load is approximately 20,000N (4,500 lbf). A 40-mm diameter guide rod is a receptacle for the disc springs. The rod is located inside a hollow cylinder. Deflection of the disc springs under minimum load should not exceed 5.5 mm (0.217 inch) including a 20 percent preload deflection. Under maximum load, the deflection is limited to 8 mm (0.315 inch) maximum. Available space for the disc spring stack inside the cylinder is 35 to 40 mm (1.38 to 1.57 inch) in length and 80 to 85 mm (3.15 to 3.54 inch) in diameter.

Select the disc spring catalog item, determine the number of springs in the stack, the spring forces, the stresses at minimum and maximum deflection, and actual disc spring fatigue life.

*Solution:* 1) Disc spring standard inside diameter is 41 mm (1.61 inch) to fit the guide rod. The outside standard diameter is 80 mm (3.15 in) to fit the cylinder inside diameter. Disc springs with such diameters are available in various thickness: 2.25, 3.0, 4.0, and 5.0 mm (0.089, 0.118, 0.157, and 0.197 inch). The 2.25- and 3.0-mm thick springs do not fit the applied loads since the maximum force values for disc springs with such thickness are 7,200N and 13,400N (1,600 lbf and 3,000 lbf) respectively. A 5.0-mm thick disc spring should not be used because its $D/t$ ratio, $80/5 = 16$, is less than 18 and is considered as unfavorable. Disc spring selection is narrowed to an 80-41-4 catalog item.

2) Checking 80 - 41 - 4 disc spring for dimensional ratios:

$$\delta = D/d = 80/41 = 1.95 \qquad h/t = 2.2/4 = 0.55 \qquad D/t = 80/4 = 20$$

Because the dimensional ratios are favorable, the 80-41-4 disc springs are selected.

3) The number of springs in the stack is found from Equation (1):

$$n = L_0 / (t + h) = 40/(4 + 2.2) = 40/6.2 = 6.45.$$

Rounding $n$ to the nearest integer gives $n = 6$. The actual length of unloaded spring stack is $L_0 = 6.2 \times 6 = 37.2$ mm (1.465 inch) and it satisfies the $L_0 < 40$ mm condition.

4) Calculating the cone angle $\alpha$ from Equation (7) and actual outside diameter $D_a$ from Equation (9) gives:

$$\alpha = \text{atan}\left(\frac{2 \times 2.2}{80 - 41}\right) = \text{atan}\,(0.11282) = 6.4°$$

$$D_a = 80 - 2 \times 4\left(\sin[\,\text{atan}(0.11282)] + \frac{1}{6}\cos[\text{atan}\,(0.11282)]\right)$$

$$D_a = 77.78 \text{ mm} \ \ (3.062 \text{ in})$$

5) Calculating constant $K_1$ from Equation (11):

$$\delta = \frac{D}{d} = 1.95122$$

$$K_1 = \frac{\left(\frac{1.95122 - 1}{1.95122}\right)^2}{\pi \cdot \left[\frac{1.95122 + 1}{1.95122 - 1} - \frac{2}{\ln(1.95122)}\right]} = 0.6841$$

6) Calculating minimum and maximum forces, $F_{min}$ and $F_{max}$ from Equation (10):

Based on the design requirements, the disc spring stack is deflecting by 5.5 mm (0.217 in) under minimum load, and each individual disc spring is deflecting by $5.5 / 6 \cong 0.92$ mm (0.036 in). A single disc spring deflection $s_{min} = 0.9$ mm (0.035 in) is used to calculate $F_{min}$. Under maximum load, the disc spring stack is permitted maximum deflection of 8 mm (0.315 in), and each individual disc spring deflects by $8 / 6 \cong 1.33$ mm (0.0524 in). A disc spring deflection $s_{max} = 1.32$ mm (0.052 in) will be used to calculate $F_{max}$. If disc springs are made of AISI 6150 alloy steel, then modulus of elasticity $E = 206{,}000$ N/mm$^2$ ($30 \times 10^6$ psi) and Poisson's ratio $\mu = 0.3$.

$$F_{min} = \frac{4 \cdot 206000}{(1 - 0.3^2)(0.6841)(77.78)^2}\left[\left(2.2 - \frac{0.9}{2}\right) \cdot (2.2 - 0.9) \cdot 4 + 4^3\right]0.9$$

$$F_{min} = 14390N \ (3235 \text{ lbf})$$

$$F_{max} = \frac{4 \cdot 206000}{(1 - 0.3^2)(0.6841)(77.78)^2}\left[\left(2.2 - \frac{1.32}{2}\right) \cdot (2.2 - 1.32) \cdot 4 + 4^3\right]1.32$$

$$F_{max} = 20050N \ (4510 \text{ lbf})$$

7) Calculating constant $K_2$, Equation (18):

$$\delta = \frac{D}{d} = \frac{80}{41} = 1.95122$$

$$K_2 = \frac{6\left(\frac{\delta - 1}{\ln\delta} - 1\right)}{\pi \cdot \ln\delta} = \frac{6\left(\frac{1.95122 - 1}{\ln(1.95122)} - 1\right)}{\pi \cdot \ln(1.95122)} = 1.2086$$

8) Calculating constant $K_3$ (Equation (19)):

$$K_3 = \frac{3 \cdot (\delta - 1)}{\pi \cdot \ln\delta} = \frac{3 \cdot (1.95122 - 1)}{\pi \cdot \ln(1.95122)} = 1.3589$$

9) Compressive stress $\sigma_0$ at point 0 due to maximum deflection, Equation (14):

$$\sigma_0 = -\frac{3}{\pi} \cdot \frac{4E \cdot t \cdot s \cdot K_4}{(1-\mu^2) \cdot K_1 \cdot D_a^2} = -\frac{3}{\pi} \cdot \frac{4 \cdot 206000 \cdot 4 \cdot 1.32 \cdot 1}{(1-0.3^2) \cdot 0.6841 \cdot 77.78^2}$$

$$\sigma_0 = 1103 \text{ N/mm}^2 = 160000 \text{ psi}$$

Because the compressive stress at point 0 does not exceed 1600 N/mm$^2$, its current value satisfies the design requirement.

10) Tensile stress $\sigma_2$ at point 2 due to minimum deflection $s = 0.9$ mm, Equation (16):

$$\sigma_{2min} = \frac{4E \cdot K_4 \cdot s \cdot \left[ K_3 \cdot t - K_2 \cdot K_4 \cdot \left(h - \frac{s}{2}\right)\right]}{(1-\mu^2) \cdot K_1 \cdot D_a^2} =$$

$$\frac{4 \cdot 206000 \cdot 1 \cdot 0.9 \left[ 1.3589 \cdot 4 - 1.2086 \cdot 1 \cdot \left(2.2 - \frac{0.9}{2}\right)\right]}{(1-0.3^2) \cdot 0.6841 \cdot 77.78^2} = 654\, N/\text{mm}^2$$

11) Tensile stress $\sigma_2$ at point 2 due to maximum deflection $s = 1.32$ mm, Equation (16):

$$\sigma_{2max} = \frac{4E \cdot K_4 \cdot s \cdot \left[ K_3 \cdot t - K_2 \cdot K_4 \cdot \left(h - \frac{s}{2}\right)\right]}{(1-\mu^2) \cdot K_1 \cdot D_a^2} =$$

$$\frac{4 \cdot 206000 \cdot 1 \cdot 1.32 \cdot \left[ 1.3589 \cdot 4 - 1.2086) \cdot 1 \cdot \left(2.2 - \frac{1.32}{2}\right)\right]}{(1-0.3^2) \cdot 0.6841 \cdot 77.78^2} = 1032 N/\text{mm}^2$$

Thus, $\sigma_{2\,min} = 654$ N/mm$^2$ (94,850 psi) and $\sigma_{2\,max} = 1032$ N/mm$^2$ (149,700 psi).

12) Tensile stress $\sigma_3$ at point 3 due to minimum deflection $s = 0.9$ mm, Equation (17):

$$\sigma_{3min} = \frac{4E \cdot K_4 \cdot s \cdot \left[ K_4 \cdot (2K_3 - K_2) \cdot \left(h - \frac{s}{2}\right) + K_3 \cdot t\right]}{(1-\mu^2) \cdot K_1 \cdot D_a^2 \cdot \delta} =$$

$$\frac{4 \cdot 206000 \cdot 1 \cdot 0.9 \cdot \left[1 \cdot (2 \cdot 1.3589 - 1.2086) \cdot \left(2.2 - \frac{0.9}{2}\right) + 1.3589 \cdot 4\right]}{(1-0.3^2) \cdot 0.6841 \cdot 77.78^2 \cdot 1.95122} = 815\, N/\text{mm}^2$$

13) Tensile stress $\sigma_3$ at point 3 due to maximum deflection $s = 1.32$ mm, Equation (17):

$$\sigma_{3max} = \frac{4E \cdot K_4 \cdot s \cdot \left[ K_4 \cdot (2K_3 - K_2) \cdot \left(h - \frac{s}{2}\right) + K_3 \cdot t\right]}{(1-\mu^2) \cdot K_1 \cdot D_a^2 \cdot \delta} =$$

$$\frac{4 \cdot 206000 \cdot 1 \cdot 1.32 \cdot \left[1 \cdot (2 \cdot 1.3589 - 1.2086) \cdot \left(2.2 - \frac{1.32}{2}\right) + 1.3589 \cdot 4\right]}{(1-0.3^2) \cdot 0.6841 \cdot 77.78^2 \cdot 1.95122} = 1149 N/\text{mm}^2$$

Thus, $\sigma_{3\,min} = 815$ N/mm$^2$ (118,200 psi) and $\sigma_{3\,max} = 1149$ N/mm$^2$ (166,600 psi).

14) Functional tensile stress range at critical points 2 and 3.

Point 2: $\sigma_{2\,max} - 0.5\sigma_{2\,min} = 1032 - 0.5 \times 654 = 705$ N/mm$^2$

Point 3: $\sigma_{3\,max} - 0.5\sigma_{3\,min} = 1149 - 0.5 \times 815 = 741.5$ N/mm$^2$

Because $\sigma_{3\,max} - 0.5\sigma_{3\,min} > \sigma_{2\,max} - 0.5\,\sigma_{2\,min}$, the tensile stresses at point 3 are used for fatigue life calculations.

15) Fatigue life of selected disc springs, Equation (21):

$N = 10^{[10.10734911 - 0.00537616(1149 - 0.5 \times 815)]} = 10^{10.10734911 - 3.98642264} = 10^{6.12092647}$

$N = 1{,}321{,}000$ cycles. Thus, the calculated actual fatigue life exceeds required minimum number of cycles by 31 percent.

In conclusion, the six 80-41-4 disc springs arranged in series stacking, satisfy the requirements and will provide a 32 percent longer fatigue life than required by the design criteria.

*Example 2:* A company wishes to use *Group 3* disc springs with contact surfaces on couplings to absorb bumping impacts between railway cars.

*Given:*

$D$ = 200 mm, disc spring outside diameter
$d$ = 102 mm, disc spring inside diameter
$t$ = 14 mm, spring standard thickness
$t'$ = 13.1 mm, spring reduced thickness
$h$ = 4.2 mm, cone height of unloaded spring
$n$ = 22, number of springs in series stacking
$S_i$ = 33.9 mm, initial deflection of the pack
$S_a$ = 36.0 mm, additional deflection of the pack

Find the fatigue life in cycles and determine if the selected springs are suitable for the application.

The calculations are performed in the following sequence:

1) Determine the minimum $s_{min}$ and maximum $s_{max}$ deflections of a single disc spring:

$$s_{\max} = \frac{(S_i + S_a)}{n} = \frac{(33.9 + 36)}{22} = 3.18 \text{ mm}$$

$$s_{\min} = \frac{S_i}{n} = \frac{33.9}{22} = 1.54 \text{ mm}$$

2) Use Equations (16) and (17) to calculate tensile stresses $\sigma_2$ and $\sigma_3$ at $s_{min}$ and $s_{max}$ deflections:

$\sigma_{2min} = 674$ N/mm$^2$, $\sigma_{2max} = 1513$ N/mm$^2$, $\sigma_{3min} = 707$ N/mm$^2$, $\sigma_{3max} = 1379$ N/mm$^2$

3) Determine critical stress points:

$\sigma_{2max} - 0.5\sigma_{2min} = 1513 - 0.5 \times 674 = 1176$ N/mm$^2$

$\sigma_{3max} - 0.5\sigma_{3min} = 1379 - 0.5 \times 707 = 1025.5$ N/mm$^2$

Because $(\sigma_{2max} - 0.5\sigma_{2min}) > (\sigma_{3max} - 0.5\sigma_{3min})$, then tensile stresses at point 2 are used to calculate fatigue life.

4) Fatigue life $N$ is calculated using Equation (22):

$N = 10^{[13.23985664 - (0.01084192 \times 1176)]} = 10^{0.49} = 3$ cycles

The selected disc springs at the above-mentioned minimum and maximum deflection values will not sustain any number of cycles. It is imperative to check the selected disc springs for dimensional ratios:

Outside-to-inside diameters ratio, 200/102 = 1.96; within recommended range.

Cone height-to-thickness ratio is 4.2/13.1 = 0.3; out of range, the minimum ratio is 0.4.

Outside diameter-to-thickness ratio is 200/13.1 = 15; out of range, the minimum ratio is 18. Thus, only one of the dimensional ratios satisfies the requirements for the best disc spring performance.

TABLE OF CONTENTS

# PROPERTIES, TREATMENT, AND TESTING OF MATERIALS

## THE ELEMENTS, ENERGY, HEAT, MASS, AND WEIGHT



## PROPERTIES OF WOOD, CERAMICS, PLASTICS, METALS



## STANDARD STEELS



## STANDARD STEELS

*(Continued)*



## TOOL STEELS



## HARDENING, TEMPERING, AND ANNEALING

TABLE OF CONTENTS

PROPERTIES, TREATMENT, AND TESTING OF MATERIALS

## HARDENING, TEMPERING, AND ANNEALING

*(Continued)*



## HARDENING, TEMPERING, AND ANNEALING

*(Continued)*



## NONFERROUS ALLOYS



PROP. OF MATERIALS

## TABLE OF CONTENTS

## PROPERTIES, TREATMENT, AND TESTING OF MATERIALS

### NONFERROUS ALLOYS

*(Continued)*



### CORROSION



### PLASTICS



### PLASTICS

*(Continued)*

## THE ELEMENTS, ENERGY, HEAT, MASS, AND WEIGHT

### Overview of Materials

Materials used in manufacturing, machining, and construction are comprised of the elements indicated in Table 1. These are made up of individual electrons, neutrons, and protons, which are comprised of subatomic particles. While most solids appear stable to our everyday senses, they are active at the atomic level, with kinetic energy of constitutive elements defining the heat of the material.

Many basic material properties are determined by the proportion and properties of their constitutive elements. Other material properties, such as modulus, strength, and thermal conductivity, are governed by both the constitutive elements and type of bonding between those elements. Materials with strong bonds between shared electrons (such as diamond) have high strength, modulus, thermal conductivity, and melting point. Materials with weak bonds (such as wax) have a low modulus, so are easily deformed and melted.

While some materials (lead, gold, and silver) are quite pure, most materials used in machinery and machine parts are purposefully designed mixtures: alloys, blends, and composites. *Alloys*, which include at least one metallic element—for example, brass contains a combination of copper and zinc—are designed for properties such as strength and corrosion resistance. *Blends* refer more broadly to mixed materials that may not be metallic (like polymers). *Composites* refer even more broadly to materials comprised of dissimilar but definable constituent materials; the resulting combined substances may be highly structured (such as laminated sheets with oriented fibers) or randomly ordered (such as concrete or glass-filled polymers).

### Material and Energy Consumption

Of the 109 elements listed in Table 1, 94 occur naturally on Earth. The prevalence and difficulty in processing constitutive elements in a material largely determines the material's cost and sustainability. Table 2 lists the most used materials with notes about each type and its properties with values given for representative comparison. As uses vary widely, more detailed data on grade-specific properties should be sought for best material choice.

Per Table 2, wood is the most consumed material by volume, while concrete is the most consumed material by mass. Both are sourced from highly prevalent elements and require relatively low energy per unit of mass or volume for primary (first-time) production. The "embodied energy" represents this energy (measured in joules; MJ=1 million joules) on a volume or mass basis; it is a significant determinant in baseline material pricing and the cost of part production.

Designers and manufacturers should strive to minimize size and mass of products, to reduce not only up-front costs and embodied energy of consumed material, but also downstream economic and environmental costs related to manufacture and transportation. Choice of manufacturing methods (e.g., CNC subtractive/extractive manufacturing, molding/forming processes, and additive manufacturing processes) drastically affect material and energy usage. Recycling and using recycled materials can help offset material and energy costs.

### Sustainability Considerations for Materials and Processes

Society is increasingly aware of the critical need for sustainable solutions across all aspects of modern life, with material selection and sustainable processes influencing manufacturing and machining decisions. In addition to consumer awareness driving growth in demand for eco-friendly products, investors increasingly are seeking sustainably-focused companies. Globally, manufacturers are having to account for sustainability impacts of products per environmental, social, and governance (ESG) standards, with best practices influenced by regulations and related incentives or penalties.

## Table 1. The Elements—Symbols, Atomic Numbers and Weights, Melting Points

| Name of Element | Sym-bol | Atomic | | Melting Point, °C | Name of Element | Sym-bol | Atomic | | Melting Point, °C |
|---|---|---|---|---|---|---|---|---|---|
| | | Num. | Weight | | | | Num. | Weight | |
| Actinium | Ac | 89 | (227) | 1050 | Mendelevium | Md | 101 | (258) | 827 |
| Aluminum | Al | 13 | 26.9815 | 660.37 | Mercury | Hg | 80 | 200.59 | −38.83 |
| Americium | Am | 95 | (243) | 1176 | Molybdenum | Mo | 42 | 95.95 | 2623 |
| Antimony | Sb | 51 | 121.76 | 630.63 | Moscovium | Mc | 115 | 290 | … |
| Argon | Ar | 18 | 39.95 | −189.3 | Neodymium | Nd | 60 | 144.24 | 1024 |
| Arsenic | As | 33 | 74.922 | 817[a] | Neon | Ne | 10 | 20.180 | −248.59 |
| Astatine | At | 85 | (210) | 302 | Neptunium | Np | 93 | (237) | 637 |
| Barium | Ba | 56 | 137.33 | 727 | Nickel | Ni | 28 | 58.693 | 1455 |
| Berkelium | Bk | 97 | (247) | 986 | Nihonium | Nh | 113 | (286) | … |
| Beryllium | Be | 4 | 9.0122 | 1287 | Niobium | Nb | 41 | 92.906 | 2477 |
| Bismuth | Bi | 83 | 208.980 | 271.3 | Nitrogen | N | 7 | 14.007 | −210.1 |
| Bohrium | Bh | 107 | (270) | … | Nobelium | No | 102 | (259) | 827 |
| Boron | B | 5 | 10.81 | 2076 | Oganesson | Og | 118 | (294) | … |
| Bromine | Br | 35 | 79.904 | −7.3 | Osmium | Os | 76 | 190.23 | 3033 |
| Cadmium | Cd | 48 | 112.41 | 321.07 | Oxygen | O | 8 | 15.999 | −218.3 |
| Calcium | Ca | 20 | 40.078 | 842 | Palladium | Pd | 46 | 106.42 | 1554 |
| Californium | Cf | 98 | (251) | 900 | Phosphorus | P | 15 | 30.974 | 44.2 |
| Carbon | C | 6 | 12.011 | 3500 | Platinum | Pt | 78 | 195.08 | 1768.3 |
| Cerium | Ce | 58 | 140.12 | 795 | Plutonium | Pu | 94 | (244) | 639.4 |
| Caesium | Cs | 55 | 132.91 | 28.44 | Polonium | Po | 84 | (209) | 254 |
| Chlorine | Cl | 17 | 35.45 | −101.5 | Potassium | K | 19 | 39.098 | 63.38 |
| Chromium | Cr | 24 | 51.996 | 1907 | Praseodymium | Pr | 59 | 140.91 | 935 |
| Cobalt | Co | 27 | 58.933 | 1495 | Promethium | Pm | 61 | (145) | 1100 |
| Copernicium | Cn | 112 | (285) | … | Protactinium | Pa | 91 | 231.04 | 1568 |
| Copper | Cu | 29 | 63.546 | 1084.62 | Radium | Ra | 88 | (226) | 700 |
| Curium | Cm | 96 | (247) | 1340 | Radon | Rn | 86 | (222) | −71 |
| Darmstadtium | Ds | 110 | (281) | … | Rhenium | Re | 75 | 186.21 | 3186 |
| Dubnium | Db | 105 | (268) | … | Rhodium | Rh | 45 | 102.91 | 1964 |
| Dysprosium | Dy | 66 | 162.5 | 1407 | Roentgenium | Rg | 111 | (282) | … |
| Einsteinium | Es | 99 | (252) | 860 | Rubidium | Rb | 37 | 85.468 | 39.31 |
| Erbium | Er | 68 | 167.26 | 1529 | Ruthenium | Ru | 44 | 101.07 | 2334 |
| Europium | Eu | 63 | 151.96 | 826 | Rutherfordium | Rf | 104 | (267) | … |
| Fermium | Fm | 100 | (257) | 1527 | Samarium | Sm | 62 | 150.36 | 1072 |
| Flerovium | Fl | 114 | (290) | … | Scandium | Sc | 21 | 44.956 | 1541 |
| Fluorine | F | 9 | 18.998 | −219.62 | Seaborgium | Sg | 106 | (269) | … |
| Francium | Fr | 87 | (223) | 30[c] | Selenium | Se | 34 | 78.971 | 221 |
| Gadolinium | Gd | 64 | 157.25 | 1312 | Silicon | Si | 14 | 28.085 | 1414 |
| Gallium | Ga | 31 | 69.723 | 29.76 | Silver | Ag | 47 | 107.87 | 961.78 |
| Germanium | Ge | 32 | 72.630 | 938.3 | Sodium | Na | 11 | 22.990 | 97.72 |
| Gold | Au | 79 | 196.97 | 1064.18 | Strontium | Sr | 38 | 87.62 | 777 |
| Hafnium | Hf | 72 | 178.49 | 2233 | Sulfur | S | 16 | 32.06 | 115.21 |
| Hassium | Hs | 108 | (269) | … | Tantalum | Ta | 73 | 180.95 | 3017 |
| Helium | He | 2 | 4.00260 | −272.2[c] | Technetium | Tc | 43 | (97) | 2157 |
| Holmium | Ho | 67 | 164.930 | 1461 | Tellurium | Te | 52 | 127.60 | 449.51 |
| Hydrogen | H | 1 | 1.0080 | −259.14 | Tennessine | Ts | 117 | (294) | … |
| Indium | In | 49 | 114.82 | 156.6 | Terbium | Tb | 65 | 158.93 | 1356 |
| Iodine | I | 53 | 126.90 | 113.7 | Thallium | Tl | 81 | 204.38 | 304 |
| Iridium | Ir | 77 | 192.22 | 2446 | Thorium | Th | 90 | 232.04 | 1750 |
| Iron | Fe | 26 | 55.845 | 1538 | Thulium | Tm | 69 | 168.934 | 1545±15 |
| Krypton | Kr | 36 | 83.798 | −157.36 | Tin | Sn | 50 | 118.71 | 231.93 |
| Lanthanum | La | 57 | 138.91 | 920 | Titanium | Ti | 22 | 47.867 | 1668 |
| Lawrencium | Lr | 103 | (262) | 1627 | Tungsten | W | 74 | 183.84 | 3422 |
| Lead | Pb | 82 | 207.2 | 327.46 | Uranium | U | 92 | 238.03 | 1132.2 |
| Lithium | Li | 3 | 6.94 | 180.54 | Vanadium | V | 23 | 50.942 | 1910 |
| Livermorium | Lv | 116 | (293) | 427 | Xenon | Xe | 54 | 131.29 | −111.7 |
| Lutetium | Lu | 71 | 174.97 | 1652 | Ytterbium | Yb | 70 | 173.05 | 824 |
| Magnesium | Mg | 12 | 24.305 | 650 | Yttrium | Y | 39 | 88.906 | 1526 |
| Manganese | Mn | 25 | 54.9380 | 1246 | Zinc | Zn | 30 | 65.38 | 419.53 |
| Meitnerium | Mt | 109 | (277) | … | Zirconium | Zr | 40 | 91.224 | 1855 |

[a] Under normal pressure.
[b] Estimated melting point.
[c] At 26 atm (2.635 MPa).

*Notes:* Values in parentheses are atomic weights of the most stable known isotopes. Melting points are at standard pressure except as noted. The melting points are based on *Macmillan's Chemical and Physical Data*, by Arthur M. James and Mary P. Lord (Pan Macmillan, 1992) and *Tables of Physical and Chemical Constants*, by G.W.C. Kaye and T.H. Laby (1911; 2019). According to other published data, melting points may vary slightly based on materials and conditions.

## Table 2. Overview of Common Materials

| Material | Material Type | Volume, tonnes/ Year | Volume, $m^3$/ Year | Density, $kg/m^3$ | Representative Modulus, GPa | Representative Strength, MPa | Thermal Conductivity, W/m-K | Embodied Energy, $MJ/m^3$ | Embodied Energy, MJ/kg | Strength: Weight $Pa/(kg/m^3)$ | Stiffness: Weight $Pa/(kg/m^3)$ | Notes |
|---|---|---|---|---|---|---|---|---|---|---|---|---|
| ABS | Plastics | 1.14E+07 | 1.12E+07 | 1.02E+03 | 2.30 | 45.00 | 0.16 | 275400 | 270 | 4.41E+04 | 2.25E+06 | Unreinforced, molded |
| Al Alloys | Metals | 6.50E+07 | 2.41E+07 | 2.70E+03 | 69.00 | 255.00 | 190.00 | 891000 | 330 | 9.44E+04 | 2.56E+07 | Al 6061-T6 |
| Asphalt | Ceramics | 1.00E+08 | 4.35E+07 | 2.30E+03 | 0.50 | 3.00 | 0.90 | 9660 | 4.20 | 1.30E+03 | 2.17E+05 | Mid-range aggregate |
| Brick | Ceramics | 1.20E+08 | 7.06E+07 | 1.70E+03 | 17.00 | 32.00 | 0.20 | 5950 | 3.50 | 1.88E+04 | 1.00E+07 | Commercial construction; compression |
| Carbon Fibers | Composites | 1.80E+05 | 1.38E+05 | 1.30E+03 | 238.00 | 3950.00 | 60.00 | 41600 | 32 | 3.04E+06 | 1.83E+08 | In fiber direction |
| Concrete | Ceramics | 4.40E+09 | 1.83E+09 | 2.40E+03 | 20.00 | 30.00 | 2.25 | 3100 | 1.29 | 1.25E+04 | 8.33E+06 | Normal concrete, compression |
| Cu Alloys | Metals | 2.10E+07 | 2.36E+06 | 8.90E+03 | 115.00 | 69.00 | 189.00 | 1958000 | 220 | 7.75E+03 | 1.29E+07 | C22000 bronze |
| Glass | Ceramics | 3.00E+08 | 1.20E+08 | 2.50E+03 | 26.00 | 212.00 | 0.80 | 70000 | 28 | 8.48E+04 | 1.04E+07 | Float glass; compression |
| Glass Fibers | Composites | 3.80E+06 | 1.52E+06 | 2.50E+03 | 79.00 | 1950.00 | 1.20 | 600000 | 240 | 7.80E+05 | 3.16E+07 | E-glass fiber |
| Gold | Metals | 3.30E+03 | 1.71E+02 | 1.93E+04 | 77.00 | 129.00 | 301.00 | 5983000000 | 310000 | 6.68E+03 | 3.99E+06 | Pure |
| Mg Alloys | Metals | 1.12E+06 | 6.44E+05 | 1.74E+03 | 45.00 | 130.00 | 159.00 | 486640 | 280 | 7.48E+04 | 2.59E+07 | AS21X cast magnesium |
| Natural Fibers | Composites | 2.50E+07 | 1.85E+07 | 1.35E+03 | 45.00 | 550.00 | 0.06 | 81000 | 60 | 4.07E+05 | 3.33E+07 | Hemp fiber |
| Ni Alloys | Metals | 2.80E+06 | 3.15E+05 | 8.90E+03 | 220.00 | 430.00 | 14.00 | 1602360 | 180 | 4.83E+04 | 2.47E+07 | Nichrome 70-30 |
| PA (Nylon) | Plastics | 5.50E+06 | 4.82E+06 | 1.14E+03 | 2.60 | 64.00 | 0.40 | 159600 | 140 | 5.61E+04 | 2.28E+06 | PA6, unreinforced, molded |
| Pb Alloys | Metals | 1.20E+07 | 1.06E+06 | 1.13E+04 | 14.00 | 14.00 | 35.00 | 225800 | 20 | 1.24E+03 | 1.24E+06 | Pure lead |
| PC | Plastics | 5.10E+06 | 4.25E+06 | 1.20E+03 | 2.40 | 62.00 | 0.22 | 228000 | 190 | 5.17E+04 | 2.00E+06 | Unreinforced, molded |
| PE | Plastics | 1.06E+08 | 1.10E+08 | 9.60E+02 | 0.97 | 25.80 | 0.35 | 124800 | 130 | 2.69E+04 | 1.01E+06 | Unreinforced HDPE, molded |
| PET | Plastics | 9.90E+07 | 7.17E+07 | 1.38E+03 | 3.20 | 49.00 | 0.26 | 115920 | 84 | 3.55E+04 | 2.32E+06 | Blow molding grade |
| Platinum | Metals | 1.90E+02 | 8.86E+00 | 2.15E+04 | 171.00 | 145.00 | 69.10 | 4075500000 | 190000 | 6.76E+03 | 7.97E+06 | Pure |
| PMMA | Plastics | 4.20E+06 | 3.56E+06 | 1.18E+03 | 2.90 | 60.00 | 0.20 | 206500 | 175 | 5.08E+04 | 2.46E+06 | Unreinforced, molded |
| PP | Plastics | 7.56E+07 | 8.35E+07 | 9.05E+02 | 1.68 | 29.20 | 0.20 | 143895 | 159 | 3.23E+04 | 1.86E+06 | Unreinforced, molded |
| PS | Plastics | 1.56E+07 | 1.53E+07 | 1.02E+03 | 0.78 | 41.00 | 0.17 | 158100 | 155 | 4.02E+04 | 7.65E+05 | Unreinforced, molded |
| PVC | Plastics | 6.00E+07 | 4.35E+07 | 1.38E+03 | 2.16 | 43.20 | 0.20 | 262200 | 190 | 3.13E+04 | 1.57E+06 | Unreinforced, molded |
| Rubber | Plastics | 2.50E+07 | 2.27E+07 | 1.10E+03 | 0.00 | 10.00 | 0.24 | 101200 | 92 | 9.09E+03 | 2.73E+03 | Nitrile rubber |
| Silver | Metals | 2.40E+04 | 2.29E+03 | 1.05E+04 | 76.00 | 140.00 | 419.00 | 15735000 | 1500 | 1.33E+04 | 7.24E+06 | Pure |
| Stainless Steel | Metals | 5.20E+07 | 6.57E+06 | 7.92E+03 | 193.00 | 505.00 | 16.20 | 1188000 | 150 | 6.38E+04 | 2.44E+07 | 304 stainless steel |
| Steel | Metals | 1.95E+09 | 2.49E+08 | 7.85E+03 | 200.00 | 500.00 | 12.40 | 431750 | 55 | 6.37E+04 | 2.55E+07 | A36 steel (construction) |
| Synthetic Fibers | Composites | 1.30E+08 | 1.04E+08 | 1.25E+03 | 2.20 | 65.00 | 0.20 | 162500 | 130 | 5.20E+04 | 1.76E+06 | Trayon fiber |
| Ti Alloys | Metals | 4.50E+06 | 1.00E+06 | 4.50E+03 | 114.00 | 880.00 | 6.70 | 3510000 | 780 | 1.96E+05 | 2.53E+07 | Ti 6Al-4V |
| Tungsten Alloys | Metals | 7.90E+04 | 4.10E+03 | 1.93E+04 | 276.00 | 605.00 | 23.00 | 11550000 | 600 | 3.14E+04 | 1.43E+07 | W-6Ni-4Cu alloy |
| Wood | Composites | 3.00E+09 | 5.00E+09 | 6.00E+02 | 8.50 | 18.00 | 0.14 | 9700 | 16.17 | 3.00E+04 | 1.42E+07 | White pine; bending |
| Zn Alloys | Metals | 1.60E+07 | 2.42E+06 | 6.60E+03 | 78.00 | 371.00 | 125.00 | 363000 | 55 | 5.62E+04 | 1.18E+07 | ZA27 die cast |

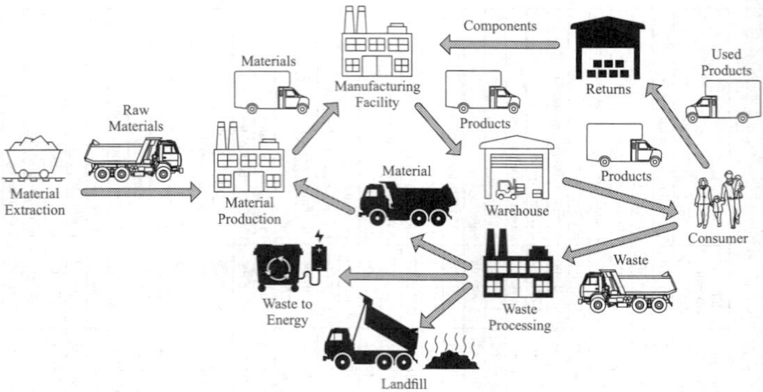


Fig. 1. Materials Lifecycle

The lifecycle of materials is depicted in Fig. 1. Raw materials are extracted or otherwise recovered from waste processing streams. Primary material production involves refinement of these raw materials, blending of alloys, and other treatments to provide feedstocks with known properties. Materials recovery from waste streams provides significant energy savings, often as much as 70 percent if recycled material sources are available.

Feedstocks are provided to manufacturing facilities where components are made and assembled into products. Significant additional energy is typically added during manufacture and assembly. Energy and cost may be reduced by minimizing material handling and shipping.

**Production Cost and Carbon Footprint.**—The amount of energy is related to the type of manufacturing process, properties of the processed material, and details of the machine design and operation. The minimum total embodied energy, E, for a component is:

$$E_{Component} = (E_{Material} = V_{Material} \times e_{Material}) + \sum (E_{Process} = P_{Process} \times t_{Process})$$

Where:

$E_{Material}$ is the embodied energy of the material

$V_{Material}$ is the volume of the material used per component

$e_{Material}$ is the embodied energy of the material as measured in MJ/m$^3$ per Table 2

$P_{process}$ is the average power of each manufacturing process

$t_{process}$ is the processing time for each manufacturing process.

While extraction and refining processes vary for different materials, material ecology research indicates a strong correlation between embodied energy and carbon footprint with a representative value of 0.5 kg $CO_2$ per MJ of embodied energy. The associated energy cost can also be estimated from the cost of energy given the economic value per kilowatt hour (kWh).

*Example 1:* Estimate the embodied energy for production of a 3/8-16 inch hex head bolt made from a 1.5 inch long, ¾ inch diameter rod of A36 steel. The bolt is produced using (1) a 75-ton cold forging press drawing 20 HP, and (2) a 50-ton threading machine drawing 12 HP, and then (3) annealed for 3 hours in an oven, drawing 10,000 BTUs per hour. The production rate is 1,000 parts per hour.

The volume of raw material, $V_{A36}$, used in the bolt is:

$$V_{A36} = \pi \frac{D^2}{4} L = 3.142 \frac{(0.75\ \text{in})^2}{4} 1.5\ in \times \left( \frac{1\ m}{39.37\ \text{in}} \right)^3 = 1.086 \times 10^{-5} m^3$$

Given that $e_{A36}$ is 431,750 MJ/m$^3$, the embodied energy of the consumed material is:

$$E_{A36} = V_{A36} \times e_{A36} = 1.086 \times 10^{-5}\text{m}^3 \times 431{,}750 \frac{\text{MJ}}{m^3} = 4.69\ \text{MJ}$$

The energy consumed by the three indicated production processes is:

$$E_{Cold\ Forging} = (20\ \text{HP} \times 746\ \text{W/HP}) \times \left( \frac{3{,}600\ \text{s/hr}}{1{,}000\ \text{bolts/hr}} \right) = 53{,}700\ \text{J}$$

$$E_{Threading} = (12\ \text{HP} \times 746\ \text{W/HP}) \times \left( \frac{3{,}600\ \text{s/hr}}{1{,}000\ \text{bolts/hr}} \right) = 32{,}200\ \text{J}$$

$$E_{Annealing} = (40{,}000\ \text{BTU/hr} \times 0.293\ \text{W/BTU/hr}) \times \left( \frac{3\ \text{hr/batch} \times 3{,}600\ \text{s/hr}}{3{,}000\ \text{bolts/batch}} \right) = 42{,}200\ \text{J}$$

The embodied energy for the component is then the sum of the input estimated energies:

$$E_{Hex\ Bolt} = 4.69\ \text{MJ} + 0.0537\ \text{MJ} + 0.0322\ \text{MJ} + 0.0422\ \text{MJ} = 4.82\ \text{MJ}$$

*Example 2:* Compute the carbon footprint and economic cost of the embodied energy for the $^3/_8$ inch-16 hex head bolt of the prior example.

Given the 4.82 MJ of embodied energy, the approximate carbon footprint is:

$$CO_{2\ Hex\ Screw} = E_{Component} \times 0.5 \frac{\text{kg}}{\text{MJ}} = 4.82\ \text{MJ} \times 0.05 \frac{\text{kg}}{\text{MJ}} = 2.4\ \text{kg}$$

Assuming an energy cost of 0.1 US dollar (USD) per kilowatt hour (kWh), the associated energy cost is:

$$Energy\ Cost_{Hex\ Screw} = 4.82\ \text{MJ} \times \frac{1\ \text{kWh}}{3.6\ \text{MJ}} \times 0.1 \frac{\text{USD}}{\text{kWh}} = 0.134\ \text{USD}$$

This energy cost sets a minimum theoretical cost for the hex head bolt. Additional costs include people, machinery, and other resources needed to design, manufacture, certify, and distribute the product.

A critical takeaway is that the embodied energy and carbon footprint of a product are largely driven by concept design, materials selected, and related manufacturing processes. Components made by subtractive machining processes will consume more feedstock and have higher environmental impact than those made by additive and net shape manufacturing processes. Natural materials like wood and stone have low embodied energy per unit volume but often are not suitable or formable for technical applications.

**Supply Chain Footprint.**—Many products are assembled from a vast supply chain involving production and assembly of components, transportation, and storage of subassemblies, and warehousing of finished goods prior to distribution (see Fig. 1). Sustainability best practices seek to minimize transportation and warehousing of components and subassemblies by co-locating and/or regional manufacturing of related products. Cargo is typically measured in 20-foot-long container equivalent units (TEU) typically holding 33 m$^3$ and 21,600 kg of goods. Given these volume and mass capacities, transportation cost can be roughly estimated based on the packaged product volume, if the density of the product is less than water, or otherwise based on mass.

*Example 3:* Estimate the transportation cost, embodied energy, and carbon footprint for the prior hex bolt if shipped from China to the US west coast. While rates for TEUs vary by route, schedule, and global demand. For this example, apply an estimated shipping cost from China to the United States of $3,000 US.

Given that the hex bolt is made of steel with a density of 7,850 kg/m$^3$, the mass of the bolt is approximately:

$$mass_{Hex\ Bolt} = Volume \times Density = 1.086 \times 10^{-5} m^3 \times 7850 \frac{\text{kg}}{\text{m}^3} = 0.085 \text{ kg}$$

Estimating the transportation cost based on mass:

$$Cost_{Transport} = 0.085 \text{ kg} \times \frac{3{,}000 \text{ USD/container}}{21{,}600 \text{ kg/container}} = 0.012 \text{ USD}$$

The transport cost can also be used to roughly estimate embodied energy and carbon footprint of the freight shipping. The associated embodied energy is approximately:

$$E_{Transportation} = Cost_{Transport} / Cost_{Energy} = 0.012 \text{ USD}/0.1 \frac{\text{USD}}{\text{kWh}} = 0.12 \text{ kWh}$$

The carbon footprint is approximately:

$$CO_{2\ Transportation} = E_{Transportation} \times \frac{3.6 \text{ MJ}}{1 \text{ kWh}} \times 0.5 \frac{\text{kg}}{\text{MJ}} = 0.21 \text{ kg}$$

These estimates represent lower bounds, as they assume the TEU is fully loaded, and partially loaded containers incur higher economic and environmental costs. Other costs to be calculated include shipping/handling beyond ports and between every node in the supply chain.

**Recycling and Material Reclamation.—**Products at end of life can be remade or partially recovered. The most sustainable approach to material reclamation involves component reuse that avoids the environmental and economic costs of reducing products to scrap for subsequent material and component production. Where not possible, manufacturers generating large-scale scrap (ton-plus batches) generally will recover the highest value from recyclers when scrap materials are clean and sorted to enable application of efficient recycling processes. The design and operation of such circular supply chains require a long-term commitment with established operations.

At the consumer end, single-stream recycling provides an opportunity to recover valuable materials for reprocessing, with manual and automatic sorting of waste facilitating return to material suppliers. For materials with lower value, the best option may be conversion back to energy; related public policies and waste-to-energy facilities are continuously evolving.

Remaining (non-recyclable) waste is typically land-filled, with a representative cost of approximately $100 US per ton. Potential consequences include potential proliferation of pests/odors, environmental contamination, and impacts on current and future land values and uses.

## Heat and Combustion

**Latent Heat.—**When a body changes from the solid to the liquid state or from the liquid to the gaseous state, a certain amount of heat is used to accomplish this change. This heat does not raise the temperature of the body and is called latent heat. When the body changes again from the gaseous to the liquid, or from the liquid to the solid state, it gives out this quantity of heat. The *latent heat of fusion* is the heat supplied to a solid body at the melting point; this heat is absorbed by the body although its temperature remains nearly stationary during the whole operation of melting. The *latent heat of evaporation* is the heat that must be supplied to a liquid at the boiling point to transform the liquid into a vapor. The latent heat is generally given in British thermal units per pound, or kilojoules per kilogram. The latent heat of evaporation of water is 966.6 Btu/pound, or 2248 kJ/kg. This means that it takes 966.6 Btu to evaporate 1 pound, or 2248 kJ to evaporate 1 kilogram, of water after it has been raised to the boiling point, 212° F or 100° C.

When a body changes from the solid to the gaseous state without passing through the liquid stage, as solid carbon dioxide does, the process is called *sublimation*.

**Table 3. Latent Heat of Fusion**

| Substance | Btu per Pound | kJ/kg | Substance | Btu per Pound | kJ/kg | Substance | Btu per Pound | kJ/kg |
|---|---|---|---|---|---|---|---|---|
| Bismuth | 22.75 | 52.92 | Paraffine | 63.27 | 147.17 | Sulfur | 16.86 | 39.22 |
| Beeswax | 76.14 | 177.10 | Phosphorus | 9.06 | 21.07 | Tin | 25.65 | 59.66 |
| Cast iron, gray | 41.40 | 96.30 | Lead | 10.00 | 23.26 | Zinc | 50.63 | 117.77 |
| Cast iron, white | 59.40 | 138.16 | Silver | 37.92 | 88.20 | Ice | 144.00 | 334.94 |

**Table 4. Latent Heat of Evaporation**

| Liquid | Btu per Pound | kJ/kg | Liquid | Btu per Pound | kJ/kg | Liquid | Btu per Pound | kJ/kg |
|---|---|---|---|---|---|---|---|---|
| Alcohol, ethyl | 371.0 | 863 | Carbon bisulfide | 160.0 | 372 | Turpentine | 133.0 | 309 |
| Alcohol, methyl | 481.0 | 1119 | Ether | 162.8 | 379 | Water | 966.6 | 2248 |
| Ammonia | 529.0 | 1230 | Sulfur dioxide | 164.0 | 381 | | | |

**Table 5. Boiling Points of Various Substances at Atmospheric Pressure**

| Substance | Boiling Point | | Substance | Boiling Point | | Substance | Boiling Point | |
|---|---|---|---|---|---|---|---|---|
| | °F | °C | | °F | °C | | °F | °C |
| Aniline | 363 | 183.9 | Chloroform | 140 | 60.0 | Saturated brine | 226 | 107.8 |
| Alcohol | 173 | 78.3 | Ether | 100 | 37.8 | Sulfur | 833 | 445.0 |
| Ammonia | −28 | −33.3 | Linseed oil | 597 | 313.9 | Sulfuric acid | 590 | 310.0 |
| Benzine | 176 | 80.0 | Mercury | 676 | 357.8 | Water, pure | 212 | 100.0 |
| Bromine | 145 | 62.8 | Napthaline | 428 | 220.0 | Water, sea | 213.2 | 100.7 |
| Carbon bisulfide | 118 | 47.8 | Nitric acid | 248 | 120.0 | Wood alcohol | 150 | 65.6 |
| | | | Oil of turpentine | 315 | 157.2 | | | |

**Specific Heat.—**The specific heat of a substance is the ratio of the heat required to raise the temperature of a certain weight of the given substance 1° F to the heat required to raise the temperature of the same weight of water 1° F. As the specific heat is not constant at all temperatures, it is generally assumed that it is determined by raising the temperature from 62 to 63° F. For most substances, however, specific heat is practically constant for temperatures up to 212° F.

**In metric units, specific heat is defined as the ratio of the heat needed to raise the temperature of a mass by 1° C to the heat needed to raise the temperature of the same mass of water by 1° C. In the metric system, heat is measured in joules (J), mass is in grams (g), and measurements are usually taken at 15° C.**

Because specific heat is a dimensionless ratio, the values given in Table 6 and Table 7 that follow are valid in both the US system and the metric system.

**Table 6. Average Specific Heats of Various Substances**

| Substance | Specific Heat | Substance | Specific Heat |
|---|---|---|---|
| Alcohol (absolute) | 0.700 | Lead | 0.031 |
| Alcohol (density 0.8) | 0.622 | Lead (fluid) | 0.037 |
| Aluminum | 0.214 | Limestone | 0.217 |
| Antimony | 0.051 | Magnesia | 0.222 |
| Benzine | 0.450 | Marble | 0.210 |
| Brass | 0.094 | Masonry, brick | 0.200 |
| Brickwork | 0.200 | Mercury | 0.033 |
| Cadmium | 0.057 | Naphtha | 0.310 |
| Carbon | 0.204 | Nickel | 0.109 |
| Charcoal | 0.200 | Oil, machine | 0.400 |
| Chalk | 0.215 | Oil, olive | 0.350 |
| Coal | 0.240 | Paper | 0.32 |
| Coke | 0.203 | Phosphorus | 0.189 |
| Copper, 32° to 212° F (0–100° C) | 0.094 | Platinum | 0.032 |
| Copper, 32° to 572° F (0–100° C) | 0.101 | Quartz | 0.188 |
| Corundum | 0.198 | Sand | 0.195 |
| Ether | 0.503 | Silica | 0.191 |
| Fusel oil | 0.564 | Silver | 0.056 |
| Glass | 0.194 | Soda | 0.231 |
| Gold | 0.031 | Steel, high carbon | 0.117 |
| Graphite | 0.201 | Steel, mild | 0.116 |
| Ice | 0.504 | Stone (generally) | 0.200 |
| Iron, cast | 0.130 | Sulfur | 0.178 |
| Iron, wrought, 32° to 212° F (0–100° C) | 0.110 | Sulfuric acid | 0.330 |
| 32° to 392° F (0–200° C) | 0.115 | Tin (solid) | 0.056 |
| 32° to 572° F (0–300° C) | 0.122 | Tin (fluid) | 0.064 |
| 32° to 662° F (0–350° C) | 0.126 | Turpentine | 0.472 |
| Iron, at high temperatures: | | Water | 1.000 |
| 1382° to 1832° F (750–1000° C) | 0.213 | Wood, fir | 0.650 |
| 1750° to 1840° F (954–1004° C) | 0.218 | Wood, oak | 0.570 |
| 1920° to 2190° F (1049–1199° C) | 0.199 | Wood, pine | 0.467 |
| Kerosene | 0.500 | Zinc | 0.095 |

**Table 7. Specific Heat of Gases**

| Gas | Constant Pressure | Constant Volume | Gas | Constant Pressure | Constant Volume |
|---|---|---|---|---|---|
| Acetic acid | 0.412 | … | Chloroform | 0.157 | … |
| Air | 0.238 | 0.168 | Ethylene | 0.404 | 0.332 |
| Alcohol | 0.453 | 0.399 | Hydrogen | 3.409 | 2.412 |
| Ammonia | 0.508 | 0.399 | Nitrogen | 0.244 | 0.173 |
| Carbonic acid | 0.217 | 0.171 | Oxygen | 0.217 | 0.155 |
| Carbonic oxide | 0.245 | 0.176 | Steam | 0.480 | 0.346 |
| Chlorine | 0.121 | … | | | |

**Heat Loss from Uncovered Steam Pipes.—**The loss of heat from a bare steam or hot-water pipe varies with the temperature difference between the inside of the pipe and the surrounding air. The loss is 2.15 Btu per hour, per square foot of pipe surface, per degree F of temperature difference when the latter is 100 degrees; for a difference of 200 degrees, the loss is 2.66 Btu; for 300 degrees, 3.26 Btu; for 400 degrees, 4.03 Btu; for 500 degrees, 5.18 Btu. Thus, if the pipe area is 1.18 square feet per foot of length, and the temperature difference 300° F, the loss per hour per foot of length = 1.18 × 300 × 3.26 = 1154 Btu. See also *Flow and Energy Loss in Pipe, Tube, and Fittings* on page 2786.

## Table 8. Values of Thermal Conductivity (*k*) and of Conductance (*C*) of Common Building and Insulating Materials

| Type of Material | Thick-ness, in. | *k* or *C*[a] | Type of Material | Thick-ness, in. | *k* or *C*[a] | Max. Temp.,°F | Density, lb per cu. ft. | *k*[a] |
|---|---|---|---|---|---|---|---|---|
| BUILDING | | | BUILDING (*Continued*) | | | | | |
| Batt: | ... | ... | Siding: | ... | ... | ... | ... | ... |
| Mineral Fiber | 2–2¾ | 0.14 | Metal[b] | Avg. | 1.61 | ... | ... | ... |
| Mineral Fiber | 3–3½ | 0.09 | Wood, Med. Density | 7⁄16 | 1.49 | ... | ... | ... |
| Mineral Fiber | 3½–6½ | 0.05 | Stone: | ... | ... | ... | ... | ... |
| Mineral Fiber | 6–7 | 0.04 | Lime or Sand | 1 | 12.50 | ... | ... | ... |
| Mineral Fiber | 8½ | 0.03 | Wall Tile: | ... | ... | ... | ... | ... |
| Block: | ... | ... | Hollow Clay, 1-Cell | 4 | 0.9 | ... | ... | ... |
| Cinder | 4 | 0.90 | Hollow Clay, 2-Cell | 8 | 0.54 | ... | ... | ... |
| Cinder | 8 | 0.58 | Hollow Clay, 3-Cell | 12 | 0.40 | ... | ... | ... |
| Cinder | 12 | 0.53 | Hollow Gypsum | Avg. | 0.7 | ... | ... | ... |
| Block: | ... | ... | INSULATING | | | | | |
| Concrete | 4 | 1.40 | Blanket, Mineral Fiber: | ... | ... | ... | ... | ... |
| Concrete | 8 | 0.90 | Felt | ... | ... | 400 | 3 to 8 | 0.26 |
| Concrete | 12 | 0.78 | Rock or Slag | ... | ... | 1200 | 6 to 12 | 0.26[c] |
| Board: | ... | ... | Glass | ... | ... | 350 | 0.65 | 0.33 |
| Asbestos Cement | ¼ | 16.5 | Textile | ... | ... | 350 | 0.65 | 0.31 |
| Plaster | ½ | 2.22 | Blanket, Hairfelt | ... | ... | 180 | 10 | 0.29 |
| Plywood | ¾ | 1.07 | Board, Block and Pipe | ... | ... | ... | ... | ... |
| Brick: | ... | ... | Insulation: | ... | ... | ... | ... | ... |
| Common | 1 | 5.0 | Amosite | ... | ... | 1500 | 15 to 18 | 0.32[c] |
| Face | 1 | 9.0 | Asbestos Paper | ... | ... | 700 | 30 | 0.40[c] |
| Concrete (poured) | 1 | 12.0 | Glass or Slag (for Pipe) | ... | ... | 350 | 3 to 4 | 0.23 |
| Floor: | ... | ... | Glass or Slag (for Pipe) | ... | ... | 1000 | 10 to 15 | 0.33[c] |
| Wood Subfloor | ¾ | 1.06 | Glass, Cellular | ... | ... | 800 | 9 | 0.40 |
| Hardwood Finish | ¾ | 1.47 | Magnesia (85%) | ... | ... | 600 | 11 to 12 | 0.35[c] |
| Tile | Avg. | 20.0 | Mineral Fiber | ... | ... | 100 | 15 | 0.29 |
| Glass: | ... | ... | Polystyrene, Beaded | ... | ... | 170 | 1 | 0.28 |
| Architectural | ... | 10.00 | Polystyrene, Rigid | ... | ... | 170 | 1.8 | 0.25 |
| Mortar: | ... | ... | Rubber, Rigid Foam | ... | ... | 150 | 4.5 | 0.22 |
| Cement | 1 | 5.0 | Wood Felt | ... | ... | 180 | 20 | 0.31 |
| Plaster: | ... | ... | Loose Fill: | ... | ... | ... | ... | ... |
| Sand | ⅜ | 13.30 | Cellulose | ... | ... | ... | 2.5 to 3 | 0.27 |
| Sand and Gypsum | ½ | 11.10 | Mineral Fiber | ... | ... | ... | 2 to 5 | 0.28 |
| Stucco | 1 | 5.0 | Perlite | ... | ... | ... | 5 to 8 | 0.37 |
| Roofing: | ... | ... | Silica Aerogel | ... | ... | ... | 7.6 | 0.17 |
| Asphalt Roll | Avg. | 6.50 | Vermiculite | ... | ... | ... | 7 to 8.2 | 0.47 |
| Shingle, asb. cem. | Avg. | 4.76 | Mineral Fiber Cement: | ... | ... | ... | ... | ... |
| Shingle, asphalt | Avg. | 2.27 | Clay Binder | ... | ... | 1800 | 24 to 30 | 0.49[c] |
| Shingle, wood | Avg. | 1.06 | Hydraulic Binder | ... | ... | 1200 | 30 to 40 | 0.75[c] |

[a] Units are in Btu/hr-ft$^2$-°F. Where thickness is given as 1 inch, the value given is thermal conductivity (k); for other thicknesses the value given is thermal conductance (*C*). All values are for a test mean temperature of 75° F, except those designated with [c], which are for 100° F.

[b] Over hollowback sheathing.

[c] Test mean temperature 100° F, see footnote [a].

*Source:* American Society of Heating, Refrigerating and Air-Conditioning Engineers, Inc.: *Handbook of Fundamentals*.

**Table 9. Typical Values of Coefficient of Linear Thermal Expansion for Thermoplastics and Other Commonly Used Materials**

| Material[a] | in/in/deg F × $10^{-5}$ | cm/cm/deg C × $10^{-5}$ | Material[a] | in/in/deg F × $10^{-5}$ | cm/cm/deg C × $10^{-5}$ |
|---|---|---|---|---|---|
| Liquid Crystal—GR | 0.3 | 0.6 | ABS—GR | 1.7 | 3.1 |
| Glass | 0.4 | 0.7 | Polypropylene—GR | 1.8 | 3.2 |
| Steel | 0.6 | 1.1 | Epoxy—GR | 2.0 | 3.6 |
| Concrete | 0.8 | 1.4 | Polyphenylene sulfide—GR | 2.0 | 3.6 |
| Copper | 0.9 | 1.6 | Acetal—GR | 2.2 | 4.0 |
| Bronze | 1.0 | 1.8 | Epoxy | 3.0 | 5.4 |
| Brass | 1.0 | 1.8 | Polycarbonate | 3.6 | 6.5 |
| Aluminum | 1.2 | 2.2 | Acrylic | 3.8 | 6.8 |
| Polycarbonate—GR | 1.2 | 2.2 | ABS | 4.0 | 7.2 |
| Nylon—GR | 1.3 | 2.3 | Nylon | 4.5 | 8.1 |
| TP polyester—GR | 1.4 | 2.5 | Acetal | 4.8 | 8.5 |
| Magnesium | 1.4 | 2.5 | Polypropylene | 4.8 | 8.6 |
| Zinc | 1.7 | 3.1 | TP Polyester | 6.9 | 12.4 |
| ABS—GR | 1.7 | 3.1 | Polyethylene | 7.2 | 13.0 |

[a] GR = Typical glass fiber-reinforced material. Other plastics materials shown are unfilled.

**Table 10. Linear Expansion of Various Substances between 32 and 212° F Expansion of Volume = 3 × Linear Expansion**

| Substance | Linear Expansion | | Substance | Linear Expansion | |
|---|---|---|---|---|---|
| | for 1° F | for 1° C | | for 1° F | for 1° C |
| Brick | 0.0000030 | 0.0000054 | Masonry, brick from | 0.0000026 | 0.0000047 |
| Cement, Portland | 0.0000060 | 0.0000108 | to | 0.0000050 | 0.0000090 |
| Concrete | 0.0000080 | 0.0000144 | Plaster | 0.0000092 | 0.0000166 |
| Ebonite | 0.0000428 | 0.0000770 | Porcelain | 0.0000020 | 0.0000036 |
| Glass, thermometer | 0.0000050 | 0.0000090 | Quartz, from | 0.0000043 | 0.0000077 |
| Glass, hard | 0.0000040 | 0.0000072 | to | 0.0000079 | 0.0000142 |
| Granite | 0.0000044 | 0.0000079 | Slate | 0.0000058 | 0.0000104 |
| Marble, from | 0.0000031 | 0.0000056 | Sandstone | 0.0000065 | 0.0000117 |
| to | 0.0000079 | 0.0000142 | Wood, pine | 0.0000028 | 0.0000050 |

**Table 11. Coefficients of Heat Transmission**

| Metal | Btu per Second | Metal | Btu per Second | Metal | Btu per Second |
|---|---|---|---|---|---|
| Aluminum | 0.00203 | German silver | 0.00050 | Steel, soft | 0.00062 |
| Antimony | 0.00022 | Iron | 0.00089 | Silver | 0.00610 |
| Brass, yellow | 0.00142 | Lead | 0.00045 | Tin | 0.00084 |
| Brass, red | 0.00157 | Mercury | 0.00011 | Zinc | 0.00170 |
| Copper | 0.00404 | Steel, hard | 0.00034 | … | … |

Heat transmitted, in British thermal units, per second, through metal 1 inch thick, per square inch of surface, for a temperature difference of 1° F.

**Table 12. Coefficients of Heat Radiation**

| Surface | Btu per Hour | Surface | Btu per Hour |
|---|---|---|---|
| Cast-iron, new | 0.6480 | Sawdust | 0.7215 |
| Cast-iron, rusted | 0.6868 | Sand, fine | 0.7400 |
| Copper, polished | 0.0327 | Silver, polished | 0.0266 |
| Glass | 0.5948 | Tin, polished | 0.0439 |
| Iron, ordinary | 0.5662 | Tinned iron, polished | 0.0858 |
| Iron, sheet-, polished | 0.0920 | Water | 1.0853 |
| Oil | 1.4800 | … | … |

Heat radiated, in British thermal units, per square foot of surface per hour, for a temperature difference of 1° F.

## Table 13. Freezing Mixtures

| Mixture | Temperature Change, °F | | Temperature Change, °C | |
|---|---|---|---|---|
| | From | To | From | To |
| Common salt (NaCl), 1 part; snow, 3 parts | 32 | ±0 | 0 | −17.8 |
| Common salt (NaCl), 1 part; snow, 1 part | 32 | −0.4 | 0 | −18 |
| Calcium chloride ($CaCl_2$), 3 parts; snow, 2 parts | 32 | −27 | 0 | −32.8 |
| Calcium chloride ($CaCl_2$), 2 parts; snow, 1 part | 32 | −44 | 0 | −42.2 |
| Sal ammoniac ($NH_4Cl$), 5 parts; saltpeter ($KNO_3$), 5 parts; water, 16 parts | 50 | +10 | 10 | −12.2 |
| Sal ammoniac ($NH_4Cl$), 1 part; saltpeter ($KNO_3$), 1 part; water, 1 part | 46 | −11 | 7.8 | −23.9 |
| Ammonium nitrate ($NH_4NO_3$), 1 part; water, 1 part | 50 | +3 | 10 | −16.1 |
| Potassium hydrate (KOH), 4 parts; snow, 3 parts | 32 | −35 | 0 | −37.2 |

**Ignition Temperatures.—**The following temperatures are required to ignite the different substances specified: Phosphorus, transparent, 120° F (49° C); bisulfide of carbon, 300° F (149° C); gun cotton, 430° F (221° C); nitroglycerine, 490° F (254° C); phosphorus, amorphous, 500° F (260° C); rifle powder, 550° F (288° C); charcoal, 660° F (349° C); dry pine wood, 800° F (427° C); dry oak wood, 900° F (482° C).

## Table 14. Typical Thermal Properties of Various Metals

| Material and Alloy Designation[a] | Density, ρ | | Melting Point, °F | | Conductivity, *k* | Specific Heat, *C* | Coeff. of Expansion, α | |
|---|---|---|---|---|---|---|---|---|
| | lb/in³ | g/cc | solidus | liquidus | Btu/hr-ft-°F | Btu/lb/°F | μin/in-°F | μm/m-°C |
| **Aluminum Alloys** | | | | | | | | |
| 2011 | 0.102 | 2.823 | 995 | 1190 | 82.5 | 0.23 | 12.8 | 23.0 |
| 2017 | 0.101 | 2.796 | 995 | 1185 | 99.4 | 0.22 | 13.1 | 23.6 |
| 2024 | 0.100 | 2.768 | 995 | 1180 | 109.2 | 0.22 | 12.9 | 23.2 |
| 3003 | 0.099 | 2.740 | 1190 | 1210 | 111 | 0.22 | 12.9 | 23.2 |
| 5052 | 0.097 | 2.685 | 1100 | 1200 | 80 | 0.22 | 13.2 | 23.8 |
| 5086 | 0.096 | 2.657 | 1085 | 1185 | 73 | 0.23 | 13.2 | 23.8 |
| 6061 | 0.098 | 2.713 | 1080 | 1200 | 104 | 0.23 | 13.0 | 23.4 |
| 7075 | 0.101 | 2.796 | 890 | 1180 | 70 | 0.23 | 13.1 | 23.6 |
| **Copper-Base Alloys** | | | | | | | | |
| Manganese Bronze | 0.302 | 8.359 | 1590 | 1630 | 61 | 0.09 | 11.8 | 21.2 |
| C11000 (Electrolytic tough pitch) | 0.321 | 8.885 | 1941 | 1981 | 226 | 0.09 | 9.8 | 17.6 |
| C14500 (Free machining Cu) | 0.323 | 8.941 | 1924 | 1967 | 205 | 0.09 | 9.9 | 17.8 |
| C17200, C17300 (Beryllium Cu) | 0.298 | 8.249 | 1590 | 1800 | 62 | 0.10 | 9.9 | 17.8 |
| C18200 (Chromium Cu) | 0.321 | 8.885 | 1958 | 1967 | 187 | 0.09 | 9.8 | 17.6 |
| C18700 (Leaded Cu) | 0.323 | 8.941 | 1750 | 1975 | 218 | 0.09 | 9.8 | 17.6 |
| C22000 (Commercial bronze, 90%) | 0.318 | 8.802 | 1870 | 1910 | 109 | 0.09 | 10.2 | 18.4 |
| C23000 (Red brass, 85%) | 0.316 | 8.747 | 1810 | 1880 | 92 | 0.09 | 10.4 | 18.7 |
| C26000 (Cartridge brass, 70%) | 0.313 | 8.664 | 1680 | 1750 | 70 | 0.09 | 11.1 | 20.0 |
| C27000 (Yellow brass) | 0.306 | 8.470 | 1660 | 1710 | 67 | 0.09 | 11.3 | 20.3 |
| C28000 (Muntz metal, 60%) | 0.303 | 8.387 | 1650 | 1660 | 71 | 0.09 | 11.6 | 20.9 |
| C33000 (Low-leaded brass tube) | 0.310 | 8.581 | 1660 | 1720 | 67 | 0.09 | 11.2 | 20.2 |
| C35300 (High-leaded brass) | 0.306 | 8.470 | 1630 | 1670 | 67 | 0.09 | 11.3 | 20.3 |
| C35600 (Extra-high-leaded brass) | 0.307 | 8.498 | 1630 | 1660 | 67 | 0.09 | 11.4 | 20.5 |
| C36000 (Free machining brass) | 0.307 | 8.498 | 1630 | 1650 | 67 | 0.09 | 11.4 | 20.5 |
| C36500 (Leaded Muntz metal) | 0.304 | 8.415 | 1630 | 1650 | 71 | 0.09 | 11.6 | 20.9 |
| C46400 (Naval brass) | 0.304 | 8.415 | 1630 | 1650 | 67 | 0.09 | 11.8 | 21.2 |
| C51000 (Phosphor bronze, 5% A) | 0.320 | 8.858 | 1750 | 1920 | 40 | 0.09 | 9.9 | 17.8 |
| C54400 (Free cutting phos. bronze) | 0.321 | 8.885 | 1700 | 1830 | 50 | 0.09 | 9.6 | 17.3 |
| C62300 (Aluminum bronze, 9%) | 0.276 | 7.640 | 1905 | 1915 | 31.4 | 0.09 | 9.0 | 16.2 |
| C62400 (Aluminum bronze, 11%) | 0.269 | 7.446 | 1880 | 1900 | 33.9 | 0.09 | 9.2 | 16.6 |
| C63000 (Ni-Al bronze) | 0.274 | 7.584 | 1895 | 1930 | 21.8 | 0.09 | 9.0 | 16.2 |
| Nickel-Silver | 0.314 | 8.691 | 1870 | 2030 | 17 | 0.09 | 9.0 | 16.2 |

**Table 13.** *(Continued)* **Typical Thermal Properties of Various Metals**

| Material and Alloy Designation[a] | Density, r | | Melting Point, °F | | Conductivity, k | Specific Heat, C | Coeff. of Expansion, α | |
|---|---|---|---|---|---|---|---|---|
| | lb/in³ | g/cc | solidus | liquidus | Btu/hr-ft-°F | Btu/lb/°F | μin/in-°F | μm/m-°C |
| **Nickel-Base Alloys** | | | | | | | | |
| Nickel 200, 201, 205 | 0.321 | 8.885 | 2615 | 2635 | 43.3 | 0.11 | 8.5 | 15.3 |
| Hastelloy C-22 | 0.314 | 8.691 | 2475 | 2550 | 7.5 | 0.10 | 6.9 | 12.4 |
| Hastelloy C-276 | 0.321 | 8.885 | 2415 | 2500 | 7.5 | 0.10 | 6.2 | 11.2 |
| Inconel 718 | 0.296 | 8.193 | 2300 | 2437 | 6.5 | 0.10 | 7.2 | 13.0 |
| Monel | 0.305 | 8.442 | 2370 | 2460 | 10 | 0.10 | 8.7 | 15.7 |
| Monel 400 | 0.319 | 8.830 | 2370 | 2460 | 12.6 | 0.10 | 7.7 | 13.9 |
| Monel K500 | 0.306 | 8.470 | 2400 | 2460 | 10.1 | 0.10 | 7.6 | 13.7 |
| Monel R405 | 0.319 | 8.830 | 2370 | 2460 | 10.1 | 0.10 | 7.6 | 13.7 |
| **Stainless Steels** | | | | | | | | |
| S30100 | 0.290 | 8.027 | 2550 | 2590 | 9.4 | 0.12 | 9.4 | 16.9 |
| S30200, S30300, S30323 | 0.290 | 8.027 | 2550 | 2590 | 9.4 | 0.12 | 9.6 | 17.3 |
| S30215 | 0.290 | 8.027 | 2500 | 2550 | 9.2 | 0.12 | 9.0 | 16.2 |
| S30400, S30500 | 0.290 | 8.027 | 2550 | 2650 | 9.4 | 0.12 | 9.6 | 17.3 |
| S30430 | 0.290 | 8.027 | 2550 | 2650 | 6.5 | 0.12 | 9.6 | 17.3 |
| S30800 | 0.290 | 8.027 | 2550 | 2650 | 8.8 | 0.12 | 9.6 | 17.3 |
| S30900, S30908 | 0.290 | 8.027 | 2550 | 2650 | 9.0 | 0.12 | 8.3 | 14.9 |
| S31000, S31008 | 0.290 | 8.027 | 2550 | 2650 | 8.2 | 0.12 | 8.8 | 15.8 |
| S31600, S31700 | 0.290 | 8.027 | 2500 | 2550 | 9.4 | 0.12 | 8.8 | 15.8 |
| S31703 | 0.290 | 8.027 | 2500 | 2550 | 8.3 | 0.12 | 9.2 | 16.6 |
| S32100 | 0.290 | 8.027 | 2550 | 2600 | 9.3 | 0.12 | 9.2 | 16.6 |
| S34700 | 0.290 | 8.027 | 2550 | 2650 | 9.3 | 0.12 | 9.2 | 16.6 |
| S34800 | 0.290 | 8.027 | 2550 | 2650 | 9.3 | 0.12 | 9.3 | 16.7 |
| S38400 | 0.290 | 8.027 | 2550 | 2650 | 9.4 | 0.12 | 9.6 | 17.3 |
| S40300, S41000, S41600, S41623 | 0.280 | 7.750 | 2700 | 2790 | 14.4 | 0.11 | 5.5 | 9.9 |
| S40500 | 0.280 | 7.750 | 2700 | 2790 | 15.6 | 0.12 | 6.0 | 10.8 |
| S41400 | 0.280 | 7.750 | 2600 | 2700 | 14.4 | 0.11 | 5.8 | 10.4 |
| S42000, S42020 | 0.280 | 7.750 | 2650 | 2750 | 14.4 | 0.11 | 5.7 | 10.3 |
| S42200 | 0.280 | 7.750 | 2675 | 2700 | 13.8 | 0.11 | 6.2 | 11.2 |
| S42900 | 0.280 | 7.750 | 2650 | 2750 | 14.8 | 0.11 | 5.7 | 10.3 |
| S43000, S43020, S43023 | 0.280 | 7.750 | 2600 | 2750 | 15.1 | 0.11 | 5.8 | 10.4 |
| S43600 | 0.280 | 7.750 | 2600 | 2750 | 13.8 | 0.11 | 5.2 | 9.4 |
| S44002, S44004 | 0.280 | 7.750 | 2500 | 2700 | 14.0 | 0.11 | 5.7 | 10.3 |
| S44003 | 0.280 | 7.750 | 2500 | 2750 | 14.0 | 0.11 | 5.6 | 10.1 |
| S44600 | 0.270 | 7.474 | 2600 | 2750 | 12.1 | 0.12 | 5.8 | 10.4 |
| S50100, S50200 | 0.280 | 7.750 | 2700 | 2800 | 21.2 | 0.11 | 6.2 | 11.2 |
| **Cast Iron and Steel** | | | | | | | | |
| Malleable Iron, A220 (50005, 60004, 80002) | 0.265 | 7.335 | | | 29.5 | 0.12 | 7.5 | 13.5 |
| Gray Cast Iron | 0.25 | 6.920 | liquidus approximately, 2100 to 2200, depending on composition | | 28.0 | 0.25 | 5.8 | 10.4 |
| Ductile Iron, A536 (120-90-02) | 0.25 | 6.920 | | | | 0.16 | 5.9–6.2 | 10.6–11.16 |
| Ductile Iron, A536 (100-70-03) | 0.25 | 6.920 | | | 20.0 | 0.16 | 5.9–6.2 | 10.6–11.16 |
| Ductile Iron, A536 (80-55-06) | 0.25 | 6.920 | | | 18.0 | 0.15 | 5.9–6.2 | 10.6–11.16 |
| Ductile Iron, A536 (65-45-120) | 0.25 | 6.920 | | | 20.8 | 0.15 | 5.9–6.2 | 10.6–11.16 |
| Ductile Iron, A536 (60-40-18) | 0.25 | 6.920 | | | | 0.12 | 5.9–6.2 | 10.6–11.16 |
| Cast Steel, 3%C | 0.25 | 6.920 | liquidus, 2640 | | 28.0 | 0.12 | 7.0 | 12.6 |
| **Titanium Alloys** | | | | | | | | |
| Commercially Pure | 0.163 | 4.512 | 3000 | 3040 | 9.0 | 0.12 | 5.1 | 9.2 |
| Ti-5Al-2.5Sn | 0.162 | 4.484 | 2820 | 3000 | 4.5 | 0.13 | 5.3 | 9.5 |
| Ti-8Mn | 0.171 | 4.733 | 2730 | 2970 | 6.3 | 0.19 | 6.0 | 10.8 |

[a] Alloy designations correspond to the Aluminum Association numbers for aluminum alloys and to the Unified Numbering System (UNS) for copper and stainless steel alloys. A220 and A536 are ASTM specified irons.

**Adjusting Lengths for Reference Temperature.—**The standard reference temperature for industrial length measurements is 20 degrees Celsius (68 degrees Fahrenheit). For other temperatures, corrections should be made in accordance with the difference in thermal expansion for the two parts, especially when the gage is made of a different material than the part to be inspected.

*Example:* An aluminum part is to be measured with a steel gage when the room temperature is 30° C. The aluminum part has a coefficient of linear thermal expansion, $\alpha_{Part} = 24.7 \times 10^{-6}$ mm/mm-°C, and for the steel gage, $\alpha_{Gage} = 10.8 \times 10^{-6}$ mm/mm-°C.

At the reference temperature, the specified length of the aluminum part is 20.021 mm. What is the length of the part at the measuring (room) temperature?

$\Delta L$, the change in the measured length due to temperature, is given by:

$$\begin{aligned}\Delta L &= L(T_R - T_0)(\alpha_{Part} - \alpha_{Gage})\\ &= 20.021(30 - 20)(24.7 - 10.8) \times 10^{-6}\text{mm}\\ &= 2782.919 \times 10^{-6} \approx 0.003 \text{ mm}\end{aligned}$$

where $L$ = length of part at reference temperature; $T_R$ = room temperature (temperature of part and gage) and $T_0$ = reference temperature.

Thus, the temperature-corrected length at 30° C is $L + \Delta L = 20.021 + 0.003 = 20.024$ mm.

**Length Changes Due to Temperature.—**Table 15 gives changes in length for variations from the standard reference temperature of 68° F (20° C) for materials of known coefficients of expansion, α. Coefficients of expansion are given in tables on pages 370, 371, 372, 384, 385, and elsewhere.

*Example:* In Table 15, for coefficients between those listed, add appropriate listed values. For example, a length change for a coefficient of 7 is the sum of values in the 5 and 2 columns. Fractional interpolation also is possible. Thus, in a steel bar with a coefficient of thermal expansion of $6.3 \times 10^{-6} = 0.0000063$ in/in = 6.3 µin/in of length/°F, the increase in length at 73° F is 25 + 5 + 1.5 = 31.5 µin/in of length. For a steel with the same coefficient of expansion, the change in length, measured in degrees C, is expressed in microns (micrometers)/meter (µm/m) of length.

Alternatively, and for temperatures beyond the scope of the table, the length difference due to a temperature change is equal to the coefficient of expansion multiplied by the change in temperature, i.e., $\Delta L = \alpha \Delta T$. Thus, for the previous example, $\Delta L = 6.3 \times (73 - 68) = 6.3 \times 5 = 31.5$ µin/in.

**Change in Radius of Thin Circular Ring Due to Temperature.—**Consider a circular ring of initial radius $r$, that undergoes a temperature change $\Delta T$. Initially, the circumference of the ring is $c = 2\pi r$. If the coefficient of expansion of the ring material is α, the change in circumference due to the temperature change is $\Delta c = 2\pi r\,\alpha\Delta T$.

The new circumference of the ring will be: $c_n = c + \Delta c = 2\pi r + 2\pi r\alpha\Delta T = 2\pi r(1 + \alpha\Delta T)$.

*Note:* An increase in temperature causes $\Delta c$ to be positive, and a decrease in temperature causes $\Delta c$ to be negative.

As the circumference increases, the radius of the circle also increases. If the new radius is $R$, the new circumference is $2\pi R$. For a given change in temperature, $\Delta T$, the change in radius of the ring is found as follows:

$$c_n = 2\pi R = 2\pi r(1 + \alpha\Delta T) \qquad R = r + r\alpha\Delta T \qquad \Delta r = R - r = r\alpha\Delta T$$

## Table 15. Differences in Length in Microinches/Inch (Microns/Meter) for Changes from the Standard Temperature of 68° F (20° C)

| Temperature Deg. | | Coefficient of Thermal Expansion of Material per Degree F (C) × $10^6$ | | | | | | | | | |
|---|---|---|---|---|---|---|---|---|---|---|---|
| | | 1 | 2 | 3 | 4 | 5 | 10 | 15 | 20 | 25 | 30 |
| | | Total Change in Length from Standard Temperature | | | | | { for °F in microinches/inch of length (µin/in) for °C or °K in microns/meter of length (µm/m) | | | | |
| F | C | | | | | | | | | | |
| 38 | −10 | −30 | −60 | −90 | −120 | −150 | −300 | −450 | −600 | −750 | −900 |
| 39 | −9 | −29 | −58 | −87 | −116 | −145 | −290 | −435 | −580 | −725 | −870 |
| 40 | −8 | −28 | −56 | −84 | −112 | −140 | −280 | −420 | −560 | −700 | −840 |
| 41 | −7 | −27 | −54 | −81 | −108 | −135 | −270 | −405 | −540 | −675 | −810 |
| 42 | −6 | −26 | −52 | −78 | −104 | −130 | −260 | −390 | −520 | −650 | −780 |
| 43 | −5 | −25 | −50 | −75 | −100 | −125 | −250 | −375 | −500 | −625 | −750 |
| 44 | −4 | −24 | −48 | −72 | −96 | −120 | −240 | −360 | −480 | −600 | −720 |
| 45 | −3 | −23 | −46 | −69 | −92 | −115 | −230 | −345 | −460 | −575 | −690 |
| 46 | −2 | −22 | −44 | −66 | −88 | −110 | −220 | −330 | −440 | −550 | −660 |
| 47 | −1 | −21 | −42 | −63 | −84 | −105 | −210 | −315 | −420 | −525 | −630 |
| 48 | 0 | −20 | −40 | −60 | −80 | −100 | −200 | −300 | −400 | −500 | −600 |
| 49 | 1 | −19 | −38 | −57 | −76 | −95 | −190 | −285 | −380 | −475 | −570 |
| 50 | 2 | −18 | −36 | −54 | −72 | −90 | −180 | −270 | −360 | −450 | −540 |
| 51 | 3 | −17 | −34 | −51 | −68 | −85 | −170 | −255 | −340 | −425 | −510 |
| 52 | 4 | −16 | −32 | −48 | −64 | −80 | −160 | −240 | −320 | −400 | −480 |
| 53 | 5 | −15 | −30 | −45 | −60 | −75 | −150 | −225 | −300 | −375 | −450 |
| 54 | 6 | −14 | −28 | −42 | −56 | −70 | −140 | −210 | −280 | −350 | −420 |
| 55 | 7 | −13 | −26 | −39 | −52 | −65 | −130 | −195 | −260 | −325 | −390 |
| 56 | 8 | −12 | −24 | −36 | −48 | −60 | −120 | −180 | −240 | −300 | −360 |
| 57 | 9 | −11 | −22 | −33 | −44 | −55 | −110 | −165 | −220 | −275 | −330 |
| 58 | 10 | −10 | −20 | −30 | −40 | −50 | −100 | −150 | −200 | −250 | −300 |
| 59 | 11 | −9 | −18 | −27 | −36 | −45 | −90 | −135 | −180 | −225 | −270 |
| 60 | 12 | −8 | −16 | −24 | −32 | −40 | −80 | −120 | −160 | −200 | −240 |
| 61 | 13 | −7 | −14 | −21 | −28 | −35 | −70 | −105 | −140 | −175 | −210 |
| 62 | 14 | −6 | −12 | −18 | −24 | −30 | −60 | −90 | −120 | −150 | −180 |
| 63 | 15 | −5 | −10 | −15 | −20 | −25 | −50 | −75 | −100 | −125 | −150 |
| 64 | 16 | −4 | −8 | −12 | −16 | −20 | −40 | −60 | −80 | −100 | −120 |
| 65 | 17 | −3 | −6 | −9 | −12 | −15 | −30 | −45 | −60 | −75 | −90 |
| 66 | 18 | −2 | −4 | −6 | −8 | −10 | −20 | −30 | −40 | −50 | −60 |
| 67 | 19 | −1 | −2 | −3 | −4 | −5 | −10 | −15 | −20 | −25 | −30 |
| 68 | 20 | 0 | 0 | 0 | 0 | 0 | 0 | 0 | 0 | 0 | 0 |
| 69 | 21 | 1 | 2 | 3 | 4 | 5 | 10 | 15 | 20 | 25 | 30 |
| 70 | 22 | 2 | 4 | 6 | 8 | 10 | 20 | 30 | 40 | 50 | 60 |
| 71 | 23 | 3 | 6 | 9 | 12 | 15 | 30 | 45 | 60 | 75 | 90 |
| 72 | 24 | 4 | 8 | 12 | 16 | 20 | 40 | 60 | 80 | 100 | 120 |
| 73 | 25 | 5 | 10 | 15 | 20 | 25 | 50 | 75 | 100 | 125 | 150 |
| 74 | 26 | 6 | 12 | 18 | 24 | 30 | 60 | 90 | 120 | 150 | 180 |
| 75 | 27 | 7 | 14 | 21 | 28 | 35 | 70 | 105 | 140 | 175 | 210 |
| 76 | 28 | 8 | 16 | 24 | 32 | 40 | 80 | 120 | 160 | 200 | 240 |
| 77 | 29 | 9 | 18 | 27 | 36 | 45 | 90 | 135 | 180 | 225 | 270 |
| 78 | 30 | 10 | 20 | 30 | 40 | 50 | 100 | 150 | 200 | 250 | 300 |
| 79 | 31 | 11 | 22 | 33 | 44 | 55 | 110 | 165 | 220 | 275 | 330 |
| 80 | 32 | 12 | 24 | 36 | 48 | 60 | 120 | 180 | 240 | 300 | 360 |
| 81 | 33 | 13 | 26 | 39 | 52 | 65 | 130 | 195 | 260 | 325 | 390 |
| 82 | 34 | 14 | 28 | 42 | 56 | 70 | 140 | 210 | 280 | 350 | 420 |
| 83 | 35 | 15 | 30 | 45 | 60 | 75 | 150 | 225 | 300 | 375 | 450 |
| 84 | 36 | 16 | 32 | 48 | 64 | 80 | 160 | 240 | 320 | 400 | 480 |
| 85 | 37 | 17 | 34 | 51 | 68 | 85 | 170 | 255 | 340 | 425 | 510 |
| 86 | 38 | 18 | 36 | 54 | 72 | 90 | 180 | 270 | 360 | 450 | 540 |
| 87 | 39 | 19 | 38 | 57 | 76 | 95 | 190 | 285 | 380 | 475 | 570 |
| 88 | 40 | 20 | 40 | 60 | 80 | 100 | 200 | 300 | 400 | 500 | 600 |
| 89 | 41 | 21 | 42 | 63 | 84 | 105 | 210 | 315 | 420 | 525 | 630 |
| 90 | 42 | 22 | 44 | 66 | 88 | 110 | 220 | 330 | 440 | 550 | 660 |
| 91 | 43 | 23 | 46 | 69 | 92 | 115 | 230 | 345 | 460 | 575 | 690 |
| 92 | 44 | 24 | 48 | 72 | 96 | 120 | 240 | 360 | 480 | 600 | 720 |
| 93 | 45 | 25 | 50 | 75 | 100 | 125 | 250 | 375 | 500 | 625 | 750 |
| 94 | 46 | 26 | 52 | 78 | 104 | 130 | 260 | 390 | 520 | 650 | 780 |
| 95 | 47 | 27 | 54 | 81 | 108 | 135 | 270 | 405 | 540 | 675 | 810 |
| 96 | 48 | 28 | 56 | 84 | 112 | 140 | 280 | 420 | 560 | 700 | 840 |
| 97 | 49 | 29 | 58 | 87 | 116 | 145 | 290 | 435 | 580 | 725 | 870 |
| 98 | 50 | 30 | 60 | 90 | 120 | 150 | 300 | 450 | 600 | 750 | 900 |

## Properties of Mass and Weight

**Density.**—The density of any solid, fluid or gaseous substance is the mass of that substance per unit volume. If weight is used in the ordinary sense as being equivalent to mass, then density may be defined as the weight per unit volume. The density depends upon the unit in which the mass or weight is expressed, and upon the unit of volume used. In engineering and scientific work, density is generally expressed in grams per cubic centimeter, without naming the units, because the density will be equal to the specific gravity.

**Specific Gravity.**—Specific gravity is a number indicating how many times a volume of material is heavier than an equal volume of water. Density of water varies slightly at different temperatures. In exacting scientific studies, a reference temperature of 4° C (39.2° F) is often used, and the weight of 1 cubic meter of pure water at 4° C is 1000 kg. In engineering practice, the usual custom is to measure specific gravity at water temperature of 60 or 62° F (15.5 or 16.6° C); 1 cubic foot of pure water at 62° F weighs 62.355 pounds.

| Given | Rule to find density | Given | Rule to find specific gravity |
|---|---|---|---|
| Specific Gravity (S.G.) | weight/cm$^3$ = S.G. | weight/cm$^3$ | S.G. = weight/cm$^3$ |
| | weight/m$^3$ = S.G. × 1000 | weight/m$^3$ | S.G. = weight/m$^3$ ÷ 1000 |
| | weight/in$^3$ = S.G. × 0.0361 | weight/in$^3$ | S.G. = weight/in$^3$ ÷ 0.0361 |
| | weight/ft$^3$ = S.G. ÷ 0.01604 | weight/ft$^3$ | S.G. = weight/ft$^3$ × 0.01604 |

When specific gravity is known, the weight per cubic centimeter is equal to its specific gravity. The weight per cubic meter equals the specific gravity × 1000. The weight per cubic inch equals the specific gravity × 0.0361. The weight of a cubic foot equals the specific gravity divided by 0.01604.

When weight per cubic centimeter is known, the specific gravity is equal to the weight per cubic centimeter. If weight per cubic meter is known, the specific gravity equals this weight divided by 1000. If density is given in lb/in$^3$, specific gravity may be determined by dividing the density by 0.0361. If weight per cubic foot is known, specific gravity equals this weight multiplied by 0.01604.

*Examples:* The specific gravity of cast iron is 7.2. The weight of 80 cm$^3$ of cast iron = 7.2 × 80 = 5.6 kg. The weight of 5 in$^3$ of cast iron = 7.2 × 0.0361 × 5 = 1.2996 pounds.

*Examples:* The weight of a cubic centimeter of gold is 19.31 grams. The specific gravity of gold = weight of a cubic centimeter of gold = 19.31. A cubic inch of gold weighs 0.697 pound. The specific gravity of gold = 0.697 ÷ 0.0361 = 19.31

**Table 16. Average Specific Gravity of Various Substances**

| Substance | Specific Gravity | [a]Weight lb/ft$^3$ | [a]Weight kg/m$^3$ | Substance | Specific Gravity | [a]Weight lb/ft$^3$ | [a]Weight kg/m$^3$ | Substance | Specific Gravity | [a]Weight lb/ft$^3$ | [a]Weight kg/m$^3$ |
|---|---|---|---|---|---|---|---|---|---|---|---|
| ABS | 1.05 | 66 | 1057 | Glass | 2.6 | 162 | 2595 | Platinum | 21.5 | 1342 | 21497 |
| Acrylic | 1.19 | 74 | 1185 | Glass, crushed | … | 74 | 1185 | Polycarbonate | 1.19 | 74 | 1185 |
| Aluminum bronze | 7.8 | 486 | 7785 | Gold, 22 carat fine | 17.5 | 1091 | 17476 | Polyethylene | 0.97 | 60 | 961 |
| Aluminum, cast | 2.6 | 160 | 2563 | Gold, pure | 19.3 | 1204 | 19286 | Polypropylene | 0.91 | 57 | 913 |
| Aluminum, wrought | 2.7 | 167 | 2675 | Granite | 2.7 | 168 | 2691 | Polyurethane | 1.05 | 66 | 1057 |
| Asbestos | 2.4 | 150 | 2403 | Gravel | … | 109 | 1746 | Quartz | 2.6 | 162 | 2595 |
| Asphaltum | 1.4 | 87 | 1394 | Gypsum | 2.4 | 150 | 2403 | Salt, common | … | 48 | 769 |
| Borax | 1.8 | 112 | 1794 | Ice | 0.9 | 56 | 897 | Sand, dry | … | 100 | 1602 |
| Brick, common | 1.8 | 112 | 1794 | Iron, cast | 7.2 | 447 | 7160 | Sand, wet | … | 125 | 2002 |
| Brick, fire | 2.3 | 143 | 2291 | Iron, wrought | 7.7 | 479 | 7673 | Sandstone | 2.3 | 143 | 2291 |
| Brick, hard | 2.0 | 125 | 2002 | Iron slag | 2.7 | 168 | 2691 | Silver | 10.5 | 656 | 10508 |
| Brick, pressed | 2.2 | 137 | 2195 | Lead | 11.4 | 711 | 11389 | Slate | 2.8 | 175 | 2803 |
| Brickwork, in cement | 1.8 | 112 | 1794 | Limestone | 2.6 | 162 | 2595 | Soapstone | 2.7 | 168 | 2691 |
| Brickwork, in mortar | 1.6 | 100 | 1602 | Marble | 2.7 | 168 | 2691 | Steel | 7.9 | 491 | 7865 |
| CPVC | 1.55 | 97 | 1554 | Masonry | 2.4 | 150 | 2403 | Sulfur | 2.0 | 125 | 2002 |
| Cement, Portland (set) | 3.1 | 193 | 3092 | Mercury | 13.56 | 845.3 | 13540 | Tar, bituminous | 1.2 | 75 | 1201 |
| Chalk | 2.3 | 143 | 2291 | Mica | 2.8 | 175 | 2803 | Tile | 1.8 | 112 | 1794 |
| Charcoal | 0.4 | 25 | 400 | Mortar | 1.5 | 94 | 1506 | Trap rock | 3.0 | 187 | 2995 |
| Coal, anthracite | 1.5 | 94 | 1506 | Nickel, cast | 8.3 | 517 | 8282 | Water at 4° C | 1.0 | 62.43 | 1000 |
| Coal, bituminous | 1.3 | 81 | 1297 | Nickel, rolled | 8.7 | 542 | 8682 | Water at 62° F | 1.0 | 62.355 | |
| Concrete | 2.2 | 137 | 2195 | Nylon 6, Cast | 1.16 | 73 | 1169 | White metal | 7.3 | 457 | 7320 |
| Earth, loose | … | 75 | 1201 | PTFE | 2.19 | 137 | 2195 | Zinc, cast | 6.9 | 429 | 6872 |
| Earth, rammed | … | 100 | 1602 | Phosphorus | 1.8 | 112 | 1794 | Zinc, sheet | 7.2 | 450 | 7208 |
| Emery | 4.0 | 249 | 3989 | Plaster of Paris | 1.8 | 112 | 1794 | … | … | … | … |

[a] The weight per cubic foot or cubic meter is calculated on the basis of the specific gravity except for those substances that occur in bulk, heaped, or loose form. In these instances, only the weights per cubic foot and cubic meter are given because the voids present in representative samples make the values of the specific gravities inaccurate.

**Average Weights and Volumes of Solid Fuels**

| Material | lb/ft³ | ft³/ton (2240 lb) | kg/m³ | lb/bushel | m³/t |
|---|---|---|---|---|---|
| Anthracite coal | 55–65 | 34–41 | 881–1041 | 67 (80[a]) | 0.96–1.14 |
| Bituminous coal | 50–55 | 41–45 | 801–881 | 78–86 (80[a]) | 1.14–1.25 |
| Charcoal | 18–18.5 | 121–124 | 288–296 | 22–23 (20[a]) | 3.37–3.47 |
| Coke | 28 | 80 | 449 | 35 (40[a]) | 2.23 |

[a] Legal commodities weight/bushel defined by statute in some states.

*Note:* t = metric ton = 1000 kg; ton = US ton of 2000 lbs; a gross or long ton = 2240 lbs.

**Specific Gravity of Gases.**—The specific gravity of gases is the number that indicates their weight in comparison with that of an equal volume of air. The specific gravity of air is 1, and the comparison is made at 32° F (0° C). Values are given in Table 17.

**Table 17. Specific Gravity of Gases At 32° F (0° C)**

| Gas | Sp. Gr. | Gas | Sp. Gr. | Gas | Sp. Gr. |
|---|---|---|---|---|---|
| Air[a] | 1.000 | Ether vapor | 2.586 | Marsh gas | 0.555 |
| Acetylene | 0.920 | Ethylene | 0.967 | Nitrogen | 0.971 |
| Alcohol vapor | 1.601 | Hydrofluoric acid | 2.370 | Nitric oxide | 1.039 |
| Ammonia | 0.592 | Hydrochloric acid | 1.261 | Nitrous oxide | 1.527 |
| Carbon dioxide | 1.520 | Hydrogen | 0.069 | Oxygen | 1.106 |
| Carbon monoxide | 0.967 | Illuminating gas | 0.400 | Sulfur dioxide | 2.250 |
| Chlorine | 2.423 | Mercury vapor | 6.940 | Water vapor | 0.623 |

[a] 1 cubic foot of air at 32° F and atmospheric pressure weighs 0.0807 pound. 1 cubic meter of air at 0° C and atmospheric pressure weighs 1.29 kg.

**Specific Gravity of Liquids.**—The specific gravity of liquids is the number that indicates how much a certain volume of the liquid weighs compared with an equal volume of water, the same as with solid bodies. Specific gravity of various liquids is given in Table 18.

The density of liquid is often expressed in degrees on the hydrometer, an instrument for determining the density of liquids that is provided with graduations made to an arbitrary scale. The hydrometer consists of a glass tube with a bulb at one end containing air and arranged with a weight at the bottom so as to float in an upright position in the liquid, the density of which is to be measured. The depth to which the hydrometer sinks in the liquid is read off on the graduated scale. The most commonly used hydrometer is the Baumé, see Table 19. The value of the degrees of the Baumé scale differs according to whether the liquid is heavier or lighter than water. The specific gravity for liquids heavier than water equals 145 ÷ (145 – degrees Baumé). For liquids lighter than water, the specific gravity equals 140 ÷ (130 + degrees Baumé).

**Table 18. Specific Gravity of Liquids**

| Liquid | Sp. Gr. | Liquid | Sp. Gr. | Liquid | Sp. Gr. | Liquid | Sp. Gr. |
|---|---|---|---|---|---|---|---|
| Acetic acid | 1.06 | Cotton-seed oil | 0.93 | Naphtha | 0.76 | Tar | 1.00 |
| Alcohol, commercial | 0.83 | Ether, sulfuric | 0.72 | Nitric acid | 1.50 | Turpentine oil | 0.87 |
| Alcohol, pure | 0.79 | Fluoric acid | 1.50 | Olive oil | 0.92 | Vinegar | 1.08 |
| Ammonia | 0.89 | Gasoline | 0.70 | Palm oil | 0.97 | Water | 1.00 |
| Benzine | 0.69 | Kerosene | 0.80 | Petroleum oil | 0.82 | Water, sea | 1.03 |
| Bromine | 2.97 | Linseed oil | 0.94 | Phosphoric acid | 1.78 | Whale oil | 0.92 |
| Carbolic acid | 0.96 | Mineral oil | 0.92 | Rape oil | 0.92 | | |
| Carbon disulfide | 1.26 | Muriatic acid | 1.20 | Sulfuric acid | 1.84 | | |

**Table 19. Degrees on Baumé's Hydrometer Converted to Specific Gravity**

| Deg. Baumé | Specific Gravity for Liquids: Heavier than Water | Specific Gravity for Liquids: Lighter than Water | Deg. Baumé | Specific Gravity for Liquids: Heavier than Water | Specific Gravity for Liquids: Lighter than Water | Deg. Baumé | Specific Gravity for Liquids: Heavier than Water | Specific Gravity for Liquids: Lighter than Water |
|---|---|---|---|---|---|---|---|---|
| 0 | 1.000 | ... | 27 | 1.229 | 0.892 | 54 | 1.593 | 0.761 |
| 1 | 1.007 | ... | 28 | 1.239 | 0.886 | 55 | 1.611 | 0.757 |
| 2 | 1.014 | ... | 29 | 1.250 | 0.881 | 56 | 1.629 | 0.753 |
| 3 | 1.021 | ... | 30 | 1.261 | 0.875 | 57 | 1.648 | 0.749 |
| 4 | 1.028 | ... | 31 | 1.272 | 0.870 | 58 | 1.667 | 0.745 |
| 5 | 1.036 | ... | 32 | 1.283 | 0.864 | 59 | 1.686 | 0.741 |
| 6 | 1.043 | ... | 33 | 1.295 | 0.859 | 60 | 1.706 | 0.737 |
| 7 | 1.051 | ... | 34 | 1.306 | 0.854 | 61 | 1.726 | 0.733 |
| 8 | 1.058 | ... | 35 | 1.318 | 0.849 | 62 | 1.747 | 0.729 |
| 9 | 1.066 | ... | 36 | 1.330 | 0.843 | 63 | 1.768 | 0.725 |
| 10 | 1.074 | 1.000 | 37 | 1.343 | 0.838 | 64 | 1.790 | 0.721 |
| 11 | 1.082 | 0.993 | 38 | 1.355 | 0.833 | 65 | 1.813 | 0.718 |
| 12 | 1.090 | 0.986 | 39 | 1.368 | 0.828 | 66 | 1.836 | 0.714 |
| 13 | 1.099 | 0.979 | 40 | 1.381 | 0.824 | 67 | 1.859 | 0.710 |
| 14 | 1.107 | 0.972 | 41 | 1.394 | 0.819 | 68 | 1.883 | 0.707 |
| 15 | 1.115 | 0.966 | 42 | 1.408 | 0.814 | 69 | 1.908 | 0.704 |
| 16 | 1.124 | 0.959 | 43 | 1.422 | 0.809 | 70 | 1.933 | 0.700 |
| 17 | 1.133 | 0.952 | 44 | 1.436 | 0.805 | 71 | 1.959 | 0.696 |
| 18 | 1.142 | 0.946 | 45 | 1.450 | 0.800 | 72 | 1.986 | 0.693 |
| 19 | 1.151 | 0.940 | 46 | 1.465 | 0.796 | 73 | 2.014 | 0.689 |
| 20 | 1.160 | 0.933 | 47 | 1.480 | 0.791 | 74 | 2.042 | 0.686 |
| 21 | 1.169 | 0.927 | 48 | 1.495 | 0.787 | 75 | 2.071 | 0.683 |
| 22 | 1.179 | 0.921 | 49 | 1.510 | 0.782 | 76 | 2.101 | 0.679 |
| 23 | 1.189 | 0.915 | 50 | 1.526 | 0.778 | 77 | 2.132 | 0.676 |
| 24 | 1.198 | 0.909 | 51 | 1.542 | 0.773 | 78 | 2.164 | 0.673 |
| 25 | 1.208 | 0.903 | 52 | 1.559 | 0.769 | 79 | 2.197 | 0.669 |
| 26 | 1.219 | 0.897 | 53 | 1.576 | 0.765 | 80 | 2.230 | 0.666 |

**How to Estimate the Weight of Natural Piles.**—To calculate the upper and lower limits of weight of a substance piled naturally on a plate, use the following:

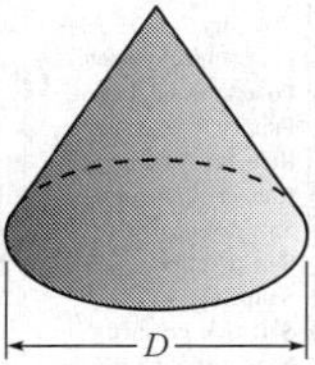


Fig. 2a. Conical Pile

For a substance piled naturally on a circular plate, forming a cone of material,

$$W = MD^3 \tag{1}$$

where $W$ = weight, lb (kg); $D$ = diameter of plate in Fig. 2a, in feet (meters); and, $M$ = a materials factor, whose upper and lower limits are given in Table 20b.

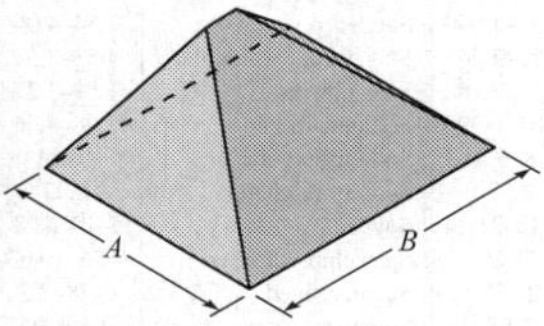


Fig. 2b. Rectangular Pile

For a substance piled naturally on a rectangular plate,

$$W = MRA^3 \tag{2}$$

where $W$ = weight, lb (kg); $A$ and $B$ = the length and width in feet (meters), respectively, of the rectangular plate in Fig. 2b, with $B \leq A$; $M$ = a materials factor, whose upper and lower limits are given in Table 20b; and, $R$ = is a factor given in Table 20a as a function of the ratio $B/A$.

*Example:* Find the upper and lower limits of the weight of dry ashes piled naturally on a plate 10 ft. in diameter.

Using Equation (1), $M = 4.58$ from Table 20b, the lower limit $W = 4.58 \times 10^3 = 4{,}580$ lb. For $M = 5.89$, the upper limit $W = 5.89 \times 10^3 = 5{,}890$ lb.

*Example:* What weight of dry ashes rests on a rectangular plate 10 ft. by 5 ft.?

For $B/A = 5/10 = 0.5$, $R = 0.39789$ from Table 20a. Using Equation (2), for $M = 4.58$, the lower limit $W = 4.58 \times 0.39789 \times 10^3 = 1{,}822$ lb. For $M = 5.89$, the upper limit $W = 5.89 \times 0.39789 \times 10^3 = 2{,}344$lb.

*Example:* What is the weight of a pile of cast iron chips resting on a rectangular plate 4 m by 2 m? For $B/A = 2/4$, $R = 0.39789$ from Table 20a. Using Equation (2), for $M = 17.02$, the lower limit $W = 17.02 \times 0.39789 \times 4^3 = 433$ kg. For $M = 26.18$, the upper limit $W = 26.18 \times 0.39789 \times 4^3 = 667$ kg.

**Table 20a. Factor *R* as a Function of *B*/*A* (*B* ≤ *A*)**

| *B*/*A* | *R* | *B*/*A* | *R* | *B*/*A* | *R* | *B*/*A* | *R* | *B*/*A* | *R* | *B*/*A* | *R* |
|---|---|---|---|---|---|---|---|---|---|---|---|
| 0.01 | 0.00019 | 0.18 | 0.05817 | 0.35 | 0.20666 | 0.52 | 0.42691 | 0.69 | 0.70015 | 0.86 | 1.00761 |
| 0.02 | 0.00076 | 0.19 | 0.06458 | 0.36 | 0.21782 | 0.53 | 0.44170 | 0.70 | 0.71747 | 0.87 | 1.02636 |
| 0.03 | 0.00170 | 0.20 | 0.07130 | 0.37 | 0.22921 | 0.54 | 0.45667 | 0.71 | 0.73491 | 0.88 | 1.04516 |
| 0.04 | 0.00302 | 0.21 | 0.07833 | 0.38 | 0.24085 | 0.55 | 0.47182 | 0.72 | 0.75245 | 0.89 | 1.06400 |
| 0.05 | 0.00470 | 0.22 | 0.08566 | 0.39 | 0.25273 | 0.56 | 0.48713 | 0.73 | 0.77011 | 0.90 | 1.08289 |
| 0.06 | 0.00674 | 0.23 | 0.09329 | 0.40 | 0.26483 | 0.57 | 0.50262 | 0.74 | 0.78787 | 0.91 | 1.10182 |
| 0.07 | 0.00914 | 0.24 | 0.10121 | 0.41 | 0.27717 | 0.58 | 0.51826 | 0.75 | 0.80572 | 0.92 | 1.12078 |
| 0.08 | 0.01190 | 0.25 | 0.10942 | 0.42 | 0.28973 | 0.59 | 0.53407 | 0.76 | 0.82367 | 0.93 | 1.13977 |
| 0.09 | 0.01501 | 0.26 | 0.11792 | 0.43 | 0.30252 | 0.60 | 0.55004 | 0.77 | 0.84172 | 0.94 | 1.15879 |
| 0.10 | 0.01846 | 0.27 | 0.12670 | 0.44 | 0.31552 | 0.61 | 0.56616 | 0.78 | 0.85985 | 0.95 | 1.17783 |
| 0.11 | 0.02226 | 0.28 | 0.13576 | 0.45 | 0.32873 | 0.62 | 0.58243 | 0.79 | 0.87807 | 0.96 | 1.19689 |
| 0.12 | 0.02640 | 0.29 | 0.14509 | 0.46 | 0.34216 | 0.63 | 0.59884 | 0.80 | 0.89636 | 0.97 | 1.21596 |
| 0.13 | 0.03088 | 0.30 | 0.15470 | 0.47 | 0.35579 | 0.64 | 0.61539 | 0.81 | 0.91473 | 0.98 | 1.23505 |
| 0.14 | 0.03569 | 0.31 | 0.16457 | 0.48 | 0.36963 | 0.65 | 0.63208 | 0.82 | 0.93318 | 0.99 | 1.25414 |
| 0.15 | 0.04082 | 0.32 | 0.17471 | 0.49 | 0.38366 | 0.66 | 0.64891 | 0.83 | 0.95169 | 1.00 | 1.27324 |
| 0.16 | 0.04628 | 0.33 | 0.18511 | 0.50 | 0.39789 | 0.67 | 0.66586 | 0.84 | 0.97027 | … | … |
| 0.17 | 0.05207 | 0.34 | 0.19576 | 0.51 | 0.41231 | 0.68 | 0.68295 | 0.85 | 0.98891 | … | … |

**Table 20b. Limits of Factor *M* for Various Materials**

| Material | Factor *M* | Material | Factor *M* | Material | Factor *M* |
|---|---|---|---|---|---|
| Almonds, whole | 2.12–3.93 | Coffee, ground | 1.89–3.27 | Peanuts, unshelled | 1.13–3.14 |
| Aluminum chips | 0.92–1.96 | Coke, pulverized | 2.21 | Peanuts, shelled | 2.65–5.89 |
| Aluminum silicate | 3.7–6.41 | Copper oxide, powdered | 20.87 | Peas, dry | 2.75–3.05 |
| Ammonium chloride | 3.93–6.81 | Cork, granulated | 1.57–1.96 | Potassium carbonate | 3.85–6.68 |
| Asbestos, shredded | 2.62–3.27 | Corn on cob | 1.29–1.33 | Potassium sulphate | 5.5–6.28 |
| Ashes, dry | 4.58–5.89 | Corn sugar | 2.34–4.06 | Pumice | 5.24–5.89 |
| Ashes, damp | 6.24–7.80 | Cottonseed, dry, de-linted | 1.66–5.24 | Rice, bran | 1.51–2.75 |
| Asphalt, crushed | 3.4–5.89 | Diatomaceous earth | 0.83–1.83 | Rubber, scrap, ground | 2.11–4.58 |
| Bakelite, powdered | 3.93–5.24 | Dicalcium phosphate | 5.63 | Salt, dry, coarse | 3.02–8.38 |
| Baking powder | 3.1–5.37 | Ebonite, crushed | 4.91–9.16 | Salt, dry, fine | 5.29–10.47 |
| Barium carbonate | 9.42 | Epsom salts | 3.02–6.54 | Saltpeter | 6.05–10.47 |
| Bauxite, mine run | 5.9–6.69 | Feldspar, ground | 8.51–9.16 | Salt rock, crushed | 4.58 |
| Beans, navy, dry | 3.63 | Fish scrap | 5.24–6.54 | Sand, very fine | 7.36–9 |
| Beets, sugar, shredded | 0.47–0.55 | Flour | 5.61–10.43 | Sawdust, dry | 0.95–2.85 |
| Bicarbonate of soda | 3.10 | Flue dust | 2.65–3.40 | Sesame seed | 2.04–4.84 |
| Borax | 3.78–9.16 | Fluorspar (Fluorite) | 10.73–14.40 | Shellac, powdered | 2.34–4.06 |
| Boric acid | 4.16–7.20 | Graphite, flake | 3.02–5.24 | Slag, furnace, granular | 4.53–8.51 |
| Bronze chips | 3.93–6.54 | Gravel | 6.8–13.18 | Soap powder | 1.51–3.27 |
| Buckwheat | 2.8–3.17 | Gypsum, calcined | 6.04–6.59 | Sodium nitrate | 3.96–4.66 |
| Calcium lactate | 3.4–3.8 | Hominy | 2.8–6.54 | Sodium sulphite | 10.54 |
| Calcium oxide (lime) | 3.30 | Hops, dry | 4.58 | Sodium sulphate | 6.92 |
| Carbon, ground | 2.51 | Kaolin clay | 12.32–21.34 | Soybeans | 3.48–6.28 |
| Casein | 2.72–4.71 | Lead silicate, granulated | 25.26 | Steel chips, crushed | 7.56–19.63 |
| Cashew nuts | 4.19–4.84 | Lead sulphate, pulverized | 24.09 | Sugar, refined | 3.78–7.2 |
| Cast iron chips | 17.02–26.18 | Lime, ground | 7.85 | Sulphur | 4.5–6.95 |
| Cement, Portland | 6.8–13.09 | Limestone, crushed | 6.42–11.78 | Talcum powder | 4.37–5.9 |
| *Cinders, coal* | 3.02–5.24 | Magnesium chloride | 4.32 | Tin oxide, ground | 9.17 |
| Clay, blended for tile | 5.89 | Malt, dry, ground | 1.66–2.88 | Tobacco stems | 1.96–3.27 |
| Coal, anthracite, chestnut | 2.43 | Manganese sulphate | 5.29–9.16 | Trisodium phosphate | 4.53–7.85 |
| Coal, bituminous, sized | 2.64–4.48 | Marble, crushed | 6.8–12.44 | Walnut shells, crushed | 2.65–5.24 |
| Coal, ground | 2.90 | Mica, ground | 1.24–1.43 | Wood chips, fir | 2.49–2.88 |
| Cocoa, powdered | 3.93–4.58 | Milk, whole, powdered | 2.62 | Zinc sulphate | 8.85–11.12 |
| Coconut, shredded | 2.62–2.88 | Oats | 1.74–2.86 | … | … |
| Coffee beans | 2.42–5.89 | Orange peel, dry | 1.96 | … | … |

**Molecular Weight.**—The smallest mass of a chemical combination which can be conceived of as existing and yet preserving its chemical properties is known as a *molecule*. The molecular weight of a chemical compound is equal to the sum of the atomic weights of the atoms contained in the molecule, and is calculated from the atomic weights, when the symbol of the compound is known. See Table 1 on page 362 for atomic weights. The atomic weight of silver is 107.88; of nitrogen, 14.01; and of oxygen, 16; hence, the molecular weight of silver-nitrate, the chemical formula of which is $AgNO_3$ equals $107.88 + 14.01 + (3 \times 16) = 169.89$.

**Mole or Mol.**—This term is used as a designation of quantity in electrochemistry to indicate the number of grams of a substance equal to its molecular weight. For example, one mol of silver-nitrate equals 169.89 grams, the molecular weight of silver-nitrate being 169.89.

**Air.**—Air is a mechanical mixture composed of 78 percent, by volume, of nitrogen, 21 percent of oxygen, and 1 percent of argon. The weight of pure air at 32° F (0° C), at an atmospheric pressure of 29.92 inches of mercury (760 mm mercury or 760 torr) or 14.70 pounds per square inch, is 0.08073 pound per cubic foot. The volume of a pound of air at the same temperature and pressure is 12.387 cubic feet. The weight of air, in pounds per cubic foot, at any other temperature or pressure may be determined by first multiplying the barometer reading (atmospheric pressure in inches of mercury) by 1.325 and then dividing the product by the absolute temperature in degrees F. The absolute zero from which all temperatures must be derived in dealing with the weight and volume of gases, is assumed to be minus 459.67° F (273.15° C). Hence, to obtain the absolute temperature, add to the temperature observed on a regular Fahrenheit thermometer the value 459.67.

**Alligation.**—Alligation or "the rule of mixtures" are names applied to several rules of arithmetical processes for determining the relation between proportions and prices of the ingredients of a mixture and the cost of the mixture per unit of weight or volume. For example, if an alloy is composed of several metals varying in price, the price per pound of the alloy can be found as in the following example: An alloy is composed of 50 pounds of copper at $1.70 a pound, 10 pounds of tin at $4.05 a pound, 20 pounds of zinc at $0.99 a pound, and 5 pounds of lead at $1.10 cents a pound. What is the cost of the alloy per pound, no account being taken of the cost of mixing it? Multiply the number of pounds of each of the ingredients by its price per pound, add these products together, and divide the sum by the total weight of all the ingredients. The quotient is the price per pound of the alloy.

*Example:* The foregoing example would be worked out numerically as follows:

Total cost of materials: $50 \times 1.70 + 10 \times 4.05 + 20 \times 0.99 + 5 \times 1.10 = \$150.80$

Total weight of metal in alloy: $50 + 10 + 20 + 5 = 85$ lbs.

Price per pound of alloy $= 150.80 \div 85 = \$1.77$, approximately.

**Earth or Soil Weight.**—Loose earth has a weight of approximately 75 lb/ft$^3$ (1200 kg/m$^3$) and rammed earth, 100 lb/ft$^3$ (1600 kg/m$^3$).

*Composition of Earth Crust:* The solid crust of the earth, according to an estimate, is composed approximately of the following elements: oxygen 44 to 48.7 percent; silicon 22.8 to 36.2 percent; aluminum 6.1 to 9.9 percent; iron 2.4 to 9.9 percent; calcium 0.9 to 6.6 percent; magnesium 0.1 to 2.7 percent; sodium 2.4 to 2.5 percent; potassium 1.7 to 3.1 percent.

*Loads on Soils and Rocks:* The bearing capacities of soils and rocks is useful in structural engineering and also of value under certain conditions in connection with the installation of very heavy machinery requiring foundations. The ultimate resistance of various soils and rocks will be given in tons per square foot: natural earth that is solid and dry, 4 to 6 tons; thick beds of absolutely dry clay, 4 tons; thick beds of moderately dry clay, 2 tons; soft clay, 1 ton; gravel that is dry, coarse, and well packed, 6 to 8 tons; soft, friable rock and shales, 5 to 10 tons; sand that is compact, dry, and well cemented, 4 tons; natural sand in a clean dry condition, 2 to 4 tons; compact bedrock, northern red sandstone, 20 tons; compact bedrock, northern sound limestone, 25 tons; compact bedrock granite, 30 tons.

# PROPERTIES OF WOOD, CERAMICS, PLASTICS, METALS

## Properties of Wood

**Mechanical Properties of Wood.**—Wood is composed of cellulose, lignin, ash-forming minerals, and extractives formed into a cellular structure. (Extractives are substances that can be removed from wood by extraction with such solvents as water, alcohol, acetone, benzene, and ether.) Variations in the characteristics and volumes of the four components and differences in the cellular structure result in some woods being heavy and some light, some stiff and some flexible, and some hard and some soft. For a single species, the properties are relatively constant within limits; therefore, selection of wood by species alone may sometimes be adequate. However, to use wood most effectively in engineering applications, the effects of physical properties or specific characteristics must be considered.

The mechanical properties listed in the accompanying Table 1 were obtained from tests on small pieces of wood termed "clear" and "straight grained" because they did not contain such characteristics as knots, cross grain, checks, and splits. However, these test pieces did contain such characteristics as growth rings that occur in consistent patterns within the piece. Since wood products may contain knots, cross grain, etc., these characteristics must be taken into account when assessing actual properties or when estimating actual performance. In addition, the methods of data collection and analysis have changed over the years during which the data in Table 1 have been collected; therefore, the appropriateness of the data should be reviewed when used for critical applications such as stress grades of lumber.

Wood is an orthotropic material; that is, its mechanical properties are unique and independent in three mutually perpendicular directions—longitudinal, radial, and tangential. These directions are illustrated in the following figure.

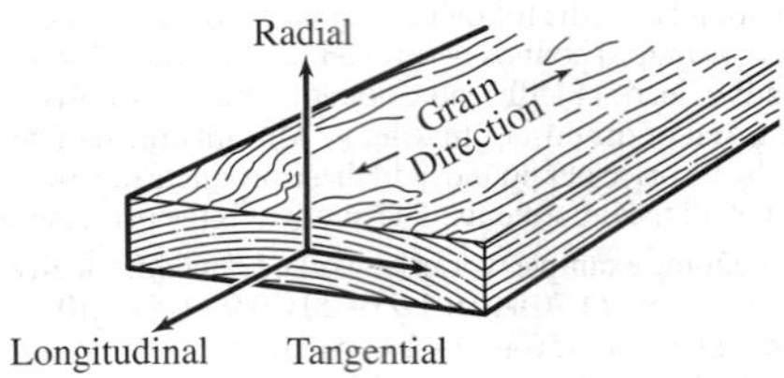


*Modulus of Rupture:* The modulus of rupture in bending reflects the maximum load-carrying capacity of a member and is proportional to the maximum moment borne by the member. The modulus is an accepted criterion of strength, although it is not a true stress because the formula used to calculate it is valid only to the proportional limit.

*Work to Maximum Load in Bending:* The work to maximum load in bending represents the ability to absorb shock with some permanent deformation and more or less injury to a specimen; it is a measure of the combined strength and toughness of the wood under bending stress.

*Maximum Crushing Strength:* The maximum crushing strength is the maximum stress sustained by a compression parallel-to-grain specimen having a ratio of length to least diameter of less than 11.

*Compression Perpendicular to Grain:* Strength in compression perpendicular to grain is reported as the stress at the proportional limit because there is no clearly defined ultimate stress for this property.

*Shear Strength Parallel to Grain:* Shear strength is a measure of the ability to resist internal slipping of one part upon another along the grain. The values listed in the table are averages of the radial and tangential shears.

*Tensile Strength Perpendicular to Grain:* The tensile strength perpendicular to the grain is a measure of the resistance of wood to forces acting across the grain that tend to split the material. Averages of radial and tangential measurements are listed.

**Table 1. Mechanical Properties of Commercially Important US Grown Woods**

| Use the first number in each column for GREEN wood; use the second number for DRY wood. | Static Bending | | | | Maximum Crushing Strength ($10^3$ psi) | | Compression Strength Perpendicular to Grain (psi) | | Shear Strength Parallel to Grain (psi) | | Tensile Strength Perp. to Grain (psi) | |
|---|---|---|---|---|---|---|---|---|---|---|---|---|
| | Modulus of Rupture ($10^3$ psi) | | Work to Max Load (in-lb/in$^3$) | | | | | | | | | |
| Basswood, American | 5.0 | 8.7 | 5.3 | 7.2 | 2.22 | 4.73 | 170 | 370 | 600 | 990 | 280 | 350 |
| Cedar, N. white | 4.2 | 6.5 | 5.7 | 4.8 | 1.90 | 3.96 | 230 | 310 | 620 | 850 | 240 | 240 |
| Cedar, W. red | 5.2 | 7.5 | 5.0 | 5.8 | 2.77 | 4.56 | 240 | 460 | 770 | 990 | 230 | 220 |
| Douglas Fir, coast[a] | 7.7 | 12.4 | 7.6 | 9.9 | 3.78 | 7.23 | 380 | 800 | 900 | 1,130 | 300 | 340 |
| Douglas Fir, interior W. | 7.7 | 12.6 | 7.2 | 10.6 | 3.87 | 7.43 | 420 | 760 | 940 | 1,290 | 290 | 350 |
| Douglas Fir, interior N. | 7.4 | 13.1 | 8.1 | 10.5 | 3.47 | 6.90 | 360 | 770 | 950 | 1,400 | 340 | 390 |
| Douglas Fir, interior S. | 6.8 | 11.9 | 8.0 | 9.0 | 3.11 | 6.23 | 340 | 740 | 950 | 1,510 | 250 | 330 |
| Fir, balsam | 5.5 | 9.2 | 4.7 | 5.1 | 2.63 | 5.28 | 190 | 404 | 662 | 944 | 180 | 180 |
| Hemlock, Eastern | 6.4 | 8.9 | 6.7 | 6.8 | 3.08 | 5.41 | 360 | 650 | 850 | 1,060 | 230 | … |
| Hemlock, Mountain | 6.3 | 11.5 | 11.0 | 10.4 | 2.88 | 6.44 | 370 | 860 | 930 | 1,540 | 330 | … |
| Hemlock, Western | 6.6 | 11.3 | 6.9 | 8.3 | 3.36 | 7.20 | 280 | 550 | 860 | 1,290 | 290 | 340 |
| Pine, E. white | 4.9 | 9.9 | 5.2 | 8.3 | 2.44 | 5.66 | 220 | 580 | 680 | 1,170 | 250 | 420 |
| Pine, Virginia | 7.3 | 13.0 | 10.9 | 13.7 | 3.42 | 6.71 | 390 | 910 | 890 | 1,350 | 400 | 380 |
| Pine, W. white | 4.7 | 9.7 | 5.0 | 8.8 | 2.43 | 5.04 | 190 | 470 | 680 | 1,040 | 260 | … |
| Redwood, old-growth | 7.5 | 10.0 | 7.4 | 6.9 | 4.20 | 6.15 | 420 | 700 | 800 | 940 | 260 | 240 |
| Redwood, young-growth | 5.9 | 7.9 | 5.7 | 5.2 | 3.11 | 5.22 | 270 | 520 | 890 | 1,110 | 300 | 250 |
| Spruce, Engelmann | 4.7 | 9.3 | 5.1 | 6.4 | 2.18 | 4.48 | 200 | 410 | 640 | 1,200 | 240 | 350 |
| Spruce, red | 6.0 | 10.8 | 6.9 | 8.4 | 2.72 | 5.54 | 260 | 550 | 750 | 1,290 | 220 | 350 |
| Spruce, white | 5.0 | 9.4 | 6.0 | 7.7 | 2.35 | 5.18 | 210 | 430 | 640 | 970 | 220 | 360 |

[a] Coast: grows west of the summit of the Cascade Mountains in OR and WA. Interior west: grows in CA and all counties in OR and WA east of but adjacent to the Cascade summit. Interior north: grows in remainder of OR and WA and ID, MT, and WY. Interior south: grows in UT, CO, AZ, and NM.

Results of tests on small, clear, straight-grained specimens. Data for dry specimens are from tests of seasoned material adjusted to a moisture content of 12%.

*Source:* US Department of Agriculture: *Wood Handbook*.

**Weight of Wood.—**The weight of seasoned wood per cord is approximately as follows, assuming about 70 cubic feet of *solid wood* per cord: beech, 3300 pounds; chestnut, 2600 pounds; elm, 2900 pounds; maple, 3100 pounds; poplar, 2200 pounds; white pine, 2200 pounds; red oak, 3300 pounds; white oak, 3500 pounds. For additional weights of green and dry woods, see Table 2.

**Weight per Foot of Wood, Board Measure.—**The following is the weight in pounds of various kinds of woods, commercially known as dry timber, per foot board measure: white oak, 4.16; white pine, 1.98; Douglas fir, 2.65; short-leaf yellow pine, 2.65; red pine, 2.60; hemlock, 2.08; spruce, 2.08; cypress, 2.39; cedar, 1.93; chestnut, 3.43; Georgia yellow pine, 3.17; California spruce, 2.08. For other woods, divide the weight/ft$^3$ from Table 2 by 12 to obtain the approximate weight per board foot.

**Effect of Pressure Treatment on Mechanical Properties of Wood.—**The strength of wood preserved with creosote, coal-tar, creosote-coal-tar mixtures, creosote-petroleum mixtures, or pentachlorophenol dissolved in petroleum oil is not reduced. However, waterborne salt preservatives contain chemicals, such as copper, arsenic, chromium, and ammonia, which have the potential of affecting mechanical properties of treated wood and causing mechanical fasteners to corrode. Preservative salt-retention levels required for marine protection may reduce bending strength by 10 percent or more.

**Density of Wood.—**The following formula can be used to find the density of wood in $lb/ft^3$ as a function of its moisture content.

$$\rho = 62.4\left(\frac{G}{1 + G \times 0.009 \times M}\right)\left(1 + \frac{M}{100}\right)$$

where ρ is the density, $G$ is the specific gravity of wood, and $M$ is the moisture content expressed in percent.

**Table 2. Weights of American Woods, in Pounds per Cubic Foot**

| Species | Green | Airdry | Species | Green | Airdry | Species | Green | Airdry |
|---|---|---|---|---|---|---|---|---|
| Alder, red | 46 | 28 | Douglas fir, Rocky Mt. region | 35 | 30 | Oak, red | 64 | 44 |
| Ash, black | 52 | 34 | Elm, American | 54 | 35 | Oak, white | 63 | 47 |
| Ash, commercial white | 48 | 41 | Elm, rock | 53 | 44 | Pine, lodgepole | 39 | 29 |
| Ash, Oregon | 46 | 38 | Elm, slippery | 56 | 37 | Pine, northern white | 36 | 25 |
| Aspen | 43 | 26 | Fir, balsam | 45 | 25 | Pine, Norway | 42 | 34 |
| Basswood | 42 | 26 | Fir, commercial white | 46 | 27 | Pine, ponderosa | 45 | 28 |
| Beech | 54 | 45 | Gum, black | 45 | 35 | Pines, southern yellow: | | |
| Birch | 57 | 44 | Gum, red | 50 | 34 | Pine, loblolly | 53 | 36 |
| Birch, paper | 50 | 38 | Hemlock, eastern | 50 | 28 | Pine, longleaf | 55 | 41 |
| Cedar, Alaska | 36 | 31 | Hemlock, western | 41 | 29 | Pine, shortleaf | 52 | 36 |
| Cedar, eastern red | 37 | 33 | Hickory, pecan | 62 | 45 | Pine, sugar | 52 | 25 |
| Cedar, northern white | 28 | 22 | Hickory, true | 63 | 51 | Pine, western white | 35 | 27 |
| Cedar, southern white | 26 | 23 | Honeylocust | 61 | … | Poplar, yellow | 38 | 28 |
| Cedar, western red | 27 | 23 | Larch, western | 48 | 36 | Redwood | 50 | 28 |
| Cherry, black | 45 | 35 | Locust, black | 58 | 48 | Spruce, eastern | 34 | 28 |
| Chestnut | 55 | 30 | Maple, bigleaf | 47 | 34 | Spruce, Engelmann | 39 | 23 |
| Cottonwood, eastern | 49 | 28 | Maple, black | 54 | 40 | Spruce, Sitka | 33 | 28 |
| Cottonwood, northern black | 46 | 24 | Maple, red | 50 | 38 | Sycamore | 52 | 34 |
| Cypress, southern | 51 | 32 | Maple, silver | 45 | 33 | Tamarack | 47 | 37 |
| Douglas fir, coast region | 38 | 34 | Maple, sugar | 56 | 44 | Walnut, black | 58 | 38 |

*Source:* United States Department of Agriculture

**Machinability of Wood.—**The ease of working wood with hand tools generally varies directly with the specific gravity of the wood; the lower the specific gravity, the easier the wood is to cut with a sharp tool. A rough idea of the specific gravity of various woods can be obtained from the preceding table by dividing the weight of wood in $lb/ft^3$ by 62.355.

A wood species that is easy to cut does not necessarily develop a smooth surface when it is machined. Three major factors, other than specific gravity, influence the smoothness of the surface obtained by machining: interlocked and variable grain, hard deposits in the grain, and reaction wood. Interlocked and variable grain is a characteristic of many tropical and some domestic species; this type of grain structure causes difficulty in planing quarter sawn boards unless careful attention is paid to feed rates, cutting angles, and sharpness of the knives. Hard deposits of calcium carbonate, silica, and other minerals in the grain tend to dull cutting edges quickly, especially in wood that has been dried to the usual in service moisture content. Reaction wood results from growth under some physical stress, such as occurs in leaning trunks and crooked branches. Generally, reaction wood occurs as tension wood in hardwoods and as compression wood in softwoods. Tension wood is particularly troublesome, often resulting in fibrous and fuzzy surfaces, especially in woods of lower density. Reaction wood may also be responsible for pinching saw blades, resulting in burning and dulling of teeth.

The Table 3 rates the suitability of various domestic hardwoods for machining. The data for each species represent the percentage of pieces machined that successfully met the listed quality requirement for the processes. For example, 62 percent of the black walnut pieces planed came out perfect, but only 34 percent of the pieces run on the shaper achieved good to excellent results.

**Table 3. Machinability and Related Properties of Various Domestic Hardwoods**

| | Planing | Shaping | Turning | Boring | Mortising | Sanding |
|---|---|---|---|---|---|---|
| | Quality Required | | | | | |
| Type of Wood | Perfect | Good to Excellent | Fair to Excellent | Good to Excellent | Fair to Excellent | Good to Excellent |
| Alder, red | 61 | 20 | 88 | 64 | 52 | … |
| Ash | 75 | 55 | 79 | 94 | 58 | 75 |
| Aspen | 26 | 7 | 65 | 78 | 60 | … |
| Basswood | 64 | 10 | 68 | 76 | 51 | 17 |
| Beech | 83 | 24 | 90 | 99 | 92 | 49 |
| Birch | 63 | 57 | 80 | 97 | 97 | 34 |
| Birch, paper | 47 | 22 | … | … | … | … |
| Cherry, black | 80 | 80 | 88 | 100 | 100 | … |
| Chestnut | 74 | 28 | 87 | 91 | 70 | 64 |
| Cottonwood | 21 | 3 | 70 | 70 | 52 | 19 |
| Elm, soft | 33 | 13 | 65 | 94 | 75 | 66 |
| Hackberry | 74 | 10 | 77 | 99 | 72 | … |
| Hickory | 76 | 20 | 84 | 100 | 98 | 80 |
| Magnolia | 65 | 27 | 79 | 71 | 32 | 37 |
| Maple, bigleaf | 52 | 56 | 8 | 100 | 80 | … |
| Maple, hard | 54 | 72 | 82 | 99 | 95 | 38 |
| Maple, soft | 41 | 25 | 76 | 80 | 34 | 37 |
| Oak, red | 91 | 28 | 84 | 99 | 95 | 81 |
| Oak, white | 87 | 35 | 85 | 95 | 99 | 83 |
| Pecan | 88 | 40 | 89 | 100 | 98 | … |
| Sweetgum | 51 | 28 | 86 | 92 | 53 | 23 |
| Sycamore | 22 | 12 | 85 | 98 | 96 | 21 |
| Tanoak | 80 | 39 | 81 | 100 | 100 | … |
| Tupelo, black | 48 | 32 | 75 | 82 | 24 | 21 |
| Tupelo, water | 55 | 52 | 79 | 62 | 33 | 34 |
| Walnut, black | 62 | 34 | 91 | 100 | 98 | … |
| Willow | 52 | 5 | 58 | 71 | 24 | 24 |
| Yellow-poplar | 70 | 13 | 81 | 87 | 63 | 19 |

The data above represent the percentage of pieces attempted that meet the quality requirement listed.

**Nominal and Minimum Sizes of Sawn Lumber**

| Type of Lumber | Thickness (inches) | | | Face Widths (inches) | | |
|---|---|---|---|---|---|---|
| | Nominal, $T_n$ | Dry | Green | Nominal, $W_n$ | Dry | Green |
| Boards | 1 | ¾ | 25⁄32 | 2 to 4 | $W_n$ − ½ | $W_n$ − 7⁄16 |
| | 1¼ | 1 | 1 1⁄32 | 5 to 7 | $W_n$ − ½ | $W_n$ − ⅜ |
| | 1½ | 1¼ | 1 9⁄32 | 8 to 16 | $W_n$ − ¾ | $W_n$ − ½ |
| Dimension Lumber | 2 | 1½ | 1 9⁄16 | 2 to 4 | $W_n$ − ½ | $W_n$ − 7⁄16 |
| | 2½ | 2 | 2 1⁄16 | 5 to 6 | $W_n$ − ½ | $W_n$ − ⅜ |
| | 3 | 2½ | 2 9⁄16 | 8 to 16 | $W_n$ − ¾ | $W_n$ − ½ |
| | 3½ | 3 | 3 1⁄16 | … | … | … |
| | 4 | 3½ | 3 9⁄16 | … | … | … |
| | 4½ | 4 | 4 1⁄16 | … | … | … |
| Timbers | 5 and up | … | $T_n$ − ½ | 5 and up | … | $W_n$ − ½ |

*Source:* National Forest Products Association: *Design Values for Wood Construction.* Moisture content: dry lumber ≤ 19%; green lumber > 19%. Dimension lumber refers to lumber 2 to 4 inches thick (nominal) and 2 inches or greater in width. Timbers refers to lumber of approximately square cross section, 5 × 5 inches or larger, and a width no more than 2 inches greater than the thickness.

## Tabulated Properties of Ceramics, Plastics, and Metals

### Typical Properties of Ceramics Materials

| Material | | Density[a] lb/in³ | Density[a] g/cm3 | Dielectric Strength (V/mil) | Coeff. of Expansion[b] 10⁻⁶ in/in-°F | Coeff. of Expansion[b] 10⁻⁶ m/m-°C | Flexural Strength (10³ psi) | Mohs's Hardness[c] | Operating Temp. (°F) | Tensile Strength 10³ psi | Tensile Strength MPa | Compressive Strength (10³ psi) | Thermal Conductivity[d] (Btu-ft/hr-ft²-°F) | Thermal Conductivity[d] W/(m-k) |
|---|---|---|---|---|---|---|---|---|---|---|---|---|---|---|
| Machinable Glass Ceramic | | 0.09 | 2.49 | 1000 | 4.1–7.0 | 7.38–12.6 | | 48 RA | 1472 | … | … | 50 | 0.85 | 1.47 |
| | | 0.11 | 3.04 | 400 | 6 | 10.8 | 15 | 5.5 | 700 | … | … | 40 | 0.24 | 0.42 |
| | | 0.10 | 2.77 | 380 | 5.2 | 9.4 | 14 | 5.0 | 1100 | … | … | 32 | 0.34 | 0.59 |
| Glass-Mica | Machining Grades | 0.09–0.10 | 2.49–2.77 | 400 | 10.5–11.2 | 18.9–20.2 | 12.5–13 | 90 RH | 750 | 6 | 41 | 40–45 | 0.24–0.29 | 0.41–0.50 |
| | | 0.10 | 2.77 | 380 | 9.4 | 16.9 | 11 | 90 RH | 1100 | 5 | 34 | 32 | 0.34 | 0.59 |
| | Molding Grades | 0.13–0.17 | 3.60–4.70 | 300–325 | 11–11.5 | 19.8–20.7 | 9–10 | 90 RH | 700–750 | 6–6.5 | 41–45 | 33–35 | 0.29–0.31 | 0.50–0.54 |
| | | 0.14 | 3.88 | 350 | 10.3 | 18.5 | 9 | 90 RH | 1300 | 6 | 41 | 30 | 0.3 | 0.52 |
| Aluminum Silicate | | 0.10 | 2.77 | 80 | 2.5 | 4.5 | 4.5 | 1–2 | 1000 | … | … | 12 | 0.92 | 1.59 |
| | | 0.08 | 2.21 | 100 | 2.9 | 5.2 | 10 | 6.0 | 2100 | … | … | 25 | 0.75 | 1.30 |
| Alumina Silicate | | 0.08 | 2.21 | 70 | … | … | … | … | 2370 | … | … | … | 0.38 | 0.66 |
| Silica Foam | | 0.03 | 0.83 | 80 | 0.3 | 0.5 | 0.4 | NA | 2000 | … | … | 1.4 | 0.10 | 0.17 |
| $TiO_2$ (Titania) | | 0.14 | 3.88 | 100 | 4.61 | 8.3 | 20 | 8 | 1800 | 7.5 | 52 | 100 | … | … |
| Lava (Grade A) | | 0.08 | 2.21 | 80 | 1.83 | 3.3 | 9 | 6 | 2000 | 2.5 | 17 | 40 | 0.92 | 1.59 |
| Zirconium Phosphate | | 0.11 | 3.04 | NA | 0.5 | 0.9 | 7.5 | NA | 2800 | … | … | 30 | 0.4 (approx.) | 0.69 |
| $ZrO_2$ | | 0.21 | 5.81 | … | 6.1 | 11.0 | 102 | 1300 HV | … | … | … | 261 | 1.69 | 2.92 |
| $ZrO_2 \cdot SiO_2$ (Zircon) | | 0.11 | 3.04 | 220 | 1.94 | 3.5 | 16 | 7.5 | 1825 | 10 | 69 | 90 | … | … |
| $2MgO \cdot SiO_2$ (Forsterite) | | 0.11 | 3.04 | 240 | 5.56 | 10.0 | 20 | 7.5 | 1825 | 10 | 69 | 85 | 4.58 | 7.93 |
| $MgO \cdot SiO_2$ (Steatite) | | 0.09–0.10 | 2.49–2.77 | 210–240 | 3.83–5.44 | 6.89–9.79 | 18–21 | 7.5 | 1825 | 8.5–10 | 59–69 | 80–90 | 3.17–3.42 | 5.49–5.92 |
| $2MgO \cdot 2Al_2O_3 \cdot 5SiO_2$ (Cordierite) | | 0.06 | 1.66 | 60 | 0.33 | 0.6 | 3.4 | 6.5 | 2000 | 2.5 | 17 | 18.5 | 1.00 | 1.73 |
| | | 0.08 | 2.21 | 100–172 | 1.22–1.28 | 2.20–2.30 | 8–12 | 7–7.5 | 2000 | 3.5–3.7 | 24–25 | 30–40 | 1.00 | 1.73 |
| | | 0.09 | 2.49 | 200 | 1.33 | 2.4 | 15 | 8 | 2000 | 4 | 28 | 50 | 1.83 | 3.17 |
| $Al_2O_3$ (Alumina) | 94% | 0.13 | 3.60 | 210 | 3.33 | 6.0 | 44 | 9 | 2700 | 20 | 138 | 315 | 16.00 | 27.69 |
| | 96% | 0.13–0.14 | 3.60–3.88 | 210 | 3.5–3.7 | 6.3–6.6 | 48–60 | 9 | 2600–2800 | 25 | 172 | 375 | 20.3–20.7 | 35.13–35.8 |
| | 99.5% | 0.14 | 3.88 | 200 | 3.72 | 6.7 | 70 | 9 | 2700 | 28 | 193 | 380 | 21.25 | 36.78 |
| | 99.9% | 0.14 | 3.88 | … | 3.75 | 6.8 | 72 | 9 | 2900 | … | … | 400 | … | … |

[a] Obtain specific gravity by dividing density in lb/in³ by 0.0361; for density in lb/ft³, multiply lb/in³ by 1728; for kg/m³, multiply density in lb/in³ by 27,679.9.

[b] To convert coefficient of expansion to $10^{-6}$ in/in-°C, multiply table value by 1.8.

[c] Mohs's Hardness scale is used unless otherwise indicated as follows: RA and RH for Rockwell A and H scales, respectively; HV for Vickers hardness.

[d] To convert conductivity from Btu-ft/hr-ft²-°F to cal-cm/sec-cm²-°C, divide by 241.9.

## Typical Properties of Plastics Materials

| Material | Density[a] | | Specific Gravity | Dielectric Strength | | Coeff. of Expansion[b] | | Tensile Modulus | | Izod Impact | | Flexural Modulus | | % Elongation | Hardness[b] | Max. Operating Temp. | |
|---|---|---|---|---|---|---|---|---|---|---|---|---|---|---|---|---|---|
| | lb/in³ | g/cm³ | | V/mil | MV/m | $10^{-6}$ in/in-°F | $10^{-6}$ m/m-°C | $10^3$ psi | MPa | ft-lb/in of notch | J/m of notch | ksi at 73° F | MPa at 23° C | | | °F | °C |
| ABS, Extrusion Grade | 0.038 | 1.052 | 1.05 | … | … | 53.0 | 95.4 | 275 | 1896 | 7 | 373.65 | 300 | 2068 | … | 105 Rr | 200 | 93 |
| ABS, High Impact | 0.037 | 1.024 | 1.03 | … | … | … | … | 200 | 1379 | … | … | 330 | 2275 | … | 105 Rr | … | … |
| Acetal, 20% Glass | 0.056 | 1.550 | 1.55 | … | … | … | … | 1000 | 6895 | 0.9 | 48.04 | 715 | 4930 | … | 94 Rm | … | … |
| Acetal, Copolymer | 0.051 | 1.412 | 1.41 | 380 | 14.96 | 47.0 | 84.6 | 437 | 3013 | 2 | 106.76 | 400 | 2758 | 13 | 94 Rm | … | … |
| Acetyl, Homopolymer | 0.051 | 1.412 | 1.41 | … | … | 58.0 | 104.4 | 310 | 2137 | … | … | 320 | 2206 | … | 94 Rm | 200 | 93 |
| Acrylic | 0.043 | 1.190 | 1.19 | 500 | 19.69 | 35.0 | 63.0 | 400 | 2758 | 0.5 | 26.69 | 400 | 2758 | 2.7 | 94 Rm | 180 | 82 |
| Azdel | 0.043 | 1.190 | 1.19 | 500 | 19.69 | 15.0 | 27.0 | 750 | 5171 | 14 | 747.30 | 800 | 5516 | 2.1 | 94 Rm | 311 | 155 |
| CPVC | 0.056 | 1.550 | 1.55 | … | … | 34.0 | 61.2 | 400 | 2758 | 3 | 160.14 | 400 | 2758 | 4 | … | 212 | 100 |
| Fiberglass Sheet | 0.067 | 1.855 | 1.87 | … | … | 11.1 | 20.0 | … | … | 8 | 427.03 | 1 | 7 | … | 101 Rm | 260 | 127 |
| Nylon 6, 30% Glass | 0.050 | 1.384 | 1.39 | … | … | … | … | 1350 | 9308 | 2.8 | 149.46 | 1400 | 9653 | … | 119 Rr | … | … |
| Nylon 6, Cast | 0.042 | 1.163 | 1.16 | 295 | 11.61 | 45.0 | 81.0 | 380 | 2620 | 1.4 | 74.73 | 450 | 3103 | 20 | 100 Rr | 210 | 99 |
| Nylon 6/6, Cast | 0.047 | 1.301 | 1.30 | … | … | … | … | … | … | … | … | … | … | … | … | … | … |
| Nylon 6/6, Extruded | 0.041 | 1.135 | 1.14 | 600 | 23.62 | 45.0 | 81.0 | 390 | 2689 | 1 | 53.38 | … | … | 240 | 118 Rr | 230 | 110 |
| Nylon 60L, Cast | 0.042 | 1.163 | 1.16 | … | … | … | … | … | … | 2.2 | 117.43 | … | … | … | … | … | … |
| PET, unfilled | 0.049 | 1.356 | 1.36 | 1300 | 51.18 | 39.0 | 70.2 | 500 | 3447 | 0.5 | 26.69 | 400 | 2758 | 70 | … | 230 | 110 |
| PTFE (Teflon) | 0.079 | 2.187 | 2.19 | 480 | 18.90 | 50.0 | 90.0 | 225 | 1551 | 3 | 160.14 | 80 | 552 | 350 | … | … | … |
| PVC | 0.050 | 1.384 | 1.39 | 500 | 19.69 | 29.5 | 53.1 | 550 | 3792 | 0.8 | 42.70 | 400 | 2758 | 31–40 | 110 Rr | 170 | 77 |
| PVDF | 0.064 | 1.772 | 1.77 | 260 | 10.24 | 60.0 | 108.0 | 320 | 2206 | 3 | 160.14 | 200 | 1379 | 80 | 100 Rr | 180 | 82 |
| Phenolics | 0.050 | 1.384 | 1.38 | … | … | 11.1 | 20.0 | … | … | 2.4 | 128.11 | 1000 | 6895 | … | 100 Rm | 248 | 120 |
| Polycarbonate | 0.043 | 1.190 | 1.19 | 380 | 14.96 | 37.5 | 67.5 | 345 | 2379 | 14 | 747.30 | 340 | 2344 | 110 | 74 Rm | 290 | 143 |
| Polyetherimide | 0.046 | 1.273 | 1.27 | 480 | 18.90 | … | … | 430 | 2965 | 1.1 | 58.72 | 480 | 3309 | … | … | … | … |
| Polyethylene, HD | 0.035 | 0.969 | 0.97 | 475 | 18.70 | 20.0 | 36.0 | 156 | 1076 | 6 | 320.27 | 160 | 1103 | 900 | … | 180 | 82 |
| Polyethylene, UHMW | 0.034 | 0.941 | 0.94 | 710 | 27.95 | 19.0 | 34.2 | 110 | 758 | No Break | … | 130 | 896 | 450 | 64 Rr | 176 | 80 |
| Polymethylpentene | 0.030 | 0.830 | 0.83 | … | … | … | … | 220 | 1517 | 2.5 | 133.45 | … | … | … | … | … | … |
| Polymid, unfilled | 0.051 | 1.412 | 1.41 | 560 | 22.05 | … | … | 300 | 2068 | 1.5 | 80.07 | … | … | … | … | … | … |
| Polyphenylene Sulfide | 0.047 | 1.301 | 1.30 | 380 | 14.96 | … | … | … | … | 0.5 | 26.69 | 550 | 3792 | … | … | … | … |
| Polypropylene | 0.033 | 0.913 | 0.91 | 600 | 23.62 | 96.0 | 172.8 | 155 | 1069 | 0.75 | 40.03 | 200 | 1379 | 120 | 92 Rr | 150 | 66 |
| Polysulfone | 0.045 | 1.246 | 1.25 | 425 | 16.73 | 31.0 | 55.8 | 360 | 2482 | 1.2 | 64.05 | 390 | 2689 | 50 | 120 Rr | 325 | 163 |
| Polyurethane | 0.038 | 1.052 | 1.05 | … | … | … | … | … | … | … | … | … | … | 465–520 | … | … | … |

[a] To obtain specific gravity, divide density in lb/in³ by 0.0361; for density in lb/ft³, multiply lb/in³ by 1728; for kg/m³, multiply density in lb/in³ by 27,679.9.

[b] To convert coefficient of expansion to $10^{-6}$ in/in-°C, multiply table value by 1.8.

[c] Hardness value scales are: RM and RR for Rockwell M and R scales, respectively.

## Mechanical Properties of Various Investment Casting Alloys

| Alloy Designation | Material Condition | Tensile Strength ($10^3$ psi) | 0.2% Yield Strength[a] ($10^3$ psi) | % Elongation | Hardness[c] |
|---|---|---|---|---|---|
| | | Aluminum | | | |
| 356 | As Cast | 32–40 | 22–30 | 3–7 | ... |
| A356 | As Cast | 38–40 | 28–36 | 3–10 | ... |
| A357 | As Cast | 33–50 | 27–40 | 3–9 | ... |
| 355, C355 | As Cast | 35–50 | 28–39 | 1–8 | ... |
| D712 (40E) | As Cast | 34–40 | 25–32 | 4–8 | ... |
| A354 | As Cast | 47–55 | 36–45 | 2–5 | ... |
| RR-350 | As Cast | 32–45 | 24–38 | 1.5–5 | ... |
| Precedent 71 | As Cast | 35–55 | 25–45 | 2–5 | ... |
| KO-1 | As Cast | 56–60 | 48–55 | 3–5 | ... |
| | | Copper-Base Alloys[a] | | | |
| Al Bronze C (954) | As Cast | 75–85 | 30–40 | 10–20 | 80–85 RB |
| | Heat-Treated | 90–105 | 45–55 | 6–10 | 91–96 RB |
| Al Bronze D (955) | As Cast | 90–100 | 40–50 | 6–10 | 91–96 RB |
| | Heat-Treated | 110–120 | 60–70 | 5–8 | 93–98 RB |
| Manganese Bronze, A | ... | 65–75 | 25–40 | 16–24 | 60–65 RB |
| Manganese Bronze, C | ... | 110–120 | 60–70 | 8–16 | 95–100 RB |
| Silicon Bronze | ... | 45 | 18 | 20 | ... |
| Tin Bronze | ... | 40–50 | 18–30 | 20–35 | 40–50 RB |
| Leaded Yellow Brass (854) | ... | 30–50 | 11–20 | 15–25 | ... |
| Red Brass | ... | 30–40 | 14–25 | 20–30 | 30–35 RB |
| Silicon Brass | ... | 70 | 32 | 24 | ... |
| Pure Copper | ... | 20–30 | ... | 4–50 | 35–42 RB |
| Beryllium Cu 10C (820) | As Cast | 45–50 | 40–45 | 15–20 | 50–55 RB |
| | Hardened | 90–100 | 90–130 | 3–8 | 90–95 RB |
| Beryllium Cu 165C (824) | ... | 70–155 | 40–140 | 1–15 | 60 RB–38 RC |
| Beryllium Cu 20C (825) | As Cast | 70–80 | 50–55 | 18–23 | 75–80 RB |
| | Hardened | 110–160 | ... | 1–4 | 25–44 RC |
| Beryllium Cu 275C (828) | As Cast | 80–90 | ... | 15–20 | 80–85 RB |
| Chrome Copper | ... | 33–50 | 20–40 | 20–30 | 70–78 RB |
| | | Carbon and Low-Alloy Steels and Iron | | | |
| IC 1010 | Annealed | 50–60 | 30–35 | 30–35 | 50–55 RB |
| IC 1020 | Annealed | 60–70 | 40–45 | 25–40 | 80 RB |
| IC 1030 | Annealed | 65–75 | 45–50 | 20–30 | 75 RB |
| | Hardened | 85–150 | 60–150 | 0–15 | 20–50 RC |
| IC 1035 | Annealed | 70–80 | 45–55 | 20–30 | 80 RB |
| | Hardened | 90–150 | 85–150 | 0–15 | 25–52 RC |
| IC 1045 | Annealed | 80–90 | 50–60 | 20–25 | 100 RB |
| | Hardened | 100–180 | 90–180 | 0–10 | 25–57 RC |
| IC 1050 | Annealed | 90–110 | 50–65 | 20–25 | 100 RB |
| | Hardened | 125–180 | 100–180 | 0–10 | 30–60 RC |
| IC 1060 | Annealed | 100–120 | 55–70 | 5–10 | 25 RC |
| | Hardened | 120–200 | 100–180 | 0–3 | 30–60 RC |
| IC 1090 | Annealed | 110–150 | 70–80 | 12–20 | 30 RC |
| | Hardened | 130–180 | 130–180 | 0–3 | 37–50 RC |
| IC 2345 | Hardened | 130–200 | 110–180 | 5–10 | 30–58 RC |
| IC 4130 | Hardened | 130–170 | 100–130 | 5–20 | 23–49 RC |
| IC 4140 | Hardened | 130–200 | 100–155 | 5–20 | 29–57 RC |
| IC 4150 | Hardened | 140–200 | 120–180 | 5–10 | 25–58 RC |
| IC 4330 | Hardened | 130–190 | 100–175 | 5–20 | 25–48 RC |
| IC 4340 | Hardened | 130–200 | 100–180 | 5–20 | 20–55 RC |
| IC 4620 | Hardened | 110–150 | 90–130 | 10–20 | 20–32 RC |
| IC 6150, IC 8740 | Hardened | 140–200 | 120–180 | 5–10 | 30–60 RC |
| IC 8620 | Hardened | 100–130 | 80–110 | 10–20 | 20–45 RC |
| IC 8630 | Hardened | 120–170 | 100–130 | 7–20 | 25–50 RC |
| IC 8640 | Hardened | 130–200 | 100–180 | 5–20 | 30–60 RC |

**Mechanical Properties of Various Investment Casting Alloys** *(Continued)*

| Alloy Designation | Material Condition | Tensile Strength ($10^3$ psi) | 0.2% Yield Strength[a] ($10^3$ psi) | % Elongation | Hardness[c] |
|---|---|---|---|---|---|
| | | Carbon and Low-Alloy Steels and Iron *(Continued)* | | | |
| IC 8665 | Hardened | 170–220 | 140–200 | 0–10 | … |
| IC 8730 | Hardened | 120–170 | 110–150 | 7–20 | … |
| IC 52100 | Hardened | 180–230 | 140–180 | 1–7 | 30–65 RC |
| IC 1722AS | Hardened | 130–170 | 100–140 | 6–12 | 25–48 RC |
| 1.2% Si Iron | … | 50–60 | 37–43 | 30–35 | 55 RB |
| Ductile Iron, Ferritic | Annealed | 60–80 | 40–50 | 18–24 | 143–200 BHN |
| Ductile Iron, Pearlitic | Normalized | 100–120 | 70–80 | 3–10 | 243–303 BHN |
| | | Hardenable Stainless Steel | | | |
| CA-15 | Hardened | 95–200 | 75–160 | 5–12 | 94 RB–45 RC |
| IC 416 | Hardened | 95–200 | 75–160 | 3–8 | 94 RB–45 RC |
| CA-40 | Hardened | 200–225 | 130–210 | 0–5 | 30–52 RC |
| IC 431 | Hardened | 110–160 | 75–105 | 5–20 | 20–40 RC |
| IC 17-4 | Hardened | 150–190 | 140–160 | 6–20 | 34–44 RC |
| Am-355 | Hardened | 200–220 | 150–165 | 6–12 | … |
| IC 15-5 | Hardened | 135–170 | 110–145 | 5–15 | 26–38 RC |
| CD-4M Cu | Annealed | 100–115 | 75–85 | 20–30 | 94–100 RB |
| | Hardened | 135–145 | 100–120 | 10–25 | 28–32 RC |
| | | Austenitic Stainless Steels | | | |
| CF-3, CF-3M, CF-8, CF-8M, IC 316F | Annealed | 70–85 | 40–50 | 35–50 | 90 RB (max) |
| CF-8C | Annealed | 70–85 | 32–36 | 30–40 | 90 RB (max) |
| CF-16F | Annealed | 65–75 | 30–35 | 35–45 | 90 RB (max) |
| CF-20 | Annealed | 65–75 | 30–45 | 35–60 | 90 RB (max) |
| CH-20 | Annealed | 70–80 | 30–40 | 30–45 | 90 RB (max) |
| CN-7M | Annealed | 65–75 | 25–35 | 35–45 | 90 RB (max) |
| IC 321, CK-20 | Annealed | 65–75 | 30–40 | 35–45 | 90 RB (max) |
| | | Nickel-Base Alloys | | | |
| Alloy B | Annealed | 75–85 | 50–60 | 8–12 | 90–100 RB |
| Alloy C | As Cast | 80–95 | 45–55 | 8–12 | 90–100 RB |
| | Annealed | 75–95 | 45–55 | 8–12 | 90 RB–25 RC |
| Alloy X[b] | AC to 24° C | 63–70 | 41–45 | 10–15 | 85–96 RB |
| | AC to 816° C | 35–45 | … | 12–20 | … |
| Invar (Fe-Ni alloy) | As Cast | 50–60 | 25–30 | 30–40 | 50–60 RB |
| In 600 (Inconel) | As Cast | 65–75 | 35–40 | 10–20 | 80–90 RB |
| In 625 (Inconel) | Annealed | 80–100 | 40–55 | 15–30 | 10–20 RC |
| Monel 410 | As Cast | 65–75 | 32–38 | 25–35 | 65–75 RB |
| S Monel | Annealed | 100–110 | 55–65 | 5–10 | 20–28 RC |
| | Hardened | 120–140 | 85–100 | 0 | 32–38 RC |
| RH Monel | As Cast | 100–110 | 60–80 | 10–20 | 20–30 RC |
| Monel E | As Cast | 65–80 | 33–40 | 25–35 | 67–78 RB |
| M-35 Monel | As Cast | 65–80 | 25–35 | 25–40 | 65–85 RB |
| | | Cobalt-Base Alloys | | | |
| Cobalt 21 | As Cast | 95–130 | 65–95 | 8–20 | 24–32 RC |
| Cobalt 25 | As Cast | 90–120 | 60–75 | 15–25 | 20–25 RC |
| Cobalt 31 | As Cast | 105–130 | 75–90 | 6–10 | 20–30 RC |
| Cobalt 36 | As Cast | 90–105 | 60–70 | 15–20 | 30–36 RC |
| F75 | As Cast | 95–110 | 70–80 | 8–15 | 25–34 RC |
| N-155 | Sol. Anneal | 90–100 | 50–60 | 15–30 | 90–100 RB |

[a] For copper alloys, yield strength is determined by 0.5% extension under load or 0.2% offset method. A number in parentheses following a copper alloy indicates the UNS designation of that alloy (for example, Al Bronze C (954) identifies the alloy as UNS C95400).

[b] AC = air cooled to temperature indicated.

[c] Hardness value scales are: RB and RC for Rockwell B and C scales, respectively, and BHN for Brinell Hardness Numbers.

*Source:* Investment Casting Institute. Mechanical properties are average values of separately cast test bars and are for reference only. For items marked … data are not available. Alloys identified by IC followed by an SAE designation number (IC 1010 steel, for example) are generally similar to the SAE material although properties and chemical composition may be different.

## Typical Properties of Compressed and Sintered Powdered Metal Alloys

| Alloy Number[a] and Nominal Composition (%) | | Density (g/cc) | Hardness[b] | Strength ($10^3$ psi) | | | % Elongation |
|---|---|---|---|---|---|---|---|
| | | | | Transverse Rupture | Ultimate Tensile | Yield | |
| Copper Base | | | | | | | |
| … | 100Cu | 7.7–7.9 | 81–82 RH | 54–68 | 24–34 | … | 10–26 |
| CZP-3002 | 70Cu, 1.5Pb, Bal. Zn | 8 | 75 RH | … | 33.9 | … | 24 |
| CNZ-1818 | 63Cu, 17.5Ni, Bal. Zn | 7.9 | 90 RH | 73 | 34 | 20 | 11 |
| CTG-1004 | 10Sn, 4.4C, Bal. Cu | 7 | 67 RH | 20 | 9.4 | 6.5 | 6 |
| CTG-1001 | 10Sn, 1C, Bal. Cu | 6.5 | 45 RH | 25.8 | 15.1 | 9.6 | 9.7 |
| Iron Base (Balance of composition, Fe) | | | | | | | |
| FC-2015 | 23.5Cu, 1.5C | 6.5 | 65 RB | 80 | 52.4 | 48.5 | 0 |
| FC-0800 | 8Cu, 0.4C | 6.3–6.8 | 39–55 RB | 75–100 | 38–54 | 32–47 | 1 or less |
| FX-2008 | 20Cu, 1C | 7.3 | 93 RB | 164.2 | 72.3 | 57.7 | 2 |
| FN-0408 | 4Ni, 1-2Cu, 0.75C | 6.3–7 | 64–84 RB | 70–107 | 37–63 | 30–47 | 1–1.6 |
| F-0000 | 100Fe | 6.5 | 26 RF | 37.7 | 15.7 | 11 | 5.7 |
| FN-0005 | 0.45C, 0.50 MnS | 6.4–6.8 | 66–78 RF | 44–61 | … | … | … |
| F-0000 | 0.02C, 0.45P | 6.6–7.2 | 35–50 RB | 90–125 | … | 29–38 | 3.9–5.5 |
| F-0008 | 0.6–0.9C | 6.2–7 | 50–70 RB | 61–100 | 35–57 | 30–40 | <0.5 to 1 |
| FC-0508 | 0.6–0.9C, 4–6Cu | 5.9–6.8 | 60–80 RB | 100–145 | 58–82 | 50–70 | <0.5 to 1 |
| FN-0405 | 4Ni, 0.5C | 6.6–7.0 | 73–82 RB | 90–100 | 47–50 | 38–40 | <1 |
| FN-0208 | 2Ni, 0.8C | 6.6–7.0 | 50–70 RB | 70–108 | 47–58 | 35–51 | <1 |
| FN-0205 | 2Ni, 0.5C | 6.6–7.0 | 51–61 RB | 72–93 | 35–45 | 27–31 | 2.0–2.5 |
| FN-0200 | 2Ni, 0.25C | 6.6 | 29 RB | 57.5 | 25.8 | 19.0 | 1.3 |
| FC-0208 | 2Cu, 0.75C | 6.5–6.7 | 68–72 RB | 95–107 | 56–61 | 51–54 | up to 1 |
| FC-2008 | 20Cu, 1C | 6.2 | 45 RB | 79.5 | 47.8 | 40.0 | 1.3 |
| … | 4Ni, 0.6C, 1.6Cu, 0.55Mo | 7.0 | 92 RB | 190.0 | 100.0 | 65.0 | 2.5 |
| FL-4605 | 1.8Ni, 0.6C, 1.6Cu, 0.55Mo | 7.0 | 87 RB | 170.0 | 80.0 | 55.0 | 2.5 |
| FL-4605 | 1.8Ni, 0.6C, 0.55Mo | 7.0 | 80 RB | 150.0 | … | … | … |
| SS-316L | 17Cr, 13Ni, 2.2Mo, 0.9Si | 6.5 | 65 RB | 94.0 | 45.0 | 30.0 | 6.0 |
| … | 17Cr, 13Ni, 2.2Mo, 0.9Si, 15-20Cu | 7.3 | 66 RB | 108.6 | 59.2 | 49.7 | 4.3 |
| SS-410 | 13Cr, 0.8Si, 0.8Mn | 6.2 | 15 RC | 85.0 | 66.7 | 56.9 | 0 |
| FL-4608 | 2Cu, 3.8Ni, 0.9C, 0.75Mo | 6.8 | 24 RC | 107.3 | 55.8 | 46.5 | 1.5 |
| SS-303N1 | 18Cr, 11Ni, 1Mn | 6.4 | 62 RB | 86.0 | 39.0 | 32.0 | 0.5 |
| SS-304N1 | 19Cr, 10Ni, 1Mn | 6.4 | 61 RB | 112.0 | 43.0 | 38.0 | 0.5 |
| Tungsten Base | | | | | | | |
| 90W, 6Ni, 4Cu | | 17.0 | 24 RC | … | 110 | 80 | 6 |
| 90W, 7Ni, 3Cu | | 17.0 | 25 RC | … | 120 | 88 | 10 |
| 92.5W, 5.25Ni, 2.25Cu | | 17.5 | 26 RC | … | 114 | 84 | 7 |
| 92.5W, Bal. Ni, Fe, and Mo | | 17.6 | 30 RC | … | 120 | 90 | 4 |
| 93W, Bal. Ni, Fe, and Mo | | 17.7 | 32 RC | … | 125 | 95 | 4 |
| 95W, 3.5Ni, 1.5Cu | | 18.0 | 27 RC | … | 110 | 85 | 7 |
| 95W, 3.5Ni, 1.5Fe | | 18.0 | 27 RC | … | 120 | 90 | 7 |
| 97W, 2.1Ni, 0.9Fe | | 18.5 | 28 RC | … | 123 | 85 | 5 |

[a] Copper- and iron-base alloy designations are Metal Powder Industries Federation (MPIF) alloy numbers.

[b] Hardness value scales are: RB, RC, RF, and RH for Rockwell B, C, F, and H scales, respectively.

## Typical Elastic Properties of Materials

| Material | Modulus of Elasticity | | Shear Modulus | | Bulk Modulus | | Poisson's Ratio |
|---|---|---|---|---|---|---|---|
| | $10^6$ psi | GPa | $10^6$ psi | GPa | $10^6$ psi | GPa | |
| Aluminum, var. alloys | 9.9–10.3 | 68–71 | 3.7–3.9 | 26–27 | 9.9–10.2 | 68–70 | 0.330–0.334 |
| Aluminum, 6061-T6 | 10 | 70 | 3.8 | 26 | … | … | 0.35 |
| Aluminum, 2024-T4 | 10.6 | 73 | 4 | 28 | … | … | 0.32 |
| Beryllium copper | 18 | 124 | 7 | 48 | … | … | 0.29 |
| Brass, 70–30 | 15.9 | 110 | 6 | 41 | 15.7 | 108 | 0.331 |
| Brass, cast | 14.5 | 100 | 5.3 | 37 | 16.8 | 116 | 0.357 |
| Bronze | 14.9 | 103 | 6.5 | 45 | … | … | 0.14 |
| Copper | 15.6 | 108 | 5.8 | 40 | 17.9 | 123 | 0.355 |
| Glass | 6.7 | 46 | 2.7 | 19 | … | … | 0.24 |
| Glass ceramic (machinable) | 9.7 | 67 | 3.7 | 26 | … | … | 0.29 |
| Inconel | 31 | 214 | 11 | 76 | … | … | 0.27–0.38 |
| Iron, cast | 13.5–21.0 | 93–145 | 5.2–8.2 | 36–57 | 8.4–15.5 | 58–107 | 0.221–0.299 |
| Iron, ductile | 23.8–25.2 | 164–174 | 9.1–9.6 | 63–66 | … | … | 0.26–0.31 |
| Iron, gray cast | 14.5 | 100 | 6 | 41 | … | … | 0.211 |
| Iron, malleable | 23.6 | 163 | 9.3 | 64 | 17.2 | 119 | 0.271 |
| Lead | 5.3 | 37 | 1.9 | 13 | … | … | 0.43 |
| Magnesium | 6.5 | 45 | 2.4 | 17 | … | … | 0.35 |
| Magnesium alloy | 6.3 | 43 | 2.5 | 17 | 4.8 | 33 | 0.281 |
| Molybdenum | 48 | 331 | 17 | 117 | … | … | 0.307 |
| Monel metal | 25 | 172 | 9.5 | 66 | 22.5 | 155 | 0.315 |
| Nickel silver | 18.5 | 128 | 7 | 48 | … | … | 0.322 |
| Nickel steel | 30 | 207 | 11.5 | 79 | … | … | 0.291 |
| Phosphor bronze | 13.8 | 95 | 5.1 | 35 | 16.3 | 112 | 0.359 |
| Stainless steel 18–8 | 27.6 | 190 | 10.6 | 73 | 23.6 | 163 | 0.305 |
| Steel, cast | 28.5 | 197 | 11.3 | 78 | 20.2 | 139 | 0.265 |
| Steel, cold-rolled | 29.5 | 203 | 11.5 | 79 | 23.1 | 159 | 0.287 |
| Steel, all others | 28.6–30.0 | 197–207 | 11.0–11.9 | 76–82 | 22.6–24.0 | 156–165 | 0.283–0.292 |
| Titanium (99.0 Ti) | 15–16 | 103–110 | 6.5 | 45 | … | … | 0.24 |
| Titanium (Ti-8Al-1Mo-1V) | 18 | 124 | 6.8 | 47 | … | … | 0.32 |
| Zinc, cast alloys | 10.9–12.4 | 75–85 | … | … | … | … | 0.33 |
| Zinc, wrought alloys | 6.2–14 | 43–97 | … | … | … | … | 0.33 |
| Z-nickel | 30 | 207 | 11 | 76 | … | … | 0.36 |

Data represent typical values, but material properties may vary widely, depending on exact composition, material condition, and processing. Symbol … indicates no data available.

## Average Ultimate Strength of Common Materials other than Metals

| Material | Compression | | Tension | | Material | Compression | | Tension | |
|---|---|---|---|---|---|---|---|---|---|
| | psi | kPa | psi | kPa | | psi | kPa | psi | kPa |
| Bricks, best hard | 12,000 | 82737 | 400 | 2758 | Concrete, Portland | 1,000 | 6895 | 200 | 1379 |
| Bricks, light red | 1,000 | 6895 | 40 | 276 | Concrete, Portland, 1 year old | 2,000 | 13790 | 400 | 2758 |
| Brickwork, common | 1,000 | 6895 | 50 | 345 | Granite | 19,000 | 131000 | 700 | 4826 |
| Brickwork, best | 2,000 | 13790 | 300 | 2068 | Limestone and sandstone | 9,000 | 62053 | 300 | 2068 |
| Cement, Portland, 1 month old | 2,000 | 13790 | 400 | 2758 | Trap rock | 20,000 | 137895 | 800 | 5516 |
| Cement, Portland, 1 year old | 3,000 | 20684 | 500 | 3447 | Slate | 14,000 | 96527 | 500 | 3447 |
| | | | | | Vulcanized fiber | 39,000 | 268896 | 13,000 | 89632 |

### Minimum Tensile Strength of Spring Wire by Diameter

| Wire Dia. | | Music Wire | | Hard-Drawn MB | | Oil Temp. MB | | Stainless Steel 18-8 | | Cr–V Alloy | | Phosphor Bronze | | Chrome Silicon | |
|---|---|---|---|---|---|---|---|---|---|---|---|---|---|---|---|
| inch | mm | kpsi | MPa | kpsi | MPa | kpsi | MPa | kpsi | MPa | kpsi | MPa | kpsi | MPa | kpsi | MPa |
| 0.004 | 0.10 | 439 | 3027 | … | … | … | … | 325 | 2241 | … | … | 140 | 965 | … | … |
| 0.008 | 0.20 | 399 | 2751 | … | … | … | … | 325 | 2241 | … | … | 140 | 965 | … | … |
| 0.012 | 0.30 | 377 | 2599 | … | … | … | … | 316 | 2179 | … | … | … | … | … | … |
| 0.020 | 0.51 | 350 | 2413 | 283 | 1951 | 288 | 1986 | 300 | 2068 | … | … | … | … | … | … |
| 0.028 | 0.71 | 333 | 2296 | 271 | 1868 | 281 | 1937 | 284 | 1958 | … | … | … | … | … | … |
| 0.032 | 0.81 | 327 | 2255 | 265 | 1827 | 275 | 1896 | 278 | 1917 | 281 | 1937 | … | … | 300 | 2068 |
| 0.035 | 0.89 | 322 | 2220 | 261 | 1800 | 268 | 1848 | 274 | 1889 | 276 | 1903 | … | … | 298 | 2055 |
| 0.041 | 1.04 | 314 | 2165 | 255 | 1758 | 261 | 1800 | 270 | 1862 | 270 | 1862 | 135 | 931 | 298 | 2055 |
| 0.047 | 1.19 | 307 | 2117 | 248 | 1710 | 254 | 1751 | 262 | 1806 | 263 | 1813 | … | … | 292 | 2013 |
| 0.054 | 1.37 | 301 | 2075 | 243 | 1675 | 248 | 1710 | 258 | 1779 | 257 | 1772 | … | … | 292 | 2013 |
| 0.063 | 1.60 | 293 | 2020 | 237 | 1634 | 242 | 1669 | 251 | 1731 | 251 | 1731 | 130 | 896 | 290 | 1999 |
| 0.072 | 1.83 | 287 | 1979 | 232 | 1600 | 236 | 1627 | 245 | 1689 | 245 | 1689 | … | … | 288 | 1986 |
| 0.080 | 2.03 | 282 | 1944 | 227 | 1565 | 230 | 1586 | 240 | 1655 | 240 | 1655 | … | … | 285 | 1965 |
| 0.092 | 2.34 | 275 | 1896 | 220 | 1517 | 225 | 1551 | 233 | 1606 | 235 | 1620 | … | … | 280 | 1931 |
| 0.105 | 2.67 | 269 | 1855 | 216 | 1489 | 220 | 1517 | 227 | 1565 | 229 | 1579 | 125 | 862 | 275 | 1896 |
| 0.120 | 3.05 | 263 | 1813 | 210 | 1448 | 215 | 1482 | 221 | 1524 | 222 | 1531 | … | … | 275 | 1896 |
| 0.135 | 3.43 | 258 | 1779 | 206 | 1420 | 210 | 1448 | 213 | 1469 | 219 | 1510 | … | … | 270 | 1862 |
| 0.148 | 3.76 | 253 | 1744 | 203 | 1400 | 205 | 1413 | 207 | 1427 | 215 | 1482 | … | … | 268 | 1848 |
| 0.162 | 4.11 | 249 | 1717 | 200 | 1379 | 200 | 1379 | 200 | 1379 | 212 | 1462 | … | … | 162 | 1117 |
| 0.177 | 4.50 | 245 | 1689 | 195 | 1344 | 195 | 1344 | 195 | 1344 | 210 | 1448 | … | … | 260 | 1793 |
| 0.192 | 4.88 | 241 | 1662 | 192 | 1324 | 190 | 1310 | 189 | 1303 | 206 | 1420 | … | … | 260 | 1793 |
| 0.207 | 5.26 | 238 | 1641 | 190 | 1310 | 185 | 1276 | 185 | 1276 | 204 | 1407 | … | … | 260 | 1793 |
| 0.225 | 5.72 | 225 | 1551 | 186 | 1282 | 183 | 1262 | 180 | 1241 | 200 | 1379 | 120 | 827 | 255 | 1758 |
| 0.250 | 6.35 | 220 | 1517 | 182 | 1255 | 180 | 1241 | 174 | 1200 | 196 | 1351 | … | … | 250 | 1724 |
| 0.312 | 7.92 | … | … | 174 | 1200 | 178 | 1227 | 160 | 1103 | 189 | 1303 | 110 | 758 | 245 | 1689 |
| 0.375 | 9.53 | … | … | 167 | 1151 | 175 | 1207 | … | … | 187 | 1289 | … | … | 240 | 1655 |
| 0.437 | 11.10 | … | … | 165 | 1138 | 170 | 1172 | … | … | 186 | 1282 | … | … | 235 | 1620 |
| 0.500 | 12.70 | … | … | 156 | 1076 | 165 | 1138 | … | … | 185 | 1276 | 100 | 689 | 230 | 1586 |

For allowable working stresses and recommended design stresses in bending related to severity of service, refer to Fig. 1 through Fig. 10 on pages 304 through 307, and for endurance limits for compression springs made from these materials refer to Fig. 11 on page 309 in the section on spring stresses.

**Effect of Temperature on Elasticity and Strength of Metals.—**Most ferrous metals have a maximum strength at approximately 400° F (204° C), whereas the strength of non-ferrous alloys is a maximum at about room temperature. The table below gives general data for variation in metal strength with temperature.

The modulus of elasticity of metals decreases regularly with increasing temperatures above room temperature until at some elevated temperature it falls off rapidly and reaches zero at the melting point.

### Influence of Temperature on the Strength of Metals

| Material | 210° F (99° C) | 400° F (204° C) | 570° F (299° C) | 750° F (399° C) | 930° F (499° C) | 1100° F (593° C) | 1300° F (704° C) | 1475° F (802° C) |
|---|---|---|---|---|---|---|---|---|
| | Strength in Percent of Strength at 70° F (21° C) | | | | | | | |
| Wrought iron | 104 | 112 | 116 | 96 | 76 | 42 | 25 | 15 |
| Cast iron | … | 100 | 99 | 92 | 76 | 42 | … | … |
| Steel castings | 109 | 125 | 121 | 97 | 57 | … | … | … |
| Structural steel | 103 | 132 | 122 | 86 | 49 | 28 | … | … |
| Copper | 95 | 85 | 73 | 59 | 42 | … | … | … |
| Bronze | 101 | 94 | 57 | 26 | 18 | … | … | … |

## STANDARD STEELS

### Properties, Compositions, and Applications

Steel is the generic term for a large family of iron-carbon alloys that are malleable, within some temperature range, immediately after solidification from the molten state. The principal raw materials used in steelmaking are iron ore, coal, and limestone. These materials are converted in a blast furnace into a product known as "pig iron," which contains considerable amounts of carbon, manganese, sulfur, phosphorus, and silicon. Pig iron is hard, brittle, and unsuitable for direct processing into wrought forms. Steelmaking is the process of refining pig iron as well as iron and steel scrap by removing undesirable elements from the melt and then adding desirable elements in predetermined amounts. A primary reaction in most steelmaking is the combination of carbon with oxygen to form a gas. If dissolved oxygen is not removed from the melt prior to or during pouring, the gaseous products continue to evolve during solidification. If the steel is strongly deoxidized by the addition of deoxidizing elements, no gas is evolved, and the steel is called "killed" because it lies quietly in the molds. Increasing degrees of gas evolution (decreased deoxidation) characterize steels called "semikilled," "capped," or "rimmed." The degree of deoxidation affects some of the properties of the steel. In addition to oxygen, liquid steel contains measurable amounts of dissolved hydrogen and nitrogen. For some critical steel applications, special deoxidation practices as well as vacuum treatments may be used to reduce and control dissolved gases.

By definition, steel is an iron-carbon alloy that contains less than 2 percent carbon and 1 percent manganese, which acts as a deoxidizer and facilitates hot-working. Silicon, phosphorus, and sulfur are also always present, if only in trace amounts. Other elements may be present, either as residuals that are not intentionally added but result from the raw materials or steelmaking practice, or as alloying elements added to effect changes in the properties of the steel.

Steels can be cast to shape, or the cast ingot or strand can be reheated and hot-worked by rolling, forging, extrusion, or other processes into a wrought mill shape. Wrought steels are the most widely used of engineering materials, offering a multitude of forms, finishes, strengths, and usable temperature ranges.

**Standard Steel Classification.—**Wrought steels may be classified into groups based on characteristics such as elemental composition and metallurgical structure; manufacturing, deoxidation, and finishing processes; product form; and so on. Chemical composition is commonly used as a basis for identifying and assigning standard designations to wrought steels. Although carbon is the principal hardening and strengthening element in steel, no single element controls the steel's characteristics. The combined effect of several elements influences microstructure, hardness, strength, response to heat treatment, formability, and corrosion resistance.

Standard steels can be divided by composition into three main groups: *plain carbon steels* (mild steels), which have no appreciable alloying element content; *low-alloy steels,* containing up to 8 percent alloying elements; and *high-alloy steels* with more than 8 percent alloys. The general term *alloy steel* also may be used for any grade to which are added any amount of alloying elements other than what is found in plain carbon steel, within specific ranges, to enhance certain attributes of steels. Thermal treatments and other processes also are used to develop specific properties.

*Carbon Steels (Mild Steels):* A steel qualifies as a carbon steel when its manganese content is limited to 1.65 percent (max), silicon to 0.60 percent (max), and copper to 0.60 percent (max). With the exception of deoxidizers and boron when specified, no other alloying elements are added intentionally, but they may be present as residuals. If any of these incidental elements are considered detrimental for special applications, maximum acceptable limits may be specified. Carbon steels are further divided into low-carbon (less than 0.2% carbon), medium-carbon (0.2% to 0.5%), and high-carbon (great than 0.5%) categories. In contrast to most alloy steels, carbon steels are most often used without a final heat treatment; however, they may be annealed, normalized, case hardened, or quenched and tempered to enhance fabrication or mechanical properties. Carbon steels may be killed, semikilled, capped, or rimmed, and, when necessary, the method of deoxidation may be specified.

*Low-Alloy Steels:* Low-alloy steels contain up to 8 percent of one or more alloying elements. Various alloy compositions are designed to enhance mechanical properties, fabricating characteristics, weathering performance, hardenability, or any other attribute of the steel.

*High-Alloy Steels:* These steels contain more than 8 percent of alloying elements, in combinations designed to achieve enhanced performance characteristics. Types of high-alloy steels include corrosion-resistant steels, heat-resistant steels, and wear-resistant steels. Stainless steels and some tool steels are important members of the high-alloy steels category.

*Stainless Steels:* These high-alloy steels owe much of their corrosion resistance to relatively large amounts of chromium. Other elements also may be added to increase corrosion resistance, but their usefulness in this respect is limited. Stainless steels generally contain at least 10 percent chromium, with or without other elements; however, it has been customary in the United States to include in this classification those steels that contain as little as 4 percent chromium. Together, these steels form a family known as *stainless and heat-resisting steels*, in which some members possess very high strength and oxidation resistance. Few, however, contain more than 30 percent chromium or less than 50 percent iron.

The most common stainless steels can be divided into groups based on their metallurgical structures: *austenitic*, *ferritic*, and *martensitic*. *Precipitation-hardening* stainless steels are classified as either semi-austenitic or martensitic. *Duplex stainless steels* contain approximately equal amounts of austenite and ferrite. In each group, the standard grades are defined by their compositional ranges, within which stainless steel will exhibit certain desired attributes. Variations in composition are made to obtain specific properties. For instance, added chromium, molybdenum, and nitrogen give *super duplex* grades enhanced corrosion resistance.

The *austenitic grades* are nonmagnetic in the annealed condition, although some may become slightly magnetic after cold-working. They can be hardened only by cold-working, and not by heat treatment, and combine outstanding corrosion and heat resistance with good mechanical properties over a wide temperature range. The austenitic grades are further classified into two subgroups: the chromium-nickel types and the less frequently used chromium-manganese-low-nickel types. The basic composition in the chromium-nickel group is widely known as 18-8 (Cr-Ni) and is the general-purpose austenitic grade. This grade is the basis for over 20 modifications that can be characterized as follows: the chromium-nickel ratio has been modified to change the forming characteristics; the carbon content has been decreased to prevent intergranular corrosion; the elements niobium or titanium have been added to stabilize the structure; or molybdenum has been added or the chromium and nickel contents have been increased to improve corrosion or oxidation resistance.

The standard *ferritic grades* are always magnetic and contain chromium but no nickel. They can be hardened to some extent by cold-working, but not by heat treatment, and they combine corrosion and heat resistance with moderate mechanical properties and decorative appeal. The ferritic grades generally are restricted to a narrower range of corrosive conditions than the austenitic grades. The basic ferritic grade contains 17 percent chromium. In this series, there are free-machining modifications and grades with increased chromium content to improve scaling resistance. Also in this ferritic group is a 12 percent chromium steel (the basic composition of the martensitic group) with other elements, such as aluminum or titanium, added to prevent hardening.

The standard *martensitic grades* are magnetic and can be hardened by quenching and tempering. They contain chromium and, with two exceptions, no nickel. The basic martensitic grade normally contains 12 percent chromium. There are more than ten standard compositions in the martensitic series; some are modified to improve machinability and others have small additions of nickel or other elements to improve the mechanical properties or their response to heat treatment. Still others have greatly increased carbon content, in the tool steel range, and are hardenable to the highest levels of all the stainless steels. The martensitic grades are excellent for service in mild environments, such as the atmosphere, freshwater, steam, and weak acids, but are not resistant to severely corrosive solutions.

**Numbering Systems for Metals and Alloys.**—Several different numbering systems have been developed for metals and alloys by various trade associations, professional engineering societies, standards organizations, and private industries for their own use. The numerical code used to identify the metal or alloy may or may not be related to a specification, which is a statement of the technical and commercial requirements that the product must meet. Numbering systems in use include those developed by the American Iron and Steel Institute (AISI), Society of Automotive Engineers (SAE), American Society for Testing and Materials (ASTM), American National Standards Institute (ANSI), Steel Founders Society of America, American Society of Mechanical Engineers (ASME), American Welding Society (AWS), Aluminum Association, Copper Development Association, US Department of Defense (Military Specifications), and the General Accounting Office (Federal Specifications).

The Unified Numbering System (UNS) was developed through a joint effort of the ASTM and the SAE to provide a means of correlating the different numbering systems for metals and alloys that have a commercial standing. This system avoids the confusion caused when more than one identification number is used to specify the same material, or when the same number is assigned to two entirely different materials. It is important to understand that a UNS number is not a specification; it is an identification number for metals and alloys for which detailed specifications are provided elsewhere. UNS numbers are shown in Table 1; each number consists of a letter prefix followed by five digits. In some, the letter is suggestive of the family of metals identified by the series, such as A for aluminum and C for copper. Whenever possible, the numbers in the UNS groups contain numbering sequences taken directly from other systems to facilitate identification of the material; e.g., the corresponding UNS number for AISI 1020 steel is G10200. The UNS numbers corresponding to the commonly used AISI-SAE numbers that are used to identify plain carbon, alloy, and tool steels are given in Table 2.

**Table 1. Unified Numbering System (UNS) for Metals and Alloys**

| UNS Series | Metal |
|---|---|
| A00001 to A99999 | Aluminum and aluminum alloys |
| C00001 to C99999 | Copper and copper alloys |
| D00001 to D99999 | Specified mechanical property steels |
| E00001 to E99999 | Rare earth and rare earthlike metals and alloys |
| F00001 to F99999 | Cast irons |
| G00001 to G99999 | AISI and SAE carbon and alloy steels (except tool steels) |
| H00001 to H99999 | AISI and SAE H-steels |
| J00001 to J99999 | Cast steels (except tool steels) |
| K00001 to K99999 | Miscellaneous steels and ferrous alloys |
| L00001 to L99999 | Low-melting metals and alloys |
| M00001 to M99999 | Miscellaneous nonferrous metals and alloys |
| N00001 to N99999 | Nickel and nickel alloys |
| P00001 to P99999 | Precious metals and alloys |
| R00001 to R99999 | Reactive and refractory metals and alloys |
| S00001 to S99999 | Heat and corrosion resistant (stainless) steels |
| T00001 to T99999 | Tool steels, wrought and cast |
| W00001 to W99999 | Welding filler metals |
| Z00001 to Z99999 | Zinc and zinc alloys |

**Identifying Metals.**—When it is necessary to sort materials, several rough methods may be used without elaborate chemical analysis. The most obvious of these is by using a magnet to pick out those materials that contain magnetic elements. To differentiate various levels of carbon and other elements in a steel bar, hold the bar in contact with a grinding wheel and observe the sparks. With high levels of carbon, for instance, sparks are produced that appear to split into several bright tracers. Patterns produced by several other elements, including small amounts of aluminum and titanium, for instance, can be identified with the aid of Data Sheet 13, issued by the American Society for Metals (ASM), Metals Park, OH.

**Standard Steel Numbering System.**—The most widely used systems for identifying wrought carbon, low-alloy, and stainless steels are based on chemical composition, and are those of the American Iron and Steel Institute (AISI) and the Society of Automotive Engineers (SAE). These systems are almost identical, but they are carefully coordinated. The standard steels so designated have been developed cooperatively by producers and users and have been found through long experience to cover most of the wrought ferrous metals used in automotive vehicles and related equipment. These designations, however, are not specifications, and should not be used for purchasing unless accompanied by supplementary information necessary to describe commercially the product desired. Engineering societies, associations, and institutes whose members make, specify, or purchase steel products publish standard specifications, many of which have become well known and respected. The most comprehensive and widely used specifications are those published by the American Society for Testing and Materials (ASTM). The US government and various companies also publish their own specification for steel products to serve their own special procurement needs. The Unified Numbering System (UNS) for metals and alloys is also used to designate steels (see pages 393 and 395).

The numerical designation system used by both AISI and SAE for wrought carbon, alloy, and stainless steels is summarized in Table 3. In Table 4 is given the compositions of the standard carbon steels; Table 5 lists the standard low-alloy steel compositions; and Table 6 includes the typical compositions of the standard stainless steels.

**Binary Alloy.**—An alloy containing two elements. When the term is used in regard to iron or steel, it refers to a material that has one alloying element in addition to iron. Since carbon is always present in steel, plain carbon steel is the typical binary iron alloy.

**Ternary Alloy.**—This is an alloy consisting of three elements. When the term refers to steel, it denotes a steel that contains two alloying elements in addition to iron; since carbon is always present, it is one of these elements. The third element may be nickel, chromium, manganese, tungsten, molybdenum, titanium, or any other element that is alloyed to give the steel some special property.

**Quarternary Alloy.**—A quarternary alloy is an alloy consisting of four elements. When applied to steel, such an alloy contains, in addition to iron, three alloying elements. Carbon is one of these, and the other two may be chromium and nickel, silicon and manganese, etc.

**Damascus Steel.**—A characteristic feature of Damascus steel is its surface patterns, which vary with the carbon content and are either in the form of wavy parallel stripes or mottled patterns. This steel represents an early development in steel making, as it was imported during the Middle Ages to Western Europe through Syria and Palestine, and is known also as Indian steel and bulat. The old Indian method of producing real damascene steel consists in using a pure ore and the best grade of charcoal. The Persian practice is to use soft iron bars and charcoal and plumbago (black lead or graphite) to supply the carbon; and a third method consists of a certain heat treatment which resembles a prolonged tempering. One investigator has concluded that the carbon, irregularly dispersed in the metal and forming two distinct combinations, is what causes the damask or characteristic pattern, and that the slower the cooling the larger the veins will be.

An imitation of Damascus steel can be obtained by etching the surface of the steel blade with acids, the parts which are not to be attacked by the acid being protected by a "resist."

**Table 2. AISI and SAE Numbers and Their Corresponding UNS Numbers for Plain Carbon, Alloy, and Tool Steels**

| AISI-SAE Numbers | UNS Numbers | AISI-SAE Numbers | UNS Numbers | AISI-SAE Numbers | UNS Numbers | AISI-SAE Numbers | UNS Numbers |
|---|---|---|---|---|---|---|---|
| Plain Carbon Steels | | | | | | | |
| 1005 | G10050 | 1030 | G10300 | 1070 | G10700 | 1566 | G15660 |
| 1006 | G10060 | 1035 | G10350 | 1078 | G10780 | 1110 | G11100 |
| 1008 | G10080 | 1037 | G10370 | 1080 | G10800 | 1117 | G11170 |
| 1010 | G10100 | 1038 | G10380 | 1084 | G10840 | 1118 | G11180 |
| 1012 | G10120 | 1039 | G10390 | 1086 | G10860 | 1137 | G11370 |
| 1015 | G10150 | 1040 | G10400 | 1090 | G10900 | 1139 | G11390 |
| 1016 | G10160 | 1042 | G10420 | 1095 | G10950 | 1140 | G11400 |
| 1017 | G10170 | 1043 | G10430 | 1513 | G15130 | 1141 | G11410 |
| 1018 | G10180 | 1044 | G10440 | 1522 | G15220 | 1144 | G11440 |
| 1019 | G10190 | 1045 | G10450 | 1524 | G15240 | 1146 | G11460 |
| 1020 | G10200 | 1046 | G10460 | 1526 | G15260 | 1151 | G11510 |
| 1021 | G10210 | 1049 | G10490 | 1527 | G15270 | 1211 | G12110 |
| 1022 | G10220 | 1050 | G10500 | 1541 | G15410 | 1212 | G12120 |
| 1023 | G10230 | 1053 | G10530 | 1548 | G15480 | 1213 | G12130 |
| 1025 | G10250 | 1055 | G10550 | 1551 | G15510 | 1215 | G12150 |
| 1026 | G10260 | 1059 | G10590 | 1552 | G15520 | 12L14 | G12144 |
| 1029 | G10290 | 1060 | G10600 | 1561 | G15610 | … | … |
| Alloy Steels | | | | | | | |
| 1330 | G13300 | 4150 | G41500 | 5140 | G51400 | 8642 | G86420 |
| 1335 | G13350 | 4161 | G41610 | 5150 | G51500 | 8645 | G86450 |
| 1340 | G13400 | 4320 | G43200 | 5155 | G51550 | 8655 | G86550 |
| 1345 | G13450 | 4340 | G43400 | 5160 | G51600 | 8720 | G87200 |
| 4023 | G40230 | E4340 | G43406 | E51100 | G51986 | 8740 | G87400 |
| 4024 | G40240 | 4615 | G46150 | E52100 | G52986 | 8822 | G88220 |
| 4027 | G40270 | 4620 | G46200 | 6118 | G61180 | 9260 | G92600 |
| 4028 | G40280 | 4626 | G46260 | 6150 | G61500 | 50B44 | G50441 |
| 4037 | G40370 | 4720 | G47200 | 8615 | G86150 | 50B46 | G50461 |
| 4047 | G40470 | 4815 | G48150 | 8617 | G86170 | 50B50 | G50501 |
| 4118 | G41180 | 4817 | G48170 | 8620 | G86200 | 50B60 | G50601 |
| 4130 | G41300 | 4820 | G48200 | 8622 | G86220 | 51B60 | G51601 |
| 4137 | G41370 | 5117 | G51170 | 8625 | G86250 | 81B45 | G81451 |
| 4140 | G41400 | 5120 | G51200 | 8627 | G86270 | 94B17 | G94171 |
| 4142 | G41420 | 5130 | G51300 | 8630 | G86300 | 94B30 | G94301 |
| 4145 | G41450 | 5132 | G51320 | 8637 | G86370 | … | … |
| 4147 | G41470 | 5135 | G51350 | 8640 | G86400 | … | … |
| Tool Steels (AISI and UNS Only) | | | | | | | |
| M1 | T11301 | T6 | T12006 | A6 | T30106 | P4 | T51604 |
| M2 | T11302 | T8 | T12008 | A7 | T30107 | P5 | T51605 |
| M4 | T11304 | T15 | T12015 | A8 | T30108 | P6 | T51606 |
| M6 | T11306 | H10 | T20810 | A9 | T30109 | P20 | T51620 |
| M7 | T11307 | H11 | T20811 | A10 | T30110 | P21 | T51621 |
| M10 | T11310 | H12 | T20812 | D2 | T30402 | F1 | T60601 |
| M3-1 | T11313 | H13 | T20813 | D3 | T30403 | F2 | T60602 |
| M3-2 | T11323 | H14 | T20814 | D4 | T30404 | L2 | T61202 |
| M30 | T11330 | H19 | T20819 | D5 | T30405 | L3 | T61203 |
| M33 | T11333 | H21 | T20821 | D7 | T30407 | L6 | T61206 |
| M34 | T11334 | H22 | T20822 | O1 | T31501 | W1 | T72301 |
| M36 | T11336 | H23 | T20823 | O2 | T31502 | W2 | T72302 |
| M41 | T11341 | H24 | T20824 | O6 | T31506 | W5 | T72305 |
| M42 | T11342 | H25 | T20825 | O7 | T31507 | CA2 | T90102 |
| M43 | T11343 | H26 | T20826 | S1 | T41901 | CD2 | T90402 |
| M44 | T11344 | H41 | T20841 | S2 | T41902 | CD5 | T90405 |
| M46 | T11346 | H42 | T20842 | S4 | T41904 | CH12 | T90812 |
| M47 | T11347 | H43 | T20843 | S5 | T41905 | CH13 | T90813 |
| T1 | T12001 | A2 | T30102 | S6 | T41906 | CO1 | T91501 |
| T2 | T12002 | A3 | T30103 | S7 | T41907 | CS5 | T91905 |
| T4 | T12004 | A4 | T30104 | P2 | T51602 | … | … |
| T5 | T12005 | A5 | T30105 | P3 | T51603 | … | … |

**Table 3. AISI and SAE System of Designating Carbon and Alloy Steels**

| AISI-SAE Designation[a] | | Type of Steel and Nominal Alloy Content (%) |
|---|---|---|
| | | Carbon Steels |
| 10xx | | Plain Carbon (Mn 1.00% max.) |
| 11xx | | Resulfurized |
| 12xx | | Resulfurized and Rephosphorized |
| 15xx | | Plain Carbon (Max. Mn range 1.00 to 1.65%) |
| | | Manganese Steels |
| 13xx | | Mn 1.75 |
| | | Nickel Steels |
| 23xx | | Ni 3.50 |
| 25xx | | Ni 5.00 |
| | | Nickel-Chromium Steels |
| 31xx | | Ni 1.25; Cr 0.65 and 0.80 |
| 32xx | | Ni 1.75; Cr 1.07 |
| 33xx | | Ni 3.50; Cr 1.50 and 1.57 |
| 34xx | | Ni 3.00; Cr 0.77 |
| | | Molybdenum Steels |
| 40xx | | Mo 0.20 and 0.25 |
| 44xx | | Mo 0.40 and 0.52 |
| | | Chromium-Molybdenum Steels |
| 41xx | | Cr 0.50, 0.80, and 0.95; Mo 0.12, 0.20, 0.25, and 0.30 |
| | | Nickel-Chromium-Molybdenum Steels |
| 43xx | | Ni 1.82; Cr 0.40 and 0.90; Mo 0.25 |
| 43BVxx | | Ni 1.82; Cr 0.50; Mo 0.12 and 0.35; V 0.03 min. |
| 47xx | | Ni 1.05; Cr 0.45; Mo 0.20 and 0.35 |
| 81xx | | Ni 0.30; Cr 0.40; Mo 0.12 |
| 86xx | | Ni 0.55; Cr 0.50; Mo 0.20 |
| 87xx | | Ni 0.55; Cr 0.50; Mo 0.25 |
| 88xx | | Ni 0.55; Cr 0.50; Mo 0.35 |
| 93xx | | Ni 3.25; Cr 1.20; Mo 0.12 |
| 94xx | | Ni 0.45; Cr 0.40; Mo 0.12 |
| 97xx | | Ni 0.55; Cr 0.20; Mo 0.20 |
| 98xx | | Ni 1.00; Cr 0.80; Mo 0.25 |
| | | Nickel-Molybdenum Steels |
| 46xx | | Ni 0.85 and 1.82; Mo 0.20 and 0.25 |
| 48xx | | Ni 3.50; Mo 0.25 |
| | | Chromium Steels |
| 50xx | | Cr 0.27, 0.40, 0.50, and 0.65 |
| 51xx | | Cr 0.80, 0.87, 0.92, 0.95, 1.00, and 1.05 |
| 50xxx | | Cr 0.50; C 1.00 min. |
| 51xxx | | Cr 1.02; C 1.00 min. |
| 52xxx | | Cr 1.45; C 1.00 min. |
| | | Chromium-Vanadium Steels |
| 61xx | | Cr 0.60, 0.80, and 0.95; V 0.10 and 0.15 min |
| | | Tungsten-Chromium Steels |
| 72xx | | W 1.75; Cr 0.75 |
| | | Silicon-Manganese Steels |
| 92xx | | Si 1.40 and 2.00; Mn 0.65, 0.82, and 0.85; Cr 0.00 and 0.65 |
| | | High-Strength Low-Alloy Steels |
| 9xx | | Various SAE grades |
| xxBxx | | B denotes boron steels |
| xxLxx | | L denotes leaded steels |
| **AISI** | **SAE** | **Stainless Steels** |
| 2xx | 302xx | Chromium-Manganese-Nickel Steels |
| 3xx | 303xx | Chromium-Nickel Steels |
| 4xx | 514xx | Chromium Steels |
| 5xx | 515xx | Chromium Steels |

[a] xx in the last two digits of the carbon and low-alloy designations (but not the stainless steels) indicates that the carbon content (in hundredths of a percent) is to be inserted.

**Table 4. Composition of AISI-SAE Standard Carbon Steels**

| AISI-SAE No. | UNS No. | Composition(%)[a] | | | |
|---|---|---|---|---|---|
| | | C | Mn | P(max)[b] | S(max)[b] |
| Nonresulfurized Grades — 1 percent Mn (max) | | | | | |
| 1005[c] | G10050 | 0.06 max | 0.35 max | 0.040 | 0.050 |
| 1006[c] | G10060 | 0.08 max | 0.25-0.40 | 0.040 | 0.050 |
| 1008 | G10080 | 0.10 max | 0.30-0.50 | 0.040 | 0.050 |
| 1010 | G10100 | 0.08-0.13 | 0.30-0.60 | 0.040 | 0.050 |
| 1012 | G10120 | 0.10-0.15 | 0.30-0.60 | 0.040 | 0.050 |
| 1015 | G10150 | 0.13-0.18 | 0.30-0.60 | 0.040 | 0.050 |
| 1016 | G10160 | 0.13-0.18 | 0.60-0.90 | 0.040 | 0.050 |
| 1017 | G10170 | 0.15-0.20 | 0.30-0.60 | 0.040 | 0.050 |
| 1018 | G10180 | 0.15-0.20 | 0.60-0.90 | 0.040 | 0.050 |
| 1019 | G10190 | 0.15-0.20 | 0.70-1.00 | 0.040 | 0.050 |
| 1020 | G10200 | 0.18-0.23 | 0.30-0.60 | 0.040 | 0.050 |
| 1021 | G10210 | 0.18-0.23 | 0.60-0.90 | 0.040 | 0.050 |
| 1022 | G10220 | 0.18-0.23 | 0.70-1.00 | 0.040 | 0.050 |
| 1023 | G10230 | 0.20-0.25 | 0.30-0.60 | 0.040 | 0.050 |
| 1025 | G10250 | 0.22-0.28 | 0.30-0.60 | 0.040 | 0.050 |
| 1026 | G10260 | 0.22-0.28 | 0.60-0.90 | 0.040 | 0.050 |
| 1029 | G10290 | 0.25-0.31 | 0.60-0.90 | 0.040 | 0.050 |
| 1030 | G10300 | 0.28-0.34 | 0.60-0.90 | 0.040 | 0.050 |
| 1035 | G10350 | 0.32-0.38 | 0.60-0.90 | 0.040 | 0.050 |
| 1037 | G10370 | 0.32-0.38 | 0.70-1.00 | 0.040 | 0.050 |
| 1038 | G10380 | 0.35-0.42 | 0.60-0.90 | 0.040 | 0.050 |
| 1039 | G10390 | 0.37-0.44 | 0.70-1.00 | 0.040 | 0.050 |
| 1040 | G10400 | 0.37-0.44 | 0.60-0.90 | 0.040 | 0.050 |
| 1042 | G10420 | 0.40-0.47 | 0.60-0.90 | 0.040 | 0.050 |
| 1043 | G10430 | 0.40-0.47 | 0.70-1.00 | 0.040 | 0.050 |
| 1044 | G10440 | 0.43-0.50 | 0.30-0.60 | 0.040 | 0.050 |
| 1045 | G10450 | 0.43-0.50 | 0.60-0.90 | 0.040 | 0.050 |
| 1046 | G10460 | 0.43-0.50 | 0.70-1.00 | 0.040 | 0.050 |
| 1049 | G10490 | 0.46-0.53 | 0.60-0.90 | 0.040 | 0.050 |
| 1050 | G10500 | 0.48-0.55 | 0.60-0.90 | 0.040 | 0.050 |
| 1053 | G10530 | 0.48-0.55 | 0.70-1.00 | 0.040 | 0.050 |
| 1055 | G10550 | 0.50-0.60 | 0.60-0.90 | 0.040 | 0.050 |
| 1059[c] | G10590 | 0.55-0.65 | 0.50-0.80 | 0.040 | 0.050 |
| 1060 | G10600 | 0.55-0.65 | 0.60-0.90 | 0.040 | 0.050 |
| 1064[c] | G10640 | 0.60-0.70 | 0.50-0.80 | 0.040 | 0.050 |
| 1065[c] | G10650 | 0.60-0.70 | 0.60-0.90 | 0.040 | 0.050 |
| 1069[c] | G10690 | 0.65-0.75 | 0.40-0.70 | 0.040 | 0.050 |
| 1070 | G10700 | 0.65-0.75 | 0.60-0.90 | 0.040 | 0.050 |
| 1078 | G10780 | 0.72-0.85 | 0.30-0.60 | 0.040 | 0.050 |
| 1080 | G10800 | 0.75-0.88 | 0.60-0.90 | 0.040 | 0.050 |
| 1084 | G10840 | 0.80-0.93 | 0.60-0.90 | 0.040 | 0.050 |
| 1086[c] | G10860 | 0.80-0.93 | 0.30-0.50 | 0.040 | 0.050 |
| 1090 | G10900 | 0.85-0.98 | 0.60-0.90 | 0.040 | 0.050 |
| 1095 | G10950 | 0.90-1.03 | 0.30-0.50 | 0.040 | 0.050 |

**Table 4.** *(Continued)* **Composition of AISI-SAE Standard Carbon Steels**

| AISI-SAE No. | UNS No. | Composition(%)[a] | | | |
|---|---|---|---|---|---|
| | | C | Mn | P(max)[b] | S(max)[b] |
| Nonresulfurized Grades — Over 1 percent Mn | | | | | |
| 1513 | G15130 | 0.10-0.16 | 1.10-1.40 | 0.040 | 0.050 |
| 1522 | G15220 | 0.18-0.24 | 1.10-1.40 | 0.040 | 0.050 |
| 1524 | G15240 | 0.19-0.25 | 1.35-1.65 | 0.040 | 0.050 |
| 1526 | G15260 | 0.22-0.29 | 1.10-1.40 | 0.040 | 0.050 |
| 1527 | G15270 | 0.22-0.29 | 1.20-1.50 | 0.040 | 0.050 |
| 1541 | G15410 | 0.36-0.44 | 1.35-1.65 | 0.040 | 0.050 |
| 1548 | G15480 | 0.44-0.52 | 1.10-1.40 | 0.040 | 0.050 |
| 1551 | G15510 | 0.45-0.56 | 0.85-1.15 | 0.040 | 0.050 |
| 1552 | G15520 | 0.47-0.55 | 1.20-1.50 | 0.040 | 0.050 |
| 1561 | G15610 | 0.55-0.65 | 0.75-1.05 | 0.040 | 0.050 |
| 1566 | G15660 | 0.60-0.71 | 0.85-1.15 | 0.040 | 0.050 |
| Free-Machining Grades — Resulfurized | | | | | |
| 1110 | G11100 | 0.08-0.13 | 0.30-0.60 | 0.040 | 0.08-0.13 |
| 1117 | G11170 | 0.14-0.20 | 1.00-1.30 | 0.040 | 0.08-0.13 |
| 1118 | G11180 | 0.14-0.20 | 1.30-1.60 | 0.040 | 0.08-0.13 |
| 1137 | G11370 | 0.32-0.39 | 1.35-1.65 | 0.040 | 0.08-0.13 |
| 1139 | G11390 | 0.35-0.43 | 1.35-1.65 | 0.040 | 0.13-0.20 |
| 1140 | G11400 | 0.37-0.44 | 0.70-1.00 | 0.040 | 0.08-0.13 |
| 1141 | G11410 | 0.37-0.45 | 1.35-1.65 | 0.040 | 0.08-0.13 |
| 1144 | G11440 | 0.40-0.48 | 1.35-1.65 | 0.040 | 0.24-0.33 |
| 1146 | G11460 | 0.42-0.49 | 0.70-1.00 | 0.040 | 0.08-0.13 |
| 1151 | G11510 | 0.48-0.55 | 0.70-1.00 | 0.040 | 0.08-0.13 |
| Free-Machining Grades — Resulfurized and Rephosphorized | | | | | |
| 1211 | G12110 | 0.13 max | 0.60-0.90 | 0.07-0.12 | 0.10-0.15 |
| 1212 | G12120 | 0.13 max | 0.70-1.00 | 0.07-0.12 | 0.16-0.23 |
| 1213 | G12130 | 0.13 max | 0.70-1.00 | 0.07-0.12 | 0.24-0.33 |
| 1215 | G12150 | 0.09 max | 0.75-1.05 | 0.04-0.09 | 0.26-0.35 |
| 12L14[d] | G12144 | 0.15 max | 0.85-1.15 | 0.04-0.09 | 0.26-0.35 |

[a] The following notes refer to boron, copper, lead, and silicon additions. Boron: Standard killed carbon steels, which are generally fine grain, may be produced with a boron treatment addition to improve hardenability. Such steels are produced to a range of 0.0005–0.003 percent B. These steels are identified by inserting the letter "B" between the second and third numerals of the AISI or SAE number, e.g., 10B46. Copper: When copper is required, 0.20 percent (min) is generally specified. Lead: Standard carbon steels can be produced with a lead range of 0.15–0.35 percent to improve machinability. Such steels are identified by inserting the letter "L" between the second and third numerals of the AISI or SAE number, e.g., 12L15 and 10L45. Silicon: It is not common practice to produce the 12XX series of resulfurized and rephosphorized steels to specified limits for silicon because of its adverse effect on machinability. When silicon ranges or limits are required for resulfurized or nonresulfurized steels, however, these values apply: a range of 0.08 percent Si for Si max up to 0.15 percent inclusive, a range of 0.10 percent Si for Si max over 0.15 to 0.20 percent inclusive, a range of 0.15 percent Si for Si max over 0.20 to 0.30 percent inclusive, and a range of 0.20 percent Si for Si max over 0.30 to 0.60 percent inclusive. Example: Si max is 0.25 percent, range is 0.10–0.25 percent.

[b] Values given are maximum percentages, except where a range of values is given.

[c] Standard grades for wire rods and wire only.

[d] 0.15–0.35 percent Pb.

## Table 5. Compositions of AISI-SAE Standard Low-Alloy Steels

| AISI-SAE No. | UNS No. | Composition (%)[a, b] C | Mn | P (max) | S (max) | Si | Ni | Cr | Mo |
|---|---|---|---|---|---|---|---|---|---|
| 1330 | G13300 | 0.28-0.33 | 1.60-1.90 | 0.035 | 0.040 | 0.15-0.35 | … | … | … |
| 1335 | G13350 | 0.33-0.38 | 1.60-1.90 | 0.035 | 0.040 | 0.15-0.35 | … | … | … |
| 1340 | G13400 | 0.38-0.43 | 1.60-1.90 | 0.035 | 0.040 | 0.15-0.35 | … | … | … |
| 1345 | G13450 | 0.43-0.48 | 1.60-1.90 | 0.035 | 0.040 | 0.15-0.35 | … | … | … |
| 4023 | G40230 | 0.20-0.25 | 0.70-0.90 | 0.035 | 0.040 | 0.15-0.35 | … | … | 0.20-0.30 |
| 4024 | G40240 | 0.20-0.25 | 0.70-0.90 | 0.035 | 0.035-0.050 | 0.15-0.35 | … | … | 0.20-0.30 |
| 4027 | G40270 | 0.25-0.30 | 0.70-0.90 | 0.035 | 0.040 | 0.15-0.35 | … | … | 0.20-0.30 |
| 4028 | G40280 | 0.25-0.30 | 0.70-0.90 | 0.035 | 0.035-0.050 | 0.15-0.35 | … | … | 0.20-0.30 |
| 4037 | G40370 | 0.35-0.40 | 0.70-0.90 | 0.035 | 0.040 | 0.15-0.35 | … | … | 0.20-0.30 |
| 4047 | G40470 | 0.45-0.50 | 0.70-0.90 | 0.035 | 0.040 | 0.15-0.35 | … | … | 0.20-0.30 |
| 4118 | G41180 | 0.18-0.23 | 0.70-0.90 | 0.035 | 0.040 | 0.15-0.35 | … | 0.40-0.60 | 0.08-0.15 |
| 4130 | G41300 | 0.28-0.33 | 0.40-0.60 | 0.035 | 0.040 | 0.15-0.35 | … | 0.80-1.10 | 0.15-0.25 |
| 4137 | G41370 | 0.35-0.40 | 0.70-0.90 | 0.035 | 0.040 | 0.15-0.35 | … | 0.80-1.10 | 0.15-0.25 |
| 4140 | G41400 | 0.38-0.43 | 0.75-1.00 | 0.035 | 0.040 | 0.15-0.35 | … | 0.80-1.10 | 0.15-0.25 |
| 4142 | G41420 | 0.40-0.45 | 0.75-1.00 | 0.035 | 0.040 | 0.15-0.35 | … | 0.80-1.10 | 0.15-0.25 |
| 4145 | G41450 | 0.43-0.48 | 0.75-1.00 | 0.035 | 0.040 | 0.15-0.35 | … | 0.80-1.10 | 0.15-0.25 |
| 4147 | G41470 | 0.45-0.50 | 0.75-1.00 | 0.035 | 0.040 | 0.15-0.35 | … | 0.80-1.10 | 0.15-0.25 |
| 4150 | G41500 | 0.48-0.53 | 0.75-1.00 | 0.035 | 0.040 | 0.15-0.35 | … | 0.80-1.10 | 0.15-0.25 |
| 4161 | G41610 | 0.56-0.64 | 0.75-1.00 | 0.035 | 0.040 | 0.15-0.35 | … | 0.70-0.90 | 0.25-0.35 |
| 4320 | G43200 | 0.17-0.22 | 0.45-0.65 | 0.035 | 0.040 | 0.15-0.35 | 1.65-2.00 | 0.40-0.60 | 0.20-0.30 |
| 4340 | G43400 | 0.38-0.43 | 0.60-0.80 | 0.035 | 0.040 | 0.15-0.35 | 1.65-2.00 | 0.70-0.90 | 0.20-0.30 |
| E4340[c] | G43406 | 0.38-0.43 | 0.65-0.85 | 0.025 | 0.025 | 0.15-0.35 | 1.65-2.00 | 0.70-0.90 | 0.20-0.30 |
| 4615 | G46150 | 0.13-0.18 | 0.45-0.65 | 0.035 | 0.040 | 0.15-0.35 | 1.65-2.00 | … | 0.20-0.30 |
| 4620 | G46200 | 0.17-0.22 | 0.45-0.65 | 0.035 | 0.040 | 0.15-0.35 | 1.65-2.00 | … | 0.20-0.30 |
| 4626 | G46260 | 0.24-0.29 | 0.45-0.65 | 0.035 | 0.040 | 0.15-0.35 | 0.70-1.00 | … | 0.15-0.25 |
| 4720 | G47200 | 0.17-0.22 | 0.50-0.70 | 0.035 | 0.040 | 0.15-0.35 | 0.90-1.20 | 0.35-0.55 | 0.15-0.25 |
| 4815 | G48150 | 0.13-0.18 | 0.40-0.60 | 0.035 | 0.040 | 0.15-0.35 | 3.25-3.75 | … | 0.20-0.30 |
| 4817 | G48170 | 0.15-0.20 | 0.40-0.60 | 0.035 | 0.040 | 0.15-0.35 | 3.25-3.75 | … | 0.20-0.30 |
| 4820 | G48200 | 0.18-0.23 | 0.50-0.70 | 0.035 | 0.040 | 0.15-0.35 | 3.25-3.75 | … | 0.20-0.30 |
| 5117 | G51170 | 0.15-0.20 | 0.70-0.90 | 0.035 | 0.040 | 0.15-0.35 | … | 070-0.90 | … |
| 5120 | G51200 | 0.17-0.22 | 0.70-0.90 | 0.035 | 0.040 | 0.15-0.35 | … | 0.70-0.90 | … |
| 5130 | G51300 | 0.28-0.33 | 0.70-0.90 | 0.035 | 0.040 | 0.15-0.35 | … | 0.80-1.10 | … |
| 5132 | G51320 | 0.30-0.35 | 0.60-0.80 | 0.035 | 0.040 | 0.15-0.35 | … | 0.75-1.00 | … |
| 5135 | G51350 | 0.33-0.38 | 0.60-0.80 | 0.035 | 0.040 | 0.15-0.35 | … | 0.80-1.05 | … |
| 5140 | G51400 | 0.38-0.43 | 0.70-0.90 | 0.035 | 0.040 | 0.15-0.35 | … | 0.70-0.90 | … |
| 5150 | G51500 | 0.48-0.53 | 0.70-0.90 | 0.035 | 0.040 | 0.15-0.35 | … | 0.70-0.90 | … |
| 5155 | G51550 | 0.51-0.59 | 0.70-0.90 | 0.035 | 0.040 | 0.15-0.35 | … | 0.70-0.90 | … |
| 5160 | G51600 | 0.56-0.64 | 0.75-1.00 | 0.035 | 0.040 | 0.15-0.35 | … | 0.70-0.90 | … |

**Table 5.** *(Continued)* **Compositions of AISI-SAE Standard Low-Alloy Steels**

| AISI-SAE | | Composition (%)[a, b] | | | | | | | |
|---|---|---|---|---|---|---|---|---|---|
| No. | UNS No. | C | Mn | P (max) | S (max) | Si | Ni | Cr | Mo |
| E51100[c] | G51986 | 0.98-1.10 | 0.25-0.45 | 0.025 | 0.025 | 0.15-0.35 | ... | 0.90-1.15 | ... |
| E52100[c] | G52986 | 0.98-1.10 | 0.25-0.45 | 0.025 | 0.025 | 0.15-0.35 | ... | 1.30-1.60 | ... |
| 6118 | G61180 | 0.16-0.21 | 0.50-0.70 | 0.035 | 0.040 | 0.15-0.35 | ... | 0.50-0.70 | 0.10-0.15 V |
| 6150 | G61500 | 0.48-0.53 | 0.70-0.90 | 0.035 | 0.040 | 0.15-0.35 | ... | 0.80-1.10 | 0.15 V min |
| 8615 | G86150 | 0.13-0.18 | 0.70-0.90 | 0.035 | 0.040 | 0.15-0.35 | 0.40-0.70 | 0.40-0.60 | 0.15-0.25 |
| 8617 | G86170 | 0.15-0.20 | 0.70-0.90 | 0.035 | 0.040 | 0.15-0.35 | 0.40-0.70 | 0.40-0.60 | 0.15-0.25 |
| 8620 | G86200 | 0.18-0.23 | 0.70-0.90 | 0.035 | 0.040 | 0.15-0.35 | 0.40-0.70 | 0.40-0.60 | 0.15-0.25 |
| 8622 | G86220 | 0.20-0.25 | 0.70-0.90 | 0.035 | 0.040 | 0.15-0.35 | 0.40-0.70 | 0.40-0.60 | 0.15-0.25 |
| 8625 | G86250 | 0.23-0.28 | 0.70-0.90 | 0.035 | 0.040 | 0.15-0.35 | 0.40-0.70 | 0.40-0.60 | 0.15-0.25 |
| 8627 | G86270 | 0.25-0.30 | 0.70-0.90 | 0.035 | 0.040 | 0.15-0.35 | 0.40-0.70 | 0.40-0.60 | 0.15-0.25 |
| 8630 | G86300 | 0.28-0.33 | 0.70-0.90 | 0.035 | 0.040 | 0.15-0.35 | 0.40-0.70 | 0.40-0.60 | 0.15-0.25 |
| 8637 | G86370 | 0.35-0.40 | 0.75-1.00 | 0.035 | 0.040 | 0.15-0.35 | 0.40-0.70 | 0.40-0.60 | 0.15-0.25 |
| 8640 | G86400 | 0.38-0.43 | 0.75-1.00 | 0.035 | 0.040 | 0.15-0.35 | 0.40-0.70 | 0.40-0.60 | 0.15-0.25 |
| 8642 | G86420 | 0.40-0.45 | 0.75-1.00 | 0.035 | 0.040 | 0.15-0.35 | 0.40-0.70 | 0.40-0.60 | 0.15-0.25 |
| 8645 | G86450 | 0.43-0.48 | 0.75-1.00 | 0.035 | 0.040 | 0.15-0.35 | 0.40-0.70 | 0.40-0.60 | 0.15-0.25 |
| 8655 | G86550 | 0.51-0.59 | 0.75-1.00 | 0.035 | 0.040 | 0.15-0.35 | 0.40-0.70 | 0.40-0.60 | 0.15-0.25 |
| 8720 | G87200 | 0.18-0.23 | 0.70-0.90 | 0.035 | 0.040 | 0.15-0.35 | 0.40-0.70 | 0.40-0.60 | 0.20-0.30 |
| 8740 | G87400 | 0.38-0.43 | 0.75-1.00 | 0.035 | 0.040 | 0.15-0.35 | 0.40-0.70 | 0.40-0.60 | 0.20-0.30 |
| 8822 | G88220 | 0.20-0.25 | 0.75-1.00 | 0.035 | 0.040 | 0.15-0.35 | 0.40-0.70 | 0.40-0.60 | 0.30-0.40 |
| 9260 | G92600 | 0.56-0.64 | 0.75-1.00 | 0.035 | 0.040 | 1.80-2.20 | ... | ... | ... |
| Standard Boron Grades[d] | | | | | | | | | |
| 50B44 | G50441 | 0.43-0.48 | 0.75-1.00 | 0.035 | 0.040 | 0.15-0.35 | ... | 0.40-0.60 | ... |
| 50B46 | G50461 | 0.44-0.49 | 0.75-1.00 | 0.035 | 0.040 | 0.15-0.35 | ... | 0.20-0.35 | ... |
| 50B50 | G50501 | 0.48-0.53 | 0.75-1.00 | 0.035 | 0.040 | 0.15-0.35 | ... | 0.40-0.60 | ... |
| 50B60 | G50601 | 0.56-0.64 | 0.75-1.00 | 0.035 | 0.040 | 0.15-0.35 | ... | 0.40-0.60 | ... |
| 51B60 | G51601 | 0.56-0.64 | 0.75-1.00 | 0.035 | 0.040 | 0.15-0.35 | ... | 0.70-0.90 | ... |
| 81B45 | G81451 | 0.43-0.48 | 0.75-1.00 | 0.035 | 0.040 | 0.15-0.35 | 0.20-0.40 | 0.35-0.55 | 0.08-0.15 |
| 94B17 | G94171 | 0.15-0.20 | 0.75-1.00 | 0.035 | 0.040 | 0.15-0.35 | 0.30-0.60 | 0.30-0.50 | 0.08-0.15 |
| 94B30 | G94301 | 0.28-0.33 | 0.75-1.00 | 0.035 | 0.040 | 0.15-0.35 | 0.30-0.60 | 0.30-0.50 | 0.08-0.15 |

[a] Small quantities of certain elements are present that are not specified or required. These incidental elements may be present to the following maximum amounts: Cu, 0.35 percent; Ni, 0.25 percent; Cr, 0.20 percent; and Mo, 0.06 percent.

[b] Standard alloy steels can also be produced with a lead range of 0.15-0.35 percent. Such steels are identified by inserting the letter "L" between the second and third numerals of the AISI or SAE number, e.g., 41L40.

[c] Electric furnace steel.

[d] 0.0005-0.003 percent B.

*Source:* American Iron and Steel Institute: *Steel Products Manual*.

**Table 6. Standard Stainless Steels — Typical Compositions**

| AISI Type (UNS) | Typical Composition (%) | AISI Type (UNS) | Typical Composition (%) |
|---|---|---|---|
| | | **Austenitic** | |
| 201 (S20100) | 16-18 Cr, 3.5-5.5 Ni, 0.15 C, 5.5-7.5 Mn, 0.75 Si, 0.060 P, 0.030 S, 0.25 N | 310 (S31000) | 24-26 Cr, 19-22 Ni, 0.25 C, 2.0 Mn, 1.5 Si, 0.045 P, 0.030 S |
| 202 (S20200) | 17-19 Cr, 4-6 Ni, 0.15 C, 7.5-10.0 Mn, 0.75 Si, 0.060 P, 0.030 S, 0.25 N | 310S (S31008) | 24-26 Cr, 19-22 Ni, 0.08 C, 2.0 Mn, 1.5 Si, 0.045 P, 0.30 S |
| 205 (S20500) | 16.5-18 Cr, 1-1.75 Ni, 0.12-0.25 C, 14-15.5 Mn, 0.75 Si, 0.060 P, 0.030 S, 0.32-0.40 N | 314 (S31400) | 23-26 Cr, 19-22 Ni, 0.25 C, 2.0 Mn, 1.5-3.0 Si, 0.045 P, 0.030 S |
| 301 (S30100) | 16-18 Cr, 6-8 Ni, 0.15 C, 2.0 Mn, 0.75 Si, 0.045 P, 0.030 S | 316 (S31600) | 16-18 Cr, 10-14 Ni, 0.08 C, 2.0 Mn, 0.75 Si, 0.045 P, 0.030 S, 2.0-3.0 Mo, 0.10 N |
| 302 (S30200) | 17-19 Cr, 8-10 Ni, 0.15 C, 2.0 Mn, 0.75 Si, 0.045 P, 0.030 S, 0.10 N | 316L (S31603) | 16-18 Cr, 10-14 Ni, 0.03 C, 2.0 Mn, 0.75 Si, 0.045 P, 0.030 S, 2.0-3.0 Mo, 0.10 N |
| 302B (S30215) | 17-19 Cr, 8-10 Ni, 0.15 C, 2.0 Mn, 2.0-3.0 Si, 0.045 P, 0.030 S | 316F (S31620) | 16-18 Cr, 10-14 Ni, 0.08 C, 2.0 Mn, 1.0 Si, 0.20 P, 0.10 S min, 1.75-2.50 Mo |
| 303 (S30300) | 17-19 Cr, 8-10 Ni, 0.15 C, 2.0 Mn, 1.0 Si, 0.20 P, 0.015 S min, 0.60 Mo (optional) | 316N (S31651) | 16-18 Cr, 10-14 Ni, 0.08 C, 2.0 Mn, 0.75 Si, 0.045 P, 0.030 S, 2-3 Mo, 0.10-0.16 N |
| 303Se (S30323) | 17-19 Cr, 8-10 Ni, 0.15 C, 2.0 Mn, 1.0 Si, 0.20 P, 0.060 S, 0.15 Se min | 317 (S31700) | 18-20 Cr, 11-15 Ni, 0.08 C, 2.0 Mn, 0.75 Si, 0.045 P, 0.030 S, 3.0-4.0 Mo, 0.10 N max |
| 304 (S30400) | 18-20 Cr, 8-10.50 Ni, 0.08 C, 2.0 Mn, 0.75 Si, 0.045 P, 0.030 S, 0.10 N | 317L (S31703) | 18-20 Cr, 11-15 Ni, 0.03 C, 2.0 Mn, 0.75 Si, 0.045 P, 0.030 S, 3-4 Mo, 0.10 N max |
| 304L (S30403) | 18-20 Cr, 8-12 Ni, 0.03 C, 2.0 Mn, 0.75 Si, 0.045 P, 0.030 S, 0.10 N | 321 (S32100) | 17-19 Cr, 9-12 Ni, 0.08 C, 2.0 Mn, 0.75 Si, 0.045 P, 0.030 S [Ti, 5(C + N) min, 0.70 max], 0.10 max |
| 304 Cu (S30430) | 17-19 Cr, 8-10 Ni, 0.08 C, 2.0 Mn, 0.75 Si, 0.045 P, 0.030 S, 3-4 Cu | 329 (S32900) | 23-28 Cr, 2.5-5 Ni, 0.08 C, 2.0 Mn, 0.75 Si, 0.040 P, 0.030 S, 1-2 Mo |
| 304N (S30451) | 18-20 Cr, 8-10.5 Ni, 0.08 C, 2.0 Mn, 0.75 Si, 0.045 P, 0.030 S, 0.10-0.16 N | 330 (N08330) | 17-20 Cr, 34-37 Ni, 0.08 C, 2.0 Mn, 0.75-1.50 Si, 0.040 P, 0.030 S |
| 305 (S30500) | 17-19 Cr, 10.50-13 Ni, 0.12 C, 2.0 Mn, 0.75 Si, 0.045 P, 0.030 S | 347 (S34700) | 17-19 Cr, 9-13 Ni, 0.08 C, 2.0 Mn, 0.75 Si, 0.045 P, 0.030 S (Nb + Ta, 10 × C min, 1 max) |
| 308 (S30800) | 19-21 Cr, 10-12 Ni, 0.08 C, 2.0 Mn, 1.0 Si, 0.045 P, 0.030 S | 348 (S34800) | 17-19 Cr, 9-13 Ni, 0.08 C, 2.0 Mn, 0.75 Si, 0.045 P, 0.030 S (Nb + Ta, 10 × C min, 1 max, but 0.10 Ta max), 0.20 Ca |
| 309 (S30900) | 22-24 Cr, 12-15 Ni, 0.20 C, 2.0 Mn, 1.0 Si, 0.045 P, 0.030 S | 384 (S38400) | 15-17 Cr, 17-19 Ni, 0.08 C, 2.0 Mn, 1.0 Si, 0.045 P, 0.030 S |
| 309S (S30908) | 22-24 Cr, 12-15 Ni, 0.08 C, 2.0 Mn, 1.0 Si, 0.045 P, 0.030 S | … | … |
| | | **Ferritic** | |
| 405 (S40500) | 11.5-14.5 Cr, 0.08 C, 1.0 Mn, 1.0 Si, 0.040 P, 0.030 S, 0.1-0.3 Al, 0.60 max | 430FSe (S43023) | 16-18 Cr, 0.12 C, 1.25 Mn, 1.0 Si, 0.060 P, 0.060 S, 0.15 Se min |
| 409 (S40900) | 10.5-11.75 Cr, 0.08 C, 1.0 Mn, 1.0 Si, 0.045 P, 0.030 S, 0.05 Ni (Ti 6 × C, but with 0.75 max) | 434 (S43400) | 16-18 Cr, 0.12 C, 1.0 Mn, 1.0 Si, 0.040 P, 0.030 S, 0.75-1.25 Mo |
| 429 (S42900) | 14-16 Cr, 0.12 C, 1.0 Mn, 1.0 Si, 0.040 P, 030 S, 0.75 Ni | 436 (S43600) | 16-18 Cr, 0.12 C, 1.0 Mn, 1.0 Si, 0.040 P, 0.030 S, 0.75-1.25 Mo (Nb + Ta 5 × C min, 0.70 max) |
| 430 (S43000) | 16-18 Cr, 0.12 C, 1.0 Mn, 1.0 Si, 0.040 P, 030 S, 0.75 Ni | 442 (S44200) | 18-23 Cr, 0.20 C, 1.0 Mn, 1.0 Si, 0.040 P, 0.030 S |
| 430F (S43020) | 16-18 Cr, 0.12 C, 1.25 Mn, 1.0 Si, 0.060 P, 0.15 S min, 0.60 Mo (optional) | 446 (S44600) | 23-27 Cr, 0.20 C, 1.5 Mn, 1.0 Si, 0.040 P, 0.030 S, 0.025 N |

**Table 6.** *(Continued)* **Standard Stainless Steels — Typical Compositions**

| AISI Type (UNS) | Typical Composition (%) | AISI Type (UNS) | Typical Composition (%) |
|---|---|---|---|
| Martensitic | | | |
| 403 (S40300) | 11.5-13.0 Cr, 1.15 C, 1.0 Mn, 0.5 Si, 0.040 P, 0.030 S, 0.60 Ni | 420F (S42020) | 12-14 Cr, over 0.15 C, 1.25 Mn, 1.0 Si, 0.060 P, 0.15 S min, 0.60 Mo max (optional) |
| 410 (S41000) | 11.5-13.5 Cr, 0.15 C, 1.0 Mn, 1.0 Si, 0.040 P, 0.030 S, 0.75 Ni | 422 (S42200) | 11-12.50 Cr, 0.50-1.0 Ni, 0.20- 0.25 C, 0.50-1.0 Mn, 0.50 Si, 0.025 P, 0.025 S, 0.90-1.25 Mo, 0.20-0.30 V, 0.90-1.25 W |
| 414 (S41400) | 11.5-13.5 Cr, 1.25-2.50 Ni, 0.15 C, 1.0 Mn, 1.0 Si, 0.040 P, 0.030 S, 1.25-2.50 Ni | 431 (S41623) | 15-17 Cr, 1.25-2.50 Ni, 0.20 C, 1.0 Mn, 1.0 Si, 0.040 P, 0.030 S |
| 416 (S41600) | 12-14 Cr, 0.15 C, 1.25 Mn, 1.0 Si, 0.060 P, 0.15 S min, 0.060 Mo (optional) | 440A (S44002) | 16-18 Cr, 0.60-0.75 C, 1.0 Mn, 1.0 Si, 0.040 P, 0.030 S, 0.75 Mo |
| 416Se (S41623) | 12-14 Cr, 0.15 C, 1.25 Mn, 1.0 Si, 0.060 P, 0.060 S, 0.15 Se min | 440B (S44003) | 16-18 Cr, 0.75-0.95 C, 1.0 Mn, 1.0 Si, 0.040 P, 0.030 S, 0.75 Mo |
| 420 (S42000) | 12-14 Cr, 0.15 C min, 1.0 Mn, 1.0 Si, 0.040 P, 0.030 S | 440C (S44004) | 16-18 Cr, 0.95-1.20 C, 1.0 Mn, 1.0 Si, 0040 P, 0.030 S, 0.75 Mo |
| Heat-Resisting | | | |
| 501 (S50100) | 4-6 Cr, 0.10 C min, 1.0 Mn, 1.0 Si, 0.040 P, 0.030 S, 0.40-0.65 Mo | 502 (S50200) | 4-6 Cr, 0.10 C. 1.0 Mn, 1.0 Si, 0.040 P, 0.030 S, 0.40-0.65 Mo |

**Thermal Treatments of Steel.—**Steel's versatility is due to its response to thermal treatment. Although most steel products are used in the as-rolled or un-heat-treated condition, thermal treatment greatly increases the number of properties that can be obtained because at certain "critical temperatures" iron changes from one type of crystal structure to another. This structural change, known as an allotropic transformation, is spontaneous and reversible and can be made to occur by simply changing the temperature of the metal.

In steel, the transformation in crystal structure occurs over a range of temperatures, bounded by lower and upper critical points. When heated, most carbon and low-alloy steels have a critical temperature range between 1300 and 1600° F (700 and 870° C). Steel above this temperature but below the melting range, has a crystalline structure known as austenite, in which the carbon and alloying elements are dissolved in a solid solution. Below this critical range, the crystal structure changes to a phase known as ferrite, which is capable of maintaining only a very small percentage of carbon in solid solution. The remaining carbon exists in the form of carbides, which are compounds of carbon and iron and certain of the other alloying elements. Depending primarily on cooling rate, the carbides may be present as thin plates alternating with the ferrite (pearlite); as spheroidal globular particles at ferrite grain boundaries or dispersed throughout the ferrite; or as a uniform distribution of extremely fine particles throughout a "ferritelike" phase, which has an acicular (needle-like) appearance, termed martensite. In some of the highly alloyed stainless steels, the addition of certain elements stabilizes the austenite structure so that it persists even at very low temperatures (austenitic grades). Other alloying elements can prevent the formation of austenite entirely up to the melting point (ferritic grades).

Fundamentally, all steel heat treatments are intended to either harden or soften the metal. They involve one or a series of operations in which the solid metal is heated and cooled under specified conditions to develop a required structure and properties. In general, there are five major forms of heat treatment for the standard steels that modify properties to suit either fabrication or end use.

*Quenching and Tempering:* The primary hardening treatment for steel, quenching and tempering, usually consists of three successive operations: heating the steel above the critical range and holding it at these temperatures for a sufficient time to approach a uniform solid solution (austenitizing); cooling the steel rapidly by quenching in oil, water, brine, salt or air to form a hard, usually brittle, metastable structure known as untempered or white martensite; tempering the steel by reheating it to a temperature below the critical range in order to obtain the required combination of hardness, strength, ductility, toughness, and structural stability (tempered martensite).

Two well-known modifications of conventional quenching and tempering are "austempering" and "martempering." They involve interrupted quenching techniques (two or more quenching media) that can be utilized for some steels to obtain desired structures and properties while minimizing distortion and cracking problems that may occur in conventional hardening.

*Normalizing:* The steel is heated to a temperature above the critical range, after which it is cooled in still air to produce a generally fine pearlite structure. The purpose is to promote uniformity of structure and properties after a hot-working operation such as forging or extrusion. Steels may be placed in service in the normalized condition, or they may be subjected to additional thermal treatment after subsequent machining or other operations.

*Annealing:* The steel is heated to a temperature above or within the critical range, then cooled at a predetermined slow rate (usually in a furnace) to produce a coarse pearlite structure. This treatment is used to soften the steel for improved machinability; to improve or restore ductility for subsequent forming operations; or to eliminate the residual stresses and microstructural effects of cold-working.

*Spheroidize Annealing:* This is a special form of annealing that requires prolonged heating at an appropriate temperature followed by slow cooling in order to produce globular carbides, a structure desirable for machining, cold forming, or cold drawing, or for the effect it will have on subsequent heat treatment.

*Stress Relieving:* This process reduces internal stresses caused by machining, cold-working, or welding. The steel is heated to a temperature below the critical range and held there long enough to equalize the temperature throughout the piece.

See the sections *HARDENING, TEMPERING, AND ANNEALING* on page 456 and *Heat Treating High-Speed Steels* on page 491 for more information about the heat treatment of steels.

**Applications.—**Many factors enter into the selection of a steel for a particular application. These factors include the mechanical and physical properties needed to satisfy the design requirements and service environment; the cost and availability of the material; the cost of processing (machining, heat treatment, welding, etc.); and the suitability of available processing equipment or the cost of any new equipment required.

These steel selection considerations require input from designers, metallurgists, manufacturing engineers, service engineers, and procurement specialists, and can be considered proper or optimum when the part is made from the lowest cost material consistent with satisfying engineering and service requirements. The factors in selection can vary widely among different organizations, so that several different steels may be used successfully for similar applications. The best choice of a steel for any application most often results from a balance or trade-offs among the various selection considerations.

The AISI/SAE designated "standard steels" provide a convenient way for engineers and metallurgists to state briefly but clearly the chemical composition and, in some instances, desired properties. These steels are widely recognized and used in the United States and in many other countries. There are, however, numerous nonstandard carbon, alloy, and stainless steel grades that are widely used for special applications.

The following sections and tables illustrate the general characteristics and typical applications of most of the standard carbon, alloy, and stainless steel grades.

*General Applications of SAE Steels:* These applications are intended as a general guide only since the selection may depend on the exact character of the service, cost of material, machinability when machining is required, or other factors. When more than one steel is recommended for a given application, information on the characteristics of each steel listed will be found in the section *Carbon Steels* starting on page 405.

Adapters, 1145
Agricultural steel, 1070, 1080
Aircraft forgings, 4140
Axles front or rear, 1040, 4140
Axle shafts, 1045, 2340, 2345, 3135, 3140, 3141, 4063, 4340
Ball-bearing races, 52100
Balls for ball bearings, 52100
Body stock for cars, rimmed*
Bolts and screws, 1035
Bolts
- anchor, 1040
- cold-headed, 4042
- connecting-rod, 3130
- heat-treated, 2330
- heavy-duty, 4815, 4820
- steering-arm, 3130

Brake levers, 1030, 1040
Bumper bars, 1085
Cams free-wheeling, 4615, 4620
Camshafts, 1020, 1040
Carburized parts, 1020, 1022, 1024, 1117, 1118, 1320, 2317, 2515, 3310, 3115, 3120, 4023, 4032
Chain pins transmission, 4320, 4815, 4820
Chains transmission, 3135, 3140
Clutch disks, 1060, 1070, 1085
Clutch springs, 1060
Coil springs, 4063
Cold-headed bolts, 4042
Cold-heading
- steel, 30905, 1070
- wire or rod, rimmed*, 1035

Cold-rolled steel, 1070
Connecting-rods, 1040, 3141
Connecting-rod bolts, 3130
Corrosion resisting, 51710, 30805
Covers transmission, rimmed*
Crankshafts, 1045, 1145, 3135, 3140, 3141
Crankshafts, Diesel engine, 4340
Cushion springs, 1060
Cutlery, stainless, 51335
Cylinder studs, 3130
Deep-drawing steel, rimmed*, 30905
Differential gears, 4023
Disks, clutch, 1070, 1060
Ductile steel, 30905
Fan blades, 1020
Fatigue resisting 4340, 4640
Fender stock for cars, rimmed*
Forgings
- aircraft, 4140
- carbon steel, 1040, 1045
- heat-treated, 3240, 5140, 6150
- high-duty, 6150
- small or medium, 1035
- large, 1036

Free-cutting steel
- carbon, 1111, 1113
- chromium-nickel steel, 30615
- manganese steel, 1132, 1137

Gears
- carburized, 1320, 2317, 3115, 3120, 3310, 4119, 4125, 4320, 4615, 4620, 4815, 4820
- heat-treated, 2345
- car and truck, 4027, 4032
- cyanide-hardening, 5140
- differential, 4023
- high duty, 4640, 6150
- oil-hardening, 3145, 3150, 4340, 5150
- ring, 1045, 3115, 3120, 4119
- transmission, 3115, 3120, 4119
- truck and bus, 3310, 4320

Gear shift levers, 1030
Harrow disks, 1080
Hay-rake teeth, 1095
Key stock, 1030, 2330, 3130
Leaf springs, 1085, 9260
Levers
- brake, 1030, 1040
- gear shift, 1030
- heat-treated, 2330

Lock washers, 1060
Mower knives, 1085
Mower sections, 1070
Music wire, 1085
Nuts, 3130
- heat-treated, 2330

Oil pans, automobile, rimmed*
Pinions, carburized, 3115, 3120, 4320
Piston pins, 3115, 3120
Plow
- beams, 1070
- disks, 1080
- shares, 1080

* The "rimmed" and "killed" steels listed are in the SAE 1008, 1010, and 1015 group. See general description of these steels.

Propeller shafts, 2340, 2345, 4140
Races, ball-bearing, 52100
Ring gears, 3115, 3120, 4119
Rings, snap, 1060, 1070, 1090
Rivets, rimmed*
Rod and wire, killed*
Rod, cold-heading, 1035
Roller bearings, 4815
Rollers for bearings, 52100
Screws and bolts, 1035
Screw stock
  Bessemer, 1111, 1112, 1113
  open-hearth, 1115
Screws, heat-treated, 2330
Seat springs, 10956
Shafts
  axle, 1045
  cyanide-hardening, 5140
  heavy-duty, 4340, 6150, 4615, 4620
  oil-hardening, 5150
  propeller, 2340, 2345, 4140
  transmission, 4140
Sheets and strips, rimmed*
Snap rings, 1060, 1070, 1090
Spline shafts, 1045, 1320, 2340, 2345, 3115, 3120, 3135, 3140, 4023
Spring clips, 1060
Springs
  coil, 1095, 4063, 6150
  clutch, 1060
  cushion, 1060
  hard-drawn coiled, 1066
  leaf, 1085, 1095, 4063, 4068, 9260, 6150
  oil-hardening, 5150
  oil-tempered wire, 1066
  seat, 1095
  valve, 1060
Spring wire, 1045
  hard-drawn, 1055
  oil-tempered, 1055
Stainless irons, 51210, 51710
Steel
  cold-rolled, 1070
  cold-heading, 30905
  free-cutting carbon, 1111, 1113
  free-cutting chrome-nickel, 30615
  free-cutting manganese, 1132
  minimum distortion, 4615, 4620, 4640
  soft ductile, 30905
Steering arms, 4042
Steering-arm bolts, 3130
Steering knuckles, 3141
Steering-knuckle pins, 4815, 4820
Tacks, rimmed*
Thrust washers, 1060
  oil-hardened, 5150
Transmission shafts, 4140
Tubing, 1040
  front axle, 4140
  seamless, 1030
  welded, 1020
Universal joints, 1145
Valve springs, 1060
Washers, lock, 1060
Welded structures, 30705
Wire and rod, killed*
Wire
  cold-heading, rimmed*
  hard-drawn spring, 1045, 1055
  music, 1085
  oil-tempered spring, 1055
Wrist-pins, automobile, 1020
Yokes, 1145

**Carbon Steels.**—*SAE Steels 1006, 1008, 1010, 1015:* These steels are the lowest carbon steels of the plain carbon type and are selected where cold formability is the primary requisite of the user. They are produced both as rimmed and killed steels. Rimmed steel is used for sheet, strip, rod, and wire where excellent surface finish or good drawing qualities are required, such as body and fender stock, hoods, lamps, oil pans, and other deep-drawn and -formed products. This steel is also used for cold-heading wire for tacks, and rivets and low carbon wire products. Killed steel (usually aluminum killed or special killed) is used for difficult stampings, or where nonaging properties are needed. Killed steels (usually silicon killed) should be used in preference to rimmed steel for forging or heat-treating applications.

These steels have relatively low tensile values and should not be selected where much strength is desired. Within the carbon range of the group, strength and hardness will rise with increases in carbon and/or with cold-work, but such increases in strength are at the sacrifice of ductility or the ability to withstand cold deformation. Where cold rolled strip is used, the proper temper designation should be specified to obtain the desired properties.

With less than 0.15 carbon, the steels are susceptible to serious grain growth, causing brittleness, which may occur as the result of a combination of critical strain (from cold-work) followed by heating to certain elevated temperatures. If cold-worked parts formed from these steels are to be later heated to temperatures in excess of 1100° F (590° C), the user should exercise care to avoid or reduce cold-working. When this condition develops, it can be overcome by heating the parts to a temperature well in excess of the upper critical point, or at least 1750° F (955° C).

Steels in this group, being nearly pure iron or ferritic in structure, do not machine freely and should be avoided for cut screws and operations requiring broaching or smooth finish on turning. The machinability of bar, rod, and wire products is improved by cold drawing. Steels in this group are readily welded.

*SAE 1016, 1017, 1018, 1019, 1020, 1021, 1022, 1023, 1024, 1025, 1026, 1027, 1030:* Steels in this group, due to the carbon range covered, have increased strength and hardness, and reduced cold formability compared to the lowest carbon group. For heat-treating purposes, they are known as carburizing or case-hardening grades. When uniform response to heat treatment is required, or for forgings, killed steel is preferred; for other uses, semi-killed or rimmed steel may be indicated, depending on the combination of properties desired. Rimmed steels can ordinarily be supplied up to 0.25 carbon.

Selection of one of these steels for carburizing applications depends on the nature of the part, the properties desired, and the processing practice preferred. Increases in carbon give greater core hardness with a given quench, or permit the use of thicker sections. Increases in manganese improve the hardenability of both the core and case; in carbon steels this is the only change in composition that will increase case hardenability. The higher manganese variants also machine much better. For carburizing applications, SAE 1016, 1018, and 1019 are widely used for thin sections or water-quenched parts. SAE 1022 and 1024 are used for heavier sections or where oil quenching is desired, and SAE 1024 is sometimes used for such parts as transmission and rear axle gears. SAE 1027 is used for parts given a light case to obtain satisfactory core properties without drastic quenching. SAE 1025 and 1030, although not usually regarded as carburizing types, are sometimes used in this manner for larger sections or where greater core hardness is needed.

For cold-formed or -headed parts, the lowest manganese grades (SAE 1017, 1020, and 1025) offer the best formability at their carbon level. SAE 1020 is used for fan blades and some frame members, and SAE 1020 and 1025 are widely used for low-strength bolts. The next higher manganese types (SAE 1018, 1021, and 1026) provide increased strength.

All steels listed may be readily welded or brazed by the common commercial methods. SAE 1020 is frequently used for welded tubing. These steels are used for numerous forged parts, the lower-carbon grades where high strength is not essential. Forgings from the lower-carbon steels usually machine better in the as-forged condition without annealing, or after normalizing.

*SAE 1030, 1033, 1034, 1035, 1036, 1038, 1039, 1040, 1041, 1042, 1043, 1045, 1046, 1049, 1050, 1052:* These steels, of the medium-carbon type, are selected for uses where higher mechanical properties are needed and are frequently further hardened and strengthened by heat treatment or by cold-work. These grades are ordinarily produced as killed steels.

Steels in this group are suitable for a wide variety of automotive-type applications. The particular carbon and manganese level selected is affected by a number of factors. Increases in the mechanical properties required in section thickness, or in depth of hardening, ordinarily indicate either higher carbon or manganese or both. The heat-treating practice preferred, particularly the quenching medium, has a great effect on the steel selected. In general, any of the grades over 0.30 carbon may be selectively hardened by induction or flame methods.

The lower-carbon and manganese steels in this group find usage for certain types of cold-formed parts. SAE 1030 is used for shift and brake levers. SAE 1034 and 1035 are used in the form of wire and rod for cold upsetting such as bolts, and SAE 1038 for bolts and studs. The parts cold-formed from these steels are usually heat treated prior to use. Stampings are generally limited to flat parts or simple bends. The higher-carbon SAE 1038, 1040, and 1042 are frequently cold drawn to specified physical properties for use without heat treatment for some applications such as cylinder head studs.

Any of this group of steels may be used for forgings, the selection being governed by the section size and the physical properties desired after heat treatment. Thus, SAE 1030 and 1035 are used for shifter forks and many small forgings where moderate properties are desired, but the deeper-hardening SAE 1036 is used for more critical parts where a higher strength level and more uniformity are essential, such as some front suspension parts. Forgings such as connecting rods, steering arms, truck front axles, axle shafts, and tractor wheels are commonly made from the SAE 1038 to 1045 group. Larger forgings at similar strength levels need more carbon and perhaps more manganese. Examples are crankshafts made from SAE 1046 and 1052. These steels are also used for small forgings where high hardness after oil quenching is desired. Suitable heat treatment is necessary on forgings from this group to provide machinability. These steels are also widely used for parts machined from bar stock, the selection following an identical pattern to that described for forgings. They are used both with and without heat treatment, depending on the application and the level of properties needed. As a class, they are considered good for normal machining operations. It is also possible to weld these steels by most commercial methods, but precautions should be taken to avoid cracking from too rapid cooling.

*SAE 1055, 1060, 1062, 1064, 1065, 1066, 1070, 1074, 1078, 1080, 1085, 1086, 1090, 1095:* Steels in this group are of the high-carbon type, having more carbon than is required to achieve maximum as quenched hardness. They are used for applications where the higher carbon is needed to improve wear characteristics for cutting edges, to make springs, and for special purposes. Selection of a particular grade is affected by the nature of the part, its end use, and the manufacturing methods available.

In general, cold-forming methods are not practical on this group of steels, being limited to flat stampings and springs coiled from small-diameter wire. Practically all parts from these steels are heat treated before use, with some variations in heat-treating methods to obtain optimum properties for the particular use to which the steel is to be put.

Uses in the spring industry include SAE 1065 for pretempered wire and SAE 1066 for cushion springs of hard-drawn wire, SAE 1064 may be used for small washers and thin stamped parts, SAE 1074 for light flat springs formed from annealed stock, and SAE 1080 and 1085 for thicker flat springs. SAE 1085 is also used for heavier coil springs. Valve spring wire and music wire are special products.

Due to good wear properties when properly heat treated, the high-carbon steels find wide usage in the farm implement industry. SAE 1070 has been used for plow beams, SAE 1074 for plowshares, and SAE 1078 for such parts as rake teeth, scrapers, cultivator shovels, and plowshares. SAE 1085 has been used for scraper blades, disks, and for spring tooth harrows. SAE 1086 and 1090 find use as mower and binder sections, twine holders, and knotter disks.

*SAE 1111, 1112, 1113:* This class of steels is intended for those uses where easy machining is the primary requirement. They are characterized by a higher sulfur content than comparable carbon steels. This composition results in some sacrifice of cold-forming properties, weldability, and forging characteristics. In general, the uses are similar to those for carbon steels of similar carbon and manganese content.

These steels are commonly known as Bessemer screw stock and are considered the best machining steels available, machinability improving within the group as sulfur increases. They are used for a wide variety of machined parts. Although of excellent strength in the cold-drawn condition, they have an unfavorable property of cold shortness and are not commonly used for vital parts. These steels may be cyanided or carburized, but when uniform response to heat treating is necessary, open-hearth steels are recommended.

*SAE 1109, 1114, 1115, 1116, 1117, 1118, 1119, 1120, 1126:* Steels in this group are used where a combination of good machinability and more uniform response to heat treatment is needed. The lower-carbon varieties are used for small parts that are to be cyanided or carbonitrided. SAE 1116, 1117, 1118, and 1119 carry more manganese for better hardenability, permitting oil quenching after case-hardening heat treatments in many instances. The higher-carbon SAE 1120 and 1126 provide more core hardness when this is needed.

*SAE 1132, 1137, 1138, 1140, 1141, 1144, 1145, 1146, 1151:* This group of steels has characteristics comparable to carbon steels of the same carbon level, except for changes due to higher sulfur as noted previously. They are widely used for parts where large amounts of machining are necessary, or where threads, splines, or other contours present special problems with tooling. SAE 1137, for example, is widely used for nuts and bolts and studs with machined threads. The higher-manganese SAE 1132, 1137, 1141, and 1144 offer greater hardenability, the higher-carbon types being suitable for oil quenching for many parts. All these steels may be selectively hardened by induction or flame heating if desired.

**Carburizing Grades of Alloy Steels.—***Properties of the Case:* The properties of carburized and hardened cases (surface layers) depend on the carbon and alloy content, the structure of the case, and the degree and distribution of residual stresses. The carbon content of the case depends on the details of the carburizing process and the response of iron and the alloying elements present to carburization. The original carbon content of the steel has little or no effect on the carbon content produced in the case. The hardenability of the case, therefore, depends on the alloy content of the steel and the final carbon content produced by carburizing, but not on the initial carbon content of the steel.

With complete carbide solution, the effect of alloying elements on the hardenability of the case is about the same as the effect of these elements on the hardenability of the core. As an exception to this statement, any element that inhibits carburizing may reduce the hardenability of the case. Some elements that raise the hardenability of the core may tend to produce more retained austenite and consequently somewhat lower hardness in the case.

Alloy steels are frequently used for case hardening because the required surface hardness can be obtained by moderate speeds of quenching. Slower quenching may mean less distortion than would be encountered with water quenching. It is usually desirable to select a steel that will attain a minimum surface hardness of 58 or 60 RC (Rockwell C scale) after carburizing and oil quenching. Where section sizes are large, a high-hardenability alloy steel may be necessary, whereas for medium and light sections, low-hardenability steels will suffice.

In general, the case-hardening alloy steels may be divided into two classes as far as the hardenability of the case is concerned. Only the general type of steel (SAE 3300-4100, etc.) is discussed. The original carbon content of the steel has no effect on the carbon content of the case, so the last two digits in the specification numbers are not meaningful as far as the case is concerned.

a) High-Hardenability Case: SAE 2500, 3300, 4300, 4800, 9300

As these are high-alloy steels, both the case and the core have high hardenability. They are used particularly for carburized parts having thick sections, such as bevel drive pinions and heavy gears. Good case properties can be obtained by oil quenching. These steels are likely to have retained austenite in the case after carburizing and quenching; consequently, special precautions or treatments, such as refrigeration, may be required.

b) Medium-Hardenability Case: SAE 1300, 2300, 4000, 4100, 4600, 5100, 8600, 8700

Carburized cases of these steels have medium hardenability, which means that their hardenability is intermediate between that of plain carbon steel and the higher-alloy carburizing steels discussed earlier. In general, these steels can be used for average-size case-hardened automotive parts, such as gears, pinions, piston pins, ball studs, universal joint crosses, crankshafts, etc. Satisfactory case hardness is usually produced by oil quenching.

*Core Properties:* The core properties of case-hardened steels depend on both carbon and alloy content of the steel. Each of the general types of alloy case-hardening steel is usually made with two or more carbon contents to permit different hardenability in the core.

The most desirable hardness for the core depends on the design and functioning of the individual part. In general, where high compressive loads are encountered, relatively high core hardness is beneficial in supporting the case. Low core hardnesses may be desirable where great toughness is essential.

The case-hardening steels may be divided into three general classes, depending on hardenability of the core.

a) Low-Hardenability Core: SAE 4017, 4023, 4024, 4027*, 4028*, 4608, 4615, 4617*, 8615*, 8617*

b) Medium-Hardenability Core: SAE 1320, 2317, 2512, 2515*, 3115, 3120, 4032, 4119, 4317, 4620, 4621, 4812, 4815*, 5115, 5120, 8620, 8622, 8720, 9420

c) High-Hardenability Core: SAE 2517, 3310, 3316, 4320, 4817, 4820, 9310, 9315, 9317

*Heat Treatments:* In general, all the alloy carburizing steels are made with fine grain and most are suitable for direct quenching from the carburizing temperature. Several other types of heat treatment involving single and double quenching are also used for most of these steels. See Table 4a on page 485 and Table 4b on page 486.

**Directly Hardenable Grades of Alloy Steels.—**These steels may be considered in five groups on the basis of approximate mean carbon content of the SAE specification. In general, the last two figures of the specification agree with the mean carbon content. Consequently the heading *0.30–0.37 Mean Carbon Content of SAE Specification* includes steels such as SAE 1330, 3135, and 4137.

It is necessary to deviate from the above plan in the classification of the carbon molybdenum steels. Because of the low alloy content of carbon molybdenum steels, it is customary to specify a higher carbon content for any given application than would be specified for other alloy steels. For example, SAE 4063 is used for the same applications as SAE 4140, 4145, and 5150. Consequently, in the following discussion, the carbon molybdenum steels have been shown in the groups where they belong on the basis of applications rather than carbon content.

| Mean Carbon Content of SAE Specification | Common Applications |
|---|---|
| (a) 0.30–0.37 percent | Heat-treated parts requiring moderate strength and great toughness. |
| (b) 0.40–0.42 percent | Heat-treated parts requiring higher strength and good toughness. |
| (c) 0.45–0.50 percent | Heat-treated parts requiring fairly high hardness and strength with moderate toughness. |
| (d) 0.50–0.62 percent | Springs and hand tools. |
| (e) 1.02 percent | Ball and roller bearings. |

For the present discussion, steels of each carbon content are divided into two or three groups on the basis of hardenability. Transformation ranges and consequently heat-treating practices vary somewhat with different alloying elements even though the hardenability is not changed.

*0.30–0.37 Mean Carbon Content of SAE Specification:* These steels are frequently used for water-quenched parts of moderate section size and for oil-quenched parts of small section size. Typical applications of these steels are connecting rods, steering arms and steering knuckles, axle shafts, bolts, studs, screws, and other parts requiring strength

* Borderline classifications might be considered in the next higher hardenability group.

and toughness where section size is small enough to permit the desired physical properties to be obtained with the customary heat treatment.

Steels falling in this classification may be subdivided into two groups on the basis of hardenability:

a) Low Hardenability: SAE 1330, 1335, 4037, 4042, 4130, 5130, 5132, 8630

b) Medium Hardenability: SAE 2330, 3130, 3135, 4137, 5135, 8632, 8635, 8637, 8735, 9437

*0.40–0.42 Mean Carbon Content of SAE Specification:* In general, these steels are used for medium and large parts requiring a high degree of strength and toughness. The choice of the proper steel depends on the section size and the mechanical properties that must be produced. The low and medium hardenabilty steels are used for average size automotive parts, such as steering knuckles, axle shafts, propeller shafts, etc. The high hardenability steels are used particularly for large axles and shafts for large aircraft parts.

These steels are usually considered as oil-quenching steels, although some large parts made of the low and medium hardenability classifications may be quenched in water under properly controlled conditions.

These steels may be divided into three groups on the basis of hardenability:

a) Low Hardenability: SAE 1340, 4047, 5140, 9440

b) Medium Hardenability: SAE 2340, 3140, 3141, 4053, 4063, 4140, 4640, 8640, 8641, 8642, 8740, 8742, 9442

c) High Hardenability: SAE 4340, 9840

*0.45–0.50 Mean Carbon Content of SAE Specification:* These steels are used primarily for gears and other parts requiring fairly high hardness as well as strength and toughness. Such parts are usually oil quenched and a minimum of 90 percent martensite in the as-quenched condition is desirable.

a) Low Hardenability: SAE 5045, 5046, 5145, 9747, 9763

b) Medium Hardenability: SAE 2345, 3145, 3150, 4145, 5147, 5150, 8645, 8647, 8650, 8745, 8747, 8750, 9445, 9845

c) High Hardenability: SAE 4150, 9850

*0.50–0.63 Mean Carbon Content of SAE Specification:* These steels are used primarily for springs and hand tools. The hardenability necessary depends on the thickness of the material and the quenching practice.

a) Medium hardenability: SAE 4068, 5150, 5152, 6150, 8650, 9254, 9255, 9260, 9261

b) High Hardenability: SAE 8653, 8655, 8660, 9262

*1.02 Mean Carbon Content of SAE Specification—SAE 50100, 51100, 52100:* These straight chromium electric furnace steels are used primarily for the races and balls or rollers of antifriction bearings. They are also used for other parts requiring high hardness and wear resistance. The compositions of the three steels are identical, except for a variation in chromium, with a corresponding variation in hardenability.

a) Low Hardenability: SAE 50100

b) Medium Hardenability: SAE 51100, 52100

*Resulfurized Steel:* Some of the alloy steels, SAE 4024, 4028, and 8641, are made resulfurized so as to give better machinability at a relatively high hardness. In general, increased sulfur results in decreased transverse ductility, notched impact toughness, and weldability.

**Characteristics and Typical Applications of Standard Stainless Steels.—**Typical applications of various stainless steel alloys are given in the following. The first number given is the AISI designation followed by the UNS number in parenthesis. (See also *Numbering Systems for Metals and Alloys* on page 393)

*201 (S20100):* High work-hardening rate; low-nickel equivalent of type 301. Flatware; automobile wheel covers, trim.

*202 (S20200):* General-purpose low-nickel equivalent of type 302. Kitchen equipment; hub caps; milk handling.

*205 (S20500):* Lower work-hardening rate than type 202; used for spinning and special drawing operations. Nonmagnetic and cryogenic parts.

*301 (S30100):* High work-hardening rate; used for structural applications where high strength plus high ductility are required. Railroad cars; trailer bodies; aircraft structurals; fasteners; automobile wheel covers, trim; pole line hardware.

*302 (S30200):* General-purpose austenitic stainless steel. Trim; food-handling equipment; aircraft cowlings; antennas; springs; cookware; building exteriors; tanks; hospital, household appliances; jewelry; oil refining equipment; signs.

*302B (S30215):* More resistant to scale than type 302. Furnace parts; still liners; heating elements; annealing covers; burner sections.

*303 (S30300):* Free-machining modification of type 302, for heavier cuts. Screw machine products; shafts; valves; bolts; bushings; nuts.

*303Se (S30323):* Free-machining modification of type 302, for lighter cuts; used where hot-working or cold-heading may be involved. Aircraft fittings; bolts; nuts; rivets; screws; studs.

*304 (S30400):* Low-carbon modification of type 302 for restriction of carbide precipitation during welding. Chemical and food processing equipment; brewing equipment; cryogenic vessels; gutters; downspouts; flashings.

*304L (S30403):* Extra-low-carbon modification of type 304 for further restriction of carbide precipitation during welding. Coal hopper linings; tanks for liquid fertilizer and tomato paste.

*304Cu (S30430):* Lower work-hardening rate than type 304. Severe cold-heading applications.

*304N (S30451):* Higher nitrogen than type 304 to increase strength with minimum effect on ductility and corrosion resistance, more resistant to increased magnetic permeability. Type 304 applications requiring higher strength.

*305 (S30500):* Low work-hardening rate; used for spin forming, severe drawing, cold heading, and forming. Coffee urn tops; mixing bowls; reflectors.

*308 (S30800):* Higher-alloy steel having high corrosion and heat resistance. Welding filler metals to compensate for alloy loss in welding; industrial furnaces.

*309 (S30900):* High-temperature strength and scale resistance. Aircraft heaters; heat-treating equipment; annealing covers; furnace parts; heat exchangers; heat-treating trays; oven linings; pump parts.

*309S (S30908):* Low-carbon modification of type 309. Welded constructions; assemblies subject to moist corrosion conditions.

*310 (S31000):* Higher elevated temperature strength and scale resistance than type 309. Heat exchangers; furnace parts; combustion chambers; welding filler metals; gas-turbine parts; incinerators; recuperators; rolls for roller hearth furnaces.

*310S (S31008):* Low-carbon modification of type 310. Welded constructions; jet engine rings.

*314 (S31400):* More resistant to scale than type 310. Severe cold-heading or -forming applications. Annealing and carburizing boxes; heat-treating fixtures; radiant tubes.

*316 (S31600):* Higher corrosion resistance than types 302 and 304; high creep strength. Chemical and pulp handling equipment; photographic equipment; brandy vats; fertilizer parts; ketchup cooking kettles; yeast tubs.

*316L (S31603):* Extra-low-carbon modification of type 316. Welded construction where intergranular carbide precipitation must be avoided. Type 316 applications requiring extensive welding.

*316F (S31620):* Higher phosphorus and sulfur than type 316 to improve machining and nonseizing characteristics. Automatic screw machine parts.

*316N (S31651):* Higher nitrogen than type 316 to increase strength with minimum effect on ductility and corrosion resistance. Type 316 applications requiring extra strength.

*317 (S31700):* Higher corrosion and creep resistance than type 316. Dyeing and ink manufacturing equipment.

*317L (S31703):* Extra-low-carbon modification of type 317 for restriction of carbide precipitation during welding. Welded assemblies.

*321 (S32100):* Stabilized for weldments under severe corrosive conditions and service from 800–1650° F (425–900° C). Aircraft exhaust manifolds; boiler shells; process equipment; expansion joints; cabin heaters; fire walls; flexible couplings; pressure vessels.

*329 (S32900):* Austenitic-ferritic type with general corrosion resistance similar to type 316 but with better resistance to stress-corrosion cracking; capable of age hardening. Valves; valve fittings; piping; pump parts.

*330 (N08330):* Good resistance to carburization and oxidation and to thermal shock. Heat-treating fixtures.

*347 (S34700):* Similar to type 321 with higher creep strength. Airplane exhaust stacks; welded tank cars for chemicals; jet engine parts.

*348 (S34800):* Similar to type 321; low retentivity. Tubes and pipes for radioactive systems; nuclear energy uses.

*384 (S38400):* Suitable for severe cold heading or cold forming; lower cold-work hardening rate than type 305. Bolts; rivets; screws; instrument parts.

*403 (S40300):* "Turbine quality" grade. Steam turbine blading and other highly stressed parts including jet engine rings.

*405 (S40500):* Nonhardenable grade for assemblies where air-hardening types such as 410 or 403 are objectionable. Annealing boxes; quenching racks; oxidation-resistant partitions.

*409 (S40900):* General-purpose construction stainless. Automotive exhaust systems; transformer and capacitor cases; dry fertilizer spreaders; tanks for agricultural sprays.

*410 (S41000):* General-purpose heat-treatable type. Machine parts; pump shafts; bolts; bushings; coal chutes; cutlery; hardware; jet engine parts; mining machinery; rifle barrels; screws; valves.

*414 (41400):* High hardenability steel. Springs; tempered rules; machine parts, bolts; mining machinery; scissors; ships' bells; spindles; valve seats.

*416 (S41600):* Free-machining modification of type 410, for heavier cuts. Aircraft fittings; bolts; nuts; fire extinguisher inserts; rivets; screws.

*416Se (S41623):* Free-machining modification of type 410, for lighter cuts. Machined parts requiring hot-working or cold-heading.

*420 (S42000):* Higher carbon modification of type 410. Cutlery; surgical instruments; valves; wear-resisting parts; glass molds; hand tools; vegetable choppers.

*420F (S42020):* Free-machining modification of type 420. Applications similar to those for type 420 requiring better machinability.

*422 (S42200):* High strength and toughness at service temperatures up to 1200° F (650° C). Steam turbine blades; fasteners.

*429 (S42900):* Improved weldability as compared to type 430. Nitric acid and nitrogen-fixation equipment.

*430 (S43000):* General-purpose nonhardenable chromium type. Decorative trim; nitric acid tanks; annealing baskets; combustion chambers; dishwashers; heaters; mufflers; range hoods; recuperators; restaurant equipment.

*430F (S43020):* Free-machining modification of type 430, for heavier cuts. Screw machine parts.

*430FSe (S43023):* Free-machining modification of type 430, for lighter cuts. Machined parts requiring light cold heading or forming.

*431 (S43100):* Special-purpose hardenable steel used where particularly high mechanical properties are required. Aircraft fittings; beater bars; paper machinery; bolts.

*434 (S43400):* Modification of type 430 designed to resist atmospheric corrosion in the presence of winter road conditioning and dust-laying compounds. Automotive trim and fasteners.

*436 (S43600):* Similar to types 430 and 434. Used where low "roping" or "ridging" required. General corrosion and heat-resistant applications, such as automobile trim.

*440A (S44002):* Hardenable to higher hardeness than type 420 with good corrosion resistance. Cutlery; bearings; surgical tools.

*440B (S44003):* Cutlery grade. Cutlery; valve parts; instrument bearings.

*440C (S44004):* Yields highest hardnesses of hardenable stainless steels. Balls; bearings; races; nozzles; balls and seats for oil well pumps; valve parts.

*442 (S44200):* High-chromium steel, principally for parts that must resist high service temperatures without scaling. Furnace parts; nozzles; combustion chambers.

*446 (S44600):* High-resistance to corrosion and scaling at high temperatures, especially for intermittent service; often used in sulfur-bearing atmosphere. Annealing boxes; combustion chambers; glass molds; heaters; pyrometer tubes; recuperators; stirring rods; valves.

*501 (S50100):* Heat resistance; good mechanical properties at moderately elevated temperatures. Heat exchangers; petroleum refining equipment.

*502 (S50200):* More ductility and less strength than type 501. Heat exchangers; petroleum refining equipment; gaskets.

**Chromium-Nickel Austenitic Steels.—**(Not capable of heat treatment) *SAE 30201:* This steel is an austenitic chromium-nickel-manganese stainless steel usually required in flat products. In the annealed condition, it exhibits higher strength values than the corresponding chromium-nickel stainless steel (SAE 30301). It is nonmagnetic in the annealed condition, but may be magnetic when cold-worked. SAE 30201 is used to obtain high strength by work-hardening and is well suited for corrosion-resistant structural members requiring high strength with low weight. It has excellent resistance to a wide variety of corrosive media, showing behavior comparable to stainless grade SAE 30301. It has high ductility and excellent forming properties. Owing to this steel's work-hardening rate and yield strength, tools for forming must be designed to allow for a higher springback or recovery rate. It is used for automotive trim, automotive wheel covers, railroad passenger car bodies and structural members, and truck trailer bodies.

*SAE 30202:* Like chromium-nickel stainless steel SAE 30302, this is a general-purpose stainless steel. It has excellent corrosion resistance and deep drawing qualities. It is nonhardenable by thermal treatments but may be cold-worked to high tensile strengths. In the annealed condition, it is nonmagnetic but slightly magnetic when cold-worked. Applications for this stainless steel are hub cap, railcar and truck trailer bodies, and spring wire.

*SAE 30301:* Capable of attaining high tensile strength and ductility by moderate or severe cold-working. It is used largely in the cold-rolled or cold-drawn condition in the form of sheet, strip, and wire. Its corrosion resistance is good but not equal to SAE 30302.

*SAE 30302:* The most widely used of the general-purpose austenitic chromium-nickel stainless steels. It is used for deep drawing largely in the annealed condition. It can be worked to high tensile strengths but with slightly lower ductility than SAE 30301.

*SAE 30303F:* A free-machining steel recommended for the manufacture of parts produced on automatic screw machines. Caution must be used in forging this steel.

*SAE 30304:* Similar to SAE 30302 but somewhat superior in corrosion resistance and having superior welding properties for certain types of equipment.

*SAE 30305:* Similar to SAE 30304 but capable of lower hardness. Has greater ductility with slower work-hardening tendency.

*SAE 30309:* A steel with high heat-resisting qualities which is resistant to oxidation at temperatures up to about 1800° F (980° C).

*SAE 30310:* This steel has the highest heat-resisting properties of the chromium nickel steels listed here and will resist oxidation at temperatures up to about 1900° F (1040° C).

*SAE 30316:* Recommended for use in parts where unusual resistance to chemical or salt water corrosion is necessary. It has superior creep strength at elevated temperatures.

*SAE 30317:* Similar to SAE 30316 but has the highest corrosion resistance of all these alloys in many environments.

*SAE 30321:* Recommended for use in the manufacture of welded structures where heat treatment after welding is not feasible. It is also recommended for use where temperatures up to 1600° F (870° C) are encountered in service.

*SAE 30325:* Used for such parts as heat control shafts.

*SAE 30347:* This steel is similar to SAE 30321. This niobium alloy is sometimes preferred to titanium because niobium is less likely to be lost in welding operations.

**Stainless Chromium Irons and Steels.—***SAE 51409:* An 11 percent chromium alloy developed especially for automotive mufflers and tailpipes. Resistance to corrosion and oxidation is very similar to SAE 51410. It is nonhardenable and has good forming and welding characteristics. This alloy is recommended for mildly corrosive applications where surface appearance is not critical.

*SAE 51410:* A general-purpose stainless steel capable of heat treatment to show good physical properties. It is used for general stainless applications, both in the annealed and heat-treated condition, but is not as resistant to corrosion as SAE 51430 in either the annealed or heat-treated condition.

*SAE 51414:* A corrosion and heat-resisting nickel-bearing chromium steel with somewhat better corrosion resistance than SAE 51410. It will attain slightly higher mechanical properties when heat treated than SAE 51410. It is used in the form of tempered strip or wire, and in bars and forgings for heat-treated parts.

*SAE 51416F:* A free-machining grade for the manufacture of parts produced in automatic screw machines.

*SAE 51420:* This steel is heat-treatable to a relatively high hardness. It will harden to a maximum of approximately 500 BHN (Brinell Hardness Number). Maximum corrosion-resisting qualities exist only in the fully hardened condition. It is used for cutlery, hardened pump shafts, etc.

*SAE 51420F:* This is similar to SAE 51420 except for its free-machining properties.

*SAE 51430:* This high-chromium steel is not capable of heat treatment and is recommended for use in shallow parts requiring moderate draw. Corrosion and heat resistance are superior to SAE 51410.

*SAE 51430F:* This steel is similar to SAE 51430 except for its free-machining properties.

*SAE 51431:* This nickel-bearing chromium steel is designed for heat treatment to high mechanical properties. Its corrosion resistance is superior to other hardenable steels.

*SAE 51440A:* A hardenable chromium steel with greater quenched hardness than SAE 51420 and greater toughness than SAE 51440B and 51440C. Maximum corrosion resistance is obtained in the fully hardened and polished condition.

*SAE 51440B:* A hardenable chromium steel with greater quenched hardness than SAE 51440A. Maximum corrosion resistance is obtained in the fully hardened and polished condition. Capable of hardening to 50–60 RC (Rockwell C scale) depending on carbon content.

*SAE 51440C:* This steel has the greatest quenched hardness and wear resistance on heat treatment of any corrosion- or heat-resistant steel.

*SAE 51440F:* The same as SAE 51440C, except for its free-machining characteristics.

*SAE 51442:* A corrosion- and heat-resisting chromium steel with corrosion-resisting properties slightly better than SAE 51430 and with good scale resistance up to 1600° F (870° C).

*SAE 51446:* A corrosion- and heat-resisting steel with maximum amount of chromium consistent with commercial malleability. Used principally for parts that must resist high temperatures in service without scaling. Resists oxidation up to 2000° F (1095° C).

*SAE 51501:* Used for its heat and corrosion resistance and good mechanical properties at temperatures up to approximately 1000° F (540° C).

**High-Strength, Low-Alloy Steels.—**High-strength, low-alloy (HSLA) steel represents a specific group of steels in which enhanced mechanical properties and, sometimes, resistance to atmospheric corrosion are obtained by the addition of moderate amounts of one or more alloying elements other than carbon. Different types are available, some of which are carbon-manganese steels and others contain further alloy additions, governed by special requirements for weldability, formability, toughness, strength, and economics. These steels may be obtained in the form of sheet, strip, plates, structural shapes, bars, and bar size sections.

HSLA steels are especially characterized by their mechanical properties, obtained in the as-rolled condition. They are not intended for quenching and tempering. For certain applications, they are sometimes annealed, normalized, or stress relieved with some influence on mechanical properties.

Where these steels are used for fabrication by welding, care must be exercised in selection of grade and in the details of the welding process. Certain grades may be welded without preheat or postheat.

Because of their high strength-to-weight ratio, abrasion resistance, and, in certain compositions, improved atmospheric corrosion resistance, these steels are adapted particularly for use in mobile equipment and other structures where substantial weight savings are generally desirable. Typical applications are truck bodies, frames, structural members, scrapers, truck wheels, cranes, shovels, booms, chutes, and conveyors.

*Grade 942X:* A niobium- or vanadium-treated carbon-manganese high-strength steel similar to 945X and 945C except for somewhat improved welding and forming properties.

*Grade 945A:* A HSLA steel with excellent welding characteristics, both arc and resistance, and the best formability, weldability, and low-temperature notch toughness of the high-strength steels. It is generally used in sheets, strip, and light plate thicknesses.

*Grade 945C:* A carbon-manganese high-strength steel with satisfactory arc welding properties if adequate precautions are observed. It is similar to grade 950C, except that lower carbon and manganese improve arc welding characteristics, formability, and low-temperature notch toughness at some sacrifice in strength.

*Grade 945X:* A niobium- or vanadium-treated carbon-manganese high-strength steel similar to 945C, except for somewhat improved welding and forming properties.

*Grade 950A:* A HSLA steel with good weldability, both arc and resistance, with good low-temperature notch toughness, and good formability. It is generally used in sheet, strip, and light plate thicknesses.

*Grade 950B:* A HSLA steel with satisfactory arc welding properties and fairly good low-temperature notch toughness and formability.

*Grade 950C:* A carbon-manganese high-strength steel that can be arc welded with special precautions, but is unsuitable for resistance welding. The formability and toughness are fair.

*Grade 950D:* A HSLA steel with good weldability, both arc and resistance, and fairly good formability. Where low-temperature properties are important, the effect of phosphorus in conjunction with other elements present should be considered.

*Grade 950X:* A niobium- or vanadium-treated carbon-manganese high-strength steel similar to 950C, except for somewhat improved welding and forming properties.

*Grades 955X, 960X, 965X, 970X, 980X:* These are steels similar to 945X and 950X with higher strength obtained by increased amounts of strengthening elements, such as carbon or manganese, or by the addition of nitrogen up to about 0.015 percent. This increased strength involves reduced formability and usually decreased weldability. Toughness will vary considerably with composition and mill practice.

The formability, composition, and minimum mechanical properties of the HSLA steel grades are shown in Table 7 through Table 9 on page 416.

**Table 7. HSLA Steel Grades in Approximate Order of Increasing Excellence**

| Weldability | Formability | Toughness |
|---|---|---|
| 980X | 980X | 980X |
| 970X | 970X | 970X |
| 965X | 965X | 965X |
| 960X | 960X | 960X |
| 955X, 950C, 942X | 955X | 955X |
| 945C | 950C | 945C, 950C, 942X |
| 950B, 950X | 950D | 945X, 950X |
| 945X | 950B, 950X, 942X | 950D |
| 950D | 945C, 945X | 950B |
| 950A | 950A | 950A |
| (Best) 945A | 945A | 945A |

*Source: SAE Handbook*, 1990. Reprinted with permission. Copyright © 1990. Society of Automotive Engineers, Inc. All rights reserved.

**Table 8. Chemical Composition Ladle Analysis of HSLA Steels (max. percent)**

| Grade | C | Mn | P | Grade | C | Mn | P |
|---|---|---|---|---|---|---|---|
| 942X | 0.21 | 1.35 | 0.04 | 950D | 0.15 | 1.00 | 0.15 |
| 945A | 0.15 | 1.00 | 0.04 | 950X | 0.23 | 1.35 | 0.04 |
| 945C | 0.23 | 1.40 | 0.04 | 955X | 0.25 | 1.35 | 0.04 |
| 945X | 0.22 | 1.35 | 0.04 | 960X | 0.26 | 1.45 | 0.04 |
| 950A | 0.15 | 1.30 | 0.04 | 965X | 0.26 | 1.45 | 0.04 |
| 950B | 0.22 | 1.30 | 0.04 | 970X | 0.26 | 1.65 | 0.04 |
| 950C | 0.25 | 1.60 | 0.04 | 980X | 0.26 | 1.65 | 0.04 |

Sulfur, 0.05 percent max; silicon, 0.90 percent max.
*Source: SAE Handbook*, 1990. Reprinted with permission. Copyright © 1990. Society of Automotive Engineers, Inc. All rights reserved.

**Table 9. Minimum Mechanical Properties of High-Strength Low-Alloy Steels**

| | | Tensile Strength[a] (psi) | | % Elongation | | | | Tensile Strength[a] (psi) | | % Elongation | |
|---|---|---|---|---|---|---|---|---|---|---|---|
| Grade | Form | Yield | Ultimate | 2 in. | 8 in. | Grade | Form | Yield | Ultimate | 2 in. | 8 in. |
| 942X | Plates, shapes, bars To 4 in. incl. | 42,000 | 60,000 | 24 | 20 | 955X | Sheet and strip | 55,000 | 70,000 | 20 | … |
| 945A, C | Sheet and strip | 45,000 | 60,000 | 22 | … | | Plates, shapes, bars | | | | |
| | Plates, shapes, bars | | | | | | To 1½ in. incl. | 55,000 | 70,000 | … | 17 |
| | To ½ in. incl. | 45,000 | 65,000 | 22 | 18 | 960X | Sheet and strip | 60,000 | 75,000 | 18 | … |
| | ½-1½ in. incl. | 42,000 | 62,000 | 24 | 19 | | Plates, shapes, bars | | | | |
| | 1½-3 in. incl. | 40,000 | 62,000 | 24 | 19 | | To 1½ in. incl. | 60,000 | 75,000 | … | 16 |
| 945X | Sheet and strip | 45,000 | 60,000 | 25 | … | 965X | Sheet and strip | 65,000 | 80,000 | 16 | … |
| | Plates, shapes, bars | | | | | | Plates, shapes, bars | | | | |
| | To 1½ in. incl. | 45,000 | 60,000 | 22 | 19 | | To ¾ in. incl. | 65,000 | 80,000 | … | 15 |
| 950A, B, C, D | Sheet and strip | 50,000 | 70,000 | 22 | … | 970X | Sheet and strip | 70,000 | 85,000 | 14 | … |
| | Plates, shapes, bars | | | | | | Plates, shapes, bars | | | | |
| | To ½ in. incl. | 50,000 | 70,000 | 22 | 18 | | To ¾ in. incl. | 70,000 | 85,000 | … | 14 |
| | ½-1½ in. incl. | 45,000 | 67,000 | 24 | 19 | 980X | Sheet and strip | 80,000 | 95,000 | 12 | … |
| | 1½-3 in. incl. | 42,000 | 63,000 | 24 | 19 | | Plates to ⅜ in. incl. | 80,000 | 95,000 | … | 10 |
| 950X | Sheet and strip | 50,000 | 65,000 | 22 | … | | | | | | |
| | Plates, shapes, bars | | | | | | | | | | |
| | To 1½ in. incl. | 50,000 | 65,000 | … | 18 | | | | | | |

[a] Yield strength to be measured at 0.2 percent offset. Mechanical properties to be determined in accordance with ASTM A370.
*Source: SAE Handbook*, 1990. Reprinted with permission. Copyright © 1990. Society of Automotive Engineers, Inc. All rights reserved.

**Typical Mechanical Properties of Steels.—**Table 10 through Table 13 provide expected minimum and/or typical mechanical properties of selected standard carbon and alloy steels and stainless steels.

## Table 10. Expected Minimum Mechanical Properties of Cold-Drawn Carbon-Steel Rounds, Squares, and Hexagons

| Size, in. | As Cold-Drawn: Tensile Strength, Ultimate (1000 lb/in²) | Tensile Strength, Yield (1000 lb/in²) | Elongation in 2 inches, Percent | Reduction in Area, Percent | Hardness, BHN | Cold-Drawn Followed by Low-Temperature Stress Relief: Tensile Strength, Ultimate (1000 lb/in²) | Tensile Strength, Yield (1000 lb/in²) | Elongation in 2 inches, Percent | Reduction in Area, Percent | Hardness, BHN | Cold-Drawn Followed by High-Temperature Stress Relief: Tensile Strength, Ultimate (1000 lb/in²) | Tensile Strength, Yield (1000 lb/in²) | Elongation in 2 inches, Percent | Reduction in Area, Percent | Hardness, BHN |
|---|---|---|---|---|---|---|---|---|---|---|---|---|---|---|---|
| AISI 1018 and 1025 Steels | | | | | | | | | | | | | | | |
| ⅝ - ⅞ | 70 | 60 | 18 | 40 | 143 | …. | …. | …. | …. | … | 65 | 45 | 20 | 45 | 131 |
| Over ⅞ - 1¼ | 65 | 55 | 16 | 40 | 131 | …. | …. | …. | …. | … | 60 | 45 | 20 | 45 | 121 |
| Over 1¼ - 2 | 60 | 50 | 15 | 35 | 121 | …. | …. | …. | …. | … | 55 | 45 | 16 | 40 | 111 |
| Over 2 - 3 | 55 | 45 | 15 | 35 | 111 | …. | …. | …. | …. | … | 50 | 40 | 15 | 40 | 101 |
| AISI 1117 and 1118 Steels | | | | | | | | | | | | | | | |
| ⅝ - ⅞ | 75 | 65 | 15 | 40 | 149 | 80 | 70 | 15 | 40 | 163 | 70 | 50 | 18 | 45 | 143 |
| Over ⅞ - 1¼ | 70 | 60 | 15 | 40 | 143 | 75 | 65 | 15 | 40 | 149 | 65 | 50 | 16 | 45 | 131 |
| Over 1¼ - 2 | 65 | 55 | 13 | 35 | 131 | 70 | 60 | 13 | 35 | 143 | 60 | 50 | 15 | 40 | 121 |
| Over 2 - 3 | 60 | 50 | 12 | 30 | 121 | 65 | 55 | 12 | 35 | 131 | 55 | 45 | 15 | 40 | 111 |
| AISI 1035 Steel | | | | | | | | | | | | | | | |
| ⅝ - ⅞ | 85 | 75 | 13 | 35 | 170 | 90 | 80 | 13 | 35 | 179 | 80 | 60 | 16 | 45 | 163 |
| Over ⅞ - 1¼ | 80 | 70 | 12 | 35 | 163 | 85 | 75 | 12 | 35 | 170 | 75 | 60 | 15 | 45 | 149 |
| Over 1¼ - 2 | 75 | 65 | 12 | 35 | 149 | 80 | 70 | 12 | 35 | 163 | 70 | 60 | 15 | 40 | 143 |
| Over 2 - 3 | 70 | 60 | 10 | 30 | 143 | 75 | 65 | 10 | 30 | 149 | 65 | 55 | 12 | 35 | 131 |
| AISI 1040 and 1140 Steels | | | | | | | | | | | | | | | |
| ⅝ - ⅞ | 90 | 80 | 12 | 35 | 179 | 95 | 85 | 12 | 35 | 187 | 85 | 65 | 15 | 45 | 170 |
| Over ⅞ - 1¼ | 85 | 75 | 12 | 35 | 170 | 90 | 80 | 12 | 35 | 179 | 80 | 65 | 15 | 45 | 163 |
| Over 1¼ - 2 | 80 | 70 | 10 | 30 | 163 | 85 | 75 | 10 | 30 | 170 | 75 | 60 | 15 | 40 | 149 |
| Over 2 - 3 | 75 | 65 | 10 | 30 | 149 | 80 | 70 | 10 | 30 | 163 | 70 | 55 | 12 | 35 | 143 |

**Table 10.** *(Continued)* **Expected Minimum Mechanical Properties of Cold-Drawn Carbon-Steel Rounds, Squares, and Hexagons**

| Size, in. | As Cold-Drawn | | | | | Cold-Drawn Followed by Low-Temperature Stress Relief | | | | | Cold-Drawn Followed by High-Temperature Stress Relief | | | | |
|---|---|---|---|---|---|---|---|---|---|---|---|---|---|---|---|
| | Tensile Strength | | Elongation in 2 inches, Percent | Reduction in Area, Percent | Hardness, BHN | Tensile Strength | | Elongation in 2 inches, Percent | Reduction in Area, Percent | Hardness, BHN | Tensile Strength | | Elongation in 2 inches, Percent | Reduction in Area, Percent | Hardness, BHN |
| | Ultimate | Yield | | | | Ultimate | Yield | | | | Ultimate | Yield | | | |
| | 1000 lb/in² | | | | | 1000 lb/in² | | | | | 1000 lb/in² | | | | |
| AISI 1045, 1145, and 1146 Steels | | | | | | | | | | | | | | | |
| ⅝-⅞ | 95 | 85 | 12 | 35 | 187 | 100 | 90 | 12 | 35 | 197 | 90 | 70 | 15 | 45 | 179 |
| Over ⅞-1¼ | 90 | 80 | 11 | 30 | 179 | 95 | 85 | 11 | 30 | 187 | 85 | 70 | 15 | 45 | 170 |
| Over 1¼-2 | 85 | 75 | 10 | 30 | 170 | 90 | 80 | 10 | 30 | 179 | 80 | 65 | 15 | 40 | 163 |
| Over 2-3 | 80 | 70 | 10 | 30 | 163 | 85 | 75 | 10 | 25 | 170 | 75 | 60 | 12 | 35 | 149 |
| AISI 1050, 1137, and 1151 Steels | | | | | | | | | | | | | | | |
| ⅝-⅞ | 100 | 90 | 11 | 35 | 197 | 105 | 95 | 11 | 35 | 212 | 95 | 75 | 15 | 45 | 187 |
| Over ⅞-1¼ | 95 | 85 | 11 | 30 | 187 | 100 | 90 | 11 | 30 | 197 | 90 | 75 | 15 | 40 | 179 |
| Over 1¼-2 | 90 | 80 | 10 | 30 | 179 | 95 | 85 | 10 | 30 | 187 | 85 | 70 | 15 | 40 | 170 |
| Over 2-3 | 85 | 75 | 10 | 30 | 170 | 90 | 80 | 10 | 25 | 179 | 80 | 65 | 12 | 35 | 163 |
| AISI 1141 Steel | | | | | | | | | | | | | | | |
| ⅝-⅞ | 105 | 95 | 11 | 30 | 212 | 110 | 100 | 11 | 30 | 223 | 100 | 80 | 15 | 40 | 197 |
| Over ⅞-1¼ | 100 | 90 | 10 | 30 | 197 | 105 | 95 | 10 | 30 | 212 | 95 | 80 | 15 | 40 | 187 |
| Over 1¼-2 | 95 | 85 | 10 | 30 | 187 | 100 | 90 | 10 | 25 | 197 | 90 | 75 | 15 | 40 | 179 |
| Over 2-3 | 90 | 80 | 10 | 20 | 179 | 95 | 85 | 10 | 20 | 187 | 85 | 70 | 12 | 30 | 170 |
| AISI 1144 Steel | | | | | | | | | | | | | | | |
| ⅝-⅞ | 110 | 100 | 10 | 30 | 223 | 115 | 105 | 10 | 30 | 229 | 105 | 85 | 15 | 40 | 212 |
| Over ⅞-1¼ | 105 | 95 | 10 | 30 | 212 | 110 | 100 | 10 | 30 | 223 | 100 | 85 | 15 | 40 | 197 |
| Over 1¼-2 | 100 | 90 | 10 | 25 | 197 | 105 | 95 | 10 | 25 | 212 | 95 | 80 | 15 | 35 | 187 |
| Over 2-3 | 95 | 85 | 10 | 20 | 187 | 100 | 90 | 10 | 20 | 197 | 90 | 75 | 12 | 30 | 179 |

*Source:* AISI Committee of Hot-Rolled and Cold-Finished Bar Producers and published in 1974 DATABOOK issue of the American Society for Metals' *Metal Progress* magazine and used with permission.

**Table 11a. Typical Mechanical Properties of Selected Carbon and Alloy Steels** (Hot Rolled, Normalized, and Annealed)

| AISI No.[a] | Treatment | Tensile Strength Ultimate | Tensile Strength Yield | Elongation, Percent | Reduction in Area, Percent | Hard-ness, BHN | Impact Strength (Izod), ft-lb |
|---|---|---|---|---|---|---|---|
| | | lb/in² | | | | | |
| 1015 | As-rolled | 61,000 | 45,500 | 39.0 | 61.0 | 126 | 81.5 |
| | Normalized (1700 F) | 61,500 | 47,000 | 37.0 | 69.6 | 121 | 85.2 |
| | Annealed (1600 F) | 56,000 | 41,250 | 37.0 | 69.7 | 111 | 84.8 |
| 1020 | As-rolled | 65,000 | 48,000 | 36.0 | 59.0 | 143 | 64.0 |
| | Normalized (1600 F) | 64,000 | 50,250 | 35.8 | 67.9 | 131 | 86.8 |
| | Annealed (1600 F) | 57,250 | 42,750 | 36.5 | 66.0 | 111 | 91.0 |
| 1022 | As-rolled | 73,000 | 52,000 | 35.0 | 67.0 | 149 | 60.0 |
| | Normalized (1700 F) | 70,000 | 52,000 | 34.0 | 67.5 | 143 | 86.5 |
| | Annealed (1600 F) | 65,250 | 46,000 | 35.0 | 63.6 | 137 | 89.0 |
| 1030 | As-rolled | 80,000 | 50,000 | 32.0 | 57.0 | 179 | 55.0 |
| | Normalized (1700 F) | 75,000 | 50,000 | 32.0 | 60.8 | 149 | 69.0 |
| | Annealed (1550 F) | 67,250 | 49,500 | 31.2 | 57.9 | 126 | 51.2 |
| 1040 | As-rolled | 90,000 | 60,000 | 25.0 | 50.0 | 201 | 36.0 |
| | Normalized (1650 F) | 85,500 | 54,250 | 28.0 | 54.9 | 170 | 48.0 |
| | Annealed (1450 F) | 75,250 | 51,250 | 30.2 | 57.2 | 149 | 32.7 |
| 1050 | As-rolled | 105,000 | 60,000 | 20.0 | 40.0 | 229 | 23.0 |
| | Normalized (1650 F) | 108,500 | 62,000 | 20.0 | 39.4 | 217 | 20.0 |
| | Annealed (1450 F) | 92,250 | 53,000 | 23.7 | 39.9 | 187 | 12.5 |
| 1060 | As-rolled | 118,000 | 70,000 | 17.0 | 34.0 | 241 | 13.0 |
| | Normalized (1650 F) | 112,500 | 61,000 | 18.0 | 37.2 | 229 | 9.7 |
| | Annealed (1450 F) | 90,750 | 54,000 | 22.5 | 38.2 | 179 | 8.3 |
| 1080 | As-rolled | 140,000 | 85,000 | 12.0 | 17.0 | 293 | 5.0 |
| | Normalized (1650 F) | 146,500 | 76,000 | 11.0 | 20.6 | 293 | 5.0 |
| | Annealed (1450 F) | 89,250 | 54,500 | 24.7 | 45.0 | 174 | 4.5 |
| 1095 | As-rolled | 140,000 | 83,000 | 9.0 | 18.0 | 293 | 3.0 |
| | Normalized (1650 F) | 147,000 | 72,500 | 9.5 | 13.5 | 293 | 4.0 |
| | Annealed (1450 F) | 95,250 | 55,000 | 13.0 | 20.6 | 192 | 2.0 |
| 1117 | As-rolled | 70,600 | 44,300 | 33.0 | 63.0 | 143 | 60.0 |
| | Normalized (1650 F) | 67,750 | 44,000 | 33.5 | 63.8 | 137 | 62.8 |
| | Annealed (1575 F) | 62,250 | 40,500 | 32.8 | 58.0 | 121 | 69.0 |
| 1118 | As-rolled | 75,600 | 45,900 | 32.0 | 70.0 | 149 | 80.0 |
| | Normalized (1700 F) | 69,250 | 46,250 | 33.5 | 65.9 | 143 | 76.3 |
| | Annealed (1450 F) | 65,250 | 41,250 | 34.5 | 66.8 | 131 | 78.5 |
| 1137 | As-rolled | 91,000 | 55,00 | 28.0 | 61.0 | 192 | 61.0 |
| | Normalized (1650 F) | 97,000 | 57,500 | 22.5 | 48.5 | 197 | 47.0 |
| | Annealed (1450 F) | 84,750 | 50,000 | 26.8 | 53.9 | 174 | 36.8 |
| 1141 | As-rolled | 98,000 | 52,000 | 22.0 | 38.0 | 192 | 8.2 |
| | Normalized (1650 F) | 102,500 | 58,750 | 22.7 | 55.5 | 201 | 38.8 |
| | Annealed (1500 F) | 86,800 | 51,200 | 25.5 | 49.3 | 163 | 25.3 |
| 1144 | As-rolled | 102,000 | 61,000 | 21.0 | 41.0 | 212 | 39.0 |
| | Normalized (1650 F) | 96,750 | 58,000 | 21.0 | 40.4 | 197 | 32.0 |
| | Annealed (1450 F) | 84,750 | 50,250 | 24.8 | 41.3 | 167 | 48.0 |

**Table 11a.** *(Continued)* **Typical Mechanical Properties of Selected Carbon and Alloy Steels** (Hot Rolled, Normalized, and Annealed)

| AISI No.[a] | Treatment | Tensile Strength Ultimate (lb/in²) | Tensile Strength Yield (lb/in²) | Elongation, Percent | Reduction in Area, Percent | Hardness, BHN | Impact Strength (Izod), ft-lb |
|---|---|---|---|---|---|---|---|
| 1340 | Normalized (1600 F) | 121,250 | 81,000 | 22.0 | 62.9 | 248 | 68.2 |
| | Annealed (1475 F) | 102,000 | 63,250 | 25.5 | 57.3 | 207 | 52.0 |
| 3140 | Normalized (1600 F) | 129,250 | 87,000 | 19.7 | 57.3 | 262 | 39.5 |
| | Annealed (1500 F) | 100,000 | 61,250 | 24.5 | 50.8 | 197 | 34.2 |
| 4130 | Normalized (1600 F) | 97,000 | 63,250 | 25.5 | 59.5 | 197 | 63.7 |
| | Annealed (1585 F) | 81,250 | 52,250 | 28.2 | 55.6 | 156 | 45.5 |
| 4140 | Normalized (1600 F) | 148,000 | 95,000 | 17.7 | 46.8 | 302 | 16.7 |
| | Annealed (1500 F) | 95,000 | 60,500 | 25.7 | 56.9 | 197 | 40.2 |
| 4150 | Normalized (1600 F) | 167,500 | 106,500 | 11.7 | 30.8 | 321 | 8.5 |
| | Annealed (1500 F) | 105,750 | 55,000 | 20.2 | 40.2 | 197 | 18.2 |
| 4320 | Normalized (1640 F) | 115,000 | 67,250 | 20.8 | 50.7 | 235 | 53.8 |
| | Annealed (1560 F) | 84,000 | 61,625 | 29.0 | 58.4 | 163 | 81.0 |
| 4340 | Normalized (1600 F) | 185,500 | 125,000 | 12.2 | 36.3 | 363 | 11.7 |
| | Annealed (1490 F) | 108,000 | 68,500 | 22.0 | 49.9 | 217 | 37.7 |
| 4620 | Normalized (1650 F) | 83,250 | 53,125 | 29.0 | 66.7 | 174 | 98.0 |
| | Annealed (1575 F) | 74,250 | 54,000 | 31.3 | 60.3 | 149 | 69.0 |
| 4820 | Normalized (1580 F) | 109,500 | 70,250 | 24.0 | 59.2 | 229 | 81.0 |
| | Annealed (1500 F) | 98,750 | 67,250 | 22.3 | 58.8 | 197 | 68.5 |
| 5140 | Normalized (1600 F) | 115,000 | 68,500 | 22.7 | 59.2 | 229 | 28.0 |
| | Annealed (1525 F) | 83,000 | 42,500 | 28.6 | 57.3 | 167 | 30.0 |
| 5150 | Normalized (1600 F) | 126,250 | 76,750 | 20.7 | 58.7 | 255 | 23.2 |
| | Annealed (1520 F) | 98,000 | 51,750 | 22.0 | 43.7 | 197 | 18.5 |
| 5160 | Normalized (1575 F) | 138,750 | 77,000 | 17.5 | 44.8 | 269 | 8.0 |
| | Annealed (1495 F) | 104,750 | 40,000 | 17.2 | 30.6 | 197 | 7.4 |
| 6150 | Normalized (1600 F) | 136,250 | 89,250 | 21.8 | 61.0 | 269 | 26.2 |
| | Annealed (1500 F) | 96,750 | 59,750 | 23.0 | 48.4 | 197 | 20.2 |
| 8620 | Normalized (1675 F) | 91,750 | 51,750 | 26.3 | 59.7 | 183 | 73.5 |
| | Annealed (1600 F) | 77,750 | 55,875 | 31.3 | 62.1 | 149 | 82.8 |
| 8630 | Normalized (1600 F) | 94,250 | 62,250 | 23.5 | 53.5 | 187 | 69.8 |
| | Annealed (1550 F) | 81,750 | 54,000 | 29.0 | 58.9 | 156 | 70.2 |
| 8650 | Normalized (1600 F) | 148,500 | 99,750 | 14.0 | 40.4 | 302 | 10.0 |
| | Annealed (1465 F) | 103,750 | 56,000 | 22.5 | 46.4 | 212 | 21.7 |
| 8740 | Normalized (1600 F) | 134,750 | 88,000 | 16.0 | 47.9 | 269 | 13.0 |
| | Annealed (1500 F) | 100,750 | 60,250 | 22.2 | 46.4 | 201 | 29.5 |
| 9255 | Normalized (1650 F) | 135,250 | 84,000 | 19.7 | 43.4 | 269 | 10.0 |
| | Annealed (1550 F) | 112,250 | 70,500 | 21.7 | 41.1 | 229 | 6.5 |
| 9310 | Normalized (1630 F) | 131,500 | 82,750 | 18.8 | 58.1 | 269 | 88.0 |
| | Annealed (1550 F) | 119,000 | 63,750 | 17.3 | 42.1 | 241 | 58.0 |

[a] All grades are fine-grained except those in the 1100 series that are coarse-grained. Austenitizing temperatures are given in parentheses. Heat-treated specimens were oil quenched unless otherwise indicated.

*Source:* Bethlehem Steel Corp. and Republic Steel Corp. as published in 1974 DATABOOK issue of the American Society for Metals' *Metal Progress* magazine and used with permission.

**Table 11b. Typical Mechanical Properties of Selected Carbon and Alloy Steels** (Quenched-and-Tempered Condition)

| AISI No.[a] | Tempering Temperature, °F | Tensile Strength Ultimate (1000 lb/in²) | Tensile Strength Yield (1000 lb/in²) | Elongation, Percent | Reduction in Area, Percent | Hardness, BHN |
|---|---|---|---|---|---|---|
| 1030[b] | 400 | 123 | 94 | 17 | 47 | 495 |
| | 600 | 116 | 90 | 19 | 53 | 401 |
| | 800 | 106 | 84 | 23 | 60 | 302 |
| | 1000 | 97 | 75 | 28 | 65 | 255 |
| | 1200 | 85 | 64 | 32 | 70 | 207 |
| 1040[b] | 400 | 130 | 96 | 16 | 45 | 514 |
| | 600 | 129 | 94 | 18 | 52 | 444 |
| | 800 | 122 | 92 | 21 | 57 | 352 |
| | 1000 | 113 | 86 | 23 | 61 | 269 |
| | 1200 | 97 | 72 | 28 | 68 | 201 |
| 1040 | 400 | 113 | 86 | 19 | 48 | 262 |
| | 600 | 113 | 86 | 20 | 53 | 255 |
| | 800 | 110 | 80 | 21 | 54 | 241 |
| | 1000 | 104 | 71 | 26 | 57 | 212 |
| | 1200 | 92 | 63 | 29 | 65 | 192 |
| 1050[b] | 400 | 163 | 117 | 9 | 27 | 514 |
| | 600 | 158 | 115 | 13 | 36 | 444 |
| | 800 | 145 | 110 | 19 | 48 | 375 |
| | 1000 | 125 | 95 | 23 | 58 | 293 |
| | 1200 | 104 | 78 | 28 | 65 | 235 |
| 1050 | 400 | … | … | … | … | … |
| | 600 | 142 | 105 | 14 | 47 | 321 |
| | 800 | 136 | 95 | 20 | 50 | 277 |
| | 1000 | 127 | 84 | 23 | 53 | 262 |
| | 1200 | 107 | 68 | 29 | 60 | 223 |
| 1060 | 400 | 160 | 113 | 13 | 40 | 321 |
| | 600 | 160 | 113 | 13 | 40 | 321 |
| | 800 | 156 | 111 | 14 | 41 | 311 |
| | 1000 | 140 | 97 | 17 | 45 | 277 |
| | 1200 | 116 | 76 | 23 | 54 | 229 |
| 1080 | 400 | 190 | 142 | 12 | 35 | 388 |
| | 600 | 189 | 142 | 12 | 35 | 388 |
| | 800 | 187 | 138 | 13 | 36 | 375 |
| | 1000 | 164 | 117 | 16 | 40 | 321 |
| | 1200 | 129 | 87 | 21 | 50 | 255 |
| 1095[b] | 400 | 216 | 152 | 10 | 31 | 601 |
| | 600 | 212 | 150 | 11 | 33 | 534 |
| | 800 | 199 | 139 | 13 | 35 | 388 |
| | 1000 | 165 | 110 | 15 | 40 | 293 |
| | 1200 | 122 | 85 | 20 | 47 | 235 |
| 1095 | 400 | 187 | 120 | 10 | 30 | 401 |
| | 600 | 183 | 118 | 10 | 30 | 375 |
| | 800 | 176 | 112 | 12 | 32 | 363 |
| | 1000 | 158 | 98 | 15 | 37 | 321 |
| | 1200 | 130 | 80 | 21 | 47 | 269 |
| 1137 | 400 | 157 | 136 | 5 | 22 | 352 |
| | 600 | 143 | 122 | 10 | 33 | 285 |
| | 800 | 127 | 106 | 15 | 48 | 262 |
| | 1000 | 110 | 88 | 24 | 62 | 229 |
| | 1200 | 95 | 70 | 28 | 69 | 197 |

**Table 11b.** *(Continued)* **Typical Mechanical Properties of Selected Carbon and Alloy Steels** (Quenched-and-Tempered Condition)

| AISI No.[a] | Tempering Temperature, °F | Tensile Strength Ultimate | Tensile Strength Yield | Elongation, Percent | Reduction in Area, Percent | Hardness, BHN |
|---|---|---|---|---|---|---|
| | | 1000 lb/in² | | | | |
| 1137[b] | 400 | 217 | 169 | 5 | 17 | 415 |
| | 600 | 199 | 163 | 9 | 25 | 375 |
| | 800 | 160 | 143 | 14 | 40 | 311 |
| | 1000 | 120 | 105 | 19 | 60 | 262 |
| | 1200 | 94 | 77 | 25 | 69 | 187 |
| 1141 | 400 | 237 | 176 | 6 | 17 | 461 |
| | 600 | 212 | 186 | 9 | 32 | 415 |
| | 800 | 169 | 150 | 12 | 47 | 331 |
| | 1000 | 130 | 111 | 18 | 57 | 262 |
| | 1200 | 103 | 86 | 23 | 62 | 217 |
| 1144 | 400 | 127 | 91 | 17 | 36 | 277 |
| | 600 | 126 | 90 | 17 | 40 | 262 |
| | 800 | 123 | 88 | 18 | 42 | 248 |
| | 1000 | 117 | 83 | 20 | 46 | 235 |
| | 1200 | 105 | 73 | 23 | 55 | 217 |
| 1330[b] | 400 | 232 | 211 | 9 | 39 | 459 |
| | 600 | 207 | 186 | 9 | 44 | 402 |
| | 800 | 168 | 150 | 15 | 53 | 335 |
| | 1000 | 127 | 112 | 18 | 60 | 263 |
| | 1200 | 106 | 83 | 23 | 63 | 216 |
| 1340 | 400 | 262 | 231 | 11 | 35 | 505 |
| | 600 | 230 | 206 | 12 | 43 | 453 |
| | 800 | 183 | 167 | 14 | 51 | 375 |
| | 1000 | 140 | 120 | 17 | 58 | 295 |
| | 1200 | 116 | 90 | 22 | 66 | 252 |
| 4037 | 400 | 149 | 110 | 6 | 38 | 310 |
| | 600 | 138 | 111 | 14 | 53 | 295 |
| | 800 | 127 | 106 | 20 | 60 | 270 |
| | 1000 | 115 | 95 | 23 | 63 | 247 |
| | 1200 | 101 | 61 | 29 | 60 | 220 |
| 4042 | 400 | 261 | 241 | 12 | 37 | 516 |
| | 600 | 234 | 211 | 13 | 42 | 455 |
| | 800 | 187 | 170 | 15 | 51 | 380 |
| | 1000 | 143 | 128 | 20 | 59 | 300 |
| | 1200 | 115 | 100 | 28 | 66 | 238 |
| 4130[b] | 400 | 236 | 212 | 10 | 41 | 467 |
| | 600 | 217 | 200 | 11 | 43 | 435 |
| | 800 | 186 | 173 | 13 | 49 | 380 |
| | 1000 | 150 | 132 | 17 | 57 | 315 |
| | 1200 | 118 | 102 | 22 | 64 | 245 |
| 4140 | 400 | 257 | 238 | 8 | 38 | 510 |
| | 600 | 225 | 208 | 9 | 43 | 445 |
| | 800 | 181 | 165 | 13 | 49 | 370 |
| | 1000 | 138 | 121 | 18 | 58 | 285 |
| | 1200 | 110 | 95 | 22 | 63 | 230 |
| 4150 | 400 | 280 | 250 | 10 | 39 | 530 |
| | 600 | 256 | 231 | 10 | 40 | 495 |
| | 800 | 220 | 200 | 12 | 45 | 440 |
| | 1000 | 175 | 160 | 15 | 52 | 370 |
| | 1200 | 139 | 122 | 19 | 60 | 290 |
| 4340 | 400 | 272 | 243 | 10 | 38 | 520 |
| | 600 | 250 | 230 | 10 | 40 | 486 |
| | 800 | 213 | 198 | 10 | 44 | 430 |
| | 1000 | 170 | 156 | 13 | 51 | 360 |
| | 1200 | 140 | 124 | 19 | 60 | 280 |

**Table 11b.** *(Continued)* **Typical Mechanical Properties of Selected Carbon and Alloy Steels** (Quenched-and-Tempered Condition)

| AISI No.[a] | Tempering Temperature, °F | Tensile Strength Ultimate (1000 lb/in²) | Tensile Strength Yield (1000 lb/in²) | Elongation, Percent | Reduction in Area, Percent | Hardness, BHN |
|---|---|---|---|---|---|---|
| 5046 | 400 | 253 | 204 | 9 | 25 | 482 |
| | 600 | 205 | 168 | 10 | 37 | 401 |
| | 800 | 165 | 135 | 13 | 50 | 336 |
| | 1000 | 136 | 111 | 18 | 61 | 282 |
| | 1200 | 114 | 95 | 24 | 66 | 235 |
| 50B46 | 400 | … | … | … | … | 560 |
| | 600 | 258 | 235 | 10 | 37 | 505 |
| | 800 | 202 | 181 | 13 | 47 | 405 |
| | 1000 | 157 | 142 | 17 | 51 | 322 |
| | 1200 | 128 | 115 | 22 | 60 | 273 |
| 50B60 | 400 | … | … | … | … | 600 |
| | 600 | 273 | 257 | 8 | 32 | 525 |
| | 800 | 219 | 201 | 11 | 34 | 435 |
| | 1000 | 163 | 145 | 15 | 38 | 350 |
| | 1200 | 130 | 113 | 19 | 50 | 290 |
| 5130 | 400 | 234 | 220 | 10 | 40 | 475 |
| | 600 | 217 | 204 | 10 | 46 | 440 |
| | 800 | 185 | 175 | 12 | 51 | 379 |
| | 1000 | 150 | 136 | 15 | 56 | 305 |
| | 1200 | 115 | 100 | 20 | 63 | 245 |
| 5140 | 400 | 260 | 238 | 9 | 38 | 490 |
| | 600 | 229 | 210 | 10 | 43 | 450 |
| | 800 | 190 | 170 | 13 | 50 | 365 |
| | 1000 | 145 | 125 | 17 | 58 | 280 |
| | 1200 | 110 | 96 | 25 | 66 | 235 |
| 5150 | 400 | 282 | 251 | 5 | 37 | 525 |
| | 600 | 252 | 230 | 6 | 40 | 475 |
| | 800 | 210 | 190 | 9 | 47 | 410 |
| | 1000 | 163 | 150 | 15 | 54 | 340 |
| | 1200 | 117 | 118 | 20 | 60 | 270 |
| 5160 | 400 | 322 | 260 | 4 | 10 | 627 |
| | 600 | 290 | 257 | 9 | 30 | 555 |
| | 800 | 233 | 212 | 10 | 37 | 461 |
| | 1000 | 169 | 151 | 12 | 47 | 341 |
| | 1200 | 130 | 116 | 20 | 56 | 269 |
| 51B60 | 400 | … | … | … | … | 600 |
| | 600 | … | … | … | … | 540 |
| | 800 | 237 | 216 | 11 | 36 | 460 |
| | 1000 | 175 | 160 | 15 | 44 | 355 |
| | 1200 | 140 | 126 | 20 | 47 | 290 |
| 6150 | 400 | 280 | 245 | 8 | 38 | 538 |
| | 600 | 250 | 228 | 8 | 39 | 483 |
| | 800 | 208 | 193 | 10 | 43 | 420 |
| | 1000 | 168 | 155 | 13 | 50 | 345 |
| | 1200 | 137 | 122 | 17 | 58 | 282 |
| 81B45 | 400 | 295 | 250 | 10 | 33 | 550 |
| | 600 | 256 | 228 | 8 | 42 | 475 |
| | 800 | 204 | 190 | 11 | 48 | 405 |
| | 1000 | 160 | 149 | 16 | 53 | 338 |
| | 1200 | 130 | 115 | 20 | 55 | 280 |

**Table 11b.** *(Continued)* **Typical Mechanical Properties of Selected Carbon and Alloy Steels** (Quenched-and-Tempered Condition)

| AISI No.[a] | Tempering Temperature, °F | Tensile Strength Ultimate | Tensile Strength Yield | Elongation, Percent | Reduction in Area, Percent | Hardness, BHN |
|---|---|---|---|---|---|---|
| | | 1000 lb/in² | | | | |
| 8630 | 400 | 238 | 218 | 9 | 38 | 465 |
| | 600 | 215 | 202 | 10 | 42 | 430 |
| | 800 | 185 | 170 | 13 | 47 | 375 |
| | 1000 | 150 | 130 | 17 | 54 | 310 |
| | 1200 | 112 | 100 | 23 | 63 | 240 |
| 8640 | 400 | 270 | 242 | 10 | 40 | 505 |
| | 600 | 240 | 220 | 10 | 41 | 460 |
| | 800 | 200 | 188 | 12 | 45 | 400 |
| | 1000 | 160 | 150 | 16 | 54 | 340 |
| | 1200 | 130 | 116 | 20 | 62 | 280 |
| 86B45 | 400 | 287 | 238 | 9 | 31 | 525 |
| | 600 | 246 | 225 | 9 | 40 | 475 |
| | 800 | 200 | 191 | 11 | 41 | 395 |
| | 1000 | 160 | 150 | 15 | 49 | 335 |
| | 1200 | 131 | 127 | 19 | 58 | 280 |
| 8650 | 400 | 281 | 243 | 10 | 38 | 525 |
| | 600 | 250 | 225 | 10 | 40 | 490 |
| | 800 | 210 | 192 | 12 | 45 | 420 |
| | 1000 | 170 | 153 | 15 | 51 | 340 |
| | 1200 | 140 | 120 | 20 | 58 | 280 |
| 8660 | 400 | … | … | … | … | 580 |
| | 600 | … | … | … | … | 535 |
| | 800 | 237 | 225 | 13 | 37 | 460 |
| | 1000 | 190 | 176 | 17 | 46 | 370 |
| | 1200 | 155 | 138 | 20 | 53 | 315 |
| 8740 | 400 | 290 | 240 | 10 | 41 | 578 |
| | 600 | 249 | 225 | 11 | 46 | 495 |
| | 800 | 208 | 197 | 13 | 50 | 415 |
| | 1000 | 175 | 165 | 15 | 55 | 363 |
| | 1200 | 143 | 131 | 20 | 60 | 302 |
| 9255 | 400 | 305 | 297 | 1 | 3 | 601 |
| | 600 | 281 | 260 | 4 | 10 | 578 |
| | 800 | 233 | 216 | 8 | 22 | 477 |
| | 1000 | 182 | 160 | 15 | 32 | 352 |
| | 1200 | 144 | 118 | 20 | 42 | 285 |
| 9260 | 400 | … | … | … | … | 600 |
| | 600 | … | … | … | … | 540 |
| | 800 | 255 | 218 | 8 | 24 | 470 |
| | 1000 | 192 | 164 | 12 | 30 | 390 |
| | 1200 | 142 | 118 | 20 | 43 | 295 |
| 94B30 | 400 | 250 | 225 | 12 | 46 | 475 |
| | 600 | 232 | 206 | 12 | 49 | 445 |
| | 800 | 195 | 175 | 13 | 57 | 382 |
| | 1000 | 145 | 135 | 16 | 65 | 307 |
| | 1200 | 120 | 105 | 21 | 69 | 250 |

[a] All grades are fine-grained except those in the 1100 series that are coarse-grained. Austenitizing temperatures are given in parentheses. Heat-treated specimens were oil quenched unless otherwise indicated.

[b] Water quenched.

*Source:* Bethlehem Steel Corp. and Republic Steel Corp. as published in 1974 DATABOOK issue of the American Society for Metals' *Metal Progress* magazine and used with permission.

**Table 12. Nominal Mechanical Properties of Standard Stainless Steels**

| Grade | Condition | Ultimate Tensile Strength (psi) | 0.2 Percent Yield Strength (psi) | Elongation in 2 in. (%) | Reduction of Area (%) | Hardness Rockwell | Hardness BHN |
|---|---|---|---|---|---|---|---|
| Austenitic Steels | | | | | | | |
| 201 | Annealed | 115,000 | 55,000 | 55 | … | 90 RB | … |
| | ¼-hard | 125,000[a] | 75,000[a] | 20[a] | … | 25 RC | … |
| | ½-hard | 150,000[a] | 110,000[a] | 10[a] | … | 32 RC | … |
| | ¾-hard | 175,000[a] | 135,000[a] | 5[a] | … | 37 RC | … |
| | Full-hard | 185,000[a] | 140,000[a] | 4[a] | … | 41 RC | … |
| 202 | Annealed | 105,000 | 55,000 | 55 | … | 90 RB | … |
| | ¼-hard | 125,000[a] | 75,000[a] | 12[a] | … | 27 RC | … |
| 301 | Annealed | 110,000 | 40,000 | 60 | … | 85 RB | 165 |
| | ¼-hard | 125,000[a] | 75,000[a] | 25[a] | … | 25 RC | … |
| | ½-hard | 150,000[a] | 110,000[a] | 15[a] | … | 32 RC | … |
| | ¾-hard | 175,000[a] | 135,000[a] | 12[a] | … | 37 RC | … |
| | Full-hard | 185,000 | 140,000[a] | 8[a] | … | 41 RC | … |
| 302 | Annealed | 90,000 | 37,000 | 55 | 65 | 82 RB | 155 |
| | 1/4-hard (sheet, strip) | 125,000[a] | 75,000[a] | 12[a] | … | 25 RC | … |
| | Cold-drawn (bar, wire)[b] | To 350,000 | … | … | … | … | … |
| 302B | Annealed | 95,000 | 40,000 | 50 | 65 | 85 RB | 165 |
| 303, 303Se | Annealed | 90,000 | 35,000 | 50 | 55 | 84 RB | 160 |
| 304 | Annealed | 85,000 | 35,000 | 55 | 65 | 80 RB | 150 |
| 304L | Annealed | 80,000 | 30,000 | 55 | 65 | 76 RB | 140 |
| 305 | Annealed | 85,000 | 37,000 | 55 | 70 | 82 RB | 156 |
| 308 | Annealed | 85,000 | 35,000 | 55 | 65 | 80 RB | 150 |
| 309, 309S | Annealed | 90,000 | 40,000 | 45 | 65 | 85 RB | 165 |
| 310, 310S | Annealed | 95,000 | 40,000 | 45 | 65 | 87 RB | 170 |
| 314 | Annealed | 100,000 | 50,000 | 45 | 60 | 87 RB | 170 |
| 316 | Annealed | 85,000 | 35,000 | 55 | 70 | 80 RB | 150 |
| | Cold-drawn (bar, wire)[b] | To 300,000 | … | … | … | … | … |
| 316L | Annealed | 78,000 | 30,000 | 55 | 65 | 76 RB | 145 |
| 317 | Annealed | 90,000 | 40,000 | 50 | 55 | 85 RB | 160 |
| 321 | Annealed | 87,000 | 35,000 | 55 | 65 | 80 RB | 150 |
| 347, 348 | Annealed | 92,000 | 35,000 | 50 | 65 | 84 RB | 160 |
| Martensitic Steels | | | | | | | |
| 403, 410, 416, 416Se | Annealed | 75,000 | 40,000 | 30 | 65 | 82 RB | 155 |
| | Hardened[c] | … | … | … | … | 43 RC | 410 |
| | Tempered at | | | | | | |
| | 400° F | 190,000 | 145,000 | 15 | 55 | 41 RC | 390 |
| | 600° F | 180,000 | 140,000 | 15 | 55 | 39 RC | 375 |
| | 800° F | 195,000 | 150,000 | 17 | 55 | 41 RC | 390 |
| | 1000° F | 145,000 | 115,000 | 20 | 65 | 31 RC | 300 |
| | 1200° F | 110,000 | 85,000 | 23 | 65 | 97 RB | 225 |
| | 1400° F | 90,000 | 60,000 | 30 | 70 | 89 RB | 180 |

**Table 12.** *(Continued)* **Nominal Mechanical Properties of Standard Stainless Steels**

| Grade | Condition | Ultimate Tensile Strength (psi) | 0.2 Percent Yield Strength (psi) | Elongation in 2 in. (%) | Reduction of Area (%) | Hardness Rockwell | Hardness BHN |
|---|---|---|---|---|---|---|---|
| | | Martensitic Steels *(Continued)* | | | | | |
| 414 | Annealed | 120,000 | 95,000 | 17 | 55 | 22 RC | 235 |
| | Hardened[c] | ... | ... | ... | ... | 44 RC | 426 |
| | Tempered at | | | | | | |
| | 400° F | 200,000 | 150,000 | 15 | 55 | 43 RC | 415 |
| | 600° F | 190,000 | 145,000 | 15 | 55 | 41 RC | 400 |
| | 800° F | 200,000 | 150,000 | 16 | 58 | 43 RC | 415 |
| | 1000° F | 145,000 | 120,000 | 20 | 60 | 34 RC | 325 |
| | 1200° F | 120,000 | 105,000 | 20 | 65 | 24 RC | 260 |
| 420, 420F | Annealed | 95,000 | 50,000 | 25 | 55 | 92 RB | 195 |
| | Hardened[d] | ... | ... | ... | ... | 54 RC | 540 |
| | Tempered at | | | | | | |
| | 600° F | 230,000 | 195,000 | 8 | 25 | 50 RC | 500 |
| 431 | Annealed | 125,000 | 95,000 | 20 | 60 | 24 RC | 260 |
| | Hardened[d] | ... | ... | ... | ... | 45 RC | 440 |
| | Tempered at | | | | | | |
| | 400° F | 205,000 | 155,000 | 15 | 55 | 43 RC | 415 |
| | 600° F | 195,000 | 150,000 | 15 | 55 | 41 RC | 400 |
| | 800° F | 205,000 | 155,000 | 15 | 60 | 43 RC | 415 |
| | 1000° F | 150,000 | 130,000 | 18 | 60 | 34 RC | 325 |
| | 1200° F | 125,000 | 95,000 | 20 | 60 | 24 RC | 260 |
| 440A | Annealed | 105,000 | 60,000 | 20 | 45 | 95 RB | 215 |
| | Hardened[d] | ... | ... | ... | ... | 56 RC | 570 |
| | Tempered | | | | | | |
| | 600° F | 260,000 | 240,000 | 5 | 20 | 51 RC | 510 |
| 440B | Annealed | 107,000 | 62,000 | 18 | 35 | 96 RB | 220 |
| | Hardened[d] | ... | ... | ... | ... | 58 RC | 590 |
| | Tempered | | | | | | |
| | 600° F | 280,000 | 270,000 | 3 | 15 | 55 RC | 555 |
| 440C, 440F | Annealed | 110,000 | 65,000 | 13 | 25 | 97 RB | 230 |
| | Hardened[d] | ... | ... | ... | ... | 60 RC | 610 |
| | Tempered | | | | | | |
| | 600° F | 285,000 | 275,000 | 2 | 10 | 57 RC | 580 |
| 501 | Annealed | 70,000 | 30,000 | 28 | 65 | ... | 160 |
| 502 | Annealed | 70,000 | 30,000 | 30 | 75 | 80 RB | 150 |
| | | Ferritic Steels | | | | | |
| 405 | Annealed | 70,000 | 40,000 | 30 | 60 | 80 RB | 150 |
| 430 | Annealed | 75,000 | 45,000 | 30 | 60 | 82 RB | 155 |
| 430F, 430FSe | Annealed | 80,000 | 55,000 | 25 | 60 | 86 RB | 170 |
| 446 | Annealed | 80,000 | 50,000 | 23 | 50 | 86 RB | 170 |

[a] Minimum.

[b] Depending on size and amount of cold reduction.

[c] Hardening temperature 1800° F (982° C), 1-inch diameter bars.

[d] Hardening temperature 1900° F (1038° C), 1-inch diameter bars.

*Source: Metals Handbook*, 8th edition, Volume 1.

## Table 13. Strength Data for Iron and Steel

| Material | Ultimate Strength | | | Yield Strength, Thousands of Pounds per Square Inch | Modulus of Elasticity | |
|---|---|---|---|---|---|---|
| | Tension, Thousands of Pounds per Square Inch, $T$ | Compression, in terms of $T$ | Shear, in terms of $T$ | | Tension, Millions of psi, $E$ | Shear,[a] in terms of $E$ |
| Cast iron, gray, class 20 | 20[b] | 3.6$T$ to 4.4$T$ | 1.6$T$ | … | 11.6 | 0.40$E$ |
| class 25 | 25[b] | 3.6$T$ to 4.4$T$ | 1.4$T$ | … | 14.2 | 0.40$E$ |
| class 30 | 30[b] | 3.7$T$ | 1.4$T$ | … | 14.5 | 0.40$E$ |
| class 35 | 35[b] | 3.2$T$ to 3.9$T$ | 1.4$T$ | … | 16.0 | 0.40$E$ |
| class 40 | 40[b] | 3.1$T$ to 3.4$T$ | 1.3$T$ | … | 17 | 0.40$E$ |
| class 50 | 50[b] | 3.0$T$ to 3.4$T$ | 1.3$T$ | … | 18 | 0.40$E$ |
| class 60 | 60[b] | 2.8$T$ | 1.0$T$ | … | 19.9 | 0.40$E$ |
| malleable | 40 to 100[c] | … | … | 30 to 80[c] | 25 | 0.43$E$ |
| nodular (ductile iron) | 60 to 120[d] | … | … | 40 to 90[d] | 23 | … |
| Cast steel, carbon | 60 to 100 | $T$ | 0.75$T$ | 30 to 70 | 30 | 0.38$E$ |
| low-alloy | 70 to 200 | $T$ | 0.75$T$ | 45 to 170 | 30 | 0.38$E$ |
| Steel, SAE 950 (low-alloy) | 65 to 70 | $T$ | 0.75$T$ | 45 to 50 | 30 | 0.38$E$ |
| 1025 (low-carbon) | 60 to 103 | $T$ | 0.75$T$ | 40 to 90 | 30 | 0.38$E$ |
| 1045 (medium-carbon) | 80 to 182 | $T$ | 0.75$T$ | 50 to 162 | 30 | 0.38$E$ |
| 1095 (high-carbon) | 90 to 213 | $T$ | 0.75$T$ | 20 to 150 | 30 | 0.39$E$ |
| 1112 (free-cutting)* | 60 to 100 | $T$ | 0.75$T$ | 30 to 95 | 30 | 0.38$E$ |
| 1212 (free-cutting) | 57 to 80 | $T$ | 0.75$T$ | 25 to 72 | 30 | 0.38$E$ |
| 1330 (alloy) | 90 to 162 | $T$ | 0.75$T$ | 27 to 149 | 30 | 0.38$E$ |
| 2517 (alloy)[e] | 88 to 190 | $T$ | 0.75$T$ | 60 to 155 | 30 | 0.38$E$ |
| 3140 (alloy) | 93 to 188 | $T$ | 0.75$T$ | 62 to 162 | 30 | 0.38$E$ |
| 3310 (alloy)[e] | 104 to 172 | $T$ | 0.75$T$ | 56 to 142 | 30 | 0.38$E$ |
| 4023 (alloy)[e] | 105 to 170 | $T$ | 0.75$T$ | 60 to 114 | 30 | 0.38$E$ |
| 4130 (alloy) | 81 to 179 | $T$ | 0.75$T$ | 46 to 161 | 30 | 0.38$E$ |
| 4340 (alloy) | 109 to 220 | $T$ | 0.75$T$ | 68 to 200 | 30 | 0.38$E$ |
| 4640 (alloy) | 98 to 192 | $T$ | 0.75$T$ | 62 to 169 | 30 | 0.38$E$ |
| 4820 (alloy)[e] | 98 to 209 | $T$ | 0.75$T$ | 68 to 184 | 30 | 0.38$E$ |
| 5150 (alloy) | 98 to 210 | $T$ | 0.75$T$ | 51 to 190 | 30 | 0.38$E$ |
| 52100 (alloy) | 100 to 238 | $T$ | 0.75$T$ | 81 to 228 | 30 | 0.38$E$ |
| 6150 (alloy) | 96 to 228 | $T$ | 0.75$T$ | 59 to 210 | 30 | 0.38$E$ |
| 8650 (alloy) | 110 to 228 | $T$ | 0.75$T$ | 69 to 206 | 30 | 0.38$E$ |
| 8740 (alloy) | 100 to 179 | $T$ | 0.75$T$ | 60 to 165 | 30 | 0.38$E$ |
| 9310 (alloy)[e] | 117 to 187 | $T$ | 0.75$T$ | 63 to 162 | 30 | 0.38$E$ |
| 9840 (alloy) | 120 to 285 | $T$ | 0.75$T$ | 45 to 50 | 30 | 0.38$E$ |
| Steel, stainless, SAE | | | | | | |
| 30302[f] | 85 to 125 | $T$ | … | 35 to 95 | 28 | 0.45$E$ |
| 30321[f] | 85 to 95 | $T$ | … | 30 to 60 | 28 | … |
| 30347[f] | 90 to 100 | $T$ | … | 35 to 65 | 28 | … |
| 51420[g] | 95 to 230 | $T$ | … | 50 to 195 | 29 | 0.40$E$ |
| 51430[h] | 75 to 85 | $T$ | … | 40 to 70 | 29 | … |
| 51446[h] | 80 to 85 | $T$ | … | 50 to 70 | 29 | … |
| 51501[g] | 70 to 175 | $T$ | … | 30 to 135 | 29 | … |
| Steel, structural | | | | | | |
| common | 60 to 75 | $T$ | 0.75$T$ | 33[b] | 29 | 0.41$E$ |
| rivet | 52 to 62 | $T$ | 0.75$T$ | 28[b] | 29 | … |
| rivet, high-strength | 68 to 82 | $T$ | 0.75$T$ | 38[b] | 29 | … |
| Wrought iron | 34 to 54 | $T$ | 0.83$T$ | 23 to 32 | 28 | … |

[a] Synonymous in other literature to the modulus of elasticity in torsion and the modulus of rigidity, $G$.

[b] Minimum specified value of the American Society for Testing and Materials. The specifications for the various materials are as follows: Cast iron, ASTM A48; structural steel for bridges and structures, ASTM A7; structural rivet steel, ASTM A141; high-strength structural rivet steel, ASTM A195.

[c] Range of minimum specified values of the ASTM (ASTM A47, A197, and A220).

[d] Range of minimum specified values of the ASTM (ASTM A339) and the Munitions Board Standards Agency (MIL-I-17166A and MIL-I-11466).

[e] Carburizing grades of steel.

[f] Nonhardenable nickel-chromium and chromium-nickel-manganese steel (austenitic).

[g] Hardenable chromium steel (martensitic).

[h] Nonhardenable chromium steel (ferritic).

# TOOL STEELS

## Overview

As the designation implies, tool steels serve primarily for making tools used in manufacturing and in the trades for the working and forming of metals, wood, plastics, and other industrial materials. Tools must withstand high specific loads, often concentrated at exposed areas, may have to operate at elevated or rapidly changing temperatures and in continual contact with abrasive types of work materials, and are often subjected to shocks, or may have to perform under other varieties of adverse conditions. Nevertheless, when employed under circumstances that are regarded as normal operating conditions, the tool should not suffer major damage, untimely wear resulting in the dulling of the edges, or be susceptible to detrimental metallurgical changes.

Tools for less demanding uses, such as ordinary handtools, including hammers, chisels, files, mining bits, etc., are often made of standard AISI steels that are not considered to belong to any of the tool steel categories.

The steel for most types of tools must be used in a heat-treated state, usually hardened and tempered, to provide properties needed for a particular application. Adaptability to heat treatment, with minimum harmful effects, dependably resulting in intended beneficial changes in material properties, is another requirement for tool steels.

To meet such varied requirements, steel types of different chemical composition, often produced by special metallurgical processes, have been developed. Due to the large number of tool steel types produced by the steel mills, which generally are made available with proprietary designations, it is rather difficult for the user to select those types that are most suitable for any specific application, unless the recommendations of a particular steel producer or producers are obtained.

Substantial clarification has resulted from the development of a classification system that is now widely accepted throughout the industry, on the part of both the producers and the users of tool steels. The tool steel classification system establishes seven basic categories of tool and die steels. These categories are associated with the predominant applicational characteristics of the tool steel types they comprise. A few of these categories are composed of several groups to distinguish between families of steel types that, while serving the same general purpose, differ with regard to one or more dominant characteristics. The classification categories are given in Table 3.

To provide an easily applicable guide for the selection of tool steel types best suited for a particular application, the subsequent discussions and tables are based on the previously mentioned application-related categories. As an introduction to the detailed surveys, a concise discussion is presented of the principal tool steel characteristics that govern the suitability for varying service purposes and operational conditions. A brief review of the major steel alloying elements and of the effect of these constituents on the significant characteristics of tool steels is also given in the following sections.

**The Properties of Tool Steels.**—Tool steels must possess certain properties to a higher than ordinary degree to make them adaptable for uses that require the ability to sustain heavy loads and perform dependably even under adverse conditions. The extent and types of loads, operating conditions, and expected performance with regard to both duration and level of consistency are the principal considerations, in combination with aspects of cost, that govern selection of tool steels for specific applications.

Although it is not possible to define and apply exact parameters for measuring significant tool steel characteristics, certain properties can be determined that may assist in appraising suitability of various tool steels for specific uses. AISI standards define the chemical composition of tool steels, but the quality of ingredients (such as element purity and initial particle size) and processing of materials (manufacturing methods and heat-treatment parameters) can result in different properties. Material testing and controlled processes help ensure that desired properties are obtained.

Because tool steels are generally heat treated to make them adaptable to the intended use by enhancing the desirable properties, *the behavior of the steel during heat treatment* is of prime importance. The behavior of the steel comprises, in this respect, both the resistance to harmful effects and the attainment of the desirable properties. The following are considered the major properties related to heat treatment:

*Safety in Hardening:* This designation expresses the ability of the steel to withstand the harmful effects of exposure to very high heat and particularly to the sudden temperature changes during quenching, without harmful effects. One way of obtaining this property is by adding alloying elements to reduce the critical speed at which quenching must be carried out, thus permitting the use of milder quenching media, such as oil, salt, or just still air.

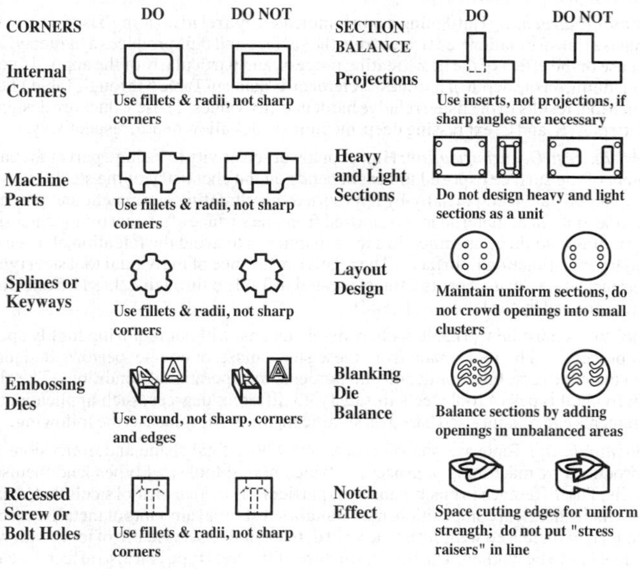


Fig. 1. Tool and Die Design Tips to Reduce Breakage in Heat Treatment
*Courtesy of Society of Automotive Engineers, Inc.*

Parts made of tool steel suffer most commonly during heat treatment from the development of cracks. In addition to the composition of the steel and the applied heat-treating process, the configuration of the part can also affect the sensitivity to cracking. The preceding figure illustrates a few design characteristics related to cracking and warpage in heat treatment; the observation of these design tips, which call for generous filleting, avoidance of sharp angles, and major changes without transition in the cross-section, is particularly advisable when using tool steel types with a low index value for safety in hardening.

In current practice, the previously mentioned property of tool steels is rated in the order of decreasing safety (i.e., increasing sensitivity) as Highest, Very High, High, Medium, and Low safety, expressed in Table 6 through Table 11 by the letters A, B, C, D, and E.

*Distortions in Heat Treating:* In parts made from tool steels, distortions are often a consequence of inadequate design (See Fig. 1) or improper heat treatment (e.g., lack of stress relieving). However, certain types of tool steels display different degrees of sensitivity to distortion. Steels that are less stable require safer design of the parts for which they are used, more careful heat treatment, including the proper support for long and slender parts, or thin sections, and possibly greater grinding allowance to permit subsequent correction of the distorted shape. Some parts made of a type of steel generally sensitive to distortions can be heat treated with very little damage when the requirements of the part call for a relatively shallow hardened layer over a soft core. However, for intricate shapes and large tools, steel types should be selected that possess superior nondeforming properties. The ratings used in Table 6 through Table 11 express the nondeforming properties (stability of shape in heat treatment) of the steel types and start with the lowest distortion (the best stability) designated as A; the greatest susceptibility to distortion is designated as E.

*Depth of Hardening:* Hardening depth is indicated by a relative rating based on how deep the phase transformation penetrates from the surface and thus produces a hardened layer. Because of the effect of the heat-treating process, and particularly of the applied quenching medium, on the depth of hardness, reference is made in Table 6 through Table 11 to the quench that results in the listed relative hardenability values. These values are designated by letters A, B, and C, expressing deep, medium, and shallow depth, respectively.

*Resistance to Decarburization:* Higher or lower sensitivity to losing a part of the carbon content of the surface exposed to heat depends on the chemistry of the steel. The sensitivity can be balanced partially by appropriate heat-treating equipment and processes. Also, the amount of material to be removed from the surface after heat treatment, usually by grinding, should be specified in such a manner as to avoid the retention of a decarburized layer on functional surfaces. The relative resistance of individual tool steel types to decarburization during heat treatment is rated in Table 6 through Table 11 from High to Low, expressed by the letters A, B, and C.

Tool steels must be workable with available means, without requiring highly specialized processes. The tools made from these steels must, of course, perform adequately, often under adverse environmental and burdensome operational conditions. The ability of individual types of tool steels to satisfy, to different degrees, such applicational requirements can be appraised based on significant properties, such as the following.

*Machinability:* Tools are precision products whose final shape and dimensions must be produced by machining, a process to which not all tool steel types lend themselves equally. The difference in machinability is particularly evident in tool steels that, depending on their chemical composition, may contain substantial amounts of metallic carbides, beneficial to increased wear resistance, yet detrimental to service life of tools with which the steel has to be worked. The microstructure of the steel type can also affect the ease of machining and, in some types, certain phase conditions, such as those due to low carbon content, may cause difficulties in achieving a fine surface finish. Machinability of certain types of tool steels may be improved by adding small amounts of sulfur or lead. Note that machinability ratings may be defined for a single cutting process not optimal for the material grades being considered.

Machinability affects the cost of making the tool, particularly for intricate tool shapes, and must be considered in selection of the steel to be used. The ratings in Table 6 through Table 11, starting with A for the greatest ease of machining to E for the lowest machinability, refer to working of the steel in an unhardened condition. Machinability is not necessarily identical with grindability, which expresses how well the steel is adapted to grinding after heat treating. The ease of grinding, however, may become an important consideration in tool steel selection, particularly for cutting tools and dies, which require regular sharpening involving extensive grinding. AVCO Bay State Abrasives Company compiled information on the relative grindability of frequently used types of tool steels. A simplified version of that information is presented in Table 1, which assigns the listed tool steel types to one of the following grindability grades: High (A), Medium (B), Low (C), and Very Low (D), expressing decreasing ratios of volume of metal removed to wheel wear.

**Table 1. Relative Grindability of Selected Types of Frequently Used Tool Steels**

| AISI Tool Steel Type | H41 | H42 | H43 | Other H | D2 | D3 | D5 | D7 | A Types | O Types | L Types | F Types |
|---|---|---|---|---|---|---|---|---|---|---|---|---|
| Relative Grindability Index | B | B | B | A | B | B | B | C | A | A | A | B |

| High-Speed Tool Steel Type | M1 | M2 | M3 (1) | M3 (2) | M4 | M7 | M8 | M10 | M15 | M36 | M43 | T1 | T2 | T3 | T5 | T6 | T15 |
|---|---|---|---|---|---|---|---|---|---|---|---|---|---|---|---|---|---|
| Relative Grindability Index | A | B | C | C | D | B | A | B | D | B | B | A | B | C | B | B | D |

*Hot Hardness:* This property designates the steel's resistance to the softening effect of elevated temperature. This characteristic is related to the tempering temperature of the type of steel, which is controlled by various alloying elements, such as tungsten, molybdenum, vanadium, cobalt, and chromium. Hot hardness testing is typically performed using a Rockwell or Vickers hardness tester. Hardness is measured over time, with the samples at high temperature (often 2100° F or 1200° C).

Hot hardness is a necessary property of tools used for hot-work, like forging, casting, and hot extrusion. Hot hardness is also important in cutting tools operated at high-speed, which generate sufficient heat to raise their temperature well above the level where ordinary steels lose their hardness; hence the designation *high-speed steels*, which refers to a family of tool steels developed for use at high cutting speeds. Frequently it is the degree of the tool steel's resistance to softening at elevated temperature that governs important process data, such as the applicable cutting speed. In the ratings of Table 6 through Table 11, tool steel types having the highest hot hardness are marked with A. Subsequent letters express gradually decreasing capacity to endure elevated temperature without losing hardness.

*Wear Resistance:* The gradual erosion of the tool's operating surface, most conspicuously occurring at the exposed edges, is known as wear. Resistance to wear prolongs the useful life of the tool by delaying the degradation of its surface through abrasive contact with the work at regular operating temperatures; these temperatures vary according to the type of process. Wear resistance is observable experimentally and measurable by comparison based on a specific and unchanging set of test parameters. As a result, these ratings must be used with caution. Certain types of metallic carbides embedded into the steel matrix are considered to be the prime contributing factors to wear resistance, besides the hardness of the heat-treated steel material. The ratings of Table 6 through Table 11, starting with A for the best to E for poor, are based on conditions thought to be normal in operations for which various types of tool materials are primarily used.

*Toughness:* Toughness is amount of energy a material can absorb without fracture. For tool steels, this is the material's ability to withstand shocks, major impacts, and sudden loading and unloading. Steels used for making tools also must be able to absorb such forces with a minimum of elastic deformation and without permanent deformation that would interfere with proper functioning of the tool. Certain types of tool steels, particularly those with high carbon content and without beneficial alloying constituents, tend to be the most sensitive to shocks. They can be made functionally tougher when they are used in tools that permit a hardened case to be supported by a soft core. Also, tempering improves toughness, while generally reducing hardness.

In Table 6 through Table 11, an A rating is for steels with the highest toughness, while an E rating is for steels that are the most sensitive to shocks. These ratings apply to tools heat-treated to hardness values normally used for the particular type of tool steel. Such relative ratings often are used when comparing materials, but toughness also is a measurable property with units of energy. Results of standardized testing methods can be affected by material variations and may not represent performance in real-world applications.

**Common Tool Faults and Failures.—**The proper selection of the steel grade used for any particular type of tool is of great importance, but it should be recognized that many of the failures experienced in common practice originate from causes other than those related to the tool material.

To permit a better appraisal of the actual causes of failure and possible corrective action, a general, although not complete, list of common tool faults, resulting failures, and corrective actions is shown in Table 2a through Table 2d. In this list, the potential failure causes are grouped into four categories. The possibility of more than a single cause being responsible for the experienced failure should not be excluded.

*Note:* Examples of tool failures from causes such as those listed in Table 2a through Table 2d may be found in "The Tool Steel Trouble Shooter" handbook, published by Bethlehem Steel Corporation.

Finally, it must be remembered that the proper usage of tools is indispensable for obtaining satisfactory performance and tool life. Using the tools properly involves, for example, the avoidance of damage to the tool, overloading, excessive speeds and feeds, the application of adequate coolant when called for, a rigid setup, proper alignment, and firm tool and work holding.

**Table 2a. Common Tool Faults, Failures, and Cures**

Improper Tool Design

| Fault Description | Probable Failure | Possible Cure |
|---|---|---|
| Drastic section changes—widely different thicknesses of adjacent wall sections or protruding elements | In liquid quenching, the thin section will cool and then harden more rapidly than the adjacent thicker section, setting up stresses that may exceed the strength of the steel. | Make such parts of two pieces or use an air-hardening tool steel that avoids the harsh action of a liquid quench. |
| Sharp corners on shoulders or in square holes | Cracking can occur, particularly in liquid quenching, due to stress concentrations. | Apply fillets to the corners and/or use an air-hardening tool steel. |
| Sharp cornered keyways | Failure may arise during service and is usually considered to be caused by fatigue. | The use of round keyways should be preferred when the general configuration of the part makes it prone to failure due to square keyways. |
| Abrupt section changes in battering tools | Due to impact in service, pneumatic tools are particularly sensitive to stress concentrations that lead to fatigue failures. | Use taper transitions, which are better than even generous fillets. |
| Functional inadequacy of tool design—e.g., insufficient guidance for a punch | Excessive wear or breakage in service may occur. | Assure solid support, avoid unnecessary play, adapt travel length to operational conditions (e.g., punch to penetrate to four-fifths of thickness in hard work material). |
| Improper tool clearance, such as in blanking and punching tools | Deformed and burred parts may be produced, excessive tool wear or breakage can result. | Adapt clearances to material conditions and dimensions to reduce tool load and to obtain clean sheared surfaces. |

**The Effect of Alloying Elements on Tool Steel Properties.—** *Carbon (C):* The presence of carbon, usually in excess of 0.60 percent for nonalloyed types, is essential for raising the hardenability of steels to the levels needed for tools. Raising the carbon content by different amounts up to a maximum of about 1.3 percent increases the hardness slightly and the wear resistance considerably. The amount of carbon in tool steels is designed to attain certain properties (such as in the water-hardening category where higher carbon content may be chosen to improve wear resistance, although to the detriment of toughness) or, in the alloyed types of tool steels, in conformance with the other constituents to produce well-balanced metallurgical and performance properties.

*Manganese (Mn):* In small amounts, to about 0.60 percent, manganese is added to reduce brittleness and to improve forgeability. Larger amounts of manganese improve hardenability, permitting oil quenching for nonalloyed carbon steels, thus reducing deformation, although, with regard to several other properties, manganese is not an equivalent replacement for the regular alloying elements.

*Silicon (Si):* In itself, silicon may not be considered an alloying element of tool steels, but it is needed as a deoxidizer and improves the hot-forming properties of the steel. In combination with certain alloying elements, the silicon content is sometimes raised to about 2 percent to increase the strength and toughness of steels used for tools that have to sustain shock loads.

**Table 2b. Common Tool Faults, Failures, and Cures**
Faulty Condition or Inadequate Grade of Tool Steel

| Fault Description | Probable Failure | Possible Cure |
|---|---|---|
| Improper tool steel grade selection | Typical failures:<br>Chipping—insufficient toughness.<br>Wear—poor abrasion resistance.<br>Softening—inadequate "red hardness." | Choose the tool steel grade by following recommendations and improve selection when needed, guided by property ratings. |
| Material defects—voids, streaks, tears, flakes, surface cooling cracks, etc. | When not recognized during material inspection, tools made of defective steel often prove to be useless. | Obtain tool steels from reliable sources and inspect tool material for detectable defects. |
| Decarburized surface layer ("bark") in rolled tool steel bars | Cracking may originate from the decarburized layer or it will not harden ("soft skin"). | Provide allowance for stock to be removed from all surfaces of hot-rolled tool steel. Recommended amounts are listed in tool steel catalogs and vary according to section size, generally about 10 percent for smaller and 5 percent for larger diameters. |
| Brittleness caused by poor carbide distribution in high-alloy tool steels | Excessive brittleness can cause chipping or breakage during service. | Bars with large diameter (above about 4 inches or 10 cm) tend to be prone to nonuniform carbide distribution. Choose upset forged discs instead of large-diameter bars. |
| Unfavorable grain flow | Improper grain flow of the steel used for milling cutters and similar tools can cause teeth to break out. | Upset forged discs made with an upset ratio of about 2 to 1 (starting to upset thickness) display radial grain flow. Highly stressed tools, such as gear-shaper cutters, may require the cross forging of blanks. |

*Tungsten (W):* Tungsten is one of the important alloying elements of tool steels, particularly because of two valuable properties: it improves "hot hardness," that is, the resistance of the steel to the softening effect of elevated temperature, and it forms hard, abrasion-resistant carbides, thus improving the wear properties of tool steels.

*Vanadium (V):* Vanadium contributes to the refinement of the carbide structure and thus improves the forgeability of alloy tool steels. Vanadium has a very strong tendency to form a hard carbide, which improves both the hardness and the wear properties of tool steels. However, a large amount of vanadium carbide makes the grinding of the tool very difficult (causing low grindability).

*Molybdenum (Mo):* In small amounts, molybdenum improves certain metallurgical properties of alloy steels, such as deep hardening and toughness. It is used often in larger amounts in certain high-speed tool steels to replace tungsten, primarily for economic reasons, often with nearly equivalent results.

*Cobalt (Co):* As an alloying element of tool steels, cobalt increases hot hardness and is used in applications where that property is needed. Substantial addition of cobalt, however, raises the critical quenching temperature of the steel with a tendency to increase the decarburization of the surface, and reduces toughness.

*Chromium (Cr):* This element is added in amounts of several percent to high-alloy tool steels and up to 12 percent to types in which chromium is the major alloying element. Chromium improves hardenability and, together with high carbon, provides both wear resistance and toughness, a combination valuable in certain tool applications. However, high chromium raises the hardening temperature of the tool steel and thus can make it prone to hardening deformations. A high percentage of chromium also affects the grindability of the tool steel.

*Nickel (Ni):* Generally in combination with other alloying elements, particularly chromium, nickel is used to improve the toughness and, to some extent, the wear resistance of tool steels.

**Table 2c. Common Tool Faults, Failures, and Cures**
Heat-Treatment Faults

| Fault Description | Probable Failure | Possible Cure |
|---|---|---|
| Improper preparation for heat treatment. Certain tools may require stress relieving or annealing, and often preheating as well | Tools highly stressed during machining or forming, unless stress-relieved, may aggravate the thermal stresses of heat treatment, thus causing cracks. Excessive temperature gradients developed in nonpreheated tools with different section thicknesses can cause warpage. | Stress relieve, when needed, before hardening. Anneal prior to heavy machining or cold forming (e.g., hobbing). Preheat tools (a) having substantial section thickness variations or (b) requiring high quenching temperatures, as those made of high-speed tool steels. |
| Overheating during hardening; quenching from too high a temperature | Causes grain coarsening and a sensitivity to cracking that is more pronounced in tools with drastic section changes. | Overheated tools have a characteristic microstructure that aids recognition of the cause of failure and indicates the need for improved temperature control. |
| Low hardening temperature | The tool may not harden at all, or in its outer portion only, thereby setting up stresses that can lead to cracks. | Controlling both the temperature of the furnace and the time of holding the tool at quenching temperature will prevent this not too frequent deficiency. |
| Inadequate composition or condition of the quenching media | Water-hardening tool steels are particularly sensitive to inadequate quenching media, which can cause soft spots or even violent cracking. | For water-hardening tool steels, use water free of dissolved air and contaminants, also assure sufficient quantity and proper agitation of the quench. |
| Improper handling during and after quenching | Cracking, particularly of tools with sharp corners, during the heat treatment can result from holding the part too long in the quench or from incorrectly applied tempering. | Following the steel producer's specifications is a safe way to assure proper heat-treatment handling. In general, the tool should be left in the quench until it reaches a temperature of 150–200° F (66–93° C), and should then be transferred promptly into a warm tempering furnace. |
| Insufficient tempering | Omission of double tempering for steel types that require it may cause early failure by heat checking in hot-work steels or make the tool abnormally sensitive to grinding checks. | Double temper highly alloyed tool steel of the high-speed, hot-work, and high-chromium categories to remove stresses caused by martensite formed during the first tempering phase. Second temper also increases hardness of most high-speed steels. |
| Decarburization and carburization | Unless hardened in a neutral atmosphere the original carbon content of the tool surface may be changed: Reduced carbon (decarburization) causes a soft layer that wears rapidly. Increased carbon (carburization) when excessive may cause brittleness. | Heating in neutral atmosphere or well-maintained salt bath and controlling the furnace temperature and the time during which the tool is subjected to heating can usually keep the carbon imbalance within acceptable limits. |

The addition of more than one element to a steel often produces what is called a synergistic effect. Thus, the combined effects of two or more alloy elements may be greater than the sum of the individual effects of each element.

**Classification of Tool Steels.**—Steels for tools must satisfy a number of different, often conflicting, requirements. The need for specific steel properties arising from widely varying applications has led to the development of many compositions of tool steels, each intended to meet a particular combination of applicational requirements. The diversity of tool steels, their number being continually expanded by the addition of new developments, makes it extremely difficult for the user to select the type best suited to his needs or to find equivalent alternatives for specific types available from particular sources.

As a cooperative industrial effort under the sponsorship of AISI and SAE, a tool classification system has been developed in which the commonly used tool steels are grouped into seven major categories. These categories, several of which contain more than a single group, are listed in Table 3 with the letter symbols used for their identification. The individual types of tool steels within each category are identified by suffix numbers following the letter symbols.

## Table 2d. Common Tool Faults, Failures, and Cures
Grinding Damages

| Fault Description | Probable Failure | Possible Cure |
|---|---|---|
| Grinding Damages | | |
| Excessive stock removal rate causing heating of the part surface beyond the applied tempering temperature | Scorched tool surface displaying temper colors varying from yellow to purple, depending on the degree of heat, causes softening of the ground surface. When coolant is used, a local rehardening can take place, often resulting in cracks. | Prevention: by reducing speed and feed, or using coarser, softer, more openstructured grinding wheel, with ample coolant. Correction: eliminate the discolored layer by subsequent light stock removal. Not always a cure, because the effects of abusive grinding may not be corrected. |
| Improper grinding wheel specifications; grain too fine or bond too hard | Intense localized heating during grinding may set up surface stresses causing grinding cracks. These cracks are either parallel but at right angles to the direction of grinding or, when more advanced, form a network. May need cold etch or magnetic particle testing to become recognizable. | Prevention: by correcting the grinding wheel specifications. Correction: in shallow (0.002–0.004-inch, or 0.05–0.1-mm) cracks, by removing the damaged layer, when permitted by the design of the tool, using very light grinding passes. |
| Incorrectly dressed or loaded grinding wheel | Heating of the work surface can cause scorching or cracking. Incorrect dressing can also cause a poor finish of the ground work surface. | Dress wheel with sharper diamond and faster diamond advance to produce coarser wheel surface. Alternate dressing methods, like crush-dressing, can improve wheel surface conditions. Dress wheel regularly to avoid loading or glazing of the wheel surface. |
| Inadequate coolant, with regard to composition, amount, distribution, and cleanliness | Introducing into the tool surface heat that is not adequately dissipated or absorbed by the coolant can cause softening or even the development of cracks. | Improve coolant supply and quality, or reduce stock removal rate to reduce generation of heat in grinding. |
| Damage caused by abusive abrasive cutoff | The intensive heat developed during this process can cause a hardening of the steel surface, or may even result in cracks. | Reduce rate of advance; adopt wheel specifications better suited for the job. Use ample coolant or, when harmful effect not eliminated, replace abrasive cutoff by some cooler-acting stock separation method (e.g., sawing or lathe cutoff) unless damaged surface is being removed by subsequent machining. |

## Table 3. Classification of Tool Steels

| Category Designation | Letter Symbol | Group Designation | Application Type |
|---|---|---|---|
| High-speed tool steels | M | Molybdenum types | Cutting Tools |
| | M50 - M52 | Intermediate types | |
| | T | Tungsten types | |
| Hot-work tool steels | H1 - H19 | Chromium types | Hot-work |
| | H20 - H39 | Tungsten types | |
| | H40 - H59 | Molybdenum types | |
| Cold-work tool steels | D | High-carbon, high-chromium types | Cold-work |
| | A | Medium-alloy, air-hardening types | |
| | O | Oil-hardening types | |
| Shock-resistant tool steels | S | — | Cold-work; hot-work (some grades) |
| Low-carbon tool and mold steels | P | — | Cold-work |
| Special-purpose tool steels | L | Low-alloy types | Cold-work |
| | F | Carbon-tungsten types | |
| Water-hardening tool steels | W | - | Cold-work |

The following detailed discussion of tool steels will be in agreement with these categories, showing for each type the percentages of the major alloying elements. However, these values are for identification only; elements in tool steels of different producers in the mean analysis of the individual types may deviate from the listed percentages.

**Table 4. Classification, Approximate Compositions, and Properties Affecting Selection of Tool and Die Steels**
(From SAE Recommended Practice)

| Type of Tool Steel | Chemical Composition[a] | | | | | | | | Non-warping Prop. | Safety in Hardening | Tough-ness | Depth of Hardening | Wear Resistance |
|---|---|---|---|---|---|---|---|---|---|---|---|---|---|
| | C | Mn | Si | Cr | V | W | Mo | Co | | | | | |
| Water Hardening | | | | | | | | | | | | | |
| 0.80 Carbon | 70-0.85 | b | b | b | … | … | … | … | Poor | Fair | Good[c] | Shallow | Fair |
| 0.90 Carbon | 0.85-0.95 | b | b | b | … | … | … | … | Poor | Fair | Good[c] | Shallow | Fair |
| 1.00 Carbon | 0.95-1.10 | b | b | b | … | … | … | … | Poor | Fair | Good[c] | Shallow | Good |
| 1.20 Carbon | 1.10-1.30 | b | b | b | … | … | … | … | Poor | Fair | Good[c] | Shallow | Good |
| 0.90 Carbon-V | 0.85-0.95 | b | b | b | 0.15-0.35 | … | … | … | Poor | Fair | Good | Shallow | Fair |
| 1.00 Carbon-V | 0.95-1.10 | b | b | b | 0.15-0.35 | … | … | … | Poor | Fair | Good | Shallow | Good |
| 1.00 Carbon-VV | 0.90-1.10 | b | b | b | 0.35-0.50 | … | … | … | Poor | Fair | Good | Shallow | Good |
| Oil Hardening | | | | | | | | | | | | | |
| Low-Manganese | 0.90 | 1.20 | 0.25 | 0.50 | 0.20[d] | 0.50 | … | … | Good | Good | Fair | Deep | Good |
| High-Manganese | 0.90 | 1.60 | 0.25 | 0.35[d] | 0.20[d] | … | 0.30[d] | … | Good | Good | Fair | Deep | Good |
| High-Carbon, High-Chromium[e] | 2.15 | 0.35 | 0.35 | 12.00 | 0.80[d] | 0.75[d] | 0.80[d] | … | Good | Good | Poor | Through | Best |
| Chromium | 1.00 | 0.35 | 0.25 | 1.40 | … | … | 0.40 | … | Fair | Good | Fair | Deep | Good |
| Molybdenum Graphitic | 1.45 | 0.75 | 1.00 | … | … | … | 0.25 | … | Fair | Good | Fair | Deep | Good |
| Nickel-Chromium[f] | 0.75 | 0.70 | 0.25 | 0.85 | 0.25[d] | … | 0.50[d] | … | Fair | Good | Fair | Deep | Fair |
| Air Hardening | | | | | | | | | | | | | |
| High-Carbon, High-Chromium | 1.50 | 0.40 | 0.40 | 12.00 | 0.80[d] | … | 0.90 | 0.60[d] | Best | Best | Fair | Through | Best |
| 5 Percent Chromium | 1.00 | 0.60 | 0.25 | 5.25 | 0.40[d] | … | 1.10 | … | Best | Best | Fair | Through | Good |
| High-Carbon, High-Chromium-Cobalt | 1.50 | 0.40 | 0.40 | 12.00 | 0.80[d] | … | 0.90 | 3.10 | Best | Best | Fair | Through | Best |
| Shock-Resisting | | | | | | | | | | | | | |
| Chromium-Tungsten | 0.50 | 0.25 | 0.35 | 1.40 | 0.20 | 2.25 | 0.40[d] | … | Fair | Good | Good | Deep | Fair |
| Silicon-Molybdenum | 0.50 | 0.40 | 1.00 | … | 0.25[d] | … | 0.50 | … | Poor[g] | Poor[h] | Best | Deep | Fair |
| Silicon-Manganese | 0.55 | 0.80 | 2.00 | 0.30[d] | 0.25[d] | … | 0.40[d] | … | Poor[g] | Poor[h] | Best | Deep | Fair |
| Hot-Work | | | | | | | | | | | | | |
| Chromium-Molybdenum-Tungsten | 0.35 | 0.30 | 1.00 | 5.00 | 0.25[d] | 1.25 | 1.50 | … | Good | Good | Good | Through | Fair |
| Chromium-Molybdenum-V | 0.35 | 0.30 | 1.00 | 5.00 | 0.40 | … | 1.50 | … | Good | Good | Good | Through | Fair |
| Chromium-Molybdenum-VV | 0.35 | 0.30 | 1.00 | 5.00 | 0.90 | … | 1.50 | … | Good | Good | Good | Through | Fair |
| Tungsten | 0.32 | 0.30 | 0.20 | 3.25 | 0.40 | 9.00 | … | … | Good | Good | Good | Through | Fair |

**Table 4.** *(Continued)* **Classification, Approximate Compositions, and Properties Affecting Selection of Tool and Die Steels**
(From SAE Recommended Practice)

| Type of Tool Steel | Chemical Composition[a] | | | | | | | | Non-warping Prop. | Safety in Hardening | Tough-ness | Depth of Hardening | Wear Resistance |
|---|---|---|---|---|---|---|---|---|---|---|---|---|---|
| | C | Mn | Si | Cr | V | W | Mo | Co | | | | | |
| High-Speed | | | | | | | | | | | | | |
| Tungsten, 18-4-1 | 0.70 | 0.30 | 0.30 | 4.10 | 1.10 | 18.00 | … | … | Good | Good | Poor | Through | Good |
| Tungsten, 18-4-2 | 0.80 | 0.30 | 0.30 | 4.10 | 2.10 | 18.50 | 0.80 | … | Good | Good | Poor | Through | Good |
| Tungsten, 18-4-3 | 1.05 | 0.30 | 0.30 | 4.10 | 3.25 | 18.50 | 0.70 | … | Good | Good | Poor | Through | Best |
| Cobalt-Tungsten, 14-4-2-5 | 0.80 | 0.30 | 0.30 | 4.10 | 2.00 | 14.00 | 0.80 | 5.00 | Good | Fair | Poor | Through | Good |
| Cobalt-Tungsten, 18-4-1-5 | 0.75 | 0.30 | 0.30 | 4.10 | 1.00 | 18.00 | 0.80 | 5.00 | Good | Fair | Poor | Through | Good |
| Cobalt-Tungsten, 18-4-2-8 | 0.80 | 0.30 | 0.30 | 4.10 | 1.75 | 18.50 | 0.80 | 8.00 | Good | Fair | Poor | Through | Good |
| Cobalt-Tungsten, 18-4-2-12 | 0.80 | 0.30 | 0.30 | 4.10 | 1.75 | 20.00 | 0.80 | 12.00 | Good | Fair | Poor | Through | Good |
| Molybdenum, 8-2-1 | 0.80 | 0.30 | 0.30 | 4.00 | 1.15 | 1.50 | 8.50 | … | Good | Fair | Poor | Through | Good |
| Molybdenum-Tungsten, 6-6-2 | 0.83 | 0.30 | 0.30 | 4.10 | 1.90 | 6.25 | 5.00 | … | Good | Fair | Poor | Through | Good |
| Molybdenum-Tungsten, 6-6-3 | 1.15 | 0.30 | 0.30 | 4.10 | 3.25 | 5.75 | 5.25 | … | Good | Fair | Poor | Through | Best |
| Molybdenum-Tungsten, 6-6-4 | 1.30 | 0.30 | 0.30 | 4.25 | 4.25 | 5.75 | 5.25 | … | Good | Fair | Poor | Through | Best |
| Cobalt-Molybdenum-Tungsten, 6-6-2-8 | 0.85 | 0.30 | 0.30 | 4.10 | 2.00 | 6.00 | 5.00 | 8.00 | Good | Fair | Poor | Through | Good |

[a] C = carbon; Mn = manganese; Si = silicon; Cr = chromium; V = vanadium; W = tungsten; Mo = molybdenum; Co = cobalt.

[b] Carbon tool steels are usually available in four grades or qualities: *Special (Grade 1)*—The highest quality water-hardening carbon tool steel, controlled for hardenability, chemistry held to closest limits, and subject to rigid tests to ensure maximum uniformity in performance; *Extra (Grade 2)*—A high-quality water-hardening carbon tool steel, controlled for hardenability, subject to tests to ensure good service; *Standard (Grade 3)*—A good-quality water-hardening carbon tool steel, not controlled for hardenability, recommended for application where some latitude with respect to uniformity is permissible; *Commercial (Grade 4)*—A commercial-quality water-hardening carbon tool steel, not controlled for hardenability, not subject to special tests. On *special* and *extra* grades, limits on manganese, silicon, and chromium are not generally required if Shepherd hardenability limits are specified. For *standard* and *commercial* grades, limits are 0.35 max. each for Mn and Si; 0.15 max. Cr for standard; 0.20 max. Cr for commercial.

[c] Toughness decreases somewhat when increasing depth of hardening.

[d] Optional element. Steels have found satisfactory application either with or without the element present. In silicon-manganese steel listed under Shock-Resisting, if chromium, vanadium, and molybdenum are not present, then hardenability will be affected.

[e] This steel may have 0.50 percent nickel as an optional element. The steel has been found to give satisfactory application either with or without the element present.

[f] Approximate nickel content of this steel is 1.50 percent.

[g] Poor when water-quenched, fair when oil quenched.

[h] Poor when water-quenched, good when oil quenched.

**The Selection of Tool Steels for Particular Applications.—**Although the advice of the specialized steel producer is often sought as a reliable source of information, the engineer is still faced with the task of selecting the tool steel. It must be realized that frequently the designation of the tool or of the process will not define the particular tool steel type best suited for the job. For that reason, tool steel selection tables naming a single type for each listed application cannot take into consideration such often conflicting work factors as ease of tool fabrication and maintenance (resharpening), productivity, product quality, and tooling cost. For certain high-speed cutting applications, alternate, advanced materials, such as carbides and ceramics, may be preferable.

When data related to past experience with tool steels for identical or similar applications are not available, a tool steel selection procedure may be followed, based on information in this *Handbook* section as follows:

Identify the AISI category that contains the type of steel sought by consulting the Quick Reference Guide, Table 5, starting on page 439. Within the defined category:

a) find from the listed applications of the most frequently used types of tool steels the particular type that corresponds to the job on hand; or

b) evaluate from the table of property ratings the best compromise between any conflicting properties (e.g., compromising on wear resistance to obtain better toughness).

For those willing to refine even further the first choice or to improve on it when there is not entirely satisfactory experience in one or more meaningful respects, the identifying analyses of the different types of tool steels within each general category may provide additional guidance. In this procedure, the general discussion of the effects of different alloying elements on the properties of tool steels, in a previous section, will probably be found useful.

The following two examples illustrate the procedure for refining an original choice with the purpose of adopting a tool steel grade best suited to a particular set of conditions:

*Example 1, Workpiece—Trimming Dies:* For the manufacture of a type of trimming die, the first choice was grade A2 because, for the planned medium rate of production, the lower material cost was considered an advantage.

A subsequent rise in the production rate indicated the use of a higher-alloy tool steel, such as D2, whose increased abrasion resistance would permit longer runs between regrinds.

A still further increase in the abrasion-resistant properties was then sought, which led to the use of D7, the high carbon and high chromium content of which provided excellent edge retainment, although at the cost of greatly reduced grindability. Finally, it became a matter of economic appraisal, whether the somewhat shorter tool regrind intervals (for D2) or the more expensive tool sharpening (for D7) constituted the lesser burden.

*Example 2, Workpiece—Circular Form Cutter Made of High-Speed Tool Steel for Use on Multiple-Spindle Automatic Turning Machines:* The first choice from Table 5 may be the classical tungsten-base high-speed tool steel T1, because of its good performance and ease of heat treatment, or its alternate in the molybdenum high-speed tool steel category, the type M2.

In practice, neither of these grades provided a tool that could hold its edge and profile over the economical tool change time because of the abrasive properties of the work material and the high cutting speeds applied in the cycle. An overrating of the problem resulted in reaching for the top of the scale, making the tool from T15, a high-alloy high-speed tool steel (high vanadium and high cobalt).

Although the performance of the tools made of T15 was excellent, the cost of this steel type was rather high, and the grinding of the tool, both for making it and in regularly needed resharpening, proved to be very time-consuming and expensive. Therefore, an intermediate tool steel type was tried, the M3 that provided added abrasion resistance (due to increased carbon and vanadium content), and was less expensive and much easier to grind than the T15.

**Table 5. Quick Reference Guide for Tool Steel Selection**

| Application Areas | Tool Steel Categories and AISI Letter Symbol | | | | | | |
|---|---|---|---|---|---|---|---|
| | High-Speed Tool Steels, M and T | Hot-Work Tool Steels, H | Cold-Work Tool Steels, D, A, and O | Shock-Resisting Tool Steels, S | Mold Steels, P | Special-Purpose Tool Steels, L and F | Water-Hardening Tool Steels, W |
| | Examples of Typical Applications | | | | | | |
| Cutting Tools<br>Single-point types (lathe, planer, boring)<br>Milling cutters<br>Drills<br>Reamers<br>Taps<br>Threading dies<br>Form cutters | General-purpose production tools: M2, T1<br>For increased abrasion resistance: M3, M4, and M10<br>Heavy-duty work calling for high hot hardness: T5, T15<br>Heavy-duty work calling for high abrasion resistance: M42, M44 | | Tools with keen edges (knives, razors)<br>Tools for operations where no high speed is involved, yet stability in heat treatment and substantial abrasion resistance are needed | Pipe cutter wheels | | | Uses that do not require hot hardness or high abrasion resistance.<br>Examples with carbon content of applicable group:<br>Taps (1.05 / 1.10% C)<br>Reamers (1.10 / 1.15% C)<br>Twist drills (1.20 / 1.25% C)<br>Files (1.35 / 1.40% C) |
| Hot Forging Tools and Dies<br>Dies and inserts<br>Forging machine plungers and pierces | For combining hot hardness with high abrasion resistance: M2, T1 | Dies for presses and hammers: H20, H21<br>For severe conditions over extended service periods: H22 to H26, also H43 | Hot-trimming dies: D2 | Hot-trimming dies<br>Blacksmith tools<br>Hot-swaging dies | | | Smith's tools (1.65 / 0.70% C)<br>Hot chisels (0.70 / 0.75% C)<br>Drop-forging dies (0.90 / 1.00% C)<br>Applications limited to short-run production |
| Hot Extrusion Tools and Dies<br>Extrusion dies and mandrels<br>Dummy blocks<br>Valve extrusion tools | Brass extrusion dies: T1 | Extrusion dies and dummy blocks: H20 to H26<br>For tools that are exposed to less heat: H10 to H19 | | Compression molding: S1 | | | |

**Table 5.** *(Continued)* **Quick Reference Guide for Tool Steel Selection**

| Application Areas | Tool Steel Categories and AISI Letter Symbol | | | | | | |
|---|---|---|---|---|---|---|---|
| | High-Speed Tool Steels, M and T | Hot-Work Tool Steels, H | Cold-Work Tool Steels, D, A, and O | Shock-Resisting Tool Steels, S | Mold Steels, P | Special-Purpose Tool Steels, L and F | Water-Hardening Tool Steels, W |
| | Examples of Typical Applications | | | | | | |
| Cold-Forming Dies<br>Bending, forming, drawing, and deep drawing dies and punches | Burnishing tools: M1, T1 | Cold heading: die-casting dies: H13 | Drawing dies: O1<br>Coining tools: O1, D2<br>Forming and bending dies: A2<br>Thread-rolling dies: D2 | Hobbing and short-run applications: S1, S7<br>Rivet sets and rivet busters | | Blanking, forming, and trimmer dies when toughness has precedence over abrasion resistance: L6 | Cold-heading dies: W1 or W2 ($C \cong 1.00\%$)<br>Bending dies: W1 ($C \cong 1.00\%$) |
| Shearing Tools<br>Dies for piercing, punching, and trimming<br>Shear blades | Special dies for cold- and hot-work: T1<br>For work requiring high abrasion resistance: M2, M3 | For shearing knives: H11, H12<br>For severe hot shearing applications: M21, M25 | Dies for medium runs: A2, A6 also O1 and O4<br>Dies for long runs: D2, D3<br>Trimming dies (also for hot trimming): A2 | Cold and hot shear blades<br>Hot punching and piercing tools<br>Boilermaker's tools | | Knives for work requiring high toughness: L6 | Trimming dies (0.90/0.95% C)<br>Cold-blanking and punching dies (1.00% C) |
| Die-Casting Dies and Plastics Molds | | For zinc and lead: H11<br>For aluminum: H13<br>For brass: H21 | A2 and A6<br><br>O1 | | Plastics molds: P2 to P4, and P20 | | |
| Structural Parts for Severe Service Conditions | Roller bearings for high-temperature environment: T1<br>Lathe centers: M2 and T1 | For aircraft components (landing gear, arrester hooks, rocket cases): H11 | Lathe centers: D2, D3<br>Arbors: O1<br>Bushings: A4<br>Gages: D2 | Pawls<br>Clutch parts | | Spindles, clutch parts (where high toughness is needed): L6 | Spring steel (1.10/1.15% C) |
| Battering Tools for Hand and Power Tool Use | | | | Pneumatic chisels for cold-work: S5<br>For higher performance: S7 | | | For intermittent use: W1 (0.80% C) |

## High-Speed Tool Steels

The primary application of high-speed steels is for tools used for the working of metals at high cutting speeds. Cutting metal at high speed generates heat, the penetration of the cutting tool edge into the work material requires great hardness and strength, and the continued frictional contact of the tool with both the parent material and the detached chips can only be sustained by an abrasion-resistant tool edge.

Accordingly, the dominant properties of high-speed steel are a) resistance to the softening effect of elevated temperature; b) great hardness penetrating to substantial depth from the surface; and c) excellent abrasion resistance.

High-speed tool steels are listed in the AISI specifications in two groups: molybdenum types and tungsten types, these designations expressing the dominant alloying element of the respective group.

**Molybdenum Type High-Speed Tool Steels.—**Unlike the traditional tungsten-base high-speed steels, the tool steels listed in this category are considered to have molybdenum as the principal alloying constituent, this element also being used in the designation of the group. Other significant elements like tungsten and cobalt might be present in equal, or even greater, amounts in several types listed in this category. The available range of types also includes high-speed tool steels with higher than usual carbon and vanadium content. Amounts of these alloying elements have been increased to obtain better abrasion resistance although such a change in composition may adversely affect the machinability and grindability of the steel. The series in whose AISI identification numbers the number 4 is the first digit was developed to attain exceptionally high hardness in heat treatment that, for these types, usually requires triple tempering rather than the double tempering generally applied for high-speed tool steels.

*Frequently Used Molybdenum Types: AISI M1:* This alloy was developed as a substitute for the classical T1 to save on the alloying element tungsten by replacing most of it with molybdenum. In most uses, this steel is an acceptable substitute, although it requires greater care or more advanced equipment for its heat treatment than the tungsten-alloyed type it replaces. The steel is often selected for cutting tools like drills, taps, milling cutters, reamers, lathe tools used for lighter cuts, and for shearing dies.

*AISI M2:* Similar to M1, yet with substantial tungsten content replacing a part of the molybdenum. This is one of the general-purpose high-speed tool steels, combining the economic advantages of the molybdenum-type steels with greater ease of hardening, excellent wear resistance, and improved toughness. It is a preferred steel type for the manufacture of general-purpose lathe tools; of most categories of multiple-edge cutting tools, like milling cutters, taps, dies, reamers, and for form tools in lathe operations.

*AISI M3:* A high-speed tool steel with increased vanadium content for improved wear resistance, yet still below the level where vanadium would interfere with the ease of grinding. This steel is preferred for cutting tools requiring improved wear resistance, like broaches, form tools, milling cutters, chasers, and reamers.

*AISI M7:* The chemical composition of this type is similar to that of M1, except for the higher carbon and vanadium content that raises the cutting efficiency without materially reducing the toughness. Because of sensitivity to decarburization, heat treatment in a salt bath or a controlled atmosphere is advisable. Used for blanking and trimming dies, shear blades, lathe tools, and thread rolling dies.

*AISI M10:* Although the relatively high vanadium content assures excellent wear and cutting properties, the only slightly increased carbon does not cause brittleness to an extent that is harmful in many applications. Form cutters and single-point lathe tools, broaches, planer tools, punches, blanking dies, and shear blades are examples of typical uses.

## Table 6. Molybdenum High-Speed Steels

| Identifying Chemical Composition and Typical Heat-Treatment Data | | | | | | | | | | | | | | | | | | | | |
|---|---|---|---|---|---|---|---|---|---|---|---|---|---|---|---|---|---|---|---|---|
| | AISI Type | | M1 | M2 | M3 Cl. 1 | M3 Cl. 2 | M4 | M6 | M7 | M10 | M30 | M33 | M34 | M36 | M41 | M42 | M43 | M44 | M46 | M47 |
| Identifying Chemical Elements in Percent | C | | 0.80 | 0.85 1.00 | 1.05 | 1.20 | 1.30 | 0.80 | 1.00 | 0.85 1.00 | 0.80 | 0.90 | 0.90 | 0.80 | 1.10 | 1.10 | 1.20 | 1.15 | 1.25 | 1.10 |
| | W | | 1.50 | 6.00 | 6.00 | 6.00 | 5.50 | 4.00 | 1.75 | … | 2.00 | 1.50 | 2.00 | 6.00 | 6.75 | 1.50 | 2.75 | 5.25 | 2.00 | 1.50 |
| | Mo | | 8.00 | 5.00 | 5.00 | 5.00 | 4.50 | 5.00 | 8.75 | 8.00 | 8.00 | 9.50 | 8.00 | 5.00 | 3.75 | 9.50 | 8.00 | 6.25 | 8.25 | 9.50 |
| | Cr | | 4.00 | 4.00 | 4.00 | 4.00 | 4.00 | 4.00 | 4.00 | 4.00 | 4.00 | 4.00 | 4.00 | 4.00 | 4.25 | 3.75 | 3.75 | 4.25 | 4.00 | 3.75 |
| | V | | 1.00 | 2.00 | 2.40 | 3.00 | 4.00 | 1.50 | 2.00 | 2.00 | 1.25 | 1.15 | 2.00 | 2.00 | 2.00 | 1.15 | 1.60 | 2.25 | 3.20 | 1.25 |
| | Co | | … | … | … | … | … | 12.00 | … | … | 5.00 | 8.00 | 8.00 | 8.00 | 5.00 | 8.00 | 8.25 | 12.00 | 8.25 | 5.00 |
| Heat-Treatment Data | Hardening Temperature Range | °F | 2150–2225 | 2175–2225 | 2200–2250 | 2200–2250 | 2200–2250 | 2150–2200 | 2150–2225 | 2150–2225 | 2200–2250 | 2200–2250 | 2200–2250 | 2225–2275 | 2175–2220 | 2175–2210 | 2175–2220 | 2190–2240 | 2175–2225 | 2150–2200 |
| | | °C | 1177–1218 | 1191–1218 | 1204–1232 | 1204–1232 | 1204–1232 | 1177–1204 | 1177–1218 | 1177–1218 | 1204–1232 | 1204–1232 | 1204–1232 | 1218–1246 | 1191–1216 | 1191–1210 | 1191–1216 | 1199–1227 | 1191–1218 | 1177–1204 |
| | Tempering Temperature Range | °F | 1000–1100 | 1000–1160 | 1000–1100 | 1000–1100 | 1000–1100 | 1000–1100 | 1000–1100 | 1000–1100 | 1000–1100 | 1000–1100 | 1000–1100 | 1000–1100 | 1000–1100 | 950–1100 | 950–1100 | 1000–1160 | 975–1050 | 975–1100 |
| | | °C | 538–593 | 538–627 | 538–593 | 538–593 | 538–593 | 538–593 | 538–593 | 538–593 | 538–593 | 538–593 | 538–593 | 538–593 | 538–593 | 510–593 | 510–593 | 538–627 | 524–566 | 524–594 |
| | Approx. Tempered Hardness, RC (Rockwell C scale) | | 65–60 | 65–60 | 66–61 | 66–61 | 66–61 | 66–61 | 66–61 | 65–60 | 65–60 | 65–60 | 65–60 | 65–60 | 70–65 | 70–65 | 70–65 | 70–62 | 69–67 | 70–65 |
| Relative Ratings of Properties (A = greatest to E = least) | | | | | | | | | | | | | | | | | | | | |
| Characteristics in Heat Treatment | Safety in Hardening | | D | D | D | D | D | D | D | D | D | D | D | D | D | D | D | D | D | D |
| | Depth of Hardening | | A | A | A | A | A | A | A | A | A | A | A | A | A | A | A | A | A | A |
| | Resistance to Decarburization | | C | B | B | B | B | C | C | C | C | C | C | C | C | C | C | C | C | C |
| | Stability of Shape in Heat Treatment — Quenching Medium | Air or Salt | C | C | C | C | C | C | C | C | C | C | C | C | C | C | C | C | C | C |
| | | Oil | D | D | D | D | D | D | D | D | D | D | D | D | D | D | D | D | D | D |
| Service Properties | Machinability | | D | D | D | D/E | D | D | D | D | D | D | D | D | D | D | D | D | D | D |
| | Hot Hardness | | B | B | B | B | B | A | B | B | A | A | A | A | A | A | A | A | A | A |
| | Wear Resistance | | B | B | B | B | A | B | B | B | B | B | B | B | B | B | B | B | B | B |
| | Toughness | | E | E | E | E | E | E | E | E | E | E | E | E | E | E | E | E | E | E |

*AISI M42:* In applications where high hardness both at regular and at elevated temperatures is needed, this type of high-speed steel with high cobalt content can provide excellent service. Typical applications are tool bits, form tools, shaving tools, fly cutters, roll turning tools, and thread rolling dies. Important uses are found for M42, and for other types of the "M40" group, in the working of "difficult-to-machine" alloys.

**Tungsten Type High-Speed Tool Steels.**—For several decades following their introduction, the tungsten-base high-speed steels were the only types available for cutting operations involving the generation of substantial heat and are still preferred by users who do not have the kind of advanced heat-treating equipment that efficient hardening of the molybdenum-type high-speed tool steels requires. Most tungsten high-speed steels display excellent resistance to decarburization and can be brought to good hardness by simple heat treatment. However, even with tungsten-type high-speed steels, heat treatment using modern methods and furnaces can appreciably improve the metallurgical qualities of the hardened material and the performance of the cutting tools made from these steels.

**Table 7. Tungsten Type High-Speed Tool Steels—Identifying Chemical Composition and Typical Heat-Treatment Data**

| AISI Type | | | T1 | T2 | T4 | T5 | T6 | T8 | T15 |
|---|---|---|---|---|---|---|---|---|---|
| Identifying Chemical Elements in Percent | | | | | | | | | |
| C | | | 0.75 | 0.80 | 0.75 | 0.80 | 0.80 | 0.75 | 1.50 |
| W | | | 18.00 | 18.00 | 18.00 | 18.00 | 20.00 | 14.00 | 12.00 |
| Cr | | | 4.00 | 4.00 | 4.00 | 4.00 | 4.50 | 4.00 | 4.00 |
| V | | | 1.00 | 2.00 | 1.00 | 2.00 | 1.50 | 2.00 | 5.00 |
| Co | | | … | … | 5.00 | … | … | 5.00 | 5.00 |
| Heat-Treatment Data | | | | | | | | | |
| Hardening Temperature Range | | °F | 2300–2375 | 2300–2375 | 2300–2375 | 2325–2375 | 2325–2375 | 2300–2375 | 2200–2300 |
| | | °C | 1260–1302 | 1260–1302 | 1260–1302 | 1274–1302 | 1274–1302 | 1260–1302 | 1204–1260 |
| Tempering Temperature Range | | °F | 1000–1100 | 1000–1100 | 1000–1100 | 1000–1100 | 1000–1100 | 1000–1100 | 1000–1200 |
| | | °C | 538–593 | 538–593 | 538–593 | 538–593 | 538–593 | 538–593 | 538–649 |
| Approx. Tempered Hardness, RC (Rockwell C scale) | | | 65–60 | 66–61 | 66–62 | 65–60 | 65–60 | 65–60 | 68–63 |
| Characteristics in Heat Treatment[a] | | | | | | | | | |
| Safety in Hardening | | | C | C | D | D | D | D | D |
| Depth of Hardening | | | A | A | A | A | A | A | A |
| Resistance to Decarburization | | | A | A | B | C | C | B | B |
| Stability of Shape in Heat Treatment | Quenching Medium | Air or Salt | C | C | C | C | C | C | C |
| | | Oil | D | D | D | D | D | D | D |
| Service Properties | | | | | | | | | |
| Machinability | | | D | D | D | D | D/E | D | D/E |
| Hot Hardness | | | B | B | A | A | A | A | A |
| Wear Resistance | | | B | B | B | B | B | B | A |
| Toughness | | | E | E | E | E | E | E | E |

[a] Relative Ratings of Properties (A = greatest to E = least)

*Frequently Used Tungsten Types: AISI T1:* Also mentioned as the 18-4-1 type with reference to the nominal percentage of its principal alloying elements (W-Cr-V), it is considered to be the classical type of high-speed tool steel. The chemical composition of T1 was developed in the early 1900s, and has changed very little since. T1 is still considered to be perhaps the best general-purpose high-speed tool steel because of the comparative ease of its machining and heat treatment. It combines a high degree of cutting ability with relative toughness. T1 steel is used for all types of multiple-edge cutting tools like drills, reamers, milling cutters, threading taps and dies, light- and medium-duty lathe tools, and is also used for punches, dies, and machine knives, as well as for structural parts that are subjected to elevated temperatures, like lathe centers and certain types of antifriction bearings.

*AISI T2:* Similar to T1 except for somewhat higher carbon content and twice the vanadium contained in the former grade. Its handling ease, both in machining and heat treating, is comparable to that of T1, although it should be held at the quenching temperature slightly longer, particularly when the heating is carried out in a controlled-atmosphere furnace. The applications are similar to those of T1. However, because of its increased wear resistance, T2 is preferred for tools required for finer cuts and where the form or size retention of the tool is particularly important, such as for form and finishing tools.

*AISI T5:* The essential characteristic of this type of high-speed steel, its superior red hardness, stems from its substantial cobalt content that, combined with the relatively high amount of vanadium, provides this steel with excellent wear resistance. In heat treatment, the tendency for decarburization must be considered, and heating in a controlled, slightly reducing atmosphere is recommended. This type of high-speed tool steel is mainly used for single-point tools and inserts; it is well adapted for working at high speeds and feeds, for cutting hard materials and those that produce discontinuous chips, also for nonferrous metals and, for all kinds of tools needed for hogging (removing great bulks of material).

*AISI T15:* The performance qualities of this high-alloy tool steel surpass most of those found in other grades of high-speed tool steels. The high vanadium content, supported by uncommonly high carbon, assures superior cutting ability and wear resistance. The addition of high cobalt increases the "hot hardness," and therefore tools made of T15 can sustain cutting speeds in excess of those commonly applicable to tools made of steel. The machining and heat treatment of T15 does not cause extraordinary problems, although for best results, heating to high temperature is often applied in its heat treatment, and double or even triple tempering is recommended. On the other hand, T15 is rather difficult to grind because of the presence of large amounts of very hard metallic carbides; therefore, it is considered to have a very low "grindability" index. The main uses are in the field of high-speed cutting and the working of hard metallic materials, T15 being often considered to represent in its application a transition from the regular high-speed tool steels to cemented carbides. Lathe tool bits, form cutters, and solid and inserted blade milling cutters are examples of uses of this steel type for cutting tools; excellent results may also be obtained with such tools as cold-work dies, punches, blanking, and forming dies, etc. The low toughness rating of the T15 steel excludes its application for operations that involve shock or sudden variations in load.

## Hot-Work Tool Steels

A family of special tool steels has been developed for tools that in their regular service are in contact with hot metals over a shorter or longer period of time, with or without cooling being applied, and are known as hot-work steels. The essential property of these steels is their ability to sustain elevated temperature without seriously affecting the usefulness of the tools made from them. Depending on the purpose of the tools for which they were developed, the particular types of hot-work tool steels have different dominant properties and are assigned to one of three groups, based primarily on their principal alloying elements.

**Hot-Work Tool Steels, Chromium Types.—**As referred to in the group designation, the chromium content is considered the characteristic element of these tool steels. Their predominant properties are high hardenability, excellent toughness, and great ductility, even at the cost of wear resistance. Some members of this family are made with the addition of tungsten, and in one type, cobalt as well. These alloying elements improve the resistance to the softening effect of elevated temperatures, but reduce ductility.

*Frequently Used Chromium Types: AISI H11:* This hot-work tool steel of the Chromium-molybdenum-vanadium type has excellent ductility, can be machined easily, and retains its strength at temperatures up to 1000° F (538° C).

These properties, combined with relatively good abrasion and shock resistance, account for the varied fields of application of H11, which include the following typical uses:

a) structural applications where high strength is needed at elevated operating temperatures, as for gas turbine engine components; and b) hot-work tools, particularly of the kind whose service involves shocks and drastic cooling of the tool, such as in extrusion tools, pierce and draw punches, bolt header dies, etc.

## Table 8. Hot-Work Tool Steels

| Identifying Chemical Composition and Typical Heat-Treatment Data | | | | | | | | | | | | | | | | | |
|---|---|---|---|---|---|---|---|---|---|---|---|---|---|---|---|---|---|
| AISI | Group | | Chromium Types | | | | | | Tungsten Types | | | | | | Molybdenum Types | | |
| | Type | | H10 | H11 | H12 | H13 | H14 | H19 | H21 | H22 | H23 | H24 | H25 | H26 | H41 | H42 | H43 |
| Identifying Chemical Elements in Percent | C | | 0.40 | 0.35 | 0.35 | 0.35 | 0.40 | 0.40 | 0.35 | 0.35 | 0.35 | 0.45 | 0.25 | 0.50 | 0.65 | 0.60 | 0.55 |
| | W | | … | … | 1.50 | … | 5.00 | 4.25 | 9.00 | 11.00 | 12.00 | 15.00 | 15.00 | 18.00 | 1.50 | 6.00 | … |
| | Mo | | 2.50 | 1.50 | 1.50 | 1.50 | … | … | … | … | … | … | … | … | 8.00 | 5.00 | 8.00 |
| | Cr | | 3.25 | 5.00 | 5.00 | 5.00 | 5.00 | 4.25 | 3.50 | 2.00 | 12.00 | 3.00 | 4.00 | 4.00 | 4.00 | 4.00 | 4.00 |
| | V | | 0.40 | 0.40 | 0.40 | 1.00 | … | 2.00 | … | … | … | … | … | 1.00 | 1.00 | 2.00 | 2.00 |
| | Co | | … | … | … | … | … | 4.25 | … | … | … | … | … | … | … | … | … |
| Heat-Treatment Data | Hardening Temperature Range | °F | 1850–1900 | 1825–1875 | 1825–1875 | 1825–1900 | 1850–1950 | 2000–2200 | 2000–2200 | 2000–2200 | 2000–2300 | 2000–2250 | 2100–2300 | 2150–2300 | 2000–2175 | 2050–2225 | 2000–2175 |
| | | °C | 1010–1038 | 996–1024 | 996–1024 | 996–1038 | 1010–1066 | 1093–1204 | 1093–1204 | 1093–1204 | 1093–1260 | 1093–1232 | 1149–1260 | 1177–1260 | 1093–1191 | 1121–1218 | 1093–1191 |
| | Tempering Temperature Range | °F | 1000–1200 | 1000–1200 | 1000–1200 | 1000–1200 | 1100–1200 | 1000–1300 | 1100–1250 | 1100–1250 | 1200–1500 | 1050–1200 | 1050–1250 | 1050–1250 | 1050–1200 | 1050–1200 | 1050–1200 |
| | | °C | 538–649 | 538–649 | 538–649 | 538–649 | 593–649 | 538–704 | 593–677 | 593–677 | 649–816 | 566–649 | 566–677 | 566–677 | 566–649 | 566–649 | 566–649 |
| | Approx. Tempered Hardness, RC (Rockwell C scale) | | 56–39 | 54–38 | 55–38 | 53–38 | 47–40 | 59–40 | 54–36 | 52–39 | 47–30 | 55–45 | 44–35 | 58–43 | 60–50 | 60–50 | 58–45 |
| Relative Ratings of Properties (A = greatest to D = least) | | | | | | | | | | | | | | | | | |
| Characteristics in Heat Treatment | Safety in Hardening | | A | A | A | A | A | B | B | B | B | B | B | B | C | C | C |
| | Depth of Hardening | | A | A | A | A | A | A | A | A | A | A | A | A | A | A | A |
| | Resistance to Decarburization | | B | B | B | B | B | B | B | B | B | B | B | B | C | B | C |
| | Stability of Shape in Heat Treatment — Quenching Medium | Air or Salt | B | B | B | B | C | C | C | C | … | C | C | C | C | C | C |
| | | Oil | … | … | … | … | … | D | D | D | D | D | D | D | D | D | D |
| Service Properties | Machinability | | C/D | C/D | C/D | C/D | D | D | D | D | D | D | D | D | D | D | D |
| | Hot Hardness | | C | C | C | C | C | C | C | C | B | B | B | B | B | B | B |
| | Wear Resistance | | D | D | D | D | D | C/D | C/D | C/D | C/D | C | D | C | C | C | C |
| | Toughness | | C | B | B | B | C | C | C | C | D | D | C | D | D | D | D |

*AISI H12:* The properties of this type of steel are comparable to those of H11, with increased abrasion resistance and hot hardness resulting from the addition of tungsten, yet in an amount that does not affect the good toughness of this steel type. The applications based on these properties are hot-work tools that often have to withstand severe impact, such as various punches, bolt header dies, trimmer dies, and hot shear blades. H12 is also used to make aluminum extrusion dies and die-casting dies.

*AISI H13:* This type of tool steel differs from the preceding ones particularly in properties related to the addition of about 1 percent vanadium, which contributes to increased hot hardness, abrasion resistance, and reduced sensitivity to heat checking. Such properties are needed in die casting, particularly of aluminum, where the tools are subjected to drastic heating and cooling at high operating temperatures. Besides die-casting dies, H13 is also widely used for extrusion dies, trimmer dies, hot gripper and header dies, and hot shear blades.

*AISI H19:* This high-alloyed hot-work tool steel containing chromium, tungsten, cobalt, and vanadium has excellent resistance to abrasion and shocks at elevated temperatures. It is particularly well adapted to severe hot-work uses where the tool, to retain its size and shape, must withstand wear and the washing-out effect of molten work material. Typical applications include brass extrusion dies and dummy blocks, inserts for forging and valve extrusion dies, press forging dies, and hot punches.

**Hot-Work Tool Steels, Tungsten Types.—**A substantial amount of tungsten and very low-carbon content characterize the hot-work tool steels of this group. These tool steels have been developed for applications where the tool is in contact with the hot-work material over extended periods of time; therefore, the resistance of the steel to the softening effect of elevated temperatures is of prime importance, even to the extent of accepting a lower degree of toughness.

*Frequently Used Tungsten Types: AISI H21:* This medium-tungsten alloyed hot-work tool steel has substantially increased abrasion resistance over the chromium-alloyed types yet possesses a degree of toughness that represents a transition between the chromium and the higher-alloyed tungsten-steel types. The principal applications are for tools subjected to continued abrasion yet to only a limited amount of shock loads, like tools for the extrusion of brass, both dies and dummy blocks, pierces for forging machines, inserts for forging tools, and hot nut tools. Another typical application is dies for the hot extrusion of automobile valves.

*AISI H24:* The comparatively high tungsten content (about 14 percent) of this steel results in good hardness, great compression strength, and excellent abrasion resistance, but makes it sensitive to shock loads. By taking these properties into account, the principal applications include extrusion dies for brass in long-run operations, hot-forming and gripper dies with shallow impressions, punches that are subjected to great wear yet only to moderate shocks, and hot shear blades.

*AISI H20:* The composition of this high-alloyed tungsten-type hot-work steel resembles the tungsten-type high-speed steel AISI T1, except for the somewhat lower carbon content for improved toughness. The high amount of tungsten provides the maximum resistance to the softening effect of elevated temperature and assures excellent wear-resistant properties, including withstanding the washing-out effect of certain processes. However, this steel is less resistant to thermal shocks than the chromium hot-work steels. Typical applications comprise extrusion dies for long production runs, extrusion mandrels operated without cooling, hot piercing punches, hot forging dies and inserts. It is also used as special structural steel for springs operating at elevated temperatures.

**Hot-Work Tool Steels, Molybdenum Types.—**These steels are closely related to certain types of molybdenum high-speed steels and possess excellent resistance to the softening effect of elevated temperature, but their ductility is rather low. These steel types are generally available on special orders only.

*Frequently Used Molybdenum Types: AISI H43:* The principal constituents of this hot-work steel, chromium, molybdenum, and vanadium, provide excellent abrasion- and wear-resistant properties at elevated temperatures. H43 has a good resistance to the development of heat checks and a toughness adequate for many different purposes. Applications include tools and operations that tend to cause surface wear in high-temperature work, like hot headers, punch and die inserts, hot heading and hot nut dies, as well as different kinds of punches operating at high temperature in service involving considerable wear.

## Cold-Work Tool Steels

Tool steels of the cold-working category are primarily intended for die work, although their use is by no means restricted to that general field. Cold-work tool steels are extensively used for tools whose regular service does not involve elevated temperatures. They are available in chemical compositions adjusted to the varying requirements of a wide range of different applications. According to their predominant properties, characterized either by the chemical composition or by the quenching medium in heat treatment, the cold-work tool steels are assigned to three different groups, as discussed below.

**Cold-Work Tool Steels, High-Carbon, High-Chromium Types.—**The chemical composition of tool steels of this family is characterized by the very high chromium content, to the order of 12 to 13 percent, and the uncommonly high carbon content, in the range of about 1.50 to 2.30 percent. Additional alloying elements that are present in different amounts in some of the steel types of this group are vanadium, molybdenum, and cobalt, each of which contributes desirable properties.

The predominant properties of the whole group are: 1) excellent dimensional stability in heat treatment, where, with one exception, air quench is used; 2) great wear resistance, particularly in the types with the highest carbon content; and 3) rather good machinability.

*Frequently Used High-Carbon, High-Chromium Types: AISI D2:* An air-hardening die steel with high-carbon, high-chromium content having several desirable tool steel properties, such as abrasion resistance, high hardness, and nondeforming characteristics. The carbon content of this type, although relatively high, is not particularly detrimental to its machining. The ease of working can be further improved by selecting the same basic type with the addition of sulfur. Several steel producers supply the sulfurized version of D2, in which the uniformly distributed sulfide particles substantially improve the machinability and the resulting surface finish. The applications comprise primarily cold-working press tools for shearing (blanking and stamping dies, punches, shear blades), for forming (bending, seaming), and for thread rolling dies, solid gages, and wear-resistant structural parts. Dies for hot trimming of forgings are also made of D2, which is then heated treated to a lower hardness for the purpose of increasing toughness.

*AISI D3:* The high carbon content of this high-chromium tool steel type results in excellent resistance to wear and abrasion and provides superior compressive strength as long as the pressure is applied gradually without exerting sudden shocks. In hardening, an oil quench is used without affecting the excellent nondeforming properties of this type. Its deep-hardening properties make it particularly suitable for tools that require repeated regrinding during their service life, such as different types of dies and punches. The more important applications comprise blanking, stamping, and trimming dies and punches for long production runs; forming, bending and drawing tools; and structural elements like plug and ring gages, and lathe centers, in applications where high wear resistance is important.

**Cold-Work Tool Steels, Oil-Hardening Types.—**With a relatively low percentage of alloying elements, yet with a substantial amount of manganese, these less expensive types of tool steels attain good depth of hardness in an oil quench, although at the cost of reduced resistance to deformation. Their good machinability supports general-purpose applications, yet, because of relatively low wear resistance, they are mostly selected for comparatively short-run work.

*Frequently Used Oil-Hardening Types: AISI O1:* A low-alloy tool steel that is hardened in oil and exhibits only a low tendency to shrinking or warping. It is used for cutting tools, the operation of which does not generate high heat, such as taps and threading dies, reamers, and broaches, and for press tools like blanking, trimming, and forming dies in short- or medium-run operations.

*AISI O2:* Manganese is the dominant alloying element in this type of oil-hardening tool steel that has good nondeforming properties, can be machined easily, and performs satisfactorily in low-volume production. The low hardening temperature results in good safety in hardening, both with regard to form stability and freedom from cracking. The combination of handling ease, including free-machining properties, with good wear resistance, makes this type of tool steel adaptable to a wide range of common applications, such as cutting tools for low- and medium-speed operations; forming tools including thread rolling dies; structural parts, such as bushings and fixed gages; and for plastics molds.

*AISI O6:* This oil-hardening type of tool steel belongs to a group often designated as graphitic because of the presence of small particles of graphitic carbon that are uniformly dispersed throughout the steel. Usually, about one-third of the total carbon is present as free graphite in nodular form, which contributes to the uncommon ease of machining. In the service of parts made of this type of steel, the free graphite acts like a lubricant, reducing wear and galling. The ease of hardening is also excellent, requiring only a comparatively low quenching temperature. Deep hardness penetration is produced and the oil quench causes very little dimensional change. The principal applications of the O6 tool steel are in the field of structural parts, like arbors, bushings, bodies for inserted tool cutters, and shanks for cutting tools, jigs, and machine parts, and fixed gages like plugs, rings, and snap gages. It is also used for blanking, forming, and trimming dies and punches, in applications where the stability of the tool material is more important than high wear resistance.

**Cold-Work Tool Steels, Medium-Alloy, Air-Hardening Types.—**The desirable nondeforming properties of the high-chromium types are approached by the members of this family, with substantially lower alloy content that, nonetheless, is sufficient to permit hardening by air quenching. The machinability is good, and the comparatively low wear resistance is balanced by relatively high toughness, a property that, in certain applications, may be considered of prime importance.

*Frequently Used Medium-Alloy, Air-Hardening Types: AISI A2:* The lower chromium content, about 5 percent, makes this air-hardening tool steel less expensive than the high-chromium types, without affecting its nondeforming properties. The somewhat reduced wear resistance is balanced by greater toughness, making this type suitable for press work where the process calls for tough tool materials. The machinability is improved by the addition of about 0.12 percent sulfur, offered as a variety of the basic composition by several steel producers. The prime uses of this tool steel type are punches for blanking and forming, cold and hot trimming dies (the latter heat treated to a lower hardness), thread rolling dies, and plastics molds.

*AISI A6:* The composition of this type of tool steel makes it adaptable to air hardening from a relatively low temperature, comparable to that of oil-hardening types, yet offers improved stability in heat treating. Its reduced tendency to heat-treatment distortions makes this tool steel type well adapted for die work, forming tools, and gages, which do not require the highest degree of wear resistance.

## Table 9. Cold-Work Tool Steels

| Identifying Chemical Composition and Typical Heat-Treatment Data | | | | | | | | | | | | | | | | | | | |
|---|---|---|---|---|---|---|---|---|---|---|---|---|---|---|---|---|---|---|---|
| AISI | Group | | High-Carbon, High-Chromium Types | | | | | Medium-Alloy, Air-Hardening Types | | | | | | | | Oil-Hardening Types | | | |
| | Types | | D2 | D3 | D4 | D5 | D7 | A2 | A3 | A4 | A6 | A7 | A8 | A9 | A10 | O1 | O2 | O6 | O7 |
| Identifying Chemical Elements in Percent | C | | 1.50 | 2.25 | 2.25 | 1.50 | 2.35 | 1.00 | 1.25 | 1.00 | 0.70 | 2.25 | 0.55 | 0.50 | 1.35 | 0.90 | 0.90 | 1.45 | 1.20 |
| | Mn | | … | … | … | … | … | … | … | 2.00 | 2.00 | … | … | … | 1.80 | 1.00 | 1.60 | … | … |
| | Si | | … | … | … | … | … | … | … | … | … | … | … | … | 1.25 | … | … | 1.00 | … |
| | W | | … | … | … | … | … | … | … | … | … | 1.00 | 1.25 | … | … | 0.50 | … | … | 1.75 |
| | Mo | | 1.00 | … | 1.00 | 1.00 | 1.00 | 1.00 | 1.00 | 1.00 | 1.25 | 1.00 | 1.25 | 1.40 | 1.50 | … | … | 0.25 | … |
| | Cr | | 12.00 | 12.00 | 12.00 | 12.00 | 12.00 | 5.00 | 5.00 | 1.00 | 1.00 | 5.25 | 5.00 | 5.00 | … | 0.50 | … | … | 0.75 |
| | V | | 1.00 | … | … | … | 4.00 | … | 1.00 | … | … | 4.75 | … | 1.00 | … | … | … | … | … |
| | Co | | … | … | … | 3.00 | … | … | … | … | … | … | … | … | … | … | … | … | … |
| Heat-Treatment Data | Ni | | … | … | … | … | … | … | … | … | … | … | … | 1.50 | 1.80 | … | … | … | … |
| | Hardening Temperature Range | °F | 1800–1875 | 1700–1800 | 1775–1850 | 1800–1875 | 1850–1950 | 1700–1800 | 1750–1850 | 1500–1600 | 1525–1600 | 1750–1800 | 1800–1850 | 1800–1875 | 1450–1500 | 1450–1500 | 1400–1475 | 1450–1500 | 1550–1525 |
| | | °C | 982–1024 | 927–982 | 968–1010 | 982–1024 | 1010–1066 | 927–982 | 954–1010 | 816–871 | 829–871 | 954–982 | 982–1010 | 982–1024 | 788–816 | 788–816 | 760–802 | 788–816 | 843–829 |
| | Quenching Medium | | Air | Oil | Air | Air | Air | Air | Air | Air | Air | Air | Air | Air | Air | Oil | Oil | Oil | Oil |
| | Tempering Temperature Range | °F | 400–1000 | 400–1000 | 400–1000 | 400–1000 | 300–1000 | 350–1000 | 350–1000 | 350–800 | 300–800 | 300–1000 | 350–1100 | 950–1150 | 350–800 | 350–500 | 350–500 | 350–600 | 350–550 |
| | | °C | 204–538 | 204–538 | 204–538 | 204–538 | 149–538 | 177–538 | 177–538 | 177–427 | 149–427 | 149–538 | 177–593 | 510–621 | 177–427 | 177–260 | 177–260 | 177–316 | 177–288 |
| | Approx. Tempered Hardness, RC (Rockwell C scale) | | 61–54 | 61–54 | 61–54 | 61–54 | 65–58 | 62–57 | 65–57 | 62–54 | 60–54 | 67–57 | 60–50 | 56–35 | 62–55 | 62–57 | 62–57 | 63–58 | 64–58 |
| Relative Ratings of Properties (A = greatest to E = least) | | | | | | | | | | | | | | | | | | | |
| Characteristics in Heat Treatment | Safety in Hardening | | A | C | A | A | A | A | A | A | A | A | A | A | A | B | B | B | B |
| | Depth of Hardening | | A | A | A | A | A | A | A | A | A | A | A | A | A | B | B | B | B |
| | Resistance to Decarburization | | B | B | B | B | B | B | B | A/B | A/B | B | B | B | A/B | A | A | A | A |
| | Stability of Shape in Heat Treatment | | A | B | A | A | A | A | A | A | A | A | A | A | A | B | B | B | B |
| Service Properties | Machinability | | E | E | E | E | E | D | D | D/E | D/E | E | D | D | C/D | C | C | B | C |
| | Hot Hardness | | C | C | C | C | C | C | C | D | D | C | C | C | D | E | E | E | E |
| | Wear Resistance | | B/C | B | B | B/C | A | C | B | C/D | C/D | A | C/D | C/D | C | D | D | D | D |
| | Toughness | | E | E | E | E | E | D | D | D | D | E | C | C | D | D | D | D | C |

## Shock-Resisting, Mold, and Special-Purpose Tool Steels

There are fields of tool application in which specific properties of the tool steels have dominant significance, determining to a great extent the performance and the service life of tools made of these materials. To meet these requirements, special types of tool steels have been developed. These individual types grew into families with members that, while similar in their major characteristics, provide related properties to different degrees. Originally developed for a specific use, the resulting particular properties of some of these tool steels made them desirable for other uses as well. In the tool steel classification system, they are shown in three groups, as discussed below.

**Shock-Resisting Tool Steels.—**These steels are made with low-carbon content for increased toughness, even at the expense of wear resistance, which is generally low. Each member of this group also contains alloying elements, different in composition and amount, selected to provide properties particularly adjusted to specific applications. Such varying properties are the degree of toughness (generally high in all members), hot hardness, abrasion resistance, and machinability.

*Properties and Applications of Frequently Used Shock-Resisting Types: AISI S1:* This Chromium-tungsten alloyed tool steel combines, in its hardened state, great toughness with high hardness and strength. Although it has a low-carbon content for reasons of good toughness, the carbon-forming alloys contribute to deep hardenability and abrasion resistance. When high wear resistance is also required, this property can be improved by carburizing the surface of the tool while still retaining its shock-resistant characteristics. Primary uses are for battering tools, including hand and pneumatic chisels. The chemical composition, particularly the silicon and tungsten content, also provides good hot hardness up to operating temperatures of about 1050° F (566° C), making this tool steel type adaptable for such hot-work tool applications involving shock loads as headers, pierces, forming tools, drop forge die inserts, and heavy shear blades.

*AISI S2:* This steel type serves primarily for hand chisels and pneumatic tools, but it also has limited applications for hot-work. Although its wear-resistance properties are only moderate, S2 is sometimes used for forming and thread rolling applications, when the resistance to rupturing is more important than extended service life. For hot-work applications, this steel requires heat treatment in a neutral atmosphere to avoid either carburization or decarburization of the surface. Such conditions make this tool steel type particularly susceptible to failure in hot-work uses.

*AISI S5:* This composition is essentially a silicon-manganese-type tool steel with small additions of chromium, molybdenum, and vanadium for the purpose of improved deep hardening and refinement of the grain structure. The most important properties of this steel are its high elastic limit and good ductility, resulting in excellent shock-resisting characteristics, when used at atmospheric temperatures. Its recommended quenching medium is oil, although a water quench may also be applied as long as the design of the tools avoids sharp corners or drastic sectional changes. Typical applications include pneumatic tools in severe service, like chipping chisels as well as shear blades, heavy-duty punches, and bending rolls. Occasionally, this steel is also used for structural applications, like shanks for carbide tools and machine parts subject to shocks.

**Mold Steels.—**These materials differ from all other types of tool steels by their very low-carbon content, generally requiring carburizing to obtain a hard operating surface. A special property of most steel types in this group is adaptability to shaping by impression (hobbing) instead of by conventional machining. They also have high resistance to decarburization in heat treatment and dimensional stability, characteristics that obviate the need for grinding following heat treatment. Molding dies for plastics materials require an excellent surface finish, even to the degree of high luster; the generally high-chromium content of these types of tool steels greatly aids in meeting this requirement.

## Table 10. Shock-Resisting, Mold, and Special-Purpose Tool Steels

| | | | S1 | S2 | S5 | S7 | P2 | P3 | P4 | P5 | P6 | P20 | P21[a] | L2[b] | L3[b] | L6 | F1 | F2 |
|---|---|---|---|---|---|---|---|---|---|---|---|---|---|---|---|---|---|---|
| Identifying Chemical Composition and Typical Heat-Treatment Data | | | | | | | | | | | | | | | | | | |
| AISI | Category | | Shock-Resisting Tool Steels | | | | Mold Steels | | | | | | | Special-Purpose Tool Steels | | | | |
| | Types | | S1 | S2 | S5 | S7 | P2 | P3 | P4 | P5 | P6 | P20 | P21[a] | L2[b] | L3[b] | L6 | F1 | F2 |
| Identifying Elements in Percent | C | | 0.50 | 0.50 | 0.55 | 0.50 | 0.07 | 0.10 | 0.07 | 0.10 | 0.10 | 0.35 | 0.20 | 0.50/1.10 | 1.00 | 0.70 | 1.00 | 1.25 |
| | Mn | | … | … | 0.80 | … | … | … | … | … | … | … | … | … | … | … | … | … |
| | Si | | … | 1.00 | 2.00 | … | … | … | … | … | … | … | … | … | … | … | … | … |
| | W | | 2.50 | … | … | … | … | … | … | … | … | … | … | … | … | … | 1.25 | 3.50 |
| | Mo | | … | 0.50 | 0.40 | 1.40 | 0.20 | … | 0.75 | … | … | 0.40 | … | … | … | 0.25 | … | … |
| | Cr | | 1.50 | … | … | 3.25 | 2.00 | 0.60 | 5.00 | 2.25 | 1.50 | 1.25 | … | 1.00 | 1.50 | 0.75 | … | … |
| | V | | … | … | … | … | … | … | … | … | … | … | … | 0.20 | 0.20 | … | … | … |
| | Ni | | … | … | … | … | 0.50 | 1.25 | … | … | 3.50 | … | 4.00 | … | … | 1.50 | … | … |
| Heat-Treatment Data | Hardening Temperature | °F | 1650–1750 | 1550–1650 | 1600–1700 | 1700–1750 | 1525–1550[c] | 1475–1525[c] | 1775–1825[c] | 1550–1600[c] | 1450–1500[c] | 1500–1600[c] | Soln. treat. | 1550–1700 | 1500–1600 | 1450–1550 | 1450–1600 | 1450–1600 |
| | | °C | 899–954 | 843–899 | 871–927 | 927–955 | 829–843[c] | 802–829[c] | 968–996[c] | 843–871[c] | 788–816[c] | 816–871[c] | Soln. treat. | 843–927 | 816–871 | 788–843 | 788–871 | 788–871 |
| | Tempering Temp. Range | °F | 400–1200 | 350–800 | 350–800 | 400–1150 | 350–500 | 350–500 | 350–900 | 350–500 | 350–450 | 900–1100 | Aged | 350–1000 | 350–600 | 350–1000 | 350–500 | 350–500 |
| | | °C | 204–649 | 177–427 | 177–427 | 204–621 | 177–260 | 177–260 | 177–482 | 177–260 | 177–232 | 482–593 | Aged | 177–538 | 177–316 | 177–538 | 177–260 | 177–260 |
| | Approx. Tempered Hardness, RC (Rockwell C scale) | | 58–40 | 60–50 | 60–50 | 57–45 | 64–58[d] | 64–58[d] | 64–58[d] | 64–58[d] | 61–58[d] | 37–28[d] | 40–30 | 63–45 | 63–56 | 62–45 | 64–60 | 65–62 |
| Relative Ratings of Properties (A = greatest to E = least) | | | | | | | | | | | | | | | | | | |
| Characteristics in Heat Treatment | Safety in Hardening | | C | E | C | B/C | C | C | C | C | C | C | A | D | D | C | E | E |
| | Depth of Hardening | | B | B | B | A | B[e] | B[e] | B[e] | B[e] | A[e] | B | A | B | B | B | C | C |
| | Resist. to Decarb. | | B | C | C | B | A | A | A | A | A | A | A | A | A | A | A | A |
| | Stability of Shape in Heat Treatment — Quench. Med. | Air | … | … | … | A | … | … | B | … | B | C | A | … | … | … | … | … |
| | | Oil | D | … | D | C | C | C | … | C | C | … | A | D | D | C | … | … |
| | | Water[f] | … | E | … | … | … | … | … | E | … | … | … | E | E | … | E | E |
| Service Properties | Machinability | | D | C/D | C/D | D | C/D | D | D/E | D | D | C/D | D | C | C | D | C | D |
| | Hot Hardness | | D | E | E | C | E | E | D | E | E | E | D | E | E | E | E | E |
| | Wear Resistance | | D/E | D/E | D/E | D/E | D | D | C | D | D | D/E | D | D/E | D | D | D | B/C |
| | Toughness | | B | A | A | B | C | C | C | C | C | C | D | B | D | B | E | E |

[a] Contains also about 1.20 percent A1. Solution treated in hardening.
[b] Quenched in oil.
[c] After carburizing.
[d] Carburized case.
[e] Core hardenability.
[f] Sometimes brine is used.

*Properties and Applications of Frequently Used Mold Steel Types: AISI P3 and P4:* Essentially, both types of tool steels were developed for the same special purpose, that is, the making of plastics molds. The application conditions of plastics molds require high core strength, good wear resistance at elevated temperature, and excellent surface finish. Both types are carburizing steels that possess good dimensional stability. Hobbing is the process of sinking a cavity by pressing a punch that is the inverse replica of the cavity into the tool material, and is the process by which many plastics mold cavities are produced. Good "hobbability" is an important requirement of tool steels used for this purpose. The different chemistry of these two types of mold steels is responsible for the high core hardness of P4, which makes it better suited for applications requiring high strength at elevated temperature.

*AISI P6:* This nickel-chromium-type plastics mold steel has exceptional core strength and develops a deep carburized case. Due to the high nickel-chromium content, the cavities of molds made of this steel type are produced by machining rather than by hobbing. An outstanding characteristic of this steel type is the high luster that is produced by polishing of the hard case surface.

*AISI P20:* This general-type mold steel is adaptable to both through hardening and carburized case hardening. In through hardening, an oil quench is used and a relatively lower, yet deeply penetrating, hardness is obtained, such as is needed for zinc die-casting dies and injection molds for plastics. After the direct quenching and tempering, carburizing produces a very hard case and comparatively high core hardness. When thus heat treated, this steel is particularly well adapted for making compression, transfer, and plunger-type plastics molds.

**Special-Purpose Tool Steels.—**These steels include several low-alloy types of tool steels that were developed to provide transitional types between the more commonly used basic types of tool steels, and thereby contribute to the balancing of certain conflicting properties, such as wear resistance and toughness. They offer intermediate depth of hardening and are less expensive than the higher-alloyed types of tool steels.

*Properties and Applications of Frequently Used Special-Purpose Types: AISI L6:* This material is a low-alloy-type special-purpose tool steel. Comparatively safe hardening and fair nondeforming properties, combined with the service advantage of good toughness in comparison to most other oil-hardening types, explain the acceptance of this steel with a rather special chemical composition. L6 is used for tools whose toughness requirements prevail over abrasion-resistant properties, such as forming rolls and forming and trimmer dies in applications where combinations of moderate shock- and wear-resistant properties are sought. The areas of use also include structural parts, like clutch members, pawls, and knuckle pins, that must withstand shock loads and still display good wear properties.

*AISI F2:* This carbon-tungsten type is one of the most abrasion-resistant of all water-hardening tool steels. However, it is sensitive to thermal changes, such as those involved in heat treatment, and it is also susceptible to distortions. Consequently, its use is limited to tools of simple shape in order to avoid cracking in hardening. The shallow hardening characteristics of F2 result in a tough core and are desirable properties for certain tool types that, at the same time, require excellent wear-resistant properties.

**Water-Hardening Tool Steels.—**Steel types in this category are made without, or with only a minimum amount of, alloying elements, and their heat-treatment needs the harsh quenching action of water or brine, hence the general designation of the category.

Water-hardening steels are usually available with different percentages of carbon, to provide properties required for different applications; the classification system lists a carbon range of 0.60 to 1.40 percent. In practice, however, the steel mills produce these steels in a few varieties of differing carbon content, often giving proprietary designations to each particular group. Typical carbon content limits of frequently used water-hardening tool steels are 0.70–0.90, 0.90–1.10, 1.05–1.20, and 1.20–1.30 percent. The appropriate group should be chosen according to the intended use, as indicated in the steel selection guide for this category, keeping in mind that whereas higher carbon content results in deeper hardness penetration, it also reduces toughness.

The general system distinguishes the following four grades, listed in the order of decreasing quality: 1) special; 2) extra; 3) standard; and 4) commercial.

The differences between these grades, which are not offered by all steel mills, are defined in principle only. The distinguishing characteristics are purity and consistency resulting from different degrees of process refinement and inspection steps applied in making the steel. Higher qualities are selected for assuring dependable uniformity and performance of the tools made from the steel.

The groups with higher carbon content are more sensitive to heat-treatment defects and are generally used for the more demanding applications, so the better grades are usually chosen for the high-carbon types and the lower grades for applications where steels with lower carbon content only are needed.

Water-hardening tool steels, although the least expensive and quite acceptable in many types of applications, have several drawbacks. Some limiting properties are the tendency to deformation in heat treatment due to harsh effects of the applied quenching medium, the sensitivity to heat during the use of the tools made of these steels, the only fair degree of toughness, and the shallow penetration of hardness. However, this last-mentioned property may prove a desirable characteristic in certain applications, such as cold-heading dies, because the relatively shallow hard case is supported by the tough, although softer, core.

The AISI designation for water-hardening tool steels is W, followed by a numeral indicating the type, primarily defined by the chemical composition, as shown in Table 11.

**Table 11. Water-Hardening Tool Steels—Identifying Chemical Composition and Heat-Treatment Data**

| Chemical Composition in Percent | AISI Types | | |
|---|---|---|---|
| | W1 | W2 | W5 |
| C | 0.60–1.40 | 0.60–1.40 | 1.10 |
| | Varying carbon content may be available | | |
| V | … | 0.25 | … |
| Cr | These elements are adjusted to satisfy the hardening requirements | | 0.50 |
| Mn | | | |
| Si | | | |

| Heat-Treatment Data | | |
|---|---|---|
| Hardening Temperature Ranges Varying with Carbon Content | 0.60–0.80% | 1450–1500° F (788–816° C) |
| | 0.85–1.05% | 1425–1550° F (774–843° C) |
| | 1.10–1.40% | 1400–1525° F (760–829° C) |
| Quenching Medium | | Brine or Water |
| Tempering Temperature Range | | 350–650° F (177–343° C) |
| Approx. Tempered Hardness, RC (Rockwell C scale) | | 64–50 |

| Relative Ratings of Properties (A = greatest to E = least) | | | | | | | |
|---|---|---|---|---|---|---|---|
| Characteristics in Heat Treatment | | | | Service Properties | | | |
| Safety in Hardening | Depth of Hardening | Resistance to Decarburization | Stability of Shape in Heat Treatment | Machinability | Hot Hardness | Wear Resistance | Toughness |
| D | C | A | E | A | E | D/E | C/D |

*Water-Hardening Type W1 (Plain Carbon) Tool Steels, Recommended Applications:*

*Group I (C 0.70 to 0.90%):* This group is relatively tough and therefore preferred for tools that are subjected to shocks or abusive treatment. Used for such applications as hand tools, chisels, screwdriver blades, cold punches, and nail sets, and fixture elements, vise jaws, anvil faces, and chuck jaws.

*Group II (C 0.90 to 1.10%):* This group combines greater hardness with fair toughness, resulting in improved cutting capacity and moderate ability to sustain shock loads. Used for such applications as hand tools, knives, center punches, pneumatic chisels, cutting tools, reamers, hand taps, and threading dies, wood augers; die parts, drawing and heading dies, shear knives, cutting and forming dies; and fixture elements, drill bushings, lathe centers, collets, and fixed gages.

*Group III (C 1.05 to 1.20%):* The higher carbon content of this group increases the depth of hardness penetrations, yet reduces toughness, thus the resistance to shock loads. Preferred for applications where wear resistance and cutting ability are the prime considerations. Used for such applications as hand tools, woodworking chisels, paper knives, cutting tools (for low-speed applications), milling cutters, reamers, planer tools, thread chasers, center drills, die parts, cold blanking, coining, bending dies.

*Group IV (C 1.20 to 1.30%):* The high carbon content of this group produces a hard case of considerable depth with improved wear resistance yet sensitivity to shock and concentrated stresses. Selected for applications where the capacity to withstand abrasive wear is needed and where the retention of a keen edge or the original shape of the tool is important. Used for such applications as cutting tools for finishing work, like cutters and reamers, and for cutting chilled cast iron and forming tools, for ferrous and nonferrous metals, and burnishing tools.

By adding small amounts of alloying elements to W-steel types 2 and 5, certain characteristics that are desirable for specific applications are improved. The vanadium in type 2 contributes to retaining a greater degree of fine-grain structure after heat treating. Chromium in type 5 improves the deep-hardening characteristics of the steel, a property needed for large sections, and assists in maintaining the keen cutting edge that is desirable in cutting tools like broaches, reamers, threading taps, and dies.

## Mill Production Forms of Tool Steels

Tool steels are produced in many different forms, but not all those listed in the following are always readily available; certain forms and shapes are made for special orders only.

*Hot-Finished Bars and Cold-Finished Bars:* These bars are the most commonly produced forms of tool steels. Bars can be furnished in many different cross sections, the round shape being the most common. Sizes can vary over a wide range, with a more limited number of standard stock sizes. Various conditions may also be available; however, technological limitations prevent all conditions applying to every size, shape, or type of steel. Tool steel bars may be supplied in one of the following conditions and surface finishes:

*Conditions:* Hot-rolled or forged (natural); hot-rolled or forged and annealed; hot-rolled or forged and heat-treated; cold- or hot-drawn (as drawn); and cold- or hot-drawn and annealed.

*Finishes:* Hot-rolled finish (scale not removed); pickled or blast-cleaned; cold-drawn; turned or machined; rough ground; centerless ground or precision flat ground; and polished (rounds only).

Other forms in which tool steels are supplied are the following:

*Rolled or Forged Special Shapes:* These shapes are usually produced on special orders only, for the purpose of reducing material loss and machining time in the large-volume manufacture of certain frequently used types of tools.

*Forgings:* All types of tool steels may be supplied in the form of forgings that are usually specified for special shapes and for dimensions that are beyond the range covered by bars.

*Wires:* Tool steel wires are produced either by hot or cold drawing and are specified when special shapes, controlled dimensional accuracy, improved surface finish, or special mechanical properties are required. Round wire is commonly produced within an approximate size range of 0.015 to 0.500 inch (0.38 to 12.7 mm), and these dimensions also indicate the limits within which other shapes of tool steel wires, like oval, square, or rectangular, may be produced.

*Drill Rods:* Rods are produced in round, rectangular, square, hexagonal, and octagonal shapes, usually with tight dimensional tolerances to eliminate subsequent machining, thereby offering manufacturing economies for the users.

*Hot-Rolled Plates and Sheets, and Cold-Rolled Strips:* Such forms of tool steel are generally specified for the high-volume production of specific tool types.

*Tool Bits:* These pieces are semifinished tools and are used by clamping in a tool holder or shank in a manner permitting ready replacement. Tool bits are commonly made of high-speed types of tool steels, mostly in square, but also in round, rectangular, and other shapes. Tool bits are made of hot-rolled bars and are commonly, yet not exclusively, supplied in hardened and ground form, ready for use after the appropriate cutting edges are ground, usually in the user's plant.

*Hollow Bars:* These bars are generally produced by trepanning, boring, or drilling of solid round rods and are used for making tools or structural parts of annular shapes, like rolls, ring gages, bushings, etc.

**Tolerances of Dimensions.**—Such tolerances have been developed and published by the American Iron and Steel Institute (AISI) as a compilation of available industry experience that, however, does not exclude the establishment of closer tolerances, particularly for hot-rolled products manufactured in large quantities. The tolerances differ for various categories of production processes (e.g., forged, hot-rolled, cold-drawn, centerless ground) and of general shapes.

**Allowances for Machining.**—These allowances provide freedom from soft spots and defects of the tool surface, thereby preventing failures in heat treatment or in service. After a layer of specific thickness, known as the allowance, has been removed, the bar or other form of tool steel material should have a surface without decarburization and other surface defects, such as scale marks or seams. The accepted industry-wide machining allowance values for tool steels in different conditions, shapes, and size ranges are spelled out in AISI specifications and are generally also listed in the tool steel catalogs of the producer companies.

**Decarburization Limits.**—Heating of steel for production operation causes the oxidation of the exposed surfaces, resulting in the loss of carbon. That condition, called decarburization, penetrates to a certain depth from the surface, depending on the applied process, the shape and the dimensions of the product. Values of tolerance for decarburization must be considered one of the factors for defining the machining allowances, which must also compensate for expected variations of size and shape, the dimensional effects of heat treatment, and so forth. Decarburization can be present not only in hot-rolled and forged, but also in rough turned and cold-drawn conditions.

**Advances in Tool Steel-Making Technology.**—Significant advances in processes for tool steel production have been made that offer more homogeneous materials of greater density and higher purity for applications where such extremely high quality is required. Two of these methods of tool steel production are of particular interest.

*Vacuum-Melted Tool Steels:* These steels are produced by the consumable electrode method, which involves remelting of the steel originally produced by conventional processes. Inside a vacuum-tight shell that has been evacuated, the electrode cast of tool steel of the desired chemical analysis is lowered into a water-cooled copper mold where it strikes a low-voltage, high-amperage arc causing the electrode to be consumed by gradual melting. The undesirable gases and volatiles are drawn off by the vacuum, and the inclusions float on the surface of the pool, accumulating on the top of the produced ingot, to be removed later by cropping. In the field of tool steels, the consumable-electrode vacuum-melting (CVM) process is applied primarily to the production of special grades of hot-work and high-speed tool steels.

*High-Speed Tool Steels Produced by Powder Metallurgy:* The steel produced by conventional methods is reduced to a fine powder by a gas atomization process. The powder is compacted by a hot isostatic method with pressures in the range of 15,000 to 17,000 psi (103 to 117 MPa). The compacted billets are hot-rolled to the final bar size, yielding a tool-steel material which has 100 percent theoretical density. High-speed tool steels produced by the PM method offer a tool material providing increased tool wear life and high impact strength, of particular advantage in interrupted cuts.

## HARDENING, TEMPERING, AND ANNEALING

### Heat Treatment of Standard Steels

**Heat Treatment Definitions.**—This glossary of terms related to heat treating has been adopted by the American Foundry Society (AFS, formerly American Foundrymen's Association), ASM International (formerly American Society for Metals), American Society for Testing and Materials (ASTM), and the Society of Automotive Engineers (SAE). Since it is not intended to be a specification but is strictly a set of definitions, temperatures have purposely been omitted.

*Aging:* Describes a time-temperature-dependent change in the properties of certain alloys. Except for strain aging and age softening, it is the result of precipitation from a solid solution of one or more compounds whose solubility decreases with decreasing temperature. For each alloy susceptible to aging, there is a unique range of time-temperature combinations to which it will respond.

*Annealing:* A term denoting a treatment, consisting of heating to and holding at a suitable temperature followed by cooling at a suitable rate, used primarily to soften but also to simultaneously produce desired changes in other properties or in microstructure. The purpose of such changes may be, but is not confined to, improvement of machinability, facilitation of cold-working, improvement of mechanical or electrical properties, or increase in stability of dimensions. The time-temperature cycles used vary widely both in maximum temperature attained and in cooling rate employed, depending on the composition of the material, its condition, and the results desired. When applicable, the following more specific process names should be used: Black Annealing, Blue Annealing, Box Annealing, Bright Annealing, Cycle Annealing, Flame Annealing, Full Annealing, Graphitizing, Intermediate Annealing, Isothermal Annealing, Process Annealing, Quench Annealing, and Spheroidizing. When the term is used without qualification, full annealing is implied. When applied only for the relief of stress, the process is properly called stress relieving.

*Black Annealing:* Box annealing or pot annealing, used mainly for sheet, strip, or wire.

*Blue Annealing:* Heating hot-rolled sheet in an open furnace to a temperature within the transformation range and then cooling in air to soften the metal. The formation of a bluish oxide on the surface is incidental.

*Box Annealing:* Annealing in a sealed container under conditions that minimize oxidation. In box annealing, the charge is usually heated slowly to a temperature below the transformation range, but sometimes above or within it, and is then cooled slowly; this process is also called "close annealing" or "pot annealing."

*Bright Annealing:* Annealing in a protective medium to prevent discoloration of the bright surface.

*Cycle Annealing:* An annealing process employing a predetermined and closely controlled time-temperature cycle to produce specific properties or microstructure.

*Flame Annealing:* Annealing in which the heat is applied directly by a flame.

*Full Annealing:* Austenitizing and then cooling at a rate such that the hardness of the product approaches a minimum.

*Graphitizing:* Annealing in such a way that some or all of the carbon is precipitated as graphite.

*Intermediate Annealing:* Annealing at one or more stages during manufacture and before final thermal treatment.

*Isothermal Annealing:* Austenitizing and then cooling to and holding at a temperature at which austenite transforms to a relatively soft ferrite-carbide aggregate.

*Process Annealing:* An imprecise term used to denote various treatments that improve workability. For the term to be meaningful, the condition of the material and the time-temperature cycle used must be stated.

*Quench Annealing:* Annealing an austenitic alloy by *Solution Heat Treatment*.

*Spheroidizing:* Heating and cooling in a cycle designed to produce a spheroidal or globular form of carbide.

*Austempering:* Quenching from a temperature above the transformation range, in a medium having a rate of heat abstraction high enough to prevent the formation of high-temperature transformation products, and then holding the alloy, until transformation is complete, at a temperature below that of pearlite formation and above that of martensite formation.

*Austenitizing:* Forming austenite by heating into the transformation range (partial austenitizing) or above the transformation range (complete austenitizing). When used without qualification, the term implies complete austenitizing.

*Baking:* Heating to a low temperature in order to remove entrained gases.

*Bluing:* A treatment of the surface of iron-base alloys, usually in the form of sheet or strip, on which, by the action of air or steam at a suitable temperature, a thin blue oxide film is formed on the initially scale-free surface, as a means of improving appearance and resistance to corrosion. This term is also used to denote a heat treatment of springs after fabrication, to reduce the internal stress created by coiling and forming.

*Carbon Potential:* A measure of the ability of an environment containing active carbon to alter or maintain, under prescribed conditions, the carbon content of the steel exposed to it. In any particular environment, the carbon level attained will depend on such factors as temperature, time, and steel composition.

*Carbon Restoration:* Replacing the carbon lost in the surface layer from previous processing by carburizing this layer to substantially restore the original carbon level.

*Carbonitriding:* A case-hardening process in which a suitable ferrous material is heated above the lower transformation temperature in a gaseous atmosphere of such composition as to cause simultaneous absorption of carbon and nitrogen by the surface and, by diffusion, create a concentration gradient. The process is completed by cooling at a rate that produces the desired properties in the workpiece.

*Carburizing:* A process in which carbon is introduced into a solid iron-base alloy by heating above the transformation temperature range while in contact with a carbonaceous material that may be a solid, liquid, or gas. Carburizing is frequently followed by quenching to produce a hardened case.

*Case:* 1) The surface layer of an iron-base alloy that has been suitably altered in composition and can be made substantially harder than the interior or core by a process of case hardening; and 2) the hardened surface layer of a piece of steel that is large enough to have a distinctly softer core or center.

*Cementation:* The process of introducing elements into the outer layer of metal objects by means of high-temperature diffusion.

*Cold Treatment:* Exposing to suitable subzero temperatures for the purpose of obtaining desired conditions or properties, such as dimensional or microstructural stability. When the treatment involves the transformation of retained austenite, it is usually followed by a tempering treatment.

*Conditioning Heat Treatment:* A preliminary heat treatment used to prepare a material for a desired reaction to a subsequent heat treatment. For the term to be meaningful, the treatment used must be specified.

*Controlled Cooling:* A term used to describe a process by which a steel object is cooled from an elevated temperature, usually from the final hot-forming operation in a predetermined manner of cooling to avoid hardening, cracking, or internal damage.

*Core:* 1) The interior portion of an iron-base alloy that after case hardening is substantially softer than the surface layer or case; and 2) the relatively soft central portion of certain hardened tool steels.

*Critical Range or Critical Temperature Range:* Synonymous with *Transformation Range*, which is preferred.

*Cyaniding:* A process of case hardening an iron-base alloy by the simultaneous absorption of carbon and nitrogen by heating in a cyanide salt. Cyaniding is usually followed by quenching to produce a hard case.

*Decarburization:* The loss of carbon from the surface of an iron-base alloy as the result of heating in a medium that reacts with the carbon.

*Drawing:* Drawing, or drawing the temper, is synonymous with *Tempering*, which is preferable.

*Eutectic Alloy:* The alloy composition that freezes at constant temperature similar to a pure metal. The lowest melting (or freezing) combination of two or more metals. The alloy structure (homogeneous) of two or more solid phases formed from the liquid eutectically.

*Hardenability:* In a ferrous alloy, the property that determines the depth and distribution of hardness induced by quenching.

*Hardening:* Any process of increasing hardness of metal by suitable treatment, usually involving heating and cooling. See also *Aging*.

*Hardening, Case:* A process of surface hardening involving a change in the composition of the outer layer of an iron-base alloy followed by appropriate thermal treatment. Typical case-hardening processes are *Carburizing*, *Cyaniding*, *Carbonitriding*, and *Nitriding*.

*Hardening, Flame:* A process of heating the surface layer of an iron-base alloy above the transformation temperature range by means of a high-temperature flame, followed by quenching.

*Hardening, Precipitation:* A process of hardening an alloy in which a constituent precipitates from a supersaturated solid solution. See also *Aging*.

*Hardening, Secondary:* An increase in hardness following the normal softening that occurs during the tempering of certain alloy steels.

*Heating, Differential:* A heating process by which the temperature is made to vary throughout the object being heated so that, on cooling, different portions may have such different physical properties as may be desired.

*Heating, Induction:* A process of local heating by electrical induction.

*Heat Treatment:* A combination of heating and cooling operations applied to a metal or alloy in the solid state to obtain desired conditions or properties. Heating for the sole purpose of hot-working is excluded from the meaning of this definition.

*Heat Treatment, Solution:* A treatment in which an alloy is heated to a suitable temperature and held at this temperature for a sufficient length of time to allow a desired constituent to enter into solid solution, followed by rapid cooling to hold the constituent in solution. The material is then in a supersaturated, unstable state, and may subsequently exhibit *Age Hardening*.

*Homogenizing:* A high-temperature heat-treatment process intended to eliminate or to decrease chemical segregation by diffusion.

*Isothermal Transformation:* A change in phase at constant temperature.

*Malleablizing:* A process of annealing white cast iron in which the combined carbon is wholly or in part transformed to graphitic or free carbon and, in some cases, part of the carbon is removed completely. See *Temper Carbon*.

*Maraging:* A precipitation hardening treatment applied to a special group of iron-base alloys to precipitate one or more intermetallic compounds in a matrix of essentially carbon-free martensite.

*Martempering:* A hardening procedure in which an austenitized ferrous workpiece is quenched into an appropriate medium whose temperature is maintained substantially at the $M_s$ of the workpiece, held in the medium until its temperature is uniform throughout but not long enough to permit bainite to form, and then cooled in air. The treatment is followed by tempering.

*Nitriding:* A process of case hardening in which an iron-base alloy of special composition is heated in an atmosphere of ammonia or in contact with nitrogenous material. Surface hardening is produced by the absorption of nitrogen without quenching.

*Normalizing:* A process in which an iron-base alloy is heated to a temperature above the transformation range and subsequently cooled in still air at room temperature.

*Overheated:* A metal is said to have been overheated if, after exposure to an unduly high temperature, it develops an undesirably coarse grain structure but is not permanently damaged. The structure damaged by overheating can be corrected by suitable heat treatment or by mechanical work or by a combination of the two. In this respect it differs from a Burnt structure.

*Patenting:* A process of heat treatment applied to medium- or high-carbon steel in wire making prior to the wire drawing or between drafts. It consists in heating to a temperature above the transformation range, followed by cooling to a temperature below that range in air or in a bath of molten lead or salt maintained at a temperature appropriate to the carbon content of the steel and the properties required of the finished product.

*Preheating:* Heating to an appropriate temperature immediately prior to austenitizing when hardening high-hardenability constructional steels, many of the tool steels, and heavy sections.

*Quenching:* Rapid cooling. When applicable, the following more specific terms should be used: Direct Quenching, Fog Quenching, Hot Quenching, Interrupted Quenching, Selective Quenching, Slack Quenching, Spray Quenching, and Time Quenching.

*Direct Quenching:* Quenching carburized parts directly from the carburizing operation.

*Fog Quenching:* Quenching in a mist.

*Hot Quenching:* An imprecise term used to cover a variety of quenching procedures in which a quenching medium is maintained at a prescribed temperature above 160° F (71° C).

*Interrupted Quenching:* A quenching procedure in which the workpiece is removed from the first quench at a temperature substantially higher than that of the quenchant and is then subjected to a second quenching system having a different cooling rate than the first.

*Selective Quenching:* Quenching only certain portions of a workpiece.

*Slack Quenching:* The incomplete hardening of steel due to quenching from the austenitizing temperature at a rate slower than the critical cooling rate for the particular steel, resulting in the formation of one or more transformation products in addition to martensite.

*Spray Quenching:* Quenching in a spray of liquid.

*Time Quenching:* Interrupted quenching in which the duration of holding in the quenching medium is controlled.

*Soaking:* Prolonged heating of a metal at a selected temperature.

*Stabilizing Treatment:* A treatment applied to stabilize the dimensions of a workpiece or the structure of a material such as 1) before finishing to final dimensions, heating a workpiece to or somewhat beyond its operating temperature and then cooling to room temperature a sufficient number of times to ensure stability of dimensions in service; 2) transforming retained austenite in those materials that retain substantial amounts when quench hardened (see cold treatment); and 3) heating a solution-treated austenitic stainless steel that contains controlled amounts of titanium or niobium plus tantalum to a temperature below the solution heat-treating temperature to cause precipitation of finely divided, uniformly distributed carbides of those elements, thereby substantially reducing the amount of carbon available for the formation of chromium carbides in the grain boundaries on subsequent exposure to temperatures in the sensitizing range.

*Stress Relieving:* A process to reduce internal residual stresses in a metal object by heating the object to a suitable temperature and holding for a proper time at that temperature. This treatment may be applied to relieve stresses induced by casting, quenching, normalizing, machining, cold-working, or welding.

*Temper Carbon:* The free or graphitic carbon that comes out of solution usually in the form of rounded nodules in the structure during *Graphitizing* or *Malleablizing*.

*Tempering:* Heating a quench-hardened or normalized ferrous alloy to a temperature below the transformation range to produce desired changes in properties.

*Double Tempering:* A treatment in which quench-hardened steel is given two complete tempering cycles at substantially the same temperature for the purpose of ensuring completion of the tempering reaction and promoting stability of the resulting microstructure.

*Snap Temper:* A precautionary interim stress-relieving treatment applied to high hardenability steels immediately after quenching to prevent cracking because of delay in tempering them at the prescribed higher temperature.

*Temper Brittleness:* Brittleness that results when certain steels are held within, or are cooled slowly through, a certain range of temperatures below the transformation range. The brittleness is revealed by notched-bar impact tests at or below room temperature.

*Transformation Ranges or Transformation Temperature Ranges:* Those ranges of temperature within which austenite forms during heating and transforms during cooling. The two ranges are distinct, sometimes overlapping but never coinciding. The limiting temperatures of the ranges depend on the composition of the alloy and on the rate of change of temperature, particularly during cooling.

*Transformation Temperature:* The temperature at which a change in phase occurs. The term is sometimes used to denote the limiting temperature of a transformation range. The following symbols are used for iron and steels:

$Ac_{cm}$ = In hypereutectoid steel, the temperature at which the solution of cementite in austenite is completed during heating

$Ac_1$ = The temperature at which austenite begins to form during heating

$Ac_3$ = The temperature at which transformation of ferrite to austenite is completed during heating

$Ac_4$ = The temperature at which austenite transforms to delta ferrite during heating

$Ae_1, Ae_3, Ae_{cm}, Ae_4$ = The temperatures of phase changes at equilibrium

$Ar_{cm}$ = In hypereutectoid steel, the temperature at which precipitation of cementite starts during cooling

$Ar_1$ = The temperature at which transformation of austenite to ferrite or to ferrite plus cementite is completed during cooling

$Ar_3$ = The temperature at which austenite begins to transform to ferrite during cooling

$Ar_4$ = The temperature at which delta ferrite transforms to austenite during cooling

$M_s$ = The temperature at which transformation of austenite to martensite starts during cooling

$M_f$ = The temperature, during cooling, at which transformation of austenite to martensite is substantially completed

All these changes, except the formation of martensite, occur at lower temperatures during cooling than during heating, and depend on the rate of change of temperature.

**Hardness and Hardenability.**—Hardenability is the property of steel that determines the *depth and distribution of hardness* induced by quenching from the austenitizing temperature. Hardenability should not be confused with hardness as such or with maximum hardness. Hardness is a measure of the ability of a metal to resist penetration as determined by any one of a number of standard tests (Brinell, Rockwell, Vickers, etc). The maximum attainable hardness of any steel depends solely on carbon content and is not significantly affected by alloy content. Maximum hardness is realized only when the cooling rate in quenching is rapid enough to ensure full transformation to martensite.

The as-quenched surface hardness of a steel part is dependent on carbon content and cooling rate, but the *depth* to which a certain hardness level is maintained with given quenching conditions is a function of its hardenability. Hardenability is largely determined by the percentage of alloying elements in the steel; however, austenite grain size, time and temperature during austenitizing, and prior microstructure also significantly affect the hardness depth. The hardenability required for a particular part depends on size, design, and service stresses. For highly stressed parts, the best combination of strength and toughness is obtained by through hardening to a martensitic structure followed by adequate tempering. There are applications, however, where through hardening is not necessary or even desirable. For parts that are stressed principally at or near the surface, or in which wear resistance or resistance to shock loading is anticipated, a shallow hardening steel with a moderately soft core may be appropriate.

For through hardening of thin sections, carbon steels may be adequate; but as section size increases, alloy steels of increasing hardenability are required. The usual practice is to select the most economical grade that can meet the desired properties consistently. It is not good practice to utilize a higher alloy grade than necessary, because excessive use of alloying elements adds little to the properties and can sometimes induce susceptibility to quenching cracks.

*Quenching Media:* The choice of quenching media is often a critical factor in the selection of steel of the proper hardenability for a particular application. Quenching severity can be varied by selection of quenching medium, agitation control, and additives that improve the cooling capability of the quenchant. Increasing the quenching severity permits the use of less expensive steels of lower hardenability; however, consideration must also be given to the amount of distortion that can be tolerated and the susceptibility to quench cracking. In general, the more severe the quenchant and the less symmetrical the part being quenched, the greater are the size and shape changes that result from quenching and the greater is the risk of quench cracking. Consequently, although water quenching is less costly than oil quenching and water-quenched steels are less expensive than those requiring oil quenching, it is important to know that the parts being hardened can withstand the resulting distortion and the possibility of cracking.

Oil, salt, and synthetic water-polymer quenchants are also used, but they often require steels of higher alloy content and hardenability. A general rule for the selection of steel and quenchant for a particular part is that the steel should have a hardenability not exceeding that required by the severity of the quenchant selected. The carbon content of the steel should also not exceed that required to meet specified hardness and strength, because quench cracking susceptibility increases with carbon content.

The choice of quenching media is important in hardening, but another factor is agitation of the quenching bath. The more rapidly the bath is agitated, the more rapidly heat is removed from the steel and the more effective is the quench.

*Hardenability Test Methods:* The most commonly used method for determining hardenability is the end-quench test developed by Jominy and Boegehold, and described in detail in both SAE J406 and ASTM A255. In this test a normalized 1-inch (25.4-mm) round, approximately 4-inch (101.6-mm) long specimen of the steel to be evaluated is heated uniformly to its austenitizing temperature. The specimen is then removed from the furnace, placed in a jig, and immediately end quenched by a jet of room-temperature water. The water is played on the end face of the specimen, without touching the sides, until the entire specimen has cooled. Longitudinal flat surfaces are ground on opposite sides of the piece, and Rockwell C scale hardness readings are taken at $\frac{1}{16}$-inch (1.6 mm) intervals from the quenched end. The resulting data are plotted on graph paper with the hardness values as ordinates ($y$-axis) and distances from the quenched end as abscissas ($x$-axis). Representative data have been accumulated for a variety of standard steel grades and are published by SAE and AISI as "H-bands." These data show graphically and in tabular form the high and low limits applicable to each grade. The suffix H following the standard AISI/SAE numerical designation indicates that the steel has been produced to specific hardenability limits.

Experiments have confirmed that the cooling rate at a given point along the Jominy bar corresponds closely to the cooling rate at various locations in round bars of various sizes. In general, when end-quench curves for different steels coincide approximately, similar treatments will produce similar properties in sections of the same size. On occasion it is necessary to predict the end-quench hardenability of a steel not available for testing, and reasonably accurate means of calculating hardness for any Jominy location on a section of steel of known analysis and grain size have been developed.

*Tempering:* As-quenched steels are in a highly stressed condition and are seldom used without tempering. Tempering imparts plasticity or toughness to the steel and is inevitably accompanied by a loss in hardness and strength. The loss in strength, however, is only incidental to the very important increase in toughness, which is due to the relief of residual stresses induced during quenching and to precipitation, coalescence, and spheroidization of iron and alloy carbides resulting in a microstructure of greater plasticity.

Alloying slows the tempering rate, so that alloy steel requires a higher tempering temperature to obtain a given hardness than carbon steel of the same carbon content. The higher tempering temperature for a given hardness permits a greater relaxation of residual stress and thereby improves the steel's mechanical properties. Tempering is done in furnaces or in oil or salt baths at temperatures varying from 300 to 1200° F (149 to 649° C). With most grades of alloy steel, the range between 500 and 700° F (260 to 371° C) is avoided because of a phenomenon known as "blue brittleness," which reduces impact properties. Tempering the martensitic stainless steels in the range of 800–1100° F (427 to 593° C) is not recommended because of the low and erratic impact properties and reduced corrosion resistance that result. Maximum toughness is achieved at higher temperatures. It is important to temper parts as soon as possible after quenching because any delay greatly increases the risk of cracking resulting from the high-stress condition in the as-quenched part.

**Surface Hardening (Case Hardening).**—Many applications require high hardness or strength primarily at the surface, and complex service stresses frequently require not only a hard, wear-resistant surface, but also core strength and toughness to withstand impact stress.

To achieve these different properties, two general processes are used: 1) The chemical composition of the surface is altered, prior to or after quenching and tempering; the processes used include carburizing, nitriding, cyaniding, and carbonitriding; and 2) Only the surface layer is hardened by the heating and quenching process; the most common processes used for surface hardening are flame hardening and induction hardening.

*Carburizing:* Carbon is diffused into the part's surface to a controlled depth by heating the part in a carbonaceous medium. The resulting depth of carburization, commonly referred to as case depth, depends on the carbon potential of the medium used and the time and temperature of the carburizing treatment. The steels most suitable for carburizing to enhance toughness are those with sufficiently low carbon contents, usually below 0.3 percent. Carburizing temperatures range from 1550 to 1750° F (843 to 954° C), with the temperature and time at temperature adjusted to obtain various case depths. Steel selection, hardenability, and type of quench are determined by section size, desired core hardness, and service requirements.

Three types of carburizing are most often used: 1) *Liquid carburizing* involves heating the steel in molten barium cyanide or sodium cyanide. The case absorbs some nitrogen in addition to carbon, thus enhancing surface hardness; 2) *Gas carburizing* involves heating the steel in a gas of controlled carbon content. When used, the carbon level in the case can be closely controlled; and 3) *Pack carburizing,* which involves sealing both the steel and solid carbonaceous material in a gas-tight container, then heating this combination.

With any of these methods, the part may be either quenched after the carburizing cycle without reheating or air cooled followed by reheating to the austenitizing temperature prior to quenching. The case depth may be varied to suit the conditions of loading in service. However, service characteristics frequently require that only selective areas of a part have to be case hardened. Covering the areas not to be cased, with copper plating or a layer of commercial paste, allows the carbon to penetrate only the exposed areas. Another method involves carburizing the entire part, then removing the case in selected areas by machining, prior to quench hardening.

*Nitriding:* The steel part is heated to a temperature of 900–1150° F (482 to 621° C) in an atmosphere of ammonia gas and dissociated ammonia for an extended period of time that depends on the case depth desired. A thin, very hard case results from the formation of nitrides. Strong nitride-forming elements (chromium and molybdenum) are required to be present in the steel, and often special nonstandard grades containing aluminum (a strong nitride former) are used. The major advantage of this process is that parts can be quenched and tempered, then machined, prior to nitriding, because only a little distortion occurs during nitriding.

*Cyaniding:* This process involves heating the part in a bath of sodium cyanide to a temperature slightly above the transformation range, followed by quenching, to obtain a thin case of high hardness.

*Carbonitriding:* This process is similar to cyaniding except that the absorption of carbon and nitrogen is accomplished by heating the part in a gaseous atmosphere containing hydrocarbons and ammonia. Temperatures of 1425–1625° F (774 to 885° C) are used for parts to be quenched, and lower temperatures, 1200–1450° F (649 to 788° C), may be used where a liquid quench is not required.

*Flame Hardening:* This process involves rapid heating with a direct high-temperature gas flame, such that the surface layer of the part is heated above the transformation range, followed by cooling at a rate that causes the desired hardening. Steels for flame hardening are usually in the range of 0.30–0.60 percent carbon, with hardenability appropriate for the case depth desired and the quenchant used. The quenchant is usually sprayed on the surface a short distance behind the heating flame. Immediate tempering is required and may be done in a conventional furnace or by a flame-tempering process, depending on part size and costs.

*Induction Hardening:* This process is similar in many respects to flame hardening except that the heating is caused by a high-frequency electric current sent through a coil or inductor surrounding the part. The depth of heating depends on the frequency, the rate of heat conduction from the surface, and the length of the heating cycle. Quenching is usually accomplished with a water spray introduced at the proper time through jets in or near the inductor block or coil. In some instances, however, parts are oil quenched by immersing them in a bath of oil after they reach the hardening temperature.

**Structure of Fully Annealed Carbon Steel.—**In carbon steel that has been fully annealed, there are normally present, apart from such impurities as phosphorus and sulfur, two constituents: the element iron in a form metallurgically known as *ferrite* and the chemical compound iron carbide in the form metallurgically known as *cementite*. This latter constituent consists of 6.67 percent carbon and 93.33 percent iron. A certain proportion of these two constituents will be present as a mechanical mixture. This mechanical mixture, the amount of which depends on the carbon content of the steel, consists of alternate bands or layers of ferrite and cementite. Under the microscope, the matrix frequently has the appearance of mother-of-pearl and hence has been named *pearlite*. Pearlite contains about 0.85 percent carbon and 99.15 percent iron, neglecting impurities. A fully annealed steel containing 0.85 percent carbon would consist entirely of pearlite. Such a steel is known as *eutectoid* steel and has a laminated structure characteristic of a eutectic alloy. Steel that has less than 0.85 percent carbon (*hypoeutectoid* steel) has an excess of ferrite above that required to mix with the cementite present to form pearlite; hence, both ferrite and pearlite are present in the fully annealed state. Steel having a carbon content greater than 0.85 percent (*hypereutectoid* steel) has an excess of cementite over that required to mix with the ferrite to form pearlite; hence, both cementite and pearlite are present in the fully annealed state. The structural constitution of carbon steel in terms of ferrite, cementite, pearlite and austenite for different carbon contents and at different temperatures is illustrated in Fig. 1.

**Effect of Heating Fully Annealed Carbon Steel.—**When carbon steel in the fully annealed state is heated above the lower critical point, which is some temperature in the range of 1335 to 1355° F (724 to 735° C) depending on the carbon content, the alternate bands or layers of ferrite and cementite that make up the pearlite begin to merge into each other. This process continues until the pearlite is thoroughly "dissolved," forming what is known as *austenite*. If the temperature of the steel continues to rise and there is present, in addition to the pearlite, any excess ferrite or cementite, this also will begin to dissolve into the austenite until finally only austenite will be present. The temperature at which the excess ferrite or cementite is completely dissolved in the austenite is called the *upper critical point*. This temperature varies with the carbon content of the steel much more widely than the lower critical point (see Fig. 1).

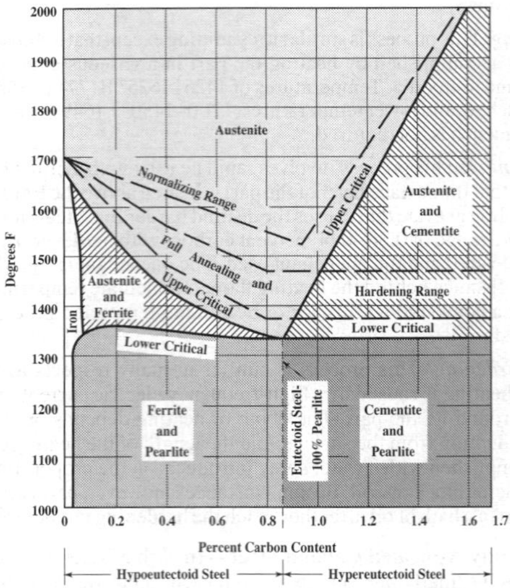


Fig. 1. Phase Diagram of Carbon Steel

**Effect of Slow Cooling on Carbon Steel.—**If carbon steel that has been heated to the point where it consists entirely of austenite is slowly cooled, the process of transformation that took place during the heating will be reversed, but the upper and lower critical points will occur at somewhat lower temperatures than they do on heating. Assuming that the steel was originally fully annealed, its structure on returning to atmospheric temperature after slow cooling will be the same as before in terms of the proportions of ferrite or cementite and pearlite present. The austenite will have entirely disappeared.

**Effect of Rapid Cooling or Quenching on Carbon Steel.—**Observations have shown that as the rate at which carbon steel is cooled from an austenitic state is increased, the temperature at which the austenite begins to change into pearlite drops more and more below the slow cooling transformation temperature of about 1300° F (704° C). For example, a 0.80 percent carbon steel that is cooled at such a rate that the temperature drops 500° F (260° C drop) in one second will show transformation of austenite beginning at 930° F (500° C). As the cooling rate is increased, the laminations of the pearlite formed by the transformation of the austenite become finer and finer up to the point where they cannot be detected under a high-power microscope, while the steel itself increases in hardness and tensile strength. As the rate of cooling is still further increased, this transformation temperature suddenly drops to around 500° F (260° C) or lower, depending on the carbon content of the steel. The cooling rate at which this sudden drop in transformation temperature takes place is called the *critical cooling rate*. When a piece of carbon steel is quenched at this rate or faster, a new structure is formed. The austenite is transformed into *martensite*, which is characterized by an angular needle-like structure and a very high hardness.

If carbon steel is subjected to a severe quench or to extremely rapid cooling, a small percentage of the austenite, instead of being transformed into martensite during the quenching operation, may be retained. Over a period of time, however, this remaining austenite tends to be gradually transformed into martensite even though the steel is not subjected to further heating or cooling. Martensite has a lower density than austenite, and such a change, or "aging" as it is called, often results in an appreciable increase in volume or "growth" and the setting up of new internal stresses in the steel.

**Steel Heat-Treating Furnaces.**—Various types of furnaces heated by gas, oil, or electricity are used for the heat treatment of steel. These furnaces include the oven or box type in various modifications for "in-and-out" or for continuous loading and unloading; the retort type; the pit type; the pot type; and the salt-bath electrode type.

*Oven or Box Furnaces:* This type of furnace has a box or oven-shaped heating chamber. The "in-and-out" oven furnaces are loaded by hand or by a track-mounted car that, when rolled into the furnace, forms the bottom of the heating chamber. The car type is used where heavy or bulky pieces must be handled. Some oven-type furnaces are provided with a full muffle or a semimuffle, which is an enclosed refractory chamber into which the parts to be heated are placed. The full-muffle, being fully enclosed, prevents any flames or burning gases from coming in contact with the work and permits a special atmosphere to be used to protect or condition the work. The semimuffle, which is open at the top, protects the work from direct impingement of the flame although it does not shut off the work from the hot gases. In the direct-heat-type oven furnace, the work is open to the flame. In the electric oven furnace, a retort is provided when gas atmospheres are to be employed to confine the gas and prevent it from attacking the heating elements. Where muffles are used, they must be replaced periodically, and a greater amount of fuel is required than in a direct-heat type of oven furnace.

For continuous loading and unloading, there are several types of furnaces, such as rotary hearth car; roller-, furnace belt-, walking-beam, or pusher-conveyor; and a continuous-kiln-type through which track-mounted cars are run. In the continuous type of furnace, the work may pass through several zones that are maintained at different temperatures for preheating, heating, soaking, and cooling.

*Retort Furnace:* This is a vertical type of furnace provided with a cylindrical metal retort into which the parts to be heat treated are suspended either individually, if large enough, or in a container of some sort. The use of a retort permits special gas atmospheres to be employed for carburizing, nitriding, etc.

*Pit-Type Furnace:* This is a vertical furnace arranged for the loading of parts in a metal basket. The parts within the basket are heated by convection, and when the basket is lowered into place, it fits into the furnace chamber in such a way as to provide a dead-air space to prevent direct heating.

*Pot-Type Furnace:* This furnace is used for the immersion method of heat treating small parts. A cast-alloy pot is employed to hold a bath of molten lead or salt in which the parts are placed for heating.

*Salt Bath Electrode Furnace:* In this type of electric furnace, heating is accomplished by means of electrodes suspended directly in the salt bath. The patented grouping and design of electrodes provide an electromagnetic action that results in an automatic stirring action. This stirring tends to produce an even temperature throughout the bath.

*Vacuum Furnace:* Vacuum heat treatment is a relatively new development in metallurgical processing, with a vacuum substituting for the more commonly used protective gas atmospheres. The most commonly used furnace is the "cold wall" type, consisting of a water-cooled vessel that is maintained near ambient temperature during operation. During quenching, the chamber is backfilled up to or above atmospheric pressure with an inert gas, which is circulated by an internal fan. When even faster cooling rates are needed, furnaces are available with capability for liquid quenching, performed in an isolated chamber.

*Fluidized-Bed Furnace:* Fluidized-bed techniques are not new; however, new furnace designs have extended the technology into the temperature ranges required for most common heat treatments. In fluidization, a bed of dry, finely divided particles, typically aluminum oxide, is made to behave like a liquid by feeding gas upward through the bed. An important characteristic of the bed is high-efficiency heat transfer. Applications include continuous or batch-type units for all general heat treatments.

**Physical Properties of Heat-Treated Steels.—**Steels that have been "fully hardened" to the same hardness when quenched will have about the same tensile and yield strengths regardless of composition and alloying elements. When the hardness of such a steel is known, it is also possible to predict its reduction of area and tempering temperature. The accompanying figures illustrating these relationships have been prepared by the Society of Automotive Engineers.

Fig. 2 gives the range of Brinell Hardness Numbers (BHN) that could be expected for any particular tensile strength, or it may be used to determine the range of tensile strengths that would correspond to any particular hardness. Fig. 3 shows the relationship between the tensile strength, hardness, and the yield point. The solid line is the normal-expectancy curve. The dotted-line curves give the range of the variation of scatter of the plotted data. Fig. 4 shows the relationship that exists between the tensile strength (or hardness) and the reduction of area. The curve to the left represents the alloy steels and that on the right the carbon steels. Both are normal-expectancy curves and the extremities of the perpendicular lines that intersect them represent the variations from the normal-expectancy curves that may be caused by quality differences and by the magnitude of parasitic stresses induced by quenching. Fig. 5 shows the relationship between the hardness (or approximately equivalent tensile strength) and the tempering temperature. Three curves are given, one for fully hardened steels with a carbon content between 0.40 and 0.55 percent, one for fully hardened steels with a carbon content between 0.30 and 0.40 percent, and one for steels that are not fully hardened.

From Fig. 2, it can be seen that for a tensile strength of, say, 200,000 pounds per square inch (1379 MPa), the Brinell Hardness Number could range between 375 and 425. By taking 400 as the mean hardness value and using Fig. 5, it can be seen that the tempering temperature of fully hardened steels of 0.40 to 0.55 percent carbon content would be 990° F (532° C) and that of fully hardened steels of 0.30 to 0.40 percent carbon would be 870° F (466° C). This chart also shows that the tempering temperature for a steel not fully hardened would approach 520° F (271° C). A yield point of $0.9 \times 200{,}000$, or 180,000 pounds per square inch (1241 MPa) is indicated (Fig. 3) for the fully hardened steel with a tensile strength of 200,000 pounds per square inch. Most alloy steels of 200,000 pounds per square inch tensile strength would probably have a reduction in area of close to 44 percent (Fig. 4), but some would have values in the range of 35 to 53 percent. Carbon steels of the same tensile strength would probably have a reduction in area of close to 24 percent but could possibly range from 17 to 31 percent.

Fig. 3 and Fig. 4 represent steel in the quenched and tempered condition, and Fig. 2 represents steel in the hardened and tempered, as-rolled, annealed, and normalized conditions. These charts give a good general indication of mechanical properties; however, more exact information when required should be obtained from tests on samples of the individual heats of steel under consideration.

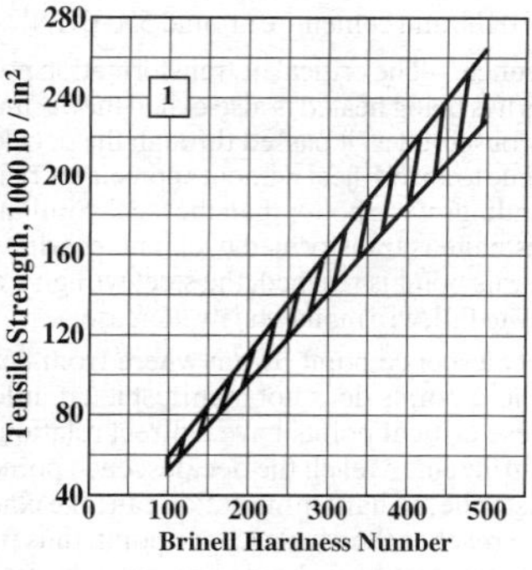


Fig. 2.

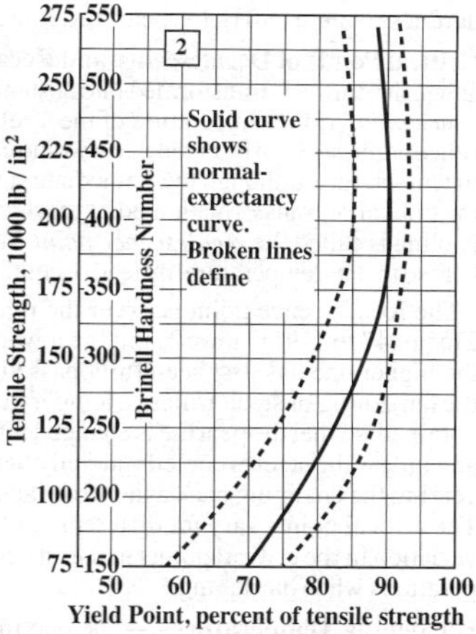


Fig. 3.

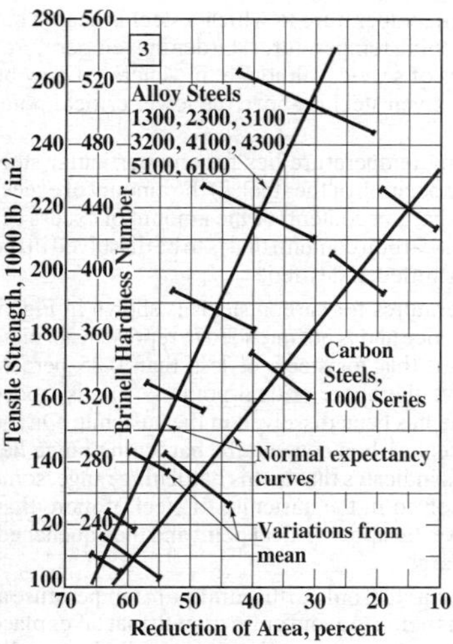


Fig. 4.

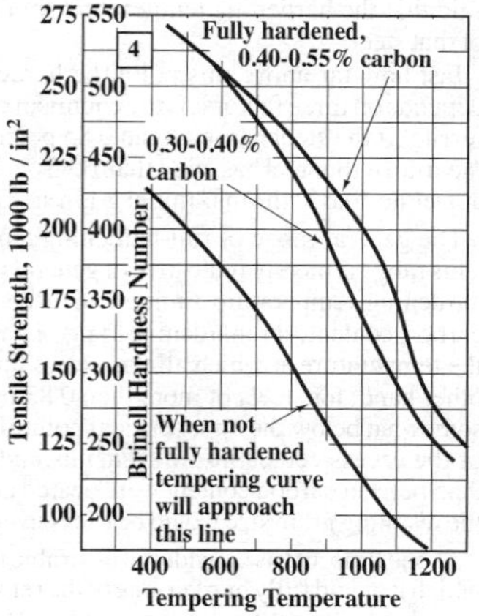


Fig. 5.

## Hardening

**Basic Steps in Hardening.**—The operation of hardening steel consists fundamentally of two steps. The first step is to heat the steel to some temperature above its transformation point (usually at least 100° F or 56° C above) so that it becomes entirely austenitic in structure. The second step is to quench the steel at some rate faster than the critical rate (which depends on the carbon content, the amounts of alloying elements present other than carbon, and the grain size of the austenite) to produce a martensitic structure. The hardness of a martensitic steel depends on its carbon content and ranges from about 460 BHN at

0.20 percent carbon to about 710 BHN above 0.50 carbon. In comparison, ferrite has a hardness of about 90 BHN, pearlite about 240 BHN, and cementite around 550 BHN.

**Critical Points of Decalescence and Recalescence.**—The critical or transformation point at which pearlite is transformed into austenite as it is being heated is also called the *decalescence point*. If the temperature of the steel was observed as it passed through the decalescence point, it would be noted that it would continue to absorb heat without appreciably rising in temperature, although the immediate surroundings were hotter than the steel. Similarly, the critical or transformation point at which austenite is transformed back into pearlite on cooling is called the *recalescence point*. When this point is reached, the steel will give out heat so that its temperature, instead of continuing to fall, will momentarily increase.

The recalescence point is lower than the decalescence point by anywhere from 85 to 215° F (47 to 119° C lower), and the lower of these points does not manifest itself unless the higher one has first been fully passed. These critical points have a direct relation to the hardening of steel. Unless a temperature sufficient to reach the decalescence point is obtained, so that the pearlite is changed into austenite, no hardening action can take place; and unless the steel is cooled suddenly before it reaches the recalescence point, thus preventing the changing back again from austenite to pearlite, no hardening can take place. The critical points vary for different kinds of steel and must be determined by tests. The variation in the critical points makes it necessary to heat different steels to different temperatures when hardening.

**Hardening Temperatures.**—The maximum temperature to which a steel is heated before quenching to harden it is called the hardening temperature. Hardening temperatures vary for different steels and different classes of service, although, in general, it may be said that the hardening temperature for any given steel is above the lower critical point of that steel.

Just how far above this point the hardening temperature lies for any particular steel depends on three factors: 1) the chemical composition of the steel; 2) the amount of excess ferrite (if the steel has less than 0.85 percent carbon content) or the amount of excess cementite (if the steel has more than 0.85 percent carbon content) that is to be dissolved in the austenite; and 3) the maximum grain size permitted, if desired.

The general range of full-hardening temperatures for carbon steels is shown in Fig. 5. This range is merely indicative of general practice and is not intended to represent absolute hardening temperature limits. It can be seen that for steels of less than 0.85 percent carbon content, the hardening range is above the upper critical point — that is, above the temperature at which all the excess ferrite has been dissolved in the austenite. On the other hand, for steels of more than 0.85 percent carbon content, the hardening range lies somewhat below the upper critical point. This indicates that in this hardening range, some of the excess cementite still remains undissolved in the austenite. If steel of more than 0.85 percent carbon content were heated above the upper critical point and then quenched, the resulting grain size would be excessively large.

At one time, it was considered desirable to heat steel only to the minimum temperature at which it would fully harden, one of the reasons being to avoid grain growth that takes place at higher temperature. It is now realized that no such rule as this can be applied generally since there are factors other than hardness that must be taken into consideration. For example, in many cases, toughness can be impaired by too low a temperature just as much as by too high a temperature. It is true, however, that excessive hardening temperatures result in warpage, distortion, increased scale, and decarburization.

**Hardening Temperatures for Carbon Tool Steels.**—The best hardening temperatures for any given tool steel are dependent on the type of tool and the intended class of service. Wherever possible, the specific recommendations of the tool steel manufacturer should be followed. General recommendations for hardening temperatures of carbon tool steels based on carbon content are as follows: For steel of 0.65 to 0.80 percent carbon content, 1450 to 1550° F (788 to 843° C); for steel of 0.80 to 0.95 percent carbon content, 1410 to

1460° F (766 to 793° C); for steel of 0.95 to 1.10 percent carbon content, 1390 to 1430° F (754 to 777° C); and for steels of 1.10 percent and over carbon content, 1380 to 1420° F (749 to 771° C). For a given hardening temperature range, the higher temperatures tend to produce deeper hardness penetration and increased compressional strength, whereas the lower temperatures tend to result in shallower hardness penetration but increased resistance to splitting or bursting stresses.

**Determining Hardening Temperatures.—**A hardening temperature can be specified directly, or it may be specified indirectly as a certain temperature rise above the lower critical point of the steel. Where the temperature is specified directly, a pyrometer of the type that indicates the furnace temperature or a pyrometer of the type that indicates the work temperature may be employed. If the pyrometer shows furnace temperature, care must be taken to allow sufficient time for the work to reach the furnace temperature after the pyrometer indicates that the required hardening temperature has been attained. If the pyrometer indicates work temperature, then, where the workpiece is large, time must be allowed for the interior of the work to reach the temperature of the surface, which is the temperature indicated by the pyrometer.

Where the hardening temperature is specified as a given temperature rise above the critical point of the steel, a pyrometer that indicates the temperature of the work should be used. The critical point, as well as the given temperature rise, can be more accurately determined with this type of pyrometer. As the work is heated, its temperature, as indicated by the pyrometer, rises steadily until the lower critical or decalescence point of the steel is reached. At this point, the temperature of the work ceases to rise and the pyrometer indicating or recording pointer remains stationary or fluctuates slightly. After a certain elapsed period, depending on the heat input rate, the internal changes in structure of the steel that take place at the lower critical point are completed and the temperature of the work again begins to rise. A small fluctuation in temperature may occur in the interval during which structural changes are taking place, so for uniform practice, the critical point may be considered as the temperature at which the pointer first becomes stationary.

**Heating Steel in Liquid Baths.—**The liquid bath commonly used for heating steel tools preparatory to hardening are molten lead, sodium cyanide, barium chloride, a mixture of barium and potassium chloride, and other metallic salts. The molten substance is retained in a crucible or pot, and the heat required may be obtained from gas, oil, or electricity. The principal advantages of heating baths are as follows: No part of the work can be heated to a temperature above that of the bath; the temperature can be easily maintained at whatever degree has proved, in practice, to give the best results; the submerged steel can be heated uniformly; and the finished surfaces are protected against oxidation.

**Salt Baths.—**Molten baths of various salt mixtures or compounds are used extensively for heat-treating operations, such as hardening and tempering; they are also utilized for annealing ferrous and nonferrous metals. Commercial salt-bath mixtures are available that meet a wide range of temperature and other metallurgical requirements. For example, there are neutral baths for heating tool and die steels without carburizing the surfaces; baths for carburizing the surfaces of low-carbon steel parts; baths adapted for the usual tempering temperatures of, say, 300 to 1100° F (approx. 150 to 595° C); and baths that may be heated to temperatures up to approximately 2400° F (1315° C) for hardening high-speed steels. Salt baths are also adapted for local or selective hardening, the type of bath being selected to suit the requirements. For example, a neutral bath may be used for annealing the ends of tubing or other parts, or an activated cyanide bath for carburizing the ends of shafts or other parts. Surfaces that are not to be carburized are protected by copper plating. When the work is immersed, the unplated surfaces are subjected to the carburizing action.

Baths may consist of a mixture of sodium, potassium, barium, and calcium chlorides or nitrates of sodium, potassium, barium, and calcium in varying proportions, to which sodium carbonate and sodium cyanide are sometimes added to prevent decarburization. Various proportions of these salts provide baths of different properties. Potassium cyanide is seldom used, as sodium cyanide costs less. The specific gravity of a salt bath is not as high as that of a lead bath; consequently, the work may be suspended in a salt bath and does not have to be held below the surface as in a lead bath.

**The Lead Bath.**—The lead bath is extensively used but is not adapted to the high temperatures required for hardening high-speed steel, as it begins to vaporize at about 1190° F (645° C). As the temperature increases, the lead volatilizes and gives off poisonous vapors; hence, lead furnaces should be equipped with hoods to carry away the fumes. Lead baths are generally used for temperatures below 1500 or 1600° F (815 or 870° C). They are often employed for heating small pieces that must be hardened in quantities. It is important to use pure lead that is free from sulfur. The work should be preheated before plunging it into the molten lead.

**Defects in Hardening.**—Uneven heating is the cause of most defects in hardening. Cracks of a circular form, from the corners or edges of a tool, indicate uneven heating in hardening. Cracks of a vertical nature and dark-colored fissures indicate that the steel has been burned and should be put on the scrap heap. Tools that have hard and soft places have been either unevenly heated, unevenly cooled, or "soaked," a term used to indicate prolonged heating. A tool not thoroughly moved about in the hardening fluid will show hard and soft places, and have a tendency to crack. Tools that are hardened by dropping them to the bottom of the tank sometimes have soft places, owing to contact with the floor or sides.

**Scale on Hardened Steel.**—The formation of scale on the surface of hardened steel is due to the contact of oxygen with the heated steel; hence, to prevent scale, the heated steel must not be exposed to the action of the air. When using an oven heating furnace, the flame should be so regulated that it is not visible in the heating chamber. The heated steel should be exposed to the air as little as possible when transferred from the furnace to the quenching bath. An old method of preventing scale and retaining a fine finish on dies used in jewelry manufacture, small taps, etc., is as follows: Fill the die impression with powdered boracic acid and place near the fire until the acid melts; then add a little more acid to ensure all the surfaces are covered. The die is then hardened in the usual way. If the boracic acid does not come off entirely in the quenching bath, immerse the work in boiling water. Dies hardened by this method are said to be as durable as those heated without the acid.

**Hardening or Quenching Baths.**—The purpose of a quenching bath is to remove heat from the steel being hardened at a rate that is faster than the critical cooling rate. Generally speaking, the more rapid the rate of heat extraction above the cooling rate, the higher will be the resulting hardness. To obtain the different rates of cooling required by different classes of work, baths of various kinds are used. These include plain or fresh water, brine, caustic soda solutions, oils of various classes, oil-water emulsions, baths of molten salt or lead for high-speed steels, and air cooling for some high-speed steel tools when a slow rate of cooling is required. To minimize distortion and cracking where such tendencies are present, without sacrificing depth-of-hardness penetration, a quenching medium should be selected that will cool rapidly at the higher temperatures and more slowly at the lower temperatures, that is below 750° F (400° C). Oil quenches in general meet this requirement.

*Oil-Quenching Baths:* Oil is used very extensively as a quenching medium, as it results in a good proportion of hardness, toughness, and freedom from warpage when used with standard steels. Oil baths are used extensively for alloy steels. Various kinds of oils are employed, such as prepared mineral oils and vegetable, animal, and fish oils, either singly or in combination. Prepared mineral quenching oils are widely used because they have good quenching characteristics, are chemically stable, do not have an objectionable odor, and are relatively inexpensive. Special compounded oils of the soluble type are used in many plants instead of such oils as fish oil, linseed oil, cottonseed oil, etc. The soluble properties enable the oil to form an emulsion with water.

Oil cools steel at a slower rate than water, but the rate is fast enough for alloy steel. Oils have different cooling rates, however, and this rate may vary through the initial and final stages of the quenching operation. Faster cooling in the initial stage and slower cooling at lower temperatures are preferable because there is less danger of cracking the steel. The temperature of quenching oil baths should range ordinarily between 90 and 130° F (32 and 55° C). A fairly constant temperature may be maintained either by circulating the oil through cooling coils or by using a tank provided with a cold-water jacket.

A good quenching oil should possess a flash and fire point sufficiently high to be safe under the conditions used. 350° F (175° C) should be about the minimum point. The specific heat of the oil regulates the hardness and toughness of the quenched steel; and the greater the specific heat, the higher will be the hardness produced. Specific heats of quenching oils vary from 0.20 to 0.75, the specific heats of fish, animal, and vegetable oils usually being from 0.2 to 0.4, and of soluble and mineral oils from 0.5 to 0.7. The efficient temperature range for quenching oil is from 90 to 140° F (32 to 60° C).

**Quenching in Water.**—Many carbon tool steels are hardened by immersing them in a bath of fresh water, but water is not an ideal quenching medium. Contact between the water and work and the cooling of the hot steel are impaired by the formation of gas bubbles or an insulating vapor film, especially in holes, cavities, or pockets. The results are uneven cooling and sometimes excessive strains, which may cause the tool to crack; in fact, there is greater danger of cracking in a fresh-water bath than in one containing salt water or brine.

In order to secure more even cooling and reduce the danger of cracking, either rock salt (8 or 9 percent) or caustic soda (3 to 5 percent) may be added to the bath to eliminate or prevent the formation of a vapor film or gas pockets, thus promoting rapid early cooling. Brine is commonly used and ¾ pound of rock salt per gallon (90 gm per liter) of water is equivalent to about 8 percent of salt. Brine is not inherently a more severe or drastic quenching medium than plain water, although it may seem to be because the brine makes better contact with the heated steel, and cooling is more effective. In still-bath quenching, a slow up-and-down movement of the tool is preferable to a violent swishing around.

The temperature of water-base quenching baths should preferably be kept around 70° F (21° C), but up to 90 or 100° F (32 to 38° C) is a safe range. The temperature of the hardening bath has a great deal to do with the hardness obtained. The higher the temperature of the quenching water, the more nearly does its effect approach that of oil; and if boiling water is used for quenching, it will have an effect even more gentle than that of oil — in fact, it would leave the steel nearly soft. Parts of irregular shape are sometimes quenched in a water bath that has been warmed somewhat to prevent sudden cooling and cracking.

When water is used, it should be "soft" because unsatisfactory results will be obtained with "hard" water. Any contamination of water-base quenching liquids by soap tends to decrease their rate of cooling. A water bath having 1 or 2 inches (2.5 to 5 cm) of oil on the top is sometimes employed to advantage for quenching tools made of high-carbon steel, as the oil through which the work first passes reduces the sudden quenching action of the water.

The bath should be amply large to dissipate the heat rapidly, and the temperature should be kept about constant so that successive pieces will be cooled at the same rate. Irregularly shaped parts should be immersed so that the heaviest or thickest section enters the bath first. After immersion, the part to be hardened should be agitated in the bath; the agitation reduces the tendency of the formation of a vapor coating on certain surfaces, and a more uniform rate of cooling is obtained. The work should never be dropped to the bottom of the bath until quite cool.

*Flush or Local Quenching by Pressure-Spraying:* When dies for cold heading, drawing, extruding, etc., or other tools, require a hard working surface and a relatively soft but tough body, the quenching may be done by spraying water under pressure against the interior or other surfaces to be hardened. Special spraying fixtures are used to hold the tool and apply the spray where the hardening is required. The pressure spray prevents the formation of gas pockets previously referred to in connection with the fresh-water quenching bath; hence, fresh water is effective for flush quenching and there is no advantage in using brine.

**Quenching in Molten Salt Bath.—**A molten salt bath may be used in preference to oil for quenching high-speed steel. The object in using a liquid salt bath for quenching (instead of an oil bath) is to obtain maximum hardness with minimum cooling stresses and distortion that might result in cracking expensive tools, especially if there are irregular sections. The temperature of the quenching bath may be around 1100 or 1200° F (595 to 650° C). Quenching is followed by cooling to room temperature and then the tool is tempered or drawn in a bath having a temperature range of 950 to 1100° F (510 to 593° C). In many cases, the tempering temperature is about 1050° F (566° C).

**Tanks for Quenching Baths.—**The main point to be considered in a quenching bath is to keep it at a uniform temperature so that successive pieces quenched will be subjected to the same heat treatment. The next consideration is to keep the bath agitated so that it will not be of different temperatures in different places; if thoroughly agitated and kept in motion, as is the case with the bath shown in Fig. 6, it is not even necessary to keep the pieces in motion in the bath, as steam will not be likely to form around the pieces quenched. Experience has proved that if a piece is held still in a thoroughly agitated bath, it will come out much straighter than if it has been moved around in an unagitated bath, an important consideration, especially when hardening long pieces. It is, besides, no easy matter to keep heavy and long pieces in motion unless by mechanical means.

Fig. 6 illustrates a water or brine tank for quenching baths. Water is forced by a pump or other means through the supply pipe into the intermediate space between the outer and inner tank. From the intermediate space, it is forced into the inner tank through holes as indicated. The water returns to the storage tank by overflowing from the inner tank into the outer one and then through the overflow pipe as indicated. Fig. 8 illustrates another water or brine tank of a more common type. In this case, the water or brine is pumped from the storage tank and continuously returned to it. If the storage tank contains a large volume of water, there is no need for a special means for cooling. Otherwise, arrangements must be made for cooling the water after it has passed through the tank. The bath is agitated by the force with which the water is pumped into it. The holes at A are drilled at an angle, so as to throw the water toward the center of the tank. Fig. 7 shows an oil-quenching tank in which water is circulated in an outer surrounding tank to keep the oil bath cool. Air is forced into the oil bath to keep it agitated. Fig. 9 shows the ordinary type of quenching tank cooled by water forced through a coil of pipe. This arrangement can be used for oil, water, or brine. Fig. 10 shows a similar type of quenching tank, but with two coils of pipe. Waterflows through one of these and steam through the other. By these means, it is possible to keep the bath at a constant temperature.

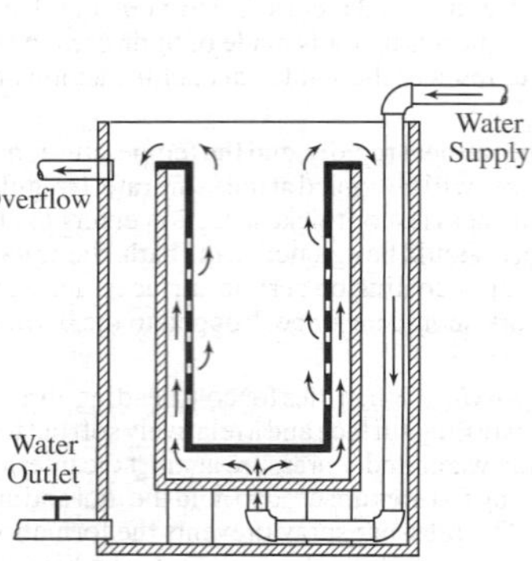

Fig. 6.

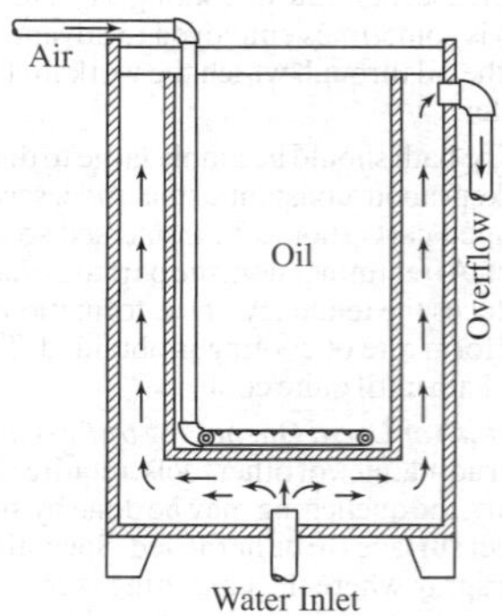

Fig. 7.

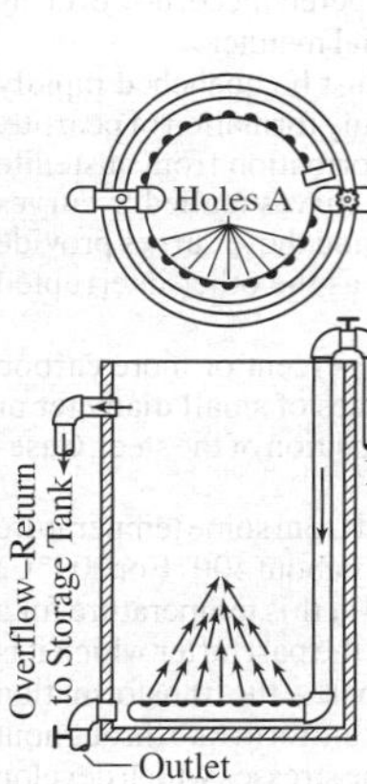


Fig. 8.

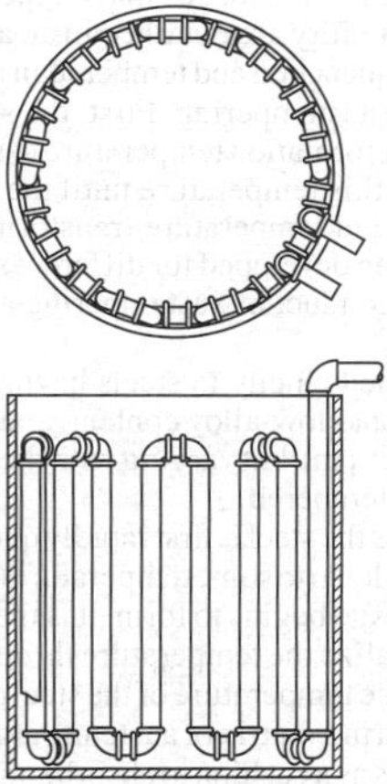

Fig. 9.

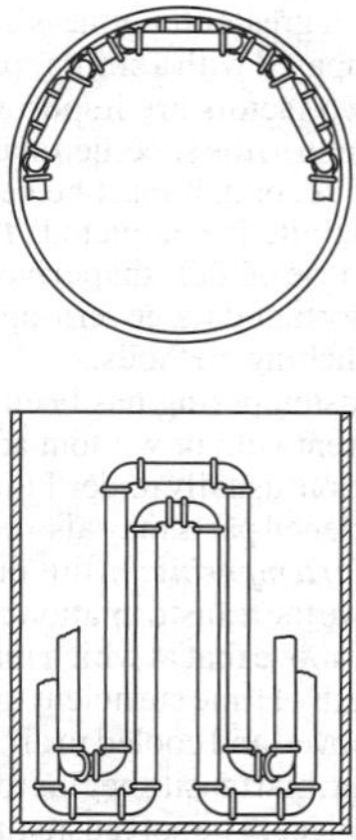

Fig. 10.

**Interrupted Quenching.—***Austempering, martempering, and isothermal quenching* are three methods of interrupted quenching that have been developed to obtain greater toughness and ductility for given hardnesses and to avoid the difficulties of quench cracks, internal stresses, and warpage, frequently experienced when the conventional method of quenching steel directly and rapidly from above the transformation point to atmospheric temperature is employed. In each of these three methods, quenching is begun when the work has reached some temperature above the transformation point and is conducted at a rate faster than the critical rate. The rapid cooling of the steel is interrupted, however, at some temperature above that at which martensite begins to form. The three methods differ in the temperature range at which interruption of the rapid quench takes place, the length of time that the steel is held at this temperature, and whether the subsequent cooling to atmospheric temperature is rapid or slow, and is or is not preceded by a tempering operation.

One of the reasons for maintaining the steel at a constant temperature for a definite period of time is to permit the inside sections of the piece to reach the same temperature as the outer sections so that when transformation of the structure does take place, it will occur at about the same rate and period of time throughout the piece. In order to maintain the constant temperature required in interrupted quenching, a quenching arrangement for absorbing and dissipating a large quantity of heat without increase in temperature is needed. Molten salt baths equipped for water spray or air cooling around the exterior of the bath container have been used for this purpose.

*Austempering:* This is a heat-treating process in which steels are quenched in a bath maintained at some constant temperature in the range of 350 to 800° F (177 to 427° C), depending on the analysis of the steel and the characteristics to be obtained. On immersion in the quenching bath, the steel is cooled more rapidly than the critical quenching rate. When the temperature of the steel reaches that of the bath, however, the quenching action is interrupted. If the steel is now held at this temperature for a predetermined length of time, say, from 10 to 60 minutes, the austenitic structure of the steel is gradually changed into a new structure, called *bainite*. The structure of bainite is acicular (needlelike) and resembles that of tempered martensite such as is usually obtained by quenching in the usual manner to atmospheric temperature and tempering at 400° F (204° C) or higher.

Hardnesses ranging up to 60 RC (Rockwell C scale), depending on the carbon and alloy content of the steel, are obtainable and compare favorably with those obtained for the respective steels by a conventional quench and tempering to above 400° F (204° C). Much greater toughness and ductility are obtained in an austempered piece, however, as compared with a similar piece quenched and tempered in the usual manner.

Two factors are important in austempering. First, the steel must be quenched rapidly enough to the specified subtransformation temperature to avoid any formation of pearlite, and, second, it must be held at this temperature until the transformation from austenite to bainite is completed. Time and temperature transformation curves (called S-curves because of their shape) have been developed for different steels, and these curves provide important data governing the conduct of austempering as well as the other interrupted quenching methods.

Austempering has been applied chiefly to steels having 0.60 percent or more carbon content with or without additional low-alloy content, and to pieces of small diameter or section, usually under 1 inch (2.5 cm), but varying with the composition of the steel. Case-hardened parts may also be austempered.

*Martempering:* In this process the steel is first rapidly quenched from some temperature above the transformation point down to some temperature (usually about 400° F or 204° C) just above that at which martensite begins to form. It is then held at this temperature for a length of time sufficient to equalize the temperature throughout the part, after which it is removed and cooled in air. As the temperature of the steel drops below the transformation point, martensite begins to form in a matrix of austenite at a fairly uniform rate throughout the piece. The soft austenite acts as a cushion to absorb some of the stresses which develop as the martensite is formed. The difficulties presented by quench cracks, internal stresses, and dimensional changes are largely avoided; thus a structure of high hardness can be obtained. If greater toughness and ductility are required, conventional tempering may follow. In general, heavier sections can be hardened more easily by the martempering process than by the austempering process. The martempering process is especially suited to the higher-alloyed steels.

*Isothermal Quenching:* This process resembles austempering in that the steel is first rapidly quenched from above the transformation point down to a temperature that is above that at which martensite begins to form and is held at this temperature until the austenite is completely transformed into bainite. The constant temperature to which the piece is quenched and then maintained is usually 450° F (232° C) or above. The process differs from austempering in that, after transformation to a bainite structure has been completed, the steel is immersed in another bath and is brought up to some higher temperature, depending on the characteristics desired. It is maintained at that temperature for a definite period of time, followed by cooling in air. Thus, tempering to obtain the desired toughness or ductility takes place immediately after the structure of the steel has changed to bainite and before it is cooled to atmospheric temperature.

**Laser and Electron-Beam Surface Hardening.**—Industrial lasers and electron-beam equipment are now available for surface hardening of steels, a process referred to as *laser peening*. The laser and electron beams can generate intense energy fluxes and steep temperature profiles in the workpiece so that external quench media are not needed. This self-quenching is due to a cold interior with sufficient mass acting as a large heat sink to rapidly cool the hot surface by conducting heat to the interior of a part. The laser beam is a beam of light and does not require a vacuum for operation. The electron beam is a stream of electrons, and processing usually takes place in a vacuum chamber or envelope. Both processes may normally be applied to finished machined or ground surfaces because little distortion results.

## Tempering

The object of *tempering* or *drawing* is to reduce the brittleness in hardened steel and to remove the internal strains caused by the sudden cooling in the quenching bath. The tempering process consists in heating the steel by various means to a certain temperature and then cooling it. When steel is in a fully hardened condition, its structure consists largely of *martensite*. On reheating to a temperature of about 300 to 750° F (150 to 400° C), a softer and tougher structure known as *troostite* is formed. If the steel is reheated to a temperature of 750 to 1290° F (400 to 700° C), a structure known as *sorbite* is formed that has somewhat less strength than troostite but much greater ductility.

**Tempering Temperatures.—**If steel is heated in an oxidizing atmosphere, a film of oxide forms on the surface that changes color as the temperature increases. These oxide colors (see Table 1) have been used extensively in the past as a means of gaging the correct amount of temper; but, since these colors are affected to some extent by the composition of the metal, the method is not dependable.

**Table 1. Temperatures as Indicated by the Color of Plain Carbon Steel**

| Degrees | | | Degrees | | |
|---|---|---|---|---|---|
| Fahrenheit | Centigrade | Color of Steel | Fahrenheit | Centigrade | Color of Steel |
| 430 | 221.1 | Very pale yellow | 510 | 265.6 | Spotted red-brown |
| 440 | 226.7 | Light yellow | 520 | 271.1 | Brown-purple |
| 450 | 232.2 | Pale straw-yellow | 530 | 276.7 | Light purple |
| 460 | 237.8 | Straw-yellow | 540 | 282.2 | Full purple |
| 470 | 243.3 | Deep straw-yellow | 550 | 287.8 | Dark purple |
| 480 | 248.9 | Dark yellow | 560 | 293.3 | Full blue |
| 490 | 254.4 | Yellow-brown | 570 | 298.9 | Dark blue |
| 500 | 260.0 | Brown-yellow | 640 | 337.8 | Light blue |

The availability of reliable pyrometers in combination with tempering baths of oil, salt, or lead makes it possible to heat the work uniformly and to a given temperature within close limits.

Suggested temperatures for tempering various tools are given in Table 2.

**Table 2. Tempering Temperatures for Various Plain Carbon Steel Tools**

| Degrees F | Degrees C | Class of Tool |
|---|---|---|
| 495 to 500 | 257 to 260 | Taps, ½ inch (1.27 cm) or over, for use on automatic screw machines |
| 495 to 500 | 257 to 260 | Nut taps, ½ inch (1.27 cm ) and under |
| 515 to 520 | 268 to 271 | Taps, ¼ inch (0.635 cm) and under, for use on automatic screw machines |
| 525 to 530 | 274 to 277 | Thread dies to cut thread close to shoulder |
| 500 to 510 | 260 to 266 | Thread dies for general work |
| 495 | 257 | Thread dies for tool steel or steel tube |
| 525 to 540 | 274 to 282 | Dies for bolt threader threading to shoulder |
| 460 to 470 | 238 to 243 | Thread rolling dies |
| 430 to 435 | 221 to 224 | Hollow mills (solid type) for roughing on automatic screw machines |
| 485 | 252 | Knurls |
| 450 | 232 | Twist drills for hard service |
| 450 | 232 | Centering tools for automatic screw machines |
| 430 | 221 | Forming tools for automatic screw machines |
| 430 to 435 | 221 to 224 | Cut-off tools for automatic screw machines |
| 440 to 450 | 227 to 232 | Profile cutters for milling machines |
| 430 | 221 | Formed milling cutters |
| 435 to 440 | 224 to 227 | Milling cutters |
| 430 to 440 | 221 to 227 | Reamers |
| 460 | 238 | Counterbores and countersinks |
| 480 | 249 | Cutters for tube- or pipe-cutting machines |
| 460 to 520 | 238 to 271 | Snaps for pneumatic hammers — harden full length, temper to 460° F (238° C), then bring point to 520° F (271° C ) |

**Tempering in Oil.—**Oil baths are extensively used for tempering tools (especially in quantity), the work being immersed in oil heated to the required temperature, as indicated by a thermometer. It is important that the oil have a uniform temperature throughout and that the work be immersed long enough to acquire this temperature. Cold steel should not be plunged into a bath heated for tempering, owing to the danger of cracking. The steel should either be preheated to about 300° F (150° C) before placing it in the bath or the bath should be at a comparatively low temperature before the steel is immersed, and then reheated to the required degree. A temperature of from 650 to 700° F (343 to 371° C) can be obtained with heavy tempering oils; for higher temperatures, either a bath of nitrate salts or a lead bath may be used.

In tempering, the best method is to immerse the pieces to be tempered before starting to heat the oil so that they are heated with the oil. After the tempered pieces are taken out of the oil bath, they should be immediately dipped in a tank of caustic soda and, after that, in a tank of hot water. This will remove all oil that might adhere to the tools. The following tempering oil has given satisfactory results: mineral oil, 94 percent; saponifiable oil, 6 percent; specific gravity, 0.920; flash point, 550° F (288° C); fire test, 625° F (329° C).

**Tempering in Salt Baths.—**Molten salt baths may be used for tempering or drawing operations. Nitrate baths are particularly adapted for the usual drawing temperature range of, say, 300 to 1100° F (150 to 595° C). Tempering in an oil bath usually is limited to temperatures of 500 to 600° F (260 to 315° C), and some heat-treating specialists recommend the use of a salt bath for temperatures above 350 or 400° F (175 or 205° C), as it is considered more efficient and economical. Tempering in a bath (salt or oil) has several advantages, such as ease in controlling the temperature range and maintenance of a uniform temperature. The work is also heated much more rapidly in a molten bath. A gas- or oil-fired muffle or semimuffle furnace may be used for tempering, but a salt bath or oil bath is preferable. A salt bath is recommended for tempering high-speed steel, although furnaces may also be used. The bath or furnace temperature should be increased gradually, say, from 300 to 400° F (150 to 205° C) up to the tempering temperature, which may range from 1050 to 1150° F (565 to 620° C) for high-speed steel.

**Tempering in a Lead Bath.—**The lead bath is commonly used for heating steel in connection with tempering, as well as for hardening. The bath is first heated to the temperature at which the steel should be tempered; the preheated work is then placed in the bath long enough to acquire this temperature, after which it is removed and cooled. As the melting temperature of pure lead is about 620° F (327° C), tin is commonly added to lower the temperature sufficiently for tempering. Reductions in temperature can be obtained by varying the proportions of lead and tin, as shown in Table 3.

**Table 3. Temperatures of Lead Bath Alloys**

| Parts Lead | Parts Tin | Melting Temperature Deg. F | Melting Temperature Deg. C | Parts Lead | Parts Tin | Melting Temperature Deg. F | Melting Temperature Deg. C | Parts Lead | Parts Tin | Melting Temperature Deg. F | Melting Temperature Deg. C |
|---|---|---|---|---|---|---|---|---|---|---|---|
| 200 | 8 | 560 | 293 | 39 | 8 | 510 | 266 | 19 | 8 | 460 | 238 |
| 100 | 8 | 550 | 288 | 33 | 8 | 500 | 260 | 17 | 8 | 450 | 232 |
| 75 | 8 | 540 | 282 | 28 | 8 | 490 | 254 | 16 | 8 | 440 | 227 |
| 60 | 8 | 530 | 277 | 24 | 8 | 480 | 249 | 15 | 8 | 430 | 221 |
| 48 | 8 | 520 | 271 | 21 | 8 | 470 | 243 | 14 | 8 | 420 | 216 |

**To Prevent Lead from Sticking to Steel.—**To prevent hot lead from sticking to parts heated in it, mix common whiting (calcium carbonate) with wood alcohol and paint the part that is to be heated. Water can be used instead of alcohol, but, in that case, the paint must be thoroughly dry, as otherwise the moisture will cause the lead to "fly." Another method is to make a thick paste according to the following formula: pulverized charred leather, 1 pound; fine wheat flour, 1½ pounds; fine table salt, 2 pounds. Coat the tool with this paste and heat slowly until dry, then proceed to harden. Still another method is to heat the work to a blue color, or about 600° F (316° C), and then dip it in a strong solution of salt water prior to heating in the lead bath. The lead is sometimes removed from parts having fine projections or teeth, by using a stiff brush just before immersing in the cooling bath. Removal of lead is necessary to prevent the formation of soft spots.

**Tempering in Sand.**—The sand bath is used for tempering certain classes of work. One method is to deposit the sand on an iron plate or in a shallow box that has burners beneath it. With this method of tempering, tools such as boiler punches, etc., can be given a varying temper by placing them endwise in the sand. As the temperature of the sand bath is higher toward the bottom, a tool can be so placed that the color of the lower end will become a deep dark blue when the middle portion is a very dark straw, and the working end or top a light straw color, the hardness gradually increasing from the bottom up.

**Double Tempering.**—In tempering high-speed steel tools, it is common practice to repeat the tempering operation or "double temper" the steel. Double tempering is done by heating the steel to tempering temperature, say 1050° F (566° C), and holding it at that temperature for 2 hours. It is then cooled to room temperature, reheated to the same temperature for another two-hour period, and again cooled to room temperature. After the first tempering operation, some untempered martensite remains in the steel. This martensite is not only tempered by a second tempering operation but is relieved of internal stresses, thus improving the steel for service conditions. The hardening temperature for the higher-alloy steels may affect the hardness after tempering. For example, molybdenum high-speed steel heated to 2100° F (1149° C) had a 61 RC (Rockwell C scale) after tempering, whereas a temperature of 2250° F (1232° C) resulted in hardness of 64.5 RC after tempering.

## Annealing, Spheroidizing, and Normalizing

Annealing of steel is a heat-treating process in which the steel is heated to some elevated temperature, usually in or near the critical range, is held at this temperature for some period of time, and is then cooled, usually at a slow rate. Spheroidizing and normalizing may be considered as special cases of annealing.

The *full annealing* of carbon steel consists in heating it slightly above the *upper* critical point for hypoeutectoid steels (steels of less than 0.85 percent carbon content) and slightly above the *lower* critical point for hypereutectoid steels (steels of more than 0.85 percent carbon content), holding it at this temperature until it is uniformly heated and then slowly cooling it to 1000° F (538° C) or below. The resulting structure is layerlike, or lamellar, in character due to the pearlite that is formed during the slow cooling.

Anealing is employed 1) to soften steel for machining, cutting, stamping, etc., or for some particular service; 2) to alter ductility, toughness, electrical or magnetic characteristics or other physical properties; 3) to refine the crystal structure; 4) to produce grain reorientation; and 5) to relieve stresses and hardness resulting from cold-working.

The *spheroidizing* of steel, according to the American Society of Metals, is "any process of heating and cooling that produces a rounded or globular form of carbide." High-carbon steels are spheroidized to improve their machinability especially in continuous cutting operations, such as are performed by lathes and screw machines. In low-carbon steels, spheroidizing may be employed to meet certain strength requirements before subsequent heat treatment. Spheroidizing also tends to increase resistance to abrasion.

The *normalizing* of steel consists in heating it to some temperature above that used for annealing, usually about 100° F (56° C) above the upper critical range, and then cooling it in still air at room temperature. Normalizing is intended to put the steel into a uniform, unstressed condition of proper grain size and refinement so that it will properly respond to further heat treatments. It is particularly important in the case of forgings that are to be later heat treated. Normalizing may or may not (depending on the composition) leave steel in a sufficiently soft state for machining with available tools. Annealing for machinability is often preceded by normalizing, and the combined treatment—frequently called a *double anneal*—produces a better result than a simple anneal.

**Annealing Practice.—**For carbon steels, the following annealing temperatures are recommended by the American Society for Testing and Materials:

| Percent Carbon | Annealing Temperature | |
|---|---|---|
| | °F | °C |
| less than 0.12% | 1600 to 1700 | 871 to 927 |
| 0.12 to 0.29% | 1550 to 1600 | 843 to 871 |
| 0.30 to 0.49% | 1500 to 1550 | 816 to 843 |
| 0.50 to 1.00% | 1450 to 1500 | 788 to 816 |

Slightly lower temperatures are satisfactory for steels having more than 0.75 percent manganese content. Heating should be uniform to avoid the formation of additional stresses. In the case of large workpieces, the heating should be slow enough so that the temperature of the interior does not lag too far behind that of the surface.

It has been found that in annealing steel, the higher the temperature to which it is heated to produce an austenitic structure, the greater the tendency of the structure to become lamellar (pearlitic) in cooling. On the other hand, the closer the austenitizing temperature to the critical temperature, the greater is the tendency of the annealed steel to become spheroidal.

*Rate of Cooling:* After the steel is heated to some temperature within the annealing range, it should be cooled slowly enough to permit the development of the desired softness and ductility. In general, the slower the cooling rate, the greater the resulting softness and ductility. Steel of a high-carbon content should be cooled more slowly than steel of a low-carbon content; and the higher the alloy content, the slower is the cooling rate usually required. Where extreme softness and ductility are not required, the steel may be cooled in the annealing furnace to some temperature well below the critical point, say, to about 1000° F (538° C) and then removed and cooled in air.

**Annealing by Constant-Temperature Transformation.—**It has been found that steel that has been heated above the critical point so that it has an austenitic structure can be transformed into a lamellar (pearlitic) or a spheroidal structure by holding it for a definite period of time at some constant subcritical temperature. In other words, it is feasible to anneal steel by means of a constant-temperature transformation as well as by the conventional continuous cooling method. When the constant-temperature transformation method is employed, the steel, after being heated to some temperature above the critical and held at this temperature until it is austenitized, is cooled as rapidly as feasible to some relatively high subcritical transformation temperature. This temperature selection is governed by the desired microstructure and hardness required, and is taken from a transformation time and temperature curve (often called a TTT curve). As drawn for a particular steel, such a curve shows the length of time required to transform that steel from an austenitic state at various subcritical temperatures. After being held at the selected sub-critical temperature for the required length of time, the steel is cooled to room temperature — again, as rapidly as feasible. This rapid cooling down to the selected transformation temperature and then down to room temperature has a negligible effect on the structure of the steel and often produces a considerable time saving over the conventional slow cooling method of annealing.

The softest condition in steel can be developed by heating it to a temperature usually less than 100° F (56° C) above the lower critical point and then cooling it to some temperature, usually less than 100 degrees (56° C), below the critical point, where it is held until the transformation is completed. Certain steels require a very lengthy period of time for transformation of the austenite when held at a constant temperature within this range. For such steels, a practical procedure is to allow most of the transformation to take place in this temperature range where a soft product is formed and then to finish the transformation at a lower temperature where the time for the completion of the transformation is short.

**Spheroidizing Practice.**—A common method of spheroidizing steel consists in heating it to or slightly below the lower critical point, holding it at this temperature for a period of time, and then cooling it slowly to about 1000° F (538° C) or below. The length of time for which the steel is held at the spheroidizing temperature largely governs the degree of spheroidization. High-carbon steel may be spheroidized by subjecting it to a temperature that alternately rises and falls between a point within and a point without the critical range. Tool steel may be spheroidized by heating to a temperature slightly above the critical range and then, after being held at this temperature for a period of time, cooling without removal from the furnace.

**Normalizing Practice.**—When using the lower-carbon steels, simple normalizing is often sufficient to place the steel in its best condition for machining and will lessen distortion in carburizing or hardening. In the medium- and higher-carbon steels, combined normalizing and annealing constitutes the best practice. For unimportant parts, the normalizing may be omitted entirely or annealing may be practiced only when the steel is otherwise difficult to machine. Both processes are recommended in the following heat treatments (for SAE steels) as representing the best metallurgical practice. The temperatures recommended for normalizing and annealing have been made indefinite in many instances because of the many different types of furnaces used in various plants and the difference in results desired.

## Case Hardening

In order to harden low-carbon steel, it is necessary to increase the carbon content of the surface of the steel so that a thin outer "case" can be hardened by heating the steel to the hardening temperature and then quenching it. The process, therefore, involves two separate operations. The first is the *carburizing* operation for impregnating the outer surface with sufficient carbon, and the second operation is that of heat treating the carburized parts so as to obtain a hard outer case and, at the same time, give the "core" the required physical properties. The term "case hardening" is ordinarily used to indicate the complete process of carburizing and hardening.

**Carburization.**—Carburization is the result of heating iron or steel to a temperature below its melting point in the presence of a solid, liquid, or gaseous material that decomposes so as to liberate carbon when heated to the temperature used. In this way, it is possible to obtain, by the gradual penetration, diffusion, or absorption of the carbon by the steel, a "zone" or "case" of higher-carbon content at the outer surfaces than that of the original object. When a carburized object is rapidly cooled or quenched in water, oil, brine, etc., from the proper temperature, this case becomes hard, leaving the inside of the piece soft but of great toughness.

**Use of Carbonaceous Mixtures.**—When carburizing materials of the solid class are used, the case-hardening process consists in packing steel articles in metal boxes or pots, with a carbonaceous compound surrounding the steel objects. The boxes or pots are sealed and placed in a carburizing oven or furnace maintained usually at a temperature of from about 1650 to 1700° F (899 to 927° C) for a length of time depending on the extent of the carburizing action desired. The carbon from the carburizing compound will then be absorbed by the steel on the surfaces desired, and the low-carbon steel is converted into high-carbon steel at these portions. The internal sections and the insulated parts of the object retain practically their original low-carbon content. The result is a steel of a dual structure, a high-carbon and a low-carbon steel in the same piece. The carburized steel may now be heat treated by heating and quenching, in much the same way as high-carbon steel is hardened, in order to develop the properties of hardness and toughness; but, as the steel is, in reality, two steels in one, one high-carbon and one low-carbon, the correct heat treatment after carburizing includes two distinct processes, one suitable for the high-carbon portion or the "case," as it is generally called, and one suitable for the low-carbon portion or core. The method of heat treatment varies according to the kind of steel used.

Usually, an initial heating and slow cooling is followed by reheating to 1400–1450° F (760 to 788° C), quenching in oil or water, and a final tempering. More definite information is given in the following section on SAE steels.

*Carburizers:* There are many commercial carburizers on the market in which the materials used as the generator may be hard and soft wood charcoal, animal charcoal, coke, coal, beans and nuts, bone and leather, or various combinations of these. The energizers may be barium, cyanogen, and ammonium compounds, various salts, soda ash, or lime and oil hydrocarbons.

**Pack-Hardening.—**When cutting tools, gages, and other parts made from high-carbon steels are heated for hardening while packed in some carbonaceous material in order to protect delicate edges, corners, or finished surfaces, the process is usually known as pack-hardening. Thus, the purpose is to protect the work, prevent scale formation, ensure uniform heating, and minimize the danger of cracking and warpage. The work is packed, as in carburizing, and in the same type of receptacle. Common hardwood charcoal often is used, especially if it has had an initial heating to eliminate shrinkage and discharge its more impure gases. The lowest temperature required for hardening should be employed for pack-hardening—usually 1400 to 1450° F (760 to 788° C) for carbon steels. Pack-hardening has also been applied to high-speed steels, but modern developments in heat-treating salts have made it possible to harden high-speed steel without decarburization, injury to sharp edges, or marring the finished surfaces. See *Salt Baths* on page 469.

**Cyanide Hardening.—**When low-carbon steel requires a very hard outer surface but does not need high shock-resisting qualities, the cyanide-hardening process may be employed to produce what is known as superficial hardness. This superficial hardening is the result of carburizing a very thin outer skin (which may be only a few thousandths inch thick) by immersing the steel in a bath containing sodium cyanide. The temperatures usually vary from 1450 to 1650° F (788 to 899° C), and the percentage of sodium cyanide in the bath extends over a wide range, depending on the steel used and properties required.

**Nitriding Process.—**Nitriding is a process for surface hardening certain alloy steels by heating the steel in an atmosphere of nitrogen (ammonia gas) at approximately 950° F (510° C). The steel is then cooled slowly. Finish machined surfaces hardened by nitriding are subject to minimum distortion. The physical properties, such as toughness, high impact strength, etc., can be imparted to the core by previous heat treatments and are unaffected by drawing temperatures up to 950° F (510° C). The "Nitralloy" steels suitable for this process may be readily machined in the heat-treated as well as in the annealed state, and they forge as easily as alloy steels of the same carbon content. Certain heat treatments must be applied prior to nitriding, the first being annealing to relieve rolling, forging, or machining strains. Parts or sections not requiring heat treating should be machined or ground to the exact dimensions required. Close tolerances must be maintained in finish machining, but allowances for growth due to adsorption of nitrogen should be made, and this usually amounts to about 0.0005 inch (0.0127 mm) for a case depth of 0.02 inch (0.508 mm). Parts requiring heat treatment for definite physical properties are forged or cut from annealed stock, heat treated for the desired physical properties, rough machined, normalized, and finish machined. If quenched and drawn parts are normalized afterwards, the drawing and normalizing temperatures should be alike. The normalizing temperature may be below but should never be above the drawing temperature.

**Ion Nitriding.—**Ion nitriding, also referred to as glow discharge nitriding, is a process for case hardening of steel parts, such as tool spindles, cutting tools, extrusion equipment, forging dies, gears, and crankshafts. An electrical potential ionizes low-pressure nitrogen gas, and the ions produced are accelerated to and impinge on the workpiece, heating it to the appropriate temperature for diffusion to take place. Therefore, there is no requirement for a supplemental heat source. The inward diffusion of the nitrogen ions forms the iron and alloy nitrides in the case. White layer formation, familiar in conventional gas nitriding, is readily controlled by this process.

**Liquid Carburizing.—**Activated liquid salt baths are now used extensively for carburizing. Sodium cyanide and other salt baths are used. The salt bath is heated by electrodes immersed in it, the bath itself acting as the conductor and resistor. One or more groups of electrodes, with two or more electrodes per group, may be used. The heating is accompanied by a stirring action to ensure uniform temperature and carburizing activity throughout the bath. The temperature may be controlled by a thermocouple immersed in the bath and connecting with a pyrometer designed to provide automatic regulation. The advantages of liquid baths include rapid action, uniform carburization, minimum distortion, and elimination of the packing and unpacking required when carbonaceous mixtures are used. In selective carburizing, the portions of the work that are not to be carburized are copperplated, and the entire piece is then immersed in an activated cyanide bath. The copper inhibits any carburizing action on the plated parts, and this method offers a practical solution for selectively carburizing any portion of a steel part.

**Gas Carburizing.—**When carburizing gases are used, the mixture varies with the type of case and quality of product desired. The gaseous hydrocarbons most widely used are methane (natural gas), propane, and butane. These carbon-bearing gases are mixed with air, with manufactured gases of several types, with flue gas, or with other specially prepared "diluent" gases. It is necessary to maintain a continuous fresh stream of carburizing gases to the carburizing retort or muffle as well as to continuously remove the spent gases from the muffle in order to obtain the correct mixture of gases inside the muffle. A slight pressure is maintained on the muffle to exclude unwanted gases.

The horizontal rotary type of gas carburizing furnace has a retort or muffle that revolves slowly. This type of furnace is adapted to small parts, such as ball and roller bearings, chain links, small axles, bolts, etc. With this type of furnace, very large pieces, such as gears, for example, may be injured by successive shocks due to tumbling within the rotor.

The vertical pit type of gas carburizer has a stationary workholder that is placed vertically in a pit. The work, instead of circulating in the gases as with the rotary type, is stationary, and the gases circulate around it. This type is applicable to long, large shafts or other parts or shapes that cannot be rolled in a rotary type of furnace.

There are three types of continuous gas furnaces that may be designated as

1) direct quench and manually operated

2) direct quench and mechanically operated

3) cooling-zone type

Where production does not warrant using a large continuous-type furnace, a horizontal muffle furnace of the batch type may be used, especially if the quantities of work are varied and the production not continuous.

**Vacuum Carburizing.—**Vacuum carburizing is a high-temperature gas carburizing process that is performed at pressures below atmospheric. The furnace atmosphere usually consists solely of an enriching gas, such as natural gas, pure methane, or propane; nitrogen is sometimes used as a carrier gas. Vacuum carburizing offers several advantages, such as combining of processing operations and reduced total processing time.

**Carburizing Steels.—**A low-carbon steel containing, say, from 0.10 to 0.20 percent of carbon is suitable for carburized case hardening. In addition to straight-carbon steels, the low-carbon alloy steels are employed. The alloys add to case-hardened parts the same advantageous properties they give to other classes of steel. Various steels suitable for case hardening will be found in the section on SAE steels.

**To Clean Work after Case Hardening.—**To clean work, especially if knurled, or if dirt is likely to stick into crevices after case hardening, wash it in caustic soda (1 part soda to 10 parts water). In making the solution, the soda should be put into hot water gradually, and the mixture stirred until the soda is thoroughly dissolved. A still more effective method of cleaning is to dip the work into a mixture of 1 part sulfuric acid and 2 parts water. Leave the pieces in this mixture about 3 minutes; then wash them immediately in a soda solution.

**Flame Hardening.—**This method of hardening is especially applicable to selective hardening of large steel forgings or castings that must be finish-machined prior to heat-treatment or, because of size or shape, cannot be heat treated by using a furnace or bath. An oxyacetylene torch is used to quickly heat the surface to be hardened, which is then quenched to secure a hardened layer that may vary in depth from a mere skin to ¼ inch (6.35 mm) and with hardness ranging from 400 to 700 BHN (Brinell Hardness Number). A multiflame torchhead may be equipped with quenching holes or a spray nozzle behind the flame. This is not a carburizing or a case-hardening process as the torch is only a heating medium. Most authorities recommend tempering or drawing of the hardened surface at temperatures between 200 and 350° F (93 and 177° C). This treatment may be performed in a standard furnace, in an oil bath, or with a gas flame. It should follow the hardening process as closely as possible. Medium-carbon and many low-alloy steels are suitable for flame hardening. Plain carbon steels ranging from 0.35 to 0.60 percent carbon will give hardnesses of from 400 to 700 BHN. Steels in the 0.40 to 0.45 percent carbon range are preferred, as they have excellent core properties and produce hardnesses of from 400 to 500 BHN without checking or cracking. Higher-carbon steels will give greater hardnesses, but extreme care must be taken to prevent cracking. Careful control of the quenching operation is required.

*Spinning Method of Flame Hardening:* This method is employed on circular objects that can be rotated or spun past a stationary flame. It may be subdivided according to the speed of rotation, where the part is rotated slowly in front of a stationary flame and the quench is applied immediately after the flame. This method is used on large circular pieces, such as track wheels and bearing surfaces. There will be a narrow band of material with lower hardness between adjacent torches if more than one path of the flame is required to harden the surface. There will also be an area of lower hardness where the flame is extinguished. A second method is applicable to small rollers or pinions. The work is spun at a speed of 50 to 150 rpm in front of the flame until the entire piece has reached the proper temperature; then it is quenched as a unit by a cooling spray or by ejecting it into a cooling bath.

*Progressive Method:* In this method the torch travels along the face of the work and the work remains stationary. It is used to harden lathe ways, gear teeth, and track rails.

*Stationary or Spot-Hardening Method:* When this method is employed, the work and torch are both stationary. When the spot to be hardened reaches the quenching temperature, the flame is removed and the quench applied.

*Combination Method:* This approach is a combination of the spinning and progressive methods, and is used for long bearing surfaces. The work rotates slowly past the torch as the torch travels longitudinally across the face of the work at the rate of the torch width per revolution of the work.

Equipment for the stationary method of flame hardening consists merely of an acetylene torch, an oxyacetylene supply, and a suitable means of quenching; but when the other methods are employed, work-handling tools are essential and specially designed torches are desirable. A lathe is ideally suited for the spinning or combination hardening method, whereas a planer is easily adapted for progressive hardening. Production jobs, such as the hardening of gears, require specially designed machines. These machines reduce handling and hardening time, as well as assuring consistent results.

**Induction Hardening.—**The hardening of steel by means of induction heating and subsequent quenching in either liquid or air is particularly applicable to parts that require localized hardening or controlled depth of hardening and to irregularly shaped parts, such as cams that require uniform surface hardening around their contour.

Advantages offered by induction hardening are: 1) a short heating cycle that may range from a fraction of a second to several seconds; heat energy can be induced in a piece of steel at the rate of 100 to 250 Btu/in$^2$/min (165 to 410 J/mm$^2$/min) by induction heating, as compared with a rate of 3 Btu/in$^2$/min (5 J/mm$^2$/min) for the same material at room temperature when placed in a furnace with a wall temperature of 2000° F (1093° C); 2) absence of tendency to produce oxidation or decarburization; 3) exact control of depth and area of hardening; 4) close regulation of degree of hardness obtained by automatic timing of heating and quenching cycles; 5) minimum warpage or distortion; and 6) possibility of substituting carbon steels for higher-cost alloy steels.

To the designer, the principal advantage of induction hardening lies in its application to localized zones. Thus, specific areas in a given part can be heat treated separately to the respective hardnesses required. Parts can be designed so that the stresses at any given point in the finished piece can be relieved by local heating. Parts can be designed in which welded or brazed assemblies are built up prior to heat treating with only internal surfaces or projections requiring hardening.

**Types of Induction Heating Equipment.**—Induction heating is secured by placing the metal part inside or close to an "applicator" coil of one or more turns through which alternating current is passed. The coil, formed to suit the general class of work to be heated, is usually made of copper tubing through which water is passed to prevent overheating of the coil itself. The workpiece is either held in a fixed position or rotated slowly within or close to the applicator coil. Where the length of work is too great to permit heating in a fixed position, progressive heating may be employed. Thus, a rod or tube of steel may be fed through an applicator coil of one or more turns so that the heating zone travels progressively along the entire length of the workpiece.

The frequency of the alternating current used and the type of generator employed to supply this current to the applicator coil depend on the character of the work to be done.

There are three types of commercial equipment used to produce high-frequency current for induction heating: 1) motor generator sets that deliver current at frequencies of approximately 1000, 2000, 3000, and 10,000 cycles; 2) spark gap oscillator units that produce frequencies ranging from 80,000 to 300,000 cycles; and 3) vacuum tube oscillator sets, which produce currents at frequencies ranging from 350,000 to 15,000,000 cycles or more.

**Depth of Heat Penetration.**—Generally speaking, the higher the frequency used, the shallower the depth of heat penetration. For heating clear through, for deep hardening, and for large workpieces, low power concentrations and low frequencies are usually used. For very shallow and closely controlled depths of heating, as in surface hardening and localized heat treating of small workpieces, currents at high frequencies are used.

For example, a ½-inch (12.7-mm) round bar of hardenable steel will be heated through its entire structure quite rapidly by an induced current of 2000 cycles. After quenching, the bar would show through hardness with a decrease in hardness from surface to center. The same piece of steel could be readily heated and surface hardened to a depth of 0.100 inch (2.5 mm) with current at 9600 cycles, and to an even shallower depth with current at 100,000 cycles. A ¼-inch (6.35-mm) bar, however, would not reach a sufficiently high temperature at 2000 cycles to permit hardening, but at 9600 cycles through hardening would be accomplished. Current at over 100,000 cycles would be needed for surface hardening such a bar.

**Types of Steel for Induction Hardening.**—Most of the standard types of steels can be hardened by induction heating, providing the carbon content is sufficient to produce the desired degree of hardness by quenching. Thus, low-carbon steels with a carburized case, medium- and high-carbon steels (both plain and alloy), and cast iron with a portion of the carbon in combined form, may be used for this purpose. Induction heating of alloy steels should be limited primarily to the shallow-hardening type, that is, those of low alloy content; otherwise the severe quench usually required may result in a highly stressed surface with consequent reduced load-carrying capacity and danger of cracking.

**Through Hardening, Annealing, and Normalizing by Induction.—**For through hardening, annealing, and normalizing by induction, low power concentrations are desirable to prevent too great a temperature differential between the surface and the interior of the work. A satisfactory rate of heating is obtained when the total power input to the work is slightly greater than the radiation losses at the desired temperature. If possible, as low a frequency should be used as is consistent with good electrical coupling. A number of applicator coils may be connected in a series so that several workpieces can be heated simultaneously, thus reducing the power input to each. Widening the spacing between work and applicator coil also will reduce the amount of power delivered to the work.

**Induction Surface Hardening.—**As indicated on page 483 in "*Depth of Heat Penetration,*" currents at much higher frequencies are required in induction surface hardening than in through hardening by induction. In general, the smaller the workpiece, the thinner the section; or the shallower the depth to be hardened, the higher the frequency required. High power concentrations are also needed to make possible a short heating period so that an undue amount of heat will not be conducted to adjacent or interior areas, where a change in hardness is not desired. Generators of large capacity and applicator coils of but a few turns, or even a single turn, provide the necessary concentration of power in the localized area to be hardened.

Induction heating of internal surfaces, such as the interior of a hollow cylindrical part or the inside of a hole, can be accomplished readily with applicator coils shaped to match the cross section of the opening, which may be round, square, elliptical or some other form. If the internal surface is of short length, a multiturn applicator coil extending along its entire length may be employed. Where the power available is insufficient to heat the entire internal surface at once, progressive heating is used. For this purpose, an applicator coil of few turns — often but a single turn — is employed, and either coil or work is moved so that the heated zone passes progressively from one end of the hole or opening to the other. For bores of small diameter, a hairpin-shaped applicator, extending the entire length of the hole, may be used and the work rotated about the axis of the hole to ensure even heating.

**Quenching After Induction Heating.—**After induction heating, quenching may be by immersion in a liquid bath (usually oil), by liquid spray (usually water), or by self-quenching. The term "self-quenching" is used when there is no quenching medium and hardening of the heated section is due chiefly to rapid absorption of heat by the mass of cool metal adjacent to it. Quenching by immersion offers the advantage of even cooling and is particularly satisfactory for through heated parts. Spray quenching may be arranged so that the quenching ring and applicator coil are in the same or adjacent units, permitting the quenching cycle to follow the heating cycle immediately without removal of the work from the holding fixture. Automatic timing to a fraction of a second may also be employed for both heating and quenching with this arrangement to secure the exact degree of hardness desired. Self-quenching is applicable only in thin-surface hardening where the mass of adjacent cool metal in the part is great enough to conduct the heat rapidly out of the surface layer being hardened. It has been recommended that for adequate self-quenching, the mass of the unheated section should be at least ten times that of the heated shell. It has been found difficult to use the self-quenching technique to produce hardened shells of much more than about 0.060 inch (1.52 mm) thickness. Close to this limit, self-quenching can only be accomplished with the easily hardenable steels. By using a combination of self-quench and liquid quench, however, it is possible to produced hardened shells on work too thin to self-quench completely. In general, self-quenching is confined chiefly to relatively small parts and simple shapes.

**Induction Hardening of Gear Teeth.—**Several advantages are claimed for the induction hardening of gear teeth. One advantage is that the gear teeth can be completely machined, including shaving, when in the soft-annealed or normalized condition, and then hardened, because when induction heating is used, distortion is held to a minimum. Another advantage claimed is that bushings and inserts can be assembled in the gears before hardening. A wide latitude in choice of built-up webs and easily machined hubs is afforded because the hardness of neither web nor hub is affected by the induction-hardening operation although slight dimensional changes may occur in certain designs. Regular carbon steels can be used in place of alloy steels for a wide variety of gears, and steel with a higher carbon content can frequently be substituted for a carburizing steel so that the carburizing operation can be eliminated. Another time-saving benefit is elimination of cleaning after hardening.

In heating spur gear teeth by induction, the gear is usually placed inside a circular unit that combines the applicator coil and quenching ring. An automatic timing device controls both the heating and quenching cycles. During the heating cycle, the gear is rotated at 25 to 35 rpm to ensure uniform heating.

In hardening bevel gears, the applicator coil is wound to conform to the face angle of the gear. In some spiral-bevel gears, there is a tendency to obtain more heat on one side of the tooth than on the other. In some sizes of spiral-bevel gears, this tendency can be overcome by applying slightly more heat to ensure hardening of the concave side. In some forms of spiral-bevel gears, it has been the practice to carburize that part of the gear surface which is to be hardened after the teeth have been rough-cut. Carburizing is followed by the finish-cutting operation, after which the teeth can be induction heated, using a long enough period to heat the entire tooth. When the gear is quenched, only the carburized surface will become hardened.

**Table 4a. Typical Heat Treatments for SAE Carbon Steels** (Carburizing Grades)

| SAE No. | Normalize, Deg. F | Carburize | | Cool[a] | Reheat | | Cool[a] | 2nd Reheat, Deg. F | Cool[a] | Temper,[b] | |
|---|---|---|---|---|---|---|---|---|---|---|---|
| | | Deg. F | Deg. C | | Deg. F | Deg. C | | | | Deg. F | Deg. C |
| 1010 to 1022 | ... | 1650–1700 | 899–927 | A | ... | ... | ... | ... | ... | 250–400 | 121–204 |
| | ... | 1650–1700 | 899–927 | B | 1400–1450 | 760–788 | A | ... | ... | 250–400 | 121–204 |
| | ... | 1650–1700 | 899–927 | C | 1400–1450 | 760–788 | A | ... | ... | 250–400 | 121–204 |
| | ... | 1650–1700 | 899–927 | C | 1650–1700 | 899–927 | B | 1400–1450 | A | 250–400 | 121–204 |
| | ... | 1500–1650[c,d] | 816–899[c,d] | B | ... | ... | ... | ... | ... | Optional | Optional |
| | ... | 1350–1575[e,d] | 732–857[e,d] | D | ... | ... | ... | ... | ... | Optional | Optional |
| 1024 | 1650–1750[f] | 1650–1700 | 899–927 | E | ...... | ...... | ... | ... | ... | 250–400 | 121–204 |
| | ... | 1350–1575[e,d] | 732–857[e,d] | D | ... | ... | ... | ... | ... | Optional | Optional |
| 1025 | ... | 1650–1700 | 899–927 | A | ... | ... | ... | ... | ... | 250–400 | 121–204 |
| 1026 | ... | 1500–1650[c,d] | 816–899[c,d] | B | ... | ... | ... | ... | ... | Optional | Optional |
| 1027 | ... | 1350–1575[e,d] | 732–857[e,d] | D | ... | ... | ... | ... | ... | Optional | Optional |
| 1030 | ... | 1500–1650[c,d] | 816–899[c,d] | B | ... | ... | ... | ... | ... | Optional | Optional |
| | ... | 1350–1575[e,d] | 732–857[e,d] | D | ... | ... | ... | ... | ... | Optional | Optional |
| 1111 1112 1113 | ... | 1500–1650[c,d] | 816–899[c,d] | B | ... | ... | ... | ... | ... | Optional | Optional |
| | ... | 1350–1575[e,d] | 732–857[e,d] | D | ... | ... | ... | ... | ... | Optional | Optional |
| 1109 to 1120 | ... | 1650–1700 | 899–927 | A | ... | ... | ... | ... | ... | 250–400 | 121–204 |
| | ... | 1650–1700 | 899–927 | B | 1400–1450 | 760–788 | A | ... | ... | 250–400 | 121–204 |
| | ... | 1650–1700 | 899–927 | C | 1400–1450 | 760–788 | A | ... | ... | 250–400 | 121–204 |
| | ... | 1650–1700 | 899–927 | C | 1650–1700 | 899–927 | B | 1400–1450 | A | 250–400 | 121–204 |
| | ... | 1500–1650[c,d] | 816–899[c,d] | B | ... | ... | ... | ... | ... | Optional | Optional |
| | ... | 1350–1575[e,d] | 732–857[e,d] | D | ... | ... | ... | ... | ... | Optional | Optional |
| 1126 | ... | 1500–1650[c,d] | 816–899[c,d] | B | ... | ... | ... | ... | ... | Optional | Optional |
| | ... | 1350–1575[e,d] | 732–857[e,d] | D | ... | ... | ... | ... | ... | Optional | Optional |

[a] Symbols: A = water or brine; B = water or oil; C = cool slowly; D = air or oil; E = oil; F = water, brine, or oil.

[b] Even where tempering temperatures are shown, tempering is not mandatory in many applications. Tempering is usually employed for partial stress relief and improved resistance to grinding cracks.

[c] Activated or cyanide baths.

[d] May be given refining heat as in other processes.

[e] Carbonitriding atmospheres.

[f] Normalizing temperatures at least 50° F (28° C) above the carburizing temperature are sometimes recommended where minimum heat-treatment distortion is of vital importance.

**Table 4b. Typical Heat Treatments for SAE Carbon Steels** (Heat-Treating Grades)

| SAE Number | Normalize, Deg. F | | Anneal, Deg. F | Harden, Deg. F | Quench[a] | Temper, Deg. F |
|---|---|---|---|---|---|---|
| 1025 & 1030 | … | | … | 1575–1650 | A | To Desired Hardness |
| 1033 to 1035 | … | | … | 1525–1575 | B | |
| 1036 | 1600–1700 | | … | 1525–1575 | B | |
| | … | | … | 1525–1575 | B | |
| 1038 to 1040 | 1600–1700 | | … | 1525–1575 | B | |
| | … | | … | 1525–1575 | B | |
| 1041 | 1600–1700 | and/or | 1400–1500 | 1475–1550 | E | |
| 1042 to 1050 | 1600–1700 | | … | 1475–1550 | B | |
| 1052 & 1055 | 1550–1650 | and/or | 1400–1500 | 1475–1550 | E | |
| 1060 to 1074 | 1550–1650 | and/or | 1400–1500 | 1475–1550 | E | |
| 1078 | … | | 1400–1500[a] | 1450–1500 | A | |
| 1080 to 1090 | 1550–1650 | and/or | 1400–1500[a] | 1450–1500 | E[b] | |
| 1095 | … | | 1400–1500[a] | 1450–1500 | F | |
| | … | | 1400–1500[a] | 1500–1600 | E | |
| 1132 & 1137 | 1600–1700 | and/or | 1400–1500 | 1525–1575 | B | |
| 1138 & 1140 | … | | … | 1500–1550 | B | |
| | 1600–1700 | | … | 1500–1550 | B | |
| 1141 & 1144 | … | | 1400–1500 | 1475–1550 | E | |
| | 1600–1700 | | 1400–1500 | 1475–1550 | E | |
| 1145 to 1151 | … | | … | 1475–1550 | B | |
| | 1600–1700 | | … | 1475–1550 | B | |

[a] Slow cooling produces a spheroidal structure in these high-carbon steels that is sometimes required for machining purposes.

[b] May be water- or brine-quenched by special techniques such as partial immersion or time-quenched; otherwise they are subject to quench cracking.

**Table 5a. Typical Heat Treatments for SAE Alloy Steels**
(Carburizing Grades)

| SAE No. | Normalize[a] | Cycle Anneal[b] | Carburized Deg. F | Carburized Deg. C | Cool[c] | Reheat Deg. F | Reheat Deg. C | Cool[c] | Temper,[d] Deg. F | Temper,[d] Deg. C |
|---|---|---|---|---|---|---|---|---|---|---|
| 1320 | yes | … | 1650–1700 | 899–927 | E | 1400–1450[e] | 760–788[e] | E | 250–350 | 121–177 |
| | yes | … | 1650–1700 | 899–927 | E | 1475–1525[f] | 802–829[f] | E | 250–350 | 121–177 |
| | yes | … | 1650–1700 | 899–927 | C | 1400–1450[e] | 760–788[e] | E | 250–350 | 121–177 |
| | yes | … | 1650–1700 | 899–927 | C | 1500–1550[f] | 816–843[f] | E | 250–350 | 121–177 |
| | yes | … | 1650–1700 | 899–927 | E[g] | … | … | … | 250–350 | 121–177 |
| | yes | … | 1500–1650[h] | 816–899[h] | E | … | … | … | 250–350 | 121–177 |
| 2317 | yes | yes | 1650–1700 | 899–927 | E | 1375–1425[e] | 746–774[e] | E | 250–350 | 121–177 |
| | yes | yes | 1650–1700 | 899–927 | E | 1450–1500[f] | 788–816[f] | E | 250–350 | 121–177 |
| | yes | yes | 1650–1700 | 899–927 | C | 1375–1425[e] | 746–774[e] | E | 250–350 | 121–177 |
| | yes | yes | 1650–1700 | 899–927 | C | 1475–1525[f] | 802–829[f] | E | 250–350 | 121–177 |
| | yes | yes | 1650–1700 | 899–927 | E[g] | … | … | … | 250–350 | 121–177 |
| | yes | yes | 1450–1650[h] | 788–899[h] | E | … | … | … | 250–350 | 121–177 |
| 2512 to 2517 | yes[i] | … | 1650–1700 | 899–927 | C | 1325–1375[e] | 718–746[e] | E | 250–350 | 121–177 |
| | yes[i] | … | 1650–1700 | 899–927 | C | 1425–1475[f] | 774–802[f] | E | 250–350 | 121–177 |
| 3115 & 3120 | yes | … | 1650–1700 | 899–927 | E | 1400–1450[e] | 760–788[e] | E | 250–350 | 121–177 |
| | yes | … | 1650–1700 | 899–927 | E | 1475–1525[f] | 802–829[f] | E | 250–350 | 121–177 |
| | yes | … | 1650–1700 | 899–927 | C | 1400–1450[e] | 760–788[e] | E | 250–350 | 121–177 |
| | yes | … | 1650–1700 | 899–927 | C | 1500–1550[f] | 816–843[f] | E | 250–350 | 121–177 |
| | yes | … | 1650–1700 | 899–927 | E[g] | … | … | … | 250–350 | 121–177 |
| | yes | …… | 1500–1650[h] | 816–899[h] | E | … | … | … | 250–350 | 121–177 |
| 3310 & 3316 | yes[i] | … | 1650–1700 | 899–927 | E | 1400–1450[e] | 760–788[e] | E | 250–350 | 121–177 |
| | yes[i] | … | 1650–1700 | 899–927 | C | 1475–1500[f] | 802–816[f] | E | 250–350 | 121–177 |
| 4017 to 4032 | yes | yes | 1650–1700 | 899–927 | E[g] | … | … | … | 250–350 | 121–177 |
| 4119 & 4125 | yes | … | 1650–1700 | 899–927 | E[g] | … | … | … | 250–350 | 121–177 |

**Table 5a.** *(Continued)* **Typical Heat Treatments for SAE Alloy Steels** (Carburizing Grades)

| SAE No. | Normalize[a] | Cycle Anneal[b] | Carburized Deg. F | Carburized Deg. C | Cool[c] | Reheat Deg. F | Reheat Deg. C | Cool[c] | Temper,[d] Deg. F | Temper,[d] Deg. C |
|---|---|---|---|---|---|---|---|---|---|---|
| 4317 & 4320 4608 to 4621 | yes | yes | 1650–1700 | 899–927 | E | 1425–1475[e] | 774–802[e] | E | 250–350 | 121–177 |
| | yes | yes | 1650–1700 | 899–927 | E | 1475–1527[f] | 802–831[f] | E | 250–350 | 121–177 |
| | yes | yes | 1650–1700 | 899–927 | C | 1425–1475[e] | 774–802[e] | E | 250–350 | 121–177 |
| | yes | yes | 1650–1700 | 899–927 | C | 1475–1525[f] | 802–829[f] | E | 250–350 | 121–177 |
| | yes | yes | 1650–1700 | 899–927 | E[g] | … | … | … | 250–350 | 121–177 |
| | yes | yes | 1650–1700 | 899–927 | E[g] | … | … | … | 250–350 | 121–177 |
| | yes | … | 1500–1650[h] | 816–899[h] | E | … | … | … | 250–350 | 121–177 |
| 4812 to 4820 | yes[i] | yes | 1650–1700 | 899–927 | E | 1375–1425[e] | 746–774[e] | E | 250–350 | 121–177 |
| | yes[i] | yes | 1650–1700 | 899–927 | E | 1450–1500[f] | 788–816[f] | E | 250–350 | 121–177 |
| | yes[i] | yes | 1650–1700 | 899–927 | C | 1375–1425[e] | 746–774[e] | E | 250–350 | 121–177 |
| | yes[i] | yes | 1650–1700 | 899–927 | C | 1450–1500[f] | 788–816[f] | E | 250–350 | 121–177 |
| | … | … | 1650–1700 | 899–927 | E[g] | … | … | … | 250–350 | 121–177 |
| 5115 & 5120 | yes | … | 1650–1700 | 899–927 | E | 1425–1475[e] | 774–802[e] | E | 250–350 | 121–177 |
| | yes | … | 1650–1700 | 899–927 | E | 1500–1550[f] | 816–843[f] | E | 250–350 | 121–177 |
| | yes | … | 1650–1700 | 899–927 | C | 1425–1475[e] | 774–802[e] | E | 250–350 | 121–177 |
| | yes | … | 1650–1700 | 899–927 | C | 1500–1550[f] | 816–843[f] | E | 250–350 | 121–177 |
| | yes | … | 1500–1650[h] | 816–899[h] | E | … | … | … | 250–350 | 121–177 |
| 8615 to 8625 8720 | yes | yes | 1650–1700 | 899–927 | E | 1475–1525[e] | 802–829[e] | E | 250–350 | 121–177 |
| | yes | yes | 1650–1700 | 899–927 | E | 1525–1575[f] | 829–857[f] | E | 250–350 | 121–177 |
| | yes | yes | 1650–1700 | 899–927 | C | 1475–1525[e] | 802–829[e] | E | 250–350 | 121–177 |
| | yes | yes | 1650–1700 | 899–927 | C | 1525–1575[f] | 829–857[f] | E | 250–350 | 121–177 |
| | yes | yes | 1650–1700 | 899–927 | E[g] | … | … | … | 250–350 | 121–177 |
| | yes | yes | 1500–1650[h] | 816–899[h] | E | … | … | … | 250–350 | 121–177 |
| 9310 to 9317 | yes[i] | … | 1650–1700 | 899–927 | E | 1400–1450[e] | 760–788[e] | E | 250–350 | 121–177 |
| | yes[i] | … | 1650–1700 | 899–927 | C | 1500–1525 | 816–829 | E | 250–350 | 121–177 |

[a] Normalizing temperatures should be not less than 50° F (28° C) higher than the carburizing temperature. Follow by air cooling.

[b] For cycle annealing, heat to normalizing temperature; hold for uniformity; cool rapidly to 1000–1250° F (538–677° C); hold 1 to 3 hours; then air or furnace cool to obtain a structure suitable for machining and finishing.

[c] Symbols: C = cool slowly; E = oil.

[d] Tempering treatment is optional and is generally employed for partial stress relief and improved resistance to cracking from grinding operations.

[e] For use when case hardness only is paramount.

[f] For use when higher core hardness is desired.

[g] Treatment is for fine-grained steels only, when a second reheat is often unnecessary.

[h] Treatment is for activated or cyanide baths. Parts may be given refining heats as indicated for other heat-treating processes.

[i] After normalizing, reheat to temperatures of 1000–1200° F (538–649° C) and hold approximately 4 hours.

**Metallography.—**The science or study of the microstructure of metal is known by most metallurgists as "metallography" or sometimes "crystallography." The examination of metals and metal alloys by the aid of the microscope is one of the most effective methods of studying their properties and is also a valuable means of controlling the quality of manufactured metallic articles and of testing the finished product. In preparing the specimen, a flat surface is first formed by filing or grinding and then given a high polish, which is later etched in order to clearly reveal the internal structure under the microscope. This process clearly shows an experienced observer the effect of variation in composition, heat treatment, etc., and in many cases it has proved a correct means of determining certain properties of industrial products that a chemical analysis has failed to reveal.

**Table 5b. Typical Heat Treatments for SAE Alloy Steels** (Directly Hardenable Grades)

| SAE No. | Normalize, °F | | Anneal, °F | Harden, °F | Quench[a] | Temper, °F |
|---|---|---|---|---|---|---|
| 1330 | … | | … | 1525-1575 | B | |
| | 1600-1700 | and/or | 1500-1600 | 1525-1575 | B | |
| 1335 & 1340 | … | | … | 1500-1550 | E | |
| | 1600-1700 | and/or | 1500-1600 | 1525-1575 | E | |
| 2330 | … | | … | 1450-1500 | E | |
| | 1600-1700 | and/or | 1400-1500 | 1450-1500 | E | |
| 2340 & 2345 | … | | … | 1425-1475 | E | To desired hardness |
| | 1600-1700 | and/or | 1400-1500 | 1425-1475 | E | |
| 3130 | 1600-1700 | | … | 1500-1550 | B | |
| 3135 to 3141 | … | | … | 1500-1550 | E | |
| | 1600-1700 | and/or | 1450-1550 | 1500-1550 | E | |
| 3145 & 3150 | … | | … | 1500-1550 | E | |
| | 1600-1700 | and/or | 1400-1500 | 1500-1550 | E | |
| 4037 & 4042 | … | | 1525-1575 | 1500-1575 | E | Gears, 350-450<br>To desired hardness |
| 4047 & 4053 | … | | 1450-1550 | 1500-1575 | E | |
| 4063 & 4068 | … | | 1450-1550 | 1475-1550 | E | |
| 4130 | 1600-1700 | and/or | 1450-1550 | 1600-1650 | B | To desired hardness |
| 4137 & 4140 | 1600-1700 | and/or | 1450-1550 | 1550-1600 | E | |
| 4145 & 4150 | 1600-1700 | and/or | 1450-1550 | 1500-1600 | E | |
| 4340 | 1600-1700 | and draw | 1100-1225 | 1475-1525 | E | |
| 4640 | 1600-1700 | and/or | 1450-1550 | 1450-1500 | E | To desired hardness |
| | 1600-1700 | and/or | 1450-1500 | 1450-1500 | E | Gears, 350-450 |
| 5045 & 5046 | 1600-1700 | and/or | 1450-1550 | 1475-1500 | E | 250-300 |
| 5130 & 5132 | 1650-1750 | and/or | 1450-1550 | 1500-1550 | G | To desired hardness |
| 5135 to 5145 | 1650-1750 | and/or | 1450-1550 | 1500-1550 | E | To desired hardness<br>Gears, 350-400 |
| 5147 to 5152 | 1650-1750 | and/or | 1450-1550 | 1475-1550 | E | To desired hardness<br>Gears, 350-400 |
| 50100<br>51100<br>52100 | …<br>… | | 1350-1450<br>1350-1450 | 1425-1475<br>1500-1600 | H<br>E | To desired hardness |
| 6150 | 1650-1750 | and/or | 1550-1650 | 1600-1650 | E | |
| 9254 to 9262 | … | | … | 1500-1650 | E | |
| 8627 to 8632 | 1600-1700 | and/or | 1450-1550 | 1550-1650 | B | |
| 8635 to 8641 | 1600-1700 | and/or | 1450-1550 | 1525-1575 | E | |
| 8642 to 8653 | 1600-1700 | and/or | 1450-1550 | 1500-1550 | E | |
| 8655 & 8660 | 1650-1750 | and/or | 1450-1550 | 1475-1550 | E | |
| 8735 & 8740 | 1600-1700 | and/or | 1450-1550 | 1525-1575 | E | To desired hardness |
| 8745 & 8750 | 1600-1700 | and/or | 1450-1500 | 1500-1550 | E | |
| 9437 & 9440 | 1600-1700 | and/or | 1450-1550 | 1550-1600 | E | |
| 9442 to 9747 | 1600-1700 | and/or | 1450-1550 | 1500-1600 | E | |
| 9840 | 1600-1700 | and/or | 1450-1550 | 1500-1550 | E | |
| 9845 & 9850 | 1600-1700 | and/or | 1450-1550 | 1500-1550 | E | |

[a] Symbols: B = water or oil; E = oil; G = water, caustic solution, or oil; H = water.

**Table 5c. Typical Heat Treatments for SAE Alloy Steels**
(Heat-Treating Grades — Chromium–Nickel Austenitic Steels)

| SAE No. | Normalize | Anneal,[a] | | Harden | Quenching Medium | Temper |
|---|---|---|---|---|---|---|
| | | Deg. F | Deg. C | °F (°C) | | |
| 30301 to 30347 | … | 1800–2100 | 982–1149 | … | Water or Air | … |

[a] Quench to produce full austenitic structure using water or air in accordance with thickness of section. Annealing temperatures given cover process and full annealing as used by industry, the lower end of the range being used for process annealing.

**Table 5d. Typical Heat Treatments for SAE Alloy Steels**
(Heat-Treating Grades — Stainless Chromium Irons and Steels)

| SAE No.[a] | Normalize | Sub-critical Anneal | | Full Anneal | | Harden | | Quenching Medium | Temper |
|---|---|---|---|---|---|---|---|---|---|
| | | Deg. F | Deg. C | Deg. F | Deg. C | Deg. F | Deg. C | | |
| 51410 | … | 1300–1350[b] | 704–732[b] | 1550–1650[c] | 843–899[c] | … | … | Oil or air | To desired hardness |
| | … | … | … | … | … | 1750–1850 | 954–1010 | | |
| 51414 | … | 1200–1250[b] | 649–677[b] | … | … | … | … | Oil or air | To desired hardness |
| | … | … | … | … | … | 1750–1850 | 954–1010 | | |
| 51416 | … | 1300–1350[b] | 704–732[b] | 1550–1650[c] | 843–899[c] | … | … | Oil or air | To desired hardness |
| | … | … | … | … | … | 1750–1850 | 954–1010 | | |
| 51420 | … | 1350–1450[b] | 732–788[b] | 1550–1650[c] | 843–899[c] | … | … | Oil or air | To desired hardness |
| 51420F | … | … | … | … | … | 1800–1850 | 982–1010 | | |
| 51430 | … | 1400–1500[d] | 760–816[d] | … | … | … | … | … | … |
| 51430F | … | 1250–1500[d] | 677–816[d] | … | … | … | … | … | … |
| 51431 | … | 1150–1225[b] | 621–663[b] | … | … | 1800–1900 | 982–1038 | Oil or air | To desired hardness |
| 51440A 51440B 51440C 51440F | … | 1350–1440[b] | 732–782[b] | 1550–1650[c] | 843–899[c] | 1850–1950 | 1010–1066 | Oil or air | To desired hardness |
| 51442 | … | 1400–1500[d] | 760–816[d] | … | … | … | … | … | … |
| 51446 | … | 1500–1650[d] | 816–899[d] | … | … | … | … | … | … |
| 51501 | … | 1325–1375[b] | 718–746[b] | 1525–1600[c] | 829–871[c] | 1600–1700 | 871–927 | Oil or air | To desired hardness |

[a] Suffixes A, B, and C denote steels differing in carbon content only. Suffix F denotes a free-machining steel.
[b] Usually air cooled, but may be furnace cooled.
[c] Cool slowly in furnace.
[d] Cool rapidly in air.

**Table 6. Typical SAE Heat Treatments for Grades of Chromium-Nickel Austenitic Steels Not Hardenable by Thermal Treatment**

| SAE Steels | AISI No. | Annealing[a] Temperature | | Quenching Medium |
|---|---|---|---|---|
| | | (degrees F) | (degrees C) | |
| 30201 | 201 | 1850-2050 | 1010-1120 | Air |
| 30202 | 202 | 1850-2050 | 1010-1120 | Air |
| 30301 | 301 | 1850-2050 | 1010-1120 | Air |
| 30302 | 302 | 1850-2050 | 1010-1120 | Air |
| 30303 | 303 | 1850-2050 | 1010-1120 | Air |
| 30304 | 304 | 1850-2050 | 1010-1120 | Air |
| 30305 | 305 | 1850-2050 | 1010-1120 | Air |
| 30309 | 309 | 1900-2050 | 1040-1120 | Air |
| 30310 | 310 | 1900-2100 | 1040-1150 | Air |
| 30316 | 316 | 1850-2050 | 1010-1120 | Air |
| 30317 | 317 | 1850-2050 | 1010-1120 | Air |
| 30321 | 321 | 1750-2050 | 955-1120 | Air |
| 30325 | 325 | 1800-2100 | 980-1150 | Air |
| 30330 | … | 1950-2150 | 1065-1175 | Air |
| 30347 | 347 | 1850-2050 | 1010-1120 | Air |

[a] Quench to produce full austenitic structure in accordance with the thickness of the section. Annealing temperatures given cover process and full annealing as already established and used by industry, the lower end of the range being used for process annealing.

*Source: SAE Handbook*, 1990. Reprinted with permission. 

**Table 7. Typical SAE Heat Treatments for Stainless Chromium Steels**

| SAE Steels | AISI No. | Normalizing Temperature | Subcritical Annealing Temperature | | Full Annealing[a] Temperature | | Hardening Temperature (degrees F) | | Quenching Medium | Temper |
|---|---|---|---|---|---|---|---|---|---|---|
| | | | degrees F | degrees C | degrees F | degrees C | degrees F | degrees C | | |
| 51409 | … | … | … | … | 1625 | 885 | … | … | Air | … |
| 51410 | 410 | | 1300–1350[b] | 705–730[b] | 1500–1650 | 845–900 | 1700–1850 | 925–1010 | Oil or air | To desired hardness |
| 51414 | 414 | … | 1200–1250[b] | 650–675[b] | … | … | 1800–1900 | 980–1040 | Oil or air | To desired hardness |
| 51416 | 416 | … | 1300–1350[c] | 705–730[c] | 1500–1650 | 845–900 | 1700–1850 | 925–1010 | Oil or air | To desired hardness |
| 51420 | 420 | … | 1350–1450[b] | 730–790[b] | 1550–1650 | 845–900 | 1800–1900 | 980–1040 | Oil or air | To desired hardness |
| 51420F[d] | … | … | 1350–1450[b] | 730–790[b] | 1550–1650 | 845–900 | 1800–1900 | 980–1040 | Oil or air | To desired hardness |
| 51430 | 430 | … | 1400–1500[c] | 760–815[c] | … | … | … | … | … | … |
| 51430F[d] | … | … | 1250–1400[c] | 675–760[c] | … | … | … | … | … | … |
| 51431 | 431 | … | 1150–1225[b] | 620–665[b] | … | … | 1800–1950 | 980–1065 | Oil or air | To desired hardness |
| 51434 | … | … | 1400–1600[c] | 760–870[c] | … | … | … | … | … | … |
| 51436 | … | | | | | | | | | |
| 51440A[d] | 440A[d] | … | 1350–1440[b] | 730–780[b] | 1550–1650 | 845–900 | 1850–1950 | 1010–1065 | Oil or air | To desired hardness |
| 51440B[d] | 440B[d] | | | | | | | | | |
| 51440C[d] | 440C[d] | | | | | | | | | |
| 51440F[d] | … | | | | | | | | | |
| 51442 | 442 | … | 1350–1500[c] | 730–815[c] | … | … | … | … | … | … |
| 51446 | 446 | … | 1450–1600[b] | 790–870[b] | … | … | … | … | … | … |
| 51501 | 501 | … | 1325–1375[c] | 720–745[c] | 1525–1600 | 829–871 | 1600–1700 | 871–927 | Oil or air | To desired hardness |

[a] Cool slowly in furnace.
[b] Usually air cooled but may be furnace cooled.
[c] Cool rapidly in air.
[d] Suffixes A, B, and C denote three types of steel differing only in carbon content. Suffix F denotes a free-machining steel.

*Source: SAE Handbook*, 1990. Reprinted with permission. 

## Heat Treating High-Speed Steels

**Cobaltcrom Steel.—**A tungstenless alloy steel or high-speed steel that contains approximately 1.5 percent carbon, 12.5 percent chromium, and 3.5 percent cobalt. Tools such as dies and milling cutters made from cobaltcrom steel can be cast to shape in suitable molds, the cutter teeth being formed so that it is necessary only to grind them.

Before the blanks can be machined, they must be annealed; this operation is performed by pack annealing at the temperature of 1800° F (982° C), for a period of from 3 to 6 hours, according to the size of the castings being annealed. The following directions are given for the hardening of blanking and trimming dies, milling cutters, and similar tools made from cobaltcrom steel: Heat slowly in a hardening furnace to about 1830° F (999° C), and hold at this temperature until the tools are thoroughly soaked. Reduce the temperature about 50° F (28° C), withdraw the tools from the furnace, and allow them to cool in the atmosphere. As soon as the red color disappears from the cooling tool, place it in quenching oil until cold. The slight drop of 50° F (28° C) in temperature while the tool is still in the hardening furnace is highly important to obtain proper results. The steel will be injured if the tool is heated above 1860° F (1016° C). In cooling milling cutters or other rotary tools, it is suggested that they be suspended on a wire to ensure a uniform rate of cooling.

Tools that are to be subjected to shocks or vibration, such as pneumatic rivet sets, shear blades, etc., should be heated slowly to 1650° F (899° C), after which the temperature should be reduced to about 1610° F (877° C), at which point the tool should be removed from the furnace and permitted to cool in the atmosphere. No appreciable scaling occurs in the hardening of cobaltcrom steel tools.

**Preheating Tungsten High-Speed Steel.—**Tungsten high-speed steel must be hardened at a very high temperature; consequently, tools made from such steel are seldom hardened without at least one preheating stage to avoid internal strain. This requirement applies especially to milling cutters, taps, and other tools having thin teeth and thick bodies and to forming tools of irregular shape and section. The tools should be heated slowly and carefully to a temperature somewhat below the critical point of the steel, usually in the range of 1500 to 1600° F (816 to 871° C). Limiting the preheating temperature prevents the operation from being unduly sensitive, and the tool may be safely left in the furnace until it reaches a uniform temperature throughout its length and cross section.

A single stage of preheating is customary for tools of simple form that are not more than from 1 to 1½ inches (25 to 38 mm) in thickness. For large, intricate tools, two stages of preheating are frequently used. The first brings the tool up to a temperature of about 1100 to 1200° F (593 to 649° C), and the second raises its temperature to 1550 to 1600° F (843 to 649° C). A preheating time of 5 minutes for each ¼ inch (6.4 mm) in tool thickness has been recommended for a furnace temperature of 1600° F (871° C). This is where a single stage of preheating is used and the furnace capacity should be sufficient to maintain practically constant temperature when the tools are changed. To prevent undue chilling, it is common practice to insert a single tool or a small lot in the hardening furnace whenever a tool or lot is removed rather than to insert a full charge of cold metal at one time.

Preheating is usually done in a simple type of oven furnace heated by gas, electricity, or oil. Atmospheric control is seldom used, although for 18-4-1 steel a slightly reducing atmosphere (2 to 6 percent carbon monoxide) has been found to produce the least amount of scale and will result in a better surface after final hardening.

**Hardening of Tungsten High-Speed Steel.—**All tungsten high-speed steels must be heated to a temperature close to their fusion point to develop their maximum efficiency as metal-cutting tools. Hardening temperatures ranging from 2200 to 2500° F (1204 to 1371° C) may be needed. The effects of changes in the hardening temperature on the cutting efficiency of several of the more common high-speed steels are shown in Table 1. The figures given are ratios, the value 1.00 for each steel being assigned to the highest observed cutting speed for that steel. The figures for different steels, therefore, cannot be directly compared with each other, except to note changes in the point of maximum cutting efficiency.

**Table 1. Relation of Hardening Temperature to Cutting Efficiency**

| Hardening Temperature | | Typical Analyses of High-Speed Steels | | | |
|---|---|---|---|---|---|
| Deg. F | Deg. C | 18 – 4 – 1 | 14 – 4 – 2 | 18 – 4 –1 Cobalt | 14 – 4 – 2 Cobalt |
| 2200 | 1204 | 0.86 | 0.83 | 0.84 | 0.85 |
| 2250 | 1232 | 0.88 | 0.88 | 0.86 | 0.88 |
| 2300 | 1260 | 0.90 | 0.93 | 0.90 | 0.91 |
| 2350 | 1288 | 0.95 | 0.98 | 0.94 | 0.94 |
| 2400 | 1316 | 0.99 | 0.98 | 0.98 | 0.98 |
| 2450 | 1343 | 1.00 | … | 0.99 | 1.00 |
| 2500 | 1371 | 0.98 | … | 1.00 | 0.97 |

The figures in the table refer to tools heated in an oven-type furnace in which a neutral atmosphere is maintained. The available data indicate that a steel reaches its best cutting qualities at a temperature approximately 50° F (28° C) lower than the figures in the table if it is hardened in a bath-type furnace. It is, however, desirable to use a hardening temperature approximately 50° F (28° C) lower than that giving maximum cutting qualities to avoid the possibility of overheating the tool.

*Length of Time for Heating:* The cutting efficiency of a tool is affected by the time that it is kept at the hardening temperature, almost as much as by the hardening temperature itself. It has been common practice to heat a tool for hardening until a "sweat" appears on its surface. This sweat is presumably a melting of the oxide film on the surface of a tool heated in an oxidizing atmosphere. It does not appear when the tool is heated in an inert atmosphere. This method of determining the proper heating time is at best an approximation and indicates only the temperature on the outside of the tool rather than the condition of the interior. As such, it cannot be relied upon to give consistent results.

The only safe method is to heat the tool for a definite predetermined time, based on the size and the thickness of metal that the heat must penetrate to reach the interior. The values given in Table 2 are based on a series of experiments to determine the relative cutting efficiency of a group of tools hardened in an identical manner except for variations in the time the tools were kept at the hardening temperature. The time given is based on that required to harden throughout a tool resting on a conducting hearth; the tool receives heat freely from three sides, on its large top surface and its smaller side surfaces. (The table does not apply to a disk lying flat on the hearth.) For a tool having a projecting cutting edge, such as a tap, the thickness or depth of the projecting portion on which the cutting edge is formed should be used when referring to the table.

**Table 2. Length of Heating Time for Through Hardening**

| High-Speed Steel Tool Thickness | | Time in Furnace at High Heat, Minutes | High-Speed Steel Tool Thickness | | Time in Furnace at High Heat, Minutes | High-Speed Steel Tool Thickness | | Time in Furnace at High Heat, Minutes |
|---|---|---|---|---|---|---|---|---|
| Inch | Centimeter | | Inch | Centimeter | | Inch | Centimeter | |
| ¼ | 0.635 | 2 | 1½ | 3.81 | 7 | 5 | 12.7 | 18 |
| ½ | 1.27 | 3 | 2 | 5.08 | 8 | 6 | 15.24 | 20 |
| ¾ | 1.905 | 4 | 3 | 7.62 | 12 | 8 | 20.32 | 25 |
| 1 | 2.54 | 5 | 4 | 10.16 | 15 | 10 | 25.4 | 30 |

The time periods given in Table 2 are based on complete penetration of the hardening effect. For very thick tools, the practical procedure is to harden to a depth sufficient to produce an adequate cutting edge, leaving the interior of the tool relatively soft.

Where atmosphere control is not provided, it often will be found impracticable to use both the temperature for maximum cutting efficiency, given in Table 1, and the heating time, given in Table 2, because abnormal scaling, grain growth, and surface decarburization of the tool will result. The principal value of an accurate control of the furnace atmosphere appears to lie in the fact that its use makes possible the particular heat treatment that produces the best structure in the tool without destruction of the tool surface or grain.

**Quenching Tungsten High-Speed Steel.—**High-speed steel is usually quenched in oil. The oil bath offers a convenient quench; it calls for no unusual care in handling and brings about a uniform and satisfactory rate of cooling, which does not vary appreciably with the temperature of the oil. Some authorities believe it desirable to withdraw the tool from the oil bath for a few seconds after it has reached a dull red. It is also believed desirable to move the tool around in the quenching oil, particularly immediately after being placed in it, to prevent the formation of a gas film on the tool. Such a film is usually a poor conductor of heat and slows the rate of cooling.

*Salt Bath:* Quenching in a lead or salt bath at from 1000 to 1200° F (538 to 649° C) has the advantage that cooling of the tool from hardening to room temperature is accomplished in two stages, thus reducing the possibility of setting up internal strains that may tend to crack the tool. The quenching temperature is sufficiently below the lower critical point for a tool so quenched to be allowed to cool to room temperature in still air. This type of quench is particularly advantageous for tools of complicated section that would easily develop hardening cracks. The salt quench has the advantage that the tool sinks and requires only a support, whereas the same tool will float in the lead bath and must be held under the surface. It is believed that the lead quench gives a somewhat higher matrix hardness, and is of advantage for tools that tend to fail by nose abrasion. Tools treated as described are brittle unless given a regular tempering treatment, because the 1000° F (538° C) quenching temperature is not a substitute for later tempering at the same temperature, after the tool has cooled to room temperature.

*Air Cooling:* Many high-speed steel tools are quenched in air, either in a stream of dry compressed air or in still air. Small sections harden satisfactorily in still air, but heavier sections should be subjected to air under pressure. One advantage of air cooling is that the tool can be kept straight and free from distortion, although it is likely that there will be more scale on a tool thus quenched than when oil, lead, or salt is used. Cooling between steel plates may help to keep thin flat tools straight and flat.

**Straightening Tools when Quenching.—**The final straightness required in a tool must be considered when it is quenched. When several similar tools are to be hardened, a jig can be used to advantage for holding the tools while quenching. When long slender tools are quenched without holders, they frequently warp and must be straightened later. The best time for this straightening is during the first few minutes after the tools have been quenched, as the steel is then quite pliable and may be straightened without difficulty. The straightening must be done at once, as the tools become hard in a few minutes.

**Anneal Before Rehardening.—**Tools that are too soft after hardening must be annealed before rehardening. A quick anneal, such as previously described, is all that is required to put such a tool into the proper condition for rehardening. This treatment is absolutely essential. For milling cutters and forming tools of irregular section, a full anneal should be used.

**Tempering or Drawing Tungsten High-Speed Steel.—**The tempering or drawing temperature for high-speed steel tools usually varies from 900 to 1200° F (482 to 649° C). This temperature is higher for turning and planing tools than for such tools as milling cutters, forming tools, etc. If the temperature is below 800° F (427° C), the tool is likely to be too brittle. The general idea is to temper tools at the highest temperature likely to occur in service. Because this temperature ordinarily is unknown, the general practice is to temper at whatever temperature experience with that particular steel and tool has proved to be the best. The furnace used for tempering usually is kept at a temperature of from 1000 to

1100° F (538 to 593° C) for ordinary high-speed steels and from 1200 to 1300° F (649 to 704° C) for steels of the cobalt type. These furnace temperatures apply to tools of the class used on lathes and planers. Such tools, in service, frequently heat to the point of visible redness. Milling cutters, forming tools, or any other tools for lighter duty may be tempered as low as 850 or 900° F (454 to 4821° C). When the tool has reached the temperature of the furnace, it should be held at this temperature for from one to several hours until it has been heated evenly throughout. It should then be allowed to cool gradually in the air and in a place that is dry and free from air drafts. In tempering, the tool should not be quenched because quenching tends to produce strains that may result later in cracks.

**Annealing Tungsten High-Speed Steel.—**The following method of annealing high-speed steel has been used extensively. Use an iron box or pipe of sufficient size to allow at least ½ inch (12.7 mm) of packing between the pieces of steel to be annealed and the sides of the box or pipe. It is not necessary that each piece of steel be kept separate from every other piece, but only that the steel be prevented from touching the sides of the annealing pipe or box. Pack carefully with powdered charcoal, fine dry lime, or mica (preferably charcoal), and cover with an airtight cap or lute with fire clay; heat slowly to 1600 to 1650° F (871 to 899° C) and keep at this heat from 2 to 8 hours, depending on the size of the pieces to be annealed. A piece measuring 2 × 1 × 8 inches (5 × 2.5 × 20.3 cm) requires about 3 hours. Cool as slowly as possible, and do not expose to the air until cold because cooling in air is likely to cause partial hardening. A good method is to allow the box or pipe to remain in the furnace until cold.

**Hardening Molybdenum High-Speed Steels.—**Table 3 gives the compositions of several molybdenum high-speed steels that are widely used for general commercial tool applications. The general method of hardening molybdenum high-speed steels resembles that used for 18-4-1 tungsten high-speed steel except that the hardening temperatures are lower and more precautions must be taken to avoid decarburization, especially on tools made from Type I or Type II steels, when the surface is not ground after hardening. Either salt baths or atmosphere-controlled furnaces are recommended for hardening molybdenum high-speed steels.

**Table 3. Compositions of Molybdenum High-Speed Steels**

| | Molybdenum-Tungsten | | Molybdenum-Vanadium | Tungsten-Molybdenum |
|---|---|---|---|---|
| Element | Type Ia (Percent) | Type Ib[a] (Percent) | Type II (Percent) | Type III (Percent) |
| Carbon | 0.70-0.85 | 0.76-0.82 | 0.70-0.90 | 0.75-0.90 |
| Tungsten | 1.25-2.00 | 1.60-2.30 | … | 5.00-6.00 |
| Chromium | 3.00-5.00 | 3.70-4.20 | 3.00-5.00 | 3.50-5.00 |
| Vanadium | 0.90-1.50 | 1.05-1.35 | 1.50-2.25 | 1.25-1.75 |
| Molybdenum | 8.00-9.50 | 8.00-9.00 | 7.50-9.50 | 3.50-5.50 |
| Cobalt | See footnote | 4.50-5.50 | See footnote | See footnote |

[a] Cobalt may be used in any of these steels in varying amounts up to 9 percent, and the vanadium content may be as high as 2.25 percent. When cobalt is used in Type III steel, the vanadium content may be as high as 2.25 percent. When cobalt is used in Type III steel, this steel becomes susceptible to decarburization. As an illustration of the use of cobalt, Type Ib steel is included. This is steel T10 in the US Navy Specification 46S37, dated November 1, 1939.

The usual method is to preheat uniformly in a separate furnace to 1250 to 1550° F (677 to 843° C), then transfer to a high-heat furnace maintained within the hardening temperature range given in Table 4. Single-point cutting tools, in general, should be hardened at the upper end of the temperature range indicated by Table 4. Slight grain coarsening is not objectionable in such tools when they are properly supported in service and are not subjected to chattering; however, when these tools are used for intermittent cuts, it is better to use the middle of the temperature range. All other cutting tools, such as drills, countersinks, taps, milling cutters, reamers, broaches, and form tools, should be hardened in the middle of the range shown. For certain tools, such as slender taps, cold punches, and blanking and trimming dies, where greater toughness to resist shocks is required, the lower end of the hardening temperature range should be used.

**Table 4. Heat Treatment of Molybdenum High-Speed Steels**

| Heat-Treating Operation | Molybdenum–Tungsten Types Ia and Ib[a] Temp. | | Molybdenum–Vanadium Type II Temp. | | Tungsten–Molybdenum Type III Temp. | |
|---|---|---|---|---|---|---|
| | Deg. F | Deg. C | Deg. F | Deg. C | Deg. F | Deg. C |
| Forging | 1850–2000 | 1010–1093 | 1850–2000 | 1010–1093 | 1900–2050 | 1038–1121 |
| Not below | 1600 | 871 | 1600 | 871 | 1600 | 871 |
| Annealing | 1450–1550 | 788–843 | 1450–1550 | 788–843 | 1450–1550 | 788–843 |
| Strain relief | 1150–1350 | 621–732 | 1150–1350 | 621–732 | 1150–1350 | 621–732 |
| Preheating | 1250–1500 | 677–816 | 1250–1500 | 677–816 | 1250–1550 | 677–843 |
| Hardening[b] | 2150–2250[a] | 1177–1232[a] | 2150–2250 | 1177–1232 | 2175–2275 | 1191–1246 |
| Salt | 2150–2225 | 1177–1218 | 2150–2225 | 1177–1218 | 2150–2250 | 1177–1232 |
| Tempering | 950–1100 | 510–593 | 950–1100 | 510–593 | 950–1100 | 510–593 |

[a] For similar working conditions, Type Ib steel requires slightly higher hardening heat than Type Ia.
[b] The higher side of the hardening range should be used for large sections, and the lower side for small sections.

Molybdenum high-speed steels can be pack-hardened following the same practice as is used for tungsten high-speed steels but keeping on the lower side of the hardening range (approximately 1850° F or 1010° C). Special surface treatments such as nitriding by immersion in molten cyanide that are used for tungsten high-speed steels are also applicable to molybdenum high-speed tools.

When heated in an open fire or in furnaces without atmosphere control, these steels do not sweat like 18-4-1 steels; consequently, determining the proper time in the high-heat chamber is a matter of experience. This time approximates that used with 18-4-1 steels, although it may be slightly longer when the lower part of the hardening range is used. Much can be learned by preliminary hardening of test pieces and checking on the hardness fracture and structure. It is difficult to give the exact heating time because it is affected by temperature, type of furnace, size and shape, and furnace atmosphere. Rate of heat transfer is most rapid in salt baths and slowest in controlled-atmosphere furnaces with high carbon monoxide content.

**Quenching and Tempering of Molybdenum High-Speed Tools.—**Quenching may be done in oil, air, or molten bath. To reduce the possibility of breakage and undue distortion of intricately shaped tools, it is advisable to quench in a molten bath at approximately 1100° F (593° C). The tool also may be quenched in oil and removed while still red, or at approximately 1100° F (593° C). The tool is then cooled in air to room temperature and tempered immediately to avoid cracking.

When straightening is necessary, it should be done after quenching and before cooling to room temperature prior to tempering.

To temper, the tools should be reheated slowly and uniformly to 950 to 1100° F (510 to 593° C). For general work, 1050° F (566° C) is most common. The tools should be held at this temperature at least 1 hour. Two hours is a safer minimum, and 4 hours is maximum. The time and temperature depend on the hardness and toughness required. Where tools are subjected to more or less shock, multiple temperings are suggested.

**Protective Coatings for Molybdenum Steels.—**To protect the surface from oxidation during heat treatment, borax may be sprinkled lightly over the steel when the latter is heated in a furnace to a low temperature, 1200 to 1400° F (649 to 760° C). Small tools may be rolled in a box of borax before heating. Another method more suitable for finished tools is to apply the borax or boric acid in the form of a supersaturated water solution. The tools are then immersed in the solution at 180 to 212° F (82 to 100° C), or the solution may be applied with a brush or spray. Pieces so treated are heated as usual and carefully handled to ensure good adherence of the coating. When properly applied, special protective coatings or paints have been found to be extremely useful. These materials do not fuse or run at the temperatures used and therefore do not affect the furnace hearth. When applying these coatings, it is necessary to have a surface free from scale or grease to ensure good adherence. Coatings may be sprayed or brushed on, and usually one thin coat is sufficient. Heavy coats tend to pit the surface of the tool and are difficult to remove. Tools covered with these coatings should be allowed to dry before they are charged into the preheat furnace. After hardening and tempering, the coating can be easily removed by light blasting with sand or steel shot. When tools are lightly ground, these coatings come off immediately. Protection may also be obtained by wrapping pieces in stainless steel foil.

**Nitriding High-Speed Steel Tools.**—Nitriding is applied to high-speed steel for the purpose of increasing tool life by producing a very hard skin or case, the thickness of which ordinarily is from 0.001 to 0.002 inch (0.025 to 0.05 mm). Nitriding is done after the tool has been fully heat treated and finish ground. (The process differs entirely from that which is applied to surface harden certain alloy steels by heating in an atmosphere of nitrogen or ammonia gas.) The temperature of the high-speed steel nitriding bath, which is a mixture of sodium and potassium cyanides, is equal to or slightly lower than the tempering temperature. For ordinary tools, this temperature usually varies from about 1025 to 1050° F (552 to 566° C); but, if the tools are exceptionally fragile, the range may be reduced to 950 or 1000° F (510 to 538° C). Accurate temperature control is essential to prevent exceeding the final tempering temperature. The nitriding time may vary from 10 or 15 minutes to 30 minutes or longer, and should be determined by experiment. The shorter periods are applied to tools for iron or steel, or any shock-resisting tools, and the longer periods are for tools used in machining nonferrous metals and plastics. This nitriding process is applied to tools, such as hobs, reamers, taps, box tools, form tools, and milling cutters. Nitriding may increase tool life 50 to 200 percent, or more, but it should always be preceded by correct heat treatment.

*Nitriding Bath Mixtures and Temperatures:* A mixture of 60 percent sodium cyanide and 40 percent potassium cyanide is commonly used for nitriding. This mixture has a melting point of 925° F (496° C), which is gradually reduced to 800° F (427° C) as the cyanate content of the bath increases. A more economical mixture of 70 percent sodium cyanide and 30 percent potassium cyanide may be used if the bath temperature is only 1050° F (566° C). Nitriding bath temperatures should not exceed 1100° F (593° C) because higher temperatures accelerate the formation of carbonate at the expense of the essential cyanide. A third mixture used for nitriding consists of 55 percent sodium cyanide, 25 percent potassium chloride, and 20 percent sodium carbonate. This mixture melts at 930° F (499° C).

**Equipment for Hardening High-Speed Steel.**—Equipment for hardening high-speed steel consists of a hardening furnace capable of maintaining a temperature of 2350 to 2450° F (1288 to 1343° C); a preheating furnace capable of maintaining a temperature of 1700 to 1800° F (927 to 982° C) and of sufficient size to hold a number of pieces of the work; a tempering (drawing) furnace capable of maintaining a temperature of 1000 to 1200° F (538 to 649° C) as a general rule; and a water-cooled tank of quenching oil.

High-speed steels usually are heated for hardening either in some type of electric furnace or in a gas-fired furnace of the muffle type. The small furnaces used for high-speed steel seldom are oil-fired. It is desirable to use automatic temperature control and, where an oven-type of furnace is employed, a controlled atmosphere is advisable because of the variations in cutting qualities caused by hardening under uncontrolled conditions. Some furnaces of both electric and fuel-fired types are equipped with a salt bath suitable for high-speed steel hardening temperatures. Salt baths have the advantage of providing protection against the atmosphere during the heating period. A type of salt developed for commercial use is water-soluble so that all deposits from the hardening bath may be removed by immersion in water after quenching in oil or salt, or after air cooling. One type of electric furnace heats the salt bath internally by electrodes immersed in it. The same type of furnace is also applied to various heat-treating operations, such as cyanide hardening, liquid carburizing, tempering, and annealing.

An open-forge fire has many disadvantages, especially in hardening cutters or other tools that cannot be ground all over after hardening. The air blast decarburizes the steel, and lack of temperature control makes it impossible to obtain uniform results. Electric and gas furnaces provide continuous uniform heat, and the temperature may be regulated accurately, especially when pyrometers are used. In shops equipped with only one furnace for carbon steel and one for high-speed steel, the tempering can be done in the furnace used for hardening carbon steel after the preheating is finished and the steel has been removed for hardening.

**Heating High-Speed Steel for Forging.—**Care should be taken not to heat high-speed steel for forging too abruptly. In winter, the steel may be extremely cold when brought into the forge shop. If the steel is put directly into the hot forge fire, it is likely to develop cracks that will show up later in the finished tool. The steel, therefore, should be warmed gradually before heating for forging.

## Subzero Treatment of Steel

Subzero treatment consists of subjecting the steel, after hardening and either before or after tempering, to a subzero temperature that usually ranges from −100 to −120° F (−73 to −84° C), and for a period of time varying with the size or volume of the tool, gage, or other part. Commercial equipment is available for obtaining these low temperatures.

The subzero treatment is employed by most gage manufacturers to stabilize precision gages and prevent subsequent changes in size or form. Subzero treatment is also applied to some high-speed steel cutting tools. The object here is to increase the durability or life of the tools; however, up to the present time, the results of tests by metallurgists and tool engineers often differ considerably and in some instances are contradictory. Methods of procedure also vary, especially with regard to the order and number of operations in the complete heat-treating and cooling cycle.

**Changes Resulting From Subzero Treatment.—**When steel is at the hardening temperature it contains a solid solution of carbon and iron known as *austenite*. When the steel is hardened by sudden cooling, most of the austenite, which is relatively soft, tough, and ductile even at room temperatures, is transformed into martensite, a hard and strong constituent. If all the austenite were changed to martensite upon reaching room temperature, this process would be an ideal hardening operation, but many steels retain some austenite. In general, the higher the carbon and alloy contents and the higher the hardening temperature, the greater the tendency to retain austenite. When steel is cooled to subzero temperatures, the stability of the retained austenite is reduced so that it is more readily transformed. To obtain more complete transformation, the subzero treatment may be repeated. The ultimate transformation of austenite to martensite may take place in carbon steel without the aid of subzero treatment, but this natural transformation might require 6 months or longer, whereas by refrigeration this change occurs in a few hours.

The thorough, uniform heating that is always recommended in heat-treating operations should be accompanied by thorough, uniform cooling when the subzero treatment is applied. To ensure uniform cooling, the subzero cooling period should be increased for the larger tools and it may range from 2 to 6 hours. The tool or other part is sometimes surrounded by one or more layers of heavy wrapping for insulating material to delay the cooling somewhat and ensure uniformity. After the cooling cycle is started, it should continue without interruption.

Subzero treatment may sometimes cause cracking. Normally, the austenite in steel provides a cushioning effect that may prevent cracking or breakage resulting from treatments involving temperature and dimensional changes; but if this cushioning effect is removed, particularly at very low temperatures as in subzero treatments, there may be danger of cracking, especially with tools having large or irregular sections and sharp corners offering relatively low resistance to stresses. This effect is one reason why subzero treatments may differ in regard to the cooling and tempering cycle.

**Stabilizing Dimensions of Gages or Precision Parts by Subzero Cooling.—** Transformation of austenite into martensite is accompanied by an increase in volume; consequently, the transformation of austenite, that may occur naturally over a period of months or years, tends to change the dimensions and form of steel parts, and such changes may be serious in the case of precision gages, close-fitting machine parts, etc. To prevent such changes, the subzero treatment has proved effective. Gage blocks, for example, may be stabilized by hardening followed by repeated cycles of chilling and tempering, to transform a large percentage of the austenite into martensite.

*Order of Operations for Stabilizing Precision Gages:* If precision gages and sine-bars are heat treated in the ordinary manner and then finished without some stabilizing treatment, dimensional changes and warpage are liable to occur. Subzero cooling provides a practical and fairly rapid method of obtaining the necessary stabilization by transforming the austenite into martensite. In stabilization treatments of this kind, tempering is the final operation. One series of treatments that has been recommended after hardening and rough-grinding is as follows:

a) Cool to −120° F (−84° C). This cooling period may require from 1 to 6 hours, depending on the size and form of the gage.

b) Place gage in boiling water for 2 hours (oil or salt bath may also be used).

Note: Steps Item a) and Item b) may be repeated from two to six times, depending on the size and form of the gage. These repeated cycles will eventually transform practically all the austenite into martensite. Two or three cooling and drawing operations usually are sufficient for such work as thread gages and gage blocks.

c) Follow with regular tempering or drawing operation and finish gage by lapping.

*Series of Stabilizing Treatments for Chromium Steel:* The following series of treatments has proved successful in stabilizing precision gage blocks made from SAE 52100 chromium steel.

a) Preheat to 600° F (316° C), then heat to 1575° F (857° C) for a period of four minutes.

b) Quench in oil at 85° F (29° C). Uniform quenching is essential.

c) Temper at 275° F (135° C) for one hour.

d) Cool in tempering furnace to room temperature.

e) Continue cooling in atmosphere of industrial refrigerator for 6 hours with temperature of atmosphere at −120° F (−84° C).

f) Allow gage blocks to return to room temperature and again temper.

*Note:* The complete treatment consists of six subzero cooling periods, each followed by a tempering operation. The transformation to martensite is believed to be complete even after the fifth cooling period. The hardness is about 66 RC (Rockwell C scale). Transformation is checked by magnetic tests based upon the magnetism of martensite and the nonmagnetic qualities of austenite.

**Stabilizing Dimensions of Close-Fitting Machine Parts.—** Subzero treatment will always cause an increase in size. Machine parts subjected to repeated and perhaps drastic changes in temperature, as in aircraft, may eventually cause trouble due to growth or warpage as the austenite gradually changes to martensite. In some instances, the sizes of close-fitting moving parts have increased sufficiently to cause seizure. Such treatment, for example, may be applied to precision bearings made from SAE 52100 or alloy carburizing steels for stabilizing or aging them. *Time* aging of 52100 steels after hardening has been found to cause changes as large as 0.0025 inch (0.064 mm) in medium size sections. A practical remedy is to apply the subzero treatment before the final grinding or other machining operation.

**Subzero Treatment of Carburized Parts to Improve Physical Properties.—**The subzero treatment has been applied to carburized machine parts. For example, the amount of retained austenite in carburized gears may be sufficient to reduce the life of the gears. In one component, the Rockwell hardness was increased from 55 to 65 RC without loss of impact resistance qualities; in fact, impact and fatigue resistance may be increased in some examples.

**Application of Subzero Treatments to High-Speed Steel.—**The subzero treatment has been applied to such tools as milling cutters, hobs, taps, broaches, and drills. It is applicable to different classes of high-speed steels, such as the 18-4-1 tungsten, 18-4-14 cobalt, and the molybdenum high-speed steels. This *cold* treatment is applied preferably in conjunction with the heat treatment, both being combined in a continuous cycle of operations. The general procedure is either to harden the steel, cool it to a subzero temperature, and then temper; or, especially if there is more than one tempering operation, the first tempering may *precede* subzero cooling. The cooling and tempering cycle may be repeated two or more times. The number and order of the operations, or the complete cycle, may be varied to suit the class of work and to minimize the danger of cracking, particularly if the tool has large or irregular sections, sharp corners or edges, or a high cobalt content. A subzero treatment of some kind with a final tempering operation for stress relief is intended to increase strength and toughness without much loss in hardness; consequently, if there is greater strength at a given hardness, tools subjected to subzero treatment can operate with a higher degree of hardness than those heat treated in the ordinary manner, or, if preferred, greater toughness can be obtained by tempering to the original degree of hardness.

**Order of Cooling and Tempering Periods for High-Speed Steel.—**The order or cycle for the cooling and tempering periods has not been standardized. The methods that follow have been applied to high-speed steel tools. They are given as examples of procedure and are subject to possible changes due to subsequent developments. The usual ranges of preheating and hardening temperatures are given, but, for a particular steel, the recommended temperatures should be obtained from the manufacturer.

1) *Double Subzero Treatment:* (For rugged simple tool forms without irregular sections, sharp corners or edges where cracks might develop during the subzero treatment).

a) Preheat between 1400 and 1600° F (760 and 871° C). Double preheating is preferable, the first preheating ranging from 700 to 1000° F (371 to 538° C).

b) Heat to the hardening temperature. *Note:* Tests indicate that the effect of subzero treatment on high-speed steel may be influenced decidedly by the hardening temperature. If this temperature is near the lower part of the range, the results are unsatisfactory. Effective temperatures for ordinary high-speed steels appear to range from 2300 to 2350° F (1260 to 1288° C).

c) Quench in oil, salt, lead, or air, down to a workpiece temperature of 150–200° F (66–93° C). *Note:* One method is to quench in oil; a second method is to quench in oil to about 200–225° F (93–107° C) and then air cool; a third method is to quench in salt bath at 1050–1100° F (566–593° C) and then air cool.

d) Cool in refrigerating unit to temperature of −100 to −120° F (−73 to −84° C) *right after quenching. Note:* Tests have shown that a delay of one hour has a detrimental effect, and in 10 hours the efficiency of the subzero treatment is reduced 50 percent. This is because the austenite becomes more and more stabilized when the subzero treatment is delayed; consequently, the austenite is more difficult to transform into martensite.) The refrigerating period usually varies from 2 to 6 hours, depending on the size of the tool. Remove the tool from the refrigerating unit and allow it to return to room temperature.

e) Temper to required hardness for a period of 2½ to 3 hours. The tempering temperature usually varies from a minimum of 1000 to 1100° F (538 to 593° C) for ordinary high-speed steels. Tests indicate that if this first tempering is less than $2^1/_2$ hours at 1050° F (566° C), there will not be sufficient precipitation of carbides at the tempering temperature to allow complete transformation of the retained austenite on cooling, whereas more than 3 hours causes some loss in room temperature hardness, hot hardness, strength, and toughness.

f) Repeat subzero treatment, step Item d).

g) Repeat the tempering operation, Item e). *Note:* The time for the second tempering operation is sometimes reduced to about ½ the time required for the first tempering.

2) *Single Subzero Treatment:* This treatment is the same as procedure Item 1) except that a second subzero cooling is omitted; hence, the cycle consists of hardening, subzero cooling, and double tempering. Procedure Item 3), which follows, also has one subzero cooling period in the cycle, but this *follows* the first tempering operation.

3) *Tempering Followed by Subzero Treatment:* This treatment is for tools having irregular sections, sharp corners, or edges where cracks might develop if the hardening operation were followed immediately by subzero cooling.

a) Preheat and heat for hardening.

b) Preheat and heat for hardening.

c) Quench as described under procedure Item 1).

d) Temper to required hardness.

e) Cool to subzero temperature −100 to −120° F (−73 to −84° C) and then allow the tool to return to room temperature.

f) Repeat tempering operation.

## Testing the Hardness of Metals

**Brinell Hardness Test.—**The Brinell test for determining the hardness of metallic materials consists in applying a known load to the surface of the material to be tested through a hardened steel ball of known diameter. The diameter of the resulting permanent impression in the metal is measured and the Brinell Hardness Number (BHN)* is then calculated from the following formula in which $D$ = diameter of ball in millimeters, $d$ = measured diameter at the rim of the impression in millimeters, and $P$ = applied load in kilograms.

$$\text{BHN} = \frac{\text{load on indenting tool in kilograms}}{\text{surface area of indentation in sq. mm.}} = \frac{P}{\frac{\pi D}{2}(D - \sqrt{D^2 - d^2})}$$

If the steel ball were not deformed under the applied load and if the impression were truly spherical, then the preceding formula would be a general one, and any combination of applied load and size of ball could be used. The impression, however, is not quite a spherical surface because there must always be some deformation of the steel ball and some recovery of form of the metal in the impression. Hence, for a standard Brinell Hardness test, the size and characteristics of the ball and the magnitude of the applied load must be standardized. In the standard test, a ball 10 millimeters in diameter and a load of 3000, 1500, or 500 kilograms is used. It is desirable, although not mandatory, that the test load be of such magnitude that the diameter of the impression be in the range of 2.50 to 4.75 millimeters. The following test loads and approximate Brinell Hardness Numbers for this range of impression diameters are: 3000 kg, 160 to 600 BHN; 1500 kg, 80 to 300 BHN; 500 kg, 26 to 100 BHN. In making a Brinell Hardness test, the load should be applied steadily and without a jerk for at least 15 seconds for iron and steel, and at least 30 seconds in testing other metals. A minimum period of 2 minutes, for example, has been recommended for magnesium and magnesium alloys. (For the softer metals, loads of 250, 125, or 100 kg are sometimes used.)

According to the American Society for Testing and Materials Standard E10-66, a steel ball may be used on material having a BHN not over 450, a Hultgren ball on material not over 500, or a carbide ball on material not over 630. The Brinell Hardness test is not recommended for material having a BHN over 630.

* Brinell Hardness Numbers also may be identified with the letters HB, HBS for hardness measured with a steel indenter, and HBW for hardness measured with a carbide indenter.

**Rockwell Hardness Test.**—The Rockwell hardness tester is essentially a machine that measures hardness by determining the depth of penetration of a penetrator into the specimen under certain fixed test conditions. The penetrator may be either a steel ball or a diamond spheroconical penetrator. The hardness number is related to the depth of indentation, and the number is higher the harder the material. A minor load of 10 kg is first applied, causing an initial penetration; the dial is set at zero on the black-figure scale, and the major load is applied. This major load is customarily 60 or 100 kg when a steel ball is used as a penetrator, but other loads may be used when necessary. The ball penetrator is $\frac{1}{16}$ inch in diameter normally, but other penetrators of larger diameter, such as $\frac{1}{8}$ inch, may be employed for soft metals. When a diamond spheroconical penetrator is employed, the load usually is 150 kg. Experience decides the best combination of load and penetrator for use. After the major load is applied and removed, according to standard procedure, the reading is taken while the minor load is still applied.

**The Rockwell Hardness Scales.**—The various Rockwell scales and their applications are shown in the following table. The type of penetrator and load used with each are shown in Table 5 and Table 6, which give comparative hardness values for different hardness scales.

| Scale | Testing Application |
|---|---|
| A | For tungsten carbide and other extremely hard materials. Also for thin, hard sheets. |
| B | For materials of medium hardness such as low- and medium-carbon steels in the annealed condition. |
| C | For materials harder than 100 RB on the Rockwell B scale. |
| D | Where a somewhat lighter load is desired than on the Rockwell C scale, as on case-hardened pieces. |
| E | For very soft materials such as bearing metals. |
| F | Same as the Rockwell E scale but using a $\frac{1}{16}$-inch ball. |
| G | For metals harder than tested on the Rockwell B scale. |
| H & K | For softer metals. |
| 15-N; 30-N; 45-N | Where a shallow impression or a small area is desired. For hardened steel and hard alloys. |
| 15-T; 30-T; 45-T | Where a shallow impression or a small area is desired for materials softer than hardened steel. |

**Shore's Scleroscope.**—The scleroscope is an instrument that measures the hardness of the work in terms of elasticity. A diamond-tipped hammer is allowed to drop from a known height on the metal to be tested. As this hammer strikes the metal, it rebounds, and the harder the metal, the greater the rebound. The extreme height of the rebound is recorded, and an average of a number of readings taken on a single piece will give a good indication of the hardness of the work. The surface smoothness of the work affects the reading of the instrument. The readings are also affected by the contour and mass of the work and the depth of the case, and, in carburized work, the soft core of light-depth carburizing, pack-hardening, or cyanide hardening, absorbing the force of the hammer fall and decreasing the rebound. The hammer weighs about 40 grains, the height of the rebound of hardened steel is in the neighborhood of 100 on the scale, or about $6\frac{1}{4}$ inches (158.8 mm), and the total fall is about 10 inches or 255 millimeters.

**Vickers Hardness Test.**—The Vickers test is similar in principle to the Brinell Hardness test. The standard Vickers penetrator is a square-based diamond pyramid having an included point angle of 136 degrees. The numerical value of the hardness number equals the applied load in kilograms divided by the area of the pyramidal impression. A smooth, firmly supported, flat surface is required. The load, which is usually applied for 30 seconds, may be 5, 10, 20, 30, 50, or 120 kilograms. The 50-kilogram load is the most usual. The hardness number is based upon the diagonal length of the square impression. The Vickers test is considered to be very accurate and may be applied to thin sheets as well as to larger sections with proper load regulation.

**Knoop Hardness Numbers.—**The Knoop hardness test is applicable to extremely thin metal, plated surfaces, exceptionally hard and brittle materials, very shallow carburized or nitrided surfaces, or whenever the applied load must be kept below 3600 grams. The Knoop indentor is a diamond ground to an elongated pyramidal form and it produces an indentation having long and short diagonals with a ratio of approximately 7 to 1. The longitudinal angle of the indentor is 172 degrees, 30 minutes, and the transverse angle 130 degrees. The Tukon Tester in which the Knoop indentor is used is fully automatic under electronic control. The Knoop hardness number equals the load in kilograms divided by the projected area of indentation in square millimeters. The indentation number corresponding to the long diagonal and for a given load may be determined from a table computed for a theoretically perfect indentor. The load, which may be varied from 25 to 3600 grams, is applied for a definite period and always normal to the surface tested. Lapped plane surfaces free from scratches are required.

**Monotron Hardness Indicator.—**With this instrument, a diamond-ball impressor point $3/4$ mm in diameter is forced into the material to a depth of $9/5000$ inch; the pressure required to produce this constant impression indicates the hardness. One of two dials shows the pressure in kilograms and pounds, and the other shows the depth of the impression in millimeters and inches. Readings in Brinell Hardness Numbers may be obtained by means of a scale designated as *M*-1.

**Keep's Test.—**With this apparatus, a standard steel drill makes a definite number of revolutions while it is pressed with standard force against the specimen to be tested. The hardness is automatically recorded on a diagram on which a dead soft material gives a horizontal line, and a material as hard as the drill itself gives a vertical line, intermediate hardness being represented by the corresponding angle between 0 and 90 degrees.

**Comparison of Hardness Scales.—**Table 5, Table 6, and Table 7 show comparisons of various hardness scales. All such tables are based on the assumption that the metal tested is homogeneous to a depth several times that of the indentation. To the extent that the metal being tested is not homogeneous, errors are introduced because different loads and different shapes of penetrators meet the resistance of metal of varying hardness, depending on the depth of indentation. Another source of error is introduced in comparing the hardness of different materials as measured on different hardness scales. This error arises from the fact that in any hardness test, metal that is severely cold-worked actually supports the penetrator, and different metals, different alloys, and different analyses of the same type of alloy have different cold-working properties. In spite of the possible inaccuracies introduced by such factors, it is of considerable value to be able to compare hardness values in a general way.

The data shown in Table 5 are based on extensive tests on carbon and alloy steels mostly in the heat-treated condition but have been found to be reliable on constructional alloy steels and tool steels in the as-forged, annealed, normalized, quenched, and tempered conditions, providing they are homogeneous. These hardness comparisons are not as accurate for special alloys such as high manganese steel, 18-8 stainless steel and other austenitic steels, nickel-base alloys, constructional alloy steels, and nickel-base alloys in the cold-worked condition.

The data shown in Table 6 are for hardness measurements of unhardened steel, steel of soft temper, gray and malleable cast iron, and most nonferrous metals. Again these hardness comparisons are not as accurate for annealed metals of high Rockwell B hardness, such as austenitic stainless steel, nickel and high nickel alloys, and cold-worked metals of low Rockwell B scale hardness, such as aluminum and the softer alloys.

Table 7 is intended to provided a general numerical comparison of the various Rockwell hardness scales; however, it should not be used for conversion between the various scales.

## Table 5. Comparative Hardness Scales for Steel

| Rockwell C Scale Hardness Number | Diamond Pyramid Hardness Number Vickers | Brinell Hardness Number 10-mm Ball, 3000-kgf Load | | | Rockwell Hardness Number | | Rockwell Superficial Hardness Number Superficial Diamond Indenter | | | Shore Scleroscope Hardness Number |
|---|---|---|---|---|---|---|---|---|---|---|
| | | Standard Ball | Hultgren Ball | Tungsten Carbide Ball | A Scale 60-kgf Load Diamond Indenter | D Scale 100-kgf Load Diamond Indenter | 15-N Scale 15-kgf Load | 30-N Scale 30-kgf Load | 45-N Scale 45-kgf Load | |
| **68** | **940** | … | … | … | **85.6** | **76.9** | **93.2** | **84.4** | **75.4** | 97 |
| **67** | **900** | … | … | … | **85.0** | **76.1** | **92.9** | **83.6** | **74.2** | 95 |
| **66** | **865** | … | … | … | **84.5** | **75.4** | **92.5** | **82.8** | **73.3** | 92 |
| **65** | **832** | … | … | **(739)** | **83.9** | **74.5** | **92.2** | **81.9** | **72.0** | 91 |
| **64** | **800** | … | … | **(722)** | **83.4** | **73.8** | **91.8** | **81.1** | **71.0** | 88 |
| **63** | **772** | … | … | **(705)** | **82.8** | **73.0** | **91.4** | **80.1** | **69.9** | 87 |
| **62** | **746** | … | … | **(688)** | **82.3** | **72.2** | **91.1** | **79.3** | **68.8** | 85 |
| **61** | **720** | … | … | **(670)** | **81.8** | **71.5** | **90.7** | **78.4** | **67.7** | 83 |
| **60** | **697** | … | **(613)** | **(654)** | **81.2** | **70.7** | **90.2** | **77.5** | **66.6** | 81 |
| **59** | **674** | … | **(599)** | **(634)** | **80.7** | **69.9** | **89.8** | **76.6** | **65.5** | 80 |
| **58** | **653** | … | **(587)** | **615** | **80.1** | **69.2** | **89.3** | **75.7** | **64.3** | 78 |
| **57** | **633** | … | **(575)** | **595** | **79.6** | **68.5** | **88.9** | **74.8** | **63.2** | 76 |
| **56** | **613** | … | **(561)** | **577** | **79.0** | **67.7** | **88.3** | **73.9** | **62.0** | 75 |
| **55** | **595** | … | **(546)** | **560** | **78.5** | **66.9** | **87.9** | **73.0** | **60.9** | 74 |
| **54** | **577** | … | **(534)** | **543** | **78.0** | **66.1** | **87.4** | **72.0** | **59.8** | 72 |
| **53** | **560** | … | **(519)** | **525** | **77.4** | **65.4** | **86.9** | **71.2** | **58.6** | 71 |
| **52** | **544** | **(500)** | **(508)** | **512** | **76.8** | **64.6** | **86.4** | **70.2** | **57.4** | 69 |
| **51** | **528** | **(487)** | **(494)** | **496** | **76.3** | **63.8** | **85.9** | **69.4** | **56.1** | 68 |
| **50** | **513** | **(475)** | **(481)** | **481** | **75.9** | **63.1** | **85.5** | **68.5** | **55.0** | 67 |
| **49** | **498** | **(464)** | **(469)** | **469** | **75.2** | **62.1** | **85.0** | **67.6** | **53.8** | 66 |
| **48** | **484** | **(451)** | **(455)** | **455** | **74.7** | **61.4** | **84.5** | **66.7** | **52.5** | 64 |
| **47** | **471** | **442** | **443** | **443** | **74.1** | **60.8** | **83.9** | **65.8** | **51.4** | 63 |
| **46** | **458** | **432** | **432** | **432** | **73.6** | **60.0** | **83.5** | **64.8** | **50.3** | 62 |
| **45** | **446** | **421** | **421** | **421** | **73.1** | **59.2** | **83.0** | **64.0** | **49.0** | 60 |
| **44** | **434** | **409** | **409** | **409** | **72.5** | **58.5** | **82.5** | **63.1** | **47.8** | 58 |
| **43** | **423** | **400** | **400** | **400** | **72.0** | **57.7** | **82.0** | **62.2** | **46.7** | 57 |
| **42** | **412** | **390** | **390** | **390** | **71.5** | **56.9** | **81.5** | **61.3** | **45.5** | 56 |
| **41** | **402** | **381** | **381** | **381** | **70.9** | **56.2** | **80.9** | **60.4** | **44.3** | 55 |
| **40** | **392** | **371** | **371** | **371** | **70.4** | **55.4** | **80.4** | **59.5** | **43.1** | 54 |
| **39** | **382** | **362** | **362** | **362** | **69.9** | **54.6** | **79.9** | **58.6** | **41.9** | 52 |
| **38** | **372** | **353** | **353** | **353** | **69.4** | **53.8** | **79.4** | **57.7** | **40.8** | 51 |
| **37** | **363** | **344** | **344** | **344** | **68.9** | **53.1** | **78.8** | **56.8** | **39.6** | 50 |
| **36** | **354** | **336** | **336** | **336** | **68.4** | **52.3** | **78.3** | **55.9** | **38.4** | 49 |
| **35** | **345** | **327** | **327** | **327** | **67.9** | **51.5** | **77.7** | **55.0** | **37.2** | 48 |
| **34** | **336** | **319** | **319** | **319** | **67.4** | **50.8** | **77.2** | **54.2** | **36.1** | 47 |
| **33** | **327** | **311** | **311** | **311** | **66.8** | **50.0** | **76.6** | **53.3** | **34.9** | 46 |
| **32** | **318** | **301** | **301** | **301** | **66.3** | **49.2** | **76.1** | **52.1** | **33.7** | 44 |
| **31** | **310** | **294** | **294** | **294** | **65.8** | **48.4** | **75.6** | **51.3** | **32.5** | 43 |
| **30** | **302** | **286** | **286** | **286** | **65.3** | **47.7** | **75.0** | **50.4** | **31.3** | 42 |
| **29** | **294** | **279** | **279** | **279** | **64.7** | **47.0** | **74.5** | **49.5** | **30.1** | 41 |
| **28** | **286** | **271** | **271** | **271** | **64.3** | **46.1** | **73.9** | **48.6** | **28.9** | 41 |
| **27** | **279** | **264** | **264** | **264** | **63.8** | **45.2** | **73.3** | **47.7** | **27.8** | 40 |
| **26** | **272** | **258** | **258** | **258** | **63.3** | **44.6** | **72.8** | **46.8** | **26.7** | 38 |
| **25** | **266** | **253** | **253** | **253** | **62.8** | **43.8** | **72.2** | **45.9** | **25.5** | 38 |
| **24** | **260** | **247** | **247** | **247** | **62.4** | **43.1** | **71.6** | **45.0** | **24.3** | 37 |
| **23** | **254** | **243** | **243** | **243** | **62.0** | **42.1** | **71.0** | **44.0** | **23.1** | 36 |
| **22** | **248** | **237** | **237** | **237** | **61.5** | **41.6** | **70.5** | **43.2** | **22.0** | 35 |

**Table 5.** *(Continued)* **Comparative Hardness Scales for Steel**

| Rockwell C Scale Hardness Number | Diamond Pyramid Hardness Number Vickers | Brinell Hardness Number 10-mm Ball, 3000-kgf Load | | | Rockwell Hardness Number | | Rockwell Superficial Hardness Number Superficial Diamond Indenter | | | Shore Scleroscope Hardness Number |
|---|---|---|---|---|---|---|---|---|---|---|
| | | Standard Ball | Hultgren Ball | Tungsten Carbide Ball | A Scale 60-kgf Load Diamond Indenter | D Scale 100-kgf Load Diamond Indenter | 15-N Scale 15-kgf Load | 30-N Scale 30-kgf Load | 45-N Scale 45-kgf Load | |
| **21** | **243** | **231** | **231** | **231** | **61.0** | **40.9** | **69.9** | **42.3** | **20.7** | 35 |
| **20** | **238** | **226** | **226** | **226** | **60.5** | **40.1** | **69.4** | **41.5** | **19.6** | 34 |
| (18) | 230 | 219 | 219 | 219 | … | … | … | … | … | 33 |
| (16) | 222 | 212 | 212 | 212 | … | … | … | … | … | 32 |
| (14) | 213 | 203 | 203 | 203 | … | … | … | … | … | 31 |
| (12) | 204 | 194 | 194 | 194 | … | … | … | … | … | 29 |
| (10) | 196 | 187 | 187 | 187 | … | … | … | … | … | 28 |
| (8) | 188 | 179 | 179 | 179 | … | … | … | … | … | 27 |
| (6) | 180 | 171 | 171 | 171 | … | … | … | … | … | 26 |
| (4) | 173 | 165 | 165 | 165 | … | … | … | … | … | 25 |
| (2) | 166 | 158 | 158 | 158 | … | … | … | … | … | 24 |
| (0) | 160 | 152 | 152 | 152 | … | … | … | … | … | 24 |

*Note:* The values in this table shown in **boldface** type correspond to those shown in American Society for Testing and Materials Specification E140-67.

Values in ( ) are beyond the normal range and are given for information only.

**Turner's Sclerometer.—**In making this test, a weighted diamond point is drawn once forward and once backward over the smooth surface of the material to be tested. The hardness number is the weight in grams required to produce a standard scratch.

**Mohs's Hardness Scale.—**Hardness, in general, is determined by what is known as Mohs's scale, a standard for hardness that is applied mainly to nonmetallic elements and minerals. In this hardness scale, there are ten degrees or steps, each designated by a mineral, the difference in hardness of the different steps being determined by the fact that any member in the series will scratch any of the preceding members. This scale is as follows:

1) talc; 2) gypsum; 3) calcite; 4) fluorspar; 5) apatite; 6) orthoclase; 7) quartz; 8) topaz; 9) sapphire or corundum; and 10) diamond.

These minerals, arbitrarily selected as standards, are successively harder, from talc, the softest of all minerals, to diamond, the hardest. This scale, which is now universally used for nonmetallic minerals, is not applied to metals.

**Relation Between Hardness and Tensile Strength.—**The approximate relationship between hardness and tensile strength is shown by the following formula:

Tensile strength = $BHN \times 515$ (for Brinell Hardness Numbers up to 175).

Tensile strength = $BHN \times 490$ (for Brinell Hardness Numbers larger than 175).

The above formulas give the tensile strength in pounds per square inch for steels. These approximate relationships between hardness and tensile strength do not apply to nonferrous metals, with the possible exception of certain aluminum alloys.

**Durometer Tests.—**The durometer is a portable hardness tester for measuring hardness of rubber, plastics, and some soft metals. The instrument is designed to apply pressure to the specimen and the hardness is read from a scale while the pressure is maintained. Various scales can be used by changing the indentor and the load applied.

## Table 6. Comparative Hardness Scales for Unhardened Steel, Soft-Temper Steel, Gray and Malleable Cast Iron, and Nonferrous Alloys

| Rockwell Hardness Number | | | Rockwell Superficial Hardness Number | | | Rockwell Hardness Number | | | | Brinell Hardness Number | |
|---|---|---|---|---|---|---|---|---|---|---|---|
| 1/16″ Ball Indenter | | | 1/16″ Ball Indenter | | | 1/8″ Ball Indenter | | | "Brale" Indenter | 10-mm Standard Ball | |
| B scale 100-kg Load | F scale 60-kg Load | G scale 150-kg Load | 15-T scale 15-kg Load | 30-T scale 30-kg Load | 45-T scale 45-kg Load | E scale 100-kg Load | H scale 60-kg Load | K scale 150-kg Load | A scale 60-kg Load | 500-kg Load | 3000-kg Load |
| 100 | … | 82.5 | 93.0 | 82.0 | 72.0 | … | … | … | 61.5 | 201 | 240 |
| 99 | … | 81.0 | 92.5 | 81.5 | 71.0 | … | … | … | 61.0 | 195 | 234 |
| 98 | … | 79.0 | … | 81.0 | 70.0 | … | … | … | 60.0 | 189 | 228 |
| 97 | … | 77.5 | 92.0 | 80.5 | 69.0 | … | … | … | 59.5 | 184 | 222 |
| 96 | … | 76.0 | … | 80.0 | 68.0 | … | … | … | 59.0 | 179 | 216 |
| 95 | … | 74.0 | 91.5 | 79.0 | 67.0 | … | … | … | 58.0 | 175 | 210 |
| 94 | … | 72.5 | … | 78.5 | 66.0 | … | … | … | 57.5 | 171 | 205 |
| 93 | … | 71.0 | 91.0 | 78.0 | 65.5 | … | … | … | 57.0 | 167 | 200 |
| 92 | … | 69.0 | 90.5 | 77.5 | 64.5 | … | … | 100 | 56.5 | 163 | 195 |
| 91 | … | 67.5 | … | 77.0 | 63.5 | … | … | 99.5 | 56.0 | 160 | 190 |
| 90 | … | 66.0 | 90.0 | 76.0 | 62.5 | … | … | 98.5 | 55.5 | 157 | 185 |
| 89 | … | 64.0 | 89.5 | 75.5 | 61.5 | … | … | 98.0 | 55.0 | 154 | 180 |
| 88 | … | 62.5 | … | 75.0 | 60.5 | … | … | 97.0 | 54.0 | 151 | 176 |
| 87 | … | 61.0 | 89.0 | 74.5 | 59.5 | … | … | 96.5 | 53.5 | 148 | 172 |
| 86 | … | 59.0 | 88.5 | 74.0 | 58.5 | … | … | 95.5 | 53.0 | 145 | 169 |
| 85 | … | 57.5 | … | 73.5 | 58.0 | … | … | 94.5 | 52.5 | 142 | 165 |
| 84 | … | 56.0 | 88.0 | 73.0 | 57.0 | … | … | 94.0 | 52.0 | 140 | 162 |
| 83 | … | 54.0 | 87.5 | 72.0 | 56.0 | … | … | 93.0 | 51.0 | 137 | 159 |
| 82 | … | 52.5 | … | 71.5 | 55.0 | … | … | 92.0 | 50.5 | 135 | 156 |
| 81 | … | 51.0 | 87.0 | 71.0 | 54.0 | … | … | 91.0 | 50.0 | 133 | 153 |
| 80 | … | 49.0 | 86.5 | 70.0 | 53.0 | … | … | 90.5 | 49.5 | 130 | 150 |
| 79 | … | 47.5 | … | 69.5 | 52.0 | … | … | 89.5 | 49.0 | 128 | 147 |
| 78 | … | 46.0 | 86.0 | 69.0 | 51.0 | … | … | 88.5 | 48.5 | 126 | 144 |
| 77 | … | 44.0 | 85.5 | 68.0 | 50.0 | … | … | 88.0 | 48.0 | 124 | 141 |
| 76 | … | 42.5 | … | 67.5 | 49.0 | … | … | 87.0 | 47.0 | 122 | 139 |
| 75 | 99.5 | 41.0 | 85.0 | 67.0 | 48.5 | … | … | 86.0 | 46.5 | 120 | 137 |
| 74 | 99.0 | 39.0 | … | 66.0 | 47.5 | … | … | 85.0 | 46.0 | 118 | 135 |
| 73 | 98.5 | 37.5 | 84.5 | 65.5 | 46.5 | … | … | 84.5 | 45.5 | 116 | 132 |
| 72 | 98.0 | 36.0 | 84.0 | 65.0 | 45.5 | … | … | 83.5 | 45.0 | 114 | 130 |
| 71 | 97.5 | 34.5 | … | 64.0 | 44.5 | 100 | … | 82.5 | 44.5 | 112 | 127 |
| 70 | 97.0 | 32.5 | 83.5 | 63.5 | 43.5 | 99.5 | … | 81.5 | 44.0 | 110 | 125 |
| 69 | 96.0 | 31.0 | 83.0 | 62.5 | 42.5 | 99.0 | … | 81.0 | 43.5 | 109 | 123 |
| 68 | 95.5 | 29.5 | … | 62.0 | 41.5 | 98.0 | … | 80.0 | 43.0 | 107 | 121 |
| 67 | 95.0 | 28.0 | 82.5 | 61.5 | 40.5 | 97.5 | … | 79.0 | 42.5 | 106 | 119 |
| 66 | 94.5 | 26.5 | 82.0 | 60.5 | 39.5 | 97.0 | … | 78.0 | 42.0 | 104 | 117 |
| 65 | 94.0 | 25.0 | … | 60.0 | 38.5 | 96.0 | … | 77.5 | … | 102 | 116 |
| 64 | 93.5 | 23.5 | 81.5 | 59.5 | 37.5 | 95.5 | … | 76.5 | 41.5 | 101 | 114 |
| 63 | 93.0 | 22.0 | 81.0 | 58.5 | 36.5 | 95.0 | … | 75.5 | 41.0 | 99 | 112 |
| 62 | 92.0 | 20.5 | … | 58.0 | 35.5 | 94.5 | … | 74.5 | 40.5 | 98 | 110 |
| 61 | 91.5 | 19.0 | 80.5 | 57.0 | 34.5 | 93.5 | … | 74.0 | 40.0 | 96 | 108 |
| 60 | 91.0 | 17.5 | … | 56.5 | 33.5 | 93.0 | … | 73.0 | 39.5 | 95 | 107 |
| 59 | 90.5 | 16.0 | 80.0 | 56.0 | 32.0 | 92.5 | … | 72.0 | 39.0 | 94 | 106 |
| 58 | 90.0 | 14.5 | 79.5 | 55.0 | 31.0 | 92.0 | … | 71.0 | 38.5 | 92 | 104 |
| 57 | 89.5 | 13.0 | … | 54.5 | 30.0 | 91.0 | … | 70.5 | 38.0 | 91 | 103 |
| 56 | 89.0 | 11.5 | 79.0 | 54.0 | 29.0 | 90.5 | … | 69.5 | … | 90 | 101 |
| 55 | 88.0 | 10.0 | 78.5 | 53.0 | 28.0 | 90.0 | … | 68.5 | 37.5 | 89 | 100 |
| 54 | 87.5 | 8.5 | … | 52.5 | 27.0 | 89.5 | … | 68.0 | 37.0 | 87 | … |
| 53 | 87.0 | 7.0 | 78.0 | 51.5 | 26.0 | 89.0 | … | 67.0 | 36.5 | 86 | … |
| 52 | 86.5 | 5.5 | 77.5 | 51.0 | 25.0 | 88.0 | … | 66.0 | 36.0 | 85 | … |
| 51 | 86.0 | 4.0 | … | 50.5 | 24.0 | 87.5 | … | 65.0 | 35.5 | 84 | … |
| 50 | 85.5 | 2.5 | 77.0 | 49.5 | 23.0 | 87.0 | … | 64.5 | 35.0 | 83 | … |
| 49 | 85.0 | 1.0 | 76.5 | 49.0 | 22.0 | 86.5 | … | 63.5 | … | 82 | … |

**Table 6.** *(Continued)* **Comparative Hardness Scales for Unhardened Steel, Soft-Temper Steel, Gray and Malleable Cast Iron, and Nonferrous Alloys**

| Rockwell Hardness Number | | | Rockwell Superficial Hardness Number | | | Rockwell Hardness Number | | | | Brinell Hardness Number | |
|---|---|---|---|---|---|---|---|---|---|---|---|
| 1/16″ Ball Indenter | | | 1/16″ Ball Indenter | | | 1/8″ Ball Indenter | | | "Brale" Indenter | 10-mm Standard Ball | |
| B scale 100-kg Load | F scale 60-kg Load | G scale 150-kg Load | 15-T scale 15-kg Load | 30-T scale 30-kg Load | 45-T scale 45-kg Load | E scale 100-kg Load | H scale 60-kg Load | K scale 150-kg Load | A scale 60-kg Load | 500-kg Load | 3000-kg Load |
| 48 | 84.5 | … | … | 48.5 | 20.5 | 85.5 | … | 62.5 | 34.5 | 81 | … |
| 47 | 84.0 | … | 76.0 | 47.5 | 19.5 | 85.0 | … | 61.5 | 34.0 | 80 | … |
| 46 | 83.0 | … | 75.5 | 47.0 | 18.5 | 84.5 | … | 61.0 | 33.5 | … | … |
| 45 | 82.5 | … | … | 46.0 | 17.5 | 84.0 | … | 60.0 | 33.0 | 79 | … |
| 44 | 82.0 | … | 75.0 | 45.5 | 16.5 | 83.5 | … | 59.0 | 32.5 | 78 | … |
| 43 | 81.5 | … | 74.5 | 45.0 | 15.5 | 82.5 | … | 58.0 | 32.0 | 77 | … |
| 42 | 81.0 | … | … | 44.0 | 14.5 | 82.0 | … | 57.5 | 31.5 | 76 | … |
| 41 | 80.5 | … | 74.0 | 43.5 | 13.5 | 81.5 | … | 56.5 | 31.0 | 75 | … |
| 40 | 79.5 | … | 73.5 | 43.0 | 12.5 | 81.0 | … | 55.5 | … | … | … |
| 39 | 79.0 | … | … | 42.0 | 11.0 | 80.0 | … | 54.5 | 30.5 | 74 | … |
| 38 | 78.5 | … | 73.0 | 41.5 | 10.0 | 79.5 | … | 54.0 | 30.0 | 73 | … |
| 37 | 78.0 | … | 72.5 | 40.5 | 9.0 | 79.0 | … | 53.0 | 29.5 | 72 | … |
| 36 | 77.5 | … | … | 40.0 | 8.0 | 78.5 | 100 | 52.0 | 29.0 | … | … |
| 35 | 77.0 | … | 72.0 | 39.5 | 7.0 | 78.0 | 99.5 | 51.5 | 28.5 | 71 | … |
| 34 | 76.5 | … | 71.5 | 38.5 | 6.0 | 77.0 | 99.0 | 50.5 | 28.0 | 70 | … |
| 33 | 75.5 | … | … | 38.0 | 5.0 | 76.5 | … | 49.5 | … | 69 | … |
| 32 | 75.0 | … | 71.0 | 37.5 | 4.0 | 76.0 | 98.5 | 48.5 | 27.5 | … | … |
| 31 | 74.5 | … | … | 36.5 | 3.0 | 75.5 | 98.0 | 48.0 | 27.0 | 68 | … |
| 30 | 74.0 | … | 70.5 | 36.0 | 2.0 | 75.0 | … | 47.0 | 26.5 | 67 | … |
| 29 | 73.5 | … | 70.0 | 35.5 | 1.0 | 74.0 | 97.5 | 46.0 | 26.0 | … | … |
| 28 | 73.0 | … | … | 34.5 | … | 73.5 | 97.0 | 45.0 | 25.5 | 66 | … |
| 27 | 72.5 | … | 69.5 | 34.0 | … | 73.0 | 96.5 | 44.5 | 25.0 | … | … |
| 26 | 72.0 | … | 69.0 | 33.0 | … | 72.5 | … | 43.5 | 24.5 | 65 | … |
| 25 | 71.0 | … | … | 32.5 | … | 72.0 | 96.0 | 42.5 | … | 64 | … |
| 24 | 70.5 | … | 68.5 | 32.0 | … | 71.0 | 95.5 | 41.5 | 24.0 | … | … |
| 23 | 70.0 | … | 68.0 | 31.0 | … | 70.5 | … | 41.0 | 23.5 | 63 | … |
| 22 | 69.5 | … | … | 30.5 | … | 70.0 | 95.0 | 40.0 | 23.0 | … | … |
| 21 | 69.0 | … | 67.5 | 29.5 | … | 69.5 | 94.5 | 39.0 | 22.5 | 62 | … |
| 20 | 68.5 | … | … | 29.0 | … | 68.5 | … | 38.0 | 22.0 | … | … |
| 19 | 68.0 | … | 67.0 | 28.5 | … | 68.0 | 94.0 | 37.5 | 21.5 | 61 | … |
| 18 | 67.0 | … | 66.5 | 27.5 | … | 67.5 | 93.5 | 36.5 | … | … | … |
| 17 | 66.5 | … | … | 27.0 | … | 67.0 | 93.0 | 35.5 | 21.0 | 60 | … |
| 16 | 66.0 | … | 66.0 | 26.0 | … | 66.5 | … | 35.0 | 20.5 | … | … |
| 15 | 65.5 | … | 65.5 | 25.5 | … | 65.5 | 92.5 | 34.0 | 20.0 | 59 | … |
| 14 | 65.0 | … | … | 25.0 | … | 65.0 | 92.0 | 33.0 | … | … | … |
| 13 | 64.5 | … | 65.0 | 24.0 | … | 64.5 | … | 32.0 | … | 58 | … |
| 12 | 64.0 | … | 64.5 | 23.5 | … | 64.0 | 91.5 | 31.5 | … | … | … |
| 11 | 63.5 | … | … | 23.0 | … | 63.5 | 91.0 | 30.5 | … | … | … |
| 10 | 63.0 | … | 64.0 | 22.0 | … | 62.5 | 90.5 | 29.5 | … | 57 | … |
| 9 | 62.0 | … | … | 21.5 | … | 62.0 | … | 29.0 | … | … | … |
| 8 | 61.5 | … | 63.5 | 20.5 | … | 61.5 | 90.0 | 28.0 | … | … | … |
| 7 | 61.0 | … | 63.0 | 20.0 | … | 61.0 | 89.5 | 27.0 | … | 56 | … |
| 6 | 60.5 | … | … | 19.5 | … | 60.5 | … | 26.0 | … | … | … |
| 5 | 60.0 | … | 62.5 | 18.5 | … | 60.0 | 89.0 | 25.5 | … | 55 | … |
| 4 | 59.5 | … | 62.0 | 18.0 | … | 59.0 | 88.5 | 24.5 | … | … | … |
| 3 | 59.0 | … | … | 17.0 | … | 58.5 | 88.0 | 23.5 | … | … | … |
| 2 | 58.0 | … | 61.5 | 16.5 | … | 58.0 | … | 23.0 | … | 54 | … |
| 1 | 57.5 | … | 61.0 | 16.0 | … | 57.5 | 87.5 | 22.0 | … | … | … |
| 0 | 57.0 | … | … | 15.0 | … | 57.0 | 87.0 | 21.0 | … | 53 | … |

Not applicable to annealed metals of high Rockwell B scale hardness, such as austenitic stainless steels, nickel and high-nickel alloys, nor to cold-worked metals of low Rockwell B scale hardness, such as aluminum and the softer alloys. (Compiled by Wilson Mechanical Instrument Co.)

**Table 7. Graphical Comparison of Rockwell Hardness Scales**

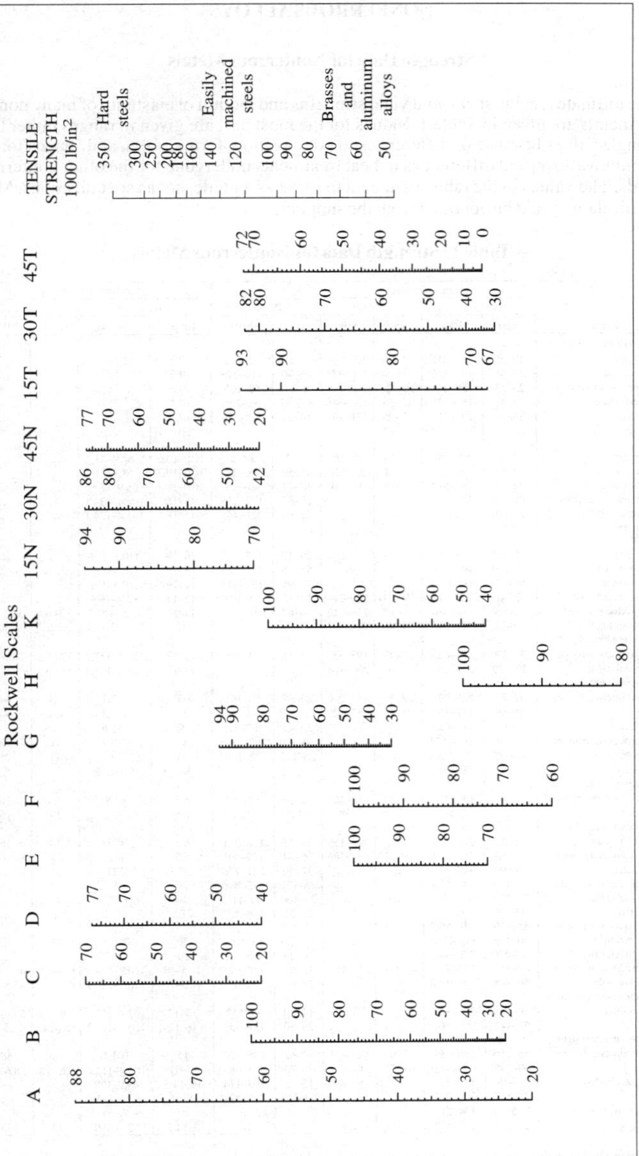


*Source:* George Dorian of GPDorian Co. *Note:* This chart is for the general comparison of Rockwell scales only.

# NONFERROUS ALLOYS

## Strength Data for Nonferrous Metals

The ultimate tensile, shear, and yield strengths and moduli of elasticity of many nonferrous metals are given in Table 1. Values for the most part are given in ranges rather than as single values because of differences in composition, forms, sizes, and shapes for the aluminum alloys, and differences in heat treatments undergone for the other nonferrous metals. The values in the table are meant to serve as a guide, not as specifications. More specific data should be obtained from the supplier.

### Table 1. Strength Data for Nonferrous Metals

| | Ultimate Strength | | | | Yield Strength (0.2% offset) | | Modulus of Elasticity | | | |
|---|---|---|---|---|---|---|---|---|---|---|
| | in Tension | | in Shear | | | | in Tension, $E$ | | in Shear, $G$ | |
| Material | kpsi | MPa | kpsi | MPa | kpsi | MPa | $10^6$ psi | $10^3$ MPa | $10^6$ psi | $10^3$ MPa |
| Aluminum alloys, cast, | | | | | | | | | | |
| sand-cast | 19–35 | 131–241 | 14–26 | 97–179 | 8–25 | 55–172 | 10.3 | 71 | … | … |
| heat-treated | 20–48 | 138–331 | 20 t–34 | 138–234 | 16–40 | 110–276 | 10.3 | 71 | … | … |
| permanent-mold-cast, | 23–35 | 159–241 | 16–27 | 110–186 | 9–24 | 62–165 | 10.3 | 71 | … | … |
| heat-treated | 23–48 | 159–331 | 15–36 | 103–248 | 8.5–43 | 59–296 | 10.3 | 71 | … | … |
| die-cast | 30–46 | 207–317 | 19–29 | 131–200 | 16–27 | 110–186 | 10.3 | 71 | … | … |
| Aluminum alloys, wrought, | | | | | | | | | | |
| annealed | 10–42 | 69–290 | 7–26 | 48–179 | 4–22 | 28–152 | 10.0–10.6 | 68.9–73 | … | … |
| cold-worked | 12–63 | 83–434 | 8–34 | 55–234 | 11–59 | 76–407 | 10.0–10.3 | 68.9–71 | … | … |
| heat-treated | 22–83 | 152–572 | 14–48 | 97–331 | 13–73 | 90–503 | 10.0–11.4 | 68.9–78.6 | … | … |
| Aluminum bronze, cast, | 62–90 | 428–621 | … | … | 25–37 | 172–255 | 15–18 | 103.4–124.1 | … | … |
| heat-treated | 80–110 | 552–758 | … | … | 32–65 | 221–448 | 15–18 | 103.4–124.1 | … | … |
| Aluminum bronze, wrought, | | | | | | | | | | |
| annealed | 55–80 | 379–552 | … | … | 20–40 | 138–276 | 16–19 | 110.3–131 | … | … |
| cold-worked | 71–110 | 490–758 | … | … | 62–66 | 427–455 | 16–19 | 110.3–131 | … | … |
| heat-treated | 101–151 | 696–1041 | … | … | 48–94 | 331–648 | 16–19 | 110.3–131 | … | … |
| Brasses, leaded, cast, | 32–40 | 221–276 | 29–31 | 200–214 | 12–15 | 83–103 | 12–14 | 82.7–96.5 | … | … |
| flat products, wrought | 46–85 | 317–586 | 31–45 | 214–310 | 14–62 | 97–427 | 14–17 | 96.5–117.2 | 5.3–6.4 | 36.5–44.1 |
| wire, wrought | 50–88 | 345–607 | 34–46 | 234–317 | … | … | 15 | 103.4 | 5.6 | 38.6 |
| Brasses, nonleaded, | | | | | | | | | | |
| flat products, wrought | 34–99 | 234–683 | 28–48 | 193–331 | 10–65 | 69–448 | 15–17 | 103.4–117.2 | 5.6–6.4 | 38.6–44.1 |
| wire, wrought | 40–130 | 276–896 | 29–60 | 200–414 | … | … | 15–17 | 103.4–117.2 | 5.6–6.4 | 38.6–44.1 |
| Copper, wrought, | | | | | | | | | | |
| flat products | 32–57 | 221–393 | 22–29 | 152–200 | 10–53 | 69–365 | 17 | 117.2 | 6.4 | 44.1 |
| wire | 35–66 | 241–455 | 24–33 | 165–228 | … | … | 17 | 117.2 | 6.4 | 44.1 |
| Inconel, cast, | 70–95 | 483–655 | … | … | 30–45 | 207–310 | 23 | 158.6 | … | … |
| flat products, wrought | 80–170 | 552–1172 | … | … | 30–160 | 207–1103 | 31 | 213.7 | 11 | 75.8 |
| wire, wrought | 80–185 | 552–1276 | … | … | 25–175 | 172–1207 | 31 | 213.7 | 11 | 75.8 |
| Lead | 2.2–4.9 | 15–34 | … | … | … | … | 0.8–2.0 | 5.5–13.8 | … | … |
| Magnesium, cast, | | | | | | | | | | |
| sand & permanent mold | 22–40 | 152–276 | 17–22 | 117–152 | 12–23 | 83–159 | 6.5 | 44.8 | 2.4 | 16.5 |
| die-cast | 33 | 228 | 20 | 138 | 22 | 152 | 6.5 | 44.8 | 2.4 | 16.5 |
| Magnesium, wrought, | | | | | | | | | | |
| sheet and plate | 35–42 | 241–290 | 21–23 | 145–159 | 20–32 | 138–221 | 6.5 | 44.8 | 2.4 | 16.5 |
| bars, rods, and shapes | 37–55 | 255–379 | 19–27 | 131–186 | 26–44 | 179–303 | 6.5 | 44.8 | 2.4 | 16.5 |
| Monel, cast, | 65–90 | 448–621 | … | … | 32–40 | 221–276 | 19 | 131 | … | … |
| flat products, wrought | 70–140 | 483–965 | … | … | 25–130 | 172–896 | 26 | 179.3 | 9.5 | 65.5 |
| wire, wrought | 70–170 | 483–1172 | … | … | 25–160 | 172–1103 | 26 | 179.3 | 9.5 | 65.5 |
| Nickel, cast, | 45–60 | 310–414 | … | … | 20–30 | 138–207 | 21.5 | 148.2 | … | … |
| flat products, wrought | 55–130 | 379–896 | … | … | 15–115 | 103–793 | 30 | 206.8 | 11 | 75.8 |
| wire, wrought | 50–165 | 345–1138 | … | … | 10–155 | 69–1069 | 30 | 206.8 | 11 | 75.8 |
| Nickel silver, cast, | 40–50 | 276–345 | … | … | 24–25 | 165–172 | … | … | … | … |
| flat products, wrought | 49–115 | 338–793 | 41–59 | 283–407 | 18–90 | 124–621 | 17.5–18 | 120.7–124.1 | 6.6–6.8 | 45.5–46.9 |
| wire, wrought | 50–145 | 345–1000 | … | … | 25–90 | 172–621 | 17.5–18 | 120.7–124.1 | 6.6–6.8 | 45.5–46.9 |
| Phosphor bronze, wrought, | | | | | | | | | | |
| flat products | 40–128 | 276–883 | … | … | 14–80 | 97–552 | 15–17 | 103.4–117.2 | 5.6–6.4 | 38.6–44.1 |
| wire | 50–147 | 345–1014 | … | … | 20–80 | 138–552 | 16–17 | 110.3–117.2 | 6–6.4 | 41.4–44.1 |
| Silicon bronze, wrought, | | | | | | | | | | |
| flat products | 56–110 | 386–758 | 42–63 | 290–434 | 21–62 | 145–427 | 15 | 103.4 | 5.6 | 38.6 |
| wire | 50–145 | 345–1000 | 36–70 | 248–483 | 25–70 | 172–483 | 15–17 | 103.4–117.2 | 5.6–6.4 | 38.6–44.1 |
| Tin bronze, leaded, cast | 21–38 | 145–262 | 23–43 | 159–296 | 15–18 | 103–124 | 10–14.5 | 68.9–100 | … | … |
| Titanium | 50–135 | 345–931 | … | … | 40–120 | 276–827 | 15.0–16.5 | 103.4–113.8 | … | … |
| Zinc, commercial rolled | 19.5–31 | 134–214 | … | … | … | … | … | … | … | … |
| Zirconium | 22–83 | 152–572 | … | … | … | … | 9–14.5 | 62.1–100 | 4.8 | 33.1 |

Consult the index for data on metals not listed and for more data on metals listed.

## Coppers and Copper Alloys

Pure copper is a reddish, highly malleable metal, and was one of the first to be found and utilized. Copper and its alloys are widely used because of their excellent electrical and thermal conductivities, outstanding resistance to corrosion, ease of fabrication, and broad ranges of obtainable strengths and special properties. Almost 400 commercial copper and copper-alloy compositions are available from mills as wrought products (rod, plate, sheet, strip, tube, pipe, extrusions, foil, forgings, and wire) and from foundries as castings.

Copper alloys are grouped into several general categories according to composition:

- coppers and high-copper alloys
- brasses
- bronzes
- copper nickels
- copper-nickel-zinc alloys (nickel silvers)
- leaded coppers
- special alloys

The designation system originally developed by the US copper and brass industry for identifying copper alloys used a three-digit number preceded by the letters CA. These designations have now been made part of the Unified Numbering System (UNS) simply by expanding the numbers to five digits preceded by the letter C. Because the old numbers are embedded in the new UNS numbers, no confusion results. UNS C10000 to C79999 are assigned to wrought compositions, and UNS C80000 to C99999 are assigned to castings. The designation system is not a specification but a method for identifying the composition of mill and foundry products. The precise technical and quality assurance requirements to be satisfied are defined in relevant standard specifications issued by the federal government, the military, and the ASTM.

**Table 1. Classification of Coppers and Copper Alloys**

| Family | Principal Alloying Element | UNS Numbers[a] |
|---|---|---|
| Coppers, high-copper alloys | | C1xxxx |
| Brasses | Zn | C2xxxx, C3xxxx, C4xxxx, C66400 to C69800 |
| Phosphor bronzes | Sn | C5xxxx |
| Aluminum bronzes | Al | C60600 to C64200 |
| Silicon bronzes | Si | C64700 to C66100 |
| Copper nickels, nickel silvers | Ni | C7xxxx |

[a] Wrought alloys.

**Cast Coppers and Copper Alloys.—** Generally, casting permits greater latitude in the use of alloying elements than in the fabrication of wrought products, which requires either hot or cold-working. The cast compositions of coppers and high-copper alloys have a designated minimum copper content and may include other elements to impart special properties. The cast brasses are comprised of copper-zinc-tin alloys (red, semired, and yellow brasses); manganese bronze alloys (high-strength yellow brasses); leaded manganese bronze alloys (leaded high-strength yellow brasses); and copper-zinc-silicon alloys (silicon brasses and bronzes).

The cast bronze alloys have four main families: copper-tin alloys (tin bronzes); copper-tin-lead alloys (leaded and high-leaded tin bronzes); copper-tin-nickel alloys (nickel-tin bronzes); and copper-aluminum alloys (aluminum bronzes).

The cast copper-nickel alloys contain nickel as the principal alloying element. The leaded coppers are cast alloys containing 20 percent or more lead.

Table 2 lists the properties and applications of common cast copper alloys.

**Table 2. Properties and Applications of Cast Coppers and Copper Alloys**

| UNS Designation | Nominal Composition (%) | Typical Mechanical Properties, as Cast or Heat-Treated[a] | | | | Typical Applications |
|---|---|---|---|---|---|---|
| | | Tensile Strength (ksi) | Yield Strength (ksi) | Elongation in 2 in. (%) | Machinability Rating[b] | |
| Copper Alloys | | | | | | |
| C80100 | 99.95 Cu + Ag min, 0.05 others max | 25 | 9 | 40 | 10 | Electrical and thermal conductors; corrosion and oxidation-resisant applications. |
| C80300 | 99.95 Cu + Ag min, 0.034 Ag min, 0.05 others max | 25 | 9 | 40 | 10 | Electrical and thermal conductors; corrosion and oxidation-resistant applications. |
| C80500 | 99.75 Cu + Ag min, 0.034 Ag min, 0.02 B max, 0.23 others max | 25 | 9 | 40 | 10 | Electrical and thermal conductors; corrosion and oxidation-resistant applications. |
| C80700 | 99.75 Cu + Ag min, 0.02 B max, 0.23 others max | 25 | 9 | 40 | 10 | Electrical and thermal conductors; corrosion and oxidation-resistant applications. |
| C80900 | 99.70 Cu + Ag min, 0.034 Ag min, 0.30 others max | 25 | 9 | 40 | 10 | Electrical and thermal conductors; corrosion and oxidation-resistant applications. |
| C81100 | 99.70 Cu + Ag min, 0.30 others max | 25 | 9 | 40 | 10 | Electrical and thermal conductors; corrosion and oxidation resstant applications. |
| High-Copper Alloys | | | | | | |
| C81300 | 98.5 Cu min, 0.06 Be, 0.80 Co, 0.40 others max | (53) | (36) | (11) | 20 | Higher hardness electrical and thermal conductors. |
| C81400 | 98.5 Cu min, 0.06 Be, 0.80 Cr, 0.40 others max | (53) | (36) | (11) | 20 | Higher hardness electrical and thermal conductors. |
| C81500 | 98.0 Cu min, 1.0 Cr, 0.50 others max | (51) | (40) | (17) | 20 | Electrical and/or thermal conductors used as structural members where strength and hardness greater than that of C80100-81100 are required. |
| C81700 | 94.2 Cu min, 1.0 Ag, 0.4 Be, 0.9 Co, 0.9 Ni | (92) | (68) | (8) | 30 | Electrical and/or thermal conductors used as structural members where strength and hardness greater than that of C80100-81100 are required. Also used in place of C81500 where electrical and/or thermal conductivities can be sacrificed for hardness and strength. |
| C81800 | 95.6 Cu min, 1.0 Ag, 0.4 Be, 1.6 Co | 50 (102) | 25 (75) | 20 (8) | 20 | Resistance-welding electrodes, dies. |
| C82000 | 96.8 Cu, 0.6 Be, 2.6 Co | 50 (100) | 20 (75) | 20 (8) | 20 | Current-carrying parts, contact and switch blades, bushings and bearings, and soldering iron and resistance-welding tips. |
| C82100 | 97.7 Cu, 0.5 Be, 0.9 Co, 0.9 Ni | (92) | (68) | (8) | 30 | Electrical and/or thermal conductors used as structural members where strength and hardness greater than that of C80100-81100 are required. Also used in place of C81500 where electrical and/or thermal conductivities can be sacrificed for hardness and strength. |
| C82200 | 96.5 Cu min, 0.6 Be, 1.5 Ni | 57 (95) | 30 (75) | 20 (8) | 20 | Clutch rings, brake drums, seam-welder electrodes, projection welding dies, spot-welding tips, beam-welder shapes, bushings, water-cooled holders. |
| C82400 | 96.4 Cu min, 1.70 Be, 0.25 Co | 72 (150) | 37 (140) | 20 (1) | 20 | Safety tools, molds for plastic parts, cams, bushings, bearings, valves, pump parts, gears. |
| C82500 | 97.2 Cu, 2.0 Be, 0.5 Co, 0.25 Si | 80 (160) | 45 | 20 (1) | 20 | Safety tools, molds for plastic parts, cams, bushings, bearings, valves, pump parts. |
| C82600 | 95.2 Cu min, 2.3 Be, 0.5 Co, 0.25 Si | 82 (165) | 47 (155) | 20 (1) | 20 | Bearings and molds for plastic parts. |
| C82700 | 96.3 Cu, 2.45 Be, 1.25 Ni | (155) | (130) | (0) | 20 | Bearings and molds for plastic parts. |
| C82800 | 96.6 Cu, 2.6 Be, 0.5 Co, 0.25 Si | 97 (165) | 55 (145) | 20 (1) | 10 | Molds for plastic parts, cams, bushings, bearings, valves, pump parts, sleeves. |

**Table 2.** *(Continued)* **Properties and Applications of Cast Coppers and Copper Alloys**

| UNS Designation | Nominal Composition (%) | Typical Mechanical Properties, as Cast or Heat-Treated[a] | | | | Typical Applications |
|---|---|---|---|---|---|---|
| | | Tensile Strength (ksi) | Yield Strength (ksi) | Elongation in 2 in. (%) | Machinability Rating[b] | |
| Red Brasses and Leaded Red Brasses | | | | | | |
| C83300 | 93 Cu, 1.5 Sn, 1.5 Pb, 4 Zn | 32 | 10 | 35 | 35 | Terminal ends for electrical cables. |
| C83400 | 90 Cu, 10 Zn | 35 | 10 | 30 | 60 | Moderate strength, moderate conductivity castings; rotating bands. |
| C83600 | 85 Cu, 5 Sn, 5 Pb, 5 Zn | 37 | 17 | 30 | 84 | Valves, flanges, pipe fittings, plumbing goods, pump castings, water pump impellers and housings, ornamental fixtures, small gears. |
| C83800 | 83 Cu, 4 Sn, 6 Pb, 7 Zn | 35 | 16 | 25 | 90 | Low-pressure valves and fittings, plumbing supplies and fittings, general hardware, air-gas-water fittings, pump components, railroad catenary fittings. |
| Semired Brasses and Leaded Semired Brasses | | | | | | |
| C84200 | 80 Cu, 5 Sn, 2.5 Pb, 12.5 Zn | 35 | 14 | 27 | 80 | Pipe fittings, elbows, T's, couplings, bushings, locknuts, plugs, unions. |
| C84400 | 81 Cu, 3 Sn, 7 Pb, 9 Zn | 34 | 15 | 26 | 90 | General hardware, ornamental castings, plumbing supplies and fixtures, low-pressure valves and fittings. |
| C84500 | 78 Cu, 3 Sn, 7 Pb, 12 Zn | 35 | 14 | 28 | 90 | Plumbing fixtures, cocks, faucets, stops, waste, air and gas fittings, low-pressure valve fittings. |
| C84800 | 76 Cu, 3 Sn, 6 Pb, 15 Zn | 36 | 14 | 30 | 90 | Plumbing fixtures, cocks, faucets, stops, waste, air, and gas, general hardware, and low-pressure valve fittings. |
| Yellow Brasses and Leaded Yellow Brasses | | | | | | |
| C85200 | 72 Cu, 1 Sn, 3 Pb, 24 Zn | 38 | 13 | 35 | 80 | Plumbing fittings and fixtures, ferrules, valves, hardware, ornamental brass, chandeliers, and-irons. |
| C85400 | 67 Cu, 1 Sn, 3 Pb, 29 Zn | 34 | 12 | 35 | 80 | General-purpose yellow casting alloy not subject to high internal pressure. Furniture hardware, ornamental castings, radiator fittings, ship trimmings, battery clamps, valves, and fittings. |
| C85500 | 61 Cu, 0.8 Al, bal Zn | 60 | 23 | 40 | 80 | Ornamental castings. |
| C85700 | 63 Cu, 1 Sn, 1 Pb, 34.7 Zn, 0.3 Al | 50 | 18 | 40 | 80 | Bushings, hardware fittings, ornamental castings. |
| C85800 | 58 Cu, 1 Sn, 1 Pb, 40 Zn | 55 | 30 | 15 | 80 | General-purpose die-casting alloy having moderate strength. |
| Manganese and Leaded Manganese Bronze Alloys | | | | | | |
| C86100 | 67 Cu, 21 Zn, 3 Fe, 5 Al, 4 Mn | 95 | 50 | 20 | 30 | Marine castings, gears, gun mounts, bushings and bearings, marine racing propellers. |
| C86200 | 64 Cu, 26 Zn, 3 Fe, 4 Al, 3 Mn | 95 | 48 | 20 | 30 | Marine castings, gears, gun mounts, bushings and bearings. |
| C86300 | 63 Cu, 25 Zn, 3 Fe, 6 Al, 3 Mn | 115 | 83 | 15 | 8 | Extra-heavy duty, high-strength alloy. Large valve stems, gears, cams, slow-speed heavy-load bearings, screwdown nuts, hydraulic cylinderparts. |
| C86400 | 59 Cu, 1 Pb, 40 Zn | 65 | 25 | 20 | 65 | Free-machining manganese bronze. Valve stems, marine fittings, lever arms, brackets, light-duty gears. |
| C86500 | 58 Cu, 0.5 Sn, 39.5 Zn, 1 Fe, 1 Al | 71 | 28 | 30 | 26 | Machinery parts requiring strength and toughness, lever arms, valve stems, gears. |
| C86700 | 58 Cu, 1 Pb, 41 Zn | 85 | 42 | 20 | 55 | High strength, free-machining manganese bronze. Valve stems. |
| C86800 | 55 Cu, 37 Zn, 3 Ni, 2 Fe, 3 Mn | 82 | 38 | 22 | 30 | Marine fittings, marine propellers. |
| Silicon Bronzes and Silicon Brasses | | | | | | |
| C87200 | 89 Cu min, 4 Si | 55 | 25 | 30 | 40 | Bearings, bells, impellers, pump and valve components, marine fittings, corrosion-resistant castings. |
| C87400 | 83 Cu, 14 Zn, 3 Si | 55 | 24 | 30 | 50 | Bearings, gears, impellers, rocker arms, valve stems, clamps. |

**Table 2.** *(Continued)* **Properties and Applications of Cast Coppers and Copper Alloys**

| UNS Designation | Nominal Composition (%) | Typical Mechanical Properties, as Cast or Heat-Treated[a] | | | | Typical Applications |
|---|---|---|---|---|---|---|
| | | Tensile Strength (ksi) | Yield Strength (ksi) | Elongation in 2 in. (%) | Machinability Rating[b] | |
| C87500 | 82 Cu, 14 Zn, 4 Si | 67 | 30 | 21 | 50 | Bearings, gears, impellers, rocker arms, valve stems, small boat propellers. |
| C87600 | 90 Cu, 5.5 Zn, 4.5 Si | 66 | 32 | 20 | 40 | Valve stems. |
| C87800 | 82 Cu, 14 Zn, 4 Si | 85 | 50 | 25 | 40 | High-strength, thin-wall die castings; brush holders, lever arms, brackets, clamps, hexagonal nuts. |
| C87900 | 65 Cu, 34 Zn, 1 Si | 70 | 35 | 25 | 80 | General-purpose die-casting alloy having moderate strength. |
| Tin Bronzes | | | | | | |
| C90200 | 93 Cu, 7 Sn | 38 | 16 | 30 | 20 | Bearings and bushings. |
| C90300 | 88 Cu, 8 Sn, 4 Zn | 45 | 21 | 30 | 30 | Bearings, bushings, pump impellers, piston rings, valve components, seal rings, steam fittings, gears. |
| C90500 | 88 Cu, 10 Sn, 2 Zn | 45 | 22 | 25 | 30 | Bearings, bushings, pump impellers, piston rings, valve components, steam fittings, gears. |
| C90700 | 89 Cu, 11 Sn | 44 (55) | 22 (30) | 20 (16) | 20 | Gears, bearings, bushings. |
| C90900 | 87 Cu, 13 Sn | 40 | 20 | 15 | 20 | Bearings and bushings. |
| C91000 | 85 Cu, 14 Sn, 1 Zn | 32 | 25 | 2 | 20 | Piston rings and bearings. |
| C91100 | 84 Cu, 16 Sn | 35 | 25 | 2 | 10 | Piston rings, bearings, bushings, bridge plates. |
| C91300 | 81 Cu, 19 Sn | 35 | 30 | 0.5 | 10 | Piston rings, bearings, bushings, bridge plates, bells. |
| C91600 | 88 Cu, 10.5 Sn, 1.5 Ni | 44 (60) | 22 (32) | 16 (16) | 20 | Gears. |
| C91700 | 86.5 Cu, 12 Sn, 1.5 Ni | 44 (60) | 22 (32) | 16 (16) | 20 | Gears. |
| Leaded Tin Bronzes | | | | | | |
| C92200 | 88 Cu, 6 Sn, 1.5 Pb, 4.5 Zn | 40 | 20 | 30 | 42 | Valves, fittings, and pressure-containing parts for use up to 550° F. |
| C92300 | 87 Cu, 8 Sn, 4 Zn | 40 | 20 | 25 | 42 | Valves, pipe fittings, and high-pressure steam castings. Superior machinability to C90300. |
| C92500 | 87 Cu, 11 Sn, 1 Pb, 1 Ni | 44 | 20 | 20 | 30 | Gears, automotive synchronizer rings. |
| C92600 | 87 Cu, 10 Sn, 1 Pb, 2 Zn | 44 | 20 | 30 | 40 | Bearings, bushings, pump impellers, piston rings, valve components, steam fittings, and gears. Superior machinability to C90500. |
| C92700 | 88 Cu, 10 Sn, 2 Pb | 42 | 21 | 20 | 45 | Bearings, bushings, pump impellers, piston rings, and gears. Superior machinability to C90500. |
| C92800 | 79 Cu, 16 Sn, 5 Pb | 40 | 30 | 1 | 70 | Piston rings. |
| C92900 | 82 Cu min, 9 Sn min, 2 Pb min, 2.8 Ni min | 47 (47) | 26 (26) | 20 (20) | 40 | Gears, wear plates, guides, cams, parts requiring machinability superior to that of C91600 or 91700. |
| High-Leaded Tin Bronzes | | | | | | |
| C93200 | 83 Cu, 6.3 Sn min, 7 Pb, 3 Zn | 35 | 18 | 20 | 70 | General-utility bearings and bushings. |
| C93400 | 84 Cu, 8 Sn, 8 Pb | 32 | 16 | 20 | 70 | Bearings and bushings. |
| C93500 | 85 Cu, 5 Sn, 9 Pb | 32 | 16 | 20 | 70 | Small bearings and bushings, bronze backing for babbit-lined automotive bearings. |
| C93700 | 80 Cu, 10 Sn, 10 Pb | 35 | 18 | 20 | 80 | Bearings for high speed and heavy pressures, pumps, impellers, corrosion-resistant applications, pressure-tight castings. |
| C93800 | 78 Cu, 7 Sn, 15 Pb | 30 | 16 | 18 | 80 | Bearings for general service and moderate pressure, pump impellers, and bodies for use in acid mine water. |
| C93900 | 79 Cu, 6 Sn, 15 Pb | 32 | 22 | 7 | 80 | Continuous castings only. Bearings for general service, pump bodies, and impellers for mine waters. |
| C94300 | 70 Cu, 5 Sn, 25 Pb | 27 | 13 | 15 | 80 | High-speed bearings for light loads. |
| C94400 | 81 Cu, 8 Sn, 11 Pb, 0.35 P | 32 | 16 | 18 | 80 | General-utility alloy for bushings and bearings. |
| C94500 | 73 Cu, 7 Sn, 20 Pb | 25 | 12 | 12 | 80 | Locomotive wearing parts; high-low, low-speed bearings. |

**Table 2.** *(Continued)* **Properties and Applications of Cast Coppers and Copper Alloys**

| UNS Designation | Nominal Composition (%) | Typical Mechanical Properties, as Cast or Heat-Treated[a] Tensile Strength (ksi) | Yield Strength (ksi) | Elongation in 2 in. (%) | Machinability Rating[b] | Typical Applications |
|---|---|---|---|---|---|---|
| | | | Nickel-Tin Bronzes | | | |
| C94700 | 88 Cu, 5 Sn, 2 Zn, 5 Ni | 50 (85) | 23 (60) | 35 (10) | 30 (20) | Valve stems and bodies, bearings, wear guides, shift forks, feeding mechanisms, circuit breaker parts, gears, piston cylinders, nozzles. |
| C94800 | 87 Cu, 5 Sn, 5 Ni | 45 (60) | 23 (30) | 35 (8) | 50 (40) | Structural castings, gear components, motion-translation devices, machinery parts, bearings. |
| | | | Aluminum Bronzes | | | |
| C95200 | 88 Cu, 3 Fe, 9 Al | 80 | 27 | 35 | 50 | Acid-resisting pumps, bearings, gears, valve seats, guides, plungers, pump rods, bushings. |
| C95300 | 89 Cu, 1 Fe, 10 Al | 75 (85) | 27 (42) | 25 (15) | 55 | Pickling baskets, nuts, gears, steel-mill slippers, marine equipment, welding jaws. |
| C95400 | 85 Cu, 4 Fe, 11 Al | 85 (105) | 35 (54) | 18 (8) | 60 | Bearings, gears, worms, bushings, valve seats and guides, pickling hooks. |
| C95500 | 81 Cu, 4 Ni, 4 Fe, 11 Al | 100 (120) | 44 (68) | 12 (10) | 50 | Valve guides and seats in aircraft engines, corrosion-resistant parts, bushings, gears, worms, pickling hooks and baskets, agitators. |
| C95600 | 91 Cu, 7 Al, 2 Si | 75 | 34 | 18 | 60 | Cable connectors, terminals, valve stems, marine hardware, gears, worms, pole-line hardware. |
| C95700 | 75 Cu, 2 Ni, 3 Fe, 8 Al, 12 Mn | 95 | 45 | 26 | 50 | Propellers, impellers, stator clamp segments, safety tools, welding rods, valves, pump casings. |
| C95800 | 81 Cu, 5 Ni, 4 Fe, 9 Al, 1 Mn | 95 | 38 | 25 | 50 | Propeller hubs, blades, and other parts in contact with salt water. |
| | | | Copper-Nickels | | | |
| C96200 | 88.6 Cu, 10 Ni, 1.4 Fe | 45 min | 25 min | 20 min | 10 | Components of items being used for seawater corrosion resistance. |
| C96300 | 79.3 Cu, 20 Ni, 0.7 Fe | 75 min | 55 min | 10 min | 15 | Centrifugally cast tailshaft sleeves. |
| C96400 | 69.1 Cu, 30 Ni, 0.9 Fe | 68 | 37 | 28 | 20 | Valves, pump bodies, flanges, elbows used for seawater corrosion resistance. |
| C96600 | 68.5 Cu, 30 Ni, 1 Fe, 0.5 Be | (110) | (70) | (7) | 20 | High-strength constructional parts for seawater corrosion resistance. |
| | | | Nickel Silvers | | | |
| C97300 | 56 Cu, 2 Sn, 10 Pb, 12 Ni, 20 Zn | 35 | 17 | 20 | 70 | Hardware fittings, valves and valve trim, statuary, ornamental castings. |
| C97400 | 59 Cu, 3 Sn, 5 Pb, 17 Ni, 16 Zn | 38 | 17 | 20 | 60 | Valves, hardware, fittings, ornamental castings. |
| C97600 | 64 Cu, 4 Sn, 4 Pb, 20 Ni, 8 Zn | 45 | 24 | 20 | 70 | Marine castings, sanitary fittings, ornamental hardware, valves, pumps. |
| C97800 | 66 Cu, 5 Sn, 2 Pb, 25 Ni, 2 Zn | 55 | 30 | 15 | 60 | Ornamental and sanitary castings, valves and valve seats, musical instrument components. |
| | | | Special Alloys | | | |
| C99300 | 71.8 Cu, 15 Ni, 0.7 Fe, 11 Al, 1.5 Co | 95 | 55 | 2 | 20 | Glass-making molds, plate glass rolls, marine hardware. |
| C99400 | 90.4 Cu, 2.2 Ni, 2.0 Fe, 1.2 Al, 1.2 Si, 3.0 Zn | 66 (79) | 34 (54) | 25 | 50 | Valve stems, marine and other uses requiring resistance to dezincification and dealuminification, propeller wheels, electrical parts, mining equipment gears. |
| C99500 | 87.9 Cu, 4.5 Ni, 4.0 Fe, 1.2 Al, 1.2 Si, 1.2 Zn | 70 min | 40 min | 12 min | 50 | Same as C99400, but where higher yield strength is required. |
| C99700 | 56.5 Cu, Al, 1.5 Pb, 12 Mn, 5 Ni, 24 Zn | 55 | 25 | 25 | 80 | |
| C99750 | 58 Cu, 1 Al, 1 Pb, 20 Mn, 20 Zn | 65 (75) | 32 (40) | 30 (20) | | |

[a] Values in parentheses are for heat-treated condition.
[b] Free cutting brass = 100.

*Source:* Copper Development Association, New York.

**Wrought Coppers and Copper Alloys.**—Wrought copper alloys can be utilized in the annealed, cold-worked, stress-relieved, or hardened-by-heat-treatment conditions, depending on composition and end use. The "temper designation" for copper alloys is defined in ASTM Standard Recommended Practice B601, which is applicable to all product forms.

Wrought coppers and high-copper alloys, like cast alloys, have a designated minimum copper content and may include other elements to impart special properties. Wrought brasses have zinc as the principal alloying element and may have other designated elements. They comprise the copper-zinc alloys; copper-zinc-lead alloys (leaded brasses); and copper-zinc-tin alloys (tin brasses).

Four main groups comprise the wrought bronzes; copper-tin-phosphorus alloys (phosphor bronze); copper-tin-lead-phosphorus alloys (leaded phosphor bronze); copper-aluminum alloys (aluminum bronzes); and copper-silicon alloys (silicon bronze).

Wrought copper-nickel alloys, like the cast alloys, have nickel as the principal alloying element. The wrought copper-nickel-zinc alloys are known as "nickel silvers" because of their color.

Table 3 lists the nominal composition, properties, and applications of common wrought copper alloys.

**Table 3. Properties and Applications of Wrought Coppers and Copper Alloys**

| Name and Number | Nominal Composition (%) | Strength (ksi) | | Elongation in 2 in. (%) | Machinability Rating[a] | Fabricating Characteristics and Typical Applications |
|---|---|---|---|---|---|---|
| | | Tensile | Yield | | | |
| C10100 Oxygen-free electronic | 99.99 Cu | 32 -66 | 10 -53 | 55 | 20 | Excellent hot and cold workability; good forgeability. Fabricated by blanking, coining, coppersmithing, drawing and upsetting, hot forging and pressing, spinning, swaging, stamping. Uses: busbars, bus conductors, waveguides, hollow conductors, lead-in wires and anodes for vacuum tubes, vacuum seals, transistor components, glass to metal seals, coaxial cables and tubes, klystrons, microwave tubes, rectifiers. |
| C10200 Oxygen-free copper | 99.95 Cu | 32-66 | 10-53 | 55 | 20 | Fabricating characteristics same as C10100. Uses: busbars, waveguides. |
| C10300 Oxygen-free, extra-low phosphorus | 99.95 Cu, 0.003 P | 32-55 | 10-50 | 50 | 20 | Fabricating characteristics same as C10100. Uses: busbars, electrical conductors, tubular bus, and applications requiring good conductivity and welding or brazing properties. |
| C10400, C10500, C10700 Oxygen-free, silver-bearing | 99.95 Cu | 32-66 | 10-53 | 55 | 20 | Fabricating characteristics same as C10100. Uses: auto gaskets, radiators, busbars, conductivity wire, contacts, radio parts, winding, switches, terminals, commutator segments; chemical process equipment, printing rolls, clad metals, printed-circuit foil. |
| C10800 Oxygen-free, low phosphorus | 99.95 Cu, 0.009 P | 32-55 | 10-50 | 50 | 20 | Fabricating characteristics same as C10100. Uses: refrigerators, air conditioners, gas and heater lines, oil burner tubes, plumbing pipe and tube, brewery tubes, condenser and heat-exchanger tubes, dairy and distiller tubes, pulp and paper lines, tanks; air, gasoline, and hydraulic lines. |
| C11000 Electrolytic tough pitch copper | 99.90 Cu, 0.04 O | 32-66 | 10-53 | 55 | 20 | Fabricating characteristics same as C10100. Uses: downspouts, gutters, roofing, gaskets, auto radiators, busbars, nails, printing rolls, rivets, radio parts. |
| C11000 Electrolytic tough pitch, anneal-resistant | 99.90 Cu, 0.04 O, 0.01 Cd | 66 | | | 20 | Fabricated by drawing and stranding, stamping. Uses: electrical power transmission where resistance to softening under overloads is desired. |
| C11300, C11400, C11500, C11600 Silver-bearing tough pitch copper | 99.90 Cu, 0.04 O, Ag | 32-66 | 10-53 | 55 | 20 | Fabricating characteristics same as C10100. Uses: gaskets, radiators, busbars, windings, switches, chemical process equipment, clad metals, printed-circuit foil. |

**Table 3.** ***(Continued)*** **Properties and Applications of Wrought Coppers and Copper Alloys**

| Name and Number | Nominal Composition (%) | Strength (ksi) | | Elongation in 2 in. (%) | Machinability Rating[a] | Fabricating Characteristics and Typical Applications |
|---|---|---|---|---|---|---|
| | | Tensile | Yield | | | |
| C12000, C12100 Phosphorus deoxidized, low residual phosphorus | 99.9 Cu | 32-57 | 10-53 | 55 | 20 | Fabricating characteristics same as C10100. Uses: busbars, electrical conductors, tubular bus, and applications requiring welding or brazing. |
| C12200, C12210 Phosphorus deoxidized copper, high residual phosphorus | 99.90 Cu, 0.02 P | 32-55 | 10-53 | 55 | 20 | Fabricating characteristics same as C10100. Uses: gas and heater lines; oil burner tubing; plumbing pipe and tubing; condenser, evaporator, heat exchanger, dairy, and distiller tubing; steam and water lines; air, gasoline, and hydraulic lines. |
| C12500, C12700, C12800, C12900, C13000 Fire-refined tough pitch with silver | 99.88 Cu | 32-66 | 10-53 | 55 | 20 | Fabricating characteristics same as C10100. Uses: same as C11000, electrolytic tough pitch copper. |
| C14200 Phosphorus deoxidized, arsenical | 99.68 Cu, 0.3 As, 0.02 P | 32-55 | 10-50 | 45 | 20 | Fabricating characteristics same as C10100. Uses: staybolts, heat-exchanger and condenser tubes. |
| C14300, C14310 Cadmium copper, deoxidized | 99.9 Cu, 0.1 Cd | 32-58 | 11-56 | 42 | 20 | Fabricating characteristics same as C10100. Uses: anneal-resistant electrical applications requiring thermal softening and embrittlement resistance, lead frames, contacts, terminals, solder-coated and solder-fabricated parts, furnace-brazed assemblies and welded components, cable wrap. |
| C14500, C14510, C14520 Tellurium bearing | 99.5 Cu, 0.50 Te, 0.008 P | 32-56 | 10-51 | 50 | 85 | Fabricating characteristics same as C10100. Uses: Forgings and screw-machine products, and parts requiring high conductivity, extensive machining, corrosion resistance, copper color, or a combination of these; electrical connectors, motor and switch parts, plumbing fittings, soldering coppers, welding torch tips, transistor bases, and furnace-brazed articles. |
| C14700, C14710, C14720 Sulfur bearing | 99.6 Cu, 0.40 S | 32-57 | 10-55 | 52 | 85 | Fabricating characteristics same as C10100. Uses: screw-machine products and parts requiring high conductivity, extensive machining, corrosion resistance, copper color, or a combination of these; electrical connectors, motor and switch components, plumbing fittings, cold-headed and machined parts, cold forgings, furnace-brazed articles, screws, soldering coppers, rivets and welding torch tips. |
| C15000 Zirconium copper | 99.8 Cu, 0.15 Zr | 29-76 | 6-72 | 54 | 20 | Fabricating characteristics same as C10100. Uses: switches, high-temperature circuit breakers, commutators, stud bases for power transmitters, rectifiers, soldering welding tips. |
| C15500 | 99.75 Cu, 0.06 P, 0.11 Mg, Ag | 40-80 | 18-72 | 40 | 20 | Fabricating characteristics same as C10100. Uses: high-conductivity light-duty springs, electrical contacts, fittings, clamps, connectors, diahragms, electronic components, resistance-welding electrodes. |
| C15715 | 99.6 Cu, 0.13 $Al_2O_3$ | 52-88 | 44-84 | 27 | 20 | Excellent cold workability. Fabricated by extrusion, drawing, rolling, heading, swaging, machining, blanking, roll threading. Uses: integrated-circuit lead frames, diode leads; vacuum, microwave, and x-ray tube components; electrical components; brush springs; commutators, electric generator and motor components. |
| C15720 | 99.5 Cu, 0.18 $Al_2O_3$ | 64-98 | 54-96 | 25 | | Excellent cold workability. Fabricated by extrusion, drawing, rolling, heading, swaging, machining, blanking. Uses: relay and switch springs, lead frames, contact supports, heat sinks, circuit breaker parts, rotor bars, resistance-welding electrodes and wheels, connectors, soldering gun tips. |

**Table 3.** *(Continued)* **Properties and Applications of Wrought Coppers and Copper Alloys**

| Name and Number | Nominal Composition (%) | Strength (ksi) Tensile | Strength (ksi) Yield | Elongation in 2 in. (%) | Machinability Rating[a] | Fabricating Characteristics and Typical Applications |
|---|---|---|---|---|---|---|
| C15760 | 98.8 Cu, 0.58 $Al_2O_3$ | 70-90 | 65-87 | 22 | | Excellent cold workability. Fabricated by extrusion and drawing. Uses: resistance-welding electrodes, soldering gun tips, MIG welding contact tips, continuous-casting molds. |
| C16200, C16210 Cadmium copper | 99.0 Cu, 1.0 Cd | 35-100 | 7-69 | 57 | 20 | Excellent cold workability; good hot formability. Uses: trolley wires, heating pads, electric-blanket elements, spring contacts, railbands, high-strength transmission lines, connectors, cable wrap, switch-gear components, and waveguide cavities. |
| C16500 | 98.6 Cu, 0.8 Cd, 0.6 Sn | 40-95 | 14-71 | 53 | 20 | Fabricating characteristics same as C16200. Uses: electrical springs and contacts, trolley wire, clips, flat cable, resistance-welding electrodes. |
| C17000 Beryllium copper | 98.3 Cu, 1.7 Be, 0.20 Co | 70-190 | 32-170 | 45 | 20 | Fabricating characteristics same as C16200. Commonly fabricated by blanking, forming and bending, turning, drilling, tapping. Uses: bellows, Bourdon tubing, diaphragms, fuse clips, fasteners, lock-washers, springs, switch parts, roll pins, valves, welding equipment. |
| C17200 Beryllium copper | 98.1 Cu, 1.9 Be, 0.20 Co | 68-212 | 25-195 | 48 | 20 | Similar to C17000, particularly for its nonsparking characteristics. |
| C17300 Beryllium copper | 98.1 Cu, 1.9 Be, 0.40 Pb | 68-212 | 25-195 | 48 | 50 | Combines superior machinability with good fabricating characteristics of C17200. |
| C17500, C17510 Beryllium copper | 96.9 Cu, 2.5 Co, 0.6 Be | 45-115 | 25-110 | 28 | | Fabricating characteristics same as C16200. Uses: fuse clips, fasteners, springs, switch and relay parts, electrical conductors, welding equipment. |
| C18200, C18400, C18500 Chromium copper | 99.2 Cu | 34-86 | 14-77 | 40 | 20 | Excellent cold workability, good hot workability. Uses: resistance-welding electrodes, seam-welding wheels, switch gear, electrode-holder jaws, cable connectors, current-carrying arms and shafts, circuit-breaker parts, molds, spot-welding tips, flash-welding electrodes, electrical and thermal conductors requiring strength, switch contacts. |
| C18700 Leaded copper | 99.0 Cu, 1.0 Pb | 32-55 | 10-50 | 45 | 85 | Good cold workability; poor hot formability. Uses: connectors, motor and switch parts, screw-machine parts requiring high conductivity. |
| C18900 | 98.7 Cu, 0.8 Sn, 0.3 Si, 0.20 Mn | 38-95 | 9-52 | 48 | 20 | Fabricating characteristics same as C10100. Uses: welding rod and wire for inert gas tungsten arc and metal arc welding, and oxyacetylene welding of copper. |
| C19000 Copper-nickel-phosphorus alloy | 98.6 Cu, 1.1 Ni, 0.3 P | 38-115 | 20-81 | 50 | 30 | Fabricating characteristics same as C10100. Uses: springs, clips, electrical connectors, power tube and electron tube components, high-strength electrical conductors, bolts, nails, screws, cotter pins, and parts requiring some combination of high strength, high electrical or thermal conductivity, high resistance to fatigue and creep, and good workability. |
| C19100 Copper-nickel-phosphorus-tellurium alloy | 98.2 Cu, 1.1 Ni, 0.5 Te, 0.2 P | 36-104 | 10-92 | 27 | 75 | Good hot and cold workability. Uses: forgings and screw-machine parts requiring high strength, hardenability, extensive machining, corrosion resistance, copper color, good conductivity, or a combination of these; bolts, bushings, electrical connectors, gears, marine hardware, nuts, pinions, tie rods, turnbuckle barrels, welding torch tips. |
| C19200 | 99 Cu, 1.0 Fe, 0.03 P | 37-77 | 11-74 | 40 | 20 | Excellent hot and cold workability. Uses: automotive hydraulic brake lines, flexible hose, electrical terminals, fuse clips, gaskets, gift hollow ware, applications requiring resistance to softening and stress corrosion, air-conditioning and heat-exchanger tubing. |

## Table 3. *(Continued)* Properties and Applications of Wrought Coppers and Copper Alloys

| Name and Number | Nominal Composition (%) | Strength (ksi) Tensile | Strength (ksi) Yield | Elongation in 2 in. (%) | Machinability Rating[a] | Fabricating Characteristics and Typical Applications |
|---|---|---|---|---|---|---|
| C19400 | 97.4 Cu, 2.4 Fe, 0.13 Zn, 0.04 P | 45-76 | 24-73 | 32 | 20 | Excellent hot and cold workability. Uses: circuit-breaker components, contact springs, electrical clamps, electrical springs, electrical terminals, flexible hose, fuse clips, gaskets, gift hollow ware, plug contacts, rivets, and welded condenser tubes. |
| C19500 | 97.0 Cu, 1.5 Fe, 0.6 Sn, 0.10 P, 0.80 Co | 80-97 | 65-95 | 15 | 20 | Excellent hot and cold workability. Uses: electrical springs, sockets, terminals, connectors, clips, and other current-carrying parts requiring strength. |
| C21000 Gilding, 95% | 95.0 Cu, 5.0 Zn | 34-64 | 10-58 | 45 | 20 | Excellent cold workability, good hot workability for blanking, coining, drawing, piercing and punching, shearing, spinning, squeezing and swaging, stamping. Uses: coins, medals, bullet jackets, fuse caps, primers, plaques, jewelry base for gold plate. |
| C22000 Commercial bronze, 90% | 90.0 Cu, 10.0 Zn | 37-72 | 10-62 | 50 | 20 | Fabricating characteristics same as C21000, plus heading and upsetting, roll threading and knurling, hot forging and pressing. Uses: etching bronze, grillwork, screen cloth, weatherstripping, lipstick cases, compacts, marine hardware, screws, rivets. |
| C22600 Jewelry bronze, 87.5% | 87.5 Cu, 12.5 Zn | 39-97 | 11-62 | 46 | 30 | Fabricating characteristics same as C21000, plus heading and up-setting, roll threading and knurling. Uses: angles, channels, chain, fasteners, costume jewelry, lipstick cases, powder compacts, base for gold plate. |
| C23000 Red brass, 85% | 85.0 Cu, 15.0 Zn | 39-105 | 10-63 | 55 | 30 | Excellent cold workability; good hot formability. Uses: weather-stripping, conduit, sockets, fasteners, fire extinguishers, condenser and heat-exchanger tubing, plumbing pipe, radiator cores. |
| C24000 Low brass, 80% | 80.0 Cu, 20.0 Zn | 42-125 | 12-65 | 55 | 30 | Excellent cold workability. Fabricating characteristics same as C23000. Uses: battery caps, bellows, musical instruments, clock dials, pump lines, flexible hose. |
| C26000, C26100, C26130, C26200 Cartridge brass, 70% | 70.0 Cu, 30.0 Zn | 44-130 | 11-65 | 66 | | Excellent cold workability. Uses: radiator cores and tanks, flashlight shells, lamp fixtures, fasteners, screws, springs, grillwork, stencils, plumbing accessories, plumbing brass goods, locks, hinges, ammunition components, plumbing accessories, pins, rivets. |
| C26800, C27000 Yellow brass | 65.0 Cu, 35.0 Zn | 46-128 | 14-62 | 65 | 30 | Excellent cold workability. Fabricating characteristics same as C23000. Uses: same as C26000 except not used for ammunition. |
| C28000 Muntz metal, 60% | 60.0 Cu, 40.0 Zn | 54-74 | 21-55 | 52 | 40 | Excellent hot formability and forgeability for blanking, forming and bending, hot forging and pressing, hot heading and upsetting, shearing. Uses: architectural, large nuts and bolts, brazing rod, condenser plates, heat-exchanger and condenser tubing, hot forgings. |
| C31400 Leaded commercial bronze | 89.0 Cu, 1.9 Pb, 0.1 Zn | 37-60 | 12-55 | 45 | 80 | Excellent machinability. Uses: screws, machine parts, pickling crates. |
| C31600 Leaded commercial bronze, nickel-bearing | 89.0 Cu, 1.9 Pb, 1.0 Ni, 8.1 Zn | 37-67 | 12-59 | 45 | 80 | Good cold workability; poor hot formability. Uses: electrical connectors, fasteners, hardware, nuts, screws, screw-machine parts. |
| C33000 Low-leaded brass tube | 66.0 Cu, 0.5 Pb, 33.5 Zn | 47-75 | 15-60 | 60 | 60 | Combines good machinability and excellent cold workability. Fabricated by forming and bending, machining, piercing and punching. Uses: pump and power cylinders and liners, ammunition primers, plumbing accessories. |

**Table 3.** *(Continued)* **Properties and Applications of Wrought Coppers and Copper Alloys**

| Name and Number | Nominal Composi-tion (%) | Strength (ksi) | | Elonga-tion in 2 in. (%) | Machin-ability Rating[a] | Fabricating Characteristics and Typical Applications |
|---|---|---|---|---|---|---|
| | | Tensile | Yield | | | |
| C33200 High-leaded brass tube | 66.0 Cu, 2.0 Pb, 32.0 Zn | 47-75 | 15-60 | 50 | 80 | Excellent machinability. Fabricated by piercing, punching, and machining. Uses: general-purpose screw-machine parts. |
| C33500 Low-leaded brass | 63.5 Cu, 0.5 Pb, 36 Zn | 46-74 | 14-60 | 65 | 60 | Similar to C33200. Commonly fabricated by blanking, drawing, machining, piercing and punching, stamping. Uses: butts, hinges, watch backs. |
| C34000 Medium-leaded brass | 63.5 Cu, 1.0 Pb, 35.5 Zn | 47-88 | 15-60 | 60 | 70 | Similar to C33200. Fabricated by blanking, heading and upsetting, machining, piercing and punching, roll threading and knurling, stamping. Uses: butts, gears, nuts, rivets, screws, dials, engravings, instrument plates. |
| C34200 High-leaded brass | 63.5 Cu, 2.0 Pb, 34.5 Zn | 49-85 | 17-62 | 52 | 90 | Combines excellent machinability with moderate cold workability. Uses: clock plates and nuts, clock and watch backs, gears, wheels and channel plate. |
| C35000 Medium-leaded brass | 62.5 Cu, 1.1 Pb, 36.4 Zn | 45-95 | 13-70 | 66 | 70 | Fair cold workability; poor hot formability. Uses: bearing cages, book dies, clock plates, gears, hinges, hose couplings, keys, lock parts, lock tumblers, meter parts, nuts, sink strainers, strike plates, templates, type characters, washers, wear plates. |
| C35300 High-leaded brass | 61.5 Cu, 2.8 Pb, 36.5 Zn | 49-85 | 17-62 | 52 | 90 | Similar to C34200. |
| C35600 Extra-high-leaded brass | 61.5 Cu, 2.5 Pb, 36 Zn | 47-97 | 17-87 | 60 | 100 | Excellent machinability. Fabricated by blanking, machining, piercing and punching, stamping. Uses: clock plates and nuts, clock and watch backs, gears, wheels, and channel plate. |
| C36000 Free-cutting brass | 61.5 Cu, 3.1 Pb, 35.4 Zn | 49-68 | 18-45 | 53 | 100 | Excellent machinability. Fabricated by machining, roll threading, and knurling. Uses: gears, pinions, automatic high-speed screw-machine parts. |
| C36500 to C36800 Leaded Muntz metal | 59.5 Cu, 0.5 Pb, 40.0 Zn | 54 (As hot rolled) | 20 | 45 | 60 | Combines good machinability with excellent hot formability. Uses: condenser-tube plates. |
| C37000 Free-cutting Muntz metal | 60.0 Cu, 1.0 Pb, 39.0 Zn | 54-80 | 20-60 | 40 | 70 | Fabricating characteristics similar to C36500 to 36800. Uses: automatic screw-machine parts. |
| C37700 Forging brass | 59.5 Cu, 2.0 Pb, 38.0 Zn | 52 (As ex-truded) | 20 | 45 | 80 | Excellent hot workability. Fabricated by heading and upsetting, hot forging and pressing, hot heading and upsetting, machining. Uses: forgings and pressings of all kinds. |
| C38500 Architectural bronze | 57.0 Cu, 3.0 Pb, 40.0 Zn | 60 (As ex-truded) | 20 | 30 | 90 | Excellent machinability and hot workability. Fabricated by hot forging and pressing, forming, bending, and machining. Uses: architectural extrusions, store fronts, thresholds, trim, butts, hinges, lock bodies, and forgings. |
| C40500 | 95 Cu, 1 Sn, 4 Zn | 39-78 | 12-70 | 49 | 20 | Excellent cold workability. Fabricated by blanking, forming, and drawing. Uses: meter clips, terminals, fuse clips, contact and relay springs, washers. |
| C40800 | 95 Cu, 2 Sn, 3 Zn | 42-79 | 13-75 | 43 | 20 | Excellent cold workability. Fabricated by blanking, stamping, and shearing. Uses: electrical connectors. |
| C41100 | 91 Cu, 0.5 Sn, 8.5 Zn | 39-106 | 11-72 | 43 | 20 | Excellent cold workability, good hot formability. Fabricated by blanking, forming and bending, drawing, piercing and punching, shearing, spinning, and stamping. Uses: bushings, bearing sleeves, thrust washers, flexible metal hose. |
| C41300 | 90.0 Cu, 1.0 Sn, 9.0 Zn | 41-105 | 12-82 | 45 | 20 | Excellent cold workability; good hot formability. Uses: plater bar for jewelry products, flat springs for electrical switchgear. |

**Table 3.** *(Continued)* **Properties and Applications of Wrought Coppers and Copper Alloys**

| Name and Number | Nominal Composition (%) | Strength (ksi) Tensile | Strength (ksi) Yield | Elongation in 2 in. (%) | Machinability Rating[a] | Fabricating Characteristics and Typical Applications |
|---|---|---|---|---|---|---|
| C41500 | 91 Cu, 1.8 Sn, 7.2 Zn | 46-81 | 17-75 | 44 | 30 | Excellent cold workability. Fabricated by blanking, drawing, bending, forming, shearing, and stamping. Uses: spring applications for electrical switches. |
| C42200 | 87.5 Cu, 1.1 Sn, 11.4 Zn | 43-88 | 15-75 | 46 | 30 | Excellent cold workability; good hot formability. Fabricated by blanking, piercing, forming, and drawing. Uses: sash chains, fuse clips, terminals, spring washers, contact springs, electrical connectors. |
| C42500 | 88.5 Cu, 2.0 Sn, 9.5 Zn | 45-92 | 18-76 | 49 | 30 | Excellent cold workability. Fabricated by blanking, piercing, forming, and drawing. Uses: electrical switches, springs, terminals, connectors, fuse clips, pen clips, weather stripping. |
| C43000 | 87.0 Cu, 2.2 Sn, 10.8 Zn | 46-94 | 18-73 | 55 | 30 | Excellent cold workability; good hot formability. Fabricated by blanking, coining, drawing, forming, bending, heading, and upsetting. Uses: same as C42500. |
| C43400 | 85.0 Cu, 0.7 Sn, 14.3 Zn | 45-90 | 15-75 | 49 | 30 | Excellent cold workability. Fabricated by blanking, drawing, bonding, forming, stamping, and shearing. Uses: electrical switch parts, blades, relay springs, contacts. |
| C43500 | 81.0 Cu, 0.9 Sn, 18.1 Zn | 46-80 | 16-68 | 46 | 30 | Excellent cold workability for fabrication by forming and bending. Uses: Bourdon tubing and musical instruments. |
| C44300, C44400, C44500 Inhibited admiralty | 71.0 Cu, 28.0 Zn, 1.0 Sn | 48-55 | 18-22 | 65 | 30 | Excellent cold workability for forming and bending. Uses: condenser, evaporator and heat-exchanger tubing, condenser tubing plates, distiller tubing, ferrules. |
| C46400 to C46700 Naval brass | 60.0 Cu, 39.2 Zn, 0.8 Sn | 55-88 | 25-66 | 50 | 30 | Excellent hot workability and hot forgeability. Fabricated by blanking, drawing, bending, heading and upsetting, hot forging, pressing. Uses: aircraft turnbuckle barrels, balls, bolts, marine hardware, nuts, propeller shafts, rivets, valve stems, condenser plates, welding rod. |
| C48200 Naval brass, medium-leaded | 60.5 Cu, 0.7 Pb, 0.8 Sn, 38.0 Zn | 56-75 | 25-53 | 43 | 50 | Good hot workability for hot forging, pressing, and machining operations. Uses: marine hardware, screw-machine products, valve stems. |
| C48500 Leaded naval brass | 60.0 Cu, 1.8 Pb, 37.5 Zn, 0.7 Sn | 57-75 | 25-53 | 40 | 70 | Combines good hot forgeability and machinability. Fabricated by hot forging and pressing, machining. Uses: marine hardware, screw-machine parts, valve stems. |
| C50500 Phosphor bronze, 1.25% E | 98.7 Cu, 1.3 Sn, trace P | 40-79 | 14-50 | 48 | 20 | Excellent cold workability; good hot formability. Fabricated by blanking, bending, heading and upsetting, shearing and swaging. Uses: electrical contacts, flexible hose, pole-line hardware. |
| C51000 Phosphor bronze, 5% A | 94.8 Cu, 5.0 Sn, trace P | 47-140 | 19-80 | 64 | 20 | Excellent cold workability. Fabricated by blanking, drawing, bending, heading and upsetting, roll threading and knurling, shearing, stamping. Uses: bellows, Bourdon tubing, clutch discs, cotter pins, diaphragms, fasteners, lock washers, wire brushes, chemical hardware, textile machinery, welding rod. |
| C51100 | 95.6 Cu, 4.2 Sn, 0.2 P | 46-103 | 50-80 | 48 | 20 | Excellent cold workability. Uses: bridge bearing plates, locator bars, fuse clips, sleeve bushings, springs, switch parts, truss wire, wire brushes, chemical hardware, perforated sheets, textile machinery. |

**Table 3.** ***(Continued)*** **Properties and Applications of Wrought Coppers and Copper Alloys**

| Name and Number | Nominal Composition (%) | Strength (ksi) | | Elongation in 2 in. (%) | Machinability Rating[a] | Fabricating Characteristics and Typical Applications |
|---|---|---|---|---|---|---|
| | | Tensile | Yield | | | |
| C52100 Phosphor bronze, 8% C | 92.0 Cu, 8.0 Sn, trace P | 55-140 | 24-80 | 70 | 20 | Good cold workability for blanking, drawing, forming and bending, shearing, stamping. Uses: generally for more severe service conditions than C51000. |
| C52400 Phosphor bronze, 10% D | 90.0 Cu, 10.0 Sn, trace P | 66-147 | 28 | 70 | 20 | Good cold workability for blanking, forming and bending, shearing. Uses: heavy bars and plates for severe compression, bridge and expansion plates and fittings, articles requiring good spring qualities, resilience, fatigue resistance, good wear and corrosion resistance. |
| C54400 | 88.0 Cu, 4.0 Pb, 4.0 Zn, 4.0 Sn | 44-75 | 19-63 | 50 | 80 | Excellent machinability; good cold workability. Fabricated by blanking, drawing, bending, machining, shearing, stamping. Uses: bearings, bushings, gears, pinions, shafts, thrust washers, valve parts. |
| C60800 | 95.0 Cu, 5.0 Al | 60 | 27 | 55 | 20 | Good cold workability; fair hot formability. Uses: condenser, evaporator and heat-exchanger tubes, distiller tubes, ferrules. |
| C61000 | 92.0 Cu, 8.0 Al | 52-60 | 17-27 | 45 | 20 | Good hot and cold workability. Uses: bolts, pump parts, shafts, tie rods, overlay on steel for wearing surfaces. |
| C61300 | 90.3 Cu, 0.35 Sn, 6.8 Al, 0.35 Sn | 70-85 | 30-58 | 42 | 30 | Good hot and cold formability. Uses: nuts, bolts, corrosion resistant vessels and tanks, structural components, machine parts, condenser tube and piping systems, marine protective sheathing and fasteners, munitions mixing troughs and blending chambers. |
| C61400 Aluminum bronze, D | 91.0 Cu, 7.0 Al, 2.0 Fe | 76-89 | 33-60 | 45 | 20 | Similar to C61300. |
| C61500 | 90.0 Cu, 8.0 Al, 2.0 Ni | 70-145 | 22-140 | 55 | 30 | Good hot and cold workability. Fabricating characteristics similar to C52100. Uses: hardware, decorative metal trim, interior furnishings and other articles requiring high tarnish resistance. |
| C61800 | 89.0 Cu, 1.0 Fe, 10.0 Al | 80-85 | 39-42.5 | 28 | 40 | Fabricated by hot forging and hot pressing. Uses: bushings, bearings, corrosion-resistant applications, welding rods. |
| C61900 | 86.5 Cu, 4.0 Fe, 9.5 Al | 92-152 | 49-145 | 30 | | Excellent hot formability for fabricating by blanking, forming, bending, shearing, and stamping. Uses: springs, contacts, and switch components. |
| C62300 | 87.0 Cu, 3.0 Fe, 10.0 Al | 75-98 | 35-52 | 35 | 50 | Good hot and cold formability. Fabricated by bending, hot forging, hot pressing, forming, and welding. Uses: bearings, bushings, valve guides, gears, valve seats, nuts, bolts, pump rods, worm gears, and cams. |
| C62400 | 86.0 Cu, 3.0 Fe, 11.0 Al | 90-105 | 40-52 | 18 | 50 | Excellent hot formability for fabrication by hot forging and hot bending. Uses: bushings, gears, cams, wear strips, nuts, drift pins, tie rods. |
| C62500 | 82.7 Cu, 4.3 Fe, 13.0 Al | 100 (As extruded) | 55 | 1 | 20 | Excellent hot formability for fabrication by hot forging and machining. Uses: guide bushings, wear strips, cams, dies, forming rolls. |
| C63000 | 82.0 Cu, 3.0 Fe, 10.0 Al, 5.0 Ni | 90-118 | 50-75 | 20 | 30 | Good hot formability. Fabricated by hot forming and forging. Uses: nuts, bolts, valve seats, plunger tips, marine shafts, valve guides, aircraft parts, pump shafts, structural members. |
| C63200 | 82.0 Cu, 4.0 Fe, 9.0 Al, 5.0 Ni | 90-105 | 45-53 | 25 | 30 | Good hot formability. Fabricated by hot forming and welding. Uses: nuts, bolts, structural pump parts, shafting requiring corrosion resistance. |

**Table 3.** *(Continued)* **Properties and Applications of Wrought Coppers and Copper Alloys**

| Name and Number | Nominal Composition (%) | Strength (ksi) Tensile | Strength (ksi) Yield | Elongation in 2 in. (%) | Machinability Rating[a] | Fabricating Characteristics and Typical Applications |
|---|---|---|---|---|---|---|
| C63600 | 95.5 Cu, 3.5 Al, 1.0 Si | 60-84 | | 64 | 40 | Excellent cold workability; fair hot formability. Fabricated by cold heading. Uses: components for pole-line hardware, cold-headed nuts for wire and cable connectors, bolts and screw products. |
| C63800 | 95.0 Cu, 2.8 Al, 1.8 Si, 0.40 Co | 82-130 | 54-114 | 36 | | Excellent cold workability and hot formability. Uses: springs, switch parts, contacts, relay springs, glass sealing, and porcelain enameling. |
| C64200 | 91.2 Cu, 7.0 Al, 1.8 Si | 75-102 | 35-68 | 32 | 60 | Excellent hot formability. Fabricated by hot forming, forging, machining. Uses: valve stems, gears, marine hardware, pole-line hardware, bolts, nuts, valve bodies, and components. |
| C65100 Low-silicon bronze, B | 98.5 Cu, 1.5 Si | 40-105 | 15-71 | 55 | 30 | Excellent hot and cold workability. Fabricated by forming and bending, heading and upsetting, hot forging and pressing, roll threading and knurling, squeezing and swaging. Uses: hydraulic pressure lines, anchor screws, bolts, cable clamps, cap screws, machine screws, marine hardware, nuts, pole-line hardware, rivets, U-bolts, electrical conduits, heat-exchanger tubing, welding rod. |
| C65500 High-silicon bronze, A | 97.0 Cu, 3.0 Si | 56-145 | 21-71 | 63 | 30 | Excellent hot and cold workability. Fabricated by blanking, drawing, forming and bending, heading and upsetting, hot forging and pressing, roll threading and knurling, shearing, squeezing and swaging. Uses: similar to C65100 including propeller shafts. |
| C66700 Manganese brass | 70.0 Cu, 28.8 Zn, 1.2 Mn | 45.8-100 | 12-92.5 | 60 | 30 | Excellent cold formability. Fabricated by blanking, bending, forming, stamping, welding. Uses: brass products resistance welded by spot, seam, and butt welding. |
| C67400 | 58.5 Cu, 36.5 Zn, 1.2 Al, 2.8 Mn, 1.0 Sn | 70-92 | 34-55 | 28 | 25 | Excellent hot formability. Fabricated by hot forging and pressing, machining. Uses: bushings, gears, connecting rods, shafts, wear plates. |
| C67500 Manganese bronze, A | 58.5 Cu, 1.4 Fe, 39.0 Zn, 1.0 Sn, 0.1 Mn | 65-84 | 30-60 | 33 | 30 | Excellent hot workability. Fabricated by hot forging and pressing, hot heading and upsetting. Uses: clutch discs, pump rods, shafting, balls, valve stems and bodies. |
| C68700 Aluminum brass, arsenical | 77.5 Cu, 20.5 Zn, 2.0 Al, trace As | 60 | 27 | 55 | 30 | Excellent cold workability for forming and bending. Uses: condenser, evaporator- and heat-exchanger tubing, condenser tubing plates, distiller tubing, ferrules. |
| C68800 | 73.5 Cu, 22.7 Zn, 3.4 Al, 0.40 Co | 82-129 | 55-114 | 36 | | Excellent hot and cold formability. Fabricated by blanking, drawing, forming and bending, shearing and stamping. Uses: springs, switches, contacts, relays, drawn parts. |
| C69000 | 73.3 Cu, 3.4 Al, 0.6 Ni, 22.7 Zn | 82-130 | 52-117 | 35 | | Fabricating characteristics same as C68800. Uses: contacts, relays, switches, springs, drawn parts. |
| C69400 Silicon red brass | 81.5 Cu, 14.5 Zn, 4.0 Si | 80-100 | 40-57 | 25 | 30 | Excellent hot formability for fabrication by forging, screw-machine operations. Uses: valve stems where corrosion resistance and high strength are critical. |
| C70400 Copper nickel, 5% | 92.4 Cu, 1.5 Fe, 5.5 Ni, 0.6 Mn | 38-77 | 40-76 | 46 | 20 | Excellent cold workability; good hot formability. Fabricated by forming, bending, and welding. Uses: condensers, evaporators, heat exchangers, ferrules, salt water piping, lithium bromide absorption tubing, shipboard condenser intake systems. |

**Table 3.** *(Continued)* **Properties and Applications of Wrought Coppers and Copper Alloys**

| Name and Number | Nominal Composition (%) | Strength (ksi) | | Elongation in 2 in. (%) | Machinability Rating[a] | Fabricating Characteristics and Typical Applications |
|---|---|---|---|---|---|---|
| | | Tensile | Yield | | | |
| C70600 Copper nickel, 10% | 88.6 Cu, 1.4 Fe, 10.0 Ni | 44-60 | 16-57 | 42 | 20 | Good hot and cold workability. Fabricated by forming and bending, welding. Uses: condensers, condenser plates, distiller tubing, evaporator and heat-exchanger tubing, ferrules. |
| C71000 Copper nickel, 20% | 79.0 Cu, 21.0 Ni | 49-95 | 13-85 | 40 | 20 | Good hot and cold formability. Fabricated by blanking, forming and bending, welding. Uses: communication relays, condensers, condenser plates, electrical springs, evaporator and heat-exchanger tubes, ferrules, resistors. |
| C71500 Copper nickel, 30% | 69.5 Cu, 30.0 Ni, 0.5 Fe | 54-75 | 20-70 | 45 | 20 | Similar to C70600. |
| C72200 | 82.2 Cu, 16.5 Ni, 0.8 Fe, 0.5 Cr | 46-70 | 18-66 | 46 | | Good hot and cold formability. Fabricated by forming, bending, and welding. Uses: condenser tubing, heat-exchanger tubing, saltwater piping. |
| C72500 | 88.2 Cu, 9.5 Ni, 2.3 Sn | 55-120 | 22-108 | 35 | 20 | Excellent cold and hot formability. Fabricated by blanking, brazing, coining, drawing, etching, forming and bending, heading and upsetting, roll threading and knurling, shearing, spinning, squeezing, stamping, and swaging. Uses: relay and switch springs, connectors, brazing alloy, lead frames, control and sensing bellows. |
| C73500 | 72.0 Cu, 10.0 Zn, 18.0 Ni | 50-100 | 15-84 | 37 | 20 | Fabricating characteristics same as C74500. Uses: hollow ware, medallions, jewelry, base for silverplate, cosmetic cases, musical instruments, name plates, contacts. |
| C74500 Nickel silver, 65-10 | 65.0 Cu, 25.0 Zn, 10.0 Ni | 49-130 | 18-76 | 50 | 20 | Excellent cold workability. Fabricated by blanking, drawing, etching, forming and bending, heading and upsetting, roll threading and knurling, shearing, spinning, squeezing, and swaging. Uses: rivets, screws, slide fasteners, optical parts, etching stock, hollow ware, nameplates, platers' bars. |
| C75200 Nickel silver, 65-18 | 65.0 Cu, 17.0 Zn, 18.0 Ni | 56-103 | 25-90 | 45 | 20 | Fabricating characteristics similar to C74500. Uses: rivets, screws, table flatware, truss wire, zippers, bows, camera parts, core bars, temples, base for silverplate, costume jewelry, etching stock, hollow ware, nameplates, radio dials. |
| C75400 Nickel silver, 65-15 | 65.0 Cu, 20.0 Zn, 15.0 Ni | 53-92 | 18-79 | 43 | 20 | Fabricating characteristics similar to C74500. Uses: camera parts, optical equipment, etching stock, jewelry. |
| C75700 Nickel silver, 65-12 | 65.0 Cu, 23.0 Zn, 12.0 Ni | 52-93 | 18-79 | 48 | 20 | Fabricating characteristics similar to C74500. Uses: slide fasteners, camera parts, optical parts, etching stock, name plates. |
| C76390 | 61 Cu, 13 Zn, 24.5 Ni, 1 Pb, 0.5 Sn | 90 | 85 | 6 | 40 | Fabricated by machining, roll threading, and knurling. Uses: hardware, fasteners, connectors for electronic applications. |
| C77000 Nickel silver, 55-18 | 55.0 Cu, 27.0 Zn, 18.0 Ni | 60-145 | 27-90 | 40 | 30 | Good cold workability. Fabricated by blanking, forming and bending, and shearing. Uses: optical goods, springs, and resistance wire. |
| C78200 | 65.0 Cu, 2.0 Pb, 25.0 Zn, 8.0 Ni | 53-91 | 23-76 | 40 | 60 | Good cold formability. Fabricated by blanking, milling, and drilling. Uses: key blanks, watch plates, watch parts. |

[a] Free-cutting brass = 100.

*Source:* Copper Development Association, New York.

**Strength of Copper-Zinc-Tin Alloys (US Government Tests)**

| Percentage of | | | Tensile Strength | | Percentage of | | | Tensile Strength | | Percentage of | | | Tensile Strength | |
|---|---|---|---|---|---|---|---|---|---|---|---|---|---|---|
| Copper | Zinc | Tin | lb/in² | MPa | Copper | Zinc | Tin | lb/in² | MPa | Copper | Zinc | Tin | lb/in² | MPa |
| 45 | 50 | 5 | 15,000 | 103.4 | 60 | 20 | 20 | 10,000 | 68.9 | 75 | 20 | 5 | 45,000 | 310.3 |
| 50 | 45 | 5 | 50,000 | 344.7 | 65 | 30 | 5 | 50,000 | 344.7 | 75 | 15 | 10 | 45,000 | 310.3 |
| 50 | 40 | 10 | 15,000 | 103.4 | 65 | 25 | 10 | 42,000 | 289.6 | 75 | 10 | 15 | 43,000 | 296.5 |
| 55 | 43 | 2 | 65,000 | 488.2 | 65 | 20 | 15 | 30,000 | 206.8 | 75 | 5 | 20 | 41,000 | 282.7 |
| 55 | 40 | 5 | 62,000 | 427.5 | 65 | 15 | 20 | 18,000 | 124.1 | 80 | 15 | 5 | 45,000 | 310.3 |
| 55 | 35 | 10 | 32,500 | 224.0 | 65 | 10 | 25 | 12,000 | 82.7 | 80 | 10 | 10 | 45,000 | 310.3 |
| 55 | 30 | 15 | 15,000 | 103.4 | 70 | 25 | 5 | 45,000 | 310.3 | 80 | 5 | 15 | 47,500 | 327.5 |
| 60 | 37 | 3 | 60,000 | 413.7 | 70 | 20 | 10 | 44,000 | 303.4 | 85 | 10 | 5 | 43,500 | 299.9 |
| 60 | 35 | 5 | 52,500 | 362.0 | 70 | 15 | 15 | 37,000 | 255.1 | 85 | 5 | 10 | 46,500 | 320.6 |
| 60 | 30 | 10 | 40,000 | 275.8 | 70 | 10 | 20 | 30,000 | 206.8 | 90 | 5 | 5 | 42,000 | 289.6 |

## Copper-Silicon and Copper-Beryllium Alloys

**Everdur.—**This copper-silicon alloy is available in five slightly different nominal compositions for applications that require high strength, good fabricating and fusing qualities, immunity to rust, free-machining and a corrosion resistance equivalent to copper. The following table gives the nominal compositions and tensile strengths, yield strengths, and percent elongations for various tempers and forms.

**Table 4. Nominal Composition and Properties of Everdur**

| Desig. No. | Nominal Composition | | | | | Temper[a] | Strength | | | | Elongation (%) |
|---|---|---|---|---|---|---|---|---|---|---|---|
| | | | | | | | Tensile | | Yield | | |
| | Cu | Si | Mn | Pb | Al | | ksi | MPa | ksi | MPa | |
| 655 | 95.80 | 3.10 | 1.10 | | | A | 52 | 359 | 15 | 103 | 35[b] |
| | | | | | | HRA | 50 | 345 | 18 | 124 | 40 |
| | | | | | | CRA | 52 | 359 | 18 | 124 | 35 |
| | | | | | | CRHH | 71 | 490 | 40 | 276 | 10 |
| | | | | | | CRH | 87 | 600 | 60 | 414 | 3 |
| | | | | | | H | 70–85 | 483–586 | 38–50 | 262–345 | 17 to 8[b] |
| 651 | 98.25 | 1.50 | 0.25 | | | AP | 38 | 262 | 10 | 69 | 35 |
| | | | | | | HP | 50 | 345 | 40 | 276 | 7 |
| | | | | | | XHB | 75–85 | 517–586 | 45–55 | 310–379 | 8 to 6[b] |
| 661 | 95.60 | 3.00 | 1.00 | 0.40 | | A | 52 | 359 | 15 | 103 | 35[b] |
| | | | | | | H | 85 | 586 | 50 | 345 | 13 to 8[b] |
| 6552 | 94.90 | 4.00 | 1.10 | | | AC | 45 | 310 | | | 15 |
| 637 | 90.75 | 2.00 | | | 7.25 | A | 75–90 | 517–621 | 37.5–45 | 259–310 | 12 to 9[b] |

[a] Symbols used are HRA for hot-rolled and annealed tank plates; CRA for cold-rolled sheets and strips; CRHH for cold-rolled half hard strips; abd CRH for cold-rolled hard strips. For round, square, hexagonal, and octagonal rods: A for annealed; H for hard; and XHB for extra-hard bolt temper (in coils for cold-heading). For pipe and tube: AP for annealed; and HP for hard. For castings: AC for as cast.

[b] Percent elongation in 4 times the diameter or thickness of the specimen. All other values are percent elongation in 2 inches.

Designation numbers are those of the American Brass Co.

The values given for tensile srength, yield strength, and elongation are all minimum values. Where ranges are shown, the first values given are for the largest diameter or largest size specimens. Yield strength values were determined at 0.50 percent elongation under load.

**Copper-Beryllium Alloys.—**Alloys of copper and beryllium present health hazards. Particles produced by machining may be absorbed into the body through the skin, the mouth, the nose, or an open wound, resulting in a condition requiring immediate medical attention. Working of these alloys requires protective clothing or other shielding in a monitored environment. Copper-beryllium alloys involved in a fire give off profuse toxic fumes that must not be inhaled.

These alloys contain copper, beryllium, cobalt, and silver, and fall into two groups. One group, whose beryllium content is greater than one percent, is characterized by its high strength and hardness, and the other, whose beryllium content is less than one percent, by its high electrical and thermal conductivity. The alloys have many applications in the electrical and aircraft industries or wherever strength, corrosion resistance, conductivity, non-magnetic and nonsparking properties are essential. Beryllium copper is obtainable in the form of strips, rods and bars, wire, platers, bars, billets, tubes, and casting ingots.

*Composition and Properties:* Table 5 lists some of the more common wrought alloys and gives some of their mechanical properties.

**Table 5. Wrought Copper–Beryllium Properties**

| Alloy[a] | Form | Temper[b] | Heat Treatment | Tensile Strength | | Yield Strength 0.2% Offset | | Elongation % in 2 inch (5.08 cm) |
|---|---|---|---|---|---|---|---|---|
| | | | | ksi | MPa | ksi | MPa | |
| 25 | Rod, Bar, and Plate | A | | 60–85 | 414–586 | 20–30 | 138–207 | 35–60 |
| | | ½ H or H | | 85–130 | 586–896 | 75–105 | 517–724 | 10–20 |
| | | AT | 3 hr at 600° F (316° C) or mill heat-treated | 165–190 | 1138–1310 | 145–175 | 1000–1207 | 3–10 |
| | | ½ HT or HT | 2 hr at 600° F (316° C) or mill heat-treated | 175–215 | 1207–1482 | 150–200 | 1034–1379 | 2–5 |
| | Wire | A | | 58–78 | 400–538 | 20–35 | 138–241 | 35–55 |
| | | ¼ H | | 90–115 | 621–793 | 70–95 | 483–655 | 10–35 |
| | | ½ H | | 110–135 | 758–931 | 90–110 | 621–758 | 5–10 |
| | | ¾ H | | 130–155 | 896–1069 | 110–135 | 758–931 | 2–8 |
| | | AT | 3 hr at 600° F (316° C) | 165–190 | 1138–1310 | 145–175 | 1000–1207 | 3–8 |
| | | ¼ HT | 2 hr at 600° F (316° C) | 175–205 | 1207–1413 | 160–190 | 1103–1310 | 2–5 |
| | | ½ HT | 2 hr at 600° F (316° C) | 190–215 | 1310–1482 | 175–200 | 1207–1379 | 1–3 |
| | | ¾ HT | 2 hr at 600° F (316° C) | 195–220 | 1344–1517 | 180–205 | 1241–1413 | 1–3 |
| | | XHT | Mill heat-treated | 115–165 | 793–1138 | 95–145 | 655–1000 | 2–8 |
| 165 | Rod, Bar, and Plate | A | | 60–85 | 414–586 | 20–30 | 138–207 | 35–60 |
| | | ½ H or H | | 85–130 | 586–896 | 75–105 | 517–724 | 10–20 |
| | | AT | 3 hr at 650° F (343° C) or mill heat-treated | 150–180 | 1034–1241 | 125–155 | 862–1069 | 4–10 |
| | | ½ HT or HT | 2 hr at 650° F (343° C) or mill heat-treated | 165–200 | 1138–1379 | 135–165 | 931–1138 | 2–5 |
| 10 | Rod, Bar, and Plate | A | | 35–55 | 241–379 | 20–30 | 138–207 | 20–35 |
| | | ½ H or H | | 65–80 | 448–552 | 55–75 | 379–517 | 10–15 |
| | | AT | 3 hr at 900° F (482° C) or mill heat-treated | 100–120 | 689–827 | 80–100 | 552–689 | 10–25 |
| | | ½ HT or HT | 2 hr at 900° F (482° C) or mill heat-treated | 110–130 | 758–896 | 100–120 | 689–827 | 8–20 |
| 50 | Rod, Bar, and Plate | A | | 35–55 | 241–379 | 20–30 | 138–207 | 20–35 |
| | | ½ H or H | | 65–80 | 448–552 | 55–75 | 379–517 | 10–15 |
| | | AT | 3 hr at 900° F (482° C) or mill heat-treated | 100–120 | 689–827 | 80–100 | 552–689 | 10–25 |
| | | ½ HT or HT | 2 hr at 900° F (482° C) or mill heat-treated | 110–130 | 758–896 | 100–120 | 689–827 | 8–20 |
| 35 | Rod, Bar, and Plate | A | | 35–55 | 241–379 | 20–30 | 138–207 | 20–35 |
| | | ½ H or H | | 65–80 | 448–552 | 55–75 | 379–517 | 10–15 |
| | | AT | 3 hr at 900° F (482° C) or mill heat-treated | 100–120 | 689–827 | 80–100 | 552–689 | 10–25 |
| | | ½ HT or HT | 2 hr at 900° F (482° C) or mill heat-treated | 110–130 | 758–896 | 100–120 | 689–827 | 8–20 |

[a] Composition (in percent) of alloys is as follows: alloy 25: 1.80–2.05 Be, 0.20–0.35 Co, balance Cu; alloy 165: 1.6–1.8 Be, 0.20–0.35 Co, balance Cu; alloy 10: 0.4–0.7 Be, 2.35–2.70 Co, balance Cu; alloy 50, 0.25–0.50 Be, 1.4–1.7 Co, 0.9–1.1 Ag, balance Cu; alloy 35, 0.25–0.50 Be, 1.4–1.6 Ni, balance Cu.

[b] Temper symbol designations: A, solution annealed; H, hard; HT, heat treated from hard; At, heat treated from solution annealed.

## Aluminum and Aluminum Alloys

Pure aluminum is a silver-white metal characterized by a slightly bluish cast. It has a specific gravity of 2.70, resists the corrosive effects of many chemicals, and has a malleability approaching that of gold. When alloyed with other metals, numerous properties are obtained that make these alloys useful over a wide range of applications.

Aluminum alloys are light in weight compared with steel, brass, nickel, or copper; can be fabricated by all common processes; are available in a wide range of sizes, shapes, and forms; resist corrosion; readily accept a wide range of surface finishes; have good electrical and thermal conductivities; and are highly reflective to both heat and light.

**Characteristics of Aluminum and Aluminum Alloys.—**Aluminum and its alloys lose part of their strength at elevated temperatures, although some alloys retain good strength at temperatures from 400 to 500° F (204 to 260° C). At subzero temperatures, however, their strength increases without loss of ductility so that aluminum is a particularly useful metal for low-temperature applications.

When aluminum surfaces are exposed to the atmosphere, a thin invisible oxide skin forms immediately that protects the metal from further oxidation. This self-protecting characteristic gives aluminum its high resistance to corrosion. Unless exposed to some substance or condition that destroys this protective oxide coating, the metal remains protected against corrosion. Aluminum is highly resistant to weathering, even in industrial atmospheres. It is also corrosion resistant to many acids. Alkalis are among the few substances that attack the oxide skin and therefore are corrosive to aluminum. Although the metal can safely be used in the presence of certain mild alkalis with the aid of inhibitors, in general, direct contact with alkaline substances should be avoided. Direct contact with certain other metals should be avoided in the presence of an electrolyte; otherwise, galvanic corrosion of the aluminum may take place in the contact area. Where other metals must be fastened to aluminum, the use of a bituminous paint coating or insulating tape is recommended.

Aluminum is one of the two common metals having an electrical conductivity high enough for use as an electric conductor. The conductivity of electric-conductor (EC) grade is about 62 percent that of the International Annealed Copper Standard. Because aluminum has less than one-third the specific gravity of copper, however, a pound of aluminum will go almost twice as far as a pound of copper when used as a conductor. Alloying lowers the conductivity somewhat so that wherever possible the EC grade is used in electric conductor applications. However, aluminum takes a set, which often results in loosening of screwed connectors, leading to arcing and fires. Special clamping designs are therefore required when aluminum is used for electrical wiring, especially in buildings.

Aluminum has nonsparking and nonmagnetic characteristics that make the metal useful for electrical shielding purposes, such as in bus bar housings or enclosures for other electrical equipment, and for use around inflammable or explosive substances.

Aluminum can be cast by any method known. It can be rolled to any desired thickness down to foil thinner than paper and in sheet form can be stamped, drawn, spun, or roll-formed. The metal also may be hammered or forged. Aluminum wire, drawn from rolled rod, may be stranded into cable of any desired size and type. The metal may be extruded into a variety of shapes. It may be turned, milled, bored, or otherwise machined in equipment often operating at their maximum speeds. Aluminum rod and bar may readily be employed in the high-speed manufacture of parts made on automatic screw-machine.

Almost any method of joining is applicable to aluminum—riveting, welding, or brazing. A wide variety of mechanical aluminum fasteners simplifies the assembly of many products. Resin bonding of aluminum parts has been successfully employed, particularly in aircraft components.

For the majority of applications, aluminum needs no protective coating. Mechanical finishes such as polishing, sandblasting, or wire brushing meet the majority of needs. When additional protection is desired, chemical, electrochemical, and paint finishes are all used. Vitreous enamels have been developed for aluminum, and the metal may also be electroplated.

**Temper Designations for Aluminum Alloys.—**The temper designation system adopted by the Aluminum Association and used in industry pertains to all forms of wrought and cast aluminum and aluminum alloys except ingot. It is based on the sequences of basic treatments used to produce the various tempers. The temper designation follows the alloy designation, being separated by a dash.

Basic temper designations consist of letters. Subdivisions of the basic tempers, where required, are indicated by one or more digits following the letter. These digits designate specific sequences of basic treatments, but only operations recognized as significantly influencing the characteristics of the product are indicated. Should some other variation of the same sequence of basic operations be applied to the same alloy, resulting in different characteristics, then additional digits are added.

The basic temper designations and subdivisions are as follows:

*-F, as fabricated:* Applies to products that acquire some temper from shaping processes not having special control over the amount of strain-hardening or thermal treatment. For wrought products, there are no mechanical property limits.

*-O, annealed, recrystallized (wrought products only):* Applies to the softest temper of wrought products.

*-H, strain-hardened (wrought products only):* Applies to products that have their strength increased by strain-hardening with or without supplementary thermal treatments to produce partial softening.

The -H is always followed by two or more digits. The first digit indicates the specific combination of basic operations, as follows:

*-H1, strain-hardened only:* Applies to products that are strain-hardened to obtain the desired mechanical properties without supplementary thermal treatment. The number following this designation indicates the degree of strain-hardening.

*-H2, strain-hardened and then partially annealed:* Applies to products that are strain-hardened more than the desired final amount and then reduced in strength to the desired level by partial annealing. For alloys that age-soften at room temperature, the -H2 tempers have approximately the same ultimate strength as the corresponding -H3 tempers. For other alloys, the -H2 tempers have approximately the same ultimate strengths as the corresponding -H1 tempers and slightly higher elongations. The number following this designation indicates the degree of strain-hardening remaining after the product has been partially annealed.

*-H3, strain-hardened and then stabililized:* Applies to products which are strain-hardened and then stabilized by a low-temperature heating to slightly lower their strength and increase ductility. This designation applies only to the magnesium-containing alloys that, unless stabilized, gradually age-soften at room temperature.The number following this designation indicates the degree of strain-hardening remaining after the product has been strain-hardened a specific amount and then stabilized.

The second digit following the designations -H1, -H2, and -H3 indicates the final degree of strain-hardening. Numeral 8 has been assigned to indicate tempers having a final degree of strain-hardening equivalent to that resulting from approximately 75 percent reduction of area. Tempers between -O (annealed) and 8 (full hard) are designated by numerals 1 through 7. Material having an ultimate strength about midway between that of the -O temper and that of the 8 temper is designated by the numeral 4 (half hard); between -O and 4 by the numeral 2 (quarter hard); and between 4 and 8 by the numeral 6 (three-quarter hard). (*Note:* For two-digit -H tempers whose second figure is odd, the standard limits for ultimate strength are exactly midway between those for the adjacent two-digit -H tempers whose second figures are even.) Numeral 9 designates extra-hard tempers.

The third digit, when used, indicates a variation of a two-digit -H temper, and is used when the degree of control of temper or the mechanical properties are different from but close to those for the two-digit -H temper designation to which it is added. (*Note:* The minimum ultimate strength of a three-digit -H temper is at least as close to that of the corresponding two-digit -H temper as it is to the adjacent two-digit -H tempers.) Numerals 1 through 9 may be arbitrarily assigned and registered with the Aluminum Association for an alloy and product to indicate a specific degree of control of temper or specific mechanical property limits. Zero has been assigned to indicate degrees of control of temper or mechanical property limits negotiated between the manufacturer and purchaser that are not used widely enough to justify registration with the Aluminum Association.

The following three-digit -H temper designations have been assigned for wrought products in all alloys:

*-H111:* Applies to products that are strain-hardened less than the amount required for a controlled H11 temper.

*-H112:* Applies to products that acquire some temper from shaping processes not having special control over the amount of strain-hardening or thermal treatment, but for which there are mechanical property limits, or require mechanical property testing.

The following three-digit H temper designations have been assigned for wrought products in alloys containing more than a normal 4 percent magnesium.

*-H311:* Applies to products that are strain-hardened less than the amount required for a controlled H31 temper.

*-H321:* Applies to products that are strain-hardened less than the amount required for a controlled H32 temper.

*-H323:* Applies to products that are specially fabricated to have acceptable resistance to stress-corrosion cracking.

*-H343:* Applies to products that are specially fabricated to have acceptable resistance to stress-corrosion cracking.

The following three-digit -H temper designations have been assigned for

| Patterned or Embossed Sheet | Fabricated Form |
|---|---|
| -H114 | -O temper |
| -H124, -H224, -H324 | -H11, -H21, -H31 temper, respectively |
| -H134, -H234, -H334 | -H12, -H22, -H32 temper, respectively |
| -H144, -H244, -H344 | -H13, -H23, -H33 temper, respectively |
| -H154, -H254, -H354 | -H14, -H24, -H34 temper, respectively |
| -H164, -H264, -H364 | -H15, -H25, -H35 temper, respectively |
| -H174, -H274, -H374 | -H16, -H26, -H36 temper, respectively |
| -H184, -H284, -H384 | -H17, -H27, -H37 temper, respectively |
| -H194, -H294, -H394 | -H18, -H28, -H38 temper, respectively |
| -H195, -H395 | -H19, -H39 temper, respectively |

*-W, solution heat treated:* An unstable temper applicable only to alloys that spontaneously age at room temperature after solution heat treatment. This designation is specific only when the period of natural aging is indicated.

*-T, thermally treated to produce stable tempers other than -F, -O, or -H:* Applies to products that are thermally treated, with or without supplementary strain-hardening, to produce stable tempers.The -T is always followed by one or more digits. Numerals 2 through 10 have been assigned to indicate specific sequences of basic treatments, as follows:

*-T1, naturally aged to a substantially stable condition:* Applies to products for which the rate of cooling from an elevated temperature-shaping process, such as casting or extrusion, is such that their strength is increased by room-temperature aging.

*-T2, annealed (cast products only):* Designates a type of annealing treatment used to improve ductility and increase dimensional stability of castings.

*-T3, solution heat treated and then cold-worked:* Applies to products that are cold-worked to improve strength, or in which the effect of cold-work in flattening or straightening is recognized in applicable specifications.

*-T4, solution heat treated and naturally aged to a substantially stable condition:* Applies to products that are not cold-worked after solution heat treatment, or in which the effect of cold-work in flattening or straightening may not be recognized in applicable specifications.

*-T5, artificially aged only:* Applies to products that are artificially aged after an elevated-temperature rapid-cool fabrication process, such as casting or extrusion, to improve mechanical properties or dimensional stability, or both.

*-T6, solution heat treated and then artificially aged:* Applies to products that are not cold-worked after solution heat treatment, or in which the effect of cold-work in flattening or straightening may not be recognized in applicable specifications.

*-T7, solution heat treated and then stabilized:* Applies to products that are stabilized to carry them beyond the point of maximum hardness, providing control of growth or residual stress or both.

*-T8, solution heat treated, cold-worked, and then artificially aged:* Applies to products that are cold-worked to improve strength, or in which the effect of cold-work in flattening or straightening is recognized in applicable specifications.

*-T9, solution heat treated, artificially aged, and then cold-worked:* Applies to products that are cold-worked to improve strength.

*-T10, artificially aged and then cold-worked:* Applies to products that are artificially aged after an elevated-temperature rapid-cool fabrication process, such as casting or extrusion, and then cold-worked to improve strength.

Additional digits may be added to designations -T1 through -T10 to indicate a variation in treatment that significantly alters the characteristics of the product. These may be arbitrarily assigned and registered with The Aluminum Association for an alloy and product to indicate a specific treatment or specific mechanical property limits.

These additional digits have been assigned for wrought products in all alloys:

*-T__51, stress-relieved by stretching:* Applies to products that are stress-relieved by stretching the following amounts after solution heat treatment:

| | |
|---|---|
| Plate | 1½ to 3 percent permanent set |
| Rod, Bar and Shapes | 1 to 3 percent permanent set |
| Drawn tube | 0.5 to 3 percent permanent set |

Applies directly to plate and rolled or cold-finished rod and bar.

These products receive no further straightening after stretching.

Applies to extruded rod and bar shapes and tube when designated as follows:

*-T__510 :* Products that receive no further straightening after stretching.

*-T__511:* Products that receive minor straightening after stretching to comply with standard tolerances.

*-T__52, stress-relieved by compressing:* Applies to products that are stress-relieved by compressing after solution heat treatment to produce a nominal permanent set of 2½ percent.

*-T__54, stress-relieved by combined stretching and compressing:* Applies to die forgings that are stress relieved by restriking cold in the finish die.

The following two-digit -T temper designations have been assigned for wrought products in all alloys:

*-T42:* Applies to products solution heat treated and naturally aged that attain mechanical properties different from those of the -T4 temper.

*-T62:* Applies to products solution heat treated and artificially aged that attain mechanical properties different from those of the -T6 temper.

**Aluminum Alloy Designation Systems.—**Aluminum casting alloys are listed in many specifications of various standardizing agencies. The numbering systems used by each differ and do not always correlate. Casting alloys are available from producers who use a commercial numbering system; this system is used in the tables of aluminum casting alloys given in this section.

Table 6a lists the nominal composition of commonly used aluminum casting alloys, and Table 6b and Table 6c list the typical tensile properties of separately cast bars.

**Table 6a. Nominal Compositions (in percent) of Common Aluminum Casting Alloys (AA/ANSI)**

| Alloy | Product[a] | Si | Fe | Cu | Mn | Mg | Cr | Ni | Zn | Ti | Others | |
|---|---|---|---|---|---|---|---|---|---|---|---|---|
| | | | | | | | | | | | Each | Total |
| 201.0 | S | 0.10 | 0.15 | 4.0-5.2 | 0.20-0.50 | 0.15-0.55 | | | | 0.15-0.35 | 0.05[b] | 0.10 |
| 204.0 | S&P | 0.20 | 0.35 | 4.2-5.0 | 0.10 | 0.15-0.35 | | 0.05 | 0.10 | 0.15-0.30 | 0.05[c] | 0.15 |
| 208.0 | S&P | 2.5-3.5 | 1.2 | 3.5-4.5 | 0.50 | 0.10 | | 0.35 | 1.0 | 0.25 | | 0.50 |
| 222.0 | S&P | 2.0 | 1.5 | 9.2-10.7 | 0.50 | 0.15-0.35 | | 0.50 | 0.8 | 0.25 | | 0.35 |
| 242.0 | S&P | 0.7 | 1.0 | 3.5-4.5 | 0.35 | 1.2-1.8 | 0.25 | 1.7-2.3 | 0.35 | 0.25 | 0.05 | 0.15 |
| 295.0 | S | 0.7-1.5 | 1.0 | 4.0-5.0 | 0.35 | 0.03 | | | 0.35 | 0.25 | 0.05 | 0.15 |
| 308.0 | P | 5.0-6.0 | 1.0 | 4.0-5.0 | 0.50 | 0.10 | | | 1.0 | 0.25 | | 0.50 |
| 319.0 | S&P | 5.5-6.5 | 1.0 | 3.0-4.0 | 0.50 | 0.10 | | 0.35 | 1.0 | 0.25 | | 0.50 |
| 328.0 | S | 7.5-8.5 | 1.0 | 1.0-2.0 | 0.20-0.6 | 0.20-0.6 | 0.35 | 0.25 | 1.5 | 0.25 | | 0.50 |
| 332.0 | P | 8.5-10.5 | 1.2 | 2.0-4.0 | 0.50 | 0.50-1.5 | | 0.50 | 1.0 | 0.25 | | 0.50 |
| 333.0 | P | 8.0-10.0 | 1.0 | 3.0-4.0 | 0.50 | 0.05-0.50 | | 0.50 | 1.0 | 0.25 | | 0.50 |
| 336.0 | P | 11.0-13.0 | 1.2 | 0.50-1.5 | 0.35 | 0.7-1.3 | | 2.0-3.0 | 0.35 | 0.25 | 0.05 | |
| 355.0 | S&P | 4.5-5.5 | 0.6[d] | 1.0-1.5 | 0.50[d] | 0.40-0.6 | 0.25 | | 0.35 | 0.25 | 0.05 | 0.15 |
| C355.0 | S&P | 4.5-5.5 | 0.20 | 1.0-1.5 | 0.10 | 0.40-0.6 | | | 0.10 | 0.20 | 0.05 | 0.15 |
| 356.0 | S&P | 6.5-7.5 | 0.6[d] | 0.25 | 0.35[d] | 0.20-0.45 | | | 0.35 | 0.25 | 0.05 | 0.15 |
| 356.0 | S&P | 6.5-7.5 | 0.20 | 0.20 | 0.10 | 0.25-0.45 | | | 0.10 | 0.20 | 0.05 | 0.15 |
| 357.0 | S&P | 6.5-7.5 | 0.15 | 0.05 | 0.03 | 0.45-0.6 | | | 0.05 | 0.20 | 0.05 | 0.15 |
| A357.0 | S&P | 6.5-7.5 | 0.20 | 0.20 | 0.10 | 0.40-0.7 | | | 0.10 | 0.04-0.20 | 0.05 | 0.15[e] |
| 443.0 | S&P | 4.5-6.0 | 0.8 | 0.6 | 0.50 | 0.05 | 0.25 | | 0.50 | 0.25 | | 0.35 |
| B443.0 | S&P | 4.5-6.0 | 0.8 | 0.15 | 0.35 | 0.05 | | | 0.35 | 0.25 | 0.05 | 0.15 |
| A444.0 | P | 6.5-7.5 | 0.20 | 0.10 | 0.10 | 0.05 | | | 0.10 | 0.20 | 0.05 | 0.15 |
| 512.0 | S | 1.4-2.2 | 0.6 | 0.35 | 0.8 | 3.5-4.5 | 0.25 | | 0.35 | 0.25 | 0.05 | 0.15 |
| 513.0 | P | 0.30 | 0.40 | 0.10 | 0.30 | 3.5-4.5 | | | 1.4-2.2 | 0.20 | 0.05 | 0.15 |
| 514.0 | S | 0.35 | 0.50 | 0.15 | 0.35 | 3.5-4.5 | | | 0.15 | 0.25 | 0.05 | 0.15 |
| 520.0 | S | 0.25 | 0.30 | 0.25 | 0.15 | 9.5-10.6 | | | 0.15 | 0.25 | 0.05 | 0.15 |
| 705.0 | S&P | 0.20 | 0.8 | 0.20 | 0.40-0.6 | 1.4-1.8 | 0.20-0.40 | | 2.7-3.3 | 0.25 | 0.05 | 0.15 |
| 707.0 | S&P | 0.20 | 0.8 | 0.20 | 0.40-0.6 | 1.8-2.4 | 0.20-0.40 | | 4.0-4.5 | 0.25 | 0.05 | 0.15 |
| 710.0 | S | 0.15 | 0.50 | 0.35-0.65 | 0.05 | 0.6-0.8 | | | 6.0-7.0 | 0.25 | 0.05 | 0.15 |
| 711.0 | P | 0.30 | 0.7-1.4 | 0.35-0.65 | 0.05 | 0.25-0.45 | | | 6.0-7.0 | 0.20 | 0.05 | 0.15 |
| 712.0 | S | 0.30 | 0.50 | 0.25 | 0.10 | 0.50-0.65 | 0.40-0.6 | | 5.0-6.5 | 0.15-0.25 | 0.05 | 0.20 |
| 850.0 | S&P | 0.7 | 0.7 | 0.7-1.3 | 0.10 | 0.10 | | 0.7-1.3 | | 0.20 | —[f] | 0.30 |
| 851.0 | S&P | 2.0-3.0 | 0.7 | 0.7-1.3 | 0.10 | 0.10 | | 0.3-0.7 | | 0.20 | —[f] | 0.30 |

[a] S = sand cast; P = permanent mold cast. The sum of those "Others," metallic elements 0.010 percent or more each, is expressed to the second decimal before determining the sum. *Source: Standards for Aluminum Sand and Permanent Mold Castings,* courtesy of the Aluminum Association.

[b] Also contains 0.40–1.0 percent silver.

[c] Also contains 0.05 max. percent tin.

[d] If iron exceeds 0.45 percent, manganese content should not be less than one-half the iron content.

[e] Also contains 0.04–0.07 percent beryllium.

[f] Also contains 5.5–7.0 percent tin.

## Table 6b. Mechanical Property Limits for Separately Cast Test Bars of Commonly Used Aluminum Sand Casting Alloys

| Alloy | Temper[a] | Minimum Properties | | | Typical Brinell Hardness Number (500 kgf load, 10-mm ball) |
|---|---|---|---|---|---|
| | | Tensile Strength (ksi) | | Elongation In 2 inches (%) | |
| | | Ultimate | Yield | | |
| 201.0 | T7 | 60.0 | 50.0 | 3.0 | 110-140 |
| 204.0 | T4 | 45.0 | 28.0 | 6.0 | |
| 208.0 | F | 19.0 | 12.0 | 1.5 | 40-70 |
| 222.0 | O | 23.0 | | | 65-95 |
| 222.0 | T61 | 30.0 | | | 100-130 |
| 242.0 | O | 23.0 | | | 55-85 |
| 242.0 | T571 | 29.0 | | | 70-100 |
| 242.0 | T61 | 32.0 | 20.0 | | 90-120 |
| 242.0 | T77 | 24.0 | 13.0 | 1.0 | 60-90 |
| 295.0 | T4 | 29.0 | 13.0 | 6.0 | 45-75 |
| 295.0 | T6 | 32.0 | 20.0 | 3.0 | 60-90 |
| 295.0 | T62 | 36.0 | 28.0 | | 80-110 |
| 295.0 | T7 | 29.0 | 16.0 | 3.0 | 55-85 |
| 319.0 | F | 23.0 | 13.0 | 1.5 | 55-85 |
| 319.0 | T5 | 25.0 | | | 65-95 |
| 319.0 | T6 | 31.0 | 20.0 | 1.5 | 65-95 |
| 328.0 | F | 25.0 | 14.0 | 1.0 | 45-75 |
| 328.0 | T6 | 34.0 | 21.0 | 1.0 | 65-95 |
| 354.0 | [b] | | | | |
| 355.0 | T51 | 25.0 | 18.0 | | 50-80 |
| 355.0 | T6 | 32.0 | 20.0 | 2.0 | 70-105 |
| 355.0 | T7 | 35.0 | | | 70-100 |
| 355.0 | T71 | 30.0 | 22.0 | | 60-95 |
| C355.0 | T6 | 36.0 | 25.0 | 2.5 | 75-105 |
| 356.0 | F | 19.0 | | 2.0 | 40-70 |
| 356.0 | T51 | 23.0 | 16.0 | | 45-75 |
| 356.0 | T6 | 30.0 | 20.0 | 3.0 | 55-90 |
| 356.0 | T7 | 31.0 | 29.0 | | 60-90 |
| 356.0 | T71 | 25.0 | 18.0 | 3.0 | 45-75 |
| A356.0 | T6 | 34.0 | 24.0 | 3.5 | 70-105 |
| 443.0 | F | 17.0 | 7.0 | 3.0 | 25-55 |
| B443.0 | F | 17.0 | 6.0 | 3.0 | 25-55 |
| 512.0 | F | 17.0 | 10.0 | | 35-65 |
| 514.0 | F | 22.0 | 9.0 | 6.0 | 35-65 |
| 520.0 | T4[c] | 42.0 | 22.0 | 12.0 | 60-90 |
| 535.0 | F or T5 | 35.0 | 18.0 | 9.0 | 60-90 |
| 705.0 | F or T5 | 30.0 | 17.0 | 5.0 | 50-80 |
| 707.0 | T5 | 33.0 | 22.0 | 2.0 | 70-100 |
| 707.0 | T7 | 37.0 | 30.0 | 1.0 | 65-95 |
| 710.0 | F or T5 | 32.0 | 20.0 | 2.0 | 60-90 |
| 712.0 | F or T5 | 34.0 | 25.0 | 4.0 | 60-90 |
| 713.0 | F or T5 | 32.0 | 22.0 | 3.0 | 60-90 |
| 771.0 | T5 | 42.0 | 38.0 | 1.5 | 85-115 |
| 771.0 | T51 | 32.0 | 27.0 | 3.0 | 70-100 |
| 771.0 | T52 | 36.0 | 30.0 | 1.5 | 70-100 |
| 771.0 | T53 | 36.0 | 27.0 | 1.5 | |
| 771.0 | T6 | 42.0 | 35.0 | 5.0 | 75-105 |
| 771.0 | T71 | 48.0 | 45.0 | 2.0 | 105-135 |
| 850.0 | T5 | 16.0 | | 5.0 | 30-60 |
| 851.0 | T5 | 17.0 | | 3.0 | 30-60 |
| 852.0 | T5 | 24.0 | 18.0 | | 45-75 |

*Source: Standards for Aluminum Sand and Permanent Mold Castings*, courtesy of the Aluminum Association.

[a] F indicates "as cast" condition.

[b] Mechanical properties for these alloys depend on the casting process. For further information consult the individual foundries.

[c] The T4 temper of Alloy 520.0 is unstable; significant room temperature aging occurs within life expectancy of most castings. Elongation may decrease by as much as 80 percent.

## Table 6c. Mechanical Property Limits for Separately Cast Test Bars of Commonly Used Aluminum Permanent Mold Casting Alloys

| Alloy | Temper[a] | Minimum Properties | | | Typical Brinell Hardness Number (500 kgf load, 10-mm ball) |
|---|---|---|---|---|---|
| | | Tensile Strength (ksi) | | Elongation In 2 inches (%) | |
| | | Ultimate | Yield | | |
| 204.0 | T4 | 48.0 | 29.0 | 8.0 | |
| 208.0 | T4 | 33.0 | 15.0 | 4.5 | 60-90 |
| 208.0 | T6 | 35.0 | 22.0 | 2.0 | 75-105 |
| 208.0 | T7 | 33.0 | 16.0 | 3.0 | 65-95 |
| 222.0 | T551 | 30.0 | | | 100-130 |
| 222.0 | T65 | 40.0 | | | 125-155 |
| 242.0 | T571 | 34.0 | | | 90-120 |
| 242.0 | T61 | 40.0 | | | 95-125 |
| 296.0 | T6 | 35.0 | | 2.0 | 75-105 |
| 308.0 | F | 24.0 | | | 55-85 |
| 319.0 | F | 28.0 | 14.0 | 1.5 | 70-100 |
| 319.0 | T6 | 34.0 | | 2.0 | 75-105 |
| 332.0 | T5 | 31.0 | | | 90-120 |
| 333.0 | F | 28.0 | | | 65-100 |
| 333.0 | T5 | 30.0 | | | 70-105 |
| 333.0 | T6 | 35.0 | | | 85-115 |
| 333.0 | T7 | 31.0 | | | 75-105 |
| 336.0 | T551 | 31.0 | | | 90-120 |
| 336.0 | T65 | 40.0 | | | 110-140 |
| 354.0 | T61 | 48.0 | 37.0 | 3.0 | |
| 354.0 | T62 | 52.0 | 42.0 | 2.0 | |
| 355.0 | T51 | 27.0 | | | 60-90 |
| 355.0 | T6 | 37.0 | | 1.5 | 75-105 |
| 355.0 | T62 | 42.0 | | | 90-120 |
| 355.0 | T7 | 36.0 | | | 70-100 |
| 355.0 | T71 | 34.0 | 27.0 | | 65-95 |
| C355.0 | T61 | 40.0 | 30.0 | 3.0 | 75-105 |
| 356.0 | F | 21.0 | | 3.0 | 40-70 |
| 356.0 | T51 | 25.0 | | | 55-85 |
| 356.0 | T6 | 33.0 | 22.0 | 3.0 | 65-95 |
| 356.0 | T7 | 25.0 | | 3.0 | 60-90 |
| 356.0 | T71 | 25.0 | | 3.0 | 60-90 |
| A356.0 | T61 | 37.0 | 26.0 | 5.0 | 70-100 |
| 357.0 | T6 | 45.0 | | 3.0 | 75-105 |
| A357.0 | T61 | 45.0 | 36.0 | 3.0 | 85-115 |
| 359.0 | T61 | 45.0 | 34.0 | 4.0 | 75-105 |
| 359.0 | T62 | 47.0 | 38.0 | 3.0 | 85-115 |
| 443.0 | F | 21.0 | 7.0 | 2.0 | 30-60 |
| B443.0 | F | 21.0 | 6.0 | 2.5 | 30-60 |
| A444.0 | T4 | 20.0 | | 20.0 | |
| 513.0 | F | 22.0 | 12.0 | 2.5 | 45-75 |
| 535.0 | F | 35.0 | 18.0 | 8.0 | 60-90 |
| 705.0 | T5 | 37.0 | 17.0 | 10.0 | 55-85 |
| 707.0 | T7 | 45.0 | 35.0 | 3.0 | 80-110 |
| 711.0 | T1 | 28.0 | 18.0 | 7.0 | 55-85 |
| 713.0 | T5 | 32.0 | 22.0 | 4.0 | 60-90 |
| 850.0 | T5 | 18.0 | | 8.0 | 30-60 |
| 851.0 | T5 | 17.0 | | 3.0 | 30-60 |
| 851.0 | T6 | 18.0 | | 8.0 | |
| 852.0 | T5 | 27.0 | | 3.0 | 55-85 |

*Source: Standards for Aluminum Sand and Permanent Mold Castings*. Courtesy of the Aluminum Association.

[a] F indicates "as cast" condition.

*Wrought Aluminum and Wrought Aluminum Alloys:* A system of four-digit numerical designations for wrought aluminum and wrought aluminum alloys was adopted by the Aluminum Association in 1954. This system is used by the commercial producers and is similar to that used by the SAE, the difference being the addition of two prefix letters.

The first digit of the designation identifies the alloy type: 1) indicating an aluminum of 99.00 percent or greater purity; 2) copper; 3) manganese; 4) silicon; 5) magnesium; 6) magnesium and silicon; 7) zinc; 8) some element other than those aforementioned; and 9) unused (not assigned at present).

If the second digit in the designation is zero, it indicates that there is no special control on individual impurities; integers 1 through 9 indicate special control on one or more individual impurities.

In the 1000 series group for aluminum of 99.00 percent or greater purity, the last two of the four digits indicate to the nearest hundredth the amount of aluminum above 99.00 percent. Thus designation 1030 indicates 99.30 percent minimum aluminum. In the 2000 to 8000 series groups, the last two of the four digits have no significance but are used to identify different alloys in the group. At the time of adoption of this designation system, most of the existing commercial designation numbers were used for these last two digits; for example, 14S became 2014, 3S became 3003, and 75S became 7075. When new alloys are developed and are commercially used, these last two digits are assigned consecutively beginning with -01, skipping any numbers previously assigned at the time of initial adoption.

Experimental alloys are also designated in accordance with this system, but they are indicated by the prefix X, which is dropped upon standardization.

Table 7a shows the product forms and nominal compositions of common wrought aluminum alloys, and Table 7b lists typical mechanical properties of wrought aluminum alloys.

**Table 7a. Nominal Compositions of Common Wrought Aluminum Alloys**

| Alloy | Alloying Elements — Aluminum and Normal Impurities Constitute Remainder | | | | | | | | | | | | |
|---|---|---|---|---|---|---|---|---|---|---|---|---|---|
| | Si | Cu | Mn | Mg | Cr | Ni | Zn | Ti | Pb | Bi | V | Z | Fe |
| 1050 | | | 99.50 percent minimum aluminum | | | | | | | | | | |
| 1060 | | | 99.60 percent minimum aluminum | | | | | | | | | | |
| 1100 | | 0.12 | 99.00 percent minimum aluminum | | | | | | | | | | |
| 1145 | | | 99.45 percent minimum aluminum | | | | | | | | | | |
| 1175 | | | 99.75 percent minimum aluminum | | | | | | | | | | |
| 1200 | | | 99.00 percent minimum aluminum | | | | | | | | | | |
| 1230 | | | 99.30 percent minimum aluminum | | | | | | | | | | |
| 1235 | | | 99.35 percent minimum aluminum | | | | | | | | | | |
| 1345 | | | 99.45 percent minimum aluminum | | | | | | | | | | |
| 1350[a] | | | 99.50 percent minimum aluminum | | | | | | | | | | |
| 2011 | | 5.5 | | | | | | | 0.4 | 0.4 | | | |
| 2014 | 0.8 | 4.4 | 0.8 | 0.50 | | | | | | | | | |
| 2017 | 0.50 | 4.0 | 0.7 | 0.6 | | | | | | | | | |
| 2018 | | 4.0 | | 0.7 | | 2.0 | | | | | | | |
| 2024 | | 4.4 | 0.6 | 1.5 | | | | | | | | | |
| 2025 | 0.8 | 4.4 | 0.8 | | | | | | | | | | |
| 2036 | | 2.6 | 0.25 | 0.45 | | | | | | | | | |
| 2117 | | 2.6 | | 0.35 | | | | | | | | | |
| 2124 | | 4.4 | 0.6 | 1.5 | | | | | | | | | |
| 2218 | | 4.0 | | 1.5 | | 2.0 | | | | | | | |
| 2219 | | 6.3 | 0.30 | | | | | 0.06 | | | 0.10 | 0.18 | |
| 2319 | | 6.3 | 0.30 | | | | | 0.15 | | | 0.10 | 0.18 | |
| 2618 | 0.18 | 2.3 | | 1.6 | | 1.0 | | 0.07 | | | | | 1.1 |

**Table 7a.** *(Continued)* **Nominal Compositions of Common Wrought Aluminum Alloys**

| Alloy | Alloying Elements — Aluminum and Normal Impurities Constitute Remainder | | | | | | | | | | | | |
|---|---|---|---|---|---|---|---|---|---|---|---|---|---|
| | Si | Cu | Mn | Mg | Cr | Ni | Zn | Ti | Pb | Bi | V | Z | Fe |
| 3003 | | 0.12 | 1.2 | | | | | | | | | | |
| 3004 | | | 1.2 | 1.0 | | | | | | | | | |
| 3005 | | | 1.2 | 0.40 | | | | | | | | | |
| 4032 | 12.2 | 0.9 | | 1.0 | | 0.9 | | | | | | | |
| 4043 | 5.2 | | | | | | | | | | | | |
| 4045 | 10.0 | | | | | | | | | | | | |
| 4047 | 12.0 | | | | | | | | | | | | |
| 4145 | 10.0 | 4.0 | | | | | | | | | | | |
| 5005 | | | | 0.8 | | | | | | | | | |
| 5050 | | | | 1.4 | | | | | | | | | |
| 5052 | | | | 2.5 | 0.25 | | | | | | | | |
| 5056 | | | 0.12 | 5.0 | 0.12 | | | | | | | | |
| 5083 | | | 0.7 | 4.4 | 0.15 | | | | | | | | |
| 5086 | | | 0.45 | 4.0 | 0.15 | | | | | | | | |
| 5183 | | | 0.8 | 4.8 | 0.15 | | | | | | | | |
| 5252 | | | | 2.5 | | | | | | | | | |
| 5254 | | | | 3.5 | 0.25 | | | | | | | | |
| 5356 | | | 0.12 | 5.0 | 0.12 | | | 0.13 | | | | | |
| 5456 | | | 0.8 | 5.1 | 0.12 | | | | | | | | |
| 5457 | | | 0.30 | 1.0 | | | | | | | | | |
| 5554 | | | 0.8 | 2.7 | 0.12 | | | 0.12 | | | | | |
| 5556 | | | 0.8 | 5.1 | 0.12 | | | 0.12 | | | | | |
| 5652 | | | | 2.5 | 0.25 | | | | | | | | |
| 5654 | | | | 3.5 | 0.25 | | | 0.10 | | | | | |
| 6003 | 0.7 | | | 1.2 | | | | | | | | | |
| 6005 | 0.8 | | | 0.50 | | | | | | | | | |
| 6053 | 0.7 | | | 1.2 | 0.25 | | | | | | | | |
| 6061 | 0.6 | 0.28 | | 1.0 | 0.20 | | | | | | | | |
| 6066 | 1.4 | 1.0 | 0.8 | 1.1 | | | | | | | | | |
| 6070 | 1.4 | 0.28 | 0.7 | 0.8 | | | | | | | | | |
| 6101 | 0.50 | | | 0.6 | | | | | | | | | |
| 6105 | 0.8 | | | 0.62 | | | | | | | | | |
| 6151 | 0.9 | | | 0.6 | 0.25 | | | | | | | | |
| 6201 | 0.7 | | | 0.8 | | | | | | | | | |
| 6253 | 0.7 | | | 1.2 | 0.25 | | 2.0 | | | | | | |
| 6262 | 0.6 | 0.28 | | 1.0 | 0.09 | | | | 0.6 | 0.6 | | | |
| 6351 | 1.0 | | 0.6 | 0.6 | | | | | | | | | |
| 6463 | 0.40 | | | 0.7 | | | | | | | | | |
| 7005 | | | 0.45 | 1.4 | 0.13 | | 4.5 | 0.04 | | | | 0.14 | |
| 7008 | | | | 1.0 | 0.18 | | 5.0 | | | | | | |
| 7049 | | 1.6 | | 2.4 | 0.16 | | 7.7 | | | | | | |
| 7050 | | 2.3 | | 2.2 | 6.2 | | | | | | | 0.12 | |
| 7072 | | | | | | | 1.0 | | | | | | |
| 7075 | | 1.6 | | 2.5 | 0.23 | | 5.6 | | | | | | |
| 7108 | | | | 1.0 | | | 5.0 | | | | | 0.18 | |
| 7178 | | 2.0 | | 2.8 | 0.23 | | 6.8 | | | | | | |
| 8017 | | 0.15 | | 0.03 | | | | | | | | | 0.7 |
| 8030 [b] | | 0.22 | | | | | | | | | | | |
| 8177 | | | | 0.08 | | | | | | | | | 0.35 |

[a] Formerly designated EC.

[b] Boron 0.02 percent.

*Source: Aluminum Standards and Data*. Courtesy of the Aluminum Association.

**Table 7b. Typical Mechanical Properties of Wrought Aluminum Alloys**

| Alloy and Temper | Tension | | | | Brinell Hardness Number (500 kg load, 10-mm ball) | Ultimate Shearing Strength (ksi) | Endurance Limit[a] (ksi) |
|---|---|---|---|---|---|---|---|
| | Strength (ksi) | | Elongation in 2 inches (%) | | | | |
| | Ultimate | Yield | 1/16-inch Thick Specimen | 1/2-inch Diameter Specimen | | | |
| 1060-O | 10 | 4 | 43 | | 19 | 7 | 3 |
| 1060-H12 | 12 | 11 | 16 | | 23 | 8 | 4 |
| 1060-H14 | 14 | 13 | 12 | | 26 | 9 | 5 |
| 1060-H16 | 16 | 15 | 8 | | 30 | 10 | 6.5 |
| 1060-H18 | 19 | 18 | 6 | | 35 | 11 | 6.5 |
| 1100-O | 13 | 5 | 35 | 45 | 23 | 9 | 5 |
| 1100-H12 | 16 | 15 | 12 | 25 | 28 | 10 | 6 |
| 1100-H14 | 18 | 17 | 9 | 20 | 32 | 11 | 7 |
| 1100-H16 | 21 | 20 | 6 | 17 | 38 | 12 | 9 |
| 1100-H18 | 24 | 22 | 5 | 15 | 44 | 13 | 9 |
| 1350-O | 12 | 4 | | [b] | | 8 | |
| 1350-H12 | 14 | 12 | | | | 9 | |
| 1350-H14 | 16 | 14 | | | | 10 | |
| 1350-H16 | 18 | 16 | | | | 11 | |
| 1350-H19 | 27 | 24 | | [c] | | 15 | 7 |
| 2011-T3 | 55 | 43 | | 15 | 95 | 32 | 18 |
| 2011-T8 | 59 | 45 | | 12 | 100 | 35 | 18 |
| 2014-O | 27 | 14 | | 18 | 45 | 18 | 13 |
| 2014-T4, T451 | 62 | 42 | | 20 | 105 | 38 | 20 |
| 2014-T6, T651 | 70 | 60 | | 13 | 135 | 42 | 18 |
| Alclad 2014-O | 25 | 10 | 21 | | | 18 | |
| Alclad 2014-T3 | 63 | 40 | 20 | | | 37 | |
| Alclad 2014-T4, T451 | 61 | 37 | 22 | | | 37 | |
| Alclad 2014-T6, T651 | 68 | 60 | 10 | | | 41 | |
| 2017-O | 26 | 10 | | 22 | 45 | 18 | 13 |
| 2017-T4, T451 | 62 | 40 | | 22 | 105 | 38 | 18 |
| 2018-T61 | 61 | 46 | | 12 | 120 | 39 | 17 |
| 2024-O | 27 | 11 | 20 | 22 | 47 | 18 | 13 |
| 2024-T3 | 70 | 50 | 18 | | 120 | 41 | 20 |
| 2024-T4, T351 | 68 | 47 | 20 | 19 | 120 | 41 | 20 |
| 2024-T361[d] | 72 | 57 | 13 | | 130 | 42 | 18 |
| Alclad 2024-O | 26 | 11 | 20 | | | 18 | |
| Alclad 2024-T3 | 65 | 45 | 18 | | | 40 | |
| Alclad 2024-T4, T351 | 64 | 42 | 19 | | | 40 | |
| Alclad 2024-T361[d] | 67 | 53 | 11 | | | 41 | |
| Alclad 2024-T81, T851 | 65 | 60 | 6 | | | 40 | |
| Alclad 2024-T861[d] | 70 | 66 | 6 | | | 42 | |
| 2025-T6 | 58 | 37 | | 19 | 110 | 35 | 18 |
| 2036-T4 | 49 | 28 | 24 | | | | 18[e] |
| 2117-T4 | 43 | 24 | | 27 | 70 | 28 | 14 |
| 2218-T72 | 48 | 37 | | 11 | 95 | 30 | |
| 2219-O | 25 | 11 | 18 | | | | |
| 2219-T42 | 52 | 27 | 20 | | | | |
| 2219-T31, T351 | 52 | 36 | 17 | | | | |
| 2219-T37 | 57 | 46 | 11 | | | | |
| 2219-T62 | 60 | 42 | 10 | | | | 15 |
| 2219-T81, T851 | 66 | 51 | 10 | | | | 15 |
| 2219-T87 | 69 | 57 | 10 | | | | 15 |

**Table 7b.** *(Continued)* **Typical Mechanical Properties of Wrought Aluminum Alloys**

| Alloy and Temper | Tension | | | | Brinell Hardness Number (500 kg load, 10-mm ball) | Ultimate Shearing Strength (ksi) | Endurance Limit[a] (ksi) |
|---|---|---|---|---|---|---|---|
| | Strength (ksi) | | Elongation in 2 inches (%) | | | | |
| | Ultimate | Yield | 1⁄16-inch Thick Specimen | ½-inch Diameter Specimen | | | |
| 3003-O | 16 | 6 | 30 | 40 | 28 | 11 | 7 |
| 3003-H12 | 19 | 18 | 10 | 20 | 35 | 12 | 8 |
| 3003-H14 | 22 | 21 | 8 | 16 | 40 | 14 | 9 |
| 3003-H16 | 26 | 25 | 5 | 14 | 47 | 15 | 10 |
| 3003-H18 | 29 | 27 | 4 | 10 | 55 | 16 | 10 |
| Alclad 3003-O | 16 | 6 | 30 | 40 | | 11 | |
| Alclad 3003-H12 | 19 | 18 | 10 | 20 | | 12 | |
| Alclad 3003-H14 | 22 | 21 | 8 | 16 | | 14 | |
| Alclad 3003-H16 | 26 | 25 | 5 | 14 | | 15 | |
| Alclad 3003-H18 | 29 | 27 | 4 | 10 | | 16 | |
| 3004-O | 26 | 10 | 20 | 25 | 45 | 16 | 14 |
| 3004-H32 | 31 | 25 | 10 | 17 | 52 | 17 | 15 |
| 3004-H34 | 35 | 29 | 9 | 12 | 63 | 18 | 15 |
| 3004-H36 | 38 | 33 | 5 | 9 | 70 | 20 | 16 |
| 3004-H38 | 41 | 36 | 5 | 6 | 77 | 21 | 16 |
| Alclad 3004-O | 26 | 10 | 20 | 25 | | 16 | |
| Alclad 3004-H32 | 31 | 25 | 10 | 17 | | 17 | |
| Alclad 3004-H34 | 35 | 29 | 9 | 12 | | 18 | |
| Alclad 3004-H36 | 38 | 33 | 5 | 9 | | 20 | |
| Alclad 3004-H38 | 41 | 36 | 5 | 6 | | 21 | |
| 3105-O | 17 | 8 | 24 | | | 12 | |
| 3105-H12 | 22 | 19 | 7 | | | 14 | |
| 3105-H14 | 25 | 22 | 5 | | | 15 | |
| 3105-H16 | 28 | 25 | 4 | | | 16 | |
| 3105-H18 | 31 | 28 | 3 | | | 17 | |
| 3105-H25 | 26 | 23 | 8 | | | 15 | |
| 4032-T6 | 55 | 46 | | 9 | 120 | 38 | 16 |
| 5005-O | 18 | 6 | 25 | | 28 | 11 | |
| 5005-H12 | 20 | 19 | 10 | | | 14 | |
| 5005-H14 | 23 | 22 | 6 | | | 14 | |
| 5005-H16 | 26 | 25 | 5 | | | 15 | |
| 5005-H18 | 29 | 28 | 4 | | | 16 | |
| 5005-H32 | 20 | 17 | 11 | | 36 | 14 | |
| 5005-H34 | 23 | 20 | 8 | | 41 | 14 | |
| 5005-H36 | 26 | 24 | 6 | | 46 | 15 | |
| 5005-H38 | 29 | 27 | 5 | | 51 | 16 | |
| 5050-O | 21 | 8 | 24 | | 36 | 15 | 12 |
| 5050-H32 | 25 | 21 | 9 | | 46 | 17 | 13 |
| 5050-H34 | 28 | 24 | 8 | | 53 | 18 | 13 |
| 5050-H36 | 30 | 26 | 7 | | 58 | 19 | 14 |
| 5050-H38 | 32 | 29 | 6 | | 63 | 20 | 14 |
| 5052-O | 28 | 13 | 25 | 30 | 47 | 18 | 16 |
| 5052-H32 | 33 | 28 | 12 | 18 | 60 | 20 | 17 |
| 5052-H34 | 38 | 31 | 10 | 14 | 68 | 21 | 18 |
| 5052-H36 | 40 | 35 | 8 | 10 | 73 | 23 | 19 |
| 5052-H38 | 42 | 37 | 7 | 8 | 77 | 24 | 20 |
| 5056-O | 42 | 22 | | 35 | 65 | 26 | 20 |
| 5056-H18 | 63 | 59 | | 10 | 105 | 34 | 22 |
| 5056-H38 | 60 | 50 | | 15 | 100 | 32 | 22 |
| 5083-O | 42 | 21 | | 22 | | 25 | |
| 5083-H321, H116 | 46 | 33 | | 16 | | | 23 |

**Table 7b.** *(Continued)* **Typical Mechanical Properties of Wrought Aluminum Alloys**

| Alloy and Temper | Tension | | | | Brinell Hardness Number (500 kg load, 10-mm ball) | Ultimate Shearing Strength (ksi) | Endurance Limit[a] (ksi) |
|---|---|---|---|---|---|---|---|
| | Strength (ksi) | | Elongation in 2 inches (%) | | | | |
| | Ultimate | Yield | 1⁄16-inch Thick Specimen | ½-inch Diameter Specimen | | | |
| 5086-O | 38 | 17 | 22 | | | 23 | |
| 5086-H32, H116 | 42 | 30 | 12 | | | | |
| 5086-H34 | 47 | 37 | 10 | | | 27 | |
| 5086-H112 | 39 | 19 | 14 | | | | |
| 5154-O | 35 | 17 | 27 | | 58 | 22 | 17 |
| 5154-H32 | 39 | 30 | 15 | | 67 | 22 | 18 |
| 5154-H34 | 42 | 33 | 13 | | 73 | 24 | 19 |
| 5154-H36 | 45 | 36 | 12 | | 78 | 26 | 20 |
| 5154-H38 | 48 | 39 | 10 | | 80 | 28 | 21 |
| 5154-H112 | 35 | 17 | 25 | | 63 | | 17 |
| 5252-H25 | 34 | 25 | 11 | | 68 | 21 | |
| 5252-H38, H28 | 41 | 35 | 5 | | 75 | 23 | |
| 5254-O | 35 | 17 | 27 | | 58 | 22 | 17 |
| 5254-H32 | 39 | 30 | 15 | | 67 | 22 | 18 |
| 5254-H34 | 42 | 33 | 13 | | 73 | 24 | 19 |
| 5254-H36 | 45 | 36 | 12 | | 78 | 26 | 20 |
| 5254-H38 | 48 | 39 | 10 | | 80 | 28 | 21 |
| 5254-H112 | 35 | 17 | 25 | | 63 | | 17 |
| 5454-O | 36 | 17 | 22 | | 62 | 23 | |
| 5454-H32 | 40 | 30 | 10 | | 73 | 24 | |
| 5454-H34 | 44 | 35 | 10 | | 81 | 26 | |
| 5454-H111 | 38 | 26 | 14 | | 70 | 23 | |
| 5454-H112 | 36 | 18 | 18 | | 62 | 23 | |
| 5456-O | 45 | 23 | | 24 | | | |
| 5456-H112 | 45 | 24 | | 22 | | | |
| 5456-H321, H116 | 51 | 37 | | 16 | 90 | 30 | |
| 5457-O | 19 | 7 | 22 | | 32 | 12 | |
| 5457-H25 | 26 | 23 | 12 | | 48 | 16 | |
| 5457-H38, H28 | 30 | 27 | 6 | | 55 | 18 | |
| 5652-O | 28 | 13 | 25 | 30 | 47 | 18 | 16 |
| 5652-H32 | 33 | 28 | 12 | 18 | 60 | 20 | 17 |
| 5652-H34 | 38 | 31 | 10 | 14 | 68 | 21 | 18 |
| 5652-H36 | 40 | 35 | 8 | 10 | 73 | 23 | 19 |
| 5652-H38 | 42 | 37 | 7 | 8 | 77 | 24 | 20 |
| 5657-H25 | 23 | 20 | 12 | | 40 | 14 | |
| 5657-H38, H28 | 28 | 24 | 7 | | 50 | 15 | |
| 6061-O | 18 | 8 | 25 | 30 | 30 | 12 | 9 |
| 6061-T4, T451 | 35 | 21 | 22 | 25 | 65 | 24 | 14 |
| 6061-T6, T651 | 45 | 40 | 12 | 17 | 95 | 30 | 14 |
| Alclad 6061-O | 17 | 7 | 25 | | | 11 | |
| Alclad 6061-T4, T451 | 33 | 19 | 22 | | | 22 | |
| Alclad 6061-T6, T651 | 42 | 37 | 12 | | | 27 | |
| 6063-O | 13 | 7 | | | 25 | 10 | 8 |
| 6063-T1 | 22 | 13 | 20 | | 42 | 14 | 9 |
| 6063-T4 | 25 | 13 | 22 | | | | |
| 6063-T5 | 27 | 21 | 12 | | 60 | 17 | 10 |
| 6063-T6 | 35 | 31 | 12 | | 73 | 22 | 10 |
| 6063-T83 | 37 | 35 | 9 | | 82 | 22 | |
| 6063-T831 | 30 | 27 | 10 | | 70 | 18 | |
| 6063-T832 | 42 | 39 | 12 | | 95 | 27 | |

**Table 7b.** *(Continued)* **Typical Mechanical Properties of Wrought Aluminum Alloys**

| Alloy and Temper | Tension | | | | Brinell Hardness Number (500 kg load, 10-mm ball) | Ultimate Shearing Strength (ksi) | Endurance Limit[a] (ksi) |
|---|---|---|---|---|---|---|---|
| | Strength (ksi) | | Elongation in 2 inches (%) | | | | |
| | Ultimate | Yield | 1/16-inch Thick Specimen | 1/2-inch Diameter Specimen | | | |
| 6066-O | 22 | 12 | | 18 | 43 | 14 | |
| 6066-T4, T451 | 52 | 30 | | 18 | 90 | 29 | |
| 6066-T6, T651ß | 57 | 52 | | 12 | 120 | 34 | 16 |
| 6070-T6 | 55 | 51 | 10 | | | 34 | 14 |
| 6101-H111 | 14 | 11 | | | | | |
| 6101-T6 | 32 | 28 | 15 | | 71 | 20 | |
| 6262-T9 | 58 | 55 | | 10 | 120 | 35 | 13 |
| 6351-T4 | 36 | 22 | 20 | | | | |
| 6351-T6 | 45 | 41 | 14 | | 95 | 29 | 13 |
| 6463-T1 | 22 | 13 | 20 | | 42 | 14 | 10 |
| 6463-T5 | 27 | 21 | 12 | | 60 | 17 | 10 |
| 6463-T6 | 35 | 31 | 12 | | 74 | 22 | 10 |
| 7049-T73 | 75 | 65 | | 12 | 135 | 44 | |
| 7049-T7352 | 75 | 63 | | 11 | 135 | 43 | |
| 7050-T73510, T73511 | 72 | 63 | | 12 | | | |
| 7050-T7451[f] | 76 | 68 | | 11 | | 44 | |
| 7050-T7651 | 80 | 71 | | 11 | | 47 | |
| 7075-O | 33 | 15 | 17 | 16 | 60 | 22 | |
| 7075-T6, T651 | 83 | 73 | 11 | 11 | 150 | 48 | 23 |
| Alclad 7075-O | 32 | 14 | 17 | | | 22 | |
| Alclad 7075-T6, T651 | 76 | 67 | 11 | | | 46 | |
| 7178-O | 33 | 15 | 15 | 16 | | | |
| 7178-T6, T651 | 88 | 78 | 10 | 11 | | | |
| 7178-T76, T7651 | 83 | 73 | | 11 | | | |
| Alclad 7178-O | 32 | 14 | 16 | | | | |
| Alclad 7178-T6, T651 | 81 | 71 | 10 | | | | |
| 8176-H24 | 17 | 14 | 15 | | | 10 | |

[a] Based on 500,000,000 cycles of completely reversed stress using the R. R. Moore type of machine and specimen.

[b] 1350-O wire should have an elongation of approximately 23 percent in 10 inches.

[c] 1350-H19 wire should have an elongation of approximately 1.5 percent in 10 inches.

[d] Tempers T361 and T861 were formerly designated T36 and T86, respectively.

[e] Based on $10^7$ cycles using flexural type testing of sheet specimens.

[f] T7451, although not previously registered, has appeared in the literature and in some specifications as T73651.

The data given in this table are intended only as a basis for comparing alloys and tempers, and should not be specified as engineering requirements or used for design purposes. The indicated typical mechanical properties for all except O temper material are higher than the specified minimum properties. For O temper products, typical ultimate and yield values are slightly lower than specified (maximum) values.

*Source: Aluminum Standards and Data.* Courtesy of the Aluminum Association.

**Heat-Treatability of Wrought Aluminum Alloys.**—In high-purity form, aluminum is soft and ductile. Most commercial uses, however, require greater strength than pure aluminum affords. This extra strength is achieved in aluminum first by the addition of other elements to produce various alloys, which singly or in combination impart strength to the metal. Further strengthening is possible by means that classify the alloys roughly into two categories, non-heat-treatable and heat-treatable.

*Non-Heat-Treatable Alloys:* The initial strength of alloys in this group depends upon the hardening effect of elements such as manganese, silicon, iron and magnesium, singly or in various combinations. The non-heat-treatable alloys are usually designated, therefore, in the 1000, 3000, 4000, or 5000 series. These alloys are work-hardenable, so further strengthening is made possible by various degrees of cold-working, denoted by the "H" series of tempers. Alloys containing appreciable amounts of magnesium when supplied in strain-hardened tempers are usually given a final elevated-temperature treatment called *stabilizing* for improving property stability.

*Heat-Treatable Alloys:* The initial strength of alloys in this group is enhanced by the addition of alloying elements such as copper, magnesium, zinc, and silicon. These elements singly or in various combinations show increasing solid solubility in aluminum with increasing temperature, so it is possible to subject them to thermal treatments that will impart pronounced strengthening.

The first step, called *heat treatment* or *solution heat treatment,* is an elevated-temperature process designed to put the soluble element in solid solution. This step is followed by rapid quenching, usually in water, which momentarily "freezes" the structure and for a short time renders the alloy very workable. Some fabricators retain this more workable structure by storing the alloys at below freezing temperatures until they can be formed. At room or elevated temperatures the alloys are not stable after quenching, however, and precipitation of the constituents from the supersaturated solution begins. After a period of several days at room temperature, termed *aging* or *room-temperature precipitation,* the alloy is considerably stronger. Many alloys approach a stable condition at room temperature, but some alloys, particularly those containing magnesium and silicon or magnesium and zinc, continue to age-harden for long periods of time at room temperature.

Heating for a controlled time at slightly elevated temperatures provides even further strengthening and property stabilization. This process is called *artificial aging* or *precipitation hardening*. By application of the proper combination of solution heat treatment, quenching, cold-working and artificial aging, the highest strengths are obtained.

**Clad Aluminum Alloys.—**The heat-treatable alloys in which copper or zinc are major alloying constituents are less resistant to corrosive attack than the majority of non-heat-treatable alloys. To increase their corrosion resistance in sheet and plate form, these alloys are often clad with high-purity aluminum, a low magnesium-silicon alloy, or an alloy containing 1 percent zinc. The cladding, usually from 2½ to 5 percent of the total thickness on each side, not only protects the composite due to its own inherently excellent corrosion resistance but also exerts a galvanic effect that further protects the core material.

Special composites, such as clad non-heat-treatable alloys for extra corrosion protection, for brazing purposes, or for special surface finishes, may be obtained. Some alloys in wire and tubular form are clad for similar reasons, and on an experimental basis extrusions also have been clad.

**Aluminum Alloys, Wrought, Sheet.—***Physical Properties:* In sheet form, tensile strength varies from 35,000 psi (241 MPa) for soft temper to 62,000 (427 MPa) psi for heat-treated sheets, and the elongation in 2 inches ( 5.08 cm) from 12 to 18 percent. The yield strength of a heat-treated sheet is about 40,000 psi (276 MPa) minimum.

**Characteristics of Principal Aluminum Alloy Series Groups.—***1000 series:* These alloys are characterized by high corrosion resistance, high thermal and electrical conductivity, low mechanical properties and good workability. Moderate increases in strength may be obtained by strain-hardening. Iron and silicon are the major impurities.

*2000 Series:* Copper is the principal alloying element in this group. These alloys require solution heat treatment to obtain optimum properties; in the heat-treated condition mechanical properties are similar to, and sometimes exceed, those of mild steel. In some instances artificial aging is employed to further increase the mechanical properties. This treatment materially increases yield strength, with attendant loss in elongation; its effect on tensile (ultimate) strength is not as great. Alloys in the 2000 series have reduced corrosion resistance compared to most other aluminum alloys, and under certain conditions they may be subject to intergranular corrosion. Therefore, these alloys in the form of sheet are usually clad with a high-purity alloy or a magnesium-silicon alloy of the 6000 series, which provides galvanic protection to the core material and thus greatly increases resistance to corrosion. Alloy 2024 is perhaps the best known and most widely used aircraft alloy.

*3000 Series:* Manganese is the major alloying element of alloys in this group, which are generally non-heat-treatable. Because only a limited percentage of manganese, up to about 1.5 percent, can be effectively added to aluminum, it is used as a major element in only a few instances. One of these, however, is the popular 3003, used for moderate-strength applications requiring good workability.

*4000 Series:* The major alloying element of this group is silicon, which can be added in sufficient quantities to cause substantial lowering of the melting point without producing brittleness in the resulting alloys. For these reasons aluminum-silicon alloys are used in welding wire and as brazing alloys where a lower melting point than that of the parent metal is required. Most alloys in this series are non-heat-treatable, but, when used in welding heat-treatable alloys, they will pick up some of the alloying constituents of the latter and so respond to heat treatment to a limited extent. The alloys containing appreciable amounts of silicon become dark gray when anodic oxide finishes are applied and hence are in demand for architectural applications.

*5000 Series:* Magnesium is one of the most effective and widely used alloying elements for aluminum. When it is used as the major alloying element or with manganese, the result is a moderate to high strength non-heat-treatable alloy. Magnesium is considerably more effective than manganese as a hardener, about 0.8 percent magnesium being equal to 1.25 percent manganese, and it can be added in considerably higher quantities. Alloys in this series possess good welding characteristics and good resistance to corrosion in marine atmospheres. However, certain limitations should be placed on the amount of cold-work and on safe operating temperatures permissible for the higher magnesium content alloys (over about $3\frac{1}{2}$ percent for operating temperatures over about 150° F or 66° C) to avoid susceptibility to stress corrosion.

*6000 Series:* Alloys in this group contain silicon and magnesium in approximate proportions to form magnesium silicide, thus making them capable of being heat treated. The major alloy in this series is 6061, one of the most versatile of the heat-treatable alloys. Though not as strong as most of the 2000 or 7000 alloys, the magnesium-silicon (or magnesium-silicide) alloys possess good formability and corrosion resistance, with medium strength. Alloys in this heat-treatable group may be formed in the -T4 temper (solution heat treated but not artificially aged) and then reach full -T6 properties by artificial aging.

*7000 Series:* Zinc is the major alloying element in this group, and when coupled with a smaller percentage of magnesium results in heat-treatable alloys of very high strength. Other elements, such as copper and chromium, are usually added in small quantities. A notable member of this group is 7075, which is among the highest strength aluminum alloys available and is used in air-frame structures and for highly stressed parts.

**Type Metal.**—Antimony gives to metals the property of expansion on solidification. Hence, it is used in type metal for casting type for the printing trades to ensure the molds are completely filled. Type metals are generally made with from 5 to 25 percent of antimony, and with lead, tin and sometimes a small percentage of copper as the other alloying metals.

The compositions of a number of type metal alloys are as follows (figures given are percentages): lead 77.5, tin 6.5, antimony 16; lead 70, tin 10, antimony 18, copper 2; lead 63.2, tin 12, antimony 24, copper 0.8 ; lead 60.5, tin 14.5, antimony 24–25, copper 0.75; lead 60, tin 35, antimony 5; and lead 55.5, tin 40, antimony 4.5.

A high grade of type metal is composed of the following percentages: lead 50; tin 25; and antimony 25.

## Magnesium Alloys

**Magnesium Alloys.**—Magnesium is the lightest of all structural metals. Silver-white in color, pure magnesium is relatively soft, so it is rarely used for structural purposes in the pure state. Principal metallurgical uses for pure magnesium are as an alloying element for aluminum and other metals; as a reducing agent in the extraction of such metals as titanium, zirconium, hafnium, and uranium; as a nodularizing agent in the manufacture of ductile iron; and as a sulfur removal agent in steel manufacture. Magnesium alloys are made by alloying up to about 10 percent of other metals and have low density and an excellent combination of mechanical properties, as shown in Table 8a, resulting in high strength-to-weight ratios.

Magnesium alloys are the easiest of all the structural metals to machine, and these alloys have very high weld efficiencies. Magnesium is readily processed by all the standard casting and fabrication techniques used in metalworking, especially by pressure die casting. Because the metal work hardens rapidly, cold forming is limited to mild deformation, but magnesium alloys have excellent working characteristics at temperatures between 300 and 500° F (149 and 260° C).

These alloys have relatively low elastic moduli, so they will absorb energy with good resistance to dents and high damping capacities. Fatigue strength also is good, particularly in the low-stress, high-cycle range. The alloys can be precipitation hardened, so mechanical properties can be improved by solution heat treatment and aging. Corrosion resistance was greatly improved recently, when methods were found to limit heavy metal impurities to "parts per million."

**Applications of Magnesium Alloys.**—Magnesium alloys are used in a wide variety of structural applications including industrial, materials handling, automotive, consumer-durable, and aerospace equipment. In industrial machinery, the alloys are used for parts that operate at high speeds, which must have light weight to allow rapid acceleration and minimize inertial forces. Materials handling equipment applications include hand trucks, dockboards, grain shovels, and gravity conveyors. Automotive applications include wheels, gearboxes, clutch housings, valve covers, and brake pedal and other brackets. Consumer durables include luggage, softball bats, tennis rackets, and housings for cameras and projectors. Their high strength-to-weight ratio suits magnesium alloys to use in a variety of aircraft structures, particularly helicopters. Very intricate shapes that are uneconomical to produce in other materials are often cast in magnesium, sometimes without draft. Wrought magnesium alloys are made in the form of bars, forgings, extrusions, wire, sheet, and plate.

**Alloy and Temper Designation.**—Magnesium alloys are designated by a standard four-part system established by the ASTM, and now also used by the SAE, that indicates both chemical composition and temper. Designations begin with two letters representing the two alloying elements that are specified in the greatest amount; these letters are arranged in order of decreasing percentage of alloying elements or alphabetically if they are present in equal amounts. The letters are followed by digits representing the respective composition percentages, rounded off to whole numbers, and then by a serial letter indicating some variation in composition of minor constituents. The final part, separated by a hyphen, consists of a letter followed by a number, indicating the temper condition. The letters that designate the more common alloying elements are A, aluminum; E, rare earths; H, thorium; K, zirconium; M, manganese; Q, silver; S, silicon; T, tin; Z, zinc.

The letters and numbers that indicate the temper designation are F, as fabricated; O, annealed; H10, H11, strain hardened; H23, H24, H26, strain hardened and annealed; T4, solution heat treated; T5, artificially aged; T6, solution heat treated and artificially aged; and T8, solution heat treated, cold-worked, and artificially aged.

The nominal composition and typical properties of magnesium alloys are listed in Table 8a and Table 8b.

## Table 8a. Nominal Composition of Magnesium Alloys

| Alloy | Al | Zn | Mn[a] | Si | Zr | Ag | Th | Y | Rare Earth |
|---|---|---|---|---|---|---|---|---|---|
| Sand and Permanent Mold (Gravity Die) Castings | | | | | | | | | |
| AM100A-T61 | 10.0 | | 0.10 | | | | | | |
| AZ63A-T6 | 6.0 | 3.0 | 0.15 | | | | | | |
| AZ81A-T4 | 7.6 | 0.7 | 0.13 | | | | | | |
| AZ91C-T6 | 8.7 | 0.7 | 0.13 | | | | | | |
| AZ91E-T6[b] | 8.7 | 0.7 | 0.17 | | | | | | |
| AZ92A-T6 | 9.0 | 2.0 | 0.10 | | | | | | |
| EZ33A-T5 | | 2.6 | | | 0.8 | | | | 3.3 |
| HK31A-T6 | | 0.3 | | | 0.7 | | 3.3 | | |
| HZ32A-T6 | | 2.1 | | | 0.8 | | 3.3 | | 0.1 |
| K1A-F | | | | | 0.7 | | | | |
| QE22A-T6 | | | | | 0.7 | 2.5 | | | 2.2 |
| QH21A-T6 | | 0.2 | | | 0.7 | | 1.1 | | 1.1 |
| ZE41A-T5 | | 4.3 | 0.15 | | 0.7 | | | | 1.3 |
| ZE63A-T6 | | 5.8 | | | 0.7 | | | | 2.6 |
| ZH62A-T5 | | 5.7 | | | 0.8 | | 1.8 | | |
| ZK51A-T5 | | 4.6 | | | 0.8 | | | | |
| ZK61A-T6 | | 6.0 | | | 0.8 | | | | |
| WE54A-F | | | | | 0.5 | | | 5.3 | 3.5 |
| Pressure Die Castings | | | | | | | | | |
| AZ91A-F | 9.0 | 0.7 | 0.13 | | | | | | |
| AZ91B-F[c] | 9.0 | 0.7 | 0.13 | | | | | | |
| AZ91D-F[b] | 9.0 | 0.7 | 0.15 | | | | | | |
| AM60A-F | 6.0 | | 0.13 | | | | | | |
| AM60B-F[b] | 6.0 | | 0.25 | | | | | | |
| AS41A-F[d] | 4.3 | | 0.35 | 1.0 | | | | | |
| Extruded Bars and Shapes | | | | | | | | | |
| AZ10A-F | 1.3 | 0.4 | 0.20 | | | | | | |
| AZ31B-F | 3.0 | 1.0 | 0.20 | | | | | | |
| AZ31C-F | 3.0 | 1.0 | 0.15 | | | | | | |
| AZ61A-F | 6.5 | 1.0 | 0.15 | | | | | | |
| AZ80A-T5 | 8.5 | 0.5 | 0.12 | | | | | | |
| HM31A-F | | | 1.20 | | | | 3.0 | | |
| M1A-F | | | 1.20 | | | | | | |
| ZK40A-T5 | | 4.0 | | | 0.45 | | | | |
| ZK60A-F | | 5.5 | | | 0.45 | | | | |
| Sheet and Plate | | | | | | | | | |
| AZ31B-H24 | 3.0 | 1.0 | 0.20 | | | | | | |
| AZ31C-H24 | 3.0 | 1.0 | 0.15 | | | | | | |
| HK31A-H24 | | | | | 0.7 | | 3.3 | | |
| HM21A-T8 | | | 0.80 | | | | 2.0 | | |

[a] All manganese values are minimum.
[b] High-purity alloy, Ni, Fe, and Cu severely restricted.
[c] 0.30 percent maximum residual copper is allowed.
[d] For battery applications.

*Source: Metals Handbook*, 9th edition, Vol. 2, American Society for Metals.

**Table 8b. Typical Room-Temperature Mechanical Properties of Magnesium Alloys**

| Alloy | Tensile Strength (ksi) | Yield Strength | | | Elongation in 2 in. (%) | Shear Strength (ksi) | Rockwell Hardness B[a] |
|---|---|---|---|---|---|---|---|
| | | Tensile (ksi) | Compressive (ksi) | Bearing (ksi) | | | |
| Sand and Permanent Mold (Gravity Die) Castings | | | | | | | |
| AM100A-T61 | 40 | 22 | 22 | 68 | 1 | | 69 |
| AZ63A-T6 | 40 | 14 | 14 | 44 | 12 | 18 | 55 |
| AZ81A-T4 | 40 | 12 | 12 | 35 | 15 | 21 | 55 |
| AZ91C-T6 | 40 | 21 | 21 | 52 | 6 | 21 | 70 |
| AZ91E-T6[b] | 40 | 21 | 21 | 52 | 6 | 21 | 70 |
| AZ92A-T6 | 40 | 22 | 22 | 65 | 3 | 21 | 81 |
| EZ33A-T5 | 23 | 16 | 16 | 40 | 3 | 20 | 50 |
| HK31A-T6 | 32 | 15 | 15 | 40 | 8 | 21 | 55 |
| HZ32A-T6 | 27 | 13 | 13 | 37 | 4 | 20 | 55 |
| K1A-F | 26 | 8 | 8 | 18 | 19 | 8 | |
| QE22A-T6 | 38 | 28 | 28 | | 3 | | 80 |
| QH21A-T6 | 40 | 30 | 30 | | 4 | 22 | |
| ZE41A-T5 | 30 | 20 | 20 | 51 | 4 | 23 | 62 |
| ZE63A-T6 | 44 | 28 | 28 | | 10 | | 60-85 |
| ZH62A-T5 | 35 | 22 | 22 | 49 | 4 | 23 | 70 |
| ZK51A-T5 | 30 | 20 | 20 | 51 | 4 | 22 | 62 |
| ZK61A-T6 | 45 | 28 | 28 | | 10 | | |
| WE54A-F | 40 | 29 | 29 | | 4 | | |
| Pressure Die Castings | | | | | | | |
| AZ91A-F | 34 | 23 | 23 | | 3 | 20 | 63 |
| AZ91B-FAZ91B-F[c] | 34 | 23 | 23 | | 3 | 20 | 63 |
| AZ91D-F[b] | 34 | 23 | 23 | | 3 | 20 | 63 |
| AM60A-F | 32 | 19 | 19 | | 8 | | |
| AM60B-F[b] | 32 | 19 | 19 | | 8 | | |
| AS41A-F[d] | 31 | 20 | 20 | | 6 | | |
| Extruded Bars and Shapes | | | | | | | |
| AZ10A-F | 35 | 21 | 10 | | 10 | | |
| AZ31B-F | 38 | 29 | 14 | 33 | 15 | 19 | 49 |
| AZ31C-F | 38 | 29 | 14 | 33 | 15 | 19 | 49 |
| AZ61A-F | 45 | 33 | 19 | 41 | 16 | 20 | 60 |
| AZ80A-T5 | 55 | 40 | 35 | | 7 | 24 | 82 |
| HM31A-F | 42 | 33 | 27 | 50 | 10 | 22 | |
| M1A-F | 37 | 26 | 12 | 28 | 12 | 18 | 44 |
| ZK40A-T5 | 40 | 37 | 20 | | 4 | | |
| ZK60A-F | 51 | 41 | 36 | 59 | 11 | 26 | 88 |
| Sheet and Plate | | | | | | | |
| AZ31B-H24 | 42 | 32 | 26 | 47 | 15 | 23 | 73 |
| AZ31C-H24 | 42 | 32 | 26 | 47 | 15 | 23 | 73 |
| HK31A-H24 | 38 | 30 | 23 | 41 | 9 | | 68 |
| HM21A-T8 | 34 | 25 | 19 | 39 | 11 | 18 | |

[a] 500 kg load, 10-mm ball.
[b] High-purity alloy, Ni, Fe, and Cu severely restricted.
[c] 0.30 percent maximum residual copper is allowed.
[d] For battery applications.

*Source: Metals Handbook,* 9th edition, Vol. 2, American Society for Metals.

## Nickel and Nickel Alloys

**Characteristics of Nickel and Nickel Alloys.**—Nickel is a white metal, similar in some respects to iron but with good oxidation and corrosion resistances. Nickel and its alloys are used in a variety of applications, usually requiring specific corrosion resistance or high strength at high temperature. Some nickel alloys exhibit very high toughness; others have very high strength, high proportional limits, and high moduli compared with steel. Commercially, pure nickel has good electrical, magnetic, and magnetostrictive properties. Nickel alloys are strong, tough, and ductile at cryogenic temperatures, and several of the nickel-based superalloys have good strength at temperatures up to 2000° F (1095° C).

Most wrought nickel alloys can be hot- and cold-worked, machined, and welded successfully; an exception is the most highly alloyed nickel compound—forged nickel-based superalloys—in which these operations are more difficult. The casting alloys can be machined or ground, and many can be welded and brazed.

There are five categories into which the common nickel-based metals and alloys can be separated:the pure nickel and high nickel (over 94 percent Ni) alloys; the nickel-molybdenum and nickel-molybdenum-chromium superalloys, which are specifically for corrosive or high-temperature, high-strength service; the nickel-molybdenum-chromium-copper alloys, which are also specified for corrosion applications; the nickel-copper (Monel) alloys, which are used in actively corrosive environments; and the nickel-chromium and nickel-chromium-iron superalloys, which are noted for their strength and corrosion resistance at high temperatures.

Descriptions and compositions of some commonly used nickel and high nickel alloys are shown in Table 9.

## Titanium and Titanium Alloys

Titanium is a gray, light metal with a better strength-to-weight ratio than any other metal at room temperature and is used in corrosive environments or in applications that take advantage of its light weight, good strength, and nonmagnetic properties. Titanium is available commercially in many alloys, but multiple requirements can be met by a single grade of the commercially pure metal. The alloys of titanium are of three metallurgical types: alpha, alpha-beta, and beta, with these designations referring to the predominant phases present in the microstructure.

Titanium has a strong affinity for hydrogen, oxygen, and nitrogen gases, which tend to embrittle the material; carbon is another embrittling agent. Titanium is outstanding in its resistance to strongly oxidizing acids, aqueous chloride solutions, moist chlorine gas, sodium hypochlorite, and seawater and brine solutions. Nearly all nonaircraft applications take advantage of this corrosion resistance. Its uses in aircraft engine compressors and in airframe structures are based on both its high corrosion resistance and high strength-to-weight ratio.

Procedures for forming titanium are similar to those for forming stainless steel. Titanium and its alloys can be machined and abrasive ground; however, sharp tools and continuous feed are required to prevent work hardening. Tapping is difficult because the metal galls.

Titanium castings can be produced by investment or graphite mold methods; however, because of the highly reactive nature of the metal in the presence of oxygen, casting must be done in a vacuum.

Generally, titanium is welded by gas-tungsten arc or plasma arc techniques, and the key to successful welding lies in proper cleaning and shielding. The alpha-beta titanium alloys can be heat treated for higher strength, but they are not easily welded. Beta and alpha-beta alloys are designed for formability; they are formed in the soft state, and then heat treated for high strength.

The properties of some wrought titanium alloys are shown in Table 10.

**Table 9. Common Cast and Wrought Nickel and High Nickel Alloys — Designations, Compositions, Typical Properties, and Uses**

| UNS Designation | Description and Common Name | Nominal Composition (Weight %) | Typical Room-Temperature Properties: Tensile (ksi) | 0.2% Yield (ksi) | Elong. (%) | Form | Typical Uses |
|---|---|---|---|---|---|---|---|
| N02200 | Commercially pure Ni (Nickel 200) | 99.5 Ni | 67 | 22 | 47 | Wrought | Food processing and chemical equipment. |
| N04400 | Nickel-copper alloy (Monel 400) | 65 Ni, 32 Cu, 2 Fe | 79 | 30 | 48 | Wrought | Valves, pumps, shafts, marine fixtures and fasteners, electrical and petroleum refining equipment. |
| N05500 | Age-hardened Ni-Cu alloy (Monel K 500) | 65 Ni, 30 Cu, 2 Fe, 3 Al + Ti | 160 | 111 | 24 | Wrought | Pump shafts, impellers, springs, fasteners, and electronic and oil well components. |
| N06002 | Ni-Cr Alloy (Hastelloy X) | 60 Ni, 22 Cr, 19 Fe, 9 Mo, 0.6 W | 114 | 52 | 43 | Wrought | Turbine and furnace parts, petrochemical equipment. |
| N06003 | Ni-Cr alloy (Nichrome V) | 80 Ni, 20 Cr | 100 | 60 | 30 | Wrought | Heating elements, resistors, electronic parts. |
| N06333 | Ni-Cr alloy (RA 333) | 48 Ni, 25 Cr, 18 Fe, 3 Mo, 3 W, 3 Co | 100 | 50 | 50 | Wrought | Turbine and furnace parts. |
| N06600 | Ni-Cr alloy (Inconel 600) | 75 Ni, 15 Cr, 10 Fe | 90 | 36 | 47 | Wrought | Chemical, electronic, food processing and heat-treating equipment; nuclear steam generator tubing. |
| N06625 | Ni-Cr alloy (Inconel 625) | 61 Ni, 21 Cr, 2 Fe, 9 Mo, 4 Nb | 142 | 86 | 42 | Wrought | Turbine parts, marine and chemical equipment. |
| N07001 | Age-hardened Ni-Cr alloy (Waspaloy) | 58 Ni, 20 Cr, 14 Co, 4 Mo, 3 Al, 1.3 Ti, B, Zr | 185 | 115 | 25 | Wrought | Turbine parts. |
| N07500 | Age-hardened Ni-Cr alloy (Udimet 500) | 52 Ni, 18 Cr, 19 Co, 4 Mo, 3 Al, 3 Ti, B, Zr | 176 | 110 | 16 | Wrought & Cast | Turbine parts. |
| N07750 | Age-hardened Ni-Cr alloy (Inconel X-750) | 73 Ni, 16 Cr, 7 Fe, 2.5 Ti, 1 Al, 1 Nb | 185 | 130 | 20 | Wrought | Turbine parts, nuclear reactor springs, bolts, extrusion dies, forming tools. |
| N08800 | Ni-Cr-Fe alloy (Incoloy 800) | 32 Ni, 21 Cr, 46 Fe, 0.4 Ti, 0.4 Al | 87 | 42 | 44 | Wrought | Heat exchangers, furnace parts, chemical and power plant piping. |
| N08825 | Ni-Cr-Fe alloy (Incoloy 825) | 42 Ni, 22 Cr, 30 Fe, 3 Mo, 2 Cu, 1 Ti, Al | 91 | 35 | 50 | Wrought | Heat-treating and chemical handling equipment. |
| N09901 | Age-hardened Ni-Cr-Fe alloy (Incoloy 901) | 43 Ni, 12 Cr, 36 Fe, 6 Mo, 3 Ti + Al, B | 175 | 130 | 14 | Wrought | Turbine parts. |
| N10001 | Ni-Mo alloy (Hastelloy B) | 67 Ni, 28 Mo, 5 Fe | 121 | 57 | 63 | Wrought | Chemical handling equipment. |
| N10004 | Ni-Cr-Mo alloy (Hastelloy W) | 59 Ni, 5 Cr, 25 Mo, 5 Fe, 0.6 V | 123 | 53 | 55 | Wrought | Weld wire for joining dissimilar metals, engine repair and maintenance. |
| N10276 | Ni-Cr-Mo alloy (Hastelloy C-276) | 57 Ni, 15 Cr, 16 Mo, 5 Fe, 4 W, 2 Co | 116 | 52 | 60 | Wrought | Chemical handling equipment. |
| N13100 | Ni-Co alloy (IN 100) | 60 Ni, 10 Cr, 15 Co, 3 Mo, 5.5 Al, 5 Ti, 1 V. B, Zr | 147 | 123 | 9 | Cast | Turbine parts. |

For strength in MPa, multipy strength in ksi by 6.8947572

**Table 10. Mechanical Properties of Wrought Titanium Alloys**

| Nominal Composition (%) | Condition | Room Temperature Tensile Strength (ksi) | Yield Strength (ksi) | Elongation (%) | Reduction in Area (%) |
|---|---|---|---|---|---|
| Commercially Pure | | | | | |
| 99.5 Ti | Annealed | 48 | 35 | 30 | 55 |
| 99.2 Ti | Annealed | 63 | 50 | 28 | 50 |
| 99.1 Ti | Annealed | 75 | 65 | 25 | 45 |
| 99.0 Ti | Annealed | 96 | 85 | 20 | 40 |
| 99.2 Ti[a] | Annealed | 63 | 50 | 28 | 50 |
| 98.9 Ti[b] | Annealed | 75 | 65 | 25 | 42 |
| Alpha Alloys | | | | | |
| 5 Al, 2.5 Sn | Annealed | 125 | 117 | 16 | 40 |
| 5 Al, 2.5 Sn (low $O_2$) | Annealed | 117 | 108 | 16 | |
| Near Alpha Alloys | | | | | |
| 8 Al, 1 Mo, 1 V | Duplex annealed | 145 | 138 | 15 | 28 |
| 11 Sn, 1 Mo, 2.25 Al, 5.0 Zr, 1 Mo, 0.2 Si | Duplex annealed | 160 | 144 | 15 | 35 |
| 6 Al, 2 Sn, 4 Zr, 2 Mo | Duplex annealed | 142 | 130 | 15 | 35 |
| 5 Al, 5 Sn, 2 Zr, 2 Mo, 0.25 Si | 975° C (1785° F) (½ h), AC 595° C (1100° F)(2 h), AC | 152 | 140 | 13 | |
| 6 Al, 2 Nb, 1 Ta, 1 Mo | As rolled 2.5 cm (1 in.) plate | 124 | 110 | 13 | 34 |
| 6 Al, 2 Sn, 1.5 Zr, 1 Mo, 0.35 Bi, 0.1 Si | Beta forge + duplex anneal | 147 | 137 | 11 | |
| Alpha-Beta Alloys | | | | | |
| 8 Mn | Annealed | 137 | 125 | 15 | 32 |
| 3 Al, 2.5 V | Annealed | 100 | 85 | 20 | |
| 6 Al, 4 V | Annealed | 144 | 134 | 14 | 30 |
| | Solution + age | 170 | 160 | 10 | 25 |
| 6 Al, 4 V (low $O_2$) | Annealed | 130 | 120 | 15 | 35 |
| 6 Al, 6 V, 2 Sn | Annealed | 155 | 145 | 14 | 30 |
| | Solution + age | 185 | 170 | 10 | 20 |
| 7 Al, 4 Mo | Solution + age | 160 | 150 | 16 | 22 |
| 6 Al, 2 Sn, 4 Zr, 6 Mo | Solution + age | 184 | 170 | 10 | 23 |
| 6 Al, 2 Sn, 2 Zr, 2 Mo, 2 Cr, 0.25 Si | Solution + age | 185 | 165 | 11 | 33 |
| 10 V, 2 Fe, 3 Al | Solution + age | 185 | 174 | 10 | 19 |
| Beta Alloys | | | | | |
| 13 V, 11 Cr, 3 Al | Solution + age | 177 | 170 | 8 | |
| | Solution + age | 185 | 175 | 8 | |
| 8 Mo, 8 V, 2 Fe, 3 Al | Solution + age | 190 | 180 | 8 | |
| 3 Al, 8 V, 6 Cr, 4 Mo, 4 Zr | Solution + age | 210 | 200 | 7 | |
| | Annealed | 128 | 121 | 15 | |
| 11.5 Mo, 6 Zr, 4.5 Sn | Solution + age | 201 | 191 | 11 | |

[a] Also contains 0.2 Pd.
[b] Also contains 0.8 Ni and 0.3 Mo.
*Source:* Titanium Metals Corp. of America and RMI Co.
For strength in MPa, multipy strength in ksi by 6.8947572

## CORROSION

Corrosion is a chemical or electrochemical process by which a material and/or its properties deteriorate due to interaction with one or more external substances. Environmental conditions, surface conditions, and component stresses can initiate or accelerate corrosion. Effects of corrosion are many, ranging from merely aesthetic changes to interference with electrical connections and weakening of a structural or mechanical part that can lead to failure. Of particular concern are common material changes, such as pitting corrosion and stress corrosion cracking, which are difficult to detect, yet can result in sudden component failure.

### Corrosion Types and Methods of Prevention

Consideration of material properties—including chemical compatibility, galvanic potential, and reaction to anticipated environmental conditions—is important in minimizing corrosion and maximizing the design life of metal and alloy parts and assemblies. However, the mechanisms of corrosion can result in a wide variety of corrosion patterns and processes, and different types and mechanisms of corrosion may occur simultaneously. Thus, corrosion protection often requires a multipronged approach.

**Uniform (General) Corrosion.—**This process describes either a chemical or electrochemical attack that uniformly affects an entire exposed surface. The corroding part will gradually thin; corrosive substances may accumulate on the surface. Usually, uniform corrosion can be predicted and measured, so designers can provide for and monitor it to ensure safety.

Methods of preventing general corrosion include using appropriate materials, passivation, polishing, coatings and other barriers, and environmental controls, as well as reducing stresses on components. To prevent corrosive interaction between incompatible materials, parts can be separated electrically from each other or electrolyte solutions with gaskets or coatings. Gaskets or sealants also can minimize moisture intrusion into joints to further reduce corrosion, though materials interactions should be investigated. For instance, graphite, often found in carbon black rubber and gaskets, should not be used in contact with stainless steels in seawater.

**Chemical Corrosion.—**This type of corrosion generally refers to various chemical interactions. In most cases, damage results from the destructive reaction of a contacting substance directly acting on and degrading a material, for instance, an acid affecting a metal. Chemical corrosion also may result in oxide formation or deposition of other surface coatings.

**Dry and High-Temperature Corrosion.—**Dry corrosion, also called *scaling,* generally is caused by chemicals (gases, molten salts, or solids) acting on a surface in a dry atmosphere (i.e., no aqueous electrolyte), though particulate abrasion also may be involved. Material selection and effective environmental controls can significantly reduce damage; numerous specialized protective coatings also are available.

High-temperature corrosion takes place when dry corrosion occurs and/or accelerates due to a high-temperature environment. Critical temperature and rate of corrosion depend on the materials involved and environmental conditions. As this form of corrosion may become a risk at temperatures as low as 194° F (90° C) in some applications (such as oil pipelines), it may simply be identified as dry corrosion. Common results of high-temperature corrosion include oxidation (in air, this is a problem above the material's scaling temperature), carburization, chlorination, and sulfidation.

**Electrochemical Corrosion.—**This occurs when an electrically active material is exposed to an electrolyte, forming an electrical cell in which ionization occurs and electrons move from an anodic (active) material to a cathodic (noble) material. As a result, the anode corrodes through oxidation, while the cathode absorbs the free electrons through a reduction reaction. With sufficient oxygen, the cathode often will develop a protective oxide layer. Areas of oxygen deprivation can become anodic, resulting in hidden pockets of accelerated corrosion.

*Wet and damp/atmospheric corrosion* occur when water and contaminants in the environment form an electrolyte liquid or damp film on the surface of a part, facilitating electrochemical corrosion. Wet corrosion rates generally are highest, while atmospheric corrosion rates depend on the amount and type of moisture.

To minimize electrochemical corrosion, the best course of action is to avoid mixing different metals in an assembly and match the potentials of the part(s) to the environment. The next lines of defense are barriers and protective area ratios (see *Galvanic Corrosion*, below). When those methods are not enough, it is common practice to add one or more auxiliary anodes or cathodes to the system. Current from an external DC power supply may be applied to change the electrical potential of the target material and control the flow of electrons. Some systems can monitor and optimize conditions in the cell by using a reference electrode.

*Cathodic Protection:* Adding one or more sacrificial anodes (either a part or a coating) to a system can protect a part that otherwise would be the anode in an electrochemical reaction. Buried or submerged metal pipes are commonly protected passively by the addition of a sacrificial magnesium anode placed nearby and connected by a wire. An impressed current system includes a DC power supply to supply current, which improves system performance and extends anode life. Cathodic protection is broadly applicable for all metals and alloys, but it is often not suitable for corrosive environments and can accelerate hydrogen embrittlement.

*Anodic Protection:* This newer method employs additional cathodes and an applied current to shift the target material's potential into passive range. Anodic protection applies to corrosive environments and can be achieved with much lower current density than cathodic protection would require. This method is limited to materials that exhibit active-passive behavior (material surfaces can change from active to passive when exposed to oxidizers or applied current). Steel, titanium, and nickel alloys are included in this group.

**Galvanic Corrosion.**—In this electrochemical corrosion process, sometimes called *bimetallic corrosion,* dissimilar metals or alloys interact when electrically coupled by a conductive fluid. If the two metals or alloys have different electrochemical potentials, current will flow from the more active (anodic) metal to the more noble (cathodic) metal. In this galvanic couple, the anode will experience dissolution (usually oxidation), while the cathode will experience reduction. In practical terms, the anode will corrode faster or differently than it would alone in the electrolyte, and the cathode usually will be protected. The metals or alloys usually are in direct contact with each other, but this is not necessary for corrosion to occur—it can expand into the entire wetted area if the fluid is highly conductive. Parts in direct contact will experience the most corrosion; as separation between parts increases, corrosion decreases.

Intensity of the corrosion process depends on differences in electrochemical potential between two metals, distance between the metals, wetted surface areas of the parts, and conductivity of the surrounding fluid. Other factors include oxygenation, fluid pH, metal composition and variations, presence of films or deposits, passive film stability, and exposure of metals to processes such as cold-working or welding. Corrosion often will intensify as oxygen levels in the electrolyte drop, and oxygen depletion frequently occurs in crevices and other stagnant areas.

*Galvanic Compatibility:* The US military published empirical compatibility information in MIL-STD-889B, "Dissimilar Metals." Data in Table 1 represents those results. Because many factors influence galvanic corrosion, this data should be used only as a general guide.

## Table 1. Galvanic Compatibility of Metals in Select Environments

| | | | Active (Anodic) → Noble (Cathodic) Metal or Alloy | | | | | | | | | | | | | | | | | | | |
|---|---|---|---|---|---|---|---|---|---|---|---|---|---|---|---|---|---|---|---|---|---|---|
| | Metal or Alloy | | A | B | C | D | E | F | G | H | I | J | K | L | M | N | O | P | Q | R | S | T |
| *Active (Anodic)* ↓ | Magnesium | A | C,C,C | C,C,I | C,C,I | I,I,I | I,I,I | I,I,I | I,I,I | I,I,I | I,I,I | I,I,I | I,I,I | I,I,I | I,I,I | I,I,I | I,I,I | I,I,I | I,I,I | I,I,I | I,I,I | I,I,I |
| | Zinc | B | | C,C,C | C,C,C | C,C,I | C,C,I | I,I,I | I,I,I | I,I,I | I,I,I | I,I,I | I,I,I | I,I,I | I,I,I | I,I,I | I,I,I | I,I,I | I,I,I | I,I,I | I,I,I | I,I,I |
| | Cadmium, Beryllium | C | | | C,C,C | C,C,I | C,C,I | I,I,I | I,I,I | I,I,I | I,I,I | I,I,I | I,I,I | I,I,I | I,I,I | I,I,I | I,I,I | I,I,I | I,I,I | I,I,I | I,I,I | I,I,I |
| | Aluminum, Al-Zn, Al-Mg | D | | | | C,C,C | C,C,C | I,I,I | I,I,I | I,I,I | I,I,I | I,I,I | I,I,I | I,I,I | I,I,I | I,I,I | I,I,I | I,I,I | I,I,I | I,I,I | I,I,I | I,I,I |
| | Aluminum-Copper | E | | | | | C,C,C | I,I,I | I,I,I | I,I,I | I,I,I | I,I,I | I,I,I | I,I,I | I,I,I | I,I,I | I,I,I | I,I,I | I,I,I | I,I,I | I,I,I | I,I,I |
| | Carbon Steels, Low-Alloy Steels | F | | | | | | C,C,C | C,I,I | I,I,I | I,I,I | I,I,I | I,I,I | I,I,I | I,I,I | I,I,I | I,I,I | C,C,I | I,I,I | I,I,I | I,I,I | I,I,I |
| | Lead | G | | | | | | | C,C,C | C,C,C | C,C,I | C,C,I | C,C,I | C,C,I | I,I,I | C,I,I | C,I,I | C,I,I | C,I,I | C,C,I | C,I,I | C,I,I |
| | Tin, Tin-Lead, Indium | H | | | | | | | | C,C,C | I,I,I | C,I,I | C,I,I | C,C,I | C,C,I | C,C,I | C,C,I | C,C,C | C,I,C | C,C,C | I,I,I | I,I,I |
| | Martensitic Stainless Steels (includes 420), Ferritic Stainless Steels | I | | | | | | | | | C,C,C | C,I,I | I,I,I | I,I,I | C,I,I | I,I,I | I,I,I | I,I,I | I,I,I | I,I,I | I,I,I | I,I,I |
| | Chromium, Molybdenum, Tungsten | J | | | | | | | | | | C,C,C | C,C,C | C,C,I | C,I,I | I,I,I | C,C,I | C,C,I | C,C,I | C,C,I | C,C,I | C,C,I |
| | Austenitic Stainless Steels (includes 200 and 300 Series), PH Stainless Steels, Super Strength Stainless Steels, Heat-Resistant Stainless Steel | K | | | | | | | | | | | C,C,C | I,I,I | I,I,I | I,I,I | C,C,I | C,C,I | C,C,I | C,C,I | C,C,I | C,C,I |
| | Lead Brass, Bronze | L | | | | | | | | | | | | C,C,C | C,C,I | C,C,I | C,I,I | C,C I | C,I,I | C,C,I | C,C,I | C,C,I |
| | Low Copper Brass, Low Copper Bronze | M | | | | | | | | | | | | | C,C,C | C,C,I | C,C,I | C,C,I | C,I,I | C,C,I | C,I,I | C,I,I |
| | High Copper Brass, High Copper Bronze | N | | | | | | | | | | | | | | C,C,C | C,I,I | C,C,I | C,I,I | C,C,I | C,I,I | C,I,I |
| | Copper-High Nickel, Monel | O | | | | | | | | | | | | | | | C,C,C | C,C,I | C,C,I | C,C,I | C,C,I | C,C,I |
| | Nickel, Cobalt | P | | | | | | | | | | | | | | | | C,C,C | C,C,I | C,C,I | C,C,I | C,C,I |
| | Titanium | Q | | | | | | | | | | | | | | | | | C,C,C | C,C,I | C,C,I | C,C,I |
| | Silver | R | | | | | | | | | | | | | | | | | | C,C,C | C,C,C | C,C,I |
| *Noble* | Palladium, Rhodium, Gold, Platinum | S | | | | | | | | | | | | | | | | | | | C,C,C | C,C,C |
| *(Cathodic)* | Graphite | T | | | | | | | | | | | | | | | | | | | | C,C,C |

Values
I = Incompatible (risk of significant galvanic interaction)
C = Compatible (negligible galvanic interaction likely)

Position of Compatibility Values (1,2,3)
1 = Industrial atmosphere
2 = Marine atmosphere
3 = Seawater immersion

Example: C,C,I = Compatible in industrial atmosphere (1)
Compatible in marine atmosphere (2)
Incompatible in seawater immersion (3)

Selecting metals with similar electrochemical potentials usually minimizes galvanic corrosion. One method of comparing potentials involves referencing a *galvanic series*. While it should be representative of anticipated environmental conditions, this tool is not used to predict corrosion rates, but rather provides a qualitative evaluation of coupled metal behavior.

To develop a series, a reference half-cell and samples of the target metals are immersed together in an electrolyte solution chosen and circulated to match the expected environmental conditions. Over time, potentials of the target metals are measured relative to the reference half-cell. There are several standard reference half-cell compositions that will yield different values; the appropriate reference is compatible with the electrolyte. A useful standard is ASTM G82-98 (2014), "Standard Guide for Development and Use of a Galvanic Series for Predicting Galvanic Corrosion Performance."

Most published galvanic series data applies to specific flowing seawater environment conditions. While saltwater is highly conductive, freshwater has low conductivity, and dynamic electrolyte conditions will greatly affect potential measurements. Therefore, for critical applications, it is good practice to develop application-specific series data, rather than using published galvanic series information.

For examples of seawater applications, refer to Table 2, which is based on Army Missile Command Report RS-TR-67-11, "Practical Galvanic Series." Materials closer together along the arrow in the series have less corrosion-inducing potential difference between them in that environment. However, use this data with caution in predicting whether corrosion will be a risk. This series indicates which material will be the anode in a couple, though polarity reversals can occur in which a metal normally anodic to another will become cathodic to that same metal. Examples include high-temperature reversals of zinc/iron, aluminum/iron, and aluminum/zinc.

## Table 2. Sample Galvanic Series, General Seawater Environment

*Active (Anodic)*

| | | |
|---|---|---|
| | Stainless Steel 410 (active) | Stainless Steel 201 (active) |
| Magnesium | Copper (plated, cast, or wrought) | Carpenter 20 (active) |
| Mg alloy AZ-31B | Nickel (plated) | Stainless Steel 321 (active) |
| Mg alloy HK-31A | Chromium (plated) | Stainless Steel 316 (active) |
| Zinc (hot-dip, die cast, or plated) | Tantalum | Stainless Steel 309 (active) |
| Beryllium (hot-pressed) | AM350 (active) | Stainless Steel 17-7PH (passive) |
| Aluminum 7072 clad on 7075 | Stainless Steel 310 (active) | Silicon Bronze 655 |
| Aluminum 2014-T3 | Stainless Steel 301 (active) | Stainless Steel 304 (passive) |
| Aluminum 1160-H14 | Stainless Steel 304 (active) | Stainless Steel 301 (passive) |
| Aluminum 7079-T6 | Stainless Steel 430 (active) | Stainless Steel 321 (passive) |
| Cadmium (plated) | Stainless Steel 410 (active) | Stainless Steel 201 (passive) |
| Uranium | Stainless Steel 17-7PH (active) | Stainless Steel 286 (passive) |
| Aluminum 218 (die cast) | Tungsten | Stainless Steel 316L (passive) |
| Aluminum 5052-0 | Niobium (columbium) 1% Zr | AM355 (active) |
| Aluminum 5052-H12 | Brass, Yellow, 268 | Stainless Steel 202 (passive) |
| Aluminum 5456-0, H353 | Uranium 8% Mo | Carpenter 20 (passive) |
| Aluminum 5052-H32 | Brass, Naval, 464 | AM355 (passive) |
| Aluminum 1100-0 | Yellow Brass | A286 (passive) |
| Aluminum 3003-H25 | Muntz Metal 280 | Titanium 5A1, 2.5 Sn |
| Aluminum 6061-T6 | Brass (plated) | Titanium 13V, 11Cr, 3Al (annealed) |
| Aluminum A360 (die cast) | Nickel-Silver (18% Ni) | Titanium 6Al, 4V |
| Aluminum 7075-T6 | Stainless Steel 316L (active) | (solution treated and aged) |
| Aluminum 6061-0 | Bronze 220 | Titanium 6Al, 4V (anneal) |
| Indium | Copper 110 | Titanium 8Mn |
| Aluminum 2014-0 | Red Brass | Titanium 13V, 11Cr 3Al |
| Aluminum 2024-T4 | Stainless Steel 347 (active) | (solution heat treated and aged) |
| Aluminum 5052-H16 | Molybdenum (commercial pure) | Titanium 75A |
| Tin (plated) | Copper-Nickel 715 | AM350 (passive) |
| Stainless Steel 430 (active) | Admiralty Brass | Silver |
| Lead | Stainless Steel 202 (active) | Gold |
| Steel 1010 | Bronze, Phosphor 534 (B-1) | Graphite |
| Iron (cast) | Monel 400 | |

*Noble (Cathodic)*

Another galvanic compatibility guide is an *anodic index*. Table 3 is an example based on US military specification MIL-F-14072, "Finishes for Ground Based Electronic Equipment." This table shows electrochemical potentials normalized relative to a gold reference. Subtract one index value from another to determine the electrochemical potential difference between the metals. The Standard recommends differential limits for three types of non-immersed application environments (controlled, normal, and harsh). Again, use caution when referring to any anodic index, because, as with a galvanic series, actual conditions can significantly change the corrosion behavior of a galvanic couple.

**Table 3. Anodic Index (Gold Reference)**

| Metals and Alloys | Anodic Index (V) | |
|---|---|---|
| Gold, Gold-Platinum Alloys, Wrought Platinum | −0.00 | *Cathodic (Noble)* ↑ |
| Rhodium Plated on Silver-Plated Copper | −0.10 | |
| Silver, High Silver Alloys | −0.15 | |
| Nickel, Monel, High Nickel-Copper Alloys, Titanium Alloys | −0.30 | |
| Copper, Low Brasses or Bronzes, Silver Solder, High Copper-Nickel-Zinc Alloys, Nickel-Chromium Alloys | −0.35 | |
| Yellow Brasses and Bronzes | −0.40 | |
| High Brasses and Bronzes, Naval Brass, Muntz Metal | −0.45 | |
| 18% Chromium-type Stainless Steels (includes 300 Series) | −0.50 | |
| Chromium Plated, Tin Plated, 12% Chromium-type Stainless Steels (some 400 Series Steels) | −0.60 | |
| Tin Plate, Tin-Lead Solder, Terneplate (Lead-Tin Alloy) | −0.65 | |
| Lead, High Lead Alloys | −0.70 | |
| 2000 Series Wrought Aluminum | −0.75 | |
| Plain Carbon Steels, Low Alloy Steels, Wrought Iron, Gray Malleable Iron | −0.85 | |
| Wrought Aluminum Alloys other than 2000 Series, Cast Aluminum-Silicon Alloys | −0.90 | |
| Cast Aluminum Alloys other than Silicon Type, Cadmium, Cadmium Chromate | −0.95 | |
| Hot-Dip Zinc Plate, Galvanized Steel | −1.20 | |
| Wrought Zinc, Zinc Die-Casting Alloys, Zinc Plate | −1.25 | |
| Magnesium, Magnesium Alloys | −1.75 | *Anodic (Active)* |
| Beryllium | −1.85 | |

**Evaluation of Galvanic Couples in Various Environments**

| Environmental Conditions | Maximum Differential (V) |
|---|---|
| Controlled: Temperature and humidity are controlled, as in laboratory or office spaces. | 0.50 |
| Normal: Industrial environments like warehouses where temperature and humidity are not controlled. | 0.25 |
| Harsh: Outdoors, high humidity, or salt exposure environments. | 0.15 |

Note that the "Electrochemical Series," also known as the "Electromotive Force (EMF) Series," which ranks metals with respect to inherent reactivity, from active to noble, depends on theoretical, static conditions. Therefore, galvanic corrosion of metals listed is likely to vary in real-world environments.

*Protective Area Ratios for Galvanic Couples:* The wetted cathode/anode area ratio for parts in contact has a significant effect on the rate of galvanic corrosion: if the wetted cathode is large and the wetted anode is small (high ratio), corrosion is intensified; if the cathode is small and the anode is large (low ratio), corrosion slows. Protective area ratios can offset large potential differences between coupled metals. This effect makes it possible to use stainless steel screws to anchor large aluminum parts in conductive environments. By contrast, use of zinc or galvanized fasteners to secure large stainless steel parts will result in intense fastener corrosion in the presence of moisture. When choosing which part to coat in a galvanic couple, coat the more noble (cathodic) part; if only the more active metal (anodic) is coated, intense corrosion can occur at sites of small coating defects due to the resulting high area ratio.

### Effects of Corrosion

Corrosion can result in many different effects on a material. The following sections describe some common effects on materials induced at least in part or enhanced by corrosion, as well as mitigation methods.

**Crevice Corrosion.—**In small gaps, holes, and crevices, such as those between contacting parts, the fluid medium can be stagnant and undergo chemical changes with oxygen depletion. This corrosion is difficult to detect due to its location. Where possible, it helps to eliminate features that can trap fluid, use full welds instead of fasteners, and seal joints with grease or nonporous gaskets.

**Pitting Corrosion.—**Localized pitting corrosion is difficult to detect and can be destructive and unpredictable. Pits characteristic of this process may be small or large, may exist singly or in groups, and may be masked by corrosion-protection products. Pitting corrosion can be initiated by damage to a coating or oxide layer; surface contaminants or inclusions; or acid, chloride, or other caustic chemical exposure. Pits can act as stress risers and cause premature part failure due to fatigue or stress corrosion cracking.

**Microbiologically Influenced Corrosion (MIC).—**Also called *microbial corrosion* or *biocorrosion,* this process usually takes the form of crevice and/or pitting corrosion under a fouling biofilm that has formed on part surfaces. Chemistry changes due to static conditions under the film, metabolic waste from the microorganisms, and metabolic reduction of oxygen all contribute to corrosive attack. Many different biological organisms, both aerobic and anaerobic, can participate in this corrosion process. Elimination of fouling biofilms is the best defense. Also applicable are the stated methods for reducing pitting or crevices and using corrosion-resistant materials where possible. For example, in seawater applications, titanium is highly resistant to MIC, though it is not immune in all environments.

**Intergranular Corrosion.—**Localized attack along microstructural grain boundaries of a material can occur due to segregation or precipitation of elements at those boundaries. The unique material composition at the grain boundaries can preferentially corrode or cause galvanic corrosion to occur. The resulting changes in material properties can lead to intergranular stress corrosion cracking or disintegration.

This type of corrosion is a risk for stainless steel welded in multiple passes, which can cause sensitization of material. Depletion of chromium and/or formation of carbides in the heat-affected zone can lead to a type of intergranular corrosion known as *weld decay,* or *knife-line attacks* in alloys stabilized with niobium. Aluminum alloys, particularly extruded or heavily worked pieces, also are susceptible to intergranular corrosion; a subtype called *exfoliation corrosion* may occur where the surface grain is thinner than the rolled grain. Prevention methods include heat treatment of sensitized parts.

**Selective Leaching (Dealloying).—**This process involves preferential corrosion of the least noble element in an alloy. Dezincification of unstabilized brass (leaving porous copper) and iron depletion of gray cast iron (leaving graphite) are examples of selective leaching corrosion. This type of corrosion is difficult to detect visually, but structural weakening can be profound. To reduce the likelihood of selective leaching, vulnerable parts may be heat treated to improve homogeneity.

**Stress Corrosion Cracking.—**Potentially devastating cracks can start and grow slowly in metals and alloys exposed to both static tensile stress (applied and/or residual) and a corrosive environment. Rather than propagating on surfaces, such cracks usually are fine, branching, transgranular, and penetrating, and, thus, difficult to detect. Environmental conditions that induce this process are very specific and depend on materials, applied processes, and part uses. Weldments and heavily worked parts are particularly vulnerable, due to the potential for high levels of residual stress. Methods to guard against stress corrosion cracking include using stress-relieving parts with heat treatment or mechanical loading and introducing surface compressive stress through methods such as shot peening, laser peening, or grit blasting.

**Tribocorrosion.—**When a material degrades due to a combination of erosion/wear and corrosion, the process is called *tribocorrosion*. Mechanical wear often accelerates the corrosion process; in turn, corrosion attack on the material often accelerates wear. There are two commonly encountered subtypes of tribocorrosion:

*Fretting Corrosion:* This occurs when two parts, exposed to a corrosive medium, rub forcibly against each other in a repetitive manner. The movements typically are very small; vibration is a common culprit. Contact surfaces experience mechanical wear and corrosive attack simultaneously, often leading to crack formation and fatigue failure. Bolted joints are especially vulnerable, as are bearings that undergo oscillating movements. Mitigation methods include eliminating movement between parts (particularly small oscillating movements), employing protective lubricants or sealants, adjusting hardness of one or both parts, and reducing contact loading, if possible.

*Erosion Corrosion:* Parts exposed to a flowing corrosive medium may exhibit localized areas where mechanical erosion occurs and corrosion is intensified. Optimizing flow geometry and velocity can help reduce turbulence and prevent such cavitation. Other methods include filtering out abrasive particles in the fluid, smoothing surfaces to remove imperfections, and minimizing edges.

**Hydrogen Embrittlement.—**High-strength steels, titanium alloys, and aluminum alloys are particularly susceptible to this type of attack. When hydrogen is absorbed into the material, this can cause embrittlement and cracking along grain boundaries, posing a risk of sudden failure. Processes that can cause excess hydrogen exposure include arc welding (unless low-hydrogen electrodes are used), electroplating, pickling, and phosphating. Mitigation steps include limiting hydrogen exposure, baking off hydrogen after exposure, relieving residual stresses, using barrier coatings, and substituting less vulnerable materials. Note that cathodic protection accelerates hydrogen embrittlement, especially in acid environments, so it should be used with caution when working with susceptible materials and processes.

**Corrosion Fatigue.—**Fatigue failure can occur prematurely in parts exposed to a corrosive environment while experiencing cyclic or alternating loading conditions. Cracks formed in this process usually are nonbranching and transgranular, form early, and grow at an accelerated rate, leading to substantially reduced fatigue life of parts. Prevention includes reducing loads and stresses, as well as normal corrosion mitigation methods.

# PLASTICS

## Types and Properties of Plastics

### Characteristics of Common Plastics Families

| | |
|---|---|
| ABS (acrylonitrile-butadiene-styrene) | Common thermoplastic designed for balance of properties; easily machined and thermoformed. |
| Acetal, POM | Engineering thermoplastic with good strength, wear resistance, and dimensional stability. More dimensionally stable than nylon under wet and humid conditions. |
| Acrylic, PMMA | Clear, transparent, strong, break-resistant thermoplastic with very good chemical resistance and weatherability. |
| Fiberglass | Thermosetting composite with high strength-to-weight ratio and insulative properties. Highly resistant to corrosion. |
| Liquid crystal polymer (LCP) | Aromatic, highly inert polymer, with high strength and rigidity, as well as chemical, fire, and temperature resistance. |
| Nylon, PA (polyamide) | Thermoplastic with high impact resistance, ideal for wear applications such as bearings and gears. Self-lubricating under some circumstances. |
| PEEK (polyetherether-ketone) | Engineering thermoplastic with excellent flexural and tensile properties. Exceptional temperature resistance makes it suitable for continuous use above 500° F (260° C). |
| Phenolic | Thermosetting family of plastics with minimal thermal expansion, high compressive strength, very high wear and abrasion resistance, and a low coefficient of friction. Used for bearing applications and molded parts. |
| Polycarbonate, PC | Transparent, tough thermoplastic with high impact strength, chemical resistance, and electrical properties, plus good dimensional stability. |
| Polyester, PET (polyethylene-terephthalate) | Dimensionally stable thermoplastic with superior machining characteristics compared to acetal. |
| Polyethylene, PE | Most common thermoplastic, with very good chemical resistance and ductility, but relatively low stiffness. Suitable for applications in temperatures no greater than 122° F (50° C). |
| Polypropylene, PP | Good chemical resistance, combined with low moisture absorption and excellent electrical properties. Retains strength up to 250° F (120° C). |
| Polystyrene, PS | Transparent, colorless, and relatively low-cost amorphous thermoplastic. relatively rigid with good electrical properties, but brittle, with relatively poor chemical and ultraviolet resistance. |
| Polysulfone, PSU | Durable thermoplastic with good electrical properties. Stable at temperatures in excess of 300° F (150° C). |
| Polyurethane | Another durable thermoplastic, with exceptional impact and abrasion resistance. Resists sunlight and weathering. |
| Teflon, PTFE (polytetrafluoro-ethylene) | Thermoplastic, with low coefficient of friction; inert to most chemicals and solvents. Generally self-lubricating with a low thermal-expansion rate. Withstands temperatures up to 500° F (260° C). |
| PVC (polyvinyl chloride) | Thermoplastic with good stiffness; resists most corrosive solutions and gases (both acid and alkaline), with softening temperatures around 190° F (90° C). Chlorinated PVC (CPVC) has similar properties, but is more rigid with softening temperatures around 230° F (110° C) higher. |
| PVDF (polyvinylidene-fluoride) | Thermoplastic with outstanding chemical resistance, plus good mechanical strength and dielectric properties. Good substitute for PCC or PP. |

**Plastics Materials.**—Plastics encompass a wide range of materials that consist of long chains of carbon atoms with attached atoms of other elements, mainly hydrogen, and small, repeating chemical functional groups that influence the plastics' properties. Some other elements found in plastics are oxygen, nitrogen, chlorine, sulfur, and fluorine. A special class, the *silicones*, have chains of alternating silicon and oxygen atoms with small carbon-based side groups. These repeating chain segments are called *mers*, giving the term *polymers* for these materials. The pure (*neat*) materials are also known as *resins*. Commercial plastics are resins containing additives in amounts ranging from trace quantities to 50 percent or more. In any sample of resin, molecular chain lengths will generally vary widely, but the average typically exceed several thousand mer units if the resin is to possess good mechanical properties. There are two main classes of resins: *thermoplastics* (which includes *elastomers*) and *thermosets*.

*Thermoplastics:* Plastics that soften when heated are called thermoplastics. These plastics can be repeatedly melted by heating and then solidified by cooling. Multiple heat cycles may be applied in manufacturing a final product. For example, freshly polymerized polypropylene is a fine powder that is converted into pelletized feedstock by extrusion. The extruder compacts and melts the powder and pumps it through a strand die, where the emerging strands are water-chilled and sliced into $\frac{1}{3}$ -inch (10-mm) BB-sized pellets. A sheet maker may then reprocess the pellets in another extruder to produce long rolls of thin sheets or shorter slabs of thicker sheets. These can be used by a thermoformer, which again heat-softens the sheets and pulls them down over a chilled mold to make such products as packaging from the thin sheets or wastebaskets from the thicker ones.

Some examples of commodity thermoplastic families, in order of decreasing annual demand are polyethylene (PE), polypropylene ( PP), polyester ( PET), polyvinyl chloride (PVC), polystyrene (PS), and acrylic (PMMA). Engineering thermoplastics typically have superior strength, toughness, and resistance to heat and chemicals as compared to commodity grades. Examples include acrylonitrile-butadiene-styrene (ABS), nylon (polyamide), polycarbonate (PC), polyvinylidene-fluoride (PVDF), polytetrafluoroethylene (PTFE), polysulfone (PSU), and polyetherether-ketone (PEEK).

*Thermosets:* Thermosetting plastics, such as amino, epoxy, phenolic, and unsaturated polyesters, are supplied to the final processor as partially polymerized mixtures of chemically active, long-chain molecules. When they are heated during molding, the mixture components react to form chemical bonds, called *cross-links*, between the original chains, forming a large network of three-dimensional molecules. Such products generally have improved mechanical properties and heat resistance. However, if heated above their molding temperatures, they do not melt but instead decompose. They cannot be reprocessed, so opportunities for salvaging scrap for recycling are limited. Many rubbers that are processed by vulcanizing, such as butyl, latex, neoprene, nitrile, polyurethane, and silicone, also are classified as thermosets.

*Elastomers and Rubbers:* Raw natural or synthetic rubber, initially thermoplastic, is mixed with finely divided carbon and sulfur. During molding, these materials undergo cross-linking by sulfur, a process called *vulcanizing*. Elastomers typically have low elastic moduli (stiffness) but high elasticity (stretch). Many, if stretched to twice their length (or much more) and then released, will snap back resiliently to their original length and shape. Thermoplastic elastomers are often used in place of rubber, and also may be used as additives in other thermoplastic resins for improving impact strength of rigid thermoplastics.

**Thermoplastic Structures.**—Thermoplastics can be classified by their structures into three categories:

*Amorphous:* These randomly structured materials are sometimes called *noncrystalline*. Such thermoplastics include acrylates, polycarbonate (PC), polystyrene, acrylonitrile-butadiene-styrene (ABS), styrene acrylonitrile (SAN), and polyvinylchloride (PVC) and its acetate copolymers.

*Semi-Crystalline:* Half or even more of the chains in crystalline thermoplastics are organized into tiny regular regions called *crystallites*. Representative members are nylons, acetals, polyethylene (PE), polypropylene (PP), and polyethylene terephthalate (PET). The crystallites are linked by amorphous zones; the resulting microstructures make these materials stronger and stiffer than similar materials without crystallinity. They generally possess higher end-use chemical and temperature resistance, and melt at higher temperatures, with more sharply defined melting ranges than amorphous resins.

*Liquid-Crystalline Polymers (LCP):* These contain crystal-like structures even in the melted resin. They have highly ordered, rod-like microstructures with good mechanical properties and high melting temperatures. They have the lowest shrinkage and warpage of the three types of thermoplastics.

**Copolymers, Blends, Additives, and Reinforcements.**—Plastics can be manipulated to modify mechanical, chemical, electrical, and cost characteristics.

*Copolymers:* Combining two or more different polymers can create a new material with properties that are quite different from those of the individual polymers (*homopolymers*) from which they are made. The copolymers are alloyed: mechanically blended from the two or more different polymers, often with special additives to make them compatible. These "alloys" are compounded so as to retain the most desirable characteristics of each constituent, especially impact strength and flame resistance. The properties usually are between those of the constituent homopolymer materials.

Copolymers such as SAN and ABS have long been commercially available. While tens of thousands of grades have been developed, less common blends and alloys are not always readily available in sheets or in standard industry stock shapes supplied for machining.

*Additives and Fillers:* Plastics compounds with desired sets of properties are produced from "neat" resins by adding many types of other chemicals and materials. Antioxidants, lubricants, pigments and dyes, fire retardants, and ultraviolet protectants in small percentages usually do not significantly alter mechanical and physical properties. *Larger* quantities of particulate fillers, such as wood flour or fibers, silica powder, glass or phenolic microballoons, and carbon and metal powders, are often used to tune density, thermal and electrical conductivity, mechanical properties, and unit cost. The use of such additives makes available many grades of a given plastic type from each commercial plastics manufacturer. Such additives can increase value in application but also may increase difficulty in recycling.

*Plasticizers:* Plasticizers soften and toughen the base resins with which they are compounded. While they are used in several resin families to some extent, polyvinyl chloride resins (PVC) and related plastics use plasticizers and many other additives in high percentages. These modified resins produce a myriad of grades of PVC from the familiar rigid pipe material (the dominant consumer of PVC) to wire coatings, flooring, and elastomerics used in footwear.

*Fibers:* Plastics, which have much lower moduli and strengths than glass and metals, can be made stiffer and stronger by incorporating glass, graphite or other fibers. Short chopped fibers in fractions up to about 40 percent, randomly oriented in molding compounds, provide substantial increases in stiffness and strength. But more dramatic improvements are obtained by incorporating longer fibers in carefully chosen directions, as in fishing rods and golf club shafts, where content of glass or carbon fiber may be more than 50 percent by weight. Historically, sturdy composites have been and still are made by compression molding multilayer canvas, paper, or cotton-felt sheets impregnated with phenolic or other thermosetting resins. (See also *Plastics Gears* on page 607.) A relatively recent reinforcement material is *oriented aramid* (aromatic polyamide, Kevlar), whose strength and modulus rival those of metals. Its most familiar use has been in "bullet-proof" vests for law-enforcement and military personnel.

**Foamed Plastics.**—Many rigid and flexible thermoplastic and thermosetting resin families are available as foamed plastics in the form of sheets, boards, bars, and cylinders, as well as molded shapes. The internal structure of plastic foam consists of tiny gas-filled cells bounded by shared membranes of the resin.

In *closed cell* foams, each cell is a tiny polyhedral cavity adjacent to, but not opening into, other cells. Foamed polystyrene hot liquid beverage cups are an example of closed cell foams. In *open cell* foams, the cells are not completely closed but instead form continuous connecting tunnels with other cells in all directions throughout the object. Kitchen sponges are open cell foams.

Densities of plastics foams range from 90 percent of the resin's unfoamed density down to just a few percent. Foaming may be accomplished physically by injection of a gas or volatile solvent during extrusion or molding, or by mixing a chemical blowing agent with the resin. When the mix is processed, the unstable blowing agent decomposes in the melt, generating bubbles. Foams have many uses: as low-cost space filling, cushioning, thermal and electrical insulation, and as the interlayer member of sandwich materials. Polystyrene, polyethylene, and polyurethane (rigid and flexible) are important foam families.

**Application and Physical Properties.**—Selecting a plastic for a specific application requires knowledge of many different kinds of properties of candidate materials. Density and mechanical properties will be important considerations in most cases. The plastics must be able to bear tensile, compressive, or bending loads; resist or allow twisting or stretching; survive impacts; withstand thousands or even millions of repeated stress cycles; and resist abrasion. Special service environments—extremely low or high temperatures and humidities, outdoor exposure, and the presence of active solvents and chemicals—will modify the selection criteria. Thermal or electrical properties, such as coefficients of thermal expansion, thermal or electrical conductivity, heat capacity (specific heat), dielectric strength, and dielectric-loss factor, may sometimes be important.

Since 1937, Committee D-20 of ASTM (formerly known as the American Society for Testing Materials) has developed hundreds of standards for plastics and their products. For example, ASTM Test D638, "Tensile Properties," prescribes the preparation and dimensions of specimens to be tested by pulling from both ends. Several useful mechanical-property values are obtained from this test: tensile modulus, yield strength, tensile strength at break, and elongation (percent stretching) at break.

Still, due to "skin effects" found in molded plastics, ASTM Test D790, "Flexure Testing of Plastics," generally is considered more accurate in determining flexural properties. Flexural moduli, usually similar to the elastic modulus *E*, actually tend to be somewhat higher in molded thermoplastics, as are the plastics' flexural strengths. While tensile and flexural specimens are typically $^1/_8$-inch (3.2-mm) thick, properties do vary with part thickness and processing conditions, so designs should be developed with appropriate safety factors and validation.

Even with homogeneous plastics, whose properties are expected to be about the same in all principal directions, certain processing conditions, such as stretching while the heat-softened plastic is cooling, can cause alignment *orientation* of the long polymer chains or crystallites in the stretching or flow direction. This will increase tensile strength in that direction but with some reduction in the perpendicular cross directions. Even more striking anisotropic effects are achieved with long-fiber reinforcements, in which the modulus and strength may be a factor of two or greater in the fiber direction than in the cross-fiber direction.

Suppliers of unreinforced sheet, bar, and rod stock for machining can relieve stress by controlled annealing of any residual stresses developed in processing, thereby preventing distortions related to the relief of residual strains during machining. Machining or heating of un-annealed stock can release residual stresses in the material, causing significant dimensional variations and warpage that can be difficult to resolve.

Some of the more important application properties are defined here.

*Density and Specific Gravity: Density* is the mass (or weight) per unit volume of a material, used mainly to calculate the mass of material required to make a part whose volume has been calculated from its dimensions. In the plastics industry, commonly used units are $lb/in^3$ or $g/cm^3$. *Specific gravity* is the ratio of a material's density to that of water, both measured at the same temperature, at 23° C (73.4° F) for plastics. Since the density of water at typical room temperatures is very close to 1.00 $gm/cm^3$, specific gravity and density measured in metric units are numerically almost equal. In US customary units, water density equals 0.0361 $lb/in^3$. Specific gravities of neat resins range from 0.83 for polymethylpentene to 2.20 for polytetrafluoroethylene (PTFE, Teflon). Because specific gravity is a ratio of densities of like units, it is dimensionless. See also *Specific Gravity* on page 375.

Unlike specific gravity, *specific volume* is not dimensionless, but is defined as the reciprocal of density, with units of $in^3/lb$ or $cm^3/g$.

*Shrinkage:* This is the difference in measurement between a linear dimension of a mold and the corresponding dimension of a plastic article made in the mold, divided by the mold dimension, and expressed in/in, mils/in, or percent. The dominant material property here is the *coefficient of linear thermal expansion* since the shrinkage is mainly due to cooling from the high temperature of the melt as it solidifies under pressure in the mold to room temperature after its removal from the mold. Expansion coefficients of plastics are much larger, 20 to over 100 times those of metals and ceramics. For type-66 nylon, which freezes at 260° C (500° F), shrinkage may be as much as 20 mils/in (0.02 cm/cm), i.e., 2 percent. Shrinkage can also be affected by other molding conditions and by additive content. Amorphous and liquid-crystalline thermoplastics have lower shrinkages than crystalline materials. Glass- or carbon-reinforced and mineral-filled compounds have lower shrinkage than their neat resins. For accurate assessment of shrinkage, test moldings at the target part thickness are advised before mold-cavity dimensions are finalized.

*Water Absorption:* This is the increase in weight of a test specimen due to absorption of water, expressed as a percentage of the original dry weight. Standard test specimens (ASTM Method D570) are first dried for 24 hours, then weighed before and after immersion in water at 23° C (73.4° F) for a specified time. Absorption of water affects not only the dimensions of plastics parts but also mechanical and electrical properties. Parts made from materials with low water absorption tend to have greater dimensional stability.

*Transparency:* This is a measure of the percentage of incident light that is transmitted through a test specimen. It is quantified as *luminous transmittance,* the ratio of transmitted to incident light. *Haze* measures the percentage of light entering a specimen that is scattered more than 2.5 degrees from the incident beam.

*Elasticity:* Most solid materials are initially elastic, such that when small stresses causing deformation are removed, the material will recover its original shape and dimensions. A more precisely formulated definition of "elastic" is "a material that obeys Hooke's law of proportionality of strain to stress up to its proportional limit" (see page 563). The *modulus of elasticity* is Hooke's law constant $E$, the quotient of stress divided by strain in the proportional region. In a tensile test, strain is the ratio of observed displacement, relative to the original length, reached under a given stress. Because plastics generally deviate from strict linearity, the moduli reported are usually "one percent moduli," that is, the quotient of stress and strain at 1 percent strain. For most materials the limit of recoverable strain is a few percent or less. For rubbers and a handful of thermoplastic elastomers, maximum recoverable strain may be as high as 100 percent or more. Stress and strain in relation to plastics part design is discussed on page 561.

*Plasticity:* This refers to the deformation exhibited when materials are stressed beyond their elastic limits without breaking (i.e., when stress exceeds the *yield point*). In this mode, there can be considerable flow within the microstructure of the material, and permanent changes in the dimensions and shape of the deformed article occur. Plastic deformation continues with small increases in stress until finally there is rupture.

Plasticity is exhibited to some degree by most unreinforced plastics. While parts are normally designed to operate in their material's elastic ranges, and plastic deformation may render a part useless for its designed function, plastic deformation *can* be used to prevent catastrophic failure with a fail-safe design. A few plastics have such compliant plasticity that they can be cold-formed by metalworking methods—rolled, stretched, or even forged into final products. This property in metals is called *ductility* and also applies to plastics.

*Toughness:* This property governs the ability of a material to absorb mechanical energy without cracking or breaking. Often it depends on the rate at which the specimen is deformed. In a standard tensile test (ASTM D638), the strain rate is fairly low—from 0.15 to 15 (in/in)/min. In a tensile-impact test (ASTM D1822) the specimen is abruptly stretched to failure in milliseconds at a strain rate of about 20,000 (in/in)/min. The tensile tester of D638 generates a graph of stress (psi) versus strain (in/in). The area under this curve is the energy in ft-$lb_f/in^3$ absorbed by the gage volume as it is stretched and finally broken. The impact tester, on the other hand, directly determines how many foot-pounds of energy are needed to stretch and break the specimen. That value, too, can be converted to energy per unit of specimen volume, but the impact energy usually will be very different from the tensile energy to failure. A widely used test of toughness for plastics is the Izod impact test (ASTM D256), in which a falling pendulum strikes and breaks off the upper half of a cantilever test bar whose leading face has a 0.1 in. (2.54 mm) deep, V-shaped notch across it just above the clamping point. This test (the oldest in ASTM's Plastics list, first published in 1926) indicates not only toughness but also the *notch-sensitivity* of the plastic. The test result is expressed in ft-$lb_f$/in of notch width. Izod-test values for $^1/_8$-inch (3.2-mm) wide specimens of neat thermoplastics range from 0.4 for crystal polystyrene or wood-flour-filled phenolics to 14 for polycarbonate and phenolics with a high content of glass fiber. Particulate fillers may increase or decrease Izod impact strength; fibrous reinforcements usually increase it. Thus, applying tabulated toughness values to plastics part design is not a simple task.

*Brittleness:* Lack of toughness, characterized in plastics by Izod impact test values below 1.0 ft-$lb_f$/in (53 J/m) or by abrupt fracture with little or no yield in tensile tests, is referred to as brittleness.

*Notch Sensitivity:* This measurement indicates the ease with which a crack progresses through a material from an existing notch, crack, or sharp corner.

*Hardness:* The ability of a material to resist deformation by surface indentation or abrasion, hardness is often related to modulus, strength, and wear resistance. Several ASTM test methods are in use for plastics, the Rockwell hardness test (ASTM D785) being a popular one. One of three hardened steel balls is pressed into the test specimen, first with a low force, then a much higher one, while a micrometer indicates the increase in depth of impression. Five scales are available for plastics: R, L, M, E, and K, in order of increasing hardness, requiring progressively smaller balls and stronger forces. The scale *must* be identified in the report of Rockwell hardness numbers. When an article is expected to suffer abrasion in service, the abrasion loss of candidate plastics may be experimentally determined with an abrasion machine (ASTM D1242).

*Homogeneous* plastics have the same composition throughout a part down to the microstructural level, but not to the molecular level as in simple chemical compounds. A $^3/_{16}$ inch (5 mm) BB-sized pellet of a neat resin contains many long-chain molecules of various chain lengths. However, even compounds containing several additives, pigments, and powdery fillers behave in molded articles as if they were homogeneous.

*Heterogeneous (inhomogeneous)* plastics vary in composition and structure. A compound reinforced with centimeter-long, randomly oriented, chopped glass fibers will not look homogeneous under 50× magnification, yet it may be safely treated as such in the design of parts that are larger than the test samples whose properties have been measured.

*Isotropy and Anisotropy:* These properties are related to homogeneous and heterogeneous types of plastics. In *isotropic* materials, measured properties are equal in all directions. Cast metals and unfilled, amorphous plastics tend to be isotropic. Processing effects such as flow orientation that do not change homogeneity can create substantial directional differences in properties. This effect is called *anisotropy*. In plastic films made by extruding molten resin onto chilled rolls, the films are usually stronger in the "stretch" direction, weaker in the lateral or cross direction. Laminates made of laid-up reinforcing fibers or woven fabric can be nearly biaxially isotropic in the laminate plane, but their interlaminar shear is low because no fibers cross through the plane. Extremely anisotropic are unidirectional laminates in which all the fibers are parallel, as in golf club shafts reinforced with graphite fibers.

Wood (see page 380) is a natural anisotropic material, strong and stiff in the grain (growth) direction, less so in the perpendicular and tangential directions. Commercially important fibers, both natural and synthetic, are also highly anisotropic with their greatest strength and moduli aligned with length.

As the degree of anisotropy increases, the number of physical constants (moduli) required to define a material's stress-strain behavior also increases, up to a maximum of 21. Absence of such complete information on complex laminates and the inadequacy of simple analysis methods point to the need for extensive end-use testing of plastics parts before approval of critical application requirements. Computer modeling using finite element analysis (FEA) techniques can help in the successful implementation of these materials. Detailed properties data are required for reliable and accurate modeling.

In addition to mechanical properties, isotropy and anisotropy can affect manufactured plastic part dimensions. Anisotropic shrinkage caused by patterns of flow during processing is important in molding crystalline and fiber-reinforced plastics, for which different shrinkage values are experienced in the flow and cross-flow directions. These values are of most concern to the tool designer and molder; however, the existence and severity of anisotropic shrinkage must be considered when a material is chosen for a part having tight tolerances.

*Coefficient of Friction:* This property comes into play in plastics gears and bearings and other applications where low dynamic (kinetic) friction is desired between moving surfaces in contact. In general, the observed friction will be different for different pairs of materials, so a test report must identify both materials according to ASTM D3028. Coefficients of friction also change with temperature. Materials such as 66-nylon, acetal resin, and polytetrafluoroethylene containing molybdenum disulfide, which have low self-coefficients, typically enjoy long operating life in unlubricated gears and bearings.

*Significance of Elasticity, Homogeneity, and Isotropy in Plastics:* A conventional structural material is low-carbon steel, which behaves as an isotropic and elastic material at temperatures below dull red heat. For this metal, $E = 30{,}000$ kpsi (206.8 GPa) and Poisson's ratio $\omega = 0.32$. The elastic limit stress $\sigma$ is about 40 kpsi (276 MPa), from which Hooke's law (see page 563) indicates that its elongation in tension at that limit is only $\sigma/E$ or 0.13 percent. Its stress-strain graph up to the elastic limit is practically a straight line. If it is loaded up to half its elastic limit and maintained at that load for a couple of decades, when the loading is removed, the 0.065 percent elongation will be quickly and completely recovered. The rate and duration of loading in normal applications do not affect the steel's apparent properties.

Plastics, in contrast, are *viscoelastic* materials. They display properties between those of elastic solids and viscous fluids. Thus, their stress-strain graphs tend not to be linear. Even at low stresses, plastics are sensitive to rate of loading and to temperature, as shown generically in Fig. 1. Curves A and B identify the stress-strain behavior of brittle materials that will tend to fail suddenly under increasing strain. By comparison, curves C through E show how increasing temperatures both increase ductility and extend the strain to failure curve, albeit with decreased stiffness.

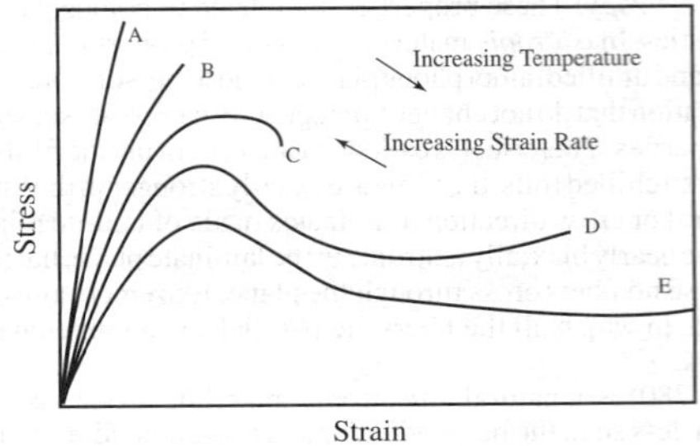


Fig. 1. Generic Tensile-Test Graphs for Plastics, Showing Typical Range of Behavior from Brittle to Highly Ductile with Increasing Temperatures

Structural analysis during design of components uses two independent constants, Young's modulus ($E$) and Poisson's ratio ($\nu$), but these two constants are sufficient only for elastic, isotropic materials that respond linearly to loads (when load is proportional to deformation). Designers often use the same values for these constants everywhere in the structure, which is correct only if the structure is homogeneous.

Assumptions of linear elasticity, isotropy, and homogeneity are reasonable for many analyses and provide a good starting point, but use of these assumptions can lead to significant design errors with plastics, particularly with glass-reinforced and liquid crystalline polymers, which are highly anisotropic. In this discussion, plastics are assumed to be linearly elastic, homogeneous, and isotropic to allow a simpler presentation of mechanical properties in line with the data provided in plastics manufacturers' marketing data sheets. The standard equations of structural analysis (bending, torsion, hoop stress, etc.) also require these assumptions.

*Creep:* When plastics are subjected to sustained high loads, they *creep*. That is, they tend to slowly deform (flow), exhibiting the viscous side of their nature. If the load is removed, the stressed object will usually not return completely to its original shape and dimensions. Creep is more rapid at higher temperatures, even at temperatures expected in normal service. Laboratory stress-strain curves obtained according to ASTM D638 at moderate strain rates and 23° C (73.4° F) or other temperatures will resemble those in Fig. 2. Most plastics have no true proportional limits, instead exhibiting curved dependence of stress versus strain even at strains less than 1 percent. Tangent moduli from D638 therefore tend to be subjective and inaccurate when used for predicting deformation. The isometric diagram of Fig. 2 shows the dependence of stress in Styron 475, a rigid modified polystyrene, at various strains and times. Note the logarithmic time scale (0.01 h = 36 s, 10,000 h = 1.14 yr). The curve on the left panel, for 0.01 h, is the stress-strain graph one might obtain at one of the slower testing rates of ASTM D638. The stress ordinate at the left edge, front, i.e., 600 psi (4.1 MPa) for a strain of 0.15 percent, corresponds to a short-time modulus for this resin of 400 psi (2.8 MPa), in agreement with values reported elsewhere. When this small strain was maintained for 1000 hours (6 weeks), the stress had decayed along the first curve (left rear to right front) to about 400 psi (2.8 MPa), two-thirds of its initial value.

Some hygroscopic resins can absorb water from the air, with stress-strain behavior changing from curve B or C of Fig. 1 when very dry to curve D or E at high moisture contents. Polyamides (nylons), classified as engineering resins with superior properties, exemplify this moisture dependence, picking up about 0.3 percent in very dry climates, and 2.5 percent or more in very humid ones.

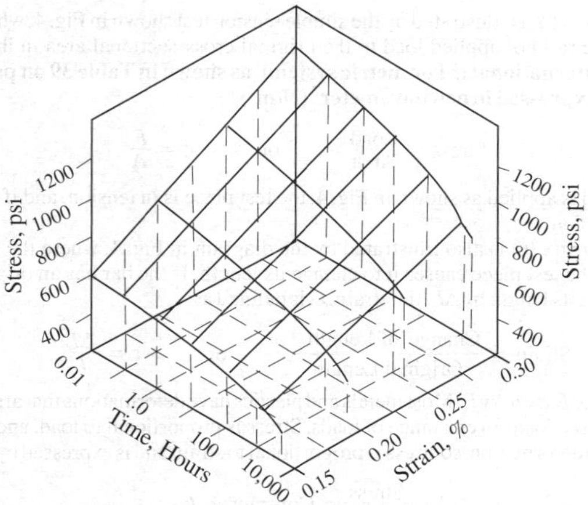


Fig. 2. Isometric Plot for High-Impact Polystyrene Showing Stress Response to Imposed Strains over Three Decades

Modulus and strength values reported from ASTM D638 testing should not be used in the design of plastics products that must sustain loads over long periods; the addition of reinforcements, such as glass fiber (GF), is observed to significantly increase the modulus and strength of both thermoplastics (e.g., PP) and thermosets (e.g., epoxy). Properties for various plastics are included in the plastics properties table on page 385.

**The Basics of Stress and Strain in Plastics.—**The mechanical behavior of plastics is similar in many ways to that of metals, but there are important differences that must be kept in mind in the design of plastics products and parts.

*Stress:* A three-dimensional body having a balanced system of external forces $F_1$ through $F_5$ acting on it such that the body is at rest is shown in Fig. 3. Such a body develops internal forces to transfer and distribute the external loads. If the body is cut at an arbitrary cross section and one part is removed, as shown at the right in Fig. 3, a new system of forces acting on the cut surface is developed to balance the remaining external forces. Similar forces (stresses) exist within the uncut body.

Stresses must be defined by both magnitude and direction. The stress $S$ acting in the direction shown in Fig. 3 on point $P$ of the cut surface has two stress components. One of these components, σ, acts perpendicular to the surface and is called a *normal stress*. The other stress, τ, acts parallel to the surface and is called a *shear stress*.

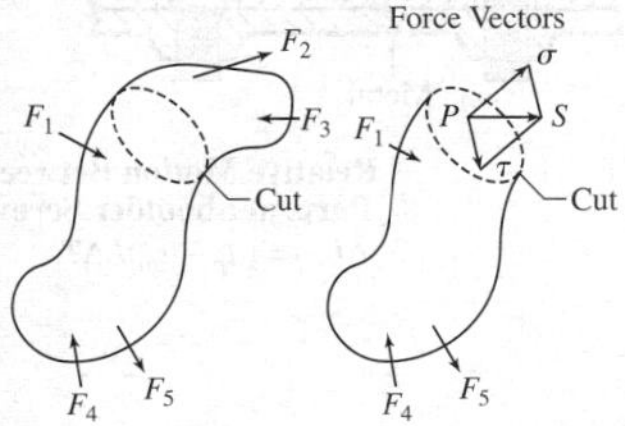


Fig. 3. Internal Forces and Stresses in a Body

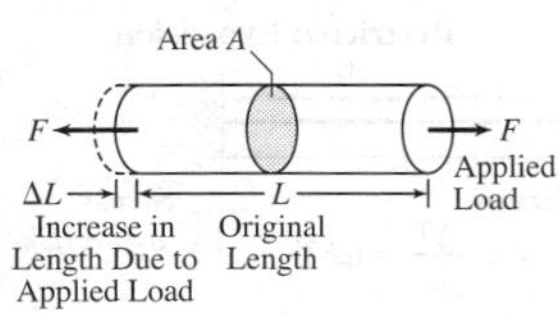


Fig. 4. Simple Tension Load

*Normal stress* (σ) is illustrated by the simple tension test shown in Fig. 4, where the direct stress is the ratio of applied load to the original cross-sectional area in $lb_f/in^2$. **In the Système International (SI or metric system), as shown in Table 39 on page 2872, the stress σ is expressed in newtons/meter$^2$ (N/m$^2$).**

$$\text{Stress} = \frac{\text{Load}}{\text{Area}} \qquad \text{or} \qquad \sigma = \frac{F}{A} \tag{1}$$

If the load is applied as shown in Fig. 4, the test piece is in tension, and if reversed, it is in compression.

*Normal strain* (ε) is also illustrated by the diagram in Fig. 4, where the load or stress applied to the test piece causes it to change its length. If the bar has an original length $L$ and changes its length by $\Delta L$, the strain, ε, is defined as

$$\text{Strain} = \frac{\text{Change of Length}}{\text{Original Length}} \qquad \text{or} \qquad \varepsilon = \frac{\Delta L}{L} \tag{2}$$

*Modulus of Elasticity E:* Most metals and plastics have deformations that are proportional to the imposed loads over a range of loads. Stress is proportional to load, and strain is proportional to deformation, so stress is proportional to strain and is expressed by Hooke's law:

$$\frac{\text{Stress}}{\text{Strain}} = \text{Constant} = E \tag{3}$$

The constant $E$ is called by various names: the *modulus of elasticity*, *Young's modulus*, or, in the plastics industry, *tensile modulus*. Referring to Fig. 4, tensile modulus is given by the formula:

$$E = \frac{\sigma}{\varepsilon} = \frac{F/A}{\Delta L/L} = \frac{FL}{A\Delta L} \tag{4}$$

Thus, the modulus is the slope of the initial portion of the stress-strain curve.

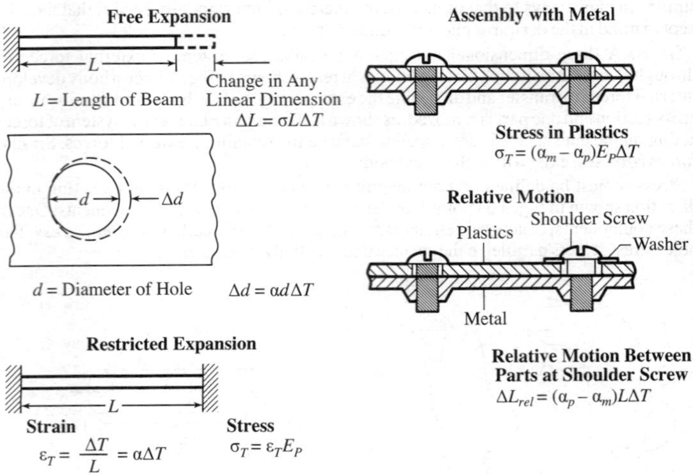


Fig. 5. Thermal Expansion Equations for Various Combinations of Materials and Situations

The linear portion of the stress-strain curve for many plastics is difficult to locate, and it is necessary to construct a straight line tangent to the initial portion of the curve to use as a modulus. The shape of a line so obtained is called the *initial modulus*. In some plastics the initial modulus can be misleading, owing to the nonlinear elasticity of the material. Some suppliers therefore provide the so-called 1 percent secant modulus, which is the ratio of stress to strain at 1 percent strain on the stress-strain curve. In the illustration of typical stress-strain curves in Fig. 6, the secant modulus at the point *E* is the slope of the line *OE*.

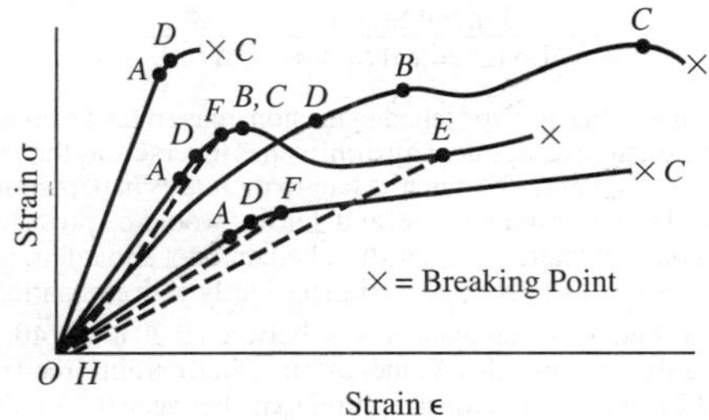


Fig. 6. Typical Stress-Strain Curves

*Secant modulus* is the ratio of stress to corresponding strain at any point on the stress-strain curve (see *Modulus of Elasticity E*).

*Proportional limit* is the greatest stress at which a material is capable of sustaining the applied load without losing the proportionality of stress to strain. This limit is the point on the stress-strain curve where the slope begins to change, as shown at *A* on each of the curves in Fig. 6. Proportional limit is expressed in $lb_f/in^2$ (MPa or GPa).

*Yield point* is the first point on the stress-strain curve where an increase in strain occurs without an increase in stress, indicated by *B* on some of the curves in Fig. 6. The slope of the curve is zero at this point; however, some materials do not have a yield point.

*Ultimate strength* is the maximum stress a material withstands when subjected to a load, and is indicated by *C* in Fig. 6. Ultimate strength is expressed in $lb_f/in^2$ (MPa or GPa).

*Elastic limit* is indicated by *D* on the stress-strain curve in Fig. 6 and is the level beyond which a material is permanently deformed when the load is removed.

Although many materials can be loaded beyond their proportional limit and still return to zero strain when the load is removed, some plastics have no proportional limit in that no region exists where the stress is proportional to strain (i.e., where the material obeys Hooke's law).

*Yield strength* is the stress at which a material shows a specified deviation from stress to strain proportionality. Some materials do not show a yield strength clearly, and it may be desirable to choose an arbitrary stress level beyond the elastic limit, especially with plastics that have a very high strain at the yield point, to establish a realistic yield strength. Such a point is seen at *F* on some of the curves in Fig. 6 and is defined by constructing a line parallel to *OA* at a specified offset strain *H*. The stress at the intersection of the line with the stress-strain curve at *F* would be the yield strength at *H* offset. If *H* were at 2 percent strain, *F* would be described as the yield strength at a 2 percent strain offset.

*Poisson's Ratio* ν is the ratio of the lateral contraction to longitudinal elongation. Under a tensile load, a rectangular bar of length *L* with sides of widths *b* and *d* lengthens by an amount $\Delta L$, producing a longitudinal strain of

$$\varepsilon = \frac{\Delta L}{L} \tag{5}$$

The bar is reduced in its lateral dimensions and the associated lateral strains will be opposite in sign, resulting in

$$\varepsilon = -\frac{\Delta b}{b} = -\frac{\Delta d}{d} \tag{6}$$

If the deformation is within the elastic range, Poisson's ratio of the lateral to the longitudinal strains will be constant. The formula is:

$$\nu = \frac{\text{Lateral Strain}}{\text{Longitudinal Strain}} = \frac{\Delta d/d}{\Delta L/L} \tag{7}$$

When a rod like that in Fig. 4 is stretched in tension, mass must be conserved; if there is no change in diameter, the average density diminishes inversely as the strain increases. If the area decreases in percentage as much as length increases in percentage, the stretched volume would equal the original volume, and density would be preserved. In fact, most real materials do contract laterally when stretched, but not enough to preserve constant density; therefore, density usually does diminish slightly with elongation.

Values of $\nu$ for most engineering materials lie between 0.20 and 0.40, and these values hold for unfilled rigid thermoplastics. Values of $\nu$ for filled or reinforced rigid thermoplastics fall between 0.10 and 0.40 and for structural foam between 0.30 and 0.40. Rigid thermoset plastics have Poisson's ratios between 0.20 and 0.40, whether filled or unfilled, and elastomers can approach 0.5.

Table 1 lists values of $\nu$ determined at 73° F (23° C) for some plastics by measuring of lateral contraction and longitudinal elongation in tensile tests.

**Table 1. Poisson's Ratio at 23° C for Resins and Compounds**

| Plastic | Poisson's Ratio, $\nu$ | Plastic | Poisson's Ratio, $\nu$ |
|---|---|---|---|
| ABS, unfilled | 0.35 | Polycarbonate | 0.37 |
| Acetal | 0.35 | Polyester, thermoplastic, gfr[a] | 0.35–0.41 |
| Acrylic (PMMA) | 0.35–0.40 | Polyester, thermoset, gfr[a] | 0.24 |
| Alkyd, mineral filled | 0.17 | Polyetherimide | 0.36 |
| Diallyl phthalate | 0.27 | Polyethylene, HD | 0.45 |
| FEP resin | 0.48 | Polyphenylene ether | 0.35 |
| PC, gfr[a] | 0.38 | Polyphenylene oxide | 0.27–0.38 |
| PCTFE | 0.33 | Polypropylene | 0.42 |
| Phenolic, cellulose-filled | 0.24 | Polystyrene, cross-linked | 0.35 |
| Phenolic, gfr[a] | 0.2 | Polyethersulfone | 0.4 |
| Polyamide, min-filled, gfr[a] | 0.32 | PTFE | 0.46 |
| Polyamide 6, dry, gfr[a] | 0.36 | PVF | 0.4 |
| Polyamide 6/6, dry, gfr[a] | 0.32 | PVDF | 0.34 |

[a] gfr = glass-fiber reinforced, 30 or 33%.

Mineral fillers and glass-fiber reinforcements in plastics reduce Poisson's ratio. In anisotropic structures such as uni- and bi-directional laminates, Poisson's ratio may be different in orthogonal (i.e., perpendicular) directions. Usually, the value perpendicular to the reinforcing fibers will be the one most likely to be relevant. Vulcanized gum rubber has exceptional behavior; its Poisson ratio = 0.5, and it conserves density. This must be kept in mind in designing items to be compressed, such as rubber cushions and bumpers. When a rubber specimen is compressed it is shortened and its sides must be free to bulge out; otherwise the incompressibility of the rubber would not allow it to deform at all.

*Shear stress* is described on page 213. Any block of material is subject to a set of equal and opposite shearing forces $Q$. If the block is envisaged as an infinite number of infinitesimally thin layers as shown diagrammatically in Fig. 7, it is easy to imagine a tendency for one layer subject to a force to slide over the next layer, producing a shear form of deformation or failure. The shear stress $\tau$ is defined as

$$\tau = \frac{\text{Shear Load}}{\text{Area Resisting Load}} = \frac{Q}{A} \tag{8}$$

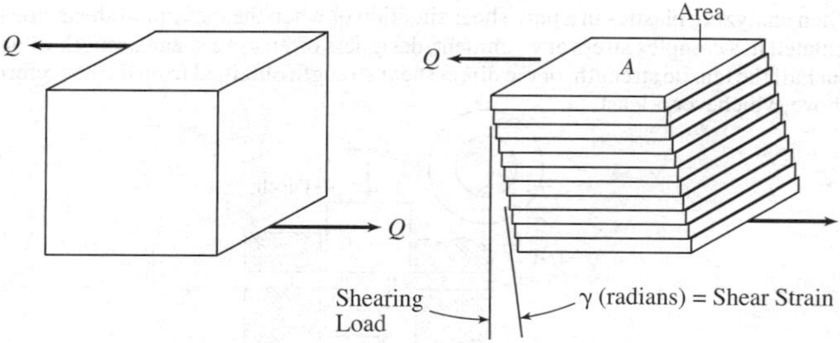


Fig. 7. Shear Stress Is Visualized as a Force $Q$ Causing Infinitely Thin Layers of a Component to Slide Past Each Other, Producing a Shear Form of Deformation

Shear stress is always tangential to the area on which it acts. Shearing strain is the angle of deformation $\gamma$ and is measured in radians.

*Example:* In Fig. 8a, shear stress in the overlapped zone equals the tensile force $F$ divided by the contact area of the adhesive. That is, $\tau = F/(WL)$. As $F$ is increased, the joint will likely fail in shear, either in the adhesive layer or in the material of one of the bars, depending on which material has the greater shear strength.

In Fig. 8b, a similar lap joint is joined with a rivet of diameter D. In this case, the area undergoing shear in the rivet is $\pi D^2/4$, a much smaller area. But what if the rivet were aluminum and much stronger than the plastic being joined? A second, very different shear area comes into play, shown as hatched cross-sections in Fig. 8b. The total shear area in this case equals $(L\text{-}D)T$. Thus, the shear stress $\tau$ exerted by the rivet and trying to push the plug out of each bar end is $F/((L\text{-}D)T)$. As the force $F$ is increased, the mode of failure in the rivet or the bar end will be decided by which of these shear stresses first exceeds the shear strength of the material on which it is acting.

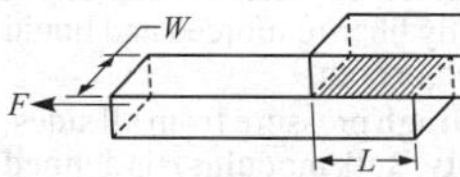


Fig. 8a. Adhesive Joint

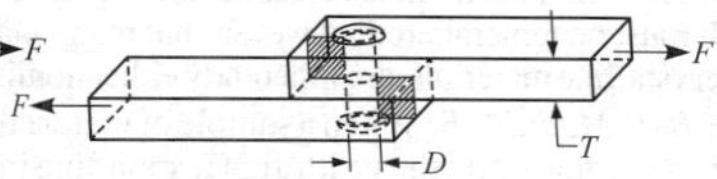


Fig. 8b. Riveted Lap Joint

Geometries like those in Fig. 8a and Fig. 8b are not useful for determining shear *modulus* because there is very little recordable shear deformation prior to failure at the shear strength. However, the very simple shear geometry shown in Fig. 9 is used in ASTM D732 test to measure the shear strengths of plastics. The washer-shaped test sample has an outer diameter of 2 inches (50.8 mm) and a central hole of about $^9/_{16}$ inch (14.3 mm) to allow the $^1/_2$-inch (12.7-mm) bolt to pass freely through the specimen. The thickness may be any measured value between 0.05 and 0.5 inch (1.27 and 12.7 mm). In this test, the shear area is given by $\pi \times$ punch diameter $\times$ disk thickness. The results of such tests are often described in manufacturers' marketing data sheets as the shear strength of the material. The shear strength reported from such a test is not a pure shear strength because a considerable part of the load is transferred by bending or compressing, or both, rather than by pure shear, and results can be affected by the susceptibility of the material to the sharpness of the load faces in the test apparatus. Thus, the test cannot be used to develop shear stress-strain curves or to determine the shear modulus. Shear strengths reported from this test have coefficients of variation of about 4 percent.

When analyzing plastics in a pure shear situation or when the maximum shear stress is calculated in a complex stress environment, designers often use a shear strength value of about half the tensile strength, or the direct shear strength obtained from the test referred to above, whichever is least.

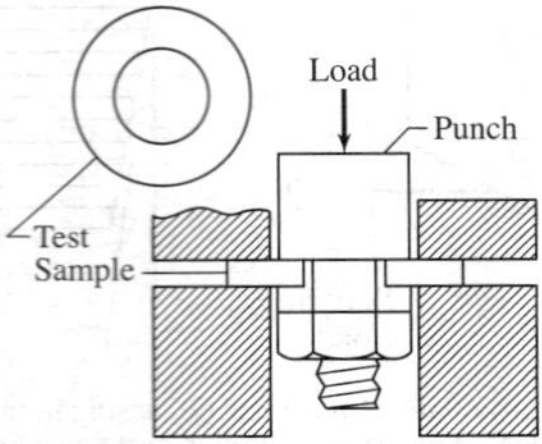


Fig. 9. Setup for Testing Shear Strength of Plastics in ASTM Test D732. Specimen may be Molded or Cut from Sheet

*Shear modulus G* is a constant otherwise called the *modulus of rigidity*. For materials that behave according to Hooke's law, it is directly comparable to the modulus of elasticity used in direct stress calculations. The constant is derived from

$$G = \frac{\text{Shear Stress}}{\text{Shear Strain}} = \frac{\tau}{\gamma} \tag{9}$$

Material suppliers can furnish values of $G$ for their materials tested according to ASTM D1043 at various temperatures by a torsion test.

*Relating Material Constants:* Although only two material constants are required to characterize a material that is linearly elastic, homogeneous, and isotropic, three such constants have been introduced here. These three constants are tensile modulus $E$, Poisson's ratio $\nu$, and shear modulus $G$, and they are based on elasticity principles:

$$\frac{E}{G} = 2(1+\nu) \tag{10}$$

This relationship holds for most metals and is generally applicable to most thermoplastics. It must be remembered, however, that many plastics, particularly fiber-reinforced and liquid crystalline materials, are inherently either nonlinear, anisotropic, or both.

*Bulk Modulus B:* When a sample of material is subjected to high pressure from all sides, its volume diminishes with a corresponding increase in density. Bulk modulus $B$ is defined as the increase in pressure divided by the fractional decrease in specific volume, $V$:

$$B = \frac{\Delta P}{(\Delta V)/V} \rightarrow \frac{dP}{d\ln V} \tag{11}$$

Since $\rho = \frac{1}{V}$, it follows that

$$\frac{1}{B} = \frac{(\Delta\rho)/\rho}{\Delta P} \rightarrow \frac{d\ln\rho}{dP} \tag{12}$$

The dimensions of $B$ are pressure per unit of fractional volume contraction. In English units $B$ is in Mpsi/(in$^3$/in$^3$), or just Mpsi; the SI units for $B$ are GPa/(cm$^3$/cm$^3$) or just GPa. The last equation indicates that what happens to volume (and specific volume) also happens in the same degree to density, but with the opposite sign, i.e., as the volume increases, the density decreases and vice versa.

For isotropic elastic materials it has been shown that bulk modulus is linked to Young's modulus and Poisson's ratio by the equation

$$B = \frac{E}{3(1-2\nu)} \tag{13}$$

Thus, all three moduli, $E$, $G$, and $B$, are linked through Poisson's ratio.

In Fig. 6 the stress, sometimes called the *engineering stress*, is calculated from an increasing load $F$, acting over a constant area $A$. Because the cross-sectional area is reduced with most materials, use of that smaller cross-sectional area in the calculation yields what is called the *true stress*. In addition, the direct strain referred to earlier, that is, the total change in length divided by the original length, is often called the *engineering strain*. The true strain would be the instantaneous deformation divided by the instantaneous length. Therefore, the shape of such a stress-strain curve would not be the same as a simple stress-strain curve. Modulus values and stress-strain curves are commonly based on engineering stress and strain.

**Other Measures of Strength and Modulus.—**Tensile and compression properties of many engineering materials, which are treated as linearly elastic, homogeneous, and isotropic, are often considered to be identical so as to eliminate the need to measure properties in compression. Also, if tension and compression properties are identical, under standard beam bending theory, there is no need to measure the properties in bending. In a concession to the nonlinear, anisotropic nature of most plastics, these properties, particularly flexural properties, are often reported on material data sheets.

*Compression Strength and Modulus:* Because of the relative simplicity of testing in tension, the elastic modulus of a material is usually measured and reported as a tension value. For design purposes it often is necessary to know the stress-strain relationship for compression loading. With most elastic materials at low stress levels, the tensile and compressive stress-strain curves are nearly equivalent. At higher stress levels, the compressive strain is less than the tensile strain. Unlike tensile loading, which usually results in a clear-cut failure, stressing in compression produces a slow and indefinite yielding that seldom leads to failure. Because of this phenomenon, compressive strength is customarily expressed as the amount of stress in $lb_f/in^2$ (or Pa) required to deform a standard plastics test specimen to a certain strain. Compression modulus is not always reported because defining a stress at a given strain is equivalent to reporting a secant modulus. If a compression modulus is given, it is usually an initial modulus.

*Bending Strength and Modulus:* When a beam of rectangular cross section is bent under a vertical load midway between the beam supports, the bottom surface is stretched in tension while the top surface is compressed. Extending from end to end along the vertical center of the beam is a plane of zero stress called the *neutral axis*. Whatever the shape of the cross section—constant along the beam—the neutral axis is located at the center of gravity (centroid) of that cross-sectional area. The theory of bending for simple beams makes the following assumptions: the beam is initially straight, unstressed, and symmetric; the material is linearly elastic, homogeneous, and isotropic; the proportional limit is not exceeded; Young's modulus for the material is the same in tension and compression; and all deflections are small so that cross sections remain planar during bending.

For a beam of length $L$ with a rectangular section of width $b$ and depth $h$ carrying a central load $F$, the formula for the maximum bending stress $\sigma$ at the bottom surface is

$$\sigma = \frac{3FL}{2bh^2} \tag{14}$$

and the formula for the maximum deflection at the center of the beam is

$$y = \frac{FL^3}{4Ebh^3} \tag{15}$$

Note that, if the deflection equation is solved for $E$, Young's modulus, the modulus can be estimated by measuring the deflection of a beam under a known central force:

$$E = \frac{FL^3}{4ybh^3} \tag{16}$$

This is the basis for ASTM Test D790 for the flexural modulus and strength of plastics. The flexural modulus reported is usually the initial modulus from the load/deflection curve. Most plastics parts undergo bending, so use of flexural values, which tend to be higher than tensile values for many molded plastics, should give more conservative and hence trustworthy results in product design.

*Rate Dependence of Mechanical Properties:* Tensile and flexural data in manufacturers' literature are measured at specific displacement rates. These rates are usually not consistent with the loading environment encountered in use of the product. The same plastics material, under differing rates or in other environmental conditions, can produce different stress-strain curves. Designers should be aware of the loading rates in specific applications and request the appropriate end-use testing.

**Time-Related Mechanical Properties.**—Mechanical properties discussed previously were related to loads applied gradually and applied for short periods. Long-term creep, cyclic fatigue, the effects of repeated high-speed impact, and other end-use scenarios should be considered.

*Creep* is defined as increasing strain over time at constant stress. The rate of creep for a given material depends on applied stress, temperature, and time.

Creep behavior of a material is an important, even crucial, issue with plastics when parts are to be subjected to loads for extended periods and the maximum allowable deflection is critical. To determine creep behavior, test samples may be loaded in tension, compression, or flexure in a constant-temperature environment. Under constant loads, deflection is recorded at regular intervals over suitable periods. results are generally obtained for four or more stress levels and recorded as creep curves of strain versus logarithmically scaled time. In general, crystalline materials have lower creep rates than amorphous plastics below glass transition temperatures. Glass reinforcement substantially improves the creep resistance. Fig. 10 shows the great differences in creep-rupture behavior that exist among different plastics at several temperatures.

*Apparent or Creep Modulus:* If the deflection of a part subjected to continuous loading is calculated using the short-test modulus of elasticity $E$, results are likely to be disappointing or even disastrous because the effects of creep have not been considered. If the stress level and temperature are known and creep curves are available for the temperature in question, an apparent, or creep, modulus $E_{app}$ can be calculated from the creep curves by the formula: $E_{app} = \sigma/\varepsilon_c$ where $\sigma$ is the calculated stress level and $\varepsilon_c$ is the strain from the creep curve at the expected time and temperature. This value $E_{app}$ can be used instead of $E$ in predicting the maximum beam deflection from the traditional equation given earlier.

Manufacturers' data often include curves of creep modulus (or log creep modulus) versus log time at either constant stress or constant strain derived from creep data. The bar graphs of Fig. 11 show 1000-hour creep moduli for several engineering resins at 73° F (23° C). This kind of information may also be provided as tables of values at constant stress and temperature for various time periods.

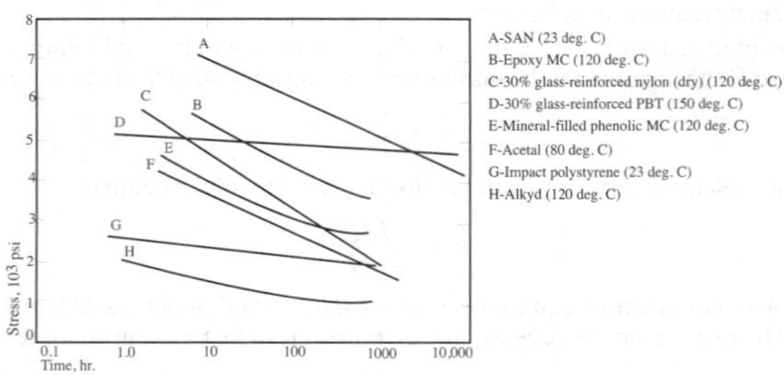


Fig. 10. Plots Showing Diminishing Strengths of Several Plastics as Creep Period is Extended (Curves are Steeper at Higher Temperatures)

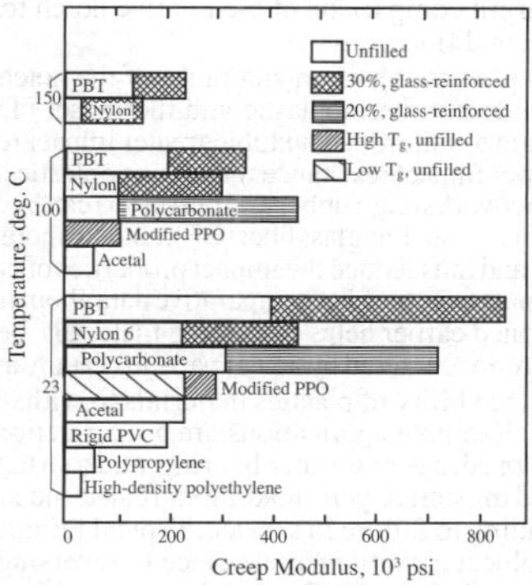


Fig. 11. 1000-Hour Creep Moduli of Several Plastics at Room Temperature and Higher

Creep rupture data are obtained in the same manner as creep data except that higher stresses are used and time is measured to failure. Such failures may be brittle or ductile with some degree of necking. Results are generally plotted as log stress versus log time to failure.

*Stress relaxation* occurs when plastics parts are assembled into a permanent deflected condition, as in a press fit, a bolted assembly, or some plastics springs. Under constant strain over a period of time the stress level decreases due to the same internal molecular movement that produces creep. Stress relaxation is important with such applications as bolt preloading and springs, where loading must be maintained. The relaxation can be assessed by applying a fixed strain to a sample and measuring the load over time. A relaxation modulus similar to the creep modulus can be derived from the relaxation data. Relaxation data are not as readily available as creep data, but the decrease in load due to stress relaxation can be approximated by using the creep modulus $E_{app}$ calculated from the creep curves.

Plastics parts often fail due to imposition of excessive fixed strains over extended periods of time, for example, a plastics tube that is a press fit over a steel shaft. No relaxation rupture equivalent to creep rupture exists, so for initial design purposes, a strain limit of 20 percent of the strain at the yield point or yield strength is suggested for high-elongation plastics. For low-elongation brittle plastics that have no yield point, 20 percent of the elongation at break is also recommended. Higher or lower property limits may also be indicated in manufacturers' data on specific materials. Still, these figures should be regarded only as guidelines for development of initial design concepts; prototype parts should be thoroughly tested under end-use conditions to confirm the suitability of the design in long-term use.

Extrapolating creep and relaxation data must be done with caution. When creep and relaxation data are plotted as log property against log time, the curves are generally less pronounced, facilitating extrapolation. While this procedure is common practice, extrapolation should not exceed one unit of log time, and the strain limit of 20 percent of the yield or ultimate strength mentioned above should not be exceeded.

*Impact loading* describes a situation in which a load is applied extremely rapidly. Any moving body has kinetic energy, and when the motion is stopped by a collision, the energy is dissipated. Ability of a plastic part to absorb energy is determined by the shape, size, thickness, and type of material. Impact-testing methods now available do not provide designers with information that can be used analytically. The tests can be used for comparing relative notch sensitivity or relative impact resistance, and so can be useful in choosing a series of materials to be evaluated for an application or in grading materials within a series. Impact testing by the Izod and Charpy (ASTM Method D6110) methods, in which a pendulum arm is swung from a certain height to impact a notched test specimen, is the most widely used for measuring impact strength. Impact with the test specimen reduces the energy remaining in the arm, and this energy loss is recorded in ft-$lb_f$ (J). The value of such tests is that they permit comparison of the relative notch toughness of two or more materials under specific conditions.

Table 2a and Table 2b provide the mean and range of the notched Izod impact strength (measured in J/cm) for different thermoplastic and thermoset (TS) materials. In general, materials with higher strain to failure will exhibit greater impact resistance. Thermosetting materials tend to have lower impact resistance than thermoplastics, though thermosets with flexible and unsaturated networks (e.g. rubber and polyester) can have high impact resistance. The addition of reinforcements, such as glass fiber (GF), tends to increase modulus and strength but lower strain to failure and thus reduce the impact properties of composite systems. While such Izod and Charpy impact tests provide comparative data about material performance, the tensile impact test mentioned earlier helps in ranking materials, because it represents more realistic conditions that are encountered by actual parts in certain applications.

*Fatigue tests* measure the ability of plastics materials to withstand repeated stresses or other cyclic phenomena. Example applications are a snap-action or snap-fit latch that is frequently opened and closed, a gear tooth, a bearing, and a structural component subject to vibration or to repeated impacts. Cyclic loading can cause mechanical deterioration and progressive fracture, leading to failure in service. Typical fatigue tests are carried out on machines designed to subject a cantilever test piece to reversing flexural loading cycles at different stress levels. ASTM Test D671 is of this type, with a cycle rate of 30 Hz. The number of cycles to failure is recorded for each stress level. Data are normally presented in plots of stress (S) versus the log of the number of cycles (N) called S-N curves for specific cycle rates and environmental temperatures.

*Tensile impact* tests mount the test specimen on the swinging arm. Attached to the test specimen is a cross-piece that is arrested by a notched anvil as the bar swings down, allowing the energy stored in the arm to break the specimen under tension as it passes through the notch. Another impact test used for plastics allows a weighted, round-ended cylindrical "dart" to fall on a flat disk of the plastics to be tested. This test is good for ranking materials because it represents conditions encountered by actual parts in certain applications.

**Thermal Properties.—**Melting temperatures of highly crystalline thermoplastics are sharp and clearly defined, but amorphous and liquid-crystalline materials soften and become more fluid over wider temperature ranges. Melting points have greater significance in molding and assembly operations than in product design, which usually deals with the product's service temperatures well beneath the melting range.

*Glass transition* is a reversible change that occurs in an amorphous polymer or in amorphous regions of a partly crystalline polymer when it is heated from a very low temperature into a certain range, peculiar to each polymer, characterized by a rather sudden change from a hard, glassy, or brittle condition to a flexible or rubbery condition. Physical properties such as coefficient of thermal expansion, specific heat, and density usually undergo distinct changes in their temperature derivatives at the same time. The symbol for the glass-transition temperature is $T_g$.

*Vicat softening point* is the temperature at which a small, circular, lightly gravity-loaded, heated probe penetrates a specific distance into a thermoplastics test specimen. This test (ASTM D1525) measures the ability of a thermoplastics material to withstand short-term contact with a heated surface and is most useful for crystalline plastics. Amorphous thermoplastics tend to creep during the test, which reduces its usefulness for such materials.

*Deflection temperature under load (DTUL)* is the temperature at which a test bar of 0.5 inch (12.7 mm) thickness, loaded to a specified bending stress, will deflect by 0.010 inch (0.25 mm). This ASTM test (D648) is run at bending stresses of 66 $lb_f/in^2$ (0.455 MPa) or 264 $lb_f/in^2$ (1.82 MPa) or both. The value obtained is referred to as the *heat deflection temperature* (HDT), also called heat distortion temperature, and is an indication of the ability of the material to perform at elevated temperatures under load. Both stress and deflection for a specific design of test bar are given, so the test may be regarded as establishing the temperature at which the flexural modulus is reduced to particular values—35,200 psi (242.7 MPa) at 66 psi stress, and 140,000 psi (965.3 MPa) at 264 psi stress. Table 2a and Table 2b provides the range of HDTs for some common thermoplastics and thermosets. The addition of reinforcements, such as glass fiber (GF), can increase the HDT, so that the parts may be used at higher service temperatures than unreinforced grades.

**Table 2a. Typical Mechanical Properties of Common Plastics (Inch)**

| Material | Yield Stress, ksi | | | Elastic Modulus, ksi | | | Heat Deflection Temperature, ° F | | | Izod Impact Strength, ft lb/in | | |
|---|---|---|---|---|---|---|---|---|---|---|---|---|
| | Minimum | Median | Maximum | Minimum | Median | Maximum | Minimum | Median | Maximum | Minimum | Median | Maximum |
| ABS | 2.9 | 6.3 | 10.7 | 112.8 | 333.5 | 884.5 | 149 | 192 | 244 | 0.19 | 4.12 | 12 |
| PC | 5.8 | 9.1 | 22.3 | 261 | 346.55 | 870 | 172 | 261 | 369 | 0.84 | 16.7 | 37.4 |
| PEEK | 9.43 | 14.1 | 16.7 | 319 | 568.4 | 939.6 | 284 | 316 | 500 | 0.39 | 1.09 | 3.18 |
| PET | 0.305 | 5.6 | 13.1 | 130.5 | 456.75 | 754 | 140 | 159 | 239 | 0.26 | 1.12 | 1.55 |
| PP | 1.3 | 3.6 | 5.1 | 108.8 | 195.75 | 507.5 | 100 | 135 | 239 | 0.5 | 4.02 | 13.5 |
| PP, GF | 3.63 | 8.7 | 13.1 | 149.4 | 609 | 949.75 | 140 | 268 | 374 | 0.26 | 1.8 | 8.42 |
| PE, HD | 1.1 | 3.1 | 6.2 | 65.3 | 134.42 | 217.5 | 100 | 118 | 187 | 0.36 | 1.44 | 37.4 |
| PE, LD | 1.12 | 1.6 | 9.4 | 16 | 33.64 | 65.11 | 100 | 153 | 214 | 4.49 | 8.42 | 37.4 |
| PMMA | 3.63 | 9.3 | 12.3 | 137.8 | 420.5 | 549.55 | 125 | 193 | 223 | 0.22 | 0.56 | 2.75 |
| PSU | 6.96 | 12.8 | 26.8 | 249.4 | 817.8 | 2798.5 | 175 | 352 | 500 | 0.5 | 1.42 | 7.86 |
| PVC | 0.21 | 2.4 | 8.6 | 0.2 | 313.2 | 469.8 | 116 | 159 | 189 | 0.39 | 11.6 | 37.4 |
| TS Phenolic | 5.95 | 7.7 | 8.4 | 594.5 | 1015 | 1252.8 | 320 | 349 | 439 | 0.39 | 0.51 | 0.6 |
| TS Epoxy | 0.1 | 3.3 | 12.3 | 14.5 | 359.6 | 870 | 149 | 295 | 649 | 0.36 | 0.67 | 1.29 |
| TS Epoxy, GF | 12.04 | 16.5 | 21.8 | 435 | 2088 | 2755 | 329 | 439 | 536 | 0.3 | 0.54 | 0.69 |
| TS Polyester | 1.45 | 7.5 | 17.8 | 145 | 584.35 | 1537 | 392 | 486 | 500 | 2.3 | 7.49 | 17 |
| TS Polyimide | 10.59 | 18.6 | 23.2 | 159.5 | 569.85 | 1566 | 356 | 707 | 752 | 0.39 | 0.66 | 0.8 |

**Table 2b. Typical Mechanical Properties of Common Plastics (Metric)**

| Material | Yield Stress, MPa | | | Elastic Modulus, GPa | | | Heat Deflection Temperature, ° C | | | Izod Impact Strength, J/cm | | |
|---|---|---|---|---|---|---|---|---|---|---|---|---|
| | Minimum | Median | Maximum | Minimum | Median | Maximum | Minimum | Median | Maximum | Minimum | Median | Maximum |
| ABS | 20 | 43.4 | 73.5 | 0.78 | 2.3 | 6.1 | 65 | 88.9 | 118 | 0.1 | 2.2 | 6.4 |
| PC | 40 | 62.7 | 154 | 1.8 | 2.39 | 6 | 77.8 | 127 | 187 | 0.45 | 8.92 | 20 |
| PEEK | 65 | 97.4 | 115 | 2.2 | 3.92 | 6.48 | 140 | 158 | 260 | 0.21 | 0.58 | 1.7 |
| PET | 2.1 | 38.8 | 90 | 0.9 | 3.15 | 5.2 | 60 | 70.3 | 115 | 0.14 | 0.6 | 0.83 |
| PP | 8.96 | 25 | 35.2 | 0.75 | 1.35 | 3.5 | 37.8 | 57 | 115 | 0.27 | 2.15 | 7.2 |
| PP, GF | 25 | 60 | 90 | 1.03 | 4.2 | 6.55 | 60 | 131 | 190 | 0.14 | 0.96 | 4.5 |
| PE, HD | 7.6 | 21.3 | 43 | 0.45 | 0.93 | 1.5 | 37.6 | 47.5 | 86.1 | 0.19 | 0.77 | 20 |
| PE, LD | 7.7 | 10.8 | 64.8 | 0.11 | 0.23 | 0.45 | 38 | 67.4 | 101 | 2.4 | 4.5 | 20 |
| PMMA | 25 | 64 | 85 | 0.95 | 2.9 | 3.79 | 51.7 | 89.4 | 106 | 0.12 | 0.3 | 1.47 |
| PSU | 48 | 88.1 | 185 | 1.72 | 5.64 | 19.3 | 79.4 | 178 | 260 | 0.27 | 0.76 | 4.2 |
| PVC | 1.47 | 16.4 | 59 | 0.002 | 2.16 | 3.24 | 46.7 | 70.8 | 87.2 | 0.21 | 6.2 | 20 |
| TS Phenolic | 41 | 53.2 | 57.9 | 4.1 | 7 | 8.64 | 160 | 176 | 226 | 0.21 | 0.27 | 0.32 |
| TS Epoxy | 0.69 | 22.8 | 85.1 | 0.1 | 2.48 | 6 | 65 | 146 | 343 | 0.19 | 0.36 | 0.69 |
| TS Epoxy, GF | 83 | 114 | 150 | 3 | 14.4 | 19 | 165 | 226 | 280 | 0.16 | 0.29 | 0.37 |
| TS Polyester | 10 | 51.8 | 123 | 1 | 4.03 | 10.6 | 200 | 252 | 260 | 1.23 | 4.0 | 9.08 |
| TS Polyimide | 73 | 128 | 160 | 1.1 | 3.93 | 10.8 | 180 | 375 | 400 | 0.21 | 0.35 | 0.43 |

Statistical summary of available grades submitted by material suppliers (data courtesy of MatWeb.com).

Like metals, thermoplastic materials expand when heated and contract when cooled. For a given temperature range, most plastics change dimensions much more than metals. The coefficient of *linear thermal expansion* ($\alpha$) is the ratio of the fractional change in a linear dimension for a unit change of temperature and is expressed as in/in-F, or cm/cm-C. Typical average values for common materials and selected plastics are shown in Table 9, page 370. ASTM Method D696 directs that samples should be prepared so as to give a minimum of anisotropy, but if anisotropy is suspected, specimens shall be cut so as to measure $\alpha$ along the principal axes of anisotropy. The values in Table 9, page 370, may be presumed to have been measured according to those instructions.

*Thermal conductivity* is the rate at which a material conducts heat energy along its length or through its thickness. For example, if a tightly covered plastic insulated cooler at 32° F (0° C) is filled with ice and chilled beverage cans and the outside is 80° F (27° C), heat from the surrounding warm air passes slowly through the box wall, melting the ice inside. A low thermal conductivity of the wall resists this heat flow. The property is defined by *Fourier's law of heat conduction* as follows:

$$Q_x = k\frac{\Delta T}{x}$$

where $Q_x$ = flux, i.e., the rate of heat flow per unit area of the wall surface,
$k$ = thermal conductivity of the box-wall material,
$\Delta T$ = temperature difference from the hot side to the cold side, and
$x$ = wall thickness.

English units for $k$ are Btu/h-in$^2$ (°F/in); SI units are J/s-m$^2$(K/m).

In sharp contrast to metals, plastics are poor heat conductors so most serve as highly efficient thermal and electrical insulators, with plastic foams even more so. Conductivities can be enhanced where needed by loading the plastics with copper or aluminum powder. In anisotropic materials, thermal conductivity may differ along major axes. Thermal conductivities for plastics are listed in the table *Typical Properties of Plastics Materials* on page 385.

*Aging at elevated temperatures* may affect physical, mechanical, electrical, or thermal properties of plastics materials. Data from tests on specimens stored at specific temperatures for suitable periods are presented as plots of properties versus aging time at various temperatures, and may be used as an indication of thermal stability of the material.

*Temperature index* is a rating by Underwriters Laboratories (UL) of electrical and mechanical properties (with and without effects of impacts) of plastics materials used in electrical equipment for certain continuous operating conditions.

*Flammability* ratings also are produced by Underwriters Laboratories. UL tests measure the ability to continue burning after a flame is removed, and the percentage of oxygen needed for the material to continue burning. Other tests measure combustibility, ignition temperatures, and smoke generation. The importance of flammability testing for plastics is reflected by ASTM's listing of ten different tests under the index subject "Flammability—Plastics." The high carbon and hydrogen content of plastics means they are usually flammable. Some exceptions are fluorocarbons and chlorofluoroplastics, which are not flammable. Silicones and rigid PVC will not sustain fire by themselves. Compounders can significantly reduce any plastic's flammability by judiciously choosing from available fire-retarding additives.

**Effect of Temperature on Mechanical Properties.**—The equivalence of high strain rates and low temperatures (and vice versa) must be kept in mind when designing with plastics materials. Stress-strain curves for tests performed with one strain rate at several temperatures are similar to those for tests with one temperature and several strain rates. That is, very high strain rates and very low temperatures produce similar responses in materials. Conversely, the effects of very low strain rates, that is, creep effects, can be determined more quickly by testing at elevated temperatures and employing time-temperature superposition methods to estimate creep at lower temperatures and longer times. End-use testing at temperatures near or above the highest values expected in everyday use of a product helps the designer estimate long-term performance of components.

Strength, modulus, and elongation behavior are similar for tensile, compressive, flexural, and shear properties. Generally, strength and modulus decrease with increasing temperature. The effect of temperature increase is shown by the curves in Fig. 12 for crystalline and amorphous materials, where a gradual drop in modulus is seen as the glass-transition temperature $T_g$ is approached. The deflection temperature marks a flexural modulus of 35,200 psi (242.7 MPa) at a test stress of 66 psi (455 kPa), giving rise to the horizontal dashed line in Fig. 12. Above the glass-transition temperature, amorphous materials suffer rapid diminution of modulus, and, even with glass-fiber and other reinforcements, modulus drops rapidly above $T_g$. Some crystalline resins maintain usable moduli at temperatures approaching the crystalline melting point ($T_m$). Glass-fiber reinforcement can substantially improve the modulus of crystalline materials between the glass-transition and melting temperatures. Generally, strength-versus-temperature curves exhibit shapes like modulus curves, while, as expected, elongations increase with rising temperatures.

As temperatures drop significantly below normal ambient values, most plastics materials become brittle and lose much of their room-temperature impact strength, although a few materials show only a gradual decrease. Concern over failures at low temperatures led to the development of ASTM D746, "Brittleness Temperature of Plastics," which reports the brittleness temperature at which half the specimens break under a falling impact load.

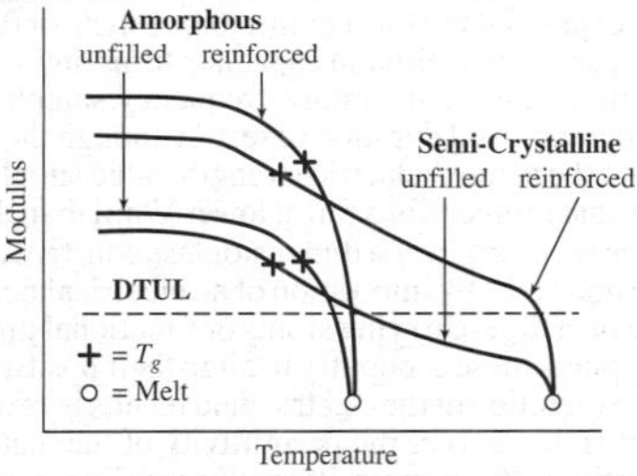


Fig. 12. Schematic of Modulus Behavior of Semi-Crystalline and Amorphous Plastics Showing $T_g$ and $T_m$ and the Beneficial Effect of Reinforcement on Deflection Temperature Under Load (DTUL)

**Electrical Properties.**—The most notable electrical property of plastics is that they are good insulators, but other electrical properties can be vital in design.

*Conductivity* in solids depends on the availability and mobility of movable charge carriers within the material. Metals are good conductors because the metal atom has a loosely held, outermost electron, and the close proximity of the atoms allows these outer electrons to break free and move within the lattice structure. These free electrons give metals the ability to conduct large currents, even at low voltages. Outer electrons in materials such as glass, porcelain, and plastics are tightly bound to the atoms or molecules so there are no free electrons. Electrical current cannot be conducted, and the materials act as insulators.

*Volume resistivity* is the electrical resistance of a material when a steady electromotive force (emf) difference is applied to a sample of the material that is made part of an electric circuit. Ohm's law governs the flow of direct current:

$$I = \frac{V}{R}$$

where $I$ = current, amperes (A)

$V$ = emf difference across the test element, volts (V)

$R$ = resistance, ohms (Ω)

$R$ is determined by the element's dimensions and the material's electrical resistivity:

$$R = \rho \frac{L}{A}$$

where $L$ = length of the element, cm
$A$ = cross-sectional area perpendicular to the current direction, $cm^2$
$\rho$ = resistivity of the conductor material, ohm-cm

Materials having resistivities above $10^8$ ohm-cm are considered insulators, and materials with values between $10^8$ and $10^3$ ohm-cm are considered partial conductors. Most plastics have volume resistivities in the range of $10^{12}$ to $10^{18}$ ohm-cm.

*Surface resistivity* is a measure of the susceptibility of an electrical insulating material to surface contamination, particularly moisture, which makes it electrically conductive. The tests use parallel electrodes that are placed on the same surface of the material. Measured resistance is multiplied by the width of the electrodes and divided by the distance between them and reported as ohms per square. Methods for determining both volume and surface resistivities of insulating materials are spelled out in ASTM D257.

The results of such measurements are sometimes reported in terms of their reciprocals, *volume conductivity* (siemens/cm) and *surface conductivity* (siemens). The unit siemens is the reciprocal of resistance, $1/R$ (with units $1/\Omega$.) The siemens was formerly known as the mho.

*Dielectric strength* is a measure of the voltage required to cause an insulator to break down and allow an electric current to pass through the resulting conductive path. It may be measured (ASTM Test D149) at any or several frequencies from 0 Hz (direct current) to 800 Hz; the results are expressed in volts per mil (0.001 inch, or 0.0254 mm) of thickness. High values of this property are desired in capacitor films and wire coverings. Variables that may affect test results include temperature, frequency, sample contamination, internal voids, rate of voltage increase, and duration of test. Although the use of volts/mil units in test results suggests that the same dielectric strength value should apply to different film thicknesses, tests show that thinner films fail at lower V/mil than thicker ones.

*Dielectric constant or permittivity* is a dimensionless constant that indicates how easily a material can become polarized by imposition of an electrical field on an insulator. Polarization occurs because of charge-carrying atoms or functional groups (dipoles) in materials. When the field is applied, these groups try to align their positive and negative ends with the field. Reversal of the direction of the electric field results in reversal of the polarization. The dielectric constant is the ratio of the permittivity of the material in normal ambient conditions to the permittivity of a vacuum. Permittivity is important when plastics are used as insulating materials in capacitors and high-frequency electrical apparatus. Changes in temperature, humidity, electrical frequency, and part thickness may affect the dielectric constant. Dielectric constants for a dozen various engineering thermoplastics are listed in Table 3, page 575. Many others are included in the main table of plastics properties, *Typical Properties of Plastics Materials* on page 385.

If a polar material is subjected to an alternating electric field, its dipoles reverse their orientation twice per cycle. This movement encounters frictional resistance at the molecular level, and at high frequencies that friction can generate considerable heat, as detected by the electrical phase angle between voltage and current. This effect is called the *heat dissipation factor*. The heat dissipation factor is a measure of heat energy dissipated by rapidly repeated reversals of polarization, as with an alternating current. The dissipation factor can also be thought of as the ratio of electrical energy lost to that transmitted at a representative frequency, such as 1 MHz ($10^6$ cycles/s). This high-frequency heat generation, undesired in many insulation uses, can be used to preheat thermosetting materials just before molding. It is also the basis for electronic heat-sealing, widely used in the fabrication of articles from plasticized-PVC films. Some heat dissipation factor values are shown in Table 3.

*Arc resistance* as determined by ASTM D495 is the length of time (seconds) required for an electric arc applied on the surface of an insulating material to develop a conductive path. Materials that resist such a development are preferred for switchgear parts and other high-voltage apparatus. This test is used mostly for thermosetting materials because, unlike thermoplastics which tend to melt in the arc, thermosets decompose and do produce conductive paths.

**Table 3. Typical Values of Dielectric Constants and Heat Dissipation Factors for Several Thermoplastics at Room Temperature**

| Material | Dielectric Constant | Heat Dissipation Factor | Material | Dielectric Constant | Heat Dissipation Factor |
|---|---|---|---|---|---|
| Acetal | 3.7-3.9 | 0.001-0.007 | Polypropylene | 2.3-2.9 | 0.003-0.014 |
| Acrylic | 2.1-3.9 | 0.001-0.060 | Polysulfone | 2.7-3.8 | 0.0008-0.009 |
| ABS | 2.9-3.4 | 0.006-0.021 | Modified PPO | 2.4-3.1 | 0.0002-0.005 |
| Nylon 6/6 | 3.1-8.3 | 0.006-0.190 | Polyphenylene sulfide | 2.9-4.5 | 0.001-0.002 |
| Polycarbonate | 2.9-3.8 | 0.0006-0.026 | Polyarylate | 2.6-3.1 | 0.001-0.022 |
| TP Polyester | 3.0-4.5 | 0.0012-0.022 | Liquid crystal | 3.7-10 | 0.010-0.060 |

The *Comparative Tracking Index (CTI)* is a UL test with an ASTM counterpart (D3638) that is similar to the arc resistance test except that the surface to be tested is precoated with an ammonium chloride electrolyte. The test measures the voltage required to cause a conductive path to form between the electrodes and indicates the arc resistance of a contaminated surface, a condition often found in electrical and electronic equipment.

**End-Use Environmental Considerations.—**The environment that will be encountered by the product is a prime consideration at the design stage. Problems with cracking, crazing, discoloration, loss of properties, melting, or dissolving can be encountered in the presence of high or low temperatures, chemical substances, energy sources, and radiation. Design also must consider that plastics components may be stressed by processing, assembly, finishing, and cleaning operations that affect their end-use properties.

The stress level in the plastics product greatly affects performance. Generally, increased stress levels resulting from injection molding, forming, assembly work, and end-use forces reduce resistance to environmental factors. Although many plastics are hygroscopic and absorption of water results in dimensional and property changes, plastics are widely accepted because of their relative compatibility with the environment compared to metals. Some chemicals attack the polymer chain directly by reaction, resulting in a progressive lowering of the molecular weight of the polymer and changes in the short-term mechanical properties; others dissolve the material, although high-molecular-weight plastics dissolve very slowly. Swelling, changes in weight and dimensions, and loss of properties are evidence of solvation.

Plasticization may result if the chemical is miscible with the polymer, resulting in loss of strength, stiffness, and creep resistance, and increased impact resistance. The material may swell and warp due to relaxation of molded-in stresses. Environmental stress cracking may cause catastrophic failure when plastics are stressed, even when the product appears to be unaffected by exposure to a chemical.

Chemical compatibility data are obtained from standard test bars exposed to or placed in the chemical of study and tested as previously described for such properties as tensile strength, flexural modulus, dimensional change, weight, and discoloration. Chemical resistances from some commonly used thermoplastics materials are shown in Table 4, but these are only general guidelines and cannot substitute for values derived from tests on the end product. More extensive tests expose samples to a chemical in the presence of fixed stress or fixed strain distribution along its length, followed by examination for the stress or strain location at which damage begins.

While the preceding tests may provide useful data about chemical compatibility, they do not generate reliable information on performance properties for design purposes. The only tests that provide such information are the creep modulus and rupture tests, conducted at appropriate temperatures in the environment that will be encountered by the product, preferably on prototype parts. Many plastics are slowly degraded to varying degrees by ultraviolet light (UV), which causes fading, chalking, and embrittlement. Acrylics and a few others are not damaged by UV. However, hundreds of additives are available that, when compounded into neat resins, will increase resistance to heat, burning, Uv, staining, etc.

## Table 4. Chemical Resistance of Various Materials by Chemical Classes

| | Acetal Copolymer | Acetal Homopolymer | Nylon 6/6 | Thermoplastic Polyester (PBT) | Thermoplastic Polyester (PET) | Polyester Elastomer | Liquid Crystal Polymer* | Polyphenylene Sulfide* | Polyethylene | Polycarbonate | Polysulfone* | Modified Polyphenylene Oxide | Polypropylene | ABS | 316 Stainless Steel | Carbon Steel | Aluminum | |
|---|---|---|---|---|---|---|---|---|---|---|---|---|---|---|---|---|---|---|
| **ACIDS AND BASES** | | | | | | | | | | | | | | | | | | |
| Acids, weak | A | B | C | A | A | A | A | A | A | A | B | A | A | A | A | A | C | Dilute mineral acids |
| Acids, strong | C | C | C | B | — | C | B | A | B | C | C | B | A | A | B | C | C | Concentrated mineral acids |
| Bases, weak | A | C | A | B | B | A | B | A | B | C | A | A | A | A | A | B | C | Dilute sodium hydroxide |
| Bases, strong | A | C | C | — | — | B | C | A | C | C | A | A | A | A | B | B | C | Concentrated sodium hydroxide |
| Acids, organic, weak | A | B | C | A | A | A | A | A | B | A | B | A | A | A | A | C | C | Acetic acid, vinegar |
| Acids, organic, strong | C | C | C | B | — | C | B | A | C | C | C | A | A | A | B | C | C | Trichloroacetic acid |
| **AUTOMOTIVE** | | | | | | | | | | | | | | | | | | |
| Automotive, fuel | A | A | A | A | A | A | A | A | C | C | A | C | C | A | A | A | A | |
| Automotive, lubricants | A | A | A | A | A | A | A | A | C | C | A | A | A | A | A | B | A | |
| Automotive, hydraulic | A | A | — | A | A | — | A | A | C | C | C | A | A | A | — | — | — | |
| **SOLVENTS** | | | | | | | | | | | | | | | | | | |
| Aliphatic hydrocarbons | A | A | A | A | A | A | A | A | C | A | A | B | C | A | A | A | A | Heptane, hexane |
| Aliphatic hydrocarbons, halogenated | A | B | C | B | B | A | A | A | C | C | C | C | C | C | B | B | B | Ethylene chloride, chloroform |
| Alcohols | A | A | B | A | A | A | A | A | B | A | A | A | A | A | A | A | B | Ethanol, cyclohexanol |
| Aldehydes | A | A | A | A | B | B | A | A | B | C | B | A | A | — | A | B | A | Acetaldehyde, formaldehyde |
| Amines | C | C | C | — | — | — | C | B | C | C | C | C | A | C | A | B | B | Aniline, triethanolamine |
| Aromatic hydrocarbons | A | B | A | A | B | B | A | A | C | C | C | C | C | C | A | A | A | Toluene, xylene, naphtha |
| Aromatic hydrocarbons, halogenated | A | A | A | — | — | C | — | A | C | C | C | C | C | C | A | A | A | Chlorobenzene |
| Aromatic, hydroxy | C | C | C | C | — | C | A | A | C | C | C | C | A | — | B | C | A | Phenol |
| Esters | B | B | A | B | B | B | A | A | C | C | C | C | C | C | B | B | B | Ethyl acetate, dioctyl phthalate |
| Ethers | B | B | A | A | — | — | — | A | C | A | B | C | C | — | A | A | A | Butyl ether, diethyl ether |
| Ketones | B | B | A | B | B | B | A | A | C | C | C | C | B | C | A | A | A | Methyl ethyl ketone, acetone |
| **MISCELLANEOUS** | | | | | | | | | | | | | | | | | | |
| Detergents | A | — | A | — | B | — | — | A | B | A | — | B | A | — | A | A | B | Laundry and dishwashing detergents, soaps |
| Inorganic salts | B | B | B | — | A | — | — | A | B | A | — | A | A | A | B | B | B | Zinc chloride, cupric sulfate |
| Oxidizing agents, strong | C | C | C | — | C | — | B | B | C | C | — | C | A | C | C | C | C | 30% hydrogen peroxide, bromine (wet) |
| Oxidizing agents, weak | C | C | C | A | — | A | A | A | B | A | — | A | A | A | B | C | A | Sodium hypochlorite solution |
| Water, ambient | A | A | B | A | A | A | A | A | A | A | A | A | A | — | A | C | B | |
| Water, hot | B | C | B | C | C | B | A | A | A | C | — | A | C | — | A | C | B | |
| Water, steam | C | C | C | C | C | C | B | A | — | C | — | — | C | — | A | C | — | |

A—minimal effect; B—some effect; C—generally not recommended.

This information is presented for instructional purposes and is not intended for design. The data were extracted from numerous sources, making consistent rating assignments difficult. Furthermore, the response of any given material to specific chemicals in any one class can vary significantly. Indeed, during the preparation of the table, the effect on one plastics of various chemicals in the same category ranged from essentially no effect to total dissolution. Therefore, an "A" rating for a particular plastics exposed to a particular class of chemicals should not be interpreted as applying to all chemicals in that class. The rating simply means that for the chemicals in that class found in the literature reviewed, the rating was generally an "A." There may be other chemicals in the same class for which the rating would be "C." Finally, the typical chemicals listed do not necessarily correspond to the ones on which the individual ratings are based.

Room temperature except for hot water, steam, and materials marked with a * ≡ 200° F (93.3° C). Generally, data covering extended exposure (more than a week) were used.

## Design Analysis

**Structural Analysis.—**Plastics parts may be subjected to stresses caused by assembly, handling, temperature variations, and other environmental effects. Simple analysis using information in *Moment of Inertia* on page 233 and *Beam Calculations* on page 250 can be used to make sure that newly designed parts can withstand these stresses. These methods may also be used for product improvement, cost reduction, and failure analysis of existing parts.

*Safety Factors:* In setting safety factors for plastics parts, there are no hard and fast rules. The most important consideration is the consequence of failure. For example, a little extra deflection in an outside wall or a crack in one of six internal screw bosses may not cause much concern, but the failure of a pressure vessel or water valve might have serious safety or product liability implications.

Tests should be run on actual parts at the most extreme operating conditions that could possibly be encountered during product life. For example, maximum working load should be applied at the maximum temperature and in the presence of any chemicals that might be encountered in service. Loads, temperatures, and chemicals to which a product may be exposed prior to its end use also should be investigated. Impact loading tests should be performed at the lowest temperature expected, taking into account temperature fluctuations during assembly and shipping. Effects of variations in resin lots and molding conditions must also be considered.

Failures in testing of preproduction lots often can be corrected by increasing the wall thickness, using ribs or gussets, and eliminating stress concentrations. Changing the material to another grade of the same resin or to a different plastics with more suitable mechanical properties is another possible solution. Reviews of product data and discussions with experienced engineers suggest the design stresses shown in Table 5 are suitable for use with the structural analysis information indicated above and the equations presented here, for preliminary design analysis and for evaluating general product dimensions. Products designed under these guidelines must be thoroughly tested before being marketed.

**Table 5. Design Stresses for Preliminary Part Designs Expressed as a Percentage of Manufacturers' Data Sheet Strength Values**

| | Failure Not Critical | Failure Critical |
|---|---|---|
| Intermittent (nonfatigue) loading | 25–50% | 10–25% |
| Continuous loading (creep or fatigue) | 10–25% | 5–10% |

*Failure Criteria:* Designers typically evaluate complex stress states using calculated Von Mises stresses (see *Three-Dimensional Stress*, page 216) relative to the stress limits expressed in Table 5 or otherwise consider the maximum shear theory of failure (otherwise known as Coulomb or Tresca theory). It is further suggested that the shear strength be taken as the manufacturer's published shear strength, or half the tensile strength, whichever is lower.

*Pressure Vessels:* The most common plastic pressure vessel is a cylindrical tube with internal pressure. In selecting a wall thickness for the tube it is convenient to use the thin-wall hoop stress equation:

$$\text{hoop stress } \sigma = \frac{Pd_i}{2t} \tag{17}$$

where $P$ = uniform internal pressure in the tube, $d_i$ = inside diameter of the tube, and $t$ = tube-wall thickness. This equation is reasonably accurate for tubes where the wall thickness is less than $0.1 \times d$. Also, it is easy to solve for the wall thickness $t$ required to support an internal pressure $P$ with an allowable hoop stress $\sigma$.

For wall thicknesses greater than $0.1d$, the maximum hoop stress on the wall surface inside the tube can be calculated from

$$\text{hoop stress } \sigma = P\frac{d_o^2 + d_i^2}{d_o^2 - d_i^2} \quad \text{or} \quad \sigma = P\frac{1+R}{1-R} \tag{18}$$

where $d_i$ and $d_o$ are the inside and outside diameters of the tube, and $R = (d_i/d_o)^2$.

By substituting $d_o = d_i + t$ into the left equation one can solve it for the required wall thickness $t$ to provide a working equation for design. The result is

$$t = \frac{d_i}{2}\left(\sqrt{\frac{\sigma + P}{\sigma - P}} - 1\right) \tag{19}$$

In the above equations, all units must be consistent.

*Press Fits:* Press fits are used widely in assembly work for speed and convenience, although they sometimes are unsatisfactory with thermoplastics parts. Common applications are to a plastics hub or boss accepting a plastics or metal shaft or pin. Forcing the pin into the hole expands the hub, creating a tensile or hoop stress.

If the interference is too great, very high strain and stress develop, and the plastics part may a) fail immediately by developing a crack parallel to the hub axis to relieve the stress, a typical hoop stress failure; b) survive assembly but fail prematurely due to creep rupture caused by the high induced-stress levels; or c) undergo stress relaxation sufficient to reduce the stress to a level that can be sustained, which can result in slippage during end use.

For a typical press fit, the allowable design stress depends on the particular plastics material, temperature, and other environmental considerations. Hoop stress equations for such a design make use of a geometry factor $\gamma$:

$$\gamma = \frac{1 + (d_s/d_o)^2}{1 - (d_s/d_o)^2} \tag{20}$$

where $d_s$ = diameter of the pin to be inserted and $d_o$ = outside diameter of the boss.

When both the shaft and the hub are of the same or essentially the same materials, the hoop stress $\sigma$ given the diametral interference $i = d_s - d_i$ is

$$\sigma = \frac{i}{d_s} E_p \frac{\gamma}{\gamma + 1} \tag{21}$$

and the allowable interference $i_a$ given the permissible design stress $\sigma_a$ is

$$i_a = d_s \frac{\sigma_a}{E_p} \frac{\gamma + 1}{\gamma} \tag{22}$$

When the shaft is metal and the hub is plastic, the hoop stress given $i$ is obtained from

$$\sigma = \frac{i}{d_s} E_p \frac{\gamma}{(\gamma + \nu_p)} \tag{23}$$

and the allowable interference $i_a$ given the permissible design stress for plastic's $\sigma_a$ is

$$i_a = d_s \frac{\sigma_a}{E_p} \frac{\gamma + \nu_p}{\gamma} \tag{24}$$

where $E_p$ = modulus of elasticity of plastics and $\nu_p$ = Poisson's ratio for plastics.

*Pipe Threads:* Pipe threads on plastics pipes and other parts used in plastics plumbing and pneumatic assemblies typically require only hand tight assembly to effect a good seat, especially if a compatible sealant tape or compound is used. Assembling a tapered male pipe thread into a mating female thread in a plastics part is analogous to driving a cone into a round hole and may result in a split boss. Sometimes straight threads and an O-ring seal can avoid the need for pipe threads. When pipe threads must be used, torque control is essential.

When mating metal to plastics pipe threads, the threaded plastics component should be the male member, so that the plastics are in compression. If torque can be controlled during assembly, fluoroplastics tape should be used on female plastics pipe threads. If torque cannot be controlled, an external hoop ring, either pressed on or molded in, should be considered. Flats should not be designed into plastics parts for assembly purposes because they can encourage overtightening. If some provision for improved gripping must be made, wings or a textured surface may be used. An approximate formula for the hoop stress $\sigma$ produced in a plastics boss with internal pipe threads is

$$\sigma = \frac{3T}{tdL} \tag{25}$$

where $T$ = torque in in-lb, $t$ = wall thickness of the plastics boss in inches, $d$ = pipe outside diameter in inches, and $L$ = length of thread engagement in inches.

Equation (25) assumes certain geometric relationships and a coefficient of friction of 0.15. If compatible thread lubricants are used during assembly, the torque must be reduced. To ensure safety and reliability, all threaded assemblies must be subjected to long-term testing under operating pressures, temperatures, and stresses caused by installation procedures exceeding those likely to be encountered in service.

**Thermal Stresses.—** When materials with different coefficients of thermal expansion are bolted, riveted, bonded, crimped, pressed, welded, or fastened by any method that prevents relative movement between the parts, there is potential for thermal stress to exist. Typical examples are joining of nonreinforced thermoplastics parts with materials such as metals, glass, or ceramics which usually have much lower coefficients of thermal expansion than plastics. The basic relationship for thermal expansion is

$$\Delta L = \alpha L \Delta T \tag{26}$$

where $\Delta L$ = change in length, $\alpha$ = coefficient of thermal expansion (see Table 9, page 370), $L$ = linear dimension under consideration (including hole diameters), and $\Delta T$ = temperature change.

If the plastics component is constrained so that it cannot expand or contract, the strain $\varepsilon_T$, induced by a temperature change, is calculated by

$$\varepsilon_T = \frac{\Delta L}{L} = \alpha \Delta T \tag{27}$$

The stress can then be calculated by multiplying the strain $\varepsilon_T$ by the tensile modulus of the material at the temperature involved. A typical example is of a plastics part to be mounted to a metal part, such as a window in a housing. Both components expand with changes in temperature. The plastics imposes insignificant load to the metal, but considerable stress is generated in the plastics. For such an example, the approximate thermal stress $\sigma_T$ in the plastics is given by

$$\sigma_T = (\alpha_m - \alpha_p) E_p \Delta T \tag{28}$$

where $\alpha_m$ = coefficient of thermal expansion of the metal, $\alpha_p$ = coefficient of thermal expansion of the plastics, and $E_p$ = tensile modulus of the plastics at the temperature involved. Other equations for thermal expansion in various situations are shown in Fig. 13.

Most plastics expand more than metals with temperature increase, and their modulus drops. The result is a compressive load in the plastics that often results in buckling. Conversely, as the temperature drops, the plastics shrinks more than the metal and develops an increased tensile stress. These conditions can cause tensile rupture of the plastics part. Clearances around fasteners, warpage, creep, or failure, or yield of adhesives tend to relieve the thermal stress. Allowances must be made for temperature changes, especially with large parts subjected to wide variations. Provision is often made for relative expansion $\Delta L_{rel}$ between two materials, as illustrated in Fig. 13:

$$\Delta L_{rel} = (\alpha_p - \alpha_m) L \Delta T \tag{29}$$

**Designing for Stiffness.—**The designer must take full advantage of the ability of plastics to be easily formed into a desired shape, in contrast to woods where every shape must be machined or assembled from planks or blocks. To a lesser degree, the same is often true of metals, although die-castable metals have a considerable shape versatility. Though the modulus of elasticity is much less in plastics than in metals, *stiffness* is a property not just of the material but also of the *structure* that can be made from it. When the product must support bending loads, as in a bookshelf, a simple beam of rectangular cross section would be the logical approach if the shelf material were pine or mahogany. If the material is plastic, other cross sections of higher stiffness per unit weight can be considered to take advantage of plastics' ease of formability and low density. Fig. 13 shows some alternatives to the traditional rectangular cross section that achieve greater stiffness with less material.

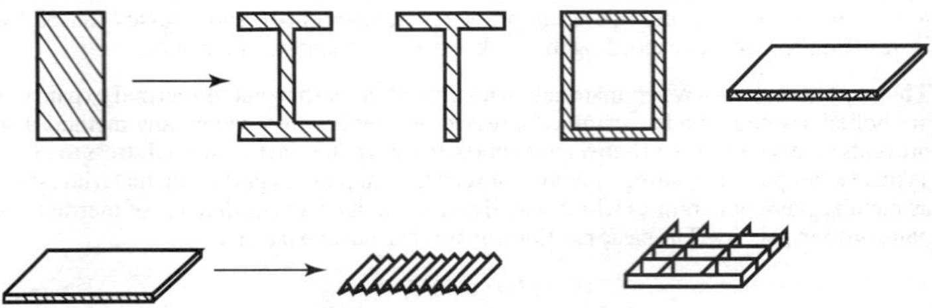

Fig. 13. Placing Most of the Material Far from the Centroid Achieves Far Higher Bending Stiffness per Unit Weight of Plastics

## Manufacture of Plastics Products

The main processes for manufacturing large quantities of plastics products are extrusion, injection molding, blow molding, and sheet thermoforming for thermoplastic materials; compression molding, transfer molding, prepreg molding, and pultrusion for thermosetting materials.

Most thermoplastics pass through one or more extrusion machines. The dominant type of extruder is the single-screw extruder. A feed hopper supplies small plastic granules to a thick-barreled, heated cylinder, inside which turns a sturdy, precisely contoured screw. A powerful variable-speed motor drives the screw at speeds typically between 10 and 1000 rpm. At the delivery end is attached a die, a block of metal with a specially shaped orifice. The screw has two main jobs: 1) to melt by friction and thoroughly mix the feed, transforming it from cold, solid particles into a homogeneous hot, viscous melt; 2) to develop the high pressure (1,000–8,000 psi, or 14–55 MPa) needed to pump the melt through the die. This produces a continuous length of product whose cross section resembles that of the die opening. Extruders are widely employed in intermediate processing of material in the plastics industry. Some final consumer products made by this process are pipes, sheets and films, garden hoses, electrical wires, netting, fishing lines and leaders, and house sidings. Many extruded products are parts of other industrial or consumer products. For example, huge quantities of ABS sheet and flexible vinyl gasketing go into refrigerators and freezers each year. Mop handles are extruded tubing cut to the length needed. Produce bags are made from extruded blown polyethylene film that is laid flat, printed, heat-sealed across and perforated for tear-off at intervals, then wound onto roll cores for delivery to supermarkets. An important class of extruded products includes rods, bars, and tubes from which machinists fabricate smaller quantities of objects, such as prototypes for testing prior to mass production.

**Sheet Thermoforming.**—Also known as *vacuum* or *pressure forming*, these widely used molding processes respectively apply negative or positive pressure to form sheet feedstock. Mounted in a frame, the sheet to be formed is preheated in an oven between

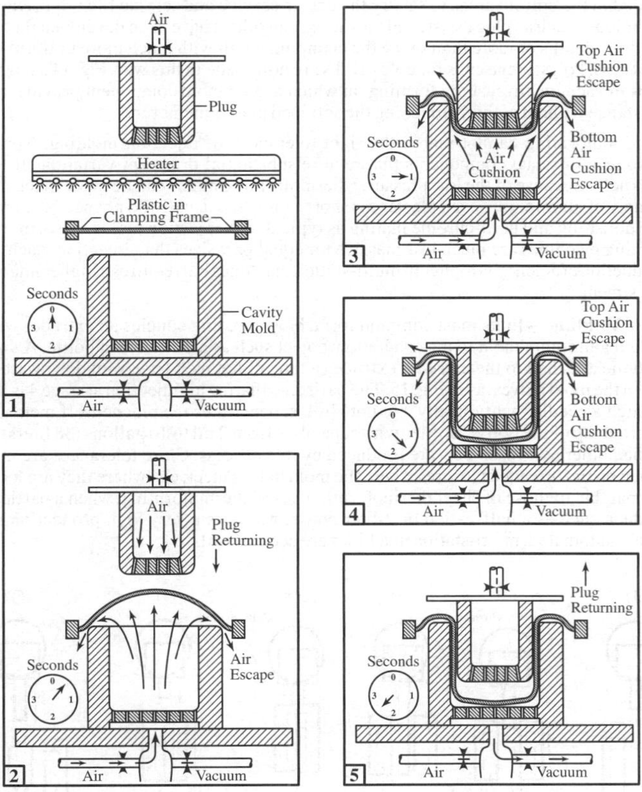


Fig. 14. Five Stages in the Cycle of Air-Cushion Vacuum Forming with Controlled Air Pressures at Stages 3 and 4. Removal and Trimming Follow Stage 5.

radiant heaters at the thermoforming press until it begins to sag. The temperature is still well below that at which the same resin would be injection molded. The sheet is then positioned over the mold, which, in the original and simplest variant of the process, is a female mold made porous by drilling many small holes through it. The mold is enclosed in a chamber (vacuum or pressure box) so that a vacuum can be drawn from below (or a pressure applied from above). The softened sheet is clamped to the mold around its edges, and the pressure difference is applied to press the sheet against the mold surface. Cooling and solidification of the part is done with fans, water mist, or cooling channels in the mold. The unclamped molding is then indexed to a die-cutting table where the edges are trimmed and the part(s) cut out. Trim scrap, which may be as much as 20 percent of the sheet area, is recycled to the sheet extruder.

Because the drawn sheet contacts the upper areas of the mold early in the mold cycle and the bottom last, the wall thickness of the finished part is thinner at the bottom than the upper areas when using a female mold. Often, a specified minimum thickness is needed in the bottom areas to satisfy the design service and service life requirements. Techniques such as plug assist, male forming, and others have been developed that permit draw depths of more than twice the clamping width with much more uniform wall thickness. Polystyrene cups for cold drinks are now made in this way. Fig. 14 shows the steps in vacuum air-cushion forming, in which a male mold component provides plug assistance without actually touching the softened plastic membrane.

Thermoforming cannot match the tight tolerances of injection molding. Nor can it produce intricate components that require substantial or abrupt variations in wall thickness. Because of the low pressure, thermoforming molds can be made much more cheaply of aluminum—even plaster for short runs—and usually do not need the extensive polishing and hard-chrome plating as typical with injection molds. Also, thermoforming machines are much less massive for equal part sizes than injection machines. Production efficiency is higher in multi-station machines but requires a higher financial investment.

**Blow Molding.**—In its most common form, blow molding couples an extruder with a chilled aluminum mold to make a hollow object such as a 2-liter soda bottle. A crosshead die attached to the extruder extrudes a tube of hot melt (*parison*) downward between the mold halves, as in Fig. 15. The parison-inflating air injection at stage 4 is done through a needle that typically, but not always, penetrates the pinched-off melt at the inverted top of the bottle. Bottles with capacities from 2 ml to 10 gallons (38 liters) and myriad other hollow objects are produced by this process. Close tolerances are maintained on the threaded closures where the mold halves meet; elsewhere they are looser, comparable to those of sheet thermoforming. As in thermoforming, when a particular product, such as a half-gallon or 2-liter bottle, warrants a very high production rate, highly automated, multi-station machines are economical.

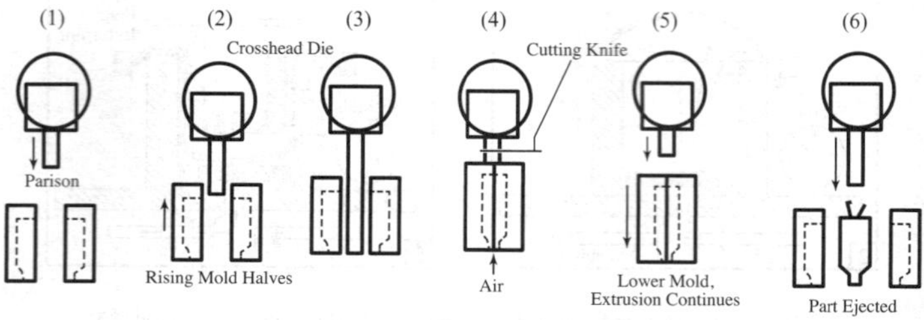


Fig. 15. Six Stages of Extrusion/Blow Molding

**Rotomolding.**—Also known as *rotational molding*, this process enables repeated production of hollow parts with complex details and low residual stresses, often more cost-effectively than other manufacturing processes. As shown in Fig. 16, rotomolding involves melting material in a rotating mold that is heated and then cooled while the rotation forces the material to freeze against the cavity walls, producing the final molded shape.

*Production Process:* Rotomolding may be performed as a single- or multi-station process, using shuttle or carousel machines. In the single-station process, a mold is heated, cooled, the part unloaded, and then the process is repeated. Multiple station machines speed up production, as mold heating, cooling, and loading/unloading steps can be performed in parallel; these machines also are more energy efficient, as heating and cooling chambers can be operated at near constant temperatures, reducing energy costs and time to dynamically reheat and again cool much of the machinery over and again. Both single- and multi-cavity molds can be mounted on a rotomolding frame to produce several parts per cycle; the size and thickness of the parts should be similar to avoid extended processing times or unintended defects.

Regardless of the machine and mold layout, rotomolding relies on rotation of the mold (usually around two different axes) to evenly distribute the molten plastic within the mold. This rotation is performed during both heating (so the polymer melt adheres to all mold surfaces) and cooling (to develop uniform wall thickness). Depending on the application, internal pressure may be applied to ensure the plastic replicates mold surface details and to minimize shrinkage. Modern machines monitor the internal air temperature and adjust mold temperature and cycle time as needed. Quality control measures include visual inspection of the product, dimensional measurements, and pressure/strength testing.

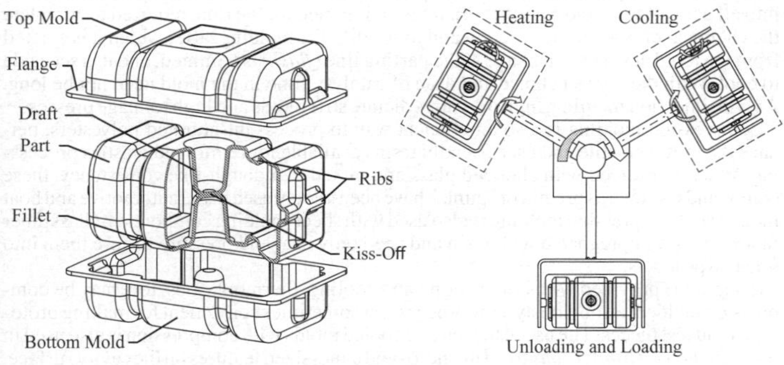


Fig. 16. Basic Rotomold Design Features (left);
Three-Station Carousel Process (right)

*Mold Design:* Even when internal pressure is added, it is relatively low. Molds can be constructed from steel or aluminum and often are comprised of a series of plates mounted to an outer frame. Molded parts can be finish machined, assembled, and welded to provide integral assemblies with functional tolerances.

Molds should feature generous fillets on corners to ensure the molten material is evenly distributed. Draft is typically employed on deeper cavities, with 5 degrees recommended for deep cores and textured surfaces. Mold inserts can be used to provide attachment points, add aesthetic surfacing, or form intentional threads.

If heat is evenly distributed across the mold, wall thickness will tend to be uniform, though internal corners will be slightly thicker than external corners. To increase rigidity, corrugated ribs can be periodically placed along a wall to increase the moment of inertia. When even more rigidity is desired without increasing wall thickness, one or more kiss-offs can be formed as shown in Fig. 16, bringing opposing mold surfaces within between 1.5 and

2 times the nominal wall thickness. To avoid unwanted kiss-offs and internal webs, opposing mold surfaces should be separated by at least 4 times the nominal wall thickness.

*Materials and Uses:* Various types of plastics can be rotomolded, though polyethylene (PE) is the most common material used for at least two reasons. First, as a semi-crystalline material, polyethylene has a sharp transition between solid and liquid phases, so it can be processed at a relatively low temperature and exhibit a suitably low viscosity for flowing within the mold cavity. Second, different types of polyethylene (HDPE, LDPE, etc.) are available at low cost and may be customized with various fillers, UV protection, and colorants; incorporation of recycled materials is common. In production, extrusion compounding of the polymer, fillers, and colorants ensures uniform material properties; feedstock should be in the form of molten shot, micro-pellets, or pulverized pellets (powder) for efficiency and consistency.

Rotomolding is versatile for use across industries such as agricultural, automotive, and medical, and in consumer products of various production volumes. For larger runs, twin-sheet thermoforming and blow molding also should be considered, due to faster cycle times, though drawbacks typically include higher tooling costs and more wall thickness variation.

**Processing of Thermosets.—**While some thermosetting materials have been adapted to injection molding processes, most are still processed by compression molding, transfer molding, prepreg molding, sprayup, and hand layup. Commonly used *compression molding* is a process in which precompounded and catalyzed resins, still thermoplastically formable, are preformed into buns or bricks of known weight. One or more bricks are preheated in convection or electronic ovens and quickly positioned in the female half of a horizontal mold. It and the hovering male half are at the molding temperature. The mold is closed and brought to high pressure, squeezing the softened bricks, which, ideally, flow into all parts of the mold cavity. Pressure is maintained for the time required to complete the chemical cross-linking reaction and to solidify the molding, after which it is ejected from the mold, any slight leakage at the parting line (*flash*) is trimmed, and it is set aside to cool. With the correct choice and dose of catalyst, time in the mold need not be long. As with injection molding, high pressures dictate stiff molds and high tonnage presses.

Compression molding is also an efficient way to process unsaturated polyesters, perhaps the most versatile and user-friendly resins available for reinforced plastics processing. When reinforced with chopped glass and mixed to a dough-like consistency, these compounds, known as premix or "gunk," have been widely used in the automotive and boat industries. Compression molding is also used with sheet-molding compounds. Glass-fiber fabrics are preimpregnated with resin and need only heat and pressure to cure them into finished products.

If a mold is provided with both heating and cooling, thermoplastics, too, may be compression molded, an obviously slow process, but sometimes convenient for making prototypes and test pieces. The use of heated and cooled molds with compression is also used in *thermal imprint lithography* to form micro- and nano-sized features on the cavity surface.

In *transfer molding*, also called *plunger molding*, the thermosetting compound is preheated in a separate cylinder (the pot) until it is fully fluid, then forced by a plunger through a runner into the final mold. This process is capable of molding complex parts that cannot be made by compression molding, and also offers closer dimensional tolerances and glossier surfaces than the latter. Multi-cavity molds and short cycle times are feasible with transfer molding. Transfer molding has been used widely in the electronics industry and in molding and overmolding of silicone products, including medical devices.

Another popular method of processing thermosets into unidirectionally reinforced products is *pultrusion*, in which continuous glass roving or mat is pulled through a resin bath or impregnator and then into a preforming fixture where the cross section is partly shaped. The preliminary profile continues through dies where curing mechanisms are applied, and finally to a cut-off saw to produce shapes such as channels, rods, and tubes. Ladder rails, fishing rods, tool handles, tent poles, and numerous other parts are made by this process.

**Polyurethanes.—**Polyurethanes (PUR) are among the most versatile of polymers, both as thermoplastic and thermosetting, as well as the most ubiquitous in everyday products. For example, the flexible thermoplastic members of the family are mainly elastomers that are favorites for athletic shoes, while the rigid, transparent thermoplastics are vying with polycarbonate for the high-value market in eyeglass lenses. Cast rigid PUR is available in rod, tube, and bar stock form.

The thermosets are formed by linking two reactive components, an isocyanate and a polyol. Toluene diisocyanate and diphenylmethane diisocyanate are the most used isocyanates, while a variety of polyols are commonly used. Short-chain, low-molecular-weight types result in rigid plastics, and longer-chain types form flexible materials. About 70 percent of all PURs worldwide go into automotive, building and construction, mattresses, and furniture including frame members produced by reaction injection molding. Densities of rigid PUR foams range from 1 to 8 lb/ft$^3$ (16–128 kg/m$^3$) and much higher in furniture members. Flexible PUR foams range from near that of the parent plastic down to 1 lb/ft$^3$ (16 kg/m$^3$). Since the reactive components must be mixed instantly and thoroughly at the moment of application, a special industry of PUR processing machinery has developed to deal with that requirement.

**Polymer Composites.**—For superior strength and modulus with high strength-to-weight ratios, products may be engineered with oriented composite plies or other reinforced polymer materials. Common applications include automotive and aircraft beams and panels, wind turbine and helicopter rotor blades, boat hulls, sports equipment, ballistic armor, prosthetics, and many others.

Fig. 17 provides an overview of composites, including types of matrix reinforcements and composite layouts. Common polymer matrices include thermoplastics, such as ABS, PA (nylon), PEEK, and PP, as well as thermosets such as epoxy, phenolic, polyester, and silicones. Generally, thermoplastic matrices tend to provide higher strain to failure and impact toughness, while thermoset matrices provide higher stiffness and end-use temperature tolerances. Notable exceptions are PEEK, an engineering thermoplastic with high strength and temperature resistance, and silicone, a thermosetting elastomer with low modulus but high strain to failure.

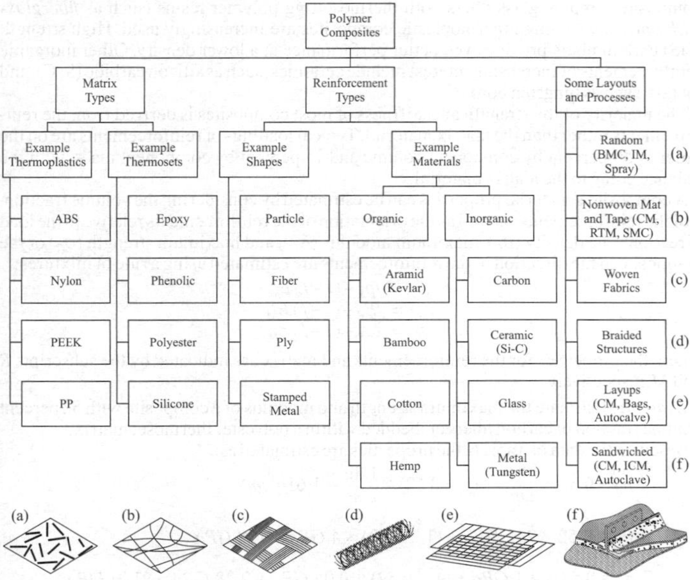


Fig. 17. Overview of Composite Materials and Structures

**Table 6. Properties of Common Polymer Matrix and Composite Reinforcements**

| Material | Specific Gravity | Tensile Strength ksi (GPa) | Tensile Modulus ksi (GPa) | Elongation to Failure (%) | Heat Resistance °F (°C) |
|---|---|---|---|---|---|
| Matrix, Thermoplastic | | | | | |
| ABS | 1.04 | 5.8 (0.04) | 261 (1.8) | 48 | 194 (90) |
| Nylon 66 | 1.14 | 10.2 (0.07) | 522 (3.6) | 47 | 275 (135) |
| PEEK | 1.31 | 14 (0.1) | 580 (4) | 50 | 480 (250) |
| Polypropylene | 0.91 | 4.4 (0.03) | 188 (1.3) | 250 | 226 (108) |
| Matrix, Thermoset | | | | | |
| Epoxy | 1.35 | 81 (0.56) | 5,100 (35) | 3 | 277 (136) |
| Phenolic | 1.45 | 7.3 (0.05) | 1,000 (7) | 0.7 | 365 (185) |
| Polyester | 1.40 | 7.3 (0.05) | 580 (4) | 1.2 | 250 (121) |
| Silicone | 1.27 | 0.9 (0.006) | 29 (0.2) | 300 | 415 (213) |
| Reinforcing Fiber, Organic | | | | | |
| Aramid (Kevlar) | 1.44 | 435 (3) | 16,200 (112) | 2.4 | 320 (160) |
| Bamboo | 1.25 | 28 (0.19) | 2,000 (14) | 2 | 284 (140) |
| Cotton | 1.55 | 58 (0.4) | 1,200 (8) | 7 | 248 (120) |
| Hemp | 1.48 | 100 (0.69) | 10,200 (70) | 2.8 | 284 (140) |
| Reinforcing Fiber, Inorganic | | | | | |
| Carbon HS | 1.81 | 600 (4.1) | 35,100 (242) | 1.5 | 1472 (800) |
| Ceramic (Si C) | 3.08 | 856 (5.9) | 60,200 (415) | 1.4 | 2500 (1370) |
| Glass (Type E) | 2.56 | 755 (0.52) | 10,500 (72.4) | 4.8 | 1544 (840) |
| Metal (Tungsten-Boron) | 2.54 | 522 (3.6) | 58,000 (400) | 0.9 | 932 (500) |

Table 6 provides comparative data for some common organic and inorganic reinforcements used in polymer matrices. Organic reinforcements contain carbon and usually are less dense and expensive than inorganic reinforcements. Kevlar is an example of an inorganic, highly oriented aramid polymer that provides outstanding properties compared to organic materials such as bamboo, cotton, and hemp.

A common and low-cost inorganic reinforcement, glass provides high strength and temperature resistance compared to organic reinforcements. The majority of polymer composites employ glass fibers with thermosetting polymer resins (such as *fiberglass with polyester*), though thermoplastic composites are increasingly used. High strength (HS) carbon fibers provide even better performance at a lower density. Other inorganic reinforcements of increasing interest include ceramics, such as silicon carbide (Si C), and metals, such as tungsten boron.

The majority of the strength and stiffness of most composites is derived from the reinforcements rather than the matrix material. Typical loadings of reinforcements are on the order of 50 percent by component volume and 75 percent by component mass, with the balance going to the matrix material.

A composite's material properties can be estimated by considering the volume fraction, $f$, of the reinforcements, as well as the orientation of the reinforcements relative to the load direction. The density ($\rho_C$), maximum modulus ($E_C$), and maximum strength ($\sigma_C$) of the composite in the direction of the reinforcements are estimated using a rule of mixtures:

$$\rho_C = f\rho_R + (1-f)\rho_M$$
$$E_C = fE_R + (1-f)E_M$$
$$\sigma_C = f\sigma_R + (1-f)\sigma_M$$

where the properties for the reinforcements and matrix are indicated by the subscripts $R$ and $M$, respectively.

*Example:* Estimate the maximum strength and modulus of a composite with 52 percent volume fraction of carbon fibers embedded within a polyester thermoset matrix.

Based on the data of Table 6, the properties are estimated as:

$$\rho_C = 0.52 \times \frac{1.81g}{cm^3} + (1-0.52) \times \frac{1.4g}{cm^3} = 1.61g/cm^3$$

$$E_C = 0.52 \times 242\ GPa + (1-0.52) \times 4\ GPa = 128\ GPa$$

$$\sigma_C = 0.52 \times 4.1\ GPa + (1-0.52) \times 0.05\ GPa = 2.28\ GPa = 2160\ MPa$$

While these properties seem precise, the modulus and strength are the theoretical maxima in the fiber direction. And the modulus and strength in the transverse direction to the reinforcements are substantially reduced, since they are constrained by the lesser properties of the matrix. As such, nonwoven mats and woven fabrics (as shown in Fig. 17b and Fig. 17c) that provide quasi-isotropic properties are commonly employed.

Table 7 provides some representative properties of common plastic composites in different matrices. Since polymer composites can be brittle and subject to stress concentrations, it is critical that composite products be tested according to established composite standards, including flexion (ASTM D7264), tension (ASTM D3039), shear (ASTM D3518), compression (ASTM D6641), impact (ASTM D7136), fatigue (ASTM D3479), and others.

**Table 7. General Properties of Reinforced Plastics** *(ASTM Test Procedures)*

| Material | Specific Gravity | Tensile Strength $10^3$ psi | Tensile Modulus $10^6$ psi | Compressive Strength $10^3$ psi | Flexural Strength $10^3$ psi | Izod Impact Strength ft-lb/in notch | Heat Resistance Continuous °F | Arc Resistance seconds |
|---|---|---|---|---|---|---|---|---|
| **Polyester** | | | | | | | | |
| Glass cloth | 1.5–2.1 | 30–70 | 1–3 | 25–50 | 40–90 | 5–30 | 300–350 | 60–120 |
| Glass mat | 1.3–2.3 | 20–25 | 0.5–2 | 15–50 | 25–40 | 2–10 | 300–350 | 120–180 |
| Asbestos | 1.6–1.9 | 30–60 | 1–3 | 30–50 | 50–70 | 2–8 | 300–450 | 100–140 |
| Paper | 1.2–1.5 | 6–14 | 0.5–1.5 | 20–25 | 13–28 | 1–2 | 220–250 | 28–75 |
| Cotton cloth | 1.2–1.4 | 7–9 | 0.5–1.5 | 23–24 | 13–18 | 1–4 | 230–250 | 70–85 |
| **Epoxy** | | | | | | | | |
| Glass cloth | 1.9–2.0 | 20–60 | 2–4 | 50–70 | 70–100 | 11–26 | 330–500 | 100–110 |
| Glass mat | 1.8–2.0 | 14–30 | 1–3 | 30–38 | 20–26 | 8–15 | 330–500 | 110–125 |
| Paper | 1.4–1.5 | 10–19 | 0.5–1 | 20–28 | 19–24 | 0.5–1 | 260–300 | 30–100 |
| **Phenolic** | | | | | | | | |
| Glass cloth | 1.8–2.0 | 40–60 | 1–3 | 35–40 | 65–95 | 10–35 | 350–500 | 20–130 |
| Glass mat | 1.7–1.9 | 5–20 | | 17–26 | 10–60 | 8–16 | 350–500 | 40–150 |
| Asbestos | 1.7–1.9 | 40–65 | 2–5 | 45–55 | 50–90 | 1–6 | 350–600 | 120–200 |
| Paper | 1.3–1.4 | 8–20 | 1–2 | 20–40 | 10–30 | 0.3–1 | 225–250 | Tracks |
| Cotton cloth | 1.3–1.4 | 7–16 | 0.5–1.5 | 30–44 | 14–30 | 0.5–3 | 225–250 | Tracks |
| Nylon cloth | 1.1–1.2 | 5–10 | 0.3–0.5 | 28–36 | 9–22 | 2–4 | 150–165 | Tracks |
| **Silicone** | | | | | | | | |
| Glass cloth | 1.6–1.9 | 10–35 | 1–2 | 25–46 | 10–38 | 5–13 | 400–700 | 150–250 |
| Asbestos cloth | 1.7–1.8 | 10–25 | 1–2 | 40–50 | 12–20 | 6–9 | 450–730 | 150–300 |

**Polymer Composite Production Methods.—**Many different layouts and methods for production of composite materials are commonly in practice (Fig. 17), and composite design and processing must be matched with end-use application requirements. The lowest cost approaches generally are random short fiber reinforcements and nonwoven mats, with more complex composites providing higher performance at higher cost.

*Random Fibers:* Randomly distributed and oriented short fiber reinforcements are dispersed in a polymer matrix, such as the "gunk" or bulk molding compound (BMC) usually used for compression molding of thermosets. The same mixing approach can be applied to create compounds for injection molding or deposition via nozzle spraying onto a form. The resulting products (Fig. 17a) tend to have properties approximately averaging those of the matrix and reinforcing materials.

*Nonwoven Mats:* Using longer fibers (often greater than 1 inch), nonwoven mats (Fig. 17b) tend to offer better performance than shorter fibers, as increasing length of the reinforcing material enhances the load-carrying capability of the resulting structure. The nonwoven fibers in a mat tend to be randomly distributed and oriented; multiple fibers may cross through the thickness of the loaded structure. Nonwoven mats can be conformed to male or female molds and set with adhesive matrices and surface coatings. Alternatively, mats may be placed in a two-piece mold with matrix material injected, as in resin transfer molding (RTM), or in sheets of matrix material, as in compression molding (CM). Sheet molding composite (SMC) is another option, with nonwoven fibers already embedded in a polymer matrix for compression molding.

*Woven and Braided:* In these more highly structured composites (Fig. 17c and Fig. 17d), the reinforcements carry a higher portion of the load, providing improved properties for designed load directions. Woven composite fabrics are formed by interlacing 0-degree fibers (the *warp*) with 90-degree fibers (the *weft*) in a repeating pattern. Varying repeating patterns results in different weaves featuring varied appearance, surface smoothness, and drape. *Plain* weaves, in which each warp and weft fiber crosses on a 1:1 over:under pattern, are efficient but prone to crimping with larger fibers. For this reason, many composite fabrics use *twill* weaves, in which the warp and weft fibers cross over at every second or more intersection. Woven composite fabrics and tapes are mass produced and typically provided on rolls for incorporation into laminate plies.

Providing an even higher level of performance, braided composites are comprised of a series of continuous, mechanically interlocked fibers, with varying orientations. The position and orientation of the individual braids can be designed with support mandrels to create seamless tubes and fabrics with fibers crossing the thickness direction.

**Forming and Curing Composite Structures.**—Many composite products are made by manually assembling composite layers (knows as *layups*) in a female mold or on a form. With layups (Fig. 17e), the reinforcing layers can be directly distributed in accordance with the stress distribution expected when the product is placed in service.

*Spray Layup:* Many large composite products are made by feeding the fibers, resin, and catalyst through a *chopper gun* that directs a spray of chopped fibers and resin onto the desired form. Such spray layup (Fig. 18a) is very common and economical for achieving high deposition rates in large structures, plus deposition can locally control orientation and thickness of the developed composite. Yet spray layup often results in only average properties and stiffness-to-weight ratios, due to relatively short fiber lengths and higher resin content than other processes. Also, airborne fibers and resin particulates cause environmental and safety issues, making automation desirable; after initial investment, it can reduce costs and improve composite properties.

*Hand Layup:* As an alternative to spray layup, woven fabrics, nonwoven mats, reinforcing tapes, and aesthetic films may be placed on forms, wetted, and adhered with resin applied with brushes and rollers. Such *hand layups* (also referred to as *wet layups*) are broadly used with varied reinforcements and resins. Typically, composites can then be cured at room temperature and pressure, with gel coats applied for glossy surface finishes. Compared to spray layups, hand layups can achieve improved mechanical properties, due to higher strength reinforcing fabric and reduced internal voids. But final properties are dependent on the skill and care of the operator.

Fig. 18a shows increasingly high performance (but more costly) ways of curing hand layups. With some thermosetting catalysts, resin can be cured at room temperature, without added pressure, facilitating manufacture of large structures, such as boats and storage tanks for liquids.

*Vacuum Curing:* A vacuum can be used to further reduce voids and increase density and strength of composites. Following hand layup, the positioned material is placed in a vacuum bag or otherwise sealed with film. When vacuum is pulled from within, the resin tends to fully wet the reinforcements, reducing the resin and void content, as well as removing volatiles emitted during cure. After the product is cured, sealing materials are removed, along with any release layers. Vacuum equipment and sealing and release materials add to costs, and again final properties are operator dependent.

*Autoclaves and Prepegs:* In lieu of vacuum bagging, autoclaves, providing positive pressure and temperatures higher than environmental conditions, may be used with a broader set of resin chemistries to produce composites with higher end-use temperatures. Autoclaves also are operable with *prepregs* (tapes and sheets of reinforcing fabrics pre-impregnated with resin). The prepregs are placed directly on the form without applied resin. After prepreg plies are built up, they may be warmed and debulked to minimize entrapped air. The form with the applied prepregs is then sealed and cured in the autoclave. Composite end-use performance can justify investment in the autoclave and associated tooling.

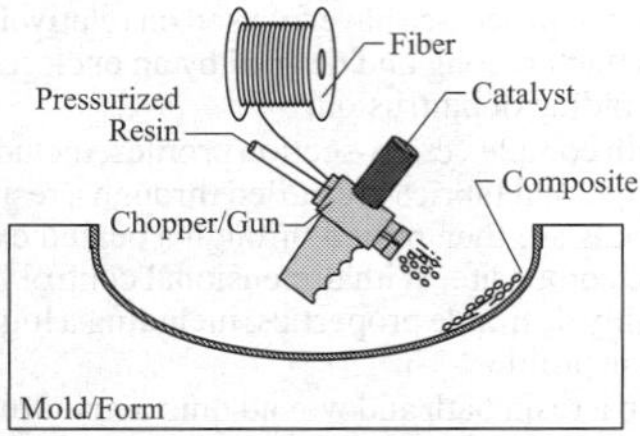


**Spray Layup**
Fibers, resin, and catalyst are fed through a chopper and sprayed onto a form, depositing composite polymer in consistent or variable thicknesses.

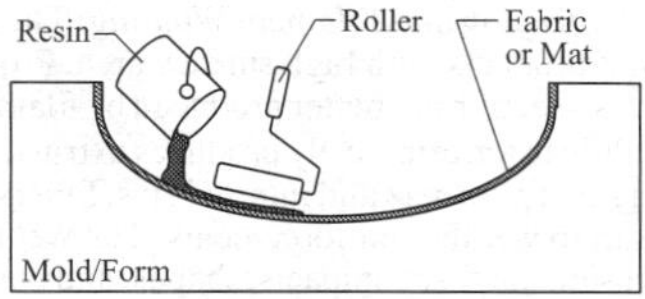


**Hand Layup**
Also known as *wet lay-up*, fabrics, mats, reinforcing tapes, and/or films are hand-placed on forms, wetted, and adhered with resin.

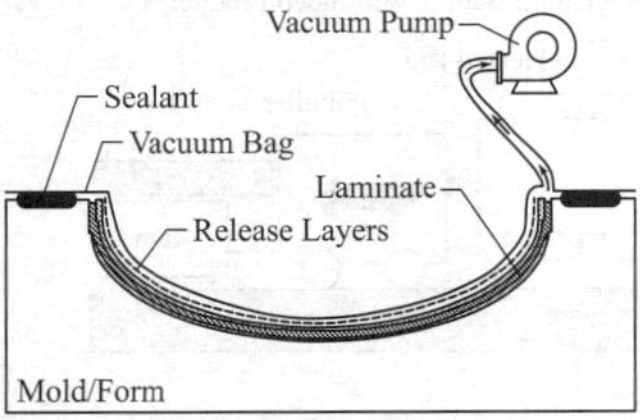


**Wet Layup with Vacuum Bag**
Use of sealing materials, release layers, and vacuum reduce voids and increase composite density.

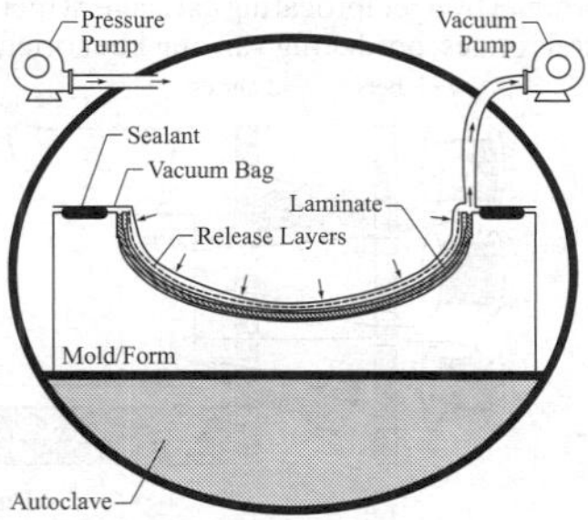


**Wet Layup with Autoclave**
Prepeg plies are built up and then sealed and cured by an autoclave providing positive pressure and increased temperatures.

Fig. 18a. Spray and Hand Layup Methods of Forming Reinforced Plastic Composites

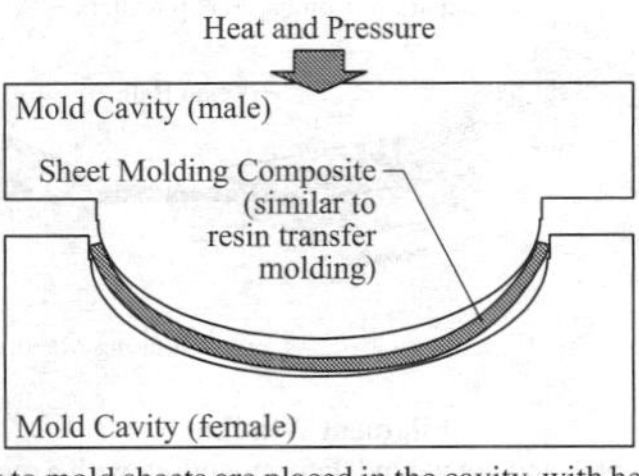


Ready to mold sheets are placed in the cavity, with heat and pressure applied to remove air and cure the product.

Fig. 18b. Compression Molding of Plastic Composites

*Resin Transfer and Compression Molding:* In resin transfer molding, woven fabrics and nonwoven mats are placed in a female mold cavity. A shot of resin is placed or otherwise injected into the mold cavity. Heat and pressure are applied with the closing of a matched male core, removing entrapped air, wetting reinforcements, and curing the composite product. The applied pressure and temperature are greater than those of an autoclave, resulting in composites with even higher fiber and lowest void contents. Compression molding (Fig. 18b) of sheet molding composites (oversized prepregs supplied as sheets ready to mold with fibers and resin) works similarly for thermoplastic matrices.

*Pultrusion and Filament Winding:* The discussed processes have focused on relatively thin products with high surface area. Products that are long and defined by an enclosed cross-section are better produced by filament winding or pultrusion.

Pultrusion effectively produces extrusions with complex cross-section profiles, including sharp corners and internal ribs. Fibers and optional fabrics are pulled through a resin bath to wet the reinforcements. The wet materials are then pulled through a heated extrusion die that compacts, shapes, and cures the composite. With dimensional control of high fiber content, pultruded products offer highly desirable properties, including a high stiffness-to-weight ratio given their cross-section profiles.

In filament winding, fibers are passed through a resin bath and wound onto a mandrel. Orientation of the fibers is controlled along the length and through the thickness of the product by a reciprocating carriage, which traverses the length of the mandrel as the mandrel rotates, producing varying fiber orientations and thicknesses.

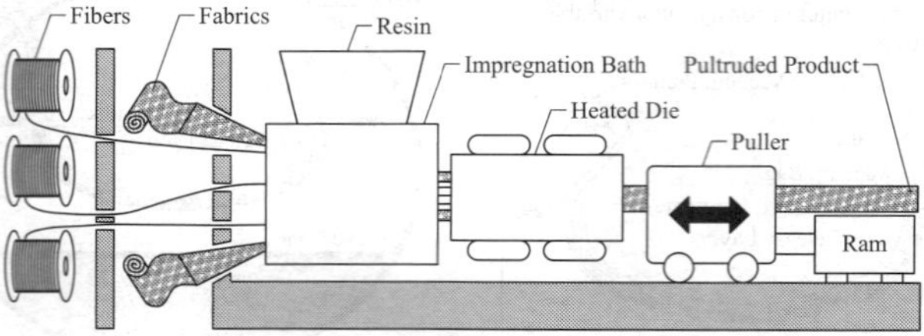


**Pultrusion**

Fibers (and sometimes fabrics) are pulled through a resin bath and heated extrusion die to compact, shape, and cure the composites.

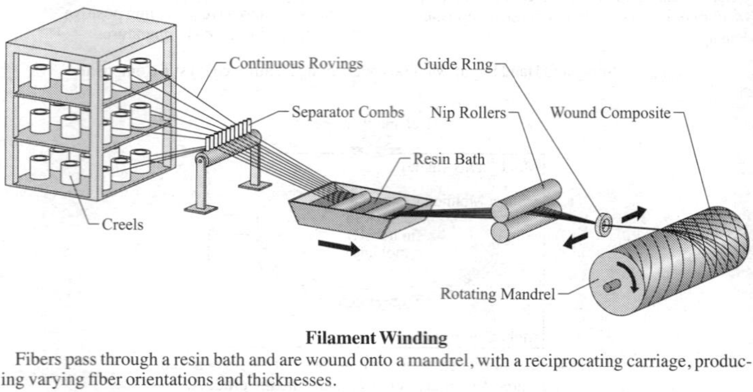


**Filament Winding**

Fibers pass through a resin bath and are wound onto a mandrel, with a reciprocating carriage, producing varying fiber orientations and thicknesses.

Fig. 19. (a) and (b). Two Methods for Creating Long Composite Products with Enclosed Cross-Sections

*Sandwich Polymer Composites:* Given increased design and automation capabilities, composites are being increasingly implemented in the form of complex, 3D sandwich structures (see Fig. 17 on page 585). This approach often uses molds into which reinforcing composite tapes, stamped or machined metal parts, or other composite structures are placed. Resin or foam is then introduced into the mold—by a shot of thermosetting resin in resin transfer molding, layer of thermosetting resin in an autoclave, sheet of thermoplastic in compression molding, shot of thermoplastic melt in injection molding—such that the polymer encapsulates and bonds with the structural members to form an integrated structure. Fig. 17f shows an example where two outer walls are formed by a carbon fiber composite supported by urethane foam with an internal metal mounting flange having external mounting points. The combination of components provides appropriate attachment points with high stiffness to weight ratio.

While sandwiched structures offer long life and superior performance, their combined materials can reduce end of life sustainability. As possible, designers should strive to make the composite parts easily separable, so the higher value components (metal and other structural reinforcements) may be recovered, reused, or recycled.

**Injection Molding.—**This process produces the majority of plastics consumer and industrial products. Two halves of a mold include a female cavity and male core that define the product features. The mold halves are clamped together with high force by a hydraulic piston or a toggle mechanism driven by hydraulic or electric means. Within the mold is a network of *runners*, designed to distribute melted material from the nozzle of the injection unit throughout the mold cavities. Coolant channels passing near the cavities keep the mold cold, below the freezing range of the plastic.

A screw or ram within the injection unit forces the molten plastic through the runners and into the cavities, holding the melt pressure (5,000–15,000 psi, 34–103 MPa) while the moldings cool and solidify within the mold cavities. The screw then rotates as it retracts, generating more hot melt for the next shot. When enough material has been fed to complete the part, the moving mold half with the core(s) and moldings are withdrawn, activating ejector pins that push the moldings and runners off the mold core(s).

Injection machines are rated by melting capacity, shot size, platen size, and clamping force. They range from machines for forming micron-sized gears up to meter-sized garbage dumpsters. Capable of making identical items with very close dimension tolerances, injection molding is ideal for production of interchangeable industrial parts destined for complex assemblies. Undercut shapes can be molded with the aid of side-draw mold sections. For example, millions of bottle caps are molded in automatic unscrewing molds.

Successful product and process design requires input from the product designer, mold designer, mold maker, molder, and customer. Producing female mold cavities and their male halves in fine tool steels is costly, so it is essential to be certain that the part design is accurate before finishing the mold cavities. Because of the investment in the mold, injection molding may not be preferable for very short runs, especially considering available 3D printing technologies. (For more on this topic, see *Additive Manufacturing Plastics* on page 612. Inexpensive molds for making a few hundred parts can be made from metal-filled epoxy casting compounds. But CNC machining of aluminum and other rapidly produced metal molds is widely used for production of tens of thousands of parts with tolerances and surface finishes that still exceed the capabilities of most current 3D printing processes.

**Design for Molding.—**Product analysis is a good starting point in the design process for plastics molding. Make a detailed checklist of the functions the new part will have to perform and the operating conditions it will encounter. If the part will replace an older design that required fabrication from other materials and, perhaps, subassembly with fasteners or adhesives, the new design may be able to take advantage of injection molding's versatility to meld the several functions into a single new part design.

*Relative Plastics Costs:* The list of some comparative material costs in Table 8 should serve to narrow materials choices, but current quantity quotations from various suppliers must be obtained. It is important to account for material density when calculating material costs, since the density of thermoplastics may vary significantly. For injection molding of a large production run, a crude estimate of part cost is double the material cost to account for amortized mold cost, as well as the molder's processing costs.

**Table 8. Costs of Plastics**

| Type/Material | Price (US $) | | Type/Material | Price (US $) | |
|---|---|---|---|---|---|
| | Per Pound | Per Kilogram | | Per Pound | Per Kilogram |
| **Thermoplastics** | | | PSU | 5.50 | 12.13 |
| ABS | 1.24 | 2.73 | PTFE | 7.35 | 16.20 |
| Acetal, POM | 1.41 | 3.11 | PVC | 0.89 | 1.96 |
| HDPE | 0.70 | 1.54 | TPU | 2.06 | 4.54 |
| LCP | 6.40 | 14.11 | **Recycled Thermoplastics** | | |
| LDPE | 0.78 | 1.72 | HDPE | 0.68 | 1.50 |
| PA66, Nylon | 1.64 | 3.62 | Recycled LDPE | 0.39 | 0.86 |
| PC | 1.77 | 3.90 | Recycled PET | 0.61 | 1.34 |
| PEEK | 45.00 | 99.21 | Recycled PP | 0.34 | 0.75 |
| PEI | 8.80 | 19.40 | **Thermoset** | | |
| PET, bottle | 0.78 | 1.72 | Epoxy | 1.12 | 2.47 |
| PMMA | 1.28 | 2.82 | Thermoset Melamine | 1.55 | 3.42 |
| Polyester | 1.23 | 2.71 | Thermoset Phenolic | 0.80 | 1.76 |
| PP | 0.85 | 1.87 | Thermoset Polyester | 1.84 | 4.06 |
| PPS, 30%GF | 3.85 | 8.49 | Thermoset Polyurethane | 1.10 | 2.43 |
| PS | 1.10 | 2.43 | Thermoset Urea | 1.07 | 2.36 |

Based on 2018 market prices. Recycled material prices are for pelletized feed stock and vary with quality.

*Wall Thicknesses:* Wall thickness throughout the part, referred to as the nominal wall thickness, is one of the most critical design parameters. All critical dimensions and tolerances should be calculated and decided before the mold is made, since modifications are costly. In general, wall thicknesses should be kept as thin as needed to meet the part's design requirements and as uniform as possible. Ideally, the flow of molten plastic should be arranged to flow through thicker sections into thinner ones rather than the reverse. This is accomplished—in consultation with the mold builder—by gating into the thick sections of cavities. The mold builder will help, too, with the decisions on how many cavities are needed, how to arrange them in the mold for optimal temperature control, the best runner layout and sizing, and what size mold and injection machine will be economical for the production desired. Most injection-molded plastics parts have wall thicknesses in the range from 0.02 to 0.25 inch (0.5 to 6 mm). For compounds containing ½-inch (12.7-mm) and longer glass fibers, and for thermoplastic polyurethanes, the range shifts upward by a factor of 2 or more.

When the plastics selection for the part has been finalized, the suppliers of the plastic resin can provide recommendations of grades and wall thicknesses appropriate to those grades. Table 9 shows typical wall thicknesses for various thermoplastics.

**Table 9. Typical Nominal Wall Thicknesses for Various Classes of Thermoplastics**

| Thermoplastics Group | Typical Working Range | | Thermoplastics Group | Typical Working Range | |
|---|---|---|---|---|---|
| | inch | mm | | inch | mm |
| Acrylonitrile-butadiene-styrene (ABS) | 0.045–0.140 | 1.14–3.56 | Polyester elastomer | 0.025–0.125 | 0.64–3.18 |
| Acetal | 0.030–0.120 | 0.76–3.05 | Polyethylene | 0.030–0.200 | 0.76–5.08 |
| Acrylic | 0.025–0.150 | 0.64–3.81 | Polyphenylene sulfide | 0.020–0.180 | 0.51–4.57 |
| Liquid-crystal polymer | 0.008–0.120 | 0.20–3.05 | Polypropylene | 0.025–0.150 | 0.64–3.81 |
| Long-fiber-reinforced plastics | 0.075–1.000 | 1.91–25.4 | Polystyrene | 0.035–0.150 | 0.89–3.81 |
| Modified polyphenylene ether | 0.045–0.140 | 1.14–3.56 | Polysulfone | 0.050–0.150 | 1.27–3.81 |
| Nylon | 0.010–0.115 | 0.25–2.92 | Polyurethane | 0.080–0.750 | 2.03–19.05 |
| Polyarylate | 0.045–0.150 | 1.14–3.81 | Polyvinyl chloride (PVC) | 0.040–0.150 | 1.02–3.81 |
| Polycarbonate | 0.040–0.150 | 1.02–3.81 | Styrene-acrylonitrile (SAN) | 0.035–0.150 | 0.89–3.81 |
| Polyester | 0.025–0.125 | 0.64–3.18 | … | … | … |

**Stresses and Deflections in Load-Bearing Parts.—**If the plastic part is to carry loads, load-bearing areas should be analyzed for stress and deflection according to the principles presented earlier. If stress or deflection is found to be too high, solutions include using ribs or contours to increase section modulus; using a higher-strength or higher-modulus (i.e., fiber-reinforced) material; or increasing the wall thickness if it is not already excessive. Where space allows, adding or thickening ribs can increase structural integrity without making walls thicker.

Formulas given in *Beam Calculations* starting on page 250, which presume elastic behavior, can be used if care is taken to use moduli and strengths related to anticipated service times and temperatures.

*Example:* Consider a simple case of a beam with rectangular cross section $b$ wide and $h$ deep, of length $l$ centrally loaded and freely supported at its ends. This is an example of Case 2 on page 251 with a beam cross section depicted generically in the first diagram of *Moments of Inertia, Section Moduli, and Radii of Gyration* on page 235. Applicable formulas for the moment of inertia $I$ of the beam cross section, the section modulus $Z$, and the maximum tensile stress $s$ beneath the central load are

$$I = \frac{bh^3}{12} \qquad Z = \frac{I}{(h/2)} = \frac{bh^2}{6} \qquad \sigma = \frac{Wl}{4Z} = \frac{6Wl}{4bh^2}$$

From the last box for Case 2, the maximum deflection at the beam center is given by

$$y = \frac{Wl^3}{48EI} = \frac{Wl^3}{4Ebh^3}$$

Consider the example of a small beam whose width = 19 mm, depth (wall thickness) = 2.0 mm, and length between supports = 102 mm. The beam is to support a central load of 22.24 N (5 lb). This beam's cross section is depicted in Fig. 21(b). As it has very little depth, hence a low sectional modulus, it may not handle the job. The plastic's flex modulus is 2069 MPa (300,000 psi). Using the formulas above, the four quantities to be calculated are

$$I = \frac{19(2^3)}{12} = 12.67 \text{ mm}^4$$

$$Z = \frac{(12.666667)2}{2} = 12.67 \text{ mm}^3$$

$$\sigma = \frac{6(22.24)102}{4(19)2^2} = 44.77 \text{ MPa}$$

$$y = \frac{(22.24)(102^3)}{(4)(2068.5)(19)(2^3)} = 18.77 \text{ mm}$$

Both the stress and the deflection are too high. The wall thickness could be increased, but instead a rib will be added that will run along the underside and center of the beam, changing its cross section to that shown in Fig. 21(c). The rib will be 10 mm deep but only 1.0 mm thick, with a slight draft of 0.5 degree per side. Following the formulas given in the table *Moments of Inertia, Section Moduli, and Radii of Gyration* on page 235, see the third row of the T-sections for the one with the tapered rib; the new cross-sectional area is 47.548 mm$^2$, and the location of the neutral axis is 9.87 mm above the most extreme material at the bottom of the rib. The new cross-sectional moment of inertia $I$ is 339.6 mm$^4$, 27 times greater than the original $I$. The section modulus $Z = 339.6/9.87 = 34.4$ mm$^3$. The recomputed maximum stress is 16.5 MPa, and the maximum deflection is 0.70 mm, acceptable for both the material and the application. To achieve the same result from a heavier rectangular-section beam would require a thickness (depth) of 5.99 mm, tripling the beam's weight and increasing molding difficulties, whereas the new rib adds only 25 percent to the total section weight. As shown in Fig. 21(c), a fillet should be added at the bottom of the rib with a radius equal to half the nominal wall thickness to assist molding and avoid stress concentrations.

*Draft:* Molded features are often cut into the mold perpendicular to the parting plane. Removal of these parts from the mold is easier if they are tapered in the direction of mold opening. This taper is called *draft* in the line of draw or mold movement, and it allows the part to break free of the mold by creating a clearance as soon as the mold starts to open. Fig. 20 defines draft and the dimensions associated with it. Plastics materials shrink as they cool, so they grip mold projections very tightly and ejection can be difficult without sufficient draft.

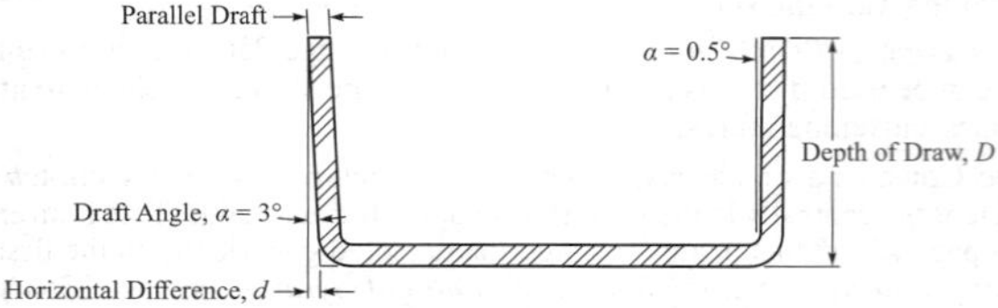


Fig. 20. Defining Dimensions Related to Draft and Draft Angle

The amount of draft also depends on the surface finish of the mold walls. A draft of 0.5 degree on each side of a projection on the part (as shown at right in Fig. 20) is generally considered the minimum, although a 3 degree draft (as shown at left in Fig. 20) is often used, especially for textured surfaces or glass-filled resins. Any texturing of the surface will increase the draft requirement by at least 1 degree per side for every 0.001 inch (0.025 mm) of texture depth. For uniform wall thickness, parallel draft can be applied to standing walls as shown in Fig. 20, though draft is primarily needed on the surfaces of the part that will contact the mold core.

The exact relationship between the draft angle α and the dimensional difference from the top to the bottom of the draw depth *D* is given by

$$d = D \tan \alpha \tag{30}$$

where tan is the tangent of the draft angle in degrees. An good approximation is

$$d = 0.0175 \alpha D \tag{31}$$

This is accurate to the nearest 0.001 inch (0.025 mm) for draft angles up to 5 degrees and draw depths up to 10 inches (254 mm).

Most plastics parts are so designed that they can be ejected parallel with the direction of mold parting. Complex parts with undercuts may require mold designs with cavity-forming projections that must move at an angle to the direction of opening. Between these two extremes lie such items as "windows," or simple openings in the side of a molding, which can be produced by the normal interaction of the two main parts of the mold.

*Fillets, Radii, and Undercuts:* Sharp corners are always to be avoided in injection-molded part designs because they represent points of stress concentration. Sharp inside corners are particularly to be avoided in moldings because severe molded-in stresses are generated as the material shrinks onto the mold corner. Sharp corners also cause poor material flow patterns, reduced mechanical properties, and increased tool wear. Therefore, inside corner radii should be equal to half the nominal wall thickness (as shown in Fig. 21(c) for a 0.04 inch or 1 mm rib thickness) with a minimum of 0.020 inch (0.51 mm) for parts subject to stress, and a 0.005 inch (0.127 mm) radius for stress-free parts. Outside corner radius should equal the inside corner radius plus the wall thickness.

With an inside radius of half the wall thickness, a stress concentration of 1.5 is a reasonable assumption, and for radii down to 0.1 times the wall thickness, a stress concentration of 3 is likely. More information on stress concentrations is found in *Working Stress* on page 206. A suitable value for *q* in Equation (8) on page 206, for plastics materials, is 1. That is, the actual stress concentration factor equals the theoretical value.

*Impact Resistance:* The impact resistance of a plastics part is directly related to its ability to absorb mechanical energy without fracture or deformation, and this ability depends on

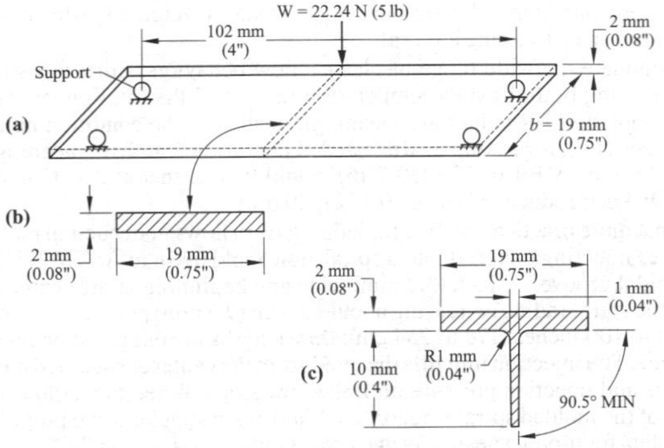


Fig. 21. (a) Original Beam; (b) Original Central Section; (c) Cross Section of Center with Rib Added to Reinforce the Structure
Note: *Dimensions in parentheses are for example on page 593.*

the material properties and the part geometry. Increasing wall thickness may improve impact resistance but may also hurt impact resistance by making the part too stiff so that it is unable to deflect and distribute the force. Validation is required in critical applications.

Other ways to improve section properties include use of top-hat and corrugated sections, crowning or doming of some areas, and reinforcement with metal or other inserts that are placed in the mold before it is closed. To keep molded parts uniform in wall thickness, cores or projections may be provided in the mold to prevent a space being filled with molding material. Blind holes can be cored by pins that are supported on only one side of the mold and through holes by pins that pass through both sides. The length-to-width ratio should be kept as low as possible to prevent bending or breakage under the high pressures used in the injection molding process.

Agency approvals for resistance to flammability or heat, electrical properties, or other characteristics are usually based on specific wall thicknesses. These restrictions sometimes necessitate thicker walls than are required for structural strength purposes.

**Melt Flow in the Mold.—**When determining wall thickness, design engineers should have knowledge of mold design and consider the ability of plastics to flow into narrow mold channels. Thermoplastic melts have very high viscosities—thousands to hundreds of thousands times that of room-temperature water—requiring pressures ($P$) of 5,000–30,000 psi (35–206 MPa) to move the melt quickly from the injection unit through the mold runners and into the cavities. Viscosity is strongly dependent on temperature, so transfer to the chilled mold must be quick to obtain fully filled cavities and accurately dimensioned products. Due to the high viscosities and rapid flow, frictional heat is dissipated within the melt, helping keep it warm during its travel.

Notably, plastic melts are non-Newtonian. That is, viscosity depends not only on temperature, but also on the flow rate and pressure drop $\Delta P$. So doubling the injection pressure may triple or quadruple flow rate through the mold passages, but unpredictable results may occur if the melt temperature is too near the freezing range.

If the product design calls for thin wall sections, *Melt Flow Rate* (*MFR*), formerly known as *Melt Flow Index* (*MFI*), can be used as a selection characteristic in choosing among grades of candidate plastics. The higher the melt flow rate, the easier it likely will be to fill thin sections during molding. Flow rate values usually are obtained by testing (ASTM Test D 1238) in a simple extrusion plastometer. In this device, a small resin sample is loaded into a 0.375 inch (9.55-mm) diameter well in a heated cylinder and brought to a test temperature above the plastic's melting range. At the bottom of the well is a die 0.083 inch

(2.1 mm) in bore diameter and 0.315 inch (8 mm) long. A weighted piston drives the melt through the die for a 10-minute interval.

To accommodate the wide range of plastics flow behaviors, thirty-six standard conditions, specifying both the melt temperature in °C and the total piston weight in kg, have been adopted. MFR values are meaningless without the condition numbers. Most polypropylenes (PP), for example, are tested at condition 230/2.16, meaning a PP resin reported to have an MFR of 12 (230/2.16) would be one that at 230°C under a piston weighing 2.16 $kg_f$, produces a flow rate of 12 g/10 min.

A somewhat more practical method for judging resin flows under actual molding conditions involves injecting material into a spiral-flow mold (akin to ASTM D3123). A shallow-trapezoidal groove, $^1/_8$ inch (3.2 mm) wide and beginning at the center, circles outward, with the spiral radius increasing about $^3/_8$ inch (9.5 mm) per turn, to a total channel length of 70 to 100 inches (178 to 254 cm); flow lengths are engraved on one side of the cavity surface. The injection unit fills the mold from the center at specified melt and mold temperatures and injection pressure until the flow stops. When the mold is opened, the flow length of the molded spiral is recorded. Many resin suppliers can provide MFR and spiral-flow data for product designers and processors.

The achievable flow length depends on temperature, pressure, wall thickness, and material properties, with representative behavior shown in Fig. 22. Spiral flow tests generally provide much more accurate information about achievable flow lengths for a given wall thickness than melt flow rate (MFR), since they are performed with shear rates similar to those of molded parts. However, spiral flow tests tend to overestimate flow lengths relative to real applications, as they calculate the maximum effect of the moving melt's heat convection relative to radial flow fronts that often occur in mold filling. Fortunately, molders can adjust melt temperature and pressure to compensate for material viscosity differences.

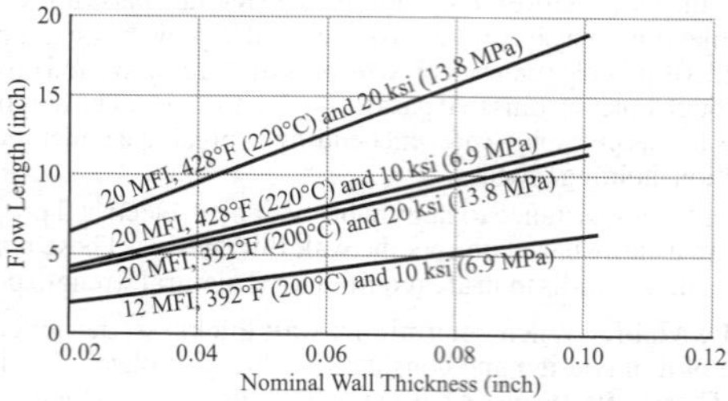


Fig. 22. Representative Spiral-Flow Curves for Plastics Materials

**Design for Assembly.**—An advantage of the flexibility of plastics parts compared to other materials is that they can often be designed for assembly by means of molded-in snap-fit, press-fit, pop-on, and thread fasteners. With careful design, no additional fasteners, adhesives, solvents, or special equipment is required, significantly reducing assembly time and costs. Improper assembly can be minimized, but tooling is often made more complex and disassembly may be difficult with these methods.

*Mechanical fasteners* designed for metals are often usable with plastics, but there are better fasteners designed specifically for plastics. Typical are bolts, self-tapping and thread-forming screws, rivets, threaded inserts, and spring clips. Care must be taken to avoid overstressing the parts. Creep can result in loss of preload in poorly designed systems.

*Snap-fit* designs are widely used, a typical application being to battery compartment covers. All snap-fit designs have a molded part that must flex like a spring, usually past a designed-in interference, then return to its unflexed position to hold the parts together. The design must have sufficient holding power without exceeding the elastic or fatigue limits of the material. With the typical snap-fit designs in Fig. 23, beam equations can be used to calculate the maximum strain during assembly. If the stress is kept below the yield point of the material, the flexing finger returns to its original position.

With some materials the calculated bending stress can exceed the yield point stress considerably if the movement is done rapidly. In other words, the flexing finger passes through its maximum deflection or strain and the material does not plasticly deform even though the yield stress has been exceeded. It is common to evaluate snap-ins by calculating the strain instead of stress. Dynamic strain ε for the straight beam, is calculated from

$$\varepsilon = \frac{3Yh_o}{2L^2}$$

and for the tapered beam, from

$$\varepsilon = \frac{3Yh_0}{2L^2K}$$

The derived values should be compared with the permissible dynamic strain limits for the material in question, if known. A tapered finger provides more uniform stress distribution and is recommended where possible. Sharp corners or structural discontinuities that will cause stress concentrations can be avoided through the use of judicious fillets.

Snap-in arrangements usually require undercuts produced by a sliding core in the mold as shown in Fig. 24(a). Sometimes the snap finger can simply be popped off when the mold is opened. An alternative to the sliding core is shown in Fig. 24(b), which requires an opening in the molding at the base of the flexing finger. Other snap-in assembly techniques that take advantage of the flexibility of plastics are shown in Fig. 25.

Molded-in threads in holes usually are formed by cores that require some type of unscrewing or collapsing mechanism, leading to tooling complications. External threads can often be molded by positioning them across the parting plane of the mold. Molding of threads finer than 28 threads per inch (0.9 mm pitch) is sometimes not practical.

*Chemical Bonding.* Chemical bonding is suited to applications that must be leak-tight or where stresses due to assembly must be minimized. However, adhesives and solvents can be hazardous, and their preparation and cure times can be prolonged. Properly applied cyanoacrylate adhesives achieve moderate strength in seconds or minutes and work well with assemblies that can be brought together very quickly after applying the bonding agent.

Chemical bonding may use solvents, solvents bodied with resin, or so-called "100% solids" adhesives, such as epoxies and cyanoacrylates. Use of solvents is limited to compatible materials that can be dissolved by the same solvent. Because most of the solvent must gradually diffuse out of the joint, solvent cementing is used mostly where a very thin film of the solvent can be applied to the accurately mating surface(s) to be joined, followed by rapid assembly, usually with clamping. Bodied solvent cements permit more flexibility in such operations. Safety precautions must be observed in handling the solvents to protect workers and in solvent recovery.

Adhesives that set up or cure with little or no loss of volatiles include epoxies, acrylics, polyurethanes, phenolics, rubbers, polyesters, and vinyls. Cyanoacrylates are often used because of their rapid setting with many materials, both similar and dissimilar, and because they are naturally catalyzed by atmospheric humidity. Their downside is the very short working time they allow for assembly.

Recommendations of both plastics suppliers and adhesives suppliers should be sought because many adhesives contain active components that can partly dissolve the plastics' surfaces, giving improved adhesion. However, some adhesives attack certain plastics, causing crazing and even joint failure. The main disadvantages of adhesives are that their bonds require time to reach full strength, may need long clamp times, may require fixtures and may involve special ovens or curing conditions. Also, surface preparation may be difficult because of the presence of grease or mold-release compounds; even a fingerprint can spoil a bond. Some adhesion-resistant materials such as polyolefins, nylons, and fluorocarbons usually need pretreatment with surface-activating primers or flame or plasma treatment prior to applying an adhesive. Some materials may need mechanical roughening to improve joint strength.

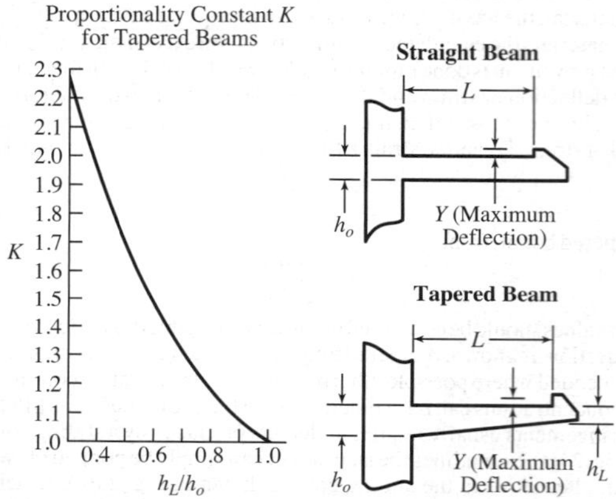


Fig. 23. Snap-Fit Designs for Cantilever Beams with Rectangular Cross Sections

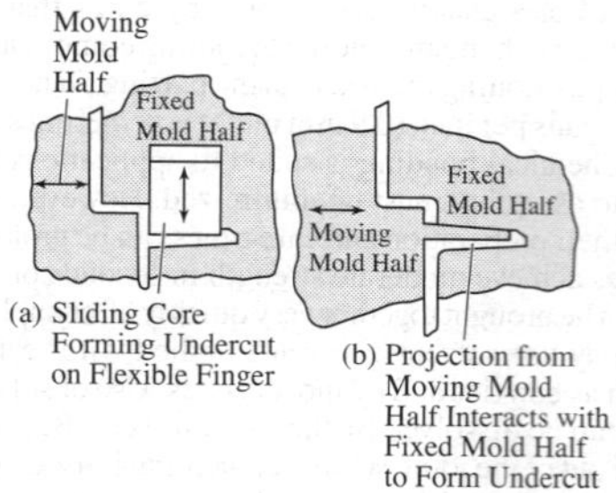


Fig. 24. (a) Arrangement for Molding an Undercut on the End of a Flexible Finger Using a Sliding Core; (b) With the Undercut Formed by a Mold Projection, the Sliding Core is Eliminated

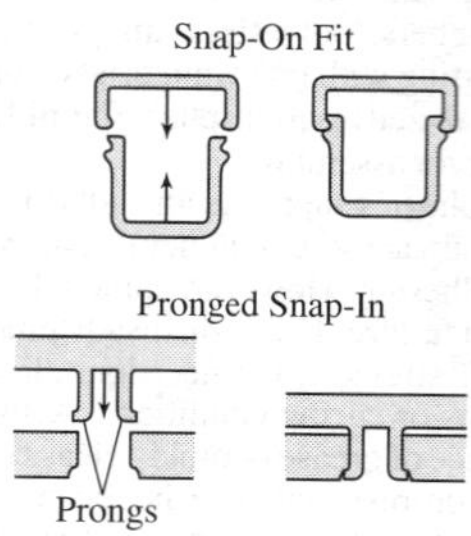


Fig. 25. Examples of Snap-In and Snap-On Arrangements

*Welding Plastics:* Techniques include ultrasonic, laser, vibration, spin, hot gas, dielectric, and induction welding, among others. Thermal methods are also used for staking, swaging, and other post-forming procedures. Materials must be melt-compatible and have similar melting temperatures to form a melt film at their interfaces.

*Ultrasonic welding* is frequently used for joining parts of a single material or melt-compatible materials that share the same melting-temperature range. High-frequency (20 to 40 kHz) acoustic vibrational energy is briefly applied to the surfaces to be joined, creating localized molecular excitation that causes a thin layer to melt. The surfaces are pressed and held together while the melt freezes. The lower part of the assembly is supported in a rigid nest fixture and the upper part is aligned, usually by the joint design. This upper part has the freedom to couple acoustically when it is in contact with the horn through which the ultrasonic energy is transmitted. Fig. 26 shows two typical joint designs for this process. The example on the left shows a simple butt-type, energy-director design that works well with amorphous materials. The inverted-cone projection concentrates the energy in a small area on both faces of the joint. This area melts quickly and the material flows as the parts are pressed together. Shown at the right in Fig. 26 is a shear-interference joint. Melting of both components starts in the small initial contact area, and flow continues along the near- vertical wall as the parts are pressed together, creating a continuous, leakproof joint with a strength that often equals that of the parts joined.

With hygroscopic plastics, welding should be done as soon as possible after molding because even small amounts of moisture can weaken bonds. Otherwise, drying immediately before welding is advisable. Also, filled and fiber-reinforced materials are more difficult to weld to achieve full part strength. It shoud be noted that horn frequencies in ultrasonic welding are not audible to the ear, but sounds that cause discomfort may be generated when plastics parts vibrate at lower frequencies, necessitating sound-proofing.

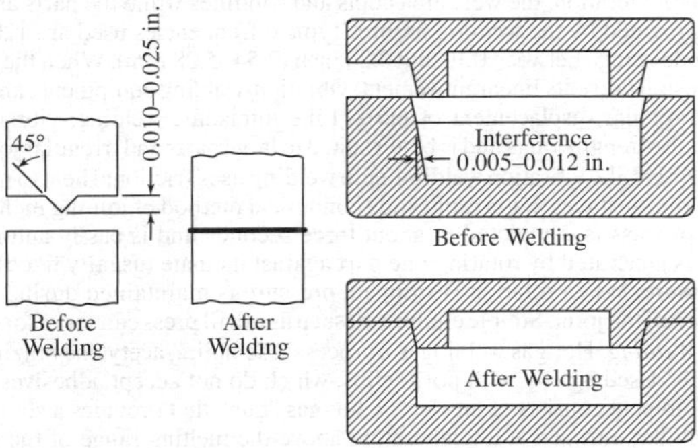


Fig. 26. (left) Energy-Director Type of Ultrasonic Weld Joint for Molded-Part Assembly and (right) Typical Shear-Interference Joint

*Laser Welding:* Laser welding has become a popular joining technique given its flexibility and fast welding velocities, and because internal welding leaves smooth outer part surfaces. However, this method is not suitable for plastics with high crystallinity or fillers, due to laser refraction and absorption.

*Direct laser welding* (Fig. 27a) is performed with a $CO_2$ laser (wavelength of 10.6 microns) that traverses a thin plastic layer located above a substrate. At this wavelength, the top layer quickly absorbs the laser energy to locally heat and weld it to the substrate. $CO_2$ lasers are available with sufficient lasing power (2 kW) such that the speed of laser welding is primarily limited by material handling and tracing velocity of the laser. Speeds of up to 45 mph (20 m/s) can be achieved with spot widths of 0.04 inch (1 mm) for layer thicknesses up to 0.02 inch (0.5 mm). Above this thickness, the top layer absorbs the laser energy and is likely to be cut without providing an adequate weld.

Transmission laser welding (Fig. 27b) is performed with a laser that passes through a transparent layer of the first part that is clamped to an absorbing layer of the second part. The infrared laser (typical wavelengths around 900 nm with a Nd:YAG or diode) passes through the first part to locally heat the top surface of the absorbing layer. Weld geometry and velocity is determined by processing parameters, including laser power, spot diameter, laser duty cycle, and material properties. While use of transparent and absorbing materials as shown is typical as shown in Fig. 27b, both materials may be transparent with an infrared-absorbing coating applied to their interface or the transparent layers can be died black to appear opaque.

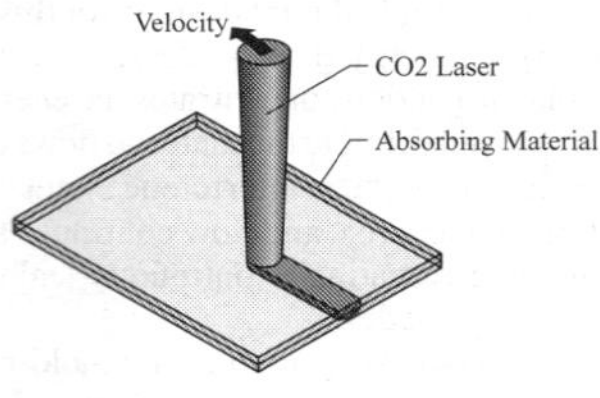


Fig. 27a. Direct Laser Welding

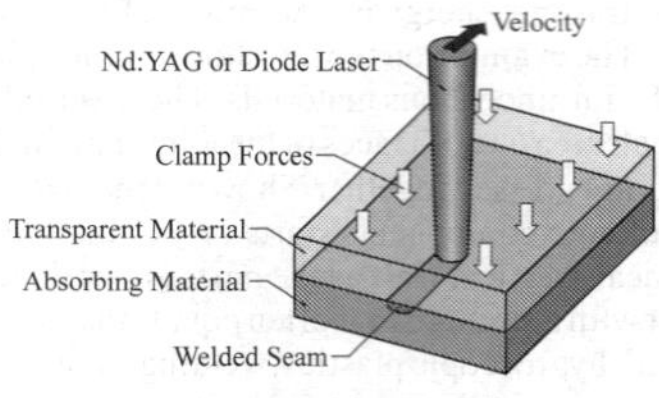


Fig. 27b. Transmission Laser Welding

*Vibration Welding:* Vibration welding resembles ultrasonic welding except that the parts to be joined are rubbed together to produce frictional heat that melts the faces. The energy is transferred in the form of high-amplitude, low-frequency, reciprocating motion. After a short period of vibration, the weld area cools and solidifies while the parts are held in the alignment provided by the welding fixture. Typical frequencies used are 120 to 240 Hz, and amplitudes range between 0.10 and 0.20 inch (2.54–5.08 mm). When the geometry or assembly design prevents linear movement, vibration-welding equipment can be designed to produce angular displacement of parts. Like ultrasonic welding, vibration welding produces high-strength joints and is better suited to large parts and irregular joint faces.

*Spin Welding:* Like vibration welding, spin welding uses frictional heat to melt the faces of circular joint surfaces. It is a rapid and economical method of joining melt-compatible parts. The process is completed in about three seconds and is easily automated. Frictional heat is generated by rotating one part against its mate (usually fixed) with a controlled pressure. When rotation is stopped, pressure is maintained during cooling and solidification of the joint. Simple equipment such as a drill press can serve for this process.

*Hot-Gas Welding:* Hot-gas welding is a process resembling acetylene brazing and welding of metals. Used mostly with polyolefins, which do not accept adhesives, the hot-gas welding equipment consists of an electric hot-gas "gun" that provides a stream of hot air or preferably hot nitrogen at a temperature above the melting range of the plastic to be worked. A welding rod of the same material provides the melt that forms the joint to be made between two parts. The parts to be joined are set up in their final orientation in a jig. A groove will usually be machined in one or both to accept the molten plastic. The operator directs the hot gas stream along the joint surfaces while feeding and melting the welding rod into the joint. The welded structure is then allowed to cool several hours. Some huge tanks have been fabricated from thick polyethylene sheets by this technique. It is also useful for repairing broken polyolefin parts. Hot-gas welding can produce joints with 85 percent of the strength of the parent material, but it is not recommended for use with walls less than $\frac{1}{16}$ inch (1.6 mm) thick.

*Dielectric Welding:* Dielectric welding is also known as *radio-frequency welding.* This process utilizes the dielectric loss that occurs in many plastics when they are placed in a rapidly reversing electric field. It is widely used with flexible thermoplastics films and sheets of materials—typically plasticized PVC and polyurethane—and for joining some injection-molded parts. The parts to be joined are brought together between two electrodes that also supply joining pressure. The field passes through the material, generating heat throughout the volume between the electrodes until it reaches melting range. The field is then turned off and pressure is maintained until the joined parts freeze. Polyolefins and some other plastics are so nonpolar that their dielectric loss is too small to permit welding them by this method.

*Induction Welding:* This form of electromagnetic welding uses inductive heating to generate fusion temperatures in thermoplastics as shown at the top in Fig. 28. Fine iron particles embedded in a gasket, preform, filament, ribbon, adhesive, or molded part are excited by the rapidly alternating magnetic field, generating eddy currents that, through resistive loss, cause distributed local heating. The heated parts are pressed together, and, as the temperature rises, the material of the particle carrier flows under pressure through the joint interface, filling voids and cavities and becoming an integral part of the weld. Ideally, the melted material should be contained and subjected to internal pressure by the surrounding component surfaces. Proper joint design is essential to successful induction welding.

Requirements of the preform often add cost to this welding method, but the cost is offset by low reject rates resulting from good reliability of the welds. Structural, hermetic welds can be produced in most thermoplastics materials, and automation can be used for large-volume production. The process also offers great latitude in joint size, configuration, tolerance requirements, and ability to bond some dissimilar materials. A limitation is that no ferrous metal can be near the joint line during the energizing of the inductor coil. All components of an assembly (except the weld medium) to be induction-welded must therefore be nonferrous, or ferrous components must be placed where they will not be subjected to the high-frequency field from the inductor.

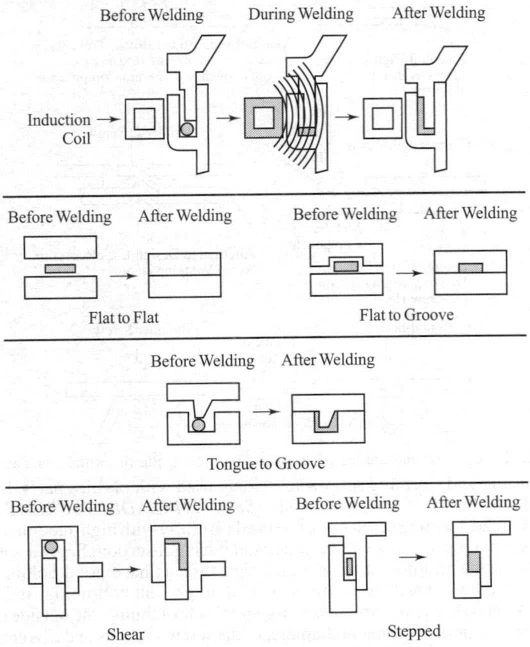


Fig. 28. Typical Joint Designs Used in Induction-Welding Plastics Assembly

**Assembly with Fasteners.**—Metal fasteners of high strength can overstress plastics parts, so torque-controlled tightening or special design provisions are needed. Examples of poor and preferred designs are shown in Fig. 29. The shoulder of the metal screw shown in the lower right portion of Fig. 29 meets the metal component when it is tightened, sparing the plastic part from over-compression. Where force cannot be satisfactorily controlled, even with a shoulder screw, various types of washers can be used to spread the compression force over wider areas.

Metal inserts are available in a wide range of shapes and sizes for permanent installation of metal threads or bushings in plastics parts. Inserts are typically installed in molded bosses, designed with holes to suit the insert to be used. Some inserts are pressed into place and others are installed by methods designed to limit stress and increase strength. Generally, the outside of the insert is knurled or otherwise provided with projections of various configurations that penetrate the plastics and prevent movement under normal torques exerted during assembly. "Helicoil" inserts have large standard threads on the outside and receive a smaller standard screw inside, thus lowering the assembly torque on the plastic by distributing it to the larger outside contact area. Inserts can also be installed with equipment similar to that used for ultrasonic welding, the plastics being melted to enhance contact with the metal and reduce insertion stresses.

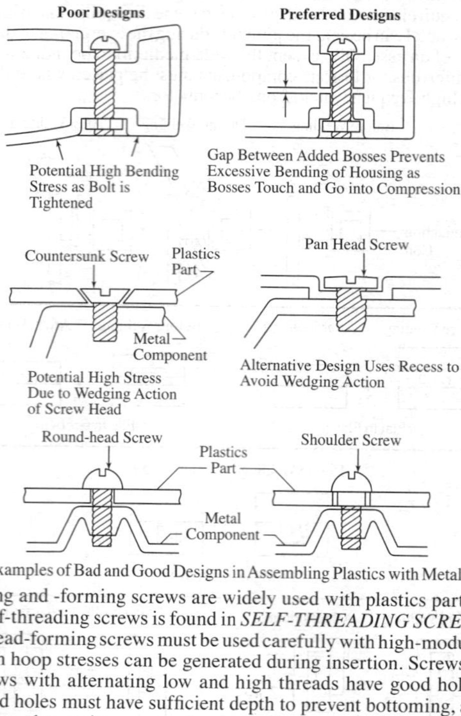


Fig. 29. Examples of Bad and Good Designs in Assembling Plastics with Metal Fasteners

Thread-cutting and -forming screws are widely used with plastics parts. Information on standard self-threading screws is found in *SELF-THREADING SCREWS* starting on page 1836. Thread-forming screws must be used carefully with high-modulus, low-creep plastics, as high hoop stresses can be generated during insertion. Screws with multiple lobes and screws with alternating low and high threads have good holding power in plastics. Molded holes must have sufficient depth to prevent bottoming, and boss walls must be thick enough to resist stresses. As a general rule of thumb, the outside diameter of the boss should be double the major diameter of the screw. Gussets and ribs can be used to increase the torsional strength of bosses.

Hollow aluminum or other metal rivets are often used in plastics assembly, as are stamped sheet metal components, especially push-on or -in designs. Molded plastics fasteners are also frequently used.

**Machining Plastics.—**Plastics can be molded into complex shapes and so do not usually need to be machined. However, machining is sometimes more cost-effective than making a complex tool, especially when requirements are for prototype development, low-volume production, undercuts, angular holes, or other openings that are difficult to produce in a mold. Special methods for development of prototypes are discussed on page 612. All machining of plastics requires dust control, adequate ventilation, safety guards, and eye protection.

Like some metals, plastics may need to be annealed before machining to avoid warpage. Some commercially available bar and rod stock are sold already annealed. If annealing is necessary, instructions can be obtained from plastics suppliers. Plastics moduli are small fractions—2 to 10 percent—of those of metals, and this lower stiffness permits much greater deflection of the work material during cutting. Thermoplastics materials must be held and supported firmly to prevent distortion, and sharp tools are essential to minimize normal forces and avoid excess heat generation.

Plastics recover elastically during and after machining so that drilled or tapped holes often end up tapered or of smaller diameter than the tool. Turned diameters also can end up larger or smaller than the dimensions measured immediately after the finishing cut. The low thermal conductivity of plastics causes much of the heat generated in cutting to be absorbed by the tool. Heat in the plastics tends to stay near the surface. The heat must be removed by an air blast or a liquid coolant for optimum surface finishes and close tolerances. A non-aromatic, water-soluble misting or flood coolant is suggested.

Plastics have thermal expansion coefficients some 10 times higher than those of metals so that even though actual heat generation during machining may be less than with metals there can easily be more expansion. Adequate tool clearances must be provided to minimize heating. Compared with most structural metals, temperatures at which plastics soften, deform and flow are quite low. Allowing frictional heat to build up causes gumming, discoloration, poor tolerance control, and rough finishes. These effects are more pronounced with plastics such as polystyrene and polyvinyl chloride that have low melting points than with plastics that have higher melting points, such as nylons, fluoroplastics, and polyphenylene sulfide. Sufficient clearances must be provided on cutting tools to prevent rubbing contact between the tool and the work. Tool surfaces that will come into contact with plastics during machining should be polished to reduce frictional drag and resulting temperature increases. Proper rake angles depend on depth of cut, cutting speed, and the type of plastic being cut. Large rake angles should be used to produce continuous-type cuttings, but they should not be so large as to cause brittle fracture of the work and resulting discontinuous chips.

*Turning and Cutting Off:* High-speed steel and carbide tools are commonly used with cutting speeds of 300–600 and 600–1200 ft/min (91–183 and 183–366 m/min), respectively. Water-soluble coolants can be used to keep down temperatures at the shear zone and improve the finish, except when they react with the work material. Chatter may result from the low modulus of elasticity and can be reduced by close chucking and follow rests. Box tools are good for long, thin parts. Tools for cutting off plastics require greater front and side clearances than are needed for metal. Cutting speeds should be about half those used for turning operations.

*Drilling:* This is a common plastics machining operation since small-diameter holes are more easily drilled than molded. However, plastics can be difficult to drill without some damage. Difficulties not encountered in drilling metals can occur, such as gumming, burning in the drilled hole, cracks around the edges, or growth of cracks after drilling. Two reasons for these difficulties are that swarf flow (chip removal) in drilling is poor and cutting speeds vary from the center to the periphery of the drill, so that drilling imposes severe loading on the workpiece. Some drill types used with plastics are shown in Fig. 30.

Drills of high-speed steel or premium high-speed steel (T15, M33, or M41-M47) are recommended, with low helix angles, point angles of 70–120 degrees, and wide, highly polished flutes to ease chip exit. Normal feed rates are in the range of 0.001–0.012 in/rev

(0.025–0.3 mm/rev) for holes of 1/16–2-inch (1.6–51-mm) diameter, and speeds of 100–250 ft/min (30.5–76.2 m/min), with lower speeds used for deep and blind holes. Point angles of 60 to 90 degrees (included) are used for many plastics, but an angle of 120 degrees should be used for rigid polyvinyl chloride and acrylic (polymethyl methacrylate).

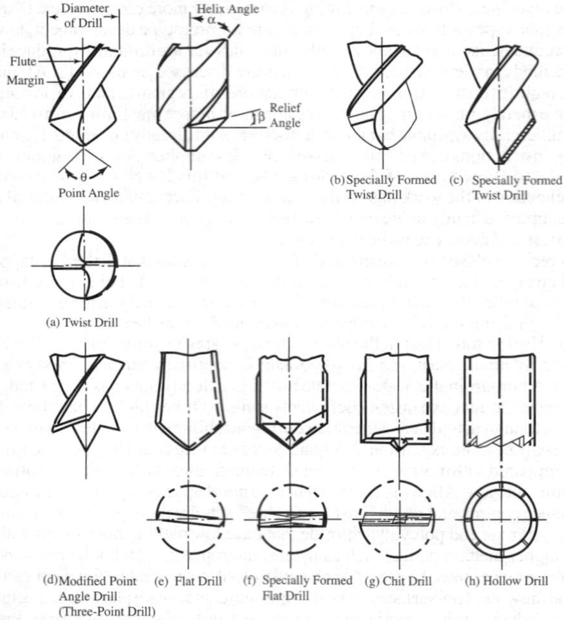


Fig. 30. Drill Designs Used for Drilling Plastics

Clearance angles of 9–15 degrees are usually sufficient to prevent the drill flanks from rubbing in the bottom of the hole, but acrylic materials require angles of 12–20 degrees. Tests may be needed to determine the drill diameter for accurately sized holes, allowing for thermal expansion and elastic recovery. Reaming may be used to size holes accurately, but even those diameters produced may also be affected by thermal expansion of the plastics. Close-fitting bushings in drill jigs may increase friction on the drill and cause swarf to plug up the drill flutes. For positioning accuracy, removable templates may be used to spot the hole position, then set aside for the drilling to be completed. Pilot holes are not necessary, except when the hole is to be reamed or counterbored. Frequent pull-out (peck drilling) to remove chips and compressed air cooling may be needed, especially for deep holes.

Drilling and reaming speed and feed suggestions for various materials are shown in Table 10. These speeds and feeds are approximate, and can be increased where there is no melting, burning, discoloration, or poor surface finish. Drills must be kept sharp and cool, and carbide tools may be needed in high production, especially with glass-reinforced materials. They must be cooled with clean compressed air to avoid contamination. Aqueous solutions are used for deep drilling because metal cutting fluids and oils may degrade or attack the plastics and may cause a cleaning problem. Plastics parts must be held firmly during drilling to counter the tendency for the tooling to grab and spin the work. Circle cutters often are preferred for holes in thin materials.

**Table 10. Suggested Approximate Speeds and Feeds for Drilling Various Thermoplastics[a]**

| Material | 1/16 in. Diameter | | 1/4 in. Diameter | | 1 in. Diameter | |
|---|---|---|---|---|---|---|
| | Speed (RPM) | Feed (in/rev) | Speed (RPM) | Feed (in/rev) | Speed (RPM) | Feed (in/rev) |
| ABS | 6000 | 0.015 | 2000 | 0.040 | 500 | 0.080 |
| Acetal (POM) | 12000 | 0.010 | 4000 | 0.030 | 1000 | 0.060 |
| Acrylic (PMMA) | 9000 | 0.010 | 3000 | 0.030 | 750 | 0.060 |
| Polyamide, Nylon 6/6 and PA6 | 6000 | 0.010 | 2000 | 0.030 | 500 | 0.060 |
| Polycarbonate (PC) | 9000 | 0.010 | 3000 | 0.030 | 750 | 0.060 |
| Polyester (PET) | 9000 | 0.007 | 3000 | 0.020 | 750 | 0.040 |
| Polyether ether ketone (PEEK) | 12000 | 0.005 | 4000 | 0.015 | 1000 | 0.030 |
| Polyethylene (PE) | 12000 | 0.015 | 4000 | 0.040 | 1000 | 0.080 |
| Polyphenylene sulfide (PPS) | 3000 | 0.010 | 1000 | 0.030 | 250 | 0.060 |
| Polypropylene (PP) | 12000 | 0.015 | 4000 | 0.040 | 1000 | 0.080 |
| Polystyrene (PS) | 6000 | 0.007 | 2000 | 0.020 | 500 | 0.040 |
| Polytetrafluoroethylene (PTFE) | 9000 | 0.007 | 3000 | 0.020 | 750 | 0.040 |
| Polyvinyl chloride (PVC) | 12000 | 0.007 | 4000 | 0.020 | 1000 | 0.040 |
| Ultem polyetherimide (PEI) | 12000 | 0.010 | 4000 | 0.030 | 1000 | 0.040 |

[a] Using a two-fluted drill.

*Tapping and Threading of Plastics:* Many different threaded fasteners can be used with plastics, including thread-tapping and -forming screws, threaded metal inserts, and molded-in threads, but threads must sometimes be machined after molding. For tapping of through holes in thin-cast, molded, or extruded thermoplastics and thermosets, a speed of 50 ft/min (15.2 m/min) is appropriate. Tapping of filled materials is done at 25 ft/min (7.6 m/min). These speeds should be reduced for deep or blind holes and when the percentage of thread is greater than 65–75 percent. Taps should be of M10, M7, or M1 molybdenum high-speed steel, with finish-ground and -polished flutes. Two-flute taps are recommended for holes up to 0.125 inch (3.2 mm) diameter. Oversize taps may be required to make up for elastic recovery of the plastics. The danger of retapping on the return stroke can be reduced by blunting the withdrawal edges of the tool.

*Sawing Thermoset Cast or Molded Plastics:* Circular or band saws may be used for sawing. Circular saws provide smoother cut faces than band saws, but, because band saws run cooler, they are often preferred even for straight cuts. Projection of the circular saw above the table should be minimized. Saws should have skip teeth or buttress teeth with zero front rake and a raker set. Precision-tooth saw blades should be used for thicknesses up to 1 inch (25.4 mm), and saws with buttress teeth are recommended for thicknesses above 1 inch (25.4 mm). Dull edges to the teeth cause chipping of the plastics and may cause breakage of the saw. Sawing speeds and other recommendations for using blades of high-carbon steel are shown in Table 11.

**Table 11. Speeds and Numbers of Teeth for Sawing Plastics Materials with High-Carbon Steel Saw Blades**

| Material Thickness | | Number of Teeth/inch on Blade | Peripheral Speed | | | |
|---|---|---|---|---|---|---|
| | | | Thermoset Cast or Molded Plastics | | Thermoplastics (and Epoxy, Melamine, Phenolic and Allyl Thermosets) | |
| (inch) | (mm) | | (ft/min) | (m/min) | (ft/min) | (m/min) |
| 0–0.5 | 0–13 | 8–14 | 2000–3000 | 607–914 | 4000–5000 | 1219–1524 |
| 0.5–1 | 13–25 | 6–8 | 1800–2200 | 549–671 | 3500–4300 | 1067–1311 |
| 1–3 | 25–76 | 3 | 1500–2200 | 475–671 | 3000–3500 | 914–1067 |
| >3 | >76 | >3 | 1200–1800 | 366–549 | 2500–3000 | 762–914 |

**Table 12. Suggested Approximate Speeds and Feeds for Milling Various Thermoplastics[a]**

| Material | Roughing 0.250 in. Cut Depth | | Finishing 0.050 in. Cut Depth | |
|---|---|---|---|---|
| | Speed (feet/min) | Feed (in/rev) | Speed (feet/min) | Feed (in/rev) |
| ABS | 600 | 0.018 | 480 | 0.008 |
| Acetal (POM) | 600 | 0.012 | 480 | 0.005 |
| Acrylic (PMMA) | 450 | 0.012 | 360 | 0.005 |
| Polyamide, Nylon 6 / 6 (PA6) | 800 | 0.012 | 480 | 0.005 |
| Polycarbonate (PC) | 500 | 0.012 | 450 | 0.005 |
| Polyester (PET) | 600 | 0.010 | 480 | 0.004 |
| Polyether ether ketone (PEEK) | 450 | 0.010 | 360 | 0.004 |
| Polyethylene (PE) | 800 | 0.018 | 620 | 0.008 |
| Polyphenylene sulfide (PPS) | 450 | 0.015 | 360 | 0.006 |
| Polypropylene (PP) | 800 | 0.018 | 620 | 0.008 |
| Polystyrene (PS) | 450 | 0.010 | 360 | 0.004 |
| Polytetrafluoroethylene (PTFE) | 600 | 0.012 | 480 | 0.005 |
| Polyvinyl chloride (PVC) | 600 | 0.010 | 480 | 0.004 |
| Ultem polyetherimide (PEI) | 450 | 0.012 | 360 | 0.005 |

[a] 2-fluted, 1 in. end mill

*Milling of Plastics:* Peripheral cutting with end mills is used for edge preparation, slotting and similar milling operations; end cutting can also be used for facing operations. Some recommended approximate speeds and feeds are provided in Table 12 for peripheral end milling of common thermoplastics; as per standard machining practice, speeds and feeds must be adjusted with observation of the machining relative to the quality requirements. Slower speeds are generally used for face and other milling operations, with some thermoplastics being machined at 300–500 ft/min (91–152 m/min), and some thermosets at 150–300 ft/min (46–91 m/min). Adequate support and suitable feed rates are very important. A table feed that is too low will generate excessive heat and cause surface cracks, loss of dimensional accuracy, and poor surface finish. Too high a feed rate will produce a rough surface. High-speed steel tools (M2, M3, M7, or T15) are generally used, but for glass-reinforced nylon, silicone, polyimide, and allyl, carbide (C2) is recommended.

*Turning of Plastics:* Plastics can be processed on commercially available lathes. A fine-grained, C2 carbide cutter, with a small cutting radius and broad-nosed finishing cutting edge, is recommended to avoid material build-up. Recommended approximate feed speeds are provided in Table 13. When feasible, turning at higher cutting speeds is recommended, using a cutting depth of at least 0.004 inch. Compressed air is a common and acceptable method for cooling, though misted and flooded coolants can be used if available. If the diameter of the plastic workpiece is small or workpiece rigidity is an issue, then a live center or lunette is recommended to stabilize the component and provide improved tolerances and surface finish.

*Other Machining Techniques:* Lasers can be used for machining plastics, especially sheet laminates, although their use may generate internal stresses. Ultrasonic machining has no thermal, chemical, or electrical reaction with the workpiece and can produce holes down to 0.003 inch (0.0762 mm) diameter; tight tolerances, 0.0005 inch (0.0127 mm); and very smooth finishes, 0.15 μinch (0.381 μm) with No. 600 boron carbide abrasive powder. Water-jet cutting using pressures up to 60,000 lb/in$^2$ (414 N/mm$^2$) is widely used for plastics and does not introduce stresses into the material. Tolerances of ± 0.004 inch (± 0.102 mm) can be held, depending on the equipment available. Process variables, pressures, feed rates, and the nozzle diameter depend on the material being cut. This method does not work reliably with hollow parts unless they can be filled with a solid core.

*Machining Hard Rubber:* Hard rubber, or ebonite, is the oldest rigid manmade plastic, and has been used since the 1840s. The base material is a highly vulcanized natural or, today, synthetic rubber, empirical formula C5H8S, containing 32 percent sulfur. A range of compositions containing various amounts of fillers is available. Unfilled ebonite has a specific gravity = 1.18, tensile strength = 9,000 psi (62 N/mm$^2$), elongation = 3 percent,

**Table 13. Suggested Approximate Speeds and Feeds for Turning Various Thermoplastics**

| Material | Roughing 0.150 in. Cut Depth | | Finishing 0.030 in. Cut Depth | |
|---|---|---|---|---|
| | Speed (feet/min) | Feed (in/rev) | Speed (feet/min) | Feed (in/rev) |
| ABS | 1200 | 0.018 | 800 | 0.007 |
| Acetal (POM) | 1200 | 0.012 | 800 | 0.006 |
| Acrylic (PMMA) | 1000 | 0.012 | 700 | 0.005 |
| Polyamide, Nylon 6 / 6 (PA6) | 1200 | 0.012 | 800 | 0.006 |
| Polycarbonate (PC) | 1000 | 0.012 | 700 | 0.005 |
| Polyester (PET) | 1200 | 0.010 | 800 | 0.005 |
| Polyether ether ketone (PEEK) | 1000 | 0.010 | 700 | 0.005 |
| Polyethylene (PE) | 1500 | 0.018 | 1200 | 0.008 |
| Polyphenylene sulfide (PPS) | 800 | 0.015 | 800 | 0.007 |
| Polypropylene (PP) | 1500 | 0.018 | 1200 | 0.008 |
| Polystyrene (PS) | 900 | 0.010 | 800 | 0.005 |
| Polytetrafluoroethylene (PTFE) | 1200 | 0.012 | 700 | 0.005 |
| Polyvinyl chloride (PVC) | 1200 | 0.010 | 800 | 0.005 |
| Ultem polyetherimide (PEI) | 800 | 0.012 | 600 | 0.005 |

and Izod impact strength = 0.5 ft-lb$_f$/in (26.7 J/m) of notch. Its continued use is due to its good insulating characteristics, good chemical resistance, and ease of machining. Tools suitable for steel, high-speed steel, or tungsten carbide may be used with hard rubber with no top or side rake angles and 10 to 20 degrees clearance angles.

Without coolant, surface speeds of 200 ft/min (60 m/min) are recommended for turning, boring and facing; they may be increased to 300 surface ft/min (90 m/min) with coolant.

Drilling of hard rubber requires high-speed steel drills of 35 to 40 degrees helix angle to obtain maximum cutting speeds and drill life. Feed rates for drilling range up to 0.015 in/rev (0.38 mm/rev). Deep-fluted taps are best for threading hard rubber and should be 0.002 to 0.005 inch (0.05-0.13 mm) oversize if close tolerances are to be held. Machine oil is used as the lubricant. Hard rubber may be sawed with band saws having 5 to 10 teeth per inch, running at about 3000 ft/min (915 m/min), or it may be cut with abrasive wheels. Use of coolant in grinding ebonite gives a smoother finish.

Piercing and blanking of hard-rubber sheet is best performed with the rubber and dies warmed. Straightening of the often-distorted blanks may be accomplished by dropping them into a pan of hot water.

**Plastics Gears.—**Plastics gears may be cut from blanks, as are metal gears, or molded to shape in an injection-molding machine for lower production costs, though tooling may cost more. Cut plastics gears may be of similar design to their metal counterparts, but molded gears are usually of modified form to suit material characteristics, including sensitivity to temperature and humidity. Plastics materials also may be preferred for gears because of superior sliding properties with reduced noise and less need for lubrication, chemical or electrical properties, or resistance to wear. However, plastics gear teeth slide more smoothly and easily against metal teeth than do plastics against plastics, and wear is less. For power transmission, plastics gear teeth are usually of involute form. See also *Nonmetallic Gearing* on page 2326.

Most plastics gears are made from nylons and acetals, although acrylonitrile-butadiene-styrenes (ABS), polycarbonates, polysulfones, phenylene oxides, polyurethanes, liquid crystal polymers, and thermoplastic polyesters have also been used. Gears can be made from virtually any plastic, including polypropylene and polyethylene. But care must be taken to manage deflection and stresses, even in low-load and low-speed applications. Additives used in plastics for gears include glass and carbon fiber for added strength and stiffness, and fibers, beads, and powders for reduced thermal expansion and better dimensional stability. Other materials, such as molybdenum disulfide, polytetrafluoroethylene powder (PTFE), and silicones may be incorporated as integral lubricants to reduce friction and wear.

Choice of plastics gear material depends on requirements for size and nature of loads to be transmitted, designated speeds, required life, working environment, type of cooling, lubrication, and operating precision. Because of cost, plastics gears are sometimes not enclosed in sealed housings and are often given only a single coating of lubricant grease. Overloading of lubricated plastics gear teeth will usually cause tooth fracture, and unlubricated teeth often suffer excessive wear. Thermoplastics strength varies with temperature, with higher temperatures reducing root stress and permitting tooth deformation. In calculating power to be transmitted by spur, helical, and straight bevel gearing, the following formulas should be used with the factors given in Table 14, Table 15, and Table 16:

| | US Customary Units | | Metric Units | |
|---|---|---|---|---|
| Internal and External Spur Gears | $HP = \dfrac{S_s FYV}{55(600+V)PC_s}$ | (32a) | $KW = \dfrac{FYmS_sV}{327(3.05+V)C_s}$ | (32b) |
| Internal and External Helical Gears | $HP = \dfrac{S_s FYV}{423(78+\sqrt{V})P_nC_s}$ | (33a) | $KW = \dfrac{FYmS_sV}{179(5.56+\sqrt{V})C_s}$ | (33b) |
| Straight Bevel Gears | $HP = \dfrac{S_s FYV(C-F)}{55(600+V)PCC_s}$ | (34a) | $KW = \dfrac{FYmS_sV(C-F)}{327(3.05+V)C_sC}$ | (34b) |

$S_s$ = safe stress in bending (from Table 15); $F$ = face width in inches (mm); $Y$ = tooth form factor (from Table 14); $m$ = module, mm; $C$ = pitch cone distance in inches (mm); $C_s$ = service factor (from Table 16); $P$ = diametral pitch; $P_n$ = normal diametral pitch; and $V$ = velocity at pitch circle diameter in ft/min (m/s).

*Example:* As an example, assume that a material is to be selected for a spur gear that must transmit ⅛ hp at 350 rpm for 8 hours/day under a steady load. The gear is to have 75 teeth, 32 diametral pitch, 20 degree pressure angle, 0.375 inch face width, and a pitch diameter of 2.3438 inches. Using Equation (32a),

$$HP = \frac{S_s FYV}{55(600+V)PC_s} \quad \text{or} \quad S_s = \frac{55(600+V)PC_sHP}{FYV}$$

$$HP = 0.125 \qquad Y = 0.434 \quad \text{and}$$

$$V = \frac{rpm \times \pi \times D}{12} = \frac{350 \times 3.1416 \times 2.3438}{12} = 215 \text{ ft/min}$$

therefore, $$S_s = \frac{55(600+215)32 \times 1.00 \times 0.125}{0.375 \times 0.434 \times 215} = 5{,}124 \text{ lb/in}^2$$

From Table 15 it is apparent that gears can be molded from several materials. Available physical and chemical characteristics must now be considered in relation to the operating environment for the gear. Strengths of plastics materials decrease with increasing temperatures, and not all plastics resist the effects of some liquids, including some lubricants. Some plastics deteriorate when in sunlight for long periods; some are more dimensionally stable than others; and wear resistance varies from one to another. Manufacturers' data sheets will answer some of these questions.

*Backlash:* Plastics gears should be so dimensioned that they will provide sufficient backlash at the highest temperatures likely to be encountered in service. Dimensional allowances must also be made for gears made of hygroscopic plastics that may be exposed to damp service conditions. Teeth of heavily loaded gears usually have tip relief to reduce effects of deflection, and have full fillet radii to reduce stress concentrations. Such modifications to tooth form are also desirable in plastics gears. If the pinion in a pair of gears has a

**Table 14. Tooth Form Factors *Y* for Plastics Gears**

| Number of Teeth | 14½-degree Involute or Cycloidal | 20-degree Full Depth Involute | 20-degree Stub Tooth Involute | 20-deg Internal Full Depth Pinion | 20-deg Internal Full Depth Gear |
|---|---|---|---|---|---|
| 12 | 0.210 | 0.245 | 0.311 | 0.327 | … |
| 13 | 0.220 | 0.261 | 0.324 | 0.327 | … |
| 14 | 0.226 | 0.276 | 0.339 | 0.330 | … |
| 15 | 0.236 | 0.289 | 0.348 | 0.330 | … |
| 16 | 0.242 | 0.259 | 0.361 | 0.333 | … |
| 17 | 0.251 | 0.302 | 0.367 | 0.342 | … |
| 18 | 0.261 | 0.308 | 0.377 | 0.349 | … |
| 19 | 0.273 | 0.314 | 0.386 | 0.358 | … |
| 20 | 0.283 | 0.320 | 0.393 | 0.364 | … |
| 21 | 0.289 | 0.327 | 0.399 | 0.371 | … |
| 22 | 0.292 | 0.330 | 0.405 | 0.374 | … |
| 24 | 0.298 | 0.336 | 0.415 | 0.383 | … |
| 26 | 0.307 | 0.346 | 0.424 | 0.393 | … |
| 28 | 0.314 | 0.352 | 0.430 | 0.399 | 0.691 |
| 30 | 0.320 | 0.358 | 0.437 | 0.405 | 0.679 |
| 34 | 0.327 | 0.371 | 0.446 | 0.415 | 0.660 |
| 38 | 0.336 | 0.383 | 0.456 | 0.424 | 0.644 |
| 43 | 0.346 | 0.396 | 0.462 | 0.430 | 0.628 |
| 50 | 0.352 | 0.480 | 0.474 | 0.437 | 0.613 |
| 60 | 0.358 | 0.421 | 0.484 | 0.446 | 0.597 |
| 75 | 0.364 | 0.434 | 0.496 | 0.452 | 0.581 |
| 100 | 0.371 | 0.446 | 0.506 | 0.462 | 0.565 |
| 150 | 0.377 | 0.459 | 0.518 | 0.468 | 0.550 |
| 300 | 0.383 | 0.471 | 0.534 | 0.478 | 0.534 |
| Rack | 0.390 | 0.484 | 0.550 | … | … |

These values assume a moderate temperature increase and some initial lubrication. With bevel gearing, divide the number of teeth by the cosine of the pitch angle and use the data in the table. For example, if a 20-degree PA bevel gear has 40 teeth and a pitch angle of 58 degrees, 40 divided by the cosine of 58 degrees = 40 ÷ 0.529919 ≈ 75, and $Y = 0.434$.

**Table 15. Safe Bending Stress Values for Plastics Gears**

| Plastics Type | Safe Stress: Unfilled lb/in² | Unfilled MPa | Glass-Filled lb/in² | Glass-Filled MPa |
|---|---|---|---|---|
| ABS | 3,000 | 20.68 | 6,000 | 41.37 |
| Acetal | 5,000 | 34.47 | 7,000 | 48.26 |
| Nylon | 6,000 | 41.37 | 12,000 | 82.74 |
| Polycarbonate | 6,000 | 41.37 | 9,000 | 62.05 |
| Polyester | 3,500 | 24.13 | 8,000 | 55.16 |
| Polyurethane | 2,500 | 17.24 | … | … |

**Table 16. Service Factors for Plastics Gears**

| Type of Load | 8–10 hr/day | 24 hr/day | Intermittent, 3 hr/day | Occasional, ½ hr/day |
|---|---|---|---|---|
| Steady | 1.00 | 1.25 | 0.80 | 0.50 |
| Light shock | 1.25 | 1.5 | 1.00 | 0.80 |
| Medium shock | 1.5 | 1.75 | 1.25 | 1.00 |
| Heavy shock | 1.75 | 2.00 | 1.5 | 1.25 |

small number of teeth, undercutting may result. Undercutting weakens teeth, causes undue wear, and may affect continuity of action. The undercutting can be reduced by using the long-short addendum system, which involves increasing the addendum of the pinion teeth and reducing that of the gear teeth. The modified addendum method will also reduce the amount of initial wear that takes place during initial stages of contact between the teeth.

*Accuracy:* "The Gear Handbook," AGMA 390-03a-1980, Part 2, Gear Classification, provides a system whereby results of gear accuracy measurements are expressed in terms of maximum tooth-to-tooth and composite tolerances. This system uses AGMA quality numbers related to maximum tolerances, by pitch and diameter, and is equally applicable to plastics and metal gears. AGMA quality numbers must be chosen for a pair of mating gears early in the design process, and the finished gears must be inspected by being run in close mesh with a master gear in a center-distance measuring instrument to make sure that the errors do not exceed the specified tolerances.

To prevent failure from fatigue and wear caused by excessive flexing of the teeth, plastics gears must be made to similar standards of accuracy as metal gears. Solidification shrinkage of plastics requires that dimensions of molds for gears be larger than the dimensions of the parts to be produced from them. The amount of the shrinkage is usually added to the mold dimension (with the mold at operating temperature). However, this procedure cannot be followed for the tooth profile as it would introduce large errors in the pressure angle. Increases in pressure angle cause gear teeth to become wider at the root and more pointed. Sliding conditions are improved and the teeth are stronger, so that higher loading values can be used.

Shrinkage allowances have the greatest effect on the accuracy of the molded gears, so tooth profiles must be calculated extremely carefully in terms of mold profile. If a tooth is merely made larger by using a standard hobbing cutter to cut the tool, whereby the teeth in the mold are electroeroded, differential shrinkage caused by the molded tooth being thicker at the root than at the tip will distort the shape of the molded tooth, making it thinner at the tip and thicker at the root. With two mating gears these faulty shapes will affect the pressure angle, resulting in binding, wear, and general malfunction. If tooth thickness limits for a molded gear are to be held to +0.000 inch, −0.001 inch (−0.025 mm), the outside diameter must be permitted to vary up to 0.0027 inch (0.069 mm) for 20 degree and 0.0039 inch (0.099 mm) for $14\frac{1}{2}$-deg pressure angle gears. All high-accuracy gears should be specified with AGMA quality numbers and inspected with center-distance measuring machines if the required accuracy is to be achieved.

**Polishing Plastics.—**Plastics can be highly polished for optical and tooling applications, low-friction surfaces, aesthetic purposes, and other uses. This can be achieved with buffing methods similar to those used on metals. But experiments typically are required to determine appropriate speed and forces, based on effects of various polishing media and frictional heat.

As with metals, polishing operations proceed from coarser to finer surfaces. If molding, additive manufacturing, or machining processes left a rough surface, sanding may be needed prior to polishing. Larger defects, including parting lines, imperfections, scratches, saw lines, and scars resulting from fabrication, also can be treated with abrasives prior to buffing. Wet or dry abrasives, such as silicon carbide or aluminum oxide, generally are used in grain sizes from 60 to as fine as 1000. Surfaces are ready for fine polishing when a satin finish is achieved.

Polishing with buffer wheels is common, with speeds for polishing comparable to those for turning (see Table 12), typically 800 to 1,500 feet (360 to 580 meters) per minute, with rotation speeds varying with the buffer wheel size. For example, a wheel measuring 6 inch (150 mm) operating at 750 rpm will provide a linear velocity of 1,180 feet/minute (360 meters/minute). For most plastics, soft cotton buffs are best, using a wet or greasy tripoli or silica compound. If a cleaning solvent is used, it should be checked to ensure it does not dissolve the plastic and used in a well-ventilated area. Each round of polishing should be performed with a separate, clean polishing bonnet.

To minimize surface heating during machine polishing, avoid lingering at any one location and keep speeds and forces low. Otherwise, surfaces can quickly heat to temperatures of 400° F (205° C), and many plastics will soften and melt. Heating also can cause plastics to give off toxic gases. So machine polishing should be performed in vented enclosures, and masks worn during manual polishing to filter dust and gases.

If the plastic parts are sufficiently robust, tumble polishing can be used with abrasive media in a vibratory bowl. Tumbling processes can operate as a batch process, with high loadings of parts. Cycle times are typically a few hours, with the usual media to parts ratio approximately 2 or 3 to 1 by volume. A wide variety of abrasive media are available, varying in shape, density, hardness, and grit. For burnishing and polishing of plastics parts, angle-cut cylinders, with finer grit and lower density and hardness, are recommended to avoid excessive material removal and loss of feature detail. Tumbling of plastics parts in ceramic media usually works best when used as a wet process, with intermittent or continuous filtering of the generated particulates.

Economical polishing of edges—for example, of acrylic sheets—may be accomplished by flame polishing. If the surface is rough, edges should first be sanded with 320 grit paper. Quickly passing an oxy-acetylene torch over the edge surface melts it with the local flow and solidification results in a polished appearance. To minimize rounding of melted edges, a small strip of plastic or other material can be clamped on each side. Note that polycarbonate and many other plastics may not be easily flame polished, given their propensity to oxidative degradation and flammability.

**Finishing and Decorating Plastics.—**There are a number of other methods for finishing and decorating the surfaces of plastic parts, such as spray painting, vacuum metallizing, hot stamping, silk screening, and even plating. Conductive coatings may be applied to inside surfaces, usually by flame- or arc-spraying, to dissipate static electricity and provide electromagnetic shielding. Thorough cleaning is essential. Materials such as polyethylene, polypropylene, and acetal have waxlike surfaces that may not be painted easily or may need pretreatment or special primers. Many amorphous plastics are easy to paint. Suitable coatings include polyurethane-, epoxy-, acrylic-, alkyd-, and vinyl-based paints. Oven curing must be carried out at temperatures well below the deflection temperatures of the plastics in order to avoid distortion of parts. See *Deflection temperature under load (DTUL)* on page 570.

*Vacuum Metallizing and Sputter Coating* require application of a special base lacquer and protective clear top coat before and after treatment. Resistance heating or an electron beam can be used to melt the coating metals such as aluminum, silver, copper, and gold, which usually are pure elements. Wires of the coating metal are hung from tungsten heating elements in an array that gives them direct views of the surfaces to be coated. Sputter plating uses a plasma to produce the metallic vapor, requires line-of-sight setup, and can use brass as well as the other metals mentioned. The appearance of gold or brass is often simulated with vacuum-metallized aluminum by tinting the overcoating lacquer. Chromium plating requires etched surfaces to ensure good adhesion.

**Development of Plastics Prototypes.—**Model prototypes are made for testing of properties, such as stress and fatigue resistance, to find ways to improve quality and reliability, to improve tooling design, and to reduce time to market. Prototyping may answer questions about finish, sink marks that result from contraction, witness lines from mold joints, ejector pin marks, knit or weld lines, texturing, moldability, shrinkage, mechanical strength, insert pull-out resistance electrical properties, and problems of mating with other parts.

Prototypes may be made by conventional machining, additive manufacturing (AM, also known as 3D printing), or standard molding processes using molds made of wood, plaster, plastics (by machining), or metal (often aluminum). The majority of pre-production molds utilize CAD/CAM methods that enable dimensional tolerances of prototypes to be held to 1 or 2 percent of design specifications. (See *CAD/CAM* on page 1377.)

Ongoing advancements in AM processes support creation of complex plastic prototypes, as well as many production parts. After the initial equipment investment or cost of a 3D printing service, using AM processes to produce plastics prototypes can be less expensive than producing prototypes with CNC machined molds, freeing up more costly CNC machines for other production. CNC systems generally require more floor space than 3D printers and, as a subtractive process, use more material, producing added scrap that must be recycled or otherwise disposed of, whereas AM machines use only the amount of material needed.

## Additive Manufacturing Plastics

Plastics prototyping was the motivation for development of the first AM process, known as stereolithography (SLA). This process patent was filed in 1984, granted in 1986, and commercialized in 1987 by 3D Systems. Since then, other AM processes for making plastics prototypes and production parts have been developed and refined, including selective laser sintering, fused deposition modeling, binder jetting, material jetting, and multijet fusion. In all of these processes, a part is built up one layer at a time, adding material to produce a shape.

In all AM methods, as in CNC machining and other computerized manufacturing systems, the objective shape is represented in a three-dimensional (3D) CAD file. The CAD file may be made by a designer or by scanning an object into a solid modeling package. Software embedded within the printer "slices" the CAD model into layers; it then determines the toolpath for fabricating each layer. The "tool" in this case is an extruder, laser, or printhead, as described in the following sections.

Unlike CNC systems, AM does not require manual programming of the toolpath. With the new programs and interfaces, the skill level required to program and operate a 3D printer and create unique plastic models in-house is less. Alternatively, parts designers can take advantage of 3D printing services to produce affordable prototypes. Thus, prototypes can be produced more frequently, providing efficiencies in product design and development time. At the same time, production parts are increasingly being manufactured with this technology.

Operating principles and features of various plastic AM processes follows. Table 17 summarizes attributes of plastic 3D printing processes. For more on the plastics mentioned here, see *Plastics Materials* on page 554.

**Table 17. Plastics Processes and Materials for 3D Printing**

| Process | Materials | Feedstock Form | Accuracy | | Resolution: Diameter of Hole | | Resolution: Thickness of Wall | | Layer Thickness | | Surface Finish |
|---|---|---|---|---|---|---|---|---|---|---|---|
| | | | inch | mm | inch | mm | inch | mm | inch | mm | Roughness Average (Ra) |
| Stereolithography (SLA) | Thermoset plastics: PP-like, ABS-like, PC-like, heat-resistant, flexible | Photosensitive liquid resin | 0.002 | 0.05 | 0.02 | 0.5 | 0.02 | 0.5 | 0.001–0.004 | 0.03–0.10 | 3–5 |
| Selective Laser Sintering (SLS) | Thermoplastics: nylon, carbon-filled nylon, glass-filled nylon, aluminum-filled nylon | Polymer powders | 0.012 | 0.30 | 0.06 | 1.5 | 0.03 | 0.8 | 0.001–0.004 | 0.03–0.10 | 10–20 |
| Fused Deposition Modeling (FDM) | Thermoplastics: ABS, PC, PLA, Ultem high-temperature | Polymer filament | 0.02 | 0.5 | 0.08 | 2.0 | 0.03 | 0.8 | 0.004–0.012 | 0.10–0.30 | 20–50 |
| Binder Jetting (BJ) | Plaster, starch | Plaster, starch powders; water-based binders | 0.02 | 0.5 | 0.04 | 1.0 | 0.04 | 1.0 | 0.004–0.012 | 0.10–0.30 | 10–20 |
| Material Jetting (MJ) | Thermoset plastics: ABS-like, PP-like, rubber | Photosensitive polymer resin | 0.004 | 0.10 | 0.02 | 0.5 | 0.04 | 1.0 | 0.0006–0.002 | 0.015–0.05 | 5–10 |
| Multijet Fusion (MJF) | Nylon (PA), glass-filled PA, PP, and TPU | Polymer particles; liquid binder and detail agents | 0.012 | 0.30 | 0.02 | 0.5 | 0.02 | 0.5 | 0.002–0.005 | 0.05–0.13 | 5–10 |

**Stereolithography (SLA).**—Under the action of light energy supplied by a laser beam, long-chain molecules in a photosensitive liquid resin cross-link to form a solid polymer. Motion of the laser spot is controlled by galvanometer-driven mirrors to trace a two-dimensional (2D) slice of the three-dimensional (3D) CAD model.

Commonly, the part is built from bottom up, as shown in Table 18a. First, the bottom slice of the part is produced on a platform that moves in the vertical direction. The platform is then lowered into a vat of photosensitive liquid resin, by one layer thickness (see

Table 17). An arm sweeps across the surface to smooth it, and the next slice is traced by the laser. This action sequence is repeated until the part specified by the CAD file is produced.

In increasingly common variations of this process, the platform is moved upward, while the emitted light projects from below onto the bottom of a transparent tray containing photosensitive liquid resin, as shown in Table 18b. In this case, the top slice of the part is produced first, the build platform moves upward, in one layer thickness, and each new slice is added in a top-down approach. This method initially requires much less resin than the bottom-up approach.

Newer versions of the top-down SLA method project an entire cross-section of light to increase the production rate of the process. Printers with *digital light processors* (DLP) convert each layer of liquid resin into a solid 2D slice with a projected flash of light that is controlled by a matrix of digital micromirrors. By comparison, printers with *liquid crystal displays* (LCD) directly project light using the LCD display to mask the pixels that are not to be cured in the current layer.

**Table 18. Three Configurations of the Stereolithography (SLA) Process**

| | |
|---|---|
| 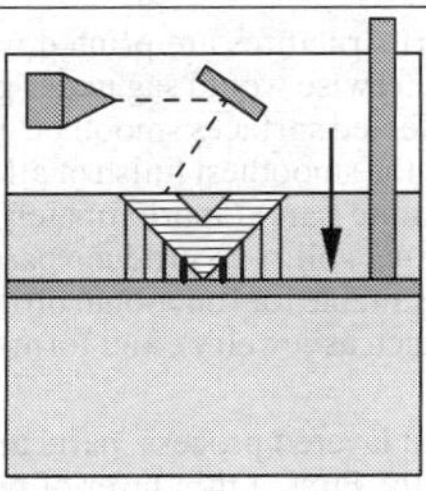 A | (a) *Bottom-up build by SLA with ultraviolet (UV) laser projecting from above:* The build platform descends into a vat of photosensitive polymer, one layer at a time. A UV laser scans the cross section of the part on the top surface. As the laser moves across this surface, it solidifies the polymer, incrementally transforming each slice of the model into a layer of the 3D part. |
| 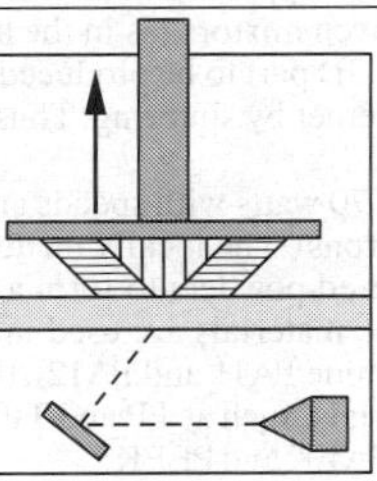 B | (b) *Top-down build by SLA with UV laser projecting from below:* The build platform moves upward, one layer at a time, over a build tray containing photosensitive polymer. A UV laser scans the cross section of the part through the transparent base of the build tray. As the laser moves across and projects through the build tray, it solidifies the polymer, incrementally transforming each slice of the model into a layer of the 3D part. |
| 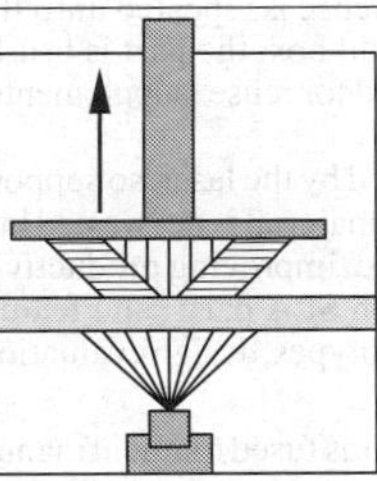 C | (c) *Top-down build by SLA with digital light processor (DLP) projecting from below:* The build platform moves upward, one layer at a time, over a build tray containing photosensitive polymer. A DLP projects through the transparent base of the build tray. As the DLP projects the complete shape of each slice on the build tray, it solidifies the polymer, transforming the slice of the model into a layer of the 3D part all at once. |

Both DLP and LCP printers rely on components derived from high-definition displays. As such, resolution and frequency continue to advance and enable higher resolution and faster 3D printing. Ultra-high definition "4K" resolutions provide a matrix of 3840 by 2160 pixels with a typical pixel providing 72 micron resolution in a 10.9 by 6.1 inch (276 by 156 mm) build area. Both types of printers typically cure photopolymers with UV light at standard wavelengths of 405 nm. Print qualities can be very similar, though the DLP printers currently provide slightly better sharpness and longevity.

Another type of photopolymer printer is *continuous liquid interface production* (CLIP). CLIP extends the capability of DLP printing by using an oxygen-permeable membrane as a window between the resin and the DLP projector. The local diffusion of oxygen at the interface between the resin and window prevents the resin from attaching to the window and results in a so-called "dead zone" with a persistent liquid interface. As a result, UV projection of cross-sections and upward translation of the build platform proceeds continuously without requiring delays to mechanically peel the developing print from the window (a time-consuming step in DLP and LCD printing). CLIP thus provides much higher production rates but may incur added costs for licensing and materials associated with newly patented technologies.

In all these photopolymerization approaches, support structures are printed with the part to support overhanging sections of the part that otherwise would sag under gravity. Supports are removed from the finished part and the affected surfaces smoothed, though these processes provide high resolution and so produce the smoothest finish of all plastic 3D printing processes. It also produces the most accurate parts (approximately 0.002 in. or 0.05 mm) of the 3D printing processes. With SLA's ability to produce parts with high resolution and good surface finish, this process is suitable for functional prototypes of mechanisms, patterns for fine investment casting (such as jewelry), and form-and-fit models.

**Selective Laser Sintering (SLS).**—In this powder-bed layered process, parts are built within a build box on a build plate that moves vertically. First, a thin layer of polymer powder is spread on the build plate by a roller, blade, or hopper. Then an infrared laser travels across the surface, guided by galvanometer-driven mirrors (as in the bottom-up SLA process in Table 18) and traces the 2D layer of the 3D part to be produced. Polymer powders within the laser spot are heated and fused together by sintering. Thus, thermoplastic materials are used in SLS processes.

SLS typically uses a pulsed $CO_2$ laser on the order of 70 watts with speeds around 400 in/s (10 m/s) and laser spot sizes of 0.002 inch (50 microns). The locally melted plastics fuse with the surrounding unfused and formerly sintered powders to form a relatively dense component. As a welding process, thermoplastic materials are used in SLS processes with the dominant material being nylon (polyamide PA11 and PA12), though the process also is commonly performed with softer materials, such as PP and TPU (elastomer), as well as higher performance materials, such as PAEK and PEEK.

The build plate is moved down by one layer thickness (see Table 17) and a new layer of powder is spread, followed by the laser trace. This sequence is repeated until the 3D part is completed. Unused powder is removed from the build box, the part is brushed off to remove residual powder, and the unused powder is saved for reuse (supplemented by new powder) in future builds.

Unprinted powder provides support for powder bonded by the laser, so supports are not added to parts produced using SLS printers. As a result, material is not wasted in supports, and multiple layers of parts can be produced in one build, improving productivity of SLS machines. SLS parts have a rougher surface finish than SLA parts, and feature resolution is less. Applications of SLS include functional prototypes, design evaluation models, short-run bridge production, and casting patterns.

**Fused Deposition Modeling (FDM).**—Also referred to as fused filament manufacturing and *material extrusion*, this process involves feeding a thermoplastic filament from a coil into a heated nozzle, which melts the plastic and extrudes it into a thin road, typically

0.016 in. (0.4 mm) in diameter (Fig. 31). The nozzle is attached to a precision XYZ motion system that traces the first 2D layer, as defined by the sliced CAD file, onto a platform. The platform then moves down one layer thickness, and the next layer is traced; this process is repeated until the defined model is completed.

Since the extruded plastic is soft, some features require supports extending from the build platform upward to any overhangs of the part; after the part is completed and separated from the build platform, supports are easily broken off. Because the part is made of extruded filaments with rounded sides, the surface will be rougher than SLA. Surface finishing can be accomplished by sanding or exposing the part to boiling acetone, which dissolves the surface slightly and allows it to reflow to a smooth finish.

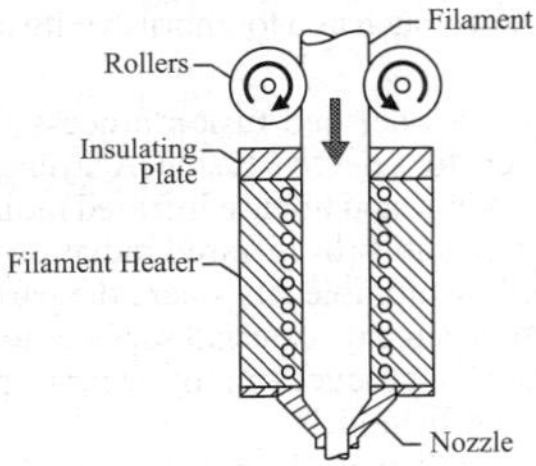


Fig. 31. Fused Deposition Modeling (FDM) Process

FDM machines are the least expensive plastics prototyping machines because they do not involve a laser and the heated extrusion head is widely available. Filaments for FDM are readily made by extrusion of thermoplastic feedstocks. Some feedstocks include reinforcing particles of glass or chopped graphite fiber, adding strength to printed parts.

One FDM method uses two nozzles and can produce multicolored parts. Or one of the nozzles can extrude a water-soluble support material that can be removed easily by soaking parts in water, reducing post-processing labor.

Another FDM process includes a separate nozzle for inserting continuous fibers of graphite, fiberglass, or Kevlar. This type of machine produces parts with selective reinforcement in high-stress areas.

Applications for FDM are based on its low cost but limited resolution. These include proof-of-concept models, low-cost prototyping, sand casting patterns, and jigs and fixtures. Because of its low cost, FDM has brought AM to a large number of nontechnical users, such as hobbyists, experimenters, and model-makers. In addition, FDM machines have provided a convenient and low-cost vehicle for introduction of students at all levels to the basic concepts of design and manufacturing.

**Binder Jetting (BJ).**—In this process, a layer of powder is spread on a build plate in a build box similar to SLS. However, instead of bonding the powder by sintering with the heat of an infrared laser, droplets of binder are jetted from a printhead containing an array of jets, as it passes over the powder layer, creating a 2D layer. After each pass, the build platform is lowered by one layer thickness, a new layer of powder is spread, and the print head passes over the new layer. This process is repeated until the complete 3D part is produced.

Binder droplets can be thought of as dots that combine to produce each slice of the CAD file. The binder not only bonds particles within each droplet, it also bonds each dot to its neighbors and penetrates to the slice below to ensure strength for subsequent handling. Because the droplet mass is relatively small, the bonded powder regions have similar density as the unbonded powder, so supports are not required. The finished part is removed from the build box, and the unbonded powder is removed and reused. Binder jetting of powders with adhesives works with a variety of materials. While earlier used with plaster and starch, the approach has been widely used for ceramics, metals, and plastics (see "Multijet Fusion," below).

**Material Jetting (MJ).**—Something of a hybrid process, material jetting involves printing fine droplets of light-curable resin onto a build plate. Immediately after the printhead deposits the resin droplets, a light bar passes over the build plate, solidifying the resin in a polymerization process similar to SLA, producing thermoset plastics. Overhangs require supports to prevent sagging of solidified material.

Some advanced versions of MJ involve multiple printheads; each one can be set up to jet a different resin. Various resins harden into solid polymers of different color and hardness. As a result, material jetting can be used to produce product prototypes and parts with varying material properties, such as a toothbrush with a rigid handle, soft grip, and stiff bristles.

**Multijet Fusion (MJF).**—A powder bed fusion process like binder jetting, multijet fusion uses two types of binders for selective fusion. A layer of plastic powder is spread, a printhead deposits the two binders, and intense infrared radiation is applied to the layer. One binder is a fusing agent that absorbs infrared radiation and fuses the polymer particles together; it is deposited by multiple jets where the part is to be made. The second binder, a detailing agent, is deposited on external surfaces to provide sharp definition of part boundaries and features. MJF produces highly accurate parts and is useful for prototype development and form-and-fit models.

The MJF process has some significant advantages over other AM processes. As a powder-bed process, structural supports are not required, processing speed is very high, and material costs are relatively low. MJF also supports multiple types of materials such as PP, TPU, PA12, and glass bead–filled PA12. Moreover, binders can be tinted to provide "full color" parts with varying colored surfaces as defined in CAD.

## AM Considerations for Tooling and Parts

All plastic AM processes build parts layer by layer. This characteristic enables production of more complex shapes than can be made by conventional machining, casting and molding, or other forming operations. Because AM processes do not involve forming tools, overhangs and re-entrant corners can be produced readily. In addition, internal cavities and channels that are neither round nor straight are feasible—overcoming limitations of conventional drilling.

On the other hand, the per-part cost of AM parts is greater than that of conventional mass production processes. The cost of powder or filament feedstocks is higher than the same materials in conventional form, and the processes are much slower than conventional processes. These advantages and limitations are illustrated as cost break-even points in Fig. 30a and Fig. 30b in *Additive Manufacturing Workflow* on page 1555.

A major application that fits within these limits is production of tooling for conventional processes. Applications include jigs and fixtures for machining, grinding, and welding; dimension-checking fixtures and gages for quality control; masking and non-marring of critical surfaces; organizing tools; and robotic end effectors. In addition, high-strength AM plastic dies are used for short-run thermoforming and injection molding. Composite layup tools and sand casting patterns also are produced economically by plastic AM processes.

Probably the most widely used AM tooling application is in fixtures for dimensioning first production articles and/or for periodic checks. Such fixtures normally are CNC machined from aluminum or assembled from machined aluminum segments, but in most cases, the strength of aluminum is not required. Furthermore, fixtures can be produced by the FDM process in a day or two, rather than the far longer lead time required for CNC programming and production or assembly. For example, a fixture needed for dimension and distortion measurement of an investment casting would require delivery time of four weeks and cost $450 if it were made from aluminum. Producing the same fixture from

ABS plastic by FDM would cost $50 and be completed in one day. Similar savings of time and money are realized in AM production of jigs, such as those used for emblem or medallion attachment on automotive assembly lines.

While AM processes and procedures currently are not capable of economically producing the large numbers of components needed for full-scale automotive production, AM processes are continuously evolving with respect to precision and speed. Already, AM is used in the automotive industry to produce customized features, such as whimsical front-end "face" coverings, with shark skin, dimple, and whisker details. race car and other custom shops have used AM for many years to produce custom spoilers for wind tunnel testing, vents to enhance engine performance, and other specialized parts. AM also is used to make parts such as air intakes and fender vents for luxury automobiles. for vintage vehicles, from cars to airplanes, AM software and systems can scan, analyze, and replicate outdated replacement parts. For this purpose, in addition to making plastic parts directly, plastic-based AM processes also are used to produce tooling for other manufacturing processes.

Plastic AM is used to make patterns and molds for metal and composite casting and molding, as well as dies for sheet metal forming processes. In many applications, AM processes can produce thermoforming dies with less time and cost than machining aluminum (see *Sheet Thermoforming* on page 581). Plastic AM processes also are used to produce injection molding tools for short runs between 10 and 1,000 parts. A major advantage of making injection molding tools using AM is that conformal internal cooling channels can be printed within molds. Because the cooling channels have the same profile as the cavity surfaces, heat is removed from molded parts rapidly and uniformly, reducing injection molding cycle time and warpage.

Plastic tooling made by AM processes has enough strength to be used for forming simple shapes in sheet metal. In one case, a nose gear failure during landing on an aircraft carrier at sea damaged two sheet metal parts in the nose of the plane, removing it from service. Replacement parts were not available, so two forming dies, used for forming such replacement sheet metal parts, were made by FDM at a seaside base and then flown to the carrier at sea. The aircraft was repaired and in service within a week, instead of waiting longer to receive replacement parts from the original equipment manufacturer or aftermarket parts supplier.

More patterns for sandcasting are being made by AM to shorten time to mold production. While such patterns commonly have been made by machining aluminum or composition, using AM can reduce delivery time from weeks to days. The pattern materials must resist the heavy pressures and abrasion of sand impact during mold compaction and be impervious to the binding agents and mold-release chemicals used in casting (see *Patterns* on page 1498). Printed patterns may be sanded, smoothed by acetone or other chemical treatment, or sealed with epoxy to provide a smoother finish.

As an alternative to printing patterns, AM can also print the sand mold itself, eliminating the need for a pattern. Printed sand molds are produced with the binder jetting (BJ) process, using conventional foundry sand as the printed material instead of metal powder and a no-bake binder. (Both materials are common in conventional mold making foundries, facilitating foundries printing sand molds.) This patternless casting method can produce the entire mold, including the internal cavity for the part shape; the metal feed system, consisting of sprue, runners, risers, and gates; and the vent holes that allow air and gases to escape as the molten liquid is poured into the mold (see *Sand Casting* on page 1497). As with other AM processes, patternless casting also is useful for part prototyping, as changes to the liquid feed system and cast shape can be made quickly by modifying the CAD file.

Plastic AM is also used to produce patterns for investment casting (see *Precision Investment Casting* on page 1504). Patterns are produced by AM processes, and then used in the conventional investment casting sequence of coating the pattern with plaster or refractory materials, melting or burning out the pattern material, and pouring liquid metal into the mold. Plastics have been developed for this purpose that leave very small amounts of ash to contaminate the investment casting. In addition, because loads on patterns for investment casting are low, the patterns are made in the form of a shell with a honeycomb internal structure, reducing the amount of material in the pattern.

Tooling for composite layups is increasingly being made by FDM because of their rapid turnaround. Various approaches are used to heat and pressurize the composite layers into a finished part, such as a blade for a wind turbine. (See *Polymer Composites* on page 585 for curing practices used in plastic composite forming.) ABS and PC can withstand curing process temperatures up to 270 °F (132 °C) and pressures up to 100 psi (690 kPa). For higher temperatures and increased pressures, tools can be made of higher temperature polymers including nylon, Ultem, and PEEK. The use of materials with chopped glass or graphite fibers can drastically reduce thermal distortion and sagging during production of large parts. The largest of these 3D printers has a production envelope of length, width, and height, respectively, of 20 by 7.5 by 6 feet (6 by 2.3 by 1.8 meters). Segmented printing enables printing of even larger vehicle and building parts.

Production of parts by AM is extending to other applications. One of the highest production and most visible applications is the use of SLA-type processes to print custom molds for thermoforming of personalized dental braces. Many companies are working towards a digital inventory in which production and replacement parts are made by AM as needed. Such an approach supports rapid product changes, enabling short runs and mass customization to gain market share without having to invest in conventional tooling and machinery. AM is especially promising for spare parts, particularly for systems near the end of their lifecycle, where parts inventory has been depleted. This capability is important for many types of older machinery still in operation. Production of legacy and other on-demand parts on location for defense systems represents a major opportunity.

## TABLE OF CONTENTS

## DIMENSIONING, GAGING, AND MEASURING

### DRAFTING PRACTICES



### ALLOWANCES AND TOLERANCES FOR FITS



### ALLOWANCES AND TOLERANCES FOR FITS

*(Continued)*



### MEASURING, INSTRUMENTS, AND INSPECTION METHODS

## TABLE OF CONTENTS

## DIMENSIONING, GAGING, AND MEASURING

### MEASURING, INSTRUMENTS, AND INSPECTION METHODS

*(Continued)*



### MICROMETERS AND CALIPERS



### MICROMETERS AND CALIPERS

*(Continued)*



### SURFACE TEXTURE

## DRAFTING PRACTICES

While there are many national standards-based drafting practices throughout the world, three sets of standards practices predominate internationally. These sets of practices include the American National Standards Institute (ANSI); American Society of Mechanical Engineers (ASME); those modeled on standards developed by the International Organization for Standardization (ISO); and to a lesser extent, those based on Japanese Industry Standards (JIS). However since JIS has committed to adopting ISO practices, coverage will be limited to ANSI/ASME practice and select ISO practices where appropriate. In general, there are several ANSI/ASME Standards for use in preparing engineering drawings and related documents.

### Standard Drafting Practices

**Sizes of Drawing Sheets.—**Recommended trimmed sheet sizes are shown in Table 1. Customary inch sizes are documented in US ANSI/ASME Y14.1-2020, are based on commercial letterhead paper ($8\frac{1}{2} \times 11$ in.), and are in general use throughout the United States. Each successive size beginning with $8\frac{1}{2} \times 11$ in. (A) is double the size of the previous (except size F). Using sizes based on commercial letterhead paper size, and its multiples, permits filing small prints and folded larger prints in the same commercial standard letter files. For drawings requiring trimmed metric sheets, sizes are based on ANSI/ASME Y14.1M-2012 in the United States and ISO 5457 in other countries. Metric sizes are consistent across US and ISO standards. As with the customary inch sheet sizes, each successive metric size sheet beginning with commercial letterhead paper size A4 is double the previous size also for ease of folding and filing.

Metric sizes are based on the width-to-length ratio of 1 to $\sqrt{2}$. Note that the metric size designators increase as the paper decreases in area. Virtually all CAD systems include both inch and metric size paper templates. Most countries outside the United States use metric paper sizes, so for foreign correspondence it is recommended that the metric sheet sizes be used. For additional sizes and details see the respective standards.

**Table 1. Drawing Sheet Sizes**

| US Customary Size, inches *ANSI/ASME Y14.1-2020* | | | | Metric Size, millimeters *ANSI/ASME Y14.1M-2020* | | | |
|---|---|---|---|---|---|---|---|
| A | $8\frac{1}{2} \times 11$ | D | $22 \times 34$ | A4 | $210 \times 297$ | A1 | $594 \times 841$ |
| B | $11 \times 17$ | E | $34 \times 44$ | A3 | $297 \times 420$ | A0 | $841 \times 1189$ |
| C | $17 \times 22$ | F | $28 \times 40$ | A2 | $420 \times 594$ | | |

**Line Conventions and Drawings.—**American National Standard ANSI/ASME Y14.2-2014 (R2020) establishes line and lettering practices for engineering drawings. The line conventions and the symbols for section lining are as shown in Table 2 and Table 3.

Approximate width of THICK lines for metric drawings is 0.6 mm, and for inch drawings, 0.032 inch. Approximate width of THIN lines for metric drawings is 0.3 mm, and for inch drawings, 0.016 inch. These approximate line widths are intended to differentiate between THICK and THIN lines and are not values influencing acceptance or rejection of drawings. Basic line conventions in ISO are established in ISO 128-20: 1996. Lines intended to comply with ISO standards use widths of 0.13 mm, 0.18 mm, 0.25 mm, 0.35 mm, 0.5 mm, 0.7 mm, 1 mm, 1.4 mm and 2 mm, but again may deviate as long as lines are consistent in width and it is possible to differentiate "unambiguously" between two adjacent lines with different widths.

**Table 2. American National Standard Lines for Engineering Drawings**
*ANSI/ASME Y14.2-2014 (R2020)*

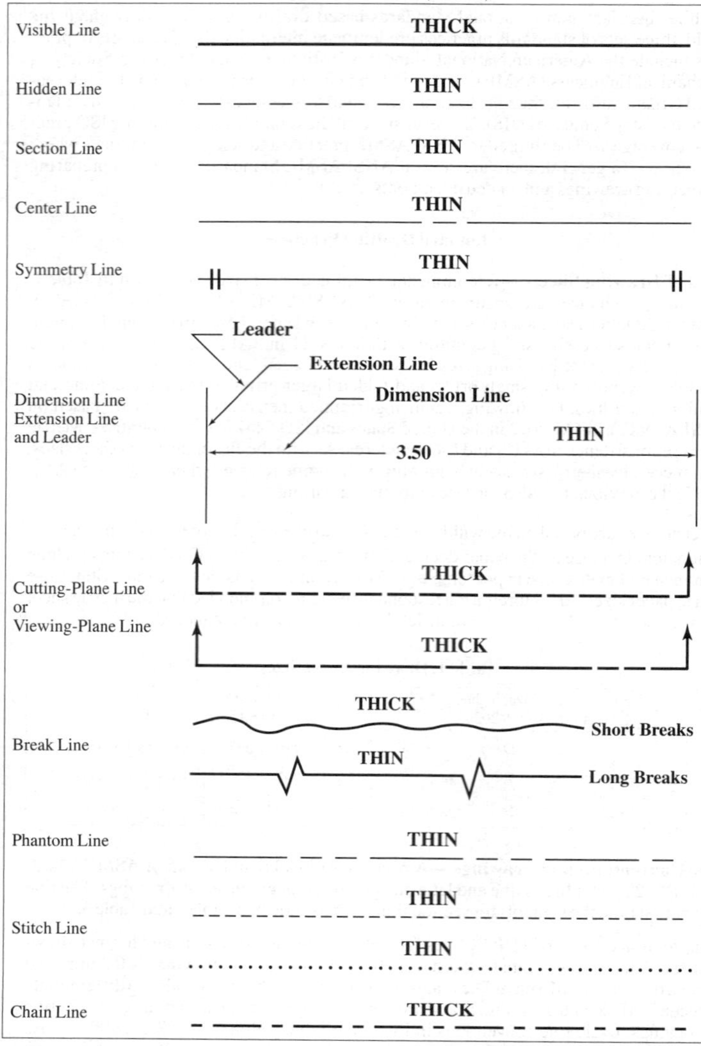

**Table 3. American National Standard Symbols for Section Lining**
*ANSI Y14.2M-1979 (R1987)*[a]

| | | | |
|---|---|---|---|
| 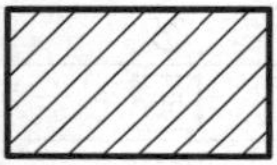 | Cast and malleable iron (Also for general use of all materials) | 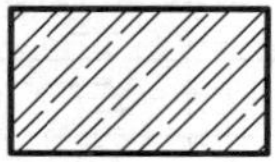 | Titanium and refractory material |
| 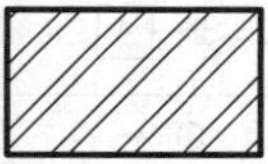 | Steel | 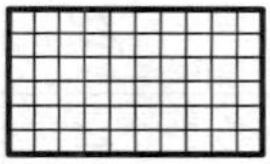 | Electric windings, electromagnets, resistance, etc. |
| 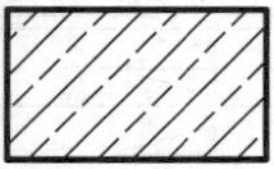 | Bronze, brass, copper, and compositions | 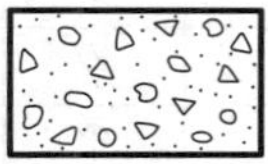 | Concrete |
| 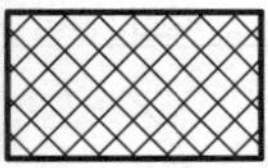 | White metal, zinc, lead, babbitt, and alloys | 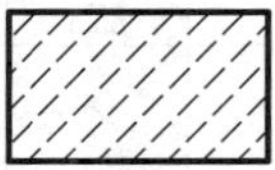 | Marble, slate, glass, porcelain, etc. |
| 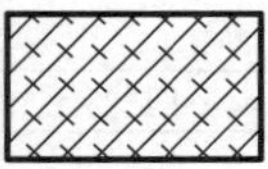 | Magnesium, aluminum, and aluminum alloys | 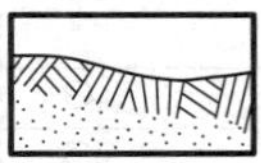 | Earth |
| 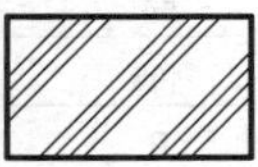 | Rubber, plastic electrical insulation |  | Rock |
| 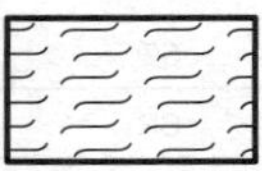 | Cork, felt, fabric, leather, fiber | 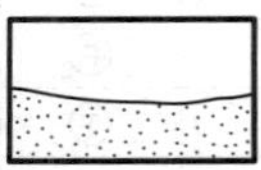 | Sand |
| 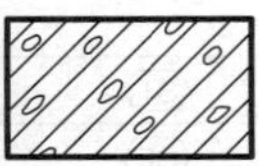 | Sound insulation | 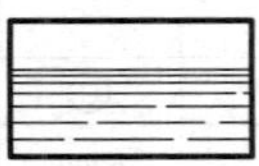 | Water and other liquids |
| 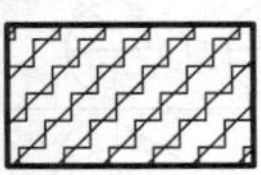 | Thermal insulation | 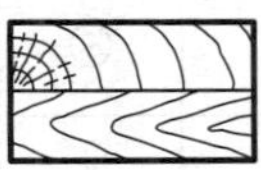 | Wood-across grain<br>Wood-with grain |

[a] This table appeared in the 2008 reaffirmed version of this Standard but has been removed from the current version, ANSI/ASME Y14.2-2014 (R2020). It has been retained here for reference.

**Table 4. ASME Geometric Symbols** *ASME Y14.5-2018*

| Symbol for | | Symbol for | |
|---|---|---|---|
| Straightness | — | Diameter | Ø |
| Flatness | ⏥ | Basic Dimension | [50] |
| Circularity | ○ | Reference Dimension | (50) |
| Cylindricity | ⌭ | Datum Feature (triangle may be filled in or not) |  |
| Angularity | ∠ | Dimension Origin |  |
| Perpendicularity | ⊥ | Feature Control Frame | ⌖ \| Ø0.5 Ⓜ \| A \| B \| C |
| Parallelism | // | Conical Taper | ⌲ |
| Position | ⌖ | Slope | ⌳ |
| Profile of a Line | ⌒ | Counterbore | ⌴ |
| Profile of a Surface | ⌓ | Spotface | ⌴SF⌴ |
| Circular Runout (arrowhead may be filled in or not) | ↗ | Countersink | ⌵ |
| Total Runout (arrowheads may be filled in or not) | ⌰ | Depth/Deep | ↧ |
| All Around |  | Square | □ |
| All Over |  | Dimension Not to Scale | 15 (underlined) |
| At Maximum Material Condition | Ⓜ | Number of Places | 8X |
| At Maximum Material Boundary | Ⓜ | Arc Length | ⌒105 |
| At Least Material Condition | Ⓛ | Radius | R |
| At Least Material Boundary | Ⓛ | Spherical Radius | SR |
| Projected Tolerance Zone | Ⓟ | Spherical Diameter | SØ |
| Tangent Plane | Ⓣ | Controlled Radius | CR |
| Free State | Ⓕ | Between (arrowheads may be filled in or not) | ↔ |
| Unequally Disposed Profile | Ⓤ | Statistical Tolerance | ⟨ST⟩ |
| Envelope Principle | Default | Continuous Feature | ⟨CF⟩ |
| Independency | Ⓘ | Datum Target | 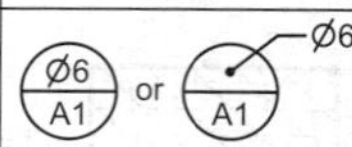 |
| Dynamic Profile Tolerance | Δ | Movable Datum Target | 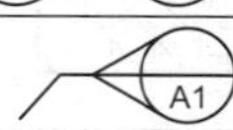 |
| Translation | ▷ | Target Point | ✕ |
| From / To | → | | |

**Table 5. ISO Geometric Symbols** *ISO 1101:2017*

| Symbol for | | Symbol for | |
|---|---|---|---|
| Straightness | — | Independency | Default |
| Flatness | ▱ | Dynamic Profile Tolerance | Δ |
| Circularity | ○ | From / To | → |
| Cylindricity | ⌭ | Diameter | ϕ |
| Angularity | ∠ | Theoretically Exact Dimension | [50] |
| Perpendicularity | ⊥ | Auxiliary Dimension | (50) |
| Parallelism | // | Datum Feature (triangle may be filled in or not) | [A] |
| Position | ⌖ | Dimension Origin | ⌀→ |
| Concentricity and Coaxiality | ◎ | Feature Control Frame | ⌖ \| ϕ 0.5 Ⓜ \| A \| B \| C |
| Symmetry | ⌯ | Conical Taper | ⌲ |
| Profile of a Line | ⌒ | Slope | ⌳ |
| Profile of a Surface | ⌓ | Square | □ |
| Circular Runout | ↗ | Dimension Not to Scale | 15 (underlined) |
| Total Runout | ⌰ | Number of Places | 8x |
| All Around | | Arc Length | ⌒105 |
| All Over | | Radius | R |
| At Maximum Material Condition | Ⓜ | Spherical Radius | SR |
| At Least Material Condition | Ⓛ | Spherical Diameter | Sϕ |
| Projected Tolerance Zone | Ⓟ | Between (arrowheads may be filled in or not) | ↔ |
| Tangent Plane | Ⓣ | Datum Target | ϕ6 / A1 or ϕ6 A1 |
| Free State | Ⓕ | Movable Datum Target | A1 |
| Unequally Disposed Profile | UZ | Target Point | ✕ |
| Envelope Principle | Ⓔ | | |

Note: This table includes the most commonly used symbols; for additional symbols, refer to the Standard.

**Surface Texture Symbols.—**A detailed explanation of the use of surface-texture symbols from American National Standard ANSI/ASME Y14.36 begins on page 804, while ISO surface-texture techniques, standardized in ISO 1302, are explained beginning on page 811.

**Geometric Dimensioning and Tolerancing (GD&T).—**ASME Y14.5-2018, Dimensioning and Tolerancing, covers dimensioning, tolerancing, and similar practices for engineering drawings and related documentation. The mathematical definitions of dimensioning and tolerancing principles are given in this Standard. ISO 8015, ISO 1101, and ISO 26921 contain a detailed explanation of ISO geometric dimensioning and tolerancing practices. Those ISO practices involving GD&T and other selected ISO standard practices are shown in contrast to ANSI/ASME practices where applicable.

Geometric dimensioning and tolerancing provides a comprehensive system for symbolically defining the geometrical tolerance zone within which features must be contained. It provides an accurate transmission of design specifications among the three primary users of engineering drawings: design, manufacturing, and quality assurance groups. Some techniques introduced in ASME Y14.5-2018 have been accepted by ISO. These techniques include projected tolerance zone, the three-plane datum concept, total runout tolerance, multiple datums, and datum targets. Although Y14.5 follows ISO practice closely, there are still differences between ISO and US practice. Summaries of commonly used symbols in ASME Y14.5 and ISO standards are given in Table 4 and Table 5, respectively.

One major area of disagreement is the ISO "principle of independency" versus the "Taylor principle." Y14.5 and standard US practice both follow the Taylor principle, in which a geometric tolerancing zone may not extend beyond the boundary (or envelope) of perfect form at MMC (maximum material condition). This boundary is prescribed to control variations as well as the size of individual features. The US definition of independency further defines features of size as being independent and not required to maintain a perfect relationship with other features. The "envelope principle" is optional in treatment of these principles. A summary of the application of ASME geometric control symbols and their use with basic dimensions and modifiers is given in Table 6.

ASME Y14.5 features metric SI units (the International System of Units), but customary units may be used without violating any principles. On drawings where all dimensions are either in millimeters or in inches, individual identification of linear units is not required. However, the drawing should contain a note stating *"Unless otherwise specified, all dimensions are in millimeters"* (or *in inches*, as applicable). According to Y14.5, all dimensions are applicable at a temperature of 20° C (68° F) unless otherwise specified. Compensation may be made for measurements taken at other temperatures.

Angular units are expressed in degrees and decimals of a degree (35.4) or in degrees (°), minutes (′), and seconds (″), as in 35° 25′ 10″. Where decimal degrees less than one are specified, a zero shall precede the decimal value. A 90-degree angle is implied where center lines and depicting features are shown on a drawing at right angles and no angle is specified. A 90-degree BASIC angle applies where center lines of features in a pattern or surface shown at right angles on a drawing are located or defined by basic dimensions and no angle is specified.

Basic to all US practice is that orthographic drawing views are arranged in 3rd angle projection while Europe and Asia generally follow ISO practices that default to 1st angle projection. In 1st angle projection, the orientation of the views is reversed from US practice. See Fig. 1a for 3rd angle projection and Fig. 1b for 1st angle projection. Note the graphical symbol at the bottom of each, which indicates the projection system in use for the drawing. ANSI, ASME and ISO practice allow the other projection system, but the use of non-default practice must be declared by use of the graphical symbol.

**Table 6. ASME Geometric Control Symbols** *ASME Y14.5-2018*

| Type [a] | Geometric Characteristics | | Pertains To | Basic Dimensions | Feature Modifier | Datum Modifier |
|---|---|---|---|---|---|---|
| Form | ⏤ | Straightness | Only individual feature | Not applicable | See *Note* 1 | No datum |
| | ▱ | Flatness | | | Modifier not applicable | |
| | ○ | Circularity | | | | |
| | ⌭ | Cylindricity | | | | |
| Profile | ⌒ | Profile of a Line | Individual or related | Yes if related | | See *Note* 1 |
| | ⌓ | Profile of a Surface | | | | |
| Orientation | ∠ | Angularity | Always related feature(s) | Yes | See *Note* 1 | |
| | ⊥ | Perpendicularity | | Not applicable | | |
| | // | Parallelism | | | | |
| Location | ⌖ | Position | | Yes | | |
| Runout | ↗ | Circular Runout [b] | | Not applicable | Only RFS | |
| | ⌰ | Total Runout [b] | | | | |

*Note* 1: Default RFS unless MMC or LMC explicitly stated.

[a] Five types of geometric control, when datums are indicated, when basic dimensions are required, and when MMC and LMC modifiers may be used.
[b] Arrowheads may be filled in or not.

**ANSI/ASME and ISO Orthographic Projections**

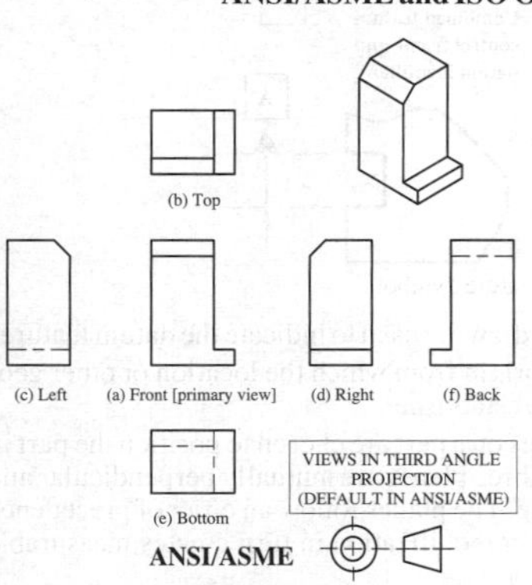


Fig. 1a. ANSI/ASME–Orthographic Projection Following Third Angle Projection

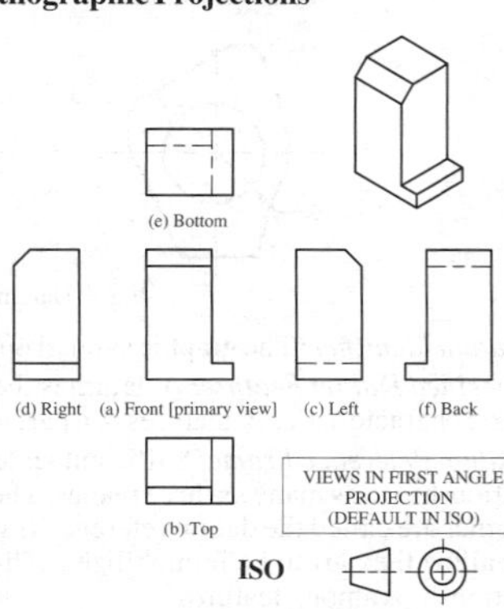


Fig. 1b. ISO–Orthographic Projection Following First Angle Projection

US ASME Y14.5-2018 and ISO practice use a preceding 0 before millimeter dimension values less than 1. Unlike US practice, ISO requires a comma as a decimal placeholder where US practice would use a decimal point, compare Fig. 2a and Fig. 2c. For decimal inch dimensioning, a zero is NOT used before the decimal for values less than 1 inch, Fig. 2b.

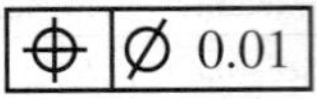


Fig. 2a. US Practice, Metric Dimensions

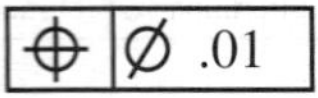


Fig. 2b. US Practice, Inch Dimensions

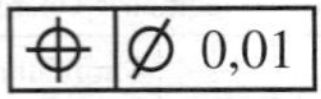


Fig. 2c. ISO Practice

When dimensioning per US default practice, unless otherwise specified, dimension lines are broken and dimension text remains horizontal, hence read from the bottom of the drawing, see Fig. 3a. ISO dimensioning practices default to aligned text with dimensions placed above an unbroken dimension line, see Fig. 3b.

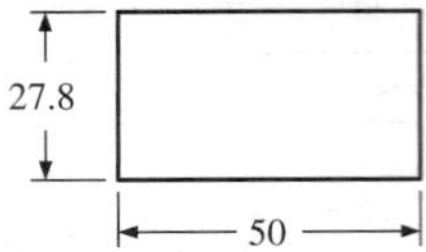


Fig. 3a. Default US Practice

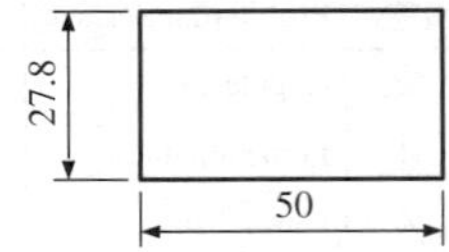


Fig. 3b. Default ISO Practice

**Standard Drafting Definitions**

The following terms are defined as their use applies to ASME Y14.5. Where ISO practice in actual use differs, the definitions are expanded to clarify the contrasting practices.

*Datum Feature:* The feature of a part that is used to establish a datum, identified with either a datum feature or datum target symbols, see Fig. 4.

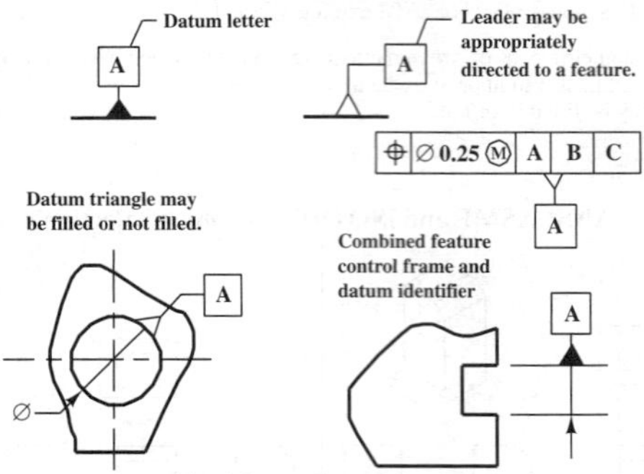


Fig. 4. Datum Feature Symbol

*Datum Identifier:* The graphic symbol on a drawing used to indicate the datum feature.

*Specified Datum Feature:* A datum is the origin from which the location or other geometric characteristics of features of a part are established.

*Datum Reference Frame:* Sufficient features on a part are chosen to position the part in relationship to as many as three planes. The three planes are mutually perpendicular and together are called the datum reference frame. The planes follow an order of precedence and allow the part to be immobilized. This immobilization in turn creates measurable relationships among features.

*Datum Simulator:* Formed by the datum feature contacting a precision surface such as a surface plate, gage surface or by a mandrel contacting the datum. Thus, the plane formed by contact restricts motion and constitutes the specific reference surface from which measurements are taken and dimensions verified. The datum simulator is the practical embodiment of the datum feature during manufacturing and quality assurance.

*Datum Target:* A specified point, line, or area on a part, used to establish a datum; see page 632.

*Degrees of Freedom:* The six directions of movement or translation are called degrees of freedom in a three-dimensional environment. They are up-down, left-right, fore-aft, roll, pitch, and yaw, see Fig. 5.

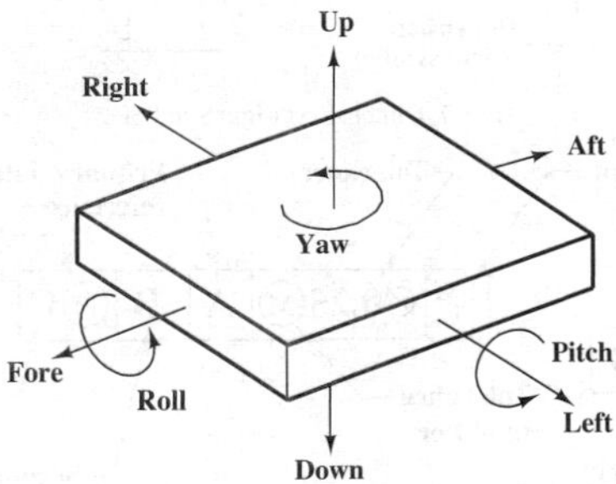


Fig. 5. Degrees of Freedom (Movement) that Must be Controlled, Depending on the Design Requirements

*Dimension, Basic:* A numerical value used to describe the theoretically exact size, orientation, location, or optionally, the profile, of a feature or datum or datum target. Basic dimensions are indicated by a rectangle around the dimension and are not toleranced directly or by default, see Fig. 6. The specific dimensional limits are determined by the permissible variations as established by the tolerance zone specified in the feature control frame. A dimension is only considered basic for the geometric control to which it is related.

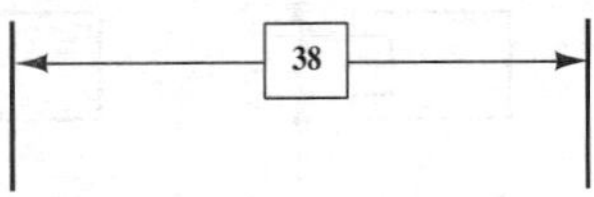


Fig. 6. Basic Dimensions

*Dimension, Origin:* A symbol used to indicate the origin and direction of a dimension between two features. The dimension originates from the symbol with the dimension tolerance zone being applied at the other feature, see Fig. 7.

*Dimension, Reference:* A dimension, usually without tolerance, used for information purposes only. Considered to be auxiliary information and not governing production or inspection operations. A reference dimension is a repeat of a dimension or is derived from a calculation or combination of other values shown on the drawing or on related drawings.

*Feature Control Frame [Tolerance Frame]:* Specification on a drawing that indicates the type of geometric control for the feature, the tolerance for the control, and the related datums, if applicable, see Fig. 8.

*Feature:* The general term applied to a physical portion of a part, such as a surface, hole, pin, tab, or slot. In ISO practice, depending on how the tolerance frame leader line is attached to the feature, different interpretations may be invoked as to whether the reference is to a line or surface, or an axis or median planer.

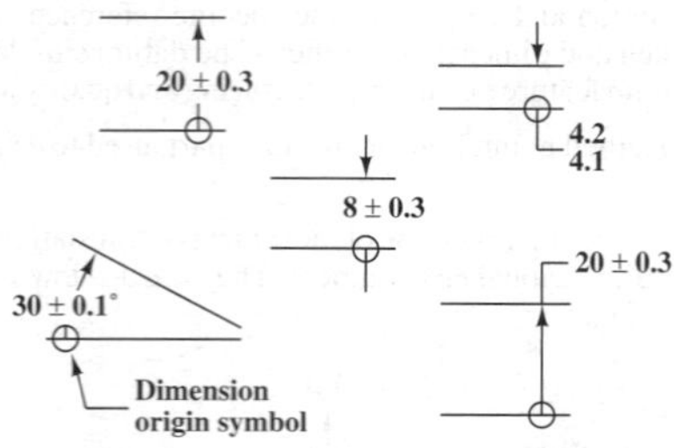


Fig. 7. Dimension Origin Symbol

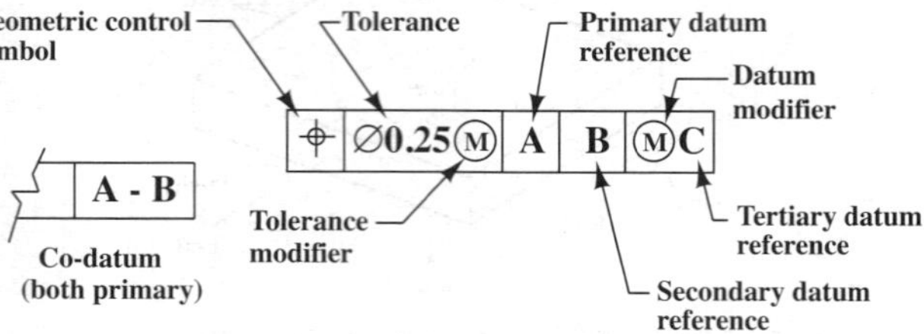


Fig. 8. Feature Control Frame and Datum Order of Precedence

When an ISO tolerance frame leader line terminates on the outline of the feature, it indicates that the control is a line or the surface itself (Fig. 9a.) When an ISO tolerance frame leader line terminates on a dimension, the axis or medium plane of the dimensioned feature is being controlled. Either inside or outside dimension lines may be used (Fig. 9b.)

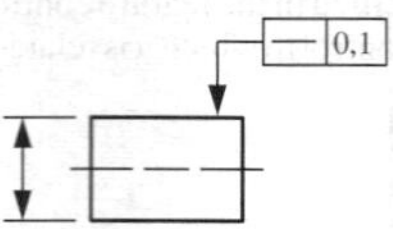


Fig. 9a. ISO Tolerance Frame Leader Line Terminating on the Outline of the Feature

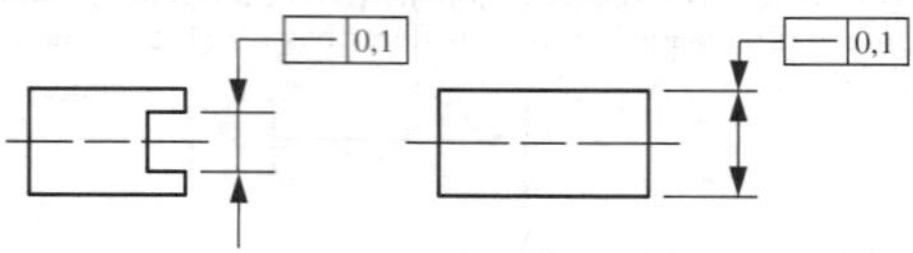


Fig. 9b. ISO Tolerance Frame Leader Line Terminating on a Dimension

*Feature of Size, Regular:* One cylindrical or spherical surface, a circular element, and a set of two opposed parallel elements or opposed parallel surfaces, each of which is associated with a directly toleranced dimension.

*Feature of Size, Irregular:* A directly toleranced feature or collection of features that may contain or be contained by an actual mating envelope that is a) a sphere, cylinder, or pair of parallel planes; or b) other than a sphere, cylinder, or pair of parallel planes.

*Least Material Boundary (LMB):* The limit defined by a tolerance or combination of tolerances that exist on or inside the material of a feature or features.

*Least Material Condition (LMC):* The condition in which a feature of size contains the least amount of material within the stated limits of size, for example, upper limit or maximum hole diameter and lower limit or minimum shaft diameter.

*Limits, Upper and Lower (UL and LL):* The arithmetic values representing the maximum and minimum size allowable for a dimension or tolerance. The upper limit represents the maximum size allowable. The lower limit represents the minimum size allowable.

*Maximum Material Boundary (MMB):* The limit defined by a tolerance or combination of tolerances that exist on or outside the material of a feature or features.

*Maximum Material Condition (MMC):* The condition in which a feature of size contains the maximum amount of material within the stated limits of size. For example, the lower limit of a hole is the minimum hole diameter. The upper limit of a shaft is the maximum shaft diameter.

*Position:* Formerly called true position, position is the theoretically exact location of a feature established by basic dimensions. In ISO practice a basic dimension is called a theoretically exact dimension (TED). A positional tolerance is indicated by the position symbol, a tolerance value, applicable material condition modifiers, and appropriate datum references placed in a feature control frame.

*Regardless of Feature Size (RFS):* The term used to indicate that a geometric tolerance or datum reference applies at any increment of size of the feature within its tolerance limits. RFS is the default condition unless MMC or LMC is specified. The concept is now the default in ASME Y14.5-2018, unless specifically stated otherwise. Thus the symbol for RFS is no longer supported in ASME Y14.5-2018.

*Regardless of Material Boundary (RMB):* indicates that a datum feature simulator progresses from MMB toward LMB until it makes maximum contact with the extremities of a feature(s). See *Datum Simulator* on page 628.

*Size, Actual:* The term indicating the size of a feature as produced.

*Tolerance Zone Symmetry:* In geometric tolerancing, the tolerance value stated in the feature control frame is always a single value. Unless otherwise specified, it is assumed that boundaries created by the stated tolerance are bilateral and equidistant about the perfect form control specified. See Fig. 10a for default zone. If desired, the tolerance may be specified as unilateral or unequally bilateral. See Fig. 10b and Fig. 10c for external and internal unilateral zones, and Fig. 10d for an example of a bilateral asymmetrical zone.

**Tolerance Zone Symmetry Examples**

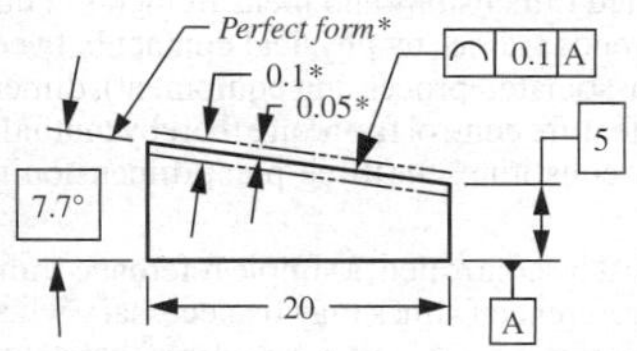


Fig. 10a. Default Symmetrical Tolerance Zone About Perfect Form

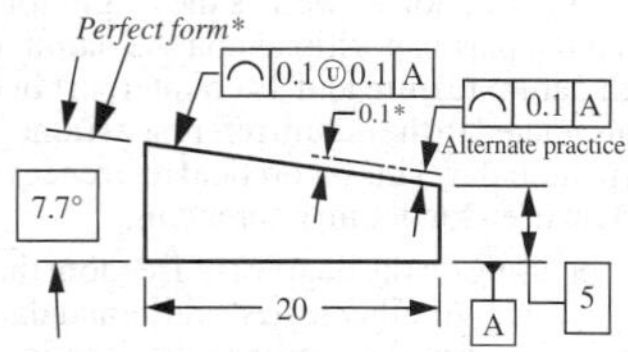


Fig. 10b. External Unilateral Tolerance Zone About Perfect Form

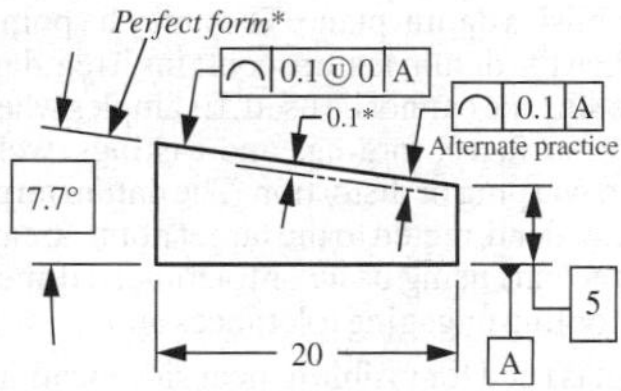


Fig. 10c. Internal Unilateral Tolerance Zone About Perfect Form

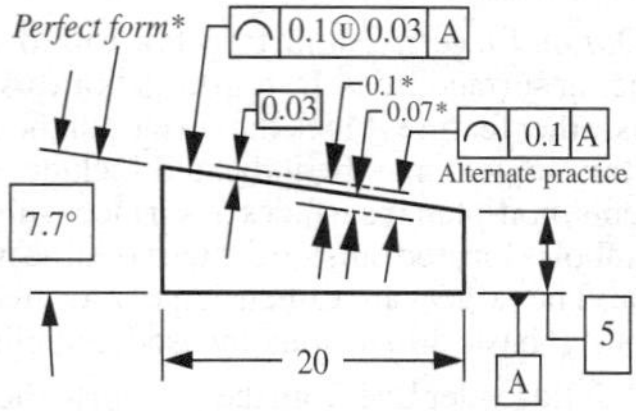


Fig. 10d. Bilateral Asymmetrical Tolerance Zone About Perfect Form

** Added for clarification and is not part of the specification.*

*Tolerance, Bilateral:* A tolerance where variation is permitted in both directions from the specified dimension. Bilateral tolerances may be equal or unequal.

*Tolerance, Geometric:* The general term applied to the category of tolerances used to control form, profile, orientation, location, and runout.

*Tolerance, Unilateral:* A tolerance where variation is permitted in only one direction from the specified dimension.

*True Geometric Counterpart:* Theoretically perfect plane of a specified datum feature.

*Virtual Condition:* A constant boundary generated by the collective effects of the feature size, its specified MMC or LMC material condition, and the geometric tolerance for that condition.

**Datum Referencing.**—A datum indicates the origin of a dimensional relationship between a toleranced feature and a designated feature or features on a part. The designated feature serves as a datum feature, whereas its true geometric counterpart establishes the datum plane. Because measurements cannot be made from a true geometric counterpart, which is theoretical, a datum is assumed to exist in and be simulated by the associated processing equipment.

For example, machine tables and surface plates, although not true planes, are of such quality that they are used to simulate the datums from which measurements are taken and dimensions are verified. When magnified, flat surfaces of manufactured parts are seen to have irregularities, so that contact is made with a datum plane formed at a number of surface extremities or high points. Sufficient datum features, those most important to the design of the part, are chosen to position the part in relation to a set of three mutually perpendicular planes, the datum reference frame. This reference frame exists only in theory and not on the part. Therefore, it is necessary to establish a method for simulating the theoretical reference frame from existing features of the part. This simulation is accomplished by positioning the part on appropriate datum features to adequately relate the part to the reference frame and to restrict the degrees of freedom of the part in relation to it.

These reference frame planes are simulated in a mutually perpendicular relationship to provide direction as well as the origin for related dimensions and measurements. Thus, when the part is positioned on the datum reference frame (by physical contact between each datum feature and its counterpart in the associated processing equipment), dimensions related to the datum reference frame by a feature control frame are thereby mutually perpendicular. This theoretical reference frame constitutes the three-plane dimensioning system used for datum referencing.

Depending on the degrees of freedom that must be controlled, a simple reference frame may suffice. At other times, additional datum reference frames may be necessary where physical separation occurs or the functional relationship of features requires that datum reference frames be applied at specific locations on the part. Each feature control frame must contain the datum feature references that are applicable.

*Datum Targets:* Datum targets are used to establish a datum plane. They may be points, lines or surface areas. Datum targets are used when the datum feature contains irregularities, other features block the surface or the entire surface cannot be used. Examples where datum targets may be indicated include uneven surfaces, forgings and castings, weldments, non-planar surfaces or surfaces subject to warping or distortion. The datum target symbol is located outside the part outline with a leader directed to the target point, area or line. The targets are dimensionally located on the part using basic or toleranced dimensions. If basic dimensions are used, established tooling or gaging tolerances apply.

A solid leader line from the symbol to the target is used for visible or near side locations, with a dashed leader line used for hidden or far side locations. The datum target symbol is divided horizontally into two halves. The top half contains the target point area if applicable; the bottom half contains a datum feature identifying letter and target number. Target numbers indicate the quantity required to define a primary, secondary, or tertiary datum. If a target point or target line is indicated, the top half is left blank. Datum targets and datum features may be combined to form the datum reference frame, see Fig. 11.

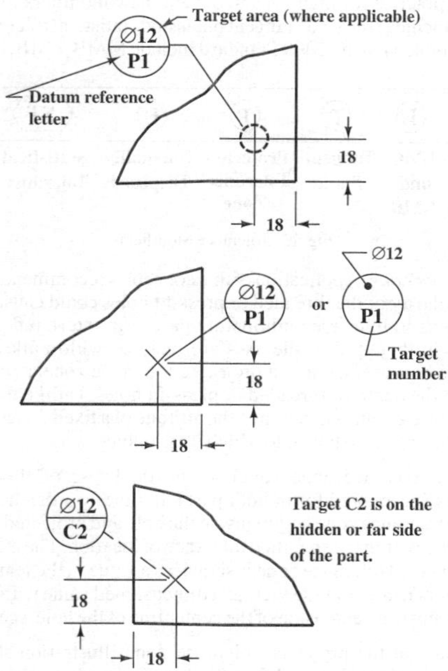


Fig. 11. Datum Target Symbols

*Datum Target Points:* A datum target point is indicated by the symbol "X," which is dimensionally located on a direct view of the surface. Where there is no direct view, the point location is dimensioned on multiple views.

*Datum Target Lines:* A datum target line is dimensionally located on an edge view of the surface using a phantom line on the direct view. Where there is no direct view, the location is dimensioned on multiple views. Where the length of the datum target line must be controlled, its length and location are dimensioned.

*Datum Target Areas:* Where it is determined that an area or areas of flat contact are necessary to ensure establishment of the datum, and where spherical or pointed pins would be inadequate, a target area of the desired shape is specified. Examples include the need to span holes, finishing irregularities, or rough surface conditions. The datum target area may be indicated with the "X" symbol as with a datum point, but the area of contact is specified in the upper half of the datum target symbol. Datum target areas may additionally be specified by defining controlling dimensions and drawing the contact area on the feature with section lines inside a phantom outline of the desired shape.

**Positional Tolerance.**—A positional tolerance defines a zone within which the center, axis, or center plane of a feature of size is permitted to vary from true (theoretically exact) position. Basic dimensions establish the true position from specified datum features and between interrelated features. A positional tolerance is indicated by the position symbol, a tolerance, and appropriate datum references placed in a feature control frame.

*Modifiers:* In certain geometric tolerances, modifiers in the form of additional symbols may be used to further refine the level of control. The use of the MMC and LMC modifiers has been common practice for many years. Several new modifiers introduced with the 1994 US Standard include free state, tangent plane and statistical tolerancing, see Fig. 12. New modifiers introduced in the 2009 Standard include MMB, LMB, and RMB.

| Ⓕ | Ⓜ | Ⓛ | Ⓣ | Ⓟ | Ⓤ | ⟨ST⟩ | ⟨CF⟩ |
|---|---|---|---|---|---|---|---|
| **Free State** | **MMC and MMB** | **LMC and LMB** | **Tangent Plane** | **Projected Tolerance Zone** | **Unequally Disposed** | **Statistical Tolerance** | **Continuous Feature** |

Fig. 12. Tolerance Modifiers

*Projected Tolerance Zone:* Application of this concept is recommended where any variation in perpendicularity of the threaded or press-fit holes could cause fasteners such as screws, studs, or pins to interfere with mating parts. An interference with subsequent parts can occur even though the hole axes are inclined within allowable limits. This interference occurs because, without a projected tolerance zone, a positional tolerance is applied solely to the depth of threaded or press-fit holes. Unlike the floating fastener application involving clearance holes only, the attitude of a fixed fastener is restrained by the inclination of the produced hole into which it assembles.

With a projected tolerance zone equal to the thickness of the mating part, the inclinational error is accounted for in both parts. In other words, the projected zone is an extension of the tolerance zone begun inside the hole and projected out, in this case 14 mm, on the same axis within the specified tolerance of location. The minimum extent and direction of the projected tolerance zone is shown as a value in the feature control frame. The zone may be shown in a drawing view as a dimensioned value with a heavy chain line drawn closely adjacent to an extension of the center line of the hole, see Fig. 13.

It has been noted that the projected tolerance zone illustration shown here differs slightly from that shown in ASME Y14.5-2018. However, it is believed that the illustration shown here is a more realistic portrayal of the intent of the text in the Standard.

*Statistical Tolerance:* The statistical tolerancing symbol is a modifier that may be used to indicate that a tolerance is controlled statistically as opposed to being controlled arithmetically. With arithmetic control, assembly tolerances are typically divided arithmetically among the individual components of the assembly. This division results in the assumption that assemblies based on "worst case" conditions would be guaranteed to fit because the worst case set of parts fits — so that anything better would fit as well. When this technique is restrictive, statistical tolerancing, via the symbol, may be specified in the feature control frame as a method of increasing tolerances for individual parts. This procedure may reduce manufacturing costs because its use changes the assumption that statistical process control may make a statistically significant quantity of parts fit, but not absolutely all. The technique should only be used when sound statistical methods are employed.

*Tangent Plane:* When it is desirable to control the surface of a feature by the contacting or high points of the surface, a tangent plane symbol is added as a modifier to the tolerance in the feature control frame, see Fig. 14.

*Free State:* The free state modifier symbol is used when the geometric tolerance applies to the feature in its "free state," or after removal of any forces used in the manufacturing process. With removal of forces, the part may distort due to gravity, flexibility, springback, or other release of internal stresses developed during fabrication. Typical applications include parts with extremely thin walls and non-rigid parts made of rubber or plastics. The modifier is placed in the tolerance portion of the feature control frame and follows any other modifier.

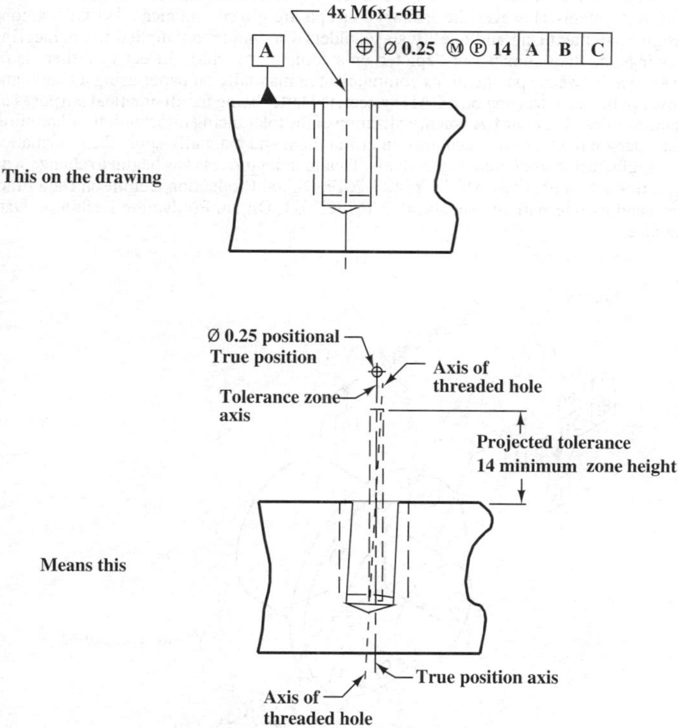


Fig. 13. Projected Tolerance Zone Application

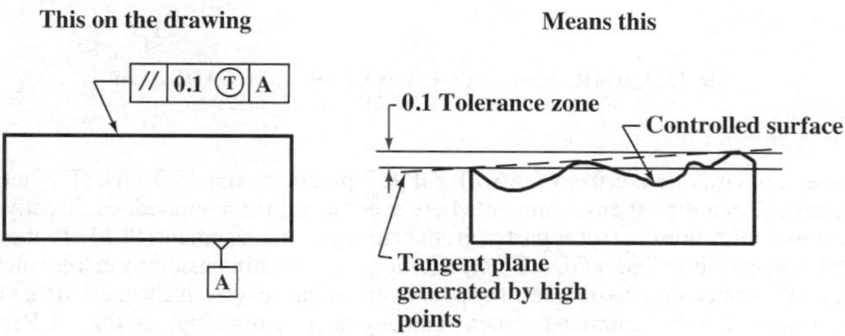


Fig. 14. Tangent Plane Modifier

The above examples are just a few of the numerous concepts and related symbols covered by ASME Y14.5-2018. Refer to the Standard for a complete discussion with further examples of the application of geometric dimensioning and tolerancing principles.

**GD&T and CAD Models.**—ASME Y14.5-2018 and all earlier editions are usable on modern CAD systems. However, the primary concepts are still communicated visually among design, manufacturing and quality stakeholders via annotation applied to engineering drawings, be they on paper or depicted on a computer monitor. Effectively, there is no difference between specifications communicated manually on paper using a pencil and those applied to a drawing on a CAD system, the latter being for all practical purposes an electronic drawing board. As such, utilization of the tolerancing information still requires that a person read the specifications, interpret them and manually apply the information in manufacturing and/or quality systems. That manual process has begun to change with the introduction of ANSI/ASME Y14.41-2019, Digital Production Definition Data Practices, and its international version ISO 16792:2021, Digital Production Definition Data Practices.

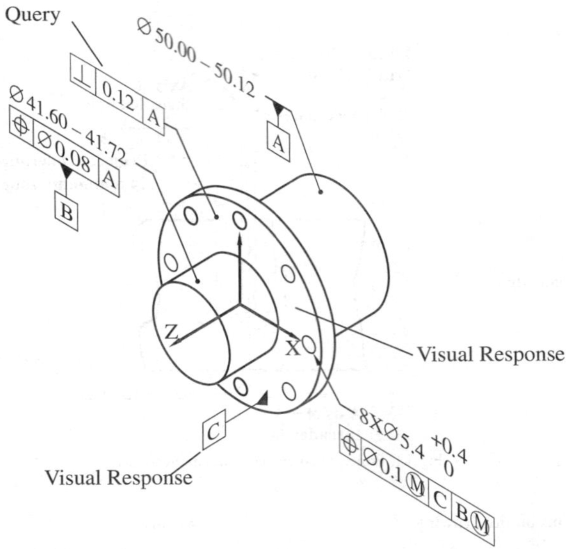


Fig. 15. Visual Response to a Tolerance Query on a 3D CAD System Compliant with ANSI/ASME Y14.41-2019

These standards define use of ASME GD&T practices and ISO GD&T practices, respectively, in a digital environment where specifications are embedded directly into the product definition data set as part of product lifecycle management (PLM). Embedded GD&T is generally called "eGD&T" (electronic geometric dimensioning & tolerancing). Once incorporated into the data set, the specifications can be selectively queried for visual display as needed on a computer screen or an engineering drawing, see Fig. 15. Perhaps more importantly, the embedded tolerancing information can be directly accessed by software for tolerance analysis, process planning, or any other applications designed to utilize the geometric and tolerancing information as part of product lifecycle management. As these standards are adopted by CAD vendors and fully implemented, their promise is for a more expedient, more accurate design and analysis process.

Both standards define how current GD&T concepts are applied to digital models. Both standards also explain where rules have changed to account for differing requirements between 2D-based annotation and 3D annotations.

**Digital Product Definition Data Practices Terms.**—The following terms are defined in both ANSI/ASME Y14.41 and ISO 16792. In ISO 16792, rules are modified to recognize ISO drawing practices where those practices differ from predominately US practice as illustrated in ANSI/ASME Y14.41-2019. Rules in both standards cover not only data sets with models only, but also those data sets with models and drawings. See the respective standards for complete rules and definitions.

*Annotation:* Dimensions, tolerances, notes, text, or symbols visible without any manual or external manipulation.

*Attribute:* A dimension, tolerance, note, text, or symbol required to complete the product definition or feature of the product that is not visible but available upon interrogating the model.

*Design Model:* The portion of the data set that contains model and supplemental geometry.

*Geometric Element:* A graphic entity used in a data set. For example: point, line, plane, surface, solid, model coordinate system or crosshatching.

*Management Data:* The data required for the release, control, and storage of product definition data as well as other relevant engineering data.

*Model:* A combination of design model, annotation and attributes that describes a product, see Fig. 16.

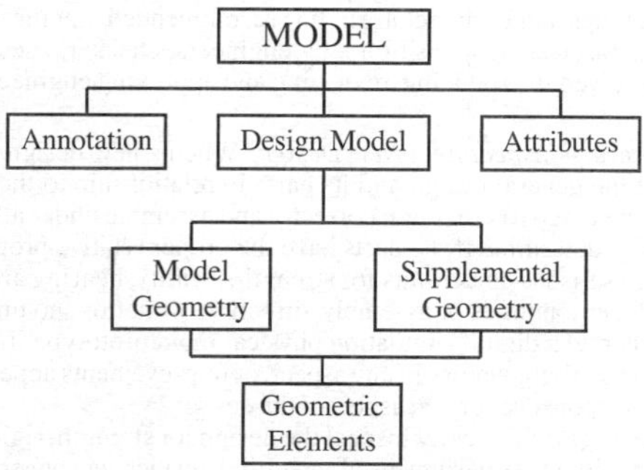


Fig. 16. Content of a Digital Model (ANSI/ASME Y14.41-2012)

*Model Value:* The numerical value derived by interrogating the model that quantifies the form and spatial relationships of the geometry composing a design model or assembly of models to the precision (number of decimal places) of the computer system.

*Model Geometry:* Geometric elements in product definition data which represent a designed product.

*Product Definition Data:* Data elements required to completely define a product.

*Product Definition Data Set:* A collection of one or more computer files that discloses (directly or by reference), by means of graphic or textual presentations, or combinations of both, the physical and functional requirements of an item, see Fig. 17.

*Query:* A means of interrogating a digital element or the relationship between digital elements, see Fig. 17.

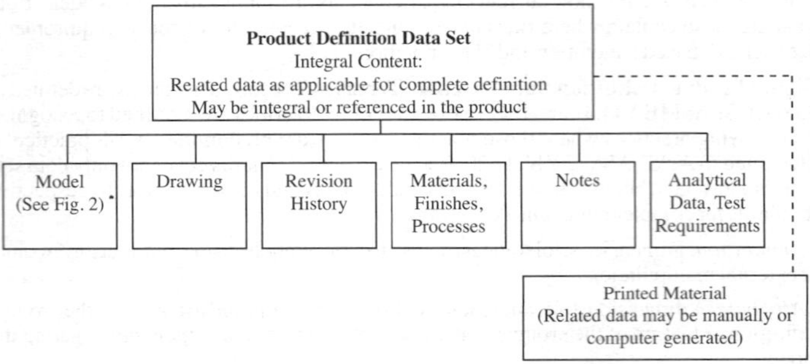


Fig. 17. Product Definition Data Set, ANSI/ASME Y14.41-2019

*Supplemental Geometry:* Geometric elements included in product definition data to communicate design requirements but not intended to represent a portion of the manufactured product.

## Checking Designs and Drawings

In order for production designs to be maintained at the highest level of consistency, a set of suggestions for checking models and drawings is provided below. The suggestions first concentrate on a review of the design parameters in a global sense, then on a detailed review of the drawings and/or model itself. It is recommended that these suggestions be followed in parts based on responsibility, by engineers, checkers, designers, detailers and any others involved in producing, reviewing and approving engineering models and drawings.

**Design Parameters.**—*Inspecting a New Design:* When a new design is involved, first carefully inspect the general design and its parts in relationship to the assemblies as a whole, ascertaining that parts function correctly and assemble under all material conditions. Additionally, determine if the parts have the proper relative proportions and that the general design satisfies constraints for strength, rigidity, bearing areas, appearance, manufacturability, ergonomics of assembly, direction of motion, no unnecessary interferences, etc. Generate a digital simulation physical rapid prototype if necessary. If the design appears to be unsatisfactory in any aspect, or improvements appear to be possible, notify the person responsible for that aspect of the design.

*Checking for Strength:* Physically inspect the design for strength, rigidity, and appearance. Compare it with other designs used in similar service whenever possible, giving preference to the known working designs in such a comparison, unless the known designs are examples of unsatisfactory parts. If there is any question, perform applicable analysis on the model or, if required, manually compute the stresses and deformations. If the new design is for a larger version of an existing device, ascertain that any standard parts necessarily increased in size will continue to be sufficiently strong to bear increased service loads.

*Materials Specified:* For the design, consider the material specification in relationship to the various manufacturing process options available, such as molding, forging, a weldment, or other method with which to form the rough shape. Then consider the machining operations to see whether changes in form or design will reduce the number of operations or reduce the machining cost. Parts should be designed with reference to the economical use of material, and whenever possible, utilize standard stock sizes and material readily obtainable from local sources. In the case of alloy steel, special bronze, and similar or exotic materials, confirm that the material can be obtained in the required size.

*Checking Casting Designs:* In checking castings, the form of the pattern is studied, the methods of molding, the method of supporting and venting the cores, and the effect of draft and rough molding on clearances. Undue metal thickness should be avoided, as well as extreme differences between thick and thin sections in the same casting.

All metal thicknesses should be specified, so that appropriate chaplets can be selected for supporting cores. Ample fillets and rounds should be provided and properly dimensioned directly or by note. Cores should be designed so that they can be secured in the mold without crushing or causing interference.

Allowances should be made for swelling, shrinkage, or misalignment of cores to ensure adequate machineability. Material should be added to surfaces destined to be finished.

On large castings, sufficient extra material should be provided for finishing to permit "clean up" to net size in case of warpage.

**Checking Drawings.**—The following are some rules-of-thumb and general guidelines to keep in mind when creating and reviewing engineering drawings and data models.

*Checking the Technique Used in Making the Drawing:* Inspect the drawing to see that all regular, auxiliary and section views are made in such a way as to illustrate the most descriptive views of the form of the piece and its relationship to other parts. Selection of type and quantity of views on a drawing serve as both a carrier of the dimensional values and to represent accurately the visual nature of the part. Ultimately, drawings must communicate design intent, both numerically and visually.

*Checking Dimensions in General:* In general, all dimensions should be checked for correctness. If manually drawn, dimensions should also be scaled to determine if the drawing is to scale. Where any dimension is "not-to-scale," it is indicated by underlining the offending dimension. This is the standard technique for indicating not-to-scale. While most CAD systems automatically indicate not-to-scale dimensions, some do not. Not-to-scale dimensions are problematic on CAD drawings and models as they are indicative of dimension values not matching the geometry. In the case of a CAD drawing dimension being not-to-scale, it too should be underlined. However, this may be a serious concern, and may demand immediate correction. More and more, CAD geometry drives downstream applications such as numerical control or tolerance analysis. Hence, errors in CAD geometry cannot be tolerated if accurate downstream results are expected.

**ASME and ISO Drawing Checklists.**—The following checklists represent questions based on industry practice and standards for engineering drawings and, unless noted, apply to both inch and metric drawings and models and both ASME Y14.5 and ISO methodology.

*Model Checklist:* Are features built within dimensional limits and to appropriate material conditions (MMC, LMC, nominal)?

Are features built to sufficient numerical accuracy?

Are fillets and rounds added to all non-finished corners of castings and forgings?

On as-cast models, is machining allowance added to surfaces destined to be finished?

Has allowance for wrench clearance been built into locations of bolts and other fasteners requiring tools?

*Drawing Checklist:* Are all dimensions indicated only once to avoid potentially misinterpreted redundant dimensioning?

Are all auxiliary and section views labeled?

Are general notes added to address default conditions such as fillets and rounds, numerical significance, default tolerances?

Are all finish marks indicated where appropriate?

Are clearance hole diameters clearly specified to avoid potential look-up errors?

Are there sufficient views so that no assumptions need be made about any feature or dimension?

If following US practice, are all dimensions, notes and other text readable from the bottom of the drawing?

If following ISO practice, is all dimensional text aligned with each dimension on the drawing?

When dimensioning in metric, are preceding 0s shown for values less than 1, and trailing 0s not shown for integers? (e.g. 0.11 and 10)

When heat treatment is required, the heat treatment should be specified.

When dimension in inches, are preceding 0s not shown for values less than 1, and trailing 0s added for consistent significant digits? (e.g. 11 and 10.00)

If dimensioning symbols are used, such as depth, diameter, radius, etc., are the symbols placed before the dimensional value?

Are quantities specified for all dimensions or features occurring in multiple locations (via 2X or TWO TIMES)?

Are threaded holes specified by tap drill and thread specifications to avoid potential look-up errors?

Are all occurrences of dimensional and non-dimensional text to consistent size or per company policy?

Are dimensions to hidden lines and those directly to object lines minimized unless impossible to avoid?

Are dimensions shown in a view depicting the true shape projection of the feature?

Do all dimensions account for tolerance, either directly or indirectly, via note or block tolerance?

Does the selection of paper size match the complexity of the part to avoid over-crowding via a size too small or wasted space via a size too large?

Are dimensions shown in a form that minimizes unnecessary calculations by readers of the drawing?

Are views aligned left-to-right and bottom-to-top, and arranged properly for 3rd angle projection (or 1st angle projection)?

Is the primary view the most descriptive view possible?

Is chain dimensioning minimized to avoid "stacking" of tolerances?

Are holes and other round features and parts located by their centers?

Are witness lines (extension lines) used with dimensions rather than dimensioning directly to features?

Is the drawing title correct and the drawing number or file name in the correct format?

**Arrangement of Views.—**A drawing may show one or more views of an object arranged according to a definite plan. In the United States, the general practice is to place the top view above the front view, and the end view next to whatever end it represents. For example, if a view of the left-hand end is considered important, the end view is placed to the left of the front view. If a right-hand view is also necessary, it is placed to the right of the front view. If a bottom view is needed, it is placed below the front view.

The view obtained by looking at an object from above is a *plan* view. The view obtained by looking at the object from its side and showing a vertical face is an elevation, which may be either a *front elevation* or an *end elevation* (also known as side elevation).

*Third-angle and First-angle Projection:* If the views are placed with the plan above the front elevation, the right-hand end view to the right and the left-hand end view (when drawn) to the left, this is known as *third-angle projection*. In ISO practice, it is the custom to use what is known as *first-angle projection*. With this method, the front elevation is placed at the top, the plan view at the bottom, the right-hand end view at the left, and the left-hand end view at the right. The first-angle projection is also generally employed in architectural and structural work, as in drawings of bridges, etc. See Fig. 1a and Fig. 1b on page 627.

# ALLOWANCES AND TOLERANCES FOR FITS

## Limits and Fits

Fits between cylindrical parts, i.e., cylindrical fits, govern the proper assembly and performance of many mechanisms. Clearance fits permit relative freedom of motion between a shaft and a hole—axially, radially, or both. Interference fits secure a certain amount of tightness between parts, whether these are meant to remain permanently assembled or to be taken apart from time to time. Or again, two parts may be required to fit together snugly—without apparent tightness or looseness. The designer's problem is to specify these different types of fits in such a way that the shop can produce them. Establishing the specifications requires the adoption of two manufacturing limits for the hole and two for the shaft, and, hence, the adoption of a manufacturing tolerance on each part.

In selecting and specifying limits and fits for various applications, it is essential in the interests of interchangeable manufacturing that 1) standard definitions of terms relating to limits and fits be used; 2) preferred basic sizes be selected wherever possible to reduce material and tooling costs; 3) limits be based upon a series of preferred tolerances and allowances; and 4) a uniform system of applying tolerances (preferably unilateral) be used. These principles have been incorporated in both the American and British Standards for limits and fits. Information about these Standards is given beginning on page 647.

**Basic Size.—**The basic size of a screw thread or machine part is the theoretical or nominal standard size from which variations are made. For example, a shaft may have a *basic* diameter of 2 inches, but a maximum variation of minus 0.010 inch may be permitted. The minimum hole should be of basic size wherever the use of standard tools represents the greatest economy. The maximum shaft should be of basic size wherever the use of standard purchased material, without further machining, represents the greatest economy, even though special tools are required to machine the mating part.

**Tolerances.—**Tolerance is the amount of variation permitted on dimensions or surfaces of machine parts. The tolerance is equal to the difference between the maximum and minimum limits of any specified dimension. For example, if the maximum limit for the diameter of a shaft is 2.000 inches and its minimum limit 1.990 inches, the tolerance for this diameter is 0.010 inch. The extent of these tolerances is established by determining the maximum and minimum clearances required on operating surfaces. As applied to the fitting of machine parts, the word tolerance means the amount that duplicate parts are allowed to vary in size in connection with manufacturing operations, owing to unavoidable imperfections of workmanship. Tolerance may also be defined as the amount that duplicate parts are permitted to vary in size to secure sufficient accuracy without unnecessary refinement. The terms "tolerance" and "allowance" are often used interchangeably, but, according to common usage, *allowance* is a difference in dimensions prescribed to secure various classes of fits between different parts.

For information on factors involved in and various methods for calculating tolerances, see *Tolerance Analysis and Assignment* on page 685.

**Unilateral and Bilateral Tolerances.—**The term "unilateral tolerance" means that the total tolerance, as related to a basic dimension, is in *one* direction only. For example, if the basic dimension were 1 inch and the tolerance were expressed as 1.000 – 0.002, or as 1.000 + 0.002, these would be unilateral tolerances because the total tolerance in each is in one direction. On the contrary, if the tolerance were divided, so as to be partly plus and partly minus, it would be classed as "bilateral."

$$\text{Thus,} \qquad 1.000^{+0.001}_{-0.001}$$

is an example of bilateral tolerance, because the total tolerance of 0.002 is given in two directions—plus and minus.

When unilateral tolerances are used, one of the following three methods should be used to express them:

1) Specify, limiting dimensions only as

Diameter of hole: 2.250, 2.252

Diameter of shaft: 2.249, 2.247

2) One limiting size may be specified with its tolerances as

Diameter of hole: 2.250 + 0.002, –0.000

Diameter of shaft: 2.249 + 0.000, –0.002

3) The nominal size may be specified for both parts, with a notation showing both allowance and tolerance, as

Diameter of hole: $2\frac{1}{4}$ + 0.002, – 0.000

Diameter of shaft: $2\frac{1}{4}$ – 0.001, – 0.003

Bilateral tolerances should be specified as such, usually with plus and minus tolerances of equal amount. An example of the expression of bilateral tolerances is

$$2 \pm 0.001 \quad \text{or} \quad 2\,{}^{+0.001}_{-0.001}$$

**Application of Tolerances.—**According to common practice, tolerances are applied in such a way as to show the permissible amount of dimensional variation in the direction that is less dangerous. When a variation in either direction is equally dangerous, a bilateral tolerance should be given. When a variation in one direction is more dangerous than a variation in another, a unilateral tolerance should be given in the less dangerous direction.

For nonmating surfaces, or atmospheric fits, the tolerances may be bilateral or unilateral, depending entirely upon the nature of the variations that develop in manufacture. On mating surfaces, with few exceptions, the tolerances should be unilateral.

Where tolerances are required on the distances between holes, usually they should be bilateral, as variation in either direction is normally equally dangerous. The variation in the distance between shafts carrying gears, however, should always be unilateral and plus; otherwise, the gears might run too tight. A slight increase in the backlash between gears is seldom of much importance.

One exception to the use of unilateral tolerances on mating surfaces occurs when tapers are involved; either bilateral or unilateral tolerances may then prove advisable, depending upon conditions. These tolerances should be determined in the same manner as the tolerances on the distances between holes. When a variation either in or out of the position of the mating taper surfaces is equally dangerous, the tolerances should be bilateral. When a variation in one direction is of less danger than a variation in the opposite direction, the tolerance should be unilateral and in the less dangerous direction.

**Locating Tolerance Dimensions.—**Only one dimension in the same straight line can be controlled within fixed limits. That dimension is the distance between the cutting surface of the tool and the locating or registering surface of the part being machined. Therefore, it is incorrect to locate any point or surface with tolerances from more than one point in the same straight line.

Every part of a mechanism must be located in each plane. Every operating part must be located with proper operating allowances. After such requirements of location are met, all other surfaces should have liberal clearances. Dimensions should be given between those points or surfaces that it is essential to hold in a specific relation to each other. This restriction applies particularly to those surfaces in each plane that control the location of other component parts. Many dimensions are relatively unimportant in this respect. It is good practice to establish a common locating point in each plane and give, as far as possible, all such dimensions from these common locating points. The locating points on the drawing, the locating or registering points used for machining the surfaces and the locating points for measuring should all be identical.

The initial dimensions placed on component drawings should be the exact dimensions that would be used if it were possible to work without tolerances. Tolerances should be given in that direction in which variations will cause the least harm or danger. When a variation in either direction is equally dangerous, the tolerances should be of equal amount in both directions, or bilateral. The initial clearance, or allowance, between operating parts should be as small as the operation of the mechanism will permit. The maximum clearance should be as great as the proper functioning of the mechanism will permit.

**Direction of Tolerances on Gages.**—The extreme sizes for all plain limit gages shall not exceed the extreme limits of the part to be gaged. All variations in the gages, whatever their cause or purpose, shall bring these gages within these extreme limits.

The data for gage tolerances on page 674 cover gages to inspect workpieces held to tolerances in the American National Standard ANSI B4.4M-1981 (Withdrawn).

**Allowance for Forced Fits.**—The allowance per inch of diameter usually ranges from 0.001 inch to 0.0025 inch (0.0254-0.0635 mm), 0.0015 inch (0.0381 mm) being a fair average. Ordinarily the allowance per inch decreases as the diameter increases; thus the total allowance for a diameter of 2 inches (50.8 mm) might be 0.004 inch (0.102 mm), whereas for a diameter of 8 inches (203.2 mm) the total allowance might not be over 0.009 or 0.010 inch (0.23 or 0.25 mm). The parts to be assembled by forced fits are usually made cylindrical, although sometimes they are slightly tapered. Advantages of the taper form are: the possibility of abrasion of the fitted surfaces is reduced; less pressure is required in assembling; and parts are more readily separated when renewal is required. On the other hand, the taper fit is less reliable because, if it loosens, the entire fit is free with but little axial movement. Some lubricant, such as white lead and lard oil mixed to the consistency of paint, should be applied to the pin and bore before assembling, to reduce the tendency toward abrasion.

**Pressure for Forced Fits.**—The pressure required for assembling cylindrical parts depends not only upon the allowance for the fit, but also upon the area of the fitted surfaces, the pressure increasing in proportion to the distance that the inner member is forced in. The approximate ultimate pressure in tons can be determined by the use of the following formula in conjunction with the accompanying table of *Pressure Factors for Forced Fits*. Assuming that $A$ = area of surface in contact in "fit"; $a$ = total allowance in inches; $P$ = ultimate pressure required, in tons; $F$ = pressure factor based upon assumption that the diameter of the hub is twice the diameter of the bore, that the shaft is of machine steel, and that the hub is of cast iron:

$$P = \frac{A \times a \times F}{2}$$

**Pressure Factors for Forced Fits**

| Diameter, Inches | Pressure Factor | Diameter, Inches | Pressure Factor | Diameter, Inches | Pressure Factor | Diameter, Inches | Pressure Factor | Diameter, Inches | Pressure Factor |
|---|---|---|---|---|---|---|---|---|---|
| 1 | 500 | 3½ | 132 | 6 | 75 | 9 | 48.7 | 14 | 30.5 |
| 1¼ | 395 | 3¾ | 123 | 6¼ | 72 | 9½ | 46.0 | 14½ | 29.4 |
| 1½ | 325 | 4 | 115 | 6½ | 69 | 10 | 43.5 | 15 | 28.3 |
| 1¾ | 276 | 4¼ | 108 | 6¾ | 66 | 10½ | 41.3 | 15½ | 27.4 |
| 2 | 240 | 4½ | 101 | 7 | 64 | 11 | 39.3 | 16 | 26.5 |
| 2¼ | 212 | 4¾ | 96 | 7¼ | 61 | 11½ | 37.5 | 16½ | 25.6 |
| 2½ | 189 | 5 | 91 | 7½ | 59 | 12 | 35.9 | 17 | 24.8 |
| 2¾ | 171 | 5¼ | 86 | 7¾ | 57 | 12½ | 34.4 | 17½ | 24.1 |
| 3 | 156 | 5½ | 82 | 8 | 55 | 13 | 33.0 | 18 | 23.4 |
| 3¼ | 143 | 5¾ | 78 | 8½ | 52 | 13½ | 31.7 | … | … |

**Allowance for Given Pressure.**—By transposing the preceding formula, the approximate allowance for a required ultimate tonnage can be determined. Thus, $a = \frac{2P}{AF}$. The average ultimate pressure in tons commonly used ranges from 7 to 10 times the diameter in inches.

**Expansion Fits.—**In assembling certain classes of work requiring a very tight fit, the inner member is contracted by sub-zero cooling to permit insertion into the outer member and a tight fit is obtained as the temperature rises and the inner part expands. To obtain the sub-zero temperature, solid carbon dioxide or "dry ice" has been used, but its temperature of about −109° F (−78° C) below zero will not contract some parts sufficiently to permit insertion in holes or recesses. Greater contraction may be obtained by using high purity liquid nitrogen which has a temperature of about −320° F (−196° C) below zero. During a temperature reduction from 75° F to −321° F (220° C difference), the shrinkage per inch of diameter varies from about 0.002 to 0.003 inch for steel; 0.0042 inch for aluminum alloys; 0.0046 inch for magnesium alloys; 0.0033 inch for copper alloys; 0.0023 inch for monel metal; and 0.0017 inch for cast iron (not alloyed). The cooling equipment may vary from an insulated bucket to a special automatic unit, depending upon the kind and quantity of work. One type of unit is so arranged that parts are precooled by vapors from the liquid nitrogen before immersion. With another type, cooling is entirely by the vapor method.

**Shrinkage Fits.—**General practice seems to favor a smaller allowance for shrinkage fits than for forced fits, although in many shops the allowances are practically the same for each, and, for some classes of work, shrinkage allowances exceed those for forced fits. The shrinkage allowance also varies to a great extent with the form and construction of the part that has to be shrunk into place. The thickness or amount of metal around the hole is the most important factor. The way in which the metal is distributed also has an influence on the results. Shrinkage allowances for locomotive driving wheel tires adopted by the American Railway Master Mechanics Association are as follows:

| | | | | | | |
|---|---|---|---|---|---|---|
| Center diameter, inches | 38 | 44 | 50 | 56 | 62 | 66 |
| Allowances, inches | 0.040 | 0.047 | 0.053 | 0.060 | 0.066 | 0.070 |

Whether parts are to be assembled by forced or shrinkage fits depends upon conditions. For example, to press a tire over its wheel center, without heating, would ordinarily be a rather difficult job. On the other hand, pins, etc., are easily and quickly forced into place with a hydraulic press, and there is the additional advantage of knowing the exact pressure required in assembling, whereas there is more or less uncertainty connected with a shrinkage fit, unless the stresses are calculated. Tests to determine the difference in the quality of shrinkage and forced fits showed that the resistance of a shrinkage fit to slippage for an axial pull was 3.66 times greater than that of a forced fit, and, in rotation or torsion, 3.2 times greater. In each comparative test, dimensions and allowances were equal.

**Allowances for Shrinkage Fits.—**The most important point to consider when calculating shrinkage fits is the stress in the hub at the bore, which depends chiefly upon the shrinkage allowance. If the allowance is excessive, the elastic limit of the material will be exceeded, and permanent set will occur, or, in extreme conditions, the ultimate strength of the metal will be exceeded and the hub will burst. The intensity of the grip of the fit and the resistance to slippage depend mainly upon the thickness of the hub; the greater the thickness, the stronger the grip, and *vice versa*. Assuming the modulus of elasticity for steel to be 30,000,000 psi ($206.8 \times 10^3$ MPa) and for cast iron 15,000,000 psi ($103.4 \times 10^3$ MPa), the shrinkage allowance per inch (mm) of nominal diameter can be determined by the following formula, in which $A$ = allowance per inch (mm) of diameter; $T$ = true tangential tensile stress at inner surface of outer member, psi (MPa); $C$ = factor taken from one of the accompanying Table 1, Table 2, and Table 3.

For a cast-iron hub and steel shaft:

$$\text{US} \quad A = \frac{T(2+C)}{30{,}000{,}000} \quad (1a) \qquad \textbf{metric} \quad A = 25.4\frac{T(2+C)}{206.843 \times 10^9} \quad (1b)$$

When both hub and shaft are of steel:

$$\text{US} \quad A = \frac{T(1+C)}{30{,}000{,}000} \qquad (2a) \qquad \textbf{metric} \quad A = 25.4\frac{T(1+C)}{206.843 \times 10^9} \quad (2b)$$

If the shaft is solid, the factor *C* is taken from Table 1; if it is hollow and the hub is of steel, factor *C* is taken from Table 2; if it is hollow and the hub is of cast iron, the factor is taken from Table 3.

**Table 1. Factors for Calculating Shrinkage Fit Allowances for Steel Shafts and Steel or Cast-Iron Hubs**

| Ratio of Diameters $\frac{D_2}{D_1}$ | Steel Hub | Cast-iron Hub | Ratio of Diameters $\frac{D_2}{D_1}$ | Steel Hub | Cast-iron Hub |
|---|---|---|---|---|---|
| | *C* | | | *C* | |
| 1.5 | 0.227 | 0.234 | 2.8 | 0.410 | 0.432 |
| 1.6 | 0.255 | 0.263 | 3.0 | 0.421 | 0.444 |
| 1.8 | 0.299 | 0.311 | 3.2 | 0.430 | 0.455 |
| 2.0 | 0.333 | 0.348 | 3.4 | 0.438 | 0.463 |
| 2.2 | 0.359 | 0.377 | 3.6 | 0.444 | 0.471 |
| 2.4 | 0.380 | 0.399 | 3.8 | 0.450 | 0.477 |
| 2.6 | 0.397 | 0.417 | 4.0 | 0.455 | 0.482 |

Values of factor *C* for solid steel shafts of nominal diameter $D_1$, and hubs of steel or cast iron of nominal external and internal diameters $D_2$ and $D_1$, respectively.

*Example 1:* A steel crank web 375 mm outside diameter is to be shrunk on a 250 mm solid steel shaft. Required is the allowance per mm of shaft diameter to produce a maximum tensile stress in the crank of 170 MPa, assuming the stresses in the crank to be equivalent to those in a ring of the diameter given.

The ratio of the external to the internal diameters equals 375 ÷ 250 = 1.5; $T = 170$ MPa; from Table 1, $C = 0.227$. Substituting in Formula (2b):

$$A = 25.4\frac{170 \times (1 + 0.227)}{206843} = 0.026 \text{ mm}$$

*Example 2:* Find the allowance per mm of diameter for a 250 mm shaft having a 125 mm axial through hole, other conditions being the same as in Example 1.

The ratio of external to internal diameters of the hub equals 375 ÷ 250 = 1.5, as before, and the ratio of external to internal diameters of the shaft equals 250 ÷ 125 = 2. From Table 2, we find that factor $C = 0.455$; $T = 170$ MPa. Substituting these values in Formula (2b):

$$A = 25.4\frac{170(1 + 0.455)}{206843} = 0.030 \text{ mm}$$

The allowance is increased, as compared with Example 1, because the hollow shaft is more compressible.

*Example 3:* If the crank web in Example 1 is of cast iron and 28 MPa is the maximum tensile stress in the hub, what is the allowance per mm of diameter?

$$\frac{D_2}{D_1} = 1.5 \qquad T = 28$$

In Table 1, we find that $C = 0.234$. Substituting in Formula (1b), for cast-iron hubs, $A = 0.0076$ mm, which, owing to the lower tensile strength of cast iron, is about one-third the shrinkage allowance in Example 1, although the stress is two-thirds of the elastic limit.

**Table 2. Factors for Calculating Shrinkage Fit Allowances for Hollow Steel Shafts and Steel Hubs**

| $\frac{D_2}{D_1}$ | $\frac{D_1}{D_0}$ | $C$[a] | $\frac{D_2}{D_1}$ | $\frac{D_1}{D_0}$ | $C$[a] | $\frac{D_2}{D_1}$ | $\frac{D_1}{D_0}$ | $C$[a] |
|---|---|---|---|---|---|---|---|---|
| | 2.0 | 0.455 | | 2.0 | 0.760 | | 2.0 | 0.876 |
| 1.5 | 2.5 | 0.357 | 2.4 | 2.5 | 0.597 | 3.4 | 2.5 | 0.689 |
| | 3.0 | 0.313 | | 3.0 | 0.523 | | 3.0 | 0.602 |
| | 3.5 | 0.288 | | 3.5 | 0.481 | | 3.5 | 0.555 |
| | 2.0 | 0.509 | | 2.0 | 0.793 | | 2.0 | 0.888 |
| 1.6 | 2.5 | 0.400 | 2.6 | 2.5 | 0.624 | 3.6 | 2.5 | 0.698 |
| | 3.0 | 0.350 | | 3.0 | 0.546 | | 3.0 | 0.611 |
| | 3.5 | 0.322 | | 3.5 | 0.502 | | 3.5 | 0.562 |
| | 2.0 | 0.599 | | 2.0 | 0.820 | | 2.0 | 0.900 |
| 1.8 | 2.5 | 0.471 | 2.8 | 2.5 | 0.645 | 3.8 | 2.5 | 0.707 |
| | 3.0 | 0.412 | | 3.0 | 0.564 | | 3.0 | 0.619 |
| | 3.5 | 0.379 | | 3.5 | 0.519 | | 3.5 | 0.570 |
| | 2.0 | 0.667 | | 2.0 | 0.842 | | 2.0 | 0.909 |
| 2.0 | 2.5 | 0.524 | 3.0 | 2.5 | 0.662 | 4.0 | 2.5 | 0.715 |
| | 3.0 | 0.459 | | 3.0 | 0.580 | | 3.0 | 0.625 |
| | 3.5 | 0.422 | | 3.5 | 0.533 | | 3.5 | 0.576 |
| | 2.0 | 0.718 | | 2.0 | 0.860 | | … | … |
| 2.2 | 2.5 | 0.565 | 3.2 | 2.5 | 0.676 | … | … | … |
| | 3.0 | 0.494 | | 3.0 | 0.591 | | … | … |
| | 3.5 | 0.455 | | 3.5 | 0.544 | | … | … |

[a] Values of factor $C$ for hollow steel shafts of external and internal diameters $D_1$ and $D_0$, respectively, and steel hubs of nominal external diameter $D_2$.

**Table 3. Factors for Calculating Shrinkage Fit Allowances for Hollow Steel Shafts and Cast-Iron Hubs**

| $\frac{D_2}{D_1}$ | $\frac{D_1}{D_0}$ | $C$[a] | $\frac{D_2}{D_1}$ | $\frac{D_1}{D_0}$ | $C$[a] | $\frac{D_2}{D_1}$ | $\frac{D_1}{D_0}$ | $C$[a] |
|---|---|---|---|---|---|---|---|---|
| | 2.0 | 0.468 | | 2.0 | 0.798 | | 2.0 | 0.926 |
| 1.5 | 2.5 | 0.368 | 2.4 | 2.5 | 0.628 | 3.4 | 2.5 | 0.728 |
| | 3.0 | 0.322 | | 3.0 | 0.549 | | 3.0 | 0.637 |
| | 3.5 | 0.296 | | 3.5 | 0.506 | | 3.5 | 0.587 |
| | 2.0 | 0.527 | | 2.0 | 0.834 | | 2.0 | 0.941 |
| 1.6 | 2.5 | 0.414 | 2.6 | 2.5 | 0.656 | 3.6 | 2.5 | 0.740 |
| | 3.0 | 0.362 | | 3.0 | 0.574 | | 3.0 | 0.647 |
| | 3.5 | 0.333 | | 3.5 | 0.528 | | 3.5 | 0.596 |
| | 2.0 | 0.621 | | 2.0 | 0.864 | | 2.0 | 0.953 |
| 1.8 | 2.5 | 0.488 | 2.8 | 2.5 | 0.679 | 3.8 | 2.5 | 0.749 |
| | 3.0 | 0.427 | | 3.0 | 0.594 | | 3.0 | 0.656 |
| | 3.5 | 0.393 | | 3.5 | 0.547 | | 3.5 | 0.603 |
| | 2.0 | 0.696 | | 2.0 | 0.888 | | 2.0 | 0.964 |
| 2.0 | 2.5 | 0.547 | 3.0 | 2.5 | 0.698 | 4.0 | 2.5 | 0.758 |
| | 3.0 | 0.479 | | 3.0 | 0.611 | | 3.0 | 0.663 |
| | 3.5 | 0.441 | | 3.5 | 0.562 | | 3.5 | 0.610 |
| | 2.0 | 0.753 | | 2.0 | 0.909 | | … | … |
| 2.2 | 2.5 | 0.592 | 3.2 | 2.5 | 0.715 | … | … | … |
| | 3.0 | 0.518 | | 3.0 | 0.625 | | … | … |
| | 3.5 | 0.477 | | 3.5 | 0.576 | | … | … |

[a] Values of factor $C$ for hollow steel shafts and cast-iron hubs. Notation as in Table 2.

**Temperatures for Shrinkage Fits.**—The temperature to which the outer member in a shrinkage fit should be heated for clearance in assembling the parts depends on the total expansion required and on the coefficient $\alpha$ of linear expansion of the metal (i.e., the increase in length of any section of the metal in any direction for an increase in temperature of 1 degree F). The total expansion in diameter that is required consists of the total allowance for shrinkage and an added amount for clearance. The value of the coefficient $\alpha$ is, for nickel-steel, 0.000007; for steel in general, 0.0000065; for cast iron, 0.0000062. As an example, take an outer member of steel to be expanded 0.005 inch per inch of internal diameter, 0.001 being the shrinkage allowance and the remainder for clearance. Then

$$\alpha \times t^{\circ} = 0.005$$

$$t = \frac{0.005}{0.0000065} = 769 \text{ degrees F}$$

The value $t$ is the number of degrees F that the temperature of the member must be raised above that of the room temperature.

## ANSI/ASME Standard Limits and Fits

This American National Standard for Preferred Limits and Fits for Cylindrical Parts, ANSI/ASME B4.1-1967 (R2020), presents definitions of terms applying to fits between plain (non threaded) cylindrical parts and makes recommendations on preferred sizes, allowances, tolerances, and fits for use wherever they are applicable. As of 2009, this Standard was in accord with the recommendations of American-British-Canadian (ABC) conferences up to a diameter of 20 inches, with experimental work thereafter carried on with the objective of reaching agreement in the range above 20 inches. The recommendations in the Standard are presented for guidance and for use where they might serve to improve and simplify products, practices, and facilities. They have practical application for a wide range of products.

**Factors Affecting Selection of Fits.**—Many factors, such as length of engagement, bearing load, speed, lubrication, temperature, humidity, and materials must be taken into consideration in the selection of fits for a particular application, and modifications in the ANSI/ASME recommendations may be required to satisfy extreme conditions. Subsequent adjustments may also be found desirable as a result of experience in a particular application to suit critical functional requirements or to permit optimum manufacturing economy.

**Definitions.**—The following terms are defined in this Standard:

*Nominal Size:* The nominal size is the designation used for the purpose of general identification.

*Dimension:* A dimension is a geometrical characteristic such as diameter, length, angle, or center distance.

*Size:* Size is a designation of magnitude. When a value is assigned to a dimension, it is referred to as the size of that dimension. (It is recognized that the words "dimension" and "size" are both used at times to convey the meaning of magnitude.)

*Allowance:* An allowance is a prescribed difference between the maximum material limits of mating parts. (See definition of *Fit*). It is a minimum clearance (positive allowance) or maximum interference (negative allowance) between such parts.

*Tolerance:* A tolerance is the total permissible variation of a size. The tolerance is the difference between the limits of size.

*Basic Size:* The basic size is that size from which the limits of size are derived by the application of allowances and tolerances.

*Design Size:* The design size is the basic size with allowance applied, from which the limits of size are derived by the application of tolerances. Where there is no allowance, the design size is the same as the basic size.

*Actual Size:* An actual size is a measured size.

*Limits of Size:* The limits of size are the applicable maximum and minimum sizes.

*Maximum Material Limit:* A maximum material limit is that limit of size that provides the maximum amount of material for the part. Normally it is the maximum limit of size of an external dimension or the minimum limit of size of an internal dimension.*

*Minimum Material Limit:* A minimum material limit is that limit of size that provides the minimum amount of material for the part. Normally it is the minimum limit of size of an external dimension or the maximum limit of size of an internal dimension.*

*Tolerance Limit:* A tolerance limit is the variation, positive or negative, by which a size is permitted to depart from the design size.

*Unilateral Tolerance:* A unilateral tolerance is a tolerance in which variation is permitted in only one direction from the design size.

*Bilateral Tolerance:* A bilateral tolerance is a tolerance in which variation is permitted in both directions from the design size.

*Unilateral Tolerance System:* A design plan that uses only unilateral tolerances is known as a Unilateral Tolerance System.

*Bilateral Tolerance System:* A design plan that uses only bilateral tolerances is known as a Bilateral Tolerance System.

**Fits.—***Fit:* Fit is the general term used to signify the range of tightness that may result from the application of a specific combination of allowances and tolerances in the design of mating parts.

*Actual Fit:* The actual fit between two mating parts is the relation existing between them with respect to the amount of clearance or interference that is present when they are assembled. (Fits are of three general types: clearance, transition, and interference.)

*Clearance Fit:* A clearance fit is one having limits of size so specified that a clearance always results when mating parts are assembled.

*Interference Fit:* An interference fit is one having limits of size so specified that an interference always results when mating parts are assembled.

*Transition Fit:* A transition fit is one having limits of size so specified that either a clearance or an interference may result when mating parts are assembled.

*Basic Hole System:* A basic hole system is a system of fits in which the design size of the hole is the basic size and the allowance, if any, is applied to the shaft.

*Basic Shaft System:* A basic shaft system is a system of fits in which the design size of the shaft is the basic size and the allowance, if any, is applied to the hole.

**Preferred Basic Sizes.—**In specifying fits, the basic size of mating parts shall be chosen from the decimal series or the fractional series in Table 4.

**Preferred Series for Tolerances and Allowances.—**All fundamental tolerances and allowances of all shafts and holes have been taken from the series given in Table 5.

**Standard Tolerances.—**The series of standard tolerances shown in Table 6 are so arranged that for any one grade they represent approximately similar production difficulties throughout the range of sizes. This table provides a suitable range from which appropriate tolerances for holes and shafts can be selected and enables standard gages to be used. The tolerances shown in Table 6 have been used in the succeeding tables for different classes of fits.

Table 7 graphically illustrates the range of tolerance grades that various machining processes may produce under normal conditions.

* An example of exceptions: an exterior corner radius where the maximum radius is the minimum material limit and the minimum radius is the maximum material limit.

**Table 4. Preferred Basic Sizes** *ANSI/ASME B4.1-1967 (R2020)*

| Decimal[a] | | | Fractional[a] | | | | | | | |
|---|---|---|---|---|---|---|---|---|---|---|
| 0.010 | 2.00 | 8.50 | $\frac{1}{64}$ | 0.015625 | $2\frac{1}{4}$ | 2.2500 | $9\frac{1}{2}$ | 9.5000 |
| 0.012 | 2.20 | 9.00 | $\frac{1}{32}$ | 0.03125 | $2\frac{1}{2}$ | 2.5000 | 10 | 10.0000 |
| 0.016 | 2.40 | 9.50 | $\frac{1}{16}$ | 0.0625 | $2\frac{3}{4}$ | 2.7500 | $10\frac{1}{2}$ | 10.5000 |
| 0.020 | 2.60 | 10.00 | $\frac{3}{32}$ | 0.09375 | 3 | 3.0000 | 11 | 11.0000 |
| 0.025 | 2.80 | 10.50 | $\frac{1}{8}$ | 0.1250 | $3\frac{1}{4}$ | 3.2500 | $11\frac{1}{2}$ | 11.5000 |
| 0.032 | 3.00 | 11.00 | $\frac{5}{32}$ | 0.15625 | $3\frac{1}{2}$ | 3.5000 | 12 | 12.0000 |
| 0.040 | 3.20 | 11.50 | $\frac{3}{16}$ | 0.1875 | $3\frac{3}{4}$ | 3.7500 | $12\frac{1}{2}$ | 12.5000 |
| 0.05 | 3.40 | 12.00 | $\frac{1}{4}$ | 0.2500 | 4 | 4.0000 | 13 | 13.0000 |
| 0.06 | 3.60 | 12.50 | $\frac{5}{16}$ | 0.3125 | $4\frac{1}{4}$ | 4.2500 | $13\frac{1}{2}$ | 13.5000 |
| 0.08 | 3.80 | 13.00 | $\frac{3}{8}$ | 0.3750 | $4\frac{1}{2}$ | 4.5000 | 14 | 14.0000 |
| 0.10 | 4.00 | 13.50 | $\frac{7}{16}$ | 0.4375 | $4\frac{3}{4}$ | 4.7500 | $14\frac{1}{2}$ | 14.5000 |
| 0.12 | 4.20 | 14.00 | $\frac{1}{2}$ | 0.5000 | 5 | 5.0000 | 15 | 15.0000 |
| 0.16 | 4.40 | 14.50 | $\frac{9}{16}$ | 0.5625 | $5\frac{1}{4}$ | 5.2500 | $15\frac{1}{2}$ | 15.5000 |
| 0.20 | 4.60 | 15.00 | $\frac{5}{8}$ | 0.6250 | $5\frac{1}{2}$ | 5.5000 | 16 | 16.0000 |
| 0.24 | 4.80 | 15.50 | $\frac{11}{16}$ | 0.6875 | $5\frac{3}{4}$ | 5.7500 | $16\frac{1}{2}$ | 16.5000 |
| 0.30 | 5.00 | 16.00 | $\frac{3}{4}$ | 0.7500 | 6 | 6.0000 | 17 | 17.0000 |
| 0.40 | 5.20 | 16.50 | $\frac{7}{8}$ | 0.8750 | $6\frac{1}{2}$ | 6.5000 | $17\frac{1}{2}$ | 17.5000 |
| 0.50 | 5.40 | 17.00 | 1 | 1.0000 | 7 | 7.0000 | 18 | 18.0000 |
| 0.60 | 5.60 | 17.50 | $1\frac{1}{4}$ | 1.2500 | $7\frac{1}{2}$ | 7.5000 | $18\frac{1}{2}$ | 18.5000 |
| 0.80 | 5.80 | 18.00 | $1\frac{1}{2}$ | 1.5000 | 8 | 8.0000 | 19 | 19.0000 |
| 1.00 | 6.00 | 18.50 | $1\frac{3}{4}$ | 1.7500 | $8\frac{1}{2}$ | 8.5000 | $19\frac{1}{2}$ | 19.5000 |
| 1.20 | 6.50 | 19.00 | 2 | 2.0000 | 9 | 9.0000 | 20 | 20.0000 |
| 1.40 | 7.00 | 19.50 | … | … | … | … | … | … |
| 1.60 | 7.50 | 20.00 | … | … | … | … | … | … |
| 1.80 | 8.00 | … | … | … | … | … | … | … |

[a] All dimensions are in inches.

**Table 5. Preferred Series of Tolerances and Allowances**[a]
*ANSI/ASME B4.1-1967 (R2020)*

| | | | | | | | |
|---|---|---|---|---|---|---|---|
| 0.1 | 1 | 10 | 100 | 0.3 | 3 | 30 | … |
| … | 1.2 | 12 | 125 | … | 3.5 | 35 | … |
| 0.15 | 1.4 | 14 | … | 0.4 | 4 | 40 | … |
| … | 1.6 | 16 | 160 | … | 4.5 | 45 | … |
| … | 1.8 | 18 | … | 0.5 | 5 | 50 | … |
| 0.2 | 2 | 20 | 200 | 0.6 | 6 | 60 | … |
| … | 2.2 | 22 | … | 0.7 | 7 | 70 | … |
| 0.25 | 2.5 | 25 | 250 | 0.8 | 8 | 80 | … |
| … | 2.8 | 28 | … | 0.9 | 9 | … | … |

[a] All values in thousandths of an inch.

**ANSI/ASME Standard Fits.**—Table 8a through Table 12 inclusive show a series of standard types and classes of fits on a unilateral hole basis, such that the fit produced by mating parts in any one class will produce approximately similar performance throughout the range of sizes. These tables prescribe the fit for any given size, or type of fit; they also prescribe the standard limits for the mating parts that will produce the fit. The fits listed in these tables contain all those that appear in the approved American-British-Canadian proposal.

*Selection of Fits:* In selecting limits of size for any application, the type of fit is determined first, based on the use or service required from the equipment being designed; then the limits of size of the mating parts are established, to ensure that the desired fit will be produced.

Theoretically, an infinite number of fits could be chosen, but the number of standard fits shown in the accompanying tables should cover most applications.

*Designation of Standard Fits:* Standard fits are designated by means of the following symbols which, facilitate reference to classes of fit for educational purposes. The symbols are not intended to be shown on manufacturing drawings; instead, sizes should be specified on drawings.

**Table 6. Standard Tolerances** *ANSI/ASME B4.1-1967 (R2020)*

| Nominal Size, Inches | | Grade | | | | | | | | | |
|---|---|---|---|---|---|---|---|---|---|---|---|
| | | 4 | 5 | 6 | 7 | 8 | 9 | 10 | 11 | 12 | 13 |
| Over | To | Tolerances in thousandths of an inch[a] | | | | | | | | | |
| 0 | 0.12 | 0.12 | 0.15 | 0.25 | 0.4 | 0.6 | 1.0 | 1.6 | 2.5 | 4 | 6 |
| 0.12 | 0.24 | 0.15 | 0.20 | 0.3 | 0.5 | 0.7 | 1.2 | 1.8 | 3.0 | 5 | 7 |
| 0.24 | 0.40 | 0.15 | 0.25 | 0.4 | 0.6 | 0.9 | 1.4 | 2.2 | 3.5 | 6 | 9 |
| 0.40 | 0.71 | 0.2 | 0.3 | 0.4 | 0.7 | 1.0 | 1.6 | 2.8 | 4.0 | 7 | 10 |
| 0.71 | 1.19 | 0.25 | 0.4 | 0.5 | 0.8 | 1.2 | 2.0 | 3.5 | 5.0 | 8 | 12 |
| 1.19 | 1.97 | 0.3 | 0.4 | 0.6 | 1.0 | 1.6 | 2.5 | 4.0 | 6 | 10 | 16 |
| 1.97 | 3.15 | 0.3 | 0.5 | 0.7 | 1.2 | 1.8 | 3.0 | 4.5 | 7 | 12 | 18 |
| 3.15 | 4.73 | 0.4 | 0.6 | 0.9 | 1.4 | 2.2 | 3.5 | 5 | 9 | 14 | 22 |
| 4.73 | 7.09 | 0.5 | 0.7 | 1.0 | 1.6 | 2.5 | 4.0 | 6 | 10 | 16 | 25 |
| 7.09 | 9.85 | 0.6 | 0.8 | 1.2 | 1.8 | 2.8 | 4.5 | 7 | 12 | 18 | 28 |
| 9.85 | 12.41 | 0.6 | 0.9 | 1.2 | 2.0 | 3.0 | 5.0 | 8 | 12 | 20 | 30 |
| 12.41 | 15.75 | 0.7 | 1.0 | 1.4 | 2.2 | 3.5 | 6 | 9 | 14 | 22 | 35 |
| 15.75 | 19.69 | 0.8 | 1.0 | 1.6 | 2.5 | 4 | 6 | 10 | 16 | 25 | 40 |
| 19.69 | 30.09 | 0.9 | 1.2 | 2.0 | 3 | 5 | 8 | 12 | 20 | 30 | 50 |
| 30.09 | 41.49 | 1.0 | 1.6 | 2.5 | 4 | 6 | 10 | 16 | 25 | 40 | 60 |
| 41.49 | 56.19 | 1.2 | 2.0 | 3 | 5 | 8 | 12 | 20 | 30 | 50 | 80 |
| 56.19 | 76.39 | 1.6 | 2.5 | 4 | 6 | 10 | 16 | 25 | 40 | 60 | 100 |
| 76.39 | 100.9 | 2.0 | 3 | 5 | 8 | 12 | 20 | 30 | 50 | 80 | 125 |
| 100.9 | 131.9 | 2.5 | 4 | 6 | 10 | 16 | 25 | 40 | 60 | 100 | 160 |
| 131.9 | 171.9 | 3 | 5 | 8 | 12 | 20 | 30 | 50 | 80 | 125 | 200 |
| 171.9 | 200 | 4 | 6 | 10 | 16 | 25 | 40 | 60 | 100 | 160 | 250 |

[a] All tolerances above the heavy line are in accordance with American-British-Canadian (ABC) agreements.

**Table 7. Relation of Machining Processes to Tolerance Grades**
*ANSI/ASME B4.1-1967 (R2020)*

This chart may be used as a general guide to determine the machining processes that will, under normal conditions, produce work within the tolerance grades indicated.

(See also *Relation of Surface Roughness to Tolerances* starting on page 802.)

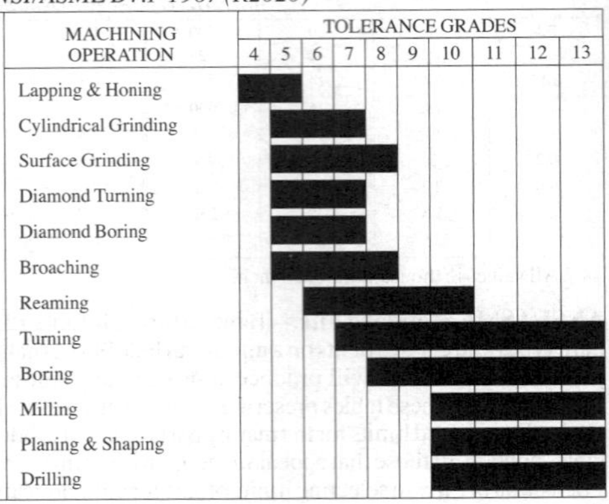


The letter symbols used to designate standard fits are as follows:

*RC* = Running or Sliding Clearance Fit *LC* = Locational Clearance Fit

*LT* = Transition Clearance or Interference Fit

*LN* = Locational Interference Fit

*FN* = Force or Shrink Fit

These letter symbols are used in conjunction with numbers representing the class of fit; thus FN 4 represents a Class 4 force fit.

Each of these symbols (two letters and a number) represents a complete fit for which the minimum and maximum clearance or interference and the limits of size for the mating parts are given directly in the tables.

**Description of Fits.—**The classes of fits are arranged in three general groups: running and sliding fits, locational fits, and force fits.

*Running and Sliding Fits (RC):* Running and sliding fits, for which limits of clearance are given in Table 8a, are intended to provide a similar running performance, with suitable lubrication allowance, throughout the range of sizes. The clearances for the first two classes, used chiefly as slide fits, increase more slowly with the diameter than for the other classes so that accurate location is maintained even at the expense of free relative motion.

These fits may be described as follows:

RC 1 *Close sliding fits* are intended for the accurate location of parts that must assemble without perceptible play.

RC 2 *Sliding fits* are intended for accurate location, but with greater maximum clearance than class RC 1. Parts made to this fit move and turn easily but are not intended to run freely, and in the larger sizes may seize with small temperature changes.

RC 3 *Precision running fits* are about the closest fits that can be expected to run freely and are intended for precision work at slow speeds and light journal pressures, but they are not suitable where appreciable temperature differences are likely to be encountered.

RC 4 *Close running fits* are intended chiefly for running fits on accurate machinery with moderate surface speeds and journal pressures, where accurate location and minimum play are desired.

RC 5 and RC 6 *Medium running fits* are intended for higher running speeds, or heavy journal pressures, or both.

RC 7 *Free running fits* are intended for use where accuracy is not essential, or where large temperature variations are likely to be encountered, or under both these conditions.

RC 8 and RC 9 *Loose running fits* are intended for use where wide commercial tolerances may be necessary, together with an allowance, on the external member.

*Locational Fits (LC, LT, and LN):* Locational fits are intended to determine only the location of the mating parts; they may provide rigid or accurate location, as with interference fits, or provide some freedom of location, as with clearance fits. Accordingly, they are divided into three groups: clearance fits (LC), transition fits (LT), and interference fits (LN).

These are described as follows:

LC *Locational clearance fits* are intended for parts that are normally stationary but that can be freely assembled or disassembled. They range from snug fits for parts requiring accuracy of location, through the medium clearance fits for parts such as spigots, to the looser fastener fits where freedom of assembly is of prime importance.

LT *Locational transition fits* are a compromise between clearance and interference fits for applications where accuracy of location is important but either a small amount of clearance or interference is permissible.

LN *Locational interference fits* are used where accuracy of location is of prime importance, and for parts requiring rigidity and alignment with no special requirements for bore pressure. Such fits are not intended for parts designed to transmit frictional loads from one part to another by virtue of the tightness of fit. These conditions are covered by force fits.

*Force Fits (FN):* Force or shrink fits constitute a special type of interference fit, normally characterized by maintenance of constant bore pressures throughout the range of sizes. The interference, therefore, varies almost directly with diameter, and the difference between its minimum and maximum value is small to maintain the resulting pressures within reasonable limits.

These fits are described as follows:

FN 1 *Light drive fits* are those requiring light assembly pressures and produce more or less permanent assemblies. They are suitable for thin sections or long fits, or in cast-iron external members.

FN 2 *Medium drive fits* are suitable for ordinary steel parts or for shrink fits on light sections. They are about the tightest fits that can be used with high-grade cast-iron external members.

FN 3 *Heavy drive fits* are suitable for heavier steel parts or for shrink fits in medium sections.

FN 4 and FN 5 *Force fits* are suitable for parts that can be highly stressed or for shrink fits where the heavy pressing forces required are impractical.

**Graphical Representation of Limits and Fits.—**A visual comparison of the hole and shaft tolerances and the clearances or interferences provided by the various types and classes of fits can be obtained from the diagrams on page 653. These diagrams have been drawn to scale for a nominal diameter of 1 inch.

**Use of Standard Fit Tables.—***Example 1:* A Class RC 1 fit is to be used in assembling a mating hole and shaft of 2-inch nominal diameter. This class of fit was selected because the application required accurate location of the parts with no perceptible play (see *Description of Fits*, RC 1 close sliding fits). From the data in Table 8a, establish the limits of size and clearance of the hole and shaft.

Maximum hole = 2 + 0.0005 = 2.0005; minimum hole = 2 inches

Maximum shaft = 2 − 0.0004 = 1.9996; minimum shaft = 2 − 0.0007 = 1.9993 inches

Minimum clearance = 0.0004; maximum clearance = 0.0012 inch

**Modified Standard Fits.—**Fits having the same limits of clearance or interference as those shown in Table 8a to Table 12 may sometimes have to be produced by using holes or shafts having limits of size other than those shown in these tables. These modifications may be accomplished by using either a *Bilateral Hole System* (*Symbol B*) or a *Basic Shaft System* (*Symbol S*). Both methods will result in nonstandard holes and shafts.

*Bilateral Hole Fits (Symbol B):* The common situation is where holes are produced with fixed tools such as drills or reamers; to provide a longer wear life for such tools, a bilateral tolerance is desired.

The symbols used for these fits are identical with those used for standard fits except that they are followed by the letter B. Thus, LC 4B is a clearance locational fit Class 4, except that it is produced with a bilateral hole.

The limits of clearance or interference are identical with those shown in Table 8a to Table 12 for the corresponding fits.

The hole tolerance, however, is changed so that the plus limit is that for one grade finer than the value shown in the tables and the minus limit equals the amount by which the plus limit was lowered. The shaft limits are both lowered by the same amount as the lower limit of size of the hole. The finer grade of tolerance required to make these modifications may be obtained from Table 6. For example, an LC 4B fit for a 6-inch diameter hole would have tolerance limits of +4.0, −2.0 ( +0.0040 inch, −0.0020 inch); the shaft would have tolerance limits of −2.0, −6.0 (−0.0020 inch, −0.0060 inch).

*Basic Shaft Fits (Symbol S):* For these fits, the maximum size of the shaft is basic. The limits of clearance or interference are identical with those shown in Table 8a to Table 12 for the corresponding fits, and the symbols used for these fits are identical with those used for standard fits except that they are followed by the letter S. Thus, LC 4S is a clearance locational fit Class 4, except that it is produced on a basic shaft basis.

The limits for hole and shaft as given in Table 8a to Table 12 are increased for clearance fits (*decreased* for transition or interference fits) by the value of the upper shaft limit, that is, by the amount required to change the maximum shaft to the basic size.

**Graphical Representation of Standard Limits and Fits**
*ANSI/ASME B4.1-1967 (R2020)*

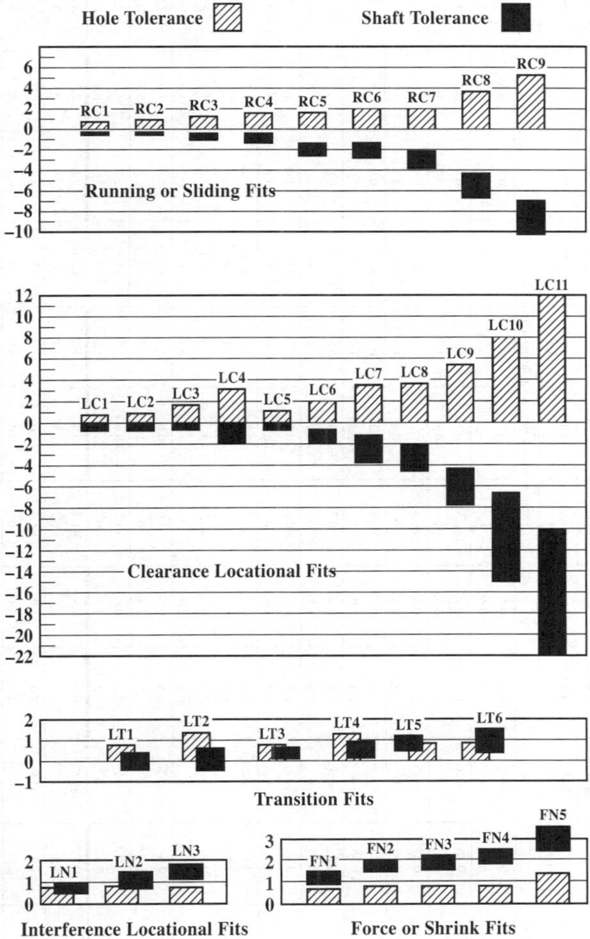


Diagrams show disposition of hole and shaft tolerances (in thousandths of an inch) with respect to basic size (0) for a diameter of 1 inch.

## Table 8a. American National Standard Running and Sliding Fits *ANSI/ASME B4.1-1967 (R2020)*

| Nominal Size Range, Inches | Class RC 1 | | | Class RC 2 | | | Class RC 3 | | | Class RC 4 | | |
|---|---|---|---|---|---|---|---|---|---|---|---|---|
| | | Standard Tolerance Limits | | | Standard Tolerance Limits | | | Standard Tolerance Limits | | | Standard Tolerance Limits | |
| | Clearance[a] | Hole H5 | Shaft g4 | Clearance[a] | Hole H6 | Shaft g5 | Clearance[a] | Hole H7 | Shaft f6 | Clearance[a] | Hole H8 | Shaft f7 |
| Over To | Values shown below are in thousandths of an inch | | | | | | | | | | | |
| 0 - 0.12 | 0.1<br>0.45 | +0.2<br>0 | −0.1<br>−0.25 | 0.1<br>0.55 | +0.25<br>0 | −0.1<br>−0.3 | 0.3<br>0.95 | +0.4<br>0 | −0.3<br>−0.55 | 0.3<br>1.3 | +0.6<br>0 | −0.3<br>−0.7 |
| 0.12 - 0.24 | 0.15<br>0.5 | +0.2<br>0 | −0.15<br>−0.3 | 0.15<br>0.65 | +0.3<br>0 | −0.15<br>−0.35 | 0.4<br>1.2 | +0.5<br>0 | −0.4<br>−0.7 | 0.4<br>1.6 | +0.7<br>0 | −0.4<br>−0.9 |
| 0.24 - 0.40 | 0.2<br>0.6 | +0.25<br>0 | −0.2<br>−0.35 | 0.2<br>0.85 | +0.4<br>0 | −0.2<br>−0.45 | 0.5<br>1.5 | +0.6<br>0 | −0.5<br>−0.9 | 0.5<br>2.0 | +0.9<br>0 | −0.5<br>−1.1 |
| 0.40 - 0.71 | 0.25<br>0.75 | +0.3<br>0 | −0.25<br>−0.45 | 0.25<br>0.95 | +0.4<br>0 | −0.25<br>−0.55 | 0.6<br>1.7 | +0.7<br>0 | −0.6<br>−1.0 | 0.6<br>2.3 | +1.0<br>0 | −0.6<br>−1.3 |
| 0.71 - 1.19 | 0.3<br>0.95 | +0.4<br>0 | −0.3<br>−0.55 | 0.3<br>1.2 | +0.5<br>0 | −0.3<br>−0.7 | 0.8<br>2.1 | +0.8<br>0 | −0.8<br>−1.3 | 0.8<br>2.8 | +1.2<br>0 | −0.8<br>−1.6 |
| 1.19 - 1.97 | 0.4<br>1.1 | +0.4<br>0 | −0.4<br>−0.7 | 0.4<br>1.4 | +0.6<br>0 | −0.4<br>−0.8 | 1.0<br>2.6 | +1.0<br>0 | −1.0<br>−1.6 | 1.0<br>3.6 | +1.6<br>0 | −1.0<br>−2.0 |
| 1.97 - 3.15 | 0.4<br>1.2 | +0.5<br>0 | −0.4<br>−0.7 | 0.4<br>1.6 | +0.7<br>0 | −0.4<br>−0.9 | 1.2<br>3.1 | +1.2<br>0 | −1.2<br>−1.9 | 1.2<br>4.2 | +1.8<br>0 | −1.2<br>−2.4 |
| 3.15 - 4.73 | 0.5<br>1.5 | +0.6<br>0 | −0.5<br>−0.9 | 0.5<br>2.0 | +0.9<br>0 | −0.5<br>−1.1 | 1.4<br>3.7 | +1.4<br>0 | −1.4<br>−2.3 | 1.4<br>5.0 | +2.2<br>0 | −1.4<br>−2.8 |
| 4.73 - 7.09 | 0.6<br>1.8 | +0.7<br>0 | −0.6<br>−1.1 | 0.6<br>2.3 | +1.0<br>0 | −0.6<br>−1.3 | 1.6<br>4.2 | +1.6<br>0 | −1.6<br>−2.6 | 1.6<br>5.7 | +2.5<br>0 | −1.6<br>−3.2 |
| 7.09 - 9.85 | 0.6<br>2.0 | +0.8<br>0 | −0.6<br>−1.2 | 0.6<br>2.6 | +1.2<br>0 | −0.6<br>−1.4 | 2.0<br>5.0 | +1.8<br>0 | −2.0<br>−3.2 | 2.0<br>6.6 | +2.8<br>0 | −2.0<br>−3.8 |
| 9.85 - 12.41 | 0.8<br>2.3 | +0.9<br>0 | −0.8<br>−1.4 | 0.8<br>2.9 | +1.2<br>0 | −0.8<br>−1.7 | 2.5<br>5.7 | +2.0<br>0 | −2.5<br>−3.7 | 2.5<br>7.5 | +3.0<br>0 | −2.5<br>−4.5 |
| 12.41 - 15.75 | 1.0<br>2.7 | +1.0<br>0 | −1.0<br>−1.7 | 1.0<br>3.4 | +1.4<br>0 | −1.0<br>−2.0 | 3.0<br>6.6 | +2.2<br>0 | −3.0<br>−4.4 | 3.0<br>8.7 | +3.5<br>0 | −3.0<br>−5.2 |
| 15.75 - 19.69 | 1.2<br>3.0 | +1.0<br>0 | −1.2<br>−2.0 | 1.2<br>3.8 | +1.6<br>0 | −1.2<br>−2.2 | 4.0<br>8.1 | +2.5<br>0 | −4.0<br>−5.6 | 4.0<br>10.5 | +4.0<br>0 | −4.0<br>−6.5 |

**Table 8b. American National Standard Running and Sliding Fits** *ANSI/ASME B4.1-1967 (R2020)*

| Nominal Size Range, Inches | Class RC 5 | | | Class RC 6 | | | Class RC 7 | | | Class RC 8 | | | Class RC 9 | | |
|---|---|---|---|---|---|---|---|---|---|---|---|---|---|---|---|
| | | Standard Tolerance Limits | | | Standard Tolerance Limits | | | Standard Tolerance Limits | | | Standard Tolerance Limits | | | Standard Tolerance Limits | |
| | Clearance[a] | Hole H8 | Shaft e7 | Clearance[a] | Hole H9 | Shaft e8 | Clearance[a] | Hole H9 | Shaft d8 | Clearance[a] | Hole H10 | Shaft c9 | Clearance[a] | Hole H11 | Shaft |
| Over To | Values shown below are in thousandths of an inch | | | | | | | | | | | | | | |
| 0 - 0.12 | 0.6 | +0.6 | − 0.6 | 0.6 | +1.0 | − 0.6 | 1.0 | +1.0 | − 1.0 | 2.5 | +1.6 | − 2.5 | 4.0 | +2.5 | − 4.0 |
| | 1.6 | 0 | − 1.0 | 2.2 | 0 | − 1.2 | 2.6 | 0 | − 1.6 | 5.1 | 0 | − 3.5 | 8.1 | 0 | − 5.6 |
| 0.12 - 0.24 | 0.8 | +0.7 | − 0.8 | 0.8 | +1.2 | − 0.8 | 1.2 | +1.2 | − 1.2 | 2.8 | +1.8 | − 2.8 | 4.5 | +3.0 | − 4.5 |
| | 2.0 | 0 | − 1.3 | 2.7 | 0 | − 1.5 | 3.1 | 0 | − 1.9 | 5.8 | 0 | − 4.0 | 9.0 | 0 | − 6.0 |
| 0.24 - 0.40 | 1.0 | +0.9 | − 1.0 | 1.0 | +1.4 | − 1.0 | 1.6 | +1.4 | − 1.6 | 3.0 | +2.2 | − 3.0 | 5.0 | +3.5 | − 5.0 |
| | 2.5 | 0 | − 1.6 | 3.3 | 0 | − 1.9 | 3.9 | 0 | − 2.5 | 6.6 | 0 | − 4.4 | 10.7 | 0 | − 7.2 |
| 0.40 - 0.71 | 1.2 | +1.0 | − 1.2 | 1.2 | +1.6 | − 1.2 | 2.0 | +1.6 | − 2.0 | 3.5 | +2.8 | − 3.5 | 6.0 | +4.0 | − 6.0 |
| | 2.9 | 0 | − 1.9 | 3.8 | 0 | − 2.2 | 4.6 | 0 | − 3.0 | 7.9 | 0 | − 5.1 | 12.8 | 0 | − 8.8 |
| 0.71 - 1.19 | 1.6 | +1.2 | − 1.6 | 1.6 | +2.0 | − 1.6 | 2.5 | +2.0 | − 2.5 | 4.5 | +3.5 | − 4.5 | 7.0 | +5.0 | − 7.0 |
| | 3.6 | 0 | − 2.4 | 4.8 | 0 | − 2.8 | 5.7 | 0 | − 3.7 | 10.0 | 0 | − 6.5 | 15.5 | 0 | −10.5 |
| 1.19 - 1.97 | 2.0 | +1.6 | − 2.0 | 2.0 | +2.5 | − 2.0 | 3.0 | +2.5 | − 3.0 | 5.0 | +4.0 | − 5.0 | 8.0 | +6.0 | − 8.0 |
| | 4.6 | 0 | − 3.0 | 6.1 | 0 | − 3.6 | 7.1 | 0 | − 4.6 | 11.5 | 0 | − 7.5 | 18.0 | 0 | −12.0 |
| 1.97 - 3.15 | 2.5 | +1.8 | − 2.5 | 2.5 | +3.0 | − 2.5 | 4.0 | +3.0 | − 4.0 | 6.0 | +4.5 | − 6.0 | 9.0 | +7.0 | − 9.0 |
| | 5.5 | 0 | − 3.7 | 7.3 | 0 | − 4.3 | 8.8 | 0 | − 5.8 | 13.5 | 0 | − 9.0 | 20.5 | 0 | −13.5 |
| 3.15 - 4.73 | 3.0 | +2.2 | − 3.0 | 3.0 | +3.5 | − 3.0 | 5.0 | +3.5 | − 5.0 | 7.0 | +5.0 | − 7.0 | 10.0 | +9.0 | −10.0 |
| | 6.6 | 0 | − 4.4 | 8.7 | 0 | − 5.2 | 10.7 | 0 | − 7.2 | 15.5 | 0 | −10.5 | 24.0 | 0 | −15.0 |
| 4.73 - 7.09 | 3.5 | +2.5 | − 3.5 | 3.5 | +4.0 | − 3.5 | 6.0 | +4.0 | − 6.0 | 8.0 | +6.0 | − 8.0 | 12.0 | +10.0 | −12.0 |
| | 7.6 | 0 | − 5.1 | 10.0 | 0 | − 6.0 | 12.5 | 0 | − 8.5 | 18.0 | 0 | −12.0 | 28.0 | 0 | −18.0 |
| 7.09 - 9.85 | 4.0 | +2.8 | − 4.0 | 4.0 | +4.5 | − 4.0 | 7.0 | +4.5 | − 7.0 | 10.0 | +7.0 | −10.0 | 15.0 | +12.0 | −15.0 |
| | 8.6 | 0 | − 5.8 | 11.3 | 0 | − 6.8 | 14.3 | 0 | − 9.8 | 21.5 | 0 | −14.5 | 34.0 | 0 | −22.0 |
| 9.85 - 12.41 | 5.0 | +3.0 | − 5.0 | 5.0 | +5.0 | − 5.0 | 8.0 | +5.0 | − 8.0 | 12.0 | +8.0 | −12.0 | 18.0 | +12.0 | −18.0 |
| | 10.0 | 0 | − 7.0 | 13.0 | 0 | − 8.0 | 16.0 | 0 | −11.0 | 25.0 | 0 | −17.0 | 38.0 | 0 | −26.0 |
| 12.41 - 15.75 | 6.0 | +3.5 | − 6.0 | 6.0 | +6.0 | − 6.0 | 10.0 | +6.0 | −10.0 | 14.0 | +9.0 | −14.0 | 22.0 | +14.0 | −22.0 |
| | 11.7 | 0 | − 8.2 | 15.5 | 0 | − 9.5 | 19.5 | 0 | −13.5 | 29.0 | 0 | −20.0 | 45.0 | 0 | −31.0 |
| 15.75 - 19.69 | 8.0 | +4.0 | − 8.0 | 8.0 | +6.0 | − 8.0 | 12.0 | +6.0 | −12.0 | 16.0 | +10.0 | −16.0 | 25.0 | +16.0 | −25.0 |
| | 14.5 | 0 | −10.5 | 18.0 | 0 | −12.0 | 22.0 | 0 | −16.0 | 32.0 | 0 | −22.0 | 51.0 | 0 | −35.0 |

[a] Pairs of values shown represent minimum and maximum amounts of clearance resulting from application of standard tolerance limits.

Tolerance limits in Table 8a and Table 8b are added to or subtracted from basic size (as indicated by + or − sign) to obtain maximum and minimum sizes of mating parts.

All data above heavy lines are in accord with ABC agreements. Symbols H5, g4, etc., are hole and shaft designations in ABC system. Limits for sizes above 19.69 inches are also given in the ANSI/ASME Standard.

**Table 9a. American National Standard Clearance Locational Fits** *ANSI/ASME B4.1-1967 (R2020)*

| Nominal Size Range, Inches | Class LC 1 | | | Class LC 2 | | | Class LC 3 | | | Class LC 4 | | | Class LC 5 | | |
|---|---|---|---|---|---|---|---|---|---|---|---|---|---|---|---|
| | Clearance[a] | Standard Tolerance Limits | | Clearance[a] | Standard Tolerance Limits | | Clearance[a] | Standard Tolerance Limits | | Clearance[a] | Standard Tolerance Limits | | Clearance[a] | Standard Tolerance Limits | |
| | | Hole H6 | Shaft h5 | | Hole H7 | Shaft h6 | | Hole H8 | Shaft h7 | | Hole H10 | Shaft h9 | | Hole H7 | Shaft g6 |
| Over To | Values shown below are in thousandths of an inch | | | | | | | | | | | | | | |
| 0- 0.12 | 0<br>0.45 | +0.25<br>0 | 0<br>−0.2 | 0<br>0.65 | +0.4<br>0 | 0<br>−0.25 | 0<br>1 | +0.6<br>0 | 0<br>−0.4 | 0<br>2.6 | +1.6<br>0 | 0<br>−1.0 | 0.1<br>0.75 | +0.4<br>0 | −0.1<br>−0.35 |
| 0.12- 0.24 | 0<br>0.5 | +0.3<br>0 | 0<br>−0.2 | 0<br>0.8 | +0.5<br>0 | 0<br>−0.3 | 0<br>1.2 | +0.7<br>0 | 0<br>−0.5 | 0<br>3.0 | +1.8<br>0 | 0<br>−1.2 | 0.15<br>0.95 | +0.5<br>0 | −0.15<br>−0.45 |
| 0.24- 0.40 | 0<br>0.65 | +0.4<br>0 | 0<br>−0.25 | 0<br>1.0 | +0.6<br>0 | 0<br>−0.4 | 0<br>1.5 | +0.9<br>0 | 0<br>−0.6 | 0<br>3.6 | +2.2<br>0 | 0<br>−1.4 | 0.2<br>1.2 | +0.6<br>0 | −0.2<br>−0.6 |
| 0.40- 0.71 | 0<br>0.7 | +0.4<br>0 | 0<br>−0.3 | 0<br>1.1 | +0.7<br>0 | 0<br>−0.4 | 0<br>1.7 | +1.0<br>0 | 0<br>−0.7 | 0<br>4.4 | +2.8<br>0 | 0<br>−1.6 | 0.25<br>1.35 | +0.7<br>0 | −0.25<br>−0.65 |
| 0.71- 1.19 | 0<br>0.9 | +0.5<br>0 | 0<br>−0.4 | 0<br>1.3 | +0.8<br>0 | 0<br>−0.5 | 0<br>2 | +1.2<br>0 | 0<br>−0.8 | 0<br>5.5 | +3.5<br>0 | 0<br>−2.0 | 0.3<br>1.6 | +0.8<br>0 | −0.3<br>−0.8 |
| 1.19- 1.97 | 0<br>1.0 | +0.6<br>0 | 0<br>−0.4 | 0<br>1.6 | +1.0<br>0 | 0<br>−0.6 | 0<br>2.6 | +1.6<br>0 | 0<br>−1 | 0<br>6.5 | +4.0<br>0 | 0<br>−2.5 | 0.4<br>2.0 | +1.0<br>0 | −0.4<br>−1.0 |
| 1.97- 3.15 | 0<br>1.2 | +0.7<br>0 | 0<br>−0.5 | 0<br>1.9 | +1.2<br>0 | 0<br>−0.7 | 0<br>3 | +1.8<br>0 | 0<br>−1.2 | 0<br>7.5 | +4.5<br>0 | 0<br>−3 | 0.4<br>2.3 | +1.2<br>0 | −0.4<br>−1.1 |
| 3.15- 4.73 | 0<br>1.5 | +0.9<br>0 | 0<br>−0.6 | 0<br>2.3 | +1.4<br>0 | 0<br>−0.9 | 0<br>3.6 | +2.2<br>0 | 0<br>−1.4 | 0<br>8.5 | +5.0<br>0 | 0<br>−3.5 | 0.5<br>2.8 | +1.4<br>0 | −0.5<br>−1.4 |
| 4.73- 7.09 | 0<br>1.7 | +1.0<br>0 | 0<br>−0.7 | 0<br>2.6 | +1.6<br>0 | 0<br>−1.0 | 0<br>4.1 | +2.5<br>0 | 0<br>−1.6 | 0<br>10.0 | +6.0<br>0 | 0<br>−4 | 0.6<br>3.2 | +1.6<br>0 | −0.6<br>−1.6 |
| 7.09- 9.85 | 0<br>2.0 | +1.2<br>0 | 0<br>−0.8 | 0<br>3.0 | +1.8<br>0 | 0<br>−1.2 | 0<br>4.6 | +2.8<br>0 | 0<br>−1.8 | 0<br>11.5 | +7.0<br>0 | 0<br>−4.5 | 0.6<br>3.6 | +1.8<br>0 | −0.6<br>−1.8 |
| 9.85- 12.41 | 0<br>2.1 | +1.2<br>0 | 0<br>−0.9 | 0<br>3.2 | +2.0<br>0 | 0<br>−1.2 | 0<br>5 | +3.0<br>0 | 0<br>−2.0 | 0<br>13.0 | +8.0<br>0 | 0<br>−5 | 0.7<br>3.9 | +2.0<br>0 | −0.7<br>−1.9 |
| 12.41- 15.75 | 0<br>2.4 | +1.4<br>0 | 0<br>−1.0 | 0<br>3.6 | +2.2<br>0 | 0<br>−1.4 | 0<br>5.7 | +3.5<br>0 | 0<br>−2.2 | 0<br>15.0 | +9.0<br>0 | 0<br>−6 | 0.7<br>4.3 | +2.2<br>0 | −0.7<br>−2.1 |
| 15.75- 19.69 | 0<br>2.6 | +1.6<br>0 | 0<br>−1.0 | 0<br>4.1 | +2.5<br>0 | 0<br>−1.6 | 0<br>6.5 | +4<br>0 | 0<br>−2.5 | 0<br>16.0 | +10.0<br>0 | 0<br>−6 | 0.8<br>4.9 | +2.5<br>0 | −0.8<br>−2.4 |

Tolerance limits given in body of Table 9a and Table 9b are added or subtracted to basic size (as indicated by + or − sign) to obtain maximum and minimum sizes of mating parts. All data above heavy lines are in accordance with American-British-Canadian (ABC) agreements. Symbols H6, H7, s6, etc., are hole and shaft designations in ABC system.

**Table 9b. American National Standard Clearance Locational Fits** *ANSI/ASME B4.1-1967 (R2020)*

| Nominal Size Range, Inches | Class LC 6 | | | Class LC 7 | | | Class LC 8 | | | Class LC 9 | | | Class LC 10 | | | Class LC 11 | | |
|---|---|---|---|---|---|---|---|---|---|---|---|---|---|---|---|---|---|---|
| | | Std. Tolerance Limits | | | Std. Tolerance Limits | | | Std. Tolerance Limits | | | Std. Tolerance Limits | | | Std. Tolerance Limits | | | Std. Tolerance Limits | |
| | Clearance[a] | Hole H9 | Shaft f8 | Clearance[a] | Hole H10 | Shaft e9 | Clearance[a] | Hole H10 | Shaft d9 | Clearance[a] | Hole H11 | Shaft c10 | Clearance[a] | Hole H12 | Shaft | Clearance[a] | Hole H13 | Shaft |
| Over To | Values shown below are in thousandths of an inch | | | | | | | | | | | | | | | | | |
| 0 - 0.12 | 0.3 | +1.0 | −0.3 | 0.6 | +1.6 | − 0.6 | 1.0 | +1.6 | − 1.0 | 2.5 | +2.5 | − 2.5 | 4 | +4 | − 4 | 5 | +6 | − 5 |
| | 1.9 | 0 | −0.9 | 3.2 | 0 | − 1.6 | 2.0 | 0 | − 2.0 | 6.6 | 0 | − 4.1 | 12 | 0 | − 8 | 17 | 0 | − 11 |
| 0.12 – 0.24 | 0.4 | +1.2 | −0.4 | 0.8 | +1.8 | − 0.8 | 1.2 | +1.8 | − 1.2 | 2.8 | +3.0 | − 2.8 | 4.5 | +5 | − 4.5 | 6 | +7 | − 6 |
| | 2.3 | 0 | −1.1 | 3.8 | 0 | − 2.0 | 4.2 | 0 | − 2.4 | 7.6 | 0 | − 4.6 | 14.5 | 0 | − 9.5 | 20 | 0 | −13 |
| 0.24 – 0.40 | 0.5 | +1.4 | −0.5 | 1.0 | +2.2 | − 1.0 | 1.6 | +2.2 | − 1.6 | 3.0 | +3.5 | − 3.0 | 5 | +6 | − 5 | 7 | +9 | − 7 |
| | 2.8 | 0 | −1.4 | 4.6 | 0 | − 2.4 | 5.2 | 0 | − 3.0 | 8.7 | 0 | − 5.2 | 17 | 0 | −11 | 25 | 0 | −16 |
| 0.40 - 0.71 | 0.6 | +1.6 | −0.6 | 1.2 | +2.8 | − 1.2 | 2.0 | +2.8 | − 2.0 | 3.5 | +4.0 | − 3.5 | 6 | +7 | − 6 | 8 | +10 | − 8 |
| | 3.2 | 0 | −1.6 | 5.6 | 0 | − 2.8 | 6.4 | 0 | − 3.6 | 10.3 | 0 | − 6.3 | 20 | 0 | −13 | 28 | 0 | −18 |
| 0.71 - 1.19 | 0.8 | +2.0 | −0.8 | 1.6 | +3.5 | − 1.6 | 2.5 | +3.5 | − 2.5 | 4.5 | +5.0 | − 4.5 | 7 | +8 | − 7 | 10 | +12 | −10 |
| | 4.0 | 0 | −2.0 | 7.1 | 0 | − 3.6 | 8.0 | 0 | − 4.5 | 13.0 | 0 | − 8.0 | 23 | 0 | −15 | 34 | 0 | −22 |
| 1.19 - 1.97 | 1.0 | +2.5 | −1.0 | 2.0 | +4.0 | − 2.0 | 3.6 | +4.0 | − 3.0 | 5.0 | +6 | − 5.0 | 8 | +10 | − 8 | 12 | +16 | −12 |
| | 5.1 | 0 | −2.6 | 8.5 | 0 | − 4.5 | 9.5 | 0 | − 5.5 | 15.0 | 0 | − 9.0 | 28 | 0 | −18 | 44 | 0 | −28 |
| 1.97 - 3.15 | 1.2 | +3.0 | −1.2 | 2.5 | +4.5 | − 2.5 | 4.0 | +4.5 | − 4.0 | 6.0 | +7 | − 6.0 | 10 | +12 | −10 | 14 | +18 | −14 |
| | 6.0 | 0 | −3.0 | 10.0 | 0 | − 5.5 | 11.5 | 0 | − 7.0 | 17.5 | 0 | −10.5 | 34 | 0 | −22 | 50 | 0 | −32 |
| 3.15 - 4.73 | 1.4 | +3.5 | −1.4 | 3.0 | +5.0 | − 3.0 | 5.0 | +5.0 | − 5.0 | 7 | +9 | − 7 | 11 | +14 | −11 | 16 | +22 | −16 |
| | 7.1 | 0 | −3.6 | 11.5 | 0 | − 6.5 | 13.5 | 0 | − 8.5 | 21 | 0 | −12 | 39 | 0 | −25 | 60 | 0 | −38 |
| 4.73 - 7.09 | 1.6 | +4.0 | −1.6 | 3.5 | +6.0 | − 3.5 | 6 | +6 | − 6 | 8 | +10 | − 8 | 12 | +16 | −12 | 18 | +25 | −18 |
| | 8.1 | 0 | −4.1 | 13.5 | 0 | − 7.5 | 16 | 0 | −10 | 24 | 0 | −14 | 44 | 0 | −28 | 68 | 0 | −43 |
| 7.09 – 9.85 | 2.0 | +4.5 | −2.0 | 4.0 | +7.0 | − 4.0 | 7 | +7 | − 7 | 10 | +12 | −10 | 16 | +18 | −16 | 22 | +28 | −22 |
| | 9.3 | 0 | −4.8 | 15.5 | 0 | − 8.5 | 18.5 | 0 | −11.5 | 29 | 0 | −17 | 52 | 0 | −34 | 78 | 0 | −50 |
| 9.85 - 12.41 | 2.2 | +5.0 | −2.2 | 4.5 | +8.0 | − 4.5 | 7 | +8 | − 7 | 12 | +12 | −12 | 20 | +20 | −20 | 28 | +30 | −28 |
| | 10.2 | 0 | −5.2 | 17.5 | 0 | − 9.5 | 20 | 0 | −12 | 32 | 0 | −20 | 60 | 0 | −40 | 88 | 0 | −58 |
| 12.41 - 15.75 | 2.5 | +6.0 | −2.5 | 5.0 | +9.0 | − 5 | 8 | +9 | − 8 | 14 | +14 | −14 | 22 | +22 | −22 | 30 | +35 | −30 |
| | 12.0 | 0 | −6.0 | 20.0 | 0 | −11 | 23 | 0 | −14 | 37 | 0 | −23 | 66 | 0 | −44 | 100 | 0 | −65 |
| 15.75- 19.69 | 2.8 | +6.0 | −2.8 | 5.0 | +10.0 | − 5 | 9 | +10 | − 9 | 16 | +16 | −16 | 25 | +25 | −25 | 35 | +40 | −35 |
| | 12.8 | 0 | −6.8 | 21.0 | 0 | −11 | 25 | 0 | −15 | 42 | 0 | −26 | 75 | 0 | −50 | 115 | 0 | −75 |

[a] Pairs of values shown represent minimum and maximum amounts of interference resulting from application of standard tolerance limits.

Limits for sizes above 19.69 inches are not covered by American-British-Canadian (ABC) agreements but are given in the ANSI/ASME Standard.

## Table 10. Standard Transition Locational Fits *ANSI/ASME B4.1-1967 (R2020)*

| Nominal Size Range, Inches | Class LT 1 | | | Class LT 2 | | | Class LT 3 | | | Class LT 4 | | | Class LT 5 | | | Class LT 6 | | |
|---|---|---|---|---|---|---|---|---|---|---|---|---|---|---|---|---|---|---|
| | | Std. Tolerance Limits | | | Std. Tolerance Limits | | | Std. Tolerance Limits | | | Std. Tolerance Limits | | | Std. Tolerance Limits | | | Std. Tolerance Limits | |
| Over To | Fit[a] | Hole H7 | Shaft js6 | Fit[a] | Hole H8 | Shaft js7 | Fit[a] | Hole H7 | Shaft k6 | Fit[a] | Hole H8 | Shaft k7 | Fit[a] | Hole H7 | Shaft n6 | Fit[a] | Hole H7 | Shaft n7 |
| | Values shown below are in thousandths of an inch | | | | | | | | | | | | | | | | | |
| 0 - 0.12 | −0.12 | +0.4 | +0.12 | −0.2 | +0.6 | +0.2 | | | | | | | −0.5 | +0.4 | +0.5 | −0.65 | +0.4 | +0.65 |
| | +0.52 | 0 | −0.12 | +0.8 | 0 | −0.2 | | | | | | | +0.15 | 0 | +0.25 | +0.15 | 0 | +0.25 |
| 0.12 - 0.24 | −0.15 | +0.5 | +0.15 | −0.25 | +0.7 | +0.25 | | | | | | | −0.6 | +0.5 | +0.6 | −0.8 | +0.5 | +0.8 |
| | +0.65 | 0 | −0.15 | +0.95 | 0 | −0.25 | | | | | | | +0.2 | 0 | +0.3 | +0.2 | 0 | +0.3 |
| 0.24 - 0.40 | −0.2 | +0.6 | +0.2 | −0.3 | +0.9 | +0.3 | −0.5 | +0.6 | +0.5 | −0.7 | +0.9 | +0.7 | −0.8 | +0.6 | +0.8 | −1.0 | +0.6 | +1.0 |
| | +0.8 | 0 | −0.2 | +1.2 | 0 | −0.3 | +0.5 | 0 | +0.1 | +0.8 | 0 | +0.1 | +0.2 | 0 | +0.4 | +0.2 | 0 | +0.4 |
| 0.40 - 0.71 | −0.2 | +0.7 | +0.2 | −0.35 | +1.0 | +0.35 | −0.5 | +0.7 | +0.5 | −0.8 | +1.0 | +0.8 | −0.9 | +0.7 | +0.9 | −1.2 | +0.7 | +1.2 |
| | +0.9 | 0 | −0.2 | +1.35 | 0 | −0.35 | +0.6 | 0 | +0.1 | +0.9 | 0 | +0.1 | +0.2 | 0 | +0.5 | +0.2 | 0 | +0.5 |
| 0.71 - 1.19 | −0.25 | +0.8 | +0.25 | −0.4 | +1.2 | +0.4 | −0.6 | +0.8 | +0.6 | −0.9 | +1.2 | +0.9 | −1.1 | +0.8 | +1.1 | −1.4 | +0.8 | +1.4 |
| | +1.05 | 0 | −0.25 | +1.6 | 0 | −0.4 | +0.7 | 0 | +0.1 | +1.1 | 0 | +0.1 | +0.2 | 0 | +0.6 | +0.2 | 0 | +0.6 |
| 1.19 - 1.97 | −0.3 | +1.0 | +0.3 | −0.5 | +1.6 | +0.5 | −0.7 | +1.0 | +0.7 | −1.1 | +1.6 | +1.1 | −1.3 | +1.0 | +1.3 | −1.7 | +1.0 | +1.7 |
| | +1.3 | 0 | −0.3 | +2.1 | 0 | −0.5 | +0.9 | 0 | +0.1 | +1.5 | 0 | +0.1 | +0.3 | 0 | +0.7 | +0.3 | 0 | +0.7 |
| 1.97 - 3.15 | −0.3 | +1.2 | +0.3 | −0.6 | +1.8 | +0.6 | −0.8 | +1.2 | +0.8 | −1.3 | +1.8 | +1.3 | −1.5 | +1.2 | +1.5 | −2.0 | +1.2 | +2.0 |
| | +1.5 | 0 | −0.3 | +2.4 | 0 | −0.6 | +1.1 | 0 | +0.1 | +1.7 | 0 | +0.1 | +0.4 | 0 | +0.8 | +0.4 | 0 | +0.8 |
| 3.15 - 4.73 | −0.4 | +1.4 | +0.4 | −0.7 | +2.2 | +0.7 | −1.0 | +1.4 | +1.0 | −1.5 | +2.2 | +1.5 | −1.9 | +1.4 | +1.9 | −2.4 | +1.4 | +2.4 |
| | +1.8 | 0 | −0.4 | +2.9 | 0 | −0.7 | +1.3 | 0 | +0.1 | +2.1 | 0 | +0.1 | +0.4 | 0 | +1.0 | +0.4 | 0 | +1.0 |
| 4.73 - 7.09 | −0.5 | +1.6 | +0.5 | −0.8 | +2.5 | +0.8 | −1.1 | +1.6 | +1.1 | −1.7 | +2.5 | +1.7 | −2.2 | +1.6 | +2.2 | −2.8 | +1.6 | +2.8 |
| | +2.1 | 0 | −0.5 | +3.3 | 0 | −0.8 | +1.5 | 0 | +0.1 | +2.4 | 0 | +0.1 | +0.4 | 0 | +1.2 | +0.4 | 0 | +1.2 |
| 7.09 - 9.85 | −0.6 | +1.8 | +0.6 | −0.9 | +2.8 | +0.9 | −1.4 | +1.8 | +1.4 | −2.0 | +2.8 | +2.0 | −2.6 | +1.8 | +2.6 | −3.2 | +1.8 | +3.2 |
| | +2.4 | 0 | −0.6 | +3.7 | 0 | −0.9 | +1.6 | 0 | +0.2 | +2.6 | 0 | +0.2 | +0.4 | 0 | +1.4 | +0.4 | 0 | +1.4 |
| 9.85 - 12.41 | −0.6 | +2.0 | +0.6 | −1.0 | +3.0 | +1.0 | −1.4 | +2.0 | +1.4 | −2.2 | +3.0 | +2.2 | −2.6 | +2.0 | +2.6 | −3.4 | +2.0 | +3.4 |
| | +2.6 | 0 | −0.6 | +4.0 | 0 | −1.0 | +1.8 | 0 | +0.2 | +2.8 | 0 | +0.2 | +0.6 | 0 | +1.4 | +0.6 | 0 | +1.4 |
| 12.41 - 15.75 | −0.7 | +2.2 | +0.7 | −1.0 | +3.5 | +1.0 | −1.6 | +2.2 | +1.6 | −2.4 | +3.5 | +2.4 | −3.0 | +2.2 | +3.0 | −3.8 | +2.2 | +3.8 |
| | +2.9 | 0 | −0.7 | +4.5 | 0 | −1.0 | +2.0 | 0 | +0.2 | +3.3 | 0 | +0.2 | +0.6 | 0 | +1.6 | +0.6 | 0 | +1.6 |
| 15.75 - 19.69 | −0.8 | +2.5 | +0.8 | −1.2 | +4.0 | +1.2 | −1.8 | +2.5 | +1.8 | −2.7 | +4.0 | +2.7 | −3.4 | +2.5 | +3.4 | −4.3 | +2.5 | +4.3 |
| | +3.3 | 0 | −0.8 | +5.2 | 0 | −1.2 | +2.3 | 0 | +0.2 | +3.8 | 0 | +0.2 | +0.7 | 0 | +1.8 | +0.7 | 0 | +1.8 |

[a] Pairs of values shown represent maximum amount of interference (–) and maximum amount of clearance (+) resulting from application of standard tolerance limits.

All data above heavy lines are in accord with American-British-Canadian (ABC) agreements. Symbols H7, js6, etc., are hole and shaft designations in the ABC system.

## Table 11. Standard Force and Shrink Fits *ANSI/ASME B4.1-1967 (R2020)*

| Nominal Size Range, Inches | Class FN 1 | | | Class FN 2 | | | Class FN 3 | | | Class FN 4 | | | Class FN 5 | | |
|---|---|---|---|---|---|---|---|---|---|---|---|---|---|---|---|
| | | Standard Tolerance Limits | | | Standard Tolerance Limits | | | Standard Tolerance Limits | | | Standard Tolerance Limits | | | Standard Tolerance Limits | |
| | Interference[a] | Hole H6 | Shaft | Interference[a] | Hole H7 | Shaft s6 | Interference[a] | Hole H7 | Shaft t6 | Interference[a] | Hole H7 | Shaft u6 | Interference[a] | Hole H8 | Shaft x7 |
| Over To | Values shown below are in thousandths of an inch | | | | | | | | | | | | | | |
| 0- 0.12 | 0.05 | +0.25 | +0.5 | 0.2 | +0.4 | +0.85 | | | | 0.3 | +0.4 | +0.95 | 0.3 | +0.6 | +1.3 |
| | 0.5 | 0 | +0.3 | 0.85 | 0 | +0.6 | | | | 0.95 | 0 | +0.7 | 1.3 | 0 | +0.9 |
| 0.12- 0.24 | 0.1 | +0.3 | +0.6 | 0.2 | +0.5 | +1.0 | | | | 0.4 | +0.5 | +1.2 | 0.5 | +0.7 | +1.7 |
| | 0.6 | 0 | +0.4 | 1.0 | 0 | +0.7 | | | | 1.2 | 0 | +0.9 | 1.7 | 0 | +1.2 |
| 0.24- 0.40 | 0.1 | +0.4 | +0.75 | 0.4 | +0.6 | +1.4 | | | | 0.6 | +0.6 | +1.6 | 0.5 | +0.9 | +2.0 |
| | 0.75 | 0 | +0.5 | 1.4 | 0 | +1.0 | | | | 1.6 | 0 | +1.2 | 2.0 | 0 | +1.4 |
| 0.40- 0.56 | 0.1 | +0.4 | +0.8 | 0.5 | +0.7 | +1.6 | | | | 0.7 | +0.7 | +1.8 | 0.6 | +1.0 | +2.3 |
| | 0.8 | 0 | +0.5 | 1.6 | 0 | +1.2 | | | | 1.8 | 0 | +1.4 | 2.3 | 0 | +1.6 |
| 0.56- 0.71 | 0.2 | +0.4 | +0.9 | 0.5 | +0.7 | +1.6 | | | | 0.7 | +0.7 | +1.8 | 0.8 | +1.0 | +2.5 |
| | 0.9 | 0 | +0.6 | 1.6 | 0 | +1.2 | | | | 1.8 | 0 | +1.4 | 2.5 | 0 | +1.8 |
| 0.71- 0.95 | 0.2 | +0.5 | +1.1 | 0.6 | +0.8 | +1.9 | | | | 0.8 | +0.8 | +2.1 | 1.0 | +1.2 | +3.0 |
| | 1.1 | 0 | +0.7 | 1.9 | 0 | +1.4 | | | | 2.1 | 0 | +1.6 | 3.0 | 0 | +2.2 |
| 0.95- 1.19 | 0.3 | +0.5 | +1.2 | 0.6 | +0.8 | +1.9 | 0.8 | +0.8 | +2.1 | +1.0 | +0.8 | +2.3 | 1.3 | +1.2 | +3.3 |
| | 1.2 | 0 | +0.8 | 1.9 | 0 | +1.4 | 2.1 | 0 | +1.6 | 2.3 | 0 | +1.8 | 3.3 | 0 | +2.5 |
| 1.19- 1.58 | 0.3 | +0.6 | +1.3 | 0.8 | +1.0 | +2.4 | 1.0 | +1.0 | +2.6 | 1.5 | +1.0 | +3.1 | 1.4 | +1.6 | +4.0 |
| | 1.3 | 0 | +0.9 | 2.4 | 0 | +1.8 | 2.6 | 0 | +2.0 | 3.1 | 0 | +2.5 | 4.0 | 0 | +3.0 |
| 1.58- 1.97 | 0.4 | +0.6 | +1.4 | 0.8 | +1.0 | +2.4 | 1.2 | +1.0 | +2.8 | 1.8 | +1.0 | +3.4 | 2.4 | +1.6 | +5.0 |
| | 1.4 | 0 | +1.0 | 2.4 | 0 | +1.8 | 2.8 | 0 | +2.2 | 3.4 | 0 | +2.8 | 5.0 | 0 | +4.0 |
| 1.97- 2.56 | 0.6 | +0.7 | +1.8 | 0.8 | +1.2 | +2.7 | 1.3 | +1.2 | +3.2 | 2.3 | +1.2 | +4.2 | 3.2 | +1.8 | +6.2 |
| | 1.8 | 0 | +1.3 | 2.7 | 0 | +2.0 | 3.2 | 0 | +2.5 | 4.2 | 0 | +3.5 | 6.2 | 0 | +5.0 |
| 2.56- 3.15 | 0.7 | +0.7 | +1.9 | 1.0 | +1.2 | +2.9 | 1.8 | +1.2 | +3.7 | 2.8 | +1.2 | +4.7 | 4.2 | +1.8 | +7.2 |
| | 1.9 | 0 | +1.4 | 2.9 | 0 | +2.2 | 3.7 | 0 | +3.0 | 4.7 | 0 | +4.0 | 7.2 | 0 | +6.0 |
| 3.15- 3.94 | 0.9 | +0.9 | +2.4 | 1.4 | +1.4 | +3.7 | 2.1 | +1.4 | +4.4 | 3.6 | +1.4 | +5.9 | 4.8 | +2.2 | +8.4 |
| | 2.4 | 0 | +1.8 | 3.7 | 0 | +2.8 | 4.4 | 0 | +3.5 | 5.9 | 0 | +5.0 | 8.4 | 0 | +7.0 |
| 3.94- 4.73 | 1.1 | +0.9 | +2.6 | 1.6 | +1.4 | +3.9 | 2.6 | +1.4 | +4.9 | 4.6 | +1.4 | +6.9 | 5.8 | +2.2 | +9.4 |
| | 2.6 | 0 | +2.0 | 3.9 | 0 | +3.0 | 4.9 | 0 | +4.0 | 6.9 | 0 | +6.0 | 9.4 | 0 | +8.0 |

**Table 11.** *(Continued)* **Standard Force and Shrink Fits** *ANSI/ASME B4.1-1967 (R2020)*

| Nominal Size Range, Inches Over – To | Class FN 1 Interference[a] | Class FN 1 Hole H6 | Class FN 1 Shaft | Class FN 2 Interference[a] | Class FN 2 Hole H7 | Class FN 2 Shaft s6 | Class FN 3 Interference[a] | Class FN 3 Hole H7 | Class FN 3 Shaft t6 | Class FN 4 Interference[a] | Class FN 4 Hole H7 | Class FN 4 Shaft u6 | Class FN 5 Interference[a] | Class FN 5 Hole H8 | Class FN 5 Shaft x7 |
|---|---|---|---|---|---|---|---|---|---|---|---|---|---|---|---|
| | Values shown below are in thousandths of an inch | | | | | | | | | | | | | | |
| 4.73- 5.52 | 1.2 | +1.0 | +2.9 | 1.9 | +1.6 | +4.5 | 3.4 | +1.6 | +6.0 | 5.4 | +1.6 | +8.0 | 7.5 | +2.5 | +11.6 |
| | 2.9 | 0 | +2.2 | 4.5 | 0 | +3.5 | 6.0 | 0 | +5.0 | 8.0 | 0 | +7.0 | 11.6 | 0 | +10.0 |
| 5.52- 6.30 | 1.5 | +1.0 | +3.2 | 2.4 | +1.6 | +5.0 | 3.4 | +1.6 | +6.0 | 5.4 | +1.6 | +8.0 | 9.5 | +2.5 | +13.6 |
| | 3.2 | 0 | +2.5 | 5.0 | 0 | +4.0 | 6.0 | 0 | +5.0 | 8.0 | 0 | +7.0 | 13.6 | 0 | +12.0 |
| 6.30- 7.09 | 1.8 | +1.0 | +3.5 | 2.9 | +1.6 | +5.5 | 4.4 | +1.6 | +7.0 | 6.4 | +1.6 | +9.0 | 9.5 | +2.5 | +13.6 |
| | 3.5 | 0 | +2.8 | 5.5 | 0 | +4.5 | 7.0 | 0 | +6.0 | 9.0 | 0 | +8.0 | 13.6 | 0 | +12.0 |
| 7.09- 7.88 | 1.8 | +1.2 | +3.8 | 3.2 | +1.8 | +6.2 | 5.2 | +1.8 | +8.2 | 7.2 | +1.8 | +10.2 | 11.2 | +2.8 | +15.8 |
| | 3.8 | 0 | +3.0 | 6.2 | 0 | +5.0 | 8.2 | 0 | +7.0 | 10.2 | 0 | +9.0 | 15.8 | 0 | +14.0 |
| 7.88- 8.86 | 2.3 | +1.2 | +4.3 | 3.2 | +1.8 | +6.2 | 5.2 | +1.8 | +8.2 | 8.2 | +1.8 | +11.2 | 13.2 | +2.8 | +17.8 |
| | 4.3 | 0 | +3.5 | 6.2 | 0 | +5.0 | 8.2 | 0 | +7.0 | 11.2 | 0 | +10.0 | 17.8 | 0 | +16.0 |
| 8.86- 9.85 | 2.3 | +1.2 | +4.3 | 4.2 | +1.8 | +7.2 | 6.2 | +1.8 | +9.2 | 10.2 | +1.8 | +13.2 | 13.2 | +2.8 | +17.8 |
| | 4.3 | 0 | +3.5 | 7.2 | 0 | +6.0 | 9.2 | 0 | +8.0 | 13.2 | 0 | +12.0 | 17.8 | 0 | +16.0 |
| 9.85- 11.03 | 2.8 | +1.2 | +4.9 | 4.0 | +2.0 | +7.2 | 7.0 | +2.0 | +10.2 | 10.0 | +2.0 | +13.2 | 15.0 | +3.0 | +20.0 |
| | 4.9 | 0 | +4.0 | 7.2 | 0 | +6.0 | 10.2 | 0 | +9.0 | 13.2 | 0 | +12.0 | 20.0 | 0 | +18.0 |
| 11.03- 12.41 | 2.8 | +1.2 | +4.9 | 5.0 | +2.0 | +8.2 | 7.0 | +2.0 | +10.2 | 12.0 | +2.0 | +15.2 | 17.0 | +3.0 | +22.0 |
| | 4.9 | 0 | +4.0 | 8.2 | 0 | +7.0 | 10.2 | 0 | +9.0 | 15.2 | 0 | +14.0 | 22.0 | 0 | +20.0 |
| 12.41- 13.98 | 3.1 | +1.4 | +5.5 | 5.8 | +2.2 | +9.4 | 7.8 | +2.2 | +11.4 | 13.8 | +2.2 | +17.4 | 18.5 | +3.5 | +24.2 |
| | 5.5 | 0 | +4.5 | 9.4 | 0 | +8.0 | 11.4 | 0 | +10.0 | 17.4 | 0 | +16.0 | 24.2 | 0 | +22.0 |
| 13.98- 15.75 | 3.6 | +1.4 | +6.1 | 5.8 | +2.2 | +9.4 | 9.8 | +2.2 | +13.4 | 15.8 | +2.2 | +19.4 | 21.5 | +3.5 | +27.2 |
| | 6.1 | 0 | +5.0 | 9.4 | 0 | +8.0 | 13.4 | 0 | +12.0 | 19.4 | 0 | +18.0 | 27.2 | 0 | +25.0 |
| 15.75- 17.72 | 4.4 | +1.6 | +7.0 | 6.5 | +2.5 | +10.6 | +9.5 | +2.5 | +13.6 | 17.5 | +2.5 | +21.6 | 24.0 | +4.0 | +30.5 |
| | 7.0 | 0 | +6.0 | 10.6 | 0 | +9.0 | 13.6 | 0 | +12.0 | 21.6 | 0 | +20.0 | 30.5 | 0 | +28.0 |
| 17.72- 19.69 | 4.4 | +1.6 | +7.0 | 7.5 | +2.5 | +11.6 | 11.5 | +2.5 | +15.6 | 19.5 | +2.5 | +23.6 | 26.0 | +4.0 | +32.5 |
| | 7.0 | 0 | +6.0 | 11.6 | 0 | +10.0 | 15.6 | 0 | +14.0 | 23.6 | 0 | +22.0 | 32.5 | 0 | +30.0 |

[a] Pairs of values shown represent minimum and maximum amounts of interference resulting from application of standard tolerance limits.

All data above heavy lines are in accordance with American-British-Canadian (ABC) agreements. Symbols H6, H7, s6, etc., are hole and shaft designations in the ABC system. Limits for sizes above 19.69 inches are not covered by ABC agreements but are given in the ANSI/ASME Standard.

**Table 12. Standard Interference Locational Fits**
*ANSI/ASME B4.1-1967 (R2020)*

| Nominal Size Range, Inches | Class LN 1 | | | Class LN 2 | | | Class LN 3 | | |
|---|---|---|---|---|---|---|---|---|---|
| | Limits of Interference | Standard Limits | | Limits of Interference | Standard Limits | | Limits of Interference | Standard Limits | |
| | | Hole H6 | Shaft n5 | | Hole H7 | Shaft p6 | | Hole H7 | Shaft r6 |
| Over To | Values shown below are given in thousandths of an inch | | | | | | | | |
| 0- 0.12 | 0 | +0.25 | +0.45 | 0 | +0.4 | +0.65 | 0.1 | +0.4 | +0.75 |
| | 0.45 | 0 | +0.25 | 0.65 | 0 | +0.4 | 0.75 | 0 | +0.5 |
| 0.12- 0.24 | 0 | +0.3 | +0.5 | 0 | +0.5 | +0.8 | 0.1 | +0.5 | +0.9 |
| | 0.5 | 0 | +0.3 | 0.8 | 0 | +0.5 | 0.9 | 0 | +0.6 |
| 0.24- 0.40 | 0 | +0.4 | +0.65 | 0 | +0.6 | +1.0 | 0.2 | +0.6 | +1.2 |
| | 0.65 | 0 | +0.4 | 1.0 | 0 | +0.6 | 1.2 | 0 | +0.8 |
| 0.40- 0.71 | 0 | +0.4 | +0.8 | 0 | +0.7 | +1.1 | 0.3 | +0.7 | +1.4 |
| | 0.8 | 0 | +0.4 | 1.1 | 0 | +0.7 | 1.4 | 0 | +1.0 |
| 0.71- 1.19 | 0 | +0.5 | +1.0 | 0 | +0.8 | +1.3 | 0.4 | +0.8 | +1.7 |
| | 1.0 | 0 | +0.5 | 1.3 | 0 | +0.8 | 1.7 | 0 | +1.2 |
| 1.19- 1.97 | 0 | +0.6 | +1.1 | 0 | +1.0 | +1.6 | 0.4 | +1.0 | +2.0 |
| | 1.1 | 0 | +0.6 | 1.6 | 0 | +1.0 | 2.0 | 0 | +1.4 |
| 1.97- 3.15 | 0.1 | +0.7 | +1.3 | 0.2 | +1.2 | +2.1 | 0.4 | +1.2 | +2.3 |
| | 1.3 | 0 | +0.8 | 2.1 | 0 | +1.4 | 2.3 | 0 | +1.6 |
| 3.15- 4.73 | 0.1 | +0.9 | +1.6 | 0.2 | +1.4 | +2.5 | 0.6 | +1.4 | +2.9 |
| | 1.6 | 0 | +1.0 | 2.5 | 0 | +1.6 | 2.9 | 0 | +2.0 |
| 4.73- 7.09 | 0.2 | +1.0 | +1.9 | 0.2 | +1.6 | +2.8 | 0.9 | +1.6 | +3.5 |
| | 1.9 | 0 | +1.2 | 2.8 | 0 | +1.8 | 3.5 | 0 | +2.5 |
| 7.09- 9.85 | 0.2 | +1.2 | +2.2 | 0.2 | +1.8 | +3.2 | 1.2 | +1.8 | +4.2 |
| | 2.2 | 0 | +1.4 | 3.2 | 0 | +2.0 | 4.2 | 0 | +3.0 |
| 9.85- 12.41 | 0.2 | +1.2 | +2.3 | 0.2 | +2.0 | +3.4 | 1.5 | +2.0 | +4.7 |
| | 2.3 | 0 | +1.4 | 3.4 | 0 | +2.2 | 4.7 | 0 | +3.5 |
| 12.41- 15.75 | 0.2 | +1.4 | +2.6 | 0.3 | +2.2 | +3.9 | 2.3 | +2.2 | +5.9 |
| | 2.6 | 0 | +1.6 | 3.9 | 0 | +2.5 | 5.9 | 0 | +4.5 |
| 15.75- 19.69 | 0.2 | +1.6 | +2.8 | 0.3 | +2.5 | +4.4 | 2.5 | +2.5 | +6.6 |
| | 2.8 | 0 | +1.8 | 4.4 | 0 | +2.8 | 6.6 | 0 | +5.0 |

All data in this table are in accordance with American-British-Canadian (ABC) agreements.

Limits for sizes above 19.69 inches are not covered by ABC agreements but are given in the ANSI/ASME Standard.

Symbols H7, p6, etc., are hole and shaft designations in the ABC system.

Tolerance limits given in body of table are added or subtracted to basic size (as indicated by + or – sign) to obtain maximum and minimum sizes of mating parts.

## ANSI/ASME Preferred Metric Limits and Fits

ANSI/ASME B4.2-1978 (R2020) describes the ISO system of metric limits and fits for mating parts as approved for general engineering usage in the United States.

It establishes 1) the designation symbols used to define dimensional limits on drawings, material stock, related tools, gages, etc.; 2) the preferred basic sizes (first and second choices); 3) the preferred tolerance zones (first, second, and third choices); 4) the preferred limits and fits for sizes (first choice only) up to and including 500 millimeters; and 5) the definitions of related terms.

The general terms "hole" and "shaft" can also be taken to refer to the space containing or contained by two parallel faces of any part, such as the width of a slot or the thickness of a key.

**Definitions.—**The most important terms relating to limits and fits are shown in Fig. 1 and are defined as follows:

*Basic Size:* The size to which limits of deviation are assigned. The basic size is the same for both members of a fit. For example, it is designated by the numbers 40 in 40H7.

*Deviation:* The algebraic difference between a size and the corresponding basic size.

*Upper Deviation:* The algebraic difference between the maximum limit of size and the corresponding basic size.

*Lower Deviation:* The algebraic difference between the minimum limit of size and the corresponding basic size.

*Fundamental Deviation:* That one of the two deviations closest to the basic size. For example, it is designated by the letter H in 40H7.

*Tolerance:* The difference between the maximum and minimum size limits on a part.

*Tolerance Zone:* A zone representing the tolerance and its position in relation to the basic size.

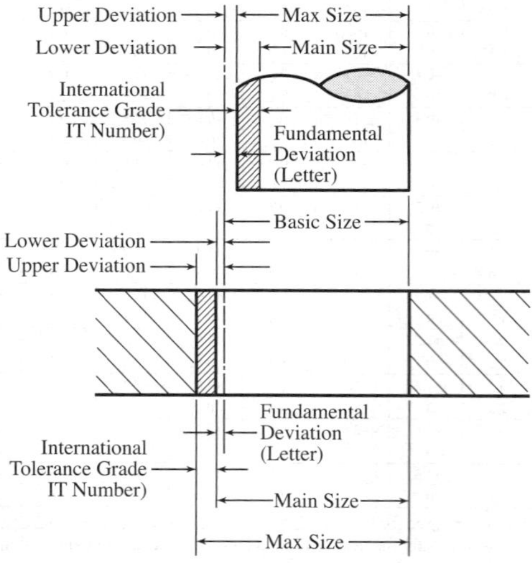


Fig. 1. Illustration of Definitions

*International Tolerance Grade (IT)*: A group of tolerances that vary depending on the basic size, but that provide the same relative level of accuracy within a given grade. For example, International Tolerance Grade is designated by the number 7 in 40H7 or as IT7.

*Hole Basis:* The system of fits where the minimum hole size is basic. The fundamental deviation for a hole basis system is H.

*Shaft Basis:* The system of fits where the maximum shaft size is basic. The fundamental deviation for a shaft basis system is h.

*Clearance Fit:* The relationship between assembled parts when clearance occurs under all tolerance conditions.

*Interference Fit:* The relationship between assembled parts when interference occurs under all tolerance conditions.

*Transition Fit:* The relationship between assembled parts when either a clearance or an interference fit can result, depending on the tolerance conditions of the mating parts.

**Tolerances Designation.**—An "International Tolerance Grade" establishes the magnitude of the tolerance zone or the amount of part size variation allowed for external and internal dimensions alike (see Fig. 1). Tolerances are expressed in grade numbers that are consistent with International Tolerance Grades identified by the prefix IT, such as IT6, IT11, etc. A smaller grade number provides a smaller tolerance zone.

A fundamental deviation establishes the position of the tolerance zone with respect to the basic size (see Fig. 1). Fundamental deviations are expressed by tolerance position letters. Capital letters are used for internal dimensions and lowercase or small letters for external dimensions.

*Symbols:* By combining the IT grade number and the tolerance position letter, the tolerance symbol is established that identifies the actual maximum and minimum limits of the part. The toleranced size is thus defined by the basic size of the part followed by a symbol composed of a letter and a number, such as 40H7, 40f7, etc.

A fit is indicated by the basic size common to both components, followed by a symbol corresponding to each component, the internal part symbol preceding the external part symbol, such as 40H8/f7.

Some methods of designating tolerances on drawings are:

40H8 $\qquad$ 40H8 $\begin{pmatrix}40.039\\40.000\end{pmatrix}$ $\qquad$ $\begin{pmatrix}40.039\\40.000\end{pmatrix}$ 40H8

The values in parentheses indicate reference only.

**Preferred Metric Fits.—**First-choice tolerance zones are used to establish preferred fits in ANSI/ASME B4.2, *Preferred Metric Limits and Fits*, as shown in Fig. 2 and Fig. 3. A complete listing of first-, second-, and third- choice tolerance zones is given in the Standard.

Hole basis fits have a fundamental deviation of H on the hole, and shaft basis fits have a fundamental deviation of *h* on the shaft; they are shown in Fig. 2 for hole basis and Fig. 3 for shaft basis fits. A description of both types of fits, which have the same relative fit condition, is given in Table 1. Normally, the hole basis system is preferred; however, when a common shaft mates with several holes, the shaft basis system should be used.

The hole basis and shaft basis fits shown in the table *Description of Preferred Fits* on page 665 are combined with the first-choice preferred metric sizes from Table 1, page 689, to form Table 2, Table 3, Table 4, and Table 5, in which specific limits as well as the resultant fits are tabulated.

If the required size is not found tabulated in Table 2 through Table 5, then the preferred fit can be calculated from numerical values given in an appendix of ANSI/ASME B4.2-1978 (R2020). It is anticipated that other fit conditions may be necessary to meet special requirements, and a preferred fit can be loosened or tightened simply by selecting a standard tolerance zone as given in the Standard. Information on how to calculate limit dimensions, clearances, and interferences for nonpreferred fits and sizes can be found in an appendix of this Standard.

*Conversion of Fits:* It may sometimes be necessary or desirable to modify the tolerance zone on one or both of two mating parts yet still keep the total tolerance and fit condition the same. Examples of this appear in Table 1 on page 665 when converting from a hole basis fit to a shaft basis fit. The corresponding fits are identical, yet the individual tolerance zones are different.

To convert from one type of fit to another, reverse the fundamental deviations between the shaft and hole, keeping the IT grade the same on each individual part. The examples below represent preferred fits from Table 1 for a 60-mm basic size. These fits have the same maximum clearance (0.520) and the same minimum clearance (0.140).

Hole basis, loose running fit, values from Table 2

Hole 60H11 $\begin{pmatrix}60.190\\60.000\end{pmatrix}$ $\qquad$ Shaft 60c11 $\begin{pmatrix}59.860\\59.670\end{pmatrix}$ $\qquad$ Fit 60H11/c11 $\begin{pmatrix}0.520\\0.140\end{pmatrix}$

Hole basis, loose running fit, values from Table 4

Hole 60C11 $\begin{pmatrix}60.330\\60.140\end{pmatrix}$ $\qquad$ Shaft 60h11 $\begin{pmatrix}60.000\\59.810\end{pmatrix}$ $\qquad$ Fit 60C11/h11 $\begin{pmatrix}0.520\\0.140\end{pmatrix}$

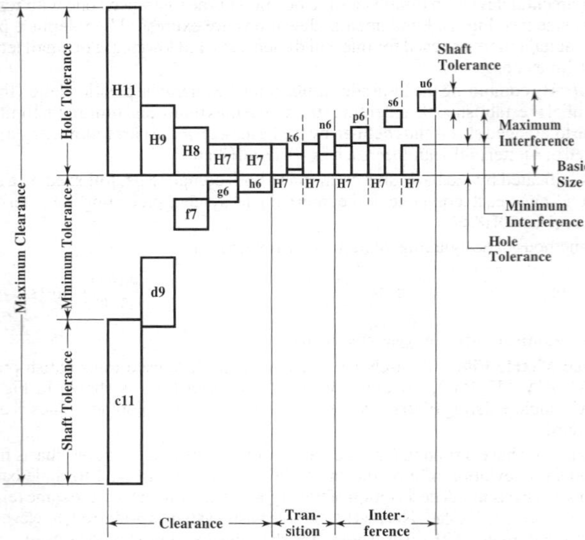


Fig. 2. Preferred Hole Basis Fits

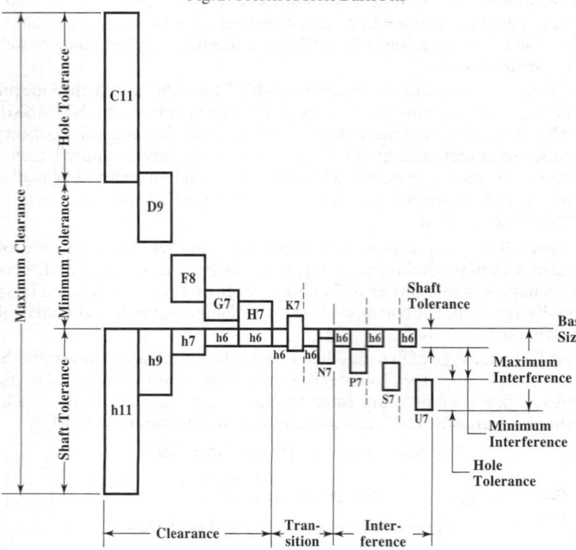


Fig. 3. Preferred Shaft Basis Fits

**Table 1. Description of Preferred Fits**

| | ISO SYMBOL | | DESCRIPTION | |
|---|---|---|---|---|
| | Hole Basis | Shaft Basis | | |
| Clearance Fits | H11/c11 | C11/h11 | *Loose running* fit for wide commercial tolerances or allowances on external members. | ↑ More Clearance |
| | H9/d9 | D9/h9 | *Free running* fit not for use where accuracy is essential, but good for large temperature variations, high running speeds, or heavy journal pressures. | |
| | H8/f7 | F8/h7 | *Close Running* fit for running on accurate machines and for accurate moderate speeds and journal pressures. | |
| | H7/g6 | G7/h6 | *Sliding fit* not intended to run freely but to move and turn freely and locate accurately. | |
| | H7/h6 | H7/h6 | *Locational clearance* fit provides snug fit for locating stationary parts but can be freely assembled and disassembled. | |
| Transition Fits | H7/k6 | K7/h6 | *Locational transition* fit for accurate location, a compromise between clearance and Interference. | |
| | H7/n6 | N7/h6 | *Locational transition* fit for more accurate location where greater interference is permissible. | |
| Interference Fits | H7/p6[a] | P7/h6 | *Locational interference* fit for parts requiring rigidity and alignment with prime accuracy of location but without special bore pressure requirements. | More Interference ↓ |
| | H7/s6 | S7/h6 | *Medium drive* fit for ordinary steel parts or shrink fits on light sections, the tightest fit usable with cast iron. | |
| | H7/u6 | U7/h6 | *Force* fit suitable for parts which can be highly stressed or for shrink fits where the heavy pressing forces required are impractical. | |

[a] Transition fit for basic sizes in range from 0 through 3 mm.

**Table 2. American National Standard Preferred Hole Basis Metric Clearance Fits** *ANSI/ASME B4.2-1978 (R2020)*

| Basic Size[a] | | Loose Running | | | Free Running | | | Close Running | | | Sliding | | | Locational Clearance | | |
|---|---|---|---|---|---|---|---|---|---|---|---|---|---|---|---|---|
| | | Hole H11 | Shaft c11 | Fit[b] | Hole H9 | Shaft d9 | Fit[b] | Hole H8 | Shaft f7 | Fit[b] | Hole H7 | Shaft g6 | Fit[b] | Hole H7 | Shaft h6 | Fit[b] |
| 1 | Max | 1.060 | 0.940 | 0.180 | 1.025 | 0.980 | 0.070 | 1.014 | 0.994 | 0.030 | 1.010 | 0.998 | 0.018 | 1.010 | 1.000 | 0.016 |
| | Min | 1.000 | 0.880 | 0.060 | 1.000 | 0.955 | 0.020 | 1.000 | 0.984 | 0.006 | 1.000 | 0.992 | 0.002 | 1.000 | 0.994 | 0.000 |
| 1.2 | Max | 1.260 | 1.140 | 0.180 | 1.225 | 1.180 | 0.070 | 1.214 | 1.194 | 0.030 | 1.210 | 1.198 | 0.018 | 1.210 | 1.200 | 0.016 |
| | Min | 1.200 | 1.080 | 0.060 | 1.200 | 1.155 | 0.020 | 1.200 | 1.184 | 0.006 | 1.200 | 1.192 | 0.002 | 1.200 | 1.194 | 0.000 |
| 1.6 | Max | 1.660 | 1.540 | 0.180 | 1.625 | 1.580 | 0.070 | 1.614 | 1.594 | 0.030 | 1.610 | 1.598 | 0.018 | 1.610 | 1.600 | 0.016 |
| | Min | 1.600 | 1.480 | 0.060 | 1.600 | 1.555 | 0.020 | 1.600 | 1.584 | 0.006 | 1.600 | 1.592 | 0.002 | 1.600 | 1.594 | 0.000 |
| 2 | Max | 2.060 | 1.940 | 0.180 | 2.025 | 1.980 | 0.070 | 2.014 | 1.994 | 0.030 | 2.010 | 1.998 | 0.018 | 2.010 | 2.000 | 0.016 |
| | Min | 2.000 | 1.880 | 0.060 | 2.000 | 1.955 | 0.020 | 2.000 | 1.984 | 0.006 | 2.000 | 1.992 | 0.002 | 2.000 | 1.994 | 0.000 |
| 2.5 | Max | 2.560 | 2.440 | 0.180 | 2.525 | 2.480 | 0.070 | 2.514 | 2.494 | 0.030 | 2.510 | 2.498 | 0.018 | 2.510 | 2.500 | 0.016 |
| | Min | 2.500 | 2.380 | 0.060 | 2.500 | 2.455 | 0.020 | 2.500 | 2.484 | 0.006 | 2.500 | 2.492 | 0.002 | 2.500 | 2.494 | 0.000 |
| 3 | Max | 3.060 | 2.940 | 0.180 | 3.025 | 2.980 | 0.070 | 3.014 | 2.994 | 0.030 | 3.010 | 2.998 | 0.018 | 3.010 | 3.000 | 0.016 |
| | Min | 3.000 | 2.880 | 0.060 | 3.000 | 2.955 | 0.020 | 3.000 | 2.984 | 0.006 | 3.000 | 2.992 | 0.002 | 3.000 | 2.994 | 0.000 |
| 4 | Max | 4.075 | 3.930 | 0.220 | 4.030 | 3.970 | 0.090 | 4.018 | 3.990 | 0.040 | 4.012 | 3.996 | 0.024 | 4.012 | 4.000 | 0.020 |
| | Min | 4.000 | 3.855 | 0.070 | 4.000 | 3.940 | 0.030 | 4.000 | 3.978 | 0.010 | 4.000 | 3.988 | 0.004 | 4.000 | 3.992 | 0.000 |
| 5 | Max | 5.075 | 4.930 | 0.220 | 5.030 | 4.970 | 0.090 | 5.018 | 4.990 | 0.040 | 5.012 | 4.996 | 0.024 | 5.012 | 5.000 | 0.020 |
| | Min | 5.000 | 4.855 | 0.070 | 5.000 | 4.940 | 0.030 | 5.000 | 4.978 | 0.010 | 5.000 | 4.988 | 0.004 | 5.000 | 4.992 | 0.000 |
| 6 | Max | 6.075 | 5.930 | 0.220 | 6.030 | 5.970 | 0.090 | 6.018 | 5.990 | 0.040 | 6.012 | 5.996 | 0.024 | 6.012 | 6.000 | 0.020 |
| | Min | 6.000 | 5.855 | 0.070 | 6.000 | 5.940 | 0.030 | 6.000 | 5.978 | 0.010 | 6.000 | 5.988 | 0.004 | 6.000 | 5.992 | 0.000 |
| 8 | Max | 8.090 | 7.920 | 0.260 | 8.036 | 7.960 | 0.112 | 8.022 | 7.987 | 0.050 | 8.015 | 7.995 | 0.029 | 8.015 | 8.000 | 0.024 |
| | Min | 8.000 | 7.830 | 0.080 | 8.000 | 7.924 | 0.040 | 8.000 | 7.972 | 0.013 | 8.000 | 7.986 | 0.005 | 8.000 | 7.991 | 0.000 |
| 10 | Max | 10.090 | 9.920 | 0.260 | 10.036 | 9.960 | 0.112 | 10.022 | 9.987 | 0.050 | 10.015 | 9.995 | 0.029 | 10.015 | 10.000 | 0.024 |
| | Min | 10.000 | 9.830 | 0.080 | 10.000 | 9.924 | 0.040 | 10.000 | 9.972 | 0.013 | 10.000 | 9.986 | 0.005 | 10.000 | 9.991 | 0.000 |
| 12 | Max | 12.110 | 11.905 | 0.315 | 12.043 | 11.956 | 0.136 | 12.027 | 11.984 | 0.061 | 12.018 | 11.994 | 0.035 | 12.018 | 12.000 | 0.029 |
| | Min | 12.000 | 11.795 | 0.095 | 12.000 | 11.907 | 0.050 | 12.000 | 11.966 | 0.016 | 12.000 | 11.983 | 0.006 | 12.000 | 11.989 | 0.000 |
| 16 | Max | 16.110 | 15.905 | 0.315 | 16.043 | 15.950 | 0.136 | 16.027 | 15.984 | 0.061 | 16.018 | 15.994 | 0.035 | 16.018 | 16.000 | 0.029 |
| | Min | 16.000 | 15.795 | 0.095 | 16.000 | 15.907 | 0.050 | 16.000 | 15.966 | 0.016 | 16.000 | 15.983 | 0.006 | 16.000 | 15.989 | 0.000 |
| 20 | Max | 20.130 | 19.890 | 0.370 | 20.052 | 19.935 | 0.169 | 20.033 | 19.980 | 0.074 | 20.021 | 19.993 | 0.041 | 20.021 | 20.000 | 0.034 |
| | Min | 20.000 | 19.760 | 0.110 | 20.000 | 19.883 | 0.065 | 20.000 | 19.959 | 0.020 | 20.000 | 19.980 | 0.007 | 20.000 | 19.987 | 0.000 |
| 25 | Max | 25.130 | 24.890 | 0.370 | 25.052 | 24.935 | 0.169 | 25.033 | 24.980 | 0.074 | 25.021 | 24.993 | 0.041 | 25.021 | 25.000 | 0.034 |
| | Min | 25.000 | 24.760 | 0.110 | 25.000 | 24.883 | 0.065 | 25.000 | 24.959 | 0.020 | 25.000 | 24.980 | 0.007 | 25.000 | 24.987 | 0.000 |

**Table 2.** *(Continued)* **American National Standard Preferred Hole Basis Metric Clearance Fits** *ANSI/ASME B4.2-1978 (R2020)*

| Basic Size[a] | | Loose Running | | | Free Running | | | Close Running | | | Sliding | | | Locational Clearance | | |
|---|---|---|---|---|---|---|---|---|---|---|---|---|---|---|---|---|
| | | Hole H11 | Shaft c11 | Fit[b] | Hole H9 | Shaft d9 | Fit[b] | Hole H8 | Shaft f7 | Fit[b] | Hole H7 | Shaft g6 | Fit[b] | Hole H7 | Shaft h6 | Fit[b] |
| 30 | Max | 30.130 | 29.890 | 0.370 | 30.052 | 29.935 | 0.169 | 30.033 | 29.980 | 0.074 | 30.021 | 29.993 | 0.041 | 30.021 | 30.000 | 0.034 |
| | Min | 30.000 | 29.760 | 0.110 | 30.000 | 29.883 | 0.065 | 30.000 | 29.959 | 0.020 | 30.000 | 29.980 | 0.007 | 30.000 | 29.987 | 0.000 |
| 40 | Max | 40.160 | 39.880 | 0.440 | 40.062 | 39.920 | 0.204 | 40.039 | 39.975 | 0.089 | 40.025 | 39.991 | 0.050 | 40.025 | 40.000 | 0.041 |
| | Min | 40.000 | 39.720 | 0.120 | 40.000 | 39.858 | 0.080 | 40.000 | 39.950 | 0.025 | 40.000 | 39.975 | 0.009 | 40.000 | 39.984 | 0.000 |
| 50 | Max | 50.160 | 49.870 | 0.450 | 50.062 | 49.920 | 0.204 | 50.039 | 49.975 | 0.089 | 50.025 | 49.991 | 0.050 | 50.025 | 50.000 | 0.041 |
| | Min | 50.000 | 49.710 | 0.130 | 50.000 | 49.858 | 0.080 | 50.000 | 49.950 | 0.025 | 50.000 | 49.975 | 0.009 | 50.000 | 49.984 | 0.000 |
| 60 | Max | 60.190 | 59.860 | 0.520 | 60.074 | 59.900 | 0.248 | 60.046 | 59.970 | 0.106 | 60.030 | 59.990 | 0.059 | 60.030 | 60.000 | 0.049 |
| | Min | 60.000 | 59.670 | 0.140 | 60.000 | 59.826 | 0.100 | 60.000 | 59.940 | 0.030 | 60.000 | 59.971 | 0.010 | 60.000 | 59.981 | 0.000 |
| 80 | Max | 80.190 | 79.850 | 0.530 | 80.074 | 79.900 | 0.248 | 80.046 | 79.970 | 0.106 | 80.030 | 79.990 | 0.059 | 80.030 | 80.000 | 0.049 |
| | Min | 80.000 | 79.660 | 0.150 | 80.000 | 79.826 | 0.100 | 80.000 | 79.940 | 0.030 | 80.000 | 79.971 | 0.010 | 80.000 | 79.981 | 0.000 |
| 100 | Max | 100.220 | 99.830 | 0.610 | 100.087 | 99.880 | 0.294 | 100.054 | 99.964 | 0.125 | 100.035 | 99.988 | 0.069 | 100.035 | 100.000 | 0.057 |
| | Min | 100.000 | 99.610 | 0.170 | 100.000 | 99.793 | 0.120 | 100.000 | 99.929 | 0.036 | 100.000 | 99.966 | 0.012 | 100.000 | 99.978 | 0.000 |
| 120 | Max | 120.220 | 119.820 | 0.620 | 120.087 | 119.880 | 0.294 | 120.054 | 119.964 | 0.125 | 120.035 | 119.988 | 0.069 | 120.035 | 120.000 | 0.057 |
| | Min | 120.000 | 119.600 | 0.180 | 120.000 | 119.793 | 0.120 | 120.000 | 119.929 | 0.036 | 120.000 | 119.966 | 0.012 | 120.000 | 119.978 | 0.000 |
| 160 | Max | 160.250 | 159.790 | 0.710 | 160.100 | 159.855 | 0.345 | 160.063 | 159.957 | 0.146 | 160.040 | 159.986 | 0.079 | 160.040 | 160.000 | 0.065 |
| | Min | 160.000 | 159.540 | 0.210 | 160.000 | 159.755 | 0.145 | 160.000 | 159.917 | 0.043 | 160.000 | 159.961 | 0.014 | 160.000 | 159.975 | 0.000 |
| 200 | Max | 200.290 | 199.760 | 0.820 | 200.115 | 199.830 | 0.400 | 200.072 | 199.950 | 0.168 | 200.046 | 199.985 | 0.090 | 200.046 | 200.000 | 0.075 |
| | Min | 200.000 | 199.470 | 0.240 | 200.000 | 199.715 | 0.170 | 200.000 | 199.904 | 0.050 | 200.000 | 199.956 | 0.015 | 200.000 | 199.971 | 0.000 |
| 250 | Max | 250.290 | 249.720 | 0.860 | 250.115 | 249.830 | 0.400 | 250.072 | 249.950 | 0.168 | 250.046 | 249.985 | 0.090 | 250.046 | 250.000 | 0.075 |
| | Min | 250.000 | 249.430 | 0.280 | 250.000 | 249.715 | 0.170 | 250.000 | 249.904 | 0.050 | 250.000 | 249.956 | 0.015 | 250.000 | 249.971 | 0.000 |
| 300 | Max | 300.320 | 299.670 | 0.970 | 300.130 | 299.810 | 0.450 | 300.081 | 299.944 | 0.189 | 300.052 | 299.983 | 0.101 | 300.052 | 300.000 | 0.084 |
| | Min | 300.000 | 299.350 | 0.330 | 300.000 | 299.680 | 0.190 | 300.000 | 299.892 | 0.056 | 300.000 | 299.951 | 0.017 | 300.000 | 299.968 | 0.000 |
| 400 | Max | 400.360 | 399.600 | 1.120 | 400.140 | 399.790 | 0.490 | 400.089 | 399.938 | 0.208 | 400.057 | 399.982 | 0.111 | 400.057 | 400.000 | 0.093 |
| | Min | 400.000 | 399.240 | 0.400 | 400.000 | 399.650 | 0.210 | 400.000 | 399.881 | 0.062 | 400.000 | 399.946 | 0.018 | 400.000 | 399.964 | 0.000 |
| 500 | Max | 500.400 | 499.520 | 1.280 | 500.155 | 499.770 | 0.540 | 500.097 | 499.932 | 0.228 | 500.063 | 499.980 | 0.123 | 500.063 | 500.000 | 0.103 |
| | Min | 500.000 | 499.120 | 0.480 | 500.000 | 499.615 | 0.230 | 500.000 | 499.869 | 0.068 | 500.000 | 499.940 | 0.020 | 500.000 | 499.960 | 0.000 |

[a] The sizes shown are first-choice basic sizes (see Table 1, page 689). Preferred fits for other sizes can be calculated from data given in ANSI/ASME B4.2-1978 (R2020).

[b] All fits shown in this table have clearance.

All dimensions are in millimeters.

**Table 3. American National Standard Preferred Hole Basis Metric Transition and Interference Fits** *ANSI/ASME B4.2-1978 (R2020)*

| Basic Size[a] | | Locational Transition | | | Locational Transition | | | Locational Interference | | | Medium Drive | | | Force | | |
|---|---|---|---|---|---|---|---|---|---|---|---|---|---|---|---|---|
| | | Hole H7 | Shaft k6 | Fit[b] | Hole H7 | Shaft n6 | Fit[b] | Hole H7 | Shaft p6 | Fit[b] | Hole H7 | Shaft s6 | Fit[b] | Hole H7 | Shaft u6 | Fit[b] |
| 1 | Max | 1.010 | 1.006 | +0.010 | 1.010 | 1.010 | +0.006 | 1.010 | 1.012 | +0.004 | 1.010 | 1.020 | −0.004 | 1.010 | 1.024 | −0.008 |
| | Min | 1.000 | 1.000 | −0.006 | 1.000 | 1.004 | −0.010 | 1.000 | 1.006 | −0.012 | 1.000 | 1.014 | −0.020 | 1.000 | 1.018 | −0.024 |
| 1.2 | Max | 1.210 | 1.206 | +0.010 | 1.210 | 1.210 | +0.006 | 1.210 | 1.212 | +0.004 | 1.210 | 1.220 | −0.004 | 1.210 | 1.224 | −0.008 |
| | Min | 1.200 | 1.200 | −0.006 | 1.200 | 1.204 | −0.010 | 1.200 | 1.206 | −0.012 | 1.200 | 1.214 | −0.020 | 1.200 | 1.218 | −0.024 |
| 1.6 | Max | 1.610 | 1.606 | +0.010 | 1.610 | 1.610 | +0.006 | 1.610 | 1.612 | +0.004 | 1.610 | 1.620 | −0.004 | 1.610 | 1.624 | −0.008 |
| | Min | 1.600 | 1.600 | −0.006 | 1.600 | 1.604 | −0.010 | 1.600 | 1.606 | −0.012 | 1.600 | 1.614 | −0.020 | 1.600 | 1.618 | −0.024 |
| 2 | Max | 2.010 | 2.006 | +0.010 | 2.010 | 2.010 | +0.006 | 2.010 | 2.012 | +0.004 | 2.010 | 2.020 | −0.004 | 2.010 | 2.024 | −0.008 |
| | Min | 2.000 | 2.000 | −0.006 | 2.000 | 2.004 | −0.010 | 2.000 | 2.006 | −0.012 | 2.000 | 2.014 | −0.020 | 2.000 | 2.018 | −0.024 |
| 2.5 | Max | 2.510 | 2.506 | +0.010 | 2.510 | 2.510 | +0.006 | 2.510 | 2.512 | +0.004 | 2.510 | 2.520 | −0.004 | 2.510 | 2.524 | −0.008 |
| | Min | 2.500 | 2.500 | −0.006 | 2.500 | 2.504 | −0.010 | 2.500 | 2.506 | −0.012 | 2.500 | 2.514 | −0.020 | 2.500 | 2.518 | −0.024 |
| 3 | Max | 3.010 | 3.006 | +0.010 | 3.010 | 3.010 | +0.006 | 3.010 | 3.012 | +0.004 | 3.010 | 3.020 | −0.004 | 3.010 | 3.024 | −0.008 |
| | Min | 3.000 | 3.000 | −0.006 | 3.000 | 3.004 | −0.010 | 3.000 | 3.006 | −0.012 | 3.000 | 3.014 | −0.020 | 3.000 | 3.018 | −0.024 |
| 4 | Max | 4.012 | 4.009 | +0.011 | 4.012 | 4.016 | +0.004 | 4.012 | 4.020 | 0.000 | 4.012 | 4.027 | −0.007 | 4.012 | 4.031 | −0.011 |
| | Min | 4.000 | 4.001 | −0.009 | 4.000 | 4.008 | −0.016 | 4.000 | 4.012 | −0.020 | 4.000 | 4.019 | −0.027 | 4.000 | 4.023 | −0.031 |
| 5 | Max | 5.012 | 5.009 | +0.011 | 5.012 | 5.016 | +0.004 | 5.012 | 5.020 | 0.000 | 5.012 | 5.027 | −0.007 | 5.012 | 5.031 | −0.011 |
| | Min | 5.000 | 5.001 | −0.009 | 5.000 | 5.008 | −0.016 | 5.000 | 5.012 | −0.020 | 5.000 | 5.019 | −0.027 | 5.000 | 5.023 | −0.031 |
| 6 | Max | 6.012 | 6.009 | +0.011 | 6.012 | 6.016 | +0.004 | 6.012 | 6.020 | 0.000 | 6.012 | 6.027 | −0.007 | 6.012 | 6.031 | −0.011 |
| | Min | 6.000 | 6.001 | −0.009 | 6.000 | 6.008 | −0.016 | 6.000 | 6.012 | −0.020 | 6.000 | 6.019 | −0.027 | 6.000 | 6.023 | −0.031 |
| 8 | Max | 8.015 | 8.010 | +0.014 | 8.015 | 8.019 | +0.005 | 8.015 | 8.024 | 0.000 | 8.015 | 8.032 | −0.008 | 8.015 | 8.037 | −0.013 |
| | Min | 8.000 | 8.001 | −0.010 | 8.000 | 8.010 | −0.019 | 8.000 | 8.015 | −0.024 | 8.000 | 8.023 | −0.032 | 8.000 | 8.028 | −0.037 |
| 10 | Max | 10.015 | 10.010 | +0.014 | 10.015 | 10.019 | +0.005 | 10.015 | 10.024 | 0.000 | 10.015 | 10.032 | −0.008 | 10.015 | 10.034 | −0.013 |
| | Min | 10.000 | 10.001 | −0.010 | 10.000 | 10.010 | −0.019 | 10.000 | 10.015 | −0.024 | 10.000 | 10.023 | −0.032 | 10.000 | 10.028 | −0.037 |
| 12 | Max | 12.018 | 12.012 | +0.017 | 12.018 | 12.023 | +0.006 | 12.018 | 12.029 | 0.000 | 12.018 | 12.039 | −0.010 | 12.018 | 12.044 | −0.015 |
| | Min | 12.000 | 12.001 | −0.012 | 12.000 | 12.012 | −0.023 | 12.000 | 12.018 | −0.029 | 12.000 | 12.028 | −0.039 | 12.000 | 12.033 | −0.044 |
| 16 | Max | 16.018 | 16.012 | +0.017 | 16.018 | 16.023 | +0.006 | 16.018 | 16.029 | 0.000 | 16.018 | 16.039 | −0.010 | 16.018 | 16.044 | −0.015 |
| | Min | 16.000 | 16.001 | −0.012 | 16.000 | 16.012 | −0.023 | 16.000 | 16.018 | −0.029 | 16.000 | 16.028 | −0.039 | 16.000 | 16.033 | −0.044 |
| 20 | Max | 20.021 | 20.015 | +0.019 | 20.021 | 20.028 | +0.006 | 20.021 | 20.035 | −0.001 | 20.021 | 20.048 | −0.014 | 20.021 | 20.054 | −0.020 |
| | Min | 20.000 | 20.002 | −0.015 | 20.000 | 20.015 | −0.028 | 20.000 | 20.022 | −0.035 | 20.000 | 20.035 | −0.048 | 20.000 | 20.041 | −0.054 |
| 25 | Max | 25.021 | 25.015 | +0.019 | 25.021 | 25.028 | +0.006 | 25.021 | 25.035 | −0.001 | 25.021 | 25.048 | −0.014 | 25.021 | 25.061 | −0.027 |
| | Min | 25.000 | 25.002 | −0.015 | 25.000 | 25.015 | −0.028 | 25.000 | 25.022 | −0.035 | 25.000 | 25.035 | −0.048 | 25.000 | 25.048 | −0.061 |

**Table 3.** *(Continued)* **American National Standard Preferred Hole Basis Metric Transition and Interference Fits** *ANSI/ASME B4.2-1978 (R2020)*

| Basic Size[a] | | Locational Transition | | | Locational Transition | | | Locational Interference | | | Medium Drive | | | Force | | |
|---|---|---|---|---|---|---|---|---|---|---|---|---|---|---|---|---|
| | | Hole H7 | Shaft k6 | Fit[b] | Hole H7 | Shaft n6 | Fit[b] | Hole H7 | Shaft p6 | Fit[b] | Hole H7 | Shaft s6 | Fit[b] | Hole H7 | Shaft u6 | Fit[b] |
| 30 | Max | 30.021 | 30.015 | +0.019 | 30.021 | 30.028 | +0.006 | 30.021 | 30.035 | –0.001 | 30.021 | 30.048 | –0.014 | 30.021 | 30.061 | –0.027 |
| | Min | 30.000 | 30.002 | –0.015 | 30.000 | 30.015 | –0.028 | 30.000 | 30.022 | –0.035 | 30.000 | 30.035 | –0.048 | 30.000 | 30.048 | –0.061 |
| 40 | Max | 40.025 | 40.018 | +0.023 | 40.025 | 40.033 | +0.008 | 40.025 | 40.042 | –0.001 | 40.025 | 40.059 | –0.018 | 40.025 | 40.076 | –0.035 |
| | Min | 40.000 | 40.002 | –0.018 | 40.000 | 40.017 | –0.033 | 40.000 | 40.026 | –0.042 | 40.000 | 40.043 | –0.059 | 40.000 | 40.060 | –0.076 |
| 50 | Max | 50.025 | 50.018 | +0.023 | 50.025 | 50.033 | +0.008 | 50.025 | 50.042 | –0.001 | 50.025 | 50.059 | –0.018 | 50.025 | 50.086 | –0.045 |
| | Min | 50.000 | 50.002 | –0.018 | 50.000 | 50.017 | –0.033 | 50.000 | 50.026 | –0.042 | 50.000 | 50.043 | –0.059 | 50.000 | 50.070 | –0.086 |
| 60 | Max | 60.030 | 60.021 | +0.028 | 60.030 | 60.039 | +0.010 | 60.030 | 60.051 | –0.002 | 60.030 | 60.072 | –0.023 | 60.030 | 60.106 | –0.057 |
| | Min | 60.000 | 60.002 | –0.021 | 60.000 | 60.020 | –0.039 | 60.000 | 60.032 | –0.051 | 60.000 | 60.053 | –0.072 | 60.000 | 60.087 | –0.106 |
| 80 | Max | 80.030 | 80.021 | +0.028 | 80.030 | 80.039 | +0.010 | 80.030 | 80.051 | –0.002 | 80.030 | 80.078 | –0.029 | 80.030 | 80.121 | –0.072 |
| | Min | 80.000 | 80.002 | –0.021 | 80.000 | 80.020 | –0.039 | 80.000 | 80.032 | –0.051 | 80.000 | 80.059 | –0.078 | 80.000 | 80.102 | –0.121 |
| 100 | Max | 100.035 | 100.025 | +0.032 | 100.035 | 100.045 | +0.012 | 100.035 | 100.059 | –0.002 | 100.035 | 100.093 | –0.036 | 100.035 | 100.146 | –0.089 |
| | Min | 100.000 | 100.003 | –0.025 | 100.000 | 100.023 | –0.045 | 100.000 | 100.037 | –0.059 | 100.000 | 100.071 | –0.093 | 100.000 | 100.124 | –0.146 |
| 120 | Max | 120.035 | 120.025 | +0.032 | 120.035 | 120.045 | +0.012 | 120.035 | 120.059 | –0.002 | 120.035 | 120.101 | –0.044 | 120.035 | 120.166 | –0.109 |
| | Min | 120.000 | 120.003 | –0.025 | 120.000 | 120.023 | –0.045 | 120.000 | 120.037 | –0.059 | 120.000 | 120.079 | –0.101 | 120.000 | 120.144 | –0.166 |
| 160 | Max | 160.040 | 160.028 | +0.037 | 160.040 | 160.052 | +0.013 | 160.040 | 160.068 | –0.003 | 160.040 | 160.125 | –0.060 | 160.040 | 160.215 | –0.150 |
| | Min | 160.000 | 160.003 | –0.028 | 160.000 | 160.027 | –0.052 | 160.000 | 160.043 | –0.068 | 160.000 | 160.100 | –0.125 | 160.000 | 160.190 | –0.215 |
| 200 | Max | 200.046 | 200.033 | +0.042 | 200.046 | 200.060 | +0.015 | 200.046 | 200.079 | –0.004 | 200.046 | 200.151 | –0.076 | 200.046 | 200.265 | –0.190 |
| | Min | 200.000 | 200.004 | –0.033 | 200.000 | 200.031 | –0.060 | 200.000 | 200.050 | –0.079 | 200.000 | 200.122 | –0.151 | 200.000 | 200.236 | –0.265 |
| 250 | Max | 250.046 | 250.033 | +0.042 | 250.046 | 250.060 | +0.015 | 250.046 | 250.079 | –0.004 | 250.046 | 250.169 | –0.094 | 250.046 | 250.313 | –0.238 |
| | Min | 250.000 | 250.004 | –0.033 | 250.000 | 250.031 | –0.060 | 250.000 | 250.050 | –0.079 | 250.000 | 250.140 | –0.169 | 250.000 | 250.284 | –0.313 |
| 300 | Max | 300.052 | 300.036 | +0.048 | 300.052 | 300.066 | +0.018 | 300.052 | 300.088 | –0.004 | 300.052 | 300.202 | –0.118 | 300.052 | 300.382 | –0.298 |
| | Min | 300.000 | 300.004 | –0.036 | 300.000 | 300.034 | –0.066 | 300.000 | 300.056 | –0.088 | 300.000 | 300.170 | –0.202 | 300.000 | 300.350 | –0.382 |
| 400 | Max | 400.057 | 400.040 | +0.053 | 400.057 | 400.073 | +0.020 | 400.057 | 400.098 | –0.005 | 400.057 | 400.244 | –0.151 | 400.057 | 400.471 | –0.378 |
| | Min | 400.000 | 400.004 | –0.040 | 400.000 | 400.037 | –0.073 | 400.000 | 400.062 | –0.098 | 400.000 | 400.208 | –0.244 | 400.000 | 400.435 | –0.471 |
| 500 | Max | 500.063 | 500.045 | +0.058 | 500.063 | 500.080 | +0.023 | 500.063 | 500.108 | –0.005 | 500.063 | 500.292 | –0.189 | 500.063 | 500.580 | –0.477 |
| | Min | 500.000 | 500.005 | –0.045 | 500.000 | 500.040 | –0.080 | 500.000 | 500.068 | –0.108 | 500.000 | 500.252 | –0.292 | 500.000 | 500.540 | –0.580 |

[a] The sizes shown are first-choice basic sizes (see Table 1, page 689). Preferred fits for other sizes can be calculated from data given in ANSI/ASME B4.2-1978 (R2020).

[b] A plus sign indicates clearance; a minus sign indicates interference.

All dimensions are in millimeters.

**Table 4. American National Standard Preferred Shaft Basis Metric Clearance Fits** *ANSI/ASME B4.2-1978 (R2020)*

| Basic Size[a] | | Loose Running | | | Free Running | | | Close Running | | | Sliding | | | Locational Clearance | | |
|---|---|---|---|---|---|---|---|---|---|---|---|---|---|---|---|---|
| | | Hole C11 | Shaft h11 | Fit[b] | Hole D9 | Shaft h9 | Fit[b] | Hole F8 | Shaft h7 | Fit[b] | Hole G7 | Shaft h6 | Fit[b] | Hole H7 | Shaft h6 | Fit[b] |
| 1 | Max | 1.120 | 1.000 | 0.180 | 1.045 | 1.000 | 0.070 | 1.020 | 1.000 | 0.030 | 1.012 | 1.000 | 0.018 | 1.010 | 1.000 | 0.016 |
| | Min | 1.060 | 0.940 | 0.060 | 1.020 | 0.975 | 0.020 | 1.006 | 0.990 | 0.006 | 1.002 | 0.994 | 0.002 | 1.000 | 0.994 | 0.000 |
| 1.2 | Max | 1.320 | 1.200 | 0.180 | 1.245 | 1.200 | 0.070 | 1.220 | 1.200 | 0.030 | 1.212 | 1.200 | 0.018 | 1.210 | 1.200 | 0.016 |
| | Min | 1.260 | 1.140 | 0.060 | 1.220 | 1.175 | 0.020 | 1.206 | 1.190 | 0.006 | 1.202 | 1.194 | 0.002 | 1.200 | 1.194 | 0.000 |
| 1.6 | Max | 1.720 | 1.600 | 0.180 | 1.645 | 1.600 | 0.070 | 1.620 | 1.600 | 0.030 | 1.612 | 1.600 | 0.018 | 1.610 | 1.600 | 0.016 |
| | Min | 1.660 | 1.540 | 0.060 | 1.620 | 1.575 | 0.020 | 1.606 | 1.590 | 0.006 | 1.602 | 1.594 | 0.002 | 1.600 | 1.594 | 0.000 |
| 2 | Max | 2.120 | 2.000 | 0.180 | 2.045 | 2.000 | 0.070 | 2.020 | 2.000 | 0.030 | 2.012 | 2.000 | 0.018 | 2.010 | 2.000 | 0.016 |
| | Min | 2.060 | 1.940 | 0.060 | 2.020 | 1.975 | 0.020 | 2.006 | 1.990 | 0.006 | 2.002 | 1.994 | 0.002 | 2.000 | 1.994 | 0.000 |
| 2.5 | Max | 2.620 | 2.500 | 0.180 | 2.545 | 2.500 | 0.070 | 2.520 | 2.500 | 0.030 | 2.512 | 2.500 | 0.018 | 2.510 | 2.500 | 0.016 |
| | Min | 2.560 | 2.440 | 0.060 | 2.520 | 2.475 | 0.020 | 2.506 | 2.490 | 0.006 | 2.502 | 2.494 | 0.002 | 2.500 | 2.494 | 0.000 |
| 3 | Max | 3.120 | 3.000 | 0.180 | 3.045 | 3.000 | 0.070 | 3.020 | 3.000 | 0.030 | 3.012 | 3.000 | 0.018 | 3.010 | 3.000 | 0.016 |
| | Min | 3.060 | 2.940 | 0.060 | 3.020 | 2.975 | 0.020 | 3.006 | 2.990 | 0.006 | 3.002 | 2.994 | 0.002 | 3.000 | 2.994 | 0.000 |
| 4 | Max | 4.145 | 4.000 | 0.220 | 4.060 | 4.000 | 0.090 | 4.028 | 4.000 | 0.040 | 4.016 | 4.000 | 0.024 | 4.012 | 4.000 | 0.020 |
| | Min | 4.070 | 3.925 | 0.070 | 4.030 | 3.970 | 0.030 | 4.010 | 3.988 | 0.010 | 4.004 | 3.992 | 0.004 | 4.000 | 3.992 | 0.000 |
| 5 | Max | 5.145 | 5.000 | 0.220 | 5.060 | 5.000 | 0.090 | 5.028 | 5.000 | 0.040 | 5.016 | 5.000 | 0.024 | 5.012 | 5.000 | 0.020 |
| | Min | 5.070 | 4.925 | 0.070 | 5.030 | 4.970 | 0.030 | 5.010 | 4.988 | 0.010 | 5.004 | 4.992 | 0.004 | 5.000 | 4.992 | 0.000 |
| 6 | Max | 6.145 | 6.000 | 0.220 | 6.060 | 6.000 | 0.090 | 6.028 | 6.000 | 0.040 | 6.016 | 6.000 | 0.024 | 6.012 | 6.000 | 0.020 |
| | Min | 6.070 | 5.925 | 0.070 | 6.030 | 5.970 | 0.030 | 6.010 | 5.988 | 0.010 | 6.004 | 5.992 | 0.004 | 6.000 | 5.992 | 0.000 |
| 8 | Max | 8.170 | 8.000 | 0.260 | 8.076 | 8.000 | 0.112 | 8.035 | 8.000 | 0.050 | 8.020 | 8.000 | 0.029 | 8.015 | 8.000 | 0.024 |
| | Min | 8.080 | 7.910 | 0.080 | 8.040 | 7.964 | 0.040 | 8.013 | 7.985 | 0.013 | 8.005 | 7.991 | 0.005 | 8.000 | 7.991 | 0.000 |
| 10 | Max | 10.170 | 10.000 | 0.260 | 10.076 | 10.000 | 0.112 | 10.035 | 10.000 | 0.050 | 10.020 | 10.000 | 0.029 | 10.015 | 10.000 | 0.024 |
| | Min | 10.080 | 9.910 | 0.080 | 10.040 | 9.964 | 0.040 | 10.013 | 9.985 | 0.013 | 10.005 | 9.991 | 0.005 | 10.000 | 9.991 | 0.000 |
| 12 | Max | 12.205 | 12.000 | 0.315 | 12.093 | 12.000 | 0.136 | 12.043 | 12.000 | 0.061 | 12.024 | 12.000 | 0.035 | 12.018 | 12.000 | 0.029 |
| | Min | 12.095 | 11.890 | 0.095 | 12.050 | 11.957 | 0.050 | 12.016 | 11.982 | 0.016 | 12.006 | 11.989 | 0.006 | 12.000 | 11.989 | 0.000 |
| 16 | Max | 16.205 | 16.000 | 0.315 | 16.093 | 16.000 | 0.136 | 16.043 | 16.000 | 0.061 | 16.024 | 16.000 | 0.035 | 16.018 | 16.000 | 0.029 |
| | Min | 16.095 | 15.890 | 0.095 | 16.050 | 15.957 | 0.050 | 16.016 | 15.982 | 0.016 | 16.006 | 15.989 | 0.006 | 16.000 | 15.989 | 0.000 |
| 20 | Max | 20.240 | 20.000 | 0.370 | 20.117 | 20.000 | 0.169 | 20.053 | 20.000 | 0.074 | 20.028 | 20.000 | 0.041 | 20.021 | 20.000 | 0.034 |
| | Min | 20.110 | 19.870 | 0.110 | 20.065 | 19.948 | 0.065 | 20.020 | 19.979 | 0.020 | 20.007 | 19.987 | 0.007 | 20.000 | 19.987 | 0.000 |
| 25 | Max | 25.240 | 25.000 | 0.370 | 25.117 | 25.000 | 0.169 | 25.053 | 25.000 | 0.074 | 25.028 | 25.000 | 0.041 | 25.021 | 25.000 | 0.034 |
| | Min | 25.110 | 24.870 | 0.110 | 25.065 | 24.948 | 0.065 | 25.020 | 24.979 | 0.020 | 25.007 | 24.987 | 0.007 | 25.000 | 24.987 | 0.000 |

**Table 4.** *(Continued)* **American National Standard Preferred Shaft Basis Metric Clearance Fits** *ANSI/ASME B4.2-1978 (R2020)*

| Basic Size[a] | | Loose Running | | | Free Running | | | Close Running | | | Sliding | | | Locational Clearance | | |
|---|---|---|---|---|---|---|---|---|---|---|---|---|---|---|---|---|
| | | Hole C11 | Shaft h11 | Fit[b] | Hole D9 | Shaft h9 | Fit[b] | Hole F8 | Shaft h7 | Fit[b] | Hole G7 | Shaft h6 | Fit[b] | Hole H7 | Shaft h6 | Fit[b] |
| 30 | Max | 30.240 | 30.000 | 0.370 | 30.117 | 30.000 | 0.169 | 30.053 | 30.000 | 0.074 | 30.028 | 30.000 | 0.041 | 30.021 | 30.000 | 0.034 |
| | Min | 30.110 | 29.870 | 0.110 | 30.065 | 29.948 | 0.065 | 30.020 | 29.979 | 0.020 | 30.007 | 29.987 | 0.007 | 30.000 | 29.987 | 0.000 |
| 40 | Max | 40.280 | 40.000 | 0.440 | 40.142 | 40.000 | 0.204 | 40.064 | 40.000 | 0.089 | 40.034 | 40.000 | 0.050 | 40.025 | 40.000 | 0.041 |
| | Min | 40.120 | 39.840 | 0.120 | 40.080 | 39.938 | 0.080 | 40.025 | 39.975 | 0.025 | 40.009 | 39.984 | 0.009 | 40.000 | 39.984 | 0.000 |
| 50 | Max | 50.290 | 50.000 | 0.450 | 50.142 | 50.000 | 0.204 | 50.064 | 50.000 | 0.089 | 50.034 | 50.000 | 0.050 | 50.025 | 50.000 | 0.041 |
| | Min | 50.130 | 49.840 | 0.130 | 50.080 | 49.938 | 0.080 | 50.025 | 49.975 | 0.025 | 50.009 | 49.984 | 0.009 | 50.000 | 49.984 | 0.000 |
| 60 | Max | 60.330 | 60.000 | 0.520 | 60.174 | 60.000 | 0.248 | 60.076 | 60.000 | 0.106 | 60.040 | 60.000 | 0.059 | 60.030 | 60.000 | 0.049 |
| | Min | 60.140 | 59.810 | 0.140 | 60.100 | 59.926 | 0.100 | 60.030 | 59.970 | 0.030 | 60.010 | 59.981 | 0.010 | 60.000 | 59.981 | 0.000 |
| 80 | Max | 80.340 | 80.000 | 0.530 | 80.174 | 80.000 | 0.248 | 80.076 | 80.000 | 0.106 | 80.040 | 80.000 | 0.059 | 80.030 | 80.000 | 0.049 |
| | Min | 80.150 | 79.810 | 0.150 | 80.100 | 79.926 | 0.100 | 80.030 | 79.970 | 0.030 | 80.010 | 79.981 | 0.010 | 80.000 | 79.981 | 0.000 |
| 100 | Max | 100.390 | 100.000 | 0.610 | 100.207 | 100.000 | 0.294 | 100.090 | 100.000 | 0.125 | 100.047 | 100.000 | 0.069 | 100.035 | 100.000 | 0.057 |
| | Min | 100.170 | 99.780 | 0.170 | 100.120 | 99.913 | 0.120 | 100.036 | 99.965 | 0.036 | 100.012 | 99.978 | 0.012 | 100.000 | 99.978 | 0.000 |
| 120 | Max | 120.400 | 120.000 | 0.620 | 120.207 | 120.000 | 0.294 | 120.090 | 120.000 | 0.125 | 120.047 | 120.000 | 0.069 | 120.035 | 120.000 | 0.057 |
| | Min | 120.180 | 119.780 | 0.180 | 120.120 | 119.913 | 0.120 | 120.036 | 119.965 | 0.036 | 120.012 | 119.978 | 0.012 | 120.000 | 119.978 | 0.000 |
| 160 | Max | 160.460 | 160.000 | 0.710 | 160.245 | 160.000 | 0.345 | 160.106 | 160.000 | 0.146 | 160.054 | 160.000 | 0.079 | 160.040 | 160.000 | 0.065 |
| | Min | 160.210 | 159.750 | 0.210 | 160.145 | 159.900 | 0.145 | 160.043 | 159.960 | 0.043 | 160.014 | 159.975 | 0.014 | 160.000 | 159.975 | 0.000 |
| 200 | Max | 200.530 | 200.000 | 0.820 | 200.285 | 200.000 | 0.400 | 200.122 | 200.000 | 0.168 | 200.061 | 200.000 | 0.090 | 200.046 | 200.000 | 0.075 |
| | Min | 200.240 | 199.710 | 0.240 | 200.170 | 199.885 | 0.170 | 200.050 | 199.954 | 0.050 | 200.015 | 199.971 | 0.015 | 200.000 | 199.971 | 0.000 |
| 250 | Max | 250.570 | 250.000 | 0.860 | 250.285 | 250.000 | 0.400 | 250.122 | 250.000 | 0.168 | 250.061 | 250.000 | 0.090 | 250.046 | 250.000 | 0.075 |
| | Min | 250.280 | 249.710 | 0.280 | 250.170 | 249.885 | 0.170 | 250.050 | 249.954 | 0.050 | 250.015 | 249.971 | 0.015 | 250.000 | 249.971 | 0.000 |
| 300 | Max | 300.650 | 300.000 | 0.970 | 300.320 | 300.000 | 0.450 | 300.137 | 300.000 | 0.189 | 300.069 | 300.000 | 0.101 | 300.052 | 300.000 | 0.084 |
| | Min | 300.330 | 299.680 | 0.330 | 300.190 | 299.870 | 0.190 | 300.056 | 299.948 | 0.056 | 300.017 | 299.968 | 0.017 | 300.000 | 299.968 | 0.000 |
| 400 | Max | 400.760 | 400.000 | 1.120 | 400.350 | 400.000 | 0.490 | 400.151 | 400.000 | 0.208 | 400.075 | 400.000 | 0.111 | 400.057 | 400.000 | 0.093 |
| | Min | 400.400 | 399.640 | 0.400 | 400.210 | 399.860 | 0.210 | 400.062 | 399.943 | 0.062 | 400.018 | 399.964 | 0.018 | 400.000 | 399.964 | 0.000 |
| 500 | Max | 500.880 | 500.000 | 1.280 | 500.385 | 500.000 | 0.540 | 500.165 | 500.000 | 0.228 | 500.083 | 500.000 | 0.123 | 500.063 | 500.000 | 0.103 |
| | Min | 500.480 | 499.600 | 0.480 | 500.230 | 499.845 | 0.230 | 500.068 | 499.937 | 0.068 | 500.020 | 499.960 | 0.020 | 500.000 | 499.960 | 0.000 |

[a] The sizes shown are first-choice basic sizes (see Table 1, page 689). Preferred fits for other sizes can be calculated from data given in ANSI/ASME B4.2-1978 (R2020).

[b] All fits shown in this table have clearance.

All dimensions are in millimeters.

**Table 5. American National Standard Preferred Shaft Basis Metric Transition and Interference Fits** *ANSI/ASME B4.2-1978 (R2020)*

| Basic Size[a] | | Locational Transition | | | Locational Transition | | | Locational Interference | | | Medium Drive | | | Force | | |
|---|---|---|---|---|---|---|---|---|---|---|---|---|---|---|---|---|
| | | Hole K7 | Shaft h6 | Fit[b] | Hole N7 | Shaft h6 | Fit[b] | Hole P7 | Shaft h6 | Fit[b] | Hole S7 | Shaft h6 | Fit[b] | Hole U7 | Shaft h6 | Fit[b] |
| 1 | Max | 1.000 | 1.000 | +0.006 | 0.996 | 1.000 | +0.002 | 0.994 | 1.000 | 0.000 | 0.986 | 1.000 | −0.008 | 0.982 | 1.000 | −0.012 |
| | Min | 0.990 | 0.994 | −0.010 | 0.986 | 0.994 | −0.014 | 0.984 | 0.994 | −0.016 | 0.976 | 0.994 | −0.024 | 0.972 | 0.994 | −0.028 |
| 1.2 | Max | 1.200 | 1.200 | +0.006 | 1.196 | 1.200 | +0.002 | 1.194 | 1.200 | 0.000 | 1.186 | 1.200 | −0.008 | 1.182 | 1.200 | −0.012 |
| | Min | 1.190 | 1.194 | −0.010 | 1.186 | 1.194 | −0.014 | 1.184 | 1.194 | −0.016 | 1.176 | 1.194 | −0.024 | 1.172 | 1.194 | −0.028 |
| 1.6 | Max | 1.600 | 1.600 | +0.006 | 1.596 | 1.600 | +0.002 | 1.594 | 1.600 | 0.000 | 1.586 | 1.600 | −0.008 | 1.582 | 1.600 | −0.012 |
| | Min | 1.590 | 1.594 | −0.010 | 1.586 | 1.594 | −0.014 | 1.584 | 1.594 | −0.016 | 1.576 | 1.594 | −0.024 | 1.572 | 1.594 | −0.028 |
| 2 | Max | 2.000 | 2.000 | +0.006 | 1.996 | 2.000 | +0.002 | 1.994 | 2.000 | 0.000 | 1.986 | 2.000 | −0.008 | 1.982 | 2.000 | −0.012 |
| | Min | 1.990 | 1.994 | −0.010 | 1.986 | 1.994 | −0.014 | 1.984 | 1.994 | −0.016 | 1.976 | 1.994 | −0.024 | 1.972 | 1.994 | −0.028 |
| 2.5 | Max | 2.500 | 2.500 | +0.006 | 2.496 | 2.500 | +0.002 | 2.494 | 2.500 | 0.000 | 2.486 | 2.500 | −0.008 | 2.482 | 2.500 | −0.012 |
| | Min | 2.490 | 2.494 | −0.010 | 2.486 | 2.494 | −0.014 | 2.484 | 2.494 | −0.016 | 2.476 | 2.494 | −0.024 | 2.472 | 2.494 | −0.028 |
| 3 | Max | 3.000 | 3.000 | +0.006 | 2.996 | 3.000 | +0.002 | 2.994 | 3.000 | 0.000 | 2.986 | 3.000 | −0.008 | 2.982 | 3.000 | −0.012 |
| | Min | 2.990 | 2.994 | −0.010 | 2.986 | 2.994 | −0.014 | 2.984 | 2.994 | −0.016 | 2.976 | 2.994 | −0.024 | 2.972 | 2.994 | −0.028 |
| 4 | Max | 4.003 | 4.000 | +0.011 | 3.996 | 4.000 | +0.004 | 3.992 | 4.000 | 0.000 | 3.985 | 4.000 | −0.007 | 3.981 | 4.000 | −0.011 |
| | Min | 3.991 | 3.992 | −0.009 | 3.984 | 3.992 | −0.016 | 3.980 | 3.992 | −0.020 | 3.973 | 3.992 | −0.027 | 3.969 | 3.992 | −0.031 |
| 5 | Max | 5.003 | 5.000 | +0.011 | 4.996 | 5.000 | +0.004 | 4.992 | 5.000 | 0.000 | 4.985 | 5.000 | −0.007 | 4.981 | 5.000 | −0.011 |
| | Min | 4.991 | 4.992 | −0.009 | 4.984 | 4.992 | −0.016 | 4.980 | 4.992 | −0.020 | 4.973 | 4.992 | −0.027 | 4.969 | 4.992 | −0.031 |
| 6 | Max | 6.003 | 6.000 | +0.011 | 5.996 | 6.000 | +0.004 | 5.992 | 6.000 | 0.000 | 5.985 | 6.000 | −0.007 | 5.981 | 6.000 | −0.011 |
| | Min | 5.991 | 5.992 | −0.009 | 5.984 | 5.992 | −0.016 | 5.980 | 5.992 | −0.020 | 5.973 | 5.992 | −0.027 | 5.969 | 5.992 | −0.031 |
| 8 | Max | 8.005 | 8.000 | +0.014 | 7.996 | 8.000 | +0.005 | 7.991 | 8.000 | 0.000 | 7.983 | 8.000 | −0.008 | 7.978 | 8.000 | −0.013 |
| | Min | 7.990 | 7.991 | −0.010 | 7.981 | 7.991 | −0.019 | 7.976 | 7.991 | −0.024 | 7.968 | 7.991 | −0.032 | 7.963 | 7.991 | −0.037 |
| 10 | Max | 10.005 | 10.000 | +0.014 | 9.996 | 10.000 | +0.005 | 9.991 | 10.000 | 0.000 | 9.983 | 10.000 | −0.008 | 9.978 | 10.000 | −0.013 |
| | Min | 9.990 | 9.991 | −0.010 | 9.981 | 9.991 | −0.019 | 9.976 | 9.991 | −0.024 | 9.968 | 9.991 | −0.032 | 9.963 | 9.991 | −0.037 |
| 12 | Max | 12.006 | 12.000 | +0.017 | 11.995 | 12.000 | +0.006 | 11.989 | 12.000 | 0.000 | 11.979 | 12.000 | −0.010 | 11.974 | 12.000 | −0.015 |
| | Min | 11.988 | 11.989 | −0.012 | 11.977 | 11.989 | −0.023 | 11.971 | 11.989 | −0.029 | 11.961 | 11.989 | −0.039 | 11.956 | 11.989 | −0.044 |
| 16 | Max | 16.006 | 16.000 | +0.017 | 15.995 | 16.000 | +0.006 | 15.989 | 16.000 | 0.000 | 15.979 | 16.000 | −0.010 | 15.974 | 16.000 | −0.015 |
| | Min | 15.988 | 15.989 | −0.012 | 15.977 | 15.989 | −0.023 | 15.971 | 15.989 | −0.029 | 15.961 | 15.989 | −0.039 | 15.956 | 15.989 | −0.044 |
| 20 | Max | 20.006 | 20.000 | +0.019 | 19.993 | 20.000 | +0.006 | 19.986 | 20.000 | −0.001 | 19.973 | 20.000 | −0.014 | 19.967 | 20.000 | −0.020 |
| | Min | 19.985 | 19.987 | −0.015 | 19.972 | 19.987 | −0.028 | 19.965 | 19.987 | −0.035 | 19.952 | 19.987 | −0.048 | 19.946 | 19.987 | −0.054 |
| 25 | Max | 25.006 | 25.000 | +0.019 | 24.993 | 25.000 | +0.006 | 24.986 | 25.000 | −0.001 | 24.973 | 25.000 | −0.014 | 24.960 | 25.000 | −0.027 |
| | Min | 24.985 | 24.987 | −0.015 | 24.972 | 24.987 | −0.028 | 24.965 | 24.987 | −0.035 | 24.952 | 24.987 | −0.048 | 24.939 | 24.987 | −0.061 |

**Table 5.** *(Continued)* **American National Standard Preferred Shaft Basis Metric Transition and Interference Fits** *ANSI/ASME B4.2-1978 (R2020)*

| Basic Size[a] | | Locational Transition | | | Locational Transition | | | Locational Interference | | | Medium Drive | | | Force | | |
|---|---|---|---|---|---|---|---|---|---|---|---|---|---|---|---|---|
| | | Hole K7 | Shaft h6 | Fit[b] | Hole N7 | Shaft h6 | Fit[b] | Hole P7 | Shaft h6 | Fit[b] | Hole S7 | Shaft h6 | Fit[b] | Hole U7 | Shaft h6 | Fit[b] |
| 30 | Max | 30.006 | 30.000 | +0.019 | 29.993 | 30.000 | +0.006 | 29.986 | 30.000 | –0.001 | 29.973 | 30.000 | –0.014 | 29.960 | 30.000 | –0.027 |
| | Min | 29.985 | 29.987 | –0.015 | 29.972 | 29.987 | –0.028 | 29.965 | 29.987 | –0.035 | 29.952 | 29.987 | –0.048 | 29.939 | 29.987 | –0.061 |
| 40 | Max | 40.007 | 40.000 | +0.023 | 39.992 | 40.000 | +0.008 | 39.983 | 40.000 | –0.001 | 39.966 | 40.000 | –0.018 | 39.949 | 40.000 | –0.035 |
| | Min | 39.982 | 39.984 | –0.018 | 39.967 | 39.984 | –0.033 | 39.958 | 39.984 | –0.042 | 39.941 | 39.984 | –0.059 | 39.924 | 39.984 | –0.076 |
| 50 | Max | 50.007 | 50.000 | +0.023 | 49.992 | 50.000 | +0.008 | 49.983 | 50.000 | –0.001 | 49.966 | 50.000 | –0.018 | 49.939 | 50.000 | –0.045 |
| | Min | 49.982 | 49.984 | –0.018 | 49.967 | 49.984 | –0.033 | 49.958 | 49.984 | –0.042 | 49.941 | 49.984 | –0.059 | 49.914 | 49.984 | –0.086 |
| 60 | Max | 60.009 | 60.000 | +0.028 | 59.991 | 60.000 | +0.010 | 59.979 | 60.000 | –0.002 | 59.958 | 60.000 | –0.023 | 59.924 | 60.000 | –0.087 |
| | Min | 59.979 | 59.981 | –0.021 | 59.961 | 59.981 | –0.039 | 59.949 | 59.981 | –0.051 | 59.928 | 59.981 | –0.072 | 59.894 | 59.981 | –0.106 |
| 80 | Max | 80.009 | 80.000 | +0.028 | 79.991 | 80.000 | +0.010 | 79.979 | 80.000 | –0.002 | 79.952 | 80.000 | –0.029 | 79.909 | 80.000 | –0.072 |
| | Min | 79.979 | 79.981 | –0.021 | 79.961 | 79.981 | –0.039 | 79.949 | 79.981 | –0.051 | 79.922 | 79.981 | –0.078 | 79.879 | 79.981 | –0.121 |
| 100 | Max | 100.010 | 100.000 | +0.032 | 99.990 | 100.000 | +0.012 | 99.976 | 100.000 | –0.002 | 99.942 | 100.000 | –0.036 | 99.889 | 100.000 | –0.089 |
| | Min | 99.975 | 99.978 | –0.025 | 99.955 | 99.978 | –0.045 | 99.941 | 99.978 | –0.059 | 99.907 | 99.978 | –0.093 | 99.854 | 99.978 | –0.146 |
| 120 | Max | 120.010 | 120.000 | +0.032 | 119.990 | 120.000 | +0.012 | 119.976 | 120.000 | –0.002 | 119.934 | 120.000 | –0.044 | 119.869 | 120.000 | –0.109 |
| | Min | 119.975 | 119.978 | –0.025 | 119.955 | 119.978 | –0.045 | 119.941 | 119.978 | –0.059 | 119.899 | 119.978 | –0.101 | 119.834 | 119.978 | –0.166 |
| 160 | Max | 160.012 | 160.000 | +0.037 | 159.988 | 160.000 | +0.013 | 159.972 | 160.000 | –0.003 | 159.915 | 160.000 | –0.060 | 159.825 | 160.000 | –0.150 |
| | Min | 159.972 | 159.975 | –0.028 | 159.948 | 159.975 | –0.052 | 159.932 | 159.975 | –0.068 | 159.875 | 159.975 | –0.125 | 159.785 | 159.975 | –0.215 |
| 200 | Max | 200.013 | 200.00 | +0.042 | 199.986 | 200.000 | +0.015 | 199.967 | 200.000 | –0.004 | 199.895 | 200.000 | –0.076 | 199.781 | 200.000 | –0.190 |
| | Min | 199.967 | 199.971 | –0.033 | 199.940 | 199.971 | –0.060 | 199.921 | 199.971 | –0.079 | 199.849 | 199.971 | –0.151 | 199.735 | 199.971 | –0.265 |
| 250 | Max | 250.013 | 250.000 | +0.042 | 249.986 | 250.000 | +0.015 | 249.967 | 250.000 | –0.004 | 249.877 | 250.000 | –0.094 | 249.733 | 250.000 | –0.238 |
| | Min | 249.967 | 249.971 | –0.033 | 249.940 | 249.971 | –0.060 | 249.921 | 249.971 | –0.079 | 249.831 | 249.971 | –0.169 | 249.687 | 249.971 | –0.313 |
| 300 | Max | 300.016 | 300.000 | +0.048 | 299.986 | 300.000 | +0.018 | 299.964 | 300.000 | –0.004 | 299.850 | 300.000 | –0.118 | 299.670 | 300.000 | –0.298 |
| | Min | 299.964 | 299.968 | –0.036 | 299.934 | 299.968 | –0.066 | 299.912 | 299.968 | –0.088 | 299.798 | 299.968 | –0.202 | 299.618 | 299.968 | –0.382 |
| 400 | Max | 400.017 | 400.000 | +0.053 | 399.984 | 400.000 | +0.020 | 399.959 | 400.000 | –0.005 | 399.813 | 400.000 | –0.151 | 399.586 | 400.000 | –0.378 |
| | Min | 399.960 | 399.964 | –0.040 | 399.927 | 399.964 | –0.073 | 399.902 | 399.964 | –0.098 | 399.756 | 399.964 | –0.244 | 399.529 | 399.964 | –0.471 |
| 500 | Max | 500.018 | 500.000 | +0.058 | 499.983 | 500.000 | +0.023 | 499.955 | 500.000 | –0.005 | 499.771 | 500.000 | –0.189 | 499.483 | 500.000 | –0.477 |
| | Min | 499.955 | 499.960 | –0.045 | 499.920 | 499.960 | –0.080 | 499.892 | 499.960 | –0.108 | 499.708 | 499.960 | –0.292 | 499.420 | 499.960 | –0.580 |

[a] The sizes shown are first-choice basic sizes (see Table 1, page 689). Preferred fits for other sizes can be calculated from data given in ANSI/ASME B4.2-1978 (R2020).

[b] A plus sign indicates clearance; a minus sign indicates interference.

All dimensions are in millimeters.

**Table 6. American National Standard Gagemakers Tolerances**
*ANSI B4.4M-1981 (Withdrawn)*

| Gagemakers Tolerance | | | Workpiece Tolerance | |
|---|---|---|---|---|
| | Class | ISO Symbol[a] | IT Grade | Recommended Gage Usage |
| Rejection of Good Parts Increase ↑ | ZM | 0.05 IT11 | IT11 | Low-precision gages recommended to be used to inspect workpieces held to internal (hole) tolerances C11 and H11 and to external (shaft) tolerances c11 and h11. |
| | YM | 0.05 IT9 | IT9 | Gages recommended to be used to inspect workpieces held to internal (hole) tolerances D9 and H9 and to external (shaft) tolerances d9 and h9. |
| | XM | 0.05 IT8 | IT8 | Precision gages recommended to be used to inspect workpieces held to internal (hole) tolerances F8 and H8. |
| Gage Cost Increase ↓ | XXM | 0.05 IT7 | IT7 | Recommended to be used for gages to inspect workpieces held to internal (hole) tolerances G7, H7, K7, N7, P7, S7, and U7, and to external (shaft) tolerances f7 and h7. |
| | XXXM | 0.05 IT6 | IT6 | High-precision gages recommended to be used to inspect workpieces held to external (shaft) tolerances g6, h6, k6, n6, p6, s6, and u6. |

[a] Gagemakers tolerance is equal to 5 percent of workpiece tolerance or 5 percent of applicable IT grade value. See Table 7.

For workpiece tolerance class values, see previous Table 2 through Table 5, inclusive.

**Table 7. American National Standard Gagemakers Tolerances**
*ANSI B4.4M-1981 (Withdrawn)*

| Basic Size | | Class ZM | Class YM | Class XM | Class XXM | Class XXXM |
|---|---|---|---|---|---|---|
| Over | To | 0.05 IT11 | 0.05 IT9 | 0.05 IT8 | 0.05 IT7 | 0.05 IT6 |
| 0 | 3 | 0.0030 | 0.0012 | 0.0007 | 0.0005 | 0.0003 |
| 3 | 6 | 0.0037 | 0.0015 | 0.0009 | 0.0006 | 0.0004 |
| 6 | 10 | 0.0045 | 0.0018 | 0.0011 | 0.0007 | 0.0005 |
| 10 | 18 | 0.0055 | 0.0021 | 0.0013 | 0.0009 | 0.0006 |
| 18 | 30 | 0.0065 | 0.0026 | 0.0016 | 0.0010 | 0.0007 |
| 30 | 50 | 0.0080 | 0.0031 | 0.0019 | 0.0012 | 0.0008 |
| 50 | 80 | 0.0095 | 0.0037 | 0.0023 | 0.0015 | 0.0010 |
| 80 | 120 | 0.0110 | 0.0043 | 0.0027 | 0.0017 | 0.0011 |
| 120 | 180 | 0.0125 | 0.0050 | 0.0031 | 0.0020 | 0.0013 |
| 180 | 250 | 0.0145 | 0.0057 | 0.0036 | 0.0023 | 0.0015 |
| 250 | 315 | 0.0160 | 0.0065 | 0.0040 | 0.0026 | 0.0016 |
| 315 | 400 | 0.0180 | 0.0070 | 0.0044 | 0.0028 | 0.0018 |
| 400 | 500 | 0.0200 | 0.0077 | 0.0048 | 0.0031 | 0.0020 |

All dimensions are in millimeters. For closer gagemakers tolerance classes than Class XXXM, specify 5 percent of IT5, IT4, or IT3 and use the designation 0.05 IT5, 0.05 IT4, etc.

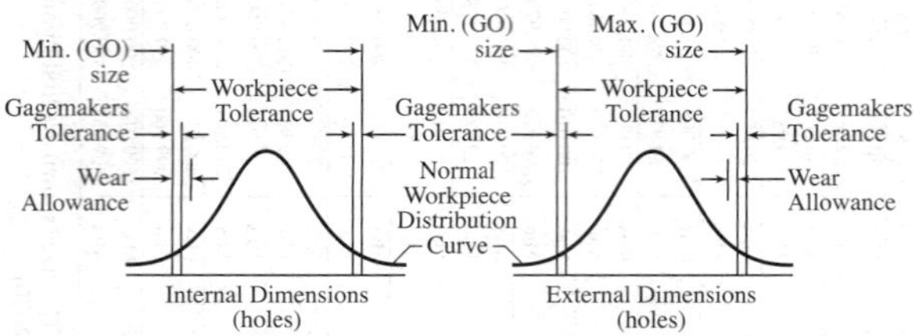


Fig. 4. Relationship between Gagemakers Tolerance, Wear Allowance and Workpiece Tolerance

## Applications

Many factors such as length of engagement, bearing load, speed, lubrication, operating temperatures, humidity, surface texture, and materials must be taken into account in fit selections for a particular application.

Choice of other than the preferred fits might be considered necessary to satisfy extreme conditions. Subsequent adjustments might also be desired as the result of experience in a particular application to suit critical functional requirements or to permit optimum manufacturing economy. Selection of a departure from these recommendations will depend upon consideration of the engineering and economic factors that might be involved; however, the benefits to be derived from the use of preferred fits should not be overlooked.

A general guide to machining processes that may normally be expected to produce work within the tolerances indicated by the IT grades given in ANSI/ASME B4.2-1978 (R2020) is shown in Table 8. Practical usage of the various IT tolerance grades is shown in Table 9.

**Table 8. Relation of Machining Processes to IT Tolerance Grades**

| | IT Grades | | | | | | | |
|---|---|---|---|---|---|---|---|---|
| | 4 | 5 | 6 | 7 | 8 | 9 | 10 | 11 |
| Lapping & Honing | ■ | ■ | | | | | | |
| Cylindrical Grinding | | ■ | ■ | ■ | | | | |
| Surface Grinding | | ■ | ■ | ■ | ■ | | | |
| Diamond Turning | | ■ | ■ | ■ | | | | |
| Diamond Boring | | ■ | ■ | ■ | | | | |
| Broaching | | ■ | ■ | ■ | ■ | | | |
| Powder Metal sizes | | | ■ | ■ | ■ | | | |
| Reaming | | | ■ | ■ | ■ | ■ | ■ | |
| Turning | | | | ■ | ■ | ■ | ■ | ■ |
| Powder Metal sintered | | | | ■ | ■ | ■ | ■ | |
| Boring | | | | ■ | ■ | ■ | ■ | ■ |
| Milling | | | | | | | ■ | ■ |
| Planing & Shaping | | | | | | | ■ | ■ |
| Drilling | | | | | | | ■ | ■ |
| Punching | | | | | | | ■ | ■ |
| Die Casting | | | | | | | | ■ |

**Table 9. Practical Use of International Tolerance Grades**

| | For Measuring Tools | | | | | | | | | For Material | | | | | | | | |
|---|---|---|---|---|---|---|---|---|---|---|---|---|---|---|---|---|---|---|
| IT Grades | 01 | 0 | 1 | 2 | 3 | 4 | 5 | 6 | 7 | 8 | 9 | 10 | 11 | 12 | 13 | 14 | 15 | 16 |
| | | | | | | | For Fits | | | | | | | For Large Manufacturing Tolerances | | | | |

## British and ISO Metric Limits and Fits

Based on ISO Recommendation R286, the British Standard BS EN ISO 286-1:2010 is intended to provide a comprehensive range of metric limits and fits for engineering purposes. Sizes up to 3,150 mm are covered by the Standard, but the condensed information presented here embraces dimensions up to 500 mm only. The system is based on a series of tolerances graded to suit all classes of work from the finest to the most coarse, and the different types of fits that can be obtained range from coarse clearance to heavy interference. In the standard, only cylindrical parts, designated holes and shafts are referred to explicitly, but it is emphasized that the recommendations apply equally well to other sections, and the general terms *hole* and *shaft* can be taken to mean the space contained by or containing two parallel faces or tangent planes of any part, such as the width of a slot or the thickness of a key. It is also strongly emphasized that the grades series of tolerances are intended for the most general application and should be used wherever possible whether the features of the component involved are members of a fit or not.

**Definitions.—**The definitions given in the Standard include the following:

*Limits of Size:* The maximum and minimum sizes permitted for a feature.

*Nominal Size:* Size of a feature of perfect form as defined by the drawing specification. Used for the location of the limits of size by the application of the upper and lower limit deviations.

*Upper Limit Deviation:* The algebraic difference between the maximum limit of size and the corresponding nominal size. It is designated as "ES" for a hole and as "es" for a shaft, which stands for the French term *écart supérieur.*

*Lower Limit Deviation:* The algebraic difference between the minimum limit of size and the corresponding nominal size. It is designated as "EI" for a hole and as "ei" for a shaft, which stands for the French term *écart inférieur.*

*Zero Line:* In a graphical representation of limits and fits, the straight line to which the deviations are referred. The zero line is the line of zero deviation and represents the nominal size.

*Tolerance:* The difference between the maximum limit of size and the minimum limit of size. It is an absolute value without sign.

*Tolerance Zone:* In a graphical representation of tolerances, the zone comprised between the two lines representing the limits of tolerance and defined by its magnitude (tolerance) and by its position in relation to the zero line.

*Fundamental Deviation:* That one of the two deviations, being the one nearest to the zero line, which is conventionally chosen to define the position of the tolerance zone in relation to the zero line.

*Shaft-Basis System of Fits:* A system of fits in which the different clearances and interferences are obtained by associating various holes with a single shaft. In the ISO system, the basic shaft is the shaft the upper deviation of which is zero.

*Hole-Basis System of Fits:* A system of fits in which the different clearances and interferences are obtained by associating various shafts with a single hole. In the ISO system, the basic hole is the hole the lower deviation of which is zero.

**Selected Limits of Tolerance and Fits.—**The number of fit combinations that can be built up with the ISO system is very large. However, experience shows that the majority of fits required for usual engineering products can be provided by a limited selection of tolerances. Limits of tolerance for selected holes are shown in Table 1 and for shafts in Table 2. Selected fits, based on combinations of the selected hole and shaft tolerances, are given in Table 3.

**Tolerances and Fundamental Deviations.—**There are 18 tolerance grades intended to meet the requirements of different classes of work, and they are designated IT01, IT0, and IT1 to IT16. (IT stands for ISO series of tolerances.) Table 4 shows the standardized numerical values for the 18 tolerance grades, which are known as standard tolerances. The system provides 27 fundamental deviations for sizes up to and including 500 mm, and Table 5a and Table 5b contain the values for shafts and Table 6a and Table 6b for holes. Uppercase (capital) letters designate hole deviations, and the same letters in lower case designate shaft deviations. The deviation $j_s$ ($J_s$ for holes) is provided to meet the need for symmetrical bilateral tolerances. In this instance, there is no fundamental deviation, and the tolerance zone, of whatever magnitude, is equally disposed about the zero line.

**Calculated Limits of Tolerance.—**The deviations and fundamental tolerances provided by the ISO system can be combined in any way that appears necessary to give a required fit. Thus, for example, the deviations H (basic hole) and f (clearance shaft) could be associated, and with each of these deviations any one of the tolerance grades IT01 to IT16 could be used. All the limits of tolerance that the system is capable of providing for sizes up to and including 500 mm can be calculated from the standard tolerances given in Table 4, and the fundamental deviations given in Table 5a, Table 5b, Table 6a and Table 6b. The range includes limits of tolerance for shafts and holes used in small high-precision work and horology.

The system provides for the use of either hole-basis or shaft-basis fits, and the Standard includes details of procedures for converting from one type of fit to the other.

The limits of tolerance for a shaft or hole are designated by the appropriate letter indicating the fundamental deviation, followed by a suffix number denoting the tolerance grade. This suffix number is the numerical part of the tolerance grade designation. Thus, a hole tolerance with deviation H and tolerance grade IT7 is designated H7. Likewise, a shaft with deviation p and tolerance grade IT6 is designated p6. The limits of size of a component feature are defined by the basic size, say, 45 mm, followed by the appropriate tolerance designation, for example, 45 H7 or 45 p6. A fit is indicated by combining the basic size common to both features with the designation appropriate to each of them, for example, 45 H7-p6 or 45 H7/p6.

When calculating the limits of size for a shaft, the upper deviation es, or the lower deviation ei, is first obtained from Table 5a or Table 5b, depending on the particular letter designation, and nominal dimension. If an upper deviation has been determined, the lower deviation ei = es − IT. The IT value is obtained from Table 4 for the particular tolerance grade being applied. If a lower deviation has been obtained from Table 5a or Table 5b, the upper deviation es = ei + IT. When the upper deviation ES has been determined for a hole from Table 6a or Table 6b, the lower deviation EI = ES − IT. If a lower deviation EI has been obtained from Table 6a, then the upper deviation ES = EI + IT.

The upper deviations for holes K, M, and N with tolerance grades up to and including IT8, and for holes P to ZC with tolerance grades up to and including IT7 must be calculated by adding the delta (Δ) values given in Table 6b as indicated.

*Example 1:* The limits of size for a part of 133 mm basic size with a tolerance designation g9 are derived as follows:

From Table 5a, the upper deviation (es) is − 0.014 mm. From Table 4, the tolerance grade (IT9) is 0.100 mm. The lower deviation (ei) = es − IT = 0.114 mm, and the limits of size are thus 132.986 and 132.886 mm.

*Example 2:* The limits of size for a part 20 mm in size, with tolerance designation D3, are derived as follows: From Table 6a, the lower deviation (EI) is + 0.065 mm. From Table 4, the tolerance grade (IT3) is 0.004 mm. The upper deviation (ES) = EI + IT = 0.069 mm, and thus the limits of size for the part are 20.069 and 20.065 mm.

*Example 3:* The limits of size for a part 32 mm in size, with tolerance designation M5, which involves a delta value, are obtained as follows: From Table 6a, the upper deviation ES is − 0.009 mm + Δ = −0.005 mm. (The delta value given at the end of Table 6b for this size and grade IT5 is 0.004 mm.) From Table 4, the tolerance grade (IT5) is 0.011 mm. The lower deviation (EI) = ES − IT = − 0.016 mm, and thus the limits of size for the part are 31.995 and 31.984 mm.

Where the designations h and H or $j_s$ and $J_s$ are used, it is only necessary to refer to Table 4. For h and H, the fundamental deviation is always zero, and the disposition of the tolerance is always negative ( − ) for a shaft and positive ( + ) for a hole.

*Example 4:* The limits for a part 40 mm in size, designated h8 are derived as follows: From Table 4, the tolerance grade (IT8) is 0.039 mm, and the limits are therefore 40.000 and 39.961 mm.

*Example 5:* The limits for a part 60 mm in size, designated $j_s$7 or $J_s$7 are derived as follows: From Table 4, the tolerance grade (IT7) is 0.030 mm, and this value is divided equally about the basic size to give limits of 60.015 and 59.985 mm.

**Table 1. British and ISO Standard Limits of Tolerance for Selected Holes (Upper and Lower Deviations)** *BS EN ISO 286-1:2010*

| Nominal Sizes, mm | | H7 | | H8 | | H9 | | H11 | |
|---|---|---|---|---|---|---|---|---|---|
| Over | Up to and Including | ES + | EI | ES + | EI | ES + | EI | ES + | EI |
| … | 3 | 10 | 0 | 14 | 0 | 25 | 0 | 60 | 0 |
| 3 | 6 | 12 | 0 | 18 | 0 | 30 | 0 | 75 | 0 |
| 6 | 10 | 15 | 0 | 22 | 0 | 36 | 0 | 90 | 0 |
| 10 | 18 | 18 | 0 | 27 | 0 | 43 | 0 | 110 | 0 |
| 18 | 30 | 21 | 0 | 33 | 0 | 52 | 0 | 130 | 0 |
| 30 | 50 | 25 | 0 | 39 | 0 | 62 | 0 | 160 | 0 |
| 50 | 80 | 30 | 0 | 46 | 0 | 74 | 0 | 190 | 0 |
| 80 | 120 | 35 | 0 | 54 | 0 | 87 | 0 | 220 | 0 |
| 120 | 180 | 40 | 0 | 63 | 0 | 100 | 0 | 250 | 0 |
| 180 | 250 | 46 | 0 | 72 | 0 | 115 | 0 | 290 | 0 |
| 250 | 315 | 52 | 0 | 81 | 0 | 130 | 0 | 320 | 0 |
| 315 | 400 | 57 | 0 | 89 | 0 | 140 | 0 | 360 | 0 |
| 400 | 500 | 63 | 0 | 97 | 0 | 155 | 0 | 400 | 0 |

ES = Upper deviation, EI = Lower deviation.

The dimensions are given in 0.001 mm, except for the nominal sizes, which are in millimeters.

**Table 2. British and ISO Standard Limits of Tolerance for Selected Shafts (Upper and Lower Deviations)** *BS EN ISO 286-1:2010*

| Nominal Sizes, mm | | c11 | | d10 | | e9 | | f7 | | g6 | | h6 | | k6 | | n6 | | p6 | | s6 | |
|---|---|---|---|---|---|---|---|---|---|---|---|---|---|---|---|---|---|---|---|---|---|
| Over | Up to and Incl. | es − | ei − | es − | ei − | es − | ei − | es − | ei − | es − | ei − | es − | ei − | es + | ei + | es + | ei + | es + | ei + | es + | ei + |
| … | 3 | 60 | 120 | 20 | 60 | 14 | 39 | 6 | 16 | 2 | 8 | 0 | 6 | 6 | 0 | 10 | 4 | 12 | 6 | 20 | 14 |
| 3 | 6 | 70 | 145 | 30 | 78 | 20 | 50 | 10 | 22 | 4 | 12 | 0 | 8 | 9 | 1 | 16 | 8 | 20 | 12 | 27 | 19 |
| 6 | 10 | 80 | 170 | 40 | 98 | 25 | 61 | 13 | 28 | 5 | 14 | 0 | 9 | 10 | 1 | 19 | 10 | 24 | 15 | 32 | 23 |
| 10 | 18 | 95 | 205 | 50 | 120 | 32 | 75 | 16 | 34 | 6 | 17 | 0 | 11 | 12 | 1 | 23 | 12 | 29 | 18 | 39 | 28 |
| 18 | 30 | 110 | 240 | 65 | 149 | 40 | 92 | 20 | 41 | 7 | 20 | 0 | 13 | 15 | 2 | 28 | 15 | 35 | 22 | 48 | 35 |
| 30 | 40 | 120 | 280 | 80 | 180 | 50 | 112 | 25 | 50 | 9 | 25 | 0 | 16 | 18 | 2 | 33 | 17 | 42 | 26 | 59 | 43 |
| 40 | 50 | 130 | 290 | 80 | 180 | 50 | 112 | 25 | 50 | 9 | 25 | 0 | 16 | 18 | 2 | 33 | 17 | 42 | 26 | 59 | 43 |
| 50 | 65 | 140 | 330 | 100 | 220 | 60 | 134 | 30 | 60 | 10 | 29 | 0 | 19 | 21 | 2 | 39 | 20 | 51 | 32 | 72 | 53 |
| 65 | 80 | 150 | 340 | 100 | 220 | 60 | 134 | 30 | 60 | 10 | 29 | 0 | 19 | 21 | 2 | 39 | 20 | 51 | 32 | 78 | 59 |
| 80 | 100 | 170 | 390 | 120 | 260 | 72 | 159 | 36 | 71 | 12 | 34 | 0 | 22 | 25 | 3 | 45 | 23 | 59 | 37 | 93 | 71 |
| 100 | 120 | 180 | 400 | 120 | 260 | 72 | 159 | 36 | 71 | 12 | 34 | 0 | 22 | 25 | 3 | 45 | 23 | 59 | 37 | 101 | 79 |
| 120 | 140 | 200 | 450 | 145 | 305 | 85 | 185 | 43 | 83 | 14 | 39 | 0 | 25 | 28 | 3 | 52 | 27 | 68 | 43 | 117 | 92 |
| 140 | 160 | 210 | 460 | 145 | 305 | 85 | 185 | 43 | 83 | 14 | 39 | 0 | 25 | 28 | 3 | 52 | 27 | 68 | 43 | 125 | 100 |
| 160 | 180 | 230 | 480 | 145 | 305 | 85 | 185 | 43 | 83 | 14 | 39 | 0 | 25 | 28 | 3 | 52 | 27 | 68 | 43 | 133 | 108 |
| 180 | 200 | 240 | 530 | 170 | 355 | 100 | 215 | 50 | 96 | 15 | 44 | 0 | 29 | 33 | 4 | 60 | 31 | 79 | 50 | 151 | 122 |
| 200 | 225 | 260 | 550 | 170 | 355 | 100 | 215 | 50 | 96 | 15 | 44 | 0 | 29 | 33 | 4 | 60 | 31 | 79 | 50 | 159 | 130 |
| 225 | 250 | 280 | 570 | 170 | 355 | 100 | 215 | 50 | 96 | 15 | 44 | 0 | 29 | 33 | 4 | 60 | 31 | 79 | 50 | 169 | 140 |
| 250 | 280 | 300 | 620 | 190 | 400 | 110 | 240 | 56 | 108 | 17 | 49 | 0 | 32 | 36 | 4 | 66 | 34 | 88 | 56 | 190 | 158 |
| 280 | 315 | 330 | 650 | 190 | 400 | 110 | 240 | 56 | 108 | 17 | 49 | 0 | 32 | 36 | 4 | 66 | 34 | 88 | 56 | 202 | 170 |
| 315 | 355 | 360 | 720 | 210 | 440 | 125 | 265 | 62 | 119 | 18 | 54 | 0 | 36 | 40 | 4 | 73 | 37 | 98 | 62 | 226 | 190 |
| 355 | 400 | 400 | 760 | 210 | 440 | 125 | 265 | 62 | 119 | 18 | 54 | 0 | 36 | 40 | 4 | 73 | 37 | 98 | 62 | 244 | 208 |
| 400 | 450 | 440 | 840 | 230 | 480 | 135 | 290 | 68 | 131 | 20 | 60 | 0 | 40 | 45 | 5 | 80 | 40 | 108 | 68 | 272 | 232 |
| 450 | 500 | 480 | 880 | 230 | 480 | 135 | 290 | 68 | 131 | 20 | 60 | 0 | 40 | 45 | 5 | 80 | 40 | 108 | 68 | 292 | 252 |

es = Upper deviation, ei = Lower deviation.

The dimensions are given in 0.001 mm, except for the nominal sizes, which are in millimeters.

**Table 3. British and ISO Standard Selected Fits, Minimum and Maximum Clearances** *BS EN ISO 286-1:2010*

| Nominal Sizes, mm | | H11—c11 | | H9—d10 | | H9—e9 | | H8—f7 | | H7—g6 | | H7—h6 | | H7—k6 | | H7—n6 | | H7—p6 | | H7—s6 | |
|---|---|---|---|---|---|---|---|---|---|---|---|---|---|---|---|---|---|---|---|---|---|
| Over | Up to and Including | Min | Max | Min | Max | Min | Max | Min | Max | Min | Max | Min | Max | Min | Max | Min | Max | Min | Max | Min | Max |
| … | 3 | 60 | 180 | 20 | 85 | 14 | 64 | 6 | 30 | 2 | 18 | 0 | 16 | −6 | +10 | −10 | +6 | −12 | +4 | −20 | −4 |
| 3 | 6 | 70 | 220 | 30 | 108 | 20 | 80 | 10 | 40 | 4 | 24 | 0 | 20 | −9 | +11 | −16 | +4 | −20 | 0 | −27 | −7 |
| 6 | 10 | 80 | 260 | 40 | 134 | 25 | 97 | 13 | 50 | 5 | 29 | 0 | 24 | −10 | +14 | −19 | +5 | −24 | 0 | −32 | −8 |
| 10 | 18 | 95 | 315 | 50 | 163 | 32 | 118 | 16 | 61 | 6 | 35 | 0 | 29 | −12 | +17 | −23 | +6 | −29 | 0 | −39 | −10 |
| 18 | 30 | 110 | 370 | 65 | 201 | 40 | 144 | 20 | 74 | 7 | 41 | 0 | 34 | −15 | +19 | −28 | +6 | −35 | −1 | −48 | −14 |
| 30 | 40 | 120 | 440 | 80 | 242 | 50 | 174 | 25 | 89 | 9 | 50 | 0 | 41 | −18 | +23 | −33 | +8 | −42 | −1 | −59 | −18 |
| 40 | 50 | 130 | 450 | 80 | 242 | 50 | 174 | 25 | 89 | 9 | 50 | 0 | 41 | −18 | +23 | −33 | +8 | −42 | −1 | −59 | −18 |
| 50 | 65 | 140 | 520 | 100 | 294 | 60 | 208 | 30 | 106 | 10 | 59 | 0 | 49 | −21 | +28 | −39 | +10 | −51 | −2 | −72 | −23 |
| 65 | 80 | 150 | 530 | 100 | 294 | 60 | 208 | 30 | 106 | 10 | 59 | 0 | 49 | −21 | +28 | −39 | +10 | −51 | −2 | −78 | −29 |
| 80 | 100 | 170 | 610 | 120 | 347 | 72 | 246 | 36 | 125 | 12 | 69 | 0 | 57 | −25 | +32 | −45 | +12 | −59 | −2 | −93 | −36 |
| 100 | 120 | 180 | 620 | 120 | 347 | 72 | 246 | 36 | 125 | 12 | 69 | 0 | 57 | −25 | +32 | −45 | +12 | −59 | −2 | −101 | −44 |
| 120 | 140 | 200 | 700 | 145 | 405 | 85 | 285 | 43 | 146 | 14 | 79 | 0 | 65 | −28 | +37 | −52 | +13 | −68 | −3 | −117 | −52 |
| 140 | 160 | 210 | 710 | 145 | 405 | 85 | 285 | 43 | 146 | 14 | 79 | 0 | 65 | −28 | +37 | −52 | +13 | −68 | −3 | −125 | −60 |
| 160 | 180 | 230 | 730 | 145 | 405 | 85 | 285 | 43 | 146 | 14 | 79 | 0 | 65 | −28 | +37 | −52 | +13 | −68 | −3 | −133 | −68 |
| 180 | 200 | 240 | 820 | 170 | 470 | 100 | 330 | 50 | 168 | 15 | 90 | 0 | 75 | −33 | +42 | −60 | +15 | −79 | −4 | −151 | −76 |
| 200 | 225 | 260 | 840 | 170 | 470 | 100 | 330 | 50 | 168 | 15 | 90 | 0 | 75 | −33 | +42 | −60 | +15 | −79 | −4 | −159 | −84 |
| 225 | 250 | 280 | 860 | 170 | 470 | 100 | 330 | 50 | 168 | 15 | 90 | 0 | 75 | −33 | +42 | −60 | +15 | −79 | −4 | −169 | −94 |
| 250 | 280 | 300 | 940 | 190 | 530 | 110 | 370 | 56 | 189 | 17 | 101 | 0 | 84 | −36 | +48 | −66 | +18 | −88 | −4 | −190 | −126 |
| 280 | 315 | 330 | 970 | 190 | 530 | 110 | 370 | 56 | 189 | 17 | 101 | 0 | 84 | −36 | +48 | −66 | +18 | −88 | −4 | −202 | −112 |
| 315 | 355 | 360 | 1080 | 210 | 580 | 125 | 405 | 62 | 208 | 18 | 111 | 0 | 93 | −40 | −53 | −73 | +20 | −98 | −5 | −226 | −133 |
| 355 | 400 | 400 | 1120 | 210 | 580 | 125 | 405 | 62 | 208 | 18 | 111 | 0 | 93 | −40 | −53 | −73 | +20 | −98 | −5 | −244 | −151 |
| 400 | 450 | 440 | 1240 | 230 | 635 | 135 | 445 | 68 | 228 | 20 | 123 | 0 | 103 | −45 | +58 | −80 | +23 | −108 | −5 | −272 | −169 |
| 450 | 500 | 480 | 1280 | 230 | 635 | 135 | 445 | 68 | 228 | 20 | 123 | 0 | 103 | −45 | +58 | −80 | +23 | −108 | −5 | −292 | −189 |

The dimensions are given in 0.001 mm, except for the nominal sizes, which are in millimeters.

Minus (–) sign indicates negative clearance, i.e., interference.

**Table 4. British and ISO Standard Limits and Fits** ***BS EN ISO 286-1:2010***

| Nominal Sizes, mm | | Tolerance Grades | | | | | | | | | | | | | | | | | |
|---|---|---|---|---|---|---|---|---|---|---|---|---|---|---|---|---|---|---|---|
| Over | To | IT01 | IT 0 | IT 1 | IT 2 | IT 3 | IT 4 | IT 5 | IT 6 | IT 7 | IT 8 | T9 | IT 10 | IT 11 | IT 12 | IT 13 | IT 14 [a] | IT 15 [a] | IT 16 [a] |
| … | 3 | 0.3 | 0.5 | 0.8 | 1.2 | 2 | 3 | 4 | 6 | 10 | 14 | 25 | 40 | 60 | 100 | 140 | 250 | 400 | 600 |
| 3 | 6 | 0.4 | 0.6 | 1 | 1.5 | 2.5 | 4 | 5 | 8 | 12 | 18 | 30 | 48 | 75 | 120 | 180 | 300 | 480 | 750 |
| 6 | 10 | 0.4 | 0.6 | 1 | 1.5 | 2.5 | 4 | 6 | 9 | 15 | 22 | 36 | 58 | 90 | 150 | 220 | 360 | 580 | 900 |
| 10 | 18 | 0.5 | 0.8 | 1.2 | 2 | 3 | 5 | 8 | 11 | 18 | 27 | 43 | 70 | 110 | 180 | 270 | 430 | 700 | 1100 |
| 18 | 30 | 0.6 | 1 | 1.5 | 2.5 | 4 | 6 | 9 | 13 | 21 | 33 | 52 | 84 | 130 | 210 | 330 | 520 | 840 | 1300 |
| 30 | 50 | 0.6 | 1 | 1.5 | 2.5 | 4 | 7 | 11 | 16 | 25 | 39 | 62 | 100 | 160 | 250 | 390 | 620 | 1000 | 1600 |
| 50 | 80 | 0.8 | 1.2 | 2 | 3 | 5 | 8 | 13 | 19 | 30 | 46 | 74 | 120 | 190 | 300 | 460 | 740 | 1200 | 1900 |
| 80 | 120 | 1 | 1.5 | 2.5 | 4 | 6 | 10 | 15 | 22 | 35 | 54 | 87 | 140 | 220 | 350 | 540 | 870 | 1400 | 2200 |
| 120 | 180 | 1.2 | 2 | 3.5 | 5 | 8 | 12 | 18 | 25 | 40 | 63 | 100 | 160 | 250 | 400 | 630 | 1000 | 1600 | 2500 |
| 180 | 250 | 2 | 3 | 4.5 | 7 | 10 | 14 | 20 | 29 | 46 | 72 | 115 | 185 | 290 | 460 | 720 | 1150 | 1850 | 2900 |
| 250 | 315 | 2.5 | 4 | 6 | 8 | 12 | 16 | 23 | 32 | 52 | 81 | 130 | 210 | 320 | 520 | 810 | 1300 | 2100 | 3200 |
| 315 | 400 | 3 | 5 | 7 | 9 | 13 | 18 | 25 | 36 | 57 | 89 | 140 | 230 | 360 | 570 | 890 | 1400 | 2300 | 3600 |
| 400 | 500 | 4 | 6 | 8 | 10 | 15 | 20 | 27 | 40 | 63 | 97 | 155 | 250 | 400 | 630 | 970 | 1550 | 2500 | 4000 |

[a] Not applicable to sizes below 1 mm.

The dimensions are given in 0.001 mm, except for the nominal sizes, which are in millimeters.

## Table 5a. British and ISO Standard Fundamental Deviations for Shafts *BS EN ISO 286-1:2010*

| Nominal Sizes, mm | | Grade | | | | | | | | | | | | | | | | |
|---|---|---|---|---|---|---|---|---|---|---|---|---|---|---|---|---|---|---|
| | | 01 to 16 | | | | | | | | | | | | 5–6 | 7 | 8 | 4–7 | ≤3>7 |
| | | Fundamental (Upper) Deviation es | | | | | | | | | | | | Fundamental (Lower) Deviation ei | | | | |
| Over | To | a[a] | b[a] | c | cd | d | e | ef | f | fg | g | h | js[b] | j | | | k | |
| … | 3 | −270 | −140 | −60 | −34 | −20 | −14 | −10 | −6 | −4 | −2 | 0 | | −2 | −4 | −6 | 0 | 0 |
| 3 | 6 | −270 | −140 | −70 | −46 | −30 | −20 | −14 | −10 | −6 | −4 | 0 | | −2 | −4 | … | +1 | 0 |
| 6 | 10 | −280 | −150 | −80 | −56 | −40 | −25 | −18 | −13 | −8 | −5 | 0 | | −2 | −5 | … | +1 | 0 |
| 10 | 14 | −290 | −150 | −95 | −70 | −50 | −32 | −23 | −16 | −10 | −6 | 0 | | −3 | −6 | … | +1 | 0 |
| 14 | 18 | −290 | −150 | −95 | −70 | −50 | −32 | −23 | −16 | −10 | −6 | 0 | | −3 | −6 | … | +1 | 0 |
| 18 | 24 | −300 | −160 | −110 | −85 | −65 | −40 | −28 | −20 | −12 | −7 | 0 | | −4 | −8 | … | +2 | 0 |
| 24 | 30 | −300 | −160 | −110 | −85 | −65 | −40 | −28 | −20 | −12 | −7 | 0 | | −4 | −8 | … | +2 | 0 |
| 30 | 40 | −310 | −170 | −120 | −100 | −80 | −50 | −35 | −25 | −15 | −9 | 0 | | −5 | −10 | … | +2 | 0 |
| 40 | 50 | −320 | −180 | −130 | −100 | −80 | −50 | −35 | −25 | −15 | −9 | 0 | | −5 | −10 | … | +2 | 0 |
| 50 | 65 | −340 | −190 | −140 | … | −100 | −60 | … | −30 | … | −10 | 0 | | −7 | −12 | … | +2 | 0 |
| 65 | 80 | −360 | −200 | −150 | … | −100 | −60 | … | −30 | … | −10 | 0 | | −7 | −12 | … | +2 | 0 |
| 80 | 100 | −380 | −220 | −170 | … | −120 | −72 | … | −36 | … | −12 | 0 | | −9 | −15 | … | +3 | 0 |
| 100 | 120 | −410 | −240 | −180 | … | −120 | −72 | … | −36 | … | −12 | 0 | ±IT/2 | −9 | −15 | … | +3 | 0 |
| 120 | 140 | −460 | −260 | −200 | … | −145 | −85 | … | −43 | … | −14 | 0 | | −11 | −18 | … | +3 | 0 |
| 140 | 160 | −520 | −280 | −210 | … | −145 | −85 | … | −43 | … | −14 | 0 | | −11 | −18 | … | +3 | 0 |
| 160 | 180 | −580 | −310 | −230 | … | −145 | −85 | … | −43 | … | −14 | 0 | | −11 | −18 | … | +3 | 0 |
| 180 | 200 | −660 | −340 | −240 | … | −170 | −100 | … | −50 | … | −15 | 0 | | −13 | −21 | … | +4 | 0 |
| 200 | 225 | −740 | −380 | −260 | … | −170 | −100 | … | −50 | … | −15 | 0 | | −13 | −21 | … | +4 | 0 |
| 225 | 250 | −820 | −420 | −280 | … | −170 | −100 | … | −50 | … | −15 | 0 | | −13 | −21 | … | +4 | 0 |
| 250 | 280 | −920 | −480 | −300 | … | −190 | −110 | … | −56 | … | −17 | 0 | | −16 | −26 | … | +4 | 0 |
| 280 | 315 | −1050 | −540 | −330 | … | −190 | −110 | … | −56 | … | −17 | 0 | | −16 | −26 | … | +4 | 0 |
| 315 | 355 | −1200 | −600 | −360 | … | −210 | −125 | … | −62 | … | −18 | 0 | | −18 | −28 | … | +4 | 0 |
| 355 | 400 | −1350 | −680 | −400 | … | −210 | −125 | … | −62 | … | −18 | 0 | | −18 | −28 | … | +4 | 0 |
| 400 | 450 | −1500 | −760 | −440 | … | −230 | −135 | … | −68 | … | −20 | 0 | | −20 | −32 | … | +5 | 0 |
| 450 | 500 | −1650 | −840 | −480 | … | −230 | −135 | … | −68 | … | −20 | 0 | | −20 | −32 | … | +5 | 0 |

[a] Not applicable to sizes up to 1 mm.

[b] In grades 7 to 11, the two symmetrical deviations ±IT/2 should be rounded if the IT value in micrometers is an odd value by replacing it with the even value immediately below. For example, if IT = 175, replace it by 174.

**Table 5b. British and ISO Standard Fundamental Deviations for Shafts** *BS EN ISO 286-1:2010*

| Nominal Sizes, mm | | Grade | | | | | | | | | | | | | |
|---|---|---|---|---|---|---|---|---|---|---|---|---|---|---|---|
| | | 01 to 16 | | | | | | | | | | | | | |
| | | Fundamental (Lower) Deviation ei | | | | | | | | | | | | | |
| Over | To | m | n | p | r | s | t | u | v | x | y | z | za | zb | zc |
| … | 3 | +2 | +4 | +6 | +10 | +14 | … | +18 | … | +20 | … | +26 | +32 | +40 | +60 |
| 3 | 6 | +4 | +8 | +12 | +15 | +19 | … | +23 | … | +28 | … | +35 | +42 | +50 | +80 |
| 6 | 10 | +6 | +10 | +15 | +19 | +23 | … | +28 | … | +34 | … | +42 | +52 | +67 | +97 |
| 10 | 14 | +7 | +12 | +18 | +23 | +28 | … | +33 | … | +40 | … | +50 | +64 | +90 | +130 |
| 14 | 18 | +7 | +12 | +18 | +23 | +28 | … | +33 | +39 | +45 | … | +60 | +77 | +108 | +150 |
| 18 | 24 | +8 | +15 | +22 | +28 | +35 | … | +41 | +47 | +54 | +63 | +73 | +98 | +136 | +188 |
| 24 | 30 | +8 | +15 | +22 | +28 | +35 | +41 | +48 | +55 | +64 | +75 | +88 | +118 | +160 | +218 |
| 30 | 40 | +9 | +17 | +26 | +34 | +43 | +48 | +60 | +68 | +80 | +94 | +112 | +148 | +200 | +274 |
| 40 | 50 | +9 | +17 | +26 | +34 | +43 | +54 | +70 | +81 | +97 | +114 | +136 | +180 | +242 | +325 |
| 50 | 65 | +11 | +20 | +32 | +41 | +53 | +66 | +87 | +102 | +122 | +144 | +172 | +226 | +300 | +405 |
| 65 | 80 | +11 | +20 | +32 | +43 | +59 | +75 | +102 | +120 | +146 | +174 | +210 | +274 | +360 | +480 |
| 80 | 100 | +13 | +23 | +37 | +51 | +71 | +91 | +124 | +146 | +178 | +214 | +258 | +335 | +445 | +585 |
| 100 | 120 | +13 | +23 | +37 | +54 | +79 | +104 | +144 | +172 | +210 | +254 | +310 | +400 | +525 | +690 |
| 120 | 140 | +15 | +27 | +43 | +63 | +92 | +122 | +170 | +202 | +248 | +300 | +365 | +470 | +620 | +800 |
| 140 | 160 | +15 | +27 | +43 | +65 | +100 | +134 | +190 | +228 | +280 | +340 | +415 | +535 | +700 | +900 |
| 160 | 180 | +15 | +27 | +43 | +68 | +108 | +146 | +210 | +252 | +310 | +380 | +465 | +600 | +780 | +1000 |
| 180 | 200 | +17 | +31 | +50 | +77 | +122 | +166 | +236 | +284 | +350 | +425 | +520 | +670 | +880 | +1150 |
| 200 | 225 | +17 | +31 | +50 | +80 | +130 | +180 | +258 | +310 | +385 | +470 | +575 | +740 | +960 | +1250 |
| 225 | 250 | +17 | +31 | +50 | +84 | +140 | +196 | +284 | +340 | +425 | +520 | +640 | +820 | +1050 | +1350 |
| 250 | 280 | +20 | +34 | +56 | +94 | +158 | +218 | +315 | +385 | +475 | +580 | +710 | +920 | +1200 | +1550 |
| 280 | 315 | +20 | +34 | +56 | +98 | +170 | +240 | +350 | +425 | +525 | +650 | +790 | +1000 | +1300 | +1700 |
| 315 | 355 | +21 | +37 | +62 | +108 | +190 | +268 | +390 | +475 | +590 | +730 | +900 | +1150 | +1500 | +1900 |
| 355 | 400 | +21 | +37 | +62 | +114 | +208 | +294 | +435 | +530 | +660 | +820 | +1000 | +1300 | +1650 | +2100 |
| 400 | 450 | +23 | +40 | +68 | +126 | +232 | +330 | +490 | +595 | +740 | +920 | +1100 | +1450 | +1850 | +2400 |
| 450 | 500 | +23 | +40 | +68 | +132 | +252 | +360 | +540 | +660 | +820 | +1000 | +1250 | +1600 | +2100 | +2600 |

The dimensions are in 0.001 mm, except the nominal sizes, which are in millimeters.

**Table 6a. British and ISO Standard Fundamental Deviations for Holes** *BS EN ISO 286-1:2010*

| Nominal Sizes, mm | | Grade | | | | | | | | | | | | | | | | | | | | |
|---|---|---|---|---|---|---|---|---|---|---|---|---|---|---|---|---|---|---|---|---|---|---|
| | | 01 to 16 | | | | | | | | | | | | 6 | 7 | 8 | ≤8 | >8 | ≤8[a] | >8 | ≤8 | >8[b] |
| | | Fundamental (Lower) Deviation EI | | | | | | | | | | | | Fundamental (Upper) Deviation ES | | | | | | | | |
| Over | To | A[b] | B[b] | C | CD | D | E | EF | F | FG | G | H | Js[c] | J | | | K[d] | | M[d] | | N[d] | |
| … | 3 | +270 | +140 | +60 | +34 | +20 | +14 | +10 | +6 | +4 | +2 | 0 | | +2 | +4 | +6 | 0 | 0 | −2 | −2 | −4 | −4 |
| 3 | 6 | +270 | +140 | +70 | +46 | +30 | +20 | +14 | +10 | +6 | +4 | 0 | | +5 | +6 | +10 | −1+Δ | … | −4+Δ | −4 | −8+Δ | 0 |
| 6 | 10 | +280 | +150 | +80 | +56 | +40 | +25 | +18 | +13 | +8 | +5 | 0 | | +5 | +8 | +12 | −1+Δ | … | −6+Δ | −6 | −10+Δ | 0 |
| 10 | 14 | +290 | +150 | +95 | +70 | +50 | +32 | … | +16 | +10 | +6 | 0 | | +6 | +10 | +15 | −1+Δ | … | −7+Δ | −7 | −12+Δ | 0 |
| 14 | 18 | +290 | +150 | +95 | +70 | +50 | +32 | … | +16 | +10 | +6 | 0 | | +6 | +10 | +15 | −1+Δ | … | −7+Δ | −7 | −12+Δ | 0 |
| 18 | 24 | +300 | +160 | +110 | +85 | +65 | +40 | … | +20 | +12 | +7 | 0 | | +8 | +12 | +20 | −2+Δ | … | −8+Δ | −8 | −15+Δ | 0 |
| 24 | 30 | +300 | +160 | +110 | +85 | +65 | +40 | … | +20 | +12 | +7 | 0 | | +8 | +12 | +20 | −2+Δ | … | −8+Δ | −8 | −15+Δ | 0 |
| 30 | 40 | +310 | +170 | +120 | +100 | +80 | +50 | … | +25 | +15 | +9 | 0 | | +10 | +14 | +24 | −2+Δ | … | −9+Δ | −9 | −17+Δ | 0 |
| 40 | 50 | +320 | +180 | +130 | +100 | +80 | +50 | … | +25 | +15 | +9 | 0 | | +10 | +14 | +24 | −2+Δ | … | −9+Δ | −9 | −17+Δ | 0 |
| 50 | 65 | +340 | +190 | +140 | … | +100 | +60 | … | +30 | … | +10 | 0 | | +13 | +18 | +28 | −2+Δ | … | −11+Δ | −11 | −20+Δ | 0 |
| 65 | 80 | +360 | +200 | +150 | … | +100 | +60 | … | +30 | … | +10 | 0 | | +13 | +18 | +28 | −2+Δ | … | −11+Δ | −11 | −20+Δ | 0 |
| 80 | 100 | +380 | +220 | +170 | … | +120 | +72 | … | +36 | … | +12 | 0 | | +16 | +22 | +34 | −3+Δ | … | −13+Δ | −13 | −23+Δ | 0 |
| 100 | 120 | +410 | +240 | +180 | … | +120 | +72 | … | +36 | … | +12 | 0 | ±IT/2 | +16 | +22 | +34 | −3+Δ | … | −13+Δ | −13 | −23+Δ | 0 |
| 120 | 140 | +460 | +260 | +200 | … | +145 | +85 | … | +43 | … | +14 | 0 | | +18 | +26 | +41 | −3+Δ | … | −15+Δ | −15 | −27+Δ | 0 |
| 140 | 160 | +520 | +280 | +210 | … | +145 | +85 | … | +43 | … | +14 | 0 | | +18 | +26 | +41 | −3+Δ | … | −15+Δ | −15 | −27+Δ | 0 |
| 160 | 180 | +580 | +310 | +230 | … | +145 | +85 | … | +43 | … | +14 | 0 | | +18 | +26 | +41 | −3+Δ | … | −15+Δ | −15 | −27+Δ | 0 |
| 180 | 200 | +660 | +340 | +240 | … | +170 | +100 | … | +50 | … | +15 | 0 | | +22 | +30 | +47 | −4+Δ | … | −17+Δ | −17 | −31+Δ | 0 |
| 200 | 225 | +740 | +380 | +260 | … | +170 | +100 | … | +50 | … | +15 | 0 | | +22 | +30 | +47 | −4+Δ | … | −17−Δ | −17 | −31+Δ | 0 |
| 225 | 250 | +820 | +420 | +280 | … | +170 | +100 | … | +50 | … | +15 | 0 | | +22 | +30 | +47 | −4+Δ | … | −17+Δ | −17 | −31+Δ | 0 |
| 250 | 280 | +920 | +480 | +300 | … | +190 | +110 | … | +56 | … | +17 | 0 | | +25 | +36 | +55 | −4+Δ | … | −20+Δ | −20 | −34+Δ | 0 |
| 280 | 315 | +1050 | +540 | +330 | … | +190 | +110 | … | +56 | … | +17 | 0 | | +25 | +36 | +55 | −4+Δ | … | −20+Δ | −20 | −34+Δ | 0 |
| 315 | 355 | +1200 | +600 | +360 | … | +210 | +125 | … | +62 | … | +18 | 0 | | +29 | +39 | +60 | −4+Δ | … | −21+Δ | −21 | −37+Δ | 0 |
| 355 | 400 | +1350 | +680 | +400 | … | +210 | +125 | … | +62 | … | +18 | 0 | | +29 | +39 | +60 | −4+Δ | … | −21+Δ | −21 | −37+Δ | 0 |
| 400 | 450 | +1500 | +760 | +440 | … | +230 | +135 | … | +68 | … | +20 | 0 | | +33 | +43 | +66 | −5+4 | … | −23+Δ | −23 | −40+Δ | 0 |
| 450 | 500 | +1650 | +840 | +480 | … | +230 | +135 | … | +68 | … | +20 | 0 | | +33 | +43 | +66 | −5+4 | … | −23+Δ | −23 | −40+Δ | 0 |

[a] Special case: for M6, ES = −9 for sizes from 250 to 315 mm, instead of −11.

[b] Not applicable to sizes up to 1 mm.

[c] In grades 7 to 11, the two symmetrical deviations ±IT/2 should be rounded if the IT value in micrometers is an odd value by replacing it with the even value below. For example, if IT = 175, replace it by 174.

[d] When calculating deviations for holes K, M, and N with tolerance grades up to and including IT8, and holes P to ZC with tolerance grades up to and including IT7, the delta (Δ) values are added to the upper deviation ES. For example, for 25 P7, ES = −0.022 + 0.008 = −0.014 mm.

**Table 6b. British and ISO Standard Fundamental Deviations for Holes *BS EN ISO 286-1:2010***

| Nominal Sizes, mm | | Grade ≤7 | Grade >7 | | | | | | | | | | | | Values for delta (Δ)[d] | | | | | |
|---|---|---|---|---|---|---|---|---|---|---|---|---|---|---|---|---|---|---|---|---|
| | | Fundamental (Upper) Deviation ES | | | | | | | | | | | | | Grade | | | | | |
| Over | To | P to ZC | P | R | S | T | U | V | X | Y | Z | ZA | ZB | ZC | 3 | 4 | 5 | 6 | 7 | 8 |
| … | 3 | Same deviation as for grades above 7 increased by Δ | −6 | −10 | −14 | … | −18 | … | −20 | … | −26 | −32 | −40 | −60 | 0 | 0 | 0 | 0 | 0 | 0 |
| 3 | 6 | | −12 | −15 | −19 | … | −23 | … | −28 | … | −35 | −42 | −50 | −80 | 1 | 1.5 | 1 | 3 | 4 | 6 |
| 6 | 10 | | −15 | −19 | −23 | … | −28 | … | −34 | … | −42 | −52 | −67 | −97 | 1 | 1.5 | 2 | 3 | 6 | 7 |
| 10 | 14 | | −18 | −23 | −28 | … | −33 | … | −40 | … | −50 | −64 | −90 | −130 | 1 | 2 | 3 | 3 | 7 | 9 |
| 14 | 18 | | −18 | −23 | −28 | … | −33 | −39 | −45 | … | −60 | −77 | −108 | −150 | 1 | 2 | 3 | 3 | 7 | 9 |
| 18 | 24 | | −22 | −28 | −35 | … | −41 | −47 | −54 | −63 | −73 | −98 | −136 | −188 | 1.5 | 2 | 3 | 4 | 8 | 12 |
| 24 | 30 | | −22 | −28 | −35 | −41 | −48 | −55 | −64 | −75 | −88 | −118 | −160 | −218 | 1.5 | 2 | 3 | 4 | 8 | 12 |
| 30 | 40 | | −26 | −34 | −43 | −48 | −60 | −68 | −80 | −94 | −112 | −148 | −200 | −274 | 1.5 | 3 | 4 | 5 | 9 | 14 |
| 40 | 50 | | −26 | −34 | −43 | −54 | −70 | −81 | −97 | −114 | −136 | −180 | −242 | −325 | 1.5 | 3 | 4 | 5 | 9 | 14 |
| 50 | 65 | | −32 | −41 | −53 | −66 | −87 | −102 | −122 | −144 | −172 | −226 | −300 | −405 | 2 | 3 | 5 | 6 | 11 | 16 |
| 65 | 80 | | −32 | −43 | −59 | −75 | −102 | −120 | −146 | −174 | −210 | −274 | −360 | −480 | 2 | 3 | 5 | 6 | 11 | 16 |
| 80 | 100 | | −37 | −51 | −71 | −91 | −124 | −146 | −178 | −214 | −258 | −335 | −445 | −585 | 2 | 4 | 5 | 7 | 13 | 19 |
| 100 | 120 | | −37 | −54 | −79 | −104 | −144 | −172 | −210 | −254 | −310 | −400 | −525 | −690 | 2 | 4 | 5 | 7 | 13 | 19 |
| 120 | 140 | | −43 | −63 | −92 | −122 | −170 | −202 | −248 | −300 | −365 | −470 | −620 | −800 | 3 | 4 | 6 | 7 | 15 | 23 |
| 140 | 160 | | −43 | −65 | −100 | −134 | −190 | −228 | −280 | −340 | −415 | −535 | −700 | −900 | 3 | 4 | 6 | 7 | 15 | 23 |
| 160 | 180 | | −43 | −68 | −108 | −146 | −210 | −252 | −310 | −380 | −465 | −600 | −780 | −1000 | 3 | 4 | 6 | 7 | 15 | 23 |
| 180 | 200 | | −50 | −77 | −122 | −166 | −226 | −284 | −350 | −425 | −520 | −670 | −880 | −1150 | 3 | 4 | 6 | 9 | 17 | 26 |
| 200 | 225 | | −50 | −80 | −130 | −180 | −258 | −310 | −385 | −470 | −575 | −740 | −960 | −1250 | 3 | 4 | 6 | 9 | 17 | 26 |
| 225 | 250 | | −50 | −84 | −140 | −196 | −284 | −340 | −425 | −520 | −640 | −820 | −1050 | −1350 | 3 | 4 | 6 | 9 | 17 | 26 |
| 250 | 280 | | −56 | −94 | −158 | −218 | −315 | −385 | −475 | −580 | −710 | −920 | −1200 | −1550 | 4 | 4 | 7 | 9 | 20 | 29 |
| 280 | 315 | | −56 | −98 | −170 | −240 | −350 | −425 | −525 | −650 | −790 | −1000 | −1300 | −1700 | 4 | 4 | 7 | 9 | 20 | 29 |
| 315 | 355 | | −62 | −108 | −190 | −268 | −390 | −475 | −590 | −730 | −900 | −1150 | −1500 | −1800 | 4 | 5 | 7 | 11 | 21 | 32 |
| 355 | 400 | | −62 | −114 | −208 | −294 | −435 | −530 | −660 | −820 | −1000 | −1300 | −1650 | −2100 | 4 | 5 | 7 | 11 | 21 | 32 |
| 400 | 450 | | −68 | −126 | −232 | −330 | −490 | −595 | −740 | −920 | −1100 | −1450 | −1850 | −2400 | 5 | 5 | 7 | 13 | 23 | 34 |
| 450 | 500 | | −68 | −132 | −252 | −360 | −540 | −660 | −820 | −1000 | −1250 | −1600 | −2100 | −2600 | 5 | 5 | 7 | 13 | 23 | 34 |

The dimensions are given in 0.001 mm, except the nominal sizes, which are in millimeters.

## Tolerance Analysis and Assignment

All parts are produced with some amount of variation, or difference, between the intended geometry and features of the part produced. While variation is unavoidable and expected, the challenge is determining an acceptable amount. The amount and type of variation is called *tolerance*. Part feature tolerances are not assigned arbitrarily, but rather are specified by designers with good understanding of manufacturing process capabilities, cost limitations, and assembly and functional requirements. Good design allows for the broadest tolerances possible that meet all requirements, resulting in parts that are easier and less costly to produce. Overly restrictive tolerances limit manufacturers to unnecessarily precise methods, driving up costs, which can be impacted dramatically by small differences in tolerances.

Tolerance analysis and tolerance assignment are bottom-up and top-down approaches, respectively, to calculating tolerances, and typically they are used in iterative fashion. Tolerance assignment is done at the design stage, to meet assembly and functional requirements. Tolerance analysis is performed later, on preliminary or final part tolerances, to detect any needed design improvements and changes in tolerance assignments. Predicting the results of variations may be an end in itself, but more often, the goal is to select the best tolerances throughout an assembly to meet design requirements.

**The Tolerance Stack-Up Chain.—**Tolerance stack-up calculations analyze the effects that tolerances have on function and guide tolerance assignment. A tolerance stack-up chain diagram is created first to clearly identify the assembly dimension to be solved and each part dimension contributing to the stack-up, as is shown in Fig. 5.

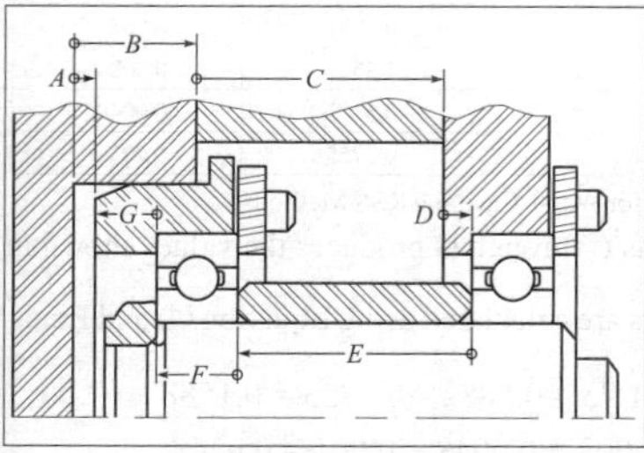


| Dim | Max | Min | Mean (μ) (from mfg. data) | Standard Deviation (σ) (from mfg. data) |
|---|---|---|---|---|
| *B* | 0.9395 | 0.9355 | 0.937331 | 0.000517 |
| *C* | 1.755 | 1.745 | 1.750748 | 0.001200 |
| *D* | 0.1875 | 0.1825 | 0.185147 | 0.000874 |
| *E* | 1.751 | 1.749 | 1.750128 | 0.000203 |
| *F* | 0.5906 | 0.5858 | 0.588126 | 0.001004 |
| *G* | 0.376 | 0.375 | 0.375464 | 0.000106 |

Fig. 5. Tolerance Stack-Up Chain Diagram, Related Specifications, and Dimension Data

From the diagram, a tolerance stack-up equation is derived. In dimension labeling, a sign convention denotes direction: positive for up or right; negative for down or left. The assembly dimension can then be solved, as in the following example, written in general form:

$$A = B + C + D - E - F - G \tag{1}$$

Clear, unambiguous part definition, as provided by geometric dimensioning and tolerancing, is required for accurate tolerance analysis. Care must be taken to ensure the effects of geometric variation are included in dimensions expressed numerically. Any sources of variation allowed by referenced drawings must be included in the stack-up equation.

**Tolerance Analysis Calculation Methods.—**For solving analysis or assignment problems, four methods of tolerance stack-up calculation are common.

*Worst-Case:* During analysis, the worst-case method predicts the largest variation in the assembly dimension and requires the smallest tolerances be assigned. Worst-case calculations are appropriate when every feature is inspected to screen from use any out-of-specification parts, and/or when there are few (approximately seven or fewer) dimensions in the chain. Benefits of this analytical method are its simplicity and speed.

Each contributing dimension is modeled as a mean (μ) and total tolerance (*t*) calculated from the maximum and minimum values.

Each contributing dimension mean (μ) is calculated as:

$$\mu = \frac{\text{Max} + \text{Min}}{2} \tag{2}$$

Each contributing dimension's total tolerance ($t$) is calculated as:

$$t = \text{Max} - \text{Min} \tag{3}$$

The assembly dimension $A$ is modeled as an assembly mean ($\mu_A$) and assembly total tolerance ($t_A$). Following the signs used in the general stack-up equation:

$$\mu_A = \mu_B + \mu_C + \mu_D - \mu_E - \mu_F - \mu_G \tag{4}$$

$$t_A = t_B + t_C + t_D + t_E + t_F + t_G \tag{5}$$

the assembly dimension is then predicted to fall between the limits:

$$A_{\text{max}} = \mu_A + \frac{t_A}{2} \tag{6a}$$

$$A_{\text{min}} = \mu_A - \frac{t_A}{2} \tag{6b}$$

*Example:* Predict the minimum and maximum assembly dimensions for the stack-up diagram and dimensions given in Fig. 5 using the worst-case method.

*Solution:* Using Equation (2) and Equation (3), calculate a mean and total tolerance to use as model parameters for each of the contributing dimensions in Fig. 5. Starting with $B$:

$$\mu = \frac{0.9395 + 0.9355}{2} = 0.9375$$

$$t = 0.9395 - 0.9355 = 0.004$$

| Dim | Mean (μ) | Total Tolerance ($t$) |
|---|---|---|
| $B$ | 0.9375 | 0.004 |
| $C$ | 1.75 | 0.01 |
| $D$ | 0.185 | 0.005 |
| $E$ | 1.75 | 0.002 |
| $F$ | 0.5882 | 0.0048 |
| $G$ | 0.3755 | 0.001 |

Fig. 6. Tabulated Calculations for Worst-Case and RSS Methods

Repeating this calculation for dimensions $C$ through $G$ produces the values shown in Fig. 6.

Next, the assembly dimension parameters are calculated using Equation (4) and Equation (5):

$$\mu_A = 0.9375 + 1.75 + 0.185 - 1.75 - 0.5882 - 0.3755 = 0.1588$$

$$t_A = 0.004 + 0.01 + 0.005 + 0.002 + 0.0048 + 0.001 = 0.0268$$

Finally, the assembly limits are calculated using Equation (6a) and Equation (6b):

$$A_{max} = 0.1588 + \frac{0.0268}{2} = 0.1722$$

$$A_{min} = 0.1588 - \frac{0.0268}{2} = 0.1454$$

Therefore, assemblies are expected to measure between 0.1454 and 0.1722.

*Root Sum of Squares (RSS):* While technically a statistical method with several simplifying assumptions applied, RSS belongs somewhere between the worst-case and statistical methods, with use depending on desired characteristics of the result. Each contributing dimension is modeled as a mean (μ) and total tolerance ($t$) calculated from the maximum and minimum values, using Equation (2) and Equation (3).

The calculation for assembly mean ($\mu_A$) is the same as in the worst-case and statistical methods; see Equation (4). The calculation for assembly total tolerance ($t_A$) is:

$$t_A = \sqrt{t_B^2 + t_C^2 + t_D^2 + t_E^2 + t_F^2 + t_G^2} \tag{7}$$

Equation (6a) and Equation (6b) are used to predict the minimum and maximum assembly dimension.

*Example:* Predict the minimum and maximum assembly dimension for the stack-up diagram and dimensions given in Fig. 5. Assume manufacturing data are not available for this example and apply the RSS method.

*Solution:* Using Equation (2) and Equation (3), calculate a mean and total tolerance to use as model parameters for each of the contributing dimensions in Fig. 5. The results are tabulated in Fig. 6.

Next, the assembly dimension parameters are calculated using Equation (4) and Equation (5):

$$\mu_A = 0.9375 + 1.75 + 0.185 - 1.75 - 0.5882 - 0.3755 = 0.1588$$

$$t_A = \sqrt{0.004^2 + 0.01^2 + 0.005^2 + 0.002^2 + 0.0048^2 + 0.001^2} = 0.013002$$

Finally, the assembly limits are calculated using Equation (6a) and Equation (6b):

$$A_{max} = 0.1588 + \frac{0.013002}{2} = 0.165301$$

$$A_{min} = 0.1588 - \frac{0.013002}{2} = 0.152299$$

Therefore, assemblies are expected to measure between 0.152299 and 0.165301.

*Statistical:* This approach produces likely values and ranges, predicts smaller assembly variations, and permits larger tolerances to be assigned. While it does not guarantee conformity to requirements, conformance and defect rates can be predicted. Statistical calculations are appropriate when many dimensions are contributing to a stack-up chain or when it can be assumed that the contributing dimensions follow normal distributions, such as when statistical process control (SPC) is used (see *STATISTICAL ANALYSIS OF MANUFACTURING DATA* on page 131 for statistics terminology used here).

The assembly dimension A is modeled as a normal distribution, with assembly mean ($\mu_A$) calculated using actual mean measurement values and applying Equation (4). The assembly standard deviation ($\sigma_A$) is calculated using:

$$\sigma_A = \sqrt{\sigma_B^2 + \sigma_C^2 + \sigma_D^2 + \sigma_E^2 + \sigma_F^2 + \sigma_G^2} \tag{8}$$

*Example:* Use the statistical method to determine the range of assembly dimension values, between which 99 percent of assembly dimensions are expected to fall, for the stack-up diagram and dimensions given in Fig. 5. Manufacturing data show the contributing dimensions are well modeled as normal distributions using the parameters included in the figure.

*Solution:* Model assembly dimension *A* as a normal distribution with parameters calculated using Equation (3) and Equation (7):

$$\mu_A = 0.937331 + 1.750748 + 0.185147 - 1.750128 - 0.588126 - 0.375464 = 0.159508$$

$$\sigma_A = \sqrt{0.000517^2 + 0.001200^2 + 0.000874^2 + 0.000203^2 + 0.001004^2 + 0.000106^2} = 0.001879$$

99 percent of the area under the normal distribution curve falls within the range of $\mu \pm 2.5758\sigma$. Upper and lower limits of this range for assembly dimension *A* are calculated as:

$$A_{upper} = 0.159508 + (2.5758)(0.001879) = 0.164348$$

$$A_{lower} = 0.159508 - (2.5758)(0.001879) = 0.154668$$

Therefore, 99 percent of assembly dimensions are expected to measure between 0.154668 and 0.164348.

*Monte Carlo Simulation:* This random-number-based calculation method attempts to model variation realistically. The simulation runs numerous cases and builds a probability distribution for the stack-up, which is used to predict results, including defect rates. For each case, contributing dimensions in the stack-up chain are randomly chosen according to their statistical distribution (normal, uniform, or otherwise). The dimensions are combined according to the stack-up equation, which may be nonlinear.

**Tolerance Assignment.**—This is the process of distributing the "tolerance budget" among contributing dimensions of the stack-up chain to manage costs and meet target quality levels. As explained above, analysis produces one solution (assembly dimension) from several inputs (dimensions of contributing parts), whereas assignment attempts to produce several contributing part dimensions from one assembly dimension. There are infinite ways to assign part tolerances to achieve a desired assembly tolerance (thus, analytical solutions usually are not possible), with several methods for assigning tolerances being most common.

*Even Assignment:* Part feature tolerances are assigned equally among assignable tolerances to stay within the required assembly tolerance.

*Proportional Scaling:* Preliminary tolerances are assigned and analyzed. If the assembly tolerance exceeds the allowed amount, assignable part feature tolerances are scaled down proportionally as needed.

*Weight Factors:* Weight factors, determined on the basis of manufacturing difficulty, are allocated to each assignable tolerance. The assembly tolerance budget is divided across each tolerance proportional to its weight factor.

*Cost Minimization:* A cost-versus-tolerance relationship is determined for each contributing dimension. The optimization algorithm varies and compares the tolerances of each dimension to arrive at the combination that minimizes total cost.

*Total Cost Minimization:* This method loosens tolerances to reduce cost while considering the cost of lower yields due to production of some number of defective parts.

## Preferred Numbers

Preferred numbers are series of numbers selected to be used for standardization purposes in preference to any other numbers. Their use will lead to simplified practice, and they should be employed whenever possible for individual standard sizes and ratings, or for a series, in applications similar to the following:

1) Important or characteristic linear dimensions such as diameters and lengths, areas, volume, weights, capacities.

2) Ratings of machinery and apparatus in horsepower, kilowatts, kilovolt-amperes, voltages, currents, speeds, power-factors, pressures, heat units, temperatures, gas or liquid-flow units, weight-handling capacities, etc.

3) Characteristic ratios of figures for all kinds of units.

**American National Standard for Preferred Numbers.—**ANSI Z17.1-1973 (Withdrawn) covers basic series of preferred numbers which are independent of any measurement system and therefore can be used with metric or customary units. This Standard has been withdrawn with no superseding version specified.

The numbers are rounded values of the following five geometric series of numbers: $10^{N/5}$, $10^{N/10}$, $10^{N/20}$, $10^{N/40}$, and $10^{N/80}$, where $N$ is an integer in the series 0, 1, 2, 3, etc. The designations used for the five series are respectively R5, R10, R20, R40, and R80, where R stands for Renard (Charles Renard, originator of the first preferred number system) and the number indicates the root of 10 on which the particular series is based.

The R5 series gives 5 numbers approximately 60 percent apart, the R10 series gives 10 numbers approximately 25 percent apart, the R20 series gives 20 numbers approximately 12 percent apart, the R40 series gives 40 numbers approximately 6 percent apart, and the R80 series gives 80 numbers approximately 3 percent apart. The number of sizes for a given purpose can be minimized by using the R5 series first and then adding sizes from the R10 and R20 series as needed. The R40 and R80 series are used principally for expressing tolerances in sizes based on preferred numbers. Preferred numbers below 1 are formed by dividing the given numbers by 10, 100, etc., and numbers above 10 are obtained by multiplying the given numbers by 10, 100, etc. Sizes graded according to the system may not be exactly proportional to one another because preferred numbers may differ from calculated values by +1.26 percent to −1.01 percent. Deviations from preferred numbers are used in some instances — for example, where whole numbers are needed, such as 32 instead of 31.5 for the number of teeth in a gear.

**Basic Series of Preferred Numbers** *ANSI Z17.1-1973 (Withdrawn)*

| Series Designation | | | | | | | | |
|---|---|---|---|---|---|---|---|---|
| R5 | R10 | R20 | R40 | R40 | R80 | R80 | R80 | R80 |
| Preferred Numbers | | | | | | | | |
| 1.00 | 1.00 | 1.00 | 1.00 | 3.15 | 1.00 | 1.80 | 3.15 | 5.60 |
| 1.60 | 1.25 | 1.12 | 1.06 | 3.35 | 1.03 | 1.85 | 3.25 | 5.80 |
| 2.50 | 1.60 | 1.25 | 1.12 | 3.55 | 1.06 | 1.90 | 3.35 | 6.00 |
| 4.00 | 2.00 | 1.40 | 1.18 | 3.75 | 1.09 | 1.95 | 3.45 | 6.15 |
| 6.30 | 2.50 | 1.60 | 1.25 | 4.00 | 1.12 | 2.00 | 3.55 | 6.30 |
| … | 3.15 | 1.80 | 1.32 | 4.25 | 1.15 | 2.06 | 3.65 | 6.50 |
| … | 4.00 | 2.00 | 1.40 | 4.50 | 1.18 | 2.12 | 3.75 | 6.70 |
| … | 5.00 | 2.24 | 1.50 | 4.75 | 1.22 | 2.18 | 3.87 | 6.90 |
| … | 6.30 | 2.50 | 1.60 | 5.00 | 1.25 | 2.24 | 4.00 | 7.10 |
| … | 8.00 | 2.80 | 1.70 | 5.30 | 1.28 | 2.30 | 4.12 | 7.30 |
| … | … | 3.15 | 1.80 | 5.60 | 1.32 | 2.36 | 4.25 | 7.50 |
| … | … | 3.55 | 1.90 | 6.00 | 1.36 | 2.43 | 4.37 | 7.75 |
| … | … | 4.00 | 2.00 | 6.30 | 1.40 | 2.50 | 4.50 | 8.00 |
| … | … | 4.50 | 2.12 | 6.70 | 1.45 | 2.58 | 4.62 | 8.25 |
| … | … | 5.00 | 2.24 | 7.10 | 1.50 | 2.65 | 4.75 | 8.50 |
| … | … | 5.60 | 2.36 | 7.50 | 1.55 | 2.72 | 4.87 | 8.75 |
| … | … | 6.30 | 2.50 | 8.00 | 1.60 | 2.80 | 5.00 | 9.00 |
| … | … | 7.10 | 2.65 | 8.50 | 1.65 | 2.90 | 5.15 | 9.25 |
| … | … | 8.00 | 2.80 | 9.00 | 1.70 | 3.00 | 5.20 | 9.50 |
| … | … | 9.00 | 3.00 | 9.50 | 1.75 | 3.07 | 5.45 | 9.75 |

**Preferred Metric Sizes.—**American National Standard ANSI/ASME B32.100-2016, presents series of preferred metric sizes for round, square, rectangular, and hexagonal metal products. Table 1 gives preferred metric diameters from 1 to 320 millimeters for round metal products. Wherever possible, sizes should be selected from the Preferred Series shown in the table. A Second Preference Series is also shown. A Third Preference Series not shown in the table is: 1.3, 2.1, 2.4, 2.6, 3.2, 3.8, 4.2, 4.8, 7.5, 8.5, 9.5, 36, 85, and 95. This Standard has now been consolidated into ANSI/ASME B32.100-2016, see *Metric Sizes for Flat Metal Products* on page 2714.

ANSI/ASME B4.2-1978 (R2020) states that the basic size of mating parts should be chosen from the first choice sizes listed in Table 1. Most of the Preferred Series sizes are derived from the American National Standard "10 series" of preferred numbers (see *American National Standard for Preferred Numbers* on page 688). Most of the Second Preference Series are derived from the "20 series" of preferred numbers. Third Preference sizes are generally from the "40 series" of preferred numbers.

For preferred metric diameters less than 1 millimeter, preferred across flat metric sizes of square and hexagon metal products, preferred across flat metric sizes of rectangular metal products, and preferred metric lengths of metal products, reference should be made to the Standard.

**Table 1. American National Standard Preferred Metric Sizes**
*ANSI/ASME B4.2-1978 (R2020)*

| Basic Size, mm | | Basic Size, mm | | Basic Size, mm | | Basic Size, mm | |
|---|---|---|---|---|---|---|---|
| 1st Choice | 2nd Choice | 1st Choice | 2nd Choice | 1st Choice | 2nd Choice | 1st Choice | 2nd Choice |
| 1 | … | 6 | … | 40 | … | 250 | … |
| … | 1.1 | … | 7 | … | 45 | … | 280 |
| 1.2 | … | 8 | … | 50 | … | 300 | … |
| … | 1.4 | … | 9 | … | 55 | … | 350 |
| 1.6 | … | 10 | … | 60 | … | 400 | … |
| … | 1.8 | … | 11 | … | 70 | … | 450 |
| 2 | … | 12 | … | 80 | … | 500 | … |
| … | 2.2 | … | 14 | … | 90 | … | 550 |
| 2.5 | … | 16 | … | 100 | … | 600 | … |
| … | 2.8 | … | 18 | … | 110 | … | 700 |
| 3 | … | 20 | … | 120 | … | 800 | … |
| … | 3.5 | … | 22 | … | 140 | … | 900 |
| 4 | … | 25 | … | 160 | … | 1000 | … |
| … | 4.5 | … | 28 | … | 180 | … | … |
| 5 | … | 30 | … | 200 | … | … | … |
| … | 5.5 | … | 35 | … | 220 | … | … |

**Preferred Metric Sizes for Metal Products.—**See *Metric Sizes for Flat Metal Products* on page 2714.

**British Standard Preferred Numbers and Preferred Sizes.**—This British Standard, PD 6481:1977 (Withdrawn), gives recommendations for the use of preferred numbers and preferred sizes for functional characteristics and dimensions of various products.

The preferred number system is internationally standardized in ISO 3. It is also referred to as the Renard, or R, series (see *American National Standard for Preferred Numbers* on page 688).

The series in the preferred number system are geometric series, that is, there is a constant ratio between each figure and the succeeding one, within a decimal framework. Thus, the R5 series has five steps between 1 and 10, the R10 series has 10 steps between 1 and 10, the R20 series, 20 steps, and the R40 series, 40 steps, giving increases between steps of approximately 60, 25, 12, and 6 percent, respectively.

The preferred size series have been developed from the preferred number series by rounding off the inconvenient numbers in the basic series and adjusting for linear measurement in millimeters. These series are shown in Table 2.

After taking all normal considerations into account, it is recommended that (a) for ranges of values of the primary *functional* characteristics (outputs and capacities) of a series of products, the preferred number series R5 to R40 (see page 688) should be used, and (b) whenever linear sizes are concerned, the preferred sizes as given in the following table should be used. The presentation of preferred sizes gives designers and users a logical selection and the benefits of rational variety reduction.

The second-choice size given should only be used when it is not possible to use the first choice, and the third choice should be applied only if a size from the second choice cannot be selected. With this procedure, common usage will tend to be concentrated on a limited range of sizes, and a contribution is thus made to variety reduction. However, the decision to use a particular size cannot be taken on the basis that one is first choice and the other not. Account must be taken of the effect on the design, the availability of tools, and other relevant factors.

**Table 2. British Standard Preferred Sizes,** ***PD 6481 : 1977 (1983)***

| Choice | | | Choice | | | Choice | | | Choice | | | Choice | | | Choice | | |
|---|---|---|---|---|---|---|---|---|---|---|---|---|---|---|---|---|---|
| 1st | 2nd | 3rd | 1st | 2nd | 3rd | 1st | 2nd | 3rd | 1st | 2nd | 3rd | 1st | 2nd | 3rd | 1st | 2nd | 3rd |
| 1 | | | | | 5.2 | | | 23 | 65 | | | | | 122 | | | 188 |
| | 1.1 | | | 5.5 | | | | 24 | | | 66 | | 125 | | 190 | | |
| 1.2 | | | | | 5.8 | 25 | | | | 68 | | | | 128 | | | 192 |
| | | 1.3 | 6 | | | | | 26 | 70 | | | 130 | | | | 195 | |
| | 1.4 | | | | 6.2 | | 28 | | | 72 | | | | 132 | | | 198 |
| | | 1.5 | | 6.5 | | 30 | | | | | 74 | | 135 | | 200 | | |
| 1.6 | | | | | 6.8 | | 32 | | 75 | | | | | 138 | | | 205 |
| | | 1.7 | | 7 | | | | 34 | | | 76 | 140 | | | | 210 | |
| | 1.8 | | | | 7.5 | 35 | | | | 78 | | | | 142 | | | 215 |
| | | 1.9 | 8 | | | | | 36 | 80 | | | | 145 | | 220 | | |
| 2 | | | | | 8.5 | | 38 | | | | 82 | | | 148 | | | 225 |
| | | 2.1 | | 9 | | 40 | | | | 85 | | 150 | | | | 230 | |
| | 2.2 | | | | 9.5 | | 42 | | | | 88 | | | 152 | | | 235 |
| | | 2.4 | 10 | | | | | 44 | 90 | | | | 155 | | 240 | | |
| 2.5 | | | | 11 | | 45 | | | | | 92 | | | 158 | | | 245 |
| | | 2.6 | 12 | | | | | 46 | | 95 | | 160 | | | | 250 | |
| 2.8 | | | | | 13 | | 48 | | | | 98 | | | 162 | | | 255 |
| 3 | | | | 14 | | 50 | | | 100 | | | | 165 | | 260 | | |
| | | 3.2 | | | 15 | | 52 | | | | 102 | | | 168 | | | 265 |
| | 3.5 | | 16 | | | | | 54 | 105 | | | 170 | | | | 270 | |
| | | 3.8 | | | 17 | 55 | | | | | 108 | | | 172 | | | 275 |
| 4 | | | | 18 | | | | 56 | 110 | | | | 175 | | 280 | | |
| | | 4.2 | | | 19 | | 58 | | | | 112 | | | 178 | | | 285 |
| | 4.5 | | 20 | | | 60 | | | | 115 | | 180 | | | | 290 | |
| | | 4.8 | | | 21 | | 62 | | | | 118 | | | 182 | | | 295 |
| 5 | | | | 22 | | | | 64 | 120 | | | | 185 | | 300 | | |

For dimensions above 300, each series continues in a similar manner, i.e., the intervals between each series number are the same as between 200 and 300.

## MEASURING, INSTRUMENTS, AND INSPECTION METHODS

### Reading Verniers and Micrometers

**Reading a Vernier.—**The general procedure for determining a vernier's unit of measurement (that is, what it reads to) is first noting the value represented by the smallest subdivision on the main scale, and then dividing that value by the number of divisions on the vernier scale. Reading the indicated value is a two-part process: (1) note the value on the main scale indicated by the vernier zero line, and (2) look for coincidence between other lines on the vernier and lines on the main scale.

Fig. 1 shows an example of a dual vernier caliper with both inch and metric scales. On the upper scale, the smallest subdivision is 0.025 inch. There are 25 divisions on the vernier scale, so the inch resolution is 0.025 inch/25 = 0.001 inch. The vernier zero is just to the right of the third subdivision (0.075 inch) after 1 inch, so the indicated true value is 1.075 inch, plus a small amount indicated by vernier coincidence, in this case at line 4, making the total 1 inch + 0.075 inch + (4 × 0.001 inch) = 1.079 inch.

On the lower main scale, there are 1 mm subdivisions. There are 50 divisions on the vernier scale, so the metric resolution is 1/50 = 0.02 mm. The vernier zero is about midway between 27 mm and 28 mm. There is coincidence at 21 vernier divisions from zero (at line 42), so the indicated total value is 27 mm + 21 × 0.02 mm = 27.42 mm.

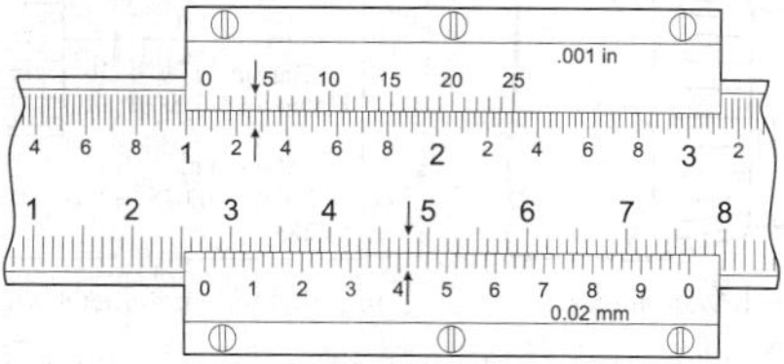


Fig. 1. Dual Inch-Metric Vernier

*Dial and Digital Vernier Calipers:* Vernier calipers are still widely used, especially for large dimensions, but the 6-inch vernier used in most shop applications has been largely superseded by dial and digital calipers, described elsewhere in this *Handbook* (see *MICROMETERS AND CALIPERS* on page 766). These have no "vernier uncertainty" and are easier for untrained users to read. A digital caliper offers added advantages: (1) it displays inches or metric; (2) it can be zeroed when the jaws are fully closed, or at any separation of the jaws (enabling measurements to be made relative to an arbitrary value instead of just zero); and (3) it usually can output measured values electronically for use in quality control and data collection applications.

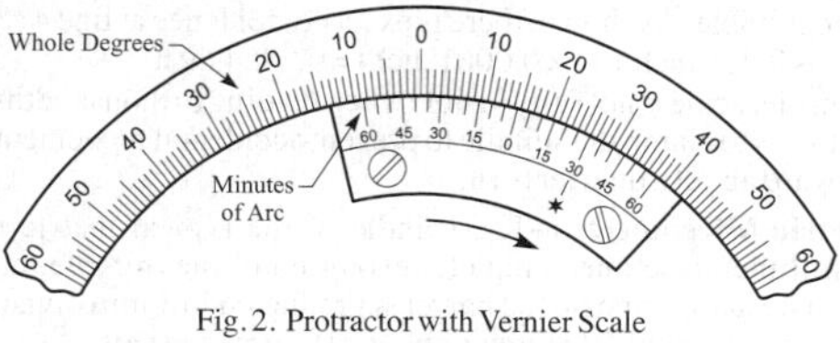


Fig. 2. Protractor with Vernier Scale

*Vernier Protractor:* The vernier on a protractor has mirror-image scales on each side of the vernier zero, graduated in minutes of arc (0-15-30-45-60) as shown in Fig. 2. Because the scales are identical to the left and the right, it is not always obvious which should be referred to. One way to avoid confusion is to start with the main scale and vernier zeros coincident, then open the protractor arms to the desired angle, noting at the same time which direction the scales move. If the vernier moves clockwise relative to the main scale, use the right-hand vernier; if counter-clockwise, use the left-hand vernier.

To measure the angle, note the main scale value indicated by the vernier 0, here between 14 and 15 degrees. Then look for coincidence between main scale and right-hand vernier graduations. This occurs at line 30, so the indicated angle is 14 degrees + 30 minutes (expressed 14°30′).

**Reading an Inch Micrometer.—**A micrometer is a highly sensitive magnifier, measuring and displaying infinitesimal spindle movements in readable increments. The barrel, or stem, encloses a precisely machined nut that mates with the threaded portion of the spindle. The spindle is rotated by a knurled thimble, usually fitted with a ratcheting device for consistent pressure between the spindle and the measured surface.

In the inch-system micrometer, the thread is 40 threads per inch, so one full turn of the thimble moves the spindle axially 1/40 of an inch (0.025 inch), which is the distance between graduations on the barrel. The thimble has 25 graduations around its circumference, so turning it through one division moves the spindle axially 1/1000 of an inch (0.001 inch).

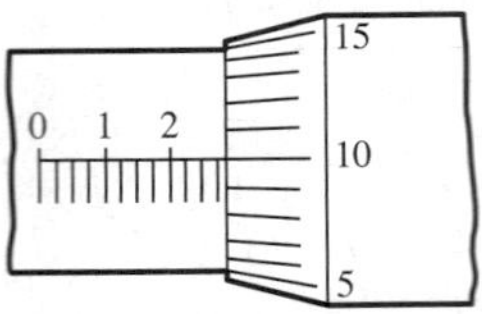


Fig. 3a. Inch Micrometer

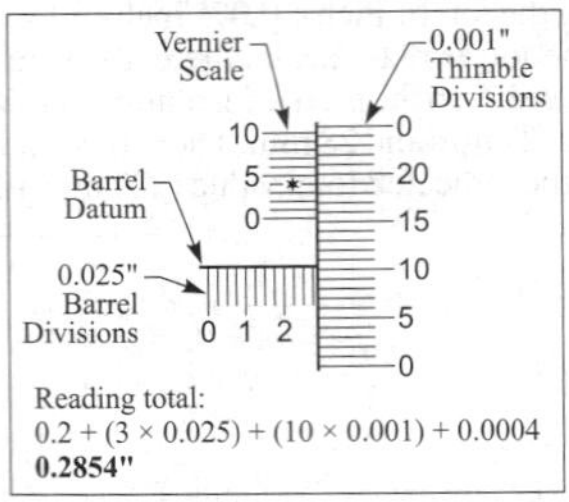


Fig. 3b. Inch Micrometer with Vernier Scale

To read an inch micrometer, as shown in Fig. 3a, note the number on the barrel (0.2 inch), then add the number of divisions (3) visible between 0.2 inch and the thimble: 0.2 inch + (3 × 0.025 inch) = 0.275 inch. Finally, add the value of the thimble graduation coinciding with the barrel datum line (10), for the total value: 0.275 inch + (10 × 0.001 inch) = 0.285 inch.

In addition to the regular graduations, some micrometers have a vernier scale reading to 0.0001 inch. The vernier has 10 divisions spanning 9 divisions on the thimble. This is "unrolled" in Fig. 3b, showing a slightly different case, where the barrel datum does not coincide exactly with 10 on the thimble, but at a fractionally greater value between 10 and 11. It usually is possible to estimate that value by interpolation, but with the vernier, there is no need for guesswork. Locate the line on the vernier scale that most closely coincides with a line on the thimble. As shown, there is exact coincidence at line 4 (*) on the vernier, so the true value is 0.285 inch + (4 × 0.0001 inch) = 0.2854 inch.

Because the vernier scale is not fully visible when reading to thousandths in the ordinary way, usual practice is to clamp the spindle to prevent accidental movement, then rotate the micrometer forward about a quarter turn.

**Reading a Metric Micrometer.—**The spindle of the typical metric micrometer has 2 threads per millimeter, so one complete revolution of the thimble moves the spindle axially 0.5 mm. The datum line on the barrel is graduated in 1 mm divisions and 0.5 mm subdivisions. The thimble has 50 graduations at 0.01 mm intervals.

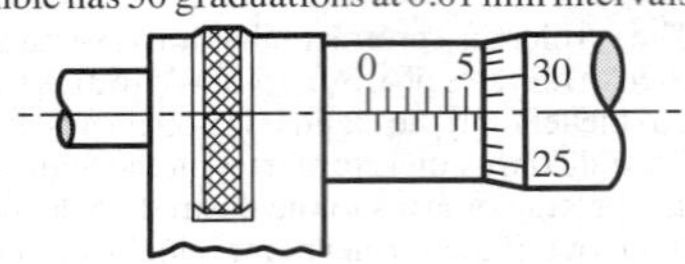


Fig. 4. Metric Micrometer

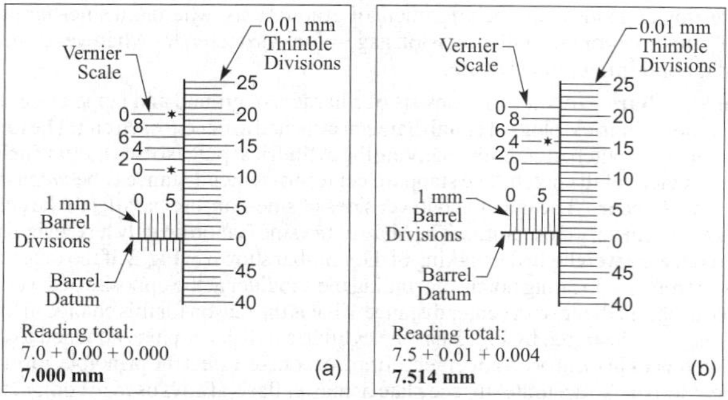


Fig. 5a and 5b. Examples of Metric Micrometer Vernier Readings

To read a metric micrometer, note the number of millimeter divisions and subdivisions visible on the barrel, then add to that total the thimble reading alongside the barrel datum. In Fig. 4, the total reading is 5.00 mm + 0.5 mm + 0.28 mm = 5.78 mm.

In addition to the regular graduations, some metric micrometers have a vernier scale, as shown "unrolled" in Figs. 5a and 5b. Unlike the inch micrometer vernier, with its 10 divisions, the metric vernier has only 5 divisions, spanning 9 divisions on the thimble. This allows readings to 0.002 mm (a little better resolution than the inch vernier micrometer's 0.0001 inch).

In Fig. 5a, the thimble has been rotated to bring the thimble edge to the 7 mm line. This is exactly on the line, because the spindle zero coincides with the barrel datum, and the vernier shows coincidence at 0 and 0 (see the *s). Therefore the reading is 7 mm.

In Fig. 5b, the thimble has been rotated fractionally more than one full turn, just exposing the 7.5 line on the barrel. The datum now coincides with the thimble between 0.01 mm and 0.02 mm. The vernier shows coincidence at line 4 (marked by *), meaning that the exact value is 0.010 mm + 0.004 mm = 0.014 mm beyond the 7.5 mm indicated by the barrel.

As with the inch micrometer, because the vernier scale is not fully visible when reading to hundredths of a millimeter, usual practice is to clamp the spindle to prevent accidental movement, then rotate the micrometer forward about a quarter turn.

**Digital Micrometers.—**Digital micrometers, also discussed elsewhere in this *Handbook* (see page 770), are becoming increasing popular, though not to the point of completely displacing conventional mechanical micrometers, which have been used since 1850. One reason for relatively slow acceptance of digital micrometers has been cost, though the price difference between digital and mechanical versions is steadily decreasing. All digital micrometers display inches or metric; many read directly in "half-tenths" (0.00005 inch) or 0.001 millimeters.

Like the digital calipers described above, digital micrometers can be zeroed when the spindle is touching the anvil (the conventional starting point), as well as at any spindle position. This enables measurements to be made relative to an arbitrary value anywhere in the 0 to 1 inch (25.4 mm) range. Most digital micrometers also can output measured values electronically for use in quality control and data collection applications, plus they feature a push-button HOLD function that saves for later viewing a displayed value otherwise unreadable due to awkward positioning.

## Sine-Bar Use

The sine-bar is used either for very accurate angular measurements or for locating work at a given angle as, for example, in surface grinding templets, gages, etc. The sine-bar is especially useful in measuring or checking angles when the limit of accuracy is 5 minutes or less. Some bevel protractors are equipped with verniers which read to 5 minutes, but the setting depends upon the alignment of graduations, whereas a sine-bar usually is located by positive contact with precision gage-blocks selected for whatever dimension is required for obtaining a given angle.

**Types of Sine-Bars.—**A sine-bar consists of a hardened, ground and lapped steel bar with very accurate cylindrical plugs of equal diameter attached to or near each end. The form illustrated by Fig. 1 has notched ends for receiving the cylindrical plugs so that they are held firmly against both faces of the notch. The standard center-to-center distance *C* between the plugs is either 5 or 10 inches. The upper and lower sides of sine-bars are parallel to the center line of the plugs within very close limits. The body of the sine-bar ordinarily has several through holes to reduce the weight. In the making of the sine-bar shown in Fig. 2, if too much material is removed from one locating notch, regrinding the shoulder at the opposite end would make it possible to obtain the correct center distance. That is the reason for this change in form. The type of sine-bar illustrated by Fig. 3 has the cylindrical disks or plugs attached to one side. These differences in form or arrangement do not, of course, affect the principle governing the use of the sine-bar. An accurate surface plate or master flat is always used in conjunction with a sine-bar in order to form the base from which the vertical measurements are made.

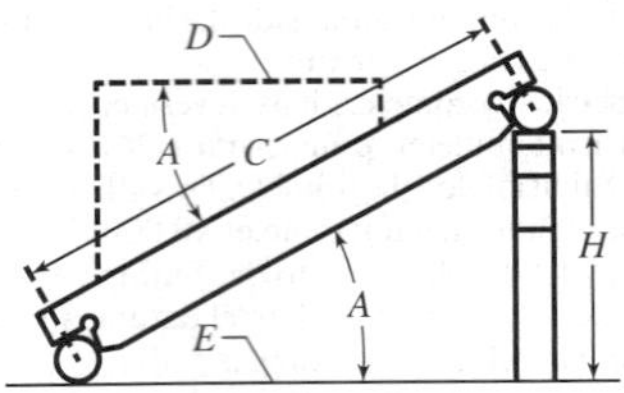

Fig. 1.

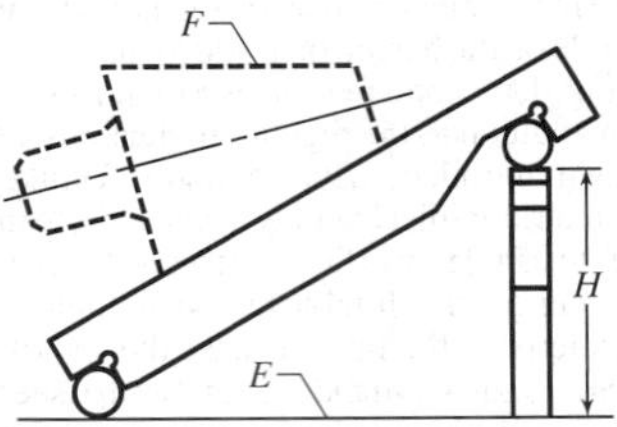

Fig. 2.

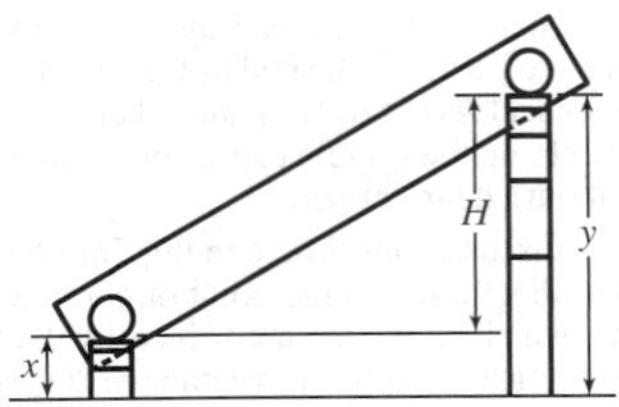

Fig. 3.

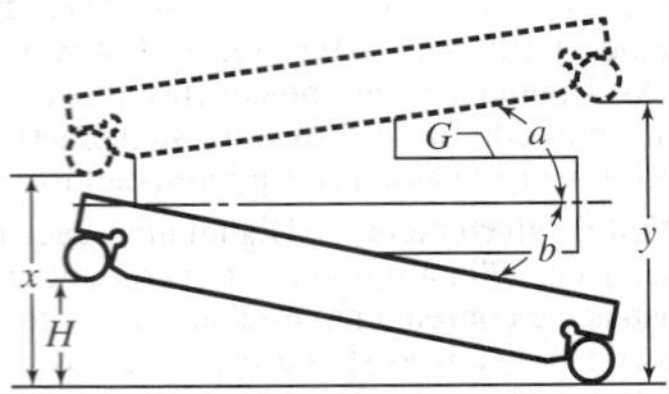

Fig. 4.

**Setting a Sine-Bar to a Given Angle.—**To find the vertical distance *H* for setting a sine-bar to the required angle, convert the angle to decimal form on a pocket calculator, take the sine of that angle, and multiply by the distance between the cylinders. For example, if an angle of 31 degrees, 30 minutes is required, the equivalent angle is 31 degrees plus $^{30}/_{60} = 31 + 0.5$, or 31.5 degrees. (For conversions from minutes and seconds to decimals of degrees and vice versa, see page 98). The sine of 31.5 degrees is 0.5225, and multiplying this value by the sine-bar length gives 2.613 inches for the height *H*, Fig. 1 and Fig. 3, of the gage blocks.

**Finding Angle when Height *H* of Sine-Bar is Known.—**To find the angle equivalent to a given height *H*, reverse the above procedure. Thus, if the height *H* is 1.4061 in., dividing by 5 gives a sine of 0.28122, which corresponds to an angle of 16.333 degrees, or 16 degrees 20 minutes.

**Checking Angle of Templet or Gage by Using Sine-Bar.—**Place templet or gage on sine-bar as indicated by dotted lines, Fig. 1. Clamps may be used to hold work in place. Place upper end of sine-bar on gage blocks having total height *H* corresponding to the required angle. If upper edge *D* of work is parallel with surface plate *E*, then angle *A* of work equals angle *A* to which sine-bar is set. Parallelism between edge *D* and surface plate may be tested by checking the height at each end with a dial gage or some type of indicating comparator.

**Measuring Angle of Templet or Gage with Sine-Bar.**—To measure such an angle, adjust height of gage blocks and sine-bar until edge *D*, Fig. 1, is parallel with surface plate *E*; then find angle corresponding to height *H* of gage blocks. For example, if height *H* is 2.5939 inches when *D* and *E* are parallel, the calculator will show that the angle *A* of the work is 31 degrees, 15 minutes.

**Checking Taper per Foot with Sine-Bar.**—As an example, assume that the plug gage in Fig. 2 is supposed to have a taper of $6\frac{1}{8}$ inches per foot and taper is to be checked by using a 5-inch sine-bar. The table of *Tapers per Foot and Corresponding Angles* on page 698 shows that the included angle for a taper of $6\frac{1}{8}$ inches per foot is 28 degrees 38 minutes 1 second, or 28.6336 degrees from the calculator. For a 5-inch sine-bar, the calculator gives a value of 2.396 inch for the height *H* of the gage blocks. Using this height, if the upper surface *F* of the plug gage is parallel to the surface plate, the angle corresponds to a taper of $6\frac{1}{8}$ inches per foot.

**Setting Sine-Bar Having Plugs Attached to Side.**—If the lower plug does not rest directly on the surface plate, as in Fig. 3, the height *H* for the sine-bar is the difference between heights *x* and *y*, or the difference between the heights of the plugs; otherwise, the procedure in setting the sine-bar and checking angles is the same as previously described.

**Checking Templets Having Two Angles.**—Assume that angle *a* of templet, Fig. 4, is 9 degrees, angle *b* is 12 degrees, and that edge *G* is parallel to the surface plate. For an angle *b* of 12 degrees, the calculator shows that the height *H* is 1.03956 inches. For an angle *a* of 9 degrees, the difference between measurements *x* and *y* when the sine-bar is in contact with the upper edge of the templet is 0.78217 inch.

**Sine-Bar Tables to Set Sine-Bars to Given Angle.**—*Machinery's Handbook 32 Digital Edition* contains tables that give constants for sine-bars of 2.5 to 10 inches and 75 to 150 mm length. These constants represent the vertical height *H* for setting a sine-bar of the corresponding length to the required angle.

**Using Sine-Bar Tables with Sine-Bars of Other Lengths.**—A sine-bar may sometimes be preferred that is longer (or shorter) than that given in available tables because of its longer working surface or because the longer center distance is conducive to greater precision. To use the sine-bar tables with a sine-bar of another length to obtain the vertical distances *H*, multiply the value obtained from the table by the fraction (length of sine-bar used ÷ length of sine-bar specified in table).

*Example:* Use the 5-inch sine-bar table to obtain the vertical height *H* for setting a 10-inch sine-bar to an angle of 39 degrees. The sine of 39 degrees is 0.62932, hence the vertical height *H* for setting a 10-inch sine-bar is 6.2932 inches.

*Solution:* The height *H* given for 39 degrees in the 5-inch sine-bar table (as shown in *Constants for 5-Inch Sine-Bar* in the *ADDITIONAL MATERIAL* in the *Machinery's Handbook 32 Digital Edition*) is 3.14660. The corresponding height for a 10-inch sine-bar is $\frac{10}{5} \times 3.14660 = 6.2932$ inches.

**Using a Calculator to Determine Sine-Bar Constants for a Given Angle.**—The constant required to set a given angle for a sine-bar of any length can be quickly determined by using a scientific calculator. The required formulas are as follows:

a) angle *A* given in degrees and calculator is set to measure angles in radian

$$H = L \times \sin\left(A \times \frac{\pi}{180}\right)$$

or

a) angle *A* is given in radian, or
b) angle *A* is given in degrees and calculator is set to measure angles in degrees

$$H = L \times \sin(A)$$

where $L$ = length of the sine-bar  $A$ = angle to which the sine-bar is to be set
$H$ = vertical height to which one end of sine-bar must be set to obtain angle $A$
$\pi = 3.141592654$

In the previous formulas, the height $H$ and length $L$ must be given in the same units but may be in either metric or US units. Thus, if $L$ is given in mm, then $H$ is in mm; and, if $L$ is given in inches, then $H$ is in inches.

**Measuring Tapers with V-Block and Sine-Bar.—**The taper on a conical part may be checked or found by placing the part in a V-block that rests on the surface of a sine-plate or sine-bar as shown in the accompanying diagram. The advantage of this method is that the axis of the V-block may be aligned with the sides of the sine-bar. Thus, when the tapered part is placed in the V-block, it will be aligned perpendicular to the transverse axis of the sine-bar.

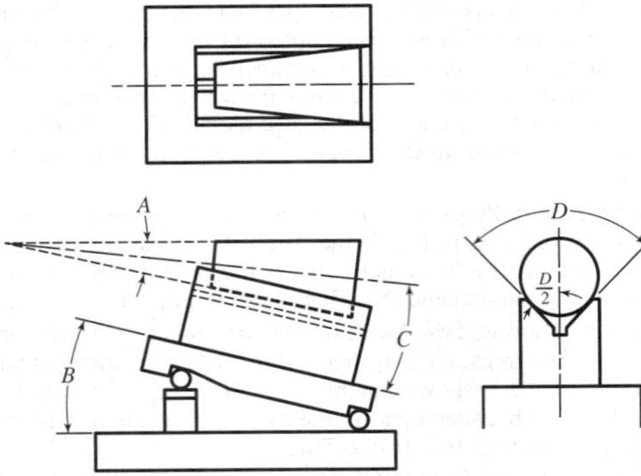


The sine-bar is set to angle $B = (C + A/2)$ where $A/2$ is one-half the included angle of the tapered part. If $D$ is the included angle of the precision V-block, the angle $C$ is calculated from the formula:

$$\sin C = \frac{\sin (A/2)}{\sin (D/2)}$$

If dial indicator readings show no change across all points along the top of the taper surface, then this checks that the angle $A$ of the taper is correct.

If the indicator readings vary, proceed as follows to find the actual angle of taper: 1) Adjust the angle of the sine-bar until the indicator reading is constant. Then find the new angle $B'$ as explained in the paragraph *Measuring Angle of Templet or Gage with Sine-Bar* on page 695; and Using the angle $B'$ calculate the actual half-angle $A'/2$ of the taper from the formula:

$$\tan\frac{A'}{2} = \frac{\sin B'}{\csc\frac{D}{2} + \cos B'}$$

The taper per foot corresponding to certain half-angles of taper may be found in the table on page 698.

**Dimensioning Tapers.—**At least three methods of dimensioning tapers are in use.

*Standard Tapers:* Give one diameter or width, the length, and insert note on drawing designating the taper by number.

*Special Tapers:* In dimensioning a taper when the slope is specified, the length and only one diameter should be given or the diameters at both ends of the taper should be given and length omitted.

*Precision Work:* In certain cases where very precise measurements are necessary, the taper surface, either external or internal, is specified by giving a diameter at a certain distance from a surface and the slope of the taper.

## Accurate Measurement of Angles and Tapers

When great accuracy is required in the measurement of angles, or when originating tapers, disks are commonly used. The principle of the disk method of taper measurement is that if two disks of unequal diameters are placed either in contact or a certain distance apart, lines tangent to their peripheries will represent an angle or taper, the degree of which depends upon the diameters of the two disks and the distance between them.

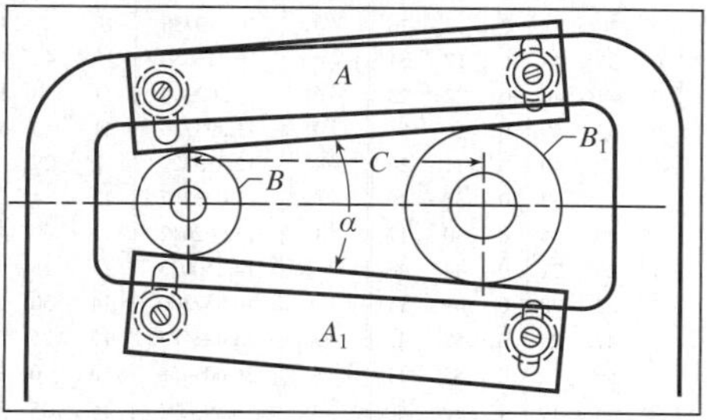


The gage shown in the accompanying illustration, which is a form commonly used for originating tapers or measuring angles accurately, is set by means of disks. This gage consists of two adjustable straight edges $A$ and $A_1$, which are in contact with disks $B$ and $B_1$. The angle α or the taper between the straight edges depends, of course, upon the diameters of the disks and the center distance $C$, and, as these three dimensions can be measured accurately, it is possible to set the gage to a given angle within very close limits. Moreover, if a record of the three dimensions is kept, the exact setting of the gage can be reproduced quickly at any time. The following rules may be used for adjusting a gage of this type and cover all problems likely to arise in practice. Disks are also occasionally used for the setting of parts in angular positions when they are to be machined accurately to a given angle: the rules are applicable to these conditions also.

**Measuring Dovetail Slides.—**Dovetail slides that must be machined accurately to a given width are commonly gaged by using pieces of cylindrical rod or wire and measuring as indicated by the dimensions *x* and *y* of the accompanying illustrations.

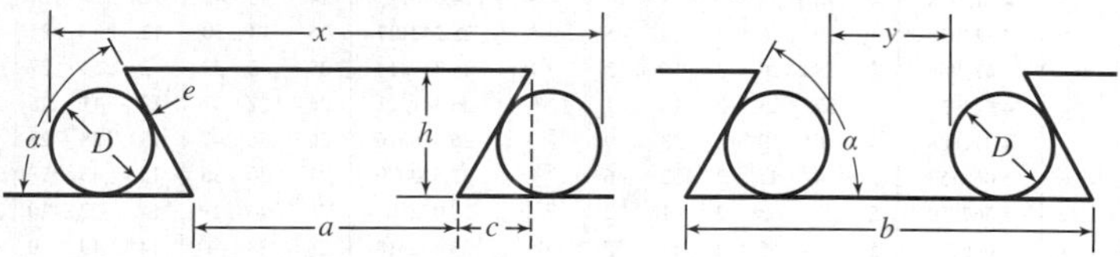


The rod or wire used should be small enough so that the point of contact *e* is somewhat below the corner or edge of the dovetail.

To obtain dimension *x* for measuring male dovetails, add 1 to the cotangent of one-half the dovetail angle α, multiply by diameter *D* of the rods used, and add the product to dimension α.

$$x = D\left(1 + \cot\frac{1}{2}\alpha\right) + a \qquad c = h \times \cot\alpha$$

To obtain dimension *y* for measuring a female dovetail, add 1 to the cotangent of one-half the dovetail angle α, multiply by diameter *D* of the rod used, and subtract the result from dimension *b*. Expressing these rules as formulas:

$$y = b - D\left(1 + \cot\frac{1}{2}\alpha\right)$$

**Tapers per Foot and Corresponding Angles**

| Taper per Foot | Included Angle | | | | Angle with Center Line | | | Taper per Foot | Included Angle | | | | Angle with Center Line | | |
|---|---|---|---|---|---|---|---|---|---|---|---|---|---|---|---|
| 1/64 | 0.074604° | 0° | 4′ | 29″ | 0° | 2′ | 14″ | 1 7/8 | 8.934318° | 8° | 56′ | 4″ | 4° | 28′ | 2″ |
| 1/32 | 0.149208° | 0 | 8 | 57 | 0 | 4 | 29 | 1 15/16 | 9.230863° | 9 | 13 | 51 | 4 | 36 | 56 |
| 1/16 | 0.298415 | 0 | 17 | 54 | 0 | 8 | 57 | 2 | 9.527283 | 9 | 31 | 38 | 4 | 45 | 49 |
| 3/32 | 0.447621 | 0 | 26 | 51 | 0 | 13 | 26 | 2 1/8 | 10.119738 | 10 | 7 | 11 | 5 | 3 | 36 |
| 1/8 | 0.596826 | 0 | 35 | 49 | 0 | 17 | 54 | 2 1/4 | 10.711650 | 10 | 42 | 42 | 5 | 21 | 21 |
| 5/32 | 0.746028 | 0 | 44 | 46 | 0 | 22 | 23 | 2 3/8 | 11.302990 | 11 | 18 | 11 | 5 | 39 | 5 |
| 3/16 | 0.895228 | 0 | 53 | 43 | 0 | 26 | 51 | 2 1/2 | 11.893726 | 11 | 53 | 37 | 5 | 56 | 49 |
| 7/32 | 1.044425 | 1 | 2 | 40 | 0 | 31 | 20 | 2 5/8 | 12.483829 | 12 | 29 | 2 | 6 | 14 | 31 |
| 1/4 | 1.193619 | 1 | 11 | 37 | 0 | 35 | 49 | 2 3/4 | 13.073267 | 13 | 4 | 24 | 6 | 32 | 12 |
| 9/32 | 1.342808 | 1 | 20 | 34 | 0 | 40 | 17 | 2 7/8 | 13.662012 | 13 | 39 | 43 | 6 | 49 | 52 |
| 5/16 | 1.491993 | 1 | 29 | 31 | 0 | 44 | 46 | 3 | 14.250033 | 14 | 15 | 0 | 7 | 7 | 30 |
| 11/32 | 1.641173 | 1 | 38 | 28 | 0 | 49 | 14 | 3 1/8 | 14.837300 | 14 | 50 | 14 | 7 | 25 | 7 |
| 3/8 | 1.790347 | 1 | 47 | 25 | 0 | 53 | 43 | 3 1/4 | 15.423785 | 15 | 25 | 26 | 7 | 42 | 43 |
| 13/32 | 1.939516 | 1 | 56 | 22 | 0 | 58 | 11 | 3 3/8 | 16.009458 | 16 | 0 | 34 | 8 | 0 | 17 |
| 7/16 | 2.088677 | 2 | 5 | 19 | 1 | 2 | 40 | 3 1/2 | 16.594290 | 16 | 35 | 39 | 8 | 17 | 50 |
| 15/32 | 2.237832 | 2 | 14 | 16 | 1 | 7 | 8 | 3 5/8 | 17.178253 | 17 | 10 | 42 | 8 | 35 | 21 |
| 1/2 | 2.386979 | 2 | 23 | 13 | 1 | 11 | 37 | 3 3/4 | 17.761318 | 17 | 45 | 41 | 8 | 52 | 50 |
| 17/32 | 2.536118 | 2 | 32 | 10 | 1 | 16 | 5 | 3 7/8 | 18.343458 | 18 | 20 | 36 | 9 | 10 | 18 |
| 9/16 | 2.685248 | 2 | 41 | 7 | 1 | 20 | 33 | 4 | 18.924644 | 18 | 55 | 29 | 9 | 27 | 44 |
| 19/32 | 2.834369 | 2 | 50 | 4 | 1 | 25 | 2 | 4 1/8 | 19.504850 | 19 | 30 | 17 | 9 | 45 | 9 |
| 5/8 | 2.983481 | 2 | 59 | 1 | 1 | 29 | 30 | 4 1/4 | 20.084047 | 20 | 5 | 3 | 10 | 2 | 31 |
| 21/32 | 3.132582 | 3 | 7 | 57 | 1 | 33 | 59 | 4 3/8 | 20.662210 | 20 | 39 | 44 | 10 | 19 | 52 |
| 11/16 | 3.281673 | 3 | 16 | 54 | 1 | 38 | 27 | 4 1/2 | 21.239311 | 21 | 14 | 22 | 10 | 37 | 11 |
| 23/32 | 3.430753 | 3 | 25 | 51 | 1 | 42 | 55 | 4 5/8 | 21.815324 | 21 | 48 | 55 | 10 | 54 | 28 |
| 3/4 | 3.579821 | 3 | 34 | 47 | 1 | 47 | 24 | 4 3/4 | 22.390223 | 22 | 23 | 25 | 11 | 11 | 42 |
| 25/32 | 3.728877 | 3 | 43 | 44 | 1 | 51 | 52 | 4 7/8 | 22.963983 | 22 | 57 | 50 | 11 | 28 | 55 |
| 13/16 | 3.877921 | 3 | 52 | 41 | 1 | 56 | 20 | 5 | 23.536578 | 23 | 32 | 12 | 11 | 46 | 6 |
| 27/32 | 4.026951 | 4 | 1 | 37 | 2 | 0 | 49 | 5 1/8 | 24.107983 | 24 | 6 | 29 | 12 | 3 | 14 |
| 7/8 | 4.175968 | 4 | 10 | 33 | 2 | 5 | 17 | 5 1/4 | 24.678175 | 24 | 40 | 41 | 12 | 20 | 21 |
| 29/32 | 4.324970 | 4 | 19 | 30 | 2 | 9 | 45 | 5 3/8 | 25.247127 | 25 | 14 | 50 | 12 | 37 | 25 |
| 15/16 | 4.473958 | 4 | 28 | 26 | 2 | 14 | 13 | 5 1/2 | 25.814817 | 25 | 48 | 53 | 12 | 54 | 27 |
| 31/32 | 4.622931 | 4 | 37 | 23 | 2 | 18 | 41 | 5 5/8 | 26.381221 | 26 | 22 | 52 | 13 | 11 | 26 |
| 1 | 4.771888 | 4 | 46 | 19 | 2 | 23 | 9 | 5 3/4 | 26.946316 | 26 | 56 | 47 | 13 | 28 | 23 |
| 1 1/16 | 5.069753 | 5 | 4 | 11 | 2 | 32 | 6 | 5 7/8 | 27.510079 | 27 | 30 | 36 | 13 | 45 | 18 |
| 1 1/8 | 5.367550 | 5 | 22 | 3 | 2 | 41 | 2 | 6 | 28.072487 | 28 | 4 | 21 | 14 | 2 | 10 |
| 1 3/16 | 5.665275 | 5 | 39 | 55 | 2 | 49 | 57 | 6 1/8 | 28.633518 | 28 | 38 | 1 | 14 | 19 | 0 |
| 1 1/4 | 5.962922 | 5 | 57 | 47 | 2 | 58 | 53 | 6 1/4 | 29.193151 | 29 | 11 | 35 | 14 | 35 | 48 |
| 1 5/16 | 6.260490 | 6 | 15 | 38 | 3 | 7 | 49 | 6 3/8 | 29.751364 | 29 | 45 | 5 | 14 | 52 | 32 |
| 1 3/8 | 6.557973 | 6 | 33 | 29 | 3 | 16 | 44 | 6 1/2 | 30.308136 | 30 | 18 | 29 | 15 | 9 | 15 |
| 1 7/16 | 6.855367 | 6 | 51 | 19 | 3 | 25 | 40 | 6 5/8 | 30.863447 | 30 | 51 | 48 | 15 | 25 | 54 |
| 1 1/2 | 7.152669 | 7 | 9 | 10 | 3 | 34 | 35 | 6 3/4 | 31.417276 | 31 | 25 | 2 | 15 | 42 | 31 |
| 1 9/16 | 7.449874 | 7 | 27 | 0 | 3 | 43 | 30 | 6 7/8 | 31.969603 | 31 | 58 | 11 | 15 | 59 | 5 |
| 1 5/8 | 7.746979 | 7 | 44 | 49 | 3 | 52 | 25 | 7 | 32.520409 | 32 | 31 | 13 | 16 | 15 | 37 |
| 1 11/16 | 8.043980 | 8 | 2 | 38 | 4 | 1 | 19 | 7 1/8 | 33.069676 | 33 | 4 | 11 | 16 | 32 | 5 |
| 1 3/4 | 8.340873 | 8 | 20 | 27 | 4 | 10 | 14 | 7 1/4 | 33.617383 | 33 | 37 | 3 | 16 | 48 | 31 |
| 1 13/16 | 8.637654 | 8 | 38 | 16 | 4 | 19 | 8 | 7 3/8 | 34.163514 | 34 | 9 | 49 | 17 | 4 | 54 |

Taper per foot represents inches of taper per foot of length. For conversions into decimal degrees and radians see *Conversion Tables of Angular Measure* on page 98.

**Rules for Figuring Tapers**

$D_L$ = diameter of taper, large end
$D_S$ = diameter of taper, small end
$L$ = length of taper (inches)
$TPF$ = taper per foot
$TPI$ = taper per inch
$T$ = taper in a certain length, in inches

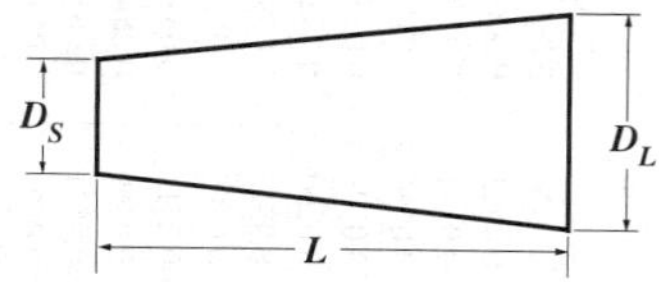


| Given | To Find | Rule |
|---|---|---|
| $TPF$ | $TPI$ | $TPI = TPF/12$ |
| $TPI$ | $TPF$ | $TPF = TPI \times 12$ |
| $D_L, D_S, L$ | $TPF$ | $TPF = 12\dfrac{D_L - D_S}{L}$ |
| $D_L, L, TPF$ | $D_S$ | $D_S = D_L - \dfrac{TPF}{12} L$ |
| $D_S, L, TPF$ | $D_L$ | $D_L = D_S + \dfrac{TPF}{12} L$ |
| $D_L, D_S, TPF$ | $L$ | $L = (D_L - D_S)\dfrac{12}{TPF}$ |
| $L, TPF$ | $T$ | $T = \dfrac{TPF}{12} \times L$ |

**Measuring Tapers and Angles.—**

*To Find Angle α for Given Taper* T *in Inches per Foot:*

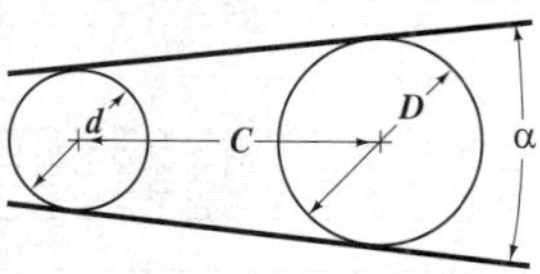


$$\alpha = 2\ \arctan(T/24)$$

*Example:* What angle α is equivalent to a taper of 1.5 inches per foot?

$$\alpha = 2 \times \arctan\ (1.5/24) = 7.153°$$

*To Find Taper per Foot* T *Given Angle α in Degrees:*

$$T = 24\tan\left(\alpha/2\right) \text{ inches per foot}$$

*Example:* What taper $T$ is equivalent to an angle of 7.153°?

$$T = 24\tan\left(7.153/2\right) = 1.5 \text{ inches per foot}$$

*To Find Angle α Given Dimensions* D, d, *and* C*:* Let $K$ be the difference in the disk diameters divided by twice the center distance. $K = (D - d)/(2C)$, then $\alpha = 2\arcsin K$.

*Example:* If the disk diameters $d$ and $D$ are 1 and 1.5 inches, respectively, and the center distance $C$ is 5 inches, find the included angle α.

$$K = (1.5 - 1) / (2 \times 5) = 0.05 \qquad \alpha = 2 \times \arcsin 0.05 = 5.732°$$

## Amount of Taper in a Given Length, When the Taper per Foot is Known

| Length of Taper | Taper per Foot | | | | | | | | | | |
|---|---|---|---|---|---|---|---|---|---|---|---|
| | 1/16 | 3/32 | 1/8 | 1/4 | 3/8 | 1/2 | 0.600 | 5/8 | 3/4 | 1 | 1 1/4 |
| 1/32 | 0.0002 | 0.0002 | 0.0003 | 0.0007 | 0.0010 | 0.0013 | 0.0016 | 0.0016 | 0.0020 | 0.0026 | 0.0033 |
| 1/16 | 0.0003 | 0.0005 | 0.0007 | 0.0013 | 0.0020 | 0.0026 | 0.0031 | 0.0033 | 0.0039 | 0.0052 | 0.0065 |
| 1/8 | 0.0007 | 0.0010 | 0.0013 | 0.0026 | 0.0039 | 0.0052 | 0.0063 | 0.0065 | 0.0078 | 0.0104 | 0.0130 |
| 3/16 | 0.0010 | 0.0015 | 0.0020 | 0.0039 | 0.0059 | 0.0078 | 0.0094 | 0.0098 | 0.0117 | 0.0156 | 0.0195 |
| 1/4 | 0.0013 | 0.0020 | 0.0026 | 0.0052 | 0.0078 | 0.0104 | 0.0125 | 0.0130 | 0.0156 | 0.0208 | 0.0260 |
| 5/16 | 0.0016 | 0.0024 | 0.0033 | 0.0065 | 0.0098 | 0.0130 | 0.0156 | 0.0163 | 0.0195 | 0.0260 | 0.0326 |
| 3/8 | 0.0020 | 0.0029 | 0.0039 | 0.0078 | 0.0117 | 0.0156 | 0.0188 | 0.0195 | 0.0234 | 0.0313 | 0.0391 |
| 7/16 | 0.0023 | 0.0034 | 0.0046 | 0.0091 | 0.0137 | 0.0182 | 0.0219 | 0.0228 | 0.0273 | 0.0365 | 0.0456 |
| 1/2 | 0.0026 | 0.0039 | 0.0052 | 0.0104 | 0.0156 | 0.0208 | 0.0250 | 0.0260 | 0.0313 | 0.0417 | 0.0521 |
| 9/16 | 0.0029 | 0.0044 | 0.0059 | 0.0117 | 0.0176 | 0.0234 | 0.0281 | 0.0293 | 0.0352 | 0.0469 | 0.0586 |
| 5/8 | 0.0033 | 0.0049 | 0.0065 | 0.0130 | 0.0195 | 0.0260 | 0.0313 | 0.0326 | 0.0391 | 0.0521 | 0.0651 |
| 11/16 | 0.0036 | 0.0054 | 0.0072 | 0.0143 | 0.0215 | 0.0286 | 0.0344 | 0.0358 | 0.0430 | 0.0573 | 0.0716 |
| 3/4 | 0.0039 | 0.0059 | 0.0078 | 0.0156 | 0.0234 | 0.0313 | 0.0375 | 0.0391 | 0.0469 | 0.0625 | 0.0781 |
| 13/16 | 0.0042 | 0.0063 | 0.0085 | 0.0169 | 0.0254 | 0.0339 | 0.0406 | 0.0423 | 0.0508 | 0.0677 | 0.0846 |
| 7/8 | 0.0046 | 0.0068 | 0.0091 | 0.0182 | 0.0273 | 0.0365 | 0.0438 | 0.0456 | 0.0547 | 0.0729 | 0.0911 |
| 15/16 | 0.0049 | 0.0073 | 0.0098 | 0.0195 | 0.0293 | 0.0391 | 0.0469 | 0.0488 | 0.0586 | 0.0781 | 0.0977 |
| 1 | 0.0052 | 0.0078 | 0.0104 | 0.0208 | 0.0313 | 0.0417 | 0.0500 | 0.0521 | 0.0625 | 0.0833 | 0.1042 |
| 2 | 0.0104 | 0.0156 | 0.0208 | 0.0417 | 0.0625 | 0.0833 | 0.1000 | 0.1042 | 0.1250 | 0.1667 | 0.2083 |
| 3 | 0.0156 | 0.0234 | 0.0313 | 0.0625 | 0.0938 | 0.1250 | 0.1500 | 0.1563 | 0.1875 | 0.2500 | 0.3125 |
| 4 | 0.0208 | 0.0313 | 0.0417 | 0.0833 | 0.1250 | 0.1667 | 0.2000 | 0.2083 | 0.2500 | 0.3333 | 0.4167 |
| 5 | 0.0260 | 0.0391 | 0.0521 | 0.1042 | 0.1563 | 0.2083 | 0.2500 | 0.2604 | 0.3125 | 0.4167 | 0.5208 |
| 6 | 0.0313 | 0.0469 | 0.0625 | 0.1250 | 0.1875 | 0.2500 | 0.3000 | 0.3125 | 0.3750 | 0.5000 | 0.6250 |
| 7 | 0.0365 | 0.0547 | 0.0729 | 0.1458 | 0.2188 | 0.2917 | 0.3500 | 0.3646 | 0.4375 | 0.5833 | 0.7292 |
| 8 | 0.0417 | 0.0625 | 0.0833 | 0.1667 | 0.2500 | 0.3333 | 0.4000 | 0.4167 | 0.5000 | 0.6667 | 0.8333 |
| 9 | 0.0469 | 0.0703 | 0.0938 | 0.1875 | 0.2813 | 0.3750 | 0.4500 | 0.4688 | 0.5625 | 0.7500 | 0.9375 |
| 10 | 0.0521 | 0.0781 | 0.1042 | 0.2083 | 0.3125 | 0.4167 | 0.5000 | 0.5208 | 0.6250 | 0.8333 | 1.0417 |
| 11 | 0.0573 | 0.0859 | 0.1146 | 0.2292 | 0.3438 | 0.4583 | 0.5500 | 0.5729 | 0.6875 | 0.9167 | 1.1458 |
| 12 | 0.0625 | 0.0938 | 0.1250 | 0.2500 | 0.3750 | 0.5000 | 0.6000 | 0.6250 | 0.7500 | 1.0000 | 1.2500 |

The amount of taper in a certain length of inches is equal to the taper per foot divided by 12 then multiplied by the given length of the tapered part.

*To Find Taper* T *Measured at Right Angles to a Line Through the Disk Centers Given Dimensions* D, d, *and Distance* C*:* Find $K$ using the formula in the previous example, then $T = 24K/\sqrt{1 - K^2}$ inches per foot.

*Example:* If disk diameters $d$ and $D$ are 1 and 1.5 inches, respectively, and the center distance $C$ is 5 inches, find the taper per foot.

$$K = (1.5 - 1)/(2 \times 5) = 0.05 \qquad T = \frac{24 \times 0.05}{\sqrt{1 - (0.05)^2}} = 1.2015 \text{ inches per foot}$$

*To Find Center Distance* C *for a Given Taper* T *in Inches per Foot:*

$$C = \frac{D - d}{2} \times \frac{\sqrt{1 + (T/24)^2}}{T/24} \text{ inches}$$

*Example:* Gage is to be set to ¾ inch per foot, and disk diameters are 1.25 and 1.5 inches, respectively. Find the required center distance for the disks.

$$C = \frac{1.5 - 1.25}{2} \times \frac{\sqrt{1 + (0.75/24)^2}}{0.75/24} = 4.002 \text{ inches}$$

*To Find Center Distance* C *for a Given Angle* α *and Dimensions* D *and* d*:*

$$C = (D - d)/2 \ \sin(\alpha/2) \text{ inches}$$

*Example:* If an angle α of 20° is required, and the disks are 1 and 3 inches in diameter, respectively, find the required center distance $C$.

$$C = (3 - 1)/(2 \times \sin 10°) = 5.759 \text{ inches}$$

*To Find Taper* T *Measured at Right Angles to One Side:* When one side is taken as a base line and the taper is measured at right angles to that side, calculate $K$ as explained above and use the following formula for determining the taper $T$:

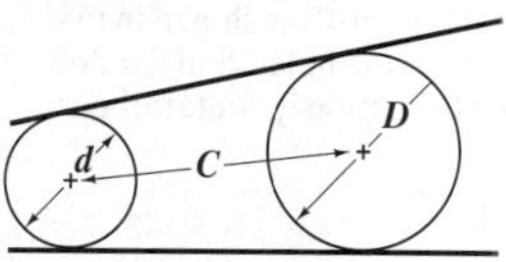


$$T = 24K\frac{\sqrt{1 - K^2}}{1 - 2K^2} \text{ inches per foot}$$

*Example:* If the disk diameters are 2 and 3 inches, respectively, and the center distance is 5 inches, what is the taper per foot measured at right angles to one side?

$$K = \frac{3 - 2}{2 \times 5} = 0.1 \qquad T = 24 \times 0.1 \times \frac{\sqrt{1 - (0.1)^2}}{1 - [2 \times (0.1)^2]} = 2.4367 \text{ in. per ft.}$$

*To Find Center Distance* C *When Taper* T *is Measured from One Side:*

$$C = \frac{D - d}{\sqrt{2 - 2/\sqrt{1 + (T/12)^2}}} \text{ inches}$$

*Example:* If the taper measured at right angles to one side is 6.9 inches per foot, and the disks are 2 and 5 inches in diameter, respectively, what is center distance $C$?

$$C = \frac{5 - 2}{\sqrt{2 - 2\sqrt{1 = (6.9/12)^2}}} = 5.815 \text{ inches}.$$

*To Find Diameter* D *of a Large Disk in Contact with a Small Disk of Diameter* d *Given Angle* $\alpha$*:*

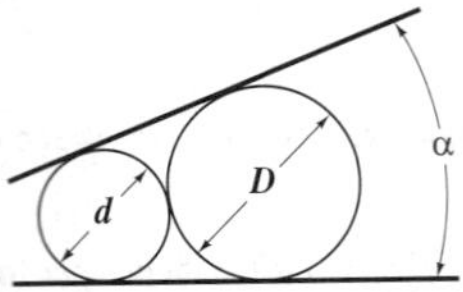


$$D = d \times \frac{1 + \sin(\alpha/2)}{1 - \sin(\alpha/2)} \text{ inches}$$

*Example:* The required angle $\alpha$ is 15°. Find diameter $D$ of a large disk that is in contact with a standard 1-inch reference disk.

$$D = 1 \times \frac{1 + \sin 7.5°}{1 - \sin 7.5°} = 1.3002 \text{ inches}$$

## Measurement Over Pins and Rolls

**Measurement Over Pins.—**When the distance across a bolt circle is too large to measure using ordinary measuring tools, then the required distance may be found from the distance across adjacent or alternate holes using one of the methods that follow:

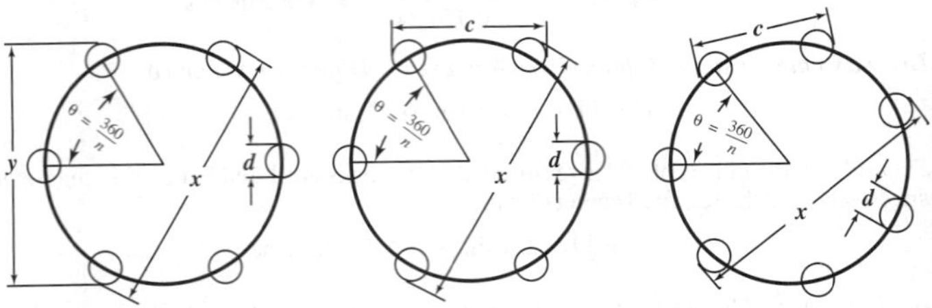


Fig. 1a. Fig. 1b. Fig. 1c.

*Even Number of Holes in Circle:* To measure the unknown distance $x$ over opposite plugs in a bolt circle of $n$ holes ($n$ is even and greater than 4), as shown in Fig. 1a, where $y$ is the distance over alternate plugs, $d$ is the diameter of the holes, and $\theta = 360/n$ is the angle between adjacent holes, use the following general equation for obtaining $x$:

$$x = \frac{y - d}{\sin\theta} + d$$

*Example:* In a die that has six 3/4-inch diameter holes equally spaced on a circle, where the distance $y$ over alternate holes is $4\frac{1}{2}$ inches, and the angle $\theta$ between adjacent holes is 60, then

$$x = \frac{4.500 - 0.7500}{\sin 60°} + 0.7500 = 5.0801$$

In a similar problem, the distance $c$ over adjacent plugs is given, as shown in Fig. 1b. If the number of holes is even and greater than 4, the distance $x$ over opposite plugs is given in the following formula:

$$x = 2(c - d)\left(\frac{\sin\left(\frac{180 - \theta}{2}\right)}{\sin\theta}\right) + d$$

where $d$ and $\theta$ are as defined above.

*Odd Number of Holes in Circle:* In a circle as shown in Fig. 1c, where the number of holes $n$ is odd and greater than 3, and the distance $c$ over adjacent holes is given, then θ equals $360/n$ and the distance $x$ across the most widely spaced holes is given by:

$$x = \frac{\frac{c-d}{2}}{\sin\frac{\theta}{4}} + d$$

**Checking a V-Shaped Groove by Measurement Over Pins.—** In checking a groove of the shape shown in Fig. 2, it is necessary to measure the dimension $X$ over the pins of radius $R$. If values for the radius $R$, dimension $Z$, and the angles α and β are known, the problem is to determine the distance $Y$, to arrive at the required overall dimension for $X$. If a line $AC$ is drawn from the bottom of the V to the center of the pin at the left in Fig. 2, and a line $CB$ from the center of this pin to its point of tangency with the side of the V, a right-angle triangle is formed in which one side, $CB$, is known and one angle, $CAB$, can be determined. A line drawn from the center of a circle to the point of intersection of two tangents to the circle bisects the angle made by the tangent lines, and angle $CAB$ therefore equals $\frac{1}{2}(\alpha + \beta)$. The length $AC$ and the angle $DAC$ can now be found, and with $AC$ known in the right-angle triangle $ADC$, $AD$, which is equal to $Y$, can be found.

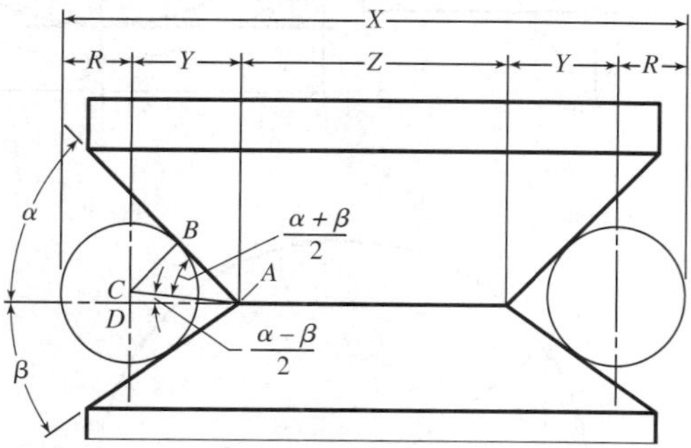


Fig. 2.

The value for $X$ can be obtained from the formula

$$X = Z + 2R\left(\csc\frac{\alpha+\beta}{2}\cos\frac{\alpha-\beta}{2} + 1\right)$$

For example, if $R = 0.500$, $Z = 1.824$, $\alpha = 45$ degrees, and $\beta = 35$ degrees,

$$X = 1.824 + (2 \times 0.5)\left(\csc\frac{45° + 35°}{2}\cos\frac{45° - 35°}{2} + 1\right)$$

$$X = 1.824 + \csc 40° \cos 5° + 1$$

$$X = 1.824 + 1.5557 \times 0.99619 + 1$$

$$X = 1.824 + 1.550 + 1 = 4.374$$

**Checking Radius of Arc by Measurement Over Rolls.—** The radius $R$ of large-radius concave and convex gages of the type shown in Fig. 3a, Fig. 3b and Fig. 3c can be checked by measurement $L$ over two rolls with the gage resting on the rolls as shown. If the diameter of the rolls $D$, the length $L$, and the height $H$ of the top of the arc above the surface plate (for the concave gage, Fig. 3a) are known or can be measured, the radius $R$ of the workpiece to be checked can be calculated trigonometrically, as follows.

Referring to Fig. 3a for the concave gage, if $L$ and $D$ are known, $cb$ can be found, and if $H$ and $D$ are known, $ce$ can be found. With $cb$ and $ce$ known, $ab$ can be found by means of a diagram as shown in Fig. 3c.

In diagram Fig. 3c, $cb$ and $ce$ are shown at right angles as in Fig. 3a. A line is drawn connecting points $b$ and $e$, and line $ce$ is extended to the right. A line is now drawn from point $b$ perpendicular to $be$ and intersecting the extension of $ce$ at point $f$. A semicircle can now be drawn through points $b$, $e$, and $f$ with point $a$ as the center. Triangles $bce$ and $bcf$ are similar and have a common side. Thus $ce$:$bc$::$bc$:$cf$. With $ce$ and $bc$ known, $cf$ can be found from this proportion and, hence, $ef$, which is the diameter of the semicircle and radius $ab$. Then $R = ab + D/2$.

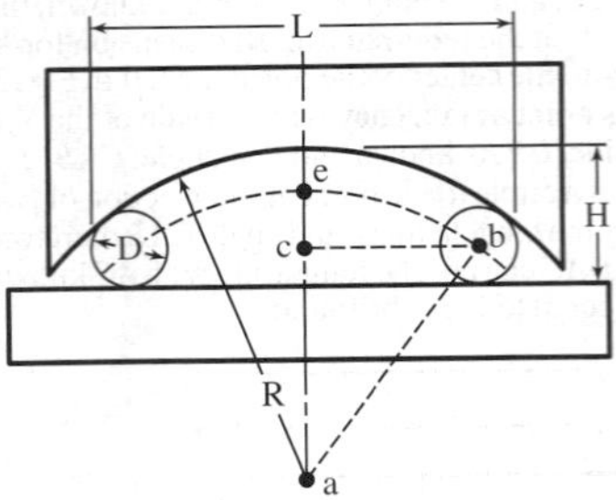


Fig. 3a.

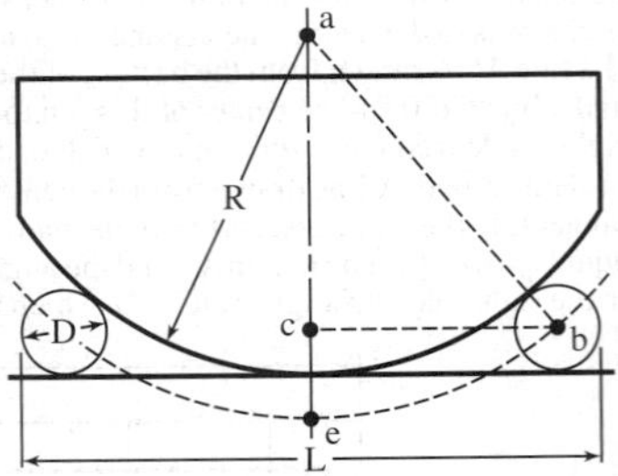


Fig. 3b.

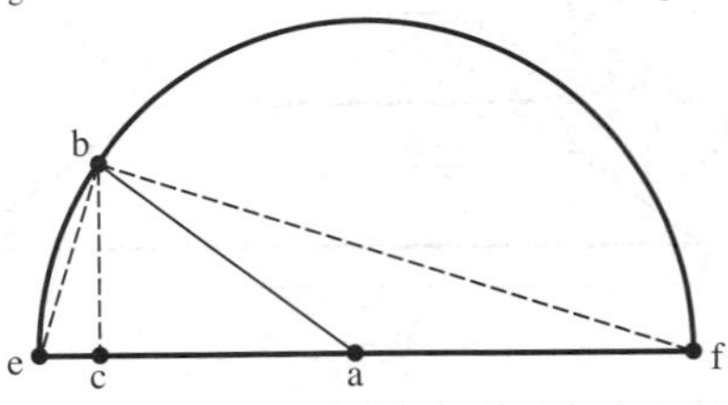


Fig. 3c.

The procedure for the convex gage is similar. The distances $cb$ and $ce$ are readily found, and from these two distances $ab$ is computed on the basis of similar triangles as before. Radius $R$ is then readily found.

The derived formulas for concave and convex gages are as follows:

Formulas:

$$R = \frac{(L-D)^2}{8(H-D)} + \frac{H}{2} \qquad \text{(Concave gage Fig. 3a)}$$

$$R = \frac{(L-D)^2}{8D} \qquad \text{(Convex gage Fig. 3b)}$$

For example: For Fig. 3a, let $L = 17.8$, $D = 3.20$, and $H = 5.72$, then

$$R = \frac{(17.8 - 3.20)^2}{8(5.72 - 3.20)} + \frac{5.72}{2} = \frac{(14.60)^2}{8 \times 2.52} + 2.86$$

$$R = \frac{213.16}{20.16} + 2.86 = 13.43$$

For Fig. 3b, let $L = 22.28$ and $D = 3.40$, then

$$R = \frac{(22.28 - 3.40)^2}{8 \times 3.40} = \frac{356.45}{27.20} = 13.1$$

## Checking Shaft Conditions

**Checking for Various Shaft Conditions.—**An indicating height gage, together with V-blocks can be used to check shafts for ovality, taper, straightness (bending or curving), and concentricity of features (as shown exaggerated in Fig. 4). If a shaft on which work has been completed shows lack of concentricity, it may be due to the shaft having become bent or bowed because of mishandling or oval or tapered due to poor machine conditions. In checking for concentricity, the first step is to check for ovality, or out-of-roundness, as in Fig. 4a. The shaft is supported in a suitable V-block on a surface table, and the dial indicator plunger is placed over the workpiece, which is then rotated beneath the plunger to obtain readings of the amount of eccentricity.

This procedure (sometimes called clocking, owing to the resemblance of the dial indicator to a clock face) is repeated for other shaft diameters as necessary, and, in addition to making a written record of the measurements, the positions of extreme conditions should be marked on the workpiece for later reference.

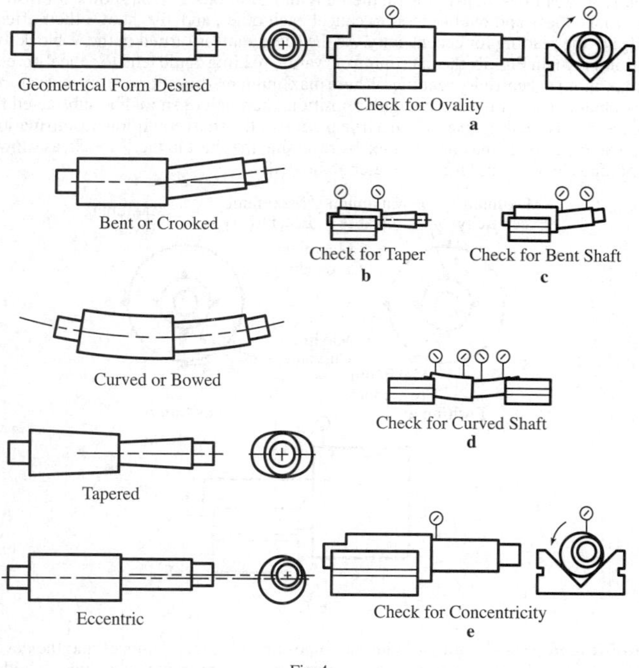


Fig. 4.

To check for taper, the shaft is supported in the V-block and the dial indicator is used to measure the maximum height over the shaft at various positions along its length, as shown in Fig. 4b, without turning the workpiece. Again, the shaft should be marked with the reading positions and values, and with the direction of the taper, and a written record should be made of the amount and direction of any taper discovered.

Checking for a bent shaft requires that the shaft be clocked at the shoulder and at the farther end, as shown in Fig. 4c. For a second check, the shaft is rotated only 90° or a quarter turn. When the recorded readings are compared with those from the ovality and taper checks, the three conditions can be distinguished.

To detect a curved or bowed condition, the shaft should be suspended in two V-blocks with only about ⅛ inch of each end in each V. Alternatively, the shaft can be placed between centers. The shaft is then clocked at several points, as shown in Fig. 4d, but preferably not at those locations used for the ovality, taper, or crookedness checks. If the single element due to curvature is to be distinguished from the effects of ovality, taper, and crookedness, and its value assessed, great care must be taken to differentiate between the conditions detected by the measurements.

Finally, the amount of eccentricity between one shaft diameter and another may be tested by the setup shown in Fig. 4e. With the indicator plunger in contact with the smaller diameter, close to the shoulder, the shaft is rotated in the V-block and the indicator needle position is monitored to find the maximum and minimum readings.

Curvature, ovality, or crookedness conditions may tend to cancel each other, as shown in Fig. 5, and one or more of these degrees of defectiveness may add themselves to the true eccentricity readings, depending on their angular positions. Fig. 5a shows, for instance, how crookedness and ovality tend to cancel each other, and also shows their effect in falsifying the reading for eccentricity. As the same shaft is turned in the V-block to the position shown in Fig. 5b, the maximum curvature reading could tend to cancel or reduce the maximum eccentricity reading. Where maximum readings for ovality, curvature, or crookedness occur at the same angular position, their values should be subtracted from the eccentricity reading to arrive at a true picture of the shaft condition. Confirmation of eccentricity readings may be obtained by reversing the shaft in the V-block, as shown in Fig. 5c, and clocking the larger diameter of the shaft.

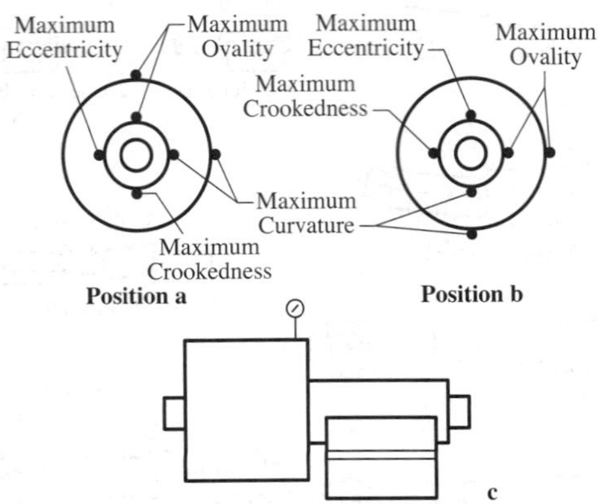


Fig. 5.

**Out-of-Roundness—Lobing.**—With the imposition of finer tolerances and the development of improved measurement methods, it has become apparent that no hole, cylinder, or sphere can be produced with a perfectly symmetrical round shape. Some of the conditions are diagrammed in Fig. 6, where Fig. 6a shows simple ovality and Fig. 6b shows ovality occurring in two directions. From the observation of such conditions have come the terms lobe and lobing. Fig. 6c shows the three-lobed shape common with centerless-ground components, and Fig. 6d is typical of multi-lobed shapes. In Fig. 6e are shown surface waviness, surface roughness, and out-of-roundness, which often are combined with lobing.

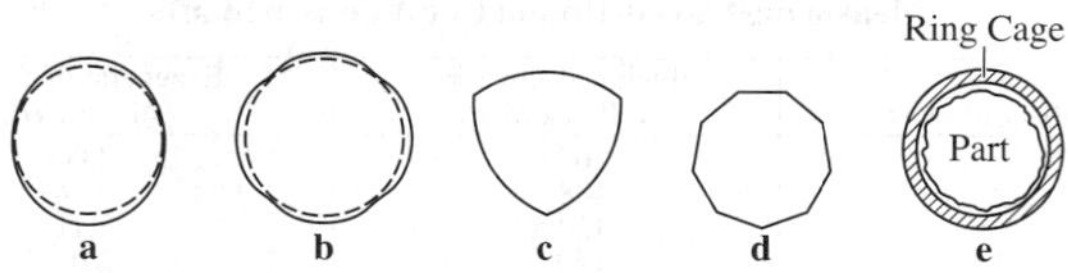


Fig. 6.

In Fig. 6a through Fig. 6d, the cylinder (or hole) diameters are shown at full size, but the lobes are magnified some 10,000 times to make them visible. In precision parts, the deviation from the round condition is usually only in the range of millionths of an inch, although it occasionally can be 0.0001 inch, 0.0002 inch, or more. For instance, a 3-inch diameter part may have a lobing condition amounting to an inaccuracy of only 30 millionths (0.000030 inch). Even if the distortion (ovality, waviness, roughness) is small, it may cause hum, vibration, heat buildup, and wear, possibly leading to eventual failure of the component or assembly.

Plain elliptical out-of-roundness (two lobes), or any even number of lobes, can be detected by rotating the part on a surface plate under a dial indicator of adequate resolution, or by using an indicating caliper or snap gage. However, supporting such a part in a V-block during measurement will tend to conceal roundness errors. Ovality in a hole can be detected by a dial-type bore gage or internal measuring machine. Parts with odd numbers of lobes require an instrument that can measure the envelope or complete circumference. Plug and ring gages will tell whether a shaft can be assembled into a bearing but not whether there will be a good fit, as illustrated in Fig. 6e.

A standard, 90-degree included-angle V-block can be used to detect and count the number of lobes, but to measure the exact amount of lobing indicated by $R$-$r$ in Fig. 7 requires a V-block with an angle $\alpha$, which is related to the number of lobes. This angle $\alpha$ can be calculated from the formula $2\alpha = 180° - 360°/N$, where $N$ is the number of lobes. Thus, for a three-lobe form, $\alpha$ becomes 30 degrees, and the V-block used should have a 60-degree included angle. The distance $M$, which is obtained by rotating the part under the comparator plunger, is converted to a value for the radial variation in cylinder contour by the formula $M = (R - r)(1 + \csc\alpha)$.

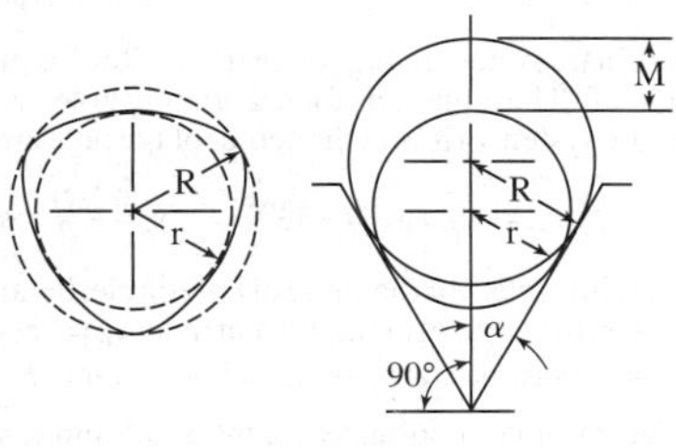


Fig. 7.

Using a V-block (even of appropriate angle) for parts with odd numbers of lobes will give exaggerated readings when the distance $R - r$ (Fig. 7) is used as the measure of the amount of out-of-roundness. The accompanying table shows the appropriate V-block angles for various odd numbers of lobes and the factors ($1 + \csc\alpha$) by which the readings are increased over the actual out-of-roundness values.

**Table of Lobes, V-Block Angles and Exaggeration Factors in Measuring Out-of-Round Conditions in Shafts**

| Number of Lobes | Included Angle of V-Block (deg) | Exaggeration Factor ($1 + \csc\alpha$) |
|---|---|---|
| 3 | 60 | 3.00 |
| 5 | 108 | 2.24 |
| 7 | 128.57 | 2.11 |
| 9 | 140 | 2.06 |

Measurement of a complete circumference requires special equipment, often incorporating a precision spindle running true within two millionths (0.000002) inch. A stylus attached to the spindle is caused to traverse the internal or external cylinder being inspected, and its divergences are processed electronically to produce a polar chart similar to the wavy outline in Fig. 6e. Electronic circuits provide for the variations due to surface effects to be separated from those of lobing and other departures from the "true" cylinder traced out by the spindle.

## Coordinates for Hole Circles

**Type "A" Hole Circles.—**Type "A" hole circles can be identified by hole number 1 at the top of the hole circle, as shown in Fig. 1a and Fig. 1b. The $x$, $y$ coordinates for hole circles of from 3 to 33 holes corresponding to the geometry of Fig. 1a are given in Table 1a, and corresponding to the geometry of Fig. 1b in Table 1b. Holes are numbered in a counterclockwise direction as shown. Coordinates given are based upon a hole circle of (1) unit diameter. For other diameters, multiply the $x$ and $y$ coordinates from the table by the hole circle diameter. For example, for a 3-inch or 3-centimeter hole circle diameter, multiply table values by 3. Coordinates are valid in any unit system.

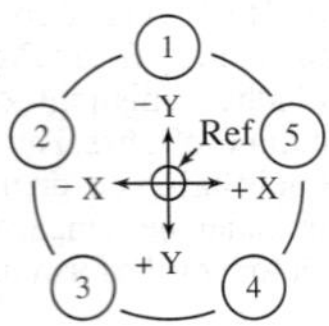

Fig. 1a. Type "A" Circle

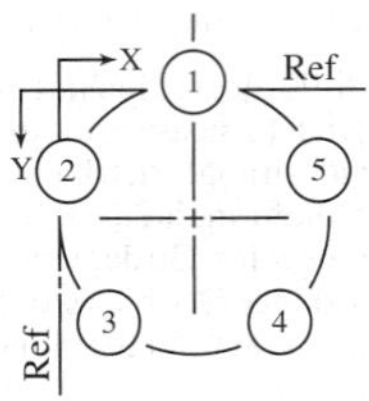

Fig. 1b. Type "A" Circle

The origin of the coordinate system in Fig. 1a, marked "Ref", is at the center of the hole circle at position $x = 0$, $y = 0$. The equations for calculating hole coordinates for type "A" circles with the coordinate system origin at the center of the hole circle are as follows:

$$\theta = \frac{360}{n} = \frac{2\pi}{n} \qquad x_H = -\frac{D}{2}\sin((H-1)\theta) \qquad y_H = -\frac{D}{2}\cos((H-1)\theta) \qquad (1a)$$

where $n$ = number of holes in circle; $D$ = diameter of hole circle; $\theta$ = angle between adjacent holes in circle; $H$ = number (from 1 to $n$) of the current hole; $x_H$ = $x$ coordinate at position of hole number $H$; and, $y_H$ = $y$ coordinate at position of hole number $H$.

*Example 1(a):* Calculate the hole coordinates for the 5-hole circle shown in Fig. 1a when circle diameter = 1. Compare the results to the data tabulated in Table 1a.

| Hole | $\theta = 360/5 = 72°$ $D = 1$ | |
|---|---|---|
| 1 | $x_H = x_1 = -½ \times \sin(0) = 0.00000$ | $y_1 = -½ \times \cos(0) = -0.50000$ |
| 2 | $x_2 = -½ \times \sin(72) = -0.47553$ | $y_2 = -½ \times \cos(72) = -0.15451$ |
| 3 | $x_3 = -½ \times \sin(144) = -0.29389$ | $y_3 = -½ \times \cos(144) = 0.40451$ |
| 4 | $x_4 = -½ \times \sin(216) = 0.29389$ | $y_4 = -½ \times \cos(216) = 0.40451$ |
| 5 | $x_5 = -½ \times \sin(288) = 0.47553$ | $y_5 = -½ \times \cos(288) = -0.15451$ |

In Fig. 1b, the origin of the coordinate system (point 0,0) is located at the top left of the figure at the intersection of the two lines labeled "Ref." The center of the hole circle is offset from the coordinate system origin by distance $X_O$ in the $+x$ direction, and by distance $Y_O$ in the $+y$ direction. In practice the origin of the coordinate system can be located at any convenient distance from the center of the hole circle. In Fig. 1b, it can be determined by inspection that the distances $X_O = Y_O = D/2$. The equations for calculating hole positions of type "A" circles of the Fig. 1b type are almost the same as in Equation (1a), but with the addition of $X_O$ and $Y_O$ terms, as follows:

$$\theta = \frac{360}{n} = \frac{2\pi}{n} \qquad x_H = -\frac{D}{2}\sin((H-1)\theta) + X_O \qquad y_H = -\frac{D}{2}\cos((H-1)\theta) + Y_O \quad \text{(1b)}$$

*Example 1(b):* Use results of Example 1 to determine hole coordinates of Fig. 1b for circle diameter = 1, and compare results with Table 1b.

| Hole | $\theta = 360/5 = 72°$ $D = 1$ $X_O = D/2 = 0.50000$ | $Y_O = D/2 = 0.50000$ |
|---|---|---|
| 1 | $x_1 = 0.00000 + X_O = 0.50000$ | $y_1 = -0.50000 + Y_O = 0.00000$ |
| 2 | $x_2 = -0.47553 + X_O = 0.02447$ | $y_2 = -0.15451 + Y_O = 0.34549$ |
| 3 | $x_3 = -0.29389 + X_O = 0.20611$ | $y_3 = 0.40451 + Y_O = 0.90451$ |
| 4 | $x_4 = 0.29389 + X_O = 0.79389$ | $y_4 = 0.40451 + Y_O = 0.90451$ |
| 5 | $x_5 = 0.47553 + X_O = 0.97553$ | $y_5 = -0.15451 + Y_O = 0.34549$ |

**Type "B" Hole Circles.**—Compared to type "A" hole circles, type "B" hole circles, Fig. 2a and Fig. 2b, are arranged such that the circle of holes is rotated about the center of the circle by $\theta/2$ degrees, that is, ½ of the angle between adjacent holes. The $x,y$ coordinates for type "B" hole circles of from 3 to 33 holes are given in Table 2a for geometry corresponding to Fig. 2a, and in Table 2b for geometry corresponding to Fig. 2b. Holes are numbered in a counterclockwise direction as shown. Coordinates given are based upon a hole circle of (1) unit diameter. For other diameters, multiply the $x$ and $y$ coordinates from the table by the hole circle diameter. For example, for a 3-inch or 3-centimeter hole circle diameter, multiply table values by 3. Coordinates are valid in any unit system.

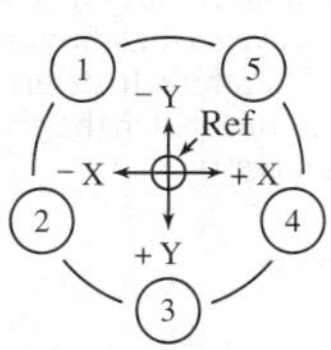


Fig. 2a. Type "B" Circle

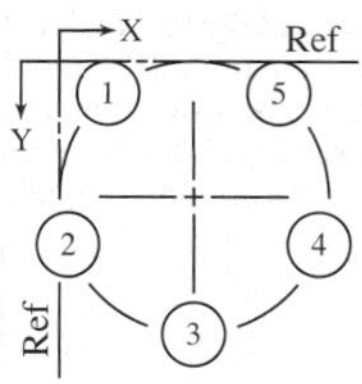


Fig. 2b. Type "B" Circle

In Fig. 2a, the coordinate system origin, marked "Ref", is at the center of the hole circle at position $x = 0, y = 0$. Equations for calculating hole coordinates for type "B" circles with the coordinate system origin at the center of the hole circle as in Fig. 2a are as follows:

$$\theta = \frac{360}{n} = \frac{2\pi}{n} \qquad x_H = -\frac{D}{2}\sin\left((H-1)\theta + \frac{\theta}{2}\right) \qquad y_H = -\frac{D}{2}\cos\left((H-1)\theta + \frac{\theta}{2}\right) \quad \text{(2a)}$$

where $n$ = number of holes in circle; $D$ = diameter of hole circle; $\theta$ = angle between adjacent holes; $x_H$ = $x$ coordinate at position of hole number $H$; and, $y_H$ = $y$ coordinate at position of hole number $H$.

*Example 2(a):* Calculate the hole coordinates for the 5-hole circle shown in Fig. 2a when circle diameter = 1. Compare the results to the data in Table 2a.

| Hole | $\theta = 360/5 = 72°$ $\quad \theta/2 = 36°$ $\quad D = 1$ | |
|---|---|---|
| 1 | $x_1 = -½ \times \sin(36) = -0.29389$ | $y_1 = -½ \times \cos(36) = -0.40451$ |
| 2 | $x_2 = -½ \times \sin(108) = -0.47553$ | $y_2 = -½ \times \cos(108) = 0.15451$ |
| 3 | $x_3 = -½ \times \sin(180) = 0.00000$ | $y_3 = -½ \times \cos(180) = 0.50000$ |
| 4 | $x_4 = -½ \times \sin(252) = 0.47553$ | $y_4 = -½ \times \cos(252) = 0.15451$ |
| 5 | $x_5 = -½ \times \sin(324) = 0.29389$ | $y_5 = -½ \times \cos(324) = -0.40451$ |

In Fig. 2b, the origin of the coordinate system (point 0,0) is located at the top left of the figure at the intersection of the two lines labeled "Ref." The center of the hole circle is offset from the coordinate system origin by distance $X_O$ in the $+x$ direction and by distance $Y_O$ in the $+y$ direction. In practice, the origin of the coordinate system can be chosen at any convenient distance from the hole circle origin. In Fig. 2b, it can be determined by inspection that distance $X_O = Y_O = D/2$. The equations for calculating hole positions of type "B" circles of the Fig. 2b type are the same as in Equation (2a) but with the addition of $X_O$ and $Y_O$ terms, as follows:

$$\theta = \frac{360}{n} = \frac{2\pi}{n} \qquad x_H = -\frac{D}{2}\sin\left((H-1)\theta + \frac{\theta}{2}\right) + X_O \qquad y_H = -\frac{D}{2}\cos\left((H-1)\theta + \frac{\theta}{2}\right) + Y_O \quad (2b)$$

*Example 2(b):* Use the coordinates obtained in Example 2(a) to determine the hole coordinates of a 5-hole circle shown in Fig. 1b with circle diameter = 1. Compare the results to the data in Table 2b.

| Hole | $\theta = 360/5 = 72°$ $\quad \theta/2 = 36°$ $\quad D = 1$ $\quad X_O = D/2 = 0.50000$ | $Y_O = D/2 = 0.50000$ |
|---|---|---|
| 1 | $x_1 = -½ \times \sin(36) + 0.50000 = 0.20611$ | $y_1 = -½ \times \cos(36) + 0.50000 = 0.09549$ |
| 2 | $x_2 = -½ \times \sin(108) + 0.50000 = 0.02447$ | $y_2 = -½ \times \cos(108) + 0.50000 = 0.65451$ |
| 3 | $x_3 = -½ \times \sin(180) + 0.50000 = 0.50000$ | $y_3 = -½ \times \cos(180) + 0.50000 = 1.00000$ |
| 4 | $x_4 = -½ \times \sin(252) + 0.50000 = 0.97553$ | $y_4 = -½ \times \cos(252) + 0.50000 = 0.65451$ |
| 5 | $x_5 = -½ \times \sin(324) + 0.50000 = 0.79389$ | $y_5 = -½ \times \cos(324) + 0.50000 = 0.09549$ |

**Adapting Hole Coordinate Equations for Different Geometry.—**Hole coordinate values in Table 1a through Table 2b are obtained using the equations given previously, along with the geometry of the corresponding figures. If the geometry does not match that given in one of the previous figures, hole coordinate values from the tables or equations will be incorrect. Fig. 3 illustrates such a case. Fig. 3 resembles a type "A" hole circle (Fig. 1b) with hole number 2 at the top, and it also resembles a type "B" hole circle (Fig. 2b) in which all holes have been rotated 90° clockwise. A closer look also reveals that the positive $y$ direction in Fig. 3 is opposite that used in Fig. 1b and Fig. 2b. Therefore, to determine the hole coordinates of Fig. 3 it is necessary to create new equations that match the given geometry or to modify the previous equations to match the Fig. 3 geometry.

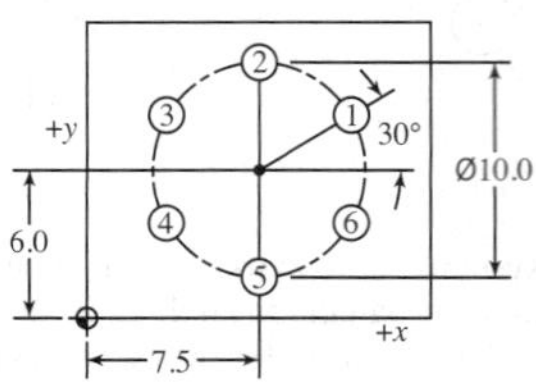


Fig. 3.

*Example 3(a), Determining Hole Coordinates for* Fig. 3*:* Write equations for the coordinates of holes 1, 2, and 3 of Fig. 3.

$$x_1 = 7.5 + 5\cos(30°) = 11.8301 \qquad y_1 = 6.0 + 5\sin(30°) = 8.5000$$
$$x_2 = 7.5 + 5\cos(30° + 60°) = 7.5 \qquad y_2 = 6.0 + 5\sin(30° + 60°) = 11.0000$$
$$x_3 = 7.5 + 5\cos(30° + 120°) = 3.1699 \qquad y_3 = 6.0 + 5\sin(30° + 120°) = 8.5000$$

*Example 3(b), Modify* Equation (2b) *for* Fig. 3*:* In Fig. 3, hole numbering is rotated 90° (π/2 radian) in the clockwise (negative) direction relative to Fig. 2b, and the direction of the $+y$ coordinate axis is the reverse, or negative, of that given in Fig. 2b. Equations for Fig. 3 can be obtained from Equation (2b) by

1) subtracting 90° from the angle of each hole in the $x$ and $y$ equations of Equation (2b)

2) multiplying the $y_H$ equation by –1 to reverse the orientation of the $y$-axis

$$x_H = -\frac{D}{2}\sin\left((H-1)\theta + \frac{\theta}{2} - 90\right) + X_0 \qquad y_H = -\left(-\frac{D}{2}\cos\left((H-1)\theta + \frac{\theta}{2} - 90\right) + Y_0\right)$$

In Fig. 3, $\theta = 360/n = 60°$ for 6 holes, $X_O = 7.5$, and $Y_O = -6.0$.

$$x_1 = -5\sin(30 - 90) + 7.5 = 11.8301 \qquad y_1 = 5\cos(30 - 90) + 6 = 8.5000$$
$$x_2 = -5\sin(60° + 30° - 90°) + 7.5 = 7.5 \qquad y_2 = 5\cos(60° + 30° - 90°) + 6 = 11.0000$$
$$x_3 = -5\sin(120° + 30° - 90°) + 7.5 = 3.1699 \qquad y_3 = 5\cos(120° + 30° - 90°) + 6 = 8.5000$$

**Lengths of Chords on Hole Circle Circumference.—**Table 3 on page 720 gives the lengths of chords for spacing off the circumferences of circles. The object of this table is to make possible the division of the periphery into a number of equal parts without trials with the dividers. Table 3 is calculated for circles having a diameter equal to 1. For circles of other diameters, the length of chord given in the table should be multiplied by the diameter of the circle. Table 3 may be used by toolmakers when setting "buttons" in circular formation and with inch or metric dimensions. See *Coordinates for Hole Circles* on page 708 for more information on this topic.

*Example:* Assume that it is required to divide the periphery of a circle of 20 inches diameter into thirty-two equal parts.

*Solution:* From the table the length of the chord is found to be 0.098017 inch, if the diameter of the circle were 1 inch. With a diameter of 20 inches the length of the chord for one division would be $20 \times 0.098017 = 1.9603$ inches.

**Another example, in metric units: For a 100 millimeter diameter requiring 5 equal divisions, the length of the chord for one division would be $100 \times 0.587785 = 58.7785$ millimeters.**

***Example:* Assume that it is required to divide a circle of $6\frac{1}{2}$ millimeter diameter into seven equal parts. Find the length of the chord required for spacing off the circumference.**

***Solution:* In Table 3, the length of the chord for dividing a circle of 1 millimeter diameter into 7 equal parts is 0.433884 mm. The length of chord for a circle of $6\frac{1}{2}$ mm diameter is $6\frac{1}{2} \times 0.433884 = 2.820246$ mm.**

*Example:* Assume that it is required to divide a circle having a diameter of $9\frac{23}{32}$ inches into 15 equal divisions.

*Solution:* In Table 3, the length of the chord for dividing a circle of 1 inch diameter into 15 equal parts is 0.207912 inch. The length of chord for a circle of 9 inches diameter is $9\frac{23}{32} \times 0.207912 = 2.020645$ inches.

**Coordinates for Locating Equally Spaced Holes.—**Table 4 on page 721 is a quick reference for calculating the coordinates of between 3 and 12 equally spaced holes using the center point and diameter of the pitch circle.

## Table 1a. Hole Coordinate Dimension Factors for Type "A" Hole Circles

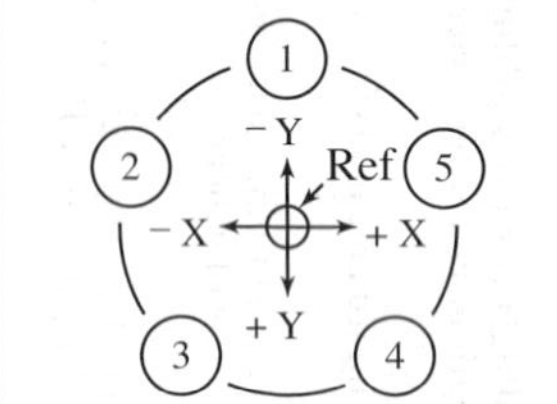


**3 holes**

| # | x | y |
|---|---|---|
| 1 | 0.00000 | −0.50000 |
| 2 | −0.43301 | 0.25000 |
| 3 | 0.43301 | 0.25000 |

**4 holes**

| # | x | y |
|---|---|---|
| 1 | 0.00000 | −0.50000 |
| 2 | −0.50000 | 0.00000 |
| 3 | 0.00000 | 0.50000 |
| 4 | 0.50000 | 0.00000 |

**5 holes**

| # | x | y |
|---|---|---|
| 1 | 0.00000 | −0.50000 |
| 2 | −0.47553 | −0.15451 |
| 3 | −0.29389 | 0.40451 |
| 4 | 0.29389 | 0.40451 |
| 5 | 0.47553 | −0.15451 |

**6 holes**

| # | x | y |
|---|---|---|
| 1 | 0.00000 | −0.50000 |
| 2 | −0.43301 | −0.25000 |
| 3 | −0.43301 | 0.25000 |
| 4 | 0.00000 | 0.50000 |
| 5 | 0.43301 | 0.25000 |
| 6 | 0.43301 | −0.25000 |

**7 holes**

| # | x | y |
|---|---|---|
| 1 | 0.00000 | −0.50000 |
| 2 | −0.39092 | −0.31174 |
| 3 | −0.48746 | 0.11126 |
| 4 | −0.21694 | 0.45048 |
| 5 | 0.21694 | 0.45048 |
| 6 | 0.48746 | 0.11126 |
| 7 | 0.39092 | −0.31174 |

**8 holes**

| # | x | y |
|---|---|---|
| 1 | 0.00000 | −0.50000 |
| 2 | −0.35355 | −0.35355 |
| 3 | −0.50000 | 0.00000 |
| 4 | −0.35355 | 0.35355 |
| 5 | 0.00000 | 0.50000 |
| 6 | 0.35355 | 0.35355 |
| 7 | 0.50000 | 0.00000 |
| 8 | 0.35355 | −0.35355 |

**9 holes**

| # | x | y |
|---|---|---|
| 1 | 0.00000 | −0.50000 |
| 2 | −0.32139 | −0.38302 |
| 3 | −0.49240 | −0.08682 |
| 4 | −0.43301 | 0.25000 |
| 5 | −0.17101 | 0.46985 |
| 6 | 0.17101 | 0.46985 |
| 7 | 0.43301 | 0.25000 |
| 8 | 0.49240 | −0.08682 |
| 9 | 0.32139 | −0.38302 |

**10 holes**

| # | x | y |
|---|---|---|
| 1 | 0.00000 | −0.50000 |
| 2 | −0.29389 | −0.40451 |
| 3 | −0.47553 | −0.15451 |
| 4 | −0.47553 | 0.15451 |
| 5 | −0.29389 | 0.40451 |
| 6 | 0.00000 | 0.50000 |
| 7 | 0.29389 | 0.40451 |
| 8 | 0.47553 | 0.15451 |
| 9 | 0.47553 | −0.15451 |
| 10 | 0.29389 | −0.40451 |

**11 holes**

| # | x | y |
|---|---|---|
| 1 | 0.00000 | −0.50000 |
| 2 | −0.27032 | −0.42063 |
| 3 | −0.45482 | −0.20771 |
| 4 | −0.49491 | 0.07116 |
| 5 | −0.37787 | 0.32743 |
| 6 | −0.14087 | 0.47975 |
| 7 | 0.14087 | 0.47975 |
| 8 | 0.37787 | 0.32743 |
| 9 | 0.49491 | 0.07116 |
| 10 | 0.45482 | −0.20771 |
| 11 | 0.27032 | −0.42063 |

**12 holes**

| # | x | y |
|---|---|---|
| 1 | 0.00000 | −0.50000 |
| 2 | −0.25000 | −0.43301 |
| 3 | −0.43301 | −0.25000 |
| 4 | −0.50000 | 0.00000 |
| 5 | −0.43301 | 0.25000 |
| 6 | −0.25000 | 0.43301 |
| 7 | 0.00000 | 0.50000 |
| 8 | 0.25000 | 0.43301 |
| 9 | 0.43301 | 0.25000 |
| 10 | 0.50000 | 0.00000 |
| 11 | 0.43301 | −0.25000 |
| 12 | 0.25000 | −0.43301 |

**13 holes**

| # | x | y |
|---|---|---|
| 1 | 0.00000 | −0.50000 |
| 2 | −0.23236 | −0.44273 |
| 3 | −0.41149 | −0.28403 |
| 4 | −0.49635 | −0.06027 |
| 5 | −0.46751 | 0.17730 |
| 6 | −0.33156 | 0.37426 |
| 7 | −0.11966 | 0.48547 |
| 8 | 0.11966 | 0.48547 |
| 9 | 0.33156 | 0.37426 |
| 10 | 0.46751 | 0.17730 |
| 11 | 0.49635 | −0.06027 |
| 12 | 0.41149 | −0.28403 |
| 13 | 0.23236 | −0.44273 |

**14 holes**

| # | x | y |
|---|---|---|
| 1 | 0.00000 | −0.50000 |
| 2 | −0.21694 | −0.45048 |
| 3 | −0.39092 | −0.31174 |
| 4 | −0.48746 | −0.11126 |
| 5 | −0.48746 | 0.11126 |
| 6 | −0.39092 | 0.31174 |
| 7 | −0.21694 | 0.45048 |
| 8 | 0.00000 | 0.50000 |
| 9 | 0.21694 | 0.45048 |
| 10 | 0.39092 | 0.31174 |
| 11 | 0.48746 | 0.11126 |
| 12 | 0.48746 | −0.11126 |
| 13 | 0.39092 | −0.31174 |
| 14 | 0.21694 | −0.45048 |

**15 holes**

| # | x | y |
|---|---|---|
| 1 | 0.00000 | −0.50000 |
| 2 | −0.20337 | −0.45677 |
| 3 | −0.37157 | −0.33457 |
| 4 | −0.47553 | −0.15451 |
| 5 | −0.49726 | 0.05226 |
| 6 | −0.43301 | 0.25000 |
| 7 | −0.29389 | 0.40451 |
| 8 | −0.10396 | 0.48907 |
| 9 | 0.10396 | 0.48907 |
| 10 | 0.29389 | 0.40451 |
| 11 | 0.43301 | 0.25000 |
| 12 | 0.49726 | 0.05226 |
| 13 | 0.47553 | −0.15451 |
| 14 | 0.37157 | −0.33457 |
| 15 | 0.20337 | −0.45677 |

**16 holes**

| # | x | y |
|---|---|---|
| 1 | 0.00000 | −0.50000 |
| 2 | −0.19134 | −0.46194 |
| 3 | −0.35355 | −0.35355 |
| 4 | −0.46194 | −0.19134 |
| 5 | −0.50000 | 0.00000 |
| 6 | −0.46194 | 0.19134 |
| 7 | −0.35355 | 0.35355 |
| 8 | −0.19134 | 0.46194 |
| 9 | 0.00000 | 0.50000 |
| 10 | 0.19134 | 0.46194 |
| 11 | 0.35355 | 0.35355 |
| 12 | 0.46194 | 0.19134 |
| 13 | 0.50000 | 0.00000 |
| 14 | 0.46194 | −0.19134 |
| 15 | 0.35355 | −0.35355 |
| 16 | 0.19134 | −0.46194 |

**17 holes**

| # | x | y |
|---|---|---|
| 1 | 0.00000 | −0.50000 |
| 2 | −0.18062 | −0.46624 |
| 3 | −0.33685 | −0.36950 |
| 4 | −0.44758 | −0.22287 |
| 5 | −0.49787 | −0.04613 |
| 6 | −0.48091 | 0.13683 |
| 7 | −0.39901 | 0.30132 |
| 8 | −0.26322 | 0.42511 |
| 9 | −0.09187 | 0.49149 |
| 10 | 0.09187 | 0.49149 |
| 11 | 0.26322 | 0.42511 |
| 12 | 0.39901 | 0.30132 |
| 13 | 0.48091 | 0.13683 |
| 14 | 0.49787 | −0.04613 |
| 15 | 0.44758 | −0.22287 |
| 16 | 0.33685 | −0.36950 |
| 17 | 0.18062 | −0.46624 |

**18 holes**

| # | x | y |
|---|---|---|
| 1 | 0.00000 | −0.50000 |
| 2 | −0.17101 | −0.46985 |
| 3 | −0.32139 | −0.38302 |
| 4 | −0.43301 | −0.25000 |
| 5 | −0.49240 | −0.08682 |
| 6 | −0.49240 | 0.08682 |
| 7 | −0.43301 | 0.25000 |
| 8 | −0.32139 | 0.38302 |
| 9 | −0.17101 | 0.46985 |
| 10 | 0.00000 | 0.50000 |
| 11 | 0.17101 | 0.46985 |
| 12 | 0.32139 | 0.38302 |
| 13 | 0.43301 | 0.25000 |
| 14 | 0.49240 | 0.08682 |
| 15 | 0.49240 | −0.08682 |
| 16 | 0.43301 | −0.25000 |
| 17 | 0.32139 | −0.38302 |
| 18 | 0.17101 | −0.46985 |

**19 holes**

| # | x | y |
|---|---|---|
| 1 | 0.00000 | −0.50000 |
| 2 | −0.16235 | −0.47291 |
| 3 | −0.30711 | −0.39457 |
| 4 | −0.41858 | −0.27347 |
| 5 | −0.48470 | −0.12274 |
| 6 | −0.49829 | 0.04129 |
| 7 | −0.45789 | 0.20085 |
| 8 | −0.36786 | 0.33864 |
| 9 | −0.23797 | 0.43974 |
| 10 | −0.08230 | 0.49318 |
| 11 | 0.08230 | 0.49318 |
| 12 | 0.23797 | 0.43974 |
| 13 | 0.36786 | 0.33864 |
| 14 | 0.45789 | 0.20085 |
| 15 | 0.49829 | 0.04129 |
| 16 | 0.48470 | −0.12274 |
| 17 | 0.41858 | −0.27347 |
| 18 | 0.30711 | −0.39457 |
| 19 | 0.16235 | −0.47291 |

**20 holes**

| # | x | y |
|---|---|---|
| 1 | 0.00000 | −0.50000 |
| 2 | −0.15451 | −0.47553 |
| 3 | −0.29389 | −0.40451 |
| 4 | −0.40451 | −0.29389 |
| 5 | −0.47553 | −0.15451 |
| 6 | −0.50000 | 0.00000 |
| 7 | −0.47553 | 0.15451 |
| 8 | −0.40451 | 0.29389 |
| 9 | −0.29389 | 0.40451 |
| 10 | −0.15451 | 0.47553 |
| 11 | 0.00000 | 0.50000 |
| 12 | 0.15451 | 0.47553 |
| 13 | 0.29389 | 0.40451 |
| 14 | 0.40451 | 0.29389 |
| 15 | 0.47553 | 0.15451 |
| 16 | 0.50000 | 0.00000 |
| 17 | 0.47553 | −0.15451 |
| 18 | 0.40451 | −0.29389 |
| 19 | 0.29389 | −0.40451 |
| 20 | 0.15451 | −0.47553 |

**21 holes**

| # | x | y |
|---|---|---|
| 1 | 0.00000 | −0.50000 |
| 2 | −0.14738 | −0.47779 |
| 3 | −0.28166 | −0.41312 |
| 4 | −0.39092 | −0.31174 |
| 5 | −0.46544 | −0.18267 |
| 6 | −0.49860 | −0.03737 |
| 7 | −0.48746 | 0.11126 |
| 8 | −0.43301 | 0.25000 |
| 9 | −0.34009 | 0.36653 |
| 10 | −0.21694 | 0.45048 |
| 11 | −0.07452 | 0.49442 |
| 12 | 0.07452 | 0.49442 |
| 13 | 0.21694 | 0.45048 |
| 14 | 0.34009 | 0.36653 |
| 15 | 0.43301 | 0.25000 |
| 16 | 0.48746 | 0.11126 |
| 17 | 0.49860 | −0.03737 |
| 18 | 0.46544 | −0.18267 |
| 19 | 0.39092 | −0.31174 |
| 20 | 0.28166 | −0.41312 |
| 21 | 0.14738 | −0.47779 |

**22 holes**

| # | x | y |
|---|---|---|
| 1 | 0.00000 | −0.50000 |
| 2 | −0.14087 | −0.47975 |
| 3 | −0.27032 | −0.42063 |
| 4 | −0.37787 | −0.32743 |
| 5 | −0.45482 | −0.20771 |
| 6 | −0.49491 | −0.07116 |
| 7 | −0.49491 | 0.07116 |
| 8 | −0.45482 | 0.20771 |
| 9 | −0.37787 | 0.32743 |
| 10 | −0.27032 | 0.42063 |
| 11 | −0.14087 | 0.47975 |
| 12 | 0.00000 | 0.50000 |
| 13 | 0.14087 | 0.47975 |
| 14 | 0.27032 | 0.42063 |
| 15 | 0.37787 | 0.32743 |
| 16 | 0.45482 | 0.20771 |
| 17 | 0.49491 | 0.07116 |
| 18 | 0.49491 | −0.07116 |
| 19 | 0.45482 | −0.20771 |
| 20 | 0.37787 | −0.32743 |
| 21 | 0.27032 | −0.42063 |
| 22 | 0.14087 | −0.47975 |

**23 holes**

| # | x | y |
|---|---|---|
| 1 | 0.00000 | −0.50000 |
| 2 | −0.13490 | −0.48146 |
| 3 | −0.25979 | −0.42721 |
| 4 | −0.36542 | −0.34128 |
| 5 | −0.44394 | −0.23003 |
| 6 | −0.48954 | −0.10173 |
| 7 | −0.49883 | 0.03412 |
| 8 | −0.47113 | 0.16744 |
| 9 | −0.40848 | 0.28834 |
| 10 | −0.31554 | 0.38786 |
| 11 | −0.19920 | 0.45861 |
| 12 | −0.06808 | 0.49534 |
| 13 | 0.06808 | 0.49534 |
| 14 | 0.19920 | 0.45861 |
| 15 | 0.31554 | 0.38786 |
| 16 | 0.40848 | 0.28834 |
| 17 | 0.47113 | 0.16744 |
| 18 | 0.49883 | 0.03412 |
| 19 | 0.48954 | −0.10173 |
| 20 | 0.44394 | −0.23003 |
| 21 | 0.36542 | −0.34128 |
| 22 | 0.25979 | −0.42721 |
| 23 | 0.13490 | −0.48146 |

**24 holes**

| # | x | y |
|---|---|---|
| 1 | 0.00000 | −0.50000 |
| 2 | −0.12941 | −0.48296 |
| 3 | −0.25000 | −0.43301 |
| 4 | −0.35355 | −0.35355 |

## Table 1a. *(Continued)* Hole Coordinate Dimension Factors for Type "A" Hole Circles

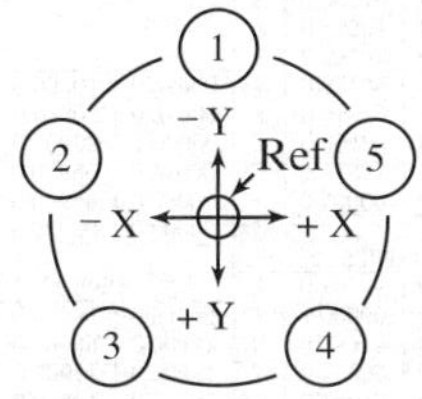


**24 holes** *(Continued)*

| # | x | y |
|---|---|---|
| 5 | −0.43301 | −0.25000 |
| 6 | −0.48296 | −0.12941 |
| 7 | −0.50000 | 0.00000 |
| 8 | −0.48296 | 0.12941 |
| 9 | −0.43301 | 0.25000 |
| 10 | −0.35355 | 0.35355 |
| 11 | −0.25000 | 0.43301 |
| 12 | −0.12941 | 0.48296 |
| 13 | 0.00000 | 0.50000 |
| 14 | 0.12941 | 0.48296 |
| 15 | 0.25000 | 0.43301 |
| 16 | 0.35355 | 0.35355 |
| 17 | 0.43301 | 0.25000 |
| 18 | 0.48296 | 0.12941 |
| 19 | 0.50000 | 0.00000 |
| 20 | 0.48296 | −0.12941 |
| 21 | 0.43301 | −0.25000 |
| 22 | 0.35355 | −0.35355 |
| 23 | 0.25000 | −0.43301 |
| 24 | 0.12941 | −0.48296 |

**25 holes**

| # | x | y |
|---|---|---|
| 1 | 0.00000 | −0.50000 |
| 2 | −0.12434 | −0.48429 |
| 3 | −0.24088 | −0.43815 |
| 4 | −0.34227 | −0.36448 |
| 5 | −0.42216 | −0.26791 |
| 6 | −0.47553 | −0.15451 |

**25 holes** *(Continued)*

| # | x | y |
|---|---|---|
| 7 | −0.49901 | −0.03140 |
| 8 | −0.49114 | 0.09369 |
| 9 | −0.45241 | 0.21289 |
| 10 | −0.38526 | 0.31871 |
| 11 | −0.29389 | 0.40451 |
| 12 | −0.18406 | 0.46489 |
| 13 | −0.06267 | 0.49606 |
| 14 | 0.06267 | 0.49606 |
| 15 | 0.18406 | 0.46489 |
| 16 | 0.29389 | 0.40451 |
| 17 | 0.38526 | 0.31871 |
| 18 | 0.45241 | 0.21289 |
| 19 | 0.49114 | 0.09369 |
| 20 | 0.49901 | −0.03140 |
| 21 | 0.47553 | −0.15451 |
| 22 | 0.42216 | −0.26791 |
| 23 | 0.34227 | −0.36448 |
| 24 | 0.24088 | −0.43815 |
| 25 | 0.12434 | −0.48429 |

**26 holes**

| # | x | y |
|---|---|---|
| 1 | 0.00000 | −0.50000 |
| 2 | −0.11966 | −0.48547 |
| 3 | −0.23236 | −0.44273 |
| 4 | −0.33156 | −0.37426 |
| 5 | −0.41149 | −0.28403 |
| 6 | −0.46751 | −0.17730 |
| 7 | −0.49635 | −0.06027 |

**26 holes** *(Continued)*

| # | x | y |
|---|---|---|
| 8 | −0.49635 | 0.06027 |
| 9 | −0.46751 | 0.17730 |
| 10 | −0.41149 | 0.28403 |
| 11 | −0.33156 | 0.37426 |
| 12 | −0.23236 | 0.44273 |
| 13 | −0.11966 | 0.48547 |
| 14 | 0.00000 | 0.50000 |
| 15 | 0.11966 | 0.48547 |
| 16 | 0.23236 | 0.44273 |
| 17 | 0.33156 | 0.37426 |
| 18 | 0.41149 | 0.28403 |
| 19 | 0.46751 | 0.17730 |
| 20 | 0.49635 | 0.06027 |
| 21 | 0.49635 | −0.06027 |
| 22 | 0.46751 | −0.17730 |
| 23 | 0.41149 | −0.28403 |
| 24 | 0.33156 | −0.37426 |
| 25 | 0.23236 | −0.44273 |
| 26 | 0.11966 | −0.48547 |

**27 holes**

| # | x | y |
|---|---|---|
| 1 | 0.00000 | −0.50000 |
| 2 | −0.11531 | −0.48652 |
| 3 | −0.22440 | −0.44682 |
| 4 | −0.32139 | −0.38302 |
| 5 | −0.40106 | −0.29858 |
| 6 | −0.45911 | −0.19804 |
| 7 | −0.49240 | −0.08682 |
| 8 | −0.49915 | 0.02907 |
| 9 | −0.47899 | 0.14340 |
| 10 | −0.43301 | 0.25000 |
| 11 | −0.36369 | 0.34312 |
| 12 | −0.27475 | 0.41774 |
| 13 | −0.17101 | 0.46985 |
| 14 | −0.05805 | 0.49662 |
| 15 | 0.05805 | 0.49662 |
| 16 | 0.17101 | 0.46985 |
| 17 | 0.27475 | 0.41774 |
| 18 | 0.36369 | 0.34312 |
| 19 | 0.43301 | 0.25000 |
| 20 | 0.47899 | 0.14340 |
| 21 | 0.49915 | 0.02907 |
| 22 | 0.49240 | −0.08682 |
| 23 | 0.45911 | −0.19804 |

**27 holes** *(Continued)*

| # | x | y |
|---|---|---|
| 24 | 0.40106 | −0.29858 |
| 25 | 0.32139 | −0.38302 |
| 26 | 0.22440 | −0.44682 |
| 27 | 0.11531 | −0.48652 |

**28 holes**

| # | x | y |
|---|---|---|
| 1 | 0.00000 | −0.50000 |
| 2 | −0.11126 | −0.48746 |
| 3 | −0.21694 | −0.45048 |
| 4 | −0.31174 | −0.39092 |
| 5 | −0.39092 | −0.31174 |
| 6 | −0.45048 | −0.21694 |
| 7 | −0.48746 | −0.11126 |
| 8 | −0.50000 | 0.00000 |
| 9 | −0.48746 | 0.11126 |
| 10 | −0.45048 | 0.21694 |
| 11 | −0.39092 | 0.31174 |
| 12 | −0.31174 | 0.39092 |
| 13 | −0.21694 | 0.45048 |
| 14 | −0.11126 | 0.48746 |
| 15 | 0.00000 | 0.50000 |
| 16 | 0.11126 | 0.48746 |
| 17 | 0.21694 | 0.45048 |
| 18 | 0.31174 | 0.39092 |
| 19 | 0.39092 | 0.31174 |
| 20 | 0.45048 | 0.21694 |
| 21 | 0.48746 | 0.11126 |
| 22 | 0.50000 | 0.00000 |
| 23 | 0.48746 | −0.11126 |
| 24 | 0.45048 | −0.21694 |
| 25 | 0.39092 | −0.31174 |
| 26 | 0.31174 | −0.39092 |
| 27 | 0.21694 | −0.45048 |
| 28 | 0.11126 | −0.48746 |

**29 holes**

| # | x | y |
|---|---|---|
| 1 | 0.00000 | −0.50000 |
| 2 | −0.10749 | −0.48831 |
| 3 | −0.20994 | −0.45379 |
| 4 | −0.30259 | −0.39805 |
| 5 | −0.38108 | −0.32369 |
| 6 | −0.44176 | −0.23420 |
| 7 | −0.48177 | −0.13376 |
| 8 | −0.49927 | −0.02707 |

**29 holes** *(Continued)*

| # | x | y |
|---|---|---|
| 9 | −0.49341 | 0.08089 |
| 10 | −0.46449 | 0.18507 |
| 11 | −0.41384 | 0.28059 |
| 12 | −0.34385 | 0.36300 |
| 13 | −0.25778 | 0.42843 |
| 14 | −0.15965 | 0.47383 |
| 15 | −0.05406 | 0.49707 |
| 16 | 0.05406 | 0.49707 |
| 17 | 0.15965 | 0.47383 |
| 18 | 0.25778 | 0.42843 |
| 19 | 0.34385 | 0.36300 |
| 20 | 0.41384 | 0.28059 |
| 21 | 0.46449 | 0.18507 |
| 22 | 0.49341 | 0.08089 |
| 23 | 0.49927 | −0.02707 |
| 24 | 0.48177 | −0.13376 |
| 25 | 0.44176 | −0.23420 |
| 26 | 0.38108 | −0.32369 |
| 27 | 0.30259 | −0.39805 |
| 28 | 0.20994 | −0.45379 |
| 29 | 0.10749 | −0.48831 |

**30 holes**

| # | x | y |
|---|---|---|
| 1 | 0.00000 | −0.50000 |
| 2 | −0.10396 | −0.48907 |
| 3 | −0.20337 | −0.45677 |
| 4 | −0.29389 | −0.40451 |
| 5 | −0.37157 | −0.33457 |
| 6 | −0.43301 | −0.25000 |
| 7 | −0.47553 | −0.15451 |
| 8 | −0.49726 | −0.05226 |
| 9 | −0.49726 | 0.05226 |
| 10 | −0.47553 | 0.15451 |
| 11 | −0.43301 | 0.25000 |
| 12 | −0.37157 | 0.33457 |
| 13 | −0.29389 | 0.40451 |
| 14 | −0.20337 | 0.45677 |
| 15 | −0.10396 | 0.48907 |
| 16 | 0.00000 | 0.50000 |
| 17 | 0.10396 | 0.48907 |
| 18 | 0.20337 | 0.45677 |
| 19 | 0.29389 | 0.40451 |
| 20 | 0.37157 | 0.33457 |
| 21 | 0.43301 | 0.25000 |

**30 holes** *(Continued)*

| # | x | y |
|---|---|---|
| 22 | 0.47553 | 0.15451 |
| 23 | 0.49726 | 0.05226 |
| 24 | 0.49726 | −0.05226 |
| 25 | 0.47553 | −0.15451 |
| 26 | 0.43301 | −0.25000 |
| 27 | 0.37157 | −0.33457 |
| 28 | 0.29389 | −0.40451 |
| 29 | 0.20337 | −0.45677 |
| 30 | 0.10396 | −0.48907 |

**31 holes**

| # | x | y |
|---|---|---|
| 1 | 0.00000 | −0.50000 |
| 2 | −0.10065 | −0.48976 |
| 3 | −0.19718 | −0.45948 |
| 4 | −0.28563 | −0.41038 |
| 5 | −0.36240 | −0.34448 |
| 6 | −0.42432 | −0.26448 |
| 7 | −0.46888 | −0.17365 |
| 8 | −0.49423 | −0.07571 |
| 9 | −0.49936 | 0.02532 |
| 10 | −0.48404 | 0.12533 |
| 11 | −0.44890 | 0.22020 |
| 12 | −0.39539 | 0.30605 |
| 13 | −0.32569 | 0.37938 |
| 14 | −0.24265 | 0.43717 |
| 15 | −0.14968 | 0.47707 |
| 16 | −0.05058 | 0.49743 |
| 17 | 0.05058 | 0.49743 |
| 18 | 0.14968 | 0.47707 |
| 19 | 0.24265 | 0.43717 |
| 20 | 0.32569 | 0.37938 |
| 21 | 0.39539 | 0.30605 |
| 22 | 0.44890 | 0.22020 |
| 23 | 0.48404 | 0.12533 |
| 24 | 0.49936 | 0.02532 |
| 25 | 0.49423 | −0.07571 |
| 26 | 0.46888 | −0.17365 |
| 27 | 0.42432 | −0.26448 |
| 28 | 0.36240 | −0.34448 |
| 29 | 0.28563 | −0.41038 |
| 30 | 0.19718 | −0.45948 |
| 31 | 0.10065 | −0.48976 |

**32 holes**

| # | x | y |
|---|---|---|
| 1 | 0.00000 | −0.50000 |
| 2 | −0.09755 | −0.49039 |
| 3 | −0.19134 | −0.46194 |
| 4 | −0.27779 | −0.41573 |
| 5 | −0.35355 | −0.35355 |
| 6 | −0.41573 | −0.27779 |
| 7 | −0.46194 | −0.19134 |
| 8 | −0.49039 | −0.09755 |
| 9 | −0.50000 | 0.00000 |
| 10 | −0.49039 | 0.09755 |
| 11 | −0.46194 | 0.19134 |
| 12 | −0.41573 | 0.27779 |
| 13 | −0.35355 | 0.35355 |
| 14 | −0.27779 | 0.41573 |
| 15 | −0.19134 | 0.46194 |
| 16 | −0.09755 | 0.49039 |
| 17 | 0.00000 | 0.50000 |
| 18 | 0.09755 | 0.49039 |
| 19 | 0.19134 | 0.46194 |
| 20 | 0.27779 | 0.41573 |
| 21 | 0.35355 | 0.35355 |
| 22 | 0.41573 | 0.27779 |
| 23 | 0.46194 | 0.19134 |
| 24 | 0.49039 | 0.09755 |
| 25 | 0.50000 | 0.00000 |
| 26 | 0.49039 | −0.09755 |
| 27 | 0.46194 | −0.19134 |
| 28 | 0.41573 | −0.27779 |
| 29 | 0.35355 | −0.35355 |
| 30 | 0.27779 | −0.41573 |
| 31 | 0.19134 | −0.46194 |
| 32 | 0.09755 | −0.49039 |

**33 holes**

| # | x | y |
|---|---|---|
| 1 | 0.00000 | −0.50000 |
| 2 | −0.09463 | −0.49096 |
| 3 | −0.18583 | −0.46418 |
| 4 | −0.27032 | −0.42063 |
| 5 | −0.34504 | −0.36187 |
| 6 | −0.40729 | −0.29003 |
| 7 | −0.45482 | −0.20771 |
| 8 | −0.48591 | −0.11788 |
| 9 | −0.49943 | −0.02379 |
| 10 | −0.49491 | 0.07116 |

**33 holes** *(Continued)*

| # | x | y |
|---|---|---|
| 11 | −0.47250 | 0.16353 |
| 12 | −0.43301 | 0.25000 |
| 13 | −0.37787 | 0.32743 |
| 14 | −0.30908 | 0.39303 |
| 15 | −0.22911 | 0.44442 |
| 16 | −0.14087 | 0.47975 |
| 17 | −0.04753 | 0.49774 |
| 18 | 0.04753 | 0.49774 |
| 19 | 0.14087 | 0.47975 |
| 20 | 0.22911 | 0.44442 |
| 21 | 0.30908 | 0.39303 |
| 22 | 0.37787 | 0.32743 |
| 23 | 0.43301 | 0.25000 |
| 24 | 0.47250 | 0.16353 |
| 25 | 0.49491 | 0.07116 |
| 26 | 0.49943 | −0.02379 |
| 27 | 0.48591 | −0.11788 |
| 28 | 0.45482 | −0.20771 |
| 29 | 0.40729 | −0.29003 |
| 30 | 0.34504 | −0.36187 |
| 31 | 0.27032 | −0.42063 |
| 32 | 0.18583 | −0.46418 |
| 33 | 0.09463 | −0.49096 |

## Table 1b. Hole Coordinate Dimension Factors for Type "A" Hole Circles

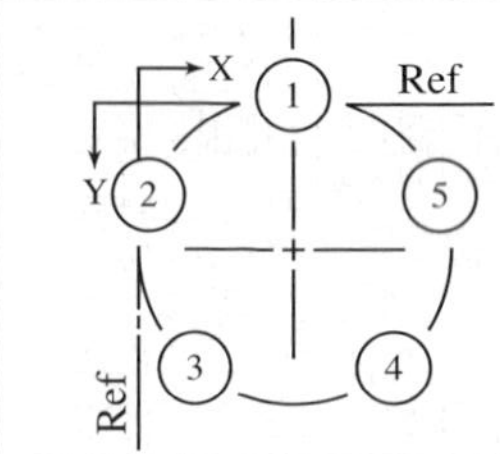


**3 holes**

| # | x | y |
|---|---|---|
| 1 | 0.50000 | 0.00000 |
| 2 | 0.06699 | 0.75000 |
| 3 | 0.93301 | 0.75000 |

**4 holes**

| # | x | y |
|---|---|---|
| 1 | 0.50000 | 0.00000 |
| 2 | 0.00000 | 0.50000 |
| 3 | 0.50000 | 1.00000 |
| 4 | 1.00000 | 0.50000 |

**5 holes**

| # | x | y |
|---|---|---|
| 1 | 0.50000 | 0.00000 |
| 2 | 0.02447 | 0.34549 |
| 3 | 0.20611 | 0.90451 |
| 4 | 0.79389 | 0.90451 |
| 5 | 0.97553 | 0.34549 |

**6 holes**

| # | x | y |
|---|---|---|
| 1 | 0.50000 | 0.00000 |
| 2 | 0.06699 | 0.25000 |
| 3 | 0.06699 | 0.75000 |
| 4 | 0.50000 | 1.00000 |
| 5 | 0.93301 | 0.75000 |
| 6 | 0.93301 | 0.25000 |

**7 holes**

| # | x | y |
|---|---|---|
| 1 | 0.50000 | 0.00000 |
| 2 | 0.10908 | 0.18826 |
| 3 | 0.01254 | 0.61126 |
| 4 | 0.28306 | 0.95048 |
| 5 | 0.71694 | 0.95048 |
| 6 | 0.98746 | 0.61126 |
| 7 | 0.89092 | 0.18826 |

**8 holes**

| # | x | y |
|---|---|---|
| 1 | 0.50000 | 0.00000 |
| 2 | 0.14645 | 0.14645 |
| 3 | 0.00000 | 0.50000 |
| 4 | 0.14645 | 0.85355 |
| 5 | 0.50000 | 1.00000 |
| 6 | 0.85355 | 0.85355 |
| 7 | 1.00000 | 0.50000 |
| 8 | 0.85355 | 0.14645 |

**9 holes**

| # | x | y |
|---|---|---|
| 1 | 0.50000 | 0.00000 |
| 2 | 0.17861 | 0.11698 |
| 3 | 0.00760 | 0.41318 |
| 4 | 0.06699 | 0.75000 |
| 5 | 0.32899 | 0.96985 |
| 6 | 0.67101 | 0.96985 |
| 7 | 0.93301 | 0.75000 |
| 8 | 0.99240 | 0.41318 |
| 9 | 0.82139 | 0.11698 |

**10 holes**

| # | x | y |
|---|---|---|
| 1 | 0.50000 | 0.00000 |
| 2 | 0.20611 | 0.09549 |
| 3 | 0.02447 | 0.34549 |
| 4 | 0.02447 | 0.65451 |
| 5 | 0.20611 | 0.90451 |
| 6 | 0.50000 | 1.00000 |
| 7 | 0.79389 | 0.90451 |
| 8 | 0.97553 | 0.65451 |
| 9 | 0.97553 | 0.34549 |
| 10 | 0.79389 | 0.09549 |

**11 holes**

| # | x | y |
|---|---|---|
| 1 | 0.50000 | 0.00000 |
| 2 | 0.22968 | 0.07937 |
| 3 | 0.04518 | 0.29229 |
| 4 | 0.00509 | 0.57116 |
| 5 | 0.12213 | 0.82743 |
| 6 | 0.35913 | 0.97975 |
| 7 | 0.64087 | 0.97975 |
| 8 | 0.87787 | 0.82743 |
| 9 | 0.99491 | 0.57116 |
| 10 | 0.95482 | 0.29229 |
| 11 | 0.77032 | 0.07937 |

**12 holes**

| # | x | y |
|---|---|---|
| 1 | 0.50000 | 0.00000 |
| 2 | 0.25000 | 0.06699 |
| 3 | 0.06699 | 0.25000 |
| 4 | 0.00000 | 0.50000 |
| 5 | 0.06699 | 0.75000 |
| 6 | 0.25000 | 0.93301 |
| 7 | 0.50000 | 1.00000 |
| 8 | 0.75000 | 0.93301 |
| 9 | 0.93301 | 0.75000 |
| 10 | 1.00000 | 0.50000 |
| 11 | 0.93301 | 0.25000 |
| 12 | 0.75000 | 0.06699 |

**13 holes**

| # | x | y |
|---|---|---|
| 1 | 0.50000 | 0.00000 |
| 2 | 0.26764 | 0.05727 |
| 3 | 0.08851 | 0.21597 |
| 4 | 0.00365 | 0.43973 |
| 5 | 0.03249 | 0.67730 |

**13 holes** *(Continued)*

| # | x | y |
|---|---|---|
| 6 | 0.16844 | 0.87426 |
| 7 | 0.38034 | 0.98547 |
| 8 | 0.61966 | 0.98547 |
| 9 | 0.83156 | 0.87426 |
| 10 | 0.96751 | 0.67730 |
| 11 | 0.99635 | 0.43973 |
| 12 | 0.91149 | 0.21597 |
| 13 | 0.73236 | 0.05727 |

**14 holes**

| # | x | y |
|---|---|---|
| 1 | 0.50000 | 0.00000 |
| 2 | 0.28306 | 0.04952 |
| 3 | 0.10908 | 0.18826 |
| 4 | 0.01254 | 0.38874 |
| 5 | 0.01254 | 0.61126 |
| 6 | 0.10908 | 0.81174 |
| 7 | 0.28306 | 0.95048 |
| 8 | 0.50000 | 1.00000 |
| 9 | 0.71694 | 0.95048 |
| 10 | 0.89092 | 0.81174 |
| 11 | 0.98746 | 0.61126 |
| 12 | 0.98746 | 0.38874 |
| 13 | 0.89092 | 0.18826 |
| 14 | 0.71694 | 0.04952 |

**15 holes**

| # | x | y |
|---|---|---|
| 1 | 0.50000 | 0.00000 |
| 2 | 0.29663 | 0.04323 |
| 3 | 0.12843 | 0.16543 |
| 4 | 0.02447 | 0.34549 |
| 5 | 0.00274 | 0.55226 |
| 6 | 0.06699 | 0.75000 |
| 7 | 0.20611 | 0.90451 |
| 8 | 0.39604 | 0.98907 |
| 9 | 0.60396 | 0.98907 |
| 10 | 0.79389 | 0.90451 |
| 11 | 0.93301 | 0.75000 |
| 12 | 0.99726 | 0.55226 |
| 13 | 0.97553 | 0.34549 |
| 14 | 0.87157 | 0.16543 |
| 15 | 0.70337 | 0.04323 |

**16 holes**

| # | x | y |
|---|---|---|
| 1 | 0.50000 | 0.00000 |
| 2 | 0.30866 | 0.03806 |
| 3 | 0.14645 | 0.14645 |
| 4 | 0.03806 | 0.30866 |
| 5 | 0.00000 | 0.50000 |
| 6 | 0.03806 | 0.69134 |
| 7 | 0.14645 | 0.85355 |
| 8 | 0.30866 | 0.96194 |
| 9 | 0.50000 | 1.00000 |
| 10 | 0.69134 | 0.96194 |
| 11 | 0.85355 | 0.85355 |
| 12 | 0.96194 | 0.69134 |
| 13 | 1.00000 | 0.50000 |
| 14 | 0.96194 | 0.30866 |
| 15 | 0.85355 | 0.14645 |
| 16 | 0.69134 | 0.03806 |

**17 holes**

| # | x | y |
|---|---|---|
| 1 | 0.50000 | 0.00000 |
| 2 | 0.31938 | 0.03376 |
| 3 | 0.16315 | 0.13050 |
| 4 | 0.05242 | 0.27713 |
| 5 | 0.00213 | 0.45387 |
| 6 | 0.01909 | 0.63683 |
| 7 | 0.10099 | 0.80132 |
| 8 | 0.23678 | 0.92511 |
| 9 | 0.40813 | 0.99149 |
| 10 | 0.59187 | 0.99149 |
| 11 | 0.76322 | 0.92511 |
| 12 | 0.89901 | 0.80132 |
| 13 | 0.98091 | 0.63683 |
| 14 | 0.99787 | 0.45387 |
| 15 | 0.94758 | 0.27713 |
| 16 | 0.83685 | 0.13050 |
| 17 | 0.68062 | 0.03376 |

**18 holes**

| # | x | y |
|---|---|---|
| 1 | 0.50000 | 0.00000 |
| 2 | 0.32899 | 0.03015 |
| 3 | 0.17861 | 0.11698 |
| 4 | 0.06699 | 0.25000 |
| 5 | 0.00760 | 0.41318 |
| 6 | 0.00760 | 0.58682 |
| 7 | 0.06699 | 0.75000 |

**18 holes** *(Continued)*

| # | x | y |
|---|---|---|
| 8 | 0.17861 | 0.88302 |
| 9 | 0.32899 | 0.96985 |
| 10 | 0.50000 | 1.00000 |
| 11 | 0.67101 | 0.96985 |
| 12 | 0.82139 | 0.88302 |
| 13 | 0.93301 | 0.75000 |
| 14 | 0.99240 | 0.58682 |
| 15 | 0.99240 | 0.41318 |
| 16 | 0.93301 | 0.25000 |
| 17 | 0.82139 | 0.11698 |
| 18 | 0.67101 | 0.03015 |

**19 holes**

| # | x | y |
|---|---|---|
| 1 | 0.50000 | 0.00000 |
| 2 | 0.33765 | 0.02709 |
| 3 | 0.19289 | 0.10543 |
| 4 | 0.08142 | 0.22653 |
| 5 | 0.01530 | 0.37726 |
| 6 | 0.00171 | 0.54129 |
| 7 | 0.04211 | 0.70085 |
| 8 | 0.13214 | 0.83864 |
| 9 | 0.26203 | 0.93974 |
| 10 | 0.41770 | 0.99318 |
| 11 | 0.58230 | 0.99318 |
| 12 | 0.73797 | 0.93974 |
| 13 | 0.86786 | 0.83864 |
| 14 | 0.95789 | 0.70085 |
| 15 | 0.99829 | 0.54129 |
| 16 | 0.98470 | 0.37726 |
| 17 | 0.91858 | 0.22653 |
| 18 | 0.80711 | 0.10543 |
| 19 | 0.66235 | 0.02709 |

**20 holes**

| # | x | y |
|---|---|---|
| 1 | 0.50000 | 0.00000 |
| 2 | 0.34549 | 0.02447 |
| 3 | 0.20611 | 0.09549 |
| 4 | 0.09549 | 0.20611 |
| 5 | 0.02447 | 0.34549 |
| 6 | 0.00000 | 0.50000 |
| 7 | 0.02447 | 0.65451 |
| 8 | 0.09549 | 0.79389 |
| 9 | 0.20611 | 0.90451 |
| 10 | 0.34549 | 0.97553 |

**20 holes** *(Continued)*

| # | x | y |
|---|---|---|
| 11 | 0.50000 | 1.00000 |
| 12 | 0.65451 | 0.97553 |
| 13 | 0.79389 | 0.90451 |
| 14 | 0.90451 | 0.79389 |
| 15 | 0.97553 | 0.65451 |
| 16 | 1.00000 | 0.50000 |
| 17 | 0.97553 | 0.34549 |
| 18 | 0.90451 | 0.20611 |
| 19 | 0.79389 | 0.09549 |
| 20 | 0.65451 | 0.02447 |

**21 holes**

| # | x | y |
|---|---|---|
| 1 | 0.50000 | 0.00000 |
| 2 | 0.35262 | 0.02221 |
| 3 | 0.21834 | 0.08688 |
| 4 | 0.10908 | 0.18826 |
| 5 | 0.03456 | 0.31733 |
| 6 | 0.00140 | 0.46263 |
| 7 | 0.01254 | 0.61126 |
| 8 | 0.06699 | 0.75000 |
| 9 | 0.15991 | 0.86653 |
| 10 | 0.28306 | 0.95048 |
| 11 | 0.42548 | 0.99442 |
| 12 | 0.57452 | 0.99442 |
| 13 | 0.71694 | 0.95048 |
| 14 | 0.84009 | 0.86653 |
| 15 | 0.93301 | 0.75000 |
| 16 | 0.98746 | 0.61126 |
| 17 | 0.99860 | 0.46263 |
| 18 | 0.96544 | 0.31733 |
| 19 | 0.89092 | 0.18826 |
| 20 | 0.78166 | 0.08688 |
| 21 | 0.64738 | 0.02221 |

**22 holes**

| # | x | y |
|---|---|---|
| 1 | 0.50000 | 0.00000 |
| 2 | 0.35913 | 0.02025 |
| 3 | 0.22968 | 0.07937 |
| 4 | 0.12213 | 0.17257 |
| 5 | 0.04518 | 0.29229 |
| 6 | 0.00509 | 0.42884 |
| 7 | 0.00509 | 0.57116 |
| 8 | 0.04518 | 0.70771 |
| 9 | 0.12213 | 0.82743 |

**22 holes** *(Continued)*

| # | x | y |
|---|---|---|
| 10 | 0.22968 | 0.92063 |
| 11 | 0.35913 | 0.97975 |
| 12 | 0.50000 | 1.00000 |
| 13 | 0.64087 | 0.97975 |
| 14 | 0.77032 | 0.92063 |
| 15 | 0.87787 | 0.82743 |
| 16 | 0.95482 | 0.70771 |
| 17 | 0.99491 | 0.57116 |
| 18 | 0.99491 | 0.42884 |
| 19 | 0.95482 | 0.29229 |
| 20 | 0.87787 | 0.17257 |
| 21 | 0.77032 | 0.07937 |
| 22 | 0.64087 | 0.02025 |

**23 holes**

| # | x | y |
|---|---|---|
| 1 | 0.50000 | 0.00000 |
| 2 | 0.36510 | 0.01854 |
| 3 | 0.24021 | 0.07279 |
| 4 | 0.13458 | 0.15872 |
| 5 | 0.05606 | 0.26997 |
| 6 | 0.01046 | 0.39827 |
| 7 | 0.00117 | 0.53412 |
| 8 | 0.02887 | 0.66744 |
| 9 | 0.09152 | 0.78834 |
| 10 | 0.18446 | 0.88786 |
| 11 | 0.30080 | 0.95861 |
| 12 | 0.43192 | 0.99534 |
| 13 | 0.56808 | 0.99534 |
| 14 | 0.69920 | 0.95861 |
| 15 | 0.81554 | 0.88786 |
| 16 | 0.90848 | 0.78834 |
| 17 | 0.97113 | 0.66744 |
| 18 | 0.99883 | 0.53412 |
| 19 | 0.98954 | 0.39827 |
| 20 | 0.94394 | 0.26997 |
| 21 | 0.86542 | 0.15872 |
| 22 | 0.75979 | 0.07279 |
| 23 | 0.63490 | 0.01854 |

**24 holes**

| # | x | y |
|---|---|---|
| 1 | 0.50000 | 0.00000 |
| 2 | 0.37059 | 0.01704 |
| 3 | 0.25000 | 0.06699 |
| 4 | 0.14645 | 0.14645 |

**Table 1b.** *(Continued)* **Hole Coordinate Dimension Factors for Type "A" Hole Circles**

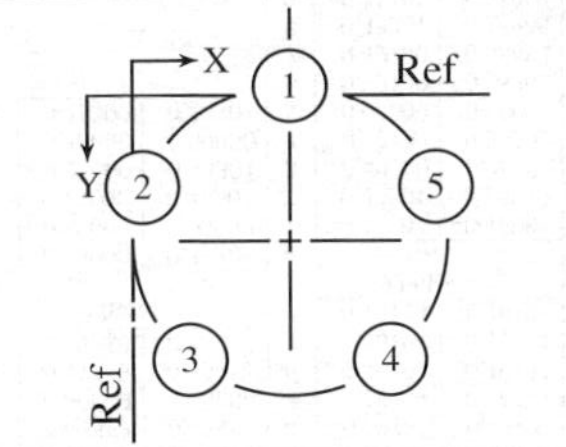


**24 holes** *(Continued)*

| # | x | y |
|---|---|---|
| 5 | 0.06699 | 0.25000 |
| 6 | 0.01704 | 0.37059 |
| 7 | 0.00000 | 0.50000 |
| 8 | 0.01704 | 0.62941 |
| 9 | 0.06699 | 0.75000 |
| 10 | 0.14645 | 0.85355 |
| 11 | 0.25000 | 0.93301 |
| 12 | 0.37059 | 0.98296 |
| 13 | 0.50000 | 1.00000 |
| 14 | 0.62941 | 0.98296 |
| 15 | 0.75000 | 0.93301 |
| 16 | 0.85355 | 0.85355 |
| 17 | 0.93301 | 0.75000 |
| 18 | 0.98296 | 0.62941 |
| 19 | 1.00000 | 0.50000 |
| 20 | 0.98296 | 0.37059 |
| 21 | 0.93301 | 0.25000 |
| 22 | 0.85355 | 0.14645 |
| 23 | 0.75000 | 0.06699 |
| 24 | 0.62941 | 0.01704 |

**25 holes**

| # | x | y |
|---|---|---|
| 1 | 0.50000 | 0.00000 |
| 2 | 0.37566 | 0.01571 |
| 3 | 0.25912 | 0.06185 |
| 4 | 0.15773 | 0.13552 |
| 5 | 0.07784 | 0.23209 |
| 6 | 0.02447 | 0.34549 |

**25 holes** *(Continued)*

| # | x | y |
|---|---|---|
| 7 | 0.00099 | 0.46860 |
| 8 | 0.00886 | 0.59369 |
| 9 | 0.04759 | 0.71289 |
| 10 | 0.11474 | 0.81871 |
| 11 | 0.20611 | 0.90451 |
| 12 | 0.31594 | 0.96489 |
| 13 | 0.43733 | 0.99606 |
| 14 | 0.56267 | 0.99606 |
| 15 | 0.68406 | 0.96489 |
| 16 | 0.79389 | 0.90451 |
| 17 | 0.88526 | 0.81871 |
| 18 | 0.95241 | 0.71289 |
| 19 | 0.99114 | 0.59369 |
| 20 | 0.99901 | 0.46860 |
| 21 | 0.97553 | 0.34549 |
| 22 | 0.92216 | 0.23209 |
| 23 | 0.84227 | 0.13552 |
| 24 | 0.74088 | 0.06185 |
| 25 | 0.62434 | 0.01571 |

**26 holes**

| # | x | y |
|---|---|---|
| 1 | 0.50000 | 0.00000 |
| 2 | 0.38034 | 0.01453 |
| 3 | 0.26764 | 0.05727 |
| 4 | 0.16844 | 0.12574 |
| 5 | 0.08851 | 0.21597 |
| 6 | 0.03249 | 0.32270 |
| 7 | 0.00365 | 0.43973 |

**26 holes** *(Continued)*

| # | x | y |
|---|---|---|
| 8 | 0.00365 | 0.56027 |
| 9 | 0.03249 | 0.67730 |
| 10 | 0.08851 | 0.78403 |
| 11 | 0.16844 | 0.87426 |
| 12 | 0.26764 | 0.94273 |
| 13 | 0.38034 | 0.98547 |
| 14 | 0.50000 | 1.00000 |
| 15 | 0.61966 | 0.98547 |
| 16 | 0.73236 | 0.94273 |
| 17 | 0.83156 | 0.87426 |
| 18 | 0.91149 | 0.78403 |
| 19 | 0.96751 | 0.67730 |
| 20 | 0.99635 | 0.56027 |
| 21 | 0.99635 | 0.43973 |
| 22 | 0.96751 | 0.32270 |
| 23 | 0.91149 | 0.21597 |
| 24 | 0.83156 | 0.12574 |
| 25 | 0.73236 | 0.05727 |
| 26 | 0.61966 | 0.01453 |

**27 holes**

| # | x | y |
|---|---|---|
| 1 | 0.50000 | 0.00000 |
| 2 | 0.38469 | 0.01348 |
| 3 | 0.27560 | 0.05318 |
| 4 | 0.17861 | 0.11698 |
| 5 | 0.09894 | 0.20142 |
| 6 | 0.04089 | 0.30196 |
| 7 | 0.00760 | 0.41318 |
| 8 | 0.00085 | 0.52907 |
| 9 | 0.02101 | 0.64340 |
| 10 | 0.06699 | 0.75000 |
| 11 | 0.13631 | 0.84312 |
| 12 | 0.22525 | 0.91774 |
| 13 | 0.32899 | 0.96985 |
| 14 | 0.44195 | 0.99662 |
| 15 | 0.55805 | 0.99662 |
| 16 | 0.67101 | 0.96985 |
| 17 | 0.77475 | 0.91774 |
| 18 | 0.86369 | 0.84312 |
| 19 | 0.93301 | 0.75000 |
| 20 | 0.97899 | 0.64340 |
| 21 | 0.99915 | 0.52907 |
| 22 | 0.99240 | 0.41318 |
| 23 | 0.95911 | 0.30196 |

**27 holes** *(Continued)*

| # | x | y |
|---|---|---|
| 24 | 0.90106 | 0.20142 |
| 25 | 0.82139 | 0.11698 |
| 26 | 0.72440 | 0.05318 |
| 27 | 0.61531 | 0.01348 |

**28 holes**

| # | x | y |
|---|---|---|
| 1 | 0.50000 | 0.00000 |
| 2 | 0.38874 | 0.01254 |
| 3 | 0.28306 | 0.04952 |
| 4 | 0.18826 | 0.10908 |
| 5 | 0.10908 | 0.18826 |
| 6 | 0.04952 | 0.28306 |
| 7 | 0.01254 | 0.38874 |
| 8 | 0.00000 | 0.50000 |
| 9 | 0.01254 | 0.61126 |
| 10 | 0.04952 | 0.71694 |
| 11 | 0.10908 | 0.81174 |
| 12 | 0.18826 | 0.89092 |
| 13 | 0.28306 | 0.95048 |
| 14 | 0.38874 | 0.98746 |
| 15 | 0.50000 | 1.00000 |
| 16 | 0.61126 | 0.98746 |
| 17 | 0.71694 | 0.95048 |
| 18 | 0.81174 | 0.89092 |
| 19 | 0.89092 | 0.81174 |
| 20 | 0.95048 | 0.71694 |
| 21 | 0.98746 | 0.61126 |
| 22 | 1.00000 | 0.50000 |
| 23 | 0.98746 | 0.38874 |
| 24 | 0.95048 | 0.28306 |
| 25 | 0.89092 | 0.18826 |
| 26 | 0.81174 | 0.10908 |
| 27 | 0.71694 | 0.04952 |
| 28 | 0.61126 | 0.01254 |

**29 holes**

| # | x | y |
|---|---|---|
| 1 | 0.50000 | 0.00000 |
| 2 | 0.39251 | 0.01169 |
| 3 | 0.29006 | 0.04621 |
| 4 | 0.19741 | 0.10195 |
| 5 | 0.11892 | 0.17631 |
| 6 | 0.05824 | 0.26580 |
| 7 | 0.01823 | 0.36624 |
| 8 | 0.00073 | 0.47293 |

**29 holes** *(Continued)*

| # | x | y |
|---|---|---|
| 9 | 0.00659 | 0.58089 |
| 10 | 0.03551 | 0.68507 |
| 11 | 0.08616 | 0.78059 |
| 12 | 0.15615 | 0.86300 |
| 13 | 0.24222 | 0.92843 |
| 14 | 0.34035 | 0.97383 |
| 15 | 0.44594 | 0.99707 |
| 16 | 0.55406 | 0.99707 |
| 17 | 0.65965 | 0.97383 |
| 18 | 0.75778 | 0.92843 |
| 19 | 0.84385 | 0.86300 |
| 20 | 0.91384 | 0.78059 |
| 21 | 0.96449 | 0.68507 |
| 22 | 0.99341 | 0.58089 |
| 23 | 0.99927 | 0.47293 |
| 24 | 0.98177 | 0.36624 |
| 25 | 0.94176 | 0.26580 |
| 26 | 0.88108 | 0.17631 |
| 27 | 0.80259 | 0.10195 |
| 28 | 0.70994 | 0.04621 |
| 29 | 0.60749 | 0.01169 |

**30 holes**

| # | x | y |
|---|---|---|
| 1 | 0.50000 | 0.00000 |
| 2 | 0.39604 | 0.01093 |
| 3 | 0.29663 | 0.04323 |
| 4 | 0.20611 | 0.09549 |
| 5 | 0.12843 | 0.16543 |
| 6 | 0.06699 | 0.25000 |
| 7 | 0.02447 | 0.34549 |
| 8 | 0.00274 | 0.44774 |
| 9 | 0.00274 | 0.55226 |
| 10 | 0.02447 | 0.65451 |
| 11 | 0.06699 | 0.75000 |
| 12 | 0.12843 | 0.83457 |
| 13 | 0.20611 | 0.90451 |
| 14 | 0.29663 | 0.95677 |
| 15 | 0.39604 | 0.98907 |
| 16 | 0.50000 | 1.00000 |
| 17 | 0.60396 | 0.98907 |
| 18 | 0.70337 | 0.95677 |
| 19 | 0.79389 | 0.90451 |
| 20 | 0.87157 | 0.83457 |
| 21 | 0.93301 | 0.75000 |

**30 holes** *(Continued)*

| # | x | y |
|---|---|---|
| 22 | 0.97553 | 0.65451 |
| 23 | 0.99726 | 0.55226 |
| 24 | 0.99726 | 0.44774 |
| 25 | 0.97553 | 0.34549 |
| 26 | 0.93301 | 0.25000 |
| 27 | 0.87157 | 0.16543 |
| 28 | 0.79389 | 0.09549 |
| 29 | 0.70337 | 0.04323 |
| 30 | 0.60396 | 0.01093 |

**31 holes**

| # | x | y |
|---|---|---|
| 1 | 0.50000 | 0.00000 |
| 2 | 0.39935 | 0.01024 |
| 3 | 0.30282 | 0.04052 |
| 4 | 0.21437 | 0.08962 |
| 5 | 0.13760 | 0.15552 |
| 6 | 0.07568 | 0.23552 |
| 7 | 0.03112 | 0.32635 |
| 8 | 0.00577 | 0.42429 |
| 9 | 0.00064 | 0.52532 |
| 10 | 0.01596 | 0.62533 |
| 11 | 0.05110 | 0.72020 |
| 12 | 0.10461 | 0.80605 |
| 13 | 0.17431 | 0.87938 |
| 14 | 0.25735 | 0.93717 |
| 15 | 0.35032 | 0.97707 |
| 16 | 0.44942 | 0.99743 |
| 17 | 0.55058 | 0.99743 |
| 18 | 0.64968 | 0.97707 |
| 19 | 0.74265 | 0.93717 |
| 20 | 0.82569 | 0.87938 |
| 21 | 0.89539 | 0.80605 |
| 22 | 0.94890 | 0.72020 |
| 23 | 0.98404 | 0.62533 |
| 24 | 0.99936 | 0.52532 |
| 25 | 0.99423 | 0.42429 |
| 26 | 0.96888 | 0.32635 |
| 27 | 0.92432 | 0.23552 |
| 28 | 0.86240 | 0.15552 |
| 29 | 0.78563 | 0.08962 |
| 30 | 0.69718 | 0.04052 |
| 31 | 0.60065 | 0.01024 |

**32 holes**

| # | x | y |
|---|---|---|
| 1 | 0.50000 | 0.00000 |
| 2 | 0.40245 | 0.00961 |
| 3 | 0.30866 | 0.03806 |
| 4 | 0.22221 | 0.08427 |
| 5 | 0.14645 | 0.14645 |
| 6 | 0.08427 | 0.22221 |
| 7 | 0.03806 | 0.30866 |
| 8 | 0.00961 | 0.40245 |
| 9 | 0.00000 | 0.50000 |
| 10 | 0.00961 | 0.59755 |
| 11 | 0.03806 | 0.69134 |
| 12 | 0.08427 | 0.77779 |
| 13 | 0.14645 | 0.85355 |
| 14 | 0.22221 | 0.91573 |
| 15 | 0.30866 | 0.96194 |
| 16 | 0.40245 | 0.99039 |
| 17 | 0.50000 | 1.00000 |
| 18 | 0.59755 | 0.99039 |
| 19 | 0.69134 | 0.96194 |
| 20 | 0.77779 | 0.91573 |
| 21 | 0.85355 | 0.85355 |
| 22 | 0.91573 | 0.77779 |
| 23 | 0.96194 | 0.69134 |
| 24 | 0.99039 | 0.59755 |
| 25 | 1.00000 | 0.50000 |
| 26 | 0.99039 | 0.40245 |
| 27 | 0.96194 | 0.30866 |
| 28 | 0.91573 | 0.22221 |
| 29 | 0.85355 | 0.14645 |
| 30 | 0.77779 | 0.08427 |
| 31 | 0.69134 | 0.03806 |
| 32 | 0.59755 | 0.00961 |

**33 holes**

| # | x | y |
|---|---|---|
| 1 | 0.50000 | 0.00000 |
| 2 | 0.40537 | 0.00904 |
| 3 | 0.31417 | 0.03582 |
| 4 | 0.22968 | 0.07937 |
| 5 | 0.15496 | 0.13813 |
| 6 | 0.09271 | 0.20997 |
| 7 | 0.04518 | 0.29229 |
| 8 | 0.01409 | 0.38212 |
| 9 | 0.00057 | 0.47621 |
| 10 | 0.00509 | 0.57116 |

**33 holes** *(Continued)*

| # | x | y |
|---|---|---|
| 11 | 0.02750 | 0.66353 |
| 12 | 0.06699 | 0.75000 |
| 13 | 0.12213 | 0.82743 |
| 14 | 0.19092 | 0.89303 |
| 15 | 0.27089 | 0.94442 |
| 16 | 0.35913 | 0.97975 |
| 17 | 0.45247 | 0.99774 |
| 18 | 0.54753 | 0.99774 |
| 19 | 0.64087 | 0.97975 |
| 20 | 0.72911 | 0.94442 |
| 21 | 0.80908 | 0.89303 |
| 22 | 0.87787 | 0.82743 |
| 23 | 0.93301 | 0.75000 |
| 24 | 0.97250 | 0.66353 |
| 25 | 0.99491 | 0.57116 |
| 26 | 0.99943 | 0.47621 |
| 27 | 0.98591 | 0.38212 |
| 28 | 0.95482 | 0.29229 |
| 29 | 0.90729 | 0.20997 |
| 30 | 0.84504 | 0.13813 |
| 31 | 0.77032 | 0.07937 |
| 32 | 0.68583 | 0.03582 |
| 33 | 0.59463 | 0.00904 |

## Table 2a. Hole Coordinate Dimension Factors for Type "B" Hole Circles

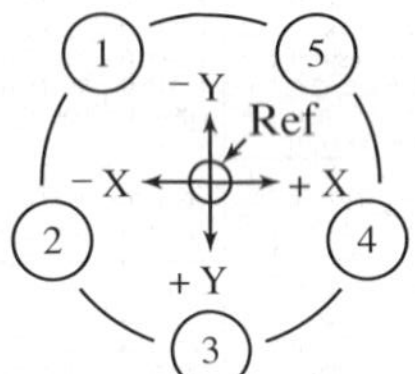


**3 holes**

| # | x | y |
|---|---|---|
| 1 | −0.43301 | −0.25000 |
| 2 | 0.00000 | 0.50000 |
| 3 | 0.43301 | −0.25000 |

**4 holes**

| # | x | y |
|---|---|---|
| 1 | −0.35355 | −0.35355 |
| 2 | −0.35355 | 0.35355 |
| 3 | 0.35355 | 0.35355 |
| 4 | 0.35355 | −0.35355 |

**5 holes**

| # | x | y |
|---|---|---|
| 1 | −0.29389 | −0.40451 |
| 2 | −0.47553 | 0.15451 |
| 3 | 0.00000 | 0.50000 |
| 4 | 0.47553 | 0.15451 |
| 5 | 0.29389 | −0.40451 |

**6 holes**

| # | x | y |
|---|---|---|
| 1 | −0.25000 | −0.43301 |
| 2 | −0.50000 | 0.00000 |
| 3 | −0.25000 | 0.43301 |
| 4 | 0.25000 | 0.43301 |
| 5 | 0.50000 | 0.00000 |
| 6 | 0.25000 | −0.43301 |

**7 holes**

| # | x | y |
|---|---|---|
| 1 | −0.21694 | −0.45048 |
| 2 | −0.48746 | −0.11126 |
| 3 | −0.39092 | 0.31174 |
| 4 | 0.00000 | 0.50000 |
| 5 | 0.39092 | 0.31174 |
| 6 | 0.48746 | −0.11126 |
| 7 | 0.21694 | −0.45048 |

**8 holes**

| # | x | y |
|---|---|---|
| 1 | −0.19134 | −0.46194 |
| 2 | −0.46194 | −0.19134 |
| 3 | −0.46194 | 0.19134 |
| 4 | −0.19134 | 0.46194 |
| 5 | 0.19134 | 0.46194 |
| 6 | 0.46194 | 0.19134 |
| 7 | 0.46194 | −0.19134 |
| 8 | 0.19134 | −0.46194 |

**9 holes**

| # | x | y |
|---|---|---|
| 1 | −0.17101 | −0.46985 |
| 2 | −0.43301 | −0.25000 |
| 3 | −0.49240 | 0.08682 |
| 4 | −0.32139 | 0.38302 |
| 5 | 0.00000 | 0.50000 |
| 6 | 0.32139 | 0.38302 |
| 7 | 0.49240 | 0.08682 |
| 8 | 0.43301 | −0.25000 |
| 9 | 0.17101 | −0.46985 |

**10 holes**

| # | x | y |
|---|---|---|
| 1 | −0.15451 | −0.47553 |
| 2 | −0.40451 | −0.29389 |
| 3 | −0.50000 | 0.00000 |
| 4 | −0.40451 | 0.29389 |
| 5 | −0.15451 | 0.47553 |
| 6 | 0.15451 | 0.47553 |
| 7 | 0.40451 | 0.29389 |
| 8 | 0.50000 | 0.00000 |
| 9 | 0.40451 | −0.29389 |
| 10 | 0.15451 | −0.47553 |

**11 holes**

| # | x | y |
|---|---|---|
| 1 | −0.14087 | −0.47975 |
| 2 | −0.37787 | −0.32743 |
| 3 | −0.49491 | −0.07116 |
| 4 | −0.45482 | 0.20771 |
| 5 | −0.27032 | 0.42063 |
| 6 | 0.00000 | 0.50000 |
| 7 | 0.27032 | 0.42063 |
| 8 | 0.45482 | 0.20771 |
| 9 | 0.49491 | −0.07116 |
| 10 | 0.37787 | −0.32743 |
| 11 | 0.14087 | −0.47975 |

**12 holes**

| # | x | y |
|---|---|---|
| 1 | −0.12941 | −0.48296 |
| 2 | −0.35355 | −0.35355 |
| 3 | −0.48296 | −0.12941 |
| 4 | −0.48296 | 0.12941 |
| 5 | −0.35355 | 0.35355 |
| 6 | −0.12941 | 0.48296 |
| 7 | 0.12941 | 0.48296 |
| 8 | 0.35355 | 0.35355 |
| 9 | 0.48296 | 0.12941 |
| 10 | 0.48296 | −0.12941 |
| 11 | 0.35355 | −0.35355 |
| 12 | 0.12941 | −0.48296 |

**13 holes**

| # | x | y |
|---|---|---|
| 1 | −0.11966 | −0.48547 |
| 2 | −0.33156 | −0.37426 |
| 3 | −0.46751 | −0.17730 |
| 4 | −0.49635 | 0.06027 |
| 5 | −0.41149 | 0.28403 |

**13 holes** *(Continued)*

| # | x | y |
|---|---|---|
| 6 | −0.23236 | 0.44273 |
| 7 | 0.00000 | 0.50000 |
| 8 | 0.23236 | 0.44273 |
| 9 | 0.41149 | 0.28403 |
| 10 | 0.49635 | 0.06027 |
| 11 | 0.46751 | −0.17730 |
| 12 | 0.33156 | −0.37426 |
| 13 | 0.11966 | −0.48547 |

**14 holes**

| # | x | y |
|---|---|---|
| 1 | −0.11126 | −0.48746 |
| 2 | −0.31174 | −0.39092 |
| 3 | −0.45048 | −0.21694 |
| 4 | −0.50000 | 0.00000 |
| 5 | −0.45048 | 0.21694 |
| 6 | −0.31174 | 0.39092 |
| 7 | −0.11126 | 0.48746 |
| 8 | 0.11126 | 0.48746 |
| 9 | 0.31174 | 0.39092 |
| 10 | 0.45048 | 0.21694 |
| 11 | 0.50000 | 0.00000 |
| 12 | 0.45048 | −0.21694 |
| 13 | 0.31174 | −0.39092 |
| 14 | 0.11126 | −0.48746 |

**15 holes**

| # | x | y |
|---|---|---|
| 1 | −0.10396 | −0.48907 |
| 2 | −0.29389 | −0.40451 |
| 3 | −0.43301 | −0.25000 |
| 4 | −0.49726 | −0.05226 |
| 5 | −0.47553 | 0.15451 |
| 6 | −0.37157 | 0.33457 |
| 7 | −0.20337 | 0.45677 |
| 8 | 0.00000 | 0.50000 |
| 9 | 0.20337 | 0.45677 |
| 10 | 0.37157 | 0.33457 |
| 11 | 0.47553 | 0.15451 |
| 12 | 0.49726 | −0.05226 |
| 13 | 0.43301 | −0.25000 |
| 14 | 0.29389 | −0.40451 |
| 15 | 0.10396 | −0.48907 |

**16 holes**

| # | x | y |
|---|---|---|
| 1 | −0.09755 | −0.49039 |
| 2 | −0.27779 | −0.41573 |
| 3 | −0.41573 | −0.27779 |
| 4 | −0.49039 | −0.09755 |
| 5 | −0.49039 | 0.09755 |
| 6 | −0.41573 | 0.27779 |
| 7 | −0.27779 | 0.41573 |
| 8 | −0.09755 | 0.49039 |
| 9 | 0.09755 | 0.49039 |
| 10 | 0.27779 | 0.41573 |
| 11 | 0.41573 | 0.27779 |
| 12 | 0.49039 | 0.09755 |
| 13 | 0.49039 | −0.09755 |
| 14 | 0.41573 | −0.27779 |
| 15 | 0.27779 | −0.41573 |
| 16 | 0.09755 | −0.49039 |

**17 holes**

| # | x | y |
|---|---|---|
| 1 | −0.09187 | −0.49149 |
| 2 | −0.26322 | −0.42511 |
| 3 | −0.39901 | −0.30132 |
| 4 | −0.48091 | −0.13683 |
| 5 | −0.49787 | 0.04613 |
| 6 | −0.44758 | 0.22287 |
| 7 | −0.33685 | 0.36950 |
| 8 | −0.18062 | 0.46624 |
| 9 | 0.00000 | 0.50000 |
| 10 | 0.18062 | 0.46624 |
| 11 | 0.33685 | 0.36950 |
| 12 | 0.44758 | 0.22287 |
| 13 | 0.49787 | 0.04613 |
| 14 | 0.48091 | −0.13683 |
| 15 | 0.39901 | −0.30132 |
| 16 | 0.26322 | −0.42511 |
| 17 | 0.09187 | −0.49149 |

**18 holes**

| # | x | y |
|---|---|---|
| 1 | −0.08682 | −0.49240 |
| 2 | −0.25000 | −0.43301 |
| 3 | −0.38302 | −0.32139 |
| 4 | −0.46985 | −0.17101 |
| 5 | −0.50000 | 0.00000 |
| 6 | −0.46985 | 0.17101 |
| 7 | −0.38302 | 0.32139 |

**18 holes** *(Continued)*

| # | x | y |
|---|---|---|
| 8 | −0.25000 | 0.43301 |
| 9 | −0.08682 | 0.49240 |
| 10 | 0.08682 | 0.49240 |
| 11 | 0.25000 | 0.43301 |
| 12 | 0.38302 | 0.32139 |
| 13 | 0.46985 | 0.17101 |
| 14 | 0.50000 | 0.00000 |
| 15 | 0.46985 | −0.17101 |
| 16 | 0.38302 | −0.32139 |
| 17 | 0.25000 | −0.43301 |
| 18 | 0.08682 | −0.49240 |

**19 holes**

| # | x | y |
|---|---|---|
| 1 | −0.08230 | −0.49318 |
| 2 | −0.23797 | −0.43974 |
| 3 | −0.36786 | −0.33864 |
| 4 | −0.45789 | −0.20085 |
| 5 | −0.49829 | −0.04129 |
| 6 | −0.48470 | 0.12274 |
| 7 | −0.41858 | 0.27347 |
| 8 | −0.30711 | 0.39457 |
| 9 | −0.16235 | 0.47291 |
| 10 | 0.00000 | 0.50000 |
| 11 | 0.16235 | 0.47291 |
| 12 | 0.30711 | 0.39457 |
| 13 | 0.41858 | 0.27347 |
| 14 | 0.48470 | 0.12274 |
| 15 | 0.49829 | −0.04129 |
| 16 | 0.45789 | −0.20085 |
| 17 | 0.36786 | −0.33864 |
| 18 | 0.23797 | −0.43974 |
| 19 | 0.08230 | −0.49318 |

**20 holes**

| # | x | y |
|---|---|---|
| 1 | −0.07822 | −0.49384 |
| 2 | −0.22700 | −0.44550 |
| 3 | −0.35355 | −0.35355 |
| 4 | −0.44550 | −0.22700 |
| 5 | −0.49384 | −0.07822 |
| 6 | −0.49384 | 0.07822 |
| 7 | −0.44550 | 0.22700 |
| 8 | −0.35355 | 0.35355 |
| 9 | −0.22700 | 0.44550 |
| 10 | −0.07822 | 0.49384 |

**20 holes** *(Continued)*

| # | x | y |
|---|---|---|
| 11 | 0.07822 | 0.49384 |
| 12 | 0.22700 | 0.44550 |
| 13 | 0.35355 | 0.35355 |
| 14 | 0.44550 | 0.22700 |
| 15 | 0.49384 | 0.07822 |
| 16 | 0.49384 | −0.07822 |
| 17 | 0.44550 | −0.22700 |
| 18 | 0.35355 | −0.35355 |
| 19 | 0.22700 | −0.44550 |
| 20 | 0.07822 | −0.49384 |

**21 holes**

| # | x | y |
|---|---|---|
| 1 | −0.07452 | −0.49442 |
| 2 | −0.21694 | −0.45048 |
| 3 | −0.34009 | −0.36653 |
| 4 | −0.43301 | −0.25000 |
| 5 | −0.48746 | −0.11126 |
| 6 | −0.49860 | 0.03737 |
| 7 | −0.46544 | 0.18267 |
| 8 | −0.39092 | 0.31174 |
| 9 | −0.28166 | 0.41312 |
| 10 | −0.14738 | 0.47779 |
| 11 | 0.00000 | 0.50000 |
| 12 | 0.14738 | 0.47779 |
| 13 | 0.28166 | 0.41312 |
| 14 | 0.39092 | 0.31174 |
| 15 | 0.46544 | 0.18267 |
| 16 | 0.49860 | 0.03737 |
| 17 | 0.48746 | −0.11126 |
| 18 | 0.43301 | −0.25000 |
| 19 | 0.34009 | −0.36653 |
| 20 | 0.21694 | −0.45048 |
| 21 | 0.07452 | −0.49442 |

**22 holes**

| # | x | y |
|---|---|---|
| 1 | −0.07116 | −0.49491 |
| 2 | −0.20771 | −0.45482 |
| 3 | −0.32743 | −0.37787 |
| 4 | −0.42063 | −0.27032 |
| 5 | −0.47975 | −0.14087 |
| 6 | −0.50000 | 0.00000 |
| 7 | −0.47975 | 0.14087 |
| 8 | −0.42063 | 0.27032 |
| 9 | −0.32743 | 0.37787 |

**22 holes** *(Continued)*

| # | x | y |
|---|---|---|
| 10 | −0.20771 | 0.45482 |
| 11 | −0.07116 | 0.49491 |
| 12 | 0.07116 | 0.49491 |
| 13 | 0.20771 | 0.45482 |
| 14 | 0.32743 | 0.37787 |
| 15 | 0.42063 | 0.27032 |
| 16 | 0.47975 | 0.14087 |
| 17 | 0.50000 | 0.00000 |
| 18 | 0.47975 | −0.14087 |
| 19 | 0.42063 | −0.27032 |
| 20 | 0.32743 | −0.37787 |
| 21 | 0.20771 | −0.45482 |
| 22 | 0.07116 | −0.49491 |

**23 holes**

| # | x | y |
|---|---|---|
| 1 | −0.06808 | −0.49534 |
| 2 | −0.19920 | −0.45861 |
| 3 | −0.31554 | −0.38786 |
| 4 | −0.40848 | −0.28834 |
| 5 | −0.47113 | −0.16744 |
| 6 | −0.49883 | −0.03412 |
| 7 | −0.48954 | 0.10173 |
| 8 | −0.44394 | 0.23003 |
| 9 | −0.36542 | 0.34128 |
| 10 | −0.25979 | 0.42721 |
| 11 | −0.13490 | 0.48146 |
| 12 | 0.00000 | 0.50000 |
| 13 | 0.13490 | 0.48146 |
| 14 | 0.25979 | 0.42721 |
| 15 | 0.36542 | 0.34128 |
| 16 | 0.44394 | 0.23003 |
| 17 | 0.48954 | 0.10173 |
| 18 | 0.49883 | −0.03412 |
| 19 | 0.47113 | −0.16744 |
| 20 | 0.40848 | −0.28834 |
| 21 | 0.31554 | −0.38786 |
| 22 | 0.19920 | −0.45861 |
| 23 | 0.06808 | −0.49534 |

**24 holes**

| # | x | y |
|---|---|---|
| 1 | −0.06526 | −0.49572 |
| 2 | −0.19134 | −0.46194 |
| 3 | −0.30438 | −0.39668 |
| 4 | −0.39668 | −0.30438 |

## Table 2a. *(Continued)* Hole Coordinate Dimension Factors for Type "B" Hole Circles

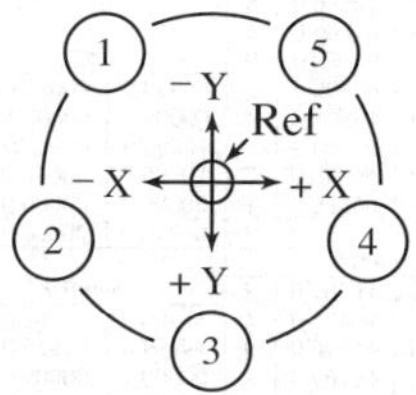


**24 holes** *(Continued)*

| # | x | y |
|---|---|---|
| 5 | −0.46194 | −0.19134 |
| 6 | −0.49572 | −0.06526 |
| 7 | −0.49572 | 0.06526 |
| 8 | −0.46194 | 0.19134 |
| 9 | −0.39668 | 0.30438 |
| 10 | −0.30438 | 0.39668 |
| 11 | −0.19134 | 0.46194 |
| 12 | −0.06526 | 0.49572 |
| 13 | 0.06526 | 0.49572 |
| 14 | 0.19134 | 0.46194 |
| 15 | 0.30438 | 0.39668 |
| 16 | 0.39668 | 0.30438 |
| 17 | 0.46194 | 0.19134 |
| 18 | 0.49572 | 0.06526 |
| 19 | 0.49572 | −0.06526 |
| 20 | 0.46194 | −0.19134 |
| 21 | 0.39668 | −0.30438 |
| 22 | 0.30438 | −0.39668 |
| 23 | 0.19134 | −0.46194 |
| 24 | 0.06526 | −0.49572 |

**25 holes**

| # | x | y |
|---|---|---|
| 1 | −0.06267 | −0.49606 |
| 2 | −0.18406 | −0.46489 |
| 3 | −0.29389 | −0.40451 |
| 4 | −0.38526 | −0.31871 |
| 5 | −0.45241 | −0.21289 |
| 6 | −0.49114 | −0.09369 |

**25 holes** *(Continued)*

| # | x | y |
|---|---|---|
| 7 | −0.49901 | 0.03140 |
| 8 | −0.47553 | 0.15451 |
| 9 | −0.42216 | 0.26791 |
| 10 | −0.34227 | 0.36448 |
| 11 | −0.24088 | 0.43815 |
| 12 | −0.12434 | 0.48429 |
| 13 | 0.00000 | 0.50000 |
| 14 | 0.12434 | 0.48429 |
| 15 | 0.24088 | 0.43815 |
| 16 | 0.34227 | 0.36448 |
| 17 | 0.42216 | 0.26791 |
| 18 | 0.47553 | 0.15451 |
| 19 | 0.49901 | 0.03140 |
| 20 | 0.49114 | −0.09369 |
| 21 | 0.45241 | −0.21289 |
| 22 | 0.38526 | −0.31871 |
| 23 | 0.29389 | −0.40451 |
| 24 | 0.18406 | −0.46489 |
| 25 | 0.06267 | −0.49606 |

**26 holes**

| # | x | y |
|---|---|---|
| 1 | −0.06027 | −0.49635 |
| 2 | −0.17730 | −0.46751 |
| 3 | −0.28403 | −0.41149 |
| 4 | −0.37426 | −0.33156 |
| 5 | −0.44273 | −0.23236 |
| 6 | −0.48547 | −0.11966 |
| 7 | −0.50000 | 0.00000 |

**26 holes** *(Continued)*

| # | x | y |
|---|---|---|
| 8 | −0.48547 | 0.11966 |
| 9 | −0.44273 | 0.23236 |
| 10 | −0.37426 | 0.33156 |
| 11 | −0.28403 | 0.41149 |
| 12 | −0.17730 | 0.46751 |
| 13 | −0.06027 | 0.49635 |
| 14 | 0.06027 | 0.49635 |
| 15 | 0.17730 | 0.46751 |
| 16 | 0.28403 | 0.41149 |
| 17 | 0.37426 | 0.33156 |
| 18 | 0.44273 | 0.23236 |
| 19 | 0.48547 | 0.11966 |
| 20 | 0.50000 | 0.00000 |
| 21 | 0.48547 | −0.11966 |
| 22 | 0.44273 | −0.23236 |
| 23 | 0.37426 | −0.33156 |
| 24 | 0.28403 | −0.41149 |
| 25 | 0.17730 | −0.46751 |
| 26 | 0.06027 | −0.49635 |

**27 holes**

| # | x | y |
|---|---|---|
| 1 | −0.05805 | −0.49662 |
| 2 | −0.17101 | −0.46985 |
| 3 | −0.27475 | −0.41774 |
| 4 | −0.36369 | −0.34312 |
| 5 | −0.43301 | −0.25000 |
| 6 | −0.47899 | −0.14340 |
| 7 | −0.49915 | −0.02907 |
| 8 | −0.49240 | 0.08682 |
| 9 | −0.45911 | 0.19804 |
| 10 | −0.40106 | 0.29858 |
| 11 | −0.32139 | 0.38302 |
| 12 | −0.22440 | 0.44682 |
| 13 | −0.11531 | 0.48652 |
| 14 | 0.00000 | 0.50000 |
| 15 | 0.11531 | 0.48652 |
| 16 | 0.22440 | 0.44682 |
| 17 | 0.32139 | 0.38302 |
| 18 | 0.40106 | 0.29858 |
| 19 | 0.45911 | 0.19804 |
| 20 | 0.49240 | 0.08682 |
| 21 | 0.49915 | −0.02907 |
| 22 | 0.47899 | −0.14340 |
| 23 | 0.43301 | −0.25000 |

**27 holes** *(Continued)*

| # | x | y |
|---|---|---|
| 24 | 0.36369 | −0.34312 |
| 25 | 0.27475 | −0.41774 |
| 26 | 0.17101 | −0.46985 |
| 27 | 0.05805 | −0.49662 |

**28 holes**

| # | x | y |
|---|---|---|
| 1 | −0.05598 | −0.49686 |
| 2 | −0.16514 | −0.47194 |
| 3 | −0.26602 | −0.42336 |
| 4 | −0.35355 | −0.35355 |
| 5 | −0.42336 | −0.26602 |
| 6 | −0.47194 | −0.16514 |
| 7 | −0.49686 | −0.05598 |
| 8 | −0.49686 | 0.05598 |
| 9 | −0.47194 | 0.16514 |
| 10 | −0.42336 | 0.26602 |
| 11 | −0.35355 | 0.35355 |
| 12 | −0.26602 | 0.42336 |
| 13 | −0.16514 | 0.47194 |
| 14 | −0.05598 | 0.49686 |
| 15 | 0.05598 | 0.49686 |
| 16 | 0.16514 | 0.47194 |
| 17 | 0.26602 | 0.42336 |
| 18 | 0.35355 | 0.35355 |
| 19 | 0.42336 | 0.26602 |
| 20 | 0.47194 | 0.16514 |
| 21 | 0.49686 | 0.05598 |
| 22 | 0.49686 | −0.05598 |
| 23 | 0.47194 | −0.16514 |
| 24 | 0.42336 | −0.26602 |
| 25 | 0.35355 | −0.35355 |
| 26 | 0.26602 | −0.42336 |
| 27 | 0.16514 | −0.47194 |
| 28 | 0.05598 | −0.49686 |

**29 holes**

| # | x | y |
|---|---|---|
| 1 | −0.05406 | −0.49707 |
| 2 | −0.15965 | −0.47383 |
| 3 | −0.25778 | −0.42843 |
| 4 | −0.34385 | −0.36300 |
| 5 | −0.41384 | −0.28059 |
| 6 | −0.46449 | −0.18507 |
| 7 | −0.49341 | −0.08089 |
| 8 | −0.49927 | 0.02707 |

**29 holes** *(Continued)*

| # | x | y |
|---|---|---|
| 9 | −0.48177 | 0.13376 |
| 10 | −0.44176 | 0.23420 |
| 11 | −0.38108 | 0.32369 |
| 12 | −0.30259 | 0.39805 |
| 13 | −0.20994 | 0.45379 |
| 14 | −0.10749 | 0.48831 |
| 15 | 0.00000 | 0.50000 |
| 16 | 0.10749 | 0.48831 |
| 17 | 0.20994 | 0.45379 |
| 18 | 0.30259 | 0.39805 |
| 19 | 0.38108 | 0.32369 |
| 20 | 0.44176 | 0.23420 |
| 21 | 0.48177 | 0.13376 |
| 22 | 0.49927 | 0.02707 |
| 23 | 0.49341 | −0.08089 |
| 24 | 0.46449 | −0.18507 |
| 25 | 0.41384 | −0.28059 |
| 26 | 0.34385 | −0.36300 |
| 27 | 0.25778 | −0.42843 |
| 28 | 0.15965 | −0.47383 |
| 29 | 0.05406 | −0.49707 |

**30 holes**

| # | x | y |
|---|---|---|
| 1 | −0.05226 | −0.49726 |
| 2 | −0.15451 | −0.47553 |
| 3 | −0.25000 | −0.43301 |
| 4 | −0.33457 | −0.37157 |
| 5 | −0.40451 | −0.29389 |
| 6 | −0.45677 | −0.20337 |
| 7 | −0.48907 | −0.10396 |
| 8 | −0.50000 | 0.00000 |
| 9 | −0.48907 | 0.10396 |
| 10 | −0.45677 | 0.20337 |
| 11 | −0.40451 | 0.29389 |
| 12 | −0.33457 | 0.37157 |
| 13 | −0.25000 | 0.43301 |
| 14 | −0.15451 | 0.47553 |
| 15 | −0.05226 | 0.49726 |
| 16 | 0.05226 | 0.49726 |
| 17 | 0.15451 | 0.47553 |
| 18 | 0.25000 | 0.43301 |
| 19 | 0.33457 | 0.37157 |
| 20 | 0.40451 | 0.29389 |
| 21 | 0.45677 | 0.20337 |

**30 holes** *(Continued)*

| # | x | y |
|---|---|---|
| 22 | 0.48907 | 0.10396 |
| 23 | 0.50000 | 0.00000 |
| 24 | 0.48907 | −0.10396 |
| 25 | 0.45677 | −0.20337 |
| 26 | 0.40451 | −0.29389 |
| 27 | 0.33457 | −0.37157 |
| 28 | 0.25000 | −0.43301 |
| 29 | 0.15451 | −0.47553 |
| 30 | 0.05226 | −0.49726 |

**31 holes**

| # | x | y |
|---|---|---|
| 1 | −0.05058 | −0.49743 |
| 2 | −0.14968 | −0.47707 |
| 3 | −0.24265 | −0.43717 |
| 4 | −0.32569 | −0.37938 |
| 5 | −0.39539 | −0.30605 |
| 6 | −0.44890 | −0.22020 |
| 7 | −0.48404 | −0.12533 |
| 8 | −0.49936 | −0.02532 |
| 9 | −0.49423 | 0.07571 |
| 10 | −0.46888 | 0.17365 |
| 11 | −0.42432 | 0.26448 |
| 12 | −0.36240 | 0.34448 |
| 13 | −0.28563 | 0.41038 |
| 14 | −0.19718 | 0.45948 |
| 15 | −0.10065 | 0.48976 |
| 16 | 0.00000 | 0.50000 |
| 17 | 0.10065 | 0.48976 |
| 18 | 0.19718 | 0.45948 |
| 19 | 0.28563 | 0.41038 |
| 20 | 0.36240 | 0.34448 |
| 21 | 0.42432 | 0.26448 |
| 22 | 0.46888 | 0.17365 |
| 23 | 0.49423 | 0.07571 |
| 24 | 0.49936 | −0.02532 |
| 25 | 0.48404 | −0.12533 |
| 26 | 0.44890 | −0.22020 |
| 27 | 0.39539 | −0.30605 |
| 28 | 0.32569 | −0.37938 |
| 29 | 0.24265 | −0.43717 |
| 30 | 0.14968 | −0.47707 |
| 31 | 0.05058 | −0.49743 |

**32 holes**

| # | x | y |
|---|---|---|
| 1 | −0.04901 | −0.49759 |
| 2 | −0.14514 | −0.47847 |
| 3 | −0.23570 | −0.44096 |
| 4 | −0.31720 | −0.38651 |
| 5 | −0.38651 | −0.31720 |
| 6 | −0.44096 | −0.23570 |
| 7 | −0.47847 | −0.14514 |
| 8 | −0.49759 | −0.04901 |
| 9 | −0.49759 | 0.04901 |
| 10 | −0.47847 | 0.14514 |
| 11 | −0.44096 | 0.23570 |
| 12 | −0.38651 | 0.31720 |
| 13 | −0.31720 | 0.38651 |
| 14 | −0.23570 | 0.44096 |
| 15 | −0.14514 | 0.47847 |
| 16 | −0.04901 | 0.49759 |
| 17 | 0.04901 | 0.49759 |
| 18 | 0.14514 | 0.47847 |
| 19 | 0.23570 | 0.44096 |
| 20 | 0.31720 | 0.38651 |
| 21 | 0.38651 | 0.31720 |
| 22 | 0.44096 | 0.23570 |
| 23 | 0.47847 | 0.14514 |
| 24 | 0.49759 | 0.04901 |
| 25 | 0.49759 | −0.04901 |
| 26 | 0.47847 | −0.14514 |
| 27 | 0.44096 | −0.23570 |
| 28 | 0.38651 | −0.31720 |
| 29 | 0.31720 | −0.38651 |
| 30 | 0.23570 | −0.44096 |
| 31 | 0.14514 | −0.47847 |
| 32 | 0.04901 | −0.49759 |

**33 holes**

| # | x | y |
|---|---|---|
| 1 | −0.04753 | −0.49774 |
| 2 | −0.14087 | −0.47975 |
| 3 | −0.22911 | −0.44442 |
| 4 | −0.30908 | −0.39303 |
| 5 | −0.37787 | −0.32743 |
| 6 | −0.43301 | −0.25000 |
| 7 | −0.47250 | −0.16353 |
| 8 | −0.49491 | −0.07116 |
| 9 | −0.49943 | 0.02379 |
| 10 | −0.48591 | 0.11788 |

**33 holes** *(Continued)*

| # | x | y |
|---|---|---|
| 11 | −0.45482 | 0.20771 |
| 12 | −0.40729 | 0.29003 |
| 13 | −0.34504 | 0.36187 |
| 14 | −0.27032 | 0.42063 |
| 15 | −0.18583 | 0.46418 |
| 16 | −0.09463 | 0.49096 |
| 17 | 0.00000 | 0.50000 |
| 18 | 0.09463 | 0.49096 |
| 19 | 0.18583 | 0.46418 |
| 20 | 0.27032 | 0.42063 |
| 21 | 0.34504 | 0.36187 |
| 22 | 0.40729 | 0.29003 |
| 23 | 0.45482 | 0.20771 |
| 24 | 0.48591 | 0.11788 |
| 25 | 0.49943 | 0.02379 |
| 26 | 0.49491 | −0.07116 |
| 27 | 0.47250 | −0.16353 |
| 28 | 0.43301 | −0.25000 |
| 29 | 0.37787 | −0.32743 |
| 30 | 0.30908 | −0.39303 |
| 31 | 0.22911 | −0.44442 |
| 32 | 0.14087 | −0.47975 |
| 33 | 0.04753 | −0.49774 |

## Table 2b. Hole Coordinate Dimension Factors for Type "B" Hole Circles

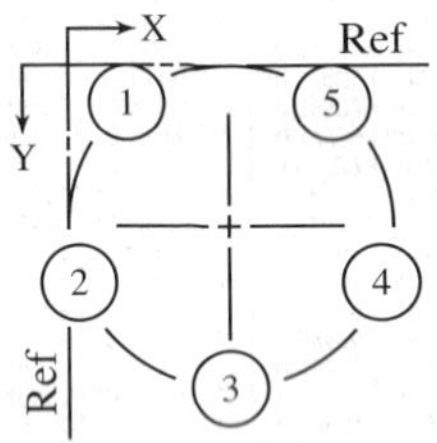


**3 holes**

| # | x | y |
|---|---|---|
| 1 | 0.06699 | 0.25000 |
| 2 | 0.50000 | 1.00000 |
| 3 | 0.93301 | 0.25000 |

**4 holes**

| # | x | y |
|---|---|---|
| 1 | 0.14645 | 0.14645 |
| 2 | 0.14645 | 0.85355 |
| 3 | 0.85355 | 0.85355 |
| 4 | 0.85355 | 0.14645 |

**5 holes**

| # | x | y |
|---|---|---|
| 1 | 0.20611 | 0.09549 |
| 2 | 0.02447 | 0.65451 |
| 3 | 0.50000 | 1.00000 |
| 4 | 0.97553 | 0.65451 |
| 5 | 0.79389 | 0.09549 |

**6 holes**

| # | x | y |
|---|---|---|
| 1 | 0.25000 | 0.06699 |
| 2 | 0.00000 | 0.50000 |
| 3 | 0.25000 | 0.93301 |
| 4 | 0.75000 | 0.93301 |
| 5 | 1.00000 | 0.50000 |
| 6 | 0.75000 | 0.06699 |

**7 holes**

| # | x | y |
|---|---|---|
| 1 | 0.28306 | 0.04952 |
| 2 | 0.01254 | 0.38874 |
| 3 | 0.10908 | 0.81174 |
| 4 | 0.50000 | 1.00000 |
| 5 | 0.89092 | 0.81174 |
| 6 | 0.98746 | 0.38874 |
| 7 | 0.71694 | 0.04952 |

**8 holes**

| # | x | y |
|---|---|---|
| 1 | 0.30866 | 0.03806 |
| 2 | 0.03806 | 0.30866 |
| 3 | 0.03806 | 0.69134 |
| 4 | 0.30866 | 0.96194 |
| 5 | 0.69134 | 0.96194 |
| 6 | 0.96194 | 0.69134 |
| 7 | 0.96194 | 0.30866 |
| 8 | 0.69134 | 0.03806 |

**9 holes**

| # | x | y |
|---|---|---|
| 1 | 0.32899 | 0.03015 |
| 2 | 0.06699 | 0.25000 |
| 3 | 0.00760 | 0.58682 |
| 4 | 0.17861 | 0.88302 |
| 5 | 0.50000 | 1.00000 |
| 6 | 0.82139 | 0.88302 |
| 7 | 0.99240 | 0.58682 |
| 8 | 0.93301 | 0.25000 |
| 9 | 0.67101 | 0.03015 |

**10 holes**

| # | x | y |
|---|---|---|
| 1 | 0.34549 | 0.02447 |
| 2 | 0.09549 | 0.20611 |
| 3 | 0.00000 | 0.50000 |
| 4 | 0.09549 | 0.79389 |
| 5 | 0.34549 | 0.97553 |
| 6 | 0.65451 | 0.97553 |
| 7 | 0.90451 | 0.79389 |
| 8 | 1.00000 | 0.50000 |
| 9 | 0.90451 | 0.20611 |
| 10 | 0.65451 | 0.02447 |

**11 holes**

| # | x | y |
|---|---|---|
| 1 | 0.35913 | 0.02025 |
| 2 | 0.12213 | 0.17257 |
| 3 | 0.00509 | 0.42884 |
| 4 | 0.04518 | 0.70771 |
| 5 | 0.22968 | 0.92063 |
| 6 | 0.50000 | 1.00000 |
| 7 | 0.77032 | 0.92063 |
| 8 | 0.95482 | 0.70771 |
| 9 | 0.99491 | 0.42884 |
| 10 | 0.87787 | 0.17257 |
| 11 | 0.64087 | 0.02025 |

**12 holes**

| # | x | y |
|---|---|---|
| 1 | 0.37059 | 0.01704 |
| 2 | 0.14645 | 0.14645 |
| 3 | 0.01704 | 0.37059 |
| 4 | 0.01704 | 0.62941 |
| 5 | 0.14645 | 0.85355 |
| 6 | 0.37059 | 0.98296 |
| 7 | 0.62941 | 0.98296 |
| 8 | 0.85355 | 0.85355 |
| 9 | 0.98296 | 0.62941 |
| 10 | 0.98296 | 0.37059 |
| 11 | 0.85355 | 0.14645 |
| 12 | 0.62941 | 0.01704 |

**13 holes**

| # | x | y |
|---|---|---|
| 1 | 0.38034 | 0.01453 |
| 2 | 0.16844 | 0.12574 |
| 3 | 0.03249 | 0.32270 |
| 4 | 0.00365 | 0.56027 |
| 5 | 0.08851 | 0.78403 |

**13 holes** (*Continued*)

| # | x | y |
|---|---|---|
| 6 | 0.26764 | 0.94273 |
| 7 | 0.50000 | 1.00000 |
| 8 | 0.73236 | 0.94273 |
| 9 | 0.91149 | 0.78403 |
| 10 | 0.99635 | 0.56027 |
| 11 | 0.96751 | 0.32270 |
| 12 | 0.83156 | 0.12574 |
| 13 | 0.61966 | 0.01453 |

**14 holes**

| # | x | y |
|---|---|---|
| 1 | 0.38874 | 0.01254 |
| 2 | 0.18826 | 0.10908 |
| 3 | 0.04952 | 0.28306 |
| 4 | 0.00000 | 0.50000 |
| 5 | 0.04952 | 0.71694 |
| 6 | 0.18826 | 0.89092 |
| 7 | 0.38874 | 0.98746 |
| 8 | 0.61126 | 0.98746 |
| 9 | 0.81174 | 0.89092 |
| 10 | 0.95048 | 0.71694 |
| 11 | 1.00000 | 0.50000 |
| 12 | 0.95048 | 0.28306 |
| 13 | 0.81174 | 0.10908 |
| 14 | 0.61126 | 0.01254 |

**15 holes**

| # | x | y |
|---|---|---|
| 1 | 0.39604 | 0.01093 |
| 2 | 0.20611 | 0.09549 |
| 3 | 0.06699 | 0.25000 |
| 4 | 0.00274 | 0.44774 |
| 5 | 0.02447 | 0.65451 |
| 6 | 0.12843 | 0.83457 |
| 7 | 0.29663 | 0.95677 |
| 8 | 0.50000 | 1.00000 |
| 9 | 0.70337 | 0.95677 |
| 10 | 0.87157 | 0.83457 |
| 11 | 0.97553 | 0.65451 |
| 12 | 0.99726 | 0.44774 |
| 13 | 0.93301 | 0.25000 |
| 14 | 0.79389 | 0.09549 |
| 15 | 0.60396 | 0.01093 |

**16 holes**

| # | x | y |
|---|---|---|
| 1 | 0.40245 | 0.00961 |
| 2 | 0.22221 | 0.08427 |
| 3 | 0.08427 | 0.22221 |
| 4 | 0.00961 | 0.40245 |
| 5 | 0.00961 | 0.59755 |
| 6 | 0.08427 | 0.77779 |
| 7 | 0.22221 | 0.91573 |
| 8 | 0.40245 | 0.99039 |
| 9 | 0.59755 | 0.99039 |
| 10 | 0.77779 | 0.91573 |
| 11 | 0.91573 | 0.77779 |
| 12 | 0.99039 | 0.59755 |
| 13 | 0.99039 | 0.40245 |
| 14 | 0.91573 | 0.22221 |
| 15 | 0.77779 | 0.08427 |
| 16 | 0.59755 | 0.00961 |

**17 holes**

| # | x | y |
|---|---|---|
| 1 | 0.40813 | 0.00851 |
| 2 | 0.23678 | 0.07489 |
| 3 | 0.10099 | 0.19868 |
| 4 | 0.01909 | 0.36317 |
| 5 | 0.00213 | 0.54613 |
| 6 | 0.05242 | 0.72287 |
| 7 | 0.16315 | 0.86950 |
| 8 | 0.31938 | 0.96624 |
| 9 | 0.50000 | 1.00000 |
| 10 | 0.68062 | 0.96624 |
| 11 | 0.83685 | 0.86950 |
| 12 | 0.94758 | 0.72287 |
| 13 | 0.99787 | 0.54613 |
| 14 | 0.98091 | 0.36317 |
| 15 | 0.89901 | 0.19868 |
| 16 | 0.76322 | 0.07489 |
| 17 | 0.59187 | 0.00851 |

**18 holes**

| # | x | y |
|---|---|---|
| 1 | 0.41318 | 0.00760 |
| 2 | 0.25000 | 0.06699 |
| 3 | 0.11698 | 0.17861 |
| 4 | 0.03015 | 0.32899 |
| 5 | 0.00000 | 0.50000 |
| 6 | 0.03015 | 0.67101 |
| 7 | 0.11698 | 0.82139 |

**18 holes** (*Continued*)

| # | x | y |
|---|---|---|
| 8 | 0.25000 | 0.93301 |
| 9 | 0.41318 | 0.99240 |
| 10 | 0.58682 | 0.99240 |
| 11 | 0.75000 | 0.93301 |
| 12 | 0.88302 | 0.82139 |
| 13 | 0.96985 | 0.67101 |
| 14 | 1.00000 | 0.50000 |
| 15 | 0.96985 | 0.32899 |
| 16 | 0.88302 | 0.17861 |
| 17 | 0.75000 | 0.06699 |
| 18 | 0.58682 | 0.00760 |

**19 holes**

| # | x | y |
|---|---|---|
| 1 | 0.41770 | 0.00682 |
| 2 | 0.26203 | 0.06026 |
| 3 | 0.13214 | 0.16136 |
| 4 | 0.04211 | 0.29915 |
| 5 | 0.00171 | 0.45871 |
| 6 | 0.01530 | 0.62274 |
| 7 | 0.08142 | 0.77347 |
| 8 | 0.19289 | 0.89457 |
| 9 | 0.33765 | 0.97291 |
| 10 | 0.50000 | 1.00000 |
| 11 | 0.66235 | 0.97291 |
| 12 | 0.80711 | 0.89457 |
| 13 | 0.91858 | 0.77347 |
| 14 | 0.98470 | 0.62274 |
| 15 | 0.99829 | 0.45871 |
| 16 | 0.95789 | 0.29915 |
| 17 | 0.86786 | 0.16136 |
| 18 | 0.73797 | 0.06026 |
| 19 | 0.58230 | 0.00682 |

**20 holes**

| # | x | y |
|---|---|---|
| 1 | 0.42178 | 0.00616 |
| 2 | 0.27300 | 0.05450 |
| 3 | 0.14645 | 0.14645 |
| 4 | 0.05450 | 0.27300 |
| 5 | 0.00616 | 0.42178 |
| 6 | 0.00616 | 0.57822 |
| 7 | 0.05450 | 0.72700 |
| 8 | 0.14645 | 0.85355 |
| 9 | 0.27300 | 0.94550 |
| 10 | 0.42178 | 0.99384 |

**20 holes** (*Continued*)

| # | x | y |
|---|---|---|
| 11 | 0.57822 | 0.99384 |
| 12 | 0.72700 | 0.94550 |
| 13 | 0.85355 | 0.85355 |
| 14 | 0.94550 | 0.72700 |
| 15 | 0.99384 | 0.57822 |
| 16 | 0.99384 | 0.42178 |
| 17 | 0.94550 | 0.27300 |
| 18 | 0.85355 | 0.14645 |
| 19 | 0.72700 | 0.05450 |
| 20 | 0.57822 | 0.00616 |

**21 holes**

| # | x | y |
|---|---|---|
| 1 | 0.42548 | 0.00558 |
| 2 | 0.28306 | 0.04952 |
| 3 | 0.15991 | 0.13347 |
| 4 | 0.06699 | 0.25000 |
| 5 | 0.01254 | 0.38874 |
| 6 | 0.00140 | 0.53737 |
| 7 | 0.03456 | 0.68267 |
| 8 | 0.10908 | 0.81174 |
| 9 | 0.21834 | 0.91312 |
| 10 | 0.35262 | 0.97779 |
| 11 | 0.50000 | 1.00000 |
| 12 | 0.64738 | 0.97779 |
| 13 | 0.78166 | 0.91312 |
| 14 | 0.89092 | 0.81174 |
| 15 | 0.96544 | 0.68267 |
| 16 | 0.99860 | 0.53737 |
| 17 | 0.98746 | 0.38874 |
| 18 | 0.93301 | 0.25000 |
| 19 | 0.84009 | 0.13347 |
| 20 | 0.71694 | 0.04952 |
| 21 | 0.57452 | 0.00558 |

**22 holes**

| # | x | y |
|---|---|---|
| 1 | 0.42884 | 0.00509 |
| 2 | 0.29229 | 0.04518 |
| 3 | 0.17257 | 0.12213 |
| 4 | 0.07937 | 0.22968 |
| 5 | 0.02025 | 0.35913 |
| 6 | 0.00000 | 0.50000 |
| 7 | 0.02025 | 0.64087 |
| 8 | 0.07937 | 0.77032 |
| 9 | 0.17257 | 0.87787 |

**22 holes** (*Continued*)

| # | x | y |
|---|---|---|
| 10 | 0.29229 | 0.95482 |
| 11 | 0.42884 | 0.99491 |
| 12 | 0.57116 | 0.99491 |
| 13 | 0.70771 | 0.95482 |
| 14 | 0.82743 | 0.87787 |
| 15 | 0.92063 | 0.77032 |
| 16 | 0.97975 | 0.64087 |
| 17 | 1.00000 | 0.50000 |
| 18 | 0.97975 | 0.35913 |
| 19 | 0.92063 | 0.22968 |
| 20 | 0.82743 | 0.12213 |
| 21 | 0.70771 | 0.04518 |
| 22 | 0.57116 | 0.00509 |

**23 holes**

| # | x | y |
|---|---|---|
| 1 | 0.43192 | 0.00466 |
| 2 | 0.30080 | 0.04139 |
| 3 | 0.18446 | 0.11214 |
| 4 | 0.09152 | 0.21166 |
| 5 | 0.02887 | 0.33256 |
| 6 | 0.00117 | 0.46588 |
| 7 | 0.01046 | 0.60173 |
| 8 | 0.05606 | 0.73003 |
| 9 | 0.13458 | 0.84128 |
| 10 | 0.24021 | 0.92721 |
| 11 | 0.36510 | 0.98146 |
| 12 | 0.50000 | 1.00000 |
| 13 | 0.63490 | 0.98146 |
| 14 | 0.75979 | 0.92721 |
| 15 | 0.86542 | 0.84128 |
| 16 | 0.94394 | 0.73003 |
| 17 | 0.98954 | 0.60173 |
| 18 | 0.99883 | 0.46588 |
| 19 | 0.97113 | 0.33256 |
| 20 | 0.90848 | 0.21166 |
| 21 | 0.81554 | 0.11214 |
| 22 | 0.69920 | 0.04139 |
| 23 | 0.56808 | 0.00466 |

**24 holes**

| # | x | y |
|---|---|---|
| 1 | 0.43474 | 0.00428 |
| 2 | 0.30866 | 0.03806 |
| 3 | 0.19562 | 0.10332 |
| 4 | 0.10332 | 0.19562 |

## Table 2b. *(Continued)* Hole Coordinate Dimension Factors for Type "B" Hole Circles

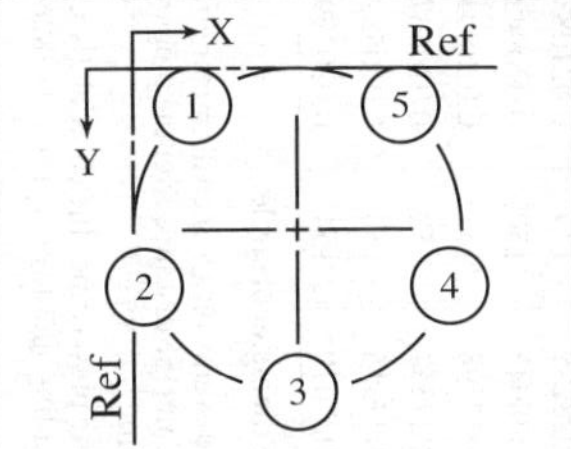


**24 holes** *(Continued)*

| # | x | y |
|---|---|---|
| 5 | 0.03806 | 0.30866 |
| 6 | 0.00428 | 0.43474 |
| 7 | 0.00428 | 0.56526 |
| 8 | 0.03806 | 0.69134 |
| 9 | 0.10332 | 0.80438 |
| 10 | 0.19562 | 0.89668 |
| 11 | 0.30866 | 0.96194 |
| 12 | 0.43474 | 0.99572 |
| 13 | 0.56526 | 0.99572 |
| 14 | 0.69134 | 0.96194 |
| 15 | 0.80438 | 0.89668 |
| 16 | 0.89668 | 0.80438 |
| 17 | 0.96194 | 0.69134 |
| 18 | 0.99572 | 0.56526 |
| 19 | 0.99572 | 0.43474 |
| 20 | 0.96194 | 0.30866 |
| 21 | 0.89668 | 0.19562 |
| 22 | 0.80438 | 0.10332 |
| 23 | 0.69134 | 0.03806 |
| 24 | 0.56526 | 0.00428 |

**25 holes**

| # | x | y |
|---|---|---|
| 1 | 0.43733 | 0.00394 |
| 2 | 0.31594 | 0.03511 |
| 3 | 0.20611 | 0.09549 |
| 4 | 0.11474 | 0.18129 |
| 5 | 0.04759 | 0.28711 |
| 6 | 0.00886 | 0.40631 |

**25 holes** *(Continued)*

| # | x | y |
|---|---|---|
| 7 | 0.00099 | 0.53140 |
| 8 | 0.02447 | 0.65451 |
| 9 | 0.07784 | 0.76791 |
| 10 | 0.15773 | 0.86448 |
| 11 | 0.25912 | 0.93815 |
| 12 | 0.37566 | 0.98429 |
| 13 | 0.50000 | 1.00000 |
| 14 | 0.62434 | 0.98429 |
| 15 | 0.74088 | 0.93815 |
| 16 | 0.84227 | 0.86448 |
| 17 | 0.92216 | 0.76791 |
| 18 | 0.97553 | 0.65451 |
| 19 | 0.99901 | 0.53140 |
| 20 | 0.99114 | 0.40631 |
| 21 | 0.95241 | 0.28711 |
| 22 | 0.88526 | 0.18129 |
| 23 | 0.79389 | 0.09549 |
| 24 | 0.68406 | 0.03511 |
| 25 | 0.56267 | 0.00394 |

**26 holes**

| # | x | y |
|---|---|---|
| 1 | 0.43973 | 0.00365 |
| 2 | 0.32270 | 0.03249 |
| 3 | 0.21597 | 0.08851 |
| 4 | 0.12574 | 0.16844 |
| 5 | 0.05727 | 0.26764 |
| 6 | 0.01453 | 0.38034 |
| 7 | 0.00000 | 0.50000 |

**26 holes** *(Continued)*

| # | x | y |
|---|---|---|
| 8 | 0.01453 | 0.61966 |
| 9 | 0.05727 | 0.73236 |
| 10 | 0.12574 | 0.83156 |
| 11 | 0.21597 | 0.91149 |
| 12 | 0.32270 | 0.96751 |
| 13 | 0.43973 | 0.99635 |
| 14 | 0.56027 | 0.99635 |
| 15 | 0.67730 | 0.96751 |
| 16 | 0.78403 | 0.91149 |
| 17 | 0.87426 | 0.83156 |
| 18 | 0.94273 | 0.73236 |
| 19 | 0.98547 | 0.61966 |
| 20 | 1.00000 | 0.50000 |
| 21 | 0.98547 | 0.38034 |
| 22 | 0.94273 | 0.26764 |
| 23 | 0.87426 | 0.16844 |
| 24 | 0.78403 | 0.08851 |
| 25 | 0.67730 | 0.03249 |
| 26 | 0.56027 | 0.00365 |

**27 holes**

| # | x | y |
|---|---|---|
| 1 | 0.44195 | 0.00338 |
| 2 | 0.32899 | 0.03015 |
| 3 | 0.22525 | 0.08226 |
| 4 | 0.13631 | 0.15688 |
| 5 | 0.06699 | 0.25000 |
| 6 | 0.02101 | 0.35660 |
| 7 | 0.00085 | 0.47093 |
| 8 | 0.00760 | 0.58682 |
| 9 | 0.04089 | 0.69804 |
| 10 | 0.09894 | 0.79858 |
| 11 | 0.17861 | 0.88302 |
| 12 | 0.27560 | 0.94682 |
| 13 | 0.38469 | 0.98652 |
| 14 | 0.50000 | 1.00000 |
| 15 | 0.61531 | 0.98652 |
| 16 | 0.72440 | 0.94682 |
| 17 | 0.82139 | 0.88302 |
| 18 | 0.90106 | 0.79858 |
| 19 | 0.95911 | 0.69804 |
| 20 | 0.99240 | 0.58682 |
| 21 | 0.99915 | 0.47093 |
| 22 | 0.97899 | 0.35660 |
| 23 | 0.93301 | 0.25000 |

**27 holes** *(Continued)*

| # | x | y |
|---|---|---|
| 24 | 0.86369 | 0.15688 |
| 25 | 0.77475 | 0.08226 |
| 26 | 0.67101 | 0.03015 |
| 27 | 0.55805 | 0.00338 |

**28 holes**

| # | x | y |
|---|---|---|
| 1 | 0.44402 | 0.00314 |
| 2 | 0.33486 | 0.02806 |
| 3 | 0.23398 | 0.07664 |
| 4 | 0.14645 | 0.14645 |
| 5 | 0.07664 | 0.23398 |
| 6 | 0.02806 | 0.33486 |
| 7 | 0.00314 | 0.44402 |
| 8 | 0.00314 | 0.55598 |
| 9 | 0.02806 | 0.66514 |
| 10 | 0.07664 | 0.76602 |
| 11 | 0.14645 | 0.85355 |
| 12 | 0.23398 | 0.92336 |
| 13 | 0.33486 | 0.97194 |
| 14 | 0.44402 | 0.99686 |
| 15 | 0.55598 | 0.99686 |
| 16 | 0.66514 | 0.97194 |
| 17 | 0.76602 | 0.92336 |
| 18 | 0.85355 | 0.85355 |
| 19 | 0.92336 | 0.76602 |
| 20 | 0.97194 | 0.66514 |
| 21 | 0.99686 | 0.55598 |
| 22 | 0.99686 | 0.44402 |
| 23 | 0.97194 | 0.33486 |
| 24 | 0.92336 | 0.23398 |
| 25 | 0.85355 | 0.14645 |
| 26 | 0.76602 | 0.07664 |
| 27 | 0.66514 | 0.02806 |
| 28 | 0.55598 | 0.00314 |

**29 holes**

| # | x | y |
|---|---|---|
| 1 | 0.44594 | 0.00293 |
| 2 | 0.34035 | 0.02617 |
| 3 | 0.24222 | 0.07157 |
| 4 | 0.15615 | 0.13700 |
| 5 | 0.08616 | 0.21941 |
| 6 | 0.03551 | 0.31493 |
| 7 | 0.00659 | 0.41911 |
| 8 | 0.00073 | 0.52707 |

**29 holes** *(Continued)*

| # | x | y |
|---|---|---|
| 9 | 0.01823 | 0.63376 |
| 10 | 0.05824 | 0.73420 |
| 11 | 0.11892 | 0.82369 |
| 12 | 0.19741 | 0.89805 |
| 13 | 0.29006 | 0.95379 |
| 14 | 0.39251 | 0.98831 |
| 15 | 0.50000 | 1.00000 |
| 16 | 0.60749 | 0.98831 |
| 17 | 0.70994 | 0.95379 |
| 18 | 0.80259 | 0.89805 |
| 19 | 0.88108 | 0.82369 |
| 20 | 0.94176 | 0.73420 |
| 21 | 0.98177 | 0.63376 |
| 22 | 0.99927 | 0.52707 |
| 23 | 0.99341 | 0.41911 |
| 24 | 0.96449 | 0.31493 |
| 25 | 0.91384 | 0.21941 |
| 26 | 0.84385 | 0.13700 |
| 27 | 0.75778 | 0.07157 |
| 28 | 0.65965 | 0.02617 |
| 29 | 0.55406 | 0.00293 |

**30 holes**

| # | x | y |
|---|---|---|
| 1 | 0.44774 | 0.00274 |
| 2 | 0.34549 | 0.02447 |
| 3 | 0.25000 | 0.06699 |
| 4 | 0.16543 | 0.12843 |
| 5 | 0.09549 | 0.20611 |
| 6 | 0.04323 | 0.29663 |
| 7 | 0.01093 | 0.39604 |
| 8 | 0.00000 | 0.50000 |
| 9 | 0.01093 | 0.60396 |
| 10 | 0.04323 | 0.70337 |
| 11 | 0.09549 | 0.79389 |
| 12 | 0.16543 | 0.87157 |
| 13 | 0.25000 | 0.93301 |
| 14 | 0.34549 | 0.97553 |
| 15 | 0.44774 | 0.99726 |
| 16 | 0.55226 | 0.99726 |
| 17 | 0.65451 | 0.97553 |
| 18 | 0.75000 | 0.93301 |
| 19 | 0.83457 | 0.87157 |
| 20 | 0.90451 | 0.79389 |
| 21 | 0.95677 | 0.70337 |

**30 holes** *(Continued)*

| # | x | y |
|---|---|---|
| 22 | 0.98907 | 0.60396 |
| 23 | 1.00000 | 0.50000 |
| 24 | 0.98907 | 0.39604 |
| 25 | 0.95677 | 0.29663 |
| 26 | 0.90451 | 0.20611 |
| 27 | 0.83457 | 0.12843 |
| 28 | 0.75000 | 0.06699 |
| 29 | 0.65451 | 0.02447 |
| 30 | 0.55226 | 0.00274 |

**31 holes**

| # | x | y |
|---|---|---|
| 1 | 0.44942 | 0.00257 |
| 2 | 0.35032 | 0.02293 |
| 3 | 0.25735 | 0.06283 |
| 4 | 0.17431 | 0.12062 |
| 5 | 0.10461 | 0.19395 |
| 6 | 0.05110 | 0.27980 |
| 7 | 0.01596 | 0.37467 |
| 8 | 0.00064 | 0.47468 |
| 9 | 0.00577 | 0.57571 |
| 10 | 0.03112 | 0.67365 |
| 11 | 0.07568 | 0.76448 |
| 12 | 0.13760 | 0.84448 |
| 13 | 0.21437 | 0.91038 |
| 14 | 0.30282 | 0.95948 |
| 15 | 0.39935 | 0.98976 |
| 16 | 0.50000 | 1.00000 |
| 17 | 0.60065 | 0.98976 |
| 18 | 0.69718 | 0.95948 |
| 19 | 0.78563 | 0.91038 |
| 20 | 0.86240 | 0.84448 |
| 21 | 0.92432 | 0.76448 |
| 22 | 0.96888 | 0.67365 |
| 23 | 0.99423 | 0.57571 |
| 24 | 0.99936 | 0.47468 |
| 25 | 0.98404 | 0.37467 |
| 26 | 0.94890 | 0.27980 |
| 27 | 0.89539 | 0.19395 |
| 28 | 0.82569 | 0.12062 |
| 29 | 0.74265 | 0.06283 |
| 30 | 0.64968 | 0.02293 |
| 31 | 0.55058 | 0.00257 |

**32 holes**

| # | x | y |
|---|---|---|
| 1 | 0.45099 | 0.00241 |
| 2 | 0.35486 | 0.02153 |
| 3 | 0.26430 | 0.05904 |
| 4 | 0.18280 | 0.11349 |
| 5 | 0.11349 | 0.18280 |
| 6 | 0.05904 | 0.26430 |
| 7 | 0.02153 | 0.35486 |
| 8 | 0.00241 | 0.45099 |
| 9 | 0.00241 | 0.54901 |
| 10 | 0.02153 | 0.64514 |
| 11 | 0.05904 | 0.73570 |
| 12 | 0.11349 | 0.81720 |
| 13 | 0.18280 | 0.88651 |
| 14 | 0.26430 | 0.94096 |
| 15 | 0.35486 | 0.97847 |
| 16 | 0.45099 | 0.99759 |
| 17 | 0.54901 | 0.99759 |
| 18 | 0.64514 | 0.97847 |
| 19 | 0.73570 | 0.94096 |
| 20 | 0.81720 | 0.88651 |
| 21 | 0.88651 | 0.81720 |
| 22 | 0.94096 | 0.73570 |
| 23 | 0.97847 | 0.64514 |
| 24 | 0.99759 | 0.54901 |
| 25 | 0.99759 | 0.45099 |
| 26 | 0.97847 | 0.35486 |
| 27 | 0.94096 | 0.26430 |
| 28 | 0.88651 | 0.18280 |
| 29 | 0.81720 | 0.11349 |
| 30 | 0.73570 | 0.05904 |
| 31 | 0.64514 | 0.02153 |
| 32 | 0.54901 | 0.00241 |

**33 holes**

| # | x | y |
|---|---|---|
| 1 | 0.45247 | 0.00226 |
| 2 | 0.35913 | 0.02025 |
| 3 | 0.27089 | 0.05558 |
| 4 | 0.19092 | 0.10697 |
| 5 | 0.12213 | 0.17257 |
| 6 | 0.06699 | 0.25000 |
| 7 | 0.02750 | 0.33647 |
| 8 | 0.00509 | 0.42884 |
| 9 | 0.00057 | 0.52379 |
| 10 | 0.01409 | 0.61788 |

**33 holes** *(Continued)*

| # | x | y |
|---|---|---|
| 11 | 0.04518 | 0.70771 |
| 12 | 0.09271 | 0.79003 |
| 13 | 0.15496 | 0.86187 |
| 14 | 0.22968 | 0.92063 |
| 15 | 0.31417 | 0.96418 |
| 16 | 0.40537 | 0.99096 |
| 17 | 0.50000 | 1.00000 |
| 18 | 0.59463 | 0.99096 |
| 19 | 0.68583 | 0.96418 |
| 20 | 0.77032 | 0.92063 |
| 21 | 0.84504 | 0.86187 |
| 22 | 0.90729 | 0.79003 |
| 23 | 0.95482 | 0.70771 |
| 24 | 0.98591 | 0.61788 |
| 25 | 0.99943 | 0.52379 |
| 26 | 0.99491 | 0.42884 |
| 27 | 0.97250 | 0.33647 |
| 28 | 0.93301 | 0.25000 |
| 29 | 0.87787 | 0.17257 |
| 30 | 0.80908 | 0.10697 |
| 31 | 0.72911 | 0.05558 |
| 32 | 0.64087 | 0.02025 |
| 33 | 0.54753 | 0.00226 |

**Table 3. Lengths of Chords for Spacing Off the Circumferences of Circles with a Diameter Equal to 1 (English or Metric units)**

| No. of Spaces | Length of Chord | No. of Spaces | Length of Chord | No. of Spaces | Length of Chord | No. of Spaces | Length of Chord |
|---|---|---|---|---|---|---|---|
| 3 | 0.866025 | 41 | 0.076549 | 79 | 0.039757 | 117 | 0.026848 |
| 4 | 0.707107 | 42 | 0.074730 | 80 | 0.039260 | 118 | 0.026621 |
| 5 | 0.587785 | 43 | 0.072995 | 81 | 0.038775 | 119 | 0.026397 |
| 6 | 0.500000 | 44 | 0.071339 | 82 | 0.038303 | 120 | 0.026177 |
| 7 | 0.433884 | 45 | 0.069756 | 83 | 0.037841 | 121 | 0.025961 |
| 8 | 0.382683 | 46 | 0.068242 | 84 | 0.037391 | 122 | 0.025748 |
| 9 | 0.342020 | 47 | 0.066793 | 85 | 0.036951 | 123 | 0.025539 |
| 10 | 0.309017 | 48 | 0.065403 | 86 | 0.036522 | 124 | 0.025333 |
| 11 | 0.281733 | 49 | 0.064070 | 87 | 0.036102 | 125 | 0.025130 |
| 12 | 0.258819 | 50 | 0.062791 | 88 | 0.035692 | 126 | 0.024931 |
| 13 | 0.239316 | 51 | 0.061561 | 89 | 0.035291 | 127 | 0.024734 |
| 14 | 0.222521 | 52 | 0.060378 | 90 | 0.034899 | 128 | 0.024541 |
| 15 | 0.207912 | 53 | 0.059241 | 91 | 0.034516 | 129 | 0.024351 |
| 16 | 0.195090 | 54 | 0.058145 | 92 | 0.034141 | 130 | 0.024164 |
| 17 | 0.183750 | 55 | 0.057089 | 93 | 0.033774 | 131 | 0.023979 |
| 18 | 0.173648 | 56 | 0.056070 | 94 | 0.033415 | 132 | 0.023798 |
| 19 | 0.164595 | 57 | 0.055088 | 95 | 0.033063 | 133 | 0.023619 |
| 20 | 0.156434 | 58 | 0.054139 | 96 | 0.032719 | 134 | 0.023443 |
| 21 | 0.149042 | 59 | 0.053222 | 97 | 0.032382 | 135 | 0.023269 |
| 22 | 0.142315 | 60 | 0.052336 | 98 | 0.032052 | 136 | 0.023098 |
| 23 | 0.136167 | 61 | 0.051479 | 99 | 0.031728 | 137 | 0.022929 |
| 24 | 0.130526 | 62 | 0.050649 | 100 | 0.031411 | 138 | 0.022763 |
| 25 | 0.125333 | 63 | 0.049846 | 101 | 0.031100 | 139 | 0.022599 |
| 26 | 0.120537 | 64 | 0.049068 | 102 | 0.030795 | 140 | 0.022438 |
| 27 | 0.116093 | 65 | 0.048313 | 103 | 0.030496 | 141 | 0.022279 |
| 28 | 0.111964 | 66 | 0.047582 | 104 | 0.030203 | 142 | 0.022122 |
| 29 | 0.108119 | 67 | 0.046872 | 105 | 0.029915 | 143 | 0.021967 |
| 30 | 0.104528 | 68 | 0.046183 | 106 | 0.029633 | 144 | 0.021815 |
| 31 | 0.101168 | 69 | 0.045515 | 107 | 0.029356 | 145 | 0.021664 |
| 32 | 0.098017 | 70 | 0.044865 | 108 | 0.029085 | 146 | 0.021516 |
| 33 | 0.095056 | 71 | 0.044233 | 109 | 0.028818 | 147 | 0.021370 |
| 34 | 0.092268 | 72 | 0.043619 | 110 | 0.028556 | 148 | 0.021225 |
| 35 | 0.089639 | 73 | 0.043022 | 111 | 0.028299 | 149 | 0.021083 |
| 36 | 0.087156 | 74 | 0.042441 | 112 | 0.028046 | 150 | 0.020942 |
| 37 | 0.084806 | 75 | 0.041876 | 113 | 0.027798 | 151 | 0.020804 |
| 38 | 0.082579 | 76 | 0.041325 | 114 | 0.027554 | 152 | 0.020667 |
| 39 | 0.080467 | 77 | 0.040789 | 115 | 0.027315 | 153 | 0.020532 |
| 40 | 0.078459 | 78 | 0.040266 | 116 | 0.027079 | 154 | 0.020399 |

For circles of other diameters, multiply length given in table by diameter of circle.

*Example:* In a drill jig, 8 holes, each ½ inch diameter, were spaced evenly on a 6-inch diameter circle. To test the accuracy of the jig, plugs were placed in adjacent holes, and the distance over the plugs was measured with a micrometer. What should be the micrometer reading?

*Solution:* The micrometer reading equals the diameter of one plug plus 6 times the chordal distance between adjacent hole centers given in the table above. Thus, the reading should be $\frac{1}{2} + (6 \times 0.382683) = 2.796098$ inches.

**Table 4. Coordinates for Locating Equally Spaced Holes**

The table below is a quick reference for calculating the coordinates of between 3 and 12 equally spaced holes using the center point and diameter of the pitch circle. Multiply the values shown by the pitch circle diameter (inch or metric units) to get the measurements for placing the holes.

*Example:* Locate three holes on a pitch circle diameter of 3 inches.
A = 0.25 x 3 = 0.75, B = 0.43301 x 3 = 1.299, C = 0.86603 x 3 = 2.598

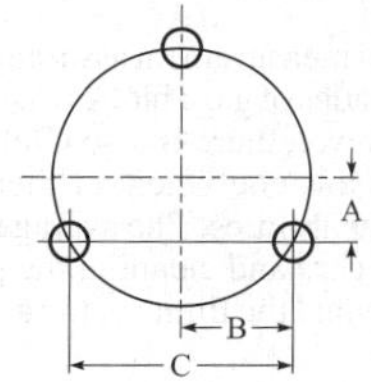


**3-Hole**
A = 0.25000
B = 0.43301
C = 0.86603

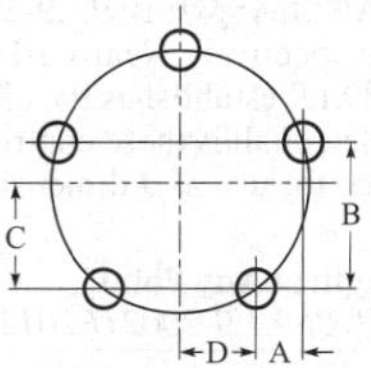


**5-Hole**
A = 0.18164
B = 0.55902
C = 0.40451
D = 0.29389

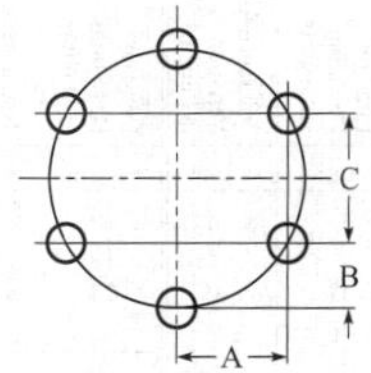


**6-Hole**
A = 0.43301
B = 0.25000
C = 0.50000

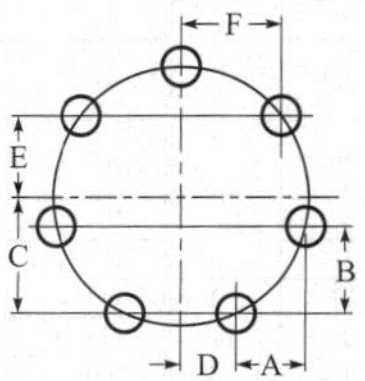


**7-Hole**
A = 0.27052
B = 0.33922
C = 0.45049
D = 0.21694
E = 0.31175
F = 0.39092

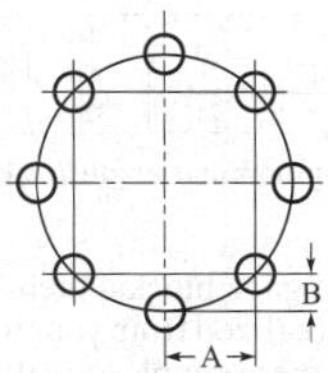


**8-Hole**
A = 0.35355
B = 0.14645

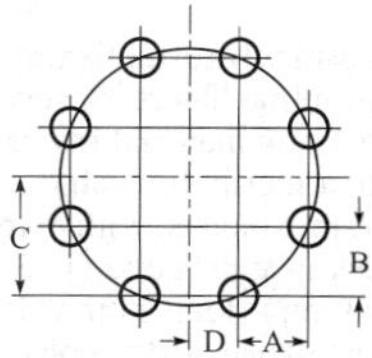


**8-Hole**
A = 0.27059
B = 0.27059
C = 0.46914
D = 0.19134

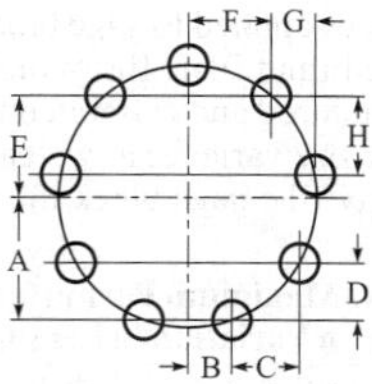


**9-Hole**
A = 0.46985
B = 0.17101
C = 0.26201
D = 0.21985
E = 0.38302
F = 0.32139
G = 0.17101
H = 0.29620

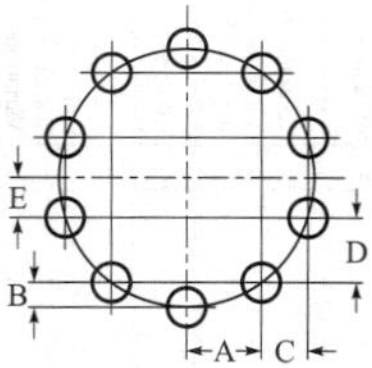


**10-Hole**
A = 0.29389
B = 0.09549
C = 0.18164
D = 0.25000
E = 0.15451

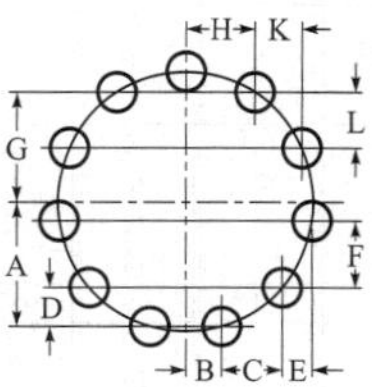


**11-Hole**
A = 0.47975
B = 0.14087
C = 0.23700
D = 0.15231
E = 0.11704
F = 0.25627
G = 0.42063
H = 0.27032
K = 0.18449
L = 0.21291

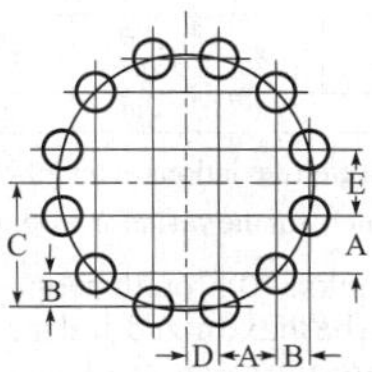


**12-Hole**
A = 0.22415
B = 0.12941
C = 0.48296
D = 0.12941
E = 0.25882

## Gage Blocks

The primary standard for linear measurement is the *gage block*. Gage blocks were originally called Jo-Blocks after their Swedish inventor, Carl Edvard Johansson. These precision lapped blocks are the primary means of establishing measurement traceability to the prime standards located in the national laboratories of every country. The primary length standards are themselves defined by the distance traveled by light in a vacuum over a fixed time period.

The gage block is critical in establishing true traceability and measurement assurance in the dimensional discipline. There are several materials and grades of gage blocks to select from. The most common material in use today is steel. However, there is also Croblox, made by Mitutoyo, a thermally stable material with a very low coefficient of thermal expansion, and ceramic blocks, with extremely good wear capabilities. The average life span of a gage block is approximately 3 years. With proper care and cleaning the gage block may last many years longer but will eventually wear beyond the limits of the allowable tolerances and will need to be replaced.

The size tolerances applied to gage blocks, defined in the ANSI/ASME B89.1.9-2002, are shown in Table 1a and Table 1b, for inch and metric units, respectively. Nearly all gage blocks are manufactured and calibrated to this Standard. B89.1.9 establishes the allowable deviations for size variance as well as flatness and parallelism. It is these controlled dimensions that give the gage block the properties necessary for use as a dimensional standard.

**Table 1a. Maximum Permitted Deviations of Length at Any Point[a] and Tolerance on Variation in Length, Inch[b]** *ANSI/ASME B89.1.9-2002 (R2012)*

| Nominal Length Range, $l_n$ inches | Calibration Grade K | | Grade 00 | | Grade 0 | | Grade AS-1 | | Grade AS-2 | |
|---|---|---|---|---|---|---|---|---|---|---|
| | Limit on Deviations of Length[a] | Tolerance for Variation in Length[b] | Limit on Deviations of Length[a] | Tolerance for Variation in Length[b] | Limit on Deviations of Length[a] | Tolerance for Variation in Length[b] | Limit on Deviations of Length[a] | Tolerance for Variation in Length[b] | Limit on Deviations of Length[a] | Tolerance for Variation in Length[b] |
| | $\pm t_e$ µin. | $t_v$ µin. | $\pm t_e$ µin. | $t_v$ µin. | $\pm t_e$ µin. | $t_v$ µin. | $\pm t_e$ µin. | $t_v$ µin. | $\pm t_e$ µin. | $t_v$ µin. |
| $\le 0.05$ | 12 | | 4 | | 6 | | 12 | | 24 | |
| $0.05 < l_n \le 0.4$ | 10 | | 3 | 2 | 5 | | 8 | | 18 | 12 |
| $0.55 < l_n \le 1$ | 12 | 2 | 3 | | 6 | 4 | 12 | 6 | 24 | |
| $1 < l_n \le 2$ | 16 | | 4 | | 8 | | 16 | | 32 | |
| $2 < l_n \le 3$ | 20 | | 5 | | 10 | | 20 | | 40 | 14 |
| $3 < l_n \le 4$ | 24 | | 6 | 3 | 12 | | 24 | | 48 | |
| $4 < l_n \le 5$ | 32 | 3 | 8 | | 16 | 5 | 32 | 8 | 64 | |
| $5 < l_n \le 6$ | 32 | | 8 | | 16 | | | | | 16 |
| $6 < l_n \le 7$ | 40 | | 10 | | 20 | | 40 | | 80 | |
| $7 < l_n \le 8$ | 40 | 4 | 10 | 4 | 20 | 6 | | 10 | | |
| $8 < l_n \le 10$ | 48 | | 12 | | 24 | | 48 | | 104 | 18 |
| $10 < l_n \le 12$ | 56 | | 14 | | 28 | 7 | 56 | | 112 | 20 |
| $12 < l_n \le 16$ | 72 | 5 | 18 | 5 | 36 | 8 | 72 | 12 | 144 | |
| $16 < l_n \le 20$ | 88 | 6 | 20 | 6 | 44 | 10 | 88 | 14 | 176 | 24 |
| $20 < l_n \le 24$ | 104 | | 25 | | 52 | | 104 | 16 | 200 | 28 |
| $24 < l_n \le 28$ | 120 | 7 | 30 | 7 | 60 | 12 | 120 | 18 | 240 | |
| $28 < l_n \le 32$ | 136 | 8 | 34 | 8 | 68 | | 136 | 20 | 260 | 32 |
| $32 < l_n \le 36$ | 152 | | 38 | | 76 | 14 | 152 | | 300 | 36 |
| $36 < l_n \le 40$ | 160 | 10 | 40 | 10 | 80 | 16 | 168 | 24 | 320 | 40 |

[a] Maximum permitted deviations of length at any point, $\pm t_e$ µinch, from nominal length, $l_e$ inches.

[b] Tolerance, $t_v$ µinch, for the variation in length.

**Care of Gage Blocks.—**Through proper care and handling of gage blocks, their functional life span can be maximized and many years of use can be realized from your investment. The basic care and cleaning of gage blocks should follow these simple guidelines.

**Table 1b. Maximum Permitted Deviations of Length at Any Point[a] and Tolerance on Variation in Length, Metric[b]** *ANSI/ASME B89.1.9-2002 (R2012)*

| Nominal Length Range, $l_n$ mm | Calibration Grade K | | Grade 00 | | Grade 0 | | Grade AS-1 | | Grade AS-2 | |
|---|---|---|---|---|---|---|---|---|---|---|
| | Limit on Deviations of Length[a] | Tolerance for Variation in Length[b] | Limit on Deviations of Length[a] | Tolerance for Variation in Length[b] | Limit on Deviations of Length[a] | Tolerance for Variation in Length[b] | Limit on Deviations of Length[a] | Tolerance for Variation in Length[b] | Limit on Deviations of Length[a] | Tolerance for Variation in Length[b] |
| | $\pm t_e$ µm | $t_v$ µm | $\pm t_e$ µm | $t_v$ µm | $\pm t_e$ µm | $t_v$ µm | $\pm t_e$ µm | $t_v$ µm | $\pm t_e$ µm | $t_v$ µm |
| $\leq 0.5$ | 0.30 | 0.05 | 0.10 | 0.05 | 0.14 | 0.10 | 0.30 | 0.16 | 0.60 | 0.30 |
| $0.5 < l_n \leq 10$ | 0.20 | | 0.07 | | 0.12 | | 0.20 | | 0.45 | |
| $10 < l_n \leq 25$ | 0.30 | | | | 0.14 | | 0.30 | | 0.60 | |
| $25 < l_n \leq 50$ | 0.40 | 0.06 | 0.10 | 0.06 | 0.20 | | 0.40 | 0.18 | 0.80 | |
| $50 < l_n \leq 75$ | 0.50 | | 0.12 | 0.07 | 0.25 | 0.12 | 0.50 | | 1.00 | 0.35 |
| $75 < l_n \leq 100$ | 0.60 | 0.07 | 0.15 | | 0.30 | | 0.60 | 0.20 | 1.20 | |
| $100 < l_n \leq 150$ | 0.80 | 0.08 | 0.20 | 0.08 | 0.40 | 0.14 | 0.80 | | 1.60 | 0.40 |
| $150 < l_n \leq 200$ | 1.00 | 0.09 | 0.25 | 0.09 | 0.50 | 0.16 | 1.00 | 0.25 | 2.00 | |
| $200 < l_n \leq 250$ | 1.20 | 0.10 | 0.30 | 0.10 | 0.60 | | 1.20 | | 2.40 | 0.45 |
| $250 < l_n \leq 300$ | 1.40 | | 0.35 | | 0.70 | 0.18 | 1.40 | | 2.80 | 0.50 |
| $300 < l_n \leq 400$ | 1.80 | 0.12 | 0.45 | 0.12 | 0.90 | 0.20 | 1.80 | 0.30 | 3.60 | |
| $400 < l_n \leq 500$ | 2.20 | 0.14 | 0.50 | 0.14 | 1.10 | 0.25 | 2.20 | 0.35 | 4.40 | 0.60 |
| $500 < l_n \leq 600$ | 2.60 | 0.16 | 0.65 | 0.16 | 1.30 | | 2.60 | 0.40 | 5.00 | 0.70 |
| $600 < l_n \leq 700$ | 3.00 | 0.18 | 0.75 | 0.18 | 1.50 | 0.30 | 3.00 | 0.45 | 6.00 | |
| $700 < l_n \leq 800$ | 3.40 | 0.20 | 0.85 | 0.20 | 1.70 | | 3.40 | 0.50 | 6.50 | 0.80 |
| $800 < l_n \leq 900$ | 3.80 | | 0.95 | | 1.90 | 0.35 | 3.80 | | 7.50 | 0.90 |
| $900 < l_n \leq 1000$ | 4.20 | 0.25 | 1.00 | 0.25 | 2.00 | 0.40 | 4.20 | 0.60 | 8.00 | 1.00 |

[a] Maximum permitted deviations of length at any point, $\pm t_e$ µm, from nominal length, $l_e$ mm.
[b] Tolerance, $t_v$ µm, for the variation in length.

1) Always keep gage blocks clean and well oiled when not in direct use. Alcohol is acceptable as a cleaner, but it is always advisable to coat the gage block with a rust inhibitor when placing it back in the case. A very light machine oil is recommended.

2) Take great care when removing gage blocks from the case so as not to nick or damage the working surface. Clean with a soft cloth or chamois and isopropanol. Never touch gage blocks with bare hands. Oil from fingers will cause corrosion on the bare metal surface.

3) Always keep the gage blocks over a soft cloth or chamois when handling or wringing them together. Dropping the gage blocks onto a hard surface or other gage blocks will damage the working surface and cause an error beyond the limits of the tolerance. Always treat gage blocks as highly accurate precision instruments.

4) Should the gage block surface show signs of degradation and the wringing together of blocks become difficult, the surface may need to be deburred. The use of a serrated sintered aluminum oxide deburring stone is recommended to recondition the surface and renew the ability to join the gage blocks through wringing. Caution must be exercised when deburring the surface of gage blocks so it is not damaged instead of repaired.

5) Gently place the gage block flat on the serrated block. With two fingers (using gloves), press down firmly, but not hard, on the gage block and slide it lengthwise over the serrations on the block for three or four strokes until the surface feels very smooth. Turn the gage block over and repeat the movement. Remove and clean the gage block thoroughly.

6) It is important that the serrated sintered aluminum oxide deburring stone is cleaned as well, and metal deposits, oils, and dirt are not allowed to build up on the surface. A cotton packing impregnated with a metal solvent will clean the serrated sintered aluminum oxide deburring stone.

**Calibration and Verification of Gage Blocks.—**The calibration and verification of gage blocks should be completed on a regular basis. This is done to maintain measurement assurance in every good quality program. The quality assurance program will determine the optimal interval for recalibration of the gage block sets to maintain the appropriate level of measurement assurance. Calibration of the gage blocks should be done by an approved and, if necessary, certified calibration laboratory that provides impartial third party confirmation of the calibrated features of the gage block set.

**Precision Gage Blocks.—**Precision gage blocks are usually purchased in sets containing a specific number of blocks of different sizes. The nominal gage lengths of individual blocks in a set are determined mathematically so that particular desired lengths can be obtained by combining selected blocks. They are made to several different tolerance grades that categorize them as master blocks, calibration blocks, inspection blocks, and workshop blocks. *Master blocks* are employed as basic reference standards; *calibration blocks* are used for high precision gaging work and calibrating inspection blocks; *inspection blocks* are used as toolroom standards and for checking and setting limit and comparator gages, for example. The *workshop blocks* are working gages used as shop standards for direct precision measurements and gaging applications, including sine-bars.

Federal Specification GGG-G-15C, Gage Blocks (see below), lists typical sets and gives details of materials, design, and manufacturing requirements, and tolerance grades. When there is in a set no single block of the exact size wanted, two or more blocks are combined by "wringing" them together. Wringing is achieved by first placing one block crosswise on the other and applying some pressure. Then a swiveling motion is used to twist the blocks to a parallel position, causing them to adhere firmly to one another.

When combining blocks for a given dimension, the object is to use as few blocks as possible to obtain the dimension. The procedure for selecting blocks is based on successively eliminating the right-hand figure of the desired dimension.

*Example:* Referring to inch-size gage block Set Number 1 below, determine the blocks required to obtain 3.6742 inches. *Step 1:* Eliminate 0.0002 by selecting a 0.1002 block. Subtract 0.1002 from 3.6742 = 3.5740. *Step 2:* Eliminate 0.004 by selecting a 0.124 block. Subtract 0.124 from 3.5740 = 3.450. *Step 3:* Eliminate 0.450 with a block this size. Subtract 0.450 from 3.450 = 3.000. *Step 4:* Select a 3.000 inch block. The combined blocks are 0.1002 + 0.124 + 0.450 + 3.000 = 3.6742 inches.

**Gage Block Sets, Inch Sizes (Federal Specification GGG-G-15C).—***Set Number 1 (81 Blocks):* First Series: 0.0001-Inch Increments (9 Blocks), 0.1001 to 0.1009; Second Series: 0.001-Inch Increments (49 Blocks), 0.101 to 0.149; Third Series: 0.050-Inch Increments (19 Blocks), 0.050 to 0.950; Fourth Series: 1.000-Inch Increments (4 Blocks), 1.000- to 4.000-inch.

*Set Numbers 2, 3, and 4:* The specification does not list a set 2 or 3. Gage block *Set Number 4* (88 Blocks), listed in the Specification, is not given here; it is the same as Set Number 1 (81 Blocks) but contains seven additional blocks measuring 0.0625, 0.078125, 0.093750, 0.100025, 0.100050, 0.100075, and 0.109375 inch.

*Set Number 5 (21 Blocks):* First Series: 0.0001-Inch Increments (9 Blocks), 0.0101 to 0.0109; Second Series: 0.001-Inch Increments (11 Blocks), 0.010 to 0.020; One Block 0.01005 inch.

*Set Number 6 (28 Blocks):* First Series: 0.0001-Inch Increments (9 Blocks), 0.0201 to 0.0209; Second Series: 0.001-Inch Increments (9 Blocks). 0.021 to 0.029; Third Series: 0.010-Inch Increments (9 Blocks), 0.010 to 0.090; One Block 0.02005 Inch.

*Long Gage Block Set Number 7 (8 Blocks):* Whole-Inch Series (8 Blocks), 5, 6, 7, 8, 10, 12, 16, 20 inches.

*Set Number 8 (36 Blocks):* First Series: 0.0001-Inch Increments (9 Blocks), 0.1001 to 0.1009; Second Series: 0.001-Inch Increments (11 Blocks), 0.100 to 0.110; Third Series: 0.010-Inch Increments (8 Blocks), 0.120 to 0.190; Fourth Series: 0.100-Inch Increments (4 Blocks), 0.200 to 0.500; Whole-Inch Series (3 Blocks), 1, 2, 4 Inches; One Block 0.050 inch.

*Set Number 9 (20 Blocks):* First Series: 0.0001-Inch Increments (9 Blocks), 0.0501 to 0.0509; Second Series: 0.001-Inch Increments (10 Blocks), 0.050 to 0.059; One Block 0.05005 inch.

**Gage Block Sets, Metric Sizes (Federal Specification GGG-G-15C).—***Set Number 1M (45 Blocks):* First Series: 0.001-Millimeter Increments (9 Blocks), 1.001 to 1.009; Second Series: 0.01-Millimeter Increments (9 Blocks), 1.01 to 1.09; Third Series: 0.10-Millimeter Increments (9 Blocks), 1.10 to 1.90; Fourth Series: 1.0-Millimeter Increments (9 Blocks), 1.0 to 9.0; Fifth Series: 10-Millimeter Increments (9 Blocks), 10 to 90 mm.

*Set Number 2M (88 Blocks):* First Series: 0.001-Millimeter Increments (9 Blocks), 1.001 to 1.009; Second Series: 0.01-Millimeter Increments (49 Blocks), 1.01 to 1.49; Third Series: 0.50-Millimeter Increments (19 Blocks), 0.5 to 9.5; Fourth Series: 10-Millimeter Increments (10 Blocks), 10 to 100; One Block 1.0005 mm.

*Set Number 3M:* Gage block Set Number 3M (112 Blocks) is not given here. It is similar to Set Number 2M (88 Blocks), and the chief difference is the inclusion of a larger number of blocks in the 0.5-millimeter increment series up to 24.5 mm.

*Set Number 4M (45 Blocks):* First Series: 0.001-Millimeter Increments (9 Blocks), 2.001 to 2.009; Second Series: 0.01-Millimeter Increments (9 Blocks), 2.01 to 2.09; Third Series: 0.10-Millimeter Increments (9 Blocks), 2.1 to 2.9; Fourth Series: 1-Millimeter Increments (9 Blocks), 1.0 to 9.0; Fifth Series: 10-Millimeter Increments (9 Blocks), 10 to 90 mm.

*Set Numbers 5M, 6M, 7M:* Set Numbers 5M (88 Blocks), 6M (112 Blocks), and 7M (17 Blocks) are not listed here.

*Long Gage Block Set Number 8M (8 Blocks):* Whole-Millimeter Series (8 Blocks), 125, 150, 175, 200, 250, 300, 400, 500 mm.

## Surface Plates

The surface plate is the primary plane from which all vertical measurements are made. The quality and dependability of this surface is one of the most critical elements in dimensional inspection measurement. Originally made from cast iron, the present-day granite plate was first developed during World War II because most metal was being used in the war effort. Faced with a need to check precision parts, Mr. Wallace Herman, a metal working and monument shop owner, decided to investigate the use of granite as a suitable replacement for the then-common cast-iron surface plate and manufactured the first granite surface plate in his shop in Dayton, Ohio.

Although surface plates have changed in their design and materials, the basic concept has remained the same. The stability and precision that can be achieved with granite is actually far superior to cast iron and is much easier to maintain. With the proper care and maintenance, a well-made surface plate can last for generations and always remain within the parameters of the grade to which it was originally made, or even better.

**Materials and Grades of Surface Plates.—**The selection of a surface plate is driven directly by the specific application the plate will be used for. A plate, for instance, that will be used in a very large machining facility would be primarily concerned with the load-bearing properties and secondarily in surface flatness accuracy, although both are important concerns. A surface plate that will be used in a metrology laboratory or high precision inspection department with a high volume of work would be concerned with high accuracy and surface wear properties. In each case the material and design would be considered for the application in mind before a selection is made and a purchase initiated.

The material properties of the granite is what makes the difference in the performance of surface plates. The differences in the various types of granite are considered in Table 1, based on Federal Specification GGG-P-463c, Plate, Surface (Granite) (Inch and Metric).

**Table 1. Granite Rock Types, Physical Properties, and Mineral Components** *GGG-P-463c*

| Rock Type | Natural Color | Texture | Mineral Constituents, Descending Order Of Abundance | Modulus of Elasticity | |
|---|---|---|---|---|---|
| | | | | $10^6$ psi | GPa |
| Biotite granite | Bluish gray | Fine-grained | Orthoclase, smokey quartz,[a] oligoclase, albite, biotite, muscovite, magnetite and zircon | 3.5–7.0 | 24.1–48.2 |
| | Light gray | Medium-grained | Oligoclase, orthoclase and microcline, quartz, biotite, apatite and zircon | 3.5–7.0 | 24.1–48.2 |
| | Pink | | Orthoclase with a small amount of microcline, plagioclase, quartz[a], biotite, magnetite, and garnet | 5.0–9.0 | 34.4–62.0 |
| Biotite hornblende granite | Reddish brown | Fine-grained | Orthoclase and microcline, quartz[a], hornblende, biotite, plagioclase and magnetite | 6.0–9.0 | 41.3–62.0 |
| Biotite-muscovite | Light gray | Medium- to fine-grained | Microcline, quartz, plagioclase, biotite, muscovite and magnetite | 5.0–7.0 | 34.4–48.2 |
| Diabase | Dark gray | Fine-grained | Plagioclase, pyroxene and magnetite | 9.0–12.0 | 62.0–82.7 |
| Hypersthene Gabbro | | | Plagioclase, pryoxene, hornblende, magnetite and biotite | 10.0–12.0 | 68.9–82.7 |
| Muscovite-biotite granite-gneiss | Light gray | Medium-grained | Microcline and orthoclase, oligoclase, quartz, rutile, muscovite | 3.5–8.0 | 24.1–55.1 |

[a] 28 to 32% quartz by volume. In certain conditions, high quartz content tends to increase wear life.

As indicated in Table 1, fine-grained pink granite containing a small amount of quartz has a lower modulus of elasticity and therefore a lower load bearing per square foot capacity. The presence of large quartz crystals, however, results in a high degree of wearability by providing an ultra-smooth surface finish with increased surface hardness that resists wear on the granite and the precision ground and lapped instruments used on it. Although the load-bearing properties are lower than those of black or dark gray granite, this can be compensated for by increasing the thickness of the plate.

As a result of the increased wearability of pink granite, the interval between lapping compared to fine-grained black granite can be as much as five times as long. This is an important consideration when planning the maintenance costs and downtime involved in maintaining a production schedule.

The precision lapped and calibrated granite surface plate is a high-precision piece of equipment and must be maintained as such. Great care should be taken at all times to protect the surface, and attention to cleanliness is critical in the life span of the surface plate. It is essential that the surface be protected from the buildup of dirt, grease, airborne grime and oils. The plate should be covered when not in use to avoid accidentally dropping objects on the surface and chipping or cracking the precision finished surface.

All surface plates should be installed and supported according to the manufacturer's design and recommendations. Plates up to and including 6–12 feet are supported in a three-point non-distortable support system of hard rubber pads that are installed during manufacturing and remain in place during lapping and finishing. These pads are critical in the correct support of the surface plate and must never be removed or repositioned. Always make sure the surface plate is resting on these pads, and never support the plate by its ledges or under the four corners as this will cause deformation of the surface and introduce errors beyond the tolerance limits.

Surface plates made to meet standard guidelines and accuracy parameters established over the last fifty years and published in documents such as the Federal Specification GGG-P-463c will be manufactured in certain pre-designed sizes. These designs have been analyzed for dimensional stability and dependability and will, with proper care, provide dependable measurement assurance for many years. There are manufacturers that will special order surface plates in a wide variety of sizes and configurations to meet the needs of specialized applications, and have engineering staffs that will design a surface plate to meet those needs. However, most applications and manufacturers' products fall under the design guidelines of the tables in the GGG-P-463c Federal Specification.

**Table 2a. Standard Sizes for Rectangular Granite Surface Plates** *GGG-P-463c*

| Inch | | | | Metric | | | |
|---|---|---|---|---|---|---|---|
| Width (inch) | Length (inch) | Calculated Diagonal (inch) | Area ($ft^2$) | Width (mm) | Length (mm) | Calculated Diagonal (mm) | Area ($m^2$) |
| 12 | 12 | 17.0 | 1 | 300 | 300 | 424 | 0.090 |
| | 18 | 21.6 | 1.5 | | 350 | 541 | 0.135 |
| 18 | 18 | 25.5 | 2.25 | 450 | 450 | 636 | 0.202 |
| | 24 | 30.0 | 3 | | 600 | 750 | 0.270 |
| 24 | 24 | 33.9 | 4 | 600 | 600 | 849 | 0.360 |
| | 36 | 43.3 | 6 | | 900 | 1082 | 0.540 |
| | 48 | 53.7 | 8 | | 1200 | 1342 | 0.720 |
| 36 | 36 | 50.9 | 9 | 900 | 900 | 1273 | 0.810 |
| | 48 | 60.0 | 12 | | 1200 | 1500 | 1.080 |
| | 60 | 70.0 | 15 | | 1500 | 1749 | 1.350 |
| | 72 | 80.5 | 18 | | 1800 | 2012 | 1.620 |
| 48 | 48 | 67.9 | 16 | 1200 | 1200 | 1697 | 1.440 |
| | 60 | 76.9 | 20 | | 1500 | 1921 | 1.800 |
| | 72 | 86.5 | 24 | | 1800 | 2163 | 2.160 |
| | 96 | 107.3 | 32 | | 2400 | 2683 | 2.880 |
| | 120 | 129.2 | 40 | | 3000 | 3231 | 3.600 |
| 60 | 120 | 134.2 | 50 | 1500 | 300 | 3304 | 4.500 |
| 72 | 96 | 120.0 | 48 | 1800 | 2400 | 3000 | 4.320 |
| | 144 | 161.0 | 72 | | 3600 | 4025 | 6.480 |

**Table 2b. Standard Sizes for Round Granite Surface Plates** *GGG-P-463c*

| Inch | | Millimeter | |
|---|---|---|---|
| Diameter (inch) | Area ($ft^2$) | Diameter (mm) | Area ($m^2$) |
| 12 | 0.8 | 300 | 0.071 |
| 18 | 1.8 | 450 | 0.159 |
| 24 | 3.1 | 600 | 0.283 |
| 36 | 7.1 | 900 | 0.636 |
| 48 | 48 | 1200 | 1.131 |

*Thickness*: For rectangular and round surface plates, specify thickness only if essential; see Appendix 30 and inch and metric versions of Tables XI and XII of GGG-P-463c.

Surface plate grades are established in the Federal Specifications and are the guidelines by which the plates are calibrated. The flatness tolerances in microinches for standard inch-dimension plates, listed in Table 3a, are obtained through the standard formula:

$$\text{Total flatness tolerance for inch} - \text{dimension grade AA plates} = 40 + \frac{D^2}{25}\,\mu\text{in.}$$

where $D$ = diagonal or diameter of the plate in inches. The calculated flatness tolerance for grade AA is rounded to the nearest 25μin.

**For metric plate sizes, the total flatness tolerance of grade AA plates in micrometers is:**

$$\textbf{Total flatness tolerance for metric grade AA plates} = 1 + 1.62D^2 10^{-6}\mu\text{m}$$

**where $D$ = diagonal or diameter of the plate in millimeters.**

For both the inch and metric plates, the tolerances of the A and B grades are 2 and 4 times, respectively, those for grade AA.

Table 3a and Table 3b, adapted from Federal Specification GGG-P-463c, contain the calculated tolerances for the standard size and grades of rectangular and round surface plates.

**Table 3a. Total Flatness Tolerance, Rectangular Surface Plates** *GGG-P-463c*

| Rectangular Plates | | | | | | | | | |
|---|---|---|---|---|---|---|---|---|---|
| Inch Sizes, Tolerances in Microinches | | | | | Millimeter Sizes, Tolerances in Micrometers | | | | |
| Width (inch) | Length (inch) | Grade AA | Grade A | Grade B | Width (mm) | Length (mm) | Grade AA | Grade A | Grade B |
| 12 | 12 | 50 | 100 | 200 | 300 | 300 | 1.3 | 2.6 | 5.2 |
| | 18 | 50 | 100 | 200 | | 450 | 1.5 | 2.9 | 5.9 |
| 18 | 18 | 50 | 100 | 200 | 450 | 450 | 1.6 | 3.3 | 6.6 |
| | 24 | 75 | 150 | 300 | | 600 | 1.9 | 3.8 | 7.6 |
| 24 | 24 | 75 | 150 | 300 | 600 | 600 | 2.2 | 4.3 | 8.6 |
| | 36 | 100 | 200 | 400 | | 900 | 2.9 | 5.7 | 11.5 |
| | 48 | 150 | 300 | 600 | | 1200 | 3.9 | 7.8 | 15.5 |
| 36 | 36 | 150 | 300 | 600 | 900 | 900 | 3.6 | 7.2 | 14.4 |
| | 48 | 200 | 400 | 800 | | 1200 | 4.6 | 9.2 | 18.4 |
| | 60 | 250 | 500 | 1000 | | 1500 | 5.9 | 11.8 | 23.6 |
| | 72 | 300 | 600 | 1200 | | 1800 | 7.5 | 15.0 | 29.9 |
| 48 | 48 | 200 | 400 | 800 | 1200 | 1200 | 5.6 | 11.2 | 22.4 |
| | 60 | 300 | 600 | 1200 | | 1500 | 6.9 | 13.8 | 17.6 |
| | 72 | 350 | 700 | 1400 | | 1800 | 8.5 | 17.0 | 33.9 |
| | 96 | 500 | 1000 | 2000 | | 2400 | 12.5 | 25.0 | 50.0 |
| | 120 | 700 | 1400 | 2800 | | 3000 | 17.7 | 35.4 | 70.8 |
| 60 | 120 | 750 | 1500 | 3000 | 1500 | 3000 | 18.5 | 36.9 | 73.9 |
| 72 | 96 | 600 | 1200 | 2400 | 1800 | 2400 | 15.4 | 30.8 | 61.6 |
| | 144 | 1100 | 2200 | 4400 | | 3600 | 26.9 | 53.8 | 107.7 |

**Table 3b. Total Flatness Tolerance, Round Surface Plates** *GGG-P-463c*

| Round Plates | | | | | | | |
|---|---|---|---|---|---|---|---|
| Inch Sizes, Tolerances in Microinches | | | | Millimeter Sizes, Tolerances in Micrometers | | | |
| Diameter (inch) | Grade AA | Grade A | Grade B | Diameter (mm) | Grade AA | Grade A | Grade B |
| 12 | 50 | 100 | 200 | 300 | 1.1 | 2.3 | 4.6 |
| 18 | 50 | 100 | 200 | 450 | 1.3 | 2.5 | 5.3 |
| 24 | 75 | 150 | 300 | 600 | 1.6 | 3.2 | 6.3 |
| 36 | 100 | 200 | 400 | 900 | 2.3 | 4.6 | 9.2 |
| 48 | 125 | 250 | 500 | 1200 | 3.3 | 6.6 | 13.2 |

**Calibration of Surface Plates.—**Surface plates, like other precision instruments, will drift out of tolerance in time and need to be periodically checked and even adjusted to maintain accuracy. For a surface plate, this adjustment involves lapping the surface and physically removing material until the entire surface is once again flat to within the limits of the grade to which it was made. This is a labor-intensive adjustment that should only be attempted by a trained technician with the appropriate tools. An untrained technician, even with the proper lapping tools and compounds, may cause more harm than good when attempting such an adjustment.

However, with proper care and cleaning, a surface plate may not need lapping at every calibration. Completing the calibration process will give the quality assurance program the data needed to identify wear patterns on a surface plate and have the proper maintenance completed before an out of tolerance condition occurs, thereby avoiding a costly failure impact analysis and reinspection of parts, or even a recall of finished parts from a customer. The calibration process is a cost-effective alternative to these undesirable effects.

Fig. 1 illustrates the "Union Jack" or eight line pattern that is used to analyze the overall flatness of the surface plate working area. Each line is measured independently, and the positive and negative elevations are recorded at predetermined intervals. The peak to valley flatness is calculated from the recorded data. This method was developed by JC Moody while working for Sandia Corporation, and the resulting plot of the data points is referred to as a Moody Plot. There are many software packages available that simplify the process of calculating this plot and provide a detailed printout of the data in a graphical format.

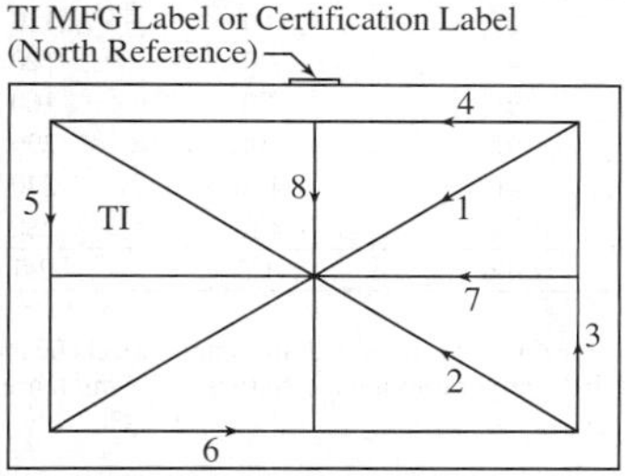


Fig. 1. Union Jack Pattern

**Repeat-o-Meter Method (Fed. Spec. GGG-P-463c)** The easiest and quickest method of monitoring for surface plate wear and tolerance adherence is the Repeat-o-Meter method. A *repeat reading gage* similar in design to the one shown in Fig. 2 is used to establish the variation in flatness of the surface plate.

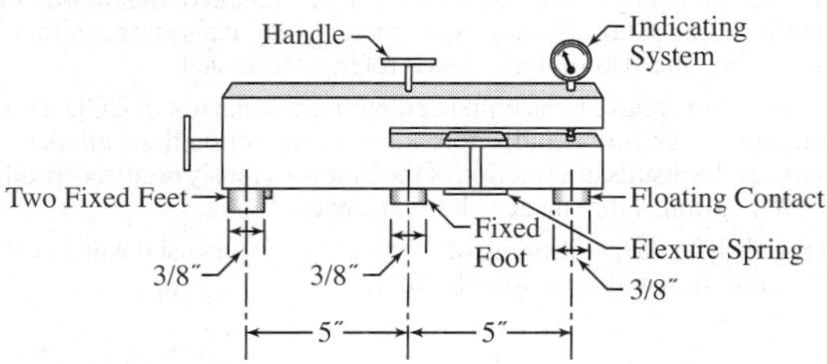


Fig. 2. Repeat Reading Gage from Federal Specification GGG-P-463c

The gage is placed at the center of the surface plate, and the indicator is set to zero. The eight-line Union Jack pattern, Fig. 1, is then scanned in the sequence defined in the Federal Specification GGG-P-463c. The result of the maximum reading minus the minimum reading shall not exceed the flatness tolerance expressed in Table 4a and Table 4b for inch- and millimeter-sized plates, respectively.

**Table 4a. Tolerance for Repeat Reading of Measurement (Microinches)**

| Diagonal / Diameter Range | | Grade AA | Grade A | Grade B | Obtained |
|---|---|---|---|---|---|
| Over | Thru | Full Indicator Movement (FIM), microinches | | | |
| … | 30 | 35 | 60 | 110 | When not specified |
| 30 | 60 | 45 | 70 | 120 | |
| 60 | 90 | 60 | 80 | 160 | |
| 90 | 120 | 75 | 100 | 200 | |
| 120 | 150 | 90 | 120 | 240 | |
| 150 | … | 100 | 140 | 280 | |
| All Sizes | | 25 | 50 | 100 | When specified |

**Table 4b. Tolerance for Repeat Reading of Measurement (Micrometers)**

| Diagonal / Diameter Range (mm) | | Grade AA | Grade A | Grade B | Obtained |
|---|---|---|---|---|---|
| Over | Thru | Full Indicator Movement (FIM), micrometers | | | |
| … | 800 | 35 | 60 | 110 | When not specified |
| 800 | 1500 | 45 | 70 | 120 | |
| 1500 | 2200 | 60 | 80 | 160 | |
| 2200 | 3000 | 75 | 100 | 200 | |
| 3000 | 3800 | 90 | 120 | 240 | |
| 3800 | … | 100 | 140 | 280 | |
| All Sizes | | 0.6 | 1.3 | 2.5 | When specified |

*Note for* Table 4a *and* Table 4b*:* If it is intended that small objects be measured on large surface plates, it should be noted that a larger tolerance in flatness over small areas is permitted on larger plates.

All points on the work surface shall be contained between two parallel planes, the roof plane and the base plane. The distance between these planes shall be no greater than that specified in the tolerance table for the respective grades.

**Autocollimator Calibration (Ref NAVAIR 17-20MD-14, 1 MAY 1995).—**Calibration of a surface plate by autocollimator is one of the most accurate and relatively quickest methods in use today. The method has been in use since the advent of the autocollimator in the 1940s but was refined by JC Moody with the development of the Moody Plot analysis method that utilized the data collected on perimeters, diagonals and bisectors. The resulting pattern resembles the "Union Jack" and is referred to as such.

Autocollimators today have been enhanced with the inclusion of CCD devices, digital readouts, computer interfacing, and automatic data collection that calculates the deviations and analyzes the results in a fraction of the time previously required, producing a full color graphical diagram of the surface plate variances.

Fig. 3 and Fig. 4 show the positioning of the turning mirrors used when performing the surface plate calibration with an autocollimator.

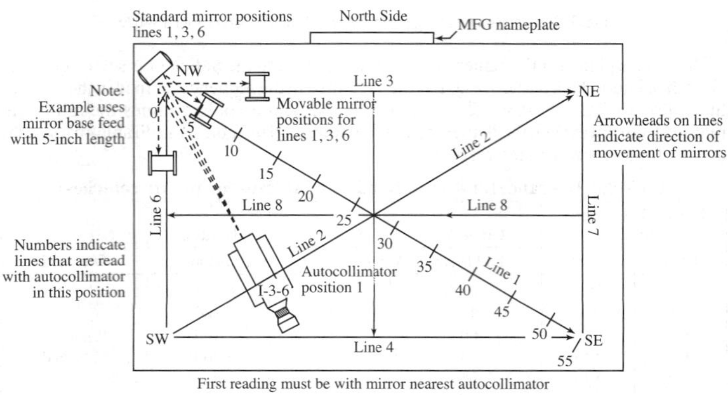


Fig. 3. Placement of the First Turning Mirror and Retroreflector: TO 33K6-4-137-1

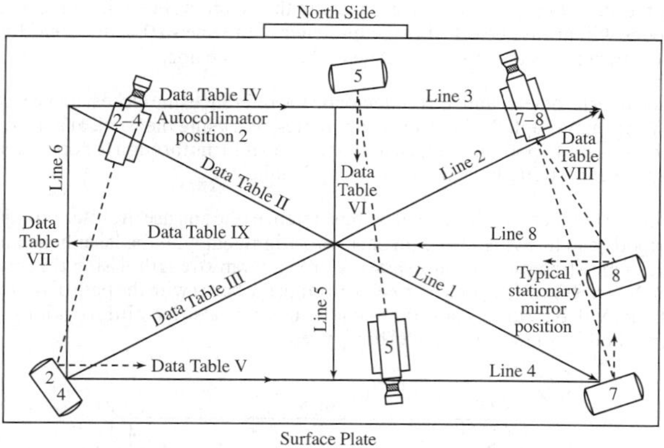


Fig. 4. Turning Mirror Placement for Remaining Lines: TO 33K6-4-137-1

**Interferometer Calibration (Ref TO 33K6-4-10-1 30 April 2006).**—Calibration of a surface plate can also be accomplished using a laser interferometer. The laser offers the absolute lowest uncertainty in the calibration process, and, although it takes more time than either of the other methods, it is sometimes required by some customers doing military or government contract work. There are several models and methods available on the market today. The general process will be covered, but familiarity with the particular model being used is necessary to complete the full calibration procedure.

The process is much the same as the autocollimator calibration method using turning mirrors. The laser tripod and laser head are located as close to the line being shot as possible. Move the tripod and realign the laser interferometer for each line during calibration. If available, use the turning mirrors to facilitate the alignment of the beam. As with any high-precision physical measurements, air currents and vibrations will affect the indications. Take precautions to minimize these affects. Clean the surface plate twice with surface plate cleaner and once again with ethyl alcohol prior to beginning the calibration process.

Mark the surface with lines as shown in Fig. 5 below. The lines will be either 3 or 4 inches away from the edge of the surface plate, depending on the size of the plate. The North side of the plate will always be the side bearing the manufacturer's label or the previous calibration label.

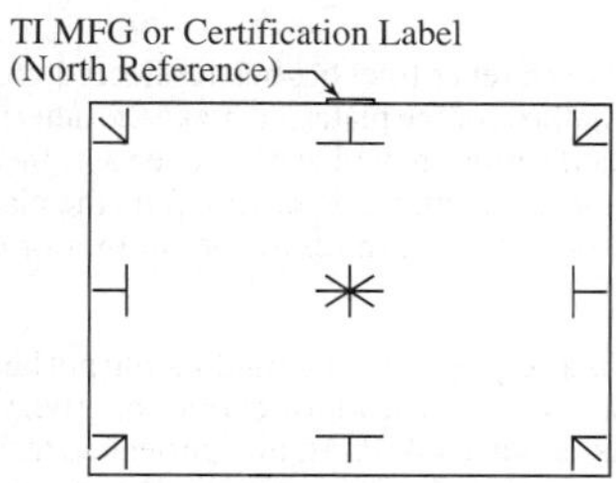


Fig. 5. Reference Lines for Union Jack 1

Clamp the straight edge to the surface plate in the position shown in Fig. 6 below. The straightedge will always be offset from the center of the line so that the centerline of the mirror will travel directly above the center of the reference line.

Place the interferometer and retroreflector on the reference line closest to the laser head. Direct the beam through the interferometer optics and, using the vertical and horizontal adjustments on the tripod head, adjust until the beam returns through the optics and registers a strong beam strength on the beam strength indicator.

Slide the retroreflector to the far end of the reference line and use the rotation adjustment on the tripod head to swing the beam back into alignment with the laser head until once again the beam strength indicator shows a strong beam strength. Using this method of Translate Near / Rotate Far you will be able to align the beam with the path of the reference line. Patience will be required, and the manufacturer will also have information regarding the best method for the specific model being used.

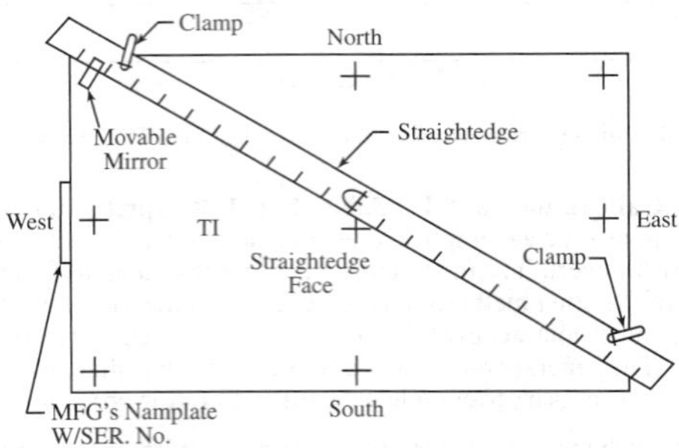


Fig. 6. Placement of the Straightedge for Measurement of the First Reference Line

Once the alignment of the first reference line has been completed, take the measurements at the stations recommended by the software calculations. The number of stations will change according to the size of the surface plate and the length of the footpad used for the retroreflector. Record the values at each station along the first reference line then return to the first position and verify that the reading repeats before proceeding to the next line. Great care must be taken to ensure a clean plate and footpad, free of residue or buildup. A very small amount of dirt or dust can cause an out of tolerance condition and result in many wasted hours of failure impact analysis and possibly recall of parts to be reinspected unnecessarily.

When the data from all the reference lines has been collected, the interferometer software will calculate the overall flatness of the plate and make the determination as to whether or not the plate meets the specific grade to which it has been assigned or needs to be lapped to meet tolerance limits. If there is a failure determination, the plate will need to be lapped and reshot to collect the "As Left" data and close the calibration event with an acceptance of the calibration data.

Lapping a surface plate is a very specific skill and should not be attempted by an amateur or enthusiastic technician however good the intentions may be. Lapping requires specialized tools and compounds as well as skill. An inexperienced technician attempting to lap a surface plate may well cause more damage than is already present. Always consult an experienced and proven professional in this field.

# V-Blocks

The V-block is a precision holding jig or fixture devised as a tool or to enable the precision and repeatable positioning of a cylindrical work piece. The V-block enables drilling, grinding, or milling operations on a robust and precision platform as compared to the vise or other simple holding devices.

In 1902, Elmer Cobb and Eugene Spaulding of Portland, Maine patented a Machinists V-block to expand their capability to hold and drill extended workpieces. The V-blocks were machined with a large channel on one side and a pair of smaller channels on the other to accommodate the various diameters that were encountered in the basic work done in the machine shop of the era. Each block had a hole in the center to allow the passage of a drill through the work piece. The blocks were a matched set that were secured together by two rods that aligned the blocks together and allowed them to slide open or to be closed together to support work of varying lengths.

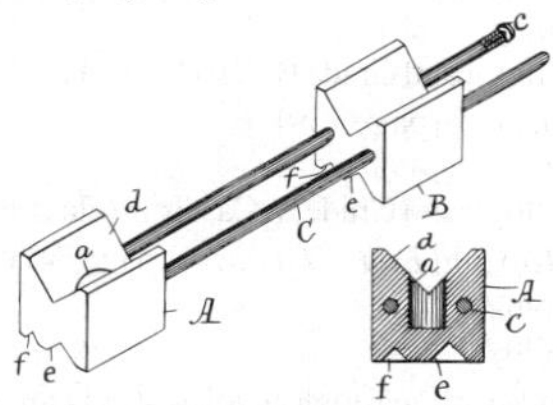


Fig. 1. Machinists' V-Block, Patented 1902 by Cobb and Spaulding

In 1923, Gustave Hines invented a device that incorporated a scale along the linear side of the work piece, and locked the work into a scaled collar that could be rotated accurately throughout 180°. The device provided a drill guide that would accurately measure distances between centerlines of holes, and provide rotational accuracy.

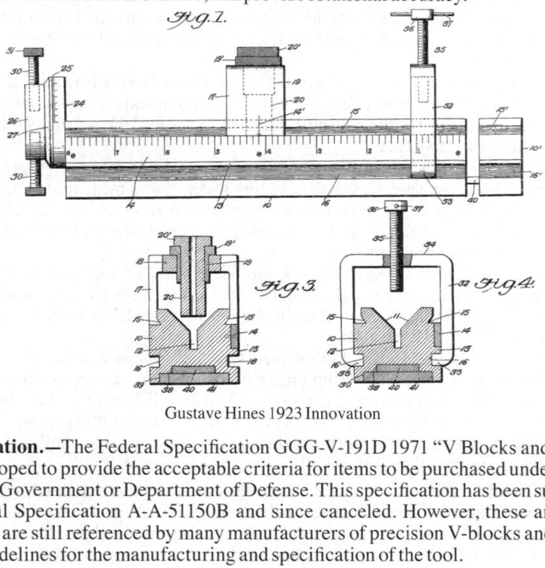


Gustave Hines 1923 Innovation

**Classification.—**The Federal Specification GGG-V-191D 1971 "V Blocks and Clamps" was developed to provide the acceptable criteria for items to be purchased under contract by the US Government or Department of Defense. This specification has been superseded by Federal Specification A-A-51150B and since canceled. However, these and related standards are still referenced by many manufacturers of precision V-blocks and were the initial guidelines for the manufacturing and specification of the tool.

*Federal Specification Coverage:* This specification covers only the types, styles and grades generally purchased by the Federal Government, and does not include all types, styles, and grades which are commercially available.

**Type, Style, and Grade**

*Type I – Single V-Groove* (page 735)

*Style 2:* Double level clamping

*Style 3:* Reversible clamping yoke

*Style 4:* Magnetic clamping (mild steel and brass laminated)

*Style 5:* Ribbed, without clamping

*Style 6:* Plain without clamping

*Grades:* Grade A: Steel (alloy); Grade B: Mild Steel (and brass laminated for style 4); and Grade C: Cast Steel.

*Type II – Double V-Groove* (page 738)

*Style 1:* Single level clamping

*Grades:* Grade A: Steel (alloy); and Grade B: Cast Iron (class 40 iron).

*Type III – Quadruple V-Groove* (page 739)

*Style 1:* Single level clamping

*Grades:* Grade A: Steel (alloy); and Grade B: Cast Iron (class 40 iron).

*Type IV – Combination, Top and End V-Grooves* (page 741)

*Style 1:* Single level clamping

*Grades:* Grade A: Steel (alloy)

**Material.—***Note:* Grade and surface finish specifications are a representation of all governing standards in use today.

*Grade A Blocks:* Grade A blocks shall be steel or steel alloy and shall have a surface hardness of 58 to 65 RC on the Rockwell C scale.

*Grade B Blocks:* Grade B blocks shall be a good quality mild steel, free from porosity and other injurious defects and have a surface hardness of 40 to 45 on the Rockwell C scale.

*Grade C Blocks:* Grade C blocks shall be a good quality cast steel free of porosity or other injurious defects and shall have a surface hardness of 150 to 180 BHN (Brinell Hardness Number).

*Precision Granite Blocks:* Precision granite V-blocks shall be made from a close grained uniform texture granite free from flaws or fissures and inclusions of softer material. Granite V-blocks shall adhere to either Grade A or B accuracy. Grade A and matched sets are to be provided in a protective case.

*Finished Surfaces:* Surfaces of the V-grooves, tops sides, bottom, V-groove relief slots, and clamp engaging grooves shall have finished surfaces measured in micrometers arithmetical average (A.A.) or roughness average (Ra) in accordance with ANSI/ASME B46.1-2019. All other surfaces shall be finished in accordance with normal manufacturer's standard commercial practice.

*Surface Finish, Grades A and B, Style 4, Blocks:* Surface of V-groove relief slots and clamp engaging grooves (when applicable) on blocks shall have a surface roughness not greater than 125 μin (3.175 μm). All other finished surfaces shall have a surface roughness not greater than 32 μin (0.81 μm) Ra.

*Surface Finish, Grade B (except Style 4), and Grade C Blocks:* Surface of V-groove(s). The V-grooves relief slots and clamp engaging grooves (when applicable) on blocks shall have a surface roughness not greater than 125 μin (3.175 μm) Ra. All other finished surfaces shall have a surface roughness not greater than 63 μin (1.600 μm) Ra.

*Surface Edges:* Edges of all exterior surfaces, including corners, shall be rounded or beveled not less than 0.010 inch and shall be free from fins and burrs.

**Tolerance.—**All Grade A; and Style 4, Grade B, mild steel laminated with brass, shall have a ground finish on the ends, sides, top, bottom, and the V-groove(s). The V-grooves shall be centered with the sides, square with the sides and ends, and parallel to the bottom within 0.0003 in. All surfaces shall be square with each other within 0.0003 in.

Grade B (except Style 4) and Grade C shall have a machined or ground finish. The V-groove(s) shall be parallel and centered with the sides to within 0.002 in. The ends shall be square with the sides within 0.002 in. and the sides shall be square with the top and the bottom within 0.002 in. The V-grooves shall be 45° each side of the center perpendicular plane for all grades. When specified, Grade B or C blocks shall be furnished with tolerances as specified for Grade A blocks.

When the V-blocks are provided in matched sets, they shall be clearly marked to identify them as such and shall be identical to each other within 0.0002 in. on all critical features.

**V-Block Design.—**Each V-block shall have a 90° V-groove 45° from center perpendicular plane cut lengthwise on the top, or top and end, or top and bottom, for the type specified. A slot shall be provided at the apex of the V-groove for chip relief. The exception being with the Type III V-blocks which shall be furnished without the chip relief unless otherwise specified.

When V-blocks are furnished in "matched pairs" they shall be marked to readily identify them as a pair. Clamping grooves shall be provided along the full length of the sides of Styles 1 and 2 blocks for engaging the clamps.

**Clamps.—**The clamps shall consist essentially of a yoke and one or two screws for the style specified, and shall be of a size to accommodate the maximum size round stock specified in the respective tables. The Clamp shall be made of good quality steel or iron and sufficiently robust to provide adequate support and security of the work piece in the V-block during drilling and grinding operations. When engaged in the clamping grooves, the clamp should not extend past the side faces of the block. When provided with matched sets the clamp shall be applicable to either member of the set. The screws for the clamping device shall be right-handed thread conforming to ANSI/ASME B1.2 or FED-STD-H28.

**Yokes.—**Yokes shall be made of good quality steel or iron and sufficiently robust to provide adequate support and security of the work piece in the V-block during drilling and grinding operations and should not extend past the side faces of the block. V-block and yoke screws for Type 1 Style 3 shall be threaded in accordance with ANSI/ASME B1.2 or FED-STD-H28 or the equivalent and shall be a Class 2 fit. When the yoke is in place it shall bisect the 90° V-groove. The yokes when engaged in the clamping grooves shall stand in an upright position.

Yokes for the style 3 blocks shall have a slot on each end to engage the two adjusting screws. The yoke shall be designed so that the pads bisect the 90° V-groove to allow the yoke to secure stock that does not extend above the V-groove. When inverted the yoke shall be able to secure stock of the maximum size specified.

**Screws.—**All screws shall be made of steel of such quality as to withstand hard usage. Screw heads shall be knurled and provided with a hole through the center for tightening with a pin or rod. All screws shall conform to the ANSI/ASME B1.2 or FED-STD-H28 or equivalent and shall be a class 2 fit.

**Type 1, Style 2.—**Type 1 V-blocks are of the single V type. The Style 2 V-block shall have two clamping grooves on each side to accommodate clamping from minimum to the full range of the capacity of the V-groove. The block shall be similar to Fig. 2 and conform to the dimensions shown in Table 1. Style 2 V-blocks can be furnished in matched pairs and the clamps shall be suitable for use with either V-block.

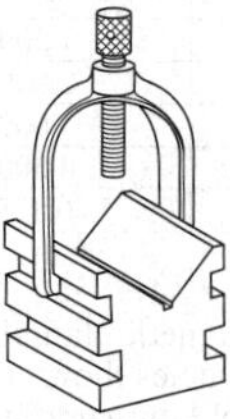

Fig. 2. Type 1, Style 2

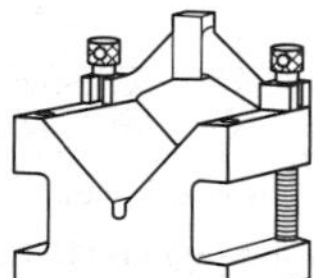

Fig. 3. Type 1, Style 3
Machinists' Table Block with Yoke

**Table 1. Accuracy and Dimensions, Type 1, Style 2**

| Grade and Designating Size | | Permitted Deviations (inch) | | | | | |
|---|---|---|---|---|---|---|---|
| | | Flatness of Working Faces | End/Side Square to Base | Parallelism of Opposite Faces | Parallelism of "V" to Base and Sides | "V" Centrality | Matching Tolerance over "V" Length |
| Grade A | 1 | 0.00015 | 0.0002 | 0.0002 | 0.0003 | 0.0003 | 0.0002 |
| | 2 | 0.0001 | 0.0002 | 0.0002 | 0.0003 | 0.0003 | 0.0002 |
| Grade B | 1 | 0.0004 | 0.002 | 0.001 | 0.002 | 0.002 | 0.001 |
| | 2 | 0.0004 | 0.002 | 0.001 | 0.002 | 0.002 | 0.001 |

| Grades A & B Designating Size | Nominal Dimensions (inch) | | | |
|---|---|---|---|---|
| | Length | Width | Height | Maximum Capacity |
| 1 | 3.000 | 2.500 | 2.500 | 2.500 |
| 2 | 2.500 | 3.000 | 2.000 | 2.000 |

All dimensions are in inches.

**Type 1, Style 3.—**Style 3, reversible clamping yoke. Style 3 blocks shall be similar to Fig. 3, conform to dimensions and accuracies shown in Table 2 for the sizes specified and be designed to facilitate clamping to a machine table or plate. The top surface shall have 4 drilled and tapped holes, one near each corner, to accommodate the yoke screws. If the blocks are furnished in matched pairs the yokes and screws shall be suitable for use with either block of the matched pair.

**Table 2. Accuracy and Dimensions, Type 1, Style 3 Machinists' Table Block**

| Designating Size | Permitted Deviations (inch) | | | | | |
|---|---|---|---|---|---|---|
| | Flatness of Working Faces | End/Side Square to Base | Parallelism of Opposite faces | Parallelism of "V" to Base and Side | "V" Centrality | Matching Tolerance |
| Grade A | 0.0001 | 0.0002 | 0.0002 | 0.0003 | 0.0003 | 0.0002 |
| Grade B | 0.0004 | 0.002 | 0.001 | 0.002 | 0.002 | 0.001 |
| Grade Characteristics | | | | | | |
| Grade A | • Up to 6.375″<br>• Surface finish NTE 32 µin | | | • Steel alloy or mild steel<br>• Material hardness 40 to 62 RC | | |
| Grade B | • Up to 6.375″<br>• Surface finish NTE 63 µin<br>• Material hardness 40–45 RC steel / 200–280 BHN cast steel | | | • Mild steel or fine grained cast steel | | |

| Grades A & B Designating Size | Nominal Dimensions (inch) | | | |
|---|---|---|---|---|
| | Length | Width | Height | Maximum Capacity |
| 1 | 1.375 | 1.375 | 1.125 | 0.75 |
| 2 | 2.125 | 2.375 | 1.875 | 1.500 |
| 3 | 2.250 | 2.9375 | 1.9375 | 2.000 |
| 4 | 2.875 | 4.000 | 2.875 | 3.000 |
| 5 | 6.000 | 6.000 | 6.000 | 4.500 |
| 6 | 8.000 | 8.000 | 8.000 | 6.375 |

All dimensions are in inches.

**Type 1, Style 4, Magnetic Clamping Blocks.—**Style 4, magnetic clamping blocks shall be similar to Fig. 4 and conform to the dimensions and accuracies shown in Table 3. Magnetic clamping V-blocks shall be either made of mild steel laminated with brass or an assembly of mild steel with an embedded magnetic core that can be disengaged to allow positioning of work piece. Style 4 magnetic clamping blocks shall have a ground finish on the sides, top, bottom, and V-grooves.

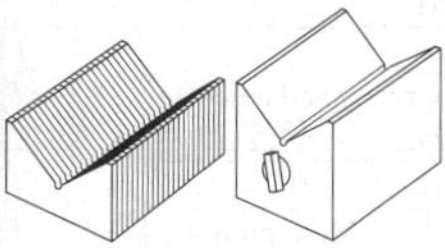

Fig. 4. Type 1, Style 4 Magnetic Clamping V Blocks

Fig. 5. Type 1, Style 5 Ribbed without Clamps

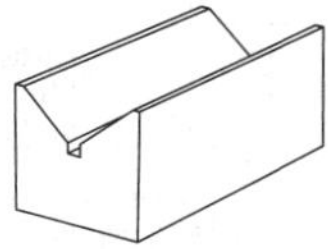

Fig. 6. Type 1, Style 6 General Machine Shop Support V-Block

**Table 3. Accuracy and Dimensions for Type 1, Style 4 Magnetic Clamping V-Block**

| Grade | Permitted Deviations (inch) | | | | | |
|---|---|---|---|---|---|---|
| | Flatness of Working Faces | End/Side Square to Base | Parallelism of opposite faces | Parallelism of "V" to Base and Side | "V" Centrality | Matching tolerance |
| Grades A, B | 0.0001 | 0.0002 | 0.0002 | 0.0003 | 0.0003 | 0.0002 |
| Grade C | 0.0004 | 0.002 | 0.001 | 0.002 | 0.002 | 0.001 |
| Grade Characteristics | | | | | | |
| Grade A, B | • Up to 2.125″ • Mild steel with embedded magnetic core, or laminated mild steel with brass • Surface finish NTE 32μin • Material hardness for mild steel 50–62 RC | | | | | |
| Grade C | • Up to 2.125″ • Mild steel with embedded magnetic core, or laminated mild steel with brass • Surface finish NTE 63μin • Material hardness 40–45 RC Steel | | | | | |

| All Grades, Designating Size | Nominal Dimensions | | | |
|---|---|---|---|---|
| | Length | Width | Height | Maximum Capacity |
| 1 | 1.750 | 2.375 | 1.875 | 2.125 |
| 2 | 2.375 | 1.875 | 1.750 | 1.625 |
| 3 | 2.500 | 2.500 | 1.750 | 2.250 |
| 4 | 3.750 | 2.375 | 1.750 | 2.125 |

All dimensions are in inches.

**Type 1, Style 5, Ribbed without Clamps.**—The Style 5 block shall be designed without clamping capability for the work piece. The block shall be similar in design to Fig. 5 and conform to the dimensions and accuracies shown in Table 4. The Style 5 block shall be of ribbed design to provide strength with light weight. There may be lightening holes in the webbing section to reduce weight without affecting the performance capabilities. The Style 5 block shall be of a design as to facilitate clamping to a machine table or plate.

Due to the nature of the material, special consideration has been given to the granite V-blocks and the general recommended dimensions are given in Table 4. Higher accuracy granite V-blocks are commercially available and manufacturers' tolerances apply.

**Table 4. Accuracy and Dimensions for Type 1, Style 5 Ribbed without Clamps**

| Grade | Permitted Deviations (inch) | | | | | |
|---|---|---|---|---|---|---|
| | Flatness of Working Faces | End/Side Square to Base | Parallelism of Opposite Faces | Parallelism of "V" to Base and Side | "V" Centrality | Matching Tolerance |
| Grades A, B | 0.0001 | 0.0002 | 0.0002 | 0.0003 | 0.0003 | 0.0002 |
| Grade C | 0.0004 | 0.002 | 0.001 | 0.002 | 0.002 | 0.001 |
| Grade Characteristics | | | | | | |
| Grades A, B | • Up to 10.500″ • Mild Steel or close grained cast steel free from porosity and defects • Surface finish NTE 63 μin • Material hardness 40–45 RC steel | | | | | |

**Table 4. Accuracy and Dimensions for Type 1, Style 5 Ribbed without Clamps** *(Continued)*

| | |
|---|---|
| Grade C | • Up to 10.500″<br>• Mild Steel or close grained cast steel free from porosity and defects<br>• Surface finish NTE 63 μin • Material hardness 40–45 RC steel |

| All Grades, Designating Size | Nominal Dimensions (inch) | | | |
|---|---|---|---|---|
| | Length | Width | Height | Maximum Capacity |
| 1 | 2.250 | 4.000 | 4.000 | 3.000 |
| 2 | 3.500 | 4.000 | 4.000 | 3.000 |
| 3 | 5.000 | 6.000 | 6.000 | 4.875 |
| 4 | 8.000 | 8.000 | 8.000 | 6.500 |
| 5 | 10.000 | 10.000 | 10.000 | 8.500 |
| 6 | 12.000 | 12.000 | 12.000 | 10.500 |

All dimensions are in inches.

**Type 1, Style 6, Plain, without Clamps.—**Type 1, Style 6 V-block shall be designed similar to the style shown in Fig. 6 and conform to the dimensions and accuracies described in Table 5. The Style 6 block shall be designed plain with a chip relief at the apex of the V-groove. The block shall be furnished in a set of three suitable for general machine shop use.

**Table 5. Type 1, Style 6 General Machine Shop Support V-Block**

| Grade and Designating Size | Permitted Deviations (inch) | | | | | |
|---|---|---|---|---|---|---|
| | Flatness of Working Faces | End/Side Square to Base | Parallelism of Opposite Faces | Parallelism of "V" to Base and Sides | "V" Centrality | Matching Tolerance over "V" Length |
| Grade A | | | | | | |
| 1 | 0.0001 | 0.0002 | 0.0002 | 0.0003 | 0.0003 | 0.0002 |
| 2 | 0.0001 | 0.0002 | 0.0002 | 0.0003 | 0.0003 | 0.0002 |
| Grade B | | | | | | |
| 1 | 0.0004 | 0.002 | 0.001 | 0.002 | 0.002 | 0.001 |
| 2 | 0.0004 | 0.002 | 0.001 | 0.002 | 0.002 | 0.001 |

| All Grades, Designating Size | Nominal Dimensions (inch) | | | |
|---|---|---|---|---|
| | Length | Width | Height | Maximum Capacity |
| 1 | 2.92 | 1.57 | 1.57 | 0.160 to 1.50 |
| 2 | 3.93 | 1.77 | 1.77 | 0.236 to 1.70 |
| 3 | 5.90 | 2.16 | 2.16 | 0.315 to 2.10 |
| 4 | 7.87 | 2.75 | 2.75 | 0.315 to 2.70 |
| 5 | 9.84 | 3.34 | 3.34 | 0.475 to 3.30 |
| 6 | 11.81 | 4.13 | 4.13 | 0.600 to 4.10 |

All dimensions are in inches.

**Type II, Double V-Blocks.—**The Type II, Double V-Blocks shall be designed to incorporate two V-grooves, as in Fig. 7a, Fig. 7b, and Fig. 7c, on one block machined on opposite sides, and both conforming to the dimensions and accuracies shown in Table 6. Nominal dimensions shown in Table 6 do not encompass all configurations available but define the tolerances that need to be adhered to and define the design parameters. In the case that the design furnishes two V-grooves of different sizes, the accuracy tolerances of the larger size shall apply to both.

The Type II, Double V-Block shall be furnished individually or in matched pairs and shall conform to Grade A or Grade B accuracy tolerances and shall come furnished with a clamp or clamps sufficiently robust to enable adequate security of work piece during average drilling or grinding operations. If furnished in matched pairs the clamps shall be usable on either member of the matched set. The clamps shall be designed to engage a single clamping groove along the outer face of the block.

**Type II, Style 1, Double Groove Examples**

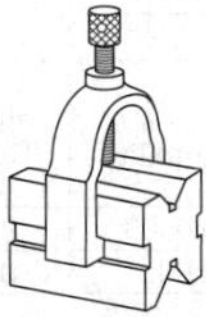

Fig. 7a. Double Groove Single Level Clamping

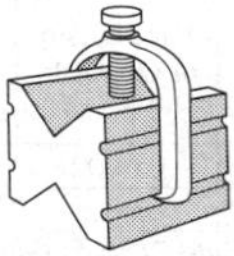

Fig. 7b. Double Groove Double Level Clamping

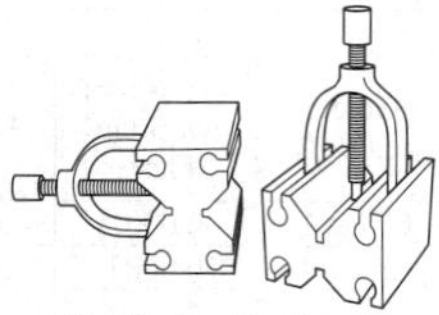

Fig. 7c. Double Groove Single Level Clamping

**Table 6. Type II, Style 1, Double Groove Single Level Clamping**

| Grade | Permitted Deviations (inch) | | | | | |
|---|---|---|---|---|---|---|
| | Flatness of Working Faces | End/Side Square to Base | Parallelism of Opposite Faces | Parallelism of "V" to Base and Side | "V" Centrality | Matching Tolerance |
| Grade A | 0.0001 | 0.0002 | 0.0002 | 0.0003 | 0.0003 | 0.0002 |
| Grade B | 0.0004 | 0.002 | 0.001 | 0.002 | 0.002 | 0.001 |

| Grade Characteristics | |
|---|---|
| Grade A | • Steel alloy<br>• Surface finish NTE 32μin<br>• Material hardness 50–62 RC |
| Grade B | • Mild steel or high grade cast steel<br>• Surface finish NTE 63μin<br>• Material hardness 40–45 RC steel / 200–280 BHN cast steel |

| Designating Size | Nominal Dimensions (inch) | | | |
|---|---|---|---|---|
| | Length | Width | Height | Maximum Capacity |
| 1 | 1.625 | 1.250 | 1.250 | 1.000 |
| 2 | 2.000 | 1.500 | 1.500 | 1.500 |

All dimensions are in inches.

**Type III, Quadruple V-Groove Block.—**The Type III, Quadruple Groove Block shall have four V-grooves, one on each of the four sides. This type shall be furnished in Grade A or B. Grade A shall be of a steel alloy and conform to the dimensions and accuracy tolerances defined in Table 7.

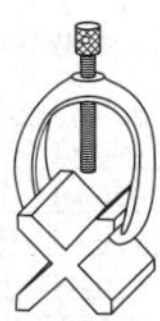

Fig. 8. Type III, Style 1 Quadruple V-Block

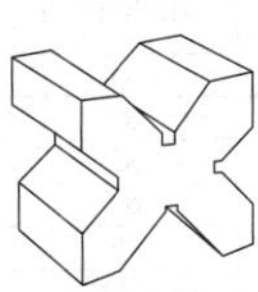

Fig. 9. Type III, Style 2 Quadruple V-Block

Fig. 10. Type IV, Style 1 Combination Top and End V-Block

The Grade B Type III shall be of fine grained cast steel and conform to the dimensions and tolerances defined in Table 8. The nominal dimensions referenced in Table 7 and Table 8 do not represent the full extent of options available. Tolerances are defined as the minimum acceptable. Higher accuracy V-blocks of the Type III Style 1 and 2 may be available.

**Table 7. Type III, Style 1, Quadruple V-Groove Block**

| Grade | Permitted Deviations (inch) | | | | | |
|---|---|---|---|---|---|---|
| | Flatness of Working Faces | End/Side Square to Base | Parallelism of Opposite Faces | Parallelism of "V" to Base and Side | "V" Centrality | Matching Tolerance |
| Grade A | 0.0001 | 0.0002 | 0.0002 | 0.0003 | 0.0003 | 0.0002 |
| Grade B | 0.0004 | 0.002 | 0.001 | 0.002 | 0.002 | 0.001 |
| Grade Characteristics | | | | | | |
| Grade A | • Steel alloy • Surface finish NTE 32µin • Material hardness 50–62 RC | | | | | |
| Grade B | • Mild steel or high grade cast steel • Surface finish NTE 63µin<br>• Material hardness 40–45 RC steel / 200–280 BHN cast steel | | | | | |

| Designating Size | Nominal Dimensions (inch) | | | |
|---|---|---|---|---|
| | Length | Width | Height | Maximum Capacity |
| 1 | 2.000 | 1.500 | 1.500 | 1.125 |

All dimensions are in inches

**Type III, Style 2, Quadruple V-Block.—**Type III, Style 2, Quadruple V-Block is designed for a robust precision support of the work piece. This style V-block is not furnished with clamp or yoke and may be made of either mild steel or fine grain cast steel. The Type III, Style 2 block may be designed with uniform V-groove capability or a range of capacities. The tolerance of the Style 2 V-block will conform to those shown in Table 8, however the nominal dimensions and accuracies referenced in Table 8 do not represent the full extent of options available. Manufacturers' tolerances are always applied.

**Table 8. Type III, Style 2, Quadruple V-Block**

| Grade | Permitted Deviations (inch) | | | | | |
|---|---|---|---|---|---|---|
| | Flatness of Working Faces | End/Side Square to Base | Parallelism of Opposite Faces | Parallelism of "V" to Base and Side | "V" Centrality | Matching Tolerance |
| Grade A | 0.0003 | 0.0005 | 0.0005 | 0.0003 | 0.0003 | 0.0002 |
| Grade B | 0.0005 | 0.001 | 0.001 | 0.0012 | 0.0012 | 0.001 |
| Grade Characteristics | | | | | | |
| Grade A | • Steel alloy • Surface finish NTE 32µin<br>• Material hardness 50–62 RC | | | | | |
| Grade B | • Mild steel or high grade cast steel • Surface finish NTE 63 µin<br>• Material hardness 40–45 RC steel / 200–280 BHN cast steel | | | | | |

| All Grades, Designating Size | Nominal Dimensions | | | |
|---|---|---|---|---|
| | Length (mm) | Height (mm) | Width (mm) | Maximum Capacity (mm) |
| 1 | 95 | 70 | 35 | 20 × 1 – 30 × 1 – 40 × 1 – 50 × 1 |
| 2 | 100 | 100 | 30 | 20 × 1 – 30 × 1 – 40 × 1 – 50 × 1 |
| 3 | 120 | 120 | 35 | 30 × 1 – 40 × 1 – 50 × 1 – 60 × 1 |
| 4 | 120 | 120 | 60 | 30 × 1 – 40 × 1 – 50 × 1 – 60 × 1 |
| 5 | 150 | 150 | 75 | 40 × 1 – 50 × 1 – 60 × 1 – 70 × 1 |
| 6 | 200 | 200 | 100 | 60 × 1 – 70 × 1 – 80 × 1 – 100 × 1 |

Dimensions of deviations are in inches. Nominal dimensions are in mm.

**Type IV, Combination Top and End V-Groove Block.—**The Type IV, Combination Top and End V-Groove Block shall be furnished in individual blocks consisting of one block, one clamp and one securing screw which is similar to the design illustrated in Fig. 10. The V-block shall be so designed that when turned on side or end that there shall be no interference from the sliding clamp. One end of the block shall be stepped and this end shall contain a V-groove and clamping grooves at right angle to the base. The blocks shall be designed to facilitate clamping to a machine bed or plate and have two tapped holes in the base and one on each side for mounting purposes.

**Style 1, Single Level Clamping.—**The Type IV, Style 1 Block shall be similar in design to the one shown in Fig. 10 and shall conform to the dimensions and tolerances as defined in Table 9. The Type IV, Style 1 shall be offered in Grade A configuration. Grade A tolerance shall be applied to the perpendicular secondary V-groove as it is to the primary V-groove.

**Table 9. Combination Top and End V-Groove Block**

| Grade | Permitted Deviations (inch) | | | | | |
|---|---|---|---|---|---|---|
| | Flatness of Working Faces | End/Side Square to Base | Parallelism of Opposite Faces | Parallelism of "V" to Base and Side | "V" Centrality | Matching Tolerance |
| Grade A | 0.0001 | 0.0002 | 0.0002 | 0.0003 | 0.0003 | N/A |
| Characteristics | • Steel alloy • Surface finish NTE 32μin<br>• Material hardness 50–62 RC | | | | | |

| Designating Size | Nominal Dimensions (inch) | | | |
|---|---|---|---|---|
| | Length | Width | Height | Maximum Capacity |
| 1 | 3.625 | 1.875 | 1.875 | 1.3125 |

All dimensions are in inches

**Metric Products.—**Products manufactured to metric dimensions (SI units) shall be considered on an equal basis with those manufactured to the inch units, providing they fall within the applicable tolerance conversions. The British Standard BS 3731:1987 provides specific requirements for hollow or solid, single or double V-groove blocks made from cast iron or steel, and solid V-blocks made from granite. Two grades of accuracy are defined, Grade 1 and Grade 2. Provisions are made for matched sets as well.

**British Standard V-Block Designs,** *BS 3731*

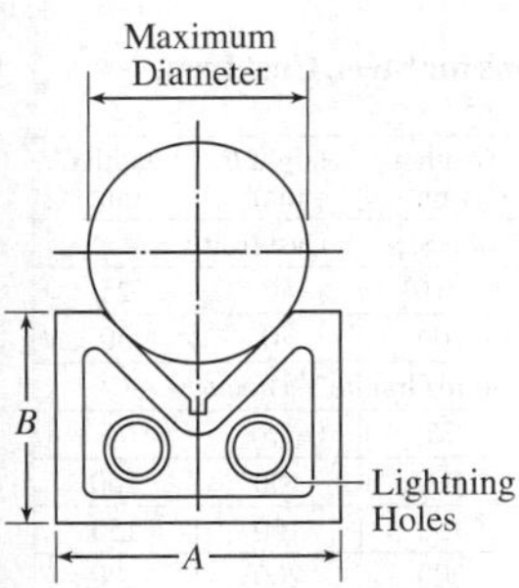


Fig. 11. Cast Iron Hollow V-Block[a]

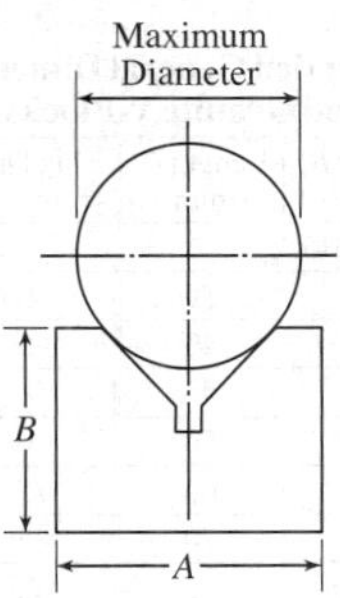


Fig. 12. Solid V-Block[a]

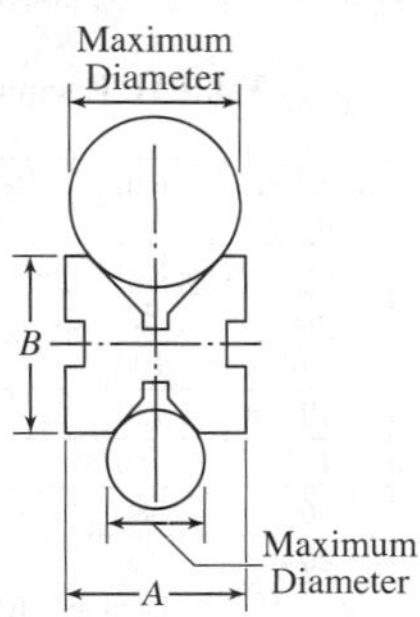


Fig. 13. Double V-Block[a]

[a] *Note:* Length C (side view) not shown.

**Table 10. Accuracy of Grade 1 and Grade 2 V-Blocks,** *BS 3731:1987*

| Designating Size, mm | Permitted Deviations | | | | | | | |
|---|---|---|---|---|---|---|---|---|
| | Flatness of Working Faces, µm | End and Side Faces, Squareness to the Base, µm | Parallelism of Opposite Faces, µm | Parallelism of Axis "V" to Base and Side Faces, µm | Squareness of Axis "V" to End Face, µm | Equality of "V" Flank Angle, minutes | "V" Centrality, µm | Matching Tolerance over "V" Length, µm |
| Grade 1 | | | | | | | | |
| 20 | 2 | 4 | 4 | 2 | 2 | 1 | 4 | 2 |
| 25 | 2 | 4 | 4 | 2 | 2 | 1 | 4 | 2 |
| 40 | 3 | 6 | 6 | 4 | 4 | 1 | 6 | 3 |
| 50 | 3 | 6 | 6 | 4 | 4 | 1 | 6 | 3 |
| 63 | 4 | 8 | 8 | 6 | 6 | 1 | 8 | 4 |
| 75 | 5 | 10 | 10 | 8 | 8 | 1 | 10 | 5 |
| 80 | 5 | 10 | 10 | 8 | 8 | 1 | 10 | 5 |
| 85 | 5 | 10 | 10 | 8 | 8 | 1 | 10 | 5 |
| 100 | 6 | 12 | 12 | 10 | 10 | 1 | 12 | 6 |
| 125 | 8 | 16 | 16 | 14 | 14 | 1 | 16 | 8 |
| 160 | 10 | 20 | 20 | 18 | 18 | 1 | 20 | 10 |
| 200 | 12 | 24 | 24 | 22 | 22 | 1 | 24 | 12 |
| Grade 2 | | | | | | | | |
| 20 | 6 | 8 | 8 | 8 | 8 | 2 | 6 | 4 |
| 25 | 6 | 10 | 10 | 10 | 10 | 2 | 8 | 4 |
| 40 | 8 | 14 | 14 | 14 | 14 | 2 | 10 | 6 |
| 50 | 8 | 14 | 14 | 14 | 14 | 2 | 10 | 6 |
| 63 | 12 | 26 | 26 | 26 | 26 | 2 | 18 | 8 |
| 75 | 16 | 34 | 34 | 34 | 34 | 2 | 22 | 10 |
| 80 | 16 | 34 | 34 | 34 | 34 | 2 | 22 | 10 |
| 85 | 16 | 34 | 34 | 34 | 34 | 2 | 22 | 10 |
| 100 | 18 | 38 | 38 | 38 | 38 | 2 | 24 | 12 |
| 125 | 24 | 40 | 40 | 40 | 40 | 2 | 24 | 16 |
| 160 | 26 | 42 | 42 | 42 | 42 | 2 | 26 | 20 |
| 200 | 28 | 46 | 46 | 46 | 46 | 2 | 28 | 24 |

*Note:* Intermediate sizes should be made to the accuracy specified for the next smaller designated size. Double V-blocks should be made to the accuracy specified for the larger designating size.

**Table 11. Recommended General Dimensions for Steel, Cast Iron, and Granite V-Blocks** *BS 3731*

| Max Dia., mm | Width *A*, mm | Height *B*, mm | Length *C*, mm | Max Dia., mm | Width *A*, mm | Height *B*, mm | Length *C*, mm |
|---|---|---|---|---|---|---|---|
| Cast Iron Hollow V-Block | | | | Solid Steel or Cast Iron | | | |
| 63 | 80 | 60 | 35 | 40 | 50 | 40 | 25 |
| 80 | 100 | 75 | 40 | 85 | 100 | 70 | 50 |
| 100 | 130 | 90 | 45 | Solid Granite V-Block | | | |
| 125 | 150 | 100 | 50 | 50 | 75 | 75 | 75 |
| 160 | 180 | 130 | 60 | 75 | 100 | 100 | 100 |
| 200 | 220 | 160 | 70 | 85 | 150 | 150 | 150 |
| Double V-Block | | | | 125 | 200 | 200 | 200 |
| 20 and 25 | 32 | 32 | 42 | | | | |
| 20 and 40 | 40 | 40 | 50 | | | | |

## Calibration of V-Blocks

**Calibration Description.**—Calibration of the V-block shall be accomplished at intervals determined by either National Conference of Standards Laboratories (NCSL) Recommended Practices found in RP-1 "Establishment and Adjustment of Calibration Intervals" or the responsible facilities Quality Assurance Manual. Calibration intervals are designed to provide an acceptable level of assurance to adherence to original manufacturer's specifications. Calibration of all precision measurement equipment is critical to providing measurement assurance to the quality system and maintaining a sound, robust manufacturing process and quality system.

**Table 12. Tolerance Limits per Grade**

| Test Instrument (TI) Characteristics | Range | Performance Specification for Accuracy | Test Method |
|---|---|---|---|
| Ends/sides Squareness to Each Other | Up to 5 in. | Grade A – 200 μin<br>Grade B – 0.002 in. | Verified with Electronic Indicator |
| | Over 5 in. | Grade A – 500 μin<br>Grade B – 0.002 in. | |
| Parallelism of V to Base | Up to 5 in. | Grade A – 300 μin<br>Grade B – 0.002 in. | |
| | Over 5 in. | Grade A – 500 μin<br>Grade B – 0.002 in. | |
| V Centrality | Up to 5 in. | Grade A – 300 μin<br>Grade B – 0.002 in. | |
| | Over 5 in. | Grade A – 500 μin<br>Grade B – 0.002 in. | |
| Flatness of Working Faces | | Grade A – 100 μin<br>Grade B – 500 μin | |
| Matching Tolerance | Less than 6.00 in. | Grade A – 200 μin | |
| | Over 6.00 in. | Grade A – 300 μin<br>Grade B – 500 μin | |

*Note:* Tolerances listed in Table 12 are a representation of the standard tolerances recommended by the GGG-V-191D-1971, AFMETCAL (Air Force Metrology & Calibration Program) and general commercial manufacturing. Be advised that not all features are calibrated on all types and styles of V-blocks. Manufacturer's tolerances shall be applied when provided. The actual use of the V-block should be taken into consideration when selecting the features to be calibrated.

**Table 13. Equipment Requirements**

| Item | Minimum Use Specifications | Calibration Equipment Examples |
|---|---|---|
| Surface Plate | Range: 24 × 36 in., Grade AA, 100 μin overall Flatness | Per Fed Spec GGG-P-463 |
| Electronic Indicator | Range: ±0.010, Accuracy: ±0.00001 in. | Mahr-Federal, Model 832 Digital Electronic Amplifier w/ EHE-2056 LVDT |
| Precision Cylindrical Gage | Range: 0.5 in. diameter × 3 in. L; Accuracy: Class XX | Van Keuren, Extra Length Reversible Plug Gage |
| | Range: 1.00 in. diameter × 2 in. L; Accuracy: Class XX | Van Keuren, Reversible Plug Gage |
| Granite Master Square | Range: 9 in. L × 12 in. H × 3 in. W Accuracy: Grade AA 25μin/6 in. | L.S.Starrett Granite Tri-Square |
| Squarol | Range: 0.375 in. to 5.00 in. | Squarol Squareness Checker Model #700 |
| Sine Plate | Range: 6 × 6 in.; Accuracy: ±200 μin | Taft Pierce Model 9123-1 |

**Table 13.** *(Continued)* **Equipment Requirements**

| Item | Minimum Use Specifications | Calibration Equipment Examples |
|---|---|---|
| Alternative Equipment Requirements | | |
| SquareMaster | Range: 0 in. to 12 in.; Accuracy ±50 μin over 12.00 in. | PMC/Mercury SquareMaster Model VSQ-12 |
| Master Square | Range 3 in. W × 5 in. L × 12 in. H | SquareMaster Master Square |
| Cylindrical Square | Range to fit max diameter of UUT. (*Example:* 5 in. × 12 in.); Accuracy: 100 μin | Taft Pierce Model 9143 |

**Preliminary Operations.—**Review and become familiar with the entire procedure before proceeding with the calibration process. Alternative standards may be substituted for the ones listed in Table 13. Thorough consideration must be paid to the *Accuracy Ratio, Total Accuracy Ratio (TAR)*, and *Measurement Uncertainties* introduced when substituting standards. Some standards represented do not offer a 4:1 TAR. If this is an unacceptable condition according to your particular quality assurance system then measures must be taken to calculate the uncertainty and introduce this into your uncertainty budget, or determine the acceptable tolerance limits through *Guard Banding* or similar method to satisfy the requirements of your quality system.

- Allow all the components to stabilize a minimum of 4 hours to the temperature of the calibration area before beginning the procedure.
- Clean and closely inspect all critical surfaces for nicks and burrs. Carefully stone with fine hard Arkansas stone to remove any incursions.
- Turn on electronic indicator and allow to warm up for a minimum of 30 minutes before proceeding.

## Calibration Procedure

*Note:* Document the results of all measurements and tests for future reference, Trend Analysis, Root Cause Analysis or Calibration Interval Adjustment.

**Sides Square to Each Other.—**1) Set up equipment as shown in Fig. 14.

a) Special attention must be paid to establishing ZERO indication at the height of the V-block (UUT or Unit Under Test) sides to be measured. The SQUAROL (squareness checker) is rotated so the crest of the radius is indicated by the lowest reading on the Electronic Indicator. Plus and Minus travel must be available on the indicator head.

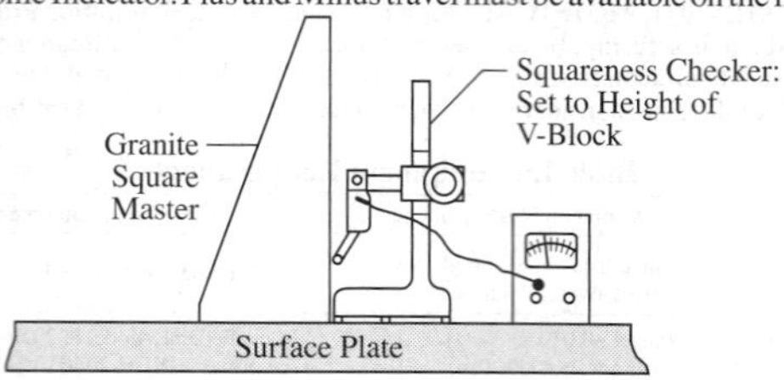


Fig. 14.

2) Set the electronic indicator to the appropriate range to measure the Squareness of the Side A of the UUT.

a) ±0.0002 in. for Grade A. (±0.0005 over 5 in.)

b) ±0.002 in. for Grade B (or C)

3) Set the gage head of the electronic indicator against the granite square master at the appropriate height to measure side A of the UUT at approximately 90% of it vertical height. Rotate the Squarol to seek the lowest indication. Using the fine adjust of the Squarol to bring the reading close to the zero indication. Assure that plus and minus travel can be realized.

4) Zero the indicator with the electronic fine adjust to set the final zero indication.

5) Remove the granite square master and replace with the UUT (Side A) as shown in Fig. 15. Side A is determined arbitrarily, but once established, must be maintained throughout the procedure to determine the relationship of all sides to each other.

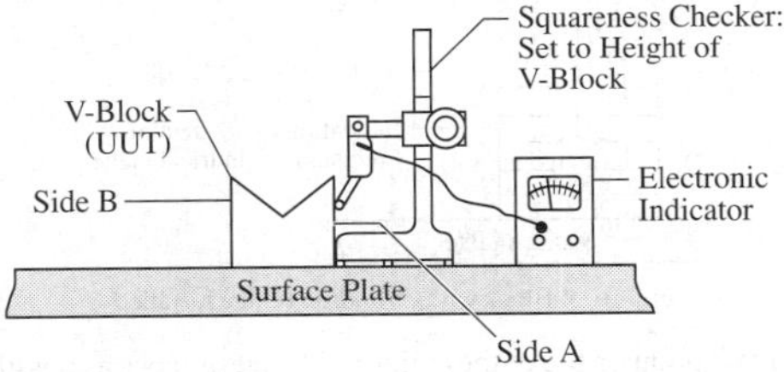


Fig. 15.

6) Rotate the Squarol to seek the lowest indication. Verify that the indication is within the tolerance limits for the UUT as defined in Table 12. Record the indication as Side A to Base Squareness.

7) It may be necessary to place a parallel bar or right-angle iron behind the UUT to secure the UUT in place while obtaining indication. This will not affect the accuracy of the measurement as long as the measurement is repeatable and consistent.

8) Reverse the UUT and repeat Items 1) through 6) for Side B. Record indication as Side B to Base Squareness.

9) Rotate UUT 90° clockwise and repeat Item 6) and record indication as End A to Base Squareness.

10) Rotate UUT 180° clockwise and repeat Item 6) and record indication as End B to Base Squareness.

11) The End measurements will not be applicable to all styles and types due to design features. In that case disregard Items 8) and 9).

12) Rotate the UUT so that Side A is down and the indication is of Base to Side A.

13) Repeat Items 6) through 9) for remaining three sides (End A, End B, Top)

14) Rotate the UUT so that Side B is down and the indication is of Base to Side B.

15) Rotate the UUT so End A is down and the indication is Base to End A.

16) Repeat Items 6) through 9) for remaining three sides. (Side A, Side B Top)

17) Repeat process for End B if applicable.

18) All readings must be within limits defined in Table 12.

**Parallelism of V to Base.**—1) Select the appropriate size precision cylindrical gage or cylindrical square for the UUT being calibrated. The appropriate size gage should contact the V-groove near the center of the working surface. Leave approximately half the length of the cylindrical gage extended over one end of the UUT as shown in Fig. 16.

*Note:* Larger V-blocks will require the use of cylindrical squares of various diameters to accomplish the engagement at the center of the working surface. Special care should be taken in handling these precision standards to avoid nicking or damaging the fine lapped surfaces. Adaptation of the methods used herein can be made to accomplish the calibration of larger V-blocks without reducing confidence levels associated with the readings taken.

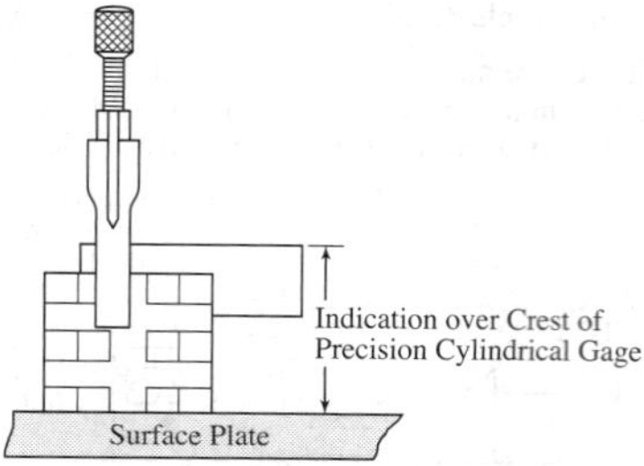


Fig. 16. V-Block with Cylindrical Gage Installed

2) Place the contact point on top of the cylindrical gage as close as possible and directly above the end of the V-block. Seek the top of the cylinder by sweeping across the radius and observing the maximum indication.

3) Zero the electronic indicator at the high point of the sweep.

4) Move the gage head to the end of the cylindrical gage. Indicate the peak of the radius and note the reading.

5) Rotate the cylindrical gage or square 180° and repeat Items 2) through 4).

6) Do not adjust the zero setting of the electronic indicator from Item 5).

*Note:* If the cylindrical gage or cylindrical square being used covers $\frac{3}{4}$ of the V-groove then Gage or Square does not have to be moved to the other end.

7) Remove the cylindrical gage or cylindrical square being used and place it on the opposite end of the V-groove with the same end facing out.

8) Place the contact point on top of the cylindrical gage as close as possible and directly above the end of the V-block. Seek the top of the Cylinder by sweeping across the radius and observing the maximum indication. Note this reading.

9) Move the gage head to the end of the cylindrical gage. Indicate the peak of the radius and note the reading.

10) Rotate the cylindrical gage or square 180° and repeat Items 8) through 9) retaining the original zero setting position.

11) Record the largest reading found as the maximum error of Parallelism of V to Base. Reading shall fall within the Tolerance Limits defined in Table 12.

12) Lay UUT on Side A and perform Items 1) through 11) to calibrate Parallelism of V-Groove to Side A.

13) Lay UUT on Side B and perform Items 1) through 11) to calibrate Parallelism of V-Groove to Side B.

14) Repeat Items 1) through 13) for V-groove opposite and all remaining V-grooves if unit is Type 2, Style 1; Type 3, Style 1 or 2; or Type 4, Style 1.

a) Double V-Groove Block Tolerance is to be derived from the tolerance of the larger of the two V-grooves.

b) The Type III Style 2 Quadruple V-Block is generally produced in the B Grade and the applicable tolerance shall be applied to all V-grooves inclusive. Grade A tolerances shall apply to all V-grooves in a Grade A Quadruple Type III Style 2 regardless of the size of the V-groove unless otherwise specified by the manufacturer.

15) If V-blocks are a matched pair the maximum deviation between the members may not exceed the tolerance limits as defined in Table 12.

16) All dimensions must be indicated in matched sets.

17) Place both members of the set side by side on the surface plate with the V-groove up as shown in Fig. 17. The clamp or yoke is not installed for this measurement.

18) Place the contact point at the upper left corner of V-Block 1 and zero the indicator at that point.

19) Indicate the top surface of the V-block and confirm the flatness of the block. If there are any erroneous readings, immediately check the V-block base for nicks or burrs that could be affecting the reading. Gently remove nicks or burrs with a hard Arkansas stone.

20) Move the contact point to V-Block 2. Do not change the zero setting and indicate the top surface of V-Block 2 and confirm that the readings are within the tolerance limits as defined in Table 12 for matched sets.

21) Rotate the UUT set so that Side A is now on the surface plate and place the contact point in the same relative position as with the TOP measurement.

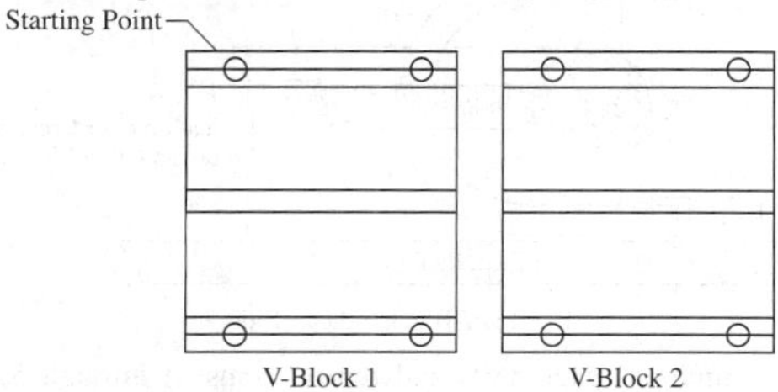


Fig. 17. Matched V-Block Setup TOP VIEW

22) Indicate the entire surface of V-Block 1, once again observing for any abnormalities that may indicate a nick or burr on the side opposite.

23) Move the contact point to the starting point on V-Block 2 and observe the deviation from the zero setting. Indicate the entire surface of V-Block 2 and note the maximum deviation from zero.

24) Verify that all readings are within the tolerance limits as defined in Table 12.

25) Rotate the UUT set so that the TOP is now on the surface plate.

26) Place the contact point in the same relative starting position as previously used. If the V-Block Set has a lapped or flat ground surface, indicate the entire surface observing any abnormalities and deviations in accordance with the tolerance table. If the UUT has a double rail supporting system, only indicate the sections that come into contact with the surface plate under normal use.

27) Move the contact point to V-Block 2 starting point and observe the initial deviation from zero. Indicate the applicable surface area and verify that all deviations are within the tolerance limits as defined in Table 12.

28) Rotate the UUT set to the final side and repeat Items 21) through 24).

29) Place UUT set on End A and repeat Items 21) through 24).

30) Place UUT set on End B and repeat Items 21) through 24).

**V-Groove Centrality.—**1) The V-Groove Centrality is not calibrated in all types and styles. Centrality of the V-Groove is only applicable if the V-block can be set on its side with no interference when the clamp or yoke is installed.

a) Precision granite V-blocks are generally produced in laboratory grade accuracies and require special attention in handling to avoid damage to the standards and V-blocks.

b) Type 1 Style 5 are sometimes large and require special care in handling due to the weight of the UUT and the size of the standard use in this calibration. Care must be taken when securing a large cylindrical square into the V-groove so no nicks or burrs are raised in the lapped surface.

2) Using the same cylindrical gage or cylindrical square as in *Parallelism of V to Base*: Items through 13), lay the UUT on Side A as shown in Fig. 18.

3) Place the contact point on top of the cylindrical gage close to the end of the V-block. Seek the top of the Cylinder by sweeping across the radius and observing the maximum indication. Zero the electronic indicator.

4) Without disturbing the position of the cylindrical gage mounted in the V-groove, rotate the setup over and lay the UUT on Side B.

5) Place the contact point on top of the cylindrical gage as close as possible to the same distance from the end of the V-block as in Item 3). Seek the top of the cylinder by sweeping across the radius and observing the maximum indication. This reading must be within the tolerance limits defined in Table 12.

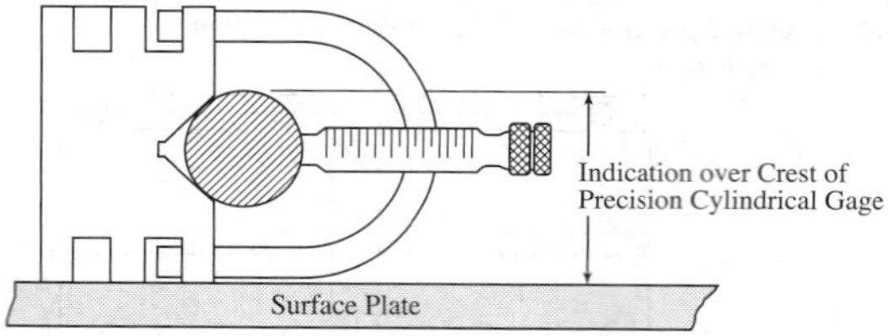


Fig. 18. V-Block Indicate Side A

6) Rotate the cylindrical gage 180° and repeat Items 2) through 5). This reading must be within the tolerance limits defined in Table 12.

7) Repeat the process for the opposite end, and for each remaining V-groove on the UUT.

8) If the UUT is a matched set, the maximum deviation must not exceed the limits for matched sets as defined in Table 12.

**Flatness of Working Surfaces.—**1) Place UUT onto sine plate as shown in Fig. 19 and seat well against the end bar. Assure that no nicks or burrs are present on either the plate or end bar that could affect the reading.

2) Set up the Sine Plate to an angle of 45°

3) Place the contact point of the electronic indicator into the V-groove at the upper left position of the first line as shown in the **Top View of Surface** of Fig. 19 and zero the electronic indicator at this position.

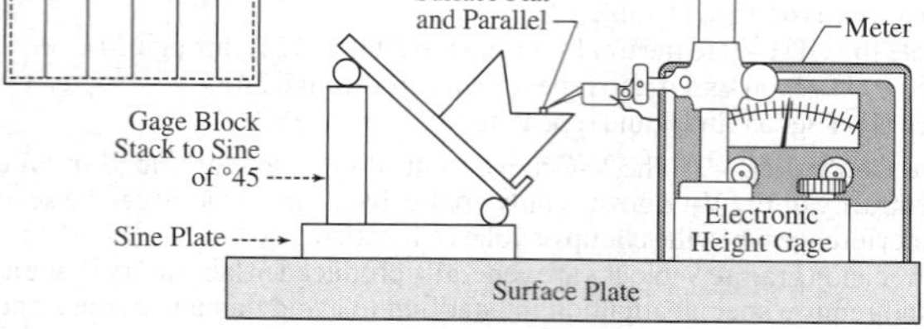


Fig. 19. Flatness of Working Surface Set Up

4) Sweep the surface along the lines shown from the throat of the V-groove to the outer edge Left to Right (or End A to End B)

5) Verify that all readings are within the tolerance limits as defined in Table 12.

6) Rotate the V-block so the side opposite is now parallel to the surface plate and repeat Items 3) through 5), working from End B to End A.

7) Record all readings on a permanent record and retain for use by the QA system.

8) Remove all standards. Clean and return to protected storage condition. Remove UUT from set up and store in appropriate protective containers or stations.

## Parallel Bars

*Parallel bars* are used for workpiece support during layout, machining, and inspection operations. Parallel bars are made from either steel, cast iron, or granite, come in a wide range of sizes and are either used alone or in matched pairs. In general, there are four types of parallel bars: Type I, Solid; Type II, Ribbed; Type III, Box; Type IV, Adjustable. Although the parallel bar is a relatively simple tool, it is still considered a precision instrument and must be handled with the same attention to care and handling as other precision ground and finished supporting gages. Most precision parallels are made from heat-treated steel and hardened to 55 to 60 on the C Rockwell scale. Generally, they have a finish of 8 μin Ra with a fine finish free from all grind marks, chatter, or cracks. Granite parallel bars are also very desirable in an environment where thermal expansion is a consideration or where steel or cast iron would not be acceptable. Granite parallel bars are made to a very high degree of accuracy and can be used in a precision inspection setup.

**Type I, Solid.—**Type I parallel bars are designed to be used independently as an individual bar to aid in inspection setups. They can be used as a reference surface or to establish a vertical plane perpendicular to a base plane. They can also be used as an extended support plane, when used in matched pairs, to establish an elevated plane that is parallel to the base plane. Type I parallel bars have four finished sides that shall not vary from a true plane by more than 0.0002 inch per foot (0.005mm/300mm). The adjacent sides will be square to each other within 0.0005 inch (0.012mm). Type I parallel bars will appear similar to Fig. 1 and will meet the tolerance specifications stated in Table 1.

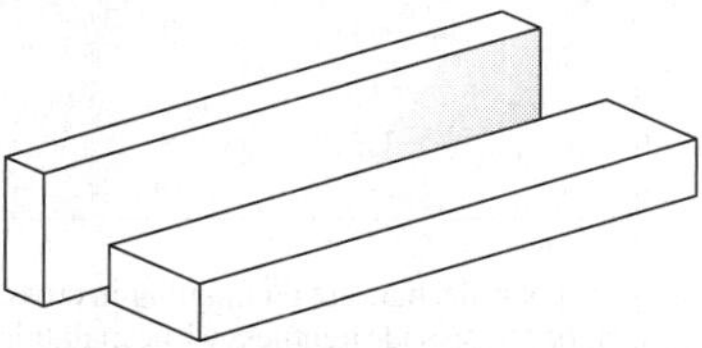

Fig. 1. Parallel Bars, Type 1, Solid

**Table 1. Steel and Cast-Iron Parallel Bar Tolerances, Inch Sizes**

| Size[a] | Tolerance Limits: Straightness (in.) | Parallelism (in.) | Width/Height (in.) | Width/Height Variance Matched Pairs (in.) |
|---|---|---|---|---|
| Inch Sizes | | | | |
| (in.) | (in.) | (in.) | (in.) | (in.) |
| ⅛ to 3⁄16 | 0.002 | 0.0001 | 0.0002 | 0.0002 |
| ¼ to ¾ | 0.0005 | 0.0001 | 0.0002 | 0.0002 |
| ¾ to 1⅛ | 0.0002 | 0.0001 | 0.0002 | 0.0002 |
| 1⅛ to 1½ | 0.0002 | 0.00015 | 0.0003 | 0.0003 |
| 1½ to 3 | 0.0002 | 0.0002 | 0.0004 | 0.0004 |
| Millimeter Sizes | | | | |
| (mm) | (mm) | (mm) | (mm) | (mm) |
| 3 to 5 | 0.0500 | 0.0025 | 0.0050 | 0.0050 |
| 6 to 20 | 0.0130 | 0.0025 | 0.0050 | 0.0050 |
| 20 to 25 | 0.0050 | 0.0025 | 0.0050 | 0.0050 |
| 25 to 35 | 0.0050 | 0.0040 | 0.0080 | 0.0080 |
| 35 to 75 | 0.0050 | 0.0050 | 0.0100 | 0.0100 |

[a] Size as applicable to either width or height dimension specified by the manufacturer.

Some common size configurations available from most manufacturers of Type I precision parallel bars appear in Table 2. This is a representation of sizes that are available; however, it is possible to have parallel bars custom made to a special order for a particular application. In that case, either the tolerance specifications of the engineering drawing or the specifications defined in the Federal Specification GGG-P-61a shall apply. It is at the discretion of the designing engineer which tolerances will apply in this case.

**Table 2. Common Sizes of Parallel Bars**

| Width and Height (in.) | Length ± 0.002 (in.) | Width and Height (in.) | Length ± 0.002 (in.) | Width and Height (in.) | Length ± 0.002 (in.) |
|---|---|---|---|---|---|
| 1/8 – 1 | 6 | 7/16 – 7/8 | 6 | 1/2 – 1 1/2 | 12 |
| 1/8 – 1 3/16 | 6 | 1/2 – 5/8 | 6 | 1/2 – 2 | 12 |
| 3/16 – 7/8 | 6 | 1/2 – 3/4 | 6 | 11/16 – 1 1/4 | 12 |
| 3/16 – 1 1/8 | 6 | 1/2 – 13/16 | 6 | 3/4 – 1 | 12 |
| 1/4 – 3/8 | 6 | 1/2 – 1 | 6 | 3/4 – 2 | 12 |
| 1/4 – 1/2 | 6 | 1/4 – 3/8 | 9 | 1 – 1 1/2 | 12 |
| 1/4 – 5/8 | 6 | 1/2 – 5/8 | 9 | 1 – 2 | 12 |
| 1/4 – 3/4 | 6 | 1/2 – 1 | 9 | 1 – 3 | 12 |
| 1/4 – 1 | 6 | 1/2 – 1 1/4 | 9 | 1 1/4 – 1 3/4 | 12 |
| 3/8 – 1/2 | 6 | 1/2 – 1 1/2 | 9 | 1 1/4 – 2 1/2 | 12 |
| 3/8 – 3/4 | 6 | 3/4 – 1 | 9 | 1 1/2 – 2 | 12 |
| 3/8 – 7/8 | 6 | 3/4 – 1 1/2 | 9 | 1 1/2 – 3 | 12 |
| 3/8 – 1 | 6 | 3/8 – 1 | 12 | | |

**Type II, Ribbed.—**The Type II parallel bars are rectangular in cross section and made from a cast gray iron. The casting is ribbed to provide lightness while maintaining strength and rigidity. The ribbing on the Type II parallel bar extends to the outer surfaces of the bar and becomes an integral part of the working surface. The working surfaces of the Type II have a fine ground surface that will not exceed a value of 16 μin Ra. All the sharp edges are removed, and the surface is free from all machining marks such as grind chatter or burn. The Type II parallel bar is finished on the four sides of its length. The ends not being considered working surfaces are not finished for work applications. The Type II parallel bar is a much more rugged design that the Type I and is designed to support workpieces of larger dimensions. Type II parallel bars are commonly available in sizes up to 4 by 8 inches with a length of 36 inches. Although larger in overall size, the Type II still maintains a very close tolerance in straightness, parallelism, and squareness.

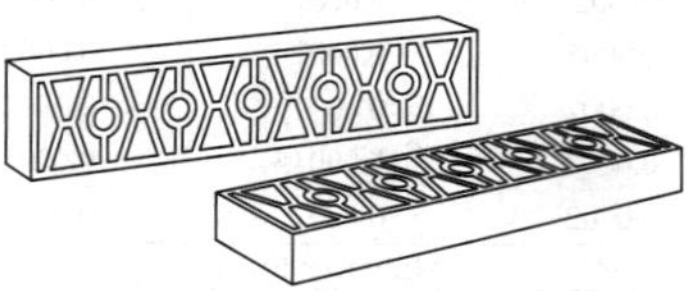

Fig. 2. Parallel Bars, Type II, Ribbed

After each bar is cast and rough ground, it is subject to a seasoning or aging process. This process can either be natural or artificial. The aging process is necessary to provide long term stability after finish grinding. Once the bar is aged and seasoned, the final dimensions will remain stable and not warp or twist for many years. Well-made and seasoned castings have been known to be in service for nearly 100 years and still hold their original geometry. Type II parallel bars will appear similar to Fig. 2 and will conform to the tolerances defined in Table 3. Squareness and parallelism tolerances are Total Indicated Readings (TIR) over the full length of the bar.

The following table shows some common sizes that are readily available from most manufacturers of precision parallel bars. These are representative of the sizes that may be available in the general market, and almost any size or combination may be specially made for a specific application.

**Table 3. Tolerance Limits for Type II, Ribbed Parallel Bars**

| Size | Tolerance Limits | | | Width/Height Variance Matched Pairs |
|---|---|---|---|---|
| | Straightness | Parallelism | Squareness | |
| Inch Sizes | | | | |
| (inch) | (inch) | (inch) | (inch) | (inch) |
| 1½ × 3 × 24 | 0.0002 | 0.0005 | 0.0005 | 0.0005 |
| 2 × 4 × 24 | 0.0002 | 0.0005 | 0.0005 | 0.0005 |
| 2½ × 5 × 24 | 0.0002 | 0.0005 | 0.0005 | 0.0005 |
| 3 × 6 × 36 | 0.0002 | 0.0005 | 0.0005 | 0.0005 |
| 4 × 8 × 36 | 0.0002 | 0.0005 | 0.0005 | 0.0005 |
| Millimeter Sizes | | | | |
| (mm) | (mm) | (mm) | (mm) | (mm) |
| 35 × 75 × 600 | 0.005 | 0.013 | 0.013 | 0.013 |
| 50 × 100 × 600 | 0.005 | 0.013 | 0.013 | 0.013 |
| 50 × 152 × 600 | 0.005 | 0.013 | 0.013 | 0.013 |
| 75 × 150 × 1000 | 0.005 | 0.013 | 0.013 | 0.013 |
| 100 × 200 × 1000 | 0.005 | 0.013 | 0.013 | 0.013 |

**Type III, Box Parallel.—**The Type III "Box" Parallel is designed to provide a wide working surface. The Type III can either be square or rectangular in cross section. The rectangular design will have a rib running through the middle of the block the entire length. This provides support to the two larger surfaces. The surfaces of all six sides will be finish ground to an average of 16 μinch Ra and all sharp edges will be removed. The ground surfaces will all be free from grind chatter marks and grind burn marks.

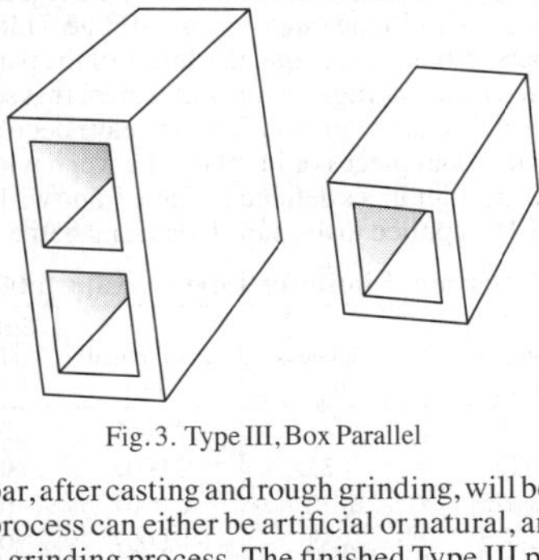

Fig. 3. Type III, Box Parallel

The Type III parallel bar, after casting and rough grinding, will be subject to a seasoning or aging process. This process can either be artificial or natural, and ensures the stability of the material after the grinding process. The finished Type III parallel bars will have a material hardness of 180 BHN (Brinell Hardness Number) checked with a 10mm ball and a 3000 Kg load.

Each of the six working surfaces will not vary from a true plane (straightness) by more than 0.0002 inch per foot. The opposite sides and ends will be parallel to each other within 0.0005 inch. All adjacent sides will be square to each other within 0.0005 inch as well. The size as specified will also be within 0.0005 inch. The Type III box parallel will appear similar to Fig. 3. Tolerances are given in Table 4.

**Table 4. Tolerance Limits for Type III, Box Parallel**

| Size (in.) | Straightness (in.) | Squareness (in.) | Parallelism (in.) |
|---|---|---|---|
| 4×4×6 | 0.0001 | 0.0005 | 0.0005 |
| 4×6×6 | 0.0001 | 0.0005 | 0.0005 |
| 5×8×12 | 0.0002 | 0.0005 | 0.0005 |
| 10×10×10 | 0.0001 | 0.0005 | 0.0005 |

**Type IV, Adjustable Parallel.—**The adjustable parallel is a precision parallel that is adjustable to any height within a specified range. This design allows a flexibility in the use of the parallel that the other designs do not possess. The adjustable parallel is made of close-grained, seasoned cast iron. After initial casting and rough machining, the members of the parallel are subject to an aging or seasoning process that can either be natural or artificial. When complete, the member will have a hardness value of no less than 87 RB on the Rockwell B scale. The final machining on the adjustable parallel shall have a finish of 16 μin Ra on the working surfaces and a 32 μin Ra on the sides. The sides of the adjustable parallel are not designed to be working surfaces.

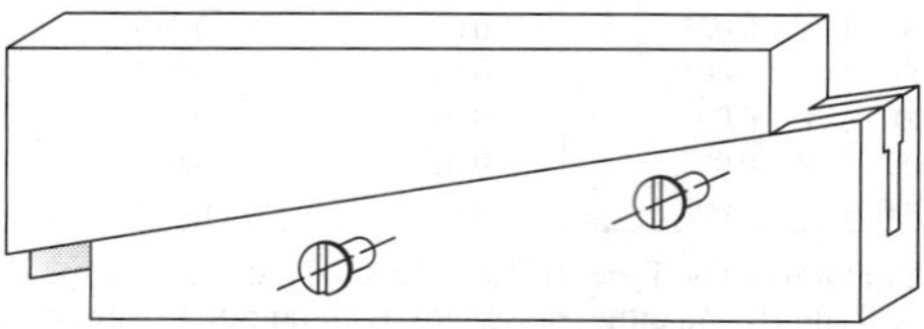

Fig. 4. Type IV, Adjustable Parallel

The adjustable parallel is made of two pieces. One piece is a fixed member machined with a dovetail slot, and the other a sliding member machined with a dovetail that fits into the slot and slides smoothly. It is this feature that gives the adjustable parallel the ability to be adjusted to any height within the range specified. The width of the bearing surface between the two members will be $\frac{1}{2}$ the total thickness of the parallel. The two members are held in place and fixed by a locking screw arrangement that secures the two members and prevents them from moving or slipping after they have been set. Adjustable parallels are supplied in either individual pieces or in sets that cover a wide range. The adjustable parallel will conform to the tolerances defined in Table 5 and will appear similar to Fig. 4. A bilateral tolerance will be applied to the parallelism and flatness

**Table 5. Tolerance Limits for Type IV, Adjustable Parallel**

| Size (range) (in.) | Length (in.) | Thickness (in.) | Parallelism (in.) | Straightness (flatness) (in.) | Lock Screws |
|---|---|---|---|---|---|
| 0.375–0.500 | 1.750 | 9/32 | 0.0005 | 0.00005 | 1 |
| 0.500–0.6875 | 2.125 | 9/32 | 0.0005 | 0.00005 | 1 |
| 0.6875–0.9375 | 2.6875 | 9/32 | 0.0007 | 0.00007 | 1 |
| 0.9375–1.3125 | 3.5625 | 9/32 | 0.0007 | 0.00009 | 2 |
| 1.3125–1.750 | 4.1875 | 9/32 | 0.0009 | 0.00010 | 2 |
| 1.750–2.250 | 5.0625 | 9/32 | 0.0009 | 0.00012 | 2 |

**Granite Parallel, High-Precision.—**The granite parallel is designed for the very high precision applications where thermal or magnetic properties must be taken into consideration. The granite parallel is made with a much tighter tolerance than the cast-iron or steel parallels. The configuration is basically the same as the Type I solid parallel; the only difference is in the material and the tolerance limits. The granite parallel is supplied in either individual members or matched sets. Granite parallels must meet the tolerances defined in the table below and will appear similar to Fig. 1.

**Table 6. Tolerance Limits, Granite Parallel**

| Size (in.) | Grade AA Flatness and Parallelism | Grade A Flatness and Parallelism | Grade B Flatness and Parallelism |
|---|---|---|---|
| 0.5 × 1 × 6 | 0.000025 | 0.00005 | 0.00010 |
| 0.750 × 1 × 6 | 0.00003 | 0.00006 | N/A |
| 0.750 × 1.5 × 9 | 0.00004 | 0.00008 | 0.00010 |
| 1 × 2 × 12 | 0.00006 | 0.00010 | N/A |
| 1.5 × 3 × 18 | 0.00015 | 0.00030 | N/A |
| 2 × 4 × 24 | 0.00020 | 0.00040 | N/A |
| 1.5 × 4 × 30 | 0.00025 | 0.00040 | N/A |

**Calibration, Precision Parallel Bars, Naval Air Systems and Air Force Metrology.—** It is commonly understood that wear and naturally occurring damage will affect the performance of all precision measuring equipment. Identifying an out of tolerance condition before it can impact a critical measurement is the primary goal of calibration. Parallel bars are no exception to this rule. A precision-ground steel or cast-iron parallel bar may become worn or even deformed from daily wear or excessive forces applied. A granite parallel can also show evidence of wear, and, due to the extremely close tolerances applied to granite parallel bars, the monitoring of these instruments is even more critical. It is not extremely difficult to perform the calibration of the parallel bar, and it can be achieved in a relatively short period of time in a temperature-controlled environment on a clean surface plate with an electronic indicator, a height transfer standard, and a few gage blocks. The features that will be observed are flatness, parallelism, and height and width of matched pairs. The exact height and width of individual parallel bars is not a critical feature, but in this status they can only be used independently.

As an additional note, a Pratt & Whitney Supermicrometer can be used to measure the exact height and width of a set of matched parallel bars, but, in the absence of this instrument, the same results can be achieved with the instruments listed above. The uncertainty of the measurement is effectively the same, and the confidence in the results is just as high with either method.

**Flatness (Straightness) Calibration Method.—** In any calibration, the first steps are to ensure a clean working surface and a proper setup of the standards and measuring instruments. To that end, the first step is to clean the surface plate thoroughly with an approved surface plate cleaner and place the electronic indicator next to the working surface. The indicator should be powered up and allowed to warm up. The surface plate should be of an accuracy grade to provide a surface flatness deviation no greater than 0.00005 inch within the working surface that will be used. The *UUT* or *Unit Under Test* should be cleaned and, as a preliminary step to calibration, deburred with a clean ultra-fine hard Arkansas stone or gage block deburring stone. Only the edges should be stoned as this is where most burrs will occur. However, all edges, corners and surfaces should be observed to ensure that no damage or burring is present before proceeding.

For precision cast-iron or steel parallel bars, the UUT will be set up on two gage blocks of the same size and accuracy grade. A 2.000 inch gage block is recommended as a minimum. This will allow access to both sides of the parallel bar when taking data to determine the flatness of each side. It is important that the gage blocks are placed at the correct points to support the UUT without any sagging effect on the bar. These are called *airy points* and are calculated by a simple formula:

Airy point separation distance = $0.554L$, where $L$ is the length of the UUT.

*Example:* For a 12.00 inch UUT, the gage blocks should be placed 0.554 × 12.00 = 6.650 inches apart and located (12.00 – 6.650)/2 = 2.675 inches from each end, as shown in Fig. 5. This distance should be very close to the calculated distance, but it does not have to be exact. Establishing the distance using a machinists' rule within 0.100 inch is sufficient.

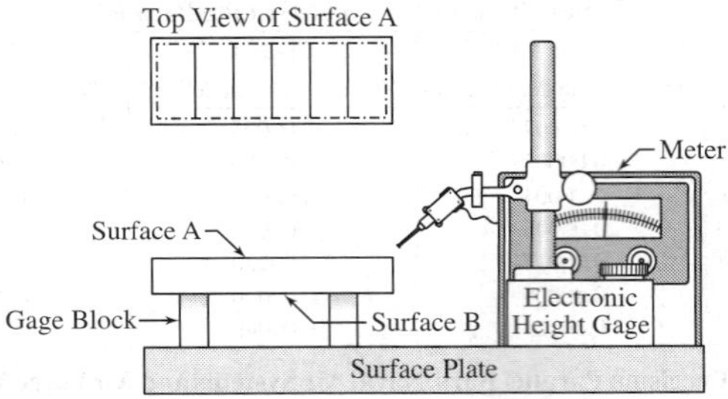


Fig. 5. Parallel Bar Flatness Setup

The UUT will be marked at five points equally spaced along the distance of each surface. The marks should be done with a felt-tipped marker or other suitable method. By general convention, the largest parallel surfaces are designated A and B. The two smaller surfaces are C and D. The markings A1 through D1 and A2 through D2 should be aligned with one another. A datasheet similar to the one shown in Table 7 should be created to collect the data and aid in the calculations of the deviations.

With the UUT placed on the gage blocks as shown, configure the electronic indicator to measure "over" and set the indicator amplifier range to ±0.0002 with a resolution of 0.00001 inch. It is recommended, however, to begin the readings at a slightly higher range and dial the scale in to a more sensitive resolution to determine the magnitude of the error. This will keep the readings on scale. Take the final readings in the highest sensitivity scale.

Place the indicator contact point on the UUT at the A1 position and zero the amplifier. All readings will have the x1 position as the zero point (A1, B1, C1, D1). All deviations will be from this point.

Sweep the surface of the UUT, recording on the datasheet the deviation as indicated at each of the calibration points. Ensure that each point does not exceed the documented tolerance for the type, style, grade, and size of the UUT.

When Side A has been scanned, adjust the electronic indicator contact point for "Under" measurements, place the contact point at the B1 calibration point and zero the indicator amplifier. This measurement is taken from the underside of the UUT, and the data collected will be used to calculate the parallelism of the two sides A to B. Once again, sweep the surface and record the deviations of the surface at each of the designated calibration points. Verify that all readings for sides A and B are within the tolerance limits assigned to the UUT.

Parallelism will be calculated from the sum of the deviations of each side. Calculate the sum of the deviations at each calibration point for surfaces A and B. The sum of the deviations will reveal the parallelism of the two surfaces. Verify that the values calculated are within the assigned tolerances for the UUT.

Carefully move the UUT to the adjacent sides (C and D) with side C up and repeat the sequence to determine the flatness and parallelism for sides C and D. Once complete and the values have been determined to be within the assigned tolerance limits, remove the UUT from the gage blocks and place on the surface pate. If this is a matched set of parallel bars, place the second member on the gage blocks as for the first member and complete the sequence for the second member.

**Table 7. Calibration of Parallel Bar and Matched Parallel Bar Sets**

| Straightness | | | | |
|---|---|---|---|---|
| Surface A | Nominal | As Found | As Left | Tolerance |
| Zero Setting (Point 1) | | | | |
| Cal Point 2 | | | | |
| Cal Point 3 | | | | |
| Cal Point 4 | | | | |
| Cal Point 5 | | | | |
| Surface B (under) | Nominal | As Found | As Left | Tolerance |
| Zero Setting (Point 1) | | | | |
| Cal Point 2 | | | | |
| Cal Point 3 | | | | |
| Cal Point 4 | | | | |
| Cal Point 5 | | | | |
| Surface C | Nominal | As Found | As Left | Tolerance |
| Zero Setting (Point 1) | | | | |
| Cal Point 2 | | | | |
| Cal Point 3 | | | | |
| Cal Point 4 | | | | |
| Cal Point 5 | | | | |
| Surface D (under) | Nominal | As Found | As Left | Tolerance |
| Zero Setting (Point 1) | | | | |
| Cal Point 2 | | | | |
| Cal Point 3 | | | | |
| Cal Point 4 | | | | |
| Cal Point 5 | | | | |
| Parallelism Calculations | | | | |
| Sum Surfaces A + B | Nominal | As Found | As Left | Tolerance |
| A2 + B2 | | | | |
| A3 + B3 | | | | |
| A4 + B4 | | | | |
| A5 + B5 | | | | |
| Sum Surfaces C + D | Nominal | As Found | As Left | Tolerance |
| C2 + D2 | | | | |
| C3 + D3 | | | | |
| C4 + D4 | | | | |
| C5 + D5 | | | | |

**Height and Width, Matched Sets.—**The calibration of height and width of matched sets of steel or cast-iron parallel bars is a relatively simple process. Place the two parallel bars side by side on the surface plate in the same orientation, that is, with the same side up and the predefined calibration points adjacent to each other. One parallel bar will be designated as "A" and the other as "B".

Position the height transfer gage next to the parallel bars and dial the transfer gage to the dimension of the side under test. Zero the electronic indicator on the transfer gage. Pull back the indicator probe off the land of the transfer gage and then replace it onto the land and ensure the zero setting repeats. If it does not repeat, re-zero the indicator amplifier and repeat the reading. Verify the zero setting in this manner until the reading repeats three times in a row with no variation. Then move the indicator probe to the surface of the "A" parallel bar and observe the deviation from zero reading. Sweep the entire surface and verify that all readings fall within the tolerance limits defined for the type and grade under test. Return the indicator probe to the transfer gage land and confirm that the zero point has not shifted before accepting the readings observed. If the zero point has shifted, it will be necessary to repeat the zero point setup and take the readings from the parallel bar again. If the readings are acceptable, proceed to the "B" parallel bar and verify that the deviation from zero reading is within the tolerance limits defined and the deviation from the "A" block is not beyond the limits for the type and grade under test.

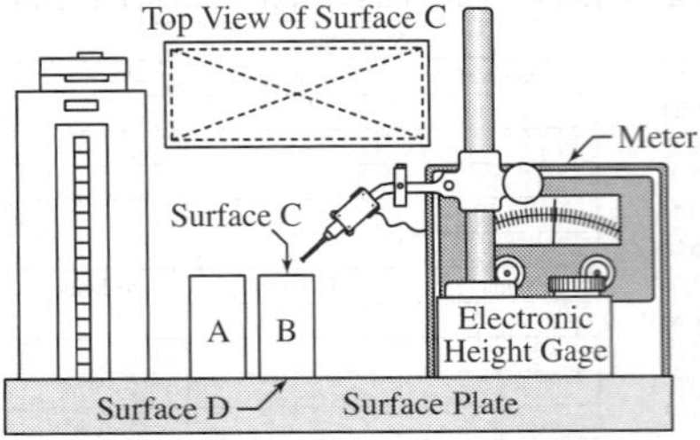


Fig. 6. Parallel Bar Matched Set, Height and Width Calibration

Size limitations for granite parallel bars are not as rigid as those for individual bars, acceptable limits for size being ±$\frac{1}{32}$ inch. However, the tolerance for size of matched sets is the same as that for steel and cast iron. The calibration procedure for granite parallel bars is the same as that for steel and cast iron.

As a final note, it should be mentioned that a Pratt & Whitney Supermicrometer can also be used to calibrate the height and width of a parallel bar or a matched set of parallel bars. The Supermicrometer is set up using a gage block or gage block stack to the precise dimension of either the height or width of the parallel bar and zeroed at that point. The gage block is then removed, and the parallel bar is set on the elevating table between the anvils of the Supermicrometer and supported by two flatted rounds. The parallel bar is then measured at the first calibration point and the reading compared to the tolerance limits to assure compliance. The bar is then measured at each remaining calibration point, verifying at each point the compliance to the tolerance limits. This same process is then completed for the "B" parallel bar in the matched set, and the values obtained in the measurement of the "A" bar are compared to assure that deviations in the matched sizes are within the tolerance limits. This method is very good for smaller parallel bar matched sets if a Supermicrometer is available.

## Right-Angle Plates

Right-angle plates or knees, as they are more generally referred to, are divided into six different types as defined in Federal Specification GGG-P-441A. Right-angle plates are used in both machining applications and the setup and inspection of machined parts and assemblies. The plate material is usually high grade, fine-grained, controlled-process iron or iron alloy castings. These castings are always of uniform quality, free from blow holes, porosity or other material inconsistencies, and defects. All castings after being rough cut are stress-relieved to assure dimensional stability and become more stable as they age. With proper care and surface maintenance, angle plates can perform for many years. Fine angle plates over 50 years old are still in service.

**Right-Angle Plate Grades.**—Most right-angle plates are manufactured to meet the guidelines and tolerances laid out in the GGG-P-441A or IS 2554:1971 (India Standards Bureau). While these two governing documents may differ slightly, they both achieve the goal of establishing dimensional guidelines and tolerances that enable the manufacturer to produce a dependably consistent product with values and accuracies that customers require to maintain a high degree of measurement confidence in their quality systems.

Angle plates come in grades according to the surface finish that is applied and the dimensional accuracies that they are manufactured to. The assigned grade is in direct correlation to the quality and accuracy of the angle plate. Federal Specification GGG-P-441A specifies three *Grades* of surface finish applied to right-angle plates. In addition, there are six distinct *Types* of angle plates that have unique properties according to their specifically designed target uses in the manufacturing process.

*Grade A* angle plates are the highest grade and are used in precise applications where the smallest deviation from square (90 degrees) is required. These plates are used in various industries for clamping and holding work in a vertical position. The scraped surface is beneficial for supporting work in a very precise vertical plane and facilitating the dispersion of lubricants and coolants during the manufacturing process. Grade A plates are required to have working surfaces that are fine precision-scraped with relief spots to prevent sticking of gage blocks or precision-lapped instruments. These relief spots allow the surface to be controlled to a high degree of flatness. Each square inch of surface on a Grade A plate has 15 to 18 spotting cavities, from 0.0002 inch to 0.0005 inch deep, and a bearing surface of 20 to 40 percent. The bearing area does not deviate from the mean plane by more than 0.0001 inch in over 24 inches, and no more than 0.0002 inch in up to 60 inches, as per Table 1. No square inch may vary from the adjacent area by more than 0.0001 inch per square foot. The adjacent working surfaces on any Grade A angle plate shall not vary from square by more than 0.0001 inch in every 8 inches. This equates to 2.5 arc seconds. These angle plates make excellent fixturing for high precision shaping, milling, grinding, drilling, or boring operations.

**Table 1. Maximum Permissible Deviation of Bearing Area from Main Plane (*GGG-P-441A*)**

<table>
<tr><th colspan="2">Maximum dimension of working surface, inches</th><th colspan="2">Maximum permissible deviation of bearing areas from mean plane, inch</th></tr>
<tr><th>Above</th><th>To and including</th><th>Grades A and B</th><th>Grade C</th></tr>
<tr><td>2½</td><td>12</td><td rowspan="2">0.0001</td><td rowspan="2">0.0002</td></tr>
<tr><td>12</td><td>24</td></tr>
<tr><td>24</td><td>36</td><td rowspan="3">0.0002</td><td>0.0004</td></tr>
<tr><td>36</td><td>48</td><td>0.0006</td></tr>
<tr><td>48</td><td>60</td><td>0.0008</td></tr>
</table>

*Grade B* plates are made to meet requirements of precision inspection and calibration operations. Although the style of the plate can accommodate uses in any of the manufacturing practices, the precision-ground surface finish lends itself to inspection applications on the surface plate and use in calibration laboratories, high precision inspection stations, or situations where a fine precision-ground finish is desired and the highest degree of accuracy necessary. The Grade B angle plate has a working surface that is precision ground to a roughness average value of not more that 16 microinches for plates less than 64 square inches and 32 microinches for larger plates. The precision-ground surface of the Grade B plate is well suited to the applications involved in the inspection process of high accuracy, tight tolerance, precision-made parts. The fine-ground finish offers a precise surface that does not deviate from the mean plane by more than 0.0001 inch in over 24 inches, and no more than 0.0002 inch in up to 60 inches, as per Table 1. No adjacent areas shall vary with each other by more than 0.0001 inch per square foot. The aspect that is most critical about the angle plate is the control of squareness and the ability to pass on the accuracy to the parts being machined or inspected. All adjacent working surfaces will not vary from square by more than 0.0001 inch in every 8 inches. This high degree of accuracy in the Grade A and Grade B angle plates provides the measurement assurance necessary to inspect precision parts and calibrate precision measuring instruments.

*Grade C* angle plates are the workhorses of the plate grades. They are designed to have a smooth machined finish and controlled to not exceed a surface roughness of 32 r.m.s. The bearing area of the working surface shall not exceed a deviation from the mean plane any greater than between 0.0002 to 0.0008 inch depending on the size of the bearing area, as per Table 1. Furthermore, no adjacent square foot may deviate from another by more than

0.0003 inch per square foot. This ensures that the bearing surface is tightly controlled in the graduation of allowable error and will not deviate from the mean plane beyond the allowable limits. Deviation beyond the allowable graduation limits will introduce an error in the surface flatness beyond the functional limits. This flatness deviation control enables the surface of the angle plate to support the workpiece in such a manner as not to introduce errors due to nonuniform support beneath the workpiece or squareness errors from the adjacent sides.

*IS 2554:1971 Grades:* Standard sizes and accuracy of angle plates specified by IS 2554:1971 (India Standards Bureau) are given in Table 2a and Table 2b.

**Table 2a. Sizes and Accuracy of Slotted Angle Plates Metric Sizes (*IS 2554:1971*)**

| Size (mm) $L \times B \times H$ | Accuracy in Microns, µm | | | |
|---|---|---|---|---|
| | Flatness of Working Faces | Squareness of Working Faces over Dimension $H$ | Parallelism of Opposite Faces & Edges over their Total Length | Squareness of End Faces with Respect to Exterior Faces as Measured over Dimension $L$ |
| **Grade 1** | | | | |
| 150 × 100 × 125 | 5 | 10 | 13 | 13 |
| 150 × 150 × 150 | 5 | 10 | 13 | 13 |
| 175 × 100 × 125 | 5 | 13 | 15 | 15 |
| 200 × 150 × 125 | 8 | 15 | 18 | 18 |
| 250 × 150 × 175 | 8 | 15 | 18 | 18 |
| 300 × 200 × 225 | 8 | 18 | 20 | 20 |
| 300 × 300 × 300 | 8 | 18 | 20 | 20 |
| 350 × 200 × 250 | 8 | 18 | 20 | 20 |
| 450 × 300 × 350 | 10 | 18 | 20 | 20 |
| 400 × 400 × 400 | 10 | 18 | 20 | 20 |
| 600 × 400 × 450 | 10 | 20 | 23 | 23 |
| **Grade 2** | | | | |
| 700 × 420 × 700 | 50 | 140 | 140 | 140 |
| 900 × 600 × 700 | | | | |
| 1000 × 700 × 500 | | | | |

**Table 2b. Sizes and Accuracy of Slotted Angle Plates Inch Sizes (*IS 2554:1971*)**

| Size (in.) $L \times B \times H$ | Accuracy in Inches | | | |
|---|---|---|---|---|
| | Flatness of Working Faces | Squareness of Working Faces over Dimension H | Parallelism of Opposite Faces & Edges over their Total Length | Squareness of End Faces with Respect to Exterior Faces as Measured over Dimension L |
| **Grade 1** | | | | |
| 6 × 4 × 5 | 0.0002 | 0.0004 | 0.0005 | 0.0005 |
| 6 × 6 × 6 | 0.0002 | 0.0004 | 0.0005 | 0.0005 |
| 7 × 6 × 5 | 0.0002 | 0.0005 | 0.0006 | 0.0006 |
| 8 × 6 × 5 | 0.0003 | 0.0006 | 0.0007 | 0.0007 |
| 10 × 6 × 7 | 0.0003 | 0.0006 | 0.0007 | 0.0007 |
| 12 × 8 × 9 | 0.0003 | 0.0007 | 0.0008 | 0.0008 |
| 12 × 12 × 12 | 0.0003 | 0.0007 | 0.0008 | 0.0008 |
| 14 × 8 × 10 | 0.0003 | 0.0007 | 0.0008 | 0.0008 |
| 18 × 12 × 14 | 0.0004 | 0.0007 | 0.0008 | 0.0008 |
| 16 × 16 × 16 | 0.0004 | 0.0007 | 0.0008 | 0.0008 |
| 24 × 16 × 18 | 0.0004 | 0.0008 | 0.0009 | 0.0009 |
| **Grade 2** | | | | |
| 28 × 16 × 28 | 0.002 | 0.0055 | 0.0055 | 0.0055 |
| 36 × 24 × 28 | | | | |
| 24 × 16 × 18 | | | | |

**Angle Plate Types.—**The right-angle plate is further divided into *Types* and *Classes*, each type having unique characteristics and uses. Types I, II, and III angle plates or knees are intended to be used by craftsmen for ordinary machine shop operations; Types IV, V, and VI are intended to be used principally by toolmakers and for precision type work.

*Type I: Class 1-Plain Right-Angle Plate and Class 2-Slotted Right-Angle Plate:* The Type I Class 1 angle plate has two members with the outside surfaces at right angles to each other. Both the working surfaces and the ends are machined square to within the tolerance limits assigned to their grade, (Fig. 1a). Type I Class 2 plates are similar to Class 1 with the vertical and horizontal slots provided on one or both faces, (Fig. 1b).

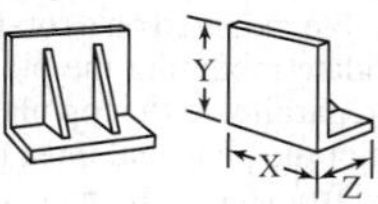

Fig. 1a. Type 1 Class 1-Plain

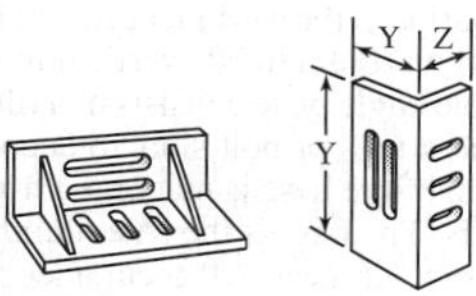

Fig. 1b. Type 1 Class 2-Slotted

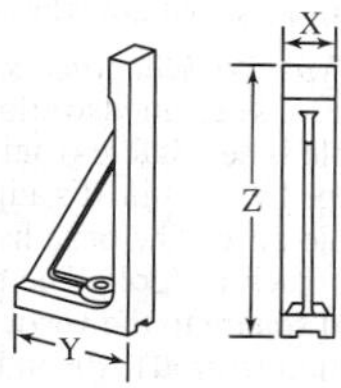

Fig. 1c. Type 1 Class 3, Measuring-Plain

All the right-angle plates of the various types have ribbing of sufficient thickness to assure maximum support for flatness, squareness, and stability. Class 1 plates of approximately 10 to 12 inches will have at least 2 ribs, those of 16 to 24 inches at least 3 ribs, and those of 36 inches and above a minimum of 4 ribs in their design.

Plates weighing over 75 pounds are designed with lifting holes in the ribbing or threaded holes for the placement of lifting eyes.

*Type I, Class 3, Measuring Plane Right-Angle Plate:* The Type I Class 3 angle plate, Fig. 1c, is designed to establish a measuring plane for production or inspection functions. It has two outside working surfaces that are at right angles to each other. The working surfaces are finished according to the grade specified and are the only two faces designed for functional operation. Unless specifically ordered otherwise, the sides and ends will be finish ground only, and these surfaces are not to be used to determine or establish squareness. The rib provides support between the ends of the plate and is cored out to provide a good hand gripping surface without sacrificing the stabilizing effect of the rib. The short base member "Y" is provided with a suitable hole so the plate may be bolted to the table of a machine or a tapped hole in a surface plate.

*Type II, Inside Right-Angle Plate:* The Type II right-angle plate is an inside right angle that is constructed with the same adherence to squareness and finish as the outside angle plates, Fig. 2. The inside angle plate is an inverted T-shaped casting formed by a base and a perpendicular member of approximately the same thickness. One side of the perpendicular, the adjacent top side of the base, and the bottom of the base are the three working surfaces and are all finished in accordance to the grade specified. The opposite ends and opposite sides are machined parallel and square to their respective working surfaces.

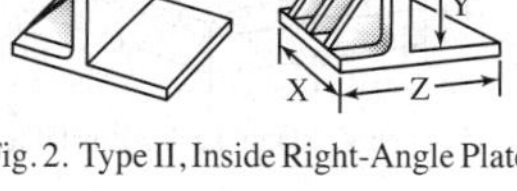

Fig. 2. Type II, Inside Right-Angle Plate

Fig. 3. Type III, Universal Right-Angle Plates

*Type III, Universal Right-Angle Plate:* By far the most versatile and useful right-angle plate in the machine shop is the Type III universal right-angle plate, Fig. 3. The Type III has two outside working surfaces at right angles and all adjacent sides are ground and finished to the same tolerance assigned to the main working surfaces. This provides a full range of positioning capabilities and makes the universal right angle just as accurate on its side as it is resting on its main working surface. All opposite sides are parallel within 0.0002 inch per foot. The universal right angle has a rib located between the ends that is square in configuration and provides support and stability to the main working surfaces while providing additional working surfaces that can be utilized for work support due to the accuracy assigned to the entire plate. The rib is hollowed out to provide good hand grip without sacrificing the stabilizing characteristics.

*Type IV: Machinist's Adjustable Angle Plate:* The Type IV adjustable angle plate, Fig. 4, is designed to allow rotation of the workpiece through 360°. The workpiece support table is adjustable from zero horizontal to 90° vertical in 10° increments. Basically, the Type IV machinist's adjustable angle plate consists of a tilting table mounted on a rotary table base. The base has heavy lugs or bolt slots to accommodate mounting the plate on machine tools. The bottom of the base is machined true flat, parallel to the top plate and square to the vertical axis. The base is the true foundation of the plate and must be manufactured to the tightest of tolerances. All accuracies extend from the craftsmanship that is put into the base work of the Type IV plate. The top of the rotary base is graduated in one degree increments for the full 360° and is marked at least every 10° for a minimum of plus and minus 90° from zero (0°). Most plates are marked around the full 360° in 10° major divisions with 1× minor divisions, permanently engraved or etched on a chrome or brightly finished face for ease of reading and maintenance. The standard rotary accuracy to which all Type IV plates are made is ±2.5 minutes, noncumulative. This tolerance maintains the accuracy ratio necessary to maintain a good measurement assurance level in your quality system.

The tilting table consists of two plates joined together at one end by a precision hinge and provided with an adjustable and positive locking mechanism. The bottom plate is attached to the base at the vertical axis. This plate will swivel through 360° freely and will also have a positive locking mechanism. The bottom of the plate is marked with a permanent line for accurate positioning in relation to the graduations on the base. The top plate is made in one piece as a casting and has T-slots milled in from the solid plate, as shown in Fig. 4.

The top plate is adjustable up to 90° and the positive locking mechanism is robust enough to secure the workpiece into position during milling, drilling, or boring operations without movement. The top plate is made with a protractor integrated into the design that is marked from 0° (horizontal) to 90° (vertical) with major scale indications at every 10°. Minor scale graduations may be indicated at 1° increments. The noncumulative error between graduations of the protractor is no greater than ±2.5 minutes, which allows for a high level of confidence in the measurements and operations made with this plate.

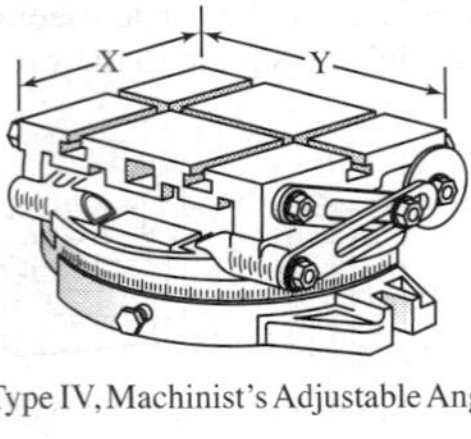


Fig. 4. Type IV, Machinist's Adjustable Angle Plate

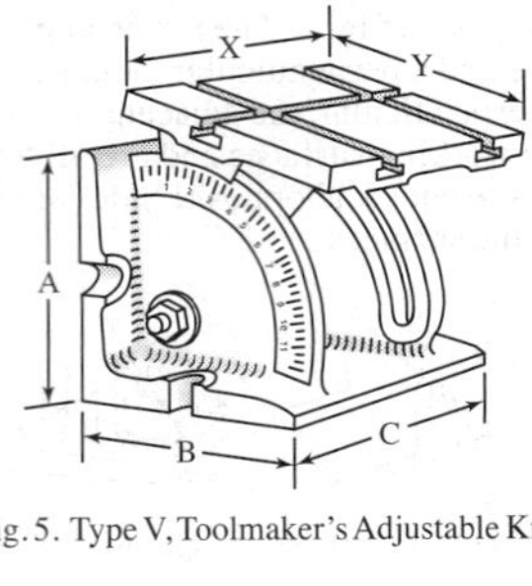


Fig. 5. Type V, Toolmaker's Adjustable Knee

*Type V, Toolmaker's Adjustable Angle Plate:* The toolmaker's adjustable knee or adjustable angle plate extends the capabilities of the right-angle plate and allows for precise adjustment of the support table or bearing surface from 0° (horizontal) through 90° (vertical), Fig. 5. The same tolerances in regard to flatness and squareness apply to the bearing surface and the base.

The toolmaker's adjustable knee is constructed of a plate, finished to the tolerances determined by the class specified, and a right-angle iron with two ribs. The right-angle iron and the plate are joined together by a precision hinge that has the capability to be secured at any position within the arc from 0° to 90°. The two ribs are of sufficient thickness to provide stability and support enough to withstand the forces applied during boring, milling, drilling, or grinding operations. The locking mechanism must be capable of maintaining its locked position under the same circumstances. The table is provided with T-slots for the mounting of the workpiece.

The quadrant scale of the adjustable angle plate is divided into 90 divisions and a vernier scale. The vernier scale is graduated into 5 minute divisions from 0 to 60 minutes on either side of the zero index. The 0 (zero index) and every 15 minute index thereafter on both sides of 0 is clearly marked with the corresponding number, and the 5 minute divisions are clearly readable. The quadrant scale has 1 degree divisions from 0° to 90°. The zero index and every 10° index following will be clearly marked with the corresponding number. A tolerance not to exceed plus or minus 2 minutes, noncumulative, is given for the error between any two graduations.

Due to its specialized nature, the toolmaker's adjustable angle plate usually is only manufactured in one size according to the GGG-P-441A Federal Specification.

*Type VI: Toolmaker's Non-Adjustable Angle Plate:* The Type VI toolmaker's nonadjustable right-angle plate is different from the other types in that it has two members. These members have two working surfaces at right angles to each other as well as a finished pad on the back of both working surfaces. The design is of sufficient thickness along with the ribbing to ensure rigidity and stability in all operations and applications.

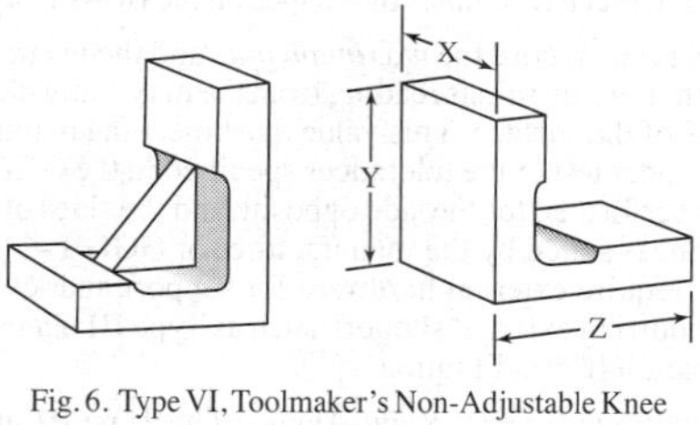


Fig. 6. Type VI, Toolmaker's Non-Adjustable Knee

Both front sides and the back side pads are considered functional working surfaces and will be manufactured according to the grade specified. The working surfaces, the opposite sides, and the opposite ends are all machined parallel to each other and square to their respective working surfaces. All machined surfaces will meet the tolerance limits assigned to the specified grade for finish, flatness, squareness, and parallelism. The Type VI toolmaker's nonadjustable right-angle plate will appear similar to Fig. 6.

**Calibration and Maintenance of Right-Angle Plates.**—In many machine shops and quality assurance programs, right-angle plates are not calibrated and are considered "Calibration Not Required" or "Reference Only" instruments that are not placed into the calibration recall system. Inevitably, this practice may lead to unknown error in a manufacturing operation or inspection procedure. However, calibration of a right-angle plate may be an investment in quality if it becomes evident that an angle plate that has worn out of squareness or flatness is introducing error in a machining operation or inspection process. There are several calibration procedures and methods for verifying the accuracy of the right-angle plate.

*Flatness Calibration:* The first feature that should be calibrated is the flatness of the working surfaces. The tolerances for the calibration of any individual right-angle plate is defined by the manufacturer and in their absence the GGG-P-441A shall provide an acceptable calibration tolerance. For Type I, II and VI angle plates, the setup for the flatness calibration will appear similar to Fig. 7.

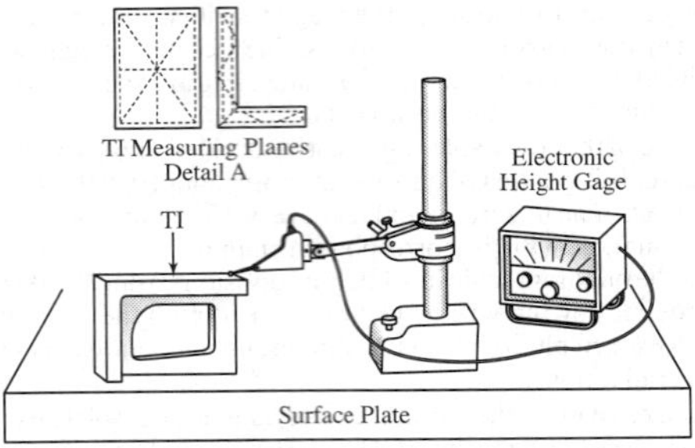


Fig. 7. Calibration of Surface Flatness

Place the angle plate (the *Test Instrument*, or *TI*) on the leveling plate, as shown, and use a jack stand or similar device to support the TI in position. A bubble level can assist in leveling the plate. When the plate has been leveled, use the electronic indicator to measure the surface following the pattern shown in the Detail A of Fig. 7. This pattern is generally called the Union Jack in that it resembles the stripes on the British flag.

Observe the readings and record the *maximum plus* and the *maximum minus* readings overall. Subtract the maximum minus reading from the maximum plus reading; the result is the overall flatness of the surface. This value must meet the manufacturer's tolerance limits for the model under test or the tolerances specified in the GGG-P-441A Standard. The same process is performed for the side opposite and the sides of the TI. All readings must meet the tolerances stated by the manufacturer or GGG-P-441A. Flatness is only checked on TIs that require external hardware for support and leveling on the surface plate. On units not requiring external support, such as Type III plates, flatness is checked as a function of the parallelism calibration.

*Parallelism Calibration (Type III Right-Angle):* The Type III universal right-angle plate has a parallelism tolerance due to the unique design of this unit. As discussed previously, the Type III is finished on all sides and is square and parallel to all working surfaces and sides opposite.

| General Parallelism Tolerances | |
|---|---|
| Size Range | Total Indicator Reading |
| > 2.5 to 24 inches | 0.0004 inch TIR |
| > 24 to 36 inches | 0.0008 inch TIR |
| > 36 to 48 inches | 0.0012 inch TIR |
| > 48 to 60 inches | 0.0016 inch TIR |

Before beginning the calibration process, attention must be paid to the surface condition to be certain that no burrs or damage to the edges or surface is present that might affect the outcome of the measurements. A super fine-grained "hard Arkansas stone" may be used to remove any burrs on the edges or surfaces prior to beginning the calibration process. After the right-angle plate is clean and free from burrs, the TI is placed on the surface plate as shown in Fig. 8. There is no need for a leveling plate as the side opposite is finished and parallel to the working surface, and the parallelism feature is what is being indicated.

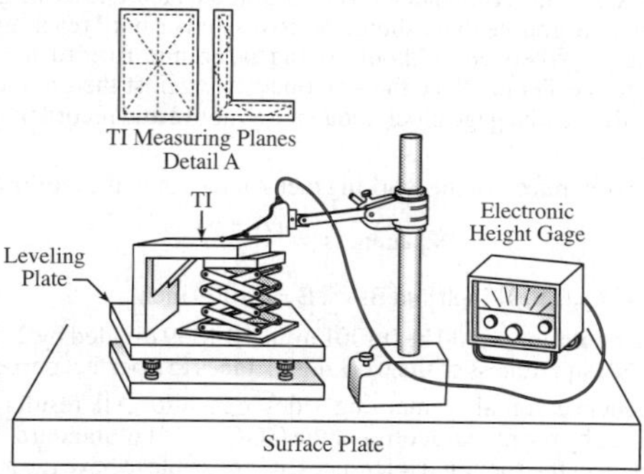


Fig. 8. Measuring Flatness/Parallelism of Type III Right-Angle Plate

Using the electronic indicator as before, indicate the surface in the Union Jack pattern and record the maximum plus and the maximum minus readings overall. In general, it is a good idea to zero the electronic indicator in the center of the Union Jack and take the readings from that point. This provides an easily repeated starting point and an overall numerical picture of the parallelism of the surface. As before, subtract the minimum reading from the maximum; the result must meet the tolerance limits specified by the manufacturer or GGG-P-441A. Repeat this process for the remaining sides. All results must meet the stated tolerances.

*Squareness Calibration (Preferred Method):* The simplest method that can be used to verify the squareness of the Type III universal right-angle plate, as well as the squareness of the other types, is quite sufficient when done properly to meet the accuracy ratio that is required for a high confidence level measurement. This method provides quantifiable data that can be assigned an uncertainty should the *Quality System* determine this necessary. This method provides a good measurement to verify the squareness value and identify any out of tolerance conditions before the error is passed on into the manufacturing or inspection process.

Place the angle plate (TI) on the surface plate with two 0.1005-inch gage blocks between the face of the angle plate and the face of the granite angle block, as shown in Fig. 9a.

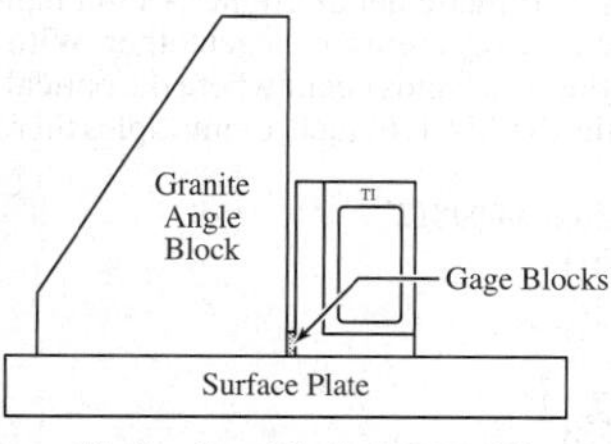


Fig. 9a. Gage Blocks Between TI and Granite Angle Block

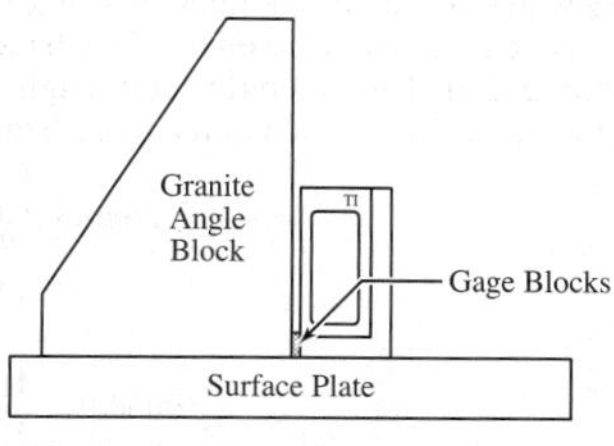


Fig. 9b. Gage Blocks Between Rotated TI and Granite Angle Block

Then, beginning with a 0.1000-inch gage block, gently insert the gage block between the TI and the granite angle block at the top edge of the TI. Continue placing increasingly larger gage blocks between the TI and the granite angle block until the next larger gage block cannot be inserted without moving the TI. A gentle force should be applied to make sure the gage block is in direct contact with the granite surface, and as the gage block slides between the TI and the granite there should be no resistance until reaching the gage block size that is too large to go between without moving the granite. Record the size of this block as Block A. Rotate the TI and place the side opposite against the granite angle block as shown in Fig. 9b. Repeat the gage block sequence as above and record the size of the gage block as Block B.

To calculate the squareness of the working faces of the rotated TI follow the formula:

$$\text{Squareness} = \frac{|A - B|}{2}$$

*Example:* Block A = 0.1001 inch and Block B = 0.1009 inch.

*Solution:* The calculation would be 0.1001 minus 0.1009 divided by 2. The result being (–)0.0004 in overall squareness of the working surfaces to a perfect perpendicular plane.

Repeat the sequence for all remaining sides opposite. All results must meet the tolerances specified by the manufacturer or the GGG-P-441A Standard. If for any reason the results do not meet the specified tolerances, it is possible to have the right-angle plates refurbished at a fraction of the cost of purchasing a new one. It is also possible that the deviations may be acceptable to the quality system, and, if the results are discussed with the Quality Engineer, the decision may be made to accept the unit "As Is" and annotate the deviation for future reference. The unit may also be downgraded to a less critical operation and the previous application be taken over by a new unit of acceptable condition. These decisions must be made by the quality engineers or quality managers and follow the guidelines of the quality system in place.

## Measurements Using Light

**Measuring by Light-Wave Interference Bands.—**Surface variations as small as two millionths (0.000002) inch can be detected by light-wave interference methods, using an optical flat. An optical flat is a transparent block, usually of plate glass, clear fused quartz, or borosilicate glass, the faces of which are finished to extremely fine limits (of the order of 1 to 8 millionths [0.000001 to 0.000008] inch, depending on the application) for flatness. When an optical flat is placed on a "flat" surface, as shown in Fig. 1, any small departure from flatness will result in formation of a wedge-shaped layer of air between the work surface and the underside of the flat.

Light rays reflected from the work surface and underside of the flat either interfere with or reinforce each other. Interference of two reflections results when the air gap measures exactly half a wavelength of the light used and produces a dark band across the work surface when viewed perpendicularly, under monochromatic helium light. A light band is produced halfway between the dark bands when the rays reinforce each other. With the 0.0000232-inch wavelength helium light used, the dark bands occur where the optical flat and work surface are separated by 11.6 millionths (0.0000116) inch, or multiples thereof.

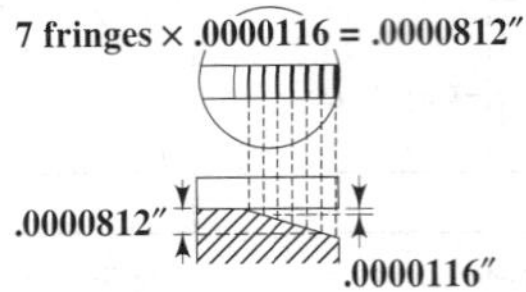


Fig. 1.

For instance, at a distance of seven dark bands from the point of contact, as shown in Fig. 1, the underface of the optical flat is separated from the work surface by a distance of $7 \times 0.0000116$ inch or 0.0000812 inch. The bands are separated more widely and the indications become increasingly distorted as the viewing angle departs from the perpendicular. If the bands appear straight, equally spaced and parallel with each other, the work surface is flat. Convex or concave surfaces cause the bands to curve correspondingly, and a cylindrical tendency in the work surface will produce unevenly spaced, straight bands.

See also *Interpreting Optical Flat Fringe Patterns* on page 781.

**Interferometer.**—The interferometer is an instrument of great precision for measuring exceedingly small movements, distances, or displacements, by means of the interference of two beams of light. Instruments of this type are used by physicists and by the makers of astronomical instruments requiring great accuracy. Prior to the introduction of the interferometer, the compound microscope had to be used in connection with very delicate measurements of length. The microscope, however, could not be used for objects smaller than one-half a wavelength of light. Two physicists (Professors Michelson and Morley) developed the *interferometer* for accomplishing in the laboratory what was beyond the range of the compound microscope. This instrument consisted principally of a system of optical mirrors arranged in such a way as to let the waves of light from a suitable source pass between and through them, the waves in the course of their travel being divided and reflected a certain number of times, thus making it possible to measure objects ten times smaller than was possible with the best compound microscope obtainable. Professor C.W. Chamberlain of Denison University invented another instrument known as the *compound interferometer,* which is much more sensitive than the one previously referred to; in fact, it is claimed that it will measure a distance as small as one twenty-millionth of an inch. These compound interferometers have been constructed in several different forms.

An important practical application of the interferometer is in measuring precision gages by a fundamental method of measurement. The use of this optical apparatus is a scientific undertaking, requiring considerable time and involving complex calculations. For this reason all commercial methods of checking accuracy must be comparative, and the taking of fundamental measurements is necessarily confined to the basic or primary standards, such as are used to a very limited extent for checking working masters, where the greatest possible degree of accuracy is required. The interferometer is used to assist in determining the number of light waves of known wavelength (or color) which at a given instant are between two planes coinciding with the opposite faces of a gage block or whatever part is to be measured. When this number is known, the thickness can be computed because the lengths of the light waves used have been determined with almost absolute precision. The light, therefore, becomes a scale with divisions — approximately two hundred-thousandths inch apart.

## MICROMETERS AND CALIPERS

The micrometer, sometimes called *micrometer caliper*, is a means of examining dimensional differences as small as one ten-thousandth of an inch (0.0001 inch) or, when using a metric micrometer, one thousandth of a millimeter (0.001 mm). Various types of micrometers and calipers are used in engineering labs and manufacturing operations for precise measurement of dimensions such as length, thickness, and diameter.

### Micrometers

**Classification and Federal Specifications.—**Micrometers are described in Federal Specification GGG-C-105 (1987) published by the US Federal Supply Service, General Services Administration (GSA). While this Federal Specification does not describe all commercially available product variations, it was created to cover those sold to the GSA, a major user of micrometers. Officially canceled in 1999, this document includes key data for each size and class of micrometer that still applies to current devices: flatness and parallelism of the measuring faces, permissible flexure of the frame, and potential error in anvil/spindle alignment and indicated measurement.

**Types, Classes, and Styles**

*Type I, Outside Micrometers* (page 767)
- *Class 1: Enameled Frame (I-Beam or Hollow Tubular)*
  - *Style A:* Solid Anvil
  - *Style B:* Interchangeable Anvil
  - *Style C:* Tubing, Ball Anvil
  - *Style D:* Tubing, Upright Anvil
- *Class 2: Finished Frame*
  - *Style A:* Solid Anvil
  - *Style B:* Paper Gage
  - *Style C:* Tubing, Ball Anvil
  - *Style D:* Thread

*Type II, Inside Micrometers* (page 769)
- *Class 1: With Jaws*
- *Class 2: Rod and Sleeve*
- *Class 3: Tubular*
  - *Style A:* Interchangeable Head
  - *Style B:* Fixed Head

*Type III, Micrometer Depth* (page 769)
- *Class 1: With Interchangeable Measuring Rods*
  - *Style A:* Full base
  - *Style B:* Half base

*Type IV, Micrometer Heads* (not covered in this *Handbook*)

**Manufacturers' Specifications and Key Features.—** Manufacturers' production tolerance tables and other data for both inch and metric micrometers can be found in ASME B89.1.13-2013 (R2022). While there are various kinds of micrometers, certain features apply to all.

*Resolution:* Micrometers using the inch system read directly to 0.001 inch and can be read to 0.0005 inch (or finer), with visual interpolation; if fitted with a vernier scale, an inch micrometer reads directly to 0.0001 inch. Micrometers using the metric system (SI) read directly to 0.01 mm (or finer), with visual interpolation; if fitted with a vernier scale, a metric micrometer reads directly to 0.001 mm.

*Identification:* All micrometers should be permanently and legibly marked with the measuring range, manufacturer's name or trademark, manufacturer's part number, and country of origin.

*Adjustments:* Each type, style, and class of micrometer provides a method for adjustment due to wear of the measuring faces and/or wear between the spindle screw and the nut within the barrel.

*Micrometer Screw:* The micrometer's accuracy depends on the precision of the threaded (screwed) section of the spindle and its mating parts. Spindles are precision ground from stabilized, high-grade tool steel or alloy steel with a hardness of not less than 62 on the Rockwell C scale. The exposed spindle section is usually mirror-finished for free-turning in the frame bearing, with no perceptible side play. All outside micrometers have some means of locking the spindle to prevent accidental movement.

*Measuring Faces:* Because of its greater resistance to wear, tungsten carbide is the preferred face material for the spindle and anvil of outside micrometers. The working surfaces are typically mirror-finished.

*Graduations:* Graduations on the barrel and thimble are etched, or otherwise incised for permanence, and usually are inked for enhanced contrast and readability.

**Type I, Outside Micrometers.—**In its traditional form (Fig. 1), the outside micrometer is used to measure dimensions of an object held between a fixed anvil and a screw-adjusted spindle. This device is functionally similar to a conventional screw clamp, but with two key differences. First, the micrometer spindle is threaded only on the thimble end; the unthreaded inner end of the spindle, about 1/4 inch in diameter, slides freely in the frame. Second, the spindle screw is of the highest quality and precisely ground.

In inch-measuring micrometers, the screw has 40 threads per inch at a pitch of 0.025 inch. The spindle screw runs inside the micrometer barrel, which is internally threaded to a precision comparable to the screw. Every turn of the screw, together with its attached thimble, advances or retracts the spindle exactly 0.025 inch per revolution.

Each of the 25 graduated divisions around the thimble represents a movement of the spindle relative to the anvil of just 0.001 inch, even though the thimble divisions are widely separated by as much as 1/10th inch. This makes the micrometer a 100:1 magnifier, displaying infinitesimal spindle movements in highly readable increments. In some micrometers, the resolution is increased from 0.001 inch to 0.0001 inch (another factor of 10) by a vernier scale on the barrel.

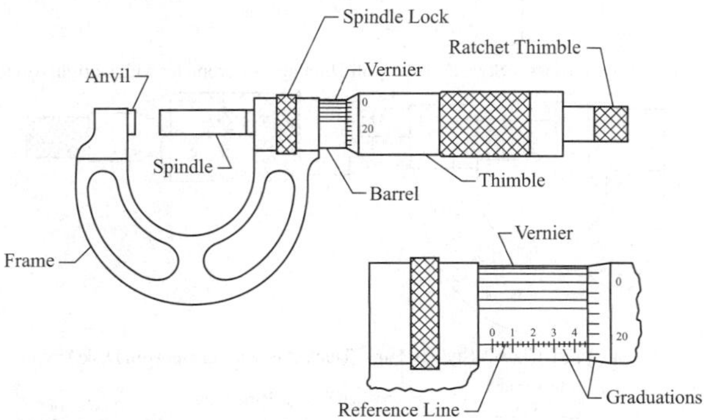


Fig. 1. Design Features of Type I, Class 1, Style A: Outside Micrometer

Outside micrometers are available in frame sizes ranging from 1 inch to at least 24 inch, but in all nondigital versions, spindle travel is limited to 1 inch. It is possible to adapt large micrometers for smaller measurements by inserting a precisely dimensioned rod, or *length standard*, between the anvil and the measured object. For example, to read a measurement of between 3 and 4 inches with a 5-inch micrometer, a 1-inch length standard is used.

For nondigital metric micrometers, the range of measurement is typically 25 mm, with a spindle screw pitch of 0.5 mm, two revolutions per millimeter. There are 50 thimble divisions, each representing a spindle motion of 0.01 mm. A vernier scale may increase resolution to either 0.002 mm or 0.001 mm, depending on the model.

Most outside micrometers have a ratchet thimble that helps ensure repeatable clamping pressure on the measured item, as well as a lock ring or lever to prevent accidental spindle movements. This is especially useful when reading the vernier scale, which is generally not wholly visible when the micrometer is held the usual way for routine measurements. For easier vernier reading, clamp the spindle, then rotate the micrometer forward for a better view.

Dedicated outside micrometers are available for many special measuring applications, including tube wall thickness (Fig. 2 and Fig. 3), paper thickness (Fig. 4), and screw pitch diameter (Fig. 5).

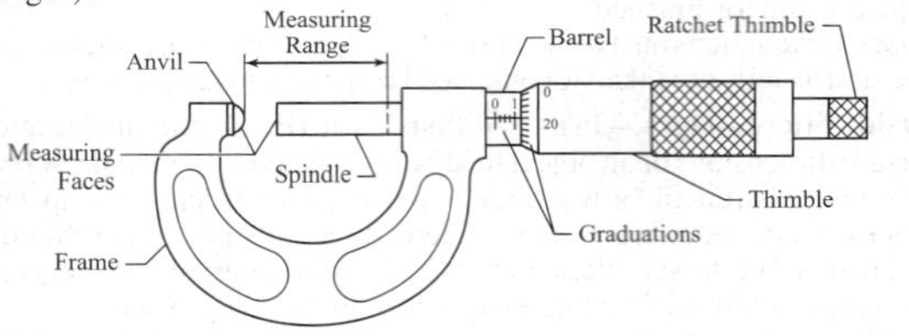


Fig. 2. Type I, Class 1, Style C: Tubing Wall-Thickness Micrometer with Ball Anvil

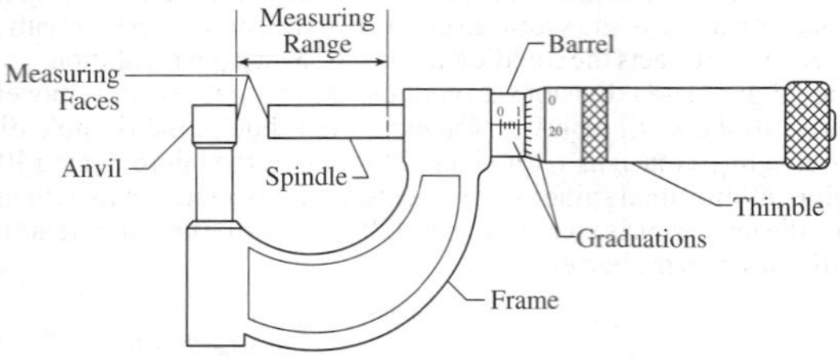


Fig. 3. Type I, Class 1, Style D: Tube Wall Thickness Micrometer with Upright Anvil

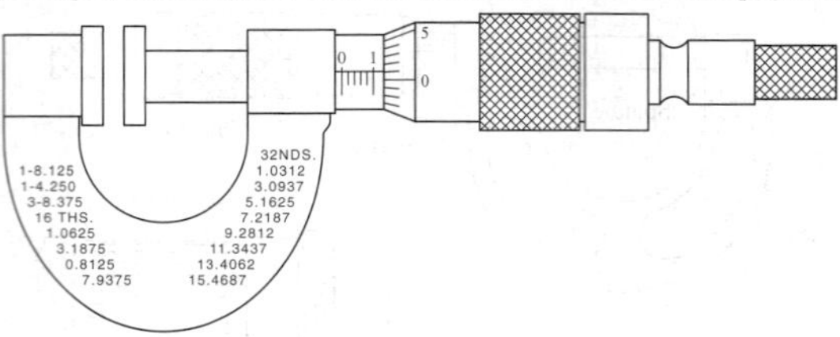


Fig. 4. Type I, Class 2, Style B: Paper Thickness Micrometer with Disk Anvil

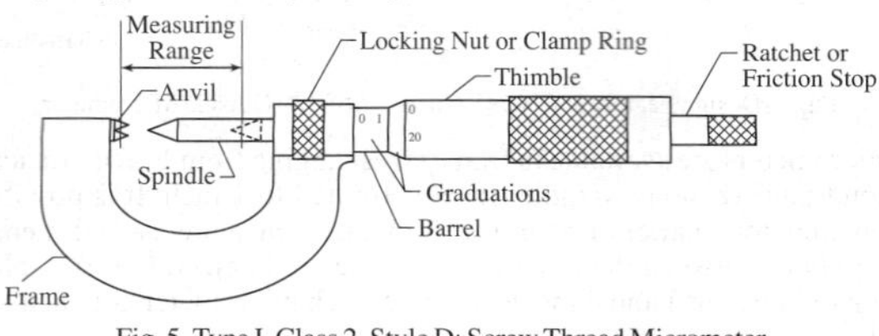


Fig. 5. Type I, Class 2, Style D: Screw Thread Micrometer

*Screw Thread Micrometers:* These devices are available for measuring inch-based or metric threads. There are two distinct types, dedicated and interchangeable. Dedicated micrometers have a single anvil and spindle, and they are limited to one or two thread pitches (example: 0.0357 inch/0.0333 inch for 28 threads/30 threads per inch). Interchangeable micrometers come with several anvil/spindle pairs, covering a wide range of threads (inch-based example: 64 to 5 threads per inch; metric example: 0.4 to 5 threads per millimeter). An interchangeable micrometer costs less than a set of dedicated micrometers covering the same range, but it is less convenient because it must be re-zeroed each time a different anvil/spindle pair is installed.

**Type II, Inside Micrometers.—**Inside micrometers are available in many different styles and sizes, with various designs enabling measurements of bore sizes from 1.5 inch to more than 100 inch. These usually comprise a micrometer head similar to the 1-inch style used in the outside micrometer, to which can be attached one of several length standards in 1-inch increments (Fig. 6). Special adaptations of inside micrometers are jaws (like those in vernier calipers) for bore measurements down to 0.2 inch (Fig. 7). Unlike outside micrometers, inside micrometer heads usually have no vernier scale.

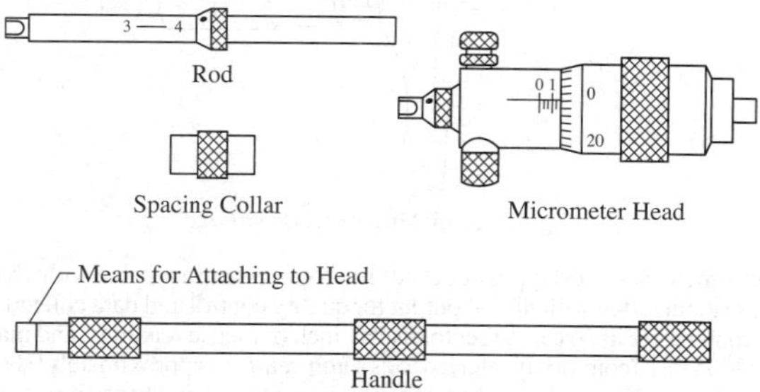


Fig. 6. Type II, Class 2: Extendable Inside Micrometer, Rod and Sleeve Type

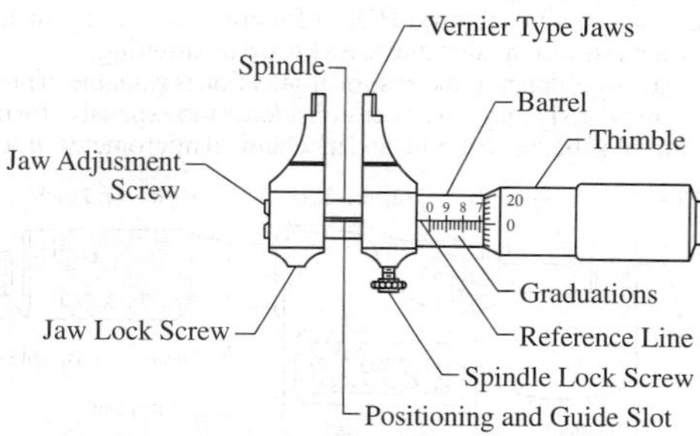


Fig. 7. Type II, Class 1: Inside Micrometer with Caliper-Style Jaws

**Type III, Micrometer Depth Gages.—**The depth micrometer has a 0- to 1-inch head, with a choice of interchangeable measuring rods (*depth rods*) in 1-inch increments, typically up to 6 inch (in special cases, 12 inch). Each of the various measuring rods, contained within the barrel, is for a specific range of depth and is individually adjustable for accuracy. However, there is no provision for extending a rod by interposing a length standard, as with other types of micrometers.

Fig. 8 shows a standard depth micrometer kit for 0- to 2-inch measurements, with a 0- to 1-inch rod installed; shown alongside it is a 1- to 2-inch rod. (Models with three or more extra measuring rods extend the depth range.) A variant of this version is the "half-base" depth micrometer, which has one wing of the base removed.

There are two types of measuring rods, conventional and nonrotating. Similar to the outside micrometer, a conventional measuring rod (as shown in Fig. 8) rotates along with the thimble. Nonrotating rods usually are thin, flat blades and may be preferred when measuring smaller components.

Compared to the outside micrometer, the micrometer barrel on the depth micrometer is "reverse reading." This means the measured value increases as the thimble rotates clockwise. As with inside micrometers, the depth micrometer typically has no vernier scale.

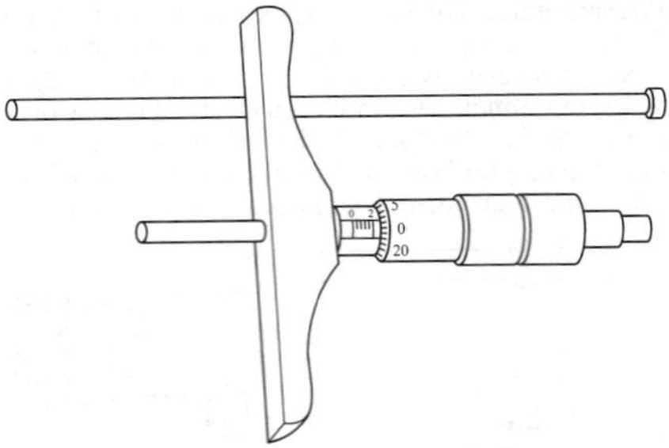

Fig. 8. Type III: Micrometer Depth Gage

**Digital Micrometers.—**Most types of outside micrometers are available with a battery-powered digital readout, often with digital output for quality control and data collection systems. Digital micrometers (Fig. 9) can be set to display inch or metric readings, and many read directly in 0.00005 inch (commonly referred to as a *half-tenth*) or approximately 0.001 mm.

A digital micrometer can be zeroed when the spindle is touching the anvil (the conventional starting point), as well as at any spindle position. This enables measurements relative to an arbitrary value anywhere in the 0- to 1-inch or 0- to 25-mm range. Digital micrometers often feature a pushbutton HOLD function that saves for later viewing a displayed value, otherwise unreadable due to awkward positioning.

Though their measuring efficiency and ease of application is valuable in production work, digital micrometers previously may have been considered too expensive for occasional use. Today, the cost differential between digital and mechanical micrometers is less significant.

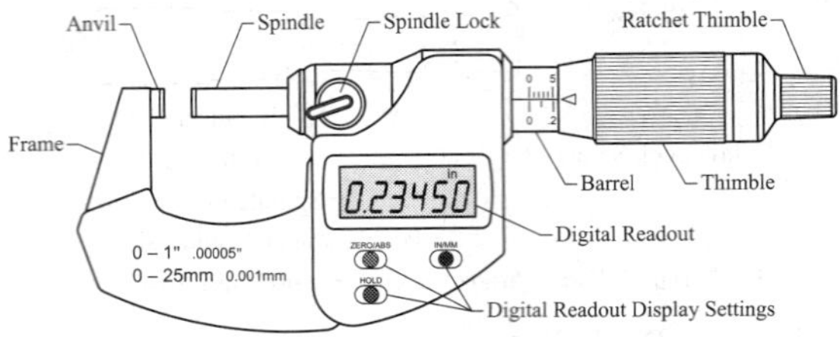


Fig. 9. Inch/Metric Digital Micrometer

## Manufacturers' Production Tolerance Tables

The *Manufacturers' Production Tolerance Tables,* Table 1a and Table 1b, represent the parameters that are utilized by the original manufacturers of the instruments. These values were defined by the Project Team B89.1.13 of the ASME Standards Committee on Dimensional Metrology. This committee is a representative group of the manufacturing industry that establishes the standards used in the production of micrometers. These standards address the requirements of the American industry and the ISO efforts in the support of international commerce. While these tolerance tables and values are utilized in the manufacturing of the instruments, the calibration of the individual instruments is slightly different and does not always include all the geometrical parameters that are taken into consideration during the manufacturing process.

**Table 1a. Manufacturers' Production Tolerance Tables, Inch** *ASME B89.1.13-2001*

| Range (in.) | Permissible Frame Flexure (2 lbf) (in.) | Maximum Parallelism Fixed Anvil (in.) | Maximum Parallelism Interchange Anvil (in.) | Indicated Measurement Error (max.) (in.) | Spindle/Anvil Alignment Error (max.) (in.) |
|---|---|---|---|---|---|
| 0–1 | 0.00010 | 0.00005 | 0.00040 | 0.00010 | 0.0020 |
| 1–2 | 0.00010 | 0.00010 | 0.00040 | 0.00020 | 0.0030 |
| 2–3 | 0.00010 | 0.00020 | 0.00040 | 0.00020 | 0.0045 |
| 3–4 | 0.00015 | 0.00020 | 0.00040 | 0.00020 | 0.0060 |
| 4–5 | 0.00015 | 0.00020 | 0.00040 | 0.00020 | 0.0080 |
| 5–6 | 0.00015 | 0.00020 | 0.00040 | 0.00020 | 0.0100 |
| 6–7 | 0.00020 | 0.00020 | 0.00040 | 0.00020 | 0.0100 |
| 7–9 | 0.00020 | 0.00025 | 0.00060 | 0.00020 | 0.0100 |
| 9–12 | 0.00030 | 0.00030 | 0.00060 | 0.00030 | 0.0100 |
| 12–18 | 0.00040 | 0.00040 | 0.00080 | 0.00040 | 0.0150 |
| 18–24 | 0.00050 | 0.00050 | 0.00080 | 0.00050 | 0.0150 |
| 24–30 | 0.00060 | 0.00060 | 0.00100 | 0.00060 | 0.0150 |
| 30–36 | 0.00070 | 0.00070 | 0.00010 | 0.00070 | 0.0150 |

**Table 1b. Manufacturers' Production Tolerance Tables, Metric** *ASME B89.1.13*

| Range (mm) | Permissible Frame Flexure (10N) (mm) | Maximum Parallelism Fixed Anvil (mm) | Maximum Parallelism Interchange Anvil (mm) | Indicated Measurement Error (max.) (mm) | Spindle/Anvil Alignment Error (max.) (mm) |
|---|---|---|---|---|---|
| 0–25 | 0.002 | 0.002 | 0.010 | 0.004 | 0.05 |
| 25–50 | 0.002 | 0.002 | 0.010 | 0.004 | 0.10 |
| 50–75 | 0.003 | 0.003 | 0.010 | 0.005 | 0.10 |
| 75–100 | 0.003 | 0.003 | 0.010 | 0.005 | 0.15 |
| 100–125 | 0.004 | 0.004 | 0.010 | 0.006 | 0.20 |
| 125–150 | 0.005 | 0.004 | 0.015 | 0.006 | 0.25 |
| 150–175 | 0.006 | 0.005 | 0.015 | 0.007 | 0.25 |
| 175–200 | 0.006 | 0.005 | 0.015 | 0.007 | 0.25 |
| 200–225 | 0.007 | 0.006 | 0.015 | 0.008 | 0.25 |
| 225–250 | 0.008 | 0.006 | 0.015 | 0.008 | 0.25 |
| 250–275 | 0.008 | 0.007 | 0.015 | 0.009 | 0.25 |
| 275–300 | 0.009 | 0.007 | 0.020 | 0.009 | 0.25 |
| 300–350 | 0.010 | 0.008 | 0.020 | 0.010 | 0.25 |
| 350–375 | 0.011 | 0.009 | 0.020 | 0.011 | 0.25 |
| 375–400 | 0.012 | 0.009 | 0.020 | 0.011 | 0.25 |
| 400–425 | 0.012 | 0.010 | 0.020 | 0.012 | 0.25 |
| 425–450 | 0.013 | 0.010 | 0.020 | 0.012 | 0.25 |
| 450–475 | 0.014 | 0.011 | 0.025 | 0.013 | 0.38 |
| 475–500 | 0.015 | 0.011 | 0.025 | 0.013 | 0.38 |
| 500–600 | 0.017 | 0.013 | 0.030 | 0.015 | 0.38 |
| 600–700 | 0.020 | 0.015 | 0.030 | 0.017 | 0.38 |
| 700–800 | 0.022 | 0.017 | 0.035 | 0.019 | 0.38 |
| 800–900 | 0.025 | 0.019 | 0.035 | 0.021 | 0.38 |

## CALIPERS

The caliper, sometimes referred to as a *slide caliper*, measures dimensional differences as small as one thousandth of an inch (0.001 inch) or, in a metric caliper, two hundredths of a millimeter (0.02 mm). Unlike micrometers, which have a 1-inch range of motion, calipers need no extension accessories and are available in jaw openings up to 60 inches.

**Classification and Federal Specifications.—**Calipers with vernier scales are described in the Federal Specification GGG-C-111 (1987; 1999), published by the US Federal Supply Service, General Services Administration (GSA). Intended to cover calipers sold to the GSA, a major user of calipers, this specification is officially inactive and does not describe all commercially available product variations. Still, the data provided is applicable to current devices.

**Type and Classes**

*Type 1 Calipers, vernier*

*Class 1:* Inch measure

*Class 2:* Metric measure

*Class 3:* Inch-metric measure

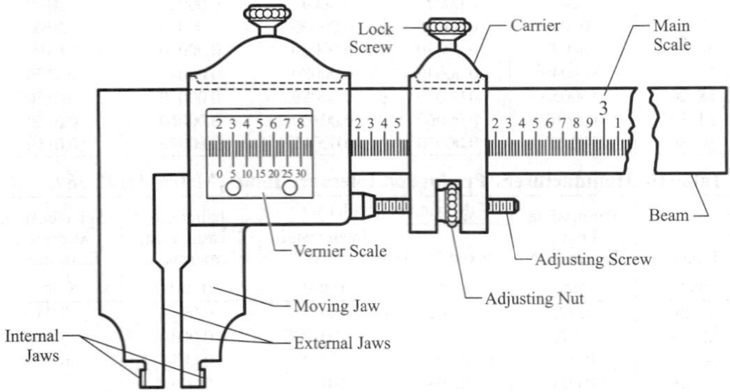


Fig. 10. Early Design of Type 1, Class 1: Inch Caliper, with Fine Adjustment

**Caliper Features.—**A desirable feature of early style calipers (Fig. 10) is screw adjustment of the moving jaw. This is helpful in achieving a repeatable degree of pressure between the moving jaw and the measured object's surface; it also enables minute adjustments to determine the exact point of coincidence between vernier and main scale graduations.

Less desirable is the caliper's ability to make inside measurements. This is because the outfacing "internal jaws" (for inside measurements) are some distance apart when the external jaws (the inner faces) are together, at which point, the caliper reads zero. This calls for the displayed value to be reduced by a set amount when an inside measurement is taken.

A more recent caliper design (Fig. 11) deals with this drawback by providing separate internal jaws exactly coincident when the displayed value is zero. Usually, the jaw is moved directly by a thumb rest as shown (some styles feature screw adjustments). An important feature of this style is the depth measuring rod attached to the moving jaw. However, the right-hand end of the rod, which serves as the base for depth measurements, has a narrow stance compared to conventional depth gauges and must be carefully seated on the reference surface. Also, the internal jaw points are fragile and prone to damage.

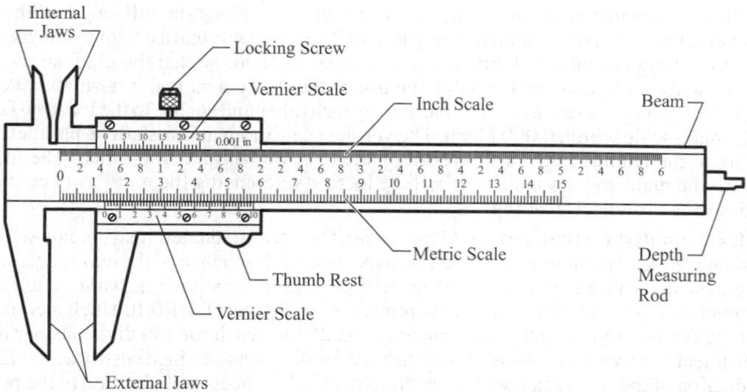


Fig. 11. Current Design of Type 1, Class 3: Inch/Metric Caliper, with Separate Internal Jaws

**Construction Requirements.—***Materials:* The beam and all measuring surfaces having direct contact with the work shall be made of high grade tool steel. The measuring faces may be made of Tool Steel with Carbide inserts. All materials shall be sound and free of injurious porosity, soft spots, hard spots and other defects. All parts subject to wear, breakage or distortion shall be replaceable.

*Main-Scale Member (Beam):* The main-scale member, or beam, of the Type 1 caliper shall be graduated as defined herein for each design variant. The main-scale shall be of a cross sectional area as to provide stability without undue weight when the instrument is used to make measurements at the maximum range. The measuring face and slide locating edge shall be straight within 0.0003 inch for each linear foot of the beam length. The total tolerance for the instrument (if over 12 inch) shall not be accumulated in any 12 inch increment and all deviation in straightness shall be evenly distributed along the entire length of the instrument.

*Slide:* The slide or moveable jaw shall be machined to receive the main-scale beam and shall have a fit and surface finish so as to allow the slide to move through the entire length of the beam without lateral shake or bind when not locked into position. The locating side of the slot shall be square with the measuring plane within 0.0003 inch along the full length of the contact surface. The opposite side of the moveable jaw shall be fitted with a nib or knife edge inside measuring jaw and a locking screw.

*Vernier Plate:* The vernier plate shall either overlap or be flush with the edge of the main scale. The overlap of the main scale shall be no less than 1/64 inch and straight with 0.002 inch in the full length. The vernier plate shall be beveled to prevent parallax error. If the scale is installed flush along the edge of the main scale, the edge of the Vernier Plate shall slide easily along the main scale. The edges of both the main scale and the vernier plate shall extend fully to the edges. The Plate shall be held in place with two screws through elongated mounting holes to allow no less than 0.010 inch longitudinal adjustment.

*Measuring Contacts:* The measuring contacts shall have a surface finish of 8 μin Ra or better and have a hardness value of no less than 60 to 65 RC (Rockwell C scale).

**Scale and Graduations.—**All calipers regardless of Type or Class will have a *main scale* and a *Vernier scale* or dial. The vernier scale shall be a direct reading scale so that when aligned with the zero point of the data scale, the first division of the vernier scale will appear below the first mark on the main scale. All vernier scale graduation spacing is a precise fraction smaller than the marks on the main scale. All graduations shall be machine cut or etched between 0.003 to 0.006 in width and from 0.001 to 0.003 inch in depth and shall not vary in width on the main scale or the vernier scale by more than 0.001 inch. The edges of the graduations shall be straight and not vary more than

0.0002 inch throughout the full length of the graduations along the full length of the instrument. The graduations shall be sharp and well defined and contrast with the surrounding area through chemical etching or other similar methods so that the graduations are distinct and easily readable to enable the use of the instrument under normal working conditions. The graduations will be parallel to each other and square to the locating edge of the main scale within 0.0004 inch. The vernier scale graduations shall be parallel and square to the reading edge of the vernier plate or scale within 0.0002 inch. The index of both the main and vernier scales will be located referencing the measuring contacts planes so as to be direct reading.

**Inch Graduated Instruments.**—*Main Scale:* The inch graduated instruments will be graduated to read in thousandths of an inch increments. Each inch of the main scale shall be represented with a graduation mark representing the inch incremental value. Each inch segment shall be subdivided into ten increments of 0.100 inch. Each 0.100 inch increment shall be further divided into 4 increments of 0.025 inch each for 100 divisions per inch instruments or two divisions of 0.050 inch for 50 divisions per inch instruments. Each graduation of the main scale within the first foot shall be accurately located to the reference zero index within 0.0003 inch. With each additional foot length of the main scale and additional 0.0003 inch shall be acceptable with no total tolerance accumulated within any single foot of main scale length. All graduations within any 1 inch increment shall be accurately located in that inch increment within 0.0002 inch.

*Vernier Scale:* The inch vernier scale shall be divided into either 25 minor divisions each equivalent to one twenty-fifth of the main scale minor division or twenty five thousandths of an inch or 50 divisions each equivalent to one fiftieth of the main scale minor division or fifty thousandths of an inch. On either scale each alignment of the vernier scale mark to the main scale is equivalent to one thousandths of an inch. The graduated length for the 25 division inch graduated vernier scale shall be either 0.6000 or 1.2250 inch ±0.0002 and are divided into 25 equal increments accurately spaced to within ±0.0002 inch. All graduations shall be of a length so as to be easily discernible with the graduations for 0 (zero) and each 0.005 inch having extra length to enable them to be easily distinguished. The 0 (zero) index and each 0.005 inch increment shall be clearly marked with the incremental value. The vernier scales that are divided into 50 increments shall have a length of either 1.225 or 2.450 ±0.0002 with each increment accurately spaced to within 0.0002 inch. Each instrument shall be marked with the scaled value.

**Metric Graduated Instruments.**—*Main Scale:* The metric graduated instruments will be graduated to read in 1/50 mm (0.02mm) increments. Each centimeter of the main scale shall be represented with a graduation mark representing the incremental value. For instruments with 25 increment vernier scales, the main scale shall be graduated in centimeters, millimeters and half-millimeters. Each centimeter segment shall be subdivided into ten increments of 1 millimeter each. Each 1 millimeter increment shall be further divided into 2 increments of ½ millimeter each for 20 divisions per centimeter. Centimeters shall be numbered from zero with easily read numerals. The graduation lines for cm, ½ cm, mm, and ½ mm shall have length of descending order. All graduation lines shall be clearly marked as described in the previous *Scale and Graduations* and *Inch Graduated Instruments* sections, and be easily readable under normal operating conditions.

Instruments with 50 increment vernier scales shall have the main scale graduated in centimeters and millimeters with half-millimeters at the manufacturers' option.

Each graduation of the main scale within the first 30 cm shall be accurately located to the reference zero index within 0.008 mm. With each additional 30 cm length of the main scale and additional 0.008 mm shall be acceptable with no total tolerance accumulated within any single 30 cm of main scale length. All graduations within any 1 cm increment shall be accurately located in that centimeter increment within 0.005 mm.

*Vernier Scale:* Vernier scales for metric instruments divided into 25 increments shall have a graduated length of 12 mm or 24.5 mm, ±0.005 mm and scales divided into 50 increments shall have a graduated length of 24.5 mm or 49 mm, ±0.005 mm.

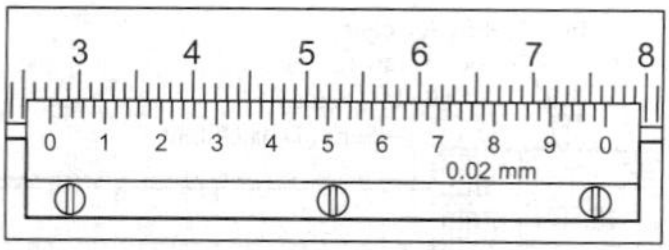


Fig. 12. Common Metric Vernier Caliper Scale Markings

The metric vernier scale most generally found in the current market shall be divided into 10 divisions or 0.1 mm, each division further divided into 5 minor divisions equivalent to a resolution of 0.02 mm. The spacing and location from zero shall be accurate to within 0.005 mm. All graduations shall be of a length so as to be easily discernible with the graduations for 0 (zero) and each 0.1 mm having extra length to enable them to be easily distinguished. The 0 (zero) index and each major division shall be clearly marked with the numerical incremental value.

**Dial Caliper.**—The dial caliper uses a precise rack and pinion movement that drives an indicating hand to provide a direct reading on a circular dial. This precision movement eliminates the necessity to visually read the vernier scale and negates the effects of subjective observation.

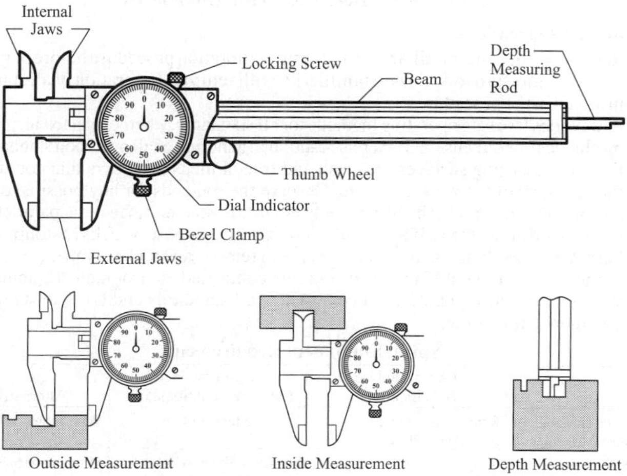


Fig. 13. Typical Dial Caliper

The movement, when adjusted properly and calibrated will provide a measurement with a higher level of confidence than the vernier scale. Typically the indicating hand rotates once every 0.100 in or 1 mm and provides a resolution of 0.001 inch or 0.02 millimeter. The main scale still indicates the major divisions that are added to the reading taken from the dial indication. An additional feature of the dial caliper is that the dial is designed to be rotatable beneath the pointer allowing the operator to "set" the caliper at a known value and zero the indicator and make "differential measurements" from the known value. This function was not possible with the original vernier caliper.

**Digital Caliper.**—The most current refinement in the instrument is in the replacement of the dial movement or vernier scale with a digital display. The dial rack and pinion movement or vernier scale has been replaced with the application of a linear encoder that allows switching between millimeters and inches on demand, zeroing the scale at

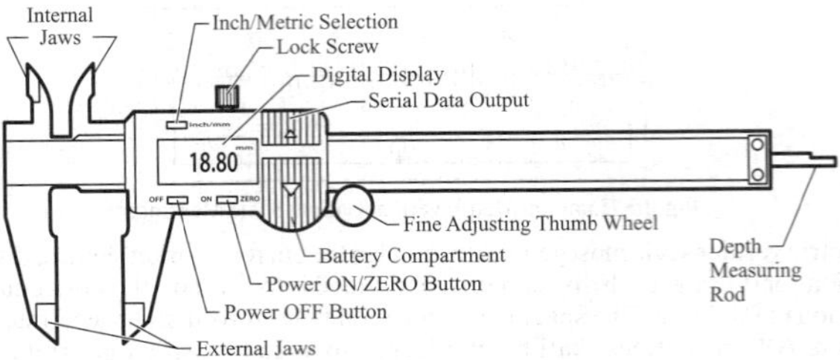


Fig. 14. Typical Features of the Digital Caliper

any point along the scale for differential measurements and electronically "holding" a measurement possibly taken in an awkward position where the digital display may not be visible. Many digital calipers offer a *serial data output* function that allows the operator to capture collected data in a spreadsheet or dedicated controller for analysis and more accurate records collection.

## Calibration of Reference Setting Standards

### Preliminary Operations

1) Review and become familiar with the entire calibration procedure before beginning the calibration process. Ensure familiarity with setup and operation of all standards involved in the procedure.
2) Always observe safe operating methods for all standards, whether electrical, physical, mechanical, or dimensional; exercise caution in handling of all precision standards.
3) Ensure measuring surfaces are clean and free of nicks and burrs that could affect the accuracy of the measurements. Observe the rod ends for any possible damage and observe the unit for bending or any condition that may affect the parallelism of the measuring surfaces. If any condition of concern is observed, light stoning with a hard Arkansas stone can be used with care to remove any nicks or burrs.
4) Bring the UUT (Unit Under Test) into the controlled environment 12 hours prior to beginning the calibration process. Allow all standards and UUT to stabilize to common temperature.

**Typical Equipment Requirements**

| Name | Minimum Use Specifications | Recommended Measurement Standards | Alternative |
|---|---|---|---|
| Universal Horizontal Measuring Machine | Range: 0 to 24 inch<br>Accuracy: ±30μin | Mahr-Federal 828 | Mahr Linear 1200 |
| Gage Block Set | Range:<br>0.050 to 4.000 inch<br>2.0 to 450 mm | Starrett Webber SC 81.A1<br>Starrett Webber S2C 88.MA1 | Mitutoyo 516-401-16<br>Mitutoyo 516-442-10 |
| Gage Block Set (Long Block) | 5 to 84 inch<br>125 to 500 mm | Starrett Webber SS 8.A1X<br>Starrett Webber SS 8.MA1X | Mitutoyo 516-762-16<br>Mitutoyo 516-753-10 |
| Height Transfer Gage | Range: 1 to 48 inch | Starrett DIGI-CHECK II DHG 49 | Mitutoyo Universal Height Master Model 515-513 |
| Electronic Height Indicator | Range: ±0.032 in. (min.)<br>Accuracy:<br>±50 μin<br>±0.001 mm | Mahr Federal Model 832 w/ LVDT | Mitutoyo Analog MU-Checker 519-420A with 519-327 Lever Head |
| Precision V-Block | Range: Adequate to support UUT in vertical position.<br>Accuracy: Calibrated to specifications of GGG-V-191D (or equivalent) | | |
| Surface Plate | Range: 24 x 36<br>Accuracy: Grade A | Starrett Crystal Pink EDP 80655 | Mitutoyo 517-808 |

**Reference Setting Standard Calibration Process.**—*Note:* Special care should be taken to clean and apply a protective coating when handling gage blocks or other precision standards with bare hands. The use of clean cotton or vinyl gloves is recommended to prevent corrosive oils from damaging precision surfaces.

The temperature of the calibration environment must be maintained at approximately 23° C (73.4° F) with a fluctuation of no more than ±0.5° C (±0.9° F) per hour.

1) Calibration of UUT up to 4.000 inch shall be performed on the horizontal measuring machine. All other measurements shall be performed in the vertical position as described in this procedure.

2) Select gage block equal to the size of the UUT. Wringing blocks together to achieve the desired size should be avoided if possible as this method increases possibility of error.

3) Insert the gage block between the anvil and spindle of the universal horizontal measuring machine resting against the anvil. Bring the head (spindle) into contact with the gage block and gently bring the indicator to the zero indication. Lock the head into position. Seat the gage block, re-zero, and assure the measurement is repeatable.

4) Remove the gage block and replace with the UUT. Assure that the UUT is seated and the measurement is being read at the peak (highest reading) for radius rod ends, or the parallel point (lowest reading) for flat rod ends. Assure radius is centered on the anvil and spindle. Verify the reading is within the tolerance limits as defined in Table 2.

5) Setup the universal horizontal measuring machine for the next size UUT as described in Items 2) and 3).

6) Repeat measurement as described in Item 4).

7) Continue the process through the range of the universal horizontal measuring machine or until all UUT in the set have been measured. If the UUT length in the set exceeds the length of the universal horizontal measuring machine, proceed to Item 8) for the process to measure the remainder of the set up to 40 inch (1000 mm).

8) Using a precision V-block, clamp the UUT into the V-block ensuring the rod end is in contact with the surface plate and the UUT is secured in a vertical orientation.

9) Using either a height transfer gage or a gage block, set the electronic indicator to a zero indication. Set the range and resolution of the indicator so as to adequately indicate the deviation from zero in relation to the tolerance limits assigned to the UUT.

10) Repeat the zero indication several times to assure no drift or error has occurred.

11) Move the indicator from the standard to UUT and seek the highest point (for radius ends) or sweep the flat rod end for deviation from nominal and indication of parallelism.

12) Confirm that the indicated deviation is within the tolerance limits as assigned to the UUT in Table 2.

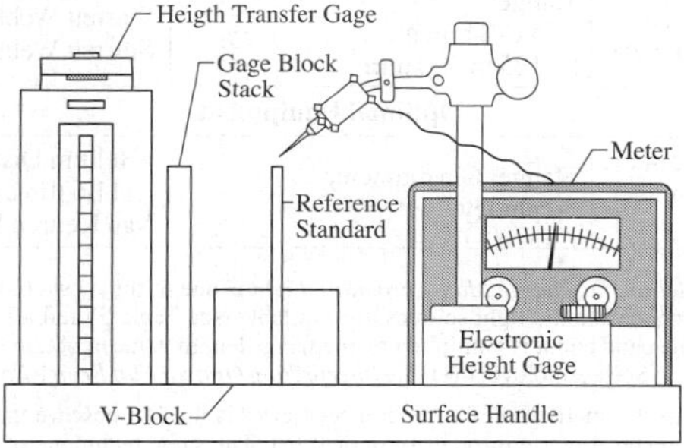


Fig. 15. Vertical Measurement of Reference Setting Standard

**Table 2. Size and Tolerance for Inch and Metric Reference Standards**
(*ASME B89.1.13-2001*)

| English | | Metric | |
|---|---|---|---|
| Length (in.) | Tolerance (in.) | Length (mm) | Tolerance (mm) |
| 1 | ±0.00005 | 25 | ±0.0012 |
| 2 to 4 | ±0.0001 | 50 to 125 | ±0.002 |
| 5 to 8 | ±0.00015 | 150 to 200 | ±0.003 |
| 9 to 11 | ±0.0002 | 225 to 275 | ±0.004 |
| 12 to 18 | ±0.00025 | 300 to 425 | ±0.005 |
| 19 to 24 | ±0.0003 | 450 to 575 | ±0.006 |
| 25 to 36 | ±0.00035 | 600 to 750 | ±0.007 |
| 37 to 40 | ±0.0004 | 775 to 900 | ±0.008 |

13) Repeat Items 8) through 12) for the remaining reference setting standards in the set.

## Calibration of Outside Micrometers

Typical equipment requirements for calibration procedures are shown in Table 3; a description of the calibration is given in Table 4, and various types of monochromatic light sources are given in Table 5.

**Table 3. Calibration Equipment Standards**

| Item | Minimum Use Specifications | Recommended Equipment |
|---|---|---|
| Monochromatic Light | Range: N/A<br>Accuracy: N/A | [a]Helium Discharge:<br>11.6 μin Length/Fringe<br>Van Keuren L-1A or L-1 |
| Optical Parallel Set | Range: Two optical parallels differing in thickness by 0.0125 in.<br>Accuracy:<br>10 μin (flat and parallel) | Van Keuren: OP-2B |
| Gage Block Set | Range:<br>0.05 to 4 inch<br>2.0 to 450 mm | Starrett Webber SC 81.A1<br>Starrett Webber S2C 88. MA1X |
| Gage Block Set (Long Block Set) | Range:<br>5 to 84 inch<br>125 to 500 mm | Starrett Webber SS 8.A1X<br>Starrett Webber SS 8.MA1X |
| **Optional Equipment** | | |
| [b]U6 Monochromatic Reflex Interference Viewer | Range: 6 in diameter<br>Accuracy: N/A | [a]Helium Discharge:<br>11.6 μin Length/Fringe<br>Van Keuren V760300 |

[a] While the *Helium Discharge Monochromatic Light* is one of the most common frequencies in use, other monochromatic light sources are available (see Table 5) and all will produce the same results. The band count is multiplied by the fringe length value to obtain the flatness of the measured surface. See examples in the table *Interpreting Optical Flat Fringe Patterns*.

[b] The U6 Monochromatic Reflex Interference Viewer is used to observe the base flatness of the depth micrometer directly through an optical flat. The same technique of fringe count can be utilized to obtain the flatness as is used for the outside diameter micrometer anvil and spindle.

**Table 4. Calibration Description**

| UUT Characteristics | Performance Specifications | Test Method |
|---|---|---|
| Flatness | Range:<br>0 to 54 in.<br>0 to 300 mm<br>Tolerance: See Table 14a and Table 14b | Measured with *Optical Flat/Parallel* converting fringe curvature to an equivalent linear value. |
| Parallelism | Range:<br>0 to 1 inch<br>0 to 25 mm<br>Tolerance:<br>50 μin.<br>0.0013 mm | For UUT with 0–1 inch (0–25 mm) range, measured at two positions with optical parallel set. |
| Length and Linearity | Range:<br>0 to 54 inch<br>0 to 1400 mm<br>Tolerance: See Table 14a and Table 14b | Comparative measurement of UUT reading to gage block test setup to test the basic length and micrometer head linearity. |

*Note*: The *Minimum Use Specifications* column of Table 3 represents the principal parameters required for performance of the calibration. References to specific equipment are recommendations only, and in no way constitute an endorsement other than as examples of the appropriate accuracy required to maintain the correct *Test Accuracy Ratio* (*TAR*) and confidence level in the calibration procedure. It is understood that equivalent equipment produced by other manufacturers is capable of equally satisfactory performance of this procedure.

*Note*: If the UUT uses the metric scale, use test points equivalent to those defined for the English scale so as to test the micrometer at five places in the circumference of the drum and five equally spaced intervals throughout the range of travel of the micrometer spindle screw.

**Table 5. Types of Monochromatic Light Source**

| Light Source | Wave Length | | Length/Fringe | |
|---|---|---|---|---|
| | μin | nm | μin | nm |
| Helium-Neon Laser (bright red) | 24.9 | 632.8 | 12.5 | 316.4 |
| Sodium Vapor (yellow) | 23.2 | 589.3 | 11.6 | 294.7 |
| Mercury Vapor (green) | 21.5 | 546.1 | 10.8 | 273.1 |
| Helium Discharge (yellow-orange) | 23.1 | 587.6 | 11.6 | 293.8 |

**Preliminary Operations.—**Ensure a clean and well illuminated work area, free from drafts, excessive humidity with a stable temperature that does not fluctuate more than ±2° per hour. It is recommended that a stable temperature of 68° F is maintained in the calibration work area.

*Note:* The calibration of outside diameter micrometers and similar precision measurement devices is accomplished through the use of gage blocks as the length standard. Due to the thermal reactivity of gage blocks, environmental stability is a major consideration in maintaining the confidence in the accuracy of the measurements taken during the calibration.

Calibration of UUT with an accuracy of ±0.001 inch or greater can be accurately calibrated in an environment with a temperature fluctuation of ±4° F. Calibration of UUT with a higher accuracy, such as ±0.0001 inch, must be accomplished in a stable environment of no greater than 68° F with a temperature fluctuation of no more than ± 2° F per hour.

Ensure that the UUT and all associated standards have been allowed to stabilize in the controlled environment for a minimum of 8 hours prior beginning the calibration procedure. This is especially critical for UUT in excess of 10 inches.

**Table 6. Anvil/Spindle Flatness Tolerances per *USAF***

| Range | | Tolerance | |
|---|---|---|---|
| Inches | Millimeters | English UUT (inch) | Metric UUT (millimeters) |
| 0 to 3 | 0 to 75 | 0.00005 | 0.001 |
| 4 to 9 | 100 to 225 | 0.00008 | 0.002 |
| 10 and above | 250 and above | 0.0001 | 0.0025 |
| All with a resolution of 50 in. or 0.001 mm | | 0.000025 | 0.0006 |

**Parallelism Test (Outside Micrometers).**—The anvil and spindle parallelism test is performed on 0–1 inch / 0–25mm micrometers where a high degree of accuracy is required to be maintained. Prior to the parallelism test, the flatness test must be performed to the satisfaction of the tolerance of the UUT.

- It is critical to the accurate measurement that no dirt or dust particles are present. It is highly recommended that several measurements be taken to verify the results of the measurement. Cleaning the anvil and spindle and optical parallel with lint free lens tissue will aid in assuring the accuracy of the measurement.
- Position the thinner optical parallel between the anvil and spindle of the UUT and adjust the UUT until both surfaces are in contact with the optical parallel. Do not exert greater than necessary force in attempting to minimize interference band count. Adjust the optical flat on the anvil end of the UUT to obtain the minimum band image. The optimum number of bands anticipated here is four. This provides an easily read image to calculate the parallelism of the anvil and spindle. Note the number of bands on the anvil face as *count A*.
- Without changing the position of the UUT, carefully turn and observe the number of bands on the spindle face. Note this as *count B*.
- Compute the parallelism of the anvil and spindle over the smaller optical flat as follows. Use the band width of the monochromatic light that is being used;

  $$11.6 \times (A + B) = Parallelism \text{ (microinches)}$$

  where $A$ = *count A*; $B$ = *count B*; and 11.6 is the fringe/length of the light source in microinches (Table 5).
- Remove the thinner optical flat and replace with the 0.0125 inch thicker one and repeat the measurement process. To calculate the anvil to spindle parallelism inclusive of the rotation of the spindle, calculate the difference between the two optical flat measurements.
- Verify that the difference between the two sets of measurements is within ±50 μinch or ±12 μm.
- If only one optical flat is available the same measurement can be achieved by using the measurement technique described above but calculated as $11.6 \times (A - B)$ = Parallelism (microinches). The exception being that this method will not take into consideration the rotation of the spindle.

**Linearity Test (Outside Micrometers).**—1) Set the UUT to indicate zero reading using the ratchet or friction thimble, and assure that the reading meets the applicable tolerance limits. If an *Out of Tolerance (OOT)* reading is identified at this time, assure there is no foreign particle or burr on the anvil or spindle. An accurate zero indication must be established at this time as all other measurements will be affected by any error at zero.

*Note:* The micrometer anvil and spindle can be closed over a sheet of clean white paper damp with isopropyl alcohol, which can then be drawn out slowly to remove any obstructing matter. Repeat with a dry paper and observe any smear or smudge that would indicate dirt or foreign matter. Clean until paper comes through with no residue.

2) For *Interchangeable Anvil Micrometers* the appropriate length gage block will be used to establish the zero indication.

3) The recommended size gage blocks used to test the micrometer spindle thread are very important. The recommended test points achieve the testing of the accuracy of the micrometer throughout the entire length and at points evenly distributed around the diameter of the spindle thread. This verifies the accuracy of the lead error of the thread as well

Closely examine the UUT for any burrs or nicks on the measuring surfaces that could in any way damage the highly polished precision surface finishes of the gage blocks and optical flats. Damage to these surfaces could degrade or potentially ruin the standard.

Thoroughly clean the UUT and rotate the micrometer through the full range of travel, ensuring smooth movement with no binding or indication of drag or misalignment of the spindle through the support bearing.

*Note:* It is recommended that well-fitting cotton or nitrile gloves be used to reduce thermal transfer and protect from corrosive elements of direct contact.

**Calibration Procedure, Flatness Test (Outside Micrometers).**—UUT used in precision work or any re-lapped anvil or spindle require a flatness test. If the instrument is known to be for general purpose; i.e. tool box issue, diameter measurement where flatness is not a factor, measurement is a non-quantified value, etc., a flatness test would not be required. Calibration records should reflect "flatness not calibrated."

Ball anvil and tubing micrometers require a flatness test on the spindle only.

- Connect the *Monochromatic Light* to a power source and turn on. Allow a few moments to warm up.
- Position the UUT in the monochromatic light field and place the optical flat on the anvil. Apply slight pressure and gently seat the optical flat until 3 to 5 bands are displayed on the surface. It is important to not apply greater than necessary force or to cause damage to the optical flat by pressing down if a burr is detected. Scratching the surface of the optical flat can ruin the flat. Caution is required.
- Using the optical flat and monochromatic light, observe interference bands on anvil face.
- Count the number of bands between the lowest ends and the crest of 1 band. Multiply this number by the frequency of the monochromatic light being used. Confirm that the calculated deviation is within the tolerance limits defined in Table 6.

  1 band = 11.6 millionths (μin) using helium discharge light source (Table 5).

Record the band count, frequency and result.

**Interpreting Optical Flat Fringe Patterns**

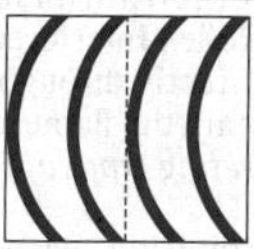
Fig. 16a.

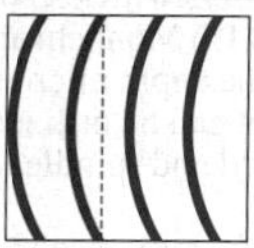
Fig. 16b.

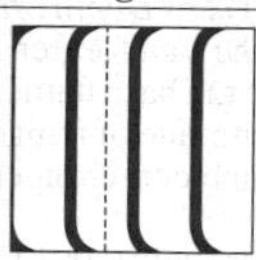
Fig. 16c.

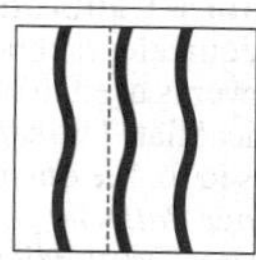
Fig. 16d.

*Example, Optical Flat Fringe Patterns:* Fig. 16a through Fig. 16d illustrate fringe patterns produced by an optical flat in contact with several test surfaces.

The following illustrates how to interpret the fringe patterns observed to determine deviations in flatness of the unit under test (UUT).

a) Convex surface (Fig. 16a): side edges low 11.6 μin

b) Convex surface (Fig. 16b): side edges low 5.7 μin.

c) Nearly flat (Fig. 16c): side edges low 5.7 μin.

d) Complex surface (Fig. 16d) both convex and concave: hollow in center, higher each side of center and lower at side edges: error 3 μin.

- Turn over the UUT and repeat preceding steps for the spindle. If this is inconvenient due to size of UUT, remove the spindle to take the reading. Calculate in the same manner as above. Record the reading.
- If UUT is an interchangeable anvil micrometer, repeat measurement for each anvil.
- *Note:* If anvil or spindle do not create an image, or are too worn to produce an image, they will need to be sent out to the appropriate vendor for lapping.

as identifying any sections that may have experienced excessive wear or damage. The recommended sizes are as follows:

a) *X*.210, *X*.420, *X*.605, *X*.815 and 1(*X*).000 (inch) and 5.10, 10.20, 15.30, 20.40, and 25.00 (mm). The micrometer should be tested at zero and full scale. *X* represents the applicable length for the UUT range.

b) When wringing gage blocks together to achieve the desired size, the stack must soak for 1 hour when calibrating UUT with 0.0001 inch (0.002 mm) resolution in a range over 8.000 inches (200 mm). The soak time does not apply to sizes over 8.000 inch (200 mm) that do not require wringing.

4) Select the first gage block stack and measure it using the ratchet or friction thimble. Do not rely on feel as the correct and controlled loading of measuring force by the ratchet or friction thimble will give the most accurate and repeatable reading. All other methods are subjective and open to operator influence. Several clicks of the ratchet or applying force until the friction thimble slips and assuring the micrometer is gently seated on the gage block stack will give an accurate reading. Assure the reading is repeatable. The reading must meet the tolerance limits as defined in Table 14a and Table 14b for *Indicated Measurement Error.*

5) Repeat these steps for each remaining gage block stack and record readings for evaluation or trend analysis. Follow all guidelines of the quality assurance programs in place in regards to recording and retaining all data collected and reporting OOT (out of tolerance) conditions.

6) For *Interchangeable Anvil Micrometers*, after calibration of the micrometer spindle thread, repeat Items 1) and 2) for each anvil in the set. Once the micrometer spindle thread accuracy has been confirmed, only the zero setting of each interchangeable anvil needs to be calibrated. Verify that each interchangeable anvil meets the tolerance limits assigned and record any OOT conditions before adjusting to meet tolerance limits if necessary.

## Calibration of Depth Micrometers

*Note:* Do not lock the micrometer head without the depth rod installed. This could result in damage to the split ring locking mechanism causing the rods to bind.

**Flatness Calibration.—** 1) For *Depth Micrometers* with a resolution of 100 μin (or metric equivalent) or better, the Van Keuren model U6 Monochromatic Reflex Interference Viewer is used to observe the base flatness of the depth micrometer directly through an optical flat. The same technique of fringe count can be utilized to obtain the flatness as is used for the outside diameter micrometer anvil and spindle (*Interpreting Optical Flat Fringe Patterns*).

2) For *Depth Micrometers* with 0.001 inch (0.01 mm) resolution per division, place the UUT on an individual 1.000 inch (25 mm) gage block with the depth rod extended through the center relief hole as shown in Fig. 17. Bring the depth rod into contact using the ratchet stop or friction thimble and observe the reading. Repeat this measurement three times ensuring repeatability of the reading. Record the reading as reading *A*.

3) Remove the UUT from the setup and place it on two gage blocks (or stacks) of 1.000 inch (25 mm) as shown in Fig. 17. Bring the depth rod into contact using the ratchet stop or friction thimble as in Item 2) and observe the reading. Repeat this measurement three times ensuring repeatability of the reading. Record the reading as reading *B*.

Subtract reading *B* from reading *A* and ensure that the result is within the tolerance limits as listed in Table 7.

**Table 7. Depth Micrometer Base Flatness**

| Range | | Tolerance | |
|---|---|---|---|
| Inches | Millimeters | (inch) | (millimeters) |
| All with a resolution of 0.0001 in. or 50μin | All with a resolution of 0.0025mm or 0.001mm | 0.0001 | 0.0025 |
| All with a resolution of 0.001 in. | All with a resolution of 0.01 mm | 0.001 | 0.01 |

*Note:* The *Linearity* of the UUT is checked using the 0–1 inch (0–25 mm) depth rod only. Once the accuracy of the *Micrometer Head* has been verified, the remaining depth rod extensions are calibrated at the zero position only. It is critical to the outcome of the calibration of the entire set that the base zero setting for the micrometer sleeve and thimble be accurate and without error or deviation. Any error in the base zero setting will cause errors to be translated to the depth rod extensions.

**Linearity Calibration (Depth Micrometers).—** 1) Select the appropriate gage blocks to make stacks to measure 0.210, 0.420, 0.605, and 0.815 inch, or 5.10, 10.20, 15.30, 20.40, and 25.00 mm. The 1.000 inch (25.00 mm) block will also be used as the micrometer head will be calibrated at zero to full scale. Clean and wring gage blocks assuring no dirt or foreign matter is present.

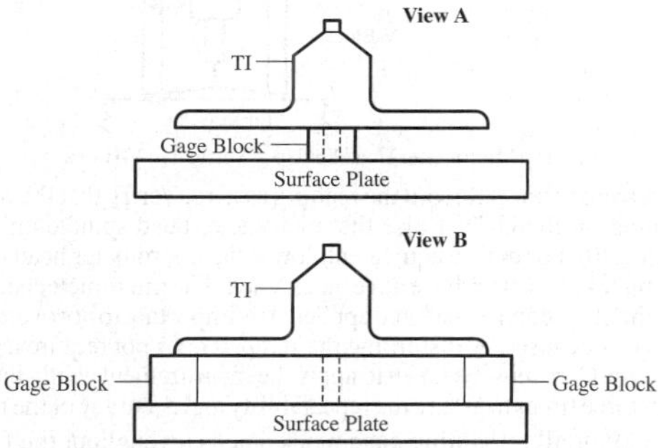


Fig. 17. Figure 4: Depth Micrometer Base Flatness Measurement

2) Place the UUT on the surface plate and zero against the plate. A calibrated surface plate must be used as it is used to establish the base zero setting and is considered as one of the standards used in the calibration procedure. Record any deviation and assure the reading is within the tolerance limits as defined.

3) Place the first gage block stack (0.210 inch or metric equivalent) on the surface plate and place the UUT on top of the stack. Measure the stack through the center relief hole in the gage block. Observe the reading. Repeat the measurement three times and assure repeatability of the measurement. Verify that the reading is within the tolerance limits.

4) Repeat Item 3) for all remaining gage block stacks. Record all readings. If an OOT condition is discovered, adjust the UUT and repeat the linearity calibration to verify the adjusted accuracy.

5) Repeat Item 2) for all remaining depth rod extensions. For UUT with resolution of 0.0001 in or better, allow gage block stacks that require wringing to assemble to stabilize for one hour for lengths over 8.00inch. Stabilization is not required for single block lengths.

## Calibration of Micrometer Heads

**Linearity Calibration.—** 1) Select the appropriate gage blocks as defined in the *Linearity Test (Outside Micrometers)*. As previously the micrometer head will be calibrated at zero to full scale. Clean and wring gage blocks, as appropriate, assuring no dirt or foreign matter is present.

2) Mount the micrometer head vertically by the shank into a precision V-block as shown in Fig. 18, and adjust so that the zero position can be achieved making contact with the calibrated surface plate. If a zero position cannot be achieved with the V-block available, use an intermediary gage block of the necessary size so as to achieve contact with the spindle face and set zero on that intermediary block. This is the zero indication and will not be included in the linear measurement. Repeat the zero measurement three time to assure repeatability of

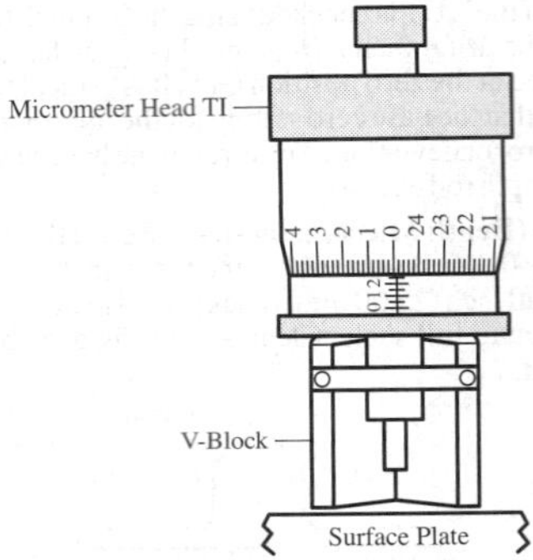


Fig. 18. Micrometer Head Set Up—Vertical in V-Block

the measurement (zero setting) Record the reading as zero. Verify that the reading is within the tolerance limits for the UUT. Raise the micrometer head spindle and insert the first step gage block (0.210 in or metric equivalent) lower the micrometer head until the spindle measuring face makes contact with the gage block stack. The micrometer head has no ratchet stop or friction thimble to control the force applied. It is important to not overdrive the spindle and introduce an error caused by disturbing the setup. It does not require a great amount of force to lift the setup. Care must be taken to apply the measurement gently but firmly. Repeat the measurement three times to assure the repeatability and accuracy of the measurement.

3) Repeat Item 2) for all remaining gage block stacks throughout the full range of the UUT. Record all readings and verify that all readings are within tolerance limits. Adhere to manufacturers' tolerances if different than the tolerance limits listed in *Calibration Tolerance Tables*, Table 14a and Table 14b.

## Calibration of Thread Micrometers

*Note:* The *Thread (Pitch) Micrometer* uses a dedicated set of spindle and anvil contacts that measure the pitch diameter of a thread of a given pitch range. There are fixed anvil micrometers as well as interchangeable anvil/spindle sets. The interchangeable sets come with matched spindle/anvil sets that can be installed into the micrometer frame and then set to zero.

**Linearity Calibration (Fixed Pitch).—**The fixed anvil micrometers usually do not have the capability to reset the zero indication other than the standard barrel adjustment. The calibration of these instruments is achieved by the measurement of master setting thread plug members that are selected to cover, as close as possible, 25, 50 and 75 percent of the range of the UUT.

1) Three calibrated *Master Setting Thread Plug Gages* are selected that cover 25, 50 and 75 percent of the measuring range of the UUT in the range of pitch that the UUT is designed for. (i.e., 8–13 pitch, 16–18 pitch, 18–24 pitch).

2) Mount the pitch micrometer in a micrometer stand to enable easier operation of the UUT and to provide better control of the standard.

3) Gently bring the *Anvil* and *Spindle* together to achieve a zero indication. If zero reading is not exact, adjust the barrel so no error from zero exists. Repeat the zero indication to assure repeatability.

4) Open the micrometer anvil and spindle beyond the point of the first standard pitch diameter and seat the standard against the anvil.

5) Slowly close the micrometer until the spindle cone seats into the thread opposite the supported anvil.

6) Gently roll the standard up and down against the spindle cone to seek the highest point on the diameter. When this point is located, take a measurement of the pitch diameter.

7) Record the measurement and assure that the measurement is within the tolerance limits for the UUT.

*Note:* UUT with the resolution of ±0.001 require a Class X master thread setting plug gage. UUT with a resolution of ±0.0001 require a class W master thread setting plug gage.

8) Remove the first standard and repeat measurement throughout the remainder of the master thread setting plug gages selected for the UUT.

**Interchangeable Anvil Pitch Micrometer Sets.—**1) *Interchangeable Anvil Pitch Micrometer* sets include several sets of Anvil and Spindle Cone, and V Anvils. These sets are dedicated to a range of thread pitch. Most sets also include a set of flat anvil and spindle inserts that can be used to calibrate the micrometer exactly the same as a standard micrometer. If this is the case proceed as follows.

2) Select the flat anvil and spindle set and install.

3) Zero the anvil and spindle and assure repeatability.

4) Perform calibration as described in *Linearity Test (Outside Micrometers)*, Item 3) on page 780 using the defined test points to verify the accuracy of the micrometer head.

5) Remove the flat anvil and spindle set and install the smallest (highest pitch) dedicated Anvil V and Spindle Cone set. Close anvil and spindle and re zero the micrometer to the new set.

6) Select one master thread setting plug gage for the pitch range, preferably at midrange of travel, and measure as described in *Linearity Calibration (Fixed Pitch)* section. Verify that the measurement is within the tolerance limits of the UUT. This measurement verified the accuracy of the flank angle of the cone and V anvil. This measurement can also be performed on an optical comparator. However, the physical measurement should be performed as the final step in the calibration of the interchangeable cone and anvil sets.

7) Repeat for all remaining Anvil V and Spindle Cone sets.

## Calibration of Inside Micrometers

**Caliper Type Inside Micrometer.—**A description of the calibration and equipment requirements for the Type II, Class 1, Inside Micrometer with Jaws are given in Table 8.

**Table 8. Calibration Requirements of the Type II, Class 1, Inside Micrometer with Jaws**

| Calibration Description | | |
|---|---|---|
| UUT Characteristics | Performance Specifications | Test Method |
| UUT Length and Linearity | Range:<br>0.200 to 1.000 inch<br>0.500 to 1.500 inch<br>1.000 to 2.000 inch<br>(or metric equivalent) | Comparative measurement of UUT reading to gage block test setup to test the basic length and micrometer head linearity. |

| Equipment Requirements | | | |
|---|---|---|---|
| Description | Minimum Use Specifications | Calibration Equipment | Sub Item |
| Gage Block Set | Range: ±0.05 to 4 inch<br>Accuracy: ±12 μin from stated value | L.S. Starrett Co. SS 36.A1 | L.S. Starrett Co. SS 81.A1 |

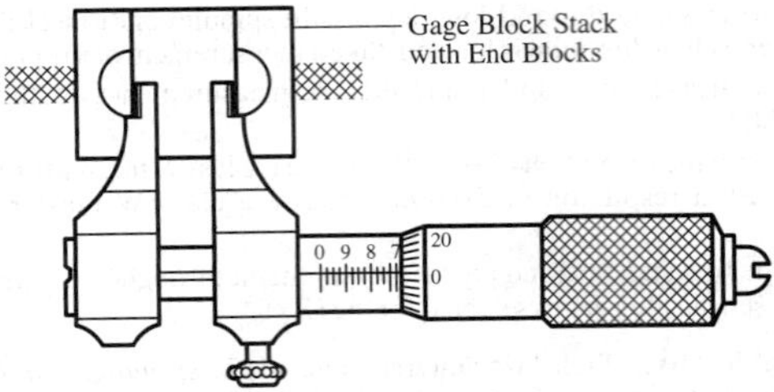


Fig. 19. Measure Gage Block Stack Assembled with End Blocks

**Calibration Process.—**1) Set the UUT to indicate the minimum reading for the UUT using the ratchet or friction thimble. Assure that the reading indicates the minimum for the UUT model and range. If the reading is observed as off the minimum at this time, assure there is no foreign particle or burr on the anvil or spindle. An accurate zero indication must be established at this time as all other measurements will be affected by any error at zero.

2) The recommended size gage blocks used to test the micrometer spindle thread are very important. The recommended test points achieve the testing of the accuracy of the micrometer throughout the entire length and at points evenly distributed around the diameter of the spindle thread. This verifies the accuracy of the lead error of the thread as well as identifying any sections that may have experienced excessive wear or damage. The recommended sizes are given in Table 9.

**Table 9. Recommended Test Points for Type II, Class 1, Inside Caliper Micrometer**

| Size | | Range | | Recommended Test Points | |
|---|---|---|---|---|---|
| (in.) | (mm) | (in.) | (mm) | (in.) | (mm) |
| 1 | 25 | 0.200 to 1 | 5 to 25 | 0.210, 0.420, 0.605, 0.815 and 1.000 | 5.10, 10.20, 15.30, 20.40 and 25.00 |
| 1½ | 30 | 0.500 to 1½ | 5 to 30 | 0.510, 0.720, 0.905, 1.115, and 1.500 | 7.10, 12.20, 17.30, 23.40, and 30.00 |
| 2 | 50 | 1 to 2 | 25 to 50 | 1.210, 1.420, 1.605, 1.815, and 2.000 | 25.10, 30.20, 35.30, 40.40, and 50.00 |

*Note:*

a) The micrometer should be tested at zero and full scale.

b) Care must be taken when wringing gage blocks together to achieve the desired size. It is critical to clean the gage blocks thoroughly before wringing and assembling the end blocks to avoid damaging the superior finish of the standard.

3) Assemble the first gage block stack and measure it using the ratchet or friction thimble as described in the *Linearity Test (Outside Micrometers)* on page 780. Do not rely on feel as the correct and controlled loading of measuring force by the ratchet or friction thimble will give the most accurate and repeatable reading. Several clicks of the ratchet or applying force until the friction thimble slips and assuring the micrometer is gently seated on the gage block stack will give an accurate reading. Gently rock the micrometer to assure that the Nibs are parallel to the gage block end block faces. Repeat the measurement several times to assure the reading is repeatable. The reading must meet the tolerance limits as defined in Table 10, *Tolerance* column.

**Table 10. Tolerance Table: Type II, Class 1 Inside Micrometer with Jaws**

| Size (in.) | Range (in.) | Tolerance (in.) |
|---|---|---|
| 1 | 0.200 to 1 | ±0.001 |
| $1\frac{1}{2}$ | 0.500 to $1\frac{1}{2}$ | ±0.001 |
| 2 | 1 to 2 | ±0.001 |

4) Repeat these steps for each remaining gage block stack and record readings for evaluation or trend analysis. Follow all guidelines of the quality assurance programs in place in regards to recording and retaining all data collected and reporting OOT conditions.

**Tubular Type Inside Micrometer.—**A description of the equipment requirements for the tubular type inside micrometer is given in Table 11.

**Table 11. Calibration Equipment Standards**

| Description | Minimum Use Specifications | Calibration Equipment | Alternate Selection |
|---|---|---|---|
| Gage Block Set | Range: 0.05 to 4 in<br>Accuracy: ±12 μin from stated value | L.S. Starrett Co.<br>SC 36.A1<br>SC 36.MA1X | L.S. Starrett Co.<br>SC 88.A1X<br>SC 88.MA1X |
| Gage Block Set | Range: 5 to 20 in<br>Accuracy: ±5 μin/in | L.S. Starrett Co.<br>SS8.A1X | |
| Supermicrometer | Range:<br>0 to 10.00 inch<br>0 to 250 mm<br><br>Accuracy: 20 μin | Pratt & Whitney<br>Model PC-250 | |
| Standard Measuring Machine | Range: 0 to 148 inch<br>0 to 1200 mm<br>Accuracy: 50 μin | Pratt & Whitney<br>Model DMM-1219 | |

**Preliminary Operations.—**1) UUT shall be brought into the calibration area at least 4 hours prior to beginning the calibration process.

2) Ensure that all working surfaces are clean and free from dust and dirt.

3) The UUT must be free from all rust, nicks and burrs that would affect the results of the measurements and operation of the UUT.

4) Linear measurements are affected by temperature and all measurements must be conducted in a temperature controlled environment. The preferred environment is 68° F with no greater than ±1° per hour deviation from nominal. The temperature must have been stable and within the preferred range for a minimum of 8 hours prior to beginning the calibration soak time.

**Length Calibration Process.—***Preferred Method:* 1) Install the shortest extension rod available for the UUT set being calibrated. If the UUT has end caps (Type II Class 3 Style A) install just the end caps. This section will calibrate the micrometer head accuracy.

*Note:* If the TI being calibrated is equipped with two end caps that can be removed and interchanged (Type II Class 3 Style A), the end caps must be checked for wear and must be the same length (nominal 0.250 inch). If the end caps differ in length by more than ±0.00025 inch (as measured on the *Supermicrometer* or *Standard Measuring Machine*, both TI end caps must be replaced.

2) Select a gage block of a length close to the nominal length of the UUT when set to zero on the micrometer head and the extension rod or end cap for minimum length indication is installed.

3) For UUT with greater than 0.0001 inch (0.001 mm) resolution, gage block stacks that are wrung together to achieve the nominal setting dimension over 8 inches, must soak for one hour before establishing the setting dimension. This does not apply to single gage blocks used as setting standards.

4) Set the *Reference* of the supermicrometer or standard measuring machine to zero.

5) With the UUT set to zero, place into the measuring plane of the supermicrometer or standard measuring machine and measure the UUT. Assure that the UUT is supported on either a flatted round or adequate support to allow it to maintain alignment with the axis of measurement of the standard.

6) Verify that the measurement is within the tolerance limits for the UUT.

7) Remove the UUT from the standard and set to the next test point as defined in Table 12 for the 1 inch (25 mm) micrometer head.

8) Repeat Items 5) and 6) for all remaining test points.

*Alternate Method:* 1) Install the shortest extension rod or end caps and set UUT to zero on the micrometer head. Wring together appropriate size gage blocks to measure the zero indication assembly.

2) Wring or attach the *Caliper Jaw End Blocks* from the gage block accessories and assure the setup is square and true. If using the connecting rods, minimal force is required to secure the end blocks to the gage block stack.

*Note:* For UUT with greater than 0.0001 inch resolution, gage block assemblies longer than 8.00 in (200 mm) will require 1 hour stabilization time before taking measurement.

3) Use the UUT to measure the gage block stack. Care must be taken to assure that the measurement is taken at the shortest distance between the caliper end block faces.

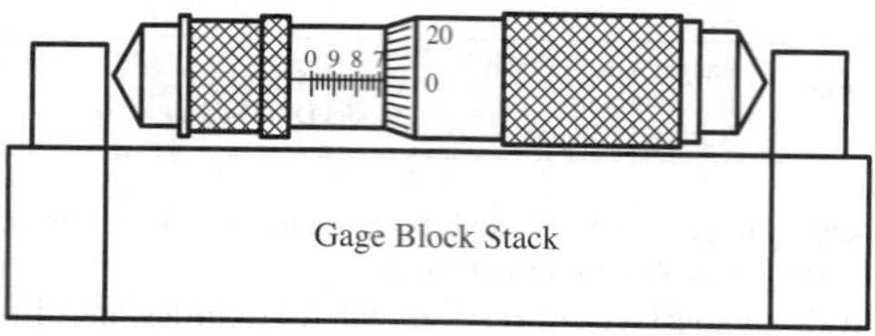


Fig. 20. Inside Micrometer Gage Block Stack Measurement

4) Repeat Items to 3) for each remaining test point. Measure each gage block setup and assure that the indication is within the tolerance limits for the Type II Class 2, Type II Class 3 Style A, and Type II Class 3 Style B inside micrometers.

**Extension Rod Calibration.**—*Preferred Method:* 1) Remove the shortest extension rod or one end cap used in the linearity calibration and replace with the next UUT extension rod in the set.

2) Ensure the *Standard Measuring Machine* has retained its reference. Verify the zero setting if necessary. The *Supermicrometer* will require re-mastering every 1 inch and can only be used up to 10 inches (generally). The standard measuring machine uses a continuous measurement path but must be re-mastered every 8 inches.

3) For UUT with 0.0001 inch (0.001 mm) resolution, allow for 1 hour stabilization time for gage block stacks that are wrung to achieve nominal length over 8 inches (200 mm).

4) Using a flatted round or appropriate supporting device, mount the UUT into the standard. Assure the alignment of the measuring path and measure the UUT.

5) Confirm that the indication is within the tolerance limits for the Type II Class 2, Type II Class 3 Style A, and Type II Class 3 Style B inside micrometer.

6) Repeat Items to 5) for all remaining extension rods, sleeves and end caps in the set.

*Alternate Method:* 1) Remove the shortest extension rod or one end cap used in the linearity calibration, and replace with the next UUT extension rod in the set.

2) Wring together a gage block stack to achieve the nominal desired length of the micrometer head at zero indication and the included extension rod.

3) Wring, or attach using the accessory attachment rods, the outside caliper jaws to the gage block stack. Assure stack is square and true. Minimal force is required to secure the outside caliper jaws to the gage block stack.

4) For UUT with 0.0001 inch (0.001 mm) resolution, allow for 1 hour stabilization time for gage block stacks that are wrung to achieve nominal length over 8 inches (200 mm).

5) Use the UUT to measure the gage block stack.

6) Confirm that the measured value is within the tolerance limits for the Type II Class 2, Type II Class 3 Style A, and Type II Class 3 Style B inside micrometer.

7) Repeat Items to 6) for all remaining extension rods, sleeves and end caps in the set.

## Calibration of Vernier, Dial and Digital Calipers

**Preliminary Operations.—**1) Review and become familiar with the entire calibration procedure before beginning calibration process. Ensure familiarity with setup and operation of all standards involved in the procedure.

2) Always observe safe handling methods for all standards and exercise caution in the handling of all precision standards.

3) Ensure the work area is clean, well illuminated, and free from excessive drafts and humidity.

4) Ensure measuring surfaces are clean and free of nicks and burrs that could affect the accuracy of the measurements. Observe the Outside Measuring Jaws and Inside Measuring Jaws for any possible damage and observe the unit for wear or deformation or any condition that may affect the parallelism of the measuring surfaces. Close the measuring jaws completely and tighten the lock screw. Hold the UUT in front of a strong light source and verify that there are no burrs on the outside diameter or inside diameter measuring jaws. If any condition of concern is observed in regards to nicks or burrs along the beam, light stoning with a hard Arkansas stone can be used with care to remove these incursions.

5) Observe the condition of the rack along the full length of a dial caliper to assure there is no damage to the rack gear tooth or embedded chips or dirt that will affect or damage the movement. Look for bent or damaged gearing. Open and close the slide slowly and listen and feel for evidence or broken or damaged drive pinion gear.

6) Assure that the movement along the length of the beam does not bind or indicate the presence of any nicks or burrs along the beam. Assure the movement is not too loose. Observe the position of the gib slide screws that they have not been removed or loosened.

7) Bring the UUT (Unit Under Test) into the controlled environment a minimum of 4 hours prior to beginning the calibration process. Allow all standards and UUT to stabilize to common temperature before proceeding.

**Table 12. Equipment Requirements and Minimum Accuracy Specifications of Required Standards**

| Noun | Minimum Use Specifications | Recommended Measurement Standards | Alternative |
|---|---|---|---|
| Gage Block Set | Range:<br>0.050 to 4.000 inch<br>2.0 to 450 mm | Starrett Webber SC 81.A1<br>Starrett Webber S2C 88.MA1 | Mitutoyo 516-401-16<br>Mitutoyo 516-442-10 |
| Gage Block Set (Long Block) | 5 to 84 inch<br>125 to 500 mm | Starrett Webber SS 8.A1X<br>Starrett Webber SS 8.MA1X | Mitutoyo 516-762-16<br>Mitutoyo 516-753-10 |

**Table 12. Equipment Requirements and Minimum Accuracy Specifications of Required Standards** *(continued)*

| Noun | Minimum Use Specifications | Recommended Measurement Standards | Alternative |
|---|---|---|---|
| Precision Micrometer | Range: 0–1.0000 inch 0–25 mm | Mitutoyo Series 293 | Starrett No.3732 |
| Precision Pin or Ball | 0.500 inch / 12 mm | Van Keuren Class XX | Precision Ground Pin or Ball ±40 in / 0.001 mm |
| Surface Plate | Range: 24 x 36 Accuracy: Grade A | Starrett Crystal Pink EDP 80655 | Mitutoyo 517-808 |

*Note:* Standards listed are recommended and are determined to meet the necessary accuracy requirements of the calibration process. Substitute or alternative standards may be used as long as they have a current in tolerance calibration status and have been determined to meet or exceed the minimum accuracy specifications as listed in Table 12.

*Note:* Special care should be taken to clean and apply protective coating when handling gage blocks or other precision standards with bare hands. The use of clean cotton or vinyl gloves is recommended to prevent corrosive oils from damaging precision surfaces.

**Vernier, Dial and Digital Caliper Calibration Process.**—The temperature of the calibration environment must be maintained at approximately 23° C with a fluctuation in temperature of no more than ±0.1° per hour.

**Table 13. Range and Tolerance Limits**

| Feature | Range (inch) | Accuracy (inch) | Method |
|---|---|---|---|
| Scale Zero | Set at Zero | ±0.0005 | Determined by observation of scale |
| Jaw Parallelism | 0.00–0.250 | ±0.001 inch | Measured with Precision cylindrical plug gage |
| ID Jaws (Nibs) Set at Scale Zero | 0 to 6.000–0.250<br>0 to 9.000–0.300<br>0 to 12.000–0.300<br>0 to 24.000–0.300<br>0 to 36.000–0.500<br>0 to 48.000–0.500 | (inch)<br>+0.000<br>−0.005 | Measured with precision micrometer |
| Outside (OD) Jaws | 0–6.00 inch<br>6–24 inch<br>24–48 inch | ±0.0010 inch<br>±0.0020 inch<br>±0.0030 inch | Determined at 4 equal points along the main scale by direct comparison with gage blocks |
| ID Jaws (Knife Edge) | 0–6.00 inch<br>6–24 inch<br>24–48 inch | ±0.0010 inch<br>±0.0020 inch<br>±0.0030 inch | Measured with gage blocks |
| Depth Rod | 0.000–6.000 inch | ±0.001 inch | Measured with gage blocks |

*Note:* Metric equivalents apply to all tolerance limits stated in Table 13. All tolerances based on AFMETCAL, NAVAIR and GGG-C-111guidelines and specifications. Default to manufacturers specifications if different from those represented here.

**Nib Calibration, Inside Measurement, Type 1.**—*Class 1, 2, and 3:* The *Total Nib Width* (measured to the fourth place) shall be measured over both nibs with the jaws fully closed as shown in Fig. 21. The offset value shall be marked on the *Inside Measurement* side of the sliding jaw or main beam and on both sides of the main beam of the Type 1, Class 3.

a) Alternately the vernier scale can be adjusted to directly compensate for and deviation from nominal dimension of the nib. This can be accomplished by adjusting the inside measurement vernier scale to zero on a known standard. Measurement of the nib nominal dimension must still be performed as excessive wear to the nib will result in adjustment in excess of the available range of the vernier scale.

b) Assemble a 1.000 inch gage block with accessory end blocks as shown in Fig. 23 of the *Calibration of Inside Dimension Jaws (Nibs)* section.

c) Measure the gage block stack and adjust the vernier scale to zero at this known point.

d) This method is applied to the Type 1 style caliper only.

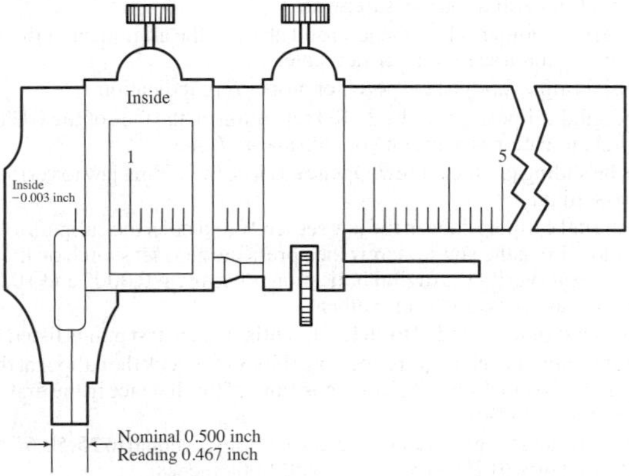


Fig. 21. Nib Dimension Measurement Location

**Calibration of Jaw Parallelism.—** 1) Slide jaws together and tighten the sliding jaw lock screw. Hold the UUT in front of a strong light source. Observe that no light is visible between the jaw measuring surfaces. Look carefully for evidence of burrs, damage to the measuring surfaces, excessive wear or bowing.

a) If any burrs or damage appears gently stone the jaws with a white hard Arkansas stone. Only use a super fine grain stone for this operation.

2) Loosen the sliding jaw lock screw and open the jaws. Place the precision cylindrical plug gage or ball in between the jaws as close as possible to the beam and measure the standard. Observe the scale, dial, or digital reading.

3) Move the plug gage or ball to the opposite end of the measuring jaws (tip) and remeasure. The difference between the two readings must be within tolerance stated in Table 13. *Range and Tolerance Limits* for "Outside Jaws."

4) If the readings are within tolerance, proceed to Calibration of Outside Jaws. If adjustment is needed, go to Items 5) through 7), in *Adjustment of Gib Slide Set Screws*.

**Adjustment of Gib Slide Set Screws.—** 5) Using a small precision screwdriver loosen the gib slide set screws then retighten evenly with a slight pressure.

a) It is important that both set screws are tightened evenly and a slight resistance is felt in the sliding jaw full range movement.

6) Operate the UUT several times over its full range to ensure that the sliding jaw can be moved smoothly over the full length of the beam.

7) Repeat steps 1 through 3 to verify the Jaw Parallelism in within tolerance as stated in Table 13 for Outside Jaws.

8) If the readings are within tolerance, proceed to *Calibration of Outside Jaws*. If readings are not within tolerance, record all readings as "As Found" and consult the end user to determine the course of action that they desire. Excessive damage or wear to the Jaws, Nibs or ID measuring contacts can result in a high degree of uncertainty in all measurements taken with the instrument and is cause for concern. Consideration of these facts should be addressed before proceeding.

**Calibration of Outside Jaws.**—1) Slide the UUT jaws together and tighten the sliding jaw set screw.

2) Set zero for this calibration measurement.

a) For vernier calipers close the jaws and observe the alignment of the zero indication on the main scale and vernier scale.

b) For dial calipers, adjust the bezel for proper zero indication.

c) For digital calipers, press the ZERO set button on the face of the UUT. Verify the digital indication is between −0.0005 and +0.0005.

3) Loosen the sliding jaw lock screw. Open and close the sliding jaw to verify repeatability of the zero setting.

a) For digital calipers, loosen the jaw set screw and open the calipers to a mid-point location. Lock the jaw set screw then press the zero set switch on the front of the instrument. Verify the digital indication is between -0.0005 and +0.0005. Close the slide jaw and zero the instrument.

4) Select the gage blocks needed to check the caliper at the test points listed below:

a) For dial calipers, select the proper gage blocks to check the caliper at the 3 o'clock, 6 o'clock, 9 o'clock and 12 o'clock positions of the dial face in the first inch. (0.125, 0.550, 0.675, 1.000)

b) For digital and vernier calipers, select gage blocks equal to 25, 50, 75, 100 percent of the first inch (0.250, 0.500, 0.750 and 1.000 inches).

5) Select the gage blocks needed to check the UUT at 25, 50, 75, and 100 percent of full scale. Place the UUT and all gage blocks on the surface plate for a minimum of 15 minutes to stabilize. The gage blocks should have both working surfaces exposed, as in Fig. 22.

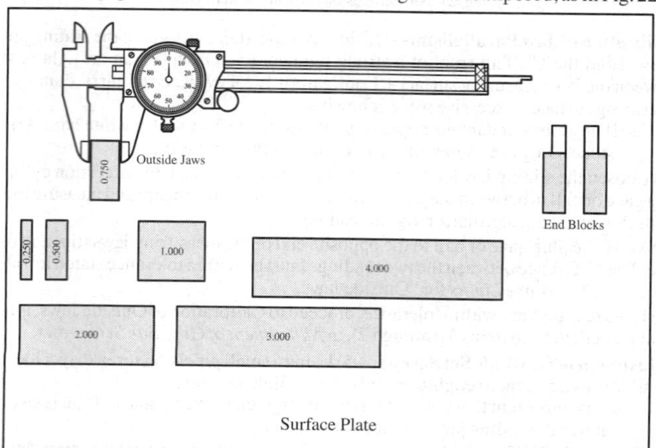


Fig. 22. Gage Blocks and End Blocks Staged for Calibration of 6-Inch Dial Caliper

6) Open the outside jaws and measure the gage block that equates to the first test point defined in step 4. Ensure that the jaws are squarely placed with a slight pressure against the gage block measuring surfaces. Use the thumb wheel or fine adjust to apply gentle but steady pressure against the gage block. Rock the caliper gently to ensure that the outside diameter measuring faces are seated squarely. Seek the smallest reading obtainable with gentle pressure. Record the value.

7) Repeat Item 6) for remaining test points defined in Item 4), parts a) and b), and Item 5).

8) If the outside jaws are within tolerance, proceed to the next section: *Calibration of Inside Dimension Jaws (Nibs)*. If the outside jaws are found Out of Tolerance, continue through each calibration section and record all data as "As Found" prior to any further adjustments. Upon completion of procedure, report any out of tolerance conditions that may have been discovered. *Any out of tolerance conditions must be addressed before returning the UUT to service.*

**Calibration of Inside Dimension Jaws (Nibs).—**1) Select the gage block needed to check the UUT at the first test point as defined above in Item 4) of *Calibration of Outside Jaws*. Clean the surface of the gage block and end blocks with lint free wipe and alcohol and assemble the Accessory End Blocks and assure they are secured in position snugly.

2) Close the caliper jaws and assure the dial or digital indication or vernier scale indicate zero.

3) Measure the dimension of the gage block as illustrated in Fig. 23. Use a gentle rocking motion to assure that the jaws are seated squarely. Seek the lowest reading as this is the true distance between the parallel faces of the end blocks.

*Note:* Excessive force must be avoided as this can cause deformation of the UUT jaws and result in erroneous readings. A gentle consistent feel must be used to provide an accurate reading. If too much force is applied or the ID measuring jaws are not in line and perpendicular to the parallel planes of the end blocks, a larger and unrepeatable reading will result. Repeat all measurements several times to assure that you have a true reading.

4) Repeat the set up and readings at the 3 o'clock, 6 o'clock, 9 o'clock and 12 o'clock positions of the dial face in the first inch. For vernier and digital calipers measure at 0.250, 0.500, 0.750 and 1.000 inches. Record all readings.

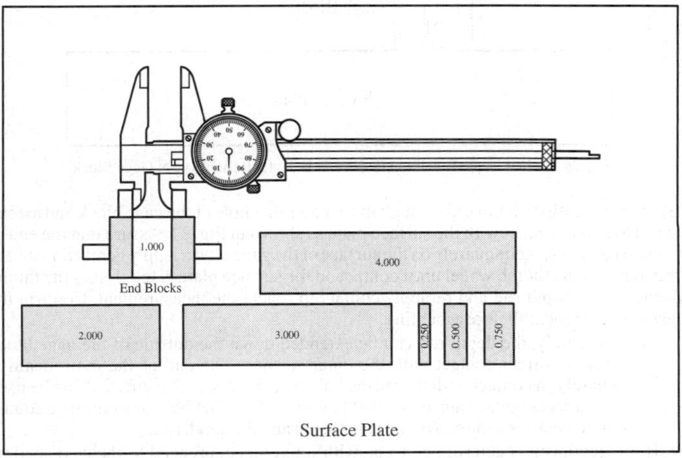


Fig. 23. Inside Dimension Measurement of Gage Blocks using Accessory End Blocks

5) Measure the remainder of the length of the UUT at 25, 50, 75, and 100 percent of full scale. Select the appropriate gage blocks and assemble as illustrated in Fig. 23. Measure each test point and record all readings.

**Calibration of Depth Rod.—**1) Select a 1.000 inch gage block and position it on the surface plate with the measuring surface against the plate.

2) Close the caliper jaws completely and place the end of the mail beam flat on the surface plate in a vertical position. Apply slight downward pressure bringing the depth rod in contact with the surface plate. Zero the dial or display at this point.

a) Type 1 Class 1 calipers do not have depth measurement capability.

b) Type 1 Class 2 have a depth rod that is directly tied to the Vernier Scale. If the Zero Reading is OOT (out of tolerance) at this point the depth rod will require service or replacement.

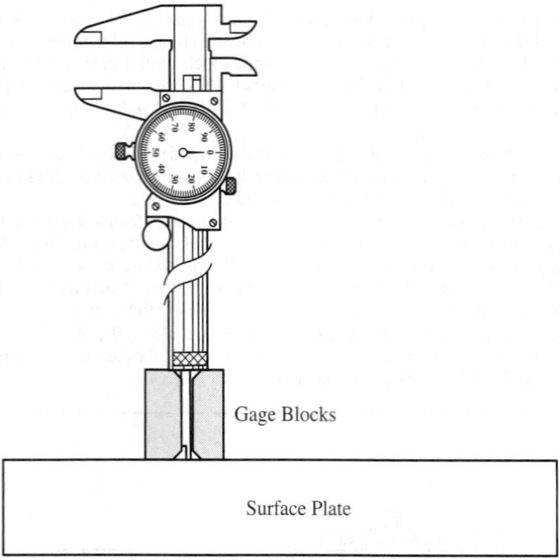


Fig. 24. Caliper Depth Rod Extended through Center Relief Hole of Gage Block.

3) Place the caliper depth rod through the center relief hole of the gage block and extend the rod to make contact with the surface plate as shown in Fig. 24. Assure that the end of the main beam is seated squarely on the surface of the gage block. Apply gently downward pressure with the thumb wheel until contact on the surface plate is felt. Using the thumb wheel, lift the depth rod and re-apply contact to repeat the measurement. Confirm the measurement repeats before accepting.

a) Alternately, the depth rod can be extended down the outside of the gage block while in direct contact with the outer surface. The end of the main beam is squarely in contact with the surface of the gage block. This method can be used with a rectangular gage block that has no center relief hole or to assure a square contact between the end of the main beam and the depth rod.

4) If the depth rod is determined to be within tolerance, proceed to the next section, *Calibration of Step Height*. If the depth rod is found out of tolerance, continue through

each calibration section and record all data as "As Found" prior to any further action. Upon completion of procedure, report any out of tolerance conditions that may have been discovered. *Any out of tolerance conditions must be addressed before returning the UUT to service.*

**Calibration of Step Height.—**1) Close the caliper jaws completely and set zero on the dial or display.

2) Place the caliper head downward on the surface plate as shown in Fig. 25.

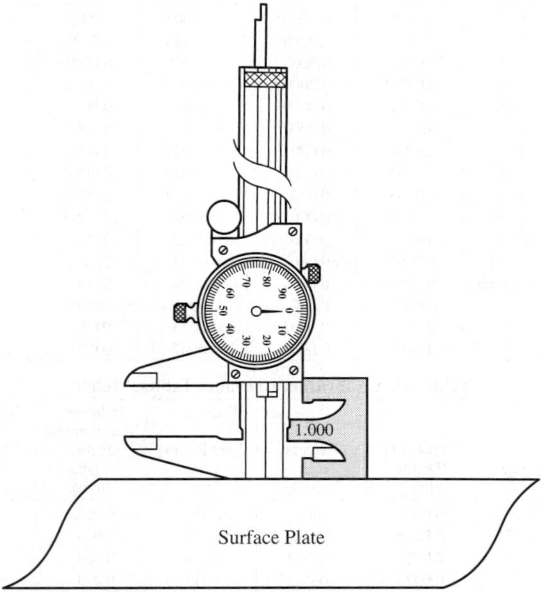


Fig. 25. Step Height Measurement with Gage Block in Position behind Main Beam Engaging Step Height Measurement Surface

3) Open the caliper jaws to accommodate the measurement of a 1.000 inch gage block by the Step Height measuring surface on the back side of the moveable jaw.

4) Bring the *Step Height* measuring surface into contact with the surface of the gage block. Assure the back of the main beam is in direct contact with the outside of the gage block to establish a good perpendicular condition.

5) Apply gently downward pressure with the thumb wheel.

6) Repeat the measurement several times to confirm repeatability before accepting the measurement.

7) Record the measurement. There is no adjustment for this feature.

8) If the Step Height is determined to be within tolerance, remove and store all standards in their appropriate containers and apply any corrosion protection methods as required. If the step height is found Out of Tolerance, record all data as "As Found" prior to any further action. Upon completion of procedure, report any out of tolerance conditions that may have been discovered. *Any out of tolerance conditions must be addressed before returning the UUT to service.*

## Calibration Tolerance Tables

### Table 14a. Calibration Tolerance Tables, English

| Size (in.) | Range (in.) | Permissible Flexure (in.) | Measuring Faces | | Indicated Measurement Error, max. (in.) | Spindle/Anvil Alignment Error max. (in.) |
|---|---|---|---|---|---|---|
| | | | Flatness, max. (in.) | Parallelism, max. (in.) | | |
| 1 | 0–1 | 0.0001 | 0.00005 | 0.00005 | 0.0001 | 0.002 |
| 2 | 1–2 | 0.0001 | 0.00005 | 0.00010 | 0.00015 | 0.003 |
| 3 | 2–3 | 0.0001 | 0.00005 | 0.00015 | 0.00015 | 0.0045 |
| 4 | 3–4 | 0.00015 | 0.00008 | 0.0002 | 0.0002 | 0.006 |
| 5 | 4–5 | 0.00015 | 0.00008 | 0.0002 | 0.0002 | 0.007 |
| 6 | 5–6 | 0.00015 | 0.00008 | 0.0002 | 0.0002 | 0.009 |
| 7 | 6–7 | 0.0002 | 0.00008 | 0.00025 | 0.00025 | 0.010 |
| 8 | 7–8 | 0.0002 | 0.00008 | 0.00025 | 0.00025 | 0.010 |
| 9 | 8–9 | 0.0003 | 0.0001 | 0.00025 | 0.00025 | 0.010 |
| 10 | 9–10 | 0.0003 | 0.0001 | 0.0003 | 0.0003 | 0.010 |
| 11 | 10–11 | 0.0003 | 0.0001 | 0.0003 | 0.0003 | 0.010 |
| 12 | 11–12 | 0.0003 | 0.0001 | 0.0003 | 0.0003 | 0.010 |
| 13 to 18 | Per 1 inch | 0.0004 | 0.0001 | 0.0004 | 0.0004 | 0.010 |
| 19 to 24 | " | 0.0005 | 0.0001 | 0.0005 | 0.0005 | 0.010 |
| 25 to 30 | " | 0.0006 | 0.0001 | 0.0006 | 0.0006 | 0.010 |
| 31 to 36 | " | 0.0007 | 0.0001 | 0.0007 | 0.0007 | 0.010 |

### Table 14b. Calibration Tolerance Tables, Metric

| Size (mm) | Range (mm) | Permissible Flexure (mm) | Measuring Faces | | Indicated Measurement Error, max. (mm) | Spindle/Anvil Alignment Error, max. (mm) |
|---|---|---|---|---|---|---|
| | | | Flatness, max. (mm) | Parallelism, max. (mm) | | |
| 13 | 0–13 | 0.0025 | 0.001 | 0.0012 | 0.003 | 0.03 |
| 25 | 0–25 | 0.0025 | 0.0012 | 0.0012 | 0.004 | 0.05 |
| 50 | 25–50 | 0.0025 | 0.0012 | 0.0025 | 0.004 | 0.08 |
| 75 | 50–75 | 0.0025 | 0.0012 | 0.004 | 0.004 | 0.11 |
| 100 | 75–100 | 0.004 | 0.002 | 0.005 | 0.005 | 0.15 |
| 125 | 100–125 | 0.004 | 0.002 | 0.005 | 0.005 | 0.19 |
| 150 | 125–150 | 0.004 | 0.002 | 0.005 | 0.005 | 0.23 |
| 175 | 150–175 | 0.005 | 0.002 | 0.006 | 0.006 | 0.25 |
| 200 | 175–200 | 0.005 | 0.002 | 0.006 | 0.006 | 0.25 |
| 225 | 200–225 | 0.005 | 0.002 | 0.006 | 0.006 | 0.25 |
| 250 | 225–250 | 0.008 | 0.0025 | 0.0075 | 0.0075 | 0.25 |
| 275 | 250–275 | 0.008 | 0.0025 | 0.0075 | 0.0075 | 0.25 |
| 300 | 275–300 | 0.008 | 0.0025 | 0.0075 | 0.0075 | 0.25 |

*Note: Indicated Measurement Error* tolerances for UUT with 0.001 inch graduations (no vernier scale) are set to ±0.001 up to 30 inch, and ±0.002 inch from 31 to 36 inch. Tolerances of UUT over 36 inch are subject to manufacturer's specification. Tolerance for UUT with a resolution of 50 μin shall be ±50μin for 0–1 inch; ±0.0001 inch for 1 through 4 inches; ±0.0002 inch for 4 three 9 inches; ±0.0003 inch for 9 three 12 inches; ±0.0004 inch for 12 to 18 inches; ±0.0005 for 18 to 24 inches; ±0.0006 for 24 to 30 inches; and ±0.0007 for 30 to 36 inches. All metric equivalents shall apply to UUT of this resolution. Metric Scale UUT with 0.01 mm graduations are set at ±0.01 mm for up to 300 mm. Tolerances above 300 mm are subject to manufacturers specifications.

## SURFACE TEXTURE

### American National Standard Surface Texture (Surface Roughness, Waviness, and Lay)

American National Standard ANSI/ASME B46.1-2019 is concerned with the geometric irregularities of surfaces of solid materials, physical specimens for gaging roughness, and the characteristics of stylus instrumentation for measuring roughness. The Standard defines surface texture and its constituents: roughness, waviness, lay, and flaws. A set of symbols for drawings, specifications, and reports is established. To ensure a uniform basis for measurements, the Standard also provides specifications for Precision Reference Specimens, and Roughness Comparison Specimens, and establishes requirements for stylus-type instruments. The Standard is not concerned with luster, appearance, color, corrosion resistance, wear resistance, hardness, subsurface microstructure, surface integrity, or many other characteristics that may govern considerations in specific applications.

The Standard is expressed in SI metric units, but US customary units may be used without prejudice. The Standard does not define the degrees of surface roughness and waviness or type of lay suitable for specific purposes, nor does it specify the means by which any degree of such irregularities may be obtained or produced. However, criteria for selection of surface qualities and information on instrument techniques and methods of producing, controlling and inspecting surfaces are included in Appendixes attached to the Standard. The Appendix sections are not considered a part of the Standard: they are included for clarification or information purposes only.

Surfaces, in general, are very complex in character. The Standard deals only with the height, width, and direction of surface irregularities because these characteristics are of practical importance in specific applications. Surface texture designations as delineated in this Standard may not be a sufficient index to performance. Other part characteristics such as dimensional and geometrical relationships, material, metallurgy, and stress must also be controlled.

**Definitions of Terms Relating to the Surfaces of Solid Materials.—**The terms and ratings in the Standard relate to surfaces produced by such means as abrading, casting, coating, cutting, etching, plastic deformation, sintering, wear, and erosion.

*Error of form* is considered to be that deviation from the nominal surface caused by errors in machine tool ways, guides, insecure clamping or incorrect alignment of the workpiece or wear, none of which are included in surface texture. Out-of-roundness and out-of-flatness are examples of errors of form. See ANSI/ASME B89.3.1-1972 (R1997) for measurement of out-of-roundness.

*Flaws* are unintentional, unexpected, and unwanted interruptions in the topography typical of a part surface and are defined as such only when agreed upon by buyer and seller. If flaws are defined, the surface should be inspected specifically to determine whether flaws are present, and rejected or accepted prior to performing final surface roughness measurements. If defined flaws are not present, or if flaws are not defined, then interruptions in the part surface may be included in roughness measurements.

*Lay* is the direction of the predominant surface pattern, ordinarily determined by the production method used.

*Roughness* consists of the finer irregularities of the surface texture, usually including those irregularities that result from the inherent action of the production process. These irregularities are considered to include traverse feed marks and other irregularities within the limits of the roughness sampling length.

*Surface* is the boundary of an object that separates that object from another object, substance or space.

*Surface, measured* is the real surface obtained by instrumental or other means.

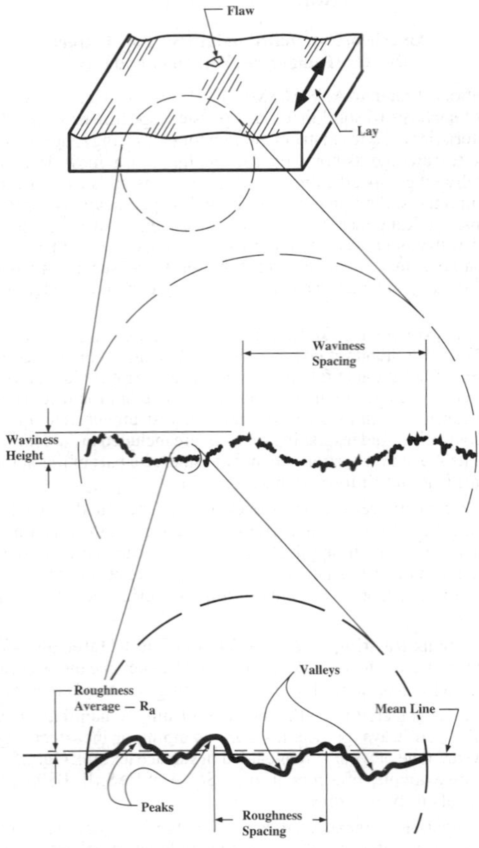


Fig. 1. Pictorial Display of Surface Characteristics

*Surface, nominal* is the intended surface contour (exclusive of any intended surface roughness), the shape and extent of which is usually shown and dimensioned on a drawing or descriptive specification.

*Surface, real* is the actual boundary of the object. Manufacturing processes determine its deviation from the nominal surface.

*Surface texture* is repetitive or random deviations from the real surface that forms the three-dimensional topography of the surface. Surface texture includes roughness, waviness, lay and flaws. Fig. 1 is an example of a unidirectional lay surface. Roughness and waviness parallel to the lay are not represented in the expanded views.

*Waviness* is the more widely spaced component of surface texture. Unless otherwise noted, waviness includes all irregularities whose spacing is greater than the roughness sampling length and less than the waviness sampling length. Waviness may result from such factors as machine or work deflections, vibration, chatter, heat treatment, or warping strains. Roughness may be considered as being superposed on a 'wavy' surface.

**Definitions of Terms Relating to the Measurement of Surface Texture.**—Terms regarding surface texture pertain to the geometric irregularities of surfaces and include roughness, waviness and lay.

*Profile* is the contour of the surface in a plane measured normal, or perpendicular, to the surface, unless another angle is specified.

*Graphical centerline*. See Mean Line.

*Height (z)* is considered to be those measurements of the profile in a direction normal, or perpendicular, to the nominal profile. For digital instruments, the profile Z(x) is approximated by a set of digitized values. Height parameters are expressed in micrometers (μm).

*Height range (z)* is the maximum peak-to-valley surface height that can be detected accurately with the instrument. It is measurement normal, or perpendicular, to the nominal profile and is another key specification.

*Mean line (M)* is the line about which deviations are measured and is parallel to the general direction of the profile within the limits of the sampling length. See Fig. 2. The mean line may be determined in one of two ways. The filtered mean line is the centerline established by the selected cutoff and its associated circuitry in an electronic roughness average measuring instrument. The least squares mean line is formed by the nominal profile, but by dividing into selected lengths the sum of the squares of the deviations minimizes the deviation from the nominal form. The form of the nominal profile could be a curve or a straight line.

*Peak* is the point of maximum height on that portion of a profile that lies above the mean line and between two intersections of the profile with the mean line.

*Profile measured* is a representation of the real profile obtained by instrumental or other means. When the measured profile is a graphical representation, it will usually be distorted through the use of different vertical and horizontal magnifications but shall otherwise be as faithful to the profile as technically possible.

*Profile, modified* is the measured profile where filter mechanisms (including the instrument datum) are used to minimize certain surface texture characteristics and emphasize others. Instrument users apply profile modifications typically to differentiate surface roughness from surface waviness.

*Profile, nominal* is the profile of the nominal surface; it is the intended profile (exclusive of any intended roughness profile). Profile is usually drawn in an x-z coordinate system. See Fig. 2.

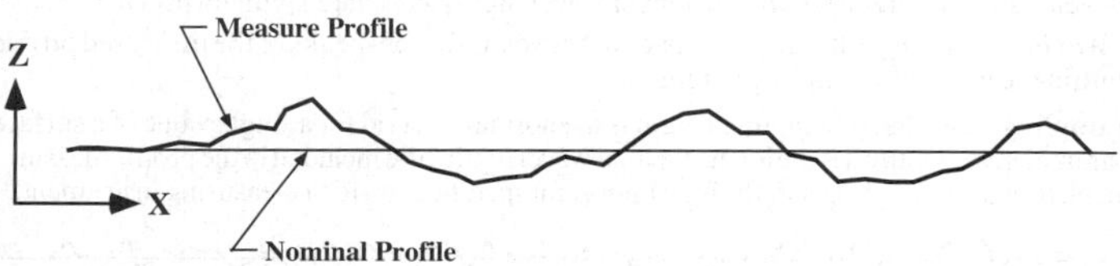


Fig. 2. Nominal and Measured Profiles

*Profile, real* is the profile of the real surface.

*Profile, total* is the measured profile where the heights and spacing may be amplified differently, but otherwise no filtering takes place.

*Roughness profile* is obtained by filtering out the longer wavelengths characteristic of waviness.

*Roughness spacing* is the average spacing between adjacent peaks of the measured profile within the roughness sampling length.

*Roughness topography* is the modified topography obtained by filtering out the longer wavelengths of waviness and form error.

*Sampling length* is the nominal spacing within which a surface characteristic is determined. The range of sampling lengths is a key specification of a measuring instrument.

*Spacing* is the distance between specified points on the profile measured parallel to the nominal profile.

*Spatial (x) resolution* is the smallest wavelength that can be resolved to 50 percent of the actual amplitude. This also is a key specification of a measuring instrument.

*System height resolution* is the minimum height that can be distinguished from background noise of the measurement instrument. Background noise values can be determined by measuring approximate rms roughness of a sample surface where actual roughness is significantly less than the background noise of the measuring instrument. It is a key instrumentation specification.

*Topography* is the three-dimensional representation of geometric surface irregularities.

*Topography, measured* is the three-dimensional representation of geometric surface irregularities obtained by measurement.

*Topography, modified* is the three-dimensional representation of geometric surface irregularities obtained by measurement but filtered to minimize certain surface characteristics and accentuate others.

*Valley* is the point of maximum depth on that portion of a profile that lies below the mean line and between two intersections of the profile with the mean line.

*Waviness, evaluation length (L),* is the length within which waviness parameters are determined.

*Waviness, long-wavelength cutoff (lcw)* is the spatial wavelength above which the undulations of waviness profile are removed to identify form parameters. A digital Gaussian filter can be used to separate form error from waviness, but its use must be specified.

*Waviness profile* is obtained by filtering out the shorter roughness wavelengths characteristic of roughness and the longer wavelengths associated with the part form parameters.

*Waviness sampling length* is a concept no longer used. See waviness long-wavelength cutoff and waviness evaluation length.

*Waviness short-wavelength cutoff (lsw)* is the spatial wavelength below which roughness parameters are removed by electrical or digital filters.

*Waviness topography* is the modified topography obtained by filtering out the shorter wavelengths of roughness and the longer wavelengths associated with form error.

*Waviness spacing* is the average spacing between adjacent peaks of the measured profile within the waviness sampling length.

**Sampling Lengths.—**Sampling length is the normal interval for a single value of a surface parameter. Generally it is the longest spatial wavelength to be included in the profile measurement. Range of sampling lengths is an important specification for a measuring instrument.

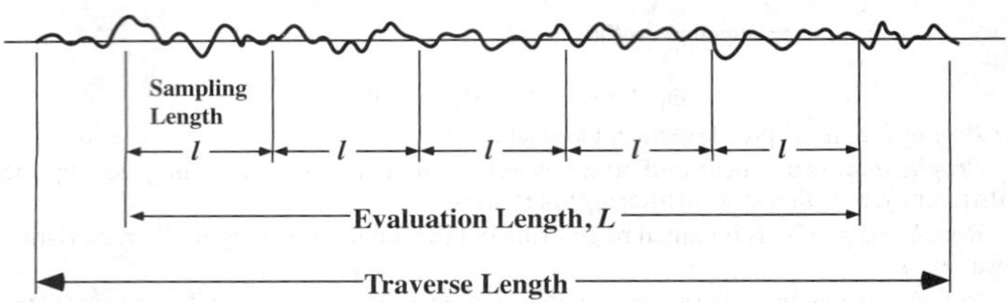


Fig. 3. Traverse Length

*Roughness sampling length (l)* is the sampling length within which the roughness average is determined. This length is chosen to separate the profile irregularities designated as roughness from those irregularities designated as waviness. It is different from evaluation length *(L)* and the traversing length. See Fig. 3.

*Evaluation length (L)* is the length the surface characteristics are evaluated. The evaluation length is a key specification of a measuring instrument.

*Traversing length* is the profile length traversed to establish a representative evaluation length. It is always longer than the evaluation length. See Section 4.4.4 of ANSI/ASME B46.1-2019 for values that should be used for different types of measurements.

*Cutoff* is the electrical response characteristic of the measuring instrument which is selected to limit the spacing of the surface irregularities to be included in the assessment of surface texture. Cutoff is rated in millimeters. In most electrical averaging instruments, the cutoff can be user-selected and is a characteristic of the instrument rather than of the surface being measured. In specifying the cutoff, care must be taken to choose a value that will include all the surface irregularities to be assessed.

*Waviness sampling length (l)* is a concept no longer used. See waviness long-wavelength cutoff and waviness evaluation length.

**Roughness Parameters.**—Roughness refers to the fine irregularities of the surface texture resulting from the production process or material condition.

*Roughness average (Ra or $R_a$)*, also known as arithmetic average (AA), is the arithmetic average of the absolute values of the measured profile height deviations divided by the evaluation length, L. This is shown as the shaded area of Fig. 4 and generally includes sampling lengths or cutoffs. For graphical determinations of roughness average, the height deviations are measured normal, or perpendicular, to the chart center line.

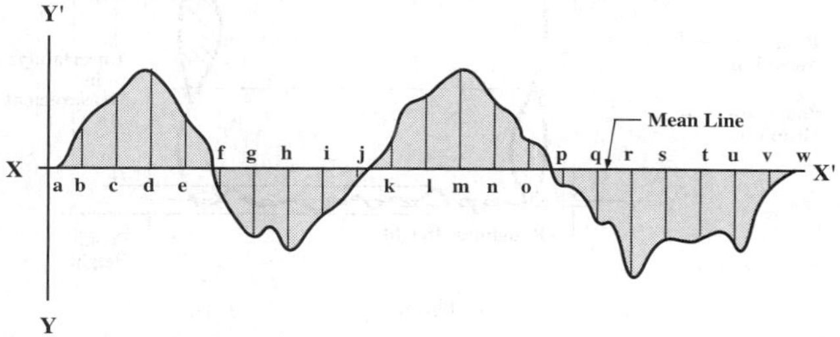


Fig. 4.

Roughness average is expressed in micrometers (μm). A micrometer is one millionth of a meter (0.000001 meter). A microinch (μin) is one millionth of an inch (0.000001 inch). One microinch equals 0.0254 micrometer (1 μin. = 0.0254 μm).

*Roughness average (Ra) value from continuously averaging meter reading* may be made of readings from stylus-type instruments of the continuously averaging type. To ensure uniform interpretation, it should be understood that the reading that is considered significant is the mean reading around which the needle tends to dwell or fluctuate with a small amplitude.

Roughness is also indicated by the root-mean-square (rms) average, which is the square root of the average value squared, within the evaluation length and measured from the mean line shown in Fig. 4, expressed in micrometers. A roughness-measuring instrument calibrated for rms average usually reads about 11 percent higher than an instrument calibrated for arithmetical average. Such instruments usually can be recalibrated to read arithmetical average. Some manufacturers consider the difference between rms and AA to be small enough that rms on a drawing may be read as AA for many purposes.

*Roughness evaluation length (L),* for statistical purposes should, whenever possible, consist of five sampling lengths (l). Use of other than five sampling lengths must be clearly indicated.

**Waviness Parameters.—**Waviness is the more widely spaced component of surface texture. Roughness may be thought of as superimposed on waviness.

*Waviness height (Wt)* is the peak-to-valley height of the modified profile with roughness and part form errors removed by filtering, smoothing or other means. This value is typically three or more times the roughness average. The measurement is taken normal, or perpendicular, to the nominal profile within the limits of the waviness sampling length.

*Waviness evaluation length (Lw)* is the evaluation length required to determine waviness parameters. For waviness, the sampling length concept is no longer used. Rather, only waviness evaluation length (*Lw*) and waviness long-wavelength cutoff (*lew*) are defined. For better statistics, the waviness evaluation length should be several times the waviness long-wavelength cutoff.

**Relation of Surface Roughness to Tolerances.—**Because the measurement of surface roughness involves the determination of the average linear deviation of the measured surface from the nominal surface, there is a direct relationship between the dimensional tolerance on a part and the permissible surface roughness. It is evident that a requirement for the accurate measurement of a dimension is that the variations introduced by surface roughness should not exceed the dimensional tolerances. If this is not the case, the measurement of the dimension will be subject to an uncertainty greater than the required tolerance, as illustrated in Fig. 5.

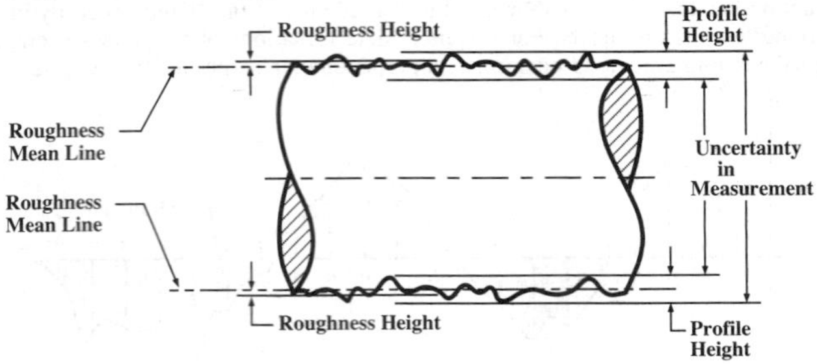


Fig. 5.

The standard method of measuring surface roughness involves the determination of the average deviation from the mean surface. On most surfaces the total profile height of the surface roughness (peak-to-valley height) will be approximately four times (4×) the measured average surface roughness. This factor will vary somewhat with the character of the surface under consideration, but the value of four may be used to establish approximate profile heights.

From these considerations it follows that if the arithmetical average value of surface roughness specified on a part exceeds one eighth of the dimensional tolerance, the whole tolerance will be taken up by the roughness height. In most cases, a smaller roughness specification than this will be found; but, on parts where very small dimensional tolerances are given, it is necessary to specify a suitably small surface roughness so useful dimensional measurements can be made. The tables on pages 649 and 675 show the relations between machining processes and working tolerances.

Values for surface roughness produced by common processing methods are shown in Table 1. The ability of a processing operation to produce a specific surface roughness depends on many factors. For example, in surface grinding, the final surface depends on the peripheral speed of the wheel, the speed of the traverse, the rate of feed, the grit size, bonding material and state of dress of the wheel, the amount and type of lubrication at the point of cutting, and the mechanical properties of the piece being ground. A small change in any of the above factors can have a marked effect on the surface produced.

**Table 1. Surface Roughness Produced by Common Production Methods**

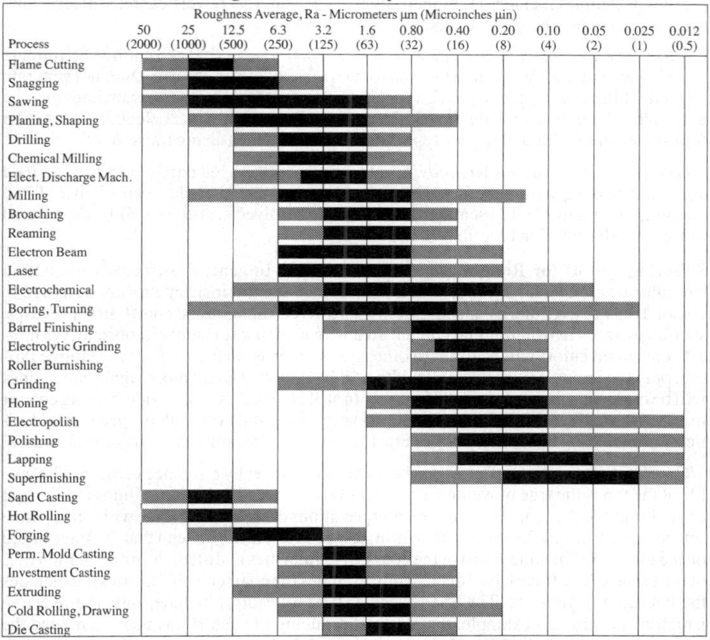


The ranges shown above are typical of the processes listed. Higher or lower values may be obtained under special conditions.

KEY: (black) Average Application; (shaded) Less Frequent Application

**Instrumentation for Surface Texture Measurement.**—Instrumentation used for measurement of surface texture, including roughness and waviness generally falls into six types. These include:

*Type I, Profiling Contact Skidless Instruments:* Used for very smooth to very rough surfaces. Used for roughness and may measure waviness. Can generate filtered or unfiltered profiles and may have a selection of filters and parameters for data analysis. Examples include: 1) skidless stylus-type with LVDT (linear variable differential transformer) vertical transducers; 2) skidless-type using an interferometric transducer; 3) skidless stylus-type using capacitance transducer.

*Type II, Profiling Non-Contact Instruments:* Capable of full profiling or topographical analysis. Non-contact operation may be advantageous for softness but may vary with sample type and reflectivity. Can generate filtered or unfiltered profiles but may have difficulty with steeply inclined surfaces. Examples include: 1) interferometric microscope; 2) optical focus sending; 3) Nomarski differential profiling; 4) laser triangulation; 5) scanning electron microscope (SEM) stereoscopy; 6) confocal optical microscope.

*Type III, Scanned Probe Microscope:* Features high spatial resolution (at or near the atomic scale) but area of measurement may be limited. Examples include: 1) scanning tunneling microscope (STM) and 2) atomic force microscope (AFM).

*Type IV, Profiling Contact Skidded Instruments:* Uses a skid as a datum to eliminate longer wavelengths; thus cannot be used for waviness or errors of form. May have a selection of filters and parameters and generates an output recording of filtered and skid-modified profiles. Examples include: 1) skidded, stylus-type with LVDT vertical measuring transducer and 2) fringe-field capacitance (FFC) transducer.

*Type V, Skidded Instruments with Parameters Only:* Uses a skid as a datum to eliminate longer wavelengths; thus cannot be used for waviness or errors of form. Does not generate a profile. Filters are typically 2RC type and generate Ra, but other parameters may be available. Examples include: 1) skidded, stylus-type with piezoelectric measuring transducer and 2) skidded, stylus-type with moving coil measuring transducer.

*Type VI, Area Averaging Methods:* Used to measure averaged parameters over defined areas, but do not generate profiles. Examples include: 1) parallel plate capacitance (PPC) method; 2) total integrated scatter (TIS); 3) angle resolved scatter (ARS)/bi-directional reflectance distribution function (BRDF).

**Selecting Cutoff for Roughness Measurements.—**In general, surfaces will contain irregularities with a large range of widths. Surface texture instruments are designed to respond only to irregularity spacings less than a given value, called cutoff. In some cases, such as surfaces in which actual contact area with a mating surface is important, the largest convenient cutoff will be used. In other cases, such as surfaces subject to fatigue failure, only the irregularities of small width will be important, and more significant values will be obtained when a short cutoff is used. In still other cases, such as identifying chatter marks on machined surfaces, information is needed on only the widely space irregularities. For such measurements, a large cutoff value and a larger radius stylus should be used.

The effect of variation in cutoff can be understood better by reference to Fig. 6. The profile at the top is the true movement of a stylus on a surface having a roughness spacing of about 1 mm and the profiles below are interpretations of the same surface with cutoff value settings of 0.8 mm, 0.25 mm and 0.08 mm, respectively. It can be seen that the trace based on 0.8 mm cutoff includes most of the coarse irregularities and all of the fine irregularities of the surface. The trace based on 0.25 mm excludes the coarser irregularities but includes the fine and medium fine. The trace based on 0.08 mm cutoff includes only the very fine irregularities. In this example, the effect of reducing the cutoff has been to reduce the roughness average indication. However, had the surface been made up only of irregularities as fine as those of the bottom trace, the roughness average values would have been the same for all three cutoff settings.

In other words, all irregularities having a spacing less than the value of the cutoff used are included in a measurement. Obviously, if the cutoff value is too small to include coarser irregularities of a surface, the measurements will not agree with those taken with a larger cutoff. For this reason, care must be taken to choose a cutoff value which will include all of the surface irregularities it is desired to assess.

To become proficient in the use of continuously averaging stylus-type instruments, the inspector or machine operator must realize that for uniform interpretation the reading considered significant is the mean reading around which the needle tends to dwell or fluctuate under small amplitude.

**Drawing Practices for Surface Texture Symbols.—**American National Standard ANSI/ASME Y14.36-2018 establishes the method to designate symbolic controls for surface texture of solid materials. It includes methods for controlling roughness, waviness, and lay, and provides a set of symbols for use on drawings, specifications, or other documents. The information in the Standard is expressed in SI metric units, but US customary units may be used without prejudice. Units used (metric or non-metric) should be consistent with the other units used on the drawing or documents. Approximate non-metric equivalents are shown for reference.

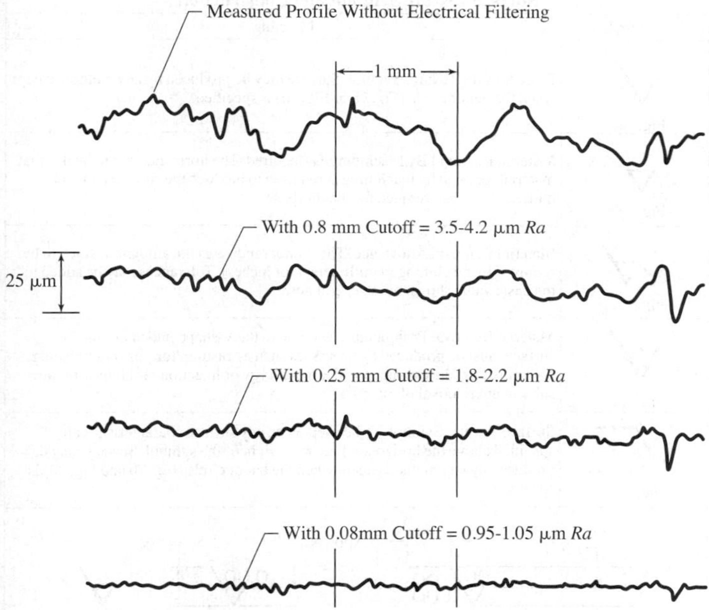


Fig. 6. Effects of Various Cutoff Values

**Surface Texture Symbol.—**The symbol used to designate control of surface irregularities is shown in Fig. 7b and Fig. 7d. Where surface texture values other than roughness average are specified, the symbol must be drawn with the horizontal extension as shown in Fig. 7f.

*Use of Surface Texture Symbols:* When required from a functional standpoint, the desired surface characteristics should be specified. Where no surface texture control is specified, the surface produced by normal manufacturing methods is satisfactory provided it is within the limits of size (and form) specified in accordance with ASME Y14.5-2018, Dimensioning and Tolerancing. It is considered good practice to always specify some maximum value, either specifically or by default (for example, in the manner of the note shown in Fig. 8 on page 807).

*Material Removal Required or Prohibited:* The surface texture symbol is modified when necessary to require or prohibit removal of material. When it is necessary to indicate that a surface must be produced by removal of material by machining, specify the symbol shown in Fig. 7b. When required, the amount of material to be removed is specified, as shown in Fig. 7c, in millimeters for metric drawings and in inches for non-metric drawings. Tolerance for material removal may be added to the basic value shown or specified in a general note. When it is necessary to indicate that a surface must be produced without material removal, specify the machining prohibited symbol as shown in Fig. 7d.

*Proportions of Surface Texture Symbols:* The recommended proportions for drawing the surface texture symbol are shown in Fig. 7f. The letter height and line width should be the same as those for dimensions and dimension lines.

**Surface Texture Symbols and Construction**

| Symbol | Meaning |
|---|---|
|  Fig. 7a. | Basic Surface Texture Symbol. Surface may be produced by any method except when the bar or circle (Fig. 7b or Fig. 7d) is specified. |
|  Fig. 7b. | Material Removal By Machining Is Required. The horizontal bar indicates that material removal by machining is required to produce the surface and that material must be provided for that purpose. |
|  Fig. 7c. | Material Removal Allowance. The number indicates the amount of stock to be removed by machining in millimeters (or inches). Tolerances may be added to the basic value shown or in general note. |
|  Fig. 7d. | Material Removal Prohibited. The circle in the V-shape indicates that the surface must be produced by processes such as casting, forging, hot finishing, cold finishing, die casting, powder metallurgy or injection molding without subsequent removal of material. |
|  Fig. 7e. | Surface Texture Symbol. To be used when any surface characteristics are specified above the horizontal line or the right of the symbol. Surface may be produced by any method except when the bar or circle (Fig. 7b and Fig. 7d) is specified. |

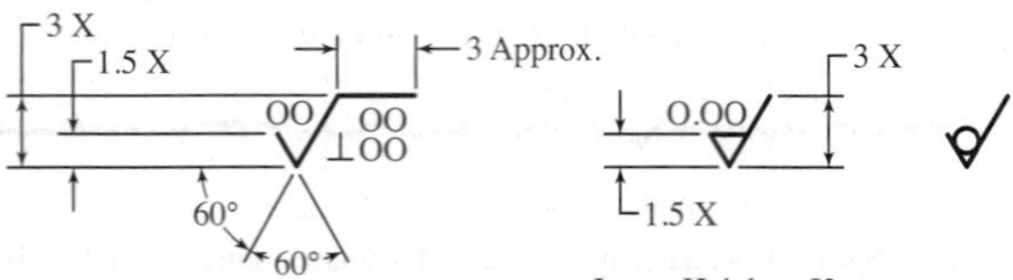


Fig. 7f.

**Applying Surface Texture Symbols.—**The point of the symbol should be on a line representing the surface, an extension line of the surface, or a leader line directed to the surface, or to an extension line. The symbol may be specified following a diameter dimension. Although ASME Y14.5-2018, "Dimensioning and Tolerancing," specifies that normally all textual dimensions and notes should be read from the bottom of the drawing, the surface texture symbol itself with its textual values may be rotated as required. Regardless, the long leg (and extension) must be to the right as the symbol is read. For parts requiring extensive and uniform surface roughness control, a general note may be added to the drawing that applies to each surface texture symbol specified without values as shown in Fig. 8.

When the symbol is used with a dimension, it affects the entire surface defined by the dimension. Areas of transition, such as chamfers and fillets, shall conform with the roughest adjacent finished area unless otherwise indicated.

Surface texture values, unless otherwise specified, apply to the complete surface. Drawings or specifications for plated or coated parts shall indicate whether the surface texture values apply before plating, after plating, or both before and after plating.

Only those values required to specify and verify the required texture characteristics should be included in the symbol. Values should be in metric units for metric drawing and non-metric units for non-metric drawings. Minority units on dual dimensioned drawings are enclosed in brackets.

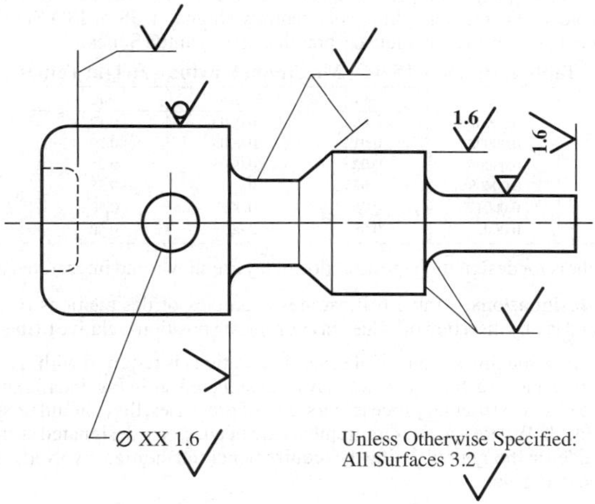


Fig. 8. Application of Surface Texture Symbols

Roughness and waviness measurements, unless otherwise specified, apply in a direction which gives the maximum reading, generally across the lay.

*Cutoff or Roughness Sampling Length, (l):* Standard values are listed in Table 2. When no value is specified, the value 0.8 mm (0.030 in.) applies.

**Table 2. Standard Roughness Sampling Length (Cutoff) Values**

| mm | in. | mm | in. |
|---|---|---|---|
| 0.08 | 0.003 | 2.5 | 0.1 |
| 0.25 | 0.010 | 8.0 | 0.3 |
| 0.80 | 0.030 | 25.0 | 1.0 |

*Roughness Average (Ra or $R_a$):* The preferred series of specified roughness average values is given in Table 3.

**Table 3. Preferred Series Roughness Average (Ra) Values**

| μm | μin | μm | μin | μm | μin |
|---|---|---|---|---|---|
| 0.012 | 0.5 | 0.40[a] | 16[a] | 4.0 | 160 |
| 0.025[a] | 1[a] | 0.50 | 20 | 5.0 | 200 |
| 0.050[a] | 2[a] | 0.63 | 25 | 6.3[a] | 250[a] |
| 0.075[a] | 3 | 0.80[a] | 32[a] | 8.0 | 320 |
| 0.10[a] | 4[a] | 1.00 | 40 | 10.0 | 400 |
| 0.125 | 5 | 1.25 | 50 | 12.5[a] | 500[a] |
| 0.15 | 6 | 1.60[a] | 63[a] | 15 | 600 |
| 0.20[a] | 8[a] | 2.0 | 80 | 20 | 800 |
| 0.25 | 10 | 2.5 | 100 | 25[a] | 1000[a] |
| 0.32 | 13 | 3.2[a] | 125[a] | … | … |

[a] Recommended

*Waviness Height (Wt):* The preferred series of maximum waviness height values is listed in Table 4. Waviness height is not currently shown in US or ISO Standards. It is included here to follow present industry practice in the United States.

**Table 4. Preferred Series Maximum Waviness Height Values**

| mm | inch | mm | inch | mm | inch |
|---|---|---|---|---|---|
| 0.0005 | 0.00002 | 0.008 | 0.0003 | 0.12 | 0.005 |
| 0.0008 | 0.00003 | 0.012 | 0.0005 | 0.20 | 0.008 |
| 0.0012 | 0.00005 | 0.020 | 0.0008 | 0.25 | 0.010 |
| 0.0020 | 0.00008 | 0.025 | 0.001 | 0.38 | 0.015 |
| 0.0025 | 0.0001 | 0.05 | 0.002 | 0.50 | 0.020 |
| 0.005 | 0.0002 | 0.08 | 0.003 | 0.80 | 0.030 |

*Lay:* Symbols for designating the direction of lay are shown and interpreted in Table 5.

**Example Designations.—**Table 6 illustrates examples of designations of roughness, waviness, and lay by insertion of values in appropriate positions relative to the symbol.

Where surface roughness control of several operations is required within a given area or on a given surface, surface qualities may be designated, as in Fig. 9a. If a surface must be produced by one particular process or a series of processes, they should be specified as shown in Fig. 9b. Where special requirements are needed on a designated surface, a note should be added at the symbol giving the requirements and the area involved. An example is illustrated in Fig. 9c.

**Surface Texture of Castings.—**Surface characteristics should not be controlled on a drawing or specification unless such control is essential to functional performance or appearance of the product. Imposition of such restrictions when unnecessary may increase production costs and in any event will serve to lessen the emphasis on the control specified for important surfaces. Surface characteristics of castings should never be considered on the same basis as machined surfaces. Castings are characterized by random distribution of nondirectional deviations from the nominal surface.

Surfaces of castings rarely need control beyond that provided by the production method necessary to meet dimensional requirements. Comparison specimens are frequently used for evaluating surfaces having specific functional requirements. Surface texture control should not be specified unless required for appearance or function of the surface. Specification of such requirements may increase cost to the user.

Engineers should recognize that different areas of the same castings may have different surface textures. It is recommended that specifications of the surface be limited to defined areas of the casting. Practicality of and methods of determining that a casting's surface texture meets the specification shall be coordinated with the producer. The Society of Automotive Engineers Standard J435 "Automotive Steel Castings" describes methods of evaluating steel casting surface texture used in the automotive and related industries.

**Metric Dimensions on Drawings.—**The length units of the metric system most generally used in connection with any work relating to mechanical engineering are the meter (39.37 inches) and the millimeter (0.03937 inch). One meter equals 1000 millimeters. On mechanical drawings, all dimensions are generally given in millimeters, no matter how large the dimensions may be. In fact, dimensions of such machines as locomotives and large electrical apparatus are given exclusively in millimeters. This practice is adopted to avoid mistakes due to misplacing decimal points, or misreading dimensions as when other units are used as well. When dimensions are given in millimeters, many of them can be given without resorting to decimal points, as a millimeter is only a little more than $\frac{1}{32}$ inch. Only dimensions of precision need be given in decimals of a millimeter; such dimensions are generally given in hundredths of a millimeter—for example, 0.02 millimeter, which is equal to 0.0008 inch. As 0.01 millimeter is equal to 0.0004 inch, dimensions are seldom given with greater accuracy than to hundredths of a millimeter.

**Table 5. Lay Symbols**

| Lay Symbol | Meaning | Example Showing Direction of Tool Marks |
|---|---|---|
| = | Lay approximately parallel to the line representing the surface to which the symbol is applied. | 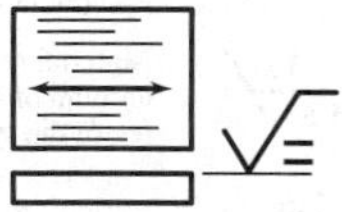 |
| ⊥ | Lay approximately perpendicular to the line representing the surface to which the symbol is applied. | 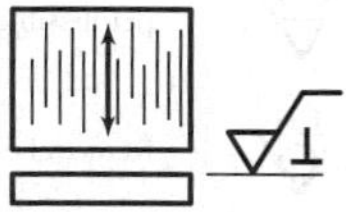 |
| **X** | Lay angular in both directions to line representing the surface to which the symbol is applied. | 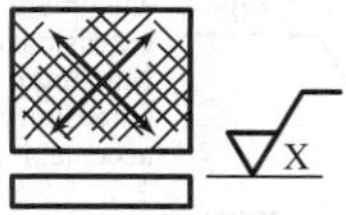  |
| **M** | Lay multidirectional. | 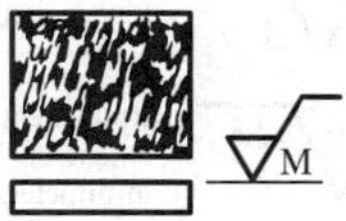  |
| **C** | Lay approximately circular relative to the center of the surface to which the symbol is applied. | 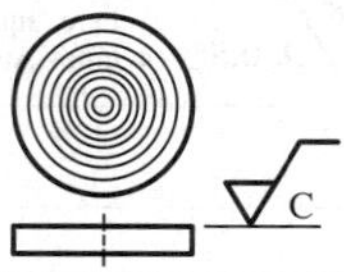  |
| **R** | Lay approximately radial relative to the center of the surface to which the symbol is applied. | 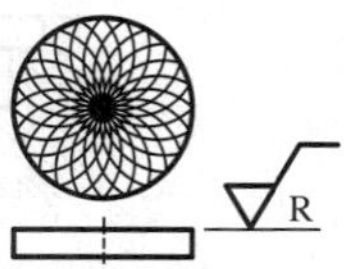  |
| **P** | Lay particulate, nondirectional, or protuberant. | 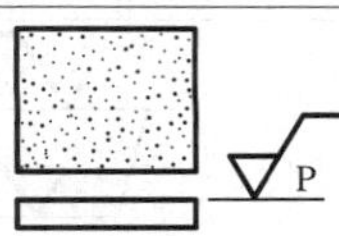  |

*Scales of Metric Drawings:* Drawings made to the metric system are not made to scales of $\frac{1}{2}$, $\frac{1}{4}$, $\frac{1}{8}$, etc., as with drawings made to the English system. If the object cannot be drawn full size, it may be drawn $\frac{1}{2}$, $\frac{1}{5}$, $\frac{1}{10}$, $\frac{1}{20}$, $\frac{1}{50}$, $\frac{1}{100}$, $\frac{1}{200}$, $\frac{1}{500}$, or $\frac{1}{1000}$ size. If the object is too small and has to be drawn larger, it is drawn 2, 5, or 10 times its actual size.

**Table 6. Application of Surface Texture Values to Symbol** *ANSI B46.1-1978*

| | |
|---|---|
|  | Roughness average rating is placed at the left of the long leg. The specification of only one rating shall indicate the maximum value, and any lesser value shall be acceptable. Specify in micrometers (microinch). |
| 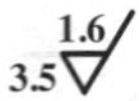 | Material removal by machining is required to produce the surface. The basic amount of stock provided for material removal is specified at the left of the short leg of the symbol. Specify in millimeters (inch). |
|  | The specification of maximum and minimum roughness average values indicates permissible range of roughness. Specify in micrometers (microinch). |
|  | Removal of material is prohibited. |
| 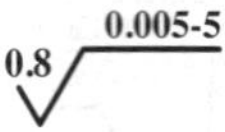 | Maximum waviness height rating is the first rating place above the horizontal extension. Any lesser rating shall be acceptable. Specify in millimeters (inch). Maximum waviness spacing rating is the second rating placed above the horizontal extension and to the right of the waviness height rating. Any lesser rating shall be acceptable. Specify in millimeters (inch). |
|  | Lay designation is indicated by the lay symbol placed at the right of the long leg. |
| 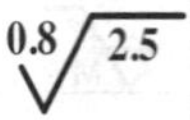 | Roughness sampling length or cutoff rating is placed below the horizontal extension. When no value is shown, 0.80 mm (0.030 inch) applies. Specify in millimeters (inch). |
| 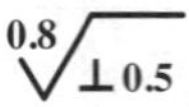 | Where required, maximum roughness spacing shall be placed at the right of the lay symbol. Any lesser rating shall be acceptable. Specify in millimeters (inch). |

## Examples of Special Designations

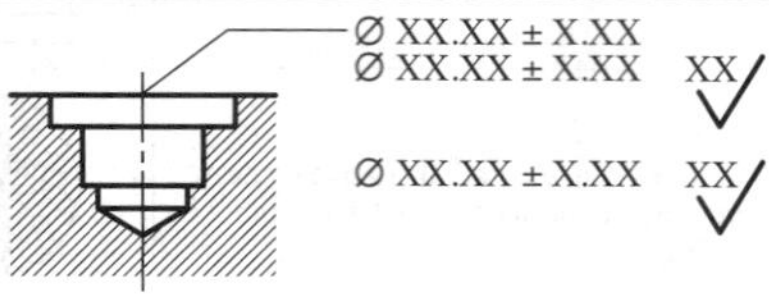


Fig. 9a.

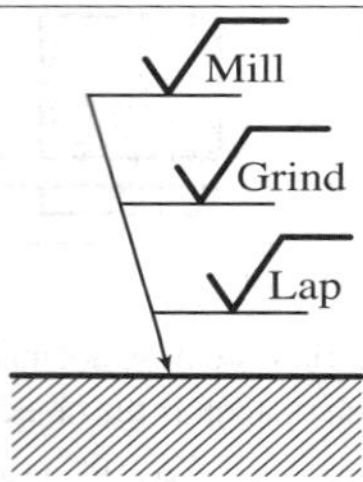


Fig. 9b.

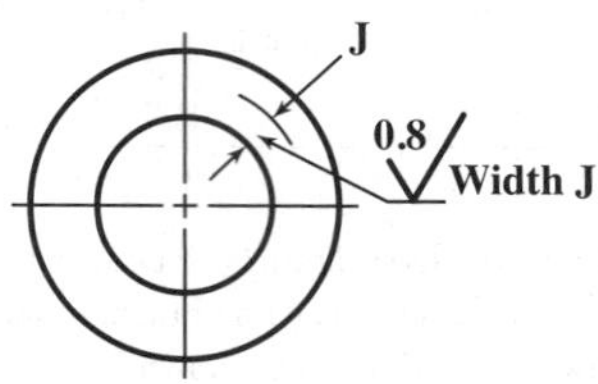


Fig. 9c.

## ISO Surface Finish Standards

ISO surface finish Standards are comprised of numerous individual Standards that, taken as a whole, form a set of Standards roughly comparable in scope to American National Standard ANSI/ASME Y14.36.

**ISO Surface Finish (ISO 1302).**—The primary Standard dealing with surface finish, ISO 1302:2002 is concerned with the methods of specifying surface texture symbology and additional indications on engineering drawings. The parameters in ISO surface finish Standards relate to surfaces produced by abrading, casting, coating, cutting, etching, plastic deformation, sintering, wear, erosion, and some other methods.

ISO 1302 defines how surface texture and its constituents, roughness, waviness, and lay, are specified on the symbology. Surface defects are specifically excluded from consideration during inspection of surface texture, but definitions of flaws and imperfections are discussed in ISO 8785.

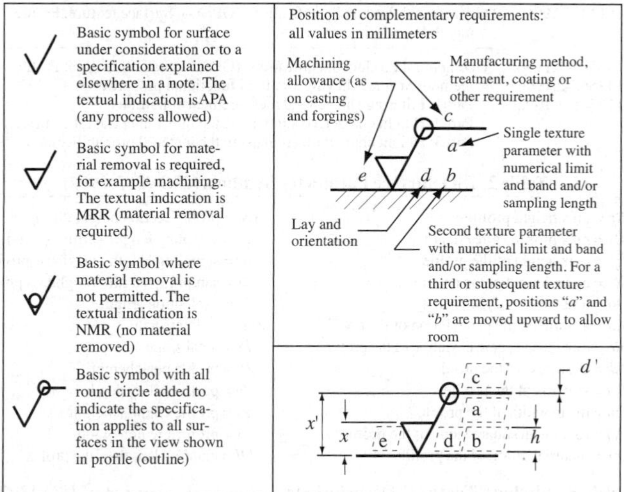


| | | | | | | | |
|---|---|---|---|---|---|---|---|
| Lettering height $h$ (ISO 3098-2) | 2.5 | 3.5 | 5 | 7 | 10 | 14 | 20 |
| Line width for symbols $d$ and $d'$ | 0.25 | 0.35 | 0.5 | 0.7 | 1 | 1.2 | 2 |
| Height for segment $x$ | 3.5 | 5 | 7 | 10 | 14 | 20 | 28 |
| Height for symbol segment $x'$ | 7.5 | 10.5 | 15 | 21 | 30 | 42 | 60 |

Fig. 1. ISO Surface Finish Symbols

*Differences Between ISO and ANSI/ASME Surface Finish Symbology:* ISO 1302, like ANSI/ASME Y14.36-2018, is not concerned with luster, appearance, color, corrosion resistance, wear resistance, hardness, sub-surface microstructure, surface integrity, or many other characteristics that may govern considerations in specific applications. Visually, ISO 1302 surface finish symbols are similar to the ANSI/ASME symbols; however, with the release of the 2002 edition, the indication of some of the parameters has changed when compared to ANSI/ASME Y14.36. The proportions of the symbol in relationship to lettering height differs in each as well. There is now less harmonization between ANSI/ASME Y14.36 and ISO 1302 than has been the case previously.

**Table 1. Other ISO Standards Related to Surface Finish**

| | |
|---|---|
| ISO 3274:1996 | "Geometrical Product Specifications (GPS) — Surface texture: Profile method; Nominal characteristics of contact (stylus) instruments." |
| ISO 4287:1997 | "Geometrical Product Specifications (GPS) — Surface texture: Profile method; Terms, definitions and surface texture parameters." |
| ISO 4288:1996 | "Geometrical Product Specifications (GPS) — Surface texture: Profile method; Rules and procedures for the assessment of surface texture." |
| ISO 8785:1998 | "Geometrical Product Specifications (GPS) — Surface imperfections — Terms, definitions and parameters." |
| ISO 12085:1996 | "Geometrical Product Specifications (GPS) — Surface texture: Profile method — Motif parameters." |
| ISO 13565-1:1996<br>ISO 13565-2:1996<br>ISO 13565-3:1998 | "Geometrical Product Specifications (GPS) — Surface texture: Profile method; Surfaces having stratified functional properties<br>Part 1: Filtering and general measurement conditions."<br>Part 2: Height characterization using the linear material ratio curve."<br>Part 3: Height characterization using the material probability curve." |

**Table 2. ISO Surface Parameter Symbols (*ISO 4287:1997*)**

| | |
|---|---|
| $Rp$ = max height profile | $R\delta c$ = profile section height difference |
| $Rv$ = max profile valley depth | $Ip$ = sampling length - primary profile |
| $Rz^*$ = max height of the profile | $lw$ = sampling length - waviness profile |
| $Rc$ = mean height of the profile | $lr$ = sampling length - roughness profile |
| $Rt$ = total height of the profile | $ln$ = evaluation length |
| $Ra$ = arithmetic mean deviation of the profile | $Z(x)$ = ordinate value |
| $Rq$ = root mean square deviation of the profile | $dZ / dX$ = local slope |
| $Rsk$ = skewness of the profile | $Zp$ = profile peak height |
| $Rku$ = kurtosis of the profile | $Zv$ = profile valley depth |
| $RSm$ = mean width of the profile | $Zt$ = profile element height |
| $R\Delta q$ = root mean square slope of the profile | $Xs$ = profile element width |
| $Rmr$ = material ration of the profile | $Ml$ = material length of the profile |

**Graphic Symbology Textural Descriptions.—**New to this version of ISO 1302:2002 is the ability to add textual descriptions of the graphic symbology used on drawing. This gives specifications writers a consistent means to describe surface texture specification from within a body of text without having to add illustrations. See Fig. 1 for textual application definitions, then Fig. 2- Fig. 6 for applications of this concept.

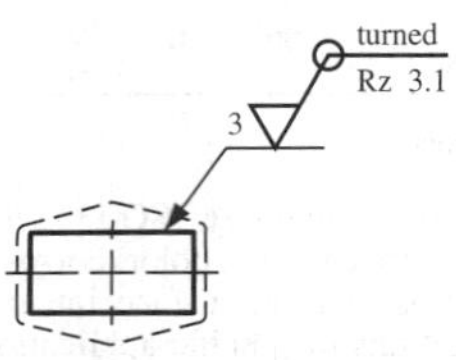

Fig. 2. Indication of Texture Requirement on a "Final" Workpiece, Reflecting a 3 mm Machining Allowance

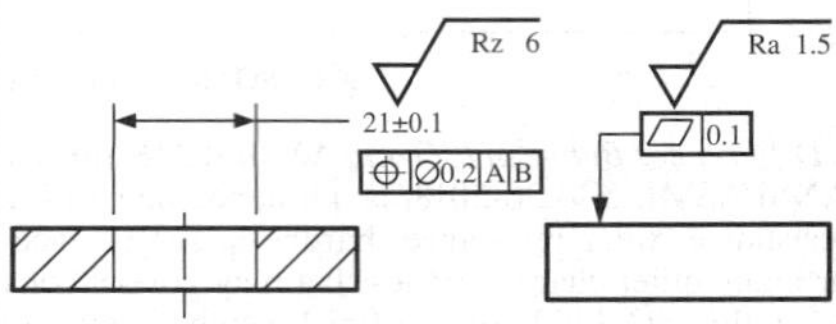

Fig. 3. Surface Texture Indications Combined with Geometric Dimensioning and Tolerancing

ISO 1302:2002 does not define the degrees of surface roughness and waviness or type of lay for specific purposes, nor does it specify the means by which any degree of such irregularities may be obtained or produced. Also, errors of form such as out-of-roundness and out-of-flatness are not addressed in the ISO surface finish standards. This edition does better illustrate how surface texture indications can be used on castings to reflect machining allowances (Fig. 2) and how symbology can be attached to geometric dimensioning and tolerancing symbology (See Fig. 3).

MRR U Rz 0.9; L Ra 0.3

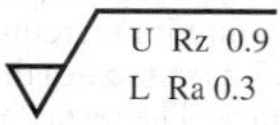


Fig. 4. Indication of Bilateral Surface Specification Shown Textually and as Indicated on a Drawing

MRR turned Rz 3.1

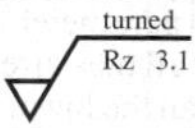


Indication of a machining process and requirement for roughness shown textually and as indicated on a drawing.

Fig. 5a. Indication of Manufacturing Processes or Related Information

NMR Fe/Ni15p Cr r; Rz 0.6

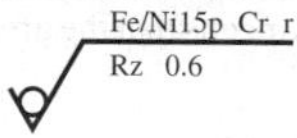


Indication of coating and roughness requirement shown textually and as indicated on a drawing.

Fig. 5b. Indication of Manufacturing Processes or Related Information

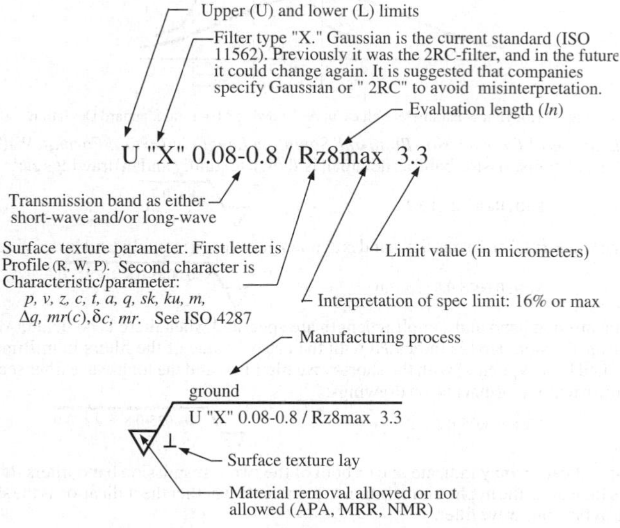


Fig. 6. Control Elements for Indication of Surface Texture Requirements on Drawings

**ISO Profiles.—**Profile parameters may be one of three types (ISO 4287). These include:

*R-profile:* Defined as the evaluation length. The ISO default length $l_n$ consists of five sampling lengths $l_r$ , thus $l_n = 5 \times l_r$

*W-profile:* This parameter indicates waviness. There is no default length.

*P-profile:* Indicates the structure parameters. The default evaluation length is defined in ISO 4288: 1996.

**Rules for Comparing Measured Values to Specified Limits.—**

*Max Rule:* When a maximum requirement is specified for a surface finish parameter on a drawing (e.g. *Rz*1.5max), none of the inspected values may extend beyond the upper limit over the entire surface. The term "max" must be added to the parametric symbol in the surface finish symbology on the drawing.

*16%-Rule:* When upper and lower limits are specified, no more than 16% of all measured values of the selected parameter within the evaluation length may exceed the upper limit. No more than 16% of all measured values of the selected parameter within the evaluation length may be less than the lower limit.

*Exceptions to the 16%-Rule:* Where the measured values of roughness profiles being inspected follow a normal distribution, the 16%-rule may be overridden. This is allowed when greater than 16% of the measured values exceeds the upper limit, but the total roughness profile conforms with the sum of the arithmetic mean and standard deviation ($\mu + \sigma$). Effectively, this means that the greater the value of $\sigma$, the further $\mu$ must be from the upper limit (see Fig. 7).

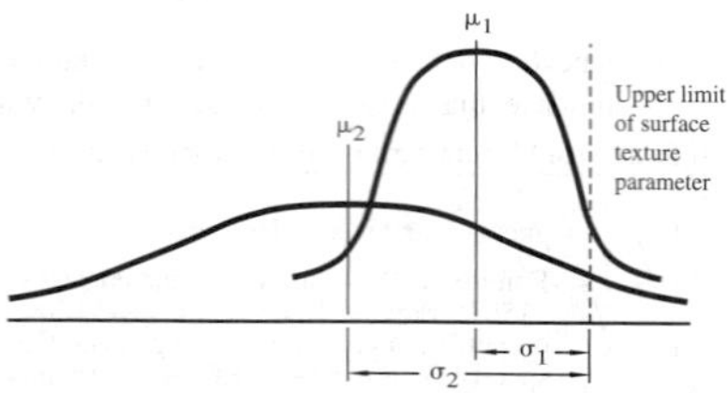


Fig. 7. Roughness Parameter Value Curves Showing Mean and Standard Deviation

*Indications of Transmission Band and Sampling Length in Textual Format:* With the "16%-rule" transmission band as default it is shown textually and in drawings as:

MRR Ra 0.7; Rz1 3.3

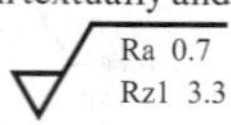


If the "max-rule" transmission band is applied, it is shown textually and in drawings as:

MRR 0.0025-0.8 / Rz 3.0

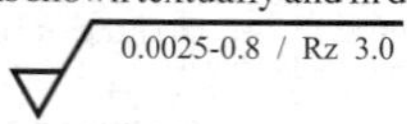


Transmission band and sampling length are specified when there is no default value. The transmission band is indicated with the cut-off value of the filters in millimeters separated by a hyphen (-) with the short-wave filter first and the long-wave filter second. Again, in textual format and on drawings:

MRR 0.0025-0.8 / Rz 3.0

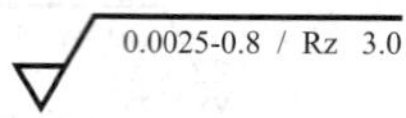


A specification may indicate one or both of the two transmission band filters. If only one is indicated, the hyphen is maintained to indicate whether the indication is the short-wave or the long-wave filter.

0.008– (short-wave filter indication) or –0.25 (long-wave filter indication)

*Determining Cut-Off Wavelength:* When the sampling length is specified on the drawing or in documentation, the cut-off wavelength $\lambda c$ is equal to the sample length. When no sampling length is specified, the cut-off wavelength is estimated using Table 3.

*Measurement of Roughness Parameters:* For non-periodic roughness, the parameter $Ra, Rz, Rz1_{max}$ or $RSm$ is first estimated using visual inspection, comparison to specimens, graphic analysis, etc. The sampling length is then selected from Table 3, based on the use of $Ra$, $Rz$, $Rz1_{max}$ or $RSm$. Then, with instrumentation, a representative sample is taken using the sampling length chosen above.

The measured values are then compared to the ranges of values in Table 3 for the particular parameter. If the value is outside the range of values for the estimated sampling length, the measuring instrument is adjusted for the next higher or lower sampling length and the measurement repeated. If the final setting corresponds to Table 3, then both the sampling length setting and $Ra, Rz, Rz1_{max}$ or $RSm$ values are correct and a representative measurement of the parameter can be taken.

For periodic roughness, the parameter $RSm$ is estimated graphically and the recommended cut-off values selected using Table 3. If the value is outside the range of values for the estimated sampling length, the measuring instrument is adjusted for the next higher or lower sampling length and the measurement repeated. If the final setting corresponds to Table 3, then both the sampling length setting and $RSm$ values are correct and a representative measurement of the parameter can be taken.

**Table 3. Sampling Lengths**

| Curves for Non-periodic Profiles such as Ground Surfaces | | Curves for Periodic and Non-periodic Profiles | Sampling length, *lr* (mm) | Evaluation length, *ln* (mm) |
|---|---|---|---|---|
| For $Ra, Rq, Rsk, Rku, R\Delta q$ | For $Rz, Rv, Rp, Rc, Rt$ | For *R*-parameters and $RSm$ | | |
| $Ra$, µm | $Rz, Rz1_{max}$, µm | $RSm$, µm | | |
| $(0.006) < Ra \leq 0.02$ | $(0.025) < Rz, Rz1_{max} \leq 0.1$ | $0.013 < RSm \leq 0.04$ | 0.08 | 0.4 |
| $0.02 < Ra \leq 0.1$ | $0.1 < Rz, Rz1_{max} \leq 0.5$ | $0.04 < RSm \leq 0.13$ | 0.25 | 1.25 |
| $0.1 < Ra \leq 2$ | $0.5 < Rz, Rz1_{max} \leq 10$ | $0.13 < RSm \leq 0.4$ | 0.8 | 4 |
| $2 < Ra \leq 10$ | $10 < Rz, Rz1_{max} \leq 50$ | $0.4 < RSm \leq 1.3$ | 2.5 | 12.5 |
| $10 < Ra \leq 80$ | $50 < Rz, Rz1_{max} \leq 200$ | $1.3 < RSm \leq 4$ | 8 | 40 |

**Table 4. Preferred Roughness Values and Roughness Grades**

| Roughness values, *Ra* | | Previous Grade Number from ISO 1302 | Roughness values, *Ra* | | Previous Grade Number from ISO 1302 |
|---|---|---|---|---|---|
| µm | µin | | µm | µin | |
| 50 | 2000 | N12 | 0.8 | 32 | N6 |
| 25 | 1000 | N11 | 0.4 | 16 | N5 |
| 12.5 | 500 | N10 | 0.2 | 8 | N4 |
| 6.3 | 250 | N9 | 0.1 | 4 | N3 |
| 3.2 | 125 | N8 | 0.05 | 2 | N2 |
| 1.6 | 63 | N7 | 0.025 | 1 | N1 |

### Examples of ISO Applications of Surface Texture Symbology

*Example 1: Surface roughness* in Fig. 8 is produced by milling with a bilateral tolerance between an upper limit of *Ra* = 55 μm and a lower limit of *Ra* = 6.2μm. Both apply the "16%-rule" default (ISO 4288). Both transmission bands are 0.008 - 4 mm, using default evaluation length (5 × 4 mm = 20 mm) (ISO 4288). The surface lay is circular about the center. *U* and *L* are omitted because it is obvious one is upper and one lower. Material removal is allowed.

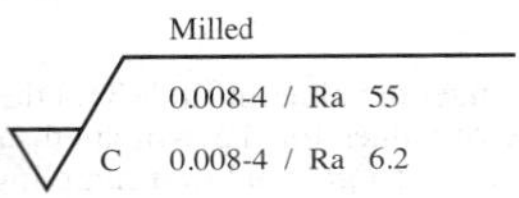


Fig. 8.

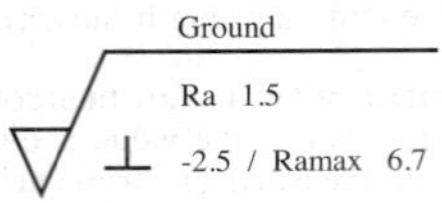


Fig. 9.

*Example 2: Surface roughness* in Fig. 9 is produced by grinding to two upper limit specifications: *Ra* = 1.5 μm and limited to *Rz* = 6.7 μm max. The default "16%-rule," default transmission band, and default evaluation length apply to the *Ra*, while the "max-rule," a −2.5 mm transmission band, and default evaluation length apply to the *Rz*. The surface lay is perpendicular relative to the plane of projection, and material removal is allowed.

*Example 3:* Fig. 10 indicates a simplified representation where *surface roughness* of *Rz* = 6.1 μm is the default for all surfaces as indicated by the *Rz* = 6.1 specification, plus basic symbol within parentheses. The default "16%-rule" applies to both as does the default transmission band (ISO 4288 and ISO 3274). Any deviating specification is called out with local notes such as the *Ra* =0.7 μm specification. There is no lay requirement, and material removal is allowed.

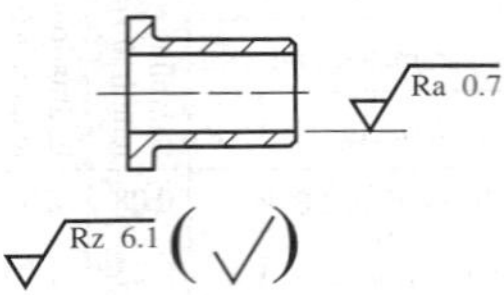


Fig. 10.

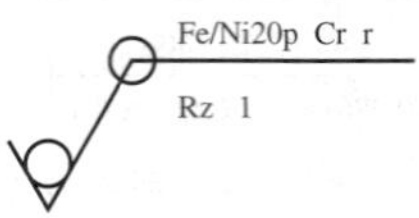


Fig. 11.

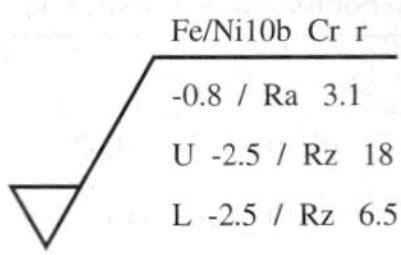


Fig. 12.

*Example 4: Surface treatment* without any material removal allowed is indicated in Fig. 11, and to a single unilateral upper limit specification of *Rz* = 1 μm. The default "16%-rule," default transmission band and default evaluation length apply. The surface treatment is nickel-chrome plated to all surfaces shown in profile (outline) in the view where the symbol is applied. There is no lay requirement.

*Example 5:* In Fig. 12, *surface roughness* is produced by any material removal process to one unilateral upper limit and one bilateral specification: the unilateral *Ra* = 3.1 is to the default "16%-rule," a transmission band of -0.8 mm and the default evaluation length (5 × 0.8 = 4 mm). The bilateral *Rz* has an upper limit of *Rz* = 18 μm and a lower limit of *Rz* = 6.5 μm. Both limits are to a transmission band of −2.5 mm with both to the default 5 × 2.5 = 12.5 mm. The symbols *U* and *L* may be indicated even if it is obvious. Surface treatment is nickel/chromium plating. There is no lay requirement.

*Example 6: Surface texture* symbology may be combined with dimension leaders and witness (extension) lines, as in Fig. 13. Surface roughness for the side surfaces of the keyway is produced by any material removal process to one unilateral upper limit specification, *Ra* = 6.5 μm. It is to the default "16%-rule," default transmission band and default evaluation length (5 × λc) (ISO 3274). There is no lay requirement.

Surface roughness for the chamfer is produced by any material removal process to one unilateral upper limit specification, *Ra* = 2.5 μm. It is to the default "16%-rule," default transmission band and default evaluation length (5 × λc) (ISO 3274). There is no lay requirement.

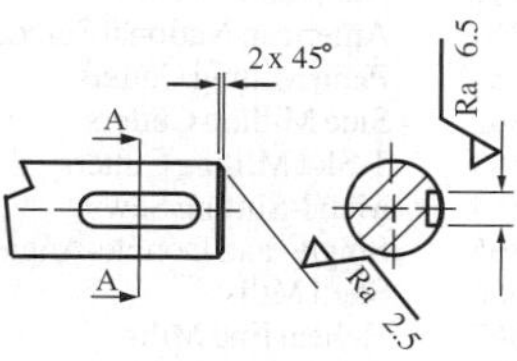


Fig. 13.

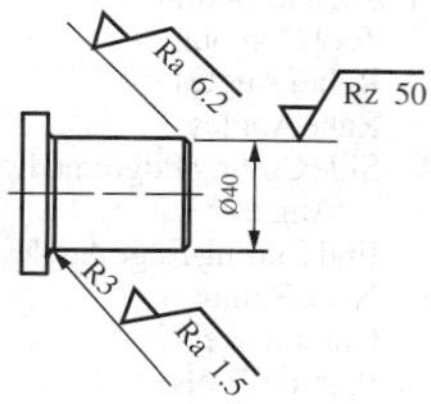


Fig. 14.

*Example 7: Surface texture* symbology may be applied to extended extension lines or on extended projection lines, Fig. 14. All feature surface roughness specifications shown are obtainable by any material removal process and are single unilateral upper limit specifications, respectively: *Ra* = 1.5 μm, *Ra* = 6.2 μm and *Rz* = 50 μm. All are to "16%-rule" default, default transmission band and default evaluation length (5 × λc). There is no lay requirement for any of the three.

*Example 8: Surface texture* symbology and dimensions may be combined on leader lines, as in Fig. 15. The feature surface roughness specifications shown are obtainable by any material removal process and are single unilateral upper limit specifications, respectively: *Rz* = 1 μm, to the default "16%-rule," default transmission band and default evaluation length (5 × λc). There is no lay requirement.

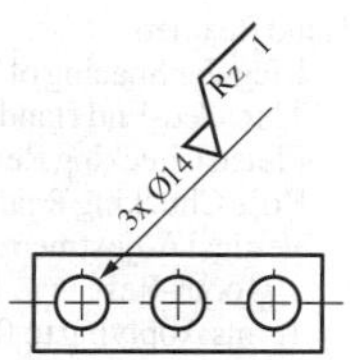


Fig. 15.

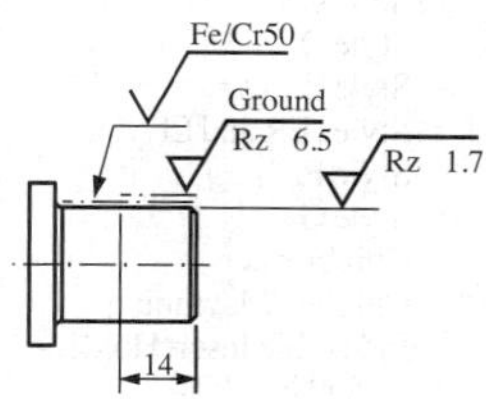


Fig. 16.

*Example 9:* Symbology can be used for *dimensional information* and *surface treatment*. Fig. 16 illustrates three successive steps of a manufacturing process.

The first step is a single unilateral upper limit *Rz* = 1.7 μm to the default "16%-rule," default evaluation length (5 × λc) and default transmission band. It is obtainable by any material removal process, with no lay characteristics specified. Step two, indicated with a phantom line over the whole length of the cylinder, has no surface texture requirement other than chromium plating.

The third step is a single unilateral upper limit of *Rz* = 6.5 μm applied only to the first 14 mm of the cylinder surface. The default "16%-rule" applies as does default evaluation length (5 × λc) and default transmission band. Material removal is to be by grinding, with no lay characteristics specified.

## TABLE OF CONTENTS

## TOOLING AND TOOLMAKING

### CUTTING TOOLS



### CEMENTED CARBIDES



### MILLING CUTTERS



### REAMERS

TABLE OF CONTENTS

# TOOLING AND TOOLMAKING

## TWIST DRILLS AND COUNTERBORES



## TAPS



## TAPS

*(Continued)*



## STANDARD TAPERS



TOOLING

## TABLE OF CONTENTS

## TOOLING AND TOOLMAKING

### ARBORS, CHUCKS, AND SPINDLES



### BROACHES AND BROACHING



### FILES AND BURS



### KNURLS AND KNURLING



### TOOL WEAR AND SHARPENING

## CUTTING TOOLS

### Terms and Definitions

**Tool Contour.**—Tools for turning, planing, etc., are made in straight, bent, offset, and other forms to place cutting edges in convenient positions for operating on differently located surfaces. The contour or shape of the cutting edge may also be varied to suit different classes of work. Tool shapes, however, are not only related to the kind of operation, but, in roughing tools particularly, the contour may have a decided effect upon the cutting efficiency of the tool. To illustrate, an increase in the side cutting edge angle of a roughing tool, or in the nose radius, tends to permit higher cutting speeds because the chip will be thinner for a given feed rate. Such changes, however, may result in chattering or vibrations unless the work and the machine are rigid; hence, the most desirable contour may be a compromise between the ideal form and one that is needed to meet practical requirements.

**Terms and Definitions.**—The terms and definitions relating to single-point tools vary somewhat in different plants, but the following are in general use.

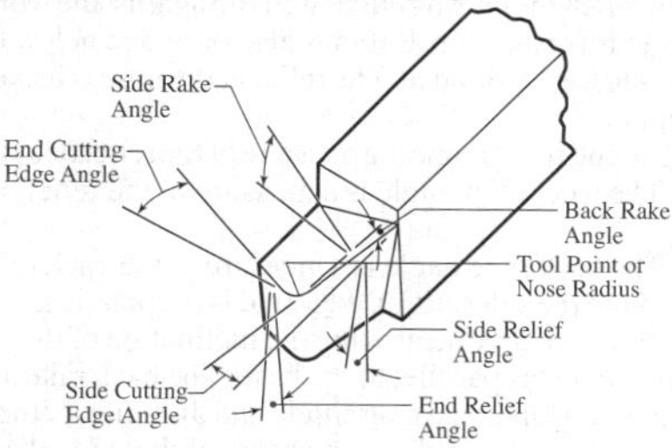


Fig. 1. Terms Applied to Single-Point Turning Tools

*Single-Point Tool:* This term is applied to tools for turning, planing, boring, etc., which have a cutting edge at one end. This cutting edge may be formed on one end of a solid piece of steel, or the cutting part of the tool may consist of an insert or tip which is held to the body of the tool by brazing, welding, or mechanical means.

*Shank:* The shank is the main body of the tool. If the tool is an inserted cutter type, the shank supports the cutter or bit. (See diagram, Fig. 1.)

*Nose:* A general term sometimes used to designate the cutting end but usually relating more particularly to the rounded tip of the cutting end.

*Face:* The surface against which the chips bear, as they are severed in turning or planing operations, is called the face.

*Flank:* The flank is that end surface adjacent to the cutting edge and below it when the tool is in a horizontal position as for turning.

*Base:* The base is the surface of the tool shank that bears against the supporting toolholder or block.

*Side Cutting Edge:* The side cutting edge is the cutting edge on the side of the tool. Tools such as that shown in Fig. 1 do the bulk of the cutting with this cutting edge and are, therefore, sometimes called side cutting edge tools.

*End Cutting Edge:* The end cutting edge is the cutting edge at the end of the tool.

On side cutting edge tools, the end cutting edge can be used for light plunging and facing cuts. Cutoff tools and similar tools have only one cutting edge located on the end. These tools and other tools that are intended to cut primarily with the end cutting edge are sometimes called end cutting edge tools.

*Rake:* A metal-cutting tool is said to have rake when the tool face or surface against which the chips bear as they are being severed is inclined for the purpose of either increasing or diminishing the keenness or bluntness of the edge. The magnitude of the rake is most conveniently measured by two angles called the back rake angle and the side rake angle. The tool shown in Fig. 1 has rake. If the face of the tool did not incline but was parallel to the base, there would be no rake; the rake angles would be zero.

*Positive Rake:* If the inclination of the tool face is such as to make the cutting edge keener or more acute than when the rake angle is zero, the rake angle is defined as positive.

*Negative Rake:* If the inclination of the tool face makes the cutting edge less keen or more blunt than when the rake angle is zero, the rake is defined as negative.

*Back Rake:* The back rake is the inclination of the face toward or away from the end or the end cutting edge of the tool. When the inclination is away from the end cutting edge, as shown in Fig. 1, the back rake is positive. If the inclination is downward toward the end cutting edge, the back rake is negative.

*Side Rake:* The side rake is the inclination of the face toward or away from the side cutting edge. When the inclination is away from the side cutting edge, as shown in Fig. 1, the side rake is positive. If the inclination is toward the side cutting edge the side rake is negative.

*Relief:* The flanks below the side cutting edge and the end cutting edge must be relieved to allow these cutting edges to penetrate into the workpiece when taking a cut. If the flanks are not provided with relief, the cutting edges will rub against the workpiece and be unable to penetrate in order to form the chip. Relief is also provided below the nose of the tool to allow it to penetrate into the workpiece. The relief at the nose is usually a blend of the side relief and the end relief.

*End Relief Angle:* The end relief angle is a measure of the relief below the end cutting edge.

*Side Relief Angle:* The side relief angle is a measure of the relief below the side cutting edge.

*Back Rake Angle:* The back rake angle is a measure of the back rake. It is measured in a plane that passes through the side cutting edge and is perpendicular to the base. Thus, the back rake angle can be defined by measuring the inclination of the side cutting edge with respect to a line or plane that is parallel to the base. The back rake angle may be positive, negative, or zero depending upon the magnitude and direction of the back rake.

*Side Rake Angle:* The side rake angle is a measure of the side rake. This angle is always measured in a plane that is perpendicular to the side cutting edge and perpendicular to the base. Thus, the side rake angle is the angle of inclination of the face perpendicular to the side cutting edge with reference to a line or a plane that is parallel to the base.

*End Cutting Edge Angle:* The end cutting edge angle is the angle made by the end cutting edge with respect to a plane perpendicular to the axis of the tool shank. It is provided to allow the end cutting edge to clear the finish machined surface on the workpiece.

*Side Cutting Edge Angle:* The side cutting edge angle is the angle made by the side cutting edge and a plane that is parallel to the side of the shank.

*Nose Radius:* The nose radius is the radius of the nose of the tool. The performance of the tool, in part, is influenced by nose radius so that it must be carefully controlled.

*Lead Angle:* The lead angle, shown in Fig. 2, is not ground on the tool. It is a tool setting angle which has a great influence on the performance of the tool. The lead angle is bounded by the side cutting edge and a plane perpendicular to the workpiece surface when the tool is in position to cut; or, more exactly, the lead angle is the angle between the side cutting edge and a plane perpendicular to the direction of the feed travel.

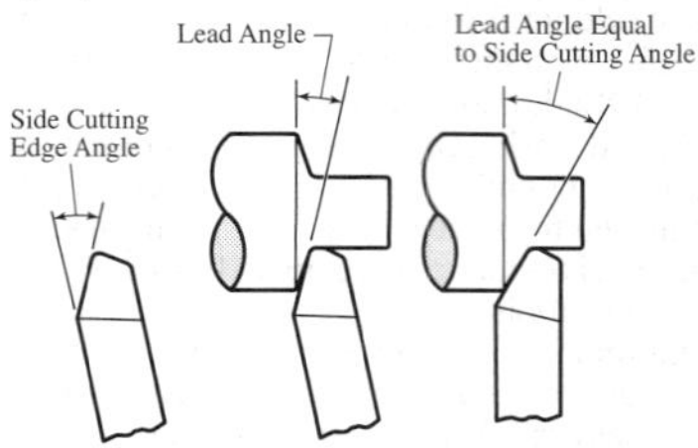


Fig. 2. Lead Angle on Single-Point Turning Tool

*Solid Tool:* A solid tool is a cutting tool made from one piece of tool material.

*Brazed Tool:* A brazed tool is a cutting tool having a blank of cutting-tool material permanently brazed to a steel shank.

*Blank:* A blank is an unground piece of cutting-tool material from which a brazed tool is made.

*Tool Bit:* A tool bit is a relatively small cutting tool that is clamped in a holder in such a way that it can readily be removed and replaced. It is intended primarily to be reground when dull and not indexed.

*Tool-Bit Blank:* The tool-bit blank is an unground piece of cutting-tool material from which a tool bit can be made by grinding. It is available in standard sizes and shapes.

*Tool-Bit Holder:* Usually made from forged steel, the tool-bit holder is used to hold the tool bit, to act as an extended shank for the tool bit, and to provide a means for clamping in the tool post.

*Straight-Shank Tool-Bit Holder:* A straight-shank tool-bit holder has a straight shank when viewed from the top. The axis of the tool bit is held parallel to the axis of the shank.

*Offset-Shank Tool-Bit Holder:* An offset-shank tool-bit holder has the shank bent to the right or left, as seen in Fig. 3. The axis of the tool bit is held at an angle with respect to the axis of the shank.

*Side Cutting Tool:* A side cutting tool has its major cutting edge on the side of the cutting part of the tool. The major cutting edge may be parallel or at an angle with respect to the axis of the tool.

*Indexable Inserts:* An indexable insert is a relatively small piece of cutting-tool material that is geometrically shaped to have two or several cutting edges that are used until dull. The insert is then indexed on the holder to apply a sharp cutting edge. When all the cutting edges have been dulled, the insert is discarded. The insert is held in a pocket or against other locating surfaces on an indexable insert holder by means of a mechanical clamping device that can be tightened or loosened easily.

*Indexable Insert Holder:* Made of steel, an indexable insert holder is used to hold indexable inserts. It is equipped with a mechanical clamping device that holds the inserts firmly in a pocket or against other seating surfaces.

*Straight-Shank Indexable Insert Holder:* A straight-shank indexable insert tool-holder is essentially straight when viewed from the top, although the cutting edge of the insert may be oriented parallel, or at an angle to, the axis of the holder.

*Offset-Shank Indexable Insert Holder:* An offset-shank indexable insert holder has the head end, or the end containing the insert pocket, offset to the right or left, as shown in Fig. 3.

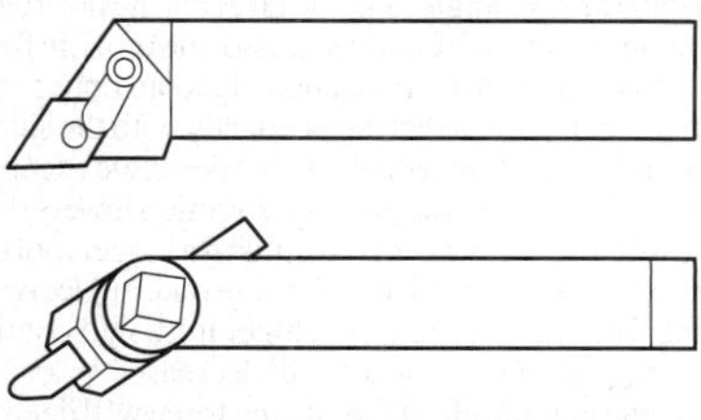

Fig. 3. Top: Right-Hand Offset-Shank, Indexable Insert Holder
Bottom: Right-Hand Offset-Shank Tool-Bit Holder

*End Cutting Tool:* An end cutting tool has its major cutting edge on the end of the cutting part of the tool. The major cutting edge may be perpendicular or at an angle, with respect to the axis of the tool.

*Curved Cutting Edge Tool:* A curved cutting edge tool has a continuously variable side cutting edge angle. The cutting edge is usually in the form of a smooth, continuous curve along its entire length, or along a large portion of its length.

*Right-Hand Tool:* A right-hand tool has the major, or working, cutting edge on the right-hand side when viewed from the cutting end with the face up. As used in a lathe, such a tool is usually fed into the work from right to left, when viewed from the shank end.

*Left-Hand Tool:* A left-hand tool has the major or working cutting edge on the left-hand side when viewed from the cutting end with the face up. As used in a lathe, the tool is usually fed into the work from left to right, when viewed from the shank end.

*Neutral-Hand Tool:* A neutral-hand tool is a tool to cut either left to right or right to left; or the cut may be parallel to the axis of the shank as when plunge cutting.

*Chipbreaker:* A groove formed in or on a shoulder on the face of a turning tool back of the cutting edge to break up the chips and prevent the formation of long, continuous chips that would be dangerous to the operator and also bulky and cumbersome to handle. A chipbreaker of the shoulder type may be formed directly on the tool face, or it may consist of a separate piece that is held either by brazing or by clamping.

**Relief Angles.—**The end relief angle and the side relief angle on single-point cutting tools are usually, though not invariably, made equal to each other. The relief angle under the nose of the tool is a blend of the side and end relief angles.

The size of the relief angles has a pronounced effect on the performance of the cutting tool. If the relief angles are too large, the cutting edge will be weakened and in danger of breaking when a heavy cutting load is placed on it by a hard and tough material. On finish cuts, rapid wear of the cutting edge may cause problems with size control on the part. Relief angles that are too small will cause the rate of wear on the flank of the tool below the cutting edge to increase, thereby significantly reducing the tool life. In general, when cutting hard and tough materials, the relief angles should be 6 to 8 degrees for high-speed steel tools and 5 to 7 degrees for carbide tools. For medium steels, mild steels, cast iron, and other average work, the recommended values of the relief angles are 8 to 12 degrees for high-speed steel tools and 5 to 10 degrees for carbides. Ductile materials having a relatively low modulus of elasticity should be cut using larger relief angles. For example, the relief angles recommended for turning copper, brass, bronze, aluminum, ferritic malleable iron, and similar metals are 12 to 16 degrees for high-speed steel tools and 8 to 14 degrees for carbides.

Larger relief angles tend to produce a better finish machined surface because less surface of the worn flank of the tool rubs against the workpiece. For this reason, single-point thread-cutting tools should be provided with relief angles that are as large as circumstances will permit. Problems encountered when machining stainless steel may be overcome by increasing the size of the relief angle. The relief angles used should not be smaller than necessary.

**Rake Angles.—**Machinability tests have confirmed that when the rake angle along which the chip slides, called the true rake angle, is made larger in the positive direction, the cutting force and the cutting temperature will decrease. Also, the tool life for a given cutting speed will increase with increases in the true rake angle up to an optimum value, after which it will decrease again. For turning tools that cut primarily with the side cutting edge, the true rake angle corresponds rather closely with the side rake angle except when taking shallow cuts. Increasing the side rake angle in the positive direction lowers the cutting force and the cutting temperature, while at the same time it results in a longer tool life or a higher permissible cutting speed up to an optimum value of the side rake angle. After the optimum value is exceeded, the cutting force and the cutting temperature will continue to drop; however, the tool life and the permissible cutting speed will decrease.

As an approximation, the magnitude of the cutting force will decrease about one percent per degree increase in the side rake angle. While not exact, this rule of thumb does correspond approximately to test results and can be used to make rough estimates. Of course, the cutting force also increases about one percent per degree decrease in the side rake angle. The limiting value of the side rake angle for optimum tool life or cutting speed depends upon the work material and the cutting tool material. In general, lower values can be used for hard and tough work materials. Cemented carbides are harder and more brittle than high-speed steel; therefore, the rake angles usually used for cemented carbides are less positive than for high-speed steel.

Negative rake angles cause the face of the tool to slope in the opposite direction from positive rake angles and, as might be expected, they have an opposite effect. For side cutting edge tools, increasing the side rake angle in a negative direction will result in an increase in the cutting force and an increase in the cutting temperature of approximately one percent per degree change in rake angle. For example, if the side rake angle is changed from 5 degrees positive to 5 degrees negative, the cutting force will be about 10 percent larger. Usually the tool life will also decrease when negative side rake angles are used, although the tool life will sometimes increase when the negative rake angle is not too large and when a fast cutting speed is used.

Negative side rake angles are usually used in combination with negative back rake angles on single-point cutting tools. The negative rake angles strengthen the cutting edges, enabling them to sustain heavier cutting loads and shock loads. They are recommended for turning very hard materials and for heavy interrupted cuts. There is also an economic advantage in favor of using negative rake indexable inserts and tool holders inasmuch as the cutting edges provided on both the top and bottom of the insert can be used.

On turning tools that cut primarily with the side cutting edge, the effect of the back rake angle alone is much less than the effect of the side rake angle, although the direction of the change in cutting force, cutting temperature, and tool life is the same. The effect that the back rake angle has can be ignored unless, of course, extremely large changes in this angle are made. A positive back rake angle does improve the performance of the nose of the tool somewhat and is helpful in taking light finishing cuts. A negative back rake angle strengthens the nose of the tool and is helpful when interrupted cuts are taken. The back rake angle has a very significant effect on the performance of end cutting edge tools, such as cut-off tools. For these tools, the effect of the back rake angle is very similar to the effect of the side rake angle on side cutting edge tools.

**Side Cutting Edge and Lead Angles.—**These angles are considered together because the side cutting edge angle is usually designed to provide the desired lead angle when the tool is being used. The side cutting edge angle and the lead angle will be equal when the shank of the cutting tool is positioned perpendicular to the workpiece, or, more correctly, perpendicular to the direction of the feed. When the shank is not perpendicular, the lead angle is determined by the side cutting edge and an imaginary line perpendicular to the feed direction.

The flow of the chips over the face of the tool is approximately perpendicular to the side cutting edge except when shallow cuts are taken. The thickness of the undeformed chip is measured perpendicular to the side cutting edge. As the lead angle is increased, the length of chip in contact with the side cutting edge is increased, and the chip will become longer and thinner. This effect is the same as increasing the depth of cut and decreasing the feed, although the actual depth of cut and feed remain the same and the same amount of metal is removed. The effect of lengthening and thinning the chip by increasing the lead angle is very beneficial as it increases the tool life for a given cutting speed or that speed can be increased. Increasing the cutting speed while the feed and the tool life remain the same leads to faster production.

However, an adverse effect must be considered. Chatter can be caused by a cutting edge that is oriented at a high lead angle when turning, and sometimes, when turning long and slender shafts, even a small lead angle can cause chatter. In fact, an unsuitable lead angle of the side cutting edge is one of the principal causes of chatter. When chatter occurs, often simply reducing the lead angle will cure it. Sometimes, very long and slender shafts can be turned successfully with a tool having a zero degree lead angle (and having a small nose radius). Boring bars, being usually somewhat long and slender, are also susceptible to chatter if a large lead angle is used. The lead angle for boring bars should be kept small, and for very long and slender boring bars a zero-degree lead angle is recommended. It is impossible to provide a rule that will determine when chatter caused by a lead angle will occur and when it will not. The first consideration is the length to diameter ratio of the part to be turned, or of the boring bar. Then the method of holding the workpiece must be considered—a part that is firmly held is less apt to chatter. Finally, the overall condition and rigidity of the machine must be considered because they may be the real cause of chatter.

Although chatter can be a problem, the advantages gained from high lead angles are such that the lead angle should be as large as possible at all times.

**End Cutting Edge Angle.—**The size of the end cutting edge angle is important when tool wear by cratering occurs. Frequently, the crater will enlarge until it breaks through the end cutting edge just behind the nose, and tool failure follows shortly. Reducing the size of the end cutting edge angle tends to delay the time of crater breakthrough. When cratering takes place, the recommended end cutting edge angle is 8 to 15 degrees. If there is no cratering, the angle can be made larger. Larger end cutting edge angles may be required to enable profile turning tools to plunge into the work without interference from the end cutting edge.

**Nose Radius.—**The tool nose is a very critical part of the cutting edge since it cuts the finished surface on the workpiece. If the nose is made to a sharp point, the finish machined surface will usually be unacceptable and the life of the tool will be short. Thus, a nose radius is required to obtain an acceptable surface finish and tool life. The surface finish obtained is determined by the feed rate and by the nose radius if other factors such as the work material, the cutting speed, and cutting fluids are not considered. A large nose radius will give a better surface finish and will permit a faster feed rate to be used.

Machinability tests have demonstrated that increasing the nose radius will also improve the tool life or allow a faster cutting speed to be used. For example, high-speed steel tools were used to turn an alloy steel in one series of tests where complete or catastrophic tool failure was used as a criterion for the end of tool life. The cutting speed for a 60-minute tool life was found to be 125 fpm (0.635 m/s) when the nose radius was $\frac{1}{16}$ inch (1.59 mm) and 160 fpm (0.8.13 m/s) when the nose radius was $\frac{1}{4}$ inch (6.35 mm).

A very large nose radius can often be used but a limit is sometimes imposed because the tendency for chatter to occur is increased as the nose radius is made larger. A nose radius that is too large can cause chatter, and, when it does, a smaller nose radius must be used on the tool. It is always good practice to make the nose radius as large as is compatible with the operation being performed.

**Chipbreakers.—**Many steel turning tools are equipped with chipbreaking devices to prevent the formation of long continuous chips in connection with the turning of steel at the high speeds made possible by high-speed steel and especially cemented carbide tools. Long steel chips are dangerous to the operator and cumbersome to handle, and they may twist around the tool and cause damage. Broken chips not only occupy less space, but permit a better flow of coolant to the cutting edge. Several different forms of chipbreakers are illustrated in Fig. 4.

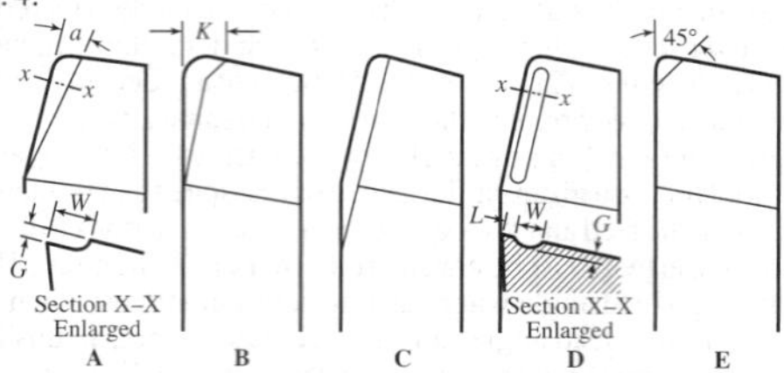


Fig. 4. Different Forms of Chipbreakers for Turning Tools

*Angular Shoulder Type:* The angular shoulder type, labeled type A in Fig. 4, is one of the commonly used forms. As the enlarged sectional view shows, the chipbreaking shoulder is located back of the cutting edge. The angle *a* between the shoulder and cutting edge may vary from 6 to 15 degrees or more, 8 degrees being a fair average. The ideal angle, width *W* and depth *G*, depend upon the speed and feed, the depth of cut, and the material. As a general rule, width *W*, at the end of the tool, varies from $\frac{3}{32}$ to $\frac{7}{32}$ inch (2.4–5.6 mm), and the depth *G* may range from $\frac{1}{64}$ to $\frac{1}{16}$ inch (0.4–1.6 mm). The shoulder radius equals depth *G*. If the tool has a large nose radius, the corner of the shoulder at the nose end may be beveled off, as illustrated as type B, to prevent it from coming into contact with the work. The width *K* for type B should equal approximately 1.5 times the nose radius.

*Parallel Shoulder Type:* Fig. 4 type C shows a design with a chipbreaking shoulder that is parallel with the cutting edge. With this form, the chips are likely to come off in short curled sections. The parallel form may also be applied to straight tools that do not have a side cutting edge angle. The tendency with this parallel shoulder form is to force the chips against the work and damage it.

*Groove Type:* This type (type D) has a groove in the face of the tool produced by grinding. Between the groove and the cutting edge, there is a land *L*. Under ideal conditions, this width *L*, the groove width *W*, and the groove depth *G*, would be varied to suit the feed, depth of cut and material. For average use, *L* and *G* are about $\frac{1}{32}$ inch (0.79 mm), and *W* $\frac{1}{16}$ inch (1.59 mm). There are differences of opinion concerning the relative merits of the groove type and the shoulder type. Both types have proved satisfactory when properly proportioned for a given class of work.

*Chipbreaker for Light Cuts:* Type E illustrates a form of chipbreaker that is sometimes used on tools for finishing cuts having a maximum depth of about $\frac{1}{32}$ inch (0.79 mm). This chipbreaker is a shoulder type, having an angle of 45 degrees and a maximum width of about $\frac{1}{16}$ inch (1.59 mm). It is important in grinding all chipbreakers to give the chip-bearing surfaces a fine finish, such as would be obtained by honing. This finish greatly increases the life of the tool.

**Planing Tools.—**Many of the principles governing the shape of turning tools also apply in the grinding of tools for planing. The amount of rake depends upon the hardness of the material, and the direction of the rake should be away from the *working part* of the cutting edge. The angle of clearance should be about 4 or 5 degrees for planer tools, which is less than for lathe tools. This small clearance is allowable because a planer tool is held about square with the platen, whereas a lathe tool, the height and inclination of which can be varied, may not always be clamped in the same position.

*Carbide Tools:* Carbide tools for planing usually have negative rake. Round-nose and square-nose end-cutting tools should have a "negative back rake" (or front rake) of 2 or 3 degrees. Side cutting tools may have a negative back rake of 10 degrees, a negative side rake of 5 degrees, and a side cutting edge angle of 8 degrees.

## Indexable Inserts

**Introduction.—**A large proportion of cemented carbide, single-point cutting tools are indexable inserts and indexable insert tool holders. Dimensional specifications for solid sintered carbide indexable inserts are given in ANSI B212.12-1991 (Withdrawn). Samples of the many insert shapes are shown in Table 3b. Most modern, cemented carbide, face milling cutters are of the indexable insert type. Larger size end milling cutters, side milling or slotting cutters, boring tools, and a wide variety of special tools are made to use indexable inserts. These inserts are primarily made from cemented carbide, although most of the cemented oxide cutting tools are also indexable inserts.

The objective of this type of tooling is to provide an insert with several cutting edges. When an edge is worn, the insert is indexed in the tool holder until all the cutting edges are used up, after which it is discarded. The insert is not intended to be reground. The advantages are that the cutting edges on the tool can be rapidly changed without removing the tool holder from the machine, tool-grinding costs are eliminated, and the cost of the insert is less than the cost of a similar, brazed carbide tool. Of course, the cost of the tool holder must be added to the cost of the insert; however, one tool holder will usually last for a long time before it, too, must be replaced.

Indexable inserts and tool holders are made with a negative or positive rake. Negative rake inserts have the advantage of having twice as many cutting edges available as comparable positive rake inserts, because the cutting edges on both the top and bottom of negative rake inserts can be used, while only the top cutting edges can be used on positive rake inserts. Positive rake inserts have a distinct advantage when machining long and slender parts, thin-walled parts, or other parts that are subject to bending or chatter when the cutting load is applied to them, because the cutting force is significantly lower as compared to that for negative rake inserts. Indexable inserts can be obtained in the following forms: utility ground, or ground on top and bottom only; precision ground, or ground on all surfaces; prehoned to produce a slight rounding of the cutting edge; and precision molded, which are unground. Positive-negative rake inserts also are available. These inserts are held on a negative-rake tool holder and have a chipbreaker groove that is formed to produce an effective positive-rake angle while cutting. Cutting edges may be available on the top surface only, or on both top and bottom surfaces. The positive-rake chipbreaker surface may be ground or precision molded on the insert.

Many materials, such as gray cast iron, form a discontinuous chip. For these materials an insert that has plain faces without chipbreaker grooves should always be used. Steels and other ductile materials form a continuous chip that must be broken into small segments when machined on lathes and planers having single-point, cemented carbide and cemented-oxide cutting tools; otherwise, the chips can cause injury to the operator. In this case a chipbreaker must be used. Some inserts are made with chipbreaker grooves molded or ground directly on the insert. When inserts with plain faces are used, a cemented-carbide plate-type chipbreaker is clamped on top of the insert.

**Identification System for Indexable Inserts.—**The size of indexable inserts is determined by the diameter of an inscribed circle (I.C.), except for rectangular and parallelogram inserts where the length and width are used. To describe an insert in its entirety, per ANSI B212.4-2002 (Withdrawn), an identification system is used where each position number designates a feature of the insert. The ANSI Standard includes items now commonly used and facilitates identification of items not in common use. Identification consists of up to ten positions; each position defines a characteristic of the insert as shown below:

| 1 | 2 | 3 | 4 | 5 | 6 | 7 | 8[a] | 9[a] | 10[a] |
|---|---|---|---|---|---|---|---|---|---|
| **T** | **N** | **M** | **G** | **5** | **4** | **3** | | | **A** |

[a] Eighth, Ninth, and Tenth Positions are used only when required.

1) *Shape:* The shape of an insert is designated by a letter: **R** for round; **S,** square; **T,** triangle; **A,** 85° parallelogram; **B,** 82° parallelogram; **C,** 80° diamond; **D,** 55° diamond; **E,** 75° diamond; **H,** hexagon; **K,** 55° parallelogram; **L,** rectangle; **M,** 86° diamond; **O,** octagon; **P,** pentagon; **V,** 35° diamond; and **W,** 80° trigon.

2) *Relief Angle (Clearances):* The second position is a letter denoting the relief angles: **N** for 0°; **A,** 3°; **B,** 5°; **C,** 7°; **P,** 11°; **D,** 15°; **E,** 20°; **F,** 25°; **G,** 30°; **H,** 0° & 11°*; **J,** 0° & 14°*; **K,** 0° & 17°*; **L,** 0° & 20°*; **M,** 11° & 14°*; **R,** 11° & 17°*; **S,** 11° & 20°*. When mounted on a holder, the actual relief angle may be different from that on the insert.

3) *Tolerances:* The third position is a letter and indicates the tolerances controlling the indexability of the insert. Tolerances specified do not imply the method of manufacture.

| Symbol | Tolerance (± from nominal) Inscribed Circle, Inch | Tolerance (± from nominal) Thickness, Inch | Symbol | Tolerance (± from nominal) Inscribed Circle, Inch | Tolerance (± from nominal) Thickness, Inch |
|---|---|---|---|---|---|
| A | 0.001 | 0.001 | H | 0.0005 | 0.001 |
| B | 0.001 | 0.005 | J | 0.002–0.005 | 0.001 |
| C | 0.001 | 0.001 | K | 0.002–0.005 | 0.001 |
| D | 0.001 | 0.005 | L | 0.002–0.005 | 0.001 |
| E | 0.001 | 0.001 | M | 0.002–0.004[a] | 0.005 |
| F | 0.0005 | 0.001 | U | 0.005–0.010[a] | 0.005 |
| G | 0.001 | 0.005 | N | 0.002–0.004[a] | 0.001 |

[a] Exact tolerance is determined by size of insert. See ANSI B212.12-1991 (Withdrawn).

4) *Type:* The type of insert is designated by a letter: **A,** with hole; **B,** with hole and countersink; **C,** with hole and two countersinks; **F,** chip grooves both surfaces, no hole; **G,** same

* Second angle is secondary facet angle, which may vary by ± 1°.

as **F** but with hole; **H,** with hole, one countersink, and chip groove on one rake surface; **J,** with hole, two countersinks and chip grooves on two rake surfaces; **M,** with hole and chip groove on one rake surface; **N,** without hole; **Q,** with hole and two countersinks; **R,** without hole but with chip groove on one rake surface; **T,** with hole, one countersink, and chip groove on one rake face; **U,** with hole, two countersinks, and chip grooves on two rake faces; and **W,** with hole and one countersink. *Note:* A dash may be used after position 4 to separate the shape-describing portion from the following dimensional description of the insert and is not to be considered a position in the Standard description.

5) *Size:* The size of the insert is designated by a one- or two-digit number. For regular polygons and diamonds, insert size is the number of eighths of an inch in the nominal size of the inscribed circle, and will be a one- or two-digit number when the number of eighths is a whole number. It will be a two-digit number, including one decimal place, when it is not a whole number. Rectangular and parallelogram inserts require two digits: the first digit indicates the number of eighths of an inch width and the second digit the number of quarters of an inch length.

6) *Thickness:* The thickness is designated by a one- or two-digit number, which indicates the number of sixteenths of an inch in the thickness of the insert. Thickness is a one-digit number when the number of sixteenths is a whole number; it is a two-digit number carried to one decimal place when the number of sixteenths of an inch is not a whole number.

7) *Cutting Point Configuration:* The cutting point, or nose radius, is designated by a number representing $\frac{1}{64}$ths of an inch; a flat at the cutting point or nose, is designated by a letter: **0** for sharp corner, 0.002 inch max. radius; **0.2** for 0.004 radius; **0.3** for 0.008 radius; **1,** $\frac{1}{64}$ inch radius; **2,** $\frac{1}{32}$ inch radius; **3,** $\frac{3}{64}$ inch radius; **4,** $\frac{1}{16}$ inch radius; **5,** $\frac{5}{64}$ inch radius; **6,** $\frac{3}{32}$ inch radius; **7,** $\frac{7}{64}$ inch radius; **8,** $\frac{1}{8}$ inch radius; **X**, any other radius; **A,** square insert with 45° chamfer; **D,** square insert with 30° chamfer; **E,** square insert with 15° chamfer; **F,** square insert with 3° chamfer; **K,** square insert with 30° double chamfer; **L,** square insert with 15° double chamfer; **M,** square insert with 3° double chamfer; **N,** truncated triangle insert; and **P,** flatted corner triangle insert.

8) *Special Cutting Point Definition:* The eighth position, if it follows a letter in the 7th position, is a number indicating the number of $\frac{1}{64}$ths of an inch in the primary facet length measured parallel to the edge of the facet.

9) *Hand:* **R,** right; **L,** left; to be used when required in ninth position.

10) *Other Conditions:* Position ten defines special conditions (such as edge treatment, surface finish): **A,** honed, 0.0005 to less than 0.003 inch (0.0127 to 0.0762 mm); **B,** honed, 0.003 to less than 0.005 inch (0.0762 to 0.127 mm); **C,** honed, 0.005 to less than 0.007 inch (0.127 to 0.178 mm); **J,** polished, 4 μinch (0.1016 μm) arithmetic average (AA) on rake surfaces only; **T,** chamfered, manufacturer's standard negative land, rake face only.

**Indexable Insert Tool Holders.—**Indexable insert tool holders are made from a good grade of steel which is heat treated to a hardness of 44 to 48 RC (Rockwell C scale) for most normal applications. Accurate pockets that serve to locate the insert in position and to provide surfaces against which the insert can be clamped are machined in the ends of tool holders. A cemented carbide seat usually is provided and is held in the bottom of the pocket by a screw or by the clamping pin, if one is used. The seat is necessary to provide a flat bearing surface upon which the insert can rest and, in so doing, it adds materially to the ability of the insert to withstand the cutting load. The seating surface of the holder may provide a positive-, negative-, or a neutral-rake orientation to the insert when it is in position on the holder. Holders, therefore, are classified as positive, negative, or neutral rake.

Four basic methods are used to clamp the insert on the holder: 1) Clamping, usually top clamping; 2) Pin-lock clamping; 3) Multiple clamping using a clamp, usually a top clamp, and a pin lock; and 4) Clamping the insert with a machine screw.

All top clamps are actuated by a screw that forces the clamp directly against the insert. When required, a cemented carbide, plate-type chipbreaker is placed between the clamp and the insert. Pin-lock clamps require an insert having a hole: the pin acts against the walls of the hole to clamp the insert firmly against the seating surfaces of the holder. Multiple or combination clamping, simultaneously using both a pin-lock and a top clamp, is recommended when taking heavier or interrupted cuts. Holders are available on which all the above-mentioned methods of clamping may be used. Other holders are made with only a top clamp or a pin lock. Screw-on type holders use a machine screw to hold the insert in the pocket. Most standard indexable insert holders are either straight-shank or offset-shank, although special holders are made having a wide variety of configurations.

The common shank sizes of indexable insert tool holders are shown in Table 1. Not all styles are available in every shank size. Positive- and negative-rake tools are also not available in every style or shank size. Some manufacturers provide additional shank sizes for certain tool holder styles. For more complete details, the manufacturers' catalogs must be consulted.

**Table 1. Standard Shank Sizes for Indexable Insert Holders**

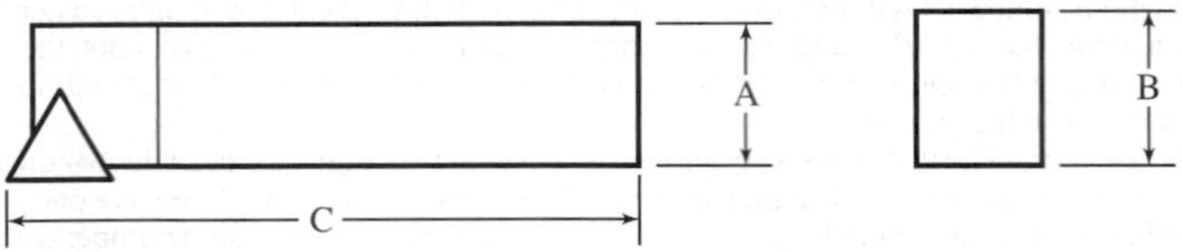


| Basic Shank Size | Shank Dimensions for Indexable Insert Holders | | | | | |
|---|---|---|---|---|---|---|
| | A | | B | | C[a] | |
| | inch | mm | inch | mm | inch | mm |
| ½ × ½ × 4½ | 0.500 | 12.70 | 0.500 | 12.70 | 4.500 | 114.30 |
| ⅝ × ⅝ × 4½ | 0.625 | 15.87 | 0.625 | 15.87 | 4.500 | 114.30 |
| ⅝ × 1¼ × 6 | 0.625 | 15.87 | 1.250 | 31.75 | 6.000 | 152.40 |
| ¾ × ¾ × 4½ | 0.750 | 19.05 | 0.750 | 19.05 | 4.500 | 114.30 |
| ¾ × 1 × 6 | 0.750 | 19.05 | 1.000 | 25.40 | 6.000 | 152.40 |
| ¾ × 1¼ × 6 | 0.750 | 19.05 | 1.250 | 31.75 | 6.000 | 152.40 |
| 1 × 1 × 6 | 1.000 | 25.40 | 1.000 | 25.40 | 6.000 | 152.40 |
| 1 × 1¼ × 6 | 1.000 | 25.40 | 1.250 | 31.75 | 6.000 | 152.40 |
| 1 × 1½ × 6 | 1.000 | 25.40 | 1.500 | 38.10 | 6.000 | 152.40 |
| 1¼ × 1¼ × 7 | 1.250 | 31.75 | 1.250 | 31.75 | 7.000 | 177.80 |
| 1¼ × 1½ × 8 | 1.250 | 31.75 | 1.500 | 38.10 | 8.000 | 203.20 |
| 1⅜ × 2¹⁄₁₆ × 6⅜ | 1.375 | 34.92 | 2.062 | 52.37 | 6.380 | 162.05 |
| 1½ × 1½ × 7 | 1.500 | 38.10 | 1.500 | 38.10 | 7.000 | 177.80 |
| 1¾ × 1¾ × 9½ | 1.750 | 44.45 | 1.750 | 44.45 | 9.500 | 241.30 |
| 2 × 2 × 8 | 2.000 | 50.80 | 2.000 | 50.80 | 8.000 | 203.20 |

[a] Holder length; may vary by manufacturer. Actual shank length depends on holder style.

**Identification System for Indexable Insert Holders.—**The following identification system conforms to the American National Standard, ANSI B212.5-2002 (Withdrawn), Metric Holders for Indexable Inserts.

Each position in the system designates a feature of the holder in the following sequence:

| 1 | 2 | 3 | 4 | 5 | — | 6 | — | 7 | — | 8[a] | — | 9 | — | 10[a] |
|---|---|---|---|---|---|---|---|---|---|---|---|---|---|---|
| **C** | **T** | **N** | **A** | **R** | — | **85** | — | **25** | — | **D** | — | **16** | — | **Q** |

1) *Method of Holding Horizontally Mounted Insert:* The method of holding or clamping is designated by a letter: **C,** top clamping, insert without hole; **M,** top and hole clamping, insert with hole; **P,** hole clamping, insert with hole; **S,** screw clamping through hole, insert with hole; **W,** wedge clamping.

2) *Insert Shape:* The insert shape is identified by a letter: **H,** hexagonal; **O,** octagonal; **P,** pentagonal; **S,** square; **T,** triangular; **C,** rhombic, 80° included angle; **D,** rhombic, 55° included angle; **E,** rhombic, 75° included angle; **M,** rhombic, 86° included angle; **V,** rhombic, 35° included angle; **W,** hexagonal, 80° included angle; **L,** rectangular; **A,** parallelogram, 85° included angle; **B,** parallelogram, 82° included angle; **K,** parallelogram, 55° included angle; **R,** round. The included angle is always the smaller angle.

3) *Holder Style:* The holder style designates the shank style and the side cutting edge angle, or end cutting edge angle, or the purpose for which the holder is used. It is designated by a letter: **A** for straight shank with 0° side cutting edge angle; **B,** straight shank with 15° side cutting edge angle; **C,** straight-shank end cutting tool with 0° end cutting edge angle; **D,** straight shank with 45° side cutting edge angle; **E,** straight shank with 30° side cutting edge angle; **F,** offset shank with 0° end cutting edge angle; **G,** offset shank with 0° side cutting edge angle; **J,** offset shank with negative 3° side cutting edge angle; **K,** offset shank with 15° end cutting edge angle; **L,** offset shank with negative 5° side cutting edge angle and 5° end cutting edge angle; **M,** straight shank with 40° side cutting edge angle; **N,** straight shank with 27° side cutting edge angle; **R,** offset shank with 15° side cutting edge angle; **S,** offset shank with 45° side cutting edge angle; **T,** offset shank with 30° side cutting edge angle; **U,** offset shank with negative 3° end cutting edge angle; **V,** straight shank with $17\frac{1}{2}$° side cutting edge angle; **W,** offset shank with 30° end cutting edge angle; **Y,** offset shank with 5° end cutting edge angle.

4) *Normal Clearances:* The normal clearances of inserts are identified by letters: **A,** 3°; **B,** 5°; **C,** 7°; **D,** 15°; **E,** 20°; **F,** 25°; **G,** 30°; **N,** 0°; **P,** 11°.

5) *Hand of tool:* The hand of the tool is designated by a letter: **R** for right-hand; **L,** left-hand; and **N,** neutral, or either hand.

6) *Tool Height for Rectangular Shank Cross Sections:* The tool height for tool holders with a rectangular shank cross section and the height of cutting edge equal to shank height are given as two-digit numbers representing these values in millimeters. For example, a height of 32 mm would be encoded as 32; 8 mm would be encoded as 08, where the one-digit value is preceded by a zero.

7) *Tool Width for Rectangular Shank Cross Sections:* The tool width for tool holders with a rectangular shank cross section is given as a two-digit number representing this value in millimeters. For example, a width of 25 mm would be encoded as 25; 8 mm would be encoded as 08, where the one-digit value is preceded by a zero.

8) *Tool Length:* The tool length is designated by a letter: **A,** 32 mm; **B,** 40 mm; **C,** 50 mm; **D,** 60 mm; **E,** 70 mm; **F,** 80 mm; **G,** 90 mm; **H,** 100 mm; **J,** 110 mm; **K,** 125 mm; **L,** 140 mm; **M,** 150 mm; **N,** 160 mm; **P,** 170 mm; **Q,** 180 mm; **R,** 200 mm; **S,** 250 mm; **T,** 300 mm; **U,** 350 mm; **V,** 400 mm; **W,** 450 mm; **X,** special length to be specified; **Y,** 500 mm.

9) *Indexable Insert Size:* The size of indexable inserts is encoded as follows: For insert shapes **C, D, E, H, M, O, P, R, S, T, V,** the side length (the diameter for **R** inserts) in millimeters is used as a two-digit number, with decimals being disregarded. For example, the symbol for a side length of 16.5 mm is 16. For insert shapes **A, B, K, L,** the length of the main cutting edge or of the longer cutting edge in millimeters is encoded as a two-digit number, disregarding decimals. If the symbol obtained has only one digit, then it should be preceded by a zero. For example, the symbol for a main cutting edge of 19.5 mm is 19; for an edge of 9.5 mm, the symbol is 09.

10) *Special Tolerances:* Special tolerances are indicated by a letter: **Q,** back and end qualified tool; **F,** front and end qualified tool; **B,** back, front, and end qualified tool. A qualified tool is one that has tolerances of ±0.08 mm for dimensions *F, G,* and *C.* (See Table 2.)

**Table 2. Letter Symbols for Qualification of Tool Holders Position 10** *ANSI B212.5-2002 (Withdrawn)*

<table>
<tr><td rowspan="3">Qualification of Tool Holder</td><td colspan="3">Letter Symbol</td></tr>
<tr><td>Q</td><td>F</td><td>B</td></tr>
<tr><td>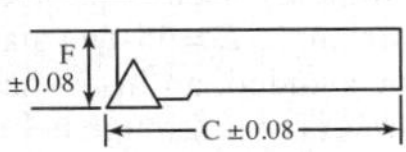

Back and end qualified tool</td><td>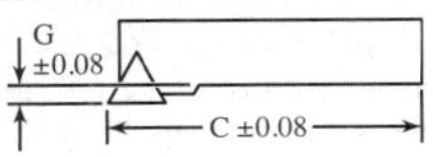

Front and end qualified tool</td><td>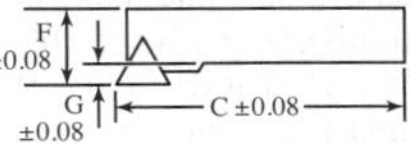

Back, front, and end qualified tool</td></tr>
</table>

**Selecting Indexable Insert Holders.—**A guide for selecting indexable insert holders is provided by Table 3b. Some operations such as deep grooving, cut-off, and threading are not given in this table. However, tool holders designed specifically for these operations are available. The boring operations listed in Table 3b refer primarily to larger holes into which the holders will fit. Smaller holes are bored using boring bars. An examination of this table shows that several tool-holder styles can be and frequently are used for each operation. Selection of the best holder for a given job depends largely on the job, and there are certain basic facts that should be considered in making the selection.

*Rake Angle:* A negative-rake insert has twice as many cutting edges available as a comparable positive-rake insert. Sometimes the tool life obtained when using the second face may be less than that obtained on the first face because the tool wear on the cutting edges of the first face may reduce the insert strength. Nevertheless, the advantage of negative-rake inserts and holders is such that they should be considered first in making any choice. Positive-rake holders should be used where lower cutting forces are required, as when machining slender or small-diameter parts, when chatter may occur, and for machining some materials, such as aluminum, copper, and certain grades of stainless steel, when positive-negative rake inserts can sometimes be used to advantage. These inserts are held on negative-rake holders that have their rake surfaces ground or molded to form a positive-rake angle.

*Insert Shape:* The configuration of the workpiece, the operation to be performed, and the lead angle required often determine the insert shape. When these factors need not be considered, the insert shape should be selected on the basis of insert strength and the maximum number of cutting edges available. Thus, a round insert is the strongest and has a maximum number of available cutting edges. It can be used with heavier feeds while producing a good surface finish. Round inserts are limited by their tendency to cause chatter, which may preclude their use.

The square insert is the next most effective shape, providing good corner strength and more cutting edges than all other inserts except the round insert. The only limitation of this insert shape is that it must be used with a lead angle. Therefore, the square insert cannot be used for turning square shoulders or for back-facing.

Triangle inserts are the most versatile and can be used to perform more operations than any other insert shape. The 80-degree diamond insert is designed primarily for heavy turning and facing operations using the 100-degree corners, and, for turning and back-facing square shoulders, using the 80-degree corners. The 55- and 35-degree diamond inserts are intended primarily for tracing.

*Lead Angle:* Tool holders should be selected to provide the largest possible lead angle, although limitations are sometimes imposed by the nature of the job. For example, when tuning and back-facing a shoulder, a negative lead angle must be used. Slender or small-diameter parts may deflect, causing difficulties in holding size, or chatter when the lead angle is too large.

*End Cutting Edge Angle:* When tracing or contour turning, the plunge angle is determined by the end cutting edge angle. A 2-deg minimum clearance angle should be provided between the workpiece surface and the end cutting edge of the insert. Table 3a provides the maximum plunge angle for holders commonly used to plunge when tracing where insert shape identifiers are $S$ = square, $T$ = triangle, $D$ = 55-deg diamond, $V$ = 35-deg diamond. When severe cratering cannot be avoided, an insert having a small, end cutting edge angle is desirable to delay the crater breakthrough behind the nose. For very heavy cuts, a small, end cutting edge angle will strengthen the corner of the tool. Tool holders for numerical control machines are discussed beginning page 844.

**Table 3a. Maximum Plunge Angle for Tracing or Contour Turning**

| Tool Holder Style | Insert Shape | Maximum Plunge Angle | Tool Holder Style | Insert Shape | Maximum Plunge Angle |
|---|---|---|---|---|---|
| E | T | 58° | J | D | 30° |
| D and S | S | 43° | J | V | 50° |
| H | D | 71° | N | T | 55° |
| J | T | 25° | N | D | 58° to 60° |

**Table 3b. Indexable Insert Holder Application Guide**

| Tool | Tool Holder Style | Insert Shape | Rake N-Negative P-Positive | Application | | | | | | | | |
|---|---|---|---|---|---|---|---|---|---|---|---|---|
| | | | | Turn | Face | Turn and Face | Turn and Backface | Trace | Groove | Chamfer | Bore | Plane |
|  | A | T | N | ● | ● | | | | | | ● | |
| | | | P | ● | ● | | | | | | ● | |
| 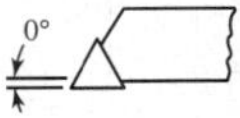 | A | T | N | ● | ● | | | ● | | | | |
| | | | P | ● | ● | | | ● | | | | |
| 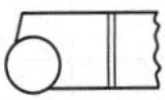 | A | R | N | ● | ● | ● | | | | | | ● |
| | | | | | | | | | | | | |
|  | A | R | N | ● | ● | ● | | ● | | | | ● |
| | | | | | | | | | | | | |
| 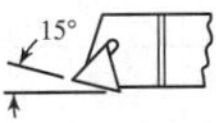 | B | T | N | ● | ● | | | | | | ● | |
| | | | P | ● | ● | | | | | | ● | |
| 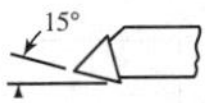 | B | T | N | ● | ● | | | ● | | | ● | |
| | | | P | ● | ● | | | ● | | | ● | |
|  | B | S | N | ● | ● | | | | | | ● | |
| | | | P | ● | ● | | | | | | ● | |
| 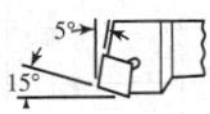 | B | C | N | ● | ● | ● | | | | | ● | ● |
| | | | | | | | | | | | | |
| 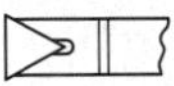 | C | T | N | ● | ● | | | | ● | ● | | |
| | | | P | ● | ● | | | | ● | ● | | |

**Table 3b.** *(Continued)* **Indexable Insert Holder Application Guide**

| Tool | Tool Holder Style | Insert Shape | Rake N-Negative P-Positive | Application: Turn | Face | Turn and Face | Turn and Backface | Trace | Groove | Chamfer | Bore | Plane |
|---|---|---|---|---|---|---|---|---|---|---|---|---|
| 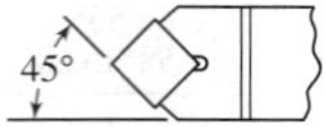  | D | S | N | ● | ● | ● | | ● | | ● | ● | ● |
| | | | P | ● | ● | ● | | ● | | ● | ● | ● |
| 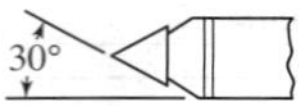  | E | T | N | ● | ● | | | ● | ● | ● | | |
| | | | P | ● | ● | | | ● | ● | ● | | |
| 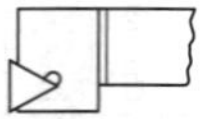 | F | T | N | ● | ● | | | | | | ● | |
| | | | P | ● | ● | | | | | | ● | |
| 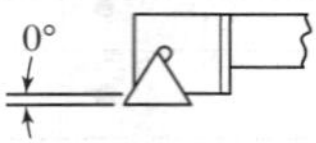  | G | T | N | ● | ● | | | | | | ● | |
| | | | P | ● | ● | | | | | | ● | |
|  | G | R | N | ● | ● | ● | | | | | | |
| | | | | | | | | | | | | |
| 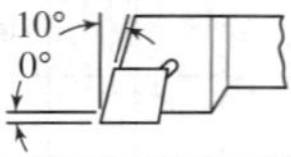  | G | C | N | ● | ● | ● | | | | | | |
| | | | P | ● | ● | ● | | | | | | |
| 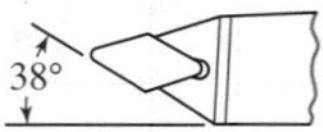  | H | D | N | ● | ● | | | ● | | | | |
| | | | | | | | | | | | | |
| 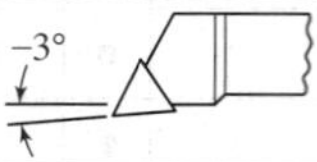  | J | T | N | | | | ● | ● | | | | |
| | | | P | | | | ● | ● | | | | |
| 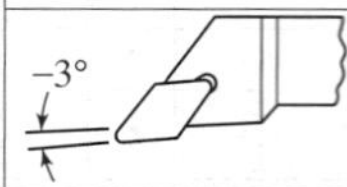  | J | D | N | | | | ● | ● | | | | |
| | | | | | | | | | | | | |
| 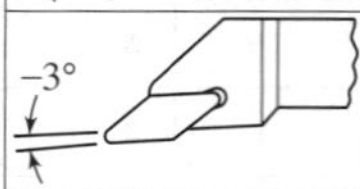  | J | V | N | | | | ● | ● | | | | |
| | | | | | | | | | | | | |
| 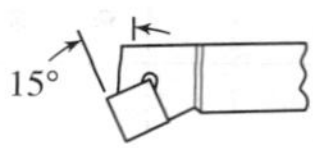  | K | S | N | ● | ● | | | | | | ● | |
| | | | P | ● | ● | | | | | | ● | |

**Table 3b.** *(Continued)* **Indexable Insert Holder Application Guide**

| Tool | Tool Holder Style | Insert Shape | Rake N-Negative P-Positive | Application: Turn | Face | Turn and Face | Turn and Backface | Trace | Groove | Chamfer | Bore | Plane |
|---|---|---|---|---|---|---|---|---|---|---|---|---|
| 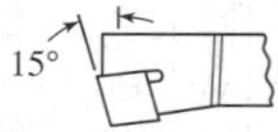 | K | C | N | • | • | | | | | | • | |
| | | | | | | | | | | | | |
| 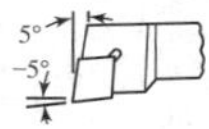 | L | C | N | | | • | • | | | | | |
| | | | | | | | | | | | | |
| 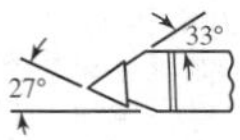 | N | T | N | • | • | | | • | | | | |
| | | | P | • | • | | | • | | | | |
| 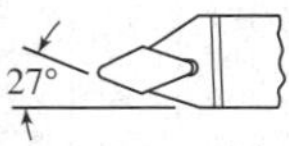 | N | D | N | • | • | | | • | | | | |
| | | | | | | | | | | | | |
| 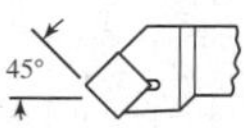 | S | S | N | • | • | • | | • | | • | • | • |
| | | | P | • | • | • | | • | | • | • | • |
| 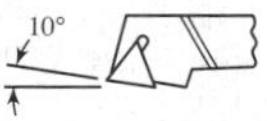 | W | T | N | • | • | | | | | | | |
| | | | | | | | | | | | | |

## Sintered Carbide Blanks and Cutting Tools

**Sintered Carbide Blanks.—**As shown in Table 4, American National Standard ANSI B212.1-2002 (Withdrawn) provides standard sizes and designations for eight styles of sintered carbide blanks. These blanks are the unground solid carbide from which either solid or tipped cutting tools are made. Tipped cutting tools are made by brazing a blank onto a shank to produce the cutting tool; these tools differ from carbide *insert* cutting tools which consist of a carbide insert held mechanically in a tool holder. A typical single-point carbide-tipped cutting tool is shown in Fig. 1 on page 837.

**Single-Point, Sintered Carbide-Tipped Tools.—**American National Standard ANSI B212.1-2002 (Withdrawn) covers eight different styles of single-point, carbide-tipped general purpose tools. These styles are designated by the letters A to G inclusive. Styles A, B, F, G, and E with offset point are either right- or left-hand cutting as indicated by the letters R or L. Dimensions of tips and shanks are given in Table 5 to Table 12. For dimensions and tolerances not shown, and for the identification system, dimensions, and tolerances of sintered carbide boring tools, see the Standard.

A number follows the letters of the tool style and hand designation, and, for square shank tools, represents the number of sixteenths of an inch of width, *W*, and height, *H*. With rectangular shanks, the first digit of the number indicates the number of eighths of an inch in the shank width, *W*, and the second digit the number of quarters of an inch in the shank height, *H*. One exception is the $1\frac{1}{2} \times 2$-inch size, which has been arbitrarily assigned the number 90.

**Table 4. American National Standard Sizes and Designations for Carbide Blanks**
*ANSI B212.1-2002 (Withdrawn)*

| Blank Dimensions[a] | | | Style[b] 1000 | Style[b] 2000 |
|---|---|---|---|---|
| *T* | *W* | *L* | Blank Designation | |
| 1/16 | 1/8 | 5/8 | 1010 | 2010 |
| 1/16 | 5/32 | 1/4 | 1015 | 2015 |
| 1/16 | 3/16 | 1/4 | 1020 | 2020 |
| 1/16 | 1/4 | 1/4 | 1025 | 2025 |
| 1/16 | 1/4 | 5/16 | 1030 | 2030 |
| 3/32 | 1/8 | 3/4 | 1035 | 2035 |
| 3/32 | 3/16 | 5/16 | 1040 | 2040 |
| 3/32 | 3/16 | 1/2 | 1050 | 2050 |
| 3/32 | 1/4 | 3/8 | 1060 | 2060 |
| 3/32 | 1/4 | 1/2 | 1070 | 2070 |
| 3/32 | 5/16 | 3/8 | 1080 | 2080 |
| 3/32 | 3/8 | 3/8 | 1090 | 2090 |
| 3/32 | 3/8 | 1/2 | 1100 | 2100 |
| 3/32 | 7/16 | 1/2 | 1105 | 2105 |
| 3/32 | 5/16 | 3/8 | 1080 | 2080 |
| 1/8 | 3/16 | 3/4 | 1110 | 2110 |
| 1/8 | 1/4 | 1/2 | 1120 | 2120 |
| 1/8 | 1/4 | 5/8 | 1130 | 2130 |
| 1/8 | 1/4 | 3/4 | 1140 | 2140 |
| 1/8 | 5/16 | 7/16 | 1150 | 2150 |
| 1/8 | 5/16 | 1/2 | 1160 | 2160 |
| 1/8 | 3/16 | 3/4 | 1110 | 2110 |
| 1/8 | 5/16 | 5/8 | 1170 | 2170 |
| 1/8 | 3/8 | 1/2 | 1180 | 2180 |
| 1/8 | 3/8 | 3/4 | 1190 | 2190 |
| 1/8 | 1/2 | 1/2 | 1200 | 2200 |
| 1/8 | 1/2 | 3/4 | 1210 | 2210 |
| 1/8 | 3/4 | 3/4 | 1215 | 2215 |
| 5/32 | 3/8 | 9/16 | 1220 | 2220 |
| 5/32 | 3/8 | 3/4 | 1230 | 2230 |
| 5/32 | 5/8 | 5/8 | 1240 | 2240 |
| 3/16 | 5/16 | 7/16 | 1250 | 2250 |
| 3/16 | 5/16 | 5/8 | 1260 | 2260 |
| 3/16 | 3/8 | 1/2 | 1270 | 2270 |
| 3/16 | 3/8 | 5/8 | 1280 | 2280 |
| 3/16 | 3/8 | 3/4 | 1290 | 2290 |
| 3/16 | 7/16 | 5/8 | 1300 | 2300 |
| 3/16 | 7/16 | 13/16 | 1310 | 2310 |
| 3/16 | 1/2 | 1/2 | 1320 | 2320 |
| 3/16 | 1/2 | 3/4 | 1330 | 2330 |
| 3/16 | 3/4 | 3/4 | 1340 | 2340 |

| Blank Dimensions[a] | | | Style[b] 0000 | Style[b] 1000 | Style[b] 3000 | Style[b] 4000 |
|---|---|---|---|---|---|---|
| *T* | *W* | *L* | Blank Designation | | | |
| 1/4 | 3/8 | 9/16 | 0350 | 1350 | 3350 | 4350 |
| 1/4 | 3/8 | 3/4 | 0360 | 1360 | 3360 | 4360 |
| 1/4 | 7/16 | 5/8 | 0370 | 1370 | 3370 | 4370 |
| 1/4 | 1/2 | 3/4 | 0380 | 1380 | 3380 | 4380 |
| 1/4 | 9/16 | 1 | 0390 | 1390 | 3390 | 4390 |
| 1/4 | 5/8 | 5/8 | 0400 | 1400 | 3400 | 4400 |
| 1/4 | 3/4 | 3/4 | 0405 | 1405 | 3405 | 4405 |
| 1/4 | 3/4 | 1 | 0410 | 1410 | 3410 | 4410 |
| 1/4 | 1 | 1 | 0415 | 1415 | 3415 | 4415 |
| 5/16 | 7/16 | 5/8 | 0420 | 1420 | 3420 | 4420 |
| 5/16 | 7/16 | 15/16 | 0430 | 1430 | 3430 | 4430 |
| 5/16 | 1/2 | 3/4 | 0440 | 1440 | 3440 | 4440 |
| 5/16 | 1/2 | 1 | 0450 | 1450 | 3450 | 4450 |
| 5/16 | 5/8 | 1 | 0460 | 1460 | 3460 | 4460 |
| 5/16 | 3/4 | 3/4 | 0470 | 1470 | 3470 | 4470 |
| 5/16 | 3/4 | 1 | 0475 | 1475 | 3475 | 4475 |
| 5/16 | 3/4 | 1 1/4 | 0480 | 1480 | 3480 | 4480 |
| 3/8 | 1/2 | 3/4 | 0490 | 1490 | 3490 | 4490 |
| 3/8 | 1/2 | 1 | 0500 | 1500 | 3500 | 4500 |
| 3/8 | 5/8 | 1 | 0510 | 1510 | 3510 | 4510 |
| 3/8 | 5/8 | 1 1/4 | 0515 | 1515 | 3515 | 4515 |
| 3/8 | 3/4 | 1 1/4 | 0520 | 1520 | 3520 | 4520 |
| 3/8 | 3/4 | 1 1/2 | 0525 | 1525 | 3525 | 4525 |
| 1/2 | 3/4 | 1 | 0530 | 1530 | 3530 | 4530 |
| 1/2 | 3/4 | 1 1/4 | 0540 | 1540 | 3540 | 4540 |
| 3/8 | 1/2 | 3/4 | 0490 | 1490 | 3490 | 4490 |
| 1/2 | 3/4 | 1 1/2 | 0550 | 1550 | 3550 | 4550 |

| *T* | *W* | *L* | *F* | Style[b] 5000 | Style[b] 6000 | Style[b] 70000 |
|---|---|---|---|---|---|---|
| 1/16 | 1/4 | 5/16 | … | 5030 | … | … |
| 3/32 | 1/4 | 3/8 | 1/16 | … | … | 7060 |
| 3/32 | 5/16 | 3/8 | … | 5080 | 6080 | … |
| 3/32 | 3/8 | 1/2 | … | 5100 | 6100 | … |
| 3/32 | 7/16 | 1/2 | … | 5105 | … | … |
| 1/8 | 5/16 | 5/8 | 3/32 | … | … | 7170 |
| 3/32 | 1/4 | 3/8 | 1/16 | … | … | 7060 |
| 1/8 | 1/2 | 1/2 | … | 5200 | 6200 | … |
| 5/32 | 3/8 | 3/4 | 1/8 | … | … | 7230 |
| 5/32 | 5/8 | 5/8 | … | 5240 | 6240 | … |
| 3/16 | 3/4 | 3/4 | … | 5340 | 6340 | … |
| 1/4 | 1 | 3/4 | … | 5410 | … | … |

[a] All dimensions are in inches.

[b] See Fig. 1 on page 837 for a description of styles.

A typical single-point carbide-tipped cutting tool is shown in Fig. 2. The side rake, side relief, and the clearance angles are normal to the side-cutting edge, rather than the shank, to facilitate its being ground on a tilting-table grinder. The end-relief and clearance angles are *normal* to the end-cutting edge. The back-rake angle is parallel to the side-cutting edge.

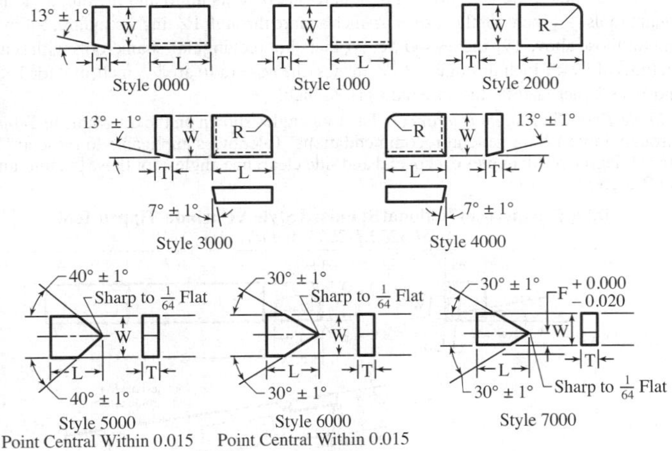


Fig. 1. Eight styles of Sintered Carbide Blanks (see Table 4)

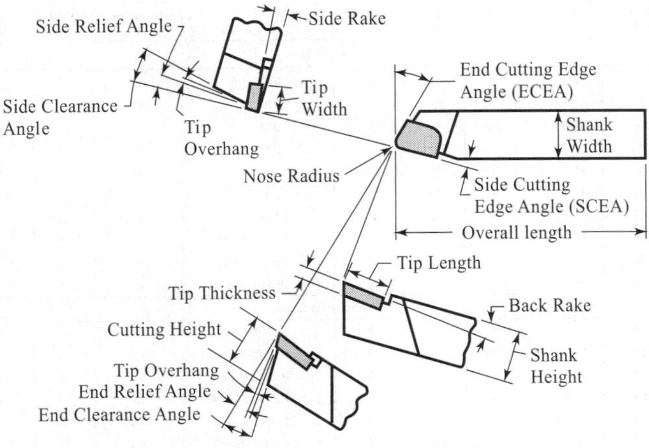


Fig. 2. A Typical Single-Point Carbide-Tipped Cutting Tool

The tip of the brazed carbide blank overhangs the shank of the tool by either $^1/_{32}$ or $^1/_{16}$ inch, depending on the size of the tool. For tools in Table 5, Table 6, Table 7, Table 8, Table 11 and Table 12, the maximum overhang is $^1/_{32}$ inch for shank sizes 4, 5, 6, 7, 8, 10, 12 and 44; for other shank sizes in these tables, the maximum overhang is $^1/_{16}$ inch. In Table 9 and Table 10, all tools have maximum overhang of $^1/_{32}$ inch.

*Single-Point Tool Nose Radii:* The tool nose radii recommended in the American National Standard are as follows: For square-shank tools up to and including 3⁄8-inch square tools, 1⁄64 inch; for those over 3⁄8-inch square through 1 1⁄4-inches square, 1⁄32 inch; and for those above 1 1⁄4-inches square, 1⁄16 inch. For rectangular-shank tools with shank sections of 1⁄2 × 1 inch through 1 × 1 1⁄2 inches, the nose radii are 1⁄32 inch, and for 1 × 2- and 1 1⁄2 × 2-inch shanks, the nose radius is 1⁄16 inch.

*Single-Point Tool Angle Tolerances:* The tool angles shown on the diagrams in Table 5 through Table 12 are general recommendations. Tolerances applicable to these angles are ± 1 degree on all angles except end and side clearance angles; for these the tolerance is ± 2 degrees.

**Table 5. American National Standard Style A Carbide-Tipped Tools**
*ANSI B212.1-2002 (Withdrawn)*

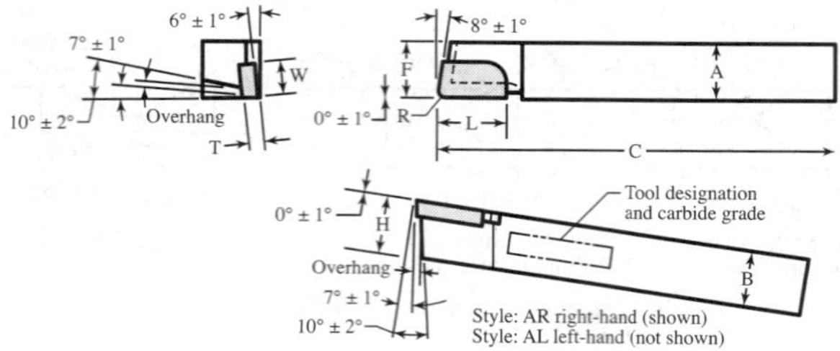


| Designation | | Shank Dimensions | | | | Tip Dimensions | | |
|---|---|---|---|---|---|---|---|---|
| Style AR[a] | Style AL[a] | Width *A* | Height *B* | Length *C* | Tip Designation[a] | Thickness *T* | Width *W* | Length *L* |
| Square Shank | | | | | | | | |
| AR 4 | AL 4 | 1⁄4 | 1⁄4 | 2 | 2040 | 3⁄32 | 3⁄16 | 5⁄16 |
| AR 5 | AL 5 | 5⁄16 | 5⁄16 | 2 1⁄4 | 2070 | 3⁄32 | 1⁄4 | 1⁄2 |
| AR 6 | AL 6 | 3⁄8 | 3⁄8 | 2 1⁄2 | 2070 | 3⁄32 | 1⁄4 | 1⁄2 |
| AR 7 | AL 7 | 7⁄16 | 7⁄16 | 3 | 2070 | 3⁄32 | 1⁄4 | 1⁄2 |
| AR 8 | AL 8 | 1⁄2 | 1⁄2 | 3 1⁄2 | 2170 | 1⁄8 | 5⁄16 | 5⁄8 |
| AR 10 | AL 10 | 5⁄8 | 5⁄8 | 4 | 2230 | 5⁄32 | 3⁄8 | 3⁄4 |
| AR 12 | AL 12 | 3⁄4 | 3⁄4 | 4 1⁄2 | 2310 | 3⁄16 | 7⁄16 | 13⁄16 |
| AR 16 | AL 16 | 1 | 1 | 6 | { P3390, P4390 | 1⁄4 | 9⁄16 | 1 |
| AR 20 | AL 20 | 1 1⁄4 | 1 1⁄4 | 7 | { P3460, P4460 | 5⁄16 | 5⁄8 | 1 |
| AR 24 | AL 24 | 1 1⁄2 | 1 1⁄2 | 8 | { P3510, P4510 | 3⁄8 | 5⁄8 | 1 |
| Rectangular Shank | | | | | | | | |
| AR 44 | AL 44 | 1⁄2 | 1 | 6 | P2260 | 3⁄16 | 5⁄16 | 5⁄8 |
| AR 54 | AL 54 | 5⁄8 | 1 | 6 | { P3360, P4360 | 1⁄4 | 3⁄8 | 3⁄4 |
| AR 55 | AL 55 | 5⁄8 | 1 1⁄4 | 7 | { P3360, P4360 | 1⁄4 | 3⁄8 | 3⁄4 |
| AR 64 | AL 64 | 3⁄4 | 1 | 6 | { P3380, P4380 | 1⁄4 | 1⁄2 | 3⁄4 |
| AR 66 | AL 66 | 3⁄4 | 1 1⁄2 | 8 | { P3430, P4430 | 5⁄16 | 7⁄16 | 15⁄16 |
| AR 85 | AL 85 | 1 | 1 1⁄4 | 7 | { P3460, P4460 | 5⁄16 | 5⁄8 | 1 |
| AR 86 | AL 86 | 1 | 1 1⁄2 | 8 | { P3510, P4510 | 3⁄8 | 5⁄8 | 1 |
| AR 88 | AL 88 | 1 | 2 | 10 | { P3510, P4510 | 3⁄8 | 5⁄8 | 1 |
| AR 90 | AL 90 | 1 1⁄2 | 2 | 10 | { P3540, P4540 | 1⁄2 | 3⁄4 | 1 1⁄4 |

[a] "A" is straight shank, 0 deg., SCEA (side cutting edge angle). "R" is right-cut. "L" is left-cut. Where a pair of tip numbers is shown, the upper number applies to AR tools, the lower to AL tools. All dimensions are in inches.

**Table 6. American National Standard Style B Carbide-Tipped Tools with 15-Degree Side Cutting Edge Angle** *ANSI B212.1-2002 (Withdrawn)*

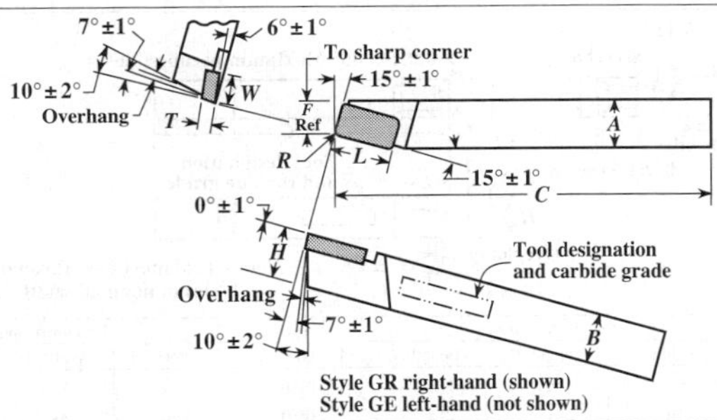


| Designation | | Shank Dimensions | | | Tip Designation[a] | Tip Dimensions | | |
|---|---|---|---|---|---|---|---|---|
| Style BR | Style BL | Width *A* | Height *B* | Length *C* | | Thickness *T* | Width *W* | Length *L* |
| Square Shank | | | | | | | | |
| BR 4 | BL 4 | 1/4 | 1/4 | 2 | 2015 | 1/16 | 5/32 | 1/4 |
| BR 5 | BL 5 | 5/16 | 5/16 | 2 1/4 | 2040 | 3/32 | 3/16 | 5/16 |
| BR 6 | BL 6 | 3/8 | 3/8 | 2 1/2 | 2070 | 3/32 | 1/4 | 1/2 |
| BR 7 | BL 7 | 7/16 | 7/16 | 3 | 2070 | 3/32 | 1/4 | 1/2 |
| BR 8 | BL 8 | 1/2 | 1/2 | 3 1/2 | 2170 | 1/8 | 5/16 | 5/8 |
| BR 10 | BL 10 | 5/8 | 5/8 | 4 | 2230 | 5/32 | 3/8 | 3/4 |
| BR 12 | BL 12 | 3/4 | 3/4 | 4 1/2 | 2310 | 3/16 | 7/16 | 13/16 |
| BR 16 | BL 16 | 1 | 1 | 6 | { 3390, 4390 | 1/4 | 9/16 | 1 |
| BR 20 | BL 20 | 1 1/4 | 1 1/4 | 7 | { 3460, 4460 | 5/16 | 5/8 | 1 |
| BR 24 | BL 24 | 1 1/2 | 1 1/2 | 8 | { 3510, 4510 | 3/8 | 5/8 | 1 |
| Rectangular Shank | | | | | | | | |
| BR 44 | BL 44 | 1/2 | 1 | 6 | 2260 | 3/16 | 5/16 | 5/8 |
| BR 54 | BL 54 | 5/8 | 1 | 6 | { 3360, 4360 | 1/4 | 3/8 | 3/4 |
| BR 55 | BL 55 | 5/8 | 1 1/4 | 7 | { 3360, 4360 | 1/4 | 3/8 | 3/4 |
| BR 64 | BL 64 | 3/4 | 1 | 6 | { 3380, 4380 | 1/4 | 1/2 | 3/4 |
| BR 66 | BL 66 | 3/4 | 1 1/2 | 8 | { 3430, 4430 | 5/16 | 7/16 | 15/16 |
| BR 85 | BL 85 | 1 | 1 1/4 | 7 | { 3460, 4460 | 5/16 | 5/8 | 1 |
| BR 86 | BL 86 | 1 | 1 1/2 | 8 | { 3510, 4510 | 3/8 | 5/8 | 1 |
| BR 88 | BL 88 | 1 | 2 | 10 | { 3510, 4510 | 3/8 | 5/8 | 1 |
| BR 90 | BL 90 | 1 1/2 | 2 | 10 | { 3540, 4540 | 1/2 | 3/4 | 1 1/4 |

[a] Where a pair of tip numbers is shown, the upper number applies to BR tools, the lower to BL tools. All dimensions are in inches.

**Brazing Carbide Tips to Steel Shanks.—**Sintered carbide tips or blanks are attached to steel shanks by brazing. Shanks usually are made of low-alloy steels having carbon contents ranging from 0.40 to 0.60 percent. Shank Preparation: The carbide tip usually is inserted into a milled recess or seat. When a recess is used, the bottom should be flat to provide a firm even support for the tip. The corner radius of the seat should be somewhat smaller than the radius on the tip to avoid contact and ensure support along each side of the recess. Cleaning: All surfaces to be brazed must be absolutely clean. Surfaces of the tip may be cleaned by grinding lightly or by sand-blasting. Brazing Materials and Equipment: The brazing metal may be copper, naval brass such as Tobin bronze, or silver solder. A flux such as borax is used to protect the clean surfaces and prevent oxidation. Heating may be done in a furnace or by oxy-acetylene torch or an oxy-hydrogen torch. Copper brazing usually is done in a furnace, although an oxy-hydrogen torch with excess hydrogen is sometimes used. Brazing Procedure: One method using a torch is to place a thin sheet material, such as copper foil, around and beneath the carbide tip, the top of which is covered with flux. The flame is applied to the underside of the tool shank, and, when the materials melt, the tip is pressed firmly into its seat with tongs or with the end of a rod. Brazing material in the form of wire or rod may be used to coat or tin the surfaces of the recess after the flux

**Table 7. American National Standard Style C Carbide-Tipped Tools** *ANSI B212.1-2002 (Withdrawn)*

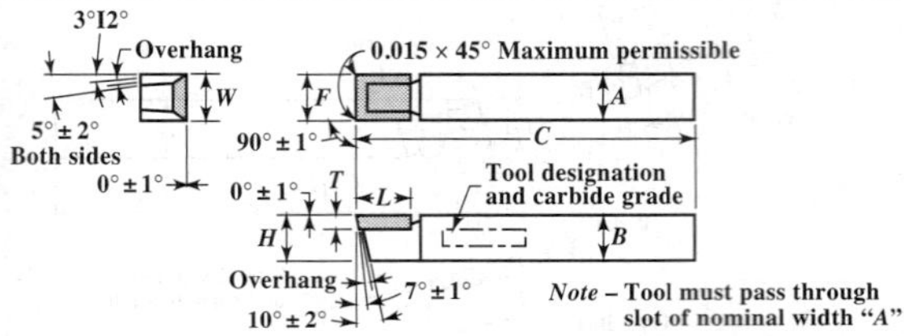


| Designation | Shank Dimensions | | | Tip Designation | Tip Dimensions | | |
|---|---|---|---|---|---|---|---|
| | Width, *A* | Height, *B* | Length, *C* | | Thickness, *T* | Width, *W* | Length, *L* |
| C 4 | ¼ | ¼ | 2 | 1030 | 1⁄16 | ¼ | 5⁄16 |
| C 5 | 5⁄16 | 5⁄16 | 2¼ | 1080 | 3⁄32 | 5⁄16 | ⅜ |
| C 6 | ⅜ | ⅜ | 2½ | 1090 | 3⁄32 | ⅜ | ⅜ |
| C 7 | 7⁄16 | 7⁄16 | 3 | 1105 | 3⁄32 | 7⁄16 | ½ |
| C 8 | ½ | ½ | 3½ | 1200 | ⅛ | ½ | ½ |
| C 10 | ⅝ | ⅝ | 4 | 1240 | 5⁄32 | ⅝ | ⅝ |
| C 12 | ¾ | ¾ | 4½ | 1340 | 3⁄16 | ¾ | ¾ |
| C 16 | 1 | 1 | 6 | 1410 | ¼ | 1 | ¾ |
| C 20 | 1¼ | 1¼ | 7 | 1480 | 5⁄16 | 1¼ | ¾ |
| C 44 | ½ | 1 | 6 | 1320 | 3⁄16 | ½ | ½ |
| C 54 | ⅝ | 1 | 6 | 1400 | ¼ | ⅝ | ⅝ |
| C 55 | ⅝ | 1¼ | 7 | 1400 | ¼ | ⅝ | ⅝ |
| C 64 | ¾ | 1 | 6 | 1405 | ¼ | ¾ | ¾ |
| C 66 | ¾ | 1½ | 8 | 1470 | 5⁄16 | ¾ | ¾ |
| C 86 | 1 | 1½ | 8 | 1475 | 5⁄16 | 1 | ¾ |

All dimensions are in inches. Square shanks above horizontal line, rectangular below.

**Table 8. American National Standard Style D, 80-Degree Nose-Angle Carbide-Tipped Tools** *ANSI B212.1-2002 (Withdrawn)*

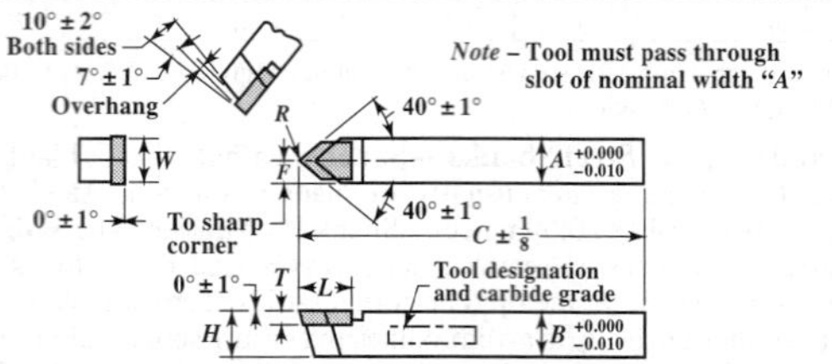


| Designation | Shank Dimensions | | | Tip Designation | Tip Dimensions | | |
|---|---|---|---|---|---|---|---|
| | Width, *A* | Height, *B* | Length, *C* | | Thickness, *T* | Width, *W* | Length, *L* |
| D 4 | ¼ | ¼ | 2 | 5030 | 1⁄16 | ¼ | 5⁄16 |
| D 5 | 5⁄16 | 5⁄16 | 2¼ | 5080 | 3⁄32 | 5⁄16 | ⅜ |
| D 6 | ⅜ | ⅜ | 2½ | 5100 | 3⁄32 | ⅜ | ½ |
| D 7 | 7⁄16 | 7⁄16 | 3 | 5105 | 3⁄32 | 7⁄16 | ½ |
| D 8 | ½ | ½ | 3½ | 5200 | ⅛ | ½ | ½ |
| D 10 | ⅝ | ⅝ | 4 | 5240 | 5⁄32 | ⅝ | ⅝ |
| D 12 | ¾ | ¾ | 4½ | 5340 | 3⁄16 | ¾ | ¾ |
| D 16 | 1 | 1 | 6 | 5410 | ¼ | 1 | ¾ |

All dimensions are in inches.

**Table 9. American National Standard Style E, 60-Degree Nose-Angle, Carbide-Tipped Tools** *ANSI B212.1-2002 (Withdrawn)*

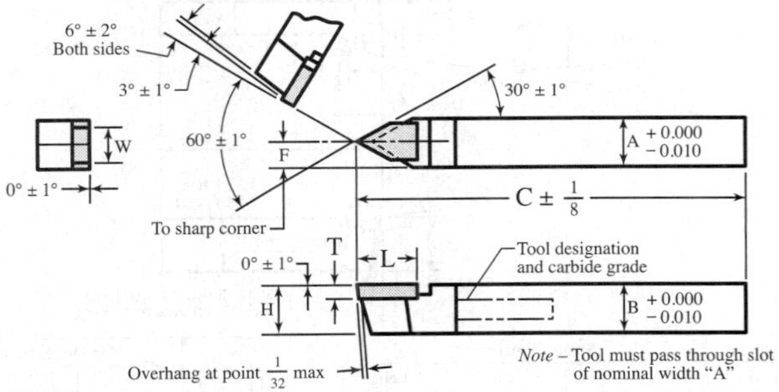


| Designation | Shank Dimensions | | | Tip Designation | Tip Dimensions | | |
|---|---|---|---|---|---|---|---|
| | Width A | Height B | Length C | | Thickness T | Width W | Length L |
| E 4 | ¼ | ¼ | 2 | 6030 | 1⁄16 | ¼ | 5⁄16 |
| E 5 | 5⁄16 | 5⁄16 | 2¼ | 6080 | 3⁄32 | 5⁄16 | ⅜ |
| E 6 | ⅜ | ⅜ | 2½ | 6100 | 3⁄32 | ⅜ | ½ |
| E 8 | ½ | ½ | 3½ | 6200 | ⅛ | ½ | ½ |
| E 10 | ⅝ | ⅝ | 4 | 6240 | 5⁄32 | ⅝ | ⅝ |
| E 12 | ¾ | ¾ | 4½ | 6340 | 3⁄16 | ¾ | ¾ |

All dimensions are in inches.

**Table 10. American National Standard Styles ER and EL, 60-Degree Nose-Angle, Carbide-Tipped Tools with Offset Point** *ANSI B212.1-2002 (Withdrawn)*

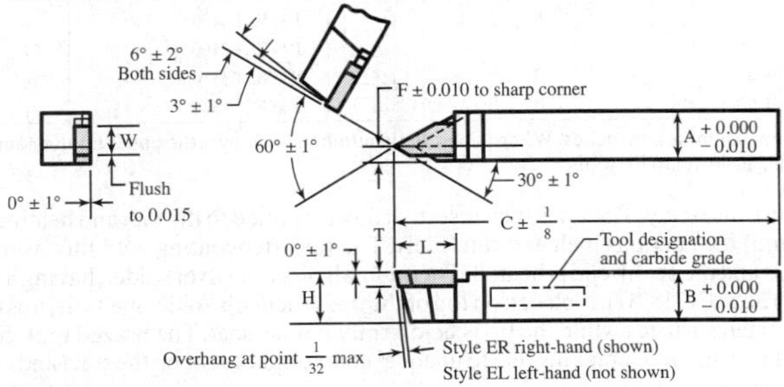


| Designation | | Shank Dimensions | | | Tip Designation | Tip Dimensions | | |
|---|---|---|---|---|---|---|---|---|
| Style ER | Style EL | Width A | Height B | Length C | | Thickness T | Width W | Length L |
| ER 4 | EL 4 | ¼ | ¼ | 2 | 1020 | 1⁄16 | 3⁄16 | ¼ |
| ER 5 | EL 5 | 5⁄16 | 5⁄16 | 2¼ | 7060 | 3⁄32 | ¼ | ⅜ |
| ER 6 | EL 6 | ⅜ | ⅜ | 2½ | 7060 | 3⁄32 | ¼ | ⅜ |
| ER 8 | EL 8 | ½ | ½ | 3½ | 7170 | ⅛ | 5⁄16 | ⅝ |
| ER 10 | EL 10 | ⅝ | ⅝ | 4 | 7170 | ⅛ | 5⁄16 | ⅝ |
| ER 12 | EL 12 | ¾ | ¾ | 4½ | 7230 | 5⁄32 | ⅜ | ¾ |

All dimensions are in inches.

**Table 11. American National Standard Style F, Offset, End-Cutting Carbide-Tipped Tools** *ANSI B212.1-2002 (Withdrawn)*

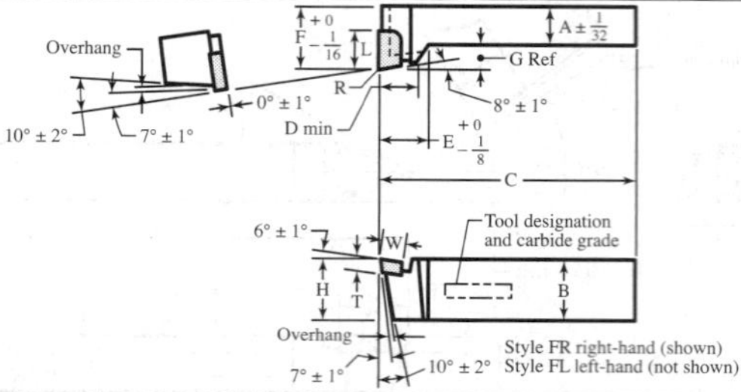


| Designation | | Shank Dimensions | | | | | Tip Designation | Tip Dimensions | | |
|---|---|---|---|---|---|---|---|---|---|---|
| Style FR | Style FL | Width *A* | Height *B* | Length *C* | Offset *G* | Length of Offset *E* | | Thickness *T* | Width *W* | Length *L* |
| | | | | | Square Shank | | | | | |
| FR 8 | FL 8 | ½ | ½ | 3½ | ¼ | ¾ | { P4170, P3170 | ⅛ | 5⁄16 | ⅝ |
| FR 10 | FL 10 | ⅝ | ⅝ | 4 | ⅜ | 1 | { P1230, P3230 | 5⁄32 | ⅜ | ¾ |
| FR 12 | FL 12 | ¾ | ¾ | 4½ | ⅝ | 1⅛ | { P4310, P3310 | 3⁄16 | 7⁄16 | 13⁄16 |
| FR 16 | FL 16 | 1 | 1 | 6 | ¾ | 1⅜ | { P4390, P3390 | ¼ | 9⁄16 | 1 |
| FR 20 | FL 20 | 1¼ | 1¼ | 7 | ¾ | 1½ | { P4460, P3460 | 5⁄16 | ⅝ | 1 |
| FR 24 | FL 24 | 1½ | 1½ | 8 | ¾ | 1½ | { P4510, P3510 | ⅜ | ⅝ | 1 |
| | | | | | Rectangular Shank | | | | | |
| FR 44 | FL 44 | ½ | 1 | 6 | ½ | ⅞ | { P4260, P1260 | 3⁄16 | 5⁄16 | ⅝ |
| FR 55 | FL 55 | ⅝ | 1¼ | 7 | ⅝ | 1⅛ | { P4360, P3360 | ¼ | ⅜ | ¾ |
| FR 64 | FL 64 | ¾ | 1 | 6 | ⅝ | 13⁄16 | { P4380, P3380 | ¼ | ½ | ¾ |
| FR 66 | FL 66 | ¾ | 1½ | 8 | ¾ | 1¼ | { P4430, P3430 | 5⁄16 | 7⁄16 | 15⁄16 |
| FR 85 | FL 85 | 1 | 1¼ | 7 | ¾ | 1½ | { P4460, P3460 | 5⁄16 | ⅝ | 1 |
| FR 86 | FL 86 | 1 | 1½ | 8 | ¾ | 1½ | { P4510, P3510 | ⅜ | ⅝ | 1 |
| FR 90 | FL 90 | 1½ | 2 | 10 | ¾ | 1⅝ | { P4540, P3540 | ½ | ¾ | 1¼ |

All dimensions are in inches. Where a pair of tip numbers is shown, the upper number applies to FR tools, the lower to FL tools.

melts and runs freely. The tip is then inserted, flux is applied to the top, and heating is continued until the coatings melt and run freely. The tip, after coating with flux, is placed in the recess and the shank end is heated. Then a small piece of silver solder, having a melting point of 1325° F (718° C), is placed on top of the tip. When this solder melts, it runs over the nickel-coated surfaces while the tip is held firmly into its seat. The brazed tool should be cooled slowly to avoid cracking due to unequal contraction between the steel and carbide.

**Carbide Tools.—**Cemented or sintered carbides are used in the machine-building and various other industries, chiefly for cutting tools but also for certain other tools or parts subject to considerable abrasion or wear. Carbide cutting tools, when properly selected to obtain the right combination of strength and hardness, are very effective in machining all classes of iron and steel, nonferrous alloys, nonmetallic materials, hard rubber, synthetic resins, slate, marble, and other materials which would quickly dull steel tools either because of hardness or abrasive action. Carbide cutting tools are not only durable, but capable of exceptionally high cutting speeds. See *CEMENTED CARBIDES* starting on page 848 for more on these materials.

*Tungsten carbide* is used extensively in cutting cast iron, nonferrous metals that form short chips in cutting, plastics and various other nonmetallic materials. A grade having

**Table 12. American National Standard Style G, Offset, Side-Cutting, Carbide-Tipped Tools** *ANSI B212.1-2002 (Withdrawn)*

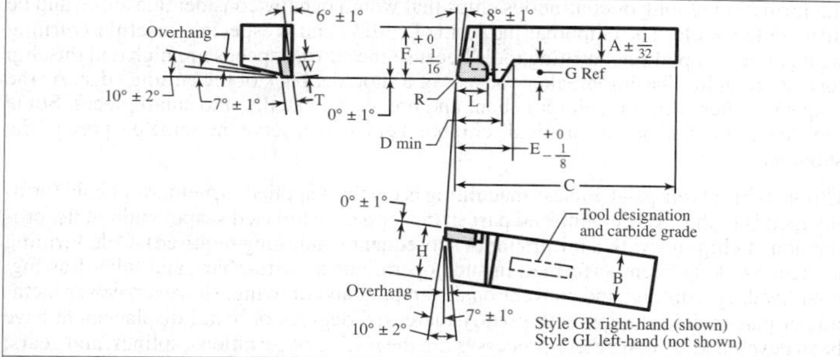


| Designation | | Shank Dimensions | | | | | Tip Designation | Tip Dimensions | | |
|---|---|---|---|---|---|---|---|---|---|---|
| Style GR | Style GL | Width *A* | Height *B* | Length *C* | Offset *G* | Length of Offset *E* | | Thickness *T* | Width *W* | Length *L* |
| Square Shank | | | | | | | | | | |
| GR 8 | GL 8 | ½ | ½ | 3½ | ¼ | 1 1/16 | { P3170, P4170 | ⅛ | 5/16 | ⅝ |
| GR 10 | GL 10 | ⅝ | ⅝ | 4 | ⅜ | 1⅜ | { P3230, P4230 | 5/32 | ⅜ | ¾ |
| GR 12 | GL 12 | ¾ | ¾ | 4½ | ⅜ | 1½ | { P3310, P2310 | 3/16 | 7/16 | 13/16 |
| GR 16 | GL 16 | 1 | 1 | 6 | ½ | 1 11/16 | { P3390, P4390 | ¼ | 9/16 | 1 |
| GR 20 | GL 20 | 1¼ | 1¼ | 7 | ¾ | 1 13/16 | { P3460, P4460 | 5/16 | ⅝ | 1 |
| GR 24 | GL 24 | 1½ | 1½ | 8 | ¾ | 1 13/16 | { P3510, P4510 | ⅜ | ⅝ | 1 |
| Rectangular Shank | | | | | | | | | | |
| GR 44 | GL 44 | ½ | 1 | 6 | ¼ | 1 1/16 | { P3260, P4260 | 3/16 | 5/16 | ⅝ |
| GR 55 | GL 55 | ⅝ | 1¼ | 7 | ⅜ | 1⅜ | { P3360, P4360 | ¼ | ⅜ | ¾ |
| GR 64 | GL 64 | ¾ | 1 | 6 | ½ | 1 7/16 | { P3380, P4380 | ¼ | ½ | ¾ |
| GR 66 | GL 66 | ¾ | 1½ | 8 | ½ | 1⅝ | { P3430, P4430 | 5/16 | 7/16 | 15/16 |
| GR 85 | GL 85 | 1 | 1¼ | 7 | ½ | 1 11/16 | { P3460, P4460 | 5/16 | ⅝ | 1 |
| GR 86 | GL 86 | 1 | 1½ | 8 | ½ | 1 11/16 | { P3510, P4510 | ⅜ | ⅝ | 1 |
| GR 90 | GL 90 | 1½ | 2 | 10 | ¾ | 2 1/16 | { P3540, P4540 | ½ | ¾ | 1¼ |

All dimensions are in inches. Where a pair of tip numbers is shown, the upper number applies to GR tools, the lower to GL tools.

a hardness of 87.5 RA on the Rockwell A scale might be used where a strong grade is required, as for roughing cuts, whereas for light high-speed finishing or other cuts, a hardness of about 92 RA might be preferable. When tungsten carbide is applied to steel, craters or chip cavities are formed back of the cutting edge; hence other carbides have been developed that offer greater resistance to abrasion.

*Tungsten-titanium carbide* (often called "titanium carbide") is adapted to cutting either heat-treated or non-heat-treated steels, cast steel, or any tough material that might form chip cavities. It is also applicable to bronzes, monel metal, aluminum alloys, etc.

*Tungsten-tantalum carbide* or "tantalum carbide" cutting tools are also applicable to steels, bronzes or other tough materials. A hardness of 86.8 RA is recommended by one manufacturer for roughing steel, whereas a grade for finishing might have a hardness ranging from 88.8 to 91.5 RA.

**Chipbreaker.—**The term "chipbreaker" indicates a method of forming or grinding turning tools which will cause the chips to break up into short pieces, thus preventing the formation of long or continuous chips that would occupy considerable space and be difficult to handle. The chipbreaking form of cutting end is especially useful in turning with carbide-tipped steel turning tools because the cutting speeds are high and the chip formation rapid. The chipbreaker consists of a shoulder back of the cutting edge. As the chip encounters this shoulder it is bent and broken repeatedly into small pieces. Some tools have attached or "mechanical" chipbreakers, which serve the same purpose as the shoulder.

**Chipless Machining.—**Chipless machining is the term applied to methods of cold forming metals to the required finished part shape (or nearly finished shape) without the production of chips (or with a minimum of subsequent machining required). Cold forming of steel has long been performed in such operations as wire-, bar-, and tube-drawing; cold-heading; coining; and conventional stamping and drawing. However, newer methods of plastic deformation with greatly increased degrees of metal displacement have been developed. Among these processes are the rolling of serrations, splines, and gears; power spinning; internal swaging; radial forging; the cold forming of multiple-diameter shafts; cold extrusion; and high-energy-rate forming, which includes explosive forming. The processes of cold heading, thread rolling and rotary swaging are also considered chipless machining processes.

**Indexable Insert Holders for NC.—**Indexable insert holders for numerical control lathes are usually made to more precise standards than ordinary holders. Where applicable, reference should be made to ANSI B212.3-2002 (Withdrawn), Precision Holders for Indexable Inserts. This Standard covers the dimensional specifications, styles, and designations of precision holders for indexable inserts, which are defined as tool holders that locate the gage insert (a combination of shim and insert thicknesses) from the back or front and end surfaces to a specified dimension with a ± 0.003 inch (± 0.08 mm) tolerance. In NC programming, the programmed path is that followed by the center of the tool tip, which is the center of the point, or nose radius, of the insert. The surfaces produced are the result of the path of the nose and the major cutting edge, so it is necessary to compensate for the nose or point radius and the lead angle when writing the program. Table 1, from B212.3, gives the compensating dimensions for different holder styles. The reference point is determined by the intersection of extensions from the major and minor cutting edges, which would be the location of the point of a sharp pointed tool. The distances from this point to the nose radius are $L1$ and $D1$; $L2$ and $D2$ are the distances from the sharp point to the center of the nose radius. Threading tools have sharp corners and do not require a radius compensation. Other dimensions of importance in programming threading tools are also given in Table 2; the data were developed by Kennametal, Inc.

The *C* and *F* characters are tool holder dimensions other than the shank size. In all instances, the *C* dimension is parallel to the length of the shank and the *F* dimension is parallel to the side dimension; actual dimensions must be obtained from the manufacturer. For all *K* style holders, the *C* dimension is the distance from the end of the shank to the tangent point of the nose radius and the end cutting edge of the insert. For all other holders, the *C* dimension is from the end of the shank to a tangent to the nose radius of the insert. The *F* dimension on all B, D, E, M, P, and V style holders is measured from the back side of the shank to the tangent point of the nose radius and the side cutting edge of the insert. For all A, F, G, J, K, and L style holders, the *F* dimension is the distance from the back side of the shank to the tangent of the nose radius of the insert. In all these designs, the nose radius is the standard radius corresponding to those given in the paragraph *Cutting Point Configuration* on page 829.

**Table 1. Insert Radius Compensation** *ANSI B212.3-2002 (Withdrawn)*

**Square Profile**

B Style[a] Also applies to R Style

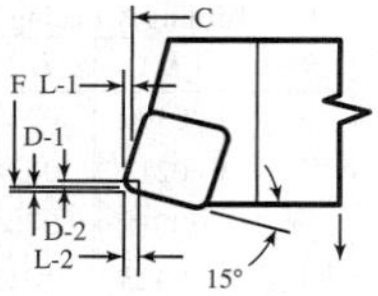


| Turning 15° Lead Angle | | | | |
|---|---|---|---|---|
| Rad. | *L*-1 | *L*-2 | *D*-1 | *D*-2 |
| 1/64 | .0035 | .0191 | .0009 | .0110 |
| 1/32 | .0070 | .0383 | .0019 | .0221 |
| 3/64 | .0105 | .0574 | .0028 | .0331 |
| 1/16 | .0140 | .0765 | .0038 | .0442 |

D Style[a] Also applies to S Style

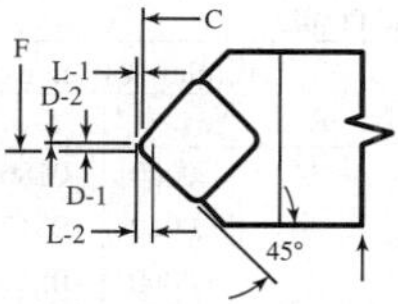


| Turning 45° Lead Angle | | | | |
|---|---|---|---|---|
| Rad. | *L*-1 | *L*-2 | *D*-1 | *D*-2 |
| 1/64 | .0065 | .0221 | .0065 | 0 |
| 1/32 | .0129 | .0442 | .0129 | 0 |
| 3/64 | .0194 | .0663 | .0194 | 0 |
| 1/16 | .0259 | .0884 | .0259 | 0 |

K Style[a]

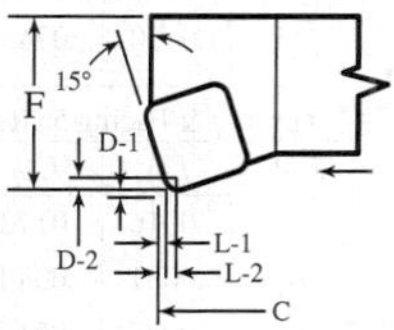


| Facing 15° Lead Angle | | | | |
|---|---|---|---|---|
| Rad. | *L*-1 | *L*-2 | *D*-1 | *D*-2 |
| 1/64 | .0009 | .0110 | .0035 | .0191 |
| 1/32 | .0019 | .0221 | .0070 | .0383 |
| 3/64 | .0028 | .0331 | .0105 | .0574 |
| 1/16 | .0038 | .0442 | .0140 | .0765 |

**Triangle Profile**

G Style[a]

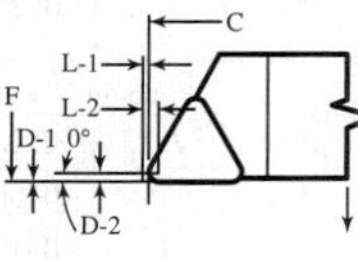


| Turning 0° Lead Angle | | | | |
|---|---|---|---|---|
| Rad. | *L*-1 | *L*-2 | *D*-1 | *D*-2 |
| 1/64 | .0114 | .0271 | 0 | .0156 |
| 1/32 | .0229 | .0541 | 0 | .0312 |
| 3/64 | .0343 | .0812 | 0 | .0469 |
| 1/16 | .0458 | .1082 | 0 | .0625 |

B Style[a] Also applies to R Style

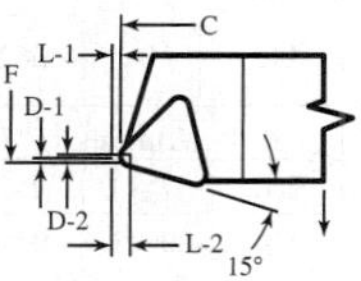


| Turning and Facing 15° Lead Angle | | | | |
|---|---|---|---|---|
| Rad. | *L*-1 | *L*-2 | *D*-1 | *D*-2 |
| 1/64 | .0146 | .0302 | .0039 | .0081 |
| 1/32 | .0291 | .0604 | .0078 | .0162 |
| 3/64 | .0437 | .0906 | .0117 | .0243 |
| 1/16 | .0582 | .1207 | .0156 | .0324 |

F Style[a]

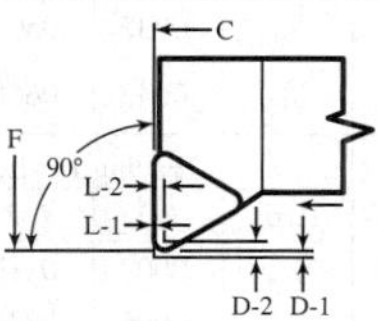


| Facing 90° Lead Angle | | | | |
|---|---|---|---|---|
| Rad. | *L*-1 | *L*-2 | *D*-1 | *D*-2 |
| 1/64 | 0 | .0156 | .0114 | .0271 |
| 1/32 | 0 | .0312 | .0229 | .0541 |
| 3/64 | 0 | .0469 | .0343 | .0812 |
| 1/16 | 0 | .0625 | .0458 | .1082 |

**Table 1.** *(Continued)* **Insert Radius Compensation** *ANSI B212.3-2002 (Withdrawn)*

| | | | | | | |
|---|---|---|---|---|---|---|
| Triangle Profile *(continued)* | | | | | | |
| J Style[a] | 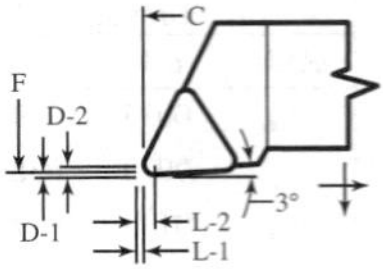 | Turning & Facing 3° Lead Angle | | | | |
| | | Rad. | *L*-1 | *L*-2 | *D*-1 | *D*-2 |
| | | 1⁄64 | .0106 | .0262 | .0014 | .0170 |
| | | 1⁄32 | .0212 | .0524 | .0028 | .0340 |
| | | 3⁄64 | .0318 | .0786 | .0042 | .0511 |
| | | 1⁄16 | .0423 | .1048 | .0056 | .0681 |
| 80° Diamond Profile | | | | | | |
| G Style[a] | 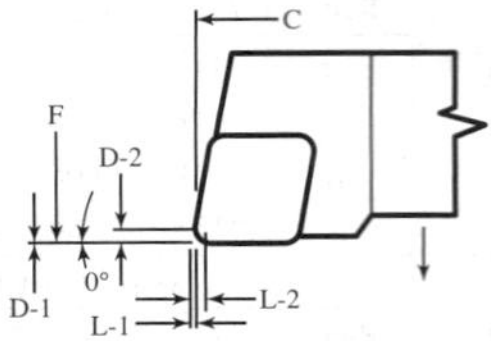 | Turning & Facing 0° Lead Angle | | | | |
| | | Rad. | *L*-1 | *L*-2 | *D*-1 | *D*-2 |
| | | 1⁄64 | .0030 | .0186 | 0 | .0156 |
| | | 1⁄32 | .0060 | .0312 | 0 | .0312 |
| | | 3⁄64 | .0090 | .0559 | 0 | .0469 |
| | | 1⁄16 | .0120 | .0745 | 0 | .0625 |
| L Style[a] | 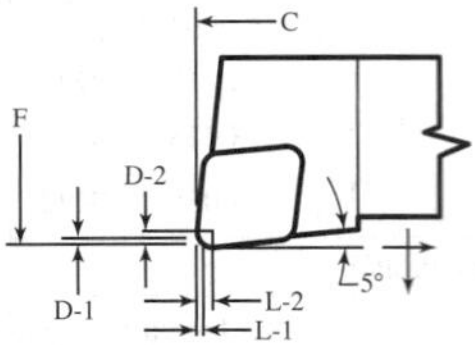 | Turning & Facing 5° Reverse Lead Angle | | | | |
| | | Rad. | *L*-1 | *L*-2 | *D*-1 | *D*-2 |
| | | 1⁄64 | .0016 | .0172 | .0016 | .0172 |
| | | 1⁄32 | .0031 | .0344 | .0031 | .0344 |
| | | 3⁄64 | .0047 | .0516 | .0047 | .0516 |
| | | 1⁄16 | .0062 | .0688 | .0062 | .0688 |
| F Style[a] | 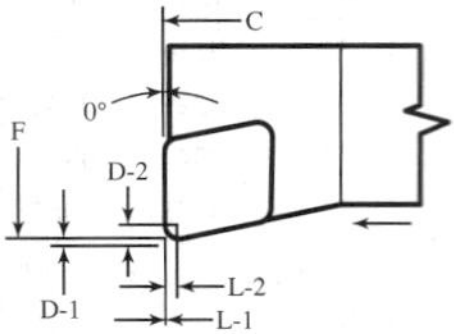 | Facing 0° Lead Angle | | | | |
| | | Rad. | *L*-1 | *L*-2 | *D*-1 | *D*-2 |
| | | 1⁄64 | 0 | .0156 | .0030 | .0186 |
| | | 1⁄32 | 0 | .0312 | .0060 | .0372 |
| | | 3⁄64 | 0 | .0469 | .0090 | .0559 |
| | | 1⁄16 | 0 | .0625 | .0120 | .0745 |
| R Style[a] | 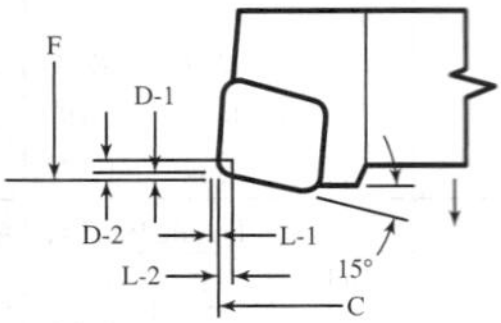 | Turning 15° Lead Angle | | | | |
| | | Rad. | *L*-1 | *L*-2 | *D*-1 | *D*-2 |
| | | 1⁄64 | .0011 | .0167 | .0003 | .0117 |
| | | 1⁄32 | .0022 | .0384 | .0006 | .0234 |
| | | 3⁄64 | .0032 | .0501 | .0009 | .0351 |
| | | 1⁄16 | .0043 | .0668 | .0012 | .0468 |
| K Style[a] | 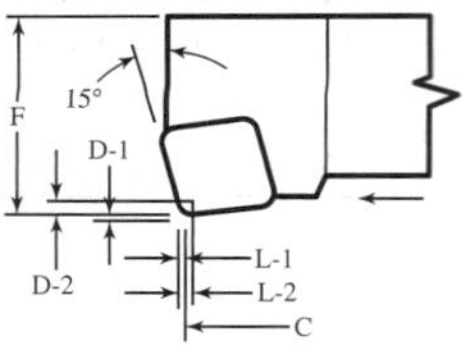 | Facing 15° Lead Angle | | | | |
| | | Rad. | *L*-1 | *L*-2 | *D*-1 | *D*-2 |
| | | 1⁄64 | .0003 | .0117 | .0011 | .0167 |
| | | 1⁄32 | .0006 | .0234 | .0022 | .0334 |
| | | 3⁄64 | .0009 | .0351 | .0032 | .0501 |
| | | 1⁄16 | .0012 | .0468 | .0043 | .0668 |

**Table 1.** *(Continued)* **Insert Radius Compensation** *ANSI B212.3-2002 (Withdrawn)*

| | | | | | | |
|---|---|---|---|---|---|---|
| 55° Profile | | | | | | |
| J Style[a] | 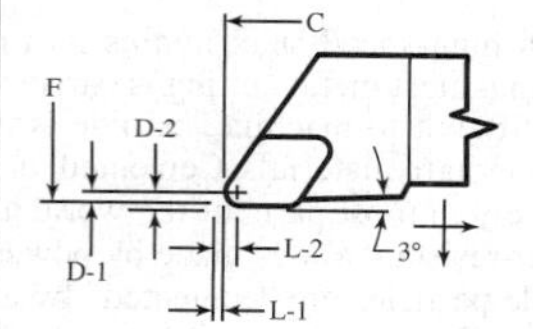 | Profiling 3° Reverse Lead Angle | | | | |
| | | Rad. | *L*-1 | *L*-2 | *D*-1 | *D*-2 |
| | | 1/64 | .0135 | .0292 | .0015 | .0172 |
| | | 1/32 | .0271 | .0583 | .0031 | .0343 |
| | | 3/64 | .0406 | .0875 | .0046 | .0519 |
| | | 1/16 | .0541 | .1166 | .0062 | .0687 |
| 35° Profile | | | | | | |
| J Style[a] Negative rake holders have 6° back rake and 6° side rake | 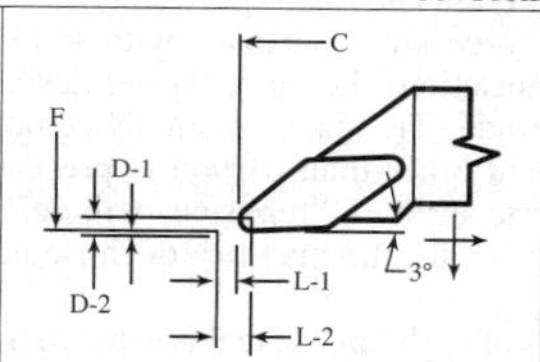 | Profiling 3° Reverse Lead Angle | | | | |
| | | Rad. | *L*-1 | *L*-2 | *D*-1 | *D*-2 |
| | | 1/64 | .0330 | .0487 | .0026 | .0182 |
| | | 1/32 | .0661 | .0973 | .0051 | .0364 |
| | | 3/64 | .0991 | .1460 | .0077 | .0546 |
| | | 1/16 | .1322 | .1947 | .0103 | .0728 |
| L Style[a] | 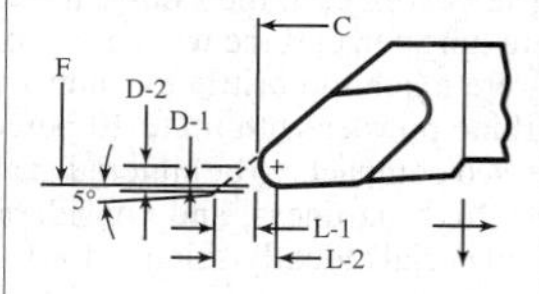 | Profiling 5° Lead Angle | | | | |
| | | Rad. | *L*-1 | *L*-2 | *D*-1 | *D*-2 |
| | | 1/64 | .0324 | .0480 | .0042 | .0198 |
| | | 1/32 | .0648 | .0360 | .0086 | .0398 |
| | | 3/64 | .0971 | .1440 | .0128 | .0597 |
| | | 1/16 | .1205 | .1920 | .0170 | .0795 |

[a] *L*-1 and *D*-1 over sharp point to nose radius; and *L*-2 and *D*-2 over sharp point to center of nose radius. The *D*-1 dimension for the *B, E, D, M, P, S, T,* and *V* style tools are over the sharp point of insert to a sharp point at the intersection of a line on the lead angle on the cutting edge of the insert and the *C* dimension. The *L*-1 dimensions on *K* style tools are over the sharp point of insert to sharp point intersection of lead angle and *F* dimensions.

All dimensions are in inches.

**Table 2. Threading Tool Insert Radius Compensation for NC Programming**

| Threading | | | | | | |
|---|---|---|---|---|---|---|
| Insert Size | T | R | U | Y | X | Z |
| 2 | 5/32 Wide | .040 | .075 | .040 | .024 | .140 |
| 3 | 3/16 Wide | .046 | .098 | .054 | .031 | .183 |
| 4 | 1/4 Wide | .053 | .128 | .054 | .049 | .239 |
| 5 | 3/8 Wide | .099 | .190 | … | … | … |

Buttress Threading | 29° Acme | 60° V-Threading

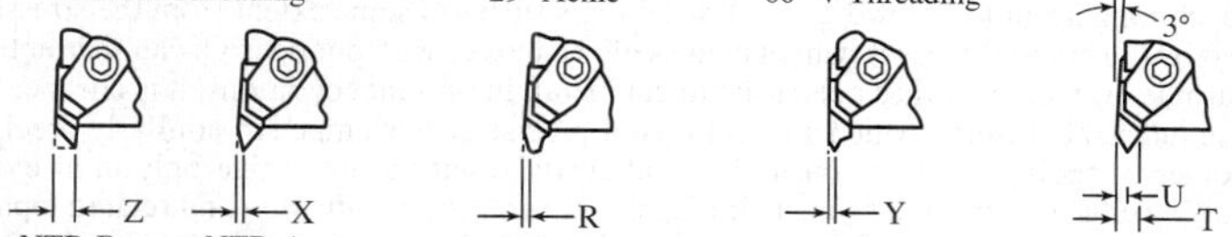


NTB-B | NTB-A | NA | NTF | NT

All dimensions are given in inches. Courtesy of Kennametal, Inc.

## CEMENTED CARBIDES

### Cemented Carbides and Other Hard Materials

**Carbides and Carbonitrides.—**Though high-speed steel retains its importance for such applications as drilling and broaching, most metal cutting is carried out with carbide tools. For materials that are very difficult to machine, carbide is now being replaced by carbonitrides, ceramics, and superhard materials. Cemented (or sintered) carbides and carbonitrides, known collectively in most parts of the world as hardmetals, are a range of very hard, refractory, wear-resistant alloys made by powder metallurgy techniques. The minute carbide or nitride particles are "cemented" by a binder metal that is liquid at the sintering temperature. Compositions and properties of individual hardmetals can be as different as those of brass and high-speed steel.

All hardmetals are *cermets*, combining *cer*amic particles with a *met*allic binder. It is unfortunate that (owing to a mistranslation) the term *cermet* has come to mean either all hardmetals with a titanium carbide (TiC) base or simply cemented titanium carbonitrides. Although no single element other than carbon is present in all hardmetals, it is no accident that the generic term is "tungsten carbide." The earliest successful grades were based on carbon, as are the majority of those made today, as listed in Table 1.

The outstanding machining capabilities of high-speed steel are due to the presence of very hard carbide particles, notably tungsten carbide, in the iron-rich matrix. Modern methods of making cutting tools from pure tungsten carbide were based on this knowledge. Early pieces of cemented carbide were much too brittle for industrial use, but it was soon found that mixing tungsten carbide powder with up to 10 percent of metals such as iron, nickel, or cobalt, allowed pressed compacts to be sintered at about 1500° C to give a product with low porosity, very high hardness, and considerable strength. This combination of properties made the materials ideally suitable for use as tools for cutting metal.

Cemented carbides for cutting tools were introduced commercially in 1927, and, although the key discoveries were made in Germany, many of the later developments have taken place in the United States, Austria, Sweden, and other countries. Recent years have seen two "revolutions" in carbide cutting tools, one led by the United States and the other by Europe. These were the change from brazed to clamped carbide inserts and the rapid development of coating technology.

When indexable tips were first introduced, it was found that so little carbide was worn away before they were discarded that a minor industry began to develop, regrinding the so-called "throwaway" tips and selling them for reuse in adapted toolholders. Hardmetal consumption, which had grown dramatically when indexable inserts were introduced, leveled off and began to decline. This situation was changed by the advent and rapid acceptance of carbide, nitride, and oxide coatings. Application of an even harder, more wear-resistant surface to a tougher, more shock-resistant substrate allowed production of new generations of longer-lasting inserts. Regrinding destroyed the enhanced properties of the coatings and was, therefore, abandoned for coated tooling.

Brazed tools have the advantage that they can be reground over and over again, until almost no carbide is left, but the tools must always be reset after grinding to maintain machining accuracy. However, all brazed tools suffer to some extent from the stresses left by the brazing process, which in unskilled hands or with poor design can shatter the carbide even before it has been used to cut metal. In present conditions, it is cheaper to use indexable inserts, which are tool tips of precise size, clamped in similarly precise holders, needing no time-consuming and costly resetting but usable only until each cutting edge or corner has lost its initial sharpness (see *Introduction* and related topics starting on page 827 and *Indexable Insert Holders for NC* on page 844). The absence of brazing stresses and the "one-use" concept also means that harder, longer-lasting grades can be used.

**Table 1. Typical Properties of Tungsten Carbide-Based Cutting Tool Hardmetals**

| ISO Application Code | Composition (%) | | | | Density ($g/cm^3$) | Hardness (Vickers) | Transverse Rupture Strength ($N/mm^2$) |
|---|---|---|---|---|---|---|---|
| | WC | TiC | TaC | Co | | | |
| P01 | 50 | 35 | 7 | 6 | 8.5 | 1900 | 1100 |
| P05 | 78 | 16 | | 6 | 11.4 | 1820 | 1300 |
| P10 | 69 | 15 | 8 | 8 | 11.5 | 1740 | 1400 |
| P15 | 78 | 12 | 3 | 7 | 11.7 | 1660 | 1500 |
| P20 | 79 | 8 | 5 | 8 | 12.1 | 1580 | 1600 |
| P25 | 82 | 6 | 4 | 8 | 12.9 | 1530 | 1700 |
| P30 | 84 | 5 | 2 | 9 | 13.3 | 1490 | 1850 |
| P40 | 85 | 5 | | 10 | 13.4 | 1420 | 1950 |
| P50 | 78 | 3 | 3 | 16 | 13.1 | 1250 | 2300 |
| M10 | 85 | 5 | 4 | 6 | 13.4 | 1590 | 1800 |
| M20 | 82 | 5 | 5 | 8 | 13.3 | 1540 | 1900 |
| M30 | 86 | 4 | | 10 | 13.6 | 1440 | 2000 |
| M40 | 84 | 4 | 2 | 10 | 14.0 | 1380 | 2100 |
| K01 | 97 | | | 3 | 15.2 | 1850 | 1450 |
| K05 | 95 | | 1 | 4 | 15.0 | 1790 | 1550 |
| K10 | 92 | | 2 | 6 | 14.9 | 1730 | 1700 |
| K20 | 94 | | | 6 | 14.8 | 1650 | 1950 |
| K30 | 91 | | | 9 | 14.4 | 1400 | 2250 |
| K40 | 89 | | | 11 | 14.1 | 1320 | 2500 |

A complementary development was the introduction of ever more complex chipbreakers derived from computer-aided design and pressed and sintered to precise shapes and dimensions. Another advance was the application of hot isostatic pressing (HIP), which has moved hardmetals into applications that were formerly uneconomical. This method allows virtually all residual porosity to be squeezed out of the carbide by means of inert gas at high pressure applied at about the sintering temperature. Toughness, rupture strength, and shock resistance can be doubled or tripled by this method, and the reject rates of very large sintered components are reduced to a fraction of their previous levels.

Further research has produced a substantial number of excellent cutting-tool materials based on titanium carbonitride. Generally called "cermets," as noted previously, carbonitride-based cutting inserts offer excellent performance and considerable prospects for the future.

*Compositions and Structures:* Properties of hardmetals are profoundly influenced by microstructure. The microstructure, in turn, depends on many factors, including basic chemical composition of the carbide and matrix phases; size, shape, and distribution of carbide particles; relative proportions of carbide and matrix phases; degree of intersolubility of carbides; excess or deficiency of carbon; variations in composition and structure caused by diffusion or segregation; production methods generally, but especially milling, carburizing, and sintering methods, and the types of raw materials; post sintering treatments such as hot isostatic pressing; and coatings or diffusion layers applied after initial sintering.

*Tungsten Carbide/Cobalt (WC/Co):* The first commercially available cemented carbides consisted of fine angular particles of tungsten carbide bonded with metallic cobalt. Intended initially for wire-drawing dies, this composition type is still considered to have the greatest resistance to simple abrasive wear and, therefore, to have many applications in machining.

For maximum hardness to be obtained from closeness of packing, the tungsten carbide grains should be as small as possible, preferably below 1 μm (0.00004 inch) and considerably less for special purposes. Hardness and abrasion resistance increase as the cobalt content is lowered, provided that a minimum of cobalt is present (2 percent can be enough, although 3 percent is the realistic minimum) to ensure complete sintering. In general, as carbide grain size or cobalt content or both are increased—frequently in unison—tougher and less hard grades are obtained. No porosity should be visible, even under the highest optical magnification.

WC/Co compositions used for cutting tools range from about 2 to 13 percent cobalt, and from less than 0.5 to more than 5 μm (0.00002–0.0002 in.) in grain size. For stamping tools, swaying dies, and other wear applications for parts subjected to moderate or severe shock, cobalt content can be as much as 30 percent, and grain size a maximum of about 10 μm (0.0004 in.). In recent years, "micrograin" carbides, combining submicron (less than 0.00004 in.) carbide grains with relatively high cobalt content have found increasing use for machining at low speeds and high feed rates. An early use was in high-speed woodworking cutters such as are used for planing.

For optimum properties, porosity should be at a minimum, carbide grain size as regular as possible, and carbon content of the tungsten carbide phase close to the theoretical (stoichiometric) value. Many tungsten carbide/cobalt compositions are modified by small but important additions—from 0.5 to perhaps 3 percent of tantalum, niobium, chromium, vanadium, titanium, hafnium, or other carbides. The basic purpose of these additions is generally inhibition of grain growth so that a consistently fine structure is maintained.

*Tungsten-Titanium Carbide/Cobalt (WC/TiC/Co):* These grades are used for tools to cut steels and other ferrous alloys, the purpose of the TiC content being to resist the high-temperature diffusive attack that causes chemical breakdown and cratering. Tungsten carbide diffuses readily into the chip surface, but titanium carbide is extremely resistant to such diffusion. A solid solution or "mixed crystal" of WC in TiC retains the anticratering property to a great extent.

Unfortunately, titanium carbide and TiC-based solid solutions are considerably more brittle and less abrasion resistant than tungsten carbide. TiC content, therefore, is kept as low as possible, only sufficient TiC being provided to avoid severe cratering wear. Even 2 or 3 percent of titanium carbide has a noticeable effect, and, as the relative content is substantially increased, the cratering tendency becomes more severe.

In the limiting formulation, the carbide is tungsten-free and based entirely on TiC, but, generally, TiC content extends to no more than about 18 percent. Above this figure, the carbide becomes excessively brittle and is very difficult to braze, although this drawback is not a problem with throwaway inserts.

WC/TiC/Co grades generally have two distinct carbide phases, angular crystals of almost pure WC and rounded TiC/WC mixed crystals. Among progressive manufacturers, although WC/TiC/Co hardmetals are very widely used, in certain important respects they are obsolescent, having been superseded by the WC/TiC/Ta(Nb)C/Co series in the many applications where higher strength combined with crater resistance is an advantage. TiC, TiN, and other coatings on tough substrates have also diminished the attractions of high-TiC grades for high-speed machining of steels and ferrous alloys.

*Tungsten-Titanium-Tantalum (-Niobium) Carbide/Cobalt:* Except for coated carbides, tungsten-titanium-tantalum (-niobium) grades could be the most popular class of hardmetals. Used mainly for cutting steel, they combine and improve upon most of the best features of the longer-established WC/TiC/Co compositions. These carbides compete directly with carbonitrides and silicon nitride ceramics, and the best cemented carbides of this class can undertake very heavy cuts at high speeds on all types of steels, including austenitic stainless varieties. These tools also operate well on ductile cast irons and nickel-base superalloys, where great heat and high pressures are generated at the cutting edge. However, they do not have the resistance to abrasive wear possessed by micrograin straight tungsten carbide grades nor the good resistance to cratering of coated grades and titanium carbide-based cermets.

*Titanium Carbide/Molybdenum/Nickel (TiC/Mo/Ni):* The extreme indentation hardness and crater resistance of titanium carbide, allied to the cheapness and availability of its main raw material (titanium dioxide, $TiO_2$), provide a strong inducement to use grades based on this carbide alone. Although developed early in the history of hardmetals, these carbides were difficult to braze satisfactorily and consequently were little used until the advent of clamped, throwaway inserts. Moreover, the carbides were notoriously brittle and could take only fine cuts in minimal-shock conditions.

Titanium carbide-based grades again came into prominence about 1960, when nickel-molybdenum began to be used as a binder instead of nickel. The new grades were able to perform a wider range of tasks, including interrupted cutting and cutting under shock conditions.

The very high indentation hardness values recorded for titanium carbide grades are not accompanied by correspondingly greater resistance to abrasive wear, the apparently less hard tungsten carbide being considerably superior in this property. Moreover, carbonitrides, advanced tantalum-containing multicarbides, and coated variants generally provide better all-around cutting performances.

*Titanium-Base Carbonitrides:* Development of titanium-carbonitride-based cutting-tool materials predates the use of coatings of this type on more conventional hardmetals by many years. Appreciable, though uncontrolled, amounts of carbonitride were often present, if only by accident, when cracked ammonia was used as a less expensive substitute for hydrogen in some stages of the production process in the 1950s and perhaps for two decades earlier.

Much of the recent, more scientific development of this class of materials has taken place in the United States, particularly by Teledyne Firth Sterling with its $SD_3$ grade and in Japan by several companies. Many of the compositions currently in use are extremely complex, and their structures—even with apparently similar compositions—can vary enormously. For instance, Mitsubishi characterizes its Himet NX series of cermets as $TiC/WC/Ta(Nb)C/Mo_2C/TiN/Ni/Co/Al$, with a structure comprised of both large and medium-size carbide particles (mainly TiC according to the quoted density) in a superalloy-type matrix containing an aluminum-bearing intermetallic compound.

*Steel- and Alloy-Bonded Titanium Carbide:* The class of material exemplified by Ferro-Tic, as it is known, consists primarily of titanium carbide bonded with heat-treatable steel, but some grades also contain tungsten carbide or are bonded with nickel- or copper-base alloys. These cemented carbides are characterized by high binder contents (typically 50–60 percent by volume) and lower hardnesses, compared with the more usual hardmetals, and by the great variation in properties obtained by heat treatment.

In the annealed condition, steel-bonded carbides have a relatively soft matrix and can be machined with little difficulty, especially by CBN (superhard cubic boron nitride) tools. After heat treatment, the degree of hardness and wear resistance achieved is considerably greater than that of normal tool steels, although understandably much less than that of traditional sintered carbides. Microstructures are extremely varied, being composed of 40–50 percent TiC by volume and a matrix appropriate to the alloy composition and the stage of heat treatment. Applications include stamping, blanking and drawing dies, machine components, and similar items where the ability to machine before hardening reduces production costs substantially.

*Coating:* As a final stage in carbide manufacture, coatings of various kinds are applied mainly to cutting tools, where for cutting steel in particular it is advantageous to give the rank and clearance surfaces characteristics that are quite different from those of the body of the insert. Coatings of titanium carbide, nitride, or carbonitride; of aluminum oxide; and of other refractory compounds are applied to a variety of hardmetal substrates by chemical or physical vapor deposition (CVD or PVD) or by newer plasma methods.

The most recent types of coatings include hafnium, tantalum, and zirconium carbides and nitrides; alumina/titanium oxide; and multiple carbide/carbonitride/nitride/oxide, oxynitride or oxycarbonitride combinations. Greatly improved properties have been claimed for variants with as many as 13 distinct CVD coatings. A markedly sharper cutting edge compared with other CVD-coated hardmetals is claimed, permitting finer cuts and the successful machining of soft but abrasive alloys.

The keenest edges on coated carbides are achieved by the techniques of physical vapor deposition. In this process, ions are deposited directionally from the electrodes, rather than evenly on all surfaces, so the sharpness of cutting edges is maintained and may even be enhanced. PVD coatings currently available include titanium nitride and carbonitride, their distinctive gold color having become familiar throughout the world on high-speed steel tooling. The high temperatures required for normal CVD tend to soften heat-treated high-speed steel. PVD-coated hardmetals have been produced commercially for several years, especially for precision milling inserts.

Recent developments in extremely hard coatings, generally involving exotic techniques, include boron carbide, cubic boron nitride, and pure diamond. Almost the ultimate in wear resistance, the commercial applications of thin plasma-generated diamond surfaces at present are mainly in the manufacture of semiconductors, where other special properties are important.

For cutting tools, the substrate is of equal importance to the coating in many respects, its critical properties including fracture toughness (resistance to crack propagation), elastic modulus, resistance to heat and abrasion, and expansion coefficient. Some manufacturers are now producing inserts with graded composition so that structures and properties are optimized at both surface and interior, and coatings are less likely to crack or break away.

*Specifications:* Compared with other standardized materials, the world of sintered hardmetals is peculiar. For instance, an engineer who seeks a carbide grade for the finish-machining of a steel component may be told to use *ISO Standard Grade P10* or *Industry Code C7.* If the composition and nominal properties of the designated tool material are then requested, the surprising answer is that, in basic composition alone, the tungsten carbide content of P10 (or of the now superseded C7) can vary from zero to about 75, titanium carbide from 8 to 80, cobalt 0 to 10, and nickel 0 to 15 percent. There are other possible constituents, also, in this so-called standard alloy, and many basic properties can vary as much as the composition. All that these dissimilar materials have in common, and all that the so-called standards mean, is that their suppliers—and sometimes their suppliers alone—consider them suitable for one particular and ill-defined machining application (which for P10 or C7 is the finish machining of steel).

This peculiar situation arose because the production of cemented carbides in occupied Europe during World War II was controlled by the German Hartmetallzentrale, and no factory other than Krupp was permitted to produce more than one grade. By the end of the war, all German-controlled producers were equipped to make the G, S, H, and F series to German standards. In the postwar years, this series of carbides formed the basis of unofficial European standardization. With the advent of the newer multicarbides, the previous identities of grades were gradually lost. The applications relating to the old grades were retained, however, as a new German DIN Standard, eventually being adopted, in somewhat modified form, by the International Organization for Standardization (ISO) and by the American National Standards Institute (ANSI).

The American cemented carbides industry developed under diverse ownership and solid competition. The major companies actively and independently developed new varieties of hardmetals, and there was little or no standardization, although there were many attempts to compile equivalent charts as a substitute for true standardization. Around 1942, the Buick division of GMC produced a simple classification code that arranged nearly 100 grades derived from 10 manufacturers under only 14 symbols (TC-1 to TC-14). In spite of serious deficiencies, this system remained in use for many years as an American industry standard; that is, Buick TC-1 was equivalent to industry code C1. Buick itself went much further, using the tremendous influence, research facilities, and purchasing potential of its parent company to standardize the products of each carbide manufacturer by properties that could be tested, rather than by the indeterminate recommended applications. Many large-scale carbide users have developed similar systems in attempts to exert some degree of in-house standardization and quality control. Small and medium-sized users, however, still suffer from so-called industry standards, which only provide a starting point for grade selection.

ISO 513, summarized in Table 2, divides all machining grades into three color-coded groups: straight tungsten carbide grades (letter K, color red) for cutting gray cast iron, nonferrous metals, and nonmetallics; highly alloyed grades (letter P, color blue) for machining steel; and less alloyed grades (letter M, color yellow, generally with less TiC than the corresponding P series), which are multipurpose and may be used on steels, nickel-base superalloys, ductile cast irons, and so on. Each grade within a group is also given a number to represent its position in a range from maximum hardness to maximum toughness (shock resistance). Typical applications are described for grades at more or less regular numerical intervals. Although coated grades scarcely existed when the ISO Standard was prepared, it is easy to classify coated as uncoated carbides—or carbonitrides, ceramics, and superhard materials—according to this system.

In this situation, it is easy to see how one plant will prefer one manufacturer's carbide and a second plant will prefer that of another. Each has found the carbide most nearly ideal for the particular conditions involved. In these circumstances it pays each manufacturer to make grades that differ in hardness, toughness, and crater resistance, so that they can provide a product that is near the optimum for a specific customer's application.

Although not classified as a hardmetal, new particle or powder metallurgical methods of manufacture, coupled with new coating technology, have led in recent years to something of an upsurge in the use of high-speed steel. Lower cost is a big factor, and the development of such coatings as titanium nitride, cubic boron nitride, and pure diamond, has enabled some high-speed steel tools to rival tools made from tungsten and other carbides in their ability to maintain cutting accuracy and prolong tool life. Multiple layers may be used to produce optimum properties in the coating, with adhesive strength where there is contact with the substrate, combined with hardness at the cutting surface to resist abrasion. Total thickness of such coating, even with multiple layers, is seldom more than 15 microns (0.00059 in.).

*Importance of Correct Grades:* A great diversity of hardmetal types is required to cope with all possible combinations of metals and alloys, machining operations, and working conditions. Tough, shock-resistant grades are needed for slow speeds and interrupted cutting, harder grades for high-speed finishing, heat-resisting alloyed grades for machining superalloys, and crater-resistant compositions, including most of the many coated varieties, for machining steels and ductile iron.

**Ceramics.**—Moving up the hardness scale, ceramics provide increasing competition for cemented carbides, both in performance and in cost-effectiveness, though not yet in reliability. Hardmetals themselves consist of ceramics—nonmetallic refractory compounds, usually carbides or carbonitrides—with a metallic binder of much lower melting point. In such systems, densification generally takes place by liquid-phase sintering. Pure ceramics have no metallic binder, but may contain lower-melting-point compounds or ceramic mixtures that permit liquid-phase sintering to take place. Where this condition is not possible, hot pressing or hot isostatic pressing can often be used to make a strong, relatively pore-free component or cutting insert. This section is restricted to those ceramics that compete directly with hardmetals, mainly in the cutting-tool category as shown in Table 3.

Ceramics are hard, completely nonmetallic substances that resist heat and abrasive wear. Increasingly used as clamped indexable tool inserts, ceramics differ significantly from tool steels, which are completely metallic. Ceramics also differ from cermets such as cemented carbides and carbonitrides, which comprise minute ceramic particles held together by metallic binders.

## Table 2. ISO Classifications of Hardmetals (Cemented Carbides and Carbonitrides) by Application

| Main Types of Chip Removal | | Groups of Applications | | | Direction of Decrease in Characteristic | |
|---|---|---|---|---|---|---|
| Symbol and Color | Broad Categories of Materials to be Machined | Designation (Grade) | Specific Material to be Machined | Use and Working Conditions | of cut | of carbide |
| P Blue | Ferrous with long chips | P01 | Steel, steel castings | Finish turning and boring; high cutting speeds, small chip sections, accurate dimensions, fine finish, vibration-free operations | ↑ speed | |
| | | P10 | Steel, steel castings | Turning, copying, threading, milling; high cutting speeds; small or medium chip sections | | ↑ wear resistance |
| | | P20 | Steel, steel castings, ductile cast iron with long chips | Turning, copying, milling; medium cutting speeds and chip sections, planing with small chip sections | | |
| | | P30 | Steel, steel castings, ductile cast iron with long chips | Turning, milling, planing; medium or large chip sections, unfavorable machining conditions | | |
| | | P40 | Steel, steel castings with sand inclusions and cavities | Turning, planing, slotting; low cutting speeds, large chip sections, with possible large cutting angles, unfavorable cutting conditions, and work on automatic machines | | |
| | | P50 | Steel, steel castings of medium- or low-tensile strength, with sand inclusions and cavities | Operations demanding very tough carbides; turning, planing, slotting; low cutting speeds, large chip sections, with possible large cutting angles, unfavorable conditions and work on automatic machines | | |
| M Yellow | Ferrous metals with long or short chips, and nonferrous metals | M10 | Steel, steel castings, manganese steel, gray cast iron, alloy cast iron | Turning; medium or high cutting speeds, small or medium chip sections | | |
| | | M20 | Steel, steel castings, austenitic or manganese steel, gray cast iron | Turning, milling; medium cutting speeds and chip sections | | |
| | | M30 | Steel, steel castings, austenitic steel, gray cast iron, high-temperature-resistant alloys | Turning, milling, planing; medium cutting speeds, medium or large chip sections | | |
| | | M40 | Mild, free-cutting steel, low-tensile steel, nonferrous metals and light alloys | Turning, parting off; particularly on automatic machines | | |
| K Red | Ferrous metals with short chips, nonferrous metals and nonmetallic materials | K01 | Very hard gray cast iron, chilled castings over 85 Shore, high-silicon aluminum alloys, hardened steel, highly abrasive plastics, hard cardboard, ceramics | Turning, finish turning, boring, milling, scraping | | |
| | | K10 | Gray cast iron over 220 BHN (Brinell), malleable cast iron with short chips, hardened steel, silicon-aluminum and copper alloys, plastics, glass, hard rubber, hard cardboard, porcelain, stone | Turning, milling, drilling, boring, broaching, scraping | | |
| | | K20 | Gray cast iron up to 220 BHN, nonferrous metals, copper, brass, aluminum | Turning, milling, planing, boring, broaching, demanding very tough carbide | | |
| | | K30 | Low-hardness gray cast iron, low-tensile steel, compressed wood | Turning, milling, planing, slotting, unfavorable conditions, and possibility of large cutting angles | ↓ feed | |
| | | K40 | Softwood or hardwood, nonferrous metals | Turning, milling, planing, slotting, unfavorable conditions, and possibility of large cutting angles | | ↓ toughness |

**Table 3. Typical Properties of Cutting Tool Ceramics**

| Group | Alumina | Alumina/TiC | Silicon Nitride | PCD | PCBN |
|---|---|---|---|---|---|
| Typical composition types | $Al_2O_3$ or $Al_2O_3/ZrO_2$ | 70/30 $Al_2O_3/TiC$ | $Si_3N_4/Y_2O_3$ plus | | |
| Density (g/cm$^3$) | 4.0 | 4.25 | 3.27 | 3.4 | 3.1 |
| Transverse rupture strength (N/mm$^2$) | 700 | 750 | 800 | | 800 |
| Compressive strength (kN/mm$^2$) | 4.0 | 4.5 | 4.0 | 4.7 | 3.8 |
| Hardness (HV) | 1750 | 1800 | 1600 | | |
| Hardness HK (kN/mm$^2$) | | | | 50 | 28 |
| Young's modulus (kN/mm$^2$) | 380 | 370 | 300 | 925 | 680 |
| Modulus of rigidity (kN/mm$^2$) | 150 | 160 | 150 | 430 | 280 |
| Poisson's ratio | 0.24 | 0.22 | 0.20 | 0.09 | 0.22 |
| Thermal expansion coefficient ($10^{-6}$/K) | 8.5 | 7.8 | 3.2 | 3.8 | 4.9 |
| Thermal conductivity (W/m K) | 23 | 17 | 22 | 120 | 100 |
| Fracture toughness ($K_{1c}$MN/m$^{3/2}$) | 2.3 | 3.3 | 5.0 | 7.9 | 10 |

Alumina-based ceramics were introduced as cutting inserts during World War II and were for many years considered too brittle for regular machine-shop use. Improved machine tools and finer-grain, tougher compositions incorporating zirconia or silicon carbide "whiskers" now permit their use in a wide range of applications. Silicon nitride, often combined with alumina (aluminum oxide), yttria (yttrium oxide), and other oxides and nitrides, is used for much of the high-speed machining of superalloys, and newer grades have been formulated specifically for cast iron—potentially a far larger market.

In addition to improvements in toolholders, great advances have been made in machine tools, many of which now feature the higher powers and speeds required for the efficient use of ceramic tooling. Brittleness at the cutting edge is no longer a disadvantage, with the improvements made to the ceramics themselves, mainly in toughness, but also in other critical properties.

Although very large numbers of useful ceramic materials are now available, only a few combinations have been found to combine such properties as minimum porosity, hardness, wear resistance, chemical stability, and resistance to shock to the extent necessary for cutting-tool inserts. Most ceramics used for machining are still based on high-purity, fine-grained alumina (aluminum oxide) but embody property-enhancing additions of other ceramics such as zirconia (zirconium oxide), titania (titanium oxide), titanium carbide, tungsten carbide, and titanium nitride. For commercial purposes, those more commonly used are often termed "white" (alumina with or without zirconia) or "black" (roughly 70¤30 alumina/titanium carbide). More recent developments are the distinctively green alumina ceramics strengthened with silicon carbide whiskers and the brown-tinged silicon nitride types.

Ceramics benefit from hot isostatic pressing to remove the last vestiges of porosity and raise substantially the material's shock resistance, even more than carbide-based hardmetals. Significant improvements are derived by even small parts such as tool inserts, although, in principle, they should not need such treatment if raw materials and manufacturing methods are properly controlled.

*Oxide Ceramics:* Alumina cutting tips have extreme hardness—more than HV 2000 or HRA 94—and give excellent service in their limited but important range of uses such as the machining of chilled iron rolls and brake drums. A substantial family of alumina-based materials has been developed, and fine-grained alumina-based composites now have sufficient strength for milling cast iron at speeds up to 2500 ft/min (800 m/min). Resistance to cratering when machining steel is exceptional.

*Oxide/Carbide Ceramics:* A second important class of alumina-based cutting ceramics combines aluminum oxide or alumina-zirconia with a refractory carbide or carbides, nearly always 30 percent TiC. The compound is black and normally hot pressed or hot isostatically pressed (HIPed). As shown in Table 3, the physical and mechanical properties of this material are generally similar to those of the pure alumina ceramics, but strength and shock resistance are generally higher, being comparable with those of higher-toughness simple alumina-zirconia grades. Current commercial grades are even more complex, combining alumina, zirconia, and titanium carbide with the further addition of titanium nitride.

*Silicon Nitride Base:* One of the most effective ceramic cutting-tool materials developed in the United Kingdom is Syalon (from SiAlON or silicon-aluminum-oxynitride) though it incorporates a substantial amount of yttria for efficient liquid-phase sintering). The material combines high strength with hot hardness, shock resistance, and other vital properties. Syalon cutting inserts are made by Kennametal and Sandvik and sold as Kyon 2000 and CC680, respectively. The brown Kyon 200 is suitable for machining high-nickel alloys and cast iron, but a later development, Kyon 3000, has good potential for machining cast iron.

Resistance to thermal stress and thermal shock of Kyon 2000 are comparable to those of sintered carbides. Toughness is substantially less than that of carbides, but roughly twice that of oxide-based cutting-tool materials at temperatures up to 850° C. Syon 200 can cut at high edge temperatures and is harder than carbide and some other ceramics at over 700° C, although softer than most at room temperature.

*Whisker-Reinforced Ceramics:* To improve toughness, Greenleaf Corporation has reinforced alumina ceramics with silicon carbide single-crystal "whiskers" that impart a distinctive green color to the material, marketed as WG300. Typically as thin as human hairs, the immensely strong whiskers improve tool life under arduous conditions. Whisker-reinforced ceramics and perhaps hardmetals are likely to become increasingly important as cutting and wear-resistant materials. Their only drawback seems to be the carcinogenic nature of the included fibers, which requires stringent precautions during manufacture.

**Superhard Materials.**—Polycrystalline synthetic diamond (PCD) and cubic boron nitride (PCBN), in the two columns at the right in Table 3, are almost the only cutting-insert materials in the "superhard" category. Both PCD and PCBN are usually made with the highest practicable concentration of the hard constituent, although ceramic or metallic binders can be almost equally important in providing overall strength and optimizing other properties. Variations in grain size are another critical factor in determining cutting characteristics and edge stability. Some manufacturers treat CBN in similar fashion to tungsten carbide, varying the composition and amount of binder within exceptionally wide limits to influence the physical and mechanical properties of the sintered compact.

In comparing these materials, users should note that some inserts comprise solid polycrystalline diamond or CBN and are double-sized to provide twice the number of cutting edges. Others consist of a layer, from 0.020- to 0.040-inch (0.5- to 1-mm) thick, on a tough carbide backing. A third type is produced with a solid superhard material almost surrounded by sintered carbide. A fourth type, used mainly for cutting inserts, is comprised of solid hardmetal with a tiny superhard insert at one or more (usually only one) cutting corners or edges. Superhard cutting inserts are expensive—up to 30 times the cost of equivalent shapes or sizes in ceramic or cemented carbide—but their outstanding properties, exceptional performance and extremely long life can make them by far the most cost-effective for certain applications.

*Diamond:* Diamond is the hardest material found or made. As harder, more abrasive ceramics and other materials came into widespread use, diamond began to be used for grinding wheel grits. Cemented carbide tools virtually demanded diamond grinding wheels for fine edge finishing. Solid single-crystal diamond tools were and are used to a small extent for special purposes, such as microtomes, for machining of hard materials, and for exceptionally fine finishes. These diamonds are made from comparatively large, high-quality gem-type diamonds, have isotropic properties, and are very expensive. By comparison, diamond abrasive grits cost only a few dollars a carat.

Synthetic diamonds are produced from graphite using high temperatures and extremely high pressures. The fine diamond particles produced are sintered together in the presence of a metal "catalyst" to produce high-efficiency anisotropic cutting-tool inserts. These tools comprise either a solid diamond compact or a layer of sintered diamond on a carbide backing, and are made under conditions similar to, though less severe than, those used in diamond synthesis. Both natural and synthetic diamond can be sintered in this way, although the latter method is the most frequently used.

Polycrystalline diamond (PCD) compacts are immensely hard and can be used to machine many substances, from highly abrasive hardwoods and glass fiber to nonferrous metals, hardmetals, and tough ceramics. Important classes of tools that are also available with cubic boron nitride inserts include brazed-tip drills, single-point turning tools, and face-milling cutters.

*Boron Nitride:* Polycrystalline diamond has one big limitation: it cannot be used to machine steel or any other ferrous material without rapid chemical breakdown. Boron nitride does not have this limitation. Normally soft and slippery like graphite, the soft hexagonal crystals (HBN) become cubic boron nitride (CBN) when subjected to ultrahigh pressures and temperatures, with a structure similar to and hardness second only to diamond. As a solid insert of polycrystalline cubic boron nitride (PCBN), the compound machines even the hardest steel with relative immunity from chemical breakdown or cratering.

Backed by sintered carbide, inserts of PCBN can readily be brazed, increasing the usefulness of the material and the range of tooling in which it can be used. With great hardness and abrasion resistance, coupled with extreme chemical stability when in contact with ferrous alloys at high temperatures, PCBN has the ability to machine both steels and cast irons at high speeds for long operating cycles. Only its currently high cost in relation to hardmetals prevents its wider use in mass-production machining.

Similar in general properties to PCBN, the recently developed "Wurbon" consists of a mixture of ultrafine (0.02 μm grain size) hexagonal and cubic boron nitride with a "wurtzite" structure, and is produced from soft hexagonal boron nitride in a microsecond by an explosive shockwave.

*Basic Machining Data:* Most mass-production metal-cutting operations are carried out with carbide-tipped tools, but their correct application is not simple. Even apparently similar batches of the same material vary greatly in their machining characteristics and may require different tool settings to attain optimum performance. Depth of cut, feed, surface speed, cutting rate, desired surface finish, and target tool life often need to be modified to suit the requirements of a particular component.

For the same downtime, the life of an insert between indexings can be less than that of an equivalent brazed tool between regrinds, so a much higher rate of metal removal is possible with the indexable or throwaway insert. It is commonplace for the claims for a new coating to include increases in surface-speed rates of 200–300 percent, and for a new insert design to offer similar improvements. Many operations are run at metal removal rates that are far from optimum for tool life because the rates used maximize productivity and cost-effectiveness.

Thus any recommendations for cutting speeds and feeds must be oversimplified or extremely complex, and must be hedged with many provisos, dependent on the technical and economic conditions in the manufacturing plant concerned. A preliminary grade selection should be made from the ISO-based tables and manufacturers' literature consulted for recommendations on the chosen grades and tool designs. If tool life is much greater than that desired under the suggested conditions, speeds, feeds, or depths of cut may be increased. If tools fail by edge breakage, a tougher (more shock-resistant) grade should be selected, with a numerically higher ISO code.

Alternatively, increasing the surface speed and decreasing the feed may be tried. If tools fail prematurely from what appears to be abrasive wear, a harder grade with numerically lower ISO designation should be tried. If cratering is severe, use a grade with higher titanium carbide content; that is, switch from an ISO K to M or M to P grade, use a P grade with lower numerical value, change to a coated grade, or use a coated grade with a (claimed) more-resistant surface layer.

*Built-Up Edge and Cratering:* The big problem in cutting steel with carbide tools is associated with the built-up edge and the familiar phenomenon called cratering. Research has shown that the built-up edge is continuous with the chip itself during normal cutting. Additions of titanium, tantalum, and niobium to the basic carbide mixture have a remarkable effect on the nature and degree of cratering, which is related to adhesion between the tool and the chip.

**Hardmetal Tooling for Wood and and Ceramics.—**Carbide-tipped circular saws are now conventional for cutting wood, wood products such as chipboard, and plastics, and tipped band saws of large size are also gaining in popularity. Tipped handsaws and mechanical equivalents are seldom needed for wood, but they are extremely useful for cutting abrasive building boards, glass-reinforced plastics, and similar material. Like the hardmetal tips used on most other woodworking tools, saw tips generally make use of straight (unalloyed) tungsten carbide/cobalt grades. However, where excessive heat is generated as with the cutting of high-silica hardwoods and particularly abrasive chipboards, the very hard but tough tungsten-titanium-tantalum-niobium carbide solid-solution grades, normally reserved for steel finishing, may be preferred. Saw tips are usually brazed and reground a number of times during service, so coated grades appear to have little immediate potential in this field.

*Cutting Blades and Plane Irons:* These tools comprise long, thin, comparatively wide slabs of carbide on a minimal-thickness steel backing. Compositions are straight tungsten carbide, preferably micrograin (to maintain a keen cutting edge with an included angle of 30° or less), but with relatively high amounts of cobalt, 11–13 percent, for toughness. Considerable expertise is necessary to braze and grind these cutters without inducing or failing to relieve the excessive stresses that cause distortion or cracking.

*Other Woodworking Cutters:* Routers and other cutters are generally similar to those used on metals and include many indexable-insert designs. The main difference with wood is that rotational and surface speeds can be the maximum available on the machine. High-speed routing of aluminum and magnesium alloys was developed largely from machines and techniques originally designed for work on wood.

*Cutting Other Materials:* The machining of plastics, fiber-reinforced plastics, graphite, asbestos, and other hard and abrasive constructional materials mainly requires abrasion resistance. Cutting pressures and power requirements are generally low. With thermoplastics and some other materials, particular attention must be given to cooling because of softening or degradation of the work material that might be caused by the heat generated in cutting. An important application of cemented carbides is the drilling and routing of printed circuit boards. Solid tungsten carbide drills of extremely small sizes are used for this work.

# MILLING CUTTERS

## Selection of Milling Cutters

The most suitable type of milling cutter for a particular milling operation depends on such factors as the kind of cut to be made, the material to be cut, the number of parts to be machined, and the type of milling machine available. Solid cutters of small size will usually cost less, initially, than inserted blade types; for long-run production, inserted-blade cutters will probably have a lower overall cost. Depending on either the material to be cut or the amount of production involved, the use of carbide-tipped cutters in preference to high-speed steel or other cutting tool materials may be justified.

Rake angles depend on both the cutter material and the work material. Carbide and cast alloy cutting tool materials generally have smaller rake angles than high-speed steel tool materials because of their lower edge strength and greater abrasion resistance. Soft work materials permit higher radial rake angles than hard materials; thin cutters permit zero or practically zero axial rake angles; and wide cutters operate more smoothly with high axial rake angles. See *Rake Angles for Milling Cutters* on page 889.

Cutting edge relief or clearance angles are usually from 3 to 6 degrees for hard or tough materials, 4 to 7 degrees for average materials, and 6 to 12 degrees for easily machined materials. See *Clearance Angles for Milling Cutter Teeth* on page 888.

The number of teeth in the milling cutter is also a factor that should be given consideration, as explained in the next paragraph.

**Number of Teeth in Milling Cutters.—**In determining the number of teeth a milling cutter should have for optimum performance, there is no universal rule.

There are, however, two factors that should be considered in making a choice: 1) The number of teeth should never be so great as to reduce the chip space between the teeth to a point where a free flow of chips is prevented; and 2) The chip space should be smooth and without sharp corners that would cause clogging of the chips in the space.

For milling ductile materials that produce a continuous and curled chip, a cutter with large chip spaces is preferable. Such coarse tooth cutters permit an easier flow of the chips through the chip space than would be obtained with fine tooth cutters and help to eliminate cutter "chatter." For cutting operations in thin materials, fine tooth cutters reduce cutter and workpiece vibration and the tendency for the cutter teeth to "straddle" the workpiece and dig in. For slitting copper and other soft nonferrous materials, teeth that are either chamfered or alternately flat and V-shaped are best.

As a general rule, to give satisfactory performance the number of teeth in milling cutters should be such that no more than two teeth at a time are engaged in the cut. Based on this rule, the following formulas (valid in both SI and English system of units) are recommended:

For face milling cutters,

$$T = \frac{6.3D}{W} \qquad (1)$$

For peripheral milling cutters,

$$T = \frac{12.6D\cos A}{D + 4d} \qquad (2)$$

where $T$ = number of teeth in cutter; $D$ = cutter diameter in inches (mm); $W$ = width of cut in inches (mm); $d$ = depth of cut in inches (mm); and $A$ = helix angle of cutter.

To find the number of teeth that a cutter should have when other than two teeth are in the cut at the same time, Formulas (1) and (2) should be divided by 2 and the result multiplied by the number of teeth desired in the cut.

*Example:* Determine the required number of teeth in a face mill where $D = 6$ inches and $W = 4$ inches. Using Formula (1),

$$T = \frac{6.3 \times 6}{4} = 10 \text{ teeth, approximately}$$

*Example:* Determine the required number of teeth in a plain milling cutter where $D =$ 4 inches and $d = 1/4$ inch. Using Formula (2),

$$T = \frac{12.6 \times 4 \times \cos 0°}{4 + \left(4 \times \frac{1}{4}\right)} = 10 \text{ teeth, approximately}$$

In *high-speed milling* with sintered carbide, high-speed steel, and cast nonferrous cutting tool materials, a formula that permits full use of the power available at the cutter but prevents overloading of the motor driving the milling machine is:

$$T = \frac{K \times H}{F \times N \times d \times W} \qquad (3)$$

where $T$ = number of cutter teeth; $H$ = horsepower (kilowatts) available at the cutter; $F$ = feed per tooth in inches (mm); $N$ = revolutions per minute of cutter; $d$ = depth of cut in inches (mm); $W$ = width of cut in inches (mm); and $K$ = a constant that may be taken as 0.65 for average steel, 1.5 for cast iron, and 2.5 for aluminum. **For metric units, $K$ = 14278 for average steel, 32949 for cast iron, and 54915 for aluminum.** These values are conservative and take into account dulling of the cutter in service.

*Example:* Determine the required number of teeth in a sintered carbide-tipped face mill for high-speed milling of 200 BHN (Brinell Hardness Number) alloy steel if $H$ = 7.5 kilowatt; $F$ = 0.2032 mm; $N$ = 272 rpm; $d$ = 3.2 mm; $W$ = 152.4 mm; and $K$ for alloy steel is 14278. Using Formula (3),

$$T = \frac{14278 \times 7.5}{0.2032 \times 272 \times 3.2 \times 152.4} = 4 \text{ teeth, approximately}$$

**American National Standard Milling Cutters.—**According to American National Standard ANSI/ASME B94.19-1997 (R2019), milling cutters may be classified in two general ways, given as follows:

*By Type of Relief on Cutting Edges:* Milling cutters may be described on the basis of one of two methods of providing relief for the cutting edges. *Profile sharpened* cutters are those on which relief is obtained and which are resharpened by grinding a narrow land in back of the cutting edges. Profile sharpened cutters may produce flat, curved, or irregular surfaces. *Form relieved* cutters are those which are so relieved that by grinding only the faces of the teeth the original form is maintained throughout the life of the cutters. Form relieved cutters may produce flat, curved or irregular surfaces.

*By Method of Mounting:* Milling cutters may be described by one of two methods used to mount the cutter. *Arbor type* cutters are those which have a hole for mounting on an arbor and usually have a keyway to receive a driving key. These are sometimes called *Shell type*. *Shank type* cutters are those which have a straight or tapered shank to fit the machine tool spindle or adapter.

**Explanation of the "Hand" of Milling Cutters.—**In the ANSI/ASME Standard, the terms "right-hand" and "left-hand" are used to describe hand of rotation, hand of cutter and hand of flute helix.

*Hand of Rotation or Hand of Cut* is described as either "right-hand" if the cutter revolves counterclockwise as it cuts when viewed from a position in front of a horizontal milling machine and facing the spindle, or "left-hand" if the cutter revolves clockwise as it cuts when viewed from the same position.

**American National Standard Plain Milling Cutters**
*ANSI/ASME B94.19-1997 (R2019)*

| Cutter Diameter | | | Range of Face Widths, Nom.[a] | Hole Diameter | | |
|---|---|---|---|---|---|---|
| Nom. | Max. | Min. | | Nom. | Max. | Min. |
| Light-duty Cutters[b] | | | | | | |
| 2½ | 2.515 | 2.485 | 3⁄16, ¼, 5⁄16, ⅜, ½, ⅝, ¾, 1, 1½, 2 and 3 | 1 | 1.00075 | 1.0000 |
| 3 | 3.015 | 2.985 | 3⁄16, ¼, 5⁄16, ⅜, ⅝, ¾, and 1½ | 1 | 1.00075 | 1.0000 |
| 3 | 3.015 | 2.985 | ½, ⅝, ¾, 1, 1¼, 1½, 2 and 3 | 1¼ | 1.2510 | 1.2500 |
| 4 | 4.015 | 3.985 | ¼, 5⁄16 and ⅜ | 1 | 1.00075 | 1.0000 |
| 4 | 4.015 | 3.985 | ⅜, ½, ⅝, ¾, 1, 1½, 2, 3 and 4 | 1¼ | 1.2510 | 1.2500 |
| Heavy-duty Cutters[c] | | | | | | |
| 2½ | 2.515 | 2.485 | 2 | 1 | 1.00075 | 1.0000 |
| 2½ | 2.515 | 2.485 | 4 | 1 | 1.0010 | 1.0000 |
| 3 | 3.015 | 2.985 | 2, 2½, 3, 4 and 6 | 1¼ | 1.2510 | 1.2500 |
| 4 | 4.015 | 3.985 | 2, 3, 4 and 6 | 1½ | 1.5010 | 1.5000 |
| High-helix Cutters[d] | | | | | | |
| 3 | 3.015 | 2.985 | 4 and 6 | 1¼ | 1.2510 | 1.2500 |
| 4 | 4.015 | 3.985 | 8 | 1½ | 1.5010 | 1.5000 |

[a] *Tolerances on Face Widths:* Up to 1 inch, inclusive, ± 0.001 inch; over 1 to 2 inches, inclusive, +0.010, −0.000 inch; over 2 inches, +0.020, −0.000 inch.

[b] Light-duty plain milling cutters with face widths under ¾ inch have straight teeth. Cutters with ¾- inch face and wider have helix angles of not less than 15 degrees nor greater than 25 degrees.

[c] Heavy-duty plain milling cutters have a helix angle of not less than 25 degrees nor greater than 45 degrees.

[d] High-helix plain milling cutters have a helix angle of not less than 45 degrees nor greater than 52 degrees.

All dimensions are in inches. All cutters are high-speed steel. Plain milling cutters are of cylindrical shape, having teeth on the peripheral surface only.

*Hand of Cutter:* Some types of cutters require special consideration when referring to their hand. These are principally cutters with unsymmetrical forms, face-type cutters, or cutters with threaded holes. *Symmetrical* cutters may be reversed on the arbor in the same axial position and rotated in the cutting direction without altering the contour produced on the work-piece, and may be considered as either right-hand or left-hand. *Unsymmetrical* cutters reverse the contour produced on the work-piece when reversed on the arbor in the same axial position and rotated in the cutting direction. A *single-angle* cutter is considered to be a right-hand cutter if it revolves counterclockwise, or a left-hand cutter if it revolves clockwise, when cutting as viewed from the side of the larger diameter. The *hand of rotation* of a single angle milling cutter need not necessarily be the same as its *hand of cutter.* A *single corner rounding* cutter is considered to be a right-hand cutter if it revolves counterclockwise, or a left-hand cutter if it revolves clockwise, when cutting as viewed from the side of the smaller diameter.

**American National Standard Side Milling Cutters**
*ANSI/ASME B94.19-1997 (R2019)*

| Cutter Diameter | | | Range of Face Widths Nom.[a] | Hole Diameter | | |
|---|---|---|---|---|---|---|
| Nom. | Max. | Min. | | Nom. | Max. | Min. |
| Side Cutters[b] | | | | | | |
| 2 | 2.015 | 1.985 | 3⁄16, 1⁄4, 3⁄8 | 5⁄8 | 0.62575 | 0.6250 |
| 2½ | 2.515 | 2.485 | 1⁄4, 3⁄8, 1⁄2 | 7⁄8 | 0.87575 | 0.8750 |
| 3 | 3.015 | 2.985 | 1⁄4, 5⁄16, 3⁄8, 7⁄16, 1⁄2 | 1 | 1.00075 | 1.0000 |
| 4 | 4.015 | 3.985 | 1⁄4, 3⁄8, 1⁄2, 5⁄8, 3⁄4, 7⁄8 | 1 | 1.00075 | 1.0000 |
| 4 | 4.015 | 3.985 | 1⁄2, 5⁄8, 3⁄4 | 1¼ | 1.2510 | 1.2500 |
| 5 | 5.015 | 4.985 | 1⁄2, 5⁄8, 3⁄4 | 1 | 1.00075 | 1.0000 |
| 5 | 5.015 | 4.985 | 1⁄2, 5⁄8, 3⁄4, 1 | 1¼ | 1.2510 | 1.2500 |
| 6 | 6.015 | 5.985 | 1⁄2 | 1 | 1.00075 | 1.0000 |
| 6 | 6.015 | 5.985 | 1⁄2, 5⁄8, 3⁄4, 1 | 1¼ | 1.2510 | 1.2500 |
| 7 | 7.015 | 6.985 | 3⁄4 | 1¼ | 1.2510 | 1.2500 |
| 7 | 7.015 | 6.985 | 3⁄4 | 1½ | 1.5010 | 1.5000 |
| 8 | 8.015 | 7.985 | 3⁄4, 1 | 1¼ | 1.2510 | 1.2500 |
| 8 | 8.015 | 7.985 | 3⁄4, 1 | 1½ | 1.5010 | 1.5000 |
| Staggered-Tooth Side Cutters[c] | | | | | | |
| 2½ | 2.515 | 2.485 | 1⁄4, 5⁄16, 3⁄8, 1⁄2 | 7⁄8 | 0.87575 | 0.8750 |
| 3 | 3.015 | 2.985 | 3⁄16, 1⁄4, 5⁄16, 3⁄8 | 1 | 1.00075 | 1.0000 |
| 3 | 3.015 | 2.985 | 1⁄2, 5⁄8, 3⁄4 | 1¼ | 1.2510 | 1.2500 |
| 4 | 4.015 | 3.985 | 1⁄4, 5⁄16, 3⁄8, 7⁄16, 1⁄2, 5⁄8, 3⁄4 and 7⁄8 | 1¼ | 1.2510 | 1.2500 |
| 5 | 5.015 | 4.985 | 1⁄2, 5⁄8, 3⁄4 | 1¼ | 1.2510 | 1.2500 |
| 6 | 6.015 | 5.985 | 3⁄8, 1⁄2, 5⁄8, 3⁄4, 7⁄8, 1 | 1¼ | 1.2510 | 1.2500 |
| 8 | 8.015 | 7.985 | 3⁄8, 1⁄2, 5⁄8, 3⁄4, 1 | 1½ | 1.5010 | 1.5000 |
| Half Side Cutters[d] | | | | | | |
| 4 | 4.015 | 3.985 | 3⁄4 | 1¼ | 1.2510 | 1.2500 |
| 5 | 5.015 | 4.985 | 3⁄4 | 1¼ | 1.2510 | 1.2500 |
| 6 | 6.015 | 5.985 | 3⁄4 | 1¼ | 1.2510 | 1.2500 |

[a] *Tolerances on Face Widths:* For side cutters, +0.002, −0.001 inch; for staggered-tooth side cutters up to 3⁄4-inch face width, inclusive, +0.000 −0.0005 inch, and over 3⁄4 to 1 inch, inclusive, +0.000 − 0.0010 inch; and for half side cutters, +0.015, −0.000 inch.

[b] Side milling cutters have straight peripheral teeth and side teeth on both sides.

[c] Staggered-tooth side milling cutters have peripheral teeth of alternate right- and left-hand helix and alternate side teeth.

[d] Half side milling cutters have side teeth on one side only. The peripheral teeth are helical of the same hand as the cut. Made either with right-hand or left-hand cut.

All dimensions are in inches. All cutters are high-speed steel. Side milling cutters are of cylindrical shape, having teeth on the periphery and on one or both sides.

*Hand of Flute Helix:* Milling cutters may have *straight flutes,* which means that their cutting edges are in planes parallel to the cutter axis. Milling cutters with flute helix in one direction only are described as having a right-hand helix if the flutes twist away from the observer in a clockwise direction when viewed from either end of the cutter or as having a left-hand helix if the flutes twist away from the observer in a counterclockwise direction when viewed from either end of the cutter. *Staggered tooth cutters* are milling cutters with every other flute of opposite (right- and left-hand) helix.

An illustration describing the various milling cutter elements of both a profile cutter and a form-relieved cutter is given on page 864.

## American National Standard Staggered Teeth, T-Slot Milling Cutters with Brown & Sharpe Taper and Weldon Shanks *ANSI/ASME B94.19-1997 (R2019)*

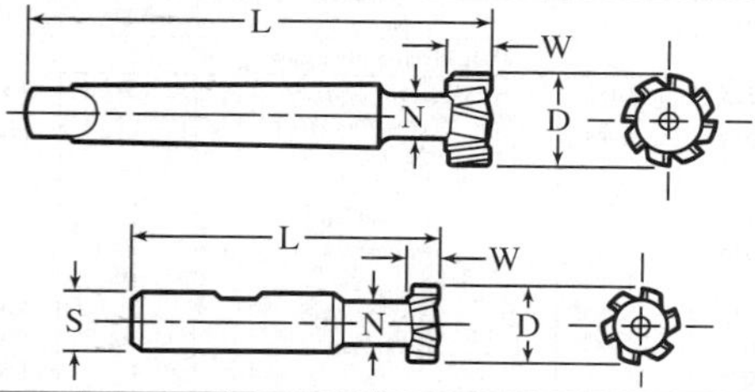


| Bolt Size | Cutter Dia., *D* | Face Width, *W* | Neck Dia., *N* | With B. & S. Taper[a,b] Length, *L* | With B. & S. Taper[a,b] Taper No. | With Weldon Shank Length, *L* | With Weldon Shank Dia., *S* |
|---|---|---|---|---|---|---|---|
| 1/4 | 9/16 | 15/64 | 17/64 | … | … | 2 19/32 | 1/2 |
| 5/16 | 21/32 | 17/64 | 21/64 | … | … | 2 11/16 | 1/2 |
| 3/8 | 25/32 | 21/64 | 13/32 | … | … | 3 1/4 | 3/4 |
| 1/2 | 31/32 | 25/64 | 17/32 | 5 | 7 | 3 7/16 | 3/4 |
| 5/8 | 1 1/4 | 31/64 | 21/32 | 5 1/4 | 7 | 3 15/16 | 1 |
| 3/4 | 1 15/32 | 5/8 | 25/32 | 6 7/8 | 9 | 4 7/16 | 1 |
| 1 | 1 27/32 | 53/64 | 1 1/32 | 7 1/4 | 9 | 4 13/16 | 1 1/4 |

[a] For dimensions of Brown & Sharpe taper shanks, see information given on page 1011.

[b] Brown & Sharpe taper shanks have been removed from ANSI/ASME B94.19; they are included for reference only.

All dimensions are in inches. All cutters are high-speed steel, and only right-hand cutters are standard.

*Tolerances:* On *D*, +0.000, −0.010 inch; on *W*, +0.000, −0.005 inch; on *N*, +0.000, −0.005 inch; on *L*, ± 1/16 inch; on *S*, −00001 to −0.0005 inch.

## American National Standard Form Relieved Corner Rounding Cutters with Weldon Shanks *ANSI/ASME B94.19-1997 (R2019)*

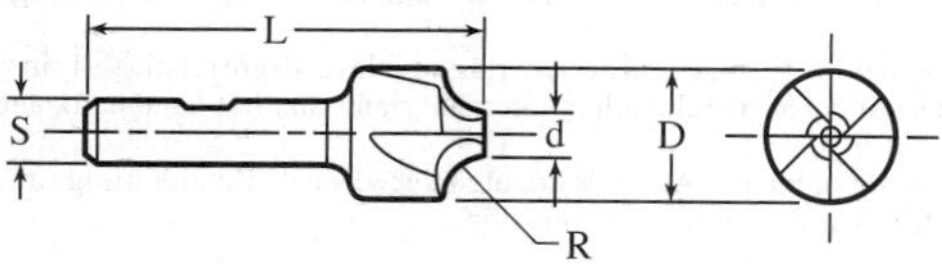


| Rad., *R* | Dia., *D* | Dia., *d* | *S* | *L* | Rad., *R* | Dia., *D* | Dia., *d* | *S* | *L* |
|---|---|---|---|---|---|---|---|---|---|
| 1/16 | 7/16 | 1/4 | 3/8 | 2 1/2 | 3/8 | 1 1/4 | 3/8 | 1/2 | 3 1/2 |
| 3/32 | 1/2 | 1/4 | 3/8 | 2 1/2 | 3/16 | 7/8 | 5/16 | 3/4 | 3 1/8 |
| 1/8 | 5/8 | 1/4 | 1/2 | 3 | 1/4 | 1 | 3/8 | 3/4 | 3 1/4 |
| 5/32 | 3/4 | 5/16 | 1/2 | 3 | 5/16 | 1 1/8 | 3/8 | 7/8 | 3 1/2 |
| 3/16 | 7/8 | 5/16 | 1/2 | 3 | 3/8 | 1 1/4 | 3/8 | 7/8 | 3 3/4 |
| 1/4 | 1 | 3/8 | 1/2 | 3 | 7/16 | 1 3/8 | 3/8 | 1 | 4 |
| 5/16 | 1 1/8 | 3/8 | 1/2 | 3 1/4 | 1/2 | 1 1/2 | 3/8 | 1 | 4 1/8 |

All dimensions are in inches. All cutters are high-speed steel. Right-hand cutters are standard.

*Tolerances:* On *D*, ±0.010 inch; on diameter of circle, 2*R*, ±0.001 inch for cutters up to and including 1/8-inch radius, +0.002, −0.001 inch for cutters over 1/8-inch radius; on *S*, −0.0001 to −0.0005 inch; and on *L*, ± 1/16 inch.

## American National Standard Metal-Slitting Saws *ANSI/ASME B94.19-1997 (R2019)*

| Cutter Diameter | | | Range of Face Widths Nom.[a] | Hole Diameter | | |
|---|---|---|---|---|---|---|
| Nom. | Max. | Min. | | Nom. | Max. | Min. |
| Plain Metal-Slitting Saws [b] | | | | | | |
| 2½ | 2.515 | 2.485 | 1⁄32, 3⁄64, 1⁄16, 3⁄32, 1⁄8 | 7⁄8 | 0.87575 | 0.8750 |
| 3 | 3.015 | 2.985 | 1⁄32, 3⁄64, 1⁄16, 3⁄32, 1⁄8 and 5⁄32 | 1 | 1.00075 | 1.0000 |
| 4 | 4.015 | 3.985 | 1⁄32, 3⁄64, 1⁄16, 3⁄32, 1⁄8, 5⁄32 and 3⁄16 | 1 | 1.00075 | 1.0000 |
| 5 | 5.015 | 4.985 | 1⁄16, 3⁄32, 1⁄8 | 1 | 1.00075 | 1.0000 |
| 5 | 5.015 | 4.985 | 1⁄8 | 1¼ | 1.2510 | 1.2500 |
| 6 | 6.015 | 5.985 | 1⁄16, 3⁄32, 1⁄8 | 1 | 1.00075 | 1.0000 |
| 6 | 6.015 | 5.985 | 1⁄8, 3⁄16 | 1¼ | 1.2510 | 1.2500 |
| 8 | 8.015 | 7.985 | 1⁄8 | 1 | 1.00075 | 1.0000 |
| 8 | 8.015 | 7.985 | 1⁄8 | 1¼ | 1.2510 | 1.2500 |
| Metal-Slitting Saws with Side Teeth [c] | | | | | | |
| 2½ | 2.515 | 2.485 | 1⁄16, 3⁄32, 1⁄8 | 7⁄8 | 0.87575 | 0.8750 |
| 3 | 3.015 | 2.985 | 1⁄16, 3⁄32, 1⁄8, 5⁄32 | 1 | 1.00075 | 1.0000 |
| 4 | 4.015 | 3.985 | 1⁄16, 3⁄32, 1⁄8, 5⁄32, 3⁄16 | 1 | 1.00075 | 1.0000 |
| 5 | 5.015 | 4.985 | 1⁄16, 3⁄32, 1⁄8, 5⁄32, 3⁄16 | 1 | 1.00075 | 1.0000 |
| 5 | 5.015 | 4.985 | 1⁄8 | 1¼ | 1.2510 | 1.2500 |
| 6 | 6.015 | 5.985 | 1⁄16, 3⁄32, 1⁄8, 3⁄16 | 1 | 1.00075 | 1.0000 |
| 6 | 6.015 | 5.985 | 1⁄8, 3⁄16 | 1¼ | 1.2510 | 1.2500 |
| 8 | 8.015 | 7.985 | 1⁄8 | 1 | 1.00075 | 1.0000 |
| 8 | 8.015 | 7.985 | 1⁄8, 3⁄16 | 1¼ | 1.2510 | 1.2500 |
| Metal-Slitting Saws with Staggered Peripheral and Side Teeth [d] | | | | | | |
| 3 | 3.015 | 2.985 | 3⁄16 | 1 | 1.00075 | 1.0000 |
| 4 | 4.015 | 3.985 | 3⁄16 | 1 | 1.00075 | 1.0000 |
| 5 | 5.015 | 4.985 | 3⁄16, 1⁄4 | 1 | 1.00075 | 1.0000 |
| 6 | 6.015 | 5.985 | 3⁄16, 1⁄4 | 1 | 1.00075 | 1.0000 |
| 6 | 6.015 | 5.985 | 3⁄16, 1⁄4 | 1¼ | 1.2510 | 1.2500 |
| 8 | 8.015 | 7.985 | 3⁄16, 1⁄4 | 1¼ | 1.2510 | 1.2500 |
| 10 | 10.015 | 9.985 | 3⁄16, 1⁄4 | 1¼ | 1.2510 | 1.2500 |
| 12 | 12.015 | 11.985 | 1⁄4, 5⁄16 | 1½ | 1.5010 | 1.5000 |

[a] Tolerances on face widths are plus or minus 0.001 inch.

[b] Plain metal-slitting saws are relatively thin plain milling cutters having peripheral teeth only. They are furnished with or without hub, and their sides are concaved to the arbor hole or hub.

[c] Metal-slitting saws with side teeth are relatively thin side milling cutters having both peripheral and side teeth.

[d] Metal-slitting saws with staggered peripheral and side teeth are relatively thin staggered-tooth milling cutters having peripheral teeth of alternate right- and left-hand helix and alternate side teeth.

All dimensions are in inches. All saws are high-speed steel. Metal-slitting saws are similar to plain or side milling cutters but are relatively thin.

### Milling Cutter Terms

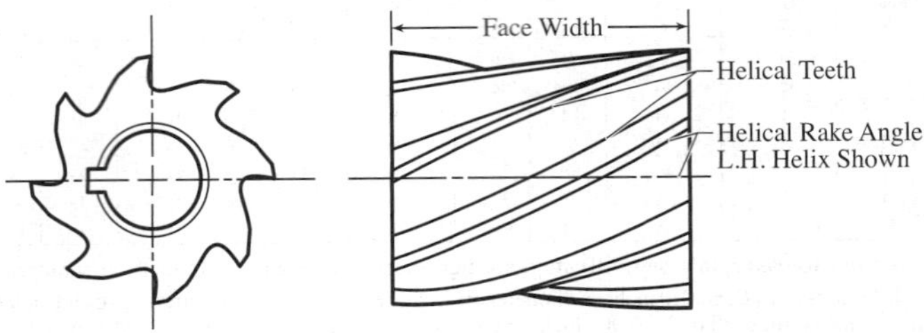

### Milling Cutter Terms *(Continued)*

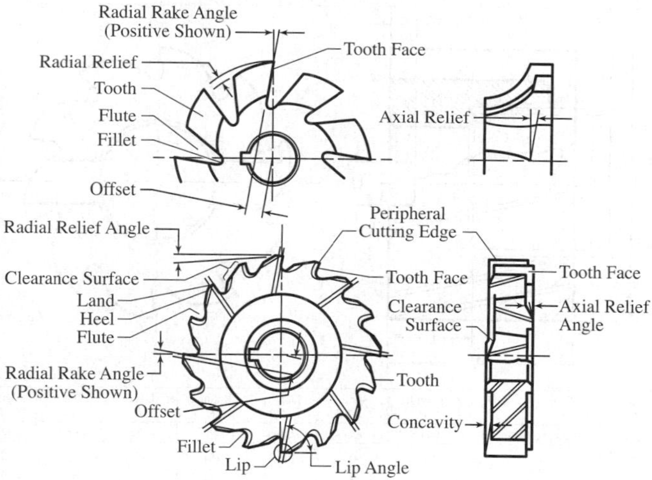


## American National Standard Single- and Double-Angle Milling Cutters *ANSI/ASME B94.19-1997 (R2019)*

| Cutter Diameter | | | Nominal Face Width[a] | Hole Diameter | | |
|---|---|---|---|---|---|---|
| Nom. | Max. | Min. | | Nom. | Max. | Min. |
| Single-Angle Cutters[b] | | | | | | |
| [c]1¼ | 1.265 | 1.235 | 7⁄16 | 3⁄8-24 UNF-2B RH<br>3⁄8-24 UNF-2B LH | | |
| [c]1⅝ | 1.640 | 1.610 | 9⁄16 | ½-20 UNF-2B RH | | |
| 2¾ | 2.765 | 2.735 | ½ | 1 | 1.00075 | 1.0000 |
| 3 | 3.015 | 2.985 | ½ | 1¼ | 1.2510 | 1.2500 |
| Double-Angle Cutters[d] | | | | | | |
| 2¾ | 2.765 | 2.735 | ½ | 1 | 1.00075 | 1.0000 |

[a] Face width tolerances are plus or minus 0.015 inch.

[b] Single-angle milling cutters have peripheral teeth, one cutting edge of which lies in a conical surface and the other in the plane perpendicular to the cutter axis. There are two types: one has a plain keywayed hole and has an included tooth angle of either 45 or 60 degrees ±10 minutes; the other has a threaded hole and an included tooth angle of 60 degrees ±10 minutes. Cutters with a right-hand threaded hole have a right-hand hand of rotation and a right-hand hand of cutter. Cutters with a left-hand threaded hole have a left-hand hand of rotation and a left-hand hand of cutter. Cutters with plain keywayed holes are standard as either right-hand or left-hand cutters.

[c] These cutters have threaded holes, the sizes of which are given under "Hole Diameter."

[d] Double-angle milling cutters have symmetrical peripheral teeth, both sides of which lie in conical surfaces. They are designated by the included angle, which may be 45, 60 or 90 degrees. Tolerances are ±10 minutes for the half angle on each side of the center.

All dimensions are in inches. All cutters are high-speed steel.

## American National Standard Shell Mills *ANSI/ASME B94.19-1997 (R2019)*

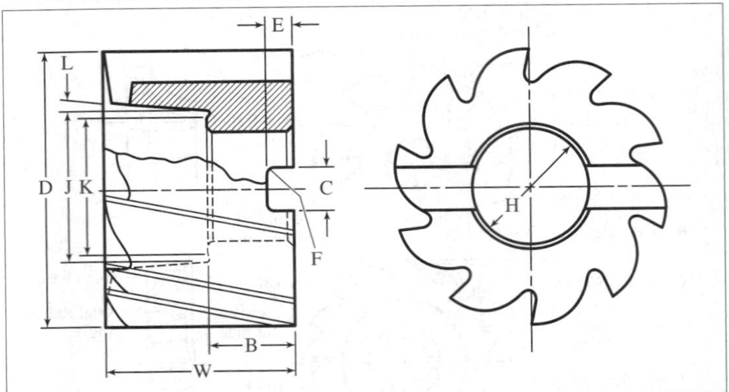


| Dia., *D* | Width, *W* | Dia., *H* | Length, *B* | Width, *C* | Depth, *E* | Radius, *F* | Dia., *J* | Dia., *K* | Angle, *L* |
|---|---|---|---|---|---|---|---|---|---|
| inches | inches | inches | inches | inches | inches | inches | inches | degrees | inches |
| 1¼ | 1 | ½ | ⅝ | ¼ | 5⁄32 | 1⁄64 | 11⁄16 | ⅝ | 0 |
| 1½ | 1⅛ | ½ | ⅝ | ¼ | 5⁄32 | 1⁄64 | 11⁄16 | ⅝ | 0 |
| 1¾ | 1¼ | ¾ | ¾ | 5⁄16 | 3⁄16 | 1⁄32 | 15⁄16 | ⅞ | 0 |
| 2 | 1⅜ | ¾ | ¾ | 5⁄16 | 3⁄16 | 1⁄32 | 15⁄16 | ⅞ | 0 |
| 2¼ | 1½ | 1 | ¾ | ⅜ | 7⁄32 | 1⁄32 | 1¼ | 1 3⁄16 | 0 |
| 2½ | 1⅝ | 1 | ¾ | ⅜ | 7⁄32 | 1⁄32 | 1⅜ | 1 3⁄16 | 0 |
| 2¾ | 1⅝ | 1 | ¾ | ⅜ | 7⁄32 | 1⁄32 | 1½ | 1 3⁄16 | 5 |
| 3 | 1¾ | 1¼ | ¾ | ½ | 9⁄32 | 1⁄32 | 1 21⁄32 | 1½ | 5 |
| 3½ | 1⅞ | 1¼ | ¾ | ½ | 9⁄32 | 1⁄32 | 1 11⁄16 | 1½ | 5 |
| 4 | 2¼ | 1½ | 1 | ⅝ | ⅜ | 1⁄16 | 2 1⁄32 | 1⅞ | 5 |
| 4½ | 2¼ | 1½ | 1 | ⅝ | ⅜ | 1⁄16 | 2 1⁄16 | 1⅞ | 10 |
| 5 | 2¼ | 1½ | 1 | ⅝ | ⅜ | 1⁄16 | 2 9⁄16 | 1⅞ | 10 |
| 6 | 2¼ | 2 | 1 | ¾ | 7⁄16 | 1⁄16 | 2 13⁄16 | 2½ | 15 |

All cutters are high-speed steel. Right-hand cutters with right-hand helix and square corners are standard.

*Tolerances:* On $D$, $+1/64$ inch; on $W$, $\pm 1/64$ inch; on $H$, $+0.0005$ inch; on $B$, $+1/64$ inch; on $C$, at least $+0.008$ but not more than $+0.012$ inch; on $E$, $+1/64$ inch; on $J$, $\pm 1/64$ inch; on $K$, $\pm 1/64$ inch.

### End Mill Terms

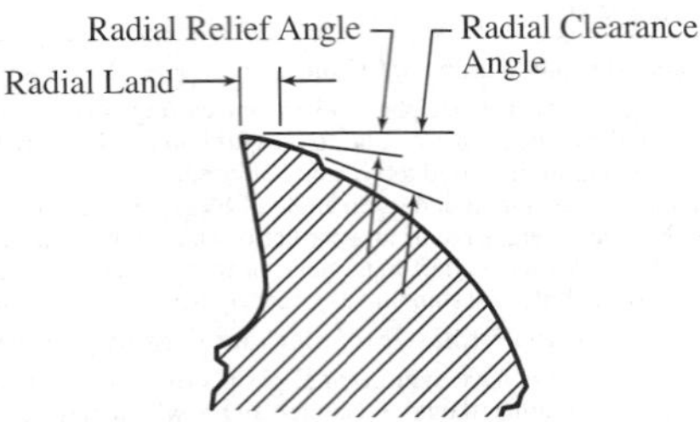


Enlarged Section of End Mill Tooth

**End Mill Terms** *(Continued)*

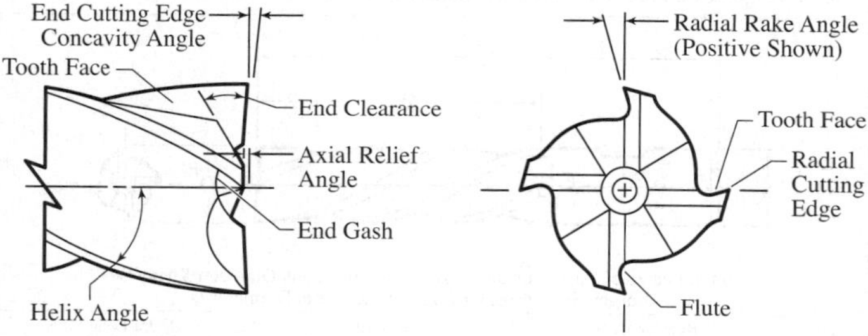


Enlarged Section of End Mill

## American National Standard Multiple- and Two-Flute Single-End Helical End Mills with Plain Straight and Weldon Shanks *ANSI/ASME B94.19-1997 (R2019)*

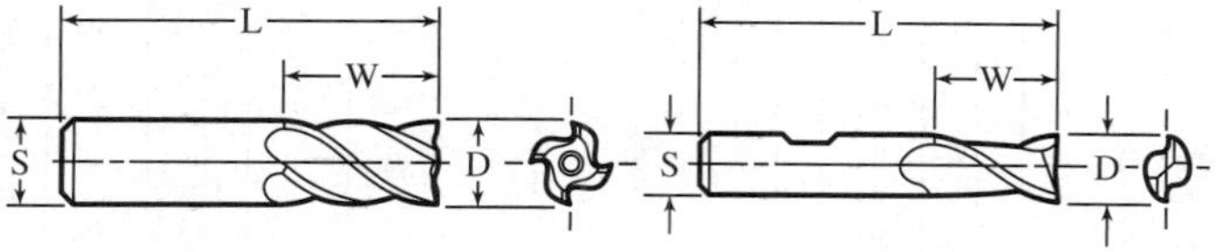


| Cutter Diameter, D | | | Shank Diameter, S | | Length of Cut, W | Length Overall, L |
|---|---|---|---|---|---|---|
| Nom. | Max. | Min. | Max. | Min. | | |
| Multiple-Flute with Plain Straight Shanks | | | | | | |
| 1⁄8 | .130 | .125 | .125 | .1245 | 5⁄16 | 1 1⁄4 |
| 3⁄16 | .1925 | .1875 | .1875 | .1870 | 1⁄2 | 1 3⁄8 |
| 1⁄4 | .255 | .250 | .250 | .2495 | 5⁄8 | 1 11⁄16 |
| 3⁄8 | .380 | .375 | .375 | .3745 | 3⁄4 | 1 13⁄16 |
| 1⁄2 | .505 | .500 | .500 | .4995 | 15⁄16 | 2 1⁄4 |
| 3⁄4 | .755 | .750 | .750 | .7495 | 1 1⁄4 | 2 5⁄8 |
| Two-Flute for Keyway Cutting with Weldon Shanks | | | | | | |
| 1⁄8 | .125 | .1235 | .375 | .3745 | 3⁄8 | 2 5⁄16 |
| 3⁄16 | .1875 | .1860 | .375 | .3745 | 7⁄16 | 2 5⁄16 |
| 1⁄4 | .250 | .2485 | .375 | .3745 | 1⁄2 | 2 5⁄16 |
| 5⁄16 | .3125 | .3110 | .375 | .3745 | 9⁄16 | 2 5⁄16 |
| 3⁄8 | .375 | .3735 | .375 | .3745 | 9⁄16 | 2 5⁄16 |
| 1⁄2 | .500 | .4985 | .500 | .4995 | 1 | 3 |
| 5⁄8 | .625 | .6235 | .625 | .6245 | 1 5⁄16 | 3 7⁄16 |
| 3⁄4 | .750 | .7485 | .750 | .7495 | 1 5⁄16 | 3 9⁄16 |
| 7⁄8 | .875 | .8735 | .875 | .8745 | 1 1⁄2 | 3 3⁄4 |
| 1 | 1.000 | .9985 | 1.000 | .9995 | 1 5⁄8 | 4 1⁄8 |
| 1 1⁄4 | 1.250 | 1.2485 | 1.250 | 1.2495 | 1 5⁄8 | 4 1⁄8 |
| 1 1⁄2 | 1.500 | 1.4985 | 1.250 | 1.2495 | 1 5⁄8 | 4 1⁄8 |

All dimensions are in inches. All cutters are high-speed steel. Right-hand cutters with right-hand helix are standard.

The helix angle is not less than 10 degrees for multiple-flute cutters with plain straight shanks; the helix angle is optional with the manufacturer for two-flute cutters with Weldon shanks.

*Tolerances:* On $W$, $\pm 1/32$ inch; on $L$, $\pm 1/16$ inch.

## ANSI Regular-, Long-, and Extra-Long-Length, Multiple-Flute Medium Helix Single-End End Mills with Weldon Shanks *ANSI/ASME B94.19-1997 (R2019)*

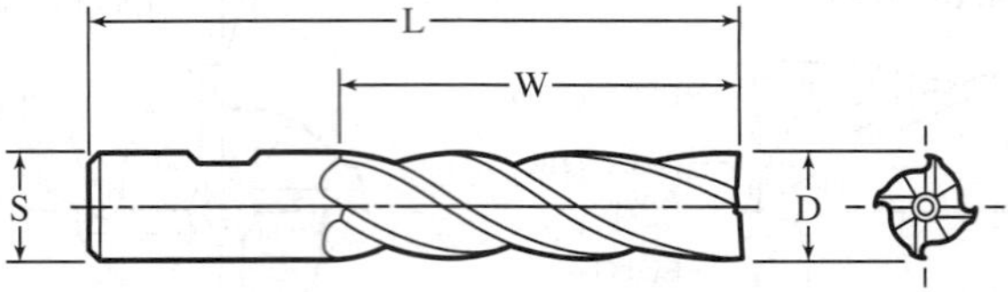


As Indicated by the Dimensions Given Below, Shank Diameter *S* May Be Larger, Smaller, or the Same as the Cutter Diameter *D*

| Cutter Dia., *D* | Regular Mills | | | | Long Mills | | | | Extra-Long Mills | | | |
|---|---|---|---|---|---|---|---|---|---|---|---|---|
| | *S* | *W* | *L* | *N*[a] | *S* | *W* | *L* | *N*[a] | *S* | *W* | *L* | *N*[a] |
| 1/8 [b] | 3/8 | 3/8 | 2 5/16 | 4 | … | … | … | … | … | … | … | … |
| 3/16 [b] | 3/8 | 1/2 | 2 3/8 | 4 | … | … | … | … | … | … | … | … |
| 1/4 [b] | 3/8 | 5/8 | 2 7/16 | 4 | 3/8 | 1 1/4 | 3 1/16 | 4 | 3/8 | 1 3/4 | 3 9/16 | 4 |
| 5/16 [b] | 3/8 | 3/4 | 2 1/2 | 4 | 3/8 | 1 3/8 | 3 1/8 | 4 | 3/8 | 2 | 3 3/4 | 4 |
| 3/8 [b] | 3/8 | 3/4 | 2 1/2 | 4 | 3/8 | 1 1/2 | 3 1/4 | 4 | 3/8 | 2 1/2 | 4 1/4 | 4 |
| 7/16 | 3/8 | 1 | 2 11/16 | 4 | 1/2 | 1 3/4 | 3 3/4 | 4 | … | … | … | … |
| 1/2 | 3/8 | 1 | 2 11/16 | 4 | 1/2 | 2 | 4 | 4 | 1/2 | 3 | 5 | 4 |
| 1/2 [b] | 1/2 | 1 1/4 | 3 1/4 | 4 | … | … | … | … | … | … | … | … |
| 9/16 | 1/2 | 1 3/8 | 3 3/8 | 4 | … | … | … | … | … | … | … | … |
| 5/8 | 1/2 | 1 3/8 | 3 3/8 | 4 | 5/8 | 2 1/2 | 4 5/8 | 4 | 5/8 | 4 | 6 1/8 | 4 |
| 11/16 | 1/2 | 1 5/8 | 3 5/8 | 4 | … | … | … | … | … | … | … | … |
| 3/4 | 1/2 | 1 5/8 | 3 5/8 | 4 | 3/4 | 3 | 5 1/4 | 4 | 3/4 | 4 | 6 1/4 | 4 |
| 5/8 [b] | 5/8 | 1 5/8 | 3 3/4 | 4 | … | … | … | … | … | … | … | … |
| 11/16 | 5/8 | 1 5/8 | 3 3/4 | 4 | … | … | … | … | … | … | … | … |
| 3/4 [b] | 5/8 | 1 5/8 | 3 3/4 | 4 | … | … | … | … | … | … | … | … |
| 13/16 | 5/8 | 1 7/8 | 4 | 6 | … | … | … | … | … | … | … | … |
| 7/8 | 5/8 | 1 7/8 | 4 | 6 | 7/8 | 3 1/2 | 5 3/4 | 4 | 7/8 | 5 | 7 1/4 | 4 |
| 1 | 5/8 | 1 7/8 | 4 | 6 | 1 | 4 | 6 1/2 | 4 | 1 | 6 | 8 1/2 | 4 |
| 7/8 | 7/8 | 1 7/8 | 4 1/8 | 4 | … | … | … | … | … | … | … | … |
| 1 | 7/8 | 1 7/8 | 4 1/8 | 4 | … | … | … | … | … | … | … | … |
| 1 1/8 | 7/8 | 2 | 4 1/4 | 6 | 1 | 4 | 6 1/2 | 6 | … | … | … | … |
| 1 1/4 | 7/8 | 2 | 4 1/4 | 6 | 1 | 4 | 6 1/2 | 6 | 1 1/4 | 6 | 8 1/2 | 6 |
| 1 | 1 | 2 | 4 1/2 | 4 | … | … | … | … | … | … | … | … |
| 1 1/8 | 1 | 2 | 4 1/2 | 6 | … | … | … | … | … | … | … | … |
| 1 1/4 | 1 | 2 | 4 1/2 | 6 | … | … | … | … | … | … | … | … |
| 1 3/8 | 1 | 2 | 4 1/2 | 6 | … | … | … | … | … | … | … | … |
| 1 1/2 | 1 | 2 | 4 1/2 | 6 | 1 | 4 | 6 1/2 | 6 | … | … | … | … |
| 1 1/4 | 1 1/4 | 2 | 4 1/2 | 6 | 1 1/4 | 4 | 6 1/2 | 6 | … | … | … | … |
| 1 1/2 | 1 1/4 | 2 | 4 1/2 | 6 | 1 1/4 | 4 | 6 1/2 | 6 | 1 1/4 | 8 | 10 1/2 | 6 |
| 1 3/4 | 1 1/4 | 2 | 4 1/2 | 6 | 1 1/4 | 4 | 6 1/2 | 6 | … | … | … | … |
| 2 | 1 1/4 | 2 | 4 1/2 | 8 | 1 1/4 | 4 | 6 1/2 | 8 | … | … | … | … |

[a] *N* = Number of flutes.

[b] In this size of regular mill, a left-hand cutter with left-hand helix is also standard.

All dimensions are in inches. All cutters are high-speed steel. Helix angle is greater than 19 degrees but not more than 39 degrees. Right-hand cutters with right-hand helix are standard.

*Tolerances:* On *D*, +0.003 inch; on *S*, −0.0001 to −0.0005 inch; on *W*, $\pm 1/32$ inch; on *L*, $\pm 1/16$ inch.

## ANSI Two-Flute, High Helix, Regular-, Long-, and Extra-Long-Length, Single-End End Mills with Weldon Shanks *ANSI/ASME B94.19-1997 (R2019)*

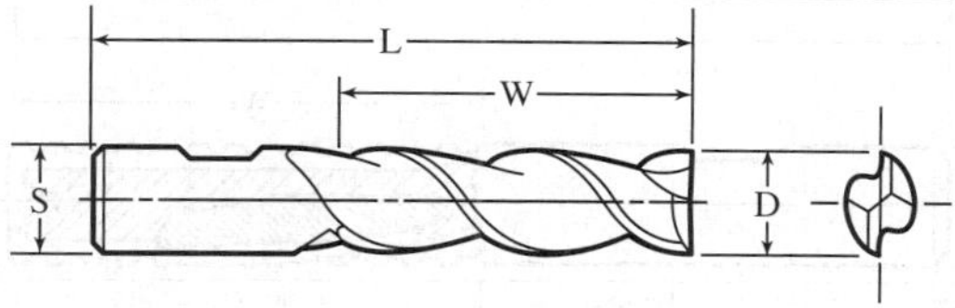


| Cutter Dia., *D* | Regular Mill *S* | Regular Mill *W* | Regular Mill *L* | Long Mill *S* | Long Mill *W* | Long Mill *L* | Extra-Long Mill *S* | Extra-Long Mill *W* | Extra-Long Mill *L* |
|---|---|---|---|---|---|---|---|---|---|
| ¼ | ⅜ | ⅝ | 2 7/16 | ⅜ | 1¼ | 3 1/16 | ⅜ | 1¾ | 3 9/16 |
| 5/16 | ⅜ | ¾ | 2½ | ⅜ | 1⅜ | 3⅛ | ⅜ | 2 | 3¾ |
| ⅜ | ⅜ | ¾ | 2½ | ⅜ | 1½ | 3¼ | ⅜ | 2½ | 4¼ |
| 7/16 | ⅜ | 1 | 2 11/16 | ½ | 1¾ | 3¾ | … | … | … |
| ½ | ½ | 1¼ | 3¼ | ½ | 2 | 4 | ½ | 3 | 5 |
| ⅝ | ⅝ | 1⅝ | 3¾ | ⅝ | 2½ | 4⅝ | ⅝ | 4 | 6⅛ |
| ¾ | ¾ | 1⅝ | 3⅞ | ¾ | 3 | 5¼ | ¾ | 4 | 6¼ |
| ⅞ | ⅞ | 1⅞ | 4⅛ | … | … | … | … | … | … |
| 1 | 1 | 2 | 4½ | 1 | 4 | 6½ | 1 | 6 | 8½ |
| 1¼ | 1¼ | 2 | 4½ | 1¼ | 4 | 6½ | 1¼ | 6 | 8½ |
| 1½ | 1¼ | 2 | 4½ | 1¼ | 4 | 6½ | 1¼ | 8 | 10½ |
| 2 | 1¼ | 2 | 4½ | 1¼ | 4 | 6½ | … | … | … |

All dimensions are in inches. All cutters are high-speed steel. Right-hand cutters with right-hand helix are standard. Helix angle is greater than 39 degrees.

*Tolerances:* On *D*, +0.003 inch; on *S*, −0.0001 to −0.0005 inch; on *W*, $\pm 1/32$ inch; and on *L*, $\pm 1/16$ inch.

## Combination Shanks for End Mills *ANSI/ASME B94.19-1997 (R2019)*

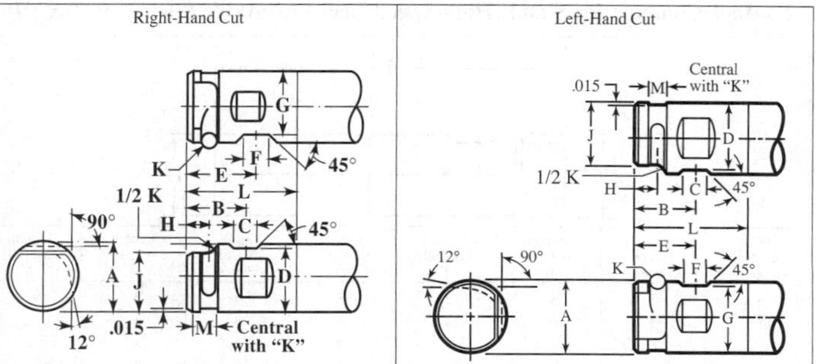


| Dia. *A* | *L*[a] | *B* | *C* | *D* | *E* | *F* | *G* | *H* | *J* | *K* | *M* |
|---|---|---|---|---|---|---|---|---|---|---|---|
| 1½ | 2 11/16 | 1 3/16 | .515 | 1.406 | 1½ | .515 | 1.371 | 9/16 | 1.302 | .377 | 7/16 |
| 2 | 3¼ | 1 23/32 | .700 | 1.900 | 1¾ | .700 | 1.809 | ⅝ | 1.772 | .440 | ½ |
| 2½ | 3½ | 1 15/16 | .700 | 2.400 | 2 | .700 | 2.312 | ¾ | 2.245 | .503 | 9/16 |

[a] Length of shank.

All dimensions are in inches.

Modified for use as Weldon or Pin Drive shank.

## ANSI Roughing, Single-End End Mills with Weldon Shanks, High-Speed Steel *ANSI/ASME B94.19-1997 (R2019)*

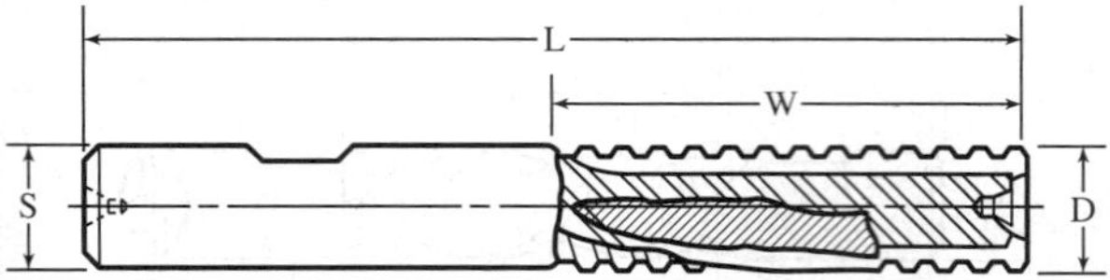


| Diameter | | Length | | Diameter | | Length | |
|---|---|---|---|---|---|---|---|
| Cutter | Shank | Cut | Overall | Cutter | Shank | Cut | Overall |
| *D* | *S* | *W* | *L* | *D* | *S* | *W* | *L* |
| ½ | ½ | 1 | 3 | 2 | 2 | 2 | 5¾ |
| ½ | ½ | 1¼ | 3¼ | 2 | 2 | 3 | 6¾ |
| ½ | ½ | 2 | 4 | 2 | 2 | 4 | 7¾ |
| ⅝ | ⅝ | 1¼ | 3⅜ | 2 | 2 | 5 | 8¾ |
| ⅝ | ⅝ | 1⅝ | 3¾ | 2 | 2 | 6 | 9¾ |
| ⅝ | ⅝ | 2½ | 4⅝ | 2 | 2 | 7 | 10¾ |
| ¾ | ¾ | 1½ | 3¾ | 2 | 2 | 8 | 11¾ |
| ¾ | ¾ | 1⅝ | 3⅞ | 2 | 2 | 10 | 13¾ |
| ¾ | ¾ | 3 | 5¼ | 2 | 2 | 12 | 15¾ |
| 1 | 1 | 2 | 4½ | 2½ | 2 | 4 | 7¾ |
| 1 | 1 | 4 | 6½ | 2½ | 2 | 6 | 9¾ |
| 1¼ | 1¼ | 2 | 4½ | 2½ | 2 | 8 | 11¾ |
| 1¼ | 1¼ | 4 | 6½ | 2½ | 2 | 10 | 13¾ |
| 1½ | 1¼ | 2 | 4½ | 3 | 2½ | 4 | 7¾ |
| 1½ | 1¼ | 4 | 6½ | 3 | 2½ | 6 | 9¾ |
| 1¾ | 1¼ | 2 | 4½ | 3 | 2½ | 8 | 11¾ |
| 1¾ | 1¼ | 4 | 6½ | 3 | 2½ | 10 | 13¾ |

All dimensions are in inches. Right-hand cutters with right-hand helix are standard.
*Tolerances:* Outside diameter, +0.025, −0.005 inch; length of cut, +⅛, −1⁄32 inch.

## American National Standard Heavy Duty, Medium Helix Single-End End Mills, 2½-Inch Combination Shank, High-Speed Steel *ANSI/ASME B94.19-1997 (R2019)*

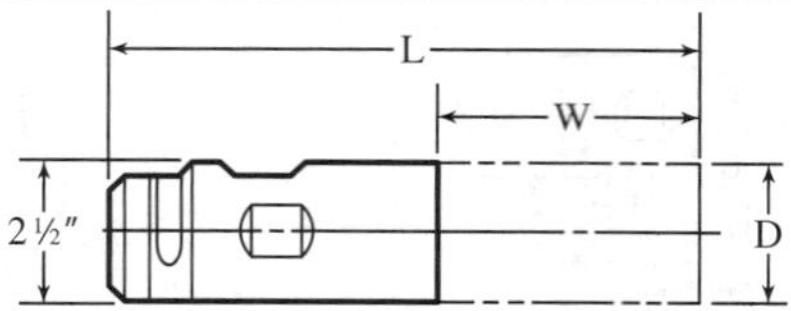


| Dia. of Cutter, *D* | No. of Flutes | Length of Cut, *W* | Length Overall, *L* | Dia. of Cutter, *D* | No. of Flutes | Length of Cut, *W* | Length Overall, *L* |
|---|---|---|---|---|---|---|---|
| 2½ | 3 | 8 | 12 | 3 | 3 | 4 | 7¾ |
| 2½ | 3 | 10 | 14 | 3 | 3 | 6 | 9¾ |
| 2½ | 6 | 4 | 8 | 3 | 3 | 8 | 11¾ |
| 2½ | 6 | 6 | 10 | 3 | 8 | 4 | 7¾ |
| 2½ | 6 | 8 | 12 | 3 | 8 | 6 | 9¾ |
| 2½ | 6 | 10 | 14 | 3 | 8 | 8 | 11¾ |
| 2½ | 6 | 12 | 16 | 3 | 8 | 10 | 13¾ |
| 3 | 2 | 4 | 7¾ | 3 | 8 | 12 | 15¾ |
| 3 | 2 | 6 | 9¾ | … | … | … | … |

All dimensions are in inches. For shank dimensions, see page 869. Right-hand cutters with right-hand helix are standard. Helix angle is greater than 19 degrees but not more than 39 degrees.
*Tolerances:* On *D*, +0.005 inch; on *W*, ±1⁄32 inch; on *L*, ±1⁄16 inch.

**ANSI Stub-, Regular-, and Long-Length, Four-Flute, Medium Helix, Plain-End, Double-End Miniature End Mills with 3⁄16-Inch Diameter Straight Shanks** *ANSI/ASME B94.19-1997 (R2019)*

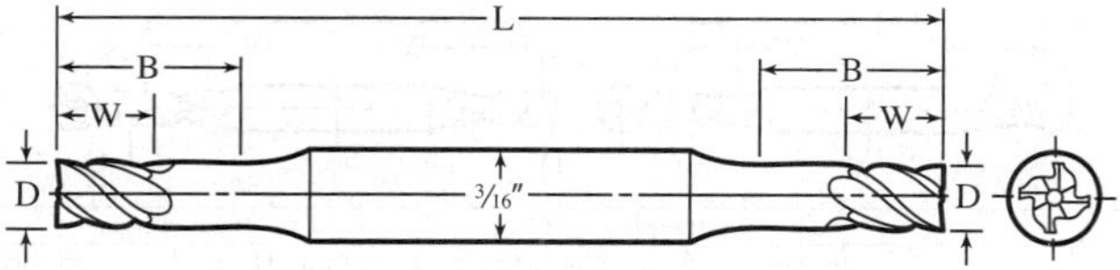


| Dia. D | Stub Length | | Regular Length | |
|---|---|---|---|---|
| | W | L | W | L |
| 1⁄16 | 3⁄32 | 2 | 3⁄16 | 2¼ |
| 3⁄32 | 9⁄64 | 2 | 9⁄32 | 2¼ |
| ⅛ | 3⁄16 | 2 | ⅜ | 2¼ |
| 5⁄32 | 15⁄64 | 2 | 7⁄16 | 2¼ |
| 3⁄16 | 9⁄32 | 2 | ½ | 2¼ |

| Dia. D | Long Length | | |
|---|---|---|---|
| | B | W | L |
| 1⁄16 | ⅜ | 7⁄32 | 2½ |
| 3⁄32 | ½ | 9⁄32 | 2⅝ |
| ⅛ | ¾ | ¾ | 3⅛ |
| 5⁄32 | ⅞ | ⅞ | 3¼ |
| 3⁄16 | 1 | 1 | 3⅜ |

All dimensions are in inches. All cutters are high-speed steel. Right-hand cutters with right-hand helix are standard. Helix angle is greater than 19 degrees but not more than 39 degrees.

*Tolerances:* On *D*, + 0.003 inch (if the shank is the same diameter as the cutting portion, however, then the tolerance on the cutting diameter is − 0.0025 inch.); on *W*, + 1⁄32, − 1⁄64 inch; and on *L*, ±1⁄16 inch.

**American National Standard 60-Degree Single-Angle Milling Cutters with Weldon Shanks** *ANSI/ASME B94.19-1997 (R2019)*

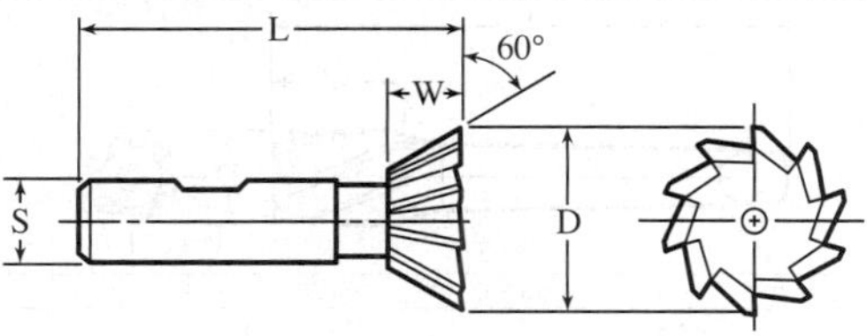


| Dia., D | S | W | L | Dia., D | S | W | L |
|---|---|---|---|---|---|---|---|
| ¾ | ⅜ | 5⁄16 | 2⅛ | 1⅞ | ⅞ | 13⁄16 | 3¼ |
| 1⅜ | ⅝ | 9⁄16 | 2⅞ | 2¼ | 1 | 1 1⁄16 | 3¾ |

All dimensions are in inches. All cutters are high-speed steel. Right-hand cutters are standard.

*Tolerances:* On *D*, ± 0.015 inch; on *S*, − 0.0001 to − 0.0005 inch; on *W*, ± 0.015 inch; and on *L*, ±1⁄16 inch.

## American National Standard Stub-, Regular-, and Long-Length, Two-Flute, Medium Helix, Plain- and Ball-End, Double-End Miniature End Mills with 3⁄16-Inch Diameter Straight Shanks *ANSI/ASME B94.19-1997 (R2019)*

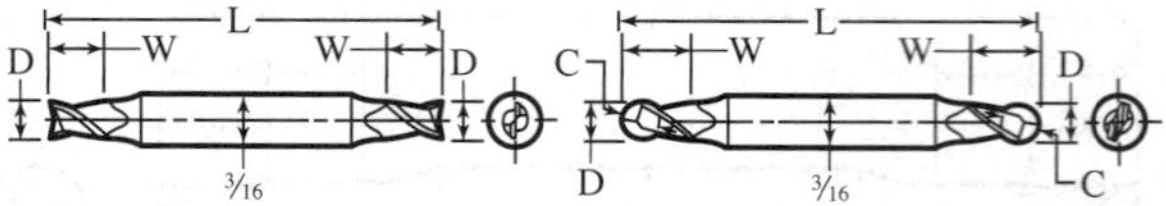


| Dia., *C* and *D* | Stub Length | | | | Regular Length | | | |
|---|---|---|---|---|---|---|---|---|
| | Plain-End | | Ball-End | | Plain-End | | Ball-End | |
| | *W* | *L* | *W* | *L* | *W* | *L* | *W* | *L* |
| 1⁄32 | 3⁄64 | 2 | … | … | 3⁄32 | 2¼ | … | … |
| 3⁄64 | 1⁄16 | 2 | … | … | 9⁄64 | 2¼ | … | … |
| 1⁄16 | 3⁄32 | 2 | 3⁄32 | 2 | 3⁄16 | 2¼ | 3⁄16 | 2¼ |
| 5⁄64 | 1⁄8 | 2 | … | … | 15⁄64 | 2¼ | … | … |
| 3⁄32 | 9⁄64 | 2 | 9⁄64 | 2 | 9⁄32 | 2¼ | 9⁄32 | 2¼ |
| 7⁄64 | 5⁄32 | 2 | … | … | 21⁄64 | 2¼ | … | … |
| 1⁄8 | 3⁄16 | 2 | 3⁄16 | 2 | 3⁄8 | 2¼ | 3⁄8 | 2¼ |
| 9⁄64 | 7⁄32 | 2 | … | … | 13⁄32 | 2¼ | … | … |
| 5⁄32 | 15⁄64 | 2 | 15⁄64 | 2 | 7⁄16 | 2¼ | 7⁄16 | 2¼ |
| 11⁄64 | 1⁄4 | 2 | … | … | ½ | 2¼ | … | … |
| 3⁄16 | 9⁄32 | 2 | 9⁄32 | 2 | ½ | 2¼ | ½ | 2¼ |

| Dia., *D* | Long-Length, Plain-End | | | Dia., *D* | Long-Length, Plain-End | | |
|---|---|---|---|---|---|---|---|
| | *B* [a] | *W* | *L* | | *B* [a] | *W* | *L* |
| 1⁄16 | 3⁄8 | 7⁄32 | 2½ | 5⁄32 | 7⁄8 | 7⁄8 | 3¼ |
| 3⁄32 | ½ | 9⁄32 | 2⅝ | 3⁄16 | 1 | 1 | 3⅜ |
| 1⁄8 | ¾ | ¾ | 3⅛ | | | | |

[a] *B* is the length below the shank.

All dimensions are in inches. All cutters are high-speed steel. Right-hand cutters with right-hand helix are standard. Helix angle is greater than 19 degrees but not more than 39 degrees.

*Tolerances:* On *C* and *D*, − 0.0015 inch for stub and regular length; + 0.003 inch for long length (if the shank is the same diameter as the cutting portion, however, then the tolerance on the cutting diameter is − 0.0025 inch.); on *W*, + 1⁄32, − 1⁄64 inch; and on *L*, ± 1⁄16 inch.

## American National Standard Multiple-Flute, Helical Series End Mills with Brown & Sharpe Taper Shanks

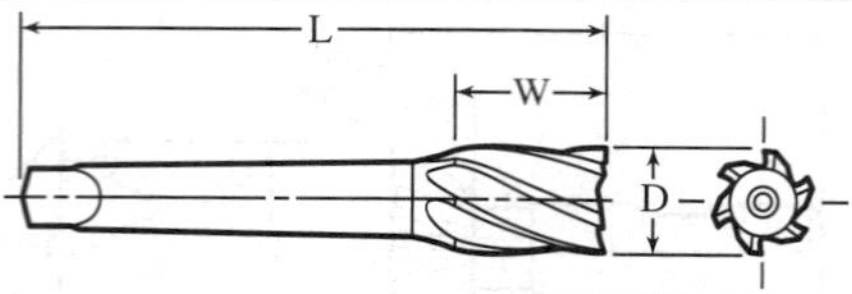


| Dia., *D* | *W* | *L* | Taper No. | Dia., *D* | *W* | *L* | Taper No. |
|---|---|---|---|---|---|---|---|
| ½ | 15⁄16 | 4 15⁄16 | 7 | 1¼ | 2 | 7¼ | 9 |
| ¾ | 1¼ | 5¼ | 7 | 1½ | 2¼ | 7½ | 9 |
| 1 | 1⅝ | 5⅝ | 7 | 2 | 2¾ | 8 | 9 |

All dimensions are in inches. All cutters are high-speed steel. Right-hand cutters with right-hand helix are standard. Helix angle is not less than 10 degrees.

No. 5 taper is standard without tang; Nos. 7 and 9 are standard with tang only.

*Tolerances:* On *D*, +0.005 inch; on *W*, ±1⁄32 inch; and on *L* ±1⁄16 inch.

For dimensions of B & S taper shanks, see information given on page 1011.

## American National Standard Stub- and Regular-Length, Two-Flute, Medium Helix, Plain- and Ball-End, Single-End End Mills with Weldon Shanks

*ANSI/ASME B94.19-1997 (R2019)*

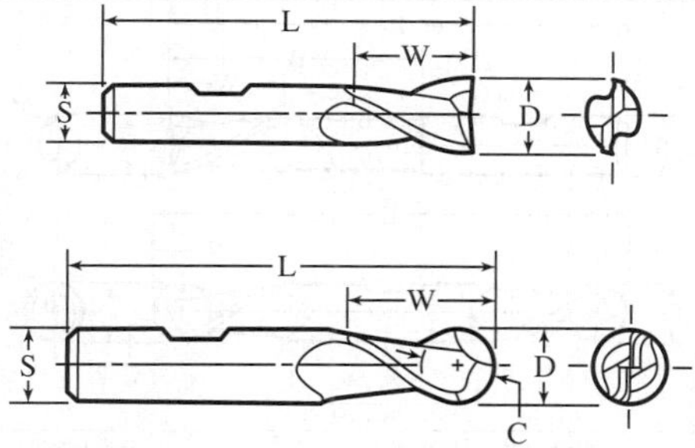


**Regular Length — Plain End**

| Dia., D | S | W | L |
|---|---|---|---|
| 1/8 | 3/8 | 3/8 | 2 5/16 |
| 3/16 | 3/8 | 7/16 | 2 5/16 |
| 1/4 | 3/8 | 1/2 | 2 5/16 |
| 5/16 | 3/8 | 9/16 | 2 5/16 |
| 3/8 | 3/8 | 9/16 | 2 5/16 |
| 7/16 | 3/8 | 13/16 | 2 1/2 |
| 1/2 | 3/8 | 13/16 | 2 1/2 |
| 1/2 | 1/2 | 1 | 3 |
| 9/16 | 1/2 | 1 1/8 | 3 1/8 |
| 5/8 | 1/2 | 1 1/8 | 3 1/8 |
| 11/16 | 1/2 | 1 5/16 | 3 5/16 |
| 3/4 | 1/2 | 1 5/16 | 3 5/16 |
| 5/8 | 5/8 | 1 5/16 | 3 7/16 |
| 11/16 | 5/8 | 1 5/16 | 3 7/16 |
| 3/4 | 5/8 | 1 5/16 | 3 7/16 |
| 13/16 | 5/8 | 1 1/2 | 3 5/8 |
| 7/8 | 5/8 | 1 1/2 | 3 5/8 |
| 1 | 5/8 | 1 1/2 | 3 5/8 |
| 7/8 | 7/8 | 1 1/2 | 3 3/4 |
| 1 | 7/8 | 1 1/2 | 3 3/4 |
| 1 1/8 | 7/8 | 1 5/8 | 3 7/8 |
| 1 1/4 | 7/8 | 1 5/8 | 3 7/8 |
| 1 | 1 | 1 5/8 | 4 1/8 |
| 1 1/8 | 1 | 1 5/8 | 4 1/8 |
| 1 1/4 | 1 | 1 5/8 | 4 1/8 |
| 1 3/8 | 1 | 1 5/8 | 4 1/8 |
| 1 1/2 | 1 | 1 5/8 | 4 1/8 |
| 1 1/4 | 1 1/4 | 1 5/8 | 4 1/8 |
| 1 1/2 | 1 1/4 | 1 5/8 | 4 1/8 |
| 1 3/4 | 1 1/4 | 1 5/8 | 4 1/8 |
| 2 | 1 1/4 | 1 5/8 | 4 1/8 |

**Stub Length — Plain End**

| Cutter Dia., D | Shank Dia., S | Length of Cut. W | Length Overall. L |
|---|---|---|---|
| 1/8 | 3/8 | 3/16 | 2 1/8 |
| 3/16 | 3/8 | 9/32 | 2 3/16 |
| 1/4 | 3/8 | 3/8 | 2 1/4 |

**Regular Length — Ball End**

| Dia., C and D | Shank Dia., S | Length of Cut. W | Length Overall. L |
|---|---|---|---|
| 1/8 | 3/8 | 3/8 | 2 5/16 |
| 3/16 | 3/8 | 1/2 | 2 3/8 |
| 1/4 | 3/8 | 5/8 | 2 7/16 |
| 5/16 | 3/8 | 3/4 | 2 1/2 |
| 3/8 | 3/8 | 3/4 | 2 1/2 |
| 7/16 | 1/2 | 1 | 3 |
| 1/2 | 1/2 | 1 | 3 |
| 9/16 | 1/2 | 1 1/8 | 3 1/8 |
| 5/8 | 1/2 | 1 1/8 | 3 1/8 |
| 5/8 | 5/8 | 1 3/8 | 3 1/2 |
| 3/4 | 1/2 | 1 5/16 | 3 5/16 |
| 3/4 | 3/4 | 1 5/8 | 3 7/8 |
| 7/8 | 7/8 | 2 | 4 1/4 |
| 1 | 1 | 2 1/4 | 4 3/4 |
| 1 1/8 | 1 | 2 1/4 | 4 3/4 |
| 1 1/4 | 1 1/4 | 2 1/2 | 5 |
| 1 1/2 | 1 1/4 | 2 1/2 | 5 |

All dimensions are in inches. All cutters are high-speed steel. Right-hand cutters with right-hand helix are standard. Helix angle is greater than 19 degrees but not more than 39 degrees.

*Tolerances:* On $C$ and $D$, −0.0015 inch for stub-length mills, + 0.003 inch for regular-length mills; on $S$, −0.0001 to −0.0005 inch; on $W$, ± 1/32 inch; and on $L$, ± 1/16 inch.

The following single-end end mills are available in premium high-speed steel: ball end, two flute, with $D$ ranging from 1/8 to 1 1/2 inches; ball end, multiple flute, with $D$ ranging from 1/8 to 1 inch; and plain end, two flute, with $D$ ranging from 1/8 to 1 1/2 inches.

**American National Standard Long-Length Single-End and Stub-, and Regular-Length, Double-End, Plain- and Ball-End, Medium Helix, Two-Flute End Mills with Weldon Shanks** *ANSI/ASME B94.19-1997 (R2019)*

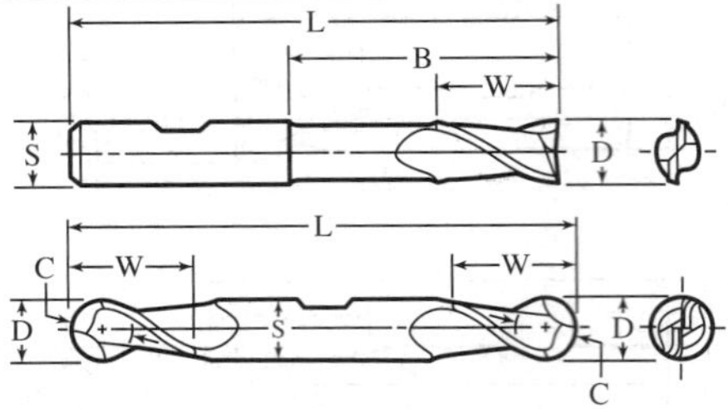


| Single End | | | | | | | | |
|---|---|---|---|---|---|---|---|---|
| Dia., *C* and *D* | Long Length — Plain End | | | | Long Length — Ball End | | | |
| | *S* | *B*[a] | *W* | *L* | *S* | *B*[a] | *W* | *L* |
| 1/8 | … | … | … | … | 3/8 | 13/16 | 3/8 | 2 3/8 |
| 3/16 | … | … | … | … | 3/8 | 1 1/8 | 1/2 | 2 11/16 |
| 1/4 | 3/8 | 1 1/2 | 5/8 | 3 1/16 | 3/8 | 1 1/2 | 5/8 | 3 1/16 |
| 5/16 | 3/8 | 1 3/4 | 3/4 | 3 5/16 | 3/8 | 1 3/4 | 3/4 | 3 5/16 |
| 3/8 | 3/8 | 1 3/4 | 3/4 | 3 5/16 | 3/8 | 1 3/4 | 3/4 | 3 5/16 |
| 7/16 | … | … | … | … | 1/2 | 1 7/8 | 1 | 3 11/16 |
| 1/2 | 1/2 | 2 7/32 | 1 | 4 | 1/2 | 2 1/4 | 1 | 4 |
| 5/8 | 5/8 | 2 23/32 | 1 3/8 | 4 5/8 | 5/8 | 2 3/4 | 1 3/8 | 4 5/8 |
| 3/4 | 3/4 | 3 11/32 | 1 5/8 | 5 3/8 | 3/4 | 3 3/8 | 1 5/8 | 5 3/8 |
| 1 | 1 | 4 31/32 | 2 1/2 | 7 1/4 | 1 | 5 | 2 1/2 | 7 1/4 |
| 1 1/4 | 1 1/4 | 4 31/32 | 3 | 7 1/4 | … | … | … | … |

[a] *B* is the length below the shank.

| Double End | | | | | | | | | |
|---|---|---|---|---|---|---|---|---|---|
| Dia., *C* and *D* | Stub Length — Plain End | | | Regular Length — Plain End | | | Regular Length — Ball End | | |
| | *S* | *W* | *L* | *S* | *W* | *L* | *S* | *W* | *L* |
| 1/8 | 3/8 | 3/16 | 2 3/4 | 3/8 | 3/8 | 3 1/16 | 3/8 | 3/8 | 3 1/16 |
| 5/32 | 3/8 | 15/64 | 2 3/4 | 3/8 | 7/16 | 3 1/8 | … | … | … |
| 3/16 | 3/8 | 9/32 | 2 3/4 | 3/8 | 7/16 | 3 1/8 | 3/8 | 7/16 | 3 1/8 |
| 7/32 | 3/8 | 21/64 | 2 7/8 | 3/8 | 1/2 | 3 1/8 | … | … | … |
| 1/4 | 3/8 | 3/8 | 2 7/8 | 3/8 | 1/2 | 3 1/8 | 3/8 | 1/2 | 3 1/8 |
| 9/32 | … | … | … | 3/8 | 9/16 | 3 1/8 | … | … | … |
| 5/16 | … | … | … | 3/8 | 9/16 | 3 1/8 | 3/8 | 9/16 | 3 1/8 |
| 11/32 | … | … | … | 3/8 | 9/16 | 3 1/8 | … | … | … |
| 3/8 | … | … | … | 3/8 | 9/16 | 3 1/8 | 3/8 | 9/16 | 3 1/8 |
| 13/32 | … | … | … | 1/2 | 13/16 | 3 3/4 | … | … | … |
| 7/16 | … | … | … | 1/2 | 13/16 | 3 3/4 | 1/2 | 13/16 | 3 3/4 |
| 15/32 | … | … | … | 1/2 | 13/16 | 3 3/4 | … | … | … |
| 1/2 | … | … | … | 1/2 | 13/16 | 3 3/4 | 1/2 | 13/16 | 3 3/4 |
| 9/16 | … | … | … | 5/8 | 1 1/8 | 4 1/2 | … | … | … |
| 5/8 | … | … | … | 5/8 | 1 1/8 | 4 1/2 | 5/8 | 1 1/8 | 4 1/2 |
| 11/16 | … | … | … | 3/4 | 1 5/16 | 5 | … | … | … |
| 3/4 | … | … | … | 3/4 | 1 5/16 | 5 | 3/4 | 1 5/16 | 5 |
| 7/8 | … | … | … | 7/8 | 1 9/16 | 5 1/2 | … | … | … |
| 1 | … | … | … | 1 | 1 5/8 | 5 7/8 | 1 | 1 5/8 | 5 7/8 |

All dimensions are in inches. All cutters are high-speed steel. Right-hand cutters with right-hand helix are standard. Helix angle is greater than 19 degrees but not more than 39 degrees.

*Tolerances:* On *C* and *D*, + 0.003 inch for single-end mills, −0.0015 inch for double-end mills; on *S*, −0.0001 to −0.0005 inch; on *W*, $\pm\frac{1}{32}$ inch; and on *L*, $\pm\frac{1}{16}$ inch.

## American National Standard Regular-, Long-, and Extra-Long-Length, Three- and Four-Flute, Medium Helix, Center Cutting, Single-End End Mills with Weldon Shanks *ANSI/ASME B94.19-1997 (R2019)*

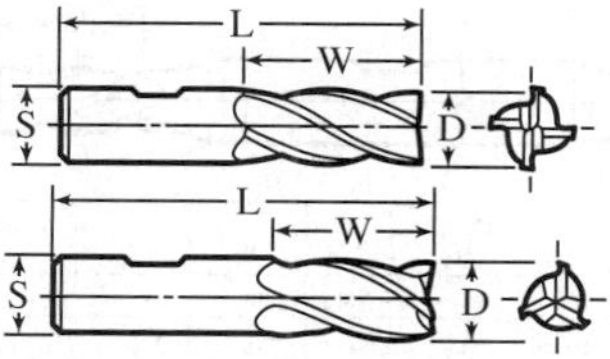


**Four Flute**

| Dia., *D* | Regular Length *S* | Regular Length *W* | Regular Length *L* | Long Length *S* | Long Length *W* | Long Length *L* | Extra-Long Length *S* | Extra-Long Length *W* | Extra-Long Length *L* |
|---|---|---|---|---|---|---|---|---|---|
| 1/8 | 3/8 | 3/8 | 2 5/16 | … | … | … | … | … | … |
| 3/16 | 3/8 | 1/2 | 2 3/8 | … | … | … | … | … | … |
| 1/4 | 3/8 | 5/8 | 2 7/16 | 3/8 | 1 1/4 | 3 3/16 | 3/8 | 1 3/4 | 3 9/16 |
| 5/16 | 3/8 | 3/4 | 2 1/2 | 3/8 | 1 3/8 | 3 1/8 | 3/8 | 2 | 3 3/4 |
| 3/8 | 3/8 | 3/4 | 2 1/2 | 3/8 | 1 1/2 | 3 1/4 | 3/8 | 2 1/2 | 4 1/4 |
| 1/2 | 1/2 | 1 1/4 | 3 1/4 | 1/2 | 2 | 4 | 1/2 | 3 | 5 |
| 5/8 | 5/8 | 1 5/8 | 3 3/4 | 5/8 | 2 1/2 | 4 5/8 | 5/8 | 4 | 6 1/8 |
| 11/16 | 5/8 | 1 5/8 | 3 3/4 | … | … | … | … | … | … |
| 3/4 | 3/4 | 1 5/8 | 3 7/8 | 3/4 | 3 | 5 1/4 | 3/4 | 4 | 6 1/4 |
| 7/8 | 7/8 | 1 7/8 | 4 1/8 | 7/8 | 3 1/2 | 5 3/4 | 7/8 | 5 | 7 1/4 |
| 1 | 1 | 2 | 4 1/2 | 1 | 4 | 6 1/2 | 1 | 6 | 8 1/2 |
| 1 1/8 | 1 | 2 | 4 1/2 | … | … | … | … | … | … |
| 1 1/4 | 1 1/4 | 2 | 4 1/2 | 1 1/4 | 4 | 6 1/2 | 1 1/4 | 6 | 8 1/2 |
| 1 1/2 | 1 1/4 | 2 | 4 1/2 | … | … | … | … | … | … |

**Three Flute**

| Dia., *D* | *S* | *W* | *L* | Dia., *D* | *S* | *W* | *L* |
|---|---|---|---|---|---|---|---|
| Regular Length | | | | Regular Length (*cont.*) | | | |
| 1/8 | 3/8 | 3/8 | 2 5/16 | 1 1/8 | 1 | 2 | 4 1/2 |
| 3/16 | 3/8 | 1/2 | 2 3/8 | 1 1/4 | 1 | 2 | 4 1/2 |
| 1/4 | 3/8 | 5/8 | 2 7/16 | 1 1/2 | 1 | 2 | 4 1/2 |
| 5/16 | 3/8 | 3/4 | 2 1/2 | 1 1/4 | 1 1/4 | 2 | 4 1/2 |
| 3/8 | 3/8 | 3/4 | 2 1/2 | 1 1/2 | 1 1/4 | 2 | 4 1/2 |
| 7/16 | 3/8 | 1 | 2 11/16 | 1 3/4 | 1 1/4 | 2 | 4 1/2 |
| 1/2 | 3/8 | 1 | 2 11/16 | 2 | 1 1/4 | 2 | 4 1/2 |
| 1/2 | 1/2 | 1 1/4 | 3 1/4 | Long Length | | | |
| 9/16 | 1/2 | 1 3/8 | 3 3/8 | | | | |
| 9/16 | 1/2 | 1 3/8 | 3 3/8 | 1/4 | 3/8 | 1 1/4 | 3 11/16 |
| 5/8 | 1/2 | 1 3/8 | 3 3/8 | 5/16 | 3/8 | 1 3/8 | 3 1/8 |
| 3/4 | 1/2 | 1 5/8 | 3 5/8 | 3/8 | 3/8 | 1 1/2 | 3 1/4 |
| 5/8 | 5/8 | 1 5/8 | 3 3/4 | 7/16 | 1/2 | 1 3/4 | 3 3/4 |
| 3/4 | 5/8 | 1 5/8 | 3 3/4 | 1/2 | 1/2 | 2 | 4 |
| 7/8 | 5/8 | 1 7/8 | 4 | 5/8 | 5/8 | 2 1/2 | 4 5/8 |
| 1 | 5/8 | 1 7/8 | 4 | 3/4 | 3/4 | 3 | 5 1/4 |
| 3/4 | 3/4 | 1 5/8 | 3 7/8 | 1 | 1 | 4 | 6 1/2 |
| 7/8 | 3/4 | 1 7/8 | 4 1/8 | 1 1/4 | 1 1/4 | 4 | 6 1/2 |
| 1 | 3/4 | 1 7/8 | 4 1/8 | 1 1/2 | 1 1/4 | 4 | 6 1/2 |
| 1 | 7/8 | 1 7/8 | 4 1/8 | 1 3/4 | 1 1/4 | 4 | 6 1/2 |
| 1 | 1 | 2 | 4 1/2 | 2 | 1 1/4 | 4 | 6 1/2 |

All dimensions are in inches. All cutters are high-speed steel. Right-hand cutters with right-hand helix are standard. Helix angle is greater than 19 degrees but not more than 39 degrees.

*Tolerances:* On *D*, +0.003 inch; on *S*, −0.0001 to −0.0005 inch; on *W*, $\pm\frac{1}{32}$ inch; and on *L*, $\pm\frac{1}{16}$ inch.

The following center-cutting, single-end end mills are available in premium high-speed steel: regular-length, multiple flute, with *D* ranging from 1/8 to 1 1/2 inches; long length, multiple flute, with *D* ranging from 3/8 to 1 1/4 inches; and extra-long-length, multiple flute, with *D* ranging from 3/8 to 1 1/4 inches.

**American National Standard Stub- and Regular-Length, Four-Flute, Medium Helix, Double-End End Mills with Weldon Shanks** *ANSI/ASME B94.19-1997 (R2019)*

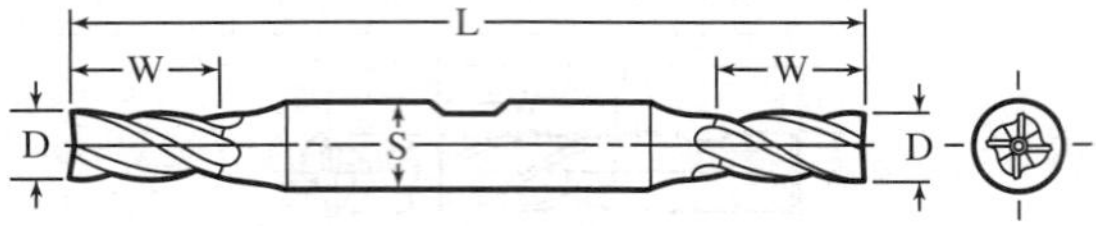


| Dia., *D* | *S* | *W* | *L* | Dia., *D* | *S* | *W* | *L* | Dia., *D* | *S* | *W* | *L* |
|---|---|---|---|---|---|---|---|---|---|---|---|
| Stub Length | | | | | | | | | | | |
| 1/8 | 3/8 | 3/16 | 2 3/4 | 3/16 | 3/8 | 9/32 | 2 3/4 | 1/4 | 3/8 | 3/8 | 2 7/8 |
| 5/32 | 3/8 | 15/64 | 2 3/4 | 7/32 | 3/8 | 21/64 | 2 7/8 | … | … | … | … |
| Regular Length | | | | | | | | | | | |
| 1/8 [a] | 3/8 | 3/8 | 3 1/16 | 11/32 | 3/8 | 3/4 | 3 1/2 | 5/8 [a] | 5/8 | 1 3/8 | 5 |
| 5/32 [a] | 3/8 | 7/16 | 3 1/8 | 3/8 [a] | 3/8 | 3/4 | 3 1/2 | 11/16 | 3/4 | 1 5/8 | 5 5/8 |
| 3/16 [a] | 3/8 | 1/2 | 3 1/4 | 13/32 | 1/2 | 1 | 4 1/8 | 3/4 [a] | 3/4 | 1 5/8 | 5 5/8 |
| 7/32 | 3/8 | 9/16 | 3 1/4 | 7/16 | 1/2 | 1 | 4 1/8 | 13/16 | 7/8 | 1 7/8 | 6 1/8 |
| 1/4 [a] | 3/8 | 5/8 | 3 3/8 | 15/32 | 1/2 | 1 | 4 1/8 | 7/8 | 7/8 | 1 7/8 | 6 1/8 |
| 9/32 | 3/8 | 11/16 | 3 3/8 | 1/2 [a] | 1/2 | 1 | 4 1/8 | 1 | 1 | 1 7/8 | 6 3/8 |
| 5/16 [a] | 3/8 | 3/4 | 3 1/2 | 9/16 | 5/8 | 1 3/8 | 5 | … | … | … | … |

[a] In this size of regular mill, a left-hand cutter with a left-hand helix is also standard.

All dimensions are in inches. All cutters are high-speed steel. Right-hand cutters with right-hand helix are standard. Helix angle is greater than 19 degrees but not more than 39 degrees.

*Tolerances:* On *D*, +0.003 inch (if the shank is the same diameter as the cutting portion, however, then the tolerance on the cutting diameter is −0.0025 inch); on *S*, −0.0001 to −0.0005 inch; on *W*, ±$^1/_{32}$ inch; and on *L*, ±$^1/_{16}$ inch.

**American National Standard Regular-Length, Three- and Four-Flute, Medium Helix, Double-End End Mills with Weldon Shanks** *ANSI/ASME B94.19-1997 (R2019)*

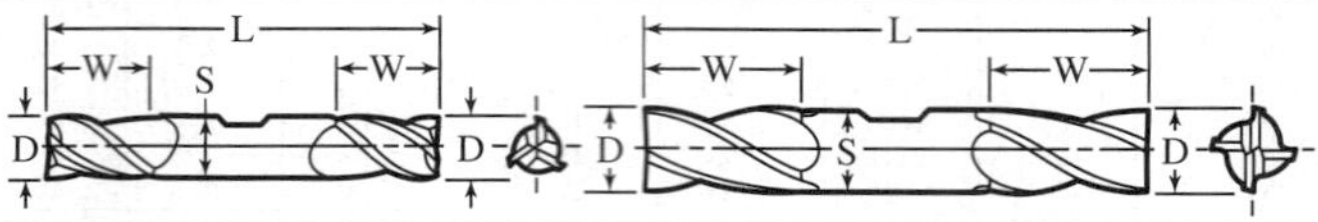


| Dia., *D* | *S* | *W* | *L* | Dia., *D* | *S* | *W* | *L* |
|---|---|---|---|---|---|---|---|
| Three Flute | | | | Four Flute | | | |
| 1/8 | 3/8 | 3/8 | 3 1/16 | 1/8 | 3/8 | 3/8 | 3 1/16 |
| 3/16 | 3/8 | 1/2 | 3 1/4 | 3/16 | 3/8 | 1/2 | 3 1/4 |
| 1/4 | 3/8 | 5/8 | 3 3/8 | 1/4 | 3/8 | 5/8 | 3 3/8 |
| 5/16 | 3/8 | 3/4 | 3 1/2 | 5/16 | 3/8 | 3/4 | 3 1/2 |
| 3/8 | 3/8 | 3/4 | 3 1/2 | 3/8 | 3/8 | 3/4 | 3 1/2 |
| 7/16 | 1/2 | 1 | 4 1/8 | 1/2 | 1/2 | 1 | 4 1/8 |
| 1/2 | 1/2 | 1 | 4 1/8 | 5/8 | 5/8 | 1 3/8 | 5 |
| 9/16 | 5/8 | 1 3/8 | 5 | 3/4 | 3/4 | 1 5/8 | 5 5/8 |
| 5/8 | 5/8 | 1 3/8 | 5 | 7/8 | 7/8 | 1 7/8 | 6 1/8 |
| 3/4 | 3/4 | 1 5/8 | 5 5/8 | 1 | 1 | 1 7/8 | 6 3/8 |
| 1 | 1 | 1 7/8 | 6 3/8 | … | … | … | … |

All dimensions are in inches. All cutters are high-speed steel. Right-hand cutters with right-hand helix are standard. Helix angle is greater than 19 degrees but not more than 39 degrees.

*Tolerances:* On *D*, +0.0015 inch; on *S*, −0.0001 to −0.0005 inch; on *W*, ±$^1/_{32}$ inch; and on *L*, ±$^1/_{16}$ inch.

## American National Standard Plain- and Ball-End, Heavy-Duty, Medium Helix, Single-End End Mills with 2-Inch Diameter Shanks *ANSI/ASME B94.19-1997 (R2019)*

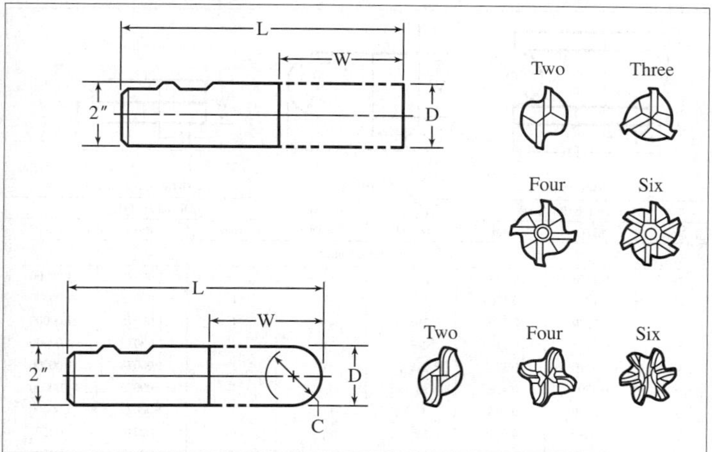


| Dia., *C* and *D* | Plain End | | | Ball End | | |
|---|---|---|---|---|---|---|
| | *W* | *L* | No. of Flutes | *W* | *L* | No. of Flutes |
| 2 | 2 | 5¾ | 2, 4, 6 | … | … | … |
| 2 | 3 | 6¾ | 2, 3 | … | … | … |
| 2 | 4 | 7¾ | 2, 3, 4, 6 | 4 | 7¾ | 6 |
| 2 | … | … | … | 5 | 8¾ | 2, 4 |
| 2 | 6 | 9¾ | 2, 3, 4, 6 | 6 | 9¾ | 6 |
| 2 | 8 | 11¾ | 6 | 8 | 11¾ | 6 |
| 2½ | 4 | 7¾ | 2, 3, 4, 6 | … | … | … |
| 2½ | … | … | … | 5 | 8¾ | 4 |
| 2½ | 6 | 9¾ | 2, 4, 6 | … | … | … |
| 2½ | 8 | 11¾ | 6 | … | … | … |

All dimensions are in inches. All cutters are high-speed steel. Right-hand cutters with right-hand helix are standard. Helix angle is greater than 19 degrees but not more than 39 degrees.

*Tolerances:* On C and D, + 0.005 inch for 2, 3, 4 and 6 flutes: on *W*, ± 1/16 inch; and on *L*, ± 1/16 inch.

## Dimensions of American National Standard Weldon Shanks *ANSI/ASME B94.19-1997 (R2019)*

| Shank | | Flat | | Shank | | Flat | |
|---|---|---|---|---|---|---|---|
| Dia. | Length | *X*[a] | Length[b] | Dia. | Length | *X*[a] | Length[b] |
| ⅜ | 1 9/16 | 0.325 | 0.280 | 1 | 2 9/32 | 0.925 | 0.515 |
| ½ | 1 25/32 | 0.440 | 0.330 | 1¼ | 2 9/32 | 1.156 | 0.515 |
| ⅝ | 1 29/32 | 0.560 | 0.400 | 1½ | 2 11/16 | 1.406 | 0.515 |
| ¾ | 2 1/32 | 0.675 | 0.455 | 2 | 3¼ | 1.900 | 0.700 |
| ⅞ | 2 1/32 | 0.810 | 0.455 | 2½ | 3½ | 2.400 | 0.700 |

[a] *X* is distance from bottom of flat to opposite side of shank.

[b] Minimum.

All dimensions are in inches.

Centerline of flat is at half-length of shank except for 1½-, 2- and 2½-inch shanks where it is 1 3/16, 1 27/32 and 1 15/16 from shank end, respectively.

Tolerance on shank diameter, − 0.0001 to − 0.0005 inch.

## Amerian National Standard Relieved, Concave, Convex, and Corner-Rounding Arbor-Type Cutters *ANSI/ASME B94.19-1997 (R2019)*

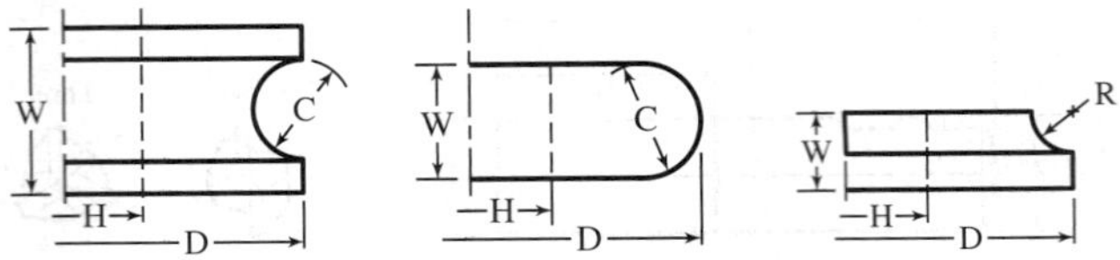


Concave Convex Corner-Rounding

| Diameter C or Radius R | | | Cutter Dia. D[a] | Width W ± .010[b] | Diameter of Hole H | | |
|---|---|---|---|---|---|---|---|
| Nom. | Max. | Min. | | | Nom. | Max. | Min. |
| **Concave Cutters[c]** | | | | | | | |
| 1/8 | 0.1270 | 0.1240 | 2 1/4 | 1/4 | 1 | 1.00075 | 1.00000 |
| 3/16 | 0.1895 | 0.1865 | 2 1/4 | 3/8 | 1 | 1.00075 | 1.00000 |
| 1/4 | 0.2520 | 0.2490 | 2 1/2 | 7/16 | 1 | 1.00075 | 1.00000 |
| 5/16 | 0.3145 | 0.3115 | 2 3/4 | 9/16 | 1 | 1.00075 | 1.00000 |
| 3/8 | 0.3770 | 0.3740 | 2 3/4 | 5/8 | 1 | 1.00075 | 1.00000 |
| 7/16 | 0.4395 | 0.4365 | 3 | 3/4 | 1 | 1.00075 | 1.00000 |
| 1/2 | 0.5040 | 0.4980 | 3 | 13/16 | 1 | 1.00075 | 1.00000 |
| 5/8 | 0.6290 | 0.6230 | 3 1/2 | 1 | 1 1/4 | 1.251 | 1.250 |
| 3/4 | 0.7540 | 0.7480 | 3 3/4 | 1 3/16 | 1 1/4 | 1.251 | 1.250 |
| 7/8 | 0.8790 | 0.8730 | 4 | 1 3/8 | 1 1/4 | 1.251 | 1.250 |
| 1 | 1.0040 | 0.9980 | 4 1/4 | 1 9/16 | 1 1/4 | 1.251 | 1.250 |
| **Convex Cutters[c]** | | | | | | | |
| 1/8 | 0.1270 | 0.1230 | 2 1/4 | 1/8 | 1 | 1.00075 | 1.00000 |
| 3/16 | 0.1895 | 0.1855 | 2 1/4 | 3/16 | 1 | 1.00075 | 1.00000 |
| 1/4 | 0.2520 | 0.2480 | 2 1/2 | 1/4 | 1 | 1.00075 | 1.00000 |
| 5/16 | 0.3145 | 0.3105 | 2 3/4 | 5/16 | 1 | 1.00075 | 1.00000 |
| 3/8 | 0.3770 | 0.3730 | 2 3/4 | 3/8 | 1 | 1.00075 | 1.00000 |
| 7/16 | 0.4395 | 0.4355 | 3 | 7/16 | 1 | 1.00075 | 1.00000 |
| 1/2 | 0.5020 | 0.4980 | 3 | 1/2 | 1 | 1.00075 | 1.00000 |
| 5/8 | 0.6270 | 0.6230 | 3 1/2 | 5/8 | 1 1/4 | 1.251 | 1.250 |
| 3/4 | 0.7520 | 0.7480 | 3 3/4 | 3/4 | 1 1/4 | 1.251 | 1.250 |
| 7/8 | 0.8770 | 0.8730 | 4 | 7/8 | 1 1/4 | 1.251 | 1.250 |
| 1 | 1.0020 | 0.9980 | 4 1/4 | 1 | 1 1/4 | 1.251 | 1.250 |
| **Corner-Rounding Cutters[d]** | | | | | | | |
| 1/8 | 0.1260 | 0.1240 | 2 1/2 | 1/4 | 1 | 1.00075 | 1.00000 |
| 1/4 | 0.2520 | 0.2490 | 3 | 13/32 | 1 | 1.00075 | 1.00000 |
| 3/8 | 0.3770 | 0.3740 | 3 3/4 | 9/16 | 1 1/4 | 1.251 | 1.250 |
| 1/2 | 0.5020 | 0.4990 | 4 1/4 | 3/4 | 1 1/4 | 1.251 | 1.250 |
| 5/8 | 0.6270 | 0.6240 | 4 1/4 | 15/16 | 1 1/4 | 1.251 | 1.250 |

[a] Tolerances on cutter diameter are $+\frac{1}{16}, -\frac{1}{16}$ inch for all sizes.

[b] Tolerance does not apply to convex cutters.

[c] *Size of cutter is designated by specifying diameter C of circular form.*

[d] Size of cutter is designated by specifying radius *R* of circular form.

All dimensions in inches. All cutters are high-speed steel and are form relieved.

Right-hand corner rounding cutters are standard, but left-hand cutter for 1/4-inch size is also standard.

For key and keyway dimensions for these cutters, see page 882.

**American National Standard Roughing and Finishing Gear Milling Cutters for Gears with $14\frac{1}{2}$-Degree Pressure Angles** *ANSI/ASME B94.19-1997 (R2019)*

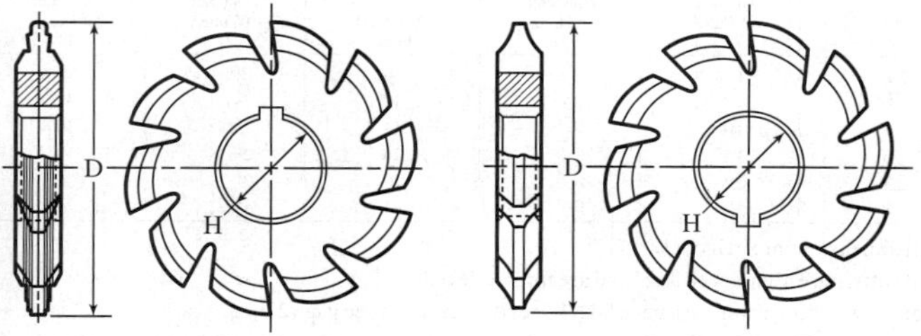


ROUGHING FINISHING

| Diametral Pitch | Dia. of Cutter, *D* | Dia. of Hole, *H* | Diametral Pitch | Dia. of Cutter, *D* | Dia. of Hole, *H* | Diametral Pitch | Dia. of Cutter, *D* | Dia. of Hole, *H* |
|---|---|---|---|---|---|---|---|---|
| Roughing Gear Milling Cutters | | | | | | | | |
| 1 | 8½ | 2 | 3 | 5¼ | 1½ | 5 | 3⅜ | 1 |
| 1¼ | 7¾ | 2 | 3 | 4¾ | 1¼ | 6 | 3⅞ | 1½ |
| 1½ | 7 | 1¾ | 4 | 4¾ | 1¾ | 6 | 3½ | 1¼ |
| 1¾ | 6½ | 1¾ | 4 | 4½ | 1½ | 6 | 3⅛ | 1 |
| 2 | 6½ | 1¾ | 4 | 4¼ | 1¼ | 7 | 3⅜ | 1¼ |
| 2 | 5¾ | 1½ | 4 | 3⅝ | 1 | 7 | 2⅞ | 1 |
| 2½ | 6⅛ | 1¾ | 5 | 4⅜ | 1¾ | 8 | 3¼ | 1¼ |
| 2½ | 5¾ | 1½ | 5 | 4¼ | 1½ | 8 | 2⅞ | 1 |
| 3 | 5⅝ | 1¾ | 5 | 3¾ | 1¼ | … | … | … |
| Finishing Gear Milling Cutters | | | | | | | | |
| 1 | 8½ | 2 | 6 | 3⅞ | 1½ | 14 | 2⅛ | ⅞ |
| 1¼ | 7¾ | 2 | 6 | 3½ | 1¼ | 16 | 2½ | 1 |
| 1½ | 7 | 1¾ | 6 | 3⅛ | 1 | 16 | 2⅛ | ⅞ |
| 1¾ | 6½ | 1¾ | 7 | 3⅝ | 1½ | 18 | 2⅜ | 1 |
| 2 | 6½ | 1¾ | 7 | 3⅜ | 1¼ | 18 | 2 | ⅞ |
| 2 | 5¾ | 1½ | 7 | 2⅞ | 1 | 20 | 2⅜ | 1 |
| 2½ | 6⅛ | 1¾ | 8 | 3½ | 1½ | 20 | 2 | ⅞ |
| 2½ | 5¾ | 1½ | 8 | 3¼ | 1¼ | 22 | 2¼ | 1 |
| 3 | 5⅝ | 1¾ | 8 | 2⅞ | 1 | 22 | 2 | ⅞ |
| 3 | 5¼ | 1½ | 9 | 3⅛ | 1¼ | 24 | 2¼ | 1 |
| 3 | 4¾ | 1¼ | 9 | 2¾ | 1 | 24 | 1¾ | ⅞ |
| 4 | 4¾ | 1¾ | 10 | 3 | 1¼ | 26 | 1¾ | ⅞ |
| 4 | 4½ | 1½ | 10 | 2¾ | 1 | 28 | 1¾ | ⅞ |
| 4 | 4¼ | 1¼ | 10 | 2⅜ | ⅞ | 30 | 1¾ | ⅞ |
| 4 | 3⅝ | 1 | 11 | 2⅝ | 1 | 32 | 1¾ | ⅞ |
| 5 | 4⅜ | 1¾ | 11 | 2⅜ | ⅞ | 36 | 1¾ | ⅞ |
| 5 | 4¼ | 1½ | 12 | 2⅞ | 1¼ | 40 | 1¾ | ⅞ |
| 5 | 3¾ | 1¼ | 12 | 2⅝ | 1 | 48 | 1¾ | ⅞ |
| 5 | 3⅜ | 1 | 12 | 2¼ | ⅞ | … | … | … |
| 6 | 4¼ | 1¾ | 14 | 2½ | 1 | … | … | … |

All dimensions are in inches.

All gear milling cutters are high-speed steel and are form relieved.

For keyway dimensions see page 882.

*Tolerances:* On outside diameter, $+\frac{1}{16}$, $-\frac{1}{16}$ inch; on hole diameter, through 1-inch hole diameter, +0.00075 inch, over 1-inch and through 2-inch hole diameter, +0.0010 inch.

For cutter number relative to numbers of gear teeth, see page 2229. Roughing cutters are made with No. 1 cutter form only.

## American National Standard Gear Milling Cutters for Mitre and Bevel Gears with 14½-Degree Pressure Angles *ANSI/ASME B94.19-1997 (R2019)*

| Diametral Pitch | Diameter of Cutter, *D* | Diameter of Hole, *H* | Diametral Pitch | Diameter of Cutter, *D* | Diameter of Hole, *H* |
|---|---|---|---|---|---|
| 3 | 4 | 1¼ | 10 | 2⅜ | ⅞ |
| 4 | 3⅝ | 1¼ | 12 | 2¼ | ⅞ |
| 5 | 3⅜ | 1¼ | 14 | 2⅛ | ⅞ |
| 6 | 3⅛ | 1 | 16 | 2⅛ | ⅞ |
| 7 | 2⅞ | 1 | 20 | 2 | ⅞ |
| 8 | 2⅞ | 1 | 24 | 1¾ | ⅞ |

All dimensions are in inches.

All cutters are high-speed steel and are form relieved.

For keyway dimensions see page 882. For cutter selection see page 2268.

*Tolerances:* On outside diameter, +1⁄16, −1⁄16 inch; on hole diameter, through 1-inch hole diameter, +0.00075 inch, for 1¼-inch hole diameter, +0.0010 inch.

To select the cutter number for bevel gears with the axis at any angle, double the back cone radius and multiply the result by the diametral pitch. This procedure gives the number of equivalent spur gear teeth and is the basis for selecting the cutter number from the table on page 2231.

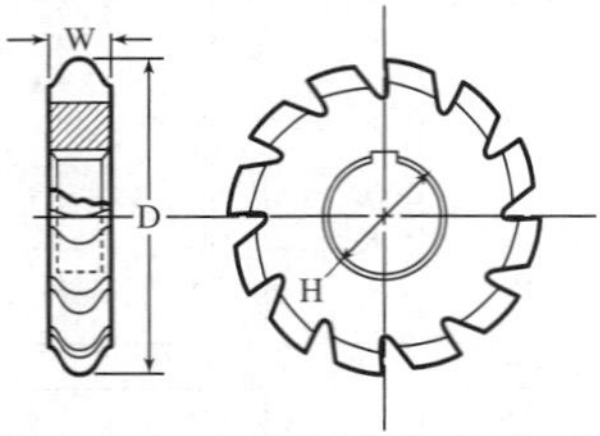


**American National Standard Roller Chain Sprocket Milling Cutters**

## American National Standard Roller Chain Sprocket Milling Cutters *ANSI/ASME B94.19-1997 (R2019)*

| Chain Pitch | Dia. of Roll | No. of Teeth in Sprocket | Dia. of Cutter, *D* | Width of Cutter, *W* | Dia. of Hole, *H* |
|---|---|---|---|---|---|
| ¼ | 0.130 | 6 | 2¾ | 5⁄16 | 1 |
| ¼ | 0.130 | 7-8 | 2¾ | 5⁄16 | 1 |
| ¼ | 0.130 | 9-11 | 2¾ | 5⁄16 | 1 |
| ¼ | 0.130 | 12-17 | 2¾ | 5⁄16 | 1 |
| ¼ | 0.130 | 18-34 | 2¾ | 9⁄32 | 1 |
| ¼ | 0.130 | 35 and over | 2¾ | 9⁄32 | 1 |
| ⅜ | 0.200 | 6 | 2¾ | 15⁄32 | 1 |
| ⅜ | 0.200 | 7-8 | 2¾ | 15⁄32 | 1 |
| ⅜ | 0.200 | 9-11 | 2¾ | 15⁄32 | 1 |
| ⅜ | 0.200 | 12-17 | 2¾ | 7⁄16 | 1 |
| ⅜ | 0.200 | 18-34 | 2¾ | 7⁄16 | 1 |
| ⅜ | 0.200 | 35 and over | 2¾ | 13⁄32 | 1 |
| ½ | 0.313 | 6 | 3 | ¾ | 1 |
| ½ | 0.313 | 7-8 | 3 | ¾ | 1 |
| ½ | 0.313 | 9-11 | 3⅛ | ¾ | 1 |
| ½ | 0.313 | 12-17 | 3⅛ | ¾ | 1 |
| ½ | 0.313 | 18-34 | 3⅛ | 23⁄32 | 1 |
| ½ | 0.313 | 35 and over | 3⅛ | 11⁄16 | 1 |
| ⅝ | 0.400 | 6 | 3⅛ | ¾ | 1 |
| ⅝ | 0.400 | 7-8 | 3⅛ | ¾ | 1 |
| ⅝ | 0.400 | 9-11 | 3¼ | ¾ | 1 |
| ⅝ | 0.400 | 12-17 | 3¼ | ¾ | 1 |
| ⅝ | 0.400 | 18-34 | 3¼ | 23⁄32 | 1 |
| ⅝ | 0.400 | 35 and over | 3¼ | 11⁄16 | 1 |

**American National Standard Roller Chain Sprocket Milling Cutters** *ANSI/ASME B94.19-1997 (R2019) (Continued)*

| Chain Pitch | Dia. of Roll | No. of Teeth in Sprocket | Dia. of Cutter, *D* | Width of Cutter, *W* | Dia. of Hole, *H* |
|---|---|---|---|---|---|
| ¾ | 0.469 | 6 | 3¼ | 29⁄32 | 1 |
| ¾ | 0.469 | 7-8 | 3¼ | 29⁄32 | 1 |
| ¾ | 0.469 | 9-11 | 3⅜ | 29⁄32 | 1 |
| ¾ | 0.469 | 12-17 | 3⅜ | ⅞ | 1 |
| ¾ | 0.469 | 18-34 | 3⅜ | 27⁄32 | 1 |
| ¾ | 0.469 | 35 and over | 3⅜ | 13⁄16 | 1 |
| 1 | 0.625 | 6 | 3⅞ | 1½ | 1¼ |
| 1 | 0.625 | 7-8 | 4 | 1½ | 1¼ |
| 1 | 0.625 | 9-11 | 4⅛ | 1 15⁄32 | 1¼ |
| 1 | 0.625 | 18-34 | 4¼ | 1 13⁄32 | 1¼ |
| 1 | 0.625 | 35 and over | 4¼ | 1 11⁄32 | 1¼ |
| 1¼ | 0.750 | 6 | 4¼ | 1 13⁄16 | 1¼ |
| 1¼ | 0.750 | 7-8 | 4⅜ | 1 13⁄16 | 1¼ |
| 1¼ | 0.750 | 9-11 | 4½ | 1 25⁄32 | 1¼ |
| 1¼ | 0.750 | 18-34 | 4⅝ | 1 11⁄16 | 1¼ |
| 1¼ | 0.750 | 35 and over | 4⅝ | 1⅝ | 1¼ |
| 1½ | 0.875 | 6 | 4⅜ | 1 13⁄16 | 1¼ |
| 1½ | 0.875 | 7-8 | 4½ | 1 13⁄16 | 1¼ |
| 1½ | 0.875 | 9-11 | 4⅝ | 1 25⁄32 | 1¼ |
| 1½ | 0.875 | 12-17 | 4⅝ | 1¾ | 1¼ |
| 1½ | 0.875 | 18-34 | 4¾ | 1 11⁄16 | 1¼ |
| 1½ | 0.875 | 35 and over | 4¾ | 1⅝ | 1¼ |
| 1¾ | 1.000 | 6 | 5 | 2 3⁄32 | 1½ |
| 1¾ | 1.000 | 7-8 | 5⅛ | 2 3⁄32 | 1½ |
| 1¾ | 1.000 | 9-11 | 5¼ | 2 1⁄16 | 1½ |
| 1¾ | 1.000 | 12-17 | 5⅜ | 2 1⁄32 | 1½ |
| 1¾ | 1.000 | 18-34 | 5½ | 1 31⁄32 | 1½ |
| 1¾ | 1.000 | 35 and over | 5½ | 1⅞ | 1½ |
| 2 | 1.125 | 6 | 5⅜ | 2 13⁄32 | 1½ |
| 2 | 1.125 | 7-8 | 5½ | 2 13⁄32 | 1½ |
| 2 | 1.125 | 9-11 | 5⅝ | 2⅜ | 1½ |
| 2 | 1.125 | 12-17 | 5¾ | 2 5⁄16 | 1½ |
| 2 | 1.125 | 18-34 | 5⅞ | 2¼ | 1½ |
| 2 | 1.125 | 35 and over | 5⅞ | 2 5⁄32 | 1½ |
| 2¼ | 1.406 | 6 | 5⅞ | 2 11⁄16 | 1½ |
| 2¼ | 1.406 | 7-8 | 6 | 2 11⁄16 | 1½ |
| 2¼ | 1.406 | 9-11 | 6¼ | 2 21⁄32 | 1½ |
| 2¼ | 1.406 | 12-17 | 6⅜ | 2 19⁄32 | 1½ |
| 2¼ | 1.406 | 18-34 | 6½ | 2 15⁄32 | 1½ |
| 2¼ | 1.406 | 35 and over | 6½ | 2 13⁄32 | 1½ |
| 2½ | 1.563 | 6 | 6⅜ | 3 | 1¾ |
| 2½ | 1.563 | 7-8 | 6⅝ | 3 | 1¾ |
| 2½ | 1.563 | 9-11 | 6¾ | 2 15⁄16 | 1¾ |
| 2½ | 1.563 | 12-17 | 6⅞ | 2 29⁄32 | 1¾ |
| 2½ | 1.563 | 18-34 | 7 | 2¾ | 1¾ |
| 2½ | 1.563 | 35 and over | 7⅛ | 2 11⁄16 | 1¾ |
| 3 | 1.875 | 6 | 7½ | 3 19⁄32 | 2 |
| 3 | 1.875 | 7-8 | 7¾ | 3 19⁄32 | 2 |
| 3 | 1.875 | 9-11 | 7⅞ | 3 17⁄32 | 2 |
| 3 | 1.875 | 12-17 | 8 | 3 15⁄32 | 2 |
| 3 | 1.875 | 18-34 | 8 | 3 11⁄32 | 2 |
| 3 | 1.875 | 35 and over | 8¼ | 3 7⁄32 | 2 |

All dimensions are in inches.

All cutters are high-speed steel and are form relieved.

For keyway dimensions see page 882.

*Tolerances:* Outside diameter, $+{}^1/_{16}$, $-{}^1/_{16}$ inch; hole diameter, through 1-inch diameter, +0.00075 inch, above 1-inch diameter and through 2-inch diameter, +0.0010 inch.

For tooth form, see ANSI/ASME sprocket tooth form table on page 2639.

**American National Standard Keys and Keyways for Milling Cutters and Arbors** *ANSI/ASME B94.19-1997 (R2019)*

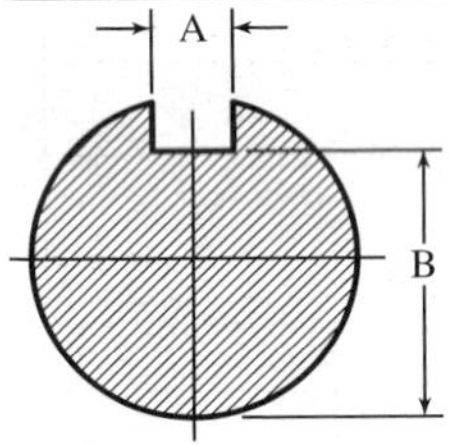


ARBOR AND KEYSEAT

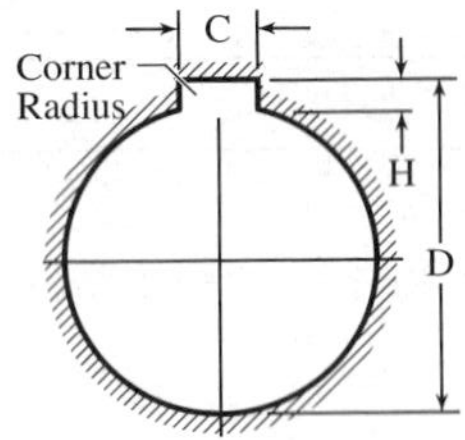


CUTTER HOLE AND KEYWAY

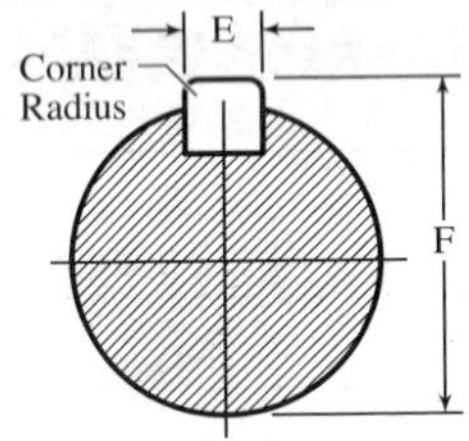


ARBOR AND KEY

| Nom. Arbor and Cutter Hole Dia. | Nom. Size Key (Square) | Arbor and Keyseat | | | | Hole and Keyway | | | | | Arbor and Key | | | |
|---|---|---|---|---|---|---|---|---|---|---|---|---|---|---|
| | | *A* Max. | *A* Min. | *B* Max. | *B* Min. | *C* Max. | *C* Min. | *D*[a] Min. | *H* Nom. | Corner Radius | *E* Max. | *E* Min. | *F* Max. | *F* Min. |
| ½ | 3⁄32 | 0.0947 | 0.0937 | 0.4531 | 0.4481 | 0.106 | 0.099 | 0.5578 | 3⁄64 | 0.020 | 0.0932 | 0.0927 | 0.5468 | 0.5408 |
| ⅝ | ⅛ | 0.1260 | 0.1250 | 0.5625 | 0.5575 | 0.137 | 0.130 | 0.6985 | 1⁄16 | 1⁄32 | 0.1245 | 0.1240 | 0.6875 | 0.6815 |
| ¾ | ⅛ | 0.1260 | 0.1250 | 0.6875 | 0.6825 | 0.137 | 0.130 | 0.8225 | 1⁄16 | 1⁄32 | 0.1245 | 0.1240 | 0.8125 | 0.8065 |
| ⅞ | ⅛ | 0.1260 | 0.1250 | 0.8125 | 0.8075 | 0.137 | 0.130 | 0.9475 | 1⁄16 | 1⁄32 | 0.1245 | 0.1240 | 0.9375 | 0.9315 |
| 1 | ¼ | 0.2510 | 0.2500 | 0.8438 | 0.8388 | 0.262 | 0.255 | 1.1040 | 3⁄32 | 3⁄64 | 0.2495 | 0.2490 | 1.0940 | 1.0880 |
| 1¼ | 5⁄16 | 0.3135 | 0.3125 | 1.0630 | 1.0580 | 0.343 | 0.318 | 1.3850 | ⅛ | 1⁄16 | 0.3120 | 0.3115 | 1.3750 | 1.3690 |
| 1½ | ⅜ | 0.3760 | 0.3750 | 1.2810 | 1.2760 | 0.410 | 0.385 | 1.6660 | 5⁄32 | 1⁄16 | 0.3745 | 0.3740 | 1.6560 | 1.6500 |
| 1¾ | 7⁄16 | 0.4385 | 0.4375 | 1.5000 | 1.4950 | 0.473 | 0.448 | 1.9480 | 3⁄16 | 1⁄16 | 0.4370 | 0.4365 | 1.9380 | 1.9320 |
| 2 | ½ | 0.5010 | 0.5000 | 1.6870 | 1.6820 | 0.535 | 0.510 | 2.1980 | 3⁄16 | 1⁄16 | 0.4995 | 0.4990 | 2.1880 | 2.1820 |
| 2½ | ⅝ | 0.6260 | 0.6250 | 2.0940 | 2.0890 | 0.660 | 0.635 | 2.7330 | 7⁄32 | 1⁄16 | 0.6245 | 0.6240 | 2.7180 | 2.7120 |
| 3 | ¾ | 0.7510 | 0.7500 | 2.5000 | 2.4950 | 0.785 | 0.760 | 3.2650 | ¼ | 3⁄32 | 0.7495 | 0.7490 | 3.2500 | 3.2440 |
| 3½ | ⅞ | 0.8760 | 0.8750 | 3.0000 | 2.9950 | 0.910 | 0.885 | 3.8900 | ⅜ | 3⁄32 | 0.8745 | 0.8740 | 3.8750 | 3.8690 |
| 4 | 1 | 1.0010 | 1.0000 | 3.3750 | 3.3700 | 1.035 | 1.010 | 4.3900 | ⅜ | 3⁄32 | 0.9995 | 0.9990 | 4.3750 | 4.3690 |
| 4½ | 1⅛ | 1.1260 | 1.1250 | 3.8130 | 3.8080 | 1.160 | 1.135 | 4.9530 | 7⁄16 | ⅛ | 1.1245 | 1.1240 | 4.9380 | 4.9320 |
| 5 | 1¼ | 1.2510 | 1.2500 | 4.2500 | 4.2450 | 1.285 | 1.260 | 5.5150 | ½ | ⅛ | 1.2495 | 1.2490 | 5.5000 | 5.4940 |

[a] *D* max. is 0.010 inch larger than *D* min.
All dimensions given in inches.

## American National Standard Woodruff Keyseat Cutters—Shank-Type Straight-Teeth and Arbor-Type Staggered-Teeth *ANSI/ASME B94.19-1997 (R2019)*

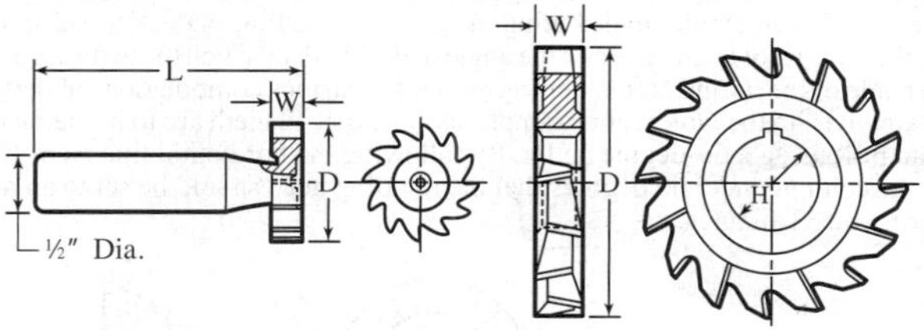


| Shank-type Cutters | | | | | | | | | | | |
|---|---|---|---|---|---|---|---|---|---|---|---|
| Cutter Number | Nom. Dia. of Cutter, *D* | Width of Face, *W* | Length Over-all, *L* | Cutter Number | Nom. Dia. of Cutter, *D* | Width of Face, *W* | Length Over-all, *L* | Cutter Number | Nom. Dia. of Cutter, *D* | Width of Face, *W* | Length Over-all, *L* |
| 202 | 1/4 | 1/16 | 2 1/16 | 506 | 3/4 | 5/32 | 2 5/32 | 809 | 1 1/8 | 1/4 | 2 1/4 |
| 202 1/2 | 5/16 | 1/16 | 2 1/16 | 606 | 3/4 | 3/16 | 2 3/16 | 1009 | 1 1/8 | 5/16 | 2 5/16 |
| 302 1/2 | 5/16 | 3/32 | 2 3/32 | 806 | 3/4 | 1/4 | 2 1/4 | 610 | 1 1/4 | 3/16 | 2 3/16 |
| 203 | 3/8 | 1/16 | 2 1/16 | 507 | 7/8 | 5/32 | 2 5/32 | 710 | 1 1/4 | 7/32 | 2 7/32 |
| 303 | 3/8 | 3/32 | 2 3/32 | 607 | 7/8 | 3/16 | 2 3/16 | 810 | 1 1/4 | 1/4 | 2 1/4 |
| 403 | 3/8 | 1/8 | 2 1/8 | 707 | 7/8 | 7/32 | 2 7/32 | 1010 | 1 1/4 | 5/16 | 2 5/16 |
| 204 | 1/2 | 1/16 | 2 1/16 | 807 | 7/8 | 1/4 | 2 1/4 | 1210 | 1 1/4 | 3/8 | 2 3/8 |
| 304 | 1/2 | 3/32 | 2 3/32 | 608 | 1 | 3/16 | 2 3/16 | 811 | 1 3/8 | 1/4 | 2 1/4 |
| 404 | 1/2 | 1/8 | 2 1/8 | 708 | 1 | 7/32 | 2 7/32 | 1011 | 1 3/8 | 5/16 | 2 5/16 |
| 305 | 5/8 | 3/32 | 2 3/32 | 808 | 1 | 1/4 | 2 1/4 | 1211 | 1 3/8 | 3/8 | 2 3/8 |
| 405 | 5/8 | 1/8 | 2 1/8 | 1008 | 1 | 5/16 | 2 5/16 | 812 | 1 1/2 | 1/4 | 2 1/4 |
| 505 | 5/8 | 5/32 | 2 5/32 | 1208 | 1 | 3/8 | 2 3/8 | 1012 | 1 1/2 | 5/16 | 2 5/16 |
| 605 | 5/8 | 3/16 | 2 3/16 | 609 | 1 1/8 | 3/16 | 2 3/16 | 1212 | 1 1/2 | 3/8 | 2 3/8 |
| 406 | 3/4 | 1/8 | 2 1/8 | 709 | 1 1/8 | 7/32 | 2 7/32 | … | … | … | … |
| Arbor-type Cutters | | | | | | | | | | | |
| Cutter Number | Nom. Dia. of Cutter, *D* | Width of Face, *W* | Dia. of Hole, *H* | Cutter Number | Nom. Dia. of Cutter, *D* | Width of Face, *W* | Dia. of Hole, *H* | Cutter Number | Nom. Dia. of Cutter, *D* | Width of Face, *W* | Dia. of Hole, *H* |
| 617 | 2 1/8 | 3/16 | 3/4 | 1022 | 2 3/4 | 5/16 | 1 | 1628 | 3 1/2 | 1/2 | 1 |
| 817 | 2 1/8 | 1/4 | 3/4 | 1222 | 2 3/4 | 3/8 | 1 | 1828 | 3 1/2 | 9/16 | 1 |
| 1017 | 2 1/8 | 5/16 | 3/4 | 1422 | 2 3/4 | 7/16 | 1 | 2028 | 3 1/2 | 5/8 | 1 |
| 1217 | 2 1/8 | 3/8 | 3/4 | 1622 | 2 3/4 | 1/2 | 1 | 2428 | 3 1/2 | 3/4 | 1 |
| 822 | 2 3/4 | 1/4 | 1 | 1228 | 3 1/2 | 3/8 | 1 | … | … | … | … |

All dimensions are given in inches. All cutters are high-speed steel.

Shank type cutters are standard with right-hand cut and straight teeth. All sizes have 1/2-inch diameter straight shank.

Arbor type cutters have staggered teeth.

For Woodruff key and key-slot dimensions, see pages 2563 through 2565.

*Tolerances:* Face with *W* for shank-type cutters: 1/16- to 5/32-inch face, + 0.0000, −0.0005; 3/16 to 7/32, − 0.0002, − 0.0007; 1/4, −0.0003, −0.0008; 5/16, −0.0004, −0.0009; 3/8, − 0.0005, −0.0010 inch. Face width *W* for arbor-type cutters; 3/16 inch face, −0.0002, −0.0007; 1/4, −0.0003, −0.0008; 5/16, −0.0004, −0.0009; 3/8 and over, −0.0005, −0.0010 inch. Hole size *H:* +0.00075, −0.0000 inch. Diameter *D* for shank type cutters: 1/4- through 3/4-inch diameter, +0.010, +0.015, 7/8 through 1 1/8, +0.012, +0.017; 1 1/4 through 1 1/2, +0.015, +0.020 inch. These tolerances include an allowance for sharpening. For arbor-type cutters, diameter *D* is furnished 1/32 inch larger than listed, and a tolerance of ±0.002 inch applies to the oversize diameter.

**Setting Angles for Milling Straight Teeth of Uniform Land Width in End Mills, Angular Cutters, and Taper Reamers.—**The accompanying tables give setting angles for the dividing head when straight teeth, having a land of uniform width throughout their length, are to be milled using single-angle fluting cutters. These setting angles depend upon three factors: the number of teeth to be cut; the angle of the blank in which the teeth are to be cut; and the angle of the fluting cutter. Setting angles for various combinations of these three factors are given in the tables. For example, assume that 12 teeth are to be cut on the end of an end mill using a 60-degree cutter. By following the horizontal line from 12 teeth, read in the column under 60 degrees that the dividing head should be set to an angle of 70 degrees and 32 minutes.

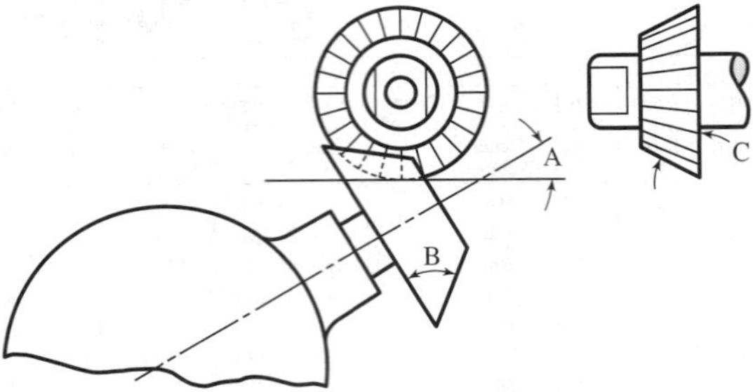


The following formulas, which were used to compile these tables, may be used to calculate the setting-angles for combinations of number of teeth, blank angle, and cutter angle not covered by the tables. In these formulas, $A$ = setting-angle for dividing head, $B$ = angle of blank in which teeth are to be cut, $C$ = angle of fluting cutter, $N$ = number of teeth to be cut, and $D$ and $E$ are angles not shown on the accompanying diagram that are used only to simplify calculations.

$$\tan D = \cos(360°/N) \times \cot B \quad (1)$$

$$\sin E = \tan(360°/N) \times \cot C \times \sin D \quad (2)$$

$$\text{Setting-angle } A = D - E \quad (3)$$

*Example:* Suppose 9 teeth are to be cut in a 35-degree blank using a 55-degree single-angle fluting cutter. Then, $N = 9$, $B = 35°$, and $C = 55°$.

$$\tan D = \cos(360°/9) \times \cot 35° = 0.76604 \times 1.4281 = 1.0940; \text{ and } D = 47°34'$$

$$\sin E = \tan(360°/9) \times \cot 55° \times \sin 47°34' = 0.83910 \times 0.70021 \times 0.73806$$
$$= 0.43365; \text{ and } E = 25°42'$$

$$\text{Setting angle } A = 47°34' - 25°42' = 21°52'$$

For end mills and side mills, the angle of the blank $B$ is 0 degrees, and the following simplified formula may be used to find the setting angle $A$:

$$\cos A = \tan\left(360°/N\right) \times \cot C \quad (4)$$

*Example:* If in the previous example the blank angle was 0 degrees,

$$\cos A = \tan(360°/9) \times \cot 55° = 0.83910 \times 0.70021 = 0.58755$$

and setting-angle $A = 54°1'$.

## Angles of Elevation for Milling Straight Teeth in 0-, 5-, 10-, 15-, 20-, 25-, 30-, and 35-Degree Blanks Using Single-Angle Fluting Cutters

| No. of Teeth | Angle of Fluting Cutter | | | | | | | | | |
|---|---|---|---|---|---|---|---|---|---|---|
| | 90° | 80° | 70° | 60° | 50° | 90° | 80° | 70° | 60° | 50° |
| | 0° Blank (End Mill) | | | | | 5° Blank | | | | |
| 6 | … | 72° 13′ | 50° 55′ | … | … | 80° 4′ | 62° 34′ | 41° 41′ | … | … |
| 8 | … | 79 51 | 68 39 | 54° 44′ | 32° 57′ | 82 57 | 72 52 | 61 47 | 48° 0′ | 25° 40′ |
| 10 | … | 82 38 | 74 40 | 65 12 | 52 26 | 83 50 | 76 31 | 68 35 | 59 11 | 46 4 |
| 12 | … | 84 9 | 77 52 | 70 32 | 61 2 | 84 14 | 78 25 | 72 10 | 64 52 | 55 5 |
| 14 | … | 85 8 | 79 54 | 73 51 | 66 10 | 84 27 | 79 36 | 74 24 | 68 23 | 60 28 |
| 16 | … | 85 49 | 81 20 | 76 10 | 69 40 | 84 35 | 80 25 | 75 57 | 70 49 | 64 7 |
| 18 | … | 86 19 | 82 23 | 77 52 | 72 13 | 84 41 | 81 1 | 77 6 | 72 36 | 66 47 |
| 20 | … | 86 43 | 83 13 | 79 11 | 74 11 | 84 45 | 81 29 | 77 59 | 73 59 | 68 50 |
| 22 | … | 87 2 | 83 52 | 80 14 | 75 44 | 84 47 | 81 50 | 78 40 | 75 4 | 70 26 |
| 24 | … | 87 18 | 84 24 | 81 6 | 77 0 | 84 49 | 82 7 | 79 15 | 75 57 | 71 44 |
| | 10° Blank | | | | | 15° Blank | | | | |
| 6 | 70° 34′ | 53° 50′ | 34° 5′ | … | … | 61° 49′ | 46° 12′ | 28° 4′ | … | … |
| 8 | 76 0 | 66 9 | 55 19 | 41° 56′ | 20° 39′ | 69 15 | 59 46 | 49 21 | 36° 34′ | 17° 34′ |
| 10 | 77 42 | 70 31 | 62 44 | 53 30 | 40 42 | 71 40 | 64 41 | 57 8 | 48 12 | 36 18 |
| 12 | 78 30 | 72 46 | 66 37 | 59 26 | 49 50 | 72 48 | 67 13 | 61 13 | 54 14 | 45 13 |
| 14 | 78 56 | 74 9 | 69 2 | 63 6 | 55 19 | 73 26 | 68 46 | 63 46 | 57 59 | 50 38 |
| 16 | 79 12 | 75 5 | 70 41 | 65 37 | 59 1 | 73 50 | 69 49 | 65 30 | 60 33 | 54 20 |
| 18 | 79 22 | 75 45 | 71 53 | 67 27 | 61 43 | 74 5 | 70 33 | 66 46 | 62 26 | 57 0 |
| 20 | 79 30 | 76 16 | 72 44 | 68 52 | 63 47 | 74 16 | 71 6 | 67 44 | 63 52 | 59 3 |
| 22 | 79 35 | 76 40 | 73 33 | 69 59 | 65 25 | 74 24 | 71 32 | 68 29 | 65 0 | 60 40 |
| 24 | 79 39 | 76 59 | 74 9 | 70 54 | 66 44 | 74 30 | 71 53 | 69 6 | 65 56 | 61 59 |
| | 20° Blank | | | | | 25° Blank | | | | |
| 6 | 53° 57′ | 39° 39′ | 23° 18′ | … | … | 47° 0′ | 34° 6′ | 19° 33′ | … | … |
| 8 | 62 46 | 53 45 | 43 53 | 31° 53′ | 14° 31′ | 56 36 | 48 8 | 38 55 | 27° 47′ | 11° 33′ |
| 10 | 65 47 | 59 4 | 51 50 | 43 18 | 32 1 | 60 2 | 53 40 | 46 47 | 38 43 | 27 47 |
| 12 | 67 12 | 61 49 | 56 2 | 49 18 | 40 40 | 61 42 | 56 33 | 51 2 | 44 38 | 36 10 |
| 14 | 68 0 | 63 29 | 58 39 | 53 4 | 46 0 | 62 38 | 58 19 | 53 41 | 48 20 | 41 22 |
| 16 | 68 30 | 64 36 | 60 26 | 55 39 | 49 38 | 63 13 | 59 29 | 55 29 | 50 53 | 44 57 |
| 18 | 68 50 | 65 24 | 61 44 | 57 32 | 52 17 | 63 37 | 60 19 | 56 48 | 52 46 | 47 34 |
| 20 | 69 3 | 65 59 | 62 43 | 58 58 | 54 18 | 63 53 | 60 56 | 57 47 | 54 11 | 49 33 |
| 22 | 69 14 | 66 28 | 63 30 | 60 7 | 55 55 | 64 5 | 61 25 | 58 34 | 55 19 | 51 9 |
| 24 | 69 21 | 66 49 | 64 7 | 61 2 | 57 12 | 64 14 | 61 47 | 59 12 | 56 13 | 52 26 |
| | 30° Blank | | | | | 35° Blank | | | | |
| 6 | 40° 54′ | 29° 22′ | 16° 32′ | … | … | 35° 32′ | 25° 19′ | 14° 3′ | … | … |
| 8 | 50 46 | 42 55 | 34 24 | 24° 12′ | 10° 14′ | 45 17 | 38 5 | 30 18 | 21° 4′ | 8° 41′ |
| 10 | 54 29 | 48 30 | 42 3 | 34 31 | 24 44 | 49 7 | 43 33 | 37 35 | 30 38 | 21 40 |
| 12 | 56 18 | 51 26 | 46 14 | 40 12 | 32 32 | 51 3 | 46 30 | 41 39 | 36 2 | 28 55 |
| 14 | 57 21 | 53 15 | 48 52 | 43 49 | 37 27 | 52 9 | 48 19 | 44 12 | 39 28 | 33 33 |
| 16 | 58 0 | 54 27 | 50 39 | 46 19 | 40 52 | 52 50 | 49 20 | 45 56 | 41 51 | 36 45 |
| 18 | 58 26 | 55 18 | 51 57 | 48 7 | 43 20 | 53 18 | 50 21 | 47 12 | 43 36 | 39 8 |
| 20 | 58 44 | 55 55 | 52 56 | 49 30 | 45 15 | 53 38 | 50 59 | 48 10 | 44 57 | 40 57 |
| 22 | 58 57 | 56 24 | 53 42 | 50 36 | 46 46 | 53 53 | 51 29 | 48 56 | 46 1 | 42 24 |
| 24 | 59 8 | 56 48 | 54 20 | 51 30 | 48 0 | 54 4 | 51 53 | 49 32 | 46 52 | 43 35 |

## Angles of Elevation for Milling Straight Teeth in 40-, 45-, 50-, 55-, 60-, 65-, 70-, and 75-Degree Blanks Using Single-Angle Fluting Cutters

| No. of Teeth | Angle of Fluting Cutter | | | | | | | | | |
|---|---|---|---|---|---|---|---|---|---|---|
| | 90° | 80° | 70° | 60° | 50° | 90° | 80° | 70° | 60° | 50° |
| | 40° Blank | | | | | 45° Blank | | | | |
| 6 | 30° 48′ | 21° 48′ | 11° 58′ | … | … | 26° 34′ | 18° 43′ | 10° 11′ | … | … |
| 8 | 40 7 | 33 36 | 26 33 | 18° 16′ | 7° 23′ | 35 16 | 29 25 | 23 8 | 15° 48′ | 5° 58′ |
| 10 | 43 57 | 38 51 | 33 32 | 27 3 | 18 55 | 38 58 | 34 21 | 29 24 | 23 40 | 16 10 |
| 12 | 45 54 | 41 43 | 37 14 | 32 3 | 25 33 | 40 54 | 37 5 | 33 0 | 28 18 | 22 13 |
| 14 | 47 3 | 43 29 | 39 41 | 35 19 | 29 51 | 42 1 | 38 46 | 35 17 | 31 18 | 26 9 |
| 16 | 47 45 | 44 39 | 41 21 | 37 33 | 32 50 | 42 44 | 39 54 | 36 52 | 33 24 | 28 57 |
| 18 | 48 14 | 45 29 | 42 34 | 39 13 | 35 5 | 43 13 | 40 42 | 38 1 | 34 56 | 30 1 |
| 20 | 48 35 | 46 7 | 43 30 | 40 30 | 36 47 | 43 34 | 41 18 | 38 53 | 36 8 | 32 37 |
| 22 | 48 50 | 46 36 | 44 13 | 41 30 | 38 8 | 43 49 | 41 46 | 39 34 | 37 5 | 34 53 |
| 24 | 49 1 | 46 58 | 44 48 | 42 19 | 39 15 | 44 0 | 42 7 | 40 7 | 37 50 | 35 55 |
| | 50° Blank | | | | | 55° Blank | | | | |
| 6 | 22° 45′ | 15° 58′ | 8° 38′ | … | … | 19° 17′ | 13° 30′ | 7° 15′ | … | … |
| 8 | 30 41 | 25 31 | 19 59 | 13° 33′ | 5° 20′ | 26 21 | 21 52 | 17 3 | 11° 30′ | 4° 17′ |
| 10 | 34 10 | 30 2 | 25 39 | 20 32 | 14 9 | 29 32 | 25 55 | 22 3 | 17 36 | 11 52 |
| 12 | 36 0 | 32 34 | 28 53 | 24 42 | 19 27 | 31 14 | 28 12 | 24 59 | 21 17 | 16 32 |
| 14 | 37 5 | 34 9 | 31 1 | 27 26 | 22 58 | 32 15 | 29 39 | 26 53 | 23 43 | 19 40 |
| 16 | 37 47 | 35 13 | 32 29 | 29 22 | 25 30 | 32 54 | 30 38 | 28 12 | 25 26 | 21 54 |
| 18 | 38 15 | 35 58 | 33 33 | 30 46 | 27 21 | 33 21 | 31 20 | 29 10 | 26 43 | 23 35 |
| 20 | 38 35 | 36 32 | 34 21 | 31 52 | 28 47 | 33 40 | 31 51 | 29 54 | 27 42 | 24 53 |
| 22 | 38 50 | 36 58 | 34 59 | 32 44 | 29 57 | 33 54 | 32 15 | 30 29 | 28 28 | 25 55 |
| 24 | 39 1 | 37 19 | 35 30 | 33 25 | 30 52 | 34 5 | 32 34 | 30 57 | 29 7 | 26 46 |
| | 60° Blank | | | | | 65° Blank | | | | |
| 6 | 16° 6′ | 11° 12′ | 6° 2′ | … | … | 13° 7′ | 9° 8′ | 4° 53′ | … | … |
| 8 | 22 13 | 18 24 | 14 19 | 9° 37′ | 3° 44′ | 18 15 | 15 6 | 11 42 | 7° 50′ | 3° 1′ |
| 10 | 25 2 | 21 56 | 18 37 | 14 49 | 10 5 | 20 40 | 18 4 | 15 19 | 12 9 | 8 15 |
| 12 | 26 34 | 23 57 | 21 10 | 17 59 | 14 13 | 21 59 | 19 48 | 17 28 | 14 49 | 11 32 |
| 14 | 27 29 | 25 14 | 22 51 | 20 6 | 16 44 | 22 48 | 20 55 | 18 54 | 16 37 | 13 48 |
| 16 | 28 5 | 26 7 | 24 1 | 21 37 | 18 40 | 23 18 | 21 39 | 19 53 | 17 53 | 15 24 |
| 18 | 28 29 | 26 44 | 24 52 | 22 44 | 20 6 | 23 40 | 22 11 | 20 37 | 18 50 | 16 37 |
| 20 | 28 46 | 27 11 | 25 30 | 23 35 | 21 14 | 23 55 | 22 35 | 21 10 | 19 33 | 17 34 |
| 22 | 29 0 | 27 34 | 26 2 | 24 17 | 22 8 | 24 6 | 22 53 | 21 36 | 20 8 | 18 20 |
| 24 | 29 9 | 27 50 | 26 26 | 24 50 | 22 52 | 24 15 | 23 8 | 21 57 | 20 36 | 18 57 |
| | 70° Blank | | | | | 75° Blank | | | | |
| 6 | 10° 18′ | 7° 9′ | 3° 48′ | … | … | 7° 38′ | 5° 19′ | 2° 50′ | … | … |
| 8 | 14 26 | 11 55 | 9 14 | 6° 9′ | 2° 21′ | 10 44 | 8 51 | 6 51 | 4° 34′ | 1° 45′ |
| 10 | 16 25 | 14 21 | 12 8 | 9 37 | 6 30 | 12 14 | 10 40 | 9 1 | 7 8 | 4 49 |
| 12 | 17 30 | 15 45 | 13 53 | 11 45 | 9 8 | 13 4 | 11 45 | 10 21 | 8 45 | 6 47 |
| 14 | 18 9 | 16 38 | 15 1 | 13 11 | 10 55 | 13 34 | 12 26 | 11 13 | 9 50 | 8 7 |
| 16 | 18 35 | 17 15 | 15 50 | 14 13 | 12 13 | 13 54 | 12 54 | 11 50 | 10 37 | 9 7 |
| 18 | 18 53 | 17 42 | 16 26 | 14 59 | 13 13 | 14 8 | 13 14 | 12 17 | 11 12 | 9 51 |
| 20 | 19 6 | 18 1 | 16 53 | 15 35 | 13 59 | 14 18 | 13 29 | 12 38 | 11 39 | 10 27 |
| 22 | 19 15 | 18 16 | 17 15 | 16 3 | 14 35 | 14 25 | 13 41 | 12 53 | 12 0 | 10 54 |
| 24 | 19 22 | 18 29 | 17 33 | 16 25 | 15 5 | 14 31 | 13 50 | 13 7 | 12 18 | 11 18 |

### Angles of Elevation for Milling Straight Teeth in 80- and 85-Degree Blanks Using Single-Angle Fluting Cutters

| No.of Teeth | Angle of Fluting Cutter | | | | | | | | | |
|---|---|---|---|---|---|---|---|---|---|---|
| | 90° | 80° | 70° | 60° | 50° | 90° | 80° | 70° | 60° | 50° |
| | 80° Blank | | | | | 85° Blank | | | | |
| 6 | 5° 2′ | 3° 30′ | 1° 52′ | … | … | 2° 30′ | 1° 44′ | 0° 55′ | … | … |
| 8 | 7 6 | 5 51 | 4 31 | 3° 2′ | 1° 8′ | 3 32 | 2 55 | 2 15 | 1° 29′ | 0° 34′ |
| 10 | 8 7 | 7 5 | 5 59 | 4 44 | 3 11 | 4 3 | 3 32 | 2 59 | 2 21 | 1 35 |
| 12 | 8 41 | 7 48 | 6 52 | 5 48 | 4 29 | 4 20 | 3 53 | 3 25 | 2 53 | 2 15 |
| 14 | 9 2 | 8 16 | 7 28 | 6 32 | 5 24 | 4 30 | 4 7 | 3 43 | 3 15 | 2 42 |
| 16 | 9 15 | 8 35 | 7 51 | 7 3 | 6 3 | 4 37 | 4 17 | 3 56 | 3 30 | 3 1 |
| 18 | 9 24 | 8 48 | 8 10 | 7 26 | 6 33 | 4 42 | 4 24 | 4 5 | 3 43 | 3 16 |
| 20 | 9 31 | 8 58 | 8 24 | 7 44 | 6 56 | 4 46 | 4 29 | 4 12 | 3 52 | 3 28 |
| 22 | 9 36 | 9 6 | 8 35 | 7 59 | 7 15 | 4 48 | 4 33 | 4 18 | 3 59 | 3 37 |
| 24 | 9 40 | 9 13 | 8 43 | 8 11 | 7 30 | 4 50 | 4 36 | 4 22 | 4 5 | 3 45 |

**Spline-Shaft Milling Cutter.—**The most efficient method of forming splines on shafts is by hobbing, but special milling cutters may also be used. Since the cutter forms the space between adjacent splines, it must be made to suit the number of splines and the root diameter of the shaft. The cutter angle $B$ equals 360 degrees divided by the number of splines. The following formulas are for determining the chordal width $C$ at the root of the splines or the chordal width across the concave edge of the cutter. In these formulas, $A$ = angle between center line of spline and a radial line passing through the intersection of the root circle and one side of the spline; $W$ = width of spline; $d$ = root diameter of splined shaft; $C$ = chordal width at root circle between adjacent splines; $N$ = number of splines.

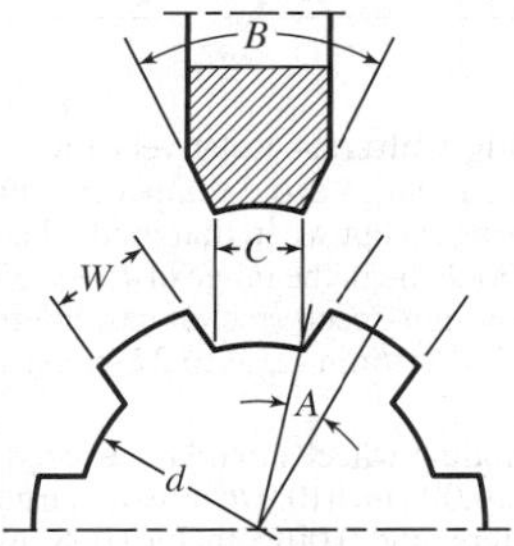


$$\sin A = \frac{W}{d} \qquad C = d \times \sin\left(\frac{180}{N} - A\right)$$

Splines of involute form are often used in preference to the straight-sided type. Dimensions of American National Standard involute splines and hobs are given in the section on *Involute Splines* on page 2333.

## Cutter Grinding

**Wheels for Sharpening Milling Cutters.—**Milling cutters may be sharpened by using either the periphery of a disk wheel or the face of a cup wheel. The latter grinds the lands of the teeth flat, whereas the periphery of a disk wheel leaves the teeth slightly concave back of the cutting edges. The concavity produced by disk wheels reduces the effective clearance angle on the teeth, the effect being more pronounced for wheels of small diameter than for wheels of large diameter. For this reason, large diameter wheels are preferred when sharpening milling cutters with disk type wheels. Irrespective of what type of wheel is used to sharpen a milling cutter, any burrs resulting from grinding should be carefully removed by a hand stoning operation. Stoning also helps to reduce the roughness of grinding marks and improves the quality of the finish produced on the surface being machined. Unless done very carefully, hand stoning may dull the cutting edge. Stoning may be avoided and a sharper cutting edge produced if the wheel rotates toward the cutting edge, which requires that the operator maintain contact between the tool and the rest while the wheel rotation is trying to move the tool away from the rest. Though slightly more difficult, this method will eliminate the burr.

**Specifications of Grinding Wheels for Sharpening Milling Cutters**

| Cutter Material | Operation | Grinding Wheel | | | |
|---|---|---|---|---|---|
| | | Abrasive Material | Grain Size | Grade | Bond |
| Carbon Tool Steel | Roughing<br>Finishing | Aluminum Oxide | 46-60<br>100 | K<br>H | Vitrified<br>Vitrified |
| High-Speed Steel: | | | | | |
| 18-4-1 { | Roughing<br>Finishing | Aluminum Oxide | 60<br>100 | K,H<br>H | Vitrified<br>Vitrified |
| 18-4-2 { | Roughing<br>Finishing | | 80<br>100 | F,G,H<br>H | Vitrified<br>Vitrified |
| Cast Nonferrous Tool Material | Roughing<br>Finishing | Aluminum Oxide | 46<br>100-120 | H,K,L,N<br>H | Vitrified<br>Vitrified |
| Sintered Carbide | Roughing after Brazing | Silicon Carbide | 60 | G | Vitrified |
| | Roughing | Diamond | 100 | [a] | Resinoid |
| | Finishing | Diamond | Up to 500 | [a] | Resinoid |
| Carbon Tool Steel and High-Speed Steel[b] | Roughing<br>Finishing | Cubic Boron Nitride | 80-100<br>100-120 | R,P<br>S,T | Resinoid<br>Resinoid |

[a] Not indicated in diamond wheel markings.

[b] For hardnesses above 56 RC on the Rockwell C scale.

**Wheel Speeds and Feeds for Sharpening Milling Cutters.—**Relatively low cutting speeds should be used when sharpening milling cutters to avoid tempering and heat checking. Dry grinding is recommended in all cases except when diamond wheels are employed. The surface speed of grinding wheels should be in the range of 4500–6500 ft/min (22.8 to 33 m/s) for grinding milling cutters of high-speed steel or cast nonferrous tool material. For sintered carbide cutters, 5000–5500 ft/min (25.4 to 27.9 m/s) should be used.

The maximum stock removed per pass of the grinding wheel should not exceed about 0.0004 inch (0.010 mm) for sintered carbide cutters; 0.003 inch (0.076 mm) for large high-speed steel and cast nonferrous tool material cutters; and 0.0015 inch (0.038 mm) for narrow saws and slotting cutters of high-speed steel or cast nonferrous tool material. The stock removed per pass of the wheel may be increased for backing-off operations such as the grinding of secondary clearance behind the teeth since there is usually a sufficient body of metal to carry off the heat.

**Clearance Angles for Milling Cutter Teeth.—**The clearance angle provided on the cutting edges of milling cutters has an important bearing on cutter performance, cutting efficiency, and cutter life between sharpenings. It is desirable in all cases to use a clearance angle as small as possible so as to leave more metal back of the cutting edges for better heat dissipation and to provide maximum support. Excessive clearance angles not only weaken the cutting edges, but also increase the likelihood of "chatter," which will result in poor finish on the machined surface and reduce the life of the cutter. According to The Cincinnati Milling Machine Co., milling cutters used for general purpose work and having diameters from $\frac{1}{8}$ to 3 inches (3.18–76.2 mm) should have clearance angles from 13 to 5 degrees, respectively, decreasing proportionately as the diameter increases. General purpose cutters over 3 inches (76.2 mm) in diameter should be provided with a clearance angle of 4 to 5 degrees. The land width is usually $\frac{1}{64}$, $\frac{1}{32}$, and $\frac{1}{16}$ inch (0.4, 0.8, and 1.6 mm), respectively, for small, medium, and large cutters.

The primary clearance or relief angle for best results varies according to the material being milled, as follows: low-carbon, high-carbon, and alloy steels, 3 to 5 degrees; cast iron and medium and hard bronze, 4 to 7 degrees; brass, soft bronze, aluminum, magnesium, plastics, etc., 10 to 12 degrees. When milling cutters are resharpened, it is customary to grind a secondary clearance angle of 3 to 5 degrees behind the primary clearance angle to reduce the land width to its original value and thus avoid interference with the surface to be milled. The following is a general formula for plain milling cutters, face mills, and form relieved cutters, which gives the clearance angle $C$, in degrees, necessitated by the feed per revolution $F$, in inches; the width of land $L$, in inches; the depth of cut d, in inches (mm); the cutter diameter $D$, in inches; and the Brinell Hardness Number $B$ of the work being cut:

$$C = \frac{45860}{DB}\left(1.5L + \frac{F}{\pi D}\sqrt{d(D-d)}\right)$$

**Rake Angles for Milling Cutters.—**In peripheral milling cutters, the rake angle is generally defined as the angle in degrees that the tooth face deviates from a radial line to the cutting edge. In face milling cutters, the teeth are inclined with respect to both the radial and axial lines. These angles are called *radial* and *axial* rake, respectively. The radial and axial rake angles may be positive, zero, or negative.

Positive rake angles should be used whenever possible for all types of high-speed steel milling cutters. For sintered carbide-tipped cutters, zero and negative rake angles are frequently employed to provide more material back of the cutting edge to resist shock loads.

*Rake Angles for High-Speed Steel Cutters:* Positive rake angles of 10 to 15 degrees are satisfactory for milling steels of various compositions with plain milling cutters. For softer materials, such as magnesium and aluminum alloys, the rake angle may be 25 degrees or more. Metal-slitting saws for cutting alloy steel usually have rake angles from 5 to 10 degrees, whereas zero and sometimes negative rake angles are used for saws to cut copper and other soft nonferrous metals to reduce the tendency to "hog in." Form relieved cutters usually have rake angles of 0, 5, or 10 degrees. Commercial face milling cutters usually have 10 degrees positive radial and axial rake angles for general use in milling cast iron, forged and alloy steel, brass, and bronze. For milling castings and forgings of magnesium and free-cutting aluminum and their alloys, the rake angles may be increased to 25 degrees positive or more, depending on the operating conditions; a smaller rake angle is used for abrasive or difficult to machine aluminum alloys.

*Cast Nonferrous Tool Material Milling Cutters:* Positive rake angles are generally provided on milling cutters using cast nonferrous tool materials, although negative rake angles may be used advantageously for some operations such as those where shock loads are encountered or where it is necessary to eliminate vibration when milling thin sections.

*Sintered Carbide Milling Cutters:* Peripheral milling cutters, such as slab mills, slotting cutters, saws, etc., tipped with sintered carbide, generally have negative radial rake angles of 5 degrees for soft low-carbon steel and 10 degrees or more for alloy steels. Positive axial rake angles of 5 and 10 degrees, respectively, may be provided, and for slotting saws and cutters, 0 degree axial rake may be used. On soft materials, such as free-cutting aluminum alloys, positive rake angles of 10 to so degrees are used. For milling abrasive or difficult to machine aluminum alloys, small positive or even negative rake angles are used.

**Eccentric-Type Radial Relief.—**When the radial relief angles on peripheral teeth of milling cutters are ground with a disc-type grinding wheel in the conventional manner, the ground surfaces on the lands are slightly concave, conforming approximately to the radius of the wheel. A flat land is produced when the radial relief angle is ground with a cup wheel. Another entirely different method of grinding the radial angle is by the eccentric method, which produces a slightly convex surface on the land. If the radial relief angle at the cutting edge is equal for all of the three types of land mentioned, it will be found that the land with the eccentric relief will drop away from the cutting edge a somewhat greater distance for a given distance around the land than will the others. This is evident from a study of Table 1, *Indicator Drops for Checking the Radial Relief Angle on Peripheral Teeth*. This feature is an advantage of the eccentric-type relief, which also produces an excellent finish.

**Table 1. Indicator Drops for Checking the Radial Relief Angle on Peripheral Teeth**

| Cutter Diameter, Inch | Rec. Range of Radial Relief Angles, Degrees | Checking Distance, Inch | Indicator Drops, Inches | | | | Rec. Max. Primary Land Width, Inch |
|---|---|---|---|---|---|---|---|
| | | | For Flat and Concave Relief | | For Eccentric Relief | | |
| | | | Min. | Max. | Min. | Max. | |
| 1⁄16 | 20-25 | .005 | .0014 | .0019 | .0020 | .0026 | .007 |
| 3⁄32 | 16-20 | .005 | .0012 | .0015 | .0015 | .0019 | .007 |
| 1⁄8 | 15-19 | .010 | .0018 | .0026 | .0028 | .0037 | .015 |
| 5⁄32 | 13-17 | .010 | .0017 | .0024 | .0024 | .0032 | .015 |
| 3⁄16 | 12-16 | .010 | .0016 | .0023 | .0022 | .0030 | .015 |
| 7⁄32 | 11-15 | .010 | .0015 | .0022 | .0020 | .0028 | .015 |
| 1⁄4 | 10-14 | .015 | .0017 | .0028 | .0027 | .0039 | .020 |
| 9⁄32 | 10-14 | .015 | .0018 | .0029 | .0027 | .0039 | .020 |
| 5⁄16 | 10-13 | .015 | .0019 | .0027 | .0027 | .0035 | .020 |
| 11⁄32 | 10-13 | .015 | .0020 | .0028 | .0027 | .0035 | .020 |
| 3⁄8 | 10-13 | .015 | .0020 | .0029 | .0027 | .0035 | .020 |
| 13⁄32 | 9-12 | .020 | .0022 | .0032 | .0032 | .0044 | .025 |
| 7⁄16 | 9-12 | .020 | .0022 | .0033 | .0032 | .0043 | .025 |
| 15⁄32 | 9-12 | .020 | .0023 | .0034 | .0032 | .0043 | .025 |
| 1⁄2 | 9-12 | .020 | .0024 | .0034 | .0032 | .0043 | .025 |
| 9⁄16 | 9-12 | .020 | .0024 | .0035 | .0032 | .0043 | .025 |
| 5⁄8 | 8-11 | .020 | .0022 | .0032 | .0028 | .0039 | .025 |
| 11⁄16 | 8-11 | .030 | .0029 | .0045 | .0043 | .0059 | .035 |
| 3⁄4 | 8-11 | .030 | .0030 | .0046 | .0043 | .0059 | .035 |
| 13⁄16 | 8-11 | .030 | .0031 | .0047 | .0043 | .0059 | .035 |
| 7⁄8 | 8-11 | .030 | .0032 | .0048 | .0043 | .0059 | .035 |
| 15⁄16 | 7-10 | .030 | .0027 | .0043 | .0037 | .0054 | .035 |
| 1 | 7-10 | .030 | .0028 | .0044 | .0037 | .0054 | .035 |
| 1 1⁄8 | 7-10 | .030 | .0029 | .0045 | .0037 | .0053 | .035 |
| 1 1⁄4 | 6-9 | .030 | .0024 | .0040 | .0032 | .0048 | .035 |
| 1 3⁄8 | 6-9 | .030 | .0025 | .0041 | .0032 | .0048 | .035 |
| 1 1⁄2 | 6-9 | .030 | .0026 | .0041 | .0032 | .0048 | .035 |
| 1 5⁄8 | 6-9 | .030 | .0026 | .0042 | .0032 | .0048 | .035 |
| 1 3⁄4 | 6-9 | .030 | .0026 | .0042 | .0032 | .0048 | .035 |
| 1 7⁄8 | 6-9 | .030 | .0027 | .0043 | .0032 | .0048 | .035 |
| 2 | 6-9 | .030 | .0027 | .0043 | .0032 | .0048 | .035 |
| 2 1⁄4 | 5-8 | .030 | .0022 | .0038 | .0026 | .0042 | .040 |
| 2 1⁄2 | 5-8 | .030 | .0023 | .0039 | .0026 | .0042 | .040 |
| 2 3⁄4 | 5-8 | .030 | .0023 | .0039 | .0026 | .0042 | .040 |
| 3 | 5-8 | .030 | .0023 | .0039 | .0026 | .0042 | .040 |
| 3 1⁄2 | 5-8 | .030 | .0024 | .0040 | .0026 | .0042 | .047 |
| 4 | 5-8 | .030 | .0024 | .0040 | .0026 | .0042 | .047 |
| 5 | 4-7 | .030 | .0019 | .0035 | .0021 | .0037 | .047 |
| 6 | 4-7 | .030 | .0019 | .0035 | .0021 | .0037 | .047 |
| 7 | 4-7 | .030 | .0020 | .0036 | .0021 | .0037 | .060 |
| 8 | 4-7 | .030 | .0020 | .0036 | .0021 | .0037 | .060 |
| 10 | 4-7 | .030 | .0020 | .0036 | .0021 | .0037 | .060 |
| 12 | 4-7 | .030 | .0020 | .0036 | .0021 | .0037 | .060 |

The setup for grinding an eccentric relief is shown in Fig. 1. In this setup, the point of contact between the cutter and the tooth rest must be in the same plane as the centers, or axes, of the grinding wheel and the cutter. A wide face is used on the grinding wheel, which is trued and dressed at an angle with respect to the axis of the cutter. An alternate method is to tilt the wheel at this angle. Then as the cutter is traversed and rotated past the grinding wheel while in contact with the tooth rest, an eccentric relief will be generated by the angular face of the wheel. This type of relief can only be ground on the peripheral teeth on milling cutters having helical flutes because the combination of the angular wheel face and the twisting motion of the cutter is required to generate the eccentric relief. Therefore, an eccentric relief cannot be ground on the peripheral teeth of straight-fluted cutters.

Table 2 is a table of wheel angles for grinding an eccentric relief for different combinations of relief angles and helix angles. When angles are required that cannot be found in this table, the wheel angle, *W*, can be calculated by using the following formula, in which *R* is the radial relief angle and *H* is the helix angle of the flutes on the cutter.

$$\tan W = \tan R \times \tan H$$

**Table 2. Grinding Wheel Angles for Grinding Eccentric-Type Radial Relief Angle**

| Radial Relief Angle, *R*, Degrees | Helix Angle of Cutter Flutes, *H*, Degrees | | | | | | | |
|---|---|---|---|---|---|---|---|---|
| | 12 | 18 | 20 | 30 | 40 | 45 | 50 | 52 |
| | Wheel Angle, *W*, Degrees | | | | | | | |
| 1 | 0°13′ | 0°19′ | 0°22′ | 0°35′ | 0°50′ | 1°00′ | 1°12′ | 1°17′ |
| 2 | 0°26′ | 0°39′ | 0°44′ | 1°09′ | 1°41′ | 2°00′ | 2°23′ | 2°34′ |
| 3 | 0°38′ | 0°59′ | 1°06′ | 1°44′ | 2°31′ | 3°00′ | 3°34′ | 3°50′ |
| 4 | 0°51′ | 1°18′ | 1°27′ | 2°19′ | 3°21′ | 4°00′ | 4°46′ | 5°07′ |
| 5 | 1°04′ | 1°38′ | 1°49′ | 2°53′ | 4°12′ | 5°00′ | 5°57′ | 6°23′ |
| 6 | 1°17′ | 1°57′ | 2°11′ | 3°28′ | 5°02′ | 6°00′ | 7°08′ | 7°40′ |
| 7 | 1°30′ | 2°17′ | 2°34′ | 4°03′ | 5°53′ | 7°00′ | 8°19′ | 8°56′ |
| 8 | 1°43′ | 2°37′ | 2°56′ | 4°38′ | 6°44′ | 8°00′ | 9°30′ | 10°12′ |
| 9 | 1°56′ | 2°57′ | 3°18′ | 5°13′ | 7°34′ | 9°00′ | 10°41′ | 11°28′ |
| 10 | 2°09′ | 3°17′ | 3°40′ | 5°49′ | 8°25′ | 10°00′ | 11°52′ | 12°43′ |
| 11 | 2°22′ | 3°37′ | 4°03′ | 6°24′ | 9°16′ | 11°00′ | 13°03′ | 13°58′ |
| 12 | 2°35′ | 3°57′ | 4°25′ | 7°00′ | 10°07′ | 12°00′ | 14°13′ | 15°13′ |
| 13 | 2°49′ | 4°17′ | 4°48′ | 7°36′ | 10°58′ | 13°00′ | 15°23′ | 16°28′ |
| 14 | 3°02′ | 4°38′ | 5°11′ | 8°11′ | 11°49′ | 14°00′ | 16°33′ | 17°42′ |
| 15 | 3°16′ | 4°59′ | 5°34′ | 8°48′ | 12°40′ | 15°00′ | 17°43′ | 18°56′ |
| 16 | 3°29′ | 5°19′ | 5°57′ | 9°24′ | 13°32′ | 16°00′ | 18°52′ | 20°09′ |
| 17 | 3°43′ | 5°40′ | 6°21′ | 10°01′ | 14°23′ | 17°00′ | 20°01′ | 21°22′ |
| 18 | 3°57′ | 6°02′ | 6°45′ | 10°37′ | 15°15′ | 18°00′ | 21°10′ | 22°35′ |
| 19 | 4°11′ | 6°23′ | 7°09′ | 11°15′ | 16°07′ | 19°00′ | 22°19′ | 23°47′ |
| 20 | 4°25′ | 6°45′ | 7°33′ | 11°52′ | 16°59′ | 20°00′ | 23°27′ | 24°59′ |
| 21 | 4°40′ | 7°07′ | 7°57′ | 12°30′ | 17°51′ | 21°00′ | 24°35′ | 26°10′ |
| 22 | 4°55′ | 7°29′ | 8°22′ | 13°08′ | 18°44′ | 22°00′ | 25°43′ | 27°21′ |
| 23 | 5°09′ | 7°51′ | 8°47′ | 13°46′ | 19°36′ | 23°00′ | 26°50′ | 28°31′ |
| 24 | 5°24′ | 8°14′ | 9°12′ | 14°25′ | 20°29′ | 24°00′ | 27°57′ | 29°41′ |
| 25 | 5°40′ | 8°37′ | 9°38′ | 15°04′ | 21°22′ | 25°00′ | 29°04′ | 30°50′ |

**Indicator Drop Method of Checking Relief and Rake Angles.—**The most convenient and inexpensive method of checking the relief and rake angles on milling cutters is by the indicator drop method. Three tables, Table 1, Table 3 and Table 4, of indicator drops are provided in this section for checking radial relief angles on the peripheral teeth, relief angles on side and end teeth, and rake angles on the tooth faces.

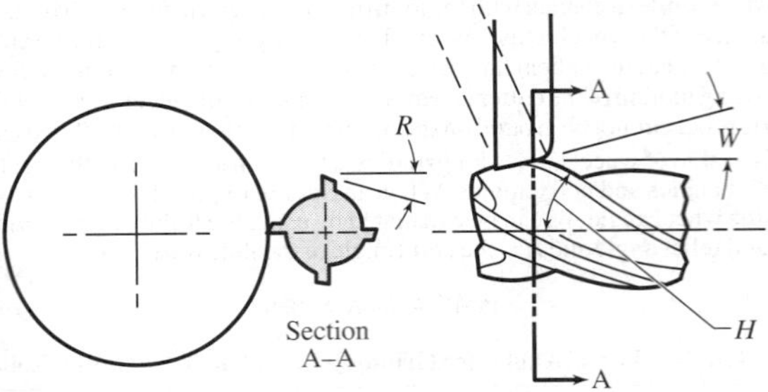


Fig. 1. Setup for Grinding Eccentric-Type Radial Relief Angle

**Table 3. Indicator Drops for Checking Relief Angles on Side Teeth and End Teeth**

| Checking Distance, Inch | Given Relief Angle | | | | | | | | |
|---|---|---|---|---|---|---|---|---|---|
| | 1° | 2° | 3° | 4° | 5° | 6° | 7° | 8° | 9° |
| | Indicator Drop, inch | | | | | | | | |
| .005 | .00009 | .00017 | .00026 | .00035 | .0004 | .0005 | .0006 | .0007 | .0008 |
| .010 | .00017 | .00035 | .00052 | .0007 | .0009 | .0011 | .0012 | .0014 | .0016 |
| .015 | .00026 | .0005 | .00079 | .0010 | .0013 | .0016 | .0018 | .0021 | .0024 |
| .031 | .00054 | .0011 | .0016 | .0022 | .0027 | .0033 | .0038 | .0044 | .0049 |
| .047 | .00082 | .0016 | .0025 | .0033 | .0041 | .0049 | .0058 | .0066 | .0074 |
| .062 | .00108 | .0022 | .0032 | .0043 | .0054 | .0065 | .0076 | .0087 | .0098 |

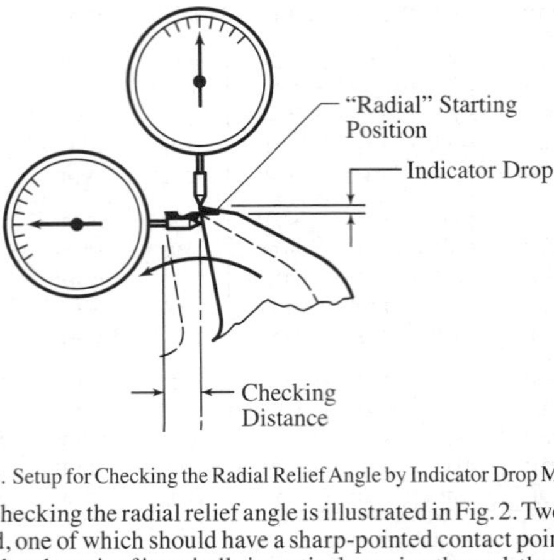


Fig. 2. Setup for Checking the Radial Relief Angle by Indicator Drop Method

The setup for checking the radial relief angle is illustrated in Fig. 2. Two dial test indicators are required, one of which should have a sharp-pointed contact point. This indicator is positioned so that the axis of its spindle is vertical, passing through the axis of the cutter. The cutter may be held by its shank in the spindle of a tool and cutter grinder workhead, or between centers while mounted on a mandrel. The cutter is rotated to the position where the vertical indicator contacts a cutting edge. The second indicator is positioned with its spindle axis horizontal and with the contact point touching the tool face just below the cutting edge. With both indicators adjusted to read zero, the cutter is rotated a distance equal to the checking distance, as determined by the reading on the second indicator. Then the indicator drop is read on the vertical indicator and checked against the values in the tables. The indicator drops for radial relief angles ground by a disc-type grinding wheel and those ground with a cup wheel are so nearly equal that the values are listed together; values for the eccentric-type relief are listed separately, since they are larger. A similar procedure is used to check the relief angles on the side and end teeth of milling cutters; however, only one indicator is used. Also, instead of rotating the cutter, the indicator or the cutter must be moved a distance equal to the checking distance in a straight line.

**Table 4. Indicator Drops for Checking Rake Angles on Milling Cutter Face**

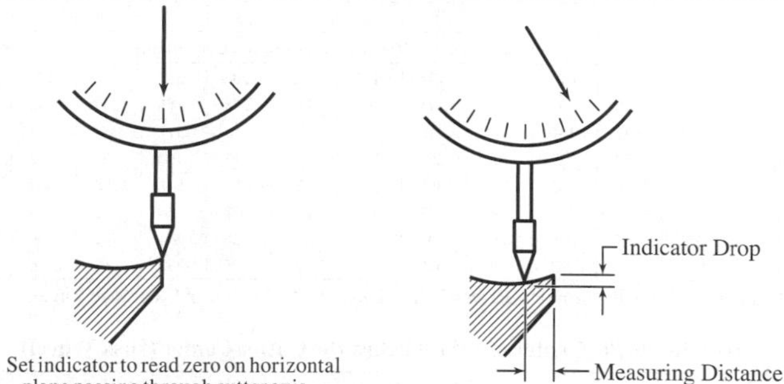


Set indicator to read zero on horizontal plane passing through cutter axis. Zero cutting edge against indicator.

Move cutter or indicator measuring distance.

| Rate Angle, Deg. | Measuring Distance, inch | | | | Rate Angle, Deg. | Measuring Distance, inch | | | |
|---|---|---|---|---|---|---|---|---|---|
| | .031 | .062 | .094 | .125 | | .031 | .062 | .094 | .125 |
| | Indicator Drop, inch | | | | | Indicator Drop, inch | | | |
| 1 | .0005 | .0011 | .0016 | .0022 | 11 | .0060 | .0121 | .0183 | .0243 |
| 2 | .0011 | .0022 | .0033 | .0044 | 12 | .0066 | .0132 | .0200 | .0266 |
| 3 | .0016 | .0032 | .0049 | .0066 | 13 | .0072 | .0143 | .0217 | .0289 |
| 4 | .0022 | .0043 | .0066 | .0087 | 14 | .0077 | .0155 | .0234 | .0312 |
| 5 | .0027 | .0054 | .0082 | .0109 | 15 | .0083 | .0166 | .0252 | .0335 |
| 6 | .0033 | .0065 | .0099 | .0131 | 16 | .0089 | .0178 | .0270 | .0358 |
| 7 | .0038 | .0076 | .0115 | .0153 | 17 | .0095 | .0190 | .0287 | .0382 |
| 8 | .0044 | .0087 | .0132 | .0176 | 18 | .0101 | .0201 | .0305 | .0406 |
| 9 | .0049 | .0098 | .0149 | .0198 | 19 | .0107 | .0213 | .0324 | .0430 |
| 10 | .0055 | .0109 | .0166 | .0220 | 20 | .0113 | .0226 | .0342 | .0455 |

**Relieving Attachments.—**A relieving attachment is a device applied to lathes (especially those used in tool-rooms) for imparting a reciprocating motion to the tool-slide and tool, in order to provide relief or clearance for the cutting edges of milling cutters, taps, hobs, etc. For example, in making a milling cutter of the formed type, such as is used for cutting gears, it is essential to provide clearance for the teeth and so form them that they may be ground repeatedly without changing the contour or shape of the cutting edge. This may be accomplished by using a relieving attachment. The tool for "backing off" or giving clearance to the teeth corresponds to the shape required, and it is given a certain amount of reciprocating movement, so that it forms a surface back of each cutting edge, which is of uniform cross section on a radial plane but eccentric to the axis of the cutter sufficiently to provide the necessary clearance for the cutting edges.

**Various Set-ups Used in Grinding the Clearance Angle on Milling Cutter Teeth**

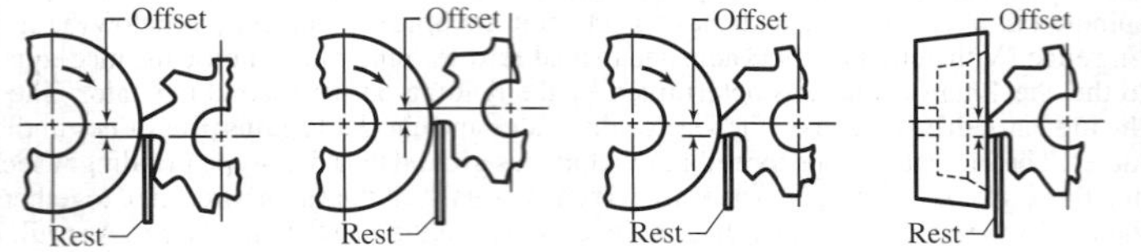

Wheel Above Center — Wheel Below Center — In-Line Centers — Cup Wheel

**Distance to Set Center of Wheel Above the Cutter Center (Disk Wheel)**

| Dia. of Wheel, Inches | Desired Clearance Angle, Degrees | | | | | | | | | | | |
|---|---|---|---|---|---|---|---|---|---|---|---|---|
| | 1 | 2 | 3 | 4 | 5 | 6 | 7 | 8 | 9 | 10 | 11 | 12 |
| | [a]Distance to Offset Wheel Center Above Cutter Center, Inches | | | | | | | | | | | |
| 3 | .026 | .052 | .079 | .105 | .131 | .157 | .183 | .209 | .235 | .260 | .286 | .312 |
| 4 | .035 | .070 | .105 | .140 | .174 | .209 | .244 | .278 | .313 | .347 | .382 | .416 |
| 5 | .044 | .087 | .131 | .174 | .218 | .261 | .305 | .348 | .391 | .434 | .477 | .520 |
| 6 | .052 | .105 | .157 | .209 | .261 | .314 | .366 | .417 | .469 | .521 | .572 | .624 |
| 7 | .061 | .122 | .183 | .244 | .305 | .366 | .427 | .487 | .547 | .608 | .668 | .728 |
| 8 | .070 | .140 | .209 | .279 | .349 | .418 | .488 | .557 | .626 | .695 | .763 | .832 |
| 9 | .079 | .157 | .236 | .314 | .392 | .470 | .548 | .626 | .704 | .781 | .859 | .936 |
| 10 | .087 | .175 | .262 | .349 | .436 | .523 | .609 | .696 | .782 | .868 | .954 | 1.040 |

[a] Calculated from the formula: Offset = Wheel Diameter × ½ × Sine of Clearance Angle.

**Distance to Set Center of Wheel Below the Cutter Center (Disk Wheel)**

| Dia. of Cutter, Inches | Desired Clearance Angle, Degrees | | | | | | | | | | | |
|---|---|---|---|---|---|---|---|---|---|---|---|---|
| | 1 | 2 | 3 | 4 | 5 | 6 | 7 | 8 | 9 | 10 | 11 | 12 |
| | [a]Distance to Offset Wheel Center Below Cutter Center, Inches | | | | | | | | | | | |
| 2 | .017 | .035 | .052 | .070 | .087 | .105 | .122 | .139 | .156 | .174 | .191 | .208 |
| 3 | .026 | .052 | .079 | .105 | .131 | .157 | .183 | .209 | .235 | .260 | .286 | .312 |
| 4 | .035 | .070 | .105 | .140 | .174 | .209 | .244 | .278 | .313 | .347 | .382 | .416 |
| 5 | .044 | .087 | .131 | .174 | .218 | .261 | .305 | .348 | .391 | .434 | .477 | .520 |
| 6 | .052 | .105 | .157 | .209 | .261 | .314 | .366 | .417 | .469 | .521 | .572 | .624 |
| 7 | .061 | .122 | .183 | .244 | .305 | .366 | .427 | .487 | .547 | .608 | .668 | .728 |
| 8 | .070 | .140 | .209 | .279 | .349 | .418 | .488 | .557 | .626 | .695 | .763 | .832 |
| 9 | .079 | .157 | .236 | .314 | .392 | .470 | .548 | .626 | .704 | .781 | .859 | .936 |
| 10 | .087 | .175 | .262 | .349 | .436 | .523 | .609 | .696 | .782 | .868 | .954 | 1.040 |

[a] Calculated from the formula: Offset = Cutter Diameter × ½ × Sine of Clearance Angle.

**Distance to Set Tooth Rest Below Center Line of Wheel and Cutter.—**When the clearance angle is ground with a disc-type wheel by keeping the center line of the wheel in line with the center line of the cutter, the tooth rest should be lowered by an amount given by the following formula:

$$\text{Offset} = \frac{\text{Wheel Dia.} \times \text{Cutter Dia.} \times \text{Sine of One-half the Clearance Angle}}{\text{Wheel Dia.} + \text{Cutter Dia.}}$$

**Distance to Set Tooth Rest Below Cutter Center When Cup Wheel is Used.—**When the clearance is ground with a cup wheel, the tooth rest is set below the center of the cutter the same amount as given in the table for *Distance to Set Center of Wheel Below the Cutter Center (Disk Wheel).*

**Counter Milling.**—Changing the direction of a linear milling operation by a specific angle requires a linear offset before changing the angle of cut. This compensates for the radius of the milling cutters, as illustrated in Fig. 1a and Fig. 1b.

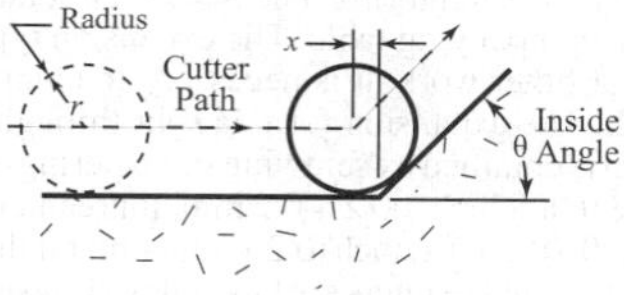

Fig. 1a. Inside Milling

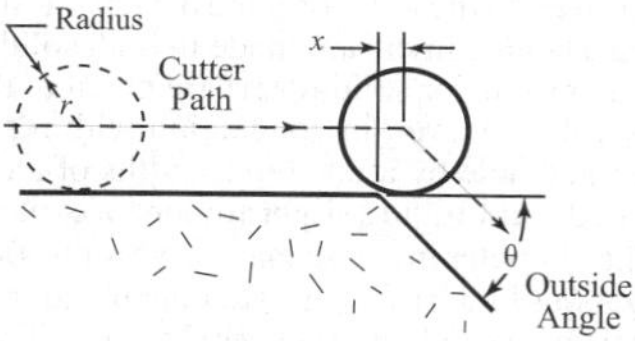

Fig. 1b. Outside Milling

For inside cuts the offset is subtracted from the point at which the cutting direction changes (Fig. 1a), and for outside cuts the offset is added to the point at which the cutting direction changes (Fig. 1b). The formula for the offset is

$$x = rM$$

where $x$ = offset distance; $r$ = radius of the milling cutter; and, $M$ = the multiplication factor ($M = \tan \theta/2$). The value of $M$ for certain angles can be found in Table 5.

**Table 5. Offset Multiplication Factors**

| Deg° | *M* | Deg° | *M* | Deg° | *M* | Deg° | *M* | Deg° | *M* |
|---|---|---|---|---|---|---|---|---|---|
| 1° | 0.00873 | 19° | 0.16734 | 37° | 0.33460 | 55° | 0.52057 | 73° | 0.73996 |
| 2° | 0.01746 | 20° | 0.17633 | 38° | 0.34433 | 56° | 0.53171 | 74° | 0.75355 |
| 3° | 0.02619 | 21° | 0.18534 | 39° | 0.35412 | 57° | 0.54296 | 75° | 0.76733 |
| 4° | 0.03492 | 22° | 0.19438 | 40° | 0.36397 | 58° | 0.55431 | 76° | 0.78129 |
| 5° | 0.04366 | 23° | 0.20345 | 41° | 0.37388 | 59° | 0.56577 | 77° | 0.79544 |
| 6° | 0.05241 | 24° | 0.21256 | 42° | 0.38386 | 60° | 0.57735 | 78° | 0.80978 |
| 7° | 0.06116 | 25° | 0.22169 | 43° | 0.39391 | 61° | 0.58905 | 79° | 0.82434 |
| 8° | 0.06993 | 26° | 0.23087 | 44° | 0.40403 | 62° | 0.60086 | 80° | 0.83910 |
| 9° | 0.07870 | 27° | 0.24008 | 45° | 0.41421 | 63° | 0.61280 | 81° | 0.85408 |
| 10° | 0.08749 | 28° | 0.24933 | 46° | 0.42447 | 64° | 0.62487 | 82° | 0.86929 |
| 11° | 0.09629 | 29° | 0.25862 | 47° | 0.43481 | 65° | 0.63707 | 83° | 0.88473 |
| 12° | 0.10510 | 30° | 0.26795 | 48° | 0.44523 | 66° | 0.64941 | 84° | 0.90040 |
| 13° | 0.11394 | 31° | 0.27732 | 49° | 0.45573 | 67° | 0.66189 | 85° | 0.91633 |
| 14° | 0.12278 | 32° | 0.28675 | 50° | 0.46631 | 68° | 0.67451 | 86° | 0.93252 |
| 15° | 0.13165 | 33° | 0.29621 | 51° | 0.47698 | 69° | 0.68728 | 87° | 0.94896 |
| 16° | 0.14054 | 34° | 0.30573 | 52° | 0.48773 | 70° | 0.70021 | 88° | 0.96569 |
| 17° | 0.14945 | 35° | 0.31530 | 53° | 0.49858 | 71° | 0.71329 | 89° | 0.98270 |
| 18° | 0.15838 | 36° | 0.32492 | 54° | 0.50953 | 72° | 0.72654 | 90° | 1.00000 |

Multiply factor *M* by the tool radius *r* to determine the offset dimension

## REAMERS

**Hand Reamers.—**Hand reamers are made with both straight and helical flutes. Helical flutes provide a shearing cut and are especially useful in reaming holes having keyways or grooves, as these are bridged over by the helical flutes, thus preventing binding or chattering. Hand reamers are made in both solid and expansion forms. The American standard dimensions for solid forms are given in the accompanying table. The expansion type is useful whenever, in connection with repair or other work, it is necessary to enlarge a reamed hole by a few thousandths of an inch. The expansion form is split through the fluted section and a slight amount of expansion is obtained by screwing in a tapering plug. The diameter increase may vary from 0.005 to 0.008 inch (0.127–0.2 mm) for reamers up to about 1 inch (25.4 mm) diameter and from 0.010 to 0.012 inch (0.25–0.3 mm) for diameters between 1 and 2 inches (25.4 and 50.8 mm). Hand reamers are tapered slightly on the end to facilitate starting them properly. The actual diameter of the shanks of commercial reamers may be from 0.002 to 0.005 inch (0.05–0.13 mm) under the reamer size. That part of the shank that is squared should be turned smaller in diameter than the shank itself, so that, when applying a wrench, no burr may be raised that may mar the reamed hole if the reamer is passed clear through it.

When fluting reamers, the cutter is so set with relation to the center of the reamer blank that the tooth gets a slight negative rake; that is, the cutter should be set *ahead* of the center, as shown in the illustration accompanying the table giving the amount to set the cutter ahead of the radial line. The amount is so selected that a tangent to the circumference of the reamer at the cutting point makes an angle of approximately 95 degrees with the front face of the cutting edge.

**Amount to Set Cutter Ahead of Radial Line to Obtain Negative Front Rake**

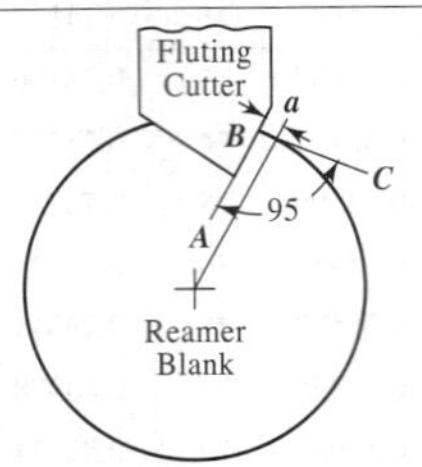


| Size of Reamer | *a*, Inches | Size of Reamer | *a*, Inches | Size of Reamer | *a*, Inches |
|---|---|---|---|---|---|
| ¼ | 0.011 | ⅞ | 0.038 | 2 | 0.087 |
| ⅜ | 0.016 | 1 | 0.044 | 2¼ | 0.098 |
| ½ | 0.022 | 1¼ | 0.055 | 2½ | 0.109 |
| ⅝ | 0.027 | 1½ | 0.066 | 2¾ | 0.120 |
| ¾ | 0.033 | 1¾ | 0.076 | 3 | 0.131 |

When fluting reamers, it is necessary to "break up the flutes," that is, to space the cutting edges unevenly around the reamer. The difference in spacing should be very slight and need not exceed two degrees one way or the other. The manner in which the breaking up of the flutes is usually done is to move the index head to which the reamer is fixed a certain amount more or less than it would be moved if the spacing were regular. A table is given showing the amount of this additional movement of the index crank for reamers with different numbers of flutes. When a reamer is provided with helical flutes, the angle of spiral should be such that the cutting edges make an angle of about 10 or at most 15 degrees with the axis of the reamer.

The relief of the cutting edges should be comparatively slight. An eccentric relief, that is, one where the land back of the cutting edge is convex, rather than flat, is used by one or two manufacturers and is preferable for finishing reamers, as the reamer will hold its size longer. When hand reamers are used merely for removing stock, or simply for enlarging holes, the flat relief is better because the reamer has a keener cutting edge. The width of the land of the cutting edges should be about $^1/_{32}$ inch (0.79 mm) for a $^1/_4$-inch (6.35-mm), $^1/_{16}$ inch (1.59 mm) for a 1-inch (25.4-mm), and $^3/_{32}$ inch (2.38 mm) for a 3-inch (76.2-mm) reamer.

**Irregular Spacing of Teeth in Reamers**

| Number of Flutes in Reamer | 4 | 6 | 8 | 10 | 12 | 14 | 16 |
|---|---|---|---|---|---|---|---|
| Index Circle to Use | 39 | 39 | 39 | 39 | 39 | 49 | 20 |
| Before Cutting | Move Spindle the Number of Holes Below More or Less than for Regular Spacing | | | | | | |
| 2d flute | 8 less | 4 less | 3 less | 2 less | 4 less | 3 less | 2 less |
| 3d flute | 4 more | 5 more | 5 more | 3 more | 4 more | 2 more | 2 more |
| 4th flute | 6 less | 7 less | 2 less | 5 less | 1 less | 2 less | 1 less |
| 5th flute | … | 6 more | 4 more | 2 more | 3 more | 4 more | 2 more |
| 6th flute | … | 5 less | 6 less | 2 less | 4 less | 1 less | 2 less |
| 7th flute | … | … | 2 more | 3 more | 4 more | 3 more | 1 more |
| 8th flute | … | … | 3 less | 2 less | 3 less | 2 less | 2 less |
| 9th flute | … | … | … | 5 more | 2 more | 1 more | 2 more |
| 10th flute | … | … | … | 1 less | 2 less | 3 less | 2 less |
| 11th flute | … | … | … | … | 3 more | 3 more | 1 more |
| 12th flute | … | … | … | … | 4 less | 2 less | 2 less |
| 13th flute | … | … | … | … | … | 2 more | 2 more |
| 14th flute | … | … | … | … | … | 3 less | 1 less |
| 15th flute | … | … | … | … | … | … | 2 more |
| 16th flute | … | … | … | … | … | … | 2 less |

**Threaded-End Hand Reamers.—**Hand reamers are sometimes provided with a thread at the extreme point in order to give them a uniform feed when reaming. The diameter on the top of this thread at the point of the reamer is slightly smaller than the reamer itself, and the thread tapers upward until it reaches a dimension of from 0.003 to 0.008 inch (0.076–0.2 mm), according to size, below the size of the reamer; at this point, the thread stops and a short neck about $^1/_{16}$-inch (1.59-mm) wide separates the threaded portion from the actual reamer, which is provided with a short taper from $^3/_{16}$ to $^7/_{16}$ inch (4.76–11.1 mm) long up to where the standard diameter is reached. The length of the threaded portion and the number of threads per inch for reamers of this kind are given in the accompanying table. The thread employed is a sharp V-thread.

**Dimensions for Threaded-End Hand Reamers**

| Sizes of Reamers | Length of Threaded Part | No. of Threads per Inch | Dia. of Thread at Point of Reamer | Sizes of Reamers | Length of Threaded Part | No. of Threads per Inch | Dia. of Thread at Point of Reamer |
|---|---|---|---|---|---|---|---|
| | | | Full diameter | | | | Full diameter |
| 1⁄8-5⁄16 | 3⁄8 | 32 | −0.006 | 1 1⁄32-1 1⁄2 | 9⁄16 | 18 | −0.010 |
| 11⁄32-1⁄2 | 7⁄16 | 28 | −0.006 | 1 17⁄32-2 | 9⁄16 | 18 | −0.012 |
| 17⁄32-3⁄4 | 1⁄2 | 24 | −0.008 | 2 1⁄32-2 1⁄2 | 9⁄16 | 18 | −0.015 |
| 25⁄32-1 | 9⁄16 | 18 | −0.008 | 2 17⁄32-3 | 9⁄16 | 18 | −0.020 |

**Fluted Chucking Reamers.—**Reamers of this type are used in turret lathes, screw machines, etc., for enlarging holes and finishing them smooth and to the required size. The best results are obtained with a floating type of holder that permits a reamer to align itself with the hole being reamed. These reamers are intended for removing a small amount of metal, 0.005 to 0.010 inch (0.127–0.25 mm) being common allowances. Fluted chucking reamers are provided either with a straight shank or a standard taper shank. (See the table on page 898 for standard dimensions.)

**Fluting Cutters for Reamers**

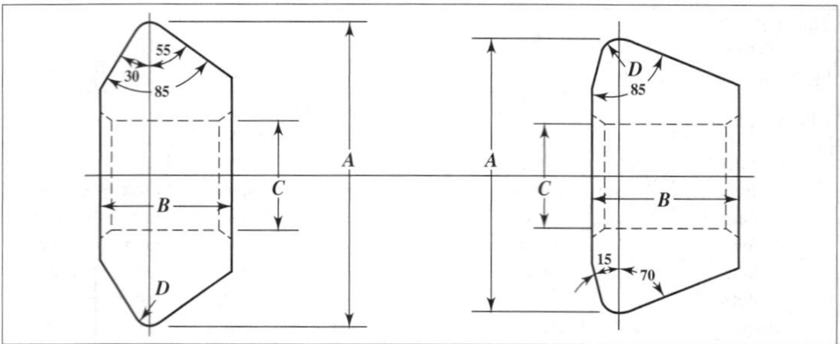


| Reamer Dia. | Fluting Cutter Dia. A | Fluting Cutter Thickness B | Hole Dia. in Cutter C | Radius between Cutting Faces D | Reamer Dia. | Fluting Cutter Dia. A | Fluting Cutter Thickness B | Hole Dia. in Cutter C | Radius between Cutting Faces D |
|---|---|---|---|---|---|---|---|---|---|
| 1/8 | 1 3/4 | 3/16 | 3/4 | none[a] | 1 1/4 | 2 1/4 | 9/16 | 1 | 1/16 |
| 3/16 | 1 3/4 | 3/16 | 3/4 | none[a] | 1 1/2 | 2 1/4 | 5/8 | 1 | 1/16 |
| 1/4 | 1 3/4 | 3/16 | 3/4 | 1/64 | 1 3/4 | 2 1/4 | 5/8 | 1 | 5/64 |
| 3/8 | 2 | 1/4 | 3/4 | 1/64 | 2 | 2 1/2 | 3/4 | 1 | 5/64 |
| 1/2 | 2 | 5/16 | 3/4 | 1/32 | 2 1/4 | 2 1/2 | 3/4 | 1 | 5/64 |
| 5/8 | 2 | 3/8 | 3/4 | 1/32 | 2 1/2 | 2 1/2 | 7/8 | 1 | 3/16 |
| 3/4 | 2 | 7/16 | 3/4 | 3/64 | 2 3/4 | 2 1/2 | 7/8 | 1 | 3/16 |
| 1 | 2 1/4 | 1/2 | 1 | 3/64 | 3 | 2 1/2 | 1 | 1 | 3/16 |

[a] Sharp corner, no radius

**Rose Chucking Reamers.—**The rose type of reamer is used for enlarging cored or other holes. The cutting edges at the end are ground to a 45-degree bevel. This type of reamer will remove considerable metal in one cut. The cylindrical part of the reamer has no cutting edges but merely grooves cut for the full length of the reamer body, providing a way for the chips to escape and a channel for lubricant to reach the cutting edges. There is no relief on the cylindrical surface of the body part, but it is slightly back-tapered so that the diameter at the point with the beveled cutting edges is slightly larger than the diameter farther back. The back-taper should not exceed 0.001 inch per inch (or mm/mm). This form of reamer usually produces holes slightly larger than its size; it is, therefore, always made from 0.005 to 0.010 inch (0.127-0.25 mm) smaller than its nominal size, so that it may be followed by a fluted reamer for finishing. The grooves on the cylindrical portion are cut by a convex cutter having a width equal to from one-fifth to one-fourth the diameter of the rose reamer itself. The depth of the groove should be from one-eighth to one-sixth the diameter of the reamer. The teeth at the end of the reamer are milled with a 75-degree angular cutter; the width of the land of the cutting edge should be about one-fifth the distance from tooth to tooth. If an angular cutter is preferred to a convex cutter for milling the grooves on the cylindrical portion because of the higher cutting speed possible when milling, an 80-degree angular cutter slightly rounded at the point may be used.

**Cutters for Fluting Rose Chucking Reamers.—**The cutters used for fluting rose chucking reamers on the end are 80-degree angular cutters for 1/4- and 5/16-inch diameter reamers; 75-degree angular cutters for 3/8- and 7/16-inch reamers; and 70-degree angular cutters for all larger sizes. The grooves on the cylindrical portion are milled with convex cutters of approximately the following sizes for given diameters of reamers: 5/32-inch convex cutter for 1/2-inch reamers; 5/16-inch cutter for 1-inch reamers; 3/8-inch cutter for 1 1/2-inch reamers; 13/32-inch cutters for 2-inch reamers; and 15/32-inch cutters for 2 1/2-inch reamers.

## Dimensions of Formed Reamer Fluting Cutters

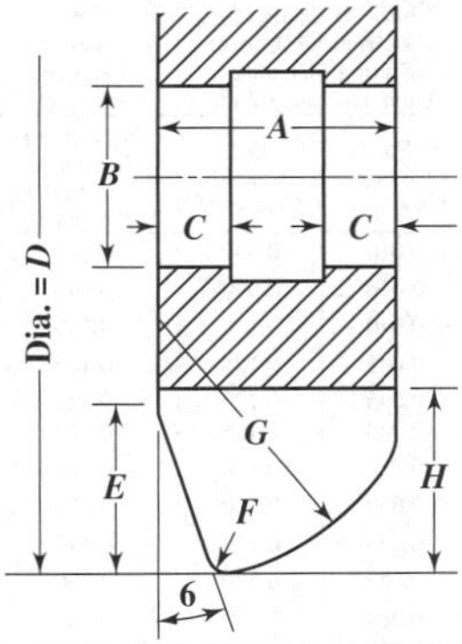


The making and maintenance of cutters of the formed type involve greater expense than the use of angular cutters, dimensions of which are given on the previous page; but the form of flute produced by the formed type of cutter is preferred by many reamer users. The claims made for the formed type of flute are that the chips can be more readily removed from the reamer and that the reamer has greater strength and is less likely to crack or spring out of shape in hardening.

| Reamer Size | No. of Teeth in Reamer | Cutter Dia. *D* | Cutter Width *A* | Hole Dia. *B* | Bearing Width *C* | Bevel Length *E* | Radius *F* | Radius *F* | Tooth Depth *H* | No. of Cutter Teeth |
|---|---|---|---|---|---|---|---|---|---|---|
| 1/8-3/16 | 6 | 1 3/4 | 3/16 | 7/8 | … | 0.125 | 0.016 | 7/32 | 0.21 | 14 |
| 1/4-5/16 | 6 | 1 3/4 | 1/4 | 7/8 | … | 0.152 | 0.022 | 9/32 | 0.25 | 13 |
| 3/8-7/16 | 6 | 1 7/8 | 3/8 | 7/8 | 1/8 | 0.178 | 0.029 | 1/2 | 0.28 | 12 |
| 1/2-11/16 | 6-8 | 2 | 7/16 | 7/8 | 1/8 | 0.205 | 0.036 | 9/16 | 0.30 | 12 |
| 3/4-1 | 8 | 2 1/8 | 1/2 | 7/8 | 5/32 | 0.232 | 0.042 | 11/16 | 0.32 | 12 |
| 1 1/16-1 1/2 | 10 | 2 1/4 | 9/16 | 7/8 | 5/32 | 0.258 | 0.049 | 3/4 | 0.38 | 11 |
| 1 9/16-2 1/8 | 12 | 2 3/8 | 5/8 | 7/8 | 3/16 | 0.285 | 0.056 | 27/32 | 0.40 | 11 |
| 2 1/4-3 | 14 | 2 5/8 | 11/16 | 7/8 | 3/16 | 0.312 | 0.062 | 7/8 | 0.44 | 10 |

The smaller sizes of reamers, from 1/4 to 3/8 inch in diameter, are often milled, with regular double-angle reamer fluting cutters having a radius of 1/64 inch for 1/4-inch reamer, and 1/32 inch for 5/16- and 3/8-inch sizes.

**Reamer Terms and Definitions.—***Reamer:* A rotary cutting tool with one or more cutting elements used for enlarging to size and contour a previously formed hole. Its principal support during the cutting action is obtained from the workpiece. (See Fig. 1a.)

*Actual Size:* The actual measured diameter of a reamer, usually slightly larger than the nominal size to allow for wear.

*Angle of Taper:* The included angle of taper on a taper tool or taper shank.

*Arbor Hole:* The central mounting hole in a shell reamer.

*Axis:* The imaginary straight line that forms the longitudinal centerline of a reamer, usually established by rotating the reamer between centers.

*Back Taper:* A slight decrease in diameter, from front to back, in the flute length of reamers.

*Bevel:* An unrelieved angular surface of revolution (not to be confused with chamfer).

*Body:* The fluted full-diameter portion of a reamer, inclusive of the chamfer, starting taper, and bevel.

*Chamfer:* The angular cutting portion at the entering end of a reamer (see also *Secondary Chamfer*).

### Vertical Adjustment of Tooth-Rest for Grinding Clearance on Reamers

| Size of Reamer | Hand Reamer for Steel. Cutting Clearance Land 0.006 inch Wide | | Hand Reamer for Cast Iron and Bronze. Cutting Clearance Land 0.025 inch Wide | | Chucking Reamer for Cast Iron and Bronze. Cutting Clearance Land 0.025 inch Wide | | Rose Chucking Reamers for Steel |
|---|---|---|---|---|---|---|---|
| | For Cutting Clearance | For Second Clearance | For Cutting Clearance | For Second Clearance | For Cutting Clearance | For Second Clearance | For Cutting Clearance on Angular Edge at End |
| ½ | 0.012 | 0.052 | 0.032 | 0.072 | 0.040 | 0.080 | 0.080 |
| ⅝ | 0.012 | 0.062 | 0.032 | 0.072 | 0.040 | 0.090 | 0.090 |
| ¾ | 0.012 | 0.072 | 0.035 | 0.095 | 0.040 | 0.100 | 0.100 |
| ⅞ | 0.012 | 0.082 | 0.040 | 0.120 | 0.045 | 0.125 | 0.125 |
| 1 | 0.012 | 0.092 | 0.040 | 0.120 | 0.045 | 0.125 | 0.125 |
| 1⅛ | 0.012 | 0.102 | 0.040 | 0.120 | 0.045 | 0.125 | 0.125 |
| 1¼ | 0.012 | 0.112 | 0.045 | 0.145 | 0.050 | 0.160 | 0.160 |
| 1⅜ | 0.012 | 0.122 | 0.045 | 0.145 | 0.050 | 0.160 | 0.175 |
| 1½ | 0.012 | 0.132 | 0.048 | 0.168 | 0.055 | 0.175 | 0.175 |
| 1⅝ | 0.012 | 0.142 | 0.050 | 0.170 | 0.060 | 0.200 | 0.200 |
| 1¾ | 0.012 | 0.152 | 0.052 | 0.192 | 0.060 | 0.200 | 0.200 |
| 1⅞ | 0.012 | 0.162 | 0.056 | 0.196 | 0.060 | 0.200 | 0.200 |
| 2 | 0.012 | 0.172 | 0.056 | 0.216 | 0.064 | 0.224 | 0.225 |
| 2⅛ | 0.012 | 0.172 | 0.059 | 0.219 | 0.064 | 0.224 | 0.225 |
| 2¼ | 0.012 | 0.172 | 0.063 | 0.223 | 0.064 | 0.224 | 0.225 |
| 2⅜ | 0.012 | 0.172 | 0.063 | 0.223 | 0.068 | 0.228 | 0.230 |
| 2½ | 0.012 | 0.172 | 0.065 | 0.225 | 0.072 | 0.232 | 0.230 |
| 2⅝ | 0.012 | 0.172 | 0.065 | 0.225 | 0.075 | 0.235 | 0.235 |
| 2¾ | 0.012 | 0.172 | 0.065 | 0.225 | 0.077 | 0.237 | 0.240 |
| 2⅞ | 0.012 | 0.172 | 0.070 | 0.230 | 0.080 | 0.240 | 0.240 |
| 3 | 0.012 | 0.172 | 0.072 | 0.232 | 0.080 | 0.240 | 0.240 |
| 3⅛ | 0.012 | 0.172 | 0.075 | 0.235 | 0.083 | 0.240 | 0.240 |
| 3¼ | 0.012 | 0.172 | 0.078 | 0.238 | 0.083 | 0.243 | 0.245 |
| 3⅜ | 0.012 | 0.172 | 0.081 | 0.241 | 0.087 | 0.247 | 0.245 |
| 3½ | 0.012 | 0.172 | 0.084 | 0.244 | 0.090 | 0.250 | 0.250 |
| 3⅝ | 0.012 | 0.172 | 0.087 | 0.247 | 0.093 | 0.253 | 0.250 |
| 3¾ | 0.012 | 0.172 | 0.090 | 0.250 | 0.097 | 0.257 | 0.255 |
| 3⅞ | 0.012 | 0.172 | 0.093 | 0.253 | 0.100 | 0.260 | 0.255 |
| 4 | 0.012 | 0.172 | 0.096 | 0.256 | 0.104 | 0.264 | 0.260 |
| 4⅛ | 0.012 | 0.172 | 0.096 | 0.256 | 0.104 | 0.264 | 0.260 |
| 4¼ | 0.012 | 0.172 | 0.096 | 0.256 | 0.106 | 0.266 | 0.265 |
| 4⅜ | 0.012 | 0.172 | 0.096 | 0.256 | 0.108 | 0.268 | 0.265 |
| 4½ | 0.012 | 0.172 | 0.100 | 0.260 | 0.108 | 0.268 | 0.265 |
| 4⅝ | 0.012 | 0.172 | 0.100 | 0.260 | 0.110 | 0.270 | 0.270 |
| 4¾ | 0.012 | 0.172 | 0.104 | 0.264 | 0.114 | 0.274 | 0.275 |
| 4⅞ | 0.012 | 0.172 | 0.106 | 0.266 | 0.116 | 0.276 | 0.275 |
| 5 | 0.012 | 0.172 | 0.110 | 0.270 | 0.118 | 0.278 | 0.275 |

*Chamfer Angle:* The angle between the axis and the cutting edge of the chamfer measured in an axial plane at the cutting edge.

*Chamfer Length:* The length of the chamfer measured parallel to the axis at the cutting edge.

*Chamfer Relief Angle:* See under *Relief.*

*Chamfer Relief:* See under *Relief.*

*Chipbreakers:* Notches or grooves in the cutting edges of some taper reamers designed to break the continuity of the chips.

*Circular Land:* See preferred term *Margin.*

*Clearance:* The space created by the relief behind the cutting edge or margin of a reamer.

## Illustration of Terms Applying to Reamers

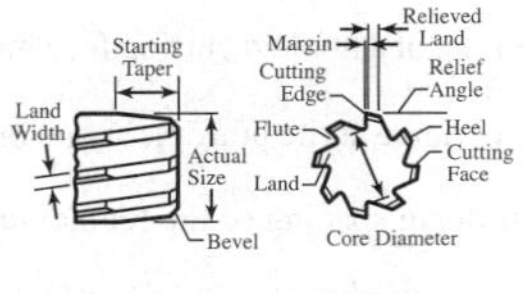


**Helical Flutes Left-Hand Helix Shown**

**Radial Zero Degrees Rake Angle and Right-Hand Rotation Shown**

Hand Reamer

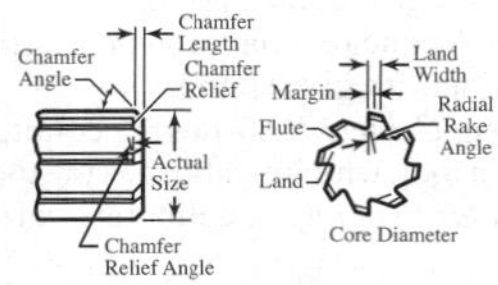


**Straight Flutes Shown**

**Positive Radial Rake Angle and Right-Hand Rotation Shown**

Machine Reamer

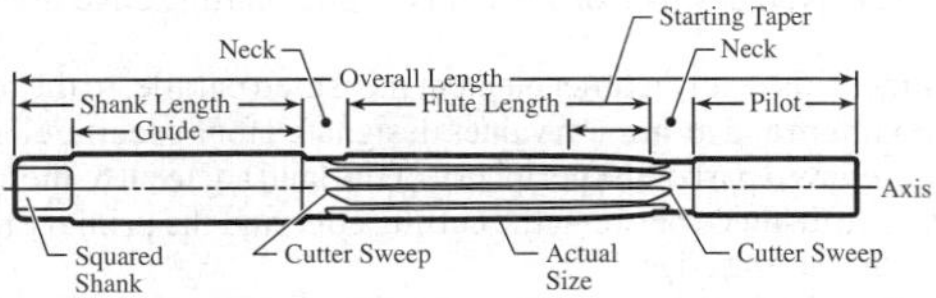


**Straight Flutes**

Hand Reamer, Pilot and Guide

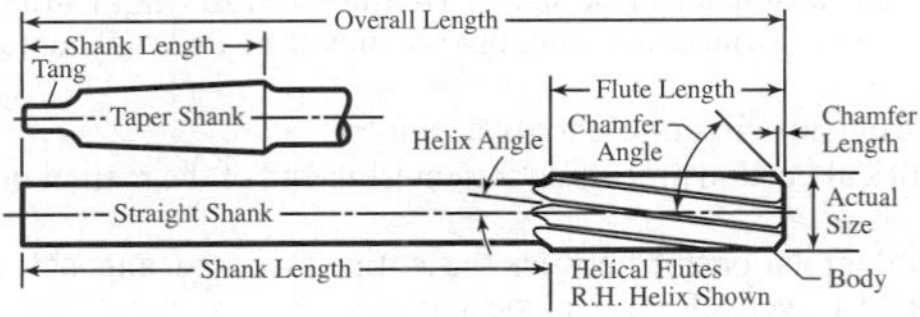


Chucking Reamer, Straight and Taper Shank

*Core:* The central portion of a reamer that joins the lands below the flutes.

*Core Diameter:* The diameter at a given point along the axis of the largest circle which does not project into the flutes.

*Cutter Sweep:* The section removed by the milling cutter or grinding wheel in entering or leaving a flute.

*Cutting Edge:* The leading edge of the relieved land in the direction of rotation for cutting.

*Cutting Face:* The leading side of the relieved land in the direction of rotation for cutting on which the chip impinges.

*External Center:* The pointed end of a reamer. The included angle varies with manufacturing practice.

*Flutes:* Longitudinal channels formed in the body of the reamer to provide cutting edges, permit passage of chips, and allow cutting fluid to reach the cutting edges.

*Angular Flute:* A flute that forms a cutting face lying in a plane intersecting the reamer axis at an angle. It is unlike a helical flute in that it forms a cutting face that lies in a single plane.

*Helical Flute:* Sometimes called a spiral flute, a flute formed in a helical path around the axis of a reamer.

*Spiral Flute:* 1) On a taper reamer, a flute of constant lead; or 2) in reference to a straight reamer, see preferred term *Helical Flute*.

*Straight Flute:* A flute that forms a cutting edge lying in an axial plane.

*Flute Length:* The length of the flutes, not including the cutter sweep.

*Guide:* A cylindrical portion following the flutes of a reamer to maintain alignment.

*Heel:* The trailing edge of the land in the direction of rotation for cutting.

*Helix Angle:* The angle which a helical cutting edge at a given point makes with an axial plane through the same point.

*Hook:* A concave condition of a cutting face. The rake of a hooked cutting face must be determined at a given point.

*Internal Center:* A 60-degree countersink with clearance at the bottom, in one or both ends of a tool, which establishes the tool axis.

*Irregular Spacing:* A deliberate variation from uniform spacing of the reamer cutting edges.

*Land:* The section of the reamer between adjacent flutes.

*Land Width:* The distance between the leading edge of the land and the heel, measured at a right angle to the leading edge.

*Lead of Flute:* The axial advance of a helical or spiral cutting edge in one turn around the reamer axis.

*Length:* The dimension of any reamer element measured parallel to the reamer axis.

*Limits:* The maximum and minimum values designated for a specific element.

*Margin:* The unrelieved part of the periphery of the land adjacent to the cutting edge.

*Margin Width:* The distance between the cutting edge and the primary relief measured at a right angle to the cutting edge.

*Neck:* The section of reduced diameter connecting shank to body, or connecting other portions of the reamer.

*Nominal Size:* The designated basic size of a reamer overall length—the extreme length of the complete reamer from end to end, but not including external centers or expansion screws.

*Periphery:* The outside circumference of a reamer.

*Pilot:* A cylindrical portion preceding the entering end of the reamer body to maintain alignment.

*Rake:* The angular relationship between the cutting face, or a tangent to the cutting face at a given point and a given reference plane or line.

*Axial Rake:* Applies to angular (not helical or spiral) cutting faces. It is the angle between a plane containing the cutting face, or tangent to the cutting face at a given point, and the reamer axis.

*Helical Rake:* Applies only to helical and spiral cutting faces (not angular). It is the angle between a plane, tangent to the cutting face at a given point on the cutting edge, and the reamer axis.

*Negative Rake:* Describes a cutting face in rotation whose cutting edge lags the surface of the cutting face.

*Positive Rake:* Describes a cutting face in rotation whose cutting edge leads the surface of the cutting face.

*Radial Rake Angle:* The angle in a transverse plane between a straight cutting face and a radial line passing through the cutting edge.

*Relief:* The result of the removal of tool material behind or adjacent to the cutting edge to provide clearance and prevent rubbing (heel drag).

*Axial Relief:* The relief measured in the axial direction between a plane perpendicular to the axis and the relieved surface. It can be measured by the amount of indicator drop at a given radius in a given amount of angular rotation.

*Cam Relief:* The relief from the cutting edge to the heel of the land produced by a cam action.

*Chamfer Relief Angle:* The axial relief angle at the outer corner of the chamfer. It is measured by projection into a plane tangent to the periphery at the outer corner of the chamfer.

*Chamfer Relief:* The axial relief on the chamfer of the reamer.

*Eccentric Relief:* A convex relieved surface behind the cutting edge.

*Flat Relief:* A relieved surface behind the cutting edge which is essentially flat.

*Radial Relief:* Relief in a radial direction measured in the plane of rotation. It can be measured by the amount of indicator drop at a given radius in a given amount of angular rotation.

*Primary Relief:* The relief immediately behind the cutting edge or margin. Properly called relief.

*Secondary Relief:* An additional relief behind the primary relief.

*Relief Angle:* The angle, measured in a transverse plane, between the relieved surface and a plane tangent to the periphery at the cutting edge.

*Secondary Chamfer:* A slight relieved chamfer adjacent to and following the initial chamfer on a reamer.

*Shank:* The portion of the reamer by which it is held and driven.

*Squared Shank:* A cylindrical shank having a driving square on the back end.

*Starting Radius:* A relieved radius at the entering end of a reamer in place of a chamfer.

*Starting Taper:* A slight relieved taper on the front end of a reamer.

*Straight Shank:* A cylindrical shank.

*Tang:* The flatted end of a taper shank which fits a slot in the socket.

*Taper per Foot:* The difference in diameter between two points 12 inches apart measured along the axis.

*Taper Shank:* A shank made to fit a specific (conical) taper socket.

**Direction of Rotation and Helix.**—The terms "right-hand" and "left-hand" are used to describe both direction of rotation and direction of flute helix or reamers.

*Hand of Rotation (or Hand of Cut): Right-Hand Rotation (or Right-Hand Cut):* When viewed from the cutting end, the reamer must revolve counterclockwise to cut

*Left-hand Rotation (or Left-Hand Cut):* When viewed from the cutting end, the reamer must revolve clockwise to cut

*Hand of Flute Helix: Right-Hand Helix:* When the flutes twist away from the observer in a clockwise direction when viewed from either end of the reamer.

*Left-Hand Helix:* When the flutes twist away from the observer in a counterclockwise direction when viewed from either end of the reamer. The standard reamers on the tables that follow are all right-hand rotation.

**Dimensions of Centers for Reamers and Arbors**

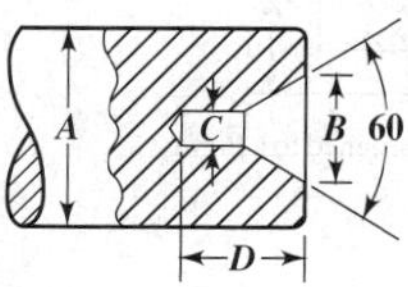


| Arbor Dia. *A* | Large Center Dia. *B* | Drill No. *C* | Hole Depth *D* |
|---|---|---|---|
| 1/4 | 1/8 | 55 | 5/32 |
| 5/16 | 5/32 | 52 | 3/16 |
| 3/8 | 3/16 | 48 | 7/32 |
| 7/16 | 7/32 | 43 | 1/4 |
| 1/2 | 1/4 | 39 | 5/16 |
| 9/16 | 9/32 | 33 | 11/32 |
| 5/8 | 5/16 | 30 | 3/8 |
| 11/16 | 11/32 | 29 | 13/32 |
| 3/4 | 3/8 | 25 | 7/16 |
| 13/16 | 13/32 | 20 | 1/2 |
| 7/8 | 7/16 | 17 | 17/32 |
| 15/16 | 15/32 | 12 | 9/16 |
| 1 | 1/2 | 8 | 19/32 |
| 1 1/8 | 33/64 | 5 | 5/8 |
| 1 1/4 | 17/32 | 3 | 21/32 |
| 1 3/8 | 35/64 | 2 | 21/32 |
| 1 1/2 | 9/16 | 1 | 11/16 |
| … | … | Letter | … |
| 1 5/8 | 37/64 | **A** | 23/32 |
| 1 3/4 | 19/32 | **B** | 23/32 |
| 1 7/8 | 39/64 | **C** | 3/4 |
| 2 | 5/8 | **E** | 3/4 |
| 2 1/8 | 41/64 | **F** | 25/32 |
| 2 1/4 | 21/32 | **G** | 13/16 |
| 2 3/8 | 43/64 | **H** | 27/32 |
| 2 1/2 | 11/16 | **J** | 27/32 |
| 2 5/8 | 45/64 | **K** | 7/8 |
| 2 3/4 | 23/32 | **L** | 29/32 |
| 2 7/8 | 47/64 | **M** | 29/32 |
| 3 | 3/4 | **N** | 15/16 |
| 3 1/8 | 49/64 | **N** | 31/32 |
| 3 1/4 | 25/32 | **O** | 31/32 |
| 3 3/8 | 51/64 | **O** | 1 |
| 3 1/2 | 13/16 | **P** | 1 |
| 3 5/8 | 53/64 | **Q** | 1 1/16 |
| 3 3/4 | 27/32 | **R** | 1 1/16 |
| 3 7/8 | 55/64 | **R** | 1 1/16 |
| 4 | 7/8 | **S** | 1 1/8 |
| 4 1/4 | 29/32 | **T** | 1 1/8 |
| 4 1/2 | 15/16 | **V** | 1 3/16 |
| 4 3/4 | 31/32 | **W** | 1 1/4 |
| 5 | 1 | **X** | 1 1/4 |

## Straight-Shank Center Reamers and Machine Countersinks

*ANSI/ASME B94.2-1995 (R2020)*

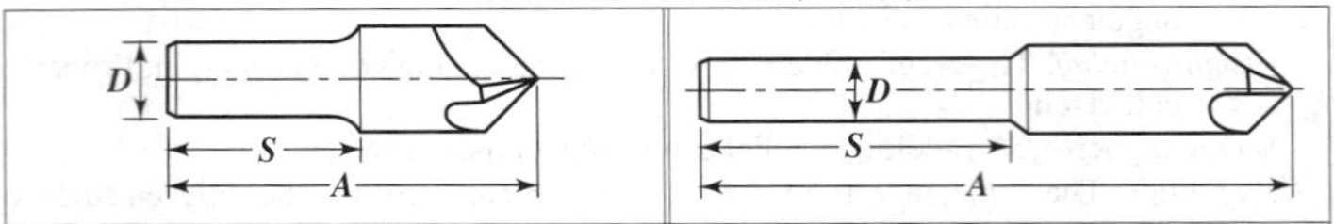


| Center Reamers (Short Countersinks) | | | | Machine Countersinks | | | |
|---|---|---|---|---|---|---|---|
| Dia. of Cut | Approx. Length Overall, *A* | Length of Shank, *S* | Dia. of Shank, *D* | Dia. of Cut | Approx. Length Overall, *A* | Length of Shank, *S* | Dia. of Shank, *D* |
| ¼ | 1½ | ¾ | 3⁄16 | ½ | 3⅞ | 2¼ | ½ |
| ⅜ | 1¾ | ⅞ | ¼ | ⅝ | 4 | 2¼ | ½ |
| ½ | 2 | 1 | ⅜ | ¾ | 4⅛ | 2¼ | ½ |
| ⅝ | 2¼ | 1 | ⅜ | ⅞ | 4¼ | 2¼ | ½ |
| ¾ | 2⅝ | 1¼ | ½ | 1 | 4⅜ | 2¼ | ½ |

All dimensions are given in inches. Material is high-speed steel. Reamers and countersinks have 3 or 4 flutes. Center reamers are standard with 60, 82, 90, or 100 degrees included angle. Machine countersinks are standard with either 60 or 82 degrees included angle.

*Tolerances:* On overall length *A*, the tolerance is ±⅛ inch for center reamers in a size range of from ¼ to ⅜ inch, incl., and machine countersinks in a size range of from ½ to ⅝ inch. incl.; ± 3⁄16 inch for center reamers, ½ to ¾ inch, incl.; and machine countersinks, ¾ to 1 inch, incl. On shank diameter *D*, the tolerance is −0.0005 to −0.002 inch. On shank length *S*, the tolerance is ±1⁄16 inch.

**Calculating Countersink and Spot Drill Depths.**—The following formulas are for countersinks and spot drills that have been sharpened to a point, allowances on tool depth will have to be made if the countersink or spot drill is not sharpened to a point.

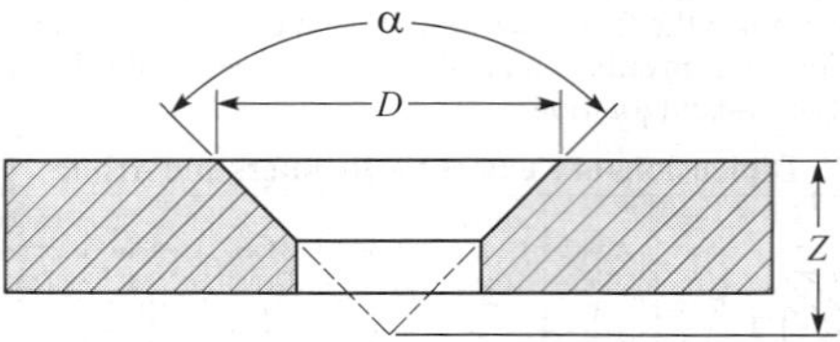


Fig. 1. For Countersinks and Spot Drills that have been Sharpened to a Point

$\alpha$ = countersink angle

$D$ = countersink diameter

$Z$ = depth of tool travel required from surface of part

To find the tool travel "Z" required for any sharp-point countersink or spot drill angle, use either of the following formulas:

$$Z = \frac{D}{2} \cdot \frac{1}{\tan\frac{\alpha}{2}} \qquad \text{or} \qquad Z = \frac{D}{2} \cdot \cot\frac{\alpha}{2}$$

To find the tool travel "*Z-a*" required for any truncated countersink or spot drill angle, use either of the following formulas:

$$Z - a = \frac{D - T}{2} \cdot \frac{1}{\tan\frac{\alpha}{2}} \quad \text{or} \quad Z - a = \frac{D - T}{2} \cdot \cot\frac{\alpha}{2}$$

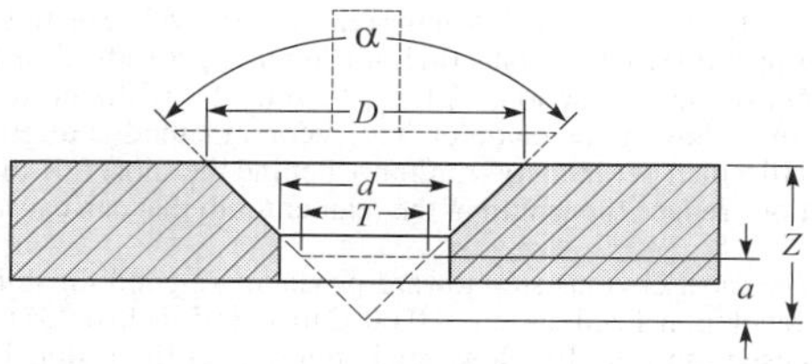


Fig. 2. For Countersinks and Spot Drills that are a Truncated Cone

$a$ = allowance for truncated countersink tool
$d$ = diameter of hole
$T$ = diameter of small end of truncated tool

The table below is a quick reference for some predefined angles on countersinks and spot drills that have been sharpened to a point. To calculate the tool travel "Z", multiply the factor by the countersink diameter.

| Countersink Angle α, degrees | Factor (Imperial or Metric Units) | Countersink Angle α, degrees | Factor (Imperial or Metric Units) |
|---|---|---|---|
| 60 | 0.8660 | 118 | 0.3004 |
| 82 | 0.5752 | 120 | 0.2887 |
| 90 | 0.5000 | 135 | 0.2071 |
| 100 | 0.4195 | 142 | 0.1722 |

**Reamer Difficulties.—**Certain frequently occurring problems in reaming require remedial measures. These difficulties include the production of oversize holes, bellmouth holes, and holes with a poor finish. The following is taken from suggestions for correction of these difficulties by the National Twist Drill and Tool Co. and Winter Brothers Co.*

*Oversize Holes:* The cutting of a hole oversize from the start of the reaming operations usually indicates a mechanical defect in the setup or reamer. Thus, the wrong reamer for the workpiece material may have been used or there may be inadequate workpiece support; inadequate or worn guide bushings; misalignment of the spindles, bushings, or workpiece; or runout of the spindle or reamer holder. The reamer itself may be defective due to chamfer runout or runout of the cutting end due to a bent or nonconcentric shank.

When reamers gradually start to cut oversize, it is due to pickup or galling, principally on the reamer margins. This condition is partly due to the workpiece material. Mild steels, certain cast irons, and some aluminum alloys are particularly troublesome in this respect.

Corrective measures include reducing the reamer margin widths to about 0.005 to 0.010 inch (0.127-0.25 mm), use of hard case surface treatments on high-speed steel reamers, either alone or in combination with black oxide treatments, and the use of a high-grade finish on the reamer faces, margins, and chamfer relief surfaces.

*Bellmouth Holes:* The cutting of a hole that becomes oversize at the entry end with the oversize decreasing gradually along its length always reflects misalignment of the cutting portion of the reamer with respect to the hole. The obvious solution is to provide improved guiding of the reamer by the use of accurate bushings and pilot surfaces. If this solution is not feasible and the reamer is cutting in a vertical position, a flexible element may be employed to hold the reamer in such a way that it has both radial and axial float, with the hope that the reamer will follow the original hole and prevent the bellmouth condition.

In horizontal setups where the reamer is held fixed and the workpiece rotated, any misalignment exerts a sideways force on the reamer as it is fed to depth, resulting in the formation of a tapered hole. This type of bellmouthing can frequently be reduced by shortening

* "Some Aspects of Reamer Design and Operation," *Metal Cuttings*, April 1963.

the bearing length of the cutting portion of the reamer. One way to do this is to reduce the reamer diameter by 0.010 to 0.030 inch (0.25–0.76 mm), depending on size and length, behind a short full-diameter section, 1/8 to 1/2 inch (3.18 to 12.7 mm) long according to length and size, following the chamfer. The second method is to grind a high back taper, 0.008 to 0.015 inch per inch (or mm/mm), behind the short full-diameter section. These modifications reduce the length of the reamer tooth that can cause the bellmouth condition.

*Poor Finish:* The most obvious step toward producing a good finish is to reduce the reamer feed per revolution. Feeds as low as 0.0002 to 0.0005 inch (0.005 to 0.013 mm) per tooth have been used successfully. However, better reamer life results if the maximum feasible feed is used.

The minimum practical amount of reaming stock allowance will often improve finish by reducing the volume of chips and the resulting heat generated on the cutting portion of the chamfer. Too little reamer stock, however, can be troublesome in that the reamer teeth may not cut freely but will deflect or push the work material out of the way. When this happens, excessive heat, poor finish, and rapid reamer wear can occur.

Because of superior abrasion resistance, carbide reamers are often used when fine finishes are required. When properly conditioned, carbide reamers can produce a large number of good-quality holes. Careful honing of the carbide reamer edges is very important.

## Types and Sizes of Reamers

### American National Standard Fluted Taper Shank Chucking Reamers—Straight and Helical Flutes, Fractional Sizes *ANSI/ASME B94.2-1995 (R2020)*

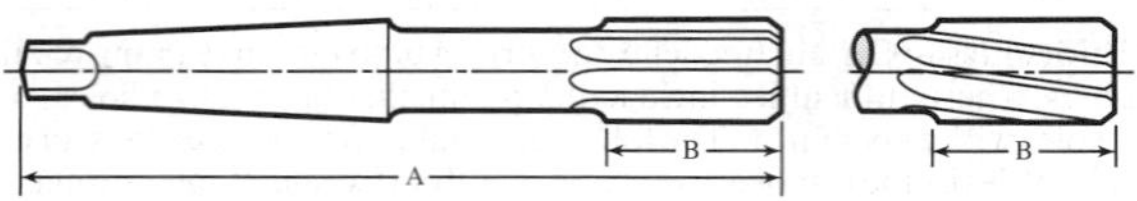


| Reamer Dia. | Length Overall *A* | Flute Length *B* | No. of Morse Taper Shank[a] | No. of Flutes | Reamer Dia. | Length Overall *A* | Flute Length *B* | No. of Morse Taper Shank[a] | No. of Flutes |
|---|---|---|---|---|---|---|---|---|---|
| 1/4 | 6 | 1 1/2 | 1 | 4 to 6 | 27/32 | 9 1/2 | 2 1/2 | 2 | 8 to 10 |
| 5/16 | 6 | 1 1/2 | 1 | 4 to 6 | 7/8 | 10 | 2 5/8 | 2 | 8 to 10 |
| 3/8 | 7 | 1 3/4 | 1 | 4 to 6 | 29/32 | 10 | 2 5/8 | 2 | 8 to 10 |
| 7/16 | 7 | 1 3/4 | 1 | 6 to 8 | 15/16 | 10 | 2 5/8 | 3 | 8 to 10 |
| 1/2 | 8 | 2 | 1 | 6 to 8 | 31/32 | 10 | 2 5/8 | 3 | 8 to 10 |
| 17/32 | 8 | 2 | 1 | 6 to 8 | 1 | 10 1/2 | 2 3/4 | 3 | 8 to 12 |
| 9/16 | 8 | 2 | 1 | 6 to 8 | 1 1/16 | 10 1/2 | 2 3/4 | 3 | 8 to 12 |
| 19/32 | 8 | 2 | 1 | 6 to 8 | 1 1/8 | 11 | 2 7/8 | 3 | 8 to 12 |
| 5/8 | 9 | 2 1/4 | 2 | 6 to 8 | 1 3/16 | 11 | 2 7/8 | 3 | 8 to 12 |
| 21/32 | 9 | 2 1/4 | 2 | 6 to 8 | 1 1/4 | 11 1/2 | 3 | 4 | 8 to 12 |
| 11/16 | 9 | 2 1/4 | 2 | 6 to 8 | 1 5/16 | 11 1/2 | 3 | 4 | 8 to 12 |
| 23/32 | 9 | 2 1/4 | 2 | 6 to 8 | 1 3/8 | 12 | 3 1/4 | 4 | 10 to 12 |
| 3/4 | 9 1/2 | 2 1/2 | 2 | 6 to 8 | 1 7/16 | 12 | 3 1/4 | 4 | 10 to 12 |
| 25/32 | 9 1/2 | 2 1/2 | 2 | 8 to 10 | 1 1/2 | 12 1/2 | 3 1/2 | 4 | 10 to 12 |
| 13/16 | 9 1/2 | 2 1/2 | 2 | 8 to 10 | … | … | … | … | … |

[a] American National Standard self-holding tapers (see Table 7a on page 1008).

All dimensions are given in inches. Material is high-speed steel.

Helical flute reamers with right-hand helical flutes are standard.

*Tolerances:* On reamer diameter, 1/4-inch size, +.0001 to +.0004 inch; over 1/4- to 1-inch size, +.0001 to +.0005 inch; over 1-inch size, +.0002 to +.0006 inch. On length overall *A* and flute length *B*, 1/4- to 1-inch size, incl., ±1/16 inch; 1 1/16- to 1 1/2-inch size, incl., 3/32 inch.

## Expansion Chucking Reamers—Straight and Taper Shanks
*ANSI/ASME B94.2-1995 (R2020)*

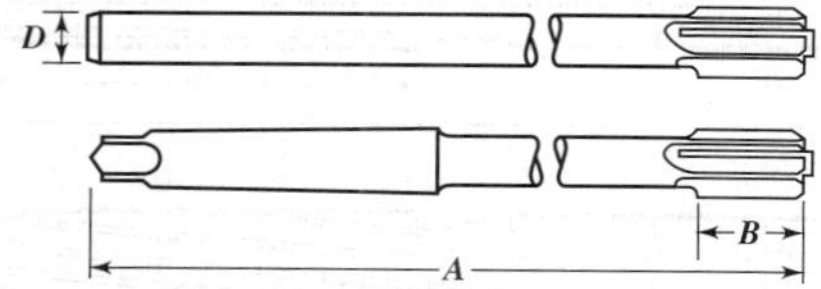


| Dia of Reamer | Length, *A* | Flute Length, *B* | Shank Dia., *D* | | Dia. of Reamer | Length, *A* | Flute Length, *B* | Shank Dia., *D* | |
|---|---|---|---|---|---|---|---|---|---|
| | | | Max. | Min. | | | | Max. | Min. |
| 3/8 | 7 | 3/4 | 0.3105 | 0.3095 | 1 3/32 | 10 1/2 | 1 5/8 | 0.8745 | 0.8730 |
| 13/32 | 7 | 3/4 | 0.3105 | 0.3095 | 1 1/8 | 11 | 1 3/4 | 0.8745 | 0.8730 |
| 7/16 | 7 | 7/8 | 0.3730 | 0.3720 | 1 5/32 | 11 | 1 3/4 | 0.8745 | 0.8730 |
| 15/32 | 7 | 7/8 | 0.3730 | 0.3720 | 1 3/16 | 11 | 1 3/4 | 0.9995 | 0.9980 |
| 1/2 | 8 | 1 | 0.4355 | 0.4345 | 1 7/32 | 11 | 1 3/4 | 0.9995 | 0.9980 |
| 17/32 | 8 | 1 | 0.4355 | 0.4345 | 1 1/4 | 11 1/2 | 1 7/8 | 0.9995 | 0.9980 |
| 9/16 | 8 | 1 1/8 | 0.4355 | 0.4345 | 1 5/16 | 11 1/2 | 1 7/8 | 0.9995 | 0.9980 |
| 19/32 | 8 | 1 1/8 | 0.4355 | 0.4345 | 1 3/8 | 12 | 2 | 0.9995 | 0.9980 |
| 5/8 | 9 | 1 1/4 | 0.5620 | 0.5605 | 1 7/16 | 12 | 2 | 1.2495 | 1.2480 |
| 21/32 | 9 | 1 1/4 | 0.5620 | 0.5605 | 1 1/2 | 12 1/2 | 2 1/8 | 1.2495 | 1.2480 |
| 11/16 | 9 | 1 1/4 | 0.5620 | 0.5605 | 1 9/16[a] | 12 1/2 | 2 1/8 | 1.2495 | 1.2480 |
| 23/32 | 9 | 1 1/4 | 0.5620 | 0.5605 | 1 5/8 | 13 | 2 1/4 | 1.2495 | 1.2480 |
| 3/4 | 9 1/2 | 1 3/8 | 0.6245 | 0.6230 | 1 11/16[a] | 13 | 2 1/4 | 1.2495 | 1.2480 |
| 25/32 | 9 1/2 | 1 3/8 | 0.6245 | 0.6230 | 1 3/4 | 13 1/2 | 2 3/8 | 1.2495 | 1.2480 |
| 13/16 | 9 1/2 | 1 3/8 | 0.6245 | 0.6230 | 1 13/16[a] | 13 1/2 | 2 3/8 | 1.4995 | 1.4980 |
| 27/32 | 9 1/2 | 1 3/8 | 0.6245 | 0.6230 | 1 7/8 | 14 | 2 1/2 | 1.4995 | 1.4980 |
| 7/8 | 10 | 1 1/2 | 0.7495 | 0.7480 | 1 15/16[a] | 14 | 2 1/2 | 1.4995 | 1.4980 |
| 29/32 | 10 | 1 1/2 | 0.7495 | 0.7480 | 2 | 14 | 2 1/2 | 1.4995 | 1.4980 |
| 15/16 | 10 | 1 1/2 | 0.7495 | 0.7480 | 2 1/8[b] | 14 1/2 | 2 3/4 | … | … |
| 31/32 | 10 | 1 1/2 | 0.7495 | 0.7480 | 2 1/4[b] | 14 1/2 | 2 3/4 | … | … |
| 1 | 10 1/2 | 1 5/8 | 0.8745 | 0.8730 | 2 3/8[b] | 15 | 3 | … | … |
| 1 1/32 | 10 1/2 | 1 5/8 | 0.8745 | 0.8730 | 2 1/2[b] | 15 | 3 | … | … |
| 1 1/16 | 10 1/2 | 1 5/8 | 0.8745 | 0.8730 | … | … | … | … | … |

[a] Straight shank only.
[b] Taper shank only.

All dimensions in inches. Material is high-speed steel. The number of flutes is as follows: 3/8- to 15/32-inch sizes, 4 to 6; 1/2- to 31/32-inch sizes, 6 to 8; 1- to 1 11/16-inch sizes, 8 to 10; 1 3/4- to 1 15/16-inch sizes, 8 to 12; 2- to 2 1/4-inch sizes, 10 to 12; 2 3/8- and 2 1/2-inch sizes, 10 to 14. The expansion feature of these reamers provides a means of adjustment that is important in reaming holes to close tolerances. When worn undersize, they may be expanded and reground to the original size.

*Tolerances:* On reamer diameter, 3/8- to 1-inch sizes, incl., +0.0001 to +0.0005 inch; over 1-inch size, + 0.0002 to + 0.0006 inch. On length *A* and flute length *B*, 3/8- to 1-inch sizes, incl., ±1/16 inch; 1 1/32- to 2-inch sizes, incl., ±3/32 inch; over 2-inch sizes, ±1/8 inch.

Taper is Morse taper: No. 1 for sizes 3/8 to 19/32 inch, incl.; No. 2 for sizes 5/8 to 29/32 incl.; No. 3 for sizes 15/16 to 1 7/32, incl.; No. 4 for sizes 1 1/4 to 1 5/8, incl.; and No. 5 for sizes 1 3/4 to 2 1/2, incl. For amount of taper, see Table on page 1001.

**Hand Reamers—Straight and Helical Flutes** *ANSI/ASME B94.2-1995 (R2020)*

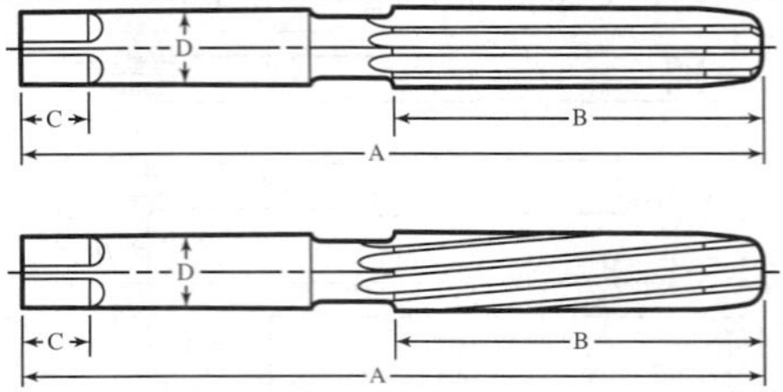


| Reamer Diameter | | | Length Overall | Flute Length | Square Length | Size of | No. of |
|---|---|---|---|---|---|---|---|
| Straight Flutes | Helical Flutes | Decimal Equivalent | *A* | *B* | *C* | Square | Flutes |
| 1/8 | ... | 0.1250 | 3 | 1 1/2 | 5/32 | 0.095 | 4 to 6 |
| 9/64 | ... | 0.1406 | 3 1/4 | 1 5/8 | 5/32 | 0.105 | 4 to 6 |
| 5/32 | ... | 0.1562 | 3 1/4 | 1 5/8 | 7/32 | 0.115 | 4 to 6 |
| 11/16 | ... | 0.1719 | 3 1/2 | 1 3/4 | 7/32 | 0.130 | 4 to 6 |
| 3/16 | ... | 0.1875 | 3 1/2 | 1 3/4 | 7/32 | 0.140 | 4 to 6 |
| 13/64 | ... | 0.2031 | 3 3/4 | 1 7/8 | 1/4 | 0.150 | 4 to 6 |
| 7/32 | ... | 0.2188 | 3 3/4 | 1 7/8 | 1/4 | 0.165 | 4 to 6 |
| 15/64 | ... | 0.2344 | 4 | 2 | 1/4 | 0.175 | 4 to 6 |
| 1/4 | 1/4 | 0.2500 | 4 | 2 | 1/4 | 0.185 | 4 to 6 |
| 17/64 | ... | 0.2969 | 4 1/4 | 2 1/8 | 1/4 | 0.200 | 4 to 6 |
| 9/32 | ... | 0.2812 | 4 1/4 | 2 1/8 | 1/4 | 0.210 | 4 to 6 |
| 19/64 | ... | 0.2969 | 4 1/2 | 2 1/4 | 5/16 | 0.220 | 4 to 6 |
| 5/16 | 5/16 | 0.3125 | 4 1/2 | 2 1/4 | 5/16 | 0.235 | 4 to 6 |
| 21/64 | ... | 0.3281 | 4 3/4 | 2 3/8 | 5/16 | 0.245 | 4 to 6 |
| 11/32 | ... | 0.3438 | 4 3/4 | 2 3/8 | 5/16 | 0.255 | 4 to 6 |
| 23/64 | ... | 0.3594 | 5 | 2 1/2 | 3/8 | 0.270 | 4 to 6 |
| 3/8 | 3/8 | 0.3750 | 5 | 2 1/2 | 3/8 | 0.280 | 4 to 6 |
| 25/64 | ... | 0.3906 | 5 1/4 | 2 5/8 | 3/8 | 0.290 | 6 to 8 |
| 13/32 | ... | 0.4062 | 5 1/4 | 2 5/8 | 3/8 | 0.305 | 6 to 8 |
| 27/64 | ... | 0.4219 | 5 1/2 | 2 3/4 | 7/16 | 0.315 | 6 to 8 |
| 7/16 | 7/16 | 0.4375 | 5 1/2 | 2 3/4 | 7/16 | 0.330 | 6 to 8 |
| 29/64 | ... | 0.4531 | 5 3/4 | 2 7/8 | 7/16 | 0.340 | 6 to 8 |
| 15/32 | ... | 0.4688 | 5 3/4 | 2 7/8 | 7/16 | 0.350 | 6 to 8 |
| 31/64 | ... | 0.4844 | 6 | 3 | 1/2 | 0.365 | 6 to 8 |
| 1/2 | 1/2 | 0.5000 | 6 | 3 | 1/2 | 0.375 | 6 to 8 |
| 17/32 | ... | 0.5312 | 6 1/4 | 3 1/8 | 1/2 | 0.400 | 6 to 8 |
| 9/16 | 9/16 | 0.5625 | 6 1/2 | 3 1/4 | 9/16 | 0.420 | 6 to 8 |
| 19/32 | ... | 0.5938 | 6 3/4 | 3 3/8 | 9/16 | 0.445 | 6 to 8 |
| 5/8 | 5/8 | 0.6250 | 7 | 3 1/2 | 5/8 | 0.470 | 6 to 8 |
| 21/32 | ... | 0.6562 | 7 3/8 | 3 11/16 | 5/8 | 0.490 | 6 to 8 |
| 11/16 | 11/16 | 0.6875 | 7 3/4 | 3 7/8 | 11/16 | 0.515 | 6 to 8 |
| 23/32 | ... | 0.7188 | 8 1/8 | 4 1/16 | 11/16 | 0.540 | 6 to 8 |
| 3/4 | 3/4 | 0.7500 | 8 3/8 | 4 3/16 | 3/4 | 0.560 | 6 to 8 |
| ... | 13/16 | 0.8125 | 9 1/8 | 4 9/16 | 13/16 | 0.610 | 8 to 10 |
| 7/8 | 7/8 | 0.8750 | 9 3/4 | 4 7/8 | 7/8 | 0.655 | 8 to 10 |
| ... | 15/16 | 0.9375 | 10 1/4 | 5 1/8 | 15/16 | 0.705 | 8 to 10 |
| 1 | 1 | 1.0000 | 10 7/8 | 5 7/16 | 1 | 0.750 | 8 to 10 |
| 1 1/8 | 1 1/8 | 1.1250 | 11 5/8 | 5 13/16 | 1 | 0.845 | 8 to 10 |
| 1 1/4 | 1 1/4 | 1.2500 | 12 1/4 | 6 1/8 | 1 | 0.935 | 8 to 12 |
| 1 3/8 | 1 3/8 | 1.3750 | 12 5/8 | 6 5/16 | 1 | 1.030 | 10 to 12 |
| 1 1/2 | 1 1/2 | 1.5000 | 13 | 6 1/2 | 1 1/8 | 1.125 | 10 to 14 |

All dimensions in inches. Material is high-speed steel. The nominal shank diameter *D* is the same as the reamer diameter. Helical-flute hand reamers with left-hand helical flutes are standard. Reamers are tapered slightly on the end to facilitate proper starting.

*Tolerances:* On diameter of reamer, up to 1/4-inch size, incl., + .0001 to + .0004 inch; over 1/4- to 1-inch size, incl., +.0001 to + .0005 inch; over 1-inch size, +.0002 to +.0006 inch. On length overall *A* and flute length *B*, 1/8- to 1-inch size, incl., ± 1/16 inch; 1 1/8- to 1 1/2-inch size, incl., ± 3/32 inch. On length of square *C*, 1/8- to 1 inch size, incl., ± 1/32 inch; 1 1/8- to 1 1/2-inch size, incl., ± 1/16 inch. On shank diameter *D*, 1/8- to 1-inch size, incl., −.001 to −.005 inch; 1 1/8- to 1 1/2-inch size, incl., −.0015 to −.006 inch. On size of square, 1/8- to 1/2-inch size, incl., −.004 inch; 17/32- to 1-inch size, incl., −.006 inch; 1 1/8- to 1 1/2-inch size, incl., −.008 inch.

**American National Standard Expansion Hand Reamers—Straight and Helical Flutes, Squared Shank** *ANSI/ASME B94.2-1995 (R2020)*

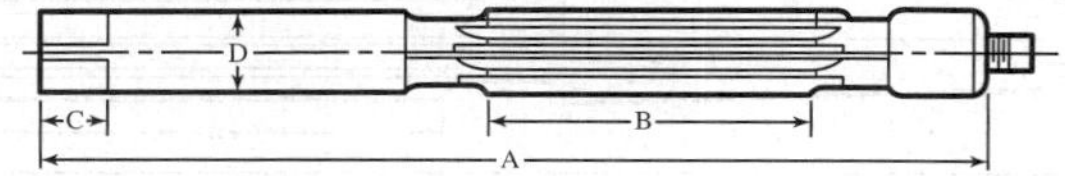


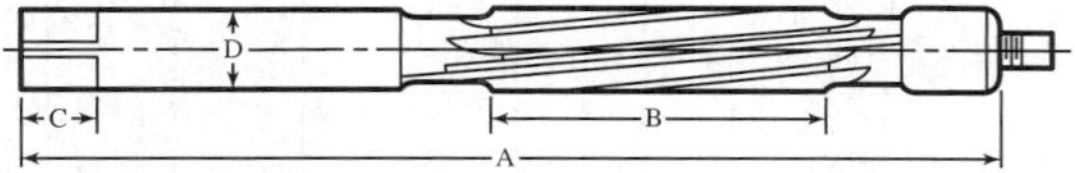


| Reamer Dia. | Length Overall A | | Flute Length B | | Length of Square C | Shank Dia. D | Size of Square | Number of Flutes |
|---|---|---|---|---|---|---|---|---|
| | Max. | Min. | Max. | Min. | | | | |
| Straight Flutes | | | | | | | | |
| ¼ | 4⅜ | 3¾ | 1¾ | 1½ | ¼ | ¼ | 0.185 | 6 to 8 |
| 5⁄16 | 4⅜ | 4 | 1⅞ | 1½ | 5⁄16 | 5⁄16 | 0.235 | 6 to 8 |
| ⅜ | 5⅜ | 4¼ | 2 | 1¾ | ⅜ | ⅜ | 0.280 | 6 to 9 |
| 7⁄16 | 5⅜ | 4½ | 2 | 1¾ | 7⁄16 | 7⁄16 | 0.330 | 6 to 9 |
| ½ | 6½ | 5 | 2½ | 1¾ | ½ | ½ | 0.375 | 6 to 9 |
| 9⁄16 | 6½ | 5⅜ | 2½ | 1⅞ | 9⁄16 | 9⁄16 | 0.420 | 6 to 9 |
| ⅝ | 7 | 5¾ | 3 | 2¼ | ⅝ | ⅝ | 0.470 | 6 to 9 |
| 11⁄16 | 7⅝ | 6¼ | 3 | 2½ | 11⁄16 | 11⁄16 | 0.515 | 6 to 10 |
| ¾ | 8 | 6½ | 3½ | 2⅝ | ¾ | ¾ | 0.560 | 6 to 10 |
| ⅞ | 9 | 7½ | 4 | 3⅛ | ⅞ | ⅞ | 0.655 | 8 to 10 |
| 1 | 10 | 8⅜ | 4½ | 3⅛ | 1 | 1 | 0.750 | 8 to 10 |
| 1⅛ | 10½ | 9 | 4¾ | 3½ | 1 | 1⅛ | 0.845 | 8 to 12 |
| 1¼ | 11 | 9¾ | 5 | 4¼ | 1 | 1¼ | 0.935 | 8 to 12 |
| Helical Flutes | | | | | | | | |
| ¼ | 4⅜ | 3⅞ | 1¾ | 1½ | ¼ | ¼ | 0.185 | 6 to 8 |
| 5⁄16 | 4⅜ | 4 | 1¾ | 1½ | 5⁄16 | 5⁄16 | 0.235 | 6 to 8 |
| ⅜ | 6⅛ | 4¼ | 2 | 1¾ | ⅜ | ⅜ | 0.280 | 6 to 9 |
| 7⁄16 | 6¼ | 4½ | 2 | 1¾ | 7⁄16 | 7⁄16 | 0.330 | 6 to 9 |
| ½ | 6½ | 5 | 2½ | 1¾ | ½ | ½ | 0.375 | 6 to 9 |
| ⅝ | 8 | 6 | 3 | 2¼ | ⅝ | ⅝ | 0.470 | 6 to 9 |
| ¾ | 8⅝ | 6½ | 3½ | 2⅝ | ¾ | ¾ | 0.560 | 6 to 10 |
| ⅞ | 9⅜ | 7½ | 4 | 3⅛ | ⅞ | ⅞ | 0.655 | 6 to 10 |
| 1 | 10¼ | 8⅜ | 4½ | 3⅛ | 1 | 1 | 0.750 | 6 to 10 |
| 1¼ | 11⅜ | 9¾ | 5 | 4¼ | 1 | 1¼ | 0.935 | 8 to 12 |

All dimensions are given in inches. Material is carbon steel. Reamers with helical flutes that are left-hand are standard. Expansion hand reamers are primarily designed for work where it is necessary to enlarge reamed holes by a few thousandths. The pilots and guides on these reamers are ground undersize for clearance. The maximum expansion on these reamers is as follows: .006 inch for the ¼- to 7⁄16-inch sizes. .010 inch for the ½- to ⅞-inch sizes and .012 inch for the 1- to 1¼-inch sizes.

*Tolerances:* On length overall *A* and flute length *B*, ±1⁄16 inch for ¼- to 1-inch sizes, ± 3⁄32 inch for 1⅛-to 1¼-inch sizes; on length of square *C*, ±1⁄32 inch for ¼- to 1-inch sizes, ± 1⁄16 inch for 1⅛-to 1¼-inch sizes; on shank diameter *D* −.001 to −.005 inch for ¼- to 1-inch sizes, −.0015 to −.006 inch for 1⅛- to 1¼-inch sizes; on size of square, −.004 inch for ¼- to ½-inch sizes. −.006 inch for 9⁄16- to 1-inch sizes, and −.008 inch for 1⅛- to 1¼-inch sizes.

## Taper Shank Jobbers Reamers—Straight Flutes *ANSI/ASME B94.2-1995 (R2020)*

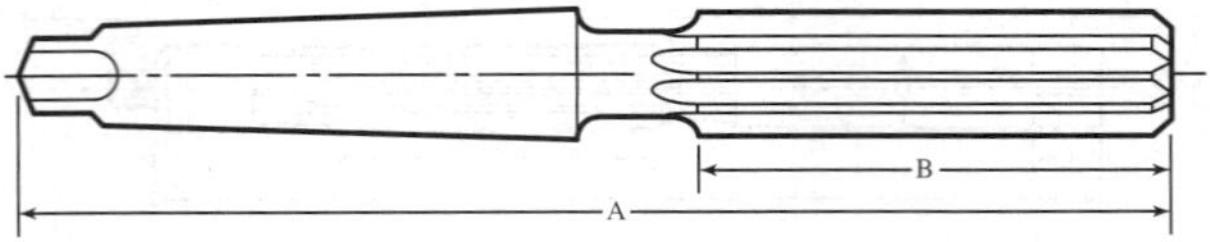


| Reamer Diameter | | Length Overall *A* | Length of Flute *B* | No. of Morse Taper Shank[a] | No. of Flutes |
|---|---|---|---|---|---|
| Fractional | Dec. Equiv. | | | | |
| 1/4 | 0.2500 | 5 3/16 | 2 | 1 | 6 to 8 |
| 5/16 | 0.3125 | 5 1/2 | 2 1/4 | 1 | 6 to 8 |
| 3/8 | 0.3750 | 5 13/16 | 2 1/2 | 1 | 6 to 8 |
| 7/16 | 0.4375 | 6 1/8 | 2 3/4 | 1 | 6 to 8 |
| 1/2 | 0.5000 | 6 7/16 | 3 | 1 | 6 to 8 |
| 9/16 | 0.5625 | 6 3/4 | 3 1/4 | 1 | 6 to 8 |
| 5/8 | 0.6250 | 7 9/16 | 3 1/2 | 2 | 6 to 8 |
| 11/16 | 0.6875 | 8 | 3 7/8 | 2 | 8 to 10 |
| 3/4 | 0.7500 | 8 3/8 | 4 3/16 | 2 | 8 to 10 |
| 13/16 | 0.8125 | 8 13/16 | 4 9/16 | 2 | 8 to 10 |
| 7/8 | 0.8750 | 9 3/16 | 4 7/8 | 2 | 8 to 10 |
| 15/16 | 0.9375 | 10 | 5 1/8 | 3 | 8 to 10 |
| 1 | 1.0000 | 10 3/8 | 5 7/16 | 3 | 8 to 10 |
| 1 1/16 | 1.0625 | 10 5/8 | 5 5/8 | 3 | 8 to 10 |
| 1 1/8 | 1.1250 | 10 7/8 | 5 13/16 | 3 | 8 to 10 |
| 1 3/16 | 1.1875 | 11 1/8 | 6 | 3 | 8 to 12 |
| 1 1/4 | 1.2500 | 12 9/16 | 6 1/8 | 4 | 8 to 12 |
| 1 3/8 | 1.3750 | 12 13/16 | 6 5/16 | 4 | 10 to 12 |
| 1 1/2 | 1.5000 | 13 1/8 | 6 1/2 | 4 | 10 to 12 |

[a] American National Standard self-holding tapers (Table 7a on page 1008).

All dimensions in inches. Material is high-speed steel.

*Tolerances:* On reamer diameter, 1/4-inch size, +.0001 to +.0004 inch; over 1/4- to 1-inch size, incl., +.0001 to +.0005 inch; over 1-inch size, +.0002 to +.0006 inch. On overall length *A* and length of flute *B*, 1/4- to 1-inch size, incl., ±1/16 inch; and 1 1/16- to 1 1/2-inch size, incl., ±3/32 inch.

## American National Standard Driving Slots and Lugs for Shell Reamers or Shell Reamer Arbors *ANSI/ASME B94.2-1995 (R2020)*

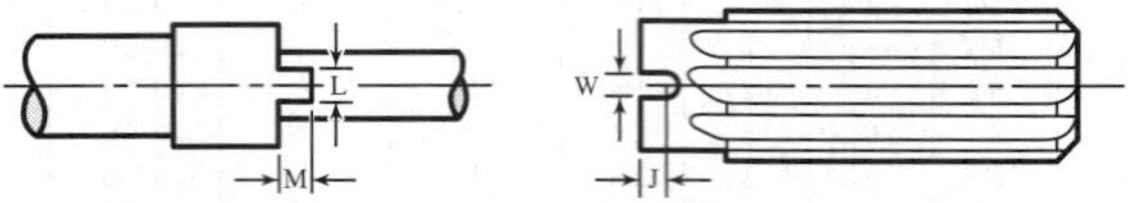


| Arbor Size No. | Fitting Reamer Sizes | Driving Slot | | Lug on Arbor | | Reamer Hole Dia. at Large End |
|---|---|---|---|---|---|---|
| | | Width *W* | Depth *J* | Width *L* | Depth *M* | |
| 4 | 3/4 | 5/32 | 3/16 | 9/64 | 5/32 | 0.375 |
| 5 | 13/16 to 1 | 3/16 | 1/4 | 11/64 | 7/32 | 0.500 |
| 6 | 1 1/16 to 1 1/4 | 3/16 | 1/4 | 11/64 | 7/32 | 0.625 |
| 7 | 1 5/16 to 1 5/8 | 1/4 | 5/16 | 15/64 | 9/32 | 0.750 |
| 8 | 1 11/16 to 2 | 1/4 | 5/16 | 15/64 | 9/32 | 1.000 |
| 9 | 2 1/16 to 2 1/2 | 5/16 | 3/8 | 19/64 | 11/32 | 1.250 |

All dimension are given in inches. The hole in shell reamers has a taper of 1/8 inch per foot, with arbors tapered to correspond. Shell reamer arbor tapers are made to permit a driving fit with the reamer.

## Straight-Shank Chucking Reamers—Straight Flutes, Wire Gage Sizes
*ANSI/ASME B94.2-1995 (R2020)*

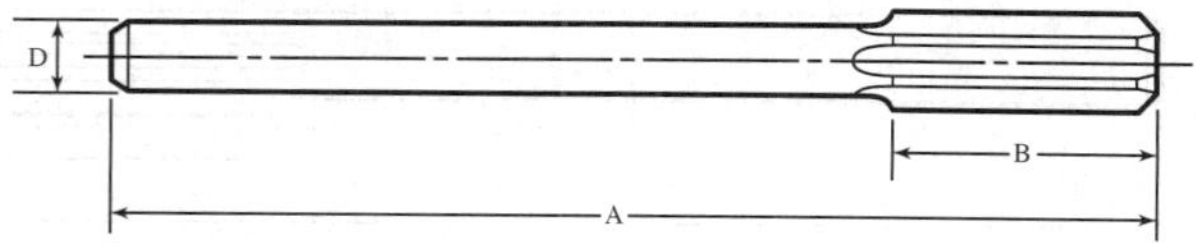


| Reamer Diameter: Wire Gage | Reamer Diameter: Inch | Lgth. Overall *A* | Lgth. of Flute *B* | Shank Dia. *D*: Max. | Shank Dia. *D*: Min. | No. of Flutes | Reamer Diameter: Wire Gage | Reamer Diameter: Inch | Lgth. Overall *A* | Lgth. of Flute *B* | Shank Dia. *D*: Max. | Shank Dia. *D*: Min. | No. of Flutes |
|---|---|---|---|---|---|---|---|---|---|---|---|---|---|
| 60 | .0400 | 2½ | ½ | .0390 | .0380 | 4 | 49 | .0730 | 3 | ¾ | .0660 | .0650 | 4 |
| 59 | .0410 | 2½ | ½ | .0390 | .0380 | 4 | 48 | .0760 | 3 | ¾ | .0720 | .0710 | 4 |
| 58 | .0420 | 2½ | ½ | .0390 | .0380 | 4 | 47 | .0785 | 3 | ¾ | .0720 | .0710 | 4 |
| 57 | .0430 | 2½ | ½ | .0390 | .0380 | 4 | 46 | .0810 | 3 | ¾ | .0771 | .0701 | 4 |
| 56 | .0465 | 2½ | ½ | .0455 | .0445 | 4 | 45 | .0820 | 3 | ¾ | .0771 | .0761 | 4 |
| 55 | .0520 | 2½ | ½ | .0510 | .0500 | 4 | 44 | .0860 | 3 | ¾ | .0810 | .0800 | 4 |
| 54 | .0550 | 2½ | ½ | .0510 | .0500 | 4 | 43 | .0890 | 3 | ¾ | .0810 | .0800 | 4 |
| 53 | .0595 | 2½ | ½ | .0585 | .0575 | 4 | 42 | .0935 | 3 | ¾ | .0880 | .0870 | 4 |
| 52 | .0635 | 2½ | ½ | .0585 | .0575 | 4 | 41 | .0960 | 3½ | ⅞ | .0928 | .0918 | 4 to 6 |
| 51 | .0670 | 3 | ¾ | .0660 | .0650 | 4 | 40 | .0980 | 3½ | ⅞ | .0928 | .0918 | 4 to 6 |
| 50 | .0700 | 3 | ¾ | .0660 | .0650 | 4 | 39 | .0995 | 3½ | ⅞ | .0928 | .0918 | 4 to 6 |
| 38 | .1015 | 3½ | ⅞ | .0950 | .0940 | 4 to 6 | 19 | .1660 | 4½ | 1⅛ | .1595 | .1585 | 4 to 6 |
| 37 | .1040 | 3½ | ⅞ | .0950 | .0940 | 4 to 6 | 18 | .1695 | 4½ | 1⅛ | .1595 | .1585 | 4 to 6 |
| 36 | .1065 | 3½ | ⅞ | .1030 | .1020 | 4 to 6 | 17 | .1730 | 4½ | 1⅛ | .1645 | .1635 | 4 to 6 |
| 35 | .1100 | 3½ | ⅞ | .1030 | .1020 | 4 to 6 | 16 | .1770 | 4½ | 1⅛ | .1704 | .1694 | 4 to 6 |
| 34 | .1110 | 3½ | ⅞ | .1055 | .1045 | 4 to 6 | 15 | .1800 | 4½ | 1⅛ | .1755 | .1745 | 4 to 6 |
| 33 | .1130 | 3½ | ⅞ | .1055 | .1045 | 4 to 6 | 14 | .1820 | 4½ | 1⅛ | .1755 | .1745 | 4 to 6 |
| 32 | .1160 | 3½ | ⅞ | .1120 | .1110 | 4 to 6 | 13 | .1850 | 4½ | 1⅛ | .1805 | .1795 | 4 to 6 |
| 31 | .1200 | 3½ | ⅞ | .1120 | .1110 | 4 to 6 | 12 | .1890 | 4½ | 1⅛ | .1805 | .1795 | 4 to 6 |
| 30 | .1285 | 3½ | ⅞ | .1190 | .1180 | 4 to 6 | 11 | .1910 | 5 | 1¼ | .1860 | .1850 | 4 to 6 |
| 29 | .1360 | 4 | 1 | .1275 | .1265 | 4 to 6 | 10 | .1935 | 5 | 1¼ | .1860 | .1850 | 4 to 6 |
| 28 | .1405 | 4 | 1 | .1350 | .1340 | 4 to 6 | 9 | .1960 | 5 | 1¼ | .1895 | .1885 | 4 to 6 |
| 27 | .1440 | 4 | 1 | .1350 | .1340 | 4 to 6 | 8 | .1990 | 5 | 1¼ | .1895 | .1885 | 4 to 6 |
| 26 | .1470 | 4 | 1 | .1430 | .1420 | 4 to 6 | 7 | .2010 | 5 | 1¼ | .1945 | .1935 | 4 to 6 |
| 25 | .1495 | 4 | 1 | .1430 | .1420 | 4 to 6 | 6 | .2040 | 5 | 1¼ | .1945 | .1935 | 4 to 6 |
| 24 | .1520 | 4 | 1 | .1460 | .1450 | 4 to 6 | 5 | .2055 | 5 | 1¼ | .2016 | .2006 | 4 to 6 |
| 23 | .1540 | 4 | 1 | .1460 | .1450 | 4 to 6 | 4 | .2090 | 5 | 1¼ | .2016 | .2006 | 4 to 6 |
| 22 | .1570 | 4 | 1 | .1510 | .1500 | 4 to 6 | 3 | .2130 | 5 | 1¼ | .2075 | .2065 | 4 to 6 |
| 21 | .1590 | 4½ | 1⅛ | .1530 | .1520 | 4 to 6 | 2 | 2210 | 6 | 1½ | .2173 | .2163 | 4 to 6 |
| 20 | .1610 | 4½ | 1⅛ | .1530 | .1520 | 4 to 6 | 1 | .2280 | 6 | 1½ | .2173 | .2163 | 4 to 6 |

All dimensions in inches. Material is high-speed steel.

*Tolerances:* On diameter of reamer, plus .0001 to plus .0004 inch. On overall length *A*, plus or minus $^{1}/_{16}$ inch. On length of flute *B*, plus or minus $^{1}/_{16}$ inch.

## Straight-Shank Chucking Reamers—Straight Flutes, Letter Sizes

*ANSI/ASME B94.2-1995 (R2020)*

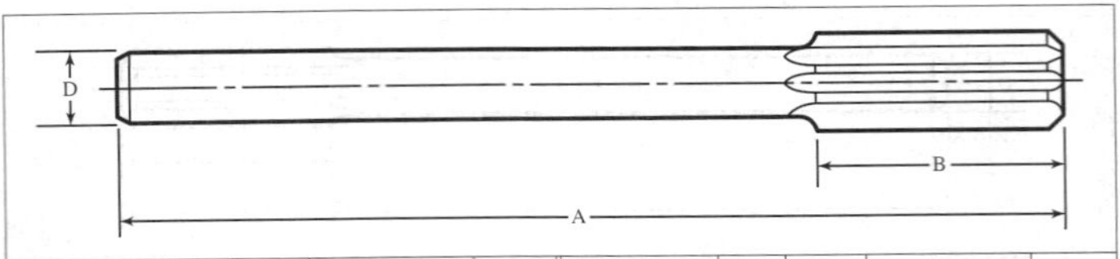


| Reamer Diameter | | Lgth. Over-all *A* | Lgth. of Flute *B* | Shank Dia. *D* | | No. of Flutes | Reamer Diameter | | Lgth. Over-all *A* | Lgth. of Flute *B* | Shank Dia. *D* | | No. of Flutes |
|---|---|---|---|---|---|---|---|---|---|---|---|---|---|
| Letter | Inch | | | Max. | Min. | | Letter | Inch | | | Max. | Min. | |
| A | 0.2340 | 6 | 1½ | 0.2265 | .2255 | 4 to 6 | N | 0.3020 | 6 | 1½ | 0.2792 | 0.2782 | 4 to 6 |
| B | 0.2380 | 6 | 1½ | 0.2329 | .2319 | 4 to 6 | O | 0.3160 | 6 | 1½ | 0.2792 | 0.2782 | 4 to 6 |
| C | 0.2420 | 6 | 1½ | 0.2329 | .2319 | 4 to 6 | P | 0.3230 | 6 | 1½ | 0.2792 | 0.2782 | 4 to 6 |
| D | 0.2460 | 6 | 1½ | 0.2329 | .2319 | 4 to 6 | Q | 0.3320 | 6 | 1½ | 0.2792 | 0.2782 | 4 to 6 |
| E | 0.2500 | 6 | 1½ | 0.2405 | .2395 | 4 to 6 | R | 0.3390 | 6 | 1½ | 0.2792 | 0.2782 | 4 to 6 |
| F | 0.2570 | 6 | 1½ | 0.2485 | .2475 | 4 to 6 | S | 0.3480 | 7 | 1¾ | 0.3105 | 0.3095 | 4 to 6 |
| G | 0.2610 | 6 | 1½ | 0.2485 | .2475 | 4 to 6 | T | 0.3580 | 7 | 1¾ | 0.3105 | 0.3095 | 4 to 6 |
| H | 0.2660 | 6 | 1½ | 0.2485 | .2475 | 4 to 6 | U | 0.3680 | 7 | 1¾ | 0.3105 | 0.3095 | 4 to 6 |
| I | 0.2720 | 6 | 1½ | 0.2485 | .2475 | 4 to 6 | V | 0.3770 | 7 | 1¾ | 0.3105 | 0.3095 | 4 to 6 |
| J | 0.2770 | 6 | 1½ | 0.2485 | .2475 | 4 to 6 | W | 0.3860 | 7 | 1¾ | 0.3105 | 0.3095 | 4 to 6 |
| K | 0.2810 | 6 | 1½ | 0.2485 | .2475 | 4 to 6 | X | 0.3970 | 7 | 1¾ | 0.3105 | 0.3095 | 4 to 6 |
| L | 0.2900 | 6 | 1½ | 0.2792 | .2782 | 4 to 6 | Y | 0.4040 | 7 | 1¾ | 0.3105 | 0.3095 | 4 to 6 |
| M | 0.2950 | 6 | 1½ | 0.2792 | .2782 | 4 to 6 | Z | 0.4130 | 7 | 1¾ | 0.3730 | 0.3720 | 6 to 8 |

All dimensions in inches. Material is high-speed steel.

*Tolerances:* On diameter of reamer, for sizes A to E, incl., plus .0001 to plus .0004 inch and for sizes F to Z, incl., plus .0001 to plus .0005 inch. On overall length *A*, plus or minus 1⁄16 inch. On length of flute *B*, plus or minus 1⁄16 inch.

## Straight-Shank Chucking Reamers—Straight Flutes, Decimal Sizes

*ANSI/ASME B94.2-1995 (R2020)*

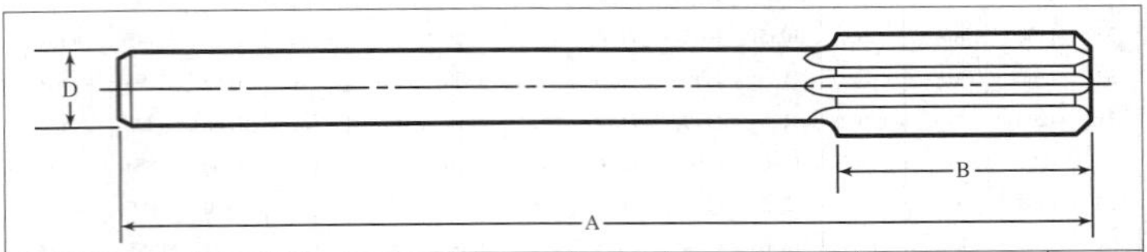


| Reamer Dia. | Lgth. Overall *A* | Lgth. of Flute *B* | Shank Diameter *D* | | No. of Flutes | Reamer Dia. | Lgth. Overall *A* | Lgth. of Flute *B* | Shank Diameter *D* | | No. of Flutes |
|---|---|---|---|---|---|---|---|---|---|---|---|
| | | | Max. | Min. | | | | | Max. | Min. | |
| 0.1240 | 3½ | ⅞ | 0.1190 | 0.1180 | 4 to 6 | 0.3135 | 6 | 1½ | 0.2792 | 0.2782 | 4 to 6 |
| 0.1260 | 3½ | ⅞ | 0.1190 | 0.1180 | 4 to 6 | 0.3740 | 7 | 1¾ | 0.3105 | 0.3095 | 6 to 8 |
| 0.1865 | 4½ | 1⅛ | 0.1805 | 0.1795 | 4 to 6 | 0.3760 | 7 | 1¾ | 0.3105 | 0.3095 | 6 to 8 |
| 0.1885 | 4½ | 1⅛ | 0.1805 | 0.1795 | 4 to 6 | 0.4365 | 7 | 1¾ | 0.3730 | 0.3720 | 6 to 8 |
| 0.2490 | 6 | 1½ | 0.2405 | 0.2395 | 4 to 6 | 0.4385 | 7 | 1¾ | 0.3730 | 0.3720 | 6 to 8 |
| 0.2510 | 6 | 1½ | 0.2405 | 0.2395 | 4 to 6 | 0.4990 | 8 | 2 | 0.4355 | 0.4345 | 6 to 8 |
| 0.3115 | 6 | 1½ | 0.2792 | 0.2782 | 4 to 6 | 0.5010 | 8 | 2 | 0.4355 | 0.4345 | 6 to 8 |

All dimensions in inches. Material is high-speed steel.

*Tolerances:* On diameter of reamer, for 0.124 to 0.249-inch sizes, plus .0001 to plus .0004 inch and for 0.251 to 0.501-inch sizes, plus .0001 to plus .0005 inch. On overall length *A*, plus or minus 1⁄16 inch. On length of flute *B*, plus or minus 1⁄16 inch.

## American National Standard Straight-Shank Rose Chucking and Chucking Reamers— Straight and Helical Flutes, Fractional Sizes *ANSI/ASME B94.2-1995 (R2020)*

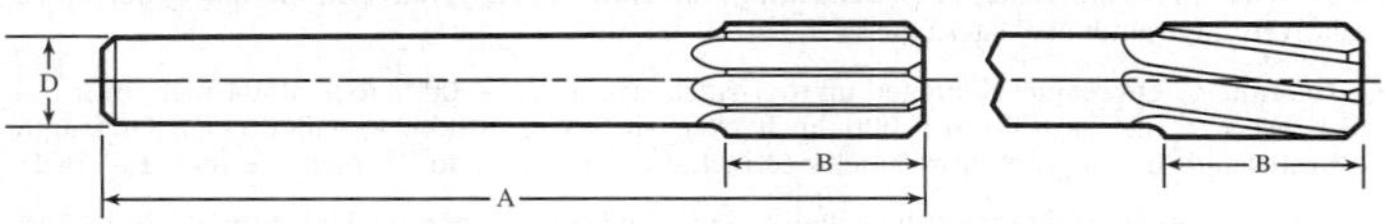


| Reamer Diameter | | Length | Flute | Shank Dia. *D* | | No. of |
|---|---|---|---|---|---|---|
| Chucking | Rose Chucking | Overall *A* | Length *B* | Max. | Min. | Flutes |
| 3/64[a] | … | 2½ | ½ | 0.0455 | 0.0445 | 4 |
| 1/16 | … | 2½ | ½ | 0.0585 | 0.0575 | 4 |
| 5/64 | … | 3 | ¾ | 0.0720 | 0.0710 | 4 |
| 3/32 | … | 3 | ¾ | 0.0880 | 0.0870 | 4 |
| 7/64 | … | 3½ | ⅞ | 0.1030 | 0.1020 | 4 to 6 |
| ⅛ | ⅛[a] | 3½ | ⅞ | 0.1190 | 0.1180 | 4 to 6 |
| 9/64 | … | 4 | 1 | 0.1350 | 0.1340 | 4 to 6 |
| 5/32 | … | 4 | 1 | 0.1510 | 0.1500 | 4 to 6 |
| 11/64 | … | 4½ | 1⅛ | 0.1645 | 0.1635 | 4 to 6 |
| 3/16 | 3/16[a] | 4½ | 1⅛ | 0.1805 | 0.1795 | 4 to 6 |
| 13/64 | … | 5 | 1¼ | 0.1945 | 0.1935 | 4 to 6 |
| 7/32 | … | 5 | 1¼ | 0.2075 | 0.2065 | 4 to 6 |
| 15/64 | … | 6 | 1½ | 0.2265 | 0.2255 | 4 to 6 |
| ¼ | ¼[a] | 6 | 1½ | 0.2405 | 0.2395 | 4 to 6 |
| 17/64 | … | 6 | 1½ | 0.2485 | 0.2475 | 4 to 6 |
| 9/32 | … | 6 | 1½ | 0.2485 | 0.2475 | 4 to 6 |
| 19/64 | … | 6 | 1½ | 0.2792 | 0.2782 | 4 to 6 |
| 5/16 | 5/16[a] | 6 | 1½ | 0.2792 | 0.2782 | 4 to 6 |
| 21/64 | … | 6 | 1½ | 0.2792 | 0.2782 | 4 to 6 |
| 11/32 | … | 6 | 1½ | 0.2792 | 0.2782 | 4 to 6 |
| 23/64 | … | 7 | 1¾ | 0.3105 | 0.3095 | 4 to 6 |
| ⅜ | ⅜[a] | 7 | 1¾ | 0.3105 | 0.3095 | 4 to 6 |
| 25/64 | … | 7 | 1¾ | 0.3105 | 0.3095 | 4 to 6 |
| 13/32 | … | 7 | 1¾ | 0.3105 | 0.3095 | 4 to 6 |
| 27/64 | … | 7 | 1¾ | 0.3730 | 0.3720 | 6 to 8 |
| 7/16 | 7/16[a] | 7 | 1¾ | 0.3730 | 0.3720 | 6 to 8 |
| 29/64 | … | 7 | 1¾ | 0.3730 | 0.3720 | 6 to 8 |
| 15/32 | … | 7 | 1¾ | 0.3730 | 0.3720 | 6 to 8 |
| 31/64 | … | 8 | 2 | 0.4355 | 0.4345 | 6 to 8 |
| ½ | ½[a] | 8 | 2 | 0.4355 | 0.4345 | 6 to 8 |
| 17/32 | … | 8 | 2 | 0.4355 | 0.4345 | 6 to 8 |
| 9/16 | … | 8 | 2 | 0.4355 | 0.4345 | 6 to 8 |
| 19/32 | … | 8 | 2 | 0.4355 | 0.4345 | 6 to 8 |
| ⅝ | … | 9 | 2¼ | 0.5620 | 0.5605 | 6 to 8 |
| 21/32 | … | 9 | 2¼ | 0.5620 | 0.5605 | 6 to 8 |
| 11/16 | … | 9 | 2¼ | 0.5620 | 0.5605 | 6 to 8 |
| 23/32 | … | 9 | 2¼ | 0.5620 | 0.5605 | 6 to 8 |
| ¾ | … | 9½ | 2½ | 0.6245 | 0.6230 | 6 to 8 |
| 25/32 | … | 9½ | 2½ | 0.6245 | 0.6230 | 8 to 10 |
| 13/16 | … | 9½ | 2½ | 0.6245 | 0.6230 | 8 to 10 |
| 27/32 | … | 9½ | 2½ | 0.6245 | 0.6230 | 8 to 10 |
| ⅞ | … | 10 | 2⅝ | 0.7495 | 0.7480 | 8 to 10 |
| 29/32 | … | 10 | 2⅝ | 0.7495 | 0.7480 | 8 to 10 |
| 15/16 | … | 10 | 2⅝ | 0.7495 | 0.7480 | 8 to 10 |
| 31/32 | … | 10 | 2⅝ | 0.7495 | 0.7480 | 8 to 10 |
| 1 | … | 10½ | 2¾ | 0.8745 | 0.8730 | 8 to 12 |
| 1 1/16 | … | 10½ | 2¾ | 0.8745 | 0.8730 | 8 to 12 |
| 1⅛ | … | 11 | 2⅞ | 0.8745 | 0.8730 | 8 to 12 |
| 1 3/16 | … | 11 | 2⅞ | 0.9995 | 0.9980 | 8 to 12 |
| 1¼ | … | 11½ | 3 | 0.9995 | 0.9980 | 8 to 12 |
| 1 5/16[b] | … | 11½ | 3 | 0.9995 | 0.9980 | 10 to 12 |
| 1⅜ | … | 12 | 3¼ | 0.9995 | 0.9980 | 10 to 12 |
| 1 7/16[b] | … | 12 | 3¼ | 1.2495 | 1.2480 | 10 to 12 |
| 1½ | … | 12½ | 3½ | 1.2495 | 1.2480 | 10 to 12 |

[a] Reamer with straight flutes is standard only.

[b] Reamer with helical flutes is standard only.

All dimensions are given in inches. Material is high-speed steel. Chucking reamers are end cutting on the chamfer and the relief for the outside diameter is ground in back of the margin for the full length of land. Lands of rose chucking reamers are not relieved on the periphery but have a relatively large amount of back taper.

*Tolerances:* On reamer diameter, up to ¼-inch size, incl., + .0001 to + .0004 inch; over ¼- to 1-inch size, incl., + .0001 to + .0005 inch; over 1-inch size, + .0002 to + .0006 inch. On length overall *A* and flute length *B*, up to 1-inch size, incl., ±1⁄16 inch; 1 1⁄16- to 1½-inch size, incl., ±3⁄32 inch.

Helical flutes are right- or left-hand helix, right-hand cut, except sizes 1 1⁄16 through 1½ inches, which are right-hand helix only.

**Shell Reamers—Straight and Helical Flutes** *ANSI/ASME B94.2-1995 (R2020)*

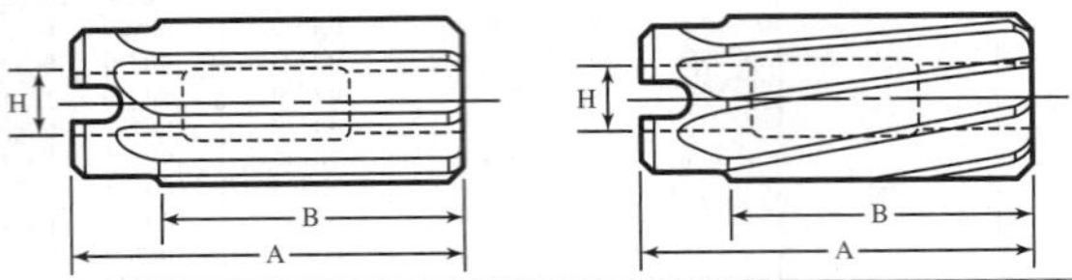


| Diameter of Reamer | Length Overall *A* | Flute Length *B* | Hole Diameter Large End *H* | Fitting Arbor No. | Number of Flutes |
|---|---|---|---|---|---|
| ¾ | 2¼ | 1½ | 0.375 | 4 | 8 to 10 |
| ⅞ | 2½ | 1¾ | 0.500 | 5 | 8 to 10 |
| 15⁄16[a] | 2½ | 1¾ | 0.500 | 5 | 8 to 10 |
| 1 | 2½ | 1¾ | 0.500 | 5 | 8 to 10 |
| 1 1⁄16 | 2¾ | 2 | 0.625 | 6 | 8 to 12 |
| 1⅛ | 2¾ | 2 | 0.625 | 6 | 8 to 12 |
| 1 3⁄16 | 2¾ | 2 | 0.625 | 6 | 8 to 12 |
| 1¼ | 2¾ | 2 | 0.625 | 6 | 8 to 12 |
| 1 5⁄16 | 3 | 2¼ | 0.750 | 7 | 8 to 12 |
| 1⅜ | 3 | 2¼ | 0.750 | 7 | 8 to 12 |
| 1 7⁄16 | 3 | 2¼ | 0.750 | 7 | 8 to 12 |
| 1½ | 3 | 2¼ | 0.750 | 7 | 10 to 14 |
| 1 9⁄16 | 3 | 2¼ | 0.750 | 7 | 10 to 14 |
| 1⅝ | 3 | 2¼ | 0.750 | 7 | 10 to 14 |
| 1 11⁄16 | 3½ | 2½ | 1.000 | 8 | 10 to 14 |
| 1¾ | 3½ | 2½ | 1.000 | 8 | 12 to 14 |
| 1 13⁄16 | 3½ | 2½ | 1.000 | 8 | 12 to 14 |
| 1⅞ | 3½ | 2½ | 1.000 | 8 | 12 to 14 |
| 1 15⁄16 | 3½ | 2½ | 1.000 | 8 | 12 to 14 |
| 2 | 3½ | 2½ | 1.000 | 8 | 12 to 14 |
| 2 1⁄16[a] | 3¾ | 2¾ | 1.250 | 9 | 12 to 16 |
| 2⅛ | 3¾ | 2¾ | 1.250 | 9 | 12 to 16 |
| 2 3⁄16[a] | 3¾ | 2¾ | 1.250 | 9 | 12 to 16 |
| 2¼ | 3¾ | 2¾ | 1.250 | 9 | 12 to 16 |
| 2⅜[a] | 3¾ | 2¾ | 1.250 | 9 | 14 to 16 |
| 2½[a] | 3¾ | 2¾ | 1.250 | 9 | 14 to 16 |

[a] Helical flutes only.

All dimensions are given in inches. Material is high-speed steel. Helical flute shell reamers with left-hand helical flutes are standard. Shell reamers are designed as a sizing or finishing reamer and are held on an arbor provided with driving lugs. The holes in these reamers are ground with a taper of ⅛ inch per foot.

*Tolerances:* On diameter of reamer, ¾- to 1-inch size, incl., + .0001 to + .0005 inch; over 1-inch size, + .0002 to + .0006 inch. On length overall *A* and flute length *B*, ¾- to 1-inch size, incl., ± 1⁄16 inch; 1 1⁄16- to 2-inch size, incl., ± 3⁄32 inch; 2 1⁄16- to 2½-inch size, incl., ± ⅛ inch.

## American National Standard Arbors for Shell Reamers— Straight and Taper Shanks *ANSI/ASME B94.2-1995 (R2020)*

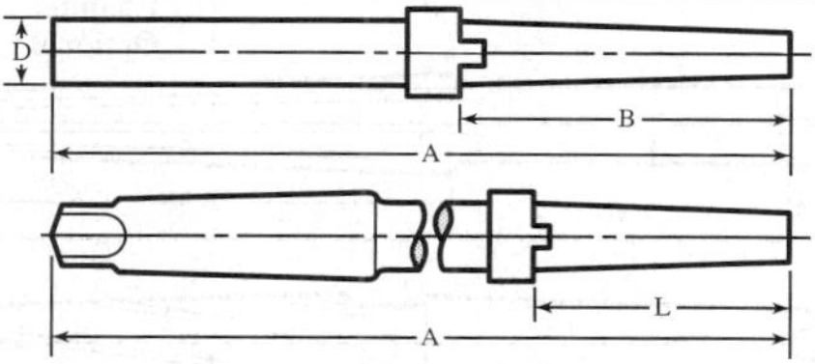


| Arbor Size No. | Overall Length *A* | Approx. Length of Taper *L* | Reamer Size | Taper Shank No.[a] | Straight Shank Dia. *D* | Arbor Size No. | Overall Length *A* | Approx. Length of Taper *L* | Reamer Size | Taper Shank No.[a] | Straight Shank Dia. *D* |
|---|---|---|---|---|---|---|---|---|---|---|---|
| 4 | 9 | 2¼ | ¾ | 2 | ½ | 7 | 11 | 3 | 1 5/16 to 1⅝ | 3 | ⅞ |
| 5 | 9½ | 2½ | 13/16 to 1 | 2 | ⅝ | 8 | 12 | 3½ | 1 11/16 to 2 | 4 | 1⅛ |
| 6 | 10 | 2¾ | 1 1/16 to 1¼ | 3 | ¾ | 9 | 13 | 3¾ | 2 1/16 to 2½ | 4 | 1⅜ |

[a] American National Standard self-holding tapers (see Table 7a on page 1008).

All dimensions are given in inches. These arbors are designed to fit standard shell reamers (see table). End that fits reamer has taper of $\frac{1}{8}$ inch per foot.

## Stub Screw Machine Reamers—Helical Flutes *ANSI/ASME B94.2-1995 (R2020)*

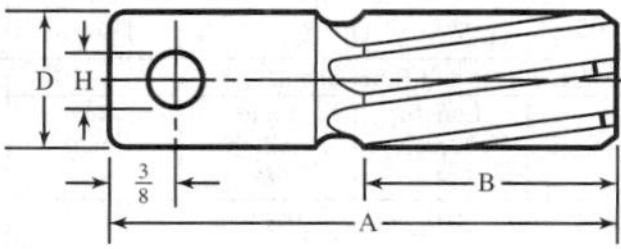


| Series No. | Diameter Range | Length Overall *A* | Length of Flute *B* | Dia. of Shank *D* | Size of Hole *H* | Flute No. | Series No. | Diameter Range | Length Overall *A* | Length of Flute *B* | Dia. of Shank *D* | Size of Hole *H* | Flute No. |
|---|---|---|---|---|---|---|---|---|---|---|---|---|---|
| 00 | .0600-.066 | 1¾ | ½ | ⅛ | 1/16 | 4 | 12 | .3761-.407 | 2½ | 1¼ | ½ | 3/16 | 6 |
| 0 | .0661-.074 | 1¾ | ½ | ⅛ | 1/16 | 4 | 13 | .4071-.439 | 2½ | 1¼ | ½ | 3/16 | 6 |
| 1 | .0741-.084 | 1¾ | ½ | ⅛ | 1/16 | 4 | 14 | .4391-.470 | 2½ | 1¼ | ½ | 3/16 | 6 |
| 2 | .0841-.096 | 1¾ | ½ | ⅛ | 1/16 | 4 | 15 | .4701-.505 | 2½ | 1¼ | ½ | 3/16 | 6 |
| 3 | .0961-.126 | 2 | ¾ | ⅛ | 1/16 | 4 | 16 | .5051-.567 | 3 | 1½ | ⅝ | ¼ | 6 |
| 4 | .1261-.158 | 2¼ | 1 | ¼ | 3/32 | 4 | 17 | .5671-.630 | 3 | 1½ | ⅝ | ¼ | 6 |
| 5 | .1581-.188 | 2¼ | 1 | ¼ | 3/32 | 4 | 18 | .6301-.692 | 3 | 1½ | ⅝ | ¼ | 6 |
| 6 | .1881-.219 | 2¼ | 1 | ¼ | 3/32 | 6 | 19 | .6921-.755 | 3 | 1½ | ¾ | 5/16 | 8 |
| 7 | .2191-.251 | 2¼ | 1 | ¼ | 3/32 | 6 | 20 | .7551-.817 | 3 | 1½ | ¾ | 5/16 | 8 |
| 8 | .2511-.282 | 2¼ | 1 | ⅜ | ⅛ | 6 | 21 | .8171-.880 | 3 | 1½ | ¾ | 5/16 | 8 |
| 9 | .2821-.313 | 2¼ | 1 | ⅜ | ⅛ | 6 | 22 | .8801-.942 | 3 | 1½ | ¾ | 5/16 | 8 |
| 10 | .3131-.344 | 2½ | 1¼ | ⅜ | ⅛ | 6 | 23 | .9421-1.010 | 3 | 1½ | ¾ | 5/16 | 8 |
| 11 | .3441-.376 | 2½ | 1¼ | ⅜ | ⅛ | 6 | … | … | … | … | … | … | … |

All dimensions in inches. Material is high-speed steel.

These reamers are standard with right-hand cut and left-hand helical flutes within the size ranges shown.

*Tolerances:* On diameter of reamer, for sizes 00 to 7, incl., plus .0001 to plus .0004 inch and for sizes 8 to 23, incl., plus .0001 to plus .0005 inch. On overall length *A*, plus or minus $\frac{1}{16}$ inch. On length of flute *B*, plus or minus $\frac{1}{16}$ inch. On diameter of shank *D*, minus .0005 to minus .002 inch.

## American National Standard Morse Taper Finishing Reamers

*ANSI/ASME B94.2-1995 (R2020)*

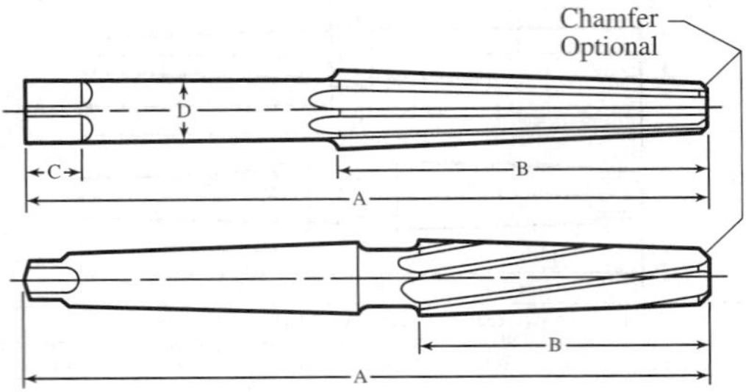


| Straight Flutes and Squared Shank | | | | | | | |
|---|---|---|---|---|---|---|---|
| Taper No.[a] | Small End Dia. (Ref.) | Large End Dia. (Ref.) | Length Overall *A* | Flute Length *B* | Square Length *C* | Shank Dia. *D* | Square Size |
| 0 | 0.2503 | 0.3674 | 3¾ | 2¼ | 5/16 | 5/16 | 0.235 |
| 1 | 0.3674 | 0.5170 | 5 | 3 | 7/16 | 7/16 | 0.330 |
| 2 | 0.5696 | 0.7444 | 6 | 3½ | 5/8 | 5/8 | 0.470 |
| 3 | 0.7748 | 0.9881 | 7¼ | 4¼ | 7/8 | 7/8 | 0.655 |
| 4 | 1.0167 | 1.2893 | 8½ | 5¼ | 1 | 1⅛ | 0.845 |
| 5 | 1.4717 | 1.8005 | 9¾ | 6¼ | 1⅛ | 1½ | 1.125 |

| Straight and Spiral Flutes and Taper Shank | | | | | | |
|---|---|---|---|---|---|---|
| Taper No.[a] | Small End Dia. (Ref.) | Large End Dia. (Ref.) | Length Overall *A* | Flute Length *B* | Taper Shank No.[a] | Squared and Taper Shank Number of Flutes |
| 0 | 0.2503 | 0.3674 | 5 11/32 | 2¼ | 0 | 4 to 6 incl. |
| 1 | 0.3674 | 0.5170 | 6 5/16 | 3 | 1 | 6 to 8 incl. |
| 2 | 0.5696 | 0.7444 | 7⅜ | 3½ | 2 | 6 to 8 incl. |
| 3 | 0.7748 | 0.9881 | 8⅞ | 4¼ | 3 | 8 to 10 incl. |
| 4 | 1.0167 | 1.2893 | 10⅞ | 5¼ | 4 | 8 to 10 incl. |
| 5 | 1.4717 | 1.8005 | 13⅛ | 6¼ | 5 | 10 to 12 incl. |

[a] Morse. For amount of taper see Table on page 1001.

All dimension are given in inches. Material is high-speed steel. The chamfer on the cutting end of the reamer is optional. Squared shank reamers are standard with straight flutes. Tapered shank reamers are standard with straight or spiral flutes. Spiral flute reamers are standard with left-hand spiral flutes.

*Tolerances:* On overall length *A* and flute length *B*, in taper numbers 0 to 3, incl., ±1/16 inch, in taper numbers 4 and 5, ±3/32 inch. On length of square *C*, in taper numbers 0 to 3, incl., ±1/32 inch; in taper numbers 4 and 5, ±1/16 inch. On shank diameter *D*, −.0005 to −.002 inch. On size of square, in taper numbers 0 and 1, −.004 inch; in taper numbers 2 and 3, −.006 inch; in taper numbers 4 and 5, −.008 inch.

**Center Reamers.—**A "center reamer" is a reamer the teeth of which meet in a point. By their use, small conical holes may be reamed in the ends of parts to be machined as on lathe centers. When large holes—usually cored—must be center-reamed, a large reamer is ordinarily used in which the teeth do not meet in a point, the reamer forming the frustum of a cone. Center reamers for such work are called "bull" or "pipe" center reamers.

*Bull Center Reamer:* A conical reamer used for reaming the ends of large holes—usually cored—so that they will fit on a lathe center. The cutting part of the reamer is generally in the shape of a frustum of a cone. It is also known as a pipe center reamer.

**Taper Pipe Reamers—Spiral Flutes** *ANSI/ASME B94.2-1995 (R2020)*

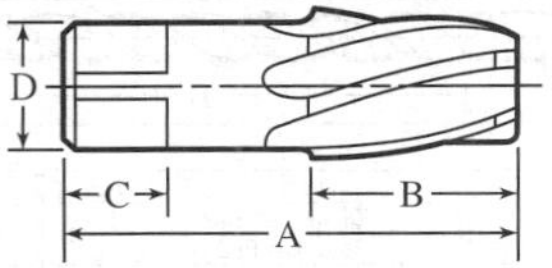


| Nom. Size | Diameter Large End | Diameter Small End | Length Overall *A* | Flute Length *B* | Square Length *C* | Shank Diameter *D* | Size of Square | No. of Flutes |
|---|---|---|---|---|---|---|---|---|
| 1/8 | 0.362 | 0.316 | 2 1/8 | 3/4 | 3/8 | 0.4375 | 0.328 | 4 to 6 |
| 1/4 | 0.472 | 0.406 | 2 7/16 | 1 1/16 | 7/16 | 0.5625 | 0.421 | 4 to 6 |
| 3/8 | 0.606 | 0.540 | 2 9/16 | 1 1/16 | 1/2 | 0.7000 | 0.531 | 4 to 6 |
| 1/2 | 0.751 | 0.665 | 3 1/8 | 1 3/8 | 5/8 | 0.6875 | 0.515 | 4 to 6 |
| 3/4 | 0.962 | 0.876 | 3 1/4 | 1 3/8 | 11/16 | 0.9063 | 0.679 | 6 to 10 |
| 1 | 1.212 | 1.103 | 3 3/4 | 1 3/4 | 13/16 | 1.1250 | 0.843 | 6 to 10 |
| 1 1/4 | 1.553 | 1.444 | 4 | 1 3/4 | 15/16 | 1.3125 | 0.984 | 6 to 10 |
| 1 1/2 | 1.793 | 1.684 | 4 1/4 | 1 3/4 | 1 | 1.5000 | 1.125 | 6 to 10 |
| 2 | 2.268 | 2.159 | 4 1/2 | 1 3/4 | 1 1/8 | 1.8750 | 1.406 | 8 to 12 |

All dimensions are given in inches. These reamers are tapered 3/4 inch per foot and are intended for reaming holes to be tapped with American National Standard Taper Pipe Thread taps. Material is high-speed steel. Reamers are standard with left-hand spiral flutes.

*Tolerances:* On length overall *A* and flute length *B*, 1/8- to 3/4-inch size, incl., ±1/16 inch; 1- to 1 1/2-inch size, incl., ±3/32 inch; 2-inch size, ±1/8 inch. On length of square *C*, 1/8- to 3/4-inch size, incl., ±1/32 inch; 1- to 2-inch size, incl., ±1/16 inch. On shank diameter *D*, 1/8-inch size, −.0015 inch; 1/4- to 1-inch size, incl., −.002 inch; 1 1/4- to 2-inch size, incl., −.003 inch. On size of square, 1/8-inch size, −.004 inch; 1/4- to 3/4-inch size, incl., −.006 inch; 1- to 2-inch size, incl., −.008 inch.

**B & S Taper Reamers—Straight and Spiral Flutes, Squared Shank**

| Taper No.[a] | Dia., Small End | Dia., Large End | Overall Length | Square Length | Flute Length | Dia. of Shank | Size of Square | No. of Flutes |
|---|---|---|---|---|---|---|---|---|
| 1 | 0.1974 | 0.3176 | 4 3/4 | 1/4 | 2 7/8 | 9/32 | 0.210 | 4 to 6 |
| 2 | 0.2474 | 0.3781 | 5 1/8 | 5/16 | 3 1/8 | 11/32 | 0.255 | 4 to 6 |
| 3 | 0.3099 | 0.4510 | 5 1/2 | 3/8 | 3 3/8 | 13/32 | 0.305 | 4 to 6 |
| 4 | 0.3474 | 0.5017 | 5 7/8 | 7/16 | 3 11/16 | 7/16 | 0.330 | 4 to 6 |
| 5 | 0.4474 | 0.6145 | 6 3/8 | 1/2 | 4 | 9/16 | 0.420 | 4 to 6 |
| 6 | 0.4974 | 0.6808 | 6 7/8 | 5/8 | 4 3/8 | 5/8 | 0.470 | 4 to 6 |
| 7 | 0.5974 | 0.8011 | 7 1/2 | 3/4 | 4 7/8 | 3/4 | 0.560 | 6 to 8 |
| 8 | 0.7474 | 0.9770 | 8 1/8 | 13/16 | 5 1/2 | 13/16 | 0.610 | 6 to 8 |
| 9 | 0.8974 | 1.1530 | 8 7/8 | 7/8 | 6 1/8 | 1 | 0.750 | 6 to 8 |
| 10 | 1.0420 | 1.3376 | 9 3/4 | 1 | 6 7/8 | 1 1/8 | 0.845 | 6 to 8 |

[a] For taper per foot, see Table 10 on page 1011.

These reamers are no longer ANSI/ASME standard.

All dimensions are given in inches. Material is high-speed steel. The chamfer on the cutting end of the reamer is optional. All reamers are finishing reamers. Spiral flute reamers are standard with left-hand spiral flutes. (Tapered reamers, especially those with left-hand spirals, should not have circular lands because cutting must take place on the outer diameter of the tool.) B & S taper reamers are designed for use in reaming out Brown & Sharpe standard taper sockets.

*Tolerances:* On length overall *A* and flute length *B*, taper nos. 1 to 7, incl., ±1/16 inch; taper nos. 8 to 10, incl., ±3/32 inch. On length of square *C*, taper nos. 1 to 9, incl., ±1/32 inch; taper no. 10, ±1/16 inch. On shank diameter *D*, −.0005 to −.002 inch. On size of square, taper nos. 1 to 3, incl., −.004 inch; taper nos. 4 to 9, incl., −.006 inch; taper no. 10, −.008 inch.

## American National Standard Die-Makers' Reamers *ANSI/ASME B94.2-1995 (R2020)*

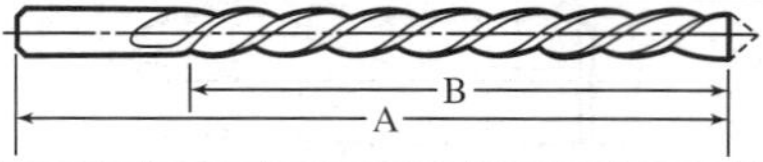


| Letter Size | Diameter | | Length | | Letter Size | Diameter | | Length | | Letter Size | Diameter | | Length | |
|---|---|---|---|---|---|---|---|---|---|---|---|---|---|---|
| | Small End | Large End | *A* | *B* | | Small End | Large End | *A* | *B* | | Small End | Large End | *A* | *B* |
| AAA | 0.055 | 0.070 | 2 1/4 | 1 1/8 | G | 0.135 | 0.158 | 3 | 1 3/4 | O | 0.250 | 0.296 | 5 | 3 1/2 |
| AA | 0.065 | 0.080 | 2 1/4 | 1 1/8 | H | 0.145 | 0.169 | 3 1/4 | 1 7/8 | P | 0.275 | 0.327 | 5 1/2 | 4 |
| A | 0.075 | 0.090 | 2 1/4 | 1 1/8 | I | 0.160 | 0.184 | 3 1/4 | 1 7/8 | Q | 0.300 | 0.358 | 6 | 4 1/2 |
| B | 0.085 | 0.103 | 2 3/8 | 1 3/8 | J | 0.175 | 0.199 | 3 1/4 | 1 7/8 | R | 0.335 | 0.397 | 6 1/2 | 4 3/4 |
| C | 0.095 | 0.113 | 2 1/2 | 1 3/8 | K | 0.190 | 0.219 | 3 1/2 | 2 1/4 | S | 0.370 | 0.435 | 6 3/4 | 5 |
| D | 0.105 | 0.126 | 2 5/8 | 1 5/8 | L | 0.205 | 0.234 | 3 1/2 | 2 1/4 | T | 0.405 | 0.473 | 7 | 5 1/4 |
| E | 0.115 | 0.136 | 2 3/4 | 1 5/8 | M | 0.220 | 0.252 | 4 | 2 1/2 | U | 0.440 | 0.511 | 7 1/4 | 5 1/2 |
| F | 0.125 | 0.148 | 3 | 1 3/4 | N | 0.235 | 0.274 | 4 1/2 | 3 | … | … | … | … | … |

All dimensions in inches. Material is high-speed steel. These reamers are designed for use in diemaking, have a taper of 3/4 degree included angle or 0.013 inch per inch, and have 2 or 3 flutes. Reamers are standard with left-hand spiral flutes.

Tip of reamer may have conical end.

*Tolerances:* On length overall *A* and flute length *B*, ±1/16 inch.

## Taper Pin Reamers — Straight and Left-Hand Spiral Flutes, Squared Shank; Left-Hand High-Spiral Flutes, Round Shank *ANSI/ASME B94.2-1995 (R2020)*

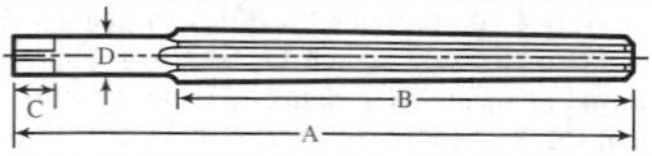


| No. of Taper Pin Reamer | Diameter at Large End of Reamer (Ref.) | Diameter at Small End of Reamer (Ref.) | Overall Length of Reamer *A* | Length of Flute *B* | Length of Square *C*[a] | Diameter of Shank *D* | Size of Square[a] |
|---|---|---|---|---|---|---|---|
| 8/0[b] | 0.0514 | 0.0351 | 1 5/8 | 25/32 | … | 1/16 | … |
| 7/0 | 0.0666 | 0.0497 | 1 13/16 | 13/16 | 5/32 | 5/64 | 0.060 |
| 6/0 | 0.0806 | 0.0611 | 1 15/16 | 15/16 | 5/32 | 3/32 | 0.070 |
| 5/0 | 0.0966 | 0.0719 | 2 3/16 | 1 3/16 | 5/32 | 7/64 | 0.080 |
| 4/0 | 0.1142 | 0.0869 | 2 5/16 | 1 5/16 | 5/32 | 1/8 | 0.095 |
| 3/0 | 0.1302 | 0.1029 | 2 5/16 | 1 5/16 | 5/32 | 9/64 | 0.105 |
| 2/0 | 0.1462 | 0.1137 | 2 9/16 | 1 9/16 | 7/32 | 5/32 | 0.115 |
| 0 | 0.1638 | 0.1287 | 2 15/16 | 1 11/16 | 7/32 | 11/64 | 0.130 |
| 1 | 0.1798 | 0.1447 | 2 15/16 | 1 11/16 | 7/32 | 3/16 | 0.140 |
| 2 | 0.2008 | 0.1605 | 3 3/16 | 1 15/16 | 1/4 | 13/64 | 0.150 |
| 3 | 0.2294 | 0.1813 | 3 11/16 | 2 5/16 | 1/4 | 15/64 | 0.175 |
| 4 | 0.2604 | 0.2071 | 4 1/16 | 2 9/16 | 1/4 | 17/64 | 0.200 |
| 5 | 0.2994 | 0.2409 | 4 5/16 | 2 13/16 | 5/16 | 5/16 | 0.235 |
| 6 | 0.3540 | 0.2773 | 5 7/16 | 3 11/16 | 3/8 | 23/64 | 0.270 |
| 7 | 0.4220 | 0.3297 | 6 5/16 | 4 7/16 | 3/8 | 13/32 | 0.305 |
| 8 | 0.5050 | 0.3971 | 7 3/16 | 5 3/16 | 7/16 | 7/16 | 0.330 |
| 9 | 0.6066 | 0.4805 | 8 5/16 | 6 1/16 | 9/16 | 9/16 | 0.420 |
| 10 | 0.7216 | 0.5799 | 9 5/16 | 6 13/16 | 5/8 | 5/8 | 0.470 |

[a] Not applicable to high-spiral flute reamers.

[b] Not applicable to straight and left-hand spiral-fluted, squared-shank reamers.

All dimensions in inches. Reamers have a taper of 1/4 inch per foot and are made of high-speed steel. Straight flute reamers of carbon steel are also standard. The number of flutes is as follows; 3 or 4, for 7/0 to 4/0 sizes; 4 to 6, for 3/0 to 0 sizes; 5 or 6, for 1 to 5 sizes; 6 to 8, for 6 to 9 sizes; 7 or 8, for the 10 size in the case of straight- and spiral-flute reamers; and 2 or 3, for 8/0 to 8 sizes; 2 to 4, for the 9 and 10 sizes in the case of high-spiral flute reamers.

*Tolerances:* On length overall *A* and flute length *B*, ±1/16 inch. On length of square *C*, ±1/32 inch. On shank diameter *D*, −.001 to −.005 inch for straight- and spiral-flute reamers and −.0005 to −.002 inch for high-spiral flute reamers. On size of square, −.004 inch for 7/0 to 7 sizes and −.006 inch for 8 to 10 sizes.

## TWIST DRILLS AND COUNTERBORES

Twist drills are rotary end-cutting tools having one or more cutting lips and one or more straight or helical flutes for the passage of chips and cutting fluids. Twist drills are made with straight or tapered shanks, but most have straight shanks. All but the smaller sizes are ground with "back taper," reducing the diameter from the point toward the shank, to prevent binding in the hole when the drill is worn.

*Straight-Shank Drills:* Straight-shank drills have cylindrical shanks that may be of the same or of a different diameter than the body diameter of the drill and may be made with or without driving flats, tang, or grooves.

*Taper-Shank Drills:* Taper-shank drills are preferable to the straight-shank type for drilling medium- and large-size holes. The taper on the shank conforms to one of the tapers in the American Standard (Morse) Series.

**American National Standard.—**American National Standard B94.11M-1993 covers nomenclature, definitions, sizes and tolerances for High-Speed Steel Straight- and Taper-Shank Drills and Combined Drills and Countersinks, Plain and Bell types. It covers both inch and metric sizes. Dimensional tables from the Standard will be found on the following pages.

**Definitions of Twist Drill Terms.—**The following definitions are included in the Standard.

*Axis:* The imaginary straight line that forms the longitudinal center of the drill.

*Back Taper:* A slight decrease in diameter from point to back in the body of the drill.

*Body:* The portion of the drill extending from the shank or neck to the outer corners of the cutting lips.

*Body Diameter Clearance:* That portion of the land that has been cut away so it will not rub against the wall of the hole.

*Chisel Edge:* The edge at the ends of the web that connects the cutting lips.

*Chisel Edge Angle:* The angle included between the chisel edge and the cutting lip as viewed from the end of the drill.

*Clearance Diameter:* The diameter over the cutaway portion of the drill lands.

*Drill Diameter:* The diameter over the margins of the drill measured at the point.

*Flutes:* Helical or straight grooves cut or formed in the body of the drill to provide cutting lips, to permit removal of chips, and to allow cutting fluid to reach the cutting lips.

*Helix Angle:* The angle made by the leading edge of the land with a plane containing the axis of the drill.

*Land:* The peripheral portion of the drill body between adjacent flutes.

*Land Width:* The distance between the leading edge and the heel of the land measured at a right angle to the leading edge.

*Lips—Two-Flute Drill:* The cutting edges extending from the chisel edge to the periphery.

*Lips—Three- or Four-Flute Drill (Core Drill):* The cutting edges extending from the bottom of the chamfer to the periphery.

*Lip Relief:* The axial relief on the drill point.

*Lip Relief Angle:* The axial relief angle at the outer corner of the lip. It is measured by projection into a plane tangent to the periphery at the outer corner of the lip. (Lip relief angle is usually measured across the margin of the twist drill.)

*Margin:* The cylindrical portion of the land that is not cut away to provide clearance.

*Neck:* The section of reduced diameter between the body and the shank of a drill.

*Overall Length:* The length from the extreme end of the shank to the outer corners of the cutting lips. It does not include the conical-shank end often used on straight-shank drills, nor does it include the conical cutting point used on both straight- and taper-shank drills. (For core drills with an external center on the cutting end, the overall length is the same as for two-flute drills. For core drills with an internal center on the cutting end, the overall length is to the extreme ends of the tool.)

*Point:* The cutting end of a drill made up of the ends of the lands, the web, and the lips. In form, it resembles a cone but departs from a true cone to furnish clearance behind the cutting lips.

*Point Angle:* The angle included between the lips projected upon a plane parallel to the drill axis and parallel to the cutting lips.

*Shank:* The part of the drill by which it is held and driven.

*Tang:* The flattened end of a taper shank, intended to fit into a driving slot in the socket.

*Tang Drive:* Two opposite parallel driving flats on the end of a straight shank.

*Web:* The central portion of the body that joins the end of the lands. The end of the web forms the chisel edge on a two-flute drill.

*Web Thickness:* The thickness of the web at the point unless another specific location is indicated.

*Web Thinning:* The operation of reducing the web thickness at the point to reduce drilling thrust.

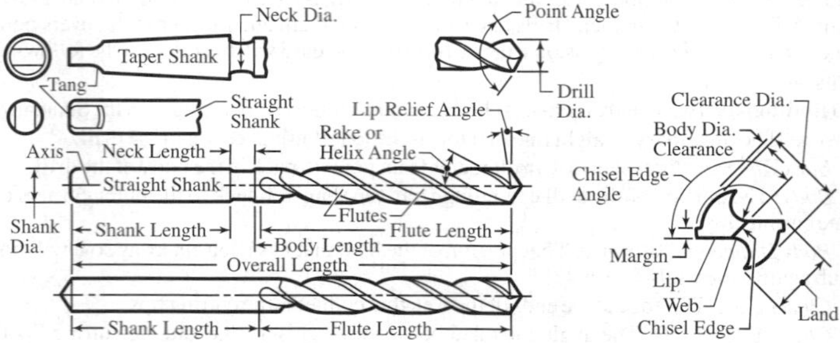


American National Standard Twist Drill Nomenclature

**Types of Drills.**—Drills may be classified based on the type of shank, number of flutes or hand of cut.

*Straight-Shank Drills:* Those having cylindrical shanks that may be the same or a different diameter than the body of the drill. The shank may be with or without driving flats, tang, grooves, or threads.

*Taper-Shank Drills:* Those having conical shanks suitable for direct fitting into tapered holes in machine spindles, driving sleeves, or sockets. Tapered shanks generally have a driving tang.

*Two-Flute Drills:* The conventional type of drill used for originating holes.

*Three-Flute Drills (Core Drills):* Drills commonly used for enlarging and finishing drilled, cast or punched holes. They will not produce original holes.

*Four-Flute Drills (Core Drills):* Used interchangeably with three-flute drills. They are of similar construction except for the number of flutes.

*Right-Hand Cut:* When viewed from the cutting point, the counterclockwise rotation of a drill in order to cut.

*Left-Hand Cut:* When viewed from the cutting point, the clockwise rotation of a drill in order to cut.

*Teat Drill:* The cutting edges of a teat drill are at right angles to the axis, and in the center there is a small teat of pyramid shape which leads the drill and holds it in position. This form is used for squaring the bottoms of holes made by ordinary twist drills or for drilling the entire hole, especially if it is not very deep and a square bottom is required. For instance, when drilling holes to form clearance spaces at the end of a keyseat, preparatory to cutting it out by planing or chipping, the teat drill is commonly used.

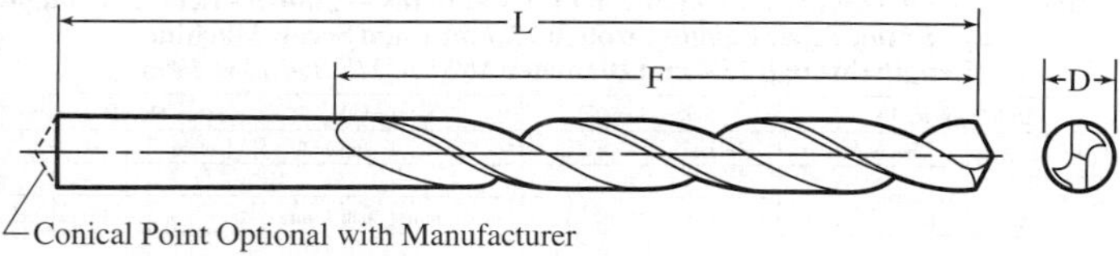


**Table 1. ANSI Straight-Shank Twist Drills — Jobbers Length through 17.5 mm, Taper Length through 12.7 mm, and Screw Machine Length through 25.4 mm Diameter** *ANSI/ASME B94.11M-1993*

| Drill Diameter, $D^a$ | | | | Jobbers Length | | | | Taper Length | | | | Screw Machine Length | | | |
|---|---|---|---|---|---|---|---|---|---|---|---|---|---|---|---|
| Fraction No. or Ltr. | mm | Equivalent | | Flute | | Overall | | Flute | | Overall | | Flute | | Overall | |
| | | Decimal Inch | mm | F | | L | | F | | L | | F | | L | |
| | | | | Inch | mm | Inch | mm | Inch | mm | Inch | mm | Inch | mm | Inch | mm |
| 97 | 0.15 | 0.0059 | 0.150 | 1⁄16 | 1.6 | 3⁄4 | 19 | … | … | … | … | … | … | … | … |
| 96 | 0.16 | 0.0063 | 0.160 | 1⁄16 | 1.6 | 3⁄4 | 19 | … | … | … | … | … | … | … | … |
| 95 | 0.17 | 0.0067 | 0.170 | 1⁄16 | 1.6 | 3⁄4 | 19 | … | … | … | … | … | … | … | … |
| 94 | 0.18 | 0.0071 | 0.180 | 1⁄16 | 1.6 | 3⁄4 | 19 | … | … | … | … | … | … | … | … |
| 93 | 0.19 | 0.0075 | 0.190 | 1⁄16 | 1.6 | 3⁄4 | 19 | … | … | … | … | … | … | … | … |
| 92 | 0.20 | 0.0079 | 0.200 | 1⁄16 | 1.6 | 3⁄4 | 19 | … | … | … | … | … | … | … | … |
| 91 | | 0.0083 | 0.211 | 5⁄64 | 2.0 | 3⁄4 | 19 | … | … | … | … | … | … | … | … |
| 90 | 0.22 | 0.0087 | 0.221 | 5⁄64 | 2.0 | 3⁄4 | 19 | … | … | … | … | … | … | … | … |
| 89 | | 0.0091 | 0.231 | 5⁄64 | 2.0 | 3⁄4 | 19 | … | … | … | … | … | … | … | … |
| 88 | | 0.0095 | 0.241 | 5⁄64 | 2.0 | 3⁄4 | 19 | … | … | … | … | … | … | … | … |
| | 0.25 | 0.0098 | 0.250 | 5⁄64 | 2.0 | 3⁄4 | 19 | … | … | … | … | … | … | … | … |
| 87 | | 0.0100 | 0.254 | 5⁄64 | 2.0 | 3⁄4 | 19 | … | … | … | … | … | … | … | … |
| 86 | | 0.0105 | 0.267 | 3⁄32 | 2.4 | 3⁄4 | 19 | … | … | … | … | … | … | … | … |
| 85 | 0.28 | 0.0110 | 0.280 | 3⁄32 | 2.4 | 3⁄4 | 19 | … | … | … | … | … | … | … | … |
| 84 | | 0.0115 | 0.292 | 3⁄32 | 2.4 | 3⁄4 | 19 | … | … | … | … | … | … | … | … |
| | 0.30 | 0.0118 | 0.300 | 3⁄32 | 2.4 | 3⁄4 | 19 | … | … | … | … | … | … | … | … |
| 83 | | 0.0120 | 0.305 | 3⁄32 | 2.4 | 3⁄4 | 19 | … | … | … | … | … | … | … | … |
| 82 | | 0.0125 | 0.318 | 3⁄32 | 2.4 | 3⁄4 | 19 | … | … | … | … | … | … | … | … |
| | 0.32 | 0.0126 | 0.320 | 3⁄32 | 2.4 | 3⁄4 | 19 | … | … | … | … | … | … | … | … |
| 81 | | 0.0130 | 0.330 | 3⁄32 | 2.4 | 3⁄4 | 19 | … | … | … | … | … | … | … | … |
| 80 | | 0.0135 | 0.343 | 1⁄8 | 3 | 3⁄4 | 19 | … | … | … | … | … | … | … | … |
| | 0.35 | 0.0138 | 0.350 | 1⁄8 | 3 | 3⁄4 | 19 | … | … | … | … | … | … | … | … |
| 79 | | 0.0145 | 0.368 | 1⁄8 | 3 | 3⁄4 | 19 | … | … | … | … | … | … | … | … |
| | 0.38 | 0.0150 | 0.380 | 3⁄16 | 5 | 3⁄4 | 19 | … | … | … | … | … | … | … | … |
| 1⁄64 | | 0.0156 | 0.396 | 3⁄16 | 5 | 3⁄4 | 19 | … | … | … | … | … | … | … | … |
| | 0.40 | 0.0157 | 0.400 | 3⁄16 | 5 | 3⁄4 | 19 | … | … | … | … | … | … | … | … |
| 78 | | 0.0160 | 0.406 | 3⁄16 | 5 | 7⁄8 | 22 | … | … | … | … | … | … | … | … |
| | 0.42 | 0.0165 | 0.420 | 3⁄16 | 5 | 7⁄8 | 22 | … | … | … | … | … | … | … | … |
| | 0.45 | 0.0177 | 0.450 | 3⁄16 | 5 | 7⁄8 | 22 | … | … | … | … | … | … | … | … |
| 77 | | 0.0180 | 0.457 | 3⁄16 | 5 | 7⁄8 | 22 | … | … | … | … | … | … | … | … |
| | 0.48 | 0.0189 | 0.480 | 3⁄16 | 5 | 7⁄8 | 22 | … | … | … | … | … | … | … | … |
| | 0.50 | 0.0197 | 0.500 | 3⁄16 | 5 | 7⁄8 | 22 | … | … | … | … | … | … | … | … |
| 76 | | 0.0200 | 0.508 | 3⁄16 | 5 | 7⁄8 | 22 | … | … | … | … | … | … | … | … |
| 75 | | 0.0210 | 0.533 | 1⁄4 | 6 | 1 | 25 | … | … | … | … | … | … | … | … |
| | 0.55 | 0.0217 | 0.550 | 1⁄4 | 6 | 1 | 25 | … | … | … | … | … | … | … | … |
| 74 | | 0.0225 | 0.572 | 1⁄4 | 6 | 1 | 25 | … | … | … | … | … | … | … | … |
| | 0.60 | 0.0236 | 0.600 | 5⁄16 | 8 | 1 1⁄8 | 29 | … | … | … | … | … | … | … | … |

**Table 1.** *(Continued)* **ANSI Straight-Shank Twist Drills — Jobbers Length through 17.5 mm, Taper Length through 12.7 mm, and Screw Machine Length through 25.4 mm Diameter** *ANSI/ASME B94.11M-1993*

| Drill Diameter, $D^a$ | | | | Jobbers Length | | | | Taper Length | | | | Screw Machine Length | | | |
|---|---|---|---|---|---|---|---|---|---|---|---|---|---|---|---|
| Frac-tion No. or Ltr. | mm | Equivalent | | Flute | | Overall | | Flute | | Overall | | Flute | | Overall | |
| | | Decimal Inch | mm | F | | L | | F | | L | | F | | L | |
| | | | | Inch | mm | Inch | mm | Inch | mm | Inch | mm | Inch | mm | Inch | mm |
| 73 | | 0.0240 | 0.610 | 5⁄16 | 8 | 1 1⁄8 | 29 | … | … | … | … | … | … | … | … |
| 72 | | 0.0250 | 0.635 | 5⁄16 | 8 | 1 1⁄8 | 29 | … | … | … | … | … | … | … | … |
| | 0.65 | 0.0256 | 0.650 | 3⁄8 | 10 | 1 1⁄4 | 32 | … | … | … | … | … | … | … | … |
| 71 | | 0.0260 | 0.660 | 3⁄8 | 10 | 1 1⁄4 | 32 | … | … | … | … | … | … | … | … |
| | 0.70 | 0.0276 | 0.700 | 3⁄8 | 10 | 1 1⁄4 | 32 | … | … | … | … | … | … | … | … |
| 70 | | 0.0280 | 0.711 | 3⁄8 | 10 | 1 1⁄4 | 32 | … | … | … | … | … | … | … | … |
| 69 | | 0.0292 | 0.742 | 1⁄2 | 13 | 1 3⁄8 | 35 | … | … | … | … | … | … | … | … |
| | 0.75 | 0.0295 | 0.750 | 1⁄2 | 13 | 1 3⁄8 | 35 | … | … | … | … | … | … | … | … |
| 68 | | 0.0310 | 0.787 | 1⁄2 | 13 | 1 3⁄8 | 35 | … | … | … | … | … | … | … | … |
| 1⁄32 | | 0.0312 | 0.792 | 1⁄2 | 13 | 1 3⁄8 | 35 | … | … | … | … | … | … | … | … |
| | 0.80 | 0.0315 | 0.800 | 1⁄2 | 13 | 1 3⁄8 | 35 | … | … | … | … | … | … | … | … |
| 67 | | 0.0320 | 0.813 | 1⁄2 | 13 | 1 3⁄8 | 35 | … | … | … | … | … | … | … | … |
| 66 | | 0.0330 | 0.838 | 1⁄2 | 13 | 1 3⁄8 | 35 | … | … | … | … | … | … | … | … |
| | 0.85 | 0.0335 | 0.850 | 5⁄8 | 16 | 1 1⁄2 | 38 | … | … | … | … | … | … | … | … |
| 65 | | 0.0350 | 0.889 | 5⁄8 | 16 | 1 1⁄2 | 38 | … | … | … | … | … | … | … | … |
| | 0.90 | 0.0354 | 0.899 | 5⁄8 | 16 | 1 1⁄2 | 38 | … | … | … | … | … | … | … | … |
| 64 | | 0.0360 | 0.914 | 5⁄8 | 16 | 1 1⁄2 | 38 | … | … | … | … | … | … | … | … |
| 63 | | 0.0370 | 0.940 | 5⁄8 | 16 | 1 1⁄2 | 38 | … | … | … | … | … | … | … | … |
| | 0.95 | 0.0374 | 0.950 | 5⁄8 | 16 | 1 1⁄2 | 38 | … | … | … | … | … | … | … | … |
| 62 | | 0.0380 | 0.965 | 5⁄8 | 16 | 1 1⁄2 | 38 | … | … | … | … | … | … | … | … |
| 61 | | 0.0390 | 0.991 | 11⁄16 | 17 | 1 5⁄8 | 41 | … | … | … | … | … | … | … | … |
| | 1.00 | 0.0394 | 1.000 | 11⁄16 | 17 | 1 5⁄8 | 41 | 1 1⁄8 | 29 | 2 1⁄4 | 57 | 1⁄2 | 13 | 1 3⁄8 | 35 |
| 60 | | 0.0400 | 1.016 | 11⁄16 | 17 | 1 5⁄8 | 41 | 1 1⁄8 | 29 | 2 1⁄4 | 57 | 1⁄2 | 13 | 1 3⁄8 | 35 |
| 59 | | 0.0410 | 1.041 | 11⁄16 | 17 | 1 5⁄8 | 41 | 1 1⁄8 | 29 | 2 1⁄4 | 57 | 1⁄2 | 13 | 1 3⁄8 | 35 |
| | 1.05 | 0.0413 | 1.050 | 11⁄16 | 17 | 1 5⁄8 | 41 | 1 1⁄8 | 29 | 2 1⁄4 | 57 | 1⁄2 | 13 | 1 3⁄8 | 35 |
| 58 | | 0.0420 | 1.067 | 11⁄16 | 17 | 1 5⁄8 | 41 | 1 1⁄8 | 29 | 2 1⁄4 | 57 | 1⁄2 | 13 | 1 3⁄8 | 35 |
| 57 | | 0.0430 | 1.092 | 3⁄4 | 19 | 1 3⁄4 | 44 | 1 1⁄8 | 29 | 2 1⁄4 | 57 | 1⁄2 | 13 | 1 3⁄8 | 35 |
| | 1.10 | 0.0433 | 1.100 | 3⁄4 | 19 | 1 3⁄4 | 44 | 1 1⁄8 | 29 | 2 1⁄4 | 57 | 1⁄2 | 13 | 1 3⁄8 | 35 |
| | 1.15 | 0.0453 | 1.150 | 3⁄4 | 19 | 1 3⁄4 | 44 | 1 1⁄8 | 29 | 2 1⁄4 | 57 | 1⁄2 | 13 | 1 3⁄8 | 35 |
| 56 | | 0.0465 | 1.181 | 3⁄4 | 19 | 1 3⁄4 | 44 | 1 1⁄8 | 29 | 2 1⁄4 | 57 | 1⁄2 | 13 | 1 3⁄8 | 35 |
| 3⁄64 | | 0.0469 | 1.191 | 3⁄4 | 19 | 1 3⁄4 | 44 | 1 1⁄8 | 29 | 2 1⁄4 | 57 | 1⁄2 | 13 | 1 3⁄8 | 35 |
| | 1.20 | 0.0472 | 1.200 | 7⁄8 | 22 | 1 7⁄8 | 48 | 1 3⁄4 | 44 | 3 | 76 | 5⁄8 | 16 | 1 5⁄8 | 41 |
| | 1.25 | 0.0492 | 1.250 | 7⁄8 | 22 | 1 7⁄8 | 48 | 1 3⁄4 | 44 | 3 | 76 | 5⁄8 | 16 | 1 5⁄8 | 41 |
| | 1.30 | 0.0512 | 1.300 | 7⁄8 | 22 | 1 7⁄8 | 48 | 1 3⁄4 | 44 | 3 | 76 | 5⁄8 | 16 | 1 5⁄8 | 41 |
| 55 | | 0.0520 | 1.321 | 7⁄8 | 22 | 1 7⁄8 | 48 | 1 3⁄4 | 44 | 3 | 76 | 5⁄8 | 16 | 1 5⁄8 | 41 |
| | 1.35 | 0.0531 | 1.350 | 7⁄8 | 22 | 1 7⁄8 | 48 | 1 3⁄4 | 44 | 3 | 76 | 5⁄8 | 16 | 1 5⁄8 | 41 |
| 54 | | 0.0550 | 1.397 | 7⁄8 | 22 | 1 7⁄8 | 48 | 1 3⁄4 | 44 | 3 | 76 | 5⁄8 | 16 | 1 5⁄8 | 41 |
| | 1.40 | 0.0551 | 1.400 | 7⁄8 | 22 | 1 7⁄8 | 48 | 1 3⁄4 | 44 | 3 | 76 | 5⁄8 | 16 | 1 5⁄8 | 41 |
| | 1.45 | 0.0571 | 1.450 | 7⁄8 | 22 | 1 7⁄8 | 48 | 1 3⁄4 | 44 | 3 | 76 | 5⁄8 | 16 | 1 5⁄8 | 41 |
| | 1.50 | 0.0591 | 1.500 | 7⁄8 | 22 | 1 7⁄8 | 48 | 1 3⁄4 | 44 | 3 | 76 | 5⁄8 | 16 | 1 5⁄8 | 41 |
| 53 | | 0.0595 | 1.511 | 7⁄8 | 22 | 1 7⁄8 | 48 | 1 3⁄4 | 44 | 3 | 76 | 5⁄8 | 16 | 1 5⁄8 | 41 |
| | 1.55 | 0.0610 | 1.550 | 7⁄8 | 22 | 1 7⁄8 | 48 | 1 3⁄4 | 44 | 3 | 76 | 5⁄8 | 16 | 1 5⁄8 | 41 |
| 1⁄16 | | 0.0625 | 1.588 | 7⁄8 | 22 | 1 7⁄8 | 48 | 1 3⁄4 | 44 | 3 | 76 | 5⁄8 | 16 | 1 5⁄8 | 41 |
| | 1.60 | 0.0630 | 1.600 | 7⁄8 | 22 | 1 7⁄8 | 48 | 2 | 51 | 3 3⁄4 | 95 | 11⁄16 | 17 | 1 11⁄16 | 43 |
| 52 | | 0.0635 | 1.613 | 7⁄8 | 22 | 1 7⁄8 | 48 | 2 | 51 | 3 3⁄4 | 95 | 11⁄16 | 17 | 1 11⁄16 | 43 |
| | 1.65 | 0.0650 | 1.650 | 1 | 25 | 2 | 51 | 2 | 51 | 3 3⁄4 | 95 | 11⁄16 | 17 | 1 11⁄16 | 43 |

**Table 1.** *(Continued)* **ANSI Straight-Shank Twist Drills — Jobbers Length through 17.5 mm, Taper Length through 12.7 mm, and Screw Machine Length through 25.4 mm Diameter** *ANSI/ASME B94.11M-1993*

| Drill Diameter, *D*[a] | | | | Jobbers Length | | | | Taper Length | | | | Screw Machine Length | | | |
|---|---|---|---|---|---|---|---|---|---|---|---|---|---|---|---|
| Frac- tion No. or Ltr. | mm | Equivalent | | Flute | | Overall | | Flute | | Overall | | Flute | | Overall | |
| | | Decimal Inch | mm | *F* | | *L* | | *F* | | *L* | | *F* | | *L* | |
| | | | | Inch | mm | Inch | mm | Inch | mm | Inch | mm | Inch | mm | Inch | mm |
| | 1.70 | 0.0669 | 1.700 | 1 | 25 | 2 | 51 | 2 | 51 | 3¾ | 95 | 11/16 | 17 | 1 11/16 | 43 |
| 51 | | 0.0670 | 1.702 | 1 | 25 | 2 | 51 | 2 | 51 | 3¾ | 95 | 11/16 | 17 | 1 11/16 | 43 |
| | 1.75 | 0.0689 | 1.750 | 1 | 25 | 2 | 51 | 2 | 51 | 3¾ | 95 | 11/16 | 17 | 1 11/16 | 43 |
| 50 | | 0.0700 | 1.778 | 1 | 25 | 2 | 51 | 2 | 51 | 3¾ | 95 | 11/16 | 17 | 1 11/16 | 43 |
| | 1.80 | 0.0709 | 1.800 | 1 | 25 | 2 | 51 | 2 | 51 | 3¾ | 95 | 11/16 | 17 | 1 11/16 | 43 |
| | 1.85 | 0.0728 | 1.850 | 1 | 25 | 2 | 51 | 2 | 51 | 3¾ | 95 | 11/16 | 17 | 1 11/16 | 43 |
| 49 | | 0.0730 | 1.854 | 1 | 25 | 2 | 51 | 2 | 51 | 3¾ | 95 | 11/16 | 17 | 1 11/16 | 43 |
| | 1.90 | 0.0748 | 1.900 | 1 | 25 | 2 | 51 | 2 | 51 | 3¾ | 95 | 11/16 | 17 | 1 11/16 | 43 |
| 48 | | 0.0760 | 1.930 | 1 | 25 | 2 | 51 | 2 | 51 | 3¾ | 95 | 11/16 | 17 | 1 11/16 | 43 |
| | 1.95 | 0.0768 | 1.950 | 1 | 25 | 2 | 51 | 2 | 51 | 3¾ | 95 | 11/16 | 17 | 1 11/16 | 43 |
| 5/64 | | 0.0781 | 1.984 | 1 | 25 | 2 | 51 | 2 | 51 | 3¾ | 95 | 11/16 | 17 | 1 11/16 | 43 |
| 47 | | 0.0785 | 1.994 | 1 | 25 | 2 | 51 | 2¼ | 57 | 4¼ | 108 | 11/16 | 17 | 1 11/16 | 43 |
| | 2.00 | 0.0787 | 2.000 | 1 | 25 | 2 | 51 | 2¼ | 57 | 4¼ | 108 | 11/16 | 17 | 1 11/16 | 43 |
| | 2.05 | 0.0807 | 2.050 | 1⅛ | 29 | 2⅛ | 54 | 2¼ | 57 | 4¼ | 108 | ¾ | 19 | 1¾ | 44 |
| 46 | | 0.0810 | 2.057 | 1⅛ | 29 | 2⅛ | 54 | 2¼ | 57 | 4¼ | 108 | ¾ | 19 | 1¾ | 44 |
| 45 | | 0.0820 | 2.083 | 1⅛ | 29 | 2⅛ | 54 | 2¼ | 57 | 4¼ | 108 | ¾ | 19 | 1¾ | 44 |
| | 2.10 | 0.0827 | 2.100 | 1⅛ | 29 | 2⅛ | 54 | 2¼ | 57 | 4¼ | 108 | ¾ | 19 | 1¾ | 44 |
| | 2.15 | 0.0846 | 2.150 | 1⅛ | 29 | 2⅛ | 54 | 2¼ | 57 | 4¼ | 108 | ¾ | 19 | 1¾ | 44 |
| 44 | | 0.0860 | 2.184 | 1⅛ | 29 | 2⅛ | 54 | 2¼ | 57 | 4¼ | 108 | ¾ | 19 | 1¾ | 44 |
| | 2.20 | 0.0866 | 2.200 | 1¼ | 32 | 2¼ | 57 | 2¼ | 57 | 4¼ | 108 | ¾ | 19 | 1¾ | 44 |
| | 2.25 | 0.0886 | 2.250 | 1¼ | 32 | 2¼ | 57 | 2¼ | 57 | 4¼ | 108 | ¾ | 19 | 1¾ | 44 |
| 43 | | 0.0890 | 2.261 | 1¼ | 32 | 2¼ | 57 | 2¼ | 57 | 4¼ | 108 | ¾ | 19 | 1¾ | 44 |
| | 2.30 | 0.0906 | 2.300 | 1¼ | 32 | 2¼ | 57 | 2¼ | 57 | 4¼ | 108 | ¾ | 19 | 1¾ | 44 |
| | 2.35 | 0.0925 | 2.350 | 1¼ | 32 | 2¼ | 57 | 2¼ | 57 | 4¼ | 108 | ¾ | 19 | 1¾ | 44 |
| 42 | | 0.0935 | 2.375 | 1¼ | 32 | 2¼ | 57 | 2¼ | 57 | 4¼ | 108 | ¾ | 19 | 1¾ | 44 |
| 3/32 | | 0.0938 | 2.383 | 1¼ | 32 | 2¼ | 57 | 2¼ | 57 | 4¼ | 108 | ¾ | 19 | 1¾ | 44 |
| | 2.40 | 0.0945 | 2.400 | 1⅜ | 35 | 2⅜ | 60 | 2½ | 64 | 4⅝ | 117 | 13/16 | 21 | 1 13/16 | 46 |
| 41 | | 0.0960 | 2.438 | 1⅜ | 35 | 2⅜ | 60 | 2½ | 64 | 4⅝ | 117 | 13/16 | 21 | 1 13/16 | 46 |
| | 2.46 | 0.0965 | 2.450 | 1⅜ | 35 | 2⅜ | 60 | 2½ | 64 | 4⅝ | 117 | 13/16 | 21 | 1 13/16 | 46 |
| 40 | | 0.0980 | 2.489 | 1⅜ | 35 | 2⅜ | 60 | 2½ | 64 | 4⅝ | 117 | 13/16 | 21 | 1 13/16 | 46 |
| | 2.50 | 0.0984 | 2.500 | 1⅜ | 35 | 2⅜ | 60 | 2½ | 64 | 4⅝ | 117 | 13/16 | 21 | 1 13/16 | 46 |
| 39 | | 0.0995 | 2.527 | 1⅜ | 35 | 2⅜ | 60 | 2½ | 64 | 4⅝ | 117 | 13/16 | 21 | 1 13/16 | 46 |
| 38 | | 0.1015 | 2.578 | 1 7/16 | 37 | 2½ | 64 | 2½ | 64 | 4⅝ | 117 | 13/16 | 21 | 1 13/16 | 46 |
| | 2.60 | 0.1024 | 2.600 | 1 7/16 | 37 | 2½ | 64 | 2½ | 64 | 4⅝ | 117 | 13/16 | 21 | 1 13/16 | 46 |
| 37 | | 0.1040 | 2.642 | 1 7/16 | 37 | 2½ | 64 | 2½ | 64 | 4⅝ | 117 | 13/16 | 21 | 1 13/16 | 46 |
| | 2.70 | 0.1063 | 2.700 | 1 7/16 | 37 | 2½ | 64 | 2½ | 64 | 4⅝ | 117 | 13/16 | 21 | 1 13/16 | 46 |
| 36 | | 0.1065 | 2.705 | 1 7/16 | 37 | 2½ | 64 | 2½ | 64 | 4⅝ | 117 | 13/16 | 21 | 1 13/16 | 46 |
| 7/64 | | 0.1094 | 2.779 | 1½ | 38 | 2⅝ | 67 | 2½ | 64 | 4⅝ | 117 | 13/16 | 21 | 1 13/16 | 46 |
| 35 | | 0.1100 | 2.794 | 1½ | 38 | 2⅝ | 67 | 2¾ | 70 | 5⅛ | 130 | ⅞ | 22 | 1⅞ | 48 |
| | 2.80 | 0.1102 | 2.800 | 1½ | 38 | 2⅝ | 67 | 2¾ | 70 | 5⅛ | 130 | ⅞ | 22 | 1⅞ | 48 |
| 34 | | 0.1110 | 2.819 | 1½ | 38 | 2⅝ | 67 | 2¾ | 70 | 5⅛ | 130 | ⅞ | 22 | 1⅞ | 48 |
| 33 | | 0.1130 | 2.870 | 1½ | 38 | 2⅝ | 67 | 2¾ | 70 | 5⅛ | 130 | ⅞ | 22 | 1⅞ | 48 |
| | 2.90 | 0.1142 | 2.900 | 1⅝ | 41 | 2¾ | 70 | 2¾ | 70 | 5⅛ | 130 | ⅞ | 22 | 1⅞ | 48 |
| 32 | | 0.1160 | 2.946 | 1⅝ | 41 | 2¾ | 70 | 2¾ | 70 | 5⅛ | 130 | ⅞ | 22 | 1⅞ | 48 |
| | 3.00 | 0.1181 | 3.000 | 1⅝ | 41 | 2¾ | 70 | 2¾ | 70 | 5⅛ | 130 | ⅞ | 22 | 1⅞ | 48 |
| 31 | | 0.1200 | 3.048 | 1⅝ | 41 | 2¾ | 70 | 2¾ | 70 | 5⅛ | 130 | ⅞ | 22 | 1⅞ | 48 |

**Table 1.** *(Continued)* **ANSI Straight-Shank Twist Drills — Jobbers Length through 17.5 mm, Taper Length through 12.7 mm, and Screw Machine Length through 25.4 mm Diameter** *ANSI/ASME B94.11M-1993*

| Drill Diameter, D[a] | | | | Jobbers Length | | | | Taper Length | | | | Screw Machine Length | | | |
|---|---|---|---|---|---|---|---|---|---|---|---|---|---|---|---|
| Frac-tion No. or Ltr. | | Equivalent | | Flute | | Overall | | Flute | | Overall | | Flute | | Overall | |
| | | | | F | | L | | F | | L | | F | | L | |
| | mm | Decimal Inch | mm | Inch | mm | Inch | mm | Inch | mm | Inch | mm | Inch | mm | Inch | mm |
| | 3.10 | 0.1220 | 3.100 | 1 5/8 | 41 | 2 3/4 | 70 | 2 3/4 | 70 | 5 1/8 | 130 | 7/8 | 22 | 1 7/8 | 48 |
| 1/8 | | 0.1250 | 3.175 | 1 5/8 | 41 | 2 3/4 | 70 | 2 3/4 | 70 | 5 1/8 | 130 | 7/8 | 22 | 1 7/8 | 48 |
| | 3.20 | 0.1260 | 3.200 | 1 5/8 | 41 | 2 3/4 | 70 | 3 | 76 | 5 3/8 | 137 | 15/16 | 24 | 1 15/16 | 49 |
| 30 | | 0.1285 | 3.264 | 1 5/8 | 41 | 2 3/4 | 70 | 3 | 76 | 5 3/8 | 137 | 15/16 | 24 | 1 15/16 | 49 |
| | 3.30 | 0.1299 | 3.300 | 1 3/4 | 44 | 2 7/8 | 73 | 3 | 76 | 5 3/8 | 137 | 15/16 | 24 | 1 15/16 | 49 |
| | 3.40 | 0.1339 | 3.400 | 1 3/4 | 44 | 2 7/8 | 73 | 3 | 76 | 5 3/8 | 137 | 15/16 | 24 | 1 15/16 | 49 |
| 29 | | 0.1360 | 3.454 | 1 3/4 | 44 | 2 7/8 | 73 | 3 | 76 | 5 3/8 | 137 | 15/16 | 24 | 1 15/16 | 49 |
| | 3.50 | 0.1378 | 3.500 | 1 3/4 | 44 | 2 7/8 | 73 | 3 | 76 | 5 3/8 | 137 | 15/16 | 24 | 1 15/16 | 49 |
| 28 | | 0.1405 | 3.569 | 1 3/4 | 44 | 2 7/8 | 73 | 3 | 76 | 5 3/8 | 137 | 15/16 | 24 | 1 15/16 | 49 |
| 9/64 | | 0.1406 | 3.571 | 1 3/4 | 44 | 2 7/8 | 73 | 3 | 76 | 5 3/8 | 137 | 15/16 | 24 | 1 15/16 | 49 |
| | 3.60 | 0.1417 | 3.600 | 1 7/8 | 48 | 3 | 76 | 3 | 76 | 5 3/8 | 137 | 1 | 25 | 2 1/16 | 52 |
| 27 | | 0.1440 | 3.658 | 1 7/8 | 48 | 3 | 76 | 3 | 76 | 5 3/8 | 137 | 1 | 25 | 2 1/16 | 52 |
| | 3.70 | 0.1457 | 3.700 | 1 7/8 | 48 | 3 | 76 | 3 | 76 | 5 3/8 | 137 | 1 | 25 | 2 1/16 | 52 |
| 26 | | 0.1470 | 3.734 | 1 7/8 | 48 | 3 | 76 | 3 | 76 | 5 3/8 | 137 | 1 | 25 | 2 1/16 | 52 |
| 25 | | 0.1495 | 3.797 | 1 7/8 | 48 | 3 | 76 | 3 | 76 | 5 3/8 | 137 | 1 | 25 | 2 1/16 | 52 |
| | 3.80 | 0.1496 | 3.800 | 1 7/8 | 48 | 3 | 76 | 3 | 76 | 5 3/8 | 137 | 1 | 25 | 2 1/16 | 52 |
| 24 | | 0.1520 | 3.861 | 2 | 51 | 3 1/8 | 79 | 3 | 76 | 5 3/8 | 137 | 1 | 25 | 2 1/16 | 52 |
| | 3.90 | 0.1535 | 3.900 | 2 | 51 | 3 1/8 | 79 | 3 | 76 | 5 3/8 | 137 | 1 | 25 | 2 1/16 | 52 |
| 23 | | 0.1540 | 3.912 | 2 | 51 | 3 1/8 | 79 | 3 | 76 | 5 3/8 | 137 | 1 | 25 | 2 1/16 | 52 |
| 5/32 | | 0.1562 | 3.967 | 2 | 51 | 3 1/8 | 79 | 3 | 76 | 5 3/8 | 137 | 1 | 25 | 2 1/16 | 52 |
| 22 | | 0.1570 | 3.988 | 2 | 51 | 3 1/8 | 79 | 3 3/8 | 86 | 5 3/4 | 146 | 1 1/16 | 27 | 2 1/8 | 54 |
| | 4.00 | 0.1575 | 4.000 | 2 1/8 | 54 | 3 1/4 | 83 | 3 3/8 | 86 | 5 3/4 | 146 | 1 1/16 | 27 | 2 1/8 | 54 |
| 21 | | 0.1590 | 4.039 | 2 1/8 | 54 | 3 1/4 | 83 | 3 3/8 | 86 | 5 3/4 | 146 | 1 1/16 | 27 | 2 1/8 | 54 |
| 20 | | 0.1610 | 4.089 | 2 1/8 | 54 | 3 1/4 | 83 | 3 3/8 | 86 | 5 3/4 | 146 | 1 1/16 | 27 | 2 1/8 | 54 |
| | 4.10 | 0.1614 | 4.100 | 2 1/8 | 54 | 3 1/4 | 83 | 3 3/8 | 86 | 5 3/4 | 146 | 1 1/16 | 27 | 2 1/8 | 54 |
| | 4.20 | 0.1654 | 4.200 | 2 1/8 | 54 | 3 1/4 | 83 | 3 3/8 | 86 | 5 3/4 | 146 | 1 1/16 | 27 | 2 1/8 | 54 |
| 19 | | 0.1660 | 4.216 | 2 1/8 | 54 | 3 1/4 | 83 | 3 3/8 | 86 | 5 3/4 | 146 | 1 1/16 | 27 | 2 1/8 | 54 |
| | 4.30 | 0.1693 | 4.300 | 2 1/8 | 54 | 3 1/4 | 83 | 3 3/8 | 86 | 5 3/4 | 146 | 1 1/16 | 27 | 2 1/8 | 54 |
| 18 | | 0.1695 | 4.305 | 2 1/8 | 54 | 3 1/4 | 83 | 3 3/8 | 86 | 5 3/4 | 146 | 1 1/16 | 27 | 2 1/8 | 54 |
| 11/64 | | 0.1719 | 4.366 | 2 1/8 | 54 | 3 1/4 | 83 | 3 3/8 | 86 | 5 3/4 | 146 | 1 1/16 | 27 | 2 1/8 | 54 |
| 17 | | 0.1730 | 4.394 | 2 3/16 | 56 | 3 3/8 | 86 | 3 3/8 | 86 | 5 3/4 | 146 | 1 1/8 | 29 | 2 3/16 | 56 |
| | 4.40 | 0.1732 | 4.400 | 2 3/16 | 56 | 3 3/8 | 86 | 3 3/8 | 86 | 5 3/4 | 146 | 1 1/8 | 29 | 2 3/16 | 56 |
| 16 | | 0.1770 | 4.496 | 2 3/16 | 56 | 3 3/8 | 86 | 3 3/8 | 86 | 5 3/4 | 146 | 1 1/8 | 29 | 2 3/16 | 56 |
| | 4.50 | 0.1772 | 4.500 | 2 3/16 | 56 | 3 3/8 | 86 | 3 3/8 | 86 | 5 3/4 | 146 | 1 1/8 | 29 | 2 3/16 | 56 |
| 15 | | 0.1800 | 4.572 | 2 3/16 | 56 | 3 3/8 | 86 | 3 3/8 | 86 | 5 3/4 | 146 | 1 1/8 | 29 | 2 3/16 | 56 |
| | 4.60 | 0.1811 | 4.600 | 2 3/16 | 56 | 3 3/8 | 86 | 3 3/8 | 86 | 5 3/4 | 146 | 1 1/8 | 29 | 2 3/16 | 56 |
| 14 | | 0.1820 | 4.623 | 2 3/16 | 56 | 3 3/8 | 86 | 3 3/8 | 86 | 5 3/4 | 146 | 1 1/8 | 29 | 2 3/16 | 56 |
| 13 | 4.70 | 0.1850 | 4.700 | 2 5/16 | 59 | 3 1/2 | 89 | 3 3/8 | 86 | 5 3/4 | 146 | 1 1/8 | 29 | 2 3/16 | 56 |
| 3/16 | | 0.1875 | 4.762 | 2 5/16 | 59 | 3 1/2 | 89 | 3 3/8 | 86 | 5 3/4 | 146 | 1 1/8 | 29 | 2 3/16 | 56 |
| 12 | 4.80 | 0.1890 | 4.800 | 2 5/16 | 59 | 3 1/2 | 89 | 3 5/8 | 92 | 6 | 152 | 1 3/16 | 30 | 2 1/4 | 57 |
| 11 | | 0.1910 | 4.851 | 2 5/16 | 59 | 3 1/2 | 89 | 3 5/8 | 92 | 6 | 152 | 1 3/16 | 30 | 2 1/4 | 57 |
| | 4.90 | 0.1929 | 4.900 | 2 7/16 | 62 | 3 5/8 | 92 | 3 5/8 | 92 | 6 | 152 | 1 3/16 | 30 | 2 1/4 | 57 |
| 10 | | 0.1935 | 4.915 | 2 7/16 | 62 | 3 5/8 | 92 | 3 5/8 | 92 | 6 | 152 | 1 3/16 | 30 | 2 1/4 | 57 |
| 9 | | 0.1960 | 4.978 | 2 7/16 | 62 | 3 5/8 | 92 | 3 5/8 | 92 | 6 | 152 | 1 3/16 | 30 | 2 1/4 | 57 |
| | 5.00 | 0.1969 | 5.000 | 2 7/16 | 62 | 3 5/8 | 92 | 3 5/8 | 92 | 6 | 152 | 1 3/16 | 30 | 2 1/4 | 57 |
| 8 | | 0.1990 | 5.054 | 2 7/16 | 62 | 3 5/8 | 92 | 3 5/8 | 92 | 6 | 152 | 1 3/16 | 30 | 2 1/4 | 57 |

**Table 1.** *(Continued)* **ANSI Straight-Shank Twist Drills — Jobbers Length through 17.5 mm, Taper Length through 12.7 mm, and Screw Machine Length through 25.4 mm Diameter** *ANSI/ASME B94.11M-1993*

| Drill Diameter, $D^a$ | | | | Jobbers Length | | | | Taper Length | | | | Screw Machine Length | | | |
|---|---|---|---|---|---|---|---|---|---|---|---|---|---|---|---|
| Frac-tion No. or Ltr. | mm | Equivalent | | Flute | | Overall | | Flute | | Overall | | Flute | | Overall | |
| | | Decimal Inch | mm | F | | L | | F | | L | | F | | L | |
| | | | | Inch | mm | Inch | mm | Inch | mm | Inch | mm | Inch | mm | Inch | mm |
| | 5.10 | 0.2008 | 5.100 | 2 7/16 | 62 | 3 5/8 | 92 | 3 5/8 | 92 | 6 | 152 | 1 3/16 | 30 | 2 1/4 | 57 |
| 7 | | 0.2010 | 5.105 | 2 7/16 | 62 | 3 5/8 | 92 | 3 5/8 | 92 | 6 | 152 | 1 3/16 | 30 | 2 1/4 | 57 |
| 13/64 | | 0.2031 | 5.159 | 2 7/16 | 62 | 3 5/8 | 92 | 3 5/8 | 92 | 6 | 152 | 1 3/16 | 30 | 2 1/4 | 57 |
| 6 | | 0.2040 | 5.182 | 2 1/2 | 64 | 3 3/4 | 95 | 3 5/8 | 92 | 6 | 152 | 1 1/4 | 32 | 2 3/8 | 60 |
| | 5.20 | 0.2047 | 5.200 | 2 1/2 | 64 | 3 3/4 | 95 | 3 5/8 | 92 | 6 | 152 | 1 1/4 | 32 | 2 3/8 | 60 |
| 5 | | 0.2055 | 5.220 | 2 1/2 | 64 | 3 3/4 | 95 | 3 5/8 | 92 | 6 | 152 | 1 1/4 | 32 | 2 3/8 | 60 |
| | 5.30 | 0.2087 | 5.300 | 2 1/2 | 64 | 3 3/4 | 95 | 3 5/8 | 92 | 6 | 152 | 1 1/4 | 32 | 2 3/8 | 60 |
| 4 | | 0.2090 | 5.309 | 2 1/2 | 64 | 3 3/4 | 95 | 3 5/8 | 92 | 6 | 152 | 1 1/4 | 32 | 2 3/8 | 60 |
| | 5.40 | 0.2126 | 5.400 | 2 1/2 | 64 | 3 3/4 | 95 | 3 5/8 | 92 | 6 | 152 | 1 1/4 | 32 | 2 3/8 | 60 |
| 3 | | 0.2130 | 5.410 | 2 1/2 | 64 | 3 3/4 | 95 | 3 5/8 | 92 | 6 | 152 | 1 1/4 | 32 | 2 3/8 | 60 |
| | 5.50 | 0.2165 | 5.500 | 2 1/2 | 64 | 3 3/4 | 95 | 3 5/8 | 92 | 6 | 152 | 1 1/4 | 32 | 2 3/8 | 60 |
| 7/32 | | 0.2188 | 5.558 | 2 1/2 | 64 | 3 3/4 | 95 | 3 5/8 | 92 | 6 | 152 | 1 1/4 | 32 | 2 3/8 | 60 |
| | 5.60 | 0.2205 | 5.600 | 2 5/8 | 67 | 3 7/8 | 98 | 3 3/4 | 95 | 6 1/8 | 156 | 1 5/16 | 33 | 2 7/16 | 62 |
| 2 | | 0.2210 | 5.613 | 2 5/8 | 67 | 3 7/8 | 98 | 3 3/4 | 95 | 6 1/8 | 156 | 1 5/16 | 33 | 2 7/16 | 62 |
| | 5.70 | 0.2244 | 5.700 | 2 5/8 | 67 | 3 7/8 | 98 | 3 3/4 | 95 | 6 1/8 | 156 | 1 5/16 | 33 | 2 7/16 | 62 |
| 1 | | 0.2280 | 5.791 | 2 5/8 | 67 | 3 7/8 | 98 | 3 3/4 | 95 | 6 1/8 | 156 | 1 5/16 | 33 | 2 7/16 | 62 |
| | 5.80 | 0.2283 | 5.800 | 2 5/8 | 67 | 3 7/8 | 98 | 3 3/4 | 95 | 6 1/8 | 156 | 1 5/16 | 33 | 2 7/16 | 62 |
| | 5.90 | 0.2323 | 5.900 | 2 5/8 | 67 | 3 7/8 | 98 | 3 3/4 | 95 | 6 1/8 | 156 | 1 5/16 | 33 | 2 7/16 | 62 |
| A | | 0.2340 | 5.944 | 2 5/8 | 67 | 3 7/8 | 98 | … | … | … | … | 1 5/16 | 33 | 2 7/16 | 62 |
| 15/64 | | 0.2344 | 5.954 | 2 5/8 | 67 | 3 7/8 | 98 | 3 3/4 | 95 | 6 1/8 | 156 | 1 5/16 | 33 | 2 7/16 | 62 |
| | 6.00 | 0.2362 | 6.000 | 2 3/4 | 70 | 4 | 102 | 3 3/4 | 95 | 6 1/8 | 156 | 1 3/8 | 35 | 2 1/2 | 64 |
| B | | 0.2380 | 6.045 | 2 3/4 | 70 | 4 | 102 | … | … | … | … | 1 3/8 | 35 | 2 1/2 | 64 |
| | 6.10 | 0.2402 | 6.100 | 2 3/4 | 70 | 4 | 102 | 3 3/4 | 95 | 6 1/8 | 156 | 1 3/8 | 35 | 2 1/2 | 64 |
| C | | 0.2420 | 6.147 | 2 3/4 | 70 | 4 | 102 | … | … | … | … | 1 3/8 | 35 | 2 1/2 | 64 |
| | 6.20 | 0.2441 | 6.200 | 2 3/4 | 70 | 4 | 102 | 3 3/4 | 95 | 6 1/8 | 156 | 1 3/8 | 35 | 2 1/2 | 64 |
| D | | 0.2460 | 6.248 | 2 3/4 | 70 | 4 | 102 | … | … | … | … | 1 3/8 | 35 | 2 1/2 | 64 |
| | 6.30 | 0.2480 | 6.300 | 2 3/4 | 70 | 4 | 102 | 3 3/4 | 95 | 6 1/8 | 156 | 1 3/8 | 35 | 2 1/2 | 64 |
| E, 1/4 | | 0.2500 | 6.350 | 2 3/4 | 70 | 4 | 102 | 3 3/4 | 95 | 6 1/8 | 156 | 1 3/8 | 35 | 2 1/2 | 64 |
| | 6.40 | 0.2520 | 6.400 | 2 7/8 | 73 | 4 1/8 | 105 | 3 7/8 | 98 | 6 1/4 | 159 | 1 7/16 | 37 | 2 5/8 | 67 |
| | 6.50 | 0.2559 | 6.500 | 2 7/8 | 73 | 4 1/8 | 105 | 3 7/8 | 98 | 6 1/4 | 159 | 1 7/16 | 37 | 2 5/8 | 67 |
| F | | 0.2570 | 6.528 | 2 7/8 | 73 | 4 1/8 | 105 | … | … | … | … | 1 7/16 | 37 | 2 5/8 | 67 |
| | 6.60 | 0.2598 | 6.600 | 2 7/8 | 73 | 4 1/8 | 105 | … | … | … | … | 1 7/16 | 37 | 2 5/8 | 67 |
| G | | 0.2610 | 6.629 | 2 7/8 | 73 | 4 1/8 | 105 | … | … | … | … | 1 7/16 | 37 | 2 5/8 | 67 |
| | 6.70 | 0.2638 | 6.700 | 2 7/8 | 73 | 4 1/8 | 105 | … | … | … | … | 1 7/16 | 37 | 2 5/8 | 67 |
| 17/64 | | 0.2656 | 6.746 | 2 7/8 | 73 | 4 1/8 | 105 | 3 7/8 | 98 | 6 1/4 | 159 | 1 7/16 | 37 | 2 5/8 | 67 |
| H | | 0.2660 | 6.756 | 2 7/8 | 73 | 4 1/8 | 105 | … | … | … | … | 1 1/2 | 38 | 2 11/16 | 68 |
| | 6.80 | 0.2677 | 6.800 | 2 7/8 | 73 | 4 1/8 | 105 | 3 7/8 | 98 | 6 1/4 | 159 | 1 1/2 | 38 | 2 11/16 | 68 |
| | 6.90 | 0.2717 | 6.900 | 2 7/8 | 73 | 4 1/8 | 105 | … | … | … | … | 1 1/2 | 38 | 2 11/16 | 68 |
| I | | 0.2720 | 6.909 | 2 7/8 | 73 | 4 1/8 | 105 | … | … | … | … | 1 1/2 | 38 | 2 11/16 | 68 |
| | 7.00 | 0.2756 | 7.000 | 2 7/8 | 73 | 4 1/8 | 105 | 3 7/8 | 98 | 6 1/4 | 159 | 1 1/2 | 38 | 2 11/16 | 68 |
| J | | 0.2770 | 7.036 | 2 7/8 | 73 | 4 1/8 | 105 | … | … | … | … | 1 1/2 | 38 | 2 11/16 | 68 |
| | 7.10 | 0.2795 | 7.100 | 2 15/16 | 75 | 4 1/4 | 108 | … | … | … | … | 1 1/2 | 38 | 2 11/16 | 68 |
| K | | 0.2810 | 7.137 | 2 15/16 | 75 | 4 1/4 | 108 | … | … | … | … | 1 1/2 | 38 | 2 11/16 | 68 |
| 9/32 | | 0.2812 | 7.142 | 2 15/16 | 75 | 4 1/4 | 108 | 3 7/8 | 98 | 6 1/4 | 159 | 1 1/2 | 38 | 2 11/16 | 68 |
| | 7.20 | 0.2835 | 7.200 | 2 15/16 | 75 | 4 1/4 | 108 | 4 | 102 | 6 3/8 | 162 | 1 9/16 | 40 | 2 3/4 | 70 |
| | 7.30 | 0.2874 | 7.300 | 2 15/16 | 75 | 4 1/4 | 108 | … | … | … | … | 1 9/16 | 40 | 2 3/4 | 70 |

**Table 1.** *(Continued)* **ANSI Straight-Shank Twist Drills — Jobbers Length through 17.5 mm, Taper Length through 12.7 mm, and Screw Machine Length through 25.4 mm Diameter** *ANSI/ASME B94.11M-1993*

| Drill Diameter, $D^a$ | | | | Jobbers Length | | | | Taper Length | | | | Screw Machine Length | | | |
|---|---|---|---|---|---|---|---|---|---|---|---|---|---|---|---|
| Frac-tion No. or Ltr. | mm | Equivalent | | Flute | | Overall | | Flute | | Overall | | Flute | | Overall | |
| | | | | F | | L | | F | | L | | F | | L | |
| | | Decimal Inch | mm | Inch | mm | Inch | mm | Inch | mm | Inch | mm | Inch | mm | Inch | mm |
| L | | 0.2900 | 7.366 | 2 15/16 | 75 | 4 1/4 | 108 | … | … | … | … | 1 9/16 | 40 | 2 3/4 | 70 |
| | 7.40 | 0.2913 | 7.400 | 3 1/16 | 78 | 4 3/8 | 111 | … | … | … | … | 1 9/16 | 40 | 2 3/4 | 70 |
| M | | 0.2950 | 7.493 | 3 1/16 | 78 | 4 3/8 | 111 | … | … | … | … | 1 9/16 | 40 | 2 3/4 | 70 |
| | 7.50 | 0.2953 | 7.500 | 3 1/16 | 78 | 4 3/8 | 111 | 4 | 102 | 6 3/8 | 162 | 1 9/16 | 40 | 2 3/4 | 70 |
| 19/64 | | 0.2969 | 7.541 | 3 1/16 | 78 | 4 3/8 | 111 | 4 | 102 | 6 3/8 | 162 | 1 9/16 | 40 | 2 3/4 | 70 |
| | 7.60 | 0.2992 | 7.600 | 3 1/16 | 78 | 4 3/8 | 111 | … | … | … | … | 1 5/8 | 41 | 2 13/16 | 71 |
| N | | 0.3020 | 7.671 | 3 1/16 | 78 | 4 3/8 | 111 | … | … | … | … | 1 5/8 | 41 | 2 13/16 | 71 |
| | 7.70 | 0.3031 | 7.700 | 3 3/16 | 81 | 4 1/2 | 114 | … | … | … | … | 1 5/8 | 41 | 2 13/16 | 71 |
| | 7.80 | 0.3071 | 7.800 | 3 3/16 | 81 | 4 1/2 | 114 | 4 | 102 | 6 3/8 | 162 | 1 5/8 | 41 | 2 13/16 | 71 |
| | 7.90 | 0.3110 | 7.900 | 3 3/16 | 81 | 4 1/2 | 114 | … | … | … | … | 1 5/8 | 41 | 2 13/16 | 71 |
| 5/16 | | 0.3125 | 7.938 | 3 3/16 | 81 | 4 1/2 | 114 | 4 | 102 | 6 3/8 | 162 | 1 5/8 | 41 | 2 13/16 | 71 |
| | 8.00 | 0.3150 | 8.000 | 3 3/16 | 81 | 4 1/2 | 114 | 4 1/8 | 105 | 6 1/2 | 165 | 1 11/16 | 43 | 2 15/16 | 75 |
| O | | 0.3160 | 8.026 | 3 3/16 | 81 | 4 1/2 | 114 | … | … | … | … | 1 11/16 | 43 | 2 15/16 | 75 |
| | 8.10 | 0.3189 | 8.100 | 3 5/16 | 84 | 4 5/8 | 117 | … | … | … | … | 1 11/16 | 43 | 2 15/16 | 75 |
| | 8.20 | 0.3228 | 8.200 | 3 5/16 | 84 | 4 5/8 | 117 | 4 1/8 | 105 | 6 1/2 | 165 | 1 11/16 | 43 | 2 15/16 | 75 |
| P | | 0.3230 | 8.204 | 3 5/16 | 84 | 4 5/8 | 117 | … | … | … | … | 1 11/16 | 43 | 2 15/16 | 75 |
| | 8.30 | 0.3268 | 8.300 | 3 5/16 | 84 | 4 5/8 | 117 | … | … | … | … | 1 11/16 | 43 | 2 15/16 | 75 |
| 21/64 | | 0.3281 | 8.334 | 3 5/16 | 84 | 4 5/8 | 117 | 4 1/8 | 105 | 6 1/2 | 165 | 1 11/16 | 43 | 2 15/16 | 75 |
| | 8.40 | 0.3307 | 8.400 | 3 7/16 | 87 | 4 3/4 | 121 | … | … | … | … | 1 11/16 | 43 | 3 | 76 |
| Q | | 0.3320 | 8.433 | 3 7/16 | 87 | 4 3/4 | 121 | … | … | … | … | 1 11/16 | 43 | 3 | 76 |
| | 8.50 | 0.3346 | 8.500 | 3 7/16 | 87 | 4 3/4 | 121 | 4 1/8 | 105 | 6 1/2 | 165 | 1 11/16 | 43 | 3 | 76 |
| | 8.60 | 0.3386 | 8.600 | 3 7/16 | 87 | 4 3/4 | 121 | … | … | … | … | 1 11/16 | 43 | 3 | 76 |
| R | | 0.3390 | 8.611 | 3 7/16 | 87 | 4 3/4 | 121 | … | … | … | … | 1 11/16 | 43 | 3 | 76 |
| | 8.70 | 0.3425 | 8.700 | 3 7/16 | 87 | 4 3/4 | 121 | … | … | … | … | 1 11/16 | 43 | 3 | 76 |
| 11/32 | | 0.3438 | 8.733 | 3 7/16 | 87 | 4 3/4 | 121 | 4 1/8 | 105 | 6 1/2 | 165 | 1 11/16 | 43 | 3 | 76 |
| | 8.80 | 0.3465 | 8.800 | 3 1/2 | 89 | 4 7/8 | 124 | 4 1/4 | 108 | 6 3/4 | 171 | 1 3/4 | 44 | 3 1/16 | 78 |
| S | | 0.3480 | 8.839 | 3 1/2 | 89 | 4 7/8 | 124 | … | … | … | … | 1 3/4 | 44 | 3 1/16 | 78 |
| | 8.90 | 0.3504 | 8.900 | 3 1/2 | 89 | 4 7/8 | 124 | … | … | … | … | 1 3/4 | 44 | 3 1/16 | 78 |
| | 9.00 | 0.3543 | 9.000 | 3 1/2 | 89 | 4 7/8 | 124 | 4 1/4 | 108 | 6 3/4 | 171 | 1 3/4 | 44 | 3 1/16 | 78 |
| T | | 0.3580 | 9.093 | 3 1/2 | 89 | 4 7/8 | 124 | … | … | … | … | 1 3/4 | 44 | 3 1/16 | 78 |
| | 9.10 | 0.3583 | 9.100 | 3 1/2 | 89 | 4 7/8 | 124 | … | … | … | … | 1 3/4 | 44 | 3 1/16 | 78 |
| 23/64 | | 0.3594 | 9.129 | 3 1/2 | 89 | 4 7/8 | 124 | 4 1/4 | 108 | 6 3/4 | 171 | 1 3/4 | 44 | 3 1/16 | 78 |
| | 9.20 | 0.3622 | 9.200 | 3 5/8 | 92 | 5 | 127 | 4 1/4 | 108 | 6 3/4 | 171 | 1 13/16 | 46 | 3 1/8 | 79 |
| | 9.30 | 0.3661 | 9.300 | 3 5/8 | 92 | 5 | 127 | … | … | … | … | 1 13/16 | 46 | 3 1/8 | 79 |
| U | | 0.3680 | 9.347 | 3 5/8 | 92 | 5 | 127 | … | … | … | … | 1 13/16 | 46 | 3 1/8 | 79 |
| | 9.40 | 0.3701 | 9.400 | 3 5/8 | 92 | 5 | 127 | … | … | … | … | 1 13/16 | 46 | 3 1/8 | 79 |
| | 9.50 | 0.3740 | 9.500 | 3 5/8 | 92 | 5 | 127 | 4 1/4 | 108 | 6 3/4 | 171 | 1 13/16 | 46 | 3 1/8 | 79 |
| 3/8 | | 0.3750 | 9.525 | 3 5/8 | 92 | 5 | 127 | 4 1/4 | 108 | 6 3/4 | 171 | 1 13/16 | 46 | 3 1/8 | 79 |
| V | | 0.3770 | 9.576 | 3 5/8 | 92 | 5 | 127 | … | … | … | … | 1 7/8 | 48 | 3 1/4 | 83 |
| | 9.60 | 0.3780 | 9.600 | 3 3/4 | 95 | 5 1/8 | 130 | … | … | … | … | 1 7/8 | 48 | 3 1/4 | 83 |
| | 9.70 | 0.3819 | 9.700 | 3 3/4 | 95 | 5 1/8 | 130 | … | … | … | … | 1 7/8 | 48 | 3 1/4 | 83 |
| | 9.80 | 0.3858 | 9.800 | 3 3/4 | 95 | 5 1/8 | 130 | 4 3/8 | 111 | 7 | 178 | 1 7/8 | 48 | 3 1/4 | 83 |
| W | | 0.3860 | 9.804 | 3 3/4 | 95 | 5 1/8 | 130 | … | … | … | … | 1 7/8 | 48 | 3 1/4 | 83 |
| | 9.90 | 0.3898 | 9.900 | 3 3/4 | 95 | 5 1/8 | 130 | … | … | … | … | 1 7/8 | 48 | 3 1/4 | 83 |
| 25/64 | | 0.3906 | 9.921 | 3 3/4 | 95 | 5 1/8 | 130 | 4 3/8 | 111 | 7 | 178 | 1 7/8 | 48 | 3 1/4 | 83 |
| | 10.00 | 0.3937 | 10.000 | 3 3/4 | 95 | 5 1/8 | 130 | 4 3/8 | 111 | 7 | 178 | 1 15/16 | 49 | 3 5/16 | 84 |

**Table 1.** *(Continued)* **ANSI Straight-Shank Twist Drills — Jobbers Length through 17.5 mm, Taper Length through 12.7 mm, and Screw Machine Length through 25.4 mm Diameter** *ANSI/ASME B94.11M-1993*

| Drill Diameter, D[a] | | | | Jobbers Length | | | | Taper Length | | | | Screw Machine Length | | | |
|---|---|---|---|---|---|---|---|---|---|---|---|---|---|---|---|
| Fraction No. or Ltr. | mm | Equivalent Decimal Inch | Equivalent mm | Flute F Inch | Flute F mm | Overall L Inch | Overall L mm | Flute F Inch | Flute F mm | Overall L Inch | Overall L mm | Flute F Inch | Flute F mm | Overall L Inch | Overall L mm |
| X | | 0.3970 | 10.084 | 3¾ | 95 | 5⅛ | 130 | … | … | … | … | 1 15/16 | 49 | 3 5/16 | 84 |
| | 10.20 | 0.4016 | 10.200 | 3⅞ | 98 | 5¼ | 133 | 4⅜ | 111 | 7 | 178 | 1 15/16 | 49 | 3 5/16 | 84 |
| Y | | 0.4040 | 10.262 | 3⅞ | 98 | 5¼ | 133 | … | … | … | … | 1 15/16 | 49 | 3 5/16 | 84 |
| 13/32 | | 0.4062 | 10.317 | 3⅞ | 98 | 5¼ | 133 | 4⅜ | 111 | 7 | 178 | 1 15/16 | 49 | 3 5/16 | 84 |
| Z | | 0.4130 | 10.490 | 3⅞ | 98 | 5¼ | 133 | … | … | … | … | 2 | 51 | 3⅜ | 86 |
| | 10.50 | 0.4134 | 10.500 | 3⅞ | 98 | 5¼ | 133 | 4⅝ | 117 | 7¼ | 184 | 2 | 51 | 3⅜ | 86 |
| 27/64 | | 0.4219 | 10.716 | 3 15/16 | 100 | 5⅜ | 137 | 4⅝ | 117 | 7¼ | 184 | 2 | 51 | 3⅜ | 86 |
| | 10.80 | 0.4252 | 10.800 | 4 1/16 | 103 | 5½ | 140 | 4⅝ | 117 | 7¼ | 184 | 2 1/16 | 52 | 3 7/16 | 87 |
| | 11.00 | 0.4331 | 11.000 | 4 1/16 | 103 | 5½ | 140 | 4⅝ | 117 | 7¼ | 184 | 2 1/16 | 52 | 3 7/16 | 87 |
| 7/16 | | 0.4375 | 11.112 | 4 1/16 | 103 | 5½ | 140 | 4⅝ | 117 | 7¼ | 184 | 2 1/16 | 52 | 3 7/16 | 87 |
| | 11.20 | 0.4409 | 11.200 | 4 3/16 | 106 | 5⅝ | 143 | 4¾ | 121 | 7½ | 190 | 2⅛ | 54 | 3 9/16 | 90 |
| | 11.50 | 0.4528 | 11.500 | 4 3/16 | 106 | 5⅝ | 143 | 4¾ | 121 | 7½ | 190 | 2⅛ | 54 | 3 9/16 | 90 |
| 29/64 | | 0.4531 | 11.509 | 4 3/16 | 106 | 5⅝ | 143 | 4¾ | 121 | 7½ | 190 | 2⅛ | 54 | 3 9/16 | 90 |
| | 11.80 | 0.4646 | 11.800 | 4 5/16 | 110 | 5¾ | 146 | 4¾ | 121 | 7½ | 190 | 2⅛ | 54 | 3⅝ | 92 |
| 15/32 | | 0.4688 | 11.908 | 4 5/16 | 110 | 5¾ | 146 | 4¾ | 121 | 7½ | 190 | 2⅛ | 54 | 3⅝ | 92 |
| | 12.00 | 0.4724 | 12.000 | 4⅜ | 111 | 5⅞ | 149 | 4¾ | 121 | 7¾ | 197 | 2 3/16 | 56 | 3 11/16 | 94 |
| | 12.20 | 0.4803 | 12.200 | 4⅜ | 111 | 5⅞ | 149 | 4¾ | 121 | 7¾ | 197 | 2 3/16 | 56 | 3 11/16 | 94 |
| 31/64 | | 0.4844 | 12.304 | 4⅜ | 111 | 5⅞ | 149 | 4¾ | 121 | 7¾ | 197 | 2 3/16 | 56 | 3 11/16 | 94 |
| | 12.50 | 0.4921 | 12.500 | 4½ | 114 | 6 | 152 | 4¾ | 121 | 7¾ | 197 | 2¼ | 57 | 3¾ | 95 |
| ½ | | 0.5000 | 12.700 | 4½ | 114 | 6 | 152 | 4¾ | 121 | 7¾ | 197 | 2¼ | 57 | 3¾ | 95 |
| | 12.80 | 0.5039 | 12.800 | 4½ | 114 | 6 | 152 | … | … | … | … | 2⅜ | 60 | 3⅞ | 98 |
| | 13.00 | 0.5118 | 13.000 | 4½ | 114 | 6 | 152 | … | … | … | … | 2⅜ | 60 | 3⅞ | 98 |
| 33/64 | | 0.5156 | 13.096 | 4 13/16 | 122 | 6⅝ | 168 | … | … | … | … | 2⅜ | 60 | 3⅞ | 98 |
| | 13.20 | 0.5197 | 13.200 | 4 13/16 | 122 | 6⅝ | 168 | … | … | … | … | 2⅜ | 60 | 3⅞ | 98 |
| 17/32 | | 0.5312 | 13.492 | 4 13/16 | 122 | 6⅝ | 168 | … | … | … | … | 2⅜ | 60 | 3⅞ | 98 |
| | 13.50 | 0.5315 | 13.500 | 4 13/16 | 122 | 6⅝ | 168 | … | … | … | … | 2⅜ | 60 | 3⅞ | 98 |
| | 13.80 | 0.5433 | 13.800 | 4 13/16 | 122 | 6⅝ | 168 | … | … | … | … | 2½ | 64 | 4 | 102 |
| 35/64 | | 0.5469 | 13.891 | 4 13/16 | 122 | 6⅝ | 168 | … | … | … | … | 2½ | 64 | 4 | 102 |
| | 14.00 | 0.5512 | 14.000 | 4 13/16 | 122 | 6⅝ | 168 | … | … | … | … | 2½ | 64 | 4 | 102 |
| | 14.25 | 0.5610 | 14.250 | 4 13/16 | 122 | 6⅝ | 168 | … | … | … | … | 2½ | 64 | 4 | 102 |
| 9/16 | | 0.5625 | 14.288 | 4 13/16 | 122 | 6⅝ | 168 | … | … | … | … | 2½ | 64 | 4 | 102 |
| | 14.50 | 0.5709 | 14.500 | 4 13/16 | 122 | 6⅝ | 168 | … | … | … | … | 2⅝ | 67 | 4⅛ | 105 |
| 37/64 | | 0.5781 | 14.684 | 4 13/16 | 122 | 6⅝ | 168 | … | … | … | … | 2⅝ | 67 | 4⅛ | 105 |
| | 14.75 | 0.5807 | 14.750 | 5 3/16 | 132 | 7⅛ | 181 | … | … | … | … | 2⅝ | 67 | 4⅛ | 105 |
| | 15.00 | 0.5906 | 15.000 | 5 3/16 | 132 | 7⅛ | 181 | … | … | … | … | 2⅝ | 67 | 4⅛ | 105 |
| 19/32 | | 0.5938 | 15.083 | 5 3/16 | 132 | 7⅛ | 181 | … | … | … | … | 2⅝ | 67 | 4⅛ | 105 |
| | 15.25 | 0.6004 | 15.250 | 5 3/16 | 132 | 7⅛ | 181 | … | … | … | … | 2¾ | 70 | 4¼ | 108 |
| 39/64 | | 0.6094 | 15.479 | 5 3/16 | 132 | 7⅛ | 181 | … | … | … | … | 2¾ | 70 | 4¼ | 108 |
| | 15.50 | 0.6102 | 15.500 | 5 3/16 | 132 | 7⅛ | 181 | … | … | … | … | 2¾ | 70 | 4¼ | 108 |
| | 15.75 | 0.6201 | 15.750 | 5 3/16 | 132 | 7⅛ | 181 | … | … | … | … | 2¾ | 70 | 4¼ | 108 |
| ⅝ | | 0.6250 | 15.875 | 5 3/16 | 132 | 7⅛ | 181 | … | … | … | … | 2¾ | 70 | 4¼ | 108 |
| | 16.00 | 0.6299 | 16.000 | 5 3/16 | 132 | 7⅛ | 181 | … | … | … | … | 2⅞ | 73 | 4½ | 114 |
| | 16.25 | 0.6398 | 16.250 | 5 3/16 | 132 | 7⅛ | 181 | … | … | … | … | 2⅞ | 73 | 4½ | 114 |
| 41/64 | | 0.6406 | 16.271 | 5 3/16 | 132 | 7⅛ | 181 | … | … | … | … | 2⅞ | 73 | 4½ | 144 |
| | 16.50 | 0.6496 | 16.500 | 5 3/16 | 132 | 7⅛ | 181 | … | … | … | … | 2⅞ | 73 | 4½ | 114 |
| 21/32 | | 0.6562 | 16.669 | 5 3/16 | 132 | 7⅛ | 181 | … | … | … | … | 2⅞ | 73 | 4½ | 114 |

**Table 1.** *(Continued)* **ANSI Straight-Shank Twist Drills — Jobbers Length through 17.5 mm, Taper Length through 12.7 mm, and Screw Machine Length through 25.4 mm Diameter** *ANSI/ASME B94.11M-1993*

| Drill Diameter, D[a] | | | | Jobbers Length | | | | Taper Length | | | | Screw Machine Length | | | |
|---|---|---|---|---|---|---|---|---|---|---|---|---|---|---|---|
| | | Equivalent | | Flute | | Overall | | Flute | | Overall | | Flute | | Overall | |
| | | | | F | | L | | F | | L | | F | | L | |
| Fraction No. or Ltr. | mm | Decimal Inch | mm | Inch | mm | Inch | mm | Inch | mm | Inch | mm | Inch | mm | Inch | mm |
| | 16.75 | 0.6594 | 16.750 | 5⅝ | 143 | 7⅝ | 194 | … | … | … | … | 2⅞ | 73 | 4½ | 114 |
| | 17.00 | 0.6693 | 17.000 | 5⅝ | 143 | 7⅝ | 194 | … | … | … | … | 2⅞ | 73 | 4½ | 114 |
| 43/64 | | 0.6719 | 17.066 | 5⅝ | 143 | 7⅝ | 194 | … | … | … | … | 2⅞ | 73 | 4½ | 114 |
| | 17.25 | 0.6791 | 17.250 | 5⅝ | 143 | 7⅝ | 194 | … | … | … | … | 2⅞ | 73 | 4½ | 114 |
| 11/16 | | 0.6875 | 17.462 | 5⅝ | 143 | 7⅝ | 194 | … | … | … | … | 2⅞ | 73 | 4½ | 114 |
| | 17.50 | 0.6890 | 17.500 | 5⅝ | 143 | 7⅝ | 194 | … | … | … | … | 3 | 76 | 4¾ | 121 |
| 45/64 | | 0.7031 | 17.859 | … | … | … | … | … | … | … | … | 3 | 76 | 4¾ | 121 |
| | 18.00 | 0.7087 | 18.000 | … | … | … | … | … | … | … | … | 3 | 76 | 4¾ | 121 |
| 23/32 | | 0.7188 | 18.258 | … | … | … | … | … | … | … | … | 3 | 76 | 4¾ | 121 |
| | 18.50 | 0.7283 | 18.500 | … | … | … | … | … | … | … | … | 3⅛ | 79 | 5 | 127 |
| 47/64 | | 0.7344 | 18.654 | … | … | … | … | … | … | … | … | 3⅛ | 79 | 5 | 127 |
| | 19.00 | 0.7480 | 19.000 | … | … | … | … | … | … | … | … | 3⅛ | 79 | 5 | 127 |
| 3/4 | | 0.7500 | 19.050 | … | … | … | … | … | … | … | … | 3⅛ | 79 | 5 | 127 |
| 49/64 | | 0.7656 | 19.446 | … | … | … | … | … | … | … | … | 3¼ | 83 | 5⅛ | 130 |
| | 19.50 | 0.7677 | 19.500 | … | … | … | … | … | … | … | … | 3¼ | 83 | 5⅛ | 130 |
| 25/32 | | 0.7812 | 19.845 | … | … | … | … | … | … | … | … | 3¼ | 83 | 5⅛ | 130 |
| | 20.00 | 0.7879 | 20.000 | … | … | … | … | … | … | … | … | 3⅜ | 86 | 5¼ | 133 |
| 51/64 | | 0.7969 | 20.241 | … | … | … | … | … | … | … | … | 3⅜ | 86 | 5¼ | 133 |
| | 20.50 | 0.8071 | 20.500 | … | … | … | … | … | … | … | … | 3⅜ | 86 | 5¼ | 133 |
| 13/16 | | 0.8125 | 20.638 | … | … | … | … | … | … | … | … | 3⅜ | 86 | 5¼ | 133 |
| | 21.00 | 0.8268 | 21.000 | … | … | … | … | … | … | … | … | 3½ | 89 | 5⅜ | 137 |
| 53/64 | | 0.8281 | 21.034 | … | … | … | … | … | … | … | … | 3½ | 89 | 5⅜ | 137 |
| 27/32 | | 0.8438 | 21.433 | … | … | … | … | … | … | … | … | 3½ | 89 | 5⅜ | 137 |
| | 21.50 | 0.8465 | 21.500 | … | … | … | … | … | … | … | … | 3½ | 89 | 5⅜ | 137 |
| 55/64 | | 0.8594 | 21.829 | … | … | … | … | … | … | … | … | 3½ | 89 | 5⅜ | 137 |
| | 22.00 | 0.8661 | 22.000 | … | … | … | … | … | … | … | … | 3½ | 89 | 5⅜ | 137 |
| 7/8 | | 0.8750 | 22.225 | … | … | … | … | … | … | … | … | 3½ | 89 | 5⅜ | 137 |
| | 22.50 | 0.8858 | 22.500 | … | … | … | … | … | … | … | … | 3⅝ | 92 | 5⅝ | 143 |
| 57/64 | | 0.8906 | 22.621 | … | … | … | … | … | … | … | … | 3⅝ | 92 | 5⅝ | 143 |
| | 23.00 | 0.9055 | 23.000 | … | … | … | … | … | … | … | … | 3⅝ | 92 | 5⅝ | 143 |
| 29/32 | | 0.9062 | 23.017 | … | … | … | … | … | … | … | … | 3⅝ | 92 | 5⅝ | 143 |
| 59/64 | | 0.9219 | 23.416 | … | … | … | … | … | … | … | … | 3¾ | 95 | 5¾ | 146 |
| | 23.50 | 0.9252 | 23.500 | … | … | … | … | … | … | … | … | 3¾ | 95 | 5¾ | 146 |
| 15/16 | | 0.9375 | 23.812 | … | … | … | … | … | … | … | … | 3¾ | 95 | 5¾ | 146 |
| | 24.00 | 0.9449 | 24.000 | … | … | … | … | … | … | … | … | 3⅞ | 98 | 5⅞ | 149 |
| 61/64 | | 0.9531 | 24.209 | … | … | … | … | … | … | … | … | 3⅞ | 98 | 5⅞ | 149 |
| | 24.50 | 0.9646 | 24.500 | … | … | … | … | … | … | … | … | 3⅞ | 98 | 5⅞ | 149 |
| 31/32 | | 0.9688 | 24.608 | … | … | … | … | … | … | … | … | 3⅞ | 98 | 5⅞ | 149 |
| | 25.00 | 0.9843 | 25.000 | … | … | … | … | … | … | … | … | 4 | 102 | 6 | 152 |
| 63/64 | | 0.9844 | 25.004 | … | … | … | … | … | … | … | … | 4 | 102 | 6 | 152 |
| 1 | | 1.0000 | 25.400 | … | … | … | … | … | … | … | … | 4 | 102 | 6 | 152 |

[a] Fractional inch, number, letter, and metric sizes.

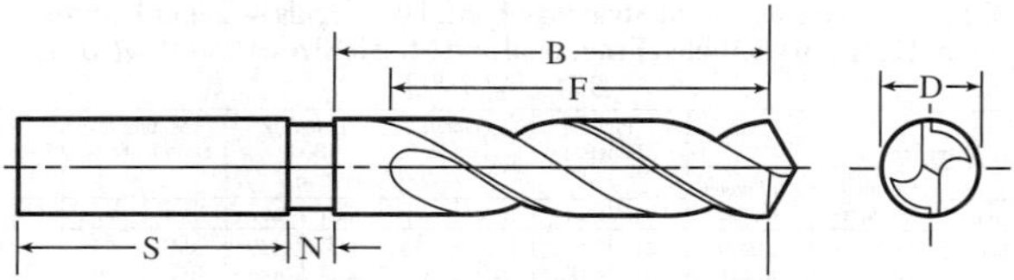


Nominal Shank Size is Same as Nominal Drill Size

## Table 2. ANSI Straight-Shank Twist Drills — Taper Length — Over $^1\!/_2$ Inch (12.7 mm) Diameter, Fractional and Metric Sizes *ANSI/ASME B94.11M-1993*

| Diameter of Drill | | | | Flute Length | | Overall Length | | Length of Body | | Minimum Length of Shk. | | Maximum Length of Neck | |
|---|---|---|---|---|---|---|---|---|---|---|---|---|---|
| *D* | | Decimal Inch Equiv. | Millimeter Equiv. | *F* | | *L* | | *B* | | *S* | | *N* | |
| Frac. | mm | | | Inch | mm | Inch | mm | Inch | mm | Inch | mm | Inch | mm |
| | 12.80 | 0.5039 | 12.800 | 4¾ | 121 | 8 | 203 | 4⅞ | 124 | 2⅝ | 66 | ½ | 13 |
| | 13.00 | 0.5117 | 13.000 | 4¾ | 121 | 8 | 203 | 4⅞ | 124 | 2⅝ | 66 | ½ | 13 |
| 33/64 | | 0.5156 | 13.096 | 4¾ | 121 | 8 | 203 | 4⅞ | 124 | 2⅝ | 66 | ½ | 13 |
| | 13.20 | 0.5197 | 13.200 | 4¾ | 121 | 8 | 203 | 4⅞ | 124 | 2⅝ | 66 | ½ | 13 |
| 17/32 | | 0.5312 | 13.492 | 4¾ | 121 | 8 | 203 | 4⅞ | 124 | 2⅝ | 66 | ½ | 13 |
| | 13.50 | 0.5315 | 13.500 | 4¾ | 121 | 8 | 203 | 4⅞ | 124 | 2⅝ | 66 | ½ | 13 |
| | 13.80 | 0.5433 | 13.800 | 4⅞ | 124 | 8¼ | 210 | 5 | 127 | 2¾ | 70 | ½ | 13 |
| 35/64 | | 0.5419 | 13.891 | 4⅞ | 124 | 8¼ | 210 | 5 | 127 | 2¾ | 70 | ½ | 13 |
| | 14.00 | 0.5512 | 14.000 | 4⅞ | 124 | 8¼ | 210 | 5 | 127 | 2¾ | 70 | ½ | 13 |
| | 14.25 | 0.5610 | 14.250 | 4⅞ | 124 | 8¼ | 210 | 5 | 127 | 2¾ | 70 | ½ | 13 |
| 9/16 | | 0.5625 | 14.288 | 4⅞ | 124 | 8¼ | 210 | 5 | 127 | 2¾ | 70 | ½ | 13 |
| | 14.50 | 0.5709 | 14.500 | 4⅞ | 124 | 8¾ | 222 | 5 | 127 | 3⅛ | 79 | ⅝ | 16 |
| 37/64 | | 0.5781 | 14.684 | 4⅞ | 124 | 8¾ | 222 | 5 | 127 | 3⅛ | 79 | ⅝ | 16 |
| | 14.75 | 0.5807 | 14.750 | 4⅞ | 124 | 8¾ | 222 | 5 | 127 | 3⅛ | 79 | ⅝ | 16 |
| | 15.00 | 0.5906 | 15.000 | 4⅞ | 124 | 8¾ | 222 | 5 | 127 | 3⅛ | 79 | ⅝ | 16 |
| 19/32 | | 0.5938 | 15.083 | 4⅞ | 124 | 8¾ | 222 | 5 | 127 | 3⅛ | 79 | ⅝ | 16 |
| | 15.25 | 0.6004 | 15.250 | 4⅞ | 124 | 8¾ | 222 | 5 | 127 | 3⅛ | 79 | ⅝ | 16 |
| 39/64 | | 0.6094 | 15.479 | 4⅞ | 124 | 8¾ | 222 | 5 | 127 | 3⅛ | 79 | ⅝ | 16 |
| | 15.50 | 0.6102 | 15.500 | 4⅞ | 124 | 8¾ | 222 | 5 | 127 | 3⅛ | 79 | ⅝ | 16 |
| | 15.75 | 0.6201 | 15.750 | 4⅞ | 124 | 8¾ | 222 | 5 | 127 | 3⅛ | 79 | ⅝ | 16 |
| 5/8 | | 0.6250 | 15.875 | 4⅞ | 124 | 8¾ | 222 | 5 | 127 | 3⅛ | 79 | ⅝ | 16 |
| | 16.00 | 0.6299 | 16.000 | 5⅛ | 130 | 9 | 228 | 5¼ | 133 | 3⅛ | 79 | ⅝ | 16 |
| | 16.25 | 0.6398 | 16.250 | 5⅛ | 130 | 9 | 228 | 5¼ | 133 | 3⅛ | 79 | ⅝ | 16 |
| 41/64 | | 0.6406 | 16.271 | 5⅛ | 130 | 9 | 228 | 5¼ | 133 | 3⅛ | 79 | ⅝ | 16 |
| | 16.50 | 0.6496 | 16.500 | 5⅛ | 130 | 9 | 228 | 5¼ | 133 | 3⅛ | 79 | ⅝ | 16 |
| 21/32 | | 0.6562 | 16.667 | 5⅛ | 130 | 9 | 228 | 5¼ | 133 | 3⅛ | 79 | ⅝ | 16 |
| | 16.75 | 0.6594 | 16.750 | 5⅜ | 137 | 9¼ | 235 | 5½ | 140 | 3⅛ | 79 | ⅝ | 16 |
| | 17.00 | 0.6693 | 17.000 | 5⅜ | 137 | 9¼ | 235 | 5½ | 140 | 3⅛ | 79 | ⅝ | 16 |
| 43/64 | | 0.6719 | 17.066 | 5⅜ | 137 | 9¼ | 235 | 5½ | 140 | 3⅛ | 79 | ⅝ | 16 |
| | 17.25 | 0.6791 | 17.250 | 5⅜ | 137 | 9¼ | 235 | 5½ | 140 | 3⅛ | 79 | ⅝ | 16 |
| 11/16 | | 0.6875 | 17.462 | 5⅜ | 137 | 9¼ | 235 | 5½ | 140 | 3⅛ | 79 | ⅝ | 16 |
| | 17.50 | 0.6890 | 17.500 | 5⅝ | 143 | 9½ | 241 | 5¾ | 146 | 3⅛ | 79 | ⅝ | 16 |
| 45/64 | | 0.7031 | 17.859 | 5⅝ | 143 | 9½ | 241 | 5¾ | 146 | 3⅛ | 79 | ⅝ | 16 |
| | 18.00 | 0.7087 | 18.000 | 5⅝ | 143 | 9½ | 241 | 5¾ | 146 | 3⅛ | 79 | ⅝ | 16 |
| 23/32 | | 0.7188 | 18.258 | 5⅝ | 143 | 9½ | 241 | 5¾ | 146 | 3⅛ | 79 | ⅝ | 16 |
| | 18.50 | 0.7283 | 18.500 | 5⅞ | 149 | 9¾ | 247 | 6 | 152 | 3⅛ | 79 | ⅝ | 16 |
| 47/64 | | 0.7344 | 18.654 | 5⅞ | 149 | 9¾ | 247 | 6 | 152 | 3⅛ | 79 | ⅝ | 16 |
| | 19.00 | 0.7480 | 19.000 | 5⅞ | 149 | 9¾ | 247 | 6 | 152 | 3⅛ | 79 | ⅝ | 16 |
| 3/4 | | 0.7500 | 19.050 | 5⅞ | 149 | 9¾ | 247 | 6 | 152 | 3⅛ | 79 | ⅝ | 16 |
| 49/64 | | 0.7656 | 19.446 | 6 | 152 | 9⅞ | 251 | 6⅛ | 156 | 3⅛ | 79 | ⅝ | 16 |
| | 19.50 | 0.7677 | 19.500 | 6 | 152 | 9⅞ | 251 | 6⅛ | 156 | 3⅛ | 79 | ⅝ | 16 |
| 25/32 | | 0.7812 | 19.842 | 6 | 152 | 9⅞ | 251 | 6⅛ | 156 | 3⅛ | 79 | ⅝ | 16 |

**Table 2.** *(Continued)* **ANSI Straight-Shank Twist Drills — Taper Length — Over ½ Inch (12.7 mm) Diameter, Fractional and Metric Sizes** *ANSI/ASME B94.11M-1993*

*B94.11M-1993*

| Diameter of Drill | | | | Flute Length | | Overall Length | | Length of Body | | Minimum Length of Shk. | | Maximum Length of Neck | |
|---|---|---|---|---|---|---|---|---|---|---|---|---|---|
| *D* | | Decimal Inch Equiv. | Millimeter Equiv. | *F* | | *L* | | *B* | | *S* | | *N* | |
| Frac. | mm | | | Inch | mm | Inch | mm | Inch | mm | Inch | mm | Inch | mm |
| | 20.00 | 0.7874 | 20.000 | 6⅛ | 156 | 10 | 254 | 6¼ | 159 | 3⅛ | 79 | ⅝ | 16 |
| 51/64 | | 0.7969 | 20.241 | 6⅛ | 156 | 10 | 254 | 6¼ | 159 | 3⅛ | 79 | ⅝ | 16 |
| | 20.50 | 0.8071 | 20.500 | 6⅛ | 156 | 10 | 254 | 6¼ | 159 | 3⅛ | 79 | ⅝ | 16 |
| 13/16 | | 0.8125 | 20.638 | 6⅛ | 156 | 10 | 254 | 6¼ | 159 | 3⅛ | 79 | ⅝ | 16 |
| | 21.00 | 0.8268 | 21.000 | 6⅛ | 156 | 10 | 254 | 6¼ | 159 | 3⅛ | 79 | ⅝ | 16 |
| 53/64 | | 0.8281 | 21.034 | 6⅛ | 156 | 10 | 254 | 6¼ | 159 | 3⅛ | 79 | ⅝ | 16 |
| 27/32 | | 0.8438 | 21.433 | 6⅛ | 156 | 10 | 254 | 6¼ | 159 | 3⅛ | 79 | ⅝ | 16 |
| | 21.50 | 0.8465 | 21.500 | 6⅛ | 156 | 10 | 254 | 6¼ | 159 | 3⅛ | 79 | ⅝ | 16 |
| 55/64 | | 0.8594 | 21.829 | 6⅛ | 156 | 10 | 254 | 6¼ | 159 | 3⅛ | 79 | ⅝ | 16 |
| | 22.00 | 0.8661 | 22.000 | 6⅛ | 156 | 10 | 254 | 6¼ | 159 | 3⅛ | 79 | ⅝ | 16 |
| 7/8 | | 0.8750 | 22.225 | 6⅛ | 156 | 10 | 254 | 6¼ | 159 | 3⅛ | 79 | ⅝ | 16 |
| | 22.50 | 0.8858 | 22.500 | 6⅛ | 156 | 10 | 254 | 6¼ | 159 | 3⅛ | 79 | ⅝ | 16 |
| 57/64 | | 0.8906 | 22.621 | 6⅛ | 156 | 10 | 254 | 6¼ | 159 | 3⅛ | 79 | ⅝ | 16 |
| | 23.00 | 0.9055 | 23.000 | 6⅛ | 156 | 10 | 254 | 6¼ | 159 | 3⅛ | 79 | ⅝ | 16 |
| 29/32 | | 0.9062 | 23.017 | 6⅛ | 156 | 10 | 254 | 6¼ | 159 | 3⅛ | 79 | ⅝ | 16 |
| 59/64 | | 0.9219 | 23.416 | 6⅛ | 156 | 10¾ | 273 | 6¼ | 159 | 3⅞ | 98 | ⅝ | 16 |
| | 23.50 | 0.9252 | 23.500 | 6⅛ | 156 | 10¾ | 273 | 6¼ | 159 | 3⅞ | 98 | ⅝ | 16 |
| 15/16 | | 0.9375 | 23.812 | 6⅛ | 156 | 10¾ | 273 | 6¼ | 159 | 3⅞ | 98 | ⅝ | 16 |
| | 24.00 | 0.9449 | 24.000 | 6⅜ | 162 | 11 | 279 | 6½ | 165 | 3⅞ | 98 | ⅝ | 16 |
| 61/64 | | 0.9531 | 24.209 | 6⅜ | 162 | 11 | 279 | 6½ | 165 | 3⅞ | 98 | ⅝ | 16 |
| | 24.50 | 0.9646 | 24.500 | 6⅜ | 162 | 11 | 279 | 6½ | 165 | 3⅞ | 98 | ⅝ | 16 |
| 31/32 | | 0.9688 | 24.608 | 6⅜ | 162 | 11 | 279 | 6½ | 165 | 3⅞ | 98 | ⅝ | 16 |
| | 25.00 | 0.9843 | 25.000 | 6⅜ | 162 | 11 | 279 | 6½ | 165 | 3⅞ | 98 | ⅝ | 16 |
| 63/64 | | 0.9844 | 25.004 | 6⅜ | 162 | 11 | 279 | 6½ | 165 | 3⅞ | 98 | ⅝ | 16 |
| 1 | | 1.0000 | 25.400 | 6⅜ | 162 | 11 | 279 | 6½ | 165 | 3⅞ | 98 | ⅝ | 16 |
| | 25.50 | 1.0039 | 25.500 | 6½ | 165 | 11⅛ | 282 | 6⅝ | 168 | 3⅞ | 98 | ⅝ | 16 |
| 1 1/64 | | 1.0156 | 25.796 | 6½ | 165 | 11⅛ | 282 | 6⅝ | 168 | 3⅞ | 98 | ⅝ | 16 |
| | 26.00 | 1.0236 | 26.000 | 6½ | 165 | 11⅛ | 282 | 6⅝ | 168 | 3⅞ | 98 | ⅝ | 16 |
| 1 1/32 | | 1.0312 | 26.192 | 6½ | 165 | 11⅛ | 282 | 6⅝ | 168 | 3⅞ | 98 | ⅝ | 16 |
| | 26.50 | 1.0433 | 26.560 | 6⅝ | 168 | 11¼ | 286 | 6¾ | 172 | 3⅞ | 98 | ⅝ | 16 |
| 1 3/64 | | 1.0469 | 26.591 | 6⅝ | 168 | 11¼ | 286 | 6¾ | 172 | 3⅞ | 98 | ⅝ | 16 |
| 1 1/16 | | 1.0625 | 26.988 | 6⅝ | 168 | 11¼ | 286 | 6¾ | 172 | 3⅞ | 98 | ⅝ | 16 |
| | 27.00 | 1.0630 | 27.000 | 6⅝ | 168 | 11¼ | 286 | 6¾ | 172 | 3⅞ | 98 | ⅝ | 16 |
| 1 5/64 | | 1.0781 | 27.384 | 6⅞ | 175 | 11½ | 292 | 7 | 178 | 3⅞ | 98 | ⅝ | 16 |
| | 27.50 | 1.0827 | 27.500 | 6⅞ | 175 | 11½ | 292 | 7 | 178 | 3⅞ | 98 | ⅝ | 16 |
| 1 3/32 | | 1.0938 | 27.783 | 6⅞ | 175 | 11½ | 292 | 7 | 178 | 3⅞ | 98 | ⅝ | 16 |
| | 28.00 | 1.1024 | 28.000 | 7⅛ | 181 | 11¾ | 298 | 7¼ | 184 | 3⅞ | 98 | ⅝ | 16 |
| 1 7/64 | | 1.1094 | 28.179 | 7⅛ | 181 | 11¾ | 298 | 7¼ | 184 | 3⅞ | 98 | ⅝ | 16 |
| | 28.50 | 1.1220 | 28.500 | 7⅛ | 181 | 11¾ | 298 | 7¼ | 184 | 3⅞ | 98 | ⅝ | 16 |
| 1⅛ | | 1.1250 | 28.575 | 7⅛ | 181 | 11¾ | 298 | 7¼ | 184 | 3⅞ | 98 | ⅝ | 16 |
| 1 9/64 | | 1.1406 | 28.971 | 7¼ | 184 | 11⅞ | 301 | 7⅜ | 187 | 3⅞ | 98 | ⅝ | 16 |
| | 29.00 | 1.1417 | 29.000 | 7¼ | 184 | 11⅞ | 301 | 7⅜ | 187 | 3⅞ | 98 | ⅝ | 16 |
| 1 5/32 | | 1.1562 | 29.367 | 7¼ | 184 | 11⅞ | 301 | 7⅜ | 187 | 3⅞ | 98 | ⅝ | 16 |
| | 29.50 | 1.1614 | 29.500 | 7⅜ | 187 | 12 | 305 | 7½ | 191 | 3⅞ | 98 | ⅝ | 16 |
| 1 11/64 | | 1.1719 | 29.766 | 7⅜ | 187 | 12 | 305 | 7½ | 191 | 3⅞ | 98 | ⅝ | 16 |
| | 30.00 | 1.1811 | 30.000 | 7⅜ | 187 | 12 | 305 | 7½ | 191 | 3⅞ | 98 | ⅝ | 16 |
| 1 3/16 | | 1.1875 | 30.162 | 7⅜ | 187 | 12 | 305 | 7½ | 191 | 3⅞ | 98 | ⅝ | 16 |
| | 30.50 | 1.2008 | 30.500 | 7½ | 190 | 12⅛ | 308 | 7⅝ | 194 | 3⅞ | 98 | ⅝ | 16 |
| 1 13/64 | | 1.2031 | 30.559 | 7½ | 190 | 12⅛ | 308 | 7⅝ | 194 | 3⅞ | 98 | ⅝ | 16 |
| 1 7/32 | | 1.2188 | 30.958 | 7½ | 190 | 12⅛ | 308 | 7⅝ | 194 | 3⅞ | 98 | ⅝ | 16 |
| | 31.00 | 1.2205 | 31.000 | 7⅞ | 200 | 12½ | 317 | 8 | 203 | 3⅞ | 98 | ⅝ | 16 |
| 1 15/64 | | 1.2344 | 31.354 | 7⅞ | 200 | 12½ | 317 | 8 | 203 | 3⅞ | 98 | ⅝ | 16 |
| | 31.50 | 1.2402 | 31.500 | 7⅞ | 200 | 12½ | 317 | 8 | 203 | 3⅞ | 98 | ⅝ | 16 |

**Table 2.** *(Continued)* **ANSI Straight-Shank Twist Drills — Taper Length — Over ½ Inch (12.7 mm) Diameter, Fractional and Metric Sizes** *ANSI/ASME B94.11M-1993*
*B94.11M-1993*

| Diameter of Drill | | | | Flute Length | | Overall Length | | Length of Body | | Minimum Length of Shk. | | Maximum Length of Neck | |
|---|---|---|---|---|---|---|---|---|---|---|---|---|---|
| *D* | | Decimal Inch Equiv. | Millimeter Equiv. | *F* | | *L* | | *B* | | *S* | | *N* | |
| Frac. | mm | | | Inch | mm | Inch | mm | Inch | mm | Inch | mm | Inch | mm |
| 1¼ | | 1.2500 | 31.750 | 7⅞ | 200 | 12½ | 317 | 8 | 203 | 3⅞ | 98 | ⅝ | 16 |
| | 32.00 | 1.2598 | 32.000 | 8½ | 216 | 14⅛ | 359 | 8⅝ | 219 | 4⅞ | 124 | ⅝ | 16 |
| | 32.50 | 1.2795 | 32.500 | 8½ | 216 | 14⅛ | 359 | 8⅝ | 219 | 4⅞ | 124 | ⅝ | 16 |
| 1 9/32 | | 1.2812 | 32.542 | 8½ | 216 | 14⅛ | 359 | 8⅝ | 219 | 4⅞ | 124 | ⅝ | 16 |
| | 33.00 | 1.2992 | 33.000 | 8⅝ | 219 | 14¼ | 362 | 8¾ | 222 | 4⅞ | 124 | ⅝ | 16 |
| 1 5/16 | | 1.3125 | 33.338 | 8⅝ | 219 | 14¼ | 362 | 8¾ | 222 | 4⅞ | 124 | ⅝ | 16 |
| | 33.50 | 1.3189 | 33.500 | 8¾ | 222 | 14⅜ | 365 | 8⅞ | 225 | 4⅞ | 124 | ⅝ | 16 |
| | 34.00 | 1.3386 | 34.000 | 8¾ | 222 | 14⅜ | 365 | 8⅞ | 225 | 4⅞ | 124 | ⅝ | 16 |
| 1 11/32 | | 1.3438 | 34.133 | 8¾ | 222 | 14⅜ | 365 | 8⅞ | 225 | 4⅞ | 124 | ⅝ | 16 |
| | 34.50 | 1.3583 | 34.500 | 8⅞ | 225 | 14½ | 368 | 9 | 229 | 4⅞ | 124 | ⅝ | 16 |
| 1⅜ | | 1.3750 | 34.925 | 8⅞ | 225 | 14½ | 368 | 9 | 229 | 4⅞ | 124 | ⅝ | 16 |
| | 35.00 | 1.3780 | 35.000 | 9 | 229 | 14⅝ | 372 | 9⅛ | 232 | 4⅞ | 124 | ⅝ | 16 |
| | 35.50 | 1.3976 | 35.500 | 9 | 229 | 14⅝ | 372 | 9⅛ | 232 | 4⅞ | 124 | ⅝ | 16 |
| 1 13/32 | | 1.4062 | 35.717 | 9 | 229 | 14⅝ | 372 | 9⅛ | 232 | 4⅞ | 124 | ⅝ | 16 |
| | 36.00 | 1.4173 | 36.000 | 9⅛ | 232 | 14¾ | 375 | 9¼ | 235 | 4⅞ | 124 | ⅝ | 16 |
| | 36.50 | 1.4370 | 36.500 | 9⅛ | 232 | 14¾ | 375 | 9¼ | 235 | 4⅞ | 124 | ⅝ | 16 |
| 1 7/16 | | 1.4375 | 36.512 | 9⅛ | 232 | 14¾ | 375 | 9¼ | 235 | 4⅞ | 124 | ⅝ | 16 |
| | 37.00 | 1.4567 | 37.000 | 9¼ | 235 | 14⅞ | 378 | 9⅜ | 238 | 4⅞ | 124 | ⅝ | 16 |
| 1 15/32 | | 1.4688 | 37.308 | 9¼ | 235 | 14⅞ | 378 | 9⅜ | 238 | 4⅞ | 124 | ⅝ | 16 |
| | 37.50 | 1.4764 | 37.500 | 9⅜ | 238 | 15 | 381 | 9½ | 241 | 4⅞ | 124 | ⅝ | 16 |
| | 38.00 | 1.4961 | 38.000 | 9⅜ | 238 | 15 | 381 | 9½ | 241 | 4⅞ | 124 | ⅝ | 16 |
| 1½ | | 1.5000 | 38.100 | 9⅜ | 238 | 15 | 381 | 9½ | 241 | 4⅞ | 124 | ⅝ | 16 |
| 1 9/16 | | 1.5625 | 39.688 | 9⅝ | 244 | 15¼ | 387 | 9¾ | 247 | 4⅞ | 124 | ⅝ | 16 |
| 1⅝ | | 1.6250 | 41.275 | 9⅞ | 251 | 15⅝ | 397 | 10 | 254 | 4⅞ | 124 | ¾ | 19 |
| 1¾ | | 1.7500 | 44.450 | 10½ | 267 | 16¼ | 413 | 10⅝ | 270 | 4⅞ | 124 | ¾ | 19 |

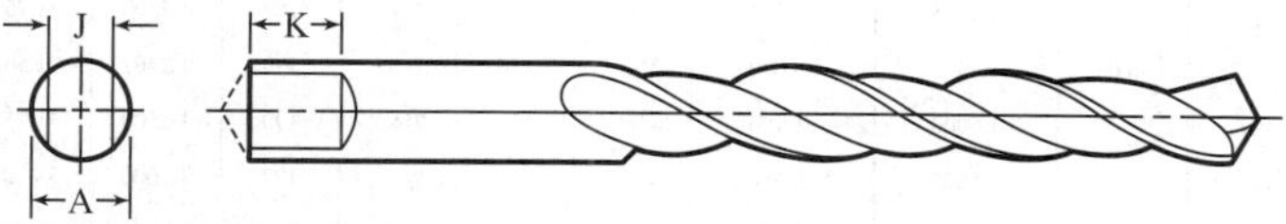


**Table 3. American National Standard Tangs for Straight-Shank Drills**
*ANSI/ASME B94.11M-1993*

| Nominal Diameter of Drill Shank, *A* | | Thickness of Tang, *J* | | | | Length of Tang, *K* | |
|---|---|---|---|---|---|---|---|
| | | Inches | | Millimeters | | | |
| Inches | Millimeters | Max. | Min. | Max. | Min. | Inches | Millimeters |
| ⅛ thru 3/16 | 3.18 thru 4.76 | 0.094 | 0.090 | 2.39 | 2.29 | 9/32 | 7.0 |
| over 3/16 thru ¼ | over 4.76 thru 6.35 | 0.122 | 0.118 | 3.10 | 3.00 | 5/16 | 8.0 |
| over ¼ thru 5/16 | over 6.35 thru 7.94 | 0.162 | 0.158 | 4.11 | 4.01 | 11/32 | 8.5 |
| over 5/16 thru ⅜ | over 7.94 thru 9.53 | 0.203 | 0.199 | 5.16 | 5.06 | ⅜ | 9.5 |
| over ⅜ thru 15/32 | over 9.53 thru 11.91 | 0.243 | 0.239 | 6.17 | 6.07 | 7/16 | 11.0 |
| over 15/32 thru 9/16 | over 11.91 thru 14.29 | 0.303 | 0.297 | 7.70 | 7.55 | ½ | 12.5 |
| over 9/16 thru 21/32 | over 14.29 thru 16.67 | 0.373 | 0.367 | 9.47 | 9.32 | 9/16 | 14.5 |
| over 21/32 thru ¾ | over 16.67 thru 19.05 | 0.443 | 0.437 | 11.25 | 11.10 | ⅝ | 16.0 |
| over ¾ thru ⅞ | over 19.05 thru 22.23 | 0.514 | 0.508 | 13.05 | 12.90 | 11/16 | 17.5 |
| over ⅞ thru 1 | over 22.23 thru 25.40 | 0.609 | 0.601 | 15.47 | 15.27 | ¾ | 19.0 |
| over 1 thru 1 3/16 | over 25.40 thru 30.16 | 0.700 | 0.692 | 17.78 | 17.58 | 13/16 | 20.5 |
| over 1 3/16 thru 1⅜ | over 30.16 thru 34.93 | 0.817 | 0.809 | 20.75 | 20.55 | ⅞ | 22.0 |

To fit split-sleeve collet-type drill drivers. See page 943.

**Table 4. American National Standard Straight-Shank Twist Drills — Screw Machine Length — Over 1 Inch (25.4 mm) Diameter** *ANSI/ASME B94.11M-1993*

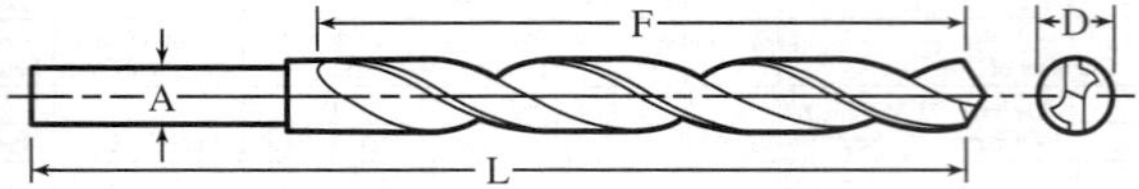


| Diameter of Drill | | | | Flute Length | | Overall Length | | Shank Diameter | |
|---|---|---|---|---|---|---|---|---|---|
| *D* | | Decimal Inch Equivalent | Millimeter Equivalent | *F* | | *L* | | *A* | |
| Frac. | mm | | | Inch | mm | Inch | mm | Inch | mm |
| | 25.50 | 1.0039 | 25.500 | 4 | 102 | 6 | 152 | 0.9843 | 25.00 |
| | 26.00 | 1.0236 | 26.000 | 4 | 102 | 6 | 152 | 0.9843 | 25.00 |
| 1 1/16 | | 1.0625 | 26.988 | 4 | 102 | 6 | 152 | 1.0000 | 25.40 |
| | 28.00 | 1.1024 | 28.000 | 4 | 102 | 6 | 152 | 0.9843 | 25.00 |
| 1 1/8 | | 1.1250 | 28.575 | 4 | 102 | 6 | 152 | 1.0000 | 25.40 |
| | 30.00 | 1.1811 | 30.000 | 4 1/4 | 108 | 6 5/8 | 168 | 0.9843 | 25.00 |
| 1 3/16 | | 1.1875 | 30.162 | 4 1/4 | 108 | 6 5/8 | 168 | 1.0000 | 25.40 |
| 1 1/4 | | 1.2500 | 31.750 | 4 3/8 | 111 | 6 3/4 | 171 | 1.0000 | 25.40 |
| | 32.00 | 1.2598 | 32.000 | 4 3/8 | 111 | 7 | 178 | 1.2402 | 31.50 |
| 1 5/16 | | 1.3125 | 33.338 | 4 3/8 | 111 | 7 | 178 | 1.2500 | 31.75 |
| | 34.00 | 1.3386 | 34.000 | 4 1/2 | 114 | 7 1/8 | 181 | 1.2402 | 31.50 |
| 1 3/8 | | 1.3750 | 34.925 | 4 1/2 | 114 | 7 1/8 | 181 | 1.2500 | 31.75 |
| | 36.00 | 1.4173 | 36.000 | 4 3/4 | 121 | 7 3/8 | 187 | 1.2402 | 31.50 |
| 1 7/16 | | 1.4375 | 36.512 | 4 3/4 | 121 | 7 3/8 | 187 | 1.2500 | 31.75 |
| | 38.00 | 1.4961 | 38.000 | 4 7/8 | 124 | 7 1/2 | 190 | 1.2402 | 31.50 |
| 1 1/2 | | 1.5000 | 38.100 | 4 7/8 | 124 | 7 1/2 | 190 | 1.2500 | 31.75 |
| 1 9/16 | | 1.5625 | 39.688 | 4 7/8 | 124 | 7 3/4 | 197 | 1.5000 | 38.10 |
| | 40.00 | 1.5748 | 40.000 | 4 7/8 | 124 | 7 3/4 | 197 | 1.4961 | 38.00 |
| 1 5/8 | | 1.6250 | 41.275 | 4 7/8 | 124 | 7 3/4 | 197 | 1.5000 | 38.10 |
| | 42.00 | 1.6535 | 42.000 | 5 1/8 | 130 | 8 | 203 | 1.4961 | 38.00 |
| 1 11/16 | | 1.6875 | 42.862 | 5 1/8 | 130 | 8 | 203 | 1.5000 | 38.10 |
| | 44.00 | 1.7323 | 44.000 | 5 1/8 | 130 | 8 | 203 | 1.4961 | 38.00 |
| 1 3/4 | | 1.7500 | 44.450 | 5 1/8 | 130 | 8 | 203 | 1.5000 | 38.10 |
| | 46.00 | 1.8110 | 46.000 | 5 3/8 | 137 | 8 1/4 | 210 | 1.4961 | 38.00 |
| 1 13/16 | | 1.8125 | 46.038 | 5 3/8 | 137 | 8 1/4 | 210 | 1.5000 | 38.10 |
| 1 7/8 | | 1.8750 | 47.625 | 5 3/8 | 137 | 8 1/4 | 210 | 1.5000 | 38.10 |
| | 48.00 | 1.8898 | 48.000 | 5 5/8 | 143 | 8 1/2 | 216 | 1.4961 | 38.00 |
| 1 15/16 | | 1.9375 | 49.212 | 5 5/8 | 143 | 8 1/2 | 216 | 1.5000 | 38.10 |
| | 50.00 | 1.9685 | 50.000 | 5 5/8 | 143 | 8 1/2 | 216 | 1.4961 | 38.00 |
| 2 | | 2.0000 | 50.800 | 5 5/8 | 143 | 8 1/2 | 216 | 1.5000 | 38.10 |

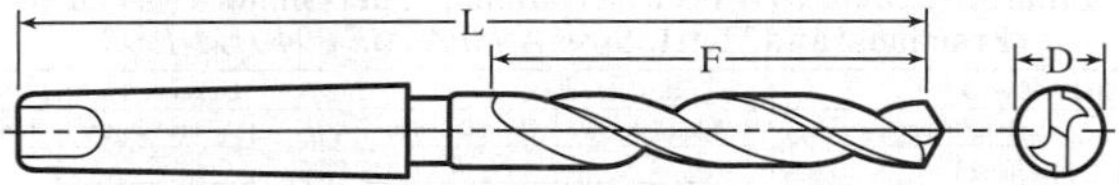


**Table 5. American National Taper-Shank Twist Drills Fractional and Metric Sizes** *ANSI/ASME B94.11M-1993*

| Drill Diameter, *D* | | | | Regular Shank | | | | | Larger or Smaller Shank[a] | | | | |
|---|---|---|---|---|---|---|---|---|---|---|---|---|---|
| | | Equivalent | | Morse Taper No. | Flute Length *F* | | Overall Length *L* | | Morse Taper No. | Flute Length *F* | | Overall Length *L* | |
| Fraction | mm | Decimal Inch | mm | | Inch | mm | Inch | mm | | Inch | mm | Inch | mm |
| | 3.00 | 0.1181 | 3.000 | 1 | 1 7/8 | 48 | 5 1/8 | 130 | … | … | … | … | … |
| 1/8 | | 0.1250 | 3.175 | 1 | 1 7/8 | 48 | 5 1/8 | 130 | … | … | … | … | … |
| | 3.20 | 0.1260 | 3.200 | 1 | 2 1/8 | 54 | 5 3/8 | 137 | … | … | … | … | … |
| | 3.50 | 0.1378 | 3.500 | 1 | 2 1/8 | 54 | 5 3/8 | 137 | … | … | … | … | … |
| 9/64 | | 0.1406 | 3.571 | 1 | 2 1/8 | 54 | 5 3/8 | 137 | … | … | … | … | … |
| | 3.80 | 0.1496 | 3.800 | 1 | 2 1/8 | 54 | 5 3/8 | 137 | … | … | … | … | … |
| 5/32 | | 0.1562 | 3.967 | 1 | 2 1/8 | 54 | 5 3/8 | 137 | … | … | … | … | … |
| | 4.00 | 0.1575 | 4.000 | 1 | 2 1/2 | 64 | 5 3/4 | 146 | … | … | … | … | … |
| | 4.20 | 0.1654 | 4.200 | 1 | 2 1/2 | 64 | 5 3/4 | 146 | … | … | … | … | … |
| 11/64 | | 0.1719 | 4.366 | 1 | 2 1/2 | 64 | 5 3/4 | 146 | … | … | … | … | … |
| | 4.50 | 0.1772 | 4.500 | 1 | 2 1/2 | 64 | 5 3/4 | 146 | … | … | … | … | … |
| 3/16 | | 0.1875 | 4.762 | 1 | 2 1/2 | 64 | 5 3/4 | 146 | … | … | … | … | … |
| | 4.80 | 0.1890 | 4.800 | 1 | 2 3/4 | 70 | 6 | 152 | … | … | … | … | … |
| | 5.00 | 0.1969 | 5.000 | 1 | 2 3/4 | 70 | 6 | 152 | … | … | … | … | … |
| 13/64 | | 0.2031 | 5.159 | 1 | 2 3/4 | 70 | 6 | 152 | … | … | … | … | … |
| | 5.20 | 0.2047 | 5.200 | 1 | 2 3/4 | 70 | 6 | 152 | … | … | … | … | … |
| | 5.50 | 0.2165 | 5.500 | 1 | 2 3/4 | 70 | 6 | 152 | … | … | … | … | … |
| 7/32 | | 0.2183 | 5.558 | 1 | 2 3/4 | 70 | 6 | 152 | … | … | … | … | … |
| | 5.80 | 0.2223 | 5.800 | 1 | 2 7/8 | 73 | 6 1/8 | 156 | … | … | … | … | … |
| 15/64 | | 0.2344 | 5.954 | 1 | 2 7/8 | 73 | 6 1/8 | 156 | … | … | … | … | … |
| | 6.00 | 0.2362 | 6.000 | 1 | 2 7/8 | 73 | 6 1/8 | 156 | … | … | … | … | … |
| | 6.20 | 0.2441 | 6.200 | 1 | 2 7/8 | 73 | 6 1/8 | 156 | … | … | … | … | … |
| 1/4 | | 0.2500 | 6.350 | 1 | 2 7/8 | 73 | 6 1/8 | 156 | … | … | … | … | … |
| | 6.50 | 0.2559 | 6.500 | 1 | 3 | 76 | 6 1/4 | 159 | … | … | … | … | … |
| 17/64 | | 0.2656 | 6.746 | 1 | 3 | 76 | 6 1/4 | 159 | … | … | … | … | … |
| | 6.80 | 0.2677 | 6.800 | 1 | 3 | 76 | 6 1/4 | 159 | … | … | … | … | … |
| | 7.00 | 0.2756 | 7.000 | 1 | 3 | 76 | 6 1/4 | 159 | … | … | … | … | … |
| 9/32 | | 0.2812 | 7.142 | 1 | 3 | 76 | 6 1/4 | 159 | … | … | … | … | … |
| | 7.20 | 0.2835 | 7.200 | 1 | 3 1/8 | 79 | 6 3/8 | 162 | … | … | … | … | … |
| | 7.50 | 0.2953 | 7.500 | 1 | 3 1/8 | 79 | 6 3/8 | 162 | … | … | … | … | … |
| 19/64 | | 0.2969 | 7.541 | 1 | 3 1/8 | 79 | 6 3/8 | 162 | … | … | … | … | … |
| | 7.80 | 0.3071 | 7.800 | 1 | 3 1/8 | 79 | 6 3/8 | 162 | … | … | … | … | … |
| 5/16 | | 0.3125 | 7.938 | 1 | 3 1/8 | 79 | 6 3/8 | 162 | … | … | … | … | … |
| | 8.00 | 0.3150 | 8.000 | 1 | 3 1/4 | 83 | 6 1/2 | 165 | … | … | … | … | … |
| | 8.20 | 0.3228 | 8.200 | 1 | 3 1/4 | 83 | 6 1/2 | 165 | … | … | … | … | … |
| 21/64 | | 0.3281 | 8.334 | 1 | 3 1/4 | 83 | 6 1/2 | 165 | … | … | … | … | … |
| | 8.50 | 0.3346 | 8.500 | 1 | 3 1/4 | 83 | 6 1/2 | 165 | … | … | … | … | … |
| 11/32 | | 0.3438 | 8.733 | 1 | 3 1/4 | 83 | 6 1/2 | 165 | … | … | … | … | … |
| | 8.80 | 0.3465 | 8.800 | 1 | 3 1/2 | 89 | 6 3/4 | 171 | … | … | … | … | … |
| | 9.00 | 0.3543 | 9.000 | 1 | 3 1/2 | 89 | 6 3/4 | 171 | … | … | … | … | … |
| 23/64 | | 0.3594 | 9.129 | 1 | 3 1/2 | 89 | 6 3/4 | 171 | … | … | … | … | … |
| | 9.20 | 0.3622 | 9.200 | 1 | 3 1/2 | 89 | 6 3/4 | 171 | … | … | … | … | … |
| | 9.50 | 0.3740 | 9.500 | 1 | 3 1/2 | 89 | 6 3/4 | 171 | … | … | … | … | … |
| 3/8 | | 0.3750 | 9.525 | 1 | 3 1/2 | 89 | 6 3/4 | 171 | 2 | 3 1/2 | 89 | 7 3/8 | 187 |
| | 9.80 | 0.3858 | 9.800 | 1 | 3 5/8 | 92 | 7 | 178 | … | … | … | … | … |
| 25/64 | | 0.3906 | 9.921 | 1 | 3 5/8 | 92 | 7 | 178 | 2 | 3 5/8 | 92 | 7 1/2 | 190 |
| | 10.00 | 0.3937 | 10.000 | 1 | 3 5/8 | 92 | 7 | 178 | … | … | … | … | … |

## Table 5. *(Continued)* American National Taper-Shank Twist Drills Fractional and Metric Sizes *ANSI/ASME B94.11M-1993*

| Drill Diameter, *D* | | | | Regular Shank | | | | | Larger or Smaller Shank[a] | | | | |
|---|---|---|---|---|---|---|---|---|---|---|---|---|---|
| | | Equivalent | | Morse Taper No. | Flute Length *F* | | Overall Length *L* | | Morse Taper No. | Flute Length *F* | | Overall Length *L* | |
| Fraction | mm | Decimal Inch | mm | | Inch | mm | Inch | mm | | Inch | mm | Inch | mm |
| | 10.20 | 0.4016 | 10.200 | 1 | 3⅝ | 92 | 7 | 178 | … | … | … | … | … |
| 13/32 | | 0.4062 | 10.320 | 1 | 3⅝ | 92 | 7 | 178 | 2 | 3⅝ | 92 | 7½ | 190 |
| | 10.50 | 0.4134 | 10.500 | 1 | 3⅞ | 98 | 7¼ | 184 | … | … | … | … | … |
| 27/64 | | 0.4219 | 10.716 | 1 | 3⅞ | 98 | 7¼ | 184 | 2 | 3⅞ | 98 | 7¾ | 197 |
| | 10.80 | 0.4252 | 10.800 | 1 | 3⅞ | 98 | 7¼ | 184 | … | … | … | … | … |
| | 11.00 | 0.4331 | 11.000 | 1 | 3⅞ | 98 | 7¼ | 184 | … | … | … | … | … |
| 7/16 | | 0.4375 | 11.112 | 1 | 3⅞ | 98 | 7¼ | 184 | 2 | 3⅞ | 98 | 7¾ | 197 |
| | 11.20 | 0.4409 | 11.200 | 1 | 4⅛ | 105 | 7½ | 190 | … | … | … | … | … |
| | 11.50 | 0.4528 | 11.500 | 1 | 4⅛ | 105 | 7½ | 190 | … | … | … | … | … |
| 29/64 | | 0.4531 | 11.509 | 1 | 4⅛ | 105 | 7½ | 190 | 2 | 4⅛ | 105 | 8 | 203 |
| | 11.80 | 0.4646 | 11.800 | 1 | 4⅛ | 105 | 7½ | 190 | … | … | … | … | … |
| 15/32 | | 0.4688 | 11.906 | 1 | 4⅛ | 105 | 7½ | 190 | 2 | 4⅛ | 105 | 8 | 203 |
| | 12.00 | 0.4724 | 12.000 | 2 | 4⅜ | 111 | 8¼ | 210 | 1 | 4⅜ | 111 | 7¾ | 197 |
| | 12.20 | 0.4803 | 12.200 | 2 | 4⅜ | 111 | 8¼ | 210 | 1 | 4⅜ | 111 | 7¾ | 197 |
| 31/64 | | 0.4844 | 12.304 | 2 | 4⅜ | 111 | 8¼ | 210 | 1 | 4⅜ | 111 | 7¾ | 197 |
| | 12.50 | 0.4921 | 12.500 | 2 | 4⅜ | 111 | 8¼ | 210 | 1 | 4⅜ | 111 | 7¾ | 197 |
| ½ | | 0.5000 | 12.700 | 2 | 4⅜ | 111 | 8¼ | 210 | 1 | 4⅜ | 111 | 7¾ | 197 |
| | 12.80 | 0.5034 | 12.800 | 2 | 4⅝ | 117 | 8½ | 216 | 1 | 4⅝ | 117 | 8 | 203 |
| | 13.00 | 0.5118 | 13.000 | 2 | 4⅝ | 117 | 8½ | 216 | 1 | 4⅝ | 117 | 8 | 203 |
| 33/64 | | 0.5156 | 13.096 | 2 | 4⅝ | 117 | 8½ | 216 | 1 | 4⅝ | 117 | 8 | 203 |
| | 13.20 | 0.5197 | 13.200 | 2 | 4⅝ | 117 | 8½ | 216 | 1 | 4⅝ | 117 | 8 | 203 |
| 17/32 | | 0.5312 | 13.492 | 2 | 4⅝ | 117 | 8½ | 216 | 1 | 4⅝ | 117 | 8 | 203 |
| | 13.50 | 0.5315 | 13.500 | 2 | 4⅝ | 117 | 8½ | 216 | 1 | 4⅝ | 117 | 8 | 203 |
| | 13.80 | 0.5433 | 13.800 | 2 | 4⅞ | 124 | 8¾ | 222 | 1 | 4⅞ | 124 | 8¼ | 210 |
| 35/64 | | 0.5469 | 13.891 | 2 | 4⅞ | 124 | 8¾ | 222 | 1 | 4⅞ | 124 | 8¼ | 210 |
| | 14.00 | 0.5572 | 14.000 | 2 | 4⅞ | 124 | 8¾ | 222 | 1 | 4⅞ | 124 | 8¼ | 210 |
| | 14.25 | 0.5610 | 14.250 | 2 | 4⅞ | 124 | 8¾ | 222 | 1 | 4⅞ | 124 | 8¼ | 210 |
| 9/16 | | 0.5625 | 14.288 | 2 | 4⅞ | 124 | 8¾ | 222 | 1 | 4⅞ | 124 | 8¼ | 210 |
| | 14.50 | 0.5709 | 14.500 | 2 | 4⅞ | 124 | 8¾ | 222 | … | … | … | … | … |
| 37/64 | | 0.5781 | 14.684 | 2 | 4⅞ | 124 | 8¾ | 222 | … | … | … | … | … |
| | 14.75 | 0.5807 | 14.750 | 2 | 4⅞ | 124 | 8¾ | 222 | … | … | … | … | … |
| | 15.00 | 0.5906 | 15.000 | 2 | 4⅞ | 124 | 8¾ | 222 | … | … | … | … | … |
| 19/32 | | 0.5938 | 15.083 | 2 | 4⅞ | 124 | 8¾ | 222 | … | … | … | … | … |
| | 15.25 | 0.6004 | 15.250 | 2 | 4⅞ | 124 | 8¾ | 222 | … | … | … | … | … |
| 39/64 | | 0.6094 | 15.479 | 2 | 4⅞ | 124 | 8¾ | 222 | … | … | … | … | … |
| | 15.50 | 0.6102 | 15.500 | 2 | 4⅞ | 124 | 8¾ | 222 | … | … | … | … | … |
| | 15.75 | 0.6201 | 15.750 | 2 | 4⅞ | 124 | 8¾ | 222 | … | … | … | … | … |
| ⅝ | | 0.6250 | 15.875 | 2 | 4⅞ | 124 | 8¾ | 222 | … | … | … | … | … |
| | 16.00 | 0.6299 | 16.000 | 2 | 5⅛ | 130 | 9 | 229 | … | … | … | … | … |
| | 16.25 | 0.6398 | 16.250 | 2 | 5⅛ | 130 | 9 | 229 | … | … | … | … | … |
| 41/64 | | 0.6406 | 16.271 | 2 | 5⅛ | 130 | 9 | 229 | 3 | 5⅛ | 130 | 9¾ | 248 |
| | 16.50 | 0.6496 | 16.500 | 2 | 5⅛ | 130 | 9 | 229 | … | … | … | … | … |
| 21/32 | | 0.6562 | 16.667 | 2 | 5⅛ | 130 | 9 | 229 | 3 | 5⅛ | 130 | 9¾ | 248 |
| | 16.75 | 0.6594 | 16.750 | 2 | 5⅜ | 137 | 9¼ | 235 | … | … | … | … | … |
| | 17.00 | 0.6693 | 17.000 | 2 | 5⅜ | 137 | 9¼ | 235 | … | … | … | … | … |
| 43/64 | | 0.6719 | 17.066 | 2 | 5⅜ | 137 | 9¼ | 235 | 3 | 5⅜ | 137 | 10 | 254 |
| | 17.25 | 0.6791 | 17.250 | 2 | 5⅜ | 137 | 9¼ | 235 | … | … | … | … | … |
| 11/16 | | 0.6875 | 17.462 | 2 | 5⅜ | 137 | 9¼ | 235 | 3 | 5⅜ | 137 | 10 | 254 |
| | 17.50 | 0.6880 | 17.500 | 2 | 5⅝ | 143 | 9½ | 241 | … | … | … | … | … |
| 45/64 | | 0.7031 | 17.859 | 2 | 5⅝ | 143 | 9½ | 241 | 3 | 5⅝ | 143 | 10¼ | 260 |
| | 18.00 | 0.7087 | 18.000 | 2 | 5⅝ | 143 | 9½ | 241 | … | … | … | … | … |
| 23/32 | | 0.7188 | 18.258 | 2 | 5⅝ | 143 | 9½ | 241 | 3 | 5⅝ | 143 | 10¼ | 260 |
| | 18.50 | 0.7283 | 18.500 | 2 | 5⅞ | 149 | 9¾ | 248 | … | … | … | … | … |
| 47/64 | | 0.7344 | 18.654 | 2 | 5⅞ | 149 | 9¾ | 248 | 3 | 5⅞ | 149 | 10½ | 267 |

**Table 5.** *(Continued)* **American National Taper-Shank Twist Drills Fractional and Metric Sizes** *ANSI/ASME B94.11M-1993*

| Drill Diameter, *D* | | | | Regular Shank | | | | | Larger or Smaller Shank[a] | | | | |
|---|---|---|---|---|---|---|---|---|---|---|---|---|---|
| | | Equivalent | | Morse Taper No. | Flute Length *F* | | Overall Length *L* | | Morse Taper No. | Flute Length *F* | | Overall Length *L* | |
| Frac-tion | mm | Decimal Inch | mm | | Inch | mm | Inch | mm | | Inch | mm | Inch | mm |
| | 19.00 | 0.7480 | 19.000 | 2 | 5⅞ | 149 | 9¾ | 248 | … | … | … | … | … |
| ¾ | | 0.7500 | 19.050 | 2 | 5⅞ | 149 | 9¾ | 248 | 3 | 5⅞ | 149 | 10½ | 267 |
| 49/64 | | 0.7656 | 19.446 | 2 | 6 | 152 | 9⅞ | 251 | 3 | 6 | 152 | 10⅝ | 270 |
| | 19.50 | 0.7677 | 19.500 | 2 | 6 | 152 | 9⅞ | 251 | … | … | … | … | … |
| 25/32 | | 0.7812 | 19.843 | 2 | 6 | 152 | 9⅞ | 251 | 3 | 6 | 152 | 10⅝ | 270 |
| | 20.00 | 0.7821 | 20.000 | 3 | 6⅛ | 156 | 10¾ | 273 | 2 | 6⅛ | 156 | 10 | 254 |
| 51/64 | | 0.7969 | 20.241 | 3 | 6⅛ | 156 | 10¾ | 273 | 2 | 6⅛ | 156 | 10 | 254 |
| | 20.50 | 0.8071 | 20.500 | 3 | 6⅛ | 156 | 10¾ | 273 | 2 | 6⅛ | 156 | 10 | 254 |
| 13/16 | | 0.8125 | 20.638 | 3 | 6⅛ | 156 | 10¾ | 273 | 2 | 6⅛ | 156 | 10 | 254 |
| | 21.00 | 0.8268 | 21.000 | 3 | 6⅛ | 156 | 10¾ | 273 | 2 | 6⅛ | 156 | 10 | 254 |
| 53/64 | | 0.8281 | 21.034 | 3 | 6⅛ | 156 | 10¾ | 273 | 2 | 6⅛ | 156 | 10 | 254 |
| 27/32 | | 0.8438 | 21.433 | 3 | 6⅛ | 156 | 10¾ | 273 | 2 | 6⅛ | 156 | 10 | 254 |
| | 21.50 | 0.8465 | 21.500 | 3 | 6⅛ | 156 | 10¾ | 273 | 2 | 6⅛ | 156 | 10 | 254 |
| 55/64 | | 0.8594 | 21.829 | 3 | 6⅛ | 156 | 10¾ | 273 | 2 | 6⅛ | 156 | 10 | 254 |
| | 22.00 | 0.8661 | 22.000 | 3 | 6⅛ | 156 | 10¾ | 273 | 2 | 6⅛ | 156 | 10 | 254 |
| ⅞ | | 0.8750 | 22.225 | 3 | 6⅛ | 156 | 10¾ | 273 | 2 | 6⅛ | 156 | 10 | 254 |
| | 22.50 | 0.8858 | 22.500 | 3 | 6⅛ | 156 | 10¾ | 273 | 2 | 6⅛ | 156 | 10 | 254 |
| 57/64 | | 0.8906 | 22.621 | 3 | 6⅛ | 156 | 10¾ | 273 | 2 | 6⅛ | 156 | 10 | 254 |
| | 23.00 | 0.9055 | 23.000 | 3 | 6⅛ | 156 | 10¾ | 273 | 2 | 6⅛ | 156 | 10 | 254 |
| 29/32 | | 0.9062 | 23.017 | 3 | 6⅛ | 156 | 10¾ | 273 | 2 | 6⅛ | 156 | 10 | 254 |
| 59/64 | | 0.9219 | 23.416 | 3 | 6⅛ | 156 | 10¾ | 273 | … | … | … | … | … |
| | 23.50 | 0.9252 | 23.500 | 3 | 6⅛ | 156 | 10¾ | 273 | … | … | … | … | … |
| 15/16 | | 0.9375 | 23.813 | 3 | 6⅛ | 156 | 10¾ | 273 | … | … | … | … | … |
| | 24.00 | 0.9449 | 24.000 | 3 | 6⅜ | 162 | 11 | 279 | … | … | … | … | … |
| 61/64 | | 0.9531 | 24.209 | 3 | 6⅜ | 162 | 11 | 279 | … | … | … | … | … |
| | 24.50 | 0.9646 | 24.500 | 3 | 6⅜ | 162 | 11 | 279 | … | … | … | … | … |
| 31/32 | | 0.9688 | 24.608 | 3 | 6⅜ | 162 | 11 | 279 | … | … | … | … | … |
| | 25.00 | 0.9843 | 25.000 | 3 | 6⅜ | 162 | 11 | 279 | … | … | … | … | … |
| 63/64 | | 0.9844 | 25.004 | 3 | 6⅜ | 162 | 11 | 279 | … | … | … | … | … |
| 1 | | 1.0000 | 25.400 | 3 | 6⅜ | 162 | 11 | 279 | 4 | 6⅜ | 162 | 12 | 305 |
| | 25.50 | 1.0039 | 25.500 | 3 | 6½ | 165 | 11⅛ | 283 | … | … | … | … | … |
| 1 1/64 | | 1.0156 | 25.796 | 3 | 6½ | 165 | 11⅛ | 283 | … | … | … | … | … |
| | 26.00 | 1.0236 | 26.000 | 3 | 6½ | 165 | 11⅛ | 283 | … | … | … | … | … |
| 1 1/32 | | 1.0312 | 26.192 | 3 | 6½ | 165 | 11⅛ | 283 | 4 | 6½ | 165 | 12⅛ | 308 |
| | 26.50 | 1.0433 | 26.500 | 3 | 6⅝ | 168 | 11¼ | 286 | … | … | … | … | … |
| 1 3/64 | | 1.0469 | 26.591 | 3 | 6⅝ | 168 | 11¼ | 286 | … | … | … | … | … |
| 1 1/16 | | 1.0625 | 26.988 | 3 | 6⅝ | 168 | 11¼ | 286 | 4 | 6⅝ | 168 | 12¼ | 311 |
| | 27.00 | 1.0630 | 27.000 | 3 | 6⅝ | 168 | 11¼ | 286 | … | … | … | … | … |
| 1 5/64 | | 1.0781 | 27.384 | 4 | 6⅞ | 175 | 12½ | 318 | 3 | 6⅞ | 175 | 11½ | 292 |
| | 27.50 | 1.0827 | 27.500 | 4 | 6⅞ | 175 | 12½ | 318 | 3 | 6⅞ | 175 | 11½ | 292 |
| 1 3/32 | | 1.0938 | 27.783 | 4 | 6⅞ | 175 | 12½ | 318 | 3 | 6⅞ | 175 | 11½ | 292 |
| | 28.00 | 1.1024 | 28.000 | 4 | 7⅛ | 181 | 12¾ | 324 | 3 | 7⅛ | 181 | 11¾ | 298 |
| 1 7/64 | | 1.1094 | 28.179 | 4 | 7⅛ | 181 | 12¾ | 324 | 3 | 7⅛ | 181 | 11¾ | 298 |
| | 28.50 | 1.1220 | 28.500 | 4 | 7⅛ | 181 | 12¾ | 324 | 3 | 7⅛ | 181 | 11¾ | 298 |
| 1⅛ | | 1.1250 | 28.575 | 4 | 7⅛ | 181 | 12¾ | 324 | 3 | 7⅛ | 181 | 11¾ | 298 |
| 1 9/64 | | 1.1406 | 28.971 | 4 | 7¼ | 184 | 12⅞ | 327 | 3 | 7¼ | 184 | 11⅞ | 302 |
| | 29.00 | 1.1417 | 29.000 | 4 | 7¼ | 184 | 12⅞ | 327 | 3 | 7¼ | 184 | 11⅞ | 302 |
| 1 5/32 | | 1.1562 | 29.367 | 4 | 7¼ | 184 | 12⅞ | 327 | 3 | 7¼ | 184 | 11⅞ | 302 |
| | 29.50 | 1.1614 | 29.500 | 4 | 7⅜ | 187 | 13 | 330 | 3 | 7⅜ | 187 | 12 | 305 |
| 1 11/64 | | 1.1719 | 29.797 | 4 | 7⅜ | 187 | 13 | 330 | 3 | 7⅜ | 187 | 12 | 305 |
| | 30.00 | 1.1811 | 30.000 | 4 | 7⅜ | 187 | 13 | 330 | 3 | 7⅜ | 187 | 12 | 305 |
| 1 3/16 | | 1.1875 | 30.162 | 4 | 7⅜ | 187 | 13 | 330 | 3 | 7⅜ | 187 | 12 | 305 |
| | 30.50 | 1.2008 | 30.500 | 4 | 7½ | 190 | 13⅛ | 333 | 3 | 7½ | 190 | 12⅛ | 308 |
| 1 13/64 | | 1.2031 | 30.559 | 4 | 7½ | 190 | 13⅛ | 333 | 3 | 7½ | 190 | 12⅛ | 308 |

**Table 5.** *(Continued)* **American National Taper-Shank Twist Drills Fractional and Metric Sizes** *ANSI/ASME B94.11M-1993*

| Drill Diameter, *D* | | | | Regular Shank | | | | | Larger or Smaller Shank[a] | | | | |
|---|---|---|---|---|---|---|---|---|---|---|---|---|---|
| | | Equivalent | | Morse Taper No. | Flute Length *F* | | Overall Length *L* | | Morse Taper No. | Flute Length *F* | | Overall Length *L* | |
| Fraction | mm | Decimal Inch | mm | | Inch | mm | Inch | mm | | Inch | mm | Inch | mm |
| 1 7/32 | | 1.2188 | 30.958 | 4 | 7½ | 190 | 13⅛ | 333 | 3 | 7½ | 190 | 12⅛ | 308 |
| | 31.00 | 1.2205 | 31.000 | 4 | 7⅞ | 200 | 13½ | 343 | 3 | 7⅞ | 200 | 12½ | 318 |
| 1 15/64 | | 1.2344 | 31.354 | 4 | 7⅞ | 200 | 13½ | 343 | 3 | 7⅞ | 200 | 12½ | 318 |
| | 31.50 | 1.2402 | 31.500 | 4 | 7⅞ | 200 | 13½ | 343 | 3 | 7⅞ | 200 | 12½ | 318 |
| 1¼ | | 1.2500 | 31.750 | 4 | 7⅞ | 200 | 13½ | 343 | 3 | 7⅞ | 200 | 12½ | 318 |
| | 32.00 | 1.2598 | 32.000 | 4 | 8½ | 216 | 14⅛ | 359 | … | … | … | … | … |
| 1 17/64 | | 1.2656 | 32.146 | 4 | 8½ | 216 | 14⅛ | 359 | … | … | … | … | … |
| | 32.50 | 1.2795 | 32.500 | 4 | 8½ | 216 | 14⅛ | 359 | … | … | … | … | … |
| 1 9/32 | | 1.2812 | 32.542 | 4 | 8½ | 216 | 14⅛ | 359 | … | … | … | … | … |
| 1 19/64 | | 1.2969 | 32.941 | 4 | 8⅝ | 219 | 14¼ | 362 | … | … | … | … | … |
| | 33.00 | 1.2992 | 33.000 | 4 | 8⅝ | 219 | 14¼ | 362 | … | … | … | … | … |
| 1 5/16 | | 1.3125 | 33.338 | 4 | 8⅝ | 219 | 14¼ | 362 | … | … | … | … | … |
| | 33.50 | 1.3189 | 33.500 | 4 | 8¾ | 222 | 14⅜ | 365 | … | … | … | … | … |
| 1 21/64 | | 1.3281 | 33.734 | 4 | 8¾ | 222 | 14⅜ | 365 | … | … | … | … | … |
| | 34.00 | 1.3386 | 34.000 | 4 | 8¾ | 222 | 14⅜ | 365 | … | … | … | … | … |
| 1 11/32 | | 1.3438 | 34.133 | 4 | 8¾ | 222 | 14⅜ | 365 | … | … | … | … | … |
| | 34.50 | 1.3583 | 34.500 | 4 | 8⅞ | 225 | 14½ | 368 | … | … | … | … | … |
| 1 23/64 | | 1.3594 | 34.529 | 4 | 8⅞ | 225 | 14½ | 368 | … | … | … | … | … |
| 1⅜ | | 1.3750 | 34.925 | 4 | 8⅞ | 225 | 14½ | 368 | … | … | … | … | … |
| | 35.00 | 1.3780 | 35.000 | 4 | 9 | 229 | 14⅝ | 371 | … | … | … | … | … |
| 1 25/64 | | 1.3906 | 35.321 | 4 | 9 | 229 | 14⅝ | 371 | … | … | … | … | … |
| | 35.50 | 1.3976 | 35.500 | 4 | 9 | 229 | 14⅝ | 371 | … | … | … | … | … |
| 1 13/32 | | 1.4062 | 35.717 | 4 | 9 | 229 | 14⅝ | 371 | … | … | … | … | … |
| | 36.00 | 1.4173 | 36.000 | 4 | 9⅛ | 232 | 14¾ | 375 | … | … | … | … | … |
| 1 27/64 | | 1.4219 | 36.116 | 4 | 9⅛ | 232 | 14¾ | 375 | … | … | … | … | … |
| | 36.50 | 1.4370 | 36.500 | 4 | 9⅛ | 232 | 14¾ | 375 | … | … | … | … | … |
| 1 7/16 | | 1.4375 | 36.512 | 4 | 9⅛ | 232 | 14¾ | 375 | … | … | … | … | … |
| 1 29/64 | | 1.4531 | 36.909 | 4 | 9¼ | 235 | 14⅞ | 378 | … | … | … | … | … |
| | 37.00 | 1.4567 | 37.000 | 4 | 9¼ | 235 | 14⅞ | 378 | … | … | … | … | … |
| 1 15/32 | | 1.4688 | 37.308 | 4 | 9¼ | 235 | 14⅞ | 378 | … | … | … | … | … |
| | 37.50 | 1.4764 | 37.500 | 4 | 9⅜ | 238 | 15 | 381 | … | … | … | … | … |
| 1 31/64 | | 1.4844 | 37.704 | 4 | 9⅜ | 238 | 15 | 381 | … | … | … | … | … |
| | 38.00 | 1.4961 | 38.000 | 4 | 9⅜ | 238 | 15 | 381 | … | … | … | … | … |
| 1½ | | 1.5000 | 38.100 | 4 | 9⅜ | 238 | 15 | 381 | … | … | … | … | … |
| 1 33/64 | | 1.5156 | 38.496 | … | … | … | … | … | 4 | 9¾ | 238 | 15 | 381 |
| 1 17/32 | | 1.5312 | 38.892 | 5 | 9⅜ | 238 | 16⅜ | 416 | 4 | 9⅜ | 238 | 15 | 381 |
| | 39.00 | 1.5354 | 39.000 | 5 | 9⅝ | 244 | 16⅝ | 422 | 4 | 9⅝ | 244 | 15¼ | 387 |
| 1 35/64 | | 1.5469 | 39.291 | … | … | … | … | … | 4 | 9⅝ | 244 | 15¼ | 387 |
| 1 9/16 | | 1.5625 | 39.688 | 5 | 9⅝ | 244 | 16⅝ | 422 | 4 | 9⅝ | 244 | 15¼ | 387 |
| | 40.00 | 1.5748 | 40.000 | 5 | 9⅞ | 251 | 16⅞ | 429 | 4 | 9⅞ | 251 | 15½ | 394 |
| 1 37/64 | | 1.5781 | 40.084 | … | … | … | … | … | 4 | 9⅞ | 251 | 15½ | 394 |
| 1 19/32 | | 1.5938 | 40.483 | 5 | 9⅞ | 251 | 16⅞ | 429 | 4 | 9⅞ | 251 | 15½ | 394 |
| 1 39/64 | | 1.6094 | 40.879 | … | … | … | … | … | 4 | 10 | 254 | 15⅝ | 397 |
| | 41.00 | 1.6142 | 41.000 | 5 | 10 | 254 | 17 | 432 | 4 | 10 | 254 | 15⅝ | 397 |
| 1⅝ | | 1.6250 | 41.275 | 5 | 10 | 254 | 17 | 432 | 4 | 10 | 254 | 15⅝ | 397 |
| 1 41/64 | | 1.6406 | 41.671 | … | … | … | … | … | 4 | 10⅛ | 257 | 15¾ | 400 |
| | 42.00 | 1.6535 | 42.000 | 5 | 10⅛ | 257 | 17⅛ | 435 | 4 | 10⅛ | 257 | 15¾ | 400 |
| 1 21/32 | | 1.6562 | 42.067 | 5 | 10⅛ | 257 | 17⅛ | 435 | 4 | 10⅛ | 257 | 15¾ | 400 |
| 1 43/64 | | 1.6719 | 42.466 | … | … | … | … | … | 4 | 10⅛ | 257 | 15¾ | 400 |
| 1 11/16 | | 1.6875 | 42.862 | 5 | 10⅛ | 257 | 17⅛ | 435 | 4 | 10⅛ | 257 | 15¾ | 400 |
| | 43.00 | 1.6929 | 43.000 | 5 | 10⅛ | 257 | 17⅛ | 435 | 4 | 10⅛ | 257 | 15¾ | 400 |
| 1 45/64 | | 1.7031 | 43.259 | … | … | … | … | … | 4 | 10⅛ | 257 | 15¾ | 400 |
| 1 23/32 | | 1.7188 | 43.658 | 5 | 10⅛ | 257 | 17⅛ | 435 | 4 | 10⅛ | 257 | 15¾ | 400 |
| | 44.00 | 1.7323 | 44.000 | 5 | 10⅛ | 257 | 17⅛ | 435 | 4 | 10⅜ | 264 | 16¼ | 413 |

**Table 5.** *(Continued)* **American National Taper-Shank Twist Drills Fractional and Metric Sizes** *ANSI/ASME B94.11M-1993*

| Drill Diameter, *D* | | | | Regular Shank | | | | | Larger or Smaller Shank[a] | | | | |
|---|---|---|---|---|---|---|---|---|---|---|---|---|---|
| | | Equivalent | | Morse Taper No. | Flute Length *F* | | Overall Length *L* | | Morse Taper No. | Flute Length *F* | | Overall Length *L* | |
| Fraction | mm | Decimal Inch | mm | | Inch | mm | Inch | mm | | Inch | mm | Inch | mm |
| 1 47/64 | | 1.7344 | 44.054 | … | … | … | … | … | 4 | 10⅜ | 264 | 16¼ | 413 |
| 1¾ | | 1.7500 | 44.450 | 5 | 10⅛ | 257 | 17⅛ | 435 | 4 | 10¾ | 264 | 16¼ | 413 |
| | 45.00 | 1.7717 | 45.000 | 5 | 10⅛ | 257 | 17⅛ | 435 | 4 | 10⅜ | 264 | 16¼ | 413 |
| 1 25/32 | | 1.7812 | 45.242 | 5 | 10⅛ | 257 | 17⅛ | 435 | 4 | 10⅜ | 264 | 16¼ | 413 |
| | 46.00 | 1.8110 | 46.000 | 5 | 10⅛ | 257 | 17⅛ | 435 | 4 | 10⅜ | 264 | 16¼ | 413 |
| 1 13/16 | | 1.8125 | 46.038 | 5 | 10⅛ | 257 | 17⅛ | 435 | 4 | 10⅜ | 264 | 16¼ | 413 |
| 1 27/32 | | 1.8438 | 46.833 | 5 | 10⅛ | 257 | 17⅛ | 435 | 4 | 10⅜ | 264 | 16¼ | 413 |
| | 47.00 | 1.8504 | 47.000 | 5 | 10⅜ | 264 | 17⅜ | 441 | 4 | 10½ | 267 | 16½ | 419 |
| 1⅞ | | 1.8750 | 47.625 | 5 | 10⅜ | 264 | 17⅜ | 441 | 4 | 10½ | 267 | 16½ | 419 |
| | 48.00 | 1.8898 | 48.000 | 5 | 10⅜ | 264 | 17⅜ | 441 | 4 | 10½ | 267 | 16½ | 419 |
| 1 29/32 | | 1.9062 | 48.417 | 5 | 10⅜ | 264 | 17⅜ | 441 | 4 | 10½ | 267 | 16½ | 419 |
| | 49.00 | 1.9291 | 49.000 | 5 | 10⅜ | 264 | 17⅜ | 441 | 4 | 10⅝ | 270 | 16⅝ | 422 |
| 1 15/16 | | 1.9375 | 49.212 | 5 | 10⅜ | 264 | 17⅜ | 441 | 4 | 10⅝ | 270 | 16⅝ | 422 |
| | 50.00 | 1.9625 | 50.000 | 5 | 10⅜ | 264 | 17⅜ | 441 | 4 | 10⅝ | 270 | 16⅝ | 422 |
| 1 31/32 | | 1.9688 | 50.008 | 5 | 10⅜ | 264 | 17⅜ | 441 | 4 | 10⅝ | 270 | 16⅝ | 422 |
| 2 | | 2.0000 | 50.800 | 5 | 10⅜ | 264 | 17⅜ | 441 | 4 | 10⅝ | 270 | 16⅝ | 422 |
| | 51.00 | 2.0079 | 51.000 | 5 | 10⅜ | 264 | 17⅜ | 441 | … | … | … | … | … |
| 2 1/32 | | 2.0312 | 51.592 | 5 | 10⅜ | 264 | 17⅜ | 441 | … | … | … | … | … |
| | 52.00 | 2.0472 | 52.000 | 5 | 10¼ | 260 | 17⅜ | 441 | … | … | … | … | … |
| 2 1/16 | | 2.0625 | 52.388 | 5 | 10¼ | 260 | 17⅜ | 441 | … | … | … | … | … |
| | 53.00 | 2.0866 | 53.000 | 5 | 10¼ | 260 | 17⅜ | 441 | … | … | … | … | … |
| 2 3/32 | | 2.0938 | 53.183 | 5 | 10¼ | 260 | 17⅜ | 441 | … | … | … | … | … |
| 2⅛ | | 2.1250 | 53.975 | 5 | 10¼ | 260 | 17⅜ | 441 | … | … | … | … | … |
| | 54.00 | 2.1260 | 54.000 | 5 | 10¼ | 260 | 17⅜ | 441 | … | … | … | … | … |
| 2 5/32 | | 2.1562 | 54.767 | 5 | 10¼ | 260 | 17⅜ | 441 | … | … | … | … | … |
| | 55.00 | 2.1654 | 55.000 | 5 | 10¼ | 260 | 17⅜ | 441 | … | … | … | … | … |
| 2 3/16 | | 2.1875 | 55.563 | 5 | 10¼ | 260 | 17¾ | 441 | … | … | … | … | … |
| | 56.00 | 2.2000 | 56.000 | 5 | 10⅛ | 257 | 17⅜ | 441 | … | … | … | … | … |
| 2 7/32 | | 2.2188 | 56.358 | 5 | 10⅛ | 257 | 17⅜ | 441 | … | … | … | … | … |
| | 57.00 | 2.2441 | 57.000 | 5 | 10⅛ | 257 | 17⅜ | 441 | … | … | … | … | … |
| 2¼ | | 2.2500 | 57.150 | 5 | 10⅛ | 257 | 17⅜ | 441 | … | … | … | … | … |
| | 58.00 | 2.2835 | 58.000 | 5 | 10⅛ | 257 | 17⅜ | 441 | … | … | … | … | … |
| 2 5/16 | | 2.3125 | 58.738 | 5 | 10⅛ | 257 | 17⅜ | 441 | … | … | … | … | … |
| | 59.00 | 2.3228 | 59.000 | 5 | 10⅛ | 257 | 17⅜ | 441 | … | … | … | … | … |
| | 60.00 | 2.3622 | 60.000 | 5 | 10⅛ | 257 | 17⅜ | 441 | … | … | … | … | … |
| 2⅜ | | 2.3750 | 60.325 | 5 | 10⅛ | 257 | 17⅜ | 441 | … | … | … | … | … |
| | 61.00 | 2.4016 | 61.000 | 5 | 11¼ | 286 | 18¾ | 476 | … | … | … | … | … |
| 2 7/16 | | 2.4375 | 61.912 | 5 | 11¼ | 286 | 18¾ | 476 | … | … | … | … | … |
| | 62.00 | 2.4409 | 62.000 | 5 | 11¼ | 286 | 18¾ | 476 | … | … | … | … | … |
| | 63.00 | 2.4803 | 63.000 | 5 | 11¼ | 286 | 18¾ | 476 | … | … | … | … | … |
| 2½ | | 2.5000 | 63.500 | 5 | 11¼ | 286 | 18¾ | 476 | … | … | … | … | … |
| | 64.00 | 2.5197 | 64.000 | 5 | 11⅞ | 302 | 19½ | 495 | … | … | … | … | … |
| | 65.00 | 2.5591 | 65.000 | 5 | 11⅞ | 302 | 19½ | 495 | … | … | … | … | … |
| 2 9/16 | | 2.5625 | 65.088 | 5 | 11⅞ | 302 | 19½ | 495 | … | … | … | … | … |
| | 66.00 | 2.5984 | 66.000 | 5 | 11⅞ | 302 | 19½ | 495 | … | … | … | … | … |
| 2⅝ | | 2.6250 | 66.675 | 5 | 11⅞ | 302 | 19½ | 495 | … | … | … | … | … |
| | 67.00 | 2.6378 | 67.000 | 5 | 12¾ | 324 | 20⅜ | 518 | … | … | … | … | … |
| | 68.00 | 2.6772 | 68.000 | 5 | 12¾ | 324 | 20⅜ | 518 | … | … | … | … | … |
| 2 11/16 | | 2.6875 | 68.262 | 5 | 12¾ | 324 | 20⅜ | 518 | … | … | … | … | … |
| | 69.00 | 2.7165 | 69.000 | 5 | 12¾ | 324 | 20⅜ | 518 | … | … | … | … | … |
| 2¾ | | 2.7500 | 69.850 | 5 | 12¾ | 324 | 20⅜ | 518 | … | … | … | … | … |
| | 70.00 | 2.7559 | 70.000 | 5 | 13⅜ | 340 | 21⅛ | 537 | … | … | … | … | … |
| | 71.00 | 2.7953 | 71.000 | 5 | 13⅜ | 340 | 21⅛ | 537 | … | … | … | … | … |
| 2 13/16 | | 2.8125 | 71.438 | 5 | 13⅜ | 340 | 21⅛ | 537 | … | … | … | … | … |

**Table 5.** *(Continued)* **American National Taper-Shank Twist Drills Fractional and Metric Sizes** *ANSI/ASME B94.11M-1993*

| Drill Diameter, D | | | | Regular Shank | | | | | Larger or Smaller Shank[a] | | | | |
|---|---|---|---|---|---|---|---|---|---|---|---|---|---|
| | | Equivalent | | Morse Taper No. | Flute Length F | | Overall Length L | | Morse Taper No. | Flute Length F | | Overall Length L | |
| Fraction | mm | Decimal Inch | mm | | Inch | mm | Inch | mm | | Inch | mm | Inch | mm |
| | 72.00 | 2.8346 | 72.000 | 5 | 13⅜ | 340 | 21⅛ | 537 | … | … | … | … | … |
| | 73.00 | 2.8740 | 73.000 | 5 | 13⅜ | 340 | 21⅛ | 537 | … | … | … | … | … |
| 2⅞ | | 2.8750 | 73.025 | 5 | 13⅜ | 340 | 21⅛ | 537 | … | … | … | … | … |
| | 74.00 | 2.9134 | 74.000 | 5 | 14 | 356 | 21¾ | 552 | … | … | … | … | … |
| 2 15/16 | | 2.9375 | 74.612 | 5 | 14 | 356 | 21¾ | 552 | … | … | … | … | … |
| | 75.00 | 2.9528 | 75.000 | 5 | 14 | 356 | 21¾ | 552 | … | … | … | … | … |
| | 76.00 | 2.9921 | 76.000 | 5 | 14 | 356 | 21¾ | 552 | … | … | … | … | … |
| 3 | | 3.0000 | 76.200 | 5 | 14 | 356 | 21¾ | 552 | … | … | … | … | … |
| | 77.00 | 3.0315 | 77.000 | 6 | 14⅝ | 371 | 24½ | 622 | 5 | 14¼ | 362 | 22 | 559 |
| | 78.00 | 3.0709 | 78.000 | 6 | 14⅝ | 371 | 24½ | 622 | 5 | 14¼ | 362 | 22 | 559 |
| 3⅛ | | 3.1250 | 79.375 | 6 | 14⅝ | 371 | 24½ | 622 | 5 | 14¼ | 362 | 22 | 559 |
| 3¼ | | 3.2500 | 82.550 | 6 | 15½ | 394 | 25½ | 648 | 5 | 15¼ | 387 | 23 | 584 |
| 3½ | | 3.5000 | 88.900 | … | … | … | … | … | 5 | 16¼ | 413 | 24 | 610 |

[a] Larger or smaller than regular shank.

**Table 6. American National Standard Combined Drills and Countersinks — Plain and Bell Types** *ANSI/ASME B94.11M-1993*

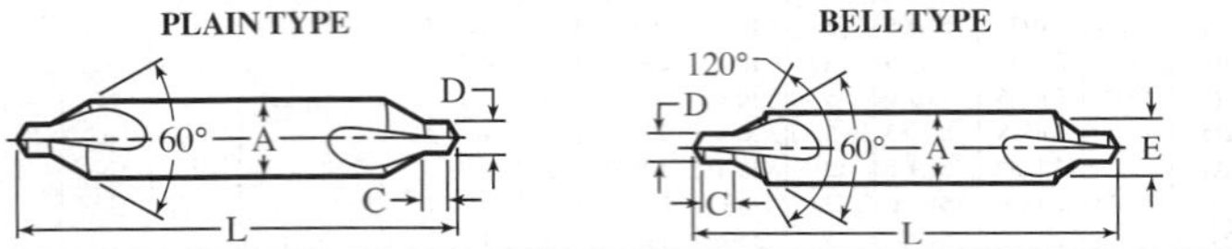


| Plain Type | | | | | | | | |
|---|---|---|---|---|---|---|---|---|
| Size Designation | Body Diameter A | | Drill Diameter D | | Drill Length C | | Overall Length L | |
| | Inches | Millimeters | Inches | Millimeters | Inches | Millimeters | Inches | Millimeters |
| 00 | ⅛ | 3.18 | .025 | 0.64 | .030 | 0.76 | 1⅛ | 29 |
| 0 | ⅛ | 3.18 | 1/32 | 0.79 | .038 | 0.97 | 1⅛ | 29 |
| 1 | ⅛ | 3.18 | 3/64 | 1.19 | 3/64 | 1.19 | 1¼ | 32 |
| 2 | 3/16 | 4.76 | 5/64 | 1.98 | 5/64 | 1.98 | 1⅞ | 48 |
| 3 | ¼ | 6.35 | 7/64 | 2.78 | 7/64 | 2.78 | 2 | 51 |
| 4 | 5/16 | 7.94 | ⅛ | 3.18 | ⅛ | 3.18 | 2⅛ | 54 |
| 5 | 7/16 | 11.11 | 3/16 | 4.76 | 3/16 | 4.76 | 2¾ | 70 |
| 6 | ½ | 12.70 | 7/32 | 5.56 | 7/32 | 5.56 | 3 | 76 |
| 7 | ⅝ | 15.88 | ¼ | 6.35 | ¼ | 6.35 | 3¼ | 83 |
| 8 | ¾ | 19.05 | 5/16 | 7.94 | 5/16 | 7.94 | 3½ | 89 |

| Bell Type | | | | | | | | | | |
|---|---|---|---|---|---|---|---|---|---|---|
| Size Designation | Body Diameter A | | Drill Diameter D | | Bell Diameter E | | Drill Length C | | Overall Length L | |
| | Inches | mm | Inches | mm | Inches | mm | Inches | mm | Inches | mm |
| 11 | ⅛ | 3.18 | 3/64 | 1.19 | 0.10 | 2.5 | 3/64 | 1.19 | 1¼ | 32 |
| 12 | 3/16 | 4.76 | 1/16 | 1.59 | 0.15 | 3.8 | 1/16 | 1.59 | 1⅞ | 48 |
| 13 | ¼ | 6.35 | 3/32 | 2.38 | 0.20 | 5.1 | 3/32 | 2.38 | 2 | 51 |
| 14 | 5/16 | 7.94 | 7/64 | 2.78 | 0.25 | 6.4 | 7/64 | 2.78 | 2⅛ | 54 |
| 15 | 7/16 | 11.11 | 5/32 | 3.97 | 0.35 | 8.9 | 5/32 | 3.97 | 2¾ | 70 |
| 16 | ½ | 12.70 | 3/16 | 4.76 | 0.40 | 10.2 | 3/16 | 4.76 | 3 | 76 |
| 17 | ⅝ | 15.88 | 7/32 | 5.56 | 0.50 | 12.7 | 7/32 | 5.56 | 3¼ | 83 |
| 18 | ¾ | 19.05 | ¼ | 6.35 | 0.60 | 15.2 | ¼ | 6.35 | 3½ | 89 |

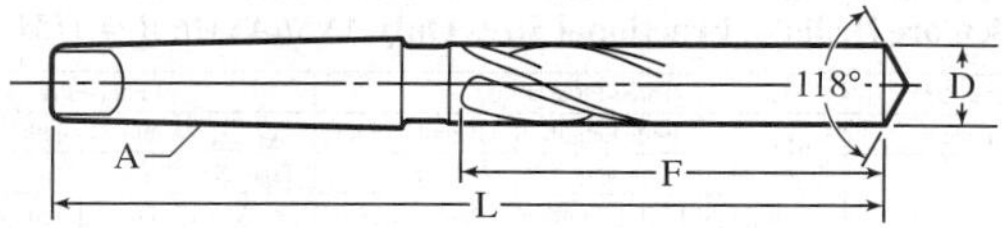


Drill Diameter 11/32″ (8.737 mm and Smaller

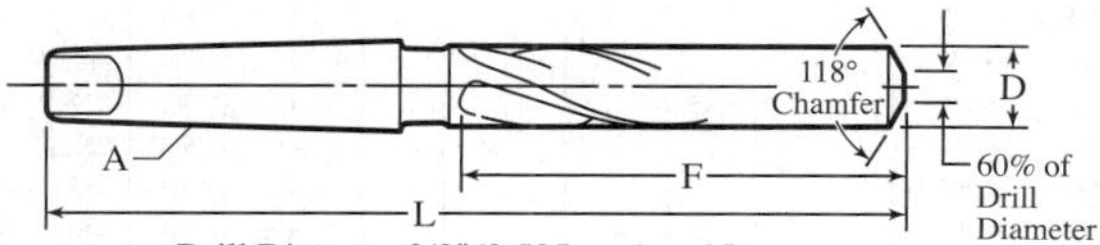


Drill Diameter 3/8″ (9.525 mm) and Larger

**Table 7. American National Standard Three- and Four-Flute Taper-Shank Core Drills — Fractional Sizes Only** *ANSI/ASME B94.11M-1993*

| Drill Diameter, *D* | | | Three-Flute Drills | | | | | Four-Flute Drills | | | | |
|---|---|---|---|---|---|---|---|---|---|---|---|---|
| | Equivalent | | Morse Taper No. | Flute Length *F* | | Overall Length *L* | | Morse Taper No. | Flute Length *F* | | Overall Length *L* | |
| Inch | Decimal Inch | mm | A | Inch | mm | Inch | mm | A | Inch | mm | Inch | mm |
| ¼ | 0.2500 | 6.350 | 1 | 2⅞ | 73 | 6⅛ | 156 | … | … | … | … | … |
| 9⁄32 | 0.2812 | 7.142 | 1 | 3 | 76 | 6¼ | 159 | … | … | … | … | … |
| 5⁄16 | 0.3175 | 7.938 | 1 | 3⅛ | 79 | 6⅜ | 162 | … | … | … | … | … |
| 11⁄32 | 0.3438 | 8.733 | 1 | 3¼ | 83 | 6½ | 165 | … | … | … | … | … |
| ⅜ | 0.3750 | 9.525 | 1 | 3½ | 89 | 6¾ | 171 | … | … | … | … | … |
| 13⁄32 | 0.4062 | 10.319 | 1 | 3⅝ | 92 | 7 | 178 | … | … | … | … | … |
| 7⁄16 | 0.4375 | 11.112 | 1 | 3⅞ | 98 | 7¼ | 184 | … | … | … | … | … |
| 15⁄32 | 0.4688 | 11.908 | 1 | 4⅛ | 105 | 7½ | 190 | … | … | … | … | … |
| ½ | 0.5000 | 12.700 | 2 | 4⅜ | 111 | 8¼ | 210 | 2 | 4⅜ | 111 | 8¼ | 210 |
| 17⁄32 | 0.5312 | 13.492 | 2 | 4⅝ | 117 | 8½ | 216 | 2 | 4⅝ | 117 | 8½ | 216 |
| 9⁄16 | 0.5625 | 14.288 | 2 | 4⅞ | 124 | 8¾ | 222 | 2 | 4⅞ | 124 | 8¾ | 222 |
| 19⁄32 | 0.5938 | 15.083 | 2 | 4⅞ | 124 | 8¾ | 222 | 2 | 4⅞ | 124 | 8¾ | 222 |
| ⅝ | 0.6250 | 15.815 | 2 | 4⅞ | 124 | 8¾ | 222 | 2 | 4⅞ | 124 | 8¾ | 222 |
| 21⁄32 | 0.6562 | 16.668 | 2 | 5⅛ | 130 | 9 | 229 | 2 | 5⅛ | 130 | 9 | 229 |
| 11⁄16 | 0.6875 | 17.462 | 2 | 5⅜ | 137 | 9¼ | 235 | 2 | 5⅜ | 137 | 9¼ | 235 |
| 23⁄32 | 0.7188 | 18.258 | 2 | 5⅝ | 143 | 9½ | 241 | 2 | 5⅝ | 143 | 9½ | 241 |
| ¾ | 0.7500 | 19.050 | 2 | 5⅞ | 149 | 9¾ | 248 | 2 | 5⅞ | 149 | 9¾ | 248 |
| 25⁄32 | 0.7812 | 19.842 | 2 | 6 | 152 | 9⅞ | 251 | 2 | 6 | 152 | 9⅞ | 251 |
| 13⁄16 | 0.8125 | 20.638 | 3 | 6⅛ | 156 | 10¾ | 273 | 3 | 6⅛ | 156 | 10¾ | 273 |
| 27⁄32 | 0.8438 | 21.433 | 3 | 6⅛ | 156 | 10¾ | 273 | 3 | 6⅛ | 156 | 10¾ | 273 |
| ⅞ | 0.8750 | 22.225 | 3 | 6⅛ | 156 | 10¾ | 273 | 3 | 6⅛ | 156 | 10¾ | 273 |
| 29⁄32 | 0.9062 | 23.019 | 3 | 6⅛ | 156 | 10¾ | 273 | 3 | 6⅛ | 156 | 10¾ | 273 |
| 15⁄16 | 0.9375 | 23.812 | 3 | 6⅛ | 156 | 10¾ | 273 | 3 | 6⅛ | 156 | 10¾ | 273 |
| 31⁄32 | 0.9688 | 24.608 | 3 | 6⅜ | 162 | 11 | 279 | 3 | 6⅜ | 162 | 11 | 279 |
| 1 | 1.0000 | 25.400 | 3 | 6⅜ | 162 | 11 | 279 | 3 | 6⅜ | 162 | 11 | 279 |
| 1 1⁄32 | 1.0312 | 26.192 | 3 | 6½ | 165 | 11⅛ | 283 | 3 | 6½ | 165 | 11⅛ | 283 |
| 1 1⁄16 | 1.0625 | 26.988 | 3 | 6⅝ | 168 | 11¼ | 286 | 3 | 6⅝ | 168 | 11¼ | 286 |
| 1 3⁄32 | 1.0938 | 27.783 | 4 | 6⅞ | 175 | 12½ | 318 | 4 | 6⅞ | 175 | 12½ | 318 |
| 1⅛ | 1.1250 | 28.575 | 4 | 7⅛ | 181 | 12¾ | 324 | 4 | 7⅛ | 181 | 12¾ | 324 |
| 1 5⁄32 | 1.1562 | 29.367 | 4 | 7¼ | 184 | 12⅞ | 327 | 4 | 7¼ | 184 | 12⅞ | 327 |
| 1 3⁄16 | 1.1875 | 30.162 | 4 | 7⅜ | 187 | 13 | 330 | 4 | 7⅜ | 187 | 13 | 330 |
| 1 7⁄32 | 1.2188 | 30.958 | 4 | 7½ | 190 | 13⅛ | 333 | 4 | 7½ | 190 | 13⅛ | 333 |
| 1¼ | 1.2500 | 31.750 | 4 | 7⅞ | 200 | 13½ | 343 | 4 | 7⅞ | 200 | 13½ | 343 |
| 1 9⁄32 | 1.2812 | 32.542 | … | … | … | … | … | 4 | 8½ | 216 | 14⅛ | 359 |

**Table 7.** *(Continued)* **American National Standard Three- and Four-Flute Taper-Shank Core Drills — Fractional Sizes Only** *ANSI/ASME B94.11M-1993*

| Drill Diameter, *D* | | | Three-Flute Drills | | | | | Four-Flute Drills | | | | |
|---|---|---|---|---|---|---|---|---|---|---|---|---|
| | Equivalent | | Morse Taper No. | Flute Length *F* | | Overall Length *L* | | Morse Taper No. | Flute Length *F* | | Overall Length *L* | |
| Inch | Decimal Inch | mm | A | Inch | mm | Inch | mm | A | Inch | mm | Inch | mm |
| 1 5/16 | 1.3125 | 33.338 | … | … | … | … | … | 4 | 8 5/8 | 219 | 14 1/4 | 362 |
| 1 11/32 | 1.3438 | 34.133 | … | … | … | … | … | 4 | 8 3/4 | 222 | 14 3/8 | 365 |
| 1 3/8 | 1.3750 | 34.925 | … | … | … | … | … | 4 | 8 7/8 | 225 | 14 1/2 | 368 |
| 1 13/32 | 1.4062 | 35.717 | … | … | … | … | … | 4 | 9 | 229 | 14 5/8 | 371 |
| 1 7/16 | 1.4375 | 36.512 | … | … | … | … | … | 4 | 9 1/8 | 232 | 14 3/4 | 375 |
| 1 15/32 | 1.4688 | 37.306 | … | … | … | … | … | 4 | 9 1/4 | 235 | 14 7/8 | 378 |
| 1 1/2 | 1.5000 | 38.100 | … | … | … | … | … | 4 | 9 3/8 | 238 | 15 | 381 |
| 1 17/32 | 1.5312 | 38.892 | … | … | … | … | … | 5 | 9 3/8 | 238 | 16 3/8 | 416 |
| 1 9/16 | 1.5675 | 39.688 | … | … | … | … | … | 5 | 9 5/8 | 244 | 16 5/8 | 422 |
| 1 19/32 | 1.5938 | 40.483 | … | … | … | … | … | 5 | 9 7/8 | 251 | 16 7/8 | 429 |
| 1 5/8 | 1.6250 | 41.275 | … | … | … | … | … | 5 | 10 | 254 | 17 | 432 |
| 1 21/32 | 1.6562 | 42.067 | … | … | … | … | … | 5 | 10 1/8 | 257 | 17 1/8 | 435 |
| 1 11/16 | 1.6875 | 42.862 | … | … | … | … | … | 5 | 10 1/8 | 257 | 17 1/8 | 435 |
| 1 23/32 | 1.7188 | 43.658 | … | … | … | … | … | 5 | 10 1/8 | 257 | 17 1/8 | 435 |
| 1 3/4 | 1.7500 | 44.450 | … | … | … | … | … | 5 | 10 1/8 | 257 | 17 1/8 | 435 |
| 1 25/32 | 1.7812 | 45.244 | … | … | … | … | … | 5 | 10 1/8 | 257 | 17 1/8 | 435 |
| 1 13/16 | 1.8125 | 46.038 | … | … | … | … | … | 5 | 10 1/8 | 257 | 17 1/8 | 435 |
| 1 27/32 | 1.8438 | 46.833 | … | … | … | … | … | 5 | 10 1/8 | 257 | 17 1/8 | 435 |
| 1 7/8 | 1.8750 | 47.625 | … | … | … | … | … | 5 | 10 3/8 | 264 | 17 3/8 | 441 |
| 1 29/32 | 1.9062 | 48.417 | … | … | … | … | … | 5 | 10 3/8 | 264 | 17 3/8 | 441 |
| 1 15/16 | 1.9375 | 49.212 | … | … | … | … | … | 5 | 10 3/8 | 264 | 17 3/8 | 441 |
| 1 31/32 | 1.9688 | 50.008 | … | … | … | … | … | 5 | 10 3/8 | 264 | 17 3/8 | 441 |
| 2 | 2.0000 | 50.800 | … | … | … | … | … | 5 | 10 3/8 | 264 | 17 3/8 | 441 |
| 2 1/8 | 2.1250 | 53.975 | … | … | … | … | … | 5 | 10 1/4 | 260 | 17 3/8 | 441 |
| 2 1/4 | 2.2500 | 57.150 | … | … | … | … | … | 5 | 10 1/8 | 257 | 17 3/8 | 441 |
| 2 3/8 | 2.3750 | 60.325 | … | … | … | … | … | 5 | 10 1/8 | 257 | 17 3/8 | 441 |
| 2 1/2 | 2.5000 | 63.500 | … | … | … | … | … | 5 | 11 1/4 | 286 | 18 3/4 | 476 |

**Table 8. American National Standard Drill Drivers — Split-Sleeve, Collet-Type** *ANSI/ASME B94.35-1972 (R2020)*

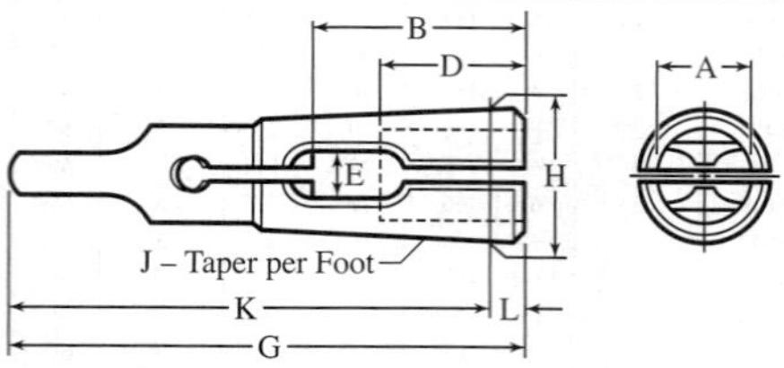


| Taper Number | *G* Overall Length | *H* Diameter at Gage Line | *J* Taper per Foot[a] | *K* Length to Gage Line | *L* Driver Projection |
|---|---|---|---|---|---|
| 0[b] | 2.38 | 0.356 | 0.62460 | 2.22 | 0.16 |
| 1 | 2.62 | 0.475 | 0.59858 | 2.44 | 0.19 |
| 2 | 3.19 | 0.700 | 0.59941 | 2.94 | 0.25 |

[a] Taper rate in accordance with ANSI/ASME B5.10-1994 (R2013), Machine Tapers.

[b] Size 0 is not an American National Standard but is included here to meet special needs.

All dimensions are in inches.

**Table 9. ANSI Three- and Four-Flute Straight-Shank Core Drills — Fractional Sizes Only** *ANSI/ASME B94.11M-1993*

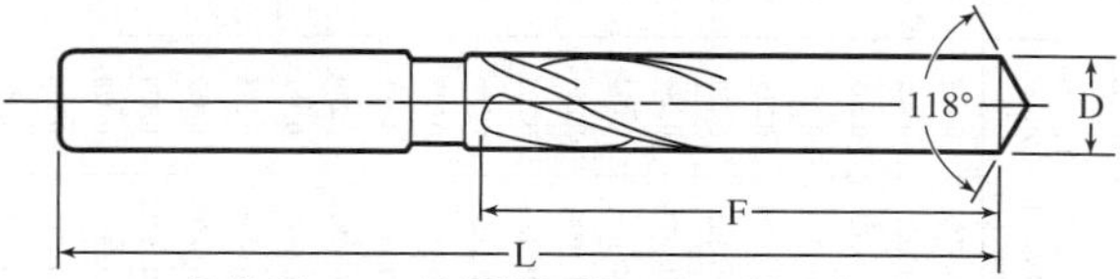


Drill Diameter 11/32″ (8.733 mm) and Smaller

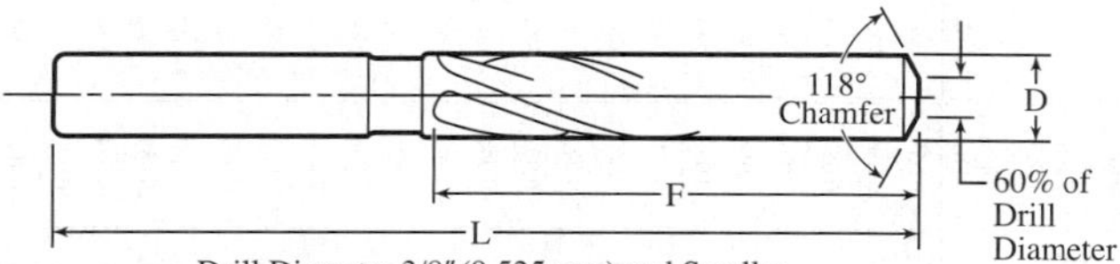


Drill Diameter 3/8″ (9.525 mm) and Smaller

Nominal Shank Size is Same Size as Nominal Drill Size

| Drill Diameter, *D* | | | Three-Flute Drills | | | | Four-Flute Drills | | | |
|---|---|---|---|---|---|---|---|---|---|---|
| | Equivalent | | Flute Length | | Overall Length | | Flute Length | | Overall Length | |
| | | | *F* | | *L* | | *F* | | *L* | |
| Inch | Decimal Inch | mm | Inch | mm | Inch | mm | Inch | mm | Inch | mm |
| ¼ | 0.2500 | 6.350 | 3¾ | 95 | 6⅛ | 156 | … | … | … | … |
| 9/32 | 0.2812 | 7.142 | 3⅞ | 98 | 6¼ | 159 | … | … | … | … |
| 5/16 | 0.3125 | 7.938 | 4 | 102 | 6⅜ | 162 | … | … | … | … |
| 11/32 | 0.3438 | 8.733 | 4⅛ | 105 | 6½ | 165 | … | … | … | … |
| ⅜ | 0.3750 | 9.525 | 4⅛ | 105 | 6¾ | 171 | … | … | … | … |
| 13/32 | 0.4062 | 10.317 | 4⅜ | 111 | 7 | 178 | … | … | … | … |
| 7/16 | 0.4375 | 11.112 | 4⅝ | 117 | 7¼ | 184 | … | … | … | … |
| 15/32 | 0.4688 | 11.908 | 4¾ | 121 | 7½ | 190 | … | … | … | … |
| ½ | 0.5000 | 12.700 | 4¾ | 121 | 7¾ | 197 | 4¾ | 121 | 7¾ | 197 |
| 17/32 | 0.5312 | 13.492 | 4¾ | 121 | 8 | 203 | 4¾ | 121 | 8 | 203 |
| 9/16 | 0.5625 | 14.288 | 4⅞ | 124 | 8¼ | 210 | 4⅞ | 124 | 8¼ | 210 |
| 19/32 | 0.5938 | 15.083 | 4⅞ | 124 | 8¾ | 222 | 4⅞ | 124 | 8¾ | 222 |
| ⅝ | 0.6250 | 15.875 | 4⅞ | 124 | 8¾ | 222 | 4⅞ | 124 | 8¾ | 222 |
| 21/32 | 0.6562 | 16.667 | 5⅛ | 130 | 9 | 229 | 5⅛ | 130 | 9 | 229 |
| 11/16 | 0.6875 | 17.462 | 5⅜ | 137 | 9¼ | 235 | 5⅜ | 137 | 9¼ | 235 |
| 23/32 | 0.7188 | 18.258 | … | … | … | … | 5⅝ | 143 | 9½ | 241 |
| ¾ | 0.7500 | 19.050 | 5⅞ | 149 | 9¾ | 248 | 5⅞ | 149 | 9¾ | 248 |
| 25/32 | 0.7812 | 19.842 | … | … | … | … | 6 | 152 | 9⅞ | 251 |
| 13/16 | 0.8125 | 20.638 | … | … | … | … | 6⅛ | 156 | 10 | 254 |
| 27/32 | 0.8438 | 21.433 | … | … | … | … | 6⅛ | 156 | 10 | 254 |
| ⅞ | 0.8750 | 22.225 | … | … | … | … | 6⅛ | 156 | 10 | 254 |
| 29/32 | 0.9062 | 23.017 | … | … | … | … | 6⅛ | 156 | 10 | 254 |
| 15/16 | 0.9375 | 23.812 | … | … | … | … | 6⅛ | 156 | 10¾ | 273 |
| 31/32 | 0.9688 | 24.608 | … | … | … | … | 6⅜ | 162 | 11 | 279 |
| 1 | 1.0000 | 25.400 | … | … | … | … | 6⅜ | 162 | 11 | 279 |
| 1 1/32 | 1.0312 | 26.192 | … | … | … | … | 6½ | 165 | 11⅛ | 283 |
| 1 1/16 | 1.0625 | 26.988 | … | … | … | … | 6⅝ | 168 | 11¼ | 286 |
| 1 3/32 | 1.0938 | 27.783 | … | … | … | … | 6⅞ | 175 | 11½ | 292 |
| 1⅛ | 1.1250 | 28.575 | … | … | … | … | 7⅛ | 181 | 11¾ | 298 |
| 1¼ | 1.2500 | 31.750 | … | … | … | … | 7⅞ | 200 | 12½ | 318 |

## Table 10. Length of Point on Twist Drills and Centering Tools

| Size of Drill | Decimal Equivalent | Length of Point when Included Angle = 90° | Length of Point when Included Angle = 118° | Size of Drill | Decimal Equivalent | Length of Point when Included Angle = 90° | Length of Point when Included Angle = 118° | Size or Dia. of Drill | Decimal Equivalent | Length of Point when Included Angle = 90° | Length of Point when Included Angle = 118° | Dia. of Drill | Decimal Equivalent | Length of Point when Included Angle = 90° | Length of Point when Included Angle = 118° |
|---|---|---|---|---|---|---|---|---|---|---|---|---|---|---|---|
| 60 | 0.0400 | 0.020 | 0.012 | 37 | 0.1040 | 0.052 | 0.031 | 14 | 0.1820 | 0.091 | 0.055 | 3/8 | 0.3750 | 0.188 | 0.113 |
| 59 | 0.0410 | 0.021 | 0.012 | 36 | 0.1065 | 0.054 | 0.032 | 13 | 0.1850 | 0.093 | 0.056 | 25/64 | 0.3906 | 0.195 | 0.117 |
| 58 | 0.0420 | 0.021 | 0.013 | 35 | 0.1100 | 0.055 | 0.033 | 12 | 0.1890 | 0.095 | 0.057 | 13/32 | 0.4063 | 0.203 | 0.122 |
| 57 | 0.0430 | 0.022 | 0.013 | 34 | 0.1110 | 0.056 | 0.033 | 11 | 0.1910 | 0.096 | 0.057 | 27/64 | 0.4219 | 0.211 | 0.127 |
| 56 | 0.0465 | 0.023 | 0.014 | 33 | 0.1130 | 0.057 | 0.034 | 10 | 0.1935 | 0.097 | 0.058 | 7/16 | 0.4375 | 0.219 | 0.131 |
| 55 | 0.0520 | 0.026 | 0.016 | 32 | 0.1160 | 0.058 | 0.035 | 9 | 0.1960 | 0.098 | 0.059 | 29/64 | 0.4531 | 0.227 | 0.136 |
| 54 | 0.0550 | 0.028 | 0.017 | 31 | 0.1200 | 0.060 | 0.036 | 8 | 0.1990 | 0.100 | 0.060 | 15/32 | 0.4688 | 0.234 | 0.141 |
| 53 | 0.0595 | 0.030 | 0.018 | 30 | 0.1285 | 0.065 | 0.039 | 7 | 0.2010 | 0.101 | 0.060 | 31/64 | 0.4844 | 0.242 | 0.145 |
| 52 | 0.0635 | 0.032 | 0.019 | 29 | 0.1360 | 0.068 | 0.041 | 6 | 0.2040 | 0.102 | 0.061 | 1/2 | 0.5000 | 0.250 | 0.150 |
| 51 | 0.0670 | 0.034 | 0.020 | 28 | 0.1405 | 0.070 | 0.042 | 5 | 0.2055 | 0.103 | 0.062 | 33/64 | 0.5156 | 0.258 | 0.155 |
| 50 | 0.0700 | 0.035 | 0.021 | 27 | 0.1440 | 0.072 | 0.043 | 4 | 0.2090 | 0.105 | 0.063 | 17/32 | 0.5313 | 0.266 | 0.159 |
| 49 | 0.0730 | 0.037 | 0.022 | 26 | 0.1470 | 0.074 | 0.044 | 3 | 0.2130 | 0.107 | 0.064 | 35/64 | 0.5469 | 0.273 | 0.164 |
| 48 | 0.0760 | 0.038 | 0.023 | 25 | 0.1495 | 0.075 | 0.045 | 2 | 0.2210 | 0.111 | 0.067 | 9/16 | 0.5625 | 0.281 | 0.169 |
| 47 | 0.0785 | 0.040 | 0.024 | 24 | 0.1520 | 0.076 | 0.046 | 1 | 0.2280 | 0.114 | 0.068 | 37/64 | 0.5781 | 0.289 | 0.173 |
| 46 | 0.0810 | 0.041 | 0.024 | 23 | 0.1540 | 0.077 | 0.046 | 15/64 | 0.2344 | 0.117 | 0.070 | 19/32 | 0.5938 | 0.297 | 0.178 |
| 45 | 0.0820 | 0.041 | 0.025 | 22 | 0.1570 | 0.079 | 0.047 | 1/4 | 0.2500 | 0.125 | 0.075 | 39/64 | 0.6094 | 0.305 | 0.183 |
| 44 | 0.0860 | 0.043 | 0.026 | 21 | 0.1590 | 0.080 | 0.048 | 17/64 | 0.2656 | 0.133 | 0.080 | 5/8 | 0.6250 | 0.313 | 0.188 |
| 43 | 0.0890 | 0.045 | 0.027 | 20 | 0.1610 | 0.081 | 0.048 | 9/32 | 0.2813 | 0.141 | 0.084 | 41/64 | 0.6406 | 0.320 | 0.192 |
| 42 | 0.0935 | 0.047 | 0.028 | 19 | 0.1660 | 0.083 | 0.050 | 19/64 | 0.2969 | 0.148 | 0.089 | 21/32 | 0.6563 | 0.328 | 0.197 |
| 41 | 0.0960 | 0.048 | 0.029 | 18 | 0.1695 | 0.085 | 0.051 | 5/16 | 0.3125 | 0.156 | 0.094 | 43/64 | 0.6719 | 0.336 | 0.202 |
| 40 | 0.0980 | 0.049 | 0.029 | 17 | 0.1730 | 0.087 | 0.052 | 21/64 | 0.3281 | 0.164 | 0.098 | 11/16 | 0.6875 | 0.344 | 0.206 |
| 39 | 0.0995 | 0.050 | 0.030 | 16 | 0.1770 | 0.089 | 0.053 | 11/32 | 0.3438 | 0.171 | 0.103 | 23/32 | 0.7188 | 0.359 | 0.216 |
| 38 | 0.1015 | 0.051 | 0.030 | 15 | 0.1800 | 0.090 | 0.054 | 23/64 | 0.3594 | 0.180 | 0.108 | 3/4 | 0.7500 | 0.375 | 0.225 |

**British Standard Combined Drills and Countersinks (Center Drills).**—BS 328: Part 2: 1972 provides dimensions of combined drills and countersinks for center holes. Three types of drill and countersink combinations are shown in this Standard but are not given here. These three types will produce center holes without protecting chamfers, with protecting chamfers, and with protecting chamfers of radius form.

**Drill Drivers—Split-Sleeve, Collet-Type.**—American National Standard ANSI/ASME B94.35-1972 (R2020) covers split-sleeve, collet-type drivers for driving straight shank drills, reamers, and similar tools, without tangs from 0.0390-inch through 0.1220-inch diameter, and with tangs from 0.1250-inch through 0.7500-inch diameter, including metric sizes.

For sizes 0.0390 through 0.0595 inch, the standard taper number is 1, and the optional taper number is 0. For sizes 0.0610 through 0.1875 inch, the standard taper number is 1, first optional taper number is 0, and second optional taper number is 2. For sizes 0.1890 through 0.2520 inch, the standard taper number is 1, first optional taper number is 2, and second optional taper number is 0. For sizes 0.2570 through 0.3750 inch, the standard taper number is 1 and the optional taper number is 2. For sizes 0.3860 through 0.5625 inch, the standard taper number is 2 and the optional taper number is 3. For sizes 0.5781 through 0.7500 inch, the standard taper number is 3 and the optional taper number is 4.

The depth *B* that the drill enters the driver is 0.44 inch for sizes 0.0390 through 0.0781 inch; 0.50 inch for sizes 0.0785 through 0.0938 inch; 0.56 inch for sizes 0.0960 through 0.1094 inch; 0.62 inch for sizes 0.1100 through 0.1220 inch; 0.75 inch for sizes 0.1250 through 0.1875 inch; 0.88 inch for sizes 0.1890 through 0.2500 inch; 1.00 inch for sizes 0.2520 through 0.3125 inch; 1.12 inches for sizes 0.3160 through 0.3750 inch; 1.25 inches for sizes 0.3860 through 0.4688 inch; 1.31 inches for sizes 0.4844 through 0.5625 inch; 1.47 inches for sizes 0.5781 through 0.6562 inch; and 1.62 inches for sizes 0.6719 through 0.7500 inch.

**British Standard Metric Twist Drills.**—BS 328: Part 1:1993 (incorporating the 1950, 1959, and 1986 Standards, with amendments issued in 1960 and 1964) covers twist drills made to inch and metric dimensions that are intended for general engineering purposes. ISO recommendations are taken into account. The accompanying tables give the standard metric sizes of Morse taper-shank twist drills and core drills, parallel-shank jobbing and long-series drills, and stub drills.

All drills are right-hand cutting unless otherwise specified, and normal, slow, or quick helix angles may be provided. A "back-taper" is ground on the diameter from point to shank to provide longitudinal clearance. Core drills may have three or four flutes and are intended for opening up cast holes or enlarging machined holes, for example. The parallel-shank jobber, long-series drills, and stub drills are made without driving tenons.

Morse taper shank drills with oversize dimensions are also listed, and Table 11 shows metric drill sizes superseding gage and letter-size drills, which are now obsolete in Britain. To meet special requirements, the Standard lists nonstandard sizes for the various types of drills.

The limits of tolerance on cutting diameters, as measured across the lands at the outer corners of a drill, shall be h8, in accordance with BS 1916, Limits and Fits for Engineering (Part I, Limits and Tolerances); Table 14 shows the values common to the different types of drills mentioned before.

The drills shall be permanently and legibly marked whenever possible, preferably by rolling, showing the size, and the manufacturer's name or trademark. If they are made from high-speed steel, they shall be marked with the letters H.S. where practicable.

*Drill Elements:* The following definitions of drill elements are given.

*Axis:* The longitudinal center line.

*Body:* That portion of the drill extending from the extreme cutting end to the commencement of the shank.

*Shank:* That portion of the drill by which it is held and driven.

*Flutes:* The grooves in the body of the drill that provide lips, permit the removal of chips and allow cutting fluid to reach the lips.

*Web (Core):* The central portion of the drill situated between the roots of the flutes and extending from the point end toward the shank; the point end of the web or core forms the chisel edge.

*Lands:* The cylindrical-ground surfaces on the leading edges of the drill flutes. The width of the land is measured at right angles to the flute helix.

*Body Clearance:* The portion of the body surface that is reduced in diameter to provide diametral clearance.

*Heel:* The edge formed by the intersection of the flute surface and the body clearance.

*Point:* The sharpened end of the drill, consisting of all that part of the drill that is shaped to produce lips, faces, flanks, and chisel edge.

*Face:* That portion of the flute surface adjacent to the lip on which the chip impinges as it is cut from the work.

*Flank:* The surface on a drill point that extends behind the lip to the following flute.

*Lip (Cutting Edge):* The edge formed by the intersection of the flank and face.

*Radial Runout:* The maximum radial runout, measured on the land at the outer corner of the drill.

*Relative Lip Height:* The relative position of the lips measured at the outer corners in a direction parallel to the drill axis.

*Outer Corner:* The corner formed by the intersection of the lip and the leading edge of the land.

*Chisel Edge:* The edge formed by the intersection of the flanks.

*Chisel Edge Corner:* The corner formed by the intersection of a lip and the chisel edge.

**Table 11. British Standard Drills — Metric Sizes Superseding Gauge and Letter Sizes** *BS 328: Part 1:1986, Appendix B (Withdrawn)*

| Obsolete Drill Size | Recommended Metric Size (mm) | Obsolete Drill Size | Recommended Metric Size (mm) | Obsolete Drill Size | Recommended Metric Size (mm) | Obsolete Drill Size | Recommended Metric Size (mm) | Obsolete Drill Size | Recommended Metric Size (mm) |
|---|---|---|---|---|---|---|---|---|---|
| 80 | 0.35 | 58 | 1.05 | 36 | 2.70 | 14 | 4.60 | I | 6.90 |
| 79 | 0.38 | 57 | 1.10 | 35 | 2.80 | 13 | 4.70 | J | 7.00 |
| 78 | 0.40 | 56 | 3/64 in. | 34 | 2.80 | 12 | 4.80 | K | 9/32 in. |
| 77 | 0.45 | 55 | 1.30 | 33 | 2.85 | 11 | 4.90 | L | 7.40 |
| 76 | 0.50 | 54 | 1.40 | 32 | 2.95 | 10 | 4.90 | M | 7.50 |
| 75 | 0.52 | 53 | 1.50 | 31 | 3.00 | 9 | 5.00 | N | 7.70 |
| 74 | 0.58 | 52 | 1.60 | 30 | 3.30 | 8 | 5.10 | O | 8.00 |
| 73 | 0.60 | 51 | 1.70 | 29 | 3.50 | 7 | 5.10 | P | 8.20 |
| 72 | 0.65 | 50 | 1.80 | 28 | 9/64 in. | 6 | 5.20 | Q | 8.40 |
| 71 | 0.65 | 49 | 1.85 | 27 | 3.70 | 5 | 5.20 | R | 8.60 |
| 70 | 0.70 | 48 | 1.95 | 26 | 3.70 | 4 | 5.30 | S | 8.80 |
| 69 | 0.75 | 47 | 2.00 | 25 | 3.80 | 3 | 5.40 | T | 9.10 |
| 68 | 1/32 in. | 46 | 2.05 | 24 | 3.90 | 2 | 5.60 | U | 9.30 |
| 67 | 0.82 | 45 | 2.10 | 23 | 3.90 | 1 | 5.80 | V | 3/8 in. |
| 66 | 0.85 | 44 | 2.20 | 22 | 4.00 | A | 15/64 in. | W | 9.80 |
| 65 | 0.90 | 43 | 2.25 | 21 | 4.00 | B | 6.00 | X | 10.10 |
| 64 | 0.92 | 42 | 3/32 in. | 20 | 4.10 | C | 6.10 | Y | 10.30 |
| 63 | 0.95 | 41 | 2.45 | 19 | 4.20 | D | 6.20 | Z | 10.50 |
| 62 | 0.98 | 40 | 2.50 | 18 | 4.30 | E | 1/4 in. | … | … |
| 61 | 1.00 | 39 | 2.55 | 17 | 4.40 | F | 6.50 | … | … |
| 60 | 1.00 | 38 | 2.60 | 16 | 4.50 | G | 6.60 | … | … |
| 59 | 1.05 | 37 | 2.65 | 15 | 4.60 | H | 17/64 in. | … | … |

Gauge and letter-size drills are now obsolete in the United Kingdom and should not be used in the production of new designs. The table is given to assist users in changing over to the recommended standard sizes.

## Table 12. British Standard Morse Taper-Shank Twist Drills and Core Drills — Standard Metric Sizes *BS 328: Part 1:1993*

| Diameter | Flute Length | Overall Length | Diameter | Flute Length | Overall Length | Diameter | Flute Length | Overall Length |
|---|---|---|---|---|---|---|---|---|
| 3.00 | 33 | 114 | 16.75 | 125 | 223 | 30.25 | | |
| 3.20 | 36 | 117 | 17.00 | | | 30.50 | | |
| 3.50 | 39 | 120 | 17.25 | | | 30.75 | 180 | 301 |
| 3.80 | | | 17.50 | 130 | 228 | 31.00 | | |
| 4.00 | 43 | 123 | 17.75 | | | 31.25 | | |
| 4.20 | | | 18.00 | | | 31.50 | | |
| 4.50 | 47 | 128 | 18.25 | | | 31.75 | 185 | 306 |
| 4.80 | | | 18.50 | 135 | 233 | 32.00 | | |
| 5.00 | 52 | 133 | 18.75 | | | 32.50 | 185 | 334 |
| 5.20 | | | 19.00 | | | 33.00 | | |
| 5.50 | | | 19.25 | | | 33.50 | | |
| 5.80 | 57 | 138 | 19.50 | 140 | 238 | 34.00 | | |
| 6.00 | | | 19.75 | | | 34.50 | 190 | 339 |
| 6.20 | 63 | 144 | 20.00 | | | 35.00 | | |
| 6.50 | | | 20.25 | | | 35.50 | | |
| 6.80 | | | 20.50 | 145 | 243 | 36.00 | | |
| 7.00 | 69 | 150 | 20.75 | | | 36.50 | 195 | 344 |
| 7.20 | | | 21.00 | | | 37.00 | | |
| 7.50 | | | 21.25 | | | 37.50 | | |
| 7.80 | | | 21.50 | | | 38.00 | | |
| 8.00 | 75 | 156 | 21.75 | 150 | 248 | 38.50 | | |
| 8.20 | | | 22.00 | | | 39.00 | 200 | 349 |
| 8.50 | | | 22.25 | | | 39.50 | | |
| 8.80 | | | 22.50 | | | 40.00 | | |
| 9.00 | 81 | | 22.75 | 155 | 253 | 40.50 | | |
| 9.20 | | 162 | 23.00 | | | 41.00 | | |
| 9.50 | | | 23.25 | 155 | 276 | 41.50 | 205 | 354 |
| 9.80 | | | 23.50 | | | 42.00 | | |
| 10.00 | 87 | 168 | 23.75 | | | 42.50 | | |
| 10.20 | | | 24.00 | | | 43.00 | | |
| 10.50 | | | 24.25 | 160 | 281 | 43.50 | | |
| 10.80 | | | 24.50 | | | 44.00 | 210 | 359 |
| 11.00 | | | 24.75 | | | 44.50 | | |
| 11.20 | 94 | 175 | 25.00 | | | 45.00 | | |
| 11.50 | | | 25.25 | | | 45.50 | | |
| 11.80 | | | 25.50 | | | 46.00 | | |
| 12.00 | | | 25.75 | 165 | 286 | 46.50 | 215 | 364 |
| 12.20 | | | 26.00 | | | 47.00 | | |
| 12.50 | 101 | 182 | 26.25 | | | 47.50 | | |
| 12.80 | | | 26.50 | | | 48.00 | | |
| 13.00 | | | 26.75 | | | 48.50 | | |
| 13.20 | | | 27.00 | | | 49.00 | 220 | 369 |
| 13.50 | | | 27.25 | 170 | 291 | 49.50 | | |
| 13.80 | 108 | 189 | 27.50 | | | 50.00 | | |
| 14.00 | | | 27.75 | | | 50.50 | 225 | 374 |
| 14.25 | | | 28.00 | | | 51.00 | | |
| 14.50 | 114 | 212 | 28.25 | | | 52.00 | 225 | 412 |
| 14.75 | | | 28.50 | | | 53.00 | | |
| 15.00 | | | 28.75 | 175 | 296 | 54.00 | | |
| 15.25 | | | 29.00 | | | 55.00 | 230 | 417 |
| 15.50 | 120 | 218 | 29.25 | | | 56.00 | | |
| 15.75 | | | | | | 57.00 | | |
| 16.00 | | | 29.50 | | | 58.00 | 235 | 422 |
| 16.25 | 125 | 223 | 29.75 | 175 | 296 | 59.00 | | |
| 16.50 | | | 30.00 | | | 60.00 | | |

**Table 12.** *(Continued)* **British Standard Morse Taper-Shank Twist Drills and Core Drills — Standard Metric Sizes** *BS 328: Part 1:1993*

| Diameter | Flute Length | Overall Length | Diameter | Flute Length | Overall Length | Diameter | Flute Length | Overall Length |
|---|---|---|---|---|---|---|---|---|
| 61.00 | | | 76.00 | 260 | 447 | 91.00 | | |
| 62.00 | 240 | 427 | 77.00 | | | 92.00 | | |
| 63.00 | | | 78.00 | 260 | 514 | 93.00 | 275 | 529 |
| 64.00 | | | 79.00 | | | 94.00 | | |
| 65.00 | 245 | 432 | 80.00 | | | 95.00 | | |
| 66.00 | | | 81.00 | | | 96.00 | | |
| 67.00 | | | 82.00 | | | 97.00 | | |
| 68.00 | | | 83.00 | 265 | 519 | 98.00 | 280 | 534 |
| 69.00 | 250 | 437 | 84.00 | | | 99.00 | | |
| 70.00 | | | 85.00 | | | 100.00 | | |
| 71.00 | 250 | 437 | 86.00 | | | | | |
| 72.00 | | | 87.00 | | | | | |
| 73.00 | 255 | 442 | 88.00 | 270 | 524 | | | |
| 74.00 | | | 89.00 | | | | | |
| 75.00 | | | 90.00 | | | | | |

All dimensions are in millimeters. Tolerances on diameters are given in the table below.

Table 13 shows twist drills that may be supplied with the shank and length oversize, but they should be regarded as nonpreferred.

The Morse taper shanks of these twist and core drills are as follows: 3.00 to 14.00 mm diameter, M.T. No. 1; 14.25 to 23.00 mm diameter, M.T. No. 2; 23.25 to 31.50 mm diameter, M.T. No. 3; 31.75 to 50.50 mm diameter, M.T. No. 4; 51.00 to 76.00 mm diameter, M.T. No. 5; 77.00 to 100.00 mm diameter, M.T. No. 6.

**Table 13. British Standard Morse Taper-Shank Twist Drills — Metric Oversize Shank and Length Series** *BS 328: Part 1:1993*

| Dia. Range | Overall Length | M.T. No. | Dia. Range | Overall Length | M.T. No. | Dia. Range | Overall Length | M.T. No. |
|---|---|---|---|---|---|---|---|---|
| 12.00 to 13.20 | 199 | 2 | 22.50 to 23.00 | 276 | 3 | 45.50 to 47.50 | 402 | 5 |
| 13.50 to 14.00 | 206 | 2 | 26.75 to 28.00 | 319 | 4 | 48.00 to 50.00 | 407 | 5 |
| 18.25 to 19.00 | 256 | 3 | 29.00 to 30.00 | 324 | 4 | 50.50 | 412 | 5 |
| 19.25 to 20.00 | 251 | 3 | 30.25 to 31.50 | 329 | 4 | 64.00 to 67.00 | 499 | 6 |
| 20.25 to 21.00 | 266 | 3 | 40.50 to 42.50 | 392 | 5 | 68.00 to 71.00 | 504 | 6 |
| 21.25 to 22.25 | 271 | 3 | 43.00 to 45.00 | 397 | 5 | 72.00 to 75.00 | 509 | 6 |

Diameters and lengths are given in millimeters. For the individual sizes within the diameter ranges given, see Table 12.

This series of drills should be regarded as non-preferred.

**Table 14. British Standard Limits of Tolerance on Diameter for Twist Drills and Core Drills — Metric Series** *BS 328: Part 1:1993*

| Drill Size (Diameter measured across lands at outer corners) | Tolerance (h8) |
|---|---|
| 0 to 1 inclusive | Plus 0.000 to Minus 0.014 |
| Over 1 to 3 inclusive | Plus 0.000 to Minus 0.014 |
| Over 3 to 6 inclusive | Plus 0.000 to Minus 0.018 |
| Over 6 to 10 inclusive | Plus 0.000 to Minus 0.022 |
| Over 10 to 18 inclusive | Plus 0.000 to Minus 0.027 |
| Over 18 to 30 inclusive | Plus 0.000 to Minus 0.033 |
| Over 30 to 50 inclusive | Plus 0.000 to Minus 0.039 |
| Over 50 to 80 inclusive | Plus 0.000 to Minus 0.046 |
| Over 80 to 120 inclusive | Plus 0.000 to Minus 0.054 |

All dimensions are given in millimeters.

**Table 15. British Standard Parallel Shank Jobber Series Twist Drills — Standard Metric Sizes** *BS 328: Part 1:1993*

| Diameter | Flute Length | Overall Length |
|---|---|---|
| 0.20<br>0.22 | 2.5 | 19 |
| 0.25<br>0.28 | 3.0 | 19 |
| 0.30<br>0.32<br>0.35<br>0.38 | 4 | 19 |
| 0.40<br>0.42<br>0.45<br>0.48 | 5 | 20 |
| 0.50<br>0.52 | 6 | 22 |
| 0.55<br>0.58<br>0.60 | 7 | 24 |
| 0.62<br>0.65 | 8 | 26 |
| 0.68<br>0.70<br>0.72<br>0.75 | 9 | 28 |
| 0.78<br>0.80<br>0.82<br>0.85 | 10 | 30 |
| 0.88<br>0.90<br>0.92<br>0.95 | 11 | 32 |
| 0.98<br>1.00<br>1.05 | 12 | 34 |
| 1.10<br>1.15 | 14 | 36 |
| 1.20<br>1.25<br>1.30 | 16 | 38 |
| 1.35<br>1.40<br>1.45<br>1.50 | 18 | 40 |
| 1.55<br>1.60<br>1.65<br>1.70 | 20 | 43 |

| Diameter | Flute Length | Overall Length |
|---|---|---|
| 1.75<br>1.80<br>1.85<br>1.90 | 22 | 46 |
| 1.95<br>2.00<br>2.05<br>2.10 | 24 | 49 |
| 2.15<br>2.20<br>2.25<br>2.30<br>2.35 | 27 | 53 |
| 2.40<br>2.45<br>2.50<br>2.55<br>2.60<br>2.65 | 30 | 57 |
| 2.70<br>2.75<br>2.80<br>2.85<br>2.90<br>2.95<br>3.00 | 33 | 61 |
| 3.10<br>3.20<br>3.30 | 36 | 65 |
| 3.40<br>3.50<br>3.60<br>3.70 | 39 | 70 |
| 3.80<br>3.90<br>4.00<br>4.10<br>4.20 | 43 | 75 |
| 4.30<br>4.40<br>4.50<br>4.60<br>4.70 | 47 | 80 |
| 4.80<br>4.90<br>5.00<br>5.10<br>5.20<br>5.30 | 52 | 86 |

| Diameter | Flute Length | Overall Length |
|---|---|---|
| 5.40<br>5.50<br>5.60<br>5.70<br>5.80<br>5.90<br>6.00 | 57 | 93 |
| 6.10<br>6.20<br>6.30<br>6.40<br>6.50<br>6.60<br>6.70 | 63 | 101 |
| 6.80<br>6.90<br>7.00<br>7.10<br>7.20<br>7.30<br>7.40<br>7.50 | 69 | 109 |
| 7.60<br>7.70<br>7.80<br>7.90<br>8.00<br>8.10<br>8.20<br>8.30<br>8.40<br>8.50 | 75 | 117 |
| 8.60<br>8.70<br>8.80<br>8.90<br>9.00<br>9.10<br>9.20<br>9.30<br>9.40<br>9.50 | 81 | 125 |
| 9.60<br>9.70<br>9.80<br>9.90<br>10.00<br>10.10 | 87 | 133 |

| Diameter | Flute Length | Overall Length |
|---|---|---|
| 10.20<br>10.30<br>10.40<br>10.50<br>10.60 | 87 | 133 |
| 10.70<br>10.80<br>10.90<br>11.00<br>11.10<br>11.20<br>11.30<br>11.40<br>11.50<br>11.60<br>11.70<br>11.80 | 94 | 142 |
| 11.90<br>12.00<br>12.10<br>12.20<br>12.30<br>12.40<br>12.50<br>12.60<br>12.70<br>12.80<br>12.90<br>13.00<br>13.10<br>13.20 | 101 | 151 |
| 13.30<br>13.40<br>13.50<br>13.60<br>13.70<br>13.80<br>13.90<br>14.00 | 108 | 160 |
| 14.25<br>14.50<br>14.75<br>15.00 | 114 | 169 |
| 15.25<br>15.50<br>15.75<br>16.00 | 120 | 178 |

All dimensions are in millimeters. Tolerances on diameters are given in Table 14.

**Table 16. British Standard Parallel Shank Long Series Twist Drills — Standard Metric Sizes** *BS 328: Part 1:1993*

| Diameter | Flute Length | Overall Length |
|---|---|---|
| 2.00, 2.05, 2.10 | 56 | 85 |
| 2.15, 2.20, 2.25, 2.30, 2.35 | 59 | 90 |
| 2.40, 2.45, 2.50, 2.55, 2.60, 2.65 | 62 | 95 |
| 2.70, 2.75, 2.80, 2.85, 2.90, 2.95, 3.00 | 66 | 100 |
| 3.10, 3.20, 3.30 | 69 | 106 |
| 3.40, 3.50, 3.60, 3.70 | 73 | 112 |
| 3.80, 3.90, 4.00, 4.10, 4.20 | 78 | 119 |
| 4.30, 4.40, 4.50, 4.60, 4.70 | 82 | 126 |
| 4.80, 4.90, 5.00, 5.10, 5.20, 5.30 | 87 | 132 |
| 5.40, 5.50, 5.60, 5.70, 5.80, 5.90, 6.00 | 91 | 139 |
| 6.10, 6.20, 6.30, 6.40, 6.50, 6.60, 6.70 | 97 | 148 |
| 6.80, 6.90, 7.00, 7.10, 7.20, 7.30, 7.40, 7.50 | 102 | 156 |
| 7.60, 7.70, 7.80, 7.90, 8.00, 8.10, 8.20, 8.30, 8.40, 8.50 | 109 | 165 |
| 8.60, 8.70, 8.80, 8.90, 9.00, 9.10, 9.20, 9.30, 9.40, 9.50 | 115 | 175 |
| 9.60, 9.70, 9.80, 9.90, 10.00, 10.10, 10.20, 10.30, 10.40, 10.50, 10.60 | 121 | 184 |
| 10.70, 10.80, 10.90, 11.00, 11.10, 11.20, 11.30, 11.40, 11.50, 11.60, 11.70, 11.80 | 128 | 195 |
| 11.90, 12.00, 12.10, 12.20, 12.30, 12.40, 12.50, 12.60 | 134 | 205 |
| 12.70, 12.80, 12.90, 13.00, 13.10, 13.20 | 134 | 205 |
| 13.30, 13.40, 13.50, 13.60, 13.70, 13.80, 13.90, 14.00 | 140 | 214 |
| 14.25, 14.50, 14.75, 15.00 | 144 | 220 |
| 15.25, 15.50, 15.75, 16.00 | 149 | 227 |
| 16.25, 16.50, 16.75, 17.00 | 154 | 235 |
| 17.25, 17.50, 17.75, 18.00 | 158 | 241 |
| 18.25, 18.50, 18.75, 19.00 | 162 | 247 |
| 19.25, 19.50, 19.75, 20.00 | 166 | 254 |
| 20.25, 20.50, 20.75, 21.00 | 171 | 261 |
| 21.25, 21.50, 21.75, 22.00, 22.25 | 176 | 268 |
| 22.50, 22.75, 23.00, 23.25, 23.50 | 180 | 275 |
| 23.75, 24.00, 24.25, 24.50, 24.75, 25.00 | 185 | 282 |

All dimensions are in millimeters. Tolerances on diameters are given in Table 14.

**Table 17. British Standard Stub Drills — Metric Sizes** *BS 328: Part 1:1993*

| Diameter | Flute Length | Overall Length | Diameter | Flute Length | Overall Length | Diameter | Flute Length | Overall Length | Diameter | Flute Length | Overall Length |
|---|---|---|---|---|---|---|---|---|---|---|---|
| 0.50 | 3 | 20 | 5.00 | 26 | 62 | 9.50 | 40 | 84 | 14.00 | 54 | 107 |
| 0.80 | 5 | 24 | 5.20 | | | 9.80 | 43 | 89 | 14.50 | 56 | 111 |
| 1.00 | 6 | 26 | 5.50 | 28 | 66 | 10.00 | | | 15.00 | | |
| 1.20 | 8 | 30 | 5.80 | | | 10.20 | | | 15.50 | 58 | 115 |
| 1.50 | 9 | 32 | 6.00 | | | 10.50 | | | 16.00 | | |
| 1.80 | 11 | 36 | 6.20 | 31 | 70 | 10.80 | 47 | 95 | 16.50 | 60 | 119 |
| 2.00 | 12 | 38 | 6.50 | | | 11.00 | | | 17.00 | | |
| 2.20 | 13 | 40 | 6.80 | 34 | 74 | 11.20 | | | 17.50 | 62 | 123 |
| 2.50 | 14 | 43 | 7.00 | | | 11.50 | | | 18.00 | | |
| 2.80 | 16 | 46 | 7.20 | | | 11.80 | | | 18.50 | 64 | 127 |
| 3.00 | | | 7.50 | | | 12.00 | 51 | 102 | 19.00 | | |
| 3.20 | 18 | 49 | 7.80 | 37 | 79 | 12.20 | | | 19.50 | 66 | 131 |
| 3.50 | 20 | 52 | 8.20 | | | 12.80 | | | 21.00 | 68 | 136 |
| 3.80 | 22 | 55 | 8.50 | | | 13.00 | | | 22.00 | 70 | 141 |
| 4.00 | | | 8.80 | 40 | 84 | 13.20 | | | 23.00 | 72 | 146 |
| 4.20 | | | 9.00 | | | 13.50 | 54 | 107 | 24.00 | 75 | 151 |
| 4.50 | 24 | 58 | 9.20 | | | 13.80 | | | 25.00 | | |
| 4.80 | 26 | 62 | | | | | | | | | |

All dimensions are given in millimeters. Tolerances on diameters are given in Table 14.

**Steels for Twist Drills.—**Twist drill steels need good toughness, abrasion resistance, and ability to resist softening due to heat generated by cutting. The amount of heat generated indicates the type of steel that should be used.

*Carbon Tool Steel* may be used where little heat is generated during drilling.

*High-Speed Steel* is preferred because of its combination of red hardness and wear resistance, which permits higher operating speeds and increased productivity. Optimum properties can be obtained by selection of alloy analysis and heat treatment.

*Cobalt High-Speed Steel* alloys have higher red hardness than standard high-speed steels, permitting drilling of materials such as heat-resistant alloys and materials with hardness greater than 38 RC (Rockwell C scale). These high-speed drills can withstand cutting speeds beyond the range of conventional high-speed steel drills and have superior resistance to abrasion but are not equal to tungsten carbide-tipped tools.

**Accuracy of Drilled Holes.—**Normally, the diameter of drilled holes is not given a tolerance; the size of the hole is expected to be as close to the drill size as can be obtained.

The accuracy of holes drilled with a two-fluted twist drill is influenced by many factors, including accuracy of the drill point; size of the drill; length and shape of the chisel edge; whether or not a bushing is used to guide the drill; work material; length of the drill; runout of the spindle and the chuck; rigidity of the machine tool, workpiece, and the setup; and cutting fluid used, if any.

The diameter of the drilled holes will be oversize in most materials. The table *Oversize Diameters in Drilling* on page 950 provides the results of tests reported by The United States Cutting Tool Institute in which the diameters of over 2800 holes drilled in steel and cast iron were measured. The values in this table indicate what might be expected under average shop conditions; however, when the drill point is accurately ground and the other machining conditions are correct, the resulting hole size is more likely to be between the mean and average minimum values given in this table. If the drill is ground and used incorrectly, holes that are even larger than the average maximum values can result.

**Oversize Diameters in Drilling**

| Drill Dia., Inch | Amount Oversize, Inch | | | Drill Dia., Inch | Amount Oversize, Inch | | |
|---|---|---|---|---|---|---|---|
| | Average Max. | Mean | Average Min. | | Average Max. | Mean | Average Min. |
| 1⁄16 | 0.002 | 0.0015 | 0.001 | ½ | 0.008 | 0.005 | 0.003 |
| ⅛ | 0.0045 | 0.003 | 0.001 | ¾ | 0.008 | 0.005 | 0.003 |
| ¼ | 0.0065 | 0.004 | 0.0025 | 1 | 0.009 | 0.007 | 0.004 |

Courtesy of The United States Cutting Tool Institute

Some conditions will cause the drilled hole to be undersize. For example, holes drilled in light metals and in other materials having a high coefficient of thermal expansion such as plastics may contract to a size that is smaller than the diameter of the drill as the material surrounding the hole is cooled after having been heated by the drilling. The elastic action of the material surrounding the hole may also cause the drilled hole to be undersize when drilling high-strength materials with a drill that is dull at its outer corner.

The accuracy of the drill point has a great effect on the accuracy of the drilled hole. An inaccurately ground twist drill will produce holes that are excessively oversize. The drill point must be symmetrical; i.e., the point angles must be equal, as well as the lip lengths and the axial height of the lips. Any alterations to the lips or to the chisel edge, such as thinning the web, must be done carefully to preserve the symmetry of the drill point. Adequate relief should be provided behind the chisel edge to prevent heel drag. On conventionally ground drill points, this relief can be estimated by the chisel edge angle.

When drilling a hole, as the drill point starts to enter the workpiece, the drill will be unstable and will tend to wander. Then, as the body of the drill enters the hole, the drill will tend to stabilize. The result of this action is a tendency to drill a bellmouth shape in the hole at the entrance and perhaps beyond. Factors contributing to bellmouthing are: an unsymmetrically ground drill point; a large chisel edge length; inadequate relief behind the chisel edge; runout of the spindle and the chuck; using a slender drill that will bend easily; and lack of rigidity of the machine tool, workpiece, or the setup. Correcting these conditions as required will reduce the tendency for bellmouthing to occur and improve the accuracy of the hole diameter and its straightness. Starting the hole with a short stiff drill, such as a center drill, will quickly stabilize the drill that follows and reduce or eliminate bellmouthing; this procedure should always be used when drilling in a lathe, where the work is rotating. Bellmouthing can also be eliminated almost entirely and the accuracy of the hole improved by using a close-fitting drill jig bushing placed close to the workpiece. Although specific recommendations cannot be made, many cutting fluids will help to increase the accuracy of the diameters of drilled holes. Double-margin twist drills, available in the smaller sizes, will drill a more accurate hole than conventional twist drills having only a single margin at the leading edge of the land. The second land, located on the trailing edge of each land, provides greater stability in the drill bushing and in the hole. These drills are especially useful in drilling intersecting off-center holes. Single and double-margin step drills, also available in the smaller sizes, will produce very accurate drilled holes, which are usually less than 0.002 inch (0.051 mm) larger than the drill size.

**Counterboring.**—Counterboring (called spot-facing if the depth is shallow) is the enlargement of a previously formed hole. Counterbores for screw holes are generally made in sets. Each set contains three counterbores: one with the body of the size of the screw head and the pilot the size of the hole to admit the body of the screw; one with the body the size of the head of the screw and the pilot the size of the tap drill; and the third with the body the size of the body of the screw and the pilot the size of the tap drill. Counterbores are usually provided with helical flutes to provide positive effective rake on the cutting edges. The four flutes are so positioned that the end teeth cut ahead of center to provide a shearing action and eliminate chatter in the cut. Three designs are most common: solid, two-piece, and three-piece. Solid designs have the body, cutter, and pilot all in one piece. Two-piece designs have an integral shank and counterbore cutter, with an interchangeable pilot, and provide true concentricity of the cutter diameter with the shank while allowing use of various pilot diameters. Three-piece counterbores have separate holder, counterbore cutter, and pilot, so that a holder will take any size of counterbore cutter. Each counterbore cutter, in turn, can be fitted with any suitable size diameter of pilot. Counterbores for brass are fluted straight.

**Counterbores with Interchangeable Cutters and Guides**

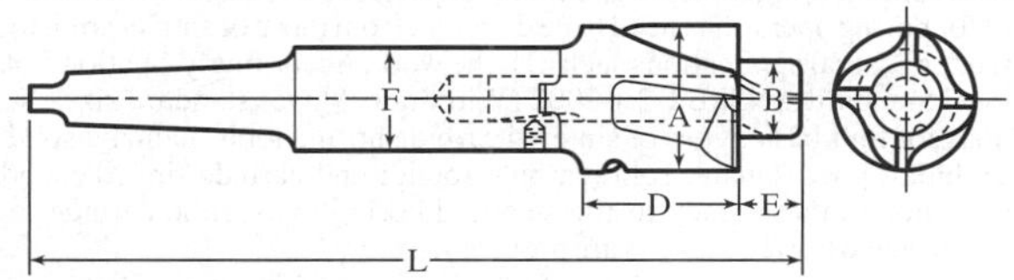


| No. of Holder | No. of Morse Taper Shank | Range of Cutter Diameters, *A* | Range of Pilot Diameters, *B* | Total Length, *C* | Length of Cutter Body, *D* | Length of Pilot, *E* | Dia. of Shank, *F* |
|---|---|---|---|---|---|---|---|
| 1 | 1 or 2 | 3/4-1 1/16 | 1/2-3/4 | 7 1/4 | 1 | 5/8 | 3/4 |
| 2 | 2 or 3 | 1 1/8-1 9/16 | 1 1/16-1 1/8 | 9 1/2 | 1 3/8 | 7/8 | 1 1/8 |
| 3 | 3 or 4 | 1 5/8-2 1/16 | 7/8-1 5/8 | 12 1/2 | 1 3/4 | 1 1/8 | 1 5/8 |
| 4 | 4 or 5 | 2 1/8-3 1/2 | 1-2 1/8 | 15 | 2 1/4 | 1 3/8 | 2 1/8 |

Small counterbores are often made with three flutes, but should then have the size plainly stamped on them before fluting, as they cannot afterwards be conveniently measured. The flutes should be deep enough to come below the surface of the pilot. The counterbore should be relieved on the end of the body only, and not on the cylindrical surface. To facilitate the relieving process, a small neck is turned between the guide and the body for clearance. The amount of clearance on the cutting edges is, for general work, from 4 to 5 degrees. The accompanying table gives dimensions for straight shank counterbores.

**Solid Counterbores with Integral Pilot**

| Counterbore Diameters | Pilot Diameters | | | Straight Shank Diameter | Overall Length | |
|---|---|---|---|---|---|---|
| | Nominal | +1/64 | +1/32 | | Short | Long |
| 0.110 | 0.060 | 0.076 | … | 7/64 | 2 1/2 | … |
| 0.133 | 0.073 | 0.089 | … | 1/8 | 2 1/2 | … |
| 0.155 | 0.086 | 0.102 | … | 5/32 | 2 1/2 | … |
| 0.176 | 0.099 | 0.115 | … | 11/64 | 2 1/2 | … |
| 0.198 | 0.112 | 0.128 | … | 3/16 | 2 1/2 | … |
| 0.220 | 0.125 | 0.141 | … | 3/16 | 2 1/2 | … |
| 0.241 | 0.138 | 0.154 | … | 7/32 | 2 1/2 | … |
| 0.285 | 0.164 | 0.180 | … | 1/4 | 2 1/2 | … |
| 0.327 | 0.190 | 0.206 | … | 9/32 | 2 3/4 | … |
| 0.372 | 0.216 | 0.232 | … | 5/16 | 2 3/4 | … |
| 13/32 | 1/4 | 17/64 | 9/32 | 3/8 | 3 1/2 | 5 1/2 |
| 1/2 | 5/16 | 21/64 | 11/32 | 3/8 | 3 1/2 | 5 1/2 |
| 19/32 | 3/8 | 25/64 | 13/32 | 1/2 | 4 | 6 |
| 11/16 | 7/16 | 29/64 | 15/32 | 1/2 | 4 | 6 |
| 25/32 | 1/2 | 33/64 | 17/32 | 1/2 | 5 | 7 |

All dimensions are in inches.

**Three-Piece Counterbores.—**Data shown for the first two styles of counterbores are for straight-shank designs. These tools are also available with taper shanks in most sizes. Sizes of taper shanks for cutter diameters of 1/4 to 9/16 in. are No. 1; for 19/32 to 7/8 in., No. 2; for 15/16 to 1 3/8 in., No. 3; for 1 1/2 to 2 in., No. 4; and for 2 1/8 to 2 1/2 in., No. 5.

**Counterbore Sizes for Hex-Head Bolts and Nuts.—**Table 3a, page 1689, shows the maximum socket wrench dimensions for standard $^1/_4$-, $^1/_2$- and $^3/_4$-inch drive socket sets. For a given socket size (nominal size equals the maximum width across the flats of nut or bolt head), the dimension *K* given in the table is the minimum counterbore diameter required to provide socket wrench clearance for access to the bolt or nut.

**Sintered Carbide Boring Tools.—**Industrial experience has shown that the shapes of tools used for boring operations need to be different from those of single-point tools ordinarily used for general applications such as lathe work. Accordingly, Section 5 of American National Standard ANSI B212.1-2002 (Withdrawn) gives standard sizes, styles and designations for four basic types of sintered carbide boring tools, namely: solid carbide square; carbide-tipped square; solid carbide round; and carbide-tipped round boring tools. In addition to these ready-to-use standard boring tools, solid carbide round and square unsharpened boring tool bits are provided.

*Style Designations for Carbide Boring Tools:* Table 1 shows designations used to specify the styles of American National Standard sintered carbide boring tools. The first letter denotes solid (S) or tipped (T). The second letter denotes square (S) or round (R). The side cutting edge angle is denoted by a third letter (A through H) to complete the style designation. Solid square and round bits with the mounting surfaces ground but the cutting edges unsharpened (Table 3) are designated using the same system except that the third letter indicating the side cutting edge angle is omitted.

**Table 1. American National Standard Sintered Carbide Boring Tools — Style Designations** *ANSI B212.1-2002 (Withdrawn)*

| Side Cutting Edge Angle *E* | | Boring Tool Styles | | | |
|---|---|---|---|---|---|
| Degrees | Designation | Solid Square (SS) | Tipped Square (TS) | Solid Round (SR) | Tipped Round (TR) |
| 0 | A | | TSA | | |
| 10 | B | | TSB | | |
| 30 | C | SSC | TSC | SRC | TRC |
| 40 | D | | TSD | | |
| 45 | E | SSE | TSE | SRE | TRE |
| 55 | F | | TSF | | |
| 90 (0° Rake) | G | | | | TRG |
| 90 (10° Rake) | H | | | | TRH |

*Size Designation of Carbide Boring Tools:* Specific sizes of boring tools are identified by the addition of numbers after the style designation. The first number denotes the diameter or square size in number of $^1/_{32}$nds for types SS and SR and in number of $^1/_{16}$ths for types TS and TR. The second number denotes length in number of 1/8ths for types SS and SR. For styles TRG and TRH, a letter "U" after the number denotes a semi-finished tool (cutting edges unsharpened). Complete designations for the various standard sizes of carbide boring tools are given in Table 2 through Table 7. In the diagrams in the tables, angles shown without tolerance are ± 1°.

*Examples of Tool Designation:* The designation TSC-8 indicates a carbide-tipped tool (T); square cross section (S); 30-degree side cutting edge angle (C); and $^8/_{16}$ or $^1/_2$ inch square size (8).

The designation SRE-66 indicates a solid carbide tool (S); round cross section (R); 45-degree side cutting edge angle (E); $^6/_{32}$ or $^3/_{16}$ inch diameter (6); and $^6/_8$ or $^3/_4$ inch long (6).

The designation SS-610 indicates a solid carbide tool (S); square cross section (S); $^6/_{32}$ or $^3/_{16}$ inch square size (6); $^{10}/_8$ or $1^1/_4$ inches long (10).

It should be noted in this last example that the absence of a third letter (from A to H) indicates that the tool has its mounting surfaces ground but that the cutting edges are unsharpened.

**Table 2. ANSI Carbide-Tipped Round General-Purpose Square-End Boring Tools Style TRG with 0° Rake and Style TRH with 10° Rake** *ANSI B212.1-2002 (Withdrawn)*

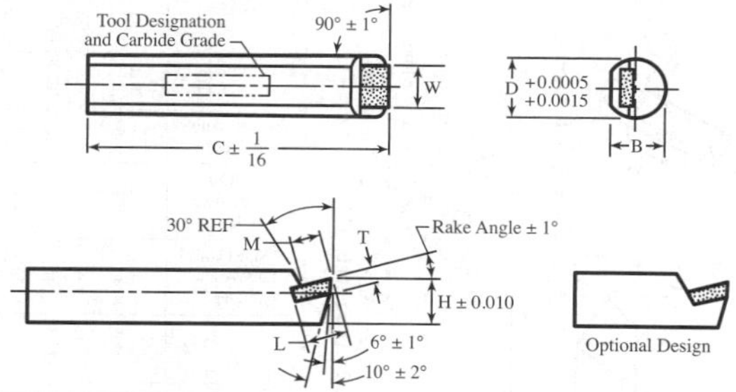


| Tool Designation | | Shank Dimensions, Inches | | | | | | | Tip Dimensions, Inches | | |
|---|---|---|---|---|---|---|---|---|---|---|---|
| Finished | Semi-finished[a] | Dia. *D* | Length *C* | Dim. Over Flat *B* | Nose Height *H* | Setback *M* (Min) | Rake Angle Deg. | Tip No. | *T* | *W* | *L* |
| TRG-5 | TRG-5U | 5/16 | 1½ | 19/64 ±.005 | 3/16 | 3/16 | 0 | 1025 | 1/16 | 1/4 | 1/4 |
| TRH-5 | TRH-5U | | | | 7/32 | 3/16 | 10 | | | | |
| TRG-6 | TRG-6U | 3/8 | 1¾ | 11/32 ±.010 | 7/32 | 3/16 | 0 | 1030 | 1/16 | 5/16 | 1/4 |
| TRH-6 | TRH-6U | | | | 1/4 | | 10 | | | | |
| TRG-7 | TRG-7U | 7/16 | 2½ | 13/32 ±.010 | 1/4 | 3/16 | 0 | 1080 | 3/32 | 5/16 | 3/8 |
| TRH-7 | TRH-7U | | | | 5/16 | | 10 | | | | |
| TRG-8 | TRG-8U | 1/2 | 2½ | 15/32 ±.010 | 9/32 | 1/4 | 0 | 1090 | 3/32 | 3/8 | 3/8 |
| TRH-8 | TRH-8U | | | | 11/32 | | 10 | | | | |

[a] Semifinished tool will be without Flat (B) and carbide unground on the end.

**Table 3. Solid Carbide Square and Round Boring Tool Bits**

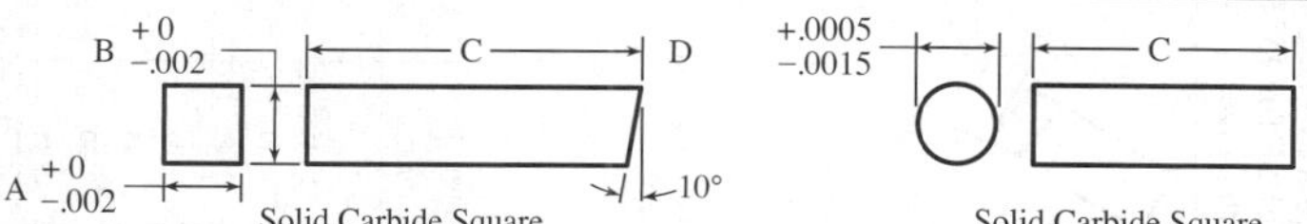


Solid Carbide Square Boring Tool Bits Style SS

Solid Carbide Square Boring Tool Bits Style SR

| Square Bits | | | | Round Bits | | | | | | | | |
|---|---|---|---|---|---|---|---|---|---|---|---|---|
| Tool Designation | *A* | *B* | *C* | Tool Designation | *D* | *C* | Tool Designation | *D* | *C* | Tool Designation | *D* | *C* |
| SS-58 | 5/32 | 5/32 | 1 | SR-33 | 3/32 | 3/8 | SR-55 | 5/32 | 5/8 | SR-88 | 1/4 | 1 |
| SS-610 | 3/16 | 3/16 | 1¼ | SR-34 | 3/32 | 1/2 | SR-64 | 3/16 | 1/2 | SR-810 | 1/4 | 1¼ |
| SS-810 | 1/4 | 1/4 | 1¼ | SR-44 | 1/8 | 1/2 | SR-66 | 3/16 | 3/4 | SR-1010 | 5/16 | 1¼ |
| SS-1012 | 5/16 | 5/16 | 1½ | SR-46 | 1/8 | 3/4 | SR-69 | 3/16 | 1⅛ | … | … | … |
| SS-1214 | 3/8 | 3/8 | 1¾ | SR-48 | 1/8 | 1 | SR-77 | 7/32 | 7/8 | … | … | … |

All dimensions are in inches.

*Tolerance on Length:* Through 1 inch, +1/32, −0; over 1 inch, +1/16, −0.

## Table 4. ANSI Solid Carbide Square Boring Tools Style SSC for 60° Boring Bar and Style SSE for 45° Boring Bar

*ANSI B212.1-2002 (Withdrawn)*

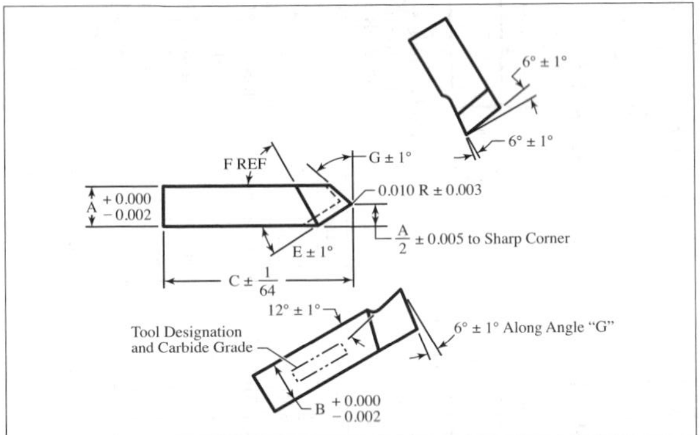


| Tool Designation | Boring Bar Angle, Deg. from Axis | Shank Dimensions, Inches: Width *A* | Height *B* | Length *C* | Side Cutting Edge Angle *E*, Deg. | End Cutting Edge Angle *G*, Deg. | Shoulder Angle *F*, Deg. |
|---|---|---|---|---|---|---|---|
| SSC-58 | 60 | 5⁄32 | 5⁄32 | 1 | 30 | 38 | 60 |
| SSE-58 | 45 | | | | 45 | 53 | 45 |
| SSC-610 | 60 | 3⁄16 | 3⁄16 | 1¼ | 30 | 38 | 60 |
| SSE-610 | 45 | | | | 45 | 53 | 45 |
| SSC-810 | 60 | ¼ | ¼ | 1¼ | 30 | 38 | 60 |
| SSE-810 | 45 | | | | 45 | 53 | 45 |
| SSC-1012 | 60 | 5⁄16 | 5⁄16 | 1½ | 30 | 38 | 60 |
| SSE-1012 | 45 | | | | 45 | 53 | 45 |

## Table 5. ANSI Carbide-Tipped Round Boring Tools Style TRC for 60° Boring Bar and Style TRE for 45° Boring Bar

*ANSI B212.1-2002 (Withdrawn)*

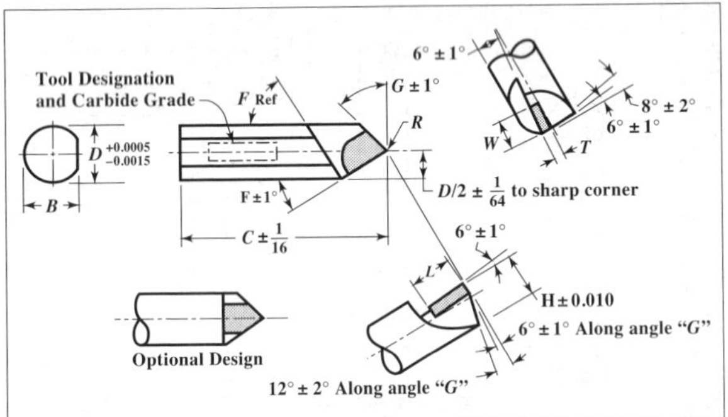


| Tool Designation | Bor. Bar Angle from Axis, Deg. | Shank Dimensions, Inches: *D* | *C* | *B* | *H* | *R* | Side Cut. Edge Angle *E*, Deg. | End Cut. Edge Angle *G*, Deg. | Shoulder Angle *F*, Deg. | Tip No. | Tip Dimensions, Inches: *T* | *W* | *L* |
|---|---|---|---|---|---|---|---|---|---|---|---|---|---|
| TRC-5 | 60 | 5⁄16 | 1½ | 19⁄64 ±.005 | 7⁄32 | 1⁄64 ±.005 | 30 | 38 | 60 | 2020 | 1⁄16 | 3⁄16 | ¼ |
| TRE-5 | 45 | | | | | | 45 | 53 | 45 | | | | |
| TRC-6 | 60 | ⅜ | 1¾ | 11⁄32 ±.010 | 9⁄32 | 1⁄64 ±.005 | 30 | 38 | 60 | 2040 | 3⁄32 | 3⁄16 | 5⁄16 |
| TRE-6 | 45 | | | | | | 45 | 53 | 45 | 2020 | 1⁄16 | 3⁄16 | ¼ |
| TRC-7 | 60 | 7⁄16 | 2½ | 13⁄32 ±.010 | 5⁄16 | 1⁄32 ±.010 | 30 | 38 | 60 | 2060 | 3⁄32 | ¼ | ⅜ |
| TRE-7 | 45 | | | | | | 45 | 53 | 45 | | | | |
| TRC-8 | 60 | ½ | 2½ | 15⁄32 ±.010 | ⅜ | 1⁄32 ±.010 | 30 | 38 | 60 | 2060 | 3⁄32 | ¼ | ⅜ |
| TRE-8 | 45 | | | | | | 45 | 53 | 45 | 2080 | 3⁄32 | 5⁄16 | ⅜ |

**Table 6. ANSI Carbide-Tipped Square Boring Tools—*ANSI B212.1-2002 (Withdrawn)* Styles TSA and TSB for 90° Boring Bar, Styles TSC and TSD for 60° Boring Bar, and Styles TSE and TSF for 45° Boring Bar**

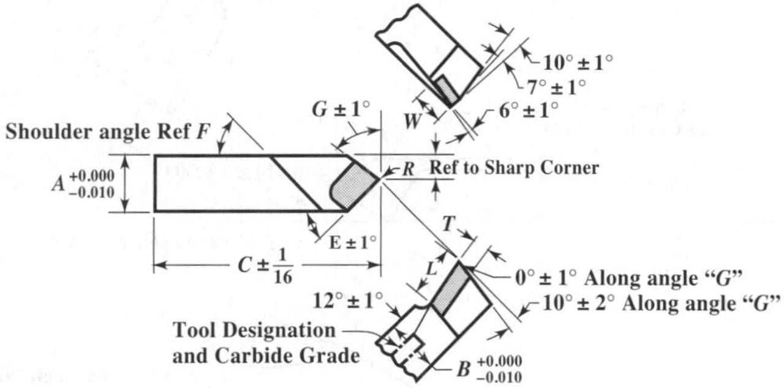


| Tool Designation | Bor. Bar Angle from Axis, Deg. | Shank Dimensions, Inches | | | | SideCut. Edge Angle E, Deg. | End Cut. Edge Angle G, Deg. | Shoulder Angle F, Deg. | Tip No. | Tip Dimensions, Inches | | |
|---|---|---|---|---|---|---|---|---|---|---|---|---|
| | | A | B | C | R | | | | | T | W | L |
| TSA-5 | 90 | 5/16 | 5/16 | 1 1/2 | 1/64 ± 0.005 | 0 | 8 | 90 | 2040 | 3/32 | 3/16 | 5/16 |
| TSB-5 | 90 | 5/16 | 5/16 | 1 1/2 | | 10 | 8 | 90 | 2040 | 3/32 | 3/16 | 5/16 |
| TSC-5 | 60 | 5/16 | 5/16 | 1 1/2 | | 30 | 38 | 60 | 2040 | 3/32 | 3/16 | 5/16 |
| TSD-5 | 60 | 5/16 | 5/16 | 1 1/2 | | 40 | 38 | 60 | 2040 | 3/32 | 3/16 | 5/16 |
| TSE-5 | 45 | 5/16 | 5/16 | 1 1/2 | | 45 | 53 | 45 | 2040 | 3/32 | 3/16 | 5/16 |
| TSF-5 | 45 | 5/16 | 5/16 | 1 1/2 | | 55 | 53 | 45 | 2040 | 3/32 | 3/16 | 5/16 |
| TSA-6 | 90 | 3/8 | 3/8 | 1 3/4 | | 0 | 8 | 90 | 2040 | 3/32 | 3/16 | 5/16 |
| TSB-6 | 90 | 3/8 | 3/8 | 1 3/4 | | 10 | 8 | 90 | 2040 | 3/32 | 3/16 | 5/16 |
| TSC-6 | 60 | 3/8 | 3/8 | 1 3/4 | | 30 | 38 | 60 | 2040 | 3/32 | 3/16 | 5/16 |
| TSD-6 | 60 | 3/8 | 3/8 | 1 3/4 | | 40 | 38 | 60 | 2040 | 3/32 | 3/16 | 5/16 |
| TSE-6 | 45 | 3/8 | 3/8 | 1 3/4 | | 45 | 53 | 45 | 2040 | 3/32 | 3/16 | 5/16 |
| TSF-6 | 45 | 3/8 | 3/8 | 1 3/4 | | 55 | 53 | 45 | 2040 | 3/32 | 3/16 | 5/16 |
| TSA-7 | 90 | 7/16 | 7/16 | 2 1/2 | 1/32 ± 0.010 | 0 | 8 | 90 | 2060 | 3/32 | 1/4 | 3/8 |
| TSB-7 | 90 | 7/16 | 7/16 | 2 1/2 | | 10 | 8 | 90 | 2060 | 3/32 | 1/4 | 3/8 |
| TSC-7 | 60 | 7/16 | 7/16 | 2 1/2 | | 30 | 38 | 60 | 2060 | 3/32 | 1/4 | 3/8 |
| TSD-7 | 60 | 7/16 | 7/16 | 2 1/2 | | 40 | 38 | 60 | 2060 | 3/32 | 1/4 | 3/8 |
| TSE-7 | 45 | 7/16 | 7/16 | 2 1/2 | | 45 | 53 | 45 | 2060 | 3/32 | 1/4 | 3/8 |
| TSF-7 | 45 | 7/16 | 7/16 | 2 1/2 | | 55 | 53 | 45 | 2060 | 3/32 | 1/4 | 3/8 |
| TSA-8 | 90 | 1/2 | 1/2 | 2 1/2 | | 0 | 8 | 90 | 2150 | 1/8 | 5/16 | 7/16 |
| TSB-8 | 90 | 1/2 | 1/2 | 2 1/2 | | 10 | 8 | 90 | 2150 | 1/8 | 5/16 | 7/16 |
| TSC-8 | 60 | 1/2 | 1/2 | 2 1/2 | | 30 | 38 | 60 | 2150 | 1/8 | 5/16 | 7/16 |
| TSD-8 | 60 | 1/2 | 1/2 | 2 1/2 | | 40 | 38 | 60 | 2150 | 1/8 | 5/16 | 7/16 |
| TSE-8 | 45 | 1/2 | 1/2 | 2 1/2 | | 45 | 53 | 45 | 2150 | 1/8 | 5/16 | 7/16 |
| TSF-8 | 45 | 1/2 | 1/2 | 2 1/2 | | 55 | 53 | 45 | 2150 | 1/8 | 5/16 | 7/16 |
| TSA-10 | 90 | 5/8 | 5/8 | 3 | 1/32 ± 0.010 | 0 | 8 | 90 | 2220 | 5/32 | 3/8 | 9/16 |
| TSB-10 | 90 | 5/8 | 5/8 | 3 | | 10 | 8 | 90 | 2220 | 5/32 | 3/8 | 9/16 |
| TSC-10 | 60 | 5/8 | 5/8 | 3 | | 30 | 38 | 60 | 2220 | 5/32 | 3/8 | 9/16 |
| TSD-10 | 60 | 5/8 | 5/8 | 3 | | 40 | 38 | 60 | 2220 | 5/32 | 3/8 | 9/16 |
| TSE-10 | 45 | 5/8 | 5/8 | 3 | | 45 | 53 | 45 | 2220 | 5/32 | 3/8 | 9/16 |
| TSF-10 | 45 | 5/8 | 5/8 | 3 | | 55 | 53 | 45 | 2220 | 5/32 | 3/8 | 9/16 |
| TSA-12 | 90 | 3/4 | 3/4 | 3 1/2 | | 0 | 8 | 90 | 2300 | 3/16 | 7/16 | 5/8 |
| TSB-12 | 90 | 3/4 | 3/4 | 3 1/2 | | 10 | 8 | 90 | 2300 | 3/16 | 7/16 | 5/8 |
| TSC-12 | 60 | 3/4 | 3/4 | 3 1/2 | | 30 | 38 | 60 | 2300 | 3/16 | 7/16 | 5/8 |
| TSD-12 | 60 | 3/4 | 3/4 | 3 1/2 | | 40 | 38 | 60 | 2300 | 3/16 | 7/16 | 5/8 |
| TSE-12 | 45 | 3/4 | 3/4 | 3 1/2 | | 45 | 53 | 45 | 2300 | 3/16 | 7/16 | 5/8 |
| TSF-12 | 45 | 3/4 | 3/4 | 3 1/2 | | 55 | 53 | 45 | 2300 | 3/16 | 7/16 | 5/8 |

**Table 7. ANSI Solid Carbide Round Boring Tools—*ANSI B212.1-2002 (Withdrawn)*
Style SRC for 60° Boring Bar and Style SRE for 45° Boring Bar**

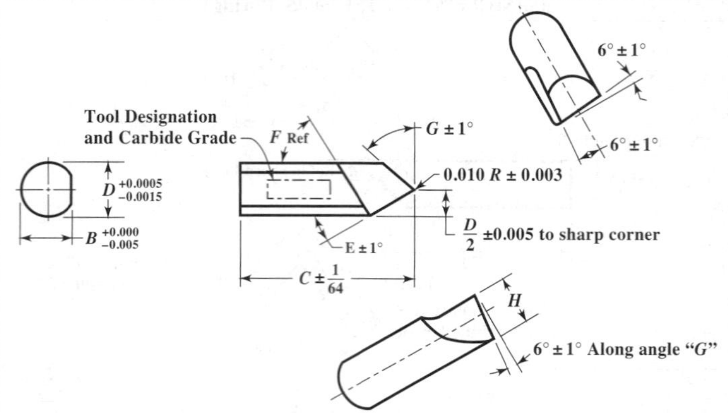


| Tool Designation | Bor. Bar Angle from Axis, Deg. | Shank Dimensions, Inches | | | | | Side Cut. Edge Angle *E*, Deg. | End Cut. Edge Angle *G*, Deg. | Shoulder Angle *F*, Deg. |
|---|---|---|---|---|---|---|---|---|---|
| | | Dia. *D* | Length *C* | Dim. Over Flat *B* | Nose Height *H* | | | | |
| SRC-33 | 60 | 3⁄32 | 3⁄8 | 0.088 | 0.070 | +0.000 −0.005 | 30 | 38 | 60 |
| SRE-33 | 45 | 3⁄32 | 3⁄8 | 0.088 | 0.070 | | 45 | 53 | 45 |
| SRC-44 | 60 | 1⁄8 | 1⁄2 | 0.118 | 0.094 | +0.000 −0.005 | 30 | 38 | 60 |
| SRE-44 | 45 | 1⁄8 | 1⁄2 | 0.118 | 0.094 | | 45 | 53 | 45 |
| SRC-55 | 60 | 5⁄32 | 5⁄8 | 0.149 | 0.117 | ±0.005 | 30 | 38 | 60 |
| SRE-55 | 45 | 5⁄32 | 5⁄8 | 0.149 | 0.117 | ±0.005 | 45 | 53 | 45 |
| SRC-66 | 60 | 3⁄16 | 3⁄4 | 0.177 | 0.140 | ±0.005 | 30 | 38 | 60 |
| SRE-66 | 45 | 3⁄16 | 3⁄4 | 0.177 | 0.140 | ±0.005 | 45 | 53 | 45 |
| SRC-88 | 60 | 1⁄4 | 1 | 0.240 | 0.187 | ±0.005 | 30 | 38 | 60 |
| SRE-88 | 45 | 1⁄4 | 1 | 0.240 | 0.187 | ±0.005 | 45 | 53 | 45 |
| SRC-1010 | 60 | 5⁄16 | 1¼ | 0.300 | 0.235 | ±0.005 | 30 | 38 | 60 |
| SRE-1010 | 45 | 5⁄16 | 1¼ | 0.300 | 0.235 | ±0.005 | 45 | 53 | 45 |

**Boring Machines, Origin.**—The first boring machine was built by John Wilkinson in 1775. John Smeaton had built one in 1769 which had a large rotary head, with inserted cutters, carried on the end of a light, overhanging shaft. The cylinder to be bored was fed forward against the cutter on a rude carriage, running on a track laid in the floor. The cutter head followed the inaccuracies of the bore, doing little more than smooth out local roughness of the surface. Watt's first steam cylinders were bored on this machine and he complained that one, 18 inches in diameter, was 3⁄8 inch out of true. Wilkinson thought of the expedient, which had escaped both Smeaton and Watt, of extending the boring-bar completely through the cylinder and giving it an out-board bearing, at the same time making it much larger and stiffer. With this machine, cylinders 57 inches in diameter were bored which were within 1⁄16 inch of true. Its importance can hardly be overestimated as it ensured the commercial success of Watt's steam engine, which, up to that time, had not passed the experimental stage.

## TAPS

A tap is a mechanical device applied to make a standard thread on a hole. A range of tap pitch diameter (PD) limits, from which the user may select to suit local conditions, is available. Taps included in ANSI/ASME B94.9-2008 (R2018) are categorized according to type, style, size and chamfer, and blank design. General dimensions and tap markings are given in ANSI/ASME B94.9 "Taps: Ground Thread With Cut Thread Appendix (Inch and Metric Sizes)" for straight-fluted taps, spiral-pointed taps, spiral-pointed-only taps, spiral-fluted taps, fast spiral-fluted taps, thread-forming taps, pulley taps, nut taps, and pipe taps. The Standard also gives the thread limits for taps with cut threads and ground threads. The tap thread limits and tolerances are given in Table 2 to Table 4, tap dimensions for cut thread and ground thread are given in Table 5a through Table 10. Pulley tap dimensions and tolerances are given in Table 12, straight and taper pipe thread tap dimensions and tolerances are given on Table 13a and Table 13b, and thread limits for cut thread and ground thread taps are given in Table 15 through Table 26a.

### Thread Form, Styles, and Types

**Thread Form.—**The basic angle of thread between the flanks of thread measured in an axial plane is 60 degrees. The line bisecting this 60° angle is perpendicular to the axis of the screw thread. The symmetrical height of the thread form, $h$, is found as follows:

$$h = 0.64951905P = \frac{0.64951905}{n} \qquad (1)$$

The basic pitch diameter (PD) is obtained by subtracting the symmetrical single thread height, $h$, from the basic major diameter as follows:

$$\text{Basic Pitch Diameter} = D_{bsc} - h \qquad (2)$$

$D_{bsc}$ = basic major diameter
$P$ = pitch of thread
$h$ = symmetrical height of thread
$n$ = number of threads per inch

**Types and Styles of Taps.—**Tap *type* is based on general dimensions such as standard straight thread, taper and straight pipe, pulley, etc., or is based on purpose, such as thread forming and screw thread inserts (STI).

Tap *style* is based on flute construction for cutting taps, such as straight, spiral, or spiral point, and on lobe style and construction for forming taps, such as straight or spiral.

*Straight-Flute Taps:* These taps have straight flutes of a number specified as either standard or optional, and are for general purpose applications. This standard applies to machine screw, fractional, metric, and STI sizes in high-speed steel ground thread, and to machine screw and fractional sizes in high speed and carbon steel cut thread, with taper, plug, semibottom, and bottom chamfer.

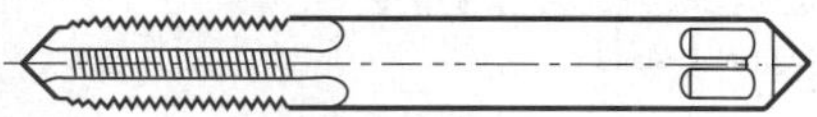

**BLANK Design 1**

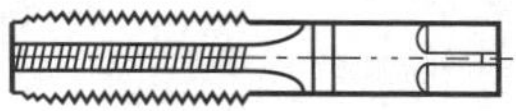

**BLANK Design 3**

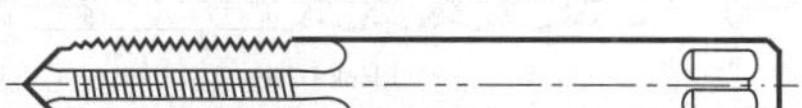

**BLANK Design 2**

*Spiral-Pointed Taps:* These taps have straight flutes, and the cutting face of the first few threads is ground at an angle to force the chips ahead and prevent clogging in the flutes. This standard applies to machine screw, fractional, metric, and STI sizes in high-speed steel ground thread, and to cut thread in machine screw and fractional sizes with plug, semibottom, and bottom chamfer.

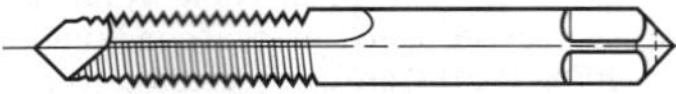

Blank Design 1

Blank Design 2

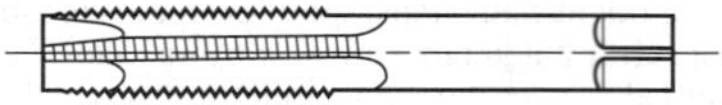

Blank Design 3

*Spiral-Pointed-Only Taps:* These taps are made with the spiral point feature only, without longitudinal flutes. These taps are especially suitable for tapping thin materials. This standard applies to machine screw and fractional sizes in high-speed steel, ground thread, with plug chamfer.

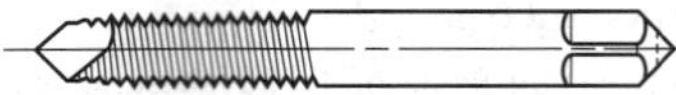

Blank Design 1

Blank Design 2

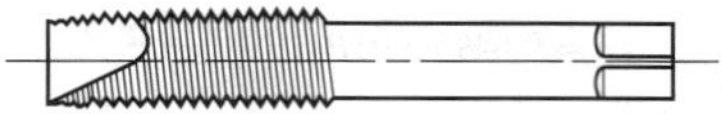

Blank Design 3

*Spiral-Fluted Taps:* These taps have right-hand helical flutes with a helix angle of 25 to 35 degrees. These features are designed to help draw chips from the hole or to bridge a keyway. This standard applies to machine screw, fractional, metric, and STI sizes in high-speed steel and to ground thread with plug, semibottom, and bottom chamfer.

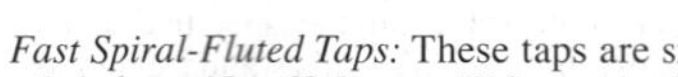

Blank Design 1

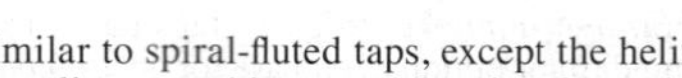

Blank Design 2

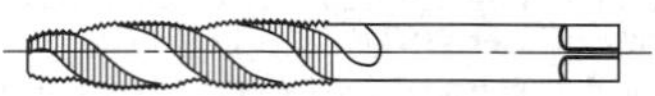

Blank Design 3

*Fast Spiral-Fluted Taps:* These taps are similar to spiral-fluted taps, except the helix angle is from 45 to 60 degrees.This standard applies to machine screw, fractional, metric, and STI sizes in high-speed steel with plug, semibottom, and bottom chamfer.

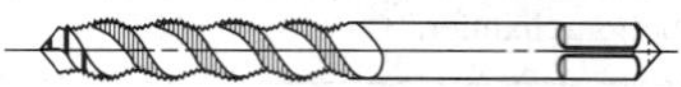

Blank Design 1

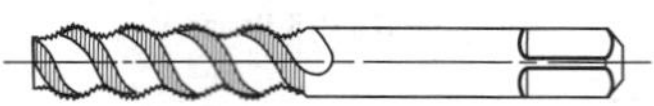

Blank Design 2

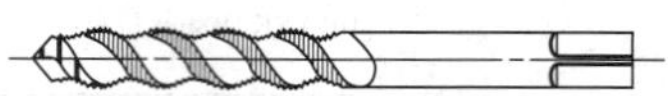

Blank Design 3

*Thread-Forming Taps:* These taps are fluteless except as optionally designed with one or more lubricating grooves. The thread form on the tap is lobed, so that there are a finite number of points contacting the work thread form. The tap does not cut, but forms the thread by extrusion. This standard applies to machine screw, fractional, and metric sizes, in high-speed steel, ground thread form, with plug, semibottom, and bottom entry taper.

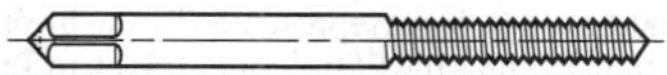

Blank Design 1

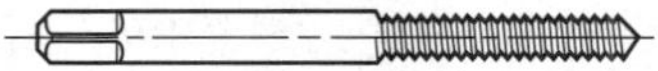

Blank Design 2

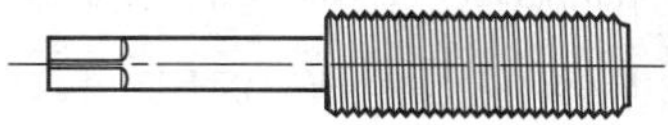

Blank Design 3

*Pulley Taps:* These taps were originally designed for tapping line-shaft pulleys by hand. Today, these taps have shanks that are extended in length by a standard amount for use where added reach is required. The shank is the same nominal diameter as the thread. This standard applies to fractional size and ground thread with plug and bottom chamfer.

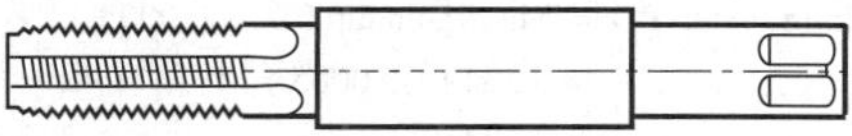

*Pipe Taps:* These taps are used to produce standard straight or tapered pipe threads. This standard applies to fractional size in high-speed steel, ground thread, to high-speed steel and carbon steel in cut thread, and to straight pipe taps having plug chamfers and taper pipe taps.

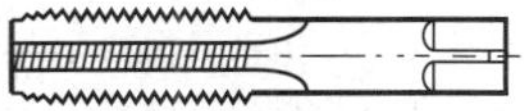

**Standard System of Tap Marking.—**Ground thread taps specified in the US customary system are marked with the nominal size, number of threads per inch, the proper symbol to identify the thread form, "HS" for high-speed steel, "G" for ground thread, and designators for tap pitch diameter and special features, such as left-hand and multi-start threads.

Cut thread taps specified in the US customary system are marked with the nominal size, number of threads per inch, and the proper symbol to identify the thread form. High-speed steel taps are marked "HS," but carbon steel taps need not be marked.

Ground thread taps made with metric screw threads (M profile) are marked with "M," followed by the nominal size and pitch in millimeters, separated by "X." Marking also includes "HS" for high-speed steel, "G" for ground thread, designators for tap pitch diameter and special features, such as left-hand and multi-start threads.

Thread symbol designators are listed in the accompanying table. Tap pitch diameter designators, systems of limits, special features, and examples for ground threads are given in the following section.

**Standard System of Tap Thread Limits and Identification for Unified Inch Screw Threads, Ground Thread.—***H or L Limits:* For Unified inch screw threads, when the maximum tap pitch diameter is over basic pitch diameter by an even multiple of 0.0005 inches, or the minimum tap pitch diameter limit is under basic pitch diameter by an even multiple of 0.0005 inches, the taps are marked "H" or "L," respectively, followed by a limit number, determined as follows:

$$\text{H Limit number} = \frac{\text{Tap PD} - \text{Basic PD}}{0.0005}$$

$$\text{L Limit number} = \frac{\text{Basic PD} - \text{Tap PD}}{0.0005}$$

The tap PD tolerances for ground threads are given in Table 2, column D; PD tolerances for cut threads are given in Table 3, column D. For standard taps, the PD limits for various H limit numbers are given in Table 20. The minimum tap PD equals the basic PD minus the number of half-thousandths (0.0005 in.) represented by the limit number. The maximum tap PD equals the minimum PD plus the PD tolerance given in Table 20.

*Tap Marking with H or L Limit Numbers*

*Example 1:* $^3\!/_8$-16 NC HS H1

$$
\begin{aligned}
\text{Maximum tap PD} &= \text{Basic PD} + 0.0005 \\
&= \frac{3}{8} - \left(0.64951904 \times \frac{1}{16}\right) + 0.0005 \\
&= 0.3344 + 0.0005 \\
&= 0.3349 \\
\text{Minimum tap PD} &= \text{Maximum tap PD} - 0.0005 \\
&= 0.3349 - 0.0005 \\
&= 0.3344
\end{aligned}
$$

*Example 2:* $^3\!/_8$-16 NC HS G L2

$$
\begin{aligned}
\text{Minimum tap PD} &= \text{Basic PD} - 0.0010 \\
&= \frac{3}{8} - \left(0.64951904 \times \frac{1}{16}\right) - 0.0010 \\
&= 0.3344 - 0.0010 \\
&= 0.3334 \\
\text{Maximum tap PD} &= \text{Minimum tap PD} + 0.0005 \\
&= 0.3334 + 0.0005 \\
&= 0.3339
\end{aligned}
$$

*Oversize or Undersize:* When the maximum tap PD over basic PD or the minimum tap PD under basic PD is not an even multiple of 0.0005, the tap PD is usually designated as an amount oversize or undersize. The amount oversize is added to the basic PD to establish the *minimum* tap PD. The amount undersize is subtracted from the basic PD to establish the *minimum* tap PD. The PD tolerance from Table 2 is added to the minimum tap PD to establish the maximum tap PD for both.

*Example:* $^7\!/_{16}$-14 NC plus 0.0017 HS G

Min. tap PD = Basic PD + 0.0017 in.

Max. tap PD = Min. tap PD + 0.0005 in.

Whenever possible for oversize or other special tap PD requirements, the maximum and minimum tap PD requirements should be specified.

*Special Tap Pitch Diameter:* Taps not made to H or L limit numbers, to the specifications in, or to the formula for oversize or undersize taps may be marked with the letter "S" enclosed by a circle or by some other special identifier. Example: $^1\!/_2$-16 NC HS G.

*Left-Hand Taps:* Taps with left-hand threads are marked "LEFT HAND" or "LH."

*Example:* $^3\!/_8$-16 NC LH HS G H3.

## Table 1. Thread Series Designations

| Standard Tap Marking | Product Thread Designation | Third Series | American National Standard References |
|---|---|---|---|
| M | M | Metric Screw Threads—M Profile, with basic ISO 68 profile | B1.13M |
| M | MJ | Metric Screw Threads: MJ Profile, with rounded root of radius 0.15011$P$ to 0.18042$P$ (external thread only) | |
| | | Class 5 interference-fit thread | |
| NC | NC5IF | Entire ferrous material range | B1.12 |
| NC | NC5INF | Entire nonferrous material range | B1.12 |
| NPS | NPSC | American National Standard straight pipe threads in pipe couplings | B1.20.1 |
| NPSF | NPSF | Dryseal American National Standard fuel internal straight pipe threads | B1.20.3 |
| NPSH | NPSH | American National Standard straight hose coupling threads for joining to American National Standard taper pipe threads | B1.20.7 |
| NPSI | NPSI | Dryseal American National Standard intermediate internal straight pipe threads | B1.20.3 |
| NPSL | NPSL | American National Standard straight pipe threads for loose-fitting mechanical joints with locknuts | B1.20.1 |
| NPSM | NPSM | American National Standard straight pipe threads for free-fitting mechanical joints for fixtures | B1.20.1 |
| ANPT | ANPT | Pipe threads, taper, aeronautical, national form | AS-71051 |
| NPT | NPT | American National Standard taper pipe threads for general use | B1.20.1 |
| NPTF | NPTF | Dryseal American National Standard taper pipe threads | B1.20.3 |
| NPTR | NPTR | American National Standard taper pipe threads for railing joints | B1.20.1 |
| PTF | PTF | Dryseal American National Standard pipe threads | B1.20.3 |
| PTF-SPL | PTF-SPL | Dryseal American National Standard pipe threads | B1.20.3 |
| STI | STI | Helical coil screw thread inserts-free running and screw locking (inch series) | B18.29.1 |
| | | **Unified Inch Screw Thread** | |
| N | UN | Constant-pitch series | B1.1 |
| NC | UNC | Coarse pitch series | B1.1 |
| NF | UNF | Fine pitch series | B1.1 |
| NEF | UNEF | Extra-fine pitch series | B1.1 |
| N | UNJ | Constant-pitch series, with rounded root of radius 0.15011$P$ to 0.18042$P$ (external thread only) | AS-8879 |
| NC | UNJC | Coarse pitch series, with rounded root of radius 0.15011$P$ to 0.18042 $P$ (external thread only) | B1.15<br>AS-8879 |
| NF | UNJF | Fine pitch series, with rounded root of radius 0.15011$P$ to 0.18042$P$ (external thread only) | B1.15<br>AS-8879 |
| NEF | UNJEF | Extra-fine pitch series, with rounded root of radius 0.15011$P$ to 0.18042$P$ (external thread only) | B1.15<br>AS-8879 |
| N | UNR | Constant-pitch series, with rounded root of radius not less than 0.108$P$ (external thread only) | B1.1 |
| NC | UNRC | Coarse thread series, with rounded root of radius not less than 0.108$P$ (external thread only) | B1.1 |
| NF | UNRF | Fine pitch series, with rounded root of radius not less than 0.108$P$ (external thread only) | B1.1 |
| NEF | UNREF | Extra-fine pitch series, with rounded root of radius not less than 0.108$P$ (external thread only) | B1.1 |
| NS | UNS | Special diameter pitch, or length of engagement | B1.1 |

## Table 2. Tap Thread Limits and Tolerances *ANSI/ASME B94.9-2008 (R2018)* Formulas for Unified Inch Screw Threads (Ground Thread)

Max. Major Diameter = Basic Diameter + $A$ Max. Pitch Diameter = Basic Diameter + $C$

Min. Major Diameter = Max. Maj. Dia. − $B$ Min. Pitch Diameter = Max. Pitch Dia. − $D$

$A$ = Constant to add = $0.130P$ for all pitches
$B$ = Major diameter tolerance= $0.087P$ for 48 to 80 TPI; $0.076P$ for 36 to 47 TPI; $0.065P$ for 4 to 35 TPI
$C$ = Amount over basic for maximum pitch diameter
$D$ = Pitch Diameter Tolerance

| Threads per Inch | *A* | *B* | *C* | | | | | *D* | | | | |
|---|---|---|---|---|---|---|---|---|---|---|---|---|
| | | | 0 to ⅝ | ⅝ to 1 | 1 to 1½ | 1½ to 2½ | Over 2½ | 0 to ⅝ | ⅝ to 1 | 1 to 1½ | 1½ to 2½ | Over 2½ |
| 80 | 0.0016 | 0.0011 | 0.0010 | 0.0015 | 0.0020 | 0.0020 | 0.0030 | 0.0005 | 0.0005 | 0.0010 | 0.0010 | 0.0015 |
| 72 | 0.0018 | 0.0012 | 0.0010 | 0.0015 | 0.0020 | 0.0020 | 0.0030 | 0.0005 | 0.0005 | 0.0010 | 0.0010 | 0.0015 |
| 64 | 0.0020 | 0.0014 | 0.0010 | 0.0015 | 0.0020 | 0.0020 | 0.0030 | 0.0005 | 0.0005 | 0.0010 | 0.0010 | 0.0015 |
| 56 | 0.0023 | 0.0016 | 0.0010 | 0.0015 | 0.0020 | 0.0020 | 0.0030 | 0.0005 | 0.0005 | 0.0010 | 0.0010 | 0.0015 |
| 48 | 0.0027 | 0.0018 | 0.0010 | 0.0015 | 0.0020 | 0.0020 | 0.0030 | 0.0005 | 0.0005 | 0.0010 | 0.0010 | 0.0015 |
| 44 | 0.0030 | 0.0017 | 0.0010 | 0.0015 | 0.0020 | 0.0020 | 0.0030 | 0.0005 | 0.0005 | 0.0010 | 0.0010 | 0.0015 |
| 40 | 0.0032 | 0.0019 | 0.0010 | 0.0015 | 0.0020 | 0.0020 | 0.0030 | 0.0005 | 0.0005 | 0.0010 | 0.0010 | 0.0015 |
| 36 | 0.0036 | 0.0021 | 0.0010 | 0.0015 | 0.0020 | 0.0020 | 0.0030 | 0.0005 | 0.0005 | 0.0010 | 0.0010 | 0.0015 |
| 32 | 0.0041 | 0.0020 | 0.0015 | 0.0015 | 0.0020 | 0.0020 | 0.0030 | 0.0005 | 0.0005 | 0.0010 | 0.0010 | 0.0015 |
| 28 | 0.0046 | 0.0023 | 0.0015 | 0.0015 | 0.0020 | 0.0020 | 0.0030 | 0.0005 | 0.0005 | 0.0010 | 0.0010 | 0.0015 |
| 24 | 0.0054 | 0.0027 | 0.0015 | 0.0015 | 0.0020 | 0.0025 | 0.0030 | 0.0005 | 0.0005 | 0.0010 | 0.0015 | 0.0015 |
| 20 | 0.0065 | 0.0032 | 0.0015 | 0.0015 | 0.0020 | 0.0025 | 0.0030 | 0.0005 | 0.0005 | 0.0010 | 0.0015 | 0.0015 |
| 18 | 0.0072 | 0.0036 | 0.0015 | 0.0015 | 0.0020 | 0.0025 | 0.0030 | 0.0005 | 0.0005 | 0.0010 | 0.0015 | 0.0015 |
| 16 | 0.0081 | 0.0041 | 0.0015 | 0.0015 | 0.0020 | 0.0025 | 0.0035 | 0.0005 | 0.0005 | 0.0010 | 0.0015 | 0.0020 |
| 14 | 0.0093 | 0.0046 | 0.0015 | 0.0020 | 0.0025 | 0.0030 | 0.0035 | 0.0005 | 0.0005 | 0.0010 | 0.0015 | 0.0020 |
| 13 | 0.0100 | 0.0050 | 0.0015 | 0.0020 | 0.0025 | 0.0030 | 0.0035 | 0.0005 | 0.0005 | 0.0010 | 0.0015 | 0.0020 |
| 12 | 0.0108 | 0.0054 | 0.0015 | 0.0020 | 0.0025 | 0.0030 | 0.0035 | 0.0005 | 0.0005 | 0.0010 | 0.0015 | 0.0020 |
| 11 | 0.0118 | 0.0059 | 0.0015 | 0.0020 | 0.0025 | 0.0030 | 0.0040 | 0.0005 | 0.0005 | 0.0010 | 0.0015 | 0.0020 |
| 10 | 0.0130 | 0.0065 | … | 0.0020 | 0.0025 | 0.0030 | 0.0040 | … | 0.0005 | 0.0010 | 0.0015 | 0.0020 |
| 9 | 0.0144 | 0.0072 | … | 0.0020 | 0.0025 | 0.0030 | 0.0040 | … | 0.0005 | 0.0010 | 0.0015 | 0.0020 |
| 8 | 0.0162 | 0.0081 | … | 0.0020 | 0.0025 | 0.0030 | 0.0040 | … | 0.0005 | 0.0010 | 0.0015 | 0.0020 |
| 7 | 0.0186 | 0.0093 | … | 0.0025 | 0.0025 | 0.0035 | 0.0045 | … | 0.0010 | 0.0010 | 0.0020 | 0.0025 |
| 6 | 0.0217 | 0.0108 | … | 0.0025 | 0.0025 | 0.0035 | 0.0045 | … | 0.0010 | 0.0010 | 0.0020 | 0.0025 |
| 5½ | 0.0236 | 0.0118 | … | 0.0025 | 0.0030 | 0.0035 | 0.0045 | … | 0.0010 | 0.0015 | 0.0020 | 0.0025 |
| 5 | 0.0260 | 0.0130 | … | 0.0025 | 0.0030 | 0.0035 | 0.0045 | … | 0.0010 | 0.0015 | 0.0020 | 0.0025 |
| 4½ | 0.0289 | 0.0144 | … | 0.0025 | 0.0030 | 0.0035 | 0.0045 | … | 0.0010 | 0.0015 | 0.0020 | 0.0025 |
| 4 | 0.0325 | 0.0162 | … | 0.0025 | 0.0030 | 0.0035 | 0.0045 | … | 0.0010 | 0.0015 | 0.0020 | 0.0025 |

Dimensions are given in inches.

The tables and formulas are used in determining the limits and tolerances for ground thread taps having a thread lead angle not in excess of 5°, unless otherwise specified.

The tap major diameter must be determined from a specified tap pitch diameter: the minimum major diameter equals the maximum specified tap pitch diameter minus constant $C$, plus $0.64951904P$ plus constant $A$.

$$\text{Minimum Major Diameter} = \text{Tap Pitch Diameter} - C + 0.64951904P + A$$

For intermediate pitches, use value of next coarser pitch; for $C$ and $D$ use formulas for $A$ and $B$.

Lead Tolerance: ± 0.0005 inch within any two threads not farther apart than 1 inch.

Angle Tolerance: ± 20′ in half angle for 4 to 5½ pitch; ± 25′ in half angle for 6 to 9 pitch; and ± 30′ in half angle for 10 to 80 pitch.

**Table 3. Tap Thread Limits and Tolerances** *ANSI/ASME B94.9-2008 (R2018)*
**Formulas for Unified Inch Screw Threads (Cut Thread)**

Min. Major Diameter = Basic Diameter + $B + C$ Min. Pitch Diameter = Basic Diameter + $B$

Max. Major Diameter = Min. Maj. Dia. + $A$ Max. Pitch Diameter = Min. Pitch Dia. + $D$

$A$ = Major diameter tolerance
$B$ = Amount over basic for minimum pitch diameter
$C$ = A constant to add for major diameter: 20% of theoretical truncation for 2 to 5.5 threads per inch and 25% for 6 to 80 threads per inch
$D$ = Pitch diameter tolerance

| Diameter of Tap (Inch) | *A* | *B* | | *D* | |
|---|---|---|---|---|---|
| | | 36 or more TPI | 34 or less TPI | Coarser than N.F. | N.F. and Finer[a] |
| 0 to 0.099 | 0.0015 | 0.0002 | 0.0005 | 0.0010 | 0.0010 |
| 0.10 to 0.249 | 0.0020 | 0.0002 | 0.0005 | 0.0015 | 0.0015 |
| 1⁄4 to 3⁄8 | 0.0025 | 0.0005 | 0.0005 | 0.0020 | 0.0015 |
| 3⁄8 to 5⁄8 | 0.0030 | 0.0005 | 0.0005 | 0.0025 | 0.0020 |
| 5⁄8 to 3⁄4 | 0.0040 | 0.0005 | 0.0005 | 0.0030 | 0.0025 |
| 3⁄4 to 1 | 0.0040 | 0.0010 | 0.0010 | 0.0030 | 0.0025 |
| 1 to 1 1⁄2 | 0.0045 | 0.0010 | 0.0010 | 0.0035 | 0.0030 |
| 1 1⁄2 to 2 | 0.0055 | 0.0015 | 0.0015 | 0.0040 | 0.0030 |
| 2 to 2 1⁄4 | 0.0060 | 0.0015 | 0.0015 | 0.0045 | 0.0035 |
| 2 1⁄4 to 2 1⁄2 | 0.0060 | 0.0020 | 0.0020 | 0.0045 | 0.0035 |
| 2 1⁄2 to 3 | 0.0070 | 0.0020 | 0.0020 | 0.0050 | 0.0035 |
| over 3 | 0.0070 | 0.0025 | 0.0025 | 0.0055 | 0.0045 |

[a] Taps over 1 1⁄2 inches with 10 or more threads per inch have tolerances for N.F. and finer.

| Threads per Inch | *C* | Threads per Inch | *C* | Threads per Inch | *C* | Threads per Inch | *C* |
|---|---|---|---|---|---|---|---|
| 2 | 0.0217 | 7 | 0.0077 | 18 | 0.0030 | 36 | 0.0015 |
| 2 1⁄2 | 0.0173 | 8 | 0.0068 | 20 | 0.0027 | 40 | 0.0014 |
| 3 | 0.0144 | 9 | 0.0060 | 22 | 0.0025 | 48 | 0.0011 |
| 3 1⁄2 | 0.0124 | 10 | 0.0054 | 24 | 0.0023 | 50 | 0.0011 |
| 4 | 0.0108 | 11 | 0.0049 | 26 | 0.0021 | 56 | 0.0010 |
| 4 1⁄2 | 0.0096 | 12 | 0.0045 | 27 | 0.0020 | 60 | 0.0009 |
| 5 | 0.0087 | 13 | 0.0042 | 28 | 0.0019 | 64 | 0.0008 |
| 5 1⁄2 | 0.0079 | 14 | 0.0039 | 30 | 0.0018 | 72 | 0.0008 |
| 6 | 0.0078 | 16 | 0.0034 | 32 | 0.0017 | 80 | 0.0007 |

| Angle Tolerance | | | | | |
|---|---|---|---|---|---|
| Threads per Inch | Deviation in Half angle | Deviation in Half angle | Threads per Inch | Deviation in Half angle | Deviation in Half angle |
| 4 and coarser | ± 30′ | ± 45′ | 10 to 28 | ± 45′ | ± 68′ |
| 4 1⁄2 to 5 1⁄2 | ± 35′ | ± 53′ | 30 and finer | ± 60′ | ± 90′ |
| 6 to 9 | ± 40′ | ± 60′ | | | |

Dimensions are given in inches.

The tables and formulas are used in determining the limits and tolerances for cut thread metric taps having special diameter, special pitch, or both.

For intermediate pitches, use value of next coarser pitch.

*Lead Tolerance:* ± 0.003 inch within any two threads not farther apart than 1 inch.

Taps over 1 1⁄2 in. with 10 or more threads per inch have tolerances for N.F. and finer.

**Standard System of Ground Thread Tap Limits and Identification for Metric Screw Threads, M Profile.—**All calculations for metric taps use millimeter values. When US customary values are needed, they are translated from the three-place millimeter tap diameters only after the calculations are completed.

**Table 4. Tap Thread Limits and Tolerances** *ANSI/ASME B94.9-2008 (R2018)*
**Formulas for Metric Thread (Ground Thread)**

Min. major diameter = Basic diameter + $W$ Max. pitch diameter = Basic diameter + $Y$

Max. major diameter = Min. maj. dia. + $X$ Min. pitch diameter = Max. pitch dia. + $Z$

$W$ = Constant to add with basic major diameter (W = 0.08P)

$X$ = Major diameter tolerance

$Y$ = Amount over basic for maximum pitch diameter

$Z$ = Pitch diameter tolerance

| | | | *Y* | | | | *Z* | | | |
|---|---|---|---|---|---|---|---|---|---|---|
| *P* Pitch (mm) | *W* (0.08*P*) | *X* | M1.6 to M6.3 | Over M6.3 to to M25 | Over M25 to M90 | Over M90 | M1.6 to M6.3 | Over M6.3 to to M25 | Over M25 to M90 | Over M90 |
| 0.30 | 0.024 | 0.025 | 0.039 | 0.039 | 0.052 | 0.052 | 0.015 | 0.015 | 0.020 | 0.020 |
| 0.35 | 0.028 | 0.025 | 0.039 | 0.039 | 0.052 | 0.052 | 0.015 | 0.015 | 0.020 | 0.020 |
| 0.40 | 0.032 | 0.025 | 0.039 | 0.052 | 0.052 | 0.052 | 0.015 | 0.015 | 0.020 | 0.025 |
| 0.45 | 0.036 | 0.025 | 0.039 | 0.052 | 0.052 | 0.052 | 0.015 | 0.020 | 0.020 | 0.025 |
| 0.50 | 0.040 | 0.025 | 0.039 | 0.052 | 0.052 | 0.065 | 0.015 | 0.020 | 0.025 | 0.025 |
| 0.60 | 0.048 | 0.025 | 0.052 | 0.052 | 0.065 | 0.065 | 0.020 | 0.020 | 0.025 | 0.025 |
| 0.70 | 0.056 | 0.041 | 0.052 | 0.052 | 0.065 | 0.065 | 0.020 | 0.020 | 0.025 | 0.025 |
| 0.75 | 0.060 | 0.041 | 0.052 | 0.065 | 0.065 | 0.078 | 0.020 | 0.025 | 0.025 | 0.031 |
| 0.80 | 0.064 | 0.041 | 0.052 | 0.065 | 0.065 | 0.078 | 0.020 | 0.025 | 0.025 | 0.031 |
| 0.90 | 0.072 | 0.041 | 0.052 | 0.065 | 0.065 | 0.078 | 0.020 | 0.025 | 0.025 | 0.031 |
| 1.00 | 0.080 | 0.041 | 0.065 | 0.065 | 0.078 | 0.091 | 0.025 | 0.025 | 0.031 | 0.031 |
| 1.25 | 0.100 | 0.064 | 0.065 | 0.065 | 0.078 | 0.091 | 0.025 | 0.031 | 0.031 | 0.041 |
| 1.50 | 0.120 | 0.064 | 0.065 | 0.078 | 0.078 | 0.091 | 0.025 | 0.031 | 0.031 | 0.041 |
| 1.75 | 0.140 | 0.064 | … | 0.078 | 0.091 | 0.104 | … | 0.031 | 0.041 | 0.041 |
| 2.00 | 0.160 | 0.064 | … | 0.091 | 0.091 | 0.104 | … | 0.041 | 0.041 | 0.041 |
| 2.50 | 0.200 | 0.063 | … | 0.091 | 0.104 | 0.117 | … | 0.041 | 0.041 | 0.052 |
| 3.00 | 0.240 | 0.100 | … | 0.104 | 0.104 | 0.130 | … | 0.041 | 0.052 | 0.052 |
| 3.50 | 0.280 | 0.100 | … | 0.104 | 0.117 | 0.130 | … | 0.041 | 0.052 | 0.052 |
| 4.00 | 0.320 | 0.100 | … | 0.104 | 0.117 | 0.143 | … | 0.052 | 0.052 | 0.064 |
| 4.50 | 0.360 | 0.100 | … | … | 0.130 | 0.143 | … | 0.052 | 0.052 | 0.064 |
| 5.00 | 0.400 | 0.100 | … | … | 0.130 | 0.156 | … | … | 0.064 | 0.064 |
| 5.50 | 0.440 | 0.100 | … | … | 0.143 | 0.156 | … | … | 0.064 | 0.064 |
| 6.00 | 0.480 | 0.100 | … | … | 0.143 | 0.156 | … | … | 0.064 | 0.064 |

Dimensions are given in millimeters.

The tables and formulas are used in determining the limits and tolerances for ground thread metric taps having a thread lead angle not in excess of 5°, unless otherwise specified. They apply only to metric thread having a 60° form with a *P*/8 flat at the major diameter of the basic thread form. All calculations for metric taps are done using millimeter values as shown. When inch values are needed, they are translated from the three-place millimeter tap diameters only after calculations are performed.

The tap major diameter must be determined from a specified tap pitch diameter: the minimum major diameter equals the maximum specified tap pitch diameter minus constant $Y$, plus $0.64951905P$ plus constant $W$.

$$\text{Minimum major diameter} = \text{Max. tap pitch diameter} - Y + 0.64951904P + W$$

For intermediate pitches, use value of next coarser pitch.

Lead Tolerance: ±0.013 mm within any two threads not farther apart than 25 mm.

Angle Tolerance: ±30′ in half angle for 0.25 to 2.5 pitch; ±25′ in half angle for 2.5 to 4 pitch; and ±20′ in half angle for 4 to 6 pitch.

*D or DU Limits:* When the maximum tap pitch diameter is over basic pitch diameter by an even multiple of 0.013 mm (0.000512 in. reference), or the minimum tap pitch diameter limit is under basic pitch diameter by an even multiple of 0.013 mm, the taps are marked with the letters "D" or "DU," respectively, followed by a limit number. The limit number is determined as follows:

$$\text{D Limit number} = \frac{\text{Tap PD} - \text{Basic PD}}{0.0013}$$

$$\text{DU Limit number} = \frac{\text{Basic PD} - \text{Tap PD}}{0.0013}$$

*Example:* M1.6×0.35 HS G D3

$$\begin{aligned}\text{Maximum tap PD} &= \text{Basic PD} + 0.0039\\ &= 1.6 - (0.64951904 \times 0.35) + 0.0039\\ &= 1.3727 + 0.039\\ &= 1.412\\ \text{Minimum tap PD} &= \text{Maximum tap PD} - 0.015\\ &= 1.412 - 0.015\\ &= 1.397\end{aligned}$$

M6×1 HS G DU4

$$\begin{aligned}\text{Minimum tap PD} &= \text{Basic PD} - 0.052\\ &= 6 - (0.64951904 \times 1.0) - 0.052\\ &= 5.350 - 0.052\\ &= 5.298\\ \text{Maximum tap PD} &= \text{Minimum tap PD} + 0.025\\ &= 5.298 + 0.025\\ &= 5.323\end{aligned}$$

**Definitions of Tap Terms.—**The definitions that follow are taken from ANSI/ASME B94.9 but include only the more important terms. Some tap terms are the same as screw thread terms; therefore, see *Definitions of Screw Threads* starting on page 1945.

*Actual size:* The measured size of an element on an individual part.

*Allowance:* A prescribed difference between the maximum material limits of mating parts. It is the minimum clearance or maximum interference between such parts.

*Basic Size:* The size from which the limits are derived by application of allowance and tolerance.

*Bottom Top:* A tap having a chamfer length of 1 to 2 pitches.

*Chamfer:* Tapering of the threads at the front end of each land or chaser of a tap by cutting away and relieving the crest of the first few teeth to distribute the cutting action over several teeth.

*Chamfer Angle:* Angle formed between the chamfer and the axis of the tap measured in an axial plane at the cutting edge.

*Chamfer Relief:* The gradual degrees in land height from cutting edge to heel on the chamfered portion of the land to provide radial clearance for the cutting edge.

*Chamfer Relief Angle:* Complement of the angle formed between a tangent to the relieved surface at the cutting edge and a radial line to the same point on the cutting edge.

*Classes of Thread:* Designation of the class that determines the specification of the size, allowance, and tolerance to which a given threaded product is to be manufactured. It is not applicable to the tools used for threading.

*Concentric:* Having a common center.

*Crest:* The surface of the thread that joins the flanks of the thread and is farthest from the cylinder or cone from which the thread projects.

*Cutter Sweep:* The section removed by the milling cutter or the grinding wheel in entering or leaving a flute.

*Cutting Edge:* The intersection of cutting edge and the major diameter in the direction of rotation for cutting, which does the actual cutting.

*Core Diameter:* The diameter of a circle that is tangent to the bottom of the flutes at a given point on the axis.

*Diameter, Major:* The major cylinder on a straight thread.

*Diameter, Minor:* The minor cylinder on a straight thread.

*Dryseal:* A thread system used for both external and internal pipe threads applications designed for use where the assembled product must withstand high fluid or gas pressure without the use of sealing compound.

*Eccentric:* Not having a common center.

*Eccentricity:* One half of the total indicator variation (TIV) with respect to the tool axis.

*Entry Taper:* The portion of the thread forming, where the thread forming is tapered toward the front to allow entry into the hole to be tapped.

*External Center:* The pointed end on a tap. On bottom-chamfered taps, the point on the front end may be removed.

*Flank:* The flank of a thread is the surface connecting the crest with the root.

*Flank Angle:* Angle between the individual flank and the perpendicular to the axis of the thread, measured in an axial plane. A flank angle of a symmetrical thread is commonly termed the "half angle of thread."

*Flank, Leading:* 1) Flank of a thread facing toward the chamfered end of a threading tool; and 2) The leading flank of a thread is the one which, when the thread is about to be assembled with a mating thread, faces the mating thread.

*Flank, Trailing:* The trailing flank of a thread is the one opposite the leading flank.

*Flutes:* Longitudinal channels formed in a tap to create cutting edges on the thread profile and to provide chip spaces and cutting fluid passages. On a parallel or straight-thread tap, they may be straight, angular or helical; on a taper-thread tap, they may be straight, angular or spiral.

*Flute Lead Angle:* Angle at which a helical- or spiral-cutting edge at a given point makes with an axial plane through the same point.

*Flute, Spiral:* A flute with uniform axial lead in a spiral path around the axis of a conical tap.

*Flute, Straight:* A flute that forms a cutting edge lying in an axial plane.

*Flute, Tapered:* A flute lying in a plane intersecting the tool axis at an angle.

*Full Indicator Movement (FIM):* The total movement of an indicator where appropriately applied to a surface to measure its surface.

*Functional Size:* The functional diameter of an external or internal thread is the PD of the enveloping thread of perfect pitch, lead, and flank angles, having full depth of engagement but clear at crests and roots, and of a specified length of engagement.

*Heel:* Edge of the land opposite the cutting edge.

*Height of Thread:* The height of a thread is the distance, measured radially, between the major and minor cylinders or cones, respectively.

*Holes, Blind:* A hole that does not pass through the work piece and is not threaded to its full depth.

*Holes, Bottom:* A blind hole that is threaded close to the bottom.

*Hook Angle:* Inclination of a concave cutting face, usually specified either as Chordal Hook or Tangential Hook.

*Hook, Chordal Angle:* Angle between the chord passing through the root and crest of a thread form at the cutting face, and a radial line through the crest at the cutting edge.

*Hook, Tangential Angle:* Angle between a line tangent to a hook cutting face at the cutting edge and a radial line to the same point.

*Internal Center:* A countersink with clearance at the bottom, in one or both ends of a tool, which establishes the tool axis.

*Interrupted Thread Tap:* A tap having an odd number of lands with alternate teeth in the thread helix removed. In some designs, alternate teeth are removed only for a portion of the thread length.

*Land:* One of the threaded sections between the flutes of a tap.

*Lead:* Distance a screw thread advances axially in one complete turn.

*Lead Error:* Deviation from prescribed limits.

*Lead Deviation:* Deviation from the basic nominal lead.

*Progressive Lead Deviation:* (1) On a straight thread, the deviation from a true helix where the thread helix advances uniformly. (2) On a taper thread, the deviation from a true spiral where the thread spiral advances uniformly.

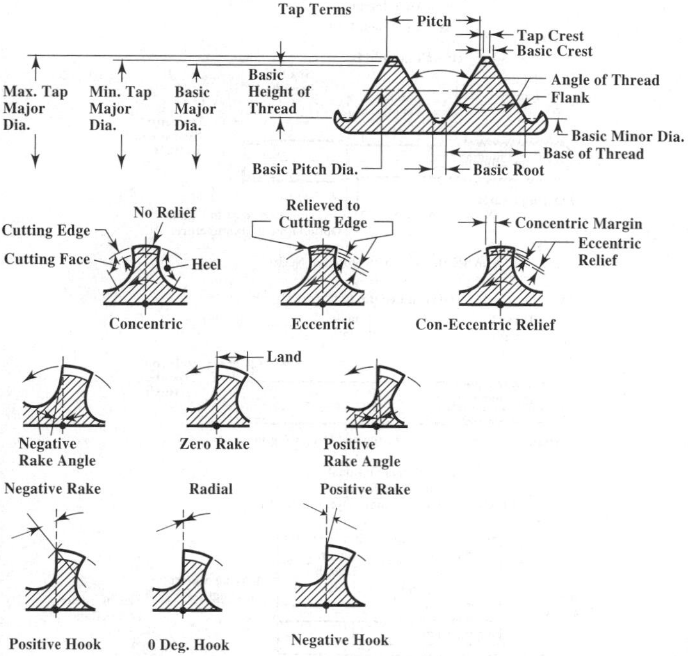


Fig. 3. Tap Terms

*Left-Hand Cut:* Rotation in a clockwise direction from cutting when viewed from the chamfered end of a tap.

*Length of Engagement:* The length of engagement of two mating threads is the axial distance over which two mating threads are designed to contact.

*Length of Thread:* The length of the thread of the tap includes the chamfered threads and the full threads, but does not include an external center. It is indicated by the letter "B" in the illustrations at the heads of the tables.

*Limits:* The limits of size are the applicable maximum and minimum sizes.

*Major Diameter:* On a straight thread, the major diameter is that of the major cylinder. On a taper thread, the major diameter at a given position on the thread axis is that of the major cone at that position.

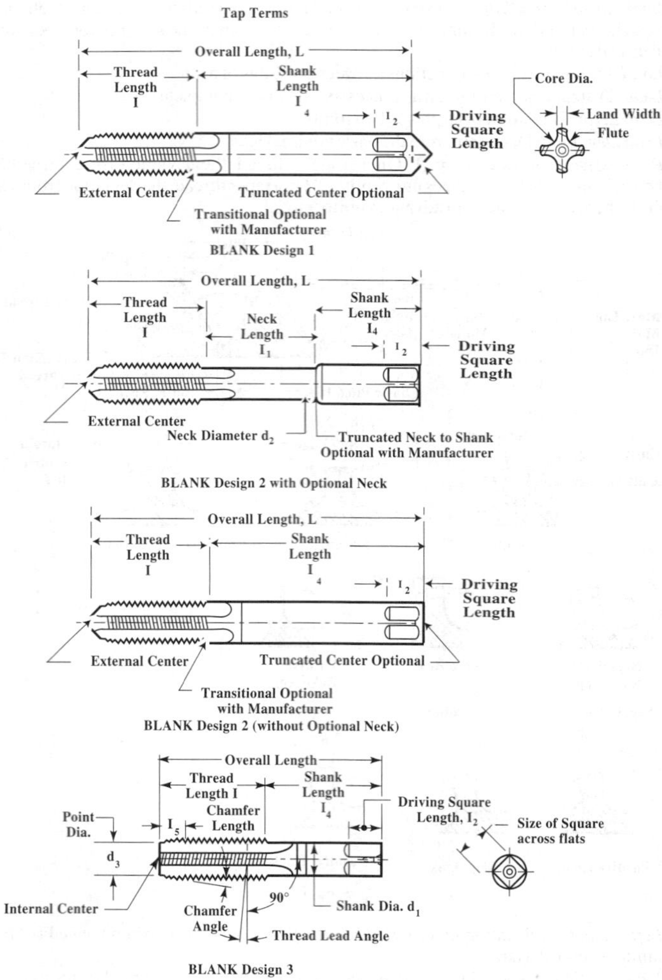


Fig. 4. Taps Terms

*Minor Diameter:* On a straight thread, the minor diameter is that of the minor cylinder. On a taper thread, the minor diameter at a given position on the thread axis is that of the minor cone at that position.

*Neck:* A section of reduced diameter between two adjacent portions of a tool.

*Pitch:* The distance from any point on a screw thread to a corresponding point in the next thread, measured parallel to the axis and on the same side of the axis.

*Pitch Diameter (Simple Effective Diameter):* On a straight thread, the pitch diameter is the diameter of the imaginary coaxial cylinder, the surface of which would pass through the thread profiles at such points as to make the width of the groove equal to one-half the basic pitch. On a perfect thread, this coincidence occurs at the point where the widths of the thread and groove are equal. On a taper thread, the pitch diameter at a given position on the thread axis is the diameter of the pitch cone at that position.

*Point Diameter:* Diameter at the cutting edge of the leading end of the chamfered section.

*Plug Tap:* A tap having a chamfer length of 3 to 5 pitches.

*Rake:* Angular relationship of the straight cutting face of a tooth with respect to a radial line through the crest of the tooth at the cutting edge. Positive rake means that the crest of the cutting face is angularly ahead of the balance of the cutting face of the tooth. Negative rake means that the crest of the cutting face is angularly behind the balance of the cutting face of the tooth. Zero rake means that the cutting face is directly on a radial line.

*Relief:* Removal of metal behind the cutting edge to provide clearance between the part being threaded and the threaded land.

*Relief, Center:* Clearance produced on a portion of the tap land by reducing the diameter of the entire thread form between cutting edge and heel.

*Relief, Chamfer:* Gradual decrease in land height from cutting edge to heel on the chamfered portion of the land on a tap to provide radial clearance for the cutting edge.

*Relief, Con-Eccentric Thread:* Radial relief in the thread form starting back of a concentric margin.

*Relief, Double Eccentric Thread:* Combination of a slight radial relief in the thread form starting at the cutting edge and continuing for a portion of the land width, and a greater radial relief for the balance of the land.

*Relief, Eccentric Thread:* Radial relief in the thread form starting at the cutting edge and continuing to the heel.

*Relief, Flatted Land:* Clearance produced on a portion of the tap land by truncating the thread between cutting edge and heel.

*Relief, Grooved Land:* Clearance produced on a tap land by forming a longitudinal groove in the center of the land.

*Relief, Radial:* Clearance produced by removal of metal from behind the cutting edge. Taps should have the chamfer relieved and should have back taper, but may or may not have relief in the angle and on the major diameter of the threads. When the thread angle is relieved, starting at the cutting edge and continuing to the heel, the tap is said to have "eccentric" relief. If the thread angle is relieved back of a concentric margin (usually one-third of land width), the tap is said to have "con-eccentric" relief.

*Right-Hand Cut:* Rotation in clockwise direction for cutting when viewed from the chamfered end of a tap or die.

*Roots:* The surface of the thread that joins the flanks of adjacent thread forms and is identical to cone from which the thread projects.

*Screw Thread:* A uniform section produced by forming a groove in the form of helix on the external or the internal surface of a cylinder.

*Screw Thread Inserts (STI):* Screw thread bushing coiled from diamond-shape cross-section wire. They are screwed into oversized tapped holes to size nominal-size internal threads.

*Screw Thread Insert (STI) Taps:* These taps are over the nominal size to the extent that the internal thread they produce will accommodate a helical coil screw insert, which at final assembly will accept a screw thread of the nominal size and pitch.

*Shank:* The portion of the tool body by which it is held and driven.

*Shaving:* The excessive removal of material from the product thread profile by the tool thread flanks caused by an axial advance per revolution less than or more than the actual lead in the tool.

*Size, Actual:* Measured size of an element on an individual part.

*Size, Basic:* That size from which the limits of size are derived by the application of allowances and tolerances.

*Size, Functional:* The functional diameter of an external or internal thread is the pitch diameter of the enveloping thread of perfect pitch, lead and flank angles, having full depth of engagement but clear at crests and roots, and of a specified length of engagement. It may be derived by adding to the pitch diameter in an external thread, or subtracting from the pitch diameter in an internal thread, the cumulative effects of deviations from specified profile, including variations in lead and flank angle over a specified length of engagement. The effects of taper, out-of-roundness, and surface defects may be positive or negative on either external or internal threads.

*Size, Nominal:* Designation used for the purpose of general identification.

*Spiral Flute:* See *Flutes*.

*Spiral Point:* Angular fluting in the cutting face of the land at the chamfered end. It is formed at an angle with respect to the tap axis of opposite hand to that of rotation. Its length is usually greater than the chamfer length and its angle with respect to the tap axis is usually made great enough to direct the chips ahead of the tap. The tap may or may not have longitudinal flutes.

*Taper, Back:* A gradual decrease in the diameter of the thread form on a tap from the chamfered end of the land towards the back, which creates a slight radial relief in the threads.

*Taper per Inch:* The difference in diameter in 1 inch measured parallel to the axis.

*Taper Tap:* A tap having a chamfer length of 7 to 10 pitches.

*Taper Thread Tap:* A tap with tapered threads for producing a tapered internal thread.

*Thread, Angle of:* The angle between the flanks of the thread measured in an axial plane.

*Thread Lead Angle:* On a straight thread, the lead angle is the angle made by the helix of the thread at the pitch line with a plane perpendicular to the axis. On a taper thread, the lead angle at a given axial position is the angle made by the conical spiral of the thread, with the plane perpendicular to the axis, at the pitch line.

*Thread per Inch:* The number of thread pitches in 1 inch of thread length.

*Tolerance:* The total permissible variation of size or difference between limits of size.

*Total Indicator Variation (TIV):* The difference between maximum and minimum indicator readings during a checking cycle.

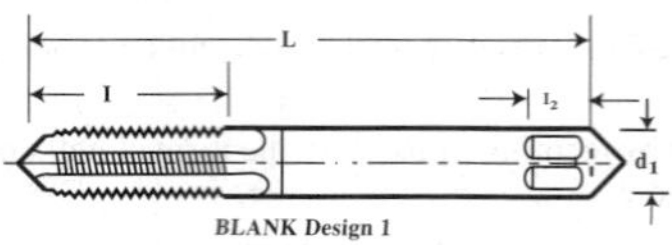


**BLANK Design 1**

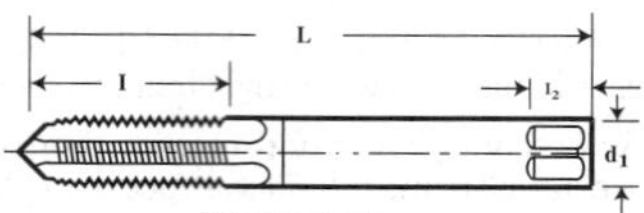


**BLANK Design 2**

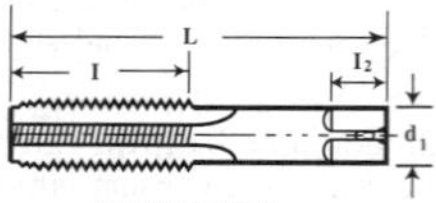


**BLANK Design 3**

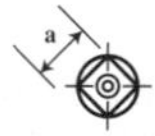

**Table 5a. Standard Tap Dimensions (Ground and Cut Thread)** *ANSI/ASME B94.9-2008 (R2018)*

| Nominal Diameter Range, inch | | Nominal Diameter, inch | | Nominal Metric Diameter | | | Tap Dimensions, inch | | | | |
|---|---|---|---|---|---|---|---|---|---|---|---|
| Over | To | Machine Screw Size No. and Fractional Sizes | Decimal Equiv. | mm | inch | Blank Design No. | Overall Length $L$ | Thread Length $I$ | Square Length $l_2$ | Shank Diameter $d_1$ | Size of Square $a$ |
| 0.052 | 0.065 | 0 | (0.0600) | M1.6 | 0.0630 | 1 | 1.63 | 0.31 | 0.19 | 0.141 | 0.110 |
| 0.065 | 0.078 | 1 | (0.0730) | M1.8 | 0.0709 | 1 | 1.69 | 0.38 | 0.19 | 0.141 | 0.110 |
| 0.078 | 0.091 | 2 | (0.0860) | M2.0 | 0.0787 | 1 | 1.75 | 0.44 | 0.19 | 0.141 | 0.110 |
| | | | | M2.2 | 0.0866 | | | | | | |
| 0.091 | 0.104 | 3 | (0.0990) | M2.5 | 0.0984 | 1 | 1.81 | 0.50 | 0.19 | 0.141 | 0.110 |
| 0.104 | 0.117 | 4 | (0.1120) | … | … | 1 | 1.88 | 0.56 | 0.19 | 0.141 | 0.110 |
| 0.117 | 0.130 | 5 | (0.1250) | M3.0 | 0.1182 | 1 | 1.94 | 0.63 | 0.19 | 0.141 | 0.110 |
| 0.130 | 0.145 | 6 | (0.1380) | M3.5 | 0.1378 | 1 | 2.00 | 0.69 | 0.19 | 0.141 | 0.110 |
| 0.145 | 0.171 | 8 | (0.1640) | M4.0 | 0.1575 | 1 | 2.13 | 0.75 | 0.25 | 0.168 | 0.131 |
| 0.171 | 0.197 | 10 | (0.1900) | M4.5 | 0.1772 | 1 | 2.38 | 0.88 | 0.25 | 0.194 | 0.152 |
| | | | | M5 | 0.1969 | | | | | | |
| 0.197 | 0.223 | 12 | (0.2160) | … | … | 1 | 2.38 | 0.94 | 0.28 | 0.220 | 0.165 |
| 0.223 | 0.260 | ¼ | (0.2500) | M6 | 0.2363 | 2 | 2.50 | 1.00 | 0.31 | 0.255 | 0.191 |
| 0.260 | 0.323 | 5⁄16 | (0.3125) | M7 | 0.2756 | 2 | 2.72 | 1.13 | 0.38 | 0.318 | 0.238 |
| | | | | M8 | 0.3150 | | | | | | |
| 0.323 | 0.395 | ⅜ | (0.3750) | M10 | 0.3937 | 2 | 2.94 | 1.25 | 0.44 | 0.381 | 0.286 |
| 0.395 | 0.448 | 7⁄16 | (0.4375) | … | … | 3 | 3.16 | 1.44 | 0.41 | 0.323 | 0.242 |
| 0.448 | 0.510 | ½ | (0.5000) | M12 | 0.4724 | 3 | 3.38 | 1.66 | 0.44 | 0.367 | 0.275 |
| 0.510 | 0.573 | 9⁄16 | (0.5625) | M14 | 0.5512 | 3 | 3.59 | 1.66 | 0.50 | 0.429 | 0.322 |
| 0.573 | 0.635 | ⅝ | (0.6250) | M16 | 0.6299 | 3 | 3.81 | 1.81 | 0.56 | 0.480 | 0.360 |
| 0.635 | 0.709 | 11⁄16 | (0.6875) | M18 | 0.7087 | 3 | 4.03 | 1.81 | 0.63 | 0.542 | 0.406 |
| 0.709 | 0.760 | ¾ | (0.7500) | … | … | 3 | 4.25 | 2.00 | 0.69 | 0.590 | 0.442 |
| 0.760 | 0.823 | 13⁄16 | (0.8125) | M20 | 0.7874 | 3 | 4.47 | 2.00 | 0.69 | 0.652 | 0.489 |
| 0.823 | 0.885 | ⅞ | (0.8750) | M22 | 0.8661 | 3 | 4.69 | 2.22 | 0.75 | 0.697 | 0.523 |

**Table 5a.** *(Continued)* **Standard Tap Dimensions (Ground and Cut Thread)** *ANSI/ASME B94.9-2008 (R2018)*

| Nominal Diameter Range, inch | | Nominal Diameter, inch | | Nominal Metric Diameter | | | Tap Dimensions, inch | | | | |
|---|---|---|---|---|---|---|---|---|---|---|---|
| Over | To | Machine Screw Size No. and Fractional Sizes | Decimal Equiv. | mm | inch | Blank Design No. | Overall Length $L$ | Thread Length $I$ | Square Length $l_2$ | Shank Diameter $d_1$ | Size of Square $a$ |
| 0.885 | 0.948 | 15⁄16 | (0.9375) | M24 | 0.9449 | 3 | 4.91 | 2.22 | 0.75 | 0.760 | 0.570 |
| 0.948 | 1.010 | 1 | (1.0000) | M25 | 0.9843 | 3 | 5.13 | 2.50 | 0.81 | 0.800 | 0.600 |
| 1.010 | 1.073 | 1 1⁄16 | (1.0625) | M27 | 1.0630 | 3 | 5.13 | 2.50 | 0.88 | 0.896 | 0.672 |
| 1.073 | 1.135 | 1 1⁄8 | (1.1250) | … | … | 3 | 5.44 | 2.56 | 0.88 | 0.896 | 0.672 |
| 1.135 | 1.198 | 1 3⁄16 | (1.1875) | M30 | 1.1811 | 3 | 5.44 | 2.56 | 1.00 | 1.021 | 0.766 |
| 1.198 | 1.260 | 1 1⁄4 | (1.2500) | … | … | 3 | 5.75 | 2.56 | 1.00 | 1.021 | 0.766 |
| 1.260 | 1.323 | 1 5⁄16 | (1.3125) | M33 | 1.2992 | 3 | 5.75 | 2.56 | 1.06 | 1.108 | 0.831 |
| 1.323 | 1.385 | 1 3⁄8 | (1.3750) | … | … | 3 | 6.06 | 3.00 | 1.06 | 1.108 | 0.831 |
| 1.358 | 1.448 | 1 7⁄16 | (1.4375) | M36 | 1.4173 | 3 | 6.06 | 3.00 | 1.13 | 1.233 | 0.925 |
| 1.448 | 1.510 | 1 1⁄2 | (1.5000) | … | … | 3 | 6.38 | 3.00 | 1.13 | 1.233 | 0.925 |
| 1.510 | 1.635 | 1 5⁄8 | (1.6250) | M39 | 1.5353 | 3 | 6.69 | 3.19 | 1.13 | 1.305 | 0.979 |
| 1.635 | 1.760 | 1 3⁄4 | (1.7500) | M42 | 1.6535 | 3 | 7.00 | 3.19 | 1.25 | 1.430 | 1.072 |
| 1.760 | 1.885 | 1 7⁄8 | (1.8750) | … | … | 3 | 7.31 | 3.56 | 1.25 | 1.519 | 1.139 |
| 1.885 | 2.010 | 2 | (2.0000) | M48 | 1.8898 | 3 | 7.63 | 3.56 | 1.38 | 1.644 | 1.233 |

Special taps greater than 1.010 inch to 1.510 inch in diameter inclusive, having 14 or more threads per inch or 1.75- mm pitch and finer, and sizes over 1.510 inch in diameter with 10 or more threads per inch or 2.5- mm pitch and finer are made to general dimensions shown in Table 10.

For standard ground thread tap limits see Table 20, and Table 21 for inch and Table 16 for metric.

For cut thread tap limits see Table 22 and Table 23.

Special ground thread tap limits are determined by using the formulas shown in Table 2 for Unified inch screw threads and Table 4 for metric M profile screw threads.

Tap sizes 0.395 inch and smaller have an external center on the thread end (may be removed on bottom taps). Sizes 0.223 inch and smaller have an external center on the shank end. Sizes 0.224 inch through 0.395 inch have truncated partial cone centers on the shank end (of diameter of shank). Sizes greater than 0.395 inch have internal centers on both the thread and shank ends.

For standard thread limits and tolerances see Table 17 for Unified inch screw threads and Table 19 for metric threads.

For runout tolerances of tap elements see Table 14.

For number of flutes see Table 11.

**Table 5b. Standard Tap Dimensions Tolerances (Ground and Cut Thread)** *ANSI/ASME B94.9-2008 (R2018)*

| Element | Nominal Diameter Range, inch | | Direction | Tolerance, inch | |
|---|---|---|---|---|---|
| | Over | To (inclusive) | | Ground Thread | Cut Thread |
| Length overall, *L* | 0.5200 | 1.0100 | ± | 0.0300 | 0.0300 |
| | 1.0100 | 2.0000 | ± | 0.0600 | 0.0600 |
| Length of thread, *I* | 0.0520 | 0.2230 | ± | 0.0500 | 0.0500 |
| | 0.2230 | 0.5100 | ± | 0.0600 | 0.0600 |
| | 0.5100 | 1.5100 | ± | 0.0900 | 0.0900 |
| | 1.5100 | 2.0000 | ± | 0.1300 | 0.1300 |
| Length of thread, $I_2$ | 0.0520 | 1.0100 | ± | 0.0300 | 0.0300 |
| | 1.0100 | 2.0000 | ± | 0.0600 | 0.0600 |
| Diameter of shank, $d_1$ | 0.0520 | 0.2230 | – | 0.0015 | 0.0040 |
| | 0.2230 | 0.6350 | – | 0.0015 | 0.0050 |
| | 0.6350 | 1.0100 | – | 0.0020 | 0.0050 |
| | 1.0100 | 1.5100 | – | 0.0020 | 0.0070 |
| | 1.5100 | 2.0000 | – | 0.0030 | 0.0070 |
| Size of square, *a* | 0.0520 | 0.5100 | – | 0.0040 | 0.0040 |
| | 0.5100 | 1.0100 | – | 0.0060 | 0.0060 |
| | 1.0100 | 2.0000 | – | 0.0080 | 0.0080 |

**Entry Taper Length.—**Entry taper length is measured on the full diameter of the thread-forming lobes and is the axial distance from the entry diameter position to the theoretical intersection of tap major diameter and entry taper angle. Beveled end threads provided on taps having internal center or incomplete threads retained when external center is removed. Whenever entry taper length is specified in terms of number of threads, this length is measured in number of pitches, *P*.

$$\text{Bottom length} = 1 \sim 2\frac{1}{2}\ \text{pitches}$$
$$\text{Plug length} = 3 \sim 5\ \text{pitches}$$

Entry diameter measured at the thread crest nearest the front of the tap is an appropriate amount smaller than the diameter of the hole drilled for tapping.

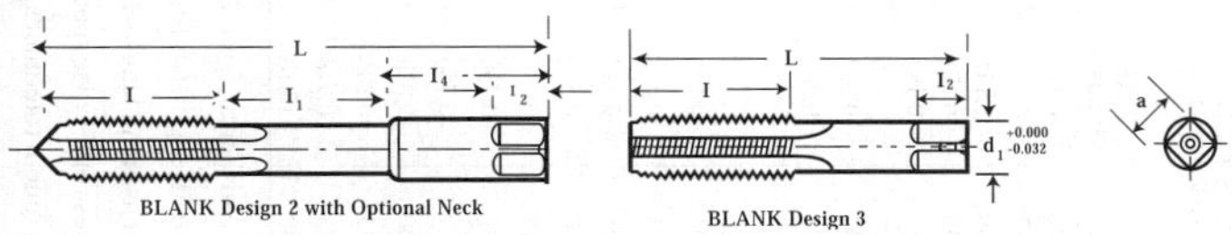


Optional Neck and Optional Shortened Thread Length, Ground and Cut Thread (Table 6)

**Table 6. Optional Neck and Optional Shortened Thread Length (Tap Dimensions, Ground and Cut Thread)** *ANSI/ASME B94.9-2008 (R2018)*

| Nominal Diameter, inch | | Nominal Diameter, inch | | Nominal Metric Diameter | | Tap Dimensions, inch | | | | | | |
|---|---|---|---|---|---|---|---|---|---|---|---|---|
| Over | To (inclusive) | Machine Screw Size No. and Fractional Sizes | Decimal Equiv. | mm | inch | Blank Design No. | Overall Length $L$ | Thread Length $I$ | Neck Length $I_1$ | Square Length $I_2$ | Shank Diameter $d_1$ | Size of Square $a$ |
| 0.104 | 0.117 | 4 | (0.1120) | | | 1 | 1.88 | 0.31 | 0.25 | 0.19 | 0.141 | 0.110 |
| 0.117 | 0.130 | 5 | (0.1250) | M3.0 | 0.1181 | 1 | 1.94 | 0.31 | 0.31 | 0.19 | 0.141 | 0.110 |
| 0.130 | 0.145 | 6 | (0.1380) | M3.5 | 0.1378 | 1 | 2.00 | 0.38 | 0.31 | 0.19 | 0.141 | 0.110 |
| 0.145 | 0.171 | 8 | (0.1640) | M4.0 | 0.1575 | 1 | 2.13 | 0.38 | 0.38 | 0.25 | 0.168 | 0.131 |
| 0.171 | 0.197 | 10 | (0.1900) | M4.5 | 0.1772 | 1 | 2.38 | 0.50 | 0.38 | 0.25 | 0.194 | 0.152 |
| … | … | … | | M5.0 | 0.1969 | … | … | … | … | … | … | … |
| 0.197 | 0.223 | 12 | (0.2160) | … | … | 1 | 2.38 | 0.50 | 0.44 | 0.28 | 0.220 | 0.165 |
| 0.223 | 0.260 | 1/4 | (0.2500) | M6.0 | 0.2362 | 2 | 2.50 | 0.63 | 0.38 | 0.31 | 0.255 | 0.191 |
| 0.260 | 0.323 | 5/16 | (0.3125) | M7.0 | 0.2756 | 2 | 2.72 | 0.69 | 0.44 | 0.38 | 0.318 | 0.238 |
| … | … | … | … | M8.0 | 0.3150 | … | … | … | … | … | … | … |
| 0.323 | 0.395 | 3/8 | (0.3750) | M10.0 | 0.3937 | 2 | 2.94 | 0.75 | 0.50 | 0.44 | 0.381 | 0.286 |
| 0.395 | 0.448 | 7/16 | (0.4375) | … | … | 3 | 3.16 | 0.88 | 0.50 | 0.41 | 0.323 | 0.242 |
| 0.448 | 0.510 | 1/2 | (0.5000) | M12.0 | 0.4724 | 3 | 3.38 | 0.94 | … | 0.44 | 0.367 | 0.275 |
| 0.510 | 0.573 | 9/16 | (0.5625) | M14.0 | 0.5512 | 3 | 3.59 | 1.00 | … | 0.50 | 0.429 | 0.322 |
| 0.573 | 0.635 | 5/8 | (0.6250) | M16.0 | 0.6299 | 3 | 3.81 | 1.09 | … | 0.56 | 0.480 | 0.360 |
| 0.635 | 0.709 | 11/16 | (0.6875) | M18.0 | 0.7087 | 3 | 4.03 | 1.09 | … | 0.63 | 0.542 | 0.406 |
| 0.709 | 0.760 | 3/4 | (0.7500) | … | … | 3 | 4.25 | 1.22 | … | 0.69 | 0.590 | 0.442 |
| 0.760 | 0.823 | 13/16 | (0.8125) | M20.0 | 0.7874 | 3 | 4.47 | 1.22 | … | 0.69 | 0.652 | 0.489 |
| 0.823 | 0.885 | 7/8 | (0.8750) | M22.0 | 0.8661 | 3 | 4.69 | 1.34 | … | 0.75 | 0.697 | 0.523 |
| 0.885 | 0.948 | 15/16 | (0.9375) | M24.0 | 0.9449 | 3 | 4.91 | 1.34 | … | 0.75 | 0.760 | 0.570 |
| 0.948 | 1.010 | 1 | (1.0000) | M25.0 | 0.9843 | 3 | 5.13 | 1.50 | … | 0.75 | 0.800 | 0.600 |

Thread length, $I$, is based on a length of 12 pitches of the UNC thread series.

Thread length, $I$, is a minimum value and has no tolerance.

When thread length, $I$, is added to neck length, $I_1$, the total shall be no less than the minimum thread length, $I$.

Unless otherwise specified, all tolerances are in accordance with Table 5b.

For runout tolerances, see Table 14.

For number of flutes see Table 11.

**Table 7. Machine Screw and Fractional Size Ground Thread Dimensions for Screw Thread Insert (STI) Taps** *ANSI/ASME B94.9-2008 (R2018)*

| Nominal Size (STI) | Threads per inch | | Blank Design No. | Tap Dimensions, inch | | | | | Table 5a Blank Equivalent (Reference) |
|---|---|---|---|---|---|---|---|---|---|
| | NC | NF | | Overall length, $L$ | Thread Length, $I$ | Square Length, $l_2$ | Shank Diameter, $d_1$ | Size of Square, $a$ | |
| 1 | 64 | … | 1 | 1.81 | 0.50 | 0.19 | 0.141 | 0.110 | No. 3 |
| 2 | 56 | 64 | 1 | 1.88 | 0.56 | 0.19 | 0.141 | 0.110 | No. 4 |
| 3 | 48 | 56 | 1 | 1.94 | 0.63 | 0.19 | 0.141 | 0.110 | No. 5 |
| 4 | 40 | 48 | 1 | 2.00 | 0.69 | 0.19 | 0.141 | 0.110 | No. 6 |
| 5 | 40 | … | 1 | 2.13 | 0.75 | 0.25 | 0.168 | 0.131 | No. 8 |
| 6 | 32 | … | 1 | 2.38 | 0.88 | 0.25 | 0.194 | 0.152 | No. 10 |
| | … | 40 | 1 | 2.13 | 0.75 | 0.25 | 0.168 | 0.131 | No. 8 |
| 8 | 32 | 36 | 1 | 2.38 | 0.94 | 0.28 | 0.220 | 0.165 | No. 12 |
| 10 | 24 | 32 | 2 | 2.50 | 1.00 | 0.31 | 0.255 | 0.191 | ¼ |
| 12 | 24 | … | 2 | 2.72 | 1.13 | 0.38 | 0.318 | 0.238 | 5⁄16 |
| ¼ | 20 | 28 | 2 | 2.72 | 1.13 | 0.38 | 0.318 | 0.238 | 5⁄16 |
| 5⁄16 | 18 | 24 | 2 | 2.94 | 1.25 | 0.44 | 0.381 | 0.286 | ⅜ |
| ⅜ | 16 | … | 3 | 3.38 | 1.66 | 0.44 | 0.367 | 0.275 | ½ |
| | … | 24 | 3 | 3.16 | 1.44 | 0.41 | 0.323 | 0.242 | 7⁄16 |
| 7⁄16 | 14 | … | 3 | 3.59 | 1.66 | 0.50 | 0.429 | 0.322 | 9⁄16 |
| | … | 20 | 3 | 3.38 | 1.66 | 0.44 | 0.367 | 0.275 | ½ |
| ½ | 13 | … | 3 | 3.81 | 1.81 | 0.56 | 0.480 | 0.360 | ⅝ |
| | … | 20 | 3 | 3.59 | 1.66 | 0.50 | 0.429 | 0.322 | 9⁄16 |
| 9⁄16 | 12 | … | 3 | 4.03 | 1.81 | 0.63 | 0.542 | 0.406 | 11⁄16 |
| | … | 18 | 3 | 3.81 | 1.81 | 0.56 | 0.480 | 0.360 | ⅝ |
| ⅝ | 11 | … | 3 | 4.25 | 2.00 | 0.69 | 0.590 | 0.442 | ¾ |
| | … | 18 | 3 | 4.03 | 1.81 | 0.63 | 0.542 | 0.406 | 11⁄16 |
| ¾ | 10 | … | 3 | 4.69 | 2.22 | 0.75 | 0.697 | 0.523 | ⅞ |
| | … | 16 | 3 | 4.47 | 2.00 | 0.69 | 0.652 | 0.489 | 13⁄16 |
| ⅞ | 9 | 14 | 3 | 5.13 | 2.50 | 0.81 | 0.800 | 0.600 | 1 |
| 1 | 8 | … | 3 | 5.75 | 2.56 | 1.00 | 1.021 | 0.766 | 1¼ |
| | … | 12, 14 NS | 3 | 5.44 | 2.56 | 0.88 | 0.896 | 0.672 | 1⅛ |
| 1⅛ | 7 | … | 3 | 6.06 | 3.00 | 1.06 | 1.108 | 0.831 | 1⅜ |
| | … | 12 | 3 | 5.75 | 2.56 | 1.00 | 1.021 | 0.766 | 1¼ |
| 1¼ | 7 | … | 3 | 6.38 | 3.00 | 1.13 | 1.233 | 0.925 | 1½ |
| | … | 12 | 3 | 6.06 | 3.00 | 1.06 | 1.108 | 0.831 | 1⅜ |
| 1⅜ | 6 | … | 3 | 6.69 | 3.19 | 1.13 | 1.305 | 0.979 | 1⅝ |
| | … | 12 | 3 | 6.38 | 3.00 | 1.13 | 1.233 | 0.925 | 1½ |
| 1½ | 6 | … | 3 | 7.00 | 3.19 | 1.25 | 1.430 | 1.072 | 1¾ |
| | … | 12 | 3 | 6.69 | 3.19 | 1.13 | 1.305 | 0.979 | 1⅝ |

These threads are larger than nominal size to the extent that the internal thread they produce will accommodate a helical coil screw insert, which at final assembly will accept a screw thread of the nominal size and pitch.

For optional necks, refer to Table 6 using dimensions for equivalent blank sizes.

*Ground Thread Taps:* STI sizes 5⁄16 inch and smaller have external center on thread end (may be removed on bottom taps); sizes 10 through 5⁄16 inch will have an external partial cone center on shank end, with the length of the cone center approximately ¼ of the diameter of shank; sizes larger than 5⁄16 inch may have internal centers on both the thread and shank ends.

For runout tolerances of tap elements, refer to Table 14 using dimensions for equivalent blank sizes.

For number of flutes, refer to Table 11 using dimensions for equivalent blank sizes.

For general dimension tolerances, refer to Table 5b using Table 5a equivalent blank size.

**Table 8. Standard Metric Size Tap Dimensions for Screw Thread Insert (STI) Taps** *ANSI/ASME B94.9-2008 (R2018)*

| Nominal Size (STI) | Thread Pitch, mm Coarse | Fine | | Blank Design No. | Tap Dimensions, inch Overall length, $L$ | Thread Length, $I$ | Square Length, $l_2$ | Shank Diameter, $d_1$ | Size of square, $a$ | Blank Diameter |
|---|---|---|---|---|---|---|---|---|---|---|
| M2.2 | 0.45 | … | | 1 | 1.88 | 0.56 | 0.19 | 0.141 | 0.110 | No.4 |
| M2.5 | 0.45 | … | | 1 | 1.94 | 0.63 | 0.19 | 0.141 | 0.110 | No.5 |
| M3 | 0.50 | … | | 1 | 2.00 | 0.69 | 0.19 | 0.141 | 0.110 | No.6 |
| M3.5 | 0.60 | … | | 1 | 2.13 | 0.75 | 0.25 | 0.168 | 0.131 | No.8 |
| M4 | 0.70 | … | | 1 | 2.38 | 0.88 | 0.25 | 0.194 | 0.152 | No.10 |
| M5 | 0.80 | … | | 2 | 2.50 | 1.00 | 0.31 | 0.255 | 0.191 | 1/4 |
| M6 | 1 | … | | 2 | 2.72 | 1.13 | 0.38 | 0.318 | 0.238 | 5/16 |
| M7 | 1 | … | | 2 | 2.94 | 1.25 | 0.44 | 0.381 | 0.286 | 3/8 |
| M8 | 1.25 | 1 | | 2 | 2.94 | 1.25 | 0.44 | 0.381 | 0.286 | 3/8 |
| M10 | 1.5 | 1.25 | | 3 | 3.38 | 1.66 | 0.44 | 0.367 | 0.275 | 1/2 |
| | … | … | | 3 | 3.16 | 1.44 | 0.41 | 0.323 | 0.242 | 7/16 |
| M12 | 1.75 | 1.5 | 1.25 | 3 | 3.59 | 1.66 | 0.50 | 0.429 | 0.322 | 9/16 |
| M14 | 2 | … | | 3 | 4.03 | 1.81 | 0.63 | 0.542 | 0.406 | 11/16 |
| | … | 1.5 | | 3 | 3.81 | 1.81 | 0.56 | 0.480 | 0.360 | 5/8 |
| M16 | 2 | … | | 3 | 4.25 | 2.00 | 0.69 | 0.590 | 0.442 | 3/4 |
| | … | 1.5 | | 3 | 4.03 | 1.81 | 0.63 | 0.542 | 0.406 | 11/16 |
| M18 | 2.5 | … | | 3 | 4.69 | 2.22 | 0.75 | 0.697 | 0.523 | 7/8 |
| | … | 2.0 | 1.25 | 3 | 4.47 | 2.00 | 0.69 | 0.652 | 0.489 | 13/16 |
| M20 | 2.5 | 2.0 | | 3 | 4.91 | 2.22 | 0.75 | 0.760 | 0.570 | 15/16 |
| | … | | 1.25 | 3 | 4.69 | 2.22 | 0.75 | 0.697 | 0.523 | 7/8 |
| M22 | 2.5 | 2.0 | | 3 | 5.13 | 2.50 | 0.81 | 0.800 | 0.600 | 1 |
| | … | 1.5 | | 3 | 4.91 | 2.22 | 0.75 | 0.760 | 0.570 | 15/16 |
| M24 | 3 | … | | 3 | 5.44 | 2.56 | 0.88 | 0.896 | 0.672 | 1 1/8 |
| | … | 2 | | 3 | 5.13 | 2.50 | 0.88 | 0.896 | 0.672 | 1 1/16 |
| M27 | 3 | … | | 3 | 5.75 | 2.56 | 1.00 | 1.021 | 0.766 | 1 1/4 |
| | … | 2 | | 3 | 5.44 | 2.56 | 0.88 | 0.896 | 0.672 | 1 1/8 |
| M30 | 3.5 | … | | 3 | 6.06 | 3.00 | 1.06 | 1.108 | 0.831 | 1 3/8 |
| | … | 2 | | 3 | 5.75 | 2.56 | 1.00 | 1.021 | 0.766 | 1 1/4 |
| M33 | 3.5 | … | | 3 | 6.38 | 3.00 | 1.13 | 1.233 | 0.925 | 1 1/2 |
| | … | 2 | | 3 | 6.06 | 3.00 | 1.06 | 1.108 | 0.831 | 1 3/8 |
| M36 | 4 | 3 | 2 | 3 | 6.69 | 3.19 | 1.13 | 1.305 | 0.979 | 1 5/8 |
| M39 | 4 | 3 | 2 | 3 | 7.00 | 3.19 | 1.25 | 1.430 | 1.072 | 1 3/4 |

These taps are larger than nominal size to the extent that the internal thread they produce will accommodate a helical coil screw insert, which at final assembly will accept a screw thread of the nominal size and pitch. For optional necks, use Table 6 and dimensions for equivalent blank sizes.

*Ground Thread Taps:* STI sizes M8 and smaller have external center on thread end (may be removed on bottom taps); STI sizes M5 through M10 will have an external partial cone center on shank end, with the length of the cone center approximately 1/4 of the diameter of shank; STI sizes larger than M10 inch may have internal centers on both the thread and shank ends.

For runout tolerances of tap elements, refer to Table 14 using dimensions for equivalent blank sizes. For number of flutes, refer to Table 11 using dimensions for equivalent blank sizes. For general dimension tolerances, refer to Table 5b using Table 5a equivalent blank sizes.

**Table 9. Special Extension Taps** *ANSI/ASME B94.9-2008 (R2018), Appendix*
**(Tap Dimensions, Ground and Cut Threads)**

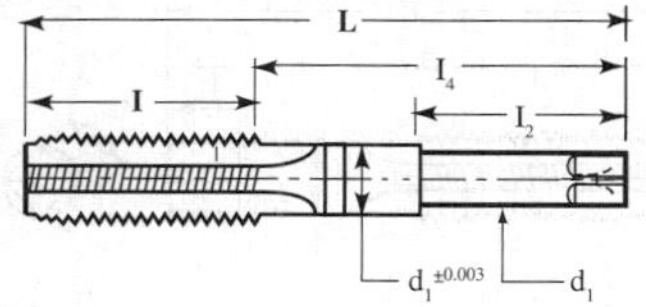


| Nominal Tap Size | | | | Nominal Tap Size | | | |
|---|---|---|---|---|---|---|---|
| Fractional | Machine Screw | Pipe | Shank Length $l_4$ | Fractional | Machine Screw | Pipe | Shank Length $l_4$ |
| … | 0-3 | … | 0.88 | 1½ | … | … | 3.00 |
| … | 4 | … | 1.00 | 1⅝ | … | 3 | 3.13 |
| … | 5-6 | … | 1.13 | 1¾ | … | … | 3.13 |
| … | 8 | … | 1.25 | 1⅞ | … | … | 3.25 |
| … | 10-12 | 1⁄16 to ¼ incl. | 1.38 | 2 | … | … | 3.25 |
| ¼ | 14 | … | 1.50 | 2⅛ | … | … | 3.38 |
| 5⁄16 | … | … | 1.56 | 2¼ | … | … | 3.38 |
| ⅜ | … | … | 1.63 | 2⅜ | … | … | 3.50 |
| 7⁄16 | … | ⅜ to ½ incl. | 1.69 | 2½ | … | … | 3.50 |
| ½ | … | … | 1.69 | 2⅝ | … | … | 3.63 |
| 9⁄16 | … | ¾ | 1.88 | 2¾ | … | … | 3.63 |
| ⅝ | … | 1 | 2.00 | 2⅞ | … | … | 3.75 |
| 11⁄16 | … | … | 2.13 | 3 | … | … | 3.75 |
| ¾ | … | 1¼ | 2.25 | 3⅛ | … | … | 3.88 |
| 13⁄16 | … | 1½ | 2.38 | 3¼ | … | … | 3.88 |
| ⅞ | … | … | 2.50 | 3⅜ | … | 4 | 4.00 |
| 15⁄16 | … | … | 2.63 | 3½ | … | … | 4.00 |
| 1 | … | … | 2.63 | 3⅝ | … | … | 4.13 |
| 1⅛ | … | 2 | 2.75 | 3¾ | … | … | 4.13 |
| 1¼ | … | 2½ | 2.88 | 3⅞ | … | … | 4.25 |
| 1⅜ | … | … | 3.00 | 4 | … | … | 4.25 |

| Tolerances | | | |
|---|---|---|---|
| For shank diameter, $d_1$ for $l_4$ length | | | |
| Fractional, Inch | Machine Screw | Pipe, Inch | Tolerances |
| ¼ to ⅝ incl. | 0 to 14 incl. | 1⁄16 to ⅛ incl. | −0.003 |
| 11⁄16 to 1½ incl. | … | ¼ to 1 incl. | −0.004 |
| 1⅝ to 4 incl. | … | 1¼ to 4 incl. | −0.006 |

Unless otherwise specified, special extension taps will be furnished with dimensions and tolerances as shown for machine screw and fractional taps in Table 5a, Table 5b, and Table 6, and for pipe taps in Table 13a. Exceptions are as follows: types of centers are optional with manufacturer. Tolerances on shank diameter $d_1$ and $l_4$ length as shown on the above Table 9. Shank runout tolerance applies only to the $l_4$ length shown on the above Table 9.

**Table 10. Special Fine Pitch Taps, Short Series** *ANSI/ASME B94.9-2008 (R2018), Appendix* **(Taps Dimensions, Ground and Cut Threads)**

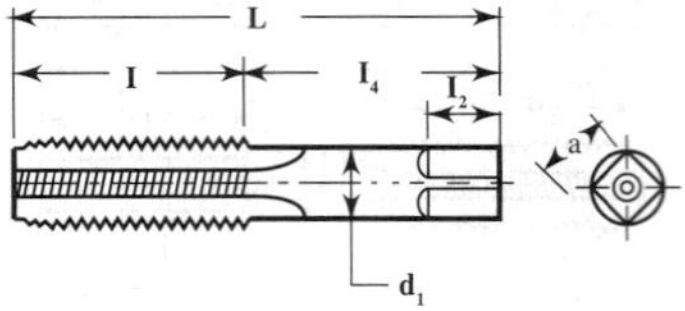


| Nominal Diameter Range, inch | | Nominal Fractional Diameter | Nominal Metric Diameter | Taps Dimensions, inches | | | | |
|---|---|---|---|---|---|---|---|---|
| | | | | Overall Length | Thread Length | Square Length | Shank Diameter | Size of Square |
| Over | To | inch | mm | $L$ | $I$ | $l_2$ | $d_1$ | $a$ |
| 1.070 | 1.073 | 1 1/16 | M27 | 4.00 | 1.50 | 0.88 | 0.8960 | 0.672 |
| 1.073 | 1.135 | 1 1/8 | … | 4.00 | 1.50 | 0.88 | 0.8960 | 0.672 |
| 1.135 | 1.198 | 1 3/16 | M30 | 4.00 | 1.50 | 1.00 | 1.0210 | 0.766 |
| 1.198 | 1.260 | 1 1/4 | … | 4.00 | 1.50 | 1.00 | 1.0210 | 0.766 |
| 1.260 | 1.323 | 1 5/16 | M33 | 4.00 | 1.50 | 1.00 | 1.1080 | 0.831 |
| 1.323 | 1.385 | 1 3/8 | … | 4.00 | 1.50 | 1.00 | 1.1080 | 0.831 |
| 1.385 | 1.448 | 1 7/16 | M36 | 4.00 | 1.50 | 1.00 | 1.2330 | 0.925 |
| 1.448 | 1.510 | 1 1/2 | … | 4.00 | 1.50 | 1.00 | 1.2330 | 0.925 |
| 1.510 | 1.635 | 1 5/8 | M39 | 5.00 | 2.00 | 1.13 | 1.3050 | 0.979 |
| 1.635 | 1.760 | 1 3/4 | M42 | 5.00 | 2.00 | 1.25 | 1.4300 | 1.072 |
| 1.760 | 1.885 | 1 7/8 | … | 5.00 | 2.00 | 1.25 | 1.5190 | 1.139 |
| 1.885 | 2.010 | 2 | M48 | 5.00 | 2.00 | 1.38 | 1.6440 | 1.233 |
| 2.010 | 2.135 | 2 1/8 | … | 5.25 | 2.00 | 1.44 | 1.7690 | 1.327 |
| 2.135 | 2.260 | 2 1/4 | M56 | 5.25 | 2.00 | 1.44 | 1.8940 | 1.420 |
| 2.260 | 2.385 | 2 3/8 | … | 5.25 | 2.00 | 1.50 | 2.0190 | 1.514 |
| 2.385 | 2.510 | 2 1/2 | … | 5.25 | 2.00 | 1.50 | 2.1000 | 1.575 |
| 2.510 | 2.635 | 2 5/8 | M64 | 5.50 | 2.00 | 1.50 | 2.1000 | 1.575 |
| 2.635 | 2.760 | 2 3/4 | … | 5.50 | 2.00 | 1.50 | 2.1000 | 1.575 |
| 2.760 | 2.885 | 2 7/8 | M72 | 5.50 | 2.00 | 1.50 | 2.1000 | 1.575 |
| 2.885 | 3.010 | 3 | … | 5.50 | 2.00 | 1.50 | 2.1000 | 1.575 |
| 3.010 | 3.135 | 3 1/8 | … | 5.75 | 2.00 | 1.50 | 2.1000 | 1.575 |
| 3.135 | 3.260 | 3 1/4 | M80 | 5.75 | 2.00 | 1.50 | 2.1000 | 1.575 |
| 3.260 | 3.385 | 3 3/8 | … | 5.75 | 2.00 | 1.50 | 2.1000 | 1.575 |
| 3.385 | 3.510 | 3 1/2 | … | 5.75 | 2.00 | 1.50 | 2.1000 | 1.575 |
| 3.510 | 3.635 | 3 5/8 | M90 | 6.00 | 2.00 | 1.75 | 2.1000 | 1.575 |
| 3.635 | 3.760 | 3 3/4 | … | 6.00 | 2.00 | 1.75 | 2.1000 | 1.575 |
| 3.760 | 3.885 | 3 7/8 | … | 6.00 | 2.00 | 1.75 | 2.1000 | 1.575 |
| 3.885 | 4.010 | 4 | M100 | 6.00 | 2.00 | 1.75 | 2.1000 | 1.575 |

Unless otherwise specified, special taps 1.010 inches to 1.510 inches in diameter, inclusive, have 14 or more threads per inch or 1.75 mm pitch and finer. Sizes greater than 1.510 inch in diameter with 10 or more threads per inch or 2.5 mm pitch and finer will be made to the general dimensions shown above.

For tolerances, see Table 5b. For runout tolerances of tap elements, see Table 14.

## Table 11. Standard Number of Flutes (Ground and Cut Thread) *ANSI/ASME B94.9-2008 (R2018)*

| Machine Screw Size, Nom. Fractional Dia. inch | | Nominal Metric Dia. | | TPI/Pitch | | | Straight Flutes | | Spiral Point | | Spiral Point Only | Reg. Spiral Flute | Fast Spiral Flute |
|---|---|---|---|---|---|---|---|---|---|---|---|---|---|
| | | mm | inch | UNC NC | UNF NF | mm | Standard | Optional | Standard | Optional | | | |
| 0 | (0.0600) | M1.6 | 0.0630 | … | 80 | 0.35 | 2 | … | 2 | … | … | … | … |
| 1 | (0.0730) | … | … | 64 | 72 | … | 2 | … | 2 | … | … | … | … |
| 2 | (0.0860) | M2.0 | 0.0787 | 56 | 64 | 0.40 | 3 | 2 | 2 | … | … | … | … |
| 3 | (0.0990) | M 2.5 | 0.0984 | 48 | 56 | 0.45 | 3 | 2 | 2 | … | … | … | 2 |
| 4 | (0.1120) | … | … | 40 | 48 | | 3 | 2 | 2 | … | 2 | 2 | 2 |
| 5 | (0.1250) | M3.0 | 0.1181 | 40 | 44 | 0.50 | 3 | 2 | 2 | … | 2 | 2 | 2 |
| 6 | (0.1380) | M3.5 | 0.1378 | 32 | 40 | 0.60 | 3 | 2 | 2 | … | 2 | 2 | 2 |
| 8 | (0.1640) | M4.0 | 0.1575 | 32 | 36 | 0.70 | 4 | 2/3 | 2 | … | 2 | 2 | 3 |
| 10 | (0.1900) | M4.5 | 0.1772 | 24 | 32 | 0.75 | 4 | 2/3 | 2 | … | 2 | 2 | 3 |
| … | | M5 | 0.1969 | … | … | 0.80 | 4 | 2/3 | 2 | … | 2 | 2 | 3 |
| 12 | (0.2160) | … | … | 24 | 28 | … | 4 | 2/3 | 2 | … | 2 | 2 | 3 |
| ¼ | (0.2500) | M6 | 0.2362 | 20 | 28 | 1.00 | 4 | 2/3 | 2 | 3 | 2 | 3 (optional) | 3 |
| … | | M7 | 0.2756 | 18 | 24 | 1.00 | 4 | 2/2 | 2 | 3 | 2 | 3 | 3 |
| 5⁄16 | (0.3125) | M8 | 0.3150 | 18 | 24 | 1.25 | 4 | 2/3 | 2 | 3 | 2 | 3 | 3 |
| ⅜ | (0.3750) | M10 | 0.3937 | 16 | 24 | 1.50 | 4 | 3 | 3 | … | 3 | 3 | 3 |
| 7⁄16 | (0.4375) | … | … | 14 | 20 | … | 4 | 3 | 3 | … | 3 | 3 | 3 |
| ½ | (0.5000) | M12 | 0.4724 | 13 | 20 | 1.75 | 4 | 3 | 3 | … | … | … | 3 |
| 9⁄16 | (0.5625) | M14 | 0.5512 | 12 | 18 | 2.00 | 4 | … | 3 | … | … | … | … |
| ⅝ | (0.6250) | M16 | 0.6299 | 11 | 18 | 2.00 | 4 | … | 3 | … | … | … | … |
| ¾ | (0.7500) | … | … | 10 | 16 | … | 4 | … | 3 | … | … | … | … |
| … | | M20 | 0.7874 | … | … | 2.5 | 4 | … | … | … | … | … | … |
| ⅞ | (0.8750) | … | … | 9 | 14 | … | 4 | … | … | … | … | … | … |
| … | | M24 | 0.9449 | … | … | 3.00 | 4 | … | … | … | … | … | … |
| 1 | (1.0000) | … | … | 8 | 12 | … | 4 | … | … | … | … | … | … |
| 1⅛ | (1.1250) | … | … | 7 | 12 | 4.00 | … | … | … | … | … | … | … |
| … | | M30 | 1.1811 | … | … | 3.50 | 4 | … | … | … | … | … | … |
| 1¼ | (1.2500) | … | … | 7 | … | … | 4 | … | … | … | … | … | … |
| … | | … | … | … | 12 | … | 6 | … | … | … | … | … | … |
| 1⅜ | (1.3750) | … | … | 6 | … | … | 6 | … | … | … | … | … | … |
| … | | … | … | … | 12 | … | 6 | … | … | … | … | … | … |
| … | | M36 | 1.4173 | … | … | 4.00 | 4 | … | … | … | … | … | … |
| 1½ | (1.5000) | … | … | 6 | … | … | 4 | … | … | … | … | … | … |
| | | … | … | … | 12 | … | 6 | … | … | … | … | … | … |
| 1¾ | (1.7500) | … | … | 5 | … | … | 6 | … | … | … | … | … | … |
| 2 | (2.0000) | … | … | 4½ | … | … | 6 | … | … | … | … | … | … |

For pulley taps, see Table 12. For taper pipe, see Table 13a. For straight pipe taps, see Table 13a. For STI taps, use number of flutes for blank size equivalent on Table 5a. For optional flutes, see Table 6.

**Table 12. Pulley Taps, Fractional Size (High-Speed Steel, Ground Thread)** *ANSI/ASME B94.9-2008 (R2018)*

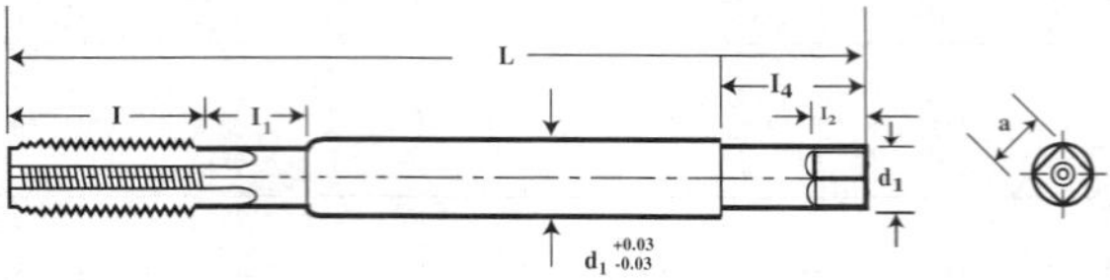


| Dia. of Tap | Threads per Inch NC UNC | Number of Flutes | Length Overall, $L$ | Thread Length, $I$ | Neck Length, $I_1$ | Square Length $I_2$ | Length[d] of Shank Close Tolerance, $I_4$ | Dia. of Shank, $d_1$ | Size[b] of Square, $a$ |
|---|---|---|---|---|---|---|---|---|---|
| ¼ | 20 | 4 | 6, 8 | 1.00 | 0.38 | 0.31 | 1.50 | 0.255 | 0.191 |
| 5⁄16 | 18 | 4 | 6, 8 | 1.13 | 0.38 | 0.38 | 1.56 | 0.318 | 0.238 |
| ⅜ | 16 | 4 | 6, 8, 10 | 1.25 | 0.38 | 0.44 | 1.63 | 0.381 | 0.286 |
| 7⁄16 | 14 | 4 | 6, 8 | 1.44 | 0.44 | 0.50 | 1.69 | 0.444 | 0.333 |
| ½ | 13 | 4 | 6, 8, 10, 12 | 1.66 | 0.50 | 0.56 | 1.69 | 0.507 | 0.380 |
| ⅝ | 11 | 4 | 6, 8,10,12 | 1.81 | 0.63 | 0.69 | 2.00 | 0.633 | 0.475 |
| ¾ | 10 | 4 | 10,12 | 2.00 | 0.75 | 0.75 | 2.25 | 0.759 | 0.569 |

| Tolerances for General Dimensions | | | | | |
|---|---|---|---|---|---|
| Element | Diameter Range | Tolerance | Element | Diameter Range | Tolerance |
| Overall length, $L$ | ¼ to ¾ | ±0.06 | Shank Diameter, $d_1$[a] | ¼ to ½ | −0.005 |
| Thread length, $I$ | ¼ to ¾ | ±0.06 | Size of Square, $a$[b] | ¼ to ½<br>⅝ to ¾ | −0.004<br>−0.006 |
| Square length, $I_2$ | ¼ to ¾ | ±0.03 | | | |
| Neck length, $I_1$ | ¼ to ¾ | c | Length of close tolerance shank, $I_4$ | ¼ to ¾ | d |

[a] Shank diameter, $d_1$, is approximately the same as the maximum major diameter for that size.
[b] Size of square, $a$, is equal to $0.75d_1$ to the nearest 0.001 in.
[c] Neck length, $I_1$, is optional with manufacturer.
[d] Length of close tolerance shank, $I_4$, is a min. length that is held to runout tolerances per Table 14.

These taps are standard with plug chamfer in H3 limit only. All dimensions are given in inches. These taps have an internal center in thread end. For standard thread limits, see Table 20. For runout tolerances of tap elements, see Table 14.

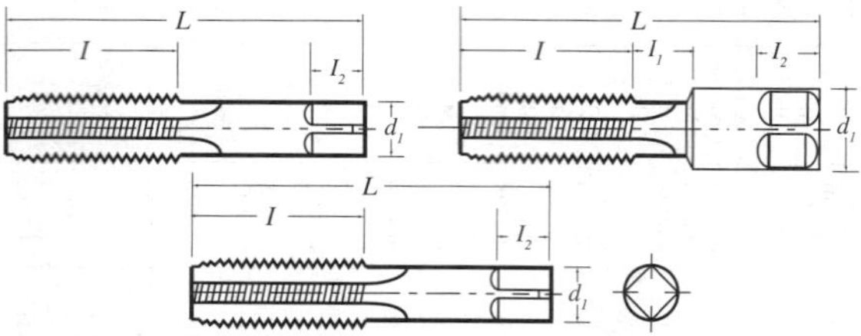


Straight and Taper Pipe Tap Dimensions, Ground and Cut Thread (Table 13a and Table 13b)

**Table 13a. Straight and Taper Pipe Tap Dimensions (Ground and Cut Thread)** *ANSI/ASME B94.9-2008 (R2018)*

| Nominal Size, Inch [a] | Threads per Inch | Number of Flutes: Regular Thread | Number of Flutes: Interrupted Thread | Length Overall, $L$ | Thread Length, $I$ | Square Length, $I_2$ | Shank Diameter, $d_1$ | Size of Square, $a$ | Length Optional Neck, $I_1$ | Ground Thread: NPT, NPTF, ANPT | Ground Thread: NPSC, NPSM, NPSF | Cut Thread Only: NPT | Cut Thread Only: NPSC, NPSM |
|---|---|---|---|---|---|---|---|---|---|---|---|---|---|
| 1/16 | 27 | 4 | … | 2.13 | 0.69 | 0.38 | 0.3125 | 0.234 | 0.375 | b | … | … | … |
| 1/8 | 27 | 4 | 5 | 2.13 | 0.75 | 0.38 | 0.3125 | 0.234 | … | b, c | d, e | f, g, h | … |
| 1/8 | 27 | 4 | 5 | 2.13 | 0.75 | 0.38 | 0.4375 | 0.328 | 0.375 | b, c | d, e | f, g, h | a |
| 1/4 | 18 | 4 | 5 | 2.44 | 1.06 | 0.44 | 0.5625 | 0.421 | 0.375 | b, c | d, e | f, g, h | a |
| 3/8 | 18 | 4 | 5 | 2.56 | 1.06 | 0.50 | 0.7000 | 0.531 | 0.375 | b, c | d, e | f, g, h | a |
| 1/2 | 14 | 4 | 5 | 3.13 | 1.38 | 0.63 | 0.6875 | 0.515 | … | b, c | d, e | f, g, h | a |
| 3/4 | 14 | 5 | 5 | 3.25 | 1.38 | 0.69 | 0.9063 | 0.679 | … | b, c | d | f, g, h | a |
| 1 | 11 1/2 | 5 | 5 | 3.75 | 1.75 | 0.81 | 1.1250 | 0.843 | … | b, c | d | f, g, h | a |
| 1 1/4 | 11 1/2 | 5 | 5 | 4.00 | 1.75 | 0.94 | 1.3125 | 0.984 | … | b, c | … | f, g, h | a |
| 1 1/2 | 11 1/2 | 7 | 7 | 4.25 | 1.75 | 1.00 | 1.5000 | 1.125 | … | b, i | … | f, h | … |
| 2 | 11 1/2 | 7 | 7 | 4.25 | 1.75 | 1.13 | 1.8750 | 1.406 | … | b, i | … | f, h | … |
| 2 1/2 | 8 | 8 | … | 5.50 | 2.56 | 1.25 | 2.2500 | 1.687 | … | … | … | h | … |
| 3 | 8 | 8 | … | 6.00 | 2.63 | 1.38 | 2.6250 | 1.968 | … | … | … | h | … |

[a] Pipe taps 1/8 inch are furnished with large-size shanks unless the small shank is specified.

[b] High-speed ground thread 1/16 to 2 inches including noninterrupted (NPT, NPTF, and ANPT).

[c] High-speed ground thread 1/8 to 1 1/4 inch including interrupted (NPT, NPTF, and ANPT).

[d] High-speed ground thread 1/8 to 1 inch including noninterrupted (NPSC and NPSM).

[e] High-speed cut thread 1/8 to 1 inch including noninterrupted (NPSC and NPSM).

[f] High-speed cut thread 1/8 to 1 inch including noninterrupted (NPT).

[g] High-speed cut thread 1/8 to 1 1/4 inch including interrupted (NPT).

[h] Carbon cut thread 1/8 to 1 1/4 inch including interrupted (NPT).

[i] High-speed ground thread 1 1/2 to 2 inches including interrupted (NPT).

**Table 13b. Straight and Taper Pipe Taps Tolerances (Ground and Cut Thread)** *ANSI/ASME B94.9-2008 (R2018)*

| Ground Thread | | | | Cut Thread | | | |
|---|---|---|---|---|---|---|---|
| Element | Nominal Diameter Range, inch: Over | To (inclusive) | Tolerances, inch | Element | Nominal Diameter Range, inch: Over | To (inclusive) | Tolerances, inch |
| Length overall, $L$ | 1/16 | 3/4 | ±0.031 | Length overall, $L$ | 1/8 | 3/4 | ±0.031 |
| | 1 | 2 | ±0.063 | | 1 | 3 | ±0.063 |
| Length of thread, $I$ | 1/16 | 3/4 | ±0.063 | Length of thread, $I$ | 1/8 | 3/4 | ±0.063 |
| | 1 | 1 1/4 | ±0.094 | | 1 | 1 1/4 | ±0.094 |
| | 1 1/2 | 2 | ±0.125 | | 1 1/2 | 3 | ±0.125 |
| Length of square, $l_2$ | 1/16 | 3/4 | ±0.031 | Length of square, $l_2$ | 1/8 | 3/4 | ±0.031 |
| | 1 | 2 | ±0.063 | | 1 | 3 | ±0.063 |
| Diameter of shank, $d_1$ | 1/16 | 1/8 | −0.002 | Diameter of shank, $d_1$ | 1/8 | 1/2 | -0.007 |
| | 1/4 | 1 | −0.002 | | 3/4 | 3 | -0.009 |
| | 1 1/4 | 2 | −0.002 | Size of square, $a$ | 1/8 | … | -0.004 |
| Size of square, $a$ | 1/16 | 1/8 | −0.004 | | 1/4 | 3/4 | -0.006 |
| | 1/4 | 3/4 | −0.006 | | 1 | 3 | -0.008 |
| | 1 | 2 | −0.008 | | | | |

All dimensions are given in inches.

The first few threads on interrupted thread pipe taps are left full. These taps have internal centers. For runout tolerances of tap elements, see Table 14. Taps marked NPS are suitable for NPSC and NPSM. These taps have 2 to 3½ threads chamfer, see Table 5a. Optional neck is for manufacturing use only. For taper pipe thread limit, see Table 24a. For straight pipe thread limits, see Table 23a, Table 23b, and Table 23d.

**Table 14. Runout and Locational Tolerance of Tap Elements** *ANSI/ASME B94.9-2008 (R2018)*

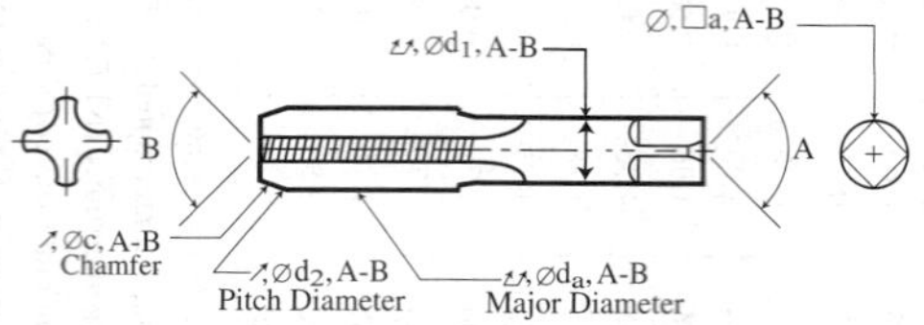


| | Range Sizes (Inclusive) | | | Total Runout FIM, Inch | | |
|---|---|---|---|---|---|---|
| | Machine Screw | Metric | Pipe, Inch | Cut Thread | Ground Thread | Location, Inch |
| Shank, $d_1$ | #0 to 5/16 | M1.6 to M8 | 1/16 | 0.0060 | 0.0010 | … |
| | 11/32 to 4 | M10 to M100 | 1/8 to 4 | 0.0080 | 0.0016 | … |
| Major diameter, $d_a$ | #0 to 5/16 | M1.6 to M8 | 1/16 | 0.0050 | 0.0010 | … |
| | 11/32 to 4 | M10 to M100 | 1/8 to 4 | 0.0080 | 0.0016 | … |
| Pitch Diameter, $d_2$ | #0 to 5/16 | M1.6 to M8 | 1/16 | 0.0050 | 0.0010 | … |
| | 11/32 to 4 | M10 to M100 | 1/8 to 4 | 0.0080 | 0.0016 | … |
| Chamfer[a], $c$ | #0 to 1/2 | M1.6 to M12 | 1/16 to 1/8 | 0.0040 | 0.0020 | … |
| | 17/32 to 4 | M14 to M100 | 1/8 to 4 | 0.0060 | 0.0030 | … |
| Square, $a$ | #0 to 1/2 | M1.6 to M12 | 1/16 to 1/8 | … | … | 0.0060 |
| | 17/32 to 4 | M14 to M100 | 1/8 to 4 | … | … | 0.0080 |

[a] Chamfer should preferably be inspected by light projection to avoid errors due to indicator contact points dropping into the thread groove.

## Table 15. Tap Thread Limits: Metric Sizes, Ground Thread (M Profile Standard Thread Limits in Inches) *ANSI/ASME B94.9-2008 (R2018)*

| Nom. Dia mm | Pitch, mm | Major Diameter (Inches) | | | Pitch Diameter (Inches) | | | | | | |
|---|---|---|---|---|---|---|---|---|---|---|---|
| | | | | | | Limit # | D # Limit | | Limit # | D # Limit | |
| | | Basic | Min. | Max. | Basic | D | Min. | Max. | D | Min. | Max. |
| 1.6 | 0.35 | 0.06299 | 0.06409 | 0.06508 | 0.05406 | | 0.05500 | 0.05559 | … | … | … |
| 2 | 0.4 | 0.07874 | 0.08000 | 0.08098 | 0.06850 | 3 | 0.06945 | 0.07004 | … | … | … |
| 2.5 | 0.45 | 0.09843 | 0.09984 | 0.10083 | 0.08693 | | 0.08787 | 0.08846 | … | … | … |
| 3 | 0.5 | 0.11811 | 0.11969 | 0.12067 | 0.10531 | | 0.10626 | 0.10685 | 5 | 0.10278[a,b] | 0.10787[a,b] |
| 3.5 | 0.6 | 0.13780 | 0.13969 | 0.14067 | 0.12244 | | 0.12370 | 0.12449 | … | | |
| 4 | 0.7 | 0.15748 | 0.15969 | 0.16130 | 0.13957 | 4 | 0.14083 | 0.14161 | 6 | 0.14185[a,b] | 0.14264[a,b] |
| 4.5 | 0.75 | 0.17717 | 0.17953 | 0.18114 | 0.15799 | | 0.15925 | 0.16004 | … | … | … |
| 5 | 0.8 | 0.19685 | 0.19937 | 0.20098 | 0.17638 | | 0.17764 | 0.17843 | 7 | 0.17917[b,c] | 0.17996[b,c] |
| 6 | 1 | 0.23622 | 0.23937 | 0.24098 | 0.21063 | | 0.21220 | 0.21319 | 8 | 0.21374[b,c] | 0.2147[b,c] |
| 7 | 1 | 0.27559 | 0.27874 | 0.28035 | 0.25000 | 5 | 0.25157 | 0.25256 | … | | |
| 8 | 1.25 | 0.31496 | 0.31890 | 0.32142 | 0.28299 | | 0.28433 | 0.28555 | 9 | 0.2864[b,d] | 0.2875[b,d] |
| 10 | 1.5 | 0.39370 | 0.39843 | 0.40094 | 0.35535 | 6 | 0.35720 | 0.35843 | 10 | 0.3593[b,e] | 0.3605[b,e] |
| 12 | 1.75 | 0.47244 | 0.47795 | 0.48047 | 0.42768 | | 0.42953 | 0.43075 | 11 | 0.43209[e] | 0.43331[e] |
| 14 | 2 | 0.55118 | 0.55748 | 0.56000 | 0.50004 | 7 | 0.50201 | 0.50362 | … | … | … |
| 14 | 1.25 | 0.55118 | 0.55500 | 0.55600 | 0.51920 | 4 | 0.52070[f] | 0.52171[f] | … | … | … |
| 16 | 2 | 0.62992 | 0.63622 | 0.63874 | 0.57878 | 7 | 0.58075 | 0.58236 | … | … | … |
| 18 | 1.5 | 0.70870 | 0.71350 | 0.71450 | 0.67030 | 4 | 0.67180[f] | 0.67230[f] | 7 | 0.58075 | 0.58236 |
| 20 | 2.5 | 0.78740 | 0.79528 | 0.79780 | 0.72346 | 7 | 0.72543 | 0.72705 | … | … | … |
| 24 | 3 | 0.94488 | 0.95433 | 0.95827 | 0.86815 | 8 | 0.87063 | 0.8722 | … | … | … |
| 30 | 3.5 | 1.18110 | 1.19213 | 1.19606 | 1.09161 | 9 | 1.0942 | 1.0962 | … | … | … |
| 36 | 4 | 1.41732 | 1.42992 | 1.43386 | 1.31504 | | 1.3176 | 1.3197 | … | … | … |
| 42 | 4.5 | 1.65354 | 1.66772 | 1.71102 | 1.53846 | 10 | 1.5415 | 1.5436 | … | … | … |
| 48 | 5 | 1.88976 | 1.90552 | 1.98819 | 1.76189 | | 1.7649 | 1.7670 | … | … | … |

[a] Minimum and maximum major diameters are 0.00102 larger than shown.

[b] Standard D limit for thread-forming taps.

[c] Minimum and maximum major diameters are 0.00154 larger than shown.

[d] Minimum and maximum major diameters are 0.00205 larger than shown.

[e] Minimum and maximum major diameters are 0.00256 larger than shown.

[f] These sizes are intended for spark plug applications; use tolerances from Table 2, column D.

All dimensions are given in inches. Not all styles of taps are available with all limits listed. For calculation of limits other than those listed, see formulas in Table 4.

## Table 16. Tap Thread Limits: Metric Sizes, Ground Thread (M Profile Standard Thread Limits in Millimeters) *ANSI/ASME B94.9-2008 (R2018)*

| Size mm | Pitch | Major Diameter | | | Pitch Diameter | | | | | | |
|---|---|---|---|---|---|---|---|---|---|---|---|
| | | | | | | | D # Limit | | | D # Limit | |
| | | Basic | Min. | Max. | Basic | D # | Min. | Max. | D # | Min. | Max. |
| 1.6 | 0.35 | 1.60 | 1.628 | 1.653 | 1.373 | | 1.397 | 1.412 | … | … | … |
| 2 | 0.4 | 2.00 | 2.032 | 2.057 | 1.740 | 3 | 1.764 | 1.779 | … | … | … |
| 2.5 | 0.45 | 2.50 | 2.536 | 2.561 | 2.208 | | 2.232 | 2.247 | … | … | … |
| 3 | 0.5 | 3.00 | 3.040 | 3.065 | 2.675 | | 2.699 | 2.714 | 5 | 2.725[a,b] | 2.740[a,b] |
| 3.5 | 0.6 | 3.50 | 3.548 | 3.573 | 3.110 | | 3.142 | 3.162 | … | … | … |
| 4 | 0.7 | 4.00 | 4.056 | 4.097 | 3.545 | 4 | 3.577 | 3.597 | 6 | 3.603[a,b] | 3.623[a,b] |
| 4.5 | 0.75 | 4.50 | 4.560 | 4.601 | 4.013 | | 4.045 | 4.065 | … | … | … |
| 5 | 0.8 | 5.00 | 5.064 | 5.105 | 4.480 | | 4.512 | 4.532 | 7 | 4.551[b,c] | 4.571[b,c] |
| 6 | 1.00 | 6.00 | 6.121 | 5.351 | 5.391 | | 5.391 | 5.416 | 8 | 5.429[b,c] | 5.454[b,c] |
| 7 | 1.00 | 7.00 | 7.121 | 6.351 | 6.391 | 5 | 6.391 | 6.416 | … | … | … |
| 8 | 1.25 | 8.00 | 8.10 | 8.164 | 7.188 | | 7.222 | 7.253 | 9 | 7.274[b,d] | 7.305[b,d] |
| 10 | 1.50 | 10.0 | 10.12 | 10.184 | 9.026 | 6 | 9.073 | 9.104 | 10 | 9.125[b,d] | 9.156[b,d] |
| 12 | 1.75 | 12.0 | 12.14 | 12.204 | 10.863 | | 10.910 | 10.941 | 11 | 10.975[b,e] | 11.006[b,e] |
| 14 | 2.00 | 14.0 | 14.01 | 14.164 | 13.188 | 4 | 7.222[f] | 7.253[f] | … | … | … |
| 14 | 1.25 | 14.0 | 14.16 | 14.224 | 12.701 | 7 | 12.751 | 12.792 | … | … | … |
| 16 | 2.00 | 16.0 | 16.16 | 16.224 | 14.701 | | 14.751 | 14.792 | … | … | … |
| 18 | 1.50 | 18.0 | 18.12 | 18.184 | 17.026 | 4 | 17.063[f] | 17.076[f] | … | … | … |

**Table 16.** *(Continued)* **Tap Thread Limits: Metric Sizes, Ground Thread (M Profile Standard Thread Limits in Millimeters)** *ANSI/ASME B94.9-2008 (R2018)*

| Size | | Major Diameter | | | Pitch Diameter | | | | | | |
|---|---|---|---|---|---|---|---|---|---|---|---|
| | | | | | | | D # Limit | | | D # Limit | |
| mm | Pitch | Basic | Min. | Max. | Basic | D # | Min. | Max. | D # | Min. | Max. |
| 20 | 2.50 | 20.0 | 20.20 | 20.263 | 18.376 | 7 | 18.426 | 18.467 | … | … | … |
| 24 | 3.00 | 24.0 | 24.24 | 24.34 | 22.051 | 8 | 22.114 | 22.155 | … | … | … |
| 30 | 3.50 | 30.0 | 30.28 | 30.38 | 27.727 | 9 | 27.792 | 27.844 | … | … | … |
| 36 | 4.00 | 36.0 | 36.32 | 36.42 | 33.402 | | 33.467 | 33.519 | … | … | … |
| 42 | 4.50 | 42.0 | 42.36 | 42.46 | 39.077 | 10 | 39.155 | 39.207 | … | … | … |
| 48 | 5.00 | 48.0 | 48.48 | 48.58 | 44.103 | | 44.182 | 44.246 | … | … | … |

[a] Minimum and maximum major diameters are 0.026 larger than shown.
[b] Standard D limit for thread-forming taps.
[c] Minimum and major diameters are 0.039 larger than shown.
[d] Minimum and major diameters are 0.052 larger than shown.
[e] Minimum and major diameters are 0.065 larger than shown.
[f] These sizes are intended for spark plug applications; use tolerances from Table 2, column D.

Notes for Table 16: Inch translations are listed in Table 15. Limits listed in Table 16 are the most commonly used in industry. Not all styles of taps are available with all limits listed. For calculations of limits other than listed, see formulas in Table 4.

**Table 17. Tap Size Recommendations for Class 6H Metric Screw Threads**

| Nominal Diameter, mm | Pitch, mm | Recommended Thread Limit Number | Internal Threads, Pitch Diameter | | | |
|---|---|---|---|---|---|---|
| | | | Min. (mm) | Max. (mm) | Min. (inch) | Max. (inch) |
| 1.6 | 0.35 | D3 | 1.373 | 1.458 | 0.05406 | 0.05740 |
| 2 | 0.4 | D3 | 1.740 | 1.830 | 0.06850 | 0.07250 |
| 2.5 | 0.45 | D3 | 2.208 | 2.303 | 0.08693 | 0.09067 |
| 3 | 0.5 | D3 | 2.675 | 2.775 | 0.10537 | 0.10925 |
| 3.5 | 0.6 | D4 | 3.110 | 3.222 | 0.12244 | 0.12685 |
| 4 | 0.7 | D4 | 3.545 | 3.663 | 0.13957 | 0.14421 |
| 4.5 | 0.75 | D4 | 4.013 | 4.131 | 0.15789 | 0.16264 |
| 5 | 0.8 | D4 | 4.480 | 4.605 | 0.17638 | 0.18130 |
| 6 | 1 | D5 | 5.350 | 5.500 | 0.201063 | 0.21654 |
| 7 | 1 | D5 | 6.350 | 6.500 | 0.2500 | 0.25591 |
| 8 | 1.25 | D5 | 7.188 | 7.348 | 0.28299 | 0.28929 |
| 10 | 1.5 | D6 | 9.206 | 9.206 | 0.35535 | 0.36244 |
| 12 | 1.75 | D6 | 10.863 | 11.063 | 0.42768 | 0.43555 |
| 14 | 2 | D7 | 12.701 | 12.913 | 0.50004 | 0.50839 |
| 16 | 2 | D7 | 14.701 | 14.913 | 0.57878 | 0.58713 |
| 20 | 2.5 | D7 | 18.376 | 18.600 | 0.72346 | 0.73228 |
| 24 | 3 | D8 | 22.051 | 22.316 | 0.86815 | 0.87858 |
| 30 | 3.5 | D9 | 27.727 | 28.007 | 1.09161 | 1.10264 |
| 36 | 4 | D9 | 33.402 | 33.702 | 1.31504 | 1.32685 |

The above recommended taps normally produce the class of thread indicated in average materials when used with reasonable care. However, if the tap specified does not give a satisfactory gage fit in the work, a choice of some other limit tap will be necessary.

**Table 18. Standard Chamfers for Thread Cutting Taps**
*ANSI/ASME B94.9-2008 (R2018)*

| Type of tap | | Chamfer length | | Type of tap | Chamfer length | |
|---|---|---|---|---|---|---|
| | | Min. | Max. | | Min. | Max. |
| Straight threads taps | Bottom | 1$P$ | 2$P$ | Taper pipe taps | 2$P$ | 3½$P$ |
| | Semibottom | 2$P$ | 3$P$ | | | |
| | Plug | 3$P$ | 5$P$ | | | |
| | Taper | 7$P$ | 10$P$ | | | |

$P$ = pitch.

The chamfered length is measured at the cutting edge and is the axial length from the point diameter to the theoretical intersection of the major diameter and the chamfer angle. Whenever chamfer length is specified in terms of threads, this length is measured in number of pitches as shown. The point diameter is approximately equal to the basic thread minor diameter.

**Table 19. Taps Sizes for Classes 2B and 3B Unified Screw Threads Machine Screw, Numbered, and Fractional Sizes** *ANSI/ASME B94.9-2008 (R2018)*

| Size | Threads per Inch | | Recommended Tap For Class of Thread[a] | | Pitch Diameter Limits For Class of Thread | | |
|---|---|---|---|---|---|---|---|
| | NC UNC | NF UNF | Class 2B[b] | Class 3B[c] | Min., All Classes (Basic) | Max. Class 2B | Max. Class 3B |
| Machine Screw Numbered-Size Taps | | | | | | | |
| 0 | … | 80 | G H2 | G H1 | 0.0519 | 0.0542 | 0.0536 |
| 1 | 64 | … | G H2 | G H1 | 0.0629 | 0.0655 | 0.0648 |
| 1 | … | 72 | G H2 | G H1 | 0.0640 | 0.0665 | 0.0659 |
| 2 | 56 | … | G H2 | G H1 | 0.0744 | 0.0772 | 0.0765 |
| 2 | … | 64 | G H2 | G H1 | 0.0759 | 0.0786 | 0.0779 |
| 3 | 48 | … | G H2 | G H1 | 0.0855 | 0.0885 | 0.0877 |
| 3 | … | 56 | G H2 | G H1 | 0.0874 | 0.0902 | 0.0895 |
| 4 | 40 | … | G H2 | G H2 | 0.0958 | 0.0991 | 0.0982 |
| 4 | … | 48 | G H2 | G H1 | 0.0985 | 0.1016 | 0.1008 |
| 5 | 40 | … | G H2 | G H2 | 0.1088 | 0.1121 | 0.1113 |
| 5 | … | 44 | G H2 | G H1 | 0.1102 | 0.1134 | 0.1126 |
| 6 | 32 | … | G H3 | G H2 | 0.1177 | 0.1214 | 0.1204 |
| 6 | … | 40 | G H2 | G H2 | 0.1218 | 0.1252 | 0.1243 |
| 8 | 32 | … | G H3 | G H2 | 0.1437 | 0.1475 | 0.1465 |
| 8 | … | 36 | G H2 | G H2 | 0.1460 | 0.1496 | 0.1487 |
| 10 | 24 | … | G H3 | G H3 | 0.1629 | 0.1672 | 0.1661 |
| 10 | … | 32 | G H3 | G H2 | 0.1697 | 0.1736 | 0.1726 |
| 12 | 24 | … | G H3 | G H3 | 0.1889 | 0.1933 | 0.1922 |
| 12 | … | 28 | G H3 | G H3 | 0.1928 | 0.1970 | 0.1959 |
| Fractional-Size Taps | | | | | | | |
| 1/4 | 20 | … | G H5 | G H3 | 0.2175 | 0.2224 | 0.2211 |
| 1/4 | … | 28 | G H4 | G H3 | 0.2268 | 0.2311 | 0.2300 |
| 5/16 | 18 | … | G H5 | G H3 | 0.2764 | 0.2817 | 0.2803 |
| 5/16 | … | 24 | G H4 | G H3 | 0.2854 | 0.2902 | 0.2890 |
| 3/8 | 16 | … | G H5 | G H3 | 0.3344 | 0.3401 | 0.3387 |
| 3/8 | … | 24 | G H4 | G H3 | 0.3479 | 0.3528 | 0.3516 |
| 7/16 | 14 | … | G H5 | G H3 | 0.3911 | 0.3972 | 0.3957 |
| 7/16 | … | 20 | G H5 | G H3 | 0.4050 | 0.4104 | 0.4091 |
| 1/2 | 13 | … | G H5 | G H3 | 0.4500 | 0.4565 | 0.4548 |
| 1/2 | … | 20 | G H5 | G H3 | 0.4675 | 0.4731 | 0.4717 |
| 9/16 | 12 | … | G H5 | G H3 | 0.5084 | 0.5152 | 0.5135 |
| 9/16 | … | 18 | G H5 | G H3 | 0.5264 | 0.5323 | 0.5308 |
| 5/8 | 11 | … | G H5 | G H3 | 0.5660 | 0.5732 | 0.5714 |
| 5/8 | … | 18 | G H5 | G H3 | 0.5889 | 0.5949 | 0.5934 |
| 3/4 | 10 | … | G H5 | G H5 | 0.6850 | 0.6927 | 0.6907 |
| 3/4 | … | 16 | G H5 | G H3 | 0.7094 | 0.7159 | 0.7143 |
| 7/8 | 9 | … | G H6 | G H4 | 0.8028 | 0.8110 | 0.8089 |
| 7/8 | … | 14 | G H6 | G H4 | 0.8286 | 0.8356 | 0.8339 |
| 1 | 8 | … | G H6 | G H4 | 0.9188 | 0.9276 | 0.9254 |
| 1 | … | 12 | G H6 | G H4 | 0.9459 | 0.9535 | 0.9516 |
| 1 | 14NS | 14NS | G H6 | G H4 | 0.9536 | 0.9609 | 0.9590 |
| 1 1/8 | 7 | … | G H8 | G H4 | 1.0322 | 1.0416 | 1.0393 |
| 1 1/8 | … | 12 | G H6 | G H4 | 1.0709 | 1.0787 | 1.0768 |
| 1 1/4 | 7 | … | G H8 | G H4 | 1.1572 | 1.1668 | 1.1644 |
| 1 1/4 | … | 12 | G H6 | G H4 | 1.1959 | 1.2039 | 1.2019 |
| 1 3/8 | 6 | … | G H8 | G H4 | 1.2667 | 1.2771 | 1.2745 |
| 1 3/8 | … | 12 | G H6 | G H4 | 1.3209 | 1.3291 | 1.3270 |
| 1 1/2 | 6 | … | G H8 | G H4 | 1.3917 | 1.4022 | 1.3996 |
| 1 1/2 | … | 12 | G H6 | G H4 | 1.4459 | 1.4542 | 1.4522 |

[a] Recommended taps are for cutting threads only and are not for roll-form threads.

[b] Cut-thread taps in sizes #3 to 1½ in. NC and NF, inclusive, may be used under all normal conditions and in average materials for producing Class 2B tapped holes.

[c] Taps suited for class 3B are satisfactory for class 2B threads.

All dimensions are given in inches.

The above recommended taps normally produce the class of thread indicated in average materials when used with reasonable care. However, if the tap specified does not give a satisfactory gage fit in the work, a choice of some other limit tap will be necessary.

**Table 20. Tap Thread Limits: Machine Screw Sizes, Ground Thread** *ANSI/ASME B94.9-2008 (R2018)*
**(Unified and American National Thread Forms, Standard Thread Limits)**

| | Threads per Inch | | | Major Diameter | | | Pitch Diameter | | | | | | | | | | | | | | | | |
|---|---|---|---|---|---|---|---|---|---|---|---|---|---|---|---|---|---|---|---|---|---|---|---|
| | | | | | | | | H1 limit | | H2 limit | | H3 limit | | H4 limit | | H5 limit | | H6 limit[a] | | H7 limit[b] | | H8 limit[c] | |
| Size | NC UNF | NF UNF | NS | Basic | Min. | Max. | Basic | Min. | Max. | Min. | Max. | Min. | Max. | Min. | Max. | Min. | Max. | Min. | Max. | Min. | Max. | Min. | Max. |
| 0 | … | 80 | … | 0.0600 | 0.0605 | 0.0616 | 0.0519 | 0.0519 | 0.0524 | 0.0524 | 0.0529 | … | … | … | … | … | … | … | … | … | … | … | … |
| 1 | 64 | … | … | 0.0730 | 0.0736 | 0.0750 | 0.0629 | 0.0629 | 0.0634 | 0.0634 | 0.0639 | … | … | … | … | … | … | … | … | … | … | … | … |
| 1 | … | 72 | … | 0.0730 | 0.0736 | 0.0748 | 0.0640 | 0.064 | 0.0645 | 0.0645 | 0.0650 | … | … | … | … | … | … | … | … | … | … | … | … |
| 2 | 56 | … | … | 0.0860 | 0.0866 | 0.0883 | 0.0744 | 0.0744 | 0.0749 | 0.0749 | 0.0754 | … | … | … | … | … | … | … | … | … | … | … | … |
| 2 | | 64 | … | 0.0860 | 0.0866 | 0.0880 | 0.0759 | … | … | 0.0764 | 0.0769 | … | … | … | … | … | … | … | … | … | … | … | … |
| 3 | 48 | | … | 0.0990 | 0.0999 | 0.1017 | 0.0855 | … | … | 0.086 | 0.0865 | … | … | … | … | … | … | … | … | … | … | … | … |
| 3 | … | 56 | … | 0.0990 | 0.0997 | 0.1013 | 0.0874 | 0.0874 | 0.0879 | 0.0879 | 0.0884 | … | … | … | … | … | … | … | … | … | … | … | … |
| 4 | 40 | … | … | 0.1120 | 0.1134 | 0.1153 | 0.0958 | 0.0958 | 0.0963 | 0.0963 | 0.0968 | … | … | … | … | 0.0978[d] | 0.0983[d] | … | … | … | … | … | … |
| 4 | … | … | 36 | 0.1120 | 0.1135 | 0.1156 | 0.0940 | 0.094 | 0.0945 | 0.0945 | 0.0950 | … | … | … | … | 0.0960[d] | 0.0965[d] | … | … | … | … | … | … |
| 4 | … | 48 | … | 0.1120 | 0.1129 | 0.1147 | 0.0985 | 0.0985 | 0.0990 | 0.0990 | 0.0995 | … | … | … | … | 0.1005[d] | 0.1010[d] | … | … | … | … | … | … |
| 5 | 40 | … | … | 0.1250 | 0.1264 | 0.1283 | 0.1088 | 0.1088 | 0.1093 | 0.1093 | 0.1098 | … | … | … | … | 0.1108[d] | 0.1113[d] | … | … | … | … | … | … |
| 5 | … | 44 | … | 0.1250 | 0.1262 | 0.1280 | 0.1102 | … | … | 0.1107 | 0.1112 | … | … | … | … | 0.1122[d] | 0.1127[d] | … | … | … | … | … | … |
| 6 | 32 | … | … | 0.1380 | 0.1400 | 0.1421 | 0.1177 | 0.1177 | 0.1182 | 0.1182 | 0.1187 | 0.1187 | 0.1192 | … | … | 0.1197[a] | 0.1202[a] | … | … | 0.1207 | 0.1212 | 0.1222 | 0.1227 |
| 6 | … | 40 | … | 0.1380 | 0.1394 | 0.1413 | 0.1218 | 0.1218 | 0.1223 | 0.1223 | 0.1228 | … | … | … | … | 0.1238[a] | 0.1243[a] | … | … | … | … | … | … |
| 8 | 32 | … | … | 0.1640 | 0.1660 | 0.1681 | 0.1437 | 0.1437 | 0.1442 | 0.1442 | 0.1447 | 0.1447 | 0.1452 | … | … | 0.1457[a] | 0.1462[a] | … | … | 0.1467 | 0.1472 | 0.1482 | 0.1487 |
| 8 | … | 36 | … | 0.1640 | 0.1655 | 0.1676 | 0.1460 | … | … | 0.1465 | 0.1470 | … | … | … | … | 0.1480[a] | 0.1485[a] | … | … | … | … | … | … |
| 10 | 24 | … | … | 0.1900 | 0.1927 | 0.1954 | 0.1629 | 0.1629 | 0.1634 | 0.1634 | 0.1639 | 0.1639 | 0.1644 | 0.1644 | 0.1649 | … | … | 0.1654 | 0.1659 | 0.1659 | 0.1664 | … | … |
| 10 | … | 32 | … | 0.1900 | 0.1920 | 0.1941 | 0.1697 | 0.1697 | 0.1702 | 0.1702 | 0.1707 | 0.1707 | 0.1712 | 0.1712 | 0.1717 | … | … | 0.1722 | 0.1727 | 0.1727 | 0.1732 | 0.1742 | 0.1747 |
| 12 | 24 | … | … | 0.2160 | 0.2187 | 0.2214 | 0.1889 | … | … | … | … | 0.1899 | 0.1904 | 0.1904 | 0.1909 | … | … | 0.1914 | 0.1919 | … | … | … | … |
| 12 | … | 28 | … | 0.2160 | 0.2183 | 0.2206 | 0.1928 | … | … | … | … | 0.1938 | 0.1943 | 0.1943 | 0.1948 | … | … | 0.1953 | 0.1958 | … | … | … | … |

[a] Minimum and maximum major diameters are 0.0010 larger than shown.

[b] Minimum and maximum major diameters are 0.0020 larger than shown.

[c] Minimum and maximum major diameters are 0.0035 larger than shown.

[d] Minimum and maximum major diameters are 0.0015 larger than shown.

General notes:

Limits listed in above table are the most commonly used in the industry.

Not all styles of taps are available with all limits listed.

For calculation of limits other than those listed, see formulas and Table 2.

## Table 21. Tap Thread Limits: Fractional Sizes, Ground Thread *ANSI/ASME B94.9-2008 (R2018)* (Unified and American National Thread Forms, Standard Thread Limits)

| Size inch | NC UNC | NF UNF | NS | Major Diameter | | | Pitch Diameter | | | | | | | | | | | | | | | | |
|---|---|---|---|---|---|---|---|---|---|---|---|---|---|---|---|---|---|---|---|---|---|---|---|
| | | | | | | | | H1 limit | | H2 limit | | H3 limit | | H4 limit | | H5 limit | | H6 limit[a] | | H7 limit | | H8 limit[b] | |
| | | | | Basic | Min. | Max. | Basic | Min. | Max. | Min. | Max. | Min. | Max. | Min. | Max. | Min. | Max. | Min. | Max. | Min. | Max. | Min. | Max. |
| 1/4 | 20 | … | … | 0.2500 | 0.2532 | 0.2565 | 0.2175 | 0.2175 | 0.2180 | 0.2180 | 0.2185 | 0.2185 | 0.2190 | … | … | 0.2195[a] | 0.2200[a] | … | … | … | … | … | … |
| 1/4 | … | 28 | … | 0.2500 | 0.2523 | 0.2546 | 0.2268 | 0.2268 | 0.2273 | 0.2273 | 0.2278 | 0.2278 | 0.2283 | 0.2283 | 0.2288 | … | … | … | … | … | … | … | … |
| 5/16 | 18 | … | … | 0.3125 | 0.3161 | 0.3197 | 0.2764 | 0.2764 | 0.2769 | 0.2769 | 02774 | 0.2774 | 0.2779 | … | … | 0.2784[a] | 0.2789[a] | … | … | 0.2794[c] | 0.2799[c] | … | … |
| 5/16 | … | 24 | … | 0.3125 | 0.3152 | 0.3179 | 0.2854 | 0.2854 | 0.2859 | 0.2859 | 0.2864 | 0.2864 | 0.2869 | 0.2869 | 0.2874 | … | … | … | … | 0.2884[c] | 0.2889[c] | … | … |
| 3/8 | 16 | … | … | 0.3750 | 0.3790 | 0.3831 | 0.3344 | 0.3344 | 0.3349 | 0.3349 | 0.3354 | 0.3354 | 0.3359 | … | … | 0.3364[a] | 0.3369[a] | … | … | 0.3374[c] | 0.3379[c] | … | … |
| 3/8 | … | 24 | … | 0.3750 | 0.3777 | 0.3804 | 0.3479 | 0.3479 | 0.3484 | 0.3484 | 0.3489 | 0.3489 | 0.3494 | 0.3494 | 0.3499 | … | … | … | … | 0.3509[c] | 0.3514[c] | … | … |
| 7/16 | 14 | … | … | 0.4375 | 0.4422 | 0.4468 | 0.3911 | … | … | 0.3916 | 0.3921 | 0.3921 | 0.3926 | … | … | 0.3931[a] | 0.3936[a] | … | … | … | … | 0.3946 | 0.3951 |
| 7/16 | … | 20 | … | 0.4375 | 0.4407 | 0.4440 | 0.4050 | … | … | … | … | 0.4060 | 0.4065 | … | … | 0.4070[a] | 0.4075[a] | … | … | … | … | 0.4085 | 0.4090 |
| 1/2 | 13 | … | … | 0.5000 | 0.5050 | 0.5100 | 0.4500 | 0.4500 | 0.4505 | 0.4505 | 0.4510 | 0.4510 | 0.4515 | … | … | 0.4520[a] | 0.4525[a] | … | … | … | … | 0.4535 | 0.4540 |
| 1/2 | … | 20 | … | 0.5000 | 0.5032 | 0.5065 | 0.4675 | 0.4675 | 0.4680 | 0.4680 | 0.4685 | 0.4685 | 0.4690 | … | … | 0.4695[a] | 0.4700[a] | … | … | … | … | 0.4710 | 0.4715 |
| 9/16 | 12 | … | … | 0.5625 | 0.5679 | 0.5733 | 0.5084 | … | … | … | … | 0.5094 | 0.5099 | … | … | 0.5104[a] | 0.5109[a] | … | … | 0.5114[c] | 0.5119[c] | … | … |
| 9/16 | … | 18 | … | 0.5625 | 0.5661 | 0.5697 | 0.5264 | … | … | 0.5269 | 0.5274 | 0.5274 | 0.5279 | … | … | 0.5284[a] | 0.5289[a] | … | … | 0.5294[c] | 0.5299[c] | … | … |
| 5/8 | 11 | … | … | 0.6250 | 0.6309 | 0.6368 | 0.566 | … | … | 0.5665 | 0.567 | 0.567 | 0.5675 | … | … | 0.5680[a] | 0.5685[a] | … | … | 0.5690[c] | 0.5695[c] | … | … |
| 5/8 | … | 18 | … | 0.6250 | 0.6286 | 0.6322 | 0.5889 | … | … | 0.5894 | 0.5899 | 0.5899 | 0.5904 | … | … | 0.5909[a] | 0.5914[a] | … | … | 0.5919[c] | 0.5924[c] | … | … |
| 11/16 | … | … | 11 | 0.6875 | 0.6934 | 0.6993 | 0.6285 | … | … | … | … | 0.6295 | 0.6300 | … | … | … | … | … | … | … | … | … | … |
| 11/16 | … | … | 16 | 0.6875 | 0.6915 | 0.6956 | 0.6469 | … | … | … | … | 0.6479 | 0.6484 | … | … | … | … | … | … | … | … | … | … |
| 3/4 | 10 | … | … | 0.7500 | 0.7565 | 0.7630 | 0.6850 | … | … | 0.6855 | 0.6860 | 0.6860 | 0.6865 | … | … | 0.6870 | 0.6875 | … | … | 0.6880[d] | 0.6885[d] | … | … |
| 3/4 | … | 16 | … | 0.7500 | 0.7540 | 0.7581 | 0.7094 | 0.7094 | 0.7099 | 0.7099 | 0.7104 | 0.7104 | 0.7109 | … | … | 0.7114[a] | 0.1119[a] | … | … | 0.7124[d] | 0.4129[d] | … | … |
| 7/8 | 9 | … | … | 0.8750 | 0.8822 | 0.8894 | 0.8028 | … | … | … | … | … | … | 0.8043 | 0.8048 | … | … | 0.8053 | 0.8058 | … | … | … | … |

**Table 21.** *(Continued)* **Tap Thread Limits: Fractional Sizes, Ground Thread** *ANSI/ASME B94.9-2008 (R2018)* **(Unified and American National Thread Forms, Standard Thread Limits)**

| | | | | Major Diameter | | | Pitch Diameter | | | | | | | | | | | | | | | | |
|---|---|---|---|---|---|---|---|---|---|---|---|---|---|---|---|---|---|---|---|---|---|---|---|
| | | | | | | | | H1 limit | | H2 limit | | H3 limit | | H4 limit | | H5 limit | | H6 limit[a] | | H7 limit | | H8 limit[b] | |
| Size inch | NC UNC | NF UNF | NS | Basic | Min. | Max. | Basic | Min. | Max. | Min. | Max. | Min. | Max. | Min. | Max. | Min. | Max. | Min. | Max. | Min. | Max. | Min. | Max. |
| ⅞ | … | 14 | … | 0.8750 | 0.8797 | 0.8843 | 0.8286 | … | … | 0.8291 | 0.8296 | … | … | 0.8301 | 0.8306 | … | … | … | … | … | … | … | … |
| 1 | 8 | … | … | 1.0000 | 1.0082 | 1.0163 | 0.9188 | … | … | … | … | … | … | 0.9203 | 0.9208 | … | … | 0.9213 | 0.9218 | … | … | … | … |
| 1 | … | 12 | … | 1.0000 | 1.0054 | 1.0108 | 0.9459 | … | … | … | … | … | … | 0.9474 | 0.9479 | … | … | … | … | … | … | … | … |
| 1 | … | … | … | 1.0000 | 1.0047 | 1.0093 | 0.9536 | … | … | … | … | … | … | 0.9551 | 0.9556 | … | … | … | … | … | … | … | … |
| 1⅛ | 7 | | … | 1.1250 | 1.1343 | 1.1436 | 1.0322 | … | … | … | … | … | … | 1.0337 | 1.0342 | … | … | … | … | … | … | … | … |
| 1⅛ | | 12 | … | 1.1250 | 1.1304 | 1.1358 | 1.0709 | … | … | … | … | … | … | 1.0724 | 1.0729 | … | … | … | … | … | … | … | … |
| 1¼ | 7 | | … | 1.2500 | 1.2593 | 1.2686 | 1.1572 | … | … | … | … | … | … | 1.1587 | 1.1592 | … | … | … | … | … | … | … | … |
| 1¼ | | 12 | … | 1.2500 | 1.2554 | 1.2608 | 1.1959 | … | … | … | … | … | … | 1.1974 | 1.1979 | … | … | … | … | … | … | … | … |
| 1⅜ | 6 | | … | 1.3750 | 1.3859 | 1.3967 | 1.2667 | … | … | … | … | … | … | 1.2682 | 1.2687 | … | … | … | … | … | … | … | … |
| 1⅜ | | 12 | … | 1.3750 | 1.3804 | 1.3858 | 1.3209 | … | … | … | … | … | … | 1.3224 | 1.3229 | … | … | … | … | … | … | … | … |
| 1½ | 6 | | … | 1.5000 | 1.5109 | 1.5217 | 1.3917 | … | … | … | … | … | … | 1.3932 | 1.3937 | … | … | … | … | … | … | … | … |
| 1½ | | 12 | … | 1.5000 | 1.5054 | 1.5108 | 1.4459 | … | … | … | … | … | … | 1.4474 | 1.4479 | … | … | … | … | … | … | … | … |
| 1¾ | | 5 | … | 1.7500 | 1.7630 | 1.7760 | 1.6201 | … | … | … | … | … | … | 1.6216 | 1.6221 | … | … | … | … | … | … | … | … |
| 2 | | 4.5 | … | 2.0000 | 2.0145 | 2.0289 | 1.8557 | … | … | … | … | … | … | 1.8572 | 1.8577 | … | … | … | … | … | … | … | … |

[a] Minimum and maximum major diameters are 0.0010 larger than shown.

[b] Minimum and maximum major diameters are 0.0035 larger than shown.

[c] Minimum and maximum major diameters are 0.0020 larger than shown.

[d] Minimum and maximum major diameters are 0.0015 larger than shown.

General notes:

Limits listed in Table 21 are the most commonly used in the industry.

Not all styles of taps are available with all limits listed.

For calculation of limits other than those listed, see formulas and Table 2.

**Table 22. Tap Thread Limits: Machine Screw Sizes, Cut Thread**
*ANSI/ASME B94.9-2008 (R2018)* **Unified and American National Thread Forms, Standard Thread Limits**

| Size | Threads per Inch | | | Major Diameter | | | Pitch Diameter | | |
|---|---|---|---|---|---|---|---|---|---|
| | NC UNC | NF UNF | NS UNS | Basic | Min. | Max. | Basic | Min. | Max. |
| 0 | … | 80 | … | 0.0600 | 0.0609 | 0.0624 | 0.0519 | 0.0521 | 0.0531 |
| 1 | 64 | … | … | 0.0730 | 0.0739 | 0.0754 | 0.0629 | 0.0631 | 0.0641 |
| 1 | … | 72 | … | 0.0730 | 0.0740 | 0.0755 | 0.0640 | 0.0642 | 0.0652 |
| 2 | 56 | … | … | 0.0860 | 0.0872 | 0.0887 | 0.0744 | 0.0746 | 0.0756 |
| 2 | … | 64 | … | 0.0860 | 0.0870 | 0.0885 | 0.0759 | 0.0761 | 0.0771 |
| 3 | 48 | … | … | 0.0990 | 0.1003 | 0.1018 | 0.0855 | 0.0857 | 0.0867 |
| 3 | … | 56 | … | 0.0990 | 0.1002 | 0.1017 | 0.0874 | 0.0876 | 0.0886 |
| 4 | … | … | 36 | 0.1120 | 0.1137 | 0.1157 | 0.0940 | 0.0942 | 0.0957 |
| 4 | 40 | … | … | 0.1120 | 0.1136 | 0.1156 | 0.0958 | 0.0960 | 0.0975 |
| 4 | … | 48 | … | 0.1120 | 0.1133 | 0.1153 | 0.0985 | 0.0987 | 0.1002 |
| 5 | 40 | … | … | 0.1250 | 0.1266 | 0.1286 | 0.1088 | 0.1090 | 0.1105 |
| 6 | 32 | … | … | 0.1380 | 0.1402 | 0.1422 | 0.1177 | 0.1182 | 0.1197 |
| 6 | … | … | 36 | 0.1380 | 0.1397 | 0.1417 | 0.1200 | 0.1202 | 0.1217 |
| 6 | … | 40 | … | 0.1380 | 0.1396 | 0.1416 | 0.1218 | 0.1220 | 0.1235 |
| 8 | 32 | … | … | 0.1640 | 0.1662 | 0.1682 | 0.1437 | 0.1442 | 0.1457 |
| 8 | … | 36 | … | 0.1640 | 0.1657 | 0.1677 | 0.1460 | 0.1462 | 0.1477 |
| 8 | … | … | 40 | 0.1640 | 0.1656 | 0.1676 | 0.1478 | 0.1480 | 0.1495 |
| 10 | 24 | … | … | 0.1900 | 0.1928 | 0.1948 | 0.1629 | 0.1634 | 0.1649 |
| 10 | … | 32 | … | 0.1900 | 0.1922 | 0.1942 | 0.1697 | 0.1702 | 0.1717 |
| 12 | 24 | … | … | 0.2160 | 0.2188 | 0.2208 | 0.1889 | 0.1894 | 0.1909 |
| 12 | … | 28 | … | 0.2160 | 0.2184 | 0.2204 | 0.1928 | 0.1933 | 0.1948 |
| 14 | … | … | 24 | 0.2420 | 0.2448 | 0.2473 | 0.2149 | 0.2154 | 0.2174 |

| Angle Tolerance | | |
|---|---|---|
| Threads per Inch | Half Angle | Full Angle |
| 20 to 28 | ±0°45′ | ±0°65′ |
| 30 and finer | ±0°60′ | ±0°90′ |

A maximum lead error of ±0.003 inch in 1 inch of thread is permitted.

All dimensions are given in inches.

Thread limits are computed from Table 3.

**Table 23. Tap Thread Limits: Fractional Sizes, Cut Thread**
*ANSI/ASME B94.9-2008 (R2018)* **(Unified and American National Thread Forms)**

| | Threads per Inch | | | Major Diameter | | | Pitch Diameter | | |
|---|---|---|---|---|---|---|---|---|---|
| Size | NC UNC | NF UNF | NS UNS | Basic | Min. | Max. | Basic | Min. | Max. |
| 1/8 | … | … | 40 | 0.1250 | 0.1266 | 0.1286 | 0.1088 | 0.1090 | 0.1105 |
| 5/32 | … | … | 32 | 0.1563 | 0.1585 | 0.1605 | 0.13595 | 0.13645 | 0.1380 |
| 3/16 | … | … | 24 | 0.1875 | 0.1903 | 0.1923 | 0.1604 | 0.1609 | 0.1624 |
| 3/16 | … | … | 32 | 0.1875 | 0.1897 | 0.1917 | 0.1672 | 0.1677 | 0.1692 |
| 1/4 | 20 | … | … | 0.2500 | 0.2532 | 0.2557 | 0.2175 | 0.2180 | 0.2200 |
| 1/4 | … | 28 | … | 0.2500 | 0.2524 | 0.2549 | 0.2268 | 0.2273 | 0.2288 |
| 5/16 | 18 | … | … | 0.3125 | 0.3160 | 0.3185 | 0.2764 | 0.2769 | 0.2789 |
| 5/16 | … | 24 | … | 0.3125 | 0.3153 | 0.3178 | 0.2854 | 0.2859 | 0.2874 |
| 3/8 | 16 | … | … | 0.3750 | 0.3789 | 0.3814 | 0.3344 | 0.3349 | 0.3369 |
| 3/8 | … | 24 | … | 0.3750 | 0.3778 | 0.3803 | 0.3479 | 0.3484 | 0.3499 |
| 7/16 | 14 | … | … | 0.4375 | 0.4419 | 0.4449 | 0.3911 | 0.3916 | 0.3941 |
| 7/16 | … | 20 | … | 0.4375 | 0.4407 | 0.4437 | 0.4050 | 0.4055 | 0.4075 |
| 1/2 | 13 | … | … | 0.5000 | 0.5047 | 0.5077 | 0.4500 | 0.4505 | 0.4530 |
| 1/2 | … | 20 | … | 0.5000 | 0.5032 | 0.5062 | 0.4675 | 0.4680 | 0.4700 |
| 9/16 | 12 | … | … | 0.5625 | 0.5675 | 0.5705 | 0.5084 | 0.5089 | 0.5114 |
| 9/16 | … | 18 | … | 0.5625 | 0.5660 | 0.5690 | 0.5264 | 0.5269 | 0.5289 |
| 5/8 | 11 | … | … | 0.6250 | 0.6304 | 0.6334 | 0.5660 | 0.5665 | 0.5690 |
| 5/8 | … | 18 | … | 0.6250 | 0.6285 | 0.6315 | 0.5889 | 0.5894 | 0.5914 |
| 3/4 | 10 | … | … | 0.7500 | 0.7559 | 0.7599 | 0.6850 | 0.6855 | 0.6885 |
| 3/4 | … | 16 | … | 0.7500 | 0.7539 | 0.7579 | 0.7094 | 0.7099 | 0.7124 |
| 7/8 | 9 | … | … | 0.8750 | 0.8820 | 0.8860 | 0.8028 | 0.8038 | 0.8068 |
| 7/8 | … | 14 | … | 0.8750 | 0.8799 | 0.8839 | 0.8286 | 0.8296 | 0.8321 |
| 1 | 8 | … | … | 1.0000 | 1.0078 | 1.0118 | 0.9188 | 0.9198 | 0.9228 |
| 1 | … | 12 | … | 1.0000 | 1.0055 | 1.0095 | 0.9459 | 0.9469 | 0.9494 |
| 1 | … | … | 14 | 1.0000 | 1.0049 | 1.0089 | 0.9536 | 0.9546 | 0.9571 |
| 1 1/8 | 7 | … | … | 1.1250 | 1.1337 | 1.1382 | 1.0322 | 1.0332 | 1.0367 |
| 1 1/8 | … | 12 | … | 1.1250 | 1.1305 | 1.1350 | 1.0709 | 1.0719 | 1.0749 |
| 1 1/4 | 7 | … | … | 1.2500 | 1.2587 | 1.2632 | 1.1572 | 1.1582 | 1.1617 |
| 1 1/4 | … | 12 | … | 1.2500 | 1.2555 | 1.2600 | 1.1959 | 1.1969 | 1.1999 |
| 1 3/8 | 6 | … | … | 1.3750 | 1.3850 | 1.3895 | 1.2667 | 1.2677 | 1.2712 |
| 1 3/8 | … | 12 | … | 1.3750 | 1.3805 | 1.3850 | 1.3209 | 1.3219 | 1.3249 |
| 1 1/2 | 6 | … | … | 1.5000 | 1.5100 | 1.5145 | 1.3917 | 1.3927 | 1.3962 |
| 1 1/2 | … | 12 | … | 1.5000 | 1.5055 | 1.5100 | 1.4459 | 1.4469 | 1.4499 |
| 1 3/4 | 5 | … | … | 1.7500 | 1.7602 | 1.7657 | 1.6201 | 1.6216 | 1.6256 |
| 2 | 4.5 | … | … | 2.0000 | 2.0111 | 2.0166 | 1.8557 | 1.8572 | 1.8612 |

| Threads per Inch | Half Angle | Full Angle |
|---|---|---|
| 4 1/2 to 5 1/2 | ±0° 35′ | ±0° 53′ |
| 6 to 9 | ±0° 40′ | ±0° 60′ |
| 10 to 28 | ±0° 45′ | ±0° 68′ |
| 30 to 64 | ±0° 60′ | ±0° 90′ |

A maximum lead error of ±0.003 inch in 1 inch of thread is permitted.

All dimensions are given in inches.

Thread limits are computed from Table 3.

**Table 23a. Straight Pipe Thread Limits: NPS, Ground Thread ANSI Straight Pipe Thread Form (NPSC, NPSM)** *ANSI/ASME B94.9-2008 (R2018)*

| Nominal Size, Inches | Threads per Inch, NPS, NPSC, NPSM | Major Diameter | | | Pitch Diameter | | |
|---|---|---|---|---|---|---|---|
| | | Plug at Gaging Notch | Min. *G* | Max. *H* | Plug at Gaging Notch *E* | Min. *K* | Max. *L* |
| 1/8 | 27 | 0.3983 | 0.4022 | 0.4032 | 0.3736 | 0.3746 | 0.3751 |
| 1/4 | 18 | 0.5286 | 0.5347 | 0.5357 | 0.4916 | 0.4933 | 0.4938 |
| 3/8 | 18 | 0.6640 | 0.6701 | 0.6711 | 0.6270 | 0.6287 | 0.6292 |
| 1/2 | 14 | 0.8260 | 0.8347 | 0.8357 | 0.7784 | 0.7806 | 0.7811 |
| 3/4 | 14 | 1.0364 | 1.0447 | 1.0457 | 0.9889 | 0.9906 | 0.9916 |
| 1 | 11½ | 1.2966 | 1.3062 | 1.3077 | 1.2386 | 1.2402 | 1.2412 |

Formulas for NPS Ground Thread Taps[a]

| Nominal Size | Major Diameter | | Minor Dia. | Threads per Inch | *A* | *B* |
|---|---|---|---|---|---|---|
| | Min. *G* | Max. *H* | Max. | 27 | 0.0296 | 0.0257 |
| 1/8 | $H - 0.0010$ | $(K + A) - 0.0010$ | $M - B$ | 18 | 0.0444 | 0.0401 |
| 1/4 to 3/4 | $H - 0.0010$ | $(K + A) - 0.0020$ | $M - B$ | 14 | 0.0571 | 0.0525 |
| 1 | $H - 0.0015$ | $(K + A) - 0.0021$ | $M - B$ | 11½ | 0.0696 | 0.0647 |

[a] In the formulas, *M* equals the actual measured pitch diameter.

All dimensions are given in inches.

*Maximum pitch diameter* of tap is based upon an allowance deducted from the maximum product pitch diameter of NPSC or NPSM, whichever is smaller.

*Minimum pitch diameter* of tap is derived by subtracting the ground thread pitch diameter tolerance for actual equivalent size.

*Lead tolerance:* A maximum lead deviation pf ± 0.0005 inch within any two threads not farther apart than 1 inch.

*Angle Tolerance:* 11½ to 27 threads per inch, plus or minus 30 min. in half angle.

Taps made to the specifications in Table 23a are to be marked NPS and used for NPSC and NPSM.

**Table 23b. Straight Pipe Thread Limits: NPSF Ground Thread ANSI Standard Straight Pipe Thread Form (NPSF)** *ANSI/ASME B94.9-2008 (R2018)*

| Nominal Size, Inches | Threads per Inch | Major Diameter | | Pitch Diameter | | | |
|---|---|---|---|---|---|---|---|
| | | Min. *G* | Max. *H* | Plug at Gaging Notch *E* | Min. *K* | Max. *L* | Minor[a] Dia. Flat, Max. |
| 1/16 | 27 | 0.3008 | 0.3018 | 0.2812 | 0.2772 | 0.2777 | 0.004 |
| 1/8 | 27 | 0.3932 | 0.3942 | 0.3736 | 0.3696 | 0.3701 | 0.004 |
| 1/4 | 18 | 0.5239 | 0.5249 | 0.4916 | 0.4859 | 0.4864 | 0.005 |
| 3/8 | 18 | 0.6593 | 0.6603 | 0.6270 | 0.6213 | 0.6218 | 0.005 |
| 1/2 | 14 | 0.8230 | 0.8240 | 0.7784 | 0.7712 | 0.7717 | 0.005 |
| 3/4 | 14 | 1.0335 | 1.0345 | 0.9889 | 0.9817 | 0.9822 | 0.005 |

[a] As specified or sharper.

All dimensions are given in inches.

## Table 23c. Standard Straight Pipe Thread Limits: NPSF Ground Thread Dryseal ANSI Standard Straight Pipe Thread Form (NPSF)

*ANSI/ASME B94.9-2008 (R2018)*

| Formulas for American Dryseal (NPSF) Ground Thread Taps | | | | | |
|---|---|---|---|---|---|
| Nominal Size, Inches | Major Diameter | | Pitch Diameter | | Max. Minor Dia. |
| | Min. $G$ | Max. $H$ | Min. $K$ | Max. $L$ | |
| 1/16 | $H-0.0010$ | $K+Q-0.0005$ | $L-0.0005$ | $E-F$ | $M-Q$ |
| 1/8 | $H-0.0010$ | $K+Q-0.0005$ | $L-0.0005$ | $E-F$ | $M-Q$ |
| 1/4 | $H-0.0010$ | $K+Q-0.0005$ | $L-0.0005$ | $E-F$ | $M-Q$ |
| 3/8 | $H-0.0010$ | $K+Q-0.0005$ | $L-0.0005$ | $E-F$ | $M-Q$ |
| 1/2 | $H-0.0010$ | $K+Q-0.0005$ | $L-0.0005$ | $E-F$ | $M-Q$ |
| 3/4 | $H-0.0010$ | $K+Q-0.0005$ | $L-0.0005$ | $E-F$ | $M-Q$ |

| Values to Use in Formulas | | | | |
|---|---|---|---|---|
| Threads per Inch | $E$ | $F$ | $M$ | $Q$ |
| 27 | Pitch diameter of plug at gaging notch | 0.0035 | Actual measured pitch diameter | 0.0251 |
| 18 | | 0.0052 | | 0.0395 |
| 14 | | 0.0067 | | 0.0533 |

All dimensions are given in inches.

*Lead Tolerance:* A maximum lead deviation of ±0.0005 inch within any two threads not farther apart than 1 inch.

*Angle Tolerance:* Plus or minus 30 min. in half angle for 14 to 27 threads per inch, inclusive.

## Table 23d. Standard Straight Pipe Tap Limits: (NPS) Cut Thread ANSI Straight Pipe Thread Form (NPSC) *ANSI/ASME B94.9-2008 (R2018)*

| Nominal Size | Threads per Inch, NPS, NPSC | Size at Gaging Notch | Pitch Diameter | | Values to Use in Formulas | | |
|---|---|---|---|---|---|---|---|
| | | | Min. | Max. | $A$ | $B$ | $C$ |
| 1/8 | 27 | 0.3736 | 0.3721 | 0.3751 | 0.0267 | 0.0296 | 0.0257 |
| 1/4 | 18 | 0.4916 | 0.4908 | 0.4938 | 0.0408 | 0.0444 | 0.0401 |
| 3/8 | 18 | 0.6270 | 0.6257 | 0.6292 | | | |
| 1/2 | 14 | 0.7784 | 0.7776 | 0.7811 | 0.0535 | 0.0571 | 0.0525 |
| 3/4 | 14 | 0.9889 | 0.9876 | 0.9916 | | | |
| 1 | 11½ | 1.2386 | 1.2372 | 1.2412 | 0.0658 | 0.0696 | 0.0647 |

The following are approximate formulas, in which $M$ = measured pitch diameter in inches:

Major dia., min. = $M + A$

Major dia., max. = $M + B$

Minor dia., max. = $M - C$

*Maximum pitch diameter* of tap is based on an allowance deducted from the maximum product pitch diameter of NPSC.

*Minimum pitch diameter* of tap equals maximum pitch diameter minus the tolerance.

All dimensions are given in inches.

*Lead Tolerance:* ± 0.003 inch per inch of thread.

*Angle Tolerance:* For all pitches, tolerance will be ± 45″ for half angle and ± 68″ for full angle. Taps made to these specifications are to be marked NPS and used for NPSC thread form.

Taps made to the specifications in Table 23a are to be marked NPS and used for NPSC.

As the American National Standard straight pipe thread form is to be maintained, the major and minor diameters vary with the pitch diameter. Either a flat or rounded form is allowable at both the crest and the root.

**Table 24a. Taper Pipe Thread Limits (Ground and Cut Thread: Ground Thread For NPS, NPTF, and ANPT; Cut Thread for NPT only)** *ANSI/ASME B94.9-2008 (R2018)*

| Nominal Size | Threads per Inch | Gage Measurement | | | Taper per Inch on Diameter[a] | | | | Reference Dimensions | |
|---|---|---|---|---|---|---|---|---|---|---|
| | | Projection Inch[b] | Tolerance ± | | Cut Thread | | Ground Thread | | $L_1$, Length[c] | Tap Drill Size NPT, ANPT, NPTF [d] |
| | | | Cut Thread | Ground Thread | Min. | Max. | Min. | Max. | | |
| 1⁄16 | 27 | 0.312 | 0.0625 | 0.0625 | 0.0599 | 0.0703 | 0.0599 | 0.0651 | 0.1600 | C |
| 1⁄8 | 27 | 0.312 | 0.0625 | 0.0625 | 0.0599 | 0.0703 | 0.0599 | 0.0651 | 0.1615 | Q |
| 1⁄4 | 18 | 0.459 | 0.0625 | 0.0625 | 0.0599 | 0.0703 | 0.0599 | 0.0651 | 0.2278 | 7⁄16 |
| 3⁄8 | 18 | 0.454 | 0.0625 | 0.0625 | 0.0599 | 0.0703 | 0.0599 | 0.0651 | 0.2400 | 9⁄16 |
| 1⁄2 | 14 | 0.579 | 0.0625 | 0.0625 | 0.0599 | 0.0677 | 0.0599 | 0.0651 | 0.3200 | 45⁄64 |
| 3⁄4 | 11.5 | 0.565 | 0.0625 | 0.0625 | 0.0599 | 0.0677 | 0.0599 | 0.0651 | 0.3390 | 29⁄32 |
| 1 | 11.5 | 0.678 | 0.0937 | 0.0937 | 0.0599 | 0.0677 | 0.0599 | 0.0651 | 0.4000 | 1 9⁄64 |
| 1 1⁄4 | 11.5 | 0.686 | 0.0937 | 0.0937 | 0.0599 | 0.0677 | 0.0599 | 0.0651 | 0.4200 | 1 31⁄64 |
| 1 1⁄2 | 11.5 | 0.699 | 0.0937 | 0.0937 | 0.0599 | 0.0677 | 0.0599 | 0.0651 | 0.4200 | 1 23⁄32 |
| 2 | 8 | 0.667 | 0.0937 | 0.0937 | 0.0599 | 0.0677 | 0.0599 | 0.0651 | 0.4360 | 2 3⁄16 |
| 2 1⁄2 | 8 | 0.925 | 0.0937 | 0.0937 | 0.0612 | 0.0664 | 0.0612 | 0.0651 | 0.6820 | 2 39⁄64 |
| 3 | 20 | 0.925 | 0.0937 | 0.0937 | 0.0612 | 0.0664 | 0.0612 | 0.0651 | 0.7660 | 3 15⁄16 |

[a] Taper is 0.0625 inch per 1.000 inch on diameter (1:16) (3⁄4 inch per 12 inches).

[b] Distance, small end of tap projects through $L_1$ taper ring gage.

[c] Dimension, $L_1$, thickness on thin ring gage; see ANSI/ASME B1.20.1 and B1.20.5.

[d] Given sizes permit direct tapping without reaming the hole, but only give full threads for approximate $L_1$ distance.

All dimensions are given in inches.

*Lead Tolerance:* ±0.003 inch per inch on cut thread, and ±0.0005 inch per inch on ground thread.

*Angle Tolerance:* ±40 min. in half angle and 60 min. in full angle for 8 cut threads per inch; ±45 min. in half angle and 68 min. in full angle for 11 1⁄2 to 27 cut threads per inch; ±25 min. in half angle for 8 ground threads per inch; and ±30 min. in half angle for 11 1⁄2 to 27 ground threads per inch.

**Table 24b. Taper Pipe Thread — Widths of Flats at Tap Crests and Roots for Cut Thread NPT and Ground Thread NPT, ANPT, and NPTF**
*ANSI/ASME B94.9-2008 (R2018)*

| Threads per Inch | Tap Flat Width at | Column I | | Column II | |
|---|---|---|---|---|---|
| | | NPT—Cut and Ground Thread[a] ANPT—Ground Thread[a] | | NPTF | Ground Thread[a] |
| | | Min.[b] | Max. | Min.[b] | Max. |
| 27 | Major diameter | 0.0014 | 0.0041 | 0.0040 | 0.0055 |
| | Minor diameter | … | 0.0041 | … | 0.0040 |
| 18 | Major diameter | 0.0021 | 0.0057 | 0.0050 | 0.0065 |
| | Minor diameter | … | 0.0057 | … | 0.0050 |
| 14 | Major diameter | 0.0027 | 0.0064 | 0.0050 | 0.0065 |
| | Minor diameter | … | 0.0064 | … | 0.0050 |
| 11 1⁄2 | Major diameter | 0.0033 | 0.0073 | 0.0060 | 0.0083 |
| | Minor diameter | … | 0.0073 | … | 0.0060 |
| 8 | Major diameter | 0.0048 | 0.0090 | 0.0080 | 0.0103 |
| | Minor diameter | … | 0.0090 | … | 0.0080 |

[a] Cut thread taps made to Column I are marked NPT but are not recommended for ANPT applications. Ground thread taps made to Column I are marked NPT and may be used for NPT and ANPT applications. Ground thread taps made to Column II are marked NPTF and used for dryseal application.

[b] Minimum minor diameter flats are not specified and may be as sharp as practicable.

All dimensions are given in inches.

**Table 25. Tap Thread Limits for Screw Thread Inserts (STI), Ground Thread, Machine Screw, and Fractional Size** *ANSI/ASME B94.9-2008 (R2018)*

| Nominal Screw Size STI | Fractional Size STI | Threads Per Inch | | Tap Major Diameter | | Pitch Diameter Limits | | | | | |
|---|---|---|---|---|---|---|---|---|---|---|---|
| | | | | | | 2B | | | 3B | | |
| | | NC | NF | Min. | Max. | H limit | Min. | Max. | H limit | Min. | Max. |
| 1 | … | 64 | … | 0.0948 | 0.0958 | H2 | 0.0837 | 0.0842 | H1 | 0.0832 | 0.0837 |
| 2 | … | 56 | … | 0.1107 | 0.1117 | H2 | 0.0981 | 0.0986 | H1 | 0.0976 | 0.0981 |
| | … | … | 64 | 0.1088 | 0.1088 | H2 | 0.0967 | 0.0972 | H1 | 0.0962 | 0.0967 |
| 3 | … | 48 | … | 0.1289 | 0.1289 | H2 | 0.1131 | 0.1136 | H1 | 0.1126 | 0.1131 |
| | … | … | 56 | 0.1237 | 0.1247 | H2 | 0.1111 | 0.1116 | H1 | 0.1106 | 0.1111 |
| 4 | … | 40 | … | 0.1463 | 0.1473 | H2 | 0.1288 | 0.1293 | H1 | 0.1283 | 0.1288 |
| | … | … | 48 | 0.1409 | 0.1419 | H2 | 0.1261 | 0.1266 | H1 | 0.1256 | 0.1261 |
| 5 | … | 40 | … | 0.1593 | 0.1603 | H2 | 0.1418 | 0.1423 | H1 | 0.1413 | 0.1418 |
| 6 | … | 32 | … | 0.1807 | 0.1817 | H3 | 0.1593 | 0.1598 | H2 | 0.1588 | 0.1593 |
| | … | … | 40 | 0.1723 | 0.1733 | H2 | 0.1548 | 0.1553 | H1 | 0.1543 | 0.1548 |
| 8 | … | 32 | … | 0.2067 | 0.2077 | H3 | 0.1853 | 0.1858 | H2 | 0.1848 | 0.1853 |
| | … | … | 36 | 0.2022 | 0.2032 | H2 | 0.1826 | 0.1831 | H1 | 0.1821 | 0.1826 |
| 10 | … | 24 | … | 0.2465 | 0.2475 | H3 | 0.2180 | 0.2185 | H2 | 0.2175 | 0.2180 |
| | … | … | 32 | 0.2327 | 0.2337 | H3 | 0.2113 | 0.2118 | H2 | 0.2108 | 0.2113 |
| 12 | … | 24 | … | 0.2725 | 0.2735 | H3 | 0.2440 | 0.2445 | H2 | 0.2435 | 0.2440 |
| … | ¼ | 20 | … | 0.3177 | 0.3187 | H3 | 0.2835 | 0.2840 | H2 | 0.2830 | 0.2835 |
| … | … | … | 28 | 0.2985 | 0.2995 | H3 | 0.2742 | 0.2747 | H2 | 0.2737 | 0.2742 |
| … | 5⁄16 | 18 | … | 0.3874 | 0.3884 | H4 | 0.3501 | 0.3506 | H3 | 0.3496 | 0.3501 |
| … | … | … | 24 | 0.3690 | 0.3700 | H3 | 0.3405 | 0.3410 | H2 | 0.3400 | 0.3405 |
| … | ⅜ | 16 | … | 0.4592 | 0.4602 | H4 | 0.4171 | 0.4176 | H3 | 0.4166 | 0.4171 |
| … | … | … | 24 | 0.4315 | 0.4325 | H3 | 0.4030 | 0.4035 | H2 | 0.4025 | 0.4030 |
| … | 7⁄16 | 14 | … | 0.5333 | 0.5343 | H4 | 0.4854 | 0.4859 | H3 | 0.4849 | 0.4854 |
| … | … | … | 20 | 0.5052 | 0.5062 | H4 | 0.4715 | 0.4720 | H3 | 0.4710 | 0.4715 |
| … | ½ | 13 | … | 0.6032 | 0.6042 | H4 | 0.5514 | 0.5519 | H3 | 0.5509 | 0.5514 |
| … | … | … | 20 | 0.5677 | 0.5687 | H4 | 0.5340 | 0.5345 | H3 | 0.5335 | 0.5340 |
| … | 9⁄16 | 12 | … | 0.6741 | 0.6751 | H4 | 0.6182 | 0.6187 | H3 | 0.6117 | 0.6182 |
| … | … | … | 18 | 06374 | 0.6384 | H4 | 0.6001 | 0.6006 | H3 | 0.5996 | 0.6001 |
| … | ⅝ | 11 | … | 0.7467 | 0.7477 | H4 | 0.6856 | 0.6861 | H3 | 0.6851 | 0.6856 |
| … | … | … | 18 | 0.6999 | 0.7009 | H4 | 0.6626 | 0.6631 | H3 | 0.6621 | 0.6626 |
| … | ¾ | 10 | … | 0.8835 | 0.8850 | H5 | 0.8169 | 0.8174 | H3 | 0.8159 | 0.8164 |
| … | … | … | 18 | 0.8342 | 0.8352 | H4 | 0.7921 | 0.7926 | H3 | 0.7916 | 0.7921 |
| … | ⅞ | 9 | … | 1.0232 | 1.0247 | H5 | 0.9491 | 0.9496 | H3 | 0.9481 | 0.9486 |
| … | … | … | 14 | 0.9708 | 0.9718 | H4 | 0.9234 | 0.9239 | H3 | 0.9224 | 0.9229 |
| … | 1 | 8 | … | 1.1666 | 1.1681 | H6 | 1.0832 | 1.0842 | H4 | 1.0822 | 1.0832 |
| … | … | … | 12 | 1.1116 | 1.1126 | H6 | 1.0562 | 1.0572 | H4 | 1.0552 | 1.0562 |
| … | … | … | 14 NS | 1.0958 | 1.0968 | H6 | 1.0484 | 1.0494 | H4 | 1.0474 | 1.0484 |
| … | 1⅛ | 7 | … | 1.3151 | 1.3171 | H6 | 1.2198 | 1.2208 | H4 | 1.2188 | 1.2198 |
| … | | | 12 | 1.2366 | 1.2376 | H6 | 1.1812 | 1.1822 | H4 | 1.1802 | 1.1812 |
| … | 1¼ | 7 | … | 1.4401 | 1.4421 | H6 | 1.3448 | 1.3458 | H4 | 1.3438 | 1.3448 |
| … | … | … | 12 | 1.3616 | 1.3626 | H6 | 1.3062 | 1.3072 | H4 | 1.3052 | 1.3062 |
| … | 1⅜ | 6 | … | 1.5962 | 1.5982 | H8 | 1.4862 | 1.4872 | H6 | 1.4852 | 1.4862 |
| … | … | … | 12 | 1.4866 | 1.4876 | H6 | 1.4312 | 1.4322 | H4 | 1.4302 | 1.4312 |
| … | 1½ | 6 | … | 1.7212 | 1.7232 | H8 | 1.6112 | 1.6122 | H6 | 1.6102 | 1.6112 |
| … | … | … | 12 | 1.6116 | 1.6126 | H6 | 1.5562 | 1.5572 | H4 | 1.5552 | 1.5562 |

These taps are over the nominal size to the extent that the internal thread they produce will accommodate a helical coil screw insert, which at final assembly will accept a screw thread of the normal size and pitch.

**Table 26a. Tap Thread Limits** *ANSI/ASME B94.9-2008 (R2018)* **for Screw Thread Inserts (STI), Ground Thread, Metric Size (Inch)**

| Metric Size STI | Pitch, mm | Tap Major Diameter, inch | | Tap Pitch Diameter Limits, inch | | | | | |
|---|---|---|---|---|---|---|---|---|---|
| | | | | Tolerance Class 4H | | | Tolerance Classes 5H and 6H | | |
| | | Min. | Max. | H limit | Min. | Max. | H limit | Min. | Max. |
| M2.5 | 0.45 | 0.1239 | 0.1229 | 1 | 0.1105 | 0.1100 | 2 | 0.1110 | 0.1105 |
| M3 | 0.5 | 0.1463 | 0.1453 | 1 | 0.1314 | 0.1309 | 2 | 0.1319 | 0.1314 |
| M3.5 | 0.6 | 0.1714 | 0.1704 | 1 | 0.1537 | 0.1532 | 2 | 0.1542 | 0.1537 |
| M4 | 0.7 | 0.1971 | 0.1955 | 2 | 0.1764 | 0.1759 | 3 | 0.1769 | 0.1764 |
| M5 | 0.8 | 0.2418 | 0.2403 | 2 | 0.2184 | 0.2179 | 3 | 0.2187 | 0.2184 |
| M6 | 1 | 0.2922 | 0.2906 | 2 | 0.2629 | 0.2624 | 3 | 0.2634 | 0.2629 |
| M7 | 1 | 0.3316 | 0.3300 | 2 | 0.3022 | 0.3017 | 3 | 0.3027 | 0.3022 |
| M8 | 1 | 0.3710 | 0.3694 | 2 | 0.3416 | 0.3411 | 3 | 0.3421 | 0.3416 |
| | 1.25 | 0.3853 | 0.3828 | 2 | 0.3480 | 0.3475 | 3 | 0.3485 | 0.3480 |
| M10 | 1 | 0.4497 | 0.4481 | 2 | 0.4203 | 0.4198 | 3 | 0.4208 | 0.4203 |
| | 1 | 0.4641 | 0.4616 | 2 | 0.4267 | 0.4262 | 3 | 0.4272 | 0.4267 |
| | 1.25 | 0.4776 | 0.4751 | 3 | 0.4336 | 0.4331 | 4 | 0.4341 | 0.4336 |
| M12 | 1.25 | 0.5428 | 0.5403 | 3 | 0.5059 | 0.5054 | 4 | 0.5064 | 0.5059 |
| | 1.5 | 0.5564 | 0.5539 | 3 | 0.5123 | 0.5118 | 4 | 0.5128 | 0.5123 |
| | 1.75 | 0.5700 | 0.5675 | 3 | 0.5187 | 0.5182 | 4 | 0.5192 | 0.5187 |
| M14 | 1.5 | 0.6351 | 0.6326 | 3 | 0.5911 | 0.5906 | 4 | 0.5916 | 0.5911 |
| | 2 | 0.6623 | 0.6598 | 3 | 0.6039 | 0.6034 | 4 | 0.6049 | 0.6044 |
| M16 | 1.5 | 0.7139 | 0.7114 | 3 | 0.6698 | 0.6693 | 4 | 0.6703 | 0.6698 |
| | 2 | 0.7410 | 0.7385 | 3 | 0.6826 | 0.6821 | 4 | 0.6836 | 0.6831 |
| M18 | 1.5 | 0.7926 | 0.7901 | 3 | 0.7485 | 0.7480 | 4 | 0.7490 | 0.7485 |
| | 2 | 0.8198 | 0.8173 | 3 | 0.7613 | 0.7608 | 4 | 0.7623 | 0.7618 |
| | 2.5 | 0.8470 | 0.8445 | 3 | 0.7741 | 0.7736 | 4 | 0.7751 | 0.7748 |
| M20 | 1.5 | 0.8713 | 0.8688 | 3 | 0.8273 | 0.8268 | 4 | 0.8278 | 0.8273 |
| | 2 | 0.8985 | 0.8960 | 3 | 0.8401 | 0.8396 | 4 | 0.8411 | 0.8406 |
| | 2.5 | 0.9257 | 0.9232 | 3 | 0.8529 | 0.8524 | 4 | 0.8539 | 0.8534 |
| M22 | 1.5 | 0.9500 | 0.9475 | 3 | 0.9060 | 0.9055 | 4 | 0.9065 | 0.9060 |
| | 2 | 0.9773 | 0.9748 | 3 | 0.9188 | 0.9183 | 5 | 0.9198 | 0.9193 |
| | 2.5 | 1.0044 | 1.0019 | 3 | 0.9316 | 0.9311 | 5 | 0.9326 | 0.9321 |
| M24 | 2 | 1.0559 | 1.0534 | 4 | 0.9981 | 0.9971 | 6 | 0.9991 | 0.9981 |
| | 3 | 1.1117 | 1.1078 | 4 | 1.0236 | 1.0226 | 6 | 1.0246 | 1.0236 |
| M27 | 2 | 1.1741 | 1.1716 | 4 | 1.1162 | 1.1152 | 6 | 1.1172 | 1.1162 |
| | 3 | 1.2298 | 1.2259 | 4 | 1.1417 | 1.1407 | 6 | 1.1427 | 1.1417 |
| M30 | 2 | 1.2922 | 1.2897 | 4 | 1.2343 | 1.2333 | 6 | 1.2353 | 1.2343 |
| | 3.5 | 1.3750 | 1.3711 | 4 | 1.2726 | 1.2716 | 6 | 1.2736 | 1.2726 |
| M33 | 2 | 1.4103 | 1.4078 | 4 | 1.3525 | 1.3515 | 6 | 1.3535 | 1.3525 |
| | 3 | 1.4931 | 1.4892 | 4 | 1.3907 | 1.3797 | 6 | 1.3917 | 1.3907 |
| M36 | 2 | 1.5284 | 1.5259 | 4 | 1.4706 | 1.4696 | 6 | 1.4716 | 1.4706 |
| | 3 | 1.5841 | 1.5802 | 6 | 1.4971 | 1.4961 | 8 | 1.4981 | 1.4971 |
| | 4 | 1.6384 | 1.6345 | 6 | 1.5226 | 1.5216 | 8 | 1.5236 | 1.5226 |
| M39 | 2 | 1.6465 | 1.6440 | 4 | 1.5887 | 1.5877 | 6 | 1.5897 | 1.5887 |
| | 3 | 1.7022 | 1.6983 | 6 | 1.6152 | 1.6142 | 8 | 1.6162 | 1.6152 |
| | 4 | 1.7565 | 1.7516 | 6 | 1.6407 | 1.6397 | 8 | 1.6417 | 1.6407 |

These taps are over the nominal size to the extent that the internal thread they produce will accommodate a helical coil screw insert, which at final assembly will accept a screw thread of the normal size and pitch.

STI basic thread dimensions are determined by adding twice the single thread height ($2 \times 0.64952P$) to the basic dimensions of the nominal thread size.

Formulas for major and pitch diameters are presented in MIL-T-21309E.

**Table 26b. Tap Thread Limits** *ANSI/ASME B94.9-2008 (R2018)* **for Screw Thread Inserts (STI), Ground Thread, Metric Size (mm)**

| Metric Size STI | Pitch, mm | Tap Major Diameter, mm | | Tap Pitch Diameter Limits, mm | | | | | |
|---|---|---|---|---|---|---|---|---|---|
| | | | | Tolerance Class 4H | | | Tolerance Classes 5H and 6H | | |
| | | Min. | Max. | H limit | Min. | Max. | H limit | Min. | Max. |
| M2.5 | 0.45 | 3.147 | 3.122 | 1 | 2.807 | 2.794 | 2 | 2.819 | 2.807 |
| M3 | 0.5 | 3.716 | 3.691 | 1 | 3.338 | 3.325 | 2 | 3.350 | 3.338 |
| M3.5 | 0.6 | 4.354 | 4.328 | 1 | 3.904 | 3.891 | 2 | 3.917 | 3.904 |
| M4 | 0.7 | 5.006 | 4.966 | 2 | 4.481 | 4.468 | 3 | 4.493 | 4.481 |
| M5 | 0.8 | 6.142 | 6.104 | 2 | 5.547 | 5.535 | 3 | 5.555 | 5.547 |
| M6 | 1 | 7.422 | 7.381 | 2 | 6.678 | 6.665 | 3 | 6.690 | 6.678 |
| M7 | 1 | 8.423 | 8.382 | 2 | 7.676 | 7.663 | 3 | 7.689 | 7.676 |
| M8 | 1 | 9.423 | 9.383 | 2 | 8.677 | 8.664 | 3 | 8.689 | 8.677 |
| | 1.25 | 9.787 | 9.723 | 2 | 8.839 | 8.827 | 3 | 8.852 | 8.839 |
| M10 | 1 | 11.422 | 11.382 | 2 | 10.676 | 10.663 | 3 | 10.688 | 10.676 |
| | 1 | 11.788 | 11.725 | 2 | 10.838 | 10.825 | 3 | 10.851 | 10.838 |
| | 1.25 | 12.131 | 12.068 | 3 | 11.013 | 11.001 | 4 | 11.026 | 11.013 |
| M12 | 1.25 | 13.787 | 13.724 | 3 | 12.850 | 12.837 | 4 | 12.863 | 12.850 |
| | 1.5 | 14.133 | 14.069 | 3 | 13.012 | 13.000 | 4 | 13.025 | 13.012 |
| | 1.75 | 14.478 | 14.415 | 3 | 13.175 | 13.162 | 4 | 13.188 | 13.175 |
| M14 | 1.5 | 16.132 | 16.068 | 3 | 15.014 | 15.001 | 4 | 15.027 | 15.014 |
| | 2 | 16.822 | 16.759 | 3 | 15.339 | 15.326 | 4 | 15.364 | 15.352 |
| M16 | 1.5 | 18.133 | 18.070 | 3 | 17.013 | 17.000 | 4 | 17.026 | 17.013 |
| | 2 | 18.821 | 18.758 | 3 | 17.338 | 17.325 | 4 | 17.363 | 17.351 |
| M18 | 1.5 | 20.132 | 20.069 | 3 | 19.012 | 18.999 | 4 | 19.025 | 19.012 |
| | 2 | 20.823 | 20.759 | 3 | 19.337 | 19.324 | 4 | 19.362 | 19.350 |
| | 2.5 | 21.514 | 21.450 | 3 | 19.662 | 19.649 | 4 | 19.688 | 19.675 |
| M20 | 1.5 | 22.131 | 22.068 | 3 | 21.013 | 21.001 | 4 | 21.026 | 21.013 |
| | 2 | 22.822 | 22.758 | 3 | 21.339 | 21.326 | 4 | 21.364 | 21.351 |
| | 2.5 | 23.513 | 23.449 | 3 | 21.664 | 21.651 | 4 | 21.689 | 21.676 |
| M22 | 1.5 | 24.130 | 24.067 | 3 | 23.012 | 23.000 | 4 | 23.025 | 23.012 |
| | 2 | 24.823 | 24.760 | 3 | 23.338 | 23.325 | 5 | 23.363 | 23.350 |
| | 2.5 | 25.512 | 25.448 | 3 | 23.663 | 23.650 | 5 | 23.688 | 23.675 |
| M24 | 2 | 26.820 | 26.756 | 4 | 25.352 | 25.352 | 6 | 25.377 | 25.352 |
| | 3 | 28.237 | 28.132 | 4 | 25.999 | 25.974 | 6 | 26.025 | 25.999 |
| M27 | 2 | 29.822 | 29.759 | 4 | 28.351 | 28.326 | 6 | 28.377 | 28.351 |
| | 3 | 31.237 | 31.138 | 4 | 28.999 | 28.974 | 6 | 29.025 | 28.999 |
| M30 | 2 | 32.822 | 32.758 | 4 | 31.351 | 31.326 | 6 | 31.377 | 31.351 |
| | 3.5 | 34.925 | 34.826 | 4 | 32.324 | 32.299 | 6 | 32.349 | 32.324 |
| M33 | 2 | 35.822 | 35.758 | 4 | 34.354 | 34.324 | 6 | 34.379 | 34.354 |
| | 3 | 37.925 | 37.826 | 4 | 35.324 | 35.298 | 6 | 35.349 | 35.324 |
| M36 | 2 | 38.821 | 38.758 | 4 | 37.353 | 37.328 | 6 | 37.379 | 37.353 |
| | 3 | 40.236 | 40.137 | 6 | 38.026 | 38.001 | 8 | 38.052 | 38.026 |
| | 4 | 41.615 | 41.516 | 6 | 38.674 | 38.649 | 8 | 38.699 | 37.674 |
| M39 | 2 | 41.821 | 41.758 | 4 | 40.353 | 40.328 | 6 | 40.378 | 40.353 |
| | 3 | 43.236 | 43.137 | 6 | 41.026 | 4 1.001 | 8 | 41.051 | 41.026 |
| | 4 | 44.615 | 44.516 | 6 | 41.674 | 41.648 | 8 | 41.699 | 41.674 |

These taps are over the nominal size to the extent that the internal thread they produce will accommodate a helical coil screw insert, which at final assembly will accept a screw thread of the normal size and pitch.

STI basic thread dimensions are determined by adding twice the single thread height ($2 \times 0.64952P$) to the basic dimensions of the nominal thread size.

Formulas for major and pitch diameters are presented in MIL-T-21309E.

## Acme and Square-Threaded Taps

These taps are usually made in sets, three taps in a set being the most common. For very fine pitches, two taps in a set will be found sufficient, whereas as many as five taps in a set are used for coarse pitches. The table on the next page gives dimensions for proportioning both Acme and square-threaded taps when made in sets. In cutting the threads of square-threaded taps, one leading tap maker uses the following rules: The width of the groove between two threads is made equal to one-half the pitch of the thread, less 0.004 inch (0.102 mm), making the width of the thread itself equal to one-half of the pitch, plus 0.004 inch (0.102 mm). The depth of the thread is made equal to 0.45 times the pitch, plus 0.0025 inch (0.064 mm). This latter rule produces a thread that for all the ordinarily used pitches for square-threaded taps has a depth less than the generally accepted standard depth, this latter depth being equal to one-half the pitch. The object of this shallow thread is to ensure that if the hole to be threaded by the tap is not bored out so as to provide clearance at the bottom of the thread, the tap will cut its own clearance. The hole should, however, always be drilled out large enough so that the cutting of the clearance is not required of the tap.

The table, *Dimensions of Acme Threads Taps in Sets of Three Taps*, may also be used for the length dimensions for Acme taps. The dimensions in this table apply to single-threaded taps. For multiple-threaded taps or taps with very coarse pitch relative to the diameter, the length of the chamfered part of the thread may be increased. Square-threaded taps are made to the same table as Acme taps, with the exception of the figures in column *K*, which for square-threaded taps should be equal to the nominal diameter of the tap, no oversize allowance being customary in these taps. The first tap in a set of Acme taps (not square-threaded taps) should be turned to a taper at the bottom of the thread for a distance of about one-quarter of the length of the threaded part. The taper should be so selected that the root diameter is about $1/32$ inch (0.794 mm) smaller at the point than the proper root diameter of the tap. The first tap should preferably be provided with a short pilot at the point. For very coarse pitches, the first tap may be provided with spiral flutes at right angles to the angle of the thread. Acme and square-threaded taps should be relieved or backed off on the top of the thread of the chamfered portion on all the taps in the set. When the taps are used as machine taps, rather than as hand taps, they should be relieved in the angle of the thread, as well as on the top, for the whole length of the chamfered portion. Acme taps should also always be relieved on the front side of the thread to within $1/32$ inch (0.794 mm) of the cutting edge.

**Adjustable Taps.—**Many adjustable taps are now used, especially for accurate work. Some taps of this class are made of a solid piece of tool steel that is split and provided with means of expanding sufficiently to compensate for wear. Most of the larger adjustable taps have inserted blades or chasers that are held rigidly but are capable of radial adjustment. The use of taps of this general class enables standard sizes to be maintained readily.

**Drill Hole Sizes for Acme Threads.—**Many tap and die manufacturers and vendors make available to their customers computer programs designed to calculate drill hole sizes for all the Acme threads in their ranges from the basic dimensions. The large variety and combination of dimensions for such tools prevent inclusion of a complete set of tables of tap drills for Acme taps in this *Handbook*. The following formulas (dimensions in inches) for calculating drill hole sizes for Acme threads are derived from the American National Standard, ANSI/ASME B1.5-1997 (R2014), Acme Screw Threads.

To select a tap drill size for an Acme thread, first calculate the maximum and minimum internal product minor diameters for the thread to be produced. (Dimensions for general purpose, centralizing, and stub Acme screw threads are given in the Threads and Threading section, starting on page 2057.) Then select a drill that will yield a finished hole somewhere between the established maximum and minimum product minor diameters. To reduce the amount of material to be removed when cutting the thread, consider staying close to the maximum product limit in selecting the hole size. If there is no standard drill size that matches the hole diameter selected, it may be necessary to drill and ream, or bore, the hole to size, to achieve the required hole diameter.

**Table 27. Dimensions of Acme Threads Taps in Sets of Three Taps**

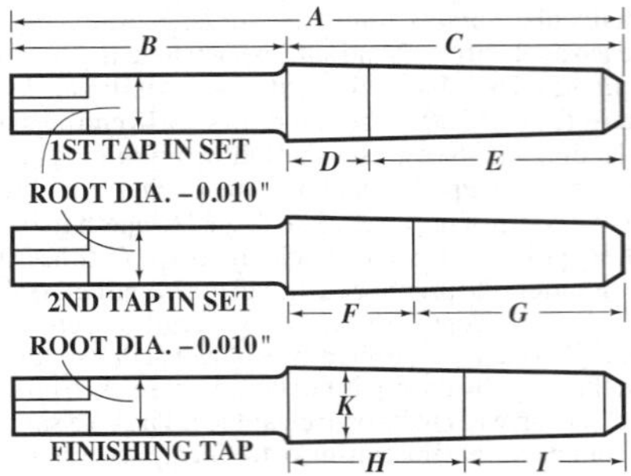


| Nominal Dia. | A | B | C | D | E | F | G | H | I | K |
|---|---|---|---|---|---|---|---|---|---|---|
| $\frac{1}{2}$ | $4\frac{1}{4}$ | $1\frac{7}{8}$ | $2\frac{3}{8}$ | $\frac{1}{2}$ | $1\frac{7}{8}$ | $\frac{5}{8}$ | $1\frac{3}{4}$ | $\frac{7}{8}$ | $1\frac{1}{2}$ | 0.520 |
| $\frac{9}{16}$ | $4\frac{7}{8}$ | $2\frac{1}{8}$ | $2\frac{3}{4}$ | $\frac{9}{16}$ | $2\frac{3}{16}$ | $\frac{3}{4}$ | 2 | 1 | $1\frac{3}{4}$ | 0.582 |
| $\frac{5}{8}$ | $5\frac{1}{2}$ | $2\frac{3}{8}$ | $3\frac{1}{8}$ | $\frac{5}{8}$ | $2\frac{1}{2}$ | $\frac{7}{8}$ | $2\frac{1}{4}$ | $1\frac{1}{8}$ | 2 | 0.645 |
| $\frac{11}{16}$ | 6 | $2\frac{1}{2}$ | $3\frac{1}{2}$ | $3\frac{13}{16}$ | $2\frac{13}{16}$ | $\frac{15}{16}$ | $2\frac{9}{16}$ | $1\frac{1}{4}$ | $2\frac{1}{4}$ | 0.707 |
| $\frac{3}{4}$ | $6\frac{1}{2}$ | $2\frac{11}{16}$ | $3\frac{13}{16}$ | $\frac{11}{16}$ | $3\frac{1}{8}$ | 1 | $2\frac{13}{16}$ | $1\frac{3}{8}$ | $2\frac{7}{16}$ | 0.770 |
| $\frac{13}{16}$ | $6\frac{7}{8}$ | $2\frac{13}{16}$ | $4\frac{1}{16}$ | $\frac{3}{4}$ | $3\frac{5}{16}$ | $1\frac{1}{16}$ | 3 | $1\frac{7}{16}$ | $2\frac{5}{8}$ | 0.832 |
| $\frac{7}{8}$ | $7\frac{1}{4}$ | 3 | $4\frac{1}{4}$ | $\frac{3}{4}$ | $3\frac{1}{2}$ | $1\frac{1}{8}$ | $3\frac{1}{8}$ | $1\frac{1}{2}$ | $2\frac{3}{4}$ | 0.895 |
| $\frac{15}{16}$ | $7\frac{9}{16}$ | $3\frac{1}{8}$ | $4\frac{7}{16}$ | $\frac{13}{16}$ | $3\frac{5}{8}$ | $1\frac{3}{16}$ | $3\frac{1}{4}$ | $1\frac{9}{16}$ | $2\frac{7}{8}$ | 0.957 |
| 1 | $7\frac{7}{8}$ | $3\frac{1}{4}$ | $4\frac{5}{8}$ | $\frac{13}{16}$ | $3\frac{13}{16}$ | $1\frac{1}{4}$ | $3\frac{3}{8}$ | $1\frac{5}{8}$ | 3 | 1.020 |
| $1\frac{1}{8}$ | $8\frac{1}{2}$ | $3\frac{9}{16}$ | $4\frac{15}{16}$ | $\frac{7}{8}$ | $4\frac{1}{16}$ | $1\frac{5}{16}$ | $3\frac{5}{8}$ | $1\frac{3}{4}$ | $3\frac{3}{16}$ | 1.145 |
| $1\frac{1}{4}$ | 9 | $3\frac{3}{4}$ | $5\frac{1}{4}$ | $\frac{15}{16}$ | $4\frac{5}{16}$ | $1\frac{3}{8}$ | $3\frac{7}{8}$ | $1\frac{7}{8}$ | $3\frac{3}{8}$ | 1.270 |
| $1\frac{3}{8}$ | $9\frac{1}{2}$ | 4 | $5\frac{1}{2}$ | 1 | $4\frac{1}{2}$ | $1\frac{7}{16}$ | $4\frac{1}{16}$ | 2 | $3\frac{1}{2}$ | 1.395 |
| $1\frac{1}{2}$ | 10 | $4\frac{1}{4}$ | $5\frac{3}{4}$ | 1 | $4\frac{3}{4}$ | $1\frac{1}{2}$ | $4\frac{1}{4}$ | $2\frac{1}{8}$ | $3\frac{5}{8}$ | 1.520 |
| $1\frac{5}{8}$ | $10\frac{1}{2}$ | $4\frac{1}{2}$ | 6 | 1 | 5 | $1\frac{1}{2}$ | $4\frac{1}{2}$ | $2\frac{1}{8}$ | $3\frac{7}{8}$ | 1.645 |
| $1\frac{3}{4}$ | 11 | $4\frac{3}{4}$ | $6\frac{1}{4}$ | $1\frac{1}{16}$ | $5\frac{3}{16}$ | $1\frac{9}{16}$ | $4\frac{11}{16}$ | $2\frac{1}{4}$ | 4 | 1.770 |
| $1\frac{7}{8}$ | $11\frac{3}{8}$ | $4\frac{7}{8}$ | $6\frac{1}{2}$ | $1\frac{1}{16}$ | $5\frac{7}{16}$ | $1\frac{9}{16}$ | $4\frac{15}{16}$ | $2\frac{1}{4}$ | $4\frac{1}{4}$ | 1.895 |
| 2 | $11\frac{3}{4}$ | 5 | $6\frac{3}{4}$ | $1\frac{1}{8}$ | $5\frac{5}{8}$ | $1\frac{5}{8}$ | $5\frac{1}{8}$ | $2\frac{3}{8}$ | $4\frac{3}{8}$ | 2.020 |
| $2\frac{1}{4}$ | $12\frac{1}{2}$ | $5\frac{1}{4}$ | $7\frac{1}{4}$ | $1\frac{1}{8}$ | $6\frac{1}{8}$ | $1\frac{3}{16}$ | $5\frac{1}{2}$ | $2\frac{1}{2}$ | $4\frac{3}{4}$ | 2.270 |
| $2\frac{1}{2}$ | $13\frac{1}{4}$ | $5\frac{1}{2}$ | $7\frac{3}{4}$ | $1\frac{3}{4}$ | $6\frac{9}{16}$ | $1\frac{7}{8}$ | $5\frac{7}{8}$ | $2\frac{5}{8}$ | $5\frac{1}{8}$ | 2.520 |
| $2\frac{3}{4}$ | 14 | $5\frac{3}{4}$ | $8\frac{1}{4}$ | $1\frac{1}{4}$ | 7 | 2 | $6\frac{1}{4}$ | $2\frac{3}{4}$ | $5\frac{1}{2}$ | 2.770 |
| 3 | 15 | $6\frac{1}{4}$ | $8\frac{3}{4}$ | $1\frac{1}{4}$ | $7\frac{1}{2}$ | 2 | $6\frac{3}{4}$ | 3 | $5\frac{3}{4}$ | 3.020 |

*Diameters of General-Purpose Acme Screw Threads of Classes 2G, 3G, and 4G* may be calculated from pitch = 1/number of threads per inch, and:

minimum diameter = basic major diameter − pitch

maximum diameter = minimum minor diameter + 0.05 × pitch

**Table 28. Proportions of Acme and Square-Threaded Taps Made in Sets**

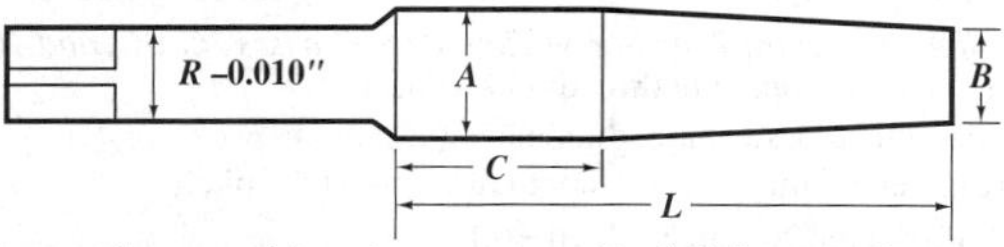


**$R$ = root diameter of thread** **$D$ = full diameter of tap**
**$T$ = double depth of full thread**

| Types of Tap | No. of Taps in Set | Order of Tap in Set | $A$ | $B$ | $C$ |
|---|---|---|---|---|---|
| Acme Thread Taps | 2 | 1st | $R + 0.65T$ | $R + 0.010$ | $\frac{1}{8}L$ to $\frac{1}{6}L$ |
| | | 2d | $D$ | $A$ on 1st tap − 0.005 | $\frac{1}{4}L$ to $\frac{1}{3}L$ |
| | 3 | 1st | $R + 0.45T$ | $R + 0.010$ | $\frac{1}{8}L$ to $\frac{1}{6}L$ |
| | | 2d | $R + 0.80T$ | $A$ on 1st tap − 0.005 | $\frac{1}{6}L$ to $\frac{1}{4}L$ |
| | | 3d | $D$ | $A$ on 2d tap − 0.005 | $\frac{1}{4}L$ to $\frac{1}{3}L$ |
| | 4 | 1st | $R + 0.40T$ | $R + 0.010$ | $\frac{1}{8}L$ |
| | | 2d | $R + 0.70T$ | $A$ on 1st tap − 0.005 | $\frac{1}{6}L$ |
| | | 3d | $R + 0.90T$ | $A$ on 2d tap − 0.005 | $\frac{1}{5}L$ |
| | | 4th | $D$ | $A$ on 3d tap − 0.005 | $\frac{1}{4}L$ to $\frac{1}{3}L$ |
| | 5 | 1st | $R + 0.37T$ | $R + 0.010$ | $\frac{1}{8}L$ |
| | | 2d | $R + 0.63T$ | $A$ on 1st tap − 0.005 | $\frac{1}{6}L$ |
| | | 3d | $R + 0.82T$ | $A$ on 2d tap − 0.005 | $\frac{1}{5}L$ |
| | | 4th | $R + 0.94T$ | $A$ on 3d tap − 0.005 | $\frac{1}{5}L$ to $\frac{1}{4}L$ |
| | | 5th | $D$ | $A$ on 4th tap − 0.005 | $\frac{1}{4}L$ to $\frac{1}{3}L$ |
| Square-Threaded Taps | 2 | 1st | $R + 0.67T$ | $R$ | $\frac{1}{8}L$ to $\frac{1}{6}L$ |
| | | 2d | $D$ | $A$ on 1st tap − 0.005 | $\frac{1}{4}L$ to $\frac{1}{3}L$ |
| | 3 | 1st | $R + 0.41T$ | $R$ | $\frac{1}{8}L$ to $\frac{1}{6}L$ |
| | | 2d | $R + 0.080T$ | $A$ on 1st tap − 0.005 | $\frac{1}{6}L$ to $\frac{1}{4}L$ |
| | | 3d | $D$ | $A$ on 2d tap − 0.005 | $\frac{1}{4}L$ to $\frac{1}{3}L$ |
| | 4 | 1st | $R + 0.32T$ | $R$ | $\frac{1}{8}L$ |
| | | 2d | $R + 0.62T$ | $A$ on 1st tap − 0.005 | $\frac{1}{6}L$ |
| | | 3d | $R + 0.90T$ | $A$ on 2d tap − 0.005 | $\frac{1}{5}L$ |
| | | 4th | $D$ | $A$ on 3d tap − 0.005 | $\frac{1}{4}L$ to $\frac{1}{3}L$ |
| | 5 | 1st | $R + 0.26T$ | $R$ | $\frac{1}{8}L$ |
| | | 2d | $R + 0.50T$ | $A$ on 1st tap − 0.005 | $\frac{1}{6}L$ |
| | | 3d | $R + 0.72T$ | $A$ on 2d tap − 0.005 | $\frac{1}{5}L$ |
| | | 4th | $R + 0.92T$ | $A$ on 3d tap − 0.005 | $\frac{1}{5}L$ to $\frac{1}{4}L$ |
| | | 5th | $D$ | $A$ on 4th tap − 0.005 | $\frac{1}{4}L$ to $\frac{1}{3}L$ |

*Example:* $\frac{1}{2}$-10 Acme 2G, pitch = 1/10 = 0.1
minimum diameter = 0.5 − 0.1 = 0.4

maximum diameter = 0.4 + (0.05 × 0.1) = 0.405

drill selected = letter X or 0.3970 + 0.0046 (probable oversize) = 0.4016

*Diameters of Acme Centralizing Screw Threads of Classes 2C, 3C, and 4C* may be calculated from pitch = 1/number of threads per inch, and:

minimum diameter = basic major diameter − 0.9 × pitch

maximum diameter = minimum minor diameter + 0.05 × pitch

*Example:* $\frac{1}{2}$-10 Acme 2C, pitch = 1/10 = 0.1

minimum diameter = 0.5 − (0.9 × 0.1) = 0.41

maximum diameter = 0.41 + (0.05 × 0.1) = 0.415

drill selected = $\frac{13}{32}$ or 0.4062 + 0.0046 (probable oversize) = 0.4108.

*Diameters for Acme Centralizing Screw Threads of Classes 5C and 6C:* These classes are not recommended for new designs, but may be calculated from:

minimum diameter = [basic major diameter − (0.025 √ basic major diameter)] − 0.9 × pitch

maximum diameter = minimum minor diameter + 0.05 × pitch

pitch = 1/number of threads per inch

*Example:* $\frac{1}{2}$-10 Acme 5C, pitch = 1/10 = 0.1

minimum diameter = [0.5 − (0.025 √0.5)] − (0.9 × 0.1) = 0.3923

maximum diameter = 0.3923 + (0.05 × 0.1) = 0.3973

drill selected = $\frac{25}{64}$ or 0.3906 + 0.0046 (probable oversize) = 0.3952

**Tapping Square Threads.—**If it is necessary to tap square threads, this should be done by using a set of taps that will form the thread by a progressive cutting action, the taps varying in size in order to distribute the work, especially for threads of comparatively coarse pitch. From three to five taps may be required in a set, depending upon the pitch. Each tap should have a pilot to steady it. The pilot of the first tap has a smooth cylindrical end from 0.003 to 0.005 inch (0.076-0.127 mm) smaller than the hole, and the pilots of following taps should have teeth.

**Collapsible Taps.—**The collapsing tap shown in the accompanying illustration is one of many different designs that are manufactured. These taps are often used in turret lathe practice in place of solid taps. When using this particular style of collapsing tap, the adjustable gage *A* is set for the length of thread required. When the tap has been fed to this depth, the gage comes into contact with the end of the work, which causes the chasers to collapse automatically. The tool is then withdrawn, after which the chasers are again expanded and locked in position by the handle seen at the side of the holder.

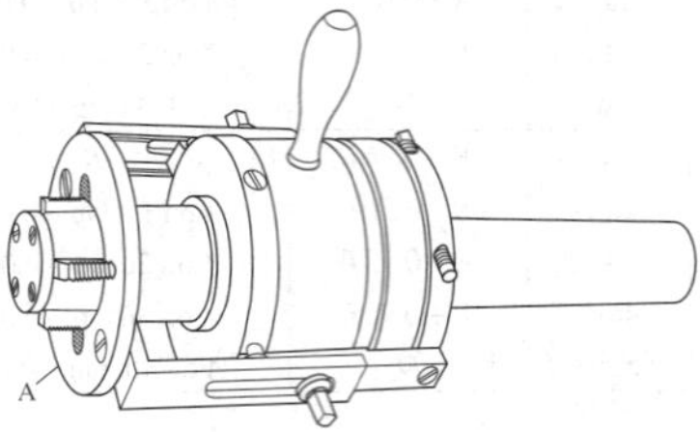


Collapsing Tap

Collapsible taps do not need to be backed out of the hole at the completion of the thread, reducing the tapping time and increasing production rates.

## STANDARD TAPERS

Certain types of small tools and machine parts, such as twist drills, end mills, arbors, lathe centers, etc., are provided with taper shanks which fit into spindles or sockets of corresponding taper, providing not only accurate alignment between the tool or other part and its supporting member, but also more or less frictional resistance for driving the tool. There are several standards for "self-holding" tapers, but the American National, the Morse, and the Brown & Sharpe are the standards most widely used by American manufacturers.

The name *self-holding* has been applied to the smaller tapers—like the Morse and the Brown & Sharpe—because, where the angle of the taper is only 2 or 3 degrees, the shank of a tool is so firmly seated in its socket that there is considerable frictional resistance to any force tending to turn or rotate the tool relative to the socket. The term "self-holding" is used to distinguish relatively small tapers from the larger or *self-releasing* type. A milling machine spindle having a taper of 3½ inches per foot is an example of a self-releasing taper. The included angle in this case is over 16 degrees and the tool or arbor requires a positive locking device to prevent slipping, but the shank may be released or removed more readily than one having a smaller taper of the self-holding type.

**Tapers for Machine Tool Spindles.—**Various standard tapers have been used for the taper holes in the spindles of machine tools, such as drilling machines, lathes, milling machines, or other types requiring a taper hole for receiving either the shank of a cutter, an arbor, a center, or any tool or accessory requiring a tapering seat. The Morse taper represents a generally accepted standard for drilling machines. See more on this subject, page 1012.

The headstock and tailstock spindles of lathes also have the Morse taper in most cases; but the Jarno, the Reed (which is the short Jarno), and the Brown & Sharpe have also been used. Milling machine spindles formerly had Brown & Sharpe tapers in most cases.

In 1927, the milling machine manufacturers of the National Machine Tool Builders' Association adopted a standard taper of 3½ inches per foot. This comparatively steep taper has the advantage of insuring instant release of arbors or adapters.

**National Machine Tool Builders' Association Tapers**

| Taper Number[a] | Large End Diameter | Taper Number[a] | Large End Diameter |
|---|---|---|---|
| 30 | 1¼ | 50 | 2¾ |
| 40 | 1¾ | 60 | 4¼ |

[a] Standard taper of 3½ inches per foot

The British Standard for milling machine spindles is also 3½ inches taper per foot and includes these large end diameters: 1⅜ inches, 1¾ inches, 2¾ inches, and 3¼ inches.

**Morse Taper.—**Dimensions relating to Morse standard taper shanks and sockets may be found in an accompanying table. The taper for different numbers of Morse tapers is slightly different, but it is approximately ⅝ inch per foot in most cases. The table gives the actual tapers, accurate to five decimal places. Morse taper shanks are used on a variety of tools, and exclusively on the shanks of twist drills. Dimensions for *Morse Stub Taper Shanks* are given in Table 1a, and for *Morse Standard Taper Shanks* in Table 1b. Also see Table 8 and Table 9 on page 1010.

**Brown & Sharpe Taper.—**This standard taper is used for taper shanks on tools such as end mills and reamers, the taper being approximately ½ inch per foot for all sizes except for taper No. 10, where the taper is 0.5161 inch per foot. Brown & Sharpe taper sockets are used for many arbors, collets, and machine tool spindles, especially milling machines and grinding machines. In many cases there are a number of different lengths of sockets corresponding to the same number of taper; all these tapers, however, are of the same diameter at the small end. See Table 10, page 1011.

## Table 1a. Morse Stub Taper Shanks

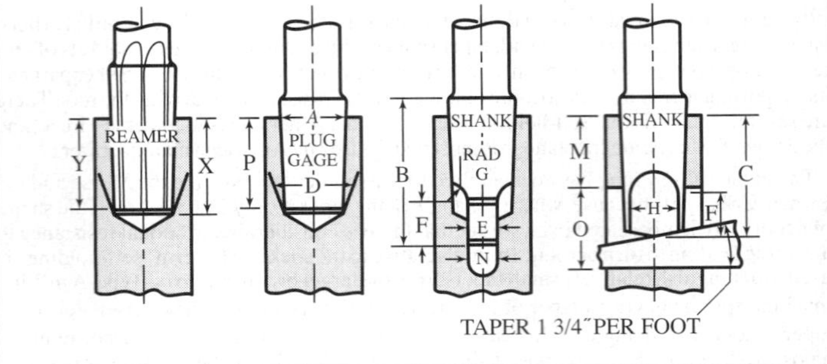


| No. of Taper | Taper per Foot[a] | Taper per Inch[b] | Small End of Plug,[b] *D* | Dia. End of Socket,[a] *A* | Shank: Total Length, *B* | Shank: Depth, *C* | Tang: Thickness, *E* | Tang: Length, *F* |
|---|---|---|---|---|---|---|---|---|
| 1 | 0.59858 | 0.049882 | 0.4314 | 0.475 | 1 5/16 | 1 1/8 | 13/64 | 5/16 |
| 2 | 0.59941 | 0.049951 | 0.6469 | 0.700 | 1 11/16 | 1 7/16 | 19/64 | 7/16 |
| 3 | 0.60235 | 0.050196 | 0.8753 | 0.938 | 2 | 1 3/4 | 25/64 | 9/16 |
| 4 | 0.62326 | 0.051938 | 1.1563 | 1.231 | 2 3/8 | 2 1/16 | 33/64 | 11/16 |
| 5 | 0.63151 | 0.052626 | 1.6526 | 1.748 | 3 | 2 11/16 | 3/4 | 15/16 |

| No. of Taper | Tang: Radius of Mill, *G* | Tang: Diameter, *H* | Socket: Plug Depth, *P* | Socket: Min. Depth of Tapered Hole, Drilled *X* | Socket: Min. Depth of Tapered Hole, Reamed *Y* | Socket: Socket End to Tang Slot, *M* | Tang Slot: Width, *N* | Tang Slot: Length, *O* |
|---|---|---|---|---|---|---|---|---|
| 1 | 3/16 | 13/32 | 7/8 | 15/16 | 29/32 | 25/32 | 7/32 | 23/32 |
| 2 | 7/32 | 39/64 | 1 1/16 | 1 5/32 | 1 7/64 | 15/16 | 5/16 | 15/16 |
| 3 | 9/32 | 13/16 | 1 1/4 | 1 3/8 | 1 5/16 | 1 1/16 | 13/32 | 1 1/8 |
| 4 | 3/8 | 1 3/32 | 1 7/16 | 1 9/16 | 1 1/2 | 1 3/16 | 17/32 | 1 3/8 |
| 5 | 9/16 | 1 19/32 | 1 13/16 | 1 15/16 | 1 7/8 | 1 7/16 | 25/32 | 1 3/4 |

All dimensions in inches.

Radius *J* is 3/64, 1/16, 5/64, 3/32, and 1/8 inch respectively for Nos. 1, 2, 3, 4, and 5 tapers.

[a] These are basic dimensions.

[b] These dimensions are calculated for reference only.

**Jarno Taper.—**The Jarno taper was originally proposed by Oscar J. Beale of the Brown & Sharpe Mfg. Co. This taper is based on such simple formulas that practically no calculations are required when the number of taper is known. The taper per foot of all Jarno taper sizes is 0.600 inch on the diameter. The diameter at the large end is as many eighths, the diameter at the small end is as many tenths, and the length as many half inches as are indicated by the number of the taper. For example, a No. 7 Jarno taper is $\frac{7}{8}$ inch in diameter at the large end; $\frac{7}{10}$, or 0.700 inch at the small end; and $\frac{7}{2}$, or $3\frac{1}{2}$ inches long; hence, diameter at large end = No. of taper ÷ 8; diameter at small end = No. of taper ÷ 10; length of taper = No. of taper ÷ 2. The Jarno taper is used on various machine tools, especially profiling machines and die-sinking machines. It has also been used for the headstock and tailstock spindles of some lathes.

**Table 1b. Morse Standard Taper Shanks**

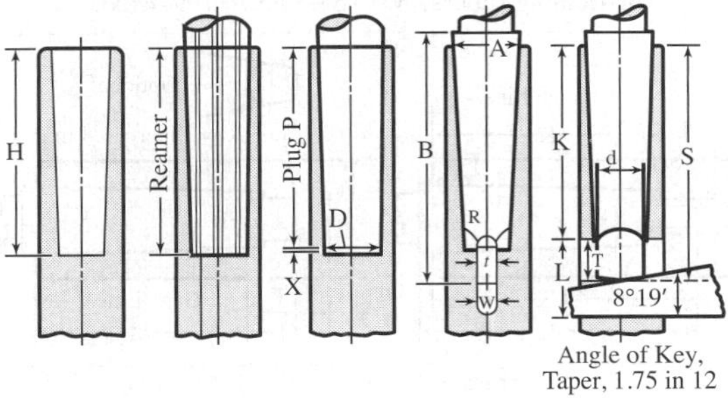


| No. of Taper | Taper per Foot | Taper per Inch | Small End of Plug $D$ | Diameter End of Socket $A$ | Shank Length $B$ | Shank Depth $S$ | Depth of Hole $H$ |
|---|---|---|---|---|---|---|---|
| 0 | 0.62460 | 0.05205 | 0.252 | 0.3561 | 2 11/32 | 2 7/32 | 2 1/32 |
| 1 | 0.59858 | 0.04988 | 0.369 | 0.475 | 2 9/16 | 2 7/16 | 2 5/32 |
| 2 | 0.59941 | 0.04995 | 0.572 | 0.700 | 3 1/8 | 2 15/16 | 2 39/64 |
| 3 | 0.60235 | 0.05019 | 0.778 | 0.938 | 3 7/8 | 3 11/16 | 3 1/4 |
| 4 | 0.62326 | 0.05193 | 1.020 | 1.231 | 4 7/8 | 4 5/8 | 4 1/8 |
| 5 | 0.63151 | 0.05262 | 1.475 | 1.748 | 6 1/8 | 5 7/8 | 5 1/4 |
| 6 | 0.62565 | 0.05213 | 2.116 | 2.494 | 8 9/16 | 8 1/4 | 7 21/64 |
| 7 | 0.62400 | 0.05200 | 2.750 | 3.270 | 11 5/8 | 11 1/4 | 10 5/64 |
| Plug Depth $P$ | Tang or Tongue Thickness $t$ | Tang or Tongue Length $T$ | Tang or Tongue Radius $R$ | Tang or Tongue Dia. $d$ | Keyway Width $W$ | Keyway Length $L$ | Keyway to End $K$ |
| 2 | 0.1562 | 1/4 | 5/32 | 0.235 | 11/64 | 9/16 | 1 15/16 |
| 2 1/8 | 0.2031 | 3/8 | 3/16 | 0.343 | 0.218 | 3/4 | 2 1/16 |
| 2 9/16 | 0.2500 | 7/16 | 1/4 | 17/32 | 0.266 | 7/8 | 2 1/2 |
| 3 3/16 | 0.3125 | 9/16 | 9/32 | 23/32 | 0.328 | 1 3/16 | 3 1/16 |
| 4 1/16 | 0.4687 | 5/8 | 5/16 | 31/32 | 0.484 | 1 1/4 | 3 7/8 |
| 5 3/16 | 0.6250 | 3/4 | 3/8 | 1 13/32 | 0.656 | 1 1/2 | 4 15/16 |
| 7 1/4 | 0.7500 | 1 1/8 | 1/2 | 2 | 0.781 | 1 3/4 | 7 |
| 10 | 1.1250 | 1 3/8 | 3/4 | 2 5/8 | 1.156 | 2 5/8 | 9 1/2 |

Tolerances on rate of taper: all sizes 0.002 in. per foot. This tolerance may be applied on shanks only in the direction that increases the rate of taper, and on sockets only in the direction that decreases the rate of taper.

**American National Standard Machine Tapers.—**This Standard includes a self-holding series (Table 2, Table 3, Table 4, Table 5 and Table 7a) and a steep taper series, Table 6. The self-holding taper series consists of 22 sizes, which are listed in Table 7a. The reference gage for the self-holding tapers is a plug gage. Table 7b gives the dimensions and tolerances for both plug and ring gages applying to this series. Table 2 through Table 5 inclusive give the dimensions for self-holding taper shanks and sockets, which are classified as to (1) means of transmitting torque from spindle to the tool shank, and (2) means of retaining the shank in the socket. The steep machine tapers consist of a preferred series (bold-face type, Table 6) and an intermediate series (light-face type). A self-holding taper is defined as "a taper with an angle small enough to hold a shank in place ordinarily by friction without holding means. (Sometimes referred to as slow taper.)" A steep taper is defined as "a taper having an angle sufficiently large to ensure the easy or self-releasing feature." The term "gage line" indicates the basic diameter at or near the large end of the taper.

**Table 2. American National Standard Taper Drive with Tang, Self-Holding Tapers** *ANSI/ASME B5.10-1994 (R2013)*

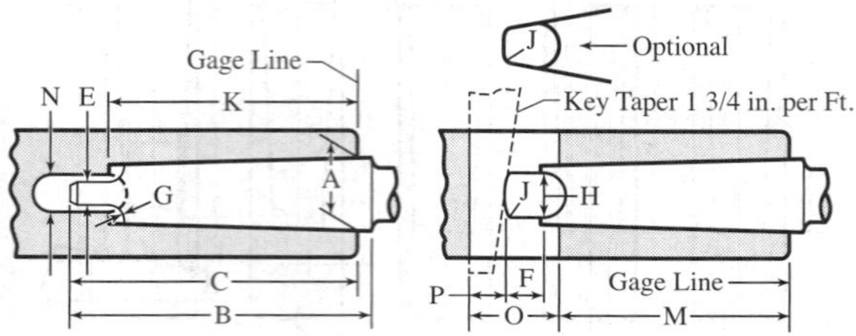


| No. of Taper | Diameter at Gage Line (1) *A* | Shank: Total Length of Shank *B* | Shank: Gage Line to End of Shank *C* | Tang: Thickness *E* | Tang: Length *F* | Tang: Radius of Mill *G* | Tang: Diameter *H* |
|---|---|---|---|---|---|---|---|
| 0.239 | 0.23922 | 1.28 | 1.19 | 0.125 | 0.19 | 0.19 | 0.18 |
| 0.299 | 0.29968 | 1.59 | 1.50 | 0.156 | 0.25 | 0.19 | 0.22 |
| 0.375 | 0.37525 | 1.97 | 1.88 | 0.188 | 0.31 | 0.19 | 0.28 |
| 1 | 0.47500 | 2.56 | 2.44 | 0.203 | 0.38 | 0.19 | 0.34 |
| 2 | 0.70000 | 3.13 | 2.94 | 0.250 | 0.44 | 0.25 | 0.53 |
| 3 | 0.93800 | 3.88 | 3.69 | 0.312 | 0.56 | 0.22 | 0.72 |
| 4 | 1.23100 | 4.88 | 4.63 | 0.469 | 0.63 | 0.31 | 0.97 |
| 4½ | 1.50000 | 5.38 | 5.13 | 0.562 | 0.69 | 0.38 | 1.20 |
| 5 | 1.74800 | 6.12 | 5.88 | 0.625 | 0.75 | 0.38 | 1.41 |
| 6 | 2.49400 | 8.25 | 8.25 | 0.750 | 1.13 | 0.50 | 2.00 |

| No. of Taper | Radius *J* | Socket: Min. Depth of Hole *K*: Drilled | Socket: Min. Depth of Hole *K*: Reamed | Socket: Gage Line to Tang Slot *M* | Tang Slot: Width *N* | Tang Slot: Length *O* | Tang Slot: Shank End to Back of Tang Slot *P* |
|---|---|---|---|---|---|---|---|
| 0.239 | 0.03 | 1.06 | 1.00 | 0.94 | 0.141 | 0.38 | 0.13 |
| 0.299 | 0.03 | 1.31 | 1.25 | 1.17 | 0.172 | 0.50 | 0.17 |
| 0.375 | 0.05 | 1.63 | 1.56 | 1.47 | 0.203 | 0.63 | 0.22 |
| 1 | 0.05 | 2.19 | 2.16 | 2.06 | 0.218 | 0.75 | 0.38 |
| 2 | 0.06 | 2.66 | 2.61 | 2.50 | 0.266 | 0.88 | 0.44 |
| 3 | 0.08 | 3.31 | 3.25 | 3.06 | 0.328 | 1.19 | 0.56 |
| 4 | 0.09 | 4.19 | 4.13 | 3.88 | 0.484 | 1.25 | 0.50 |
| 4½ | 0.13 | 4.62 | 4.56 | 4.31 | 0.578 | 1.38 | 0.56 |
| 5 | 0.13 | 5.31 | 5.25 | 4.94 | 0.656 | 1.50 | 0.56 |
| 6 | 0.16 | 7.41 | 7.33 | 7.00 | 0.781 | 1.75 | 0.50 |

All dimensions are in inches. (1) See Table 7b for plug and ring gage dimensions.

*Tolerances:* For shank diameter *A* at gage line, +0.002, −0.000; for hole diameter *A*, +0.000, −0.002. For tang thickness *E* up to No. 5 inclusive, +0.000, −0.006; No. 6, +0.000, −0.008. For width *N* of tang slot up to No. 5 inclusive, +0.006, −0.000; No. 6, +0.008, −0.000. For centrality of tang *E* with center line of taper, 0.0025 (0.005 total indicator variation). These centrality tolerances also apply to the tang slot *N*. On rate of taper, all sizes 0.002 per foot. This tolerance may be applied on *shanks* only in the direction that *increases* the rate of taper and on *sockets* only in the direction that *decreases* the rate of taper. Tolerances for two-decimal dimensions are plus or minus 0.010, unless otherwise specified.

**Table 3. American National Standard Taper Drive with Keeper Key Slot, Self-Holding Tapers** *ANSI/ASME B5.10-1994 (R2013)*

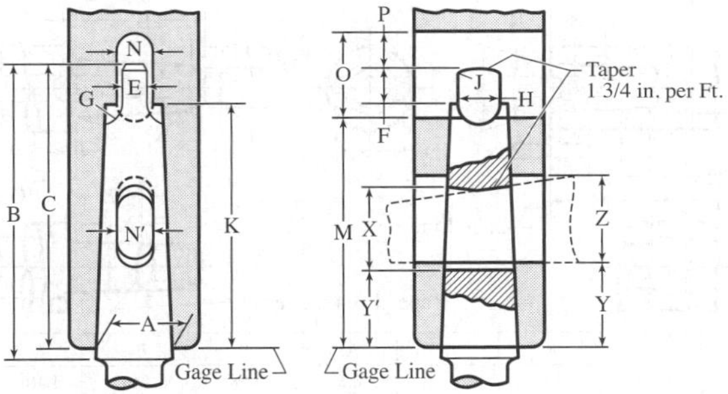


| No. of Taper | Shank: Dia. at Gage Line (1) *A* | Shank: Total Length *B* | Shank: Gage Line to End *C* | Tang: Thickness *E* | Tang: Length *F* | Tang: Radius of Mill *G* | Tang: Diameter *H* | Tang: Radius *J* | Socket: Min. Depth of Hole *K*: Drill | Socket: Min. Depth of Hole *K*: Ream | Socket: Gage Line to Tang Slot *M* |
|---|---|---|---|---|---|---|---|---|---|---|---|
| 3 | 0.938 | 3.88 | 3.69 | 0.312 | 0.56 | 0.28 | 0.78 | 0.08 | 3.31 | 3.25 | 3.06 |
| 4 | 1.231 | 4.88 | 4.63 | 0.469 | 0.63 | 0.31 | 0.97 | 0.09 | 4.19 | 4.13 | 3.88 |
| 4½ | 1.500 | 5.38 | 5.13 | 0.562 | 0.69 | 0.38 | 1.20 | 0.13 | 4.63 | 4.56 | 4.32 |
| 5 | 1.748 | 6.13 | 5.88 | 0.625 | 0.75 | 0.38 | 1.41 | 0.13 | 5.31 | 5.25 | 4.94 |
| 6 | 2.494 | 8.56 | 8.25 | 0.750 | 1.13 | 0.50 | 2.00 | 0.16 | 7.41 | 7.33 | 7.00 |
| 7 | 3.270 | 11.63 | 11.25 | 1.125 | 1.38 | 0.75 | 2.63 | 0.19 | 10.16 | 10.08 | 9.50 |

| No. of Taper | Tang Slot: Width *N* | Tang Slot: Length *O* | Tang Slot: Shank End to Back of Slot *P* | Keeper Slot in Shank: Gage Line to Bottom of Slot *Y′* | Keeper Slot in Shank: Length *X* | Keeper Slot in Shank: Width *N′* | Keeper Slot in Socket: Gage Line to Front of Slot *Y* | Keeper Slot in Socket: Length *Z* | Keeper Slot in Socket: Width *N′* |
|---|---|---|---|---|---|---|---|---|---|
| 3 | 0.328 | 1.19 | 0.56 | 1.03 | 1.13 | 0.266 | 1.13 | 1.19 | 0.266 |
| 4 | 0.484 | 1.25 | 0.50 | 1.41 | 1.19 | 0.391 | 1.50 | 1.25 | 0.391 |
| 4½ | 0.578 | 1.38 | 0.56 | 1.72 | 1.25 | 0.453 | 1.81 | 1.38 | 0.453 |
| 5 | 0.656 | 1.50 | 0.56 | 2.00 | 1.38 | 0.516 | 2.13 | 1.50 | 0.516 |
| 6 | 0.781 | 1.75 | 0.50 | 2.13 | 1.63 | 0.641 | 2.25 | 1.75 | 0.641 |
| 7 | 1.156 | 2.63 | 0.88 | 2.50 | 1.69 | 0.766 | 2.63 | 1.81 | 0.766 |

All dimensions are in inches. (1) See Table 7b for plug and ring gage dimensions.

*Tolerances:* For shank diameter *A* at gage line, +0.002, −0; for hole diameter *A*, +0, −0.002. For tang thickness *E* up to No. 5 inclusive, +0, −0.006; larger than No. 5, +0, −0.008. For width of slots *N* and *N′* up to No. 5 inclusive, +0.006, −0; larger than No. 5, +0.008, −0. For centrality of tang *E* with center line of taper 0.0025 (0.005 total indicator variation). These centrality tolerances also apply to slots *N* and *N′*. On rate of taper, see footnote in Table 2. Tolerances for two-decimal dimensions are ±0.010 unless otherwise specified.

**Table 4. American National Standard Nose Key Drive with Keeper Key Slot, Self-Holding Tapers** *ANSI/ASME B5.10-1994 (R2013)*

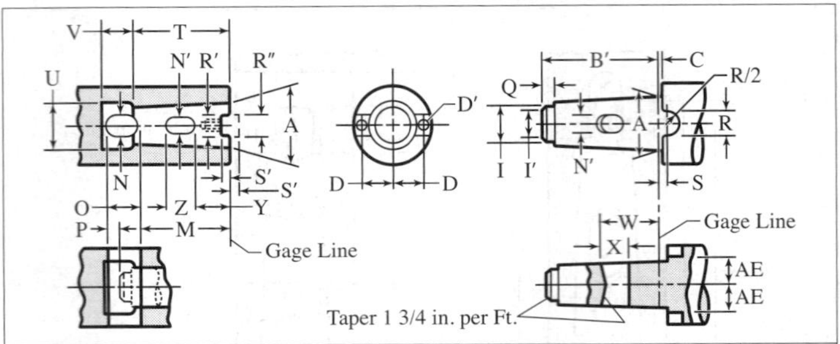


| Taper | A(1) | B′ | C | Q | I′ | I | R | S |
|---|---|---|---|---|---|---|---|---|
| 200 | 2.000 | 5.13 | Min. 0.003 Max. 0.035 for all sizes | 0.25 | 1.38 | 1.63 | 1.010 | 0.562 |
| 250 | 2.500 | 5.88 | | 0.25 | 1.38 | 2.06 | 1.010 | 0.562 |
| 300 | 3.000 | 6.63 | | 0.25 | 1.63 | 2.50 | 2.010 | 0.562 |
| 350 | 3.500 | 7.44 | | 0.31 | 2.00 | 2.94 | 2.010 | 0.562 |
| 400 | 4.000 | 8.19 | | 0.31 | 2.13 | 3.31 | 2.010 | 0.562 |
| 450 | 4.500 | 9.00 | | 0.38 | 2.38 | 3.81 | 3.010 | 0.812 |
| 500 | 5.000 | 9.75 | | 0.38 | 2.50 | 4.25 | 3.010 | 0.812 |
| 600 | 6.000 | 11.31 | | 0.44 | 3.00 | 5.19 | 3.010 | 0.812 |
| 800 | 8.000 | 14.38 | | 0.50 | 3.50 | 7.00 | 4.010 | 1.062 |
| 1000 | 10.000 | 17.44 | | 0.63 | 4.50 | 8.75 | 4.010 | 1.062 |
| 1200 | 12.000 | 20.50 | | 0.75 | 5.38 | 10.50 | 4.010 | 1.062 |

| Taper | D | D′[a] | W | X | N′ | R′ | S′ | T |
|---|---|---|---|---|---|---|---|---|
| 200 | 1.41 | 0.375 | 3.44 | 1.56 | 0.656 | 1.000 | 0.50 | 4.75 |
| 250 | 1.66 | 0.375 | 3.69 | 1.56 | 0.781 | 1.000 | 0.50 | 5.50 |
| 300 | 2.25 | 0.375 | 4.06 | 1.56 | 1.031 | 2.000 | 0.50 | 6.25 |
| 350 | 2.50 | 0.375 | 4.88 | 2.00 | 1.031 | 2.000 | 0.50 | 6.94 |
| 400 | 2.75 | 0.375 | 5.31 | 2.25 | 1.031 | 2.000 | 0.50 | 7.69 |
| 450 | 3.00 | 0.500 | 5.88 | 2.44 | 1.031 | 3.000 | 0.75 | 8.38 |
| 500 | 3.25 | 0.500 | 6.44 | 2.63 | 1.031 | 3.000 | 0.75 | 9.13 |
| 600 | 3.75 | 0.500 | 7.44 | 3.00 | 1.281 | 3.000 | 0.75 | 10.56 |
| 800 | 4.75 | 0.500 | 9.56 | 4.00 | 1.781 | 4.000 | 1.00 | 13.50 |
| 1000 | … | … | 11.50 | 4.75 | 2.031 | 4.000 | 1.00 | 16.31 |
| 1200 | … | … | 13.75 | 5.75 | 2.031 | 4.000 | 1.00 | 19.00 |

| Taper | U | V | M | N | O | P | Y | Z |
|---|---|---|---|---|---|---|---|---|
| 200 | 1.81 | 1.00 | 4.50 | 0.656 | 1.56 | 0.94 | 2.00 | 1.69 |
| 250 | 2.25 | 1.00 | 5.19 | 0.781 | 1.94 | 1.25 | 2.25 | 1.69 |
| 300 | 2.75 | 1.00 | 5.94 | 1.031 | 2.19 | 1.50 | 2.63 | 1.69 |
| 350 | 3.19 | 1.25 | 6.75 | 1.031 | 2.19 | 1.50 | 3.00 | 2.13 |
| 400 | 3.63 | 1.25 | 7.50 | 1.031 | 2.19 | 1.50 | 3.25 | 2.38 |
| 450 | 4.19 | 1.50 | 8.00 | 1.031 | 2.75 | 1.75 | 3.63 | 2.56 |
| 500 | 4.63 | 1.50 | 8.75 | 1.031 | 2.75 | 1.75 | 4.00 | 2.75 |
| 600 | 5.50 | 1.75 | 10.13 | 1.281 | 3.25 | 2.06 | 4.63 | 3.25 |
| 800 | 7.38 | 2.00 | 12.88 | 1.781 | 4.25 | 2.75 | 5.75 | 4.25 |
| 1000 | 9.19 | 2.50 | 15.75 | 2.031 | 5.00 | 3.31 | 7.00 | 5.00 |
| 1200 | 11.00 | 3.00 | 18.50 | 2.531 | 6.00 | 4.00 | 8.25 | 6.00 |

[a] Thread is UNF-2B for hole; UNF-2A for screw. (1) See Table 7b for plug and ring gage dimensions.

All dimensions are in inches. *AE* is 0.005 greater than one-half of *A*.

Width of drive key *R″* is 0.001 less than width *R″* of keyway.

*Tolerances:* For diameter *A* of hole at gage line, +0, −0.002; for diameter *A* of shank at gage line, +0.002, −0; for width of slots *N* and *N′*, +0.008, −0; for width of drive keyway *R′* in socket, +0, −0.001; for width of drive keyway *R* in shank, 0.010, −0; for centrality of slots *N* and *N′* with center line of spindle, 0.007; for centrality of keyway with spindle center line: for *R*, 0.004 and for *R′*, 0.002 T.I.V. On rate of taper, see footnote in Table 2. Two-decimal dimensions, ±0.010 unless otherwise specified.

**Table 5. American National Standard Nose Key Drive with Drawbolt, Self-Holding Tapers** *ANSI/ASME B5.10-1994 (R2013)*

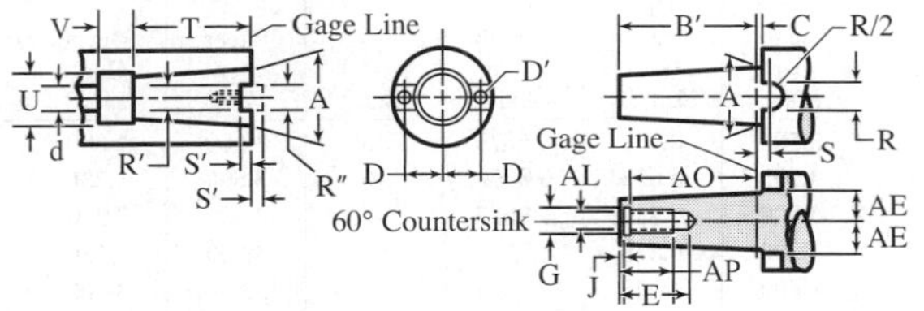


| Sockets | | | | | | | | | | |
|---|---|---|---|---|---|---|---|---|---|---|
| | | Drive Key | | | Drive Keyway | | | | | |
| | | Screw Holes | | | | | | | | |
| No. of Taper | Dia. at Gage Line $A^a$ | Center Line to Center of Screw *D* | UNF 2B Hole UNF 2A Screw *D′* | Width *R″* | Width *R′* | Depth *S′* | Gage Line to Front of Relief *T* | Dia. of Relief *U* | Depth of Relief *V* | Dia. of Draw Bolt Hole *d* |
| 200 | 2.000 | 1.41 | 0.38 | 0.999 | 1.000 | 0.50 | 4.75 | 1.81 | 1.00 | 1.00 |
| 250 | 2.500 | 1.66 | 0.38 | 0.999 | 1.000 | 0.50 | 5.50 | 2.25 | 1.00 | 1.00 |
| 300 | 3.000 | 2.25 | 0.38 | 1.999 | 2.000 | 0.50 | 6.25 | 2.75 | 1.00 | 1.13 |
| 350 | 3.500 | 2.50 | 0.38 | 1.999 | 2.000 | 0.50 | 6.94 | 3.19 | 1.25 | 1.13 |
| 400 | 4.000 | 2.75 | 0.38 | 1.999 | 2.000 | 0.50 | 7.69 | 3.63 | 1.25 | 1.63 |
| 450 | 4.500 | 3.00 | 0.50 | 2.999 | 3.000 | 0.75 | 8.38 | 4.19 | 1.50 | 1.63 |
| 500 | 5.000 | 3.25 | 0.50 | 2.999 | 3.000 | 0.75 | 9.13 | 4.63 | 1.50 | 1.63 |
| 600 | 6.000 | 3.75 | 0.50 | 2.999 | 3.000 | 0.75 | 10.56 | 5.50 | 1.75 | 2.25 |
| 800 | 8.000 | 4.75 | 0.50 | 3.999 | 4.000 | 1.00 | 13.50 | 7.38 | 2.00 | 2.25 |
| 1000 | 10.000 | … | … | 3.999 | 4.000 | 1.00 | 16.31 | 9.19 | 2.50 | 2.25 |
| 1200 | 12.000 | … | … | 3.999 | 4.000 | 1.00 | 19.00 | 11.00 | 3.00 | 2.25 |

[a] See Table 7b for plug and ring gage dimensions.

| Shanks | | | | | | | | | | |
|---|---|---|---|---|---|---|---|---|---|---|
| | | Drawbar Hole | | | | | | Drive Keyway | | |
| No. of Taper | Length from Gage Line B¢ | Dia. UNC-2B *AL* | Depth of Drilled Hole *E* | Depth of Thread *AP* | Dia. of Counter Bore *G* | Gage Line to First Thread *AO* | Depth of 60° Chamfer *J* | Width *R* | Depth *S* | Center Line to Bottom of Keyway *AE* |
| 200 | 5.13 | 7⁄8-9 | 2.44 | 1.75 | 0.91 | 4.78 | 0.13 | 1.010 | 0.562 | 1.005 |
| 250 | 5.88 | 7⁄8-9 | 2.44 | 1.75 | 0.91 | 5.53 | 0.13 | 1.010 | 0.562 | 1.255 |
| 300 | 6.63 | 1-8 | 2.75 | 2.00 | 1.03 | 6.19 | 0.19 | 2.010 | 0.562 | 1.505 |
| 350 | 7.44 | 1-8 | 2.75 | 2.00 | 1.03 | 7.00 | 0.19 | 2.010 | 0.562 | 1.755 |
| 400 | 8.19 | 1½-6 | 4.00 | 3.00 | 1.53 | 7.50 | 0.31 | 2.010 | 0.562 | 2.005 |
| 450 | 9.00 | 1½-6 | 4.00 | 3.00 | 1.53 | 8.31 | 0.31 | 3.010 | 0.812 | 2.255 |
| 500 | 9.75 | 1½-6 | 4.00 | 3.00 | 1.53 | 9.06 | 0.31 | 3.010 | 0.812 | 2.505 |
| 600 | 11.31 | 2-4½ | 5.31 | 4.00 | 2.03 | 10.38 | 0.50 | 3.010 | 0.812 | 3.005 |
| 800 | 14.38 | 2-4½ | 5.31 | 4.00 | 2.03 | 13.44 | 0.50 | 4.010 | 1.062 | 4.005 |
| 1000 | 17.44 | 2-4½ | 5.31 | 4.00 | 2.03 | 16.50 | 0.50 | 4.010 | 1.062 | 5.005 |
| 1200 | 20.50 | 2-4½ | 5.31 | 4.00 | 2.03 | 19.56 | 0.50 | 4.010 | 1.062 | 6.005 |

All dimensions in inches.

Exposed length *C* is 0.003 minimum and 0.035 maximum for all sizes.

Drive Key *D′* screw sizes are 3⁄8-24 UNF-2A up to taper No. 400 inclusive and 1⁄2-20 UNF-2A for larger tapers.

*Tolerances:* For diameter *A* of hole at gage line, +0.000, −0.002 for all sizes; for diameter *A* of shank at gage line, +0.002, −0.000; for all sizes; for width of drive keyway *R′* in socket, +0.000, −0.001; for width of drive keyway *R* in shank, +0.010, −0.000; for centrality of drive keyway *R′*, with center line of shank, 0.004 total indicator variation, and for drive keyway *R′*, with center line of spindle, 0.002. On rate of taper, see footnote in Table 2. Tolerances for two-decimal dimensions are ±0.010 unless otherwise specified.

**Table 6. American National Standard Steep Machine Tapers**
*ANSI/ASME B5.10-1994 (R2013)*

| No. of Taper | Taper per Foot[a] | Dia. at Gage Line[b] | Length Along Axis | No. of Taper | Taper per Foot[a] | Dia. at Gage Line[b] | Length Along Axis |
|---|---|---|---|---|---|---|---|
| 5 | 3.500 | 0.500 | 0.6875 | 35 | 3.500 | 1.500 | 2.2500 |
| **10** | **3.500** | **0.625** | **0.8750** | **40** | **3.500** | **1.750** | **2.5625** |
| 15 | 3.500 | 0.750 | 1.0625 | 45 | 3.500 | 2.250 | 3.3125 |
| **20** | **3.500** | **0.875** | **1.3125** | **50** | **3.500** | **2.750** | **4.0000** |
| 25 | 3.500 | 1.000 | 1.5625 | 55 | 3.500 | 3.500 | 5.1875 |
| **30** | **3.500** | **1.250** | **1.8750** | **60** | **3.500** | **4.250** | **6.3750** |

[a] This taper corresponds to an included angle of 16°, 35′, 39.4″.
[b] The basic diameter at gage line is at large end of taper.
All dimensions given in inches.

The tapers numbered 10, 20, 30, 40, 50, and 60 that are printed in heavy-faced type are designated as the "Preferred Series." The tapers numbered 5, 15, 25, 35, 45, and 55 that are printed in light-faced type are designated as the "Intermediate Series."

**Table 7a. American National Standard Self-Holding Tapers — Basic Dimensions** *ANSI/ASME B5.10-1994 (R2013)*

| No. of Taper | Taper per Foot | Dia. at Gage Line[a] A | Means of Driving and Holding[a] | Origin of Series |
|---|---|---|---|---|
| .239 | 0.50200 | 0.23922 | | Brown & Sharpe Taper Series |
| .299 | 0.50200 | 0.29968 | | |
| .375 | 0.50200 | 0.37525 | Tang Drive with Shank Held in by Friction (See Table 2) | |
| 1 | 0.59858 | 0.47500 | | Morse Taper Series |
| 2 | 0.59941 | 0.70000 | | |
| 3 | 0.60235 | 0.93800 | | |
| 4 | 0.62326 | 1.23100 | | |
| 4½ | 0.62400 | 1.50000 | Tang Drive with Shank Held in by Key (See Table 3) | |
| 5 | 0.63151 | 1.74800 | | |
| 6 | 0.62565 | 2.49400 | | |
| 7 | 0.62400 | 3.27000 | | |
| 200 | 0.750 | 2.000 | | ¾ Inch per Foot Taper Series |
| 250 | 0.750 | 2.500 | | |
| 300 | 0.750 | 3.000 | | |
| 350 | 0.750 | 3.500 | Key Drive with Shank Held in by Key (See Table 4) | |
| 400 | 0.750 | 4.000 | | |
| 450 | 0.750 | 4.500 | | |
| 500 | 0.750 | 5.000 | Key Drive with Shank Held in by Draw-bolt (See Table 5) | |
| 600 | 0.750 | 6.000 | | |
| 800 | 0.750 | 8.000 | | |
| 1000 | 0.750 | 10.000 | | |
| 1200 | 0.750 | 12.000 | | |

[a] See illustrations above Table 2 through Table 5.
All dimensions given in inches.

**Table 7b. American National Standard Plug and Ring Gages for the Self-Holding Taper Series** *ANSI/ASME B5.10-1994 (R2013)*

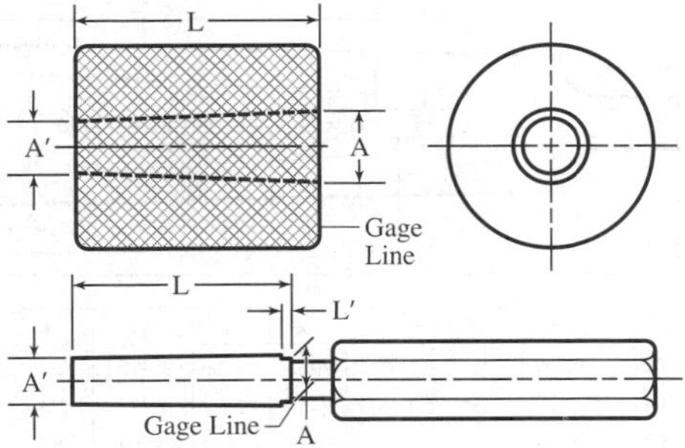


| No. of Taper | Taper[a] per Foot | Diameter[a] at Gage Line *A* | Tolerances for Diameter *A*[b] Class X Gage | Class Y Gage | Class Z Gage | Diameter at Small End *A′* | Length Gage Line to End *L* | Depth of Gaging-Notch, Plug Gage *L′* |
|---|---|---|---|---|---|---|---|---|
| 0.239 | 0.50200 | 0.23922 | 0.00004 | 0.00007 | 0.00010 | 0.20000 | 0.94 | 0.048 |
| 0.299 | 0.50200 | 0.29968 | 0.00004 | 0.00007 | 0.00010 | 0.25000 | 1.19 | 0.048 |
| 0.375 | 0.50200 | 0.37525 | 0.00004 | 0.00007 | 0.00010 | 0.31250 | 1.50 | 0.048 |
| 1 | 0.59858 | 0.47500 | 0.00004 | 0.00007 | 0.00010 | 0.36900 | 2.13 | 0.040 |
| 2 | 0.59941 | 0.70000 | 0.00004 | 0.00007 | 0.00010 | 0.57200 | 2.56 | 0.040 |
| 3 | 0.60235 | 0.93800 | 0.00006 | 0.00009 | 0.00012 | 0.77800 | 3.19 | 0.040 |
| 4 | 0.62326 | 1.23100 | 0.00006 | 0.00009 | 0.00012 | 1.02000 | 4.06 | 0.038 |
| 4½ | 0.62400 | 1.50000 | 0.00006 | 0.00009 | 0.00012 | 1.26600 | 4.50 | 0.038 |
| 5 | 0.63151 | 1.74800 | 0.00008 | 0.00012 | 0.00016 | 1.47500 | 5.19 | 0.038 |
| 6 | 0.62565 | 2.49400 | 0.00008 | 0.00012 | 0.00016 | 2.11600 | 7.25 | 0.038 |
| 7 | 0.62400 | 3.27000 | 0.00010 | 0.00015 | 0.00020 | 2.75000 | 10.00 | 0.038 |
| 200 | 0.75000 | 2.00000 | 0.00008 | 0.00012 | 0.00016 | 1.703 | 4.75 | 0.032 |
| 250 | 0.75000 | 2.50000 | 0.00008 | 0.00012 | 0.00016 | 2.156 | 5.50 | 0.032 |
| 300 | 0.75000 | 3.00000 | 0.00010 | 0.00015 | 0.00020 | 2.609 | 6.25 | 0.032 |
| 350 | 0.75000 | 3.50000 | 0.00010 | 0.00015 | 0.00020 | 3.063 | 7.00 | 0.032 |
| 400 | 0.75000 | 4.00000 | 0.00010 | 0.00015 | 0.00020 | 3.516 | 7.75 | 0.032 |
| 450 | 0.75000 | 4.50000 | 0.00010 | 0.00015 | 0.00020 | 3.969 | 8.50 | 0.032 |
| 500 | 0.75000 | 5.00000 | 0.00013 | 0.00019 | 0.00025 | 4.422 | 9.25 | 0.032 |
| 600 | 0.75000 | 6.00000 | 0.00013 | 0.00019 | 0.00025 | 5.328 | 10.75 | 0.032 |
| 800 | 0.75000 | 8.00000 | 0.00016 | 0.00024 | 0.00032 | 7.141 | 13.75 | 0.032 |
| 1000 | 0.75000 | 10.00000 | 0.00020 | 0.00030 | 0.00040 | 8.953 | 16.75 | 0.032 |
| 1200 | 0.75000 | 12.00000 | 0.00020 | 0.00030 | 0.00040 | 10.766 | 19.75 | 0.032 |

[a] The taper per foot and diameter *A* at gage line are basic dimensions. Dimensions in Column *A′* are calculated for reference only.

[b] Tolerances for diameter A are plus for plug gages and minus for ring gages.

All dimensions are in inches.

The amount of taper deviation for Class X, Class Y, and Class Z gages are the same, respectively, as the amounts shown for tolerances on diameter *A*. Taper deviation is the permissible allowance from true taper at any point of diameter in the length of the gage. On taper *plug* gages, this deviation may be applied only in the direction that *decreases* the rate of taper. On taper *ring* gages, this deviation may be applied only in the direction that *increases* the rate of taper. Tolerances on two-decimal dimensions are ±0.010.

**British Standard Tapers.—**British Standard BS 1660: 1972, Machine Tapers, Reduction Sleeves, and Extension Sockets, contains dimensions for self-holding and self-releasing tapers, reduction sleeves, extension sockets, and turret sockets for tools having Morse and metric 5 percent taper shanks. Adapters for use with $^7/_{24}$ tapers and dimensions for spindle noses and tool shanks with self-release tapers and cotter slots are included in this Standard.

**Table 8. Dimensions of Morse Taper Sleeves**

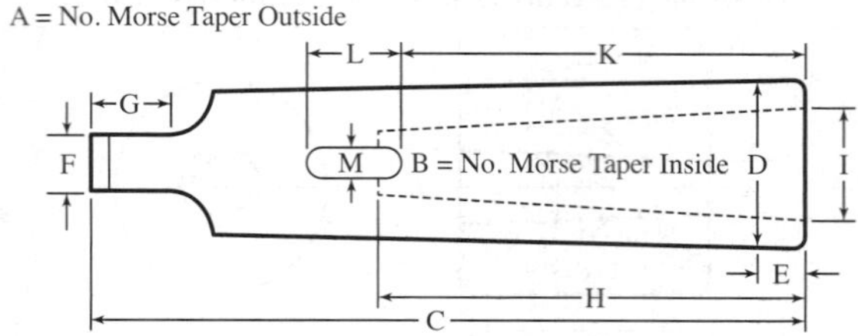


| *A* | *B* | *C* | *D* | *E* | *F* | *G* | *H* | *I* | *K* | *L* | *M* |
|---|---|---|---|---|---|---|---|---|---|---|---|
| 2 | 1 | 3 9/16 | 0.700 | 5/8 | 1/4 | 7/16 | 2 3/16 | 0.475 | 2 1/16 | 3/4 | 0.213 |
| 3 | 1 | 3 15/16 | 0.938 | 1/4 | 5/16 | 9/16 | 2 3/16 | 0.475 | 2 1/16 | 3/4 | 0.213 |
| 3 | 2 | 4 7/16 | 0.938 | 3/4 | 5/16 | 9/16 | 2 5/8 | 0.700 | 2 1/2 | 7/8 | 0.260 |
| 4 | 1 | 4 7/8 | 1.231 | 1/4 | 15/32 | 5/8 | 2 3/16 | 0.475 | 2 1/16 | 3/4 | 0.213 |
| 4 | 2 | 4 7/8 | 1.231 | 1/4 | 15/32 | 5/8 | 2 5/8 | 0.700 | 2 1/2 | 7/8 | 0.260 |
| 4 | 3 | 5 3/8 | 1.231 | 3/4 | 15/32 | 5/8 | 3 1/4 | 0.938 | 3 1/16 | 1 3/16 | 0.322 |
| 5 | 1 | 6 1/8 | 1.748 | 1/4 | 5/8 | 3/4 | 2 3/16 | 0.475 | 2 1/16 | 3/4 | 0.213 |
| 5 | 2 | 6 1/8 | 1.748 | 1/4 | 5/8 | 3/4 | 2 5/8 | 0.700 | 2 1/2 | 7/8 | 0.260 |
| 5 | 3 | 6 1/8 | 1.748 | 1/4 | 5/8 | 3/4 | 3 1/4 | 0.938 | 3 1/16 | 1 3/16 | 0.322 |
| 5 | 4 | 6 5/8 | 1.748 | 3/4 | 5/8 | 3/4 | 4 1/8 | 1.231 | 3 7/8 | 1 1/4 | 0.478 |
| 6 | 1 | 8 5/8 | 2.494 | 3/8 | 3/4 | 1 1/8 | 2 3/16 | 0.475 | 2 1/16 | 3/4 | 0.213 |
| 6 | 2 | 8 5/8 | 2.494 | 3/8 | 3/4 | 1 1/8 | 2 5/8 | 0.700 | 2 1/2 | 7/8 | 0.260 |
| 6 | 3 | 8 5/8 | 2.494 | 3/8 | 3/4 | 1 1/8 | 3 1/4 | 0.938 | 3 1/16 | 1 3/16 | 0.322 |
| 6 | 4 | 8 5/8 | 2.494 | 3/8 | 3/4 | 1 1/8 | 4 1/8 | 1.231 | 3 7/8 | 1 1/4 | 0.478 |
| 6 | 5 | 8 5/8 | 2.494 | 3/8 | 3/4 | 1 1/8 | 5 1/4 | 1.748 | 4 15/16 | 1 1/2 | 0.635 |
| 7 | 3 | 11 5/8 | 3.270 | 3/8 | 1 1/8 | 1 3/8 | 3 1/4 | 0.938 | 3 1/16 | 1 3/16 | 0.322 |
| 7 | 4 | 11 5/8 | 3.270 | 3/8 | 1 1/8 | 1 3/8 | 4 1/8 | 1.231 | 3 7/8 | 1 1/4 | 0.478 |
| 7 | 5 | 11 5/8 | 3.270 | 3/8 | 1 1/8 | 1 3/8 | 5 1/4 | 1.748 | 4 15/16 | 1 1/2 | 0.635 |
| 7 | 6 | 12 1/2 | 3.270 | 1 1/4 | 1 1/8 | 1 3/8 | 7 3/8 | 2.494 | 7 | 1 3/4 | 0.760 |

**Table 9. Morse Taper Sockets — Hole and Shank Sizes**

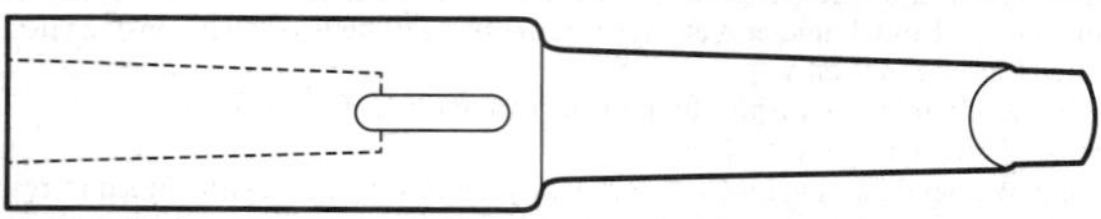

| | Morse Taper | | | Morse Taper | | | Morse Taper | |
|---|---|---|---|---|---|---|---|---|
| Size | Hole | Shank | Size | Hole | Shank | Size | Hole | Shank |
| 1 by 2 | No. 1 | No. 2 | 2 by 5 | No. 2 | No. 5 | 4 by 4 | No. 4 | No. 4 |
| 1 by 3 | No. 1 | No. 3 | 3 by 2 | No. 3 | No. 2 | 4 by 5 | No. 4 | No. 5 |
| 1 by 4 | No. 1 | No. 4 | 3 by 3 | No. 3 | No. 3 | 4 by 6 | No. 4 | No. 6 |
| 1 by 5 | No. 1 | No. 5 | 3 by 4 | No. 3 | No. 4 | 5 by 4 | No. 5 | No. 4 |
| 2 by 3 | No. 2 | No. 3 | 3 by 5 | No. 3 | No. 5 | 5 by 5 | No. 5 | No. 5 |
| 2 by 4 | No. 2 | No. 4 | 4 by 3 | No. 4 | No. 3 | 5 by 6 | No. 5 | No. 6 |

## Table 10. Brown & Sharpe Taper Shanks

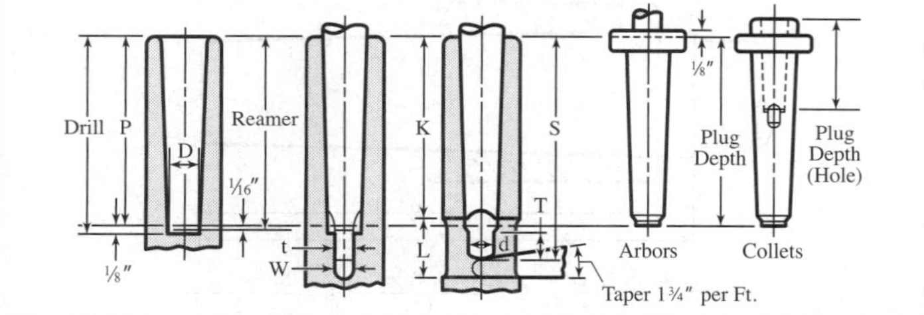


| Number of Taper | Taper per Foot (inch) | Dia. of Plug at Small End D | Plug Depth, P: B & S[b] Standard | Plug Depth, P: Mill. Mach. Standard | Plug Depth, P: Miscell. | Keyway from End of Spindle K | Shank Depth S | Length of Keyway[a] L | Width of Keyway W | Length of Arbor Tongue T | Diameter of Arbor Tongue d | Thickness of Arbor Tongue t |
|---|---|---|---|---|---|---|---|---|---|---|---|---|
| 1[c] | .50200 | .20000 | 15/16 | … | … | 15/16 | 1 3/16 | 3/8 | .135 | 3/16 | .170 | 1/8 |
| 2[c] | .50200 | .25000 | 1 3/16 | … | … | 1 11/64 | 1 1/2 | 1/2 | .166 | 1/4 | .220 | 5/32 |
| 3[c] | .50200 | .31250 | 1 1/2 | … | … | 1 15/32 | 1 7/8 | 5/8 | .197 | 5/16 | .282 | 3/16 |
| | | | … | … | 1 3/4 | 1 23/32 | 2 1/8 | 5/8 | .197 | 5/16 | .282 | 3/16 |
| | | | … | … | 2 | 1 31/32 | 2 3/8 | 5/8 | .197 | 5/16 | .282 | 3/16 |
| 4 | .50240 | .35000 | … | 1 1/4 | … | 1 13/64 | 1 21/32 | 11/16 | .228 | 11/32 | .320 | 7/32 |
| | | | 1 11/16 | … | … | 1 41/64 | 2 3/32 | 11/16 | .228 | 11/32 | .320 | 7/32 |
| 5 | .50160 | .45000 | … | 1 3/4 | … | 1 11/16 | 2 3/16 | 3/4 | .260 | 3/8 | .420 | 1/4 |
| | | | … | … | 2 | 1 15/16 | 2 7/16 | 3/4 | .260 | 3/8 | .420 | 1/4 |
| | | | 2 1/8 | … | … | 2 1/16 | 2 9/16 | 3/4 | .260 | 3/8 | .420 | 1/4 |
| 6 | .50329 | .50000 | 2 3/8 | … | … | 2 19/64 | 2 7/8 | 7/8 | .291 | 7/16 | .460 | 9/32 |
| 7 | .50147 | .60000 | … | … | 2 1/2 | 2 13/32 | 3 1/32 | 15/16 | .322 | 15/32 | .560 | 5/16 |
| | | | 2 7/8 | … | … | 2 25/32 | 3 13/32 | 15/16 | .322 | 15/32 | .560 | 5/16 |
| | | | … | 3 | … | 2 29/32 | 3 17/32 | 15/16 | .322 | 15/32 | .560 | 5/16 |
| 8 | .50100 | .75000 | 3 9/16 | … | … | 3 29/64 | 4 1/8 | 1 | .353 | 1/2 | .710 | 11/32 |
| 9 | .50085 | .90010 | … | 4 | … | 3 7/8 | 4 5/8 | 1 1/8 | .385 | 9/16 | .860 | 3/8 |
| | | | 4 1/4 | … | … | 4 1/8 | 4 7/8 | 1 1/8 | .385 | 9/16 | .860 | 3/8 |
| 10 | .51612 | 1.04465 | 5 | … | … | 4 27/32 | 5 23/32 | 1 5/16 | .447 | 21/32 | 1.010 | 7/16 |
| | | | … | 5 11/16 | … | 5 17/32 | 6 13/32 | 1 5/16 | .447 | 21/32 | 1.010 | 7/16 |
| | | | … | … | 6 7/32 | 6 1/16 | 6 15/16 | 1 5/16 | .447 | 21/32 | 1.010 | 7/16 |
| 11 | .50100 | 1.24995 | 5 15/16 | … | … | 5 25/32 | 6 21/32 | 1 5/16 | .447 | 21/32 | 1.210 | 7/16 |
| | | | … | 6 3/4 | … | 6 19/32 | 7 15/32 | 1 5/16 | .447 | 21/32 | 1.210 | 7/16 |
| 12 | .49973 | 1.50010 | 7 1/8 | 7 1/8 | … | 6 15/16 | 7 15/16 | 1 1/2 | .510 | 3/4 | 1.460 | 1/2 |
| | | | … | … | 6 1/4 | … | … | … | … | … | … | … |
| 13 | .50020 | 1.75005 | 7 3/4 | … | … | 7 9/16 | 8 9/16 | 1 1/2 | .510 | 3/4 | 1.710 | 1/2 |
| 14 | .50000 | 2.00000 | 8 1/4 | 8 1/4 | … | 8 1/32 | 9 5/32 | 1 11/16 | .572 | 27/32 | 1.960 | 9/16 |
| 15 | .5000 | 2.25000 | 8 3/4 | … | … | 8 17/32 | 9 21/32 | 1 11/16 | .572 | 27/32 | 2.210 | 9/16 |
| 16 | .50000 | 2.50000 | 9 1/4 | … | … | 9 | 10 1/4 | 1 7/8 | .635 | 15/16 | 2.450 | 5/8 |
| 17 | .50000 | 2.75000 | 9 3/4 | … | … | … | … | … | … | … | … | … |
| 18 | .50000 | 3.00000 | 10 1/4 | … | … | … | … | … | … | … | … | … |

[a] Special lengths of keyway are used instead of standard lengths in some places. Standard lengths need not be used when keyway is for driving only and not for admitting key to force out tool.
[b] "B & S Standard" Plug Depths are not used in all cases.
[c] Adopted by American Standards Association.

**Table 11. Jarno Taper Shanks**

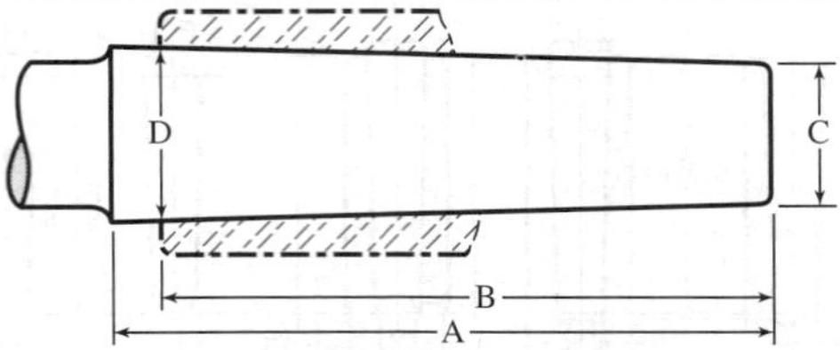


$$D = \frac{\text{No. of Taper}}{8} \qquad C = \frac{\text{No. of Taper}}{10} \qquad B = \frac{\text{No. of Taper}}{2}$$

| Number of Taper | Length *A* | Length *B* | Diameter *C* | Diameter *D* | Taper per foot |
|---|---|---|---|---|---|
| 2 | 1 1/8 | 1 | 0.20 | 0.250 | 0.600 |
| 3 | 1 5/8 | 1 1/2 | 0.30 | 0.375 | 0.600 |
| 4 | 2 3/16 | 2 | 0.40 | 0.500 | 0.600 |
| 5 | 2 11/16 | 2 1/2 | 0.50 | 0.625 | 0.600 |
| 6 | 3 3/16 | 3 | 0.60 | 0.750 | 0.600 |
| 7 | 3 11/16 | 3 1/2 | 0.70 | 0.875 | 0.600 |
| 8 | 4 3/16 | 4 | 0.80 | 1.000 | 0.600 |
| 9 | 4 11/16 | 4 1/2 | 0.90 | 1.125 | 0.600 |
| 10 | 5 1/4 | 5 | 1.00 | 1.250 | 0.600 |
| 11 | 5 3/4 | 5 1/2 | 1.10 | 1.375 | 0.600 |
| 12 | 6 1/4 | 6 | 1.20 | 1.500 | 0.600 |
| 13 | 6 3/4 | 6 1/2 | 1.30 | 1.625 | 0.600 |
| 14 | 7 1/4 | 7 | 1.40 | 1.750 | 0.600 |
| 15 | 7 3/4 | 7 1/2 | 1.50 | 1.875 | 0.600 |
| 16 | 8 5/16 | 8 | 1.60 | 2.000 | 0.600 |
| 17 | 8 13/16 | 8 1/2 | 1.70 | 2.125 | 0.600 |
| 18 | 9 5/16 | 9 | 1.80 | 2.250 | 0.600 |
| 19 | 9 13/16 | 9 1/2 | 1.90 | 2.375 | 0.600 |
| 20 | 10 5/16 | 10 | 2.00 | 2.500 | 0.600 |

**Tapers for Machine Tool Spindles.—**Most lathe spindles have Morse tapers, most milling machine spindles (particularly large vertical ones) have American National Standard tapers and almost all smaller milling machine spindles have R8 tapers (see page 1021). The spindles of drilling machines and the taper shanks of twist drills are made to fit the Morse taper. For lathes, the Morse taper is generally used, but lathes may have the Jarno, the Brown & Sharpe, or a special taper. Of 33 lathe manufacturers, 20 use the Morse taper; 5, the Jarno; 3 use special tapers of their own; 2 use modified Morse (longer than the standard size but the same taper); 2 use Reed (which is a short Jarno); 1 uses the Brown & Sharpe Standard. For grinding machine centers, Jarno, Morse, and Brown & Sharpe tapers are used. Of ten grinding machine manufacturers, 3 use Brown & Sharpe; 3 use Morse; and 4 use Jarno. The Brown & Sharpe taper is used extensively for milling machine and dividing head spindles. The standard milling machine spindle adopted in 1927 by the milling machine manufacturers of the National Machine Tool Builders' Association (now The Association for Manufacturing Technology [AMT]) has a taper of $3\frac{1}{2}$ inches per foot. This comparatively steep taper was adopted to ensure easy release of arbors.

**Table 12. American National Standard Plug and Ring Gages for Steep Machine Tapers** *ANSI/ASME B5.10-1994 (R2013)*

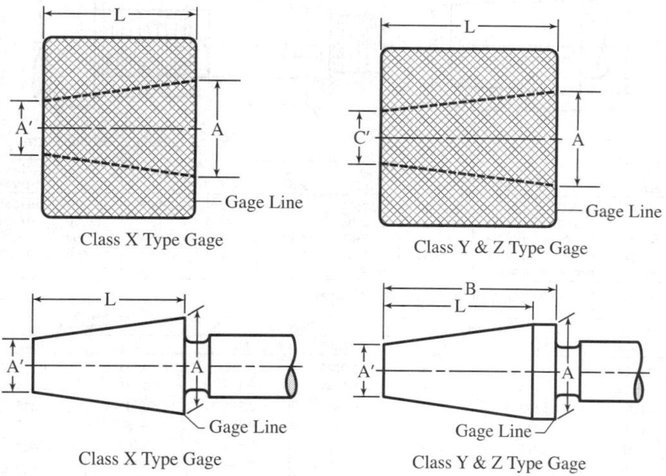


| No. of Taper | Taper per Foot[a] (Basic) | Diameter at Gage Line[a] *A* | Tolerances for Diameter A[b] Class X Gage | Class Y Gage | Class Z Gage | Diameter at Small End[a] *A′* | Length Gage Line to Small End *L* | Overall Length of Gage Body *B* | Dia. of Opening *C* |
|---|---|---|---|---|---|---|---|---|---|
| 5 | 3.500 | 0.500 | 0.00004 | 0.00007 | 0.00010 | 0.2995 | 0.6875 | 0.81 | 0.30 |
| 10 | 3.500 | 0.625 | 0.00004 | 0.00007 | 0.00010 | 0.3698 | 0.8750 | 1.00 | 0.36 |
| 15 | 3.500 | 0.750 | 0.00004 | 0.00007 | 0.00010 | 0.4401 | 1.0625 | 1.25 | 0.44 |
| 20 | 3.500 | 0.875 | 0.00006 | 0.00009 | 0.00012 | 0.4922 | 1.3125 | 1.50 | 0.48 |
| 25 | 3.500 | 1.000 | 0.00006 | 0.00009 | 0.00012 | 0.5443 | 1.5625 | 1.75 | 0.53 |
| 30 | 3.500 | 1.250 | 0.00006 | 0.00009 | 0.00012 | 0.7031 | 1.8750 | 2.06 | 0.70 |
| 35 | 3.500 | 1.500 | 0.00006 | 0.00009 | 0.00012 | 0.8438 | 2.2500 | 2.44 | 0.84 |
| 40 | 3.500 | 1.750 | 0.00008 | 0.00012 | 0.00016 | 1.0026 | 2.5625 | 2.75 | 1.00 |
| 45 | 3.500 | 2.250 | 0.00008 | 0.00012 | 0.00016 | 1.2839 | 3.3125 | 3.50 | 1.00 |
| 50 | 3.500 | 2.750 | 0.00010 | 0.00015 | 0.00020 | 1.5833 | 4.0000 | 4.25 | 1.00 |
| 55 | 3.500 | 3.500 | 0.00010 | 0.00015 | 0.00020 | 1.9870 | 5.1875 | 5.50 | 1.00 |
| 60 | 3.500 | 4.250 | 0.00010 | 0.00015 | 0.00020 | 2.3906 | 6.3750 | 6.75 | 2.00 |

[a] The taper per foot and diameter *A* at gage line are basic dimensions. Dimensions in Column *A′* are calculated for reference only.

[b] Tolerances for diameter *A* are plus for plug gages and minus for ring gages.

All dimensions are in inches.

The amounts of taper deviation for Class X, Class Y, and Class Z gages are the same, respectively, as the amounts shown for tolerances on diameter *A*. Taper deviation is the permissible allowance from true taper at any point of diameter in the length of the gage. On taper *plug* gages, this deviation may be applied only in the direction that *decreases* the rate of taper. On taper *ring* gages, this deviation may be applied only in the direction that *increases* the rate of taper. Tolerances on two-decimal dimensions are ±0.010.

**Table 13. Jacobs Tapers and Threads for Drill Chucks and Spindles**

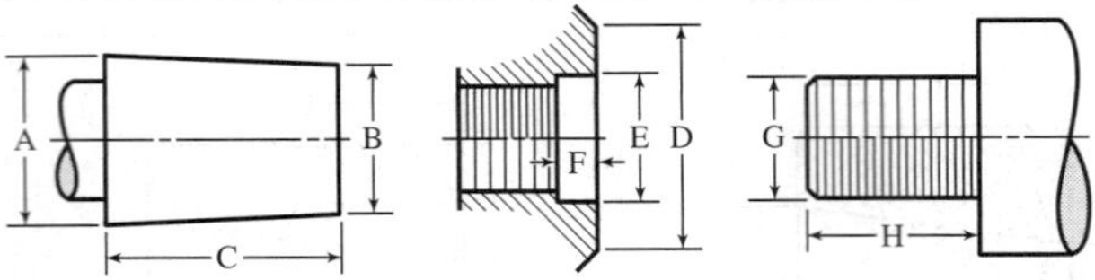


American Standard Thread Form

| Taper Series | A | B | C | Taper per Ft. | Taper Series | A | B | C | Taper per Ft. |
|---|---|---|---|---|---|---|---|---|---|
| No. 0 | 0.2500 | 0.22844 | 0.43750 | 0.59145 | No. 4 | 1.1240 | 1.0372 | 1.6563 | 0.62886 |
| No. 1 | 0.3840 | 0.33341 | 0.65625 | 0.92508 | No. 5 | 1.4130 | 1.3161 | 1.8750 | 0.62010 |
| No. 2 | 0.5590 | 0.48764 | 0.87500 | 0.97861 | No. 6 | 0.6760 | 0.6241 | 1.0000 | 0.62292 |
| No. 2[a] | 0.5488 | 0.48764 | 0.75000 | 0.97861 | No. 33 | 0.6240 | 0.5605 | 1.0000 | 0.76194 |
| No. 3 | 0.8110 | 0.74610 | 1.21875 | 0.63898 | … | … | … | … | … |

[a] These dimensions are for the No. 2 "short" taper.

| Thread Size | Diameter *D* | | Diameter *E* | | Dimension *F* | |
|---|---|---|---|---|---|---|
| | Max. | Min. | Max. | Min. | Max. | Min. |
| 5/16-24 | 0.531 | 0.516 | 0.3245 | 0.3195 | 0.135 | 0.115 |
| 5/16-24 | 0.633 | 0.618 | 0.3245 | 0.3195 | 0.135 | 0.115 |
| 3/8-24 | 0.633 | 0.618 | 0.385 | 0.380 | 0.135 | 0.115 |
| 1/2-20 | 0.860 | 0.845 | 0.510 | 0.505 | 0.135 | 0.115 |
| 5/8-11 | 1.125 | 1.110 | 0.635 | 0.630 | 0.166 | 0.146 |
| 5/8-16 | 1.125 | 1.110 | 0.635 | 0.630 | 0.166 | 0.146 |
| 45/64-16 | 1.250 | 1.235 | 0.713 | 0.708 | 0.166 | 0.146 |
| 3/4-16 | 1.250 | 1.235 | 0.760 | 0.755 | 0.166 | 0.146 |
| 1-8 | 1.437 | 1.422 | 1.036 | 1.026 | 0.281 | 0.250 |
| 1-10 | 1.437 | 1.422 | 1.036 | 1.026 | 0.281 | 0.250 |
| 1 1/2-8 | 1.871 | 1.851 | 1.536 | 1.526 | 0.343 | 0.312 |

| Thread[a] Size | *G* | | *H*[b] | Plug Gage Pitch Dia. | | Ring Gage Pitch Dia. | |
|---|---|---|---|---|---|---|---|
| | Max. | Min. | | Go | Not Go | Go | Not Go |
| 5/16-24 | 0.3114 | 0.3042 | 0.437[c] | 0.2854 | 0.2902 | 0.2843 | 0.2806 |
| 3/8-24 | 0.3739 | 0.3667 | 0.562[d] | 0.3479 | 0.3528 | 0.3468 | 0.3430 |
| 1/2-20 | 0.4987 | 0.4906 | 0.562 | 0.4675 | 0.4731 | 0.4662 | 0.4619 |
| 5/8-11 | 0.6234 | 0.6113 | 0.687 | 0.5660 | 0.5732 | 0.5644 | 0.5589 |
| 5/8-16 | 0.6236 | 0.6142 | 0.687 | 0.5844 | 0.5906 | 0.5830 | 0.5782 |
| 45/64-16 | 0.7016 | 0.6922 | 0.687 | 0.6625 | 0.6687 | 0.6610 | 0.6561 |
| 3/4-16 | 0.7485 | 0.7391 | 0.687 | 0.7094 | 0.7159 | 0.7079 | 0.7029 |
| 1-8 | 1.000 | 0.9848 | 1.000 | 0.9188 | 0.9242 | 0.9188 | 0.9134 |
| 1-10 | 1.000 | 0.9872 | 1.000 | 0.9350 | 0.9395 | 0.9350 | 0.9305 |
| 1 1/2-8 | 1.500 | 1.4848 | 1.000 | 1.4188 | 1.4242 | 1.4188 | 1.4134 |

[a] Except for 1-8, 1-10, 1 1/2-8, all threads are now manufactured to the American National Standard Unified Screw Thread System, Internal Class 2B, External Class 2A. Effective date 1976.

[b] Tolerances for dimension *H* are as follows: 0.030 inch for thread sizes 5/16-24 to 3/4-16, inclusive, and 0.125 inch for thread sizes 1-8 to 1 1/2-8, inclusive.

[c] Length for Jacobs 0B5/16 chuck is 0.375 inch, length for 1B5/16 chuck is 0.437 inch.

[d] Length for Jacobs No. 1BS chuck is 0.437 inch.

*Usual Chuck Capacities for Different Taper Series Numbers:* No. 0 taper, drill diameters, 0-5/32 inch; No. 1, 0-1/4 inch; No. 2, 0-1/2 inch; No. 2 "Short," 0-5/16 inch; No. 3, 0-1/2, 1/8-5/8, 3/16-3/4, or 1/4-13/16 inch; No. 4, 1/8-3/4 inch; No. 5, 3/8-1; No. 6, 0-1/2 inch; No. 33, 0-1/2 inch.

*Usual Chuck Capacities for Different Thread Sizes:* Size 5/16-24, drill diameters 0-1/4 inch; size 3/8-24, drill diameters 0-3/8, 1/16-3/8, or 5/64-1/2 inch; size 1/2-20, drill diameters 0-1/2, 1/16-3/8, or 5/64-1/2 inch; size 5/8-11, drill diameters 0-1/2 inch; size 5/8-16, drill diameters 0-1/2, 1/8- -5/8, or 3/16-3/4 inch; size 45/64-16, drill diameters 0-1/2 inch; size 3/4-16, drill diameters 0-1/2 or 3/16-3/4 .

**Table 1. Essential Dimensions of American National Standard Spindle Noses for Milling Machines** *ANSI/ASME B5.18-1972 (R2014)*

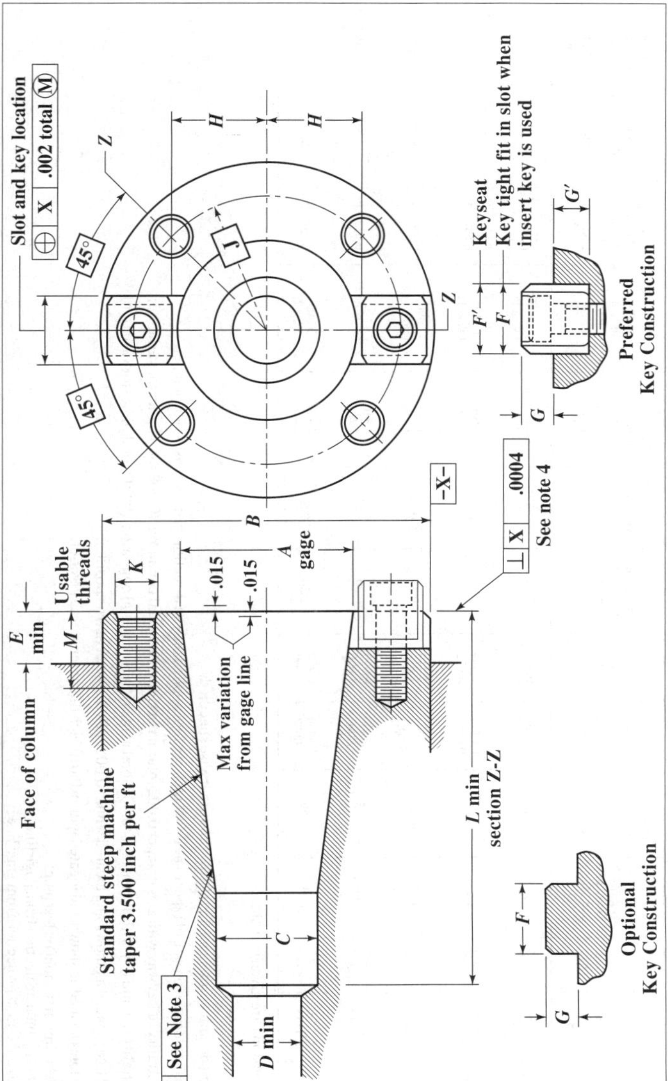

**Table 1.** *(Continued)* **Essential Dimensions of American National Standard Spindle Noses for Milling Machines** *ANSI/ASME B5.18-1972 (R2014)*

| Size No. | Gage Dia.of Taper *A* | Dia.of Spindle *B* | Pilot Dia. *C* | Clearance Hole for Draw-in Bolt Min. *D* | Minimum Dimension Spindle End to Column *E* | Width of Driving Key *F* | Width of Keyseat *F′* | Maximum Height of Driving Key *G* | Minimum Depth of Keyseat *G′* | Distance from Center to Driving Keys *H* | Radius of Bolt Hole Circle *J* | Size of Threads for Bolt Holes UNC-2B *K* | Full Depth of Arbor Hole in Spindle Min. *L* | Depth of Usable Thread for Bolt Hole *M* |
|---|---|---|---|---|---|---|---|---|---|---|---|---|---|---|
| 30 | 1.250 | 2.7493<br>2.7488 | 0.692<br>0.685 | 0.66 | 0.50 | 0.6255<br>0.6252 | 0.624<br>0.625 | 0.31 | 0.31 | 0.660<br>0.654 | 1.0625<br>(Note 1) | 0.375-16 | 2.88 | 0.62 |
| 40 | 1.750 | 3.4993<br>3.4988 | 1.005<br>0.997 | 0.66 | 0.62 | 0.6255<br>0.6252 | 0.624<br>0.625 | 0.31 | 0.31 | 0.910<br>0.904 | 1.3125<br>(Note 1) | 0.500-13 | 3.88 | 0.81 |
| 45 | 2.250 | 3.9993<br>3.9988 | 1.286<br>1.278 | 0.78 | 0.62 | 0.7505<br>0.7502 | 0.749<br>0.750 | 0.38 | 0.38 | 1.160<br>1.154 | 1.500<br>(Note 1) | 0.500-13 | 4.75 | 0.81 |
| 50 | 2.750 | 5.0618<br>5.0613 | 1.568<br>1.559 | 1.06 | 0.75 | 1.0006<br>1.0002 | 0.999<br>1.000 | 0.50 | 0.50 | 1.410<br>1.404 | 2.000(Note 2) | 0.625-11 | 5.50 | 1.00 |
| 60 | 4.250 | 8.7180<br>8.7175 | 2.381<br>2.371 | 1.38 | 1.50 | 1.0006<br>1.0002 | 0.999<br>1.000 | 0.50 | 0.50 | 2.420<br>2.414 | 3.500<br>(Note 2) | 0.750-10 | 8.62 | 1.25 |

All dimensions are given in inches.

*Tolerances:*

Two-digit decimal dimensions ±0.010 unless otherwise specified.

*A*—Taper: Tolerance on rate of taper to be 0.001 inch per foot applied only in direction that decreases rate of taper.

*F′*—Centrality of keyway with axis of taper 0.002 total at maximum material condition. (0.002 total indicator variation)

*F*—Centrality of solid key with axis of taper 0.002 total at maximum material condition. (0.002 total indicator variation)

*Note 1:* Holes spaced as shown and located within 0.006 inch diameter of true position.

*Note 2:* Holes spaced as shown and located within 0.010 inch diameter of true position.

*Note 3:* Maximum turnout on test plug:
0.0004 at 1-inch projection from gage line.
0.0010 at 12-inch projection from gage line.

*Note 4:* Squareness of mounting face measured near mounting bolt-hole circle.

**Table 2. Essential Dimensions of American National Standard Tool Shanks for Milling Machines** *ANSI/ASME B5.18-1972 (R2014)*

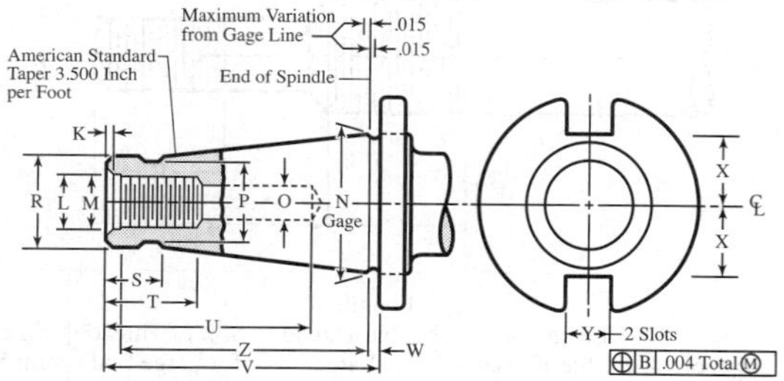


| Size No. | Gage Dia. of Taper *N* | Tap Drill Size for Draw-in Thread *O* | Dia. of Neck *P* | Size of Thread for Draw-in Bolt UNC-2B *M* | Pilot Dia. *R* | Length of Pilot *S* | Minimum Length of Usable Thread *T* | Minimum Depth of Clearance Hole *U* |
|---|---|---|---|---|---|---|---|---|
| 30 | 1.250 | 0.422<br>0.432 | 0.66<br>0.65 | 0.500-13 | 0.675<br>0.670 | 0.81 | 1.00 | 2.00 |
| 40 | 1.750 | 0.531<br>0.541 | 0.94<br>0.93 | 0.625-11 | 0.987<br>0.980 | 1.00 | 1.12 | 2.25 |
| 45 | 2.250 | 0.656<br>0.666 | 1.19<br>1.18 | 0.750-10 | 1.268<br>1.260 | 1.00 | 1.50 | 2.75 |
| 50 | 2.750 | 0.875<br>0.885 | 1.50<br>1.49 | 1.000-8 | 1.550<br>1.540 | 1.00 | 1.75 | 3.50 |
| 60 | 4.250 | 1.109<br>1.119 | 2.28<br>2.27 | 1.250-7 | 2.360<br>2.350 | 1.75 | 2.25 | 4.25 |

| Size. No. | Distance from Rear of Flange to End of Arbor *V* | Clearance of Flange from Gage Diameter *W* | Tool Shank Centerline to Driving Slot *X* | Width of Driving Slot *Y* | Distance from Gage Line to Bottom of C'bore *Z* | Depth of 60° Center *K* | Diameter of C'bore *L* |
|---|---|---|---|---|---|---|---|
| 30 | 2.75 | 0.045<br>0.075 | 0.640<br>0.625 | 0.635<br>0.645 | 2.50 | 0.05<br>0.07 | 0.525<br>0.530 |
| 40 | 3.75 | 0.045<br>0.075 | 0.890<br>0.875 | 0.635<br>0.645 | 3.50 | 0.05<br>0.07 | 0.650<br>0.655 |
| 45 | 4.38 | 0.105<br>0.135 | 1.140<br>1.125 | 0.760<br>0.770 | 4.06 | 0.05<br>0.07 | 0.775<br>0.780 |
| 50 | 5.12 | 0.105<br>0.135 | 1.390<br>1.375 | 1.010<br>1.020 | 4.75 | 0.05<br>0.12 | 1.025<br>1.030 |
| 60 | 8.25 | 0.105<br>0.135 | 2.400<br>2.385 | 1.010<br>1.020 | 7.81 | 0.05<br>0.12 | 1.307<br>1.312 |

All dimensions are given in inches.

*Tolerances:* Two-digit decimal dimensions ±0.010 inch unless otherwise specified.

*M*—Permissible for Class 2B "NoGo" gage to enter five threads before interference.

*N*—Taper tolerance on rate of taper to be 0.001 inch per foot applied only in direction that increases rate of taper.

*Y*—Centrality of drive slot with axis of taper shank 0.004 inch at maximum material condition (0.004 inch total indicator variation).

**Table 3. American National Standard Draw-In Bolt Ends**
*ANSI/ASME B5.18-1972 (R2014)*

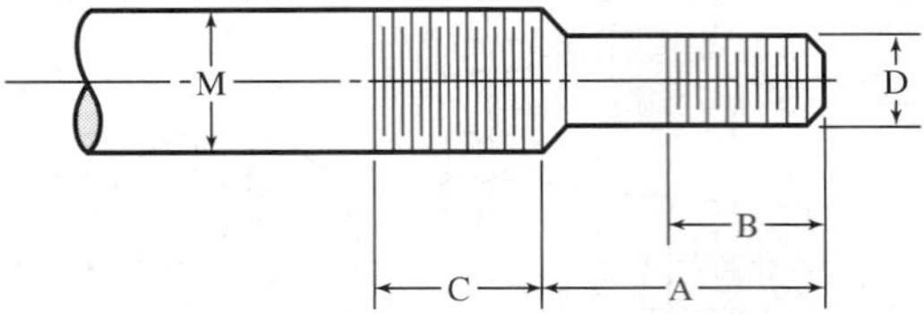


| Size No. | Length of Small End *A* | Length of Usable Thread at Small End *B* | Length of Usable Thread on Large Diameter *C* | Size of Thread for Large End UNC-2A *M* | Size of Thread for Small End UNC-2A *D* |
|---|---|---|---|---|---|
| 30 | 1.06 | 0.75 | 0.75 | 0.500-13 | 0.375-16 |
| 40 | 1.25 | 1.00 | 1.12 | 0.625-11 | 0.500-13 |
| 45 | 1.50 | 1.12 | 1.25 | 0.750-10 | 0.625-11 |
| 50 | 1.50 | 1.25 | 1.38 | 1.000-8 | 0.625-11 |
| 60 | 1.75 | 1.37 | 2.00 | 1.250-7 | 1.000-8 |

All dimensions are given in inches.

**Table 4. American National Standard Pilot Lead on Centering Plugs for Flatback Milling Cutters** *ANSI/ASME B5.18-1972 (R2014)*

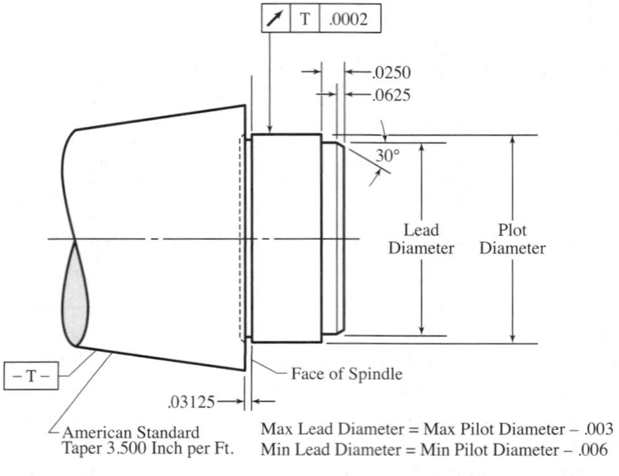

**Table 5. Essential Dimensions for American National Standard Spindle Nose with Large Flange** *ANSI/ASME B5.18-1972 (R2014)*

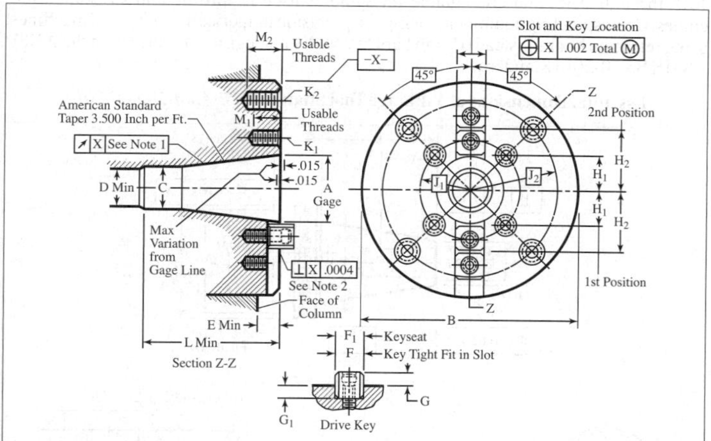


| Size No. | Gage Dia. of Taper *A* | Dia. of Spindle Flange *B* | Pilot Dia. *C* | Clearance Hole for Draw-in Bolt Min. *D* | Min. Dim. Spindle End to Column *E* | Width of Driving Key *F* | Height of Driving Key Max. *G* | Depth of Keyseat Min. *G*1 | Distance from Center to Driving Keys First Position *H*1 |
|---|---|---|---|---|---|---|---|---|---|
| 50A | 2.750 | 8.7180<br>8.7175 | 1.568<br>1.559 | 1.06 | 0.75 | 1.0006<br>1.0002 | 0.50 | 0.50 | 1.410<br>1.404 |

| Size No. | Distance from Center to Driving Keys Second Position *H*2 | Radius of Bolt Hole Circles (See Note 3) Inner *J*1 | Outer *J*2 | Size of Threads for Bolt Holes UNC-2B *K*1 | *K*2 | Full Depth of Arbor Hole in Spindle Min. *L* | Depth of Usable Thread for Bolt Holes *M*1 | *M*2 | Width of Keyseat *F*1 |
|---|---|---|---|---|---|---|---|---|---|
| 50A | 2.420<br>2.410 | 2.000 | 3.500 | 0.625-11 | 0.750-10 | 5.50 | 1.00 | 1.25 | 0.999<br>1.000 |

All dimensions are given in inches.

*Tolerances:* Two-digit decimal dimensions ±0.010 unless otherwise specified.

*A*—Tolerance on rate of taper to be 0.001 inch per foot applied only in direction that decreases rate of taper.

*F*—Centrality of solid key with axis of taper 0.002 inch total at maximum material condition. (0.002 inch total indicator variation)

$F_1$—Centrality of keyseat with axis of taper 0.002 inch total at maximum material condition. (0.002 inch total indicator variation)

*Note 1:* Maximum runout on test plug:
0.0004 at 1-inch projection from gage line.
0.0010 at 12-inch projection from gage line.

*Note 2:* Squareness of mounting face measured near mounting bolt-hole circle.

*Note 3:* Holes located as shown and within 0.010 inch diameter of true position.

**V-Flange Tool Shanks and Retention Knobs.—**Dimensions of ANSI/ASME B5.18-1972 (R2014) standard tool shanks and corresponding spindle noses are detailed on pages 1015 through 1018, and are suitable for spindles used in milling and associated machines. Corresponding equipment for higher-precision numerically controlled machines, using retention knobs instead of drawbars, is usually made in accordance with the ANSI/ASME B5.50-2015 Standard.

### Essential Dimensions of V-Flange Tool Shanks *ANSI/ASME B5.50-2015*

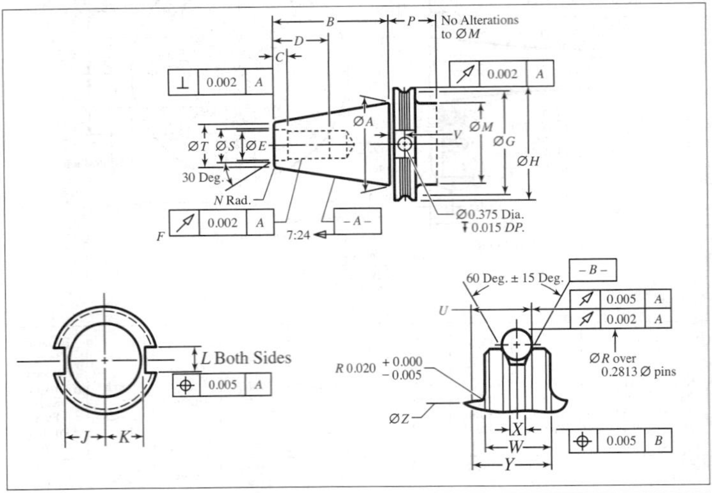


| | A | B | C | D | E | F | G | H | J | K |
|---|---|---|---|---|---|---|---|---|---|---|
| | Tolerance | ±0.005 | ±0.010 | Min. | +0.015 -0.000 | UNC 2B | ±0.010 | ±0.002 | +0.000 -0.015 | +0.000 -0.015 |
| Size | Gage Dia. | | | | | | | | | |
| 30 | 1.250 | 1.875 | 0.188 | 1.00 | 0.516 | 0.500-13 | 1.531 | 1.812 | 0.735 | 0.640 |
| 40 | 1.750 | 2.687 | 0.188 | 1.12 | 0.641 | 0.625-11 | 2.219 | 2.500 | 0.985 | 0.890 |
| 45 | 2.250 | 3.250 | 0.188 | 1.50 | 0.766 | 0.750-10 | 2.969 | 3.250 | 1.235 | 1.140 |
| 50 | 2.750 | 4.000 | 0.250 | 1.75 | 1.031 | 1.000-8 | 3.594 | 3.875 | 1.485 | 1.390 |
| 60 | 4.250 | 6.375 | 0.312 | 2.25 | 1.281 | 1.250-7 | 5.219 | 5.500 | 2.235 | 2.140 |

| | A | L | M | N | P | R | S | T | Z |
|---|---|---|---|---|---|---|---|---|---|
| | Tolerance | ±0.001 | Max. | ±0.010 | Min. | ±0.002 | ±0.010 | Min. | +0.000 -0.005 |
| Size | Gage Dia. | | | | | | | | |
| 30 | 1.250 | 0.645 | 1.812 | 0.020 | 1.38 | 2.176 | 0.590 | 0.652 | 1.250 |
| 40 | 1.750 | 0.645 | 2.500 | 0.040 | 1.38 | 2.863 | 0.720 | 0.880 | 1.750 |
| 45 | 2.250 | 0.770 | 3.250 | 0.040 | 1.38 | 3.613 | 0.850 | 1.233 | 2.250 |
| 50 | 2.750 | 1.020 | 3.875 | 0.040 | 1.38 | 4.238 | 1.125 | 1.427 | 2.750 |
| 60 | 4.250 | 1.020 | 5.500 | 0.040 | 1.500 | 5.863 | 1.375 | 2.309 | 4.250 |

Notes: Taper tolerance to be 0.001 in. in 12 in. applied in direction that increases rate of taper. Geometric dimensions symbols are to ASME Y14.5-2018. Dimensions are in inches. Deburr all sharp edges. Unspecified fillets and radii to be 0.03 ± 0.010*R*, or 0.03 ± 0.010 × 45 degrees. Data for size 60 are not part of the Standard. For all sizes, the values for dimensions *U* (tol. ± 0.005) are 0.579: for *V* (tol. ± 0.010), 0.440; for *W* (tol. ± 0.002), 0.625; for *X* (tol. ± 0.005), 0.152; and for *Y* (tol. ± 0.002), 0.750.

**Essential Dimensions of V-Flange Tool Shank Retention Knobs**
*ANSI/ASME B5.50-2015*

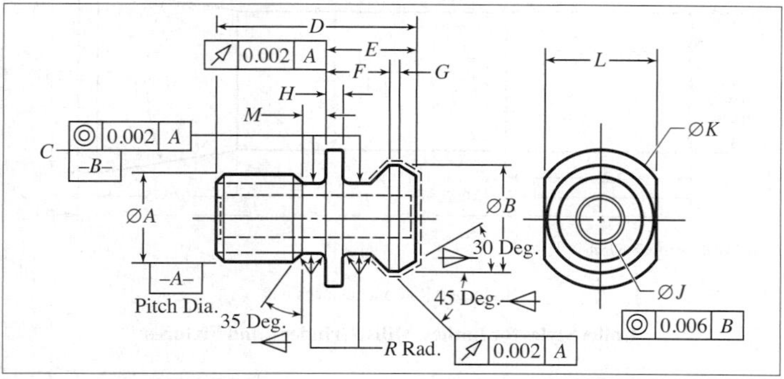


| Size | *A* | *B* | *C* | *D* | *E* | *F* |
|---|---|---|---|---|---|---|
| 30 | 0.500-13 | 0.520 | 0.385 | 1.10 | 0.460 | 0.320 |
| 40 | 0.625-11 | 0.740 | 0.490 | 1.50 | 0.640 | 0.440 |
| 45 | 0.750-10 | 0.940 | 0.605 | 1.80 | 0.820 | 0.580 |
| 50 | 1.000-8 | 1.140 | 0.820 | 2.30 | 1.000 | 0.700 |
| 60 | 1.250-7 | 1.460 | 1.045 | 3.20 | 1.500 | 1.080 |
| Tolerances | UNC-2A | ±0.005 | ±0.005 | ±0.040 | ±0.005 | ±0.005 |

| Size | *G* | *H* | *J* | *K* | *L* | *M* | *R* |
|---|---|---|---|---|---|---|---|
| 30 | 0.04 | 0.10 | 0.187 | 0.65<br>0.64 | 0.53 | 0.19 | 0.094 |
| 40 | 0.06 | 0.12 | 0.281 | 0.94<br>0.92 | 0.75 | 0.22 | 0.094 |
| 45 | 0.08 | 0.16 | 0.375 | 1.20<br>1.18 | 1.00 | 0.22 | 0.094 |
| 50 | 0.10 | 0.20 | 0.468 | 1.44<br>1.42 | 1.25 | 0.25 | 0.125 |
| 60 | 0.14 | 0.30 | 0.500 | 2.14<br>2.06 | 1.50 | 0.31 | 0.125 |
| Tolerances | ±0.010 | ±0.010 | ±0.010 | | +0.000<br>−0.010 | ±0.040 | +0.010<br>−0.005 |

Notes: Dimensions are in inches. Material: low-carbon steel. Heat treatment: carburize and harden to 0.016 to 0.028 in. effective case depth. Hardness of noted surfaces to be 56–60 RC (Rockwell C scale); core hardness 35–45 RC. Hole *J* shall not be carburized. Surfaces *C* and *R* to be free from tool marks. Deburr all sharp edges. Geometric dimension symbols are to ASME Y14.5-2018.

Data for size 60 are not part of Standard.

## Collets

**R8 Collet.—**The dimensions in this figure are believed reliable. However, there are variations among manufacturers of R8 collets, especially regarding the width and depth of the keyway. Some sources do not agree with all dimensions in this figure. R8 collets are not always interchangeable.

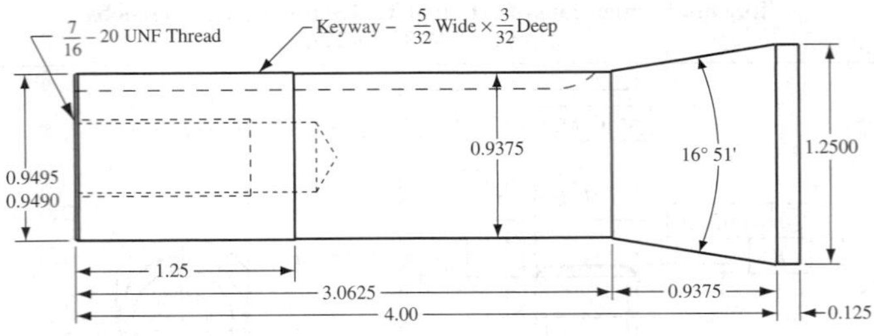


Bridgeport R8 Collet Dimensions

## Collet Styles for Lathes, Mills, Grinders, and Fixtures

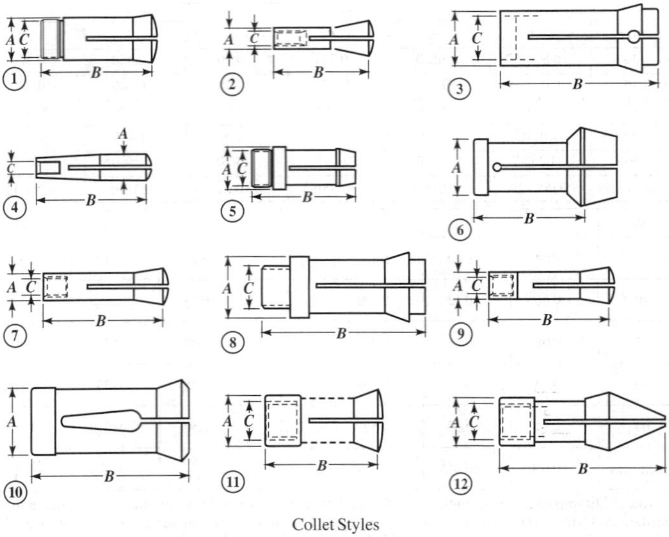


Collet Styles

## Collets for Lathes, Mills, Grinders, and Fixtures

| Collet | Style | Dimensions | | | Max. Capacity (inches) | | |
|---|---|---|---|---|---|---|---|
| | | Bearing Dia., *A* | Length, *B* | Thread, *C* | Round | Hex | Square |
| 1A | 1 | 0.650 | 2.563 | 0.640 × 26 RH | 0.500 | 0.438 | 0.344 |
| 1AM | 1 | 1.125 | 3.906 | 1.118 × 24 RH | 1.000 | 0.875 | 0.719 |
| 1B | 2 | 0.437 | 1.750 | 0.312 × 30 RH | 0.313 | 0.219 | 0.188 |
| 1C | 1 | 0.335 | 1.438 | 0.322 × 40 RH | 0.250 | 0.219 | 0.172 |
| 1J | 1 | 1.250 | 3.000 | 1.238 × 20 RH | 1.063 | 0.875 | 0.750 |
| 1K | 3 | 1.250 | 2.813 | None | 1.000 | 0.875 | 0.719 |
| 2A | 1 | 0.860 | 3.313 | 0.850 × 20 RH | 0.688 | 0.594 | 0.469 |
| 2AB | 2 | 0.750 | 2.563 | 0.500 × 20 RH | 0.625 | 0.484 | 0.391 |
| 2AM | 1 | 0.629 | 3.188 | 0.622 × 24 RH | 0.500 | 0.438 | 0.344 |

**Collets for Lathes, Mills, Grinders, and Fixtures** *(Continued)*

| Collet | Style | Dimensions | | | Max. Capacity (inches) | | |
|---|---|---|---|---|---|---|---|
| | | Bearing Dia., *A* | Length, *B* | Thread, *C* | Round | Hex | Square |
| 2B | 2 | 0.590 | 2.031 | 0.437 × 26 RH | 0.500 | 0.438 | 0.344 |
| 2C | 1 | 0.450 | 1.812 | 0.442 × 30 RH | 0.344 | 0.594 | 0.234 |
| 2H | 1 | 0.826 | 4.250 | 0.799 × 20 RH | 0.625 | 0.531 | 1.000 |
| 2J | 1 | 1.625 | 3.250 | 1.611 × 18 RH | 1.375 | 1.188 | 0.438 |
| 2L | 1 | 0.950 | 3.000 | 0.938 × 20 RH | 0.750 | 0.656 | 1.000 |
| 2M | 4 | 2 Morse | 2.875 | 0.375 × 16 RH | 0.500 | 0.438 | 0.344 |
| 2NS | 1 | 0.324 | 1.562 | 0.318 × 40 RH | 0.250 | 0.203 | 0.172 |
| 2OS | 1 | 0.299 | 1.250 | 0.263 × 40 RH | 0.188 | 0.156 | 0.125 |
| 2S | 1 | 0.750 | 3.234 | 0.745 × 18 RH | 0.563 | 0.484 | 0.391 |
| 2VB | 2 | 0.595 | 2.438 | 0.437 × 26 RH | 0.500 | 0.438 | 0.344 |
| 3AM | 1 | 0.750 | 3.188 | 0.742 × 24 RH | 0.625 | 0.531 | 0.438 |
| 3AT | 1 | 0.687 | 2.313 | 0.637 × 26 RH | 0.500 | 0.438 | 0.344 |
| 3B | 2 | 0.875 | 3.438 | 0.625 × 16 RH | 0.750 | 0.641 | 0.531 |
| 3C | 1 | 0.650 | 2.688 | 0.640 × 26 RH | 0.500 | 0.438 | 0.344 |
| 3H | 1 | 1.125 | 4.438 | 1.050 × 20 RH | 0.875 | 0.750 | 0.625 |
| 3J | 1 | 2.000 | 3.750 | 1.988 × 20 RH | 1.750 | 1.500 | 1.250 |
| 3NS | 1 | 0.687 | 2.875 | 0.647 × 20 RH | 0.500 | 0.438 | 0.344 |
| 3OS | 1 | 0.589 | 2.094 | 0.518 × 26 RH | 0.375 | 0.313 | 0.266 |
| 3PN | 1 | 0.650 | 2.063 | 0.645 × 24 RH | 0.500 | 0.438 | 0.344 |
| 3PO | 1 | 0.599 | 2.063 | 0.500 × 24 RH | 0.375 | 0.313 | 0.266 |
| 3S | 1 | 1.000 | 4.594 | 0.995 × 20 RH | 0.750 | 0.656 | 0.531 |
| 3SC | 1 | 0.350 | 1.578 | 0.293 × 36 RH | 0.188 | 0.156 | 0.125 |
| 3SS | 1 | 0.589 | 2.125 | 0.515 × 26 RH | 0.375 | 0.313 | 0.266 |
| 4C | 1 | 0.950 | 3.000 | 0.938 × 20 RH | 0.750 | 0.656 | 0.531 |
| 4NS | 1 | 0.826 | 3.500 | 0.800 × 20 RH | 0.625 | 0.531 | 0.438 |
| 4OS | 1 | 0.750 | 2.781 | 0.660 × 20 RH | 0.500 | 0.438 | 0.344 |
| 4PN | 1 | 1.000 | 2.906 | 0.995 × 16 RH | 0.750 | 0.656 | 0.531 |
| 4S | 1 | 0.998 | 3.250 | 0.982 × 20 RH | 0.750 | 0.656 | 0.531 |
| 5C | 1 | 1.250 | 3.281 | 1.238 × 20 RH[a] | 1.063 | 0.906 | 0.750 |
| 5M | 5 | 1.438 | 3.438 | 1.238 × 20 RH | 0.875 | 0.750 | 0.625 |
| 5NS | 1 | 1.062 | 4.219 | 1.050 × 20 RH | 0.875 | 0.750 | 0.625 |
| 5OS | 1 | 3.500 | 3.406 | 0.937 × 18 RH | 0.750 | 0.641 | 0.516 |
| 5P | 1 | 0.812 | 3.687 | 0.807 × 24 RH | 0.625 | 0.531 | 0.438 |
| 5PN | 1 | 1.312 | 3.406 | 1.307 × 16 RH | 1.000 | 0.875 | 0.719 |
| 5SC | 1 | 0.600 | 2.438 | 0.500 × 26 RH | 0.375 | 0.328 | 0.266 |
| 5ST | 1 | 1.250 | 3.281 | 1.238 × 20 RH | 1.063 | 0.906 | 0.750 |
| 5V | 1 | 0.850 | 3.875 | 0.775 × 18 RH | 0.563 | 0.484 | 0.391 |
| 6H | 1 | 1.375 | 4.750 | 1.300 × 10 RH | 1.125 | 0.969 | 0.797 |
| 6K | 1 | 0.842 | 3.000 | 0.762 × 26 RH | 0.625 | 0.531 | 0.438 |
| 6L | 1 | 1.250 | 4.438 | 1.178 × 20 RH | 1.000 | 0.875 | 0.719 |
| 6NS | 1 | 1.312 | 5.906 | 1.234 × 14 RH | 1.000 | 0.859 | 0.703 |
| 6R | 1 | 1.375 | 4.938 | 1.300 × 20 RH | 1.125 | 0.969 | 0.781 |
| 7B | 4 | 7 B&S | 3.125 | 0.375 × 16 RH | 0.500 | 0.406 | 0.344 |
| 7 B&S | 4 | 7 B&S | 2.875 | 0.375 × 16 RH | 0.500 | 0.406 | 0.344 |
| 7P | 1 | 1.125 | 4.750 | 1.120 × 20 RH | 0.875 | 0.750 | 0.625 |
| 7R | 6 | 1.062 | 3.500 | None | 0.875 | 0.750 | 0.625 |
| 8H | 1 | 1.500 | 4.750 | 1.425 × 20 RH | 1.250 | 1.063 | 0.875 |
| 8ST | 1 | 2.375 | 5.906 | 2.354 × 12 RH | 2.125 | 1.844 | 1.500 |
| 8WN | 1 | 1.250 | 3.875 | 1.245 × 16 RH | 1.000 | 0.875 | 0.719 |
| 9B | 4 | 9 B&S | 4.125 | 0.500 × 13 RH | 0.750 | 0.641 | 0.531 |
| 10L | 1 | 1.562 | 5.500 | 1.490 × 18 RH | 1.250 | 1.063 | 0.875 |
| 10P | 1 | 1.500 | 4.750 | 1.495 × 20 RH | 1.250 | 1.063 | 0.875 |
| 16C | 1 | 1.889 | 4.516 | 1.875 × 1.75 mm RH[b] | 1.625 | 1.406 | 1.141 |
| 20W | 1 | 0.787 | 2.719 | 0.775 × 6-1 cm | 0.563 | 0.484 | 0.391 |
| 22J | 1 | 2.562 | 4.000 | 2.550 × 18 RH | 2.250 | 1.938 | 1.563 |
| 32S | 1 | 0.703 | 2.563 | 0.690 × 24 RH | 0.500 | 0.438 | 0.344 |

## Collets for Lathes, Mills, Grinders, and Fixtures *(Continued)*

| Collet | Style | Dimensions | | | Max. Capacity (inches) | | |
|---|---|---|---|---|---|---|---|
| | | Bearing Dia., *A* | Length, *B* | Thread, *C* | Round | Hex | Square |
| 35J | 1 | 3.875 | 5.000 | 3.861 × 18 RH | 3.500 | 3.000 | 2.438 |
| 42S | 1 | 1.250 | 3.688 | 1.236 × 20 RH | 1.000 | 0.875 | 0.719 |
| 50V | 8 | 1.250 | 4.000 | 1.125 × 24 RH | 0.938 | 0.813 | 0.656 |
| 52SC | 1 | 0.800 | 3.688 | 0.795 × 20 RH | 0.625 | 0.531 | 0.438 |
| 115 | 1 | 1.344 | 3.500 | 1.307 × 20 LH | 1.125 | 0.969 | 0.797 |
| 215 | 1 | 2.030 | 4.750 | 1.990 × 18 LH | 1.750 | 1.500 | 1.219 |
| 315 | 1 | 3.687 | 5.500 | 3.622 × 16 LH | 3.250 | 2.813 | 2.250 |
| B3 | 7 | 0.650 | 3.031 | 0.437 × 20 RH | 0.500 | 0.438 | 0.344 |
| D5 | 7 | 0.780 | 3.031 | 0.500 × 20 RH | 0.625 | 0.531 | 0.438 |
| GTM | 7 | 0.625 | 2.437 | 0.437 × 20 RH | 0.500 | 0.438 | 0.344 |
| J&L | 9 | 0.999 | 4.375 | None | 0.750 | 0.641 | 0.516 |
| JC | 8 | 1.360 | 4.000 | None | 1.188 | 1.000 | 0.813 |
| LB | 10 | 0.687 | 2.000 | None | 0.500 | 0.438 | 0.344 |
| RO | 11 | 1.250 | 2.938 | 0.875 × 16 RH | 1.125 | 0.969 | 0.781 |
| RO | 12 | 1.250 | 4.437 | 0.875 × 16 RH | 0.800 | 0.688 | 0.563 |
| RO | 12 | 1.250 | 4.437 | 0.875 × 16 RH | 1.125 | 0.969 | 0.781 |
| RO | 11 | 1.250 | 2.938 | 0.875 × 16 RH | 0.800 | 0.688 | 0.563 |
| R8 | 7 | 0.950 | 4.000 | 0.437 × 20 RH | 0.750 | 0.641 | 0.531 |

[a] Internal stop thread is 1.041 × 24 RH.

[b] Internal stop thread is 1.687 × 20 RH.

Dimensions in inches unless otherwise noted. Courtesy of Hardinge Brothers, Inc.

Additional dimensions of the R8 collet are given on page 1021.

## DIN 6388, Type B, and DIN 6499, ER Type Collets

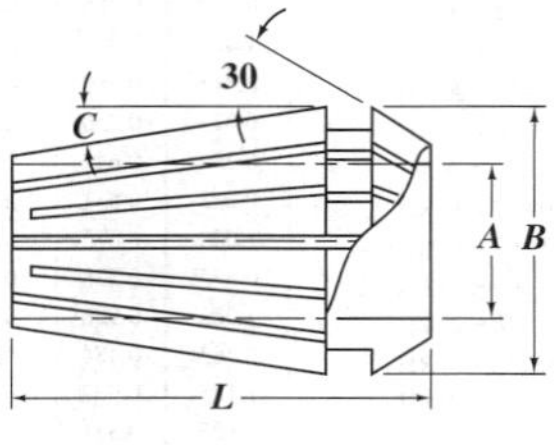


ER Type

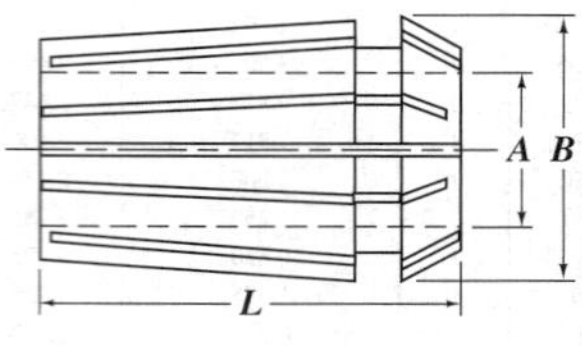

Type B

| Collet Standard | Type | Dimensions | | | |
|---|---|---|---|---|---|
| | | *B* (mm) | *L* (mm) | *A* (mm) | *C* |
| Type B, DIN 6388 | 16 | 25.50 | 40 | 4.5-16 | … |
| | 20 | 29.80 | 45 | 5.5-20 | … |
| | 25 | 35.05 | 52 | 5.5-25 | … |
| | 32 | 43.70 | 60 | 9.5-32 | … |
| ER Type, DIN 6499 | ERA8 | 8.50 | 13.5 | 0.5-5 | 8° |
| | ERA11 | 11.50 | 18 | 0.5-7 | 8° |
| | ERA16 | 17 | 27 | 0.5-10 | 8° |
| | ERA20 | 21 | 31 | 0.5-13 | 8° |
| | ERA25 | 26 | 35 | 0.5-16 | 8° |
| | ERA32 | 33 | 40 | 2-20 | 8° |
| | ERA40 | 41 | 46 | 3-26 | 8° |
| | | 41 | 39 | 26-30 | 8° |
| | ERA50 | 52 | 60 | 5-34 | 8° |

## ARBORS, CHUCKS, AND SPINDLES

### Portable Tool Spindles

**Circular Saw Arbors.—**ASME B107.4-2005 (R2011) "Driving and Spindle Ends for Portable Hand, Air, and Air Electric Tools" calls for a round arbor of 5⁄8-inch diameter for nominal saw blade diameters of 6 to 8.5 inches, inclusive, and a 3⁄4-inch diameter round arbor for saw blade diameters of 9 to 12 inches, inclusive.

**Spindles for Geared Chucks.—**Recommended threaded and tapered spindles for portable tool geared chucks of various sizes are as given in the following table:

**Recommended Spindle Sizes**

| Chuck Sizes, Inch | Recommended Spindles | |
|---|---|---|
| | Threaded | Taper[a] |
| 3⁄16 and 1⁄4 Light | 3⁄8-24 | 1 |
| 1⁄4 and 5⁄16 Medium | 3⁄8-24 or 1⁄2-20 | 2 Short |
| 3⁄8 Light | 3⁄8-24 or 1⁄2-20 | 2 |
| 3⁄8 Medium | 1⁄2-20 or 5⁄8-16 | 2 |
| 1⁄2 Light | 1⁄2-20 or 5⁄8-16 | 33 |
| 1⁄2 Medium | 5⁄8-16 or 3⁄4-16 | 6 |
| 5⁄8 and 3⁄4 Medium | 5⁄8-16 or 3⁄4-16 | 3 |

[a] Jacobs number.

**Vertical and Angle Portable Tool Grinder Spindles.—**The 5⁄8-11 spindle with a length of 1 1⁄8 inches shown on page 1028 is designed to permit the use of a jam nut with threaded cup wheels. When a revolving guard is used, the length of the spindle is measured from the wheel bearing surface of the guard. For unthreaded wheels with a 7⁄8-inch hole, a safety sleeve nut is recommended. The unthreaded wheel with 5⁄8-inch hole is not recommended because a jam nut alone may not resist the inertia effect when motor power is cut off.

**Straight Grinding Wheel Spindles for Portable Tools.—**Portable grinders with pneumatic or induction electric motors should be designed for the use of organic bond wheels rated 9500 ft per min (48.25 m/s). Light-duty electric grinders may be designed for vitrified wheels rated 6500 ft per min (33.0 m/s). Recommended maximum sizes of wheels of both types are as given in the following table:

**Recommended Maximum Grinding Wheel Sizes for Portable Tools**

| Spindle Size | Maximum Wheel Dimensions | | | |
|---|---|---|---|---|
| | 9500 fpm | | 6500 fpm | |
| | Diameter $D$ | Thickness $T$ | Diameter $D$ | Thickness $T$ |
| 3⁄8-24 × 1 1⁄8 | 2 1⁄2 | 1⁄2 | 4 | 1⁄2 |
| 1⁄2-13 × 1 3⁄4 | 4 | 3⁄4 | 5 | 3⁄4 |
| 5⁄8-11 × 2 1⁄8 | 8 | 1 | 8 | 1 |
| 5⁄8-11 × 3 1⁄8 | 6 | 2 | … | … |
| 5⁄8-11 × 3 1⁄8 | 8 | 1 1⁄2 | … | … |
| 3⁄4-10 × 3 1⁄4 | 8 | 2 | … | … |

Minimum $T$ with the first three spindles is about 1⁄8 inch to accommodate cutting off wheels. Flanges are assumed to be according to ANSI B7.1 and threads to ANSI/ASME B1.1.

## American Standard Square Drives for Portable Air and Electric Tools *ASA B5.38-1958*

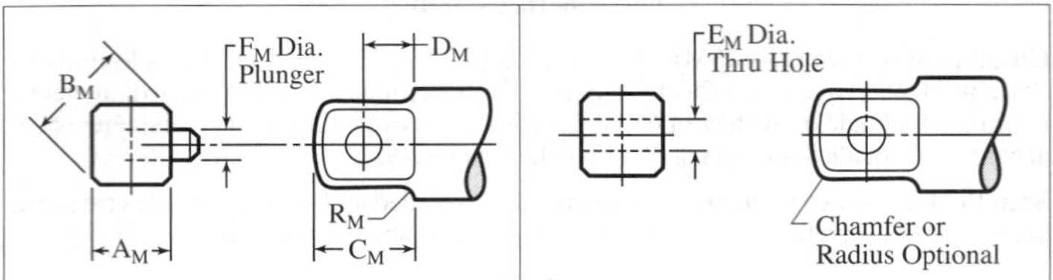


**DESIGN A** **DESIGN B**

| Drive Size | Design. | Male End | | | | | | | | | |
|---|---|---|---|---|---|---|---|---|---|---|---|
| | | $A_M$ | | $B_M$ | $C_M$ | | $D_M$ | | $E_M$ | $F_M$ | $R_M$ |
| | | Max. | Min. | Max. | Max. | Min. | Max. | Min. | Min. | Max. | Max. |
| ¼ | A | 0.252 | 0.247 | 0.330 | 0.312 | 0.265 | 0.165 | 0.153 | … | 0.078 | 0.015 |
| ⅜ | A | 0.377 | 0.372 | 0.500 | 0.438 | 0.406 | 0.227 | 0.215 | … | 0.156 | 0.031 |
| ½ | A | 0.502 | 0.497 | 0.665 | 0.625 | 0.531 | 0.321 | 0.309 | … | 0.187 | 0.031 |
| ⅝ | A | 0.627 | 0.622 | 0.834 | 0.656 | 0.594 | 0.321 | 0.309 | … | 0.187 | 0.047 |
| ¾ | B | 0.752 | 0.747 | 1.000 | 0.938 | 0.750 | 0.415 | 0.403 | 0.216 | … | 0.047 |
| 1 | B | 1.002 | 0.997 | 1.340 | 1.125 | 1.000 | 0.602 | 0.590 | 0.234 | … | 0.063 |
| 1½ | B | 1.503 | 1.498 | 1.968 | 1.625 | 1.562 | 0.653 | 0.641 | 0.310 | … | 0.094 |

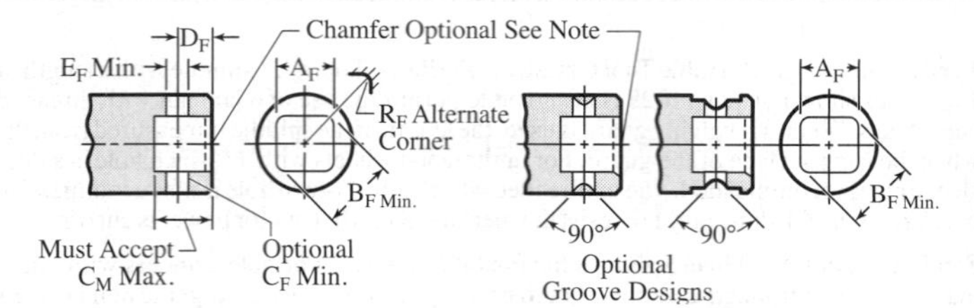


**DESIGN A** **DESIGN B**

| Drive Size | Design | Female End | | | | | | |
|---|---|---|---|---|---|---|---|---|
| | | $A_F$ | | $B_F$ | $D_F$ | | $E_F$ | $R_F$ |
| | | Max. | Min. | Min. | Max. | Min. | Min. | Max. |
| ¼ | A | 0.258 | 0.253 | 0.335 | 0.159 | 0.147 | 0.090 | … |
| ⅜ | A | 0.383 | 0.378 | 0.505 | 0.221 | 0.209 | 0.170 | … |
| ½ | A | 0.508 | 0.503 | 0.670 | 0.315 | 0.303 | 0.201 | … |
| ⅝ | A | 0.633 | 0.628 | 0.839 | 0.315 | 0.303 | 0.201 | … |
| ¾ | B | 0.758 | 0.753 | 1.005 | 0.409 | 0.397 | 0.216 | 0.047 |
| 1 | B | 1.009 | 1.004 | 1.350 | 0.596 | 0.584 | 0.234 | 0.062 |
| 1½ | B | 1.510 | 1.505 | 1.983 | 0.647 | 0.635 | 0.310 | 0.125 |

All dimensions in inches.

Incorporating fillet radius ($R_M$) at shoulder of male tang precludes use of minimum diameter cross-hole in socket ($E_F$), unless female drive end is chamfered (shown as optional).

If female drive end is not chamfered, socket cross-hole diameter ($E_F$) is increased to compensate for fillet radius $R_M$, max.

Minimum clearance across flats male to female is 0.001 inch through ¾-inch size; 0.002 inch in 1- and 1½-inch sizes. For impact wrenches, $A_M$ should be held as close to maximum as practical.

$C_F$, min., for both designs A and B should be equal to $C_M$, max.

## American Standard Threaded and Tapered Spindles for Portable Air and Electric Tools *ASA B5.38-1958*

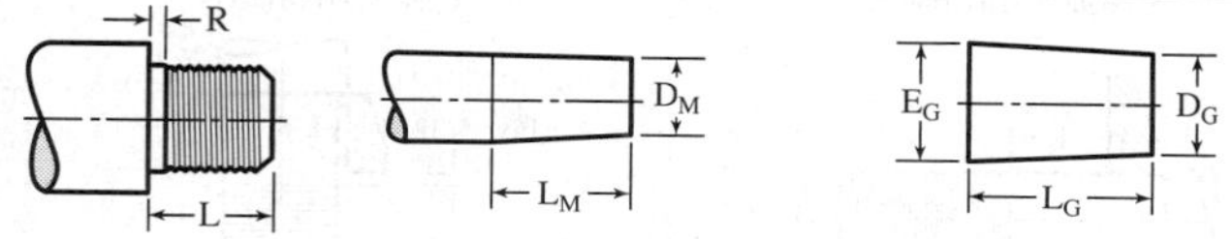


**Threaded Spindle** **Taper Spindle (Jacobs)** **Master Plug Gage**

| Nom. Dia. and Thd. | Pitch Dia. Max. | Pitch Dia. Min. | *R* | *L* | No.[a] | $D_M$ | $L_M$ | $E_G$ | $D_G$ | $L_G$ | Taper per Foot[b] |
|---|---|---|---|---|---|---|---|---|---|---|---|
| 3/8-24 | 0.3479 | 0.3455 | 1/16 | 9/16[c] | 1 | 0.335-0.333 | 0.656 | 0.38400 | 0.33341 | 0.65625 | 0.92508 |
| 1/2-20 | 0.4675 | 0.4649 | 1/16 | 9/16 | 2S[d] | 0.490-0.488 | 0.750 | 0.54880 | 0.48764 | 0.7500 | 0.97861 |
| | | | | | 2 | 0.490-0.488 | 0.875 | 0.55900 | 0.48764 | 0.87500 | 0.97861 |
| 5/8-16 | 0.5844 | 0.5812 | 3/32 | 11/16 | 33 | 0.563-0.561 | 1.000 | 0.62401 | 0.56051 | 1.000 | 0.76194 |
| | | | | | 6 | 0.626-0.624 | 1.000 | 0.67600 | 0.62409 | 1.000 | 0.62292 |
| 3/4-16 | 0.7094 | 0.7062 | 3/32 | 11/16 | 3 | 0.748-0.746 | 1.219 | 0.81100 | 0.74610 | 1.21875 | 0.63898 |

[a] Jacobs taper number.
[b] Calculated from $E_G, D_G, L_G$ for the master plug gage.
[c] Also 7/16 inch.
[d] 2S stands for 2 Short.

All dimensions in inches. Threads are per inch and right-hand. Tolerances: On *R*, plus or minus 1/64 inch; on *L*, plus 0.000, minus 0.030 inch.

## American Standard Abrasion Tool Spindles for Portable Air and Electric Tools *ASA B5.38-1958*

**Sanders and Polishers**

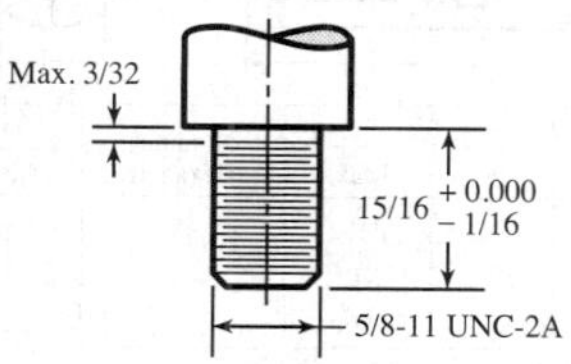


**Vertical and Angle Grinders**

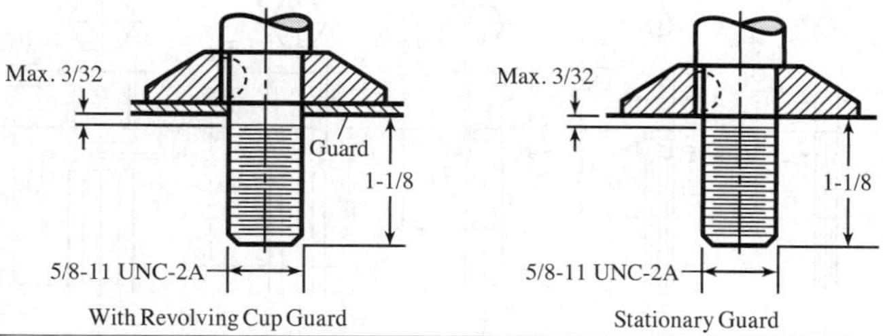


With Revolving Cup Guard    Stationary Guard

## American Standard Abrasion Tool Spindles for Portable Air and Electric Tools *ASA B5.38-1958 (Continued)*

**Straight Wheel Grinders**

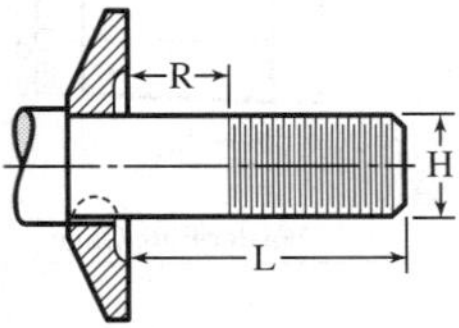


| *H* | *R* | *L* |
|---|---|---|
| 3⁄8-24 UNF-2A | 1⁄4 | 1 1⁄8 |
| 1⁄2-13 UNC-2A | 3⁄8 | 1 3⁄4 |
| 5⁄8-11 UNC-2A | 1⁄2 | 2 1⁄8 |
| 5⁄8-11 UNC-2A | 1 | 3 1⁄8 |
| 3⁄4-10 UNC-2A | 1 | 3 1⁄4 |

**Cone Wheel Grinders**

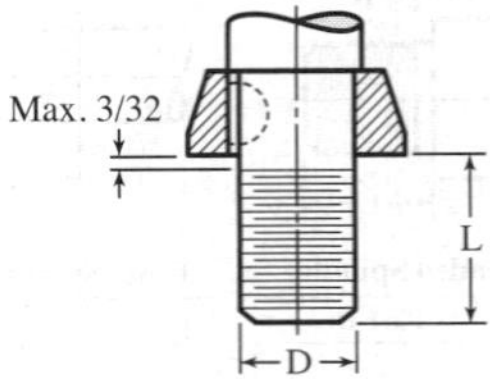


| *D* | *L* |
|---|---|
| 3⁄8-24 UNF-2A | 9⁄16 |
| 1⁄2-13 UNC-2A | 11⁄16 |
| 5⁄8-11 UNC-2A | 15⁄16 |

All dimensions in inches. Threads are right-hand.

## American Standard Hexagonal Chucks and Shanks for Portable Air and Electric Tools *ASA B5.38-1958*

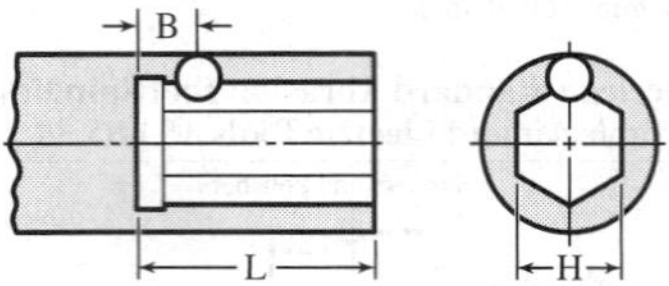


| Nominal Hexagon | *H* Min. | *H* Max. | *B* | *L* Max. | Nominal Hexagon | *H* Min. | *H* Max. | *B* | *L* Max. |
|---|---|---|---|---|---|---|---|---|---|
| 1⁄4 | 0.253 | 0.255 | 3⁄8 | 15⁄16 | 5⁄8 | 0.630 | 0.632 | 11⁄32 | 1 5⁄8 |
| 5⁄16 | 0.314 | 0.316 | 13⁄64 | 1 | 3⁄4 | 0.755 | 0.758 | 11⁄32 | 1 7⁄8 |
| 7⁄16 | 0.442 | 0.444 | 17⁄64 | 1 1⁄8 | … | … | … | … | … |

Shanks

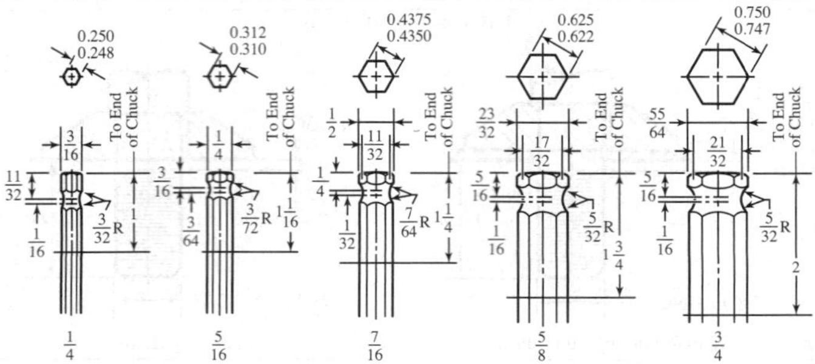


All dimensions in inches. Tolerance on *B* is plus or minus 0.005 inch.

**Mounted Wheels and Mounted Points**

These wheels and points are used in places that are hard to access and are available with a vitrified bond. The wheels are available with aluminum oxide or silicon carbide abrasive grains. The aluminum oxide wheels are used to grind tough and tempered die steels and the silicon carbide wheels, cast iron, chilled iron, bronze, and other nonferrous metals.

The illustrations on pages 1029 and 1030 give the standard shapes of mounted wheels and points as published by the Grinding Wheel Institute. A note about the maximum operating speed for these wheels is given at the bottom of the first page of illustrations. Metric sizes are given on page 1031.

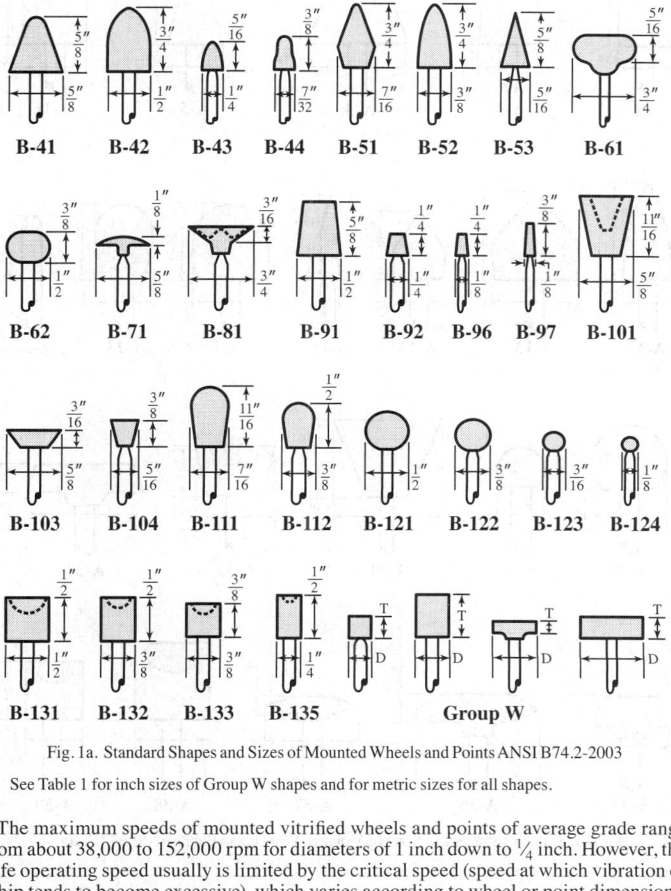


Fig. 1a. Standard Shapes and Sizes of Mounted Wheels and Points ANSI B74.2-2003

See Table 1 for inch sizes of Group W shapes and for metric sizes for all shapes.

The maximum speeds of mounted vitrified wheels and points of average grade range from about 38,000 to 152,000 rpm for diameters of 1 inch down to ¼ inch. However, the safe operating speed usually is limited by the critical speed (speed at which vibration or whip tends to become excessive), which varies according to wheel or point dimensions, spindle diameter, and overhang.

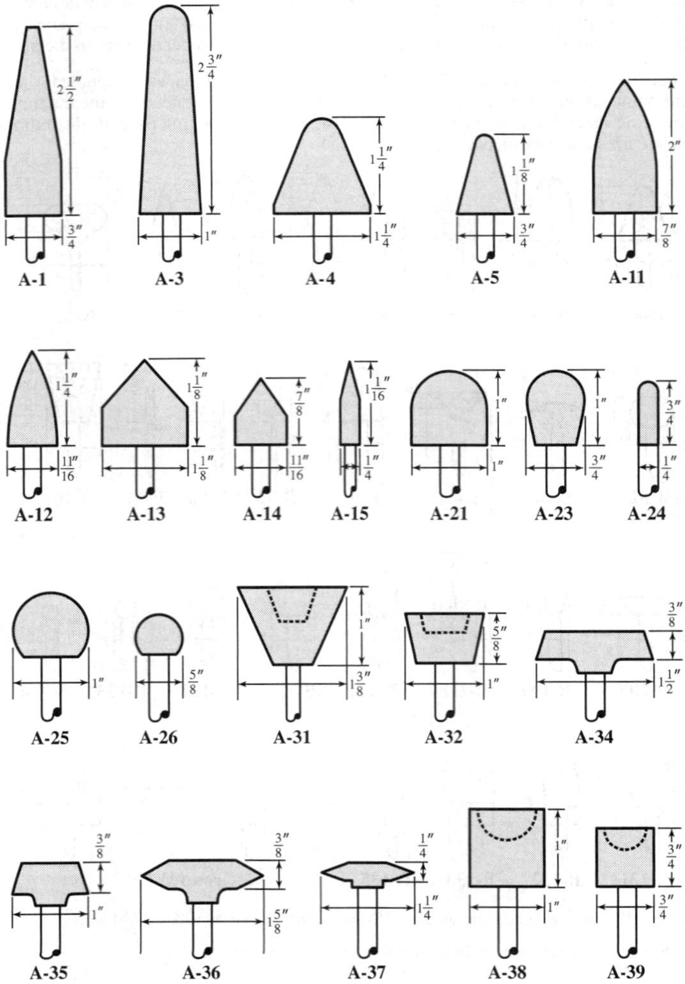


Fig. 1b. Standard Shapes and Sizes of Mounted Wheels and Points ANSI B74.2-2003

## Table 1. Shapes and Sizes of Mounted Wheels and Points *ANSI B74.2-2003*

| Abrasive Shape No.[a] | Abrasive Shape Size Diameter mm | Abrasive Shape Size Thickness mm | Abrasive Shape No.[a] | Abrasive Shape Size Diameter mm | Abrasive Shape Size Thickness mm |
|---|---|---|---|---|---|
| A 1 | 20 | 65 | A 24 | 6 | 20 |
| A 3 | 22 | 70 | A 25 | 25 | … |
| A 4 | 30 | 30 | A 26 | 16 | … |
| A 5 | 20 | 28 | A 31 | 35 | 26 |
| A 11 | 21 | 45 | A 32 | 25 | 20 |
| A 12 | 18 | 30 | A 34 | 38 | 10 |
| A 13 | 25 | 25 | A 35 | 25 | 10 |
| A 14 | 18 | 22 | A 36 | 40 | 10 |
| A 15 | 6 | 25 | A 37 | 30 | 6 |
| A 21 | 25 | 25 | A 38 | 25 | 25 |
| A 23 | 20 | 25 | A 39 | 20 | 20 |
| B 41 | 16 | 16 | B 97 | 3 | 10 |
| B 42 | 13 | 20 | B 101 | 16 | 18 |
| B 43 | 6 | 8 | B 103 | 16 | 5 |
| B 44 | 5.6 | 10 | B 104 | 8 | 10 |
| B 51 | 11 | 20 | B 111 | 11 | 18 |
| B 52 | 10 | 20 | B 112 | 10 | 13 |
| B 53 | 8 | 16 | B 121 | 13 | … |
| B 61 | 20 | 8 | B 122 | 10 | … |
| B 62 | 13 | 10 | B 123 | 5 | … |
| B 71 | 16 | 3 | B 124 | 3 | … |
| B 81 | 20 | 5 | B 131 | 13 | 13 |
| B 91 | 13 | 16 | B 132 | 10 | 13 |
| B 92 | 6 | 6 | B 133 | 10 | 10 |
| B 96 | 3 | 6 | B 135 | 6 | 13 |

| Abrasive Shape No.[a] | Abrasive Shape Size *D* mm | *T* mm | *D* inch | *T* inch | Abrasive Shape No.[a] | Abrasive Shape Size *D* mm | *T* mm | *D* inch | *T* inch |
|---|---|---|---|---|---|---|---|---|---|
| W 144 | 3 | 6 | ⅛ | ¼ | W 196 | 16 | 26 | ⅝ | 1 |
| W 145 | 3 | 10 | ⅛ | ⅜ | W 197 | 16 | 50 | ⅝ | 2 |
| W 146 | 3 | 13 | ⅛ | ½ | W 200 | 20 | 3 | ¾ | ⅛ |
| W 152 | 5 | 6 | 3/16 | ¼ | W 201 | 20 | 6 | ¾ | ¼ |
| W 153 | 5 | 10 | 3/16 | ⅜ | W 202 | 20 | 10 | ¾ | ⅜ |
| W 154 | 5 | 13 | 3/16 | ½ | W 203 | 20 | 13 | ¾ | ½ |
| W 158 | 6 | 3 | ¼ | ⅛ | W 204 | 20 | 20 | ¾ | ¾ |
| W 160 | 6 | 6 | ¼ | ¼ | W 205 | 20 | 25 | ¾ | 1 |
| W 162 | 6 | 10 | ¼ | ⅜ | W 207 | 20 | 40 | ¾ | 1½ |
| W 163 | 6 | 13 | ¼ | ½ | W 208 | 20 | 50 | ¾ | 2 |
| W 164 | 6 | 20 | ¼ | ¾ | W 215 | 25 | 3 | 1 | ⅛ |
| W 174 | 10 | 6 | ⅜ | ¼ | W 216 | 25 | 6 | 1 | ¼ |
| W 175 | 10 | 10 | ⅜ | ⅜ | W 217 | 25 | 10 | 1 | ⅜ |
| W 176 | 10 | 13 | ⅜ | ½ | W 218 | 25 | 13 | 1 | ½ |
| W 177 | 10 | 20 | ⅜ | ¾ | W 220 | 25 | 25 | 1 | 1 |
| W 178 | 10 | 25 | ⅜ | 1 | W 221 | 25 | 40 | 1 | 1½ |
| W 179 | 10 | 30 | ⅜ | 1¼ | W 222 | 25 | 50 | 1 | 2 |
| W 181 | 13 | 1.5 | ½ | 1/16 | W 225 | 30 | 6 | 1¼ | ¼ |
| W 182 | 13 | 3 | ½ | ⅛ | W 226 | 30 | 10 | 1¼ | ⅜ |
| W 183 | 13 | 6 | ½ | ¼ | W 228 | 30 | 20 | 1¼ | ¾ |
| W 184 | 13 | 10 | ½ | ⅜ | W 230 | 30 | 30 | 1¼ | 1¼ |
| W 185 | 13 | 13 | ½ | ½ | W 232 | 30 | 50 | 1¼ | 2 |
| W 186 | 13 | 20 | ½ | ¾ | W 235 | 40 | 6 | 1½ | ¼ |
| W 187 | 13 | 25 | ½ | 1 | W 236 | 40 | 13 | 1½ | ½ |
| W 188 | 13 | 40 | ½ | 1½ | W 237 | 40 | 25 | 1½ | 1 |
| W 189 | 13 | 50 | ½ | 2 | W 238 | 40 | 40 | 1½ | 1½ |
| W 195 | 16 | 20 | ⅝ | ¾ | W 242 | 50 | 25 | 2 | 1 |

[a] See shape diagrams in Fig. 1a and Fig. 1b on pages 1029 and 1030.

# BROACHES AND BROACHING

## The Broaching Process

The broaching process may be applied in machining holes or other internal surfaces and also to many flat or other external surfaces. Internal broaching is applied in forming either symmetrical or irregular holes, grooves, or slots in machine parts, especially when the size or shape of the opening, or its length in proportion to diameter or width, makes other machining processes impracticable. Broaching originally was utilized for such work as cutting keyways; machining round holes into square, hexagonal, or other shapes; forming splined holes; and for a large variety of other internal operations. The development of broaching machines and broaches finally resulted in extensive application of the process to external, flat, and other surfaces. Most external or surface broaching is done on machines of vertical design, but horizontal machines are also used for some classes of work. The broaching process is very rapid and accurate, and it leaves a finish of good quality. It is employed extensively in automotive and other plants where duplicate parts must be produced in large quantities and for dimensions within small tolerances.

**Types of Broaches.—**A number of typical broaches and the operations for which they are intended are shown by the diagrams in Fig. 1. Broach *A* produces a round-cornered, square hole. Prior to broaching square holes, it is usually the practice to drill a round hole having a diameter *d* somewhat larger than the width of the square. Hence, the sides are not completely finished, but this unfinished part is not objectionable in most cases. In fact, this clearance space is an advantage during the broaching operation in that it serves as a channel for the broaching lubricant; moreover, the broach has less metal to remove. Broach *B* is for finishing round holes. Broaching is superior to reaming for some classes of work because the broach will hold its size for a much longer period, thus insuring greater accuracy. Broaches *C* and *D* are for cutting single and double keyways, respectively. Broach *C* is of rectangular section and, when in use, slides through a guiding bushing which is inserted in the hole. Broach *E* is for forming four integral splines in a hub. The broach at *F* is for producing hexagonal holes. Rectangular holes are finished by broach *G*. The teeth on the sides of this broach are inclined in opposite directions, which has the following advantages: The broach is stronger than it would be if the teeth were opposite and parallel to each other; thin work cannot drop between the inclined teeth, as it tends to do when the teeth are at right angles, because at least two teeth are always cutting; the inclination in opposite directions neutralizes the lateral thrust. The teeth on the edges are staggered, the teeth on one side being midway between the teeth on the other edge, as shown by the dotted line. A double-cut broach is shown at *H*. This type is for finishing, simultaneously, both sides *f* of a slot, and for similar work. Broach *I* is the style used for forming the teeth in internal gears. It is practically a series of gear-shaped cutters, the outside diameters of which gradually increase toward the finishing end of the broach. Broach *J* is for round holes but differs from style *B* in that it has a continuous helical cutting edge. Some prefer this form because it gives a shearing cut. Broach *K* is for cutting a series of helical grooves in a hub or bushing. In helical broaching, either the work or the broach is rotated to form the helical grooves as the broach is pulled through.

In addition to the typical broaches shown in Fig. 1, many special designs are now in use for performing more complex operations. Two surfaces on opposite sides of a casting or forging are sometimes machined simultaneously by twin broaches and, in other cases, three or four broaches are drawn through a part at the same time, for finishing as many duplicate holes or surfaces. Notable developments have been made in the design of broaches for external or "surface" broaching.

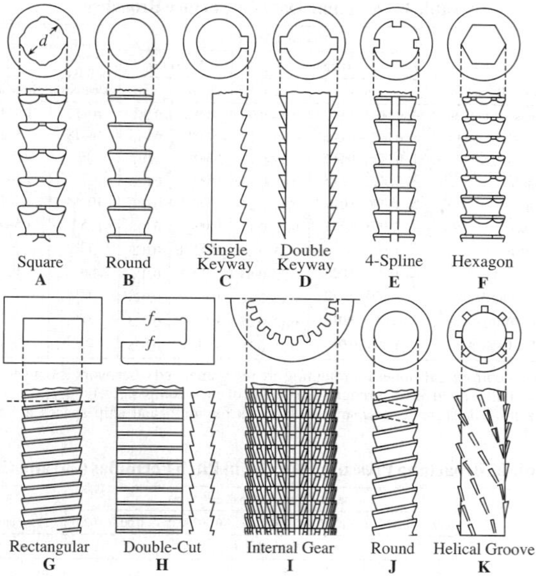


Fig. 1. Types of Broaches

*Burnishing Broach:* This is a broach having teeth or projections that are rounded on the top instead of being provided with a cutting edge, as in the ordinary type of broach. The teeth are highly polished, the tool being used for broaching bearings and for operations on other classes of work where the metal is relatively soft. The tool compresses the metal, thus making the surface hard and smooth. The amount of metal that can be displaced by a smooth-toothed burnishing broach is about the same as that removed by reaming. Such broaches are primarily intended for use on babbitt, white metal, and brass, but may also be satisfactorily used for producing a glazed surface on cast iron. This type of broach is also used when it is only required to accurately size a hole.

**Pitch of Broach Teeth.**—The pitch of broach teeth depends upon the depth of cut or chip thickness, the length of cut, the cutting force required and the power of the broaching machine. In the pitch formulas that follow

$L$ = length, in inches (mm), of layer to be removed by broaching

$d$ = depth of cut per tooth as shown by Table 1 (For internal broaches, $d$ = depth of cut as measured on one side of broach or one-half difference in diameters of successive teeth in the case of a round broach.)

$F$ = a factor (For brittle types of material, $F$ = 3 or 4 for roughing teeth and 6 for finishing teeth. For ductile types of material, $F$ = 4 to 7 for roughing teeth and 8 for finishing teeth.)

$b$ = width in inches (mm) of layer to be removed by broaching

$P$ = pressure required in tons per square inch (MPa) of an area equal to depth of cut times width of cut, in inches (mm) (Table 2)

$T$ = usable capacity, in tons (metric tons), of broaching machine = 70 percent of maximum tonnage

### Table 1. Designing Data for Surface Broaches

| Material to be Broached | Depth of Cut per Tooth: Roughing[a], inch | Roughing[a], mm | Finishing, inch | Finishing, mm | Face Angle or Rake, Degrees | Clearance Angle, Degrees: Rough | Clearance Angle, Degrees: Finish |
|---|---|---|---|---|---|---|---|
| Steel, High Tensile Strength | 0.0015–0.002 | 0.04–0.05 | 0.0005 | 0.013 | 10–12 | 1.5–3 | 0.5–1 |
| Steel, Med.Tensile Strength | 0.0025–0.005 | 0.06–0.13 | 0.0005 | 0.013 | 14–18 | 1.5–3 | 0.5–1 |
| Cast Steel | 0.0025–0.005 | 0.06–0.13 | 0.0005 | 0.013 | 10 | 1.53 | 0.5 |
| Malleable Iron | 0.0025–0.005 | 0.06–0.13 | 0.0005 | 0.013 | 7 | 1.5–3 | 0.5 |
| Cast Iron, Soft | 0.006–0.010 | 0.15–0.25 | 0.0005 | 0.013 | 10–15 | 1.5–3 | 0.5 |
| Cast Iron, Hard | 0.003–0.005 | 0.08–0.13 | 0.0005 | 0.013 | 5 | 1.5–3 | 0.5 |
| Zinc Die Castings | 0.005–0.010 | 0.13–0.25 | 0.0010 | 0.025 | 12[b] | 5 | 2 |
| Cast Bronze | 0.010–0.025 | 0.25–0.64 | 0.0005 | 0.013 | 8 | 0 | 0 |
| Wrought Aluminum Alloys | 0.005–0.010 | 0.13–0.25 | 0.0010 | 0.025 | 15[b] | 3 | 1 |
| Cast Aluminum Alloys | 0.005–0.010 | 0.13–0.25 | 0.0010 | 0.025 | 12[b] | 3 | 1 |
| Magnesium Die Castings | 0.010–0.015 | 0.25–0.38 | 0.0010 | 0.025 | 20[b] | 3 | 1 |

[a] The lower depth-of-cut values for roughing are recommended when work is not very rigid, the tolerance is small, a good finish is required, or length of cut is comparatively short.

[b] In broaching these materials, smooth surfaces for tooth and chip spaces are especially recommended.

### Table 2. Broaching Pressure P for Use in Pitch Formulas (2a) and (2b)

| Material to be Broached | Depth $d$ of cut per tooth, inch (mm): 0.024 (0.60) Pressure, $P$ Ton/in² | 0.024 (0.60) MPa | 0.01 (0.25) Ton/in² | 0.01 (0.25) MPa | 0.004 (0.10) Ton/in² | 0.004 (0.10) MPa | 0.002 (0.05) Ton/in² | 0.002 (0.05) MPa | 0.001 (0.025) Ton/in² | 0.001 (0.025) MPa | Side-cutting Broaches: Pressure, $P$ Ton/in² | Pressure, $P$ MPa | Cut, $d$ inch | Cut, $d$ mm |
|---|---|---|---|---|---|---|---|---|---|---|---|---|---|---|
| Steel, High Tensile Strength | … | … | … | … | | | 250 | 3447 | 312 | 4302 | 200 | 2758 | 0.004 | 0.10 |
| Steel, Med. Tensile Strength | … | … | … | … | 158 | 2179 | 185 | 2551 | 243 | 3351 | 143 | 1972 | 0.006 | 0.15 |
| Cast Steel | … | … | … | … | 128 | 1765 | 158 | 2179 | … | … | 115 | 1586 | 0.006 | 0.15 |
| Malleable Iron | … | … | … | … | 108 | 1489 | 128 | 1765 | … | … | 100 | 1379 | 0.006 | 0.15 |
| Cast Iron | … | … | 115 | 1586 | 115 | 1586 | 143 | 1972 | … | … | 115 | 1586 | 0.020 | 0.51 |
| Cast Brass | … | … | 50 | 689 | 50 | 689 | … | … | … | … | … | … | … | … |
| Brass, Hot-pressed | … | … | 85 | 1172 | 85 | 1172 | … | … | … | … | … | … | … | … |
| Zinc Die Castings | … | … | 70 | 965 | 70 | 965 | … | … | … | … | … | … | … | … |
| Cast Bronze | 35 | 483 | 35 | 483 | … | … | … | … | … | … | … | … | … | … |
| Wrought Aluminum | … | … | 70 | 965 | 70 | 965 | … | … | … | … | … | … | … | … |
| Cast Aluminum | … | … | 85 | 1172 | 85 | 1172 | … | … | … | … | … | … | … | … |
| Magnesium Alloy | 35 | 483 | 35 | 483 | … | … | … | … | … | … | … | … | … | … |

The minimum pitch shown by Formula (1) is based upon the receiving capacity of the chip space. The minimum pitch should not be less than 0.2 inch (5.0 mm) unless a smaller pitch is required for extremely short cuts to provide at least two teeth in contact simultaneously with the part being broached. A reduction below 0.2 inch (5.0 mm) is seldom required in surface broaching but may be necessary in connection with internal broaching.

$$\text{Minimum pitch} = 3\sqrt{LdF} \qquad (1)$$

Whether the minimum pitch may be used or not depends upon the power of the available machine. The factor $F$ in the formula provides for the increase in volume as the material is broached into chips. If a broach has adjustable inserts for the finishing teeth, the pitch of the finishing teeth may be smaller than the pitch of the roughing teeth because of the smaller depth $d$ of the cut. The higher value of $F$ for finishing teeth prevents the pitch from becoming too small, so that the spirally curled chips will not be crowded into too small a space. The pitch of the roughing and finishing teeth should be equal for broaches without separate inserts (notwithstanding the different values of $d$ and $F$) so that, after wear makes this necessary, some of the finishing teeth may be ground into roughing teeth.

| US Units | Metric Units |
| --- | --- |
| $\text{Allowable pitch} = \frac{dLbP}{T} \text{ inch}$ (2a) | $\textbf{Allowable pitch} = \frac{dLbP}{9810T} \textbf{ mm}$ (2b) |

If the pitch obtained by Formula (2a), or Formula (2b) in metric calculations, is larger than the minimum obtained by Formula (1), this larger value should be used because it is based upon the usable power of the machine. As the notation indicates, 70 percent of the maximum tonnage $T$ is taken as the usable capacity. The 30 percent reduction is to provide a margin for the increase in broaching load resulting from the gradual dulling of the cutting edges. The procedure in calculating both minimum and allowable pitches will be illustrated by an example.

***Example:* Determine pitch of broach for cast iron if $L = 220$ mm; $d = 0.1$ mm; and $F = 4$.**

$$\textbf{Minimum pitch} = 3\sqrt{220 \times 0.1 \times 4} = 28.14 \textbf{ mm}$$

**Next, apply Formula (2b). Assume that $b = 75$ mm and $T = 8$ metric ton; for cast iron and depth $d$ of 0.1 mm, $P = 1586$ MPa (Table 2). Then,**

$$\textbf{Allowable pitch} = \frac{0.1 \times 220 \times 75 = 1586}{9810(8)} = 33.34 \textbf{ mm}$$

This pitch is safely above the minimum. If in this case the usable tonnage of an available machine were, say, 7 metric tons instead of 8 metric tons, the pitch as shown by Formula (2b) might be increased to about 38.1 mm, thus reducing the number of teeth cutting simultaneously and, consequently, the load on the machine; or the cut per tooth might be reduced instead of increasing the pitch, especially if only a few teeth are in cutting contact, as might be the case with a short length of cut. If the usable tonnage in the preceding example were, say, 10 metric tons, then a pitch of 26.68 mm would be obtained by Formula (2b); hence, the pitch in this case should not be less than the minimum of approximately 28.14 mm obtained from Formula (1).

**Depth of Cut per Tooth.**—The term "depth of cut" as applied to surface or external broaches means the difference in the heights of successive teeth. This term, as applied to internal broaches for round, hexagonal or other holes, may indicate the total increase in the diameter of successive teeth; however, to avoid confusion, the term as here used means in all cases and regardless of the type of broach, the depth of cut as measured on one side.

In broaching free-cutting steel, the Broaching Tool Institute recommends 0.003 to 0.006 inch (0.076–0.15 mm) depth of cut for surface broaching; 0.002 to 0.003 inch (0.05–0.076 mm) for multispline broaching; and 0.0007 to 0.0015 inch (0.018–0.038 mm) for round hole broaching. The accompanying table contains data from a German source and applies specifically to surface broaches. All data relating to depth of cut are intended as a general guide only. While depth of cut is based primarily upon the machinability of the material, some reduction from the depth thus established may be required, particularly when the work-supporting fixture in surface broaching is not sufficiently rigid to resist the thrust from the broaching operation. In some cases, the pitch and cutting length may be increased to reduce the thrust force. Another possible remedy in surface broaching certain classes of work is to use a side-cutting broach instead of the ordinary depth-cutting type. A broach designed for side-cutting takes relatively deep narrow cuts that extend nearly to the full depth required. The side-cutting section is followed by teeth arranged for depth cutting to obtain the required size and surface finish on the work. In general, small tolerances in surface broaching require a reduced cut per tooth to minimize work deflection resulting from the pressure of the cut. See *Cutting Speed for Broaching* starting on page 1130 for broaching speeds.

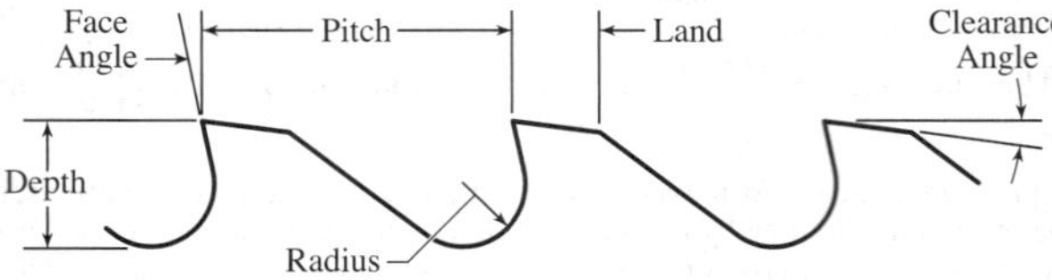


Terms Commonly Used in Broach Design

**Face Angle or Rake.—**The face angle (see diagram) of broach teeth affects the chip flow and varies considerably for different materials. While there are some variations in practice, even for the same material, the angles given in the accompanying table are believed to represent commonly used values. Some broach designers increase the rake angle for finishing teeth in order to improve the finish on the work.

**Clearance Angle.—**The clearance angle (see illustration) for roughing steel varies from 1.5 to 3 degrees and for finishing steel from 0.5 to 1 degree. Some recommend the same clearance angles for cast iron, and others larger clearance angles varying from 2 to 4 or 5 degrees. Additional data will be found in Table 1.

**Land Width.—**The width of the land usually is about 0.25 × pitch. It varies, however, from about one-fourth to one-third of the pitch. The land width is selected so as to obtain the proper balance between tooth strength and chip space.

**Depth of Broach Teeth.—**The tooth depth as established experimentally and on the basis of experience usually varies from about 0.37 to 0.40 of the pitch. This depth is measured radially from the cutting edge to the bottom of the tooth fillet.

**Radius of Tooth Fillet.—**The "gullet" or bottom of the chip space between the teeth should have a rounded fillet to strengthen the broach, facilitate curling of the chips, and safeguard against cracking in connection with the hardening operation. One rule is to make the radius equal to one-fourth the pitch. Another is to make it equal 0.4 to 0.6 the tooth depth. A third method preferred by some broach designers is to make the radius equal one-third of the sum obtained by adding together the land width, one-half the tooth depth, and one-fourth of the pitch.

**Total Length of Broach.—**After the depth of cut per tooth has been determined, the total amount of material to be removed by a broach is divided by this decimal to ascertain the number of cutting teeth required. This number of teeth multiplied by the pitch gives the length of the active portion of the broach. By adding to this dimension the distance over three or four straight teeth, the length of a pilot to be provided at the finishing end of the broach, and the length of a shank which must project through the work and the faceplate of the machine to the draw-head, the overall length of the broach is found. This calculated length is often greater than the stroke of the machine, or greater than is practical for a broach of the diameter required. In such cases, a set of broaches must be used.

**Chipbreakers.—**The teeth of broaches frequently have rounded chipbreaking grooves located at intervals along the cutting edges. These grooves break up wide curling chips and prevent them from clogging the chip spaces, thus reducing the cutting pressure and strain on the broach. These chipbreaking grooves are on the roughing teeth only. They are staggered and applied to both round and flat or surface broaches. The grooves are formed by a round-edged grinding wheel and usually vary in width from about $\frac{1}{32}$ to $\frac{3}{32}$ inch (0.79 to 2.38 mm) depending upon the size of broach. The more ductile the material, the wider the chipbreaker grooves should be and the smaller the distance between them. Narrow slotting broaches may have the right- and left-hand corners of alternate teeth beveled to obtain chipbreaking action.

**Shear Angle.**—The teeth of surface broaches ordinarily are inclined so they are not at right angles to the broaching movement. The object of this inclination is to obtain a shearing cut, which results in smoother cutting action and an improvement in surface finish. The shearing cut also tends to eliminate troublesome vibration. Shear angles for surface broaches are not suitable for broaching slots or any profiles that resist the outward movement of the chips. When the teeth are inclined, the fixture should be designed to resist the resulting thrusts unless it is practicable to incline the teeth of right- and left-hand sections in opposite directions to neutralize the thrust. The shear angle usually varies from 10 to 25 degrees.

**Types of Broaching Machines.**—Broaching machines may be divided into horizontal and vertical designs, and classified further according to the method of operation, as, for example, whether a broach in a vertical machine is pulled up or pulled down in forcing it through the work. Horizontal machines usually pull the broach through the work in internal broaching, but short rigid broaches may be pushed through. External surface broaching is also done on some machines of horizontal design, but usually vertical machines are employed for flat or other external broaching. Although parts usually are broached by traversing the broach itself, some machines are designed to hold the broach or broaches stationary during the actual broaching operation. This principle has been applied both to internal and surface broaching.

*Vertical Duplex Type:* The vertical duplex type of surface-broaching machine has two slides or rams that move in opposite directions and operate alternately. While the broach connected to one slide is moving downward on the cutting stroke, the other broach and slide is returning to the starting position, and this returning time is utilized for reloading the fixture on that side; consequently, the broaching operation is practically continuous. Each ram or slide may be equipped to perform a separate operation on the same part when two operations are required.

*Pull-Up Type:* Vertical hydraulically-operated machines that pull the broach or broaches up through the work are used for internal broaching of holes of various shapes and for broaching bushings, splined holes, small internal gears, etc. A typical machine of this kind is so designed that all broach handling is done automatically.

*Pull-Down Type:* The various movements in the operating cycle of a hydraulic pull-down type of machine equipped with an automatic broach-handling slide are the reverse of the pull-up type. The broaches for a pull-down type of machine have shanks on each end, an upper one for the broach-handling slide and a lower one for pulling through the work.

*Hydraulic Operation:* Modern broaching machines, as a general rule, are operated hydraulically rather than by mechanical means. Hydraulic operation is efficient, flexible in the matter of speed adjustments, low in maintenance cost, and the "smooth" action required for fine precision finishing may be obtained. The hydraulic pressures required, which frequently are 800 to 1000 pounds per square inch (5.5 to 6.9 MPa), are obtained from a motor-driven pump forming part of the machine. The cutting speeds of broaching machines frequently are between 20 and 30 feet per minute (6.1 to 9.1 m/min), and the return speeds often are double the cutting speed, or higher, to reduce the idle period.

**Ball-Broaching.**—Ball-broaching is a method of securing bushings, gears, or other components without the need for keys, pins, or splines. A series of axial grooves, separated by ridges, is formed in the bore of the workpiece by cold plastic deformation of the metal when a tool, having a row of three rotating balls around its periphery, is pressed through the parts. When the bushing is pressed into a broached bore, the ridges displace the softer material of the bushing into the grooves, thus securing the assembly. The balls can be made of high-carbon chromium steel or carbide, depending on the hardness of the component.

**Broaching Difficulties.—**The accompanying table has been compiled from information supplied by the National Broach and Machine Co. and presents some of the common broaching difficulties, their causes and means of correction.

**Causes of Broaching Difficulties**

| Broaching Difficulty | Possible Causes |
|---|---|
| Stuck broach | Insufficient machine capacity; dulled teeth; clogged chip gullets; failure of power during cutting stroke. To remove a stuck broach, remove workpiece and broach from the machine as a unit; never try to back out broach by reversing machine. If broach does not loosen by tapping workpiece lightly and trying to slide it off its starting end, mount workpiece and broach in a lathe and turn down workpiece to the tool surface. Workpiece may be sawed longitudinally into several sections in order to free the broach. Check broach design; perhaps tooth relief (back off) angle is too small or depth of cut per tooth is too great. |
| Galling and pickup | Lack of homogeneity of material being broached—uneven hardness, porosity; improper or insufficient coolant; poor broach design, mutilated broach; dull broach; improperly sharpened broach; improperly designed or outworn fixtures. Good broach design will do away with possible chip build-up on tooth faces and excessive heating. Grinding of teeth should be accurate so that the correct gullet contour is maintained. Contour should be fair and smooth. |
| Broach breakage | Overloading; broach dullness; improper sharpening; interrupted cutting stroke; backing up broach with workpiece in fixture; allowing broach to pass entirely through guide hole; ill fitting and/or sharp-edged key; crooked holes; untrue locating surface; excessive hardness of workpiece; insufficient clearance angle; sharp corners on pull end of broach. When grinding bevels on pull end of broach, use wheel that is not too pointed. |
| Chatter | Too few teeth in cutting contact simultaneously; excessive hardness of material being broached; loose or poorly constructed tooling; surging of ram due to load variations. Chatter can be alleviated by changing the broaching speed, by using shear cutting teeth instead of right-angle teeth, and by changing the coolant and the face and relief angles of the teeth. |
| Drifting or misalignment of tool during cutting stroke | Lack of proper alignment when broach is sharpened in grinding machine, which may be caused by dirt in the female center of the broach; inadequate support of broach during the cutting stroke, on a horizontal machine especially; body diameter too small; cutting resistance variable around I.D. due to lack of symmetry of surfaces to be cut; variations in hardness around I.D.; too few teeth in cutting contact. |
| Streaks in broached surface | Lands too wide; presence of forging, casting or annealing scale; metal pickup; presence of grinding burrs and grinding and cleaning abrasives. |
| Rings in the broached hole | Surging resulting from uniform pitch of teeth; presence of sharpening burrs on broach; tooth clearance angle too large; locating face not smooth or square; broach not supported for all cutting teeth passing through the work. The use of differential tooth spacing or shear cutting teeth helps in preventing surging. Sharpening burrs on a broach may be removed with a wood block. |

# FILES AND BURS

## Files

**Definitions of File Terms.**—The following file terms apply to hand files but not to rotary files and burs.

*Axis:* Imaginary line extending the entire length of a file equidistant from faces and edges.

*Back:* The convex side of a file having the same or similar cross section as a half-round file.

*Bastard Cut:* A grade of file coarseness between coarse and second cut of American pattern files and rasps.

*Blank:* A file in any process of manufacture before being cut.

*Blunt:* A file whose cross-sectional dimensions from point to tang remain unchanged.

*Coarse Cut:* The coarsest of all American pattern file and rasp cuts.

*Coarseness:* Term describing the relative number of teeth per unit length, the coarsest having the least number of file teeth per unit length; the smoothest, the most. American pattern files and rasps have four degrees of coarseness: coarse, bastard, second and smooth. Swiss pattern files usually have seven degrees of coarseness: 00, 0, 1, 2, 3, 4, 6 (from coarsest to smoothest). Curved-tooth files have three degrees of coarseness: standard, fine, and smooth.

*Curved Cut:* File teeth that are made in curved contour across the file blank.

*Cut:* Term used to describe file teeth with respect to their coarseness or their character (single, double, rasp, curved, special).

*Double Cut:* A file tooth arrangement formed by two series of cuts, namely the overcut followed, at an angle, by the upcut.

*Edge:* Surface joining faces of a file. May have teeth or be smooth.

*Face:* Widest cutting surface or surfaces that are used for filing.

*Heel or Shoulder:* That portion of a file that abuts the tang.

*Hopped:* A term used among file makers to represent a very wide skip or spacing between file teeth.

*Length:* The distance from the heel to the point.

*Overcut:* The first series of teeth put on a double-cut file.

*Point:* The front end of a file; the end opposite the tang.

*Rasp Cut:* A file tooth arrangement of round-topped teeth, usually not connected, that are formed individually by means of a narrow, punch-like tool.

*Re-Cut:* A worn-out file that has been re-cut and re-hardened after annealing and grinding off the old teeth.

*Safe Edge:* An edge of a file that is made smooth or uncut, so that it will not injure that portion or surface of the workplace with which it may come in contact during filing.

*Second Cut:* A grade of file coarseness between bastard and smooth of American pattern files and rasps.

*Set:* To blunt the sharp edges or corners of file blanks before and after the overcut is made, in order to prevent weakness and breakage of the teeth along such edges or corners when the file is put to use.

*Shoulder or Heel:* See *Heel or Shoulder.*

*Single Cut:* A file tooth arrangement where the file teeth are composed of single unbroken rows of parallel teeth formed by a single series of cuts.

*Smooth Cut:* An American pattern file and rasp cut that is smoother than second cut.

*Tang:* The narrowed portion of a file which engages the handle.

*Upcut:* The series of teeth superimposed on the overcut, and at an angle to it, on a double-cut file.

**File Characteristics.—**Files are classified according to their shape or cross section and according to the pitch or spacing of their teeth and the nature of the cut.

*Cross Section and Outline:* The cross section may be quadrangular, circular, triangular, or some special shape. The outline or contour may be tapered or blunt. In the former, the point is more or less reduced in width and thickness by a gradually narrowing section that extends for one-half to two-thirds of the length. In the latter the cross section remains uniform from tang to point.

*Cut:* The character of the teeth is designated as single, double, rasp or curved. The *single-cut file* (or *float* as the coarser cuts are sometimes called) has a single series of parallel teeth extending across the face of the file at an angle of from 45 to 85 degrees with the axis of the file. This angle depends upon the form of the file and the nature of the work for which it is intended. The single-cut file is customarily used with a light pressure to produce a smooth finish. The *double-cut file* has a multiplicity of small pointed teeth inclining toward the point of the file arranged in two series of diagonal rows that cross each other. For general work, the angle of the first series of rows is from 40 to 45 degrees and of the second from 70 to 80 degrees. For *double-cut finishing files,* the first series has an angle of about 30 degrees, and the second from 80 to 87 degrees. The second, or upcut, is almost always deeper than the first or overcut. Double-cut files are usually employed, under heavier pressure, for fast metal removal and where a rougher finish is permissible. The *rasp* is formed by raising a series of individual rounded teeth from the surface of the file blank with a sharp, narrow, punch-like cutting tool and is used with a relatively heavy pressure on soft substances for fast removal of material. The curved-tooth file has teeth that are in the form of parallel arcs extending across the face of the file, the middle portion of each arc being closest to the point of the file. The teeth are usually single cut and are relatively coarse. They may be formed by steel displacement but are more commonly formed by milling.

With reference to coarseness of cut, the terms *coarse, bastard, second* and *smooth cuts* are used, the coarse or bastard files for the heavier classes of work and the second- or smooth-cut files for the finishing or more exacting work. These degrees of coarseness are only comparable when files of the same length are compared, as the number of teeth per inch of length decreases as the length of the file increases. The number of teeth per inch varies considerably for different sizes and shapes and for files of different makes. The coarseness range for the curved-tooth files is given as standard, fine, and smooth. In the case of Swiss pattern files, a series of numbers is used to designate coarseness instead of names; Nos. 00, 0, 1, 2, 3, 4 and 6 being the most common with No. 00 the coarsest and No. 6 the finest.

**Classes of Files.—**There are five main classes of files: mill or saw files; machinists' files; curved-tooth files; Swiss pattern files; and rasps. The first two classes are commonly referred to as American pattern files.

*Mill or Saw Files:* These are used for sharpening mill or circular saws, and large crosscut saws; for lathe work; for draw filing; for filing brass and bronze; and for smooth filing generally. The number identifying the following files refers to the illustration in Fig. 1.

1) *Cantsaw files* have an obtuse isosceles triangular section, a blunt outline, are single cut and are used for sharpening saws having M-shaped teeth and teeth of less than 60-degree angle; 2) *Crosscut files* have a narrow triangular section with the short side rounded and a blunt outline, are single cut and are used to sharpen crosscut saws. The rounded portion is used to deepen the gullets of saw teeth, and the sides are used to sharpen the teeth themselves; 3) *Double ender files* have a triangular section, are tapered from the middle to both ends, are tangless and single cut, and are used reversibly for sharpening saws; 4) *The mill file* itself is usually single cut, tapered in width, and often has two square cutting edges in addition to the cutting sides. Either or both edges may be rounded, however, for filing the gullets of saw teeth. *The blunt mill file* has a uniform rectangular cross section from tip to tang; 5) The *triangular saw files* or *taper saw files* have an equilateral

triangular section, are tapered, are single cut and are used for filing saws with 60-degree angle teeth. They come in taper, slim taper, extra-slim taper and double-extra-slim taper thicknesses. *Blunt triangular* and *blunt hand-saw files* are without taper; and 6) *Web saw files* have a diamond-shaped section, a blunt outline, are single cut and are used for sharpening pulpwood or web saws.

*Machinists' Files:* These files are used throughout industry where metal must be removed rapidly and finish is of secondary importance. Except for certain exceptions in the round and half-round shapes, all are double cut. 7) *Flat files* have a rectangular section, are tapered in width and thickness, are cut on both sides and edges, and are used for general utility work; 8) *Half-round files* have a circular segmental section, are tapered in width and thickness, have their flat side double cut, their rounded side mostly double but sometimes single cut, and are used to file rounded holes, concave corners, etc., in general filing work; 9) *Hand files* are similar to flat files but taper in thickness only. One edge is uncut or "safe."; and 10) *Knife files* have a "knife-blade" section, are tapered in width only, are double cut, and are used by tool and die makers on work having acute angles.

*Machinist's general purpose files* have a rectangular section, are tapered and have single-cut teeth divided by angular serrations that produce short cutting edges. These edges help stock removal but still leave a smooth finish and are suitable for use on various materials, including aluminum, bronze, cast iron, malleable iron, mild steels and annealed tool steels.

11) *Pillar files* are similar to hand files but are thicker and not as wide; 12) *Round files* have a circular section, are tapered, single cut, and are generally used to file circular openings or curved surfaces; 13) *Square files* have a square section, are tapered, and are used for filing slots and keyways, and for general surface filing where a heavier section is preferred; 14) *Three-square files* have an equilateral triangular section and are tapered on all sides. They are double cut and have sharp corners, as contrasted with taper triangular files, which are single cut and have somewhat rounded corners. They are used for filing accurate internal angles, for clearing out square corners, and for filing taps and cutters; and 15) *Warding files* have a rectangular section, and taper in width to a narrow point. They are used for general narrow-space filing.

*Wood files* are made in the same sections as flat and half round files but with coarser teeth especially suited for working on wood.

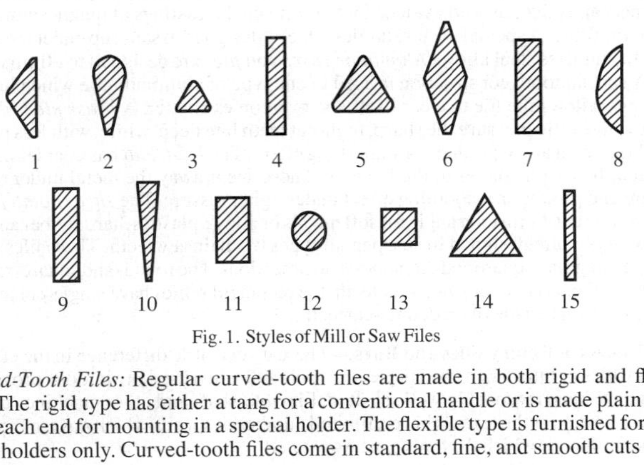


Fig. 1. Styles of Mill or Saw Files

*Curved-Tooth Files:* Regular curved-tooth files are made in both rigid and flexible forms. The rigid type has either a tang for a conventional handle or is made plain with a hole at each end for mounting in a special holder. The flexible type is furnished for use in special holders only. Curved-tooth files come in standard, fine, and smooth cuts and in parallel flat, square, pillar, pillar narrow, half-round, and shell types. A special curved-tooth file is available with teeth divided by long angular serrations. The teeth are cut in an "off center" arc. When moved across the work toward one edge of the file a fast cutting action is provided; when moved toward the other edge, a smoothing action; thus the file is made to serve a dual purpose.

*Swiss Pattern Files:* These are used by tool and die makers, model makers and delicate instrument parts finishers. They are made to closer tolerances than the conventional American pattern files although with similar cross sections. The points of the Swiss pattern files are smaller, the tapers are longer, and they are available in much finer cuts. They are primarily finishing tools for removing burrs left from previous finishing operations truing up narrow grooves, notches and keyways, cleaning out corners and smoothing small parts. For very fine work, *round-* and *square-handled needle files,* available in numerous cross-sectional shapes in overall lengths from 4 to 7 $^3/_4$ inches, are used. Die sinkers use *die sinkers' files* and *die sinkers' rifflers*. The files, also made in many different cross-sectional shapes, are $3^1/_2$ inches in length and are available in the cut Nos. 0, 1, 2, and 4. The rifflers are from $5^1/_2$ to $6^3/_4$ inches long, have cutting surfaces on either end, and come in numerous cross-sectional shapes in cut Nos. 0, 2, 3, 4 and 6. These rifflers are used by die makers for getting into corners, crevices, holes and contours of intricate dies and molds. Used in the same fashion as die sinkers' rifflers, *silversmiths' rifflers,* that have a much heavier cross section, are available in lengths from 6 $^7/_8$ to 8 inches and in cut Nos. 0, 1, 2, and 3. *Blunt machine files* in cut Nos. 00, 0, and 2 for use in ordinary and bench filing machines are available in many different cross-sectional shapes, in lengths from 3 to 8 inches.

*Rasps:* Rasps are employed for work on relatively soft substances such as wood, leather, and lead where fast removal or material is required. They come in rectangular and half round cross sections, the latter with and without a sharp edge.

*Special Purpose Files:* Falling under one of the preceding five classes of files, but modified to meet the requirements of some particular function, are a number of special-purpose files. The *long-angle lathe file* is used for filing work that is rotating in a lathe. The long tooth angle provides a clean shear, eliminates drag or tear and is self-clearing. This file has safe or uncut edges to protect shoulders of the work which are not to be filed. The *foundry file* has especially sturdy teeth with heavy-set edges for the snagging of castings—the removing of fins, sprues, and other projections. The *die-casting file* has extra-strong teeth on corners and edges as well as sides for working on die castings of magnesium, zinc, or aluminum alloys. A special file for stainless steel is designed to stand up under the abrasive action of stainless steel alloys. *Aluminum rasps* and *files* are designed to eliminate clogging. A special tooth construction is used in one type of aluminum tile which breaks up the filings, allows the file to clear itself and overcomes chatter. A *brass file* is designed so that with a little pressure the sharp, high-cut teeth bite deep, while, with less pressure, their short uncut angle produces a smoothing effect. The *lead float* has coarse, single-cut teeth at almost right angles to the file axis. These shear away the metal under ordinary pressure and produce a smoothing effect under light pressure. The *shear-tooth file* has a coarse single cut with a long angle for soft metals or alloys, plastics, hard rubber and wood. *Chainsaw files* are designed to sharpen all types of chainsaw teeth. These files come in round, rectangular, square and diamond-shaped sections. The round- and square-sectioned files have either double- or single-cut teeth, the rectangular files have single-cut teeth, and the diamond-shaped files have double-cut teeth.

**Effectiveness of Rotary Files and Burs.—**There it very little difference in the efficiency of rotary files or burs when used in electric tools and when used in air tools, provided the speeds have been reasonably well selected. Flexible-shaft and other machines used as a source of power for these tools have a limited number of speeds which govern the revolutions per minute at which the tools can be operated.

The carbide bur may be used on hard or soft materials with equally good results. The principle difference in construction of the carbide bur is that its teeth or flutes are provided with a negative rather than a radial rake. Carbide burs are relatively brittle and must be treated more carefully than ordinary burs. They should be kept cutting freely, in order to prevent too much pressure, which might result in crumbling of the cutting epics.

At the same speeds, both high-speed steel and carbide burs remove approximately the same amount of metal. However, when carbide burs are used at their most efficient speeds, the rate of stock removal may be as much as four times that of ordinary burs. In certain cases, speeds much higher than those shown in the table can be used. It has been demonstrated that a carbide bur will last up to 100 times as long as a high-speed steel bur of corresponding size and shape.

**Approximate Speeds of Rotary Files and Burs**

| Tool Diameter | | Medium Cut, High-Speed Steel Bur or File | | | | | Carbide Bur | |
|---|---|---|---|---|---|---|---|---|
| | | Mild Steel | Cast Iron | Bronze | Aluminum | Magnesium | Medium Cut | Fine Cut |
| inches | mm | Speed, Revolutions per Minute | | | | | Any Material | |
| ⅛ | 3.2 | 4600 | 7000 | 15,000 | 20,000 | 30,000 | 45,000 | 30,000 |
| ¼ | 6.4 | 3450 | 5250 | 11,250 | 15,000 | 22,500 | 30,000 | 20,000 |
| ⅜ | 9.5 | 2750 | 4200 | 9000 | 12,000 | 18,000 | 24,000 | 16,000 |
| ½ | 12.7 | 2300 | 3500 | 7500 | 10,000 | 15,000 | 20,000 | 13,350 |
| ⅝ | 15.9 | 2000 | 3100 | 6650 | 8900 | 13,350 | 18,000 | 12,000 |
| ¾ | 19.1 | 1900 | 2900 | 6200 | 8300 | 12,400 | 16,000 | 10,650 |
| ⅞ | 22.2 | 1700 | 2600 | 5600 | 7500 | 11,250 | 14,500 | 9650 |
| 1 | 25.4 | 1600 | 2400 | 5150 | 6850 | 10,300 | 13,000 | 8650 |
| 1⅛ | 28.6 | 1500 | 2300 | 4850 | 6500 | 9750 | … | … |
| 1¼ | 31.8 | 1400 | 2100 | 4500 | 6000 | 9000 | … | … |

As recommended by the Nicholson File Company.

**Steel Wool.—** Steel wool is made by shaving thin layers of steel from wire. The wire is pulled, by special machinery built for the purpose, past cutting tools or through cutting dies that shave off chips from the outside. Steel wool consists of long, relatively strong, and resilient steel shavings having sharp edges. This characteristic renders it an excellent abrasive. The fact that the cutting characteristics of steel wool vary with the size of the fiber, which is readily controlled in manufacture, has adapted it to many applications.

Metals other than steel have been made into wool by the same processes as steel, and when so manufactured have the same general characteristics. Thus wool has been made from copper, lead, aluminum, bronze, brass, monel metal, and nickel. The wire from which steel wool is made may be produced by either the Bessemer, or the basic or acid open-hearth processes. It should contain from 0.10 to 0.20 percent carbon; from 0.50 to 1.00 percent manganese; from 0.020 to 0.090 percent sulphur; from 0.050 to 0.120 percent phosphorus; and from 0.001 to 0.010 percent silicon. When drawn on a standard tensile-strength testing machine, a sample of the steel should show an ultimate strength of not less than 120,000 pounds per square inch (828 MPa).

**Steel Wool Grades**

| Description | Grade | Fiber Thickness | | Description | Grade | Fiber Thickness | |
|---|---|---|---|---|---|---|---|
| | | Inch | Millimeter | | | Inch | Millimeter |
| Super Fine | 0000 | 0.001 | 0.025 | Medium | 1 | 0.0025 | 0.06 |
| Extra Fine | 000 | 0.0015 | 0.035 | Medium Coarse | 2 | 0.003 | 0.075 |
| Very Fine | 00 | 0.0018 | 0.04 | Coarse | 3 | 0.0035 | 0.09 |
| Fine | 0 | 0.002 | 0.05 | Extra Coarse | 4 | 0.004 | 0.10 |

## KNURLS AND KNURLING

**Standard Knurls and Knurling.—**The ANSI/ASME Standard B94.6-1984 (R2014) covers knurling tools with standardized diametral pitches and their dimensional relations with respect to the work in the production of straight, diagonal, and diamond knurling on cylindrical surfaces having teeth of uniform pitch parallel to the cylinder axis or at a helix angle not exceeding 45 degrees with the work axis.

These knurling tools and the recommendations for their use are equally applicable to general purpose and precision knurling. The advantage of this ANSI/ASME Standard system is the provision by which good tracking (the ability of teeth to mesh as the tool penetrates the work blank in successive revolutions) is obtained by tools designed on the basis of diametral pitch instead of TPI (teeth per inch) when used with work blank diameters that are multiples of 1⁄64 inch for 64 and 128 diametral pitch or 1⁄32 inch for 96 and 160 diametral pitch. The use of knurls and work blank diameters that will permit good tracking should improve the uniformity and appearance of knurling, eliminate costly trial and error methods, reduce the failure of knurling tools and the production of defective work, and decrease the number of tools required. Preferred sizes for cylindrical knurls are given in Table 1, and detailed specifications appear in Table 2.

**Table 1. ANSI Standard Preferred Sizes for Cylindrical Type Knurls**
*ANSI/ASME B94.6-1984 (R2014)*

| Nominal Outside Diameter $D_{nt}$ | Width of Face $F$ | Diameter of Hole $A$ | Standard Diametral Pitches, $P$: 64 | 96 | 128 | 160 |
|---|---|---|---|---|---|---|
| | | | Number of Teeth, $N_t$, for Standard Pitches | | | |
| ½ | 3⁄16 | 3⁄16 | 32 | 48 | 64 | 80 |
| 5⁄8 | ¼ | ¼ | 40 | 60 | 80 | 100 |
| ¾ | 3⁄8 | ¼ | 48 | 72 | 96 | 120 |
| 7⁄8 | 3⁄8 | ¼ | 56 | 84 | 112 | 140 |
| Additional Sizes for Bench and Engine Lathe Tool Holders | | | | | | |
| 5⁄8 | 5⁄16 | 7⁄32 | 40 | 60 | 80 | 100 |
| ¾ | 5⁄8 | ¼ | 48 | 72 | 96 | 120 |
| 1 | 3⁄8 | 5⁄16 | 64 | 96 | 128 | 160 |

The 96 diametral pitch knurl should be given preference in the interest of tool simplification. Dimensions $D_{nt}$, $F$, and $A$ are in inches.

**Table 2. ANSI Standard Specifications for Cylindrical Knurls with Straight or Diagonal Teeth** *ANSI/ASME B94.6-1984 (R2014)*

| Diametral Pitch $P$ | Nominal Diameter, $D_{nt}$: ½ | 5⁄8 | ¾ | 7⁄8 | 1 | Tracking Correction Factor $Q$ | Tooth Depth, $h$, +0.0015, −0.0000: Straight | Diagonal | Radius at Root $R$ |
|---|---|---|---|---|---|---|---|---|---|
| | Major Diameter of Knurl, $D_{ot}$, +0.0000, −0.0015 | | | | | | | | |
| 64 | 0.4932 | 0.6165 | 0.7398 | 0.8631 | 0.9864 | 0.0006676 | 0.024 | 0.021 | 0.0070<br>0.0050 |
| 96 | 0.4960 | 0.6200 | 0.7440 | 0.8680 | 0.9920 | 0.0002618 | 0.016 | 0.014 | 0.0060<br>0.0040 |
| 128 | 0.4972 | 0.6215 | 0.7458 | 0.8701 | 0.9944 | 0.0001374 | 0.012 | 0.010 | 0.0045<br>0.0030 |
| 160 | 0.4976 | 0.6220 | 0.7464 | 0.8708 | 0.9952 | 0.00009425 | 0.009 | 0.008 | 0.0040<br>0.0025 |

All dimensions except diametral pitch are in inches.

Approximate angle of space between sides of adjacent teeth for both straight and diagonal teeth is 80 degrees. The permissible eccentricity of teeth for all knurls is 0.002 inch maximum (total indicator reading).

Number of teeth in a knurl equals diametral pitch multiplied by nominal diameter.

Diagonal teeth have 30-degree helix angle, $\psi$.

The term *Diametral Pitch* applies to the quotient obtained by dividing the total number of teeth in the circumference of the work by the basic blank diameter; in the case of the knurling tool, the diametrical pitch would be the total number of teeth in the circumference divided by the *nominal* diameter. In the Standard, the diametral pitch and number of teeth are always measured in a transverse plane that is perpendicular to the axis of rotation for diagonal as well as straight knurls and knurling.

**Cylindrical Knurling Tools.—**The cylindrical type of knurling tool is comprised of a tool holder and one or more knurls. The knurl has a centrally located mounting hole and is provided with straight or diagonal teeth on its periphery. The knurl is used to reproduce this tooth pattern on the work blank as the knurl and work blank rotate together.

**Formulas for Cylindrical Knurls*

$$P = \text{diametral pitch of knurl} = N_t \div D_{nt} \quad (1)$$

$$D_{nt} = \text{nominal diameter of knurl} = N_t \div P \quad (2)$$

$$N_t = \text{no. of teeth on knurl} = P \times D_{nt} \quad (3)$$

$$^*P_{nt} = \text{circular pitch on nominal diameter} = \pi \div P \quad (4)$$

$$^*P_{ot} = \text{circular pitch on major diameter} = \pi D_{ot} \div N_t \quad (5)$$

$$D_{ot} = \text{major diameter of knurl} = D_{nt} - (N_t Q \div \pi) \quad (6)$$

$$Q = P_{nt} - P_{ot} = \text{tracking correction factor in Formula} \quad (7)$$

*Tracking Correction Factor Q:* Use of the preferred pitches for cylindrical knurls, Table 2, results in good tracking on all fractional work-blank diameters that are multiples of $\frac{1}{64}$ inch for 64 and 128 diametral pitch, and $\frac{1}{32}$ inch for 96 and 160 diametral pitch; an indication of good tracking is evenness of marking on the work surface during the first revolution of the work.

The many variables involved in knurling practice require that an empirical correction method be used to determine what actual circular pitch is needed at the major diameter of the knurl to produce good tracking and the required circular pitch on the workpiece. The empirical tracking correction factor, $Q$, in Table 2 is used in the calculation of the major diameter of the knurl, Formula (6).

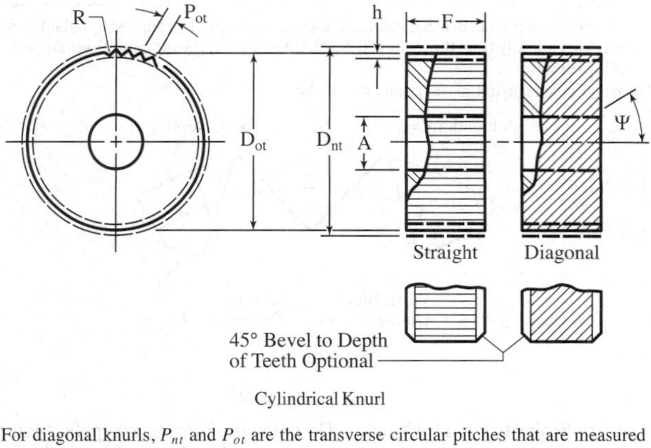


Cylindrical Knurl

* *Note:* For diagonal knurls, $P_{nt}$ and $P_{ot}$ are the transverse circular pitches that are measured in the plane perpendicular to the axis of rotation.

**Flat Knurling Tools.—**The flat type of tool is a knurling die, commonly used in reciprocating types of rolling machines. Dies may be made with either single or duplex faces having either straight or diagonal teeth. No preferred sizes are established for flat dies.

*Flat Knurling Die with Straight Teeth:*

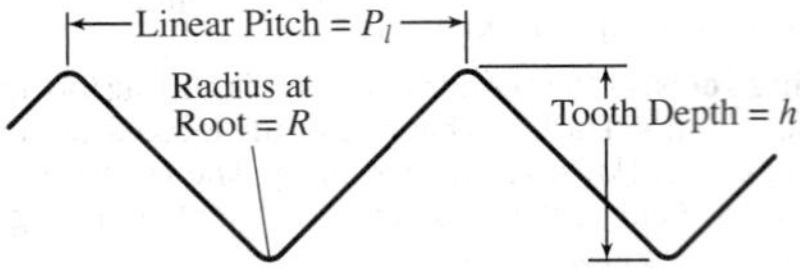


$R$ = radius at root

$P$ = diametral pitch = $N_w \div D_w$ (8)

$D_w$ = work blank (pitch) diameter = $N_w \div P$ (9)

$N_w$ = number of teeth on work = $P \times D_w$ (10)

$h$ = tooth depth

$Q$ = tracking correction factor (see Table 2)

$P_l$ = linear pitch on die

= circular pitch on work pitch diameter = $P - Q$ (11)

**Table 3. ANSI Standard Specifications for Flat Knurling Dies**

*ANSI/ASME B94.6-1984 (R2014)*

| Diametral Pitch, $P$ | Linear Pitch,[a] $P_l$ | Tooth Depth, $h$ Straight | Tooth Depth, $h$ Diagonal | Radius at Root, $R$ | Diametral Pitch, $P$ | Linear Pitch,[a] $P_l$ | Tooth Depth, $h$ Straight | Tooth Depth, $h$ Diagonal | Radius at Root, $R$ |
|---|---|---|---|---|---|---|---|---|---|
| 64 | 0.0484 | 0.024 | 0.021 | 0.0070<br>0.0050 | 128 | 0.0244 | 0.012 | 0.010 | 0.0045<br>0.0030 |
| 96 | 0.0325 | 0.016 | 0.014 | 0.0060<br>0.0040 | 160 | 0.0195 | 0.009 | 0.008 | 0.0040<br>0.0025 |

[a] The linear pitches are theoretical. The exact linear pitch produced by a flat knurling die may vary slightly from those shown, depending upon the rolling condition and the material being rolled.

All dimensions except diametral pitch are in inches.

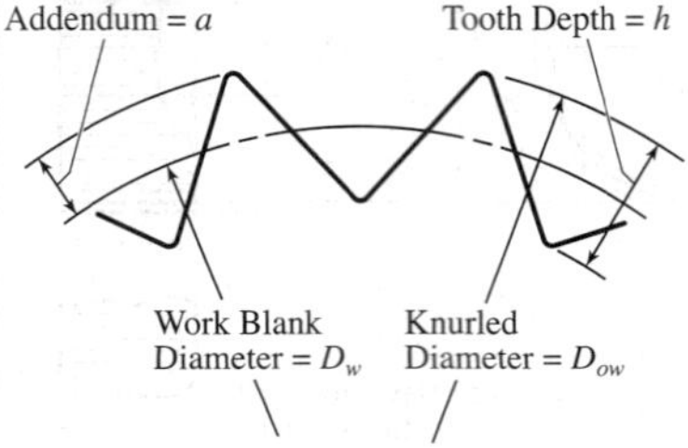


Teeth on Knurled Work

**Formulas Applicable to Knurled Work.—**The following formulas are applicable to knurled work with straight, diagonal, and diamond knurling.

*Formulas for Straight or Diagonal Knurling with Straight- or Diagonal-Tooth Cylindrical Knurling Tools Set with Knurl Axis Parallel with Work Axis:*

$$P = \text{diametral pitch} = N_w \div D_w \quad (12)$$

$$D_w = \text{work blank diameter} = N_w \div P \quad (13)$$

$$N_w = \text{no. of teeth on work} = P \times D_w \quad (14)$$

$$a = \text{“addendum” of tooth on work} = (D_{ow} - D_w) \div 2 \quad (15)$$

$h$ = tooth depth (see Table 2)

$$D_{ow} = \text{knurled diameter (outside diameter after knurling)} = D_w + 2a \quad (16)$$

*Formulas for Diagonal and Diamond Knurling with Straight-Tooth Knurling Tools Set at an Angle to the Work Axis:*

If, $\psi$ = angle between tool axis and work axis

$P$ = diametral pitch on tool

$P_\psi$ = diametral pitch produced on work blank (as measured in the transverse plane) by setting tool axis at an angle $^y$ with respect to work blank axis

$D_w$ = diameter of work blank; and

$N_w$ = number of teeth produced on work blank (as measured in the transverse plane)

then, $P_\psi = P\cos\psi$ (17)

and, $N = D_w P\cos\psi$ (18)

For example, if 30-degree diagonal knurling were to be produced on 1-inch diameter stock with a 160 pitch straight knurl:

$$N_w = D_w P\cos 30° = 1.000 \times 160 \times 0.86603 = 138.56 \text{ teeth}$$

Good tracking is theoretically possible by changing the helix angle as follows to correspond to a whole number of teeth (138):

$$\cos\psi = N_w \div D_w P = 138 \div (1 \times 160) = 0.8625$$

$$\psi = 30\tfrac{1}{2} \text{ degrees, approximately}$$

Whenever it is more practical to machine the stock, good tracking can be obtained by reducing the work blank diameter as follows to correspond to a whole number of teeth (138):

$$D_w = \frac{N_w}{P\cos\psi} = \frac{138}{160 \times 0.866} = 0.996 \text{ inch}$$

**Table 4. ANSI Standard Recommended Tolerances on Knurled Diameters**
*ANSI/ASME B94.6-1984 (R2014)*

| Tolerance Class | Diametral Pitch | | | | | | | |
|---|---|---|---|---|---|---|---|---|
| | 64 | 96 | 128 | 160 | 64 | 96 | 128 | 160 |
| | Tolerance on Knurled Outside Diameter | | | | Tolerance on Work-Blank Diameter Before Knurling | | | |
| I | +0.005<br>−0.012 | +0.004<br>−0.010 | +0.003<br>−0.008 | +0.002<br>−0.006 | ±0.0015 | ±0.0010 | ±0.0007 | ±0.0005 |
| II | +0.000<br>−0.010 | +0.000<br>−0.009 | +0.000<br>−0.008 | +0.000<br>−0.006 | ±0.0015 | ±0.0010 | ±0.0007 | ±0.0005 |
| III | +0.000<br>−0.006 | +0.000<br>−0.005 | +0.000<br>−0.004 | +0.000<br>−0.003 | +0.000<br>−0.0015 | +0.0000<br>−0.0010 | +0.000<br>−0.0007 | +0.0000<br>−0.0005 |

**Recommended Tolerances on Knurled Outside Diameters.—**The recommended applications of the tolerance classes shown in Table 4 are as follows:

*Class I:* Tolerances in this classification may be applied to straight, diagonal and raised diamond knurling where the knurled outside diameter of the work need not be held to close dimensional tolerances. Such applications include knurling for decorative effect, grip on thumb screws, and inserts for moldings and castings.

*Class II:* Tolerances in this classification may be applied to straight knurling only and are recommended for applications requiring closer dimensional control of the knurled outside diameter than provided for by Class I tolerances.

*Class III:* Tolerances in this classification may be applied to straight knurling only and are recommended for applications requiring closest possible dimensional control of the knurled outside diameter. Such applications include knurling for close fits.

*Note:* The width of the knurling should not exceed the diameter of the blank, and knurling wider than the knurling tool cannot be produced unless the knurl starts at the end of the work.

**Marking on Knurls and Dies.—**Each knurl and die should be marked as follows: *a*. when straight to indicate its diametral pitch; *b*. when diagonal, to indicate its diametral pitch, helix angle, and hand of angle.

**Concave Knurls.—**The radius of a concave knurl should not be the same as the radius of the piece to be knurled. If the knurl and the work are of the same radius, the material compressed by the knurl will be forced down on the shoulder *D* and spoil the appearance of the work. A design of concave knurl is shown in the accompanying illustration, and all the important dimensions are designated by letters. To find these dimensions, the pitch of the knurl required must be known, and also, approximately, the throat diameter *B*. This diameter must suit the knurl holder used and be such that the circumference contains an even number of teeth with the required pitch. When these dimensions have been decided upon, all the other unknown factors can be found by the following formulas: Let $R$ = radius of piece to be knurled; $r$ = radius of concave part of knurl; $C$ = radius of cutter or hob for cutting the teeth in the knurl; $B$ = diameter over concave part of knurl (throat diameter); $A$ = outside diameter of knurl; $d$ = depth of tooth in knurl; $P$ = pitch of knurl (number of teeth per inch circumference); $p$ = circular pitch of knurl; then $r = R + \frac{1}{2}d$; $C = r + d$; $A = B + 2r - (3d + 0.010 \text{ inch})$; and $d = 0.5 \times p \times \cot \alpha/2$, where $\alpha$ is the included angle of the teeth.

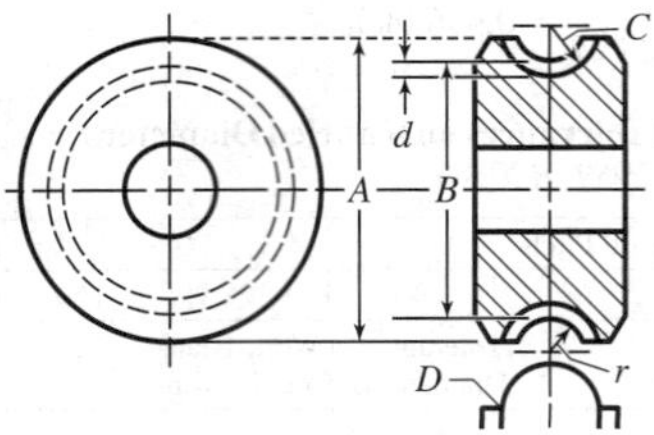


As the depth of the tooth is usually very slight, the throat diameter *B* will be accurate enough for all practical purposes for calculating the pitch, and it is not necessary to take into consideration the pitch circle. For example, assume that the pitch of a knurl is 32, that the throat diameter *B* is 0.5561 inch, that the radius *R* of the piece to be knurled is $\frac{1}{16}$ inch, and that the angle of the teeth is 90 degrees; find the dimensions of the knurl. Using the notation given:

$$p = \frac{1}{P} = \frac{1}{32} = 0.03125 \text{ inch} \qquad d = 0.5 \times 0.03125 \times \cot 45° = 0.0156 \text{ inch}$$

$$r = \frac{1}{16} + \frac{0.0156}{2} = 0.0703 \text{ inch} \qquad C = 0.0703 + 0.0156 = 0.0859 \text{ inch}$$

$$A = 0.5561 + 0.1406 - (0.0468 + 0.010) = 0.6399 \text{ inch}$$

## TOOL WEAR AND SHARPENING

Metal cutting tools wear constantly when they are being used. A normal amount of wear should not be a cause for concern until the size of the worn region indicates that the tool should be replaced. Normal wear cannot be avoided and should be differentiated from abnormal tool breakage or excessively fast wear. Tool breakage and an excessive rate of wear indicate that the tool is not operating correctly, and steps should be taken to correct this situation.

There are several basic mechanisms that cause tool wear. It is generally understood that tools wear as a result of abrasion caused by hard particles of work material plowing over the surface of the tool. Wear is also caused by diffusion or alloying between the work and tool material. In regions where the conditions of contact are favorable, work material reacts with tool material, causing an attrition of the tool material. The rate of this attrition is dependent upon the temperature in the region of contact and the reactivity of the tool and work materials with each other. Diffusion or alloying also occurs where particles of the work material are welded to the surface of the tool. These welded deposits are often quite visible in the form of a built-up edge, as particles or a layer of work material inside a crater, or as small mounds attached to the face of the tool. The diffusion or alloying occurring between these deposits and the tool weakens the tool material below the weld. Frequently these deposits are again rejoined to the chip by welding, or they are simply broken away by the force of collision with the passing chip. When this happens, a small amount of the tool material may remain attached to the deposit and be plucked from the surface of the tool, to be carried away with the chip. This mechanism can cause chips to be broken from the cutting edge and cause the formation of small craters, or pull-outs, on the tool face. It can also contribute to the enlargement of the larger crater that sometimes forms behind the cutting edge. Among the other mechanisms that can cause tool wear are severe thermal gradients and thermal shocks, which cause cracks to form near the cutting edge, ultimately leading to tool failure. This condition can be caused by improper tool grinding procedures, by heavy interrupted cuts, or by the improper application of cutting fluids when machining at high cutting speeds. Chemical reactions between the active constituents in some cutting fluids sometimes accelerate the rate of tool wear. Oxidation of the heated metal near the cutting edge also contributes to tool wear, particularly when fast cutting speeds and high cutting temperatures are encountered. Breakage of the cutting edge caused by overloading, heavy shock loads, or improper tool design is not normal wear and should be corrected.

The wear mechanisms described above bring about visible manifestations of wear on the tool, which should be understood so that the proper corrective measures can be taken, when required. These visible signs of wear are described in the following paragraphs, and the corrective measures that might be required are given in the table *Tool Troubleshooting Check List* on page 1071. The best procedure when troubleshooting is to try to correct only one condition at a time. When a correction has been made, it should be checked. After one condition has been corrected, work can then be started to correct the next condition.

**Flank Wear.**—Tool wear occurring on the flank of the tool below the cutting edge is called flank wear. Flank wear always takes place and cannot be avoided. It should not give rise to concern unless the rate of flank wear is too fast or the flank wear land becomes too large in size. The size of the flank wear can be measured as the distance between the top of the cutting edge and the bottom of the flank wear land. In practice, a visual estimate is usually made instead of a precise measurement, although in many instances flank wear is ignored and the tool wear is "measured" by the loss of size on the part. The best measure of tool wear, however, is flank wear. When it becomes too large, the rubbing action of the wear land against the workpiece increases and the cutting edge must be replaced. Because conditions vary, it is not possible to give an exact amount of flank wear at which the tool should be replaced. Although there are many exceptions, as a rough estimate, high-speed steel tools should be replaced when the width of the flank wear land reaches 0.005 to

0.010 inch (0.13–0.25 mm) for finish turning and 0.030 to 0.060 inch (0.76–1.52 mm) for rough turning; and for cemented carbides 0.005 to 0.010 inch (0.13–0.25 mm) for finish turning and 0.020 to 0.040 inch (0.51–1.02 mm) for rough turning.

Under ideal conditions, which, surprisingly, occur quite frequently, the width of the flank wear land will be very uniform along its entire length. When the depth of cut is uneven, such as when turning out-of-round stock, the bottom edge of the wear land may become somewhat slanted, the wear land being wider toward the nose. A jagged-appearing wear land usually is evidence of chipping at the cutting edge. Sometimes, only one or two sharp depressions of the lower edge of the wear land will appear to indicate that the cutting edge has chipped above these depressions. A deep notch will sometimes occur at the "depth of cut line," or that part of the cutting opposite the original surface of the work. This can be caused by a hard surface scale on the work, by a work-hardened surface layer on the work, or when machining high-temperature alloys. Often the size of the wear land is enlarged at the nose of the tool. This can be a sign of crater breakthrough near the nose or of chipping in this region. Under certain conditions, when machining with carbides, it can be an indication of deformation of the cutting edge in the region of the nose.

When a sharp tool is first used, the initial amount of flank wear is quite large in relation to the subsequent total amount. Under normal operating conditions, the width of the flank wear land will increase at a uniform rate until it reaches a critical size, after which the cutting edge breaks down completely. This is called catastrophic failure, and the cutting edge should be replaced before this occurs. When cutting at slow speeds with high-speed steel tools, there may be long periods when no increase in the flank wear can be observed. For a given work material and tool material, the rate of flank wear is primarily dependent on the cutting speed and then on the feed rate.

**Cratering.—**A deep crater will sometimes form on the face of the tool, which is easily recognizable. The crater forms at a short distance behind the side cutting edge, leaving a small shelf between the cutting edge and the edge of the crater. This shelf is sometimes covered with the built-up edge and at other times it is uncovered. Often the bottom of the crater is obscured with work material that is welded to the tool in this region. Under normal operating conditions, the crater will gradually enlarge until it breaks through a part of the cutting edge. Usually this occurs on the end cutting edge just behind the nose. When this takes place, the flank wear at the nose increases rapidly and complete tool failure follows shortly. Sometimes cratering cannot be avoided, and a slow increase in the size of the crater is considered normal. However, if the rate of crater growth is rapid, leading to a short tool life, corrective measures must be taken.

**Cutting Edge Chipping.—**Small chips are sometimes broken from the cutting edge, which accelerates tool wear but does not necessarily cause immediate tool failure. Chipping can be recognized by the appearance of the cutting edge and the flank wear land. A sharp depression in the lower edge of the wear land is a sign of chipping, and, if this edge of the wear land has a jagged appearance, it indicates that a large amount of chipping has taken place. Often the vacancy or cleft in the cutting edge that results from chipping is filled up with work material that is tightly welded in place. This occurs very rapidly when chipping is caused by a built-up edge on the face of the tool. In this manner, the damage to the cutting edge is healed; however, the width of the wear land below the chip is usually increased and the tool life is shortened.

**Deformation.—**Deformation occurs on carbide cutting tools when taking a very heavy cut using a slow cutting speed and a high feed rate. A large section of the cutting edge then becomes very hot, and the heavy cutting pressure compresses the nose of the cutting edge, thereby lowering the face of the tool in the area of the nose. This reduces the relief under the nose, increases the width of the wear land in this region, and shortens tool life.

**Surface Finish.—**The finish on the machined surface does not necessarily indicate poor cutting tool performance unless there is a rapid deterioration. A good surface finish is, however, sometimes a requirement. The principal cause of a poor surface finish is the built-up edge that forms along the edge of the cutting tool. The elimination of the built-up edge will always result in an improvement of the surface finish. The most effective way to eliminate the built-up edge is to increase cutting speed. When cutting speed is increased beyond a certain critical limit, there will be a rather sudden and large improvement in the surface finish. Cemented carbide tools can operate successfully at higher cutting speeds, where the built-up edge does not occur and where a good surface finish is obtained. Whenever possible, cemented carbide tools should be operated at cutting speeds where a good surface finish will result. There are times when such speeds are not possible. Also, high-speed tools cannot be operated at the speed where the built-up edge does not form. In these conditions, the most effective method of obtaining a good surface finish is to employ a cutting fluid that has active sulphur or chlorine additives.

Cutting tool materials that do not alloy readily with the work material are also effective in obtaining an improved surface finish. Straight titanium carbide and diamond are the two principal tool materials that fall into this category.

The presence of feed marks can mar an otherwise good surface finish, and attention must be paid to the feed rate and the nose radius of the tool if a good surface finish is desired. Changes in tool geometry can also be helpful. A small "flat," or secondary cutting edge, ground on the end cutting edge behind the nose will sometimes provide the desired surface finish. When the tool is in operation, the flank wear should not be allowed to become too large, particularly in the region of the nose where the finished surface is produced.

**Sharpening Twist Drills.—**Twist drills are cutting tools designed to perform concurrently several functions, such as penetrating directly into solid material, ejecting the removed chips outside the cutting area, maintaining the essentially straight direction of the advance movement and controlling the size of the drilled hole. The geometry needed for these multiple functions is incorporated into the design of the twist drill in such a manner that it can be retained even after repeated sharpening operations. Twist drills are resharpened many times during their service life, with the practically complete restitution of their original operational characteristics. However, in order to obtain all the benefits which the design of the twist drill is capable of providing, the surfaces generated in the sharpening process must agree with the original form of the tool's operating surfaces, unless a change of shape is required for use on a different work material.

The principal elements of tool geometry essential for the adequate cutting performance of twist drills are shown in Fig. 1. The generally used values for these dimensions are the following:

*Point angle:* Commonly 118°, except for high strength steels, 118° to 135°; aluminum alloys, 90° to 140°; and magnesium alloys, 70° to 118°.

*Helix angle:* Commonly 24° to 32°, except for magnesium and copper alloys, 10° to 30°.

*Lip relief angle:* Commonly 10° to 15°, except for high strength or tough steels, 7° to 12°. The lower values of these angle ranges are used for drills of larger diameter, the higher values for the smaller diameters. For drills of diameters less than $\frac{1}{4}$ inch (6.35 mm), the lip relief angles are increased beyond the listed maximum values up to 24°. For soft and free machining materials, 12° to 18° except for diameters less than $\frac{1}{4}$ inch (6.35 mm), 20° to 26°.

**Relief Grinding of the Tool Flanks.—**In sharpening twist drills the tool flanks containing the two cutting edges are ground. Each flank consists of a curved surface, which provides the relief needed for the easy penetration and free cutting of the tool edges. In grinding the flanks, Fig. 2, the drill is swung around the axis *A* of an imaginary cone while resting in a support that holds the drill at one-half the point angle $\beta$ with respect to the face of the grinding wheel. Feed *f* for stock removal is in the direction of the drill axis.

The relief angle is usually measured at the periphery of the twist drill and is also specified by that value. It is not a constant but should increase toward the center of the drill.

The relief grinding of the flank surfaces will generate the chisel angle on the web of the twist drill. The value of that angle, typically 55°, which can be measured, for example, with the protractor of an optical projector, is indicative of the correctness of the relief grinding.

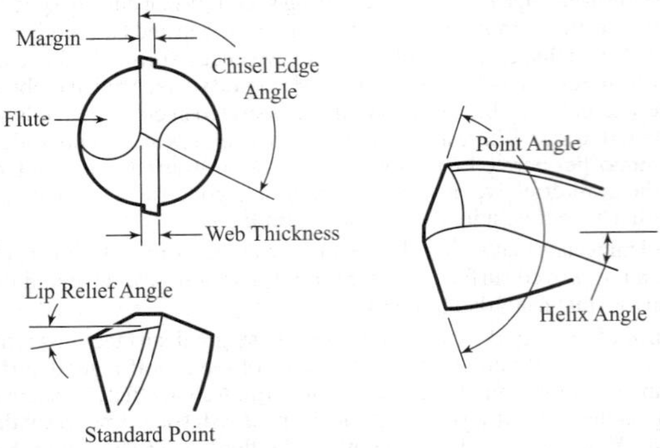


Fig. 1. The principal elements of tool geometry on twist drills.

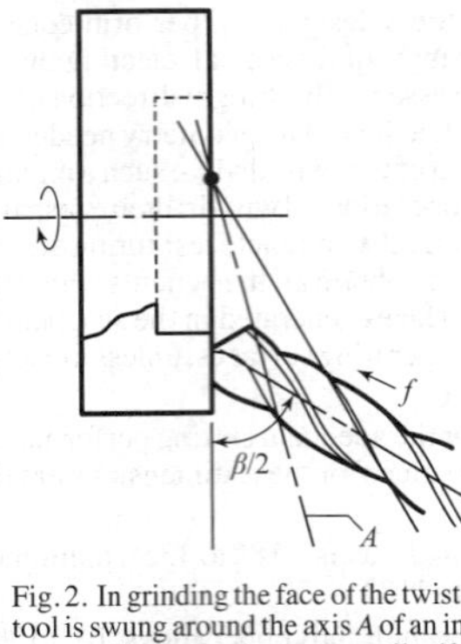


Fig. 2. In grinding the face of the twist drill, the tool is swung around the axis $A$ of an imaginary cone, while resting in a support tilted by half of the point angle $\beta$ with respect to the face of the grinding wheel. Feed $f$ for stock removal is in the direction of the drill axis.

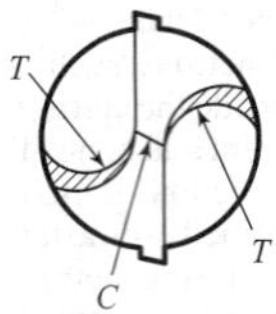


Fig. 3. The chisel edge $C$ after thinning the web on larger size or shorter drills by grinding off area $T$.

Fig. 4. Split-point or "crankshaft" type web thinning.

**Drill Point Thinning.—**The chisel edge is the least efficient operating surface element of the twist drill because it does not cut, but actually squeezes or extrudes, the work material. To improve the inefficient cutting conditions caused by the chisel edge, the point width is often reduced in a drill-point thinning operation, resulting in a condition such as that shown in Fig. 3. Point thinning is particularly desirable on larger size drills and also on those which become shorter in usage, because the thickness of the web increases toward the shaft of the twist drill, thereby adding to the length of the chisel edge. The extent of point thinning is limited by the minimum strength of the web needed to avoid splitting of the drill point under the influence of cutting forces.

Both sharpening operations—the relieved face grinding and the point thinning—should be carried out in special drill-grinding machines or with twist drill-grinding fixtures mounted on general-purpose tool-grinding machines designed to assure the essential accuracy of the required tool geometry. Offhand grinding may be used for the important web thinning when a special machine is not available; however, such operations require skill and experience.

Improperly sharpened twist drills, e.g. those with unequal edge length or asymmetrical point angle, will tend to produce holes with poor diameter and directional control.

For deep holes and for drilling into stainless steel, titanium alloys, high temperature alloys, nickel alloys, very high-strength materials and in some cases tool steels, split-point grinding, resulting in a "crankshaft" type drill point, is recommended. In this type of pointing, see Fig. 4, the chisel edge is entirely eliminated, extending the positive rake cutting edges to the center of the drill, thereby greatly reducing the required thrust in drilling. Points on modified-point drills must be restored after sharpening to maintain their increased drilling efficiency.

**Sharpening Carbide Tools.—**Cemented carbide indexable inserts are usually not re-sharpened, but sometimes they require a special grind in order to form a contour on the cutting edge to suit a special purpose. Brazed-type carbide cutting tools are resharpened after the cutting edge has become worn. On brazed carbide tools, the cutting edge wear should not be allowed to become excessive before the tool is re-sharpened. One method of determining when brazed carbide tools need resharpening is by periodic inspection of the flank wear and the condition of the face. Another method is to determine the amount of production normally obtained before excessive wear has taken place, or to determine the equivalent period of time. One disadvantage of this method is that slight variations in the work material will often cause the wear rate not to be uniform and the number of parts machined before regrinding will not be the same each time. Usually, sharpening should not require removal of more than 0.005 to 0.010 inch (0.13–0.25 mm) of carbide.

*General Procedure in Carbide-Tool Grinding:* The general procedure depends upon the kind of grinding operation required. If the operation is to resharpen a dull tool, a diamond wheel of 100- to 120-grain size is recommended, although a finer wheel—up to 150-grain size—is sometimes used to obtain a better finish. If the tool is new or is a "standard" design and changes in shape are necessary, a 100-grit diamond wheel is recommended for roughing and a finer grit diamond wheel can be used for finishing. Some shops prefer to rough-grind the carbide with a vitrified silicon carbide wheel, the finish grinding being done with a diamond wheel. A final operation commonly designated as lapping may or may not be employed for obtaining an extra-fine finish.

*Wheel Speeds:* The speed of silicon carbide wheels usually is about 5000 feet per minute (25.4 m/s). The speeds of diamond wheels generally range from 5000–6000 fpm (25.4–30.5 m/s); yet lower speeds (550–3000 fpm or 2.8–15.2 m/s) can be effective.

*Offhand Grinding:* In grinding single-point tools (excepting chipbreakers), the common practice is to hold the tool by hand, press it against the wheel face and traverse it continuously across the wheel face while the tool is supported on the machine rest or table, which is adjusted to the required angle. This is known as "offhand grinding" to distinguish it from the machine grinding of cutters as in regular cutter-grinding practice. The selection of wheels adapted to carbide-tool grinding is very important.

**Silicon Carbide Wheels.—**The green-colored silicon carbide wheels generally are preferred to the dark gray or gray-black variety, although the latter are sometimes used.

*Grain or Grit Sizes:* For roughing, a grain size of 60 is very generally used. For finish grinding with silicon carbide wheels, a finer grain size of 100 or 120 is common. A silicon carbide wheel such as C60-I-7V may be used for grinding both the steel shank and carbide tip. However, for under-cutting steel shanks up to the carbide tip, it may be advantageous to use an aluminum oxide wheel suitable for grinding softer, carbon steel.

*Grade:* According to the standard system of marking, different grades from soft to hard are indicated by letters from A to Z. For carbide-tool grinding fairly soft grades such as G, H, I, and J are used. The usual grades for roughing are I or J and for finishing H, I, and J. The grade should be such that a sharp free-cutting wheel will be maintained without excessive grinding pressure. Harder grades than those indicated tend to overheat and crack the carbide.

*Structure:* The common structure numbers for carbide-tool grinding are 7 and 8. The larger cup-wheels (10 to 14 inches or 254–356 mm) may be of the porous type and be designated as 12P. The standard structure numbers range from 1 to 15 with progressively higher numbers indicating less density and more open wheel structure.

**Diamond Wheels.—**Wheels with diamond-impregnated grinding faces are fast and cool cutting and have a very low rate of wear. They are used extensively both for resharpening and for finish grinding of carbide tools when preliminary roughing is required. Diamond wheels are also adapted for sharpening multi-tooth cutters such as milling cutters, reamers, etc., which are ground in a cutter-grinding machine.

*Resinoid bonded* wheels are commonly used for grinding chipbreakers, milling cutters, reamers or other multi-tooth cutters. They are also applicable to precision grinding of carbide dies, gages, and various external, internal and surface grinding operations. Fast, cool cutting action is characteristic of these wheels.

*Metal bonded* wheels are often used for offhand grinding of single-point tools especially when durability or long life and resistance to grooving of the cutting face are considered more important than the rate of cutting. *Vitrified bonded* wheels are used both for roughing of chipped or very dull tools and for ordinary resharpening and finishing. They provide rigidity for precision grinding, a porous structure for fast cool cutting, sharp cutting action and durability.

**Diamond Wheel Grit Sizes.—**For roughing with diamond wheels, a grit size of 100 is the most common both for offhand and machine grinding.

Grit sizes of 120 and 150 are frequently used in offhand grinding of single-point tools 1) for resharpening; 2) for a combination roughing and finishing wheel; and 3) for chipbreaker grinding.

Grit sizes of 220 or 240 are used for ordinary finish grinding all types of tools (offhand and machine) and also for cylindrical, internal and surface finish grinding. Grits of 320 and 400 are used for "lapping" to obtain very fine finishes, and for hand hones. A grit of 500 is for lapping to a mirror finish on such work as carbide gages and boring or other tools for exceptionally fine finishes.

**Diamond Wheel Grades.—**Diamond wheels are made in several different grades to better adapt them to different classes of work. The grades vary for different types and shapes of wheels. Standard Norton grades are H, J, and L for resinoid-bonded wheels, grade N for metal-bonded wheels and grades J, L, N, and P for vitrified wheels. Harder and softer grades than standard may at times be used to advantage.

**Diamond Concentration.—**The relative amount (by carat weight) of diamond in the diamond section of the wheel is known as the "diamond concentration." Concentrations of 100 (high), 50 (medium) and 25 (low) ordinarily are supplied. A concentration of 50 represents one-half the diamond content of 100 (if the depth of the diamond is the same in each case) and 25 equals one-fourth the content of 100 or one-half the content of 50 concentration.

*100 Concentration:* Generally interpreted to mean 72 carats of diamond/in$^3$ of abrasive section. (A 75 concentration indicates 54 carats/in$^3$.) Recommended (especially in grit sizes up to about 220) for general machine grinding of carbides, and for grinding cutters and chipbreakers. Vitrified and metal-bonded wheels usually have 100 concentration.

*50 Concentration:* In the finer grit sizes of 220, 240, 320, 400, and 500, a 50 concentration is recommended for offhand grinding with resinoid-bonded cup-wheels.

*25 Concentration:* A low concentration of 25 is recommended for offhand grinding with resinoid-bonded cup-wheels with grit sizes of 100, 120 and 150.

*Depth of Diamond Section:* The radial depth of the diamond section usually varies from $1/16$ to $1/4$ inch (1.6 to 6.4 mm). The depth varies somewhat according to the wheel size and type of bond.

**Dry versus Wet Grinding of Carbide Tools.**—In using silicon carbide wheels, grinding should be done either absolutely dry or with enough coolant to flood the wheel and tool. Satisfactory results may be obtained either by the wet or dry method. However, dry grinding is the most prevalent usually because, in wet grinding, operators tend to use an inadequate supply of coolant to obtain better visibility of the grinding operation and avoid getting wet; hence checking or cracking in many cases is more likely to occur in wet grinding than in dry grinding.

*Wet Grinding with Silicon Carbide Wheels:* One advantage commonly cited in connection with wet grinding is that an ample supply of coolant permits using wheels about one grade harder than in dry grinding, thus increasing the wheel life. Plenty of coolant also prevents thermal stresses and the resulting cracks, and there is less tendency for the wheel to load. A dust exhaust system also is unnecessary.

*Wet Grinding with Diamond Wheels:* In grinding with diamond wheels, the general practice is to use a coolant to keep the wheel face clean and promote free cutting. The amount of coolant may vary from a small stream to a coating applied to the wheel face by a felt pad.

**Coolants for Carbide Tool Grinding.**—In grinding either with silicon carbide or diamond wheels, a coolant that is used extensively consists of water plus a small amount either of soluble oil, sal soda, or soda ash to prevent corrosion. One prominent manufacturer recommends for silicon carbide wheels about 1 ounce of soda ash per gallon of water, and for diamond wheels kerosene. The use of kerosene is quite general for diamond wheels and is usually applied to the wheel face by a felt pad. Another coolant recommended for diamond wheels consists of 80 percent water and 20 percent soluble oil.

**Peripheral versus Flat Side Grinding.**—In grinding single-point carbide tools with silicon carbide wheels, the roughing preparatory to finishing with diamond wheels may be done either by using the flat face of a cup-shaped wheel (side grinding) or the periphery of a "straight" or disk-shaped wheel. Even where side grinding is preferred, the periphery of a straight wheel may be used for heavy roughing as in grinding back-chipped or broken tools (see left-hand diagram). Reasons for preferring peripheral grinding include faster cutting with less danger of localized heating and checking especially in grinding broad surfaces. The advantages usually claimed for side grinding are that proper rake or relief angles are easier to obtain and the relief or land is ground flat. The diamond wheels used for tool sharpening are designed for side grinding. (See right-hand diagram.)

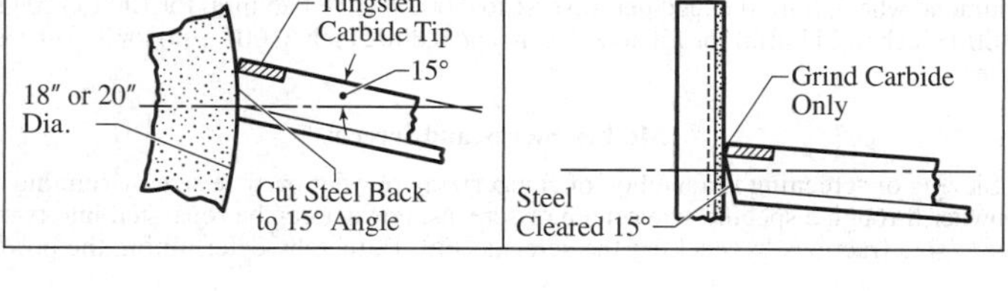

**Lapping Carbide Tools.—**Carbide tools may be finished by lapping, especially if an exceptionally fine finish is required on the work as, for example, with tools used for precision boring or turning nonferrous metals. If the finishing is done by using a diamond wheel of very fine grit (such as 240, 320, or 400), the operation is often called "lapping." A second lapping method is by means of a power-driven lapping disk charged with diamond dust, Norbide powder, or silicon carbide finishing compound. A third method is by using a hand lap or hone usually of 320 or 400 grit. In many plants, the finishes obtained with carbide tools meet requirements without a special lapping operation. In all cases, any feather edge which may be left on tools should be removed, and it is good practice to bevel the edges of roughing tools at 45 degrees to leave a chamfer 0.005 to 0.010 inch wide (0.127–0.254 mm). This is done by hand honing, and the object is to prevent crumbling or flaking off at the edges when hard scale or heavy chip pressure is encountered.

*Hand Honing:* The cutting edge of carbide tools, and tools made from other tool materials, is sometimes hand honed before it is used in order to strengthen the cutting edge. When interrupted cuts or heavy roughing cuts are to be taken, or when the grade of carbide is slightly too hard, hand honing is beneficial because it will prevent chipping or even possibly, breakage of the cutting edge. Whenever chipping is encountered, hand honing the cutting edge before use will be helpful. It is important, however, to hone the edge lightly and only when necessary. Heavy honing will always cause a reduction in tool life. Normally, removing 0.002 to 0.004 inch (0.051–0.102 mm) from the cutting edge is sufficient. When indexable inserts are used, the use of pre-honed inserts is preferred to hand honing, although sometimes an additional amount of honing is required. Hand honing of carbide tools in between cuts is sometimes done to defer grinding or to increase the life of a cutting edge on an indexable insert. If correctly done, so as not to change the relief angle, this procedure is sometimes helpful. If improperly done, it can result in a reduction in tool life.

**Chipbreaker Grinding.—**For this operation a straight diamond wheel is used on a universal tool and cutter grinder, a small surface grinder, or a special chipbreaker grinder. A resinoid-bonded wheel of grade J or N commonly is used, and the tool is held rigidly in an adjustable holder or vise. The width of the diamond wheel usually varies from $\frac{1}{8}$ to $\frac{1}{4}$ inch (3.2–6.4 mm). A vitrified bond may be used for wheels as thick as $\frac{1}{4}$ inch (6.35 mm), and a resinoid bond for relatively narrow wheels.

**Summary of Miscellaneous Points.—**In grinding a single-point carbide tool, traverse it across the wheel face continuously to avoid localized heating. This traverse movement should be quite rapid in using silicon carbide wheels and comparatively slow with diamond wheels. A hand traversing and feeding movement, whenever practicable, is generally recommended because of greater sensitivity. In grinding, maintain a constant, moderate pressure. Manipulating the tool so as to keep the contact area with the wheel as small as possible will reduce heating and increase the rate of stock removal. Never cool a hot tool by dipping it in a liquid, as this may crack the tip. Wheel rotation should preferably be *against* the cutting edge or from the front face toward the back. If the grinder is driven by a reversing motor, opposite sides of a cup wheel can be used for grinding right-and left-hand tools and with rotation against the cutting edge. If it is necessary to grind the top face of a single-point tool, this should precede the grinding of the side and front relief, and top-face grinding should be minimized to maintain the tip thickness. In machine grinding with a diamond wheel, limit the feed per traverse to 0.001 inch (0.025 mm) for 100 to 120 grit; 0.0005 inch (0.013 mm) for 150 to 240 grit; and 0.0002 inch (0.005 mm) for 320 grit and finer.

## Meshes, Sieves, and Screens

Sieving or screening is a method of categorizing powder particle size by running the powder through a specific size screen or screens. Powder can be separated into two or more size fractions by stacking the screens (Fig. 1), thereby determining the powder particle size distribution. Sieves and screens are usually used for larger particle sized materials, greater than 44 micron (325 Mesh).

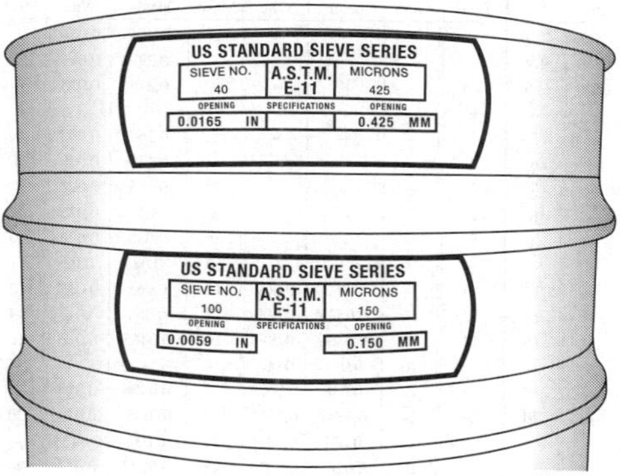


Fig. 1.

Two scales are frequently used to classify particle sizes; *US Standard Sieve Series* and *Tyler Mesh Size,* also called *Tyler Standard Sieve Series*. The larger the mesh number the smaller the particle size of the powder. The Tyler mesh number indicates the number of openings in the screen per linear inch.

- A −6 mesh powder contains particles that pass through a 6 mesh screen. A −6 mesh powder has particles that measure less than 3360 microns.
- A powder that is −325 mesh has particles that measure less than 44 micron.

*Market Grade* sieves use thicker wire than other commercial grades, and are commonly used for applications where mesh strength (and therefore screen life) is important.

*Mill Grade* sieves use a thinner wire diameter, which provide more open area for a given mesh size. Therefore, Mill Grade sieves are used when throughput is more important than durability.

*Tensile Bolting Cloth* uses very fine wire diameters, and thus provides the highest fraction of open area of all sieve types; it is often used for fine sifting and screen printing.

A comparison of standard screen and sieve sizes are indicated in Table 2 and Table 5.

*Example 1:* A tantalum powder has the following description:

Tantalum Powder, −325 Mesh, 3N8 Purity

The minus (−) sign in the designation −325 indicates that particle sizes of less than 44 micron (μm) (Table 1) are able to pass through a 325 mesh screen. Much of the powder may be considerably smaller and still qualify as a −325 mesh powder, but the designation −325 mesh simply means that the powder will pass through a 325 mesh screen.

The 3N8 Purity indicates a purity of 99.98% relative to other rare earth powders; similarily, 2N5 indicates 99.5% purity, 5N8 indicates 99.9998%, and 6N indicates 99.9999%, etc.

*Example 2*: illustrates two stacked screens. If a powder is required to fall within a specified max/min range of sizes, two mesh sizes maybe specified such as in the following:

Tantalum Powder, −40+100 Mesh, 3N8 Purity

The −40 designation indicates that the powder will pass through an 40 mesh screen, and the designation +100 indicates that particles are trapped (do not pass through) by a 100 mesh screen. Thus −40+100 indicates a range of particle sizes small enough to pass the 40 mesh and too large to pass the 100 mesh screen.

**Table 1. Commercial Sieve Mesh Dimensions**

| Sieve size | Opening | | Standard Mesh | | Tensile Bolting Cloth | | | Mill Grade | | | Market Grade | | |
|---|---|---|---|---|---|---|---|---|---|---|---|---|---|
| (mm) | (in.) | (μm) | US | Tyler | Mesh | Opening | Wire | Mesh | Opening | Wire | Mesh | Opening | Wire |
| 11.2 | 0.438 | 11200 | 7/16 | – | – | – | – | 2 | 0.466 | 0.054 | 2 | 0.437 | 0.063 |
| 6.35 | 0.25 | 6350 | 1/4 | – | – | – | – | 3 | 0.292 | 0.041 | 3 | 0.279 | 0.054 |
| 5.6 | 0.223 | 5600 | 3.5 | 3.5 | – | – | – | 4 | 0.215 | 0.035 | 4 | 0.2023 | 0.0475 |
| 4.75 | 0.187 | – | 4 | 4 | – | – | – | – | – | – | 4 | 0.187 | 0.063 |
| 4 | 0.157 | – | 5 | 5 | – | – | – | 5 | 0.168 | 0.032 | 5 | 0.159 | 0.041 |
| 3.36 | 0.132 | – | 6 | 6 | – | – | – | 6 | 0.139 | 0.028 | 6 | 0.132 | 0.0348 |
| 2.8 | 0.11 | – | 7 | 7 | – | – | – | 7 | 0.115 | 0.028 | 7 | 0.108 | 0.035 |
| 2.36 | 0.0937 | – | 8 | 8 | – | – | – | 8 | 0.1 | 0.025 | 8 | 0.0964 | 0.0286 |
| 2 | 0.0787 | – | 10 | 9 | – | – | – | 9 | 0.088 | 0.023 | 10 | 0.0742 | 0.0258 |
| 1.85 | – | – | – | – | – | – | – | 10 | 0.08 | 0.02 | 11 | 0.073 | 0.018 |
| 1.7 | 0.0661 | – | 12 | 10 | 14 | 0.062 | 0.009 | 12 | 0.065 | 0.018 | 12 | 0.0603 | 0.023 |
| 1.4 | 0.0555 | – | 14 | 12 | 16 | 0.0535 | 0.009 | 14 | 0.054 | 0.017 | 14 | 0.051 | 0.0204 |
| 1.18 | 0.0469 | – | 16 | 14 | 18 | 0.0466 | 0.009 | 16 | 0.0465 | 0.016 | 16 | 0.0445 | 0.0181 |
| 1.04 | – | – | – | – | 20 | 0.041 | 0.009 | – | – | – | – | – | – |
| 1 | 0.0394 | – | 18 | 16 | 22 | 0.038 | 0.0075 | 18 | 0.0406 | 0.015 | 18 | 0.0386 | 0.0173 |
| 0.841 | 0.0331 | 841 | 20 | 20 | 24 | 0.0342 | 0.0075 | 20 | 0.036 | 0.014 | 20 | 0.034 | 0.0162 |
| 0.787 | – | – | – | – | 26 | 0.031 | 0.0075 | 22 | 0.032 | 0.0135 | – | – | – |
| 0.71 | 0.0278 | – | 25 | 24 | 28 | 0.0282 | 0.0075 | 24 | 0.0287 | 0.013 | 24 | 0.0277 | 0.014 |
| 0.681 | – | – | – | – | 30 | 0.0268 | 0.0065 | 26 | 0.0275 | 0.011 | – | – | – |
| 0.63 | – | – | – | – | 32 | 0.0248 | 0.0065 | 28 | 0.0275 | 0.01 | – | – | – |
| 0.595 | 0.0232 | 595 | 30 | 28 | 34 | 0.0229 | 0.0065 | 30 | 0.0238 | 0.0095 | – | – | – |
| 0.541 | – | – | – | – | 36 | 0.0213 | 0.0065 | 32 | 0.0223 | 0.009 | – | – | – |
| 0.5 | 0.0197 | – | 35 | 32 | 38 | 0.0198 | 0.0065 | 34 | 0.0204 | 0.009 | 30 | 0.0203 | 0.0128 |
| 0.47 | – | – | – | – | 40 | 0.0185 | 0.0065 | 36 | 0.0188 | 0.009 | – | – | – |
| 0.465 | – | – | – | – | 42 | 0.0183 | 0.0055 | 38 | 0.0178 | 0.0085 | – | – | – |
| 0.437 | – | – | – | – | 44 | 0.0172 | 0.0055 | – | – | – | 35 | 0.0176 | 0.0118 |
| 0.4 | 0.0165 | 400 | 40 | 35 | 46 | 0.0162 | 0.0055 | 40 | 0.0165 | 0.0085 | – | – | – |
| 0.389 | – | – | – | – | 48 | 0.0153 | 0.0055 | – | – | – | 40 | 0.015 | 0.0104 |
| 0.368 | – | – | – | – | 50 | 0.0145 | 0.0055 | – | – | – | – | – | – |
| 0.355 | 0.0139 | – | 45 | 42 | 52 | 0.0137 | 0.0055 | 45 | 0.0142 | 0.008 | – | – | – |
| 0.33 | – | – | – | – | 54 | 0.013 | 0.0055 | – | – | – | – | – | – |
| 0.323 | – | – | – | – | 58 | 0.0127 | 0.0045 | – | – | – | – | – | – |
| 0.31 | – | – | – | – | 60 | 0.0122 | 0.0045 | 50 | 0.0125 | 0.0075 | – | – | – |
| 0.3 | 0.0117 | – | 50 | 48 | 62 | 0.0116 | 0.0045 | 55 | 0.0112 | 0.007 | – | – | – |
| 0.282 | – | – | – | – | 64 | 0.0111 | 0.0045 | – | – | – | 50 | 0.011 | 0.009 |
| 0.27 | – | – | – | – | 70 | 0.0106 | 0.0037 | – | – | – | – | – | – |
| 0.26 | – | – | – | – | 72 | 0.0102 | 0.0037 | – | – | – | – | – | – |
| 0.25 | 0.0098 | 250 | 60 | 60 | 74 | 0.0098 | 0.0037 | 60 | 0.0102 | 0.0065 | – | – | – |
| 0.241 | – | – | – | – | 76 | 0.0095 | 0.0037 | – | – | – | – | – | – |
| 0.231 | – | – | – | – | 78 | 0.0091 | 0.0037 | – | – | – | 60 | 0.0092 | 0.0075 |
| 0.224 | – | – | – | – | 80 | 0.0088 | 0.0037 | – | – | – | – | – | – |
| 0.21 | 0.0083 | 210 | 70 | 65 | 84 | 0.0084 | 0.0035 | – | – | – | – | – | – |
| 0.2 | – | – | – | – | 88 | 0.0079 | 0.0035 | – | – | – | – | – | – |
| 0.193 | – | – | – | – | 90 | 0.0076 | 0.0035 | – | – | – | – | – | – |
| 0.177 | 0.007 | 177 | 80 | 80 | 94 | 0.0071 | 0.0035 | – | – | – | 80 | 0.007 | 0.0055 |
| 0.165 | – | – | – | – | 105 | 0.0065 | 0.003 | – | – | – | – | – | – |
| 0.149 | 0.0059 | 149 | 100 | 100 | 120 | 0.0058 | 0.0025 | – | – | – | 100 | 0.0055 | 0.0045 |
| 0.125 | 0.0049 | 125 | 120 | 115 | 145 | 0.0047 | 0.0022 | – | – | – | 120 | 0.0046 | 0.0037 |
| 0.105 | 0.0041 | 105 | 140 | 150 | 165 | 0.0042 | 0.0019 | – | – | – | 150 | 0.0041 | 0.0026 |
| 0.088 | 0.0035 | 88 | 170 | 170 | 200 | 0.0034 | 0.0016 | – | – | – | 180 | 0.0033 | 0.0023 |
| 0.074 | 0.0029 | 74 | 200 | 200 | 230 | 0.0029 | 0.0014 | – | – | – | 200 | 0.0029 | 0.0021 |
| 0.063 | 0.0024 | 63 | 230 | 250 | – | – | – | – | – | – | 250 | 0.0024 | 0.0016 |
| 0.053 | 0.0021 | 53 | 270 | 270 | 300 | 0.0021 | 0.0012 | – | – | – | 270 | 0.0021 | 0.0016 |
| 0.044 | 0.0017 | 44 | 325 | 325 | – | – | – | – | – | – | 325 | 0.0017 | 0.0014 |
| 0.037 | 0.0015 | 37 | 400 | 400 | – | – | – | – | – | – | 400 | 0.0015 | 0.001 |

As indicated by the fragment below from Table 1, the specified particles are therefore smaller than 40 mesh and larger than 100 mesh. Particles trapped above the 100 mesh screen range in size from greater than 149 μm to 400 μm. Particles smaller than 149 micron are passed through the 100 mesh screen. Approximately 90 percent of the particles trapped will fit within the specified range.

| Sieve size (mm) | Opening (in.) | Opening (μm) | Standard Mesh US | Standard Mesh Tyler |
|---|---|---|---|---|
| … | … | … | … | … |
| 0.4 | 0.0165 | 400 | 40 | 35 |
| 0.149 | 0.0059 | 149 | 100 | 100 |

Sieves and screens are stacked with the largest opening (smallest mesh size) above, and the smallest opening (largest mesh size) below, as in Fig. 2.

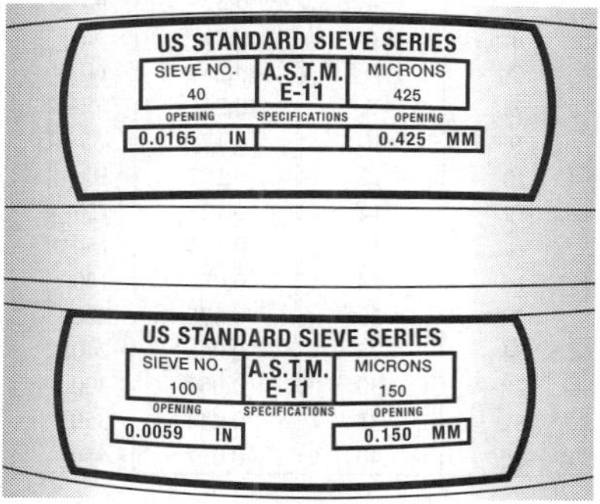


Fig. 2.

**Table 2. Standard Sieves and Mesh Sizes**

| Mesh Size μm | TYLER Mesh | ASTM E-11 No. | BS 410 Mesh | DIN 4188 mm | Mesh Size μm | TYLER Mesh | ASTM E-11 No. | BS 410 Mesh | DIN 4188 mm |
|---|---|---|---|---|---|---|---|---|---|
| 5 | 2500 | … | 2500 | 0.005 | 106 | 150 | 140 | 150 | … |
| 10 | 1250 | … | 1250 | 0.01 | 112 | … | … | … | 0.112 |
| 15 | 800 | … | 800 | 0.015 | 125 | 115 | 120 | 120 | 0.125 |
| 20 | 625 | … | 625 | 0.02 | 140 | … | … | … | 0.14 |
| 22 | … | … | … | 0.022 | 150 | 100 | 100 | 100 | … |
| 25 | 500 | … | 500 | 0.025 | 160 | … | … | … | 0.16 |
| 28 | … | … | … | 0.028 | 180 | 80 | 80 | 85 | 0.18 |
| 32 | … | … | … | 0.032 | 200 | … | … | … | 0.2 |
| 36 | … | … | … | 0.036 | 212 | 65 | 70 | 72 | … |
| 38 | 400 | 400 | 400 | … | 250 | 60 | 60 | 60 | 0.25 |
| 40 | … | … | … | 0.04 | 280 | … | … | … | 0.28 |
| 45 | 325 | 325 | 350 | 0.045 | 300 | 48 | 50 | 52 | … |
| 50 | … | … | … | 0.05 | 315 | … | … | … | 0.315 |
| 53 | 270 | 270 | 300 | … | 355 | 42 | 45 | 44 | 0.355 |
| 56 | … | … | … | 0.056 | 400 | … | … | … | 0.4 |
| 63 | 250 | 230 | 240 | 0.063 | 425 | 35 | 40 | 36 | … |
| 71 | … | … | … | 0.071 | 450 | … | … | … | 0.45 |

**Table 2.** *(Continued)* **Standard Sieves and Mesh Sizes**

| Mesh Size µm | TYLER Mesh | ASTM E-11 No. | BS 410 Mesh | DIN 4188 mm | Mesh Size µm | TYLER Mesh | ASTM E-11 No. | BS 410 Mesh | DIN 4188 mm |
|---|---|---|---|---|---|---|---|---|---|
| 75 | 200 | 200 | 200 | … | 500 | 32 | 35 | 30 | 0.5 |
| 80 | … | … | … | 0.08 | 560 | … | … | … | 0.56 |
| 90 | 170 | 170 | 170 | 0.09 | 600 | 28 | 30 | 25 | … |
| 100 | … | … | … | 0.1 | … | … | … | … | … |

**Table 3. Typical Openings in Laboratory Sieve Series**

| Sieve size (mm) | BSS | Tyler (approx) | US (approx) | Sieve size (mm) | BSS | Tyler (approx) | US (approx) |
|---|---|---|---|---|---|---|---|
| 4.75 | - | 4 | 4 | 0.354 | 44 | 42 | 45 |
| 3.35 | 5 | 6 | 6 | 0.297 | 52 | 48 | 50 |
| 2.81 | 6 | 7 | 7 | 0.251 | 60 | 60 | 60 |
| 2.38 | 7 | 8 | 8 | 0.211 | 72 | 65 | 70 |
| 2 | 8 | 9 | 10 | 0.178 | 85 | 80 | 80 |
| 1.68 | 10 | 10 | 12 | 0.152 | 100 | 100 | 100 |
| 1.4 | 12 | 12 | 14 | 0.125 | 120 | 115 | 120 |
| 1.2 | 14 | 14 | 16 | 0.104 | 150 | 150 | 140 |
| 1 | 16 | 16 | 18 | 0.089 | 170 | 170 | 170 |
| 0.853 | 18 | 20 | 20 | 0.075 | 200 | 200 | 200 |
| 0.71 | 22 | 24 | 25 | 0.066 | 240 | 250 | 230 |
| 0.599 | 25 | 28 | 30 | 0.053 | 300 | 270 | 270 |
| 0.5 | 30 | 32 | 35 | 0.044 | 350 | 325 | 325 |
| 0.422 | 36 | 35 | 40 | 0.037 | 440 | 400 | 400 |

BS 812-103.1:1985 Withdrawn replaced by BS EN 933-1:2012.

BS EN 933-1:2012 describes two methods for the determination of the particle size distribution of samples of aggregates and fillers by sieving.

**Abrasive Grit Number.—**Standard abrasive grain sizes are designated by numbers. These numbers range from number 8, which is the coarsest, to number 240, which is the finest. The allowable limits for the sizing of aluminum-oxide and silicon-carbide abrasives for grinding wheel manufacture are given in US Simplified Practice Recommendation 118. These numbers in most cases equal approximately the number of sieve openings per inch in the United States Standard Fine Sieve series. For example, a number 30 sieve has 0.0232-inch openings and a sieve wire diameter of 0.0130 inch, making the pitch equal to 0.0362 inch; hence there are 27.6 meshes per inch. The United States Standard Fine Sieve series ranges from number $3\frac{1}{2}$ to number 400.

*Grading Abrasives:* In the actual grading of abrasives, several standard sieves are used. To illustrate, take grit No. 10. All material must pass through the coarsest sieve—in this case the No. 7. Through the next to the coarsest sieve, termed the "control sieve"—in this case the No. 8—all material may pass, but not more than 15 percent may be retained on it. At least 45 percent must pass through No. 8, and be retained on No. 10 sieve, but it is permissible to have 100 percent pass through No. 8, and remain on No. 10 sieve, the requirement being that the grain passing through No. 8, and retained on No. 10 and No. 12 must add to at least 80 percent; consequently, if 45 percent passed through No. 8 sieve and was retained on No. 10 sieve, then at least 35 percent must be retained on the No. 12 sieve. Not more than 3 percent is permitted to pass through the No. 14 sieve.

## TABLE OF CONTENTS

## MACHINING OPERATIONS

### CUTTING SPEEDS AND FEEDS



### SPEEDS AND FEEDS TABLES



### ESTIMATING SPEEDS AND MACHINING POWER



### MICROMACHINING



MACHINING

# TABLE OF CONTENTS

## MACHINING OPERATIONS

### MICROMACHINING

*(Continued)*



### MACHINING ECONOMETRICS



### SCREW MACHINES, BAND SAWS, CUTTING FLUIDS



### MACHINING NONFERROUS METALS AND NONMETALLIC MATERIALS

## TABLE OF CONTENTS

## MACHINING OPERATIONS

### GRINDING FEEDS AND SPEEDS



### GRINDING AND OTHER ABRASIVE PROCESSES



### GRINDING AND OTHER ABRASIVE PROCESSES

*(Continued)*

## TABLE OF CONTENTS
## MACHINING OPERATIONS

### GRINDING AND OTHER ABRASIVE PROCESSES

*(Continued)*



### NONTRADITIONAL MACHINING AND CUTTING



### CNC PROGRAMMING AND CAD/CAM



### CNC PROGRAMMING AND CAD/CAM

*(Continued)*

## CUTTING SPEEDS AND FEEDS

**Work Materials.—**The large number of work materials that are commonly machined vary greatly in their basic structure and the ease with which they can be machined. Yet it is possible to group together certain materials having similar machining characteristics, for the purpose of recommending the cutting speed at which they can be cut. Most materials that are machined are metals and it has been found that the most important single factor influencing the ease with which a metal can be cut is its microstructure, followed by any cold-work that may have been done to the metal, which increases its hardness. Metals that have a similar, but not necessarily the same microstructure, will tend to have similar machining characteristics. Thus, the grouping of the metals in the accompanying tables has been done on the basis of their microstructure.

Except for a few soft and gummy metals, experience indicates that harder metals are more difficult to cut than softer metals. Also, any given metal is more difficult to cut when it is in a harder form than when it is softer. It is more difficult to penetrate the harder metal and more power is required. These factors in turn will generate a higher cutting temperature at any given cutting speed, thereby making it necessary to use a slower speed, for the cutting temperature must always be kept within the limits that can be sustained by the cutting tool without failure. Hardness, then, is an important property that must be considered when machining a given metal. Hardness alone, however, cannot be used as a measure of cutting speed. For example, if pieces of AISI 11L17 and AISI 1117 steel both have a hardness of 150 BHN (Brinell Hardness Number), their recommended cutting speeds for high-speed steel tools may be 140 fpm (0.71 m/s) and 130 fpm (0.66 m/s), respectively. In some metals, two entirely different microstructures can produce the same hardness. As an example, a fine pearlite microstructure and a tempered martensite microstructure can result in the same hardness in a steel. These microstructures will not machine alike. For practical purposes, however, information on hardness is usually easier to obtain than information on microstructure; thus, hardness alone is usually used to differentiate between different cutting speeds for machining a metal. In some situations, the hardness of a metal to be machined is not known. When the hardness is not known, the material condition can be used as a guide.

The surface of ferrous metal castings has a scale that is more difficult to machine than the metal below. Some scale is more difficult to machine than others, depending on the foundry sand used, the casting process, the method of cleaning the casting, and the type of metal cast. Special electrochemical treatments sometimes can be used that almost entirely eliminate the effect of the scale on machining, although castings so treated are not frequently encountered. Usually, when casting scale is encountered, the cutting speed is reduced approximately 5 or 10 percent. Difficult-to-machine surface scale can also be encountered when machining hot-rolled or forged steel bars.

Metallurgical differences that affect machining characteristics are often found within a single piece of metal. The occurrence of hard spots in castings is an example. Different microstructures and hardness levels may occur within a casting as a result of variations in the cooling rate in different parts of the casting. Such variations are less severe in castings that have been heat treated. Steel bar stock is usually harder toward the outside than toward the center of the bar. Sometimes there are slight metallurgical differences along the length of a bar that can affect its cutting characteristics.

**Cutting Tool Materials.—**The recommended cutting speeds and feeds in the accompanying tables are given for high-speed steel, coated and uncoated carbides, ceramics, cermets, and polycrystalline diamonds. More data are available for HSS and carbides because these materials are the most commonly used. Other materials that are used to make cutting tools are cemented oxides or ceramics, cermets, cast nonferrous alloys (Stellite), single-crystal diamonds, polycrystalline diamonds, and cubic boron nitride.

*Carbon Tool Steel:* It is used primarily to make the less expensive drills, taps, and reamers. It is seldom used to make single-point cutting tools. Hardening in carbon steels is very shallow, although some have a small amount of vanadium and chromium added to improve their hardening quality. The cutting speed to use for plain carbon tool steel should be approximately one-half of the recommended speed for high-speed steel.

*High-Speed Steel:* This designates a number of steels having several properties that enhance their value as cutting tool material. They can be hardened to a high initial or room-temperature hardness ranging from 63 to 65 RC (Rockwell C scale) for ordinary high-speed steels and up to 70 RC for the so-called super high-speed steels. They retain sufficient hardness at temperatures up to 1,000 or 1,100° F (573 or 593° C) to enable them to cut at cutting speeds that will generate these tool temperatures, and return to their original hardness when cooled to room temperature. They harden very deeply, enabling high-speed steels to be ground to the tool shape from solid stock and to be reground many times without sacrificing hardness at the cutting edge. High-speed steels can be made soft by annealing so that they can be machined into complex cutting tools such as drills, reamers, and milling cutters and then hardened.

The principal alloying elements of high-speed steels are tungsten (W), molybdenum (Mo), chromium (Cr), vanadium (V), together with carbon (C). There are a number of grades of high-speed steel that are divided into two types: tungsten high-speed steels and molybdenum high-speed steels. Tungsten high-speed steels are designated by the prefix T before the number that designates the grade. Molybdenum high-speed steels are designated by the prefix letter M. There is little performance difference between comparable grades of tungsten or molybdenum high-speed steel.

The addition of 5 to 12 percent cobalt to high-speed steel increases its hardness at the temperatures encountered in cutting, thereby improving its wear resistance and cutting efficiency. Cobalt slightly increases the brittleness of high-speed steel, making it susceptible to chipping at the cutting edge. For this reason, cobalt high-speed steels are primarily made into single-point cutting tools that are used to take heavy roughing cuts in abrasive materials and through rough abrasive surface scales.

The M40 series and T15 are a group of high-hardness or so-called super high-speed steels that can be hardened to 70 RC; however, they tend to be brittle and difficult to grind. For cutting applications, they are usually heat treated to 67–68 RC to reduce their brittleness and tendency to chip. The M40 series is appreciably easier to grind than T15. They are recommended for machining tough die steels and other difficult-to-cut materials; they are not recommended for applications where conventional high-speed steels perform well. High-speed steels made by the powder-metallurgy process are tougher and have an improved grindability when compared with similar grades made by the customary process. Tools made of these steels can be hardened about 1 RC higher than comparable high-speed steels made by the customary process without a sacrifice in toughness. They are particularly useful in applications involving intermittent cutting and where tool life is limited by chipping. All these steels augment rather than replace the conventional high-speed steels.

*Cemented Carbides* are also called sintered carbides or simply carbides. They are harder than high-speed steels and have excellent wear resistance. Information on these and other hardmetal tools is included in the section *CEMENTED CARBIDES* starting on page 848.

Cemented carbides retain a very high degree of hardness at temperatures up to 1400° F (760° C) and even higher; therefore, very fast cutting speeds can be used. When used at fast cutting speeds, they produce good surface finishes on the workpiece. Carbides are more brittle than high-speed steel and, therefore, must be used with more care.

There are four distinct types of carbides: 1) straight tungsten carbides; 2) crater-resistant carbides; 3) titanium carbides; and 4) coated carbides.

*Straight Tungsten Carbide:* This is the most abrasion-resistant cemented carbide and is used to machine gray cast iron, most nonferrous metals, and nonmetallic materials, where abrasion resistance is the primary criterion. Straight tungsten carbide will rapidly form a crater on the tool face when used to machine steel, which reduces the life of the tool. Titanium carbide is added to tungsten carbide in order to counteract the rapid formation of the crater. In addition, tantalum carbide is usually added to prevent the cutting edge from deforming when subjected to the intense heat and pressure generated in taking heavy cuts.

*Crater-Resistant Carbides:* These carbides, containing titanium and tantalum carbides in addition to tungsten carbide, are used to cut steels, alloy cast irons, and other materials that have a strong tendency to form a crater.

*Titanium Carbides:* These carbides are made entirely from titanium carbide and small amounts of nickel and molybdenum. They have an excellent resistance to cratering and to heat. Their high hot hardness enables them to operate at higher cutting speeds, but they are more brittle and less resistant to mechanical and thermal shock. Therefore, they are not recommended for taking heavy or interrupted cuts. Titanium carbides are less abrasion-resistant and not recommended for cutting through scale or oxide films on steel. Although the resistance to cratering of titanium carbides is excellent, failure caused by crater formation can sometimes occur because the chip tends to curl very close to the cutting edge, thereby forming a small crater in this region that may break through.

*Coated Carbides:* These are available only as indexable inserts because the coating would be removed by grinding. The principal coating materials are titanium carbide (TiC), titanium nitride (TiN), and aluminum oxide ($Al_2O_3$). A very thin layer approximately 0.0002 inch (5.08 μm) of coating material is deposited over a cemented carbide insert; the material below the coating is the substrate. The overall performance of the coated carbide is limited by the substrate, which provides the required toughness, resistance to deformation, and thermal shock. With an equal tool life, coated carbides can operate at higher cutting speeds than uncoated carbides. The increase may be 20 to 30 percent and sometimes up to 50 percent faster. Titanium carbide and titanium nitride coated carbides usually operate in the medium (200–800 fpm, 1.0–4.1 m/s) cutting speed range, and aluminum oxide coated carbides are used in the higher (800–1600 fpm, 4.1–8.1 m/s) cutting speed range.

*Carbide Grade Selection:* The selection of the best grade of carbide for a particular application is very important. An improper grade of carbide will result in a poor performance—it may even cause the cutting edge to fail before any significant amount of cutting has been done. Because of the many grades and the many variables that are involved, the carbide producers should be consulted to obtain recommendations for the application of their grades of carbide. A few general guidelines can be given that are useful to form an orientation. Metal cutting carbides usually range in hardness from about 89.5 to 93 RA (Rockwell A scale) with the exception of titanium carbide, which has a hardness range of 90.5 to 93.5 RA. Generally, the harder carbides are more wear-resistant and more brittle, whereas the softer carbides are less wear-resistant but tougher. A choice of hardness must be made to suit the given application. The very hard carbides are generally used for taking light finishing cuts. For other applications, select the carbide that has the highest hardness with sufficient strength to prevent chipping or breaking. Straight tungsten carbide grades should always be used unless cratering is encountered. Straight tungsten carbides are used to machine gray cast iron, ferritic malleable iron, austenitic stainless steel, high-temperature alloys, copper, brass, bronze, aluminum alloys, zinc alloy die castings, and plastics. Crater-resistant carbides should be used to machine plain carbon steel, alloy steel, tool steel, pearlitic malleable iron, nodular iron, other highly alloyed cast irons, ferritic stainless steel, martensitic stainless steel, and certain high-temperature alloys. Titanium carbides are recommended for taking high-speed finishing and semifinishing cuts on steel, especially the low-carbon, low-alloy steels, which are less abrasive and have a strong tendency to form a crater. They are also used to take light cuts on alloy cast iron and on some high-nickel alloys. Nonferrous materials, such as some aluminum alloys and brass, that are essentially nonabrasive may also be machined with titanium carbides. Abrasive materials and others that should not be machined with titanium carbides include gray cast iron, titanium alloys, cobalt- and nickel-base superalloys, stainless steel, bronze, many aluminum alloys, fiberglass, plastics, and graphite. The feed used should not exceed about 0.020 inch/rev (0.51 mm/rev).

Coated carbides can be used to take cuts ranging from light finishing to heavy roughing on most materials that can be cut with these carbides. The coated carbides are recommended for machining all free-machining steels, all plain carbon and alloy steels, tool steels, martensitic and ferritic stainless steels, precipitation-hardening stainless steels, alloy cast iron, pearlitic and martensitic malleable iron, and nodular iron. They are also recommended for taking light finishing and roughing cuts on austenitic stainless steels. Coated carbides should not be used to machine nickel- and cobalt-base superalloys, titanium and titanium alloys, brass, bronze, aluminum alloys, pure metals, refractory metals, and nonmetals such as fiberglass, graphite, and plastics.

*Ceramic Cutting Tool Materials:* These are made from finely powdered aluminum oxide particles sintered into a hard dense structure without a binder material. Aluminum oxide is also combined with titanium carbide to form a composite, which is called a cermet. These materials have a very high hot hardness enabling very high cutting speeds to be used. For example, ceramic cutting tools have been used to cut AISI 1040 steel at a cutting speed of 18,000 fpm (91.4 m/s) with a satisfactory tool life. However, much lower cutting speeds, in the range of 1000–4000 fpm (5.1–20.3 m/s) and lower, are more common because of limitations placed by the machine tool, cutters, and chucks. Although most applications of ceramic and cermet cutting tool materials are for turning, they have also been used successfully for milling. Ceramics and cermets are relatively brittle and a special cutting edge preparation is required to prevent chipping or edge breakage. This preparation consists of honing or grinding a narrow flat land, 0.002 to 0.006 inch (50.8–152.4 μm) wide, on the cutting edge that is made about 30 degrees with respect to the tool face. For some heavy-duty applications, a wider land is used. The setup should be as rigid as possible and the feed rate should not normally exceed 0.020 inch (508 μm), although 0.030 inch (762 μm) has been used successfully. Ceramics and cermets are recommended for roughing and finishing operations on all cast irons, plain carbon and alloy steels, and stainless steels. Materials up to a hardness of 60 RC (Rockwell C scale) can be cut with ceramic and cermet cutting tools. These tools should not be used to machine aluminum and aluminum alloys, magnesium alloys, titanium, and titanium alloys.

*Cast Nonferrous Alloy:* Cutting tools of this alloy are made from tungsten, tantalum, chromium, and cobalt plus carbon. Other alloying elements are also used to produce materials with high temperature and wear resistance. These alloys cannot be softened by heat treatment and must be cast and ground to shape. The room-temperature hardness of cast nonferrous alloys is lower than for high-speed steel, but the hardness and wear resistance is retained to a higher temperature. The alloys are generally marketed under trade names such as Stellite, Crobalt, and Tantung. The initial cutting speed for cast nonferrous tools can be 20 to 50 percent greater than the recommended cutting speed for high-speed steel.

*Diamond Cutting Tools* are available in three forms: single-crystal natural diamonds shaped to a cutting edge and mounted on a tool holder on a boring bar; polycrystalline diamond indexable inserts made from synthetic or natural diamond powders that have been compacted and sintered into a solid mass, and chemically vapor-deposited diamond. Single-crystal and polycrystalline diamond cutting tools are very wear-resistant, and recommended for machining abrasive materials that cause other cutting tool materials to wear rapidly. Typical of abrasive materials machined with single-crystal and polycrystalline diamond tools and cutting speeds used are the following: fiberglass, 300 to 1000 fpm (1.5 to 5.1 m/s); fused silica, 900 to 950 fpm (4.6 to 4.8 m/s); reinforced melamine plastics, 350 to 1000 fpm (1.8 to 5.1 m/s); reinforced phenolic plastics, 350 to 1000 fpm (1.8 to 5.1 m/s); thermosetting plastics, 300 to 2000 fpm (1.5 to 10.2 m/s); Teflon, 600 fpm

(3.0 m/s); nylon, 200 to 300 fpm (1.0 to 1.5 m/s); mica, 300 to 1000 fpm (1.5 to 5.1 m/s); graphite, 200 to 2000 fpm (1.0 to 10.1 m/s); babbitt bearing metal, 700 fpm (3.6 m/s); and aluminum-silicon alloys, 1000 to 2000 fpm (5.1 to 10.2 m/s).

Another important application of diamond cutting tools is to produce fine surface finishes on soft nonferrous metals that are difficult to finish by other methods. Surface finishes of 1 to 2 microinches (0.025 to 0.051 μm) can be readily obtained with single-crystal diamond tools, and finishes down to 10 microinches (0.25 μm) can be obtained with polycrystalline diamond tools. In addition to babbitt and the aluminum-silicon alloys, other metals finished with diamond tools include: soft aluminum, 1000 to 2000 fpm (5.1 to 10.2 m/s); all wrought and cast aluminum alloys, 600 to 1500 fpm (3.0 to 7.6 m/s); copper, 1000 fpm (5.1 m/s); brass, 500 to 1000 fpm (2.5 to 5.1 m/s); bronze, 300 to 600 fpm (1.5 to 3.0 m/s); oilite bearing metal, 500 fpm (2.5 m/s); silver, gold, and platinum, 300 to 2500 fpm (1.5 to 12.7 m/s); and zinc, 1000 fpm (5.1 m/s). Ferrous alloys, such as cast iron and steel, should not be machined with diamond cutting tools because the high cutting temperatures generated will cause the diamond to transform into carbon.

*Chemically Vapor-Deposited (CVD) Diamond:* This tool material offers performance characteristics well suited to highly abrasive or corrosive materials, and hard-to-machine composites. CVD diamond is available in two forms: thick-film tools, which are fabricated by brazing CVD diamond tips, approximately 0.020 inch (0.51 mm) thick, to carbide substrates; and thin-film tools, having a pure diamond coating over the rake and flank surfaces of a ceramic or carbide substrate.

CVD is pure diamond, made at low temperatures and pressures, with no metallic binder phase. This diamond purity gives CVD diamond tools extreme hardness, high abrasion resistance, low friction, high thermal conductivity, and chemical inertness. CVD tools are generally used as direct replacements for PCD (polycrystalline diamond) tools, primarily in finishing, semifinishing, and continuous turning applications of extremely wear-intensive materials. The small grain size of CVD diamond (ranging from less than 1 μm to 50 μm) yields superior surface finishes compared with PCD, and the higher thermal conductivity and better thermal and chemical stability of pure diamond allow CVD tools to operate at faster speeds without generating harmful levels of heat. The extreme hardness of CVD tools may also result in significantly longer tool life.

CVD diamond cutting tools are recommended for the following materials:aluminum and other ductile; nonferrous alloys such as copper, brass, and bronze; and highly abrasive composite materials such as graphite, carbon-carbon, carbon-filled phenolic, fiberglass, and honeycomb materials.

*Cubic Boron Nitride (CBN):* Next to diamond, CBN is the hardest known material. It will retain its hardness at a temperature of 1800° F and higher, making it an ideal cutting tool material for machining very hard and tough materials at cutting speeds beyond those possible with other cutting tool materials. Indexable inserts and cutting tool blanks made from this material consist of a layer, approximately 0.020 inch thick, of polycrystalline cubic boron nitride firmly bonded to the top of a cemented carbide substrate. Cubic boron nitride is recommended for rough and finish turning hardened plain carbon and alloy steels, hardened tool steels, hard cast irons, all hardness grades of gray cast iron, and superalloys. As a class, the superalloys are not as hard as hardened steel; however, their combination of high strength and tendency to deform plastically under the pressure of the cut, or gumminess, places them in the class of hard-to-machine materials. Conventional materials that can be readily machined with other cutting tool materials should not be machined with cubic boron nitride. Round indexable CBN inserts are recommended when taking severe cuts in order to provide maximum strength to the insert. When using square or triangular inserts, a large lead angle should be used, normally 15°, and whenever possible, 45°. A negative rake angle should always be used, which for most applications is negative 5°. The relief angle should be 5° to 9°. Although cubic boron nitride cutting tools can be used without a coolant, flooding the tool with a water-soluble type coolant is recommended.

**Cutting Speed, Feed, Depth of Cut, Tool Wear, and Tool Life.—**The cutting conditions that determine the rate of metal removal are the cutting speed, the feed rate, and the depth of cut. These cutting conditions and the nature of the material to be cut determine the power required to take the cut. The cutting conditions must be adjusted to stay within the power available on the machine tool to be used. Power requirements are discussed in *ESTIMATING SPEEDS AND MACHINING POWER* starting on page 1131.

Cutting conditions must also be considered in relation to the tool life. Tool life is defined as the cutting time to reach a predetermined amount of wear, usually flank wear. Tool life is determined by assessing the time—the tool life—at which a given predetermined flank wear is reached, 0.01 in. (0.25 mm), 0.015 in. (0.38 mm), 0.025 in. (0.64 mm), 0.03 in. (0.76 mm), for example. This amount of wear is called the tool wear criterion, and its size depends on the tool grade used. Usually, a tougher grade can be used with a bigger flank wear, but for finishing operations, where close tolerances are required, the wear criterion is relatively small. Other wear criteria are a predetermined value of the machined surface roughness and the depth of the crater that develops on the rake face of the tool.

ANSI/ASME B94.55M, specification for tool life testing with single-point tools, defines the end of tool life as a given amount of wear on the flank of a tool. This Standard is followed when making scientific machinability tests with single-point cutting tools in order to achieve uniformity in testing procedures so that results from different machinability laboratories can be readily compared. It is not practicable or necessary to follow this Standard in the shop; however, it should be understood that the cutting conditions and tool life are related.

Tool life is influenced most by cutting speed, then by feed rate, and least by depth of cut. When depth of cut is increased to about 10 times greater than the feed, a further increase in the depth of cut will have no significant effect on tool life. This characteristic of the cutting tool performance is very important in determining the operating or cutting conditions for machining metals. Conversely, if the cutting speed or feed is decreased, the increase in the tool life will be proportionately greater than the decrease in the cutting speed or the feed.

Tool life is reduced when either feed or cutting speed is increased. For example, the cutting speed and the feed may be increased if a shorter tool life is accepted; furthermore, the reduction in the tool life will be proportionately greater than the increase in the cutting speed or the feed. However, it is less well understood that a higher feed rate (feed/rev × speed) may result in a longer tool life if a higher feed/rev is used in combination with a lower cutting speed. This principle is well illustrated in the speed tables of this section, where two sets of speed and feed data are given (labeled *optimum* and *average*) that result in the same tool life. The *optimum* set results in a greater feed rate (i.e., increased productivity) although the feed/rev is higher and cutting speed lower than the *average* set. Complete instructions for using the speed tables and for estimating tool life are given in *How to Use the Tables* starting on page 1078.

**Selecting Cutting Conditions.—**The first step in establishing cutting conditions is to select depth of cut. The depth of cut will be limited by the amount of metal to be machined from the workpiece, by the power available on the machine tool, by the rigidity of the workpiece and cutting tool, and by the rigidity of the setup. Depth of cut has the least effect upon tool life, so the heaviest possible depth of cut should always be used.

The second step is to select the feed (feed/rev for turning, drilling, and reaming, or feed/tooth for milling). The available power must be sufficient to make the required depth of cut at the selected feed. The maximum feed possible that will produce an acceptable surface finish should be selected.

The third step is to select the cutting speed. Although the accompanying tables provide recommended cutting speeds and feeds for many materials, experience in machining a certain material may form the best basis for adjusting given cutting speeds to a particular job. In general, depth of cut should be selected first, followed by feed, and last cutting speed.

**Table 1. Tool Troubleshooting Check List**

| Problem | Tool Material | Remedy |
|---|---|---|
| Excessive flank wear—tool life too short | Carbide | 1. Change to harder, more wear-resistant grade<br>2. Reduce the cutting speed<br>3. Reduce the cutting speed and increase the feed to maintain production<br>4. Reduce the feed<br>5. For work-hardenable materials—increase the feed<br>6. Increase the lead angle<br>7. Increase the relief angles |
| | HSS | 1. Use a coolant<br>2. Reduce the cutting speed<br>3. Reduce the cutting speed and increase the feed to maintain production<br>4. Reduce the feed<br>5. For work-hardenable materials—increase the feed<br>6. Increase the lead angle<br>7. Increase the relief angle |
| Excessive cratering | Carbide | 1. Use a crater-resistant grade<br>2. Use a harder, more wear-resistant grade<br>3. Reduce the cutting speed<br>4. Reduce the feed<br>5. Widen the chip breaker groove |
| | HSS | 1. Use a coolant<br>2. Reduce the cutting speed<br>3. Reduce the feed<br>4. Widen the chip breaker groove |
| Cutting edge chipping | Carbide | 1. Increase the cutting speed<br>2. Lightly hone the cutting edge<br>3. Change to a tougher grade<br>4. Use negative-rake tools<br>5. Increase the lead angle<br>6. Reduce the feed<br>7. Reduce the depth of cut<br>8. Reduce the relief angles<br>9. If low cutting speed must be used, use a high-additive EP cutting fluid |
| | HSS | 1. Use a high additive EP cutting fluid<br>2. Lightly hone the cutting edge before using<br>3. Increase the lead angle<br>4. Reduce the feed<br>5. Reduce the depth of cut<br>6. Use a negative rake angle<br>7. Reduce the relief angles |
| | Carbide and HSS | 1. Check the setup for cause if chatter occurs<br>2. Check the grinding procedure for tool overheating<br>3. Reduce the tool overhang |
| Cutting edge deformation | Carbide | 1. Change to a grade containing more tantalum<br>2. Reduce the cutting speed<br>3. Reduce the feed |
| Poor surface finish | Carbide | 1. Increase the cutting speed<br>2. If low cutting speed must be used, use a high additive EP cutting fluid<br>4. For light cuts, use straight titanium carbide grade<br>5. Increase the nose radius<br>6. Reduce the feed<br>7. Increase the relief angles<br>8. Use positive rake tools |

**Table 1.** *(Continued)* **Tool Troubleshooting Check List**

| Problem | Tool Material | Remedy |
|---|---|---|
| Poor surface finish *(Continued)* | HSS | 1. Use a high additive EP cutting fluid<br>2. Increase the nose radius<br>3. Reduce the feed<br>4. Increase the relief angles<br>5. Increase the rake angles |
| | Diamond | 1. Use diamond tool for soft materials |
| Notching at the depth of cut line | Carbide and HSS | 1. Increase the lead angle<br>2. Reduce the feed |

## Cutting Speed Formulas

Most machining operations are conducted on machine tools having a rotating spindle. Cutting speeds are usually given in feet or meters per minute and these speeds must be converted to spindle speeds, in revolutions per minute, to operate the machine. Conversion is accomplished by use of the following formulas:

US Units:

$$N = \frac{12V}{\pi D} = \frac{12 \times 252}{\pi \times 8} = 120 \text{ rpm}$$

**Metric Units:**

$$N = \frac{1000V}{\pi D} = 318.3\frac{V}{D} \text{ rpm}$$

where $N$ is the spindle speed in revolutions per minute (rpm); $V$ is the cutting speed in feet per minute (fpm) for US units and meters per minute (m/min) for metric units. In turning, $D$ is the diameter of the workpiece; in milling, drilling, reaming, and other operations that use a rotating tool, $D$ is the cutter diameter in inches for US units and in millimeters for metric units. $\pi = 3.1416$.

*Example:* The cutting speed for turning a 4-inch (101.6-mm) diameter bar has been found to be 575 fpm (175.3 m/min). Using both the inch and metric formulas, calculate the lathe spindle speed.

US Units:

$$N = \frac{12V}{\pi D} = \frac{12 \times 575}{3.1416 \times 4} = 549 \text{ rpm}$$

**Metric Units:**

$$N = \frac{1000V}{\pi D} = \frac{1000 \times 175.3}{3.1416 \times 101.6} = 549 \text{ rpm}$$

When the cutting tool or workpiece diameter and the spindle speed in rpm are known, it is often necessary to calculate the cutting speed in feet or meters per minute. In this event, the following formulas are used.

US Units:

$$V = \frac{\pi DN}{12} \text{ fpm}$$

**Metric Units:**

$$V = \frac{\pi DN}{1000} \text{ m/min}$$

As in the previous formulas, $N$ is the rpm and $D$ is the diameter in inches for the US Units formula and in millimeters for the metric formula.

*Example:* Calculate the cutting speed in feet per minute and in meters per minute if the spindle speed of a ¾-inch (19.05-mm) drill is 400 rpm.

US Units: $$V = \frac{\pi DN}{12} = \frac{\pi \times 0.75 \times 400}{12} = 78.5 \text{ fpm}$$

**Metric Units:** $$V = \frac{\pi DN}{1000} = \frac{\pi \times 19.05 \times 400}{1000} = 24.9 \text{ m/min}$$

## Cutting Speeds and Equivalent RPM for Drills of Number and Letter Sizes

| Size No. | Cutting Speed, Feet per Minute | | | | | | | | | | |
|---|---|---|---|---|---|---|---|---|---|---|---|
| | 30′ | 40′ | 50′ | 60′ | 70′ | 80′ | 90′ | 100′ | 110′ | 130′ | 150′ |
| | Revolutions per Minute for Number Sizes | | | | | | | | | | |
| 1 | 503 | 670 | 838 | 1005 | 1173 | 1340 | 1508 | 1675 | 1843 | 2179 | 2513 |
| 2 | 518 | 691 | 864 | 1037 | 1210 | 1382 | 1555 | 1728 | 1901 | 2247 | 2593 |
| 4 | 548 | 731 | 914 | 1097 | 1280 | 1462 | 1645 | 1828 | 2010 | 2376 | 2741 |
| 6 | 562 | 749 | 936 | 1123 | 1310 | 1498 | 1685 | 1872 | 2060 | 2434 | 2809 |
| 8 | 576 | 768 | 960 | 1151 | 1343 | 1535 | 1727 | 1919 | 2111 | 2495 | 2879 |
| 10 | 592 | 790 | 987 | 1184 | 1382 | 1579 | 1777 | 1974 | 2171 | 2566 | 2961 |
| 12 | 606 | 808 | 1010 | 1213 | 1415 | 1617 | 1819 | 2021 | 2223 | 2627 | 3032 |
| 14 | 630 | 840 | 1050 | 1259 | 1469 | 1679 | 1889 | 2099 | 2309 | 2728 | 3148 |
| 16 | 647 | 863 | 1079 | 1295 | 1511 | 1726 | 1942 | 2158 | 2374 | 2806 | 3237 |
| 18 | 678 | 904 | 1130 | 1356 | 1582 | 1808 | 2034 | 2260 | 2479 | 2930 | 3380 |
| 20 | 712 | 949 | 1186 | 1423 | 1660 | 1898 | 2135 | 2372 | 2610 | 3084 | 3559 |
| 22 | 730 | 973 | 1217 | 1460 | 1703 | 1946 | 2190 | 2433 | 2676 | 3164 | 3649 |
| 24 | 754 | 1005 | 1257 | 1508 | 1759 | 2010 | 2262 | 2513 | 2764 | 3267 | 3769 |
| 26 | 779 | 1039 | 1299 | 1559 | 1819 | 2078 | 2338 | 2598 | 2858 | 3378 | 3898 |
| 28 | 816 | 1088 | 1360 | 1631 | 1903 | 2175 | 2447 | 2719 | 2990 | 3534 | 4078 |
| 30 | 892 | 1189 | 1487 | 1784 | 2081 | 2378 | 2676 | 2973 | 3270 | 3864 | 4459 |
| 32 | 988 | 1317 | 1647 | 1976 | 2305 | 2634 | 2964 | 3293 | 3622 | 4281 | 4939 |
| 34 | 1032 | 1376 | 1721 | 2065 | 2409 | 2753 | 3097 | 3442 | 3785 | 4474 | 5162 |
| 36 | 1076 | 1435 | 1794 | 2152 | 2511 | 2870 | 3228 | 3587 | 3945 | 4663 | 5380 |
| 38 | 1129 | 1505 | 1882 | 2258 | 2634 | 3010 | 3387 | 3763 | 4140 | 4892 | 5645 |
| 40 | 1169 | 1559 | 1949 | 2339 | 2729 | 3118 | 3508 | 3898 | 4287 | 5067 | 5846 |
| 42 | 1226 | 1634 | 2043 | 2451 | 2860 | 3268 | 3677 | 4085 | 4494 | 5311 | 6128 |
| 44 | 1333 | 1777 | 2221 | 2665 | 3109 | 3554 | 3999 | 4442 | 4886 | 5774 | 6662 |
| 46 | 1415 | 1886 | 2358 | 2830 | 3301 | 3773 | 4244 | 4716 | 5187 | 6130 | 7074 |
| 48 | 1508 | 2010 | 2513 | 3016 | 3518 | 4021 | 4523 | 5026 | 5528 | 6534 | 7539 |
| 50 | 1637 | 2183 | 2729 | 3274 | 3820 | 4366 | 4911 | 5457 | 6002 | 7094 | 8185 |
| 52 | 1805 | 2406 | 3008 | 3609 | 4211 | 4812 | 5414 | 6015 | 6619 | 7820 | 9023 |
| 54 | 2084 | 2778 | 3473 | 4167 | 4862 | 5556 | 6251 | 6945 | 7639 | 9028 | 10417 |
| Size | Revolutions per Minute for Letter Sizes | | | | | | | | | | |
| A | 491 | 654 | 818 | 982 | 1145 | 1309 | 1472 | 1636 | 1796 | 2122 | 2448 |
| B | 482 | 642 | 803 | 963 | 1124 | 1284 | 1445 | 1605 | 1765 | 2086 | 2407 |
| C | 473 | 631 | 789 | 947 | 1105 | 1262 | 1420 | 1578 | 1736 | 2052 | 2368 |
| D | 467 | 622 | 778 | 934 | 1089 | 1245 | 1400 | 1556 | 1708 | 2018 | 2329 |
| E | 458 | 611 | 764 | 917 | 1070 | 1222 | 1375 | 1528 | 1681 | 1968 | 2292 |
| F | 446 | 594 | 743 | 892 | 1040 | 1189 | 1337 | 1486 | 1635 | 1932 | 2229 |
| G | 440 | 585 | 732 | 878 | 1024 | 1170 | 1317 | 1463 | 1610 | 1903 | 2195 |
| H | 430 | 574 | 718 | 862 | 1005 | 1149 | 1292 | 1436 | 1580 | 1867 | 2154 |
| I | 421 | 562 | 702 | 842 | 983 | 1123 | 1264 | 1404 | 1545 | 1826 | 2106 |
| J | 414 | 552 | 690 | 827 | 965 | 1103 | 1241 | 1379 | 1517 | 1793 | 2068 |
| K | 408 | 544 | 680 | 815 | 951 | 1087 | 1223 | 1359 | 1495 | 1767 | 2039 |
| L | 395 | 527 | 659 | 790 | 922 | 1054 | 1185 | 1317 | 1449 | 1712 | 1976 |
| M | 389 | 518 | 648 | 777 | 907 | 1036 | 1166 | 1295 | 1424 | 1683 | 1942 |
| N | 380 | 506 | 633 | 759 | 886 | 1012 | 1139 | 1265 | 1391 | 1644 | 1897 |
| O | 363 | 484 | 605 | 725 | 846 | 967 | 1088 | 1209 | 1330 | 1571 | 1813 |
| P | 355 | 473 | 592 | 710 | 828 | 946 | 1065 | 1183 | 1301 | 1537 | 1774 |
| Q | 345 | 460 | 575 | 690 | 805 | 920 | 1035 | 1150 | 1266 | 1496 | 1726 |
| R | 338 | 451 | 564 | 676 | 789 | 902 | 1014 | 1127 | 1239 | 1465 | 1690 |
| S | 329 | 439 | 549 | 659 | 769 | 878 | 988 | 1098 | 1207 | 1427 | 1646 |
| T | 320 | 426 | 533 | 640 | 746 | 853 | 959 | 1066 | 1173 | 1387 | 1600 |
| U | 311 | 415 | 519 | 623 | 727 | 830 | 934 | 1038 | 1142 | 1349 | 1557 |
| V | 304 | 405 | 507 | 608 | 709 | 810 | 912 | 1013 | 1114 | 1317 | 1520 |
| W | 297 | 396 | 495 | 594 | 693 | 792 | 891 | 989 | 1088 | 1286 | 1484 |
| X | 289 | 385 | 481 | 576 | 672 | 769 | 865 | 962 | 1058 | 1251 | 1443 |
| Y | 284 | 378 | 473 | 567 | 662 | 756 | 851 | 945 | 1040 | 1229 | 1418 |
| Z | 277 | 370 | 462 | 555 | 647 | 740 | 832 | 925 | 1017 | 1202 | 1387 |

For fractional drill sizes, use the following table.

## Revolutions per Minute for Various Cutting Speeds and Diameters

| Dia., Inches | Cutting Speed, Feet per Minute | | | | | | | | | | | |
|---|---|---|---|---|---|---|---|---|---|---|---|---|
| | 40 | 50 | 60 | 70 | 80 | 90 | 100 | 120 | 140 | 160 | 180 | 200 |
| | Revolutions per Minute | | | | | | | | | | | |
| ¼ | 611 | 764 | 917 | 1070 | 1222 | 1376 | 1528 | 1834 | 2139 | 2445 | 2750 | 3056 |
| 5/16 | 489 | 611 | 733 | 856 | 978 | 1100 | 1222 | 1466 | 1711 | 1955 | 2200 | 2444 |
| ⅜ | 408 | 509 | 611 | 713 | 815 | 916 | 1018 | 1222 | 1425 | 1629 | 1832 | 2036 |
| 7/16 | 349 | 437 | 524 | 611 | 699 | 786 | 874 | 1049 | 1224 | 1398 | 1573 | 1748 |
| ½ | 306 | 382 | 459 | 535 | 611 | 688 | 764 | 917 | 1070 | 1222 | 1375 | 1528 |
| 9/16 | 272 | 340 | 407 | 475 | 543 | 611 | 679 | 813 | 951 | 1086 | 1222 | 1358 |
| ⅝ | 245 | 306 | 367 | 428 | 489 | 552 | 612 | 736 | 857 | 979 | 1102 | 1224 |
| 11/16 | 222 | 273 | 333 | 389 | 444 | 500 | 555 | 666 | 770 | 888 | 999 | 1101 |
| ¾ | 203 | 254 | 306 | 357 | 408 | 458 | 508 | 610 | 711 | 813 | 914 | 1016 |
| 13/16 | 190 | 237 | 284 | 332 | 379 | 427 | 474 | 569 | 664 | 758 | 853 | 948 |
| ⅞ | 175 | 219 | 262 | 306 | 349 | 392 | 438 | 526 | 613 | 701 | 788 | 876 |
| 15/16 | 163 | 204 | 244 | 285 | 326 | 366 | 407 | 488 | 570 | 651 | 733 | 814 |
| 1 | 153 | 191 | 229 | 267 | 306 | 344 | 382 | 458 | 535 | 611 | 688 | 764 |
| 1 1/16 | 144 | 180 | 215 | 251 | 287 | 323 | 359 | 431 | 503 | 575 | 646 | 718 |
| 1⅛ | 136 | 170 | 204 | 238 | 272 | 306 | 340 | 408 | 476 | 544 | 612 | 680 |
| 1 3/16 | 129 | 161 | 193 | 225 | 258 | 290 | 322 | 386 | 451 | 515 | 580 | 644 |
| 1¼ | 123 | 153 | 183 | 214 | 245 | 274 | 306 | 367 | 428 | 490 | 551 | 612 |
| 1 5/16 | 116 | 146 | 175 | 204 | 233 | 262 | 291 | 349 | 407 | 466 | 524 | 582 |
| 1⅜ | 111 | 139 | 167 | 195 | 222 | 250 | 278 | 334 | 389 | 445 | 500 | 556 |
| 1 7/16 | 106 | 133 | 159 | 186 | 212 | 239 | 265 | 318 | 371 | 424 | 477 | 530 |
| 1½ | 102 | 127 | 153 | 178 | 204 | 230 | 254 | 305 | 356 | 406 | 457 | 508 |
| 1 9/16 | 97.6 | 122 | 146 | 171 | 195 | 220 | 244 | 293 | 342 | 390 | 439 | 488 |
| 1⅝ | 93.9 | 117 | 141 | 165 | 188 | 212 | 234 | 281 | 328 | 374 | 421 | 468 |
| 1 11/16 | 90.4 | 113 | 136 | 158 | 181 | 203 | 226 | 271 | 316 | 362 | 407 | 452 |
| 1¾ | 87.3 | 109 | 131 | 153 | 175 | 196 | 218 | 262 | 305 | 349 | 392 | 436 |
| 1⅞ | 81.5 | 102 | 122 | 143 | 163 | 184 | 204 | 244 | 286 | 326 | 367 | 408 |
| 2 | 76.4 | 95.5 | 115 | 134 | 153 | 172 | 191 | 229 | 267 | 306 | 344 | 382 |
| 2⅛ | 72.0 | 90.0 | 108 | 126 | 144 | 162 | 180 | 216 | 252 | 288 | 324 | 360 |
| 2¼ | 68.0 | 85.5 | 102 | 119 | 136 | 153 | 170 | 204 | 238 | 272 | 306 | 340 |
| 2⅜ | 64.4 | 80.5 | 96.6 | 113 | 129 | 145 | 161 | 193 | 225 | 258 | 290 | 322 |
| 2½ | 61.2 | 76.3 | 91.7 | 107 | 122 | 138 | 153 | 184 | 213 | 245 | 275 | 306 |
| 2⅝ | 58.0 | 72.5 | 87.0 | 102 | 116 | 131 | 145 | 174 | 203 | 232 | 261 | 290 |
| 2¾ | 55.6 | 69.5 | 83.4 | 97.2 | 111 | 125 | 139 | 167 | 195 | 222 | 250 | 278 |
| 2⅞ | 52.8 | 66.0 | 79.2 | 92.4 | 106 | 119 | 132 | 158 | 185 | 211 | 238 | 264 |
| 3 | 51.0 | 63.7 | 76.4 | 89.1 | 102 | 114 | 127 | 152 | 178 | 203 | 228 | 254 |
| 3⅛ | 48.8 | 61.0 | 73.2 | 85.4 | 97.6 | 110 | 122 | 146 | 171 | 195 | 219 | 244 |
| 3¼ | 46.8 | 58.5 | 70.2 | 81.9 | 93.6 | 105 | 117 | 140 | 164 | 188 | 211 | 234 |
| 3⅜ | 45.2 | 56.5 | 67.8 | 79.1 | 90.4 | 102 | 113 | 136 | 158 | 181 | 203 | 226 |
| 3½ | 43.6 | 54.5 | 65.5 | 76.4 | 87.4 | 98.1 | 109 | 131 | 153 | 174 | 196 | 218 |
| 3⅝ | 42.0 | 52.5 | 63.0 | 73.5 | 84.0 | 94.5 | 105 | 126 | 147 | 168 | 189 | 210 |
| 3¾ | 40.8 | 51.0 | 61.2 | 71.4 | 81.6 | 91.8 | 102 | 122 | 143 | 163 | 184 | 205 |
| 3⅞ | 39.4 | 49.3 | 59.1 | 69.0 | 78.8 | 88.6 | 98.5 | 118 | 138 | 158 | 177 | 197 |
| 4 | 38.2 | 47.8 | 57.3 | 66.9 | 76.4 | 86.0 | 95.6 | 115 | 134 | 153 | 172 | 191 |
| 4¼ | 35.9 | 44.9 | 53.9 | 62.9 | 71.8 | 80.8 | 89.8 | 108 | 126 | 144 | 162 | 180 |
| 4½ | 34.0 | 42.4 | 51.0 | 59.4 | 67.9 | 76.3 | 84.8 | 102 | 119 | 136 | 153 | 170 |
| 4¾ | 32.2 | 40.2 | 48.2 | 56.3 | 64.3 | 72.4 | 80.4 | 96.9 | 113 | 129 | 145 | 161 |
| 5 | 30.6 | 38.2 | 45.9 | 53.5 | 61.1 | 68.8 | 76.4 | 91.7 | 107 | 122 | 138 | 153 |
| 5¼ | 29.1 | 36.4 | 43.6 | 50.9 | 58.2 | 65.4 | 72.7 | 87.2 | 102 | 116 | 131 | 145 |
| 5½ | 27.8 | 34.7 | 41.7 | 48.6 | 55.6 | 62.5 | 69.4 | 83.3 | 97.2 | 111 | 125 | 139 |
| 5¾ | 26.6 | 33.2 | 39.8 | 46.5 | 53.1 | 59.8 | 66.4 | 80.0 | 93.0 | 106 | 120 | 133 |
| 6 | 25.5 | 31.8 | 38.2 | 44.6 | 51.0 | 57.2 | 63.6 | 76.3 | 89.0 | 102 | 114 | 127 |
| 6¼ | 24.4 | 30.6 | 36.7 | 42.8 | 48.9 | 55.0 | 61.1 | 73.3 | 85.5 | 97.7 | 110 | 122 |
| 6½ | 23.5 | 29.4 | 35.2 | 41.1 | 47.0 | 52.8 | 58.7 | 70.4 | 82.2 | 93.9 | 106 | 117 |
| 6¾ | 22.6 | 28.3 | 34.0 | 39.6 | 45.3 | 50.9 | 56.6 | 67.9 | 79.2 | 90.6 | 102 | 113 |
| 7 | 21.8 | 27.3 | 32.7 | 38.2 | 43.7 | 49.1 | 54.6 | 65.5 | 76.4 | 87.4 | 98.3 | 109 |
| 7¼ | 21.1 | 26.4 | 31.6 | 36.9 | 42.2 | 47.4 | 52.7 | 63.2 | 73.8 | 84.3 | 94.9 | 105 |
| 7½ | 20.4 | 25.4 | 30.5 | 35.6 | 40.7 | 45.8 | 50.9 | 61.1 | 71.0 | 81.4 | 91.6 | 102 |
| 7¾ | 19.7 | 24.6 | 29.5 | 34.4 | 39.4 | 44.3 | 49.2 | 59.0 | 68.9 | 78.7 | 88.6 | 98.4 |
| 8 | 19.1 | 23.9 | 28.7 | 33.4 | 38.2 | 43.0 | 47.8 | 57.4 | 66.9 | 76.5 | 86.0 | 95.6 |

## Revolutions per Minute for Various Cutting Speeds and Diameters

| Dia., Inches | Cutting Speed, Feet per Minute | | | | | | | | | | | |
|---|---|---|---|---|---|---|---|---|---|---|---|---|
| | 225 | 250 | 275 | 300 | 325 | 350 | 375 | 400 | 425 | 450 | 500 | 550 |
| | Revolutions per Minute | | | | | | | | | | | |
| 1/4 | 3438 | 3820 | 4202 | 4584 | 4966 | 5348 | 5730 | 6112 | 6493 | 6875 | 7639 | 8403 |
| 5/16 | 2750 | 3056 | 3362 | 3667 | 3973 | 4278 | 4584 | 4889 | 5195 | 5501 | 6112 | 6723 |
| 3/8 | 2292 | 2546 | 2801 | 3056 | 3310 | 3565 | 3820 | 4074 | 4329 | 4584 | 5093 | 5602 |
| 7/16 | 1964 | 2182 | 2401 | 2619 | 2837 | 3056 | 3274 | 3492 | 3710 | 3929 | 4365 | 4802 |
| 1/2 | 1719 | 1910 | 2101 | 2292 | 2483 | 2675 | 2866 | 3057 | 3248 | 3439 | 3821 | 4203 |
| 9/16 | 1528 | 1698 | 1868 | 2037 | 2207 | 2377 | 2547 | 2717 | 2887 | 3056 | 3396 | 3736 |
| 5/8 | 1375 | 1528 | 1681 | 1834 | 1987 | 2139 | 2292 | 2445 | 2598 | 2751 | 3057 | 3362 |
| 11/16 | 1250 | 1389 | 1528 | 1667 | 1806 | 1941 | 2084 | 2223 | 2362 | 2501 | 2779 | 3056 |
| 3/4 | 1146 | 1273 | 1401 | 1528 | 1655 | 1783 | 1910 | 2038 | 2165 | 2292 | 2547 | 2802 |
| 13/16 | 1058 | 1175 | 1293 | 1410 | 1528 | 1646 | 1763 | 1881 | 1998 | 2116 | 2351 | 2586 |
| 7/8 | 982 | 1091 | 1200 | 1310 | 1419 | 1528 | 1637 | 1746 | 1855 | 1965 | 2183 | 2401 |
| 15/16 | 917 | 1019 | 1120 | 1222 | 1324 | 1426 | 1528 | 1630 | 1732 | 1834 | 2038 | 2241 |
| 1 | 859 | 955 | 1050 | 1146 | 1241 | 1337 | 1432 | 1528 | 1623 | 1719 | 1910 | 2101 |
| 1 1/16 | 809 | 899 | 988 | 1078 | 1168 | 1258 | 1348 | 1438 | 1528 | 1618 | 1798 | 1977 |
| 1 1/8 | 764 | 849 | 933 | 1018 | 1103 | 1188 | 1273 | 1358 | 1443 | 1528 | 1698 | 1867 |
| 1 3/16 | 724 | 804 | 884 | 965 | 1045 | 1126 | 1206 | 1287 | 1367 | 1448 | 1609 | 1769 |
| 1 1/4 | 687 | 764 | 840 | 917 | 993 | 1069 | 1146 | 1222 | 1299 | 1375 | 1528 | 1681 |
| 1 5/16 | 654 | 727 | 800 | 873 | 946 | 1018 | 1091 | 1164 | 1237 | 1309 | 1455 | 1601 |
| 1 3/8 | 625 | 694 | 764 | 833 | 903 | 972 | 1042 | 1111 | 1181 | 1250 | 1389 | 1528 |
| 1 7/16 | 598 | 664 | 730 | 797 | 863 | 930 | 996 | 1063 | 1129 | 1196 | 1329 | 1461 |
| 1 1/2 | 573 | 636 | 700 | 764 | 827 | 891 | 955 | 1018 | 1082 | 1146 | 1273 | 1400 |
| 1 9/16 | 550 | 611 | 672 | 733 | 794 | 855 | 916 | 978 | 1039 | 1100 | 1222 | 1344 |
| 1 5/8 | 528 | 587 | 646 | 705 | 764 | 822 | 881 | 940 | 999 | 1057 | 1175 | 1293 |
| 1 11/16 | 509 | 566 | 622 | 679 | 735 | 792 | 849 | 905 | 962 | 1018 | 1132 | 1245 |
| 1 3/4 | 491 | 545 | 600 | 654 | 709 | 764 | 818 | 873 | 927 | 982 | 1091 | 1200 |
| 1 13/16 | 474 | 527 | 579 | 632 | 685 | 737 | 790 | 843 | 895 | 948 | 1054 | 1159 |
| 1 7/8 | 458 | 509 | 560 | 611 | 662 | 713 | 764 | 815 | 866 | 917 | 1019 | 1120 |
| 1 15/16 | 443 | 493 | 542 | 591 | 640 | 690 | 739 | 788 | 838 | 887 | 986 | 1084 |
| 2 | 429 | 477 | 525 | 573 | 620 | 668 | 716 | 764 | 811 | 859 | 955 | 1050 |
| 2 1/8 | 404 | 449 | 494 | 539 | 584 | 629 | 674 | 719 | 764 | 809 | 899 | 988 |
| 2 1/4 | 382 | 424 | 468 | 509 | 551 | 594 | 636 | 679 | 721 | 764 | 849 | 933 |
| 2 3/8 | 362 | 402 | 442 | 482 | 522 | 563 | 603 | 643 | 683 | 724 | 804 | 884 |
| 2 1/2 | 343 | 382 | 420 | 458 | 496 | 534 | 573 | 611 | 649 | 687 | 764 | 840 |
| 2 5/8 | 327 | 363 | 400 | 436 | 472 | 509 | 545 | 582 | 618 | 654 | 727 | 800 |
| 2 3/4 | 312 | 347 | 381 | 416 | 451 | 486 | 520 | 555 | 590 | 625 | 694 | 763 |
| 2 7/8 | 299 | 332 | 365 | 398 | 431 | 465 | 498 | 531 | 564 | 598 | 664 | 730 |
| 3 | 286 | 318 | 350 | 381 | 413 | 445 | 477 | 509 | 541 | 572 | 636 | 700 |
| 3 1/8 | 274 | 305 | 336 | 366 | 397 | 427 | 458 | 488 | 519 | 549 | 611 | 672 |
| 3 1/4 | 264 | 293 | 323 | 352 | 381 | 411 | 440 | 470 | 499 | 528 | 587 | 646 |
| 3 3/8 | 254 | 283 | 311 | 339 | 367 | 396 | 424 | 452 | 481 | 509 | 566 | 622 |
| 3 1/2 | 245 | 272 | 300 | 327 | 354 | 381 | 409 | 436 | 463 | 490 | 545 | 600 |
| 3 5/8 | 237 | 263 | 289 | 316 | 342 | 368 | 395 | 421 | 447 | 474 | 527 | 579 |
| 3 3/4 | 229 | 254 | 280 | 305 | 331 | 356 | 382 | 407 | 433 | 458 | 509 | 560 |
| 3 7/8 | 221 | 246 | 271 | 295 | 320 | 345 | 369 | 394 | 419 | 443 | 493 | 542 |
| 4 | 214 | 238 | 262 | 286 | 310 | 334 | 358 | 382 | 405 | 429 | 477 | 525 |
| 4 1/4 | 202 | 224 | 247 | 269 | 292 | 314 | 337 | 359 | 383 | 404 | 449 | 494 |
| 4 1/2 | 191 | 212 | 233 | 254 | 275 | 297 | 318 | 339 | 360 | 382 | 424 | 466 |
| 4 3/4 | 180 | 201 | 221 | 241 | 261 | 281 | 301 | 321 | 341 | 361 | 402 | 442 |
| 5 | 171 | 191 | 210 | 229 | 248 | 267 | 286 | 305 | 324 | 343 | 382 | 420 |
| 5 1/4 | 163 | 181 | 199 | 218 | 236 | 254 | 272 | 290 | 308 | 327 | 363 | 399 |
| 5 1/2 | 156 | 173 | 190 | 208 | 225 | 242 | 260 | 277 | 294 | 312 | 347 | 381 |
| 5 3/4 | 149 | 166 | 182 | 199 | 215 | 232 | 249 | 265 | 282 | 298 | 332 | 365 |
| 6 | 143 | 159 | 174 | 190 | 206 | 222 | 238 | 254 | 270 | 286 | 318 | 349 |
| 6 1/4 | 137 | 152 | 168 | 183 | 198 | 213 | 229 | 244 | 259 | 274 | 305 | 336 |
| 6 1/2 | 132 | 146 | 161 | 176 | 190 | 205 | 220 | 234 | 249 | 264 | 293 | 322 |
| 6 3/4 | 127 | 141 | 155 | 169 | 183 | 198 | 212 | 226 | 240 | 254 | 283 | 311 |
| 7 | 122 | 136 | 149 | 163 | 177 | 190 | 204 | 218 | 231 | 245 | 272 | 299 |
| 7 1/4 | 118 | 131 | 144 | 158 | 171 | 184 | 197 | 210 | 223 | 237 | 263 | 289 |
| 7 1/2 | 114 | 127 | 139 | 152 | 165 | 178 | 190 | 203 | 216 | 229 | 254 | 279 |
| 7 3/4 | 111 | 123 | 135 | 148 | 160 | 172 | 185 | 197 | 209 | 222 | 246 | 271 |
| 8 | 107 | 119 | 131 | 143 | 155 | 167 | 179 | 191 | 203 | 215 | 238 | 262 |

**Revolutions per Minute for Various Cutting Speeds and Diameters (Metric Units)**

| Dia., mm | Cutting Speed, Meters per Minute | | | | | | | | | | | |
|---|---|---|---|---|---|---|---|---|---|---|---|---|
| | 5 | 6 | 8 | 10 | 12 | 16 | 20 | 25 | 30 | 35 | 40 | 45 |
| | Revolutions per Minute | | | | | | | | | | | |
| 5 | 318 | 382 | 509 | 637 | 764 | 1019 | 1273 | 1592 | 1910 | 2228 | 2546 | 2865 |
| 6 | 265 | 318 | 424 | 530 | 637 | 849 | 1061 | 1326 | 1592 | 1857 | 2122 | 2387 |
| 8 | 199 | 239 | 318 | 398 | 477 | 637 | 796 | 995 | 1194 | 1393 | 1592 | 1790 |
| 10 | 159 | 191 | 255 | 318 | 382 | 509 | 637 | 796 | 955 | 1114 | 1273 | 1432 |
| 12 | 133 | 159 | 212 | 265 | 318 | 424 | 531 | 663 | 796 | 928 | 1061 | 1194 |
| 16 | 99.5 | 119 | 159 | 199 | 239 | 318 | 398 | 497 | 597 | 696 | 796 | 895 |
| 20 | 79.6 | 95.5 | 127 | 159 | 191 | 255 | 318 | 398 | 477 | 557 | 637 | 716 |
| 25 | 63.7 | 76.4 | 102 | 127 | 153 | 204 | 255 | 318 | 382 | 446 | 509 | 573 |
| 30 | 53.1 | 63.7 | 84.9 | 106 | 127 | 170 | 212 | 265 | 318 | 371 | 424 | 477 |
| 35 | 45.5 | 54.6 | 72.8 | 90.9 | 109 | 145 | 182 | 227 | 273 | 318 | 364 | 409 |
| 40 | 39.8 | 47.7 | 63.7 | 79.6 | 95.5 | 127 | 159 | 199 | 239 | 279 | 318 | 358 |
| 45 | 35.4 | 42.4 | 56.6 | 70.7 | 84.9 | 113 | 141 | 177 | 212 | 248 | 283 | 318 |
| 50 | 31.8 | 38.2 | 51 | 63.7 | 76.4 | 102 | 127 | 159 | 191 | 223 | 255 | 286 |
| 55 | 28.9 | 34.7 | 46.3 | 57.9 | 69.4 | 92.6 | 116 | 145 | 174 | 203 | 231 | 260 |
| 60 | 26.6 | 31.8 | 42.4 | 53.1 | 63.7 | 84.9 | 106 | 133 | 159 | 186 | 212 | 239 |
| 65 | 24.5 | 29.4 | 39.2 | 49 | 58.8 | 78.4 | 98 | 122 | 147 | 171 | 196 | 220 |
| 70 | 22.7 | 27.3 | 36.4 | 45.5 | 54.6 | 72.8 | 90.9 | 114 | 136 | 159 | 182 | 205 |
| 75 | 21.2 | 25.5 | 34 | 42.4 | 51 | 68 | 84.9 | 106 | 127 | 149 | 170 | 191 |
| 80 | 19.9 | 23.9 | 31.8 | 39.8 | 47.7 | 63.7 | 79.6 | 99.5 | 119 | 139 | 159 | 179 |
| 90 | 17.7 | 21.2 | 28.3 | 35.4 | 42.4 | 56.6 | 70.7 | 88.4 | 106 | 124 | 141 | 159 |
| 100 | 15.9 | 19.1 | 25.5 | 31.8 | 38.2 | 51 | 63.7 | 79.6 | 95.5 | 111 | 127 | 143 |
| 110 | 14.5 | 17.4 | 23.1 | 28.9 | 34.7 | 46.2 | 57.9 | 72.3 | 86.8 | 101 | 116 | 130 |
| 120 | 13.3 | 15.9 | 21.2 | 26.5 | 31.8 | 42.4 | 53.1 | 66.3 | 79.6 | 92.8 | 106 | 119 |
| 130 | 12.2 | 14.7 | 19.6 | 24.5 | 29.4 | 39.2 | 49 | 61.2 | 73.4 | 85.7 | 97.9 | 110 |
| 140 | 11.4 | 13.6 | 18.2 | 22.7 | 27.3 | 36.4 | 45.5 | 56.8 | 68.2 | 79.6 | 90.9 | 102 |
| 150 | 10.6 | 12.7 | 17 | 21.2 | 25.5 | 34 | 42.4 | 53.1 | 63.7 | 74.3 | 84.9 | 95.5 |
| 160 | 9.9 | 11.9 | 15.9 | 19.9 | 23.9 | 31.8 | 39.8 | 49.7 | 59.7 | 69.6 | 79.6 | 89.5 |
| 170 | 9.4 | 11.2 | 15 | 18.7 | 22.5 | 30 | 37.4 | 46.8 | 56.2 | 65.5 | 74.9 | 84.2 |
| 180 | 8.8 | 10.6 | 14.1 | 17.7 | 21.2 | 28.3 | 35.4 | 44.2 | 53.1 | 61.9 | 70.7 | 79.6 |
| 190 | 8.3 | 10 | 13.4 | 16.8 | 20.1 | 26.8 | 33.5 | 41.9 | 50.3 | 58.6 | 67 | 75.4 |
| 200 | 8 | 39.5 | 12.7 | 15.9 | 19.1 | 25.5 | 31.8 | 39.8 | 47.7 | 55.7 | 63.7 | 71.6 |
| 220 | 7.2 | 8.7 | 11.6 | 14.5 | 17.4 | 23.1 | 28.9 | 36.2 | 43.4 | 50.6 | 57.9 | 65.1 |
| 240 | 6.6 | 8 | 10.6 | 13.3 | 15.9 | 21.2 | 26.5 | 33.2 | 39.8 | 46.4 | 53.1 | 59.7 |
| 260 | 6.1 | 7.3 | 9.8 | 12.2 | 14.7 | 19.6 | 24.5 | 30.6 | 36.7 | 42.8 | 49 | 55.1 |
| 280 | 5.7 | 6.8 | 9.1 | 11.4 | 13.6 | 18.2 | 22.7 | 28.4 | 34.1 | 39.8 | 45.5 | 51.1 |
| 300 | 5.3 | 6.4 | 8.5 | 10.6 | 12.7 | 17 | 21.2 | 26.5 | 31.8 | 37.1 | 42.4 | 47.7 |
| 350 | 4.5 | 5.4 | 7.3 | 9.1 | 10.9 | 14.6 | 18.2 | 22.7 | 27.3 | 31.8 | 36.4 | 40.9 |
| 400 | 4 | 4.8 | 6.4 | 8 | 9.5 | 12.7 | 15.9 | 19.9 | 23.9 | 27.9 | 31.8 | 35.8 |
| 450 | 3.5 | 4.2 | 5.7 | 7.1 | 8.5 | 11.3 | 14.1 | 17.7 | 21.2 | 24.8 | 28.3 | 31.8 |
| 500 | 3.2 | 3.8 | 5.1 | 6.4 | 7.6 | 10.2 | 12.7 | 15.9 | 19.1 | 22.3 | 25.5 | 28.6 |

## Revolutions per Minute for Various Cutting Speeds and Diameters (Metric Units)

| Dia., mm | Cutting Speed, Meters per Minute | | | | | | | | | | | |
|---|---|---|---|---|---|---|---|---|---|---|---|---|
| | 50 | 55 | 60 | 65 | 70 | 75 | 80 | 85 | 90 | 95 | 100 | 200 |
| | Revolutions per Minute | | | | | | | | | | | |
| 5 | 3183 | 3501 | 3820 | 4138 | 4456 | 4775 | 5093 | 5411 | 5730 | 6048 | 6366 | 12,732 |
| 6 | 2653 | 2918 | 3183 | 3448 | 3714 | 3979 | 4244 | 4509 | 4775 | 5039 | 5305 | 10,610 |
| 8 | 1989 | 2188 | 2387 | 2586 | 2785 | 2984 | 3183 | 3382 | 3581 | 3780 | 3979 | 7958 |
| 10 | 1592 | 1751 | 1910 | 2069 | 2228 | 2387 | 2546 | 2706 | 2865 | 3024 | 3183 | 6366 |
| 12 | 1326 | 1459 | 1592 | 1724 | 1857 | 1989 | 2122 | 2255 | 2387 | 2520 | 2653 | 5305 |
| 16 | 995 | 1094 | 1194 | 1293 | 1393 | 1492 | 1591 | 1691 | 1790 | 1890 | 1989 | 3979 |
| 20 | 796 | 875 | 955 | 1034 | 1114 | 1194 | 1273 | 1353 | 1432 | 1512 | 1592 | 3183 |
| 25 | 637 | 700 | 764 | 828 | 891 | 955 | 1019 | 1082 | 1146 | 1210 | 1273 | 2546 |
| 30 | 530 | 584 | 637 | 690 | 743 | 796 | 849 | 902 | 955 | 1008 | 1061 | 2122 |
| 35 | 455 | 500 | 546 | 591 | 637 | 682 | 728 | 773 | 819 | 864 | 909 | 1818 |
| 40 | 398 | 438 | 477 | 517 | 557 | 597 | 637 | 676 | 716 | 756 | 796 | 1592 |
| 45 | 354 | 389 | 424 | 460 | 495 | 531 | 566 | 601 | 637 | 672 | 707 | 1415 |
| 50 | 318 | 350 | 382 | 414 | 446 | 477 | 509 | 541 | 573 | 605 | 637 | 1273 |
| 55 | 289 | 318 | 347 | 376 | 405 | 434 | 463 | 492 | 521 | 550 | 579 | 1157 |
| 60 | 265 | 292 | 318 | 345 | 371 | 398 | 424 | 451 | 477 | 504 | 530 | 1061 |
| 65 | 245 | 269 | 294 | 318 | 343 | 367 | 392 | 416 | 441 | 465 | 490 | 979 |
| 70 | 227 | 250 | 273 | 296 | 318 | 341 | 364 | 387 | 409 | 432 | 455 | 909 |
| 75 | 212 | 233 | 255 | 276 | 297 | 318 | 340 | 361 | 382 | 403 | 424 | 849 |
| 80 | 199 | 219 | 239 | 259 | 279 | 298 | 318 | 338 | 358 | 378 | 398 | 796 |
| 90 | 177 | 195 | 212 | 230 | 248 | 265 | 283 | 301 | 318 | 336 | 354 | 707 |
| 100 | 159 | 175 | 191 | 207 | 223 | 239 | 255 | 271 | 286 | 302 | 318 | 637 |
| 110 | 145 | 159 | 174 | 188 | 203 | 217 | 231 | 246 | 260 | 275 | 289 | 579 |
| 120 | 133 | 146 | 159 | 172 | 186 | 199 | 212 | 225 | 239 | 252 | 265 | 530 |
| 130 | 122 | 135 | 147 | 159 | 171 | 184 | 196 | 208 | 220 | 233 | 245 | 490 |
| 140 | 114 | 125 | 136 | 148 | 159 | 171 | 182 | 193 | 205 | 216 | 227 | 455 |
| 150 | 106 | 117 | 127 | 138 | 149 | 159 | 170 | 180 | 191 | 202 | 212 | 424 |
| 160 | 99.5 | 109 | 119 | 129 | 139 | 149 | 159 | 169 | 179 | 189 | 199 | 398 |
| 170 | 93.6 | 103 | 112 | 122 | 131 | 140 | 150 | 159 | 169 | 178 | 187 | 374 |
| 180 | 88.4 | 97.3 | 106 | 115 | 124 | 133 | 141 | 150 | 159 | 168 | 177 | 354 |
| 190 | 83.8 | 92.1 | 101 | 109 | 117 | 126 | 134 | 142 | 151 | 159 | 167 | 335 |
| 200 | 79.6 | 87.5 | 95.5 | 103 | 111 | 119 | 127 | 135 | 143 | 151 | 159 | 318 |
| 220 | 72.3 | 79.6 | 86.8 | 94 | 101 | 109 | 116 | 123 | 130 | 137 | 145 | 289 |
| 240 | 66.3 | 72.9 | 79.6 | 86.2 | 92.8 | 99.5 | 106 | 113 | 119 | 126 | 132 | 265 |
| 260 | 61.2 | 67.3 | 73.4 | 79.6 | 85.7 | 91.8 | 97.9 | 104 | 110 | 116 | 122 | 245 |
| 280 | 56.8 | 62.5 | 68.2 | 73.9 | 79.6 | 85.3 | 90.9 | 96.6 | 102 | 108 | 114 | 227 |
| 300 | 53.1 | 58.3 | 63.7 | 69 | 74.3 | 79.6 | 84.9 | 90.2 | 95.5 | 101 | 106 | 212 |
| 350 | 45.5 | 50 | 54.6 | 59.1 | 63.7 | 68.2 | 72.8 | 77.3 | 81.8 | 99.1 | 91 | 182 |
| 400 | 39.8 | 43.8 | 47.7 | 51.7 | 55.7 | 59.7 | 63.7 | 67.6 | 71.6 | 75.6 | 79.6 | 159 |
| 450 | 35.4 | 38.9 | 42.4 | 46 | 49.5 | 53.1 | 56.6 | 60.1 | 63.6 | 67.2 | 70.7 | 141 |
| 500 | 31.8 | 35 | 38.2 | 41.4 | 44.6 | 47.7 | 50.9 | 54.1 | 57.3 | 60.5 | 63.6 | 127 |

## SPEEDS AND FEEDS TABLES

### How to Use the Tables

The principal tables of speed and feed values are listed in the table below. In this section, Table 1 through Table 9 give data for turning, Table 10 through Table 15e give data for milling, and Table 17 through Table 23 give data for reaming, drilling, threading.

The materials in these tables are categorized by description, and Brinell Hardness Number (BHN) range or material condition. So far as possible, work materials are grouped by similar machining characteristics. The types of cutting tools (HSS end mill, for example) are identified in one or more rows across the tops of the tables. Other important details concerning the use of the tables are contained in the footnotes to Table 1, Table 10 and Table 17. Information concerning specific cutting tool grades is given in notes at the end of each table.

### Principal Speeds and Feeds Tables

| Speeds and Feeds for Turning |
|---|
| Table 1. Cutting Speeds and Feeds for Turning Plain Carbon and Alloy Steels |
| Table 2. Cutting Speeds and Feeds for Turning Tool Steels |
| Table 3. Cutting Speeds and Feeds for Turning Stainless Steels |
| Table 4a. Cutting Speeds and Feeds for Turning Ferrous Cast Metals |
| Table 4b. Cutting Speeds and Feeds for Turning Ferrous Cast Metals |
| Table 5a. Turning Speed Adjustment Factors for Feed, Depth of Cut, and Lead Angle |
| Table 5b. Tool Life Factors for Turning with Carbides, Ceramics, Cermets, CBN, and Polycrystalline Diamond |
| Table 5c. Cutting-Speed Adjustment Factors for Turning with HSS Tools |
| Table 6. Cutting Speeds and Feeds for Turning Copper Alloys |
| Table 7. Cutting Speeds and Feeds for Turning Titanium and Titanium Alloys |
| Table 8. Cutting Speeds and Feeds for Turning Light Metals |
| Table 9. Cutting Speeds and Feeds for Turning Superalloys |
| **Speeds and Feeds for Milling** |
| Table 10. Cutting Speeds and Feeds for Milling Aluminum Alloys |
| Table 11. Cutting Speeds and Feeds for Milling Plain Carbon and Alloy Steels |
| Table 12. Cutting Speeds and Feeds for Milling Tool Steels |
| Table 13. Cutting Speeds and Feeds for Milling Stainless Steels |
| Table 14. Cutting Speeds and Feeds for Milling Ferrous Cast Metals |
| Table 15a. Recommended Feed in Inches per Tooth (ft) for Milling with High-Speed Steel Cutters |
| Table 15b. End Milling (Full Slot) Speed Adjustment Factors for Feed, Depth of Cut, and Lead Angle |
| Table 15c. End, Slit, and Side Milling Speed Adjustment Factors for Radial Depth of Cut |
| Table 15d. Face Milling Speed Adjustment Factors for Feed, Depth of Cut, and Lead Angle |
| Table 15e. Tool Life Adjustment Factors for Face Milling, End Milling, Drilling, and Reaming |
| Table 16. Cutting Tool Grade Descriptions and Common Vendor Equivalents |
| **Speeds and Feeds for Drilling, Reaming, and Threading** |
| Table 17. Speeds and Feeds for Drilling, Reaming, and Threading Plain Carbon and Alloy Steels |
| Table 18. Speeds and Feeds for Drilling, Reaming, and Threading Tool Steels |
| Table 19. Speeds and Feeds for Drilling, Reaming, and Threading Stainless Steels |
| Table 20. Speeds and Feeds for Drilling, Reaming, and Threading Ferrous Cast Metals |
| Table 21. Speeds and Feeds for Drilling, Reaming, and Threading Light Metals |
| Table 22. Diameter Speed and Feed Adjustment Factors for HSS Twist Drills and Reamers |
| Table 23. Speeds and Feeds for Drilling and Reaming Copper Alloys |

Each of the cutting speed tables in this section contains two distinct types of cutting speed data. The speed columns at the left of each table contain traditional *Handbook* cutting speeds for use with high-speed steel (HSS) tools. For many years, this extensive collection of cutting data has been used successfully as starting speed values for turning, milling, drilling, and reaming operations. Instructions and adjustment factors for use with these speeds are given in Table 5c (feed and depth-of-cut factors) for turning, and in Table 15a (feed, depth of cut, and cutter diameter) for milling. Feeds for drilling and reaming are discussed in *Using the Speeds and Feeds Tables for Drilling, Reaming, and Threading* on page 1116. With traditional speeds and feeds, tool life may vary greatly from material to material, making it very difficult to plan efficient cutting operations, in particular for setting up unattended jobs on CNC equipment where the tool life must exceed cutting time, or at least be predictable so that tool changes can be scheduled. This limitation is reduced by using the combined feed/speed data contained in the remaining columns of the speed tables.

The combined feed/speed portion of the speed tables gives two sets of speed and feed data for each material represented. These feed/speed pairs are the *optimum* and *average* data (identified by *Opt*. and *Avg*.); the *optimum* set is always on the left side of the column and the *average* set is on the right. The *optimum* feed/speed data are approximate values of speed and feed that achieve minimum-cost machining by combining a high productivity rate with low tooling cost at a fixed tool life. The *average* feed/speed data are expected to achieve approximately the same tool life and tooling costs, but productivity is usually lower, so machining costs are higher. The data in this portion of the tables are given in the form of two numbers, of which the first is the feed in thousandths of an inch per revolution (or per tooth, for milling) and the second is the cutting speed in feet per minute. For example, the feed/speed set 15⁄215 represents a feed of 0.015 in/rev (0.38 mm/rev) at a speed of 215 fpm (65.6 m/min). Blank cells in the data tables indicate that feed/speed data for these materials were not available at the time of publication.

Generally, the feed given in the *optimum* set should be interpreted as the maximum safe feed for the given work material and cutting tool grade, and the use of a greater feed may result in premature tool wear or tool failure before the end of the expected tool life. The primary exception to this rule occurs in milling, where the feed may be greater than the *optimum* feed if the radial depth of cut is less than the value established in the table footnote; this topic is covered later in the milling examples. Thus, except for milling, the speed and tool life adjustment tables, to be discussed later, do not permit feeds that are greater than the *optimum* feed. On the other hand, the speed and tool life adjustment factors often result in cutting speeds that are well outside the given *optimum* to *average* speed range.

The combined feed/speed data in this section were contributed by Dr. Colding of Colding International Corp., Ann Arbor, MI. The speed, feed, and tool life calculations were made by means of a special computer program and a large database of cutting speed and tool life testing data. The COMP computer program uses tool life equations that are extensions of the F. W. Taylor tool life equation, first proposed in the early 1900s. The Colding tool life equations use a concept called equivalent chip thickness (*ECT*), which simplifies cutting speed and tool life predictions, and the calculation of cutting forces, torque, and power requirements. *ECT* is a basic metal cutting parameter that combines the four basic turning variables (depth of cut, lead angle, nose radius, and feed per revolution) into one basic parameter. For other metal cutting operations (milling, drilling, and grinding, for example), *ECT* also includes additional variables such as the number of teeth, width of cut, and cutter diameter. The *ECT* concept was first presented in 1931 by Prof. R. Woxen, who showed that equivalent chip thickness is a basic metal cutting parameter for high-speed cutting tools. Dr. Colding later extended the theory to include other tool materials and metal cutting operations, including grinding.

The equivalent chip thickness is defined by $ECT = A/CEL$, where $A$ is the cross-sectional area of the cut (approximately equal to the feed times the depth of cut), and $CEL$ is the cutting edge length or tool contact rubbing length. *ECT* and several other terms related to tool geometry are illustrated in Fig. 1 and Fig. 2. Many combinations of feed, lead angle, nose radius and cutter diameter, axial and radial depth of cut, and numbers of teeth can give the same value of *ECT*. However, for a constant cutting speed, no matter how the depth of cut, feed, or lead angle, etc., are varied, if a constant value of *ECT* is maintained, the tool life will also remain constant. A constant value of *ECT* means that a constant cutting speed gives a constant tool life and an increase in speed results in a reduced tool life. Likewise, if *ECT* were increased and cutting speed were held constant, as illustrated in the generalized cutting speed versus *ECT* graph that follows, tool life would be reduced.

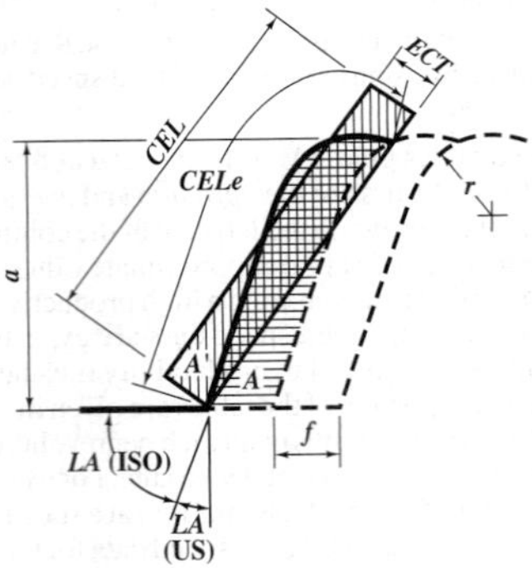


$a$ = depth of cut
$A = A'$ = chip cross-sectional area
$CEL = CELe$ = engaged cutting edge length
$ECT$ = equivalent chip thickness = $A$/CEL
$f$ = feed/rev
$r$ = nose radius
$LA$ = lead angle (US)
$LA(ISO)$ = 90–LA

Fig. 1. Cutting Geometry, Equivalent Chip Thickness, and Cutting Edge Length

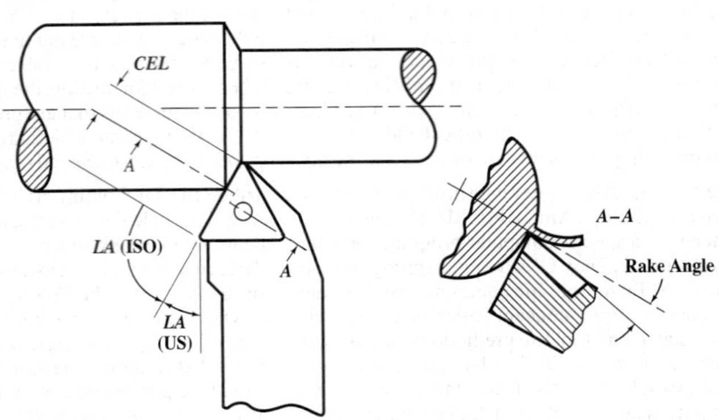


Fig. 2. Cutting Geometry for Turning

In the tables, the *optimum* feed/speed data have been calculated by COMP to achieve a fixed tool life based on the maximum *ECT* that will result in successful cutting, without premature tool wear or early tool failure. The same tool life is used to calculate the *average* feed/speed data, but these values are based on one-half of the maximum *ECT*. Because the data are not linear except over a small range of values, both *optimum* and *average* sets are required to adjust speeds for feed, lead angle, depth of cut, and other factors.

Tool life is the most important factor in a machining system, so speeds and feeds cannot be selected as simple numbers, but must be considered with respect to the many parameters that influence tool life. The accuracy of the combined feed/speed data presented is believed to be very high. However, machining is a variable and complicated process and use of the speeds and feeds tables requires the user to follow the instructions carefully to achieve good predictability. The results achieved, therefore, may vary due to material condition, tool material, machine setup, and other factors, and cannot be guaranteed.

The feed values given in the tables are valid for the standard tool geometries and fixed depths of cut that are identified in the table footnotes. If the cutting parameters and tool geometry established in the table footnotes are maintained, turning operations using either the *optimum* or *average* feed/speed data (Table 1 through Table 9) should achieve a constant tool life of approximately 15 minutes; tool life for milling, drilling, reaming, and threading data (Table 10 through Table 14 and Table 17 through Table 22) should be approximately 45 minutes. The reason for the different economic tool lives is the higher tooling cost associated with milling-drilling operations than for turning. If the cutting parameters or tool geometry are different from those established in the table footnotes, the same tool life (15 or 45 minutes) still may be maintained by applying the appropriate speed adjustment factors, or tool life may be increased or decreased using tool life adjustment factors. The use of the speed and tool life adjustment factors is described in the examples that follow.

Both the *optimum* and *average* feed/speed data given are reasonable values for effective cutting. However, the *optimum* set with its higher feed and lower speed (always the left entry in each table cell) will usually achieve greater productivity. In Table 1, for example, the two entries for turning 1212 free-machining plain carbon steel with uncoated carbide are 17/805 and 8/1075. These values indicate that a feed of 0.017 in/rev and a speed of 805 ft/min, or a feed of 0.008 in/rev and a speed of 1075 ft/min can be used for this material. The tool life, in each case, will be approximately 15 minutes. If one of these speed and feed pairs is assigned an arbitrary cutting time of 1 minute, then the relative cutting time of the second pair to the first is equal to the ratio of their respective feed × speed products. Here, the same amount of material that can be cut in 1 minute, at the higher feed and lower speed (17/805), will require 1.6 minutes at the lower feed and higher speed (8/1075) because $17 \times 805/(8 \times 1075) = 1.6$ minutes.

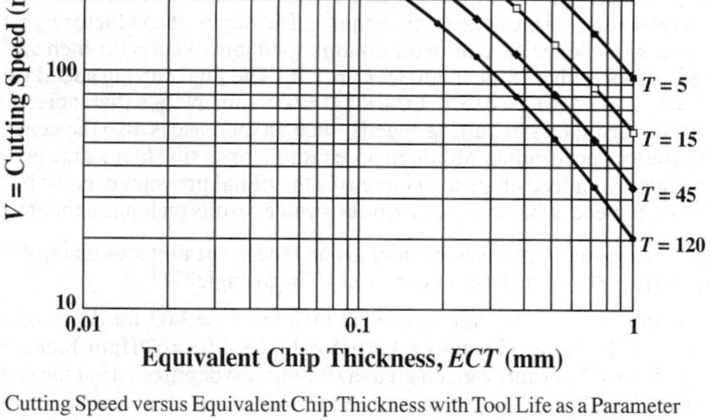


Cutting Speed versus Equivalent Chip Thickness with Tool Life as a Parameter

**Speeds and Feeds Tables for Turning.—**Speeds for HSS (high-speed steel) tools are based on a feed of 0.012 inch/rev and a depth of cut of 0.125 inch; use Table 5c to adjust the given speeds for other feeds and depths of cut. The combined feed/speed data in the remaining columns are based on a depth of cut of 0.1 inch, lead angle of 15 degrees, and nose radius of $^3\!/_{64}$ inch. Use Table 5a to adjust given speeds for other feeds, depths of cut, and lead angles; use Table 5b to adjust given speeds for increased tool life up to 180 minutes. Examples are given in the text.

*Examples Using the Speeds and Feeds Tables for Turning:* The examples that follow give instructions for determining cutting speeds for turning. In general, the same methods are also used to find cutting speeds for milling, drilling, reaming, and threading, so reading through these examples may bring some additional insight to those other metalworking processes as well. The first step in determining cutting speeds is to locate the work material in the left column of the appropriate table for turning, milling, or drilling, reaming, and threading.

*Example 1, Turning:* Find the cutting speed for turning SAE 1074 plain carbon steel of 225 to 275 BHN (Brinell Hardness Number), using an uncoated carbide insert, a feed of 0.015 in/rev, and a depth of cut of 0.1 inch.

In Table 1, speed and feed data for two types of uncoated carbide tools are given, one for hard tool grades, the other for tough tool grades. In general, use the speed data from the tool category that most closely matches the tool to be used because there are often significant differences in the speeds and feeds for different tool grades. From the uncoated carbide hard grade values, the *optimum* and *average* feed/speed data given in Table 1 are 17/615 and 8/815, or 0.017 in/rev at 615 ft/min and 0.008 in/rev at 815 ft/min. Because the selected feed (0.015 in/rev) is different from either of the feeds given in the table, the cutting speed must be adjusted to match the feed. The other cutting parameters to be used must also be compared with the general tool and cutting parameters given in the speed tables to determine if adjustments need to be made for these parameters as well. The general tool and cutting parameters for turning, given in the footnote to Table 1, are depth of cut = 0.1 inch, lead angle = 15°, and tool nose radius = $^3\!/_{64}$ inch.

Table 5a is used to adjust the cutting speeds for turning (from Table 1 through Table 9) for changes in feed, depth of cut, and lead angle. The new cutting speed $V$ is found from $V = V_{opt} \times F_f \times F_d$, where $V_{opt}$ is the *optimum* speed from the table (always the lower of the two speeds given), and $F_f$ and $F_d$ are the adjustment factors from Table 5a for feed and depth of cut, respectively.

To determine the two factors $F_f$ and $F_d$, calculate the ratio of the selected feed to the *optimum* feed, 0.015/0.017 = 0.9, and the ratio of the two given speeds $V_{avg}$ and $V_{opt}$, 815/615 = 1.35 (approximately). The feed factor $F_f$ = 1.07 is found in Table 5a at the intersection of the feed ratio row and the speed ratio column. The depth-of-cut factor $F_d$ = 1.0 is found in the same row as the feed factor in the column for depth of cut = 0.1 inch and lead angle = 15°, or for a tool with a 45° lead angle, $F_d$ = 1.18. The final cutting speed for a 15° lead angle is $V = V_{opt} \times F_f \times F_d = 615 \times 1.07 \times 1.0 = 658$ fpm. Notice that increasing the lead angle tends to permit higher cutting speeds; such an increase is also the general effect of increasing the tool nose radius, although nose radius correction factors are not included in this table. Increasing lead angle also increases the radial pressure exerted by the cutting tool on the workpiece, which may cause unfavorable results on long, slender workpieces.

*Example 2, Turning:* For the same material and feed as the previous example, what is the cutting speed for a 0.4-inch depth of cut and a 45° lead angle?

As before, the feed is 0.015 in/rev, so $F_f$ is 1.07, but $F_d$ = 1.03 for depth of cut equal to 0.4 inch and a 45° lead angle. Therefore, $V = 615 \times 1.07 \times 1.03 = 676$ fpm. Increasing the lead angle from 15° to 45° permits a much greater (four times) depth of cut, at the same feed and nearly constant speed. Tool life remains constant at 15 minutes. *(Continued on page 1092)*

## Table 1. Cutting Speeds and Feeds for Turning Plain Carbon and Alloy Steels

| Material AISI/SAE Designation | Brinell Hardness Number [a] | HSS Speed (fpm) | | Uncoated Carbide Hard Opt. | Uncoated Carbide Hard Avg. | Uncoated Carbide Tough Opt. | Uncoated Carbide Tough Avg. | Coated Carbide Hard Opt. | Coated Carbide Hard Avg. | Coated Carbide Tough Opt. | Coated Carbide Tough Avg. | Ceramic Hard Opt. | Ceramic Hard Avg. | Ceramic Tough Opt. | Ceramic Tough Avg. | Cermet Opt. | Cermet Avg. |
|---|---|---|---|---|---|---|---|---|---|---|---|---|---|---|---|---|---|
| | | | | f = feed (0.001 in/rev), s = speed (ft/min) | | | | | | *Metric Units:* f × 25.4 = mm/rev, s × 0.3048 = m/min | | | | | | | |
| Free-machining plain carbon steels (resulfurized): 1212, 1213, 1215 | 100-150 | 150 | f | 17 | 8 | 36 | 17 | 17 | 8 | 28 | 13 | 15 | 8 | 15 | 8 | 7 | 3 |
| | | | s | 805 | 1075 | 405 | 555 | 1165 | 1295 | 850 | 1200 | 3340 | 4985 | 1670 | 2500 | 1610 | 2055 |
| | 150-200 | 160 | f | 17 | 8 | 36 | 17 | 28 | 13 | 28 | 13 | 15 | 8 | 15 | 8 | 7 | 3 |
| | | | s | 745 | 935 | 345 | 470 | 915 | 1130 | 785 | 1110 | 1795 | 2680 | 1485 | 2215 | 1490 | 1815 |
| 1108, 1109, 1115, 1117, 1118, 1120, 1126, 1211 | 100-150 | 130 | | | | | | | | | | | | | | | |
| | 150-200 | 120 | f | 17 | 8 | 36 | 17 | 17 | 8 | 28 | 13 | 15 | 8 | 15 | 8 | 7 | 3 |
| | | | s | 730 | 990 | 300 | 430 | 1090 | 1410 | 780 | 1105 | 1610 | 2780 | 1345 | 2005 | 1355 | 1695 |
| 1132, 1137, 1139, 1140, 1144, 1146, 1151 | 175-225 | 120 | f | 17 | 8 | 36 | 17 | 17 | 8 | 28 | 13 | 13 | 7 | 13 | 7 | | |
| | | | s | 615 | 815 | 300 | 405 | 865 | 960 | 755 | 960 | 1400 | 1965 | 1170 | 1640 | | |
| | 275-325 | 75 | | | | | | | | | | | | | | | |
| | 325-375 | 50 | f | 17 | 8 | 36 | 17 | 17 | 8 | 28 | 13 | 10 | 5 | 10 | 5 | | |
| | | | s | 515 | 685 | 235 | 340 | 720 | 805 | 650 | 810 | 1430 | 1745 | 1070 | 1305 | | |
| | 375-425 | 40 | | | | | | | | | | | | | | | |
| (Leaded): 11L17, 11L18, 12L13, 12L14 | 100-150 | 140 | | | | | | | | | | | | | | | |
| | 150-200 | 145 | f | 17 | 8 | 36 | 17 | 28 | 13 | 28 | 13 | 15 | 8 | 15 | 8 | 7 | 3 |
| | | | s | 745 | 935 | 345 | 470 | 915 | 1130 | 785 | 1110 | 1795 | 2680 | 1485 | 2215 | 1490 | 1815 |
| | 200-250 | 110 | f | 17 | 8 | 36 | 17 | 17 | 8 | 28 | 13 | 13 | 7 | 13 | 7 | | |
| | | | s | 615 | 815 | 300 | 405 | 865 | 960 | 755 | 960 | 1400 | 1965 | 1170 | 1640 | | |
| Plain carbon steels: 1006, 1008, 1009, 1010, 1012, 1015, 1016, 1017, 1018, 1019, 1020, 1021, 1022, 1023, 1024, 1025, 1026, 1513, 1514 | 100-125 | 120 | f | 17 | 8 | 36 | 17 | 17 | 8 | 28 | 13 | 15 | 8 | 15 | 8 | 7 | 3 |
| | | | s | 805 | 1075 | 405 | 555 | 1165 | 1295 | 850 | 1200 | 3340 | 4985 | 1670 | 2500 | 1610 | 2055 |
| | 125-175 | 110 | f | 17 | 8 | 36 | 17 | 28 | 13 | 28 | 13 | 15 | 8 | 15 | 8 | 7 | 3 |
| | | | s | 745 | 935 | 345 | 470 | 915 | 1130 | 785 | 1110 | 1795 | 2680 | 1485 | 2215 | 1490 | 1815 |
| | 175-225 | 90 | | 17 | 8 | 36 | 17 | 17 | 8 | 28 | 13 | 13 | 7 | 13 | 7 | | |
| | | | | 615 | 815 | 300 | 405 | 865 | 960 | 755 | 960 | 1400 | 1965 | 1170 | 1640 | | |
| | 225-275 | 70 | f | | | | | | | | | | | | | | |
| | | | s | | | | | | | | | | | | | | |

**Table 1.** *(Continued)* **Cutting Speeds and Feeds for Turning Plain Carbon and Alloy Steels**

| Material AISI/SAE Designation | Brinell Hardness Number [a] | HSS Speed (fpm) | | Uncoated Carbide Hard Opt. | Uncoated Carbide Hard Avg. | Uncoated Carbide Tough Opt. | Uncoated Carbide Tough Avg. | Coated Carbide Hard Opt. | Coated Carbide Hard Avg. | Coated Carbide Tough Opt. | Coated Carbide Tough Avg. | Ceramic Hard Opt. | Ceramic Hard Avg. | Ceramic Tough Opt. | Ceramic Tough Avg. | Cermet Opt. | Cermet Avg. |
|---|---|---|---|---|---|---|---|---|---|---|---|---|---|---|---|---|---|
| | | | | f = feed (0.001 in/rev), s = speed (ft/min) *Metric Units:* f × 25.4 = mm/rev, s × 0.3048 = m/min | | | | | | | | | | | | | |
| Plain carbon steels *(continued)*: 1027, 1030, 1033, 1035, 1036, 1037, 1038, 1039, 1040, 1041, 1042, 1043, 1045, 1046, 1048, 1049, 1050, 1052, 1524, 1526, 1527, 1541 | 125-175 | 100 | f | 17 | 8 | 36 | 17 | 28 | 13 | 28 | 13 | 15 | 8 | 15 | 8 | 7 | 3 |
| | | | s | 745 | 935 | 345 | 470 | 915 | 1130 | 785 | 1110 | 1795 | 2680 | 1485 | 2215 | 1490 | 1815 |
| | 175-225 | 85 | | | | | | | | | | | | | | | |
| | 225-275 | 70 | f | 17 | 8 | 36 | 17 | 17 | 8 | 28 | 13 | 13 | 7 | 13 | 7 | | |
| | | | s | 615 | 815 | 300 | 405 | 865 | 960 | 755 | 960 | 1400 | 1965 | 1170 | 1640 | | |
| | 275-325 | 60 | | | | | | | | | | | | | | | |
| | 325-375 | 40 | f | 17 | 8 | 36 | 17 | 17 | 8 | 28 | 13 | 10 | 5 | 10 | 5 | | |
| | | | s | 515 | 685 | 235 | 340 | 720 | 805 | 650 | 810 | 1430 | 1745 | 1070 | 1305 | | |
| | 375-425 | 30 | | | | | | | | | | | | | | | |
| Plain carbon steels *(continued)*: 1055, 1060, 1064, 1065, 1070, 1074, 1078, 1080, 1084, 1086, 1090, 1095, 1548, 1551, 1552, 1561, 1566 | 125-175 | 100 | | | | | | | | | | | | | | | |
| | 175-225 | 80 | f | 17 | 8 | 36 | 17 | 17 | 8 | 28 | 13 | 15 | 8 | 15 | 8 | 7 | 3 |
| | | | s | 730 | 990 | 300 | 430 | 1090 | 1410 | 780 | 1105 | 1610 | 2780 | 1345 | 2005 | 1355 | 1695 |
| | 225-275 | 65 | f | 17 | 8 | 36 | 17 | 17 | 8 | 28 | 13 | 13 | 7 | 13 | 7 | 7 | 3 |
| | | | s | 615 | 815 | 300 | 405 | 865 | 960 | 755 | 960 | 1400 | 1965 | 1170 | 1640 | 1365 | 1695 |
| | 275-325 | 50 | | | | | | | | | | | | | | | |
| | 325-375 | 35 | f | 17 | 8 | 36 | 17 | 17 | 8 | 28 | 13 | 10 | 5 | 10 | 5 | | |
| | | | s | 515 | 685 | 235 | 340 | 720 | 805 | 650 | 810 | 1430 | 1745 | 1070 | 1305 | | |
| | 375-425 | 30 | | | | | | | | | | | | | | | |
| Free-machining alloy steels, (resulfurized): 4140, 4150 | 175-200 | 110 | | | | | | | | | | | | | | | |
| | 200-250 | 90 | f | 17 | 8 | 36 | 17 | 17 | 8 | 28 | 13 | 15 | 8 | 15 | 8 | 7 | 3 |
| | | | s | 525 | 705 | 235 | 320 | 505 | 525 | 685 | 960 | 1490 | 2220 | 1190 | 1780 | 1040 | 1310 |
| | 250-300 | 65 | f | 17 | 8 | 36 | 17 | 17 | 8 | 28 | 13 | 10 | 5 | 10 | 5 | 7 | 3 |
| | | | s | 355 | 445 | 140 | 200 | 630 | 850 | 455 | 650 | 1230 | 1510 | 990 | 1210 | 715 | 915 |
| | 300-375 | 50 | | | | | | | | | | | | | | | |
| | 375-425 | 40 | f | 17 | 8 | 36 | 17 | 17 | 8 | 28 | 13 | 8 | 4 | 8 | 4 | 7 | 3 |
| | | | s | 330 | 440 | 125 | 175 | 585 | 790 | 125 | 220 | 1200 | 1320 | 960 | 1060 | 575 | 740 |

**Table 1.** *(Continued)* **Cutting Speeds and Feeds for Turning Plain Carbon and Alloy Steels**

| Material AISI/SAE Designation | Brinell Hardness Number [a] | HSS Speed (fpm) | | Uncoated Carbide Hard Opt. | Uncoated Carbide Hard Avg. | Uncoated Carbide Tough Opt. | Uncoated Carbide Tough Avg. | Coated Carbide Hard Opt. | Coated Carbide Hard Avg. | Coated Carbide Tough Opt. | Coated Carbide Tough Avg. | Ceramic Hard Opt. | Ceramic Hard Avg. | Ceramic Tough Opt. | Ceramic Tough Avg. | Cermet Opt. | Cermet Avg. |
|---|---|---|---|---|---|---|---|---|---|---|---|---|---|---|---|---|---|
| | | | | f = feed (0.001 in/rev), s = speed (ft/min) *Metric Units:* f × 25.4 = mm/rev, s × 0.3048 = m/min | | | | | | | | | | | | | |
| Free-machining alloy steels: (leaded): 41L30, 41L40, 41L47, 41L50, 43L47, 51L32, 52L100, 86L20, 86L40 | 150-200 | 120 | f | 17 | 8 | 36 | 17 | 17 | 8 | 28 | 13 | 15 | 8 | 15 | 8 | 7 | 3 |
| | | | s | 730 | 990 | 300 | 430 | 1090 | 1410 | 780 | 1105 | 1610 | 2780 | 1345 | 2005 | 1355 | 1695 |
| | 200-250 | 100 | f | 17 | 8 | 36 | 17 | 17 | 8 | 28 | 13 | 13 | 7 | 13 | 7 | 7 | 3 |
| | | | s | 615 | 815 | 300 | 405 | 865 | 960 | 755 | 960 | 1400 | 1965 | 1170 | 1640 | 1355 | 1695 |
| | 250-300 | 75 | | | | | | | | | | | | | | | |
| | 300-375 | 55 | f | 17 | 8 | 36 | 17 | 17 | 8 | 28 | 13 | 10 | 5 | 10 | 5 | | |
| | | | s | 515 | 685 | 235 | 340 | 720 | 805 | 650 | 810 | 1430 | 1745 | 1070 | 1305 | | |
| | 375-425 | 50 | | | | | | | | | | | | | | | |
| Alloy steels: 4012, 4023, 4024, 4028, 4118, 4320, 4419, 4422, 4427, 4615, 4620, 4621, 4626, 4718, 4720, 4815, 4817, 4820, 5015, 5117, 5120, 6118, 8115, 8615, 8617, 8620, 8622, 8625, 8627, 8720, 8822, 94B17 | 125-175 | 100 | | | | | | | | | | | | | | | |
| | 175-225 | 90 | f | 17 | 8 | 36 | 17 | 17 | 8 | 28 | 13 | 15 | 8 | 15 | 8 | 7 | 3 |
| | | | s | 525 | 705 | 235 | 320 | 505 | 525 | 685 | 960 | 1490 | 2220 | 1190 | 1780 | 1040 | 1310 |
| | 225-275 | 70 | f | 17 | 8 | 36 | 17 | 17 | 8 | 28 | 13 | 10 | 5 | 10 | 5 | 7 | 3 |
| | | | s | 355 | 445 | 140 | 200 | 630 | 850 | 455 | 650 | 1230 | 1510 | 990 | 1210 | 715 | 915 |
| | 275-325 | 60 | f | 17 | 8 | 36 | 17 | 17 | 8 | 28 | 13 | 9 | 5 | 8 | 5 | 7 | 3 |
| | | | s | 330 | 440 | 135 | 190 | 585 | 790 | 240 | 350 | 1230 | 1430 | 990 | 1150 | 655 | 840 |
| | 325-35 | 50 | | | | | | | | | | | | | | | |
| | 375-425 | 30 (20) | f | 17 | 8 | 36 | 17 | 17 | 8 | 28 | 13 | 8 | 4 | 8 | 4 | 7 | 3 |
| | | | s | 330 | 440 | 125 | 175 | 585 | 790 | 125 | 220 | 1200 | 1320 | 960 | 1060 | 575 | 740 |
| Alloy steels: 1330, 1335, 1340, 1345, 4032, 4037, 4042, 4047, 4130, 4135, 4137, 4140, 4142, 4145, 4147, 4150, 4161, 4337, 4340, 50B44, 50B46, 50B50, 50B60, 5130, 5132, 5140, 5145, 5147, 5150, 5160, 51B60, 6150, 81B45, 8630, 8635, 8637, 8640, 8642, 8645, 8650, 8655, 8660, 8740, 9254, 9255, 9260, 9262, 94B30 E51100, E52100 use (HSS Speeds) | 175-225 | 85 (70) | f | 17 | 8 | 36 | 17 | 17 | 8 | 28 | 13 | 15 | 8 | 15 | 8 | 7 | 3 |
| | | | s | 525 | 705 | 235 | 320 | 505 | 525 | 685 | 960 | 1490 | 2220 | 1190 | 1780 | 1020 | 1310 |
| | 225-275 | 70 (65) | f | 17 | 8 | 36 | 17 | 17 | 8 | 28 | 13 | 10 | 5 | 10 | 5 | 7 | 3 |
| | | | s | 355 | 445 | 140 | 200 | 630 | 850 | 455 | 650 | 1230 | 1510 | 990 | 1210 | 715 | 915 |
| | 275-325 | 60 (50) | f | 17 | 8 | 36 | 17 | 17 | 8 | 28 | 13 | 9 | 5 | 8 | 5 | 7 | 3 |
| | | | s | 330 | 440 | 135 | 190 | 585 | 790 | 240 | 350 | 1230 | 1430 | 990 | 1150 | 655 | 840 |
| | 325-375 | 40 (30) | | | | | | | | | | | | | | | |
| | 375-425 | 30 (20) | f | 17 | 8 | 36 | 17 | 17 | 8 | 28 | 13 | 8 | 4 | 8 | 4 | 7 | 3 |
| | | | s | 330 | 440 | 125 | 175 | 585 | 790 | 125 | 220 | 1200 | 1320 | 960 | 1060 | 575 | 740 |

**Table 1.** *(Continued)* **Cutting Speeds and Feeds for Turning Plain Carbon and Alloy Steels**

| Material AISI/SAE Designation | Brinell Hardness Number [a] | HSS Speed (fpm) | | Uncoated Carbide Hard Opt. | Uncoated Carbide Hard Avg. | Uncoated Carbide Tough Opt. | Uncoated Carbide Tough Avg. | Coated Carbide Hard Opt. | Coated Carbide Hard Avg. | Coated Carbide Tough Opt. | Coated Carbide Tough Avg. | Ceramic Hard Opt. | Ceramic Hard Avg. | Ceramic Tough Opt. | Ceramic Tough Avg. | Cermet Opt. | Cermet Avg. |
|---|---|---|---|---|---|---|---|---|---|---|---|---|---|---|---|---|---|
| | | | | f = feed (0.001 in/rev), s = speed (ft/min) | | | | | *Metric Units:* f × 25.4 = mm/rev, s × 0.3048 = m/min | | | | | | | | |
| Ultra-high-strength steels (not AISI): AMS alloys 6421 (98B37 Mod.), 6422 (98BV40), 6424, 6427, 6428, 6430, 6432, 6433, 6434, 6436, and 6442; 300M and D6ac | 220-300 | 65 | | | | | | | | | | | | | | | |
| | 300-350 | 50 | f | 17 | 8 | 36 | 17 | 20 | 10 | 28 | 13 | | | 10 | 5 | 7 | 3 |
| | | | s | 220 | 295 | 100 | 150 | 355 | 525 | 600 | 865 | | | 660 | 810 | 570 | 740 |
| | 350-400 | 35 | f | 17 | 8 | 36 | 17 | 17 | 8 | 28 | 13 | | | 8 | 4 | 7 | 3 |
| | | | s | 165 | 185 | 55† | 105 | 325 | 350 | 175 | 260 | | | 660 | 730 | 445 | 560 |
| | 43-48 RC | 25 | | | | | | | | | | | | | | | |
| | 48-52 RC | 10 | f | | | 17 | 8 | | | | | 7 | 3 | 10 | 5 | | |
| | | | s | | | 55† | 90 | | | | | 385 | 645 | 270 | 500 | | |
| Maraging steels (not AISI): 18% Ni, Grades 200, 250, 300, and 350 | 250-325 | 60 | f | 17 | 8 | 36 | 17 | 20 | 10 | 28 | 13 | | | 10 | 5 | | |
| | | | s | 220 | 295 | 100 | 150 | 355 | 525 | 600 | 865 | 660 | 810 | 570 | 740 | | |
| | 50-52 RC | 10 | f | | | 17 | 8 | | | | | 7 | 3 | 10 | 5 | | |
| | | | s | | | 55 | 90 | | | | | 385‡ | 645 | 270 | 500 | | |
| Nitriding steels (not AISI): Nitralloy 125, 135, 135 Mod., 225, and 230, Nitralloy N, Nitralloy EZ, Nitrex 1 | 200-250 | 70 | f | 17 | 8 | 36 | 17 | 17 | 8 | 28 | 13 | 15 | 8 | 15 | 8 | 7 | 3 |
| | | | s | 525 | 705 | 235 | 320 | 505 | 525 | 685 | 960 | 1490 | 2220 | 1190 | 1780 | 1040 | 1310 |
| | 300-350 | 30 | f | 17 | 8 | 36 | 17 | 17 | 8 | 28 | 13 | 8 | 4 | 8 | 4 | 7 | 3 |
| | | | s | 330 | 440 | 125 | 175 | 585 | 790 | 125 | 220 | 1200 | 1320 | 960 | 1060 | 575 | 740 |

[a] Brinell Hardness Number given unless otherwise indicated by RC for hardness on the Rockwell C scale.

Speeds for HSS (high-speed steel) tools are based on a feed of 0.012 inch/rev and a depth of cut of 0.125 inch; use Table 5c to adjust the given speeds for other feeds and depths of cut. The combined feed/speed data in the remaining columns are based on a depth of cut of 0.1 inch, lead angle of 15 degrees, and nose radius of $^3/_{64}$ inch. Use Table 5a to adjust given speeds for other feeds, depths of cut, and lead angles; use Table 5b to adjust given speeds for increased tool life up to 180 minutes. Examples are given in the text.

The combined feed/speed data in this table are based on tool grades (identified in Table 16) as follows: uncoated carbides, hard = 17, tough = 19, † = 15; coated carbides, hard = 11, tough = 14; ceramics, hard = 2, tough = 3, ‡ = 4; cermet = 7.

## Table 2. Cutting Speeds and Feeds for Turning Tool Steels

| Material AISI Designation | Brinell Hardness Number[a] | Uncoated HSS Speed (fpm) | | Uncoated Carbide Hard Opt. | Uncoated Carbide Hard Avg. | Uncoated Carbide Tough Opt. | Uncoated Carbide Tough Avg. | Coated Carbide Hard Opt. | Coated Carbide Hard Avg. | Coated Carbide Tough Opt. | Coated Carbide Tough Avg. | Ceramic Hard Opt. | Ceramic Hard Avg. | Ceramic Tough Opt. | Ceramic Tough Avg. | Cermet Opt. | Cermet Avg. |
|---|---|---|---|---|---|---|---|---|---|---|---|---|---|---|---|---|---|
| | | | | f = feed (0.001 in/rev), s = speed (ft/min) | | | | *Metric Units:* f × 25.4 = mm/rev, s × 0.3048 = m/min | | | | | | | | | |
| Water hardening: W1, W2, W5 | 150-200 | 100 | | | | | | | | | | | | | | | |
| Shock resisting: S1, S2, S5, S6, S7 | 175-225 | 70 | | | | | | | | | | | | | | | |
| Cold-work, oil hardening: O1, O2, O6, O7 | 175-225 | 70 | f | 17 | 8 | 36 | 17 | 17 | 8 | 28 | 13 | 13 | 7 | 13 | 7 | 7 | 3 |
| Cold-work, high carbon, high chromium: D2, D3, D4, D5, D7 | 200-250 | 45 | s | 455 | 610 | 210 | 270 | 830 | 1110 | 575 | 805 | 935 | 1310 | 790 | 1110 | 915 | 1150 |
| Cold-work, air hardening: A2, A3, A8, A9, A10 | 200-250 | 70 | | | | | | | | | | | | | | | |
| A4, A6 | 200-250 | 55 | f | 17 | 8 | 36 | 17 | 17 | 8 | 28 | 13 | 13 | 7 | 13 | 7 | 7 | 3 |
| A7 | 225-275 | 45 | s | 445 | 490 | 170 | 235 | 705 | 940 | 515 | 770 | 660 | 925 | 750 | 1210 | 1150 | 1510 |
| | 150-200 | 80 | | | | | | | | | | | | | | | |
| | 200-250 | 65 | | | | | | | | | | | | | | | |
| Hot-work, chromium type: H10, H11, H12, H13, H14, H19 | 325-375 | 50 | f | 17 | 8 | 36 | 17 | 17 | 8 | 28 | 13 | | | 8 | 4 | 7 | 3 |
| | | | s | 165 | 185 | 55 | 105 | 325 | 350 | 175 | 260 | | | 660 | 730 | 445 | 560 |
| | 48-50 RC | 20 | | | | | | | | | | | | | | | |
| | 50-52 RC | 10 | f | | | 17 | 8 | | | | | 7 | 3 | 10 | 5 | | |
| | 52-56 RC | — | s | | | 55† | 90 | | | | | 385‡ | 645 | 270 | 500 | | |
| Hot-work, tungsten type: H21, H22, H23, H24, H25, H26 | 150-200 | 60 | | | | | | | | | | | | | | | |
| | 200-250 | 50 | f | 17 | 8 | 36 | 17 | 17 | 8 | 28 | 13 | 13 | 7 | 13 | 7 | 7 | 3 |
| Hot-work, molybdenum type: H41, H42, H43 | 150-200 | 55 | s | 445 | 490 | 170 | 235 | 705 | 940 | 515 | 770 | 660 | 925 | 750 | 1210 | 1150 | 1510 |
| | 200-250 | 45 | | | | | | | | | | | | | | | |
| Special purpose, low alloy: L2, L3, L6 | 150-200 | 75 | f | 17 | 8 | 36 | 17 | 17 | 8 | 28 | 13 | 13 | 7 | 13 | 7 | 7 | 3 |
| | | | s | 445 | 610 | 210 | 270 | 830 | 1110 | 575 | 805 | 935 | 1310 | 790 | 1110 | 915 | 1150 |
| Mold: P2, P3, P4, P5, P6, P26, P21 | 100-150 | 90 | f | 17 | 8 | 36 | 17 | 17 | 8 | 28 | 13 | 13 | 7 | 13 | 7 | 7 | 3 |
| | 150-200 | 80 | s | 445 | 610 | 210 | 270 | 830 | 1110 | 575 | 805 | 935 | 1310 | 790 | 1110 | 915 | 1150 |
| High-speed steel: M1, M2, M6, M10, T1, T2, T6 | 200-250 | 65 | | | | | | | | | | | | | | | |
| M3-1, M4 M7, M30, M33, M34, M36, M41, M42, M43, M44, M46, M47, T5, T8 | 225-275 | 55 | f | 17 | 8 | 36 | 17 | 17 | 8 | 28 | 13 | 13 | 7 | 13 | 7 | 7 | 3 |
| | | | s | 445 | 490 | 170 | 235 | 705 | 940 | 515 | 770 | 660 | 925 | 750 | 1210 | 1150 | 1510 |
| T15, M3-2 | 225-275 | 45 | | | | | | | | | | | | | | | |

[a] Brinell Hardness Number given unless otherwise indicated by RC for hardness on the Rockwell C scale.

Speeds for HSS (high-speed steel) tools are based on a feed of 0.012 inch/rev and a depth of cut of 0.125 inch; use Table 5c to adjust the given speeds for other feeds and depths of cut. The combined feed/speed data in the remaining columns are based on a depth of cut of 0.1 inch, lead angle of 15 degrees, and nose radius of 3⁄64 inch. Use Table 5a to adjust given speeds for other feeds, depths of cut, and lead angles; use Table 5b to adjust given speeds for increased tool life up to 180 minutes. Examples are given in the text. The combined feed/speed data in this table are based on tool grades (identified in Table 16) as follows: uncoated carbides, hard = 17, tough = 19, † = 15; coated carbides, hard = 11, tough = 14; ceramics, hard = 2, tough = 3, ‡ = 4; cermet = 7.

## Table 3. Cutting Speeds and Feeds for Turning Stainless Steels

| Material | Brinell Hardness Number | Uncoated HSS Speed (fpm) | | Uncoated Carbide Hard Opt. | Uncoated Carbide Hard Avg. | Uncoated Carbide Tough Opt. | Uncoated Carbide Tough Avg. | Coated Carbide Hard Opt. | Coated Carbide Hard Avg. | Coated Carbide Tough Opt. | Coated Carbide Tough Avg. | Cermet Opt. | Cermet Avg. |
|---|---|---|---|---|---|---|---|---|---|---|---|---|---|
| | | | | f = feed (0.001 in/rev), s = speed (ft/min) *Metric Units:* **f** × 25.4 = mm/rev, s × 0.3048 = m/min | | | | | | | | | |
| Free-machining stainless steel (Ferritic): 430F, 430FSe | 135-185 | 110 | **f** | 20 | 10 | 36 | 17 | 17 | 8 | 28 | 13 | 7 | 3 |
| | | | **s** | 480 | 660 | 370 | 395 | 755 | 945 | 640 | 810 | 790 | 995 |
| (Austenitic): 203EZ, 303, 303Se, 303MA, 303Pb, 303Cu, 303 Plus X | 135-185 | 100 | **f** | 13 | 7 | 36 | 17 | | | 28 | 13 | 7 | 3 |
| | 225-275 | 80 | **s** | 520 | 640 | 310 | 345 | | | 625 | 815 | 695 | 875 |
| (Martensitic): 416, 416Se, 416 Plus X, 420F, 420FSe, 440F, 440FSe | 135-185 | 110 | **f** | 13 | 7 | 36 | | | | 28 | 13 | 7 | 3 |
| | 185-240 | 100 | **s** | 520 | 640 | 310 | | | | 625 | 815 | 695 | 875 |
| | 275-325 | 60 | **f** | 13 | 7 | 36 | 17 | | | 28 | 13 | | |
| | 375-425 | 30 | **s** | 210 | 260 | 85 | 135 | | | 130 | 165 | | |
| Stainless steels (Ferritic): 405, 409 429, 430, 434, 436, 442, 446, 502 | 135-185 | 90 | **f** | 20 | 10 | 36 | 17 | 17 | 8 | 28 | 13 | 7 | 3 |
| | | | **s** | 480 | 660 | 370 | 395 | 755 | 945 | 640 | 810 | 790 | 995 |
| (Austenitic): 201, 202, 301, 302, 304, 304L, 305, 308, 321, 347, 348 | 135-185 | 75 | | | | | | | | | | | |
| | 225-275 | 65 | | | | | | | | | | | |
| (Austenitic): 302B, 309, 309S, 310, 310S, 314, 316, 316L, 317, 330 | 135-185 | 70 | **f** | 13 | 7 | 36 | 17 | | | 28 | 13 | 7 | 3 |
| | | | **s** | 520 | 640 | 310 | 345 | | | 625 | 815 | 695 | 875 |
| (Martensitic): 403, 410, 420, 501 | 135-175 | 95 | | | | | | | | | | | |
| | 175-225 | 85 | | | | | | | | | | | |
| | 275-325 | 55 | | | | | | | | | | | |
| | 375-425 | 35 | | | | | | | | | | | |
| (Martensitic): 414, 431, Greek Ascoloy, 440A, 440B, 440C | 225-275 | 55-60 | **f** | 13 | 7 | 36 | 17 | | | 28 | 13 | 13 | 7 |
| | 275-325 | 45-50 | **s** | 210 | 260 | 85 | 135 | | | 130 | 165 | 200† | 230 |
| | 375-425 | 30 | | | | | | | | | | | |
| (Precipitation hardening): 15-5PH, 17-4PH, 17-7PH, AF-71, 17-14CuMo, AFC-77, AM-350, AM-355, AM-362, Custom 455, HNM, PH13-8, PH14-8Mo, PH15-7Mo, Stainless W | 150-200 | 60 | **f** | 13 | 7 | 36 | 17 | | | 28 | 13 | 13 | 7 |
| | 275-325 | 50 | **s** | 520 | 640 | 310 | 345 | | | 625 | 815 | 695 | 875 |
| | 325-375 | 40 | **f** | 13 | 7 | 36 | 17 | | | | | | |
| | 375-450 | 25 | **s** | 195 | 240 | 85 | 155 | | | | | | |

See footnote to Table 1 for more information. The combined feed/speed data in this table are based on tool grades (identified in Table 16) as follows: uncoated carbides, hard = 17, tough = 19; coated carbides, hard = 11, tough = 14; cermet = 7, † = 18.

**Table 4a. Cutting Speeds and Feeds for Turning Ferrous Cast Metals**

| | | | Tool Material | | | | | | | | | | | | | | |
|---|---|---|---|---|---|---|---|---|---|---|---|---|---|---|---|---|---|
| | | HSS | | Uncoated Carbide | | Coated Carbide | | | | Ceramic | | | | Cermet | | CBN | |
| | Brinell | | | Tough | | Hard | | Tough | | Hard | | Tough | | | | | |
| | Hardness | Speed | | f = feed (0.001 in/rev), s = speed (ft/min) | | | | | | *Metric Units:* f × 25.4 = mm/rev, s × 0.3048 = m/min | | | | | | | |
| Material | Number | (fpm) | | *Opt.* | *Avg.* | *Opt.* | *Avg.* | *Opt.* | *Avg.* | *Opt.* | *Avg.* | *Opt.* | *Avg.* | *Opt.* | *Avg.* | *Opt.* | *Avg.* |
| Gray Cast Iron | | | | | | | | | | | | | | | | | |
| ASTM Class 20 | 120-150 | 120 | | | | | | | | | | | | | | | |
| ASTM Class 25 | 160-200 | 90 | **f** | 28 | 13 | 28 | 13 | 28 | 13 | 15 | 8 | 15 | 8 | 8 | 4 | 24 | 11 |
| ASTM Class 30, 35, and 40 | 190-220 | 80 | **s** | 240 | 365 | 665 | 1040 | 585 | 945 | 1490 | 2220 | 1180 | 1880 | 395 | 510 | 8490 | 36380 |
| ASTM Class 45 and 50 | 220-260 | 60 | **f** | 28 | 13 | 28 | 13 | 28 | 13 | 11 | 6 | 11 | 6 | 8 | 4 | 24 | 11 |
| ASTM Class 55 and 60 | 250-320 | 35 | **s** | 160 | 245 | 400 | 630 | 360 | 580 | 1440 | 1880 | 1200 | 1570 | 335 | 420 | 1590 | 2200 |
| ASTM Type 1, 1b, 5 (Ni resist) | 100-215 | 70 | | | | | | | | | | | | | | | |
| ASTM Type 2, 3, 6 (Ni resist) | 120-175 | 65 | **f** | 28 | 13 | | | 28 | 13 | 15 | 8 | 15 | 8 | 8 | 4 | | |
| ASTM Type 2b, 4 (Ni resist) | 150-250 | 50 | **s** | 110 | 175 | | | 410 | 575 | 1060 | 1590 | 885 | 1320 | 260 | 325 | | |
| Malleable Iron | | | | | | | | | | | | | | | | | |
| (Ferritic): 32510, 35018 | 110-160 | 130 | **f** | 28 | 13 | 28 | 13 | 28 | 13 | 15 | 8 | 15 | 8 | | | | |
| | | | **s** | 180 | 280 | 730 | 940 | 660 | 885 | 1640 | 2450 | 1410 | 2110 | | | | |
| (Pearlitic): 40010, 43010, 45006, 45008, 48005, 50005 | 160-200 | 95 | **f** | 28 | 13 | 28 | 13 | 28 | 13 | 13 | 7 | 13 | 7 | | | | |
| | 200-240 | 75 | **s** | 125 | 200 | 335 | 505 | 340 | 510 | 1640 | 2310 | 1400 | 1970 | | | | |
| (Martensitic): 53004, 60003, 60004 | 200-255 | 70 | | | | | | | | | | | | | | | |
| (Martensitic): 70002, 70003 | 220-260 | 60 | **f** | 28 | 13 | | | 28 | 13 | 11 | 6 | 11 | 6 | | | | |
| (Martensitic): 80002 | 240-280 | 50 | **s** | 100 | 120 | | | 205 | 250 | 1720 | 2240 | 1460 | 1910 | | | | |
| (Martensitic): 90001 | 250-320 | 30 | | | | | | | | | | | | | | | |

Speeds for HSS (high-speed steel) tools are based on a feed of 0.012 inch/rev and a depth of cut of 0.125 inch; use Table 5c to adjust the given speeds for other feeds and depths of cut. The combined feed/speed data in the remaining columns are based on a depth of cut of 0.1 inch, lead angle of 15 degrees, and nose radius of $^3/_{64}$ inch. Use Table 5a to adjust the given speeds for other feeds, depths of cut, and lead angles; use Table 5b to adjust given speeds for increased tool life up to 180 minutes. Examples are given in the text.

The combined feed/speed data in this table are based on tool grades (identified in Table 16) as follows: uncoated carbides, tough = 15; Coated carbides, hard = 11, tough = 14; ceramics, hard = 2, tough = 3; cermet = 7; CBN = 1.

## Table 4b. Cutting Speeds and Feeds for Turning Ferrous Cast Metals

| Material | Brinell Hardness Number | Uncoated HSS Speed (fpm) | | Uncoated Carbide Hard Opt. | Uncoated Carbide Hard Avg. | Uncoated Carbide Tough Opt. | Uncoated Carbide Tough Avg. | Coated Carbide Hard Opt. | Coated Carbide Hard Avg. | Coated Carbide Tough Opt. | Coated Carbide Tough Avg. | Ceramic Hard Opt. | Ceramic Hard Avg. | Ceramic Tough Opt. | Ceramic Tough Avg. | Cermet Opt. | Cermet Avg. |
|---|---|---|---|---|---|---|---|---|---|---|---|---|---|---|---|---|---|
| | | | | f = feed (0.001 in/rev), s = speed (ft/min) *Metric Units:* f × 25.4 = mm/rev, s × 0.3048 = m/min | | | | | | | | | | | | | |
| **Nodular (Ductile) Iron** | | | | | | | | | | | | | | | | | |
| (Ferritic): 60-40-18, 65-45-12 | 140-190 | 100 | f | | | 28 | 13 | 28 | 13 | 28 | 13 | 15 | 8 | 15 | 8 | 8 | 4 |
| | | | s | | | 200 | 325 | 490 | 700 | 435 | 665 | 970 | 1450 | 845 | 1260 | 365 | 480 |
| (Ferritic-Pearlitic): 80-55-06 | 190-225<br>225-260 | 80<br>65 | f | | | 28 | 13 | 28 | 13 | 28 | 13 | 11 | 6 | 11 | 6 | 8 | 4 |
| (Pearlitic-Martensitic): 100-70-03 | 240-300 | 45 | s | | | 130 | 210 | 355 | 510 | 310 | 460 | 765 | 995 | 1260 | 1640 | 355 | 445 |
| (Martensitic): 120-90-02 | 270-330<br>300-400 | 30<br>15 | f | | | 28 | 13 | | | 28 | 13 | 10 | 5 | 10 | 5 | 8 | 4 |
| | | | s | | | 40 | 65 | | | 145 | 175 | 615 | 750 | 500 | 615 | 120 | 145 |
| **Cast Steels** | | | | | | | | | | | | | | | | | |
| (Low-carbon): 1010, 1020 | 100-150<br>125-175 | 110<br>100 | f | 17 | 8 | 36 | 17 | 17 | 8 | 28 | 13 | 15 | 8 | | | 7 | 3 |
| | | | s | 370 | 490 | 230 | 285 | 665 | 815 | 495 | 675 | 2090 | 3120 | | | 625 | 790 |
| (Medium-carbon): 1030, 1040, 1050 | 175-225<br>225-300 | 9070 | | | | | | | | | | | | | | | |
| (Low-carbon alloy): 1320, 2315, 2320, 4110, 4120, 4320, 8020, 8620 | 150-200<br>200-250<br>250-300 | 90<br>80<br>60 | f | 17 | 8 | 36 | 17 | 17 | 8 | 28 | 13 | 15 | 8 | | | 7 | 3 |
| | | | s | 370 | 490 | 150 | 200 | 595 | 815 | 410 | 590 | 1460 | 2170 | | | 625 | 790 |
| (Medium-carbon alloy): 1330, 1340, 2325, 2330, 4125, 4130, 4140, 4330, 4340, 8030, 80B30, 8040, 8430, 8440, 8630, 8640, 9525, 9530, 9535 | 175-225<br>225-250 | 80<br>70 | f | 17 | 8 | 36 | 17 | 17 | 8 | | | 15 | 8 | | | | |
| | | | s | 310 | 415 | 115 | 150 | 555 | 760 | | | 830 | 1240 | | | | |
| | 250-300<br>300-350 | 55<br>45 | f | | | 28 | 13 | | | | | 15 | 8 | | | | |
| | | | s | | | 70† | 145 | | | | | 445 | 665 | | | | |
| | 350-400 | 30 | f | | | 28 | 13 | | | 28 | 13 | | | 15 | 8 | | |
| | | | s | | | 115† | 355 | | | 335 | 345 | | | 955 | 1430 | | |

The combined feed/speed data in this table are based on tool grades (identified in Table 16) as shown: uncoated carbides, hard = 17; tough = 19, † = 15; coated carbides, hard = 11; tough = 14; ceramics, hard = 2; tough = 3; cermet = 7. Also, see footnote to Table 4a.

**Table 5a. Turning Speed Adjustment Factors for Feed, Depth of Cut, and Lead Angle**

| Ratio of Chosen Feed to *Optimum* Feed | Ratio of the two cutting speeds given in the tables $V_{avg}/V_{opt}$ | | | | | | | Depth of Cut and Lead Angle | | | | | | | | | |
|---|---|---|---|---|---|---|---|---|---|---|---|---|---|---|---|---|---|
| | | | | | | | | 1 in. (25.4 mm) | | 0.4 in. (10.2 mm) | | 0.2 in. (5.1 mm) | | 0.1 in. (2.5 mm) | | 0.04 in. (1.0 mm) | |
| | 1.00 | 1.10 | 1.25 | 1.35 | 1.50 | 1.75 | 2.00 | 15° | 45° | 15° | 45° | 15° | 45° | 15° | 45° | 15° | 45° |
| | Feed Factor, $F_f$ | | | | | | | Depth of Cut and Lead Angle Factor, $F_d$ | | | | | | | | | |
| 1.00 | 1.0 | 1.0 | 1.0 | 1.0 | 1.0 | 1.0 | 1.0 | 0.74 | 1.0 | 0.79 | 1.03 | 0.85 | 1.08 | 1.0 | 1.18 | 1.29 | 1.35 |
| 0.90 | 1.00 | 1.02 | 1.05 | 1.07 | 1.09 | 1.10 | 1.12 | 0.75 | 1.0 | 0.80 | 1.03 | 0.86 | 1.08 | 1.0 | 1.17 | 1.27 | 1.34 |
| 0.80 | 1.00 | 1.03 | 1.09 | 1.10 | 1.15 | 1.20 | 1.25 | 0.77 | 1.0 | 0.81 | 1.03 | 0.87 | 1.07 | 1.0 | 1.15 | 1.25 | 1.31 |
| 0.70 | 1.00 | 1.05 | 1.13 | 1.22 | 1.22 | 1.32 | 1.43 | 0.77 | 1.0 | 0.82 | 1.03 | 0.87 | 1.08 | 1.0 | 1.15 | 1.24 | 1.30 |
| 0.60 | 1.00 | 1.08 | 1.20 | 1.25 | 1.35 | 1.50 | 1.66 | 0.78 | 1.0 | 0.82 | 1.03 | 0.88 | 1.07 | 1.0 | 1.14 | 1.23 | 1.29 |
| 0.50 | 1.00 | 1.10 | 1.25 | 1.35 | 1.50 | 1.75 | 2.00 | 0.78 | 1.0 | 0.82 | 1.03 | 0.88 | 1.07 | 1.0 | 1.14 | 1.23 | 1.28 |
| 0.40 | 1.00 | 1.09 | 1.28 | 1.44 | 1.66 | 2.03 | 2.43 | 0.78 | 1.0 | 0.84 | 1.03 | 0.89 | 1.06 | 1.0 | 1.13 | 1.21 | 1.26 |
| 0.30 | 1.00 | 1.06 | 1.32 | 1.52 | 1.85 | 2.42 | 3.05 | 0.81 | 1.0 | 0.85 | 1.02 | 0.90 | 1.06 | 1.0 | 1.12 | 1.18 | 1.23 |
| 0.20 | 1.00 | 1.00 | 1.34 | 1.60 | 2.07 | 2.96 | 4.03 | 0.84 | 1.0 | 0.89 | 1.02 | 0.91 | 1.05 | 1.0 | 1.10 | 1.15 | 1.19 |
| 0.10 | 1.00 | 0.80 | 1.20 | 1.55 | 2.24 | 3.74 | 5.84 | 0.88 | 1.0 | 0.91 | 1.01 | 0.92 | 1.03 | 1.0 | 1.06 | 1.10 | 1.12 |

Use with Table 1 through Table 9. Not for HSS tools. Table 1 through Table 9 data, except for HSS tools, are based on depth of cut = 0.1 inch, lead angle = 15 degrees, and tool life = 15 minutes. For other depths of cut, lead angles, or feeds, use the two feed/speed pairs from the tables and calculate the ratio of desired (new) feed to *optimum* feed (largest of the two feeds given in the tables), and the ratio of the two cutting speeds ($V_{avg}/V_{opt}$). Use the value of these ratios to find the feed factor $F_f$ at the intersection of the feed ratio row and the speed ratio column in the left half of the table. The depth-of-cut factor $F_d$ is found in the same row as the feed factor in the right half of the table under the column corresponding to the depth of cut and lead angle. The adjusted cutting speed can be calculated from $V = V_{opt} \times F_f \times F_d$, where $V_{opt}$ is the smaller (*optimum*) of the two speeds from the speed table (from the left side of the column containing the two feed/speed pairs). See the text for examples.

**Table 5b. Tool Life Factors for Turning with Carbides, Ceramics, Cermets, CBN, and Polycrystalline Diamond**

| Tool Life, $T$ (minutes) | Turning with Carbides: Workpiece < 300 BHN | | | Turning with Carbides: Workpiece > 300 BHN; Turning with Ceramics: Any Hardness | | | Turning with Mixed Ceramics: Any Workpiece Hardness | | |
|---|---|---|---|---|---|---|---|---|---|
| | $f_s$ | $f_m$ | $f_l$ | $f_s$ | $f_m$ | $f_l$ | $f_s$ | $f_m$ | $f_l$ |
| 15 | 1.0 | 1.0 | 1.0 | 1.0 | 1.0 | 1.0 | 1.0 | 1.0 | 1.0 |
| 45 | 0.86 | 0.81 | 0.76 | 0.80 | 0.75 | 0.70 | 0.89 | 0.87 | 0.84 |
| 90 | 0.78 | 0.71 | 0.64 | 0.70 | 0.63 | 0.56 | 0.82 | 0.79 | 0.75 |
| 180 | 0.71 | 0.63 | 0.54 | 0.61 | 0.53 | 0.45 | 0.76 | 0.72 | 0.67 |

Except for HSS speed tools, speeds and feeds given in Table 1 through Table 9 are based on 15-minute tool life. To adjust speeds for another tool life, multiply the cutting speed for 15-minute tool life $V_{15}$ by the tool life factor from this table according to the following rules: for small feeds where feed ≤ $\frac{1}{2}f_{opt}$, the cutting speed for desired tool life is $V_T = f_s \times V_{15}$; for medium feeds where $\frac{1}{2}f_{opt}$ < feed < $\frac{3}{4}f_{opt}$, $V_T = f_m \times V_{15}$; and for larger feeds where $\frac{3}{4}f_{opt}$ ≤ feed ≤ $f_{opt}$, $V_T = f_l \times V_{15}$. Here, $f_{opt}$ is the largest (*optimum*) feed of the two feed/speed values given in the speed tables.

**Table 5c. Cutting Speed Adjustment Factors for Turning with HSS Tools**

| Feed | | Feed Factor | Depth of Cut | | Depth-of-Cut Factor |
|---|---|---|---|---|---|
| in. | mm | $F_f$ | in. | mm | $F_d$ |
| 0.002 | 0.05 | 1.50 | 0.005 | 0.13 | 1.50 |
| 0.003 | 0.08 | 1.50 | 0.010 | 0.25 | 1.42 |
| 0.004 | 0.10 | 1.50 | 0.016 | 0.41 | 1.33 |
| 0.005 | 0.13 | 1.44 | 0.031 | 0.79 | 1.21 |
| 0.006 | 0.15 | 1.34 | 0.047 | 1.19 | 1.15 |
| 0.007 | 0.18 | 1.25 | 0.062 | 1.57 | 1.10 |
| 0.008 | 0.20 | 1.18 | 0.078 | 1.98 | 1.07 |
| 0.009 | 0.23 | 1.12 | 0.094 | 2.39 | 1.04 |
| 0.010 | 0.25 | 1.08 | 0.100 | 2.54 | 1.03 |
| 0.011 | 0.28 | 1.04 | 0.125 | 3.18 | 1.00 |
| 0.012 | 0.30 | 1.00 | 0.150 | 3.81 | 0.97 |
| 0.013 | 0.33 | 0.97 | 0.188 | 4.78 | 0.94 |
| 0.014 | 0.36 | 0.94 | 0.200 | 5.08 | 0.93 |
| 0.015 | 0.38 | 0.91 | 0.250 | 6.35 | 0.91 |
| 0.016 | 0.41 | 0.88 | 0.312 | 7.92 | 0.88 |
| 0.018 | 0.46 | 0.84 | 0.375 | 9.53 | 0.86 |
| 0.020 | 0.51 | 0.80 | 0.438 | 11.13 | 0.84 |
| 0.022 | 0.56 | 0.77 | 0.500 | 12.70 | 0.82 |
| 0.025 | 0.64 | 0.73 | 0.625 | 15.88 | 0.80 |
| 0.028 | 0.71 | 0.70 | 0.688 | 17.48 | 0.78 |
| 0.030 | 0.76 | 0.68 | 0.750 | 19.05 | 0.77 |
| 0.032 | 0.81 | 0.66 | 0.812 | 20.62 | 0.76 |
| 0.035 | 0.89 | 0.64 | 0.938 | 23.83 | 0.75 |
| 0.040 | 1.02 | 0.60 | 1.000 | 25.40 | 0.74 |
| 0.045 | 1.14 | 0.57 | 1.125 | 28.58 | 0.73 |
| 0.050 | 1.27 | 0.55 | 1.250 | 31.75 | 0.72 |
| 0.060 | 1.52 | 0.50 | 1.375 | 34.93 | 0.71 |

For use with HSS tool data only from Table 1 through Table 9. Adjusted cutting speed $V = V_{HSS} \times F_f \times F_d$, where $V_{HSS}$ is the tabular speed for turning with high-speed tools.

*Example 3, Turning:* Determine the cutting speed for turning 1055 steel of 175 to 225 BHN (Brinell Hardness Number) using a hard ceramic insert, a 15° lead angle, a 0.04-inch depth of cut and 0.0075 in/rev feed.

The two feed/speed combinations given in Table 1 on page 1083 for 1055 steel are 15/1610 and 8/2780, corresponding to 0.015 in/rev at 1610 fpm and 0.008 in/rev at 2780 fpm, respectively. In Table 5a, the feed factor $F_f = 1.75$ is found at the intersection of the row corresponding to feed/$f_{opt}$ = 7.5/15 = 0.5 and the column corresponding to $V_{avg}/V_{opt}$ = 2780/1610 = 1.75 (approximately). The depth-of-cut factor $F_d = 1.23$ is found in the same row, under the column heading for a depth of cut = 0.04 inch and lead angle = 15°. The adjusted cutting speed is $V = 1610 \times 1.75 \times 1.23 = 3466$ fpm.

*Example 4, Turning:* The cutting speed for 1055 steel calculated in Example 3 represents the speed required to obtain a 15-minute tool life. Estimate the cutting speed needed to obtain a tool life of 45, 90, and 180 minutes using the results of Example 3.

To estimate the cutting speed corresponding to another tool life, multiply the cutting speed for 15-minute tool life $V_{15}$ by the adjustment factor from the Table 5b, Tool Life Factors for Turning. This table gives three factors for adjusting tool life based on the feed used, $f_s$ for feeds less than or equal to $\frac{1}{2} f_{opt}$, $f_m$ for midrange feeds between $\frac{1}{2}$ and $\frac{3}{4} f_{opt}$ and $f_l$ for large feeds greater than or equal to $\frac{3}{4} f_{opt}$ and less than $f_{opt}$. In Example 3, $f_{opt}$ is 0.015 in/rev and the selected feed is 0.0075 in/rev = $\frac{1}{2} f_{opt}$. The new cutting speeds for the various tool lives are obtained by multiplying the cutting speed for 15-minute tool life $V_{15}$ by the factor for small feeds $f_s$ from the column for turning with ceramics in Table 5b. These calculations, using the cutting speed obtained in Example 3, follow.

| *Tool Life* | *Cutting Speed* |
|---|---|
| 15 min | $V_{15} = 3466$ fpm |
| 45 min | $V_{45} = V_{15} \times 0.80 = 2773$ fpm |
| 90 min | $V_{90} = V_{15} \times 0.70 = 2426$ fpm |
| 180 min | $V_{180} = V_{15} \times 0.61 = 2114$ fpm |

Depth of cut, feed, and lead angle remain the same as in Example 3. Notice, increasing the tool life from 15 to 180 minutes, a factor of 12, reduces the cutting speed by only about one-third of the $V_{15}$ speed.

**Table 6. Cutting Speeds and Feeds for Turning Copper Alloys**

| Group 1 |
|---|
| Architectural bronze (C38500); Extra-high-headed brass (C35600); Forging brass (C37700); Free-cutting phosphor bronze, B2 (C54400); Free-cutting brass (C36000); Free-cutting Muntz metal (C37000); High-leaded brass (C33200; C34200); High-leaded brass tube (C35300); Leaded commercial bronze (C31400); Leaded naval brass (C48500); Medium-leaded brass (C34000) |
| **Group 2** |
| Aluminum brass, arsenical (C68700); Cartridge brass, 70% (C26000); High-silicon bronze, B (C65500); Admiralty brass (inhibited) (C44300, C44500); Jewelry bronze, 87.5% (C22600); Leaded Muntz metal (C36500, C36800); Leaded nickel silver (C79600); Low brass, 80% (C24000); Low-leaded brass (C33500); Low-silicon bronze, B (C65100); Manganese bronze, A (C67500); Muntz metal, 60% (C28000); Nickel silver, 55-18 (C77000); Red brass, 85% (C23000); Yellow brass (C26800) |
| **Group 3** |
| Aluminum bronze, D (C61400); Beryllium copper (C17000, C17200, C17500); Commercialbronze, 90% (C22000); Copper nickel, 10% (C70600); Copper nickel, 30% (C71500); Electrolytic tough pitch copper (C11000); Guilding, 95% (C21000); Nickel silver, 65-10 (C74500); Nickel silver, 65-12 (C75700); Nickel silver, 65-15 (C75400); Nickel silver, 65-18 (C75200); Oxygen-free copper (C10200) ; Phosphor bronze, 1.25% (C50200); Phosphor bronze, 10% D (C52400) Phosphor bronze, 5% A (C51000); Phosphor bronze, 8% C (C52100); Phosphorus deoxidized copper (C12200) |

| Wrought Alloys Description and UNS Alloy Numbers | Material Condition | HSS Speed (fpm) | | Uncoated Carbide Opt. | Uncoated Carbide Avg. | Polycrystalline Diamond Opt. | Polycrystalline Diamond Avg. |
|---|---|---|---|---|---|---|---|
| | | | | f = feed (0.001 in/rev), s = speed (ft/min) *Metric Units:* f × 25.4 = mm/rev, s × 0.3048 = m/min | | | |
| Group 1 | A<br>CD | 300<br>350 | f<br>s | 28<br>1170 | 13<br>1680 | | |
| Group 2 | A<br>CD | 200<br>250 | f<br>s | 28<br>715 | 13<br>900 | | |
| Group 3 | A<br>CD | 100<br>110 | f<br>s | 28<br>440 | 13<br>610 | 7<br>1780 | 13<br>2080 |

Abbreviations designate: A, annealed; CD, cold drawn.

The combined feed/speed data in this table are based on tool grades (identified in Table 16) as follows: uncoated carbide, 15; diamond, 9. See the footnote to Table 7.

**Table 7. Cutting Speeds and Feeds for Turning Titanium and Titanium Alloys**

| Material | Brinell Hardness Number | HSS Speed (fpm) | | Uncoated Carbide (Tough) Opt. | Uncoated Carbide (Tough) Avg. |
|---|---|---|---|---|---|
| | | | | f = feed (0.001 in/rev), s = speed (ft/min) *Metric Units:* f × 25.4 = mm/rev, s × 0.3048 = m/min | |
| **Commercially Pure and Low Alloyed** | | | | | |
| 99.5Ti, 99.5Ti-0.15Pd | 110-150 | 100-105 | f<br>s | 28<br>55 | 13<br>190 |
| 99.1Ti, 99.2Ti, 99.2Ti-0.15Pd, 98.9Ti-0.8Ni-0.3Mo | 180-240 | 85-90 | f<br>s | 28<br>50 | 13<br>170 |
| 99.0 Ti | 250-275 | 70 | f<br>s | 20<br>75 | 10<br>210 |
| **Alpha Alloys and Alpha-Beta Alloys** | | | | | |
| 5Al-2.5Sn, 8Mn, 2Al-11Sn-5Zr-1Mo, 4Al-3Mo-1V, 5Al-6Sn-2Zr-1Mo, 6Al-2Sn-4Zr-2Mo, 6Al-2Sn-4Zr-6Mo, 6Al-2Sn-4Zr-2Mo-0.25Si | 300-350 | 50 | f<br>s | 17<br>95 | 8<br>250 |
| 6Al-4V | 310-350 | 40 | | | |
| 6Al-6V-2Sn, Al-4Mo, | 320-370 | 30 | | | |
| 8V-5Fe-IAl | 320-380 | 20 | | | |
| 6Al-4V, 6Al-2Sn-4Zr-2Mo, 6Al-2Sn-4Zr-6Mo, 6Al-2Sn-4Zr-2Mo-0.25Si | 320-380 | 40 | | | |
| 4Al-3Mo-1V, 6Al-6V-2Sn, 7Al-4Mo | 375-420 | 20 | | | |
| I Al-8V-5Fe | 375-440 | 20 | | | |
| **Beta Alloys** | | | | | |
| 13V-11Cr-3Al, 8Mo-8V-2Fe-3Al, 3Al-8V-6Cr-4Mo-4Zr, 11.5Mo-6Zr-4.5Sn | 275-350<br>375-440 | 25<br>20 | f<br>s | 17<br>55 | 8<br>150 |

The speed recommendations for turning with HSS (high-speed steel) tools may be used as starting speeds for milling titanium alloys, using Table 15a to estimate the feed required. Speeds for HSS (high-speed steel) tools are based on a feed of 0.012 inch/rev and a depth of cut of 0.125 inch; use Table 5c to adjust the given speeds for other feeds and depths of cut. The combined feed/speed data in the remaining columns are based on a depth of cut of 0.1 inch, lead angle of 15 degrees, and nose radius of $\frac{3}{64}$ inch. Use Table 5a to adjust given speeds for other feeds, depths of cut, and lead angles; use Table 5b to adjust given speeds for increased tool life up to 180 minutes. Examples are given in the text. The combined feed/speed data in this table are based on tool grades (identified in Table 16) as follows: uncoated carbide, 15.

**Table 8. Cutting Speeds and Feeds for Turning Light Metals**

| Material Description | Material Condition | HSS Speed (fpm) | | Uncoated Carbide (Tough) Opt. | Uncoated Carbide (Tough) Avg. | Polycrystalline Diamond Opt. | Polycrystalline Diamond Avg. |
|---|---|---|---|---|---|---|---|
| | | | | f = feed (0.001 in/rev), s = speed (ft/min) *Metric:* f × 25.4 = mm/rev, s × 0.3048 = m/min | | | |
| All wrought and cast magnesium alloys | A, CD, ST, and A | 800 | | | | | |
| All wrought aluminum alloys, including 6061-T651, 5000, 6000, and 7000 series | CD<br>ST and A | 600<br>500 | f<br>s | 36<br>2820 | 17<br>4570 | | |
| All aluminum sand and permanent mold casting alloys | AC<br>ST and A | 750<br>600 | | | | | |
| **Aluminum Die-Casting Alloys** | | | | | | | |
| Alloys 308.0 and 319.0 | — | — | f<br>s | 36<br>865 | 17<br>1280 | 11<br>5890[a] | 8<br>8270 |
| Alloys 390.0 and 392.0 | AC<br>ST and A | 80<br>60 | f<br>s | 24<br>2010 | 11<br>2760 | 8<br>4765 | 4<br>5755 |
| Alloy 413 | — | — | f<br>s | 32<br>430 | 15<br>720 | 10<br>5085 | 5<br>6570 |
| All other aluminum die-casting alloys including alloys 360.0 and 380.0 | ST and A | 100 | | | | | |
| | AC | 125 | f<br>s | 36<br>630 | 17<br>1060 | 11<br>7560 | 6<br>9930 |

[a]The speeds and feeds for turning Al alloys 308.0 and 319.0 with (polycrystalline) diamond tooling represent an expected tool life $T = 960$ minutes = 16 hours; corresponding speeds and feeds for 15-minute tool life are 11⁄28600 and 6⁄37500.

Abbreviations for material condition: A, annealed; AC, as cast; CD, cold drawn; and ST and A, solution treated and aged, respectively. Speeds for HSS (high-speed steel) tools are based on a feed of 0.012 inch/rev and a depth of cut of 0.125 inch; use Table 5c to adjust the HSS speeds for other feeds and depths of cut. The combined feed/speed data are based on a depth of cut of 0.1 inch, lead angle of 15 degrees, and nose radius of 3⁄64 inch. Use Table 5a to adjust given speeds for other feeds, depths of cut, and lead angles; use Table 5b to adjust given speeds for increased tool life up to 180 minutes. The data are based on tool grades (identified in Table 16) as follows: uncoated carbide, 15; diamond, 9.

## Table 9. Cutting Speeds and Feeds for Turning Superalloys

| | Tool Material | | | | | | | | | | |
|---|---|---|---|---|---|---|---|---|---|---|---|
| | HSS Turning | | | Uncoated Carbide | | Ceramic | | | | | |
| | Rough | Finish | | Tough | | Hard | | Tough | | CBN | |
| | | | | f = feed (0.001 in/rev), s = speed (ft/min)<br>*Metric Units:* f × 25.4 = mm/rev, s × 0.3048 = m/min | | | | | | | |
| Material Description | Speed (fpm) | | | *Opt.* | *Avg.* | *Opt.* | *Avg.* | *Opt.* | *Avg.* | *Opt.* | *Avg.* |
| T-D Nickel | 70-80 | 80-100 | | | | | | | | | |
| Discalloy | 15-35 | 35-40 | | | | | | | | | |
| 19-9DL, W-545 | 25-35 | 30-40 | f | 24 | 11 | | | | | 20 | 10 |
| 16-25-6, A-286, Incoloy 800, 801, and 802, V-57 { | 30-35 | 35-40 | s | 90 | 170 | | | | | 365 | 630 |
| Refractaloy 26 | 15-20 | 20-25 | f | 20 | 10 | | | | | 20 | 10 |
| J1300 | 15-25 | 20-30 | s | 75 | 135 | | | | | 245 | 420 |
| Inconel 700 and 702, Nimonic 90 and 95 { | 10-12 | 12-15 | | | | | | | | | |
| S-816, V-36 | 10-15 | 15-20 | | | | | | | | | |
| S-590 | 10-20 | 15-30 | | | | | | | | | |
| Udimet 630 | | 20-25 | | | | | | | | | |
| N-155 { | | 15-25 | | | | | | | | | |
| Air Resist 213; Hastelloy B, C, G and X (wrought); Haynes 25 and 188; J1570; M252 (wrought); Mar-M905 and M918; Nimonic 75 and 80 { | 15-20 | 20-25 | f<br>s | 20<br>75 | 10<br>125 | 11<br>1170 | 6<br>2590 | 11<br>405 | 6<br>900 | 20<br>230 | 10<br>400 |
| CW-12M; Hastelloy B and C (cast); N-12M { | 8-12 | 10-15 | | | | | | | | | |
| Rene 95 (Hot Isostatic Pressed) | — | — | | | | | | | | | |
| HS 6, 21, 2, 31 (X 40), 36, and 151; Haynes 36 and 151; Mar-M302, M322, and M509, WI-52 { | 10-12 | 10-15 | | | | | | | | | |
| Rene 41 | 10-15 | 12-20 | | | | | | | | | |
| Incoloy 901 | 10-20 | 20-35 | | | | | | | | | |
| Waspaloy | 10-30 | 25-35 | f | 28 | 13 | 11 | 6 | 10 | 5 | 20 | 10 |
| Inconel 625, 702, 706, 718 (wrought), 721, 722, X750, 751, 901, 600, and 604 { | 15-20 | 20-35 | s | 20 | 40 | 895 | 2230 | 345 | 815 | 185 | 315 |
| AF2-1DA, Unitemp 1753 | 8-10 | 10-15 | | | | | | | | | |
| Colmonoy, Inconel 600, 718, K-Monel, Stellite { | — | — | | | | | | | | | |
| Air Resist 13 and 215, FSH-H14, Nasa C-W-Re, X-45 | 10-12 | 10-15 | | | | | | | | | |
| Udimet 500, 700, and 710 | 10-15 | 12-20 | | | | | | | | | |
| Astroloy | 5-10 | 5-15 | | | | | | | | | |
| Mar-M200, M246, M421, and Rene 77, 80, and 95 (forged) | | 10-12<br>10-15 | f<br>s | 28<br>15 | 13<br>15 | 11<br>615 | 6<br>1720 | 10<br>290 | 5<br>700 | 20<br>165 | 10<br>280 |
| B-1900, GMR-235 and 235D, IN 100 and 738, Inconel 713C and 718 (cast), M252 (cast) { | 8-10 | 8-10 | | | | | | | | | |

The speed recommendations for rough turning may be used as starting values for milling and drilling with HSS tools. The combined feed/speed data in this table are based on tool grades (identified in Table 16) as follows: uncoated carbide = 15; ceramic, hard = 4, tough = 3; CBN = 1.

Speeds for HSS (high-speed steel) tools are based on a feed of 0.012 inch/rev and a depth of cut of 0.125 inch; use Table 5c to adjust the given speeds for other feeds and depths of cut. The combined feed/speed data in the remaining columns are based on a depth of cut of 0.1 inch, lead angle of 15 degrees, and nose radius of $\frac{3}{64}$ inch. Use Table 5a to adjust given speeds for other feeds, depths of cut, and lead angles; use Table 5b to adjust given speeds for increased tool life up to 180 minutes. Examples are given in the text.

**Speeds and Feeds Tables for Milling.—**Table 10 through Table 14 give speeds and feeds for milling. The data in the first speed column can be used with high-speed steel tools using the feeds given in Table 15a; these are the same speeds contained in previous editions of the *Handbook*. The remaining data in Table 10 through Table 14 are combined speeds and feeds for end, face, and slit, slot, and side milling that use the speed adjustment factors given in Table 15b, Table 15c, and Table 15d. Tool life for the combined feed/speed data can also be adjusted using the factors in Table 15e. Table 16 lists cutting tool grades and vendor equivalents.

*End Milling:* Table data for end milling are based on a 3-tooth, 20-degree helix angle tool with a diameter of 1.0 inch, an axial depth of cut of 0.2 inch, and a radial depth of cut of 1 inch (full slot). Use Table 15b to adjust speeds for other feeds and axial depths of cut, and Table 15c to adjust speeds if the radial depth of cut is less than the tool diameter. Speeds are valid for all tool diameters.

*Face Milling:* Table data for face milling are based on a 10-tooth, 8-inch diameter face mill, operating with a 15-degree lead angle, $\frac{3}{64}$-inch nose radius, axial depth of cut = 0.1 inch, and radial depth (width) of cut = 6 inches (i.e., width of cut to cutter diameter ratio = $\frac{3}{4}$). These speeds are valid if the cutter axis is above or close to the center line of the workpiece (eccentricity is small). Under these conditions, use Table 15d to adjust speeds for other feeds and axial and radial depths of cut. For larger eccentricity (i.e., when the cutter axis to workpiece center line offset is one half the cutter diameter or more), use the end and side milling adjustment factors (Table 15b and Table 15c) instead of the face milling factors.

*Slit and Slot Milling:* Table data for slit milling are based on an 8-tooth, 10-degree helix angle tool with a cutter width of 0.4 inch, diameter $D$ of 4.0 inch, and a depth of cut of 0.6 inch. Speeds are valid for all tool diameters and widths. See the examples in the text for adjustments to the given speeds for other feeds and depths of cut.

Tool life for all tabulated values is approximately 45 minutes; use Table 15e to adjust tool life from 15 to 180 minutes.

*Using the Speeds and Feeds Tables for Milling:* The basic feed for milling cutters is the feed per tooth ($f$), which is expressed in inches per tooth. There are many factors to consider in selecting the feed per tooth and no formula is available to resolve these factors. Among the factors to consider are the cutting tool material; the work material and its hardness; the width and the depth of the cut to be taken; the type of milling cutter to be used and its size; the surface finish to be produced; the power available on the milling machine; and the rigidity of the milling machine, the workpiece, the workpiece setup, the milling cutter, and the cutter mounting.

The cardinal principle is to always use the maximum feed that conditions will permit. Avoid, if possible, using a feed that is less than 0.001 inch per tooth because such low feeds reduce the tool life of the cutter. When milling hard materials with small-diameter end mills, such small feeds may be necessary, but otherwise use as much feed as possible. Harder materials in general will require lower feeds than softer materials. The width and the depth of cut also affect the feeds. Wider and deeper cuts must be fed somewhat more slowly than narrow and shallow cuts. A slower feed rate will result in a better surface finish; however, always use the heaviest feed that will produce the surface finish desired. Fine chips produced by fine feeds are dangerous when milling magnesium because spontaneous combustion can occur. Thus, when milling magnesium, a fast feed that will produce a relatively thick chip should be used. Cutting stainless steel produces a work-hardened layer on the surface that has been cut. Thus, when milling this material, the feed should be large enough to allow each cutting edge on the cutter to penetrate below the work-hardened layer produced by the previous cutting edge. The heavy feeds recommended for face milling cutters are to be used primarily with larger cutters on milling machines having an adequate amount of power. For smaller face milling cutters, start with smaller feeds and increase as indicated by the performance of the cutter and the machine.

When planning a milling operation that requires a high cutting speed and a fast feed, always check to determine if the power required to take the cut is within the capacity of the milling machine. Excessive power requirements are often encountered when milling with cemented carbide cutters. The large metal removal rates that can be attained require a high horsepower output. An example of this type of calculation is given in the section on Machining Power that follows this section. If the size of the cut must be reduced in order to stay within the power capacity of the machine, start by reducing the cutting speed rather than the feed in inches per tooth.

The formula for calculating the table feed rate, when the feed in inches per tooth is known, is as follows:

$$f_m = f_t n_t N$$

where $f_m$ = milling machine table feed rate in inches per minute (ipm)

$f_t$ = feed in inch per tooth (ipt)

$n_t$ = number of teeth in the milling cutter

$N$ = spindle speed of the milling machine in revolutions per minute (rpm)

*Example:* Calculate the feed rate for milling a piece of AISI 1040 steel having a hardness of 180 BHN. The cutter is a 3-inch diameter high-speed steel plain or slab milling cutter with 8 teeth. The width of the cut is 2 inches, the depth of cut is 0.062 inch, and the cutting speed from Table 11 is 85 fpm. From Table 15a, the feed rate selected is 0.008 inch per tooth.

$$N = \frac{12V}{\pi D} = \frac{12 \times 85}{3.14 \times 3} = 108 \text{ rpm}$$
$$f_m = f_t n_t N = 0.008 \times 8 \times 108$$
$$= 7 \text{ ipm (approximately)}$$

*Example 1, Face Milling:* Determine the cutting speed and machine operating speed for face milling an aluminum die casting (alloy 413) using a 4-inch polycrystalline diamond cutter, a 3-inch width of cut, a 0.10-inch depth of cut, and a feed of 0.006 inch/tooth.

Table 10 gives the speeds and feeds for milling aluminum alloys. The feed/speed pairs for face milling die-cast alloy 413 with polycrystalline diamond (PCD) are 8/2320 (0.008 in/tooth feed at 2320 fpm) and 4/4755 (0.004 in/tooth feed at 4755 fpm). These speeds are based on an axial depth of cut of 0.10 inch, an 8-inch cutter diameter *D*, a 6-inch radial depth (width) of cut *ar*, with the cutter approximately centered above the workpiece, i.e., eccentricity is low, as shown in Fig. 3. If the preceding conditions apply, the given speeds and feeds can be used without adjustment for a 45-minute tool life. The given speeds are valid for all cutter diameters if a radial depth of cut to cutter diameter ratio ($ar/D$) of $\frac{3}{4}$ is maintained (i.e., $\frac{6}{8} = \frac{3}{4}$). However, if a different feed or axial depth of cut is required, or if the $ar/D$ ratio is not equal to $\frac{3}{4}$, the cutting speed must be adjusted for the conditions. The adjusted cutting speed $V$ is calculated from $V = V_{opt} \times F_f \times F_d \times F_{ar}$, where $V_{opt}$ is the lower of the two speeds given in the speed table, and $F_f$, $F_d$, and $F_{ar}$ are adjustment factors for feed, axial depth of cut, and radial depth of cut, respectively, obtained from Table 15d (face milling); except, when cutting near the end or edge of the workpiece as in Fig. 4, Table 15c (side milling) is used to obtain $F_f$.

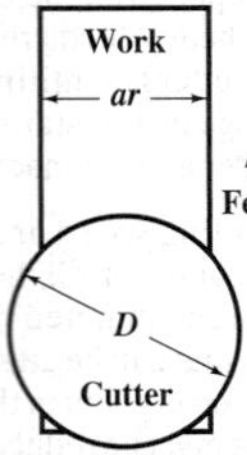


Fig. 3.

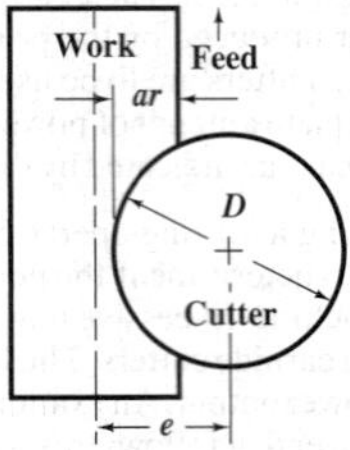


Fig. 4.

In this example, the cutting conditions match the standard conditions specified in the speed table for radial depth of cut to cutter diameter (3 in/4 in), and depth of cut (0.01 in), but the desired feed of 0.006 in/tooth does not match either of the feeds given in the speed table (0.004 or 0.008). Therefore, the cutting speed must be adjusted for this feed. As with turning, the feed factor $F_f$ is determined by calculating the ratio of the desired feed $f$ to maximum feed $f_{opt}$ from the speed table, and from the ratio $V_{avg}/V_{opt}$ of the two speeds given in the speed table. The feed factor is found at the intersection of the feed ratio row and the speed ratio column in Table 15d. The speed is then obtained using the following equation:

$$\frac{\text{Chosen feed}}{\text{Optimum feed}} = \frac{f}{f_{opt}} = \frac{0.006}{0.008} = 0.75 \qquad \frac{\text{Average speed}}{\text{Optimum speed}} = \frac{V_{avg}}{V_{opt}} = \frac{4755}{2320} \approx 2.0$$

$$F_f = (1.25 + 1.43)/2 = 1.3 \qquad F_d = 1.0 \qquad F_{ar} = 1.0$$

$$V = 2320 \times 1.34 \times 1.0 \times 1.0 = 3109 \text{ fpm, and } 3.82 \times 3109/4 = 2970 \text{ rpm}$$

*Example 2, End Milling:* What cutting speed should be used for cutting a full slot (i.e., a slot cut from the solid, in one pass, that is the same width as the cutter) in 5140 steel with hardness of 300 BHN using a 1-inch diameter coated carbide (insert) 0° lead angle end mill, a feed of 0.003 in/tooth, and a 0.2-inch axial depth of cut?

The speed and feed data for end milling 5140 steel, Brinell Hardness Number = 275–325, with a coated carbide tool are given in Table 11 as 15/80 and 8/240 for *optimum* and *average* sets, respectively. The speed adjustment factors for feed and depth of cut for full slot (end milling) are obtained from Table 15b. The calculations are the same as in the previous examples: $f/f_{opt}$ = 3/15 = 0.2 and $V_{avg}/V_{opt}$ = 240/80 = 3.0, therefore, $F_f$ = 6.86 and $F_d$ = 1.0. The cutting speed for a 45-minute tool life is $V = 80 \times 6.86 \times 1.0 = 548.8$, approximately 550 ft/min.

*Example 3, End Milling:* What cutting speed should be used in Example 2 if the radial depth of cut *ar* is 0.02 inch and axial depth of cut is 1 inch?

In end milling, when the radial depth of cut is less than the cutter diameter (as in Fig. 4), first obtain the feed factor $F_f$ from Table 15c, then the axial depth of cut and lead angle factor $F_d$ from Table 15b The radial depth of cut to cutter diameter ratio *ar*/*D* is used in Table 15c to determine the maximum and minimum feeds that guard against tool failure at high feeds and against premature tool wear caused by the tool rubbing against the work at very low feeds. The feed used should be selected so that it falls within the minimum to maximum feed range, and then the feed factor $F_f$ can be determined from the feed factors at minimum and maximum feeds, $F_{f1}$ and $F_{f2}$ as explained below.

The maximum feed $f_{max}$ is found in Table 15c by multiplying the *optimum* feed from the speed table by the maximum feed factor that corresponds to the $ar/D$ ratio, which in this instance is 0.02/1 = 0.02; the minimum feed $f_{min}$ is found by multiplying the *optimum* feed by the minimum feed factor. Thus, $f_{max} = 4.5 \times 0.015 = 0.0675$ in/tooth and $f_{min} = 3.1 \times 0.015 = 0.0465$ in/tooth. If a feed between these maximum and minimum values is selected, 0.050 in/tooth for example, then for $ar/D = 0.02$ and $V_{avg}/V_{opt} = 3.0$, the feed factors at maximum and minimum feeds are $F_{f1} = 7.90$ and $F_{f2} = 7.01$, respectively, and by interpolation, $F_f = 7.90 + (0.050 - 0.0465)/(0.0675 - 0.0465) \times (7.01 - 7.90) = 7.75$.

The depth of cut factor $F_d$ is obtained from Table 15b, using $f_{max}$ from Table 15c instead of the *optimum* feed $f_{opt}$ for calculating the feed ratio (chosen feed/*optimum* feed). In this example, the feed ratio = chosen feed/$f_{max}$ = 0.050/0.0675 = 0.74, so the feed factor is $F_d = 0.93$ for a depth of cut = 1.0 inch and 0° lead angle. Therefore, the final cutting speed is $80 \times 7.75 \times 0.93 = 577$ ft/min. Notice that $f_{max}$ obtained from Table 15c was used instead of the *optimum* feed from the speed table, in determining the feed ratio needed to find $F_d$.

**Slit Milling.—**The tabular data for slit milling is based on an 8-tooth, 10-degree helix angle cutter with a width of 0.4 inch, a diameter $D$ of 4.0 inch, and a depth of cut of 0.6 inch. The given speeds and feeds are valid for any diameters and tool widths, as long as sufficient machine power is available. Adjustments to cutting speeds for other feeds and depths of cut are made using Table 15c or Table 15d, depending on the orientation of the cutter to the work, as illustrated in Case 1 and Case 2 of Fig. 5. The situation illustrated in Case 1 is approximately equivalent to that illustrated in Fig. 3, and Case 2 is approximately equivalent to that shown in Fig. 4.

*Case 1:* If the cutter is fed directly into the workpiece, i.e., the feed is perpendicular to the surface of the workpiece, as in cutting off, then Table 15d (face milling) is used to adjust speeds for other feeds. The depth of cut portion of Table 15d is not used in this case ($F_d = 1.0$), so the adjusted cutting speed $V = V_{opt} \times F_f \times F_{ar}$. In determining the factor $F_{ar}$ from Table 15d, the radial depth of cut $ar$ is the length of cut created by the portion of the cutter engaged in the work.

*Case 2:* If the cutter feed is parallel to the surface of the workpiece, as in slotting or side milling, then Table 15c (side milling) is used to adjust the given speeds for other feeds. In Table 15c, the cutting depth (slot depth, for example) is the radial depth of cut $ar$ that is used to determine maximum and minimum allowable feed/tooth and the feed factor $F_f$. These minimum and maximum feeds are determined in the manner described previously, however, the axial depth of cut factor $F_d$ is not required. The adjusted cutting speed, valid for cutters of any thickness (width), is given by $V = V_{opt} \times F_f$.

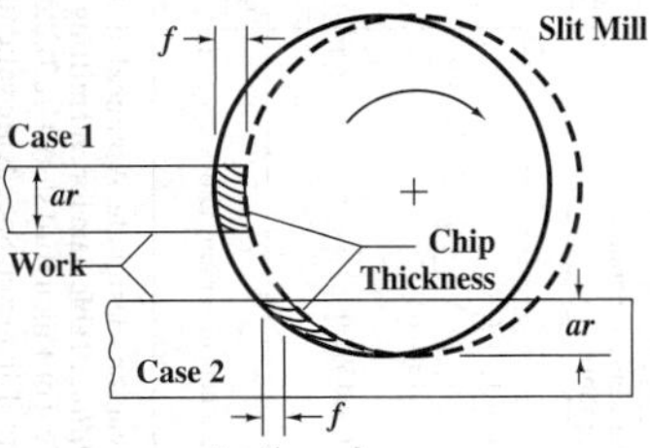


Fig. 5. Determination of Radial Depth of Cut or in Slit Milling

## Table 10. Cutting Speeds and Feeds for Milling Aluminum Alloys

| Material | Material Condition* | | End Milling HSS Opt. | End Milling HSS Avg. | End Milling Indexable Insert Uncoated Carbide Opt. | End Milling Indexable Insert Uncoated Carbide Avg. | Face Milling Indexable Insert Uncoated Carbide Opt. | Face Milling Indexable Insert Uncoated Carbide Avg. | Face Milling Polycrystalline Diamond Opt. | Face Milling Polycrystalline Diamond Avg. | Slit Milling HSS Opt. | Slit Milling HSS Avg. | Slit Milling Indexable Insert Uncoated Carbide Opt. | Slit Milling Indexable Insert Uncoated Carbide Avg. |
|---|---|---|---|---|---|---|---|---|---|---|---|---|---|---|
| | | f = feed (0.001 in/tooth), s = speed (ft/min) *Metric Units:* **f** × 25.4 = mm/rev, **s** × 0.3048 = m/min | | | | | | | | | | | | |
| All wrought aluminum alloys, 6061-T651, 5000, 6000, 7000 series | CD<br>ST and A | f | 15 | 8 | 15 | 8 | 39 | 20 | 8 | 4 | 16 | 8 | 39 | 20 |
| All aluminum sand and permanent mold casting alloys | CD<br>ST and A | s | 165 | 850 | 620 | 2020 | 755 | 1720 | 3750 | 8430 | 1600 | 4680 | 840 | 2390 |
| Aluminum Die-Casting Alloys | | | | | | | | | | | | | | |
| Alloys 308.0 and 319.0 | — | f | 15 | 8 | 15 | 8 | 39 | 20 | | | 16 | 8 | 39 | 20 |
| | | s | 30 | 100 | 620 | 2020 | 755 | 1720 | | | 160 | 375 | 840 | 2390 |
| Alloys 360.0 and 380.0 | — | f | 15 | 8 | 15 | 8 | 39 | 20 | 8 | 4 | 16 | 8 | 39 | 20 |
| | | s | 30 | 90 | 485 | 1905 | 555 | 1380 | 3105 | 7845 | 145 | 355 | 690 | 2320 |
| Alloys 390.0 and 392.0 | — | f | | | | | 39 | 20 | | | | | | |
| | | s | | | | | 220 | 370 | | | | | | |
| Alloy 413 | — | f | | | 15 | 8 | 39 | 20 | 8 | 4 | | | 39 | 20 |
| All other aluminum die-casting alloys { | ST and A | s | | | 355 | 1385 | 405 | 665 | 2320 | 4755 | | | 500 | 1680 |
| | AC | f | 15 | 8 | 15 | 8 | 39 | 20 | 8 | 4 | 16 | 8 | 39 | 20 |
| | | s | 30 | 90 | 485 | 1905 | 555 | 1380 | 3105 | 7845 | 145 | 335 | 690 | 2320 |

Abbreviations designate: A, annealed; AC, as cast; CD, cold drawn; and ST and A, solution treated and aged, respectively.

*End Milling:* Table data for end milling are based on a 3-tooth, 20-degree helix angle tool with a diameter of 1.0 inch, an axial depth of cut of 0.2 inch, and a radial depth of cut of 1 inch (full slot). Use Table 15b to adjust speeds for other feeds and axial depths of cut, and Table 15c to adjust speeds if the radial depth of cut is less than the tool diameter. Speeds are valid for all tool diameters.

*Face Milling:* Table data for face milling are based on a 10-tooth, 8-inch diameter face mill, operating with a 15-degree lead angle, $\frac{3}{64}$-inch nose radius, axial depth of cut = 0.1 inch, and radial depth (width) of cut = 6 inches (i.e., width of cut to cutter diameter ratio = $\frac{3}{4}$). These speeds are valid if the cutter axis is above or close to the center line of the workpiece (eccentricity is small). Under these conditions, use Table 15d to adjust speeds for other feeds and axial and radial depths of cut. For larger eccentricity (i.e., when the cutter axis to workpiece center line offset is one half the cutter diameter or more), use the end and side milling adjustment factors (Table 15b and Table 15c) instead of the face milling factors.

*Slit and Slot Milling:* Table data for slit milling are based on an 8-tooth, 10-degree helix angle tool with a cutter width of 0.4 inch, diameter *D* of 4.0 inch, and a depth of cut of 0.6 inch. Speeds are valid for all tool diameters and widths. See the examples in the text for adjustments to the given speeds for other feeds and depths of cut.

Tool life for all tabulated values is approximately 45 minutes; use Table 15e to adjust tool life from 15 to 180 minutes. The combined feed/speed data in this table are based on tool grades (identified in Table 16) as follows: uncoated carbide = 15; diamond = 9.

## Table 11. Cutting Speeds and Feeds for Milling Plain Carbon and Alloy Steels

| Material | Brinell Hardness Number [a] | HSS Speed (fpm) | | End Milling HSS Opt. | End Milling HSS Avg. | End Milling Uncoated Carbide Opt. | End Milling Uncoated Carbide Avg. | End Milling Coated Carbide Opt. | End Milling Coated Carbide Avg. | Face Milling Uncoated Carbide Opt. | Face Milling Uncoated Carbide Avg. | Face Milling Coated Carbide Opt. | Face Milling Coated Carbide Avg. | Slit Milling Uncoated Carbide Opt. | Slit Milling Uncoated Carbide Avg. | Slit Milling Coated Carbide Opt. | Slit Milling Coated Carbide Avg. |
|---|---|---|---|---|---|---|---|---|---|---|---|---|---|---|---|---|---|
| | | | | f = feed (0.001 in/tooth), s = speed (ft/min) *Metric Units:* f × 25.4 = mm/rev, s × 0.3048 = m/min | | | | | | | | | | | | | |
| Free-machining plain carbon steels (resulfurized): 1212, 1213, 1215 | 100-150 | 140 | f<br>s | 7<br>45 | 4<br>125 | 7<br>465 | 4<br>735 | 7<br>800 | 4<br>1050 | 39<br>225 | 20<br>335 | 39<br>415 | 20<br>685 | 39<br>265 | 20<br>495 | 39<br>525 | 20<br>830 |
| | 150-200 | 130 | f<br>s | 7<br>35 | 4<br>100 | | | | | | | 39<br>215 | 20<br>405 | | | | |
| (Resulfurized): 1108, 1109, 1115, 1117, 1118, 1120, 1126, 1211 | 100-150<br>150-200 | 130<br>115 | f<br>s | 730 | 4<br>85 | 7<br>325 | 4<br>565 | 7<br>465 | 4<br>720 | 39<br>140 | 20<br>220 | 39<br>195 | 20<br>365 | 39<br>170 | 20<br>350 | 39<br>245 | 20<br>495 |
| | 175-225 | 115 | f<br>s | 7<br>30 | 4<br>85 | | | | | | | | | 39<br>185 | 20<br>350 | | |
| (Resulfurized): 1132, 1137, 1139, 1140, 1144, 1146, 1151 | 275-325<br>325-375<br>375-425 | 70<br>45<br>35 | f<br>s | 7<br>25 | 4<br>70 | 7<br>210 | 4<br>435 | 7<br>300 | 4<br>560 | 39<br>90 | 20<br>170 | 39<br>175 | 20<br>330 | 39<br>90 | 20<br>235 | 39<br>135 | 20<br>325 |
| (Leaded): 11L17, 11L18, 12L13, 12L14 | 100-150<br>150-200 | 140<br>130 | f<br>s | 7<br>35 | 4<br>100 | | | | | | | 39<br>215 | 20<br>405 | | | | |
| | 200-250 | 110 | f<br>s | 7<br>30 | 4<br>85 | | | | | | | 39<br>185 | 20<br>350 | | | | |
| Plain carbon steels: 1006, 1008, 1009, 1010, 1012, 1015, 1016, 1017, 1018, 1019, 1020, 1021, 1022, 1023, 1024, 1025, 1026, 1513, 1514 | 100-125 | 110 | f<br>s | 7<br>45 | 4<br>125 | 7<br>465 | 4<br>735 | 7<br>800 | 4<br>1050 | 39<br>225 | 20<br>335 | 39<br>415 | 20<br>685 | 39<br>265 | 20<br>495 | 39<br>525 | 20<br>830 |
| | 125-175 | 110 | f<br>s | 7<br>35 | 4<br>100 | | | | | | | 39<br>215 | 20<br>405 | | | | |
| | 175-225<br>225-275 | 90<br>65 | f<br>s | 7<br>30 | 4<br>85 | | | | | | | 39<br>185 | 20<br>350 | | | | |

**Table 11.** *(Continued)* **Cutting Speeds and Feeds for Milling Plain Carbon and Alloy Steels**

| Material | Brinell Hardness Number [a] | HSS Speed (fpm) | | End Milling HSS Opt. | End Milling HSS Avg. | End Milling Uncoated Carbide Opt. | End Milling Uncoated Carbide Avg. | End Milling Coated Carbide Opt. | End Milling Coated Carbide Avg. | Face Milling Uncoated Carbide Opt. | Face Milling Uncoated Carbide Avg. | Face Milling Coated Carbide Opt. | Face Milling Coated Carbide Avg. | Slit Milling Uncoated Carbide Opt. | Slit Milling Uncoated Carbide Avg. | Slit Milling Coated Carbide Opt. | Slit Milling Coated Carbide Avg. |
|---|---|---|---|---|---|---|---|---|---|---|---|---|---|---|---|---|---|
| | | | | f = feed (0.001 in/tooth), s = speed (ft/min) *Metric Units:* f × 25.4 = mm/rev, s × 0.3048 = m/min | | | | | | | | | | | | | |
| | 125-175 | 100 | f | 7 | 4 | | | | | | | 39 | 20 | | | | |
| | | | s | 35 | 100 | | | | | | | 215 | 405 | | | | |
| Plain carbon steels: 1027, 1030, 1033, 1035, 1036, 1037, 1038, 1039, 1040, 1041, 1042, 1043, 1045, 1046, 1048, 1049, 1050, 1052, 1524, 1526, 1527, 1541 | 175-225<br>225-275 | 85<br>70 | f | 7 | 4 | | | | | | | 39 | 20 | | | | |
| | | | s | 30 | 85 | | | | | | | 185 | 350 | | | | |
| | 275-325<br>325-375<br>375-425 | 55<br>35<br>25 | f | 7 | 4 | 7 | 4 | 7 | 4 | 39 | 20 | 39 | 20 | 39 | 20 | 39 | 20 |
| | | | s | 25 | 70 | 210 | 435 | 300 | 560 | 90 | 170 | 175 | 330 | 90 | 235 | 135 | 325 |
| | 125-175<br>175-225 | 90<br>75 | f | 7 | 4 | 7 | 4 | 7 | 4 | 39 | 20 | 39 | 20 | 39 | 20 | 39 | 20 |
| | | | s | 30 | 85 | 325 | 565 | 465 | 720 | 140 | 220 | 195 | 365 | 170 | 350 | 245 | 495 |
| Plain carbon steels: 1055, 1060, 1064, 1065, 1070, 1074, 1078, 1080, 1084, 1086, 1090, 1095, 1548, 1551, 1552, 1561, 1566 | 225-275 | 60 | f | 7 | 4 | | | | | | | 39 | 20 | | | | |
| | | | s | 30 | 85 | | | | | | | 185 | 350 | | | | |
| | 275-325<br>325-375<br>375-425 | 45<br>30<br>15 | f | 7 | 4 | 7 | 4 | 7 | 4 | 39 | 20 | 39 | 20 | 39 | 20 | 39 | 20 |
| | | | s | 25 | 70 | 210 | 435 | 300 | 560 | 90 | 170 | 175 | 330 | 90 | 235 | 135 | 325 |
| | 175-200<br>200-250 | 100<br>90 | f | 15 | 8 | 15 | 8 | 15 | 8 | | | 39 | 20 | 39 | 20 | 7 | 4 |
| | | | s | 7 | 30 | 105 | 270 | 270 | 450 | | | 295 | 475 | 135 | 305 | 25 | 70 |
| Free-machining alloy steels (Resulfurized): 4140, 4150 | 250-300 | 60 | f | 15 | 8 | 15 | 8 | 15 | 8 | | | 39 | 20 | 39 | 20 | 7 | 4 |
| | | | s | 6 | 25 | 50 | 175 | 85 | 255 | | | 200 | 320 | 70 | 210 | 25 | 70 |
| | 300-375<br>375-425 | 45<br>35 | f | 15 | 8 | 15 | 8 | 15 | 8 | | | 39 | 20 | | | | |
| | | | s | 5 | 20 | 40 | 155 | 75 | 225 | | | 175 | 280 | | | | |

## Table 11. *(Continued)* Cutting Speeds and Feeds for Milling Plain Carbon and Alloy Steels

| Material | Brinell Hardness Number [a] | HSS Speed (fpm) | | End Milling HSS Opt. | End Milling HSS Avg. | End Milling Uncoated Carbide Opt. | End Milling Uncoated Carbide Avg. | End Milling Coated Carbide Opt. | End Milling Coated Carbide Avg. | Face Milling Uncoated Carbide Opt. | Face Milling Uncoated Carbide Avg. | Face Milling Coated Carbide Opt. | Face Milling Coated Carbide Avg. | Slit Milling Uncoated Carbide Opt. | Slit Milling Uncoated Carbide Avg. | Slit Milling Coated Carbide Opt. | Slit Milling Coated Carbide Avg. |
|---|---|---|---|---|---|---|---|---|---|---|---|---|---|---|---|---|---|
| | | | | f = feed (0.001 in/tooth), s = speed (ft/min) *Metric Units:* f × 25.4 = mm/rev, s × 0.3048 = m/min | | | | | | | | | | | | | |
| Free-machining alloy steels (Leaded): 41L30, 41L40, 41L47, 41L50, 43L47, 51L32, 52L100, 86L20, 86L40 | 150-200 | 115 | f | 7 | 4 | 7 | 4 | 7 | 4 | 39 | 20 | 39 | 20 | 39 | 20 | 39 | 20 |
| | | | s | 30 | 85 | 325 | 565 | 465 | 720 | 140 | 220 | 195 | 365 | 170 | 350 | 245 | 495 |
| | 200-250 | 95 | f | 7 | 4 | | | | | | | 39 | 20 | | | | |
| | | | s | 30 | 85 | | | | | | | 185 | 350 | | | | |
| | 250-300 | 70 | | | | | | | | | | | | | | | |
| | 300-375 | 50 | f | 7 | 4 | 7 | 4 | 7 | 4 | 39 | 20 | 39 | 20 | 39 | 20 | 39 | 20 |
| | | | s | 25 | 70 | 210 | 435 | 300 | 560 | 90 | 170 | 175 | 330 | 90 | 235 | 135 | 325 |
| | 375-425 | 40 | | | | | | | | | | | | | | | |
| Alloy steels: 4012, 4023, 4024, 4028, 4118, 4320, 4419, 4422, 4427, 4615, 4620, 4621, 4626, 4718, 4720, 4815, 4817, 4820, 5015, 5117, 5120, 6118, 8115, 8615, 8617, 8620, 8622, 8625, 8627, 8720, 8822, 94B17 | 125-175 | 100 | f | 15 | 8 | 15 | 8 | 15 | 8 | | | 39 | 20 | 39 | 20 | 39 | 20 |
| | 175-225 | 90 | s | 7 | 30 | 105 | 270 | 220 | 450 | | | 295 | 475 | 135 | 305 | 265 | 495 |
| | 225-275 | 60 | f | 15 | 8 | 15 | 8 | 15 | 8 | | | 39 | 20 | 39 | 20 | 39 | 20 |
| | | | s | 6 | 25 | 50 | 175 | 85 | 255 | | | 200 | 320 | 70 | 210 | 115 | 290 |
| | 275-325 | 50 | f | 15 | 8 | 15 | 8 | 15 | 8 | | | 39 | 20 | | | | |
| | | | s | 5 | 20 | 45 | 170 | 80 | 240 | | | 190 | 305 | | | | |
| | 325-375 | 40 | f | 15 | 8 | 15 | 8 | 15 | 8 | | | 39 | 20 | | | | |
| | 375-425 | 25 | s | 5 | 20 | 40 | 155 | 75 | 225 | | | 175 | 280 | | | | |
| Alloy steels: 1330, 1335, 1340, 1345, 4032, 4037, 4042, 4047, 4130, 4135, 4137, 4140, 4142, 4145, 4147, 4150, 4161, 4337, 4340, 50B44, 50B46, 50B50, 50B60, 5130, 5132, 5140, 5145, 5147, 5150, 5160, 51B60, 6150, 81B45, 8630, 8635, 8637, 8640, 8642, 8645, 8650, 8655, 8660, 8740, 9254, 9255, 9260, 9262, 94B30 E51100, E52100: use (HSS speeds) | 175-225 | 75 (65) | f | 15 | 8 | 15 | 8 | 15 | 8 | | | 39 | 20 | 39 | 20 | 39 | 20 |
| | | | s | 5 | 30 | 105 | 270 | 220 | 450 | | | 295 | 475 | 135 | 305 | 265 | 495 |
| | 225-275 | 60 | f | 15 | 8 | 15 | 8 | 15 | 8 | | | 39 | 20 | 39 | 20 | 39 | 20 |
| | | | s | 5 | 25 | 50 | 175 | 85 | 255 | | | 200 | 320 | 70 | 210 | 115 | 290 |
| | 275-325 | 50 (40) | f | 15 | 8 | 15 | 8 | 15 | 8 | | | 39 | 20 | | | | |
| | | | s | 5 | 25 | 45 | 170 | 80 | 240 | | | 190 | 305 | | | | |
| | 325-375 | 35 (30) | | | | | | | | | | | | | | | |
| | 375-425 | 20 | f | 15 | 8 | 15 | 8 | 15 | 8 | | | 39 | 20 | | | | |
| | | | s | 5 | 20 | 40 | 155 | 75 | 225 | | | 175 | 280 | | | | |

**Table 11.** *(Continued)* **Cutting Speeds and Feeds for Milling Plain Carbon and Alloy Steels**

| Material | Brinell Hardness Number[a] | HSS Speed (fpm) | | End Milling HSS Opt. | Avg. | End Milling Uncoated Carbide Opt. | Avg. | End Milling Coated Carbide Opt. | Avg. | Face Milling Uncoated Carbide Opt. | Avg. | Face Milling Coated Carbide Opt. | Avg. | Slit Milling Uncoated Carbide Opt. | Avg. | Slit Milling Coated Carbide Opt. | Avg. |
|---|---|---|---|---|---|---|---|---|---|---|---|---|---|---|---|---|---|
| | | | | f = feed (0.001 in/tooth), s = speed (ft/min) *Metric Units:* f × 25.4 = mm/rev, s × 0.3048 = m/min | | | | | | | | | | | | | |
| Ultra-high-strength steels (not AISI): AMS 6421 (98B37 Mod.), 6422 (98BV40), 6424, 6427, 6428, 6430, 6432, 6433, 6434, 6436, and 6442; 300M, D6ac | 220-300 | 60 | f | | | 8 | 4 | 8 | 4 | | | | | | | | |
| | 300-350 | 45 | s | | | 165 | 355 | 300 | 480 | | | | | | | | |
| | 350-400 | 20 | f | 8 | 4 | 8 | 4 | | | | | 39 | 20 | 39 | 20 | | |
| | | | s | 15 | 45 | 150 | 320 | | | | | 130 | 235 | 75 | 175 | | |
| | 43-52 RC | — | f | | | 5 | 3 | | | | | | | 39 | 20 | | |
| | | | s | | | 20† | 55 | | | | | | | 5 | 15 | | |
| Maraging steels (not AISI): 18% Ni Grades 200, 250, 300, and 350 | 250-325 | 50 | f | | | 8 | 4 | 8 | 4 | | | | | | | | |
| | | | s | | | 165 | 355 | 300 | 480 | | | | | | | | |
| | 50-52 RC | — | f | | | 5 | 3 | | | | | | | 39 | 20 | | |
| | | | s | | | 20† | 55 | | | | | | | 5 | 15 | | |
| Nitriding steels (not AISI): Nitralloy 125, 135, 135 Mod., 225, and 230, Nitralloy N, Nitralloy EZ, Nitrex 1 | 200-250 | 60 | f | 15 | 8 | 15 | 8 | 15 | 8 | | | 39 | 20 | 39 | 20 | 39 | 20 |
| | | | s | 7 | 30 | 105 | 270 | 220 | 450 | | | 295 | 475 | 135 | 305 | 265 | 495 |
| | 300-350 | 25 | f | 15 | 8 | 15 | 8 | 15 | 8 | | | 39 | 20 | | | | |
| | | | s | 5 | 20 | 40 | 155 | 75 | 225 | | | 175 | 280 | | | | |

[a] Brinell Hardness Number given unless otherwise indicated by RC for hardness on the Rockwell C scale.

For HSS (high-speed steel) tools in the first speed column only, use Table 15a for recommended feed in inches per tooth and depth of cut.

*End Milling:* Table data for end milling are based on a 3-tooth, 20-degree helix angle tool with a diameter of 1.0 inch, an axial depth of cut of 0.2 inch, and a radial depth of cut of 1 inch (full slot). Use Table 15b to adjust speeds for other feeds and axial depths of cut, and Table 15c to adjust speeds if the radial depth of cut is less than the tool diameter. Speeds are valid for all tool diameters.

*Face Milling:* Table data for face milling are based on a 10-tooth, 8-inch diameter face mill, operating with a 15-degree lead angle, $\frac{3}{64}$-inch nose radius, axial depth of cut = 0.1 inch, and radial depth (width) of cut = 6 inches (i.e., width of cut to cutter diameter ratio = $\frac{3}{4}$). These speeds are valid if the cutter axis is above or close to the center line of the workpiece (eccentricity is small). Under these conditions, use Table 15d to adjust speeds for other feeds and axial and radial depths of cut. For larger eccentricity (i.e., when the cutter axis to workpiece center line offset is one half the cutter diameter or more), use the end and side milling adjustment factors (Table 15b and Table 15c) instead of the face milling factors.

*Slit and Slot Milling:* Table data for slit milling are based on an 8-tooth, 10-degree helix angle tool with a cutter width of 0.4 inch, diameter $D$ of 4.0 inches, and a depth of cut of 0.6 inch. Speeds are valid for all tool diameters and widths. See the examples in the text for adjustments to the given speeds for other feeds and depths of cut.

Tool life for all tabulated values is approximately 45 minutes; use Table 15e to adjust tool life from 15 to 180 minutes. The combined feed/speed data in this table are based on tool grades (identified in Table 16) as follows: end and slit milling uncoated carbide = 20 except † = 15; face milling uncoated carbide = 19; end, face, and slit milling coated carbide = 10.

## Table 12. Cutting Speeds and Feeds for Milling Tool Steels

| Material | Brinell Hardness Number[a] | HSS Speed (fpm) | | End Milling: HSS Opt. | End Milling: HSS Avg. | End Milling: Uncoated Carbide Opt. | End Milling: Uncoated Carbide Avg. | End Milling: Coated Carbide Opt. | End Milling: Coated Carbide Avg. | Face Milling: Uncoated Carbide Opt. | Face Milling: Uncoated Carbide Avg. | Face Milling: CBN Opt. | Face Milling: CBN Avg. | Slit Milling: Uncoated Carbide Opt. | Slit Milling: Uncoated Carbide Avg. | Slit Milling: Coated Carbide Opt. | Slit Milling: Coated Carbide Avg. |
|---|---|---|---|---|---|---|---|---|---|---|---|---|---|---|---|---|---|
| | | | | f = feed (0.001 in/tooth), s = speed (ft/min) *Metric Units:* **f** × 25.4 = mm/rev, **s** × 0.3048 = m/min | | | | | | | | | | | | | |
| Water hardening: W1, W2, W5 | 150-200 | 85 | | | | | | | | | | | | | | | |
| Shock resisting: S1, S2, S5, S6, S7 | 175-225 | 55 | | | | | | | | | | | | | | | |
| Cold-work, oil hardening: O1, O2, O6, O7 | 175-225 | 50 | **f** | 8 | 4 | 8 | 4 | 8 | 4 | 39 | 20 | | | 39 | 20 | 39 | 20 |
| | | | **s** | 25 | 70 | 235 | 455 | 405 | 635 | 235 | 385 | | | 115 | 265 | 245 | 445 |
| Cold-work, high carbon, high chromium: D2, D3, D4, D5, D7 | 200-250 | 40 | | | | | | | | | | | | | | | |
| Cold-work, air hardening: A2, A3, A8, A9, A10 | 200-250 | 50 | | | | | | | | | | | | | | | |
| A4, A6 | 200-250 | 45 | **f** | | | | | | | 39 | 20 | | | | | | |
| A7 | 225-275 | 40 | **s** | | | | | | | 255 | 385 | | | | | | |
| Hot-work, chromium type: H10, H11, H12, H13, H14, H19 | 150-200 | 60 | | | | | | | | | | | | | | | |
| | 200-250 | 50 | | | | | | | | | | | | | | | |
| | 325-375 | 30 | **f** | 8 | 4 | 8 | 4 | | | 39 | 20 | | | 39 | 20 | | |
| | | | **s** | 15 | 45 | 150 | 320 | | | 130 | 235 | | | 75 | 175 | | |
| | 48-50 RC | — | | | | | | | | | | | | | | | |
| | 50-52 RC | — | **f** | | | 5 | 3 | | | | | 39 | 20 | 39 | 20 | | |
| | | | **s** | | | 20† | 55 | | | | | 50 | 135 | 5† | 15 | | |
| | 52-56 RC | — | | | | | | | | | | | | | | | |
| Hot-work, tungsten and molybdenum types: H21, H22, H23, H24, H25, H26, H41, H42, H43 | 150-200 | 55 | **f** | | | | | | | 39 | 20 | | | | | | |
| | 200-250 | 45 | **s** | | | | | | | 255 | 385 | | | | | | |
| Special-purpose, low alloy: L2, L3, L6 | 150-200 | 65 | **f** | 8 | 4 | 8 | 4 | 8 | 4 | 39 | 20 | | | 39 | 20 | 39 | 20 |
| Mold: P2, P3, P4, P5, P6 P20, P21 | 100-150 | 75 | **s** | 25 | 70 | 235 | 455 | 405 | 635 | 235 | 385 | | | 115 | 265 | 245 | 445 |
| | 150-200 | 60 | | | | | | | | | | | | | | | |
| High-speed steel: M1, M2, M6, M10, T1, T2, T6 | 200-250 | 50 | | | | | | | | | | | | | | | |
| M3-1, M4, M7, M30, M33, M34, M36, M41, M42, M43, M44, M46, M47, T5, T8 | 225-275 | 40 | **f** | | | | | | | 39 | 20 | | | | | | |
| | | | **s** | | | | | | | 255 | 385 | | | | | | |
| T15, M3-2 | 225-275 | 30 | | | | | | | | | | | | | | | |

[a] Brinell Hardness Number given unless otherwise indicated by RC for hardness on the Rockwell C scale.

For HSS (high-speed steel) tools in the first speed column only, use Table 15a for recommended feed in inches per tooth and depth of cut.

*End Milling:* Table data for end milling are based on a 3-tooth, 20-degree helix angle tool with a diameter of 1.0 inch, an axial depth of cut of 0.2 inch, and a radial depth of cut of 1 inch (full slot). Use Table 15b to adjust speeds for other feeds and axial depths of cut, and Table 15c to adjust speeds if the radial depth of cut is less than the tool diameter. Speeds are valid for all tool diameters.

*Face Milling:* Table data for face milling are based on a 10-tooth, 8-inch diameter face mill, operating with a 15-degree lead angle, $^3/_{64}$-inch nose radius, axial depth of cut = 0.1 inch, and radial depth (width) of cut = 6 inches (i.e., width of cut to cutter diameter ratio = $^3/_4$). These speeds are valid if the cutter axis is above or close to the center line of the workpiece (eccentricity is small). Under these conditions, use Table 15d to adjust speeds for other feeds and axial and radial depths of cut. For larger eccentricity (i.e., when the cutter axis to workpiece center line offset is one half the cutter diameter or more), use the end and side milling adjustment factors (Table 15b and Table 15c) instead of the face milling factors.

*Slit and Slot Milling:* Table data for slit milling are based on an 8-tooth, 10-degree helix angle tool with a cutter width of 0.4 inch, diameter *D* of 4.0 inches, and a depth of cut of 0.6 inch. Speeds are valid for all tool diameters and widths. See the examples in the text for adjustments to the given speeds for other feeds and depths of cut.

Tool life for all tabulated values is approximately 45 minutes; use Table 15e to adjust tool life from 15 to 180 minutes. The combined feed/speed data in this table are based on tool grades (identified in Table 16) as follows: uncoated carbide = 20, † = 15; coated carbide = 10; CBN = 1.

## Table 13. Cutting Speeds and Feeds for Milling Stainless Steels

| Material | Brinell Hardness Number | HSS Speed (fpm) | | End Milling HSS | | End Milling Uncoated Carbide | | End Milling Coated Carbide | | Face Milling Coated Carbide | | Slit Milling Uncoated Carbide | | Slit Milling Coated Carbide | |
|---|---|---|---|---|---|---|---|---|---|---|---|---|---|---|---|
| | | | | f = feed (0.001 in/tooth), s = speed (ft/min) *Metric Units:* **f** × 25.4 = mm/rev, **s** × 0.3048 = m/min | | | | | | | | | | | |
| | | | | *Opt.* | *Avg.* | *Opt.* | *Avg.* | *Opt.* | *Avg.* | *Opt.* | *Avg.* | *Opt.* | *Avg.* | *Opt.* | *Avg.* |
| Free-machining stainless steels (Ferritic): 430F, 430FSe | 135-185 | 110 | **f**<br>**s** | 7<br>30 | 4<br>80 | 7<br>305 | 4<br>780 | 7<br>420 | 4<br>1240 | 39<br>210 | 20<br>385 | 39<br>120 | 20<br>345 | 39<br>155 | 20<br>475 |
| (Austenitic): 203EZ, 303, 303Se, 303MA, 303Pb, 303Cu, 303 Plus X | 135-185<br>225-275 | 100<br>80 | **f**<br>**s** | 7<br>20 | 4<br>55 | 7<br>210 | 4<br>585 | | | | | 39<br>75 | 20<br>240 | | |
| (Martensitic): 416, 416Se, 416 Plus X, 420F, 420FSe, 440F, 440FSe | 135-185<br>185-240<br>275-325<br>375-425 | 110<br>100<br>60<br>30 | | | | | | | | | | | | | |
| Stainless steels (Ferritic): 405, 409, 429, 430, 434, 436, 442, 446, 502 | 135-185 | 90 | **f**<br>**s** | 7<br>30 | 4<br>80 | 7<br>305 | 4<br>780 | 7<br>420 | 4<br>1240 | 39<br>210 | 20<br>385 | 39<br>120 | 20<br>345 | 39<br>155 | 20<br>475 |
| (Austenitic): 201, 202, 301, 302, 304, 304L, 305, 308, 321, 347, 348 | 135-185<br>225-275 | 75<br>65 | | | | | | | | | | | | | |
| (Austenitic): 302B, 309, 309S, 310, 310S, 314, 316, 316L, 317, 330 | 135-185 | 70 | **f**<br>**s** | 7<br>20 | 4<br>55 | 7<br>210 | 4<br>585 | | | | | 39<br>75 | 20<br>240 | | |
| (Martensitic): 403, 410, 420, 501 | 135-175<br>175-225<br>275-325<br>375-425 | 95<br>85<br>55<br>35 | | | | | | | | | | | | | |

**Table 13.** *(Continued)* **Cutting Speeds and Feeds for Milling Stainless Steels**

| Material | Brinell Hardness Number | HSS Speed (fpm) | | End Milling HSS Opt. | Avg. | End Milling Uncoated Carbide Opt. | Avg. | End Milling Coated Carbide Opt. | Avg. | Face Milling Coated Carbide Opt. | Avg. | Slit Milling Uncoated Carbide Opt. | Avg. | Slit Milling Coated Carbide Opt. | Avg. |
|---|---|---|---|---|---|---|---|---|---|---|---|---|---|---|---|
| | | | **f** = feed (0.001 in/tooth), **s** = speed (ft/min) *Metric Units:* **f** × 25.4 = mm/rev, **s** × 0.3048 = m/min | | | | | | | | | | | | |
| Stainless Steels (Martensitic): 414, 431, Greek Ascoloy, 440A, 440B, 440C | 225-275 | 55-60 | | | | | | | | | | | | | |
| | 275-325 | 45-50 | | | | | | | | | | | | | |
| | 375-425 | 30 | | | | | | | | | | | | | |
| (Precipitation hardening): 15-5PH, 17-4PH, 17-7PH, AF-71, 17-14CuMo, AFC-77, AM-350, AM-355, AM-362, Custom 455, HNM, PH13-8, PH14-8Mo, PH15-7Mo, Stainless W | 150-200 | 60 | **f** | 7 | 4 | 7 | 4 | | | | | 39 | 20 | | |
| | | | **s** | 20 | 55 | 210 | 585 | | | | | 75 | 240 | | |
| | 275-325 | 50 | | | | | | | | | | | | | |
| | 325-375 | 40 | | | | | | | | | | | | | |
| | 375-450 | 25 | | | | | | | | | | | | | |

For HSS (high-speed steel) tools in the first speed column only, use Table 15a for recommended feed in inches per tooth and depth of cut.

*End Milling:* Table data for end milling are based on a 3-tooth, 20-degree helix angle tool with a diameter of 1.0 inch, an axial depth of cut of 0.2 inch, and a radial depth of cut of 1 inch (full slot). Use Table 15b to adjust speeds for other feeds and axial depths of cut, and Table 15c to adjust speeds if the radial depth of cut is less than the tool diameter. Speeds are valid for all tool diameters.

*Face Milling:* Table data for face milling are based on a 10-tooth, 8-inch diameter face mill, operating with a 15-degree lead angle, $\frac{3}{64}$-inch nose radius, axial depth of cut = 0.1 inch, and radial depth (width) of cut = 6 inches (i.e., width of cut to cutter diameter ratio = $\frac{3}{4}$). These speeds are valid if the cutter axis is above or close to the center line of the workpiece (eccentricity is small). Under these conditions, use Table 15dto adjust speeds for other feeds and axial and radial depths of cut. For larger eccentricity (i.e., when the cutter axis to workpiece center line offset is one half the cutter diameter or more), use the end and side milling adjustment factors (Table 15b and Table 15c) instead of the face milling factors.

*Slit and Slot Milling:* Table data for slit milling are based on an 8-tooth, 10-degree helix angle tool with a cutter width of 0.4 inch, diameter $D$ of 4.0 inch, and a depth of cut of 0.6 inch. Speeds are valid for all tool diameters and widths. See the examples in the text for adjustments to the given speeds for other feeds and depths of cut.

Tool life for all tabulated values is approximately 45 minutes; use Table 15e to adjust tool life from 15 to 180 minutes. The combined feed/speed data in this table are based on tool grades (identified in Table 16) as follows: uncoated carbide = 20; coated carbide = 10.

## Table 14. Cutting Speeds and Feeds for Milling Ferrous Cast Metals

| Material | Brinell Hardness Number | HSS Speed (fpm) | | End Milling HSS Opt. | Avg. | End Milling Uncoated Carbide Opt. | Avg. | End Milling Coated Carbide Opt. | Avg. | Face Milling Uncoated Carbide Opt. | Avg. | Face Milling Coated Carbide Opt. | Avg. | Face Milling Ceramic Opt. | Avg. | Face Milling CBN Opt. | Avg. | Slit Milling Uncoated Carbide Opt. | Avg. | Slit Milling Coated Carbide Opt. | Avg. |
|---|---|---|---|---|---|---|---|---|---|---|---|---|---|---|---|---|---|---|---|---|---|
| | | | | f = feed (0.001 in/tooth), s = speed (ft/min) *Metric Units:* **f** × 25.4 = mm/rev, **s** × 0.3048 = m/min | | | | | | | | | | | | | | | | | |
| | | | | Gray Cast Iron | | | | | | | | | | | | | | | | | |
| ASTM Class 20 | 120-150 | 100 | | | | | | | | | | | | | | | | | | | |
| ASTM Class 25 | 160-200 | 80 | **f** | 5 | 3 | 5 | 3 | | | 39 | 20 | 39 | 20 | 39 | 20 | 39 | 20 | 39 | 20 | | |
| ASTM Class 30, 35, and 40 | 190-220 | 70 | **s** | 35 | 90 | 520 | 855 | | | 140 | 225 | 285 | 535 | 1130 | 1630 | 200 | 530 | 205 | 420 | | |
| ASTM Class 45 and 50 | 220-260 | 50 | **f** | 5 | 3 | 5 | 3 | | | 39 | 20 | 39 | 20 | 39 | 20 | 39 | 20 | 39 | 20 | | |
| ASTM Class 55 and 60 | 250-320 | 30 | **s** | 30 | 70 | 515 | 1100 | | | 95 | 160 | 185 | 395 | 845 | 1220 | 150 | 400 | 145 | 380 | | |
| ASTM Type 1, 1b, 5 (Ni resist) | 100-215 | 50 | | | | | | | | | | | | | | | | | | | |
| ASTM Type 2, 3, 6 (Ni resist) | 120-175 | 40 | | | | | | | | | | | | | | | | | | | |
| ASTM Type 2b, 4 (Ni resist) | 150-250 | 30 | | | | | | | | | | | | | | | | | | | |
| | | | | Malleable Iron | | | | | | | | | | | | | | | | | |
| (Ferritic): 32510, 35018 | 110-160 | 110 | **f** | 5 | 3 | 5 | 3 | | | 39 | 20 | 39 | 20 | 39 | 20 | | | 39 | 20 | | |
| | | | **s** | 30 | 70 | 180 | 250 | | | 120 | 195 | 225 | 520 | 490 | 925 | | | 85 | 150 | | |
| (Pearlitic): 40010, 43010, 45006, 45008, 48005, 50005 | 160-200 | 80 | **f** | 5 | 3 | 5 | 3 | | | 39 | 20 | 39 | 20 | 39 | 20 | | | 39 | 20 | | |
| | 200-240 | 65 | **s** | 25 | 65 | 150 | 215 | | | 90 | 150 | 210 | 400 | 295 | 645 | | | 70 | 125 | | |
| (Martensitic): 53004, 60003, 60004 | 200-255 | 55 | | | | | | | | | | | | | | | | | | | |
| (Martensitic): 70002, 70003 | 220-260 | 50 | | | | | | | | | | | | | | | | | | | |
| (Martensitic): 80002 | 240-280 | 45 | | | | | | | | | | | | | | | | | | | |
| (Martensitic): 90001 | 250-320 | 25 | | | | | | | | | | | | | | | | | | | |
| | | | | Nodular (Ductile) Iron | | | | | | | | | | | | | | | | | |
| (Ferritic): 60-40-18, 65-45-12 | 140-190 | 75 | **f** | 7 | 4 | 7 | 4 | | | 39 | 20 | 39 | 20 | 39 | 20 | | | 39 | 20 | | |
| | | | **s** | 15 | 35 | 125 | 240 | | | 100 | 155 | 120 | 255 | 580 | 920 | | | 60 | 135 | | |
| (Ferritic-Pearlitic): 80-55-06 { | 190-225 | 60 | | | | | | | | | | | | | | | | | | | |
| | 225-260 | 50 | **f** | 7 | 4 | 7 | 4 | | | 39 | 20 | 39 | 20 | 39 | 20 | | | 39 | 20 | | |
| (Pearlitic-Martensitic): 100-70-03 | 240-300 | 40 | **s** | 10 | 30 | 90 | 210 | | | 95 | 145 | 150 | 275 | 170 | 415 | | | 40 | 100 | | |
| (Martensitic): 120-90-02 | 270-330 | 25 | | | | | | | | | | | | | | | | | | | |

**Table 14.** *(Continued)* **Cutting Speeds and Feeds for Milling Ferrous Cast Metals**

| Material | Brinell Hardness Number | HSS Speed (fpm) | | End Milling: HSS Opt. | End Milling: HSS Avg. | End Milling: Uncoated Carbide Opt. | End Milling: Uncoated Carbide Avg. | End Milling: Coated Carbide Opt. | End Milling: Coated Carbide Avg. | Face Milling: Uncoated Carbide Opt. | Face Milling: Uncoated Carbide Avg. | Face Milling: Coated Carbide Opt. | Face Milling: Coated Carbide Avg. | Face Milling: Ceramic Opt. | Face Milling: Ceramic Avg. | Face Milling: CBN Opt. | Face Milling: CBN Avg. | Slit Milling: Uncoated Carbide Opt. | Slit Milling: Uncoated Carbide Avg. | Slit Milling: Coated Carbide Opt. | Slit Milling: Coated Carbide Avg. |
|---|---|---|---|---|---|---|---|---|---|---|---|---|---|---|---|---|---|---|---|---|---|
| | | | | f = feed (0.001 in/tooth), s = speed (ft/min) *Metric Units:* f × 25.4 = mm/rev, s × 0.3048 = m/min | | | | | | | | | | | | | | | | | |
| Cast Steels | | | | | | | | | | | | | | | | | | | | | |
| (Low carbon): 1010, 1020 | 100-150 | 100 | f | 7 | 4 | 7 | 4 | 7 | 4 | | | 39 | 20 | | | | | 39 | 20 | 39 | 20 |
| | 125-175 | 95 | s | 25 | 70 | 245† | 410 | 420 | 650 | | | 265‡ | 430 | | | | | 135† | 260 | 245 | 450 |
| (Medium carbon): 1030, 1040 1050 | 175-225 | 80 | | | | | | | | | | | | | | | | | | | |
| | 225-300 | 60 | f | 7 | 4 | 7 | 4 | 7 | 4 | | | 39 | 20 | | | | | 39 | 20 | 39 | 20 |
| (Low-carbon alloy): 1320, 2315, 2320, 4110, 4120, 4320, 8020, 8620 | 150-200 | 85 | s | 20 | 55 | 160† | 400 | 345 | 560 | | | 205‡ | 340 | | | | | 65† | 180 | 180 | 370 |
| | 200-250 | 75 | | | | | | | | | | | | | | | | | | | |
| | 250-300 | 50 | | | | | | | | | | | | | | | | | | | |
| (Medium-carbon alloy): 1330, 1340, 2325, 2330, 4125, 4130, 4140, 4330, 4340, 8030, 80B30, 8040, 8430, 8440, 8630, 8640, 9525, 9530, 9535 | 175-225 | 70 | f | 7 | 4 | 7 | 4 | | | | | | | | | | | 39 | 20 | | |
| | 225-250 | 65 | s | 15 | 45 | 120† | 310 | | | | | | | | | | | 45† | 135 | | |
| | 250-300 | 50 | f | | | | | | | 39 | 20 | | | | | | | | | | |
| | 300-350 | 30 | s | | | | | | | 25 | 40 | | | | | | | | | | |

For HSS (high-speed steel) tools in the first speed column only, use Table 15a for recommended feed in inches per tooth and depth of cut.

*End Milling:* Table data for end milling are based on a 3-tooth, 20-degree helix angle tool with a diameter of 1.0 inch, an axial depth of cut of 0.2 inch, and a radial depth of cut of 1 inch (full slot). Use Table 15b to adjust speeds for other feeds and axial depths of cut, and Table 15c to adjust speeds if the radial depth of cut is less than the tool diameter. Speeds are valid for all tool diameters.

*Face Milling:* Table data for face milling are based on a 10-tooth, 8-inch diameter face mill, operating with a 15-degree lead angle, $\frac{3}{64}$-inch nose radius, axial depth of cut = 0.1 inch, and radial depth (width) of cut = 6 inches (i.e., width of cut to cutter diameter ratio = $\frac{3}{4}$). These speeds are valid if the cutter axis is above or close to the center line of the workpiece (eccentricity is small). Under these conditions, use Table 15d to adjust speeds for other feeds and axial and radial depths of cut. For larger eccentricity (i.e., when the cutter axis to workpiece center line offset is one half the cutter diameter or more), use the end and side milling adjustment factors (Table 15b and Table 15c) instead of the face milling factors.

*Slit and Slot Milling:* Table data for slit milling are based on an 8-tooth, 10-degree helix angle tool with a cutter width of 0.4 inch, diameter *D* of 4.0 inches, and a depth of cut of 0.6 inch. Speeds are valid for all tool diameters and widths. See the examples in the text for adjustments to the given speeds for other feeds and depths of cut.

Tool life for all tabulated values is approximately 45 minutes; use Table 15e to adjust tool life from 15 to 180 minutes. The combined feed/speed data in this table are based on tool grades (identified in Table 16) as follows: uncoated carbide = 15 except † = 20; end and slit milling coated carbide = 10; face milling coated carbide = 11 except ‡ = 10. ceramic = 6; CBN = 1.

**Table 15a. Recommended Feed in Inches per Tooth ($f_t$) for Milling with High-Speed Steel Cutters**

| Material | Brinell Hardness Number | End Mills: Depth of Cut, .250 inch (6.35 mm); Cutter Diameter, inch (mm): ½ inch (12.7 mm) | ¾ inch (25.4 mm) | 1 inch and up (19.05 mm) | End Mills: Depth of Cut, .050 inch (1.27 mm); Cutter Diameter, inch (mm): ¼ inch (6.35 mm) | ½ inch (12.7 mm) | ¾ inch (19.05 mm) | 1 inch and up (25.4 mm) | Plain or Slab Mills | Form Relieved Cutters | Face Mills and Shell End Mills | Slotting and Side Mills |
|---|---|---|---|---|---|---|---|---|---|---|---|---|
| | | $f_t$ = feed per tooth, inch; *Metric Units:* $f_t$ × 25.4 = mm | | | | | | | | | | |
| Free-machining plain carbon steels | 100-185 | .001 | .003 | .004 | .001 | .002 | .003 | .004 | .003-.008 | .005 | .004-.012 | .002-.008 |
| Plain carbon steels, AISI 1006 to 1030; 1513 to 1522 | 100-150 | .001 | .003 | .003 | .001 | .002 | .003 | .004 | .003-.008 | .004 | .004-.012 | .002-.008 |
| | 150-200 | .001 | .002 | .003 | .001 | .002 | .002 | .003 | .003-.008 | .004 | .003-.012 | .002-.008 |
| AISI 1033 to 1095; 1524 to 1566 | 120-180 | .001 | .003 | .003 | .001 | .002 | .003 | .004 | .003-.008 | .004 | .004-.012 | .002-.008 |
| | 180-220 | .001 | .002 | .003 | .001 | .002 | .002 | .003 | .003-.008 | .004 | .003-.012 | .002-.008 |
| | 220-300 | .001 | .002 | .002 | .001 | .001 | .002 | .003 | .002-.006 | .003 | .002-.008 | .002-.006 |
| Alloy steels having less than 3% carbon. Typical examples: AISI 4012, 4023, 4027, 4118, 4320 4422, 4427, 4615, 4620, 4626, 4720, 4820, 5015, 5120, 6118, 8115, 8620 8627, 8720, 8820, 8822, 9310, 93B17 | 125-175 | .001 | .003 | .003 | .001 | .002 | .003 | .004 | .003-.008 | .004 | .004-.012 | .002-.008 |
| | 175-225 | .001 | .002 | .003 | .001 | .002 | .003 | .003 | .003-.008 | .004 | .003-.012 | .002-.008 |
| | 225-275 | .001 | .002 | .003 | .001 | .001 | .002 | .003 | .002-.006 | .003 | .003-.008 | .002-.006 |
| | 275-325 | .001 | .002 | .002 | .001 | .001 | .002 | .002 | .002-.005 | .003 | .002-.008 | .002-.005 |
| Alloy steels having 3% carbon or more. Typical examples: AISI 1330, 1340, 4032, 4037, 4130, 4140, 4150, 4340, 50B40, 50B60, 5130, 51B60, 6150, 81B45, 8630, 8640, 86B45, 8660, 8740, 94B30 | 175-225 | .001 | .002 | .003 | .001 | .002 | .003 | .004 | .003-.008 | .004 | .003-.012 | .002-.008 |
| | 225-275 | .001 | .002 | .003 | .001 | .001 | .002 | .003 | .002-.006 | .003 | .003-.010 | .002-.006 |
| | 275-325 | .001 | .002 | .002 | .001 | .001 | .002 | .003 | .002-.005 | .003 | .002-.008 | .002-.005 |
| | 325-375 | .001 | .002 | .002 | .001 | .001 | .002 | .002 | .002-.004 | .002 | .002-.008 | .002-.005 |
| Tool steel | 150-200 | .001 | .002 | .002 | .001 | .002 | .003 | .003 | .003-.008 | .004 | .003-.010 | .002-.006 |
| | 200-250 | .001 | .002 | .002 | .001 | .002 | .002 | .003 | .002-.006 | .003 | .003-.008 | .002-.005 |
| Gray cast iron | 120-180 | .001 | .003 | .004 | .002 | .003 | .004 | .004 | .004-.012 | .005 | .005-.016 | .002-.010 |
| | 180-225 | .001 | .002 | .003 | .001 | .002 | .003 | .003 | .003-.010 | .004 | .004-.012 | .002-.008 |
| | 225-300 | .001 | .002 | .002 | .001 | .001 | .002 | .002 | .002-.006 | .003 | .002-.008 | .002-.005 |
| Free malleable iron | 110-160 | .001 | .003 | .004 | .002 | .003 | .004 | .004 | .003-.010 | .005 | .005-.016 | .002-.010 |

## Table 15a. *(Continued)* Recommended Feed in Inches per Tooth ($f_t$) for Milling with High-Speed Steel Cutters

| Material | Brinell Hardness Number | End Mills | | | | | | | Plain or Slab Mills | Form Relieved Cutters | Face Mills and Shell End Mills | Slotting and Side Mills |
|---|---|---|---|---|---|---|---|---|---|---|---|---|
| | | Depth of Cut, .250 inch (6.35 mm) Cutter Diameter, inch (mm) | | | Depth of Cut, .050 inch (1.27 mm) Cutter Diameter, inch (mm) | | | | | | | |
| | | ½ inch (12.7 mm) | ¾ inch (25.4 mm) | 1 inch and up (19.05 mm) | ¼ inch (6.35 mm) | ½ inch (12.7 mm) | ¾ inch (19.05 mm) | 1 inch and up (25.4 mm) | | | | |
| | | $f_t$ = feed per tooth, inch; *Metric Units:* $f_t \times 25.4$ = mm | | | | | | | | | | |
| Pearlitic-Martensitic malleable iron | 160-200 | .001 | .003 | .004 | .001 | .002 | .003 | .004 | .003-.010 | .004 | .004-.012 | .002-.018 |
| | 200-240 | .001 | .002 | .003 | .001 | .002 | .003 | .003 | .003-.007 | .004 | .003-.010 | .002-.006 |
| | 240-300 | .001 | .002 | .002 | .001 | .001 | .002 | .002 | .002-.006 | .003 | .002-.008 | .002-.005 |
| Cast steel | 100-180 | .001 | .003 | .003 | .001 | .002 | .003 | .004 | .003-.008 | .004 | .003-.012 | .002-.008 |
| | 180-240 | .001 | .002 | .003 | .001 | .002 | .003 | .003 | .003-.008 | .004 | .003-.010 | .002-.006 |
| | 240-300 | .001 | .002 | .002 | .005 | .002 | .002 | .002 | .002-.006 | .003 | .003-.008 | .002-.005 |
| Zinc alloys (die castings) | … | .002 | .003 | .004 | .001 | .003 | .004 | .006 | .003-.010 | .005 | .004-.015 | .002-.012 |
| Copper alloys (brasses & bronzes) | 100-150 | .002 | .004 | .005 | .002 | .003 | .005 | .006 | .003-.015 | .004 | .004-.020 | .002-.010 |
| | 150-250 | .002 | .003 | .004 | .001 | .003 | .004 | .005 | .003-.015 | .004 | .003-.012 | .002-.008 |
| Free cutting brasses & bronzes | 80-100 | .002 | .004 | .005 | .002 | .003 | .005 | .006 | .003-.015 | .004 | .004-.015 | .002-.010 |
| Cast aluminum alloys—as cast | … | .003 | .004 | .005 | .002 | .004 | .005 | .006 | .005-.016 | .006 | .005-.020 | .004-.012 |
| Cast aluminum alloys—hardened | … | .003 | .004 | .005 | .002 | .003 | .004 | .005 | .004-.012 | .005 | .005-.020 | .004-.012 |
| Wrought aluminum alloys— cold drawn | … | .003 | .004 | .005 | .002 | .003 | .004 | .005 | .004-.014 | .005 | .005-.020 | .004-.012 |
| Wrought aluminum alloys—hardened | … | .002 | .003 | .004 | .001 | .002 | .003 | .004 | .003-.012 | .004 | .005-.020 | .004-.012 |
| Magnesium alloys | … | .003 | .004 | .005 | .003 | .004 | .005 | .007 | .005-.016 | .006 | .008-.020 | .005-.012 |
| Ferritic stainless steel | 135-185 | .001 | .002 | .003 | .001 | .002 | .003 | .003 | .002-.006 | .004 | .004-.008 | .002-.007 |
| Austenitic stainless steel | 135-185 | .001 | .002 | .003 | .001 | .002 | .003 | .003 | .003-.007 | .004 | .005-.008 | .002-.007 |
| | 185-275 | .001 | .002 | .003 | .001 | .002 | .002 | .002 | .003-.006 | .003 | .004-.006 | .002-.007 |
| Martensitic stainless steel | 135-185 | .001 | .002 | .002 | .001 | .002 | .003 | .003 | .003-.006 | .004 | .004-.010 | .002-.007 |
| | 185-225 | .001 | .002 | .002 | .001 | .002 | .002 | .003 | .003-.006 | .004 | .003-.008 | .002-.007 |
| | 225-300 | .0005 | .002 | .002 | .0005 | .001 | .002 | .002 | .002-.005 | .003 | .002-.006 | .002-.005 |
| Monel | 100-160 | .001 | .003 | .004 | .001 | .002 | .003 | .004 | .002-.006 | .004 | .002-.008 | .002-.006 |

## Table 15b. End Milling (Full Slot) Speed Adjustment Factors for Feed, Depth of Cut, and Lead Angle

| Cutting Speed, $V = V_{opt} \times F_f \times F_d$ | | | | | | | | | | | | | | | | | |
|---|---|---|---|---|---|---|---|---|---|---|---|---|---|---|---|---|---|
| Ratio of Chosen Feed to *Optimum* Feed | Ratio of the two cutting speeds (*average*/*optimum*) given in the tables $V_{avg}/V_{opt}$ | | | | | | | Depth of Cut and Lead Angle | | | | | | | | | |
| | | | | | | | | 1 in (25.4 mm) | | 0.4 in (10.2 mm) | | 0.2 in (5.1 mm) | | 0.1 in (2.4 mm) | | 0.04 in (1.0 mm) | |
| | 1.00 | 1.25 | 1.50 | 2.00 | 2.50 | 3.00 | 4.00 | 0° | 45° | 0° | 45° | 0° | 45° | 0° | 45° | 0° | 45° |
| | Feed Factor, $F_f$ | | | | | | | Depth of Cut and Lead Angle Factor, $F_d$ | | | | | | | | | |
| 1.00 | 1.0 | 1.0 | 1.0 | 1.0 | 1.0 | 1.0 | 1.0 | 0.91 | 1.36 | 0.94 | 1.38 | 1.00 | 0.71 | 1.29 | 1.48 | 1.44 | 1.66 |
| 0.90 | 1.00 | 1.06 | 1.09 | 1.14 | 1.18 | 1.21 | 1.27 | 0.91 | 1.33 | 0.94 | 1.35 | 1.00 | 0.72 | 1.26 | 1.43 | 1.40 | 1.59 |
| 0.80 | 1.00 | 1.12 | 1.19 | 1.31 | 1.40 | 1.49 | 1.63 | 0.92 | 1.30 | 0.95 | 1.32 | 1.00 | 0.74 | 1.24 | 1.39 | 1.35 | 1.53 |
| 0.70 | 1.00 | 1.18 | 1.30 | 1.50 | 1.69 | 1.85 | 2.15 | 0.93 | 1.26 | 0.95 | 1.27 | 1.00 | 0.76 | 1.21 | 1.35 | 1.31 | 1.44 |
| 0.60 | 1.00 | 1.20 | 1.40 | 1.73 | 2.04 | 2.34 | 2.89 | 0.94 | 1.22 | 0.96 | 1.25 | 1.00 | 0.79 | 1.18 | 1.28 | 1.26 | 1.26 |
| 0.50 | 1.00 | 1.25 | 1.50 | 2.00 | 2.50 | 3.00 | 4.00 | 0.95 | 1.17 | 0.97 | 1.18 | 1.00 | 0.82 | 1.14 | 1.21 | 1.20 | 1.21 |
| 0.40 | 1.00 | 1.23 | 1.57 | 2.29 | 3.08 | 3.92 | 5.70 | 0.96 | 1.11 | 0.97 | 1.12 | 1.00 | 0.86 | 1.09 | 1.14 | 1.13 | 1.16 |
| 0.30 | 1.00 | 1.14 | 1.56 | 2.57 | 3.78 | 5.19 | 8.56 | 0.98 | 1.04 | 0.99 | 1.04 | 1.00 | 0.91 | 1.04 | 1.07 | 1.05 | 1.09 |
| 0.20 | 1.00 | 0.90 | 1.37 | 2.68 | 4.49 | 6.86 | 17.60 | 1.00 | 0.85 | 1.00 | 0.95 | 1.00 | 0.99 | 0.97 | 0.93 | 0.94 | 0.88 |
| 0.10 | 1.00 | 0.44 | 0.80 | 2.08 | 4.26 | 8.00 | 20.80 | 1.05 | 0.82 | 1.00 | 0.81 | 1.00 | 1.50 | 0.85 | 0.76 | 0.78 | 0.67 |

For HSS (high-speed steel) tool speeds in the first speed column of Table 10 through Table 14, use Table 15a to determine appropriate feeds and depths of cut.

Cutting speeds and feeds for end milling given in Table 11 through Table 14 (except those for high-speed steel in the first speed column) are based on milling a 0.20-inch deep full slot (i.e., radial depth of cut = end mill diameter) with a 1-inch diameter, 20-degree helix angle, 0-degree lead angle end mill. For other depths of cut (axial), lead angles, or feed, use the two feed/speed pairs from the tables and calculate the ratio of desired (new) feed to *optimum* feed (largest of the two feeds are given in the tables), and the ratio of the two cutting speeds ($V_{avg}/V_{opt}$). Find the feed factor $F_f$ at the intersection of the feed ratio row and the speed ratio column in the left half of the Table. The depth of cut factor $F_d$ is found in the same row as the feed factor, in the right half of the table under the column corresponding to the depth of cut and lead angle. The adjusted cutting speed can be calculated from $V = V_{opt} \times F_f \times F_d$, where $V_{opt}$ is the smaller (*optimum*) of the two speeds from the speed table (from the left side of the column containing the two feed/speed pairs). See the text for examples.

If the radial depth of cut is less than the cutter diameter (i.e., for cutting less than a full slot), the feed factor $F_f$ in the previous equation and the maximum feed $f_{max}$ must be obtained from Table 15c. The axial depth of cut factor $F_d$ can then be obtained from this table using $f_{max}$ in place of the *optimum* feed in the feed ratio. Also see the footnote to Table 15c.

## Table 15c. End, Slit, and Side Milling Speed Adjustment Factors for Radial Depth of Cut

| Ratio of Radial Depth of Cut to Diameter | Maximum Feed/Tooth Factor | Cutting Speed, $V = V_{opt} \times F_f \times F_d$ | | | | | | Minimum Feed/Tooth Factor | | | | | | |
|---|---|---|---|---|---|---|---|---|---|---|---|---|---|---|
| | | $V_{avg}/V_{opt}$ | | | | | | | $V_{avg}/V_{opt}$ | | | | | |
| | | 1.25 | 1.50 | 2.00 | 2.50 | 3.00 | 4.00 | | 1.25 | 1.50 | 2.00 | 2.50 | 3.00 | 4.00 |
| | | Feed Factor $F_f$ at Maximum Feed per Tooth, $F_{f1}$ | | | | | | | Feed Factor $F_f$ at Minimum Feed per Tooth, $F_{f2}$ | | | | | |
| 1.00 | 1.00 | 1.00 | 1.00 | 1.00 | 1.00 | 1.00 | 1.00 | 0.70 | 1.18 | 1.30 | 1.50 | 1.69 | 1.85 | 2.15 |
| 0.75 | 1.00 | 1.15 | 1.24 | 1.46 | 1.54 | 1.66 | 1.87 | 0.70 | 1.24 | 1.48 | 1.93 | 2.38 | 2.81 | 3.68 |
| 0.60 | 1.00 | 1.23 | 1.40 | 1.73 | 2.04 | 2.34 | 2.89 | 0.70 | 1.24 | 1.56 | 2.23 | 2.95 | 3.71 | 5.32 |
| 0.50 | 1.00 | 1.25 | 1.50 | 2.00 | 2.50 | 3.00 | 4.00 | 0.70 | 1.20 | 1.58 | 2.44 | 3.42 | 4.51 | 6.96 |
| 0.40 | 1.10 | 1.25 | 1.55 | 2.17 | 2.83 | 3.51 | 4.94 | 0.77 | 1.25 | 1.55 | 2.55 | 3.72 | 5.08 | 8.30 |
| 0.30 | 1.35 | 1.20 | 1.57 | 2.28 | 3.05 | 3.86 | 5.62 | 0.88 | 1.23 | 1.57 | 2.64 | 4.06 | 5.76 | 10.00 |
| 0.20 | 1.50 | 1.14 | 1.56 | 2.57 | 3.78 | 5.19 | 8.56 | 1.05 | 1.40 | 1.56 | 2.68 | 4.43 | 6.37 | 11.80 |
| 0.10 | 2.05 | 0.92 | 1.39 | 2.68 | 4.46 | 6.77 | 13.10 | 1.44 | 0.92 | 1.29 | 2.50 | 4.66 | 7.76 | 17.40 |
| 0.05 | 2.90 | 0.68 | 1.12 | 2.50 | 4.66 | 7.75 | 17.30 | 2.00 | 0.68 | 1.12 | 2.08 | 4.36 | 8.00 | 20.80 |
| 0.02 | 4.50 | 0.38 | 0.71 | 1.93 | 4.19 | 7.90 | 21.50 | 3.10 | 0.38 | 0.70 | 1.38 | 3.37 | 7.01 | 22.20 |

This table is for side milling, end milling when the radial depth of cut (width of cut) is less than the tool diameter (i.e., less than full slot milling), and slit milling when the feed is parallel to the work surface (slotting). The radial depth of cut to diameter ratio is used to determine the recommended maximum and minimum values of feed/tooth, which are found by multiplying the feed/tooth factor from the appropriate column above (maximum or minimum) by feed$_{opt}$ from the speed tables. For example, given two feed/speed pairs $^7/_{15}$ and $^4/_{45}$ for end milling cast, medium-carbon, alloy steel, and a radial depth of cut to diameter ratio $ar/D$ of 0.10 (a 0.05-inch width of cut for a ½-inch diameter end mill, for example), the maximum feed $f_{max} = 2.05 \times 0.007 = 0.014$ in/tooth and the minimum feed $f_{min} = 1.44 \times 0.007 = 0.010$ in/tooth. The feed selected should fall in the range between $f_{min}$ and $f_{max}$. The feed factor $F_d$ is determined by interpolating between the feed factors $F_{f1}$ and $F_{f2}$ corresponding to the maximum and minimum feed per tooth, at the appropriate $ar/D$ and speed ratio. In the example given, $ar/D = 0.10$ and $V_{avg}/V_{opt} = 45/15 = 3$, so the feed factor $F_{f1}$ at the maximum feed per tooth is 6.77, and the feed factor $F_{f2}$ at the minimum feed per tooth is 7.76. If a working feed of 0.012 in/tooth is chosen, the feed factor $F_f$ is half way between 6.77 and 7.76 or by formula, $F_f = F_{f2} + (\text{feed} - f_{min})/(f_{max} - f_{min}) \times (F_{f1} - F_{f2}) = 7.76 + (0.012 - 0.010)/(0.014 - 0.010) \times (6.77 - 7.76) = 7.27$. The cutting speed is $V = V_{opt} \times F_f \times F_d$, where $F_d$ is the depth of cut and lead angle factor from Table 15b that corresponds to the feed ratio (chosen feed)/$f_{max}$, not the ratio (chosen feed)/*optimum* feed. For a feed ratio = 0.012/0.014 = 0.86 (chosen feed/$f_{max}$), depth of cut = 0.2 inch and lead angle = 45°, the depth of cut factor $F_d$ in Table 15b is between 0.72 and 0.74. Therefore, the final cutting speed for this example is $V = V_{opt} \times F_f \times F_d = 15 \times 7.27 \times 0.73 = 80$ ft/min.

*Slit and Side Milling:* This table only applies when feed is parallel to the work surface, as in slotting. If feed is perpendicular to the work surface, as in cutting off, obtain the required speed-correction factor from Table 15d (face milling). The minimum and maximum feeds/tooth for slit and side milling are determined in the manner described above, however, the axial depth of cut factor $F_d$ is not required. The adjusted cutting speed, valid for cutters of any thickness (width), is given by $V = V_{opt} \times F_f$. Examples are given in the text.

## Table 15d. Face Milling Speed Adjustment Factors for Feed, Depth of Cut, and Lead Angle

| | Cutting Speed $V = V_{opt} \times F_f \times F_d \times F_{ar}$ | | | | | | | | | | | | | | | | | | | | | | | |
|---|---|---|---|---|---|---|---|---|---|---|---|---|---|---|---|---|---|---|---|---|---|---|---|---|
| Ratio of Chosen Feed to *Optimum* | Ratio of the two cutting speeds (*average/optimum*) given in the tables $V_{avg}/V_{opt}$ | | | | | | | Depth of Cut, inch (mm), and Lead Angle | | | | | | | | | | Ratio of Radial Depth of Cut/Cutter Diameter, $ar/D$ | | | | | | |
| | | | | | | | | 1 in (25.4 mm) | | 0.4 in (10.2 mm) | | 0.2 in (5.1 mm) | | 0.1 in (2.4 mm) | | 0.04 in (1.0 mm) | | | | | | | | |
| | 1.00 | 1.10 | 1.25 | 1.35 | 1.50 | 1.75 | 2.00 | 15° | 45° | 15° | 45° | 15° | 45° | 15° | 45° | 15° | 45° | 1.00 | 0.75 | 0.50 | 0.40 | 0.30 | 0.20 | 0.10 |
| Feed | Feed Factor, $F_f$ | | | | | | | Depth of Cut Factor, $F_d$ | | | | | | | | | | Radial Depth of Cut Factor, $F_{ar}$ | | | | | | |
| 1.00 | 1.0 | 1.0 | 1.0 | 1.0 | 1.0 | 1.0 | 1.0 | 0.78 | 1.11 | 0.94 | 1.16 | 0.90 | 1.10 | 1.00 | 1.29 | 1.47 | 1.66 | 0.72 | 1.00 | 1.53 | 1.89 | 2.43 | 3.32 | 5.09 |
| 0.90 | 1.00 | 1.02 | 1.05 | 1.07 | 1.09 | 1.10 | 1.12 | 0.78 | 1.10 | 0.94 | 1.16 | 0.90 | 1.09 | 1.00 | 1.27 | 1.45 | 1.58 | 0.73 | 1.00 | 1.50 | 1.84 | 2.24 | 3.16 | 4.69 |
| 0.80 | 1.00 | 1.03 | 1.09 | 1.10 | 1.15 | 1.20 | 1.25 | 0.80 | 1.10 | 0.94 | 1.14 | 0.91 | 1.08 | 1.00 | 1.25 | 1.40 | 1.52 | 0.75 | 1.00 | 1.45 | 1.73 | 2.15 | 2.79 | 3.89 |
| 0.70 | 1.00 | 1.05 | 1.13 | 1.22 | 1.22 | 1.32 | 1.43 | 0.81 | 1.09 | 0.95 | 1.14 | 0.91 | 1.08 | 1.00 | 1.24 | 1.39 | 1.50 | 0.75 | 1.00 | 1.44 | 1.72 | 2.12 | 2.73 | 3.77 |
| 0.60 | 1.00 | 1.08 | 1.20 | 1.25 | 1.35 | 1.50 | 1.66 | 0.81 | 1.09 | 0.95 | 1.13 | 0.92 | 1.08 | 1.00 | 1.23 | 1.38 | 1.48 | 0.76 | 1.00 | 1.42 | 1.68 | 2.05 | 2.61 | 3.52 |
| 0.50 | 1.00 | 1.10 | 1.25 | 1.35 | 1.50 | 1.75 | 2.00 | 0.81 | 1.09 | 0.95 | 1.13 | 0.92 | 1.08 | 1.00 | 1.23 | 1.37 | 1.47 | 0.76 | 1.00 | 1.41 | 1.66 | 2.02 | 2.54 | 3.39 |
| 0.40 | 1.00 | 1.09 | 1.28 | 1.44 | 1.66 | 2.03 | 2.43 | 0.82 | 1.08 | 0.95 | 1.12 | 0.92 | 1.07 | 1.00 | 1.21 | 1.34 | 1.43 | 0.78 | 1.00 | 1.37 | 1.60 | 1.90 | 2.34 | 2.99 |
| 0.30 | 1.00 | 1.06 | 1.32 | 1.52 | 1.85 | 2.42 | 3.05 | 0.84 | 1.07 | 0.96 | 1.11 | 0.93 | 1.06 | 1.00 | 1.18 | 1.30 | 1.37 | 0.80 | 1.00 | 1.32 | 1.51 | 1.76 | 2.10 | 2.52 |
| 0.20 | 1.00 | 1.00 | 1.34 | 1.60 | 2.07 | 2.96 | 4.03 | 0.86 | 1.06 | 0.96 | 1.09 | 0.94 | 1.05 | 1.00 | 1.15 | 1.24 | 1.29 | 0.82 | 1.00 | 1.26 | 1.40 | 1.58 | 1.79 | 1.98 |
| 0.10 | 1.00 | 0.80 | 1.20 | 1.55 | 2.24 | 3.74 | 5.84 | 0.90 | 1.04 | 0.97 | 1.06 | 0.96 | 1.04 | 1.00 | 1.10 | 1.15 | 1.18 | 0.87 | 1.00 | 1.16 | 1.24 | 1.31 | 1.37 | 1.32 |

For HSS (high-speed steel) tool speeds in the first speed column, use Table 15a to determine appropriate feeds and depths of cut.

Tabular speeds and feeds data for face milling in Table 11 through Table 14 are based on a 10-tooth, 8-inch diameter face mill, operating with a 15-degree lead angle, $\frac{3}{64}$-inch cutter insert nose radius, axial depth of cut = 0.1 inch, and radial depth (width) of cut = 6 inches (i.e., width of cut to cutter diameter ratio = $\frac{3}{4}$). For other depths of cut (radial or axial), lead angles, or feed, calculate the ratio of desired (new) feed to *optimum* feed (largest of the two feeds given in the speed table), and the ratio of the two cutting speeds ($V_{avg}/V_{opt}$). Use these ratios to find the feed factor $F_f$ at the intersection of the feed ratio row and the speed ratio column in the left third of the table. The depth of cut factor $F_d$ is found in the same row as the feed factor, in the center third of the table, in the column corresponding to the depth of cut and lead angle. The radial depth of cut factor $F_{ar}$ is found in the same row as the feed factor, in the right third of the table, in the column corresponding to the radial depth of cut to cutter diameter ratio $ar/D$. The adjusted cutting speed can be calculated from $V = V_{opt} \times F_f \times F_d \times F_{ar}$, where $V_{opt}$ is the smaller (*optimum*) of the two speeds from the speed table (from the left side of the column containing the two feed/speed pairs).

The cutting speeds as calculated above are valid if the cutter axis is centered above or close to the center line of the workpiece (eccentricity is small). For larger eccentricity (i.e., the cutter axis is offset from the center line of the workpiece by about one-half the cutter diameter or more), use the adjustment factors from Table 15b and Table 15c (end and side milling) instead of the factors from this table. Use Table 15e to adjust end and face milling speeds for increased tool life up to 180 minutes.

*Slit and Slot Milling:* Tabular speeds are valid for all tool diameters and widths. Adjustments to the given speeds for other feeds and depths of cut depend on the circumstances of the cut. *Case 1:* If the cutter is fed directly into the workpiece, i.e., the feed is perpendicular to the surface of the workpiece, as in cutting off, then this table (face milling) is used to adjust speeds for other feeds. The depth of cut factor is not used for slit milling ($F_d$ = 1.0), so the adjusted cutting speed $V = V_{opt} \times F_f \times F_{ar}$. For determining the factor $F_{ar}$, the radial depth of cut *ar* is the length of cut created by the portion of the cutter engaged in the work. *Case 2:* If the cutter is fed parallel to the surface of the workpiece, as in slotting, then Table 15b and Table 15c are used to adjust the given speeds for other feeds. See Fig. 5.

**Table 15e. Tool Life Adjustment Factors for Face Milling, End Milling, Drilling, and Reaming**

| Tool Life, $T$ (minutes) | Face Milling with Carbides and Mixed Ceramics | | | End Milling with Carbides and HSS | | | Twist Drilling and Reaming with HSS | | |
|---|---|---|---|---|---|---|---|---|---|
| | $f_s$ | $f_m$ | $f_l$ | $f_s$ | $f_m$ | $f_l$ | $f_s$ | $f_m$ | $f_l$ |
| 15 | 1.69 | 1.78 | 1.87 | 1.10 | 1.23 | 1.35 | 1.11 | 1.21 | 1.30 |
| 45 | 1.00 | 1.00 | 1.00 | 1.00 | 1.00 | 1.00 | 1.00 | 1.00 | 1.00 |
| 90 | 0.72 | 0.70 | 0.67 | 0.94 | 0.89 | 0.83 | 0.93 | 0.89 | 0.85 |
| 180 | 0.51 | 0.48 | 0.45 | 0.69 | 0.69 | 0.69 | 0.87 | 0.80 | 0.72 |

The speeds and feeds given in Table 11 through Table 14 and Table 17 through Table 23 (except for HSS speeds in the first speed column) are based on a 45-minute tool life. To adjust the given speeds to obtain another tool life, multiply the adjusted cutting speed for the 45-minute tool life $V_{45}$ by the tool life factor from this table according to the following rules: for small feeds, where feed $\leq ½ f_{opt}$, the cutting speed for the desired tool life $T$ is $V_T = f_s \times V_{15}$; for medium feeds, where $½ f_{opt} < \text{feed} < ¾ f_{opt}$, $V_T = f_m \times V_{15}$; and for larger feeds, where $¾ f_{opt} \leq \text{feed} \leq f_{opt}$, $V_T = f_l \times V_{15}$. Here, $f_{opt}$ is the largest (*optimum*) feed of the two feed/speed values given in the speed tables or the maximum feed $f_{max}$ obtained from Table 15c, if that table was used in calculating speed adjustment factors.

**Table 16. Cutting Tool Grade Descriptions and Common Vendor Equivalents**

| Grade Description | Tool Identification Code | Approximate Vendor Equivalents | | | |
|---|---|---|---|---|---|
| | | Sandvik Coromant | Kennametal | Seco | Valenite |
| Cubic boron nitride | 1 | CB50 | KD050 | CBN20 | VC721 |
| Ceramics | 2 | CC620 | K060 | 480 | — |
| | 3 | CC650 | K090 | 480 | Q32 |
| | 4 (Whiskers) | CC670 | KYON2500 | — | — |
| | 5 (Sialon) | CC680 | KYON2000 | 480 | — |
| | 6 | CC690 | KYON3000 | — | Q6 |
| Cermets | 7 | CT515 | KT125 | CM | VC605 |
| | 8 | CT525 | KT150 | CR | VC610 |
| Polycrystalline | 9 | CD10 | KD100 | PAX20 | VC727 |
| Coated carbides | 10 | GC-A | — | — | — |
| | 11 | GC3015 | KC910 | TP100 | SV310 |
| | 12 | GC235 | KC9045 | TP300 | SV235 |
| | 13 | GC4025 | KC9025 | TP200 | SV325 |
| | 14 | GC415 | KC950 | TP100 | SV315 |
| Uncoated carbides | 15 | H13A | K8, K4H | 883 | VC2 |
| | 16 | S10T | K420, K28 | CP20 | VC7 |
| | 17 | S1P | K45 | CP20 | VC7 |
| | 18 | S30T | — | CP25 | VC5 |
| | 19 | S6 | K21, K25 | CP50 | VC56 |
| | 20 | SM30 | KC710 | CP25 | VC35M |

See Table 2 on page 854 and the section *Cemented Carbides and Other Hard Materials* for more detailed information on cutting tool grades.

The identification codes in column two correspond to the grade numbers given in the footnotes to Table 1 to Table 4b, Table 6 to Table 14, and Table 17 to Table 23.

**Using the Speeds and Feeds Tables for Drilling, Reaming, and Threading.—**The first two speed columns in Table 17 through Table 23 give traditional *Handbook* speeds for drilling and reaming. The following material can be used for selecting feeds for use with the traditional speeds.

The remaining columns in Table 17 through Table 23 contain combined feed/speed data for drilling, reaming, and threading, organized in the same manner as in the turning and milling tables. Operating at the given speeds and feeds is expected to result in a tool life of approximately 45 minutes, except for indexable insert drills, which have an expected tool life of approximately 15 minutes per edge. Examples of using this data follow.

Adjustments to HSS drilling speeds for feed and diameter are made using Table 22; Table 5a is used for adjustments to indexable insert drilling speeds, where one-half the drill diameter $D$ is used for the depth of cut. Tool life for HSS drills, reamers, and thread chasers and taps may be adjusted using Table 15e and for indexable insert drills using Table 5b.

The feed for drilling is governed primarily by the size of the drill and by the material to be drilled. Other factors that also affect selection of the feed are the workpiece configuration, the rigidity of the machine tool and the workpiece setup, and the length of the chisel edge. A chisel edge that is too long will result in a very significant increase in the thrust force, which may cause large deflections to occur on the machine tool and drill breakage.

For ordinary twist drills, the feed rate used is given in the table that follows. For additional information also see the table *Approximate Cutting Speeds and Feeds for Standard Automatic Screw Machine Tools—Brown and Sharpe* on page 1223.

**Feet Rate for Twist Drills**

| Drill Size, inch (mm) | Feed Rate, inch/rev (mm/rev) |
|---|---|
| smaller than ⅛ inch (3.175 mm) | 0.001 to 0.003 in/rev (0.025–0.08 mm/rev) |
| from ⅛ to ¼ inch (3.175–6.35 mm) | 0.002 to 0.006 in/rev (0.05–0.15 mm/rev) |
| from ¼ to ½ inch (6.35–12.7 mm) | 0.004 to 0.010 in/rev (0.10–0.25 mm/rev) |
| from ½ to 1 inch (12.7–25.4 mm) | 0.007 to 0.015 in/rev (0.18–0.38 mm/rev) |
| larger than 1 inch (25.4 mm) | 0.010 to 0.025 in/rev (0.25–0.64 mm/rev) |

The lower values in the feed ranges should be used for hard materials such as tool steels, superalloys, and work-hardening stainless steels; the higher values in the feed ranges should be used to drill soft materials such as aluminum and brass.

*Example 1, Drilling:* Determine the cutting speed and feed for use with HSS drills in drilling 1120 steel.

Table 17 gives two sets of speed and feed parameters for drilling 1120 steel with HSS drills. These sets are 16⁄50 and 8⁄95, i.e., 0.016 in/rev feed at 50 ft/min and 0.008 in/rev at 95 fpm, respectively. These feed/speed sets are based on a 0.6-inch diameter drill. Tool life for either of the given feed/speed settings is expected to be approximately 45 minutes.

For different feeds or drill diameters, the cutting speeds must be adjusted and can be determined from $V = V_{opt} \times F_f \times F_d$, where $V_{opt}$ is the minimum speed for this material given in the speed table (50 fpm in this example) and $F_f$ and $F_d$ are the adjustment factors for feed and diameter, respectively, found in Table 22.

**Table 17. Speeds and Feeds for Drilling, Reaming, and Threading Plain Carbon and Alloy Steels**

| Material | Brinell Hardness Number | Drilling HSS Speed (fpm) | Reaming HSS Speed (fpm) | | Drilling HSS Opt. | Drilling HSS Avg. | Drilling Indexable Insert Coated Carbide Opt. | Drilling Indexable Insert Coated Carbide Avg. | Reaming HSS Opt. | Reaming HSS Avg. | Threading HSS Opt. | Threading HSS Avg. |
|---|---|---|---|---|---|---|---|---|---|---|---|---|
| | | | | f = feed (0.001 in/rev), s = speed (ft/min) *Metric Units:* f × 25.4 = mm/rev, s × 0.3048 = m/min | | | | | | | | |
| Free-machining plain carbon steels (Resulfurized): 1212, 1213, 1215 | 100-150 | 120 | 80 | f | 21 | 11 | 8 | 4 | 36 | 18 | 83 | 20 |
| | | | | s | 55 | 125 | 310 | 620 | 140 | 185 | 140 | 185 |
| | 150-200 | 125 | 80 | | | | | | | | | |
| (Resulfurized): 1108, 1109, 1115, 1117, 1118, 1120, 1126, 1211 | 100-150 | 110 | 75 | f | 16 | 8 | 8 | 4 | 27 | 14 | 83 | 20 |
| | 150-200 | 120 | 80 | s | 50 | 95 | 370 | 740 | 105 | 115 | 90 | 115 |
| (Resulfurized): 1132, 1137, 1139, 1140, 1144, 1146, 1151 | 175-225 | 100 | 65 | f | | | 8 | 4 | | | | |
| | | | | s | | | 365 | 735 | | | | |
| | 275-325 | 70 | 45 | | | | | | | | | |
| | 325-375 | 45 | 30 | | | | | | | | | |
| | 375-425 | 35 | 20 | | | | | | | | | |
| (Leaded): 11L17, 11L18, 12L13, 12L14 | 100-150 | 130 | 85 | | | | | | | | | |
| | 150-200 | 120 | 80 | | | | | | | | | |
| | 200-250 | 90 | 60 | f | | | 8 | 4 | | | | |
| | | | | s | | | 365 | 735 | | | | |
| Plain carbon steels: 1006, 1008, 1009, 1010, 1012, 1015, 1016, 1017, 1018, 1019, 1020, 1021, 1022, 1023, 1024, 1025, 1026, 1513, 1514 | 100-125 | 100 | 65 | f | 21 | 11 | 8 | 4 | 36 | 18 | 83 | 20 |
| | | | | s | 55 | 125 | 310 | 620 | 140 | 185 | 140 | 185 |
| | 125-175 | 90 | 60 | | | | | | | | | |
| | 175-225 | 70 | 45 | f | | | 8 | 4 | | | | |
| | 225-275 | 60 | 40 | s | | | 365 | 735 | | | | |
| Plain carbon steels: 1027, 1030, 1033, 1035, 1036, 1037, 1038, 1039, 1040, 1041, 1042, 1043, 1045, 1046, 1048, 1049, 1050, 1052, 1524, 1526, 1527, 1541 | 125-175 | 90 | 60 | | | | | | | | | |
| | 175-225 | 75 | 50 | | | | | | | | | |
| | 225-275 | 60 | 40 | f | | | 8 | 4 | | | | |
| | 275-325 | 50 | 30 | s | | | 365 | 735 | | | | |
| | 325-375 | 35 | 20 | | | | | | | | | |
| | 375-425 | 25 | 15 | | | | | | | | | |

**Table 17.** *(Continued)* **Speeds and Feeds for Drilling, Reaming, and Threading Plain Carbon and Alloy Steels**

| Material | Brinell Hardness Number | Drilling | Reaming | | Drilling | | | | Reaming | | Threading | |
|---|---|---|---|---|---|---|---|---|---|---|---|---|
| | | HSS | | | HSS | | Indexable Insert Coated Carbide | | HSS | | HSS | |
| | | Speed (fpm) | | | f = feed (0.001 in/rev), s = speed (ft/min) *Metric Units:* f × 25.4 = mm/rev, s × 0.3048 = m/min | | | | | | | |
| | | | | | *Opt.* | *Avg.* | *Opt.* | *Avg.* | *Opt.* | *Avg.* | *Opt.* | *Avg.* |
| Plain carbon steels *(Continued)*: 1055, 1060, 1064, 1065, 1070, 1074, 1078, 1080, 1084, 1086, 1090, 1095, 1548, 1551, 1552, 1561, 1566 | 125-175 | 85 | 55 | f | 16 | 8 | 8 | 4 | 27 | 14 | 83 | 20 |
| | 175-225 | 70 | 45 | s | 50 | 95 | 370 | 740 | 105 | 115 | 90 | 115 |
| | 225-275 | 50 | 30 | f<br>s | | | 8<br>365 | 4<br>735 | | | | |
| | 275-325 | 40 | 25 | | | | | | | | | |
| | 325-375 | 30 | 20 | | | | | | | | | |
| | 375-425 | 15 | 10 | | | | | | | | | |
| Free-machining alloy steels (Resulfurized): 4140, 4150 | 175-200 | 90 | 60 | f | 16 | 8 | 8 | 4 | 26 | 13 | 83 | 20 |
| | 200-250 | 80 | 50 | s | 75 | 140 | 410 | 685 | 150 | 160 | 125 | 160 |
| | 250-300 | 55 | 30 | f<br>s | | | 8<br>355 | 4<br>600 | | | | |
| | 300-375 | 40 | 25 | f | | | 8 | 4 | | | | |
| | 375-425 | 30 | 15 | s | | | 310 | 525 | | | | |
| (Leaded): 41L30, 41L40, 41L47, 41L50, 43L47, 51L32, 52L100, 86L20, 86L40 | 150-200 | 100 | 65 | f<br>s | 16<br>50 | 8<br>95 | 8<br>370 | 4<br>740 | 27<br>105 | 14<br>115 | 83<br>90 | 20<br>115 |
| | 200-250 | 90 | 60 | f<br>s | | | 8<br>365 | 4<br>735 | | | | |
| | 250-300 | 65 | 40 | | | | | | | | | |
| | 300-375 | 45 | 30 | | | | | | | | | |
| | 375-425 | 30 | 15 | | | | | | | | | |
| Alloy steels: 4012, 4023, 4024, 4028, 4118, 4320, 4419, 4422, 4427, 4615, 4620, 4621, 4626, 4718, 4720, 4815, 4817, 4820, 5015, 5117, 5120, 6118, 8115, 8615, 8617, 8620, 8622, 8625, 8627, 8720, 8822, 94B17 | 125-175 | 85 | 55 | f | 16 | 8 | 8 | 4 | 26 | 13 | 83 | 20 |
| | 175-225 | 70 | 45 | s | 75 | 140 | 410 | 685 | 150 | 160 | 125 | 160 |
| | 225-275 | 55 | 35 | f<br>s | | | 8<br>355 | 4<br>600 | | | | |
| | 275-325 | 50 | 30 | f<br>s | 11<br>50 | 6<br>85 | 8<br>335 | 4<br>570 | 19<br>95 | 10<br>135 | 83<br>60 | 20<br>95 |
| | 325-375 | 35 | 25 | f | | | 8 | 4 | | | | |
| | 375-425 | 25 | 15 | s | | | 310 | 525 | | | | |

**Table 17.** *(Continued)* **Speeds and Feeds for Drilling, Reaming, and Threading Plain Carbon and Alloy Steels**

| Material | Brinell Hardness Number | Drilling | Reaming | | Drilling | | | | Reaming | | Threading | |
|---|---|---|---|---|---|---|---|---|---|---|---|---|
| | | HSS | | | HSS | | Indexable Insert Coated Carbide | | HSS | | HSS | |
| | | Speed (fpm) | | | f = feed (0.001 in/rev), s = speed (ft/min) *Metric Units:* f × 25.4 = mm/rev, s × 0.3048 = m/min | | | | | | | |
| | | | | | *Opt.* | *Avg.* | *Opt.* | *Avg.* | *Opt.* | *Avg.* | *Opt.* | *Avg.* |
| Alloy steels: 1330, 1335, 1340, 1345, 4032, 4037, 4042, 4047, 4130, 4135, 4137, 4140, 4142, 4145, 4147, 4150, 4161, 4337, 4340, 50B44, 50B46, 50B50, 50B60, 5130, 5132, 5140, 5145, 5147, 5150, 5160, 51B60, 6150, 81B45, 8630, 8635, 8637, 8640, 8642, 8645, 8650, 8655, 8660, 8740, 9254, 9255, 9260, 9262, 94B30 | 175-225 | 75 (60) | 50 (40) | f | 16 | 8 | 8 | 4 | 26 | 13 | 83 | 20 |
| | | | | s | 75 | 140 | 410 | 685 | 150 | 160 | 125 | 160 |
| | 225-275 | 60 (50) | 40 (30) | f | | | 8 | 4 | | | | |
| | | | | s | | | 355 | 600 | | | | |
| | 275-325 | 45 (35) | 30 (25) | f | 11 | 6 | 8 | 4 | 19 | 10 | 83 | 20 |
| | | | | s | 50 | 85 | 335 | 570 | 95 | 135 | 60 | 95 |
| E51100, E52100: use (HSS speeds) | 325-375 | 30 (30) | 15 (20) | f | | | 8 | 4 | | | | |
| | 375-425 | 20 (20) | 15 (10) | s | | | 310 | 525 | | | | |
| Ultra-high-strength steels (not AISI): AMS 6421 (98B37 Mod.), 6422 (98BV40), 6424, 6427, 6428, 6430, 6432, 6433, 6434, 6436, and 6442; 300M, D6ac | 220-300 | 50 | 30 | f | | | 8 | 4 | | | | |
| | 300-350 | 35 | 20 | s | | | 325 | 545 | | | | |
| | 350-400 | 20 | 10 | f | | | 8 | 4 | | | | |
| | | | | s | | | 270 | 450 | | | | |
| Maraging steels (not AISI): 18% Ni Grade 200, 250, 300, and 350 | 250-325 | 50 | 30 | f | | | 8 | 4 | | | | |
| | | | | s | | | 325 | 545 | | | | |
| Nitriding steels (not AISI): Nitralloy 125, 135, 135 Mod., 225, and 230, Nitralloy N, Nitralloy EZ, Nitrex I | 200-250 | 60 | 40 | f | 16 | 8 | 8 | 4 | 26 | 13 | 83 | 20 |
| | | | | s | 75 | 140 | 410 | 685 | 150 | 160 | 125 | 160 |
| | 300-350 | 35 | 20 | f | | | 8 | 4 | | | | |
| | | | | s | | | 310 | 525 | | | | |

The two leftmost speed columns in this table contain traditional *Handbook* speeds for drilling and reaming with HSS steel tools. The section Feed Rates for Drilling and Reaming contains useful information concerning feeds to use in conjunction with these speeds.

*HSS Drilling and Reaming:* The combined feed/speed data for drilling are based on a 0.60-inch diameter HSS drill with standard drill point geometry (2-flute with 118° tip angle). Speed adjustment factors in Table 22 are used to adjust drilling speeds for other feeds and drill diameters. Examples of using this data are given in the text. The given speeds and feeds for reaming are based on an 8-tooth, 25⁄32-inch diameter, 30° lead angle reamer, and a 0.008-inch radial depth of cut. For other feeds, the correct speed can be obtained by interpolation using the given speeds if the desired feed lies in the recommended range (between the given values of *optimum* and *average* feed). If a feed lower than the given *average* value is chosen, the speed should be maintained at the corresponding *average* speed (i.e., the highest of the two speed values given). The cutting speeds for reaming do not require adjustment for tool diameters for standard ratios of radial depth of cut to reamer diameter (i.e., $f_d$ = 1.00). Speed adjustment factors to modify tool life are found in Table 15e.

*Indexable Insert Drilling:* The feed/speed data for indexable insert drilling are based on a tool with two cutting edges, an insert nose radius of 3⁄64 inch (1.2 mm), a 10-degree lead angle, and diameter $D = 1$ inch (2.54 mm). Adjustments to cutting speed for feed and depth of cut are made using Table 5a on page 1091 (Adjustment Factors) using a depth of cut of $D/2$, or one-half the drill diameter. Expected tool life at the given speeds and feeds is approximately 15 minutes for short hole drilling (i.e., where maximum hole depth is about $2D$ or less). Speed adjustment factors to increase tool life are found in Table 5b.

*Tapping and Threading:* The data in this column are intended for use with thread chasers and for tapping. The feed used for tapping and threading must be equal to the lead (feed = lead = pitch) of the thread being cut. The two feed/speed pairs given for each material, therefore, are representative speeds for two thread pitches, 12 and 50 threads per inch (1/0.083 = 12, and 1/0.020 = 50). Tool life is expected to be approximately 45 minutes at the given speeds and feeds. When cutting fewer than 12 threads per inch (pitch ≥ 0.08 inch or 2.1 mm), use the lower (*optimum*) speed; for cutting more than 50 threads per inch (pitch ≤ 0.02 inch or 0.51 mm), use the larger (*average*) speed; and, in the intermediate range between 12 and 50 threads per inch, interpolate between the given *average* and *optimum* speeds.

The combined feed/speed data in this table are based on tool grades (identified in Table 16) as follows: coated carbide = 10.

*Example 2, Drilling:* If the 1120 steel of Example 1 is to be drilled with a 0.60-inch drill at a feed of 0.012 in/rev, what is the cutting speed in ft/min? Also, what spindle rpm of the drilling machine is required to obtain this cutting speed?

To find the feed factor $F_d$ in Table 22, calculate the ratio of the desired feed to the *optimum* feed and the ratio of the two cutting speeds given in the speed tables. The desired feed is 0.012 in/rev and the *optimum* feed, as explained above is 0.016 in/rev, therefore, $feed/f_{opt} = 0.012/0.016 = 0.75$ and $V_{avg}/V_{opt} = 95/50 = 1.9$, approximately 2.

The feed factor $F_f$ is found at the intersection of the feed ratio row and the speed ratio column. $F_f = 1.40$ corresponds to about halfway between 1.31 and 1.50, which are the feed factors that correspond to $V_{avg}/V_{opt} = 2.0$ and $feed/f_{opt}$ ratios of 0.7 and 0.8, respectively. $F_d$, the diameter factor, is found on the same row as the feed factor (halfway between the 0.7 and 0.8 rows, for this example) under the column for drill diameter = 0.60 inch. Because the speed table values are based on a 0.60-inch drill diameter, $F_d = 1.0$ for this example, and the cutting speed is $V = V_{opt} \times F_f \times F_d = 50 \times 1.4 \times 1.0 = 70$ ft/min. The spindle speed in rpm is $N = 12 \times V/(\pi \times D) = 12 \times 70/(3.14 \times 0.6) = 445$ rpm.

*Example 3, Drilling:* Using the same material and feed as in the previous example, what cutting speeds are required for 0.079-inch and 4-inch diameter drills? What machine rpm is required for each?

Because the feed is the same as in the previous example, the feed factor is $F_f = 1.40$ and does not need to be recalculated. The diameter factors are found in Table 22 on the same row as the feed factor for the previous example (about halfway between the diameter factors corresponding to $feed/f_{opt}$ values of 0.7 and 0.8) in the column corresponding to drill diameters 0.079 and 4.0 inches, respectively. Results of the calculations are summarized below.

| *Drill diameter = 0.079 inch* | *Drill diameter = 4.0 inches* |
|---|---|
| $F_f = 1.40$ | $F_f = 1.40$ |
| $F_d = (0.34 + 0.38)/2 = 0.36$ | $F_d = (1.95 + 1.73)/2 = 1.85$ |
| $V = 50 \times 1.4 \times 0.36 = 25.2$ fpm | $V = 50 \times 1.4 \times 1.85 = 129.5$ fpm |
| $12 \times 25.2/(3.14 \times 0.079) = 1219$ rpm | $12 \times 129.5/(3.14 \times 4) = 124$ rpm |

*Drilling Difficulties:* A drill split at the web is evidence of too much feed or insufficient lip clearance at the center due to improper grinding. Rapid wearing away of the extreme outer corners of the cutting edges indicates that the speed is too high. A drill chipping or breaking out at the cutting edges indicates that either the feed is too heavy or the drill has been ground with too much lip clearance. Nothing will "check" a high-speed steel drill quicker than to turn a stream of cold water on it after it has been heated while in use. It is equally bad to plunge it in cold water after the point has been heated in grinding. The small checks or cracks resulting from this practice will eventually chip out and cause rapid wear or breakage. Insufficient speed in drilling small holes with hand feed greatly increases the risk of breakage, especially at the moment the drill is breaking through the farther side of the work, due to the operator's inability to gage the feed when the drill is running too slowly.

Small drills have heavier webs and smaller flutes in proportion to their size than do larger drills, so breakage due to clogging of chips in the flutes is more likely to occur. When drilling holes deeper than three times the diameter of the drill, it is advisable to withdraw the drill (peck feed) at intervals to remove the chips and permit coolant to reach the tip of the drill.

*Drilling Holes in Glass:* The simplest method of drilling holes in glass is to use a standard, tungsten carbide-tipped masonry drill of the appropriate diameter, in a gun-drill. The edges of the carbide in contact with the glass should be sharp. Kerosene or other liquid may be used as a lubricant, and a light force is maintained on the drill until just before the point breaks through. The hole should then be started from the other side if possible, or a very light force applied for the remainder of the operation, to prevent excessive breaking of material from the sides of the hole. As the hard particles of glass are abraded, they accumulate and act to abrade the hole, so it may be advisable to use a slightly smaller drill than the required diameter of the finished hole.

Alternatively, for holes of medium and large size, use brass or copper tubing, having an outside diameter equal to the size of hole required. Revolve the tube at a peripheral speed of about 100 feet per minute (30.5 m/min), and use carborundum (80 to 100 grit) and light machine oil between the end of the pipe and the glass. Insert the abrasive under the drill with a thin piece of soft wood, to avoid scratching the glass. The glass should be supported by a felt or rubber cushion, not much larger than the hole to be drilled. If practicable, it is advisable to drill about halfway through, then turn the glass over, and drill down to meet the first cut. Any fin that may be left in the hole can be removed with a round second-cut file wetted with turpentine.

Smaller-diameter holes may also be drilled with triangular-shaped cemented carbide drills that can be purchased in standard sizes. The end of the drill is shaped into a long tapering triangular point. The other end of the cemented carbide bit is brazed onto a steel shank. A glass drill can be made to the same shape from hardened drill rod or an old three-cornered file. The location at which the hole is to be drilled is marked on the workpiece. A dam of putty or glazing compound is built up on the work surface to contain the cutting fluid, which can be either kerosene or turpentine mixed with camphor. Chipping on the back edge of the hole can be prevented by placing a scrap plate of glass behind the area to be drilled and drilling into the backup glass. This procedure also provides additional support to the workpiece and is essential for drilling very thin plates. The hole is usually drilled with an electric hand drill. When the hole is being produced, the drill should be given a small circular motion using the point as a fulcrum, thereby providing a clearance for the drill in the hole.

Very small round or intricately shaped holes and narrow slots can be cut in glass by the ultrasonic machining process or by the abrasive jet cutting process.

## Table 18. Speeds and Feeds for Drilling, Reaming, and Threading Tool Steels

| Material | Brinell Hardness Number | Drilling | Reaming | Drilling | | | | | Reaming | | Threading | |
|---|---|---|---|---|---|---|---|---|---|---|---|---|
| | | HSS | | | HSS | | Indexable Insert Uncoated Carbide | | HSS | | HSS | |
| | | Speed (fpm) | | f = feed (0.001 in/rev), s = speed (ft/min) *Metric Units:* f × 25.4 = mm/rev, s × 0.3048 = m/min | | | | | | | | |
| | | | | | *Opt.* | *Avg.* | *Opt.* | *Avg.* | *Opt.* | *Avg.* | *Opt.* | *Avg.* |
| Water hardening: W1, W2, W5 | 150-200 | 85 | 55 | | | | | | | | | |
| Shock resisting: S1, S2, S5, S6, S7 | 175-225 | 50 | 35 | | | | | | | | | |
| Cold-work (oil hardening): O1, O2, O6, O7 | 175-225 | 45 | 30 | | | | | | | | | |
| (High carbon, high chromium): D2, D3, D4, D5, D7 | 200-250 | 30 | 20 | | | | | | | | | |
| (Air hardening): A2, A3, A8, A9, A10 | 200-250 | 50 | 35 | f<br>s | 15<br>45 | 7<br>85 | 8<br>360 | 4<br>605 | 24<br>90 | 12<br>95 | 83<br>75 | 20<br>95 |
| A4, A6 | 200-250 | 45 | 30 | | | | | | | | | |
| A7 | 225-275 | 30 | 20 | | | | | | | | | |
| Hot-work (chromium type): H10, H11, H12, H13, H14, H19 | 150-200 | 60 | 40 | | | | | | | | | |
| | 200-250 | 50 | 30 | | | | | | | | | |
| | 325-375 | 30 | 20 | f<br>s | | | 8<br>270 | 4<br>450 | | | | |
| (Tungsten type): H21, H22, H23, H24, H25, H26 | 150-200 | 55 | 35 | | | | | | | | | |
| | 200-250 | 40 | 25 | | | | | | | | | |
| (Molybdenum type): H41, H42, H43 | 150-200 | 45 | 30 | | | | | | | | | |
| | 200-250 | 35 | 20 | | | | | | | | | |
| Special-purpose, low alloy: L2, L3, L6 | 150-200 | 60 | 40 | | | | | | | | | |
| Mold steel: P2, P3, P4, P5, P6P20, P21 | 100-150 | 75 | 50 | f<br>s | 15<br>45 | 7<br>85 | 8<br>360 | 4<br>605 | 24<br>90 | 12<br>95 | 83<br>75 | 20<br>95 |
| | 150-200 | 60 | 40 | | | | | | | | | |
| High-speed steel: M1, M2, M6, M10, T1, T2, T6 | 200-250 | 45 | 30 | | | | | | | | | |
| M3-1, M4, M7, M30, M33, M34, M36, M41, M42, M43, M44, M46, M47, T5, T8 | 225-275 | 35 | 20 | | | | | | | | | |
| T15, M3-2 | 225-275 | 25 | 15 | | | | | | | | | |

See the footnote to Table 17 for instructions concerning the use of this table. The combined feed/speed data in this table are based on tool grades (identified in Table 16) as follows: coated carbide = 10.

## Table 19. Speeds and Feeds for Drilling, Reaming, and Threading Stainless Steels

| Material | Brinell Hardness Number | Drilling HSS Speed (fpm) | Reaming HSS Speed (fpm) | | Drilling HSS Opt. | Drilling HSS Avg. | Drilling Indexable Insert Coated Carbide Opt. | Drilling Indexable Insert Coated Carbide Avg. | Reaming HSS Opt. | Reaming HSS Avg. | Threading HSS Opt. | Threading HSS Avg. |
|---|---|---|---|---|---|---|---|---|---|---|---|---|
| | | | | **f** = feed (0.001 in/rev), **s** = speed (ft/min) *Metric Units:* **f** × 25.4 = mm/rev, **s** × 0.3048 = m/min | | | | | | | | |
| Free-machining stainless steels (Ferritic): 430F, 430FSe | 135-185 | 90 | 60 | **f** | 15 | 7 | 8 | 4 | 24 | 12 | 83 | 20 |
| | | | | **s** | 25 | 45 | 320 | 540 | 50 | 50 | 40 | 51 |
| (Austenitic): 203EZ, 303, 303Se, 303MA, 303Pb, 303Cu, 303 Plus X | 135-185 | 85 | 55 | | | | | | | | | |
| | 225-275 | 70 | 45 | **f** | 15 | 7 | 8 | 4 | 24 | 12 | 83 | 20 |
| (Martensitic): 416, 416Se, 416 Plus X, 420F, 420FSe, 440F, 440FSe | 135-185 | 90 | 60 | **s** | 20 | 40 | 250 | 425 | 40 | 40 | 35 | 45 |
| | 185-240 | 70 | 45 | | | | | | | | | |
| | 275-325 | 40 | 25 | | | | | | | | | |
| | 375-425 | 20 | 10 | | | | | | | | | |
| Stainless steels (Ferritic): 405, 409, 429, 430, 434 | 135-185 | 65 | 45 | **f** | 15 | 7 | 8 | 4 | 24 | 12 | 83 | 20 |
| | | | | **s** | 25 | 45 | 320 | 540 | 50 | 50 | 40 | 51 |
| (Austenitic): 201, 202, 301, 302, 304, 304L, 305, 308, 321, 347, 348 | 135-185 | 55 | 35 | | | | | | | | | |
| | 225-275 | 50 | 30 | **f** | 15 | 7 | 8 | 4 | 24 | 12 | 83 | 20 |
| (Austenitic): 302B, 309, 309S, 310, 310S, 314, 316 | 135-185 | 50 | 30 | **s** | 20 | 40 | 250 | 425 | 40 | 40 | 35 | 45 |
| (Martensitic): 403, 410, 420, 501 | 135-175 | 75 | 50 | | | | | | | | | |
| | 175-225 | 65 | 45 | | | | | | | | | |
| | 275-325 | 40 | 25 | | | | | | | | | |
| | 375-425 | 25 | 15 | | | | | | | | | |
| (Martensitic): 414, 431, Greek Ascoloy | 225-275 | 50 | 30 | | | | | | | | | |
| | 275-325 | 40 | 25 | | | | | | | | | |
| | 375-425 | 25 | 15 | | | | | | | | | |
| (Martensitic): 440A, 440B, 440C | 225-275 | 45 | 30 | | | | | | | | | |
| | 275-325 | 40 | 25 | | | | | | | | | |
| | 375-425 | 20 | 10 | | | | | | | | | |
| (Precipitation hardening): 15-5PH, 17-4PH, 17-7PH, AF-71, 17-14CuMo, AFC-77, AM-350, AM-355, AM-362, Custom 455, HNM, PH13-8, PH14-8Mo, PH15-7Mo, Stainless W | 150-200 | 50 | 30 | **f** | 15 | 7 | 8 | 4 | 24 | 12 | 83 | 20 |
| | | | | **s** | 20 | 40 | 250 | 425 | 40 | 40 | 35 | 45 |
| | 275-325 | 45 | 25 | | | | | | | | | |
| | 325-375 | 35 | 20 | | | | | | | | | |
| | 375-450 | 20 | 10 | | | | | | | | | |

See the footnote to Table 17 for instructions concerning the use of this table. The combined feed/speed data in this table are based on tool grades (identified in Table 16) as follows: coated carbide = 10.

**Table 20. Speeds and Feeds for Drilling, Reaming, and Threading Ferrous Cast Metals**

| Material | Brinell Hardness Number | Drilling HSS Speed (fpm) | Reaming HSS Speed (fpm) | | Drilling HSS Opt. | Drilling HSS Avg. | Indexable Carbide Insert Uncoated Opt. | Uncoated Avg. | Coated Opt. | Coated Avg. | Reaming HSS Opt. | Reaming HSS Avg. | Threading HSS Opt. | Threading HSS Avg. |
|---|---|---|---|---|---|---|---|---|---|---|---|---|---|---|
| | | | | | f = feed (0.001 in/rev), s = speed (ft/min) *Metric Units:* f × 25.4 = mm/rev, s × 0.3048 = m/min | | | | | | | | | |
| ASTM Class 20 | 120-150 | 100 | 65 | | | | | | | | | | | |
| ASTM Class 25 | 160-200 | 90 | 60 | f | 16 | 8 | 11 | 6 | 11 | 6 | 26 | 13 | 83 | 20 |
| ASTM Class 30, 35, and 40 | 190-220 | 80 | 55 | s | 80 | 90 | 85 | 180 | 235 | 485 | 85 | 65 | 90 | 80 |
| ASTM Class 45 and 50 | 220-260 | 60 | 40 | f | 13 | 6 | 11 | 6 | 11 | 6 | 21 | 10 | 83 | 20 |
| ASTM Class 55 and 60 | 250-320 | 30 | 20 | s | 50 | 50 | 70 | 150 | 195 | 405 | 50 | 30 | 55 | 45 |
| ASTM Type 1, 1b, 5 (Ni resist) | 100-215 | 50 | 30 | | | | | | | | | | | |
| ASTM Type 2, 3, 6 (Ni resist) | 120-175 | 40 | 25 | | | | | | | | | | | |
| ASTM Type 2b, 4 (Ni resist) | 150-250 | 30 | 20 | | | | | | | | | | | |
| Malleable Iron | | | | | | | | | | | | | | |
| (Ferritic): 32510, 35018 | 110-160 | 110 | 75 | f | 19 | 10 | | | 11 | 6 | 30 | 16 | 83 | 20 |
| | | | | s | 80 | 100 | | | 270 | 555 | 95 | 80 | 100 | 85 |
| (Pearlitic): 40010, 43010, 45006, 45008, 48005, 50005 | 160-200 | 80 | 55 | f | 14 | 7 | 11 | 6 | 11 | 6 | 22 | 11 | 83 | 20 |
| | 200-240 | 70 | 45 | s | 65 | 65 | 85 | 180 | 235 | 485 | 65 | 45 | 70 | 60 |
| (Martensitic): 53004, 60003, 60004 | 200-255 | 55 | 35 | | | | | | | | | | | |
| (Martensitic): 70002, 70003 | 220-260 | 50 | 30 | | | | | | | | | | | |
| (Martensitic): 80002 | 240-280 | 45 | 30 | | | | | | | | | | | |
| (Martensitic): 90001 | 250-320 | 25 | 15 | | | | | | | | | | | |
| Nodular (Ductile) Iron | | | | | | | | | | | | | | |
| (Ferritic): 60-40-18, 65-45-12 | 140-190 | 100 | 65 | f | 17 | 9 | 11 | 6 | 11 | 6 | 28 | 14 | 83 | 20 |
| | | | | s | 70 | 80 | 85 | 180 | 235 | 485 | 80 | 60 | 80 | 70 |

**Table 20.** *(Continued)* **Speeds and Feeds for Drilling, Reaming, and Threading Ferrous Cast Metals**

| Material | Brinell Hardness Number | Drilling | Reaming | | Drilling | | | | | | Reaming | | Threading | |
|---|---|---|---|---|---|---|---|---|---|---|---|---|---|---|
| | | HSS | | | HSS | | Indexable Carbide Insert: Uncoated | | Indexable Carbide Insert: Coated | | HSS | | HSS | |
| | | Speed (fpm) | | | f = feed (0.001 in/rev), s = speed (ft/min) *Metric Units:* f × 25.4 = mm/rev, s × 0.3048 = m/min | | | | | | | | | |
| | | | | | *Opt.* | *Avg.* | *Opt.* | *Avg.* | *Opt.* | *Avg.* | *Opt.* | *Avg.* | *Opt.* | *Avg.* |
| (Martensitic): 120-90-02 | 270-330 | 25 | 15 | | | | | | | | | | | |
| | 330-400 | 10 | 5 | | | | | | | | | | | |
| (Ferritic-Pearlitic): 80-55-06 | 190-225 | 70 | 45 | | | | | | | | | | | |
| | 225-260 | 50 | 30 | f<br>s | 13<br>60 | 6<br>60 | 11<br>70 | 6<br>150 | 11<br>195 | 6<br>405 | 21<br>55 | 11<br>40 | 83<br>60 | 20<br>55 |
| (Pearlitic-Martensitic): 100-70-03 | 240-300 | 40 | 25 | | | | | | | | | | | |
| Cast Steels | | | | | | | | | | | | | | |
| (Low carbon): 1010, 1020 | 100-150 | 100 | 65 | f<br>s | 18<br>35 | 9<br>70 | | | | | 29<br>75 | 15<br>85 | 83<br>65 | 20<br>85 |
| (Medium carbon): 1030, 1040, 1050 | 125-175 | 90 | 60 | | | | | | | | | | | |
| | 175-225 | 70 | 45 | | | | | | | | | | | |
| | 225-300 | 55 | 35 | f<br>s | 15<br>35 | 7<br>60 | | | 8<br>195† | 4<br>475 | 24<br>65 | 12<br>70 | 83<br>55 | 20<br>70 |
| (Low-carbon alloy): 1320, 2315, 2320, 4110, 4120, 4320, 8020, 8620 | 150-200 | 75 | 50 | | | | | | | | | | | |
| | 200-250 | 65 | 40 | | | | | | | | | | | |
| | 250-300 | 50 | 30 | | | | | | | | | | | |
| (Medium-carbon alloy): 1330, 1340, 2325, 2330, 4125, 4130, 4140, 4330, 4340, 8030, 80B30, 8040, 8430, 8440, 8630, 8640, 9525, 9530, 9535 | 175-225 | 70 | 45 | f<br>s | | | | | 8<br>130† | 4<br>315 | | | | |
| | 225-250 | 60 | 35 | | | | | | | | | | | |
| | 250-300 | 45 | 30 | | | | | | | | | | | |
| | 300-350 | 30 | 20 | | | | | | | | | | | |
| | 350-400 | 20 | 10 | | | | | | | | | | | |

See the footnote to Table 17 for instructions concerning the use of this table. The combined feed/speed data in this table are based on tool grades (identified in Table 16) as follows: uncoated = 15; coated carbide = 11, † = 10.

**Table 21. Speeds and Feeds for Drilling, Reaming, and Threading Light Metals**

| Material | Brinell Hardness Number | Drilling | Reaming | | Drilling | | | | Reaming | | Threading | |
|---|---|---|---|---|---|---|---|---|---|---|---|---|
| | | HSS | | | HSS | | Indexable Insert Uncoated Carbide | | HSS | | HSS | |
| | | Speed (fpm) | | | f = feed (0.001 in/rev), s = speed (ft/min) *Metric Units:* f × 25.4 = mm/rev, s × 0.3048 = m/min | | | | | | | |
| | | | | | *Opt.* | *Avg.* | *Opt.* | *Avg.* | *Opt.* | *Avg.* | *Opt.* | *Avg.* |
| All wrought aluminum alloys, 6061-T651, 5000, 6000, 7000 series | CD | 400 | 400 | | | | | | | | | |
| | ST and A | 350 | 350 | f | 31 | 16 | 11 | 6 | 52 | 26 | 83 | 20 |
| All aluminum sand and permanent mold casting alloys | AC | 500 | 500 | s | 390 | 580 | 3235 | 11370 | 610 | 615 | 635 | 565 |
| | ST and A | 350 | 350 | | | | | | | | | |
| **Aluminum Die-Casting Alloys** | | | | | | | | | | | | |
| Alloys 308.0 and 319.0 | — | — | — | f | 23 | 11 | 11 | 6 | 38 | 19 | 83 | 20 |
| | | | | s | 110 | 145 | 945 | 3325 | 145 | 130 | 145 | 130 |
| Alloys 360.0 and 380.0 | — | — | — | f | 27 | 14 | 11 | 6 | 45 | 23 | 83 | 20 |
| | | | | s | 90 | 125 | 855 | 3000 | 130 | 125 | 130 | 115 |
| Alloys 390.0 and 392.0 { | AC | 300 | 300 | | | | | | | | | |
| | ST and A | 70 | 70 | | | | | | | | | |
| Alloys 413 | — | — | — | f | 24 | 12 | 11 | 6 | 40 | 20 | 83 | 20 |
| | ST and A | 45 | 40 | s | 65 | 85 | 555 | 1955 | 85 | 80 | 85 | 80 |
| All other aluminum die-casting alloys { | AC | 125 | 100 | f | 27 | 14 | 11 | 6 | 45 | 23 | 83 | 20 |
| | | | | s | 90 | 125 | 855 | 3000 | 130 | 125 | 130 | 115 |
| **Magnesium Alloys** | | | | | | | | | | | | |
| All wrought magnesium alloys | A,CD,ST and A | 500 | 500 | | | | | | | | | |
| All cast magnesium alloys | A,AC, ST and A | 450 | 450 | | | | | | | | | |

Abbreviations designate: A, annealed; AC, as cast; CD, cold drawn; and ST and A, solution treated and aged, respectively. See the footnote to Table 17 for instructions concerning the use of this table. The combined feed/speed data in this table are based on tool grades (identified in Table 16) as follows; uncoated carbide = 15.

**Table 22. Diameter Speed and Feed Adjustment Factors for HSS Twist Drills and Reamers**

| Ratio of Chosen Feed to *Optimum* Feed | Ratio of the two cutting speeds (*average/optimum*) given in the tables $V_{avg}/V_{opt}$ | | | | | | | Cutting Speed, $V = V_{opt} \times F_f \times F_d$ — Tool Diameter | | | | | | | | |
|---|---|---|---|---|---|---|---|---|---|---|---|---|---|---|---|---|
| | 1.00 | 1.25 | 1.50 | 2.00 | 2.50 | 3.00 | 4.00 | 0.08 in (2 mm) | 0.15 in (4 mm) | 0.25 in (6 mm) | 0.40 in (10 mm) | 0.60 in (15 mm) | 1.00 in (25 mm) | 2.00 in (50 mm) | 3.00 in (75 mm) | 4.00 in (100 mm) |
| | Feed Factor, $F_f$ | | | | | | | Diameter Factor, $F_d$ | | | | | | | | |
| 1.00 | 1.00 | 1.00 | 1.00 | 1.00 | 1.00 | 1.00 | 1.00 | 0.30 | 0.44 | 0.56 | 0.78 | 1.00 | 1.32 | 1.81 | 2.11 | 2.29 |
| 0.90 | 1.00 | 1.06 | 1.09 | 1.14 | 1.18 | 1.21 | 1.27 | 0.32 | 0.46 | 0.59 | 0.79 | 1.00 | 1.30 | 1.72 | 1.97 | 2.10 |
| 0.80 | 1.00 | 1.12 | 1.19 | 1.31 | 1.40 | 1.49 | 1.63 | 0.34 | 0.48 | 0.61 | 0.80 | 1.00 | 1.27 | 1.64 | 1.89 | 1.95 |
| 0.70 | 1.00 | 1.15 | 1.30 | 1.50 | 1.69 | 1.85 | 2.15 | 0.38 | 0.52 | 0.64 | 0.82 | 1.00 | 1.25 | 1.52 | 1.67 | 1.73 |
| 0.60 | 1.00 | 1.23 | 1.40 | 1.73 | 2.04 | 2.34 | 2.89 | 0.42 | 0.55 | 0.67 | 0.84 | 1.00 | 1.20 | 1.46 | 1.51 | 1.54 |
| 0.50 | 1.00 | 1.25 | 1.50 | 2.00 | 2.50 | 3.00 | 5.00 | 0.47 | 0.60 | 0.71 | 0.87 | 1.00 | 1.15 | 1.30 | 1.34 | 1.94 |
| 0.40 | 1.00 | 1.23 | 1.57 | 2.29 | 3.08 | 3.92 | 5.70 | 0.53 | 0.67 | 0.77 | 0.90 | 1.00 | 1.10 | 1.17 | 1.16 | 1.12 |
| 0.30 | 1.00 | 1.14 | 1.56 | 2.57 | 3.78 | 5.19 | 8.56 | 0.64 | 0.76 | 0.84 | 0.94 | 1.00 | 1.04 | 1.02 | 0.96 | 0.90 |
| 0.20 | 1.00 | 0.90 | 1.37 | 2.68 | 4.49 | 6.86 | 17.60 | 0.83 | 0.92 | 0.96 | 1.00 | 1.00 | 0.96 | 0.81 | 0.73 | 0.66 |
| 0.10 | 1.00 | 1.44 | 0.80 | 2.08 | 4.36 | 8.00 | 20.80 | 1.29 | 1.26 | 1.21 | 1.11 | 1.00 | 0.84 | 0.60 | 0.46 | 0.38 |

This table is specifically for use with the combined feed/speed data for HSS twist drills in Table 17 through Table 23; use Table 5a and Table 5b to adjust speed and tool life for indexable insert drilling with carbides. The combined feed/speed data for HSS twist drilling are based on a 0.60-inch diameter HSS drill with standard drill point geometry (2-flute with 118° tip angle). To adjust the given speeds for different feeds and drill diameters, use the two feed/speed pairs from the tables and calculate the ratio of desired (new) feed to *optimum* feed (largest of the two feeds from the speed table), and the ratio of the two cutting speeds $V_{avg}/V_{opt}$. Use the values of these ratios to find the feed factor $F_f$ at the intersection of the feed ratio row and the speed ratio column in the left half of the table. The diameter factor $F_d$ is found in the same row as the feed factor, in the right half of the table, under the column corresponding to the drill diameter. For diameters not given, interpolate between the nearest available sizes. The adjusted cutting speed can be calculated from $V = V_{opt} \times F_f \times F_d$, where $V_{opt}$ is the smaller (*optimum*) of the two speeds from the speed table (from the left side of the column containing the two feed/speed pairs). Tool life using the selected feed and the adjusted speed should be approximately 45 minutes. Speed adjustment factors to modify tool life are found in Table 15e.

**Table 23. Speeds and Feeds for Drilling and Reaming Copper Alloys**

| Group 1 |
|---|
| Architectural bronze(C38500); Extra-high-leaded brass (C35600); Forging brass (C37700); Free-cutting phosphor bronze (B-2) (C54400); Free-cutting brass (C36000); Free-cutting Muntz metal (C37000); High-leaded brass (C33200, C34200); High-leaded brass tube (C35300); Leaded commercial bronze (C31400); Leaded naval brass (C48500); Medium-leaded brass (C34000) |
| **Group 2** |
| Aluminum brass, arsenical (C68700); Cartridge brass, 70% (C26000); High-silicon bronze, B (C65500); Admiralty brass (inhibited) (C44300, C44500); Jewelry bronze, 87.5% (C22600); Leaded Muntz metal (C36500, C36800); Leaded nickel silver (C79600); Low brass, 80% (C24000); Low-leaded brass (C33500); Low-silicon bronze, B (C65100); Manganese bronze, A (C67500); Muntz metal, 60% (C28000); Nickel silver, 55-18 (C77000); Red brass, 85% (C23000); Yellow brass (C26800) |
| **Group 3** |
| Aluminum bronze, D (C61400); Beryllium copper (C17000, C17200, C17500); Commercial bronze, 90% (C22000); Copper nickel, 10% (C70600); Copper nickel, 30% (C71500);Electrolytic tough-pitch copper (C11000); Gilding, 95% (C21000); Nickel silver, 65-10 (C74500); Nickel silver, 65-12 (C75700); Nickel silver, 65-15 (C75400); Nickel silver, 65-18 (C75200); Oxygen-free copper (C10200); Phosphor bronze, 1.25% (C50200); Phosphor bronze, 10% D (C52400); Phosphor bronze, 5% A (C51000); Phosphor bronze, 8% C (C52100); Phosphorus deoxidized copper (C12200) |

| Alloy Description and UNS Alloy Numbers | Material Condition | Drilling | Reaming | Drilling | | | | | Reaming | |
|---|---|---|---|---|---|---|---|---|---|---|
| | | HSS | | | HSS | | Indexable Insert Uncoated Carbide | | HSS | |
| | | Speed (fpm) | | f = feed (0.001 in/rev), s = speed (ft/min) *Metric Units:* f × 25.4 = mm/rev, s × 0.3048 = m/min | | | | | | |
| | | | | | *Opt.* | *Avg.* | *Opt.* | *Avg.* | *Opt.* | *Avg.* |
| Wrought Alloys | | | | | | | | | | |
| Group 1 | A | 160 | 160 | **f** | 21 | 11 | 11 | 6 | 36 | 18 |
| | CD | 175 | 175 | s | 210 | 265 | 405 | 915 | 265 | 230 |
| Group 2 | A | 120 | 110 | **f** | 24 | 12 | 11 | 6 | 40 | 20 |
| | CD | 140 | 120 | s | 100 | 130 | 205 | 455 | 130 | 120 |
| Group 3 | A | 60 | 50 | **f** | 23 | 11 | 11 | 6 | 38 | 19 |
| | CD | 65 | 60 | s | 155 | 195 | 150 | 340 | 100 | 175 |

Abbreviations designate: A, annealed; CD, cold drawn. The two leftmost speed columns in this table contain traditional *Handbook* speeds for HSS steel tools. The text contains information concerning feeds to use in conjunction with these speeds.

*HSS Drilling and Reaming:* The combined feed/speed data for drilling and Table 22 are used to adjust drilling speeds for other feeds and drill diameters. Examples are given in the text. The given speeds and feeds for reaming are based on an 8-tooth, $^{25}/_{32}$-inch diameter, 30° lead angle reamer, and a 0.008-inch radial depth of cut. For other feeds, the correct speed can be obtained by interpolation using the given speeds if the desired feed lies in the recommended range (between the given values of *optimum* and *average* feed). The cutting speeds for reaming do not require adjustment for tool diameter as long as the radial depth of cut does not become too large. Speed adjustment factors to modify tool life are found in Table 15e.

*Indexable Insert Drilling:* The feed/speed data for indexable insert drilling are based on a tool with two cutting edges, an insert nose radius of $^{3}/_{64}$ inch, a 10-degree lead angle, and diameter *D* of 1 inch. Adjustments for feed and depth of cut are made using Table 5a (Turning Speed Adjustment Factors) using a depth of cut of *D*/2, or one-half the drill diameter. Expected tool life at the given speeds and feeds is 15 minutes for short hole drilling (i.e., where hole depth is about 2*D* or less). Speed adjustment factors to increase tool life are found in Table 5b. The combined feed/speed data in this table are based on tool grades (identified in Table 16) as follows: uncoated carbide = 15.

**Using the Speeds and Feeds Tables for Tapping and Threading.—**The feed used in tapping and threading is always equal to the pitch of the screw thread being formed. The threading data contained in the tables for drilling, reaming, and threading (Table 17 through Table 23) are primarily for tapping and thread chasing, and do not apply to thread cutting with single-point tools.

The threading data in Table 17 through Table 23 give two sets of feed (pitch) and speed values, for 12 and 50 threads/inch, but these values can be used to obtain the cutting speed for any other thread pitches. If the desired pitch falls between the values given in the tables, i.e., between 0.020 inch (50 tpi) and 0.083 inch (12 tpi), the required cutting speed is obtained by interpolation between the given speeds. If the pitch is less than 0.020 inch (more than 50 tpi), use the *average* speed, i.e., the largest of the two given speeds. For pitches greater than 0.083 inch (fewer than 12 tpi), the *optimum* speed should be used. Tool life using the given feed/speed data is intended to be approximately 45 minutes, and should be about the same for threads between 12 and 50 threads per inch.

*Example:* Determine the cutting speed required for tapping 303 stainless steel with a ½-20 coated HSS tap.

The two feed/speed pairs for 303 stainless steel, in Table 19, are 83/35 (0.083 in/rev at 35 fpm) and 20/45 (0.020 in/rev at 45 fpm). The pitch of a ½-20 thread is 1/20 = 0.05 inch, so the required feed is 0.05 in/rev. Because 0.05 is between the two given feeds (Table 19), the cutting speed can be obtained by interpolation between the two given speeds as follows:

$$V = 35 + \frac{0.05 - 0.02}{0.083 - 0.02}(45 - 35) = 40 \text{ fpm}$$

The cutting speed for coarse-pitch taps must be lower than for fine-pitch taps with the same diameter. Usually, the difference in pitch becomes more pronounced as the diameter of the tap becomes larger and slight differences in the pitch of smaller-diameter taps have little significant effect on the cutting speed. Unlike all other cutting tools, the feed per revolution of a tap cannot be independently adjusted—it is always equal to the lead of the thread and is always greater for coarse pitches than for fine pitches. Furthermore, the thread form of a coarse-pitch thread is larger than that of a fine-pitch thread; therefore, it is necessary to remove more metal when cutting a coarse-pitch thread.

Taps with a long chamfer, such as starting or tapper taps, can cut faster in a short hole than short chamfer taps, such as plug taps. In deep holes, however, short chamfer or plug taps can run faster than long chamfer taps. Bottoming taps must be run more slowly than either starting or plug taps. The chamfer helps to start the tap in the hole. It also functions to involve more threads, or thread form cutting edges, on the tap in cutting the thread in the hole, thus reducing the cutting load on any one set of thread form cutting edges. In so doing, more chips and thinner chips are produced that are difficult to remove from deeper holes. Shortening the chamfer length causes fewer thread form cutting edges to cut, thereby producing fewer and thicker chips that can easily be disposed of. Only one or two sets of thread form cutting edges are cut on bottoming taps, causing these cutting edges to assume a heavy cutting load and produce very thick chips.

Spiral-pointed taps can operate at a faster cutting speed than taps with normal flutes. These taps are made with supplementary angular flutes on the end that push the chips ahead of the tap and prevent the tapped hole from becoming clogged with chips. They are used primarily to tap open or through holes although some are made with shorter supplementary flutes for tapping blind holes.

The tapping speed must be reduced as the percentage of full thread to be cut is increased. Experiments have shown that the torque required to cut a 100 percent thread form is more than twice that required to cut a 50 percent thread form. An increase in the percentage of full thread will also produce a greater volume of chips.

The tapping speed must be lowered as the length of the hole to be tapped is increased. More friction must be overcome in turning the tap and more chips accumulate in the hole. It will be more difficult to apply the cutting fluid at the cutting edges and to lubricate the tap to reduce friction. This problem becomes greater when the hole is being tapped in a horizontal position.

Cutting fluids have a very great effect on the cutting speed for tapping. Although other operating conditions when tapping frequently cannot be changed, a free selection of the cutting fluid usually can be made. When planning the tapping operation, the selection of a cutting fluid warrants a very careful consideration and perhaps an investigation.

Taper threaded taps, such as pipe taps, must be operated at a slower speed than straight thread taps with a comparable diameter. All the thread form cutting edges of a taper threaded tap that are engaged in the work cut and produce a chip, but only those cutting edges along the chamfer length cut on straight thread taps. Pipe taps often are required to cut the tapered thread from a straight hole, adding to the cutting burden.

The machine tool used for the tapping operation must be considered in selecting the tapping speed. Tapping machines and other machines that are able to feed the tap at a rate of advance equal to the lead of the tap, and that have provisions for quickly reversing the spindle, can be operated at high cutting speeds. On machines where the feed of the tap is controlled manually—such as on drill presses and turret lathes—the tapping speed must be reduced to allow the operator to maintain safe control of the operation.

There are other special considerations in selecting the tapping speed. Very accurate threads are usually tapped more slowly than threads with a commercial grade of accuracy. Thread forms that require deep threads for which a large amount of metal must be removed, producing a large volume of chips, require special techniques and slower cutting speeds. Acme, buttress, and square threads, therefore, are generally cut at lower speeds.

**Cutting Speed for Broaching.**—Broaching offers many advantages in manufacturing metal parts, including high production rates, excellent surface finishes, and close dimensional tolerances. These advantages are not derived from the use of high cutting speeds; they are derived from the large number of cutting teeth that can be applied consecutively in a given period of time, from their configuration and precise dimensions, and from the width or diameter of the surface that can be machined in a single stroke. Most broaching cutters are expensive in their initial cost and are expensive to sharpen. For these reasons, a long tool life is desirable, and to obtain a long tool life, relatively slow cutting speeds are used. In many instances, slower cutting speeds are used because of the limitations of the machine in accelerating and stopping heavy broaching cutters. At other times, the available power on the machine places a limit on the cutting speed that can be used; i.e., the cubic inches of metal removed per minute must be within the power capacity of the machine.

The cutting speeds for high-speed steel broaches range from 3 to 50 feet per minute, although faster speeds have been used. In general, the harder and more difficult to machine materials are cut at a slower cutting speed and those that are easier to machine are cut at a faster speed. Some typical recommendations for high-speed steel broaches are: AISI 1040, 10 to 30 fpm; AISI 1060, 10 to 25 fpm; AISI 4140, 10 to 25 fpm; AISI 41L40, 20 to 30 fpm; 201 austenitic stainless steel, 10 to 20 fpm; Class 20 gray cast iron, 20 to 30 fpm; Class 40 gray cast iron, 15 to 25 fpm; aluminum and magnesium alloys, 30 to 50 fpm; copper alloys, 20 to 30 fpm; commercially pure titanium, 20 to 25 fpm; alpha and beta titanium alloys, 5 fpm; and the superalloys, 3 to 10 fpm. Surface broaching operations on gray iron castings have been conducted at a cutting speed of 150 fpm, using *indexable insert* cemented carbide broaching cutters. In selecting the speed for broaching, the cardinal principle of the performance of all metal cutting tools should be kept in mind; i.e., increasing the cutting speed may result in a proportionately larger reduction in tool life, and reducing the cutting speed may result in a proportionately larger increase in the tool life. When broaching most materials, a suitable cutting fluid should be used to obtain a good surface finish and a better tool life. Gray cast iron can be broached without using a cutting fluid although some shops prefer to use a soluble oil.

## ESTIMATING SPEEDS AND MACHINING POWER

**Estimating Planer Cutting Speeds.—**Whereas most planers of modern design have a means of indicating the speed at which the table is traveling, or cutting, many older planers do not. The following formulas are useful for planers that do not have a means of indicating the table or cutting speed. It is not practicable to provide a formula for calculating the exact cutting speed at which a planer is operating because the time to stop and start the table when reversing varies greatly. The formulas below will provide a reasonable estimate.

$$V_c \approx S_c L \qquad \text{and} \qquad S_c \approx \frac{V_c}{L}$$

where $V_c$ = cutting speed; fpm or m/min

$S_c$ = number of cutting strokes per minute of planer table

$L$ = length of table cutting stroke; ft or m

**Cutting Speed for Planing and Shaping.—**The traditional HSS cutting tool speeds in Table 1 through Table 4b and Table 6 through Table 9, pages 1083 through 1095, can be used for planing and shaping. The feed and depth of cut factors in Table 5c should also be used, as explained previously. Very often, other factors relating to the machine or the setup will require a reduction in the cutting speed used on a specific job.

**Cutting Time for Turning, Boring, and Facing.—**The time required to turn a length of metal can be determined by the following formula in which $T$ = time in minutes, $L$ = length of cut in inches (or mm), $f$ = feed in inches per revolution (or mm/min), and $N$ = lathe spindle speed in revolutions per minute.

$$T = \frac{L}{fN}$$

When making job estimates, the time required to load and to unload the workpiece on the machine, and the machine handling time, must be added to the cutting time for each length cut to obtain the floor-to-floor time.

**Planing Time.—**The approximate time required to plane a surface can be determined from the following formula in which $T$ = time in minutes, $L$ = length of stroke in feet (or meter), $V_c$ = cutting speed in feet per minute (m/min), $V_r$ = return speed in feet per minute (m/min); $W$ = width of surface to be planed in inches (or mm), $F$ = feed in inches (or mm), and 0.025 = approximate reversal time factor per stroke in minutes for most planers:

$$T = \frac{W}{F}\left[L \times \left(\frac{1}{V_c} + \frac{1}{V_r}\right) + 0.025\right]$$

**Speeds for Metal-Cutting Saws.—**The table on page 1132 gives speeds and feeds for solid-tooth, high-speed steel, circular, metal-cutting saws as recommended by Saws International, Inc. (sfpm = surface feet per minute = 3.142 × blade diameter in inches × rpm of saw shaft ÷ 12). Also see page 1233 for band saw blade speeds.

**Speeds for Turning Unusual Materials.—***Slate*, on account of its peculiarly stratified formation, is rather difficult to turn, but if handled carefully, can be machined in an ordinary lathe. The cutting speed should be about the same as for cast iron. A sheet of fiber or pressed paper should be interposed between the chuck or steadyrest jaws and the slate, to protect the latter. Slate rolls must not be centered and run on the tailstock. A satisfactory method of supporting a slate roll having journals at the ends is to bore a piece of lignum vitae to receive the turned end of the roll, and center it for the tailstock spindle.

*Rubber* can be turned at a peripheral speed of 200 feet per minute (61 m/min), although it is much easier to grind it with an abrasive wheel that is porous and soft. For cutting a rubber roll in two, the ordinary parting tool should not be used, but a tool shaped like a knife; such a tool severs the rubber without removing any material.

**Speeds, Feeds, and Tooth Angles for Sawing Various Material with Solid-Tooth, High-Speed Steel, Circular, Metal-Cutting Saws**

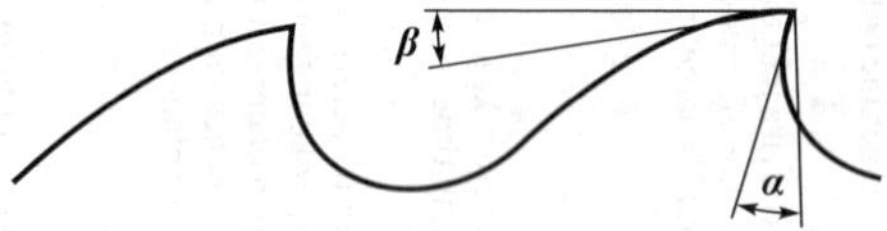


First entry is cutting speed, second entry is feed

α = Cutting angle (Front), degree

β = Relief angle (Back), degree

| Materials | Rake Angle | | Stock Diameters, inches | | | | Stock Diameters, millimeters | | | |
|---|---|---|---|---|---|---|---|---|---|---|
| | α | β | ¼–¾ | ¾–1½ | 1½–2½ | 2½–3½ | 6–19 | 19–38 | 38–63 | 63–89 |
| Aluminum | 24 | 12 | 6500 sfpm<br>100 in/min | 6200 sfpm<br>85 in/min | 6000 sfpm<br>80 in/min | 5000 sfpm,<br>75 in/min | 1981 m/min<br>2540 mm/min | 1890 m/min<br>2159 mm/min | 1829 m/min<br>2159 mm/min | 1524 m/min<br>2159 mm/min |
| Light Alloys with Cu, Mg, and Zn | 22 | 10 | 3600 sfpm<br>70 in/min | 3300 sfpm<br>65 in/min | 3000 sfpm,<br>63 in/min | 2600 sfpm<br>60 in/min | 1097 m/min<br>1778 mm/min | 1006 m/min<br>1651 mm/min | 914 m/min<br>1600 mm/min | 792 m/min<br>1524 mm/min |
| Light Alloys with High Si | 20 | 8 | 650 sfpm<br>16 in/min | 600 sfpm<br>16 in/min | 550 sfpm<br>14 in/min | 550 sfpm<br>12 in/min | 198 m/min<br>406 mm/min | 183 m/min<br>406 mm/min | 168 m/min<br>356 mm/min | 168 m/min<br>305 mm/min |
| Copper | 20 | 10 | 1300 sfpm<br>24 in/min | 1150 sfpm<br>24 in/min | 1000 sfpm<br>22 in/min | 800 sfpm<br>22 in/min | 396 m/min<br>610 mm/min | 351 m/min<br>610 mm/min | 305 m/min<br>559 mm/min | 244 m/min<br>559 mm/min |
| Bronze | 15 | 8 | 1300 sfpm<br>24 in/min | 1150 sfpm<br>24 in/min | 1000 sfpm<br>22 in/min | 800 sfpm<br>20 in/min | 396 m/min<br>610 mm/min | 351 m/min<br>610 mm/min | 305 m/min<br>559 mm/min | 244 m/min<br>508 mm/min |
| Hard Bronze | 10 | 8 | 400 sfpm<br>6.3 in/min | 360 sfpm<br>6 in/min | 325 sfpm<br>5.5 in/min | 300 sfpm<br>5.1 in/min | 122 m/min<br>160 mm/min | 110 m/min<br>152 mm/min | 99 m/min<br>140 mm/min | 91 m/min<br>130 mm/min |
| Cu-Zn Brass | 16 | 8 | 2000 sfpm<br>43 in/min | 2000 sfpm<br>43 in/min | 1800 sfpm<br>39 in/min | 1800 sfpm<br>35 in/min | 610 m/min<br>1092 mm/min | 610 m/min<br>192 mm/min | 549 m/min<br>991 mm/min | 549 m/min<br>889 mm/min |
| Gray Cast Iron | 12 | 8 | 82 sfpm<br>4 in/min | 75 sfpm<br>4 in/min | 72 sfpm<br>3.5 in/min | 66 sfpm<br>3 in/min | 25 m/min<br>102 mm/min | 23 m/min<br>102 mm/min | 22 m/min<br>89 mm/min | 20 m/min<br>76 mm/min |
| Carbon Steel | 20 | 8 | 160 sfpm<br>6.3 in/min | 150 sfpm<br>5.9 in/min | 150 sfpm<br>5.5 in/min | 130 sfpm<br>5.1 in/min | 49 m/min<br>160 mm/min | 46 m/min<br>150 mm/min | 46 m/min<br>140 mm/min | 40 m/min<br>130 mm/min |
| Medium Hard Steel | 18 | 8 | 100 sfpm<br>5.1 in/min | 100 sfpm<br>4.7 in/min | 80 sfpm<br>4.3 in/min | 80 sfpm<br>4.3 in/min | 30 m/min<br>130 mm/min | 30 m/min<br>119 mm/min | 24 m/min<br>109 mm/min | 24 m/min<br>109 mm/min |
| Hard Steel | 15 | 8 | 66 sfpm<br>4.3 in/min | 66 sfpm<br>4.3 in/min | 60 sfpm<br>4 in/min | 57 sfpm<br>3.5 in/min | 20 m/min<br>109 mm/min | 20 m/min<br>109 mm/min | 18 m/min<br>102 mm/min | 17 m/min<br>89 mm/min |
| Stainless Steel | 15 | 8 | 66 sfpm<br>2 in/min | 63 sfpm<br>1.75 in/min | 60 sfpm<br>1.75 in/min | 57 sfpm<br>1.5 in/min | 20 m/min<br>51 mm/min | 19 m/min<br>44 mm/min | 18 m/min<br>44 mm/min | 17 m/min<br>38 mm/min |

*Gutta percha* can be turned as easily as wood, but the tools must be sharp and a good soap-and-water lubricant used.

*Copper* can be turned easily at 200 feet per minute (61 m/min). See also Table 6 on page 1093.

*Limestone* such as is used in the construction of pillars for balconies, etc., can be turned at 150 feet per minute (46 m/min), and the formation of ornamental contours is quite easy. *Marble* is a treacherous material to turn. It should be cut with a tool such as would be used for brass, but at a speed suitable for cast iron. It must be handled very carefully to prevent flaws in the surface.

The foregoing speeds are for high-speed steel tools. Tools tipped with tungsten carbide are adapted for cutting various nonmetallic products which cannot be machined readily with steel tools, such as slate, marble, synthetic plastic materials, etc. In drilling slate and marble, use flat drills; and for plastic materials, tungsten carbide-tipped twist drills. Cutting speeds ranging from 75 to 150 feet per minute (23–46 m/min) have been used for drilling slate (without coolant) and a feed of 0.025 inch per revolution (0.64 mm/rev) for drills ¾ and 1 inch (19.05 and 25.4 mm) in diameter.

## Estimating Machining Power

Knowledge of the power required to perform machining operations is useful when planning new machining operations, for optimizing existing machining operations, and to develop specifications for new machine tools that are to be acquired. The available power on any machine tool places a limit on the size of the cut that it can take. When much metal must be removed from the workpiece it is advisable to estimate the cutting conditions that will utilize the maximum power on the machine. Many machining operations require only light cuts to be taken for which the machine obviously has ample power; in this event, estimating the power required is a wasteful effort. Conditions in different shops may vary and machine tools are not all designed alike, so some variations between the estimated results and those obtained on the job are to be expected. However, by using the methods provided in this section a reasonable estimate of the power required can be made, which will suffice in most practical situations.

The measure of power in customary inch units is the horsepower; in SI metric units it is the kilowatt, which is used for both mechanical and electrical power. The power required to cut a material depends upon the rate at which the material is being cut and upon an experimentally determined power constant, $K_p$, which is also called the unit horsepower, unit power, or specific power consumption. The power constant is equal to the horsepower required to cut a material at a rate of one cubic inch per minute; in SI metric units the power constant is equal to the power in kilowatts required to cut a material at a rate of one cubic centimeter per second, or 1000 cubic millimeters per second (1 $cm^3$ = 1000 $mm^3$). Different values of the power constant are required for inch and for metric units, which are related as follows: to obtain the SI metric power constant, multiply the inch power constant by 2.73; to obtain the inch power constant, divide the SI metric power constant by 2.73. Values of the power constant in Table 1a, and Table 1b can be used for all machining operations except drilling and grinding. Values given are for sharp tools.

**Table 1a. Power Constants $K_p$ Using Sharp Cutting Tools**

| Material | Brinell Hardness Number | $K_p$ Inch Units | $K_p$ Metric Units | Material | Brinell Hardness Number | $K_p$ Inch Units | $K_p$ Metric Units |
|---|---|---|---|---|---|---|---|
| **Ferrous Cast Metals** | | | | | | | |
| Gray Cast Iron | 100-120 | 0.28 | 0.76 | Malleable Iron | | | |
| | 120-140 | 0.35 | 0.96 | Ferritic | 150-175 | 0.42 | 1.15 |
| | 140-160 | 0.38 | 1.04 | | | | |
| | 160-180 | 0.52 | 1.42 | | 175-200 | 0.57 | 1.56 |
| | 180-200 | 0.60 | 1.64 | Pearlitic | 200-250 | 0.82 | 2.24 |
| | 200-220 | 0.71 | 1.94 | | 250-300 | 1.18 | 3.22 |
| | 220-240 | 0.91 | 2.48 | | | | |
| | | | | Cast Steel | 150-175 | 0.62 | 1.69 |
| Alloy Cast Iron | 150-175 | 0.30 | 0.82 | | 175-200 | 0.78 | 2.13 |
| | 175-200 | 0.63 | 1.72 | | 200-250 | 0.86 | 2.35 |
| | 200-250 | 0.92 | 2.51 | | … | … | … |

**Table 1a.** *(Continued)* **Power Constants $K_p$ Using Sharp Cutting Tools**

| Material | Brinell Hardness Number | $K_p$ Inch Units | $K_p$ Metric Units | Material | Brinell Hardness Number | $K_p$ Inch Units | $K_p$ Metric Units |
|---|---|---|---|---|---|---|---|
| **High-Temperature Alloys, Tool Steel, Stainless Steel, and Nonferrous Metals** | | | | | | | |
| High-Temperature Alloys | | | | | 150-175 | 0.60 | 1.64 |
| A286 | 165 | 0.82 | 2.24 | Stainless Steel | 175-200 | 0.72 | 1.97 |
| A286 | 285 | 0.93 | 2.54 | | 200-250 | 0.88 | 2.40 |
| Chromoloy | 200 | 0.78 | 3.22 | Zinc Die-Cast Alloys | … | 0.25 | 0.68 |
| Chromoloy | 310 | 1.18 | 3.00 | Copper (pure) | … | 0.91 | 2.48 |
| Inco 700 | 330 | 1.12 | 3.06 | Brass | | | |
| Inco 702 | 230 | 1.10 | 3.00 | Hard | … | 0.83 | 2.27 |
| Hastelloy-B | 230 | 1.10 | 3.00 | Medium | … | 0.50 | 1.36 |
| M-252 | 230 | 1.10 | 3.00 | Soft | … | 0.25 | 0.68 |
| M-252 | 310 | 1.20 | 3.28 | Leaded | … | 0.30 | 0.82 |
| Ti-150A | 340 | 0.65 | 1.77 | | | | |
| U-500 | 375 | 1.10 | 3.00 | Bronze | | | |
| | | | | Hard | … | 0.91 | 2.48 |
| Monel Metal | … | 1.00 | 2.73 | Medium | … | 0.50 | 1.36 |
| | 175-200 | 0.75 | 2.05 | Aluminum | | | |
| | 200-250 | 0.88 | 2.40 | Cast | … | 0.25 | 0.68 |
| Tool Steel | 250-300 | 0.98 | 2.68 | Rolled (hard) | … | 0.33 | 0.90 |
| | 300-350 | 1.20 | 3.28 | | | | |
| | 350-400 | 1.30 | 3.55 | Magnesium Alloys | … | 0.10 | 0.27 |

The value of the power constant is essentially unaffected by the cutting speed, the depth of cut, and the cutting tool material. Factors that do affect the value of the power constant, and thereby the power required to cut a material, include the hardness and microstructure of the work material, the feed rate, the rake angle of the cutting tool, and whether the cutting edge of the tool is sharp or dull. Values are given in the power constant tables for different material hardness levels, whenever this information is available. Feed factors for the power constant are given in Table 2. All metal cutting tools wear but a worn cutting edge requires more power to cut than a sharp cutting edge.

Factors to provide for tool wear are given in Table 3. In this table, the extra-heavy-duty category for milling and turning occurs only on operations where the tool is allowed to wear more than a normal amount before it is replaced, such as roll turning. The effect of the rake angle usually can be disregarded. The rake angle for which most of the data in the power constant tables are given is positive 14 degrees. Only when the deviation from this angle is large is it necessary to make an adjustment. Using a rake angle that is more positive reduces the power required approximately 1 percent per degree; using a rake angle that is more negative increases the power required; again approximately 1 percent per degree.

Many indexable insert cutting tools are formed with an integral chip breaker or other cutting edge modifications, which have the effect of reducing the power required to cut a material. The extent of this effect cannot be predicted without a test of each design. Cutting fluids will also usually reduce the power required, when operating in the lower range of cutting speeds. Again, the extent of this effect cannot be predicted because each cutting fluid exhibits its own characteristics.

**Table 1b. Power Constants $K_p$ Using Sharp Cutting Tools**

| Material | Brinell Hardness Number | $K_p$ Inch Units | $K_p$ Metric Units | Material | Brinell Hardness Number | $K_p$ Inch Units | $K_p$ SI Metric Units |
|---|---|---|---|---|---|---|---|
| **Wrought Steels** | | | | | | | |
| Plain Carbon Steels | | | | | | | |
| All Plain Carbon Steels | 80-100 | 0.63 | 1.72 | All Plain Carbon Steels | 220-240 | 0.89 | 2.43 |
| | 100-120 | 0.66 | 1.80 | | 240-260 | 0.92 | 2.51 |
| | 120-140 | 0.69 | 1.88 | | 260-280 | 0.95 | 2.59 |
| | 140-160 | 0.74 | 2.02 | | 280-300 | 1.00 | 2.73 |
| | 160-180 | 0.78 | 2.13 | | 300-320 | 1.03 | 2.81 |
| | 180-200 | 0.82 | 2.24 | | 320-340 | 1.06 | 2.89 |
| | 200-220 | 0.85 | 2.32 | | 340-360 | 1.14 | 3.11 |
| Free Machining Steels | | | | | | | |
| AISI 1108, 1109, 1110, 1115, 1116, 1117, 1118, 1119, 1120, 1125, 1126, 1132 | 100-120 | 0.41 | 1.12 | AISI 1137, 1138, 1139, 1140, 1141, 1144, 1145, 1146, 1148, 1151 | 180-200 | 0.51 | 1.39 |
| | 120-140 | 0.42 | 1.15 | | 200-220 | 0.55 | 1.50 |
| | 140-160 | 0.44 | 1.20 | | 220-240 | 0.57 | 1.56 |
| | 160-180 | 0.48 | 1.31 | | 240-260 | 0.62 | 1.69 |
| | 180-200 | 0.50 | 1.36 | | … | … | … |
| Alloy Steels | | | | | | | |
| AISI 4023, 4024, 4027, 4028, 4032, 4037, 4042, 4047, 4137, 4140, 4142, 4145, 4147, 4150, 4340, 4640, 4815, 4817, 4820, 5130, 5132, 5135, 5140, 5145, 5150, 6118, 6150, 8637, 8640, 8642, 8645, 8650, 8740 | 140-160 | 0.62 | 1.69 | AISI 4130, 4320, 4615, 4620, 4626, 5120, 8615, 8617, 8620, 8622, 8625, 8630, 8720 | 140-160 | 0.56 | 1.53 |
| | 160-180 | 0.65 | 1.77 | | 160-180 | 0.59 | 1.61 |
| | 180-200 | 0.69 | 1.88 | | 180-200 | 0.62 | 1.69 |
| | 200-220 | 0.72 | 1.97 | | 200-220 | 0.65 | 1.77 |
| | 220-240 | 0.76 | 2.07 | | 220-240 | 0.70 | 1.91 |
| | 240-260 | 0.80 | 2.18 | | 240-260 | 0.74 | 2.02 |
| | 260-280 | 0.84 | 2.29 | | 260-280 | 0.77 | 2.10 |
| | 280-300 | 0.87 | 2.38 | | 280-300 | 0.80 | 2.18 |
| | 300-320 | 0.91 | 2.48 | | 300-320 | 0.83 | 2.27 |
| | 320-340 | 0.96 | 2.62 | | 320-340 | 0.89 | 2.43 |
| | 340-360 | 1.00 | 2.73 | | … | … | … |
| AISI 1330, 1335, 1340, E52100 | 160-180 | 0.79 | 2.16 | | … | … | … |
| | 180-200 | 0.83 | 2.27 | | … | … | … |
| | 200-220 | 0.87 | 2.38 | | … | … | … |

The machine tool transmits the power from the driving motor to the workpiece, where it is used to cut the material. The effectiveness of this transmission is measured by the machine tool efficiency factor, $E$. Average values of this factor are given in Table 4. Formulas for calculating the metal removal rate, $Q$, for different machining operations are given in Table 5. These formulas are used together with others given below. The following formulas can be used with either customary inch or with SI metric units.

$$P_c = K_p CQW \quad (1)$$

$$P_m = \frac{P_c}{E} = \frac{K_p CQW}{E} \quad (2)$$

where $P_c$ = power at the cutting tool; hp, or kW

**Table 2. Feed Factors, *C*, for Power Constants**

| Inch Units | | | | SI Metric Units | | | |
|---|---|---|---|---|---|---|---|
| Feed in.[a] | *C* | Feed in.[a] | *C* | Feed mm[b] | *C* | Feed mm[b] | *C* |
| 0.001 | 1.60 | 0.014 | 0.97 | 0.02 | 1.70 | 0.35 | 0.97 |
| 0.002 | 1.40 | 0.015 | 0.96 | 0.05 | 1.40 | 0.38 | 0.95 |
| 0.003 | 1.30 | 0.016 | 0.94 | 0.07 | 1.30 | 0.40 | 0.94 |
| 0.004 | 1.25 | 0.018 | 0.92 | 0.10 | 1.25 | 0.45 | 0.92 |
| 0.005 | 1.19 | 0.020 | 0.90 | 0.12 | 1.20 | 0.50 | 0.90 |
| 0.006 | 1.15 | 0.022 | 0.88 | 0.15 | 1.15 | 0.55 | 0.88 |
| 0.007 | 1.11 | 0.025 | 0.86 | 0.18 | 1.11 | 0.60 | 0.87 |
| 0.008 | 1.08 | 0.028 | 0.84 | 0.20 | 1.08 | 0.70 | 0.84 |
| 0.009 | 1.06 | 0.030 | 0.83 | 0.22 | 1.06 | 0.75 | 0.83 |
| 0.010 | 1.04 | 0.032 | 0.82 | 0.25 | 1.04 | 0.80 | 0.82 |
| 0.011 | 1.02 | 0.035 | 0.80 | 0.28 | 1.01 | 0.90 | 0.80 |
| 0.012 | 1.00 | 0.040 | 0.78 | 0.30 | 1.00 | 1.00 | 0.78 |
| 0.013 | 0.98 | 0.060 | 0.72 | 0.33 | 0.98 | 1.50 | 0.72 |

[a] Turning, in/rev; milling, in/tooth; planing and shaping, in/stroke; broaching, in/tooth.
[b] Turning, mm/rev; milling, mm/tooth; planing and shaping, mm/stroke; broaching, mm/tooth.

**Table 3. Tool Wear Factors, *W***

| Type of Operation | | *W* |
|---|---|---|
| For all operations with sharp cutting tools | | 1.00 |
| Turning: | Finish turning (light cuts) | 1.10 |
| | Normal rough and semifinish turning | 1.30 |
| | Extra-heavy-duty rough turning | 1.60-2.00 |
| Milling: | Slab milling | 1.10 |
| | End milling | 1.10 |
| | Light and medium face milling | 1.10-1.25 |
| | Extra-heavy-duty face milling | 1.30-1.60 |
| Drilling: | Normal drilling | 1.30 |
| | Drilling hard-to-machine materials and drilling with a very dull drill | 1.50 |
| Broaching: | Normal broaching | 1.05-1.10 |
| | Heavy-duty surface broaching | 1.20-1.30 |
| Planing and Shaping | Use values given for turning | |

$P_m$ = power at the motor; hp, or kW
$K_p$ = power constant (see Table 1a and Table 1b)
$Q$ = metal removal rate; $in^3$/min or $cm^3$/s (see Table 5)
$C$ = feed factor for power constant (see Table 2)
$W$ = tool wear factor (see Table 3)
$E$ = machine tool efficiency factor (see Table 4)
$V$ = cutting speed, fpm, or m/min
$N$ = cutting speed, rpm
$f$ = feed rate for turning; in/rev or mm/rev
$f$ = feed rate for planing and shaping; in/stroke, or mm/stroke

$f_t$ = feed per tooth; in/tooth, or mm/tooth
$f_m$ = feed rate; in/min or mm/min
$d_t$ = maximum depth of cut per tooth: inch, or mm
$d$ = depth of cut; inch, or mm
$n_t$ = number of teeth on milling cutter
$n_c$ = number of teeth engaged in work
$w$ = width of cut; inch, or mm

**Table 4. Machine Tool Efficiency Factors, *E***

| Type of Drive | $E$ | Type of Drive | $E$ |
|---|---|---|---|
| Direct Belt Drive | 0.90 | Geared Head Drive | 0.70-0.80 |
| Back Gear Drive | 0.75 | Oil-Hydraulic Drive | 0.60-0.90 |

**Table 5. Formulas for Calculating the Metal Removal Rate, *Q***

| Operation | Metal Removal Rate For Inch Units Only $Q$ = in³/min | Metal Removal Rate For SI Metric Units Only $Q$ = cm³/s |
|---|---|---|
| Single-Point Tools (Turning, Planing, and Shaping) | $12Vfd$ | $\frac{V}{60}fd$ |
| Milling | $f_m wd$ | $\frac{f_m wd}{60{,}000}$ |
| Surface Broaching | $12Vwn_c d_t$ | $\frac{V}{60}un_c d_t$ |

*Example:* A 180-200 BHN AISI 4130 shaft is to be turned on a geared head lathe using a cutting speed of 350 fpm (107 m/min), a feed rate of 0.016 in/rev (0.40 mm/rev), and a depth of cut of 0.100 inch (2.54 mm). Estimate the power at the cutting tool and at the motor, using both the inch and metric data.

Inch units:

$K_p = 0.62$ (from Table 1b)
$C = 0.94$ (from Table 2)
$W = 1.30$ (from Table 3)
$E = 0.80$ (from Table 4)
$Q = 12Vfd = 12 \times 350 \times 0.016 \times 0.100$ (from Table 5)
$Q = 6.72$ in³/min
$P_c = K_p CQW = 0.62 \times 0.94 \times 6.72 \times 1.30 = 5.1$ hp

$$P_m = \frac{P_c}{E} = \frac{5}{0.80} = 6.4 \text{ hp}$$

**SI metric units:**

$K_p = 1.69$ **(from Table 1b)**
$C = 0.94$ **(from Table 2)**
$W = 1.30$ **(from Table 3)**
$E = 0.80$ **(from Table 4)**

$$Q = \frac{V}{60}fd = \frac{107}{60} \times 0.40 \times 2.54 = 1.81 \text{ cm}^3\text{/s (from Table 5)}$$

$$P_c = K_p CQW = 1.69 \times 0.94 \times 1.81 \times 1.30 = 3.74 \text{ kW}$$

$$P_m = \frac{P_c}{E} = \frac{3.74}{0.80} = 4.677 \text{ kW}$$

Whenever possible the maximum power available on a machine tool should be used when heavy cuts must be taken.

The cutting conditions for utilizing the maximum power should be selected in the following order: 1) select the maximum depth of cut that can be used; 2) select the maximum feed rate that can be used; and 3) estimate the cutting speed that will utilize the maximum power available on the machine. This sequence is based on obtaining the longest tool life of the cutting tool and at the same time obtaining as much production as possible from the machine.

*The life of a cutting tool is most affected by cutting speed, then by feed rate, and least of all by depth of cut.* The maximum metal removal rate that a given machine is capable of machining from a given material is used as the basis for estimating the cutting speed that will utilize all the power available on the machine.

*Example:* A 160–180 BHN gray iron casting that is 6 inches wide is to have ⅛ inch stock removed on a 10 hp milling machine, using an 8 inch diameter, 10 tooth, indexable insert cemented carbide face milling cutter. The feed rate selected for this cutter is 0.012 in/tooth, and all the stock (0.125 inch) will be removed in one cut. Estimate the cutting speed that will utilize the maximum power available on the machine.

$K_p = 0.52$ (From Table 1a)
$C = 1.00$ (From Table 2)
$W = 1.20$ (From Table 3)
$E = 0.80$ (From Table 4)

$$Q_{max} = \frac{P_m E}{K_p CW} = \frac{10 \times 0.80}{0.52 \times 1.00 \times 1.20} = 12.82 \text{ in}^3/\text{min} \qquad \left(P_m = \frac{K_p CQW}{E}\right)$$

$$f_m = \frac{Q_{max}}{wd} = \frac{12.82}{6 \times 0.125} = 17.1 \text{ in/min} \qquad (Q = f_m wd)$$

$$N = \frac{f_{max}}{f_t n_t} = \frac{17}{0.012 \times 10} = 142.4 \text{ rpm} \qquad (f_m = f_t n_t N)$$

$$V = \frac{\pi DN}{12} = \frac{\pi \times 8 \times 142}{12} = 298.3 \text{ fpm} \qquad \left(N = \frac{12V}{\pi D}\right)$$

Calculated cutting speed of 298.3 fpm is correct, but very low for available cutting tools. Today, milling of gray irons is performed, mostly, using cubic boron nitride indexable inserts. Typical range of the cutting speed for milling with large engagement (0.8) and 0.004–0.008 inch per tooth chip load is 2700–2000 sfm. Estimating maximum productivity of cutting should be based on the recommended cutting speeds versus work materials.

***Example:*** **A 3.2 mm deep cut is to be taken on a 200–210 BHN AISI 1050 steel part using a 7.5 kW geared head lathe. The feed rate selected for this job is 0.45 mm/rev. Estimate the cutting speed that will utilize the maximum power available on the lathe.**

**$K_p = 2.32$ (From Table 1b)**
**$C = 0.92$ (From Table 2)**
**$W = 1.30$ (From Table 3)**
**$E = 0.80$ (From Table 4)**

$$Q_{max} = \frac{P_m E}{K_p C W} = \frac{7.5 \times 0.80}{2.32 \times 0.92 \times 1.30} \qquad \left(P_m = \frac{K_p C Q W}{E}\right)$$

$$= 2.16 \text{cm}^3/\text{sec}$$

$$V = \frac{60 Q_{max}}{fd} = \frac{60 \times 2.16}{0.45 \times 3.2} \qquad \left(Q = \frac{V}{60} fd\right)$$

$$= 90.0 \text{m/min}$$

The calculated cutting speed of 90.0 m/min is correct, but too low for carbide cutting tools. Thirty years ago, recommended cutting speed for turning AISI 1050 steel, 200–210 BHN, depth of cut 4.0 mm, was 160 m/min. Currently, using advanced coated carbide inserts, the cutting speed is increased up to 300 m/min for the depth of cut of 2.5–5.0 mm and a 0.38 mm/rev feed rate.

**Estimating Drilling Thrust, Torque, and Power.—**Although the lips of a drill cut metal and produce a chip in the same manner as the cutting edges of other metal cutting tools, the chisel edge removes the metal by means of a very complex combination of extrusion and cutting. For this reason a separate method must be used to estimate the power required for drilling. Also, it is often desirable to know the magnitude of the thrust and the torque required to drill a hole. The formulas and tabular data provided in this section are based on information supplied by the National Twist Drill Division of Regal-Beloit Corp. The values in Table 6 through Table 9 are for sharp drills and the tool wear factors are given in Table 3. For most ordinary drilling operations 1.30 can be used as the tool wear factor. When drilling most difficult-to-machine materials and when the drill is allowed to become very dull, 1.50 should be used as the value of this factor. It is usually more convenient to measure the web thickness at the drill point than the length of the chisel edge; for this reason, the approximate *w*/*d* ratio corresponding to each *c*/*d* ratio for a correctly ground drill is provided in Table 7. For most standard twist drills the *c*/*d* ratio is 0.18, unless the drill has been ground short or the web has been thinned. The *c*/*d* ratio of split-point drills is 0.03. The formulas given below can be used for spade drills, as well as for twist drills. Separate formulas are required for use with customary inch units and for SI metric units.

**Table 6. Work Material Factor, $K_d$, for Drilling with a Sharp Drill**

| Work Material | Material Constant, $K_d$ | Work Material | Material Constant, $K_d$ |
|---|---|---|---|
| AISI 1117 (Resulfurized free machining mild steel) | 12,000 | Austenitic Stainless Steel (Type 316) | 24,000[a] for Torque |
| Steel, 200 BHN | 24,000 | Titanium Alloy Ti6Al4V 40$R_c$ | 18,000[a] for Torque<br>29,000[a] for Thrust |
| Steel, 300 BHN | 31,000 | | |
| Steel, 400 BHN | 34,000 | René 41 40$R_c$ | 40,000[ab] min. |
| Cast Iron, 150 BHN | 14,000 | | |
| Most Aluminum Alloys | 7,000 | Hastelloy-C | 30,000[a] for Torque<br>37,000[a] for Thrust |
| Most Magnesium Alloys | 4,000 | | |
| Most Brasses | 14,000 | | |
| Leaded Brass | 7,000 | | |

[a] Values based upon a limited number of tests.

[b] Will increase with rapid wear.

**Table 7. Chisel Edge Factors for Torque and Thrust**

| c/d | Approx. w/d | Torque Factor A | Thrust Factor B | Thrust Factor J | c/d | Approx. w/d | Torque Factor A | Thrust Factor B | Thrust Factor J |
|---|---|---|---|---|---|---|---|---|---|
| 0.03 | 0.025 | 1.000 | 1.100 | 0.001 | 0.18 | 0.155 | 1.085 | 1.355 | 0.030 |
| 0.05 | 0.045 | 1.005 | 1.140 | 0.003 | 0.20 | 0.175 | 1.105 | 1.380 | 0.040 |
| 0.08 | 0.070 | 1.015 | 1.200 | 0.006 | 0.25 | 0.220 | 1.155 | 1.445 | 0.065 |
| 0.10 | 0.085 | 1.020 | 1.235 | 0.010 | 0.30 | 0.260 | 1.235 | 1.500 | 0.090 |
| 0.13 | 0.110 | 1.040 | 1.270 | 0.017 | 0.35 | 0.300 | 1.310 | 1.575 | 0.120 |
| 0.15 | 0.130 | 1.080 | 1.310 | 0.022 | 0.40 | 0.350 | 1.395 | 1.620 | 0.160 |

For drills of standard design, use $c/d = 0.18$; for split-point drills, use $c/d = 0.03$
$c/d$ = Length of Chisel Edge ÷ Drill Diameter.
$w/d$ = Web Thickness at Drill Point ÷ Drill Diameter.

For inch units only:

$$T = 2K_dF_fF_TBW + K_dD^2JW \quad (1)$$

$$M = K_dF_fF_MAW \quad (2)$$

$$P_c = MN/63{,}025 \quad (3)$$

**For SI metric units only:**

$$T = 0.05\,K_dF_fF_TBW + 0.007\,K_dD^2JW \quad (4)$$

$$M = \frac{K_dF_fF_MAW}{40{,}000} = 0.000025\,K_dF_fF_MAW \quad (5)$$

$$P_c = MN/9550 \quad (6)$$

Use with either inch or metric units:

$$P_m = \frac{P_c}{E} \quad (7)$$

where $P_c$ = Power at the cutter; hp, or kW $\quad$ $P_m$ = Power at the motor; hp, or kW
$M$ = Torque; in. lb, or N.m
$T$ = Thrust; lb, or N
$K_d$ = Work material factor (See Table 6)
$F_f$ = Feed factor (See Table 8)
$F_T$ = Thrust factor for drill diameter (See Table 9)
$F_M$ = Torque factor for drill diameter (See Table 9)
$A$ = Chisel edge factor for torque (See Table 7)
$B$ = Chisel edge factor for thrust (See Table 7)
$J$ = Chisel edge factor for thrust (See Table 7)
$W$ = Tool wear factor (See Table 3)
$N$ = Spindle speed; rpm
$E$ = Machine tool efficiency factor (See Table 4)
$D$ = Drill diameter; in., or mm
$c$ = Chisel edge length; in., or mm (See Table 7)
$w$ = Web thickness at drill point; in., or mm (See Table 7)

*Example:* A standard $\frac{7}{8}$ inch drill is to drill steel parts having a hardness of 200 BHN on a drilling machine having an efficiency of 0.80. The spindle speed to be used is 350 rpm and the feed rate will be 0.008 in/rev. Calculate the thrust, torque, and power required to drill these holes:

$K_d$ = 24,000 (From Table 6) $\quad$ $F_f$ = 0.021 (From Table 8)
$F_T$ = 0.899 (From Table 9) $\quad$ $F_M$ = 0.786 (From Table 9)
$A$ = 1.085 (From Table 7) $\quad$ $B$ = 1.355 (From Table 7) $\quad$ $J$ = 0.030 (From Table 7)

### Table 8. Feed Factors $F_f$ for Drilling

| Inch Units | | | | SI Metric Units | | | |
|---|---|---|---|---|---|---|---|
| Feed, in/rev | $F_f$ | Feed, in/rev | $F_f$ | Feed, mm/rev | $F_f$ | Feed, mm/rev | $F_f$ |
| 0.0005 | 0.0023 | 0.012 | 0.029 | 0.01 | 0.025 | 0.30 | 0.382 |
| 0.001 | 0.004 | 0.013 | 0.031 | 0.03 | 0.060 | 0.35 | 0.432 |
| 0.002 | 0.007 | 0.015 | 0.035 | 0.05 | 0.091 | 0.40 | 0.480 |
| 0.003 | 0.010 | 0.018 | 0.040 | 0.08 | 0.133 | 0.45 | 0.528 |
| 0.004 | 0.012 | 0.020 | 0.044 | 0.10 | 0.158 | 0.50 | 0.574 |
| 0.005 | 0.014 | 0.022 | 0.047 | 0.12 | 0.183 | 0.55 | 0.620 |
| 0.006 | 0.017 | 0.025 | 0.052 | 0.15 | 0.219 | 0.65 | 0.708 |
| 0.007 | 0.019 | 0.030 | 0.060 | 0.18 | 0.254 | 0.75 | 0.794 |
| 0.008 | 0.021 | 0.035 | 0.068 | 0.20 | 0.276 | 0.90 | 0.919 |
| 0.009 | 0.023 | 0.040 | 0.076 | 0.22 | 0.298 | 1.00 | 1.000 |
| 0.010 | 0.025 | 0.050 | 0.091 | 0.25 | 0.330 | 1.25 | 1.195 |

### Table 9. Drill Diameter Factors: $F_T$ for Thrust, $F_M$ for Torque

| Inch Units | | | | | | SI Metric Units | | | | | |
|---|---|---|---|---|---|---|---|---|---|---|---|
| Drill Dia., in. | $F_T$ | $F_M$ | Drill Dia., in. | $F_T$ | $F_M$ | Drill Dia., mm | $F_T$ | $F_M$ | Drill Dia., mm | $F_T$ | $F_M$ |
| 0.063 | 0.110 | 0.007 | 0.875 | 0.899 | 0.786 | 1.60 | 1.46 | 2.33 | 22.00 | 11.86 | 260.8 |
| 0.094 | 0.151 | 0.014 | 0.938 | 0.950 | 0.891 | 2.40 | 2.02 | 4.84 | 24.00 | 12.71 | 305.1 |
| 0.125 | 0.189 | 0.024 | 1.000 | 1.000 | 1.000 | 3.20 | 2.54 | 8.12 | 25.50 | 13.34 | 340.2 |
| 0.156 | 0.226 | 0.035 | 1.063 | 1.050 | 1.116 | 4.00 | 3.03 | 12.12 | 27.00 | 13.97 | 377.1 |
| 0.188 | 0.263 | 0.049 | 1.125 | 1.099 | 1.236 | 4.80 | 3.51 | 16.84 | 28.50 | 14.58 | 415.6 |
| 0.219 | 0.297 | 0.065 | 1.250 | 1.195 | 1.494 | 5.60 | 3.97 | 22.22 | 32.00 | 16.00 | 512.0 |
| 0.250 | 0.330 | 0.082 | 1.375 | 1.290 | 1.774 | 6.40 | 4.42 | 28.26 | 35.00 | 17.19 | 601.6 |
| 0.281 | 0.362 | 0.102 | 1.500 | 1.383 | 2.075 | 7.20 | 4.85 | 34.93 | 38.00 | 18.36 | 697.6 |
| 0.313 | 0.395 | 0.124 | 1.625 | 1.475 | 2.396 | 8.00 | 5.28 | 42.22 | 42.00 | 19.89 | 835.3 |
| 0.344 | 0.426 | 0.146 | 1.750 | 1.565 | 2.738 | 8.80 | 5.96 | 50.13 | 45.00 | 21.02 | 945.8 |
| 0.375 | 0.456 | 0.171 | 1.875 | 1.653 | 3.100 | 9.50 | 6.06 | 57.53 | 48.00 | 22.13 | 1062 |
| 0.438 | 0.517 | 0.226 | 2.000 | 1.741 | 3.482 | 11.00 | 6.81 | 74.90 | 50.00 | 22.86 | 1143 |
| 0.500 | 0.574 | 0.287 | 2.250 | 1.913 | 4.305 | 12.50 | 7.54 | 94.28 | 58.00 | 25.75 | 1493 |
| 0.563 | 0.632 | 0.355 | 2.500 | 2.081 | 5.203 | 14.50 | 8.49 | 123.1 | 64.00 | 27.86 | 1783 |
| 0.625 | 0.687 | 0.429 | 2.750 | 2.246 | 6.177 | 16.00 | 9.19 | 147.0 | 70.00 | 29.93 | 2095 |
| 0.688 | 0.741 | 0.510 | 3.000 | 2.408 | 7.225 | 17.50 | 9.87 | 172.8 | 76.00 | 31.96 | 2429 |
| 0.750 | 0.794 | 0.596 | 3.500 | 2.724 | 9.535 | 19.00 | 10.54 | 200.3 | 90.00 | 36.53 | 3293 |
| 0.813 | 0.847 | 0.689 | 4.000 | 3.031 | 12.13 | 20.00 | 10.98 | 219.7 | 100.00 | 39.81 | 3981 |

$W = 1.30$ (From Table 3)

$$T = 2K_dF_fF_TBW + K_dd^2JW$$
$$= 2 \times 24{,}000 \times 0.021 \times 0.899 \times 1.355 \times 1.30 + 24{,}000 \times 0.875^2 \times 0.030 \times 1.30$$
$$= 2313\ lb$$

$$M = K_dF_fF_MAW$$
$$= 24{,}000 \times 0.021 \times 0.786 \times 1.085 \times 1.30 = 559 \text{ in-lb}$$

$$P_c = \frac{MN}{63{,}025} = \frac{559 \times 350}{63{,}025} = 3.1 \text{ hp} \qquad P_m = \frac{P_c}{E} = \frac{3.1}{0.80} = 3.9 \text{ hp}$$

Twist drills are generally the most highly stressed of all metal cutting tools. They must not only resist the cutting forces on the lips, but also the drill torque resulting from these forces and the very large thrust force required to push the drill through the hole. Therefore, often when drilling smaller holes, the twist drill places a limit on the power used and for very large holes, the machine may limit the power.

## Boring

Boring is an internal turning operation performed with a boring bar to enlarge a previously drilled hole to form an internal shape of required dimensions. Boring operations range from semi-roughing to finishing.

An indexable boring bar has three basic elements: indexable cutting insert, shank, and anchor. The designation system for indexable inserts is the same as for turning. The anchor is the clamping portion of the shank that is held in the tool block. The minimum clamping length is approximately three to four diameters of the shank. The distance the boring bar extends beyond the tool block is the overhang; it determines the cutting depth. The overhang is the unsupported portion of the boring bar. Long overhang causes excessive deflection of the shank and generates vibration (chatter), which worsens the surface finish of the bore.

Eliminating chatter, especially in long-bore workpieces, is one of the greatest challenges faced by manufacturers and users of boring bars. Deflection of a boring bar depends on the mechanical properties of the shank material, the length of the overhang, and the cutting conditions.

The following equation is used for calculating deflection ($y$) of a boring bar.

$$y = \frac{FL^3}{3EI} \tag{1}$$

where $F$ = the cutting force, lbf or N

$L$ = unsupported length of a boring bar (overhang), in. or mm

$E$ = the modulus of elasticity (in tension) of a boring bar material, psi or N/mm$^2$

$I$ = the moment of inertia of a boring bar cross-sectional area, in$^4$ or mm$^4$.

**Cutting Force.—**Cutting force ($F$) is calculated by Formula (2) in customary US units.

$$F = 396{,}000 dfK_p \tag{2}$$

where $d$ = depth of cut, in.

$f$ = a feed rate, ipm

$K_p$ = a power constant, hp/in$^3$/min. The power constant $K_p$ data are given on pages 1133–1135 (Table 1a and Table 1b). The feed rate factors, $C$, for power constants are given on page 1136 (Table 2).

396,000 = a conversion factor that represents a unit of power equal to 550 in-lb/min

> *Example 1(a):* A workpiece is AISI 4140 chromium-molybdenum steel, 220 HB; depth of cut $d = 0.080$ in.; feed rate $f = 0.008$ ipr. Calculate the cutting force $F$ in customary US units of measure.
>
> Power constant $K_p = 0.72$ hp/ in$^3$/min (from Table 1b, page 1135)
>
> Feed factor $C = 1.08$ (from Table 2, page 1136)
>
> Adjusted power constant, $K_{pa} = K_p \times C = 0.72 \times 1.08 \cong 0.78$.
>
> Calculating: $F = 396{,}000 dfK_{pa} = 396{,}000 \times 0.08 \times 0.008 \times 0.78 = 197.7$ lbf

Cutting force ($F$) is calculated by Formula (3) in metric units.

$$F = 60{,}000 dfK_p \tag{3}$$

where $d$ = depth of cut, mm

$f$ = feed rate, mm/min.

$K_p$ = is a power constant, kW/cm$^3$/min. The power constant $K_p$ data are given on pages 1133–1135 (Table 1a and Table 1b). The feed rate factors, $C$, for power constants are given on page 1136 (Table 2).

60,000 = is a conversion factor that represents a unit of power equal to 1 kW × mm/min.

*Example 1(b):* A workpiece is AISI 4140 chromium-molybdenum steel, 220 HB; depth of cut $d = 2.03$mm; feed rate $f = 0.20$ mm/rev. Calculate the cutting force $F$ in metric units.

Power constant $K_p = 0.0328$ kW/cm³/min. (from Table 1b, page 1135)

Feed factor $C = 1.08$ (from Table 2, page 1136)

Adjusted power constant, $K_{pa} = K_p \times C = 0.0328 \times 1.08 \cong 0.0354$ kW/cm³/min.

Calculating: $F = 60{,}000 dfK_{pa} = 60{,}000 \times 2.03 \times 0.20 \times 0.0354 = 863.6$ N

Comparing the cutting forces calculations of Example 1(a) and Example 1(b), the cutting forces are approximately equivalent within about 2 percent.

**Moduli of Elasticity ($E$) of Boring Bar Materials.—**Boring bar shanks are made of steels, tungsten-base metals, or cemented carbides. Most commonly used boring bar materials are alloy steels. Some boring bar manufacturers use AISI 1144 free machining medium-carbon steel. Regardless of grades, all carbon and alloy steels have approximately the same modulus of elasticity:

Customary US units, $E = 30 \times 10^6$ psi

Metric units, $E = 20.6 \times 10^4$ N/mm²

A common mistake is to assume that a steel shank with a higher hardness, or one made from a higher quality of steel will deflect less. As can be seen from Equation (1), the material property that determines deflection is the modulus of elasticity. Hardness does not appear in this equation.

Tungsten heavy alloys for boring bars, $E = (45–48) \times 106$ psi (customary US units) and $E = (31–33) \times 10^4$ N/mm² (metric units). Boring bars made of tungsten heavy alloys will deflect less than steel boring bars of the same diameter and overhang by 50 to 60 percent when cutting at the same depth of cut and feed rate.

Cemented carbides for boring bars, $E = (84–89) \times 10^6$ psi (customary US units) and $E = (52–61) \times 10^4$ N/mm² (metric units).

Boring bars made of cemented carbides provide minimum deflection because their moduli of elasticity are higher than those of steels and tungsten heavy alloys.

**Moment of Inertia ($I$) of a Boring Bar Cross-Sectional Area.—**Moment of inertia is a property of areas. Since boring bars are available in a variety of diameters, it is important to calculate the moment of inertia of a bar cross section using appropriate formulas. Boring bars are usually round with a solid or tubular cross section. Moments of inertia of solid or tubular cross-sectional areas are calculated by:

$$I = \frac{\pi D^4}{64} \quad (4a) \qquad\qquad I = \frac{\pi(D_o^4 - D_i^4)}{64} \quad (4b)$$

where $D$ is the diameter of the bar in inches or mm, $D_o$ is the outside diameter of the bar and $D_i$ is the inside diameter, in inches or mm.

*Example 2:* A boring bar diameter is 1 inch. Moment of inertia $I = \pi \times (1)^4/64 = 0.0491$ in⁴. Moments of inertia of cross sections are given starting on page 235.

**Deflection ($y$) of the Boring Bar.—**To calculate the deflection of the boring bar, it is necessary to enter one more data point in Equation (1). It is the unsupported length of the boring bar ($L$).

*Example, Boring Bar Deflection:* The unsupported length of a 1-inch diameter boring bar is 4 inches. Using $F$ calculated in Example 1(a) and $I$ calculated in Example 2:

$$y = \frac{FL^3}{3\,EI} = \frac{(197.7)4^3}{3\,(30 \times 10^6)\,0.0491} = 0.0028 \text{ inch}$$

# MICROMACHINING

## Introduction

Recent technological advancement and market need for product miniaturization demand three-dimensional (3D) microcomponents. Although microelectronic manufacturing techniques can produce pseudo 3D microdevices using silicon and other semiconducting materials, such materials are neither robust nor biocompatible for demanding applications in aerospace, medical, sensor, defense, petroleum, and transportation. Examples of robust applications include microdrilling holes for fuel or ink injection nozzles, electronic printed circuit boards, microfabrication of watch components, air bearings, cooling holes in turbomachinery, high aspect ratio features on tool steel molds and dies, etc. There are alternative nontraditional processes to produce microfeatures on robust engineering materials such as laser micromachining, electrical discharge microdrilling, electrochemical micromachining, chemical etching, electron/ion beam machining; however, these processes are either cost prohibitive, limited to conductive materials, or inferior when comparing resulting surface integrity, subsurface damage, high aspect ratio, or microfeature quality. Microfabrication with traditional processes such as micromilling, microdrilling, microturning… are still the preferred choice in most applications. There is no standard that defines micromachining, but most researchers uses cutting tools to produce components with key dimensions less than 1 mm (0.040 inch) or when depth of cut is comparable to tool sharpness or tool grain size in their micromachining studies.

Realizing the needs for traditional micromanufacturing, there are more commercially available machine tools and microtools in the market. However, costly equipment, lack of in-depth understanding of micromachining, and limited guidelines for effective use of microtools are still the bottleneck for full application of micromachining. Universities and research institutes worldwide have started theoretical investigation of micromachining and produced positive results from the academic point of view. Without practical guidelines on micromachining, technicians and machinists probably would make wrong and costly decisions when simply extending macroscale machining practices into microscale machining applications—a microtool simply breaks at even conservative macroscale parameters for speeds, feeds, and depth of cut.This section, while complementing other chapters in this *Handbook*, focuses on practicality, based on proven theories and published data, to help decision makers to understand the requirements for micromachining, and as a guide to people on the shop floor to quickly and confidently begin using the recommended parameters and techniques. Both US standard and SI metric units are included for convenience. Examples of how to use the data and equations are given throughout this chapter.

## Machine Tool Requirements

To obtain the same surface speed as in macromachining, a machine tool must:

a) Be capable of rotating a workpiece or tool at high speeds of 25,000 rpm or above
b) Control spindle runout to submicron level
c) Have a very robust mechanical and thermal structure that is not affected by vibration or thermal drift
d) Have high resolution tool positioning and feeding mechanisms

Success in micromachining depends on tool quality and precision of the machine tool. Machine spindle runout, tool concentricity and tool positioning accuracy must be in the neighborhood of 1/100th of tool diameter or less for successful operation. Tolerance stack up for spindle runout, tool eccentricity, and wandering of a microdrill causes cyclic bending of the tool that lead to catastrophic failure. At a low rotatational speeds, the displacement of a spindle can be monitored with a sensitive mechanical indicator. However, this option is not applicable for machines that operates at or above a few thousands rpm. Other non-contact techniques using capacitance, magnetism, or light would be more appropriate. Fig. 1a shows an example of spindle runout measuring setup. A laser beam is pointed at a rotating precision plug gage. The spindle displacement is then recorded on a computer for further analysis and display in either frequency or time domain. Commercial laser systems can provide displacement readings to ±0.1 μm resolution.

An example of spindle runout is shown in Fig. 1b; the spindle runout of a Haas OM2 machine was measured with a Keyence laser system to be ±1.25 μm. Care must be practiced to isolate vibration of the spindle or it would affect the sensor reading, and avoid direct eye contact with the reflected laser from the shiny plug gage.

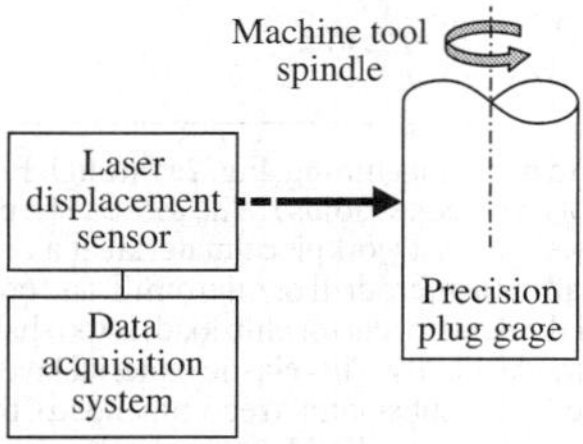


Fig. 1a. Setup for Spindle Runout Measurement.

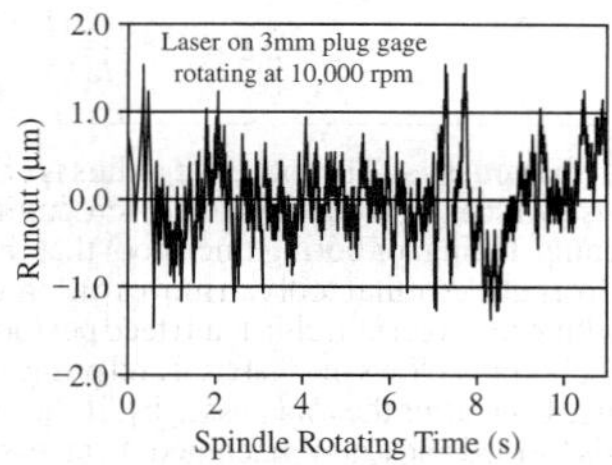


Fig. 1b. Spindle Runout of Haas OM2 CNC Micromilling Machine.

*Example 1, Spindle Speed for Macro versus Micro Machining:* The speed and feed table on page 1106 recommends a milling speed of 585 ft/min (178 m/min) and feed of 0.004 in/tooth (0.1 mm/tooth) for end milling 316L stainless steel using an uncoated carbide tool.

*Macromachining:* To have said surface speed for an Ø1/2 inch (Ø12.5 mm) end mill, the required spindle speed is

$$N = \frac{V}{\pi D} = \frac{585(\text{ft/min})}{\pi(\text{rad/rev}) \times 0.5(\text{in})} \times 12(\text{in/ft}) = 4{,}469 \text{ rpm}$$

*Micromachining:* To obtain the same surface speed for an Ø0.004 inch (Ø0.1 mm) micromill, the new spindle speed is

$$N = \frac{V}{\pi D} = \frac{585(\text{ft/min})}{\pi(\text{rad/rev}) \times 0.004(\text{in})} \times 12(\text{in/ft}) = 558{,}633 \text{ rpm}$$

To turn, face, or bore a stainless steel microshaft of Ø0.004 inch (Ø0.1 mm) at this cutting speed, a lathe spindle would need to rotate at 558,633 rpm too. A machine tool with spindle speed exceeding 500,000 rpm is rare or simply not commercially available at this time.

Applying the recommended macro feed of 0.004 in/tooth (0.1 mm/tooth) for an 0.004 inch (0.1 mm) diameter micromill would break the tool because the feed/tooth is as large as the microtool diameter.

## Microcutting Tools

**Tool Stiffness.—**It is relatively easy to have a rigid turning or facing microtool, but it requires careful planning to maintain rigidity of a micromill or a microdrill. Geometries of macroscale and microscale drilling/milling tools are the same: tool diameter, number of cutting flutes, point included angle for microdrill, helix angle, web thickness, clearance angle, flute length, shank diameter, and overall length. A careful selection of microtools must consider the intended machined features and highest possible tool stiffness. The two most important geometries that affect the microtool stiffness are the tool diameter and flute length assuming the number of flutes has been chosen. It can be shown that the torsional stiffness of a mill/drill is proportional to (tool diameter)$^4$ and (flute length)$^{-2}$. For a specific mill/drill tool dimension, the milling/drilling strategy must be adjusted accordingly to avoid tool breakage.

*Example 2, Stiffness of Microtools:* If a drill diameter of 0.8 mm is selected instead of 1.0 mm, then the 20% reduction of diameter will result in a reduction in torsional stiffness $E$ of:

$$\Delta E = \frac{(D_2)^4 - (D_1)^4}{(D_1)^4} = \frac{0.8^4 - 1.0^4}{1.0^4} = -59\%$$

Similarly, if a flute length of 1.2 mm is chosen instead of 1.0 mm, the 20% change in flute length will lead to a decrease in torsional stiffness $E$ of:

$$\Delta E = \frac{(L_2)^{-2} - (L_1)^{-2}}{(L_1)^{-2}} = \frac{1.2^{-2} - 1.0^{-2}}{1.0^{-2}} = -30\%$$

**Tool Sharpness.**—The tool edge radius is critical in micromachining. Fig. 2a through Fig. 2d shows two scenarios for the same microcutting tools with edge radius $r$. The tool can be either a turning, facing, or boring microtool that linearly engages a workpiece material at a certain depth of cut. A similar tool can move in a circular path as a microdrill or micromill, and engage a workpiece at a certain chip load (feed per tooth). If the depth of cut (or chip load) is too shallow, the tool simply plows the material and pushes it away elastically. This elastic material layer just springs back after the tool passes by. If the depth of cut is substantial (recommended), then a chip is formed and a new machined surface is generated with negligible spring back.

*Chip load* is commonly used interchangeably with feedrate for a cutting tool with multiple cutting edges (teeth) such as in milling or drilling. Chip load is defined as tool feed distance for each tooth and represents the chip size forming for each tooth. Chip load can also be interpreted as the radial depth of cut for each tooth in milling. The following equation converts chip load of a cutting edge to feedrate of a multiple-edge cutting tool:

$$f = c_L nN$$

where $f$ = feedrate of tool (mm/min, in/min)
$c_L$ = chip load of a cutting edge (mm/tooth, in/tooth)
$n$ = number of cutting flutes or cutting edges (#teeth/rev)
$N$ = rotational speed (rpm)

*Example 3:* A two-flute uncoated carbide end mill with diameter Ø1 mm (Ø0.040 in) is used for micromilling pure titanium. Table 13b suggests a chip load of 17 μm/tooth and cutting speed of 90 m/min. The rotational speed is computed as:

$$N = \frac{V}{\pi D} = \frac{90\,(\text{m/min})}{\pi\left(\frac{\text{rad}}{\text{rev}}\right) \times 0.001\,(\text{m})} = 28{,}600 \text{ rpm}$$

The feedrate for this operation is:

$$f = c_L nN = 17\left(\frac{\mu\text{m}}{\text{tooth}}\right) \times 2\left(\frac{\text{teeth}}{\text{rev}}\right) \times 28{,}600\left(\frac{\text{rev}}{\text{min}}\right) = 972{,}400\frac{\mu\text{m}}{\text{min}} \approx 972\frac{\text{mm}}{\text{min}} \approx 38\frac{\text{in}}{\text{min}}$$

Typical fine grain carbide tools are first sintered from submicron carbide particles in a cobalt matrix, and then ground and lapped to final geometry. Optimal edge radii of 1–4 μm (39–156 μinch) are typically designed for sintered tools to balance edge sharpness and edge strength. Only single crystalline diamond tools can be ground and lapped to edge radii within the nanometer range.

The threshold for minimum depth of cut has been investigated theoretically and verified experimentally by many researchers. It varies from 5 to 40 percent of the tool edge radius depending on the workpiece material and original rake angles. The threshold depth of cut or chip load, therefore, can be conservatively set to be 50 percent of the tool edge radius. When machining below this threshold, a microtool just rubs and plows the surface with negative effective rake angle and deforms it elastically during the first pass. This results in high cutting force, high specific energy, fast tool wear, rough surface finish, and significant burrs. In subsequent passes when the cumulative depth is greater than the critical depth of cut, then a tool can remove materials as chips and the cycle repeats.

**Effect of Depth of Cut (Chip Load) in Micromachining**

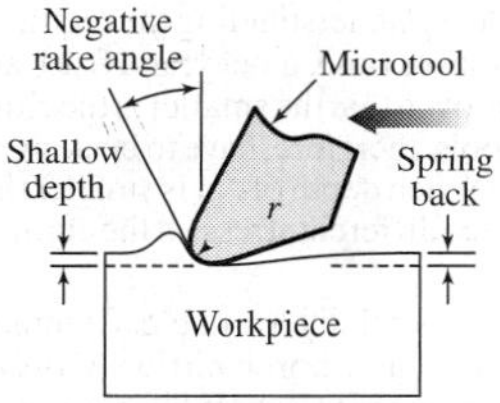


Fig. 2a. Microfacing, Depth of Cut < 0.5 *r*.

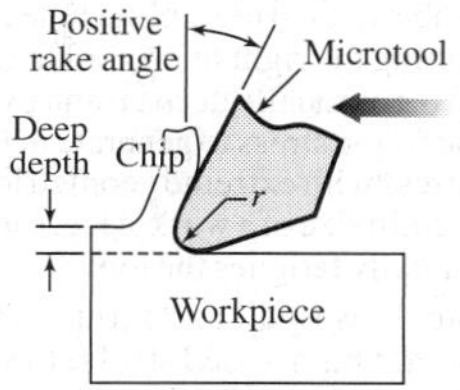


Fig. 2b. Microfacing, Depth of Cut > 0.5 *r*.

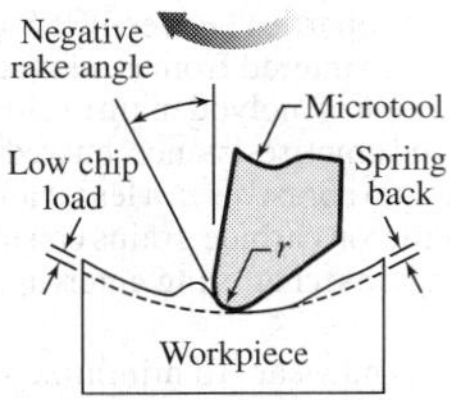


Fig. 2c. Micromilling, Chip Load < 0.5 *r*

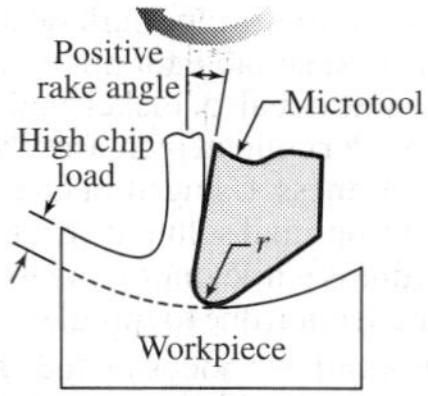


Fig. 2d. Micromilling, Chip Load > 0.5 *r*.

Fig. 2a and Fig. 2c illustrate rubbing and plowing of material with negative effective rake angle at a shallow depth of cut. Fig. 2b and Fig. 2d illustrate chip removal from material with positive effective rake angle at a deep depth of cut.

It is crucial to verify the tool edge radius before deciding on cutting parameters. Measuring of tool edge radius, however, is not trivial. A tool edge radius can be estimated from a scanning electron microscopic picture when the cutting edge is parallel to the electron beam (Fig. 6), or from a scanned image at the neighborhood of a cutting edge on an atomic force microscope (Fig. 3a and Fig. 3b), or by scanning an edge on an optical microscope profiler in different views to reconstruct a 3D image of the tool edge before finding its radius.

**Tool Edge Measurement by Atomic Force Microscopy**

Note the different vertical and horizontal scales.

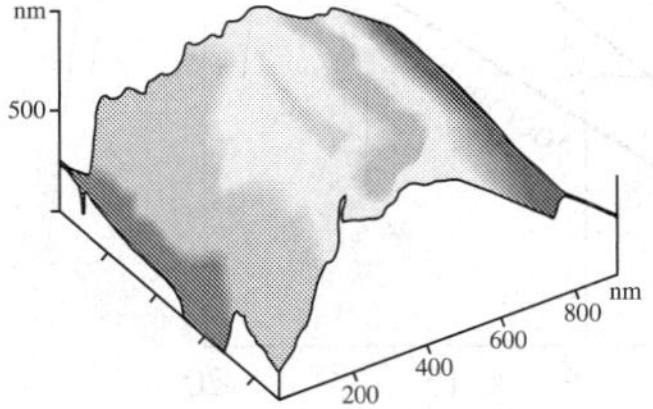


Fig. 3a. New Polycrystalline Diamond Tool with a 750 nm Edge Radius.

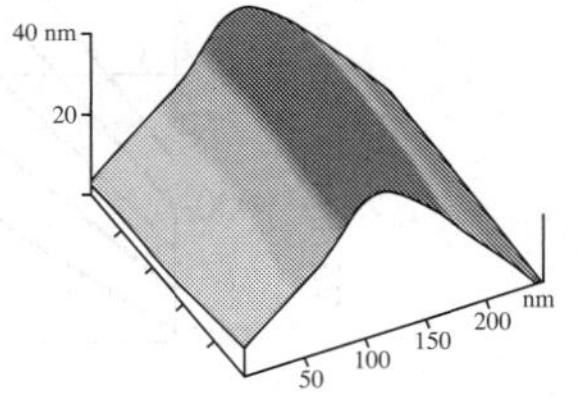


Fig. 3b. New Single Crystalline Diamond Tool with a 10 nm Edge Radius.

**Tool Materials.**—Using the right microtool is essential for micromachining. A microtool that successfully drills through holes on a plastic printed circuit board is not necessarily able to drill deep blind holes on titanium alloys. Understanding the requirements and selecting the right microtool for each condition saves time, money, and frustration. It has been theoretically derived and experimentally proven that the smaller is the chip, then the higher is the stress to generate it. Microcutting tools, therefore, have to be designed for higher stress with extreme geometrical constraints. When depth of cut is smaller than the average grain size of a workpiece, each grain generates different stress on the cutting edge and eventually fatigues the tool.

Microtools as small as 25 µm (0.001 inch) are commercially available. Common tool materials are high-speed steel (HSS), cermet, carbide, cubic boron nitride (CBN), polycrystalline diamond (PCD), and single crystalline diamond (SCD). HSS is commonly not used in micromachining of metal since it does not have required hardness and strength to resist plastic deformation. A SCD tool is available for microturning, but not for microdrilling or micromilling. Carbide and cermet, having properties between HSS and diamond, are most suitable for microcutting tools. They are sintered from random abrasive grains in either cobalt or nickel binder with a small addition of molybdenum or chromium. A higher binder content increases the tool toughness and crack resistance, but reduces the tool bulk hardness. Using ultra fine grain (submicron size) abrasives in a lesser amount of binder is the optimal solution because a tool with submicron carbide grains can maintain a high hardness while improving its crack resistance against chattering, interrupted cuts, or cyclic deflection due to spindle runout.

Microtool failure modes include shearing, chipping, and wear. To minimize shearing and catastrophic tool failure, a tool should be made from a high hardness substrate and with a geometry suitable for micromachining, i.e., large included angle and sharp cutting edge (Fig. 4). A tool with smaller than minimum included angle will be deformed and fractured in service.

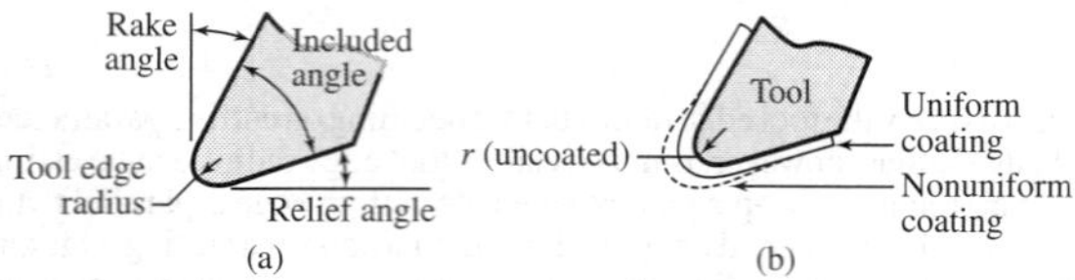


Fig. 4. (a) Tool Geometry, and (b) Change of Tool Edge Radius Due to Coating.

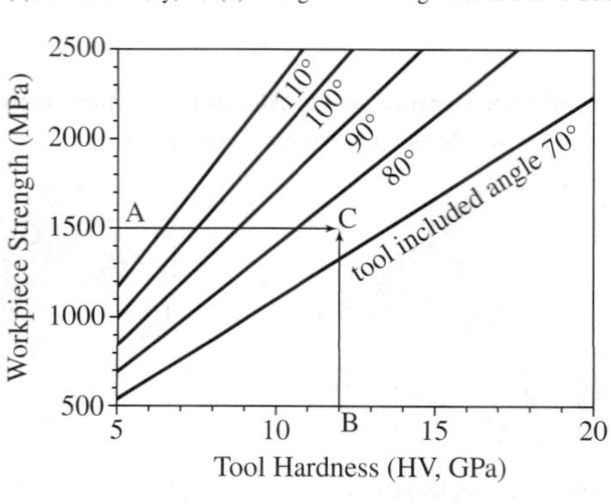


Fig. 5. Microtool Minimum Included Angles.

*Example 4, Required Included Angle:* Find the minimum included angle for an ultra-fine grain carbide tool of 12 GPa Vicker hardness to machine the F799 Co-Cr alloy of 1500 MPa (0.2 × $10^6$ psi) tensile strength.

*Solution:* Referring to Fig. 5, locate workpiece strength at point A on the vertical axis (1500 MPa). Locate point B for tool hardness on the horizontal axis (12 GPa). The intersection at C of the horizontal line from A and the vertical line from B indicates that the minimum included tool angle should be 75°.

Coating of microtools is still a technical challenge due to conflicting constraints for tool performance. Chemical or physical vapor deposition (CVD or PVD) techniques have been developed to coat cutting tools with mono/multiple layers of intermetallic or ceramic compounds (Table 1). Criteria for acceptable tool coating are numerous: uniformity, high hardness, high toughness, low friction, high wear resistance, surface smoothness, high chemical/diffusion resistance, and high temperature stability at a reasonable cost. Although a coating thickness of 2–4 μm (79–157 μin) is acceptable for a macrotool, the coating thickness on a microtool should be thinner to minimize fracture and peeling of the coating. Both CVD and PVD processes not only add the coating thickness to the edge radius, but the extra coating also increases the radius at sharp corners (Fig. 4b). This is unfortunate since the thicker coating reduces the tool sharpness by enlarging the tool edge radius and causes an unfavorable plowing effect with negative effective rake angle. An uncoated microtool might perform satisfactorily, but the same machining parameters can be devastating to an over-coated microtool (Fig. 6). A thin coating of less than 1.5 μm following by an edge sharpening process would improve the tool performance, however, at the expense of higher tool cost. Published data indicate that micrograin carbide tools with 1.5 μm TiN coating is the best for micromilling of H13 tool steel hardened to 45 RC.

**Table 1. Commercial Coatings for Microtools**

| Coating | Structure | Hardness GPa | Hardness $10^6$ psi | Coefficient of Friction | Coating Thickness μm | Coating Thickness μinch | Maximum Temperature °C | Maximum Temperature °F |
|---|---|---|---|---|---|---|---|---|
| TiN | monolayer | 24 | 3.5 | 0.55 | 1-5 | 39-197 | 600 | 1110 |
| TiCN | gradient | 37 | 5.4 | 0.20 | 1-4 | 39-157 | 400 | 750 |
| TiAlCN | gradient | 28 | 4.1 | 0.30 | 1-4 | 39-157 | 500 | 930 |
| TiAlN | multilayer | 28 | 4.1 | 0.60 | 1-4 | 39-157 | 700 | 1290 |
| AlTiN | gradient | 38 | 5.5 | 0.70 | 1-3 | 39-118 | 900 | 1650 |
| ZrN | monolayer | 20 | 2.9 | 0.40 | 1-4 | 39-157 | 550 | 1020 |
| CrN | monolayer | 18 | 2.6 | 0.30 | 1-4 | 39-157 | 700 | 1290 |
| Diamond like | gradient | 20 | 2.9 | 0.15 | 0.5-1.5 | 20-59 | 400 | 750 |
| $AlTiN/Si_3N_4$ | nanocomposite | 45 | 6.5 | 0.45 | 1-4 | 39-157 | 1200 | 2190 |
| $AlCrN/Si_3N_4$ | nanocomposite | 42 | 6.1 | 0.35 | 1-5 | 39-197 | 1100 | 2010 |

**Tool Offset and Positioning.—**Tool offset and tool positioning are crucial in micromilling and microdrilling because a tool is small and extremely fragile especially if it has a high aspect ratio (length to diameter ratio). Common shop practices to find tool offset and position often damage a tool or workpiece. Non-contact techniques using light, magnetism, capacitance, ultrasound, etc. are the preferred choice for precisely locating the relative position between tool and workpiece. Selection of a suitable sensor depends at least on following criteria:

- Better resolution compared to that of the machine tool axis
- Small working zone to cover a microtool
- Fast sampling rate for intended tool speed

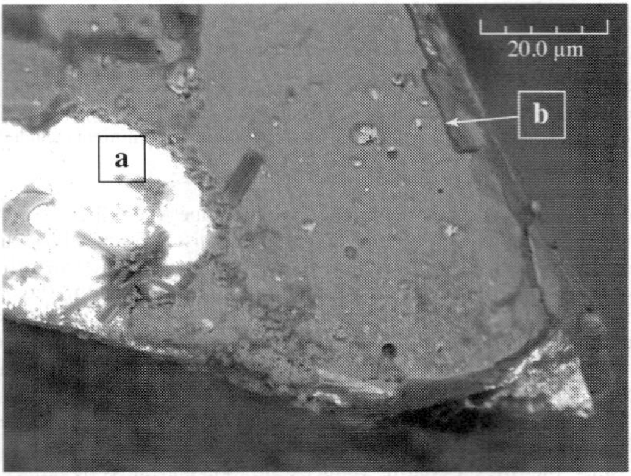


Fig. 6. Peeling (a) and cracking (b) of 4μm-thick TiN coating layer on a WC micromill. Back scattered electron technique shows high contrast of a dark TiN coating layer against the bright WC/Co substrate in the background.

*Example 5, Sensor Requirement:* Select a sensor for microdrilling using
Microdrill: 100 μm diameter, 1 mm flute length (Ø0.004 inch, 0.040 inch flute length).
Machine tool: 1 μm (40 μinch) repeatability and 500,000 rpm capability.

*Solution:* A laser displacement sensor is selected to satisfy the following specifications:
Resolution: 0.1 μm (4 μinch)

Spot size: 25–75 μm (0.001–0.003 μinch). Although most drill shanks are Ø3.175 mm (Ø0.125 inch), the working zone should be as small as possible to detect the shank center.

In order to make 6 measurements when a tool is rotating at 500,000 rpm, the time between measurements is:

$$t = \left(\frac{1\ \text{rev}}{6\ \text{measurements}}\right)\left(\frac{1\ \text{min}}{500{,}000\ \text{rev}}\right)\left(\frac{60\ \text{s}}{1\ \text{min}}\right) = 2 \times 10^{-5}\ \text{s} = 20\mu\text{s}$$

A laser with minimum 20 μs sampling rate (50 kHz) would be sufficient.

A mechanical edge finder is adequate for most macromachining setups, but it is not suitable for micromachining especially with small and pliable parts. Fig. 7a shows a non-contact technique to detect part edge or find lateral tool offset. A rotating precision plug gage, mounted on a machine spindle, is positioned between a stationary laser sensor and the workpiece. The small laser beam is aimed at the plug gage center and on the part edge when the plug gage is withdrawn away from the beam path. These two laser sensor readings allow computing the tool center offset. A precision plug gage should be used instead of a cutting tool shank for better repeatability.

*Example 6, Lateral Tool Offset Calculation:* Use a laser displacement sensor and a Ø3.175 mm (Ø1/8 inch) plug gage to detect the edge of a ground block.

*Solution:*

i) Mount the plug gage on the machine spindle and rotate it at 5000 rpm.

ii) Scan a laser beam across the plug gage and stop when the distance from the laser source to the target is minimum, i.e., the beam is at the gage center. Read $L_1 = 35$ mm.

iii) Jog the plug gage away from the beam path, read distance to the part edge $L_2 = 55$ mm.

iv) The lateral offset from the spindle center to the workpiece edge is then:

$$\text{Lateral offset} = L_2 - L_1 - D/2 = 55 - 35 - (3.175/2) = 18.412\ \text{mm}$$

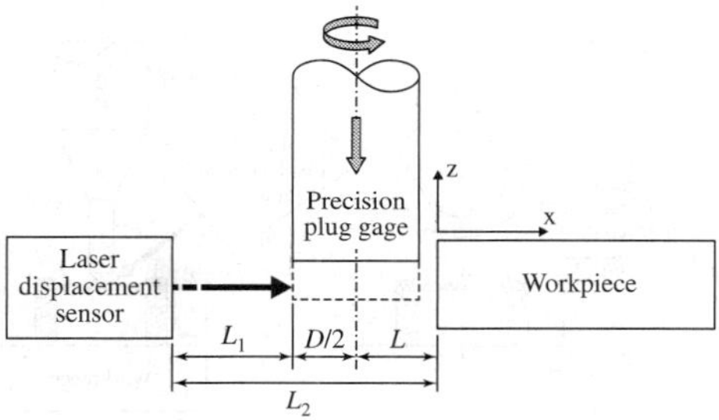


Fig. 7a. Setup for Lateral Edge Detection Using Laser Sensor.

Fig. 7b compares the accuracy and repeatability of the non-contact method shown in Fig. 7a against those of a mechanical edge finder.

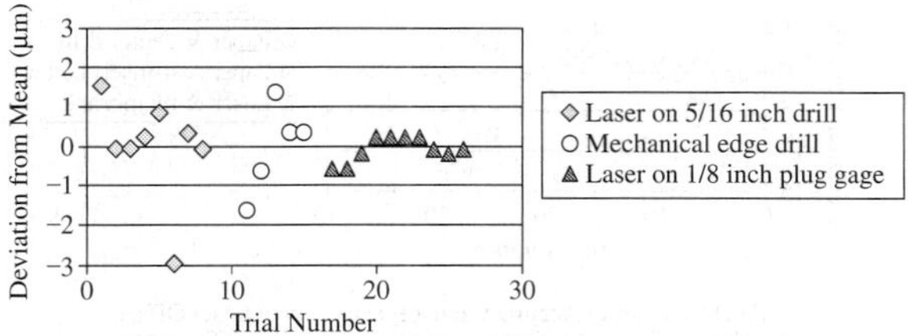


Fig. 7b. Superior Accuracy and Repeatability of Laser Edge Detection Technique Compared to Mechanical Technique.

It is rather simple to find the lateral offset as illustrated in Example 6, but it is more difficult to find the exact vertical offset for a slender microdrill or micromill without damaging it. One can attempt to use the common "paper technique" or take a risk with an available contact sensor for z-setting. In the paper technique, one would use a hand to slide a piece of paper on top of a workpiece while gradually lowering a tool. The tool stops when a resistance on the paper is felt. The paper technique is tedious, subjective, and tool dependent. Fig. 8b shows scattering of data up to ±5 µm when finding z-offset for a center drill, but it is ±15 µm for a milling cutter with 4 teeth.

A commercial contact sensor requires a tool to move down and press against a solid surface. A pressure sensor then triggers an audible or visual signal to indicate a positive contact. The pressure level on such sensor is preset for macrotool setting and cannot be adjusted for a microtool. In both cases, the tool tip is one paper thickness or one contact sensor height above the workpiece—if the tool survives.

A non-contact sensor is more practical and reliable. The same laser displacement sensor used for lateral tool offset can also be used for vertical tool offset. Fig. 8a shows a precision ring with secured circular plastic membrane that is used for indirect measurement. The membrane center is marked with a reference (e.g., crossing lines) at which the height can be measured with the laser displacement sensor. Upon placing the fixture on top of a workpiece and then lowering a tool onto the reference mark, a slight contact of the tool and the flexible membrane is precisely detected with the laser beam pointing near the contact point. When this happens, the tool tip is at the same height as the membrane. The repeatability of tool offset using this technique is well within the positioning repeatability 3 µm of the tested Haas OM2 machine tool (Fig. 8b).

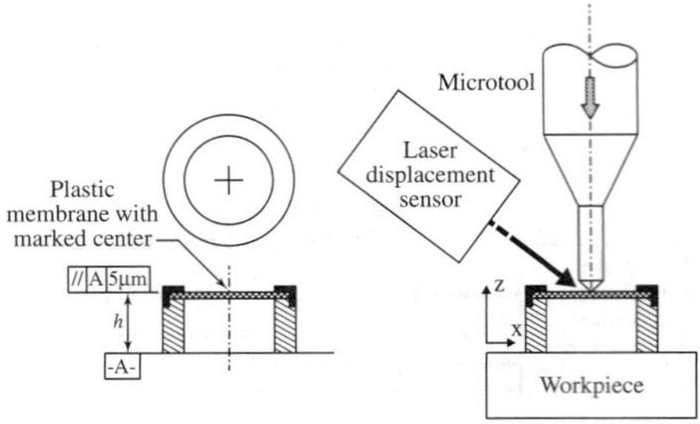


Fig. 8a. Microtool Offset and Microtool Height Detection Using Laser.

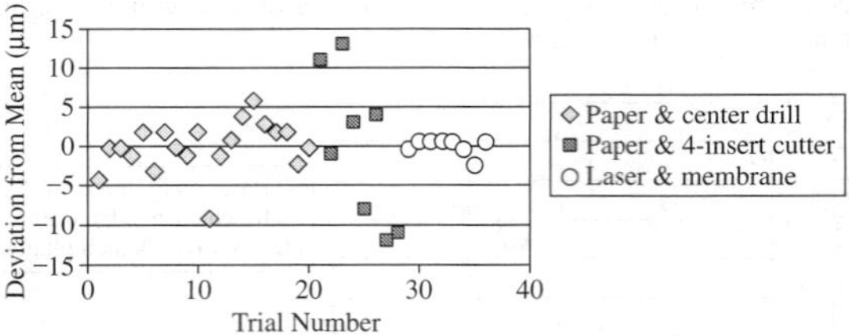


Fig. 8b. Superior Accuracy and Repeatability of Laser Offset and Height Detection Technique Compared to Paper Technique.

When a part is small or does not have a large surface for the fixture to rest on, then an indirect technique to find vertical tool offset for a microdrill or micromill is recommended. The following example illustrates this.

*Example 7, Vertical Tool Offset Calculation:* A vice or collet is used to clamp a micropart for drilling. The micropart protrudes upward a distance $h_1 = 0.1000$ inch. If the vice surface has been qualified as a reference, it can be used to find the vertical offset of a microdrill tip (Fig. 9).

i) Measure the fixture height at the reference mark using the laser sensor, $h_2 = 0.3500$ inch.

ii) Position the fixture on top of the vice.

iii) Lower the microdrill onto the reference mark of the membrane. Stop when the membrane is slightly deflected which can be detected easily with the laser sensor.

iv) Calculate the required drill vertical offset:

$$\text{Vertical offset} = h_2 - h_1 = 0.3500 - 0.1000 = 0.2500 \text{ inch}$$

**Tool Damage.—**Tool damage can be categorized by the relative size of the damage, ranging from submicron to hundreds of microns, as indicated in Table 2. The tool failure mechanisms include damages due to mechanical, thermal, and chemical effects, and adhesion. Examples of microtool damages are illustrated in Fig. 10a through Fig. 10d.

*Mechanical effect* is the most common source of tool damage. Abrasive wear is caused by sliding of hard particles from workpiece or tool against the cutting tool surface. Attrition wear is larger than abrasion wear; it occurs when one or a few grains of the tool are weakened at their grain boundaries and are dislodged from the tool. Microchipping and chipping are larger chunks of tool being removed due to mechanical or thermal shocks upon loading and unloading. Machining at optimal parameters and with a rigid setup will reduce vibration, shock, and mechanical damage to a microtool.

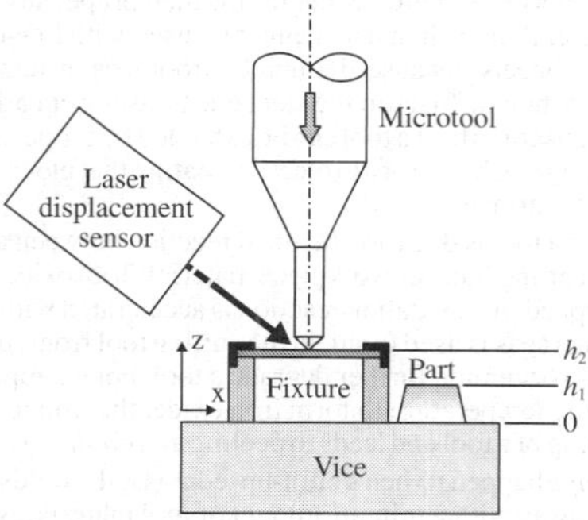


Fig. 9. Indirect Vertical Tool Offset Detection Technique.

**Microtool Damages**

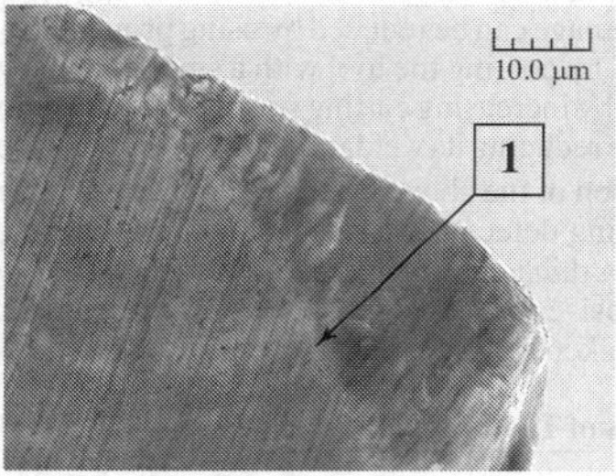


Fig. 10a. Abrasion with progressive wear from center to edge (label "1") due to different cutting speed on a microdrill.

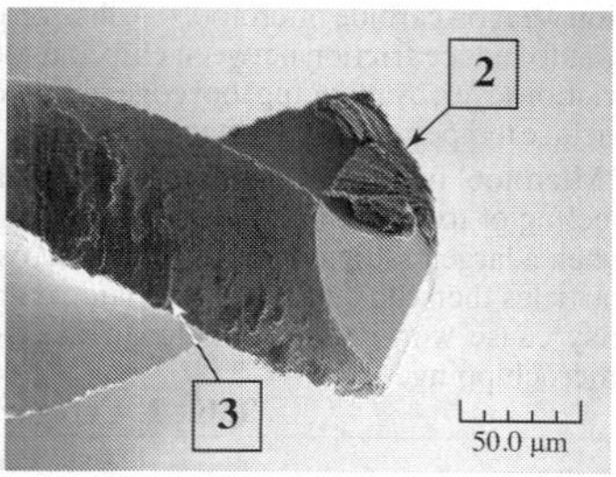


Fig. 10b. Built-up-edge at a cutting lip (label "2") and side (label "3") of a microdrill.

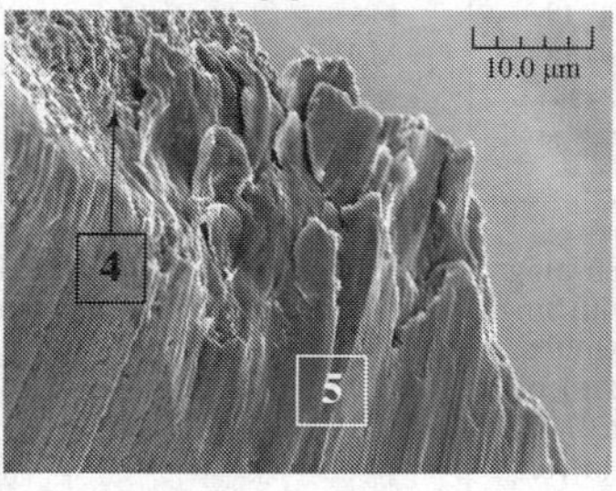


Fig. 10c. Adhesion wear (label "4") due to built-up-edge on a micromill (label "5").

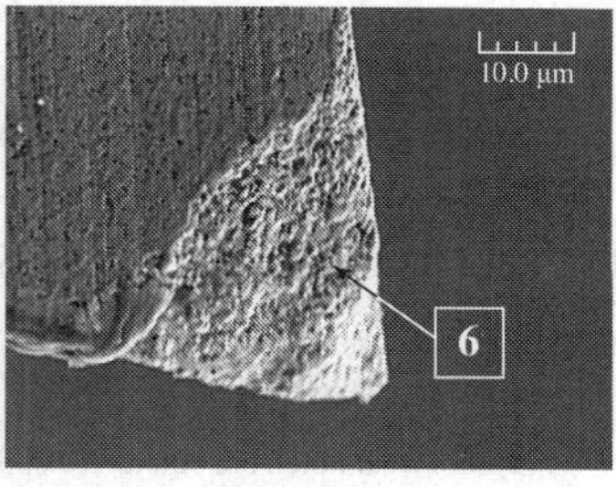


Fig. 10d. Chipping at the cutting edge of a carbide micromill due to local depletion of cobalt binder.

*Thermal effect* is the second cause of tool damage. A cutting tool edge is softened at high machining temperature, deformed plastically, and removed from the tool. Both high-speed steel tools and carbide tools with high cobalt content are vulnerable to thermal damage. High temperature also promotes diffusion, i.e., atoms from the tool and workpiece move mutually across their interfaces, therefore degrading their properties and causing diffusion wear. Diamond with a carbon-rich matrix cannot be used with low-carbon ferrous alloys like steels or stainless steels because diamond carbonizes at temperatures exceeding 1112°F (600° C) and carbon diffuses to the steel due to its lower carbon content and high affinity to carbon. The useful life of a tool can be extended by proper application of coolant to reduce thermal damage, or by use of a protective coating that blocks undesirable thermal diffusion from/to a tool surface.

*Chemical damage* of a tool is due to a chemical reaction between a tool material and its environment like air, cutting fluid, or workpiece material. Tool oxidation is common when cutting in air at high speed. An oxidation reaction is accelerated with temperature, but can be eliminated when inert gas is used to shield the cutting tool from surrounding oxygen. A chain reaction can also occur and further degrade a tool. For example, iron in steel is first oxidized at high cutting temperature to form iron oxide; the iron oxide then weakens the aluminum oxide coating of a tool and leads to peeling and chipping of the coating.

*Adhesion tool damage* happens when a built-up-edge (BUE) welds strongly to a tool surface and then breaks away with a minute amount of tool material. When machining soft materials, a chip tends to adhere to the tool and grow in size. When the BUE is large and becomes unstable, it is removed with the chip while also shearing off part of the cutting tool due to the higher adhesion strength between BUE and tool than the intergrain binding strength of the tool. Stainless steel, nickel and titanium alloys are known for causing adhesion wear on carbide microtools. Adhesion damage can be reduced by using proper lubricant to reduce friction between chip and tool, by coating the tool with a smooth and low friction layer, by reducing tool edge radius, or by increasing cutting speed to raise the tool surface temperature and soften the BUE while reducing its weldability to the tool surface.

Microtool failures occur due to a combination of the above mechanisms. For example, peeling of tool coating might be due to coating defects, or to mechanical mechanisms when a large gradient of stress exists across a thick coating layer; the loosened coating particles then rub and cause mechanical abrasive wear on a tool. Thermal mechanisms may cause workpiece atoms to diffuse, weaken, and dislodge several tool grains as microchipping.

**Table 2. Categories of Tool Damage**

| Microtool damage | Damage size | | Mechanism |
|---|---|---|---|
| | μm | μinch | |
| Abrasion | < 1 | < 39 | Mechanical, thermal |
| Attrition | 1–3 | 39–118 | Mechanical, thermal |
| Peeling | 1–3 | 39–118 | Mechanical, chemical |
| Microchipping | 3–10 | 118–394 | Mechanical, adhesion |
| Chipping | 10–30 | 394–1180 | Mechanical |
| Fracture | > 100 | > 3940 | Mechanical |

**Tool Life.—**Tool life criteria in macromachining are documented in ANSI/ASME B94.55M-1985 (R2019), "Tool Life Testing with Single-Point Turning Tools." This Standard suggests an end of tool life when a tool exhibits:

- An average flank wear of 300 μm (0.0118 in), or
- Any maximum flank wear land of 600 μm (0.0236 in), or
- Any tool wear notch of 1000 μm (0.0394 in), or
- A crater wear of 100 μm (0.0039 in).

It is obvious that such criteria for a macrotool cannot be applied to a microtool because (i) it would be cost prohibitive to continue testing until 300 μm flank wear, and (ii) the wear criteria are even larger than most tool dimensions.

In the absence of a microtool standard, researchers have set their own criteria based on direct observation and/or indirect monitoring of microtool tool wear effects. Published data varies on microtool wear thresholds: 5 μm flank/nose wear on diamond tools, or

50 μm flank/nose wear on carbide tools, or chipping dimensions relative to cutting tool grain size, or peeling of tool coating, etc. A variety of techniques have been suggested for tool monitoring; the direct techniques measure the tool conditions (e.g., flank wear, crater wear) while the indirect techniques measure the consequence of tool wear (e.g., burr size, change of microhole diameter).

**Table 3. Microtool Wear and Monitoring Techniques**

| Measurement | | Metrology Equipment / Sensor |
|---|---|---|
| Direct | Tool wear<br>Tool edge conditions | Microscope |
| | Wear particles<br>Particle radioactivity | Spectrophotometer, scintillator |
| | Tool-workpiece junction resistance | Voltmeter |
| | Workpiece features (hole diameter, slot depth…) | Microscope<br>Interferometer |
| Indirect | Cutting force, torque, power | Dynamometer<br>Strain gage, ampere meter |
| | Sound emitted from tool-workpiece friction | Acoustic emission transducer<br>Microphone |
| | Vibration | Accelerometer<br>Displacement sensor |
| | Temperature | Thermocouple<br>Pyrometer |
| | Surface roughness | Profilometer, interferometer, optical profiler… |
| | Burr dimension | Microscope, interferometer, optical profiler… |

The importance of tool life monitoring and tool life prediction is presented in the section *MACHINING ECONOMETRICS* starting on page 1184. The following material expands from that and covers relevant information for tool life of microtools.

The general Taylor equation that relates tool life and machining parameters also applies in micromachining:

$$V^a f^b d^c T = g \quad (1)$$

where $V$ = surface cutting speed (m/min, ft/min)

$f$ = tool feed (mm/rev, in/rev) or chip load (mm/tooth, in/tooth)

$d$ = depth of cut (mm, inch)

$T$ = tool life (min)

$g, a, b, c$ = constants

When thermal damage mechanism dominates then $a >> b, c$ in Equation (1). The term $a$ dominates mathematically and the effects of feed and depth of cut are insignificant compared to speed. The general Taylor equation can be rewritten as:

$$V^a T = \frac{g}{f^b d^c} \quad (2)$$

If $n = 1/a$, then this equation is the same as that in the *Econometrics* section:

$$VT^n = \left(\frac{g}{f^b d^c}\right)^n = C \quad (3)$$

When tool chipping occurs then both terms $b, c >> a$, therefore the feed and depth of cut are more important than surface speed. The general tool life reduces to

$$f^b d^c T = \frac{g}{V^a} = C' \tag{4}$$

Chipping is generally not acceptable since a chipped tool generates excessive burr and a very rough surface. By reducing depth of cut and feed, then chipping should be eliminated assuming micromachining with a quality tool and machine tool. When stable parameters are applied, then the only damage mechanism is thermal and tool life can be predicted with Equation (3).

It has been shown that flank wear due to abrasion is directly proportional to the magnitude of acoustic signal or feeding force. An increase of 300 percent in micromilling feeding force from an initial value was established as a threshold for reaching the tool life. A reduction in feeding force, however, might indicate gradual failure of a microtool due to fatigue crack propagation. Indirect monitoring of tool wear by monitoring feeding force for both micromilling and microdrilling would be a preferred technique since this does not interfere with the machining process and reduce productivity. In the absence of a sensitive commercial system that can reliably and accurately monitor tool force and tool life in micromachining, direct tool wear monitoring should still be a popular practice.

Traditional tests using the Taylor approach would machine at the same cutting speed until reaching the predetermined tool failure criteria. Such tests can be time consuming if a chosen speed is too low, and only applicable to turning since a constant cutting speed is required. In reality, a part must be machined with the same tool in different directions and speeds to obtain the final profile and surface finish. Several techniques were developed to accelerate the testing method since turning tests alone are tedious, expensive, and do not reflect actual part machining. The cumulative wear technique, assuming that the abrasion wear mechanism is the same at different cutting speeds, is more flexible and can reduce the testing time and cost. The proposed cumulative tool life testing technique:

- Is flexible. If an initial speed is too slow, testing speed can be increased and the cumulative time and tool wear recorded.
- Is simple. Manual machines can be used instead of CNC machines. The same rpm on a manual lathe can be used for the turning test until tool failure. Times and cutting speeds for all passes are used to calculate the equivalent time and speed.
- Is more cost-effective. Both turning and facing can be combined to completely consume an expensive workpiece material.
- Is order independent. The level of cutting speed is not important if providing the same tool wear mechanism. Experimental data for macromachining shows no difference of tool life if changing cutting speeds from low to high, or in reverse order.

Consider a tool that machines at cutting speed $V$ and stops after machining time $\Delta t$ before reaching its tool life $T$. The tool then cuts at different speeds and times until reaching the tool life criteria—for example, 50 μm flank wear on a carbide microendmill. The fraction of tool life when cutting at each speed and time is $\Delta t/T$, and the total tool life fraction is

$$\frac{\Delta t_1}{T_1} + \frac{\Delta t_2}{T_2} + \ldots + \frac{\Delta t_k}{T_k} = \sum_{i=1}^{k} \frac{\Delta t_i}{T_i} = Q \tag{5}$$

The theoretical value of the total tool life fraction $Q$ should be one. Experimental values for $Q$ were found to be in the range 1.2–1.5. When combined with Taylor Equation (3), then Equation (5) becomes

$$\sum_{i=1}^{k} \Delta t_i V_i^{1/n} = QC^{1/n} \tag{6}$$

After machining with a tool at different times and speeds in different conditions (e.g., different tool coatings), it is necessary to compare the tool performance by calculating its equivalent tool life and equivalent tool speed. The equivalent tool life $T_e$ is just the sum of all machining time periods:

$$T_e = \sum_{i=1}^{k} \Delta t_i \tag{7}$$

The equivalent tool speed must produce the same tool damage as a tool after cumulative machining. The total tool damage is given in Equation (6) as:

$$\frac{1}{Q}\sum_{i=1}^{k} \Delta t_i V_i^{1/n} = C^{1/n} = T_e V_e^{1/n} \tag{8a}$$

Solving for the equivalent cutting speed $V_e$

$$V_e = \left( \frac{1}{Q} \frac{\sum_{i=1}^{k} \Delta t_i V_i^{1/n}}{\sum_{i=1}^{k} \Delta t_i} \right)^n \tag{8b}$$

When $Q = 1$ then, $(V_e)^{1/n}$ is the mathematical average of all $(V_i)^{1/n}$ terms, by definition.

Mathematical models for cumulative tool wear are now derived for most popular machining operations, namely turning, drilling, facing, and milling.

- For turning with different cutting speeds, Equation (6) is applied. If turning speeds are kept the same from one pass to another, substitute $V = V_i$ into Equation (6) to obtain:

$$V^{1/n} \sum_{i=1}^{k} \Delta t_i = QC^{1/n} \tag{9}$$

- For drilling, tool wear would be most substantial at the cutting lip where cutting speed is at the highest. Since cutting speed is constant during drilling as in turning, the tool wear model for drilling is the same as in Equation (6) for variable speeds, and Equation (9) for constant speed.

- For facing, the cutting speed reduces linearly from the maximum $V_i$ at the outermost radius to zero at the spindle center. It can be shown that the cumulative tool life model for facing is:

$$\sum_{i=1}^{k} \Delta t_i V_i^{1/n} = \frac{n+1}{n} QC^{1/n} \tag{10}$$

- For milling, the actual machining time is the time during which chips are produced. The chip generating time involves geometry of a tool and milling parameters. The cumulative tool life model for face milling is

$$\sum_{i=1}^{k} \Delta t_i V_i^{1/n} = \frac{1}{\lambda} QC^{1/n} \quad \text{and} \quad \lambda = \frac{M\cos^{-1}\left(1 - 2\frac{a}{D}\right)}{360°} \tag{11}$$

where $\lambda$ = milling factor
$a$ = width of cut (radial depth) in milling
$M$ = number of teeth
$D$ = milling cutter diameter

*Example 8, Cumulative Tool Life:*

*Turning Test:* Dry turning a metal matrix composite rod (Ø18 mm, 100 mm long) at constant 256 rpm on a manual lathe, depth of cut 0.5 mm, feed 0.07 mm/rev. Carbide tool TNPR331M-H1, tool holder MTENN2020-33.

This Al-SiC composite is very abrasive and is ideal for tool life model testing since abrasive wear is the main mechanism and flank wear is clearly seen and measured on a carbide tool. In this test, a tool is turned at constant rpm until reaching 300 μm flank wear. At least two data points are required to calculate the effect of speed, or the slope $n$ in Taylor equation. From Table 4, the speed and tool life pairs are (14.48 m/min, 3.54 min) and (9.56 m/min, 5.58 min). The slope $n$ derived from Equation (3) for these two data points is

$$n = \frac{\log(V_2/V_1)}{\log(T_1/T_2)} = \frac{\log(9.56/14.48)}{\log(3.54/5.58)} = 0.91$$

When considering many data points, the averaged value of $n$ is 0.94. A spreadsheet such as Table 4 is a convenient way to tabulate cumulative values of each $\Delta t$ and $\Delta t V^{1/n}$ term and then use these to calculate the equivalent tool life $T_e$ with Equation (7), and the equivalent tool speed $V_e$ with Equation (8b). The plot for all experimental data at constant cutting speeds and cumulative speeds is shown in Fig. 11a. Having all data points fitting on the same line indicates the validity of cumulative tool life models.

*Facing Test:* The same material and cutting tools are used in the facing test. Tool wear and tool life plots are shown in Fig. 11b and Fig. 11c. There is no difference in tool life when machining at low then high speed or the other way around.

**Table 4. Spread Sheet for** Example 8**, Cumulative Tool Life in Turning**

| RPM | Δ Length (mm) | Diameter (mm) | Speed (m/min) | $\Delta t$ (min) | $\Delta t V^{1/n}$ | Cumulative $\Delta t V^{1/n}$ | Cumulative $\Delta t$ (min) | Flank Wear (μm) | Feed (mm/rev) | Equivalent $V_e$ (m/min) | Equivalent $T_e$ (min) |
|---|---|---|---|---|---|---|---|---|---|---|---|
| 256 | 26.5 | 18.0 | 14.48 | 1.48 | 25.39 | 25.39 | 1.48 | 199 | 0.07 | | |
| 256 | 16.0 | 18.0 | 14.48 | 0.89 | 15.33 | 40.72 | 2.37 | 242 | 0.07 | | |
| 256 | 21.0 | 18.0 | 14.48 | 1.17 | 20.12 | 60.84 | 3.54 | 300 | 0.07 | 14.48 | 3.54 |
| 179 | 18.5 | 17.0 | 9.56 | 1.48 | 16.30 | 16.30 | 1.48 | 160 | 0.07 | | |
| 179 | 21.5 | 17.0 | 9.56 | 1.72 | 18.95 | 35.25 | 3.19 | 233 | 0.07 | | |
| 179 | 25.0 | 17.0 | 9.56 | 2.00 | 22.03 | 57.28 | 5.19 | 289 | 0.07 | | |
| | | | | | projected | 61.61 | 5.58 | 300 | | 9.56 | 5.58 |

## Workpiece Materials

Micromachining is often utilized to fabricate components for miniaturized sensors, medical, optical, and electronic devices, etc. Common engineering materials for these applications include stainless steel, aluminum, titanium, copper, and tool steel for miniature molds and dies.

Workpiece materials must meet certain conditions for successful micromachining. Unlike macromachining, a micromachining tool is subjected to fluctuating cutting force when it encounters each grain since microtool size is comparable to material grain size. A microtool is more vulnerable to fatigue fracture and the resulting surface—if the tool survives—would be rough due to different spring-back protrusion from each grain due to different crystallographic orientations of the grains, and direction-dependent properties of the material. Homogenous workpiece materials with very fine and uniform grain sizes should be chosen for micromachining. Inclusions and large precipitates should be minimized to avoid damage to a fragile tool edge.

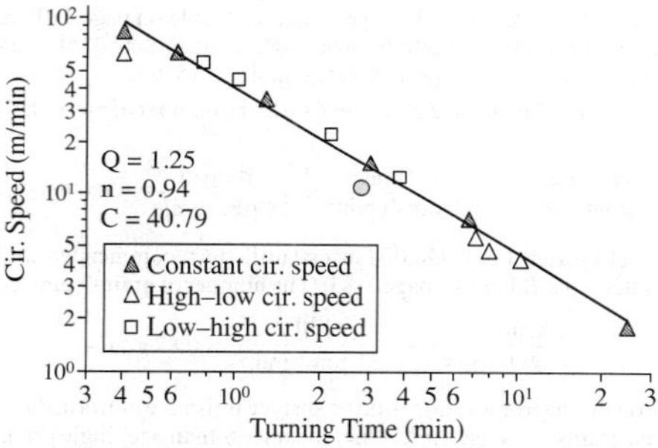


Fig. 11a. Tool life plot for turning tools. Circumference speed refers to the maximum cutting speed at the outer radius in turning.
Cast A359/SiC/20p; tool H1 WC (-8,0,9,5,60,30,0.4mm); 0.5 mm depth; 0.07 mm/rev feed; dry.

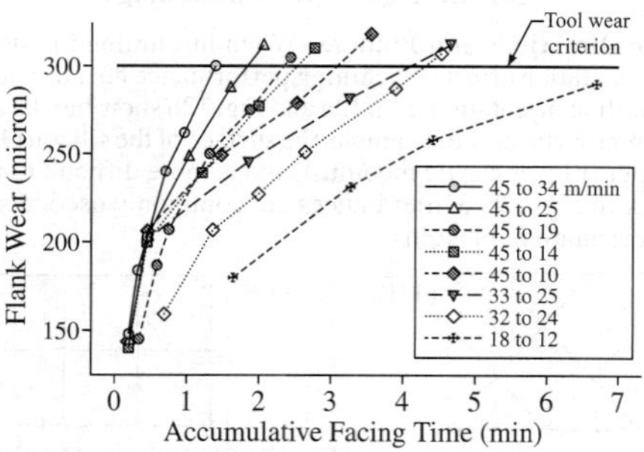


Fig. 11b. Cumulative flank wear of tool facing at high-to-low circumference speed.
Cast A359/SiC/20p; tool H1 WC (-8,0,9,5,60,30,0.4mm); 0.5 mm depth; 0.07 mm/rev feed; dry.

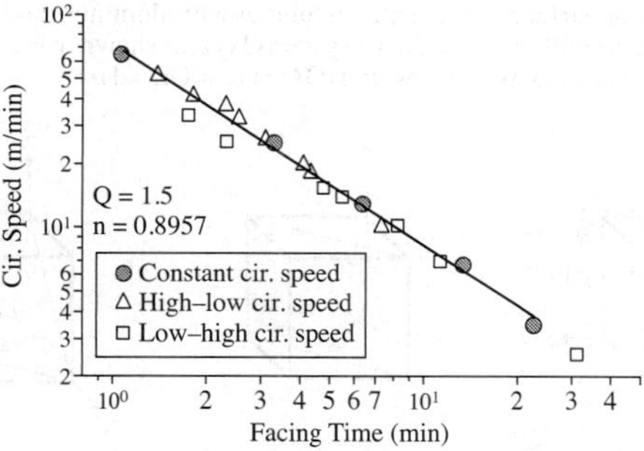


Fig. 11c. Tool life plot for facing tools. Circumference speed refers to the maximum cutting speed at the outer radius in facing.
Cast A359/SiC/20p; tool H1 WC (-8,0,9,5,60,30,0.4mm); 0.5 mm depth; 0.07 mm/rev feed; dry.

*Example 9, Grain Size Consideration:* The speed and feed table on page 1106 recommends a chip load (feed) of 0.1 mm/tooth (0.004 in/tooth) for macro-scale end milling 316L stainless steel using an uncoated carbide tool. Assume the average material grain size is 15 μm.

*Macromilling:* Using a Ø1/2 inch (Ø12.5 mm) end mill, the number of grains being cut by each tooth would be

$$\frac{\text{chip load}}{\text{grain size}} = \frac{0.1\text{mm}}{15\ \mu\text{m/grain}} = \frac{100\mu\text{m}}{15\ \mu\text{m/grain}} = 6.67\text{ grains}$$

*Micromilling:* Selecting a Ø0.1 mm (Ø0.004 in) end mill, the recommended chip load would be 13 mm for stainless steel (see Table 13b, page 1183). The number of grains being cut by each tooth is

$$\frac{\text{chip load}}{\text{grain size}} = \frac{13\mu\text{m}}{15\ \mu\text{m/grain}} = 0.87\text{ grains}$$

The cutting force on the macrotool and resulting surface finish are uniform due to the averaging effect from seven grains. Because a microtool shears less than one single grain at a time, the micromachined surface is irregular due to different spring-back amounts of each individual grain, and the cutting force on the microtool fluctuates depending on each grain orientation.

## Ductile Regime Micromachining

**Crystallographic Directions and Planes.—**When machining in micro or nano scale, workpiece atom orientation affects machining performance because material properties change with crystalline orientation. Fig. 12a and Fig. 12b show blocks of the same material but with different surfaces. For example, the surface of the silicon block shown in Fig. 12a is harder, stiffer (higher elastic modulus), and is more difficult to machine than the same silicon block in Fig. 12b. Miller indices are commonly used to specify particular crystallographic orientations of atoms.

Fig. 12a. Block of Material with (110) Surface.

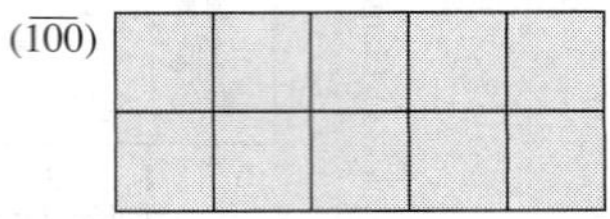


Fig. 12b. Block of Same Material with (100) Surface.

Consider a simple cubic system where atoms are located at corners (cubic as with manganese), at corners and inside (body centered cubic as with iron and chrominum), or at corners and on the surfaces (face centered cubic as with aluminum and copper) systems. For convenience, we will set a coordinate system Oxyz, as shown in Fig. 13a, and the size of the cube is set at one atomic spacing unit (OA = OC = OD = 1).

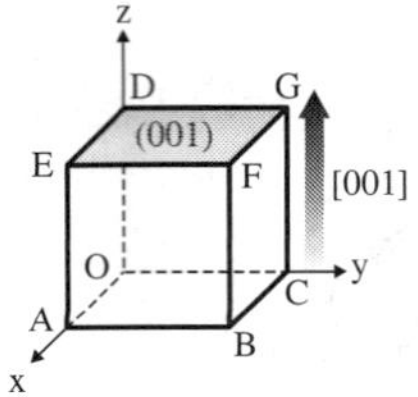


Fig. 13a.

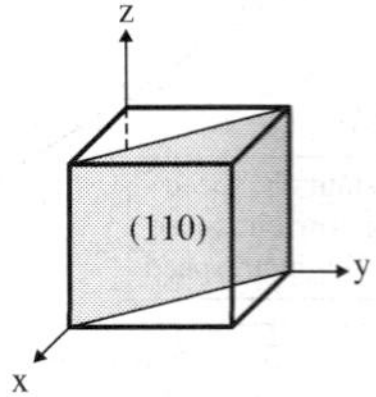


Fig. 13b.

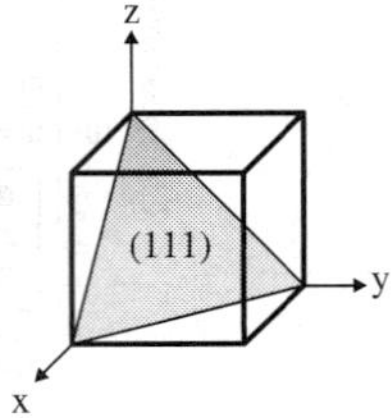


Fig. 13c.

During a micro/nano machining process, a cutting tool pushes and shears a grain. The cutting action forces some atoms to slide in certain directions and planes to form chips on the tool rake face. These specific directions and planes are called slip systems. Soft materials such as copper and aluminum have more slip systems, therefore, are easier to be machined compared to harder materials such as steel with fewer slip systems.

The directional Miller index is the coordinate of a vector representing the atom sliding direction. In Fig. 13a, if the atom slides from C to G (also the same as sliding from O to D), then this vector with corresponding coordinates are given by:

$$\overrightarrow{CG} = \overrightarrow{OD} = [x_D - x_O, y_D - y_O, z_D - z_O] = [0, 0, 1] \text{ or } [001] \text{ direction}$$

Similarly, if an atom slides from B to A (or C to O) then the Miller direction is:

$$\overrightarrow{BA} = \overrightarrow{CO} = [x_O - x_C, y_O - y_C, z_O - z_C] = [0, -1, 0] \text{ or } [0\bar{1}0] \text{ direction}$$

The Miller plane represents the planes that intersect with the coordinate system. The plane DEFG in Fig. 13a, congruent with OABC, intersects the z-axis while parallel with the x and y axes. The Miller index for this plane is represented by the inverse of the axis intersection:

$$\text{Plane DEFG} = \left(\frac{1}{x\text{-intercept}}, \frac{1}{y\text{-intercept}}, \frac{1}{z\text{-intercept}}\right) = \left(\frac{1}{\infty}, \frac{1}{\infty}, \frac{1}{1}\right) = (0,0,1) \text{ or } (001) \text{ plane}$$

The plane EGCA in Fig. 13b, congruent with DFBO, intersects the x and y axes while parallel with z-axis. The Miller index for this plane is:

$$\text{Plane EGCA} = \left(\frac{1}{x\text{-intercept}}, \frac{1}{y\text{-intercept}}, \frac{1}{z\text{-intercept}}\right) = \left(\frac{1}{1}, \frac{1}{1}, \frac{1}{\infty}\right) = (1,1,0) \text{ or } (110) \text{ plane}$$

The plane DCA in Fig. 13c, congruent with BEG, intersects all the x, y, and z axes. The Miller index for this plane is:

$$\text{Plane DCA} = \left(\frac{1}{x\text{-intercept}}, \frac{1}{y\text{-intercept}}, \frac{1}{z\text{-intercept}}\right) = \left(\frac{1}{1}, \frac{1}{1}, \frac{1}{1}\right) = (1,1,1) \text{ or } (111) \text{ plane}$$

*Miller Index Nomenclature:* In both direction and plane Miller indices, any minus sign is written on top of the number, for example $[\bar{1}\bar{1}\bar{1}]$, and all commas are omitted for simplicity.

Square brackets "[ ]" are used to indicate a specific direction, and pointed brackets "<>" are used to indicate a family of directions with similar geometries. For example, the <100> family has 12 directions similar to [100], [001], …, which are all the edges of the cube in Fig. 13a.

Parentheses "( )" are used to indicate a specific plane, and curly brackets "{ }" are used to indicate a family of planes with similar geometries. For example, the {100} family has 6 planes similar to (100), (001), …, which are all the surfaces of the cube in Fig. 13a.

**Introduction.**—The concept of ductile-regime machining has been investigated since the 1960s for amorphous brittle materials such as glasses. Silicon, germanium, and glasses have become strategic materials that are widely used to fabricate intricate components in microelectronics, optical, defense industries, and recently as micro optical-electrical-mechanical systems. Silicon and other brittle materials are known for their low machinability unless they are machined in the ductile-regime conditions. When utilized at the optimal machining conditions, only minimum effort is required for the subsequent etching, grinding, or polishing to remove the damaged subsurface. This section summarizes the theory and provides practical guidance for ductile regime machining.

**Theory.**—The mechanism of ductile-regime machining has been studied by many researchers. Using a fracture mechanics approach, it can be shown that there is a threshold below which the ductile regime prevails:

$$d_c = \frac{\text{plastic flow energy}}{\text{fracture energy}} = A\left(\frac{E}{H}\right)\left(\frac{K_c}{H}\right)^2 \qquad (12)$$

where $d_c$ = critical depth of cut (m, inch)

$A$ = constant

$E$ = Young's modulus (Pa, psi)

$K_c$ = surface fracture toughness (Pa·m$^{0.5}$, psi·in$^{0.5}$)

$H$ = surface microhardness (Pa, psi)

A shallow depth of cut, therefore, would energetically promote plastic flow rather than brittle fracture in the substrate and the chips. Table 5 tabulates properties of some brittle materials and their experimental critical depths of cut.

**Table 5. Selected Properties of Some Brittle Materials**

| Materials | Young modulus (GPa) | Fracture toughness (MPa·m$^{0.5}$) | Knoop hardness (GPa) | Critical depth of cut (μm) |
|---|---|---|---|---|
| α-$Al_2O_3$ | 275–393 | 3.85–5.90 | 19.6–20.1 | 1.0 |
| SiC | 382–475 | 2.50–3.50 | 24.5–25.0 | 0.2 |
| Si | 168 | 0.6 | 10 | 0.5 |

The constant $A$ in Equation (12) varies in the range 0.1–0.6 due to measuring uncertainty of surface toughness $K_c$, elastic modulus $E$, and microhardness $H$ in a testing environment. These properties depend on crystalline orientation of the materials, surface conditions, and tool geometry.

- The critical resolved shear stress, on a crystalline plane due to the cutting action, is directly proportional to the Schmid factor $\cos\lambda\cos\phi$, where $\phi$ and $\lambda$ are the orientations of the slip plane and slip direction. An ideal ductile mode machining would happen when the cutting shear stress is parallel to both the slip plane and the slip direction, otherwise a pseudo ductile mode with micro cleavages occurs. True ductile-regime machining happens only along certain crystalline orientations, but brittle machining occurs at other crystalline orientations. This explains why micromachining a crystalline specimen at the same speed, depth of cut, and coolant produces ductile machined surfaces in one direction but brittle machined surfaces on others.
- Cutting fluid changes the surface properties of materials ($K_c$, $E$, and $H$) and affects conditions for ductile regime micromachining. When micromachining the (100) germanium using a single crystalline diamond tool, the critical depth of cut changes from 0.13 μm (5 μin.) with distilled water as cutting fluid to 0.29 μm (11 μin.) in dry machining.
- Tool geometry also affects the results. Plowing and fracture of material occurs when depth of cut is less than approximately half of the tool cutting edge radius (see *Microcutting Tools* on page 1145). Tools with negative top rake angle are usually utilized because a negative rake causes a compressive zone in the workpiece ahead of and below the tool and suppresses microcrack formation.

*Example 10, Mirror-finish Micromachining:* Diamond tools with sharp cutting edge radii are very effective for machining brittle or ductile material with the exception of ferrous alloys such as tool steels or stainless steels. The cutting speed has minimum effect on surface finish, but a reduction of the feedrate leads to improvement of surface finish.

An optical quality surface of 1.4–1.9 nm $R_{max}$ was obtained when turning single crystalline quartz with a diamond tool (−20° rake, 0.8 mm nose radius) at <0.3 μm depth of cut, 3 m/s speed, and 8.1 μm/rev feedrate.

**Case Study.—**A study used polished (001) p-type silicon wafers of Ø100 mm (Ø4 inch). Small grooves were faced at different constant depth of cut or gradually changing depth of cut to study the ductile behavior (Fig. 14). Single crystalline diamond tools with (001) rake surface, 10–40 nm edge sharpness, +5° rake angle, and 0.51 mm or 2.00 mm nose radii were used for a facing operation. The complete tool nomenclature follows the American Standards Association (back rake angle, side rake angle, end relief angle, end clearance angle, side relief angle, side clearance angle, end cutting edge angle, side cutting edge angle, nose radius) with the addition of edge sharpness as (5°, 0°, 0°, 5°, 5°, 30°, 0°, 0.51–2.00 mm, 10–40 nm). The ultraprecision machining process was performed on a rigid system that has 9 nm positioning accuracy. Compressed air was used to blow chips away from the finish machined area. Surface finish of a machined wafer was measured with an atomic force microscope (AFM) and a phase-shift interferometer (PSI).

Surface finish measurements indicated ductile or brittle chip fracture on machined surfaces. As depth of cut reduced below 1 μm, the surface finish was also diminished due to a higher percentage of ductile machined surfaces. Perfect ductile regime machining was achieved when depth of cut was between 0.1–0.5 μm. A smaller depth of cut in the neighborhood of 0.05 μm (50 nm), however, worsened the surface finish because machining at such shallow depth of cut (close to the cutting tool edge radius of 40 nm) would plow and fracture the material surface. At the same cutting parameters, micromachining along the silicon <110> directions gave better surface finish while brittle chipping was seen when cutting along the silicon <100> directions (Fig. 15, Fig. 16a, and Fig. 16b).

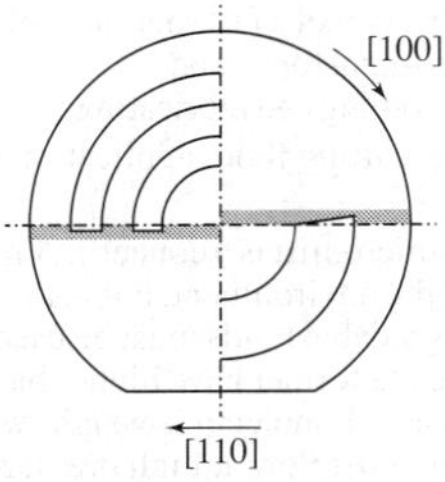


Fig. 14. Machining plan on (001) wafers. A wafer was faced at a constant depth in different zones (left) or changing cutting depth in a taper cut (right).

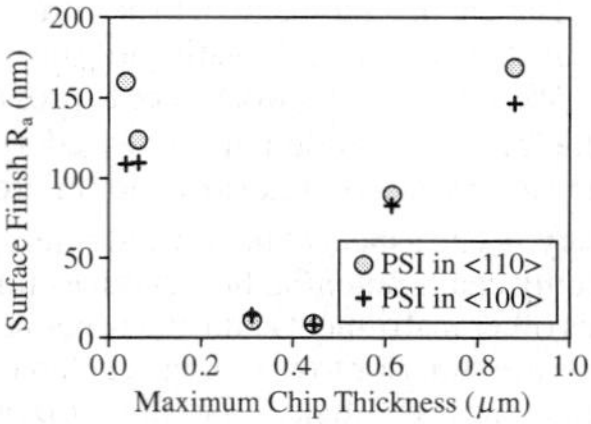


Fig. 15. Surface finish as a function of the maximum chip thickness (depth of cut) and crystalline direction of the silicon wafer. The minimum surface finish is with ductile machined surface.

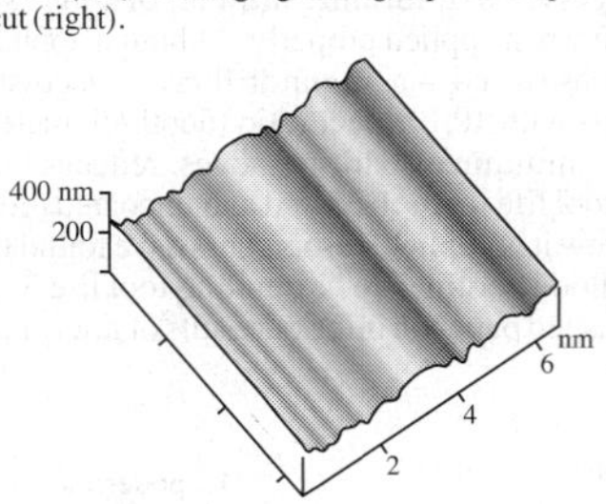


Fig. 16a. Perfect ductile regime machining of (001) silicon along [$\bar{1}$10].

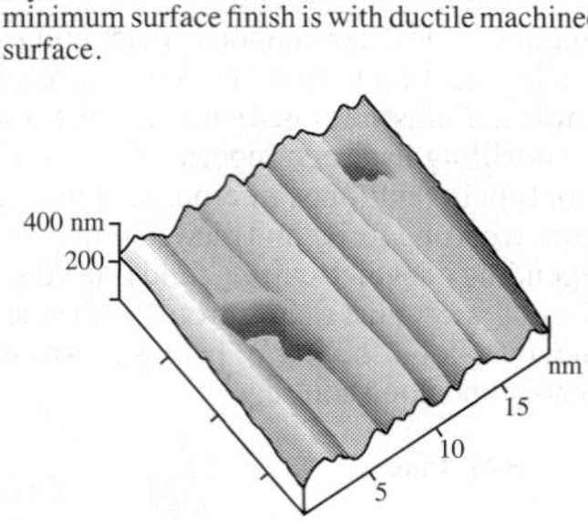


Fig. 16b. Mixed mode of ductile regime and pitting along [$\bar{1}$00].

Speed 75 m/min, feed 2.5μm/rev; depth 0.5μm; SCD tool (5°,0°,0°,5°,5°,30°, 0°, 0.5mm,10–40nm).

## Cutting Fluids in Micromachining

Micromilling and microdrilling, among the most versatile manufacturing processes, can be leveraged from existing technology to produce 3D microparts or microcavities in molds and dies for mass-replication. Although macro-scale milling and drilling technology is mature, micro-scale milling/drilling technology is yet to be fully developed. Extending common practices in macromachining to micromachining often ends up with failure. Very short tool life is experienced with micromachining, and flood cooling is not effective in microdrilling because coolant cannot flow into a partially drilled microhole. This section recommends how to select and apply a cutting fluid for effective micro milling/drilling.

Micro milling/drilling requires high rotation speed exceeding 25,000 rpm of a small tool to achieve an acceptable surface cutting speed for material removal. When drilling steel, 50 percent of the heat generated conducts into the drill, but 80 percent of heat will go to the tool when drilling titanium. A microdrill with sharp cutting edges subjected to high temperature and high stress will fail easily if cutting fluid is not adequate. When rotating a microtool at a very high rotating speed, flood coolant is not effective since it does not have enough momentum to penetrate the boundary layer (fast moving air layer) around a fast rotating tool, or wet the bottom of a deep microhole. In addition, any unfiltered chip from recycled coolant can damage a microtool or fragile workpiece. Micromist (minimum quantity lubrication, MQL) has been studied by many researchers and is proven to provide proper cooling and lubricating in micromachining. In ideal conditions, a stream of micron-size lubricant particles in micromist:

- Does not contain any chip or solid contaminant
- Has enough momentum to penetrate the boundary layer of a fast rotating tool
- Adheres to the fast rotating tool despite high centrifugal force, and
- Wets the tool and workpiece to provide effective cooling and lubricating.

The following section discusses safety, selection of cutting fluid, application method, and recommends optimal setup for micromachining.

**Safety.**—The aspect of health and safety when using micromist is a concern. A mist does not only cause potential health issues for workers in the environment, but also contaminates other instruments and machines nearby. Biodegradable fluids must be used; polyol esters are superior to common vegetable oils because the former have higher biodegradability, are less "sticky" due to oxidizing, and increase in molecular weight with time and temperature. Due to the aerosol formation during mist flow at high pressure, an air purification unit or proper ventilating fan should be installed to minimize breathing of the aerosol particles by operators, and prevent damage to adjacent equipment.

**Benefits.**—Most conventional machining processes like turning, milling, drilling, and grinding can benefit from micromist lubrication when applied properly. Although application of micromist is limited when the mist flow is obstructed — as in gun drilling — successful microdrilling has been reported for microholes with 10:1 aspect ratio (depth/diameter). At optimal conditions, micromist significantly minimizes built-up-edges, reduces burr size and cutting force, and therefore improves tool life for both coated and uncoated tools. Depending on which cutting fluid is used and how it is applied, researchers have found the effect of micromist ranges from "the same as flood cooling" to "extending tool life 3–10 times over flood cooling." There is yet any published paper on inferior results of micromist over dry and flood cooling.

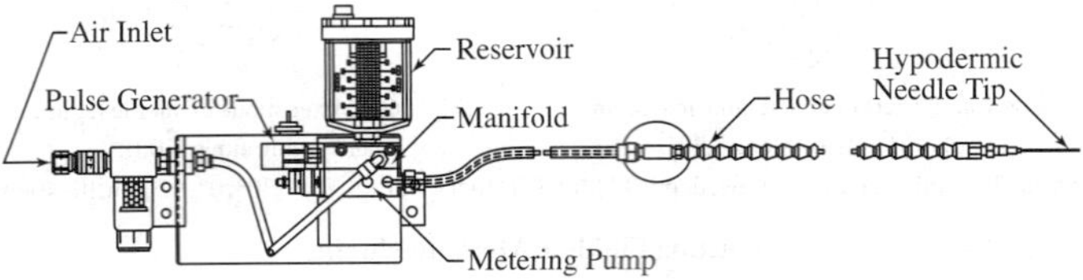


Fig. 17. Schematic of a Micromist System for Micromachining. *Courtesy of Unist, Inc.*

Systems that can generate micromist for minimum quantity lubrication machining are commercially available. A typical design (Fig. 17) includes a resevoir for biocompatible oil, feeding tubes, and an atomizing unit that mixes a compressed air flow with a controlled volume of oil. A needle is necessary to direct the mist to a predetermined location.

The resulting oil microdroplets—size and speed—should be adjustable to effectively penetrate and wet a tool/part interface. This can be done by adjusting the air pressure, type of oil, and volume of oil released into the air stream.

**Selection of Cutting Fluid.—**A cutting fluid is selected for both cooling and lubricating purposes in micromachining. It should be environmentally friendly, should not interact chemically at high temperature with tool or workpiece, and can be cleaned and disinfected from the machined parts. It must have low surface energy relative to the surface energy of the cutting tool and workpiece material, high thermal diffusivity, and lubricity. For micromist applications, a cutting fluid must be able to flow easily in a small tube (low viscosity) and to form microdroplets.

Complete wetting is desirable for a cutting fluid because it covers large surface areas of tool and workpiece and effectively removes heat from the source. Its self-spreading capability due to the differential surface energy allows cutting fluid to penetrate deep into the chip/tool interface to effectively lubricate and cool this zone. Wetting condition can be assessed by two methods:

*Pendant drop technique:* A drop of liquid is formed and suspended vertically at the end of a solid tube. The side view of a drop is analyzed to compute the liquid surface tension using a tensiometer (Fig. 18a and Fig. 18b).

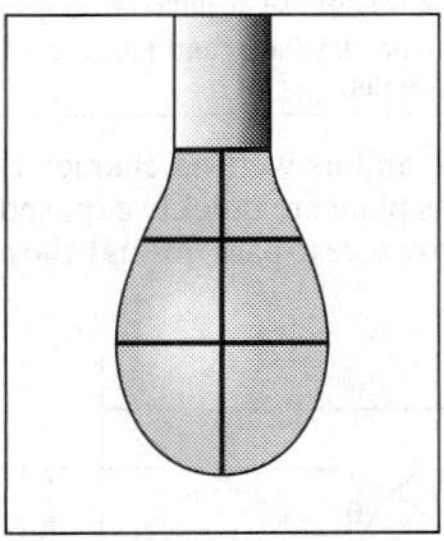

Fig. 18a. Side View of a Pendant Drop Below a Stainless Steel Tube.

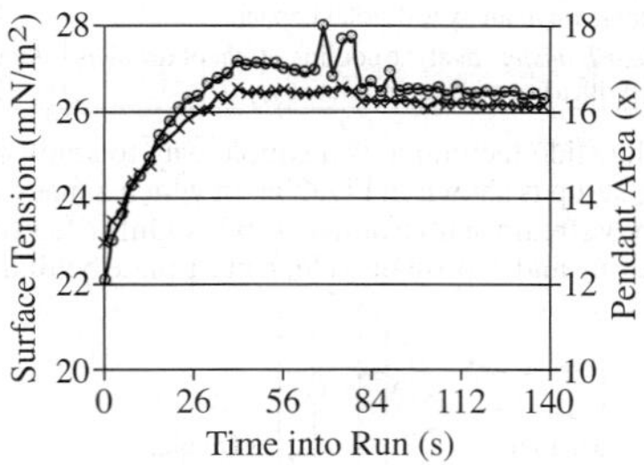


Fig. 18b. Calculated Surface Tension from Starting to Full Forming of a Droplet.

*Sessile drop technique:* A drop of liquid is placed on a horizontal surface. The side view of the drop is analyzed to calculate the liquid contact angle or measure it with a goniometer (Fig. 19).

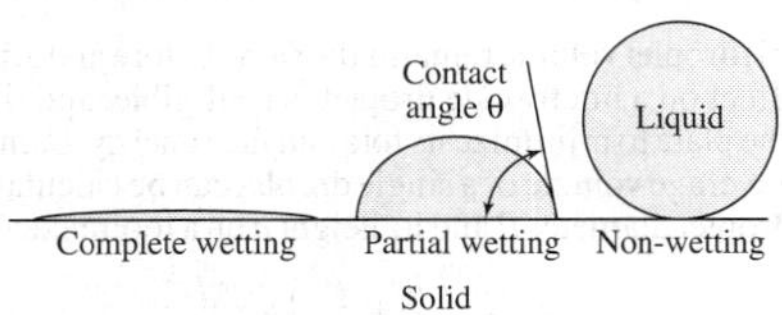


Fig. 19. Sessile Drop Technique to Assess Wetting of Cutting Fluid.

The following section presents a simpler approach to calculating the contact angle and drop size using a modified sessile drop technique. This technique uses a toolmaker's microscope, available at most manufacturing shops, to measure the top view of a drop, instead of a goniometer to measure the side view.

**Drop Size Measurement.—**A microdroplet must have sufficient momentum to penetrate the boundary air layer moving around a fast rotating microtool and to wet the tool afterward. Calculation of momentum and contact angle for wetting assessment requires the droplet dimension. Knowing the lubricant drop size allows proper calibration of a micromist system to maintain the system effectiveness. Table 6 summarizes different techniques to measure the liquid drop size. The techniques are basically intrusive and nonintrusive methods to either collect the droplets for subsequent analysis, or for in situ imaging of the in-flight droplets.

Nonintrusive techniques use dedicated laboratory research instruments to provide accurate dimensions and comprehensive statistical information of microdroplets. The effect of variables like air pressure on drop size and speed can be automatically calculated and analyzed.

Intrusive techniques use less sophisticated instruments to collect and analyze droplets directly or indirectly. These simpler techniques, however, depend on operator skills for collecting reliable data.

**Table 6. Liquid Droplet Measurement Techniques**

| Intrusive | Nonintrusive |
|---|---|
| *Slide:* collect droplets on a slide for microscopic assessment.<br>*Solidification:* transform droplets to solid for sieving or weighting.<br>*Momentum:* analyze droplet impact.<br>*Heat Transfer:* analyze cooling effect of droplets with a hot wire anemometer. | *Light shadowing:* analyze shadows of in flight droplets.<br>*Laser Doppler Anemometry:* analyze visibility, intensity and phase shift of scattered laser from a small sample of droplets.<br>*Laser Diffractometry:* scan and analyze a large group of droplets. |

The slide technique is a simple way to study drop size and its wetting characteristic. The setup is shown in Fig. 20a, in which a mask and glass plate are quickly exposed to a steady stream of micromist droplets. Only a few droplets are able to pass through the mask opening and deposit on a clean glass plate behind it.

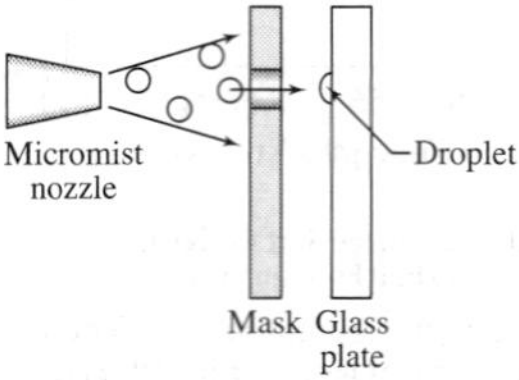


Fig. 20a. Setup for microdroplet collection.

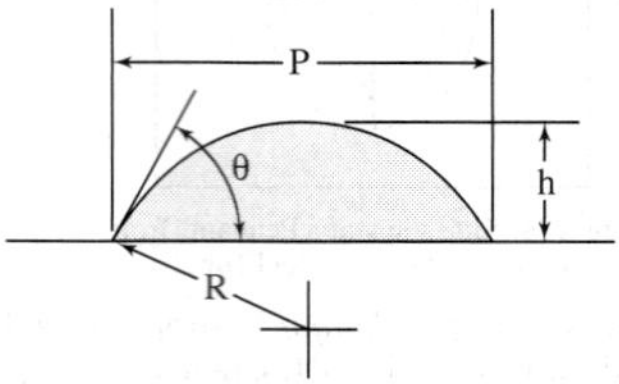


Fig. 20b. Analysis of droplet geometry.

It is assumed that (i) droplet volume remains the same before and after touching the glass plate, (ii) gravity effect on a microscale droplet is negligible, and (iii) the droplet forms part of a sphere on the plate to minimize its total surface energy. Using Equations (13) and (14) that follow, the average volume of a single droplet can be calculated by measuring the average projected droplet diameter $P$ and its height $h$ on a toolmaker's microscope:

$$V = \frac{\pi h^2}{2}\left(\frac{h}{3} + \frac{P^2}{4h}\right) = \frac{\pi D^3}{6} \tag{13}$$

$$D = \left(\frac{6V}{\pi}\right)^{1/3} = \left[3h^2\left(\frac{h}{3} + \frac{P^2}{4h}\right)\right]^{1/3} \tag{14}$$

where $V$ = volume of microdroplet (mm$^3$, in$^3$)

$P$ = projected droplet diameter (mm, in)

$h$ = height of a microdroplet (mm, in)

$D$ = air-borne diameter of microdroplet (mm, in)

Drop size varies with air pressure and volume of oil for atomization. In general, higher air pressure and velocity give more uniform and smaller drop size. Table 7 lists results of average drop sizes measured using this technique. Among different cutting fluids, the average diameter of in-flight droplets is approximately 1 μm for the CL2210EP lubricant, but it can be as large as 9 μm for other fluids.

**Table 7. Properties of Selected Lubricants**

| Lubricant | 2210EP | 2210 | 2200 | 2300HD |
|---|---|---|---|---|
| Surface tension (mN/m) | 26 | 29 | 34 | 34 |
| Droplet diameter (μm) | 0.97 | 2.3 | 6.7 | 8.4 |
| Viscosity (Pa-s) @270s$^{-1}$ shear rate | 0.016 | 0.014 | 0.023 | 0.061 |
| Contact angle on 316L (°) | 7 | 14 | 10 | 18 |
| Contact angle on WC (°) | 7 | 7 | 10 | 12 |

*Example 11, Drop Size Calculation:* Set the Unist system at 32 strokes/min, 3.6 mm stroke length, 3.78 bar pressure. Collect droplets of CL2100EP and measure with a toolmaker's microscope.

The average projected drop size is 2 μm and average drop height is 0.3 μm. The drop volume is calculated using Equation (13) to be

$$V = \frac{\pi h^2}{2}\left(\frac{h}{3} + \frac{P^2}{4h}\right) = \frac{\pi(0.3\mu\text{m})^2}{2}\left(\frac{0.3\mu\text{m}}{3} + \frac{(2\,\mu\text{m})^2}{4\times 0.3\,\mu\text{m}}\right) = 0.49\mu\text{m}^3$$

Using Equation (14), the average air-borne diameter of a droplet is

$$D = \left(\frac{6V}{\pi}\right)^{1/3} = \left(\frac{6\times 0.49\mu\text{m}^3}{\pi}\right)^{1/3} = 0.97\mu\text{m}$$

The following practical guides will assist in obtaining reliable results using the slide technique.

- The solid surface (glass plate, workpiece, or cutting tool) should be as smooth as possible, flat and polished, clean and positioned horizontally. Any surface defect such as machining marks or burrs will distort the droplet profile.
- The solid must be cleaned thoroughly before testing to avoid contamination of the tested liquid and distorted data. Ultrasonic cleaning in alcohol or degreaser following by dry air blowing should be adequate.
- For meaningful information, about 10–20 droplets should be measured. Ignore very large drops that are coalesced from smaller droplets, and very small satellite drops that are splashed off upon impact of a droplet on the glass plate.
- A measurement should be as quick as possible, using minimum light since a tiny liquid droplet might evaporate or spread when heated in bright light.

**Contact Angle Measurement.—**Droplet volume is calculated in the previous section by the sessile drop technique, and the same volume can be used for calculating the droplet contact angle. Alternatively, a predetermined droplet volume can be set and dispensed on a solid surface using a micropipette. The contact angle of a sessile droplet on a flat surface can be computed from the following equation:

$$\frac{P}{V^{1/3}} = \left[\frac{24}{\pi}\,\frac{(1 - K\cos^2\theta)^{3/2}}{2 - 3\cos\theta + \cos^3\theta}\right]^{1/3} \tag{15}$$

where $P$ = projected droplet diameter (mm, in)

$V$ = droplet volume (mm$^3$, in$^3$)

$\theta$ = contact angle (°)

$K$ = 0 for $\theta$ between 90° and 180°, $K$ = 1 for $\theta$ between 0° and 90°

The ratio $P$ over $V^{1/3}$ is called the *normalized diameter* of the droplet. Using the normalized diameter, the contact angle can be calculated using Equation (15), by looking up the value in a table (Table 8), or read from a graph (Fig. 22a and Fig. 22b).

Fig. 21 shows the top view of some liquid droplets on stainless steel sheets from which contact angles are calculated and plotted for comparison. The cutting fluids are chosen from different commercially available oil-based lubricants: CL2200, CL2210, CL2210EP, and CL2300HD. Other coolants were also included for comparison: water, KM, and water-soluble coolants RL1:15, BC1:5, and CL1:30. The nomenclature x:yy indicates dilution ratio of concentrated coolant in water for the water soluble fluids.

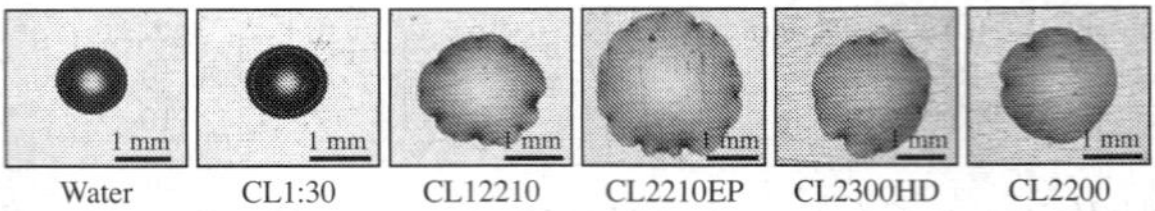


Fig. 21. Wetting of Different Cutting Fluids on 316L Stainless Steel. Constant Drop Volume = 0.25 μL.

Table 7 and Fig. 23 compare contact angles of these cutting fluids on different solids. The oil-based CL2210EP lubricant forms the lowest contact angle of 7° on 316L stainless steel, uncoated carbide, and titanium. This is in contrast to 60–70° from other water-based fluids. Surface tension of the CL2210EP lubricant is 26 mN/m, the lowest among others, thus it easily wets a carbide microtool and workpiece surfaces during machining.

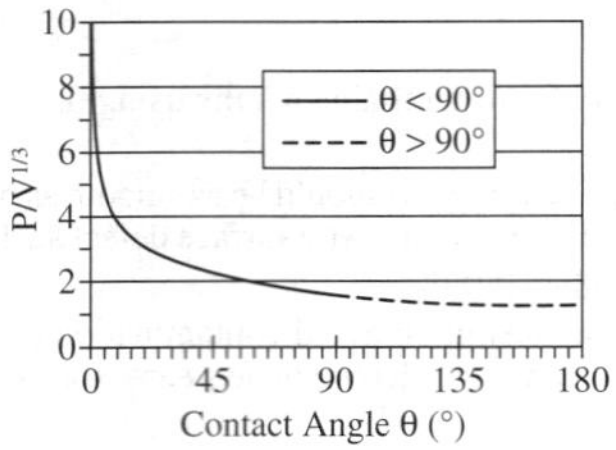


Fig. 22a. Plot of Contact Angle versus Normalized Diameter for All Angles.

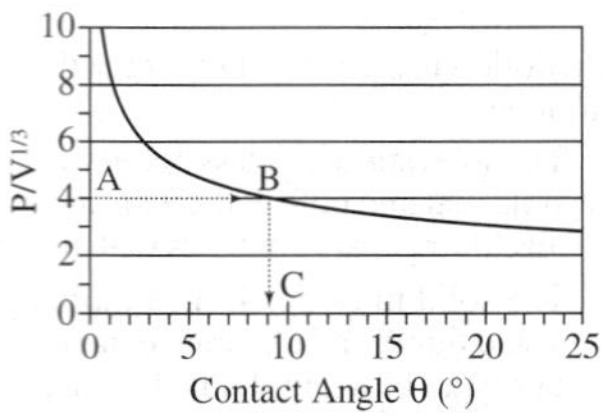


Fig. 22b. Plot of Contact Angle versus Normalized Diameter for Small Angles.

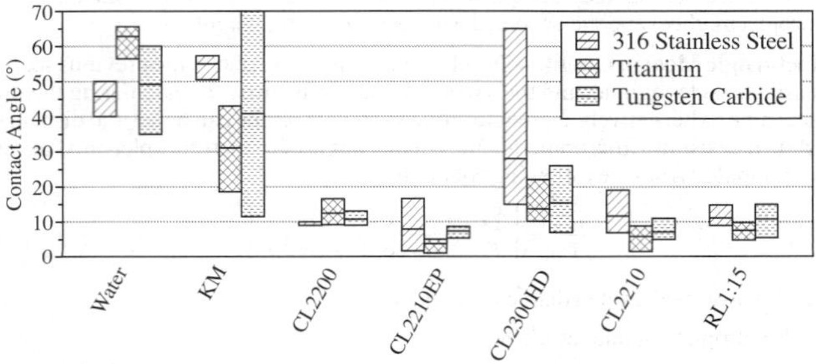


Fig. 23. Average and Range of Contact Angles for Different Cutting Fluids on 316L Stainless Steel, Tungsten Carbide, and Titanium.

**Table 8. Contact Angle (θ°) of Micromist Droplets for Practical Tests**

| Drop Size, $P$ | | Drop Volume, $V$ (μL) | | | | | | | | | | | | | | | | | | | | |
|---|---|---|---|---|---|---|---|---|---|---|---|---|---|---|---|---|---|---|---|---|---|---|
| mm | inch | 0.10 | 0.11 | 0.12 | 0.13 | 0.14 | 0.15 | 0.16 | 0.17 | 0.18 | 0.19 | 0.20 | 0.21 | 0.22 | 0.23 | 0.24 | 0.25 | 0.26 | 0.27 | 0.28 | 0.29 | 0.30 |
| 1.00 | 0.039 | 51 | 54 | 58 | 61 | 64 | 67 | 70 | 73 | 75 | 77 | 79 | 81 | 83 | 85 | 87 | 88 | 90 | 91 | 93 | 94 | 95 |
| 1.10 | 0.043 | 40 | 44 | 47 | 50 | 53 | 55 | 58 | 61 | 63 | 65 | 67 | 69 | 71 | 73 | 75 | 77 | 78 | 80 | 81 | 83 | 84 |
| 1.20 | 0.047 | 32 | 35 | 38 | 40 | 43 | 45 | 48 | 50 | 52 | 54 | 57 | 59 | 60 | 62 | 64 | 66 | 67 | 69 | 71 | 72 | 73 |
| 1.30 | 0.051 | 26 | 28 | 30 | 33 | 35 | 37 | 39 | 41 | 43 | 45 | 47 | 49 | 51 | 53 | 54 | 56 | 57 | 59 | 60 | 62 | 63 |
| 1.40 | 0.055 | 21 | 23 | 25 | 27 | 29 | 30 | 32 | 34 | 36 | 38 | 39 | 41 | 42 | 44 | 46 | 47 | 49 | 50 | 51 | 53 | 54 |
| 1.50 | 0.059 | 17 | 19 | 20 | 22 | 24 | 25 | 27 | 28 | 30 | 31 | 33 | 34 | 36 | 37 | 38 | 40 | 41 | 42 | 44 | 45 | 46 |
| 1.60 | 0.063 | 14 | 15 | 17 | 18 | 20 | 21 | 22 | 24 | 25 | 26 | 27 | 29 | 30 | 31 | 32 | 34 | 35 | 36 | 37 | 38 | 39 |
| 1.70 | 0.067 | 12 | 13 | 14 | 15 | 16 | 18 | 19 | 20 | 21 | 22 | 23 | 24 | 25 | 26 | 27 | 28 | 30 | 21 | 32 | 33 | 34 |
| 1.80 | 0.071 | 10 | 11 | 12 | 13 | 14 | 15 | 16 | 17 | 18 | 19 | 20 | 21 | 21 | 22 | 23 | 24 | 25 | 26 | 27 | 28 | 29 |
| 1.90 | 0.075 | 8.5 | 9.3 | 10 | 11 | 12 | 13 | 13 | 14 | 15 | 16 | 17 | 18 | 18 | 19 | 20 | 21 | 22 | 22 | 23 | 24 | 25 |
| 2.00 | 0.079 | 7.3 | 8.0 | 8.7 | 9.4 | 10 | 11 | 12 | 12 | 13 | 14 | 14 | 15 | 16 | 17 | 17 | 18 | 19 | 19 | 20 | 21 | 21 |
| 2.10 | 0.083 | 6.3 | 6.9 | 7.5 | 8.2 | 8.8 | 9.4 | 10 | 11 | 11 | 12 | 13 | 13 | 14 | 14 | 15 | 16 | 16 | 17 | 17 | 18 | 19 |
| 2.20 | 0.087 | 5.5 | 6.0 | 6.6 | 7.1 | 7.7 | 8.2 | 8.7 | 9.3 | 10 | 10 | 11 | 11 | 12 | 13 | 13 | 14 | 14 | 15 | 15 | 16 | 16 |
| 2.30 | 0.091 | 4.8 | 5.3 | 5.7 | 6.2 | 6.7 | 7.2 | 7.7 | 8.1 | 8.6 | 9.1 | 10 | 10 | 10 | 11 | 11 | 12 | 12 | 13 | 13 | 14 | 14 |
| 2.40 | 0.094 | 4.2 | 4.6 | 5.1 | 5.5 | 5.9 | 6.3 | 6.7 | 7.2 | 7.6 | 8.0 | 8.4 | 8.8 | 9.2 | 10 | 10 | 10 | 11 | 11 | 12 | 12 | 13 |
| 2.50 | 0.098 | 3.7 | 4.1 | 4.5 | 4.8 | 5.2 | 5.6 | 6.0 | 6.3 | 6.7 | 7.1 | 7.4 | 7.8 | 8.2 | 8.6 | 8.9 | 9.3 | 10 | 10 | 10 | 11 | 11 |
| 2.60 | 0.102 | 3.3 | 3.7 | 4.0 | 4.3 | 4.6 | 5.0 | 5.3 | 5.6 | 6.0 | 6.3 | 6.6 | 7.0 | 7.3 | 7.6 | 7.9 | 8.3 | 8.6 | 8.9 | 9.3 | 10 | 10 |
| 2.70 | 0.106 | 3.0 | 3.3 | 3.6 | 3.9 | 4.1 | 4.4 | 4.7 | 5.0 | 5.3 | 5.6 | 5.9 | 6.2 | 6.5 | 6.8 | 7.1 | 7.4 | 7.7 | 8.0 | 8.3 | 8.6 | 8.9 |
| 2.80 | 0.110 | 2.7 | 2.9 | 3.2 | 3.5 | 3.7 | 4.0 | 4.2 | 4.5 | 4.8 | 5.0 | 5.3 | 5.6 | 5.8 | 6.1 | 6.4 | 6.6 | 6.9 | 7.2 | 7.4 | 7.7 | 8.0 |
| 2.90 | 0.114 | 2.4 | 2.6 | 2.9 | 3.1 | 3.3 | 3.6 | 3.8 | 4.1 | 4.3 | 4.5 | 4.8 | 5.0 | 5.3 | 5.5 | 5.7 | 6.0 | 6.2 | 6.4 | 6.7 | 6.9 | 7.2 |
| 3.00 | 0.118 | 2.2 | 2.4 | 2.6 | 2.8 | 3.0 | 3.2 | 3.5 | 3.7 | 3.9 | 4.1 | 4.3 | 4.5 | 4.7 | 5.0 | 5.2 | 5.4 | 5.6 | 5.8 | 6.0 | 6.3 | 6.5 |

*Example 12, Contact Angle Measurement:* Set volume $V = 0.25\ \mu L$ on a micropipette and then dispense several droplets of CL2100EP on a clean titanium plate. The average size of the droplets, measured on a toolmaker's microscope, is $P = 2.520$ mm. To find wetting capability of this coolant on titanium, it is necessary to calculate its contact angle.

*Graphical Solution:* Using $1\ \mu L = 10^{-6}\ L = 1\ mm^3$, the normalized diameter is

$$\frac{P}{V^{1/3}} = \frac{2.520\ \text{mm}}{(0.25\ \text{mm}^3)^{1/3}} = 4.0$$

Starting at point A = 4.0 on the vertical Normalized Diameter ($P/V^{1/3}$) axis of Fig. 22b, draw a horizontal line until it intersects with the curve at B.

Draw a vertical line from point B until it intersects with the Contact Angle axis at C. Read the contact angle ~9° for this oil and titanium.

*Table Look-up Solution:* Locate drop volume of 0.25 μL on the first row of Table 8.

Locate projected drop size of 2.52 mm on the first column. Since 2.52 mm is not available, choose the closest number, 2.50 mm.

Read the contact angle from the intersection of row and column as 9.3°

*Example 13, Contact Angle Measurement (continued from* Example 11*):* In Example 11, droplets of CL2100EP were collected on a glass plate, the average projected drop size $P = 2\ \mu m$ was measured, and the average drop volume was calculated to be $V = 0.49\ \mu m^3$. The information obtained can be used to calculate contact angle.

a) The normalized diameter is

$$\frac{P}{V^{1/3}} = \frac{2\ \mu\text{m}}{(0.49\ \mu\text{m}^3)^{1/3}} = 2.53$$

b) Use Fig. 22b with normalized diameter of 2.53, and read the contact angle of ~18°.

The same cutting fluid Cl2100EP forms different contact angles of 18° on glass and 9° on titanium. This is due to the different surface energies of glass and titanium.

**Dynamics of Microdroplets.—**Several models are derived to study the dynamics of a microdroplet when it approaches a fast rotating tool. To effectively wet and lubricate a rotating tool, a microdroplet must (i) have enough momentum to penetrate the boundary layer around a fast rotating tool to reach the tool surface, and then (ii) adhere and wet the tool surface despite centrifugal force acting on the microdroplet.

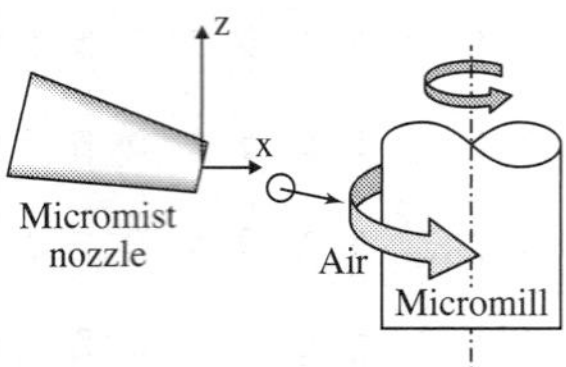


Fig. 24a. Propelling of Microdroplets Toward a Rotating Tool.

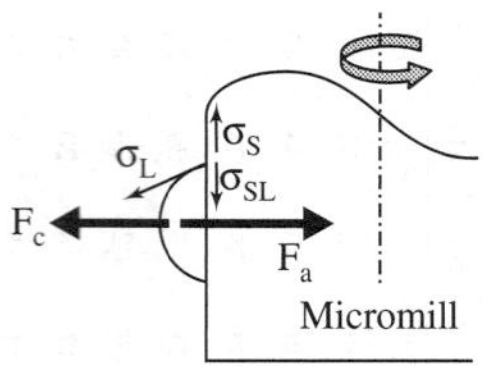


Fig. 24b. Force Balancing of a Microdroplet on the Rotating Tool.

*Propelling Microdroplets Toward a Rotating Tool:* The coordinates of a microdroplet after leaving a nozzle and moving toward a rotating tool can be expressed as

$$x_{pn} = V_f t + \frac{\alpha}{m}(V_0 \cos\beta - V_f)(1 - e^{-\alpha t/m}) \tag{16}$$

$$y_{pn} = \frac{\alpha}{m} V_0 \sin\beta (1 - e^{-\alpha t/m}) \tag{17}$$

where $x_{pn}, y_{pn}$ = coordinates of a microdroplet from the nozzle

$\alpha = 3\mu D$

$D$, $m$ = diameter and mass of a droplet

$t$ = time

$V_f$ = velocity of boundary air layer

$V_0$ = velocity of a microdroplet when leaving the mist nozzle

$\mu$ = viscosity

$\beta$ = angle between microdroplet velocity and the x-axis

The distance $y_{pn}$ away from a nozzle will reach a steady state value $(\alpha/m)V_0 \sin\beta$ after a long time ($t \to \infty$). This means if a tool and nozzle distance is closer than this steady state value, then the microdroplets will reach the tool surface.

*Force Balancing of Microdroplets on a Rotating Tool:* After reaching the tool surface, a microdroplet on a rotating tool is subject to adhesion force along lubricant/solid interface and centrifugal force. The work of adhesion is given by Young-Dupre equation:

$$W_a = \sigma_S + \sigma_L - \sigma_{SL} = (1 + \cos\theta)\sigma_L \tag{18a}$$

where $\sigma_L$ = surface tension of lubricant/air

$\sigma_S$ = surface tension of solid/air

$\sigma_{SL}$ = surface tension of solid/lubricant

$W_a$ = work of adhesion between solid/lubricant

$\theta$ = contact angle

If the adhesion force from surface tension is greater than centrifugal force, then the microdroplet will adhere and spread on the tool surface. Otherwise, the droplet will be separated from the rotating tool due to a higher centrifugal force. The condition for a microdroplet to adhere to the rotating cutting tool is given in Equation (18b) and plotted with tool surface speed and microdroplet size as two independent variables.

$$\frac{2mV_c^{\,2}}{D_t} \le P(1 + \cos\theta)\sigma_L \tag{18b}$$

where $\sigma_L$ = surface tension of lubricant/air (N/m)

$\theta$ = contact angle (°)

$m$ = mass of microdroplet (kg)

$V_c$ = cutting tool surface speed (m/s)

$P$ = projected drop size (m)

$D_t$ = tool diameter (m)

The plot in Fig. 25 is for CL2210EP lubricant and uncoated carbide tool with Ø1 mm flute diameter. The curve divides the plot into two regions: the upper region where adhesion is dominating, and the lower region where centrifugal force is stronger, i.e., a microdroplet is propelled radially away and does not wet the tool. With the droplet size of 1–10 μm for all lubricants in this study, the adhesion force is dominating and a microdroplet should adhere to the rotating tool at any rotating speed of the tested machine (0–50,000 rpm or 0–150 m/min for a Ø1 mm tool). Wetting of CL2210EP microdroplets on carbide tools was also experimentally verified.

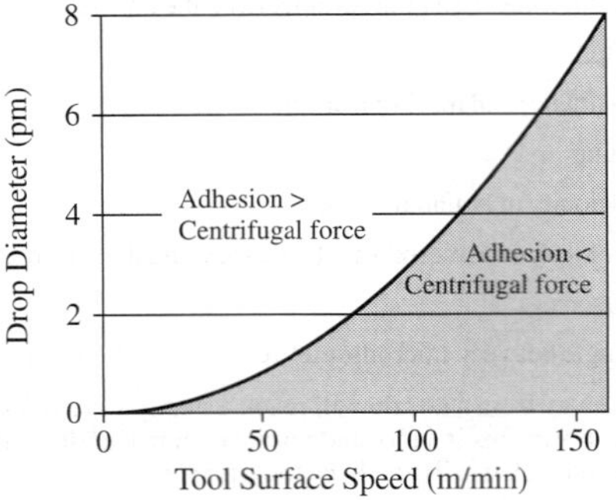


Fig. 25. Adhesion Threshold of Microdroplets on a Rotating Tool.
CL2210EP lubricant, Ø1 mm uncoated carbide tool at different speeds.
*Note:* 1 pm = 1 picometer = $10^{-12}$ meter.

Simulation using computational fluid dynamics is used to study the 2D flow of microdroplets near a rotating cutting tool. Fig. 26a shows the velocity field of microdroplets moving from left to right and around a counter-clockwise rotating cylinder. A stagnant location with zero microdroplet speed is found near the top of the cylinder (cutting tool). When micromilling in MQL condition, a workpiece should not be positioned at such a stagnant location since it would receive no lubrication. Practical setups are suggested in Fig. 26b.

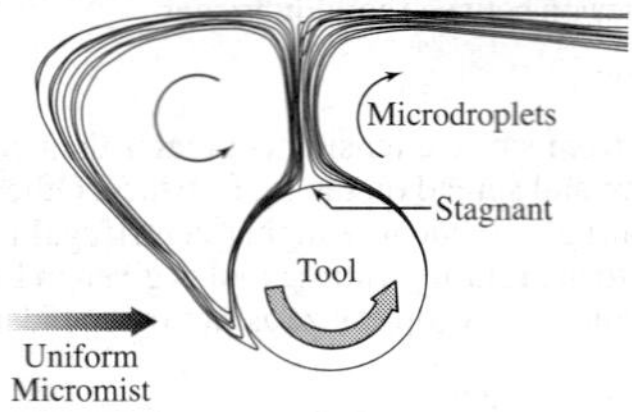


Fig. 26a. Computational Fluid Dynamic Simulation of Microdroplet Dynamics.

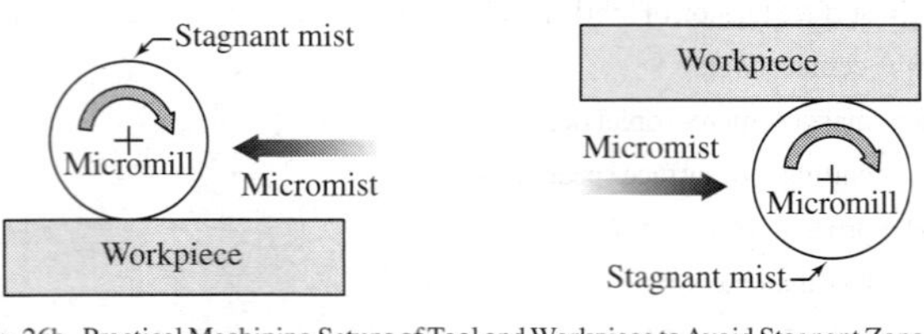


Fig. 26b. Practical Machining Setups of Tool and Workpiece to Avoid Stagnant Zone.

**Case Studies.**—Many researchers have applied micromist in their micro and macro machining studies. An example of each is given in this section.

*Use of Micromist in Micromilling of 316L Stainless Steel:* Micromilling using Ø1 mm WC tool, 10 μm/tooth chip load, 0.348 mm axial depth, 0.558 mm radial depth were performed on 316L stainless steel with different coolants. Using the cumulative tool life models (see *Tool Life* on page 1154), tool life of micromilling tools are plotted and compared in Fig. 27. Large scattering of data and low machinability is observed when dry micromilling of 316L stainless steel. The machinability improves with flood cooling, spray mist, and micromist respectively. Fig. 27 shows the data points start at lower left corner for dry machining then shift to the upper right corner of the graph for micromilling in MQL, i.e., a tool can be used at higher speed for longer time to reach the same flank wear of 50 μm. Scanning electron microscopy examination indicates significant built-up edges and attrition failure of dry cutting tools (Fig. 28a). In contrast, a well-defined abrasive tool wear is observed on a long lasting tool after micromilling with spray mist or in MQL conditions (Fig. 28b). Micromilling in MQL condition using CL2210EP extends tool life significantly over dry machining of 316L stainless steel.

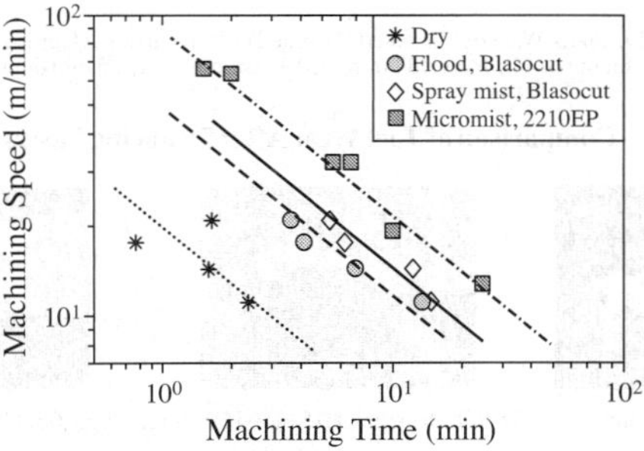


Fig. 27. Effect of Cutting Fluid Conditions on Micromilling of 316L Stainless Steel.
10 μm/tooth chip load, 0.348 mm axial depth, 0.558 mm radial depth, Ø1 mm uncoated carbide tool.

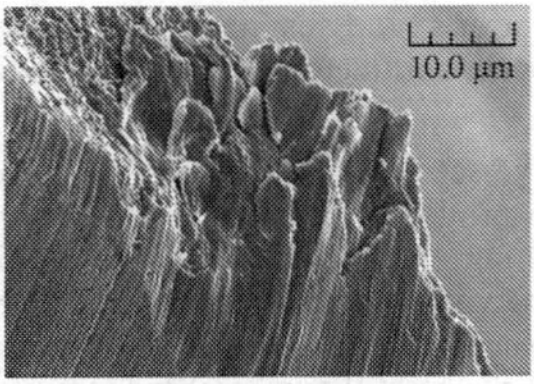


Fig. 28a. Built-Up Edge (Lower Right) and Adhesion Wear (Upper Left) on a Carbide Tool. (dry cutting, 1.4 min @ 18 m/min)

Fig. 28b. Uniform Abrasive Flank Wear on a Carbide Tool. (micromist, 21.7 min @ 15 m/min)

Micromill Ø1 mm, 10 μm/tooth chip load, 0.348 mm axial depth, 0.558 mm radial depth.

*Use of Micromist in Macrofacing of 4140 Steel:* Bars of 4140 steel, 2 inch (50 mm) diameter, 6 inch (150 mm) long were faced at 54 m/min (177 ft/min) maximum surface cutting speed, 0.1 mm/rev feed (0.004 in/rev), and 0.5 mm (0.020 in.) depth of cut. Uncoated carbide inserts TNG431 were used in the study. The operation was in dry condition, flood with Rustlick 1:15 water soluble water-based coolant, and CL2210EP oil micromist. Periodic interruption of the operation was made to remove a tool for wear assessment on a toolmaker's microscope. Fig. 29 plots flank wear of all tools and Fig. 30a, Fig. 30b, and Fig. 30c compare the tool tip conditions after 75 passes. At identical machining conditions, crater, flank and nose wear are worst for dry machining. Flood cooling improves the tool wear but nose wear is still substantial. Micromist provides the best tool protection with reduced flank and crater wear, and negligible nose wear.

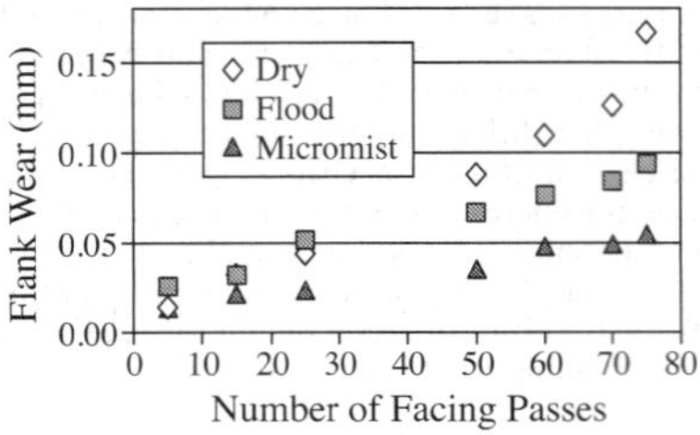


Fig. 29. Flank Wear of Uncoated Carbide Tools in Different Cutting Fluids. Macrofacing of 4140 steel, 54 m/min, 0.1 mm/rev feed, 0.5 mm depth of cut.

**Comparison of Tool Wear After 75 Facing Passes**

Fig. 30a. Dry Machining.

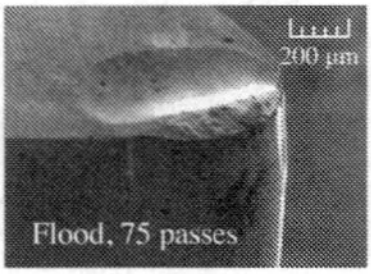


Fig. 30b. Rustlick 1:15 Flood Cooling.

Fig. 30c. CL2210EP Micromist.

There is negligible nose wear for machining in micromist. Macrofacing on 4140 steel, 54 m/min, 0.1 mm/rev feed, 0.5 mm depth of cut.

## Microfabrication Processes and Parameters

This section discusses three major microfabrication processes: micromilling, microdrilling, and microturning. Setup, tooling, and process parameters for common engineering materials are then recommended.

**Micromilling.**—Micromilling is among the most versatile of microfabrication processes. Although alternative nontraditional processes to produce microfeatures such as laser micromachining, electrical discharge micromachining, electrochemical micromachining, chemical microetching, electron/ion beam micromachining are available, these processes are either cost prohibitive, or inferior when comparing resulting surface and subsurface integrity, anisotropic aspect ratio or feature quality. Successful micromilling requires new tool geometry, tool material, machining parameters, and machining skills. It is technically incorrect and costly to perform micromilling by just scaling down a milling cutter, or parameters from macroscale milling. Commercial micromills are available for diameters of 25 μm (0.001 inch) and up, see Table 9.

- Tool material: Carbide tools should be sintered from fine grains, and ground to small cutting edge radius (see *Microcutting Tools* on page 1145).
- Milling direction: Down milling is the preferred mode since a micromill will engage a workpiece and remove a wedge shape chip with decreasing chip thickness. In contrast, a tool in upmilling would rub on the workpiece until the effective chip thickness is greater than one-half of the cutting edge radius (see *Tool Sharpness* on page 1146).
- Lubrication: Micromist should be used with all micromachining, but adequate ventilation and filtering are required to avoid inhalation of micromist. The nozzle should be as close as possible to the work and positioned to let the cutting flutes pull the mist into the cutting zone. Tool and workpiece should be arranged to avoid the stagnant zone (see Fig. 26a, Fig. 26b, Fig. 31a, and Fig. 31b).

- Tool vibration: Avoid unnecessary disengaging then engaging of microtool and workpiece in a milling program when programing the tool path. Vibration and bending of a microtool when starting and ending a cut could fatigue and shorten the tool life of the microtool (Fig. 32).

**Positioning of a Micromist Nozzle in Micromilling**

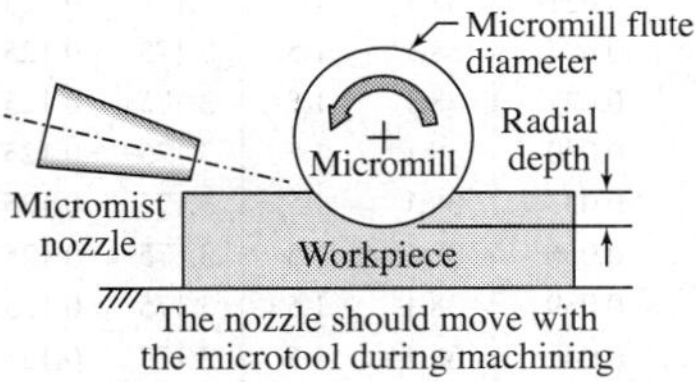


Fig. 31a. Top View.

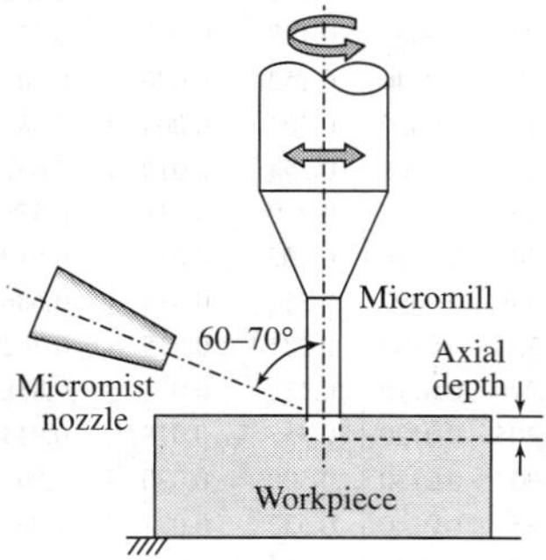


Fig. 31b. Side View.

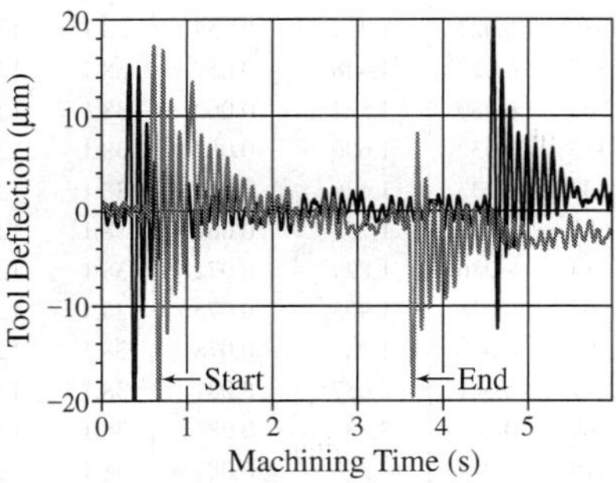


Fig. 32. Vibration of a Micromill When Engaging and Disengaging a Workpiece.
Carbide mill Ø1 mm (Ø0.040 inch), 316L stainless steel, 25000 rpm,
10 μm/tooth feed, 0.348 mm axial depth, 0.558 mm radial depth.

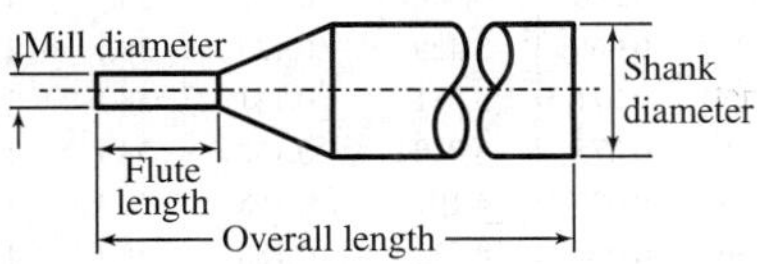


Fig. 33. Micromill Nomenclature.

**Table 9. Commercial Micromills**

(A stub flute length is about half of standard flute length)

| Mill diameter | | Stub flute length | | Standard flute length | | Overall length | | Shank diameter | |
|---|---|---|---|---|---|---|---|---|---|
| (mm) | (inch) | (mm) | (inch) | (mm) | (inch) | (mm) | (inch) | (mm) | (inch) |
| 0.025 | 0.0010 | 0.051 | 0.002 | 0.076 | 0.003 | 38.1 | 1.5 | 3.175 | 0.125 |
| 0.051 | 0.0020 | 0.076 | 0.003 | 0.152 | 0.006 | 38.1 | 1.5 | 3.175 | 0.125 |
| 0.076 | 0.0030 | 0.127 | 0.005 | 0.229 | 0.009 | 38.1 | 1.5 | 3.175 | 0.125 |
| 0.102 | 0.0040 | 0.152 | 0.006 | 0.305 | 0.012 | 38.1 | 1.5 | 3.175 | 0.125 |
| 0.127 | 0.0050 | 0.203 | 0.008 | 0.381 | 0.015 | 38.1 | 1.5 | 3.175 | 0.125 |
| 0.152 | 0.0060 | 0.254 | 0.010 | 0.457 | 0.018 | 38.1 | 1.5 | 3.175 | 0.125 |
| 0.178 | 0.0070 | 0.279 | 0.011 | 0.533 | 0.021 | 38.1 | 1.5 | 3.175 | 0.125 |
| 0.203 | 0.0080 | 0.305 | 0.012 | 0.610 | 0.024 | 38.1 | 1.5 | 3.175 | 0.125 |
| 0.229 | 0.0090 | 0.356 | 0.014 | 0.686 | 0.027 | 38.1 | 1.5 | 3.175 | 0.125 |
| 0.254 | 0.0100 | 0.381 | 0.015 | 0.762 | 0.030 | 38.1 | 1.5 | 3.175 | 0.125 |
| 0.279 | 0.0110 | 0.432 | 0.017 | 0.838 | 0.033 | 38.1 | 1.5 | 3.175 | 0.125 |
| 0.305 | 0.0120 | 0.457 | 0.018 | 0.914 | 0.036 | 38.1 | 1.5 | 3.175 | 0.125 |
| 0.330 | 0.0130 | 0.508 | 0.020 | 0.991 | 0.039 | 38.1 | 1.5 | 3.175 | 0.125 |
| 0.356 | 0.0140 | 0.533 | 0.021 | 1.067 | 0.042 | 38.1 | 1.5 | 3.175 | 0.125 |
| 0.381 | 0.0150 | 0.584 | 0.023 | 1.143 | 0.045 | 38.1 | 1.5 | 3.175 | 0.125 |
| 0.406 | 0.0160 | 0.610 | 0.024 | 1.219 | 0.048 | 38.1 | 1.5 | 3.175 | 0.125 |
| 0.432 | 0.0170 | 0.660 | 0.026 | 1.295 | 0.051 | 38.1 | 1.5 | 3.175 | 0.125 |
| 0.457 | 0.0180 | 0.686 | 0.027 | 1.372 | 0.054 | 38.1 | 1.5 | 3.175 | 0.125 |
| 0.483 | 0.0190 | 0.737 | 0.029 | 1.448 | 0.057 | 38.1 | 1.5 | 3.175 | 0.125 |
| 0.508 | 0.0200 | 0.762 | 0.030 | 1.524 | 0.060 | 38.1 | 1.5 | 3.175 | 0.125 |
| 0.533 | 0.0210 | 0.813 | 0.032 | 1.600 | 0.063 | 38.1 | 1.5 | 3.175 | 0.125 |
| 0.559 | 0.0220 | 0.838 | 0.033 | 1.676 | 0.066 | 38.1 | 1.5 | 3.175 | 0.125 |
| 0.584 | 0.0230 | 0.889 | 0.035 | 1.753 | 0.069 | 38.1 | 1.5 | 3.175 | 0.125 |
| 0.610 | 0.0240 | 0.914 | 0.036 | 1.829 | 0.072 | 38.1 | 1.5 | 3.175 | 0.125 |
| 0.635 | 0.0250 | 0.965 | 0.038 | 1.905 | 0.075 | 38.1 | 1.5 | 3.175 | 0.125 |
| 0.660 | 0.0260 | 0.991 | 0.039 | 1.981 | 0.078 | 38.1 | 1.5 | 3.175 | 0.125 |
| 0.686 | 0.0270 | 1.041 | 0.041 | 2.057 | 0.081 | 38.1 | 1.5 | 3.175 | 0.125 |
| 0.711 | 0.0280 | 1.067 | 0.042 | 2.134 | 0.084 | 38.1 | 1.5 | 3.175 | 0.125 |
| 0.737 | 0.0290 | 1.118 | 0.044 | 2.210 | 0.087 | 38.1 | 1.5 | 3.175 | 0.125 |
| 0.762 | 0.0300 | 1.143 | 0.045 | 2.286 | 0.090 | 38.1 | 1.5 | 3.175 | 0.125 |
| 0.792 | 0.0312 | 1.194 | 0.047 | 2.388 | 0.094 | 38.1 | 1.5 | 3.175 | 0.125 |
| 0.889 | 0.0350 | 1.346 | 0.053 | 2.667 | 0.105 | 38.1 | 1.5 | 3.175 | 0.125 |
| 1.016 | 0.0400 | 1.524 | 0.060 | 3.048 | 0.120 | 38.1 | 1.5 | 3.175 | 0.125 |
| 1.143 | 0.0450 | 1.727 | 0.068 | 3.429 | 0.135 | 38.1 | 1.5 | 3.175 | 0.125 |
| 1.194 | 0.0470 | 1.803 | 0.071 | 3.581 | 0.141 | 38.1 | 1.5 | 3.175 | 0.125 |
| 1.270 | 0.0500 | 1.905 | 0.075 | 3.810 | 0.150 | 38.1 | 1.5 | 3.175 | 0.125 |
| 1.397 | 0.0550 | 2.108 | 0.083 | 4.191 | 0.165 | 38.1 | 1.5 | 3.175 | 0.125 |
| 1.524 | 0.0600 | 2.286 | 0.090 | 4.572 | 0.180 | 38.1 | 1.5 | 3.175 | 0.125 |

Optional geometries include: Flute lengths: stub, standard, or optional extended length (10–80% longer); Number of flutes: 2, 3, or 4; Helix angles: 25°, 30°, 50°; End configuration: hemisphere, flat; Units: metric and US customary.

**Microdrilling.—**Microdrilling is a more complex operation compared to turning or milling. Chip removal and effective supply of cutting fluid is easy with the latter, but not with microdrilling due to extremely limited space around a microdrill.

- Tool material: As with a micromill, a carbide microdrill should be sintered from fine grains, and ground to small cutting edge radius (*Tool Sharpness* on page 1146).
- Hole quality: Spindle runout, tool eccentricity, and wandering of a microdrill causes cyclic bending of the tool that leads to catastrophic failure. To control drill wandering, precision pre-drilling of a center hole can be tried, or the drilled surface must be ground to minimize deflection of a slender drill when starting on an irregular surface.
- High aspect ratio: Pecking is essential for microhole drilling since chips have to be extracted and cutting fluid must penetrate into a small and deep microhole. The pecking depth can be deep in the beginning, but it must be reduced when drilling deeper. Start with an initial pecking depth of 2 × drill diameter and gradually reduce it to 0.5 × diameter at a depth of 10 × diameter. Pecking depth and cycles can be calculated from:

$$\frac{P}{D} = \frac{1}{9}(-1.5R + 19.5) \quad \text{for } R \le 10$$
$$\frac{P}{D} = 0.5 \quad \text{for } R > 10 \qquad (19)$$

  $P$ = incremental pecking depth (mm, in)
  $D$ = drill diameter (mm, in)
  $R$ = drill aspect ratio = hole depth / drill diameter
  See Example 14 for more on the the use of Equation (19).
- Apply micromist with a fixed nozzle pointing to the drill tip; making an angle of 60–70° with the tool axis is recommended (Fig. 34). In this way, the chip is blown away after a pecking cycle and the microdrill is re-lubricated before re-entering into the hole. Adequate ventilation and filtering are required to avoid inhaling of micromist.

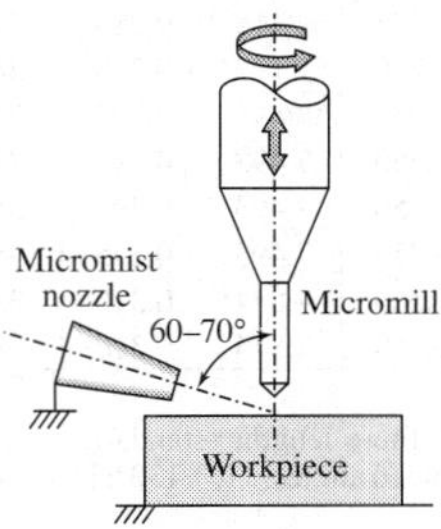


Fig. 34. Positioning of a Micromist Nozzle in Microdrilling. The Nozzle and Workpiece should be Stationary.

Commercial microdrills are available for drill diameters of 100 μm (0.004 inch) and above (Table 10).

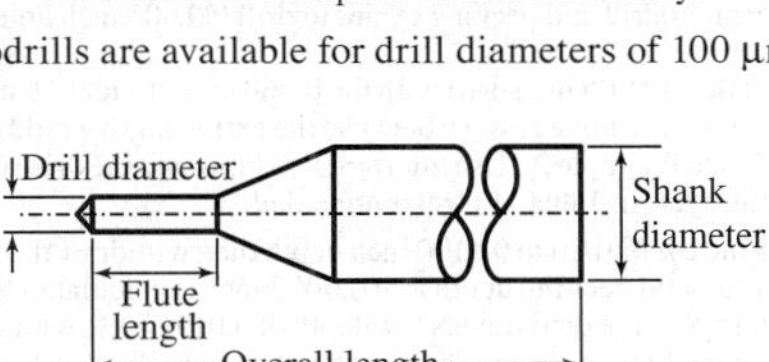


Fig. 35. Microdrill Nomenclature.

**Table 10. Commercial Microdrills**

| Drill diameter | | Standard flute length | | Extended flute length | | Overall length | | Shank diameter | |
|---|---|---|---|---|---|---|---|---|---|
| (mm) | (inch) | (mm) | (inch) | (mm) | (inch) | (mm) | (inch) | (mm) | (inch) |
| 0.102 | 0.0040 | 1.016 | 0.040 | 1.778 | 0.070 | 38.1 | 1.5 | 3.175 | 0.125 |
| 0.127 | 0.0050 | 1.524 | 0.060 | 2.286 | 0.090 | 38.1 | 1.5 | 3.175 | 0.125 |
| 0.150 | 0.0059 | 2.032 | 0.080 | 3.048 | 0.120 | 38.1 | 1.5 | 3.175 | 0.125 |
| 0.160 | 0.0063 | 2.032 | 0.080 | 3.048 | 0.120 | 38.1 | 1.5 | 3.175 | 0.125 |
| 0.170 | 0.0067 | 2.032 | 0.080 | 3.048 | 0.120 | 38.1 | 1.5 | 3.175 | 0.125 |
| 0.180 | 0.0071 | 2.540 | 0.100 | 3.810 | 0.150 | 38.1 | 1.5 | 3.175 | 0.125 |
| 0.191 | 0.0075 | 2.540 | 0.100 | 3.810 | 0.150 | 38.1 | 1.5 | 3.175 | 0.125 |
| 0.201 | 0.0079 | 2.540 | 0.100 | 3.810 | 0.150 | 38.1 | 1.5 | 3.175 | 0.125 |
| 0.211 | 0.0083 | 2.540 | 0.100 | 3.810 | 0.150 | 38.1 | 1.5 | 3.175 | 0.125 |
| 0.221 | 0.0087 | 2.540 | 0.100 | 3.810 | 0.150 | 38.1 | 1.5 | 3.175 | 0.125 |
| 0.231 | 0.0091 | 3.810 | 0.150 | 5.588 | 0.220 | 38.1 | 1.5 | 3.175 | 0.125 |
| 0.241 | 0.0095 | 3.810 | 0.150 | 5.588 | 0.220 | 38.1 | 1.5 | 3.175 | 0.125 |
| 0.249 | 0.0098 | 3.810 | 0.150 | 5.588 | 0.220 | 38.1 | 1.5 | 3.175 | 0.125 |
| 0.254 | 0.0100 | 3.810 | 0.150 | 5.588 | 0.220 | 38.1 | 1.5 | 3.175 | 0.125 |
| 0.267 | 0.0105 | 3.810 | 0.150 | 5.588 | 0.220 | 38.1 | 1.5 | 3.175 | 0.125 |
| 0.279 | 0.0110 | 3.810 | 0.150 | 5.588 | 0.220 | 38.1 | 1.5 | 3.175 | 0.125 |
| 0.292 | 0.0115 | 3.810 | 0.150 | 5.588 | 0.220 | 38.1 | 1.5 | 3.175 | 0.125 |
| 0.300 | 0.0118 | 5.715 | 0.225 | 7.112 | 0.280 | 38.1 | 1.5 | 3.175 | 0.125 |
| 0.305 | 0.0120 | 5.715 | 0.225 | 7.112 | 0.280 | 38.1 | 1.5 | 3.175 | 0.125 |
| 0.318 | 0.0125 | 5.715 | 0.225 | 7.112 | 0.280 | 38.1 | 1.5 | 3.175 | 0.125 |
| 0.330 | 0.0130 | 5.715 | 0.225 | 7.112 | 0.280 | 38.1 | 1.5 | 3.175 | 0.125 |
| 0.343 | 0.0135 | 5.715 | 0.225 | 7.112 | 0.280 | 38.1 | 1.5 | 3.175 | 0.125 |
| 0.351 | 0.0138 | 5.715 | 0.225 | 7.112 | 0.280 | 38.1 | 1.5 | 3.175 | 0.125 |
| 0.368 | 0.0145 | 5.715 | 0.225 | 7.112 | 0.280 | 38.1 | 1.5 | 3.175 | 0.125 |
| 0.396 | 0.0156 | 6.350 | 0.250 | 7.112 | 0.280 | 38.1 | 1.5 | 3.175 | 0.125 |
| 0.399 | 0.0157 | 6.350 | 0.250 | 7.493 | 0.295 | 38.1 | 1.5 | 3.175 | 0.125 |
| 0.406 | 0.0160 | 6.350 | 0.250 | 7.493 | 0.295 | 38.1 | 1.5 | 3.175 | 0.125 |
| 0.450 | 0.0177 | 6.350 | 0.250 | 7.493 | 0.295 | 38.1 | 1.5 | 3.175 | 0.125 |
| 0.457 | 0.0180 | 6.350 | 0.250 | 7.493 | 0.295 | 38.1 | 1.5 | 3.175 | 0.125 |
| 0.500 | 0.0197 | 6.604 | 0.260 | 7.874 | 0.310 | 38.1 | 1.5 | 3.175 | 0.125 |
| 0.508 | 0.0200 | 6.604 | 0.260 | 7.874 | 0.310 | 38.1 | 1.5 | 3.175 | 0.125 |

Optional geometries include: Flute lengths: standard or extended length (10–80% longer); Number of flutes: 2, 3, or 4; Included angles: 118°, 130°; Helix angles: 25°, 30°, 50°; Units: metric and US customary.

*Example 14:* Select a microdrill and pecking cycles to drill Ø0.005 inch holes to 0.050 inch depth in titanium.

From Table 10, select the Ø0.005 inch drill with the 0.060 inch standard length. A drill with extended length of 0.090 inch is unnecessary because the extra length would decrease the drill stiffness by about 56% (see Example 2). Use micromist to lubricate the drill and blow the extracted chips away. The pecking cycle in Table 11 is recommended.

During the first pecking cycle, drill to 0.0100 inch depth then withdraw the tool. For the second pecking cycle, the incremental pecking depth $P$ of 0.0092 inch is calculated from Equation (19). During the 2nd pecking cycle, the drill travels to a depth of 0.0192 inch, withdraws to remove chips, then continues down for pecking cycle #3. The incremental pecking in cycle #7 should be increased by 0.0003 inch so that the pecking cycle #8 will drill the hole to the required depth of 0.0500 inch.

**Table 11. Pecking Cycle, Drill Diameter $D = 0.005$ Inch**

| Pecking cycle # | Hole depth (inch) | Aspect ratio | $P/D$ | Pecking depth, $P$ (inch) |
|---|---|---|---|---|
| 1 | 0.0100 | 2.00 | 1.83 | 0.0092 |
| 2 | 0.0192 | 3.83 | 1.53 | 0.0076 |
| 3 | 0.0268 | 5.36 | 1.27 | 0.0064 |
| 4 | 0.0332 | 6.63 | 1.06 | 0.0053 |
| 5 | 0.0385 | 7.70 | 0.88 | 0.0044 |
| 6 | 0.0429 | 8.58 | 0.74 | 0.0037 |
| 7 | 0.0466 | 9.32 | 0.61 | 0.0031 |
| 8 | 0.0497 | 9.93 | 0.51 | 0.0026 |
| 9 | 0.0522 | 10.44 | 0.43 | 0.0021 |

**Microturning.**—Product miniaturization and the demand for ultraprecision products drives the rapid development of micro/nano turning. This technology produces polished and high quality spherical and aspherical parts from metals, ceramics, semiconductors, and polymers that cannot be economically produced by traditional grinding, lapping, or polishing processes. Micro/nano turning also produces intricate shapes with low or no subsurface damage because it operates in the ductile-regime mode (see *Ductile Regime Micromachining* on page 1160).

Diamonds are commonly used for micro/nano turning. Polycrystalline diamond (PCD) tools are sintered from microsize diamond grains. PCD tools are less expensive but with limited capability due to large edge radius (few hundred nanometers) and lower edge strength due to attrition wear. Single crystalline diamond (SCD) tools are best for micro/nano turning because they:

- Have single crystalline structure that allows a sharp cutting edge as small as 10 nm
- Have the highest thermal conductivity among all engineering materials
- Retain high strength and hardness at high temperature
- Have high elastic and shear moduli to resist plastic deformation, and
- Have a low coefficient of friction

A diamond tool, however, is costly and brittle. A tool with zero or negative rake angle (i) improves its edge strength, and (ii) forms a hydrostatic compressive stress field in the material just in front of and below the tool, and therefore, minimizes crack initiation. Single crystal diamond typically have the (110) crystal plane as the rake face and are brazed onto a steel shanks of different shapes and sizes.

Not all materials can be successfully micro/nano turned with a diamond tool. Ferrous alloys and silicon carbide (SiC) are not suitable for diamond turning because of diffusion from the highly concentrated carbon in the diamond tool to a lower concentration zone of carbon in the workpiece materials. Other materials, although machinable with diamond, should be homogeneous and contain few if no impurities. Any hard inclusions might either damage a sharp diamond edge or be sheared off and smear against the machined surface. Fig. 36a shows hard beryllides in beryllium copper CA173 that plow and smear a mirror finish surface (Fig. 36b).

**Table 12. Examples of Diamond Machinable Materials**

| Semiconductor | Metal | Ceramic | Plastics |
|---|---|---|---|
| Cadmium telluride | Aluminum alloys | Aluminum oxide | Acrylic |
| Gallium arsenide | Copper alloys | Zirconium oxide | Fluoroplastics |
| Germanium | Electroless nickel | Optical glasses | Nylon |
| Lithium niobate | Gold | Quartz | Polycarbonate |
| Silicon | Magnesium | | Polymethylmethacrylate |
| Silicon nitride | Silver | | Propylene |
| Zinc selenide | Zinc | | Styrene |
| Zinc sulphide | | | |

Micromist is required to lubricate and cool both the tool and the machined surface. The micromist nozzle should move with the tool while blowing the micro/nano chips away from the machined surface. Adequate ventilation and filtering of micromist are required to avoid environmental issues.

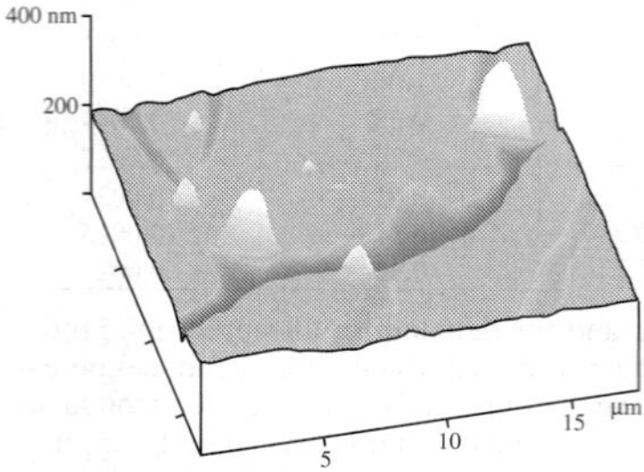


Fig. 36a. Beryllide Particles at Grain Boundaries of Diamond Turned Beryllium Copper.

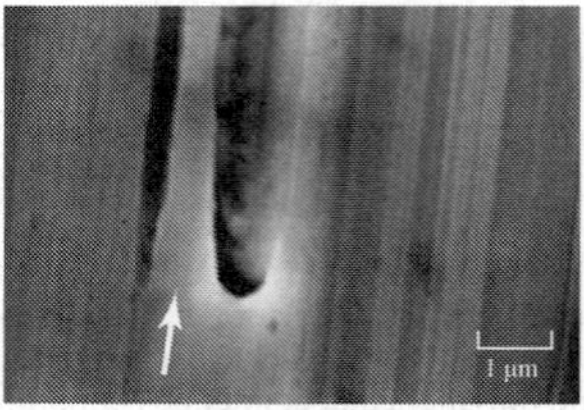


Fig. 36b. Deep Scratch on Machined Surface Caused by a Broken Beryllide. The Arrow Shows Tool Cutting Direction.

**Speeds and Feeds.**—In macromachining of engineering materials, the material grains are very small compared to the cutting tool dimensions and edge radius. Only bulk material properties such as material strength, hardness, and thermal conductivity affect machining performance. However, when the workpiece grain size is similar to the tool edge radius then the workpiece material will have more influence on micromachining performance (see *Workpiece Materials* on page 1158). Since plastically deformed materials are normally harder than the original material, machining feed must be (i) deeper than this hardened layer, and (ii) at least half of the tool edge radius (see *Microcutting Tools* on page 1145).

*Example 15, Work Hardening of Titanium Alloy:* Micromachining Ti 6Al 4V alloy with a carbide tool generates a work-hardening depth of about 1–2 µm (40–80 µinch). The microhardness increases 8–14% above the base material hardness to ~350 Vicker. Chip load for microdrilling or micromilling should be greater than this hardened zone depth.

*Microturning Parameters:* Criteria for selecting turning speed and feed are tool life and surface finish, and less concern on tool fracture since a turning tool is robust and rigid. Cutting speeds for single crystalline diamond are very high and depend mostly on the rigidity of the setup and the machine tool capability. Turning at 100–500 m/min (330–1650 ft/min) is common for most metals and metal matrix composites.

Selection of feed depends on the surface finish required because this is the main objective for using diamond for turning in the ductile regime mode. Models for a macromachined surface finish have been proposed which assume:

- The depth of cut is less than tool nose radius but more than cutting edge radius
- There is no built-up-edge on the tool, therefore, effect of cutting speed is negligible
- Workpiece of polycrystalline material with fine grains is used and grain orientation does not significantly affect surface finish of large areas
- Chips are completely removed without side-burr, and
- No error is introduced from imperfect machine kinematics (such as asynchronous spindle error motion).

The theoretical surface finish values are functions of feed and tool nose radius, and can be computed from Equations (20) and (21) and plotted as in Fig. 37a and Fig. 37b.

$$R_t = R - \sqrt{R^2 - \frac{f^2}{4}} = R - R\sqrt{1 - \left(\frac{f}{2R}\right)^2} \approx 0.125\frac{f^2}{R} \quad (20)$$

$$R_a = k(f, R)\frac{f^2}{R} \approx 0.031553\frac{f^2}{R} \text{ for } \frac{f}{R} < 0.1 \quad (21)$$

where $R_a$ = average surface finish (m, inch)

$R_t$ = peak-to-valley surface finish (m, inch)

$f$ = tool feed for each revolution (m, inch)

$R$ = tool nose radius (m, inch)

*Example 16, Turning Parameter Selection:* Select tool and parameters to achieve $R_a$ = 1 μm when turning 6061-T6 aluminum.

*Solution:* Use a single crystalline diamond tool with tool nose radius $R$ = 1 mm (0.040 inch) and cutting edge radius of 80 nm (3 μin).

*Depth of cut:* Depth must be between the nose radius and edge radius, so choose 0.5 mm.

*Cutting speed:* Speed does not affect surface finish, so choose 303 m/min (1000 ft/min).

*Feed:* Referring to Fig. 37b, from point A ($R$ = 1 mm), draw a vertical line to intersect with $R_a$ = 1 μm curve at B. Draw a horizontal line at B and intersect vertical axis at C, and read the feed = 0.18 mm/rev (0.007 in/rev). Alternatively, Equation (21) can be rearranged and used to compute the feed distance for each revolution as follows:

$$f = \sqrt{\frac{R_a R}{0.031553}} = \sqrt{\frac{1\,\mu\text{m} \times 1\,\text{mm}}{0.031553}\frac{1\text{ mm}}{1000\,\mu\text{m}}} = 0.178\text{ mm}$$

**Effect of Tool Nose Radius and Tool Feed on Surface Finish**

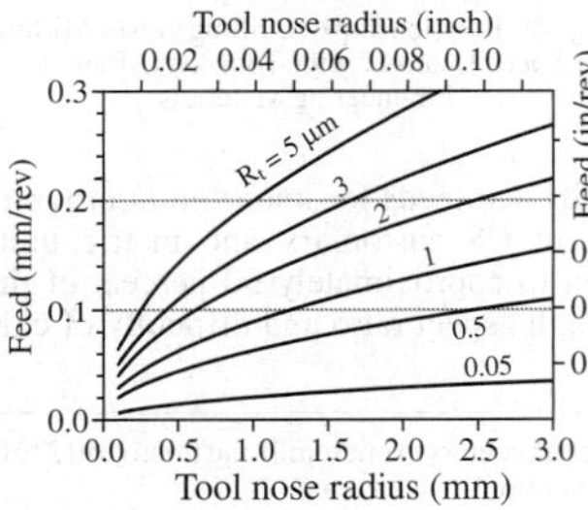


Fig. 37a. Peak-to-Valley Surface Finish $R_t$.

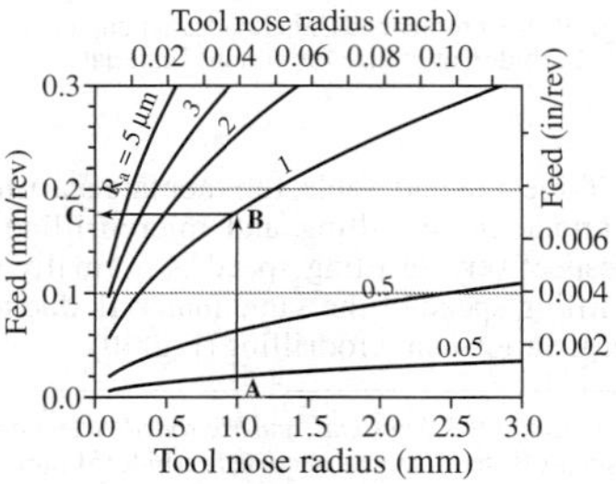


Fig. 37b. Average Surface Finish $R_a$.

*Microdrilling/Milling Parameters:* The combination of spindle runout and radial cutting force can deflect and break a tool. Finite element analysis (FEA) of a micromill shows that the critical area is the junction between the cylindrical flute and solid conical shank. Using 4.7 GPa and 93 GPa as the average values of measured flexural strength and elastic modulus of carbide, the analysis indicates that a Ø1 mm carbide micromill will break if deflected more than 34 percent of its diameter.

Fig. 38 shows crash-test and calculated data for catastrophic tool failure when micromilling 316L stainless steel. The radial depth of cut and chip load are normalized to the tool diameter for ease of comparison. The data points form a line that divides the plane into two regions: tool failure and tool safe areas. A tool will break if milling parameters are chosen to be above this threshold line. Therefore, all milling parameters should be conservatively selected below the threshold for a production run.

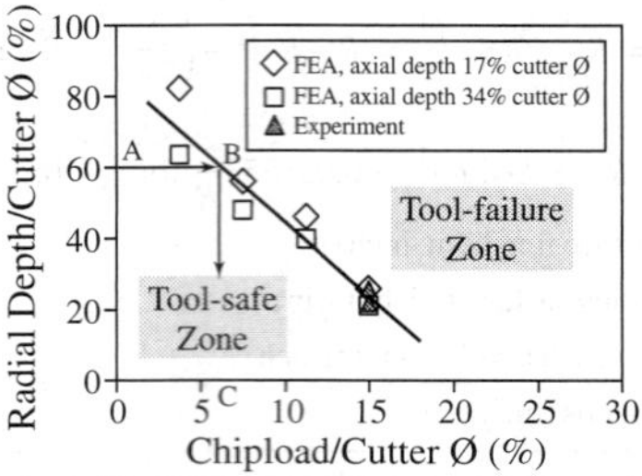


Fig. 38. Catastrophic Failure Threshold of Micro-Milling Tools as Percentage of Tool Diameter. Dry milling 316L stainless steel, Ø1 mm tool diameter, 2 flutes, 0.348 mm axial depth of cut.

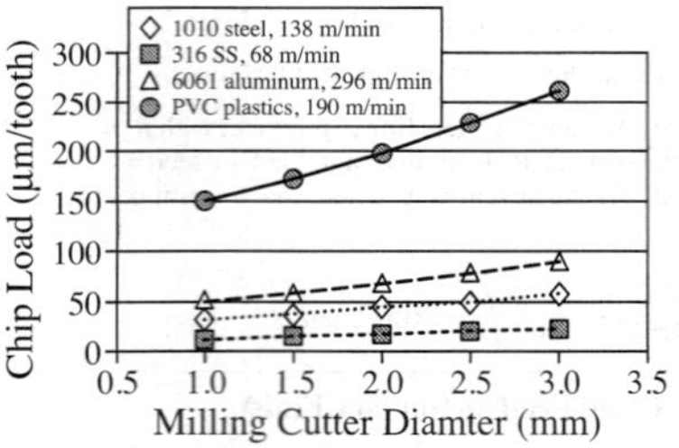


Fig. 39. Recommended Chip Load as a Function of Carbide Cutter Size for Various Materials.

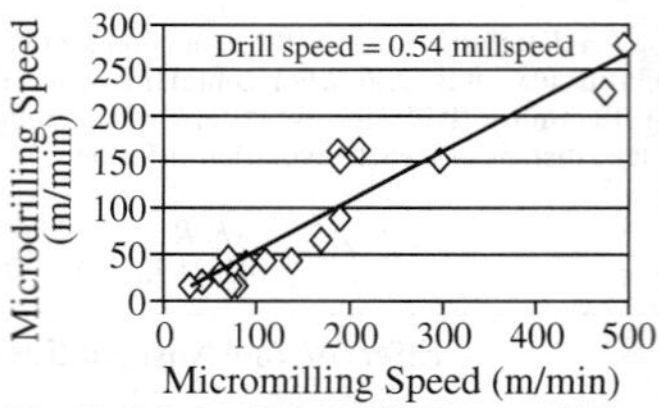


Fig. 40. Relationship of Drilling versus Milling Speed Using Carbide Tools on Different Engineering Materials.

Table 13a and Table 13b tabulate starting speeds and feeds for uncoated micrograin carbide micromilling and microdrilling tools, in US customary and metric units respectively. Drilling speed is normally reduced to approximately 50 percent of the milling speed of the same material, due to the high aspect ratio and difficulty of chip evacuation in microdrilling (Fig. 40).

*Example 17, Micromill and Microdrill Selection:* Select parameters to micromill and microdrill 316L stainless steel using an uncoated carbide Ø1 mm, 2 flute microtool.

*Micromilling:* Speed selection depends on tool life and cutting fluid. Table 13b suggests 68 m/min for micromilling. Both radial and axial depth of cut affect selection of feed. Select 35% axial depth (0.35 × tool diameter = 0.35 mm), and 60% radial depth (0.60 × tool diameter = 0.6 mm). From point A (60%) on the vertical axis of Fig. 38, draw a horizontal line that intersects the threshold line at point B, then find point C on the horizontal axis. The chip load that causes immediate tool fracture would be slightly more than 5% of tool diameter (50 μm/tooth). A conservative chip load would be 13 μm/tooth as indicated in Table 13b.

*Microdrilling:* Table 13b suggests 28 m/min speed and 13 μm/tooth feed for drilling. Reduction of drilling speed to only 50–60% (59% in this example) of milling speed is necessary to facilitate chip removal in microdrilling.

**Table 13a. Speeds and Feeds for Micro Milling/Drilling with Uncoated Carbide Tools** (US Customary Units)

| Materials | Examples | Vicker micro-hardness | Mill speed (ft/min) | Drill speed (ft/min) | Chip load (μin/tooth), *D* = drill or mill diameter | | | | |
|---|---|---|---|---|---|---|---|---|---|
| | | | | | *D* < 0.04 in. | *D* < 0.06 in. | *D* < 0.08 in. | *D* < 0.10 in. | *D* < 0.12 in. |
| Steel | 12L14 | < 120 | 558 | 213 | 1400 | 1700 | 2000 | 2300 | 2600 |
| | 1010 | < 265 | 453 | 141 | 1300 | 1500 | 1700 | 2000 | 2300 |
| | 4063 | < 208 | 361 | 141 | 1300 | 1500 | 1700 | 2000 | 2300 |
| Stainless steel | 409, 410, 446 | < 318 | 246 | 125 | 600 | 600 | 800 | 900 | 1000 |
| | 304, 316, 316L | < 265 | 223 | 92 | 500 | 600 | 700 | 800 | 900 |
| | 17-7 PH | < 318 | 230 | 148 | 400 | 450 | 600 | 700 | 900 |
| Nickel | Pure nickel | | 197 | 92 | 500 | 600 | 700 | 800 | 900 |
| | Monel 400 | | 138 | 66 | 500 | 600 | 700 | 800 | 900 |
| | Inconel 718 | | 98 | 56 | 300 | 300 | 350 | 400 | 500 |
| Titanium | Pure titanium | | 295 | 131 | 650 | 700 | 850 | 950 | 1100 |
| | Cast titanium | | 266 | 52 | 650 | 700 | 850 | 950 | 1100 |
| | Ti 6Al 4V | | 243 | 52 | 650 | 700 | 850 | 950 | 1100 |
| Aluminum | 1100 | | 1624 | 902 | 1500 | 1700 | 1950 | 2250 | 2550 |
| | A356 | | 1558 | 738 | 2100 | 2350 | 2700 | 3150 | 3600 |
| | 6061 | | 971 | 492 | 2100 | 2350 | 2700 | 3150 | 3600 |
| Copper | C17200 (soft Be-Cu) | | 617 | 531 | 1299 | 1457 | 1693 | 1969 | 2244 |
| | C85400 (annealed brass) | | 689 | 531 | 1772 | 2047 | 2362 | 2677 | 3071 |
| | C95400 (Al bronze) | | 623 | 289 | 1457 | 1732 | 1969 | 2244 | 2598 |
| Plastics | ABS, PVC thermoplastics | | 623 | 492 | 5900 | 6800 | 7800 | 9000 | 10300 |

**Table 13b. Speeds and Feeds for Micro Milling/Drilling with Uncoated Carbide Tools** (SI Metric Units)

| Materials | Examples | Vicker micro-hardness | Mill speed (m/min) | Drill speed (m/min) | Chip load (μm/tooth), *D* = drill or mill diameter | | | | |
|---|---|---|---|---|---|---|---|---|---|
| | | | | | *D* < 1.0 mm | *D* < 1.5 mm | *D* < 2.0 mm | *D* < 2.5 mm | *D* < 3.0 mm |
| Steel | 12L14 | < 120 | 170 | 65 | 38 | 43 | 50 | 57 | 65 |
| | 1010 | < 265 | 138 | 43 | 33 | 38 | 43 | 51 | 58 |
| | 4063 | < 208 | 110 | 43 | 33 | 38 | 43 | 51 | 58 |
| Stainless steel | 409, 410, 446 | < 318 | 75 | 38 | 15 | 15 | 20 | 23 | 25 |
| | 304, 316, 316L | < 265 | 68 | 28 | 13 | 15 | 18 | 20 | 23 |
| | 17-7 PH | < 318 | 70 | 45 | 10 | 11 | 15 | 18 | 23 |
| Nickel | Pure nickel | | 60 | 28 | 13 | 15 | 18 | 20 | 23 |
| | Monel 400 | | 42 | 20 | 13 | 15 | 18 | 20 | 23 |
| | Inconel 718 | | 30 | 17 | 8 | 8 | 9 | 10 | 13 |
| Titanium | Pure titanium | | 90 | 40 | 17 | 18 | 22 | 24 | 28 |
| | Cast titanium | | 81 | 16 | 17 | 18 | 22 | 24 | 28 |
| | Ti 6Al 4V | | 74 | 16 | 17 | 18 | 22 | 24 | 28 |
| Aluminum | 1100 | | 495 | 275 | 38 | 43 | 50 | 57 | 65 |
| | A356 | | 475 | 225 | 53 | 60 | 69 | 80 | 91 |
| | 6061 | | 296 | 150 | 53 | 60 | 69 | 80 | 91 |
| Copper | C17200 (soft Be-Cu) | | 188 | 162 | 33 | 37 | 43 | 50 | 57 |
| | C85400 (annealed brass) | | 210 | 162 | 45 | 52 | 60 | 68 | 78 |
| | C95400 Al bronze | | 190 | 88 | 37 | 44 | 50 | 57 | 66 |
| Plastics | ABS, PVC thermoplastics | | 190 | 150 | 150 | 173 | 198 | 229 | 262 |

# MACHINING ECONOMETRICS

## Tool Wear and Tool Life Relationships

Tool life is defined as the cutting time to reach a predetermined wear limit, called the *tool wear criterion*. The size of tool wear criterion depends on the grade used, usually a tougher grade can be used at bigger flank wear. For finishing operations, where close tolerances are required, the wear criterion is relatively small. Other alternative wear criteria are a predetermined value of the surface roughness, or a given depth of the crater which develops on the rake face of the tool. The most appropriate wear criteria depends on cutting geometry, grade, and materials.

Tool life is determined by assessing the time—the tool life—at which a given predetermined flank wear is reached, 0.25, 0.4, 0.6, 0.8 mm, etc. Fig. 1 depicts how flank wear varies with cutting time (approximately straight lines in a semi-logarithmic graph) for three combinations of cutting speeds and feeds. Alternatively, these curves may represent how variations of machinability impact on tool life, when cutting speed and feed are constant. All tool wear curves will sooner or later bend upwards abruptly and the cutting edge will break, i.e., catastrophic failure as indicated by the white arrows in Fig. 1.

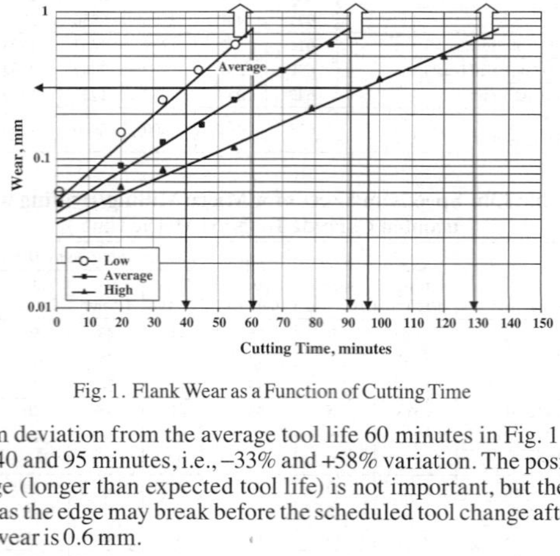


Fig. 1. Flank Wear as a Function of Cutting Time

The maximum deviation from the average tool life 60 minutes in Fig. 1 is assumed to range between 40 and 95 minutes, i.e., −33% and +58% variation. The positive deviation from the average (longer than expected tool life) is not important, but the negative one (shorter life) is, as the edge may break before the scheduled tool change after 60 minutes, when the flank wear is 0.6 mm.

It is therefore important to set the wear criterion at a safe level such that tool failures due to "normal" wear become negligible. This is the way machinability variations are mastered.

**Equivalent Chip Thickness (*ECT*).—** *ECT* combines the four basic turning variables, depth of cut, lead angle, nose radius and feed per revolution into one basic parameter. For all other metal cutting operations such as drilling, milling and grinding, additional variables such as number of teeth, width of cut, and cutter diameter are included in the parameter *ECT*. In turning, milling, and drilling, according to the *ECT* principle, when the product of feed times depth of cut is constant the tool life is constant no matter how the depth of cut or feed is selected, provided that the cutting speed and cutting edge length are maintained constant. By replacing the geometric parameters with *ECT*, the number of tool life tests to evaluate cutting parameters can be reduced considerably, by a factor of 4 in turning, and in milling by a factor of 7 because radial depth of cut, cutter diameter and number of teeth are additional parameters.

The introduction of the *ECT* concept constitutes a major simplification when predicting tool life and calculating cutting forces, torque, and power. *ECT* was first presented in 1931 by Professor R. Woxen, who both theoretically and experimentally proved that *ECT* is a basic metal cutting parameter for high-speed cutting tools. Dr. Colding later proved that the concept also holds for carbide tools, and extended the calculation of *ECT* to be valid for cutting conditions when the depth of cut is smaller than the tool nose radius, or for round inserts. Colding later extended the concept to all other metal cutting operations, including the grinding process.

The definition of *ECT* is:

$$ECT = \frac{Area}{CEL} \text{(mm or inch)}$$

where $A$ = cross sectional area of cut (approximately = feed × depth of cut), ($mm^2$ or $inch^2$)

$CEL$ = cutting edge length (tool contact rubbing length), (mm or inch), see Fig. 1 on page 1080.

An exact value of $A$ is obtained by the product of *ECT* and *CEL*. In turning, milling, and drilling, *ECT* varies between 0.05 and 1 mm, and is always less than the feed/rev or feed/tooth; its value is usually about 0.7 to 0.9 times the feed.

*Example 1:* For a feed of 0.8 mm/rev, depth of cut $a = 3$ mm, and cutting edge length $CEL = 4$ mm, the value of *ECT* is approximately $ECT = 0.8 \times 3 \div 4 = 0.6$ mm.

The product of *ECT*, *CEL*, and cutting speed $V$ (m/min or ft/min) equals the metal removal rate *MRR*, measured in terms of the volume of chips removed per minute:

$$\begin{aligned} MRR &= 1000V \times Area = 1000V \times ECT \times CEL \text{ mm}^3\text{/min} \\ &= V \times Area \text{ cm}^3\text{/min or inch}^3\text{/min} \end{aligned}$$

The specific metal removal rate *SMRR* is the metal removal rate per mm cutting edge length *CEL*, thus:

$$\begin{aligned} SMRR &= 1000V \times ECT \text{ mm}^3\text{/min/mm} \\ \text{and } MRR &= V \times ECT \text{ cm}^3\text{/min/mm or inch}^3\text{/min/inch} \end{aligned}$$

*Example 2:* Using the Example 1 data and a cutting speed of $V = 250$ m/min, the specific metal removal rate becomes $SMRR = 0.6 \times 250 = 150$ ($cm^3$/min/mm).

*ECT in Grinding:* In grinding *ECT* is defined as in the other metal cutting processes, and is approximately equal to $ECT = Vw \times ar \div V$, where $Vw$ is the work speed, $ar$ is the depth of cut, and $A = Vw \times ar$. Wheel life is constant no matter how depth $ar$, or work speed $Vw$, is selected at $V$ = constant (usually the influence of grinding contact width can be neglected). This translates into the same wheel life as long as the specific metal removal rate is constant, thus:

$$SMRR = 1000Vw \times ar \text{ mm}^3\text{/min/mm}$$

In grinding, *ECT* is much smaller than in the other cutting processes, ranging from about 0.0001 to 0.001 mm (0.000004 to 0.00004 inch). The grinding process is described in a separate chapter *GRINDING FEEDS AND SPEEDS* starting on page 1249.

**Tool Life Relationships.**—Plotting the cutting times to reach predetermined values of wear typically results in curves similar to those shown in Fig. 2 (cutting time versus cutting speed at constant feed per tooth) and Fig. 3 (cutting time versus feed per tooth at constant cutting speed). These tests were run in 1993 with mixed ceramics turn-milling 82 RC (Rockwell C scale) hard steel at the Technische Hochschule Darmstadt.

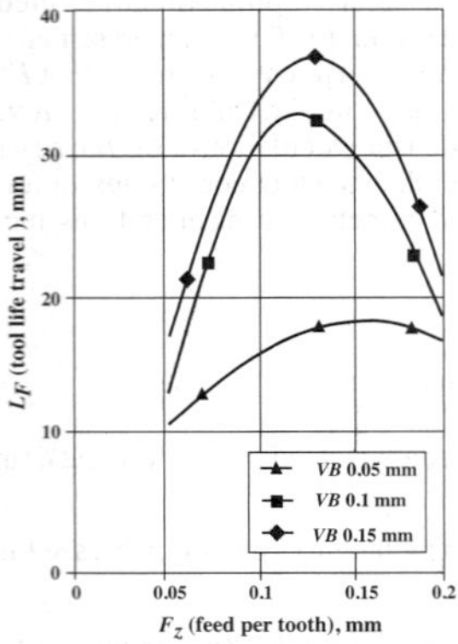


Fig. 2. Influence of Feed per Tooth on Cutting Time

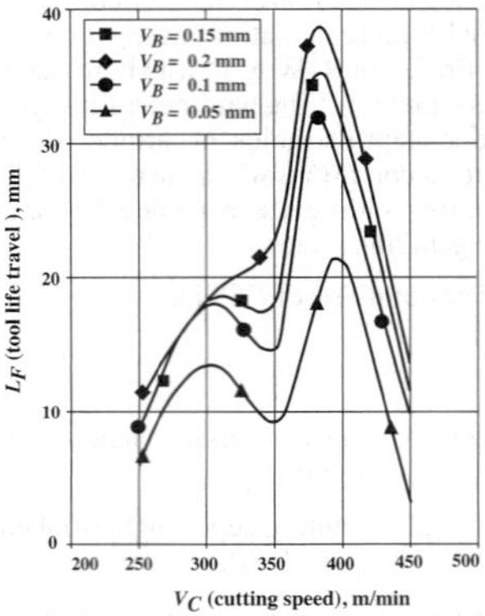


Fig. 3. Influence of Cutting Speed on Tool Life

Tool life has a maximum value at a particular setting of speed and feed. Economic and productive cutting speeds always occur on the right side of the curves in Fig. 2 and Fig. 4, which are called Taylor curves, represented by the so called Taylor's equation.

The variation of tool life with speed and feed constitute complicated relationships, illustrated in Fig. 6a, Fig. 6b, and Fig. 6c.

**Taylor's Equation.**—Taylor's equation is the most commonly used relationship between tool life *T*, and cutting speed *V*. It constitutes a straight line in a log-log plot, one line for each feed, nose radius, lead angle, or depth of cut, mathematically represented by:

$$V \times T^n = C \tag{1a}$$

where $n$ = slope of the line

$C$ = constant equal to the cutting speed for $T = 1$ minute

By transforming the equation to logarithmic axes, the Taylor lines become straight lines with slope = $n$. The constant C is the cutting speed on the horizontal *V*-axis at tool life $T = 1$ minute, expressed as follows

$$\ln V + n \times \ln T = \ln C \tag{1b}$$

For different values of feed or *ECT*, log-log plots of Equation (1a) form approximately straight lines in which the slope decreases slightly with a larger value of feed or *ECT*. In practice, the Taylor lines are usually drawn parallel to each other, i.e., the slope $n$ is assumed to be constant.

Fig. 4 illustrates the Taylor equation, tool life *T* versus cutting speed *V*, plotted in log-log coordinates, for four values of $ECT = 0.1, 0.25, 0.5$ and $0.7$ mm.

In Fig. 4, starting from the right, each *T-V* line forms a generally straight line that bends off and reaches its maximum tool life, then drops off with decreasing speed (see also Fig. 2 and Fig. 3. When operating at short tool-lives, approximately when *T* is less than 5 minutes, each line bends a little so that the cutting speed for 1 minute life becomes less than the value calculated by constant C.

The Taylor equation is a very good approximation of the right-hand side of the real tool life curve (slightly bent). The portion of the curve to the left of the maximum tool life gives shorter and shorter tool lives when decreasing the cutting speed starting from the point of maximum tool life. Operating at the maximum point of maximum tool life, or to the left of it, causes poor surface finish, high cutting forces, and sometimes vibrations.

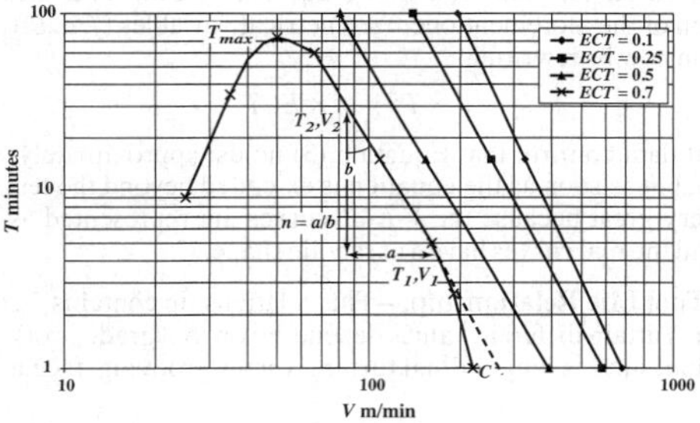


Fig. 4. Definition of Slope *n* and Constant C in Taylor's Equation

**Evaluation of Slope *n*, and Constant C.**—When evaluating the value of the Taylor slope based on wear tests, care must be taken in selecting the tool life range over which the slope is measured, as the lines are slightly curved.

The slope *n* can be found in three ways:

- Calculate *n* from the formula $n = (\ln C - \ln V)/\ln T$, reading the values of C and *V* for any value of *T* in the graph.
- Alternatively, using two points on the line, $(V_1, T_1)$ and $(V_2, T_2)$, calculate *n* using the relationship $V_1 \times T_1^n = V_2 \times T_2^n$. Then, solving for *n*,

$$n = \frac{\ln(V_1/V_2)}{\ln(T_2/T_1)}$$

- Graphically, *n* may be determined from the graph by measuring distances *a* and *b* using a mm scale, and *n* is the ratio of *a* and *b*, thus, $n = a/b$

*Example:* Using Fig. 4 and a given value of *ECT*= 0.7 mm, calculate the slope and constant of the Taylor line.

On the Taylor line for *ECT*= 0.7, locate points corresponding to tool-lives $T_1 = 15$ minutes and $T_2 = 60$ minutes. Read off the associated cutting speeds as, approximately, $V_1 = 110$ m/min and $V_2 = 65$ m/min.

The slope *n* is then found to be $n = \ln(110/65)/\ln(60/15) = 0.38$

The constant C can be then determined using the Taylor equation and either point $(T_1, V_1)$ or point $(T_2, V_2)$, with equivalent results, as follows:

$$C = V \times T^n = 110 \times 15^{0.38} = 65 \times 60^{0.38} = 308 \text{ m/min (1027 fpm)}$$

**The Generalized Taylor Equation.**—The above calculated slope and constant C define tool life at one particular value of feed *f*, depth of cut *a*, lead angle *LA*, nose radius *r*, and other relevant factors.

The generalized Taylor equation includes these parameters and is written

$$T^n = A \times f^m \times a^p \times LA^q \times r^s \qquad (2)$$

where *A* = area; and, *n*, *m*, *p*, *q*, and *s* = constants.

There are two problems with the generalized equation: 1) a great number of tests have to be run in order to establish the constants *n*, *m*, *p*, *q*, *s*, etc.; and 2) the accuracy is not very good because Equation (2) yields straight lines when plotted versus *f*, *a*, *LA*, and *r*, when in reality, they are parabolic curves.

*The Generalized Taylor Equation Using Equivalent Chip Thickness (ECT):* Due to the compression of the aforementioned geometrical variables ($f, a, LA, r$, etc.) into $ECT$, Equation (2) can now be rewritten:

$$V \times T^n = A \times ECT^m \tag{3}$$

Experimental data confirm that Equation (3) holds, approximately, within the range of the test data, but as soon as the equation is extended beyond the test results, the error can become very great because the *V-ECT* curves are represented as straight lines by Equation (3) and the real curves have a parabolic shape.

**The Colding Tool Life Relationship.—**This relationship contains 5 constants, $H, K, L, M$, and $N_0$, which attain different values depending on tool grade, work material, and the type of operation, such as longitudinal turning versus grooving, face milling versus end milling, etc.

This tool life relationship is proven to describe, with reasonable accuracy, how tool life varies with $ECT$ and cutting speed for any metal cutting and grinding operation. It is expressed mathematically as follows either as a generalized Taylor equation (4a), or, in logarithmic coordinates (4b):

$$V \times T^{(N_0 - L \times \ln ECT)} \times ECT^{\left(-\frac{H}{2M} + \frac{\ln ECT}{4M}\right)} = e^{\left(K - \frac{H}{4M}\right)} \tag{4a}$$

$$y = K - \frac{x - H}{4M} - z(N_0 - L_x) \tag{4b}$$

where $x = \ln ECT$ $\quad y = \ln V$ $\quad z = \ln T$

$M$ = vertical distance between maximum point of cutting speed $(ECT_H, V_H)$ for $T = 1$ minute and speed $V_G$ at point $(ECT_G, V_G)$, as shown in Fig. 5.

$2M$ = horizontal distance between point $(ECT_H, V_G)$ and point $(V_G, ECT_G)$

$H$ and $K$ = logarithms of coordinates of maximum speed point $(ECT_H, V_H)$ at tool life $T =$ 1 minute, thus $H = \ln(ECT_H)$ and $K = \ln(V_H)$

$N_0$ and $L$ = variation of Taylor slope $n$ with $ECT$: $n = N_0 - L \times \ln(ECT)$

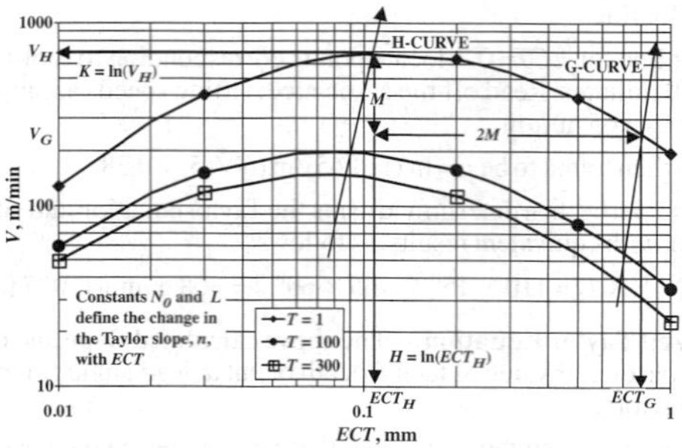


Fig. 5. Definitions of the Constants $H, K, L, M$, and $N_0$ for Tool Life Equation in the *V-ECT* Plane with Tool Life Constant

The constants $L$ and $N_0$ are determined from the slopes $n_1$ and $n_2$ of two Taylor lines at $ECT_1$ and $ECT_2$, and the constant $M$ from three *V-ECT* values at any constant tool life. Constants $H$ and $K$ are then solved using the tool life equation with the above-calculated values of $L, N_0$ and $M$.

**The *G*- and *H*-Curves.—**The *G*-curve defines the longest possible tool life for any given metal removal rate, *MRR*, or specific metal removal rate, *SMRR*. It also defines the point where the total machining cost is minimum, after the economic tool life $T_E$, or optimal tool life $T_O$, has been calculated, see *Optimization Models, Economic Tool Life when Feed is Constant* starting on page 1201.

The tool life relationship is depicted in the three planes: *T-V*, where *ECT* is the plotted parameter (the Taylor plane); *T-ECT*, where *V* is plotted; and, *V-ECT*, where *T* is a parameter. The latter plane is the most useful because the optimal cutting conditions are more readily understood when viewing in the *V-ECT* plane. Fig. 6a, Fig. 6b, and Fig. 6c show how the tool life curves look in these three planes in log-log coordinates.

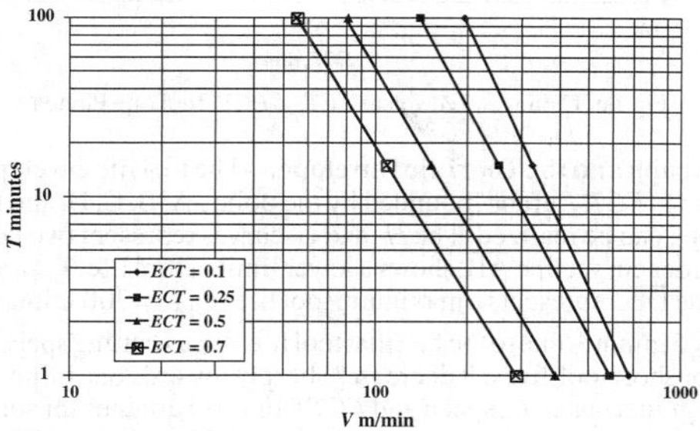


Fig. 6a. Tool Life versus Cutting Speed *T-V*, *ECT* Plotted

Fig. 6a shows the Taylor lines, and Fig. 6b illustrates how tool life varies with *ECT* at different values of cutting speed, and shows the *H*-curve. Fig. 6c illustrates how cutting speed varies with *ECT* at different values of tool life. The *H*- and *G*-curves are also drawn in Fig. 6c.

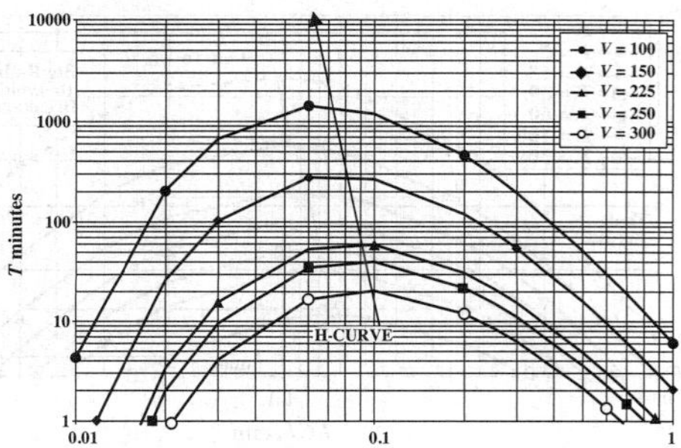


Fig. 6b. Tool Life versus *ECT*, *T-ECT*, Cutting Speed Plotted

A simple and practical method to ascertain that machining is not done to the left of the *H*-curve is to examine the chips. When *ECT* is too small, about 0.03-0.05 mm, the chips tend to become irregular and show up more or less as dust.

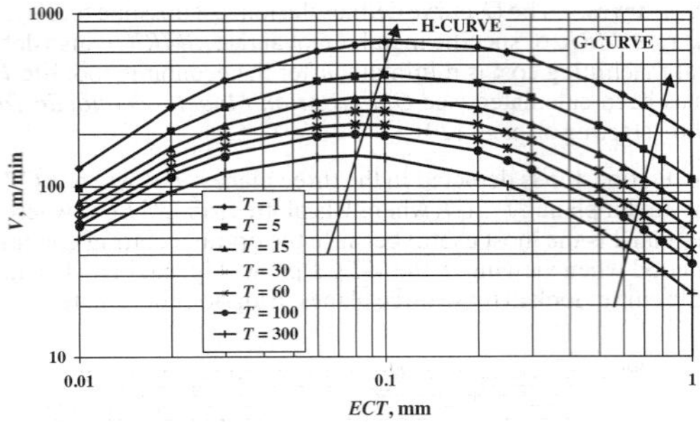


Fig. 6c. Cutting Speed versus *ECT*, *V-ECT*, Tool Life Plotted

**The *V-ECT-T* Graph and the Tool Life Envelope.—**The tool life envelope, in Fig. 7, is an area laid over the *V-ECT-T* graph, bounded by the points A, B, C, D, and E, within which successful cutting can be realized. The *H*- and *G*-curves represent two borders, lines $\overline{AE}$ and $\overline{BC}$. The border curve, line $\overline{AB}$, shows a lower limit of tool life, $T_{MIN} = 5$ minutes, and border curve, line $\overline{DE}$, represents a maximum tool life, $T_{MAX} = 300$ minutes.

$T_{MIN}$ is usually 5 minutes due to the fact that tool life versus cutting speed does not follow a straight line for short tool-lives; it decreases sharply towards one minute tool life. $T_{MAX}$ varies with tool grade, material, speed and *ECT* from 300 minutes for some carbide tools to 10,000 minutes for diamond tools or diamond grinding wheels, although systematic studies of maximum tool-lives have not been conducted.

Sometimes the metal cutting system cannot utilize the maximum values of the *V-ECT-T* envelope, that is, cutting at optimum *V-ECT* values along the *G*-curve, due to machine power or fixture constraints, or vibrations. Maximum *ECT* values, $ECT_{MAX}$, are related to the strength of the tool material and the tool geometry, and depend on the tool grade and material selection, and require a relatively large nose radius.

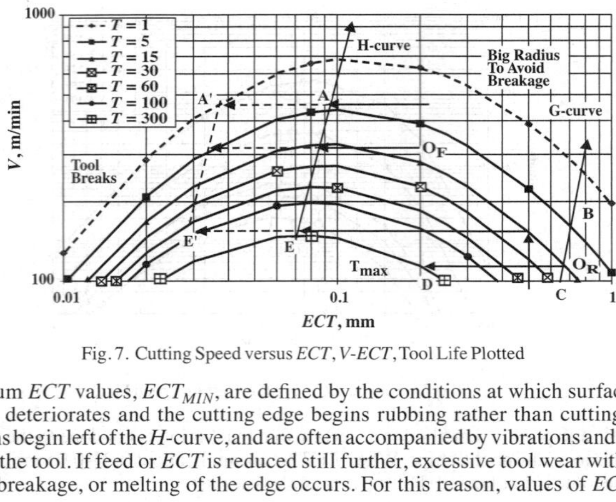


Fig. 7. Cutting Speed versus *ECT*, *V-ECT*, Tool Life Plotted

Minimum *ECT* values, $ECT_{MIN}$, are defined by the conditions at which surface finish suddenly deteriorates and the cutting edge begins rubbing rather than cutting. These conditions begin left of the *H*-curve, and are often accompanied by vibrations and built-up edges on the tool. If feed or *ECT* is reduced still further, excessive tool wear with sparks and tool breakage, or melting of the edge occurs. For this reason, values of *ECT* lower than approximately 0.03 mm should not be allowed. In Fig. 7, the $ECT_{MIN}$ boundary is indicated by contour line $\overline{A' E'}$.

In milling, the minimum feed/tooth depends on the ratio *ar*/*D*, of radial depth of cut *ar*, and cutter diameter *D*. For small *ar*/*D* ratios, the chip thickness becomes so small that it is necessary to compensate by increasing the feed/tooth. See *High-Speed Machining Econometrics* starting on page 1213 for more on this topic.

Fig. 7 demonstrates, in principle, minimum cost conditions for roughing at point $\mathbf{O_R}$, and for finishing at point $\mathbf{O_F}$, where surface finish or tolerances have set a limit. Maintaining the speed at $\mathbf{O_R}$, 125 m/min, and decreasing feed reaches a maximum tool life = 300 minutes at $ECT = 0.2$, and a further decrease of feed will result in shorter lives.

Similarly, starting at point X ($V = 150, ECT = 0.5, T = 15$) and reducing feed, the *H*-curve will be reached at point E ($ECT = 0.075$, $T = 300$). Continuing to the left, tool life will decrease and serious troubles occur at point E' ($ECT = 0.03$).

Starting at point $\mathbf{O_F}$ ($V = 300, ECT = 0.2, T = 15$) and reducing feed, the *H*-curve will be reached at point E ($ECT = 0.08, T = 15$). Continuing to the left, life will decrease and serious troubles occur at $ECT = 0.03$.

Starting at point X ($V = 400$, $ECT = 0.2$, $T = 5$) and reducing feed, the *H*-curve will be reached at point E ($ECT = 0.09, T = 7$). Continuing to the left, life will decrease and serious troubles occur at point A' ($ECT = 0.03$), where $T = 1$ minute.

**Cutting Forces and Chip Flow Angle.**—There are three cutting forces, illustrated in Fig. 8, that are associated with the cutting edge with its nose radius *r*, depth of cut *a*, lead angle *LA*, and feed per revolution *f*, or in milling feed per tooth $f_z$. There is one drawing for roughing and one for finishing operations.

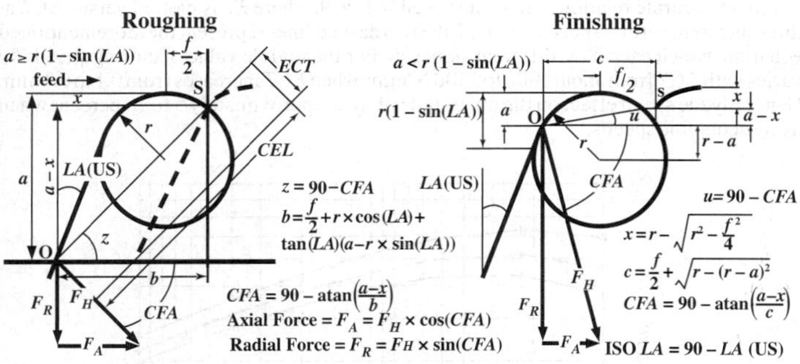


Fig. 8. Definitions of Equivalent Chip Thickness, *ECT*, and Chip Flow Angle, *CFA*.

The cutting force $F_C$, or tangential force, is perpendicular to the paper plane. The other two forces are the feed or axial force $F_A$, and the radial force $F_R$ directed towards the work piece. The resultant of $F_A$ and $F_R$ is called $F_H$. When finishing, $F_R$ is bigger than $F_A$, while in roughing $F_A$ is usually bigger than $F_R$. The direction of $F_H$, measured by the chip flow angle *CFA*, is perpendicular to the rectangle formed by the cutting edge length *CEL* and *ECT* (the product of *ECT* and CEL constitutes the cross sectional area of cut, *A*). The important task of determining the direction of $F_H$, and calculation of $F_A$ and $F_R$, are shown in the formulas given in the Fig. 8.

The method for calculating the magnitudes of $F_H$, $F_A$, and $F_R$ is described in the following. The first thing is to determine the value of the cutting force $F_C$. Approximate formulas to calculate the tangential cutting force, torque and required machining power are found in the section *ESTIMATING SPEEDS AND MACHINING POWER* starting on page 1131.

*Specific Cutting Force, Kc:* The specific cutting force, or the specific energy to cut, $Kc$, is defined as the ratio between the cutting force $F_C$ and the chip cross-sectional area $A$. Thus, $Kc = F_C \div A$ N/mm$^2$.

The value of $Kc$ decreases when $ECT$ increases, and when the cutting speed $V$ increases. Usually, $Kc$ is written in terms of its value at $ECT = 1$, called $Kc_1$, and neglecting the effect of cutting speed, thus $Kc = Kc_1 \times ECT^B$, where $B$ = slope in log-log coordinates.

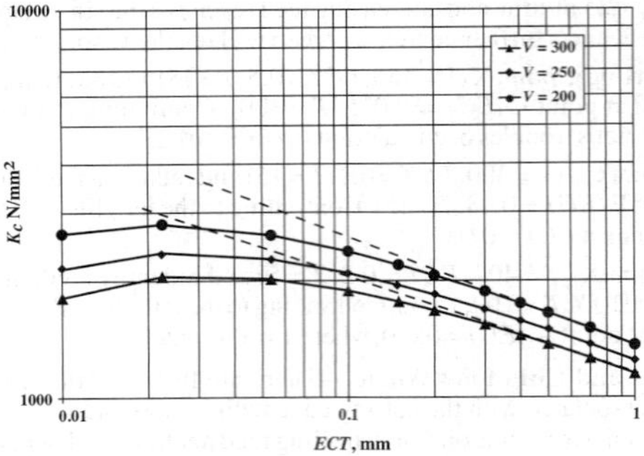


Fig. 9. *Kc* versus *ECT*, Cutting Speed Plotted

A more accurate relationship is illustrated in Fig. 9, where $Kc$ is plotted versus $ECT$ at three different cutting speeds. In Fig. 9, the two dashed lines represent the aforementioned equation, which each have different slopes, $B$. For the middle value of cutting speed, $Kc$ varies with $ECT$ from about 1900 to 1300 N/mm$^2$ when $ECT$ increases from 0.1 to 0.7 mm. Generally the speed effect on the magnitude of $Kc$ is approximately 5 to 15 percent when using economic speeds.

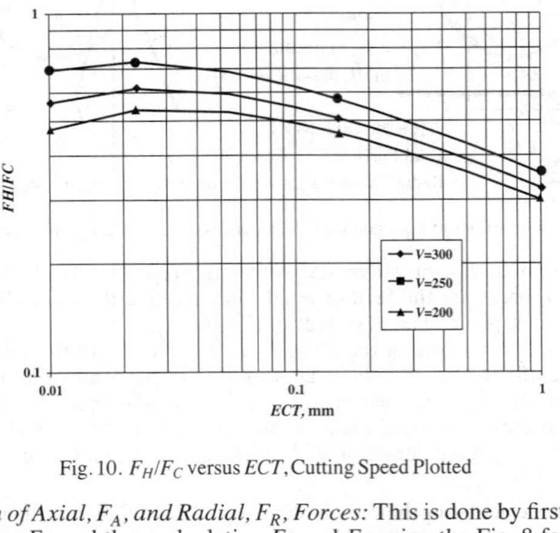


Fig. 10. $F_H/F_C$ versus *ECT*, Cutting Speed Plotted

*Determination of Axial, $F_A$, and Radial, $F_R$, Forces:* This is done by first determining the resultant force $F_H$ and then calculating $F_A$ and $F_R$ using the Fig. 8 formulas. $F_H$ is derived from the ratio $F_H/F_C$, which varies with *ECT* and speed in a fashion similar to *Kc*. Fig. 10 shows how this relationship may vary.

As seen in Fig. 10, $F_H/F_C$ is in the range 0.3 to 0.6 when *ECT* varies from 0.1 to 1 mm, and speed varies from 200 to 250 m/min using modern insert designs and grades. Hence, using reasonable large feeds $F_H/F_C$ is around 0.3–0.4 and when finishing about 0.5–0.6.

*Example:* Determine $F_A$ and $F_R$, based on the chip flow angle *CFA* and the cutting force $F_C$, in turning.

Using a value of $Kc = 1500$ N/mm$^2$ for roughing, when $ECT = 0.4$, and the cutting edge length $CEL = 5$ mm, first calculate the area $A = 0.4 \times 5 = 2$ mm$^2$. Then, determine the cutting force $F_C = 2 \times 1500 = 3000$ N, and an approximate value of $F_H = 0.5 \times 3000 = 1500$ N.

Using a value of $Kc = 1700$ N/mm$^2$ for finishing, when $ECT = 0.2$, and the cutting edge length $CEL = 2$ mm, calculate the area $A = 0.2 \times 2 = 0.4$ mm$^2$. The cutting force $F_C = 0.4 \times 1700 = 680$ N and an approximate value of $F_H = 0.35 \times 680 = 238$ N.

Fig. 8 can be used to estimate *CFA* for rough and finish turning. When the lead angle *LA* is 15 degrees and the nose radius is relatively large, an estimated value of the chip flow angle becomes about 30 degrees when roughing, and about 60 degrees in finishing. Using the formulas for $F_A$ and $F_R$ relative to $F_H$ gives:

**Roughing:**

$$F_A = F_H \times \cos(CFA) = 1500 \times \cos 30 = 1299 \text{ N}$$
$$F_R = F_H \times \sin(CFA) = 1500 \times \sin 30 = 750 \text{ N}$$

**Finishing:**

$$F_A = F_H \times \cos(CFA) = 238 \times \cos 60 = 119 \text{ N}$$
$$F_R = F_H \times \sin(CFA) = 238 \times \sin 60 = 206 \text{ N}$$

The force ratio $F_H/F_C$ also varies with the tool rake angle and increases with negative rakes. In grinding, $F_H$ is much larger than the grinding cutting force $F_C$; generally $F_H/F_C$ is approximately 2 to 4, because grinding grits have negative rakes of the order –35 to –45 degrees.

**Forces and Tool Life.—**Forces and tool life are closely linked. The ratio $F_H/F_C$ is of particular interest because of the unique relationship of $F_H/F_C$ with tool life.

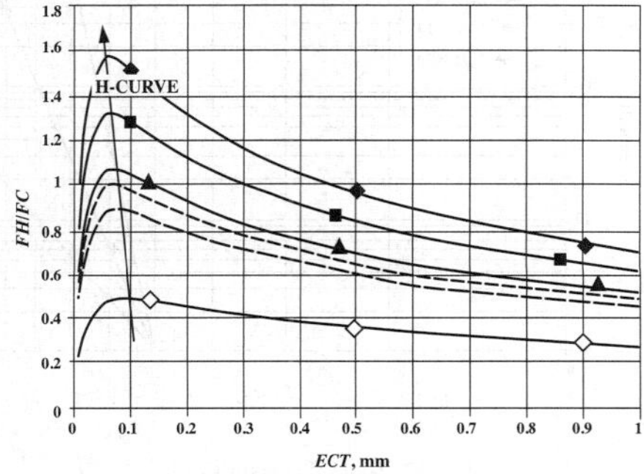


Fig. 11a. $F_H/F_C$ versus *ECT*

The results of extensive tests at Ford Motor Company are shown in Fig. 11a and Fig. 11b, where $F_H/F_C$ and tool life *T* are plotted versus *ECT* at different values of cutting speed *V*. For any constant speed, tool life has a maximum at approximately the same values of *ECT* as has the function $F_H/F_C$.

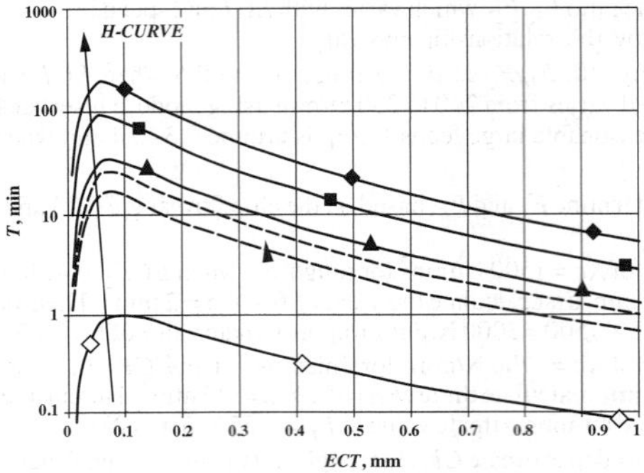


Fig. 11b. Tool Life versus *ECT*

*The Force Relationship:* Similar tests performed elsewhere confirm that the $F_H/F_C$ function can be determined using the 5 tool life constants ($H, K, M, L, N_0$) introduced previously, and a new constant ($L_F/L$).

$$\ln\left(\frac{1}{a}\cdot\frac{F_H}{F_C}\right) = \frac{K - y - \dfrac{(x-H)^2}{4M}}{\dfrac{L_F}{L}(N_0 - Lx)} \tag{5}$$

The constant $a$ depends on the rake angle; in turning, $a$ is approximately 0.25 to 0.5 and $L_F/L$ is 10 to 20. $F_C$ attains it maximum values versus *ECT* along the *H*-curve, when the tool life equation has maxima, and the relationships in the three force ratio planes look very similar to the tool life functions shown in the tool life planes in Fig. 6a, Fig. 6b, and Fig. 6c.

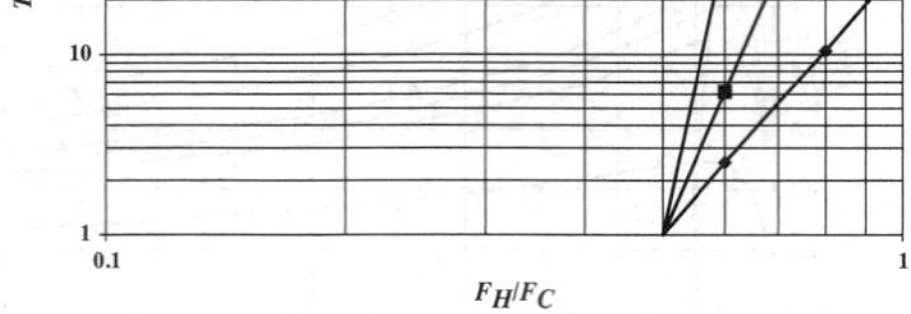


Fig. 12. Tool Life versus $F_H/F_C$

Tool life varies with $F_H/F_C$ with a simple formula according to Equation (5) as follows:

$$T = \left(\frac{F_H}{aF_C}\right)^{\frac{L_F}{L}}$$

where $L$ is the constant in the tool life equation, Equation (4a) or (4b), and $L_F$ is the corresponding constant in the force ratio equation, Equation (5). In Fig. 12 this function is plotted for $a = 0.5$ and for $L_F/L = 5, 10,$ and 20.

Accurate calculations of aforementioned relationships require elaborate laboratory tests, or better, the design of a special test and follow-up program for parts running in the ordinary production. A software machining program, such as Colding International Corp. *COMP* program can be used to generate the values of all three forces, torque, and power requirements both for sharp and worn tools.

**Surface Finish and Tool Life.**—It is well known that the surface finish in turning decreases with a bigger tool nose radius and increases with feed; usually it is assumed that the $R_a$ (or Ra, for roughness average) value increases with the square of the feed per revolution, and decreases inversely with increasing size of the nose radius. This formula, derived from simple geometry, gives rise to great errors. In reality, the relationship is more complicated because the tool geometry must be taken into account, and the work material and the cutting conditions also have a significant influence.

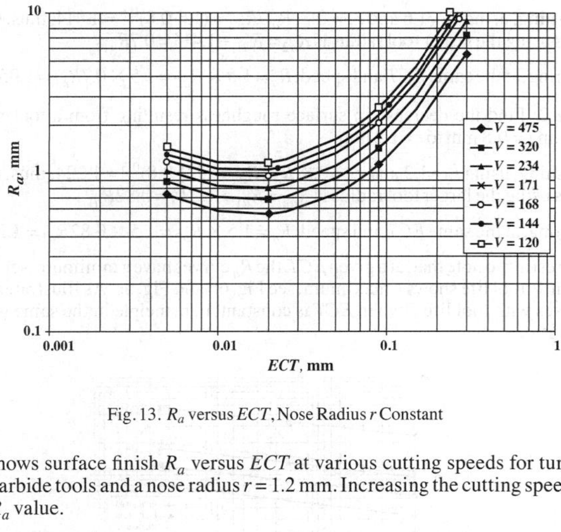


Fig. 13. $R_a$ versus *ECT*, Nose Radius *r* Constant

Fig. 13 shows surface finish $R_a$ versus *ECT* at various cutting speeds for turning cast iron with carbide tools and a nose radius $r = 1.2$ mm. Increasing the cutting speed leads to a smaller $R_a$ value.

Fig. 14 shows how the finish improves when the tool nose radius, *r*, increases at a constant cutting speed (168 m/min) in cutting nodular cast iron.

In Fig. 15, $R_a$ is plotted versus *ECT* with cutting speed *V* for turning a 4310 steel with carbide tools, for a nose radius $r = 1.2$ mm, illustrating that increasing the speed also leads to a smaller $R_a$ value for steel machining.

A simple rule of thumb for the effect of increasing nose radius *r* on decreasing surface finish $R_a$, regardless of the ranges of *ECT* or speeds used, albeit within common practical values, is as follows. In finishing,

$$\frac{R_{a1}}{R_{a2}} = \left(\frac{r_2}{r_1}\right)^{0.5} \tag{6}$$

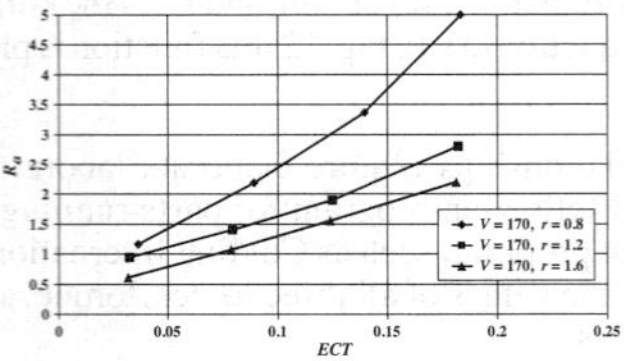


Fig. 14. $R_a$ versus *ECT*, Cutting Speed Constant, Nose Radius *r* Varies

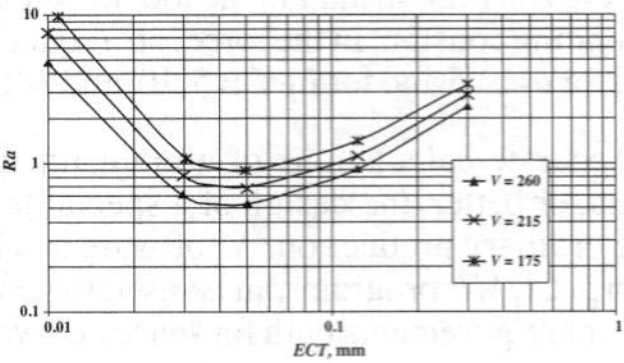


Fig. 15. $R_a$ versus *ECT*, Cutting Speed Varies, Nose Radius *r* Constant

In roughing, multiply the finishing values found using Equation (6) by 1.5, thus, $R_{a(\text{Rough})} = 1.5 \times R_{a(\text{Finish})}$ for each *ECT* and speed.

*Example 1:* Find the decrease in surface roughness resulting from a tool nose radius change from $r = 0.8$ mm to $r = 1.6$ mm in finishing. Also, find the comparable effect in roughing.

For finishing, using $r_2 = 1.6$ and $r_1 = 0.8$, $R_{a1}/R_{a2} = (1.6/0.8)^{0.5} = 1.414$, thus, the surface roughness using the larger tool radius is $R_{a2} = R_{a1} \div 1.414 = 0.7R_{a1}$

In roughing, at the same *ECT* and speed, $R_a = 1.5 \times R_{a2} = 1.5 \times 0.7R_{a1} = 1.05R_{a1}$.

*Example 2:* Find the decrease in surface roughness resulting from a tool nose radius change from $r = 0.8$ mm to $r = 1.2$ mm.

For finishing, using $r_2 = 1.2$ and $r_1 = 0.8$, $R_{a1}/R_{a2} = (1.2/0.8)^{0.5} = 1.224$, thus, the surface roughness using the larger tool radius is $R_{a2} = R_{a1} \div 1.224 = 0.82R_{a1}$.

In roughing, at the same *ECT* and speed, $R_a = 1.5 \times R_{a2} = 1.5 \times 0.82R_{a1} = 1.23R_{a1}$.

It is interesting to note that, at a given *ECT*, the $R_a$ curves have a minimum, see Fig. 13 and Fig. 15, while tool life shows a maximum, see Fig. 6b and Fig. 6c. As illustrated in Fig. 16, $R_a$ increases with tool life *T* when *ECT* is constant, in principle in the same way as does the force ratio.

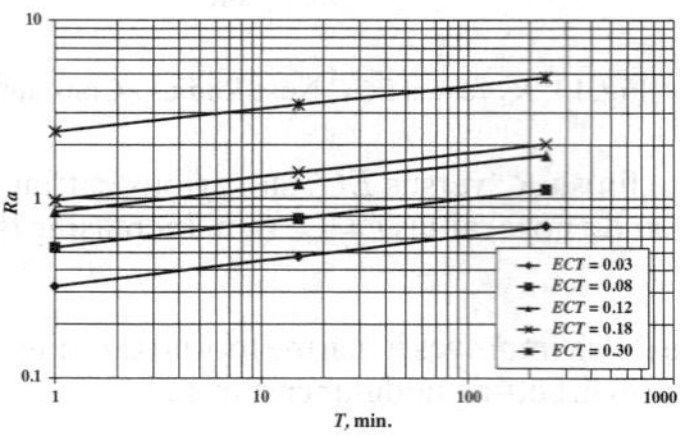


Fig. 16. $R_a$ versus *T*, Holding *ECT* Constant

*The Surface Finish Relationship:* $R_a$ is determined using the same type of mathematical relationship as tool life and force calculations:

$$y = K_{Ra} - \frac{x - H_{Ra}^2}{4M_{Ra}} - (N_{0Ra} - L_{Ra})\ln(R_a)$$

where $K_{Ra}$, $H_{Ra}$, $M_{Ra}$, $N_{0Ra}$, and $L_{Ra}$ are the five surface finish constants.

**Shape of Tool Life Relationships for Turning, Milling, Drilling and Grinding Operations—Overview.—**A summary of the general shapes of tool life curves (*V-ECT-T* graphs) for the most common machining processes, including grinding, is shown in double logarithmic coordinates in Fig. 17a through Fig. 17h.

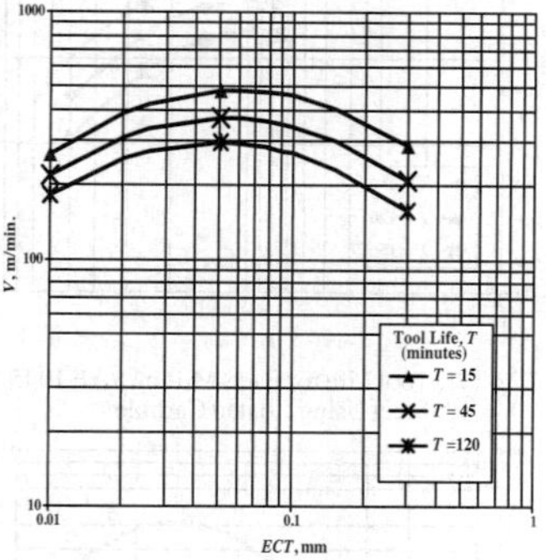


Fig. 17a. Tool Life for Turning Cast Iron Using Coated Carbide

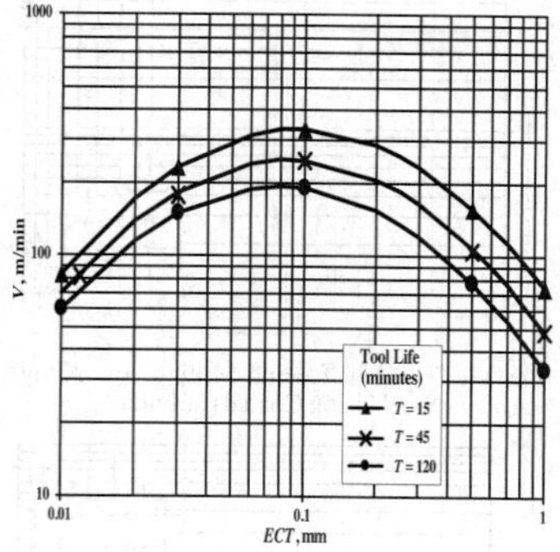


Fig. 17b. Tool Life for Turning Low-Alloy Steel Using Coated Carbide

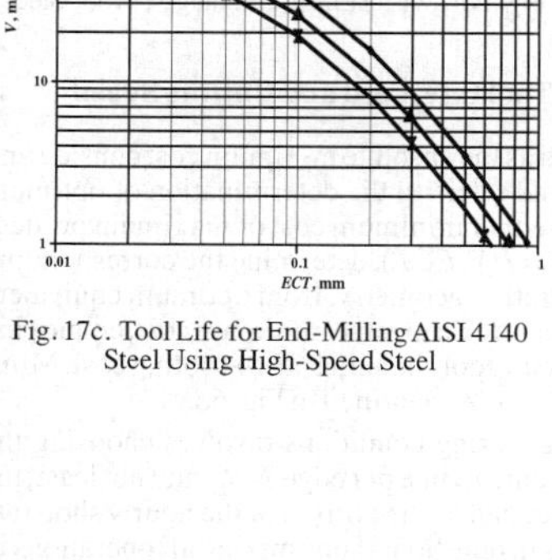


Fig. 17c. Tool Life for End-Milling AISI 4140 Steel Using High-Speed Steel

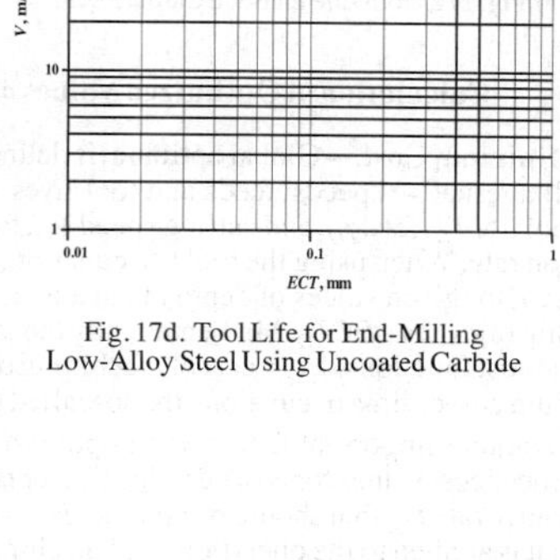


Fig. 17d. Tool Life for End-Milling Low-Alloy Steel Using Uncoated Carbide

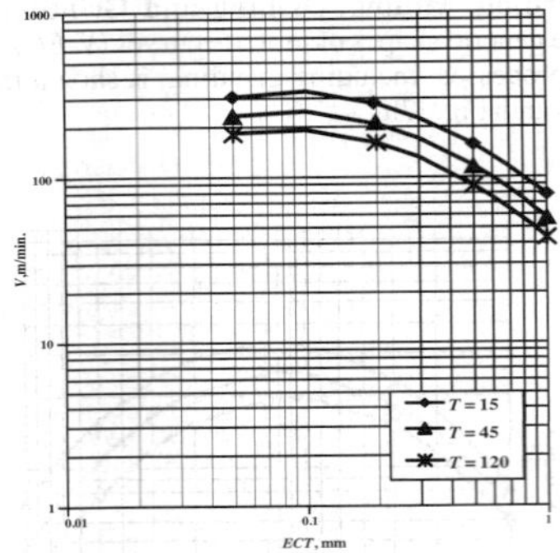


Fig. 17e. Tool Life for End-Milling Low-Alloy Steel Using Coated Carbide

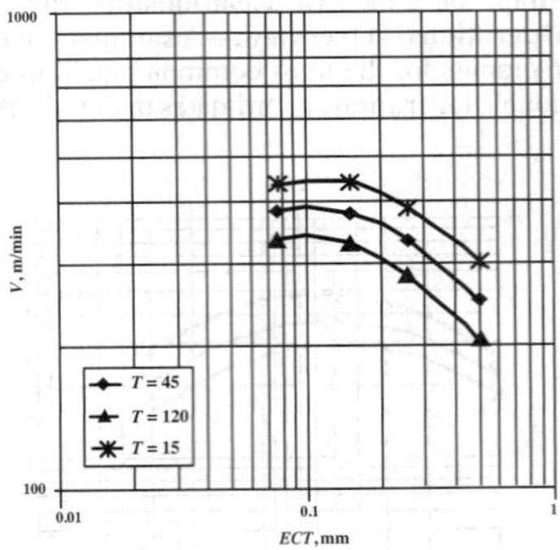


Fig. 17f. Tool Life for Face-Milling SAE 1045 Steel Using Coated Carbide

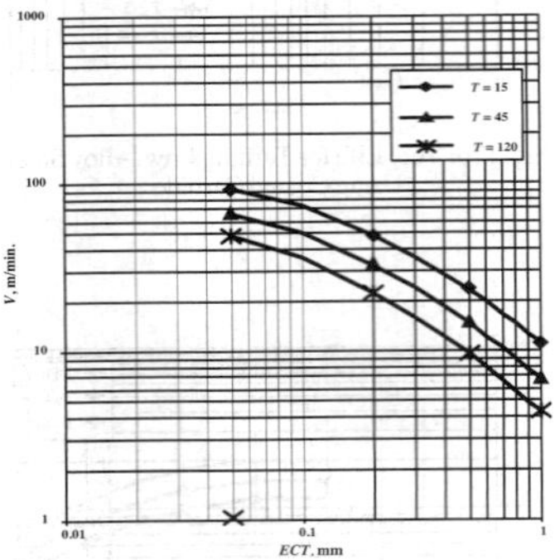


Fig. 17g. Tool Life for Solid Carbide Drill

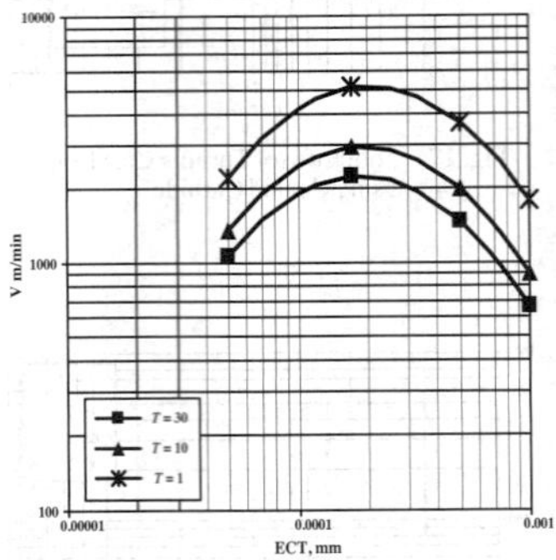


Fig. 17h. Wheel Life in Grinding M4 Tool-Steel

## Calculation of Optimized Values of Tool Life, Feed and Cutting Speed

**Minimum Cost.—** Global optimum is defined as the absolute minimum cost considering all alternative speeds, feeds and tool-lives, and refers to the determination of optimum tool life $T_O$, feed $f_O$, and cutting speed $V_O$, for either minimum cost or maximum production rate. When using the tool life equation, $T = f(V, ECT)$, determine the corresponding feed, for given values of depth of cut and operation geometry, from optimum equivalent chip thickness, $ECT_O$. Mathematically the task is to determine minimum cost, employing the cost function $C_{TOT}$ = cost of machining time + tool changing cost + tooling cost. Minimum cost optima occur along the so-called *G*-curve, identified in Fig. 6c.

Another important factor when optimizing cutting conditions involves choosing the proper cost values for cost per edge $C_E$, replacement time per edge $T_{RPL}$, and not least, the hourly rate $H_R$ that should be applied. $H_R$ is defined as the portion of the hourly shop rate that is applied to the operations and machines in question. If optimizing all operations in the portion of the shop for which $H_R$ is calculated, use the full rate; if only one machine is involved, apply a lower rate, as only a portion of the general overhead rate should be used, otherwise the optimum, and anticipated savings, are erroneous.

**Production Rate.**—The production rate is defined as the cutting time or the metal removal rate, corrected for the time required for tool changes, but neglecting the cost of tools.

The result of optimizing production rate is a shorter tool life, higher cutting speed, and a higher feed compared to minimum cost optimization, and the tooling cost is considerably higher. Production rate optima also occur along the *G*-curve.

**The Cost Function.**—There are a number of ways the total machining cost $C_{TOT}$ can be plotted, for example, versus feed, *ECT*, tool life, cutting speed or other parameter. In Fig. 18a, cost for a face milling operation is plotted versus cutting time, holding feed constant, and using a range of tool-lives, *T*, varying from 1 to 240 minutes.

| | $t_c$ | $C_{TOOL}$ | $C_{TOT}$ | $T$ | $V$ |
|---|---|---|---|---|---|
| | 5.85 | 0.487 | 0.569 | 1 | 598 |
| | 6.91 | 0.192 | 0.288 | 3 | 506 |
| | 7.47 | 0.125 | 0.228 | 5 | 468 |
| | 8.30 | 0.069 | 0.185 | 10 | 421 |
| | 8.83 | 0.049 | 0.172 | 15 | 396 |
| **Minimum cost** | **9.81** | **0.027** | **0.164** | **30** | **356** |
| | 10.91 | 0.015 | 0.167 | 60 | 321 |
| | 11.60 | 0.011 | 0.172 | 90 | 302 |
| | 12.12 | 0.008 | 0.177 | 120 | 289 |
| | 13.47 | 0.005 | 0.192 | 240 | 260 |

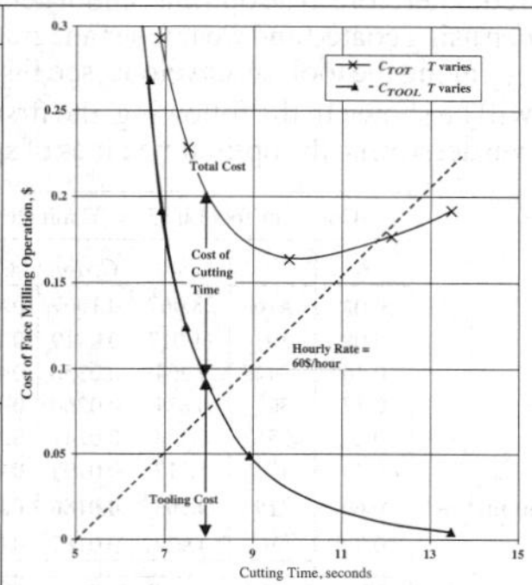


Fig. 18a. Variation of Tooling Cost $C_{TOOL}$, and Total Cost $C_C$, with Cutting Time $t_c$, Including Minimum Cost Cutting Time

The tabulated values show the corresponding cutting speeds determined from the tool life equation, and the influence of tooling on total cost. Tooling cost, $C_{TOOL}$ = sum of tool cost + cost of replacing worn tools, decreases the longer the cutting time, while the total cost, $C_{TOT}$, has a minimum at around 10 seconds of cutting time. The dashed line in the graph represents the cost of machining time: the product of hourly rate $H_R$, and the cutting time $t_c$ divided by 60. The slope of the line defines the value of $H_R$.

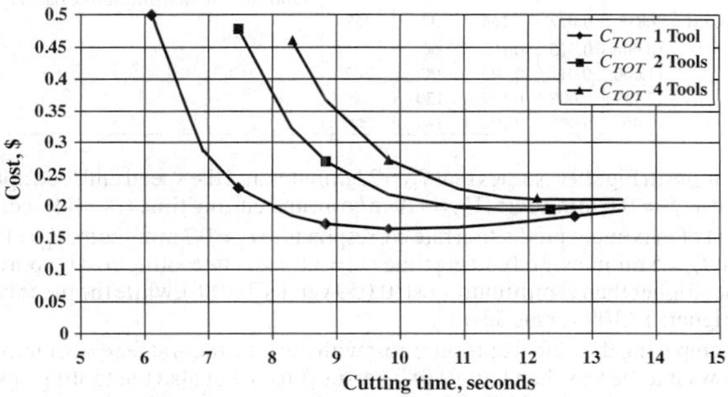


Fig. 18b. Total Cost versus Cutting Time for Simultaneously Cutting with 1, 2, and 4 Tools

The cutting time for minimum cost varies with the ratio of tooling cost and $H_R$. Minimum cost moves towards a longer cutting time (longer tool life) when either the price of the tooling increases, or when several tools cut simultaneously on the same part. In Fig. 18b, this is exemplified by running 2 and 4 cutters simultaneously on the same work piece, at the same feed and depth of cut, and with a similar tool as in Fig. 18a. As the tooling cost goes up 2 and 4 times, respectively, and $H_R$ is the same, the total costs curves move up, but also moves to the right, as do the points of minimum cost and optimal cutting times. This means that going somewhat slower, with more simultaneously cutting tools, is advantageous.

**Global Optimum.**—Usually, global optimum occurs for large values of feed, heavy roughing, and in many cases the cutting edge will break trying to apply the large feeds required. Therefore, true optima cannot generally be achieved when roughing, in particular when using coated and wear resistant grades; instead, use the maximum values of feed, $ECT_{max}$, along the tool life envelope, see Fig. 7.

As will be shown in the following, the first step is to determine the optimal tool life $T_O$, and then determine the optimum values of speeds and feeds.

| | Optimum Tool Life $T_O$ = 22 minutes | | | | |
|---|---|---|---|---|---|
| | $f_z$ | $V_{22}$ | $t_c$, sec. | $C_{TOOL}$ | $C_{TOT}$ |
| | 0.03 | 416 | 28.067 | 0.1067 | 0.4965 |
| | 0.08 | 397 | 11.017 | 0.0419 | 0.1949 |
| | 0.10 | 374 | 9.357 | 0.0356 | 0.1655 |
| | 0.17 | 301 | 6.831 | 0.0260 | 0.1208 |
| | 0.20 | 276 | 6.334 | 0.0241 | 0.1120 |
| | 0.40 | 171 | 5.117 | 0.0194 | 0.0905 |
| **Minimum Cost** | **0.60** | **119** | **4.903** | **0.0186** | **0.0867** |
| | 0.70 | 91 | 4.924 | 0.0187 | 0.0871 |

| Maximum Production Rate, $T$ = 5 minutes | | | | |
|---|---|---|---|---|
| $f_z$ | $V_5$ | $t_c$ | $C_{TOOL}$ | $C_{TOT}$ |
| 0.7 | 163 | 3.569 | 0.059 | 0.109 |

| | $T$ varies between 1 and 240 minutes | | | | |
|---|---|---|---|---|---|
| | $f_z$ = 0.10 | | $ECT$ = 0.26 | | |
| | $t_c$ sec. | $C_{TOOL}$ | $C_{TOT}$ | $T$ | $V$ |
| | 5.850 | 0.487 | 0.569 | 1 | 598 |
| | 6.914 | 0.192 | 0.288 | 3 | 506 |
| | 7.473 | 0.125 | 0.228 | 5 | 468 |
| | 8.304 | 0.069 | 0.185 | 10 | 421 |
| | 8.832 | 0.049 | 0.172 | 15 | 396 |
| **Minimum Cost** | **9.815** | **0.027** | **0.164** | **30** | **357** |
| | 10.906 | 0.015 | 0.167 | 60 | 321 |
| | 11.600 | 0.011 | 0.172 | 90 | 302 |
| | 12.119 | 0.008 | 0.177 | 120 | 289 |
| | 13.467 | 0.005 | 0.192 | 240 | 260 |

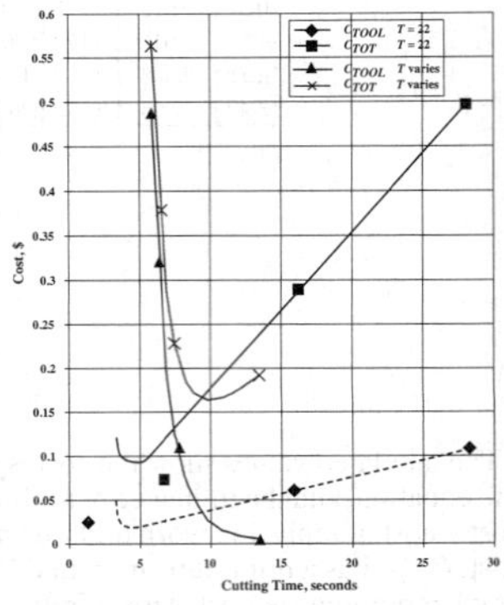


Fig. 19. Variation of Tooling and Total Cost with Cutting Time, Comparing Global Optimum with Minimum Cost at $f_z = 0.1$ mm

The example in Fig. 19 assumes that $T_O = 22$ minutes and the speed and feed optima were calculated as $f_O = 0.6$ mm/tooth, $V_O = 119$ m/min, and cutting time $t_{cO} = 4.9$ sec.

The point of maximum production rate corresponds to $f_O = 0.7$ mm/tooth, $V_O = 163$ m/min, at tool life $T_O = 5$ minutes, and cutting time $t_{cO} = 3.6$ sec. The tooling cost is approximately three times higher than at minimum cost (0.059 versus 0.0186), while the piece cost is only slightly higher: $0.109 versus $0.087.

When comparing the global optimum cost with the minimum at feed = 0.1 mm/tooth the graph shows it to be less than half (0.087 versus 0.164), but also the tooling cost is about 1/3 lower (0.0186 versus 0.027). The reason why tooling cost is lower depends on the tooling cost term $t_c \times C_E/T$ (see *Calculation of Cost of Cutting and Grinding Operations* on page 1206). In this example, cutting times $t_c$ = 4.9 and 9.81 seconds, at $T$ = 22 and 30 minutes respectively, and the ratios are proportional to 4.9/22 = 0.222 and 9.81/30 = 0.327 respectively.

The portions of the total cost curve for shorter cutting times than minimum correspond to using speeds and feeds to the right of the *G*-curve, and those on the other side are left of this curve.

**Optimization Models, Economic Tool Life when Feed is Constant.—**Usually, optimization is performed versus the parameters tool life and cutting speed, keeping feed at a constant value. The cost of cutting as function of cutting time is a straight line with the slope = $H_R$ = hourly rate. This cost is independent of the values of tool change and tooling. Adding the cost of tool change and tooling, gives the variation of total cutting cost which shows a minimum with cutting time that corresponds to an economic tool life, $T_E$. Economic tool life represents a local optima (minimum cost) at a given constant value of feed, feed/tooth, or *ECT*.

Using the Taylor Equation: $V \times T = C$ and differentiating $C_{TOT}$ with respect to $T$ yields:

**Economic tool life:**

$$T_E = T_V \times (1/n - 1), \text{minutes}$$

**Economic cutting speed:**

$$V_E = C/T_E{}^n, \text{m/min, or sfm}$$

In these equations, $n$ and C are constants in the Taylor equation for the given value of feed. Values of Taylor slopes, n, are estimated using the speed and feed Table 1 through Table 23 starting on page 1083 and Table 5b on page 1091 for turning, and Table 15e on page 1115 for milling and drilling; $T_V$ is the equivalent tooling-cost time. $T_V = T_{RPL} + 60 \times C_E \div H_R$, minutes, where $T_{RPL}$ = time for replacing a worn insert, or a set of inserts in a milling cutter or inserted drill, or a twist drill, reamer, thread chaser, or tap. $T_V$ is described in detail, later; $C_E$ = cost per edge, or set of edges, or cost per regrind including amortized price of tool; and $H_R$ = hourly shop rate, or that rate that is impacted by the changes of cutting conditions .

In two dimensions, Fig. 20a shows how economic tool life varies with feed per tooth. In this figure, the equivalent tooling-cost time $T_V$ is constant, however the Taylor constant $n$ varies with the feed per tooth.

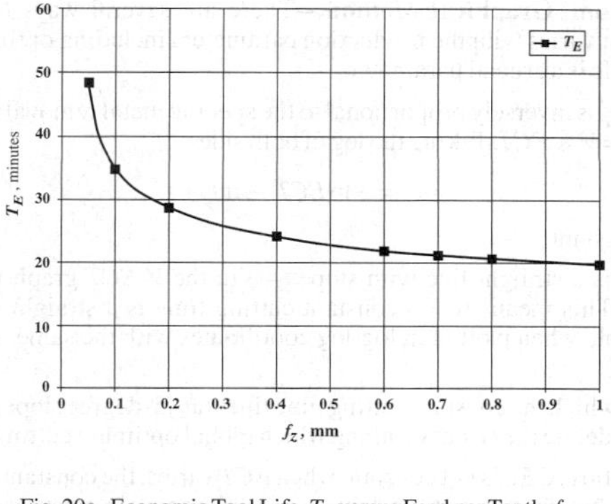


Fig. 20a. Economic Tool Life, $T_E$ versus Feed per Tooth, $f_z$

Economic tool life increases with greater values of $T_V$, either when $T_{RPL}$ is longer, or when cost per edge $C_E$ is larger for constant $H_R$, or when $H_R$ is smaller and $T_{RPL}$ and $C_E$ are unchanged. For example, when using an expensive machine (which makes $H_R$ bigger) the value of $T_V$ gets smaller, as does the economic tool life, $T_E = T_V \times (1/n - 1)$. Reducing $T_E$ results in an increase in the economic cutting speed, $V_E$. This means raising the cutting speed and illustrates the importance, in an expensive system, of utilizing the equipment better by using more aggressive machining data.

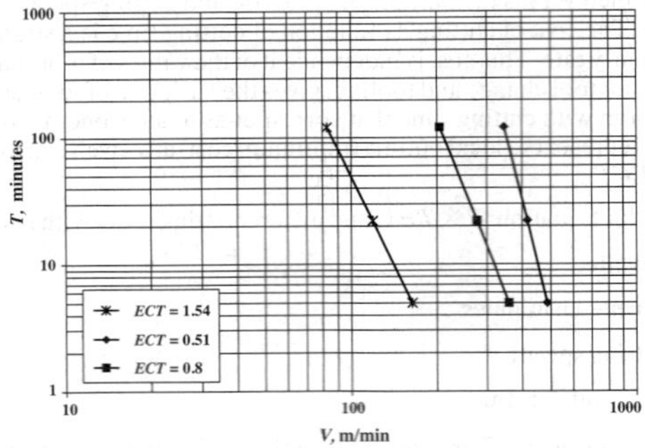


Fig. 20b. Tool Life versus Cutting Speed, Constant *ECT*

As shown in Fig. 20a for a face milling operation, economic tool life $T_E$ varies considerably with feed/tooth $f_z$, in spite of the fact that the Taylor lines have only slightly different slopes ($ECT$=0.51, 0.6, 1.54), as shown in Fig. 20b. The calculation is based on the following cost data: $T_V = 6$, hourly shop rate $H_R$ = \$60/hour, cutter diameter $D$ = 125 mm with number of teeth $z$ = 10, and radial depth of cut $ar$ = 40 mm.

The conclusion relating to the determination of economic tool life is that both hourly rate $H_R$ and slope $n$ must be evaluated with reasonable accuracy in order to arrive at good values. However, the method shown will aid in setting the trend for general machining economics evaluations.

**Global Optimum, Graphical Method.**—There are several ways to demonstrate in graphs how cost varies with the production parameters including optimal conditions. In all cases, tool life is a crucial parameter.

Cutting time $t_c$ is inversely proportional to the specific metal removal rate, $SMRR = V \times ECT$, thus, $1/t_c = V \times ECT$. Taking the log of both sides,

$$\ln V = -\ln ECT - \ln t_c + C \tag{7}$$

where $C$ is a constant.

Equation (7) is a straight line with slope (–1) in the *V-ECT* graph when plotted in a log-log graph. This means that a constant cutting time is a straight 45-degree line in the *V-ECT* graph, when plotted in log-log coordinates with the same scale on both axes (a square graph).

The points at which the constant cutting time lines (at 45-degree slope) are tangent to the tool life curves define the *G*-curve, along which global optimum cutting occurs.

*Note:* If the ratio *a*/*CEL* is not constant when *ECT* varies, the constant cutting time lines are not straight, but the cutting time deviation is quite small in most cases.

In the *V-ECT* graph, Fig. 21, 45-degree lines have been drawn tangent to each tool life curve: $T = 1, 5, 15, 30, 60, 100$ and $300$ minutes. The tangential points define the *G*-curve, and the 45-degree lines represent different constant cutting times: 1, 2, 3, 10 minutes, etc. Following one of these lines and noting the intersection points with the tool life curves $T = 1, 5$, etc., many different speed and feed combinations can be found that will give the same cutting time. As tool life gets longer (tooling cost is reduced), *ECT* (feed) increases but the cutting speed has to be reduced.

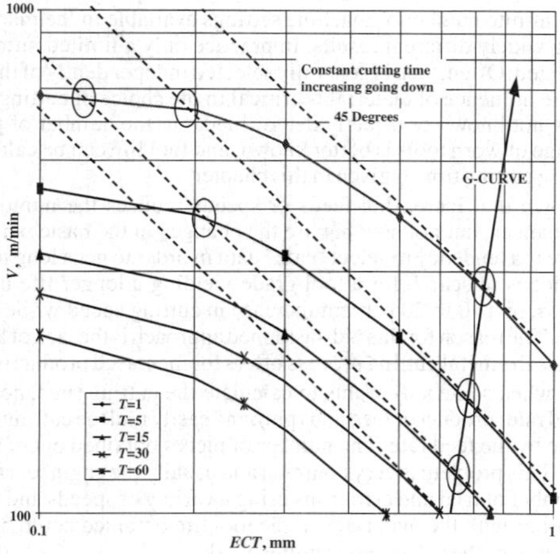


Fig. 21. Constant Cutting Time in the *V-ECT* Plane, Tool Life Constant

**Global Optimum, Mathematical Method.**—Global optimization is the search for extremum of $C_{TOT}$ for the three parameters: *T*, *ECT*, and *V*. The results, in terms of the tool life equation constants, are:

**Optimum tool life:**

$$T_O = T_V \times \left(\frac{1}{n_O} - 1\right)$$
$$n_O = 2M \times (L \times \ln T_O)^2 + 1 - N_0 + L \times (2M + H)$$

where $n_O$ = slope at optimum *ECT*.

The same approach is used when searching for maximum production rate, but without the term containing tooling cost.

**Optimum cutting speed:**

$$V_O = e^{-M + K + (H \times L - N_0) \times \ln T_O + M \times L^2 \times (\ln T_O)^2}$$

**Optimum *ECT*:**

$$ECT_O = e^{H + 2M \times (L \times \ln(T_O) + 1)}$$

Global optimum is not reached when face milling for very large feeds, and $C_{TOT}$ decreases continually with increasing feed/tooth, but can be reached for a cutter with many teeth, say 20 to 30. In end milling, global optimum can often be achieved for big feeds and for 3 to 8 teeth.

## Determination of Machine Settings and Calculation of Costs

Based on the rules and knowledge presented in Chapters 1 and 2, this chapter demonstrates, with examples, how machining times and costs are calculated.

Additional formulas are given, and the speed and feed tables given in *SPEEDS AND FEEDS TABLES* starting on page 1078 should be used. Finally the selection of feeds, speeds and tool-lives for optimized conditions are described with examples related to turning, end milling, and face milling.

There are an infinite number of machine settings available in the machine tool power train producing widely different results. In practice only a limited number of available settings are utilized. Often, feed is generally selected independently of the material being cut, however, the influence of material is critical in the choice of cutting speed. The tool life is normally not known or directly determined, but the number of pieces produced before the change of worn tools is better known, and tool life can be calculated using the formula for piece cutting time $t_c$ given in this chapter.

It is well known that increasing feeds or speeds reduces the number of pieces cut between tool changes, but not how big are the changes in the basic parameter tool life. Therefore, there is a tendency to select "safe" data in order to get a long tool life. Another common practice is to search for a tool grade yielding a longer life using the current speeds and feeds, or a 10 to 20 percent increase in cutting speed while maintaining the current tool life. The reason for this old-fashioned approach is the lack of knowledge about the opportunities the metal cutting process offers for increased productivity.

For example, when somebody wants to calculate the cutting time, he/she can select a value of the feed rate (product of feed and rpm), and easily find the cutting time by dividing cutting distance by the feed rate. The number of pieces obtained out of a tool is a guesswork, however. This problem is very common and usually the engineers find desired tool-lives after a number of trial and error runs using a variety of speeds and feeds. If the user is not well familiar with the material cut, the tool life obtained could be any number of seconds or minutes, or the cutting edge might break.

There are an infinite number of speeds and feeds, giving the same feed rate, producing equal cutting time. The same cutting time per piece $t_c$ is obtained independent of the selection of feed/rev $f$ and cutting speed $V$ (or rpm), as long as the feed rate $F_R$ remains the same: $F_R = f_1 \times \text{rpm}_1 = f_2 \times \text{rpm}_2 = f_3 \times \text{rpm}_3, \ldots$, etc. However, the number of parts before tool change $N_{ch}$ will vary considerably, including the tooling cost $c_{tool}$ and the total cutting cost $c_{tot}$.

The dilemma confronting the machining-tool engineer or the process planner is how to set speeds and feeds for either desired cycle time, or number of parts between tool changes, while balancing the process versus other operations or balancing the total times in one cell with another. These problems are addressed in this section.

### Nomenclature

$f$ = feed/rev or tooth, mm $f_E$ = economic feed $f_O$ = optimum feed

$T$ = tool life, minutes $T_E$ = economic tool life $T_O$ = optimum tool life

$V$ = cutting speed, m/min $V_E$ = economic cutting speed

$V_O$ = optimum cutting speed, m/min

Similarly, economic and optimum values of:

$c_{tool}$ = piece cost of tooling, \$ $C_{TOOL}$ = cost of tooling per batch, \$

$c_{tot}$ = piece total cost of cutting, \$ $C_{TOT}$ = total cost of cutting per batch, \$

$F_R$ = feed rate measured in the feeding direction, mm/rev

$N$ = batch size

$N_{ch}$ = number of parts before tool change

$t_c$ = piece cutting time, minutes $T_C$ = cutting time per batch, minutes

$t_{cyc}$ = piece cycle time, minutes $T_{CYC}$ = cycle time before tool change, minutes

$t_i$ = idle time (tool "air" motions during cycle), minutes

$z$ = cutter number of teeth

The following variables are used for calculating the per batch cost of cutting:

$C_C$ = cost of cutting time per batch, $

$C_{CH}$ = cost of tool changes per batch, $

$C_E$ = cost per edge, for replacing or regrinding, $

$H_R$ = hourly rate, $

$T_V$ = equivalent tooling-cost time, minutes

$T_{RPL}$ = time for replacing worn edge(s), or tool for regrinding, minutes

*Note:* In the list above, when two variables use the same name, one in capital letters and one lower case, $T_C$ and $t_c$ for example, the variable name in capital letters refers to batch processing and lowercase letters to per piece processing, such as $T_C = N_{ch} \times t_c$, $C_{TOT} = N_{ch} \times c_{tot}$, etc.

### Formulas Valid for All Operation Types Including Grinding

**Calculation of Cutting Time and Feed Rate**

**Feed Rate:**

$F_R = f \times \text{rpm}$ (mm/min), where $f$ is the feed in mm/rev along the feeding direction, rpm is defined in terms of workpiece or cutter diameter $D$ in mm, and cutting speed $V$ in m/min, as follows:

$$\text{rpm} = \frac{1000V}{\pi D} = \frac{318V}{D}$$

**Cutting Time per Piece:**

*Note:* Constant cutting time is a straight 45-degree line in the *V-ECT* graph, along which tool life varies considerably, as is shown in Fig. 21.

$$t_c = \frac{Dist}{F_R} = \frac{Dist}{f \times \text{rpm}} = \frac{Dist \times \pi D}{1000V \times f}$$

where the units of distance cut *Dist*, diameter *D*, and feed *f* are mm, and *V* is in m/min. In terms of *ECT*, cutting time per piece, $t_c$, is as follows:

$$t_c = \frac{Dist \times \pi D}{1000V} \times \frac{a}{CEL \times ECT}$$

where $a$ = depth of cut, because feed × cross sectional chip area = $f \times a = CEL \times ECT$.

*Example 3, Cutting Time:* Given $Dist$ = 105 mm, $D$ = 100 mm, $f$ = 0.3 mm, $V$ = 300 m/min, rpm = 700, $F_R$ = 210 mm/min, find the cutting time.

Cutting time = $t_c$ = 105 × 3.1416 × 100 ÷ (1000 × 300 × 0.3) = 0.366 minutes = 22 seconds

**Scheduling of Tool Changes**

**Number of Parts Before Tool Change:**

$N_{ch} = T \div t_c$

**Cycle Time Before Tool Change:**

$T_{CYC} = N_{ch} \times (t_c + t_i)$, where $t_{cyc} = t_c + t_i$, where $t_c$ = cutting time per piece, $t_i$ = idle time per piece

**Tool Life:**

$T = N_{ch} \times t_c$

*Example 4:* Given tool life $T$ = 90 minutes, cutting time $t_c$ = 3 minutes, and idle time $t_i$ = 3 minutes, find the number of parts produced before a tool change is required and the time until a tool change is required.

Number of parts before tool change = $N_{ch}$ = 90/3 = 30 parts.

Cycle time before tool change = $T_{CYC}$ = 30 × (3 + 3) = 180 minutes

*Example 5:* Given cutting time, $t_c$ = 1 minute, idle time $t_i$ = 1 minute, $N_{ch}$ = 100 parts, calculate the tool life $T$ required to complete the job without a tool change, and the cycle time before a tool change is required.

Tool life = $T = N_{ch} \times t_c$ = 100 × 1 = 100 minutes.

Cycle time before tool change = $T_{CYC}$ = 100 × (1 + 1) = 200 minutes.

**Calculation of Cost of Cutting and Grinding Operations.—**When machining data varies, the cost of cutting, tool changing, and tooling will change, but the costs of idle and slack time are considered constant.

**Cost of Cutting per Batch:**

$$C_C = H_R \times T_C/60$$

$T_C$ = cutting time per batch = (number of parts) × $t_c$, minutes, or when determining time for tool change $T_{ch} = N_{ch} \times t_c$ minutes = cutting time before tool change.

$t_c$ = cutting time/part, minutes

$H_R$ = hourly rate

**Cost of Tool Changes per Batch:**

$$C_{CH} = \frac{H_R}{60} \times T_C \times \frac{T_{RPL}}{T} \qquad \frac{\$}{min} \cdot min = \$$$

where $T$ = tool life, minutes, and $T_{RPL}$ = time for replacing a worn edge(s), or tool for regrinding, minutes

**Cost of Tooling per Batch:**

Including cutting tools and holders, but without tool changing costs,

$$C_{TOOL} = \frac{H_R}{60} \times T_C \times \frac{\dfrac{60C_E}{H_R}}{T} \qquad \frac{\$}{min} \cdot min \cdot \frac{\dfrac{min}{hr} \cdot \$ \cdot \dfrac{hr}{\$}}{min} = \$$$

**Cost of Tooling + Tool Changes per Batch:**

Including cutting tools, holders, and tool changing costs,

$$(C_{TOOL} + C_{CH}) = \frac{H_R}{60} \times T_C \times \frac{T_{RPL} + \dfrac{60C_E}{H_R}}{T}$$

**Total Cost of Cutting per Batch:**

$$C_{TOT} = \frac{H_R}{60} \times T_C \left(1 + \frac{T_{RPL} + \dfrac{60C_E}{H_R}}{T}\right)$$

**Equivalent Tooling-Cost Time, $T_V$:**

The two previous expressions can be simplified by using $T_V = T_{RPL} + \dfrac{60C_E}{H_R}$ thus:

$$(C_{TOOL} + C_{CH}) = \frac{H_R}{60} \times T_C \times \frac{T_V}{T}$$

$$C_{TOT} = \frac{H_R}{60} \times T_C \left(1+ \frac{T_V}{T}\right)$$

$C_E$ = cost per edge(s) is determined using two alternate formulas, depending on whether tools are reground or inserts are replaced:

**Cost per Edge, Tools for Regrinding:**

$$C_E = \frac{\text{cost of tool} + (\text{number of regrinds} \times \text{cost/regrind})}{1 + \text{number of regrinds}}$$

**Cost per Edge, Tools with Inserts:**

$$C_E = \frac{\text{cost of insert(s)}}{\text{number of edges per insert}} + \frac{\text{cost of cutter body}}{\text{cutter body life in number of edges}}$$

*Note:* In practice allow for insert failures by multiplying the insert cost by 4⁄3, that is, assuming only 3 out of 4 edges can be effectively used.

*Example 6, Cost per Edge-Tools for Regrinding:* Use the data in the table below to calculate the cost per edge(s) $C_E$, and the equivalent tooling-cost time $T_V$, for a drill.

| Time for cutter replacement $T_{RPL}$, minute | Cutter Price, $ | Cost per regrind, $ | Number of regrinds | Hourly shop rate, $ | Batch size | Taylor slope, $n$ | Economic cutting time, $t_{cE}$ minute |
|---|---|---|---|---|---|---|---|
| 1 | 40 | 6 | 5 | 50 | 1000 | 0.25 | 1.5 |

Using the cost per edge formula for reground tools, $C_E = (40 + 5 \times 6) \div (1 + 5) = \$6.80$

When the hourly rate is \$50/hr, $T_V = T_{RPL} + \frac{60C_E}{H_R} = 1 + \frac{60(6.8)}{50} = 9.16$ minutes

Calculate economic tool life using $T_E = T_V \times \left(\frac{1}{n} - 1\right)$ thus, $T_E = 9.17 \times (1/0.25 - 1) = 9.16 \times 3 = 27.48$ minutes.

Having determined, elsewhere, the economic cutting time per piece to be $t_{cE} = 1.5$ minutes, for a batch size = 1000 calculate:

**Cost of Tooling + Tool Change per Batch:**

$$(C_{TOOL} + C_{CH}) = \frac{H_R}{60} \times T_C \times \frac{T_V}{T} = \frac{50}{60} \times 1000 \times 1.5 \times \frac{9.16}{27.48} = \$417$$

**Total Cost of Cutting per Batch:**

$$C_{TOT} = \frac{H_R}{60} \times T_C \left(1+ \frac{T_V}{T}\right) = \frac{50}{60} \times 1000 \times 1.5 \times \left(1 + \frac{9.16}{27.48}\right) = \$1617$$

*Example 7, Cost per Edge-Tools with Inserts:* Use data from the table below to calculate the cost of tooling and tool changes, and the total cost of cutting.

For face milling, multiply insert price by safety factor 4⁄3 then calculate the cost per edge: $C_E = 10 \times (5/3) \times (4/3) + 750/500 = 23.72$ per set of edges

When the hourly rate is \$50, equivalent tooling-cost time is $T_V = 2 + 23.72 \times 60/50 = 30.466$ minutes (first line in table below). The economic tool life for Taylor slope $n = 0.333$ would be $T_E = 30.466 \times (1/0.333 - 1) = 30.466 \times 2 = 61$ minutes.

When the hourly rate is \$25, equivalent tooling-cost time is $T_V = 2 + 23.72 \times 60/25 = 58.928$ minutes (second line in table below). The economic tool life for Taylor slope $n = 0.333$ would be $T_E = 58.928 \times (1/0.333 - 1) = 58.928 \times 2 = 118$ minutes.

| Time for replacement of inserts $T_{RPL}$, minutes | Number of inserts | Price per insert | Edges per insert | Cutter Price | Edges per cutter | Cost per set of edges, $C_E$ | Hourly shop rate | $T_V$ minutes |
|---|---|---|---|---|---|---|---|---|
| Face mill | | | | | | | | |
| 2 | 10 | 5 | 3 | 750 | 500 | 23.72 | 50 | 30.466 |
| 2 | 10 | 5 | 3 | 750 | 500 | 23.72 | 25 | 58.928 |
| End mill | | | | | | | | |
| 1 | 3 | 6 | 2 | 75 | 200 | 4.375 | 50 | 6.25 |
| Turning | | | | | | | | |
| 1 | 1 | 5 | 3 | 50 | 100 | 2.72 | 30 | 6.44 |

With above data for the face mill, and after having determined the economic cutting time as $t_{cE}$ = 1.5 minutes, calculate for a batch size = 1000 and $50 per hour rate:

**Cost of Tooling + Tool Change per Batch:**

$$(C_{TOOL} + C_{CH}) = \frac{H_R}{60} \times T_C \times \frac{T_V}{T} = \frac{50}{60} \times 1000 \times 1.5 \times \frac{30.466}{61} = \$624$$

**Total Cost of Cutting per Batch:**

$$C_{TOT} = \frac{H_R}{60} \times T_C\left(1 + \frac{T_V}{T}\right) = \frac{50}{60} \times 1000 \times 1.5 \times \left(1 + \frac{30.466}{61}\right) = \$1874$$

Similarly, at the $25/hour shop rate, ($C_{TOOL}$ + $C_{CH}$) and $C_{TOT}$ are $312 and $937, respectively.

*Example 8, Turning:* Production parts were run in the shop at feed/rev = 0.25 mm. One series was run with speed $V_1$ = 200 m/min and tool life was $T_1$ = 45 minutes. Another was run with speed $V_2$ = 263 m/min and tool life was $T_2$ = 15 minutes. Given idle time $t_i$ = 1 minute, cutting distance *Dist* =1000 mm, work diameter *D* = 50 mm.

First, calculate Taylor slope, *n*, using Taylor's equation $V_1 \times T_1{}^n = V_2 \times T_2{}^n$, as follows:

$$n = \ln\frac{V_1}{V_2} \div \ln\frac{T_2}{T_1} = \ln\frac{200}{263} \div \ln\frac{15}{45} = 0.25$$

Economic tool life $T_E$ is next calculated using the equivalent tooling-cost time $T_V$, as described previously. Assuming a calculated value of $T_V$ = 4 minutes, then $T_E$ can be calculated from

$$T_E = T_V \times\left(\frac{1}{n} - 1\right) = 4 \times\left(\frac{1}{0.25} - 1\right) = 12 \text{ minutes}$$

Economic cutting speed, $V_E$ can be found using Taylor's equation again, this time using the economic tool life, as follows,

$$V_{E1} \times (T_E)^n = V_2 \times (T_2)^n$$

$$V_{E1} = V_2 \times\left(\frac{T_2}{T_E}\right)^n = 263 \times\left(\frac{15}{12}\right)^{0.25} = 278 \text{ m/min}$$

Using the process data, the remaining economic parameters can be calculated as follows:

Economic spindle rpm, $\text{rpm}_E = 1000V_E/\pi D = (1000 \times 278)/(3.1416 \times 50) =$ 1770 rpm

Economic feed rate, $F_{RE} = f \times \text{rpm}_E = 0.25 \times 1770 = 443$ mm/min

Economic cutting time, $t_{cE} = Dist/F_{RE} = 1000/443 = 2.259$ minutes

Economic number of parts before tool change, $N_{chE} = T_E \div t_{cE} = 12 \div 2.259 =$ 5.31 parts

Economic cycle time before tool change, $T_{CYCE} = N_{chE} \times (t_c + t_i) =$ $5.31 \times (2.259 + 1) = 17.3$ minutes.

### Variation of Tooling and Total Cost with the Selection of Speeds and Feeds

It is a well-known fact that tool life is reduced when either feed or cutting speed is increased. When a higher feed/rev is selected, the cutting speed must be decreased in order to maintain tool life. However, a higher feed rate (feed rate = feed/rev × rpm, mm/min) can result in a longer tool life if proper cutting data are applied. Optimized cutting data require accurate machinability databases and a computer program to analyze the options. Reasonably accurate optimized results can be obtained by selecting a large feed/rev or tooth, and then calculating the economic tool life $T_E$. Because the cost versus feed or *ECT* curve is shallow around the true minimum point, i.e., the global optimum, the error in applying a large feed is small compared with the exact solution.

Once a feed has been determined, the economic cutting speed $V_E$ can be found by calculating the Taylor slope, and the time/cost calculations can be completed using the formulas described in last section.

The remainder of this section contains examples useful for demonstrating the required procedures. Global optimum may or may not be reached, and tooling cost may or may not be reduced, compared to currently used data. However, the following examples prove that significant time and cost reductions are achievable in today's industry.

*Note:* Starting values of reasonable feeds in mm/rev can be found in the *Handbook* speed and feed tables (see *Principal Speeds and Feeds Tables* on page 1078) by using the $f_{avg}$ values converted to mm as follows: feed (mm/rev) = feed (inch/rev) × 25.4 (mm/inch), thus 0.001 inch/rev = 0.001× 25.4 = 0.0254 mm/rev. When using speed and feed Table 1 through Table 23, where feed values are given in thousandths of inch per revolution, simply multiply the given feed by 25.4/1000 = 0.0254, thus feed (mm/rev) = feed (0.001 inch/rev) × 0.0254 (mm/0.001inch).

*Example 9, Converting* Handbook *Feed Values from Inches to Millimeters:* The *Handbook* tables give feed values $f_{opt}$ and $f_{avg}$ for 4140 steel as 17 and 8 × (0.001 inch/rev) = 0.017 and 0.009 inch/rev, respectively. Convert the given feeds to mm/rev.

feed = 0.017 × 25.4 = 17 × 0.0254 = 0.4318 mm/rev
feed = 0.008 × 25.4 = 8 × 0.0254 = 0.2032 mm/rev

*Example 10, Using* Handbook *Tables to Find the Taylor Slope and Constant:* Calculate the Taylor slope and constant, using cutting speed data for 4140 steel in Table 1 starting on page 1083, and for ASTM Class 20 gray cast iron using data from Table 4a on page 1089, as follows:

For the 175–250 Brinell Hardness range, and the hard tool grade,

$$n = \frac{\ln(V_1/V_2)}{\ln(T_2/T_1)} = \frac{\ln(525/705)}{\ln(15/45)} = 0.27 \qquad C = V_1 \times (T_1)^n = 1458$$

For the 175–250 Brinell Hardness range, and the tough tool grade,

$$n = \frac{\ln(V_1/V_2)}{\ln(T_2/T_1)} = \frac{\ln(235/320)}{\ln(15/45)} = 0.28 \qquad C = V_1 \times (T_1)^n = 685$$

For the 300–425 Brinell Hardness range, and the hard tool grade,

$$n = \frac{\ln(V_1/V_2)}{\ln(T_2/T_1)} = \frac{\ln(330/440)}{\ln(15/45)} = 0.26 \qquad C = V_1 \times (T_1)^n = 894$$

For the 300–425 Brinell Hardness range, and the tough tool grade,

$$n = \frac{\ln(V_1/V_2)}{\ln(T_2/T_1)} = \frac{\ln(125/175)}{\ln(15/45)} = 0.31 \qquad C = V_1 \times (T_1)^n = 401$$

For ASTM Class 20 gray cast iron, using hard ceramic,

$$n = \frac{\ln (V_1 / V_2)}{\ln (T_2 / T_1)} = \frac{\ln (1490/2220)}{\ln (15/45)} = 0.36 \qquad C = V_1 \times T_1)^n = 5932$$

**Selection of Optimized Data.—**Fig. 22 illustrates cutting time, cycle time, number of parts before a tool change, tooling cost, and total cost, each plotted versus feed for a constant tool life. Approximate minimum cost conditions can be determined using the formulas previously given in this section.

First, select a large feed/rev or tooth, and then calculate economic tool life $T_E$, and the economic cutting speed $V_E$, and do all calculations using the time/cost formulas as described previously.

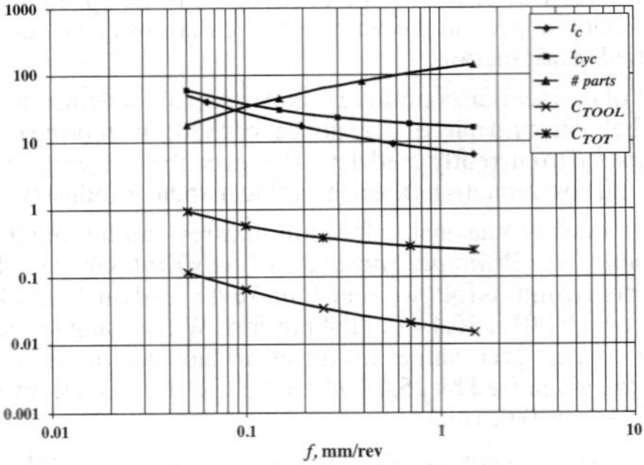


Fig. 22. Cutting Time, Cycle Time, Number of Parts Before Tool Change, Tooling Cost, and Total Cost versus Feed for Tool Life = 15 Minutes, Idle Time = 10 s, and Batch Size = 1000 Parts

*Example 11, Step by Step Procedure: Turning—Facing out:* 1) Select a big feed/rev, in this case $f = 0.9$ mm/rev (0.035 inch/rev). A Taylor slope $n$ is first determined using the *Handbook* tables and the method described in Example 10. In this example, use $n = 0.35$ and C = 280.

2) Calculate $T_V$ from the tooling cost parameters:

If cost of insert = \$7.50; edges per insert = 2; cost of tool holder = \$100; life of holder = 100 insert sets; and for tools with inserts, allowance for insert failures = cost per insert by 4/3, assuming only 3 out of 4 edges can be effectively used.

Then, cost per edge = $C_E$ is calculated as follows:

$$C_E = \frac{\text{cost of insert(s)}}{\text{number of edges per insert}} + \frac{\text{cost of cutter body}}{\text{cutter body life in number of edges}}$$

$$= \frac{7.50 \times 4/3}{2} + \frac{100}{100} = \$6.00$$

The time for replacing a worn edge of the facing insert $= T_{RPL} = 2.24$ minutes. Assuming an hourly rate $H_R = \$50$/hour, calculate the equivalent tooling-cost time $T_V$

$$T_V = T_{RPL} + 60 \times C_E / H_R = 2.24 + 60 \times 6/50 = 9.44 \text{ minutes}$$

3) Determine economic tool life $T_E$

$$T_E = T_V \times (1/n - 1) = 9.44 \times (1/0.35 - 1) = 17.5 \text{ minutes}$$

4) Determine economic cutting speed using the *Handbook* tables using the method shown in Example 10,

$$V_E = \frac{C}{T_E^n} \text{ m/min} = 280 / 17.5^{0.35} = 103 \text{ m/min}$$

5) Determine cost of tooling per batch (cutting tools, holders and tool changing) then total cost of cutting per batch:

$$C_{TOOL} = H_R \times T_C \times (C_E/T)/60$$

$$C_{TOOL} + C_{CH} = H_R \times T_C \times (T_{RPL} + C_E/T)/60$$

$$C_{TOT} = H_R \times T_C (1 + (T_{RPL} + C_E)/T)$$

*Example 12, Face Milling—Minimum Cost :* This example demonstrates how a modern firm, using the formulas previously described, can determine optimal data. It is here applied to a face mill with 10 teeth, milling a 1045 type steel, and the radial depth versus the cutter diameter is 0.8. The *V-ECT-T* curves for tool-lives 5, 22, and 120 minutes for this operation are shown in Fig. 23a.

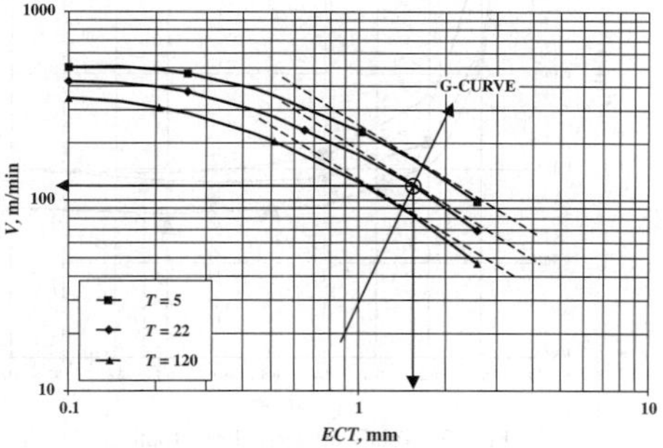


Fig. 23a. Cutting Speed versus *ECT,* Tool Life Constant

The global cost minimum occurs along the *G*-curve, see Fig. 6c and Fig. 23a, where the 45-degree lines defines this curve. Optimum *ECT* is in the range 1.5 to 2 mm.

For face and end milling operations, $ECT = z \times f_z \times ar/D \times aa/CEL \div \pi$. The ratio $aa/CEL = 0.95$ for lead angle $LA = 0$, and for $ar/D = 0.8$ and 10 teeth, using the formula to calculate the feed/tooth range gives for $ECT = 1.5, f_z = 0.62$ mm and for $ECT = 2, f_z = 0.83$ mm.

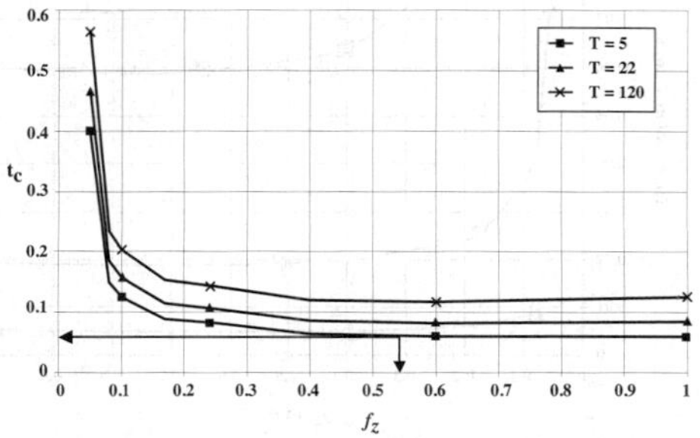


Fig. 23b. Cutting Time per Part versus Feed per Tooth

Using computer simulation, the minimum cost occurs approximately where Fig. 23a indicates it should be. Total cost has a global minimum at $f_z$ around 0.6 to 0.7 mm and a speed of around 110 m/min. *ECT* is about 1.9 mm and the optimal cutter life is $T_O$ = 22 minutes. Because it may be impossible to reach the optimum feed value due to tool breakage, the maximum practical feed $f_{max}$ is used as the optimal value. The difference in costs between a global optimum and a practical minimum cost condition is negligible, as shown in Fig. 23c and Fig. 23e. A summary of the results are shown in Fig. 23a through Fig. 23e, and Table 1.

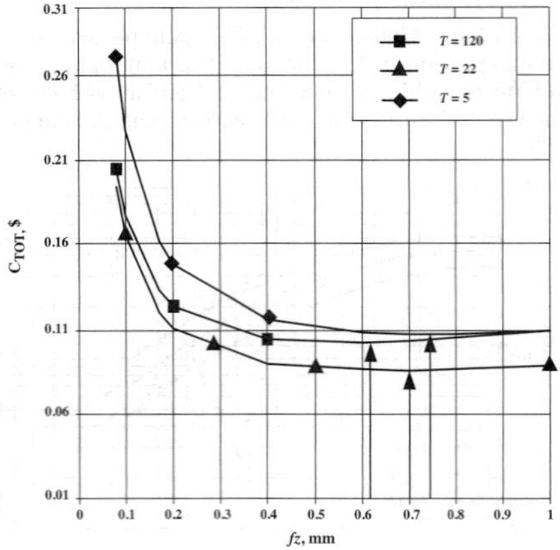


Fig. 23c. Total Cost versus Feed/Tooth

When plotting cutting time/part, $t_c$, versus feed/tooth, $f_z$, at $T = 5, 22, 120$ in Fig. 23b, tool life $T = 5$ minutes yields the shortest cutting time, but total cost is the highest; the minimum occurs for $f_z$ about 0.75 mm, see Fig. 23c. The minimum for $T = 120$ minutes is about 0.6 mm and for $T_O = 22$ minutes around 0.7 mm.

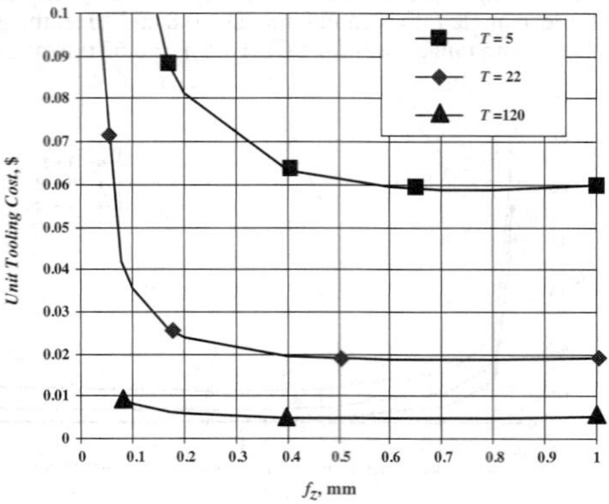


Fig. 23d. Tooling Cost versus Feed/Tooth

Fig. 23d shows that tooling cost drops off quickly when increasing feed from 0.1 to 0.3 to 0.4 mm, and then diminishes slowly and is almost constant up to 0.7 to 0.8 mm/tooth. It is generally very high at the short tool life 5 minutes, while tooling cost of optimal tool life 22 minutes is about three times higher than when going slow at $T$=120 minutes.

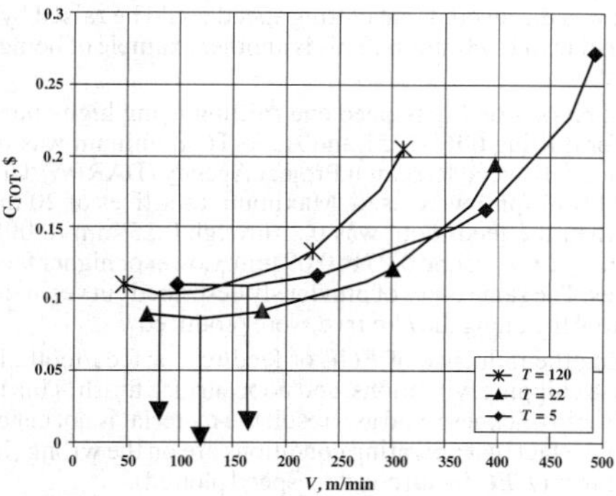


Fig. 23e. Total Cost versus Cutting Speed at 3 Constant Tool-Lives, Feed Varies

The total cost curves in Fig. 23e were obtained by varying feed and cutting speed in order to maintain constant tool-lives at 5, 22 and 120 minutes. Cost is plotted as a function of speed $V$ instead of feed/tooth. Approximate optimum speeds are $V$ = 150 m/min at $T$ = 5 minutes, $V$ = 180 m/min at $T$ = 120 minutes, and the global optimum speed is $V_O$ = 110 m/min for $T_O$ = 22 minutes.

Table 1 displays the exact numerical values of cutting speed, tooling cost and total cost for the selected tool-lives of 5, 22, and 120 minutes, obtained from the software program.

**Table 1. Face Milling, Total and Tooling Cost versus *ECT*, Feed/Tooth *fz*, and Cutting Speed *V*, at Tool-Lives 5, 22, and 120 Minutes**

| | | $T$ = 5 minutes | | | $T$ = 22 minutes | | | $T$ = 120 minutes | | |
|---|---|---|---|---|---|---|---|---|---|---|
| $f_z$ | $ECT$ | $V$ | $C_{TOT}$ | $C_{TOOL}$ | $V$ | $C_{TOT}$ | $C_{TOOL}$ | $V$ | $C_{TOT}$ | $C_{TOOL}$ |
| 0.03 | 0.08 | 489 | 0.72891 | 0.39759 | 416 | 0.49650 | 0.10667 | 344 | 0.49378 | 0.02351 |
| 0.08 | 0.21 | 492 | 0.27196 | 0.14834 | 397 | 0.19489 | 0.04187 | 311 | 0.20534 | 0.00978 |
| 0.10 | 0.26 | 469 | 0.22834 | 0.12455 | 374 | 0.16553 | 0.03556 | 289 | 0.17674 | 0.00842 |
| 0.17 | 0.44 | 388 | 0.16218 | 0.08846 | 301 | 0.12084 | 0.02596 | 225 | 0.13316 | 0.00634 |
| 0.20 | 0.51 | 359 | 0.14911 | 0.08133 | 276 | 0.11204 | 0.02407 | 205 | 0.12466 | 0.00594 |
| 0.40 | 1.03 | 230 | 0.11622 | 0.06339 | 171 | 0.09051 | 0.01945 | 122 | 0.10495 | 0.00500 |
| 0.60 | 1.54 | 164 | 0.10904 | 0.05948 | 119 | 0.08672 | 0.01863 | 83 | 0.10301 | 0.00491 |
| 0.70 | 1.80 | 141 | 0.10802 | 0.05892 | 102 | 0.08665 | 0.01862 | 70 | 0.10393 | 0.00495 |
| 0.80 | 2.06 | 124 | 0.10800 | 0.05891 | 89 | 0.08723 | 0.01874 | 60 | 0.10547 | 0.00502 |
| 1.00 | 2.57 | 98 | 0.10968 | 0.05982 | 69 | 0.08957 | 0.01924 | 47 | 0.10967 | 0.00522 |

## High-Speed Machining Econometrics

**High-Speed Machining, No Mystery.—**This section describes the theory and gives the basic formulas for any milling operation and high-speed milling in particular, followed by several examples on high-speed milling econometrics. These rules constitute the basis on which selection of milling feed factors is done. Selection of cutting speeds for general milling is done using the *Handbook* Table 10 through Table 14, starting on page 1100.

High-speed machining is no mystery to those who understand metal cutting. Machining materials with good machinability, such as low-alloyed aluminum, has for ages been performed at cutting speeds well below the speed values at which these materials should be cut. Operating at such low speeds often results in built-up edges and poor surface finish, because the operating conditions selected are on the wrong side of the Taylor curve, i.e. to the left of the $H$-curve representing maximum tool life (see Fig. 4 on page 1187).

In the 1950's it was discovered that cutting speed could be raised by a factor of 5 to 10 when hobbing steel with HSS cutters. This is another example of being on the wrong side of the Taylor curve.

One of the first reports on high-speed end milling using high-speed steel (HSS) and carbide cutters for milling 6061-T651 and A356-T6 aluminum was reported in a study funded by Defense Advanced Research Project Agency (DARPA). Cutting speeds of up to 4400 m/min (14140 fpm) were used. Maximum tool-lives of 20 through 40 minutes were obtained when the feed/tooth was 0.2 through 0.25 mm (0.008 to 0.01 inch), or measured in terms of *ECT* around 0.07 to 0.09 mm. Lower or higher feed/tooth resulted in shorter cutter lives. The same types of previously described curves, namely *T-ECT* curves with maximum tool life along the *H*-curve, were produced.

When examining the influence of *ECT*, or feed/rev, or feed/tooth, it is found that too small values cause chipping, vibrations, and poor surface finish. This is caused by inadequate (too small) chip thickness, and as a result the material is not cut but plowed away or scratched, due to the fact that operating conditions are on the wrong (left) side of the tool life versus *ECT* curve (*T-ECT* with constant speed plotted).

There is a great difference in the thickness of chips produced by a tooth traveling through the cutting arc in the milling process, depending on how the center of the cutter is placed in relation to the workpiece centerline, in the feed direction. Although end and face milling cut in the same way, from a geometry and kinematics standpoint they are in practice distinguished by the cutter center placement away from, or close to, the work centerline, respectively, because of the effect of cutter placement on chip thickness. This is the criterion used to distinguish between the end and face milling processes in the following.

*Depth of Cut/Cutter Diameter, ar/D* is the ratio of the radial depth of cut *ar* and the cutter diameter *D*. In face milling when the cutter axis points approximately to the middle of the work piece axis, eccentricity is close to zero, as illustrated in Fig. 3 and Fig. 4, page 1098, and Fig. 5 on page 1099. In end milling, $ar/D = 1$ for full slot milling.

*Mean Chip Thickness hm* is a key parameter that is used to calculate forces and power requirements in high-speed milling. If the mean chip thickness *hm* is too small, which may occur when feed/tooth is too small (this holds for all milling operations), or when *ar/D* decreases (this holds for ball nose as well as for straight end mills), then cutting occurs on the left (wrong side) of the tool life versus *ECT* curve, as illustrated in Fig. 6b and Fig. 6c.

In order to maintain a given chip thickness in end milling, the feed/tooth has to be increased, up to 10 times for very small *ar/D* values in an extreme case with no run out and otherwise perfect conditions. A 10 times increase in feed/tooth results in 10 times bigger feed rates ($F_R$) compared to data for full slot milling (valid for $ar/D = 1$), yet maintains a given chip thickness. The cutter life at any given cutting speed will not be the same, however.

Increasing the number of teeth from say 2 to 6 increases equivalent chip thickness *ECT* by a factor of 3 while the mean chip thickness *hm* remains the same, but does not increase the feed rate to 30 (3 × 10) times bigger, because the cutting speed must be reduced. However, when the *ar/D* ratio matches the number of teeth, such that one tooth enters when the second tooth leaves the cutting arc, then $ECT = hm$. Hence, *ECT* is proportional to the number of teeth. Under ideal conditions, an increase in number of teeth *z* from 2 to 6 increases the feed rate by, say, 20 times, maintaining tool life at a reduced speed. In practice about 5 times greater feed rates can be expected for small *ar/D* ratios (0.01 to 0.02), and up to 10 times with 3 times as many teeth. So, high-speed end milling is no mystery.

**Chip Geometry in End and Face Milling.**—Fig. 24 illustrates how the chip forming process develops differently in face and end milling, and how mean chip thickness *hm* varies with the angle of engagement *AE*, which depends on the *ar/D* ratio. The pertinent chip geometry formulas are given in the text that follows.

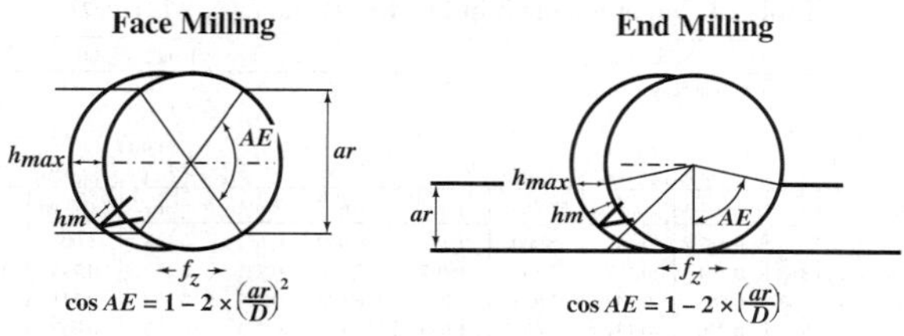


Fig. 24.

*Comparison of Face Milling and End Milling Geometry:* High-speed end milling refers to values of *ar/D* that are less than 0.5, in particular to *ar/D* ratios which are considerably smaller. When *ar/D* = 0.5 (*AE* = 90 degrees) and diminishing in end milling, the chip thickness gets so small that poor cutting action develops, including plowing or scratching. This situation is remedied by increasing the feed/tooth, as shown in Table 2a as an increasing $f_z/f_{z0}$ ratio with decreasing *ar/D*. For end milling, the $f_z/f_{z0}$ feed ratio is 1.0 for *ar/D* = 1 and also for *ar/D* = 0.5. In order to maintain the same *hm* as at *ar/D* = 1, the feed/tooth should be increased, by a factor of 6.38 when *ar/D* is 0.01 and by more than 10 when *ar/D* is less than 0.01. Hence high-speed end milling could be said to begin when *ar/D* is less than 0.5

In end milling, the ratio $f_z/f_{z0} = 1$ is set at *ar/D* = 1.0 (full slot), a common value in vendor catalogs and handbooks, for *hm* = 0.108 mm.

The face milling chip making process is exactly the same as end milling the side of a workpiece and *ar/D* = 0.5 or less. However, when face milling close to and along the work centerline (eccentricity is close to zero) chip making is quite different, as shown in Fig. 24. When *ar/D* = 0.74 (*AE* = 95 degrees) in face milling, the $f_z/f_{z0}$ ratio = 1 and increases up to 1.4 when the work width is equal to the cutter diameter (*ar/D* = 1). The face milling $f_z/f_{z0}$ ratio continues to diminish when the *ar/D* ratio decreases below *ar/D* = 0.74, but very insignificantly, only about 11 percent when *ar/D* = 0.01.

In face milling, $f_z/f_{z0} = 1$ is set at *ar/D* = 0.74, a common value recommended in vendor catalogs and handbooks, for *hm* = 0.151 mm.

Fig. 25 shows the variation of the feed/tooth-ratio in a graph for end and face milling.

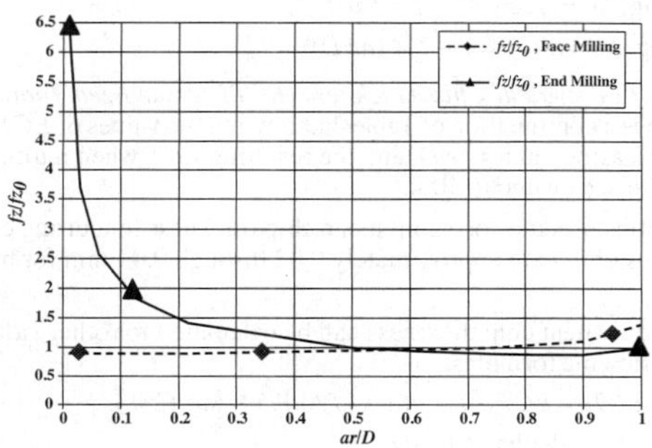


Fig. 25. Feed/Tooth versus *ar/D* for Face and End Milling

**Table 2a. Variation of Chip Thickness and $f_z/f_{z0}$ with $ar/D$**

| | Face Milling | | | | | End Milling (straight) | | | | |
|---|---|---|---|---|---|---|---|---|---|---|
| | ecentricity $e = 0$<br>$z = 8$<br>$f_{z0} = 0.17$<br>$\cos AE = 1 - 2 \times (ar/D)^2$ | | | | | $z = 2$<br>$f_{z0} = 0.17$<br>$\cos AE = 1 - 2 \times (ar/D)$ | | | | |
| $ar/D$ | $AE$ | $hm/f_z$ | $hm$ | $ECT/hm$ | $f_z/f_{z0}$ | $AE$ | $hm/f_z$ | $hm$ | $ECT/hm$ | $f_z/f_{z0}$ |
| 1.0000 | 180.000 | 0.637 | 0.108 | 5.000 | 1.398 | 180.000 | 0.637 | 0.108 | 1.000 | 1.000 |
| 0.9000 | 128.316 | 0.804 | 0.137 | 3.564 | 1.107 | 143.130 | 0.721 | 0.122 | 0.795 | 0.884 |
| 0.8000 | 106.260 | 0.863 | 0.147 | 2.952 | 1.032 | 126.870 | 0.723 | 0.123 | 0.711 | 0.881 |
| 0.7355 | 94.702 | 0.890 | 0.151 | 2.631 | 1.000 | 118.102 | 0.714 | 0.122 | 0.667 | 0.892 |
| 0.6137 | 75.715 | 0.929 | 0.158 | 1.683 | 0.958 | 103.144 | 0.682 | 0.116 | 0.573 | 0.934 |
| 0.5000 | 60.000 | 1.025 | 0.162 | 1.267 | 0.932 | 90.000 | 0.674 | 0.115 | 0.558 | 1.000 |
| 0.3930 | 46.282 | 0.973 | 0.165 | 1.028 | 0.915 | 77.643 | 0.580 | 0.099 | 0.431 | 1.098 |
| 0.2170 | 25.066 | 0.992 | 0.169 | 0.557 | 0.897 | 55.528 | 0.448 | 0.076 | 0.308 | 1.422 |
| 0.1250 | 14.361 | 0.997 | 0.170 | 0.319 | 0.892 | 41.410 | 0.346 | 0.059 | 0.230 | 1.840 |
| 0.0625 | 7.167 | 0.999 | 0.170 | 0.159 | 0.891 | 28.955 | 0.247 | 0.042 | 0.161 | 2.574 |
| 0.0300 | 3.438 | 1.000 | 0.170 | 0.076 | 0.890 | 19.948 | 0.172 | 0.029 | 0.111 | 3.694 |
| 0.0100 | 1.146 | 1.000 | 0.170 | 0.025 | 0.890 | 11.478 | 0.100 | 0.017 | 0.064 | 6.377 |
| 0.0010 | 0.115 | 1.000 | 0.000 | 0.000 | 0.890 | 3.624 | 0.000 | 0.000 | 0.000 | 20.135 |

In Table 2a, a standard value $f_{z0}$ = 0.17 mm/tooth (commonly recommended average feed) was used, but the $f_z/f_{z0}$ values are independent of the value of feed/tooth, and the previously mentioned relationships are valid whether $f_{z0}$ = 0.17 or any other value.

In both end and face milling, $hm$ = 0.108 mm for $f_{z0}$ = 0.17 mm when $ar/D$ =1. When the $f_z/f_{z0}$ ratio = 1, $hm$ = 0.15 for face milling, and 0.108 in end milling both at $ar/D$ = 1 and 0.5. The tabulated data hold for perfect milling conditions, such as, zero run-out and accurate sharpening of all teeth and edges.

**Mean Chip Thickness *hm* and Equivalent Chip Thickness *ECT*.—**The basic formula for equivalent chip thickness $ECT$ for any milling process is:

$ECT = f_z \times z/\pi \times (ar/D) \times aa/CEL$, where $f_z$ = feed/tooth, $z$ = number of teeth, $D$ = cutter diameter, $ar$ = radial depth of cut, $aa$ = axial depth of cut, and $CEL$ = cutting edge length. As a function of mean chip thickness $hm$:

$ECT = hm \times (z/2) \times (AE/180)$, where $AE$ = angle of engagement.

Both terms are exactly equal when one tooth engages as soon as the preceding tooth leaves the cutting section. Mathematically, $hm = ECT$ when $z = 360/AE$; thus:

for face milling, $AE = \arccos(1 - 2 \times (ar/D)^2)$

for end milling, $AE = \arccos(1 - 2 \times (ar/D))$

*Calculation of Equivalent Chip Thickness (ECT) versus Feed/Tooth and Number of Teeth:* Table 2b is a continuation of Table 2a, showing the values of $ECT$ for face and end milling for decreasing values $ar/D$, and the resulting $ECT$ when multiplied by the $f_z/f_{z0}$ ratio $f_{z0}$ = 0.17 (based on $hm$ = 0.108).

Small $ar/D$ ratios produce too small mean chip thickness for cutting chips. In practice, minimum values of $hm$ are approximately 0.02 through 0.04 mm for both end and face milling.

**Formulas.—**Equivalent chip thickness can be calculated for other values of $f_z$ and $z$ by means of the following formulas:

*Face milling:* $ECT_F = ECT_{0F} \times (z/8) \times (f_z/0.17) \times (aa/CEL)$

or, if $ECT_F$ is known calculate $f_z$ using:

$f_z = 0.17 \times (ECT_F/ECT_{0F}) \times (8/z) \times (CEL/aa)$

**Table 2b. Variation of $ECT$, Chip Thickness and $f_z/f_{z0}$ with $ar/D$**

| | Face Milling | | | | End Milling (straight) | | | |
|---|---|---|---|---|---|---|---|---|
| $ar/D$ | $hm$ | $f_z/f_{z0}$ | $ECT$ | $ECT_0$ corrected for $f_z/f_{z0}$ | $hm$ | $f_z/f_{z0}$ | $ECT$ | $ECT_0$ corrected for $f_z/f_{z0}$ |
| 1.0000 | 0.108 | 1.398 | 0.411 | 0.575 | 0.108 | 1.000 | 0.103 | 0.103 |
| 0.9000 | 0.137 | 1.107 | 0.370 | 0.410 | 0.122 | 0.884 | 0.093 | 0.082 |
| 0.8080 | 0.146 | 1.036 | 0.332 | 0.344 | 0.123 | 0.880 | 0.083 | 0.073 |
| 0.7360 | 0.151 | 1.000 | 0.303 | 0.303 | 0.121 | 0.892 | 0.076 | 0.067 |
| 0.6137 | 0.158 | 0.958 | 0.252 | 0.242 | 0.116 | 0.934 | 0.063 | 0.059 |
| 0.5900 | 0.159 | 0.952 | 0.243 | 0.231 | 0.115 | 0.945 | 0.061 | 0.057 |
| 0.5000 | 0.162 | 0.932 | 0.206 | 0.192 | 0.108 | 1.000 | 0.051 | 0.051 |
| 0.2170 | 0.169 | 0.897 | 0.089 | 0.080 | 0.076 | 1.422 | 0.022 | 0.032 |
| 0.1250 | 0.170 | 0.892 | 0.051 | 0.046 | 0.059 | 1.840 | 0.013 | 0.024 |
| 0.0625 | 0.170 | 0.891 | 0.026 | 0.023 | 0.042 | 2.574 | 0.006 | 0.017 |
| 0.0300 | 0.170 | 0.890 | 0.012 | 0.011 | 0.029 | 3.694 | 0.003 | 0.011 |
| 0.0100 | 0.170 | 0.890 | 0.004 | 0.004 | 0.017 | 6.377 | 0.001 | 0.007 |
| 0.0010 | 0.170 | 0.890 | 0.002 | 0.002 | 0.005 | 20.135 | 0.001 | 0.005 |

In face milling, the approximate values of $aa/CEL = 0.95$ for lead angle $LA = 0°$ (90° in the metric system); for other values of $LA$, $aa/CEL = 0.95 \times \sin(LA)$, and $0.95 \times \cos(LA)$ in the metric system.

*Example, Face Milling:* For a cutter with $D = 250$ mm and $ar = 125$ mm, calculate $ECT_F$ for $f_z = 0.1$, $z = 12$, and $LA = 30$ degrees. First calculate $ar/D = 0.5$, and then use Table 2b and find $ECT_{0F} = 0.2$.

Calculate $ECT_F$ with above formula:

$ECT_F = 0.2 \times (12/8) \times (0.1/0.17) \times 0.95 \times \sin 30 = 0.084$ mm.

*End milling:* $ECT_E = ECT_{0E} \times (z/2) \times (f_z/0.17) \times (aa/CEL)$,

or if $ECT_E$ is known calculate $f_z$ from:

$f_z = 0.17 \times (ECT_E/ECT_{0E}) \times (2/z)) \times (CEL/aa)$

The approximate values of $aa/CEL = 0.95$ for lead angle $LA = 0°$ (90° in the metric system).

*Example, High-Speed End Milling:* For a cutter with $D = 25$ mm and $ar = 3.125$ mm, calculate $ECT_E$ for $f_z = 0.1$ and $z = 6$. First calculate $ar/D = 0.125$, and then use Table 2b and find $ECT_{0E} = 0.0249$.

Calculate $ECT_E$ with above formula:

$ECT_E = 0.0249 \times (6/2) \times (0.1/0.17) \times 0.95 \times 1 = 0.042$ mm.

*Example, High-Speed End Milling:* For a cutter with $D = 25$ mm and $ar = 0.75$ mm, calculate $ECT_E$ for $f_z = 0.17$ and $z = 2$ and 6. First calculate $ar/D = 0.03$, and then use Table 2b and find $f_z/f_{z0} = 3.694$

Then, $f_z = 3.694 \times 0.17 = 0.58$ mm/tooth and $ECT_E = 0.0119 \times 0.95 = 0.0113$ mm and $0.0357 \times 0.95 = 0.0339$ mm for 2 and 6 teeth respectively. These cutters are marked HS2 and HS6 in Fig. 26a, Fig. 26d, and Fig. 26e.

*Example, High-Speed End Milling:* For a cutter with $D = 25$ mm and $ar = 0.25$ mm, calculate $ECT_E$ for $f_z = 0.17$ and $z = 2$ and 6. First calculate $ar/D = 0.01$, and then use Table 2b and find $ECT_{0E} = 0.0069$ and 0.0207 for 2 and 6 teeth respectively. When obtaining such small values of $ECT$, there is a great danger to be far on the left side of the $H$-curve, at least when there are only 2 teeth. Doubling the feed would be the solution if cutter design and material permit.

*Example, Full Slot Milling:* For a cutter with $D = 25$ mm and $ar = 25$ mm, calculate $ECT_E$ for $f_z = 0.17$ and $z = 2$ and 6. First calculate $ar/D = 1$, and then use Table 2b and find $ECT_E = 0.108 \times 0.95 = 0.103$ and $3 \times 0.108 \times 0.95 = 0.308$ for 2 and 6 teeth, respectively. These cutters are marked SL2 and SL6 in Fig. 26a, Fig. 26d, and Fig. 26e.

**Physics Behind *hm* and *ECT*, Forces and Tool Life (*T*).—**The *ECT* concept for all metal cutting and grinding operations says that the more energy put into the process, by increasing feed/rev, feed/tooth, or cutting speed, the life of the edge decreases. When increasing the number of teeth (keeping everything else constant) the work and the process are subjected to a higher energy input resulting in a higher rate of tool wear.

In high-speed milling when the angle of engagement *AE* is small the contact time is shorter compared to slot milling ($ar/D = 1$) but the chip becomes shorter as well. Maintaining the same chip thickness as in slot milling has the effect that the energy consumption to remove the chip will be different. Hence, maintaining a constant chip thickness is a good measure when calculating cutting forces (keeping speed constant), but not when determining tool wear. Depending on cutting conditions the wear rate can either increase or decrease, this depends on whether cutting occurs on the left or right side of the *H*-curve.

Fig. 26a shows an example of end milling of steel with coated carbide inserts, where cutting speed *V* is plotted versus *ECT* at 5, 15, 45 and 180 minutes tool-lives. Notice that the *ECT* values are independent of *ar*/*D* or number of teeth or feed/tooth, or whether $f_z$ or $f_{z0}$ is used, as long as the corresponding $f_z/f_{z0}$-ratio is applied to determine $ECT_E$. The result is one single curve per tool life. Had cutting speed been plotted versus $f_{z0}$, *ar*/*D*, or *z* values (number of teeth), several curves would be required at each constant tool life, one for each of these parameters This illustrates the advantage of using the basic parameter *ECT* rather than $f_z$, or *hm*, or *ar*/*D* on the horizontal axis.

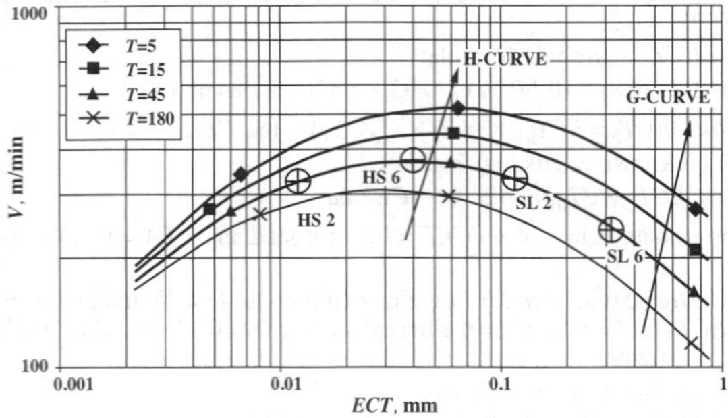


Fig. 26a. Cutting Speed versus *ECT*, Tool Life Plotted, for End Milling

*Example:* The points (HS2, HS6) and (SL2, SL6) on the 45-minute curve in Fig. 26a relate to the previous high-speed and full slot milling examples for 2 and 6 teeth, respectively.

Running a slot at $f_{z0} = 0.17$ mm/tooth ($hm = 0.108$, $ECT_E = 0.103$ mm) with 2 teeth and for a tool life 45 minutes, the cutting speed should be selected at $V = 340$ m/min at point SL2 and for six teeth ($hm = 0.108$ mm, $ECT_E = 0.308$) at $V = 240$ m/min at point SL6.

When high-speed milling for $ar/D = 0.03$ at $f_z = 3.394 \times 0.17 = 0.58$ mm/tooth = 0.58 mm/tooth, *ECT* is reduced to 0.011 mm ($hm = 0.108$) the cutting speed is 290 m/min to maintain $T = 45$ minutes, point HS2. This point is far to the left of the *H*-curve in Fig. 26b, but if the number of teeth is increased to 6 ($ECT_E = 3 \times 0.103 = 0.3090$), the cutting speed is 360 m/min at $T = 45$ minutes and is close to the *H*-curve, point HS6. Slotting data using 6 teeth are on the right of this curve at point SL6, approaching the *G*-curve, but at a lower slotting speed of 240 m/min.

Depending on the starting $f_z$ value, and on the combination of cutter grade and work material, the location of the $H$-curve plays an important role when selecting high-speed end milling data.

**Feed Rate and Tool Life in High-Speed Milling, Effect of *ECT* and Number of Teeth.—** Calculation of feed rate is done using the formulas previously given:

Feed Rate:

$F_R = z \times f_z \times \text{rpm}$, where $z \times f_z = f$ (feed/rev of cutter). Feed is measured along the feeding direction.

$\text{rpm} = 1000 \times V/3.1416/D$, where $D$ is diameter of cutter.

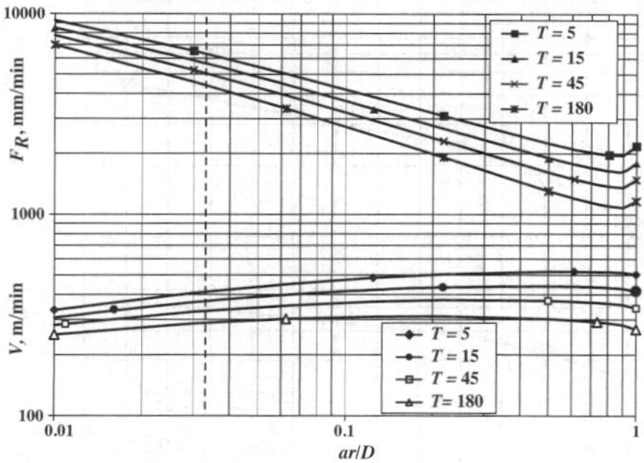


Fig. 26b. High-Speed Feed Rate and Cutting Speed versus $ar/D$ at $T = 5, 15, 45$, and 180 Minutes

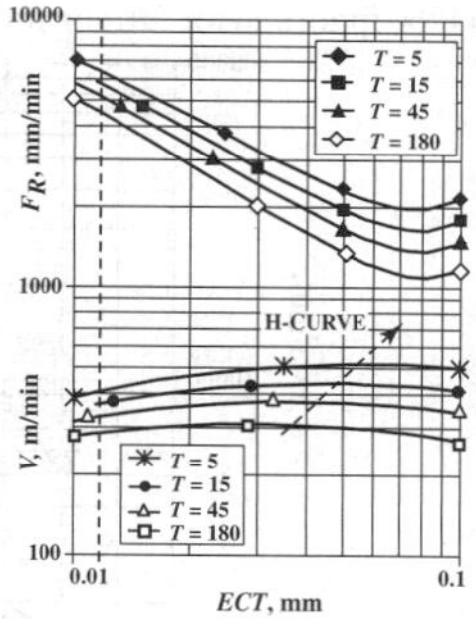


Fig. 26c. High-Speed Feed Rate and Cutting Speed versus $ECT$, $ar/D$ Plotted at $T = 5, 15, 45$, and 180 Minutes

Fig. 26b shows the variation of feed rate $F_R$ plotted versus $ar/D$ for tool-lives 5, 15, 45 and 180 minutes with a 25 mm diameter cutter and 2 teeth. Fig. 26c shows the variation of feed rate $F_R$ when plotted versus $ECT$. In both graphs the corresponding cutting speeds are also plotted. The values for $ar/D = 0.03$ in Fig. 26b correspond to $ECT = 0.011$ in Fig. 26c.

Feed rates have a minimum around values of $ar/D = 0.8$ and $ECT = 0.75$ and not along the $H$-curve. This is due to the fact that the $f_z/f_{z0}$ ratio to maintain a mean chip thickness = 0.108 mm changes $F_R$ in a different proportion than the cutting speed.

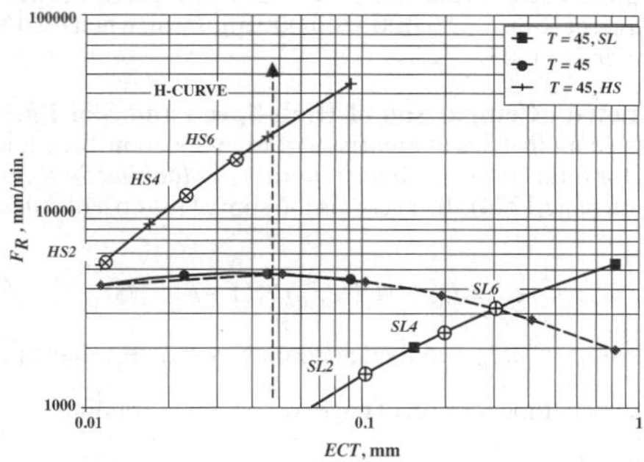


Fig. 26d. Feed Rate versus $ECT$ Comparison of Slot Milling ($ar/D = 1$) and High-Speed Milling at ($ar/D = 0.03$) for 2, 4, and 6 Teeth at $T = 45$ Minutes

A comparison of feed rates for full slot ($ar/D = 1$) and high-speed end milling ($ar/D = 0.03$ and $f_z = 3.69 \times f_{z0} = 0.628$ mm) for tool life 45 minutes is shown in Fig. 26d. The points SL2, SL4, SL6 and HS2, HS4, HS6, refer to 2, 4, and 6 teeth (2 to 6 teeth are commonly used in practice). Feed rate is also plotted versus number of teeth $z$ in Fig. 26e, for up to 16 teeth, still at $f_z = 0.628$ mm.

Comparing the effect of using 2 versus 6 teeth in high-speed milling shows that feed rates increase from 5250 mm/min (413 ipm) up to 18,000 mm/min (1417 ipm) at 45 minutes tool life. The effect of using 2 versus 6 teeth in full slot milling is that feed rate increases from 1480 mm/min (58 ipm) up to 3230 mm/min (127 ipm) at tool life 45 minutes. If 16 teeth could be used at $ar/D = 0.03$, the feed rate increases to $F_R = 44{,}700$ mm/min (1760 ipm), and for full slot milling $F_R = 5350$ mm/min (210 ipm).

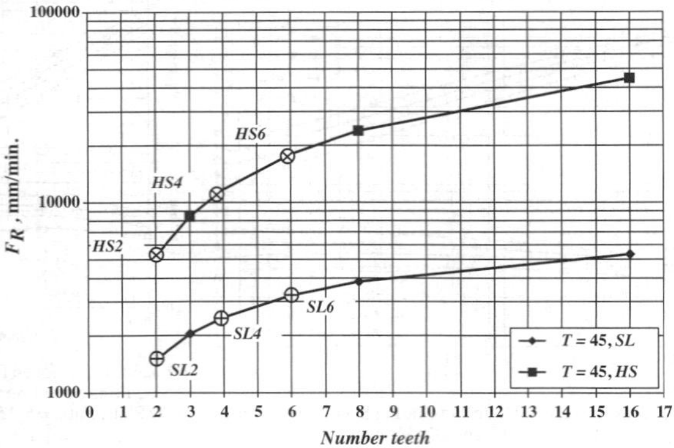


Fig. 26e. Feed Rate versus Number of Teeth Comparison of Slot Milling ($ar/D = 1$) and High-Speed Milling at ($ar/D = 0.03$) for 2, 4, and 6 Teeth at $T = 45$ Minutes

Comparing the feed rates that can be obtained in steel cutting with the one achieved in the earlier referred DARPA investigation, using HSS and carbide cutters milling 6061-T651 and A356-T6 aluminum, it is obvious that aluminum end milling can be run at 3 to 6 times higher feed rates. This requires 3 to 6 times higher spindle speeds (cutter diameter 25 mm, radial depth of cut $ar = 12.5$ mm, 2 teeth). Had these tests been run with 6 teeth, the feed rates would increase up to 150,000–300,000 mm/min, when feed/tooth = $3.4 \times 0.25 = 0.8$ mm/tooth at $ar/D = 0.03$.

**Process Econometrics Comparison of High-Speed and Slot End Milling.**—When making a process econometrics comparison of high-speed milling and slot end milling use the formulas for total cost $c_{tot}$ (*Determination of Machine Settings and Calculation of Costs* starting on page 1204). Total cost is the sum of the cost of cutting, tool changing, and tooling:

$$c_{tot} = H_R \times (Dist/F_R) \times (1 + T_V/T)/60$$

where $T_V = T_{RPL} + 60 \times C_E/H_R$ = equivalent tooling-cost time, minutes

$T_{RPL}$ = replacement time for a set of edges or tool for regrinding

$C_E$ = cost per edge(s)

$H_R$ = hourly rate, $

Fig. 27 compares total cost $c_{tot}$, using the end milling cutters of the previous examples, for full slot milling with high-speed milling at $ar/D$ =0.03, and versus $ECT$ at $T$ = 45 minutes.

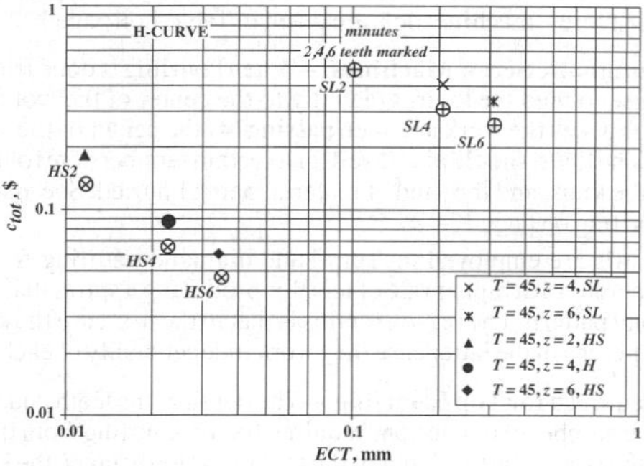


Fig. 27. Cost Comparison of Slot Milling ($ar/D = 1$) and High-Speed Milling at ($ar/D = 0.03$) for 2, 4, and 6 Teeth at $T = 45$ Minutes

The feed/tooth for slot milling is $f_{z0} = 0.17$ and for high-speed milling at $ar/D = 0.03$ the feed is $f_z = 3.69 \times f_{z0} = 0.628$ mm.

The calculations for total cost are done according to above formula using tooling cost at $T_V = 6, 10$, and 14 minutes, for $z = 2, 4$, and 6 teeth respectively. The distance cut is $Dist$ = 1000 mm. Full slot milling costs are,

at feed rate $F_R = 3230$ and $z = 6$

$$c_{tot} = 50 \times (1000/3230) \times (1 + 14/45)/60 = \$0.338 \text{ per part}$$

at feed rate $F_R$ =1480 and $z = 2$

$$c_{tot} = 50 \times (1000/1480) \times (1 + 6/45)/60 = \$0.638 \text{ per part}$$

High-speed milling costs are,

at $F_R$ =18000, $z = 6$

$$c_{tot} = 50 \times (1000/18000) \times (1 + 14/45)/60 = \$0.0606 \text{ per part}$$

at $F_R = 5250, z = 2$

$$c_{tot} = 50 \times (1000/5250) \times (1 + 6/45)/60 = \$0.180 \text{ per part}$$

The cost reduction using high-speed milling compared to slotting is enormous. The cost for high-speed milling with 2 teeth is 61 percent (0.208/0.338) of full slot milling with 6 teeth ($z = 6$). The cost for high-speed milling with 6 teeth is 19 percent (0.0638/0.338) of full slot for $z = 6$.

Aluminum end milling can be run at 3 to 6 times lower costs than when cutting steel. Costs of idle (non-machining) and slack time (waste) are not considered in the example. These data hold for perfect milling conditions such as zero run-out and accurate sharpening of all teeth and edges.

## SCREW MACHINES, BAND SAWS, CUTTING FLUIDS

### Automatic Screw Machine Tool Operations

**Speeds and Feeds for Automatic Screw Machine Tools.—**Approximate speeds and feeds for standard screw machine tools are given in Table 1 on page 1223.

**Knurling in Automatic Screw Machines.—**When knurling is done from the cross slide, it is good practice to feed the knurl gradually to the center of the work, starting to feed when the knurl touches the work and then passing off the center of the work with a quick rise of the cam. The knurl should also dwell for a certain number of revolutions, depending on the pitch of the knurl and the kind of material being knurled. See also *KNURLS AND KNURLING* starting on page 1044.

When two knurls are employed for spiral and diamond knurling from the turret, the knurls can be operated at a higher rate of feed for producing a spiral than they can for producing a diamond pattern. The reason for this is that in the first case the knurls work in the same groove, whereas in the latter case they work independently of each other.

**Revolutions Required for Top Knurling.—**The depth of the teeth and the feed per revolution govern the number of revolutions required for top knurling from the cross slide. If $R$ is the radius of the stock, $d$ is the depth of the teeth, $c$ is the distance the knurl travels from the point of contact to the center of the work at the feed required for knurling, and $r$ is the radius of the knurl. Inch or metric unts may be used in the following equation.

$$c = \sqrt{(R+r)^2 - (R+r-d)^2}$$

*Example:* If the stock radius $R$ is $\frac{5}{32}$ inch (or mm), depth of teeth $d$ is 0.0156 inch (or mm), and radius of knurl $r$ is 0.3125 inch (or mm), then

$$c = \sqrt{(0.1562 + 0.3125)^2 - (0.1562 + 0.3125 - 0.0156)^2}$$
$$= 0.120 \text{ inch (or mm)} = \text{cam rise required}$$

*Example:* Assume that it is required to find the number of revolutions to knurl a piece of brass $\frac{5}{16}$ inch in diameter using a 32 pitch knurl. The included angle of the teeth for brass is 90 degrees, the circular pitch is 0.03125 inch, and the calculated tooth depth is 0.0156 inch. The distance $c$ (as determined in the previous example) is 0.120 inch. Referring to Table 1 on page 1223, the feed for top knurling brass is 0.005 inch per revolution. The number of revolutions required for knurling is, therefore, $0.120 \div 0.005 =$ 24 revolutions. If conditions permit, the higher feed of 0.008 inch per revolution given in the table may be used, and 15 revolutions are then required for knurling.

**Cams for Threading.—**Table 2 on page 1225 gives the revolutions required for threading various lengths and pitches and the corresponding rise for the cam lobe. To illustrate the use of this table, suppose a set of cams is required for threading a screw to the length of $\frac{3}{8}$ inch in a Brown & Sharpe machine. Assume that the spindle speed is 2400 revolutions per minute; the number of revolutions to complete one piece, 400; time required to make one piece, 10 seconds; pitch of the thread, $\frac{1}{32}$ inch or 32 threads per inch. By referring to Table 2, under 32 threads per inch, and opposite $\frac{3}{8}$ inch (length of threaded part), the number of revolutions required is found to be 15 and the rise required for the cam, 0.413 inch.

**Table 1. Approximate Cutting Speeds and Feeds for Standard Automatic Screw Machine Tools—Brown and Sharpe**

| Tool | | Cut | | Material to be Machined | | | | | | | |
|---|---|---|---|---|---|---|---|---|---|---|---|
| | | | | Brass[a] | Mild or Soft Steel | | | Tool Steel, 0.80–1.00% C | | | |
| | | | | | | Surface Speed, Feet per Min. | | | Surface Speed, Feet per Min. | | |
| Tool | | Width or Depth, Inches | Dia. of Hole, Inches | Feed, Inches per Rev. | Feed, Inches per Rev. | Carbon Tools | H.S.S. Tools | Feed, Inches per Rev. | Carbon Tools | H.S.S. Tools | |
| Boring tools | | 0.005 | … | … | 0.008 | 50 | 110 | 0.004 | 30 | 60 | |
| Box tools, roller rest | Single chip finishing | 1⁄32 | … | 0.012 | 0.010 | 70 | 150 | 0.005 | 40 | 75 | |
| | | 1⁄16 | … | 0.010 | 0.008 | 70 | 150 | 0.004 | 40 | 75 | |
| | | 1⁄8 | … | 0.008 | 0.007 | 70 | 150 | 0.003 | 40 | 75 | |
| | | 3⁄16 | … | 0.008 | 0.006 | 70 | 150 | 0.002 | 40 | 75 | |
| | | 1⁄4 | … | 0.006 | 0.005 | 70 | 150 | 0.0015 | 40 | 75 | |
| | Finishing | 0.005 | … | 0.010 | 0.010 | 70 | 150 | 0.006 | 40 | 75 | |
| Center drills | | … | Under 1⁄8 | 0.003 | 0.0015 | 50 | 110 | 0.001 | 30 | 75 | |
| | | … | Over 1⁄8 | 0.006 | 0.0035 | 50 | 110 | 0.002 | 30 | 75 | |
| Cutoff tools { | Angular | … | … | 0.0015 | 0.0006 | 80 | 150 | 0.0004 | 50 | 85 | |
| | Circular | 3⁄64–1⁄8 | … | 0.0035 | 0.0015 | 80 | 150 | 0.001 | 50 | 85 | |
| | Straight | 1⁄16–1⁄8 | … | 0.0035 | 0.0015 | 80 | 150 | 0.001 | 50 | 85 | |
| Stock diameter under 1⁄8 in. | | … | … | 0.002 | 0.0008 | 80 | 150 | 0.0005 | 50 | 85 | |
| Dies { | Button | … | … | … | … | 30 | … | … | 14 | … | |
| | Chaser | … | … | … | … | 30 | 40 | … | 16 | 20 | |
| Drills, twist cut | | … | 0.02 | 0.0014 | 0.001 | 40 | 60 | 0.0006 | 30 | 45 | |
| | | … | 0.04 | 0.002 | 0.0014 | 40 | 60 | 0.0008 | 30 | 45 | |
| | | … | 1⁄16 | 0.004 | 0.002 | 40 | 60 | 0.0012 | 30 | 45 | |
| | | … | 3⁄32 | 0.006 | 0.0025 | 40 | 60 | 0.0016 | 30 | 45 | |
| | | … | 1⁄8 | 0.009 | 0.0035 | 40 | 75 | 0.002 | 30 | 60 | |
| | | … | 3⁄16 | 0.012 | 0.004 | 40 | 75 | 0.003 | 30 | 60 | |
| | | … | 1⁄4 | 0.014 | 0.005 | 40 | 75 | 0.003 | 30 | 60 | |
| | | … | 5⁄16 | 0.016 | 0.005 | 40 | 75 | 0.0035 | 30 | 60 | |
| | | … | 3⁄8–5⁄8 | 0.016 | 0.006 | 40 | 85 | 0.004 | 30 | 60 | |
| Form tools, circular | | 1⁄8 | … | 0.002 | 0.0009 | 80 | 150 | 0.0006 | 50 | 85 | |
| | | 1⁄4 | … | 0.002 | 0.0008 | 80 | 150 | 0.0005 | 50 | 85 | |
| | | 3⁄8 | … | 0.0015 | 0.0007 | 80 | 150 | 0.0004 | 50 | 85 | |
| | | 1⁄2 | … | 0.0012 | 0.0006 | 80 | 150 | 0.0004 | 50 | 85 | |
| | | 5⁄8 | … | 0.001 | 0.0005 | 80 | 150 | 0.0003 | 50 | 85 | |
| | | 3⁄4 | … | 0.001 | 0.0005 | 80 | 150 | 0.0003 | 50 | 85 | |
| | | 1 | … | 0.001 | 0.0004 | 80 | 150 | … | … | … | |

**Table 1.** *(Continued)* **Approximate Cutting Speeds and Feeds for Standard Automatic Screw Machine Tools — Brown and Sharpe**

| Tool | | Cut | | Material to be Machined | | | | | | |
|---|---|---|---|---|---|---|---|---|---|---|
| | | | | Brass[a] | Mild or Soft Steel | | | Tool Steel, 0.80–1.00% C | | |
| | | | | | | Surface Speed, Feet per Min. | | | Surface Speed, Feet per Min. | |
| | | Width or Depth, Inches | Dia. of Hole, Inches | Feed, Inches per Rev. | Feed, Inches per Rev. | Carbon Tools | H.S.S. Tools | Feed, Inches per Rev. | Carbon Tools | H.S.S. Tools |
| Hollow mills and balance turning tools | Turned Dia. under 5/32 in. | 1/32 | … | 0.012 | 0.010 | 70 | 150 | 0.008 | 40 | 85 |
| | | 1/16 | … | 0.010 | 0.009 | 70 | 150 | 0.006 | 40 | 85 |
| | Turned Dia. over 5/32 in. | 1/32 | … | 0.017 | 0.014 | 70 | 150 | 0.010 | 40 | 85 |
| | | 1/16 | … | 0.015 | 0.012 | 70 | 150 | 0.008 | 40 | 85 |
| | | 1/8 | … | 0.012 | 0.010 | 70 | 150 | 0.008 | 40 | 85 |
| | | 3/16 | … | 0.010 | 0.008 | 70 | 150 | 0.006 | 40 | 85 |
| | | 1/4 | … | 0.009 | 0.007 | 70 | 150 | 0.0045 | 40 | 85 |
| Knee tools | | 1/32 | … | … | 0.010 | 70 | 150 | 0.008 | 40 | 85 |
| Knurling tools | Turret | On | … | 0.020 | 0.015 | 150 | … | 0.010 | 105 | … |
| | | Off | … | 0.040 | 0.030 | 150 | … | 0.025 | 105 | … |
| | Side or swing | … | … | 0.004 | 0.002 | 150 | … | 0.002 | 105 | … |
| | | … | … | 0.006 | 0.004 | 150 | … | 0.003 | 105 | … |
| | Top | … | … | 0.005 | 0.003 | 150 | … | 0.002 | 105 | … |
| | | … | … | 0.008 | 0.006 | 150 | … | 0.004 | 105 | … |
| Pointing and facing tools | | … | … | 0.001 | 0.0008 | 70 | 150 | 0.0005 | 40 | 80 |
| | | … | … | 0.0025 | 0.002 | 70 | 150 | 0.0008 | 40 | 80 |
| Reamers and bits | | 0.003—0.004 | 1/8 or less | 0.010—0.007 | 0.008—0.006 | 70 | 105 | 0.006—0.004 | 40 | 60 |
| | | 0.004—0.008 | 1/8 or over | 0.010 | 0.010 | 70 | 105 | 0.006—0.008 | 40 | 60 |
| Recessing tools | End cut | … | … | 0.001 | 0.0006 | 70 | 150 | 0.0004 | 40 | 75 |
| | | … | … | 0.005 | 0.003 | 70 | 150 | 0.002 | 40 | 75 |
| | Inside cut | 1/16-1/8 | … | 0.0025 | 0.002 | 70 | 105 | 0.0015 | 40 | 60 |
| | | | … | 0.0008 | 0.0006 | 70 | 105 | 0.0004 | 40 | 60 |
| Swing tools, forming | | 1/8 | … | 0.002 | 0.0007 | 70 | 150 | 0.0005 | 40 | 85 |
| | | 1/4 | … | 0.0012 | 0.0005 | 70 | 150 | 0.0003 | 40 | 85 |
| | | 3/8 | … | 0.001 | 0.0004 | 70 | 150 | 0.0002 | 40 | 85 |
| | | 1/2 | … | 0.0008 | 0.0003 | 70 | 150 | 0.0002 | 40 | 85 |
| Turning, straight and taper[b] | | 1/32 | … | 0.008 | 0.006 | 70 | 150 | 0.0035 | 40 | 85 |
| | | 1/16 | … | 0.006 | 0.004 | 70 | 150 | 0.003 | 40 | 85 |
| | | 1/8 | … | 0.005 | 0.003 | 70 | 150 | 0.002 | 40 | 85 |
| | | 3/16 | … | 0.004 | 0.0025 | 70 | 150 | 0.0015 | 40 | 85 |
| Taps | | … | … | … | … | 25 | 30 | … | 12 | 15 |

[a] Use maximum spindle speed on machine.

[b] For taper turning use feed slow enough for greatest depth depth of cut.

## Table 2. Spindle Revolutions and Cam Rise for Threading

| Length of Threaded Portion, Inch | Number of Threads per Inch | | | | | | | | | | | | | | |
|---|---|---|---|---|---|---|---|---|---|---|---|---|---|---|---|
| | 80 | 72 | 64 | 56 | 48 | 40 | 36 | 32 | 30 | 28 | 24 | 20 | 18 | 16 | 14 |
| | First Line: Revolutions of Spindle for Threading. Second Line: Rise on Cam for Threading, Inch | | | | | | | | | | | | | | |
| 1⁄16 | 9.50 | 9.00 | 8.50 | 8.00 | 6.00 | 5.50 | 5.50 | 5.00 | 5.00 | 5.00 | 3.00 | … | … | … | … |
| | 0.107 | 0.113 | 0.120 | 0.129 | 0.110 | 0.121 | 0.134 | 0.138 | 0.147 | 0.157 | 0.106 | … | … | … | … |
| 1⁄8 | 14.50 | 13.50 | 12.50 | 11.50 | 9.00 | 8.00 | 7.00 | 7.00 | 7.00 | 6.50 | 4.50 | 4.00 | 3.50 | 3.50 | … |
| | 0.163 | 0.169 | 0.176 | 0.185 | 0.165 | 0.176 | 0.171 | 0.193 | 0.205 | 0.204 | 0.159 | 0.170 | 0.165 | 0.186 | … |
| 3⁄16 | 19.50 | 18.00 | 16.50 | 15.00 | 12.00 | 10.50 | 10.00 | 9.00 | 8.50 | 8.50 | 6.00 | 5.50 | 5.00 | 4.50 | 4.00 |
| | 0.219 | 0.225 | 0.232 | 0.241 | 0.220 | 0.231 | 0.244 | 0.248 | 0.249 | 0.267 | 0.213 | 0.234 | 0.236 | 0.239 | 0.243 |
| 1⁄4 | 24.50 | 23.508 | 20.50 | 18.50 | 15.00 | 13.00 | 12.00 | 11.00 | 10.50 | 10.00 | 7.50 | 6.50 | 6.00 | 5.50 | 5.00 |
| | 0.276 | 0.294 | 0.288 | 0.297 | 0.275 | 0.286 | 0.293 | 0.303 | 0.308 | 0.314 | 0.266 | 0.276 | 0.283 | 0.292 | 0.304 |
| 5⁄16 | 29.50 | 27.00 | 24.50 | 22.00 | 18.00 | 15.50 | 14.50 | 13.00 | 12.50 | 12.00 | 9.00 | 8.00 | 7.00 | 6.50 | 6.00 |
| | 0.332 | 0.338 | 0.345 | 0.354 | 0.340 | 0.341 | 0.354 | 0.358 | 0.367 | 0.377 | 0.319 | 0.340 | 0.330 | 0.345 | 0.364 |
| 3⁄8 | 34.50 | 31.50 | 28.50 | 25.50 | 21.00 | 18.00 | 16.50 | 15.00 | 14.50 | 13.50 | 10.50 | 9.00 | 8.50 | 7.50 | 7.00 |
| | 0.388 | 0.394 | 0.401 | 0.410 | 0.385 | 0.396 | 0.403 | 0.413 | 0.425 | 0.424 | 0.372 | 0.383 | 0.401 | 0.398 | 0.425 |
| 7⁄16 | 39.50 | 36.00 | 32.50 | 29.00 | 24.00 | 20.50 | 19.00 | 17.00 | 16.00 | 15.50 | 12.00 | 10.50 | 9.50 | 8.50 | 7.50 |
| | 0.444 | 0.450 | 0.457 | 0.466 | 0.440 | 0.451 | 0.464 | 0.468 | 0.469 | 0.487 | 0.425 | 0.446 | 0.448 | 0.451 | 0.455 |
| 1⁄2 | 44.50 | 40.50 | 36.50 | 32.50 | 27.00 | 23.00 | 21.00 | 19.00 | 18.00 | 17.00 | 13.50 | 11.50 | 10.50 | 9.50 | 8.50 |
| | 0.501 | 0.506 | 0.513 | 0.522 | 0.495 | 0.506 | 0.513 | 0.523 | 0.528 | 0.534 | 0.478 | 0.489 | 0.496 | 0.504 | 0.516 |
| 9⁄16 | 49.50 | 45.00 | 40.50 | 36.00 | 30.00 | 25.50 | 23.50 | 21.00 | 20.00 | 19.00 | 15.00 | 13.00 | 11.50 | 10.50 | 9.50 |
| | 0.559 | 0.563 | 0.570 | 0.579 | 0.550 | 0.561 | 0.574 | 0.578 | 0.587 | 0.597 | 0.531 | 0.553 | 0.543 | 0.558 | 0.577 |
| 5⁄8 | 54.50 | 49.50 | 44.50 | 39.50 | 33.00 | 28.00 | 25.50 | 23.00 | 22.00 | 20.50 | 16.50 | 14.00 | 13.00 | 11.50 | 10.50 |
| | 0.613 | 0.619 | 0.626 | 0.635 | 0.605 | 0.616 | 0.623 | 0.633 | 0.645 | 0.644 | 0.584 | 0.595 | 0.614 | 0.611 | 0.637 |
| 11⁄16 | 59.50 | 54.00 | 48.50 | 43.00 | 36.00 | 30.50 | 28.00 | 25.00 | 23.50 | 22.50 | 18.00 | 15.50 | 14.00 | 12.50 | 11.00 |
| | 0.679 | 0.675 | 0.682 | 0.691 | 0.660 | 0.671 | 0.684 | 0.688 | 0.689 | 0.707 | 0.638 | 0.659 | 0.661 | 0.664 | 0.668 |
| 3⁄4 | 64.50 | 58.50 | 52.50 | 46.50 | 39.00 | 33.00 | 30.00 | 27.00 | 25.50 | 24.00 | 19.50 | 16.50 | 15.00 | 13.50 | 12.00 |
| | 0.726 | 0.731 | 0.738 | 0.747 | 0.715 | 0.726 | 0.733 | 0.743 | 0.748 | 0.754 | 0.691 | 0.701 | 0.708 | 0.717 | 0.728 |

Threading cams are often cut on a circular milling attachment. When this method is employed, the number of minutes the attachment should be revolved for each 0.001 inch rise, is first determined. As 15 spindle revolutions are required for threading and 400 for completing one piece, that part of the cam surface required for the actual threading operation equals 15 ÷ 400 = 0.0375, which is equivalent to 810 minutes of the circumference. The total rise, through an arc of 810 minutes is 0.413 inch, so the number of minutes for each 0.001 inch rise equals 810 ÷ 413 = 1.96 or, approximately, two minutes. If the attachment is graduated to read to five minutes, the cam will be fed laterally 0.0025 inch each time it is turned through five minutes of arc.

**Practical Points on Cam and Tool Design.—**The following general rules are given to aid in designing cams and special tools for automatic screw machines, and apply particularly to Brown and Sharpe machines:

1) Use the highest speeds recommended for the material used that the various tools will stand.

2) Use the arrangement of circular tools best suited for the class of work.

3) Decide on the quickest and best method of arranging the operations before designing the cams.

4) Do not use turret tools for forming when the cross-slide tools can be used to better advantage.

5) Make the shoulder on the circular cutoff tool large enough so that the clamping screw will grip firmly.

6) Do not use too narrow a cutoff blade.

7) Allow 0.005 to 0.010 inch (0.127–0.254 mm) for the circular tools to approach the work and 0.003 to 0.005 inch (0.076–0.127 mm) for the cutoff tool to pass the center.

8) When cutting off work, the feed of the cutoff tool should be decreased near the end of the cut where the piece breaks off.

9) When a thread is cut up to a shoulder, the piece should be grooved or necked to make allowance for the lead on the die. An extra projection on the forming tool and an extra amount of rise on the cam will be needed.

10) Allow sufficient clearance for tools to pass one another.

11) Always make a diagram of the cross-slide tools in position on the work when difficult operations are to be performed; do the same for the tools held in the turret.

12) Do not drill a hole the depth of which is more than 3 times the diameter of the drill, but rather use two or more drills as required. If there are not enough turret positions for the extra drills needed, make provision for withdrawing the drill clear of the hole and then advancing it into the hole again.

13) Do not run drills at low speeds. Feeds and speeds recommended in Table 1 on page 1223 should be followed as far as is practicable.

14) When the turret tools operate farther in than the face of the chuck, see that they will clear the chuck when the turret is revolved.

15) See that the bodies of all turret tools will clear the side of the chute when the turret is revolved.

16) Use a balance turning tool or a hollow mill for roughing cuts.

17) The rise on the thread lobe should be reduced so that the spindle will reverse when the tap or die holder is drawn out.

18) When bringing another tool into position after a threading operation, allow clearance before revolving the turret.

19) Make provision to revolve the turret rapidly, especially when pieces are being made in from three to five seconds and when only a few tools are used in the turret. It is sometimes desirable to use two sets of tools.

20) When using a belt-shifting attachment for threading, clearance should be allowed, as it requires extra time to shift the belt.

21) When laying out a set of cams for operating on a piece that requires to be slotted, cross-drilled or burred, allowance should be made on the lead cam so that the transferring arm can descend and ascend to and from the work without coming in contact with any of the turret tools.

22) Always provide a vacant hole in the turret when it is necessary to use the transferring arm.

23) When designing special tools allow as much clearance as possible. Do not make them so that they will just clear each other, as a slight inaccuracy in the dimensions will often cause trouble.

24) When designing special tools having intricate movements, avoid springs as much as possible, and use positive actions.

**Stock for Screw Machine Products.—**The amount of stock required for the production of 1000 pieces on the automatic screw machine can be obtained directly from Table 3 on page 1228. To use this table, add to the length of the work the width of the cut-off tool blade; then the number of feet of material required for 1000 pieces can be found opposite the figure thus obtained, in the column headed "Feet per 1000 Parts." Screw machine stock usually comes in bars 10 feet (3.048 m) long, and in compiling this table an allowance was made for chucking on each bar.

The table can be extended by using the following formula, in which

$F$ = number of feet required for 1000 pieces

$L$ = length of piece in inches

$W$ = width of cut-off tool blade in inches

$$F = (L + W) \times 84$$

The amount to add to the length of the work, or the width of the cut-off tool, is given in the following, which is standard in a number of machine shops:

| Diameter of Stock | | Width of Cut-off Tool Blade | |
|---|---|---|---|
| in | mm | in | mm |
| 0.000–0.250 | 0.000–6.35 | 0.045 | 1.44 |
| 0.251–0.375 | 6.38–9.53 | 0.062 | 1.57 |
| 0.376–0.625 | 9.55–15.88 | 0.093 | 2.36 |
| 0.626–1.000 | 15.90–25.40 | 0.125 | 3.18 |
| 1.001–1.500 | 25.43–38.10 | 0.156 | 3.96 |

It is sometimes convenient to know the weight of a certain number of pieces, when estimating the price. The weight of round bar stock can be found by means of the following formulas, in which

$W$ = weight in pounds

$D$ = diameter of stock in inches

$F$ = length in feet

For brass stock: $W = D^2 \times 2.86 \times F$

For steel stock: $W = D^2 \times 2.675 \times F$

For iron stock: $W = D^2 \times 2.65 \times F$

**Table 3. Stock Required for Screw Machine Products**

The table gives the amount of stock, in feet, required for 1000 pieces, when the length of the finished part plus the thickness of the cut-off tool blade is known. Allowance has been made for chucking. To illustrate, if length of cut-off tool and work equals 0.140 inch, 11.8 feet of stock is required for the production of 1000 parts.

| Length of Piece and Cut-Off Tool | Feet per 1000 Parts | Length of Piece and Cut-Off Tool | Feet per 1000 Parts | Length of Piece and Cut-Off Tool | Feet per 1000 Parts | Length of Piece and Cut-Off Tool | Feet per 1000 Parts |
|---|---|---|---|---|---|---|---|
| 0.050 | 4.2 | 0.430 | 36.1 | 0.810 | 68.1 | 1.380 | 116.0 |
| 0.060 | 5.0 | 0.440 | 37.0 | 0.820 | 68.9 | 1.400 | 117.6 |
| 0.070 | 5.9 | 0.450 | 37.8 | 0.830 | 69.7 | 1.420 | 119.3 |
| 0.080 | 6.7 | 0.460 | 38.7 | 0.840 | 70.6 | 1.440 | 121.0 |
| 0.090 | 7.6 | 0.470 | 39.5 | 0.850 | 71.4 | 1.460 | 122.7 |
| 0.100 | 8.4 | 0.480 | 40.3 | 0.860 | 72.3 | 1.480 | 124.4 |
| 0.110 | 9.2 | 0.490 | 41.2 | 0.870 | 73.1 | 1.500 | 126.1 |
| 0.120 | 10.1 | 0.500 | 42.0 | 0.880 | 73.9 | 1.520 | 127.7 |
| 0.130 | 10.9 | 0.510 | 42.9 | 0.890 | 74.8 | 1.540 | 129.4 |
| 0.140 | 11.8 | 0.520 | 43.7 | 0.900 | 75.6 | 1.560 | 131.1 |
| 0.150 | 12.6 | 0.530 | 44.5 | 0.910 | 76.5 | 1.580 | 132.8 |
| 0.160 | 13.4 | 0.540 | 45.4 | 0.920 | 77.3 | 1.600 | 134.5 |
| 0.170 | 14.3 | 0.550 | 46.2 | 0.930 | 78.2 | 1.620 | 136.1 |
| 0.180 | 15.1 | 0.560 | 47.1 | 0.940 | 79.0 | 1.640 | 137.8 |
| 0.190 | 16.0 | 0.570 | 47.9 | 0.950 | 79.8 | 1.660 | 139.5 |
| 0.200 | 16.8 | 0.580 | 48.7 | 0.960 | 80.7 | 1.680 | 141.2 |
| 0.210 | 17.6 | 0.590 | 49.6 | 0.970 | 81.5 | 1.700 | 142.9 |
| 0.220 | 18.5 | 0.600 | 50.4 | 0.980 | 82.4 | 1.720 | 144.5 |
| 0.230 | 19.3 | 0.610 | 51.3 | 0.990 | 83.2 | 1.740 | 146.2 |
| 0.240 | 20.2 | 0.620 | 52.1 | 1.000 | 84.0 | 1.760 | 147.9 |
| 0.250 | 21.0 | 0.630 | 52.9 | 1.020 | 85.7 | 1.780 | 149.6 |
| 0.260 | 21.8 | 0.640 | 53.8 | 1.040 | 87.4 | 1.800 | 151.3 |
| 0.270 | 22.7 | 0.650 | 54.6 | 1.060 | 89.1 | 1.820 | 152.9 |
| 0.280 | 23.5 | 0.660 | 55.5 | 1.080 | 90.8 | 1.840 | 154.6 |
| 0.290 | 24.4 | 0.670 | 56.3 | 1.100 | 92.4 | 1.860 | 156.3 |
| 0.300 | 25.2 | 0.680 | 57.1 | 1.120 | 94.1 | 1.880 | 158.0 |
| 0.310 | 26.1 | 0.690 | 58.0 | 1.140 | 95.8 | 1.900 | 159.7 |
| 0.320 | 26.9 | 0.700 | 58.8 | 1.160 | 97.5 | 1.920 | 161.3 |
| 0.330 | 27.7 | 0.710 | 59.7 | 1.180 | 99.2 | 1.940 | 163.0 |
| 0.340 | 28.6 | 0.720 | 60.5 | 1.200 | 100.8 | 1.960 | 164.7 |
| 0.350 | 29.4 | 0.730 | 61.3 | 1.220 | 102.5 | 1.980 | 166.4 |
| 0.360 | 30.3 | 0.740 | 62.2 | 1.240 | 104.2 | 2.000 | 168.1 |
| 0.370 | 31.1 | 0.750 | 63.0 | 1.260 | 105.9 | 2.100 | 176.5 |
| 0.380 | 31.9 | 0.760 | 63.9 | 1.280 | 107.6 | 2.200 | 184.9 |
| 0.390 | 32.8 | 0.770 | 64.7 | 1.300 | 109.2 | 2.300 | 193.3 |
| 0.400 | 33.6 | 0.780 | 65.5 | 1.320 | 110.9 | 2.400 | 201.7 |
| 0.410 | 34.5 | 0.790 | 66.4 | 1.340 | 112.6 | 2.500 | 210.1 |
| 0.420 | 35.3 | 0.800 | 67.2 | 1.360 | 114.3 | 2.600 | 218.5 |

## Band Saw Blade Selection

The primary factors to consider in choosing a saw blade are: the pitch, or the number of teeth per inch of blade; the tooth form; and the blade type (material and construction). Tooth pitch selection depends on the size and shape of the work, whereas tooth form and blade type depend on material properties of the workpiece and on economic considerations of the job.

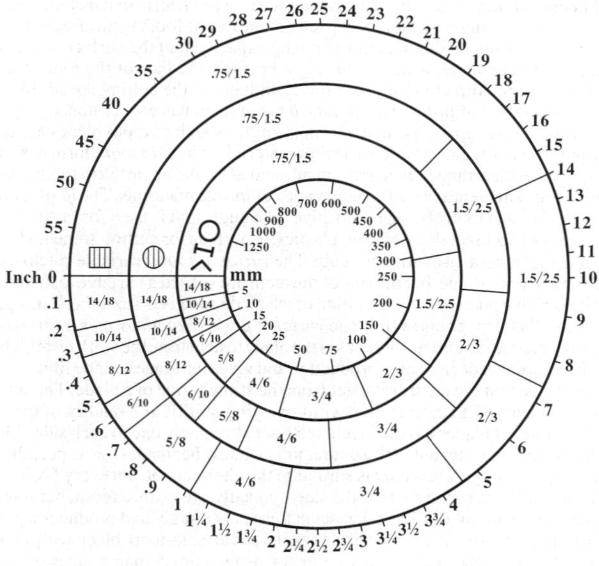


*Courtesy of American Saw and Manufacturing Company*

The tooth selection chart above is a guide to help determine the best blade pitch for a particular job. The tooth specifications in the chart are standard variable-pitch blade sizes as specified by the Hack and Band Saw Association. The variable-pitch blades listed are designated by two numbers that refer to the approximate maximum and minimum tooth pitch. A 4⁄6 blade, for example, has a maximum tooth spacing of approximately ¼ inch (6.4 mm) and a minimum tooth spacing of about ⅙ inch (4.2 mm). Blades are available, from most manufacturers, in sizes within about ±10 percent of the sizes listed.

To use the chart, locate the length of cut in inches on the outside circle of the table (for millimeters use the inside circle) and then find the tooth specification that aligns with the length, on the ring corresponding to the material shape. The length of cut is the distance that any tooth of the blade is in contact with the work as it passes once through the cut. For cutting solid round stock, use the diameter as the length of cut and select a blade from the ring with the solid circle. When cutting angles, channels, I-beams, tubular pieces, pipe, and hollow or irregular shapes, the length of cut is found by dividing the cross-sectional area of the cut by the distance the blade needs to travel to finish the cut. Locate the length of cut on the outer ring (inner ring for mm) and select a blade from the ring marked with the angle, I-beam, and pipe sections.

*Example:* A 4-inch pipe with a 3-inch inside diameter is to be cut. Select a variable pitch blade for cutting this material.

The area of the pipe is $\pi/4 \times (4^2 - 3^2) = 5.5\ \text{in}^2$. The blade has to travel 4 inches to cut through the pipe, so the average length of cut is 5.5⁄4 = 1.4 inches. On the tooth selection wheel, estimate the location of 1.4 inches on the outer ring, and read the tooth specification from the ring marked with the pipe, angle, and I-beam symbols. The chart indicates that a 4⁄6 variable-pitch blade is the preferred blade for this cut.

**Tooth Forms.—**Band saw teeth are characterized by a tooth form that includes the shape, spacing (pitch), rake angle, and gullet capacity of the tooth. Tooth form affects the cutting efficiency, noise level, blade life, chip-carrying capacity, and the surface finish quality of the cut. The rake angle, which is the angle between the face of the tooth and a line perpendicular to the direction of blade travel, influences the cutting speed. In general, positive rake angles cut faster. The *standard tooth* form has conventional shape teeth, evenly spaced with deep gullets and a 0° rake angle. Standard tooth blades are used for general-purpose cutting on a wide variety of materials. The *skip tooth* form has shallow, widely spaced teeth arranged in narrow bands and a 0° rake angle. Skip tooth blades are used for cutting soft metals, wood, plastics, and composite materials. The *hook tooth* form is similar to the skip tooth, but has a positive rake angle and is used for faster cutting of large sections of soft metal, wood, and plastics, as well as for cutting some metals, such as cast iron, that form a discontinuous chip. The *variable-tooth* (variable-pitch) form has a conventional tooth shape, but the tips of the teeth are spaced a variable distance (pitch) apart. The variable pitch reduces vibration of the blade and gives smoother cutting, better surface finish, and longer blade life. The *variable positive tooth* form is a variable-pitch tooth with a positive rake angle that causes the blade to penetrate the work faster. The variable positive tooth blade increases production and gives the longest blade life.

*Set* is the angle that the teeth are offset from the straight line of a blade. The set affects the blade efficiency (i.e., cutting rate), chip-carrying ability, and quality of the surface finish. *Alternate set* blades have adjacent teeth set alternately one to each side. Alternate set blades, which cut faster but with a poorer finish than other blades, are especially useful for rapid rough cutting. A *raker set* is similar to the alternate set, but every few teeth, one of the teeth is set to the center, not to the side (typically every third tooth, but sometimes every fifth or seventh tooth). The raker set pattern cuts rapidly and produces a good surface finish. The *vari-raker set,* or variable raker, is a variable-tooth blade with a raker set. The vari-raker is quieter and produces a better surface finish than a raker set standard tooth blade. *Wavy set* teeth are set in groups, alternately to one side, then to the other. Both wavy set and vari-raker set blades are used for cutting tubing and other interrupted cuts, but the blade efficiency and surface finish produced are better with a vari-raker set blade.

**Types of Blades.—**The most important band saw blade types are carbon steel, bimetal, carbide tooth, and grit blades made with embedded carbide or diamond. *Carbon steel blades* have the lowest initial cost, but they may wear out faster. Carbon steel blades are used for cutting a wide variety of materials, including mild steels, aluminum, brass, bronze, cast iron, copper, lead, and zinc, as well as some abrasive materials such as cork, fiberglass, graphite, and plastics. *Bimetal blades* are made with a high-speed steel cutting edge that is welded to a spring steel blade back. Bimetal blades are stronger and last longer, and they tend to produce straighter cuts because the blade can be tensioned higher than carbon steel blades. Because bimetal blades last longer, the cost per cut is frequently lower than when using carbon steel blades. Bimetal blades are used for cutting all ferrous and nonferrous metals, a wide range of shapes of easy to moderately machinable material, and solids and heavy wall tubing with moderate to difficult machinability. *Tungsten carbide blades* are similar to bimetal blades but have tungsten carbide teeth welded to the blade back. The welded teeth of carbide blades have greater wear and high-temperature resistance than either carbon steel or bimetal blades and produce less tooth vibration, while giving smoother, straighter, faster, and quieter cuts requiring less feed force. Carbide blades are used on tough alloys such as cobalt, nickel- and titanium-based alloys, and for nonferrous materials such as aluminum castings, fiberglass, and graphite. The *carbide grit blade* has tungsten carbide grit metallurgically bonded to either a gulleted (serrated) or toothless steel band. The blades are made in several styles and grit sizes. Both carbide grit and diamond grit blades are used to cut materials that conventional (carbon and bimetal) blades are unable to cut such as: fiberglass, reinforced plastics, composite materials, carbon and graphite, aramid fibers, plastics, cast iron, stellites, high-hardness tool steels, and superalloys.

**Band Saw Speed and Feed Rate.—**The band speed necessary to cut a particular material is measured in feet per minute (fpm) or in meters per minute (m/min), and depends on material characteristics and size of the workpiece. Typical speeds for a bimetal blade cutting 4-inch material with coolant are given in the speed selection table that follows. For other size materials or when cutting without coolant, adjust speeds according to the instructions at the bottom of the table.

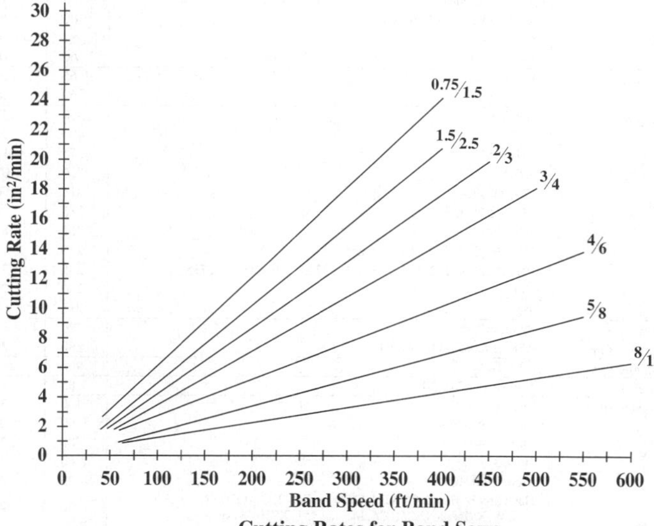


**Cutting Rates for Band Saws**

The feed or cutting rate, usually measured in square inches or square meters per minute, indicates how fast material is being removed and depends on the speed and pitch of the blade, not on the workpiece material. The graph above, based on material provided by American Saw and Mfg., gives approximate cutting rates ($in^2$/min) for various variable-pitch blades and cutting speeds. Use the value from the graph as an initial starting value and then adjust the feed based on the performance of the saw. The size and character of the chips being produced are the best indicators of the correct feed force. Chips that are curly, silvery, and warm indicate the best feed rate and band speed. If the chips appear burned and heavy, the feed is too great, so reduce the feed rate, the band speed, or both. If the chips are thin or powdery, the feed rate is too low, so increase the feed rate or reduce the band speed. The actual cutting rate achieved during a cut is equal to the area of the cut divided by the time required to finish the cut. The time required to make a cut is equal to the area of the cut divided by the cutting rate in square inches per minute.

**Bimetal Band Saw Speeds for Cutting 4-Inch Material with Coolant**

| Material | Category (AISI/SAE) | Speed (fpm) | Speed (m/min) |
|---|---|---|---|
| Aluminum Alloys | 1100, 2011, 2017, 2024, 3003, 5052, 5086, 6061, 6063, 6101, 6262, 7075 | 500 | 152 |
| Cast Iron | A536 (60-40-18) | 360 | 110 |
| | A47 | 300 | 91 |
| | A220 (50005), A536 (80-55-06) | 240 | 73 |
| | A48 (20 ksi) | 230 | 70 |
| | A536 (100-70-03) | 185 | 56 |
| | A48 (40 ksi) | 180 | 55 |
| | A220 (60004) | 170 | 52 |
| | A436 (1B) | 150 | 46 |
| | A220 (70003) | 145 | 44 |
| | A436 (2) | 140 | 43 |
| | A220 (80002), A436 (2B) | 125 | 38 |
| | A536 (120-90-02) | 120 | 37 |
| | A220 (90001), A48 (60 ksi) | 100 | 30 |
| | A439 (D-2) | 80 | 24 |
| | A439 (D-2B) | 60 | 18 |
| Cobalt | WF-11 | 65 | 20 |
| | Astroloy M | 60 | 18 |
| Copper | 356, 360 | 450 | 137 |
| | 353 | 400 | 122 |
| | 187, 1452 | 375 | 114 |
| | 380, 544 | 350 | 107 |
| | 173, 932, 934 | 315 | 96 |
| | 330, 365 | 285 | 87 |
| | 623, 624 | 265 | 81 |
| | 230, 260, 272, 280, 464, 632, 655 | 245 | 75 |
| | 101, 102, 110, 122, 172, 17510, 182, 220, 510, 625, 706, 715 | 235 | 72 |
| | 630 | 230 | 70 |
| | 811 | 215 | 66 |
| Iron Base Super Alloy | Pyromet X-15 | 120 | 37 |
| | A286, Incoloy 800 and 801 | 90 | 27 |
| Magnesium | AZ31B | 900 | 274 |
| Nickel | Nickel 200, 201, 205 | 85 | 26 |
| Nickel Alloy | Inconel 625 | 100 | 30 |
| | Incoloy 802, 804 | 90 | 27 |
| | Monel R405 | 85 | 26 |
| | 20CB3 | 80 | 24 |
| | Monel 400, 401 | 75 | 23 |
| | Hastelloy B, B2, C, C4, C22, C276, F, G, G2, G3, G30, N, S, W, X, Incoloy 825, 926, Inconel 751, X750, Waspaloy | 70 | 21 |
| | Monel K500 | 65 | 20 |
| | Incoloy 901, 903, Inconel 600, 718, Ni-Span-C902, Nimonic 263, Rene 41, Udimet 500 | 60 | 18 |
| | Nimonic 75 | 50 | 15 |
| Stainless Steel | 416, 420 | 190 | 58 |
| | 203EZ, 430, 430F, 4302 | 150 | 46 |
| | 303, 303PB, 303SE, 410, 440F, 30323 | 140 | 43 |
| | 304 | 120 | 37 |
| | 414, 30403 | 115 | 35 |
| | 347 | 110 | 34 |
| | 316, 31603 | 100 | 30 |
| | Greek Ascoloy | 95 | 29 |
| | 18-18-2, 309, Ferralium | 90 | 27 |
| | 15-5PH, 17-4PH, 17-7PH, 2205, 310, AM350, AM355, Custom 450, Custom 455, PH13-8Mo, PH14-8Mo, PH15-7Mo | 80 | 24 |
| | 22-13-5, Nitronic 50, 60 | 60 | 18 |

**Bimetal Band Saw Speeds for Cutting 4-Inch Material with Coolant** *(Continued)*

| Material | Category (AISI/SAE) | Speed (fpm) | Speed (m/min) |
|---|---|---|---|
| Steel | 12L14 | 425 | 130 |
| | 1213, 1215 | 400 | 122 |
| | 1117 | 340 | 104 |
| | 1030 | 330 | 101 |
| | 1008, 1015, 1020, 1025 | 320 | 98 |
| | 1035 | 310 | 94 |
| | 1018, 1021, 1022, 1026, 1513, A242 Cor-Ten A | 300 | 91 |
| | 1137 | 290 | 88 |
| | 1141, 1144, 1144 Hi Stress | 280 | 85 |
| | 41L40 | 275 | 84 |
| | 1040, 4130, A242 Cor-Ten B, (A36 Shapes) | 270 | 82 |
| | 1042, 1541, 4140, 4142 | 250 | 76 |
| | 8615, 8620, 8622 | 240 | 73 |
| | W-1 | 225 | 69 |
| | 1044, 1045, 1330, 4340, E4340, 5160, 8630 | 220 | 67 |
| | 1345, 4145, 6150 | 210 | 64 |
| | 1060, 4150, 8640, A-6, O-1, S-1 | 200 | 61 |
| | H-11, H-12, H-13, L-6, O-6 | 190 | 58 |
| | 1095 | 185 | 56 |
| | A-2 | 180 | 55 |
| | E9310 | 175 | 53 |
| | 300M, A-10, E52100, HY-80, HY-100 | 160 | 49 |
| | S-5 | 140 | 43 |
| | S-7 | 125 | 38 |
| | M-1 | 110 | 34 |
| | HP 9-4-20, HP 9-4-25 | 105 | 32 |
| | M-2, M-42, T1 | 100 | 30 |
| | D-2 | 90 | 27 |
| | T-15 | 70 | 21 |
| Titanium | Pure, Ti-3Al-8V-6Cr-4Mo-4Z, Ti-8Mo-8V-2Fe-3Al | 80 | 24 |
| | Ti-2Al-11Sn-5Zr-1Mo, Ti-5Al-2.5Sn, Ti-6Al-2Sn-4Zr-2Mo | 75 | 23 |
| | Ti-6Al-4V | 70 | 21 |
| | Ti-7Al-4Mo, Ti-8Al-1Mo-1V | 65 | 20 |

The speed figures given are for 4-in. material (length of cut) using a 3⁄4 variable-tooth bimetal blade and cutting fluid. For cutting dry, reduce speed 30–50%; for carbon steel band saw blades, reduce speed 50%. For other cutting lengths: increase speed 15% for ¼-in. material (10⁄14 blade); increase speed 12% for ¾-in. material (6⁄10 blade); increase speed 10% for 1¼-in. material (4⁄6 blade); decrease speed 12% for 8-in. material (2⁄3 blade).

Data are based on material provided by LENOX Blades, American Saw & Manufacturing Co.

*Example:* Find the band speed, the cutting rate, and the cutting time if the 4-inch pipe of the previous example is made of 304 stainless steel.

The preceding blade speed table gives the band speed for 4-inch 304 stainless steel as 120 fpm (feet per minute). The average length of cut for this pipe (see the previous example) is 1.4 inches, so increase the band saw speed by about 10 percent (see table footnote on page 1233) to 130 fpm to account for the size of the piece. On the cutting rate graph above, locate the point on the 4⁄6 blade line that corresponds to the band speed of 130 fpm and then read the cutting rate from the left axis of the graph. The cutting rate for this example is approximately 4 $in^2$/min. The cutting time is equal to the area of the cut divided by the cutting rate, so cutting time = 5.5⁄4 = 1.375 minutes.

**Band Saw Blade Break-In.**—A new band saw blade must be broken in gradually before it is allowed to operate at its full recommended feed rate. Break-in relieves the blade of residual stresses caused by the manufacturing process so that the blade retains its cutting ability longer. Break-in requires starting the cut at the material cutting speed with a low feed rate and then gradually increasing the feed rate over time until enough material has been cut. A blade should be broken in with the material to be cut.

To break in a new blade, first set the band saw speed at the recommended cutting speed for the material and start the first cut at the feed indicated on the starting feed rate graph below. After the saw has penetrated the work to a distance equal to the width of the blade, increase the feed slowly. When the blade is about halfway through the cut, increase the feed again slightly and finish the cut without increasing the feed again. Start the next and each successive cut with the same feed rate that ended the previous cut, and increase the feed rate slightly again before the blade reaches the center of the cut. Repeat this procedure until the area cut by the new blade is equal to the total area required as indicated on the graph below. At the end of the break-in period, the blade should be cutting at the recommended feed rate, otherwise adjusted to that rate.

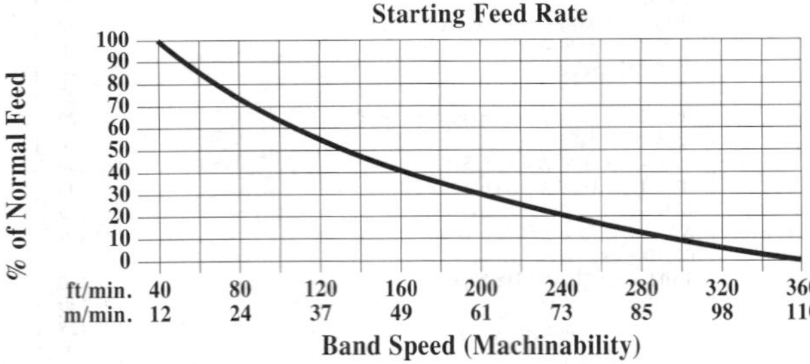


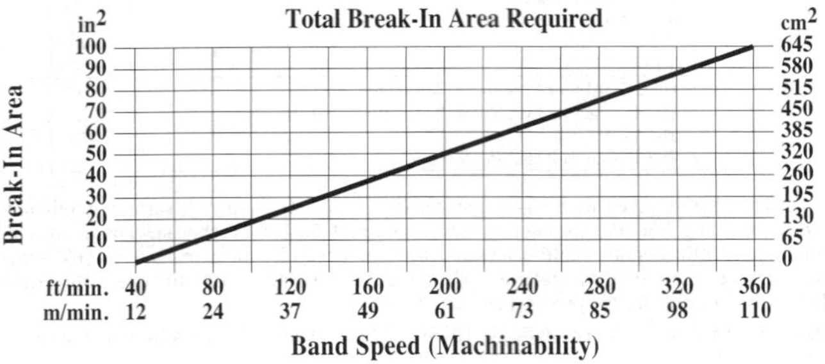


## Cutting Fluids for Machining

The goal in all conventional metal-removal operations is to raise productivity and reduce costs by machining at the highest practical speed consistent with long tool life, fewest rejects, and minimum downtime, and with the production of surfaces of satisfactory accuracy and finish. Many machining operations can be performed "dry," but the proper application of a cutting fluid generally makes possible: higher cutting speeds, higher feed rates, greater depths of cut, lengthened tool life, decreased surface roughness, increased dimensional accuracy, and reduced power consumption. Selecting the proper cutting fluid for a specific machining situation requires knowledge of fluid functions, properties, and limitations. Cutting fluid selection deserves as much attention as the choice of machine tool, tooling, speeds, and feeds.

To understand the action of a cutting fluid it is important to realize that almost all the energy expended in cutting metal is transformed into heat, primarily by the deformation of the metal into the chip and, to a lesser degree, by the friction of the chip sliding against the tool face. With these factors in mind it becomes clear that the primary functions of any cutting fluid are: cooling of the tool, workpiece, and chip; reducing friction at the sliding contacts; and reducing or preventing welding or adhesion at the contact surfaces, which forms the "built-up edge" on the tool. Two other functions of cutting fluids are flushing away chips from the cutting zone and protecting the workpiece and tool from corrosion.

The relative importance of the functions is dependent on the material being machined, the cutting tool and conditions, and the finish and accuracy required on the part. For example, cutting fluids with greater lubricity are generally used in low-speed machining and on most difficult-to-cut materials. Cutting fluids with greater cooling ability are generally used in high-speed machining on easier-to-cut materials.

**Types of Cutting and Grinding Fluids.**—In recent years a wide range of cutting fluids has been developed to satisfy the requirements of new materials of construction and new tool materials and coatings.

There are four basic types of cutting fluids; each has distinctive features, as well as advantages and limitations. Selection of the right fluid is made more complex because the dividing line between types is not always clear. Most machine shops try to use as few different fluids as possible and prefer fluids that have long life, do not require constant changing or modifying, have reasonably pleasant odors, do not smoke or fog in use, and, most important, are neither toxic nor cause irritation to the skin. Other issues in selection are the cost and ease of disposal.The major divisions and subdivisions used in classifying cutting fluids are:

*Cutting Oils*, including straight and compounded mineral oils plus additives.

*Water-Miscible Fluids*, including emulsifiable oils; chemical or synthetic fluids; and semichemical fluids.

*Gases*.

*Paste and Solid Lubricants*.

Since cutting oils and water-miscible types are the most commonly used cutting fluids in machine shops, discussion will be limited primarily to these types. It should be noted, however, that compressed air and inert gases, such as carbon dioxide, nitrogen, and Freon, are sometimes used in machining. Paste, waxes, soaps, graphite, and molybdenum disulfide may also be used, either applied directly to the workpiece or as an impregnant in the tool, such as in a grinding wheel.

**Cutting Oils.**—Cutting oils are generally compounds of mineral oil with the addition of animal, vegetable, or marine oils to improve the wetting and lubricating properties. Sulfur, chlorine, and phosphorous compounds, sometimes called extreme pressure (EP) additives, provide for even greater lubricity. In general, these cutting oils do not cool as well as water-miscible fluids.

**Water-Miscible Fluids.**—*Emulsions or soluble oils* are a suspension of oil droplets in water. These suspensions are made by blending the oil with emulsifying agents (soap and soaplike materials) and other materials. These fluids combine the lubricating and rust-prevention properties of oil with water's excellent cooling properties. Their properties are affected by the emulsion concentration, with "lean" concentrations providing better cooling but poorer lubrication, and with "rich" concentrations having the opposite effect. Additions of sulfur, chlorine, and phosphorus, as with cutting oils, yield "extreme pressure" (EP) grades.

*Chemical fluids* are true solutions composed of organic and inorganic materials dissolved in water. Inactive types are usually clear fluids combining high rust inhibition, high cooling, and low lubricity characteristics with high surface tension. Surface-active types include wetting agents and possess moderate rust inhibition, high cooling, and moderate lubricating properties with low surface tension. They may also contain chlorine and/or sulfur compounds for extreme pressure properties.

*Semichemical fluids* are combinations of chemical fluids and emulsions. These fluids have a lower oil content but a higher emulsifier and surface-active-agent content than emulsions, producing oil droplets of much smaller diameter. They possess low surface tension, moderate lubricity and cooling properties, and very good rust inhibition. Sulfur, chlorine, and phosphorus also are sometimes added.

**Selection of Cutting Fluids for Different Materials and Operations.—**The choice of a cutting fluid depends on many complex interactions including the machinability of the metal; the severity of the operation; the cutting tool material; metallurgical, chemical, and human compatibility; fluid properties, reliability, and stability; and finally cost. Other factors affect results. Some shops standardize on a few cutting fluids which have to serve all purposes. In other shops, one cutting fluid must be used for all the operations performed on a machine. Sometimes, a very severe operating condition may be alleviated by applying the "right" cutting fluid manually while the machine supplies the cutting fluid for other operations through its coolant system. Several voluminous textbooks are available with specific recommendations for the use of particular cutting fluids for almost every combination of machining operation and workpiece and tool material. In general, when experience is lacking, it is wise to consult the material supplier and/or any of the many suppliers of different cutting fluids for advice and recommendations. Another excellent source is the Machinability Data Center, one of the many information centers supported by the US Department of Defense. While the following recommendations represent good practice, they are to serve as a guide only, and it is not intended to say that other cutting fluids will not, in certain specific cases, also be effective.

*Steels:* Caution should be used when using a cutting fluid on steel that is being turned at a high cutting speed with cemented carbide cutting tools. See *Application of Cutting Fluids to Carbides* later. Frequently this operation is performed dry. If a cutting fluid is used, it should be a soluble oil mixed to a consistency of about 1 part oil to 20 to 30 parts water. A sulfurized mineral oil is recommended for reaming with carbide tipped reamers although a heavy-duty soluble oil has also been used successfully.

The cutting fluid recommended for machining steel with high-speed cutting tools depends largely on the severity of the operation. For ordinary turning, boring, drilling, and milling on medium and low strength steels, use a soluble oil having a consistency of 1 part oil to 10 to 20 parts water. For tool steels and tough alloy steels, a heavy-duty soluble oil having a consistency of 1 part oil to 10 parts water is recommended for turning and milling. For drilling and reaming these materials, a light sulfurized mineral-fatty oil is used. For tough operations such as tapping, threading, and broaching, a sulfochlorinated mineral-fatty oil is recommended for tool steels and high-strength steels, and a heavy sulfurized mineral-fatty oil or a sulfochlorinated mineral oil can be used for medium- and low-strength steels. Straight sulfurized mineral oils are often recommended for machining tough, stringy low carbon steels to reduce tearing and produce smooth surface finishes.

*Stainless Steel:* For ordinary turning and milling a heavy-duty soluble oil mixed to a consistency of 1 part oil to 5 parts water is recommended. Broaching, threading, drilling, and reaming produce best results using a sulfochlorinated mineral-fatty oil.

*Copper Alloys:* Most brasses, bronzes, and copper are stained when exposed to cutting oils containing active sulfur and chlorine; thus, sulfurized and sulfochlorinated oils should not be used. For most operations a straight soluble oil, mixed to 1 part oil and 20 to 25 parts water is satisfactory. For very severe operations and for automatic screw machine work a mineral-fatty oil is used. A typical mineral-fatty oil might contain 5 to 10 percent lard oil with the remainder mineral oil.

*Monel Metal:* When turning this material, an emulsion gives a slightly longer tool life than a sulfurized mineral oil, but the latter aids in chip breakage, which is frequently desirable.

*Aluminum Alloys:* Aluminum and aluminum alloys are frequently machined dry. When a cutting fluid is used it should be selected for its ability to act as a coolant. Soluble oils mixed to a consistency of 1 part oil to 20 to 30 parts water can be used. Mineral oil-base cutting fluids, when used to machine aluminum alloys, are frequently cut back to increase their viscosity so as to obtain good cooling characteristics and to make them flow easily to cover the tool and the work. For example, a mineral-fatty oil or a mineral plus a sulfurized fatty oil can be cut back by the addition of as much as 50 percent kerosene.

*Cast Iron:* Ordinarily, cast iron is machined dry. Some increase in tool life can be obtained or a faster cutting speed can be used with a chemical cutting fluid or a soluble oil mixed to consistency of 1 part oil and 20 to 40 parts water. A soluble oil is sometimes used to reduce the amount of dust around the machine.

*Magnesium:* Magnesium may be machined dry, or with an air blast for cooling. A light mineral oil of low acid content may be used on difficult cuts. Coolants containing water should not be used on magnesium because of the danger of releasing hydrogen caused by reaction of the chips with water. Proprietary water-soluble oil emulsions containing inhibitors that reduce the rate of hydrogen generation are available.

*Grinding:* Soluble oil emulsions or emulsions made from paste compounds are used extensively in precision grinding operations. For cylindrical grinding, 1 part oil to 40 to 50 parts water is used. Solution type fluids and translucent grinding emulsions are particularly suited for many fine finish grinding applications. Mineral oil-base grinding fluids are recommended for many applications where a fine surface finish is required on the ground surface. Mineral oils are used with vitrified wheels but are not recommended for wheels with rubber or shellac bonds. Under certain conditions the oil vapor mist caused by the action of the grinding wheel can be ignited by the grinding sparks and explode. To quench the grinding spark a secondary coolant line to direct a flow of grinding oil below the grinding wheel is recommended.

*Broaching:* For steel, a heavy mineral oil such as sulfurized oil of 300 to 500 Saybolt viscosity at 100° F (38° C) can be used to provide adequate lubricating effect and a dampening of the shock loads. Soluble oil emulsions may be used for the lighter broaching operations.

**Cutting Fluids for Turning, Milling, Drilling and Tapping.—**The following table, *Cutting Fluids Recommended for Machining Operations*, gives specific cutting oil recommendations for common machining operations.

*Soluble Oils:* Types of oil paste compounds that form emulsions when mixed with water. Soluble oils are used extensively in machining both ferrous and nonferrous metals when the cooling quality is paramount and the chip-bearing pressure is not excessive. Care should be taken in selecting the proper soluble oil for precision grinding operations. Grinding coolants should be free from fatty materials that tend to load the wheel, thus affecting the finish on the machined part. Soluble coolants should contain rust preventive constituents to prevent corrosion.

*Base Oils:* Various types of highly sulfurized and chlorinated oils containing inorganic, animal, or fatty materials. This "base stock" usually is "cut back" or blended with a lighter oil, unless the chip-bearing pressures are high, as when cutting alloy steel. Base oils usually have a viscosity range of from 300 to 900 seconds at 100° F (38° C).

*Mineral Oils:* This group includes all types of oils extracted from petroleum such as paraffin oil, mineral seal oil, and kerosene. Mineral oils are often blended with base stocks, but they are generally used in the original form for light machining operations on both free-machining steels and nonferrous metals. The coolants in this class should be of a type that has a relatively high flash point. Care should be taken to see that they are nontoxic so that they will not be injurious to the operator. The heavier mineral oils (paraffin oils) usually have a viscosity of about 100 seconds at 100° F (38° C). Mineral seal oil and kerosene have a viscosity of 35 to 60 seconds at 100° F (38° C).

## Cutting Fluids Recommended for Machining Operations

| Material to be Cut | Turning | Milling |
|---|---|---|
| Aluminum[a] | Mineral Oil with 10 Percent Fat (or) Soluble Oil | Soluble Oil (96 Percent Water) (or) Mineral Seal Oil (or) Mineral Oil |
| Alloy Steels[b] | 25 Percent Sulfur base Oil[b] with 75 Percent Mineral Oil | 10 Percent Lard Oil with 90 Percent Mineral Oil |
| Brass | Mineral Oil with 10 Percent Fat | Soluble Oil (96 Percent Water) |
| Tool Steels and Low-carbon Steels | 25 Percent Lard Oil with 75 Percent Mineral Oil | Soluble Oil |
| Copper | Soluble Oil | Soluble Oil |
| Monel Metal | Soluble Oil | Soluble Oil |
| Cast Iron[c] | Dry | Dry |
| Malleable Iron | Soluble Oil | Soluble Oil |
| Bronze | Soluble Oil | Soluble Oil |
| Magnesium[d] | 10 Percent Lard Oil with 90 Percent Mineral Oil | Mineral Seal Oil |
| **Material to be Cut** | **Drilling** | **Tapping** |
| Aluminum[a] | Soluble Oil (75 to 90 Percent Water) (or) 10 Percent Lard Oil with 90 Percent Mineral Oil | Lard Oil (or) Sperm Oil (or) Wool Grease (or) 25 Percent Sulfur-base Oil[b] Mixed with Mineral Oil |
| Alloy Steels[b] | Soluble Oil | 30 Percent Lard Oil with 70 Percent Mineral Oil |
| Brass | Soluble Oil (75 to 90 Percent Water) (or) 30 Percent Lard Oil with 70 Percent Mineral Oil | 10 to 20 Percent Lard Oil with Mineral Oil |
| Tool Steels and Low-carbon Steels | Soluble Oil | 25 to 40 Percent Lard Oil with Mineral Oil (or) 25 Percent Sulfur-base Oil[b] with 75 Percent Mineral Oil |
| Copper | Soluble Oil | Soluble Oil |
| Monel Metal | Soluble Oil | 25 to 40 Percent Lard Oil Mixed with Mineral Oil (or) Sulfur-base Oil[b] Mixed with Mineral Oil |
| Cast Iron[c] | Dry | Dry (or) 25 Percent Lard Oil with 75 Percent Mineral Oil |
| Malleable Iron | Soluble Oil | Soluble Oil |
| Bronze | Soluble Oil | 20 Percent Lard Oil with 80 Percent Mineral Oil |
| Magnesium[d] | 60-second Mineral Oil | 20 Percent Lard Oil with 80 Percent Mineral Oil |

[a] In machining aluminum, several varieties of coolants may be used. For rough machining, where the stock removal is sufficient to produce heat, water soluble mixtures can be used with good results to dissipate the heat. Other oils that may be recommended are straight mineral seal oil; a 50-50 mixture of mineral seal oil and kerosene; a mixture of 10 percent lard oil with 90 percent kerosene; and a 100-second mineral oil cut back with mineral seal oil or kerosene.

[b] The sulfur-base oil referred to contains $4\frac{1}{2}$ percent sulfur compound. Base oils are usually dark in color. As a rule, they contain sulfur compounds resulting from a thermal or catalytic refinery process. When so processed, they are more suitable for industrial coolants than when they have had such compounds as flowers of sulfur added by hand. The adding of sulfur compounds by hand to the coolant reservoir is of temporary value only, and the non-uniformity of the solution may affect the machining operation.

[c] A soluble oil or low-viscosity mineral oil may be used in machining cast iron to prevent excessive metal dust.

[d] When a cutting fluid is needed for machining magnesium, low or nonacid mineral seal or lard oils are recommended. Coolants containing water should not be used because of the fire danger when magnesium chips react with water, forming hydrogen gas.

[e] Sulfurized oils ordinarily are not recommended for tapping aluminum; however, for some tapping operations they have proved very satisfactory, although the work should be rinsed in a solvent right after machining to prevent discoloration.

**Application of Cutting Fluids to Carbides.—**Turning, boring, and similar operations on lathes using carbides are performed dry or with the help of soluble oil or chemical cutting fluids. The effectiveness of cutting fluids in improving tool life or by permitting higher cutting speeds to be used is less with carbides than with high-speed steel tools. Furthermore, the effectiveness of the cutting fluid is reduced as the cutting speed is increased. Cemented carbides are very sensitive to sudden changes in temperature and to temperature gradients within the carbide. Thermal shocks to the carbide will cause thermal cracks to form near the cutting edge, which are a prelude to tool failure. An unsteady or interrupted flow of the coolant reaching the cutting edge will generally cause these thermal cracks. The flow of the chip over the face of the tool can cause an interruption to the flow of the coolant reaching the cutting edge even though a steady stream of coolant is directed at the tool. When a cutting fluid is used and frequent tool breakage is encountered, it is often best to cut dry. When a cutting fluid must be used to keep the workpiece cool for size control or to allow it to be handled by the operator, special precautions must be used. Sometimes applying the coolant from the front and the side of the tool simultaneously is helpful. On lathes equipped with overhead shields, it is very effective to apply the coolant from below the tool into the space between the shoulder of the work and the tool flank, in addition to applying the coolant from the top. Another method is not to direct the coolant stream at the cutting tool at all but to direct it at the workpiece above or behind the cutting tool.

The danger of thermal cracking is great when milling with carbide cutters. The nature of the milling operation itself tends to promote thermal cracking because the cutting edge is constantly heated to a high temperature and rapidly cooled as it enters and leaves the workpiece. For this reason, carbide milling operations should be performed dry.

Lower cutting-edge temperatures diminish the danger of thermal cracking. The cutting-edge temperatures usually encountered when reaming with solid carbide or carbide-tipped reamers are generally such that thermal cracking is not apt to occur except when reaming certain difficult-to-machine metals. Therefore, cutting fluids are very effective when used on carbide reamers. Practically every kind of cutting fluid has been used, depending on the job material encountered. For difficult surface-finish problems in holes, heavy duty soluble oils, sulfurized mineral-fatty oils, and sulfochlorinated mineral-fatty oils have been used successfully. On some work, the grade and the hardness of the carbide also have an effect on the surface finish of the hole.

Cutting fluids should be applied where the cutting action is taking place and at the highest possible velocity without causing splashing. As a general rule, it is preferable to supply from 3 to 5 gallons per minute (11–19 liter/min) for each single-point tool on a machine such as a turret lathe or automatic. The temperature of the cutting fluid should be kept below 110° F (43° C) . If the volume of fluid used is not sufficient to maintain the proper temperature, means of cooling the fluid should be provided.

**Cutting Fluids for Machining Magnesium.—**In machining magnesium, it is the general but not invariable practice in the United States to use a cutting fluid. In other places, magnesium usually is machined dry except where heat generated by high cutting speeds would not be dissipated rapidly enough without a cutting fluid. This condition may exist when, for example, small tools without much heat-conducting capacity are employed on automatics.

The cutting fluid for magnesium should be an anhydrous oil having, at most, a very low acid content. Various mineral-oil cutting fluids are used for magnesium.

## Metalworking Fluids

The term *metalworking fluids* (MWFs) describes coolants and lubricants used during the fabrication of products from metals and metal substitutes. These fluids are used to prolong the life of machine tools, carry away debris, and protect or treat the surfaces of the material being processed. MWFs reduce friction between the cutting tool and work surfaces, reduce wear and galling, protect surface characteristics, reduce surface adhesion or welding, carry away generated heat, and flush away swarf, chips, fines, and residues. Table 1 describes the four different classes of metalworking fluids:

**Table 1. Classes of Metalworking Fluids (MWFs)**

| MWF | Description | Dilution factor |
|---|---|---|
| **Straight oil** (neat oil or cutting oil) | Highly refined petroleum oils (lubricant-base oils) or other animal, marine, vegetable, or synthetic oils used singly or in combination with or without additives. These are lubricants, or function to improve the finish on the metal cut, and prevent corrosion. | none |
| **Soluble oil** (emulsifiable oil) | Combinations of 30 to 85% highly refined, high-viscosity lubricant-base oils and emulsifiers that may include other performance additives. Soluble oils are diluted with water before use at ratios of parts water. | 1 part concentrate to 5 to 40 parts water |
| **Semisynthetic** | Contain smaller amounts of severely refined lubricant-base oil (5 to 30% in the concentrate), a higher proportion of emulsifiers that may include other performance additives, and 30 to 50% water. | 1 part concentrate to 10 to 40 parts water |
| **Synthetic**[a] | Contain no petroleum oils and may be water soluble or water dispersible. The simplest synthetics are made with organic and inorganic salts dissolved in water. Offer good rust protection and heat removal but usually have poor lubricating ability. May be formulated with other performance additives. Stable, can be made bioresistant. | 1 part concentrate to 10 to 40 parts water |

[a] Over the last several decades major changes in the US machine tool industry have increased the consumption of MWFs. Specifically, the use of synthetic MWFs increased as tool and cutting speeds increased.

**Occupational Exposures to Metalworking Fluids (MWFs).—**Workers can be exposed to MWFs by inhalation of aerosols (mists) or by skin contact resulting in an increased risk of respiratory (lung) and skin disease. Health effects vary based on the type of MWF, route of exposure, concentration, and length of exposure.

Skin contact usually occurs when the worker dips his/her hands into the fluid, floods the machine tool, or handling parts, tools, equipment or workpieces coated with the fluid, without the use of personal protective equipment such as gloves and apron. Skin contact can also result from fluid splashing onto the worker from the machine if guarding is absent or inadequate.

Inhalation exposures result from breathing MWF mist or aerosol. The amount of mist generated (and the severity of the exposure) depends on a variety of factors: the type of MWF and its application process; the MWF temperature; the specific machining or grinding operation; the presence of splash guarding; and the effectiveness of the ventilation system. In general, the exposure will be higher if the worker is in close proximity to the machine, the operation involves high tool speeds and deep cuts, the machine is not enclosed, or if ventilation equipment was improperly selected or poorly maintained. In addition, high-pressure and/or excessive fluid application, contamination of the fluid with tramp oils, and improper fluid selection and maintenance will tend to result in higher exposure.

Each MWF class consists of a wide variety of chemicals used in different combinations and the risk these chemicals pose to workers may vary because of different manufacturing processes, various degrees of refining, recycling, improperly reclaimed chemicals, different degrees of chemical purity, and potential chemical reactions between components.

Exposure to hazardous contaminants in MWFs may present health risks to workers. Contamination may occur from: process chemicals and ancillary lubricants inadvertently introduced; contaminants, metals, and alloys from parts being machined; water and cleaning agents used for routine housekeeping; and, contaminants from other environmental sources at the worksite. In addition, bacterial and fungal contaminants may metabolize and degrade the MWFs to hazardous end-products as well as produce endotoxins.

The improper use of biocides to manage microbial growth may result in potential health risks. Attempts to manage microbial growth solely with biocides may result in the emergence of biocide-resistant strains from complex interactions that may occur among different member species or groups within the population. For example, the growth of one species, or the elimination of one group of organisms may permit the overgrowth of another. Studies also suggest that exposure to certain biocides can cause either allergic or contact dermatitis.

**Fluid Selection, Use, and Application.**—The MWFs selected should be as nonirritating and nonsensitizing as possible while remaining consistent with operational requirements. Petroleum-containing MWFs should be evaluated for potential carcinogenicity using ASTM E1687-10 (R2014), "Standard Test Method for Determining Carcinogenic Potential of Virgin Base Oils in Metalworking Fluids." If soluble oil or synthetic MWFs are used, ASTM Standard E1497-17, "Standard Practice for Selection and Safe Use of Water-Miscible and Straight Oil Metal Removal Fluids," should be consulted for safe use guidelines, including those for product selection, storage, dispensing, and maintenance. To minimize the potential for nitrosamine formation, nitrate-containing materials should not be added to MWFs containing ethanolamines.

Many factors influence the generation of MWF mists, which can be minimized through the proper design and operation of the MWF delivery system. ANSI Technical Report B11 TR2-1997, "Mist Control Considerations for the Design, Installation and Use of Machine Tools Using Metalworking Fluids," provides directives for minimizing mist and vapor generation. These include minimizing fluid delivery pressure, matching the fluid to the application, using MWF formulations with low oil concentrations, avoiding contamination with tramp oils, minimizing the MWF flow rate, covering fluid reservoirs and return systems where possible, and maintaining control of the MWF chemistry. Also, proper application of MWFs can minimize splashing and mist generation. Proper application includes: applying MWFs at the lowest possible pressure and flow volume consistent with provisions for adequate part cooling, chip removal, and lubrication; applying MWFs at the tool/workpiece interface to minimize contact with other rotating equipment; ceasing fluid delivery when not performing machining; not allowing MWFs to flow over the unprotected hands of workers loading or unloading parts; and using mist collectors engineered for the operation and specific machine enclosures.

Properly maintained filtration and delivery systems provide cleaner MWFs, reduce mist, and minimize splashing and emissions. Proper maintenance of the filtration and delivery systems includes: the selection of appropriate filters; ancillary equipment such as chip handling operations, dissolved air-flotation devices, belt-skimmers, chillers or plate and frame heat exchangers, and decantation tanks; guard coolant return trenches to prevent dumping of floor wash water and other waste fluids; covering sumps or coolant tanks to prevent contamination with waste or garbage (e.g., cigarette butts, food, etc.); and, keeping the machine(s) clean of debris. Parts washing before machining can be an important part of maintaining cleaner MWFs.

Since all additives will be depleted with time, the MWF and additives concentrations should be monitored frequently so that components and additives can be made up as needed. The MWF should be maintained within the pH and concentration ranges recommended by the formulator or supplier. MWF temperature should be maintained at the lowest practical level to slow the growth of microorganisms, reduce water losses and changes in viscosity, and—in the case of straight oils—reduce fire hazards.

**Fluid Maintenance.—**Drums, tanks, or other containers of MWF concentrates should be stored appropriately to protect them from outdoor weather conditions and exposure to low or high temperatures. Extreme temperature changes may destabilize the fluid concentrates, especially in the case of concentrates mixed with water, and cause water to seep into unopened drums encouraging bacterial growth. MWFs should be maintained at as low a temperature as is practical. Low temperatures slow the growth of microorganisms, reduce water losses and change in viscosity, and in the case of straight oils, reduce the fire hazard risks.

To maintain proper MWF concentrations, neither water nor concentrate should be used to top off the system. The MWF mixture should be prepared by first adding the concentrate to the clean water (in a clean container) and then adding the emulsion to that mixture in the coolant tank. MWFs should be mixed just before use; large amounts should not be stored, as they may deteriorate before use.

*Personal Protective Clothing:* Personal protective clothing and equipment should always be worn when removing MWF concentrates from the original container, mixing and diluting concentrate, preparing additives (including biocides), and adding MWF emulsions, biocides, or other potentially hazardous ingredients to the coolant reservoir. Personal protective clothing includes eye protection or face shields, gloves, and aprons which do not react with but shed MWF ingredients and additives.

*System Service:* Coolant systems should be regularly serviced, and the machines should be rigorously maintained to prevent contamination of the fluids by tramp oils (e.g., hydraulic oils, gear box oils, and machine lubricants leaking from the machines or total loss slideway lubrication). Tramp oils can destabilize emulsions, cause pumping problems, and clog filters. Tramp oils can also float to the top of MWFs, effectively sealing the fluids from the air, allowing metabolic products such as volatile fatty acids, mercaptols, scatols, ammonia, and hydrogen sulfide produced by the anaerobic and facultative anaerobic species growing within the biofilm to accumulate in the reduced state.

When replacing the fluids, thoroughly clean all parts of the system to inhibit the growth of microorganisms growing on surfaces. Some bacteria secrete layers of slime that may grow in stringy configurations that resemble fungal growth. Many bacteria secrete polymers of polysaccharide and/or protein, forming a glycocalyx which cements cells together much as mortar holds bricks. Fungi may grow as masses of hyphae-forming mycelial mats. The attached community of microorganisms is called a biofilm and may be very difficult to remove by ordinary cleaning procedures.

*Biocide Treatment:* Biocides are used to maintain the functionality and efficacy of MWFs by preventing microbial overgrowth. These compounds are often added to the stock fluids as they are formulated, but over time the biocides are consumed by chemical and biological demands. Biocides with a wide spectrum of biocidal activity should be used to suppress the growth of the widely diverse contaminant population. Only the concentration of biocide needed to meet fluid specifications should be used since overdosing could lead to skin or respiratory irritation in workers, and under-dosing could lead to an inadequate level of microbial control.

*Ventilation Systems:* The ventilation system should be designed and operated to prevent the accumulation or recirculation of airborne contaminants in the workplace. The ventilation system should include a positive means of bringing in at least an equal volume of air from the outside, conditioning it, and evenly distributing it throughout the exhausted area.

Exhaust ventilation systems function through suction openings placed near a source of contamination. The suction opening or exhaust hood creates air motion sufficient to overcome room air currents and any airflow generated by the process. This airflow captures the contaminants and conveys them to a point where they can either be discharged or removed from the airstream. Exhaust hoods are classified by their position relative to the process as canopy, side draft, down draft or enclosure. ANSI Technical Report B11 TR 2-1997 contains guidelines for exhaust ventilation of machining and grinding operations. Enclosures are the only type of exhaust hood recommended by the ANSI committee. They consist of physical barriers between the process and the worker's environment. Enclosures can be further classified by the extent of the enclosure: close capture (enclosure of the point of operation, total enclosure (enclosure of the entire machine), or tunnel enclosure (continuous enclosure over several machines).

If no fresh make up air is introduced into the plant, air will enter the building through open doors and windows, potentially causing cross-contamination of all process areas. Ideally, all air exhausted from the building should be replaced by tempered air from an uncontaminated location. By providing a slight excess of make up air in relatively clean areas and slight deficit of make up air in dirty areas, cross-contamination can be reduced. In addition, this air can be channeled directly to operator work areas, providing the cleanest possible work environment. Ideally, this fresh air should be supplied in the form of a low-velocity air shower (<100 ft/min or <30 m/min) to prevent interference with the exhaust hoods) directly above the worker.

*Protective Clothing and Equipment:* Engineering controls are used to reduce worker exposure to MWFs. But in the event of airborne exposures that exceed the NIOSH REL or dermal contact with the MWFs, the added protection of chemical protective clothing (CPC) and respirators should be provided. Maintenance staff may also need CPC because their work requires contact with MWFs during certain operations. All workers should be trained in the proper use and care of CPC. After any item of CPC has been in routine use, it should be examined to ensure that its effectiveness has not been compromised.

Selection of the appropriate respirator depends on the operation, chemical components, and airborne concentrations in the worker's breathing zone. Table 2 lists the NIOSH-recommended respiratory protection for workers exposed to MWF aerosol.

**Table 2. Respiratory Protection for Workers Exposed to MWF Aerosols***

| Concentration of MWF aerosol (mg/m$^3$) | Minimum respiratory protection[a] |
|---|---|
| #0.5 mg/m$^3$ (1 REL)[b] | No respiratory protection required for healthy workers[c] |
| #5.0 mg/m$^3$ (10 REL) | Any air-purifying, half-mask respirator including a disposable respirator[d,e] equipped with any P- or R-series particulate filter (P95, P99, P100, R95, R99, or R100) number |
| #12.5 mg/m$^3$ (25 REL) | Any powered, air-purifying respirator equipped with a hood or helmet and a HEPA filter[f] |

[a] Respirators with higher assigned protection factors (APFs) may be substituted for those with lower APFs [NIOSH 1987a].

[b] APF times the NIOSH REL for total particulate mass. The APF [NIOSH 1987b] is the minimum anticipated level of protection provided by each type of respirator.

[c] See text for recommendations regarding workers with asthma and for other workers affected by MWF aerosols.

[d] A respirator that should be discarded after the end of the manufacturer's recommended period of use or after a noticeable increase in breathing resistance or when physical damage, hygiene considerations, or other warning indicators render the respirator unsuitable for further use.

[e] An APF of 10 is assigned to disposable particulate respirators if they have been properly fitted.

[f] High-efficiency particulate air filter. When organic vapors are a potential hazard during metalworking operations, a combination particulate and organic vapor filter is necessary.

* Only NIOSH/MSHA-approved or NIOSH-approved (effective date July 10, 1995) respiratory equipment should be used.

## MACHINING NONFERROUS METALS AND NONMETALLIC MATERIALS

### Nonferrous Metals

**Machining Aluminum.—**Some of the alloys of aluminum have been machined successfully without any lubricant or cutting compound, but some form of lubricant is desirable to obtain the best results. For many purposes, a soluble cutting oil is good.

Tools for aluminum and aluminum alloys should have larger relief and rake angles than tools for cutting steel. For high-speed steel turning tools the following angles are recommended: relief angles, 14 to 16 degrees; back rake angle, 5 to 20 degrees; side rake angle, 15 to 35 degrees. For very soft alloys even larger side rake angles are sometimes used. High silicon aluminum alloys and some others have a very abrasive effect on the cutting tool. While these alloys can be cut successfully with high-speed steel tools, cemented carbides are recommended because of their superior abrasion resistance. The tool angles recommended for cemented carbide turning tools are: relief angles, 12 to 14 degrees; back rake angle, 0 to 15 degrees; side rake angle, 8 to 30 degrees.

Cut-off tools and necking tools for machining aluminum and its alloys should have from 12 to 20 degrees back rake angle and the end relief angle should be from 8 to 12 degrees. Excellent threads can be cut with single-point tools in even the softest aluminum. Experience seems to vary somewhat regarding the rake angle for single-point thread cutting tools. Some prefer to use a rather large back and side rake angle although this requires a modification in the included angle of the tool to produce the correct thread contour. When both rake angles are zero, the included angle of the tool is ground equal to the included angle of the thread. Excellent threads have been cut in aluminum with zero rake angle thread-cutting tools using large relief angles, which are 16 to 18 degrees opposite the front side of the thread and 12 to 14 degrees opposite the back side of the thread. In either case, the cutting edges should be ground and honed to a keen edge. It is sometimes advisable to give the face of the tool a few strokes with a hone between cuts when chasing the thread to remove any built-up edge on the cutting edge.

Fine surface finishes are often difficult to obtain on aluminum and aluminum alloys, particularly the softer metals. When a fine finish is required, the cutting tool should be honed to a keen edge and the surfaces of the face and the flank will also benefit by being honed smooth. Tool wear is inevitable, but it should not be allowed to progress too far before the tool is changed or sharpened. A sulphurized mineral oil or a heavy-duty soluble oil will sometimes be helpful in obtaining a satisfactory surface finish. For best results, however, a diamond cutting tool is recommended. Excellent surface finishes can be obtained on even the softest aluminum and aluminum alloys with these tools.

Although ordinary milling cutters can be used successfully in shops where aluminum parts are only machined occasionally, the best results are obtained with coarse-tooth, large helix-angle cutters having large rake and clearance angles. Clearance angles up to 10 to 12 degrees are recommended. When slab milling and end milling a profile, using the peripheral teeth on the end mill, climb milling (also called down milling) will generally produce a better finish on the machined surface than conventional (or up) milling. Face milling cutters should have a large axial rake angle. Standard twist drills can be used without difficulty in drilling aluminum and aluminum alloys although high helix-angle drills are preferred. The wide flutes and high helix-angle in these drills helps to clear the chips. Sometimes split-point drills are preferred. Carbide tipped twist drills can be used for drilling aluminum and its alloys and may afford advantages in some production applications. Ordinary hand and machine taps can be used to tap aluminum and its alloys although spiral-fluted ground thread taps give superior results. Experience has shown that such taps should have a right-hand ground flute when intended to cut right-hand threads and the helix angle should be similar to that used in an ordinary twist drill.

**Machining Magnesium.**—Magnesium alloys are readily machined and with relatively low power consumption per cubic inch of metal removed. The usual practice is to employ high cutting speeds with relatively coarse feeds and deep cuts. Exceptionally fine finishes can be obtained so that grinding to improve the finish usually is unnecessary. The horsepower normally required in machining magnesium varies from 0.15 to 0.30 hp per cubic inch per minute (0.41-0.82 $kW/cm^3/s$). While this value is low, especially in comparison with power required for cast iron and steel, the total amount of power for machining magnesium usually is high because of the exceptionally rapid rate at which metal is removed.

Carbide tools are recommended for maximum efficiency, although high-speed steel frequently is employed. Tools should be designed so as to dispose of chips readily or without excessive friction, by employing polished chip-bearing surfaces, ample chip spaces, large clearances, and small contact areas. *Keen-edged tools should always be used.*

*Speeds and Feeds for Magnesium:* Speeds ordinarily range up to 5000 ft/min (1524 m/min) for rough- and finish-turning, up to 3000 ft/min (915 m/min) for rough-milling, and up to 9000 ft/min (2743 m/min) for finish-milling. The following combinations of speed, feed per revolution, and depth of cut are recommended:

| Process | Speed | | Feed | | Depth of Cut | |
|---|---|---|---|---|---|---|
| | ft/min | m/min | in/rev | mm/rev | inch | mm |
| Rough Turning | 300–600 | 90–185 | 0.030–0.100 | 0.76–2.5 | 0.5 | 12.7 |
| | 600–1000 | 185–305 | 0.020–0.080 | 0.50–2.0 | 0.4 | 10 |
| | 1000–1500 | 305–460 | 0.010–0.060 | 0.25–1.5 | 0.3 | 7.6 |
| | 1500–2000 | 460–610 | 0.010–0.040 | 0.25–1.0 | 0.2 | 5.0 |
| | 2000–5000 | 610–1525 | 0.010–0.030 | 0.25–0.76 | 0.1 | 3.8 |
| Rough Milling | up to 900 | up to 275 | 0.005–0.025 | 0.13–0.64 | up to 0.5 | up to 12.7 |
| | 900–1500 | 275–455 | 0.005–0.020 | 0.13–0.51 | up to 0.375 | up to 9.5 |
| | 1500–3000 | 455–915 | 0.005–0.010 | 0.13–0.25 | up to 0.2 | up to 5.1 |

*Lathe Tool Angles for Magnesium:* The true or actual rake angle resulting from back and side rakes usually varies from 10 to 15 degrees. Back rake varies from 10 to 20, and side rake from 0 to 10 degrees. Reduced back rake may be employed to obtain better chip breakage. The back rake may also be reduced to from 2 to 8 degrees on form tools or other broad tools to prevent chatter.

*Parting Tools:* For parting tools, the back rake varies from 15 to 20 degrees, the front end relief 8 to 10 degrees, the side relief measured perpendicular to the top face 8 degrees, the side relief measured in the plane of the top face from 3 to 5 degrees.

*Milling Magnesium:* In general, the coarse-tooth type of cutter is recommended. The number of teeth or cutting blades may be one-third to one-half the number normally used; however, the two-blade fly-cutter has proved to be very satisfactory. As a rule, the land relief or primary peripheral clearance is 10 degrees followed by secondary clearance of 20 degrees. The lands should be narrow, the width being about $^3/_{64}$ to $^1/_{16}$ inch (1.19 to 1.59 mm). The rake, which is positive, is about 15 degrees.

*Drilling Magnesium:* If the depth of a hole is less than five times the drill diameter, an ordinary twist drill with highly polished flutes may be used. The included angle of the point may vary from 70 degrees to the usual angle of 118 degrees. The relief angle is about 12 degrees. The drill should be kept sharp and the outer corners rounded to produce a smooth finish and prevent burr formation. For deep hole drilling, use a drill having a helix angle of 40 to 45 degrees with large polished flutes of uniform cross section throughout the drill length to facilitate the flow of chips. A pyramid-shaped "spur" or "pilot point" at the tip of the drill will reduce the "spiraling or run-off."

Drilling speeds vary from 300 to 2000 feet per minute (90–610 m/min) with feeds per revolution ranging from 0.015 to 0.050 inch (0.38–1.27 mm).

*Reaming Magnesium:* Reamers up to 1 inch in diameter should have four flutes; larger sizes, six flutes. These flutes may be either parallel with the axis or have a negative helix angle of 10 degrees. The positive rake angle varies from 5 to 8 degrees, the relief angle from 4 to 7 degrees, and the clearance angle from 15 to 20 degrees.

*Tapping Magnesium:* Standard taps may be used unless Class 3B tolerances are required, in which case the tap should be designed for use in magnesium. A high-speed steel concentric type with a ground thread is recommended. The concentric form, which eliminates the radial thread relief, prevents jamming of chips while the tap is being backed out of the hole. The positive rake angle at the front may vary from 10 to 25 degrees and the "heel rake angle" at the back of the tooth from 3 to 5 degrees. The chamfer extends over two to three threads. For holes up to $\frac{1}{4}$ inch (6.35 mm) in diameter, two-fluted taps are recommended; for sizes from $\frac{1}{2}$ to $\frac{3}{4}$ inch (12.7–19.05 mm), three flutes; and for larger holes, four flutes. Tapping speeds ordinarily range from 75 to 200 feet per minute (23–61 m/min), and mineral oil cutting fluid should be used.

*Threading Dies for Magnesium* should have about the same cutting angles as taps. Narrow lands should be used to provide ample chip space. Either solid or self-opening dies may be used. The latter type is recommended when maximum smoothness is required. Threads may be cut at speeds up to 1000 feet per minute (305 m/min).

*Grinding Magnesium:* As a general rule, magnesium is ground dry. The highly inflammable dust should be formed into a sludge by means of a spray of water or low-viscosity mineral oil. Accumulations of dust or sludge should be avoided. For surface grinding, when a fine finish is desirable, a low-viscosity mineral oil may be used.

**Machining Zinc Alloy Die Castings.—**Machining of zinc alloy die castings is mostly done without a lubricant. For particular work, especially deep drilling and tapping, a lubricant such as lard oil and kerosene (about half and half) or a 50-50 mixture of kerosene and machine oil may be used to advantage. A mixture of turpentine and kerosene has been been found effective on certain difficult jobs.

*Reaming:* In reaming, tools with six straight flutes are commonly used, although tools with eight flutes irregularly spaced have been found to yield better results by one manufacturer. Many standard reamers have a land that is too wide for best results. A land about 0.015 inch (0.38 mm) wide is recommended but this may often be ground down to around 0.007 or even 0.005 inch (0.18 or even 0.127 mm) to obtain freer cutting, less tendency to loading, and reduced heating.

*Turning:* Tools of high-speed steel are commonly employed although the application of Stellite and carbide tools, even on short runs, is feasible. For steel or Stellite, a positive top rake of from 0 to 20 degrees and an end clearance of about 15 degrees are commonly recommended. Where side cutting is involved, a side clearance of about 4 degrees minimum is recommended. With carbide tools, the end clearance should not exceed 6 to 8 degrees and the top rake should be from 5 to 10 degrees positive. For boring, facing, and other lathe operations, rake and clearance angles are about the same as for tools used in turning.

**Machining Monel and Nickel Alloys.—**These alloys are machined with high-speed steel and with cemented carbide cutting tools. High-speed steel lathe tools usually have a back rake of 6 to 8 degrees, a side rake of 10 to 15 degrees, and relief angles of 8 to 12 degrees. Broad-nose finishing tools have a back rake of 20 to 25 degrees and an end relief angle of 12 to 15 degrees. In most instances, standard commercial cemented carbide tool holders and tool shanks can be used which provide an acceptable tool geometry. Honing the cutting edge lightly will help if chipping is encountered.

The most satisfactory tool materials for machining Monel and the softer nickel alloys, such as Nickel 200 and Nickel 230, are M2 and T5 for high-speed steel and crater resistant grades of cemented carbides. For the harder nickel alloys such as K Monel, Permanickel, Duranickel, and Nitinol alloys, the recommended tool materials are T15, M41, M42, M43, and for high-speed steel, M42. For carbides, a grade of crater resistant carbide is recommended when the hardness is less than 300 BHN, and when the hardness is more than

300 BHN, a grade of straight tungsten carbide will often work best, although some crater resistant grades will also work well.

A sulfurized oil or a water-soluble oil is recommended for rough and finish turning. A sulfurized oil is also recommended for milling, threading, tapping, reaming, and broaching. Recommended cutting speeds for Monel and the softer nickel alloys are 70 to 100 fpm (20–30 m/min) for high-speed steel tools and 200 to 300 fpm (60–90 m/min) for cemented carbide tools. For the harder nickel alloys, the recommended speed for high-speed steel is 40 to 70 fpm (12–21 m/min) for a hardness up to 300 BHN and for a higher hardness, 10 to 20 fpm (3–6 m/min); for cemented carbides, 175 to 225 fpm (55–70 m/min) when the hardness is less than 300 BHN and for a higher hardness, 30 to 70 fpm (9–21 m/min).

Nickel alloys have a high tendency to work harden. To minimize work hardening caused by machining, the cutting tools should be provided with adequate relief angles and positive rake angles. Furthermore, the cutting edges should be kept sharp and replaced when dull to prevent burnishing of the work surface. The depth of cut and feed should be sufficiently large to ensure that the tool penetrates the work without rubbing.

**Machining Copper Alloys.—**Copper alloys can be machined by tooling and methods similar to those used for steel, but at higher surface speeds. Machinability of copper alloys is discussed in Table 2 on page 510 and Table 3 on page 514. Machinability is based on a rating of 100 percent for the free-cutting alloy C35000, which machines with small, easily broken chips. As with steels, copper alloys containing lead have the best machining properties, with alloys containing tin and lead having machinability ratings of 80 and 70 percent. Tellurium and sulphur are added to copper alloys to increase machinability with minimum effect on conductivity. Lead additions are made to facilitate machining, as their effect is to produce easily broken chips.

Copper alloys containing silicon, aluminum, manganese and nickel become progressively more difficult to machine, and produce long, stringy chips, the latter alloys having only 20 percent of the machinability of the free-cutting alloys. Although copper is frequently machined dry, a cooling compound is recommended. Other lubricants that have been used include tallow for drilling, gasoline for turning, and beeswax for threading.

## Machining Nonmetals

**Machining Hard Rubber.—**Tools suitable for steel may be used for hard rubber, with no top or side rake angles and 10 to 20 degree clearance angles, of high-speed steel or tungsten carbide. Without coolant, surface speeds of about 200 ft/min (60m/min) are recommended for turning, boring and facing; with coolant, increase up to 300 surface ft/min (90 m/min).

*Drilling of hard rubber* requires high-speed steel drills of 35 to 40 degree helix angle to obtain maximum cutting speeds and drill life. Feed rates for drilling range up to 0.015 in/rev (0.38 mm/rev). Deep-fluted taps are best for threading hard rubber, and should be 0.002 to 0.005 inch (0.05-0.13 mm) oversize if close tolerances are to be held. Machine oil is used for a lubricant. Hard rubber may be sawn with band saws having 5 to 10 teeth per inch (2–4 teeth/cm), running at about 3000 ft/min (914 m/min) or cut with abrasive wheels. Use of coolant in grinding rubber gives a smoother finish.

*Piercing and blanking* of sheet rubber is best performed with the rubber or dies heated. Straightening of the often-distorted blanks may be carried out by dropping them into a pan of hot water.

**Formica Machining.—**Blanks can be cut from sheets of "Formica" either by a band saw or by trepanning tools in a boring mill or a drill press. To saw blanks, first describe a circle as a guide line, then use a 21-gage $3\frac{1}{2}$-point saw running at a speed of 5000 ft/min (1525 m/min). The saw should be sharp, with a $\frac{1}{64}$-inch (0.4-mm) set on both sides. In drilling, use an ordinary high-speed drill whose point is ground to an included angle of 55 to 60 degrees. Another method is to grind the drill point slightly off center. The feed must be rapid and caution used to prevent the drill from lagging in its work, and the speed must be 1200 rpm. For all machining operations on "Formica" gear material, provision must be made in grinding for the tools to clear themselves. For reaming, the entry of the reamer and the reaming process must be rapid. There must not be a lag between the end of the reaming operation and the withdrawal of the reamer. In turning the outside diameter and the sides of blanks, the tools must be sharp and have 3 to 5 degrees more rake than is common practice for metal. A cutting speed of 750 fpm (228.6 m/min), equal to 720 rpm on a 4-inch (101.6-mm) diameter blank, is recommended. The depth of the cut can be $\frac{1}{16}$ to $\frac{1}{8}$ inch (1.6–3.2 mm), but the feed should be 0.010 inch (0.25 mm), regardless of the depth of the cut. Teeth may be cut on a hobbing machine, shaper, or milling machine. The speed of the cutter should be 150 fpm (46 m/min) and the feed from 0.023 to 0.040 inch per revolution (0.59–1.02 mm/rev). It is advisable to back up the blank to prevent fraying or breaking out of the material as the cutter comes through. The backing plates can be economically made from hard wood.

**Micarta Machining.—**In cutting blanks from sheets of "micarta" a band saw running at a speed of 350 revolutions per minute has been found suitable. The saw should be of the bevel-tooth type, seven teeth to the inch. For large quantities a trepanning tool should be used. In trepanning blanks, the tool should be fed so as to cut part way through all of the "layouts"; then the micarta plate should be turned over, and the cutting completed from the reverse side.

Turning tools should be high-speed steel cutting at speeds similar to those for bronze or cast iron. If two cuts are taken, about 0.010 inch (0.254 mm) of stock should be left for the finishing cut.

Drilling at right angles to the layers is done with a standard drill, which should be backed off sufficiently to provide plenty of clearance. When drilling parallel to layers, a "flat" or "bottom" drill should be used. In rough-drilling, the hole should preferably be drilled partly through the material from each side to prevent possible splitting as the tool protrudes. If this is impracticable, the hole can be drilled all the way through the material, provided the material is "backed up" with wood, stiff cardboard, or any other material that is sufficiently rigid to support the under surface at the point where the drill comes through.

The methods described for drilling apply as well to tapping, except that when the tapping is done parallel to the layers, it is advisable to clamp the material to equalize the stress on the layers and prevent possible splitting.

In milling, a standard tool may be used at a speed and feed corresponding to that used in working bronze or soft steel. The cutting angle of the cutter will give better results if ground with a slight rake.

While there is a wide range of practice as to speeds and feeds in cutting gears on hobbing machines, a hob speed of not less than 140 revolutions per minute has given satisfaction. In machining gear teeth on a gear shaper, a speed of about 100 to 130 strokes per minute with a fairly fine feed has given good results. Backing-up plates should be used in machining micarta gears.

**Ultrasonic Machining.—**This method of cutting and engraving hard materials such as glass, precious stones, and carbides uses a transducer (vibratory unit) to obtain the necessary mechanical vibrations needed. The transducer converts the input energy, in this case electrical, into another form of energy, in this case mechanical.

A tool of the required size and shape is made of brass or other soft material and is attached to the transducer. The tool is lowered until it just barely touches the work, and current is applied. At the same time, a slurry of water and fine abrasive, usually boron carbide, is pumped over the work. The tool does not actually touch the work, but the vibrations literally hammer the particles of abrasive into the surface and chip off tiny fragments. Some wear does take place in the tool, but it is very slight and, as it is equally distributed, it does not change the shape. The method is commonly applied to cutting designs in stones of signet rings, and also applied to cutting intricately shaped holes in carbide or hardened steel.

## GRINDING FEEDS AND SPEEDS

Grinding data are scarcely available in handbooks, which usually recommend a small range of depths and work speeds at constant wheel speed, including small variations in wheel and work material composition. Wheel life or grinding stiffness are seldom considered.

Grinding parameter recommendations typically range as follows:

- Wheel speeds are usually recommended in the 1200 to 1800 m/min (4000 to 6000 fpm) range, or in rare cases up to 3600 m/min (12000 fpm)
- Work speeds are in the range 20 to 40 m/min (70 to 140 fpm); and, depths of cut of 0.01 to 0.025 mm (0.0004 to 0.001 inch) for roughing, and around 0.005 mm (.0002 in.) for finish grinding.
- Grit sizes for roughing are around 46 to 60 for easy-to-grind materials, and for difficult-to-grind materials higher such as 80 grit. In finishing, a smaller grit size (higher grit number) is recommended. Internal grinding grit sizes for small holes are approximately 100 to 320.
- Specific metal removal rate, *SMRR*, represents the rate of material removal per unit of wheel contact width and are commonly recommended from 200 to 500 $mm^3$/mm width/min (0.3 to 0.75 $in^3$/inch width/min).
- Grinding stiffness is a major variable in determining wheel-life and spark-out time. A typical value of system stiffness in outside-diameter grinding, for 10:1 length/diameter ratio, is approximately $K_{ST}$ = 30–50 N/μm. System stiffness $K_{ST}$ is calculated from the stiffness of the part, $K_w$ and the machine and fixtures, $K_m$. Machine values can be obtained from manufacturers, or can be measured using simple equipment along with the part stiffness.
- Generally a lower wheel hardness (soft wheel) is recommended when the system stiffness is poor or when a better finish is desired.

### Basic Rules

The wheel speed *V* and *equivalent chip thickness* $ECT = SMRR \div V \div 1000$ are the primary parameters that determine wheel-life, forces and surface finish in grinding. The following general rules and recommendations, using *ECT*, are based on extensive laboratory and industry tests both in Europe and the United States. The relationships and shapes of curves pertaining to grinding tool life, grinding time, and cost are similar to those of any metal cutting operation such as turning, milling and drilling.

In turning and milling, the *ECT* theory says that if the product of feed times depth of cut is constant, the tool life is constant no matter how the depth of cut or feed is varied, provided that the cutting speed and cutting edge length are maintained constant.

In grinding, wheel-life *T* remains constant for constant cutting speed *V*, regardless of how depth of cut $a_r$ or work speed $V_w$ are selected as long as the specific metal removal rate $SMRR = V_w \times a_r$ is held constant (neglecting the influence of grinding contact width).

*ECT* is much smaller in grinding than in milling, ranging from about 0.0001 to 0.001 mm (0.000004 to 0.00004 inch). See the section *MACHINING ECONOMETRICS* starting on page 1184 for a detailed explanation of the role of *ECT* in conventional machining.

**Wheel Life and Grinding Ratio.—**A commonly used measure of relative wheel-life in grinding is the *grinding ratio*, which is used to compare grindability when varying grinding wheel composition and work material properties under otherwise constant cutting conditions.

The *grinding ratio* is defined as the slope of the wear curve versus metal removal rate: *grinding ratio* = $MRR \div W^*$, where *MRR* is the metal removal rate, and $W^*$ is the volume wheel wear at which the wheel has to be dressed. The grinding ratio is not a measure of wheel-life, but a relationship between *grinding ratio* and wheel-life *T* can be obtained from the formula *grinding ratio* = $SMRR \times T \div W^*$, where *SMRR* (specific metal removal rate) is determined from $MRR = SMRR \times T$ or from $ECT = SMRR \div V \div 1000$.

Thus, *grinding ratio* = $1000 \times ECT \times V \times T \div W^*$, and $T$ = *grinding ratio* $\times W^* \div (1000 \times ECT \times V)$, provided that the wheel wear criterion $W^*$ is valid for all data combinations.

*Example 1:* If $W^*$ in one test is found to be 500 mm$^3$ for $ECT = 0.00033$ mm and $V = 3600$ m/min, and *grinding ratio* = 10, then wheel-life will vary with measured grinding ratios, wheel speed, and *ECT* as follows: $T = 500 \times$ *grinding ratio* $\div (V \times ECT) = 4.2$ minutes.

In the remainder of this section the *grinding ratio* will not used, and wheel-life is expressed in terms of *ECT* or *SMRR* and wheel speed *V*.

***ECT* in Grinding.**—In turning and milling, *ECT* is defined as the volume of chips removed per unit cutting edge length per revolution of the work or cutter. In milling specifically, *ECT* is defined as the ratio of (number of teeth $z \times$ feed per tooth $f_z \times$ radial depth of cut $a_r \times$ and axial depth of cut $a_a$) and (cutting edge length *CEL* divided by $\pi D$), where *D* is the cutter diameter, thus,

$$ECT = \frac{\pi D z f_z a_r a_a}{CEL}$$

In grinding, the same definition of *ECT* applies if we replace the number of teeth with the average number of grits along the wheel periphery, and replace the feed per tooth by the average feed per grit. This definition is not very practical, however, and *ECT* is better defined by the ratio of the specific metal removal rate *SMRR*, and the wheel speed *V*. Thus, $ECT = 1000 \times SMRR \div V$. Keeping *ECT* constant when varying *SMRR* requires that the wheel speed must be changed proportionally.

In milling and turning *ECT* can also be redefined in terms of *SMRR* divided by the work and the cutter speeds, respectively, because *SMRR* is proportional to the feed rate $F_R$.

*Work Speed and Depth of Cut Selection:* Work speed $V_w$ is determined by dividing *SMRR* by the depth of cut $a_r$, or by using the graph in Fig. 1.

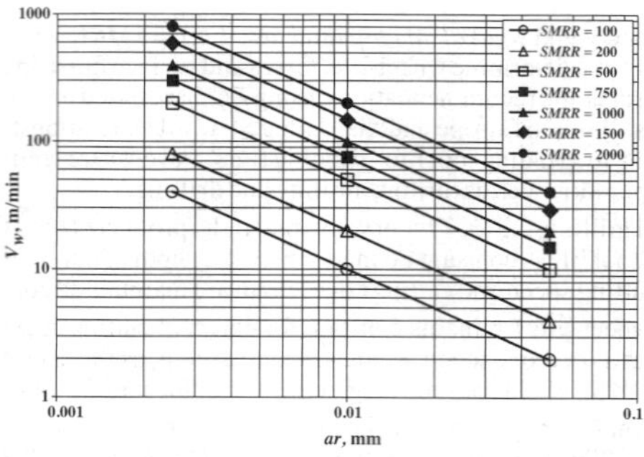


Fig. 1. Work Speed $V_w$ versus Depth of Cut $a_r$

Referring to Fig. 1, for depths of cuts of 0.01 and 0.0025 mm, a specific metal removal rate $SMRR = 1000$ mm$^3$/mm width/min is achieved at work speeds of 100 and 400 m/min, respectively, and for $SMRR = 100$ mm$^3$/mm width/min at work speeds of 10 and 40 m/min, respectively.

Unfortunately, the common use of low values of work speed (20 to 40 m/min) in finishing cause thermal surface damage, disastrous in critical parts such as aircraft components. As the grains slide across the work they generate surface heat and fatigue-type loading may cause residual tensile stresses and severe surface cracks. Proper finish grinding conditions are obtained by increasing the work speed 5 to 10 times higher than the above recommendations indicate. These higher work speeds will create compressive stresses that are not detrimental to the surface. The by-product of higher work speeds is much higher *SMRR* values and thereby much shorter grinding times. Compressive stresses are also obtained by reducing the depth of cut $a_r$.

**Wheel Life Relationships and Optimum Grinding Data.—**Fig. 2a, Fig. 2b, and Fig. 2c show, in three planes, the 3-dimensional variation of wheel-life *T* with wheel speed *V* and *ECT* when grinding a hardened tool steel. Fig. 2a depicts wheel-life versus wheel speed (the *T-V* plane) with constant *ECT* appearing as approximately straight lines when plotted in log-log coordinates.

In grinding, the wheel-life variation follows curves similar to those obtained for conventional metal cutting processes, including a bend-off of the Taylor lines (*T-V* graph) towards shorter life and lower cutting speeds when a certain maximum life is achieved for each value of *ECT*. In the two other planes (*T-ECT*, and *V-ECT*) we usually find smooth curves in which the maximum values of wheel-life are defined by points along a curve called the *H*-curve.

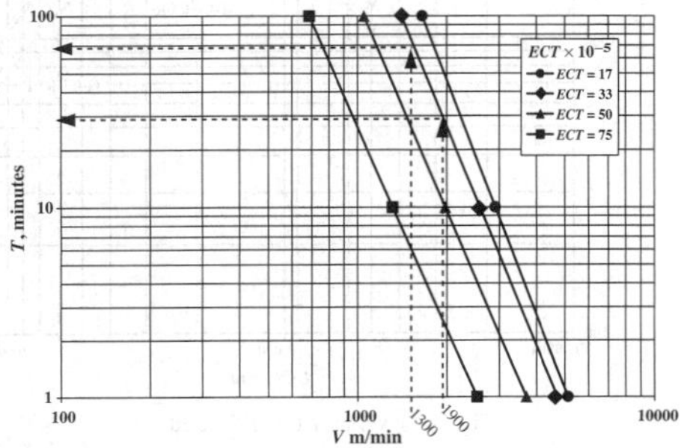


Fig. 2a. Taylor Lines: *T* versus *V*, *ECT* Plotted for Grinding M4 Tool Steel, Hardness 64 RC

*Example 2:* The variation of $SMRR = V \times ECT \times 1000$ and wheel-life at various wheel speeds can be obtained from Fig. 2a. Using sample values of $ECT = 33 \times 10^{-5}$ mm and $V = 1300$ and 1900 m/min, $SMRR = 1300 \times 33 \times 10^{-5} \times 1000 = 429$, and $1900 \times 33 \times 10^{-5} \times 1000 = 627$ mm$^3$/mm width/min, respectively; the corresponding wheel lives are read off as approximately 70 and 30 minutes, respectively.

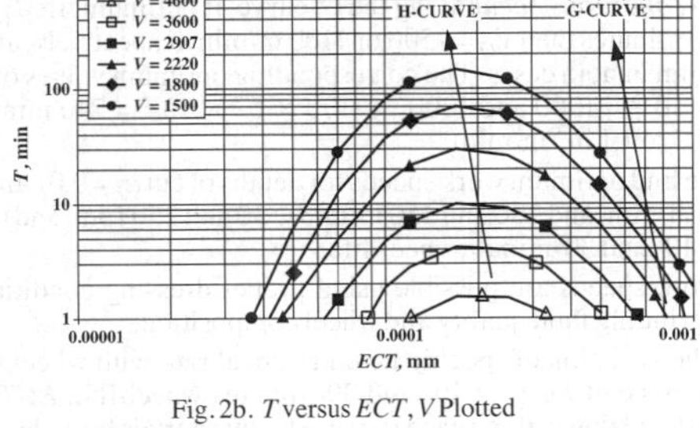


Fig. 2b. *T* versus *ECT*, *V* Plotted

Fig. 2b depicts wheel-life *T* versus *ECT* with constant wheel speed *V* shown as curves plotted in log-log coordinates, similar to those for the other cutting operations.

*Example 3:* Fig. 2b shows that maximum values of wheel-life occur along the *H*-curve. For the 3 speeds 1800, 2700, and 3600 m/min, maximum wheel lives are approximately 70, 14 and 4 minutes, respectively, at *ECT* around 17 × $10^{-5}$ through 20 × $10^{-5}$ mm along the *H*-curve. Left and right of the *H*-curve wheel-life is shorter.

Fig. 2c depicts wheel speed *V* versus *ECT* with wheel-life *T* parameter shown as curves in log-log coordinates, similar to those for the other cutting operations, with the characteristic *H*- and *G*-curves.

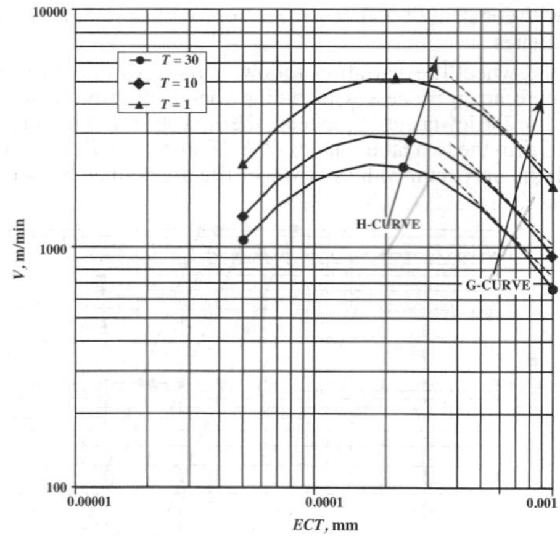


Fig. 2c. *V* versus *ECT*, *T* Plotted

*Optimum grinding data for roughing* occur along the *G*-curve, which is determined from the *V-ECT* graph by drawing 45-degree lines tangent to the *T*-curves, as shown in Fig. 2c, and drawing a line (the *G*-curve) through the points of tangency on the respective *T*-curves, thus the location and direction of the *G*-curve is determined. Globally optimum data correspond to the *T*-curve for which wheel-life is calculated using the corresponding equivalent tooling-cost time, $T_V$, calculated from $T_V = T_{RPL} + 60 \times C_E \div H_R$, minutes, where $T_{RPL}$ is the time required to replace wheel, $C_E$ = cost per wheel dressing = wheel cost + cost per dressing, and $H_R$ is the hourly rate.

*Minimum cost conditions occur* along the *G*-curve; if optimum life $T_O$ was determined at either 10 or 30 minutes then $V_O$ = 1500 or 1100 m/min, respectively, and *ECT* is around 65 – 70 × $10^{-5}$ mm in both cases. The corresponding optimum values of *SMRR* are 1000 × 1500 × 67 × $10^{-5}$ = 1000 and 1000 × 1100 × 67 × $10^{-5}$ = 740 $mm^3$/min/mm wheel contact width (1.5 to 1.1 $in^3$/in/min).

Using Fig. 1 we find optimum work speeds for depths of cut $a_r$ = 0.01 and 0.005 mm to be $V_w$ = 100 and 75 m/min, and 200 and 150 m/min (330 and 250 fpm, and 660 and 500 fpm) respectively for 10- and 30-minute wheel-life.

These high work speeds are possible using proper dressing conditions, high system stiffness, good grinding fluid quality and wheel composition.

Fig. 3 shows the variation of specific metal removal rate with wheel speed for several materials and a range of *ECT*s at 10- and 30-minutes wheel-life. *ECT* decreases when moving to the left and down along each curve. The two curves for unhardened 1020 steel have the largest values of *SMRR*, and represent the most productive grinding conditions, while the heat resistant alloy Inconel yields the least productive grinding conditions. Each branch attains a maximum *SMRR* along the *G*-curve (compare with the same curve in the V-*ECT* graph, Fig. 2c) and a maximum speed region along the *H*-curve. When the *SMRR*-values are lower than the *H*-curve the *ECT* values for each branch decrease towards the bottom of the graph, then the speed for constant wheel-life must be reduced due to the fact that the *ECT* values are to the left of their respective *H*-curves in *V-ECT* graphs.

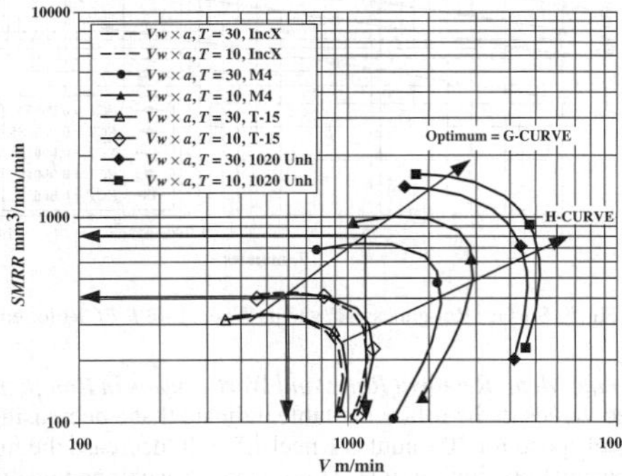


Fig. 3. Specific Metal Removal Rate versus Cutting Speed at *T*=10 and 30 Minutes Wheel Life
In the figure, IncX is Inconel; M4, and T-15 are tool steels; and 1020 Unh is unhardened 1020 steel.

**Surface Finish, *Ra*.**—The finish is improved by decreasing the value of *ECT* as shown in Fig. 4, where the *Ra* value (or $R_a$, for roughness average) is plotted versus *ECT* at 3 different wheel lives 1, 10 and 30 minutes at constant wheel speed. Because *ECT* is proportional to the depth of cut, a smaller depth of cut is favorable for reducing surface roughness when the work speed is constant.

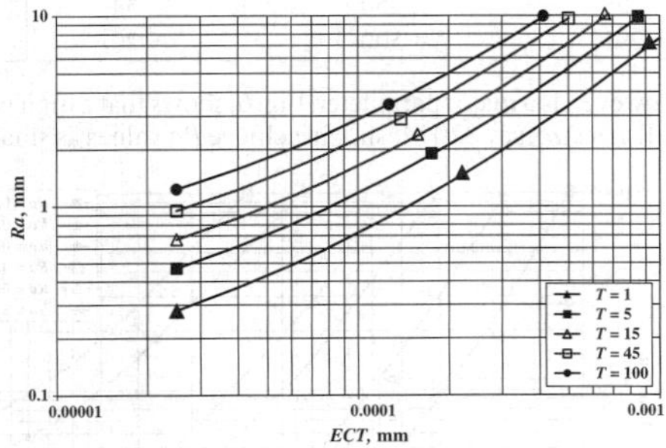


Fig. 4. Surface Finish, *Ra* versus *ECT*, Wheel-Life *T* Plotted

In Fig. 5, *Ra* is plotted versus wheel-life at 5 different *ECT*'s. Both Fig. 4 and Fig. 5 illustrate that a shorter life improves the surface finish, which means that either an increased wheel speed (wheel-life decreases) at constant *ECT*, or a smaller *ECT* at constant speed (wheel-life increases), will result in an improved finish. For a required surface finish, *ECT* and wheel-life have to be selected appropriately in order to also achieve an optimum grinding time or cost. In cylindrical grinding a reduction of side feed $f_s$ improves *Ra* as well.

In terms of specific metal removal rate, reducing *SMRR* will improve the surface finish *Ra*.

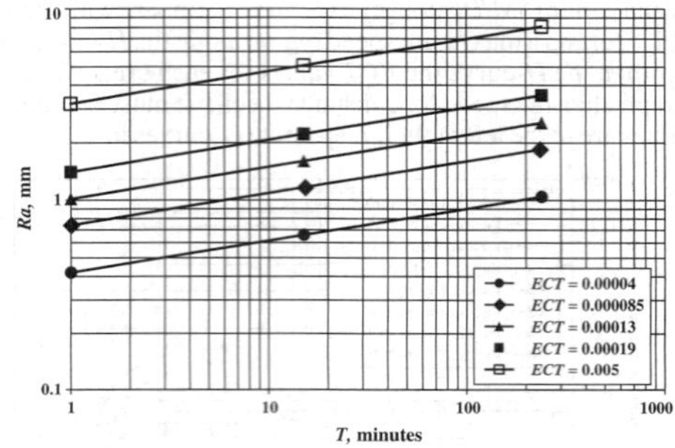


Fig. 5. Surface Roughness, *Ra* versus Wheel Life *T*, *ECT* Plotted

*Example 4, Specific Metal Removal Rates and Work Speeds in Rough and Finish Grinding:* The tabulated values in the following table indicate that a decreasing *ECT* combined with a higher wheel speed for 10 minutes wheel-life will decrease the metal removal rate and thereby increasing the grinding time. This change is accompanied by a better finish in both roughing and finishing operations. Note the high work speeds when finishing.

| | Tool Life $T = 10$ minutes | | Roughing Depth $a_r = 0.025$ mm | Finishing Depth $a_r = 0.0025$ mm |
|---|---|---|---|---|
| *ECT* mm | Wheel speed $V_{10}$ m/min | Removal Rate $SMRR_{10}$ mm³/mm/ min | Work speed $V_w$ m/min | |
| 0.00050 | 1970 | 985 | 39 | 390 |
| 0.00033 | 2580 | 850 | 34 | 340 |
| 0.00017 | 2910 | 500 | 20 | 200 |

The grit size, however, is a major parameter. Fig. 6, shows that a high wheel speed, combined with a small grit size, say 320 Mesh, can achieve *Ra* values as small as 0.03 micron.

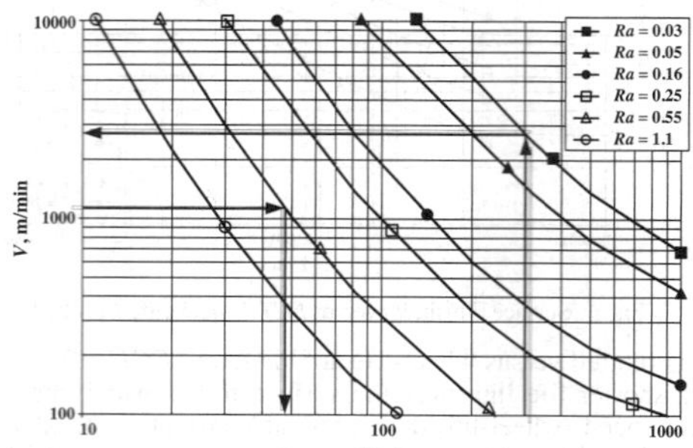


Fig. 6. Wheel Speed versus Wheel Mesh, *Ra* Plotted

**Spark-Out Time.—**Fig. 7 shows how spark-out time varies with system stiffness. As with surface finish, when wheel-life is short (high wear rate) the spark-out time decreases.

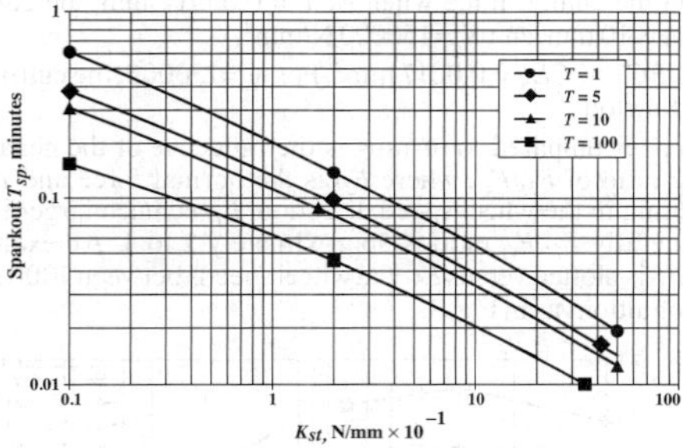


Fig. 7. Sparkout Time versus System Stiffness, Wheel-Life *T* Plotted

*Equivalent Diameter (Work Conformity) Factor:* The difference in curvature of the work and wheel in the contact region, determined by the equivalent diameter or work conformity formula, is an important factor for calculating spark-out time and forces, but has a negligible influence on wheel-life. Therefore, an equivalent diameter, $D_e = D/(1 \pm D/D_w)$, with the minus sign for internal grinding and the plus sign for external grinding operations, is used to consider the effect of conformity when using internal and external grinding with varying work and wheel diameters. $D_e$ is equal to the wheel diameter in surface grinding (work flat); in internal grinding, the wheel conforms closely to the work and $D_e$ is therefore larger than in external grinding.

**Grinding Cutting Forces, Torque and Power.—**Formulas to calculate the tangential cutting force, torque and required machining power are found in *Estimating Machining Power* on page 1133, but the values of $K_c$, specific cutting force or specific energy, are approximately 30 to 40 times higher in grinding than in turning, milling and drilling. This is primarily due to the fact that the *ECT* values in grinding are 1000 to 10000 times smaller, and also due to the negative rake angles of the grit. Average grinding rake angles are around −35 to −45 degrees. $K_c$ for grinding unhardened steel is around 50000 to 70000 N/mm$^2$ and up to 150000 to 200000 N/mm$^2$ for hardened steels and heat resistant alloys. The grinding cutting forces are relatively small because the chip area is very small.

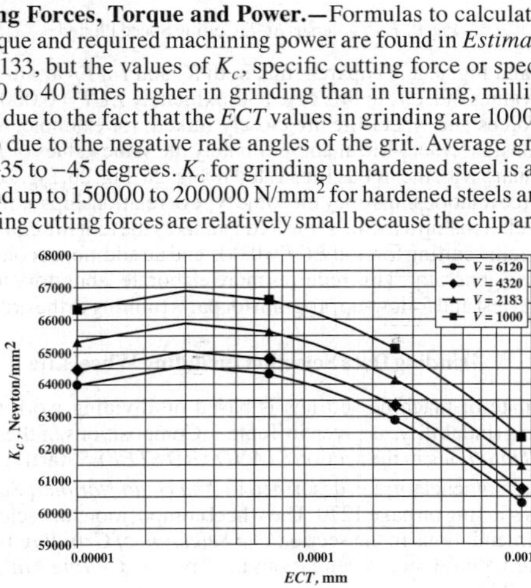


Fig. 8. Specific Grinding Force $K_c$ versus *ECT*; *V* Plotted

As in the other metal cutting operations, the forces vary with *ECT* and to a smaller extent with the wheel speed *V*. An example is shown in Fig. 8, where $K_c$, specific cutting force, is plotted versus *ECT* at wheel speeds between 1000 and 6000 m/min. The material is medium unhardened carbon steel ground by an aluminum oxide wheel. The impact of wheel speed is relatively small (2 to 5 percent lower with increasing speed).

*Example 5:* Find the cutting force when $ECT$ = 0.00017 mm, the cutting edge length (width of cut) $CEL$ is 10 mm, and $K_c$=150000 N/mm$^2$.

The chip area is $ECT \times CEL$ = 0.0017 mm$^2$. For $K_c$=150000, the cutting force is 0.0017 × 150000 = 255 Newton.

Another difference compared to turning is the influence of the negative rake angles, illustrated by the ratio of $F_H/F_C$, where $F_H$ is the normal force and $F_C$ the tangential grinding force acting in the wheel speed direction. $F_H$ is much larger than the grinding cutting force, generally $F_H/F_C$ ratio is approximately 2 to 4. An example is shown in Fig. 9, where $F_H/F_C$ is plotted versus $ECT$ at wheel speeds between 1000 and 6000 m/min, under the same conditions as in Fig. 8.

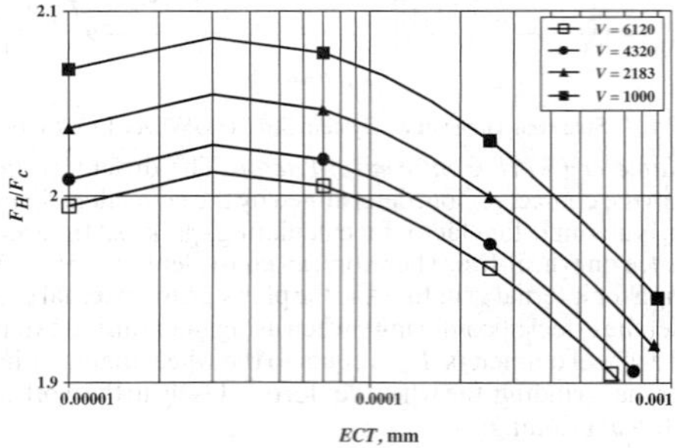


Fig. 9. $F_H/F_C$ versus $ECT$; Cutting Speed Plotted

In both Fig. 8 and Fig. 9, it is apparent that both $K_c$ and $F_H/F_C$ attain maximum values for given small values of $ECT$, in this case approximately $ECT$ = 0.00005 mm. This fact illustrates that forces and wheel-life are closely linked; for example, wheel speed has a maximum for constant wheel-life at approximately the same values of $ECT$ shown in the two graphs (compare with the trends illustrated in Fig. 2a, Fig. 2b, Fig. 2c, and Fig. 3). As a matter of fact, force relationships obey the same type of relationships as those of wheel-life. Colding's force relationship uses the same 5 constants as the tool life equation, but requires values for the specific cutting force at $ECT$=0.001 and an additional constant, obtained by a special data base generator. This requires more elaborate laboratory tests, or better, the design of a special test and follow-up program for parts running in the ordinary production.

## Grinding Data Selection Including Wheel Life

The first estimate of machine settings is based on dividing work materials into 10 groups, based on grindability, as given in Table 1. Compositions of these work materials are found in the *Handbook* in the section *STANDARD STEELS* starting on page 391.

Grinding wheel nomenclature is described in *American National Standard Grinding Wheel Markings* starting on page 1270. The wheel compositions are selected according to the grade recommendations in the section *The Selection of Grinding Wheels* starting on page 1271. Grinding fluid recommendations are given in *Cutting Fluids for Machining* starting on page 1234.

*Note:* Maximum wheel speeds should always be checked using the safety standards in the section *Safe Operating Speeds* starting on page 1300, because the recommendations will sometimes lead to speeds above safety levels.

The material in this section is based on the use of a typical standard wheel composition such as 51-A-46-L-5-V-23, with wheel grade (wheel hardness) = L or above, and mesh (grit size) = 46 or above.

**Table 1. Grindability Groups**

| Group | Examples |
|---|---|
| Group 1 Unhardened Steels | |
| Group 2 Stainless Steels | SAE 30201-30347, 51409-51501 |
| Group 3 Cast Iron | |
| Group 4 Tool Steels | M1, M8, T1, H, O, L, F, 52100 |
| Group 5 Tool Steels | M2, T2, T5, T6, D2, H41, H42, H43, M50 |
| Group 6 Tool Steels | M3, M4, T3, D7 |
| Group 7 Tool Steels | T15, M15 |
| Group 8 Heat Resistant Steels | Inconel, Rene etc. |
| Group 9 Carbide Materials | P30 Diamond Wheel |
| Group 10 Ceramic Materials | |

For each grindability group there is one table and 2 graphs (one with Taylor lines and the other with *SMRR* versus wheel speed *V*) that are used to get a first estimate of standardized machine settings, assuming a good system stiffness ($K_{ST} > 30$ N/μm). These data are then calibrated with the users own data in order to refine the estimate and optimize the grinding process, as discussed in *User Calibration of Recommendations*. The recommendations are valid for all grinding processes such as plunge grinding, cylindrical, and surface grinding with periphery or side of wheel, as well as for creep feed grinding.

The grinding data machinability system is based on the basic parameters equivalent chip thickness *ECT*, and wheel speed *V*, and is used to determine specific metal removal rates *SMRR* and wheel-life *T*, including the work speed $V_w$ after the grinding depths for roughing and finishing are specified.

For each material group, the grinding data machinability system consists of *T-V* Taylor lines in log-log coordinates for 3 wheel speeds at wheel lives of 1, 10 and 100 minutes wheel-life with 4 different values of equivalent chip thickness *ECT*. The wheel speeds are designated $V_1$, $V_{10}$, and $V_{100}$ respectively. In each table the corresponding specific metal removal rates *SMRR* are also tabulated and designated as $SMRR_1$, $SMRR_{10}$ and $SMRR_{100}$ respectively. The user can select any value of *ECT* and interpolate between the Taylor lines. These curves look the same in grinding as in the other metal cutting processes and the slope is set at n = 0.26, so each Taylor line is formulated by $V \times T^{0.26} = C$, where C is a constant tabulated at four *ECT* values, *ECT* = 17, 33, 50 and 75 × $10^{-5}$ mm, for each material group. Hence, for each value of *ECT*, $V_1 \times 1^{0.26} = V_{10} \times 10^{0.26} = V_{100} \times 100^{0.26} = C$.

**Side Feed, Roughing and Finishing.**—In cylindrical grinding, the side feed, $f_s = C \times$ Width, does not impact on the values in the tables, but on the feed rate $F_R$, where the fraction of the wheel width C is usually selected for roughing and in finishing operations, as shown in the following table.

| Work Material | Roughing, C | Finishing, C |
|---|---|---|
| Unhardened Steel | 2/3-3/4 | 1/3-3/8 |
| Stainless Steel | 1/2 | 1/4 |
| Cast Iron | 3/4 | 3/8 |
| Hardened Steel | 1/2 | 1/4 |

*Finishing:* The depth of cut in rough grinding is determined by the allowance and usually set at $a_r = 0.01$ to 0.025 mm. The depth of cut for finishing is usually set at $a_r = 0.0025$ mm and accompanied by higher wheel speeds in order to improve surface finish. However, the most important criterion for critical parts is to increase the work speed in order to avoid thermal damage and surface cracks. In cylindrical grinding, a reduction of side feed $f_s$ improves $R_a$ as well. Small grit sizes are very important when very small finishes are required. See Fig. 4, Fig. 5, and Fig. 6 for reference.

*Terms and Definitions*

$a_a$ = depth of cut

$a_r$ = radial depth of cut, mm

$C$ = fraction of grinding wheel width

$CEL$ = cutting edge length, mm

$C_U$ = Taylor constant

$D$ = wheel diameter, mm

$DIST$ = grinding distance, mm

$d_w$ = work diameter, mm

$ECT$ = equivalent chip thickness $= f(a_r, V, V_w, f_s)$, mm

$= 1 \div (V \div V_w \div a_r + 1 \div f_s) = \dfrac{V_w f_s (a_r + 1)}{V}$

$=$ approximately $V_w \times a_r \div V = SMRR \div V \div 1000$

$= z \times f_z \times a_r \times a_a \div CEL \div (\pi D)$ mm

$F_R$ = feed rate, mm/min

$= f_s \times RPM_w$ for cylindrical grinding

$= f_i \times RPM_w$ for plunge (in-feed) grinding

$f_i$ = in-feed in plunge grinding, mm/rev of work

$f_s$ = side feed or engaged wheel width in cylindrical grinding $= C \times$ *Width* $= a_a$ approximately equal to the cutting edge length $CEL$

*Grinding ratio* $= MRR \div W^* = SMRR \times T \div W^* = 1000 \times ECT \times V \times T \div W^*$

$MRR$ = metal removal rate $= SMRR \times T = 1000 \times f_s \times a_r \times V_w$ mm$^3$/min

$SMRR$ = specific metal removal rate obtained by dividing $MRR$ by the engaged wheel width ($C \times$ *Width*) $= 1000 \times a_r \times V_w$ mm$^3$/mm width/min

*Note:* 100 mm$^3$/mm/min = 0.155 in$^3$/in/min, and 1 in$^3$/in/min = 645.16 mm$^3$/mm/min

$T, T_U$ = wheel-life = *Grinding ratio* $\times W \div (1000 \times ECT \times V)$ minutes

$t_c$ = grinding time per pass $= DIST \div F_R$ min

$= DIST \div F_R + t_{sp}$ (min) when spark-out time is included

$= \#$ Strokes $\times (DIST \div F_R + t_{sp})$ (min) when spark-out time and strokes are included

$t_{sp}$ = spark-out time, minutes

$V, V_U$ = wheel speed, m/min

$V_w, V_{wU}$ = work speed $= SMRR \div 1000 \div a_r$ m/min

$W^*$ = volume wheel wear, mm$^3$

*Width* = wheel width (mm)

$RPM$ = wheel speed $= 1000 \times V \div D \div \pi$ rpm

$RPM_w$ = work speed $= 1000 \times V_w \div D_w \div \pi$ rpm

**Relative Grindability.—**An overview of grindability of the data base, which must be based on a constant wheel wear rate, or wheel-life, is demonstrated using 10 minutes wheel-life shown in Table 2.

**Table 2. Grindability Overview**

| Material Group | $ECT \times 10^{-5}$ | $V_{10}$ | $SMRR_{10}$ | $V_w$ | |
|---|---|---|---|---|---|
| | | | | Roughing Depth $a_r = 0.025$ | *Finishing Depth* $a_r = 0.0025$ |
| 1 Unhardened | 33 | 3827 | 1263 | 50 | 500 |
| 2 Stainless | 33 | 1080 | 360 | 15 | 150 |
| 3 Cast Iron | 33 | 4000 | 1320 | 53 | 530 |
| 4 Tool Steel | 33 | 3190 | 1050 | 42 | 420 |
| 5 Tool Steel | 33 | 2870 | 950 | 38 | 380 |
| 6 Tool Steel | 33 | 2580 | 850 | 35 | 350 |
| 7Tool Steel | 33 | 1080 | 360 | 15 | 150 |
| 8 Heat resistant | 33 | 1045 | 345 | 14 | 140 |
| 9 Carbide with Diamond Wheel | 5 | $V_{600} = 1200$ | $SMRR_{600} =$ 50 | 2 | 20 |
| 10 Ceramics with Diamond Wheel | 5 | $V_{600} = 411$ | $SMRR_{600} =$ 21 | 0.84 | 84 |

**Procedure to Determine Data.**—The following wheel-life recommendations are designed for 4 values of *ECT* = 0.00017, 0.00033, 0.00050 and 0.00075 mm (shown as 17, 33, 50 and 75 in the tables). Lower values of *ECT* than 0.00010 mm (0.000004 in.) are not recommended as these may lie to the left of the *H*-curve.

The user selects any one of the *ECT* values, or interpolates between these, and selects the wheel speed for 10 or 100 minutes life, denoted by $V_{10}$ and $V_{100}$, respectively. For other desired wheel lives the wheel speed can be calculated from the tabulated Taylor constants C and $n = 0.26$ as follows:

$(V \times T_{(desired)})^{0.26}$ = C, the value of which is tabulated for each *ECT* value. C is the value of cutting speed $V$ at $T = 1$ minute, hence is the same as for the speed $V_1$. $V_1 \times 1^{0.26} = \text{C}$, thus $V_1 = \text{C}$

$$V_1 = \text{C} \div 1^{0.26} = \text{C}$$

$$V_{10} = \text{C} \div 10^{0.26} = \text{C} \div 1.82$$

$$V_{100} = \text{C} \div 100^{0.26} = \text{C} \div 3.31.$$

*Example 6:* A tool steel in material group 6 with *ECT* = 0.00033, has constant C= 4690, $V_{10} = 2578$ m/min, and $V_{100} = 1417$ m/min. From this information, find the wheel speed for desired wheel-life of $T = 15$ minutes and $T = 45$ minutes.

For $T = 15$ minutes we get $V_{15} = 4690 \div 15^{0.26} = 2319$ m/min (7730 fpm) and for $T = 45$ minutes $V_{45} = 4690 \div 45^{0.26} = 1743$ m/min (5810 fpm).

**The Tables are arranged in 3 sections:**

1. Speeds $V_{10}$ and $V_1$ = Constant CST(standard) for 4 *ECT* values 0.00017, 0.00033, 0.00050 and 0.00075 mm. Values $\text{C}_\text{U}$ and $\text{V}_{10\text{U}}$ refer to user calibration of the standard values in each material group, explained in the following.

2. Speeds $V_{100}$ (first row of 3), $\text{V}_{10}$ and $V_1$ (last in row) corresponding to wheel lives 100, 10 and 1 minutes, for 4 *ECT* values 0.00017, 0.00033, 0.00050 and 0.00075 mm.

3. Specific metal removal rates $SMRR_{100}$, $SMRR_{10}$ and $SMRR_1$ corresponding to wheel lives 100, 10 and 1 minutes, for the 4 *ECT* values 0.00017, 0.00033, 0.00050, and 0.00075 mm.

The 2 Graphs show: wheel life versus wheel speed in double logarithmic coordinates (Taylor lines); and, *SMRR* versus wheel speed in double logarithmic coordinates for 4 *ECT* values: 0.00017, 0.00033, 0.00050 and 0.00075 mm.

**Table 1. Group 1 — Unhardened Steels**

| Tool Life $T$ (min) | $ECT = 0.00017$ mm | | $ECT = 0.00033$ mm | | $ECT = 0.00050$ mm | | $ECT = 0.00075$ mm | |
|---|---|---|---|---|---|---|---|---|
| | Constant C = 8925 | | Constant C = 6965 | | Constant C = 5385 | | Constant C = 3885 | |
| | $V_T$ | *SMRR* | $V_T$ | *SMRR* | $V_T$ | *SMRR* | $V_T$ | *SMRR* |
| 100 | 2695 | 460 | 2105 | 695 | 1625 | 815 | 1175 | 880 |
| **10** | **4905** | **835** | **3830** | **1265** | **2960** | **1480** | **2135** | **1600** |
| 1 | 8925 | 1520 | 6965 | 2300 | 5385 | 2695 | 3885 | 2915 |

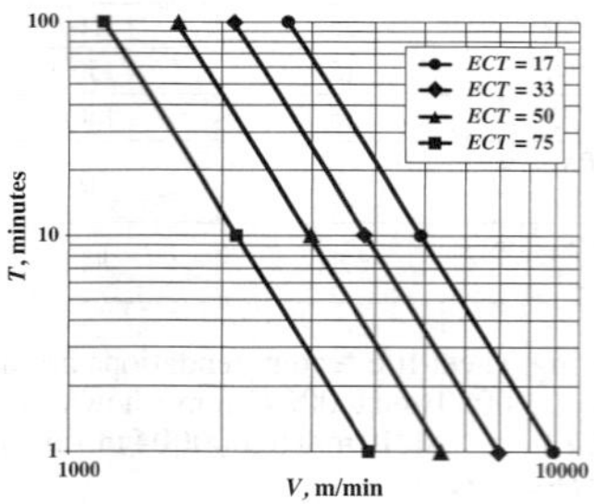


Fig. 1a. *T-V*

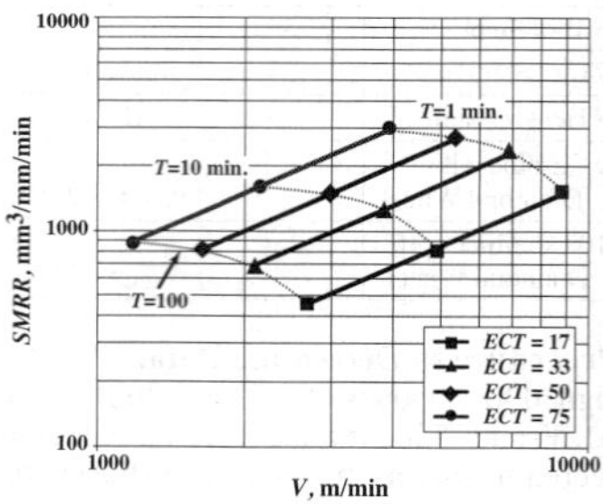


Fig. 1b. *SMRR* versus *V*, *T* = 100, 10, 1 minutes

**Table 2. Group 2 — Stainless Steels** SAE 30201 – 30347, SAE 51409 – 51501

| Tool Life $T$ (min) | $ECT = 0.00017$ mm | | $ECT = 0.00033$ mm | | $ECT = 0.00050$ mm | | $ECT = 0.00075$ mm | |
|---|---|---|---|---|---|---|---|---|
| | Constant C = 2270 | | Constant C = 1970 | | Constant C = 1505 | | Constant C = 1010 | |
| | $V_T$ | *SMRR* | $V_T$ | *SMRR* | $V_T$ | *SMRR* | $V_T$ | *SMRR* |
| 100 | 685 | 115 | 595 | 195 | 455 | 225 | 305 | 230 |
| **10** | **1250** | **210** | **1080** | **355** | **825** | **415** | **555** | **415** |
| 1 | 2270 | 385 | 1970 | 650 | 1505 | 750 | 1010 | 760 |

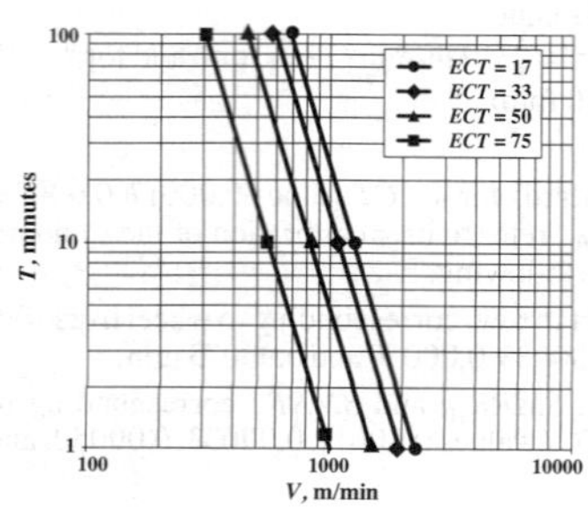


Fig. 2a. *T-V*

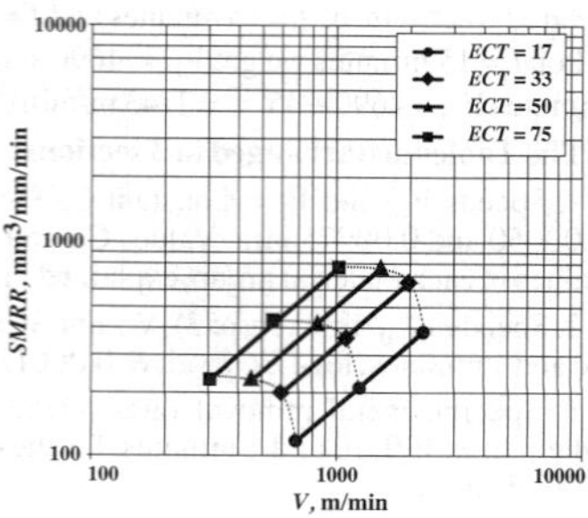


Fig. 2b. *SMRR* versus *V*, *T* = 100, 10, 1 minutes

**Table 3. Group 3—Cast Iron**

| Tool Life T (min) | ECT = 0.00017 mm | | ECT = 0.00033 mm | | ECT = 0.00050 mm | | ECT = 0.00075 mm | |
|---|---|---|---|---|---|---|---|---|
| | Constant C = 10710 | | Constant C = 8360 | | Constant C = 6465 | | Constant C = 4665 | |
| | $V_T$ | *SMRR* | $V_T$ | *SMRR* | $V_T$ | *SMRR* | $V_T$ | *SMRR* |
| 100 | 3235 | 550 | 2525 | 835 | 1950 | 975 | 1410 | 1055 |
| **10** | **5885** | **1000** | **4595** | **1515** | **3550** | **1775** | **2565** | **1920** |
| 1 | 10710 | 1820 | 8360 | 2760 | 6465 | 3230 | 4665 | 3500 |

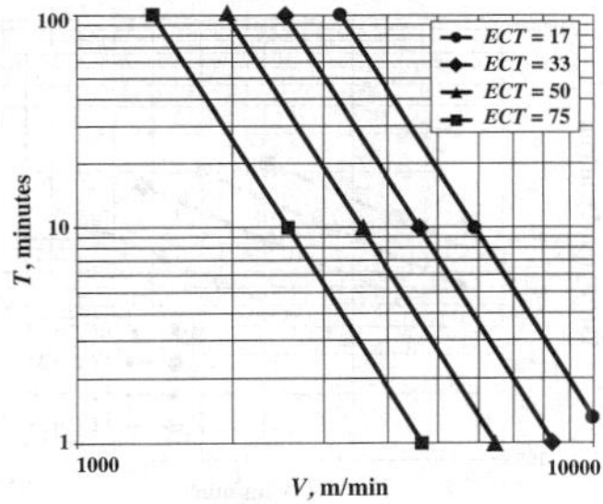


Fig. 3a. *T-V*

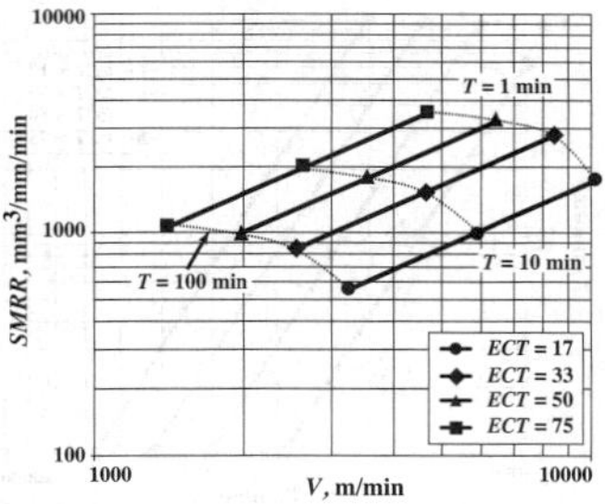


Fig. 3b. *SMRR* versus *V*, *T* = 100, 10, 1 minutes

**Table 4. Group 4—Tool Steels,** M1, M8, T1, H, O, L, F, 52100

| Tool Life T (min) | ECT = 0.00017 mm | | ECT = 0.00033 mm | | ECT = 0.00050 mm | | ECT = 0.00075 mm | |
|---|---|---|---|---|---|---|---|---|
| | Constant C = 7440 | | Constant C = 5805 | | Constant C = 4490 | | Constant C = 3240 | |
| | $V_T$ | *SMRR* | $V_T$ | *SMRR* | $V_T$ | *SMRR* | $V_T$ | *SMRR* |
| 100 | 2245 | 380 | 1755 | 580 | 1355 | 680 | 980 | 735 |
| **10** | **4090** | **695** | **3190** | **1055** | **2465** | **1235** | **1780** | **1335** |
| 1 | 7440 | 1265 | 5805 | 1915 | 4490 | 2245 | 3240 | 2430 |

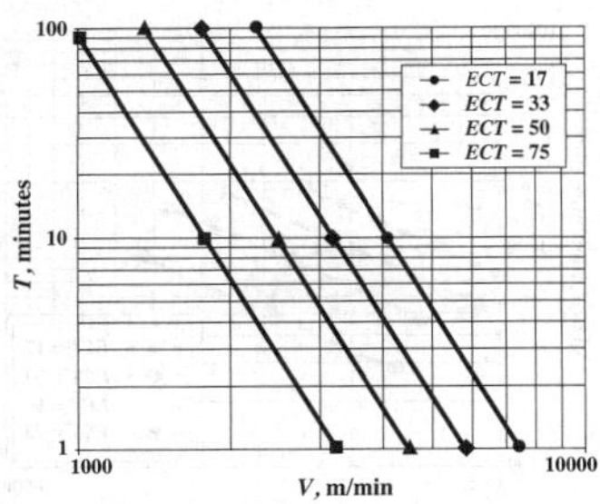


Fig. 4a. *T-V*

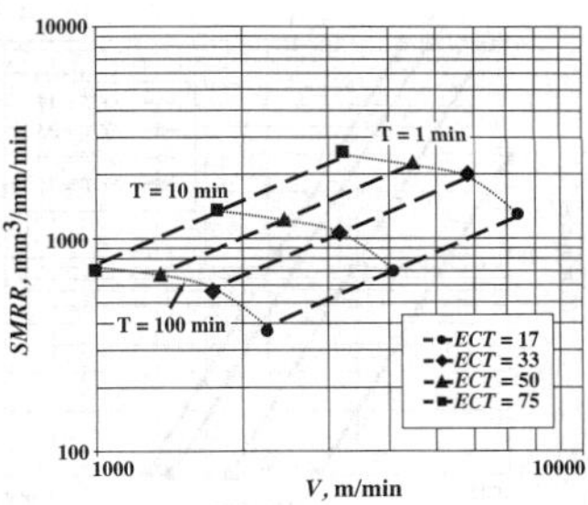


Fig. 4b. *SMRR* versus *V*, *T* = 100, 10, 1 minutes

**Table 5. Group 5—Tool Steels,** M2, T2, T5, T6, D2, D5, H41, H42, H43, M50

| Tool Life $T$ (min) | $ECT$ = 0.00017 mm | | $ECT$ = 0.00033 mm | | $ECT$ = 0.00050 mm | | $ECT$ = 0.00075 mm | |
|---|---|---|---|---|---|---|---|---|
| | Constant C = 6695 | | Constant C = 5224 | | Constant C = 4040 | | Constant C = 2915 | |
| | $V_T$ | *SMRR* | $V_T$ | *SMRR* | $V_T$ | *SMRR* | $V_T$ | *SMRR* |
| 100 | 2020 | 345 | 1580 | 520 | 1220 | 610 | 880 | 660 |
| **10** | **3680** | **625** | **2870** | **945** | **2220** | **1110** | **1600** | **1200** |
| 1 | 6695 | 1140 | 5225 | 1725 | 4040 | 2020 | 2915 | 2185 |

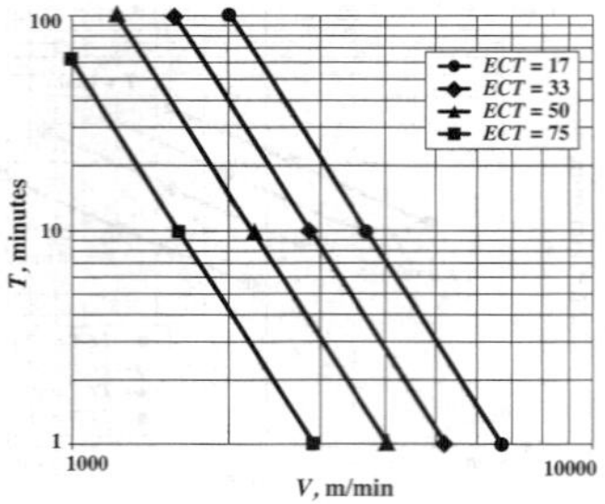


Fig. 5a. *T-V*

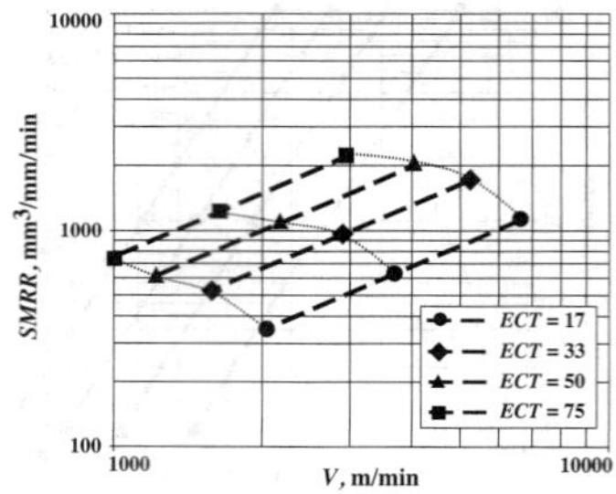


Fig. 5b. *SMRR* versus $V, T$ = 100, 10, 1 minutes

**Table 6. Group 6—Tool Steels,** M3, M4, T3, D7

| Tool Life $T$ (min) | $ECT$ = 0.00017 mm | | $ECT$ = 0.00033 mm | | $ECT$ = 0.00050 mm | | $ECT$ = 0.00075 mm | |
|---|---|---|---|---|---|---|---|---|
| | Constant C = 5290 | | Constant C = 4690 | | Constant C = 3585 | | Constant C = 2395 | |
| | $V_T$ | *SMRR* | $V_T$ | *SMRR* | $V_T$ | *SMRR* | $V_T$ | *SMRR* |
| 100 | 1600 | 270 | 1415 | 465 | 1085 | 540 | 725 | 540 |
| **10** | **2910** | **495** | **2580** | **850** | **1970** | **985** | **1315** | **985** |
| 1 | 5290 | 900 | 4690 | 1550 | 3585 | 1795 | 2395 | 1795 |

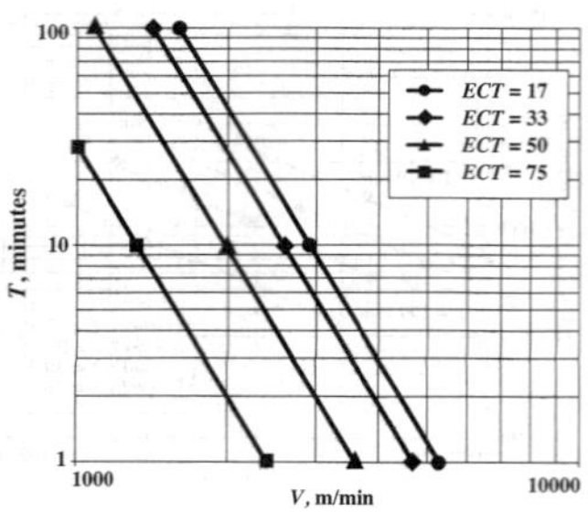


Fig. 6a. Group 6 Tool Steels *T-V*

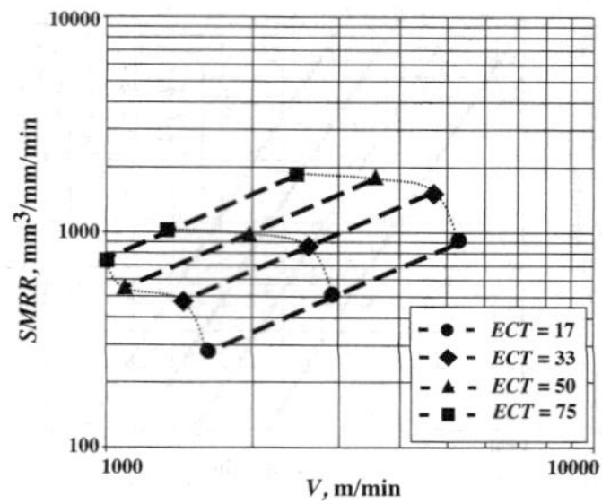


Fig. 6b. *SMRR* versus $V, T$ = 100, 10, 1 minutes

**Table 7. Group 7 — Tool Steels,** T15, M15

| Tool Life $T$ (min) | *ECT* = 0.00017 mm | | *ECT* = 0.00033 mm | | *ECT* = 0.00050 mm | | *ECT* = 0.00075 mm | |
|---|---|---|---|---|---|---|---|---|
| | Constant C = 2270 | | Constant C = 1970 | | Constant C = 1505 | | Constant C = 1010 | |
| | $V_T$ | *SMRR* | $V_T$ | *SMRR* | $V_T$ | *SMRR* | $V_T$ | *SMRR* |
| 100 | 685 | 115 | 595 | 195 | 455 | 225 | 305 | 230 |
| **10** | **1250** | **210** | **1080** | **355** | **825** | **415** | **555** | **415** |
| 1 | 2270 | 385 | 1970 | 650 | 1505 | 750 | 1010 | 760 |

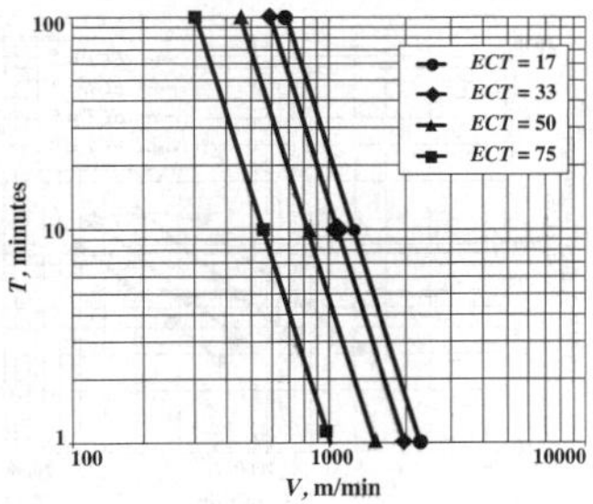


Fig. 7a. *T-V*

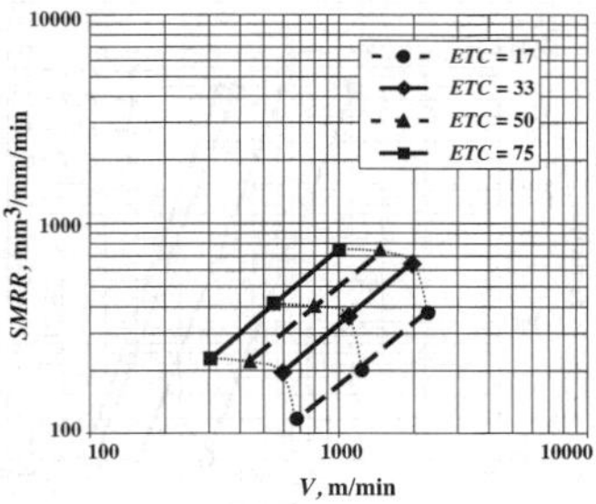


Fig. 7b. *SMRR* versus *V*, *T* = 100, 10, 1 minutes

**Table 8. Group 8 — Heat Resistant Alloys,** Inconel, Rene, etc.

| Tool Life $T$ (min) | *ECT* = 0.00017 mm | | *ECT* = 0.00033 mm | | *ECT* = 0.00050 mm | | *ECT* = 0.00075 mm | |
|---|---|---|---|---|---|---|---|---|
| | Constant C = 2150 | | Constant C = 1900 | | Constant C = 1490 | | Constant C = 1035 | |
| | $V_T$ | *SMRR* | $V_T$ | *SMRR* | $V_T$ | *SMRR* | $V_T$ | *SMRR* |
| 100 | 650 | 110 | 575 | 190 | 450 | 225 | 315 | 235 |
| **10** | **1185** | **200** | **1045** | **345** | **820** | **410** | **570** | **425** |
| 1 | 2150 | 365 | 1900 | 625 | 1490 | 745 | 1035 | 780 |

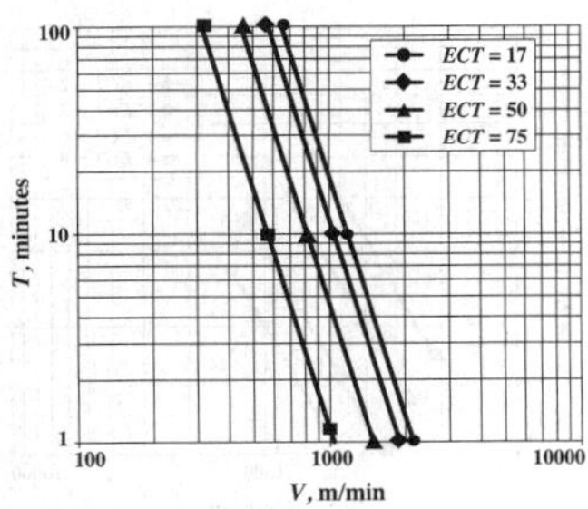


Fig. 8a. *T-V*

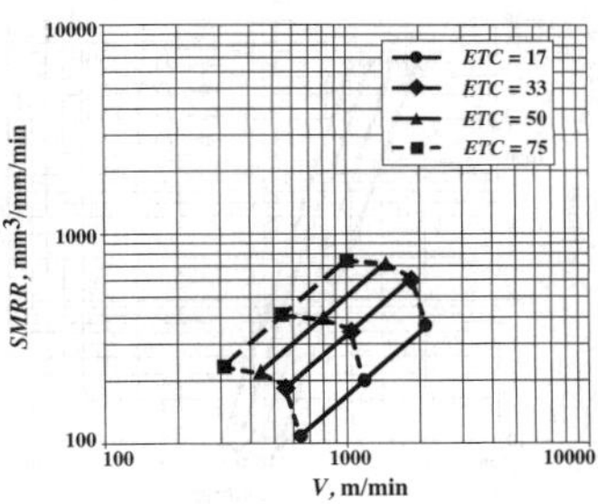


Fig. 8b. *SMRR* versus *V*, *T* = 100, 10, 1 minutes

**Table 9. Group 9 — Carbide Materials,** Diamond Wheel

| Tool Life $T$ (min) | $ECT$ = 0.00002 mm | | $ECT$ = 0.00003 mm | | $ECT$ = 0.00005 mm | | $ECT$ = 0.00008 mm | |
|---|---|---|---|---|---|---|---|---|
| | Constant C = 9030 | | Constant C = 8030 | | Constant C = 5365 | | Constant C = 2880 | |
| | $V_T$ | *SMRR* | $V_T$ | *SMRR* | $V_T$ | *SMRR* | $V_T$ | *SMRR* |
| 4800 | 1395 | 30 | 1195 | 35 | 760 | 40 | 390 | 30 |
| 600 | 2140 | 45 | 1855 | 55 | 1200 | 60 | 625 | 50 |
| 10 | 4960 | 100 | 4415 | 130 | 2950 | 145 | 1580 | 125 |

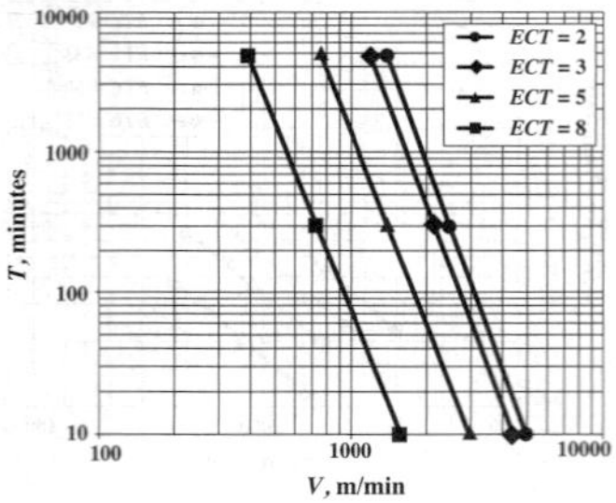


Fig. 9a. *T-V*

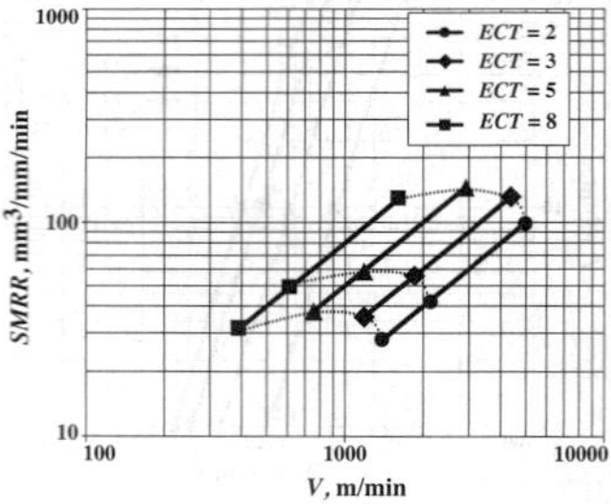


Fig. 9b. *SMRR* versus $V$, $T$ = 100, 10, 1 minutes

**Table 10. Group 10 — Ceramic Materials,** $Al_2O_3$, $ZrO_2$, SiC, $Si_3N_4$, Diamond Wheel

| Tool Life $T$ (min) | $ECT$ = 0.00002 mm | | $ECT$ = 0.00003 mm | | $ECT$ = 0.00005 mm | | $ECT$ = 0.00008 mm | |
|---|---|---|---|---|---|---|---|---|
| | Constant C = 2460 | | Constant C = 2130 | | Constant C = 1740 | | Constant C = 1420 | |
| | $V_T$ | *SMRR* | $V_T$ | *SMRR* | $V_T$ | *SMRR* | $V_T$ | *SMRR* |
| 4800 | 395 | 8 | 335 | 10 | 265 | 13 | 210 | 17 |
| 600 | 595 | 12 | 510 | 15 | 410 | 20 | 330 | 25 |
| 10 | 1355 | 25 | 1170 | 35 | 955 | 50 | 780 | 60 |

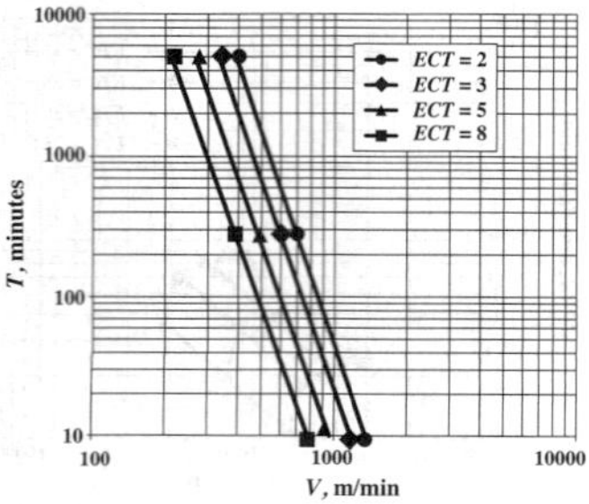


Fig. 10a. *T-V*

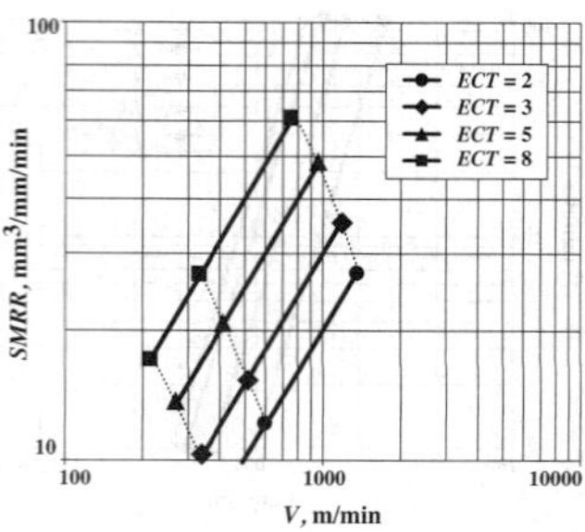


Fig. 10b. *SMRR* versus $V$, $T$ = 100, 10, 1 minutes

## User Calibration of Recommendations

It is recommended to copy or redraw the standard graph for any of the material groups before applying the data calibration method described below. The method is based on the user's own experience and data. The procedure is described in the following and illustrated in Table 11 and Fig. 12.

Only one shop data set is needed to adjust all four Taylor lines as shown below. The required shop data is the user's wheel-life $T_U$ obtained at the user's wheel speed $V_U$, the user's work speed $V_{wU}$, and depth of cut $a_r$.

1) First the user finds out which wheel-life $T_U$ was obtained in the shop, and the corresponding wheel speed $V_U$, depth of cut $a_r$ and work speed $V_{wU}$.

2) Second, calculate:

a) $ECT = V_{wU} \times ar \div V_U$

b) the user Taylor constant $C_U = V_U \times T_U^{0.26}$

c) $V_{100U} = C_U \div 100^{0.26}$

d) $V_{10U} = C_U \div 10^{0.26}$

3) Thirdly, the user Taylor line is drawn in the pertinent graph. If the user wheel-life $T_U$ is longer than that in the standard graph the speed values will be higher, or if the user wheel-life is shorter the speeds $C_U$, $V_{10U}$, $V_{100U}$ will be lower than the standard values C, $V_{10}$ and $V_{100}$.

The results are a series of lines moved to the right or to the left of the standard Taylor lines for $ECT = 17, 33, 50$ and $75 \times 10^{-5}$ mm. Each standard table contains the values $C = V_1, V_{10}, V_{100}$ and empty spaces for filling out the calculated user values: $C_U = V_U \times T_U^{0.26}$, $V_{10U} = C_U \div 10^{0.26}$ and $V_{100U} = C_U \div 100^{0.26}$.

*Example 7:* Assume the following test results on a Group 6 material: user speed is $V_U$ = 1800 m/min, wheel-life $T_U$ = 7 minutes, and $ECT$ = 0.00017 mm. The Group 6 data is repeated below for convenience.

**Standard Table Data, Group 6 Material**

| Tool Life T (min) | ECT = 0.00017 mm | | ECT = 0.00033 mm | | ECT = 0.00050 mm | | ECT = 0.00075 mm | |
|---|---|---|---|---|---|---|---|---|
| | Constant C = 5290 | | Constant C = 4690 | | Constant C = 3585 | | Constant C = 2395 | |
| | $V_T$ | *SMRR* | $V_T$ | *SMRR* | $V_T$ | *SMRR* | $V_T$ | *SMRR* |
| 100 | 1600 | 270 | 1415 | 465 | 1085 | 540 | 725 | 540 |
| **10** | **2910** | **495** | **2580** | **850** | **1970** | **985** | **1315** | **985** |
| 1 | 5290 | 900 | 4690 | 1550 | 3585 | 1795 | 2395 | 1795 |

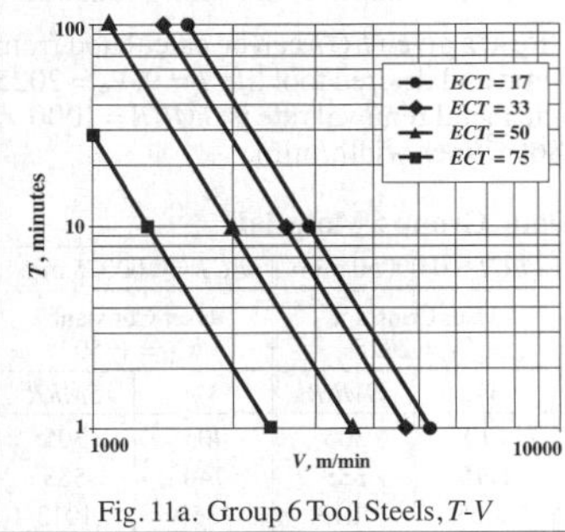


Fig. 11a. Group 6 Tool Steels, *T-V*

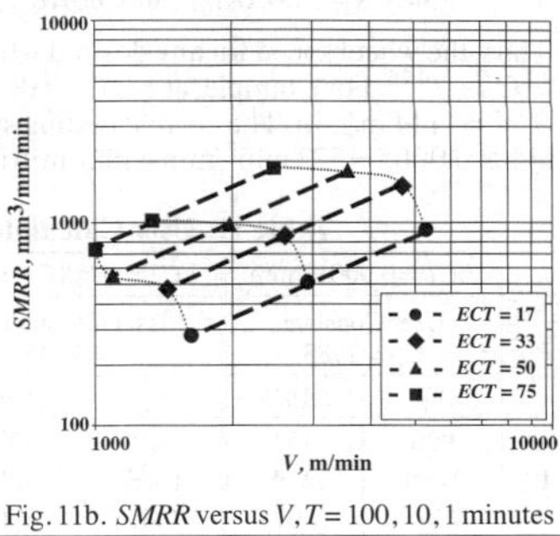


Fig. 11b. *SMRR* versus *V*, *T* = 100, 10, 1 minutes

*Calculation Procedure*

1) Calculate $V_{1U}$, $V_{10U}$, $V_{100U}$ and $SMRR_{1U}$, $SMRR_{10U}$, $SMRR_{100U}$ for $ECT = 0.00017$ mm

a) $V_{1U}$ = the user Taylor constant $C_U = V_U \times T_U^{0.26} = 1800 \times 7^{0.26} = 2985$ m/min, and $SMRR_{1U} = 1000 \times 2985 \times 0.00017 = 507$ mm³/mm width/min

b) $V_{10U} = C_U \div 10^{0.26} = 2985 \div 10^{0.26} = 1640$ m/min, and $SMRR_{10U} = 1000 \times 1640 \times 0.00017 = 279$ mm³/mm width/min

c) $V_{100U} = C_U \div 100^{0.26} = 2985 \div 100^{0.26} = 900$ m/min, and $SMRR_{100U} = 1000 \times 900 \times 0.00017 = 153$ mm³/mm width/min

2) For $ECT = 0.00017$ mm, calculate the ratio of user Taylor constant to standard Taylor constant from the tables $= C_U \div C_{ST} = C_U \div V_1 = 2985 \div 5290 = 0.564$ (see Table 6 for the value of $C_{ST} = V_1$ at $ECT = 0.00017$ mm).

3) For $ECT$ = 0.00033, 0.00050, and 0.00075 mm calculate the user Taylor constants from $C_U = C_{ST} \times$ (the ratio calculated in step 2) $= V_1 \times 0.564 = V_{1U}$. Then, calculate $V_{10U}$ and $V_{100U}$ and $SMRR_{1U}$, $SMRR_{10U}$, $SMRR_{100U}$ using the method in items 1b) and 1c) above.

a) For $ECT = 0.00033$ mm

$V_{1U} = C_U = 4690 \times 0.564 = 2645$ m/min

$V_{10U} = C_U \div 10^{0.26} = 2645 \div 10^{0.26} = 1455$ m/min

$V_{100U} = C_U \div 100^{0.26} = 2645 \div 100^{0.26} = 800$ m/min

$SMRR_{1U}$, $SMRR_{10U}$, and $SMRR_{100U}$ = 876, 480, and 264 mm³/mm width/min

b) For $ECT = 0.00050$ mm

$V_{1U} = C_U = 3590 \times 0.564 = 2025$ m/min

$V_{10U} = C_U \div 10^{0.26} = 2025 \div 10^{0.26} = 1110$ m/min

$V_{100U} = C_U \div 100^{0.26} = 2025 \div 100^{0.26} = 610$ m/min

$SMRR_{1U}$, $SMRR_{10U}$, and $SMRR_{100U}$ = 1013, 555, and 305 mm³/mm width/min

c) For $ECT = 0.00075$ mm

$V_{1U} = C_U = 2395 \times 0.564 = 1350$ m/min

$V_{10U} = C_U \div 10^{0.26} = 1350 \div 10^{0.26} = 740$ m/min

$V_{100U} = C_U \div 100^{0.26} = 1350 \div 100^{0.26} = 405$ m/min

$SMRR_{1U}$, $SMRR_{10U}$, and $SMRR_{100U}$ = 1013, 555, and 305 mm³/mm width/min

Thus, the wheel speed for any desired wheel-life at a given $ECT$ can be calculated from $V = C_U \div T^{0.26}$. For example, at $ECT = 0.00050$ mm and desired tool life $T = 9$, $V_9 = 2025 \div 9^{0.26} = 1144$ m/min. The corresponding specific metal removal rate is $SMRR = 1000 \times 1144 \times 0.0005 = 572$ mm³/mm width/min (0.886 in³/inch width/min).

**Table 11. User Calculated Data, Group 6 Material**

| Tool Life $T$ (min) | $ECT = 0.00017$ mm | | $ECT = 0.00033$ mm | | $ECT = 0.00050$ mm | | $ECT = 0.00075$ mm | |
|---|---|---|---|---|---|---|---|---|
| | User Constant $C_U = 2985$ | | User Constant $C_U = 2645$ | | User Constant $C_U = 2025$ | | User Constant $C_U = 1350$ | |
| | $V_T$ | *SMRR* | $V_T$ | *SMRR* | $V_T$ | *SMRR* | $V_T$ | *SMRR* |
| 100 | 900 | 153 | 800 | 264 | 610 | 305 | 405 | 305 |
| **10** | **1640** | **279** | **1455** | **480** | **1110** | **555** | **740** | **555** |
| 1 | 2985 | 507 | 2645 | 876 | 2025 | 1013 | 1350 | 1013 |

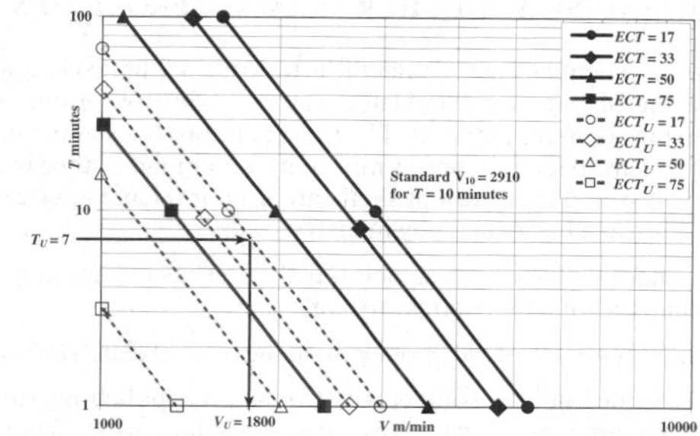


Fig. 12. Calibration of User Grinding Data to Standard Taylor Lines
User Input: $V_U = 1800$ m/min, $T_U = 7$ minutes, $ECT = 0.00017$ mm

**Optimization.—**As shown, a global optimum occurs along the *G*-curve, in selected cases for values of *ECT* around 0.00075, i.e. at high metal removal rates as in other machining operations. It is recommended to use the simple formula for economic life: $T_E = 3 \times T_V$ minutes. $T_V = T_{RPL} + 60 \times C_E \div H_R$, minutes, where $T_{RPL}$ is the time required to replace wheel, $C_E$ = cost per wheel dressing = wheel cost + cost per dressing, and $H_R$ is the hourly rate.

In grinding, values of $T_V$ range between 2 and 5 minutes in conventional grinders, which means that the economic wheel lives range between 6 and 15 minutes indicating higher metal removal rates than are commonly used. When wheels are sharpened automatically after each stroke as in internal grinding, or when grits are continually replaced as in abrasive grinding (machining), $T_V$ may be less than one minute. This translates into wheel lives around one minute in order to achieve minimum cost grinding.

*Grinding Cost, Optimization and Process Planning:* More accurate results are obtained when the firm collects and systemizes the information on wheel lives, wheel and work speeds, and depths of cut from production runs. A computer program can be used to plan the grinding process and apply the rules and formulas presented in this chapter. A complete grinding process planning program, such as that developed by Colding International Corporation, can be used to optimize machine settings for various feed-speed preferences corresponding wheel-life requirements, minimum cost or maximum production rate grinding, required surface finish and sparkout time; machine and fixture requirements based on the grinding forces, torque and power for sharp and worn grinding wheels; and, detailed time and cost analysis per part and per batch including wheel dressing and wheel changing schedules.

Table 12 summarizes the time and cost savings per batch as it relates to tool life. The sensitivity of how grinding parameters are selected is obvious. Minimum cost conditions yield a 51 percent reduction of time and 44 percent reduction of cost, while maximum production rate reduces total time by 65 percent but, at the expense of heavy wheel consumption (continuous dressing), cost by only 18 percent.

**Table 12. Wheel Life versus Cost**

| Preferences | Time per Batch, minutes | Cost per Batch, $ | | Reduction from Long Life,% | |
|---|---|---|---|---|---|
| | | Tooling | Total Cost | Time | Cost |
| Long Life | 2995 | 39 | 2412 | — | — |
| Economic Life | 2433 | 252 | 2211 | 19 | 8 |
| Minimum Cost | 1465 | 199 | 1344 | 51 | 44 |
| Max Production Rate | 1041 | 1244 | 1980 | 65 | 18 |

## GRINDING AND OTHER ABRASIVE PROCESSES

Processes and equipment discussed under this heading use abrasive grains for shaping workpieces by means of machining or related methods. Abrasive grains are hard crystals either found in nature or manufactured. The most commonly used materials are aluminum oxide, silicon carbide, cubic boron nitride and diamond. Other materials such as garnet, zirconia, glass and even walnut shells are used for some applications. Abrasive products are used in three basic forms by industry:

a) *Bonded* to form solid-shaped tools, such as disks (the basic shape of grinding wheels), cylinders, rings, cups, segments, or sticks to name a few.

b) *Coated* on backings made of paper or cloth, in the form of sheets, strips, or belts.

c) *Loose,* held in some liquid or solid carrier (for lapping, polishing, tumbling), or propelled by centrifugal force, air, or water pressure against the work surface (blast cleaning).

The applications for abrasive processes are multiple and varied. They include:

a) *Cleaning* of surfaces, also the coarse removal of excess material—such as rough offhand grinding in foundries to remove gates and risers.

b) *Shaping,* such as in form grinding and tool sharpening.

c) *Sizing,* a general objective, but of primary importance in precision grinding.

d) *Surface finish improvement,* either primarily as in lapping, honing, and polishing or as a secondary objective in other types of abrasive processes.

e) *Separating,* as in cut-off or slicing operations.

The main field of application of abrasive processes is in metalworking, because of the capacity of abrasive grains to penetrate into even the hardest metals and alloys. However, the great hardness of the abrasive grains also makes the process preferred for working other hard materials, such as stones, glass, and certain types of plastics. Abrasive processes are also chosen for working relatively soft materials, such as wood, rubber, etc., for such reasons as high stock removal rates, long-lasting cutting ability, good form control, and fine finish of the worked surface.

### Grinding Wheels

**Abrasive Materials.**—In earlier times, only natural abrasives were available. From about the beginning of this century, however, manufactured abrasives, primarily silicon carbide and aluminum oxide, have replaced the natural materials; even natural diamonds have been almost completely supplanted by synthetics. Superior and controllable properties, and dependable uniformity characterize the manufactured abrasives.

Both silicon carbide and aluminum oxide abrasives are very hard and brittle. This brittleness, called friability, is controllable for different applications. Friable abrasives break easily, thus forming sharp edges. This decreases the force needed to penetrate into the work material and the heat generated during cutting. Friable abrasives are most commonly used for precision and finish grinding. Tough abrasives resist fracture and last longer. They are used for rough grinding, snagging, and offhand grinding.

As a general rule, although subject to variation:

1) Aluminum oxide abrasives are used for grinding plain and alloyed steel in a soft or hardened condition.

2) Silicon carbide abrasives are selected for cast iron, nonferrous metals, and nonmetallic materials.

3) Diamond is the best type of abrasive for grinding cemented carbides. It is also used for grinding glass, ceramics, and hardened tool steel.

4) Cubic Boron Nitride (CBN) is known by several trade names including Borazon (General Electric Co.), ABN (De Beers), Sho-bon (Showa-Denko), and Elbor (USSR). CBN is a synthetic superabrasive used for grinding hardened steels and wear-resistant superalloys. (See *Cubic Boron Nitride (CBN)* starting on page 1069.) CBN grinding wheels have long lives and can maintain close tolerances with superior surface finishes.

**Bond Properties and Grinding Wheel Grades.—**The four main types of bonds used for grinding wheels are the vitrified, resinoid, rubber, and metal.

*Vitrified bonds* are used for more than half of all grinding wheels made, and are preferred because of their strength and other desirable qualities. Being inert, glass-like materials, vitrified bonds are not affected by water or by the chemical composition of different grinding fluids. Vitrified bonds also withstand the high temperatures generated during normal grinding operations. The structure of vitrified wheels can be controlled over a wide range of strength and porosity. Vitrified wheels, however, are more sensitive to impact than those made with organic bonds.

*Resinoid bonds* are selected for wheels subjected to impact, or sudden loads, or very high operating speeds. They are preferred for snagging, portable grinder uses, or roughing operations. The higher flexibility of this type of bond—essentially a filled thermosetting plastic—helps it withstand rough treatment.

*Rubber bonds* are even more flexible than the resinoid type, and for that reason are used for producing a high finish and for resisting sudden rises in load. Rubber bonded wheels are commonly used for wet cut-off wheels because of the nearly burr-free cuts they produce, and for centerless grinder regulating wheels to provide a stronger grip and more reliable workpiece control.

*Metal bonds* are used in CBN and diamond wheels. In metal bonds produced by electrodeposition, a single layer of superabrasive material (diamond or CBN) is bonded to a metal core by a matrix of metal, usually nickel. The process is so controlled that about 30–40 percent of each abrasive particle projects above the deposited surface, giving the wheel a very aggressive and free-cutting action. With proper use, such wheels have remarkably long lives. When dulled, or worn down, the abrasive can be stripped off and the wheel renewed by a further deposit process. These wheels are also used in electrical discharge grinding and electrochemical grinding where an electrically conductive wheel is needed.

In addition to the basic properties of the various bond materials, each can also be applied in different proportions, thereby controlling the grade of the grinding wheel.

*Grinding wheel grades* commonly associated with hardness, express the amount of bond material in a grinding wheel, and hence the strength by which the bond retains the individual grains.

During grinding, the forces generated when cutting the work material tend to dislodge the abrasive grains. As the grains get dull, if they do not fracture to resharpen themselves, the cutting forces will eventually tear the grains from their supporting bond. For a "soft" wheel the cutting forces will dislodge the abrasive grains before they have an opportunity to fracture. When a "hard" wheel is used, the situation is reversed. Because of the extra bond in the wheel the grains are so firmly held that they never break loose and the wheel becomes glazed. During most grinding operations it is desirable to have an intermediate wheel where there is a continual slow wearing process composed of both grain fracture and dislodgement.

The grades of the grinding wheels are designated by capital letters used in alphabetical order to express increasing "hardness" from A to Z.

**Grinding Wheel Structure.—**The individual grains, which are encased and held together by the bond material, do not fill the entire volume of the grinding wheel; the intermediate open space is needed for several functional purposes such as heat dissipation, coolant application, and particularly, for the temporary storage of chips. It follows that the spacing of the grains must be greater for coarse grains which cut thicker chips and for large contact areas within which the chips have to be retained on the surface of the wheel before being disposed of. On the other hand, a wide spacing reduces the number of grains that contact the work surface within a given advance distance, thereby producing a coarser finish.

In general, denser structures are specified for grinding hard materials, for high-speed grinding operations, when the contact area is narrow, and for producing fine finishes and/or accurate forms. Wheels with open structure are used for tough materials, high stock removal rates, and extended contact areas, such as grinding with the face of the wheel. There are, however, several exceptions to these basic rules, an important one being the grinding of parts made by powder metallurgy, such as cemented carbides; although they represent one of the hardest industrial materials, grinding carbides requires wheels with an open structure.

Most kinds of general grinding operations, when carried out with the periphery of the wheel, call for medium spacing of the grains. The structure of the grinding wheels is expressed by numerals from 1 to 16, ranging from dense to open. Sometimes, "induced porosity" is used with open structure wheels. This term means that the grinding wheel manufacturer has placed filler material (which later burns out when the wheel is fired to vitrify the bond) in the grinding wheel mix. These fillers create large "pores" between grain clusters without changing the total volume of the "pores" in the grinding wheel. Thus, an A46-H12V wheel and an A46H12VP wheel will contain the same amounts of bond, abrasive, and air space. In the former, a large number of relatively small pores will be distributed throughout the wheel. The latter will have a smaller number of larger pores.

**American National Standard Grinding Wheel Markings.—**ANSI B74.13-2016 "Markings for Identifying Grinding Wheels and Other Bonded Abrasives," applies to grinding wheels and other bonded abrasives, segments, bricks, sticks, hones, rubs, and other shapes that are for removing material, or producing a desired surface or dimension. It does not apply to specialities such as sharpening stones and provides only a standard system of markings. Wheels having the same standard markings but made by different wheel manufacturers may not—and probably will not—produce exactly the same grinding action. This desirable result cannot be obtained because of the impossibility of closely correlating any measurable physical properties of bonded abrasive products in terms of their grinding action.

Symbols for designating diamond and cubic boron wheel compositions are given on page 1295.

**Sequence of Markings.—**The accompanying illustration taken from ANSI B74.13-2016 shows the makeup of a typical wheel or bonded abrasive marking.

| Prefix | 1 Abrasive Type | 2 Grain Size | 3 Grade | 4 Structure | 5 Bond Type | 6 Manufacturer's Record |
|---|---|---|---|---|---|---|
| **51 –** | **A –** | **36 –** | **L –** | **5 –** | **V –** | **23** |

The meaning of each letter and number in this or other markings is indicated by the following complete list.

1) *Abrasive Letters:* The letter (A) is used for aluminum oxide, (C) for silicon carbide, and (Z) for aluminum zirconium. The manufacturer may designate some particular type in any one of these broad classes, by using his own symbol as a prefix (example, 51).

2) *Grain Size:* The grain sizes commonly used and varying from coarse to very fine are indicated by the following numbers: 8, 10, 12, 14, 16, 20, 24, 30, 36, 46, 54, 60,70, 80, 90, 100, 120, 150, 180, and 220. The following additional sizes are used occasionally: 240, 280, 320, 400, 500, and 600. The wheel manufacturer may add to the regular grain number an additional symbol to indicate a special grain combination.

3) *Grade:* Grades are indicated by letters of the alphabet from A to Z in all bonds or processes. Wheel grades from A to Z range from soft to hard.

4) *Structure:* The use of a structure symbol is optional. The structure is indicated by Nos. 1 to 16 (or higher, if necessary) with progressively higher numbers indicating a progressively wider grain spacing (more open structure).

5) *Bond or Process:* Bonds are indicated by the following letters: V, vitrified; S, silicate; E, shellac or elastic; R, rubber; RF, rubber reinforced; B, resinoid (synthetic resins); BF, resinoid reinforced; O, oxychloride; P, Epoxy/Polyester.

6) *Manufacturer's Record:* The sixth position may be used for manufacturer's private factory records; this is optional.

**Standard Shapes and Sizes of Grinding Wheels.—**ANSI B74.2-2003, which includes shapes and sizes of grinding wheels, gives a wide variety of grinding wheel shape and size combinations. These are suitable for the majority of applications. Although grinding wheels can be manufactured to shapes and dimensions different from those listed, it is advisable, for reasons of cost and inventory control, to avoid using special shapes and sizes, unless technically warranted.

Standard shapes and size ranges as given in this Standard together with typical applications are shown in Table 1a for inch dimensions and in Table 1b for metric dimensions.

The operating surface of the grinding wheel is often referred to as the wheel face. In the majority of cases it is the periphery of the grinding wheel which, when not specified otherwise, has a straight profile. However, other face shapes can also be supplied by the grinding wheel manufacturers, and also reproduced during usage by appropriate truing. ANSI B74.2-2003 offers fifteen different shapes for grinding wheel faces, which are shown in Table 2.

**The Selection of Grinding Wheels.—**In selecting a grinding wheel, the determining factors are the composition of the work material, the type of grinding machine, the size range of the wheels used, and the expected grinding results, in this approximate order.

The Norton Company has developed, as the result of extensive test series, a method of grinding wheel recommendation that is more flexible and also better adapted to taking into consideration pertinent factors of the job, than are listings based solely on workpiece categories. This approach is the basis for Table 3 through Table 6, inclusive. Tool steels and constructional steels are considered in the detailed recommendations presented in these tables.

Table 3 assigns most of the standardized tool steels to five different grindability groups. The AISI-SAE tool steel designations are used.

After having defined the grindability group of the tool steel to be ground, the operation to be carried out is found in the first column of Table 4. The second column in this table distinguishes between different grinding wheel size ranges, because wheel size is a factor in determining the contact area between wheel and work, thus affecting the apparent hardness of the grinding wheel. Distinction is also made between wet and dry grinding.

Finally, the last two columns define the essential characteristics of the recommended types of grinding wheels under the headings of first and second choice, respectively. Where letters are used *preceding* A, the standard designation for aluminum oxide, they indicate a degree of friability different from the regular, thus: SF = semi friable (Norton equivalent 16A) and F = friable (Norton equivalent 33A and 38A). The suffix P, where applied, expresses a degree of porosity that is more open than the regular.

**Table 1a. Standard Shapes and Inch Size Ranges of Grinding Wheels**
*ANSI B74.2-2003*

| Applications | Size Ranges of Principal Dimensions, Inches | | |
|---|---|---|---|
| | *D* = Dia. | *T* = Thickness | *H* = Hole |
| 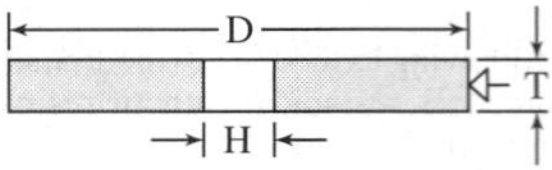 | **Type 1. Straight Wheel** For peripheral grinding. | | |
| CUTTING OFF (Organic bonds only) | 1 to 48 | 1⁄64 to 3⁄8 | 1⁄16 to 6 |
| CYLINDRICAL GRINDING<br>Between centers | 12 to 48 | ½ to 6 | 5 to 20 |
| CYLINDRICAL GRINDING<br>Centerless grinding wheels | 14 to 30 | 1 to 20 | 5 or 12 |
| CYLINDRICAL GRINDING<br>Centerless regulating wheels | 8 to 14 | 1 to 12 | 3 to 6 |
| INTERNAL GRINDING | ¼ to 4 | ¼ to 2 | 3⁄32 to 7⁄8 |
| OFFHAND GRINDING<br>Grinding on the periphery | | | |
| General purpose | 6 to 36 | ½ to 4 | ½ to 3 |
| For wet tool grinding only | 30 or 36 | 3 or 4 | 20 |
| SAW GUMMING (F-type face) | 6 to 12 | ¼ to 1½ | ½ to 1¼ |
| SNAGGING<br>Floor stand machines | 12 to 24 | 1 to 3 | 1¼ to 2½ |
| SNAGGING<br>Floor stand machines (Organic bond, wheel speed over 6500 sfpm) | 20 to 36 | 2 to 4 | 6 or 12 |
| SNAGGING<br>Mechanical grinders (Organic bond, wheel speed up to 16,500 sfpm) | 24 | 2 to 3 | 12 |
| SNAGGING<br>Portable machines | 3 to 8 | ¼ to 1 | 3⁄8 to 5⁄8 |
| SNAGGING<br>Portable machines (Reinforced organic bond, 17,000 sfpm) | 6 or 8 | ¾ or 1 | 1 |
| SNAGGING<br>Swing frame machines | 12 to 24 | 2 to 3 | 3½ to 12 |
| SURFACE GRINDING<br>Horizontal spindle machines | 6 to 24 | ½ to 6 | 1¼ to 12 |
| TOOL GRINDING<br>Broaches, cutters, mills, reamers, taps, etc. | 6 to 10 | ¼ to ½ | 5⁄8 to 5 |
| 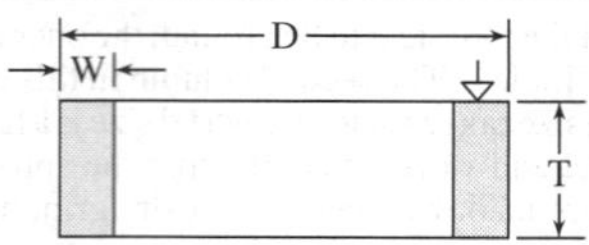 | **Type 2. Cylindrical Wheel** Side grinding wheel—mounted on the diameter; may also be mounted in a chuck or on a plate. | | |
| | | | *W* = Wall |
| SURFACE GRINDING<br>Vertical spindle machines | 8 to 20 | 4 or 5 | 1 to 4 |

**Table 1a.** *(Continued)* **Standard Shapes and Inch Size Ranges of Grinding Wheels**
*ANSI B74.2-2003*

| Applications | Size Ranges of Principal Dimensions, Inches | | |
|---|---|---|---|
| | $D$ = Dia. | $T$ = Thickness | $H$ = Hole |
| 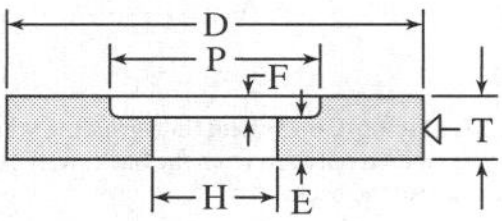  | **Type 5. Wheel,** recessed one side For peripheral grinding. Allows wider faced wheels than the available mounting thickness, also grinding clearance for the nut and flange. | | |
| CYLINDRICAL GRINDING<br>Between centers | 12 to 36 | 1½ to 4 | 5 or 12 |
| CYLINDRICAL GRINDING<br>Centerless regulating wheel | 8 to 14 | 3 to 6 | 3 or 5 |
| INTERNAL GRINDING | ⅜ to 4 | ⅜ to 2 | ⅛ to ⅞ |
| SURFACE GRINDING<br>Horizontal spindle machines | 7 to 24 | ¾ to 6 | 1¼ to 12 |
| 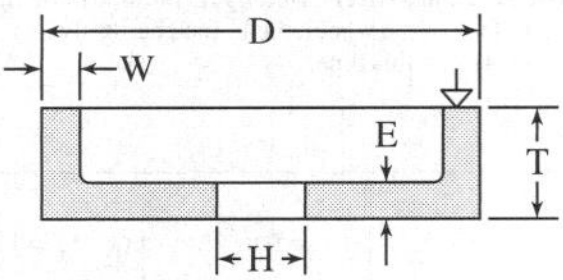  | **Type 6. Straight-Cup Wheel** Side grinding wheel, in whose dimensioning the wall thickness ($W$) takes precedence over the diameter of the recess. Hole is ⅝-11UNC-2B threaded for the snagging wheels and ½ or 1¼″ for the tool grinding wheels. | | |
| | | | $W$ = Wall |
| SNAGGING<br>Portable machines, organic bond only. | 4 to 6 | 2 | ¾ to 1½ |
| TOOL GRINDING<br>Broaches, cutters, mills, reamers, taps, etc. | 2 to 6 | 1 ¼ to 2 | 5⁄16 or ⅜ |
| 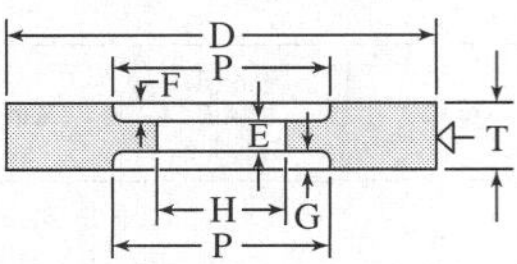  | **Type 7. Wheel,** recessed two sides Peripheral grinding. Recesses allow grinding clearance for both flanges and also narrower mounting thickness than overall thickness. | | |
| CYLINDRICAL GRINDING<br>Between centers | 12 to 36 | 1½ to 4 | 5 or 12 |
| CYLINDRICAL GRINDING<br>Centerless regulating wheel | 8 to 14 | 4 to 20 | 3 to 6 |
| SURFACE GRINDING<br>Horizontal spindle machines | 12 to 24 | 2 to 6 | 5 to 12 |

**Table 1a.** *(Continued)* **Standard Shapes and Inch Size Ranges of Grinding Wheels**
*ANSI B74.2-2003*

| Applications | Size Ranges of Principal Dimensions, Inches | | |
|---|---|---|---|
| | $D$ = Dia. | $T$ = Thickness | $H$ = Hole |
| 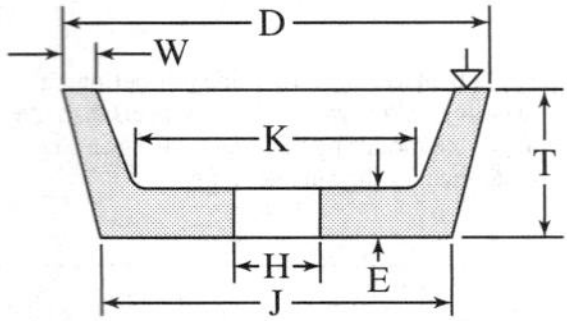 | **Type 11. Flaring-Cup Wheel** Side grinding wheel with wall tapered outward from the back; wall generally thicker in the back. | | |
| SNAGGING<br>Portable machines, organic bonds only, threaded hole | 4 to 6 | 2 | ⅝-11 UNC-2B |
| TOOL GRINDING<br>Broaches, cutters, mills, reamers, taps, etc. | 2 to 5 | 1 ¼ to 2 | ½ to 1 ¼ |
| 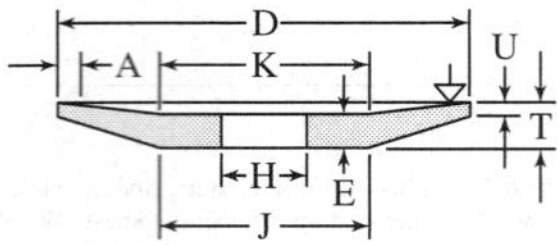 | **Type 12. Dish Wheel** Grinding on the side or on the U-face of the wheel, the U-face being always present in this type. | | |
| TOOL GRINDING<br>Broaches, cutters, mills, reamers, taps, etc. | 3 to 8 | ½ or ¾ | ½ to 1 ¼ |
| 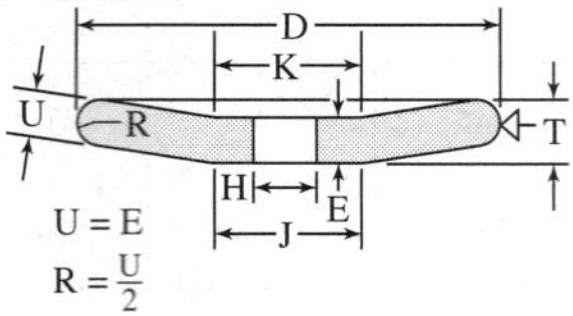 | **Type 13. Saucer Wheel** Peripheral grinding wheel, resembling the shape of a saucer, with cross section equal throughout. | | |
| SAW GUMMING<br>Saw tooth shaping and sharpening | 8 to 12 | ½ to 1 ¾<br>*U & E*<br>¼ to 1½ | ¾ to 1 ¼ |
| 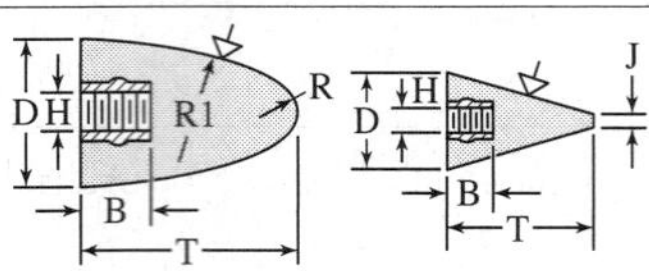 | **Type 16. Cone, Curved Side**<br>**Type 17. Cone, Straight Side, Square Tip**<br>**Type 17R. Cone, Straight Side, Round Tip** (Tip Radius $R = J/2$) | | |
| SNAGGING<br>Portable machine, threaded holes | 1¼ to 3 | 2 to 3½ | ⅜-24UNF-2B<br>to<br>⅝-11UNC-2B |

**Table 1a.** *(Continued)* **Standard Shapes and Inch Size Ranges of Grinding Wheels**
*ANSI B74.2-2003*

| Applications | Size Ranges of Principal Dimensions, Inches | | |
|---|---|---|---|
| | $D$ = Dia. | $T$ = Thickness | $H$ = Hole |
| 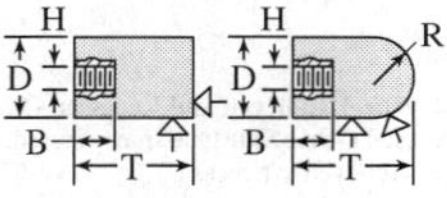 | **Type 18. Plug, Square End**<br>**Type 18R. Plug, Round End** $R = D/2$ | | |
| 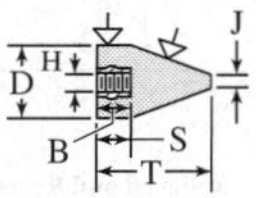 | **Type 19. Plugs, Conical End, Square Tip**<br>**Type 19R. Plugs, Conical End, Round Tip**<br>(Tip Radius $R = J/2$) | | |
| SNAGGING<br>Portable machine, threaded holes | 1¼ to 3 | 2 to 3½ | ⅜-24UNF-2B to ⅝-11UNC-2B |
| 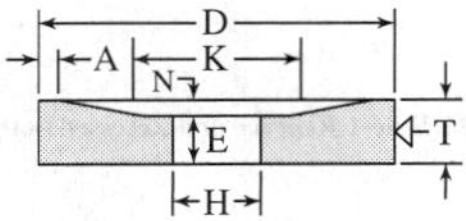 | **Type 20. Wheel, Relieved One Side** Peripheral grinding wheel, one side flat, the other side relieved to a flat. | | |
| CYLINDRICAL GRINDING<br>Between centers | 12 to 36 | ¾ to 4 | 5 to 20 |
| 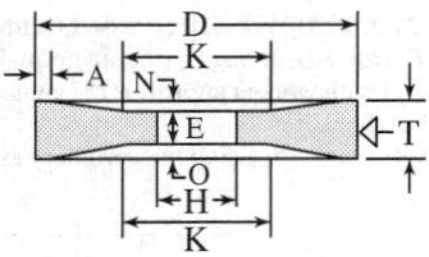 | **Type 21. Wheel, Relieved Two Sides** Both sides relieved to a flat. | | |
| 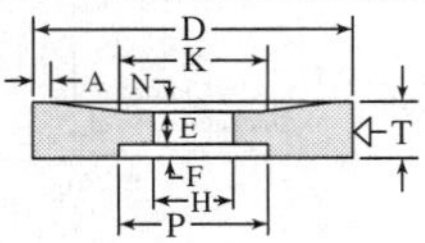 | **Type 22. Wheel, Relieved One Side, Recessed Other Side** One side relieved to a flat. | | |
| 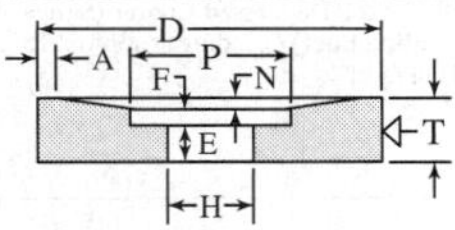 | **Type 23. Wheel, Relieved and Recessed Same Side** The other side is straight. | | |
| CYLINDRICAL GRINDING<br>Between centers, with wheel periphery | 20 to 36 | 2 to 4 | 12 or 20 |

**Table 1a.** *(Continued)* **Standard Shapes and Inch Size Ranges of Grinding Wheels**
*ANSI B74.2-2003*

| Applications | Size Ranges of Principal Dimensions, Inches | | |
|---|---|---|---|
| | $D$ = Dia. | $T$ = Thickness | $H$ = Hole |
| 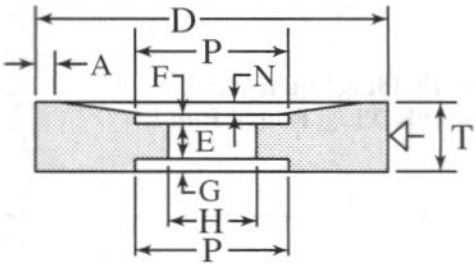  | **Type 24. Wheel, Relieved and Recessed One Side, Recessed Other Side** One side recessed, the other side is relieved to a recess. | | |
| 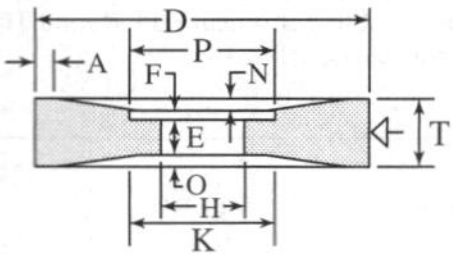  | **Type 25. Wheel, Relieved and Recessed One Side, Relieved Other Side** One side relieved to a flat, the other side relieved to a recess. | | |
| 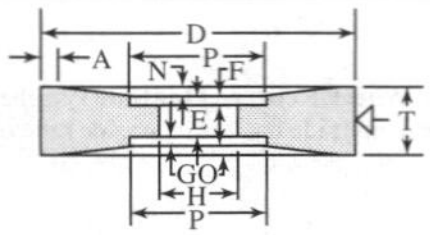  | **Type 26. Wheel, Relieved and Recessed Both Sides** | | |
| CYLINDRICAL GRINDING<br>Between centers, with the periphery of the wheel | 20 to 36 | 2 to 4 | 12 or 20 |
| 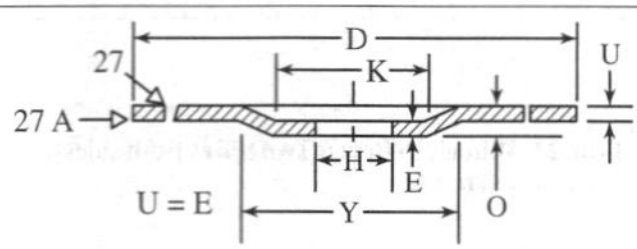  | **TYPES 27 & 27A. Wheel, Depressed Center**<br>27. *Portable Grinding:* Grinding normally done by contact with work at approx. a 15° angle with face of the wheel.<br>27A. *Cutting-off:* Using the periphery as grinding face. | | |
| CUTTING OFF<br>Reinforced organic bonds only | 16 to 30 | $U$ = $E$ = 5⁄32 to ¼ | 1 or 1 ½ |
| SNAGGING<br>Portable machine | 3 to 9 | $U$ = Uniform thickness ⅛ to ⅜ | ⅜ or ⅞ |
| 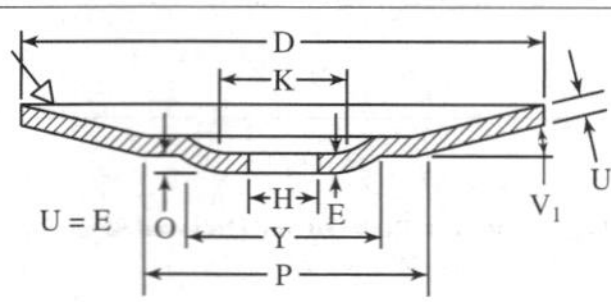  | **Type 28. Wheel, Depressed Center (Saucer-Shaped Grinding Face)** Grinding at approx. 15° angle with wheel face. | | |
| SNAGGING<br>Portable machine | 7 or 9 | $U$ = Uniform thickness ¼ | ⅞ |

Throughout table large open-head arrows indicate grinding surfaces.

**Table 1b. Standard Shapes and Metric Size Ranges of Grinding Wheels**
*ANSI B74.2-2003*

| Applications | Size Ranges of Principal Dimensions, Millimeters | | |
|---|---|---|---|
| | *D* = Dia. | *T* = Thickness | *H* = Hole |
| **Type 1. Straight Wheel[a]** | | | |
| CUTTING OFF (nonreinforced and reinforced organic bonds only) | 150 to 1250 | 0.8 to 10 | 16 to 152.4 |
| CYLINDRICAL GRINDING Between centers | 300 to 1250 | 20 to 160 | 127 to 508 |
| CYLINDRICAL GRINDING Centerless grinding wheels | 350 to 750 | 25 to 500 | 127 or 304.8 |
| CYLINDRICAL GRINDING Centerless regulating wheels | 200 to 350 | 25 to 315 | 76.2 to 152.4 |
| INTERNAL GRINDING | 6 to 100 | 6 to 50 | 2.5 to 25 |
| OFFHAND GRINDING Grinding on the periphery | | | |
| General purpose | 150 to 900 | 13 to 100 | 20 to 76.2 |
| For wet tool grinding only | 750 or 900 | 80 or 100 | 508 |
| SAW GUMMING (F-type face) | 150 to 300 | 6 to 40 | 32 |
| SNAGGING Floor stand machines | 300 to 600 | 25 to 80 | 32 to 76.2 |
| SNAGGING Floor stand machines(organic bond, wheel speed over 33 meters per second) | 500 to 900 | 50 to 100 | 152.4 or 304.8 |
| SNAGGING Mechanical grinders (organic bond, wheel speed up to 84 meters per second) | 600 | 50 to 80 | 304.8 |
| SNAGGING Portable machines | 80 to 200 | 6 to 25 | 10 to 16 |
| SNAGGING Swing frame machines (organic bond) | 300 to 600 | 50 to 80 | 88.9 to 304.8 |
| SURFACE GRINDING Horizontal spindle machines | 150 to 600 | 13 to 160 | 32 to 304.8 |
| TOOL GRINDING Broaches, cutters, mills, reamers, taps, etc. | 150 to 250 | 6 to 20 | 32 to 127 |
| **Type 2. Cylindrical Wheel[a]** | | | |
| | | | *W* = Wall |
| SURFACE GRINDING Vertical spindle machines | 200 to 500 | 100 or 125 | 25 to 100 |

**Table 1b.** *(Continued)* **Standard Shapes and Metric Size Ranges of Grinding Wheels** *ANSI B74.2-2003*

| Applications | Size Ranges of Principal Dimensions, Millimeters | | |
|---|---|---|---|
| | $D$ = Dia. | $T$ = Thickness | $H$ = Hole |
| **Type 5. Wheel, Recessed One Side**[a] | | | |
| CYLINDRICAL GRINDING<br>Between centers | 300 to 900 | 40 to 100 | 127 or 304.8 |
| CYLINDRICAL GRINDING<br>Centerless regulating wheels | 200 to 350 | 80 to 160 | 76.2 or 127 |
| INTERNAL GRINDING | 10 to 100 | 10 to 50 | 3.18 to 25 |
| **Type 6. Straight-Cup Wheel**[a] | | | |
| | | | $W$ = Wall |
| SNAGGING<br>Portable machines, organic bond only<br>(hole is ⅝-11 UNC-2B) | 100 to 150 | 50 | 20 to 40 |
| TOOL GRINDING<br>Broaches, cutters, mills, reamers, taps, etc.<br>(Hole is 13 to 32 mm) | 50 to 150 | 32 to 50 | 8 or 10 |
| **Type 7. Wheel, Recessed Two Sides**[a] | | | |
| CYLINDRICAL GRINDING<br>Between centers | 300 to 900 | 40 to 100 | 127 or 304.8 |
| CYLINDRICAL GRINDING<br>Centerless regulating wheels | 200 to 350 | 100 to 500 | 76.2 to 152.4 |
| **Type 11. Flaring-Cup Wheel**[a] | | | |
| SNAGGING<br>Portable machines, organic bonds only, threaded hole | 100 to 150 | 50 | ⅝-11 UNC-2B |
| TOOL GRINDING<br>Broaches, cutters, mills, reamers, taps, etc. | 50 to 125 | 32 to 50 | 13 to 32 |
| **Type 12. Dish Wheel**[a] | | | |
| TOOL GRINDING<br>Broaches, cutters, mills, reamers, taps, etc. | 80 to 200 | 13 or 20 | 13 to 32 |
| **Type 27 and 27A. Wheel, Depressed Center**[a] | | | |
| CUTTING OFF<br>Reinforced organic bonds only | 400 to 750 | $U = E = 6$ | 25.4 or 38.1 |
| SNAGGING<br>Portable machines | 80 to 230 | $U = E =$ 3.2 to 10 | 9.53 or 22.23 |

[a] See Table 1a for diagrams and descriptions of each wheel type.
All dimensions in millimeters.

**Table 2. Standard Shapes of Grinding Wheel Faces** *ANSI B74.2-2003*

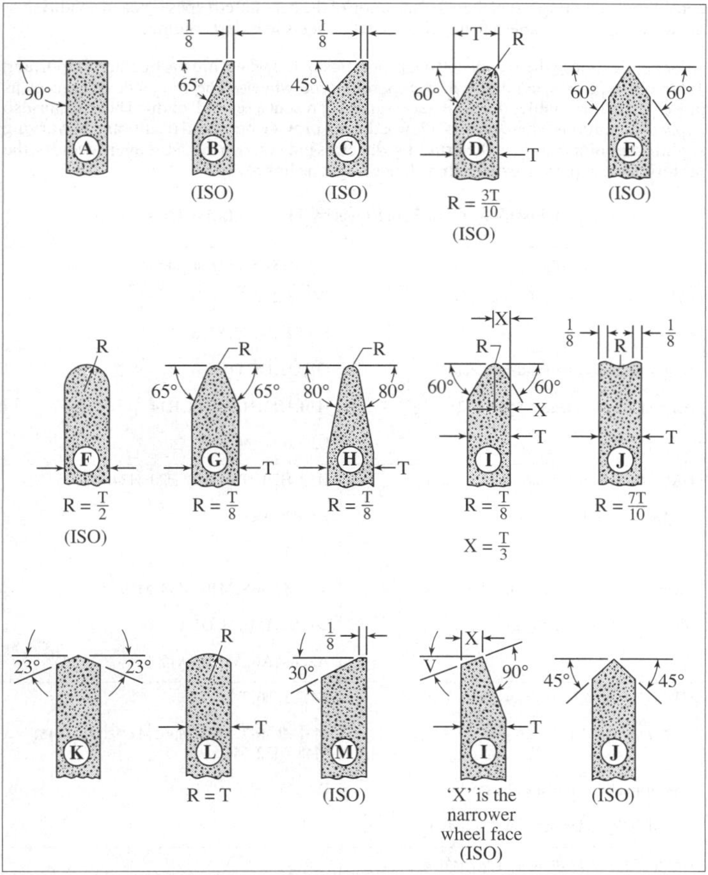


Recommendations, similar in principle, yet somewhat less discriminating have been developed by the Norton Company for *constructional steels*. These materials can be ground either in their original state (soft) or in their after-hardened state (directly or following carburization). Constructional steels must be distinguished from structural steels which are used primarily by the building industry in mill shapes, without or with a minimum of machining.

Constructional steels are either plain carbon or alloy type steels assigned in the AISI-SAE specifications to different groups, according to the predominant types of alloying elements. In the following recommendations no distinction is made because of different compositions since that factor generally, has a minor effect on grinding wheel choice in constructional steels. However, separate recommendations are made for soft (Table 5) and hardened (Table 6) constructional steels. For the relatively rare instance where the use of a single type of wheel for both soft and hardened steel materials is considered more important than the selection of the best suited types for each condition of the work materials, Table 5 lists "All Around" wheels in its last column.

For applications where cool cutting properties of the wheel are particularly important, Table 6 lists, as a second alternative, porous-type wheels. The sequence of choices as presented in these tables does not necessarily represent a second, or third best; it can also apply to conditions where the first choice did not provide optimum results and by varying slightly the composition of the grinding wheel, as indicated in the subsequent choices, the performance experience of the first choice might be improved.

**Table 3. Classification of Tool Steels by Their Relative Grindability**

| Relative Grindability Group | AISI-SAE Designation of Tool Steels |
|---|---|
| *GROUP 1—Any area of work surface* | W1, W2, W5 |
| | S1, S2, S4, S5, S6, S7 |
| High grindability tool and die steels | O1, O2, O6, O7 |
| (Grindability index greater than 12) | H10, H11, H12, H13, H14 |
| | L2, L6 |
| *GROUP 2—Small area of work surface* | H19, H20, H21, H22, H23, H24, H26 |
| *(as found in tools)* | P6, P20, P21 |
| | T1, T7, T8 |
| Medium grindability tool and die steels | M1, M2, M8, M10, M33, M50 |
| (Grindability index 3 to 12) | D1, D2, D3, D4, D5, D6 |
| | A2, A4, A6, A8, A9, A10 |
| *GROUP 3—Small area of work surface* | T4, T5, T6, T8 |
| *(as found in tools)* | M3, M6, M7, M34, M36, M41, M42, M46, M48, M52, M62 |
| Low grindability tool and die steels | D2, D5 |
| (Grindability index between 1.0 and 3) | A11 |
| *GROUP 4—Large area of work surface* | |
| *(as found in dies)* | All steels found in Groups 2 and 3 |
| Medium and low grindability tool and die steels | |
| (Grindability index between 1.0 and 12) | |
| *GROUP 5—Any area of work surface* | D3, D4, D7 |
| | M4 |
| Very low grindability tool and die steels | A7 |
| (Grindability index less than 1.0) | T15 |

**Table 4. Grinding Wheel Recommendations for Hardened Tool Steels According to Their Grindability**

| Operation | Wheel or Rim Diameter, Inches | First-Choice Specifications | Second-Choice Specifications |
|---|---|---|---|
| Group 1 Steels | | | |
| Surfacing | | | |
| Surfacing wheels | 14 and smaller | Wet FA46-I8V | SFA46-G12VP |
| | 14 and smaller | Dry FA46-H8V | FA46-F12VP |
| | Over 14 | Wet FA36-I8V | SFA36-I8V |
| Segments or Cylinders | $1\frac{1}{2}$ rim or less | Wet FA30-H8V | FA30-F12VP |
| Cups | $\frac{3}{4}$ rim or less | Wet FA36-H8V | FA46-F12VP |
| | (for rims wider than $1\frac{1}{2}$ inches, go one grade softer in available specifications) | | |
| Cutter sharpening | | | |
| Straight wheel | … | Wet FA46-K8V | FA60-K8V |
| | … | Dry FA46-J8V | FA46-H12VP |
| Dish shape | … | Dry FA60-J8V | FA60-H12VP |
| Cup shape | … | Dry FA46-L8V | FA60-H12VP |
| | … | Wet SFA46-L5V | SFA60-L5V |
| Form tool grinding | 8 and smaller | Wet FA60-L8V to FA100-M7V | |
| | 8 and smaller | Dry FA60-K8V to FA100-L8V | |
| | 10 and larger | Wet FA60-L8V to FA80-M6V | |
| Cylindrical | 14 and smaller | Wet SFA60-L5V | … |
| | 16 and larger | Wet SFA60-M5V | … |
| Centerless | … | Wet SFA60-M5V | … |
| Internal | | | |
| Production grinding | Under $\frac{1}{2}$ | Wet SPA80-N6V | SFA80-N7V |
| | $\frac{1}{2}$ to 1 | Wet SFA60-M5V | SFA60-M6V |
| | Over 1 to 3 | Wet SFA54-L5V | SFA54-L6V |
| | Over 3 | Wet SFA46-L5V | SFA46-K5V |
| Tool room grinding | Under $\frac{1}{2}$ | Dry FA80-L6V | SFA80-L7V |
| | $\frac{1}{2}$ to 1 | Dry FA70-K7V | SFA70-K7V |
| | Over 1 to 3 | Dry FA60-J8V | FA60-H12VP |
| | Over 3 | Dry FA46-J8V | FA54-H12VP |
| Group 2 Steels | | | |
| Surfacing | | | |
| Straight wheels | 14 and smaller | Wet FA46-I8V | FA46-G12VP |
| | 14 and smaller | Dry FA46-H8V | FA46-F12VP |
| | Over 14 | Wet FA46-H8V | SFA46-I8V |
| Segments or Cylinders | $1\frac{1}{2}$ rim or less | Wet FA30-G8V | FA36-E12VP |
| Cups | $\frac{3}{4}$ rim or less | Wet FA36-H8V | FA46-F12VP |
| | (for rims wider than $1\frac{1}{2}$ inches, go one grade softer in available specifications) | | |

**Table 4.** *(Continued)* **Grinding Wheel Recommendations for Hardened Tool Steels According to Their Grindability**

| Operation | Wheel or Rim Diameter, Inches | First-Choice Specifications | Second-Choice Specifications |
|---|---|---|---|
| Cutter sharpening | | | |
| Straight wheel | … | Wet FA46-L5V | FA60-K8V |
| | … | Dry FA46-J8V | FA60-H12VP |
| Dish shape | … | Dry FA60-J5V | FA60-G12VP |
| Cup shape | … | Dry FA46-K5V | FA60-G12VP |
| | … | Wet FA46-L5V | FA60-J8V |
| Form tool grinding | 8 and smaller | Wet FA60-K8V to FA120-L8V | |
| | 8 and smaller | Dry FA80-K8V to FA150-K8V | |
| | 10 and larger | Wet FA60-K8V to FA120-L8V | |
| Cylindrical | 14 and less | Wet FA60-L5V | SFA60-L5V |
| | 16 and larger | Wet FA60-K5V | SFA60-K5V |
| Centerless | … | Wet FA60-M5V | SFA60-M5V |
| Internal | | | |
| Production grinding | Under ½ | Wet FA80-L6V | SFA80-L6V |
| | ½ to 1 | Wet FA70-K5V | SFA70-K5V |
| | Over 1 to 3 | Wet FA60-J8V | SFA60-J7V |
| | Over 3 | Wet FA54-J8V | SFA54-J8V |
| Tool room grinding | Under ½ | Dry FA80-I8V | SFA80-K7V |
| | ½ to 1 | Dry FA70-J8V | SFA70-J7V |
| | Over 1 to 3 | Dry FA60-I8V | FA60-G12VP |
| | Over 3 | Dry FA54-I8V | FA54-G12VP |
| Group 3 Steels | | | |
| Surfacing | | | |
| Straight wheels | 14 and smaller | Wet FA60-I8V | FA60-G12VP |
| | 14 and smaller | Dry FA60-H8V | FA60-F12VP |
| | Over 14 | Wet FA60-H8V | SFA60-I8V |
| Segments or Cylinders | 1½ rim or less | Wet FA46-G8V | FA46-E12VP |
| Cups | ¾ rim or less | Wet FA46-G8V | FA46-E12VP |
| | (for rims wider than 1½ inches, go one grade softer in available specifications) | | |
| Cutter grinding | | | |
| Straight wheel | … | Wet FA46-J8V | FA60-J8V |
| | … | Dry FA46-I8V | FA46-G12VP |
| Dish shape | … | Dry FA60-H8V | FA60-F12VP |
| Cup shape | … | Dry FA46-I8V | FA60-F12VP |
| | … | Wet FA46-J8V | FA60-J8V |
| Form tool grinding | 8 and smaller | Wet FA80-K8V to FA150-L9V | |
| | 8 and smaller | Dry FA100-J8V to FA150-K8V | |
| | 10 and larger | Wet FA80-J8V to FA150-J8V | |

**Table 4.** *(Continued)* **Grinding Wheel Recommendations for Hardened Tool Steels According to Their Grindability**

| Operation | Wheel or Rim Diameter, Inches | First-Choice Specifications | Second-Choice Specifications |
|---|---|---|---|
| Cylindrical | 14 and less | Wet FA80-L5V | SFA80-L6V |
| | 16 and larger | Wet FA60-L6V | SFA60-K5V |
| Centerless | … | Wet FA60-L5V | SFA60-L5V |
| Internal | | | |
| Production grinding | Under ½ | Wet FA90-L6V | SFA90-L6V |
| | ½ to 1 | Wet FA80-L6V | SFA80-L6V |
| | Over 1 to 3 | Wet FA70-K5V | SFA70-K5V |
| | Over 3 | Wet FA60-J5V | SFA60-J5V |
| Tool room grinding | Under ½ | Dry FA90-K8V | SFA90-K7V |
| | ½ to 1 | Dry FA80-J8V | SFA80-J7V |
| | Over 1 to 3 | Dry FA70-I8V | SFA70-G12VP |
| | Over 3 | Dry FA60-I8V | SFA60-G12VP |
| Group 4 Steels | | | |
| Surfacing | | | |
| Straight wheels | 14 and smaller | Wet FA60-I8V | C60-JV |
| | 14 and smaller | Wet FA60-H8V | C60-IV |
| | Over 14 | Wet FA46-H8V | C60-HV |
| Segments | 1½ rim or less | Wet FA46-G8V | C46-HV |
| Cylinders | 1½ rim or less | Wet FA46-G8V | C60-HV |
| Cups | ¾ rim or less | Wet FA46-G6V | C60-IV |
| | (for rims wider than 1½ inches, go one grade softer in available specifications) | | |
| Form tool grinding | 8 and smaller | Wet FA60-J8V to FA150-K8V | |
| | 8 and smaller | Dry FA80-I8V to FA180-J8V | |
| | 10 and larger | Wet FA60-J8V to FA150-K8V | |
| Cylindrical | 14 and less | Wet FA80-K8V | C60-KV |
| | 16 and larger | Wet FA60-J8V | C60-KV |
| Internal | | | |
| Production grinding | Under ½ | Wet FA90-L8V | C90-LV |
| | ½ to 1 | Wet FA80-K5V | C80-KV |
| | Over 1 to 3 | Wet FA70-J8V | C70-JV |
| | Over 3 | Wet FA60-I8V | C60-IV |
| Tool room grinding | Under ½ | Dry FA90-K8V | C90-KV |
| | ½ to 1 | Dry FA80-J8V | C80-JV |
| | Over 1 to 3 | Dry FA70-I8V | C70-IV |
| | Over 3 | Dry FA60-H8V | C60-HV |

**Table 4.** *(Continued)* **Grinding Wheel Recommendations for Hardened Tool Steels According to Their Grindability**

| Operation | Wheel or Rim Diameter, Inches | First-Choice Specifications | Second-Choice Specifications | Third-Choice Specifications |
|---|---|---|---|---|
| Group 5 Steels | | | | |
| Surfacing | | | | |
| Straight wheels | 14 and smaller | Wet SFA60-H8V | FA60-E12VP | C60-IV |
| | 14 and smaller | Dry SFA80-H8V | FA80-E12VP | C80-HV |
| | Over 14 | Wet SFA60-H8V | FA60-E12VP | C60-HV |
| Segments or Cylinders | 1½ rim or less | Wet SFA46-G8V | FA46-E12VP | C46-GV |
| Cups | ¾ rim or less | Wet SFA60-G8V | FA60-E12VP | C60-GV |
| | (for rims wider than 1½ inches, go one grade softer in available specifications) | | | |
| Cutter grinding | | | | |
| Straight wheels | … | Wet SFA60-I8V | SFA60-G12VP | … |
| | … | Dry SFA60-H8V | SFA80-F12VP | … |
| Dish shape | … | Dry SFA80-H8V | SFA80-F12VP | … |
| Cup shape | … | Dry SFA60-I8V | SFA60-G12VP | … |
| | … | Wet SFA60-J8V | SFA60-H12VP | … |
| Form tool grinding | 8 and smaller | Wet FA80-J8V to FA180-J9V | | … |
| | 8 and smaller | Dry FA100-I8V to FA220-J9V | | … |
| | 10 and larger | Wet FA80-J8V to FA180-J9V | | … |
| Cylindrical | 14 and less | Wet FA80-J8V | C80-KV | FA80-H12VP |
| | 16 and larger | Wet FA80-I8V | C80-KV | FA80-G12VP |
| Centerless | … | Wet FA80-J5V | C80-LV | … |
| Internal | | | | |
| Production grinding | Under ½ | Wet FA100-L8V | C90-MV | … |
| | ½ to 1 | Wet FA90-K8V | C80-LV | … |
| | Over 1 to 3 | Wet FA80-J8V | C70-KV | FA80-H12VP |
| | Over 3 | Wet FA70-I8V | C60-JV | FA70-G12VP |
| Tool room grinding | Under ½ | Dry FA100-K8V | C90-KV | … |
| | ½ to 1 | Dry FA90-J8V | C80-JV | … |
| | Over 1 to 3 | Dry FA80-I8V | C70-IV | FA80-G12VP |
| | Over 3 | Dry FA70-I8V | C60-IV | FA70-G12VP |

**Table 5. Grinding Wheel Recommendations for Constructional Steels (Soft)**

| Grinding Operation | Wheel or Rim Diameter, Inches | First Choice | Alternate Choice (Porous type) | All-Around Wheel |
|---|---|---|---|---|
| Surfacing | | | | |
| Straight wheels | 14 and smaller | Wet FA46-J8V | FA46-H12VP | FA46-J8V |
| | 14 and smaller | Dry FA46-I8V | FA46-H12VP | FA46-I8V |
| | Over 14 | Wet FA36-J8V | FA36-H12VP | FA36-J8V |
| Segments | 1½ rim or less | Wet FA24-H8V | FA30-F12VP | FA24-H8V |
| Cylinders | 1½ rim or less | Wet FA24-I8V | FA30-G12VP | FA24-H8V |
| Cups | ¾ rim or less | Wet FA24-H8V | FA30-F12VP | FA30-H8V |
| | (for wider rims, go one grade softer) | | | |
| Cylindrical | 14 and smaller | Wet SFA60-M5V | … | SFA60-L5V |
| | 16 and larger | Wet SFA54-M5V | … | SFA54-L5V |
| Centerless | … | Wet SFA54-N5V | … | SFA60-M5V |
| Internal | Under ½ | Wet SFA60-M5V | … | SFA80-L6V |
| | ½ to 1 | Wet SFA60-L5V | … | SFA60-K5V |
| | Over 1 to 3 | Wet SFA54-K5V | … | SFA54-J5V |
| | Over 3 | Wet SFA46-K5V | … | SFA46-J5V |

**Table 6. Grinding Wheel Recommendations for Constructional Steels (Hardened or Carburized)**

| Grinding Operation | Wheel or Rim Diameter, Inches | First Choice | Alternate Choice (Porous Type) |
|---|---|---|---|
| Surfacing | | | |
| Straight wheels | 14 and smaller | Wet FA46-I8V | FA46-G12VP |
| | 14 and smaller | Dry FA46-H8V | FA46-F12VP |
| | Over 14 | Wet FA36-I8V | FA36-G12VP |
| Segments or Cylinders | 1½ rim or less | Wet FA30-H8V | FA36-F12VP |
| Cups | ¾ rim or less | Wet FA36-H8V | FA46-F12VP |
| | (for wider rims, go one grade softer) | | |
| Forms and Radius Grinding | 8 and smaller | Wet FA60-L7V to FA100-M8V | |
| | 8 and smaller | Dry FA60-K8V to FA100-L8V | |
| | 10 and larger | Wet FA60-L7V to FA80-M7V | |
| Cylindrical | | | |
| Work diameter | | | |
| 1 inch and smaller | 14 and smaller | Wet SFA80-L6V | … |
| Over 1 inch | 14 and smaller | Wet SFA80-K5V | … |
| 1 inch and smaller | 16 and larger | Wet SFA60-L5V | … |
| Over 1 inch | 16 and larger | Wet SFA60-L5V | … |
| Centerless | … | Wet SFA80-M6V | … |
| Internal | Under ½ | Wet SFA80-N6V | … |
| | ½ to 1 | Wet SFA60-M5V | … |
| | Over 1 to 3 | Wet SFA54-L5V | … |
| | Over 3 | Wet SFA46-K5V | … |
| | Under ½ | Dry FA80-L6V | … |
| | ½ to 1 | Dry FA70-K8V | … |
| | Over 1 to 3 | Dry FA60-J8V | FA60-H12VP |
| | Over 3 | Dry FA46-J8V | FA54-H12VP |

**Cubic Boron Nitride (CBN) Grinding Wheels.—**Although CBN is not quite as hard, strong, and wear-resistant as a diamond, it is far harder, stronger, and more resistant to wear than aluminum oxide and silicon carbide. As with diamond, CBN materials are available in different types for grinding workpieces of 50 RC (Rockwell C scale) and above, and for superalloys of 35 RC and harder. Microcrystalline CBN grinding wheels are suitable for grinding mild steels, medium-hard alloy steels, stainless steels, cast irons, and forged steels. Wheels with larger mesh size grains (up to 20⁄30), now available, provide for higher rates of metal removal.

Special types of CBN are produced for resin, vitrified, and electrodeposited bonds. Wheel standards and nomenclature generally conform to those used for diamond wheels (page 1292), except that the letter **B** instead of **D** is used to denote the type of abrasive. Grinding machines for CBN wheels are generally designed to take full advantage of the ability of CBN to operate at high surface speeds of 9,000–25,000 sfm (2745–7620 m/min). CBN is very responsive to changes in grinding conditions, and an increase in wheel speed from 5,000 to 10,000 sfm (1525–3050 m/min) can increase wheel life by a factor of 6 or more. A change from a water-based coolant to a coolant such as a sulfochlorinated or sulfurized straight grinding oil can increase wheel life by a factor of 10 or more.

Machines designed specifically for use with CBN grinding wheels generally use either electrodeposited wheels or have special truing systems for other CBN bond wheels, and are totally enclosed so they can use oil as a coolant. Numerical control systems are used, often running fully automatically, including loading and unloading. Machines designed for CBN grinding with electrodeposited wheels are extensively used for form and gear grinding, special systems being used to ensure rapid mounting to exact concentricity and truth in running, no truing or dressing being required. CBN wheels can produce workpieces having excellent accuracy and finish, with no truing or dressing for the life of the wheel, even over many hours or days of production grinding of hardened steel components.

Resin-, metal-, and vitrified-bond wheels are used extensively in production grinding, in standard and special machines. Resin-bonded wheels are used widely for dry tool and cutter resharpening on conventional hand-operated tool and cutter grinders. A typical wheel for such work would be designated 11V9 cup type, 100⁄120 mesh, 75 concentration, with a $\frac{1}{16}$ or $\frac{1}{8}$ inch (1.6 or 3.2 mm) rim section. Special shapes of resin-bonded wheels are used on dedicated machines for cutting tool manufacture. These types of wheels are usually self-dressing, and allow full machine control of the operation without the need for an operator to see, hear, or feel the action.

Metal-bonded CBN wheels are usually somewhat cheaper than those using other types of bond because only a thin layer of abrasive is present. Metal bonding is also used in manufacture of CBN honing stones. Vitrified-bond CBN wheels are a recent innovation, and high-performance bonds are still being developed. These wheels are used for grinding cams, internal diameters, and bearing components, and can be easily redressed.

An important aspect of grinding with CBN and diamond wheels is reduced heating of the workpiece, thought to result from their superior thermal conductivity compared with aluminum oxide, for instance. CBN and diamond grains also are harder, which means that they stay sharp longer than aluminum oxide grains. The superior ability to absorb heat from the workpiece during the grinding process reduces formation of untempered martensite in the ground surface, caused by overheating followed by rapid quenching. At the same time, a higher compressive residual stress is induced in the surface, giving increased fatigue resistance, compared with the tensile stresses found in surfaces ground with aluminum oxide abrasives. Increased fatigue resistance is of particular importance for gear grinding, especially in the root area.

**Variations from General Grinding Wheel Recommendations.—**Recommendations for the selection of grinding wheels are usually based on average values with regard to both operational conditions and process objectives. With variations from such average values, the composition of the grinding wheels must be adjusted to obtain optimum results. Although it is impossible to list and to appraise all possible variations and to define their effects on the selection of the best suited grinding wheels, some guidance is obtained from experience. The following tabulation indicates the general directions in which the characteristics of the initially selected grinding wheel may have to be altered in order to approach optimum performance. Variations in a sense opposite to those shown will call for wheel characteristic changes in reverse.

| Conditions or Objectives | Direction of Change |
|---|---|
| To increase cutting rate | Coarser grain, softer bond, higher porosity |
| To retain wheel size and/or form | Finer grain, harder bond |
| For small or narrow work surface | Finer grain, harder bond |
| For larger wheel diameter | Coarser grain |
| To improve finish on work | Finer grain, harder bond, or resilient bond |
| For increased work speed or feed rate | Harder bond |
| For increased wheel speed | Generally, softer bond, except for high-speed grinding, which requires a harder bond for added wheel strength |
| For interrupted or coarse work surface | Harder bond |
| For thin walled parts | Softer bond |
| To reduce load on the machine drive motor | Softer bond |

**Dressing and Truing Grinding Wheels.—**The perfect grinding wheel operating under ideal conditions will be self sharpening, i.e., as the abrasive grains become dull, they will tend to fracture and be dislodged from the wheel by the grinding forces, thereby exposing new, sharp abrasive grains. Although in precision machine grinding this ideal sometimes may be partially attained, it is almost never attained completely. Usually, the grinding wheel must be dressed and trued after mounting on the precision grinding machine spindle and periodically thereafter.

Dressing may be defined as any operation performed on the face of a grinding wheel that improves its cutting action. Truing is a dressing operation but is more precise, i.e., the face of the wheel may be made parallel to the spindle or made into a radius or special shape. Regularly applied truing is also needed for accurate size control of the work, particularly in automatic grinding. The tools and processes generally used in grinding wheel dressing and truing are listed and described in Table 1.

**Table 1. Tools and Methods for Grinding Wheel Dressing and Truing**

| Designation | Description | Application |
|---|---|---|
| Rotating Hand Dressers | Freely rotating discs, either star-shaped with protruding points or discs with corrugated or twisted perimeter, supported in a fork-type handle, the lugs of which can lean on the tool rest of the grinding machine. | Preferred for bench- or floor-type grinding machines; also for use on heavy portable grinders (snagging grinders) where free-cutting proper ties of the grinding wheel are primarily sought and the accuracy of the trued profile is not critical. |
| Abrasive Sticks | Made of silicon carbide grains with a hard bond. Applied directly or supported in a handle. Less frequently abrasive sticks are also made of boron carbide. | Usually handheld and use limited to smaller-size wheels. Because it also shears the grains of the grinding wheel, or preshaping, prior to final dressing with, e.g., a diamond. |

**Table 1.** *(Continued)* **Tools and Methods for Grinding Wheel Dressing and Truing**

| Designation | Description | Application |
|---|---|---|
| Abrasive Wheels (Rolls) | Silicon carbide grains in a hard vitrified bond are cemented on ball-bearing mounted spindles. Use either as hand tools with handles or rigidly held in a supporting member of the grinding machine. Generally freely rotating; also available with adjustable brake for diamond wheel dressing. | Preferred for large grinding wheels as a diamond saver, but also for improved control of the dressed surface characteristics. By skewing the abrasive dresser wheel by a few degrees out of parallel with the grinding wheel axis, the basic crushing action is supplemented with wiping and shearing, thus producing the desired degree of wheel surface smoothness. |
| Single-Point Diamonds | A diamond stone of selected size is mounted in a steel nib of cylindrical shape with or without head, dimensioned to fit the truing spindle of specific grinding machines. Proper orientation and retainment of the diamond point in the setting is an important requirement. | The most widely used tool for dressing and truing grinding wheels in precision grinding. Permits precisely controlled dressing action by regulating infeed and cross feed rate of the truing spindle when the latter is guided by cams or templates for accurate form truing. |
| Single-Point Form Truing Diamonds | Selected diamonds having symmetrically located natural edges with precisely lapped diamond points, controlled cone angles and vertex radius, and the axis coinciding with that of the nib. | Used for truing operations requiring very accurately controlled, and often steeply inclined wheel profiles, such as are needed for thread and gear grinding, where one or more diamond points participate in generating the resulting wheel periphery form. Dependent on specially designed and made truing diamonds and nibs. |
| Cluster-Type Diamond Dresser | Several, usually seven, smaller diamond stones are mounted in spaced relationship across the working surface of the nib. In some tools, more than a single layer of such clusters is set at parallel levels in the matrix, the deeper positioned layer becoming active after the preceding layer has worn away. | Intended for straight-face dressing and permits the utilization of smaller, less expensive diamond stones. In use, the holder is canted at a 3° to 10° angle, bringing two to five points into contact with the wheel. The multiple-point contact permits faster cross feed rates during truing than may be used with single-point diamonds for generating a specific degree of wheel-face finish. |
| Impregnated Matrix-Type Diamond Dressers | The operating surface consists of a layer of small, randomly distributed, yet rather uniformly spaced diamonds that are retained in a bond holding the points in an essentially common plane. Supplied either with straight or canted shaft, the latter being used to cancel the tilt of angular truing posts. | For the truing of wheel surfaces consisting of a single or several flat elements. The nib face should be held tangent to the grinding wheel periphery or parallel with a flat working surface. Offers economic advantages where technically applicable because of using less expensive diamond splinters presented in a manner permitting efficient utilization. |
| Form-Generating Truing Devices | Swiveling diamond holder post with adjustable pivot location, arm length, and swivel arc, mounted on angularly adjustable cross slides with controlled traverse movement, permits the generation of various straight and circular profile elements, kept in specific mutual locations. | Such devices are made in various degrees of complexity for the positionally controlled interrelation of several different profile elements. Limited to regular straight and circular sections, yet offers great flexibility of setup, very accurate adjustment, and unique versatility for handling a large variety of frequently changing profiles. |

**Table 1.** *(Continued)* **Tools and Methods for Grinding Wheel Dressing and Truing**

| Designation | Description | Application |
|---|---|---|
| Contour-Duplicating Truing Devices | The form of a master, called cam or template, shaped to match the profile to be produced on the wheel, or its magnified version, is translated into the path of the diamond point by means of mechanical linkage, a fluid actuator, or a pantograph device. | Preferred single-point truing method for profiles to be produced in quantities warranting the making of special profile bars or templates. Used also in small- and medium-volume production when the complexity of the profile to be produced excludes alternate methods of form generation. |
| Grinding Wheel Contouring by Crush Truing | A hardened steel or carbide roll, which is free to rotate and has the desired form of the workpiece, is fed gradually into the grinding wheel, which runs at slow speed. The roll will, by crushing action, produce its reverse form in the wheel. Crushing produces a free-cutting wheel face with sharp grains. | Requires grinding machines designed for crush truing, having stiff spindle bearings, rigid construction, slow wheel speed for truing, etc. Due to the cost of crush rolls and equipment, the process is used for repetitive work only. It is one of the most efficient methods for precisely duplicating complex wheel profiles that are capable of grinding in the 8-μin. (0.2032 μm) AA range. Applicable for both surface and cylindrical grinding. |
| Rotating Diamond Roll-Type Grinding Wheel Truing | Special rolls made to agree with specific profile specifications have their periphery coated with a large number of uniformly distributed diamonds, held in a matrix into which the individual stones are set by hand (for larger diamonds) or bonded by a plating process (for smaller elements). | The diamond rolls must be rotated by an air, hydraulic, or electric motor at about one-fourth of the grinding wheel surface speed and in opposite direction to the wheel rotation. Whereas the initial costs are substantially higher than for single-point diamond truing the savings in truing time warrants the method's application in large-volume production of profile-ground components. |
| Diamond Dressing Blocks | Made as flat blocks for straight wheel surfaces, are also available for radius dressing and profile truing. The working surface consists of a layer of electroplated diamond grains, uniformly distributed and capable of truing even closely toleranced profiles. | For straight wheels, dressing blocks can reduce dressing time and offer easy installation on surface grinders, where the blocks mount on the magnetic plate. Recommended for small- and medium-volume production for truing intricate profiles on regular surface grinders, because the higher pressure developed in crush dressing is avoided. |

**Guidelines for Dressing and Truing with Single-Point Diamonds.—**The diamond nib should be canted at an angle of 10 to 15 degrees in the direction of the wheel rotation and also, if possible, by the same amount in the direction of the cross feed traverse during the truing (see diagram). The dragging effect resulting from this "angling," combined with the occasional rotation of the diamond nib in its holder, will prolong the diamond life by limiting the extent of wear facets and will also tend to produce a pyramid shape of the diamond tip. The diamond may also be set to contact the wheel at about ⅛ to ¼ inch (3.2–6.4 mm) below its centerline.

*Depth of Cut:* This amount should not exceed 0.001 inch (0.025 mm) per pass for general work, and will have to be reduced to 0.0002 to 0.0004 inch (0.005-0.010 mm) per pass for wheels with fine grains used for precise finishing work.

*Diamond Crossfeed Rate:* This value may be varied to some extent depending on the required wheel surface: faster crossfeed for free cutting, and slower crossfeed for producing fine finishes. Such variations, however, must always stay within the limits set by the grain size of the wheel. Thus, the advance rate of the truing diamond per wheel revolution should not exceed the diameter of a grain or be less than half of that rate. Consequently, the diamond crossfeed must be slower for a large wheel than for a smaller wheel having the same grain size number.Typical crossfeed values for frequently used grain sizes are given in Table 2.

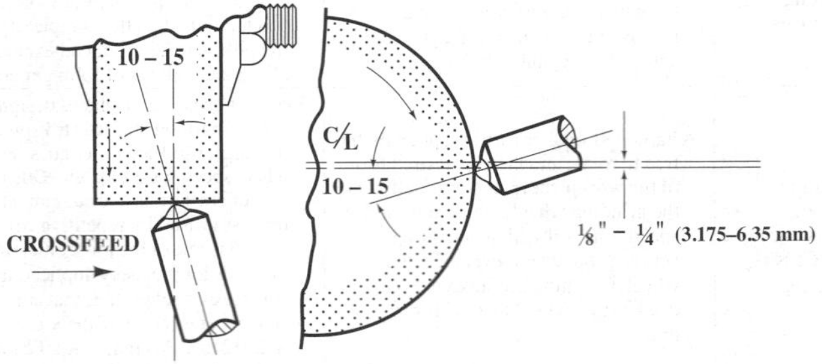


**Table 2. Typical Diamond Truing and Crossfeeds**

| Grain Size | Crossfeed per Wheel Rev. inch | Crossfeed per Wheel Rev. mm | Grain Size | Crossfeed per Wheel Rev. inch | Crossfeed per Wheel Rev. mm |
|---|---|---|---|---|---|
| 30 | 0.014–0.024 | 0.35–0.61 | 60 | 0.006–0.010 | 0.15–0.25 |
| 36 | 0.012–0.019 | 0.30–0.48 | 80 | 0.004–0.007 | 0.10–0.18 |
| 46 | 0.008–0.014 | 0.20–0.35 | 120 | 0.0025–0.004 | 0.06–0.10 |
| 50 | 0.007–0.012 | 0.18–0.30 | … | … | … |

These values can be easily converted into the more conveniently used inch-per-minute units, simply by multiplying the crossfeed rate in inches by the rpm of the grinding wheel.

*Example:* For a 20-inch diameter wheel, Grain No. 46, running at 1200 rpm: Crossfeed rate for roughing-cut truing, approximately 17 ipm; for finishing-cut truing, approximately 10 ipm.

*Coolant* should be applied before the diamond comes into contact with the wheel and must be continued in generous supply while truing.

*The speed of the grinding wheel* should be at the regular grinding rate, or not much lower. For that reason, the feed wheels of centerless grinding machines usually have an additional speed rate higher than functionally needed, that speed being provided for wheel truing only.

*The initial approach of the diamond* to the wheel surface must be carried out carefully to prevent sudden contact with the diamond, resulting in penetration in excess of the selected depth of cut. It should be noted that the highest point of a worn wheel is often in its center portion and not at the edge from which the crossfeed of the diamond starts.

*The general conditions* of the truing device are important for best truing results and for assuring extended diamond life. A rigid truing spindle, well-seated diamond nib, and firmly set diamond point are mandatory. Sensitive infeed and smooth traverse movement at uniform speed also must be maintained.

*Resetting of the diamond point.* Never let the diamond point wear to a degree where the grinding wheel is in contact with the steel nib. Such contact can damage the setting of the diamond point and result in its loss. Expert resetting of a worn diamond can repeatedly add to its useful life, even when applied to lighter work because of reduced size.

**Size Selection Guide for Single-Point Truing Diamonds.—**There are no rigid rules for determining the proper size of the diamond for any particular truing application because of the very large number of factors affecting that choice. Several of these factors are related to the condition, particularly the rigidity, of the grinding machine and truing device, as well as to such characteristics of the diamond itself as purity, crystalline structure, etc. Although these factors are difficult to evaluate in a generally applicable manner, the expected effects of several other conditions can be appraised and should be considered in the selection of the proper diamond size.

The recommended sizes in Table 3 must be considered as informative only and as representing minimum values for generally favorable conditions. Factors calling for larger diamond sizes than listed are the following:

Silicon carbide wheels (Table 3 refers to aluminum oxide wheels)

Dry truing

Grain sizes coarser than No. 46

Bonds harder than M

Wheel speed substantially higher than 6500 sfm (1981 m/min).

It is advisable to consider any single or pair of these factors as justifying the selection of one size larger diamond. As an example: for truing an SiC wheel, with grain size No. 36 and hardness P, select a diamond that is two sizes larger than that shown in Table 3 for the wheel size in use.

**Table 3. Recommended Minimum Sizes for Single-Point Truing Diamonds**

| Diamond Size in Carats[a] | Index Number (Wheel Dia. × Width, Inches) | Examples of Max. Grinding Wheel Dimensions (inches) | |
|---|---|---|---|
| | | Diameter | Width |
| 0.25 | 3 | 4 | 0.75 |
| 0.35 | 6 | 6 | 1 |
| 0.50 | 10 | 8 | 1.25 |
| 0.60 | 15 | 10 | 1.50 |
| 0.75 | 21 | 12 | 1.75 |
| 1.00 | 30 | 12 | 2.50 |
| 1.25 | 48 | 14 | 3.50 |
| 1.50 | 65 | 16 | 4.00 |
| 1.75 | 80 | 20 | 4.00 |
| 2.00 | 100 | 20 | 5.00 |
| 2.50 | 150 | 24 | 6.00 |
| 3.00 | 200 | 24 | 8.00 |
| 3.50 | 260 | 30 | 8.00 |
| 4.00 | 350 | 36 | 10.00 |

[a] One carat equals 0.2 gram.

Single-point diamonds are available as loose stones, but are preferably procured from specialized manufacturers supplying the diamonds set into steel nibs. Expert setting, comprising both the optimum orientation of the stone and its firm retainment, is mandatory for assuring adequate diamond life and satisfactory truing. Because the holding devices for truing diamonds are not yet standardized, the required nib dimensions vary depending on the make and type of different grinding machines. Some nibs are made with angular heads, usually hexagonal, to permit occasional rotation of the nib either manually, with a wrench, or automatically.

## Diamond Wheels

**Diamond Wheels.—**A diamond wheel is a special type of grinding wheel in which the abrasive elements are diamond grains held in a bond and applied to form a layer on the operating face of a non-abrasive core. Diamond wheels are used for grinding very hard or highly abrasive materials. Primary applications are the grinding of cemented carbides, such as the sharpening of carbide cutting tools; the grinding of glass, ceramics, asbestos, and cement products; and the cutting and slicing of germanium and silicon.

**Shapes of Diamond Wheels.—**The industry-wide accepted Standard ANSI B74.3-2003 (R2014) specifies ten basic diamond wheel core shapes which are shown in Table 1 with the applicable designation symbols. The applied diamond abrasive layer may have different cross-sectional shapes. Those standardized are shown in Table 2. The third aspect which is standardized is the location of the diamond section on the wheel as shown by the diagrams in Table 3. Finally, modifications of the general core shape together with pertinent designation letters are given in Table 4.

The characteristics of the wheel shape listed in these four tables make up the components of the standard designation symbol for diamond wheel shapes. An example of that symbol with arbitrarily selected components is shown in Fig. 1.

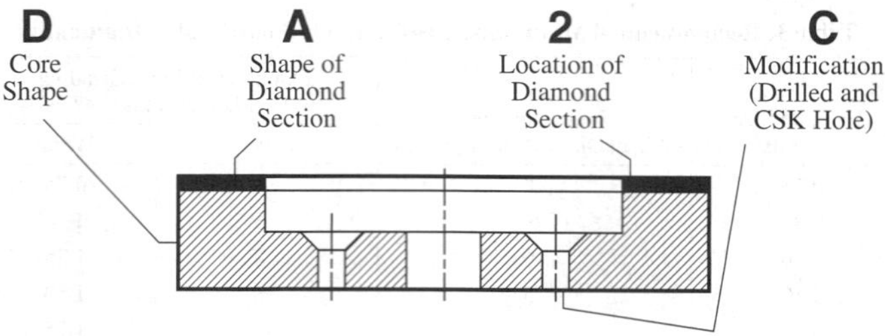


Fig. 1. A Typical Diamond Wheel Shape Designation Symbol

An explanation of these components is as follows:

*Basic Core Shape:* This portion of the symbol indicates the basic shape of the core on which the diamond abrasive section is mounted. The shape is actually designated by a number. The various core shapes and their designations are given in Table 1.

*Diamond Cross-Sectional Shape:* This, the second component, consisting of one or two letters, denotes the cross-sectional shape of the diamond abrasive section. The various shapes and their corresponding letter designations are given in Table 2.

*Diamond Section Location:* The third component of the symbol consists of a number which gives the location of the diamond section, i.e., periphery, side, corner, etc. An explanation of these numbers is shown in Table 3.

*Modification:* The fourth component of the symbol is a letter designating some modification, such as drilled and counterbored holes for mounting or special relieving of diamond section or core. This modification position of the symbol is used only when required. The modifications and their designations are given in Table 4.

**Table 1. Diamond Wheel Core Shapes and Designations** *ANSI B74.3-2003 (R2014)*

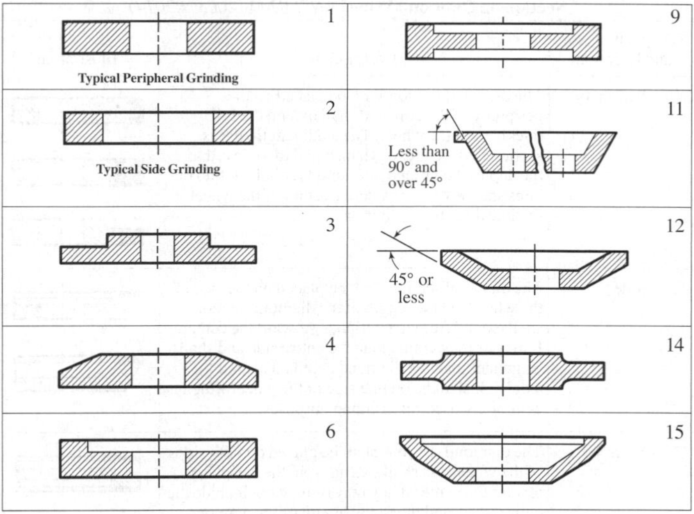


**Table 2. Diamond Cross Sections and Designations**
*ANSI B74.3-2003 (R2014)*

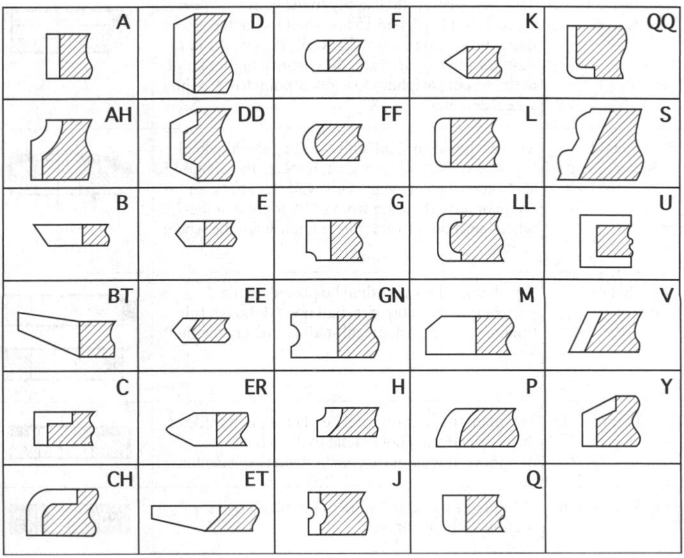

**Table 3. Designations for Location of Diamond Section on Diamond Wheel** *ANSI B74.3-2003 (R2014)*

| Designation No. and Location | Description | Illustration |
|---|---|---|
| 1 — Periphery | The diamond section shall be placed on the periphery of the core and shall extend the full thickness of the wheel. The axial length of this section may be greater than, equal to, or less than the depth of diamond, measured radially. A hub or hubs shall not be considered as part of the wheel thickness for this definition. | 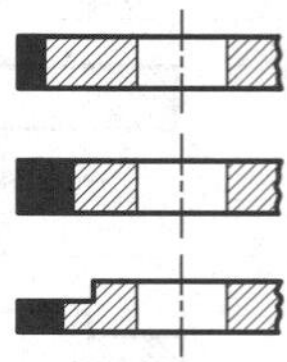 |
| 2 — Side | The diamond section shall be placed on the side of the wheel and the length of the diamond section shall extend from the periphery toward the center. It may or may not include the entire side and shall be greater than the diamond depth measured axially. It shall be on that side of the wheel which is commonly used for grinding purposes. | 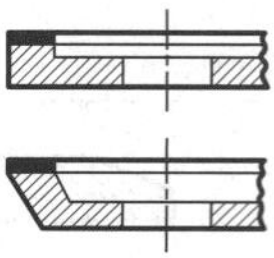 |
| 3 — Both Sides | The diamond sections shall be placed on both sides of the wheel and shall extend from the periphery toward the center. They may or may not include the entire sides, and the radial length of the diamond section shall exceed the axial diamond depth. | 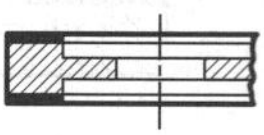 |
| 4 — Inside Bevel or Arc | This designation shall apply to the general wheel types 2, 6, 11, 12, and 15 and shall locate the diamond section on the side wall. This wall shall have an angle or arc extending from a higher point at the wheel periphery to a lower point toward the wheel center. | 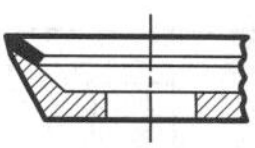 |
| 5 — Outside Bevel or Arc | This designation shall apply to the general wheel types, 2, 6, 11, and 15 and shall locate the diamond section on the side wall. This wall shall have an angle or arc extending from a lower point at the wheel periphery to a higher point toward the wheel center. | 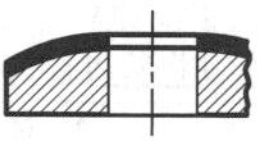 |
| 6 — Part of Periphery | The diamond section shall be placed on the periphery of the core but shall not extend the full thickness of the wheel and shall not reach to either side. | 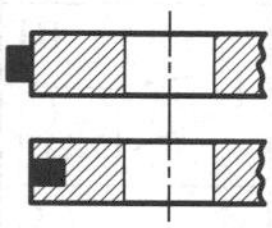 |
| 7 — Part of Side | The diamond section shall be placed on the side of the core and shall not extend to the wheel periphery. It may or may not extend to the center. | 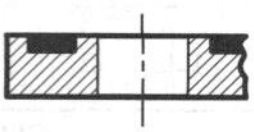 |
| 8 — Throughout | Designates wheels of solid diamond abrasive section without cores. | 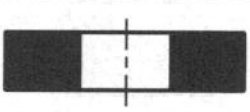 |

**Table 3.** *(Continued)* **Designations for Location of Diamond Section on Diamond Wheel** *ANSI B74.3-2003 (R2014)*

| Designation No. and Location | Description | Illustration |
|---|---|---|
| 9 — Corner | Designates a location which would commonly be considered to be on the periphery except that the diamond section shall be on the corner but shall not extend to the other corner. | 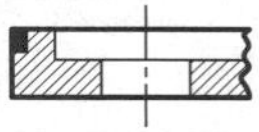 |
| 10 — Annular | Designates a location of the diamond abrasive section on the inner annular surface of the wheel. | 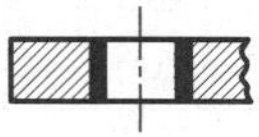 |

**Composition of Diamond and Cubic Boron Nitride Wheels.—**According to American National Standard ANSI B74.13-2016, a series of symbols is used to designate the composition of these wheels. An example is shown below.

| Prefix | Abrasive | Grain Size | Grade | Concentration | Bond Type | Bond Modification | Depth of Abrasive | Manufacturer's Identification Symbol |
|---|---|---|---|---|---|---|---|---|
| M | D | 120 | N | 100 | B | 56 | $1/8$ | * |

Fig. 2. Designation Symbols for Composition of Diamond and Cubic Boron Nitride Wheels

The meaning of each symbol is indicated by the following list:

1) *Prefix:* The prefix is a manufacturer's symbol indicating the exact kind of abrasive. Its use is optional.

2) *Abrasive Type:* The letter (B) is used for cubic boron nitride and (D) for diamond.

3) *Grain Size:* The grain sizes commonly used and varying from coarse to very fine are indicated by the following numbers: 8, 10, 12, 14, 16, 20, 24, 30, 36, 46, 54, 60, 70, 80, 90, 100, 120, 150, 180, and 220. The following additional sizes are used occasionally: 240, 280, 320, 400, 500, and 600. The wheel manufacturer may add to the regular grain number an additional symbol to indicate a special grain combination.

4) *Grade:* Grades are indicated by letters of the alphabet from A to Z in all bonds or processes. Wheel grades from A to Z range from soft to hard.

5) *Concentration:* The concentration symbol is a manufacturer's designation. It may be a number or a symbol.

6) *Bond:* Bonds are indicated by the following letters: B, resinoid; V, vitrified; M, metal.

7) *Bond Modification:* Within each bond type a manufacturer may have modifications to tailor the bond to a specific application. These modifications may be identified by either letters or numbers.

8) *Abrasive Depth:* Abrasive section depth, in inches or millimeters (inches illustrated), is indicated by a number or letter which is the amount of total dimensional wear a user may expect from the abrasive portion of the product. Most diamond and CBN wheels are made with a depth of coating on the order of $1/16$ in., $1/8$ in., (1.6 mm, 3.2 mm) or more as specified. In some cases the diamond is applied in thinner layers, as thin as one thickness of diamond grains. The L is included in the marking system to identify a layered type product.

9) *Manufacturer's Identification Symbol:* The use of this symbol is optional.

## Table 4. Designation Letters for Modifications of Diamond Wheels
*ANSI B74.3-2003 (R2014)*

| Designation Letter | Description | Illustration |
|---|---|---|
| B — Drilled and Counterbored | Holes drilled and counterbored in core. | 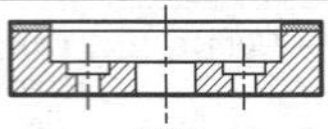 6A2B |
| C — Drilled and Countersunk | Holes drilled and countersunk in core. | 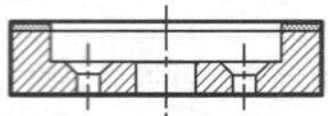 6A2C |
| F — Hub | Hub on basic wheel cover | 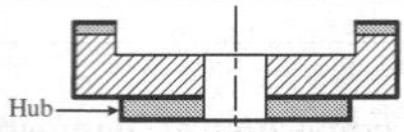  6A2F |
| H — Plain Hole | Straight hole drilled in core. | 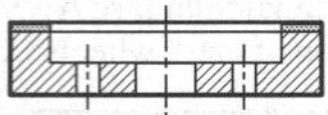 6A2H |
| J — Recess One Side | | 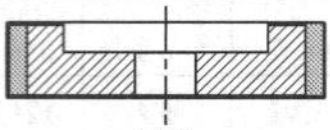 1A1J |
| JJ — Recess Two Sides | | 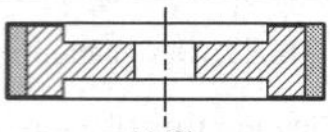 1A1JJ |
| K — Keyway | Arbor hole with keyway | 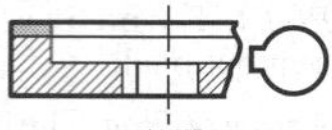 6A2K |
| M — Holes Plain and Threaded | Mixed holes, some plain, some threaded, are in core. | 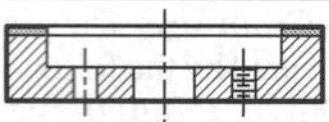 6A2M |
| N — Nonsteel Core | Nonsteel core for 1A1R and related shapes | |
| P — Relieved One Side | Core relieved on one side of wheel. Thickness of core is less than wheel thickness. | 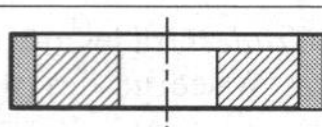 1A1P |
| R — Relieved Two Sides | Core relieved on both sides of wheel. Thickness of core is less than wheel thickness. | 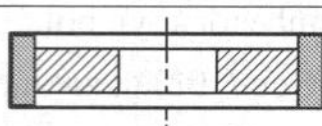 1A1R |
| S — Segmented-Diamond Section | Wheel has segmental diamond section mounted on core. (Clearance between segments has no bearing on definition.) | 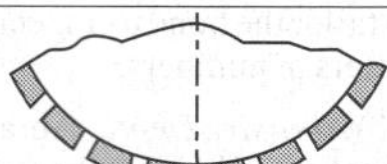 1A1S |
| SS — Segmental and Slotted | Wheel has separated segments mounted on a slotted core. | 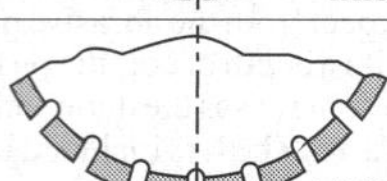 1A1SS |

**Table 4.** *(Continued)* **Designation Letters for Modifications of Diamond Wheels**
*ANSI B74.3-2003 (R2014)*

| Designation Letter | Description | Illustration |
|---|---|---|
| T — Threaded Holes | Threaded holes are in core. | 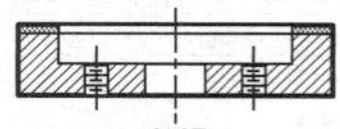 6A2T |
| Q — Diamond Inserted | Three surfaces of the diamond section are partially or completely enclosed by the core. | 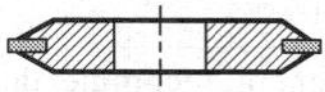 1A6Q |
| V — Diamond Inverted | Any diamond cross section that is mounted on the core so that the interior point of any angle, or the concave side of any arc, is exposed shall be considered inverted. *Exception:* Diamond cross section AH shall be placed on the core with the concave side of the arc exposed. | 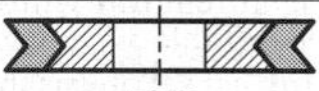 1EE1V 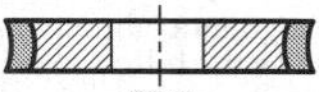 1FF1V |
| W — Wheels Affixed on a Mandrel | | 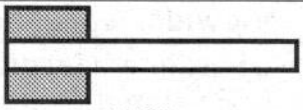 1A8W 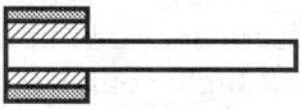 1A1W |
| Y — Diamond Inserted and Inverted | See definitions for Q and V. | 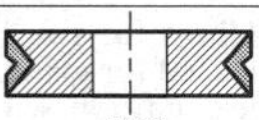 1EE6Y 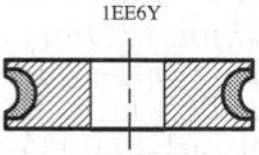 1FF6Y |

**The Selection of Diamond Wheels.—**Two general aspects must be defined: (a) The shape of the wheel, also referred to as the basic wheel type and (b) The specification of the abrasive portion.

**Table 5. General Diamond Wheel Recommendations for Wheel Type and Abrasive Specification**

| Typical Applications or Operation | Basic Wheel Type | | Abrasive Specification |
|---|---|---|---|
| Single Point Tools (offhand grinding) | D6A2C | *Rough:* | MD100-N100-B1/8 |
| | | *Finish:* | MD220-P75-B1/8 |
| Single Point Tools (machine ground) | D6A2H | *Rough:* | MD180-J100-B1/8 |
| | | *Finish:* | MD320-L75-B1/8 |
| Chip Breakers | D1A1 | | MD150-R100-B1/8 |
| Multitooth Tools and Cutters (face mills, end mills, reamers, broaches, etc.) Sharpening and Backing off | D11V9 | *Rough:* | MD100-R100-B1/8 |
| | | *Combination:* | MD150-R100-B1/8 |
| | | *Finish:* | MD220-R100-B1/8 |
| Fluting | D12A2 | | MD180-N100-B1/8 |
| Saw Sharpening | D12A2 | | MD180-R100-B1/8 |
| Surface Grinding (horizontal spindle) | D1A1 | *Rough:* | MD120-N100-B1/8 |
| | | *Finish:* | MD240-P100-B1/8 |
| Surface Grinding (vertical spindle) | D2A2T | | MD80-R75-B1/8 |

**Table 5.** *(Continued)* **General Diamond Wheel Recommendations for Wheel Type and Abrasive Specification**

| Typical Applications or Operation | Basic Wheel Type | Abrasive Specification | |
|---|---|---|---|
| Cylindrical or Centertype Grinding | D1A1 | | MD120-P100-B1/8 |
| Internal Grinding | D1A1 | | MD150-N100-B1/8 |
| Slotting and Cutoff | D1A1R | | MD150-R100-B1/4 |
| Lapping | Disc | | MD400-L50-B1/16 |
| Hand Honing | DH1, DH2 | *Rough:* | MD220-B1/16 |
| | | *Finish:* | MD320-B1/8 |

General recommendations for the dry grinding, with resin bond diamond wheels, of most grades of cemented carbides of average surface to ordinary finishes at normal rates of metal removal with average size wheels, as published by Cincinnati Milacron, are listed in Table 5.

A further set of variables are *the dimensions of the wheel,* which must be adapted to the available grinding machine and, in some cases, to the configuration of the work.

The general abrasive specifications in Table 5 may be modified to suit operating conditions by the following suggestions:

Use softer wheel grades for harder grades of carbides, for grinding larger areas or larger or wider wheel faces.

Use harder wheel grades for softer grades of carbides, for grinding smaller areas, for using smaller and narrower face wheels and for light cuts.

Use fine grit sizes for harder grades of carbides and to obtain better finishes.

Use coarser grit sizes for softer grades of carbides and for roughing cuts.

Use higher diamond concentration for harder grades of carbides, for larger diameter or wider face wheels, for heavier cuts, and for obtaining better finish.

**Guidelines for the Handling and Operation of Diamond Wheels.—**Grinding machines used for grinding with diamond wheels should be of the precision type, in good service condition, with true running spindles and smooth slide movements.

*Mounting of Diamond Wheels:* Wheel mounts should be used which permit the precise centering of the wheel, resulting in a runout of less than 0.001 inch (0.0254 mm) axially and 0.0005 inch (0.0127 mm) radially. These conditions should be checked with a 0.0001-inch (0.0025-mm) type dial indicator. Once mounted and centered, the diamond wheel should be retained on its mount and stored in that condition when temporarily removed from the machine.

*Truing and Dressing:* Resinoid bonded diamond wheels seldom require dressing, but when necessary a soft silicon carbide stick may be handheld against the wheel. Peripheral and cup type wheels may be sharpened by grinding the cutting face with a 60 to 80 grit silicon carbide wheel. This can be done with the diamond wheel mounted on the spindle of the machine, and with the silicon carbide wheel driven at a relatively slow speed by a specially designed table-mounted grinder or by a small table-mounted tool post grinder. The diamond wheel can be mounted on a special arbor and ground on a lathe with a tool post grinder; peripheral wheels can be ground on a cylindrical grinder or with a special brake-controlled truing device with the wheel mounted on the machine on which it is used. Cup and face type wheels are often lapped on a cast iron or glass plate using a 100 grit silicon carbide abrasive. Care must be used to lap the face parallel to the back, otherwise they must be ground to restore parallelism. Peripheral diamond wheels can be trued and dressed by grinding a silicon carbide block or a special diamond impregnated bronze block in a manner similar to surface grinding. Conventional diamonds must not be used for truing and dressing diamond wheels.

**Speeds and Feeds in Diamond Grinding.—**General recommendations are as follows:

*Wheel Speeds:* The generally recommended wheel speeds for diamond grinding are in the range of 5000 to 6000 surface feet per minute (1525–1830 m/min), with this upper limit as a maximum to avoid harmful "overspeeding." Exceptions from that general rule are diamond wheels with coarse grains and high concentration (100 percent) where the wheel wear in dry surface grinding can be reduced by lowering the speed to 2500–3000 sfpm (760–915 m/min). However, this lower speed range can cause rapid wheel breakdown in finer grit wheels or in those with reduced diamond concentration.

*Work Speeds:* In diamond grinding, work rotation and table traverse are usually established by experience, adjusting these values to the selected infeed so as to avoid excessive wheel wear.

*Infeed per Pass:* Often referred to as downfeed and usually a function of the grit size of the wheel. The following are general values which may be increased for raising the productivity, or lowered to improve finish or to reduce wheel wear.

| Wheel Grit Size Range | Infeed per Pass |
|---|---|
| 100 to 120 | 0.001 inch (0.025 mm) |
| 150 to 220 | 0.0005 inch (0.013 mm) |
| 250 and finer | 0.00025 inch (0.006 mm) |

## Grinding Wheel Safety

**Safety in Operating Grinding Wheels.—**Grinding wheels, although capable of exceptional cutting performance due to hardness and wear resistance, are prone to damage caused by improper handling and operation. Vitrified wheels, comprising the major part of grinding wheels used in industry, are held together by an inorganic bond which is actually a type of pottery product and therefore brittle and breakable. Although most of the organic bond types are somewhat more resistant to shocks, it must be realized that all grinding wheels are conglomerates of individual grains joined by a bond material whose strength is limited by the need of releasing the dull, abrasive grains during use.

It must also be understood that during the grinding process very substantial forces act on the grinding wheel, including the centrifugal force due to rotation, the grinding forces resulting from the resistance of the work material, and shocks caused by sudden contact with the work. To be able to resist these forces, the grinding wheel must have a substantial minimum strength throughout that is well beyond that needed to hold the wheel together under static conditions.

Finally, a damaged grinding wheel can disintegrate during grinding, liberating dormant forces which normally are constrained by the resistance of the bond, thus presenting great hazards to both operator and equipment.

To avoid breakage of the operating wheel and, should such a mishap occur, to prevent damage or injury, specific precautions must be applied. These safeguards have been formulated into rules and regulations and are set forth in the American National Standard ANSI B7.1-2017, entitled the American National Standard Safety Requirements for the Use, Care, and Protection of Abrasive Wheels.

**Handling, Storage and Inspection.—**Grinding wheels should be hand carried, or transported, with proper support, by truck or conveyor. A grinding wheel must not be rolled around on its periphery.

The storage area, positioned not far from the location of the grinding machines, should be free from excessive temperature variations and humidity. Specially built racks are recommended on which the smaller or thin wheels are stacked lying on their sides and the larger wheels in an upright position on two-point cradle supports consisting of appropriately spaced wooden bars. Partitions should separate either the individual wheels, or a small group of identical wheels. Good accessibility to the stored wheels reduces the need of undesirable handling.

Inspection will primarily be directed at detecting visible damage, mostly originating from handling and shipping. Cracks which are not obvious can usually be detected by "ring testing," which consists of suspending the wheel from its hole and tapping it with a nonmetallic implement. Heavy wheels may be allowed to rest vertically on a clean, hard floor while performing this test. A clear metallic tone, a "ring," should be heard; a dead sound being indicative of a possible crack or cracks in the wheel.

**Machine Conditions.—**The general design of the grinding machines must ensure safe operation under normal conditions. The bearings and grinding wheel spindle must be dimensioned to withstand the expected forces and ample driving power should be provided to ensure maintenance of the rated spindle speed. For the protection of the operator, stationary machines used for dry grinding should have a provision made for connection to an exhaust system and when used for offhand grinding, a work support must be available.

Wheel guards are particularly important protection elements and their material specifications, wall thicknesses and construction principles should agree with the Standard's specifications. The exposure of the wheel should be just enough to avoid interference with the grinding operation. The need for access of the work to the grinding wheel will define the boundary of guard opening, particularly in the direction of the operator.

**Grinding Wheel Mounting.**—The mass and speed of the operating grinding wheel makes it particularly sensitive to imbalance. Vibrations that result from such conditions are harmful to the machine, particularly the spindle bearings, and they also affect the ground surface, i.e., wheel imbalance causes chatter marks and interferes with size control. Grinding wheels are shipped from the manufacturer's plant in a balanced condition, but retaining the balanced state after mounting the wheel is quite uncertain. Balancing of the mounted wheel is thus required, and is particularly important for medium and large size wheels, as well as for producing accurate and smooth surfaces. The most common way of balancing mounted wheels is by using balancing flanges with adjustable weights. The wheel and balancing flanges are mounted on a short balancing arbor, the two concentric and round stub ends of which are supported in a balancing stand.

Such stands are of two types: 1) the parallel straight-edged, which must be set up precisely level; and 2) the disk type having two pairs of ball bearing mounted overlapping disks, which form a V for containing the arbor ends without hindering the free rotation of the wheel mounted on that arbor.

The wheel will then rotate only when it is out of balance and its heavy spot is not in the lowest position. Rotating the wheel by hand to different positions will move the heavy spot, should such exist, from the bottom to a higher location where it can reveal its presence by causing the wheel to turn. Having detected the presence and location of the heavy spot, its effect can be cancelled by displacing the weights in the circular groove of the flange until a balanced condition is accomplished.

Flanges are commonly used means for holding grinding wheels on the machine spindle. For that purpose, the wheel can either be mounted directly through its hole or by means of a sleeve which slips over a tapered section of the machine spindle. Either way, the flanges must be of equal diameter, usually not less than one-third of the new wheel's diameter. The purpose is to securely hold the wheel between the flanges without interfering with the grinding operation even when the wheel becomes worn down to the point where it is ready to be discarded. Blotters or flange facings of compressible material should cover the entire contact area of the flanges.

One of the flanges is usually fixed while the other is loose and can be removed and adjusted along the machine spindle. The movable flange is held against the mounted grinding wheel by means of a nut engaging a threaded section of the machine spindle. The sense of that thread should be such that the nut will tend to tighten as the spindle revolves. In other words, to remove the nut, it must be turned in the direction that the spindle revolves when the wheel is in operation.

**Safe Operating Speeds.**—Safe grinding processes are predicated on the proper use of the previously discussed equipment and procedures, and are greatly dependent on the application of adequate operating speeds.

The Standard establishes maximum speeds at which grinding wheels can be operated, assigning the various types of wheels to several classification groups. Different values are listed according to bond type and to wheel strength, distinguishing between low, medium and high strength wheels.

For the purpose of general information, the accompanying Table 1 shows an abbreviated version of the Standard's specification. However, for the governing limits, the authoritative source is the manufacturer's tag on the wheel which, particularly for wheels of lower strength, might specify speeds below those of the table.

All grinding wheels of 6 inches or greater diameter must be test run in the wheel manufacturer's plant at a speed that for all wheels having operating speeds in excess of 5000 sfpm is 1.5 times the maximum speed marked on the tag of the wheel.

The table shows the permissible wheel speeds in surface feet per minute (sfpm) units, whereas the tags on the grinding wheels state, for the convenience of the user, the maximum operating speed in revolutions per minute (rpm). The sfpm unit has the advantage of remaining valid for worn wheels whose rotational speed may be increased to the applicable sfpm value. The conversion from either one to the other of these two kinds of units is a matter of simple calculation using the formulas:

$$\text{sfpm} = \text{rpm} \times \frac{D}{12} \times \pi \qquad \text{or} \qquad \text{rpm} = \frac{\text{sfpm} \times 12}{D \times \pi}$$

where $D$ = maximum diameter of the grinding wheel, in inches. Table 2, showing the conversion values from surface speed into rotational speed, can be used for the direct reading of the rpm values corresponding to several different wheel diameters and surface speeds.

*Special Speeds:* Continuing progress in grinding methods has led to the recognition of certain advantages that can result from operating grinding wheels above, sometimes even higher than twice, the speeds considered earlier as the safe limits of grinding wheel operations. Advantages from the application of high-speed grinding are limited to specific processes, but the Standard admits, and offers code regulations for the use of wheels at special high speeds. These regulations define the structural requirements of the grinding machine and the responsibilities of the grinding wheel manufacturers, as well as of the users. High-speed grinding should not be applied unless the machines, particularly guards, spindle assemblies, and drive motors, are suitable for such methods. Also, appropriate grinding wheels expressly made for special high speeds must be used and, of course, the maximum operating speeds indicated on the wheel's tag must never be exceeded.

**Portable Grinders.—**The above discussed rules and regulations, devised primarily for stationary grinding machines apply also to portable grinders. In addition, the details of various other regulations, specially applicable to different types of portable grinders are discussed in the Standard, which should be consulted, particularly for safe applications of portable grinding machines.

**Table 1. Maximum Peripheral Speeds for Grinding Wheels** *(Based on ANSI 7.1-2017)*

| Classification Number and Types of Wheels[a] | | Maximum Operating Speeds Depending on Strength of Bond | | | |
|---|---|---|---|---|---|
| | | Inorganic Bonds | | Organic Bonds | |
| | | sfpm | m/min | sfpm | m/min |
| 1 | Straight wheels—Type 1, except classifications 6, 9, 11, and 12, 13, and 14 below; recessed wheels—Types 5 and 7 Type 20, 21, 22, 23, 24, 25, 26 | 5500–8500 | 1675–2590 | 6500–9500 | 1980–2895 |
| 2 | Taper Side Wheels—Type 4[b]; dish wheels—Type 12; Saucer wheels—Type 13; cones and plugs—Types 16, 17, 18, 19 | 5000–6500 | 1525–1980 | 6500–9500 | 1980–2895 |
| | Cylinder wheels—Type 2 segments | 5000–6000 | 1525–1830 | 5000–7000 | 1525–2135 |
| 3 | Cup shape tool grinding wheels—Types 6 and 11 (for fixed base machines) | 4500–6000 | 1370–1830 | 6000–8500 | 1830–2590 |
| 4 | Cup shape snagging wheels—Types 6 and 11 (for portable machines) | 4500–6500 | 1370–1980 | 6000–9500 | 1830–2895 |
| 5 | Abrasive disks and cylinder type discs solid and segmental | 5500–6500 | 1675–1980 | 5500–8500 | 1675–2590 |
| 6 | Reinforced wheels—except cutting-off wheels (depending on diameter and thickness) | … | … | 9500–16000 | 2895–4875 |
| 7 | Type 1 wheels for bench and pedestal grinders, Types 1, 5 and 7 also in certain sizes for surface grinders | 5500–7550 | 1675–2300 | 6500–9500 | 1980–2895 |
| 8 | Diamond and cubic boron nitride wheels | to 6500 | to 1980 | to 9500 | to 2895 |
| | Metal bond | … | to 3660 | … | … |
| | Steel centered cutting off | to 16000 | to 4875 | to 16000 | to 4875 |
| 9 | Cutting-off wheels—maximum 4-inch diameter (including reinforced organic) | … | … | 14200–20000 | 4330–6100 |
| | Cutting-off wheels—maximum 16-inch diameter (including reinforced organic) | … | … | 9500–16000 | 2895–4875 |
| | Cutting-off wheels—larger than 16-inch diameter (including reinforced organic) | … | … | 9500–14200 | 2895–4330 |
| | Non-reinforced, all diameters | … | … | 9500–14200 | 2895–4330 |
| 10 | Thread and flute grinding wheels | 8000–12000 | 2440–3660 | 10000–12000 | 3050–3660 |
| 11 | Crankshaft and camshaft grinding wheels | 8500–12000 | 2590–3660 | 6500–9500 | 1980–2895 |
| 12 | Snagging wheels 16-inch or larger (including reinforced organic)—used on swing frame grinders | … | … | 12500 | 3810 |
| | Snagging wheels 16-inch or larger (including reinforced organic)—used on semi-automatic snagging grinders | … | … | 16500 | 5030 |
| 13 | Internal wheels—Type 1 and 4, maximum diameter 6-inch | 5500–8500 | 1675–2590 | 6500–9500 | 1980–2895 |
| 14 | Mounted wheels | 10000 | 3050 | 10000 | 3050 |

[a] See Table 1a and Table 1b starting on pages 1272.

[b] Nonstandard shape. For snagging wheels, 16 inches and larger—Type 1, internal wheels—Types 1 and 5, and mounted wheels, see ANSI B7.1–2017. Under no conditions should a wheel be operated faster than the maximum operating speed established by the manufacturer.

Values in this table are for general information only.

**Table 2. Revolutions per Minute for Various Grinding Speeds and Wheel Diameters** *(Based on ANSI B7.1-2017)*

| Wheel Diameter, Inch | Peripheral (Surface) Speed, Feet per Minute | | | | | | | | | | | | | | | | | | Wheel Diameter, Inch |
|---|---|---|---|---|---|---|---|---|---|---|---|---|---|---|---|---|---|---|---|
| | 4000 | 4500 | 5000 | 5500 | 6000 | 6500 | 7000 | 7500 | 8000 | 8500 | 9000 | 9500 | 10000 | 12000 | 14200 | 16000 | 16500 | 17000 | |
| | Revolutions per Minute | | | | | | | | | | | | | | | | | | |
| 1 | 15279 | 17189 | 19099 | 21008 | 22918 | 24828 | 26738 | 28648 | 30558 | 32468 | 34377 | 36287 | 38197 | 45837 | 54240 | 61115 | 63025 | 64935 | 1 |
| 2 | 7639 | 8594 | 9549 | 10504 | 11459 | 12414 | 13369 | 14324 | 15279 | 16234 | 17189 | 18144 | 19099 | 22918 | 27120 | 30558 | 31513 | 32468 | 2 |
| 3 | 5093 | 5730 | 6366 | 7003 | 7639 | 8276 | 8913 | 9549 | 10186 | 10823 | 11459 | 12096 | 12732 | 15279 | 18080 | 20372 | 21008 | 21645 | 3 |
| 4 | 3820 | 4297 | 4775 | 5252 | 5730 | 6207 | 6685 | 7162 | 7639 | 8117 | 8594 | 9072 | 9549 | 11459 | 13560 | 15279 | 15756 | 16234 | 4 |
| 5 | 3056 | 3438 | 3820 | 4202 | 4584 | 4966 | 5348 | 5730 | 6112 | 6494 | 6875 | 7257 | 7639 | 9167 | 10848 | 12223 | 12605 | 12987 | 5 |
| 6 | 2546 | 2865 | 3183 | 3501 | 3820 | 4138 | 4456 | 4775 | 5093 | 5411 | 5730 | 6048 | 6366 | 7639 | 9040 | 10186 | 10504 | 10823 | 6 |
| 7 | 2183 | 2456 | 2728 | 3001 | 3274 | 3547 | 3820 | 4093 | 4365 | 4638 | 4911 | 5184 | 5457 | 6548 | 7749 | 8731 | 9004 | 9276 | 7 |
| 8 | 1910 | 2149 | 2387 | 2626 | 2865 | 3104 | 3342 | 3581 | 3820 | 4058 | 4297 | 4536 | 4775 | 5730 | 6780 | 7639 | 7878 | 8117 | 8 |
| 9 | 1698 | 1910 | 2122 | 2334 | 2546 | 2759 | 2971 | 3183 | 3395 | 3608 | 3820 | 4032 | 4244 | 5093 | 6027 | 6791 | 7003 | 7215 | 9 |
| 10 | 1528 | 1719 | 1910 | 2101 | 2292 | 2483 | 2674 | 2865 | 3056 | 3247 | 3438 | 3629 | 3820 | 4584 | 5424 | 6112 | 6303 | 6494 | 10 |
| 12 | 1273 | 1432 | 1592 | 1751 | 1910 | 2069 | 2228 | 2387 | 2546 | 2706 | 2865 | 3024 | 3183 | 3820 | 4520 | 5093 | 5252 | 5411 | 12 |
| 14 | 1091 | 1228 | 1364 | 1501 | 1637 | 1773 | 1910 | 2046 | 2183 | 2319 | 2456 | 2592 | 2728 | 3274 | 3874 | 4365 | 4502 | 4638 | 14 |
| 16 | 955 | 1074 | 1194 | 1313 | 1432 | 1552 | 1671 | 1790 | 1910 | 2029 | 2149 | 2268 | 2387 | 2865 | 3390 | 3820 | 3939 | 4058 | 16 |
| 18 | 849 | 955 | 1061 | 1167 | 1273 | 1379 | 1485 | 1592 | 1698 | 1804 | 1910 | 2016 | 2122 | 2546 | 3013 | 3395 | 3501 | 3608 | 18 |
| 20 | 764 | 859 | 955 | 1050 | 1146 | 1241 | 1337 | 1432 | 1528 | 1623 | 1719 | 1814 | 1910 | 2292 | 2712 | 3056 | 3151 | 3247 | 20 |
| 22 | 694 | 781 | 868 | 955 | 1042 | 1129 | 1215 | 1302 | 1389 | 1476 | 1563 | 1649 | 1736 | 2083 | 2465 | 2778 | 2865 | 2952 | 22 |
| 24 | 637 | 716 | 796 | 875 | 955 | 1035 | 1114 | 1194 | 1273 | 1353 | 1432 | 1512 | 1592 | 1910 | 2260 | 2546 | 2625 | 2706 | 24 |
| 26 | 588 | 661 | 735 | 808 | 881 | 955 | 1028 | 1102 | 1175 | 1249 | 1322 | 1396 | 1469 | 1763 | 2086 | 2351 | 2424 | 2498 | 26 |
| 28 | 546 | 614 | 682 | 750 | 819 | 887 | 955 | 1023 | 1091 | 1160 | 1228 | 1296 | 1364 | 1637 | 1937 | 2183 | 2251 | 2319 | 28 |
| 30 | 509 | 573 | 637 | 700 | 764 | 828 | 891 | 955 | 1019 | 1082 | 1146 | 1210 | 1273 | 1528 | 1808 | 2037 | 2101 | 2165 | 30 |
| 32 | 477 | 537 | 597 | 657 | 716 | 776 | 836 | 895 | 955 | 1015 | 1074 | 1134 | 1194 | 1432 | 1695 | 1910 | 1970 | 2029 | 32 |
| 34 | 449 | 506 | 562 | 618 | 674 | 730 | 786 | 843 | 899 | 955 | 1011 | 1067 | 1123 | 1348 | 1595 | 1798 | 1854 | 1910 | 34 |
| 36 | 424 | 477 | 531 | 584 | 637 | 690 | 743 | 796 | 849 | 902 | 955 | 1008 | 1061 | 1273 | 1507 | 1698 | 1751 | 1804 | 36 |
| 38 | 402 | 452 | 503 | 553 | 603 | 653 | 704 | 754 | 804 | 854 | 905 | 955 | 1005 | 1206 | 1427 | 1608 | 1659 | 1709 | 38 |
| 40 | 382 | 430 | 477 | 525 | 573 | 621 | 668 | 716 | 764 | 812 | 859 | 907 | 955 | 1146 | 1356 | 1528 | 1576 | 1623 | 40 |
| 42 | 364 | 409 | 455 | 500 | 546 | 591 | 637 | 682 | 728 | 773 | 819 | 864 | 909 | 1091 | 1291 | 1455 | 1500 | 1546 | 42 |
| 44 | 347 | 391 | 434 | 477 | 521 | 564 | 608 | 651 | 694 | 738 | 781 | 825 | 868 | 1042 | 1233 | 1389 | 1432 | 1476 | 44 |
| 46 | 332 | 374 | 415 | 457 | 498 | 540 | 581 | 623 | 664 | 706 | 747 | 789 | 830 | 996 | 1179 | 1329 | 1370 | 1412 | 46 |
| 48 | 318 | 358 | 398 | 438 | 477 | 517 | 557 | 597 | 637 | 676 | 716 | 756 | 796 | 955 | 1130 | 1273 | 1313 | 1353 | 48 |
| 53 | 288 | 324 | 360 | 396 | 432 | 468 | 504 | 541 | 577 | 613 | 649 | 685 | 721 | 865 | 1023 | 1153 | 1189 | 1225 | 53 |
| 60 | 255 | 286 | 318 | 350 | 382 | 414 | 446 | 477 | 509 | 541 | 573 | 605 | 637 | 764 | 904 | 1019 | 1050 | 1082 | 60 |
| 72 | 212 | 239 | 265 | 292 | 318 | 345 | 371 | 398 | 424 | 451 | 477 | 504 | 531 | 637 | 753 | 849 | 875 | 902 | 72 |

## Cylindrical Grinding

Cylindrical grinding designates a general category of various grinding methods that have the common characteristic of rotating the workpiece around a fixed axis while grinding outside surface sections in controlled relation to that axis of rotation.

The form of the part or section being ground in this process is frequently cylindrical, hence the designation of the general category. However, the shape of the part may be tapered or of curvilinear profile; the position of the ground surface may also be perpendicular to the axis; and it is possible to grind concurrently several surface sections, adjacent or separated, of equal or different diameters, located in parallel or mutually inclined planes, etc., as long as the condition of a common axis of rotation is satisfied.

*Size Range of Workpieces and Machines:* Cylindrical grinding is applied in the manufacture of miniature parts, such as instrument components and, at the opposite extreme, for grinding rolling mill rolls weighing several tons. Accordingly, there are cylindrical grinding machines of many different types, each adapted to a specific work-size range. Machine capacities are usually expressed by such factors as maximum work diameter, work length and weight, complemented, of course, by many other significant data.

**Plain, Universal, and Limited-Purpose Cylindrical Grinding Machines.—**The plain cylindrical grinding machine is considered the basic type of this general category, and is used for grinding parts with cylindrical or slightly tapered form.

The universal cylindrical grinder can be used, in addition to grinding the basic cylindrical forms, for the grinding of parts with steep tapers, of surfaces normal to the part axis, including the entire face of the workpiece, and for internal grinding independently or in conjunction with the grinding of the part's outer surfaces. Such variety of part configurations requiring grinding is typical of work in the tool room, which constitutes the major area of application for universal cylindrical grinding machines.

Limited-purpose cylindrical grinders are needed for special work configurations and for high-volume production, where productivity is more important than flexibility of adaptation. Examples of limited-purpose cylindrical grinding machines are crankshaft and camshaft grinders, polygonal grinding machines, roll grinders, etc.

**Traverse or Plunge Grinding.—**In traverse grinding, the machine table carrying the work performs a reciprocating movement of specific travel length for transporting the rotating workpiece along the face of the grinding wheel. At each or at alternate stroke ends, the wheel slide advances for the gradual feeding of the wheel into the work. The length of the surface that can be ground by this method is generally limited only by the stroke length of the machine table. In large roll grinders, the relative movement between work and wheel is accomplished by the traverse of the wheel slide along a stationary machine table.

In plunge grinding, the machine table, after having been set, is locked and, while the part is rotating, the wheel slide continually advances at a preset rate, until the finish size of the part is reached. The width of the grinding wheel is a limiting factor of the section length that can be ground in this process. Plunge grinding is required for profiled surfaces and for the simultaneous grinding of multiple surfaces of different diameters or located in different planes.

When the configuration of the part does not make use of either method mandatory, the choice may be made on the basis of the following general considerations: traverse grinding usually produces a better finish, and the productivity of plunge grinding is generally higher.

**Work Holding on Cylindrical Grinding Machines.—**The manner in which the work is located and held in the machine during the grinding process determines the configuration of the part that can be adapted for cylindrical grinding and affects the resulting accuracy of the ground surface. The method of work holding also affects the attainable production rate, because the mounting and dismounting of the part can represent a substantial portion of the total operating time.

Whatever method is used for holding the part on cylindrical types of grinding machines, two basic conditions must be satisfied: 1) the part should be located with respect to its correct axis of rotation; and 2) the work drive must cause the part to rotate, at a specific speed, around the established axis.

The lengthwise location of the part, although controlled, is not too critical in traverse grinding; however, in plunge grinding, particularly when shoulder sections are also involved, it must be assured with great accuracy.

Table 1 presents a listing, with brief discussions, of work-holding methods and devices that are most frequently used in cylindrical grinding.

**Table 1. Work-Holding Methods and Devices for Cylindrical Grinding**

| Designation | Description | Discussion |
|---|---|---|
| Centers, nonrotating ("dead"), with drive plate | Headstock with nonrotating spindle holds the center. Around the spindle, an independently supported sleeve carries the drive plate for rotating the work. Tailstock for opposite center. | The simplest method of holding the work between two opposite centers is also the potentially most accurate, as long as correctly prepared and located center holes are used in the work. |
| Centers, driving type | Word held between two centers obtains its rotation from the concurrently applied drive by the live headstock spindle and live tailstock spindle. | Eliminates the drawback of the common center-type grinding with driver plate, which requires a dog attached to the workpiece. Driven spindles permit the grinding of the work up to both ends. |
| Chuck, geared, or cam-actuated | Two, three, or four jaws moved radially through mechanical elements, hand-, or power-operated, exert concentrically acting clamping force on the workpiece. | Adaptable to workpieces of different configurations and within a generally wide capacity of the chuck. Flexible in uses that, however, do not include high-precision work. |
| Chuck, diaphragm | Force applied by hand or power of a flexible diaphragm causes the attached jaws to deflect temporarily for accepting the work, which is held when force is released. | Rapid action and flexible adaptation to different work configurations by means of special jaws offer varied uses for the grinding of disk-shaped and similar parts. |
| Collets | Holding devices with externally or internally acting clamping force, easily adaptable to power actuation, assuring high centering accuracy. | Limited to parts with previously machined or ground holding surfaces, because of the small range of clamping movement of the collet jaws. |
| Faceplate | Has four independently actuated jaws, any or several of which may be used, or entirely removed, using the base plate for supporting special clamps. | Used for holding bulky parts, or those of awkward shape, which are ground in small quantities not warranting special fixtures. |
| Magnetic plate | Flat plates, with pole distribution adapted to the work, are mounted on the spindle like chucks and may be used for work with the locating face normal to the axis. | Applicable for light cuts such as are frequent in tool making, where the rapid clamping action and easy access to both the outside diameter and the exposed face are sometimes of advantage. |
| Steady rests | Two basic types are used: (a) the two-jaw type supporting the work from the back (back rest), leaving access by the wheel; (b) the three-jaw type (center rest). | A complementary work-holding device, used in conjunction with primary work holders, to provide additional support, particularly to long and/or slender parts. |
| Special fixtures | Single-purpose devices, designed for a particular workpiece, primarily for providing special locating elements. | Typical workpieces requiring special fixturing are, as examples, crankshafts where the holding is combined with balancing functions; or internal gears located on the pitch circle of the teeth for outside diameter grinding. |

**Selection of Grinding Wheels for Cylindrical Grinding.**—For cylindrical grinding, as for grinding in general, the primary factor to be considered in wheel selection is the work material. Other factors are the amount of excess stock and its rate of removal (feeds and speeds), the desired accuracy and surface finish, the ratio of wheel and work diameter, wet or dry grinding, etc. In view of these many variables, it is not practical to set up a complete list of grinding wheel recommendations with general validity. Instead, examples of recommendations embracing a wide range of typical applications and assuming common practices are presented in Table 2. This is intended as a guide for the starting selection of grinding wheel specifications which, in case of a not entirely satisfactory performance, can be refined subsequently. The content of the table is a version of the grinding wheel recommendations for cylindrical grinding by the Norton Company using, however, non-proprietary designations for the abrasive types and bonds.

**Table 2. Wheel Recommendations for Cylindrical Grinding**

| Material | Wheel Marking | Material | Wheel Marking |
|---|---|---|---|
| Aluminum | SFA46-18V | Forgings | A46-M5V |
| Armatures (laminated) | SFA100-18V | Gages (plug) | SFA80-K8V |
| Axles (auto & railway) | A54-M5V | General-purpose grinding | SFA54-L5V |
| Brass | C36-KV | Glass | BFA220-011V |
| Bronze | | Gun barrels | |
| Soft | C36-KV | Spotting and outside diameter | BFA60-M5V |
| Hard | A46-M5V | Nitralloy | |
| Bushings (hardened steel) | BFA60-L5V | Before nitriding | A60-K5V |
| Bushings (cast iron) | C36-JV | After nitriding | |
| Cam lobes (cast alloy) | | Commercial finish | SFA60-18V |
| Roughing | BFA54-N5V | High finish | C100-1V |
| Finishing | A70-P6B | Reflective finish | C500-19E |
| Cam lobes (hardened steel) | | Pistons (aluminum) | SFA46-18V |
| Roughing | BFA54-L5V | (cast iron) | C36-KV |
| Finishing | BFA80-T8B | Plastics | C46-JV |
| Cast iron | C36-JV | Rubber | |
| Chromium plating | | Soft | SFA20-K5B |
| Commercial finish | SFA60-J8V | Hard | C36-KB |
| High finish | A150-K5E | Spline shafts | SFA60-N5V |
| Reflective finish | C500-I9E | Sprayed metal | C60-JV |
| Commutators (copper) | C60-M4E | Steel | |
| Crankshafts (airplane) | | Soft | |
| Pins | BFA46-K5V | 1 in. diameter and smaller | SFA60-M5V |
| Bearings | A46-L5V | over 1 in. diameter | SFA46-L5V |
| Crankshafts (automotive | | Hardened | |
| pins and bearings) | | 1 in. diameter and smaller | SFA80-L8V |
| Finishing | A54-N5V | over 1 in. diameter | SFA60-K5V |
| Roughing & finishing | A54-O5V | 300 series stainless | SFA46-K8V |
| Regrinding | A54-M5V | Stellite | BFA46-M5V |
| Regrinding, sprayed | | Titanium | C60-JV |
| metal | C60-JV | Valve stems (automative) | BFA54-N5V |
| Drills | BFA54-N5V | Valve tappets | BFA54-M5V |

*Note:* Prefixes to the standard designation "A" of aluminum oxide indicate modified abrasives as follows: BFA = Blended friable (a blend of regular and friable), SFA = Semifriable.

**Operational Data for Cylindrical Grinding.**—In cylindrical grinding, similarly to other metalcutting processes, the applied speed and feed rates must be adjusted to the operational conditions as well as to the objectives of the process. Grinding differs, however, from other types of metalcutting methods in regard to the cutting speed of the tool which, in grinding, is generally not a variable; it should be maintained at, or close to the optimum rate, commonly 6500 feet per minute (1980 m/min) peripheral speed.

In establishing the proper process values for grinding, of prime consideration are the work material, its condition (hardened or soft), and the type of operation (roughing or finishing). Other influencing factors are the characteristics of the grinding machine (stability, power), the specifications of the grinding wheel, the material allowance, the rigidity and balance of the workpiece, as well as several grinding process conditions, such as wet or dry grinding, the manner of wheel truing, etc.

Variables of the cylindrical grinding process, often referred to as *grinding data*, comprise the speed of work rotation (measured as the surface speed of the work); the infeed (in inches per pass for traverse grinding, or in inches per minute for plunge grinding); and, in the case of traverse grinding, the speed of the reciprocating table movement (expressed either in feet per minute, or as a fraction of the wheel width for each revolution of the work).

For the purpose of starting values in setting up a cylindrical grinding process, a brief listing of basic data for common cylindrical grinding conditions and involving frequently used materials, is presented in Table 3.

**Table 3. Basic Process Data for Cylindrical Grinding**

| Traverse Grinding | | | | | | | | Traverse for Each Work Revolution, In Fractions of the Wheel Width | |
|---|---|---|---|---|---|---|---|---|---|
| | | Work Surface Speed, | | Infeed per Pass | | | | | |
| | | | | inch | | mm | | | |
| Work Material | Material Condition | fpm | m/min | Rough | Finish | Rough | Finish | Rough | Finish |
| Plain Carbon Steel | Annealed | 100 | 30.5 | 0.002 | 0.0005 | 0.05 | 0.013 | ½ | ⅙ |
| | Hardened | 70 | 21 | 0.002 | 0.0003–0.0005 | 0.05 | 0.007–0.013 | ¼ | ⅛ |
| Alloy Steel | Annealed | 100 | 30.5 | 0.002 | 0.0005 | 0.05 | 0.013 | ½ | ⅙ |
| | Hardened | 70 | 21 | 0.002 | 0.0002–0.0005 | 0.05 | 0.005–0.013 | ¼ | ⅛ |
| Tool Steel | Annealed | 60 | 18 | 0.002 | 0.0005 max. | 0.05 | 0.013 max. | ½ | ⅙ |
| | Hardened | 50 | 15 | 0.002 | 0.0001–0.0005 | 0.05 | 0.003–0.013 | ¼ | ⅛ |
| Copper Alloys | Annealed or Cold Drawn | 100 | 30.5 | 0.002 | 0.0005 max. | 0.05 | 0.013 max. | ⅓ | ⅙ |
| Aluminum Alloys | Cold Drawn or Solution Treated | 150 | 45.7 | 0.002 | 0.0005 max. | 0.05 | 0.013 max. | ⅓ | ⅙ |

| Plunge Grinding, Infeed per Revolution of the Work | | | | |
|---|---|---|---|---|
| Work Material | Inch | | Millimeter | |
| | Roughing | Finishing | Roughing | Finishing |
| Steel, soft | 0.0005 | 0.0002 | 0.013 | 0.005 |
| Plain carbon steel, hardened | 0.0002 | 0.000050 | 0.005 | 0.0013 |
| Alloy and tool steel, hardened | 0.0001 | 0.000025 | 0.003 | 0.00064 |

These data, which are, in general, considered conservative, are based on average operating conditions and may be modified subsequently by: a) reducing the values in case of unsatisfactory quality of the grinding or the occurrence of failures; and b) increasing the rates for raising the productivity of the process, particularly for rigid workpieces, substantial stock allowance, etc.

**High-Speed Cylindrical Grinding.—**The maximum peripheral speed of the wheels in regular cylindrical grinding is generally 6500 feet per minute (1980 m/min); the commonly used grinding wheels and machines are designed to operate efficiently at this speed. Recently, efforts were made to raise the productivity of different grinding methods, including cylindrical grinding, by increasing the peripheral speed of the grinding wheel to a substantially higher than traditional level, such as 12,000 feet per minute (3658 m/min) or more. Such methods are designated by the distinguishing term of high-speed grinding.

For high-speed grinding, special grinding machines have been built with high dynamic stiffness and static rigidity, equipped with powerful drive motors, extra-strong spindles and bearings, reinforced wheel guards, etc., and using grinding wheels expressly made and tested for operating at high peripheral speeds. The higher stock-removal rate accomplished by high-speed grinding represents an advantage when the work configuration and material permit, and the removable stock allowance warrants its application.

*CAUTION:* High-speed grinding must *not* be applied on standard types of equipment, such as general types of grinding machines and regular grinding wheels. Operating grinding wheels, even temporarily, at higher than approved speed constitutes a grave safety hazard.

**Areas and Degrees of Automation in Cylindrical Grinding.**—Power drive for the work rotation and for the reciprocating table traverse are fundamental machine movements that, once set for a certain rate, will function without requiring additional attention. Loading and removing the work, starting and stopping the main movements, and applying infeed by hand wheel are carried out by the operator on cylindrical grinding machines in their basic degree of mechanization. Such equipment is still frequently used in tool room and jobbing-type work.

More advanced levels of automation have been developed for cylindrical grinders and are being applied in different degrees, particularly in the following principal respects:

a) *Infeed,* in which different rates are provided for rapid approach, roughing and finishing, followed by a spark-out period, with presetting of the advance rates, the cutoff points, and the duration of time-related functions.

b) *Automatic cycling* actuated by a single lever to start work rotation, table reciprocation, grinding-fluid supply, and infeed, followed at the end of the operation by wheel slide retraction, the successive stopping of the table movement, the work rotation, and the fluid supply.

c) *Table traverse dwells* (tarry) in the extreme positions of the travel, over preset periods, to assure uniform exposure to the wheel contact of the entire work section.

d) *Mechanized work loading,* clamping, and, after termination of the operation, unloading, combined with appropriate work-feeding devices such as indexing-type drums.

e) *Size control* by in-process or post-process measurements. Signals originated by the gage will control the advance movement or cause automatic compensation of size variations by adjusting the cutoff points of the infeed.

f) *Automatic wheel dressing* at preset frequency, combined with appropriate compensation in the infeed movement.

g) *Numerical control* obviates the time-consuming setups for repetitive work performed on small- or medium-size lots. As an application example: shafts with several sections of different lengths and diameters can be ground automatically in a single operation, grinding the sections in consecutive order to close dimensional limits, controlled by an in-process gage, which is also automatically set by means of the program.

The choice of the grinding machine functions to be automated and the extent of automation will generally be guided by economic considerations, after a thorough review of the available standard and optional equipment. Numerical control of partial or complete cycles is being applied to modern cylindrical and other grinding machines.

**Cylindrical Grinding Troubles and Their Correction.**—Troubles that may be encountered in cylindrical grinding may be classified as work defects (chatter, checking, burning, scratching, and inaccuracies), improperly operating machines (jumpy infeed or traverse), and wheel defects (too hard or soft action, loading, glazing, and breakage). The Landis Tool Company has listed some of these troubles, their causes, and corrections as follows:

*Chatter:* Sources of chatter include: 1) faulty coolant; 2) wheel out of balance; 3) wheel out of round; 4) wheel too hard; 5) improper dressing; 6) faulty work support or rotation; 7) improper operation; 8) faulty traverse; 9) work vibration; 10) outside vibration transmitted to machine; 11) interference; 12) wheel base; and 13) headstock.

Suggested procedures for correction of these troubles are:

1) *Faulty coolant:* Clean tanks and lines. Replace dirty or heavy coolant with correct mixture.

2) *Wheel out of balance:* Rebalance on mounting before and after dressing. Run wheel without coolant to remove excess water. Store a removed wheel on its side to keep retained water from causing a false heavy side. Tighten wheel mounting flanges. Make sure wheel center fits spindle.

3) *Wheel out of round:* True before and after balancing. True sides to face.

4) *Wheel too hard:* Use coarser grit, softer grade, more open bond. See *Wheel Defects* on page 1310.

5) *Improper dressing:* Use sharp diamond and hold rigidly close to wheel. It must not overhang excessively. Check diamond in mounting.

6) *Faulty work support or rotation:* Use sufficient number of work rests and adjust them more carefully. Use proper angles in centers of work. Clean dirt from footstock spindle and be sure spindle is tight. Make certain that work centers fit properly in spindles.

7) *Improper operation:* Reduce rate of wheel feed.

8) *Faulty traverse:* See *Uneven Traverse or Infeed of Wheel Head* on page 1310.

9) *Work vibration:* Reduce work speed. Check workpiece for balance.

10) *Outside vibration transmitted to machine:* Check and make sure that machine is level and sitting solidly on foundation. Isolate machine or foundation.

11) *Interference:* Check all guards for clearance.

12) *Wheel base:* Check spindle bearing clearance. Use belts of equal lengths or uniform cross section on motor drive. Check drive motor for unbalance. Check balance and fit of pulleys. Check wheel feed mechanism to see that all parts are tight.

13) *Headstock:* Put belts of same length and cross section on motor drive; check for correct work speeds. Check drive motor for unbalance. Make certain that headstock spindle is not loose. Check work center fit in spindle. Check wear of faceplate and jackshaft bearings.

*Spirals on Work (traverse lines with same lead on work as rate of traverse):* Sources of spirals include: 1) machine parts out of line; and 2) truing.

Suggested procedures for correction of these troubles are:

1) *Machine parts out of line:* Check wheel base, headstock, and footstock for proper alignment.

2) *Truing:* Point truing tool down 3 degrees at the workwheel contact line. Round off wheel edges.

*Check Marks on Work:* Sources of check marks include: 1) improper operation; 2) improper heat treatment; 3) improper size control; 4) improper wheel; and 5) improper dressing.

Suggested procedures for correction of these troubles are:

1) *Improper operation:* Make wheel act softer. See *Wheel Defects*. Do not force wheel into work. Use greater volume of coolant and a more even flow. Check the correct positioning of coolant nozzles to direct a copious flow of clean coolant at the proper location.

2) *Improper heat treatment:* Take corrective measures in heat-treating operations.

3) *Improper size control:* Make sure that engineering establishes reasonable size limits. See that they are maintained.

4) *Improper wheel:* Make wheel act softer. Use softer-grade wheel. Review the grain size and type of abrasive. A finer grit or more friable abrasive or both may be called for.

5) *Improper dressing:* Check that the diamond is sharp, of good quality, and well set. Increase speed of the dressing cycle. Make sure diamond is not cracked.

*Burning and Discoloration of Work:* Sources of burning and discoloration are improper operation, and improper wheel.

Suggested procedures for correction of these troubles are:

1) *Improper operation:* Decrease rate of infeed. Don't stop work while in contact with wheel.

2) *Improper wheel:* Use softer wheel or obtain softer effect. See *Wheel Defects*. Use greater volume of coolant.

*Isolated Deep Marks on Work:* Source of trouble is an unsuitable wheel. Use a finer wheel and consider a change in abrasive type.

*Fine Spiral or Thread on Work:* Sources of this trouble are: 1) improper operation; and 2) faulty wheel dressing.

Suggested procedures for corrections of these troubles are:

1) *Improper operation:* Reduce wheel pressure. Use more work rests. Reduce traverse with respect to work rotation. Use different traverse rates to break up pattern when making numerous passes. Prevent edge of wheel from penetrating by dressing wheel face parallel to work.

2) *Faulty wheel dressing:* Use slower or more even dressing traverse. Set dressing tool at least 3 degrees down and 30 degrees to the side from time to time. Tighten holder. Don't take too deep a cut. Round off wheel edges. Start dressing cut from wheel edge.

*Narrow and Deep Regular Marks on Work:* Source of trouble is that the wheel is too coarse. Use finer grain size.

*Wide, Irregular Marks of Varying Depth on Work:* Source of trouble is too soft a wheel. Use a harder grade wheel. See *Wheel Defects*.

*Widely Spaced Spots on Work:* Sources of trouble are oil spots or glazed areas on wheel face. Balance and true wheel. Keep oil from wheel face.

*Irregular "Fish-tail" Marks of Various Lengths and Widths on Work:* Source of trouble is dirty coolant. Clean tank frequently. Use filter for fine finish grinding. Flush wheel guards after dressing or when changing to finer wheel.

*Wavy Traverse Lines on Work:* Source of trouble is wheel edges. Round off. Check for loose thrust on spindle and correct if necessary.

*Irregular Marks on Work:* Cause is loose dirt. Keep machine clean.

*Deep, Irregular Marks on Work:* Source of trouble is loose wheel flanges. Tighten and make sure blotters are used.

*Isolated Deep Marks on Work:* Sources of trouble are: 1) grains pull out; 2) coolant too strong; coarse grains or foreign matter in wheel face; and 3) improper dressing.

Respective suggested procedures for corrections of these troubles are: 1) decrease soda content in coolant mixture; 2) dress wheel; and 3) use sharper dressing tool.

Brush wheel after dressing with stiff bristle brush.

*Grain Marks on Work:* Sources of trouble are: 1) improper finishing cut; 2) grain sizes of roughing and finishing wheels differ too much; 3) dressing too coarse; and 4) wheel too coarse or too soft.

Respective suggested procedures for corrections of these troubles are: start with high work and traverse speeds; finish with high work speed and slow traverse, letting wheel "spark-out" completely; finish out better with roughing wheel or use finer roughing wheel; use shallower and slower cut; and use finer grain size or harder-grade wheel.

*Inaccuracies in Work:* Work out-of-round, out-of-parallel, or tapered.

Sources of trouble are: 1) misalignment of machine parts; 2) work centers; 3) improper operation; 4) coolant; 5) wheel; 6) improper dressing; 7) spindle bearings; and 8) work.

Suggested procedures for corrections of these troubles are:

1) *Misalignment of machine parts:* Check headstock and tailstock for alignment and proper clamping.

2) *Work centers:* Centers in work must be deep enough to clear center point. Keep work centers clean and lubricated. Check play of footstock spindle and see that footstock spindle is clean and tightly seated. Regrind work centers if worn. Work centers must fit taper of work-center holes. Footstock must be checked for proper tension.

3) *Improper operation:* Don't let wheel traverse beyond end of work. Decrease wheel pressure so work won't spring. Use harder wheel or change speeds and feeds to make wheel

act harder. Allow work to "spark-out." Decrease feed rate. Use proper number of work rests. Allow proper amount of tarry. Workpiece must be balanced if it is an odd shape.

4) *Coolant:* Use greater volume of coolant.

5) *Wheel:* Rebalance wheel on mounting before and after truing.

6) *Improper dressing:* Use same positions and machine conditions for dressing as in grinding.

7) *Spindle bearings:* Check clearance.

8) *Work:* Work must come to machine in reasonably accurate form.

*Inaccurate Work Sizing (when wheel is fed to same position, it grinds one piece to correct size, another oversize, and still another undersize):* Sources of trouble are: 1) improper work support or rotation; 2) wheel out of balance; 3) loaded wheel; 4) improper infeed; 5) improper traverse; 6) coolant; 7) misalignment; and 8) work.

Suggested procedures for corrections of these troubles are:

1) *Improper work support or rotation:* Keep work centers clean and lubricated. Regrind work-center tips to proper angle. Be sure footstock spindle is tight. Use sufficient work rests, properly spaced.

2) *Wheel out of balance:* Balance wheel on mounting before and after truing.

3) *Loaded wheel:* See *Wheel Defects*.

4) *Improper infeed:* Check forward stops of rapid feed and slow feed. When readjusting position of wheel base by means of the fine feed, move the wheel base back after making the adjustment and then bring it forward again to take up backlash and relieve strain in feed-up parts. Check wheel spindle bearings. Don't let excessive lubrication of wheel base slide cause "floating." Check and tighten wheel feed mechanism. Check parts for wear. Check pressure in hydraulic system. Set infeed cushion properly. Check to see that pistons are not sticking.

5) *Improper traverse:* Check traverse hydraulic system and the operating pressure. Prevent excessive lubrication of carriage ways with resultant "floating" condition. Check to see if carriage traverse piston rods are binding. Carriage rack and driving gear must not bind. Change length of tarry period.

6) *Coolant:* Use greater volume of clean coolant.

7) *Misalignment:* Check level and alignment of machine.

8) *Work:* Workpieces may vary too much in length, permitting uneven center pressure.

*Uneven Traverse or Infeed of Wheel Head:* Sources of uneven traverse or infeed of wheel head are: carriage and wheel head, hydraulic system, interference, unbalanced conditions, and wheel out of balance. Suggested procedures for correction of these troubles are:

1) *Carriage and wheel head:* Ways may be scored. Be sure to use recommended oil for both lubrication and hydraulic system. Make sure ways are not so smooth that they press out oil film. Check lubrication of ways. Check wheel feed mechanism, traverse gear, and carriage rack clearance. Prevent binding of carriage traverse cylinder rods.

2) *Hydraulic systems:* Remove air and check pressure of hydraulic oil. Check pistons and valves for oil leakage and for gumminess caused by incorrect oil. Check worn valves or pistons that permit leakage.

3) *Interference:* Make sure guard strips do not interfere.

4) *Unbalanced conditions:* Eliminate loose pulleys, unbalanced wheel drive motor, uneven belts, or high spindle keys.

5) *Wheel out of balance:* Balance wheel on mounting before and after truing.

*Wheel Defects:* When *wheel is acting too hard,* such defects as glazing, some loading, lack of cut, chatter, and burning of work result.

Suggested procedures for correction of these faults are: 1) Increase work and traverse speeds as well as rate of in-feed; 2) decrease wheel speed, diameter, or width; 3) dress more sharply; 4) use thinner coolant; 5) don't tarry at end of traverse; 6) select softer wheel grade and coarser grain size; 7) avoid gummy coolant; and 8) on hardened work select finer grit, more fragile abrasive or both to get penetration. Use softer grade.

When *wheel is acting too soft,* such defects as wheel marks, tapered work, short wheel life, and not-holding-cut result.

Suggested procedures for correction of these faults are: 1) Decrease work and traverse speeds as well as rate of in-feed; 2) increase wheel speed, diameter, or width; 3) dress with little in-feed and slow traverse; 4) use heavier coolants; 5) don't let wheel run off work at end of traverse; and 6) select harder wheel or less fragile grain or both.

*Wheel Loading and Glazing:* Sources of the trouble of wheel loading or glazing are: 1) Incorrect wheel; 2) improper dress; 3) faulty operation; 4) faulty coolant; and 5) gummy coolant.

Suggested procedures for correction of these faults are:

1) *Incorrect wheel:* Use coarser grain size, more open bond, or softer grade.

2) *Improper dressing:* Keep wheel sharp with sharp dresser, clean wheel after dressing, use faster dressing traverse, and deeper dressing cut.

3) *Faulty operation:* Control feeds and speeds to soften action of wheel. Use less in-feed to prevent loading; more in-feed to stop glazing.

4) *Faulty coolant:* Use more, cleaner and thinner coolant, and less oily coolant.

5) *Gummy coolant:* To stop wheel glazing, increase soda content and avoid the use of soluble oils if water is hard. In using soluble oil coolant with hard water a suitable conditioner or "softener" should be added.

*Wheel Breakage:* Suggested procedures for the correction of a radial break with three or more pieces are: 1) Reduce wheel speed to or below rated speed; 2) mount wheel properly, use blotters, tight arbors, even flange pressure and be sure to keep out dirt between flange and wheel; 3) use plenty of coolant to prevent over-heating; 4) use less in-feed; and 5) don't allow wheel to become jammed on work.

A radial break with two pieces may be caused by excessive side strain. To prevent an irregular wheel break, don't let wheel become jammed on work; don't allow striking of wheel; and never use wheels that have been damaged in handling. In general, do not use a wheel that is too tight on the arbor since the wheel is apt to break when started. Prevent excessive hammering action of wheel. Follow rules of the American National Standard Safety Requirements for the Use, Care, and Protection of Abrasive Wheels (ANSI B7.1-2017).

## Centerless Grinding

In centerless grinding the work is supported on a work rest blade and is between the grinding wheel and a regulating wheel. The regulating wheel generally is a rubber bonded abrasive wheel. In the normal grinding position the grinding wheel forces the work downward against the work rest blade and also against the regulating wheel. The latter imparts a uniform rotation to the work giving it its same peripheral speed which is adjustable.

The higher the work center is placed above the line joining the centers of the grinding and regulating wheels the quicker the rounding action. Rounding action is also increased by a high work speed and a slow rate of traverse (if a through-feed operation). It is possible to have a higher work center when using softer wheels, as their use gives decreased contact pressures and the tendency of the workpiece to lift off the work rest blade is lessened.

Long rods or bars are sometimes ground with their centers below the line-of-centers of the wheels to eliminate the whipping and chattering due to slight bends or kinks in the rods or bars, as they are held more firmly down on the blade by the wheels.

There are three general methods of centerless grinding which may be described as through-feed, in-feed, and end-feed methods.

**Through-Feed Method of Grinding.—**The through-feed method is applied to straight cylindrical parts. The work is given an axial movement by the regulating wheel and passes between the grinding and regulating wheels from one side to the other. The rate of feed depends upon the diameter and speed of the regulating wheel and its inclination which is adjustable. It may be necessary to pass the work between the wheels more than once, the number of passes depending upon such factors as the amount of stock to be removed, the roundness and straightness of the unground work, and the limits of accuracy required.

The work rest fixture also contains adjustable guides on either side of the wheels that directs the work to and from the wheels in a straight line.

**In-Feed Method of Centerless Grinding.—**When parts have shoulders, heads or some part larger than the ground diameter, the in-feed method usually is employed. This method is similar to "plungecut" form grinding on a center type of grinder. The length or sections to be ground in any one operation are limited by the width of the wheel. As there is no axial feeding movement, the regulating wheel is set with its axis approximately parallel to that of the grinding wheel, there being a slight inclination to keep the work tight against the end stop.

**End-Feed Method of Grinding.—**The end-feed method is applied only to taper work. The grinding wheel, regulating wheel, and the work rest blade are set in a fixed relation to each other and the work is fed in from the front mechanically or manually to a fixed end stop. Either the grinding or regulating wheel, or both, are dressed to the proper taper.

**Automatic Centerless Grinding.—**The grinding of relatively small parts may be done automatically by equipping the machine with a magazine, gravity chute, or hopper feed, provided the shape of the part will permit using these feed mechanisms.

**Internal Centerless Grinding.—**Internal grinding machines based upon the centerless principle utilize the outside diameter of the work as a guide for grinding the bore which is concentric with the outer surface. In addition to straight and tapered bores, interrupted and "blind" holes can be ground by the centerless method. When two or more grinding operations such as roughing and finishing must be performed on the same part, the work can be rechucked in the same location as often as required.

**Centerless Grinding Troubleshooting.—**A number of troubles and some corrective measures compiled by a manufacturer are listed here for the through-feed and in-feed methods of centerless grinding.

*Chattermarks* are caused by having the work center too high above the line joining the centers of the grinding and regulating wheels; using too hard or too fine a grinding wheel; using too steep an angle on the work support blade; using too thin a work support blade; "play" in the set-up due to loosely clamped members; having the grinding wheel fit loosely on the spindle; having vibration either transmitted to the machine or caused by a defective drive in the machine; having the grinding wheel out-of-balance; using too heavy a stock removal; and having the grinding wheel or the regulating wheel spindles not properly adjusted.

*Feed lines or spiral marks* in through-feed grinding are caused by too sharp a corner on the exit side of the grinding wheel which may be alleviated by dressing the grinding wheel to a slight taper about $\frac{1}{2}$ inch (1.27 mm) from the edge, dressing the edge to a slight radius, or swiveling the regulating wheel a bit.

*Scored work* is caused by burrs, abrasive grains, or removed material being imbedded in or fused to the work support blade. This condition may be alleviated by using a coolant with increased lubricating properties and if this does not help a softer grade wheel should be used.

*Work not ground round* may be due to the work center not being high enough above the line joining the centers of the grinding and regulating wheels. Placing the work center higher and using a softer grade wheel should help to alleviate this condition.

*Work not ground straight* in through-feed grinding may be due to an incorrect setting of the guides used in introducing and removing the work from the wheels, and the existence of convex or concave faces on the regulating wheel. For example, if the work is tapered on the front end, the work guide on the entering side is deflected toward the regulating wheel. If tapered on the back end, then the work guide on the exit side is deflected toward the regulating wheel. If both ends are tapered, then both work guides are deflected toward the regulating wheel. The same barrel-shaped pieces are also obtained if the face of the regulating wheel is convex at the line of contact with the work. Conversely, the work would be ground with hollow shapes if the work guides were deflected toward the grinding wheel or if the face of the regulating wheel were concave at the line of contact with the work. The use of a warped work rest blade may also result in the work not being ground straight and the blade should be removed and checked with a straight edge.

In in-feed grinding, in order to keep the wheel faces straight which will ensure straightness of the cylindrical pieces being ground, the first item to be checked is the straightness and the angle of inclination of the work rest blade. If this is satisfactory then one of three corrective measures may be taken: the first might be to swivel the regulating wheel to compensate for the taper, the second might be to true the grinding wheel to that angle that will give a perfectly straight workpiece, and the third might be to change the inclination of the regulating wheel (this is true only for correcting very slight tapers up to 0.0005 inch or 0.0127 mm).

*Difficulties in sizing* the work in in-feed grinding are generally due to a worn in-feed mechanism and may be overcome by adjusting the in-feed nut.

*Flat spots* on the workpiece in in-feed grinding usually occur when grinding heavy work and generally when the stock removal is light. This condition is due to insufficient driving power between the work and the regulating wheel which may be alleviated by equipping the work rest with a roller that exerts a force against the workpiece; and by feeding the workpiece to the end stop using the upper slide.

## Surface Grinding

The term surface grinding implies, in current technical usage, the grinding of surfaces which are essentially flat. Several methods of surface grinding, however, are adapted and used to produce surfaces characterized by parallel straight line elements in one direction, while normal to that direction the contour of the surface may consist of several straight line sections at different angles to each other (e.g., the guideways of a lathe bed); in other cases the contour may be curved or profiled (e.g., a thread cutting chaser).

**Advantages of Surface Grinding.—**Alternate methods for machining work surfaces similar to those produced by surface grinding are milling and, to a much more limited degree, planing. Surface grinding, however, has several advantages over alternate methods that are carried out with metal-cutting tools. Examples of such potential advantages are as follows:

1) Grinding is applicable to very hard and/or abrasive work materials, without significant effect on the efficiency of the stock removal.

2) The desired form and dimensional accuracy of the work surface can be obtained to a much higher degree and in a more consistent manner.

3) Surface textures of very high finish and—when the appropriate system is utilized—with the required lay, are generally produced.

4) Tooling for surface grinding as a rule is substantially less expensive, particularly for producing profiled surfaces, the shapes of which may be dressed into the wheel, often with simple devices, in processes that are much more economical than the making and the maintenance of form cutters.

5) Fixturing for work holding is generally very simple in surface grinding, particularly when magnetic chucks are applicable, although the mechanical holding fixture can also be simpler, because of the smaller clamping force required than in milling or planing.

6) Parallel surfaces on opposite sides of the work are produced accurately, either in consecutive operations using the first ground surface as a dependable reference plane or, simultaneously, in double face grinding, which usually operates without the need for holding the parts by clamping.

7) Surface grinding is well adapted to process automation, particularly for size control, but also for mechanized work handling in the large volume production of a wide range of component parts.

**Principal Systems of Surface Grinding.—**Flat surfaces can be ground with different surface portions of the wheel, by different arrangements of the work and wheel, as well as by different interrelated movements. The various systems of surface grinding, with their respective capabilities, can best be reviewed by considering two major distinguishing characteristics:

1) *The operating surface of the grinding wheel,* which may be the periphery or the face (the side);

2) *The movement of the work during the process,* which may be traverse (generally reciprocating) or rotary (continuous), depending on the design of a particular category of surface grinders.

The accompanying Table 1 and the text that follows provides a concise review of the principal surface grinding systems, defined by the preceding characteristics. It should be noted that many surface grinders are built for specific applications, and do not fit exactly into any one of these major categories.

*Operating Surface, Periphery of Wheel: Movement of Work, Reciprocating:* Work is mounted on the horizontal machine table that is traversed in a reciprocating movement at a speed generally selected from a steplessly variable range. The transverse movement, called cross feed of the table or of the wheel slide, operates at the end of the reciprocating stroke and assures the gradual exposure of the entire work surface, which commonly exceeds the width of the wheel. The depth of the cut is controlled by the downfeed of the wheel, applied in increments at the reversal of the transverse movement.

*Operating Surface, Periphery of Wheel: Movement of Work, Rotary:* Work is mounted, usually on the full-diameter magnetic chuck of the circular machine table that rotates at a preset constant or automatically varying speed, the latter maintaining an approximately equal peripheral speed of the work surface area being ground. The wheelhead, installed on a cross slide, traverses over the table along a radial path, moving in alternating directions, toward and away from the center of the table. Infeed is by vertical movement of the saddle along the guideways of the vertical column, at the end of the radial wheelhead stroke. The saddle contains the guideways along which the wheelhead slide reciprocates.

*Operating Surface, Face of Wheel: Movement of Work, Reciprocating:* Operation is similar to the reciprocating table-type peripheral surface grinder, but grinding is with the face, usually with the rim of a cup-shaped wheel, or a segmental wheel for large machines. Capable of covering a much wider area of the work surface than the peripheral grinder, thus frequently no need for cross feed. Provides efficient stock removal, but is less adaptable than the reciprocating table-type peripheral grinder.

*Operating Surface, Face of Wheel: Movement of Work, Rotary:* The grinding wheel, usually of segmental type, is set in a position to cover either an annular area near the periphery of the table or, more commonly, to reach beyond the table center. A large circular magnetic chuck generally covers the entire table surface and facilitates the mounting of workpieces, even of fixtures, when needed. The uninterrupted passage of the work in contact with the large wheel face permits a very high rate of stock removal and the machine, with single or double wheelhead, can be adapted also to automatic operation with continuous part feed by mechanized work handling.

**Table 1. Principal Systems of Surface Grinding — Diagrams**

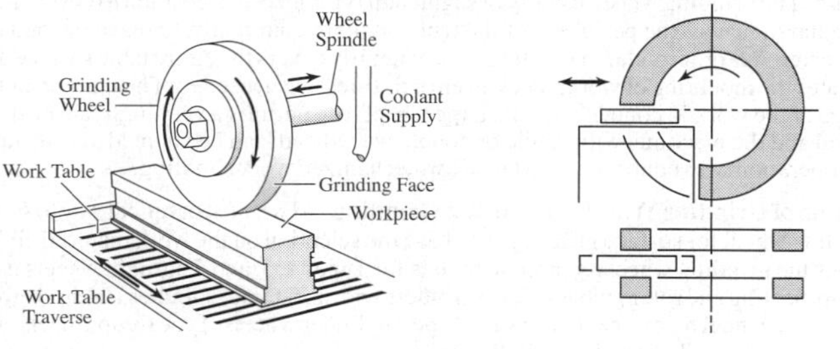


Reciprocating — Periphery of Wheel

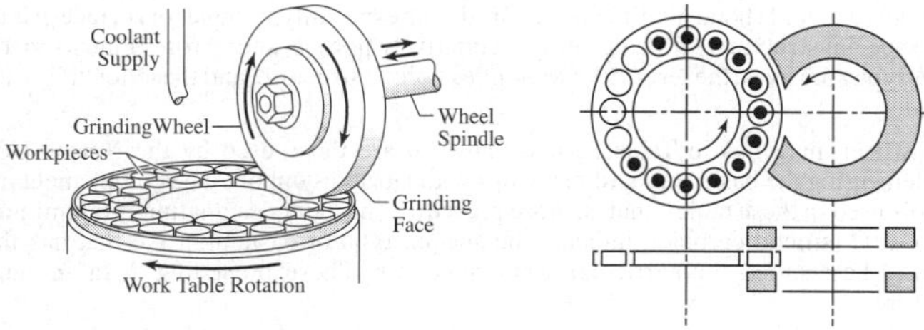


Rotary — Periphery of Wheel

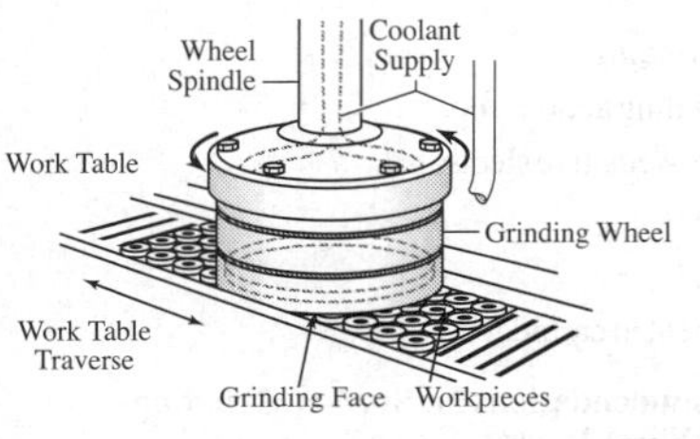


Reciprocating — Face (Side) of Wheel

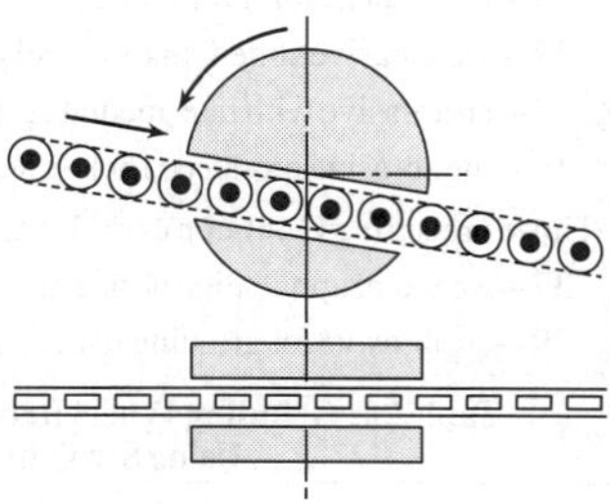

Traverse Along Straight Line or
Arcuate Path — Face (Side) of Wheel

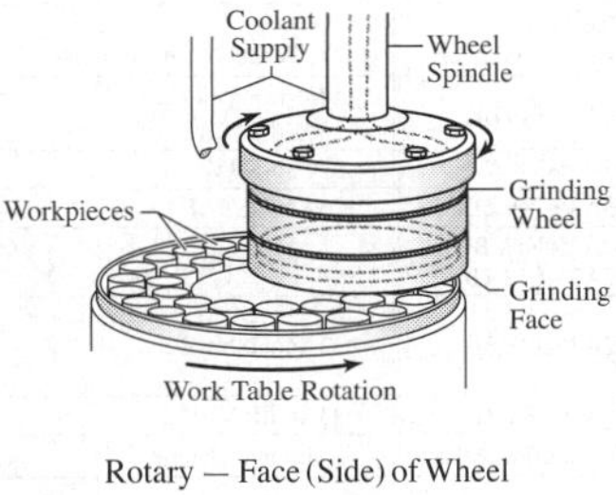


Rotary — Face (Side) of Wheel

*Operating Surface, Face of Wheel: Movement of Work, Traverse Along Straight or Arcuate Path:* The grinding wheel, usually of segmental type, is set in a position to cover either an annular area near the periphery of the table or, more commonly, to reach beyond the table center. A large circular magnetic chuck generally covers the entire table surface and facilitates the mounting of workpieces, even of fixtures, when needed. The uninterrupted passage of the work in contact with the large wheel face permits a very high rate of stock removal and the machine, with single or double wheelhead, can be adapted also to automatic operation with continuous part feed by mechanized work handling.

**Selection of Grinding Wheels for Surface Grinding.—**The most practical way to select a grinding wheel for surface grinding is to base the selection on the work material. Table 2a gives the grinding wheel recommendations for Types 1, 5, and 7 straight wheels used on reciprocating and rotary table surface grinders with horizontal spindles. Table 2b gives the grinding wheel recommendations for Type 2 cylinder wheels, Type 6 cup wheels, and wheel segments used on vertical spindle surface grinders.

The last letters (two or three) that may follow the bond designation V (vitrified) or B (resinoid) refer to: 1) bond modification, "BE" being especially suitable for surface grinding; 2) special structure, "P" type being distinctively porous; and 3) for segments made of 23A type abrasives, the term 12VSM implies porous structure, and the letter "P" is not needed.

The wheel markings in Table 2a and Table 2b are those used by the Norton Co., complementing the basic standard markings with Norton symbols. The complementary symbols used in these tables, that is, those preceding the letter designating A (aluminum oxide) or C (silicon carbide), indicate the special type of basic abrasive that has the friability best suited for particular work materials. Those preceding A (aluminum oxide) are

57—a versatile abrasive suitable for grinding steel in either a hard or soft state.

38—the most friable abrasive.

32—the abrasive suited for tool steel grinding.

23—an abrasive with intermediate grinding action, and

19—the abrasive produced for less heat-sensitive steels.

Those preceding C (silicon carbide) are

37—a general application abrasive, and

39—an abrasive for grinding hard cemented carbide.

**Table 2a. Grinding Wheel Recommendations for Surface Grinding—Using Straight Wheel Types 1, 5, and 7**

| Material | Horizontal-spindle, reciprocating-table surface grinders | | Horizontal-spindle, rotary-table surface grinders |
|---|---|---|---|
| | Wheels less than 16 inches diameter | Wheels 16 inches diameter and over | Wheels of any diameter |
| Cast iron | 37C36-K8V or 23A46-I8VBE | 23A36-I8VBE | 37C36-K8V or 23A46-I8VBE |
| Nonferrous metal | 37C36-K8V | 37C36-K8V | 37C36-K8V |
| Soft steel | 23A46-J8VBE | 23A36-J8VBE | 23A46-J8VBE |
| Hardened steel, broad contact | 32A46-H8VBE or 32A60-F12VBEP | 32A36-H8VBE or 32A36-F12VBEP | 32A46-I8VBE |
| Hardened steel, narrow contact or interrupted cut | 32A46-I8VBE | 32A36-J8VBE | 32A46-J8VBE |
| General-purpose wheel | 23A46-H8VBE | 23A36-I8VBE | 23A46-I8VBE |
| Cemented carbides | Diamond wheels[a] | Diamond wheels[a] | Diamond wheels[a] |

[a] General diamond wheel recommendations are listed in Table 5 on page 1297.

**Table 2b. Grinding Wheel Recommendations for Surface Grinding—Using Type 2 Cylinder Wheels, Type 6 Cup Wheels, and Wheel Segments**

| Material | Type 2 Cylinder Wheels | Type 6 Cup Wheels | Wheel Segments |
|---|---|---|---|
| High tensile cast iron and nonferrous metals | 37C24-HKV | 37C24-HVK | 37C24-HVK |
| Soft steel, malleable cast iron, steel castings, boiler plate | 23A24-I8VBE or 23A30-G12VBEP | 23A24-I8VBE | 23A24-I8VSM or 23A30-H12VSM |
| Hardened steel—broad contact | 32A46-G8VBE or 32A36-E12VBEP | 32A46-G8VBE or 32A60-E12VBEP | 32A36-G8VBE or 32A46-E12VBEP |
| Hardened steel—narrow contact or interrupt cut | 32A46-H8VBE | 32A60-H8VBE | 32A46-G8VBE or 32A60-G12VBEP |
| General-purpose use | 23A30-H8VBE or 23A30-E12VBEP | … | 23A30-H8VSM or 23A30-G12VSM |

**Process Data for Surface Grinding.**—In surface grinding, similarly to other metal-cutting processes, the speed and feed rates that are applied must be adjusted to the operational conditions as well as to the objectives of the process. Grinding differs, however, from other types of metal cutting methods in regard to the cutting speed of the tool; the peripheral speed of the grinding wheel is maintained within a narrow range, generally 5500 to 6500 surface feet per minute. Speed ranges different from the common one are used in particular processes which require special wheels and equipment.

**Table 3. Basic Process Data for Peripheral Surface Grinding on Reciprocating Table Surface Grinders**

| Work Material | Hardness (Brinell/ Rockwell) | Material Condition | Wheel Speed, fpm | Table Speed, fpm | Downfeed, in. per pass | | Crossfeed per pass, fraction of wheel width |
|---|---|---|---|---|---|---|---|
| | | | | | Rough | Finish, max. | |
| Plain carbon steel | 52 RC max. | Annealed, cold drawn | 5500-6500 | 50-100 | 0.003 | 0.0005 | 1/4 |
| | 52-65 RC | Carburized and/or quenched and tempered | 5500-6500 | 50-100 | 0.003 | 0.0005 | 1/10 |
| Alloy steels | 52 RC max. | Annealed or quenched and tempered | 5500-6500 | 50-100 | 0.003 | 0.001 | 1/4 |
| | 52-65 RC | Carburized and/or quenched and tempered | 5500-6500 | 50-100 | 0.003 | 0.0005 | 1/10 |
| Tool steels | 150-275 BHN | Annealed | 5500-6500 | 50-100 | 0.002 | 0.0005 | 1/5 |
| | 56-65 RC | Quenched and tempered | 5500-6500 | 50-100 | 0.002 | 0.0005 | 1/10 |
| Nitriding steels | 200-350 BHN | Normalized, annealed | 5500-6500 | 50-100 | 0.003 | 0.001 | 1/4 |
| | 60-65 RC | Nitrided | 5500-6500 | 50-100 | 0.003 | 0.0005 | 1/10 |
| Cast steels | 52 RC max. | Normalized, annealed | 5500-6500 | 50-100 | 0.003 | 0.001 | 1/4 |
| | Over 52 RC | Carburized and/or quenched and tempered | 5500-6500 | 50-100 | 0.003 | 0.0005 | 1/10 |
| Gray irons | 52 RC max. | As cast, annealed, and/or quenched and tempered | 5000-6500 | 50-100 | 0.003 | 0.001 | 1/3 |
| Ductile irons | 52 RC max. | As cast, annealed or quenched and tempered | 5500-6500 | 50-100 | 0.003 | 0.001 | 1/5 |
| Stainless steels, martensitic | 135-235 BHN | Annealed or cold drawn | 5500-6500 | 50-100 | 0.002 | 0.0005 | 1/4 |
| | Over 275 BHN | Quenched and tempered | 5500-6500 | 50-100 | 0.001 | 0.0005 | 1/8 |
| Aluminum alloys | 30-150 BHN | As cast, cold drawn or treated | 5500-6500 | 50-100 | 0.003 | 0.001 | 1/3 |

In establishing the proper process values for grinding, of prime consideration are the work material, its condition, and the type of operation (roughing or finishing). Table 3 gives basic process data for peripheral surface grinding on reciprocating table surface grinders. For different work materials and hardness ranges data are given regarding table speeds, downfeed (infeed) rates and cross feed, the latter as a function of the wheel width.

**Common Faults and Possible Causes in Surface Grinding.—**Approaching the ideal performance with regard to both the quality of the ground surface and the efficiency of surface grinding, requires the monitoring of the process and the correction of conditions adverse to the attainment of that goal.

Defective, or just not entirely satisfactory surface grinding may have any one or more of several causes. Exploring and determining the cause for eliminating its harmful effects is facilitated by knowing the possible sources of the experienced undesirable performance. Table 4, associating the common faults with their possible causes, is intended to aid in determining the actual cause, the correction of which should restore the desired performance level.

While the table lists the more common faults in surface grinding, and points out their frequent causes, other types of improper performance and/or other causes, in addition to those indicated, are not excluded.

**Vitrified Grinding Wheels.—**The term "vitrified" denotes the type of bond used in these grinding wheels. The bond in a grinding wheel is the material which holds the abrasive grains together and supports them while they cut. With a given type of bond, it is the *amount* of bond that determines the "hardness" or softness" of wheels. The abrasive itself is extremely hard in all wheels, and the terms "hard" and "soft" refer to the *strength of bonding*; the greater the percentage of bond with respect to the abrasive, the heavier the coating of bond around the abrasive grains and the stronger the bond posts, the "harder" the wheel.

Most wheels are made with a vitrified bond composed of clays and feldspar selected for their fusibility. During the "burning" process in grinding wheel manufacture, the clays are fused into a molten glass condition. Upon cooling, a span or post of this glass connects each abrasive grain to its neighbors to make a rigid, strong, grinding wheel. These wheels are porous, free cutting and unaffected by water, acids, oils, heat, or cold. Vitrified wheels are extensively used for cylindrical grinding, surface grinding, internal grinding and cutter grinding.

**Silicate Bonding Process.—**Silicate grinding wheels derive their name from the fact that silicate of soda or water glass is the principal ingredient used in the bond. These wheels are also sometimes referred to as *semi-vitrified* wheels. Ordinarily, they cut smoothly and with comparatively little heat, and for grinding operations requiring the lowest wheel wear, compatible with cool cutting, silicate wheels are often used. Their grade is also dependable and much larger wheels can be made by this bonding process than by the vitrified process. Some of the grinding operations for which silicate wheels have been found to be especially adapted are as follows: for grinding high-speed steel machine shop tools, such as reamers, milling cutters, etc.; for hand-grinding lathe and planer tools; for surface grinding with machines of the vertical ring-wheel type; and for operations requiring dish-shaped wheels and cool cutting. These wheels are unequaled for wet grinding on hardened steel and for wet tool grinding. They are easily recognized by their light gray color.

**Oilstones.—**The natural oilstones commonly used are the *Washita* and *Arkansas*. The Washita is a coarser and more rapidly cutting stone, and is generally considered the most satisfactory for sharpening woodworkers' tools. There are various grades of Washita rock, varying from the perfect crystallized and porous whetstone grit, to vitreous flint and hard sandstone. The best whetstones are porous and uniform in texture and are composed entirely of silica crystals. The poorer grades are less porous, making them vitreous or "glassy." They may also have hard spots or sand holes, or contain grains of sand among the crystals. For general work, a soft, free-grit, quick-cutting stone is required, although a fine-grit medium-hard stone is sometimes preferable. These are commonly furnished in three grits: fine, medium, and coarse, and in all required shapes.

**Table 4. Common Faults and Possible Causes in Surface Grinding**

| CAUSES | FAULTS | WORK DIMENSION | | | METALLURGICAL DEFECTS | | SURFACE QUALITY | | | | WHEEL CONDITION | | | WORK RETAINMENT | |
|---|---|---|---|---|---|---|---|---|---|---|---|---|---|---|---|
| | | Work not flat | Work not parallel | Poor size holding | Burnishing of work | Burning or checking | Feed lines | Chatter marks | Scratches on surface | Poor finish | Wheel loading | Wheel glazing | Rapid wheel wear | Not firmly seated | Work sliding on chuck |
| WORK CONDITION | Heat-treatment stresses | ● | … | … | … | … | … | … | … | … | … | … | … | … | … |
| | Work too thin | ● | ● | … | … | … | … | … | … | … | … | … | … | … | … |
| | Work warped | ● | … | … | … | … | … | … | … | … | … | … | … | ● | … |
| | Abrupt section changes | ● | ● | … | … | … | … | … | … | … | … | … | … | … | … |
| GRINDING WHEEL | Grit too fine | … | … | … | ● | ● | … | … | … | … | ● | ● | … | … | … |
| | Grit too coarse | … | … | … | … | … | … | … | … | ● | … | … | … | … | … |
| | Grade too hard | ● | … | … | ● | ● | … | ● | … | … | ● | ● | … | … | … |
| | Grade too soft | … | … | ● | … | … | … | ● | ● | … | … | … | ● | … | … |
| | Wheel not balanced | … | … | … | … | … | … | ● | … | … | … | … | … | … | … |
| | Dense structure | … | … | … | … | … | … | … | … | … | ● | ● | … | … | … |
| TOOLING AND COOLANT | Improper coolant | … | … | … | … | … | … | … | … | … | ● | … | … | … | … |
| | Insufficient coolant | ● | ● | … | ● | ● | … | … | … | … | … | ● | … | … | … |
| | Dirty coolant | … | … | … | … | … | … | … | ● | … | ● | … | … | … | … |
| | Diamond loose or chipped | ● | ● | … | … | … | … | … | ● | … | … | … | … | … | … |
| | Diamond dull | … | … | ● | … | … | … | … | … | ● | ● | ● | … | … | … |
| | No or poor magnetic force | … | … | ● | … | … | … | … | ● | … | … | … | … | ● | ● |
| | Chuck surface worn or burred | ● | ● | … | … | … | … | … | ● | … | … | … | … | ● | … |
| MACHINE AND SETUP | Chuck not aligned | ● | ● | … | … | … | … | … | … | … | … | … | … | … | … |
| | Vibrations in machine | … | … | … | … | … | … | ● | … | … | … | … | … | … | … |
| | Plane of movement out of parallel | ● | ● | … | … | … | … | … | … | … | … | … | … | … | … |
| OPERATIONAL CONDITIONS | Too low work speed | … | … | … | … | … | … | … | … | … | ● | … | … | … | … |
| | Too light feed | … | … | … | … | … | … | … | … | … | … | ● | … | … | … |
| | Too heavy cut | ● | … | … | ● | … | … | … | … | ● | … | … | … | … | … |
| | Chuck retained swarf | ● | ● | … | … | … | … | … | … | … | … | … | … | … | ● |
| | Chuck loading improper | ● | ● | … | … | … | … | … | … | … | … | … | … | … | ● |
| | Insufficient blocking of parts | … | … | … | … | … | … | … | ● | … | … | … | … | … | … |
| | Wheel runs off the work | … | ● | ● | … | ● | … | … | … | … | … | … | ● | … | … |
| | Wheel dressing too fine | ● | … | … | … | … | … | … | … | … | … | … | … | … | … |
| | Wheel edge not chamfered | … | … | … | … | … | ● | … | … | … | … | … | … | … | … |
| | Loose dirt under guard | … | … | … | … | … | … | … | ● | … | … | … | … | … | … |

## Offhand Grinding

Offhand grinding consists of holding the wheel to the work or the work to the wheel and grinding to broad tolerances and includes such operations as certain types of tool sharpening, weld grinding, snagging castings and other rough grinding. Types of machines that are used for rough grinding in foundries are floor- and bench-stand machines. Wheels for these machines vary from 6 to 30 inches in diameter. Portable grinding machines (electric, flexible shaft, or air-driven) are used for cleaning and smoothing castings.

Many rough grinding operations on castings can be best done with shaped wheels, such as cup wheels (including plate mounted) or cone wheels, and it is advisable to have a good assortment of such wheels on hand to do the odd jobs the best way.

**Floor- and Bench-Stand Grinding.—**The most common method of rough grinding is on double-end floor and bench stands. In machine shops, welding shops, and automotive repair shops, these grinders are usually provided with a fairly coarse grit wheel on one end for miscellaneous rough grinding and a finer grit wheel on the other end for sharpening tools. The pressure exerted is a very important factor in selecting the proper grinding wheel. If grinding is to be done mostly on hard sharp fins, then durable, coarse and hard wheels are required, but if grinding is mostly on large gate and riser pads, then finer and softer wheels should be used for best cutting action.

**Portable Grinding.—**Portable grinding machines are usually classified as air grinders, flexible shaft grinders, and electric grinders. The electric grinders are of two types; namely, those driven by standard 60 cycle current and so-called high-cycle grinders. Portable grinders are used for grinding down and smoothing weld seams; cleaning metal before welding; grinding out imperfections, fins and parting lines in castings and smoothing castings; grinding punch press dies and patterns to proper size and shape; and grinding manganese steel castings.

Wheels used on portable grinders are of three bond types; namely, resinoid, rubber, and vitrified. By far the largest percentage is resinoid. Rubber bond is used for relatively thin wheels and where a good finish is required. Some of the smaller wheels such as cone and plug wheels are vitrified bonded.

Grit sizes most generally used in wheels from 4 to 8 inches (102–203 mm) in diameter are 16, 20, and 24. In the still smaller diameters, finer sizes are used, such as 30, 36, and 46.

The particular grit size to use depends chiefly on the kind of grinding to be done. If the work consists of sharp fins and the machine has ample power, a coarse grain size combined with a fairly hard grade should be used. If the job is more in the nature of smoothing or surfacing and a fairly good finish is required, then finer and softer wheels are called for.

**Swing-Frame Grinding.—**This type of grinding is employed where a considerable amount of material is to be removed as on snagging large castings. It may be possible to remove 10 times as much material from steel castings using swing-frame grinders as with portable grinders; and 3 times as much material as with high-speed floor-stand grinders.

The largest field of application for swing-frame machines is on castings which are too heavy to handle on a floor stand; but often it is found that comparatively large gates and risers on smaller castings can be ground more quickly with swing-frame grinders, even if fins and parting lines have to be ground on floor stands as a second operation.

In foundries, the swing-frame machines are usually suspended from a trolley on a jib that can be swung out of the way when placing the work on the floor with the help of an overhead crane. In steel mills when grinding billets, a number of swing-frame machines are usually suspended from trolleys on a line of beams which facilitate their use as required.

The grinding wheels used on swing-frame machines are made with coarser grit sizes and harder grades than wheels used on floor stands for the same work. The reason is that greater grinding pressures can be obtained on the swing-frame machines.

## Abrasive Belt Grinding

Abrasive belts are used in the metalworking industry for removing stock, light cleaning up of metal surfaces, grinding welds, deburring, breaking and polishing hole edges, and finish grinding of sheet steel. The types of belts that are used may be coated with aluminum oxide (the most common coating) for stock removal and finishing of all alloy steels, high-carbon steel, and tough bronzes; and silicon carbide for use on hard, brittle, and low-tensile strength metals which would include aluminum and cast irons.

Table 1 is a guide to the selection of the proper abrasive belt, lubricant, and contact wheel. This table is entered on the basis of the material used and type of operation to be done and gives the abrasive belt specifications (type of bonding and abrasive grain size and material), the range of speeds at which the belt may best be operated, the type of lubricant to use, and the type and hardness of the contact wheel to use. Table 2 serves as a guide in the selection of contact wheels. This table is entered on the basis of the type of contact wheel surface and the contact wheel material. The table gives the hardness and/or density, the type of abrasive belt grinding for which the contact wheel is intended, the character of the wheel action and such comments as the uses, and hints for best use. Both tables are intended only as guides for general shop practice; selections may be altered to suit individual requirements.

There are three types of abrasive belt grinding machines. One type employs a contact wheel behind the belt at the point of contact of the workpiece to the belt and facilitates a high rate of stock removal. Another type uses an accurate parallel ground platen over which the abrasive belt passes and facilitates the finishing of precision parts. A third type which has no platens or contact wheel is used for finishing parts having uneven surfaces or contours. In this type there is no support behind the belt at the point of contact of the belt with the workpiece. Some machines are so constructed that besides grinding against a platen or a contact wheel the workpiece may be moved and ground against an unsupported portion of the belt, thereby in effect making it a dual machine.

Although abrasive belts at the time of their introduction were used dry, since the advent of the improved waterproof abrasive belts, they have been used with coolants, oil-mists, and greases to aid the cutting action. The application of a coolant to the area of contact retards loading, resulting in a cool, free cutting action, a good finish and a long belt life.

## Abrasive Cutting

Abrasive cut-off wheels are used for cutting steel, brass and aluminum bars and tubes of all shapes and hardnesses, ceramics, plastics, insulating materials, glass and cemented carbides. Originally a tool or stock room procedure, this method has developed into a high-speed production operation. While the abrasive cut-off machine and cut-off wheel can be said to have revolutionized the practice of cutting off materials, the metal saw continues to be the more economical method for cutting off large cross sections of certain materials. However, there are innumerable materials and shapes that can be cut with much greater speed and economy by the abrasive wheel method. On conventional chop-stroke abrasive cutting machines using 16-inch (406.4-mm) diameter wheels, 2-inch (50.8-mm) diameter bar stock is the maximum size that can be cut with satisfactory wheel efficiency, but bar stock up to 6 inches (152.4 mm) in diameter can be cut efficiently on oscillating-stroke machines. Tubing up to $3\frac{1}{2}$ inches (88.9 mm) in diameter can also be cut efficiently.

Abrasive wheels are commonly available in four types of bonds: Resinoid, rubber, shellac and fiber or fabric reinforced. In general, resinoid bonded cut-off wheels are used for dry cutting where burrs and some burn are not objectionable and rubber bonded wheels are used for wet cutting where cuts are to be smooth, clean and free from burrs. Shellac bonded wheels have a soft, free cutting quality which makes them particularly useful in the tool room where tool steels are to be cut without discoloration. Fiber reinforced bonded wheels are able to withstand severe flexing and side pressures and fabric reinforced bonded wheels which are highly resistant to breakage caused by extreme side pressures, are fast cutting and have a low rate of wear.

## Table 1. Guide to the Selection and Application of Abrasive Belts

| Material | Type of Operation | Abrasive Belt[a] | Grit | Belt Speed fpm | Belt Speed m/min | Type of Grease Lubricant | Contact Wheel Type | Contact Wheel Durometer Hardness |
|---|---|---|---|---|---|---|---|---|
| Hot-and Cold-Rolled Steel | Roughing | R/R $Al_2O_3$ | 24–60 | 4000–6500 | 1219–1981 | Light-body or none | Cog-tooth, serrated rubber | 70–90 |
| | Polishing | R/G or R/R $Al_2O_3$ | 80–150 | 4500–7000 | 1372–2134 | Light-body or none | Plain or serrated rubber, sectional or finger-type cloth wheel, free belt | 20–60 |
| | Fine Polishing | R/G or electrocoated $Al_2O_3$ cloth | 180–500 | 4500–7000 | 1372–2134 | Heavy or with abrasive compound | Smooth-faced rubber or cloth | 20–40 |
| Stainless Steel | Roughing | R/R $Al_2O_3$ | 50–80 | 3500–5000 | 1067–1524 | Light-body or none | Cog-tooth, serrated rubber | 70–90 |
| | Polishing | R/G or R/R $Al_2O_3$ | 80–120 | 4000–5500 | 1219–1676 | Light-body or none | Plain or serrated rubber, sectional or finger-type cloth wheel, free belt | 30–60 |
| | Fine Polishing | Closed-coat SiC | 150–280 | 4500–5500 | 1372–1676 | Heavy or oil mist | Smooth-faced rubber or cloth | 20–40 |
| Aluminum, Cast or Fabricated | Roughing | R/R SiC or $Al_2O_3$ | 24–80 | 5000–6500 | 1524–1981 | Light | Cog-tooth, serrated rubber | 70–90 |
| | Polishing | R/G SiC or $Al_2O_3$ | 100–180 | 4500–6500 | 1372–1981 | Light | Plain or serrated rubber, sectional or finger-type cloth wheel, free belt | 30–50 |
| | Fine Polishing | Closed-coat SiC or electrocoated $Al_2O_3$ | 220–320 | 4500–6500 | 1372–1981 | Heavy or with abrasive compound | Plain faced rubber, finger-type cloth or free belt | 20–50 |
| Copper Alloys or Brass | Roughing | R/R SiC or $Al_2O_3$ | 36–80 | 2200–4500 | 671–1372 | Light-body | Cog-tooth, serrated rubber | 70–90 |
| | Polishing | Closed-coat SiC or electrocoated $Al_2O_3$ or R/G SiC or $Al_2O_3$ | 100–150 | 4000–6500 | 1219–1981 | Light-body | Plain or serrated rubber, sectional or finger-type cloth wheel, free belt | 30–50 |
| | Fine Polishing | Closed-coat SiC or electrocoated $Al_2O_3$ | 180–320 | 4000–6500 | 1219–1981 | Light or with abrasive compound | Same as for polishing | 20–30 |
| Nonferrous Die-Castings | Roughing | R/R SiC or $Al_2O_3$ | 24–80 | 4500–6500 | 1372–1981 | Light-body | Hard wheel depending on application | 50–70 |
| | Polishing | R/G SiC or $Al_2O_3$ | 100–180 | 4500–6500 | 1372–1981 | Light-body | Plain rubber, cloth or free belt | 30–50 |
| | Fine Polishing | Electrocoated $Al_2O_3$ or closed-coat SiC | 220–320 | 4500–6500 | 1372–1981 | Heavy or with abrasive compound | Plain or finger-type cloth wheel, or free belt | 20–30 |
| Cast Iron | Roughing | R/R $Al_2O_3$ | 24–60 | 2000–4000 | 610–1219 | None | Cog-tooth, serrated rubber | 70–90 |
| | Polishing | R/R $Al_2O_3$ | 80–150 | 4000–5500 | 1219–1676 | None | Serrated rubber | 30–70 |
| | Fine Polishing | R/R $Al_2O_3$ | 120–240 | 4000–5500 | 1219–1676 | Light-body | Smooth-faced rubber | 30–40 |
| Titanium | Roughing | R/R SiC or $Al_2O_3$ | 36–50 | 700–1500 | 213–457 | Sulfur-chlorinated | Small-diameter, cog-tooth serrated rubber | 70–80 |
| | Polishing | R/R SiC | 60–120 | 1200–2000 | 366–610 | Light-body | Standard serrated rubber | 50 |
| | Fine Polishing | R/R SiC | 120–240 | 1200–2000 | 366–610 | Light-body | Smooth-faced rubber or cloth | 20–40 |

[a] R/R indicates that both the making and sizing bond coats are resin. R/G indicates that the making coat is glue and the sizing coat is resin. The abbreviations $Al_2O_3$ for aluminum oxide and SiC for silicon carbide are used. Almost all R/R and R/G $Al_2O_3$ and SiC belts have a heavy-drill weight cloth backing. Most electrocoated $Al_2O_3$ and closed-coat SiC belts have a jeans weight cloth backing.

**Table 2. Guide to the Selection and Application of Contact Wheels**

| Surface | Material | Hardness and Density | Purposes | Wheel Action | Comments |
|---|---|---|---|---|---|
| Cog-tooth | Rubber | 70 to 90 durometer | Roughing | Fast cutting, allows long belt life. | For cutting down projections on castings and weld beads. |
| Standard serrated | Rubber | 40 to 50 durometer, medium density | Roughing | Leaves rough- to medium-ground surface. | For smoothing projections and face defects. |
| X-shaped serrations | Rubber | 20 to 50 durometer | Roughing and polishing | Flexibility of rubber allows entry into contours. Medium polishing, light removal. | Same as for standard serrated wheels but preferred for soft nonferrous metals. |
| Plain face | Rubber | 20 to 70 durometer | Roughing and polishing | Plain wheel face allows controlled penetration of abrasive grain. Softer wheels give better finishes. | For large or small flat faces. |
| Flat flexible | Compressed canvas | About nine densities from very hard to very soft | Roughing and polishing | Hard wheels can remove metal, but not as quickly as cog-tooth rubber wheels. Softer wheels polish well. | Good for medium-range grinding and polishing. |
| Flat flexible | Solid sectional canvas | Soft, medium, and hard | Polishing | Uniform polishing. Avoids abrasive pattern on work. Adjusts to contours. Can be performed for contours. | A low-cost wheel with uniform density at the face. Handles all types of polishing. |
| Flat flexible | Buff section canvas | Soft | Contour polishing | For fine polishing and finishing. | Can be widened or narrowed by adding or removing sections. Low cost. |
| Flat flexible | Sponge rubber inserts | 5 to 10 durometer, soft | Polishing | Uniform polishing and finishing. Polishes and blends contours. | Has replaceable segments. Polishes and blends contours. Segments allow density changes. |
| Flexible | Fingers of canvas attached to hub | Soft | Polishing | Uniform polishing and finishing. | For polishing and finishing. |
| Flat flexible | Rubber segments | Varies in hardness | Roughing and polishing | Grinds or polishes depending on density and hardness of inserts. | For portable machines. Uses replaceable segments that save on wheel costs and allow density changes. |
| Flat flexible | Inflated rubber | Air pressure controls hardness | Roughing and polishing | Uniform finishing. | Adjusts to contours. |

The types of abrasives available in cut-off wheels are: Aluminum oxide, for cutting steel and most other metals; silicon carbide, for cutting nonmetallic materials such as carbon, tile, slate, ceramics, etc.; and diamond, for cutting cemented carbides. The method of denoting abrasive type, grain size, grade, structure and bond type by using a system of markings is the same as for grinding wheels (see page 1270). Maximum wheel speeds given in ANSI B7.1-2017, "Safety Requirements for The Use, Care, and Protection of Abrasive Wheels," range from 9500 to 14,200 sfpm (2896–4328 m/min) for organic bonded cut-off wheels larger than 16 inches (40.6 cm) in diameter and from 9500 to 16,000 sfpm (2896–4877 m/min) for organic bonded cut-off wheels 16 inches in diameter and smaller. Maximum wheel speeds specified by the manufacturer should never be exceeded even though they may be lower than those given in the B7.1 Standard.

There are four basic types of abrasive cutting machines: Chop-stroke, oscillating stroke, horizontal stroke and work rotating. Each of these four types may be designed for dry cutting or for wet cutting (includes submerged cutting).

The accompanying table based upon information made available by The Carborundum Co. gives some of the probable causes of cutting off difficulties that might be experienced when using abrasive cut-off wheels.

**Probable Causes of Cutting-Off Difficulties**

| Difficulty | Probable Cause |
|---|---|
| Angular Cuts and Wheel Breakage | (1) Inadequate clamping which allows movement of work while the wheel is in the cut. The work should be clamped on both sides of the cut. (2) Work vise higher on one side than the other causing wheel to be pinched. (3) Wheel vibration resulting from worn spindle bearings. (4) Too fast feeding into the cut when cutting wet. |
| Burning of Stock | (1) Insufficient power or drive allowing wheel to stall. (2) Cuts too heavy for grade of wheel being used. (3) Wheel fed through the work too slowly. This causes a heating up of the material being cut. This difficulty encountered chiefly in dry cutting. |
| Excessive Wheel Wear | (1) Too rapid cutting when cutting wet. (2) Grade of wheel too hard for work, resulting in excessive heating and burning out of bond. (3) Inadequate coolant supply in wet cutting. (4) Grade of wheel too soft for work. (5) Worn spindle bearings allowing wheel vibration. |
| Excessive Burring | (1) Feeding too slowly when cutting dry. (2) Grit size in wheel too coarse. (3) Grade of wheel too hard. (4) Wheel too thick for job. |

## Honing Process

The hone-abrading process for obtaining cylindrical forms with precise dimensions and surfaces can be applied to internal cylindrical surfaces with a wide range of diameters such as engine cylinders, bearing bores, pin holes, etc. and also to some external cylindrical surfaces. The process is used to: 1) eliminate inaccuracies resulting from previous operations by generating a true cylindrical form with respect to roundness and straightness within minimum dimensional limits; 2) generate final dimensional size accuracy within low tolerances, as may be required for interchangeability of parts; 3) provide rapid and economical stock removal consistent with accomplishment of the other results; and 4) generate surface finishes of a specified degree of surface smoothness with high surface quality.

**Amount and Rate of Stock Removal.—**Honing may be employed to increase bore diameters by as much as 0.100 inch (2.5 mm) or as little as 0.001 inch (0.25 mm). The amount of stock removed by the honing process is entirely a question of processing economy. If other operations are performed before honing then the bulk of the stock should be taken off by the operation that can do it most economically. In large diameter bores that have been distorted in heat treating, it may be necessary to remove as much as 0.030 to 0.040 inch (0.76–1.02 mm) from the diameter to make the bore round and straight. For out-of-round or tapered bores, a good "rule of thumb" is to leave twice as much stock (on the diameter) for honing as there is error in the bore. Another general rule is: For bores over 1 inch (25.4 mm) in diameter, leave 0.001 to 0.0015 inch stock per inch of diameter (0.025–0.038 mm stock/mm). For example, 0.002 to 0.003 inch (0.05–0.08 mm) of stock is left in 2-inch bores (50.8-mm) and 0.010 to 0.015 inch (0.25–0.38 mm) in 10-inch (254-mm) bores. Where parts are to be honed for finish only, the amount of metal to be left for removing tool marks may be as little as 0.0002 to 0.015 inch (0.005–0.38 mm) on the diameter.

In general, the honing process can remove stock from bore diameters at the rate of 0.009 to 0.012 inch per minute (0.23–0.30 mm/min) on cast-iron parts and from 0.005 to 0.008 inch per minute (0.13–0.20 mm/min) on steel parts having a hardness of 60 to 65 RC (Rockwell C scale). These rates are based on parts having a length equal to three or four times the diameter. Stock has been removed from long parts such as gun barrels, at the rate of 65 cubic inches per hour (17.75 cc/s). Recommended honing speeds for cast iron range from 110 to 200 surface feet per minute (34–61 m/min) of rotation and from 50 to 110 lineal feet per minute (15–34 m/min) of reciprocation. For steel, rotating surface speeds range from 50 to 110 feet per minute (15–34 m/min) and reciprocation speeds from 20 to 90 lineal feet per minute (6–27 m/min) . The rotation and reciprocation speeds to be used depend on the size of the work, the amount and characteristics of the material to be removed, and the quality of the finish desired. In general, the harder the material to be honed, the lower the speed. Interrupted bores are usually honed at faster speeds than plain bores.

**Formulas for Rotative Speeds.—**Empirical formulas for determining rotative speeds for honing have been developed by the Micromatic Hone Corp. These formulas take into consideration the type of material being honed, its hardness and its surface characteristics; the abrasive area; and the type of surface pattern and degree of surface roughness desired. Because of the wide variations in material characteristics, abrasives available, and types of finishes specified, these formulas should be considered as a guide only in determining which of the available speeds (pulley or gear combinations) should be used for any particular application.

The formula for rotative speed, $S$, in surface feet per minute is: $S = \frac{K \times D}{W \times N}$

The formula for rotative speed in revolutions per minute is: $\text{R.P.M} = \frac{R}{W \times N}$

where, $K$ and $R$ are factors taken from the table on the following page, $D$ is the diameter of the bore in inches, $W$ is the width of the abrasive stone or stock in inches, and $N$ is the number of stones.

Although the actual speed of the abrasive is the resultant of both the rotative speed and the reciprocation speed, this latter quantity is seldom solved for or used. The reciprocation speed is not determined empirically but by testing under operating conditions. Changing the reciprocation speed affects the dressing action of the abrasive stones, therefore, the reciprocation speed is adjusted to provide for a desired surface finish which is usually a well lubricated bearing surface that will not scuff.

**Table of Factors for Use in Rotative Speed Formulas**

| Character of Surface[a] | Material | Hardness[b] Soft K | Soft R | Medium K | Medium R | Hard K | Hard R |
|---|---|---|---|---|---|---|---|
| | | Factors | | | | | |
| Base Metal | Cast Iron | 110 | 420 | 80 | 300 | 60 | 230 |
| | Steel | 80 | 300 | 60 | 230 | 50 | 190 |
| Dressing Surface | Cast Iron | 150 | 570 | 110 | 420 | 80 | 300 |
| | Steel | 110 | 420 | 80 | 300 | 60 | 230 |
| Severe Dressing | Cast Iron | 200 | 760 | 150 | 570 | 110 | 420 |
| | Steel | 150 | 570 | 110 | 420 | 80 | 300 |

[a] The character of the surface is classified according to its effect on the abrasive; *Base Metal* being a honed, ground or fine bored section that has little dressing action on the grit; *Dressing Surface* being a rough bored, reamed or broached surface or any surface broken by cross holes or ports; *Severe Dressing* being a surface interrupted by keyways, undercuts or burrs that dress the stones severely. If over half of the stock is to be removed after the surface is cleaned up, the speed should be computed using the *Base Metal* factors for $K$ and $R$.

[b] Hardness designations of soft, medium, and hard cover the following ranges on the Rockwell C hardness scale, respectively: 15 to 45, 45 to 60, and 60 to 70.

**Abrasive Stones for Honing.—**Honing stones consist of aluminum oxide, silicon carbide, CBN or diamond abrasive grits, held together in stick form by a vitrified clay, resinoid or metal bond. CBN metal-bond stones are particularly suitable and widely used for honing. The grain and grade of abrasive to be used in any particular honing operation depend upon the quality of finish desired, the amount of stock to be removed, the material being honed and other factors.

The following general rules may be followed in the application of abrasive for honing: 1) Silicon-carbide abrasive is commonly used for honing cast iron, while aluminum-oxide abrasive is generally used on steel; 2) The harder the material being honed, the softer the abrasive stick used; 3) A rapid reciprocating speed will tend to make the abrasive cut fast because the dressing action on the grits will be severe; and 4) To improve the finish, use a finer abrasive grit, incorporate more multi-direction action, allow more “run-out” time after honing to size, or increase the speed of rotation.

### Possible Adjustments for Eliminating Undesirable Honing Conditions

| Undesirable Condition | Adjustment Required to Correct Condition[a] | | | | | | | | |
|---|---|---|---|---|---|---|---|---|---|
| | Abrasive[b] | | | | Other | | | | |
| | Friability | Grain Size | Hardness | Structure | Feed Pressure | Reciprocation | R.P.M. | Runout Time | Stroke Length |
| Abrasive Glazing | + | – – | – – | + | + + | + + | – – | – | 0 |
| Abrasive Loading | 0 | – – | – | – | + + | + | – – | 0 | 0 |
| Too Rough Surface Finish | 0 | + + | + + | – | – | – | + + | + | 0 |
| Too Smooth Surface Finish | 0 | – – | – – | + | + | + | – – | – | 0 |
| Poor Stone Life | – | + | + + | – | – | – | + | 0 | 0 |
| Slow Stock Removal | + | – – | – | + | + + | + + | – – | 0 | 0 |
| Taper — Large at Ends | 0 | 0 | 0 | 0 | 0 | 0 | 0 | 0 | – |
| Taper — Small at Ends | 0 | 0 | 0 | 0 | 0 | 0 | 0 | 0 | + |

[a] The + and + + symbols generally indicate that there should be an increase or addition while the – and – – symbols indicate that there should be a reduction or elimination. In each case, the double sym- bol indicates that the contemplated change would have the greatest effect. The 0 symbol means that a change would have no effect.

[b] For the abrasive adjustments the + and + + symbols indicate a more friable grain, a finer grain, a harder grade or a more open structure and the – and – – symbols just the reverse.

Compiled by Micromatic Hone Corp.

Surface roughnesses ranging from less than 1 micro-inch rms (0.0254 μm rms) to a relatively coarse roughness can be obtained by judicious choice of abrasive and honing time but the most common range is from 3 to 50 micro-inches rms (0.0762–1.27 μm rms).

**Adjustments for Eliminating Undesirable Honing Conditions.—**The accompanying table indicates adjustments that may be made to correct certain undesirable conditions encountered in honing. Only one change should be made at a time and its effect noted before making other adjustments.

**Tolerances.—**For bore diameters above 4 inches (101.6 mm) the tolerance of honed surfaces with respect to roundness and straightness ranges from 0.0005 to 0.001 inch (0.0127–0.0254 mm); for bore diameters from 1 to 4 inches (25.4–101.6 mm), 0.0003 to 0.0005 inch (0.0076–0.0127 mm); and for bore diameters below 1 inch (25.4 mm), 0.00005 to 0.0003 inch (0.0013–0.0076 mm).

## Laps and Lapping

**Material for Laps.—**Laps are usually made of soft cast iron, copper, brass or lead. In general, the best material for laps to be used on very accurate work is soft, close-grained cast iron. If the grinding, prior to lapping, is of inferior quality, or an excessive allowance has been left for lapping, copper laps may be preferable. They can be charged more easily and cut more rapidly than cast iron, but do not produce as good a finish. Whatever material is used, the lap should be softer than the work, as, otherwise, the latter will become charged with the abrasive and cut the lap, the order of the operation being reversed. A common and inexpensive form of lap for holes is made of lead which is cast around a tapering steel arbor. The arbor usually has a groove or keyway extending lengthwise, into which the lead flows, thus forming a key that prevents the lap from turning. When the lap has worn slightly smaller than the hole and ceases to cut, the lead is expanded or stretched a little by the driving in of the arbor. When this expanding operation has been repeated two or three times, the lap usually must be trued or replaced with a new one, owing to distortion.

The tendency of lead laps to lose their form is an objectionable feature. They are, however, easily molded, inexpensive, and quickly charged with the cutting abrasive. A more elaborate form for holes is composed of a steel arbor and a split cast-iron or copper shell which is sometimes prevented from turning by a small dowel pin. The lap is split so that it can be expanded to accurately fit the hole being operated upon. For hardened work, some toolmakers prefer copper to either cast iron or lead. For holes varying from $\frac{1}{4}$ to $\frac{1}{2}$ inch (6.35–12.7 mm) in diameter, copper or brass is sometimes used; cast iron is used for holes larger than $\frac{1}{2}$ inch (12.7 mm) in diameter. The arbors for these laps should have a taper of about $\frac{1}{4}$ or $\frac{3}{8}$ inch per foot (20.83 or 31.25 mm/m). The length of the lap should be somewhat greater than the length of the hole, and the thickness of the shell or lap proper should be from $\frac{1}{8}$ to $\frac{1}{6}$ its diameter.

External laps are commonly made in the form of a ring, with an outer ring or holder and an inner shell which forms the lap proper. This inner shell is made of cast iron, copper, brass, or lead. Ordinarily the lap is split and screws are provided in the holder for adjustment. The length of an external lap should at least equal the diameter of the work, and may be longer. Large ring laps usually have a handle for moving them across the work.

**Laps for Flat Surfaces.—**Laps for producing plane surfaces are made of cast iron. In order to secure accurate results, the lapping surface must be a true plane. A flat lap that is used for roughing or "blocking down" will cut better if the surface is scored by narrow grooves. These are usually located about $\frac{1}{2}$ inch (12.7 mm) apart and extend both lengthwise and crosswise, thus forming a series of squares similar to those on a checker-board. An abrasive of No. 100 or 120 emery and lard oil can be used for charging the roughing lap. For finer work, a lap having an unscored surface is used, and the lap is charged with a finer abrasive. After a lap is charged, all loose abrasive should be washed off with gasoline, for fine work, and when lapping, the surface should be kept moist, preferably with kerosene. Gasoline will cause the lap to cut a little faster, but it evaporates so rapidly that the lap soon becomes dry and the surface caked and glossy in spots. Loose emery should not be applied while lapping, for if the lap is well charged with abrasive in the beginning, is kept well moistened and not crowded too hard, it will cut for a considerable time. The pressure upon the work should be just enough to ensure constant contact. The lap can be made to cut only so fast, and if excessive pressure is applied it will become "stripped" in places. The causes of scratches are: Loose abrasive on the lap; too much pressure on the work, and poorly graded abrasive. To produce a perfectly smooth surface free from scratches, the lap should be charged with a very fine abrasive.

**Grading Abrasives for Lapping.—**For high-grade lapping, abrasives can be evenly graded as follows: A quantity of flour-emery or other abrasive is placed in a heavy cloth bag, which is gently tapped, causing very fine particles to be sifted through. When a sufficient quantity has been obtained in this way, it is placed in a dish of lard or sperm oil. The largest particles will then sink to the bottom and in about one hour the oil should be poured into another dish, care being taken not to disturb the sediment at the bottom. The oil is then allowed to stand for several hours, after which it is poured again, and so on, until the desired grade is obtained.

**Charging Laps.—**To charge a flat cast-iron lap, spread a very thin coating of the prepared abrasive over the surface and press the small cutting particles into the lap with a hard steel block. There should be as little rubbing as possible. When the entire surface is apparently charged, clean and examine for bright spots; if any are visible, continue charging until the entire surface has a uniform gray appearance. When the lap is once charged, it should be used without applying more abrasive until it ceases to cut. If a lap is over-charged and an excessive amount of abrasive is used, there is a rolling action between the work and lap which results in inaccuracy. The surface of a flat lap is usually finished true, prior to charging, by scraping and testing with a standard surface-plate, or by the well-known method of scraping-in three plates together, in order to secure a plane surface. In any case, the bearing marks or spots should be uniform and close together. These spots can be blended by covering the plates evenly with a fine abrasive and rubbing them together. While the plates are being ground in, they should be carefully tested and any high spots which may form should be reduced by rubbing them down with a smaller block.

To charge cylindrical laps for internal work, spread a thin coating of prepared abrasive over the surface of a hard steel block, preferably by rubbing lightly with a cast-iron or copper block; then insert an arbor through the lap and roll the latter over the steel block, pressing it down firmly to embed the abrasive into the surface of the lap. For external cylindrical laps, the inner surface can be charged by rolling-in the abrasive with a hard steel roller that is smaller in diameter than the lap. The taper cast-iron blocks sometimes used for lapping taper holes can also be charged by rolling-in the abrasive, as described; there is usually one roughing and one finishing lap, and when charging the former, it may be necessary to vary the charge according to any error that might exist in the taper.

**Rotary Diamond Lap.—**This style of lap is used for accurately finishing very small holes, which, because of their size, cannot be ground. While the operation is referred to as lapping, it is, in reality, a grinding process, the lap being used the same as a grinding wheel. Laps employed for this work are made of mild steel, soft material being desirable because it can be charged readily. Charging is usually done by rolling the lap between two hardened steel plates. The diamond dust and a little oil is placed on the lower plate, and as the lap revolves, the diamond is forced into its surface. After charging, the lap should be washed in benzine. The rolling plates should also be cleaned before charging with dust of a finer grade. It is very important not to force the lap when in use, especially if it is a small size. The lap should just make contact with the high spots and gradually grind them off. If a diamond lap is lubricated with kerosene, it will cut freer and faster. These small laps are run at very high speeds, the rate depending upon the lap diameter. Soft work should never be ground with diamond dust because the dust will leave the lap and charge the work.

When using a diamond lap, it should be remembered that such a lap will not produce sparks like a regular grinding wheel; hence, it is easy to crowd the lap and "strip" some of the diamond dust. To prevent this, a sound intensifier or "harker" should be used. This is placed against some stationary part of the grinder spindle, and indicates when the lap touches the work, the sound produced by the slightest contact being intensified.

**Grading Diamond Dust.—**The grades of diamond dust used for charging laps are designated by numbers, the fineness of the dust increasing as the numbers increase. The diamond, after being crushed to powder in a mortar, is thoroughly mixed with high-grade olive oil. This mixture is allowed to stand 5 minutes and then the oil is poured into another receptacle. The coarse sediment is left is removed and labeled No. 0, according to one system. The oil poured from No. 0 is again stirred and allowed to stand 10 minutes, after which it is poured into another receptacle and the sediment remaining is labeled No. 1. This operation is repeated until practically all of the dust has been recovered from the oil, the time the oil is allowed to stand being increased as shown by the following table. This is done to obtain the smaller particles that require a longer time for precipitation:

| | |
|---|---|
| To obtain No. 1 — 10 minutes | To obtain No. 4 — 2 hours |
| To obtain No. 2 — 30 minutes | To obtain No. 5 — 10 hours |
| To obtain No. 3 — 1 hour | To obtain No. 6 — until oil is clear |

The No. 0 or coarse diamond which is obtained from the first settling is usually washed in benzine, and re-crushed unless very coarse dust is required. This No. 0 grade is sometimes known as "ungraded" dust. In some places the time for settling, in order to obtain the various numbers, is greater than that given in the table.

**Cutting Properties of Laps and Abrasives.—**In order to determine the cutting properties of abrasives when used with different lapping materials and lubricants, a series of tests was conducted, the results of which were given in a paper by W. A. Knight and A. A. Case, presented before the American Society of Mechanical Engineers. In connection with these tests, a special machine was used, the construction being such that quantitative results could be obtained with various combinations of abrasive, lubricant, and lap material. These tests were confined to surface lapping.

It was not the intention to test a large variety of abrasives, three being selected as representative; namely, Naxos emery, carborundum, and alundum. Abrasive No. 150 was used in each case, and seven different lubricants, five different pressures, and three different lap materials were employed. The lubricants were lard oil, machine oil, kerosene, gasoline, turpentine, alcohol, and soda water.

These tests indicated throughout that there is, for each combination of lap and lubricant, a definite size of grain that will give the maximum amount of cutting. With all the tests, except when using the two heavier lubricants, some reduction in the size of the grain below that used in the tests (No. 150) seemed necessary before the maximum cutting rate was reached. This reduction was continuous and soon passed below that which gave the maximum cutting rate.

**Cutting Qualities with Different Laps.—**The surfaces of the steel and cast-iron laps were finished by grinding. The hardness of the different laps, as determined by the scleroscope was, for cast-iron, 28; steel, 18; copper, 5. The total amount ground from the test-pieces with each of the three laps showed that, taking the whole number of tests as a standard, there is scarcely any difference between the steel and cast iron, but that copper has somewhat better cutting qualities, although, when comparing the laps on the basis of the highest and lowest values obtained with each lap, steel and cast iron are as good for all practical purposes as copper, when the proper abrasive and lubricant are used.

**Wear of Laps.—**The wear of laps depends upon the material from which they are made and the abrasive used. The wear on all laps was about twice as fast with carborundum as with emery; with alundum the wear was about one and one-fourth times that with emery. On an average, the wear of the copper lap was about three times that of the cast-iron lap. This is not absolute wear, but wear in proportion to the amount ground from the test-pieces.

**Lapping Abrasives.—**As to the qualities of the three abrasives tested, it was found that carborundum usually began at a lower rate than the other abrasives, but, when once started, its rate was better maintained. The performance gave a curve that was more nearly a straight line. The charge or residue as the grinding proceeded remained cleaner and sharper and did not tend to become pasty or mucklike, as is so frequently the case with emery. When using a copper lap, carborundum shows but little gain over the cast-iron and steel laps, whereas, with emery and alundum, the gain is considerable.

**Effect of Different Lapping Lubricants.—**The action of the different lubricants, when tested, was found to depend upon the kind of abrasive and the lap material.

*Lard and Machine Oil:* The test showed that lard oil, without exception, gave the higher rate of cutting, and that, in general, the initial rate of cutting is higher with the lighter lubricants, but falls off more rapidly as the test continues. The lowest results were obtained with machine oil, when using an emery-charged, cast-iron lap. When using lard oil and a carborundum-charged steel lap, the highest results were obtained.

*Gasoline and Kerosene:* On the cast-iron lap, gasoline was superior to any of the lubricants tested. Considering all three abrasives, the relative value of gasoline, when applied to the different laps, is as follows: Cast iron, 127; copper, 115; steel, 106. Kerosene, like gasoline, gives the best results on cast iron and the poorest on steel. The values obtained by carborundum were invariably higher than those obtained with emery, except when using gasoline and kerosene on a copper lap.

*Turpentine and Alcohol:* Turpentine was found to do good work with carborundum on any lap. With emery, turpentine did fair work on the copper lap, but, with the emery on cast-iron and steel laps, it was distinctly inferior. Alcohol gives the lowest results with emery on the cast-iron and steel laps.

*Soda Water:* Soda water gives medium results with almost any combination of lap and abrasives, the best work being on the copper lap and the poorest on the steel lap. On the cast-iron lap, soda water is better than machine or lard oil, but not so good as gasoline or kerosene. Soda water when used with alundum on the copper lap, gave the highest results of any of the lubricants used with that particular combination.

**Lapping Pressures.—**Within the limits of the pressures used, that is, up to 25 pounds per square inch (172 kPa), the rate of cutting was found to be practically proportional to the pressure. The higher pressures of 20 and 25 pounds per square inch (138 and 172 kPa) are not so effective on the copper lap as on the other materials.

**Wet and Dry Lapping.—**With the "wet method" of using a surface lap, there is a surplus of oil and abrasive on the surface of the lap. As the specimen being lapped is moved over it, there is more or less movement or shifting of the abrasive particles. With the "dry method," the lap is first charged by rubbing or rolling the abrasive into its surface. All surplus oil and abrasive are then washed off, leaving a clean surface, but one that has embedded uniformly over it small particles of the abrasive. It is then like the surface of a very fine oilstone and will cut away hardened steel that is rubbed over it. While this has been termed the dry method, in practice, the lap surface is kept moistened with kerosene or gasoline.

Experiments on dry lapping were carried out on the cast-iron, steel, and copper laps used in the previous tests, and on one of tin made expressly for the purpose. Carborundum alone was used as the abrasive and a uniform pressure of 15 pounds per square inch (103 kPa) was applied to the specimen throughout the tests. In dry lapping, much depends upon the manner of charging the lap. The rate of cutting decreased more rapidly after the first 100 revolutions than with the wet method. Considering the amounts ground off during the first 100 revolutions, and the best result obtained with each lap taken for comparison, it was found that with a tin lap, charged by rolling No. 150 carborundum into the surface, the rate of cutting, when dry, approached that obtained with the wet method. With the other lap materials, the rate with the dry method was about one-half that of the wet method.

**Summary of Lapping Tests.—**The initial rate of cutting does not greatly differ for different abrasives. There is no advantage in using an abrasive coarser than No. 150. The rate of cutting is practically proportional to the pressure. The wear of the laps is in the following proportions: cast iron, 1.00; steel, 1.27; copper, 2.62. In general, copper and steel cut faster than cast iron, but, where permanence of form is a consideration, cast iron is the superior metal. Gasoline and kerosene are the best lubricants to use with a cast-iron lap. Machine and lard oil are the best lubricants to use with copper or steel laps. They are, however, least effective on a cast-iron lap. In general, wet lapping is from 1.2 to 6 times as fast as dry lapping, depending upon the material of the lap and the manner of charging.

## NONTRADITIONAL MACHINING AND CUTTING

**Introduction.—**Manufacturers are constantly pursuing new methods to machine complex features faster, with more accuracy, and without imposing significant cutting forces on sensitive or thin parts. In addition, use of advanced composites and harder and brittle materials is increasing. These challenges create a growing need for nontraditional machining processes, which can be categorized by material removal processes: chemical, electrochemical, mechanical erosion, and electrothermal. Such modern methods tend to be automated, with sophisticated systems and software often used to control process variables.

Shown in Table 1 are characteristics of some of the most commonly used nontraditional machining methods discussed in this section. Equipment improvements are ongoing and capabilities vary, so values given are representative rather than absolute. Table 2 contains some examples of cutting speeds for metals from industry literature. Actual speeds will vary, depending on equipment, setup, materials processed, and other factors.

**Table 1. Typical Characteristics of Some Nontraditional Machining Methods**

| | Water Jet (WJM/AWJM) | Laser (LBM) | Plasma (PAM) | Electrical Discharge (EDM/WEDM) |
|---|---|---|---|---|
| **Process Used** | Erosion | Melting/Vaporization | Melting | Vaporization/Erosion |
| **Materials Processed** | Most | Many | Conductive | Conductive |
| **Materials Not Processed** | Diamond | Some Reflective | Non-Conductive | Non-Conductive |
| | Tempered Glass | Non-Homogeneous | Non-Homogeneous | Non-Homogeneous |
| **Feature Detail Level** | High | Very High | Medium | Very High |
| **Maximum Part Thickness (Rough Cut)** | 12 in. (305 mm) | 1.5 in. (38.1 mm) | 6 in. (152.4 mm) | 12 in. (305 mm) |
| **Maximum Part Thickness (High-Quality Cut)** | 2 in. (50.8 mm) | 1 in. (25.4 mm) | 2 in. (50.8 mm) | 12 in. (305 mm) |
| **Typical Accuracy (Thin Stock)** | ±0.001 in. (0.03 mm) | ±0.001 in. (0.03 mm) | ±0.01 in. (0.25 mm) | ±0.001 in. (0.03 mm) |
| **Typical Accuracy (Thick Stock)** | ±0.005 in. (0.13 mm) | ±0.005 in. (0.13 mm) | ±0.03 in. (0.76 mm) | ±0.001 in. (0.03 mm) |
| **Minimum Kerf Width (Thin Stock)** | Pure Water: 0.003 in. (0.08 mm) | 0.001 in. (0.03 mm) | 0.02 in. (0.5 mm) | Wire EDM: 0.0008 in. (0.02 mm) |
| | Abrasive Jet: 0.015 in. (0.38 mm) | | | Typical: 0.013 in. (0.34 mm) |
| **Cut Speed** | Slow to Moderate | Fast to Very Fast | Very Fast | Slow |
| **Cut Surface Quality** | Very Good | Excellent | Good | Excellent |
| **Cut Characteristics to be Managed** | Kerf Taper, Exit Burr | Kerf Taper, Dross | Kerf Taper, Dross, Slag | True to Form |
| **Heat Affected Zone on Cut Surfaces** | None | Narrow Zone | Yes | Narrow Zone |
| **Post-Processing Needed** | Sometimes: Deburr | Sometimes: HAZ and Dross Removal | Sometimes: Stress Relieve, Straighten, Slag/Dross/HAZ Removal | None |
| | | | | Wire EDM: Pre-Processing is Required to Add Pierce Holes |

**Table 2. Examples of Cutting Speeds for Metals**

| Material | Thickness in (mm) | Laser Assist Gas | High-Quality Cut Feed Rate, in/min (mm/s): $CO_2$ Laser 4kW | Fiber Laser 4kW | Abrasive Waterjet 60 HP, 60 ksi | High-Definition Plasma, 260A | Plasma Process Gas/ Shield Gas |
|---|---|---|---|---|---|---|---|
| Aluminum | 0.04 (1.0) | Nitrogen | 420 (178) | 1457 (617) | 225 (95) | — | — |
| | 0.25 (6.4) | | 40 (17) | 100 (42) | 32 (14) | 60 (25) | Air/Air |
| | 0.5 (12.7) | | — | 21 (9) | 14 (6) | 190 (80) | H35/Nitrogen |
| Carbon Steel | 0.04 (1.0) | Oxygen | 280 (119) | 319 (135) | 80 (34) | 215 (91) | Oxygen/Air |
| | 0.25 (6.4) | | 100 (42) | 110 (47) | 10 (4) | 200 (85) | |
| | 0.5 (12.7) | | 56 (24) | 60 (25) | 3 (1) | 115 (49) | |
| | 0.75 (19.1) | | 33 (14) | 33 (14) | 2 (0.8) | 90 (38) | |
| Stainless Steel | 0.04 (1.0) | Nitrogen | 400 (170) | 1456 (616) | 75 (32) | — | — |
| | 0.25 (6.4) | | 80 (34) | 138 (58) | 11 (5) | 60 (25) | F5/Nitrogen |
| | 0.5 (12.7) | | 16 (7) | 39 (17) | 4 (2) | 65 (28) | H35/Nitrogen |
| Brass | 0.04 (1.0) | Nitrogen | N/A | 1338 (566) | 162 (69) | N/A | N/A |
| | 0.25 (6.4) | | | 42 (18) | 23 (10) | | |
| Copper | 0.04 (1.0) | Oxygen | N/A | 1181 (500) | 146 (62) | N/A | N/A |
| | 0.2 (5.0) | | | 48 (20) | 21 (9) | | |
| | 0.5 (12.7) | | | — | 5 (2) | | |

## Mechanical Erosion Processes

Many conductive and non-conductive materials can be machined or cut using mechanical erosion processes that exert relatively low force on the workpiece or part and do not introduce thermal effects. Currently, the most prevalent of these processes are water jet machining (WJM) and abrasive water jet machining (AWJM). Other mechanical erosion processes include ultrasonic machining (USM), abrasive jet machining (AJM), abrasive flow machining (AFM), and magnetic abrasive finishing (MAF).

**Water Jet Machining (WJM).—**Conductive, non-conductive, reflective, thin, and very thick materials can be cut using water jet machining technology, also known as hydrodynamic machining, which includes both abrasive water jet machining and pure water jet machining. In both processes, a pump or intensifier pressurizes water and drives it through a small orifice to form a supersonic jet that exerts force on the workpiece. A typical system pressure is 60 ksi (414 MPa); some systems are capable of up to 90 ksi (620 MPa); and with some pump and orifice combinations, jet velocity can be as high as three times the speed of sound, or 2952 fps (900 m/s). The jet orifice normally is made of jewel, often sapphire or ruby; as the orifice can crack or chip from use, diamond may be used to extend tool life.

One limitation of water jet cutting is its inability to cut into hollow sections without special precautions. Upon cutting through the first wall of the section, the jet will spread rapidly and either just gouge or cut a wide path through the far wall. A sacrificial part can be placed inside the hollow to absorb the jet force, but if the inserted part does not completely fill the space, some blasting damage will occur inside the hollow due to ricochet.

*Water Jet Cut Quality:* Water jet systems can create relatively smooth cuts that require no secondary finishing. Cut quality typically decreases with decreasing speed. Subjective cut quality designations typically used in industry range from Q1 to Q5 and refer to surface quality. Q1 cuts, suitable for rapid separations or roughing, have a heavily striated surface, with curved striations, due to stream lag induced by head travel. Q2 is appropriate for general-purpose through cuts, often producing heavy striations toward the bottom. Q3 cuts are commonly used for general-purpose clean cuts, while Q4 cuts produce a finer surface finish. Q5 cuts have the best surface finish, often targeting a roughness average (Ra) of 80 μin. (2 μm).

Additional cut characteristics, such as accuracy, kerf width, taper (bevel), entrance rounding, and exit burr, are managed through process and maintenance. (Refer to Table 1 and Table 2 for some typical performance characteristics of precision water jet systems.) Obtaining a straight cut edge may require optimizing process parameters or employing a tilting head to compensate. Head component wear affects accuracy and is of particular concern in abrasive water jet processes.

Kerf width, or the width of material removed in cutting, is a function of jet diameter, standoff distance, cutting speed, and taper. With increased standoff distance, kerf width and taper increase. Standoff distance between the nozzle and workpiece typically ranges from 0.063 in. (1.6 mm) to 0.125 in. (3.2 mm). Excessive rounding of the entrance often occurs at standoff distances beyond this range.

Taper angle grows with increasing cutting speed; it is possible to produce a cut with negligible taper, but the process is likely to be slow. High speeds also will produce a tapered cut that is wider at the entrance than it is at the exit. Taper tends to increase in workpieces less than 0.125 in. (3.2 mm) thick, so stacking parts may be used to decrease taper.

Very slow speeds or very soft workpieces can result in a reverse tapered cut that is narrower at the entrance than it is at the exit; minimizing standoff distance can reduce this taper. When the workpiece is very thick, cuts may exhibit barrel taper, where the kerf is widest halfway through the cut, and narrowest at both entrance and exit. When cutting through ductile materials, the presence of a small burr at the cut exit is likely, but can be minimized through speed and standoff reduction.

*Abrasive Water Jet Machining (AWJM):* This process is appropriate for cutting hard and/or thick materials, including certain metals, ceramics, and stone. Metal workpieces less than 2 in. (50.8 mm) thick are well suited for abrasive water jet machining, and even thicker metals in excess of 12 in. (305 mm) thick can be cut slowly. Supersonic water jets pull in metered, abrasive particles just after the orifice, and this solution travels down the mixing tube to the nozzle outlet.

The abrasive material often is garnet or similar material, with a particle size selected according to the nozzle opening, workpiece, and process. Grit sizes commonly used range from 50 mesh to 220 mesh (higher numbers indicate finer particles) in the Tyler Mesh Sizes (see *Meshes, Sieves, and Screens*, page 1056). An 80-mesh sieve size (0.007 in. or 0.177 mm) is most commonly used. The orifice and mixing tube combination determines the abrasive water jet stream diameter, which can be as small as 0.015 in. (0.38 mm), though 0.03–0.04 in. (0.76–1 mm) is more common. Nozzle mixing tubes wear rapidly with use; this can affect jet size and location.

*Pure Water Jet Machining:* Use of this process is limited to low-strength materials such as foam, fabric, and some plastics. The food industry often uses pure water jet cutting because it can be run as a sanitary process. As the jet contains no particles, very small jet sizes are possible. Typical orifices range from 0.003 in. to 0.015 in. (0.08 mm to 0.38 mm).

**Ultrasonic Machining (USM).—**Also referred to as ultrasonic vibration machining, this abrasive method makes cuts in material or changes its surface texture using a transducer that vibrates a tool at low amplitude (25 to 100 microns) and high frequency (15 to 30 kHz) and in a slurry containing fine abrasive particles. The vibrations force the particles against the material surface, removing material to the desired depth. For further information, see page 1260.

**Abrasive Flow Machining (AFM).—**Often used for deburring, smoothing, or polishing an interior part surface, but also applicable to final shaping, breakthrough, and drilling holes, this process forces a viscous fluid, containing abrasive material, through a workpiece.

**Magnetic Abrasive Finishing (MAF).—**This finishing process can be used to create a high surface quality, even on freeform parts. It involves adding micron-sized iron particulates to a flexible magnetic abrasive brush that can easily deform, based on the magnetic field strength, matching and honing a part, regardless of the complexity of the shape.

## Electrothermal Processes

These processes use melting, burning, or vaporization to remove material from a workpiece. Included are electrical discharge machining (EDM), laser beam machining (LBM), plasma arc cutting (PAM), and electron beam machining (EBM). The heat of these cutting methods can cause distortion, discoloration, changes in surface hardness or composition, and introduction of residual stresses in the workpiece.

To specify and evaluate quality and tolerances of cuts made with oxyfuel flame systems, plasma arc systems, and laser cutting systems, refer to ISO 9013, "Classification of thermal cuts – Geometrical product specification and quality tolerances." Below are the most commonly used electrothermal processes. Additional information on some of these processes is included in the *ELECTRICAL DISCHARGE MACHINING* on page 1469 and the *WELDING* section, starting on page 1576.

**Electrical Discharge Machining (EDM).—**This process vaporizes conductive materials in contact with an electrode in a dielectric fluid bath. Both through-cutting wire systems and three-dimensional die-sinking systems are widely available and highly accurate. Refer to Table 1 for some typical performance characteristics of EDM processes. For detailed information, see page 1471.

**Laser Beam Machining (LBM).—**Lasers can cut both metals and nonmetals through a process of vaporization, fusion (melt and blow), burning, or thermal stress cracking. Shallow engraving, blind features, and through cuts are all possible. Lasers can cut delicate parts, since there is no contact; however, laser processes generally use gas jets to create a reactive environment or eject material, and that may produce some pressure on the part in the area of the cut.

Material, geometry, and type of cut determine the best type of laser to use. $CO_2$ (gas) laser systems and Nd:YAG (rod crystal, solid state) laser systems have been used in industry for many years. Fiber lasers are a newer solid-state technology that uses an Yb-doped optical fiber to amplify, contain, and direct a diode-generated laser beam; when the amplified light exits the fiber, a lens system collimates and focuses the beam. Thin disk (Yb:YAG) lasers are another new solid-state technology that provides power and beam quality comparable to fiber lasers. Most laser cutting systems in industry today use either fiber lasers or $CO_2$ lasers, and thin disk lasers are gaining in popularity.

Fiber and disk laser systems have the advantage of providing efficiency typically around 30 percent, while $CO_2$ lasers typically operate at 6 to 8 percent efficiency. Additional advantages of fiber and disk lasers are that they are made of solid-state components and do not require the high-speed turbines used in $CO_2$ laser systems, reducing periodic maintenance costs. However, $CO_2$ lasers often are preferred for cutting nonmetals, such as ceramics, plastics, and various composites, due to their longer wavelength.

Most laser systems operate within an interlocked enclosure capable of blocking harmful light emissions, enclosing moving parts, and managing cutting fumes. Unlike $CO_2$ systems, fiber and disk laser light is delivered through an optical fiber, eliminating the need for periodic cleaning and alignment of the delivery system.

Wavelengths for common industrial lasers are as follows: $CO_2$ = 10.6μm, Nd:YAG = 1.06 μm, thin disk Yb:YAG = 1.03 μm, and fiber = 1.07 μm. Due to increased absorptivity of the laser beam at shorter wavelengths, high cutting speeds are possible when cutting steel less than 0.2 in. (5 mm) thick with solid state lasers. For example, fiber-delivered systems can cut up to three times faster than $CO_2$ systems at the same power level.

*Laser Cutting Reflective Materials:* Laser cutters, $CO_2$ systems in particular, are vulnerable to damage due to back-reflection when processing reflective workpieces. Back-reflections that reach the laser cavity can cause instabilities in laser output or even destroy laser components. Most high-power $CO_2$ laser manufacturers do not warranty their equipment for processing highly reflective metals, and devices available to protect $CO_2$ lasers from unwanted back-reflection are bulky, expensive, and impractical for high-powered industrial uses.

For fiber-delivered lasers, both the forward- and backward-traveling light is confined to the optical fiber. So back-reflection isolators designed to protect the laser can be compact fiber components, incorporated either near the output of the delivery fiber or at the output of the laser.

When the workpiece is reflective, some portion of laser light is reflected instead of being absorbed to supply cutting energy. Metals reflect laser light at increasing percentages with increasing laser wavelength. Increasing laser power or using a shorter wavelength can help overcome reflectivity problems.

*Laser Cutting Assist Gas:* A gas jet emitting from the laser head is used to eject molten material from the cut, protect the cutting head from splatter, and either react with the material or shield it. Oxygen assist gas reacts with the workpiece material and leaves an oxidized cut edge. The exothermic reaction adds energy to the cutting process and can enhance the depth or cutting speed; however, secondary operations may be necessary to remove an oxide layer if the laser-cut metal part is to be welded or painted.

Low-pressure oxygen assist gas can be used when cutting carbon steel. It performs this function economically due to the low flow rates and increased thickness capability afforded by the exothermic reaction. Oxygen assist can make cuts in stainless steel or aluminum possible with power levels that would be inadequate without oxygen. This facilitated initial development of industrial laser cutting, when laser power was limited and very expensive. However, use of oxygen assist to cut stainless steel has some disadvantages: cut edge discoloration (which may be unacceptable for cosmetic reasons), reduced corrosion resistance (due to the formation of chromium oxides), and dross formation (due to the higher melting point and surface tension of chromium oxide). For these reasons, and with the increased viability of higher laser power, the use of oxygen in laser cutting of stainless steel is no longer common.

In current practice, high-pressure nitrogen is the most commonly used assist gas for cutting stainless steels and aluminum with high-power lasers. This inert gas's shielding effect results in a higher-quality cut than can be obtained with oxygen. Nitrogen-assisted cut edges suffer no discoloration and are oxide-free, so they are ready for subsequent welding or painting. Since the assist gas is inert, there is no addition of energy to the process, and achievable cutting speeds are lower than with oxygen. The high viscosity of molten steel (compared to that of iron oxide) can lead to dross adhesion at the bottom of the cut. This can be addressed by using very high assist gas pressures, resulting in high consumption of pure nitrogen and increased operating costs.

Air-assisted cutting is often used with aluminum, but the air must meet high cleanliness and pressure requirements. The presence of reactive oxygen in air can enhance cutting speed, while still producing a clean cut edge in aluminum; however, this technique can require high investment costs for a setup that delivers clean air at elevated pressure and high flow rates. Instead, laser cutting with shop air, which is both inexpensive and commonly available in industrial settings, can be used to cut relatively thin steel (both carbon and stainless steel) far more economically than using pure oxygen or nitrogen.

Assist gas flow, pressure, purity, and delivery (standoff distance and jet shape) all affect cut characteristics, necessitating careful selection and monitoring of gas pressure, nozzle size, feed rate, laser power, and material thickness. For nitrogen-assisted cutting, high gas volumes and pressures are needed to successfully cut narrow kerfs, and selection of the gas nozzle size will directly affect volume capability. In many cases, a larger nozzle diameter is needed to cut a narrower kerf. For oxygen-assisted cutting, gas pressure plays a key role in determining cut quality: low pressure can result in incomplete cuts, while high pressure can lead to uncontrolled burning and a rough cut edge.

Laser cutting equipment manufacturers provide a variety of nozzle types designed to provide assist gas flow characteristics tailored for optimal performance. Performance requirements and user preferences determine system priorities, such as cut edge quality (roughness, angle), cutting speed, gas consumption, and so on. Fig. 1 shows a laser cutting head with a commonly used double-nozzle design, which can accommodate two assist gas streams: an inner, primary assist gas stream and an outer, shield gas stream. These two streams could be fed by different gases, or by different supply pressures of the same assist gas to produce the desired outcome.

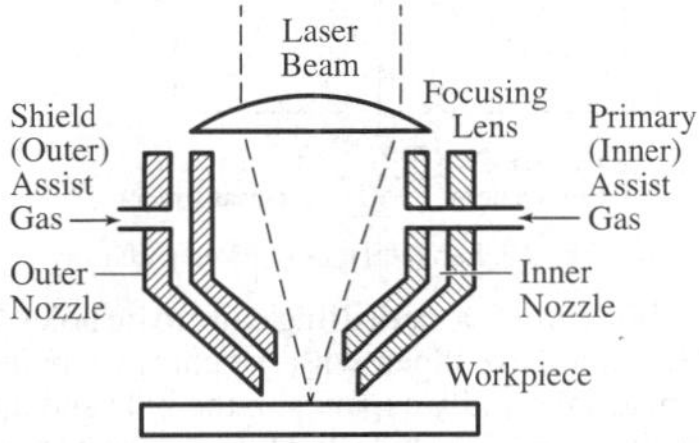


Fig. 1. Laser Cutting Nozzle with Two Assist Gas Pathways

*Laser Cutting Heat Effects:* Laser cutting and machining normally exposes workpieces to high thermal energy for a brief period. This leads to formation of a narrow heat-affected zone (HAZ) and minor heat effects in the part. Plastics and composites generally are much more vulnerable to heat effects than metals; some materials, such as PVC, will give off hazardous fumes, and others may discolor, warp, or char.

The thickness of the HAZ for a laser cut in thin material is small, often under 0.005 in. (0.13 mm), and is based on beam focus, assist gas, speed, and part material and thickness. The HAZ will be thicker around a cut in thicker material than in a thinner piece. Pulsing the laser can help to reduce HAZ thickness and other heat effects. Ultra-short pulse laser systems, sometimes called femtosecond lasers, are available that create no HAZ in applicable materials.

Efficiency loss during laser cutting occurs primarily due to heat conduction in the workpiece. In thermally conductive materials, a portion of the absorbed laser energy is conducted away from the cut zone. This diverted energy performs no useful work and only serves to heat the material. As a result, efficiency may dictate reduced cutting speeds and cut depths for highly thermally conductive metals, such as aluminum, brass, and copper.

*Laser Cutting Performance Characteristics:* Modern laser systems in the 6–8 kW range can cut through 1.5 in. (38 mm) steel, 0.75 in. (19 mm) aluminum, or 0.38 in. (9.6 mm) copper. The depth of cut is proportional to laser power and inversely proportional to speed. While fiber lasers are much more efficient and usually faster for cutting thin materials, $CO_2$ lasers usually are as fast as fiber systems when oxygen (flame) cutting materials over 0.2 in. (5 mm) thick. The cutting tolerance of a typical system on 0.5 in. (12.7 mm) thick carbon steel is ±0.005 in. (0.13 mm). Higher precision is possible in some cases. Table 1, Table 2, and Fig. 2 provide some examples of expected performance characteristics.

In thin workpieces, laser systems can create extremely small holes and fine features with a great deal of precision. Cut patterns can be as close as one beam diameter apart, if the cutting process is accurate and material thin. Kerf width will depend on wavelength, setup (spot size), process, and workpiece characteristics. A fiber laser has a wavelength ten times smaller than that of a $CO_2$ laser. As a result, a fiber laser can focus on a much smaller spot size, and kerf widths of 0.001 in. (0.025 mm) are achievable. Deeper cuts require a larger beam and wider kerf.

Beam shape is not cylindrical, so cuts will tend to have a wider entrance than exit. A 2 degree taper angle is typical. Reduction of this taper angle is possible by changing head angle, process parameters, standoff distance, or material thickness. Zero taper can be achieved in some cases.

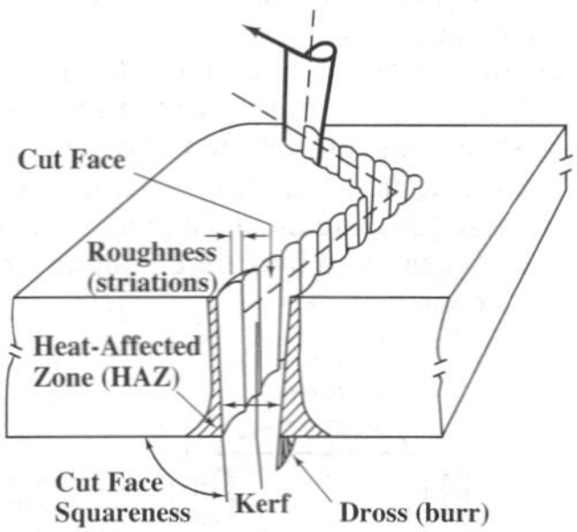


Fig. 2. Characteristics of Laser Cuts

*Laser Hole Cutting and Drilling:* Laser drilling can create holes that are blind or pierce through. The holes will tend to have taper, with the entrance being larger than the exit. When drilling, molten material rapidly expands in the hole and ejects upward. Some of this material can cling to the sides of the hole and form a recast layer.

In some cases, a single-shot direct drilling process can rapidly create a shallow drilled hole with minimal recast. A low-power pulsed beam can create holes of greater depth, smaller holes, and higher precision holes. Called percussion drilling, this method is slower than direct drilling and more prone to form a recast layer.

Helical drilling and trepanning can produce large diameter holes with high precision, minimal recast, and controllable taper. These methods require more sophisticated equipment and take longer than direct or percussion drilling. Helical drilling involves moving a pulsing laser along the circumference of the desired hole, while increasing the cut depth on each pass. Trepanning starts with percussion drilling a pilot hole, and then moving the pulsing laser in a spiral pattern, until it reaches the target diameter. Unlike the other drilling methods, trepanning begins with a through hole, allowing downward ejection of molten material.

**Plasma Arc Machining (PAM).—**Conductive metals, including expanded metal, cast iron, and rusted or painted parts, can be plasma cut. In this electrothermal process, an electric arc is generated between the plasma torch and workpiece. The plasma arc melts the material and a high velocity gas jet removes the molten material.

A plasma cutting torch consists of an electrode that serves as an attachment point of the arc and a nozzle used to confine the plasma arc and direct the high velocity gas jet toward the workpiece. This gas jet further confines or constricts the arc and protects the nozzle from the high temperature plasma. The core of the plasma arc is made of ionized gas atoms and electrons, which exit the torch nozzle at temperatures near 40,000° F (22,000° C). The plasma arc transfers energy via electrons re-entering the conductive workpiece, thermal conduction, and recombination of the plasma ions.

Some plasma torches may use a secondary or shield gas flow to protect the nozzle from spatter, electrically isolate the nozzle from the workpiece, and improve the cutting process. Other plasma torch enhancements include liquid cooling of torch components, high-precision nozzles to enhance confinement of the plasma jet via additional gases or nozzle geometry, or use of magnetic fields and the resulting Lorentz forces to constrict the arc. Efforts to constrict the arc as much as possible are meant to narrow the plasma jet to produce a narrow kerf, as well as increase the energy density of the plasma to allow a higher range of cutting speeds.

Some plasma torch components wear out and need to be replaced periodically to maintain cut quality and reliable operation. Such consumable items include the electrode, nozzle and shield.

*Plasma Cutting Performance Characteristics:* Plasma cutting is a highly flexible method of cutting metal. Small portable systems are available for large or difficult-to-access workpieces. Some plasma cutting processes can operate on materials submerged in water to reduce noise, fumes, and light emission. Plasma torches also can be used in multiple orientations, enabling three-dimensional cutting.

Plasma systems can make both through cuts and non-through gouges. It is possible to cut some materials up to 6 in. (152 mm) thick, but cut quality tends to suffer for thicknesses beyond 2 in. (50.8 mm). Since this is a thermal process, an HAZ will surround each cut. Depending on material and process parameters, an HAZ less than 0.01 in. (0.25 mm) thick is possible with little edge hardening. Cut quality characteristics, including cut smoothness, bevel angle, kerf width, and the presence of slag or dross on the bottom of the cut, are highly dependent on material and process parameters. Smoothness of the cut is generally good but can deteriorate at very high speeds or for thicknesses above 2 in. (50.8 mm).

A plasma cut usually has a positive bevel angle, where the entrance to the cut is larger than the cut exit. Plasma cuts also have two different bevel angles on the two sides of the cut kerf. This difference is a result of the plasma gas flow swirl component relative to the direction of torch motion. As a result, the plasma torch produces a "good" cut side, with a typical bevel angle of 2 degrees, and a "bad" cut side, with an angle close to 5 degrees. Cutting speed and standoff are the most direct ways to influence bevel angle. High cutting speed and standoff distances increase bevel angle, while lower speed and standoff lower the angle. Other factors include type and pressure of the gas, arc current, and nozzle characteristics. Precision systems with the right process settings can improve on this, in some cases, providing a straight cut (0 degree bevel).

While kerf width is process and workpiece dependent, it often is estimated as 1.5 times the nozzle orifice diameter. Kerf width is typically larger than 0.02 in. (0.5 mm) and often near 0.1 in. (2.5 mm). High standoff, low speed, and high amperage will widen the kerf.

Holes cut with a standard plasma torch should be larger in diameter than 1.5 times the material thickness, and larger than 3/16 in. (4.8 mm). State-of-the-art automated plasma systems can produce holes that are cylindrical and accurate enough to rival drilled holes, with a diameter to thickness ratio as low as 1:1. Some automated table systems also can tilt the torch relative to the workpiece, allowing additional control of the cut bevel angle or produce angled cuts.

Dross formation depends primarily on cutting speed. Speeds that are too slow will produce thick dross on the bottom of the cut; speeds that are too fast will produce a fine, beady dross. For a given application, there usually is a speed range for dross-free cutting, though there may be no dross-free speed for very thick workpieces.

Gouges are made with plasma systems by tilting the torch relative to the workpiece, so the molten material is blown away on the top of the plate. This technique may be used to remove material in preparation for repair or re-work.

*Plasma Cutting Assist Gas:* Plasma cutting gas selection for best cut quality depends on material type and thickness. Oxygen is the best gas for cutting mild steel, because it allows the highest cutting speed for a high-quality cut. Air also can be used to cut mild steel, but nitriding can occur and should be considered if the cut edge is to be welded.

Stainless steel can be cut with good speed and bevel angle using air, but the cut edges will be rough and black because of oxidation and may require secondary cleanup operations. Using nitrogen plasma gas on stainless steel reduces oxidation and gives a smooth cut surface but still adds dark color, especially with material thicker than 0.25 in. (6 mm). Adding a water shield flow when using nitrogen plasma gas can give an edge color that is close to the base material color and improve the taper and rounding of the cut edge.

Plasma gases containing hydrogen provide a further improvement in the appearance of the cut edge. F5 plasma gas (95 percent nitrogen, 5 percent hydrogen) can give bright silver cut edges in stainless steel up to 0.375 in. (9.5 mm) thick, while H35 (65 percent argon, 35 percent hydrogen) improves the edge color for thicknesses greater than 0.5 in. (12.7 mm). Nitrogen is the most common shield gas when cutting with hydrogen-containing plasma gases. Aluminum cutting can be done with air, nitrogen, or hydrogen-containing gases. Edge quality is often rougher with the oxygen-containing gases and smoother with the hydrogen gas processes. Some modern systems can mix gases, such as argon, hydrogen and/or nitrogen to improve results on both stainless steel and aluminum.

**Electron Beam Machining (EBM).—**This non-contact machining method employs a beam of high-velocity electrons, usually applied in a vacuum, to make cuts and gouges. As the electrons collide with the workpiece, their kinetic energy produces heat and vaporization of material. This process is often chosen to perforate thin materials at high speeds. Almost any material up to 0.4 in. (10mm) can be machined with this method, usually without burr, and with relatively minor thermal effects. Tapered holes and cuts occur when part thickness exceeds 0.005 in. (0.13 mm). The finish and speed of electron beam hole drilling are superior to most other methods, but the capital equipment cost is high. Electron beams also are used in hardening, diffusion bonding, powder bed fusion, and welding processes.

## CNC PROGRAMMING AND CAD/CAM

### CNC (Computer Numerical Control) Programming

Computer numerical control programming—better known as CNC programming—originated from the technology known as Numerical Control (NC). This technology can be defined as an operation of machine tools by means of specifically coded sequential instructions to be processed by the control system. These instructions are combinations of letters of the English alphabet "A-Z", all ten digits "0–9", and a few selected symbols.

The main purpose of a CNC program is to provide detailed instructions necessary to machine a part (also called a workpiece) for a given machine setup—with minimal human interaction. All instructions to the control system must be written in a logical order and in a specified format called the program structure. The resulting part program can be stored on various media for the future and used repeatedly to achieve identical machining results at any time.

A completed part program is processed by the CNC unit—commonly called the control system—and individual program instructions are carried out to complete desired actions of the CNC machine. These actions include all required machine activities, such as motion control, spindle speed, cutting feed rates, coolant functions, program flow control, various adjustments, etc. To develop the CNC program, a qualified person—typically called a CNC programmer—is assigned to the task.

As a key person in the production environment, the CNC programmer has the responsibility for developing safe, efficient, and error free programs for CNC machines installed in the machine shop. Many CNC programmers have machining skills and hands-on experience as machine operators working on manual or CNC machines. Typical skills of a CNC programmer include the ability to interpret drawings, knowledge of setup and tooling, and the very important understanding of "*how to machine a part*." The programmer has to be able to visualize all tool motions and recognize any restricting factors that may be involved. The programmer collects, analyzes, processes, and logically integrates all data into a single, production ready, part program. Equally important skills include mathematical aptitude and knowlege, particularly the ability to solve equations and trigonometry. The knowledge of manual programming is essential even in the age of computers and computer-based programming software. Another important quality of a professional CNC programmer is his or her ability to listen to and work with other people such as engineers, operators, managers, and customers.

## CNC Coordinate Geometry

Key to working with CNC principals is understanding the geometry of machine tools. This encompasses the relationship between the machine data, part data, and tool data, which includes setup.

As with all CAD/CAM systems (see *CAD/CAM* on page 1377), CNC is based on the same principles of a system of coordinates that define the location of a point in three dimensional space (3D). This system of coordinates is founded on the concept of two perpendicular lines (named axes "X" and "Y"), intersecting at a point called the *origin*, where both coordinates have a value of zero (X0, Y0). Both lines are divided into equal units of measurement.

A point can be defined in a plane (any two axes = 2D) or in space (three axes = 3D). For CNC work, the *rectangular coordinate system*—also known as the *Cartesian coordinate system*—is the most commonly used. It is based on three standard axes: X,Y,Z. *Points* are defined as locations with a distance from the origin defined by projecting a line at 90 degrees to each axis, forming a visual rectangle. For example, in Fig. 2, P1 is defined as X3.0, Y2.0, P2 is defined as X-4.0, Y1.0, P3 as X-2.0,Y-3.0, and P4 as X2.0, Y-2.0.

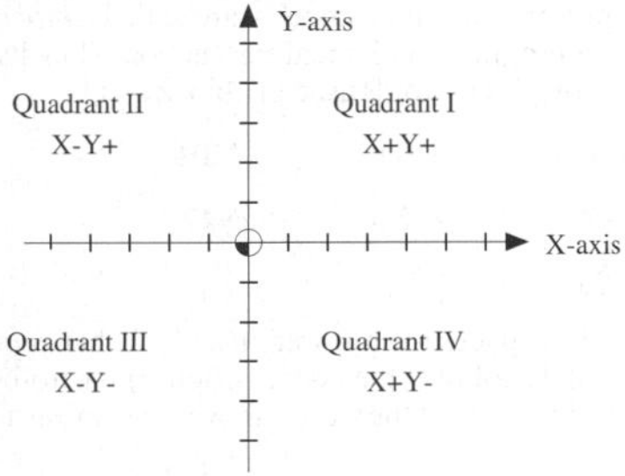


Fig. 1. Rectangular Coordinate System.

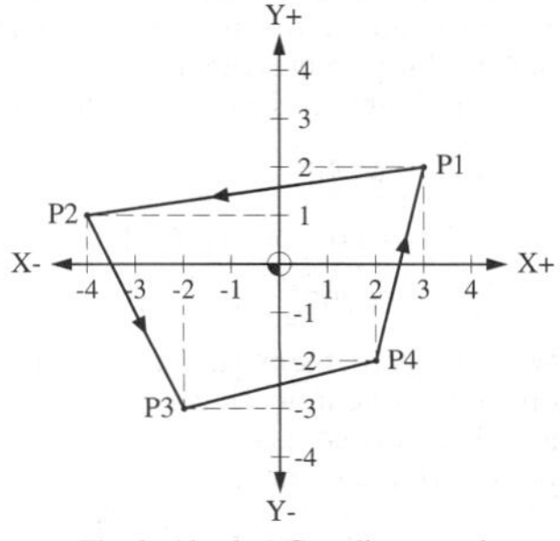


Fig. 2. Absolute Coordinates and Incremental Linear Motion.

Once the point locations are established, the programmer can use these points to specify the center of individual holes to be machined or as endpoints for a continuous contour. The points can also be connected by providing specific motion instructions, indicating a linear toolpath between given points, from P1 to P2 to P3 to P4 and back to P1 (shown by the solid lines in Fig. 2).

## CNC Programming Process

CNC program (also called *part program*) development starts with an engineering drawing and material selection. Initial decisions include tasks to be done and goals to be achieved. The part program has to provide a safe and efficient method of machining a part, it should have a consistent structure (method of data entry), and be as portable as possible. Program portability allows the same program to run on machines that have similar characteristics. The typical, logical program development process can be summarized in the short, general list shown below (in actual use, specific steps will vary):

Program Development List

a) Evaluation of engineering drawing and material.

b) Sequence of machining operations.

c) Tooling selection and arrangement of cutting tools.

d) Part setup method.

e) Technological data selection (speeds, feed rates, depths, etc.)

f) Toolpath determination.

g) Mathematical calculations.

h) Program writing and preparation for transfer to CNC.

i) Program testing.

j) Program documentation.

**Word Address Format.—**Writing any data to be processed by a computer requires a certain input format called a program. CNC program instructions are written in a special coded form following strict rules required by the programming language. Where a single letter is followed by one or more digits, the letter must always be first. Special symbols, such as a decimal point (.), negative sign (–), backslash (/), percentage sign (%), colon (:), semicolon (;), and parenthesis, may also be used, if required. Each letter, digit, or symbol represents one program character, as well as one character in the control memory. A *character* is the basic unit of a CNC program. CNC programs only use characters that are readily available on a standard computer keyboard, without any multi-key combinations. A *word* in CNC program consists of two or more characters and is a single instruction to the control system. It can be used individually or it can be grouped into a series of combined instructions (words), to be performed at the same time. A program word forms the alphanumerical *unit of instruction* to the control system. Instructions performed simultaneously are called a *block*, which is always represented by one line of the program. Individual instructions (blocks) are written in sequential order. Some examples of typical words used in a block are:

| | | | | |
|---|---|---|---|---|
| M08 | D14 | X7.875 | H03 | /M01 |
| Y0 | G01 | S850 | Z-2.5 | N49 |
| T0707 | T05 | F10.0 | | |

Within a word, there cannot be any spaces. But spaces are allowed in a block, between words only. The numeric value that immediately follows the address (letter) has to be entered in a format specific for that address. Using some of the words above, a meaningful block can be constructed:

N49 G01 Z- 2.5 F10.0 M08

where:

N49 = Block number (sequence number)

G01 = Linear motion selected

Z-2.5 = Z-axis motion only

F10.0 = Feed rate specification

M08 = Coolant ON function

An early Standard of Electronic Industry Association (EIA), RS-274 from the 1960s, defines various commands and functions for part programming. It was last updated in February 1980 to RS-274D. Yet many features of this EIA Standard, which established what is commonly referred to as *G-code*, are used to this day.

**Program Development.—**There are several methods of programming CNC equipment that will generate a working part program:

1) Manual programming (also known as G-code programming)

2) Computer-aided manufacturing (also known as CAM programming)

(a) Simulation and verification

(b) Feature-based machining

3) Shop-floor programming (also known as conversational programming)

4) Macro programming

*Manual (G-code) programming* requires the most effort, is time consuming, and prone to errors. The programmer writes every instruction separately, typically using a text editor on a computer to enter programming instructions. Usually, calculations are done with the use of a pocket calculator. In spite of its disadvantages, manual programming has two major advantages: (1) It offers complete human control over every detailed aspect of toolpath development and all related activities, and (2) it provides an important knowledge base for part programming using computer methods.

Originally, all manual programming and calculations were done by hand and written on paper. Now, computers that include text editors and software-based calculators can be used to write the program using the keyboard and display. Using today's sophisticated CAD/CAM software, the programmed toolpath can be verified in a simulation graphically presented on the computer screen.

*Computer-Aided Programming:* Computer-generated programs are based on graphical input, and typically require development of part geometry (point, lines, arcs, etc.). The final edited geometry serves as the toolpath geometry and can be assigned machining parameters. Specifying tool definition, spindle speeds and cutting feeds, and various cutting conditions is part of the programming process. A good knowledge of manual programming provides many necessary skills for computer-aided part programming.

Any CAD/CAM system that provides CNC capabilities also has to provide precise formatting of the final program output. This output is always based on a particular machine tool and control system combination. Software vendors offering program development software are very much aware of this need, yet face the extremely difficult task of covering every machine tool design with every control system available. When the toolpath is initially processed, it is generic—it has no connection to any particular control or machine specifications. The same toolpath has to be processed again (post processed) for a specific type of formatting, unique to a particular machine and control. The majority of post processors are customizable by users who have applicable knowledge of manual programming methods, as well as a various machine and control systems.

The major advantage of computer-aided program generation is elimination of tedious calculations, minimizing errors, and providing a program in much shorter time than when using the manual approach. Graphical interfaces quickly allow obvious errors to be detected visually. For more on this topic, see *CAD/CAM* on page 1377.

*Shop-Floor Programming:* Often called *conversational programming*, shop-floor programming is performed when a CNC programmer/operator uses system software to write programs directly on a machine tool controller. The operator interactively defines material(s), tools, setup, toolpath geometry, cutting conditions, etc. The resulting program is in a G-Code format, but the operator may not see the code on the display (knowledge of G-Code sub programs is still desirable). The program is saved on the control system and/or stored on a portable memory device for use.

Although relatively fast, visual, and easy to edit, conversational programming software may not offer the same flexibility in toolpath control as manual or CAM programming methods. While suitable for simple parts and production of small lot sizes, complex parts and larger jobs may benefit from CAM-generated programs; programming at the machine control also can reduce production time. Separate computer programming software may address this problem, but generally is specific to the machine tool manufacturer, lacking industry standardization.

The steps in shop-floor programming are similar to those in manual programming. First, the technical part drawing is examined for work holding and tool considerations. It is efficient to establish a tool library file, containing representations of all shop tools. On some controllers, feeds and speeds can be calculated automatically based on defined workpiece and cutting tool materials, as well as cutting condition parameters.

Conversational programming software allows definition of machined features, including patterns of holes, pocketing, automatic roughing, corner rounding and chamfering, threading, and other operations. Much like CAD/CAM, programming involves (a) recreating the part geometry, using lines, arcs, points, and other features, (b) identifying tools and cutting condition parameters, and (c) generating the toolpath code to control the machine. Typically, the machinist/programmer answers a series of questions (called "question answer format" or "prompting") to form the part program. Functional icons and buttons on the controller help identify machining operations, such as drilling, contouring, and face machining.

Controls with conversational programming enable graphically checking toolpaths as completed to verify program accuracy. If the program is flawed or missing information, the controller will not execute the toolpath; a user prompt/alarm will be displayed on the screen identifying the problem and where it occurs. Another capability is the ability to perform calculations when programming data is unknown or missing from the engineering drawing. The programmer constructs intersection points or tangency points; the software computes the desired geometry and inserts data into the program. (These automated steps resemble the formerly popular automatic programmed tool, or APT, programming method.) When the program is completed and toolpath verified by graphical simulation, machine setup can begin. The operator takes over, measuring tool and work offsets, and performing a final program test by dry run, before the first CNC machining begins.

*Macro Programming:* The highest level of manual programming, macro programming requires thorough knowledge of applicable programming concepts. In principle, the single purpose of a macro program is to develop a generic toolpath or other activity for a particular purpose. Once the macro program is stored in the control memory, it can be used many times in the future—with variable data supplied for a specific job. Typical examples include a bolt circle or a rectangular pocket. In both cases, the basic input features remain the same, but the actual dimensions change from part to part. All basic shapes of a bolt circle or pocket are retained while their sizes change—the macro itself does not change. See page 1365 for detailed information on macro programming.

**Control System.—**The control system, the actual computer controlling a CNC machine, is the main core of any CNC machine activity. It has many features that are important not only to the machine operator, but to the programmer as well. Some features are unique to a particular type of control system (mill, lathe, wire, etc.), while others are optional and may not be available on all systems.

Typical features of various CNC systems include:

*ON / OFF switch:* Power switch for the main power and the control unit.
*Cycle start:* Starts program execution or an MDI command.
*Emergency stop:* Stops all machine activity and turns off power to the control unit.
*Feedhold:* Temporarily stops motion of all axes.
*Single block:* Allows a program to run one block at a time.
*Optional stop:* Temporarily stops the program execution (requires M01).
*Block skip:* Ignores blocks preceded with a forward slash ( / ) in the program.

*Dry run:* Enables program testing at accelerated feed rates (without a mounted part).
*Spindle override:* Overrides programmed spindle speed within a percentage range.
*Feed rate override:* Overrides the programmed feed rate, within a percentage range.
*Chuck clamp:* Displays current status of chuck clamping (Outside / Inside).
*Table clamp:* Shows current status of table clamping.
*Coolant switch:* Coolant control ON / OFF / AUTO.
*Gear selection:* Shows current status of working gear range selection.
*Spindle rotation:* Spindle rotation direction indicator (CW or CCW).
*Spindle orientation:* Manually controls spindle orientation.
*Tool change:* Switch allowing a manual tool change.
*Tailstock switch:* Tailstock and/or quill switch to manually position tailstock.
*Index table switch:* Manually indexes machine table during setup.
*M-S-T lock:* Disables machine, spindle, and tool functions in the program.
*Machine lock:* Disables all machine motions.
*Z-axis neglect:* Milling Z-axis can be turned off during program verification.
*MDI mode:* Manual Data Input mode.
*AUTO mode:* Allows automatic operations.
*MEMORY mode:* Allows program execution from memory of the CNC unit.
*DNC mode:* Allows program execution from an external device.
*EDIT mode:* Allows changes to programs stored in the CNC memory.
*JOG mode:* Allows manual axes movement operation during setup using jog buttons.
*HOME mode:* Allows manual or automatic return to machine zero.
*HANDLE mode:* Manual Pulse Generator (MPG) allows manual axes movement during setup using the jog buttons.
*RAPID mode:* Selects the rapid mode for setup.
*Memory access:* This key (switch) allows access to the program stored in memory.
*Lube light:* Indicates low level of slide lubrication.
*Error light:* When red, indicates a program or other error.
*Optional stop:* With switch active, the cycle will stop when M01 is encountered in the program.
*Block skip:* With switch active, the line/lines of code preceded with the slash (/) symbol will be skipped.

*Manual Data Input (MDI):* This feature of the control panel allows quick entry of program data. An operating mode can be selected as MDI, which allows program data to be entered by the machine operator. MDI is not intended to enter complete machining programs, but enables the operator to make adjustments during setup. Typical applications include input of position data for setup, tool motion to a given location, changes to machine functions control, even some special settings of machine parameters. Knowledge of basic programming principles and language is necessary, as MDI requires the use of normal programming methods.

## CNC Program Data

The program data is a collection of all the program entries that will be accepted by the CNC system. Although the basic program principles are the same, the actual methods of inputting will vary to a certain extent with different control models.

**Program Structure.**—A CNC program structure is a method of developing a part program that is well designed and written. This means that the program should be easy to read on paper or computer screen and written in a consistent manner. Program initialization should be logical; the same tool(s) should be used for multiple operations when feasible to minimize tool changes; and the program should include important comments and be as short as possible. Finally, a part program should always be designed with ease of use for the machine operator in mind.

*Comments and Messages:* Various comments and other messages can be included in the program, provided they are enclosed with a single set of matching parentheses. Comments, when used in moderation, are an excellent way to inform the machine operator about key issues relating to the job at hand.

Typical comments should be in capital letters (for best compatibility between programs) and contain a short message, description, or caution. Comments may be used as separate blocks or at the end of a block following the program data—for example, both of the following comments are valid:

| | |
|---|---|
| Comment as a whole block: | (CHECK HOLE DEPTH) |
| Comment as part of a block: | N55 M00 (CHECK HOLE DEPTH) |

*Program Number (O-Address):* Program numbers are assigned by the programmer and accessed at the control by the operator. Program number requires the letter "O" to be followed by four or five numeric digits (O1 to O9999 or O99999), depending on the control. Macro and subprogram numbers use number series 8000 and 9000.

*Block Number (N-Address):* One or more instructions can be programmed in a single program sequence (block). A typical program may only require a very small number of blocks for a simple machining job, but, literally, hundreds or thousands of blocks for complex work. To distinguish one instruction (one block) from another, a simple numbering system can be used for each block sequence. Depending on the control system, the block numbers can be ascending, descending, or even mixed. Most control systems offer all these options, and it is up to the programmer to choose the most advantageous ones.

The address "N" is used for block numbering. The smallest block number is N1 (N0 is usually not allowed), the largest is either N9999 or N99999, depending on the control system. Block numbers can be used for each program sequence, or they can be used selectively (at each tool change), to enable more efficient search.

A forward slash symbol "/" placed in front of the N-address identifies a block that can be selectively skipped during processing. This *block skip* function is activated or deactivated from the operation panel of the CNC machine.

*Preparatory Commands (G-Address, G-Codes):* Preparatory commands are commonly known as G-codes. The address "G" identifies a preparatory command, which has one and only objective: to preset or to prepare the control system for a certain desired condition, or to a certain mode or a state of operation. For example, the address G00 presets (prepares or selects) a rapid motion mode, G01 selects a linear motion, G81 selects the drilling cycle, etc. The term *preparatory command* indicates its meaning—a G-code will prepare the control to correctly interpret the programming instructions following the G-code in a specific way.

Standard G-codes are programmed in the range of G00 to G99, although not all numbers are available. G-codes with three digits usually apply to special machine functions provided by the control manufacturers. Many G-codes are standard for various controls, and most, but not all, are *modal*. A modal G-code in one that is programmed only once and remains in effect until changed or canceled. Non-modal G-codes have to be repeated each time they are used. On the majority of controls, one or more G-codes can be programmed in a single block, providing there is no conflict between them. It is customary to program a G-code or G-codes at the beginning of the block, after the sequence number.

**Table 1. Typical Turning G-Codes**

| G-Code | Description | G -Code | Description |
|---|---|---|---|
| G00 | Rapid positioning | G54–G59 | Work coordinate offsets 1–6 |
| G01 | Linear interpolation | G61 | Exact stop mode |
| G02 | Circular interpolation clockwise | G62 | Automatic corner override mode |
| G03 | Circular interpolation counterclockwise | G64 | Cutting mode |
| G04 | Dwell (as a separate block) | G65 | Custom macro call |

**Table 1.** *(Continued)* **Typical Turning G-Codes**

| G-Code | Description | G -Code | Description |
|---|---|---|---|
| G09 | Exact stop check—one block only | G66 | Custom macro modal call |
| G10 | Programmable data input (Data setting) | G67 | Custom macro modal call—cancel |
| G11 | Data setting mode—cancel | G68 | Mirror image for double turrets |
| G20 | US customary units of input | G69 | Mirror image for double turrets—cancel |
| G21 | Metric units of input | G70 | Profile finishing cycle |
| G22 | Stored stroke check ON | G71 | Profile roughing cycle—Z axis direction |
| G23 | Stored stroke check OFF | G72 | Profile roughing cycle—X axis direction |
| G25 | Spindle speed fluctuation detection ON | G73 | Pattern repetition cycle |
| G26 | Spindle speed fluctuation detection OFF | G74 | Drilling cycle |
| G27 | Machine zero position check | G75 | Grooving cycle |
| G28 | Machine zero return (reference point 1) | G76 | Threading cycle |
| G29 | Return from machine zero | G90 | Cutting cycle A (Group type A) |
| G30 | Machine zero return (reference point 2) | G90 | Absolute command (Group type B) |
| G31 | Skip function | G91 | Incremental command (Group type B) |
| G32 | Threading—constant lead | G92 | Thread cutting cycle (Group type A) |
| G35 | Circular threading CW | G92 | Tool position register (Group type B) |
| G36 | Circular threading CCW | G94 | Cutting cycle B (Group type A) |
| G40 | Cutter radius—cancel (Tool nose radius offset cancel) | G94 | Feed rate per minute (Group type B) |
| G41 | Cutter radius—left (Tool nose radius offset left) | G95 | Feed rate per revolution (Group type B) |
| G42 | Cutter radius—right (Tool nose radius compensation right) | G96 | Constant surface speed |
| G50 | Maximum r/min preset | G97 | Constant spindle speed (r/min) |
| G52 | Local coordinate system setting | G98 | Feed rate per minute (Group type A) |
| G53 | Machine coordinate system setting | G99 | Feed rate per revolution (Group type A) |

Note: Some G-codes are optional and will vary from one control system to another. A typical list includes Group type A (most common) and Group type B, where some G-codes are different. For example, Group A uses addresses U and W to identify incremental motion, while Group B uses G91 and G92; there are other differences as well.

**Table 2. Typical Milling G-Codes**

| G-Code | Description | G-Code | Description |
|---|---|---|---|
| G00 | Rapid positioning | G52 | Local coordinate system setting |
| G01 | Linear interpolation | G53 | Machine coordinate system |
| G02 | Circular interpolation clockwise | G54–G59 | Work coordinate offsets 1–6 |
| G03 | Circular interpolation counterclockwise | G60 | Single direction positioning |
| G04 | Dwell (as a separate block) | G61 | Exact stop mode |
| G09 | Exact stop check—one block only | G62 | Automatic corner override mode |
| G10 | Programmable data input (Data setting) | G63 | Tapping mode |
| G11 | Data setting mode—cancel | G64 | Cutting mode |
| G15 | Polar coordinate command—cancel | G65 | Custom macro call |
| G16 | Polar coordinate command | G66 | Custom macro modal call |
| G17 | XY plane designation | G67 | Custom macro modal call—cancel |

**Table 2.** *(Continued)* **Typical Milling G-Codes**

| G-Code | Description | G-Code | Description |
|---|---|---|---|
| G18 | ZX plane designation | G68 | Coordinate system rotation |
| G19 | YZ plane designation | G69 | Coordinate system rotation—cancel |
| G20 | US customary units of input | G73 | High-speed peck drilling cycle (deep hole) |
| G21 | Metric units of input | G74 | Left-hand threading cycle |
| G22 | Stored stroke check ON | G76 | Fine boring cycle |
| G23 | Stored stroke check OFF | G80 | Fixed cycle—cancel |
| G25 | Spindle speed fluctuation detection ON | G81 | Drilling cycle |
| G26 | Spindle speed fluctuation detection OFF | G82 | Spot-drilling cycle |
| G27 | Machine zero position check | G83 | Peck-drilling cycle (deep hole) |
| G28 | Machine zero return (reference point 1) | G84 | Right-hand threading cycle |
| G29 | Return from machine zero | G85 | Boring cycle |
| G30 | Machine zero return (reference point 2) | G86 | Boring cycle |
| G31 | Skip function | G87 | Back boring cycle |
| G40 | Cutter radius offset—cancel | G88 | Boring cycle |
| G41 | Cutter radius offset—left | G89 | Boring cycle |
| G42 | Cutter radius offset—right | G90 | Absolute dimensioning mode |
| G43 | Tool length offset—positive | G91 | Incremental dimensioning mode |
| G44 | Tool length offset—negative | G92 | Tool position register |
| G49 | Tool length offset cancel | G98 | Return to initial level in a fixed cycle |
| G50 | Scaling function cancel | G99 | Return to R level in a fixed cycle |
| G51 | Scaling function | | |

Note: Some G-codes are optional, and some will vary from one control system to another.

A typical example of using G-codes is a simple program entry, such as X18.0. An X-address is mainly used as a dimension, but not exclusively. The amount of 18.0 could be in millimeters or inches, it can be an absolute location from part zero or an incremental distance and direction from the current point. In addition, the same address can be a dwell time in seconds (a pause). Using one or more G-codes in a program block is often required to accomplish the required action. Here are some examples:

G21 G90 G00 X18.0 … move at rapid feed to absolute location of 18 mm

G20 G91 G01 X18.0 F10.0 … feed by the incremental distance of 18 inches in positive direction

G04 X18.0 … dwell for 18 seconds

**Miscellaneous Function (M-Codes).—**Along with G-codes (preparatory commands), a CNC program requires a number of miscellaneous functions, called M-codes or M-functions. These auxiliary functions fall into two groups: those that control various machine functions, and those that control program flow. Miscellaneous functions are the least standard words in CNC applications and vary greatly between control systems. Those that are typical to most controls are:

**Table 3. M-Codes**

| M-Code | Description | M-Code | Description |
|---|---|---|---|
| M00 | Program stop | M09 | Coolant OFF |
| M01 | Optional program stop | M19 | Spindle orientation |
| M03 | Spindle rotation normal (clockwise) | M30 | Program end |
| M04 | Spindle rotation reverse (counterclockwise) | M60 | Automatic Pallet Change (APC) |
| M05 | Spindle stop | M98 | Subprogram call |
| M06 | Automatic Tool Change (ATC) | M99 | Subprogram end |
| M08 | Coolant ON | | |

**Units of Measurement (G20, G21).—**Dimension words in CNC programming are data that contain a location, distance, radius, or feed rate. In all cases, such data is dependent on the active units of measurement. For CNC work, either metric or US customary units can be selected by the program or by manual setting. US customary units can be set by the G20 (G70 for older controls) command, metric units by the G21 (G71 for older controls) command, with each command canceling out the other. US customary mode is specified in inches, metric mode in millimeters. In some cases, feet/min or m/min are also used.

Dimensions measured along an axis can be programmed up to four decimal places in inches, and three decimal places in millimeters. The minimum increment (i.e., amount of motion) possible is 0.0001 inch or 0.001 mm. The maximum programmable motion amount is typically 9999.9999 inches or 99999.999 mm. Note that there is no actual unit conversion involved, just a shift in decimal point position.

**Absolute and Incremental Programming (G90, G91).—**A motion dimension along the axes X, Y, and Z for milling and axes X and Z for turning (as well as any parallel or rotary axis) can be programmed either as an absolute location measured from part zero (origin) or as an incremental distance (U and W for turning) and direction measured from the current tool position.

The *absolute* method of programming is selected by the G90 command, the *incremental* method by the G91 command. One command cancels the other. The benefit of the absolute programming method is that one change of a point location requires only one change in the program. The same change in incremental program will require two changes. The main benefit of incremental programming method is that it can be used for toolpath repetition, typically in the form of subprograms or macros.

**Spindle Function (S-Address).—**The *spindle* function programmed with the S-address causes the machine spindle to rotate at a defined rate in revolutions per minute (r/min). Its programmable range is set by the machine manufacturer in increments of one revolution. The major difference between machining centers and some lathe operations defines the meaning of the S-address used in the program:

1) Milling—direct spindle speed in r/min (all operations)
2) Turning, Boring, Facing, and Grooving—cutting speed in ft/min or m/min
3) Threading and Centerline Operations—direct spindle speed in r/min

In all three application groups, spindle rotation direction must also be specified, using either the M03 or M04 miscellaneous functions.

*Example:* For milling, the programming is straightforward:

S800 M03—normal spindle rotation (CW) at 800 revolutions per minute
S750 M04—reverse spindle rotation (CCW) at 750 revolutions per minute

*CNC Lathes:* For CNC lathes, the situation is somewhat different, because the part itself is rotating, and cutting diameters change constantly, particularly during turning and boring operations. To select the required spindle speed mode, two G-codes are available:

G96—constant cutting speed—programmed in ft/min or m/min
G97—constant spindle speed—programmed in r/min

*Example:* G96 S300 M03 is 300 ft /min in G20 mode, or 300 m/min in G21 mode.
G97 S1200 M03 is 1200 r/min, regardless of the measuring units selected.

*Constant Surface Speed (CSS):* The purpose of constant surface speed (CSS), also known as peripheral speed or surface speed, is to increase or decrease the spindle speed with each change in cutting diameter. The speed will be slowed down for increasing diameters, and speeded up for decreasing diameters. These changes are automatically adjusted by the control system in very fine increments.

In order to limit the spindle speed from increasing to potentially dangerous levels at very small diameters, the programmer can limit the maximum spindle speed by using the G50 command.

*Example:* G50 S2500 limits r/min to 2500 in cutting speed mode

*Example:* (G96) G96 S300 M03 specifies cutting speed of 300 ft/min or 300 m/min.

G50 is only applicable in G96 mode and has no effect in G97 mode. Its setting should always be less than the maximum spindle speed of the machine.

Cutting speed (CS) can be converted to spindle speed using the following formulas:

US customary units: r/min = (ft/min × 12) / (3.14 × diameter in inches)

**Metric units: r/min = (m/min × 1000) / (3.14 × diameter in mm)**

The mathematical constant of 3.14 represents mathematical function *pi* ($\pi$), rounded to two decimal places, is more than suitable for such calculations. An even shorter formula can be used for US customary units and is also suitable for most applications, particularly without a calculator:

r/min = (ft/min × 4) / Diameter in inches

*Spindle Override:* Spindle override is a controller feature of CNC machines that allows temporary adjustments to the programmed spindle speed during program execution, without changing the program itself. For this purpose, a spindle override is provided at the operation panel. Usually, it is a rotary switch or buttons, graduated in 10 percent increments, with the range (commonly 50 to 120 percent) dependent on the machine builder. Permanent changes to the speed must be made in the program itself.

**Feed Rate Function (F-Address).—**Cutting motions are defined as movement of a cutter that is in contact with the material. Non-cutting motions are typically rapid motions without a physical contact with the material. Programming the feed rate address depends on the type of machine and the units of measurement. Milling machines and machining centers generally use feed rate per minute (inch/min or mm/min), while lathes and turning centers use feed rate per revolution (inch/rev or mm/rev). For lathes, the preparatory commands G94 and G95 determine whether feed rate per minute is selected (G94) or feed rate per revolution is selected (G95). G95 mode—feed rate per revolution—is the default for lathe work. Feed rate specification is always modal, and remains in effect until a different feed rate is selected.

Feed rate per revolution (feed/rev) can be converted to feed rate per minute (feed/min) by multiplying spindle speed in revolutions per minute (r/min) by the feed rate per revolution:

Feed/min = r/min × feed/rev

For multi-edge cutters, feed rate is also given as "per tooth" or "per flute." This feed rate per edge is also generally called "chipload," and it may be specified in inches or millimeters. Feed per tooth (fpt) can be converted to cutting feed rate:

Feed/rev = feed per tooth × number of teeth

Once the feed/rev is known, feed rate per minute can be calculated, if desired.

**Inverse Time Feed Rate.—**One type of feed rate that may be required in some special applications is the *inverse time* feed rate. This feed rate represents the reciprocal of the time in minutes required to complete a programmed motion. The feed command is indicated by a G93 command followed by an F-address value found by dividing the feed rate, in inches (millimeters) or degrees per minute, by the distance moved in the block:

Inverse time feed command = feed rate / distance = (distance / time) / distance = 1 / time

The inverse time feed rate has been used in some rotary and multi-axis applications.

**Feed Rate Override.—**The *feed rate override* is a feature of CNC machines that allows temporary adjustments to the programmed cutting feed rate during program execution, without changing the program itself. For this purpose, a feed rate override is provided at the operation panel. As with the spindle override, it usually is a rotary switch or buttons graduated in 10 percent increments, with the range (commonly 0 to 150 or 200) dependent on the machine builder. Permanent changes to the feed rate must be made in the program.

**Tool Function (T-Address).—**Both CNC milling and turning machine groups use automatic tool change (ATC). In the program, each tool has to be numbered using the T-address. On machining centers, tools are stored in a tool magazine, located on a side of the machine. On lathes, tools are stored in an indexing turret, located within the machining area. There is a difference in programming each group. In both groups, the T-address indicates the tool number of the next tool. For example, on a machining center, T06 is tool number six. Tools can be stored in any magazine position, but must be registered in the control system prior to use. On lathes, the four-digit tool number is determined by the turret station number, followed by a wear offset number. The turret station number is also equivalent to the geometry offset for the stored tool. Wear offset is used for fine-tuning. For example T0313 calls a tool located in turret station three, activates geometry offset three, as well as wear offset thirteen.

Tool offset screens (geometry and wear) are special tables visible on the control display panel that show the setting information about each tool. Such information includes the XZ setting, tool nose radius (R), and the tool tip orientation number (T) which should not be confused with tool number.

**Tool Nose Radius Compensation.—**A typical manually generated program uses both coordinate points read directly from drawings and calculated on the basis of drawing dimensions. In both cases, these calculated coordinate points represent a toolpath along the part edge suitable only for a cutter that has a radius of zero. Since a zero tool radius is not practical, the tool's actual radius has to be considered. This is achieved by using a powerful control feature called cutter radius offset (compensation) or—for lathe terms—tool nose radius compensation.

During program development, the programmer ignores the cutter radius, and uses drawing based dimensions. In order to allow for radius compensation at the machine, two preparatory commands are used—G41 for the compensation to left, and G42 for the compensation to the right. In both cases, the direction is determined by viewing along the cutter path direction.

At the control, the operator sets not only the XZ geometry and wear offsets, but also enters the tool nose radius (R-column) and the tool tip number (T-column). The purpose of the tool tip number is to establish the location of the tool nose radius center point.

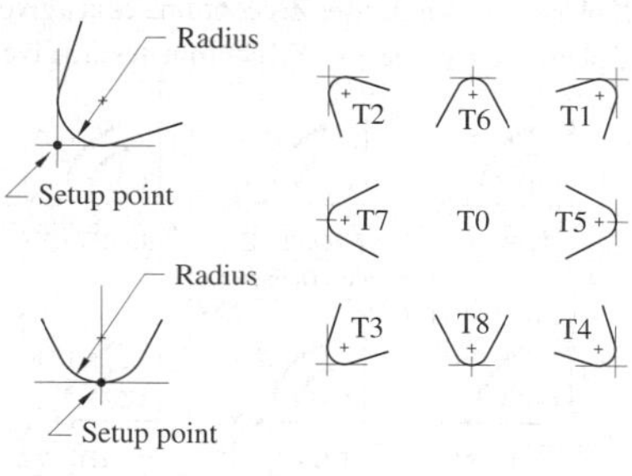


Fig. 3. Tool Tip Numbers—Rear Turret Lathe.

**Rapid Positioning (G00).**—In basic terms, machine motions are divided into two groups, non-cutting motions (positioning) and cutting motions (material removal). Rapid positioning is the only non-cutting (not intended for material removal) motion available to the CNC programmer. Its single purpose is to move the cutter from one location to another at the highest speed possible between actual cuts. CNC manufacturers determine the maximum rapid motion rate.

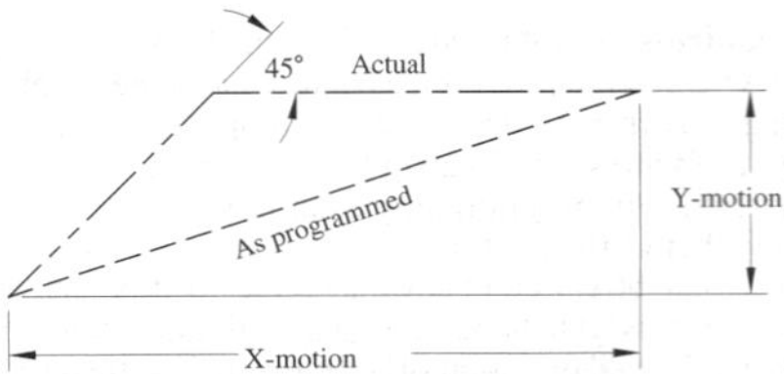


Fig. 4. Rapid Motion G00.

Fig. 4 shows the effect of programmed and actual rapid motion applied on a machining center. As both X and Y axes have the same rapid rate—and the actual toolpath is not important—the resulting motion starts at 45 degrees. When one axis reaches its destination, the other axis continues as a single axis motion. Some machines can produce rapid motion as programmed.

**Linear Interpolation (G01).**—In *linear interpolation mode* (programmed in G01 mode with a cutting feed rate in effect), the tool moves along a linear path, directly from one point to the next (traversing the shortest distance). This is a major departure from the rapid motion mode (programmed in G00 mode), where the shortest distance is disregarded and the result is the fastest motion available. The word *interpolation* defines a mathematical process that translates into synchronized feed rate for all programmed axes. This is done by the control system automatically.

The programming command for linear interpolation is G01 (modal), which cancels any other active motion command. Feed rate is required with G01, and its selection is based on cutting conditions.

**Circular Interpolation (G02, G03).**—*Circular interpolation* applies to a cutting toolpath moving along an arc or a circle. An arc (circle) can be programmed directly, without linear approximation. The cutting motion along an arc is two-dimensional and always takes place in a single plane. No arc or circle exists as a three-dimensional entity (it becomes a helix). Depending on the required outcome, the desired plane can be defined using one of three specific G-codes (Fig. 5–Fig. 7):

G17 = XY plane—arc/circle uses XY coordinates at a given Z-depth

G18 = ZX plane—arc/circle uses ZX coordinates at a given Y-depth

G19 = YZ plane—arc/circle uses YZ coordinates at a given X-depth

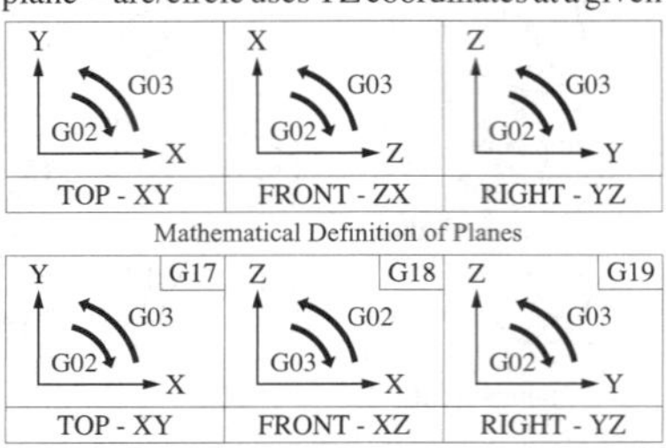


Fig. 5. Planes on a CNC Vertical Machining Center.

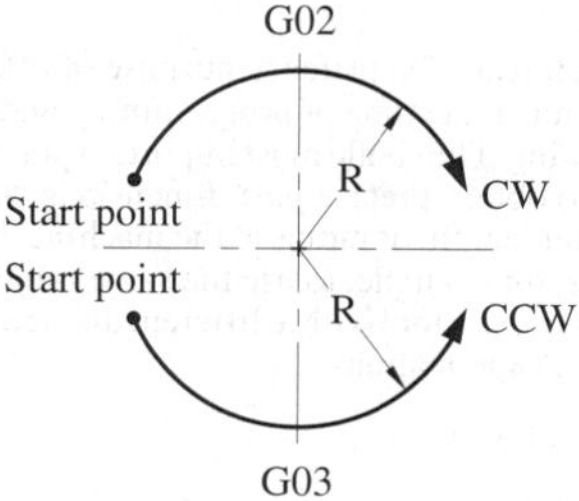


Fig. 6. Circular Interpolation.

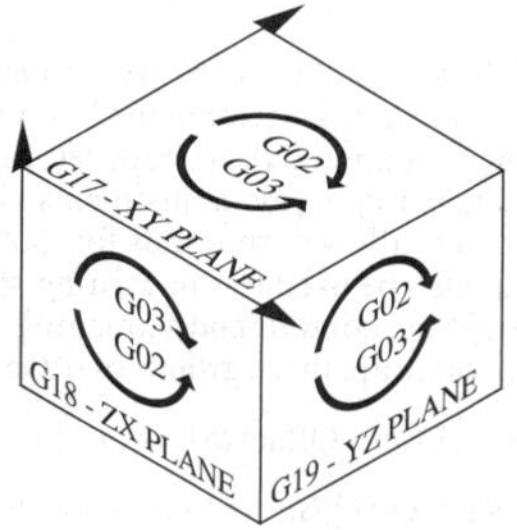


Fig. 7. Circular Interpolation in Planes.

It is important to understand that the G18 mathematical plane does not correspond to the machine axes orientation and that the arc direction appears to be incorrect. Programming an arc requires a direction of motion in the plane (G02 = CW, G03 = CCW), a start point (typically the current block), an end point (typically the target point, the end of motion), as well as the arc radius. The arc radius can be specified directly (for example R1.5), or it can be defined by special vectors, using I-J-K addresses (see Fig. 8. Arc Vectors. and definition below).

When using direct radius R, there are some limitations. Radius R (positive) can only be programmed for arcs up to and including 180 degree arc sweep angle. If the sweep angle exceeds 180 degrees but is less than 360 degrees, R– (negative) must be used. For a full circle (360 degrees arc sweep angle), the address R cannot be used at all, and a two-vector combination must be used. There is no need to break arc motion at the quadrants.

The definition of vectors I, J, and K is the same for all three axes, where X and I, Y and J, and Z and K are related (paired): The vector I-J-K is the distance and direction from the start point of the arc to the arc center, measured along axis X-Y-Z respectively.

A typical milling application in the plane XY is shown in Fig. 8. Current plane selection determines the arc vectors. Some controls use I-J-K as an absolute location. The start point of the arc is defined as the endpoint of the previous motion.

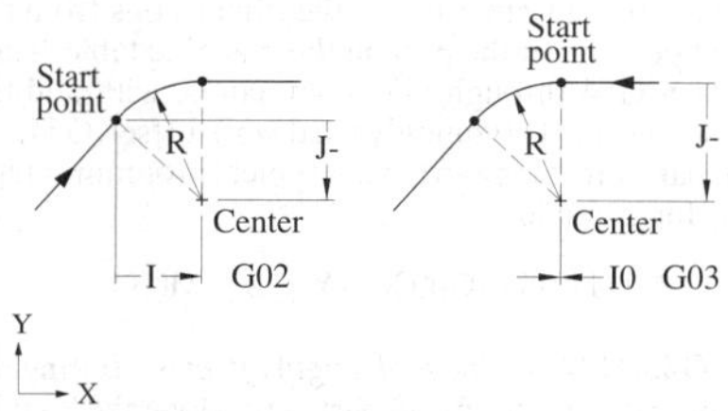


Fig. 8. Arc Vectors.

**Helical and Other Interpolation Methods.—**Most milling controls offer another motion—helical interpolation (either as a standard feature or as an option), but only very few specialized controls offer parabolic or hyperbolic interpolation. Helical interpolation produces a helix. Two axes (typically XY) form the circle, while the third axis (typically Z-axis) provides synchronized linear motion. The selected plane (G17, G18, or G19) determines the circular axes, while the remaining axis is linear motion. Helical interpolation is most commonly used for thread milling—a method of thread development that offers many advantages over tapping.

## Offsets for Milling

In CNC terms, the word *offset* means an adjustment. The primary purpose of offsets is to provide a link between the known information at the time of programming and the unknown situation at the time of setup and machining. The single most important purpose of offsets is to provide adjustments necessary to obtain precise part dimensions. Such adjustments allow corrections for tool wear, cutter length, or radius at the machine. Offsets can also be used for machining applications, for example, to use the same program for a roughing contour and a finishing contour, or to use a tool with a different diameter or length. There are three groups of offsets for milling applications:

*Work Offset* = Offset to link machine zero and part zero.

*Tool Length Offset* = Offset to link tool tip and part zero.

*Cutter Radius Offset* = Offset to link tool radius and drawing dimensions.

**Work Coordinate Offsets (G54 through G59).**—Every CNC machine has a built-in location called the machine reference point (also known as machine zero or home position). By default, this location has coordinates of X0Y0Z0. The purpose of *work offset* is to establish the difference between machine zero and part zero. All machine axes can be set, although only X and Y axes are normally used.

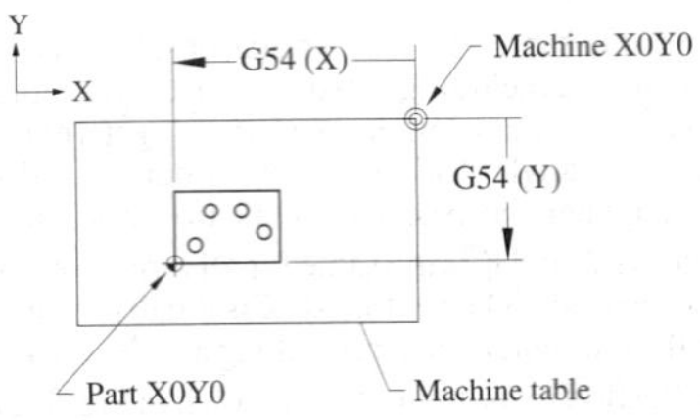


Fig. 9. Work Offset Using G54.

Work offset allows the programmer to use the dimensions from the drawing without considering the actual position of the part on the machine table. There are six standard work offsets defined by G54 through G59 commands, with additional sets available optionally. Fig. 9 shows the most commonly used work offset, G54. Note the direction is from machine zero to part zero, along an axis. A typical program entry combines the work offset with other data, for example:

N3 G90 G54 G00 X__ Y__ S__ M03 T__

**Tool Length Offset (G43, G44).**—The *tool length offset* is a setting that specifies the distance between the tool reference point and part zero, along the Z-axis. Two methods are available—the *preset method* and the *touch-off method*. The preset method requires tool measuring equipment and is used off-machine. The touch-off method is much simpler, and is done at the CNC machine. Each tool used in the program normally requires one tool length offset. A specific offset number is necessary to include with the G43 command. For example,

N5 G43 Z0.1 H01 M08

G44 is used seldomly and only for special applications. Fig. 10 shows typical tool length offset measurement using the touch-off method with G43 command.

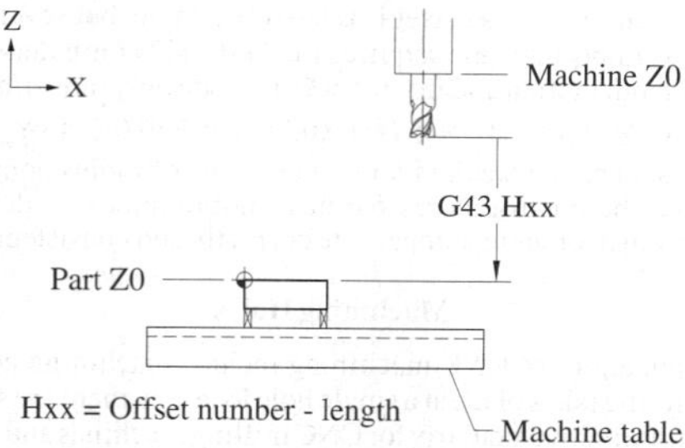


Fig. 10. Tool Length of Offset.

**Cutter Radius Offset (G41, G42).**—The *cutter radius offset* is an adjustment of the contour toolpath from the part edge to the cutter center. It is also known as *cutter radius compensation* or *tool nose radius compensation for lathes* (see page 1349).

Point calculations based on the drawing dimensions are far more convenient and practical than programming coordinates to the center of a cutter. Fig. 11 illustrates a toolpath offset by the cutter radius. The toolpath generated on the basis of drawing dimensions may only be suitable for holes where the spindle centerline and hole centerline match. Such a toolpath cannot be used on a contour, because the cutter centerline must be shifted (offset) away from the contour by the cutter radius. The purpose of the cutter radius offset is to provide automatic adjustment for the toolpath. This toolpath shifts the cutter center in such a way that the cutter edge is in constant contact with the programmed contour. This control generated toolpath is called an *equidistant toolpath*.

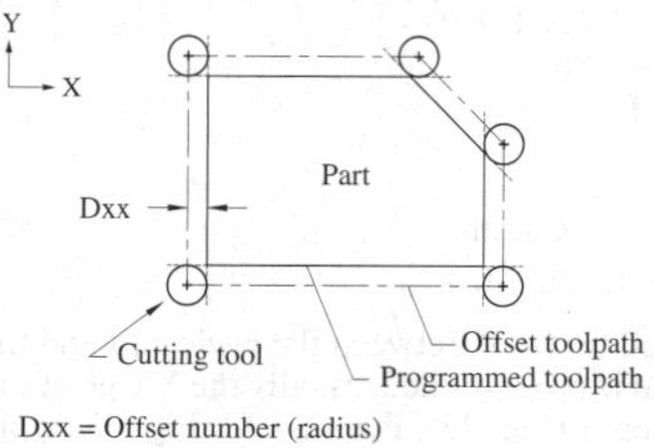


Fig. 11. Cutter Radius Offset.

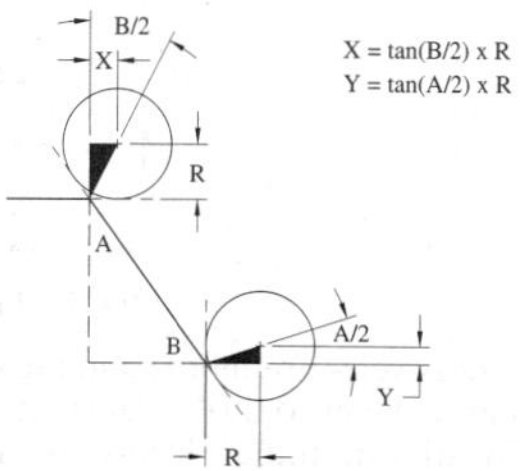


Fig. 12. Manually Calculated Radius Offset.

Programming to the center of the cutter requires tedious calculations for every point on the contour. Fig. 12 shows such a typical calculation. In addition, manually calculated center toolpaths cannot be adjusted at the machine.

To activate a cutter radius offset, two preparatory commands are available—G41 (offset to the left) and G42 (offset to the right)—both left and right offsets are determined by the cutting direction. As more than one tool can use cutter radius offset, G41 or G42 has to be programmed together with corresponding offset number, using the D-address. Many control systems use only a single offset registry for both tool length offset and cutter radius offset. In such cases, the D-address must have a number distinct from that of the tool length offset, even if the addresses are different.

*Example:* Tool 1 will be programmed as T01, using G43 H01 as the tool length offset, and G41 D01 as the cutter radius offset. G40 command cancels the cutter radius offset.

Programming the cutter radius offset is relatively simple, but several rules have to be observed. The most important rule requires the G41/G42 command to be applied after completing the Z-depth motion and in G00 or G01 mode only (not on an arc).

*Example:* N44 G01 Z-0.5 F15.0; N45 G41 X0.25 D01 F20.0; N46 Y__

Cutter radius offset uses a so called look-ahead type of radius compensation. During program execution, the control stores the next motion block in the buffer—it "looks ahead" in the program in order to compensate correctly and consistently.

### Machining Holes

Virtually all applications of CNC machining include machining holes. Regardless of what type of machining takes place at a single hole location, there are similarities important to CNC programming, particularly for CNC milling machines and machining centers. The only cutting tool motion that takes place is along the Z-axis, after a required XY location has been applied. This type of machining is commonly known as point-to-point machining and is applied to operations such as drilling, spot drilling, spot facing, counter-boring, counter-sinking, tapping, boring, and reaming.

Each toolpath step relating to a particular machining operation on a hole is always consistent and can be preset within the control system software (built-in software routine). This is the basis for shortening programming holes by using fixed cycles. As the name suggests, these cycles contain an internal set order of tool motions and cannot be changed.

**Fixed Cycles.—**"Fixed cycles" sometimes called canned cycles, are series of preset sequences of toolpath motions, allowing consecutive motions to be performed with a single block of program. Typical fixed cycles are G80–G89, as listed in Table 2 on page 1345.

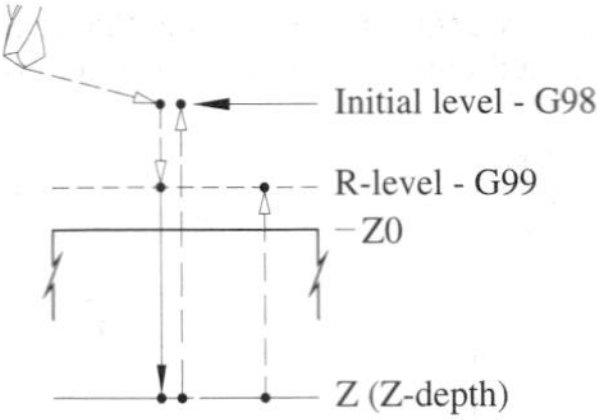


Fig. 13. Initial Level and R-Level.

All fixed cycles are modal and are canceled by G80. Between the cycle call and the cycle cancel block, only the data that changed is programmed, usually the XY position, occasionally the Initial, R-level or the Z-depth (Fig. 13). For Fig. 13–Fig. 23, rapid traverse and feed motions are indicated, respectively, by dashed lines with hollow arrowheads and by solid lines with filled arrowheads. The general and most commonly used format for all fixed cycles is:

N__G98 (G99) G__X__Y__R__Z__P__Q__F__

where:

*N*__ = block number,
*G*__ = cycle number,
*X*__*Y*__ = hole location,
*R*__ = start position from where feed rate mode starts,
*Z*__ = final depth,
*P*__ = amount time delay (dwell) for G82 and G89,
*Q*__ = either the depth of each peck for G83 and G73, or the shift amount for G76 and G87,
*F*__ = cutting feed rate in feed per minute mode.

*Drilling Cycle (G81):* The G81 code is used mainly for drilling, and has the following format see Fig. 14:

N__G99 (G98) G81 X__Y__R__Z__F__

In the G81 cycle, the tool moves at rapid traverse to the XY location first, then to the R-level (feed plane). From that position, it feeds to the depth, followed by a rapid retract to the R-level in G99 mode or the initial level, in G98 mode.

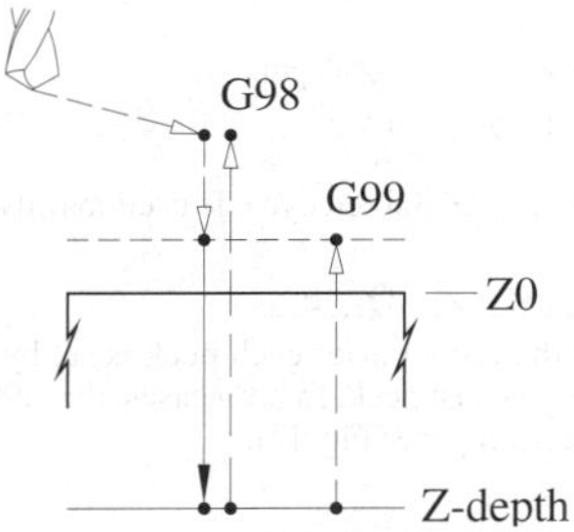

Fig. 14. G81 Drilling Cycle.

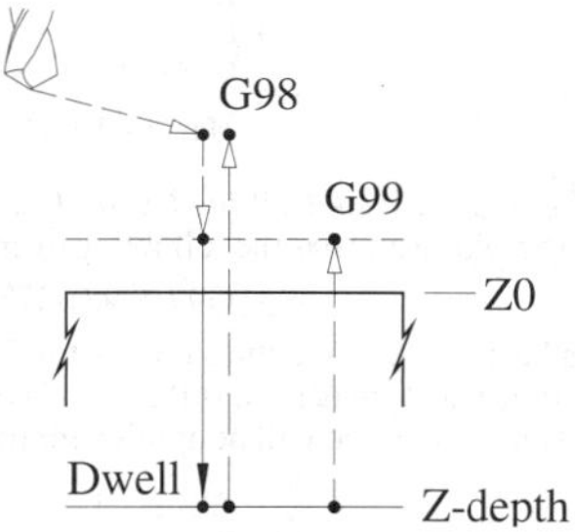

Fig. 15. G82 Spot Drilling Cycle.

*Spot Drilling (G82):* The G82 fixed cycle is used for spot drilling, spot facing, counter boring, countersinking, and any other operation where a dwell at the bottom of the hole is required. It has the following format per Fig. 15:

N__G99 (G98) G82 X__Y__R__Z__P__F__

All tool motions are identical to G81 cycle, with the added pause P__ (dwell). Pause is programmed in milliseconds (ms), without decimal a point. Its purpose is to smooth out the bottom of the hole. The dwell period should be a minimum of two spindle revolutions to allow for a full clean-up, in case 50 percent spindle speed override is in effect. To calculate minimum dwell, divide 60 by the spindle revolutions (r/min):

$$\text{Minimum Dwell (seconds)} = \frac{60}{r/min}$$

To convert to milliseconds, multiply the result by 1000.

*Example:* For 750 r/min, the minimum dwell will be 60 / 750 × 1000 = 80 ms.

*Peck-Drilling Cycle (Deep Hole) (G83):* The G83 fixed cycle is used mainly for deep hole drilling using the following format:

N__G99 (G98) G83 X__Y__R__Z__Q__F__

From the R-level, the drill will cut by the amount of Q__, which specifies the depth of each peck. When the peck depth Q__ is completed, the drill retracts to R-level, then moves rapidly back to cut at another depth. There is small non-programmable clearance between the cuts that is part of the cycle (See Fig. 16).

A general rule of peck drilling is that the greater the Q__ amount, the fewer pecks will be generated. The peck amount is calculated as two-thirds of the drill diameter—with fine adjustments made for material type and hardness. Some controls allow programming of the actual number of pecks required. No peck motion will ever exceed the final Z__ depth. For Fig. 16 and Fig. 17, the clearance distance (d) is set by the machine manufacturer in a system parameter.

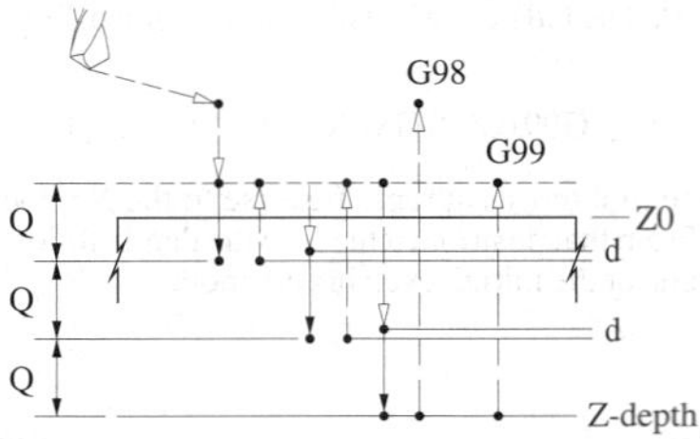


Fig. 16. Peck Drilling Cycle (Deep Hole).

*High-Speed Peck Drilling Cycle (Deep Hole):* The G73 fixed cycle is used mainly for chip breaking and has the following format

N__G99 (G98) G73 X__Y__R__Z__Q__F__

Its basic function is the same as for G83, but the retract after each peck is set by the clearance amount (d) and full retract is only after the last peck. In both cases, the G98 or G99 retract method will be applied for the last peck only (see Fig. 17).

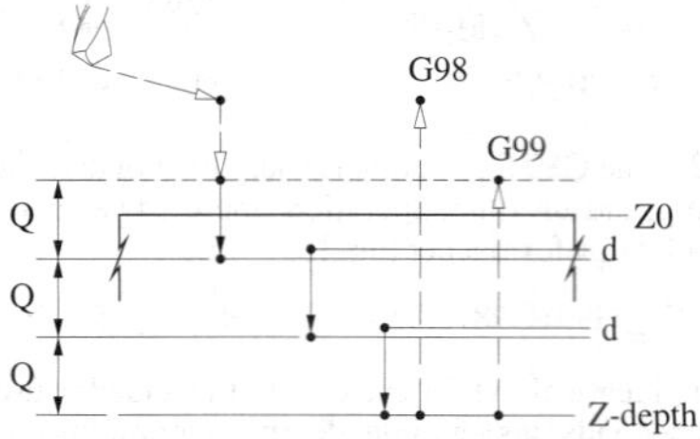


Fig. 17. G73 High-Speed Peck Drilling Cycle (Deep Hole).

*Right-Hand Threading Cycle (G84):* The G84 fixed cycle is used for cutting a right-hand thread and has the following format

N__G99 (G98) G84 X__Y__R__Z__F__

The cycle will start feeding from the R-level to the Z-depth, where the tap stops, reverses its rotation and feeds back to the R-level, where it stops again and resumes normal spindle rotation (see Fig. 18).

The G84 cycle must be programmed with spindle rotation CW (normal), typical in M03 mode.

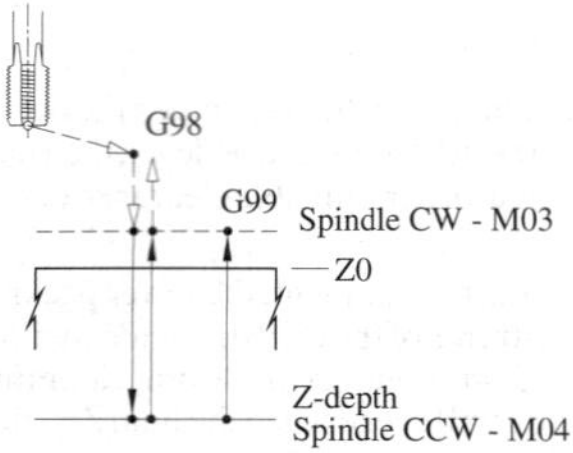


Fig. 18. G84 Right-Hand Threading Cycle.

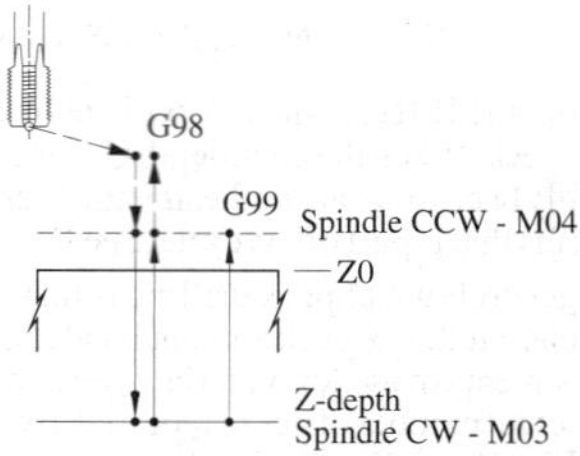


Fig. 19. G74 Left-Hand Threading Cycle.

*Left-Hand Threading Cycle (G74):* The G74 fixed cycle is used for cutting a left-hand thread and has the following format

N__G99 (G98) G74 X__Y__R__Z__F__

The G74 cycle must be programmed with spindle rotation CCW (reverse), typical in M04 mode. In all other respects, G74 is identical to G84 (see Fig. 19).

*Boring Cycle (G85):* The G85 fixed cycle is used for boring and has the following format

N__G99 (G98) G85 X__Y__R__Z__F__

From the R-level, the boring bar will feed to the Z-depth, than feeds back to R-level, both at the programmed feed rate F__. A similar cycle G89 has a dwell at the bottom of the hole. Either one of these cycles is commonly used for reaming (see Fig. 20).

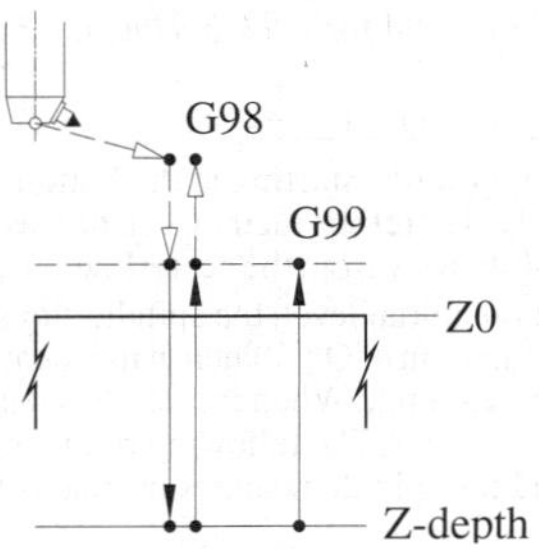


Fig. 20. G85 Fixed Cycle.

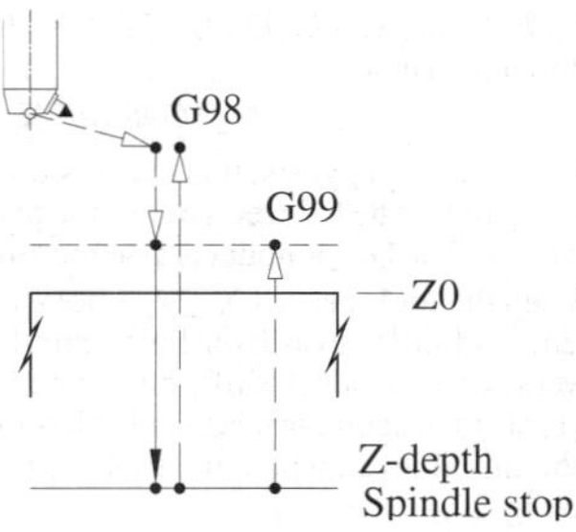


Fig. 21. G86 Fixed Cycle.

*Boring Cycle (G86):* The G86 fixed cycle is used for boring and has the following format:

N__G99 (G98) G86 X__Y__R__Z__F__

In G86 mode, the boring bar will feed from the R-level to the Z-depth, where the spindle stops rotating and the tool is retracted at a rapid rate. Note that a straight line (tool mark) will be generated on the bored hole (see Fig. 21).

*Fine Boring Cycle (G76):* The G76 fixed cycle is used for fine boring and has the following format:

N__G99 (G98) G76 X__Y__R__Z__Q__F__

This cycle is called a "*fine*" boring cycle, because it does not leave any tool marks on the bored hole. The Q__ in the block is the amount of shift away from the hole centerline that takes place before tool retraction. Typically, the amount of Q__ is very small, usually in the range of 0.005-0.010 inches or 0.125-0.250 mm.

In this internally complex cycle, the tool starts at the R-level and feeds to the Z-depth. When the depth is reached, the spindle stops and orients (so called OSS—oriented spindle stop). When oriented, the tool moves away by the Q__ amount, then retracts to the R-level. Once outside of the hole, it moves back by the Q__ amount and the spindle resumes normal rotation (see Fig. 22).

When adjusting a boring bar to size (setting) at the machine, the operator must align the cutting tip so that it is aimed in the opposite direction of the axis shift motion.

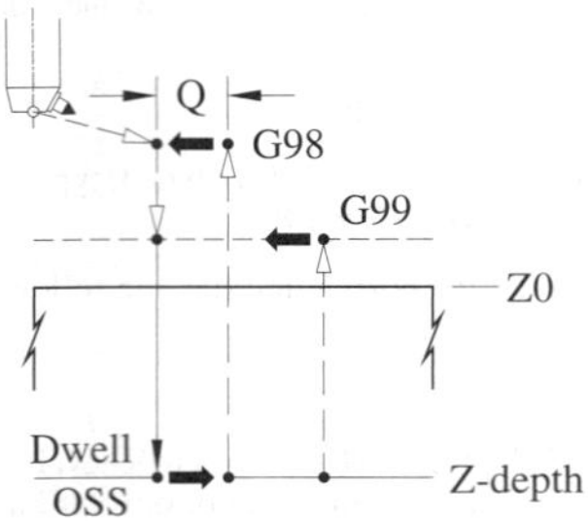

Fig. 22. G76 Fine Boring Cycle.

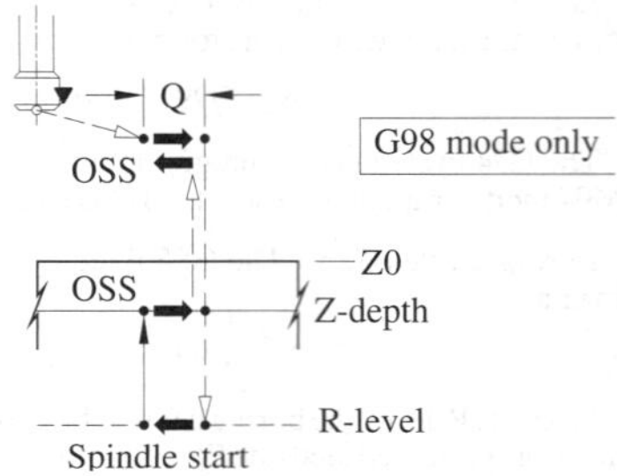

Fig. 23. G87 Back Boring Cycle.

*Back Boring Cycle (G87):* The G87 fixed cycle is used for *back boring* and has the following format:

N__G98 G87 X__Y__R__Z__Q__F__

As the name suggests, this cycle is used to bore upwards, starting at the bottom of an existing hole. The R-level is below the part, and only G98 retract method can be used with this cycle. The Q__ amount is also the amount of shift away from the centerline.

When the tool reaches X__Y__ hole location at the initial level, the spindle stops, tool orients and shifts away from hole centerline by the amount of Q__. Then, it moves at rapid traverse to the R-level, shifts back, and feeds to the Z-depth. When the depth is reached, the tool stops again, orients, and shifts by the Q__ amount. The following retract will be to the initial level, where the tool shifts back, and the spindle resumes normal rotation (see Fig. 23).

This cycle can only be used if the boring bar can move clearly through the existing hole. The shift amount Q__ must be greater that one half of the difference between the back bore diameter and the existing hole diameter. Note that G99 retract mode cannot be used.

*Manual Boring (G88):* The G88 fixed cycle is used for boring with manual intervention and has the following format:

N__G99 (G98) G88 X__Y__R__Z__P__F__

The G88 fixed cycle is rather rare. Its use is limited to boring operations with special tools that require manual interference at the bottom of a hole. When such a operation is completed, the tool is moved out of the hole for safety reasons. This cycle may be used by some tool manufactures for certain special operations (see Fig. 24).

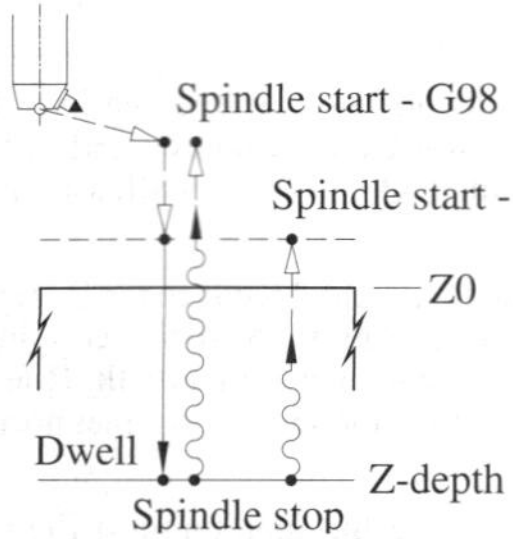

Fig. 24. G88 Fixed Cycle.

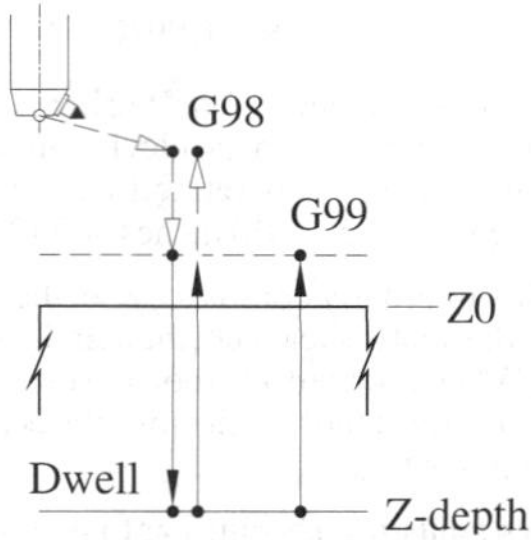

Fig. 25. G89 Boring Cycle with Dwell.

*Boring Cycle with Dwell (G89):* The G89 fixed cycle is used for boring with dwell and has the following format:

N__G99 (G98) G89 X__Y__R__Z__P__F__

From the R-level, the boring bar will feed to the Z-depth, dwells for the number of milli-seconds specified by P__, than feeds back to R-level, both at the programmed feed rate F__. G85, a similar cycle, does not have a dwell. This cycle is commonly used for reaming (see Fig. 25).

*Fixed Cycle Cancel (G80) :* The G80 fixed cycle cancel is used to cancel any fixed cycle and has the following format:

N__G80

G80 is the recommended method for fixed cycle cancellation, although programming any motion command (G00, G01, G02, G03) will also cancel any active cycle.

*General Notes on Fixed Cycles:* Except G87, all fixed cycles can retract to either the R-level (using G99) or the initial level (using G98). The R-level is often called the feed plane and is the Z-position from where the feed rate mode begins. In the cycle, the Z-address is used for depth, and the R-address is used for the feed plane to avoid a conflict of two identical addresses. The initial level is the last Z-address programmed before a fixed cycle is called. Although the initial level can be equivalent of the R-level, it usually is higher. Its purpose is to provide means for special clearance setting to bypass obstacles between holes.

Fixed cycles for boring can be applied to either single point or double point boring bars, with the exception of G76 and G87 cycles, which require a shift. In all cases, boring bars must be preset to the diameter of the hole machined.

## Calculations for Contouring

Contouring is the most common semi-finishing and finishing toolpath for both machining centers and lathes. The programmer must manually calculate every single point of the part contour based on drawing dimensions, in the order of machining and considering the cutter radius offset. For vertical and horizontal lines the calculations are quite simple, but for angular lines and partial arcs trigonometric calculations are necessary.

Fig. 26 shows typical calculations for start and end points of an arc tangent to two lines, a common situation in programming. As in any calculations, certain features must be known. For the calculations shown, point P, radius R, and angle A (as shown) must be known.

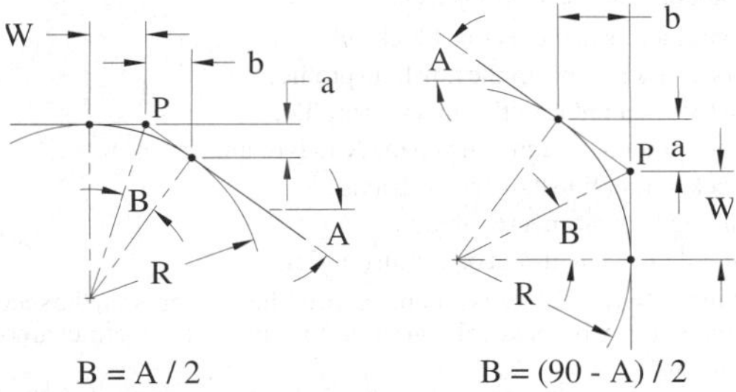


B = A / 2 B = (90 - A) / 2

W = R x tanB a = W x sinA b = W x cosA

Fig. 26. Calculation of Tangent Points.

## CNC Turning and Boring

Control manufacturers provide easy-to-program cutting cycles capable of complex lathe operations that can be readily changed and optimized at the control to improve performance. These special cycles are called *multiple repetitive cycles*, and they offer facing, turning, boring, grooving, and threading operations. Their simplified input provides excellent flexibility.

Covered next are the most common multiple repetitive cycles, G70–G76, as listed in Table 2 starting on page 1345.

For turning and boring operations, G71 cycle is the most common, along with G76 cycle for threading (described separately).

*Profile Roughing Cycle (G71):* The most common profile roughing cycle is G71. Its purpose is to remove stock by horizontal cutting along the Z-axis, typically from right to left. It is used for roughing out material out of a solid cylinder. It is available in two formats—a one-block and a two-block format, depending on the control system.

Older controls use a one-block format:

N__G71 P__Q__I__K__U__W__D__F__S__

where:

*P*__ = first block number of the finishing profile,
*Q*__ = last block number of the finishing profile,
*I*__ = distance and direction of rough semi-finishing in the X-axis—per side,
*K*__ = distance and direction of rough semi-finishing in the Z-axis,
*U*__ = stock amount for finishing on the X-axis diameter,
*W*__ = stock left for finishing on the Z-axis,
*D*__ = depth of roughing cut,
*F*__ = cutting feed rate per revolution,
*S*__ = spindle speed in ft/min or m/min (cutting speed).

The I and K parameters are not available on all machines. They control the amount of cut for semi-finishing, the last continuous cut before final roughing motions.

Newer controls require a two-block data entry for the G71 cycle, and the programming format is:

G71 U__R__

G71 P__Q__U__W__F__S__

where the parameters in the first block are:

*U*__ = depth of roughing cut,
*R*__ = amount of retract from each cut.

where the parameters in the second block are:

*P*__ = first block number of the finishing profile,
*Q*__ = last block number of the finishing profile,
*U*__ = stock amount for finishing on the X-axis diameter,
*W*__ = stock left for finishing on the Z-axis,
*F*__ = cutting feed rate per revolution,
*S*__ = spindle speed in ft/min or m/min (cutting speed).

The U-address in the first block is independent of the U-address in the second block. The I and K parameters may be used only on some controls and the retract amount R is set by a system parameter.

The G71 cycle can be used for external roughing (turning), as well as for internal roughing (boring). The control system determines the type of machining by comparing the start point SP with the point P (see Fig. 27).

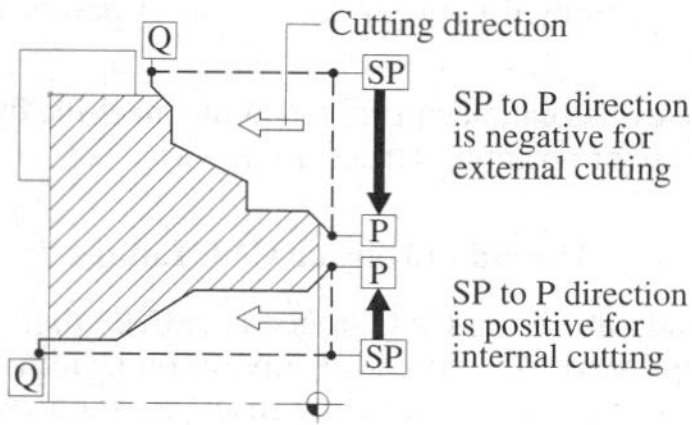


Fig. 27. G71 Roughing Cycle.

A typical program segment will have the following structure (tool 01 used):

N1 G20 T0100
N2 G96 S__M03
N3 G00 X__Z__T0101 M08 (START POINT SP)
N4 G71 U0.15 R0.04
N5 G71 P6 Q11 U0.06 W0.005 F0.015
N6 G00 X__(POINT P BLOCK)
[Block numbers N7–N10 (not shown) define the intermediate points on the profile.]
N11 U0.2 (POINT Q BLOCK)
N12 G28 U0 W0
N13 M30

G72 Profile roughing cycle is similar, but applies for vertical direction of cut (along the X-axis).

There are two types of G71 and G72—Type I and Type II. Type I does not allow for a change of direction, whereby Type II allows for a change in one axis only.

*Pattern Repeating Cycle (G73):* The G73 pattern repeating cycle is used to machine a finishing contour repeatedly, by a given distance, in specified number of steps.

*Profile Finishing Cycle (G70):* G70 is programmed after G71, G72, and G73 as a finishing cycle, and has a simple format:

N__G70 P__Q__F__S__

where:

*P*__ = the first block number of the finishing profile,
*Q*__ = the last block number of the finishing profile,
*F*__ = the cutting feed rate per revolution,
*S*__ = the spindle speed in ft/min or m/min (cutting speed).

The spindle speed and feed rate active in the G71/G72/G73 cycles will remain in effect, unless otherwise specified within the G70 cycle block.

N__G70 P6 Q11 F0.01 S400

*Basic Rules for G71, G73 Cycles:* It is important to observe some general rules, in order to make the multiple repetitive safe and efficient. Here are some more important rules:

a) Apply a tool nose radius offset before the stock removal cycle is called.

b) Cancel the tool nose radius offset after the stock removal cycle is completed.

c) Return motion to the start point is automatic and must not be programmed.

d) The P block in G71 should not include the Z-axis value for cycle Type I.

e) Change of direction is allowed only for Type II G71 cycle and along one axis only.

f) Stock amount U is programmed on a diameter (U+ for turning, U- for boring.)

g) The Feed Rate programmed between the P and Q points will be ignored during roughing.

h) Multiple repetitive cycles can be optimized at the machine by changing one or more settings. G72 and G73 are programmed in a similar way.

### Thread Cutting on CNC Lathes

The method of thread cutting on CNC lathes is called *single point threading* using an indexable threading insert. As a threading operation is both a cutting and a forming operation, the shape and size of the threading insert must correspond to the shape and size of the finished thread. By definition, single point threading is a machining process of cutting a helical groove of a specific shape with a uniform advancement per spindle revolution. The uniformity of the thread is controlled by the programmed feed rate in feed rate per revolution. Feed rate for threading is always the lead of the thread, never the pitch. For single start threads, the lead and the pitch are identical. As single point threading is a multi-pass operation, the CNC system provides spindle synchronization for each threading pass.

**Depth of Thread Calculations.—**Regardless of the threading method used, the depth of a thread will be required for various calculations. It can be calculated from these common formulas (*TPI* is the number of threads per inch):

External V-threads (60-degrees for metric or US customary units):

$$\text{Depth of Thread} = \frac{0.61343}{TPI} = 0.61343 \times Pitch \quad (1)$$

Internal V-threads (60-degrees for metric or US customary units):

$$\text{Depth of Thread} = \frac{0.54127}{TPI} = 0.54127 \times Pitch \quad (2)$$

*Thread Pitch* = the distance between two corresponding points of adjacent threads.

$$Pitch = \frac{1}{TPI} \quad (3)$$

In metric drawings, the pitch is specified as part of the thread designation.

*Thread Lead* = the distance the threading tool will advance along an axis in one revolution of the spindle

$$\text{Feed Rate} = \text{Lead} = \frac{\text{Number of Starts}}{TPI} = \text{Number of Starts} \times \text{Pitch} \quad (4)$$

Spindle speed is *always* programmed in direct r/min mode (G97), *not* in the constant surface speed mode G96.

**Infeed Methods.—**The way a threading tool enters into the material can be programmed in several ways, using two infeed methods available. Infeed is the type of motion from one pass to the next. Three basic thread infeed methods are shown in Fig. 28:

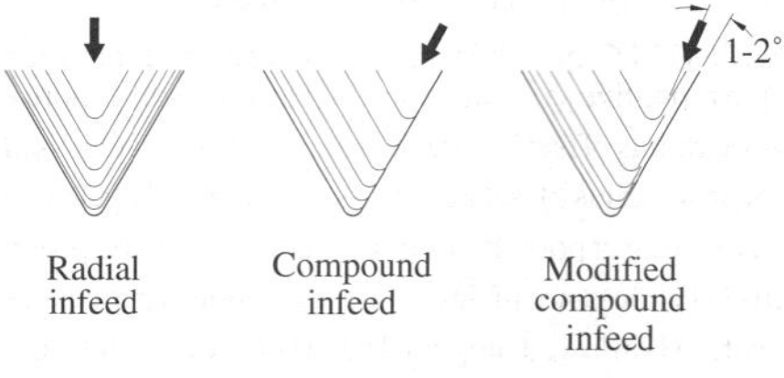


Fig. 28. Thread Infeed Methods.

1) Plunge method—also known as the radial infeed
2) Angular method—also known as the compound or flank infeed
3) Modified angular method—also known as the modified compound (flank) infeed

Specified infeed is typically selected to achieve the best cutting conditions for the insert edge in a given material. With the exception of some very fine leads and soft material, the majority of threading cuts will benefit from a compound infeed or modified compound infeed (angular method), providing the thread geometry allows this method. For example, a square thread will require a radial infeed, while an Acme thread will benefit from compound infeed. Four methods can be applied for compound infeed threading:

1) Constant cutting amount
2) Constant cutting depth
3) One edge cutting
4) Both edges cutting

**Radial Infeed.—**Radial infeed is one of the more common threading methods, if the conditions are suitable. It is applied to a cutting motion that is perpendicular to the diameter being cut. Each threading pass diameter is specified as the X-axis, while the Z-axis start point remains unchanged. This infeed method is suitable for soft materials, such as brass, certain aluminum grades, etc. In harder materials it could damage the thread integrity and is not recommended.

The inevitable outcome of a radial infeed motion is that both insert edges are working simultaneously. Since the insert edges are opposite of each other, the chips are formed at both edges at the same time, causing problems that can be traced to high temperatures, lack of coolant access, and tool wear problems. If the radial infeed results in a poor quality thread, a compound infeed method will usually solve the problem.

**Compound Infeed.—**Compound infeed method—also known as a flank infeed method—works on a different principle. Instead of feeding the threading tool perpendicularly towards the part diameter, the position for each pass is shifted by trigonometric calculation to a new Z-position. This approach results in a threading pass where most of the cutting takes place at one edge. As only one insert edge does most of the work, the generated heat can dissipate away from the tool edge while the cutting chips curl away, which extends the tool life.

Using the compound threading method, a heavier pass depth with fewer passes can be used for majority of threads. Compound infeed can be modified by providing 1 to 2 degrees of clearance on one edge to prevent rubbing. The angle of the thread will be maintained by the angle of the threading insert.

**Threading Operations.—**There are many threading operations that can be programmed for a typical CNC lathe. Several operations require a special type of threading insert and some operations can only be programmed if the control system is equipped with special (optional) features:

a) Constant lead single start threads (using G32 or G76)
b) Variable lead threads—increasing or decreasing (special option) (G34 and G35)
c) External and internal threading
d) Tapered threads (conical threads)
e) Right-hand (R/H) and left-hand (L/H) threads
f) Face threads (scroll threads)
g) Multi-start threads
h) Circular threads (special option)
i) Multi-block threads (so called long-hand threading, block-by-block) (G32)

The most common threading modes use either G32 (G33 on some controls) and G76.

**Constant Lead Threading (G32).**—The G32 command is sometimes called "long-hand threading," because each depth pass is programmed as one block. Programs using G32 can be quite long and virtually impossible to edit without major reprogramming. On the other hand, G32 method offers great flexibility and often is the only method that can be used, particularly for special threads. The programming format for G32 requires at least four blocks of input for a single thread pass from a start position:

| | |
|---|---|
| N__ G00 X__ Z__ | Position to start point |
| N__ (G00) X__ | Current threading pass diameter |
| N__ G32 Z__ F__ | Actual thread cutting (X__ Z__ for taper threads) |
| N__ G00 X__ | Retract to X-start position |
| N__ Z__ | Return to Z-start position |

**Threading Cycle (G76).**—G76 is a multiple repetitive cycle for threading, and is the most common method used to generate most of thread forms. Similar to roughing cycles, G76 is programmed in two versions, depending on the control system used. For older controls, a one-block format is used, for newer controls a two-block format is used. The two-block format offers additional settings that are not available in the one-block method.

The G76 threading cycle requires information about the thread and the cutting method.

*One-Block Cycle:* For a one-block G76 cycle, the format is:

N__ G76 X__ Z__ I__ K__ D__ A__ (P__) F__

Where:

*X* = the diameter of the last threading pass
*Z* = the end of thread along the Z-axis
*I* = the amount of taper over the total length (per side)
*K* = the single depth of the thread
*D* = the depth of the first threading pass
*A* = the included insert angle (only A0, A29, A30, A55, A60, and A80 allowed)
*P* = the infeed method adjustment—positive (not available on all controls)
*F* = the feed rate (thread lead)

*Two-Block Cycle:* The two-block G76 format includes several additional programmable features that make the cycle more flexible. For a two-block G76 cycle, the format is:

N__ G76 P__ Q__ R__
N__ G76 X__ Z__ R__ P__ Q__ F__

Where the first block:

*P* = a six-digit data entry in three pairs:
Digits 1 and 2—number of finishing cuts (01–99)
Digits 3 and 4—number of leads for gradual pull-out (0.0–9.9 times lead), no decimal point used (00–99)
Digits 5 and 6—angle of thread (00, 29, 30, 55, 60, 80 degrees only)
*Q* = the minimum cutting depth (positive radial value—no decimal point)
*R* = the fixed amount for finish allowance (decimal point allowed)

Where the second block:

*X* = the diameter of the last threading pass
*Z* = the end of thread along the Z-axis
*R* = the amount of taper over the total length (per side)
*P* = the single depth of the thread (positive radial value—no decimal point)
*Q* = the depth of the first threading pass (positive radial value—no decimal point)
*F* = the feed rate (thread lead)

The P/Q/R addresses of the first block are not related to the P/Q/R addresses of the second block. They have their own meaning, applied within each block only.

Several of parameters of the G76 cycle can be changed at the machine to optimize the threading operation.

*Example:* The code block for cutting an external thread on 2 inch diameter, 12 threads per inch, using tool number 6 to the length of 1.5 inches (thread ends in a recess) would be:

```
N31 T0600
N32 G97 S800 M03
N33 G00 X2.2 Z0.3 T0606 M08 (START POINT FOR THE THREAD)
N34 G76 P010060 Q0040 R0.002
N35 G76 X1.8978 Z-1.5 R0 P0511 Q0100 F0.083333
N36 G00 X8.0 Z5.0 T0600
N37 M01
```

**Multi-Start Threads.—**Threads with more than one start can be programmed using either G32 or G76 threading commands. The lead (and feed rate) for multi-start threads is always the number of starts times the pitch. For example, a triple start thread with the pitch of 0.0625 (16 TPI) will be 0.1875 (F0.1875). In order to achieve proper distribution of each start around the cylinder, each thread has to start at an equal angle, determined by the following calculation:

Thread cylindrical spacing angle = 360 / Number of starts

To achieve this angular spacing, the Z-axis point for each start must be one pitch further than the previous start:

Shift amount = Pitch

Number of shifts = Number of starts − 1

## Subprograms, Macros and Parametric Programming

CNC programs use several techniques to allow the programmer to create complex pieces of code using CNC code building blocks.

**Subprograms.—***Subprograms* (also known as subroutines), are specially designed programs that can be called on demand by another program and repeated as desired. Such repetition can take place in one program or in several programs. The content of standard subprograms is fixed and can be used only for the intended, specific purpose. For variable type subprograms—called macros—a special control option is required (see page 1366).

The main advantages of subprograms include transportability between programs (even machines), shorter programs, and easier editing. Any repetitive toolpath or machine activity is suitable to be developed as a subprogram.

Any subprogram has to have a number (for example O6001), in the same format as the main program (or other program of origin). The end of subprogram uses the M99 function, which transfers the processing put of the subprogram back to the program of origin. Another miscellaneous function—M98—is used to call a previously stored subprogram, for example M98 P6001. The address O is used for storing the subprogram, address P is used for calling the subprogram.

If required, a subprogram can be repeated, up to 9999 times. That is achieved in three ways, depending on the control system:

Method 1: M98 P6001 L5 will repeat subprogram O6001 five times

Method 2: M98 P6001 K5 will repeat subprogram O6001 five times

Method 3: M98 P00056001 will repeat subprogram O6001 five times

Subprograms can be nested. Nesting is a feature that allows one subprogram to call another subprogram up to four levels deep.

In the example of a simple subprogram application Fig. 29, a complete toolpath is developed for one pocket, stored as a subprogram, and called at the equivalent location of the three remaining pockets. Typical programming methods will use a incremental method of dimensioning.

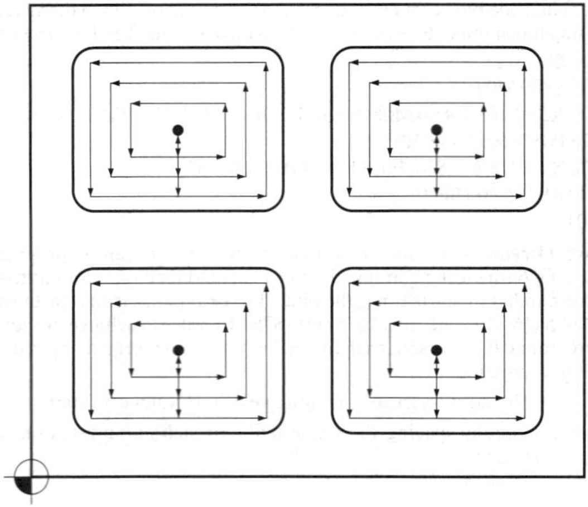

Fig. 29. Typical Subprogram Application.

**Macros and Parametric Programming.—**In this general introduction to the subject of macros and parametric programming, the most common development tool is known as Fanuc Custom Macros B. In spite of the great advancement of CAD/CAM software, macros still have their special place. In addition to manual or computer-assisted programming, macros are used to provide a special solution to a special requirement. Macro programming is a method using variable data is a special subprogram (now called a macro), with actual values defined in the main program. This method provides a single 'master' program that can be used many times with different numerical values. A typical example is a bolt circle, where only a few values change. The main purpose of a macro program is to save programming time.

**Basic Macro Skills.—**Depending on the complexity of the macro to be developed, some basic knowledge and skills are required. The following list identifies the most important items:

a) Understanding various methods of data settings
b) CNC machines and controls—operation and programming
c) Machining skills—how to machine a part
d) Basic mathematics skills—calculations, formulas
e) Program structure development skills—convenience and consistency
f) Offsets and compensations applications skills—various adjustments
g) Fixed cycles in depth—how they work, in detail
h) Subprograms in depth, including multi nesting applications
i) System parameters, their purpose and functions

In addition, if a macro is written for a particular function of the machine, knowledge of that function as it relates to the machine is also very important. If the macro is developed for probing or in-process gauging, the programmer has to understand the full process, as well as the objective of the macro development.

**Confirming Macro Capability.—**A macro is a control option, and may not be automatically available to the user. There is a simple way to find out if a macro function is available. Set the control to MDI mode (Manual Data Input) and enter this command:

#1=1

After pressing the "Cycle Start" button, the control will either accept the statement or reject it. The # symbol identifies the subsequent number as a variable number 1 with an assigned value of 1. A statement using the # symbol is unique to macros. If the control rejects it, or it is impossible to enter the symbols, that means the control does not support macro functions.

**Common Features and Applications.—**Typical features found in macros are:

1) Arithmetic and algebraic calculations
2) Trigonometric calculations
3) Variable data storage
4) Logical operations
5) Branching
6) Looping
7) Error detection
8) Alarm generation
9) Input and Output
10) Miscellaneous features

A macro program looks like a regular CNC program, but includes many other features. A macro program is structured as a subprogram—it is stored under its own program number (O), and it is usually called by the main program, using the G65 preparatory command. In a simple form, macro features can be used in the main program, without the macro call. The following list highlights some of the most common applications of macros:

1) Groups of similar parts (parametric programming)
2) Offset control
3) Custom cycles
4) Nonstandard tool motions (simulating curves)
5) Special G-codes and M-codes
6) Alarm and message generation
7) Hiding and protecting macro programs
8) Probing and gauging
9) Various shortcuts and utilities

**Macro Structure.—**There are three basic areas to understand for successful macro development. These three areas offer many special functions that are used within the macro.

*Variables* are the key element of all macros. They provide flexibility by easily storing changeable data. When a value is assigned to a variable, it is stored for future use. Stored values are called *Defined Variables*.

*Functions and Constants* are also an important part of macros. Functions calculate something—they solve a mathematical calculation or a formula. For example, a "+" (plus) function will sum two or more values together. In addition to functions, constants can be defined in a macro as well, for example the pi ($\pi$) with the value of 3.14159265359

*Logical Functions* are used for branching and looping purposes. Looping and branching provide a change in the program flow that is based on a certain condition that has been previously defined. The given condition may be tested (evaluated), using the comparison functions, such as 'greater than' (>), 'equal to' (=), 'less than or equal to' (≤), and several others, used together with the IF function. They are called Boolean or logical operators. The given condition can be tested only once, using the IF checking function or it may be evaluated repeatedly, using the loop function WHILE—while the given condition is true, the loop is repeated. The result of the evaluation will determine further flow of the program.

**Macro Definition and Call.**—Macros are defined in a similar way as subprograms, but they are called or invoked by the G65 command:

O7001 (MACRO)
...
...
M99
%

**Variable Definition (G65).**—The G65 macro command accepts variable definitions, called arguments. Arguments are actual program values required for a particular macro only. They are passed to the macro. Variable data in the macro is then replaced with the supplied arguments. A typical program sample of a G65 macro using three arguments may have the following format: G65 P7001 A6.0 B8.5 F10.0

G65… Macro call command
P7001… Program number containing the macro—stored as O7001
A6.0… Assignment of local variable A (#1) argument to be passed to macro
B8.5… Assignment of local variable B (#2) argument to be passed to macro F10.0… Assignment of a local variable F (#9) argument to be passed to macro

The illustration Fig. 30 shows graphically the flow between the main program and a macro, as well as the definition and use of the three variables (defined as arguments). In the example, a macro linear motion command will be replaced with the defined amounts:

G01 X#1 Y#2 F#9 = G01 X6.0 Y8.5 F10.0

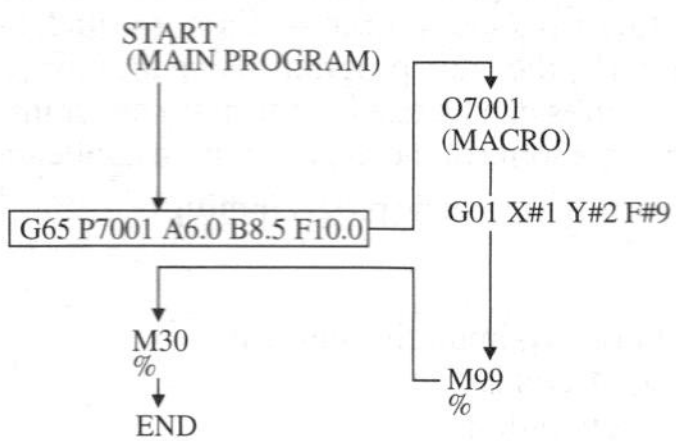


Fig. 30. Macro Call From Main Program.

Assignment of macro variables as arguments is one of the most important parts of developing any macro. Macro B has an arbitrary list of variables (called *Argument List 1*) and their meaning, shown in the following table:

**Table 4. Macro B Argument List**

| Argument List 1 | Macro Local Variable | Argument List 1 | Macro Local Variable | Argument List 1 | Macro Local Variable | Argument List 1 | Macro Local Variable |
|---|---|---|---|---|---|---|---|
| A | 1 | H | 11 | O | Not assigned | V | 22 |
| B | 2 | I | 4 | P | Not assigned | W | 23 |
| C | 3 | J | 5 | Q | 17 | X | 24 |
| D | 7 | K | 6 | R | 18 | Y | 25 |
| E | 8 | L | Not assigned | S | 19 | Z | 26 |
| F | 9 | M | 13 | T | 20 | | |
| G | Not assigned | N | Not assigned | U | 21 | | |

The above variables are LOCAL variables and can be used in arguments or in the macro.

Note that there are five letters missing (not assigned)—they are G N O P L. These letters are reserved and cannot be used as arguments. The letter L is used for the number of macro repetitions.

Variables #10, #12, #14, #15, #16, #27, #28, #29, #30, #31, #32 and #33 can only be used in a macro, not as an argument.

**Types of Variables.—**There are four types of variables available for macros:

*Null Variable:* Variable identified as #0 is a null variable. A null variable has no value—it is an empty variable, often called a vacant variable. This variable can be read by the macro program, but it cannot be assigned a value, which means data cannot be assigned to it.

*Local Variables:* Local variables are temporary only—they are used in macro body and hold certain data. When the macro is called, the local variables are set to their assigned values. When the macro is completed and exits (using the miscellaneous function M99), or the control power is turned off, all local variables are set to null values—they will no longer exist.

*Common Variables:* Common—or global—variables are still stored in the control memory when macro is completed. These variables are maintained by the system and they can be shared by several other macro programs. This group of variables is normally cleared by a specially designed macro program.

*System Variables:* System variables are used for setting and/or changing default conditions and can read and write different CNC data, for example, a current status of a G-code mode, the current work offset, etc. Their numbers are assigned by the control system manufacturer.

**Variable Declarations and Expressions.—**Variable data is not only defined in the argument section of the G65 macro call, variables can also be defined—or declared—in the macro body itself. Compare the two following declarations:

| | |
|---|---|
| #19 = 800 | A value of 800 is assigned to variable number 19—it can be spindle speed (r/min) |
| #9 = 15.0 | A value of 15.0 is assigned to variable number 9—it can be a feed rate (mm/min, m/min, ft/min, in/min, etc.) |

Both declarations store values—the value of "800" is stored into the variable #19 and the value of "15.0" into the variable #9. Both values shown in the example are numbers, but they are two different types of a number. 800 is an integer, 15.0 is a real number. In macros, this distinction is important. Variables may also use special expressions. For example, the variable definition #9 = 15.0 can be redefined later, as #9 = #9 * 1.1, which returns 16.5. If #9 defines the cutting feed rate, the new feed rate will be F16.5. Expressions can also be nested, following the standard mathematical hierarchy of the order of calculations:

#1 = 5.0

#2 = 8.5

#3 = 10.0

#4 = [#3 * #2] / #1; which returns 17.0

Note that brackets are used for nesting, not parentheses. Parentheses indicate a program comment or a message:

#19 = 800(SPINDLE SPEED DEFINED)

#9 = 15.0(CUTTING FEEDRATE DEFINED)

If a macro feature is available, variables can be defined and used in the main program directly, for example:

#19 = 800

N4 G90 G54 G00 X24.0 Y13.5 S#19 M03

Many variations and applications are possible.

**Macro Functions.—**The available macro functions can be divided into six groups:

1) ARITHMETIC functions

2) TRIGONOMETRIC functions

3) ROUNDING functions

4) MISCELLANEOUS functions
5) LOGICAL functions and operations
6) CONVERSION functions (not commonly used)

*Macro Arithmetic Functions:* There are four arithmetic functions:

| | | | |
|---|---|---|---|
| + | Addition | * | Multiplication |
| - | Subtraction | / | Division |

*Example:*

| | | |
|---|---|---|
| #1 = 3.5 | 3.5 | Returned value of variable #1 is 3.5 |
| #2 = 4.25 | 4.25 | Returned value of variable #2 is 4.25 |
| #3 = 2.0 + 5.0 | 7.0 | Returned value of variable #3 is a sum of 2+5 |
| #4 = #3 + 1 | 8.0 | Add 1 to the result of variable #3 |
| #5 = #2—0.8 | 3.45 | Subtract 0.8 from the current value of variable #2 |
| #6 = #1—#3 | -3.5 | Subtract contents of #3 from the contents of #1 |
| #7 = #2 * 6 | 25.5 | Multiply contents of #2 by 6 |
| #8 = 7.0 / 8.0 | 0.875 | Divide 7 by 8 as real numbers |
| #9 = 7 / 8 | 0.875 | Divide 7 by 8 as integer numbers |

*Trigonometric Functions:* The following trigonometric functions are generally available for macros:

| | | | | | |
|---|---|---|---|---|---|
| SIN | Sine | COS | Cosine | TAN | Tangent |
| ASIN | arcsine | ACOS | arccosine | ATAN | arctangent |

All inputs for SIN, COS, and TAN are in degrees and the output of the inverse functions ATAN, ASIN and ACOS is also in degrees. Inverse functions are usually marked as $\tan^{-1}$, $\sin^{-1}$, and $\cos^{-1}$ on the calculator.

*Example:*

| | | |
|---|---|---|
| #1 = SIN[38] | 0.6156615 | Actual value must be in brackets |
| #2 = 23.7 | 23.7 | |
| #3 = COS[#2] | 0.9156626 | Reference to a variable must be in brackets |
| #4 = TAN[12.86] | 0.2282959 | |

The inverse trigonometric functions accept the length of two sides of a triangle, both enclosed in brackets, "[ ]", and separated by the slash symbol, "/", standing alone between them. The acceptable range is within $0 \leq \text{RESULT} < 360$:

*Example:*

| | | |
|---|---|---|
| #5 = ATAN[0.25]/[0.5] | 26.5650512 | Position of the slash symbol is important |

ASIN and ACOS functions are not available on all controls.

*Rounding Functions:* Rounding functions control how decimal places are handled. There are three rounding functions:

1) ROUND
2) FIX
3) UP

*(1) ROUND:* The ROUND function rounds off a value to the nearest whole number. The function disregards fractions that are less than 0.5. For fractions that are equal to or greater than 0.5, the next whole number is the rounded value:

| | |
|---|---|
| ROUND[0.00001] | Returns 0.0 |
| ROUND[0.5] | Returns 1.0 |
| ROUND[0.99999] | Returns 1.0 |
| ROUND[1.0] | Returns 1.0 |

There is one very important difference between using the ROUND function directly in a program statement or in a variable definition. Consider the following two blocks:

#101 = 27/32

G91 G00 X[ROUND[#101]]

The incremental motion will be X0.8438 (US customary) or X0.844 (metric). However, if the ROUND function is used in a variable definition, the result will be different:

#101 = [ROUND[27/32]] Returns 1.0

G91 G00 X[#101]

The incremental motion will be X1.0

Rounding can also be performed to a given number of decimal places, but it is a two step process. In the first step, the given value must be multiplied by the factor of:

| | |
|---|---|
| 10... | to round off to one decimal place |
| 100... | to round off to two decimal places |
| 1000... | to round off to three decimal places (typical for metric system) |
| 10000... | to round off to four decimal places (typical for US customary system) |

The second step is to use the ROUND function together with the multiplier. For example, to round 123.56789 to three decimal places:

| | |
|---|---|
| #101 = 123.456789 | |
| #101 = #101*1000 | Returns 123456.789 |
| #101 = ROUND[#101] | Returns 123457.0 (nearest whole number) |
| #101 = #101/1000 | Returns 123.457 |

Note that only a single variable has been used for all four calculations. If the original value is not required later, this is a very efficient method of defining variables in the macro body.

The remaining two rounding functions FIX and FUP are used to round a given value up or down only, regardless of whether the decimal portion is over or under 0.5. The FIX and FUP functions are commonly used for counting rather than calculations.

*(2) FIX:* The FIX function is designed to round down the given value (discard fractions less than 1.0) i.e., strip all values after decimal point.

| | |
|---|---|
| ROUND[0.00001] | Returns 0.0 |
| ROUND[0.5] | Returns 0.0 |
| ROUND[0.99999] | Returns 0.0 |
| ROUND[1.0] | Returns 1.0 |

*(3) FUP:* The FUP function is designed to round up the given value (raise fractions less than 1.0).

| | |
|---|---|
| ROUND[0.00001] | Returns 1.0 |
| ROUND[0.5] | Returns 1.0 |
| ROUND[0.99999] | Returns 1.0 |
| ROUND[1.0] | Returns 1.0 |

*Miscellaneous Functions:* There are two common miscellaneous functions, SQRT and ABS. Several others exist but are not available on all control models.

*(1) SQRT:* The SQRT function calculates the square root of a number supplied between brackets:

SQRT[25.0] Returns 5.0

*(2) ABS:* The ABS function (absolute function) always returns a positive value of a given number:

| | |
|---|---|
| ABS[–1.3] | Returns 1.3 |
| ABS[1.3] | Returns 1.3 |

Using the ABS function is very useful in many applications, for example, to guarantee a positive or a negative value:

| | |
|---|---|
| #26 = 1.3 | Positive input |
| G01 Z–[ABS[#26]] | The motion will be equivalent to Z–1.3 |
| #26 = –1.3 | Negative input |
| G01 Z–[ABS[#26]] | The motion will be equivalent to Z–1.3 |

*Logical Functions:* Logical functions consist of Boolean operators and bitwise or binary functions:

Boolean functions (operators) compare two values and return a true or false condition:

| | | | | | |
|---|---|---|---|---|---|
| EQ | Equal to | GT | Greater than | GE | Greater than or equal to |
| NE | Not equal to | LT | Less than | LE | Less than or equal to |

Bitwise functions provide a means of combining comparisons.

| | | | | | |
|---|---|---|---|---|---|
| AND | And | OR | Or | XOR | Exclusive Or |

The AND, OR, and XOR functions compare two defined conditions simultaneously. The compared conditions are evaluated, and return either a TRUE value or a FALSE value. In English language, it is easy to understand the difference between the AND and the OR functions, because they follow the basic logic of everyday language. Function XOR is not used with the same frequency in machining macros. Examples of usage:

*Example 1, (OR)* #33 = [#1 EQ #2] OR [#3 EQ #4]

Variable #33 returns true (return value is 1) if #1 equals #2 OR #3 equals #4—otherwise it returns zero (false).

*Example 2, (AND) :* #33 = [#1 EQ 1] AND [#2 EQ 2]

Variable #33 returns 1 (true) if #1 equals 1 AND #2 equals 2—otherwise it returns 0 (false).

*Example 3, (XOR)* #33 = [#1 EQ 1] XOR [#2 EQ 2]

Variable #33 returns 0 (false) if both expressions are true or both expressions are false. If only one expression is true and the other is expression false, #33 returns 1 (true). XOR function is also known as the "exclusive or" function.

**Branching and Looping.**—Both branches and loops (iterations) are based on the return value of an evaluated condition. There are three macro functions available: "IF", "GOTOn", and WHILE (also IF-THEN on some controls).

*IF Function:* The "IF" function evaluates a given condition and branches to a specified block number providing the condition is true. If the condition is false, the macro processing continues in the next block. This function is commonly used when choosing between two alternatives. Effectively, the IF function bypasses one or more blocks if a certain condition is true. The programming format is:

IF [<*this condition is true*>] GOTOn

*GOTOn Function:* The "GOTOn" function is an unconditional branch and is programmed by itself in the block. The "n" is replaced by the block number to branch to. For example, GOTO901 will branch to block N901. The programming format is:

GOTO901

*WHILE Function:* The "WHILE" function is for looping—it also requires a specified condition to be true for the loop to be processed. The word "while" can be interpreted as meaning "as long as." Its programming format is:

WHILE [<*this condition is true*>] DOn

The "ENDn" is the end of loop identification. There can be up to three levels of loop nesting. Each nested loop will have is own WHILE-ENDn. For a single loop, the ENDn is END1.

Note—there is also IF-THEN method available on some control systems. If available, the programming is much more streamlined. The programming format is:

IF [<*this condition is true*>] THEN [<*argument*>]

The 'THEN' argument is a statement, normally a definition based on the condition being true:

*Example:* IF [#1 EQ 1] THEN #33 = 0

All branching and looping functions are extremely powerful and are a fundamental part of effective use of macros.

**Macro Example.—**A classic example of macro development is an equally spaced arrangement of holes along a circle—a bolt circle pattern, see Fig. 31. Many versions can be developed, depending on the actual needs

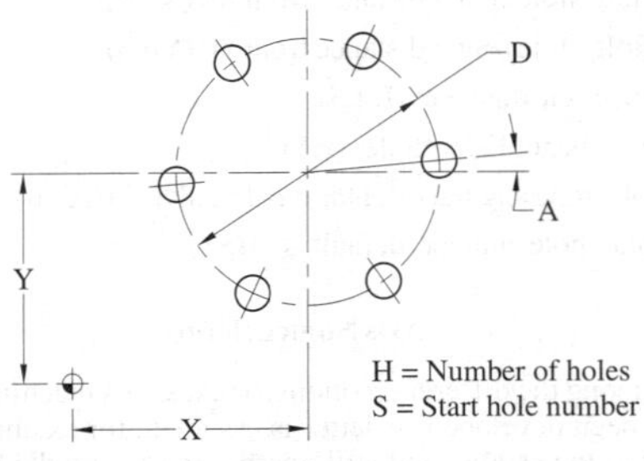


Fig. 31. Bolt Circle Macro Sample

(MAIN PROGRAM—CALLS MACRO **O7051**)

N1 G20
N2 G90 G00 G54 X0 Y0 S900 M03
N3 G43 Z1.0 H01 M08
N4 G99 G81 R0.1 Z-0.6 P250 F12.0 L0
**N5 G65 P7051 X3.0 Y6.5 D4.5 H6 A5.0 S1**
N6 G80 Z10.0 M09
N7 G28 Z10.0 M05
N8 M01
...
%

```
O7051 (BOLT HOLE CIRCLE MACRO)
#10 = #4003 (STORE CURRENT G90 OR G91)
IF [#7 LE 0] GOTO 9101
IF [#11 NE FUP[#11]] GOTO 9102
IF [#11 LE 0] GOTO 9103
IF [#19 EQ #0] THEN #19 = 1
IF [#19 NE FUP[#19]] GOTO 9102
IF [#19 LT 1] GOTO 9104
IF [#19 GT #11] GOTO 9105
#7 = #7/2 (BOLT CIRCLE RADIUS)
WHILE [#19 LE #11] DO1
 #30 = [#19−1] * 360 / #11 + #1
 X[COS[#30] * #7 + #24] Y[SIN[#30] * #7 + #25]
 #19 = #19 + 1
END1
GOTO 9999
N9101 #3000=101 (DIA MUST BE POSITIVE)
N9102 #3000=102 (ONLY INTEGER NUMBER ALLOVED)
N9103 #3000=103 (ONLY POSITIVE NUMBER OF HOLES)
N9104 #3000=104 (START HOLE MUST BE AN INTEGER)
N9105 #3000=105 (START HOLE NUMBER IS TOO HIGH)
N9999 G#10 (RESTORE ORIGINAL G90 OR G91)
M99
%
```

The macro call in block N5 includes six arguments that will be passed to the macro:

X = #24 = Bolt circle center distance from X0 (X3.0)

Y = #25 = Bolt circle center distance from Y0 (Y6.5)

D = #7 = Bolt circle diameter (D4.5)

H = #11 = Number of EQSP holes (H6)

A = #1 = Bolt circle angular orientation (default is 0) (A5.0)

S = #19 = Start hole number (default is 1) (S1)

### Axis Nomenclature

To distinguish among the different motions, or axes, of a machine tool, a system of letter addresses has been developed. A letter is assigned, for example, to the table of the machine, another to the saddle, and still another to the spindle head. These letter addresses, or axis designations, are necessary for the electronic control system to assign movement instructions to the proper machine element. The assignment of these letter addresses has been standardized on a worldwide basis and is contained in three standards, all of which are in agreement. These standards are EIA267C, issued by the Electronics Industries Association; AIA NAS-938, issued by the Aerospace Industries Association; and ISO841:2001, issued by the International Organization for Standardization.

The standards are based on a "right-hand rule," which describes the orientation of the motions, as well as whether the motions are positive or negative. If a right hand is laid palm up on the table of a vertical milling machine, as shown in Fig. 1, for example, the thumb will point in the positive *X* direction, the forefinger in the positive *Y* direction, and the erect middle finger in the positive *Z* direction, or up. The direction signs are based on the motion of the cutter relative to the workpiece. The motions are considered from the part programmer's viewpoint, which assumes that the cutter always moves around the part, regardless of whether the cutter or the part moves. The right-hand rule also holds with a horizontal-spindle machine and a vertical table, or angle plate, as shown in Fig. 2.

Here, spindle movement back and away from the angle plate, or workpiece, is a positive *Z*-motion, and movement toward the angle plate is a negative *Z*-motion.

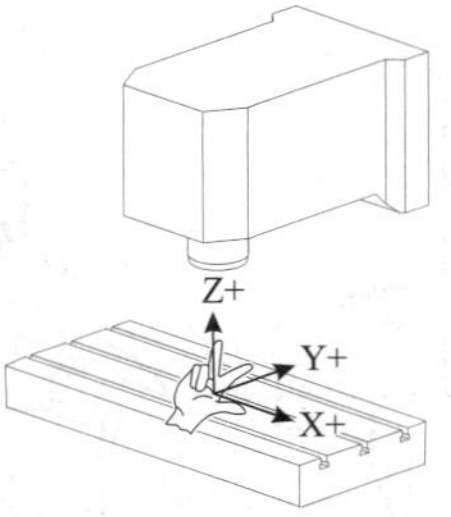


Fig. 1. Right-Hand Rule (Vertical Milling Machine)

Rotary motions also are governed by a right-hand rule, but the fingers are joined and the thumb is pointed in the positive direction of the axis. Fig. 3 shows the designations of the rotary motions about the three linear axes, *X*, *Y*, and *Z*. Rotary motion about the *X*-axis is designated as *A*; rotary motion about the *Y*-axis is *B*; and rotary motion about the *Z*-axis is *C*. The arrowhead direction indicates positive rotary movement. Movement of the rotary table around the *Y*-axis shown in Fig. 3 is a *B* motion and is common with horizontal machining centers. Here, the view is from the spindle face looking toward the rotary table. Referring, again, to linear motions, if the spindle is withdrawn axially from the work, the motion is a positive *Z*. A move toward the work is a negative *Z*.

When a second linear motion is parallel to another linear motion, as with the horizontal boring mill seen in Fig. 4, the horizontal motion of the spindle, or quill, is designated as *Z* and a parallel motion of the angle plate is *W*. A movement parallel to the *X*-axis is *U* and a movement parallel to the *Y*-axis is *V*. Corresponding motions are summarized as follows:

| Linear | Rotary | Linear and Parallel |
|---|---|---|
| *X* | *A* | *U* |
| *Y* | *B* | *V* |
| *Z* | *C* | *W* |

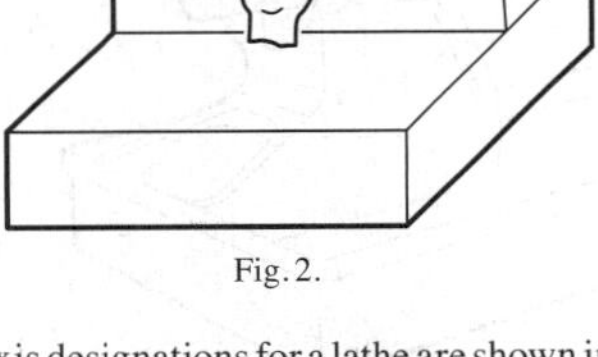


Fig. 2.

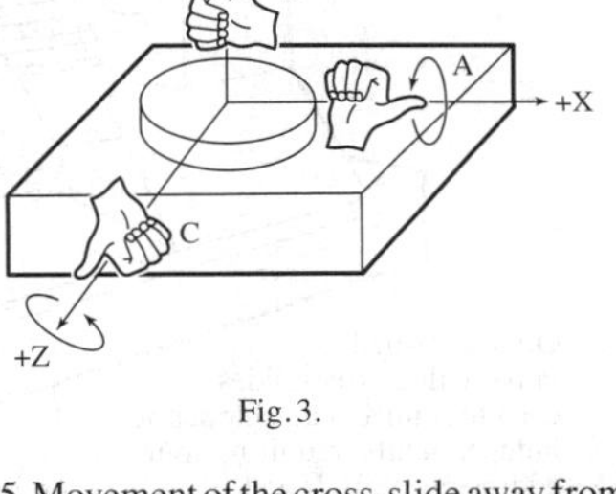


Fig. 3.

Axis designations for a lathe are shown in Fig. 5. Movement of the cross-slide away from the workpiece, or the centerline of the spindle, is noted as a plus *X*. Movement toward the workpiece is a minus *X*. The middle finger points in the positive *Z* direction; therefore, movement away from the headstock is positive and movement toward the headstock is negative. Generally, there is no *Y*-movement.

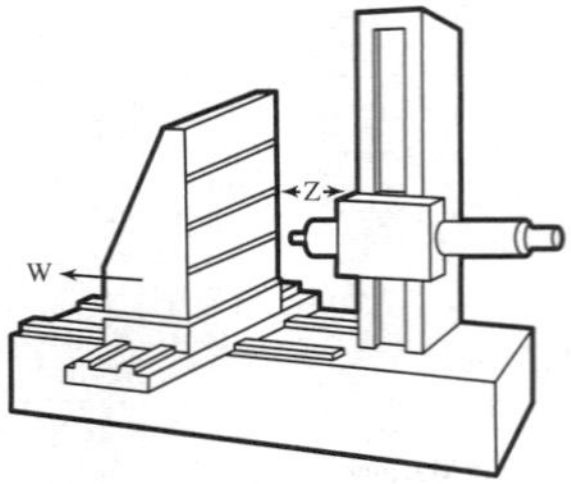


Fig. 4.

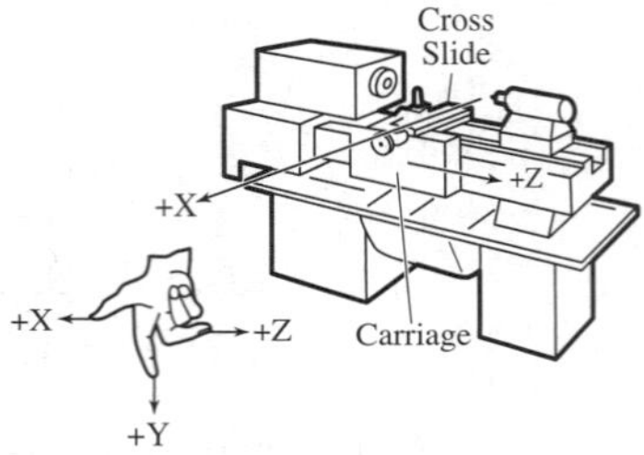


Fig. 5.

The machine shown in Fig. 5 is of conventional engine lathe design, but most NC lathes look more like that shown in Fig 6. The same right-hand rule applies to this four-axis lathe, on which each turret moves along its own two independent axes. Movement of the outside-diameter or upper turret, up and away from the workpiece, or spindle centerline, is a positive *X*-motion, and movement toward the workpiece is a negative *X*-motion. The same rules apply to the *U*-movement of the inside-diameter, or boring, turret. Movement of the lower turret parallel to the *Z*-motion of the outside-diameter turret is called the *W*-motion. A popular lathe configuration is to have both turrets on one slide, giving a two-axis system rather than the four-axis system shown. *X*-and *Z*-motions may be addressed for either of the two heads. Upward movement of the boring head therefore is a positive *X*-motion.

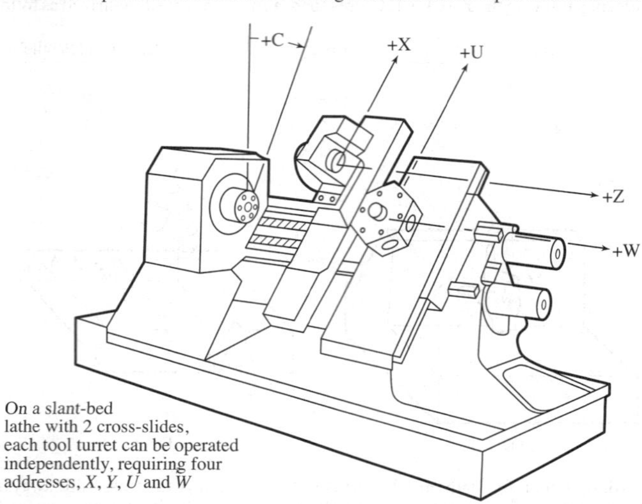


Fig. 6.

**Total Indicator Reading (TIR).—**Total indicator reading is used as a measure of the range of machine tool error. TIR is particularly useful for describing the error in a machine tool spindle, referred to as runout. As shown in Fig. 7, there are two types of runout: axial and radial, which can be measured with a dial indicator. Axial runout refers to the wobble of a spindle and is measured at the spindle face. Radial runout is the range of movement of the spindle centerline and is measured on the side of the spindle or quill.

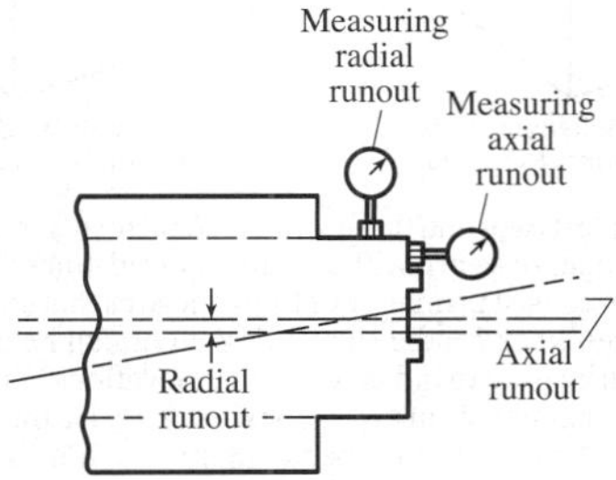


Fig. 7.

## CAD/CAM

CAD in engineering means *computer-aided design* using a computer graphics system to develop mechanical, electrical/electronic, and architectural designs. A second D (CADD) is sometimes added for *computer-aided drafting and design* to indicate use of a computerized drafting or drawing program. CAD technology is the foundation for a wide variety of engineering, design, drafting, analysis, and manufacturing activities. Often a set of drawings initially developed in the design phase of a project is also used for analyzing and optimizing the design, creating mechanical drawings of parts and assemblies and for generating CNC part programs that control machining operations.

Formerly, after a component had been designed with CAD, the design was passed to a part programmer, who developed a program for machining the components, either manually or directly on a computer (graphic) screen, but the process often required redefining and reentering part geometry. This procedure is often regarded as the CAM part of CAD/CAM, although CAM (for computer-aided manufacturing) has a much broader meaning and involves the computer in many other manufacturing activities, such as factory simulation and planning analyses. Improvements in the speed and capability of computers, operating systems, and programs have simplified the process of integrating the manufacturing process and passing drawings (revised, modified, and translated, as necessary) through the design, analysis, simulation, and manufacturing stages.

A CAD drawing is a graphic representation of part geometry data stored in a drawing database file. The drawing database generally contains the complete list of entity (line, arc, etc.) and coordinate information required to build the item represented. This may include additional information required to define solid surfaces and other model characteristics.

**Types of CAD Drawings.—**The simplest CAD drawings are two-dimensional (2D) and conform to normal engineering drafting practice showing orthographic (front, top, and side views, for example), exploded, isometric, or other views of a component. Depending on the complexity of the part and machining requirements, 2D drawings are often sufficient for use in developing CNC part programs.

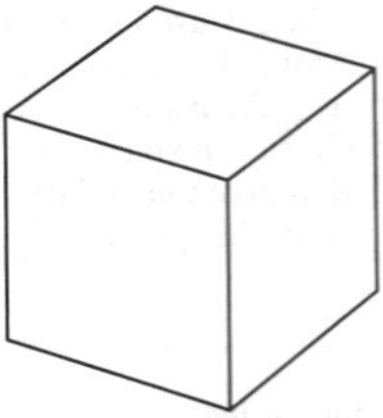

Fig. 1. Simple Wireframe Cube with Hidden Lines Removed

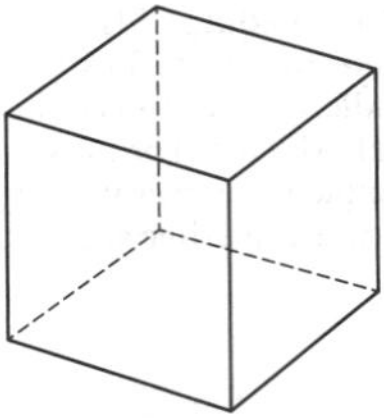

Fig. 2. Simple Wireframe Cube with Hidden Lines Shown

*Wireframes* are the simplest two- and three-dimensional forms of drawing images and are created by defining all edges of a part, with, where required, lines defining surfaces. Wireframe drawing elements consist primarily of lines and arcs that can be used in practically any combination. Hidden lines can be shown dashed or can be omitted from the view; hidden-line removal can be used to indicate relative elevations of the elements.

A wireframe drawing of a cube, as in Fig. 1, consists of twelve lines of equal length (some are hidden and thus not shown), each perpendicular to the others. Fig. 2 shows the same wireframe drawing of a cube but with the hidden lines shown. Information about the interior of the cube and the character of the surfaces is not included in the drawing. With such a system, if a 1-inch cube is drawn and a 0.5-inch cylinder is required to intersect the cube's surface at the center of one of its faces, the intersection points cannot be determined, because nothing is known about the area between the edges. A wireframe model of this type is ambiguous if the edges overlap or do not meet where they should.

Two-dimensional drawing elements, such as lines, arcs, and circles, are constructed by specifying point coordinates, usually $X$ and $Y$, to identify the location, size, and orientation of these elements. Three-dimensional (3D) drawings are also made up of a collection of lines, arcs, circles, and other drawing elements and are stored in a similar manner. A third point coordinate, $Z$, indicates the elevation of a point in 3D drawings. On the drawing screen, working in the $X$–$Y$ plane, the elevation is commonly thought of as the distance of a point or object into the viewing screen (away from the observer) or out of the viewing screen (toward the observer). Coordinate axes are orientated according to the right-hand rule. Assigning a thickness (or extruding) to objects drawn in two dimensions gives some 3D characteristics to an object and can be used to create simple prismatic 3D shapes, such as cubes and cylinders.

*Surface Images:* A surface image defines not only the edges of the part, but also the "skin" of each face or surface. Surface models are necessary for designing free-form objects such as automotive body panels and plastics injection moldings used in consumer goods. For a surface model, the computer must be provided with the $X, Y, Z$ coordinates defining each point, as in a wireframe drawing, plus much more information about the part. This may include tangent vectors, surface, and weighting information that determines how much influence one point has on another, twists, and other mathematical data that define abstract curves. For instance, Fig. 3 shows a typical 3D surface patch.

Fig. 3. A 3D Surface Patch

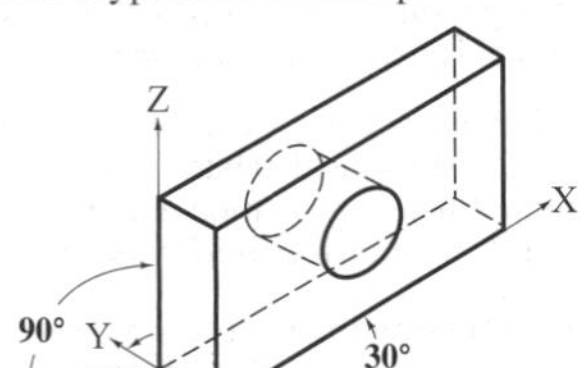


Fig. 4. Isometric Drawing Showing Orientation of Principle Drawing Axis

*Solid Models:* Solid modeling is the ultimate electronic representation of a part, with the resultant computer-generated model containing all the necessary information about edges, surfaces, and the interior. Most solid modeling programs can calculate volume, center of mass, centroid, and moment of inertia. Solid models may be sliced or sectioned on the screen to provide a view of the interior. This type of image is also useful for checking fit and assembly of parts.

One method for building a solid model is to perform Boolean operations on simple shapes such as cylinders, cones, cubes, and blocks. Boolean operations are used to union (join), difference (subtract one from another), and intersect (find the common volume between two objects). Thus, making a hole in a part requires subtracting a cylinder from a rectangular block, in a process called constructive solid geometry (CSG). The boundary representation type of imaging program uses profiles of 2D shapes that it extrudes, rotates, and otherwise translates in 3D space to create the required solid. Solid models provide complete, unambiguous representation of a part.

**Drawing Projections.—**Different techniques are used to display objects to give an accurate 3D appearance. Several methods are commonly used in CAD drawings.

*Isometric drawings,* as in Fig. 4, can be used to good effect for visualizing a part, because they give the impression of a 3D view and are often much faster to create. Isometric drawings are created in 2D space, with the $X$ and $Y$ axes being inclined at 30 degrees to the horizontal, as shown in Fig. 4, and the $Z$-axis as vertical. Holes and cylinders in isometric drawings become elliptical. Because of the orientation of the axes $X$, $Y$, and $Z$, the true length of lines may not be accurately represented in isometric drawings, so dimensions should not be taken directly from a print.

In *parallel projections*, lines that are parallel in an object, assembly, or part being portrayed remain parallel in the drawing. Parallel projections show 3D objects in a dimensionally correct manner. Though the resulting drawing may not appear as realistic as an isometric or perspective view, it can be used for accurate relative and scaled dimensions.

A characteristic of *perspective drawings* is that parallel lines converge (see Fig. 5) so that objects that are farther away from the observer appear smaller. Perspective drawing techniques are used in some 3D drawings to convey the true look of an object, or group of objects. Because objects are not drawn to scale, dimensional information cannot be extracted from a perspective drawings. Some 3D drawing packages have a true perspective drawing capability; others use a simulation technique to portray a 3D perspective.

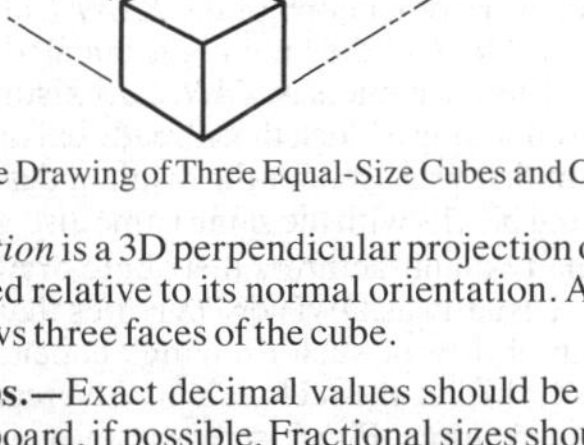

Fig. 5. Perspective Drawing of Three Equal-Size Cubes and Construction Lines

An *axonometric projection* is a 3D perpendicular projection of an object onto a surface, such that the object is tilted relative to its normal orientation. An axonometric projection of a cube, as in Fig. 1, shows three faces of the cube.

**Drawing Tips and Traps.—**Exact decimal values should be used when entering point coordinates from the keyboard, if possible. Fractional sizes should be entered as fractions, not truncated decimals. For example, $^5/_{16}$ should be entered as 0.3125 or $^5/_{16}$, not 0.313. Accumulated rounding errors and surprises later on when parts do not fit are thus reduced.

However, drawing dimensions should not have more significant digits or otherwise be more precise than necessary to avoid difficulty in production.

*Snap* and *object snap* commands make selecting lines, arcs, circles, or other drawing entities faster, easier, and more accurate when picking and placing objects on the screen. Snap permits only points that are even multiples of the snap increment to be selected by the pointer. It also is useful to reset the snap increment frequently. The snap feature can be turned off during a command to override the setting or to select points at a smaller interval than the snap increment allows. Some systems permit setting a different snap value for each coordinate axis.

The *object snap* selection mode is designed to select points on a drawing entity according to predefined characteristics of the entity. For example, if end-point snap is in effect, picking a point anywhere along a line will select the end point of the line nearest the point picked. Other object snap modes include point, intersection, midpoint, center and quadrants of circles, tangency point (allows picking a point on an arc or circle that creates a tangent to a line), and perpendicular point (picks a point that makes a perpendicular from the base point to the object selected). When two or more object snap modes are used together, the nearest point that meets the selection criteria will be chosen.

*Create Blocks of Frequently Used Objects:* Once created, complete drawings or parts of drawings can be saved and later recalled into another drawing. Such objects can be scaled, copied, stretched, mirrored, rotated, or otherwise modified without changing the original. When shapes are initially drawn in unit size (i.e., fitting within a $1 \times 1$ square) and saved, they can be inserted into any drawing and easily scaled. One or more drawing elements can be saved as a group element, or *block*. Block properties vary, depending on the design program, but are among the most powerful features of CAD.

Typically, blocks are uniquely named and, as with simple objects, may be saved in a file. (Blocks are ideal for creating libraries of frequently used drawing symbols.) Blocks can be copied, moved, scaled, rotated, arrayed, and inserted as many times as is required in a drawing and manipulated in a drawing as one element. When scaled, each object within the block is scaled to the same degree.

Once blocked, the individual components of a block (lines, arcs, circles, surfaces, and text, for example) cannot be changed or edited. To edit them, a copy (instance) of the block must be *exploded* (unblocked) to divide it into its original components. Once exploded, all of the elements of the block (except other blocks) can be edited. When the required changes have been made, the block must be redefined (redeclared as a block by giving it a name and identifying its components). If the block is redefined using the same name, previous references to the block in the drawing will be updated to match the redefined block.

For example, an assembly drawing is needed that shows a mechanical frame with 24 similar control panels attached to it. Once one of the panels is drawn and defined as a block (using the name *PANEL*, for instance), the block can be inserted (or copied) into the drawing 24 times. Later, if changes need to be made to the panel design, one instance of the block *PANEL* can be exploded, modified, and redefined with the name *PANEL*. When *PANEL* is redefined, every other copy of the *PANEL* block in the drawing is also redefined, so every copy of *PANEL* in the drawing is updated. On the other hand, if the block was redefined with a different name, say, *PANEL1*, existing copies of *PANEL* would remain unchanged. When redefining a block that already exists in the drawing, be sure to use the same insertion point that was used for the original definition of the block; otherwise, the positions of existing blocks with the same name also will be changed.

It is worth noting that many manufacturers distribute drawings of their products in libraries of CAD drawings, usually as DWG or DXF files that can be incorporated into existing drawings. The suitability of such drawings depends on the CAD program, drawing format being used, skill of the technician who created the drawings, and accuracy of the drawings. An example, Fig. 6, shows a sewage pump assembly drawing. Libraries of frequently used drawing symbols and blocks are also available from commercial sources.

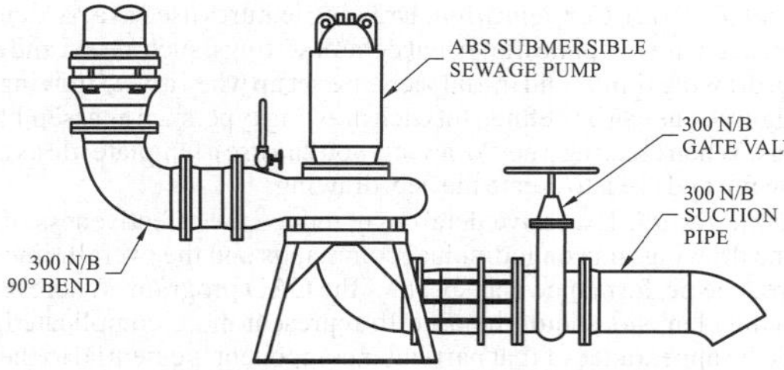


Fig. 6. Available CAD Drawing of a Sewage Pump Assembly

*Use of Text Attributes to Request Drawing Information Automatically:* Text attributes are a useful method for attaching textual information to a particular part or feature of a drawing. An attribute is basically a text variable that has a name and can be assigned a value. Attributes are created by defining attribute characteristics such as a name, location in the drawing, text size and style, and default value.

Fig. 7 shows two views of a title block for size A, B, C, and G drawing sheets. The top view includes the title block dimensions per ASME Y14.1-2019, with the names and locations of the attributes (COMPANY, TITLE1, TITLE2, etc.). The first step for creating this title block is to draw the frame of the title block and define the attributes (name, location, and default value for: company name and address, drawing titles [2 lines], drawing size, drawing number, revision number, scale, and sheet number). Then the block must be created and named with the title frame and the attribute definitions.

When the block is inserted into a drawing, the operator is asked to enter the attribute values (company name, drawing title, etc.). The bottom view in Fig. 7 shows a completed title block as it might appear in a drawing. A complete drawing sheet could include several blocks: a sheet frame, a revision block, a parts list block, and any others needed. Some might be combined into a single block for insertion into a drawing at one time.

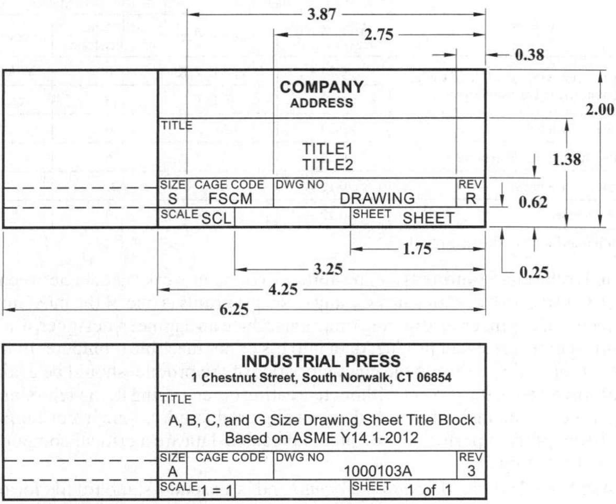


Fig. 7. Top: Title Block for A, B, C and G Drawing Sheets Showing Placement of Text Attributes with ASME Standard Dimensions for Creating this Block. Bottom: Completed Title Block

*Define a Default Drawing Configuration:* Drawing features used in a particular type of drawing can be set up in a template file so that default settings, such as text and dimension styles, text size, drawing limits, and initial view, are set up when a new drawing is started. Different configurations can be defined for each drawing type, such as assembly, parts, or circuit drawings. When creating a new drawing, you can use a template file as a pattern or open a template file and use it to create the new drawing.

*Use Appropriate Detail:* Excessive detail may reduce the effectiveness of the drawing, increase the drawing time on individual commands and the overall time spent on a drawing, and reduce performance and speed of the CAD program. Wherever possible, symbolic drawing elements should be used to represent more complicated parts of a drawing unless the appearance of that particular component is essential to the drawing.

*Scaling Text on Drawing Sheets:* It is usually desirable that text, dimensions, and a few other features on drawings stay a consistent size on each sheet, even when the drawing size is different. Following this procedure ensures that text and dimensions (other features as well, if desired) will be the same size from drawing to drawing, without resorting to scaling the drawing to fit onto the drawing sheet.

*Example:* Create a drawing sheet having the dimensions of the actual sheet to be output (A, B, C, D, or E size, for example). Use text attributes, such as the title block illustrated in Fig. 7, to include text that needs to be entered each time the drawing sheet is used. Create a block of the drawing sheet, including the text attributes, and save the block to disk. Repeat for each size drawing sheet required. Establish the text and dimension size requirements for the drawing sheet when it is plotted full size (1:1 scale). Use Table 1 as a guide to recommended text sizes. Test the sheet by placing text and dimensions on the drawing sheet. Print a copy of the sheet to check that text and dimensions are the expected size.

**Table 1. Standard Sizes of Mechanical Drawing Lettering**
*ANSI/ASME Y14.2-2014 (R2020)*

| Use For | Inch | | Metric | |
|---|---|---|---|---|
| | Min. Letter Heights, (in.) | Drawing Size | Min. Letter Heights, (mm) | Drawing Size |
| Drawing title, drawing size, CAGE Code, drawing number, and revision letter[a] | 0.24 | D, E, F, H, J, K | 6 | A0, A1 |
| | 0.12 | A, B, C, G | 3 | A2, A3, A4 |
| Section and view letters | 0.24 | All | 6 | All |
| Zone letters and numerals in borders | 0.24 | All | 6 | All |
| Drawing block headings | 0.10 | All | 2.5 | All |
| All other characters | 0.12 | All | 3 | All |

[a] When used within the title block.

**Drawing Exchange Standards.**—The ability to transfer working data between different CAD, CAD/CAM, design analysis, and CNC programs is one of the most important requirements of engineering drawing programs. Once an engineer, designer, draftsman, or machinist enters relevant product data into his or her machine (computer or machine tool), the information defining the characteristics of the product should be available to the others involved in the project without recreating or reentering it. In view of manufacturing goals of reducing lead time and increasing productivity, concurrent engineering, and improved product performance, interchangeable data are a critical component in a CAD/CAM program.

*DXF* stands for *Drawing Exchange Format* and is a pseudo-standard file format used for exchanging drawings and associated information between different CAD and design analysis programs. Nearly all 2D and 3D CAD programs support some sort of drawing exchange through the use of DXF files, and most can read and export DXF files. There are, however, differences in the drawing features supported and the manner the DXF files are handled by each program. For example, if a 3D drawing is exported in the DXF format and imported into a 2D CAD program, some loss of information results, because all the 3D features are not supported by the 2D program. Most common drawing entities (lines, arcs, etc.) will transfer, although other problems may occur. For example, entities treated as a single object (such as blocks, hatch patterns, and symbols) may be divided into hundreds of components when converted into a DXF file, making the drawing more difficult to edit.

*IGES* stands for *Initial Graphics Exchange Specification* and is a means of exchanging or converting drawings and CAD file for use in a different engineering design software. Normally, a drawing prepared on the design system used by company A would have to be redrawn before it would operate on a different design system used by company B. However, with IGES, the drawing can be passed through a software package called a preprocessor, which converts it into a standardized IGES format that can be saved on a storage device, such as a portable hard drive or memory stick. A postprocessor at company B is then used to convert the standard IGES format to that required for their design system. Most design software companies have or can assist with IGES programs for file conversion.

*STEP* stands for *Standard for the Exchange of Product* model data and is a series of existing and proposed ISO standards written to allow access to all the data surrounding a product. It extends the IGES idea of providing a geometric data transfer to include all the other data that would need to be communicated about a product over its lifetime, and facilitates the use and accessibility of the product data. Software tools have been developed for converting data between IGES and STEP formats.

*STL,* an abbreviation for *standard tessellation language,* is the format primarily used to send CAD drawings to additive manufacturing machines. Many full-featured CAD systems support translation of drawing files into STL format. CAD programs without STL file format capability can use the IGES or DXF file format. The safest route, reducing unexpected errors, is to create a CAD model and export it directly into the STL format. Reverse STL generators are available to display a file saved in STL format or convert it into a form that can be imported into a CAD program. For more on this topic, see *Metal Additive Manufacturing* on page 1555.

**CAM.—**Computer-aided design and computer-aided manufacturing (CAD/CAM) utilize computers to design part and assembly models for manufacture. CAD is limited to the generation of engineering drawings; CAD/CAM combines design and manufacturing capabilities and is much more efficient than manually writing CNC code. With CAM, the design model from CAD is used to identify specific cutting tool data and create cutting toolpaths for CNC machining. Selection of model geometry identifies part features and how they will be machined to develop the code (part program) for CNC machine tools.

*Designers* can create the models and engineering drawings needed and share them electronically with the manufacturing team. Model files are saved in native format or exported to a common file format, such as the Initial Graphics Exchange Specification (IGES), Standard for the Exchange of Product (STEP) model data, and Stereo Lithographic (STL). Most CAM systems support direct import of solid model files for toolpath creation (drag-and-drop functionality), facilitating the manufacturing engineer's/CNC programmer's assignment of cutting tool information and toolpath creation to produce the desired results.

When using CAM, the drawing may be created from scratch or imported from a CAD program; the drawing does not have to be dimensioned, but the full scale of the part is required. To make best use of CAM, it is important to fully understand all machining processes, the rectangular and polar coordinate systems, cutting tool selection, and applicable speeds and feeds. Most CAM programs will automatically develop speed and feed data based on the tool and work material selection saved in the software database (which should be matched to shop requirements), but adjustments are commonly made to fine tune and improve results.

**Simulation and Verification of CNC Programs.**—To have absolute confidence to run the CNC program, visual playback of CAM-created toolpaths is used to confirm actual program results. This *simulation* can be used for an individual tool or operation groups in continuous, step, forward, and reverse. The CAM software allows representation of the cutting tool, holder, fixturing, and raw and in-process stock to aid visualization, with rapid and feed moves displayed in different colors, depending on the software used. The programmed part can be viewed from the top, front, 3D, multiple simultaneous views, etc. This visual feedback enables adjustment and optimization of programmed toolpaths and confirmation of correct machine setup.

The simulation and final output of CNC code requires input of accurate setup data to perform effectively and has some common criteria:

*Machine Type:* To begin the programming process, the type of machine being used (mill, lathe, etc.) must be defined. Selection of the specific machine controller type directs output (post-processing) of correct G-Code.

*Stock Dimensions:* Raw material stock dimensions can be defined by overall dimensions, from boundary extents of geometry and toolpath, or imported from a solid stock model file.

*Part Material:* Selected from commonly used materials available in the CAM software libraries, part material is critical for automatic development of feeds and speeds calculations.

*Cutting Tool Data:* Detailed definitions—tool type, materials, sizes, lengths—added within the software or imported from vendor-provided data are critical to toolpath simulation and feeds and speeds calculations.

*Work Holding:* Definitions of work holding and fixturing devices (vises, chucks, clamps, etc.) again can be created within the software or imported from a solid model file and is needed for collision checks.

*Verification:* Toolpath verification functions, whether part of the CAM software, an add-on module, or separate software, is vital to ensure safe, collision-free operation of multi-axis machining programs. Solid model-based cutting (material removal) and non-cutting (rapid) connecting toolpath moves are simulated. The actual CAM-generated G-Code is processed for verification. The software checks for program errors and fixture collisions related to tool capabilities, based on the data defined in setup. This gives the programmer immediate feedback regarding setup data, program errors, warnings, and informational items. All of the *criteria* described above for simulation is required for verification.

An additional feature within verification is simulation of *machine kinematics*. Using builder-provided solid models of machine tools ensures an accurate virtual representation (digital twin or exact copy) of all moving components, including axis travel limits. When combined with cutting tool, work holding, and fixture information, every detail of the machine setup and CNC program can be collision checked.

Another powerful function in verification, *gouge detection* is the comparison of the simulated cutting toolpath to the design model. The programmer sets allowable gouge or excess limit values for material differences and during playback receives visual feedback and/or reports.

### Feature-Based Machining

The automation of CNC programming minimizes time-consuming manual programming efforts and repetitive tasks, streamlining program development by reducing the steps to complete and output CNC program code. *Feature-based machining* (*FBM*), which defines machining operations from modeled features, helps limit input and selection errors in programming for increased accuracy. Best practice machining processes are defined in a database called the *machining knowledge library* (*MKL*). This customizable system controls tool selection and cutting parameters using rules, taking dimensional tolerances into consideration. FBM is useful in creating machining programs for parts with similar features. Primary uses include simple repetitive tasks such as hole-making, 2.5D milling, drilling, reaming, and tapping.

Software for solid modeling readily facilitates design modifications. When dimensional characteristics are modified, all changes are automatically updated to the CAM and FBM program when the file is reopened.

**FBM and Solid Modeling Basics.—**Solid models are feature-based parametric designs. A *feature* is any entity of a part that has volume or area. Basic features include holes, pockets, steps, notches, and surface texture. *Parametric* refers to the parameters defining the feature's shape and size. For instance, a prismatic part may have a length of 3 inches, a width of 2 inches, and a height of 1 inch. These values are parameters that can be adjusted, resulting in a similar prismatic part. For efficient use of FBM for toolpath creation, a properly constructed solid model is required. Features must be "water tight," meaning there can be no holes or gaps in the solid.

Each machine tool being programmed has specific physical characteristics, such as tooling pockets, that must be identified as part of the programming setup. Characteristics of tools used in a shop may be retrieved from the CAM software and added for new or unique tools. When the design model is imported into the working CAM environment, it also is necessary to establish the machine coordinate system (MCS), position the workpiece in relation to it, and identify the part blank and any work holding fixtures.

**Basic Steps for Programming with Feature-Based Machining.—**As CAD/CAM software package processes differ slightly, users should consult the software documentation for specific techniques. However, the standard steps are as follows:

1) Identify the cutting tool library (database) from which tools to be used will be selected; add any special tools not included.
2) Establish the machining environment—milling, drilling, tapping, turning, wire EDM machining, and so on.
3) Specify the part and the raw material blank.
4) Use a feature finder to identify parametric features.
5) Create feature processes and operations. (As processes are completed and toolpaths are generated, the software keeps track of the finished material condition in what is called the in process workpiece.)
6) Generate toolpaths. Review and edit as needed.
7) Simulate and verify toolpaths.
8) Post process, complete setup documentation, and release the CNC program to production.

### CNC Machining of Carbon-Fiber Reinforced Polymers

Carbon Fiber-Reinforced Polymer (CFRP), a multi-layered matrix of glass cloth impregnated with epoxy resin and carbon fibers, has the desirable characteristics of a high strength-to-weight ratio and corrosion resistance. It is commonly used in aerospace, marine, auto body, and other outdoor and motorsports applications. Most components are lay-up molded and require machining of holes for fasteners and edge trimming, presenting unique machining challenges.

CNC programming techniques for CFRP are essentially the same as for metallic parts. Specific differences include selection of cutting tool type, tool material, and coatings; feeds and speeds; and the recommended use of separate tools for roughing and finishing. Common operations include drilling, edge trimming, and surface machining. When milling edge-of-part contours, the cut direction should be conventional. For large complex surfaces and part edges, a five-axis, gantry-style milling/routing machine may be required. When surface machining, ball-nose end mills can be used and tilted off-axis to effectively utilize the cutting edges at the tool tip. For all machining, spindle aligned cutting tool shank engagement should be as close to the holder/arbor as possible. When available, spindle-axis oscillation should be employed to spread out edge wear over the flute length.

*Safety Precautions:* When handling CFRP, sharp edges and splinters, especially after machining, require protective gloves to avoid injury. During machining, carbon dust and fibers can become airborne; appropriate particulate masks/respirators and regularly maintained ventilation systems with filtration are recommended. The carbon dust also is conductive, so electrical equipment, including computers must be protected. Machining of this highly abrasive material creates a great deal of heat, causing extreme tool wear that can result in surface/subsurface fracturing and delamination.

*Using Coolant for Machining CFRP:* In metal machining, chip formation is predictable. But in machining of CFRP, material removal is more like fracturing or breaking, making it difficult to control fraying, splitting, or fiber tear-out. Application of coolant can aid in reducing heat and friction and promoting chip evacuation during machining, especially when delivered through the tool. The coolant should be fully synthetic, mixed with water, and not contain oil. Using coolant engineered for machining CFRP will help optimize cutting tool life, improve part surface quality, and reduce carbon dust. Disposal of contaminated fluid and sludge must follow Hazardous Materials Management (HAZMAT) procedures.

*Work Holding for Machining:* Part vibration of any kind during machining is destructive, and CFRP only exacerbates this issue. An effective work holding method is to use vacuum fixtures, designed and machined to match component-underside (inner mold line, or IML) surfaces. For rigid support across part surfaces, vacuum channels should be machined, in a grid, across the entire mating surface, and a rubber seal groove added as close as possible to part edges. Dampening techniques, such as strategically placed sand bags, can be used to further mitigate part and/or fixture vibration.

**Cutting Tools for Machining CFRP.—**High-speed steel cutting tools designed to cut metals are not durable enough to successfully machine CFRP. A common base tool material is solid uncoated carbide, which is a good choice for roughing operations but demonstrates high wear characteristics. Polycrystalline diamond (PCD) coated carbide is better for finishing and has enhanced wear characteristics. Chemical vapor deposition (CVD) diamond-coated carbide has the best wear resistance. Polycrystalline cubic boron nitride (PCBN), a synthetic abrasive coating, is nearly as wear-resistant as diamond and is suited for finishing surfaces and edges (see *Cutting Tool Materials* on pages 1065–1074 for more information). Cutting tool manufacturers are continually developing tooling solutions for machining CRFP and should be consulted for the most up-to-date tooling and cutting data for best results.

Standard metal-cutting end mills employ a flute helix that tends to pull/push material upward/downward (conventional/climb), increasing the chance of delamination and fiber pull-out at part edges when machining CFRP. The best tool configurations for end milling are straight fluted router, compression cutter, chip breaker burr-style, low helix, and diamond cut. A straight fluted router has two flutes with no helix, nearly eliminating upward or downward pressures causing fiber pull-out. Compression cutters have an opposed (right- and left-hand) flute helix at the tool tip that centers at approximately 70 percent of the tool diameter along the tool axis (i.e., .500 diameter = .35 from the tool tip); this configuration compresses the material while machining toward the center of the part edge to decrease delamination. The chip breaker burr-style cutter is a multi-fluted end mill with serrated cutting-edge geometry that results in smaller, more easily evacuated chips. Low helix end mills, with four to six flutes, are designed for finish edge machining; the minimal helix reduces vertical lifting forces and damage to part edges.

The drill/mill tool incorporates a drill point tip enabling penetration through the material, combined with right- and left-hand cut teeth (diamond-cut) along the flute length. This tool diameter must be less than the finished hole size and requires circular interpolation after drilling to reach the finished diameter. This tool reduces deformation or delamination at the hole's bottom edge and provides a desirable surface finish from milling. PCBN abrasive-coated carbide ball nose endmills also are effective for surfacing and can be used for roughing and finishing.

PCBN abrasive-coated carbide ball nose endmills also are effective for surfacing and can be used for roughing and finishing.

**Drilling and Milling CFRP.—**Drilling tools designed for metal-cutting are not effective when machining CFRP. Also, the common method for drilling most materials penetrates the part surface with the tool with a *Z*-axis thrust that leads to edge deformation and delamination in CFRP.

*Edge Solutions:* The preferred solution is to use multi-angled diamond-coated drilling tools that minimize downward thrust pressures at entry and exit of the cut; these incorporate a 60°–118° point angle followed with a 15°–18° taper to the full drill diameter design. In some cases, pre-drilling the hole to an undersized condition and then using a second tool to bring the hole to finish size can be effective, with the downside of added tool changes. When hole size allows (where cutter diameter is less than finished hole size), circular milling with a helical ramp-in entry is an option.

*Feeds and Speeds:* PCD and diamond-coated cutting tools allow for higher cutting speeds than uncoated solid carbide and exhibit considerably better wear characteristics. Depending on tool diameter, spindle revolutions may exceed 25,000 rev/min with relative high feed rates. As spindle revolutions increase to high levels, even the smallest amount of tool holder or cutting tool runout causes vibration, affecting tool life and surface finish, and even causing premature spindle failure. Such high-performance conditions call for balanced tooling and the use of "shrink fit" tool holders and emphasize the need for fixture and part clamping rigidity. Removal of chips via coolant or other methods is key to minimizing heat buildup at cutting edges, detrimental to tool and finished part surfaces.

Starting parameters for drilling with a diamond-coated carbide taper drill for revolutions per minute and feed calculations are 200–500 surface feet per minute (SFM) and .0006–.0015 inches per revolution (IPR). Users should refer to the cutting tool manufacturer technical data for the most current speed and feed parameters. Specific conditions will dictate adjustments for fine-tuning results.

Starting parameters for milling with diamond-coated carbide for rev/min and feed calculations are 400–1500 SFM and .010–.020 IPR, depending on the tool diameter and number of cutting flutes. A smaller tool diameter starts with a lower chip load as does a higher flute count. As a general rule, the cutting tool diameter should be greater than or equal to the part thickness. As the number of cutting flutes increases, removal of chips becomes more difficult, and the need for coolant evacuation is critical to performance. Once again, it is recommended that the user refers to the cutting tool manufacturer technical data for the most current speed and feed parameters, with specific conditions dictating adjustments.

## TABLE OF CONTENTS

## MANUFACTURING PROCESSES

### SHEET METAL WORKING AND PRESSES



### SHEET METAL WORKING AND PRESSES

*(Continued)*



### ELECTRICAL DISCHARGE MACHINING



### METAL CASTING, MOLDING, AND EXTRUSION

## TABLE OF CONTENTS

## MANUFACTURING PROCESSES

### METAL CASTING, MOLDING, AND EXTRUSION

*(Continued)*



### METAL CASTING, MOLDING, AND EXTRUSION

*(Continued)*



### POWDER METALLURGY



MANUFACT.

## TABLE OF CONTENTS
## MANUFACTURING PROCESSES

### POWDER METALLURGY
*(Continued)*



### SOLDERING AND BRAZING



### WELDING



### WELDING
*(Continued)*

## TABLE OF CONTENTS

## MANUFACTURING PROCESSES

# SHEET METAL WORKING AND PRESSES

## Basic Theory of Metal Working

Metal working theory provides a background from which reasonable evaluations may be made of the deformations obtainable without instability and fracture. The fundamental principles, rules, and laws of metal working theory are used to describe the plastic flow or deformation of solid materials when subjected to external loads. For this purpose, such materials are considered as homogeneous, continuous, isotropic media. Understanding the theory underlying metal working is important for both the design and production engineer.

Solid materials may be subjected to forces that may be classified as either volume forces or surface forces. In this discussion, only surface forces acting on the surface as external forces are considered.

In analyzing design situations for either dimensioning purposes or forming processes, it is most appropriate to use force per unit area as a measure of the load rather than the total force distributed over the area.

The force per unit area is called the *stress* ($\sigma$) and is described as follows:

$$\sigma = \frac{\text{force}}{\text{area}} \tag{1}$$

Common ways of loading solid bodies include compression, tension, shear, torsion, or a combination of these stresses, such as fatigue. Fig. 1 shows a tensile specimen loaded with force $F$. The stress on a cross section $A$ perpendicular to the longitudinal axis is defined

$$\sigma = \frac{F}{A} \tag{2}$$

where $F$ = force (lb) and $A$ = cross-sectional area (in$^2$).

If a cross section is inclined at an angle to the longitudinal axis, the mean oblique stress may be defined by

$$\sigma_m = \frac{F}{A_\theta} = \frac{F}{A}\sin\theta \tag{3}$$

where $A_\theta$ = inclined cross-sectional area of specimen (in$^2$); and, $\theta$ = inclined angle cross-sectional area of specimen (°).

The mean oblique stress lies in the direction of the longitudinal axis of the specimen. Force $F$ can be divided into components $F_n$, which is perpendicular to cross section $A_\theta$ and $F_t$, which is parallel to cross section $A_\theta$, so that the state of the stresses can be defined by

$$\begin{aligned} \sigma_\theta &= \frac{F_n}{A_\theta} = \frac{F}{A}\sin^2\theta \\ \tau_\theta &= \frac{F_t}{A_\theta} = \frac{F}{2A}\sin 2\theta \end{aligned} \tag{4}$$

where $\sigma_\theta$ = normal stress—stress normal to cross section $A_\theta$ (lb/in$^2$); and $\tau_\theta$= shear stress—stress parallel to cross section $A_\theta$ (lb/in$^2$).

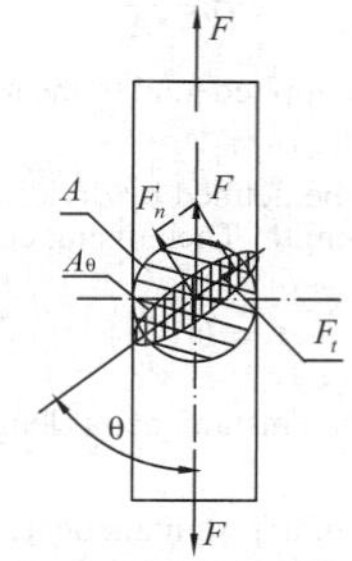


Fig. 1. A Tensile Specimen with Cross-Sectional Area $A$ Subjected to the Load $F$

**Stress-Strain Relationship.**—When solid materials are subjected to stress, they usually respond in an elastic fashion; that is, the strain produced by the stress is reversible (the strain goes back to zero when the stress is removed), and the magnitude of the strain is directly proportional to the magnitude of the stress. This relationship between stress and strain is usually referred to as *Hooke's Law* and can be written

$$\frac{\text{stress}}{\text{strain}} = E = \text{constant} \tag{5}$$

where $E$ = modulus of elasticity (lb/in$^2$).

The most common procedure for describing the various relationships between stress and strain is the tensile test, which is used to determine the modulus of elasticity, the elastic limit, the elongation, the proportional limit, the reduction area, the tensile strength, the yield point, the yield strength, and other tensile properties. The stress ($\sigma$), calculated from the load, and the strain ($\varepsilon$), calculated from the extension, can either be plotted as

1) nominal (engineering) stress-strain, or

2) true stress-strain

The first is more important in design, and the second is more important in manufacturing. The graphs used in each case will be different.

*Nominal Stress-Strain:* Nominal stress, also called engineering stress, is defined as the ratio of the applied load $F$ to the original cross-sectional area $A$ of the specimen:

$$\sigma_n = \frac{F}{A_0} \tag{6}$$

where $\sigma_n$ = nominal stress (lb/in$^2$); $F$ = applied load in the test (lb); and, $A_0$ = original cross-section area of specimen (in$^2$).

*Nominal strain* is defined as

$$\varepsilon_n = \frac{l - l_0}{l} = \frac{\Delta l}{l} \tag{7}$$

where $\varepsilon_n$ = nominal strain (in./in.); $l_0$ = original gauge length (in.); and $l$ = instantaneous length of the specimen (in.).

*True Stress-Strain:* A nominal stress-strain curve does not give a true indication of the deformation characteristic of a solid because it is based on the original cross-sectional area $A_0$ of the specimen, and this dimension changes continuously during the test. In the solution of technical problems in metalworking, true stress and true strain are much more important.

*True stress* $\sigma$ is defined as the ratio of the load $F$ to the actual stress-section area $A$ of the specimen:

$$\sigma = \frac{F}{A} \tag{8}$$

where $\sigma$ = true stress (lb/in$^2$); $F$ = applied load in the test (lb); $A$ = actual area (instantaneous) area resisting the load (in$^2$).

*True strain* in a tensile test can be defined by dividing the total elongation into small increments of actual change in length. Then, using calculus, it can be shown that true strain is defined by the equation

$$\varepsilon = \ln\left(\frac{l}{l_0}\right) \tag{9}$$

where $\varepsilon$ = true strain (in./in.); and $l$ = instantaneous length at any moment during elongation (in.).

If the true stress, based on the actual (instantaneous) cross section area of the specimen, is used, it is found that the stress-strain curve increases continuously up to fracture. If the strain measurement is also based on instantaneous measurements, the curve obtained is known as a true stress-strain curve (Fig. 2).

The stress-strain curve in Fig. 2(a) can be represented by the equation

$$\sigma = K\varepsilon^n \tag{10}$$

where $K$ = the strength coefficient (lb/in$^2$), and $n$ = strain-hardening (work-hardening) exponent.

This equation is called the *flow curve*, and it represents the behavior of metals in the plastic zone, including their capacity for cold strain hardening.

When the curve shown in Fig. 2(a) is plotted on a logarithmic graph as in Fig. 2(b), it is found that the curve is a straight line, and the slope of the line is equal to the exponent $n$. The value of constant $K$ equals the value of true stress at a true strain value to 1.

The strain-hardening exponent may have a value from $n = 0$ (perfectly plastic solid) to $n = 1$ (elastic solid). For most metals, $n$ has values between 0.10 and 0.50.

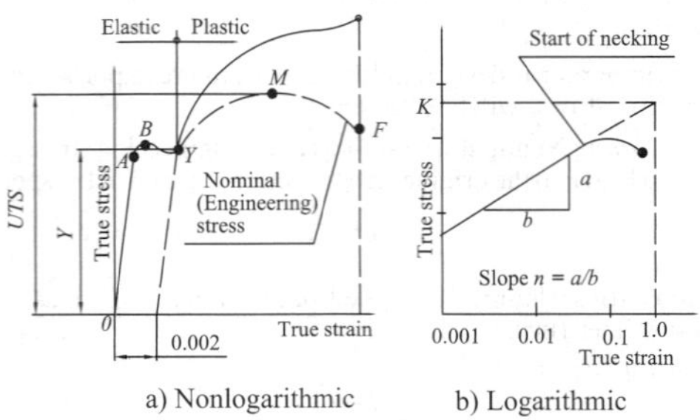


Fig. 2. True Stress-Strain Curve for Medium Steel

*Ductility* is most commonly defined as the ability of a metal to plastically deform easily upon application of a tensile force without breaking or fracturing. Ductility may be expressed as either percentage of elongation or percentage of area reduction in the specimen.

Elongation can be defined as

$$\delta = \frac{l_f - l_0}{l_0} \times 100 \tag{11}$$

Reduction can be defined as

$$\psi = \frac{A_0 - A_f}{A_0} \times 100 \quad (12)$$

where $l_f$ = length at the fracture (in.); $l_0$ = the original specimen's gauge length (in.); $A_0$ = original specimen's gauge cross-sectional area, (in$^2$); $A_f$ = cross-sectional area of the specimen at the fracture, (in$^2$). *Note:* $l_f$ length is measured between the original gauge marks after the pieces of the broken specimen are placed together.

Elongation ranges between about 8 and 60 percent for most metals, and typical measurements for reduction of area range from about 20 to 90 percent. Ductile materials such as thermoplastic and superplastic materials show large deformation before fracture, and of course, exhibit much higher ductility, but brittle materials have little or no ductility.

## Designing Sheet Metal Parts for Production

**Sheet metal parts should be designed to satisfy the following criteria:**

The parts should allow high productivity rates.

They should make highly efficient use of the materials involved.

The production machines should be easy to service.

Machines should be usable by workers with relatively basic skills.

Unfortunately, very few product designers concern themselves with suitability of production; their prime concern is usually the function of the part.

**Design rules for parts to be produced by blanking and punching:**

Avoid part design with complex configurations.

Use minimum dimensions of punched openings relative to material thickness given in Table 1.

Use minimum distances between punched opening and rounded radius relative to material thickness given in Table 2.

**Table 1. Minimal Dimensions of Punched Openings**

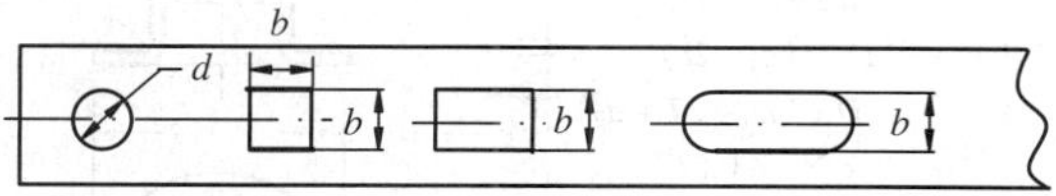


| Material | Form of opening | | | |
|---|---|---|---|---|
| | Circle $d_{min}$ = | Square $b_{min}$ = | Rectangular $b_{min}$ = | Oval $b_{min}$ = |
| Stainless steel | 1.50 $T$ | 1.40 $T$ | 1.20 $T$ | 1.10 $T$ |
| High-carbon steel | 1.20 $T$ | 1.10 $T$ | 0.90 $T$ | 0.80 $T$ |
| Medium-carbon steel | 1.00 $T$ | 0.90 $T$ | 0.70 $T$ | 0.60 $T$ |
| Low-carbon steel | 0.90 $T$ | 0.80 $T$ | 0.60 $T$ | 0.55 $T$ |
| Brass and copper | 0.80 $T$ | 0.70 $T$ | 0.60 $T$ | 0.55 $T$ |
| Magnesium alloy at 500° F | 0.25 $T$ | 0.45 $T$ | 0.35 $T$ | 0.30 $T$ |

Note: With fine punching process, minimal diameter of punched hole is (0.50 - 0.70) $T$.

**Table 2. Minimum Distance between Punched Opening and Edges and Rounded Radius**

| Part is | Form of opening | Minimum Distance | Sketch |
|---|---|---|---|
| Blanked | Circle | $c \geq T$<br>$r \geq 0.5T$ | 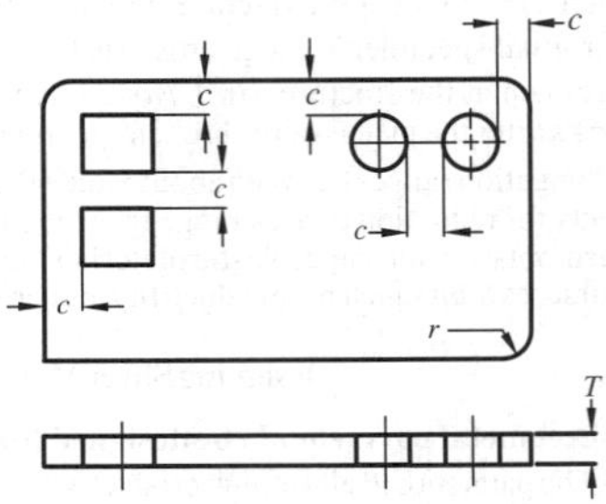 |
| | Rectangle | $c \geq 1.2T$<br>$r \geq 0.5T$ | |
| Bent | Circle | $c \geq 2T$<br>$r \geq (0.5-1.0)T$ [a]<br>$r \geq (1.0-2.0)T$ [b]<br>$r \geq (2.0-3.0)T$ [c] | 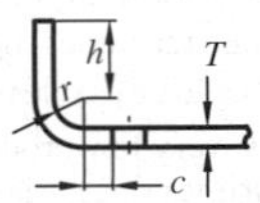 |
| Drawn | Circle | $d \leq (d_1 - 2r_1)$<br>$D_1 \geq (d_1 - 2r)$<br>$D \geq (D_1 + 3T + d_2)$<br>$c \geq r + 0.5T$ | 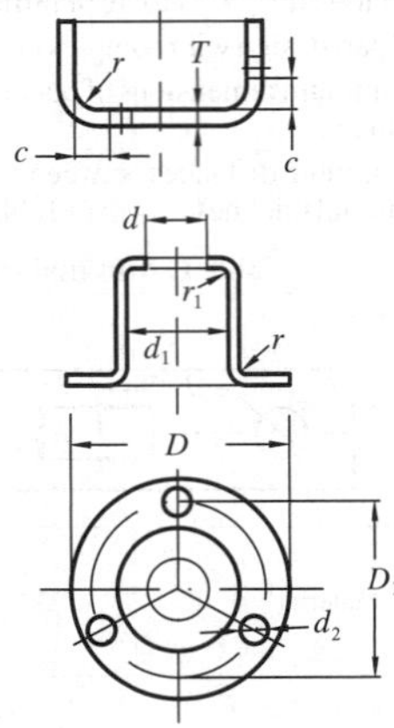 |

[a] Al and brass
[b] Steel
[c] Al-alloy 6000 series

**Design rules for parts produced by bending:**

Minimum bend radius should be used only if it is necessary for correct function of part.

The bend radius should be larger than the thickness of the material.

Use minimum distances between punched opening and bend radius given in Table 2.

Flange length as shown in Table 2 needs to be $h \geq 2T$.

If a part has more than one bend, it is necessary to define technological data.

**Design rules of parts produced by drawing:**

Avoid very complicated parts.

Make diameter of flange $D$ less than three times the diameter of shell ($D < 3d$), if height $h$ of shell is greater than twice the diameter of shell ($h > 2d$) as shown in Table 2.

Avoid design of rectangular and square shells with bottom radii less than the corner radius in the junction area.

The shortest distance between corner radii should be no less than the depth of shell.

## Shearing

Shearing involves the cutting of flat material such as metal sheets, plates, or strips. To be classified as shearing, the cutting action must be along a straight line. The piece of sheet metal sheared off may or may not be called a blank. Shearing is performed in a special machine with different types of blades or cutters. The machines may be foot-, hand-, or power-operated. The shear is equipped with long or rotary blades for cutting. The upper blade of power shears is often inclined to reduce the required cutting force.

During shearing operations, three phases (Fig. 3) may be noted:

*Phase I - Plastic Deformation:* As the upper blade begins to push into the work material, plastic deformation occurs in the surfaces of the sheet, and the stress on the material is lower than the yield stress.

*Phase II - Penetration:* As the blade moves downward, penetration occurs, in which the blade compresses the work material and cuts into the metal. In this phase, the stress on the material is higher than the yield stress but lower than the ultimate tensile strength ($UTS$).

*Phase III - Fracture:* As the blade continues to travel into the work material, fracture begins in the material at the two cutting edges. The stress on the work material is equal to the shearing stress. If the clearance between the blades is correct, the two fracture lines meet, resulting in a separation of the work material into two parts.

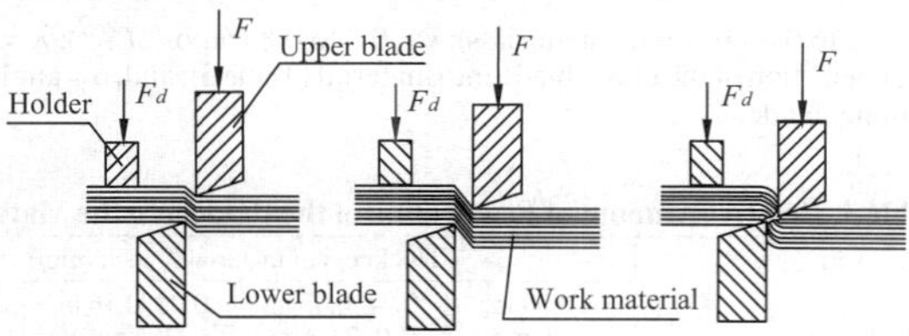


Fig. 3. Schematic Illustration of the Phases of Shearing

Shearing is the preferred way to cut blanks whenever the blank shape permits its use. In most cases, however, the limitation of straight lines in the shape of the blank eliminates the use of shears. Shearing is economical because no expensive dies have to be made for cutting out the blanks. Shearing is used for the following purposes:

1) To cut strip or coiled stock into blanks

2) To cut strip or coiled stock into smaller strips to feed into a blanking or drawing die

3) To trim large sheets, squaring the edges of the sheet.

**Shearing Forces.**—Calculating the force and power involved in shearing operations varies according to the types of blades. There are three types of blades:

a) straight parallel blades; b) straight inclined blades; and c) rotary cutters.

*Shearing with Straight Parallel Blades:* The shearing force with straight parallel cutters can be calculated approximately as

$$F = \tau A \tag{13}$$

where $F$ = shearing force (lb); $\tau$ = shear strength of the material (lb/in$^2$); and $A$ = cutting area (in$^2$). The cutting area is calculated as

$$A = bT \tag{13a}$$

where $b$ = length of the cutting material (in.), and $T$ = thickness of material (in.).

This calculated shearing force needs to be increased by 20 to 40 percent, depending on whether the following conditions exist: an enlarged clearance between the blades, variations in the thickness of the material, the obtuseness of the cutting edge angles, and other unpredictable factors.

The real force of the shearing machine is

$$F_M = 1.3F \tag{14}$$

*Shearing with Straight Inclined Blades:* Shears with straight inclined blades are used for cutting material of relatively small thickness compared with the width of cutting. Using inclined blades reduces the shearing force and increases the range of movement necessary to disjoin the material. The penetration of the upper blade into the material is gradual and as a result, the shearing force is lower.

The shearing force can be calculated as

$$F = n \cdot k \cdot UTS \cdot \lambda \frac{T^2}{\tan \phi} \tag{15}$$

where $n$ = 0.75 to 0.85 (for most materials); $k$ = 0.7 to 0.8 (ratio $UTS/\tau$); $\lambda$ = the relative amount of penetration of the upper blade into material (Table 3); and, $\phi$ = angle of inclination of the upper blade.

**Table 3. Relative Amount of Penetration of the Blade into the Material**

| Material | Thickness of material $T$, inch (mm) | | | |
|---|---|---|---|---|
| | <0.04 in. (<1.0 mm) | 0.04–0.08 in. (1.0–2.0 mm) | 0.08–0.16 in. (2–4 mm) | >0.16 in. (>4 mm) |
| Plain carbon steel | 0.75 to 0.70 | 0.70 to 0.65 | 0.65 to 0.55 | 0.50 to 0.40 |
| Medium steel | 0.65 to 0.60 | 0.60 to 0.55 | 0.55 to 0.48 | 0.45 to 0.35 |
| Hard steel | 0.50 to 0.47 | 0.47 to 0.45 | 0.44 to 0.38 | 0.35 to 0.25 |
| Aluminum and copper (annealed) | 0.80 to 0.75 | 0.75 to 0.70 | 0.70 to 0.60 | 0.60 to 0.50 |

The real force of the shearing machine is

$$F_M = 1.3F \tag{16}$$

*Shearing with Rotary Cutters:* The rotary shearing operation is much like shearing with straight inclined blades because the straight blade may be thought of as a rotary cutter with an endless radius. It is possible to make straight line cuts as well as to produce circular blanks and irregular shapes by this method. Fig. 4 illustrates the conventional arrangement of the cutters in a rotary shearing machine for the production of a perpendicular edge. Only the upper cutter is rotated by the power drive system. The upper cutter pinches the material and causes it to rotate between the two cutters.

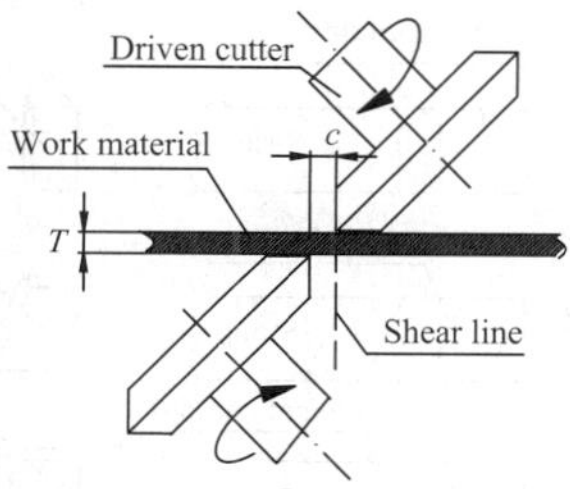


Fig. 4. Schematic Illustration of Shearing with Rotary Cutters

Shearing force with rotary cutters can be calculated approximately as

$$F = n \cdot k \cdot UTS \cdot \lambda \frac{T^2}{2\tan\phi} \qquad (17)$$

where $\phi = \operatorname{acos}\left[1 - \frac{\mu + T}{D}\right]$; $\mu$ = lap of cutters (in.); $T$ = thickness of material (in.); $D$ = diameter of cutter (in.); $n$ = 0.75 to 0.85 (for most materials); $k$ = 0.7 to 0.8 (ratio $UTS/\tau$); and, $\lambda$ = relative amount of penetration of the cutters (Table 3).

The real force of a shearing machine with rotary cutter is

$$F_M = 1.3F \qquad (18)$$

Rotary shearing machines are equipped with special holding fixtures that rotate the work material to generate the desired circle.

**Clearance.**—Clearance is defined as the space between the upper and lower blades. Without proper clearance, the cutting action no longer progresses. With too little clearance, a defect known as "secondary shear" is produced. If too much clearance is used, extreme plastic deformation will occur. Proper clearance may be defined as that clearance which causes no secondary shear and a minimum of plastic deformation.

The clearance between straight blades (parallel and inclined) is: $c$ = (0.02 to 0.05), mm.

The clearance between rotary cutters with parallel inclined axes is

$$c = (0.1 \text{ to } 0.2)T \qquad (19)$$

where $T$ = material thickness (in.).

## Cutoff and Parting

**Cutoff.**—Cutoff is a shearing operation in which the shearing action must be along a line. The pieces of sheet metal cutoff are the blanks. Fig. 5 shows several types of cutoff operations. As seen in the illustration, a cutoff is made by one or more single line cuts. The line of cutting may be straight, curved, or angular. The blanks need to be nested on the strip in such a way that scrap is avoided. Some scrap may be produced at the start of a new strip or coil of sheet metal in certain cases. This small amount is usually negligible.

The use of cutoff operations is limited by the shape of a blank. Only blanks that nest perfectly may be produced by this operation. Cutoff is performed in a die and therefore may be classified as a stamping operation. With each cut, a new part is produced. More blanks may be produced per stroke of the press ram by adding more single-line cutting edges.

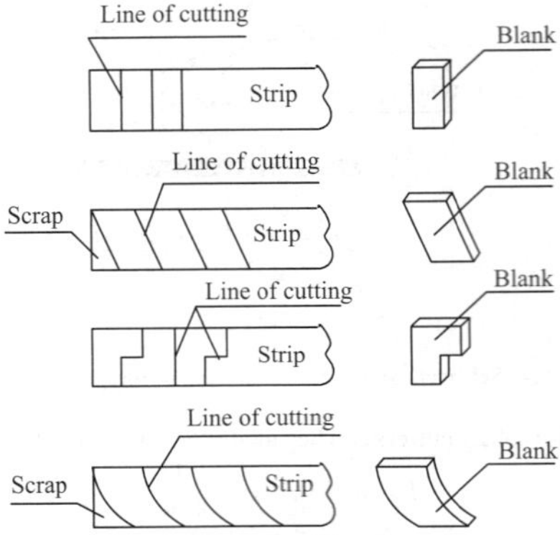


Fig. 5. Types of Cutoff

**Parting.**—Parting is a cutting operation of a sheet metal strip by a die with cutting edges on two opposite sides. During parting, some amount of scrap is produced, as shown in Fig. 6. This might be required when the blank outline is not a regular shape and is precluded from perfectly nesting on the strip. Thus, parting is not as efficient an operation as cutoff.

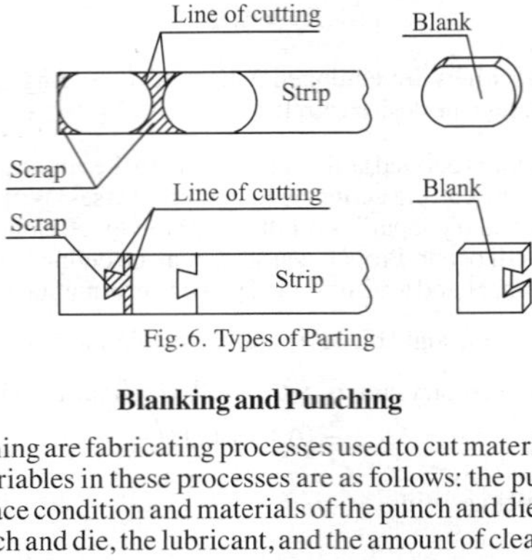


Fig. 6. Types of Parting

## Blanking and Punching

Blanking and punching are fabricating processes used to cut materials into forms by the use of a die. Major variables in these processes are as follows: the punch force, the speed of the punch, the surface condition and materials of the punch and die, the condition of the blade edge of the punch and die, the lubricant, and the amount of clearance.

In blanking, a workpiece is removed from the primary material strip or sheet when it is punched. The material that is removed is the new workpiece or blank. Punching is a fabricating process that removes a scrap slug from the workpiece each time a punch enters the punching die. This process leaves a hole in the workpiece (Fig. 7).

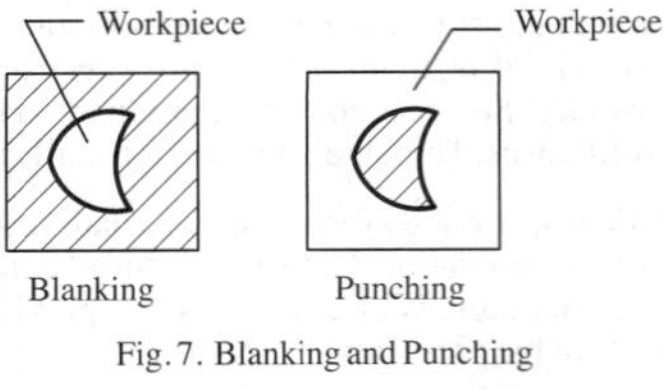


Fig. 7. Blanking and Punching

**Characteristics of the blanking process include:**

1) Ability to produce workpieces in both strip and sheet material during medium and mass production.

2) Removal of the workpiece from the primary material stock as a punch enters a die.

3) Control of the quality by the punch and die clearance.

4) Ability to produce holes of varying shapes quickly.

**Characteristics of the punching process include:**

1) Ability to produce holes in both strip and sheet material during medium and mass production.

2) Ability to produce holes of varying shapes quickly.

There are three phases in the process of shearing during blanking and punching as illustrated in Fig. 8.

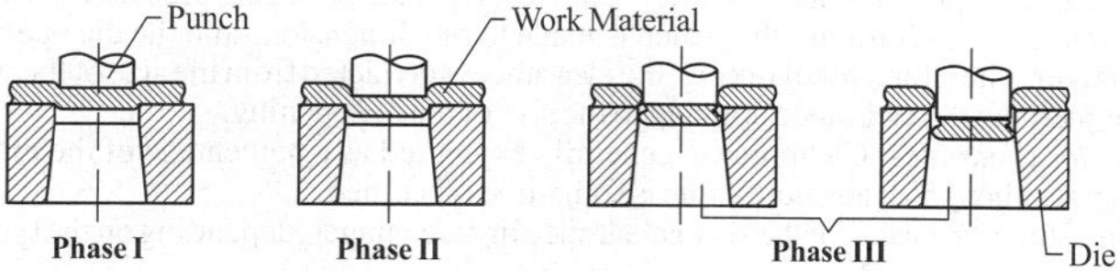


Fig. 8. Phases in the Process of Shearing

In Phase I, work material is compressed across and slightly deformed between the punch and die, but the stress and deformation in the material does not exceed the plastic limit. This phase is known as the elastic phase.

In Phase II, the work material is pushed farther into the die opening by the punch; at this point in the operation the material has been obviously deformed at the rim, between the cutting edges of the punch and the die. The concentration of outside forces causes plastic deformation at the rim of the material. At the end of this phase, the stress in the work material close to the cutting edges reaches a value corresponding to the material shear strength, but the material resists fracture. This phase is called the plastic phase.

During Phase III, the strain in the work material reaches the fracture limit, and micro-cracks appear, which turn into macro-cracks, followed by separation of the parts of the workpiece. The cracks in the material start at the cutting edge of the punch on the upper side of the work material, and at the die edge on the lower side of the material; the crack propagates along the slip planes until complete separation of the part from the sheet occurs. A slight burr is generally left at the bottom of the hole and at the top of the slug. The slug is then pushed farther into the die opening. The slug burnish zone expands and is held in the die opening. The whole burnish zone contracts and clings to the punch.

**Blanking and Punching Clearance.—***Clearance, c,* is the space (per side) between the punch and the die opening shown in Fig. 9, such that: $c = \frac{D_d - d_p}{2}$

Ideally, proper clearance between the cutting edges enables the fractures to start at the cutting edge of the punch and the die. The fractures will proceed toward each other until they meet. The fractured portion of the sheared edge then has a clean appearance. For optimum finish of a cut edge, correct clearance is necessary. This clearance is a function of the type, thickness, and temper of the material.

When clearance is not sufficient, additional layers of the material must be cut before complete separation is accomplished. With correct clearance, the angle of the fracture will permit a clean break below the burnish zone because the upper and lower fractures will extend toward one another. Excessive clearance will result in a tapered cut edge, because for any cutting operation, the opposite side of the material that the punch enters after cutting will be the same size as the die opening.

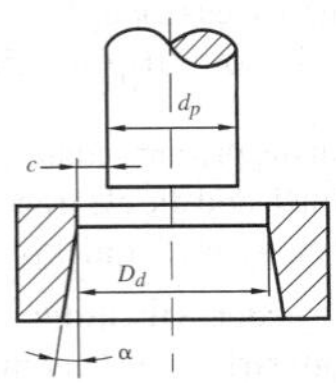


Fig. 9. Punch and Die Clearance

*Where Clearance is Applied:* Whether clearance is deducted from the dimensions of the punch or added to the dimensions of the die opening depends upon the nature of the workpiece. In the blanking process (a blank of given size is required), the die opening is made to that size and the punch is made smaller. Conversely, in the punching process (when holes of a given size are required), the punch is made to the dimensions and the die opening is made larger. Therefore, for blanking, the clearance is deducted from the size of the punch, and for piercing the clearance is added to the size of the die opening.

*Value for Clearance:* Clearance is generally expressed as a percentage of the material thickness, although an absolute value is sometimes specified.

Table 4 shows the value of the shear clearance in percentages, depending on the type and thickness of the material.

**Table 4. Values for Clearance as a Percentage of the Thickness of the Material**

| Material | Material Thickness, $T$ | | | | |
|---|---|---|---|---|---|
| | < 0.040 in. (< 1.0 mm) | 0.040-0.080 in. (1.0-2.0 mm) | 0.082-0.118 in. (2.1- 3.0 mm) | 0.122-0.197 in. (3.1-5.0 mm) | 0.200-0.275 in. (5.1-7.0 mm) |
| Low carbon steel | 5.0 | 6.0 | 7.0 | 8.0 | 9.0 |
| Copper and soft brass | 5.0 | 6.0 | 7.0 | 8.0 | 9.0 |
| Medium carbon steel 0.20% to 0.25% carbon | 6.0 | 7.0 | 8.0 | 9.0 | 10.0 |
| Hard brass | 6.0 | 7.0 | 8.0 | 9.0 | 10.0 |
| Hard steel, 0.40% to 0.60% carbon | 7.0 | 8.0 | 9.0 | 10.0 | 12.0 |

Table 5 shows absolute values for the blanking and punching clearance for high-carbon steel (0.60% to 1.0% carbon) depending on the thickness of the work material.

**Table 5. Absolute Values of Clearance for Blanking and Punching High-Carbon Steel**

| Material Thickness, $T$ | | Clearance, $c$ | | Material Thickness, $T$ | | Clearance, $c$ | |
|---|---|---|---|---|---|---|---|
| (in.) | (mm) | (in.) | (mm) | (in.) | (mm) | (in.) | (mm) |
| 0.012 | 0.3048 | 0.00006 | 0.001524 | 0.157 | 3.9878 | 0.0095 | 0.24130 |
| 0.020 | 0.5004 | 0.0009 | 0.02286 | 0.177 | 4.4958 | 0.0116 | 0.29464 |
| 0.032 | 0.8001 | 0.0013 | 0.03302 | 0.197 | 5.0038 | 0.0138 | 0.35052 |
| 0.040 | 1.0160 | 0.0016 | 0.04064 | 0.236 | 5.9944 | 0.0177 | 0.44958 |
| 0.047 | 1.1938 | 0.0020 | 0.05080 | 0.275 | 6.9850 | 0.0226 | 0.57404 |
| 0.060 | 1.5240 | 0.0026 | 0.06604 | 0.315 | 8.0010 | 0.0285 | 0.72390 |
| 0.078 | 1.9812 | 0.0035 | 0.08890 | 0.394 | 10.0076 | 0.0394 | 1.00076 |
| 0.098 | 2.4892 | 0.0047 | 0.11938 | 0.472 | 11.9888 | 0.0502 | 1.27508 |
| 0.118 | 2.9972 | 0.0059 | 0.14986 | 0.590 | 14.9860 | 0.0689 | 1.75006 |
| 0.138 | 3.5052 | 0.0077 | 0.19558 | 0.748 | 18.9992 | 0.0935 | 2.37490 |

*Effect of Clearance:* Manufacturers have performed many studies on the effect of clearance on punching and blanking. Clearance affects not only the smoothness of the fracture, but also the deformation force and work. A tighter blanking and punching clearance generates more heat on the cutting edge and the bulging area tightens around the punch. These effects produce a faster breakdown of the cutting edge. If the clearance increases, the bulging area disappears and the roll-over surface is stretched and will retract after the slug breaks free. Less heat is generated with increases in the blanking and punching clearance, and the edge breakdown rate is reduced. The deformation force is greatest when the punch diameter is small compared to the thickness of the work material. In one example, a punching force of about 142 kN was required to punch 19 mm holes into 8 mm mild steel when the clearance was about 10 percent. With a clearance of about 4.5 percent, the punching force increased to 147 kN, and a clearance of 2.75 percent resulted in a force of 153.5 kN.

**Die Opening Profile.—**Die opening profiles depend on the purpose and required tolerance of the workpiece. Two opening profiles are shown in Fig. 10a and Fig. 10b.

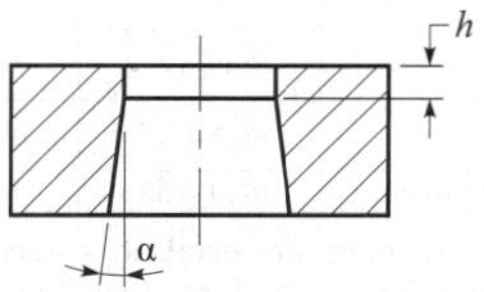


Fig. 10a. Opening Profile for High Quality Part

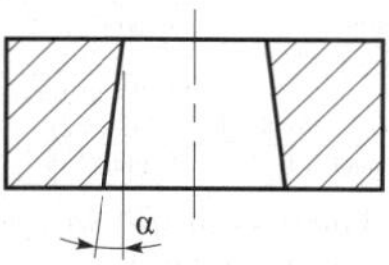


Fig. 10b. Opening for Low Accuracy Part

The profile in Fig. 10a gives the highest quality workpiece. To allow a die block to be sharpened more times, the height *h* of the block needs to be greater than the thickness of the workpiece. The value of *h* is given in Table 6. The die opening profile in Fig. 10b is used for making a small part with low accuracy from very soft material, such as soft thin brass. The angle of the cone $\alpha = 15'$ to $45'$.

**Table 6. Value of Dimension *h* Based on Material Thickness**

| | Work material thickness, *T* | | |
|---|---|---|---|
| | < 0.04 in. (< 1 mm) | > 0.04 to 0.2 in. (>1 to 5 mm) | > 0.2 to 0.4 in. (> 5 to 10 mm) |
| Height *h* | 0.14 in (3.5 mm) | 0.26 in (6.5 mm) | 0.45 in (11.5 mm) |

Angle $\alpha = 3°$ to $5°$

**Deformation Force, Deformation Work, and Force of Press.—**Deformation force *F* for punching and blanking with flat face of punch is defined by the following equation:

$$F = LT\tau_m = 0.8LT(UTS) \quad (20)$$

where $F$ = deformation force, lb (N)

$L$ = the total length of cutting, in. (mm)

$T$ = thickness of the material, in. (mm)

$\tau_m$ = shear stress, $lb/in^2$ (MPa)

$UTS$ = the ultimate tensile strength of the work material, $lb/in^2$ (MPa)

*Force of Press:* Such variables as unequal thickness of the material, friction between the punch and workpiece, or dull cutting edges, can increase the necessary force by up to 30 percent, so these variables must be considered in selecting the power requirements of the press. That is, the force requirement of the press, $F_p$ is

$$F_p = 1.3F \quad (21)$$

The blanking and punching force can be reduced if the punch or die has bevel-cut edges. In blanking operations, bevel shear angles should be used on the die to ensure workpiece remains flat. In punching operations, bevel shear angles should be used on the punch.

*Deformation Work W* for punching and blanking with flat face of punch is defined by the following equation:

$$W = kFT \quad (22)$$

where $k$ = a coefficient that depends on the shear strength of the material and the thickness of the material
$F$ = deformation force (lb)
$T$ = material thickness (in.)

**Table 7. Values for Coefficient $k$ for Some Materials**

| Material | Shear Strength lb/in² (MPa) | Material Thickness, inch (mm) | | | |
|---|---|---|---|---|---|
| | | <0.040 in. <(1.0 mm) | 0.040–0.078 in. (1.0–2.0 mm) | 0.078–0.157 in. (2.0–4.0 mm) | >0.157 in. >4.0 mm |
| Low carbon steel | 35,000–50,000 (240–345) | 0.70–0.65 | 0.64–0.60 | 0.58–0.50 | 0.45–0.35 |
| Medium carbon steel 0.20 to 0.25% carbon | 50,000–70,000 (345–483) | 0.60–0.55 | 0.54–0.50 | 0.49–0.42 | 0.40–0.30 |
| Hard steel 0.40 to 0.60% carbon | 70,000–95,000 (483–655) | 0.45–0.42 | 0.41–0.38 | 0.36–0.32 | 0.30–0.20 |
| Copper, annealed | 21,000 (145) | 0.75–0.69 | 0.70–0.65 | 0.64–0.55 | 0.50–0.40 |

**Stripper Force.—***Elastic Stripper:* When spring strippers are used, it is necessary to calculate the amount of force required to effect stripping. This force may be calculated by the following equation:

$$F_s = \frac{1}{0.00117}PT = 855PT \qquad (23a)$$

where $F_s$ = stripping force (lb)
$P$ = sum of the perimeters of all the punching or blanking faces (in.)
$T$ = thickness of material (in.)

$$F_s = 5.9PT \qquad (23b)$$

where $F_s$ = stripping force (N)
$P$ = sum of the perimeters of all the punching or blanking faces (mm)
$T$ = thickness of material (mm)

This formula has been used for many years by a number of manufacturers and has been found to be satisfactory for most punching and blanking operations.

After the total stripping force has been determined, the stripping force per spring must be found in order to establish the number and dimensions of springs required. Maximum force per spring is usually listed in the manufacturer's catalog.

The correct determined force per spring must satisfy the following relationship:

$$F_{max} > F_{so} > \frac{F_s}{n} \qquad (24)$$

where $F_{max}$ = maximum force per spring, lb (newton)
$F_{SO}$ = stripping force per spring, lb (newton)
$F_s$ = total stripping force, lb (newton)
$n$ = number of springs

**Blanking Pressure.—**When designing parts that are to be blanked in a press, it is often necessary to work out the pressure required. A press of sufficient tonnage, providing the necessary blanking force, must be used. Otherwise, the press is likely to stall. To calculate the pressure required in tons for blanking various materials, use the following formula:

$$P = A \times T \times S$$

where $P$ = pressure required (tons)
$A$ = perimeter of blank (in.)
$T$ = thickness of material (in.)
$S$ = shear strength of material (tons per in²)

*Example:* Calculate the pressure required in tons for the blank illustrated by Fig. 11. Soft sheet brass is to be used with a thickness of 0.125 inch and a shear strength of 17 tons per square inch.

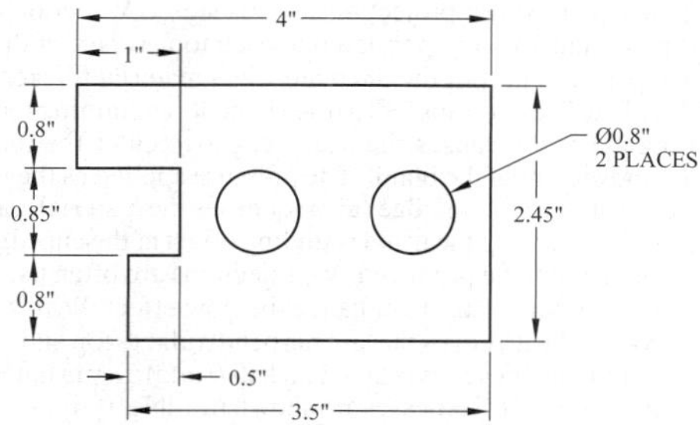


Fig. 11. Dimensioned Drawing of Part to be Blanked

Perimeter of blank = 4 + 2.45 + 3.5 + 0.8 + 0.5 + 0.85 + 1 + 0.8 = 13.9 in.
Circumference of holes = 2(π × 0.8) = 5.026 in.
Total perimeter = 13.9 + 5.026 = 18.926 in.
Pressure required = $P$ = A × T × S = 18.926 × 0.125 × 17 = 40.22 tons

When metric dimensions are used, the same formula applies:

where $P$ = pressure required (kg)
$A$ = perimeter of blank (mm)
$T$ = thickness of material (mm)
$S$ = shear strength of material (kg per mm$^2$)

Note: The general rule for punches that have a shear, where the cutting faces are ground at an angle, is to halve the tonnage for material thicknesses of up to 0.25 inch or 6 mm. For material thicknesses over 0.25 inch or 6 mm, two-thirds of the tonnage should be used.

**Fine Blanking.—**This process uses special presses and tooling to produce flat components from sheet metal or plate, with high dimensional accuracy. According to Hydrel A. G., Romanshorn, Switzerland, fine-blanking presses can be powered hydraulically or mechanically, or by a combination of these methods, but they must have three separate and distinct movements. These movements clamp the work material, perform the blanking operation, and eject the finished part from the tool. Forces of 1.5–2.5 times those used in conventional stamping are needed for fine blanking, so machines and tools must be designed accordingly. In mechanical fine-blanking presses, clamping and ejection forces are exerted hydraulically. Such presses generally are of toggle-type design and are limited to total forces of up to about 280 tons. Higher forces usually require all-hydraulic designs. These presses are also suited to embossing, coining, and impact extrusion work.

Cutting elements of tooling for fine blanking generally are made from 12 percent chromium steel, although high-speed steel and tungsten carbide also are used for long runs or improved quality. Cutting clearances between the punch and die as a percentage of the thickness of material are given in Table 8.

**Table 8. Values for Clearances Used in Fine-Blanking Tools as a Percentage of the Thickness of the Material**

| Material Thickness | | Clearance, % | |
|---|---|---|---|
| (in.) | (mm) | Inside Contour | Outside Contour |
| <0.040 | <1.016 | 2.0 | 1.0 |
| 0.040–0.063 | 1.016–1.600 | 1.5 | |
| 0.063–0.098 | 1.600–2.489 | 1.25 | |
| 0.098–0.125 | 2.489–3.175 | 1.0 | |
| 0.125–0.197 | 3.175–5.004 | 0.8 | 0.5 |
| 0.197–0.315 | 5.004–8.001 | 0.7 | |
| 0.315–0.630 | 8.001–16.002 | 0.5 | |

The clamping elements are sharp projections of 90-degree V-section that follow the outline of the workpiece and are incorporated into each tool as part of the stripper plate with thin material and also as part of the die plate when material thicker than 0.15 inch (3.81 mm) is to be blanked. Pressure applied to the elements containing the V-projections prior to the blanking operation causes the sharp edges to enter the material surface, preventing sideways movement of the blank. The pressure applied as the projections bite into the work surface near the contour edges also squeezes the material, causing it to flow toward the cutting edges, reducing the usual rounding effect at the cut edge. When small details such as gear teeth are to be produced, V-projections are often used on both sides of the work, even with thin materials, to enhance the flow effect. With suitable tooling, workpieces can be produced with edges that are perpendicular to top and bottom surfaces within 0.004 inch (0.1mm) on thicknesses of 0.2 inch (5.0 mm), for instance. V-projection dimensions for various material thicknesses are shown in Table 9.

**Table 9. Dimensions for V-projections Used in Fine-Blanking Tools**

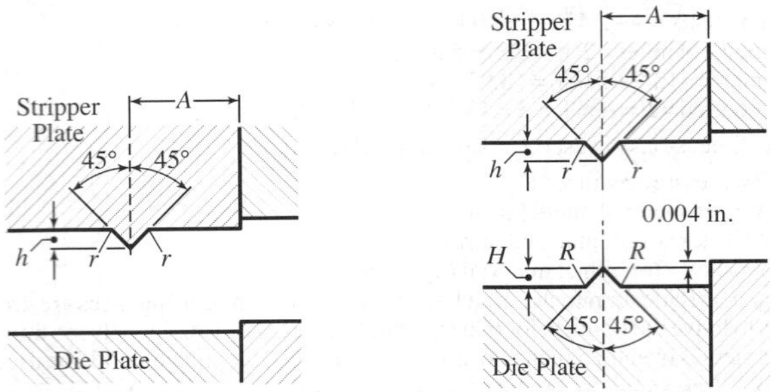


V-Projections on Stripper Plate Only V-Projections on Both Stripper and Die Plate

| Material Thickness | A | h | r | H | R |
|---|---|---|---|---|---|
| V-Projections on Stripper Plate Only | | | | | |
| 0.040–0.063 | 0.040 | 0.012 | 0.008 | … | … |
| 0.063–0.098 | 0.055 | 0.015 | 0.008 | … | … |
| 0.098–0.125 | 0.083 | 0.024 | 0.012 | … | … |
| 0.125–0.157 | 0.098 | 0.028 | 0.012 | … | … |
| 0.157–0.197 | 0.110 | 0.032 | 0.012 | … | … |
| V-Projections on Both Stripper and Die Plate | | | | | |
| 0.157–0.197 | 0.098 | 0.020 | 0.008 | 0.032 | 0.032 |
| 0.197–0.248 | 0.118 | 0.028 | 0.008 | 0.040 | 0.040 |
| 0.248–0.315 | 0.138 | 0.032 | 0.008 | 0.047 | 0.047 |
| 0.315–0.394 | 0.177 | 0.040 | 0.020 | 0.060 | 0.060 |
| 0.394–0.492 | 0.217 | 0.047 | 0.020 | 0.070 | 0.080 |
| 0.492–0.630 | 0.276 | 0.063 | 0.020 | 0.087 | 0.118 |

All units are in inches.

Fine-blanked edges are free from the fractures that result from conventional tooling and can have surface finishes down to 80 μin. (2.0 μm) Ra with suitable tooling. Close tolerances can be held on inner and outer forms and on hole center distances. Flatness of fine-blanked components is better than that of parts made by conventional methods but distortion may occur with thin materials due to release of internal stresses. Widths must be slightly greater than are required for conventional press working. Generally, the strip width must be 2–3 times the thickness, plus the width of the part measured transverse to the feed direction. Other factors are shape, material quality, size and shape of the V-projection in relation to the die outline, and spacing between adjacent blanked parts. Holes and slots can be produced with ratios of width to material thickness down to 0.7, compared with the 1:1 ratio normally specified for conventional tooling. Operations such as countersinking, coining, and bending up to 60 degrees can be incorporated in fine-blanking tooling.

The cutting force in pounds, lb (Newton, N) exerted in fine blanking is 0.9 times the length of the cut in inches (mm) times the material thickness in inches (mm), times the tensile strength in $lb_f/in^2$ (MPa). Pressure in lb (N) exerted by the clamping element(s) carrying the V-projections is calculated by multiplying the length of the V-projection, which depends on its shape, in inches (mm) by its height ($h$), times the material tensile strength in $lb_f/in^2$ (MPa), times an empirical factor $f$. Factor $f$ has been determined to be 2.4–4.4 for a tensile strength of 28,000–113,000 $lb_f/in^2$ (193–779 MPa). The clamping pressure is about 30 percent of the cutting force, as calculated. Dimensions and positioning of the V-projection(s) are related to the material thickness, quality, and tensile strength. A small V-projection close to the line of cut has about the same effect as a large V-projection spaced away from the cut. But if the V-projection is too close to the cut, it may move out of the material at the start of the cutting process, reducing its effectiveness. Positioning the V-projection at a distance from the line of cut increases both material and blanking force requirements; its location relative to the line of cut also affects tool life.

**Shaving.—**The edges of punched and blanked parts are generally rough and uneven. Shaving is the process of removing a thin layer of material, from the inside or outside contour of a workpiece or from both sides with a sharp punch and die, to achieve precise, clean parts.

*Shaving a Punched Workpiece:* It is necessary to provide a small amount of stock on the punched or blanked workpiece for subsequent shaving. This amount, δ, is the difference between diameters of the hole after shaving and before shaving.

$$\delta = d - d_o \tag{25}$$

where $d$ = diameter of hole after shaving (inch or mm)and, $d_o$ = diameter of hole before shaving (inch or mm).

The value of δ is 0.006 to 0.0098 inch (0.15 to 0.25 mm) for a previously-punched hole, and 0.004 to 0.006 inch (0.10 to 0.15 mm) for a previously-drilled hole.

The diameter of punch $d_p$ can be calculated from the formula:

$$d_p = d + \varepsilon + i \tag{26}$$

where $d$ = diameter of hole after shaving (inch or mm)

$\varepsilon$ = production tolerance of the hole (inch or mm)

$i$ = amount of compensation for tightening of the hole after shaving (inch or mm) ($i$ = 0.0002 to 0.00067 in. or 0.005-0.017 mm)

The diameter of the die $D_d$ is

$$D_d = (1.20 \text{ to } 1.30)\, d_p \tag{27}$$

*Shaving a Blanked Workpiece:* Thin layers of material can be removed from a blanked surface by a process similar to punching. If the workpiece after shaving needs to have a diameter $D$, the punch diameter for the blanking operation is

$$d_p = D + \delta \tag{28}$$

The die diameter for the blanking operation is

$$d_d = d_p + 2c = D + \delta + 2c \tag{29}$$

where $D$ = diameter of final piece (inch or mm)

$c$ = clearance between die and punch (inch or mm)

δ = amount of material for shaving (inch or mm)

## Bending

One of the most common processes for sheet metal forming is bending, which is used to form pieces such as L, U, or V-profiles, and also to improve the stiffness of a piece by increasing its moment of inertia. Bending metal is a uniform straining process that plastically deforms the material and changes its shape. The material is stressed above the yield strength but below the ultimate tensile strength. The surface area of the material changes only in the bending zone. "Bending" usually refers to linear deformation about one axis. Bending may be performed by air bending, bottoming bending, or coining.

*Air Bending:* Air bending is done with the punch touching the workpiece but not bottoming it in the lower die. The profile of a die for air bending can have a right angle or an acute angle. The edges of the die with which the workpiece is in contact are rounded, and the radius of the punch will always be smaller than the bending radius.

*Bottoming Bending and Coining:* Bottoming or coining bending is the process by which the punch and the workpiece bottom on the die. It is necessary to flatten the bottom bend area of the workpiece between the tip of the punch and bottom on the die in order to avoid springback. The tonnage required on this type of press is higher than in air bending.

**Inside Bend Radius.—**Fig. 12 shows the terminology used in the bending process.

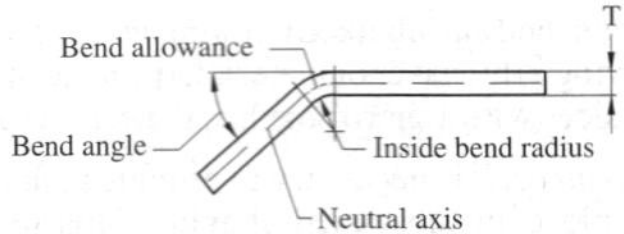


Fig. 12. Schematic Illustration of Terminology Used in the Bending Process

One of the most important factors influencing the quality of bent workpieces is the inside bend radius which must be within defined limits.

*Minimum Bend Radius:* If the bend radius is less than $R_{min}$ given in Equation (30), particularly in harder materials, the material at the outside of the bend will tend to "orange peel." If this orange peeling, or opening of the grain, is severe enough, the metal will fracture or crack off completely in extreme cases.

The minimum bend radius, $R_{min}$ is given by the following formula:

$$R_{\min} = T\left(\frac{50}{r} - 1\right) \quad (30)$$

where $T$ = material thickness (inch or mm)and, $r$ = percentage reduction in a tensile test for a given material (%).

*Maximum Bend Radius:* If the bend radius is greater than $R_{max}$ given in Equation (31), the bend will be very hard to control and will spring back erratically. The amount of springback will worsen on thinner materials. When large radius bends are required, an allowance should always be made for this in the tolerance of the part.

To achieve permanent plastic deformation in the outer fibers of the bent workpiece the maximum bend radius must be

$$R_{\max} \le \frac{TE}{2(YS)} \quad (31)$$

where $E$ = modulus of elasticity, lb/in$^2$ (N/mm$^2$); $YS$ = yield strength, lb/in$^2$ (N/mm$^2$); and, $T$ = thickness of material, inches (mm).

*Neutral Axis:* When material is formed, the deformation in the inside fibers of the material will compress during forming and the fibers of the material on the outside of the bend will expand. The material between these two regions remains neutral during forming and is referred to as the neutral axis of the material. The length of fibers along the neutral axis of the bend does not change during forming. This neutral axis is used when figuring the bend allowance for flat blank layouts.

## Table 10. Lengths of Straight Stock Required for 90-Degree Bends in Soft Copper and Soft Brass

| Radius *R* of Bend, Inches | Thickness *T* of Material, Inch | | | | | | | | | | | | |
|---|---|---|---|---|---|---|---|---|---|---|---|---|---|
| | 1/64 | 1/32 | 3/64 | 1/16 | 5/64 | 3/32 | 1/8 | 5/32 | 3/16 | 7/32 | 1/4 | 9/32 | 5/16 |
| 1/32 | 0.058 | 0.066 | 0.075 | 0.083 | 0.092 | 0.101 | 0.118 | 0.135 | 0.152 | 0.169 | 0.187 | 0.204 | 0.221 |
| 3/64 | 0.083 | 0.091 | 0.100 | 0.108 | 0.117 | 0.126 | 0.143 | 0.160 | 0.177 | 0.194 | 0.212 | 0.229 | 0.246 |
| 1/16 | 0.107 | 0.115 | 0.124 | 0.132 | 0.141 | 0.150 | 0.167 | 0.184 | 0.201 | 0.218 | 0.236 | 0.253 | 0.270 |
| 3/32 | 0.156 | 0.164 | 0.173 | 0.181 | 0.190 | 0.199 | 0.216 | 0.233 | 0.250 | 0.267 | 0.285 | 0.302 | 0.319 |
| 1/8 | 0.205 | 0.213 | 0.222 | 0.230 | 0.239 | 0.248 | 0.265 | 0.282 | 0.299 | 0.316 | 0.334 | 0.351 | 0.368 |
| 5/32 | 0.254 | 0.262 | 0.271 | 0.279 | 0.288 | 0.297 | 0.314 | 0.331 | 0.348 | 0.365 | 0.383 | 0.400 | 0.417 |
| 3/16 | 0.303 | 0.311 | 0.320 | 0.328 | 0.337 | 0.346 | 0.363 | 0.380 | 0.397 | 0.414 | 0.432 | 0.449 | 0.466 |
| 7/32 | 0.353 | 0.361 | 0.370 | 0.378 | 0.387 | 0.396 | 0.413 | 0.430 | 0.447 | 0.464 | 0.482 | 0.499 | 0.516 |
| 1/4 | 0.401 | 0.409 | 0.418 | 0.426 | 0.435 | 0.444 | 0.461 | 0.478 | 0.495 | 0.512 | 0.530 | 0.547 | 0.564 |
| 9/32 | 0.450 | 0.458 | 0.467 | 0.475 | 0.484 | 0.493 | 0.510 | 0.527 | 0.544 | 0.561 | 0.579 | 0.596 | 0.613 |
| 5/16 | 0.499 | 0.507 | 0.516 | 0.524 | 0.533 | 0.542 | 0.559 | 0.576 | 0.593 | 0.610 | 0.628 | 0.645 | 0.662 |
| 11/32 | 0.549 | 0.557 | 0.566 | 0.574 | 0.583 | 0.592 | 0.609 | 0.626 | 0.643 | 0.660 | 0.678 | 0.695 | 0.712 |
| 3/8 | 0.598 | 0.606 | 0.615 | 0.623 | 0.632 | 0.641 | 0.658 | 0.675 | 0.692 | 0.709 | 0.727 | 0.744 | 0.761 |
| 13/32 | 0.646 | 0.654 | 0.663 | 0.671 | 0.680 | 0.689 | 0.706 | 0.723 | 0.740 | 0.757 | 0.775 | 0.792 | 0.809 |
| 7/16 | 0.695 | 0.703 | 0.712 | 0.720 | 0.729 | 0.738 | 0.755 | 0.772 | 0.789 | 0.806 | 0.824 | 0.841 | 0.858 |
| 15/32 | 0.734 | 0.742 | 0.751 | 0.759 | 0.768 | 0.777 | 0.794 | 0.811 | 0.828 | 0.845 | 0.863 | 0.880 | 0.897 |
| 1/2 | 0.794 | 0.802 | 0.811 | 0.819 | 0.828 | 0.837 | 0.854 | 0.871 | 0.888 | 0.905 | 0.923 | 0.940 | 0.957 |
| 9/16 | 0.892 | 0.900 | 0.909 | 0.917 | 0.926 | 0.935 | 0.952 | 0.969 | 0.986 | 1.003 | 1.021 | 1.038 | 1.055 |
| 5/8 | 0.990 | 0.998 | 1.007 | 1.015 | 1.024 | 1.033 | 1.050 | 1.067 | 1.084 | 1.101 | 1.119 | 1.136 | 1.153 |
| 11/16 | 1.089 | 1.097 | 1.106 | 1.114 | 1.123 | 1.132 | 1.149 | 1.166 | 1.183 | 1.200 | 1.218 | 1.235 | 1.252 |
| 3/4 | 1.187 | 1.195 | 1.204 | 1.212 | 1.221 | 1.230 | 1.247 | 1.264 | 1.281 | 1.298 | 1.316 | 1.333 | 1.350 |
| 13/16 | 1.286 | 1.294 | 1.303 | 1.311 | 1.320 | 1.329 | 1.346 | 1.363 | 1.380 | 1.397 | 1.415 | 1.432 | 1.449 |
| 7/8 | 1.384 | 1.392 | 1.401 | 1.409 | 1.418 | 1.427 | 1.444 | 1.461 | 1.478 | 1.495 | 1.513 | 1.530 | 1.547 |
| 15/16 | 1.481 | 1.489 | 1.498 | 1.506 | 1.515 | 1.524 | 1.541 | 1.558 | 1.575 | 1.592 | 1.610 | 1.627 | 1.644 |
| 1 | 1.580 | 1.588 | 1.597 | 1.605 | 1.614 | 1.623 | 1.640 | 1.657 | 1.674 | 1.691 | 1.709 | 1.726 | 1.743 |
| 1 1/16 | 1.678 | 1.686 | 1.695 | 1.703 | 1.712 | 1.721 | 1.738 | 1.755 | 1.772 | 1.789 | 1.807 | 1.824 | 1.841 |
| 1 1/8 | 1.777 | 1.785 | 1.794 | 1.802 | 1.811 | 1.820 | 1.837 | 1.854 | 1.871 | 1.888 | 1.906 | 1.923 | 1.940 |
| 1 3/16 | 1.875 | 1.883 | 1.892 | 1.900 | 1.909 | 1.918 | 1.935 | 1.952 | 1.969 | 1.986 | 2.004 | 2.021 | 2.038 |
| 1 1/4 | 1.972 | 1.980 | 1.989 | 1.997 | 2.006 | 2.015 | 2.032 | 2.049 | 2.066 | 2.083 | 2.101 | 2.118 | 2.135 |

**Table 11. Lengths of Straight Stock Required for 90-Degree Bends in Half-Hard Brass and Sheet Copper, Soft Steel, and Aluminum**

| Radius *R* of Bend, Inches | Thickness *T* of Material, Inch | | | | | | | | | | | | |
|---|---|---|---|---|---|---|---|---|---|---|---|---|---|
| | 1⁄64 | 1⁄32 | 3⁄64 | 1⁄16 | 5⁄64 | 3⁄32 | 1⁄8 | 5⁄32 | 3⁄16 | 7⁄32 | 1⁄4 | 9⁄32 | 5⁄16 |
| 1⁄32 | 0.059 | 0.069 | 0.079 | 0.089 | 0.099 | 0.109 | 0.129 | 0.149 | 0.169 | 0.189 | 0.209 | 0.229 | 0.249 |
| 3⁄64 | 0.084 | 0.094 | 0.104 | 0.114 | 0.124 | 0.134 | 0.154 | 0.174 | 0.194 | 0.214 | 0.234 | 0.254 | 0.274 |
| 1⁄16 | 0.108 | 0.118 | 0.128 | 0.138 | 0.148 | 0.158 | 0.178 | 0.198 | 0.218 | 0.238 | 0.258 | 0.278 | 0.298 |
| 3⁄32 | 0.157 | 0.167 | 0.177 | 0.187 | 0.197 | 0.207 | 0.227 | 0.247 | 0.267 | 0.287 | 0.307 | 0.327 | 0.347 |
| 1⁄8 | 0.206 | 0.216 | 0.226 | 0.236 | 0.246 | 0.256 | 0.276 | 0.296 | 0.316 | 0.336 | 0.356 | 0.376 | 0.396 |
| 5⁄32 | 0.255 | 0.265 | 0.275 | 0.285 | 0.295 | 0.305 | 0.325 | 0.345 | 0.365 | 0.385 | 0.405 | 0.425 | 0.445 |
| 3⁄16 | 0.305 | 0.315 | 0.325 | 0.335 | 0.345 | 0.355 | 0.375 | 0.395 | 0.415 | 0.435 | 0.455 | 0.475 | 0.495 |
| 7⁄32 | 0.354 | 0.364 | 0.374 | 0.384 | 0.394 | 0.404 | 0.424 | 0.444 | 0.464 | 0.484 | 0.504 | 0.524 | 0.544 |
| 1⁄4 | 0.403 | 0.413 | 0.423 | 0.433 | 0.443 | 0.453 | 0.473 | 0.493 | 0.513 | 0.533 | 0.553 | 0.573 | 0.593 |
| 9⁄32 | 0.452 | 0.462 | 0.472 | 0.482 | 0.492 | 0.502 | 0.522 | 0.542 | 0.562 | 0.582 | 0.602 | 0.622 | 0.642 |
| 5⁄16 | 0.501 | 0.511 | 0.521 | 0.531 | 0.541 | 0.551 | 0.571 | 0.591 | 0.611 | 0.631 | 0.651 | 0.671 | 0.691 |
| 11⁄32 | 0.550 | 0.560 | 0.570 | 0.580 | 0.590 | 0.600 | 0.620 | 0.640 | 0.660 | 0.680 | 0.700 | 0.720 | 0.740 |
| 3⁄8 | 0.599 | 0.609 | 0.619 | 0.629 | 0.639 | 0.649 | 0.669 | 0.689 | 0.709 | 0.729 | 0.749 | 0.769 | 0.789 |
| 13⁄32 | 0.648 | 0.658 | 0.668 | 0.678 | 0.688 | 0.698 | 0.718 | 0.738 | 0.758 | 0.778 | 0.798 | 0.818 | 0.838 |
| 7⁄16 | 0.697 | 0.707 | 0.717 | 0.727 | 0.737 | 0.747 | 0.767 | 0.787 | 0.807 | 0.827 | 0.847 | 0.867 | 0.887 |
| 15⁄32 | 0.746 | 0.756 | 0.766 | 0.776 | 0.786 | 0.796 | 0.816 | 0.836 | 0.856 | 0.876 | 0.896 | 0.916 | 0.936 |
| 1⁄2 | 0.795 | 0.805 | 0.815 | 0.825 | 0.835 | 0.845 | 0.865 | 0.885 | 0.905 | 0.925 | 0.945 | 0.965 | 0.985 |
| 17⁄32 | 0.844 | 0.854 | 0.864 | 0.874 | 0.884 | 0.894 | 0.914 | 0.934 | 0.954 | 0.974 | 0.994 | 1.014 | 1.034 |
| 9⁄16 | 0.894 | 0.904 | 0.914 | 0.924 | 0.934 | 0.944 | 0.964 | 0.984 | 1.004 | 1.024 | 1.044 | 1.064 | 1.084 |
| 5⁄8 | 0.992 | 1.002 | 1.012 | 1.022 | 1.032 | 1.042 | 1.062 | 1.082 | 1.102 | 1.122 | 1.42 | 1.162 | 1.182 |
| 11⁄16 | 1.090 | 1.100 | 1.110 | 1.120 | 1.130 | 1.140 | 1.160 | 1.180 | 1.200 | 1.220 | 1.240 | 1.260 | 1.280 |
| 3⁄4 | 1.188 | 1.198 | 1.208 | 1.218 | 1.228 | 1.238 | 1.258 | 1.278 | 1.298 | 1.318 | 1.338 | 1.358 | 1.378 |
| 13⁄16 | 1.286 | 1.296 | 1.306 | 1.316 | 1.326 | 1.336 | 1.356 | 1.376 | 1.396 | 1.416 | 1.436 | 1.456 | 1.476 |
| 7⁄8 | 1.384 | 1.394 | 1.404 | 1.414 | 1.424 | 1.434 | 1.454 | 1.474 | 1.494 | 1.514 | 1.534 | 1.554 | 1.574 |
| 15⁄16 | 1.483 | 1.493 | 1.503 | 1.513 | 1.523 | 1.553 | 1.553 | 1.573 | 1.693 | 1.613 | 1.633 | 1.653 | 1.673 |
| 1 | 1.581 | 1.591 | 1.601 | 1.611 | 1.621 | 1.631 | 1.651 | 1.671 | 1.691 | 1.711 | 1.731 | 1.751 | 1.771 |
| 1 1⁄16 | 1.697 | 1.689 | 1.699 | 1.709 | 1.719 | 1.729 | 1.749 | 1.769 | 1.789 | 1.809 | 1.829 | 1.849 | 1.869 |
| 1 1⁄8 | 1.777 | 1.787 | 1.797 | 1.807 | 1.817 | 1.827 | 1.847 | 1.867 | 1.887 | 1.907 | 1.927 | 1.947 | 1.967 |
| 1 3⁄16 | 1.875 | 1.885 | 1.895 | 1.905 | 1.915 | 1.925 | 1.945 | 1.965 | 1.985 | 1.005 | 2.025 | 2.045 | 2.065 |
| 1 1⁄4 | 1.973 | 1.983 | 1.993 | 1.003 | 2.013 | 2.023 | 2.043 | 2.063 | 2.083 | 2.103 | 2.123 | 2.143 | 2.163 |

**Table 12. Lengths of Straight Stock Required for 90-Degree Bends in Hard Copper, Bronze, Cold-Rolled Steel, and Spring Steel**

| Radius $R$ of Bend, Inches | Thickness $T$ of Material, Inch | | | | | | | | | | | | |
|---|---|---|---|---|---|---|---|---|---|---|---|---|---|
| | 1/64 | 1/32 | 3/64 | 1/16 | 5/64 | 3/32 | 1/8 | 5/32 | 3/16 | 7/32 | 1/4 | 9/32 | 5/16 |
| 1/32 | 0.060 | 0.071 | 0.082 | 0.093 | 0.104 | 0.116 | 0.138 | 0.160 | 0.182 | 0.204 | 0.227 | 0.249 | 0.271 |
| 3/64 | 0.085 | 0.096 | 0.107 | 0.118 | 0.129 | 0.141 | 0.163 | 0.185 | 0.207 | 0.229 | 0.252 | 0.274 | 0.296 |
| 1/16 | 0.109 | 0.120 | 0.131 | 0.142 | 0.153 | 0.165 | 0.187 | 0.209 | 0.231 | 0.253 | 0.276 | 0.298 | 0.320 |
| 3/32 | 0.158 | 0.169 | 0.180 | 0.191 | 0.202 | 0.214 | 0.236 | 0.258 | 0.280 | 0.302 | 0.325 | 0.347 | 0.369 |
| 1/8 | 0.207 | 0.218 | 0.229 | 0.240 | 0.251 | 0.263 | 0.285 | 0.307 | 0.329 | 0.351 | 0.374 | 0.396 | 0.418 |
| 5/32 | 0.256 | 0.267 | 0.278 | 0.289 | 0.300 | 0.312 | 0.334 | 0.356 | 0.378 | 0.400 | 0.423 | 0.445 | 0.467 |
| 3/16 | 0.305 | 0.316 | 0.327 | 0.338 | 0.349 | 0.361 | 0.383 | 0.405 | 0.427 | 0.449 | 0.472 | 0.494 | 0.516 |
| 7/32 | 0.355 | 0.366 | 0.377 | 0.388 | 0.399 | 0.411 | 0.433 | 0.455 | 0.477 | 0.499 | 0.522 | 0.544 | 0.566 |
| 1/4 | 0.403 | 0.414 | 0.425 | 0.436 | 0.447 | 0.459 | 0.481 | 0.503 | 0.525 | 0.547 | 0.570 | 0.592 | 0.614 |
| 9/32 | 0.452 | 0.463 | 0.474 | 0.485 | 0.496 | 0.508 | 0.530 | 0.552 | 0.574 | 0.596 | 0.619 | 0.641 | 0.663 |
| 5/16 | 0.501 | 0.512 | 0.523 | 0.534 | 0.545 | 0.557 | 0.579 | 0.601 | 0.623 | 0.645 | 0.668 | 0.690 | 0.712 |
| 11/32 | 0.551 | 0.562 | 0.573 | 0.584 | 0.595 | 0.607 | 0.629 | 0.651 | 0.673 | 0.695 | 0.718 | 0.740 | 0.762 |
| 3/8 | 0.600 | 0.611 | 0.622 | 0.633 | 0.644 | 0.656 | 0.678 | 0.700 | 0.722 | 0.744 | 0.767 | 0.789 | 0.811 |
| 13/32 | 0.648 | 0.659 | 0.670 | 0.681 | 0.692 | 0.704 | 0.726 | 0.748 | 0.770 | 0.792 | 0.815 | 0.837 | 0.859 |
| 7/16 | 0.697 | 0.708 | 0.719 | 0.730 | 0.741 | 0.753 | 0.775 | 0.797 | 0.819 | 0.841 | 0.864 | 0.886 | 0.908 |
| 15/32 | 0.736 | 0.747 | 0.758 | 0.769 | 0.780 | 0.792 | 0.814 | 0.836 | 0.858 | 0.880 | 0.903 | 0.925 | 0.947 |
| 1/2 | 0.796 | 0.807 | 0.818 | 0.829 | 0.840 | 0.852 | 0.874 | 0.896 | 0.918 | 0.940 | 0.963 | 0.985 | 1.007 |
| 9/16 | 0.894 | 0.905 | 0.916 | 0.927 | 0.938 | 0.950 | 0.972 | 0.994 | 1.016 | 1.038 | 1.061 | 1.083 | 1.105 |
| 5/8 | 0.992 | 1.003 | 1.014 | 1.025 | 1.036 | 1.048 | 1.070 | 1.092 | 1.114 | 1.136 | 1.159 | 1.181 | 1.203 |
| 11/16 | 1.091 | 1.102 | 1.113 | 1.124 | 1.135 | 1.147 | 1.169 | 1.191 | 1.213 | 1.235 | 1.258 | 1.280 | 1.302 |
| 3/4 | 1.189 | 1.200 | 1.211 | 1.222 | 1.233 | 1.245 | 1.267 | 1.289 | 1.311 | 1.333 | 1.356 | 1.378 | 1.400 |
| 13/16 | 1.288 | 1.299 | 1.310 | 1.321 | 1.332 | 1.344 | 1.366 | 1.388 | 1.410 | 1.432 | 1.455 | 1.477 | 1.499 |
| 7/8 | 1.386 | 1.397 | 1.408 | 1.419 | 1.430 | 1.442 | 1.464 | 1.486 | 1.508 | 1.530 | 1.553 | 1.575 | 1.597 |
| 15/16 | 1.483 | 1.494 | 1.505 | 1.516 | 1.527 | 1.539 | 1.561 | 1.583 | 1.605 | 1.627 | 1.650 | 1.672 | 1.694 |
| 1 | 1.582 | 1.593 | 1.604 | 1.615 | 1.626 | 1.638 | 1.660 | 1.682 | 1.704 | 1.726 | 1.749 | 1.771 | 1.793 |
| 1 1/16 | 1.680 | 1.691 | 1.702 | 1.713 | 1.724 | 1.736 | 1.758 | 1.780 | 1.802 | 1.824 | 1.847 | 1.869 | 1.891 |
| 1 1/8 | 1.779 | 1.790 | 1.801 | 1.812 | 1.823 | 1.835 | 1.857 | 1.879 | 1.901 | 1.923 | 1.946 | 1.968 | 1.990 |
| 1 3/16 | 1.877 | 1.888 | 1.899 | 1.910 | 1.921 | 1.933 | 1.955 | 1.977 | 1.999 | 2.021 | 2.044 | 2.066 | 2.088 |
| 1 1/4 | 1.974 | 1.985 | 1.996 | 2.007 | 2.018 | 2.030 | 2.052 | 2.074 | 2.096 | 2.118 | 2.141 | 2.163 | 2.185 |

*Allowances for Bending Sheet Metal:* In bending steel, brass, bronze, or other metals, the problem is to find the length of straight stock required for each bend; these lengths are added to the straight section lengths to obtain the total length of material before bending.

If $L$ = length (inch or mm) of straight stock required before bending; $T$ = thickness (inch or mm); and $R$ = inside radius of bend (inch or mm):

For 90° bends in soft brass and soft copper see Table 10 or:

$$L = (0.55 \times T) + (1.57 \times R) \tag{32}$$

For 90° bends in half-hard copper and brass, soft steel, and aluminum see Table 11 or:

$$L = (0.64 \times T) + (1.57 \times R) \tag{33}$$

For 90° bends in bronze, hard copper, cold-rolled steel, and spring steel see Table 12 or:

$$L = (0.71 \times T) + (1.57 \times R) \tag{34}$$

*Example, Showing Application of Formulas 32–34:* Find the length before bending of the part illustrated by Fig. 13. Soft steel is to be used.

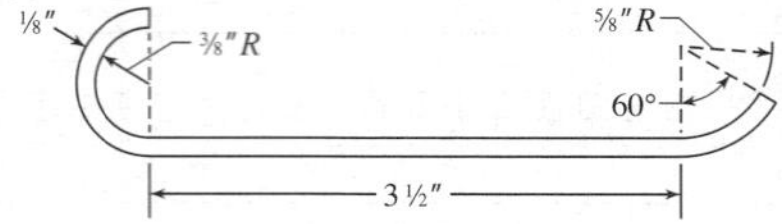


Fig. 13.

For bend at left-hand end (180-degree bend):

$$L = [(0.64 \times 0.125) + (1.57 \times 0.375)] \times \frac{180}{90} = 1.338$$

For bend at right-hand end (60-degree bend):

$$L = [(0.64 \times 0.125) + (1.57 \times 0.625)] \times \frac{60}{90} = 0.707$$

Total length before bending = 3.5 + 1.338 + 0.707 = 5.545 inches

*Angle of Bend Other Than 90 Degrees:* For angles other than 90 degrees, find length $L$, using tables or formulas, and multiply $L$ by angle of bend, in degrees, divided by 90 to find length of stock before bending. In using this rule, note that *angle of bend* is the angle through which the material has actually been bent; hence, it is not always the angle as given on a drawing. To illustrate, in Fig. 14, the angle on the drawing is 60 degrees, but the angle of bend $A$ is 120 degrees (180 − 60 = 120); in Fig. 15, the angle of bend $A$ is 60 degrees; in Fig. 16, angle $A$ is 90 − 30 = 60 degrees. Formulas (32), (33), and (34) apply to parts bent with simple tools or on the bench, where limits of $\pm\,^1\!/_{64}$ inch are specified. If a part has two or more bends of the same radius, it is, of course, only necessary to obtain the length required for one of the bends and then multiply by the number of bends, to obtain the total allowance for the bent sections.

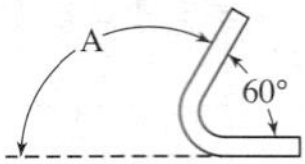


Fig. 14.

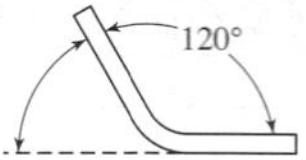


Fig. 15.

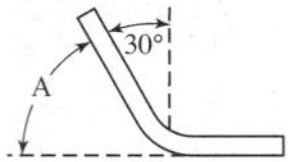


Fig. 16.

*Springback:* Every plastic deformation is followed by elastic recovery. As a consequence of this phenomenon, which occurs when a flat-rolled metal or alloy is cold-worked, upon release of the forming force, the material has a tendency to partially return to its original shape. This effect is called springback and is influenced not only by the tensile and yield strengths, but also by the thickness, bend radius, and bend angle. To estimate springback, an approximate formula in terms of the bend radius before springback $R_i$ and bend radius after springback $R_f$ is as follows

$$\frac{R_i}{R_f} = 4\left(\frac{R_i(YS)}{ET}\right)^3 - 3\left(\frac{R_i(YS)}{ET}\right) + 1 \tag{35}$$

where $R_i$ = bend radius before springback, inch (mm); $R_f$ = bend radius after springback, inch (mm); $YS$ = yield strength of the material, $lb/in^2$ (MPa); $E$ = modulus of elasticity of the material $lb/in^2$ (MPa); and, $T$ = material thickness, inch (mm).

*Other Bending Allowance Formulas:* When bending sheet steel or brass, add from $\frac{1}{3}$ to $\frac{1}{2}$ the thickness of the stock, for *each bend,* to the sum of the inside dimensions of the finished piece, to get the length of the straight blank. The harder the material the greater the allowance ($\frac{1}{3}$ of the thickness is added for soft stock and $\frac{1}{2}$ of the thickness for hard material). The data given in Table 13 refer to the bending of sheet metal for counters, bank fittings, and general office fixtures, for which purpose it is not essential to have the sections of the bends within very close limits. Absolutely accurate data for this work cannot be deduced as the hardness and other mechanical properties vary considerably. The values given in the table apply to sheet steel, aluminum, brass and bronze. Experience has demonstrated that for semi-square corners, such as those formed in a V-die, the amount to be deducted from the sum of the outside bend dimensions, shown in Fig. 17 as the sum of the letters from *a* to *e*, is as follows: $X = 1.67\,BG$, where $X$ = the amount to be deducted; $B$ = the number of bends; and $G$ = the decimal equivalent of the gage thickness of the stock. The values of $X$ for different gages and numbers of bends are given in the table.

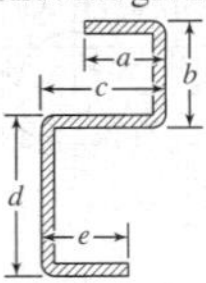


Fig. 17.

**Table 13. Allowances for Square Bends in Sheet Metal**

| Square Bends | Gage | Thickness Inches | Amount to be deducted from the sum of the outside bend dimensions, (in.) | | | | | | |
|---|---|---|---|---|---|---|---|---|---|
| | | | 1 Bend | 2 Bends | 3 Bends | 4 Bends | 5 Bends | 6 Bends | 7 Bends |
| Formed in a Press by a V-die | 18 | 0.0500 | 0.083 | 0.166 | 0.250 | 0.333 | 0.416 | 0.500 | 0.583 |
| | 16 | 0.0625 | 0.104 | 0.208 | 0.312 | 0.416 | 0.520 | 0.625 | 0.729 |
| | 14 | 0.0781 | 0.130 | 0.260 | 0.390 | 0.520 | 0.651 | 0.781 | 0.911 |
| | 13 | 0.0937 | 0.156 | 0.312 | 0.468 | 0.625 | 0.781 | 0.937 | 1.093 |
| | 12 | 0.1093 | 0.182 | 0.364 | 0.546 | 0.729 | 0.911 | 1.093 | 1.276 |
| | 11 | 0.1250 | 0.208 | 0.416 | 0.625 | 0.833 | 1.041 | 1.250 | 1.458 |
| | 10 | 0.1406 | 0.234 | 0.468 | 0.703 | 0.937 | 1.171 | 1.406 | 1.643 |
| Rolled or Drawn in a Draw-bench | 18 | 0.0500 | 0.066 | 0.133 | 0.200 | 0.266 | 0.333 | 0.400 | 0.466 |
| | 16 | 0.0625 | 0.083 | 0.166 | 0.250 | 0.333 | 0.416 | 0.500 | 0.583 |
| | 14 | 0.0781 | 0.104 | 0.208 | 0.312 | 0.416 | 0.521 | 0.625 | 0.729 |
| | 13 | 0.0937 | 0.125 | 0.250 | 0.375 | 0.500 | 0.625 | 0.750 | 0.875 |
| | 12 | 0.1093 | 0.145 | 0.291 | 0.437 | 0.583 | 0.729 | 0.875 | 1.020 |
| | 11 | 0.1250 | 0.166 | 0.333 | 0.500 | 0.666 | 0.833 | 1.000 | 1.166 |
| | 10 | 0.1406 | 0.187 | 0.375 | 0.562 | 0.750 | 0.937 | 1.125 | 1.312 |

Approximate values for sheet steel, aluminum, brass, and bronze.

The lower part of the table applies to square bends that are either drawn through a block of steel made to the required shape, or are drawn through rollers in a drawbench. The pressure applied not only gives a much sharper corner, but it also elongates the material more than in the V-die process. In this example, the deduction is $X = 1.33\,BG$.

*Example:* The following further illustrates this type of calculation. A strip having two bends is to have outside dimensions of 2, $1\frac{1}{2}$ and 2 inches, and is made of stock 0.125 inch thick. The sum of the outside dimensions is thus $5\frac{1}{2}$ inches, and from Table 13 the amount to be deducted is found to be 0.416; hence, the blank will be $5.5 - 0.416 = 5.084$ inches long.

*Bending Force:* The bending force is a function of the strength of the material, the length of the workpiece, and the die opening. A good approximation of the required force $F$ is

$$F = \frac{LT^2(UTS)}{W} \qquad (36)$$

where $F$ = bending force (lb or N)
$L$ = length of the workpiece (inch or mm)
$T$ = material thickness (inch or mm)
$UTS$ = ultimate tensile strength of the material (lb/in$^2$ or MPa)
$W$ = die opening (inch or mm)

**Three-Roll Bending.—**Many curved sheet metal parts, such as rings, cylinders, truncated cones, or segments of these shapes, are impractical to produce by press forming. Such parts are best produced by a process called roll bending. In this process, the workpiece is produced from a flat blank by passing it between three staggered rolls. Depending upon such variables as the composition of the work material, machine capability, or workpiece size, the shape may be formed in a single pass or a series of passes, with roll adjustments needed after each pass. Fig. 18 illustrates the basic setup for three-roll bending on pyramid-type machines.

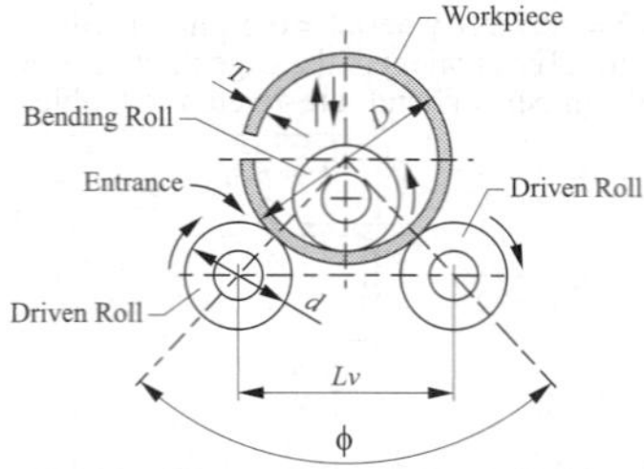


Fig. 18. Three-Roll Bending

To achieve permanent deformation in the outside and inside fibers of the material, the following relationship must apply:

$$\frac{D}{T} < \frac{E}{YS} + 1 \qquad (37)$$

where $D$ = outer diameter of the workpiece, inch (mm)
$T$ = material thickness, inch (mm)
$E$ = modulus of elasticity, lb/in$^2$ (MPa)
$YS$ = yield stress, lb/in$^2$ (MPa)

Three-roll bending is one of the bending operations most used in industrial manufacturing. With only one set of rolls and by setting only one axis, it is possible to achieve different radii or a distribution of radii along the profile length within the bending plane. In addition to the high flexibility and thus economy of this process, it can achieve high quality standards, too, if suitable measures are used.

**Two-Roll Bending.—**Traditionally, roll-bending machines have used three rolls to form the curve. As shown in Fig. 18, the blank passes between or is supported by a third roll, whose position can be adjusted to control the bend. The two-roll bending process shown in Fig. 19 is an alternate method that often is the best choice for high-volume production. Soft to hard materials can be used, thin-gauge material can be rolled to reasonably small diameters, and round workpieces can be made even if the blank has cutouts or holes.

Two-roll bending machines have two basic frame designs: vertical and horizontal. Major parts of both designs include one urethane-covered roll, another steel roll, and an overarm that can be opened and closed to facilitate removal of a finished workpieces. On horizontal machines, the lower urethane-covered roll deforms under the pressure of the upper steel roll to impact the desired shape and size to the blank in a single pass. The diameter of the rolled workpiece is controlled by the diameter of the upper steel roll; a slip-on tube is used to change the desired diameter. The operation of vertical roll-bending machines is basically the same as that of the horizontal one, except that workpieces are loaded and unloaded vertically.

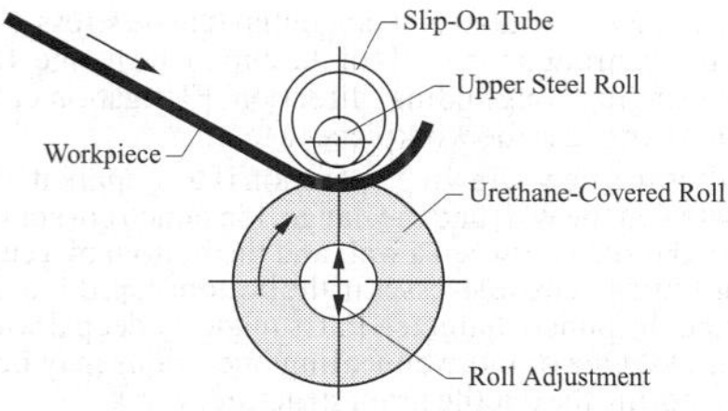

Fig. 19. Two-Roll Bending

When sheet metal is inserted between the two rolls, the urethane roll wraps the blank tightly against the steel roll. The urethane transmits applied force in all directions, as does hydraulic fluid in a hydroforming presses, exerting high pressure; since the urethane roll is driven, true curvature of the sheet metal is obtained. This radius remains uniform, whether the shape being formed is a complete circle or only a segment. Variations in diameter of the final workpiece are due to variations in and natural spring-back of the material.

Since the workpiece is completely controlled, flatting and kinking are eliminated. In most applications, there is no need for pre-forming or cutting off blank ends; flat spots on leading and trailing edges are minimal: one to four material thicknesses, depending on the material.

## Drawing

The drawing of metal, or deep drawing, is the process by which a punch is used to force sheet metal to flow between the surfaces of a punch and a die. Many products made from sheet metals are given the required shape by using a drawing operation. A blank is first cut from flat stock, and then a shell of cylindrical, conical or special shape is produced from this flat blank by means of one or more drawing dies. Most drawn parts are of cylindrical shape, but rectangular, square, and specialized shapes are sometimes produced. With this process, it is possible to get a final part, using minimal operations and generating minimal scrap, that can be assembled without further operations.

**Mechanics of Deep Drawing.—**As the material is drawn into the die by the punch, it flows into a three-dimensional shape. The blank is held in place with a blank holder using a fixed force. High compressive stresses act upon the metal, which without the offsetting effect of a blank holder, would result in a severely wrinkled workpiece.

Wrinkling is one of the major defects in deep drawing; it can damage the dies and adversely affect part assembly and function. The prediction and prevention of wrinkling is very important. There are a number of different analytical and experimental methods that can help to predict and prevent flange wrinkling, including finite element modeling (FEM).

There are many important variables in the deep drawing process, but they can be classified as either material and friction factors, or tooling and equipment factors.

Important material properties such as the strain-hardening coefficient ($n$) and normal anisotropy ($R$) affect deep-drawing operations. Friction and lubrication at the punch, die, and workpiece interfaces are very important in a successful deep drawing process.

Unlike bending operations, in which metal is plastically deformed in a relatively small area, drawing operations impose plastic deformation over large areas, and stress states are different in different regions of the part. As a starting point, consider what appear to be three zones undergoing types of deformation:

1) The flat portion of the blank that has not yet entered the die cavity (the flange)
2) The portion of the blank that is in the die cavity (the wall)
3) The zone of contact between the punch and the blank (bottom)

The radial tensile stress is due to the blank being pulled into the female die, and the compressive stress, normal in the blank sheet, is due to the blank holder pressure. The punch transmits force $F$ to the bottom of the cup, so the part of the blank that is formed into the bottom of the cup is subjected to radial and tangential tensile stress. From the bottom, the punch transmits the force through the walls of the cup to the flange. In this stressed state, the walls tend to elongate in a longitudinal direction. Elongation causes the cup wall to become thinner, which can cause the workpiece to tear.

If a drawing die radius in a deep drawing operation is too small, it will cause fracture of the cup in the zone between the wall and the flange. If a punch corner radius is too small, it may cause fracture in the zone between a wall and the bottom of a cup. Fracture can also result from high longitudinal tensile stresses in the bottom cup, due to a high ratio between the blank diameter and the punch diameter. Parts made by deep drawing usually require several successive draws. One or more annealing operations may be required to reduce work hardening by restoring the ductile grain structure.

*Number of Draws:* The number of successive draws $n$ required is a function of the ratio of the part height $h$ to the part diameter $d$, and is given by this formula:

$$n = \frac{h}{d} \qquad (38)$$

where $n$ = number of draws; $h$ = part height; and, $d$= part diameter.

The value of $n$ for the cylindrical cup draw is given in Table 14.

**Table 14. Number of Draws ($n$) for a Cylindrical Cup Draw**

| $h/d$ | <0.6 | 0.6 to1.4 | 1.4 to 2.5 | 2.5 to 4.0 | 4.0 to 7.0 | 7.0 to 12.0 |
|---|---|---|---|---|---|---|
| $n$ | 1 | 2 | 3 | 4 | 5 | 6 |

*Deep Drawability :* Deep drawability is the ability of a sheet metal to be formed, or drawn, into a cupped or cavity shape without cracking or otherwise failing. The depth to which metal can be drawn in one operation depends upon the quality and kind of material, its thickness, and the amount that the work material is thinned in drawing.

*Drawing a Cylindrical Cup without a Flange:* A general rule for determining the depth to which a cylindrical cup without a flange can be drawn in one operation is defined as the ratio of the mean diameter $d_m$ of the drawn cup to the blank diameter $D$. This relation is known as the drawing ratio $m$. The value of the drawing ratio for the first and succeeding operations is given by:

$$m_1 = \frac{d_{m_1}}{D} \qquad m_2 = \frac{d_{m_2}}{D_{m_1}} \qquad m_3 = \frac{d_{m_3}}{D_{m_2}} \qquad \ldots \qquad m_n = \frac{d_{m_n}}{D_{m_{n-1}}}$$

The magnitude of these ratios determines the following parameters:

1) the stresses and forces of the deep drawing processes
2) the number of successive draws
3) the blank holder force
4) the quality of the final drawn parts.

Table 15 shows optimal drawing ratios for cylindrical cups of sheet steel and brass without a flange.

**Table 15. Optimal Ratios $M$ for Drawing a Cylindrical Cup without Flanges**

| Drawing ratio $m$ | Relative Thickness of the Material $T_r = \frac{T}{D}100\ (\%)$ | | | | | |
|---|---|---|---|---|---|---|
| | 2.0–1.5 | 1.5–1.0 | 1.0–0.6 | 0.6–0.3 | 0.3–0.15 | 0.15–0.08 |
| $m_1$ | 0.48–0.50 | 0.50–0.53 | 0.53–0.55 | 0.55–0.58 | 0.58–0.60 | 0.60–0.63 |
| $m_2$ | 0.73–0.75 | 0.75–0.76 | 0.76–0.78 | 0.78–0.79 | 0.79–0.80 | 0.80–0.82 |
| $m_3$ | 0.76–0.78 | 0.78–0.79 | 0.79–0.80 | 0.81–0.82 | 0.81–0.82 | 0.82–0.84 |
| $m_4$ | 0.78–0.80 | 0.80–0.81 | 0.81–0.82 | 0.82–0.83 | 0.83–0.85 | 0.85–0.86 |
| $m_5$ | 0.80–0.82 | 0.82–0.84 | 0.84–0.85 | 0.85–0.86 | 0.86–0.87 | 0.78–0.90 |

Diameters of drawing workpieces for the first and succeeding operations are given by:

$$d_1 = m_1 D \qquad d_2 = m_2 d_1 \qquad \ldots \qquad d_i = m_i d_{i-1}$$

*Drawing a Cylindrical Cup with a Flange:* Table 16 gives values of the drawing ratio $m$ for the first and succeeding operations for drawing a cylindrical cup with flange.

Diameters of drawing workpiece for the first and succeeding operations are given by

$$d_1 = m_1 D \qquad d_2 = m_2 d_1 \qquad \ldots \qquad d_i = m_i d_{i-1}$$

However, diameter $D_f$ needs to be accomplished in the first drawing operation if possible.

*Diameters of Shell Blanks:* The diameters of blanks for drawing plain cylindrical shells can be obtained from Table 17, which gives a very close approximation for thin stock. The blank diameters given in this table are for sharp-cornered shells and are found by the following formula

$$D = \sqrt{d^2 + 4dh} \qquad (39)$$

where $D$ = diameter of flat blank (inch or mm); $d$ = diameter of finished shell (inch or mm); and, $h$ = height of finished shell (inch or mm).

*Example 1:* If the diameter of the finished shell $d$, is to be 1.5 inches (38.1 mm), and the height $h$, 2 inches (50.8 mm), the trial diameter of the blank $D$, would be found as follows:

**Table 16. Values of Ratio *m* for Drawing a Cylindrical Cup with Flange**

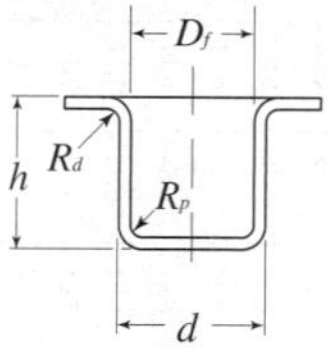


| Drawing ratio $m$ | $\frac{D_f}{d}$ | Relative thickness of the material $T_r = \frac{T}{D}100$ (%) | | | | |
|---|---|---|---|---|---|---|
| | | 2.0–1.5 | 1.5–1.0 | 1.0–0.6 | 0.6–0.3 | 0.3–0.15 |
| $m_1$ | 1.1 | 0.51 | 0.53 | 0.55 | 0.57 | 0.59 |
| | 1.3 | 0.49 | 0.51 | 0.53 | 0.54 | 0.55 |
| | 1.5 | 0.47 | 0.49 | 0.50 | 0.51 | 0.52 |
| | 1.8 | 0.45 | 0.46 | 0.47 | 0.48 | 0.48 |
| | 2.0 | 0.42 | 0.43 | 0.44 | 0.45 | 0.45 |
| | 2.5 | 0.37 | 0.38 | 0.38 | 0.38 | 0.38 |
| | 3.0 | 0.32 | 0.33 | 0.33 | 0.33 | 0.33 |
| $m_2$ | … | 0.73 | 0.75 | 0.76 | 0.78 | 0.80 |
| $m_3$ | … | 0.75 | 0.78 | 0.79 | 0.80 | 0.82 |
| $m_4$ | … | 0.78 | 0.80 | 0.82 | 0.83 | 0.84 |
| $m_5$ | … | 0.80 | 0.82 | 0.84 | 0.85 | 0.86 |

$$D = \sqrt{1.5^2 + 4 \times 1.5 \times 2} = \sqrt{14.25} = 3.78 \text{ inches (96 mm)}$$

For a round-cornered cup, the following formula, in which $r$ equals the radius of the corner, will give fairly accurate diameters, provided the radius does not exceed, say, $\frac{1}{4}$ the height of the shell:

$$D = \sqrt{d^2 + 4dh} - r \qquad (40)$$

These formulas are based on the assumption that the thickness of the drawn shell is to be the same as the original thickness of the stock and that the blank is so proportioned that its area will equal the area of the drawn shell. This method of calculating the blank diameter is quite accurate for thin material, when there is only a slight reduction in the thickness of the metal incident to drawing; but when heavy stock is drawn and the thickness of the finished shell is much less than the original thickness of the stock, the blank diameter obtained from Formulas (39) or (40) will be too large, because when the stock is drawn thinner, there is an increase in area. When an appreciable reduction in thickness is to be made, the blank diameter can be obtained by first determining the "mean height" of the drawn shell by the following formula, which is only approximately correct, but will give results sufficiently accurate for most work:

$$M = \frac{ht}{T} \qquad (41)$$

## Table 17. Diameters of Blanks for Drawn Cylindrical Shells

| Dia. of Shell, inch | Height of Shell, inch | | | | | | | | | | | | | | | | | | | |
|---|---|---|---|---|---|---|---|---|---|---|---|---|---|---|---|---|---|---|---|---|
| | 1/4 | 1/2 | 3/4 | 1 | 1 1/4 | 1 1/2 | 1 3/4 | 2 | 2 1/4 | 2 1/2 | 2 3/4 | 3 | 3 1/4 | 3 1/2 | 3 3/4 | 4 | 4 1/2 | 5 | 5 1/2 | 6 |
| 1/4 | 0.56 | 0.75 | 0.90 | 1.03 | 1.14 | 1.25 | 1.35 | 1.44 | 1.52 | 1.60 | 1.68 | 1.75 | 1.82 | 1.89 | 1.95 | 2.01 | 2.14 | 2.25 | 2.36 | 2.46 |
| 1/2 | 0.87 | 1.12 | 1.32 | 1.50 | 1.66 | 1.80 | 1.94 | 2.06 | 2.18 | 2.29 | 2.40 | 2.50 | 2.60 | 2.69 | 2.78 | 2.87 | 3.04 | 3.21 | 3.36 | 3.50 |
| 3/4 | 1.14 | 1.44 | 1.68 | 1.89 | 2.08 | 2.25 | 2.41 | 2.56 | 2.70 | 2.84 | 2.97 | 3.09 | 3.21 | 3.33 | 3.44 | 3.54 | 3.75 | 3.95 | 4.13 | 4.31 |
| 1 | 1.41 | 1.73 | 2.00 | 2.24 | 2.45 | 2.65 | 2.83 | 3.00 | 3.16 | 3.32 | 3.46 | 3.61 | 3.74 | 3.87 | 4.00 | 4.12 | 4.36 | 4.58 | 4.80 | 5.00 |
| 1 1/4 | 1.68 | 2.01 | 2.30 | 2.56 | 2.79 | 3.01 | 3.21 | 3.40 | 3.58 | 3.75 | 3.91 | 4.07 | 4.22 | 4.37 | 4.51 | 4.64 | 4.91 | 5.15 | 5.39 | 5.62 |
| 1 1/2 | 1.94 | 2.29 | 2.60 | 2.87 | 3.12 | 3.36 | 3.57 | 3.78 | 3.97 | 4.15 | 4.33 | 4.50 | 4.66 | 4.82 | 4.98 | 5.12 | 5.41 | 5.68 | 5.94 | 6.18 |
| 1 3/4 | 2.19 | 2.56 | 2.88 | 3.17 | 3.44 | 3.68 | 3.91 | 4.13 | 4.34 | 4.53 | 4.72 | 4.91 | 5.08 | 5.26 | 5.41 | 5.58 | 5.88 | 6.17 | 6.45 | 6.71 |
| 2 | 2.45 | 2.83 | 3.16 | 3.46 | 3.74 | 4.00 | 4.24 | 4.47 | 4.69 | 4.90 | 5.10 | 5.29 | 5.48 | 5.66 | 5.83 | 6.00 | 6.32 | 6.63 | 6.93 | 7.21 |
| 2 1/4 | 2.70 | 3.09 | 3.44 | 3.75 | 4.04 | 4.31 | 4.56 | 4.80 | 5.03 | 5.25 | 5.46 | 5.66 | 5.86 | 6.05 | 6.23 | 6.41 | 6.75 | 7.07 | 7.39 | 7.69 |
| 2 1/2 | 2.96 | 3.36 | 3.71 | 4.03 | 4.33 | 4.61 | 4.87 | 5.12 | 5.36 | 5.59 | 5.81 | 6.02 | 6.22 | 6.42 | 6.61 | 6.80 | 7.16 | 7.50 | 7.82 | 8.14 |
| 2 3/4 | 3.21 | 3.61 | 3.98 | 4.31 | 4.62 | 4.91 | 5.18 | 5.44 | 5.68 | 5.92 | 6.15 | 6.37 | 6.58 | 6.79 | 6.99 | 7.18 | 7.55 | 7.91 | 8.25 | 8.58 |
| 3 | 3.46 | 3.87 | 4.24 | 4.58 | 4.90 | 5.20 | 5.48 | 5.74 | 6.00 | 6.25 | 6.48 | 6.71 | 6.93 | 7.14 | 7.35 | 7.55 | 7.94 | 8.31 | 8.66 | 9.00 |
| 3 1/4 | 3.71 | 4.13 | 4.51 | 4.85 | 5.18 | 5.48 | 5.77 | 6.04 | 6.31 | 6.56 | 6.80 | 7.04 | 7.27 | 7.49 | 7.70 | 7.91 | 8.31 | 8.69 | 9.06 | 9.41 |
| 3 1/2 | 3.97 | 4.39 | 4.77 | 5.12 | 5.45 | 5.77 | 6.06 | 6.34 | 6.61 | 6.87 | 7.12 | 7.36 | 7.60 | 7.83 | 8.05 | 8.26 | 8.67 | 9.07 | 9.45 | 9.81 |
| 3 3/4 | 4.22 | 4.64 | 5.03 | 5.39 | 5.73 | 6.05 | 6.35 | 6.64 | 6.91 | 7.18 | 7.44 | 7.69 | 7.92 | 8.16 | 8.38 | 8.61 | 9.03 | 9.44 | 9.83 | 10.20 |
| 4 | 4.47 | 4.90 | 5.29 | 5.66 | 6.00 | 6.32 | 6.63 | 6.93 | 7.21 | 7.48 | 7.75 | 8.00 | 8.25 | 8.49 | 8.72 | 8.94 | 9.38 | 9.80 | 10.20 | 10.58 |
| 4 1/4 | 4.72 | 5.15 | 5.55 | 5.92 | 6.27 | 6.60 | 6.91 | 7.22 | 7.50 | 7.78 | 8.05 | 8.31 | 8.56 | 8.81 | 9.04 | 9.28 | 9.72 | 10.15 | 10.56 | 10.96 |
| 4 1/2 | 4.98 | 5.41 | 5.81 | 6.19 | 6.54 | 6.87 | 7.19 | 7.50 | 7.79 | 8.08 | 8.35 | 8.62 | 8.87 | 9.12 | 9.37 | 9.60 | 10.06 | 10.50 | 10.92 | 11.32 |
| 4 3/4 | 5.22 | 5.66 | 6.07 | 6.45 | 6.80 | 7.15 | 7.47 | 7.78 | 8.08 | 8.37 | 8.65 | 8.92 | 9.18 | 9.44 | 9.69 | 9.93 | 10.40 | 10.84 | 11.27 | 11.69 |
| 5 | 5.48 | 5.92 | 6.32 | 6.71 | 7.07 | 7.42 | 7.75 | 8.06 | 8.37 | 8.66 | 8.94 | 9.22 | 9.49 | 9.75 | 10.00 | 10.25 | 10.72 | 11.18 | 11.62 | 12.04 |
| 5 1/4 | 5.73 | 6.17 | 6.58 | 6.97 | 7.33 | 7.68 | 8.02 | 8.34 | 8.65 | 8.95 | 9.24 | 9.52 | 9.79 | 10.05 | 10.31 | 10.56 | 11.05 | 11.51 | 11.96 | 12.39 |
| 5 1/2 | 5.98 | 6.42 | 6.84 | 7.23 | 7.60 | 7.95 | 8.29 | 8.62 | 8.93 | 9.23 | 9.53 | 9.81 | 10.08 | 10.36 | 10.62 | 10.87 | 11.37 | 11.84 | 12.30 | 12.74 |
| 5 3/4 | 6.23 | 6.68 | 7.09 | 7.49 | 7.86 | 8.22 | 8.56 | 8.89 | 9.21 | 9.52 | 9.81 | 10.10 | 10.38 | 10.66 | 10.92 | 11.18 | 11.69 | 12.17 | 12.63 | 13.08 |
| 6 | 6.48 | 6.93 | 7.35 | 7.75 | 8.12 | 8.49 | 8.83 | 9.17 | 9.49 | 9.80 | 10.10 | 10.39 | 10.68 | 10.95 | 11.23 | 11.49 | 12.00 | 12.49 | 12.96 | 13.42 |

where $M$ = approximate mean height of drawn shell (inch or mm); $h$ = height of drawn shell (inch or mm); $t$ = thickness of shell (inch or mm); and $T$ = thickness of metal (inch or mm) before drawing.

After determining the mean height, the blank diameter for the required shell diameter is obtained from Table 16, the mean height being used instead of the actual height.

*Example 2:* Suppose a shell 2 inches (50.8 mm) in diameter and 3¾ inches (95.25 mm) high is to be drawn, and that the original thickness of the stock is 0.050 inch (12.7 mm) and the thickness of the drawn shell, 0.040 inch (1.02 mm). To what diameter should the blank be cut? Obtain the mean height from Formula (41):

$$M = \frac{ht}{T} = \frac{3.75 \times 0.040}{0.050} = 3 \text{ inches } (76.2 \text{ mm})$$

According to Table 16, the blank diameter for a shell 2 inches (50.8 mm) in diameter and 3 inches (76.2 mm) high is 5.29 inches (134.37 mm). Formula (41) is accurate enough for all practical purposes, unless the reduction in the thickness of the metal is greater than about one-fifth of the original thickness. When there is considerable reduction, a blank calculated by this formula produces a shell that is too long. However, the error is in the right direction, as the edges of drawn shells are ordinarily trimmed.

If the shell has a rounded corner, the radius of the corner should be deducted from the figures given in the table. For example, if the shell referred to in Example 2 had a corner of ¼-inch (6.35 mm) radius, the blank diameter would equal 5.29 − 0.25 = 5.04 inches (128.02 mm).

Another formula sometimes used for obtaining blank diameters for shells, when there is a reduction in the thickness of the stock, is as follows:

$$D = \sqrt{a^2 + (a^2 - b^2)\frac{h}{t}} \qquad (42)$$

where $D$ = blank diameter (inch or mm); $a$ = outside diameter (inch or mm); $b$ = inside diameter (inch or mm); $t$ = thickness of shell at bottom (inch or mm); and, $h$ = depth of shell (inch or mm).

Equation (42) is based on the volume of metal in the drawn shell. It is assumed that the shells are cylindrical, and no allowance is made for a rounded corner at the bottom, or for trimming the shell after drawing. To allow for trimming, add required amount to depth $h$.

When a shell is of irregular cross section, if its weight is known, the blank diameter ($D$) can be determined by the following formula:

$$D = 1.1284\sqrt{\frac{W}{wt}} \qquad (43)$$

where $D$ = blank diameter, inches (mm)
$W$ = weight of shell, lbs (kg)
$w$ = weight of metal, $lb/in^3$ ($kg/mm^3$)
$t$ = thickness of the shell, inch (mm)

In the construction of dies for producing shells, especially of irregular form, a common method to be used is to make the drawing tool first. The required blank diameter then can be determined by trial. One method is to cut a trial blank as near to size and shape as can be estimated. The outline of this blank is then scribed on a flat sheet, after which the blank is drawn. If the finished shell shows that the blank is not of the right diameter or shape, a new trial blank is cut either larger or smaller than the size indicated by the line previously scribed, this line acting as a guide. If a model shell is available, the blank diameter can also be determined as follows:

First, cut a blank somewhat large and from the same material used for making the model; then, reduce the size of the blank until its weight equals the weight of the model.

*Forces:* The punch force for drawing a cylindrical shell needs to supply the various types of work required in deep drawing, such as the work of deformation, redundant work, friction work, and the work required for ironing (if required).

*Force for the First Drawing Operation:* The calculation of the punch force for the first drawing operation (neglecting friction) is given by the following formula:

$$F_1 = \pi d_{m_1} T(UTS) \tag{44}$$

where $d_{m1}$ = mean diameter of shell after the first operation, (inch or mm)
$T$ = material thickness, (inch or mm)
$UTS$ = ultimate tensile strength of the material, (lb/in² or N/mm²)

*Force for Subsequent Drawing Operations:* Subsequent drawing operations are different from the first operation: as in the deep-drawing process, the flange diameter decreases but the zone of the plastic deformation does not change. The punch force for the next drawing operation can be calculated by the approximate empirical formula as follows:

$$F_i = \pi d_p T(UTS) \cdot \left(\frac{D}{d_p} - 0.7\right) \tag{45}$$

where $d_p$ = punch diameter (inch or mm)
$D$ = blank diameter (inch or mm)
$T$ = material thickness (inch or mm)
$UTS$ = ultimate tensile strength of the material (lb/in² or MPa)

*Shapes of Blanks for Rectangular Shells:* There is no formula for determining the shape of the blank for rectangular drawing that will produce the part as drawn to print. All corner contours must be developed. However, the following conservative procedure will get the die in the final design ballpark with a minimum of trials. When laying out a blank by this method, first draw a plan view of the finished shell or lines representing the shape of the part at the bottom, the corners being given the required radius, as shown in Fig. 20. Next, insert the sides and ends, making the length *L* and the width *W* equal to the length and width of the drawn part minus twice the radius *r* at the corners. To provide just the right amount of material for the corners, the first step is to find what blank diameter will be required to draw a cylindrical shell having a radius *r*. This diameter can be calculated by the formula for the blank diameter (*D*) of the cylindrical shell:

$$D = \sqrt{d^2 - 4dh} \tag{46}$$

where $D$ = blank diameter (inch or mm)
$d$ = diameter of drawn shell (inch or mm)
$h$ = height of shell (inch or mm)

After determining the diameter *D*, scribe arcs at each corner having radius *R* equal to one-half of diameter *D*. The outline of the blank for the rectangular part is then obtained by drawing curved lines between the ends and the sides, as shown in Fig. 20. These curves should touch the arcs *R*.

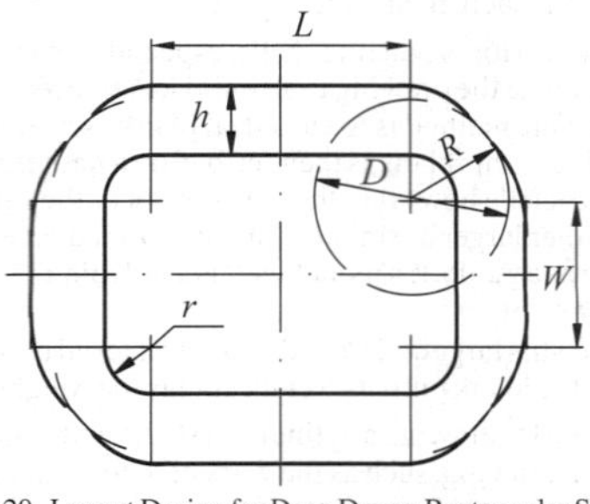


Fig. 20. Layout Design for Deep Drawn Rectangular Shell

When laying out the blank it is usually advisable to plan for a form that will produce corners a little higher than the sides. The wear of the die is at the corners, and when it occurs, the material will thicken and the drawn part will be low at the corners if no allowance for this wear has been made on the blank.

*Blank for Rectangular Flanged Shells:* The shape of the blank for a rectangular flanged shell may be determined in practically the same way as described in the foregoing, except that the width of the flange must be considered. Referring to Fig. 20, the dimension $h$ in the flat blank is made equal to the height of the drawn part plus the width of the flange; however, the blank diameter $D$ for a cylindrical shell having a flange can be determined by the formula

$$D = \sqrt{d^2 + 4dh} \qquad (47)$$

where $D$ = blank diameter (inch or mm)

$d$ = diameter of drawn shell (inch or mm)

$d_1$ = diameter measured across the flange (inch or mm)

$h$ = height of shell (inch or mm)

After determining diameter $D$ and the corresponding radius $R$, the outline of the blank is drawn the same as for a rectangular shell without the flange.

**Ironing Process.—**The ironing process is the reduction in thickness of drawn shell walls by pulling them through tight dies. Ironing is a very useful process when employed in combination with deep drawing to produce a uniform wall thickness and to increase the wall height. It is done to obtain a wall that is thin compared with the shell bottom or merely to correct natural wall thickening toward the top edge of a drawn shell.

Basically, in the ironing processes, a previously deep-drawn shell is placed on a punch and pushed through one or more ironing die rings that have a smaller inside diameter than the outside diameter of the shell (Fig. 21). Hence, the clearance between the ironing rings and the punch is less than the shell's wall thickness, so the shell after ironing has a constant wall thickness equal to the clearance. The theoretical maximum reduction in wall thickness per operation due to ironing is approximately 60 percent.

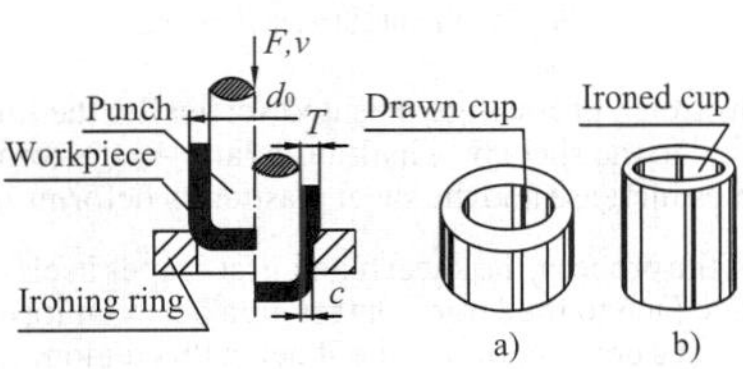


Fig. 21. Ironing of Drawn Cup: a) Cup Before Ironing; b) Cup After Ironing.

*Ironing Force:* Ironing involves the compression and additional work hardening of the metal. The stress on the sidewall can be quite severe, creating the additional possibility of cup failure.

The force required to iron a cylindrical workpiece can be calculated as

$$F_{ir} = S_c \pi d_0 (T - c) \qquad (48)$$

where $F_{ir}$ = ironing force (lb); $d_0$ = outside diameter of cylindrical cup (in.); $S_c$ = compressive strength of metal (lb/in$^2$); $T$ = thickness of material (in.); and, $c$ = clearance between punch and die (in.).

## Stretch Forming

Stretch forming is a metal forming process in which a blank of sheet metal is formed by the simultaneous application of tensile loads to the material over a die in order to form large contoured parts in the required shape.

During stretch forming, the sheet blank is subjected to both *elastic* and *plastic deformation*. The most appropriate measure of formability for stretch forming is the strain-hardening exponent, or $n$ value:

$$c = k\varepsilon^n \tag{49}$$

where $k$ = a constant.

A high value of $n$ is desired if the sheet is to show good stretch formability. In assessing the formability of sheet metals while forming a workpiece, a technique called circle grade analysis (CGA) is used to construct a forming limit diagram of the sheet metal to be used.

Two methods are used in stretch forming: simple stretch forming, also called the *block method*, and stretch-wrap forming, also called *tangential stretch forming*.

**Simple Stretch Forming.—**In the simple stretch forming process, the sheet blank to be formed is clamped between two gripping jaws located on opposite ends, and the tool moves into the clamped sheet blank as shown in Fig. 22.

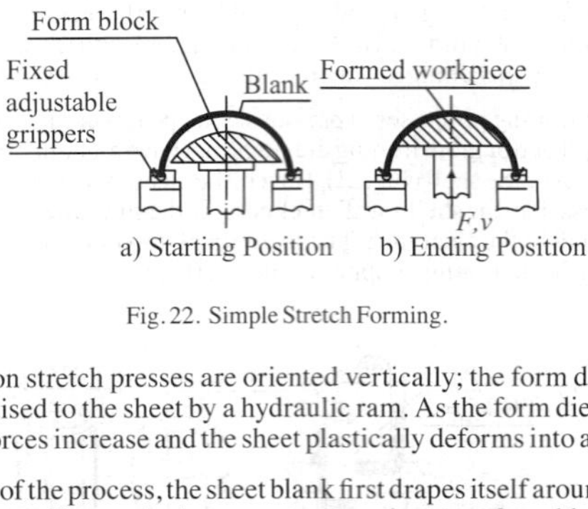


Fig. 22. Simple Stretch Forming.

The most common stretch presses are oriented vertically; the form die rests on a press table that can be raised to the sheet by a hydraulic ram. As the form die is driven into the sheet, the tensile forces increase and the sheet plastically deforms into a new shape.

At the beginning of the process, the sheet blank first drapes itself around the form block, following its contours. Due to the large contact area between form block and blank, the frictional forces prevent a deformation of the sheet in this region.

**Stretch-Wrap Forming.—**In this method also, a sheet blank is gripped from two opposite ends and stretched into the plastic region before being wrapped over a punch, so that the whole cross section of the material undergoes a uniform plastic deformation; then the die is brought down to complete the operation. The main difference from the simple stretch forming process is that both the form block and the gripping jaws are movable (Fig. 23).

Stretch formed parts are typically large and possess large radial bends. The shapes that can be produced range from a simple curved surface to complex non-uniform cross sections. Stretch forming is capable of shaping parts with very high accuracy and smooth surfaces. Ductile materials are preferable, the most commonly used being aluminum, steel, and titanium. Typical stretch formed parts are large curved panels such as door panels in cars or wing panels on aircraft.

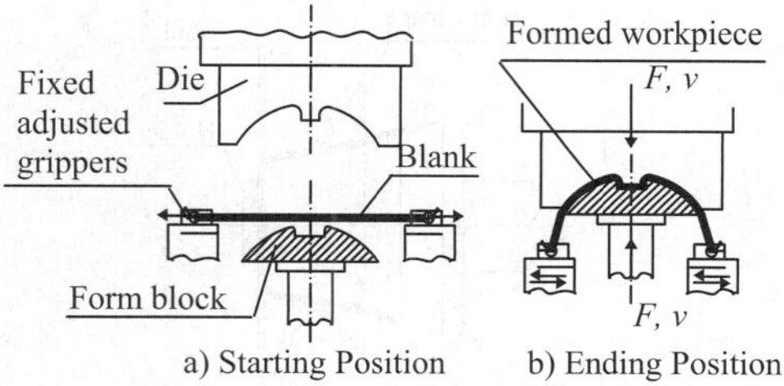


Fig. 23. Stretch Draw Forming.

The advantages of this process are the following:

a) The tensile forces applied always act tangentially to the body of the blank

b) The lack of springback in the finished part

c) Flexible low-cost tooling

d) Increase of yield stress up to 10 percent

e) Less forming pressure required

f) The die can be made of inexpensive material

The disadvantages are these:

a) More material required for gripping

b) Difficulty of adaptation to modern high-speed automated lines

c) Reduction of material thickness by 5 to 7 percent

## Spinning

Metal spinning is the process of forming various seamless and axially symmetrical parts from flat circles of sheet metal (blanks) or from a length of tubing over a mandrel with tools or rollers. There are three types of metal spinning processes: conventional spinning, shear spinning, and tube spinning.

**Conventional Spinning.—**In conventional metal spinning, a disc of metal is rotated at controlled speeds on a specialized machine similar in design to a machine lathe. Instead of the clamping chuck common on a machine lathe, a wood or metal spinning mandrel is used, the form of which corresponds with the internal contour of the part to be produced. The blank is clamped between the spinning mandrel and a follower on the tailstock spindle, as shown in Fig. 24.

The mandrel, blank, and holder are then set in rotation. Spinning tools or spinning rollers are forced against the rotating blank by hand or by a computer-controlled hydraulic mechanism. The process requires a series of crossing steps, as indicated in Fig. 24, to complete the shaping of the workpiece. With this forming technique, a material's thickness generally does not change from the blank to the finished component. If the spinning forces applied with the hydraulic mechanism are higher in comparison to hand spinning, mandrels made of a harder material have to be used, e.g., boiler plate, chilled cast iron, or hardened tool steel.

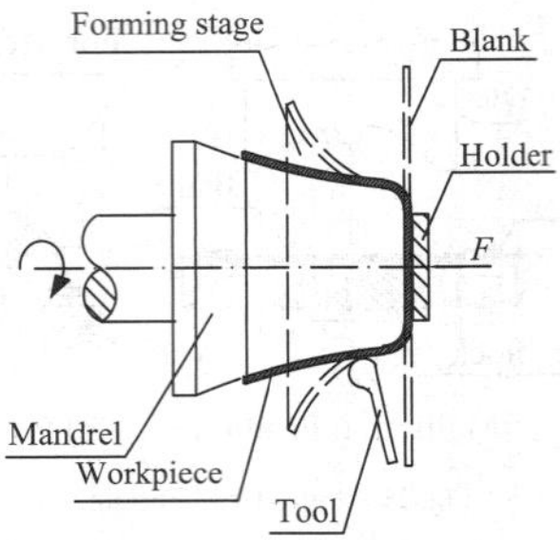


Fig. 24. Conventional Spinning

**Shear Spinning.—** Shear spinning is the process of forming complex shapes such as cones with tapering walls or symmetrical-axis curvilinear shapes such as the nose cones of missiles. This process achieves a deliberate and controlled reduction in blank thickness (Fig. 25), as opposed to a limited reduction of blank thickness in conventional spinning.

The shear forming roller or rollers achieve local metal flow, which applies a pressure on the blank against the support from the steel mandrel. Special shear forming rollers control the material's flow, and they move the free material parallel to the axis of the mandrel. The remaining portion of the blank, which does not take part in the actual deformation, remains always at right angles with respect to the axis of rotation and does not change its external diameter. The material flow takes place in the axial direction. The wall of the workpiece is produced from the reduction in blank thickness, but the diameter of the blank stays constant, as shown in Fig. 25. The thickness of the initial blank depends on the angle of the final part and the finished wall requirements and can be calculated by the formula:

$$T = \frac{T_w}{\sin \alpha} \quad (50)$$

where $T$ = thickness of the initial blank (inch or mm); $T_w$ = wall thickness of the spun part (inch or mm); and, $\alpha$ = half angle of cone.

The bottom and the flange maintain their original thickness. A particular advantage is that the surface finishes achieved are comparable to those achieved with grinding or fine turning. The accuracy of the shape and the dimensional repeatability are excellent.

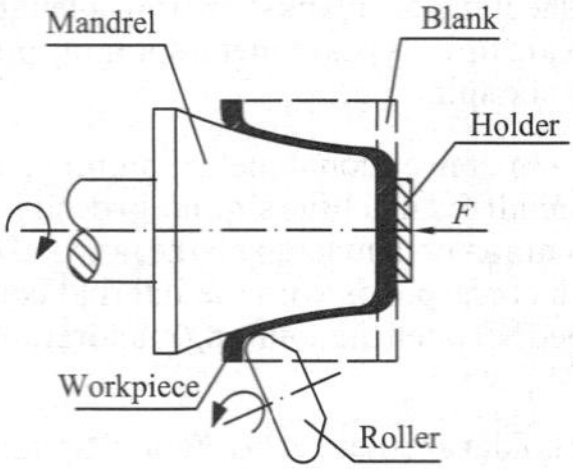


Fig. 25. Shear Spinning

**Tube Spinning.—** Tube spinning is used to reduce the wall thickness of a hollow cylindrical blank while it is spinning on a cylindrical mandrel using spinning tools (Fig. 26). Applying the spinning tool to the cylindrical blank internally or externally achieves reduction of the workpiece wall, and this reduction in turn results in an increase in the workpiece's length. Workpieces may be spun either forward or backward.

The ideal tangential force in forward tube spinning may be calculated by the following formula:

$$F_t = \sigma_{f(m)}(T - T_w)f \tag{51}$$

where $F_t$ = ideal tangential force (lb or N); $\sigma_{f(m)}$ = average flow stress of the material (lb/in² or MPa); $T$ = thickness of the initial blank (inch or mm); $T_w$ = wall thickness of the spun part (inch or mm); and, $f$ = feed (inch or mm).

Because of friction and other influencing factors, the force exerted is about twice that of the ideal force.

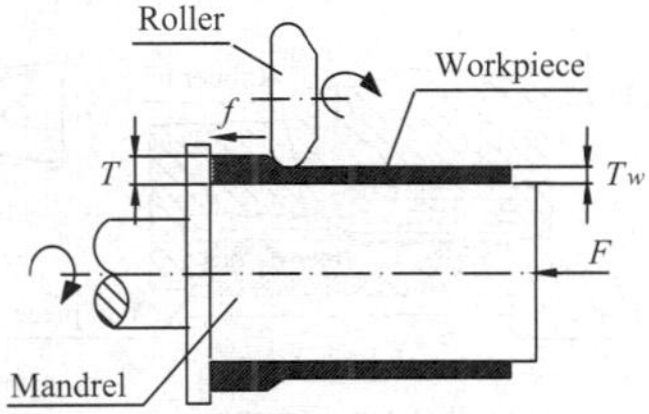


Fig. 26. Tube Spinning

## Rubber Pad and Hydroforming Processes

In the rubber pad forming process, one of the dies in a set (punch or die) is replaced with a flexible material such as rubber or polyurethane. Polyurethane is widely used because of its resistance to abrasion and its long fatigue life. Several processes utilize rubber pad forming techniques. Fig. 27 schematically illustrates the Verson-Wheelon, Marform, and hydroforming processes.

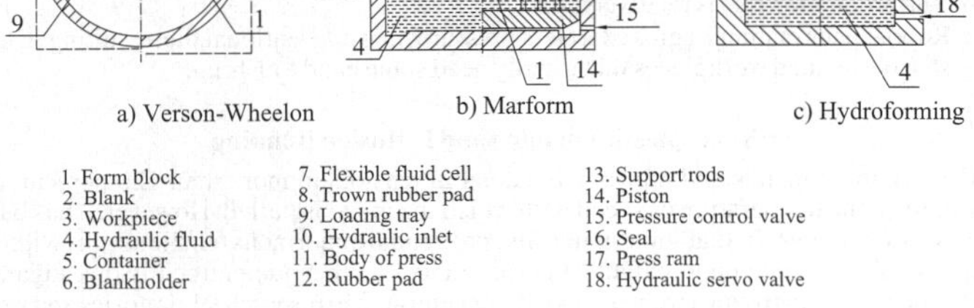


a) Verson-Wheelon b) Marform c) Hydroforming

1. Form block
2. Blank
3. Workpiece
4. Hydraulic fluid
5. Container
6. Blankholder
7. Flexible fluid cell
8. Trown rubber pad
9. Loading tray
10. Hydraulic inlet
11. Body of press
12. Rubber pad
13. Support rods
14. Piston
15. Pressure control valve
16. Seal
17. Press ram
18. Hydraulic servo valve

Fig. 27. Rubber Pad Forming Processes: a) Verson-Wheelon; b) Marform; c) Hydroforming

**Guerin Process.—**Fig. 28 shows the Guerin process, synonymous with the term "rubber pad forming," in which a rigid forming block is placed on the lower bed of the press; on top of this block, the blank is positioned and a soft die of rubber or polyurethane (hardness 50 to 70 Shore) is forced over the rigid block and blank into its required shape. The thickness of the rubber pad, which is held in a sturdy cast-iron or steel container, is usually three times the height of the formed block, but it must be a minimum of 1.5 times thicker than the height of a rigid form block. During the process cycle, the rubber pad deforms elastically over the form block and the blank, applying a strong pressure. The pressure that the soft die exerts on the blank is uniform, so that the forming process creates no thinning of the material, but the radii are more shallow than those produced in conventional dies.

The following formula can be used to determine total pad forming pressure:

$$p = \frac{F}{A} \times 2000$$

where $p$ = total rubber pad pressure (psi); $F$ = capacity of press (tons); and $A$ = area of pad (in$^2$).

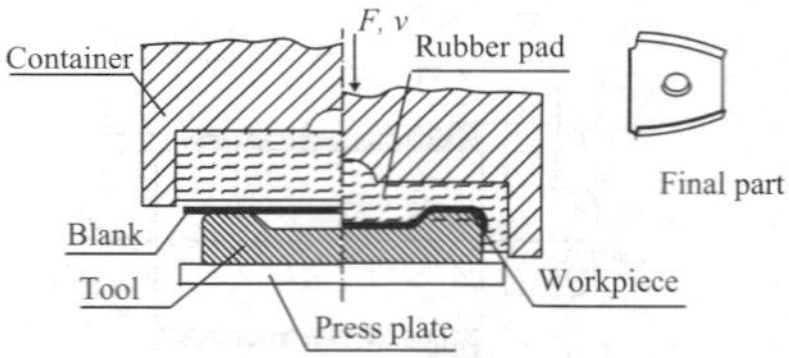


Fig. 28. Guerin Forming Process

An improvement over the Guerin process is the Marform process (see Fig. 27), which features the addition of a blankholder and die cushion to make this process suitable for deeper draws and to alleviate the wrinkling problems common to the Guerin process.

The advantages of the rubber pad forming processes compared to conventional forming processes are the following:

a) For forming a workpiece, only one part of the tool (punch or die) is necessary.
b) One rubber pad or diaphragm takes the place of many different shapes, thicknesses, and kinds of tools, returning to its original shape when the pressure is released.
c) Tool material is low cost and easy to machine.
d) No tool marks are created during forming, so parts with very fine surfaces can be formed.
e) Set-up time is usually shorter than in conventional forming operations.

However, these processes also have some disadvantages:

a) The rubber pad and diaphragm have limited lifetimes.
b) The production rate is relatively slow.
c) Rubber pads or diaphragms exert less pressure than conventional die, resulting in less sharply formed workpieces that usually need some hand finishing.

## Superplastic Forming and Diffusion Bonding

Conventional metals and alloys will extend in tension no more than 120 percent, regardless of the temperature or speed with which the metal is pulled. However, it has been known since the 1920s that some materials could endure enormous tensile strains without necking. This phenomenon, called "superplasticity," has been scientifically investigated. In the beginning, activities were primarily concentrated in research laboratories and were entirely directed towards the exploration of basic material science. The materials investigated appeared not to be sufficiently attractive for real production. But this changed with the development of supersonic aircraft with high requirements for power density and skin temperature.

A general definition of the term "superplasticity" was formulated for the first time in 1991 during the International World Conference on Superplasticity of Advanced Materials: "Superplasticity is the ability of a polycrystalline material to exhibit, in a generally isotropic manner, very high tensile elongations prior to failure."

Some materials developed for superplastic forming include

- titanium (Ti-6Al-N)
- aluminum (2004, 2419, 7475)
- aluminum-lithium (2090, 2091, 8090)
- stainless steel (2205 series).

In general, the alloys chosen for superplastic forming should have a grain size below 10 microns in diameter. The grain size must not increase if it is kept at temperatures 90 percent of melting for a few hours, and the alloys must have strain rate sensitivity parameters of $0.35 < m <= 0.85$.

High strain rate sensitivity is necessary for reducing the rate of flow localization, i.e., necking. A low rate of damage accumulation, e.g., cavitation, is necessary to allow large plastic strains to be reached.

**Superplastic Forming.**—Superplastic forming (SPF) is a metal forming process that takes advantage of the high extendability of certain materials in order to form components whose shapes might be otherwise very difficult to obtain. Today, superplastic forming of titanium and aluminum alloys is a standard industrial practice accepted worldwide. Due to the high temperatures and simultaneously relatively low gas pressure (typically less than 200 psi), superplastic forming has found widespread application. Because of titanium's (and some other alloys') high affinity with oxygen and hydrogen, inert gases are exclusively used as the pressure medium. Fig. 29 schematically illustrates the SPF process in four steps.

Typically, the closed die and the blank with created seal are heated to the same temperature in a special hydraulic hot-press (Step 1). Inert gas pressure is introduced at a controlled rate using a sophisticated gas management system until the sheet is fully formed against the die surface (Steps 2 and 3). Each workpiece's geometry is unique and requires a unique pressure/time profile to maintain the appropriate strain rate. As the sheet thins, it requires less forming pressure. However, as the workpiece radius decreases, more pressure is required.

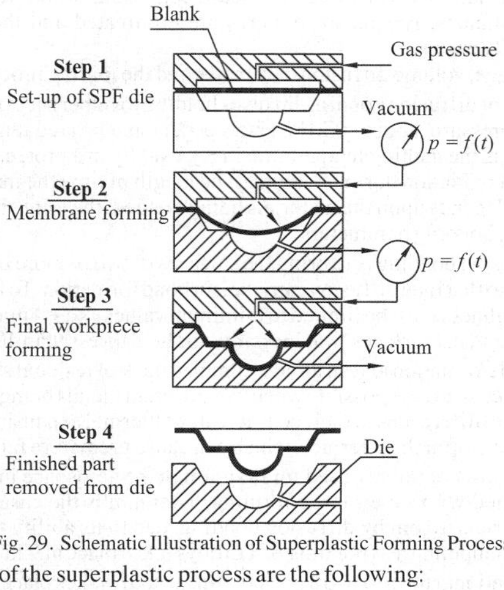


Fig. 29. Schematic Illustration of Superplastic Forming Process

The advantages of the superplastic process are the following:

a) Using one of the alloys with superplastic capabilities means getting to the finished product in fewer steps, which means lower costs.

b) Products that are usually formed in several parts separately can be integrally formed as a single part through superplastic forming. The single piece forming minimizes the number of parts and joints, and thus leads to weight savings.
c) Since the forming needs only the female die, the investment cost for the die is reduced.
d) Little or no residual stress develops in the formed parts.
e) Superior transferability of the die surface to the workpiece is provided.
f) Material waste is minimized.

Disadvantages are these:

a) There is low productivity because of low strain rates, typically $10^{-4}$ to $10^{-2}$/s
b) Material costs are high.

Although the process is increasingly being applied in the aerospace and automotive industries as a way of manufacturing very complex geometries at a fraction of the cost of conventional stamping, some practical problems are still of concern, the main ones being predicting the final thickness distribution of the formed parts, determining the optimum pressure cycles, and learning more about the microstructure of superplastic material and how it changes during such dramatic elongation.

**Diffusion Bonding.**—The International Institute of Welding (IIW) has accepted the definition of solid state diffusion bonding proposed by Kazakov. This definition is: "Diffusion bonding of materials in the solid state is a process for making a monolithic joint through the formation of bonds at atomic level, as a result of closure of plastic deformation at elevated temperature, which aids interdiffusion at the surface layers of the materials being joined."

The process is dependent on a number of parameters, such as time, applied pressure, bonding temperature, and the method of heat application. The process allows bonding of homogeneous or heterogeneous materials. Hence, structures can be manufactured from two or three metal sheets. Diffusion bonding generally occurs in three stages:

a) The deformation process results in the surfaces to be joined coming into intimate contact, but not enough to produce gross deformation.

b) Bonds are formed by diffusion-controlled mechanisms where the diffusion grain boundary predominates. At this stage, pores are eliminated and the grain boundary arrangement finally ensues.

c) In the third stage, volume diffusion dominates and the joining process is completed.

The mechanism of diffusion bonding involves holding together sheet metal components under moderate pressure, about 10 MPa (1450 psi), at an elevated temperature of (0.5–0.8) $T_m$ (where $T_m$ is the melting temperature in K), usually in a protective atmosphere or vacuum to protect oxidation during bonding. The length of time the materials are held at this temperature depends upon the materials being bonded, the joint properties required, and the remaining bonding parameters.

The aim in diffusion bonding is to bring the surfaces of two or more pieces being joined sufficiently close so that interdiffusion can result in bond formation. To form a high quality bond, surface roughness must be limited to minimum values (Ra < 4 microns); cleanliness must be absolute; and flat surfaces' waviness must be held to less than 400 microns.

A minimum of deformation and an almost complete lack of residual stresses are characteristics of the process, except possibly when two different metals being diffusion-bonded together have large differences in their coefficients of thermal expansion (CTE). This can cause strains to develop at the interface, which can cause premature failure of the bond.

The process is most commonly used for titanium in the aerospace industry, and sometimes it is combined with superplastic forming. Titanium is the easiest of all common engineering materials to join by diffusion bonding, due to its ability to dissolve its own oxide at bonding temperatures (bonding of Ti alloys takes place at 925°C).

The more conventional form of diffusion bonding usually takes place in a uniaxial loading press. Pressure and heat can be applied by different means. More complex geometries than are possible by the uniaxial process can be handled by hot isocratic pressing, which involves the application of high-temperature, high-pressure argon gas to components. A hot isostatic press consists of a furnace within a gas pressure vessel. The components must be encapsulated in a sealed can to prevent the gas from entering the site of the bond.

The advantages of diffusion bonding include the following:

a) Limited microstructural changes.
b) The ability to join dissimilar alloys.
c) The ability to fabricate very complex shapes, especially using superplastic forming.
d) Minimal deformation.
e) A highly automated process that does not require highly skilled workers
f) The ability to produce high quality joints so that neither metallurgical discontinuities nor porosity exist across the interface.
g) Diffusion bonding is free from ultraviolet radiation and gas emission, so there is no direct detrimental effect on the environment and health and safety standards are maintained.

Disadvantages:

a) Slowness of the process.
b) Protective atmosphere required.
c) Expensive equipment.
d) Smooth surface finish requirements.
e) Need for exceptional cleanliness.

*Electron beam diffusion bonding* is a variant of diffusion bonding in which only the interface region is heated, resulting in considerable energy savings. The heating source is an electron beam that is swept over the area of the joint at such a speed that fusion of the titanium or aluminum alloy is prevented. A force is applied across the joint. As the heated area is very limited, higher forces can be used without the risk of plastic collapse of the components being bonded, resulting in a significant reduction of bonding time.

**Combined DB/SPF.**—Diffusion bonding (DB) is often combined with superplastic forming (SPF) in the manufacture of complex structures in aerospace industry. This process is probably the most spectacular near-net shape process that has been developed specifically within the aerospace industry, and its industrial importance is such that it should be considered separately. The process is used commercially for titanium and its alloys.

The most common DB/SPF approaches are 2-sheet (hat-stiffened structures), 3-sheet (truss-core structures) and 4-sheet (rib-stiffed structures).

The processing conditions for superplastic forming and diffusion bonding are similar, both requiring an elevated temperature and benefiting from the fine grain size. Therefore, these two processes have been combined into one manufacturing process known as DB/SPF, which produces parts of greater complexity than sheet forming alone can. Fig. 30 schematically illustrates how combining diffusion bonding with superplastic forming creates a more complicated part shape in the same die.

The combined DB/SPF process generally occurs in two steps:

*Step 1:* The sheets have a stop-off material, such as boron nitride, placed on them in locations where no bonding is to occur in order to prevent diffusion bonding. Sheets are put down in layers, with stop-off areas into the die, heated together at an elevated temperature, and then bonded together by the use of pressure.

*Step 2:* In the same die, the SPF process is used to shape the outside of the laminated sheets. Pressure is applied by blowing gas between the sheets, usually in two phases. In the first phase, gas pressure is applied to cause a plastic stretching of the sheets, which eventually contact the die cavity and take a shape like that of the membrane. In the second phase, the pressure is increased to make the final shape of the part. This process generates a part that is very well bonded in the required locations.

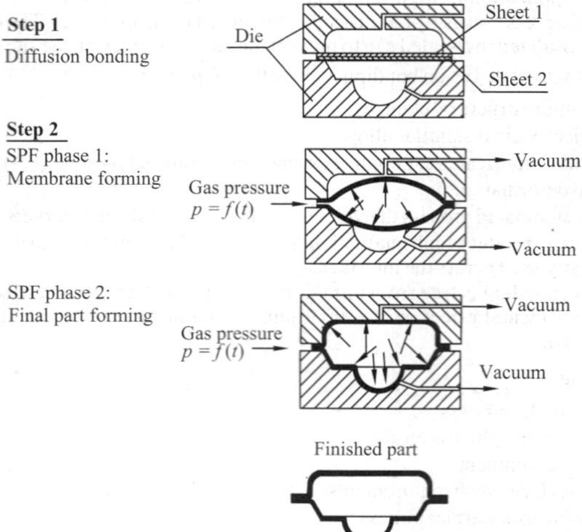


Fig. 30. Schematic Illustration of DB/SPF Process

However, this process also has its challenges. One of these involves how to apply the stop-off material in the proper location using the most cost-effective process. Historically, silk screening has been used to define the required pattern for the stop-off material. This process requires several pieces of equipment, including a wash booth, since the screen needs to be cleaned after each part. A masking paper and laser scribing process has also been developed for defining the stop-off pattern.

There are a number of commercial applications of superplastic forming and diffusion bonding, including aerospace, ground transportation, and numerous miscellaneous other uses. Examples are wing access panels in the Airbus A310 and A320; bathroom sinks in the Boeing 737; turbo fan engine-cooling duct components in the B-1, the T-38, the C-17, and the F-15E; and external window frames in the space shuttle.

## High-Energy Rate Metal Forming Processes

The term "high-energy rate forming processes" (HERF) refers to dynamic metal forming processes that form workpieces at very high velocities and extremely high pressure. HERF processes involve a short, sharp forming energy input usually of microsecond duration that is transmitted to the workpiece surface through a medium such as air or water. The resulting shockwave accelerates the workpiece to high velocity and, with its significant kinetic energy, impacts the die, which has the desired shape of the finished part.

High-energy rate metal forming was studied fairly extensively as early as the 1950s. Several processes have been developed, including explosive forming and two-capacitor, discharge-based forming methods; they are *electrohydraulic* and *electromagnetic forming*.

**Explosive Forming.—**Explosive forming is a manufacturing process that uses explosions to force sheet metal into dies. In this method, the explosive charge is located at some predetermined distance from the workpiece, and the energy is transmitted through an intervening medium such as air, oil, or water. The maximum pressure at the workpiece may range from a few thousand psi to several hundred thousand psi, depending on the parameters of the operation. Fig. 31 shows a typical explosive forming operation.

The workpiece is clamped and sealed over the die cavity. A vacuum is then created in the die cavity. The die assembly is put together at the bottom of the tank. The explosive charge is placed in the intervening medium at a certain distance above the workpiece. After the detonation of the charge, the liquid buffer is instantaneously converted from a fluid of low density, low temperature, and low pressure to a fluid of high density, high temperature, and high pressure, causing the rapid forming of the blank into the cavity die. The extremely high forming velocity of explosive forming minimizes material springback, but it does not completely eliminate it. If the relative elongation of material is more than 10 percent, it can be formed without the chance of fracturing.

The distance between the charge and the blank is called the "standoff distance." The standoff distance and the amount of charge determine the amount of pressure transmitted to the workpiece. Other factors, such as the explosive type, the explosive's shape, and the type of buffer medium, also affect the pressure.

An understanding of the compression waves and rarefactions developed in detonation is extremely important in predicting the forming metal's reaction. In the aviation and aerospace industries, the explosive method of forming has been used since the 1980s.

*Explosive:* Explosives are substances that undergo rapid chemical reaction, during which heat and large quantities of gaseous products are evolved. Explosives can be solid (TNT-trinitro toluene), liquid (nitroglycerine), or gaseous (oxygen-acetylene mixtures).

*Die Material:* Different materials are used for the explosive forming process. For instance: kirksite and fiberglass are used for low pressure and few parts; epoxy and concrete are used for low pressure and larger numbers of parts; and ductile iron, tool steel, and cast steel are used for high pressure and many parts.

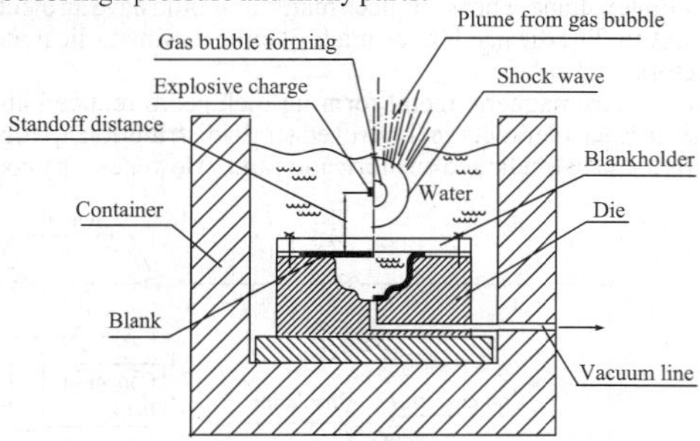


Fig. 31. Explosive Forming Set-Up

Explosive forming has the following characteristic features: a) very large sheets with relatively complex shapes, although usually axisymmetric; b) low tooling costs, but high labor costs; c) long cycle times; d) suitability for low-quantity production; e) maintenance of precise tolerances; and controllable smoothness of contours.

An exceptionally interesting development in the field of explosive method forming is the fabrication of spherical pressure vessels with diameters up to 13 ft, and thickness from 0.12 to 0.79 in. of high strength steel designed for water or oil storage in the chemical industry, and other uses.

**Electromagnetic Forming.—**Electromagnetic metal forming (EMF) is a high-energy rate cold forming process that can deform metal workpieces without contact. This process uses ultrastrong pulsed magnetic fields to form metal parts rapidly. A capacitor bank is suddenly discharged by means of a switch through a coil in which the workpiece is placed.

Basically, whenever an electrical current is rapidly sent through an electrical conductor, it will develop a magnetic field. This change in magnetic field will induce eddy currents in the workpiece that generally run in a direction opposite to the current in the coil (Lenz's law). The two antiparallel currents repel each other, supplying energy to the workpiece in the form of kinetic energy, which accelerates in the workpiece up to a certain velocity, such as 650 to 1000 ft/s. This kinetic energy drives the material into the die, causing forming on impact. This method is quite useful for general applications and suitable for any workpiece made from a good conductor, provided the current pulse is of a sufficiently high frequency.

There are two very broad ways in which this technique can be employed:

*Radial forming,* in which a round part such as a tube or ring is compressed or expanded. The forming can be done either inward or outward onto a die to give the tube a more complex shape (Fig. 32). One of the most common applications of electromagnetic forming is the compression crimp sealing and assembly of axisymmetric components such as automotive oil filter canisters.

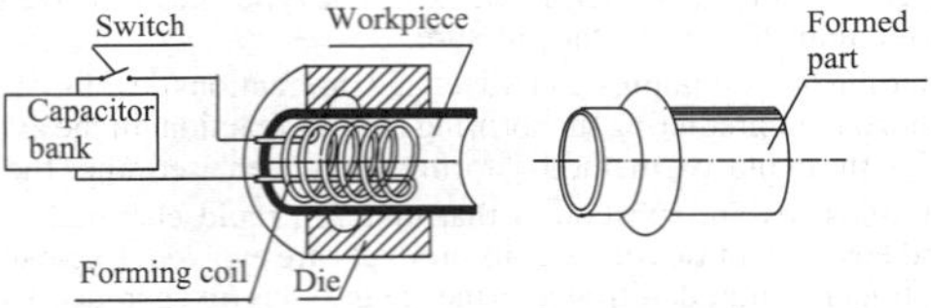


Fig. 32. Schematic Illustration of Electromagnetic Radial Forming of Tube

*Sheet metal forming,* in which using the flat coil configuration shown in Fig. 33 forms a sheet metal. The velocity of the workpiece is sufficient to cause its impact against a die to give it a more complex shape. The workpiece material should have an electrical resistance of less than 38 μΩ/in. The die needs to be made of either nonmetallic materials or of poor electric conductors.

Advantages of electromagnetic metal forming include: a) reduced number of operations needed; b) narrow tolerances; c) improved strain distribution; d) high repeatability; e) high productivity; f) less reliance on lubricants; and g) lower energy cost.

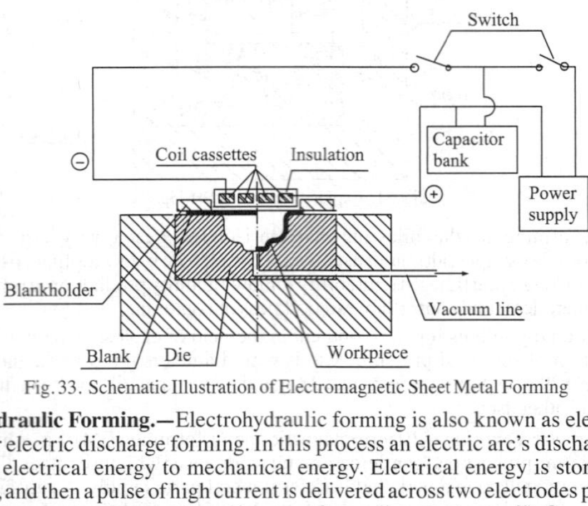


Fig. 33. Schematic Illustration of Electromagnetic Sheet Metal Forming

**Electrohydraulic Forming.—**Electrohydraulic forming is also known as electric spark forming or electric discharge forming. In this process an electric arc's discharge is used to convert electrical energy to mechanical energy. Electrical energy is stored in large capacitors, and then a pulse of high current is delivered across two electrodes positioned a short distance apart while submerged in a transfer medium (water or oil). Capacitor banks have typically stored 55 to 58 BTU at a charged voltage of 20 kV. This creates a sudden release of steam, which, along with ionization, causes the development of a high pressure shock wave within the transfer medium. The die cavity containing the blank to be formed is immersed in the tank as well. When exposed to the shock wave, the blank is forced to take on the shape of the die. A schematic illustration of the process is shown in Fig. 34.

Electrohydraulic forming is a hybrid between explosive forming and electromagnetic forming. The liquid-based shock is very similar to what would be produced by the explosive method. However, this method uses essentially the same equipment (capacitor banks) that produces the current in electromagnetic forming. Electrohydraulic forming methods are adapted for use in the production of smaller part sizes. On the other hand, these methods are more favorable for automation because of the fine control of energy discharges and the compactness of the system.

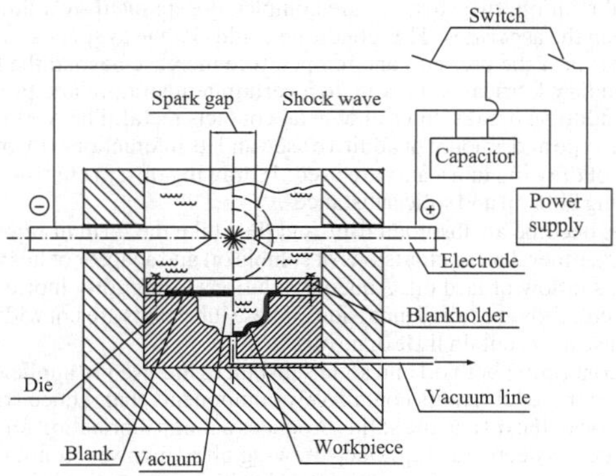


Fig. 34. Electrohydraulic Forming Set-Up

## Lubricants and Their Effects on Press Work

Most sheet metal forming operations use lubricants to protect the die and part from excessive wear caused by scratching, scoring, welding, and galling. The physical characteristics of the lubricant and metal-forming operation involved determine the application method to be used.

Methods for applying lubricant to sheet metal include dips, swabs, brushes, wipers, rollers, or recirculation. Of these, the three most common are the following: 1) manually wiping lubricant onto a surface with a rag; 2) roll coating, during which metal blanks pass through rollers that apply the compound; and 3) flooding, during which tooling and metal sheets are drenched with lubricant and the excess liquid is recovered via a filtration and recirculation system.

**Lubricants for Blanking Operations.—**Blanking dies used for carbon and low-alloy steels are often run with only mill lubricant, but will last longer if lightly oiled. Higher alloy steels and stainless steels require thicker lubricants. Kerosene is usually used with aluminum. Lubricant thickness needs to be about 0.0001 inch (0.0025 mm). During successive strokes, metal debris adheres to the punch and may accelerate wear, but damage may be reduced by application of the lubricant to the sheet or strip. High-speed blanking may require heavier applications of lubrication. For sheets thicker than 1/8 inch (3.18 mm) and for stainless steel, high-pressure lubricants containing sulfurs and chlorines are often used.

**Lubricants for Drawing Operations.—**Shallow drawing and forming of steel can be done with low-viscosity oils and soap solutions, but during deep drawing, different lubrication requirements exist, from hydrodynamic lubrication in the blank holder to boundary lubrication at the drawing radius, where breakdown of the film very often occurs. Characteristic of deep drawing is the high pressure involved in the operation, on the order of 100,000 pounds per square inch (690 MPa). To deal with such force, the choice of lubricant is critical to the success of the operation. Under such pressure, the drawing lubricant should cool the die and the workpiece, provide boundary lubrication between the die and the workpiece, prevent metal-to-metal adhesion or welding, and cushion the die during the drawing operation.

Lubricants work by forming lubricating films between two sliding surfaces in contact with each other. When these metal surfaces are viewed under magnification, peaks and valleys become apparent, even on finely-ground surfaces.

The lubricating film needs to prevent the asperities (peaks) on the two surfaces in sliding contact with each other from damaging the mating surface. Under hydrodynamic or full-film lubrication, two surfaces are completely separated by a fluid film, with no contact between the asperities. This condition could change as speeds vary during start-and-stop modes or if the pressure and temperature increase beyond the lubricant's film strength. Boundary lubricants work up to a certain temperature and pressure, and then the boundary additive breaks down and metal contacts metal. The working temperature varies with the type and amount of additive used and its interaction with other additives.

Three types of drawing lubricants are used: 1) drawing oils; 2) emulsions; and 3) lubricants containing both oil and solid substances.

Drawing oils become an absorbed film, and they take the form of light or soluble oils such as straight mineral oil or emulsions of soluble oil and soap, or of heavy oils, fats, and greases such as tallow or lard oil. Aqueous solutions of non-oily lubricants containing some suspended solids are called emulsions. These lubricants are not widely used in deep drawing because they contain little or no oil.

Lubricants containing both oil and solid substances are used in applications involving severe drawing; these lubricants contain oily components that reduce friction and heat. The combination of the oil and the solids produces enough lubrication for severe drawing applications such as deep drawing. Deep drawing often involves ironing or thinning the wall by up to 35 percent, and lubricant containing high proportions of chemically-active components. Dry soaps and polymer films are frequently used for these purposes. Aluminum can be shallow drawn with oils of low to medium viscosity, and for deep drawing, tallow may be added, as well as wax or soap suspensions for very large reductions.

**Lubricant Removal.**—Removing lubricant from a formed part after the deep drawing operation is important because any lubricant left behind can interfere with subsequent steps in the manufacturing of the part. Mineral oils, animal fat, and vegetable oils can be removed with an organic solvent by emulsification or saponification, or with an aqueous alkaline cleaner. Greases can also be removed from sheet metal with an organic solvent or an alkaline cleaner. Solids are more difficult to remove because they are not readily soluble. The presence of solids often requires that additional cleaning methods be used. Petroleum oils can raise special issues from removal through disposal. These oils require the use of alkaline cleaners for removal, which can then contaminate cleaner tanks with oil, leading to potential disposal challenges. Vegetable oils can be removed with hot water if the parts are cleaned immediately, and with mildly to moderately alkaline cleaners if the parts have been left standing for a few days.

## Joining and Edging

A duct system is an assembly whose main function is to convey air. Elements of the duct system are sheets, transverse joints, longitudinal seams, and reinforcements. The sheets must be able to withstand deflection caused by both internal pressure and vibration due to turbulent air flow. Transverse joints must be able to withstand 1.5 times the maximum operating pressure without failure. Transverse joint designs should be consistent with the static pressure class, sealing requirements, materials involved, and support interval distances. Notching, bending, folding, and fit up tolerances shall be appropriate for the proper class. Longitudinal seams also must be able to withstand 1.5 times the operating pressure without deformation. Seams must be formed and assembled with proper dimension and proportion for tight and secure fit up. Seams may be a butt, corner, plug, or spot weld design. Seam types must be selected based on material and pressure. A duct section between adjacent hangers must be able to carry its own weight and to resist the external loads for which it is constructed. Reinforcing members must be able to resist the external deflection of the sheet and their own deflection.

There is a relationship between duct width, reinforcement spacing, reinforcement size, pressure, and sheet thickness. For constant pressure and constant duct size, the thicker sheet allows more distance between reinforcements. The higher the pressure the shorter the spacing between reinforcements. Joints and intermediate reinforcements are labor-intensive and may be more costly than the savings gained by a reduction in wall thickness. Thicker duct wall and stronger joints are more cost-effective than using more reinforcement. The following material illustrates various joint designs, used both in duct work and other sheet metal assemblies.

## Sheet Metal Joints

*Plain Lap and Flush Lap:*

Fig. 35. Plain Lap

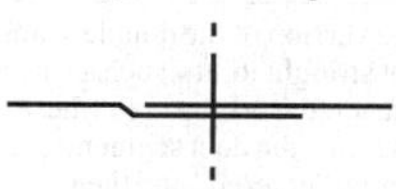

Fig. 36. Flush Lap

The *plain lap* (Fig. 35) and *flush lap* (Fig. 36) are both used for various materials such as galvanized or black iron, copper, stainless steel, aluminum, or other metals, and may be soldered, and/or riveted, as well as spot, tack, or solid-welded. Lap dimensions vary with the particular application, and since it is the duty of the drafter to specify straight joints in lengths that use full-sheet sizes, transverse lap dimensions must be known.

*Raw and Flange Corner:*

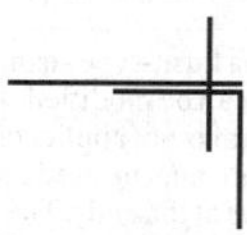

Fig. 37. Raw and Flange Corner

The *raw and flange corner* (Fig. 37) is generally spot-welded, but may be riveted or soldered. For heavy gages it is tack-welded or solid-welded.

*Flange and Flange Corner:*

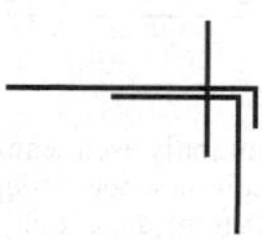

Fig. 38. Flange and Flange Corner

The *flange and flange corner* (Fig. 38) is a refinement of the raw and flange corner. It is particularly useful for heavy-gage duct sections that require flush outside corners and must be field-erected.

*Standing Seam:*

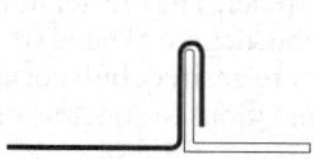

Fig. 39. Standing Seam

The *standing seam* (Fig. 39) is often used for large plenums, or casings. Before the drafter is able to lay out a casing drawing, the seam allowance measurements are needed, so that panel sizes can be detailed for economical use of standard sheets. Considering velocity levels, standing seams are considered for duct interiors: 1-inch (25.4-mm) seam is normally applied for duct widths up to 42 inches (1066.8 mm) and 1½-inch (38.1-mm) seam is applied for bigger ducts.

*Groove Seam:*

Fig. 40. Groove Seam

The *groove seam* (Fig. 40) is often used for rectangular or round duct straight joints, or to join some sheets for fittings that are too large to be cut out from standard sheets. It is also known as the pipelock, or flat lock seam.

*Corner Standing Seam:*

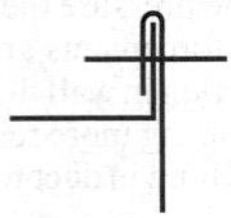

Fig. 41. Corner Standing Seam

The *corner standing seam* (Fig. 41) has applications similar to the standing seam, and also can be used for straight-duct sections. This type of seam is mostly applied at the ends at 8 inches (203 mm) intervals.

*Double Seam:*

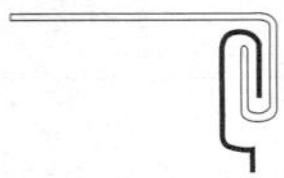

Fig. 42. Double Corner Seam

The *double corner seam* (Fig. 42) at one time was the most commonly used method for duct fabrication. However, although it is seldom used because of the hand operations required for assembly, the double seam can be used advantageously for duct fittings with compound curves. It is called the slide lock seam. Machines are available to automatically close this seam.

*Slide-Corner:*

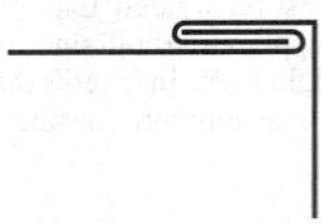

Fig. 43. Slide Corner

The *slide-corner* (Fig. 43) is a large version of the double seam. It is often used for field assembly of straight joints, such as in an existing ceiling space, or other restricted working area where ducts must be built in place. To assemble the duct segments, opposite ends of each seam are merely "entered" and then pushed into position. Ducts are sent to job sites "knocked-down" for more efficient use of shipping space.

*Button Punch Snap Lock:*

Fig. 44. Button Punch Snap Lock

The *button punch snap lock* (Fig. 44) is a flush-type seam that may be soldered or caulked. This seam can be modified slightly for use as a "snap lock." This type of seam is not applicable for aluminum or other soft metals. This seam may be used up to 4 inches (101.6 mm) w.g. by using screws at the ends. The pocket depth should not be smaller than ⅝ inch (15.88 mm) for 20-, 22- and 26-gage material (0.91- 0.45 mm).

*Pittsburgh:*

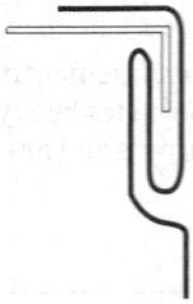

Fig. 45. Pittsburgh

The *Pittsburgh* (Fig. 45) is the most commonly used seam for standard-gage duct construction. The common pocket depths are $\frac{5}{16}$ inch (7.94 mm) and ⅝ inch (15.88 mm), depending on the thickness of the sheet.

*Flange:*

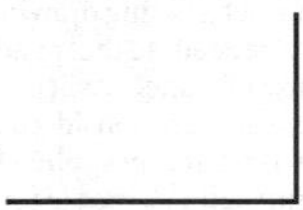

Fig. 46. Flange

The *flange* (Fig. 46) is an end edge stiffener. The drafter must indicate size of flange, direction of bend, degree of bend (if other than 90°) and when full corners are desired. Full corners are generally advisable for collar connections to concrete or masonry wall openings at louvers.

*Hem:*

Fig. 47. Hem

The *hem edge* (Fig. 47) is a flat, finished edge. As with the flange, this hem must be designated by the drafter. For example, drawing should show: ¾ inch (19 mm) hem out.

*Flat Drive Slip:*

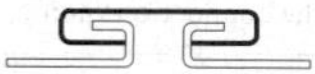

Fig. 48. Drive Slip

The drive slip is one of the simplest transverse joints. It is applicable where pressure is less than 2 inches (50.8 mm) w.g. This is a slide-type connection generally used on small ducts in combination with "S" slips, but should not be used above 2 inches w.g.

*Standing Drive Slip:*

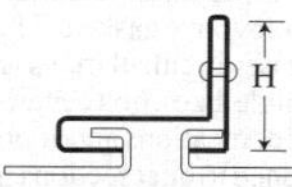


Fig. 49. Standing Drive Slip

This slip is also a slide type connection. It is made by elongating the flat drive slip and fastening standing portions 2 inches (50.8 mm) from each end. The design is applicable for any length in 2 inches w.g, 36 inches (914.4 mm) for 3 inches (76.2 mm) w.g., and 30 inches (762 mm) at 4 inches w.g. service.

*Flat Drive Slip Reinforced:*

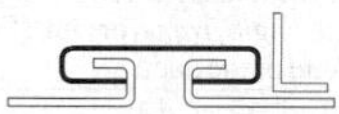

Fig. 50. Drive Slip Reinforced

This reinforcement on the flat drive slip is made by adding a transverse angle section after a fixed interval.

*Double "S" Slip Reinforced:*

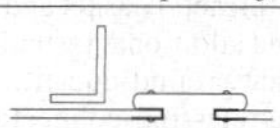

Fig. 51. Double "S" Slip

The double "S" slip is used to eliminate the problem of notching and bending, especially for large ducts. Use 24 gage sheet for 30 inches (762 mm) width or less, and 22 gage sheet over 30 inches width. (22 gage = 0.76 mm, 24 gage= 0.60 mm)

*Flat "S" Slip:*

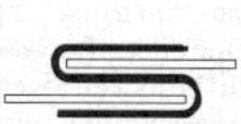

Fig. 52. Plain "S" Slip

Normally the "S" slip is used for small ducts. However, it is also useful if the connection of a large duct is tight to a beam, column or other object, and an "S" slip is substituted for the shop standard slip. Service above 2 inches (50.8 mm) w.g. is not applicable. Gage shall not be less than 24, and shall not be less than the duct gage. When it is applied on all four edges, fasten within 2 inches of the corners and at 12-inch (304.8 mm) maximum intervals.

*Hemmed "S" Slip:*

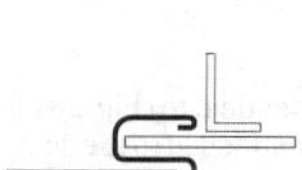

Fig. 53. Hemmed "S" Slip

This modified "S" slip is made by adding hem and an angle for reinforcing. The hem edge is a flat and finished edge. Hemmed "S" slip is mostly applied with angle. The drive is generally 16 gage (0.76 mm), forming a 1-inch height slip pocket and screws at the end. Notching and bending operations on "S" slip joints can be cumbersome and costly, especially for large sizes. Tie each section of the duct within 2 inches (50.8 mm) from the corner at maximum 6-inch (152.4-mm) intervals.

## Other Types of Duct Connections

*Clinch-Bar Slip and Flange:*

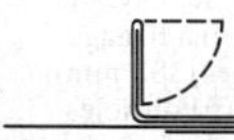

Fig. 54. Clinch-Bar Slip and Flange

The *clinch-bar slip and flange* (Fig. 54) uses the principle of the standing seam, but with a duct lap in the direction of airflow. These slips are generally assembled as a framed unit with full corners either riveted or spot-welded, which adds to the duct cross section rigidity. Reinforcement may be accomplished by spot welding the flat-bar to the flange of the large end. Accessibility to all four sides of the duct is required because the flange of the slip must be folded over the flange on the large end after the ducts are connected.

*Clinch-Bar Slip and Angle:*

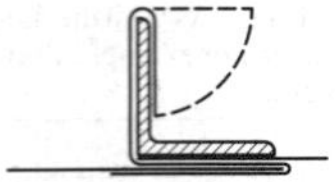

Fig. 55. Clinch-Bar Slip and Angle

The *clinch-bar slip and angle* (Fig. 55), is similar to clinch bar slip (Fig. 54), but it has a riveted or spot-welded angle on the large end. This connection can also have a raw large end that is inserted into the space between the angle and the shop-fabricated slip. Matched angles, minimum of 16 gage (1.52 mm), are riveted or spot-welded to the smaller sides of the ducts to pull the connection "home."

## Flanged Duct Connections

*Angle Frame, or Ring:*

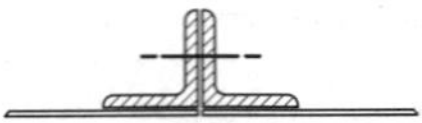

Fig. 56. Raw Ends and Matched $\angle$s

Any of the following flanged connections may have gaskets. The drafter should not allow for gasket thicknesses in calculations for running length dimensions, nor indicate angle sizes, bolt centers, etc., as these items are established in job specifications and approved shop standards. Generally, angles are fastened to duct sections in the shop. If conditions at the job site require consideration for length contingencies, the drafter should specify "loose angles" such as at a connection to equipment that may be located later. The most common matched angle connection is the *angle frame*, or *ring* (Fig. 56). The angles are fastened flush to the end of the duct.

*Flanged End and Angle:*

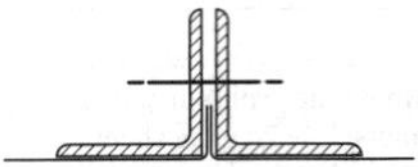

Fig. 57. Flanged Ends and Matched $\angle$s

The *flanged end and angle* (Fig. 57), is often used for ducts 16 gage or lighter, as the flange provides a metal-to-metal gasket and holds the angle frame or ring on the duct without additional fastening. The drafter may indicate in a field note that a round-duct fitting is to be "rotated as required." This type of angle-ring-connection is convenient for such a condition.

*Formed Flanges:*

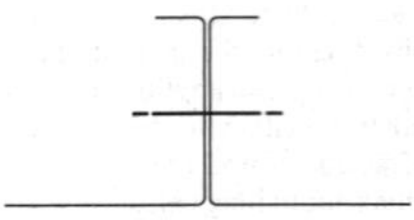

Fig. 58. Formed Flanges

*Double flanges* (Fig. 58), are similar to Fig. 46, except that the connecting flange has a series of matched bolt holes. This connection, caulked airtight, is ideal for single-wall apparatus casings or plenums. The flanges are formed at the ends of the duct; after assembly they will form a T shape. Mating flanges shall be locked together by long clips. In order to form an effective seal, gaskets with suitable density and resiliency are used. At the corners 16 gage (1.5 mm) thickness steel corners are used with $\frac{3}{8}$ inch (9.5 mm) diameter bolts.

*Double Flanges and Cleat:*

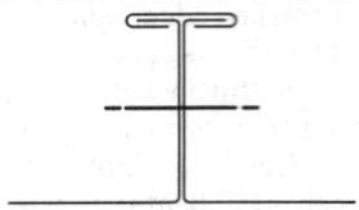

Fig. 59. Double Flanges and Cleat

*Double Flanges and Cleat* (Fig. 59) is identical to (Fig. 58), but has an air seal cleat. The reinforcements are attached to the duct wall on both sides of the joint.

*Clinch-type Flanged Connections:*

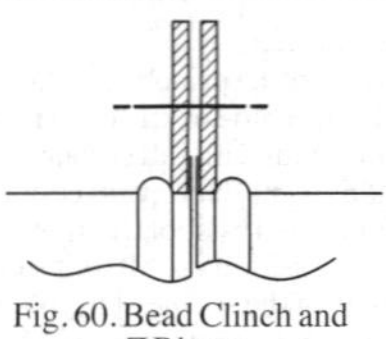

Fig. 60. Bead Clinch and Z Rings

*Clinch-type flanged connections* for round ducts, 16 gage or lighter, are shown in Fig. 60. The angles or rings can be loose, as explained in *Flanged End and Angle* (Fig. 57). The drafter should indicate flange sizes, bend direction, and type of assembly. An example such as the flange lap for a field assembly of a 10 gage (3.4 mm) casing corner would be written: $1\frac{1}{2}$ inches (38.1 mm) flange out square on side with $\frac{9}{32}$-inch (7.14-mm) ∅ bolt holes 12 inches (304.8 mm) CC. At the beginning and ending, angles are connected by rivets or welding. The bolt will be $\frac{5}{16}$ inch (7.94 mm) ∅ at 6-inch (152.4-mm) maximum spacing.

## Sheet Metal Mill Finishes

Mill finish is the natural surface texture of sheet metal after it exits a rolling mill, with no external mechanical or chemical finishing. This surface can be rough and lack sheen. It also may show oxidation or contamination with mill oil. Most mill finish surfaces are treated with polishing, industrial etching, or some other surface finishing process before being considered complete. Typical sheet metal mill finishes can be identified as:

*Unspecified:* This is the natural finish provided by hot-rolling or cold-rolling using unpolished rolls. Ranging from dull to bright, the finished surface may retain stains or a light film of oil.

*Specular as Fabricated:* Produced on one or both sides of a sheet by passing it through highly polished cold rolls, this is a mirror-like finish.

*Non-Specular as Fabricated:* Refers to a finish that is more uniform than the "unspecified" process but without the reflective properties of a "specular as fabricated" finish.

The quality and characteristics of various mill finishing processes can vary widely from one mill to another and even between lots (sets of similar parts, processed consecutively or in a short time). Hot-rolled sheets usually have dark, dull surfaces, oxidized from being hot-worked. But it is possible for a mill to influence the finish of produced stock. Carefully maintained and polished rollers can increase surface smoothness and luster; some mills will follow rolling with an annealing process to give stock a matte finish.

**Surface Finishes of Aluminum Sheets.—**A light metal, aluminum has good tensile strength over a wide range of temperatures and excellent corrosion resistance. These properties and aluminum's appearance can be improved by alloying it with other metals.

Mill finish is the natural state of the aluminum sheet as it comes from the rolling mill, with no additional finishing. When aluminum is exposed to air, a thin, glass-hard oxidized film forms on the surface. This film clings tenaciously, presenting an effective surface barrier against the elements. Generally, a rolled sheet develops a thinner oxide film than hot extruded aluminum.

With little care, aluminum sheets in mill finish state will maintain their original appearance for a long time. Surface coating can increase the effectiveness of aluminum's corrosion resistance. Types of aluminum surface treatment are shown in Table 18.

**Table 18. Types of Surface Treatment**

| Type of Treatment | Technique | Characteristics |
|---|---|---|
| Mechanical | Grinding/brushing | Produces a flat, even surface with a silk or matte appearance. Leaves fine lines in direction of grinding. |
| | Polishing and buffing | Reduces or removes grinding lines, die lines, tool marks, surface blemishes, and "stress risers." |
| | Vibration polishing | Yields a matte-to-shiny surface. Suited to small batches and mass-produced parts. |
| | High-finish polishing | Creates a mirror-like finish. Usually followed by bright anodizing. |
| Chemical | Milling | Removes surface to desired depth. Can be selectively applied for sign writing and other etched products. |
| | Etching | Leaves a matte or silky appearance. Usually a pretreatment for other finishing processes. Different etchants leave different textures that may show in anodizing sheen. |
| | Brightening | Removes surface texture for a mirror-like finish. Residual film is usually less than 0.000591 in. or 15 μm. |
| | Chromating/phosphating | Pretreatment applied prior to powder coating or other coating processes. |
| | Electrochemically plating with copper, tin, nickel, or silver | Improves corrosion resistance. Can provide a good soldering surface, conductivity, and reflectivity. |

**Table 18.** *(Continued)* **Types of Surface Treatment**

| | | |
|---|---|---|
| Electrochemical | Decorative anodizing | Deposits a hard, natural, or colored oxide layer up to 0.0009842 in. or 25 μm. Mainly for decoration, resists pitting and corrosion, while adding wear resistance and electrical insulation. |
| | Hard anodizing | Adds a gray to brown hard oxide layer of 0.0029528 to 0.0049213 in. or 75 to 125 μm. Mainly used for engineering purposes to resist abrasion and wear. |
| | Electrolytic polishing | Creates a smooth surface with high reflectivity. |
| Organic surface coating | Powder coating | Provides various degrees of protective and decorative finishes that can withstand heavy chemical attack (if surface coating remains unbroken). |
| | Screen printing | Used for printing text, decorative patterns, and so on. |
| | Coating with protective foils | Added for decoration, protection, or other properties. |
| | Wet coating | Used in capital-intensive operations for coil coating. Effective, efficient, and common in the building industry. |

*Mechanical Finishes:* Mechanical finishing is done before chemical finishing or anodizing to remove surface defects and irregularities and/or add desired texture. While the resulting surface appearance can vary, most mechanical processes result in a directional or nondirectional sanded finish, uniform matte finish, or a specular (mirror-like) finish. The mechanical finish will show through anodic oxides, because uncolored anodic oxides are transparent or translucent; anodized finishes also conform to surface textures.

Grinding requires a rotary grinder with a bonded adhesive cup wheel or canvas wheel, faced with suitable abrasives. Low-speed grinding at 59 in/min (1,500 mm/min) using aluminum oxide is preferred to avoid overheating the surface.

Machine polishing involves wheels or belts with abrasives either bonded to them or sprayed on at regular intervals. Methods include both contact and mush polishing (where only the polishing compound touches the metal surface), the latter being more common. Polishing can include final buffing to remove emery marks.

*Chemical Finishes:* Many types of chemical treatment are used on aluminum sheets as finishing processes, but few are final finishes. Most are intermediate steps in an overall finishing process, requiring subsequent application of some form of protective coating.

There are numerous methods of providing intermediate process finishes on sheet metals that involve washing or dipping fabricated product in chemical solutions; many are proprietary processes. Such treatments have various purposes but in general are used to: a) clean the surface, without otherwise affecting it; b) create a clean, matte, textured surface for applying a smooth, bright finish; or c) chemically add a film to provide a substrate for organic coatings.

*Conversion Coatings:* Although chemical conversion coatings generally are used on aluminum to prepare surfaces for painting, some can be used as a final finish. Since the natural oxide film present on aluminum surfaces does not always provide a good bond for paints, other organic coatings, and laminates, the chemical nature of the film is often converted to improve adhesion and undercoating protection. Such conversion films or coatings usually are applied using phosphate or chromate solutions, many of which are proprietary in nature.

One of the simplest methods for improving adhesion of organic coatings is etching surfaces with a phosphoric acid solution to provide a good mechanical bond. Finishes of this type, such as acid-chromate fluoride treatments, usually are governed by ASTM B449, "Specifications for Chromates on Aluminum."

*Anodic Coatings:* These economical, electrolytically formed oxides in sulfuric or phosphoric acid coatings impart excellent, comparatively thick 0.0003 to 0.002 in. (8 to 50 μm), transparent, durable, protective coatings to aluminum surfaces. Color anodic coatings are formed by: impregnating the coating with organic dyes or pigments; using alloys and processes that produce integral color; or electrolytically depositing pigments in the anodic film.

Quality control tests after anodizing are performed on representative samples taken at random from rack loads to determine whether the coating meets the specified thickness and weight and is properly sealed. Test methods for checking thickness and weight are specified in: 1) ASTM B244, "Test Method for Measurement of Thickness of Anodic

Coatings on Aluminum and of Other Nonconductive Coatings on Nonmagnetic Basis Metals with Eddy Current Instruments": This nondestructive method determines coating thickness using the lift-off effect of a probe coil, which contacts the coating and generates measurable eddy currents in the aluminum substrate; 2) ASTM B137, "Test Method for Measurement of Weight of Coating Mass per Unit Area on Anodically Coated Aluminum": This test involves weighing a small sample of known surface area on an analytical balance, dissolving the coating in a solution that does not attack aluminum, and then re-weighing the specimen and comparing weights.

**Surface Finishes of Stainless Steel Sheet.—**Finishes for stainless steel include standard mill and polished, proprietary, electropolished, blasted, rolled, textured, colored, or etched, or any combination of these.

Specific finishes for stainless steel sheet and strip products are defined in ASTM A480/ ASTM A480M, "Standard Specification for General Requirements for Flat-Rolled Stainless and Heat-Resisting Steel Plate, Sheet, and Strip" (see Table 19). These finishes are identified by a system of numbers and letters: unpolished or "mill" finishes are No. 1, No. 2D, and BA for bright annealed (usually 2BA); polished finishes are No. 3, No. 4, No. 6, No. 7, and No. 8. Temper-rolled (TR) finishes are used when strong mechanical properties are needed. Also specified are additional architectural finishes, which may be described as No. 5 and form a separate category, as there are numerous techniques and finish variations available worldwide. Other finishes, such as embossing, coining, and special rolled finishes, also are applied mechanically; non-mechanical options include coloring and etching.

**Table 19. Standard Mechanical Finishes for Stainless Steel Sheet,** *ASTM A480/A480M*

| Finish Designation | DEFINITION | NOTES |
|---|---|---|
| | **Unpolished Finishes** (Mill Finishes) | |
| No. 1 | Hot-rolled, annealed and pickled, and descaled | This is a dull, nonreflective finish. |
| No. 2D | Cold-rolled, dull finish | A smooth, nonreflective, cold-rolled, annealed, and pickled or descaled finish. This nondirectional finish is effective in retaining lubricants in deep drawing applications. |
| No. 2B | Cold-rolled, bright finish | A smooth, moderately reflective, cold-rolled and annealed and pickled or descaled finish typically produced with a final, light, cold-rolled pass using polished rolls. This general-purpose finish is more readily polished than No. 1 or No. 2D. Product with No. 2B finish is normally supplied in annealed plus lightly cold-rolled condition, unless a tensile-rolled product is specified. |
| BA (bright annealed) | Bright, cold-rolled finish, retained by final annealing in a controlled atmosphere furnace | This smooth, bright, reflective finish is usually produced by cold-rolling, followed by annealing in a protective atmosphere to prevent oxidation and scaling during annealing. |
| | **Polished Finishes** | |
| No. 3 | Intermediate polished finish, one or both sides | A linearly textured finish, produced by either mechanical polishing or rolling, with surface roughness average generally up to 40 μin. or 1.016 μm Ra. Surface roughness differs with various instruments, laboratories, and skilled operators, who may blend this finish. There can be overlap in surface roughness measurements for No. 3 and No. 4 finishes. |
| No. 4 | General purpose polished finish, one or both sides | A linearly textured finish produced by either mechanical polishing or rolling, with surface roughness average generally up to 25 μin. or 0.635 μm Ra. Surface roughness differs with various instruments, laboratories, and skilled operators, who may blend this finish. There can be overlap in surface roughness measurements for No. 3 and No. 4 finishes. |
| No. 6 | Dull, satin finish, Tampico brushed, one or both sides | Has a soft, satin appearance, typically produced by Tampico brushing a No. 4 finished sheet. |
| No. 7 | High-luster finish | Has a high degree of reflectivity, produced by buffing a finely ground surface, but grit lines are not removed. Chiefly used for architectural or ornamental purposes. |

**Table 19.** *(Continued)* **Standard Mechanical Finishes for Stainless Steel Sheet,** *ASTM A480/A480M*

| | | |
|---|---|---|
| No. 8 | Mirror-like finish | A highly reflective, smooth finish typically produced by polishing with successively finer grit abrasives, then buffing. Faint polish lines may be visible on the final product. Blending after part assembly can be done with buffing. |
| TR (temper-rolled) | Cold-worked to obtain specified properties | Results from the cold-rolling of an annealed and descaled or bright annealed product to obtain mechanical properties higher than that seen in the annealed condition. Appearance will vary, depending on the alloy, starting finish, and amount of cold work. |

*Standard Rolled Mill Finishes:* Standard rolled mill finishes are created when a stainless steel coil is passed through a set of rolls, and the finish on the rolls is transferred to the surface of the coil.

No. 1 is the standard rolled mill finish and least expensive. Produced by hot-rolling, annealing, and descaling by pickling, it is dull with uneven color and texture. Grinding marks and similar blemishes are common. Heavy sheet and plate for structural applications are supplied with this finish. When components will be visible, it is common to apply a secondary finish, such as polishing or abrasive blasting, for a uniform appearance.

No. 2D, No. 2B, and 2BA are standard mill sheet and strip finishes. They start with a cold-rolled coil that is annealed and pickled.

No. 2D is a uniform, dull, low-reflectivity sheet and strip finish. The dull appearance is produced by pickling or descaling. This is the preferred finish for deep drawing applications, as it holds lubricants well.

No. 2B finish is more reflective, resembling a smooth, cloudy mirror. The final finish is produced by a light pass through polished rolls. Heavier passes can be used if higher tensile strengths are required. This is a common starting finish for polishing and embossing.

A 2BA finish is a bright, reflective, mirror-like surface, ranging from hazy to clear. It is used as a lower cost alternative to highly polished stainless for architectural trim, panels, and fascia, though it is less uniform. When the final annealing is done in a furnace with a controlled atmosphere, it is referred to as bright annealing (BA); to produce a 2BA finish, the bright-annealed coil is given a light pass through highly polished rolls to flatten the coil and enhance finish uniformity. Buffing can produce a consistent mirror finish; it is not unusual for buffed BA to be specified for aesthetic applications where consistency is important. 2BA also can be the starting finish for embossing, coining, selective polishing, or etching. It is slightly more expensive than a No. 2D or No. 2B finish.

For all of these finishes, the rolling direction should be marked on the back of cut panels to facilitate installing them in the same direction. Or a protective strippable film can be applied to the front, with printed arrows indicating the rolling direction. As there can be differences in appearance from one end of the coil to the other, due to roller wear, numbering panels as they are cut from the coil enables adjacent installation of consecutive panels, optimizing uniformity. If more than one coil is needed for a project, ideally all coils should come from the same producer off the same line.

*Polished Finishes:* These finishes for stainless steel are produced by using successively finer polishing belts or by passing a stainless steel coil through special rollers that simulate a No. 3 or No. 4 polish. No common stainless steel specification requires that a finish meet specific surface roughness requirements, but surface roughness influences appearance and corrosion performance. Therefore, it can be important to specify a range or maximum for surface roughness. No. 7 and No. 8 finishes are produced by polishing with successively finer belts and then buffing, with cost based on the number of operations involved and degree of smoothness desired. These finishes can be applied to sheet, strip, tube, pipe, and any of the other stainless steel shapes.

No. 3 and No. 4 polished finishes are produced by belt or wheel polishing with fine grit abrasives or by rolling, and they are characterized by very fine parallel "grit lines." When a piece is mechanically finished, No. 3 involves abrasives between 80 and 150 grit; applied to sheets or strips, this finish typically produces an average surface roughness of up to 40 μin. or 1.016 μm Ra. No. 4 is commonly used as an architectural finish and can be produced by mechanical polishing or rolling. If mechanical polishing is used, the final pass can be between 120 and 320 grit; average surface roughness on sheets or strips is typically up to 25 μin. or 0.635 μm Ra. There may be overlap between the surface roughness measurements and grit ranges for the No. 3 and No. 4 finishes. Either finish can be polished wet or dry. Wet polishing produces a smoother, finer finish.

A soft, satin, directional finish, No. 6 was used for the columnar features on the exterior walls of the Empire State Building and other early stainless steel projects. It is typically produced by Tampico brushing a No. 4 finish with an oil and abrasive mixture, for a silver-white appearance; however, the high wear rate of Tampico brushes makes it difficult for workers to obtain a uniform appearance, and this finish is known for having fingerprint problems. ASTM A480/A480M allows other polishing methods to be substituted if they produce a finish with the same surface roughness, reflectivity, and an appearance that a stainless steel surface finish expert considers visually identical. Polishing lines should be soft and the resultant finish much less reflective than a No. 4. A finished surface sample that acts as a visual standard is particularly important when specifying this finish, because a wide range of visually different finishes may be called a No. 6.

Still smoother and more reflective sheet finishes are obtained by buffing polished surfaces. The semi-bright No. 7 is polished to a finely ground surface and buffed to a mirror-like finish, with residual grit lines still visible. The highly reflective, mirror-like No. 8 is progressively polished and buffed so only faint polishing grit lines show on final products.

**Surface Finishes of Stainless Steel Strips.—**Five mill finishes are defined by ASTM A480/A480M for stainless steel strips. Within each of these finish categories, some variation is expected, due to the characteristics of different alloys and differences in metal thickness and methods of processing.

No.1 strip finish is similar to a No. 2D sheet finish, which is produced by cold-rolling to the specified thickness, followed by annealing and pickling. It varies in appearance from dull gray matte to fairly reflective.

No. 2 strip finish is produced the same way as a No. 1 finish, except that it is followed by a final cold roll pass, often on highly polished rolls. It is a smoother, more reflective finish than a No. 1 strip finish. If the rolls are highly polished, it is equivalent to a No. 2B sheet finish.

BA (bright-annealed), TR (temper-rolled), and No. 3 and No. 4 polished finishes also can be used on stainless steel strips and follow the requirements for sheet finishes. No. 5 is a separate category and may be negotiated between buyer and seller.

**Surface Films.—**When stainless steel arrives from the mill or processor, it generally has a protective passive film on the surface. As long as the surface is clean and exposed to oxygen, the passive film automatically forms when the surface is abraded or machined. If the surface is protected from contamination and scratching during fabrication, transit, and installation, and there is no welding, then chemical passivation of the surface after fabrication is not required. Washing with a mild, chloride-free detergent and water solution or an ammonia and water solution still may be necessary.

*Protective Plastic Films:* These can be applied by the producer, processor, fabricator, or finisher to protect stainless steel surfaces. The use of strippable protective plastic film during forming operations that do not require heat helps protect surfaces from contamination by other metals and light scratching and also acts as a lubricant. Film also can help prevent damage from light scratching during shipping, storage, or installation. If an appropriate protective film is specified, it can be left in place until just prior to turnover, eliminating or reducing the need for final cleaning.

Commonly used protective films are porous and do not prevent corrosion from saltwater, water containing deicing salt, swimming pool atmospheres, corrosive acids, and other substances that can be corrosive to some stainless steels. Where there is exposure to such elements, a film can even accelerate corrosion by acting as a crevice, so protective strippable films should be removed immediately after installation.

If film is applied to a component that will be stored or installed in an exterior location, a UV-rated film should be specified. Film life takes into consideration the total time a film is to be used, including potential project delays. Film left in place past its warranted life or exposed to conditions for which it was not designed can be difficult to remove, leave residual adhesive, or be a factor in corrosion.

If an improper film type is selected or if film is left in place too long, the film supplier should be contacted to identify an appropriate water-based cleaner or solvent for removal of residual adhesive, based on the type of adhesive. If a solvent is needed, rubbing alcohol or citric cleaners often are tried before resorting to more aggressive solutions.

### Sheet Metal Gauge Sizes

Gauge sizes are numbers that indicate the thickness of a piece of sheet metal. The larger the gauge number, the thinner the sheet metal specification. Gauge differs between ferrous (iron-based) metals and nonferrous metals, such as aluminum or copper. Sheet metal thickness conversion for various metals is shown in Table 20.

**Table 20. Sheet Metal Gauge Conversion Cart**

| Gauge Number | Nominal Thickness | | | | | |
|---|---|---|---|---|---|---|
| | Stainless Steel | | Mild Steel | | Aluminum | |
| | Inches | Millimeters | Inches | Millimeters | Inches | Millimeters |
| 3 | 0.250 | 6.350 | 0.2391 | 6.073 | 0.2294 | 5.823 |
| 4 | 0.2343 | 5.953 | 0.2242 | 5.695 | 0.2043 | 5.182 |
| 5 | 0.2118 | 5.556 | 0.2092 | 5.314 | 0.1819 | 4.620 |
| 6 | 0.2031 | 5.159 | 0.1943 | 4.935 | 0.1620 | 4.115 |
| 7 | 0.1875 | 4.762 | 0.1793 | 4.551 | 0.1443 | 3.665 |
| 8 | 0.1719 | 4.366 | 0.1644 | 4.176 | 0.1285 | 3.264 |
| 9 | 0.1562 | 3.967 | 0.1495 | 3.787 | 0.1144 | 2.906 |
| 10 | 0.1406 | 3.572 | 0.1395 | 3.416 | 0.1019 | 2.588 |
| 11 | 0.1250 | 3.175 | 0.1196 | 3.038 | 0.0907 | 2.304 |
| 12 | 0.1094 | 2.778 | 0.1046 | 2.657 | 0.0808 | 2.052 |
| 13 | 0.0937 | 2.381 | 0.0897 | 2.278 | 0.0720 | 1.829 |
| 14 | 0.0781 | 1.984 | 0.0747 | 1.897 | 0.0641 | 1.628 |
| 15 | 0.0703 | 1.785 | 0.0673 | 1.709 | 0.0571 | 1.450 |
| 16 | 0.0625 | 1.587 | 0.0598 | 1.519 | 0.0508 | 1.290 |
| 17 | 0.0562 | 1.429 | 0.0538 | 1.367 | 0.0453 | 1.150 |
| 18 | 0.0500 | 1.270 | 0.0478 | 1.214 | 0.0403 | 1.024 |
| 19 | 0.0437 | 1.111 | 0.0418 | 1.062 | 0.0359 | 0.912 |
| 20 | 0.0375 | 0.952 | 0.0359 | 0.912 | 0.0320 | 0.813 |
| 21 | 0.0244 | 0.873 | 0.0329 | 0.836 | 0.0285 | 0.724 |
| 22 | 0.0312 | 0.793 | 0.0299 | 0.759 | 0.0253 | 0.643 |
| 23 | 0.0281 | 0.715 | 0.0269 | 0.683 | 0.0226 | 0.574 |
| 24 | 0.0250 | 0.635 | 0.0239 | 0.607 | 0.0201 | 0.510 |
| 25 | 0.0219 | 0.556 | 0.0209 | 0.531 | 0.0179 | 0.455 |
| 26 | 0.0187 | 0.476 | 0.0179 | 0.455 | 0.0159 | 0.404 |
| 27 | 0.0172 | 0.436 | 0.0164 | 0.417 | 0.0142 | 0.361 |
| 28 | 0.0156 | 0.397 | 0.0149 | 0.378 | 0.0126 | 0.320 |
| 29 | 0.0141 | 0.357 | 0.0135 | 0.343 | 0.0113 | 0.287 |
| 30 | 0.0125 | 0.317 | 0.0120 | 0.305 | 0.0100 | 0.254 |
| 31 | 0.0109 | 0.278 | 0.0105 | 0.267 | 0.0089 | 0.261 |
| 32 | 0.0102 | 0.258 | 0.0097 | 0.246 | 0.0080 | 0.203 |
| 33 | 0.0094 | 0.238 | 0.0090 | 0.229 | 0.0071 | 0.180 |
| 34 | 0.0085 | 0.218 | 0.0082 | 0.208 | 0.0063 | 0.160 |
| 35 | 0.0078 | 0.198 | 0.0075 | 0.191 | 0.0056 | 0.142 |
| 36 | 0.0070 | 0.178 | 0.0067 | 0.170 | 0.0050 | 0.127 |
| 37 | 0.0066 | 0.168 | 0.0064 | 0.163 | 0.0045 | 0.114 |
| 38 | 0.0062 | 0.158 | 0.0060 | 0.152 | 0.0040 | 0.101 |

## METALWORKING DIES

A die is a specialized tool used in manufacturing industries to cut, shape, and form a wide variety of components and products. Dies are generally customized and uniquely matched to the form they are used to create. Products made with dies range from simple paper clips to complex pieces used in the aircraft and automotive industries, as well as other advanced technologies.

### Classification of Metalworking Dies

Dies may be classified according to a variety of elements and in keeping with the diversity of die design. According to the number of stations involved, dies for sheet metal working may be classified as: a) single-station dies (either compound dies or combination dies) or b) multiple-station dies (progressive dies and transfer dies).

**Single-Station Dies.—** Single-station dies may be compound or combination dies.

*Compound Die:* This die is used when two or more cutting operations are performed to produce a part in a single press stroke. A die that produces washers is a good example (Fig. 61).

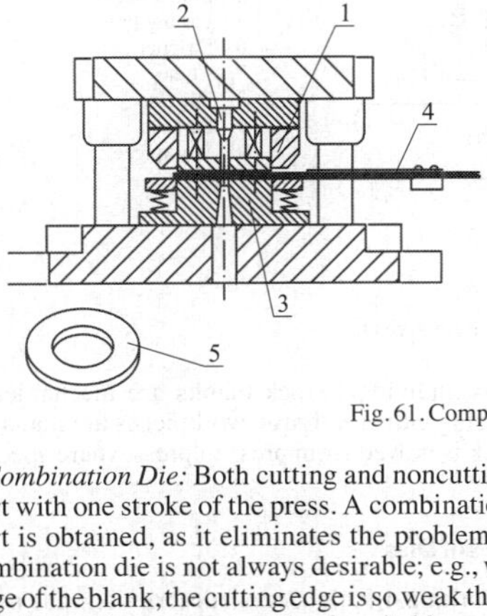


1. Blanking Die
2. Piercing Punch
3. Blanking Punch and Piercing Die
4. Sheet Metal Strip
5. Workpiece

Fig. 61. Compound Die

*Combination Die:* Both cutting and noncutting operations are performed to produce a part with one stroke of the press. A combination die is economical, and a more accurate part is obtained, as it eliminates the problem of relocating the workpiece. However, a combination die is not always desirable; e.g., when the punching hole is too close to the edge of the blank, the cutting edge is so weak that failure is likely. A die that produces cups with flanges is a good example of a suitable use of combination dies (Fig. 62).

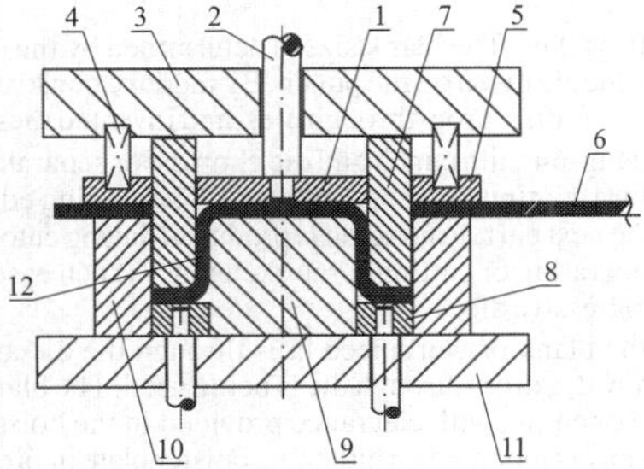


1. Knockout Plate
2. Knockout Pin
3. Shoe
4. Spring
5. Stripper Plate
6. Sheet Metal
7. Blanking Punch and Drawing Die
8. Blank Holder
9. Drawing Punch
10. Blanking Die
11. Shoe
12. Workpiece

Fig. 62. Combination Die

**Multiple-Station Dies.—**Multiple-station dies are arranged so a series of sequential operations is performed with each press stroke. The two types are progressive and transfer dies.

*Progressive Die:* A progressive die (Fig. 63) is used to transform coil stock or strips into a finished part. This transformation is performed progressively by a series of stations aligned in a row; the workpiece is fed from station to station with each stroke, by being attached to the scrap skeleton. Force for the movement is applied to the incoming sheet metal strip or coil and outgoing scrap skeleton by means of rollers. The cost of a progressive die is high, so it is best used for high-volume production.

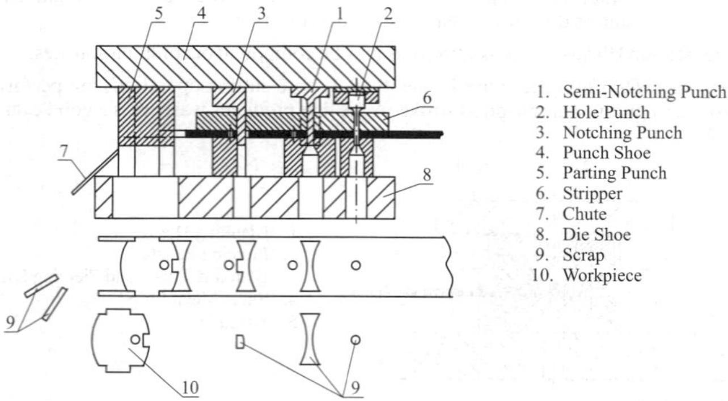


Fig. 63. Progressive Die

*Transfer Die:* In transfer die operations, individual stock blanks are mechanically moved from die station to die station in a single die set. Large workpieces are managed with tandem press lines, in which the stock is moved from press to press where specific operations are performed.

## Types of Dies

There are many types of dies for metalworking. In this section, descriptions and schematic illustrations are given of die types, with a focus on those most used for sheet metal working.

**Blanking Dies.—**Blanking dies are cutting dies. The blank size is determined by the dimension of the die, and the cutting clearance is given on the punch. By mode of construction, blanking dies can be classified as cutoff dies, drop-through dies, and invented dies.

*Cutoff Blanking Die:* In the cutoff die (Fig. 64), the punch and die chop off or separate a flat and/or square profile workpiece from a continuous supply of metal. The trailing edge of a part also forms the leading edge of the next part, so no scrap is produced during cutoff. This type of die also may be used for separating or cutting scrap skeleton. In such cases, the cutoff die is usually a section of a progressive die.

*Drop-Through Blanking Die:* When the blank or workpiece falls through the die and press bed into a pan or conveyer for removal, a drop-through die is being used. The blank drops of its own weight through the die opening, with clearance provided in the bolster plate and press bed. The die block assembly is mounted on either the bolster plate or press bed, and the punch assembly is mounted on the press slide.

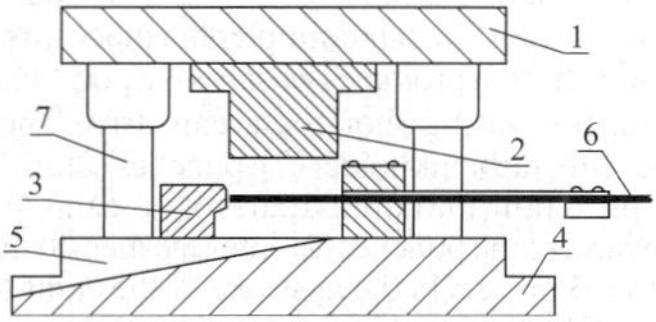

1. Punch Shoe
2. Punch
3. Die
4. Die Shoe
5. Chute
6. Sheet Metal
7. Guide Post

Fig. 64. Cutoff Blanking Die

1. Punch Shoe
2. Punch
3. Die
4. Die Shoe
5. Stripper
6. Strip
7. Blank
8. Keeper
9. Scrap Ribbon
10. Spring

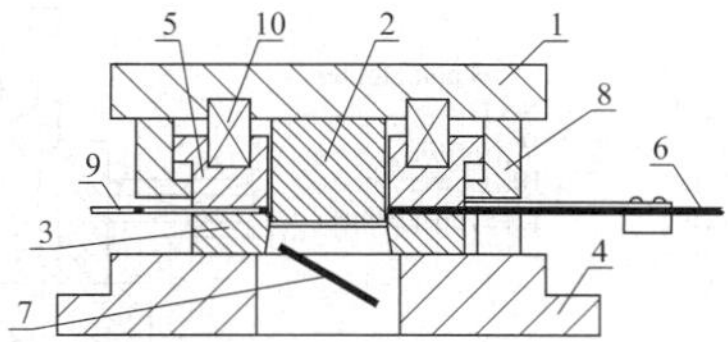

Fig. 65. Drop-Through Blanking Die

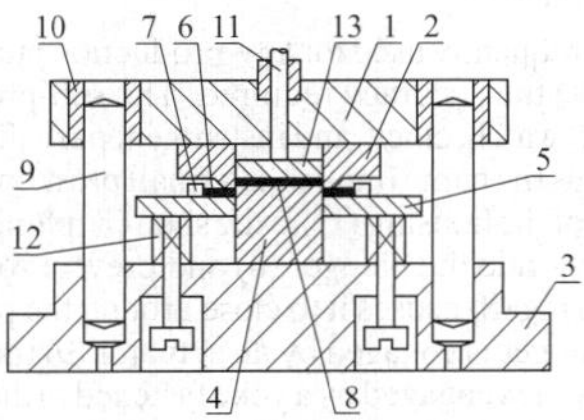

1. Upper Shoe
2. Die Block
3. Lower Shoe
4. Punch
5. Stripper
6. Scrap
7. Guide
8. Blank
9. Guide Post
10. Guide Bush
11. Knockout Pin
12. Spring
13. Knockout Plate

Fig. 66. Inverted Return Blanking Die

Fig. 65 shows a drop-through blank die with a fixed stripper. In this version, sheet metal scrap is around the punch, and the die incorporates a stripper plate. The stripper can be fixed or retained by a keeper and is actuated by a spring, rubber block, or air cylinder.

Drop-through dies are economical to build and maintain and fast in operation, but are not suitable for cases where: (a) the blank is too thin and fragile to be dropped, (b) the blank is too heavy to be dropped, (c) the blank is too awkward to be easily removed from below the press, and (d) the blank is larger than the press bed opening.

*Inverted Return Blanking Die:* In the inverted return version (Fig. 66), the punch is the lower stationary part, and the die is mounted on the ram. Scrap disposal is easier, but blank removal from the die opening requires use of special knockout devices. Complicated, costly, and slower in operation, this type of die is commonly used on presses that do not have a hole in the bed, and when large parts are being manufactured, such as automobile hoods or roofs.

The die block may be a solid single piece or assembled from sectional pieces. If the die opening is small and its contour is simple, a solid die block is suitable. In more complicated applications, sectional dies with accurately ground matching components have advantages.

**Punching Dies.—**The punching die cuts clean holes with resulting scrape slug. The punching die is often called a piercing die (though "piercing" is properly used to describe producing a hole with a tearing action—not typical of cutting operations). The punched hole size is determined by the dimension of the punch; clearance is provided in the die.

Generally, there are two main reasons for punching holes in a separate operation, rather than in combination with other operations: (a) when a subsequent bending, forming, or drawing operation would distort the previously punched hole(s), (b) when the edge of the punched hole is too close to the edge of the die section. To make simpler flat pieces, a punching die can be combined with a compound die; a die that produces washers is a good example.

Fig. 67 shows a punching die used for making holes using a horizontal cam drive. The blank cap (6) is set to the die block (1) and is pressed with the blankholder (7); punches (2) are held in sliding blocks (3). On the lower stroke of the press ram, the beveled parts of the cams (4) force the sliding blocks (3) with the punches (2) inward, so the punches (2) enter the dies (8) inserted in the die block and punch two holes in the workpiece. On the upstroke of the cams (4), the sliding blocks (3) are released, and the springs (5) return the blocks to their initial position.

1. Die Block
2. Punch
3. Slide Block
4. Cam
5. Spring
6. Workpiece
7. Blank Holder
8. Die
9. Spring
10. Lower Shoe
11. Upper Shoe

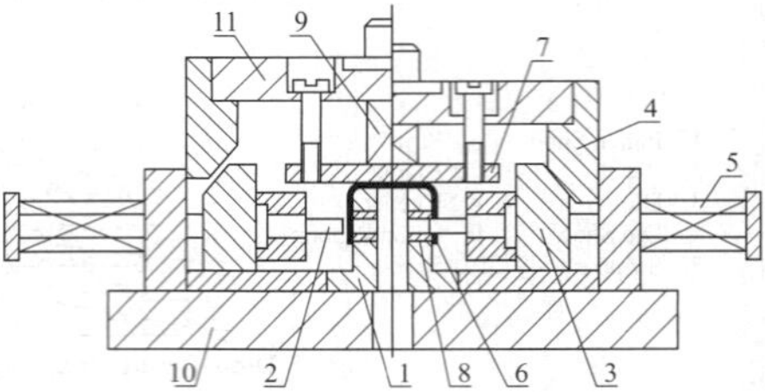

Fig. 67. Punching Die

*Sub-Press Punching Die:* Sub-press dies are frequently used for low-production punching of holes, and usually are compound dies due to the accuracy required. The sub-press die shown in Fig. 68 is used to blank and form small watch, clock, and instrument parts. The die components are retained in a sub-press, which, as the name implies, is a small press operated inside a larger one. The sub-press is composed of the housing (11), a die shoe (1), plunger (6), and bearing (8), with longitudinal key slots to guide the plunger (6) and prevent rotation. Tightening the spanner nut (7) against the bearing (8) causes it to close around the plunger to remove looseness. The top portion of the plunger is engaged by an actuator (9) threaded into the central tapped hole; the slot of the actuator is engaged by a yoke fastened to the press ram, which does not guide the sub-press. The press ram simply applies the up-down motion.

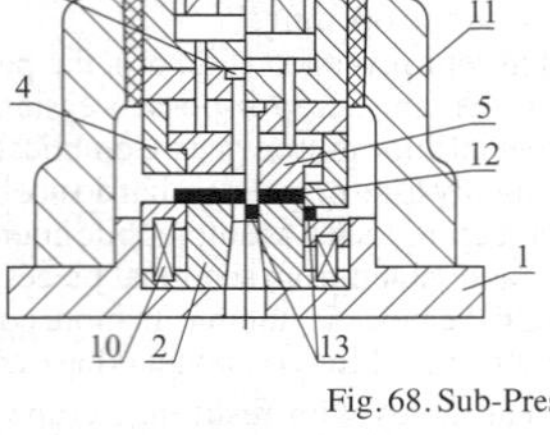

1. Die Shoe
2. Die
3. Punch
4. Keeper
5. Stripper
6. Plunger
7. Spanner Nut
8. Bearing
9. Actuator
10. Spring
11. Housing
12. Blank
13. Scrap Slug

Fig. 68. Sub-Press Punching Die

**Bending Dies.—**A bending die allows sheet metal to be bent, deforming a portion of a flat blank to some angular position and defining the resulting shape of the bent workpiece. The bending die choice depends on the parameters of the bent piece, and the line of bend is straight along its entire length. Bending dies can be classified as: profile bending dies, universal bending dies, multiple-bending dies, closing-bending dies, and special design bending dies.

*U-Profile Bending Dies:* Fig. 69 illustrates a simple die for bending a U-profile. The blank is positioned on the die (1), between stop pins (9). The die (1) is mounted on the die holder (3) in the conventional position. The punch (2) is attached to the punch holder (4), which is fitted to the press ram. The pressure pad (8) applies pressure to the blank; as the punch pushes the blank into the die, the workpiece is formed by one stroke of the press. The bent workpiece is ejected from the die by the pressure pad mechanism (A) when the ram retracts.

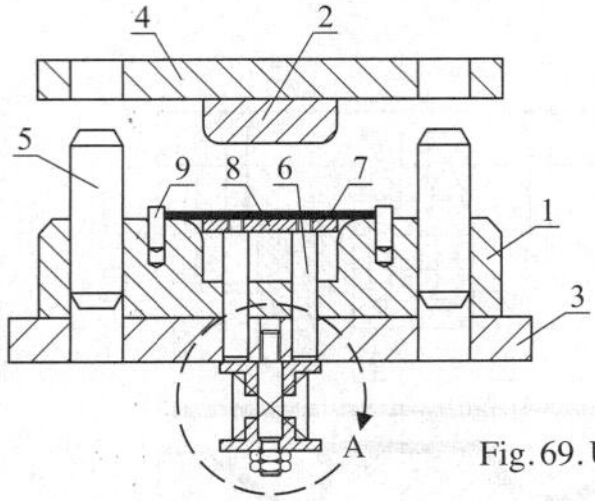


1. Die
2. Punch
3. Die Holder
4. Punch Holder
5. Guide Post
6. Cushion Pin
7. Blank
8. Pressure Pad
9. Stop Pin
A. Spring Type Pad Mechanism

Fig. 69. U-Profile Bending Die

For bending workpieces of small dimensions and thin material, the punches and dies are made from a single block of metal. For bending larger workpieces, a U-profile bending die with segments (Fig. 70) is used. The die is made of segments, generally alloy tool steel. The segments (4) are fixed to the die shoe with screws and dowels.

1. Fixed Stripper
2. Punch
3. Punch Holder
4. Die Segment
5. Cushion Pin
6. Pressure Pad Plate
7. Stop Pin
8. Die Shoe
9. Workpiece

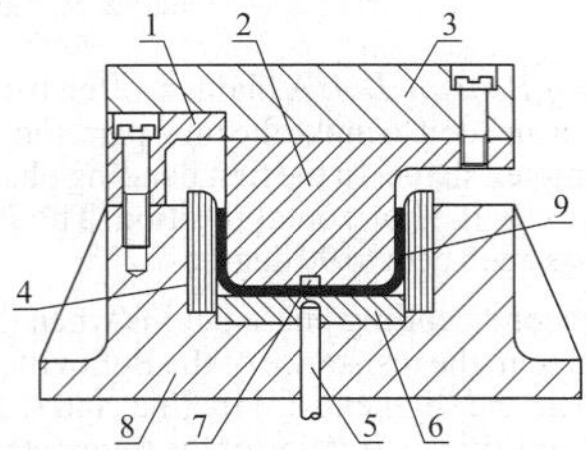


Fig. 70. U-Profile Bending Die with Segments

*Universal Bending Die:* Fig. 71 shows a typical, simple universal bending die and an example of a workpiece bent in four phases. The die, which has several slots of different shapes, consists of a die shoe (4), on which are assembled the universal die (1) and adjustable stops (3). The punch (2) is attached directly to the press ram. Selecting an appropriate profile die (1) with an appropriate punch (2) allows the workpiece to be bent into the desired shape.

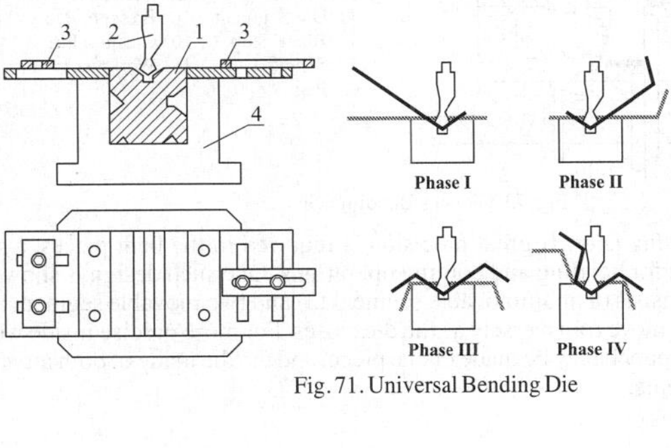


1. Die
2. Punch
3. Adjustable Stop
4. Die Shoe

Fig. 71. Universal Bending Die

*Multiple-Bending Die:* In this die (Fig. 72), a swinging arm (1) pivots at one end of the punch (3) and carries a steel roller (2). When the punch descends, the roller bends the blank down into the die (6). When downward travel of the roller is checked by the die, the arm (1) swings to the left and continues moving until it meets the lower finished workpiece. Before the punch has reached the end of its downward travel, the pressure pad (8) engages the blank and holds it in position, with tension of the two springs (9), through the final stages of the bending operation.

1. Arm
2. Roller
3. Punch
4. Punch Plate
5. Blank
6. Die
7. Workpiece
8. Pressure Pad
9. Spring
10. Base Plate

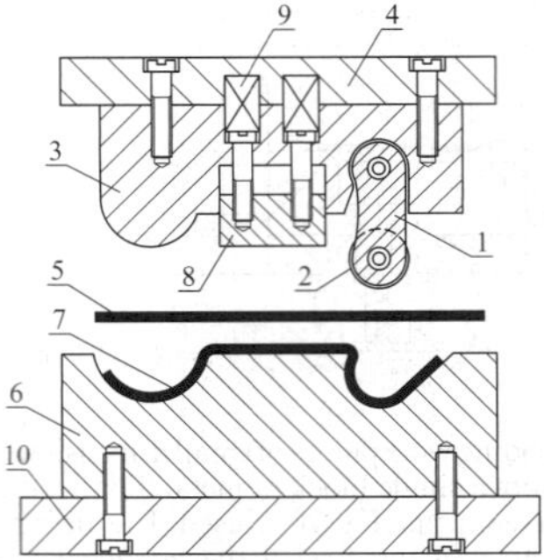


Fig. 72. Multiple-Bending Die

*Closing-Bending Die:* A blank is often bent to form the workpiece for the next bending operation needed to make the final part. Fig. 73 shows a bending die for closing a U-profile on a complex shape. In the first bending phase, the pegs (6) hold the movable segments of the die (3) in the horizontal position. The Belleville springs provide resistance until the blank has been bent to 90 degrees.

The second bending phase begins when the punch (2) touches the die segments (3). At this moment, the resistance of the Belleville springs (7) begins to be less than the force of the punch, the segments (3) incline, and the workpiece is bent past 90 degrees, with the stop (4) limiting inclination of the segments. When the press ram is moved up, the springs return the segments to the horizontal position, and the pressure pad ejects the workpiece.

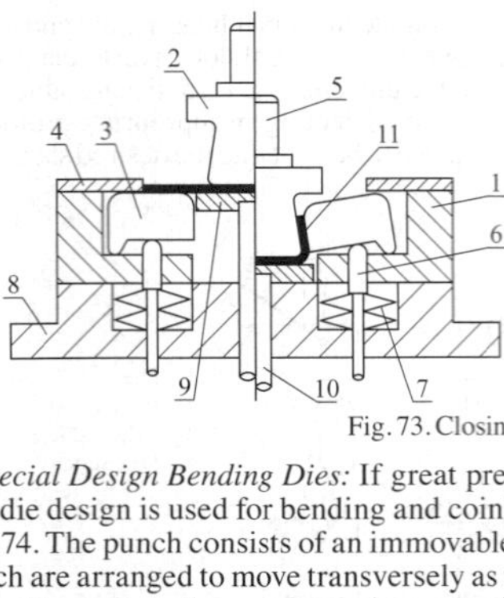


1. Die
2. Punch
3. Die Segment
4. Blank Stop
5. Shank
6. Peg
7. Belleville Spring
8. Die Shoe
9. Pressure Pad
10. Cushion Pin
11. Workpiece

Fig. 73. Closing-Bending Die

*Special Design Bending Dies:* If great precision is required in the bent pieces, a special die design is used for bending and coining operations. One such design is shown in Fig. 74. The punch consists of an immovable segment (3) and two movable segments (4), which are arranged to move transversely as the die closes. For more precise inside workpiece dimensions, the punch may be made of one piece, and the die made of both movable and immovable segments.

1. Punch Shoe
2. Die Shoe
3. Fixed Punch
4. Slide Punch
5. Spring
6. Die
7. Slide Punch Stop
8. Guide
9. Blank
10. Security Ring
11. Workpiece

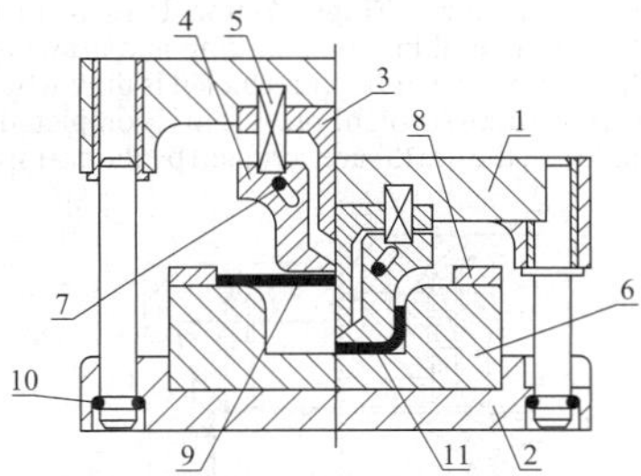

Fig. 74. Special Design of a Bending Die

**Forming Dies.—**The operation of a forming die (Fig. 75) is similar to that of bending, except the line of the bend is curved instead of straight, and the plastic deformation of the workpiece material is more severe. The blank is positioned in a nest, composed of two blanking stops mounted on the die. When the ram descends, the blank is gripped between the bottom of the punch and the surface of the pressure pad. Further descent causes the side of the blank to conform to the curved shape of the forming die and punch. When the die ascends, the part is carried up with the pressure pad.

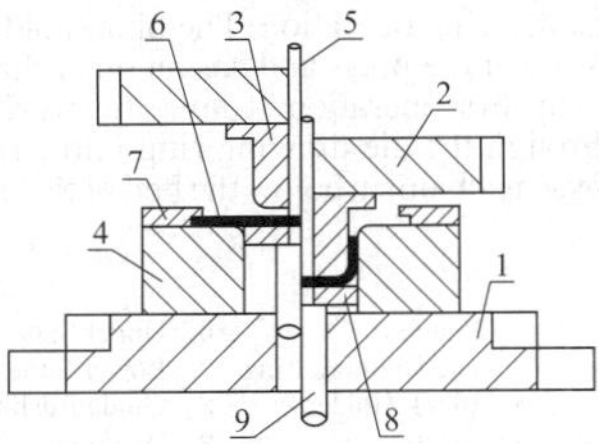

1. Die Shoe
2. Punch Shoe
3. Punch
4. Die
5. Stripper Pin
6. Blank
7. Blanking Stop
8. Pressure Pad
9. Cushion Pin

Fig. 75. Forming Die

*Rubber Pad Forming Dies:* Rubber pad forming dies are inexpensive and used for small production series, particularly in such industries as aircraft; about half of all sheet metal parts are made using these types of dies. Rubber pad forming can be accomplished in numerous ways, and as technology has advanced, so have applications for this simple process. In general, an elastic upper die, usually made of rubber or polyurethane (widely used because of its resistance to abrasion and fatigue), is connected to the ram of the hydraulic press. A rigid lower punch, often called a *form block*, provides the mold the sheeted metal is formed to.

As shown in Fig. 76, the blank is on top of the punch, and the soft die is forced over the rigid punch, partly forming the die into the required shape. During the process cycle, the rubber (or polyurethane) pad deforms elastically over the punch and the blank, applying a large amount of pressure, typically 15,000 psi. The thickness of the rubber pad should be not less than 1.5 times but preferably 3 times the height of the punch.

1. Punch
2. Container
3. Rubber Pad
4. Blank
5. Workpiece
6. Press Plate

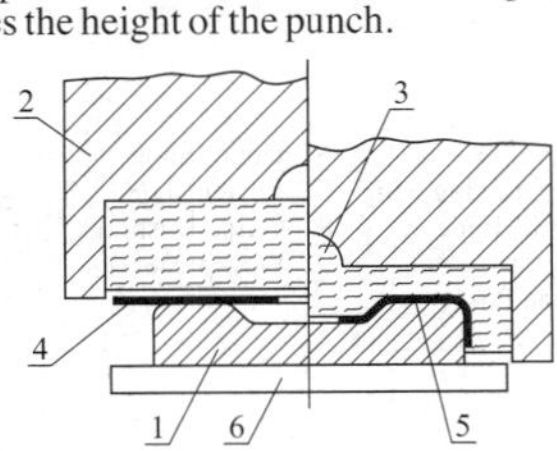

Fig. 76. Rubber Pad Forming Die

**Drawing Dies for a Single-Action Press.**—In this drawing operation (Fig. 77), the punch (1) is a combination blanking and drawing member. The blank is sheared as the punch (1) enters the die (5), and then it is drawn by the recess in the center of the punch, as the downward stroke of the press ram is completed. The ejector (3) on the punch (1) and the ejector (6) on the die (5) are actuated by the coil springs (4) and (10).

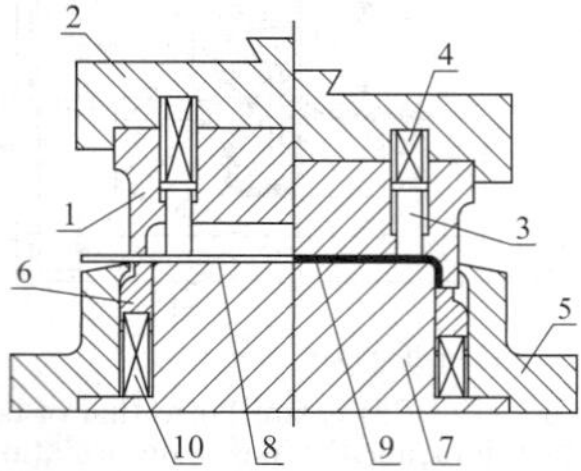


1. Punch
2. Punch Holder
3. Ejector
4. Spring
5. Die
6. Ejector
7. Block
8. Strip or Coil
9. Workpiece
10. Spring

Fig. 77. Drawing Die for Single-Action Press

**Redrawing Dies.**—There are two types of redrawing dies: dies for direct redrawing and dies for reverse redrawing.

*Die for Direct Redrawing:* A direct redrawing die (Fig. 78) is designed for use on double-action presses. The punch is attached directly to the inner slide of the press and conforms to the inner diameter of the part after the first drawing operation. The blankholder surrounding the punch is attached to the outer slide of the press and has an outer diameter about the same as that of the drawing punch in the first operation. It enters the workpiece, and then the punch pushes the workpiece through the die drawing ring into a smaller diameter. The workpiece ejector is powered by a mechanism below the bed of the press.

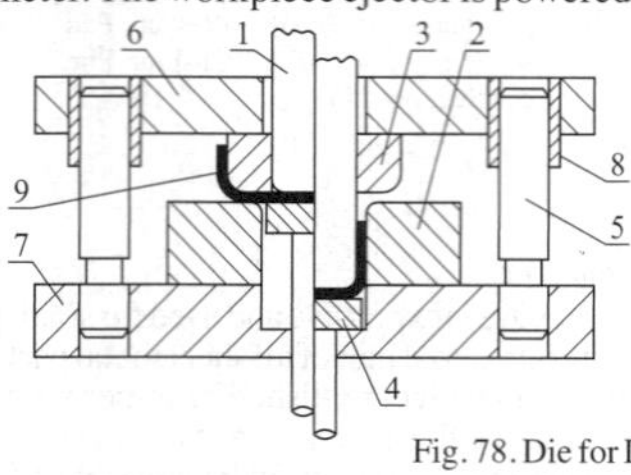


1. Punch
2. Die Drawing Ring
3. Blank Holder
4. Ejector
5. Guide Post
6. Upper Shoe
7. Lower Shoe
8. Guide Bushing
9. Workpiece

Fig. 78. Die for Direct Redrawing

*Die for Reverse Redrawing:* The advantage of reverse redrawing is that the workpiece material is not subjected to severe stress in one direction or another. In reverse redrawing (Fig. 79), the punch moves in the opposite direction to fit the shape of a hollow die, reversing the direction of material flow. Returning the inside of the cup to the outside, a second cup with new dimensions is obtained. The die drawing ring radius should not be less than four times the material's thickness.

1. Punch
2. Die Drawing Ring
3. Blank Holder
4. Ejector
5. Guide Post
6. Upper Shoe
7. Lower Shoe
8. Guide Bushing
9. Workpiece

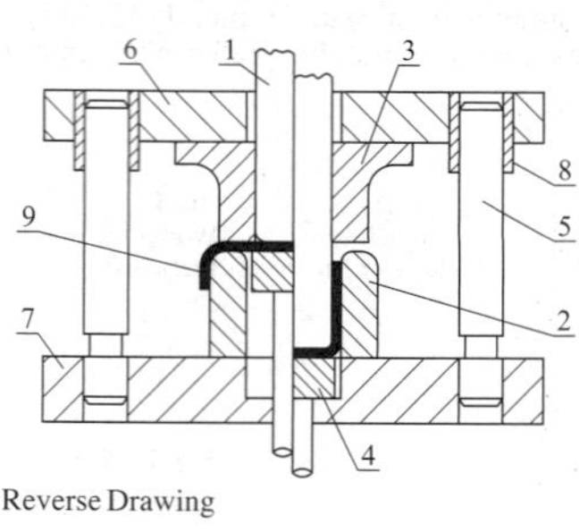


Fig. 79. Die for Reverse Drawing

**Die for Drawing a Spherical Part.—**A special triple-acting die design (Fig. 80) is used in a double-acting press for drawing a spherical part to its full depth in one operation without leaving a wrinkle. After the blank has been placed in position on the top face of the die (4), the blankholder (3) descends until it holds the blank firmly against the die face. The pressure ring (2), which is backed by heavy springs strong enough to form the first drawing operation, then acts as an inner blankholder, while the punch (1) descends to the bottom of the die, forming the central part of the blank.

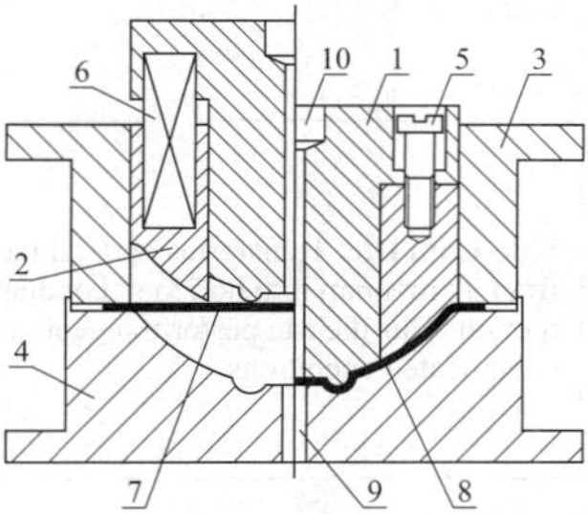


1. Punch
2. Pressure Ring
3. Blank Holder
4. Die
5. Screw
6. Spring
7. Blank
8. Workpiece
9. Vent
10. Press Shank

Fig. 80. Die for Drawing Spherical Part

**Tube-Forming Dies.—**A tube-forming die (Fig. 81), used for production of more precise tubular pieces, consists of the punch holder (10) and die shoe (11) with a guide system: the guide post (8), guide bushing (9), and cam slide die (3), with springs (6) and (7), are fixed to the die shoe. The blank (12) is located on the cam slide die and bends freely at the beginning of the bending operation. At the end of the downstroke of the press ram, the driver (5) pushes the die segment (3) to coin the workpiece around the punch (1) into its final dimensions. On the upstroke, the springs (6) and (7) return the die segments to the start point, and the tube is slipped off the punch (1).

1. Punch
2. Half Segment Die
3. Cam Slide Die
4. Die Cam
5. Driver
6. Spring
7. Spring
8. Guide Post
9. Guide Bushing
10. Punch Holder
11. Die Shoe
12. Blank
13. Workpiece

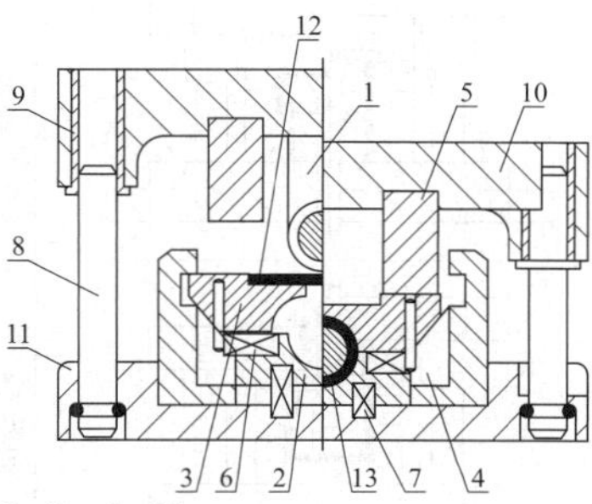


Fig. 81. Tube-Forming Die

**Table 1. Additional Die Types Commonly Used for Sheet Metalworking**

| | |
|---|---|
| 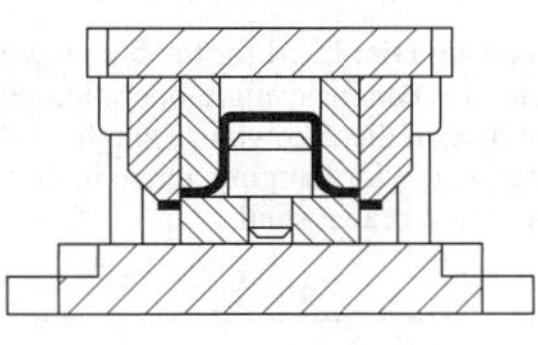 | **Trimming Die:** Cuts away portions of a formed or drawn workpiece that has become wavy and irregular. |

**Table 1.** *(Continued)* **Additional Die Types Commonly Used for Sheet Metalworking**

| | |
|---|---|
| 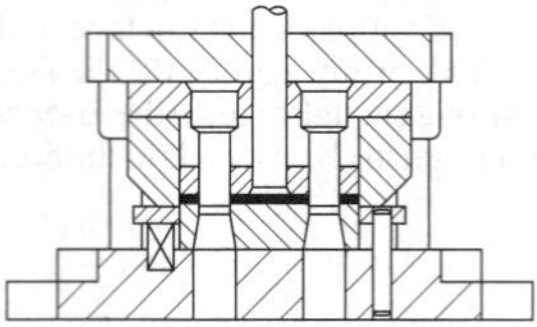 | **Shaving Die:** Removes a small amount of material from around the edges of the blank or hole to improve the surface and accuracy of the cut. |
| 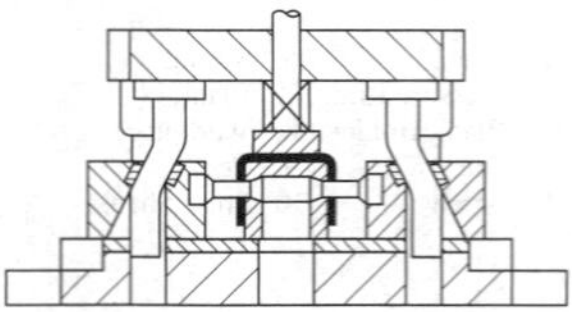 | **Side-Cam Die:** Transforms vertical motion from a press ram into horizontal or angular motion. This die can perform ingenious and complicated operations. |
| 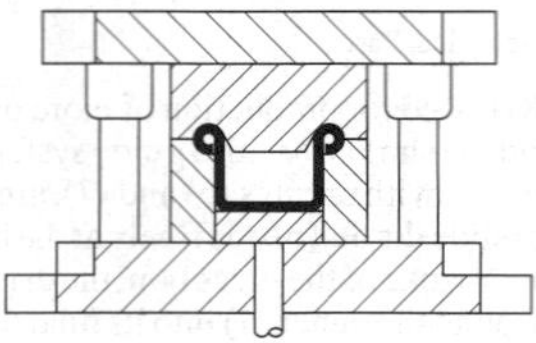 | **Curling Die:** Forms the metal at the edge of the workpiece into a circular shape or hollow ring. |
| 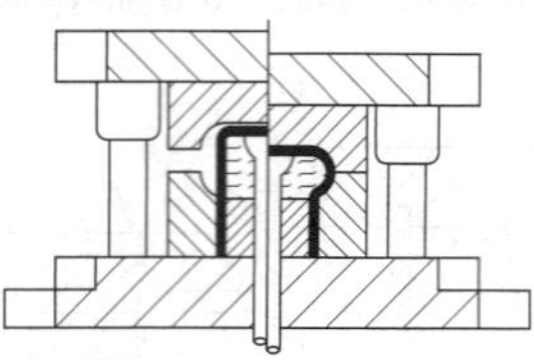 | **Bulging Die:** Expands a portion of a drawn shell, causing it to bulge. |
| 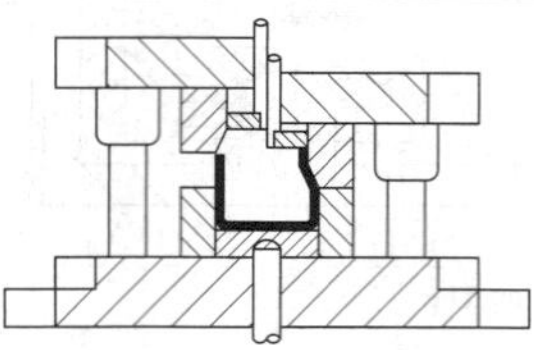 | **Necking Die:** Reduces a portion of the shell diameter and makes that portion longer. (Necking is the opposite of bulging.) |
| 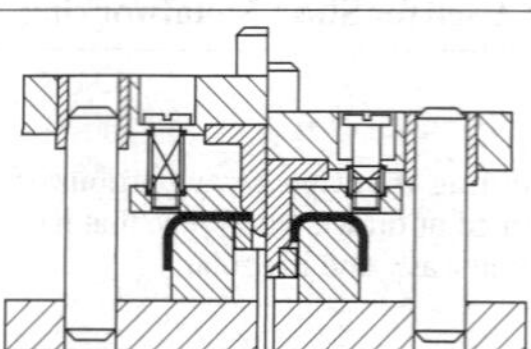 | **Extruding Die:** Used for the flanging of a hole, this can be a single-operation or combination die. Shown is a single-operation die for flanging a hole in the bottom of a drawn shell. |

**Table 1.** *(Continued)* **Additional Die Types Commonly Used for Sheet Metalworking**

| | |
|---|---|
| 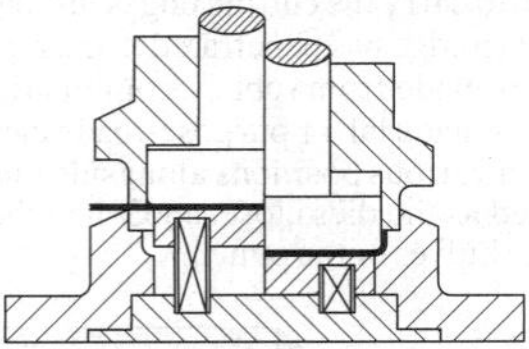 | **Drawing Die:** Used in a double-action press, cuts the blank as the cutting punch enters the die, and then holds the blank in place as the drawing punch pushes the blank into the die. |
| 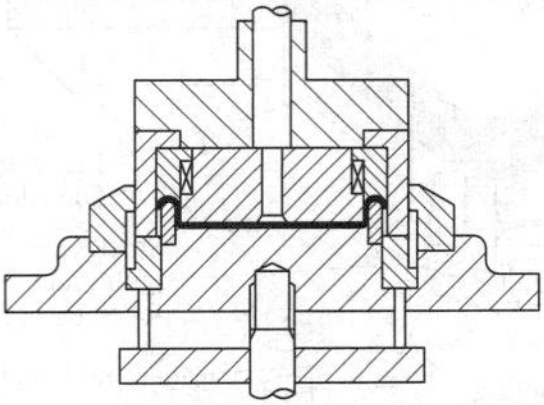 | **Inverted Drawing Die:** Cuts the blank and then performs two drawing operations: the workpiece is drawn into a cylindrical cup, and then it is curled. |

**Steel Rule Dies.—**Steel rule dies (or knife dies) were patented by Robert Gair in 1879, and, as the name implies, have cutting edges made from steel strips of about the same proportions as those used in making graduated rules for measuring purposes. According to J. A. Richards, Sr., of the J. A. Richards Co., Kalamazoo, MI, a pioneer in the field, these dies were first used in the printing and shoemaking industries for cutting out shapes in paper, cardboard, leather, rubber, cork, felt, and similar soft materials. Steel rule dies were later adopted for cutting upholstery material for the automotive and other industries, and for cutting out simple to intricate shapes in sheet metal, including copper, brass, and aluminum. A typical steel rule die, designed to cut a simple circular shape, is shown in Fig. 82, partially cut away to illustrate the construction. This die also cuts a rectangular opening in the workpiece, and pierces four holes, all in one press stroke. Such dies generally cost 25 to 35 percent of the cost of conventional blanking dies, and can be produced in much less time.

The die blocks that hold the steel strips on edge on the press platen or in the die set may be made from plaster, hot lead or type metal, or epoxy resin, all of which can be poured to shape. For light work, ¾-in. (19.05 mm) thick, five- or seven-ply maple or birch wood may be used. Narrow slots are cut in this wood with a jig saw to hold the strips vertically. Where greater forces are involved, as with operations on metal sheets, the blocks usually are made from Lignostone densified wood or from metal. In the ¾-in. thickness mostly used, metal or medium- and high-density Lignostone-laminated wood, which is available in the ¾-in. thickness, is made from about 35 plies of highly compressed lignite wood, bonded with phenolformaldehyde resin, which imparts great density and strength. The material is made in thicknesses up to 6 in. (15.24 cm) and in various widths and lengths.

Steel rule die blocks can carry punches of various shapes to pierce holes in the stock, also projections designed to form strengthening ribs and other shapes in material such as aluminum, at the same time as the die cuts the component to shape. Several dies can be combined or nested, and operated together in a large press, to produce various shapes simultaneously from one sheet of material.

As shown in Fig. 82, the die steel is held in the die block slot on its edge, usually against the flat platen of a die set attached to the moving slide of the press. The sharp, free end of the rule faces toward the workpiece, which is supported by the face of the other die half. This other die half may be flat or may have a punch attached to it, as shown, and it withstands the pressure exerted in the cutting or forming action when the press is operated.

The closed height of the die is adjusted to permit the cutting edge to penetrate into the material to the extent needed, or, if there is a punch, to carry the cutting edges just past the punch edges for the cutting operation. After the sharp edge has penetrated it, the material often clings to the sides of the knife. Ejector inserts made from rubber, combinations of cork and rubber, and specially compounded plastics material, or purpose-made ejectors, either spring- or positively actuated, are installed in various positions alongside the steel rules and the punch. These ejectors are compressed as the dies close, and when the dies open, they expand, pushing the material clear of the knives or the punch.

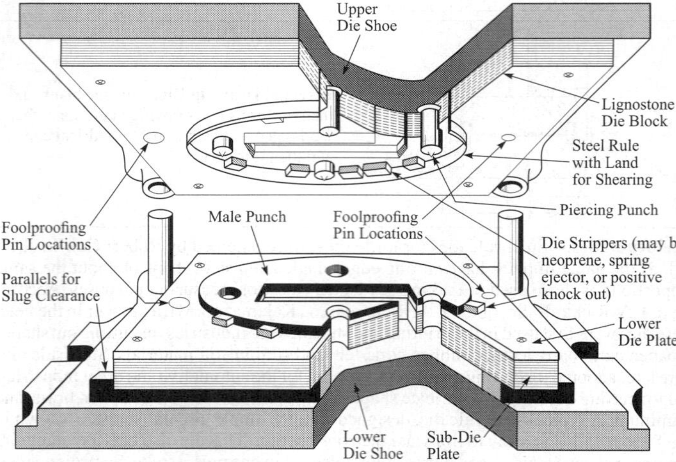


Fig. 82. Steel Rule Die for Cutting a Circular Shape, Sectioned to Show the Construction

The cutting edges of the steel rules can be of several shapes, as shown in profile in Fig. 83, to suit the material to be cut, or the type of cutting operation. Shape *A* is used for shearing in the punch in making tools for blanking and piercing operations, the sharp edge later being modified to a flat, producing a 90° cutting edge, *B*. The other shapes in Fig. 83 are used for cutting various soft materials that are pressed against a flat surface for cutting. The shape at *C* is used for thin, and the shape at *D* for thicker materials.

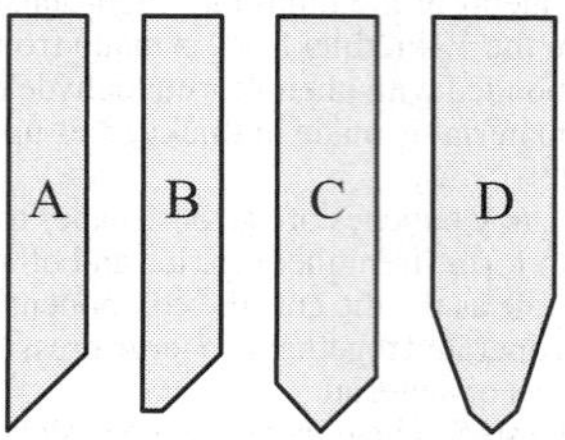


Fig. 83. Cutting Edges for Steel Rule Dies

Steel rule die steel is supplied in lengths of 30 and 50 in., or in coils of any length, with the edges ground to the desired shape, and heat treated, ready for use. The rule material width is usually referred to as the height, and material can be obtained in heights of 0.95,

1, $1\frac{1}{8}$, $1\frac{1}{4}$, and $1\frac{1}{2}$ in. Rules are available in thicknesses of 0.055, 0.083, 0.11, 0.138, 0.166, and 0.25 in. (4 to 18 points in printers' measure of 72 points = 1 in.). Generally, stock thicknesses of 0.138 or 0.166 in. (10 and 12 points) are preferred, the thinner rules being used mainly for dies requiring intricate outlines. The stock can be obtained in soft or hard temper. The standard edge bevel is 46°, but bevels of 40 to 50° can be used. Thinner rule stock is easiest to form to shape and is often used for short runs of 50 pieces or thereabouts. The thickness and hardness of the material to be blanked also must be considered when choosing rule thickness.

*Making of Steel Rule Dies:* Die making begins with a drawing of the shape required. Saw cutting lines may be marked directly on the face of the die block in a conventional layout procedure using a height gage, or a paper drawing may be pasted to or drawn on the die board. Because paper stretches and shrinks, Mylar or other nonshrink plastics sheets may be preferred for the drawing. A hole is drilled off the line to allow a jig saw to be inserted, and jig saw or circular saw cuts are then made under manual control along the drawing lines to produce the slots for the rules. Jig saw blades are available in a range of sizes to suit various thicknesses of rule and for sawing medium-density Lignostone, a speed of 300 strokes/min is recommended, the saw having a stroke of about 2 inch (50.8 mm). To make sure the rule thickness to be used will be a tight fit in the slot, trials are usually carried out on scrap pieces of die block before cuts are made on a new block.

During slot cutting, the saw blade must always be maintained vertical to the board being cut, and magnifying lenses are often used to keep the blade close to the line. Carbide or carbide-tipped saw blades are recommended for clean cuts as well as for long life. To keep any "islands" (such as the center of a circle) in position, various places in the sawn line are cut to less than full depth for lengths of $\frac{1}{4}$–$\frac{1}{2}$ in. (6.4–12.7 mm), and to heights of $\frac{5}{8}$ to $\frac{3}{4}$ in. (16–19 mm) to bridge the gaps. Slots of suitable proportions must be provided in the steel rules, on sides away from cutting edges, to accommodate these die block bridges.

Rules for steel rule dies are bent to shape to fit the contours called for on the drawing by means of small, purpose-built bending machines, fitted with suitable tooling. For bends of small radius, the tooling on these machines is arranged to perform a peening or hammering action to force the steel rule into close contact with the radius-forming component of the machine so that quite small radii, as required for jig saw puzzles, for instance, can be produced with good accuracy. Some forms are best made in two or more pieces, then joined by welding or brazing. The edges to be joined are mitered for a perfect fit, and are clamped securely in place for joining. Electrical resistance or a gas heating torch is used to heat the joint. Wet rags are applied to the steel at each side of the joint to keep the material cool and the hardness at the preset level, as long as possible.

When shapes are to be blanked from sheet metal, the steel rule die is arranged with flat, 90° edges (*B*, in Fig. 83), which cut by pushing the work past a close-fitting counter-punch. This counterpunch, shown in Fig. 82, may be simply a pad of steel or other material, and has an outline corresponding to the shape of the part to be cut. Sometimes the pad may be given a gradual, slight reduction in height to provide a shearing action as the moving tool pushes the work material past the pad edges. As shown in Fig. 82, punches can be incorporated in the die to pierce holes, cut slots, or form ribs and other details during the blanking operation. These punches are preferably made from high-carbon, high-vanadium, alloy steel, heat treated to 61 to 63 RC (Rockwell C scale), with the head tempered to 45 to 50 RC.

Heat treatment of the high-carbon-steel rules is designed to produce a hardness suited to the application. Rules in dies for cutting cartons and similar purposes, with mostly straight cuts, are hardened to 51 to 58 RC. For dies requiring many intricate bends, lower-carbon material is used, and is hardened to 38 to 45 RC. And for dies to cut very intricate shapes, a steel in dead-soft condition with hardness of about 95 RB is recommended. After the intricate bends are made, this steel must be carburized before it is hardened and tempered. For this material, heat treatment uses an automatic cycle furnace, and consists of carburizing in a liquid compound heated to 1500° F (816° C) and quenching in oil, followed by "tough" tempering at 550° F (288° C) and cooling in the furnace.

After the hardened rule has been reinstalled in the die block, the tool is loaded into the press and the sharp die is used with care to shear the sides of the pad to match the die contours exactly. A close fit, with clearances of about half those used in conventional blanking dies, is thus ensured between the steel rule and the punch. Adjustments to the clearances can be made at this point by grinding the die steel or the punch. After the adjustment work is done, the sharp edges of the rule steel are ground flat to produce a land of about $^{1}/_{64}$ in. (0.40 mm) wide (*B* in Fig. 83), for the working edges of the die. Clearances for piercing punches should be similar to those used on conventional piercing dies.

## Pipe and Tube Bending

The difference between a pipe and a tube is how they are measured, and ultimately what they are used for. A pipe is a vessel; a tube is structural. A pipe is measured by the inner diameter; a tube is measured by the outer diameter.

Generally, a tube will have a consistent outer diameter and its inner diameter may have varying wall thicknesses to increase its strength. However, a pipe will have a consistent inner diameter and its outer diameter may have varying wall thicknesses.

The terms used in tube bending are defined in Fig. 84a, and those used in pipe bending are defined in Fig. 84b.

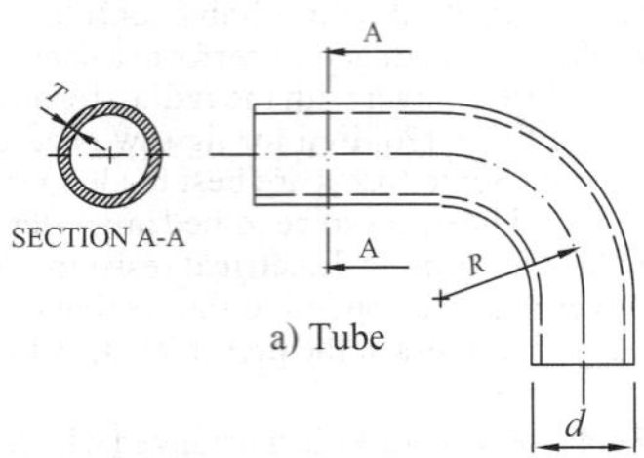


Fig. 84a. Dimension and Terms: Tube

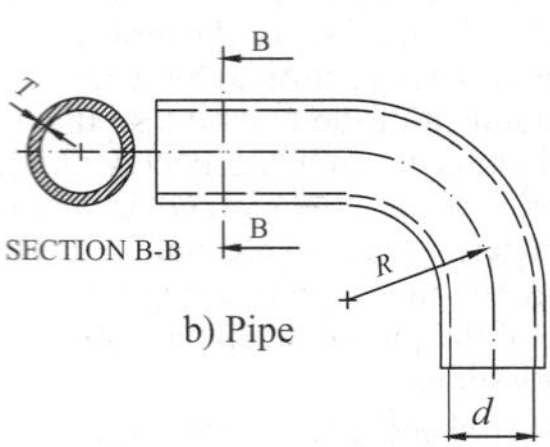


Fig. 84b. Dimension and Terms: Pipe

The radius of the bend *R* is defined with respect to the centerline of the tube or pipe. When the tube or pipe is bent, fibers at the outside wall are in tension and fibers at the wall on the inside bend are in compression. This condition of tensile stress causes thinning and elongation of the wall at the outside, and the compression stress causes thickening and shortening of the inner wall. As a result, the cross section of the bent section of the tube is flattened. The oval distortions grow stronger if thinner tube/pipe walls and smaller bending radii of the workpiece have been selected. Ovality can be calculated by the following formula:

$$u = \frac{D_{max} - D_{min}}{D} 100\% \qquad (52)$$

where $u$ = percent ovality of tube

$D_{max}$ = maximum outer tube diameter after bending (inch or mm)

$D_{min}$ = minimum outer tube diameter after bending (inch or mm)

$D$ = initial outside tube diameter (inch or mm)

When the ratio of the tube diameter to the wall thickness is small enough, the tube can be bent on a relatively small radius. The material-specific diagram in Fig. 85 may be used for a first proposition to determine whether a tube or pipe with defined dimensions (outer diameter and wall thickness) can be bent at all.

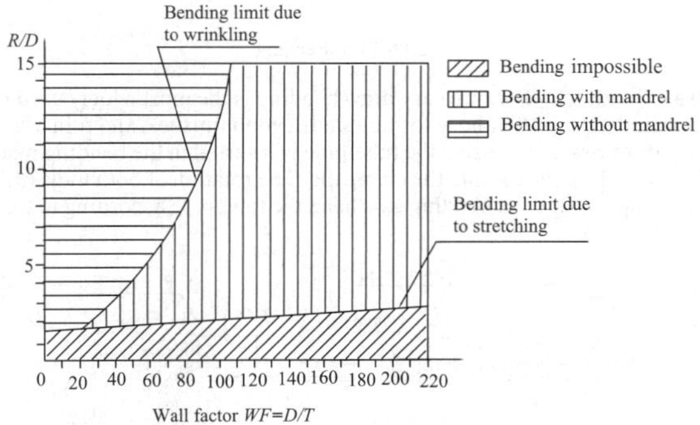


Fig. 85. Technical Limits of the Tube/Pipe Bending Process

Bending without workpiece failure is impossible below the bending limit determined by stretching. The bending limit due to wrinkling separates the range where bending with mandrel and wiper shoe is possible from the range in which the tubes/pipes can also be bent without a mandrel.

Due to the elastic-plastic behavior of metallic materials, the tube/pipe will spring back by a certain angle after every bending attempt because of the phenomenon of elasticity. While in the valid range of Hooke's law (elastic line), the shaping energy is completely given back as work of elastic strain in the form of resiliency. But after the external strain has been removed, it is partly dissipated as work of plasticity when performing the elastic-plastic shaping. In this case, the extent of springback is only caused by the elastic (reversible) part of the shaping workpiece that is stored in the tube/pipe as potential energy during the bending process. Springback is an inevitable phenomenon of bending and can only be compensated for by overbending the workpiece.

Different methods can be used for bending tubes/pipes, depending on the material in use and the required finishing precision. The most common processes for bending tubes or pipes include press bending (ram tube bending), rotary draw bending (round bending), compression bending, and 3-roll bending.

**Press Bending.—**Press bending, also called *ram tube bending*, was probably the first tube/pipe bending method used to cold-form materials. When press bending is applied, the bending tool with the inwrought bending radius is pressed against two counter rollers, either manually or by means of hydraulics. This motion forces the tube/pipe inserted between the radius block (die) and the counter-rollers to bend around the die radius (Fig. 86). The tube/pipe cannot be supported by the mandrel. This tube-bending method creates some cross-sectional ovality; therefore, this method is suitable for thick-walled pipes and large bending radii where high levels of cross-sectional ovality are acceptable, such as in furniture tubing and handrails. Large sweeping curves can be bent in small increments, moving the tube/pipe for each bend.

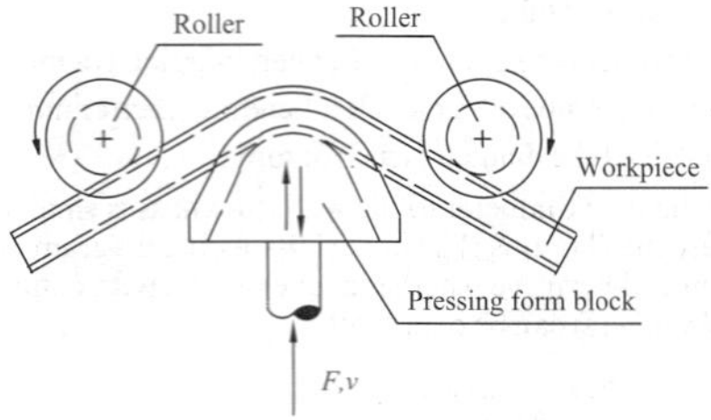


Fig. 86. Press Bending

**Rotary Draw Bending.—**Today, rotary draw bending is the most widely used method of bending tube and pipe, particularly for tight radii, good finishes, and thin-walled tubes. When rotary draw bending is used, the tube/pipe is inserted in the bending machine and fastened between the bend die and the clamp die. The rotation of both tools (the bending die and the clamp) around the bending axis bend the tube/pipe according to the radius of the bend die (Fig. 87).

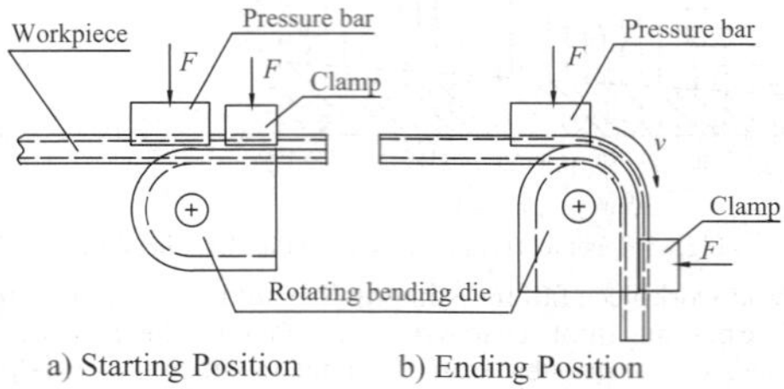


Fig. 87. Rotary Draw Bending.

The pressure bar serves the purpose of receiving the radial stress generated during the forming process, and in addition it supports the straight tube/pipe end from outside. If a mandrel and a wiper shoe are applied additionally (mandrel bending), high workpiece quality can be achieved even in thin-walled pipes and tight-bending radii.

The steel plug, or mandrel, fits inside the tube during rotary draw bending. The plug or mandrel supports the tube internally to reduce the amount of tube cross-sectional flattening during tube bending. After the tube has been bent, the operator extracts the mandrel from the tube, releases the clamp, and then removes the workpiece from the machine.

With modern mandrel bending machines, almost any kind of cold formable tube/pipe, depending on the material, can be safely bent to bending radii of approximately 1.5$D$ with the desired precision. The forming possibilities are not at all limited to bending only round pipes. Oval pipes and flat and mono-block material can be formed just as well by bending as square or other open profiles. Depending on the shape of the workpiece, the required forming tools are adapted accordingly.

This method of bending tube/pipe is used in the manufacture of exhaust pipes, custom exhaust pipes, turbocharger exhaust and intake tubing, dairy tubing and process tubing, heat exchanger tubing, and all stainless and aluminum tubing where a nondeformed diameter finish is critical.

The main differences among rotary-draw-bending machines are the maximum workable outside pipe diameter and the degree of automation of the various functions. Only the bending function of the so-called "1-axis controlled bending machines" is automatic; in them, feeding and contortion are carried out manually. For the user of CNC or bending machines with fully automatic control, however, all functions are available automatically. These bending machines can be supplemented with automatic feeding and unloading systems for the production of large series without any problems.

**Compression Bending.—**Compression bending is a process whereby pipe or tube is bent to a reasonably tight radius, usually without the use of mandrel or precision tooling. It is accomplished by clamping the tube/pipe behind the rear tangent point and then by a wiper shoe on a rotary arm rolling or compressing the material around and onto a bending die (Fig. 88).

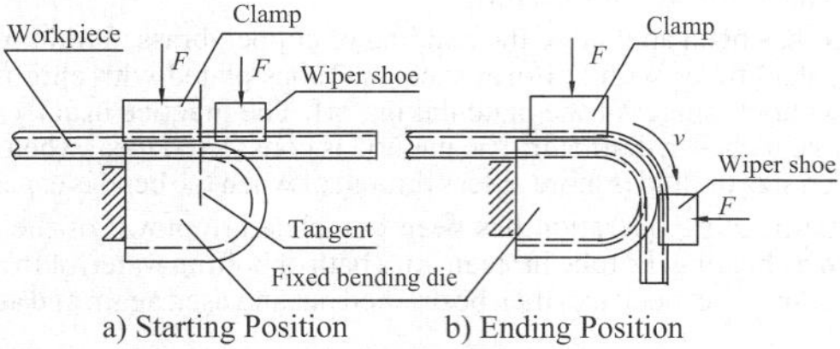


Fig. 88. Compression Bending

**Roll Bending.—**Roll bending is used for bending tube/pipe to a large radius (i.e., to a large circumference). Pipe and tube roll benders comprise three rolls on separate shafts; the tube or pipe is rolled through the rolls while the top roller exerts downward pressure on the top roll to deform the tube or pipe (Fig. 89). Roll pipe and tube benders are available in 2- or 3-driven roll machines, with either manual or hydraulic adjustment of the top roll.

This tube/pipe bending process is ideal for forming helical pipe coils for heat transfer applications, as well as for making long sweeping sections such as those used in steel construction-curved trusses and roof components for structures requiring large open spaces. Pipe cross sections are defined very little when such sweeping sections are formed.

The oldest method of bending a tube/pipe consists of first packing it with loose particles (usually sand) and then bending it into a suitable tool. The filler functions to prevent the tube from buckling inward. After the tube has been bent, the sand is shaken out.

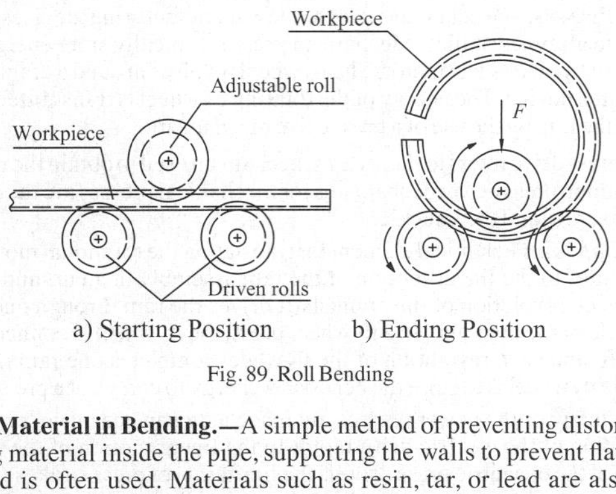


Fig. 89. Roll Bending

**Use of Filling Material in Bending.—**A simple method of preventing distortion consists of using filling material inside the pipe, supporting the walls to prevent flattening at the bend. Dry sand is often used. Materials such as resin, tar, or lead are also sometimes employed. The pipe is first filled with the molten resin, lead, or low-melting-point alloy, and then after bending, the pipe is heated to melt and remove the filling material. Resin has often been used for bending small brass and copper pipes, and lead or other alloys for small iron and steel pipes. Before bending copper or brass pipe or tubing, the latter should be annealed.

*Alloy of Low Melting Point Used as Filler:* Filling tubes with lead may result in satisfactory bends, but the comparatively high melting point of lead often negatively effects the physical properties of the tube. Commercial alloys such as "Cerrobend" and "Bendalloy" have melting points of about 160 degrees F. They are composed of bismuth, lead, tin, and cadmium. With these materials, tubes having a wall as thin as 0.007 inch have been bent to small radii. The metal filler conforms to the inside of the tube so closely that the tube can be bent just as though it were a solid rod.

This method has been applied to the bending of copper, brass, duralumin, plain steel, and stainless steel tubes with uniform success. Tubes plated with chromium or nickel can be bent without danger of the plate flaking off. The practice usually is economical for tubes up to 2 inches in diameter. The method is considered ideal when the number of tubes of a given size or kind is more or less limited or when the bend is especially severe.

When a tube-bending operation has been completed, removal of the metal filler is accomplished by heating the tube in steam, in a bath of boiling water, or in air of about the same temperature. The metal can then be drained out and used again and again.

## Presses for Sheet Metal Working

A stamping press is a metal working machine tool that utilizes the force and speed of a moving ram to transmit force, or an amount of tonnage, to a specific die in order to achieve a workpiece's final shape, often with little or no scrap, and whenever possible, with minimal operator intervention. There are two major types of press brakes, which are classified by the nature of their drive systems: mechanical and hydraulic. Mechanical and hydraulic sheet metal working presses are available in several basic designs and a wide range of sizes, tonnage capacities, and operating speeds. The moving forces of the presses are generated by either mechanical or hydraulic mechanisms that are mounted in the frame. The two most common types of frames are the gap-frame (C-frame) and the straight side frame. Each has its advantages and disadvantages. Frames may be fabricated by casting or by welding heavy rolled steel plates. Both mechanical and hydraulic sheet metal stamping presses are classified by the following main characteristics: frame, drive, action, tonnage, stroke, and strokes per minute.

**Mechanical Presses.—**Mechanical presses are manufacturing devices designed and built to operate all types of dies. Mechanical presses typically store energy in a rotating flywheel driven by an electric motor. The flywheel revolves around a crankshaft until engaged by a clutch device. The energy of the rotating flywheel is transmitted to the vertical movement of the ram by the use of a press drive mechanism.

*Drive:* The press drive refers to the style of mechanism used to obtain the ram movement. The most common press drive mechanisms for mechanical presses are the crankshaft, the eccentric, the screw, and the knuckle.

The *crankshaft* is a mechanical element for translating the rotational motion of the electrical drive motor to the linear motion of the ram assembly in an up-and-down motion. In the press, each revolution of the crankshaft drives the ram through one complete up-and-down cycle of the machine. In a flywheel press, the flywheel is connected directly to the crankshaft, and each revolution of the flywheel completes one ram movement. The function of the flywheel is to store the necessary energy to carry out a pressing operation.

*Eccentric*—many of the newer presses are made using an eccentric for translating the rotational motion of the electric drive motor to the linear motion of the ram. The main disadvantage of the eccentric press drive is the limitation of press strokes. To obtain more press strokes, the eccentric offset must be increased in diameter.

The *screw* drive press is not as widely used, but the unique characteristics of the screw press have driven an increase in its use. As the name suggests, this type of press drive uses a mechanical screw to translate rotational motion into the linear motion of the ram. Briefly, the ram acts as the nut on a rotating screw shaft, moving up or down depending on the screw rotation. Energy is either delivered from a flywheel, which is usually coupled with a torque-limiting (slipping) clutch, or by a direct drive reversing electric motor.

The *knuckle*—the knuckle press is a modified version of the crankshaft style. The crankshaft is behind the press frame in this type of press. The knuckle crankshaft runs from left to right in relation to the press frame. The knuckle is a group of three levers, one lever or connector of which joins the crankshaft throw to the other two levers. The upper lever or knuckle is fixed to a pivot point in the press crown at one end. The lower knuckle is fixed to a pivot point on the press ram at one end. The knuckle press is used for dies that coin, or when heavy sheet metal is drawn or formed. Because of the knuckle, the stroke of the press is limited. Fig. 90 schematically illustrates a knuckle drive mechanism.

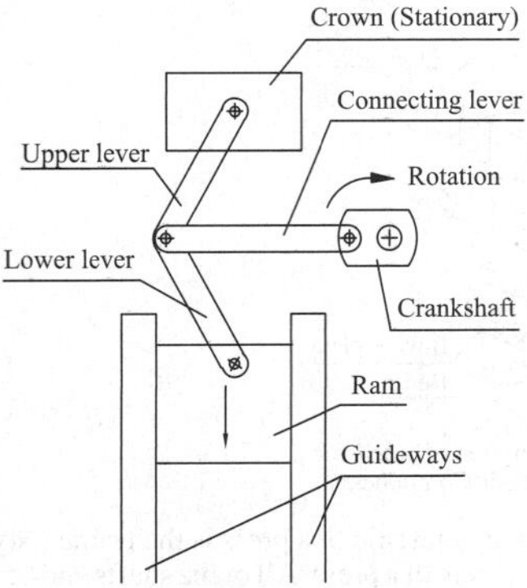

Fig. 90. Knuckle Drive Mechanism

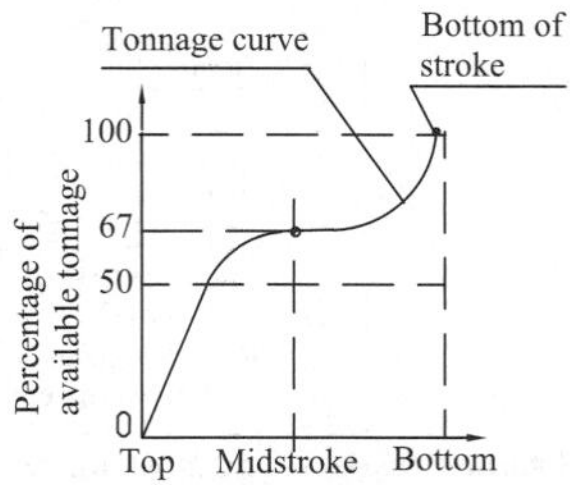

Fig. 91. Available Tonnage Curve for a Mechanical Press

*Tonnage:* The tonnage of a press is the force that a press ram is able to exert. In actual practice, press rams can and do exert forces greater than the rated tonnage. A safety factor is designed into the frame and drive mechanism. The tonnage of a mechanical press is calculated when the ram is near the bottom of its stroke. Thus, the tonnage of a mechanical press is constant and cannot be varied as in the hydraulic press.

The tonnage of a mechanical press when the ram is not near the bottom of its stroke is greatly reduced. Most press manufacturers will supply curves showing the tonnage available for each inch of stroke. Fig. 91 illustrates the available tonnage curve for a mechanical press. Mechanical presses have forces that typically range from 20 to 5,000 tons. A few specially designed large capacity presses with ratings up to 6,000 tons are in operation.

*Stroke:* The stroke of a press is the distance of ram movement from its up position to its down position. The offset on a crankshaft determines the press stroke. The stroke is constant for the crankshaft and eccentric drives. Strokes on mechanical presses range from 0.2 to 45 inches.

*Strokes per Minute:* Strokes per minute (press speed) is a self-explanatory term. The C-frame low tonnage presses (15 to 30 tons) have the highest stroke per minute of all press types. They run at up to 1,800 strokes per minute and typically involve light forming, such as electrical connectors; when cutting dies are operated, the speeds usually range from around 20 to 800 strokes per minute.

*Action:* The term "action" simply means the number of rams on the press. Types of press actions include the following:

*Single action:* One ram operated by a mechanism located in the crown or bed of a press. These presses are used for blanking, bending, forming, and other operations. They perform a single action with each cycle of the press.

*Double action:* One inner ram and one outer ram are operated by a mechanism located in the crown or bed on the press. The rams move from the crown towards the bed on the downstroke. They are used for severe forming and drawing operations.

*Triple action:* Same as double action with the addition of a third ram located in the bed of the press. The third ram moves upward in the bed soon after the other two rams lower.

Large presses (like those used in the automotive industry) have a die cushion integrated in the bolster plate to apply the blank holder force. This is necessary when a single acting press is used for deep drawing.

Fig. 92 schematically illustrates a mechanical press with its basic components.

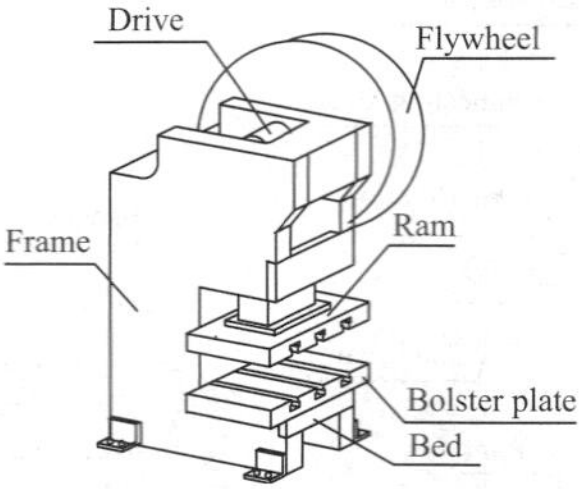


Fig. 92. Schematic Illustration of Mechanical Drive Presses with Basic Components

**Gap-Frame Presses.—**The most noticeable characteristic of a press is the frame's style of construction. The frame is the supporting element of a press. All of the shafts and bearings, as well as the ram and gearing, are mounted in the frame. Frames may be fabricated by casting or by welding heavy rolled steel plates.

The cast frame has excellent rigidity but is expensive. The welded frame is lower in cost in most cases. Due to the greater toughness of rolled steel, welded frames have more resistance to shock loading.

The C-frame press gets its name from its shape. The most common types of C-frame are the inclinable or open-back inclinable (OBI), the solid frame or open-back stationary (OBS), the open-end press, and the adjustable bed or knee frame press.

*Inclinable-Frame Press:* The frame of an inclinable press can be tilted backward to an angle of around 30°. This inclination permits finished parts, slugs, or both, to slide to the rear of the machine upon completion. As the name implies, a fixed-position, or noninclinable machine does not tilt, and parts and/or slugs must either drop vertically or be removed by some other means.

*Solid-Frame Press:* Because of the more rigid base and solid construction, solid frame presses or OBS presses are built in higher tonnages than inclinable presses. These presses have larger beds and frame openings than inclinable presses and can handle larger dies.

*Adjustable-Bed Press:* The third type of C-frame press is the adjustable bed or knee press. The lower leg of the C-frame forms an adjustable table or bed, and thus a wide range of shut heights may be obtained. Because of possible lack of rigidity, only smaller tonnages are made with this style of frame. The adjustable bed press is one of the most versatile presses.

*Open-End Press:* This press has the unusual feature of having the driveshaft located from front to back. Most presses of the C-frame type have the driveshaft from right to left.

The advantage gained in the open-end press is great accessibility. All of the large gears and flywheel are at the back and out of the way. Only smaller tonnage presses are made in this style, however.

**Straight Side Presses.**—Straight side presses are so named because of the vertical columns on either side of the machine. This design eliminates the problem of angular deflection. Also, die life and part accuracy are enhanced. Straight side presses have frames consisting of a crown member, two upright side members, a bed or fundament of the press, and the bolster, which mounts on the press bed and accommodates the die while strengthening the bed. These components are often secured in a preloaded position by four tie rods. They may also be bolted and keyed together or welded into one piece. As a result, straight side presses are stiffer vertically than gap frame units, and any deflection under load tends to be symmetrical.

Straight side presses are suitable for progressive die and transfer die applications and cover an enormous range of types, sizes, and speeds.

**Hydraulic Presses.**—Hydraulic presses are manufacturing devices designed and built to operate dies using hydraulic drive systems to deliver a controlled force.

*Drive System:* These systems are relatively less complex, and are comprised of a motor, a pump, a valving system, and a hydraulic cylinder with the piston connected to the ram. The cylinder is usually double-acting so that hydraulic fluid (oil or water) is pumped under pressure into the top, above the piston, to make the ram move down; to make it come up, fluid is pumped into the bottom, under the piston. The pressure of the fluid under the piston is generally held between 10 to 15 percent of the pressure of the fluid above the piston. This is done to keep the ram from dropping by force of gravity, and also to help control the ram throughout the stroke.

*Valving System:* The valving system ensures the desired control and direction of flow either into or out of the cylinder. Fig. 93 schematically illustrates a hydraulic press drive system with its basic components.

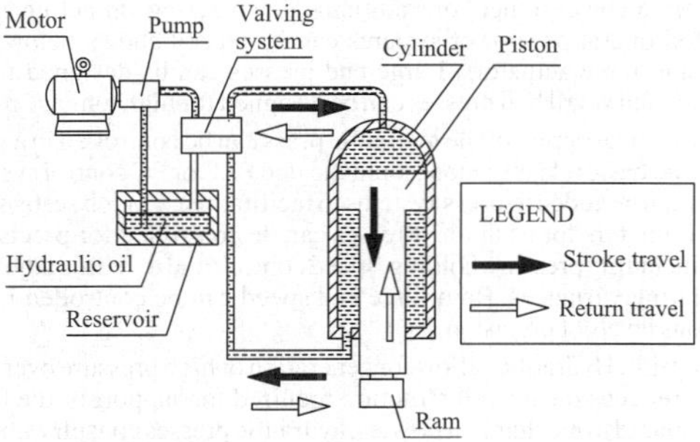


Fig. 93. Schematic Illustration of Hydraulic Press Drive System

*Frame:* Hydraulic presses have various frame types, including C-frames, straight-sided frames, H-frames, four-column frames, and other shapes depending on the application.

*Tonnage:* The tonnage of a hydraulic press is easily calculated: the piston area times the hydraulic pressure in the cylinder. By changing the pressure, the tonnage may be varied. The tonnage of a hydraulic press ranges from about 50 tons to 50,000 tons.

*Stroke:* The stroke is variable on the hydraulic press and can vary from 0.4 in. to 48 in.

*Strokes per Minute:* Hydraulic presses are faster and more reliable than ever. The technology has undergone constant change. Improvements in seals, more efficient pumps, and stronger hoses and couplings have virtually eliminated leaks and minimized maintenance. So, for operations requiring very short strokes, such as punching and blanking, a hydraulic press can achieve speeds that compare favorably with those of mechanical presses, especially when the material is hand fed. Because drawing and forming operations must be run more slowly to allow time for the metal to flow, press speeds usually range from around 5 to 100 strokes per minute, depending on the part size and the severity of the operation being performed.

**Mechanical and Hydraulic Presses, Advantages and Disadvantages.—**The mechanical press has been the first choice of many press users for years. Training of tool and die makers and tool designers has been oriented toward applying mechanical presses to sheet metal pressworking. However, modern hydraulic presses also offer good performance and reliability. Factors that may favor the use of hydraulic presses over their mechanical counterparts may include the following:

a) Full power during the stroke: Maximum power is maintained during the entire stroke of a hydraulic press. This allows for rapid movement to a position just above the part, when the stroke is slowed for the working stroke and an adjustable slowdown speed during forming. The result is more strokes per minute.

b) Overload protection: Because the pressure is pre-adjusted, if the pressure exceeds a limit, such as might occur when a part is not properly ejected, the machine shuts down, eliminating catastrophic results to tooling or to the machine.

c) Cost: Depending on the application, a hydraulic press may cost less than an equivalent mechanical press.

d) Lower operating costs: Hydraulic presses have fewer operating parts; therefore, there are fewer things to break. Automatic lubrication of moving parts helps eliminate maintenance problems.

e) Flexibility: Owing primarily to electronic controls and robotics, hydraulic presses fit well into such areas as flexible manufacturing systems (FMS) and factory automation.

f) No design limitations: The principles of hydraulic force allow for creative engineering. Presses can be designed for traditional down-acting, up-acting, side-acting and multi-action operation. Power systems can be placed above, below, or away from the press and force actuators. Large bed presses can be designed for low tonnage applications and small bed presses can be designed for high tonnage requirements.

g) Unlimited control options: The hydraulic press can be controlled in a variety of ways, ranging from basic relays to more sophisticated PLC or PC control systems. Operator interfaces can be added to press systems to facilitate ease of job set by storing individual job parameters for each die. Presses can be controlled for precise pressure and position, including pressure holding, speed control, and dynamic adjustments to real-time operating variances. Ram force and speed can be controlled in any direction with various levels of precision.

h) Small footprint: Hydraulics allow for generation of high pressure over a small surface area. This reduces the overall structure required for support of the force actuators. When compared to mechanical presses, hydraulic presses consume almost 50 percent less space for the same tonnage capability. This size advantage results in lower manufacturing costs and a faster return on investment by requiring less long-term overhead expense.

i) Hydroforming: Hydroforming is one of the areas in which hydraulic presses have no competition. The hydraulic press ram moves down to close the die and the hydroforming high-pressure fluid fills and acts onto the inside tubular parts and expands it to fill the die. The mechanical presses are not capable of carrying out the hydroforming operation, as they do not have the speed control or the ability to stay on the bottom for performing high-pressure-forming operations.

Although hydraulic presses have many advantages, they also suffer from some disadvantages when compared to mechanical presses. Some of these include:

a) Tolerances: While some hydraulic presses do maintain high tolerances, most of them are limited to approximately 0.020 inch. When closer tolerances are required on these machines, they are usually accounted for in the tooling, which increases the cost of the die or other tooling.

b) Speed: Although high speed is attainable in hydraulic presses, generally, a mechanical press will produce parts faster. This is especially true when short strokes are used.

c) Automatic feeding: Because of fewer moving parts, there is less equipment on which to attach automatic feed mechanisms. Therefore, most automatic feed equipment must be integrated into the process via electronic components, which increases cost.

Factors that favor the use of mechanical presses over their hydraulic counterparts may include these:

- The mechanical press is faster than the hydraulic press.
- The mechanical press is by far the most suitable for blanking and punching.
- Mechanical presses do not require as large a motor as hydraulic presses because they can store energy in the flywheel and then dissipate the energy throughout the press stroke.
- Mechanical presses can be easily adapted to use different rolls and transfer feeds for progressive dies.
- Mechanical presses with short strokes are more economical than hydraulic presses.

**Lubrication System.**—The lubrication systems of mechanical and hydraulic presses are very important. If the lubricating system should fail, not only will the press stop, but also many of the parts are likely to be damaged beyond repair. Therefore, when lubrication failure occurs, the press can seldom be run again without a major overhaul.

The lubricating system delivers oil to the moving parts of the press to reduce friction and to assist in keeping the parts cool. Most newer and more modern presses are equipped with a pressure re-circulating lubricating system that delivers the oil under pressure to the bearings and bushings, and other lubricant points.

After selection of the correct lubricant, the next most critical factors to long machine and lubricant life are keeping the lubricant clean and dry. Details such as machine criticality, operating environment, and component clearances as well as lubricant type, viscosity, flow rate, and economic issues must be carefully considered for optimum lubricant contamination control.

In any press application, the most important aspect for insuring maximum press and lubricant life is the selection of the correct lubricant. This process includes choosing the correct base oil, the correct oil viscosity, and the correct additives for the application. Next in importance is keeping the oil clean and dry. Particulate and water contamination can have devastating effects on machine and lubricant life. The primary source of particulate contamination in lubricants is ambient dust and dirt. While the composition can vary, in general, dust and dirt will contain materials such as silicon oxides and aluminum oxides. Elemental indicators of dirt ingression would be silicon (Si), aluminum (Al), and in some cases calcium (Ca) and magnesium (Mg). If a contaminant particle is larger than the clearance between two slide surfaces, the particle will grind against them, removing metal from the slide surfaces in a process called abrasive wear. The resultant wear particles can cause a chain reaction by increasing the total number of particulates in the lubricant. Additionally, this newly generated abrasive wear material can get broken into smaller particles and become harder due to the process of work hardening. These more numerous and harder particles combine with the original solid contaminants to increase the amount of abrasive material in the lubricant. Under conditions of high velocity or high pressure or both, small particles can impinge on a press surface and result in erosive wear. In this case, particles can be much smaller than the machine clearances and still cause extensive damage due to the velocities and pressures involved. Particles generated during erosive wear also add to the overall contamination of the system and further increase machine wear. Erosive wear most commonly occurs in hydraulic systems that contain servo or proportional devices. Two of the most commonly used lubricant systems are:

*Lost Lubricant Systems:* Lost lubricant systems are used on smaller gap presses. Older presses are usually lubricated by grease, and newer ones by oil. A hand pump is often used in lost lubricant systems. The major disadvantages of this system are that used lubricant must be cleaned periodically, and if a hand pump is used, lubrication of the machine depends on the reliability of the operator.

*Re-Circulating Lubricant Systems:* In these, oil is pumped from a reservoir, filtered, and then distributed to all lubrication points. The pump is powered by an electric motor or by a press-driven mechanical device. The circulating oil lubricates, cools, and flushes small particles from the bearing and friction surfaces.

If the failure of the lubrication system is not detected early, serious damage to the machine will occur. Thermo-chromatic indicators are often used for early detection of lubrication system failure. By the use of thermo-chromatic indicators, mechanical or electrical problems can be detected immediately by the operator or maintenance person. The machine can then be inspected and repaired before serious damage occurs. A few dollars invested in thermo-chromatic indicators can save thousands of dollars on expensive repairs and production losses.

Another sign of lubrication system failure is the presence of bronze particles around the bearing. In this case, it is usually too late for inexpensive repairs because serious damage of the bearing or friction surfaces has already occurred.

**Press Selection.**—When selecting a press, compromises must be made if more than just one type of stamping operation is desired; there is not a single universal type of press that provides productive and efficient operation. Such compromises include consideration of the following primary factors: tonnage; energy capacity; press size and frame design; speed; and control system for press.

Other factors can be considered as well, such as the number of operations to be performed, quantities and production rates, size, geometry and accuracy of workpieces, and equipment costs.

A press-rated tonnage for mechanical presses is the maximum force that should be exerted by the slide against the workpiece at a given distance above the bottom of the stroke. The higher the rating, the greater the torque capacity of its drive members and its capability of delivering more flywheel energy. Presses with flywheel-type drives are basically used for light blanking and piercing operations. The energy requirements of these machines are small and the machines operate at relatively high speeds. Single-geared presses are mostly used for shallow drawn workpieces and require more energy than flywheel types. Double-geared presses are used for deeper draw operations when a larger amount of energy is needed.

As stamping operations become more automated, the use of CNC system and various electromechanical systems to feed material to the press must also be factored in. There are mechanical blank handing systems where manual handling is not practical due to speed and size. High volume feeding is done with coil stock, which also requires an investment in additional feeding equipment.

## ELECTRICAL DISCHARGE MACHINING

Generally called EDM, electrical discharge machining uses an electrode to remove metal from a workpiece by generating electric sparks between conducting surfaces. The two main types of EDM are termed sinker or plunge, used for making mold or die cavities, and wire, used to cut shapes such as are needed for stamping dies. For die sinking, the electrode usually is made from copper or graphite and is shaped as a positive replica of the shape to be formed on or in the workpiece. A typical EDM sinker machine, shown diagrammatically in Fig. 1, resembles a vertical milling machine, with the electrode attached to the vertical slide. The slide is moved down and up by an electronic, servo-controlled drive unit that controls the spacing between the electrode and the workpiece on the table. The table can be adjusted in three directions, often under numerical control, to positions that bring a workpiece surface to within 0.0005–0.030 inch (0.013–0.76 mm) from the electrode surface, where a spark is generated.

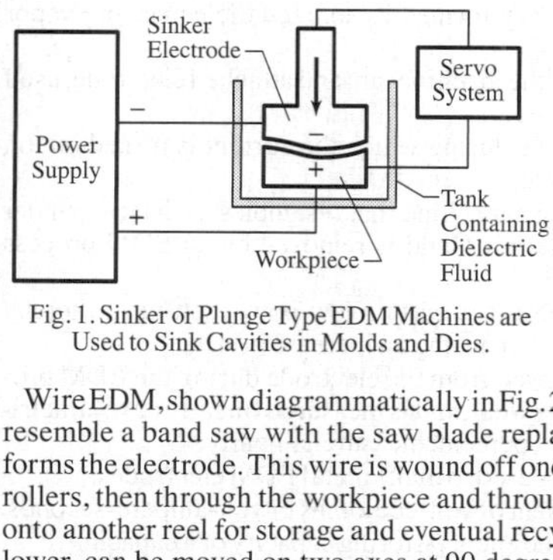


Fig. 1. Sinker or Plunge Type EDM Machines are Used to Sink Cavities in Molds and Dies.

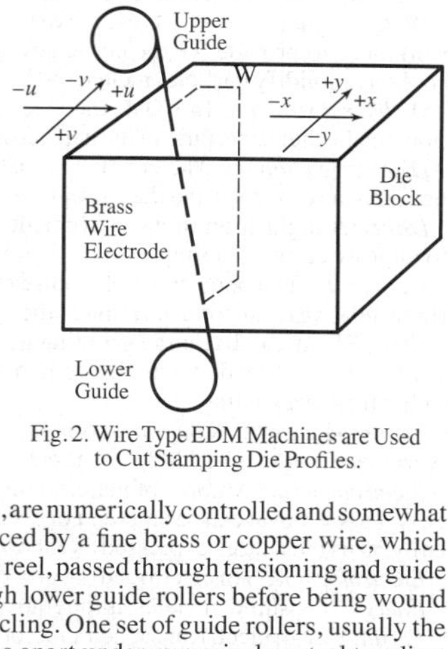


Fig. 2. Wire Type EDM Machines are Used to Cut Stamping Die Profiles.

Wire EDM, shown diagrammatically in Fig. 2, are numerically controlled and somewhat resemble a band saw with the saw blade replaced by a fine brass or copper wire, which forms the electrode. This wire is wound off one reel, passed through tensioning and guide rollers, then through the workpiece and through lower guide rollers before being wound onto another reel for storage and eventual recycling. One set of guide rollers, usually the lower, can be moved on two axes at 90 degrees apart under numerical control to adjust the angle of the wire when profiles of varying angles are to be produced. The table also is movable in two directions under numerical control to adjust the position of the workpiece relative to the wire. Provision must be made for the cut-out part to be supported when it is freed from the workpiece so that it does not pinch and break the wire.

EDM applied to grinding machines is termed EDG. The process uses a graphite wheel as an electrode, and wheels can be up to 12 inches (30.48 cm) in diameter by 6 inches (15.24 cm) wide. The wheel periphery is dressed to the profile required on the workpiece, and the wheel profile can then be transferred to the workpiece as it is traversed past the wheel, which rotates but does not touch the work. EDG machines are highly specialized and are mainly used for producing complex profiles on polycrystaline diamond cutting tools and for shaping carbide tooling such as form tools, thread chasers, dies, and crushing rolls.

**EDM Terms*.—***Anode:* The positive terminal of an electrolytic cell or battery. In EDM, incorrectly applied to the tool or electrode.

* *Source:* Hansvedt Industries

*Barrel effect:* In wire EDM, a condition where the center of the cut is wider than the entry and exit points of the wire, due to secondary discharges caused by particles being pushed to the center by flushing pressure from above and beneath the workpiece.

*Capacitor:* An electrical component that stores an electric charge. In some EDM power supplies, several capacitors are connected across the machining gap and the current for the spark comes directly from the capacitors when they are discharged.

*Cathode:* The negative terminal in an electrolytic cell or battery. In EDM incorrectly applied to the workpiece.

*Colloidal suspension:* Particles suspended in a liquid that are too fine to settle out. In EDM, the tiny particles produced in the sparking action form a colloidal suspension in the dielectric fluid.

*Craters:* Small cavities left on an EDM surface by the sparking action, also known as pits.

*Dielectric filter:* A filter that removes particles from 5 μm (0.00020 inch) down to as fine as 1 μm (0.00004 inch) in size, from dielectric fluid.

*Dielectric fluid:* The non-conductive fluid that circulates between the electrode and the workpiece to provide the dielectric strength across which an arc can occur, to act as a coolant to solidify particles melted by the arc, and to flush away the solidified particles.

*Dielectric strength:* In EDM, the electrical potential (voltage) needed to break down (ionize) the dielectric fluid in the gap between the electrode and the workpiece.

*Discharge channel:* The conductive pathway formed by ionized dielectric and vapor between the electrode and the workpiece.

*Dither:* A slight up and down movement of the machine ram and attached electrode, used to improve cutting stability.

*Duty cycle:* The percentage of a pulse cycle during which the current is turned on (on time), relative to the total duration of the cycle.

*EDG:* Electrical discharge grinding using a machine that resembles a surface grinder but has a wheel made from electrode material. Metal is removed by an EDM process rather than by grinding.

*Electrode growth:* A plating action that occurs at certain low-power settings, whereby workpiece material builds up on the electrode, causing an increase in size.

*Electrode wear:* Amount of material removed from the electrode during the EDM process. This removal can be end wear or corner wear, and is measured linearly or volumetrically but is most often expressed as end wear percent, measured linearly.

*Electroforming:* An electroplating process used to make metal EDM electrodes.

*Energy:* Measured in joules, is the equivalent of volt-coulombs or volt-ampere-seconds.

*Farad:* Unit of electrical capacitance, or the energy-storing capacity of a capacitor.

*Gap:* The closest point between the electrode and the workpiece where an electrical discharge will occur. (See *Overcut*)

*Gap current:* The average amperage flowing across the machining gap.

*Gap voltage:* The voltage across the gap while current is flowing. The voltage across the electrode/workpiece before current flows is called the open gap voltage.

*Heat-affected zone:* The layer below the recast layer, which has been subjected to elevated temperatures that have altered the properties of the workpiece metal.

*Ion:* An atom or group of atoms that has lost or gained one or more electrons and is therefore carrying a positive or negative electrical charge, and is described as being ionized.

*Ionization:* The change in the dielectric fluid that is subjected to a voltage potential whereby it becomes electrically conductive, allowing it to conduct the arc.

*Low-wear:* An EDM process in which the volume of electrode wear is between 2 and 15 percent of the volume of workpiece wear. Normal negative polarity wear ratios are 15 to 40 percent.

*Negative electrode:* The electrode voltage potential is negative relative to the workpiece.

*No-wear:* An EDM process in which electrode wear is virtually eliminated and the wear ratio is usually less than 2 percent by volume.

*Orbit:* A programmable motion between the electrode and the workpiece, produced by a feature built in to the machine, or an accessory, that produces a cavity or hole larger than the electrode. The path can be planetary (circular), vectorial, or polygonal (trace). These motions can often be performed in sequence, and combined with x-axis movement of the electrode.

*Overcut:* The distance between one side of an electrode and the adjacent wall of the workpiece cavity.

*Overcut taper:* The difference between the overcut dimensions at the top (entrance) and at the bottom of the cavity.

*Plasma:* A superheated, highly ionized gas that forms in the discharge channel due to the applied voltage.

*Positive electrode:* The electrode voltage potential is positive with respect to the workpiece. *Negative electrode* is the opposite of this condition.

*Power parameters:* A set of power supply, servo, electrode material, workpiece material, and flushing settings that are selected to produce a desired metal removal rate and surface finish.

*Quench:* The rapid cooling of the EDM surface by the dielectric fluid, which is partially responsible for metallurgical changes in the recast layer and in the heat-affected zone.

*Recast layer:* A layer created by the solidification of molten metal on the workpiece surface after it has been melted by the EDM process.

*Secondary discharge:* A discharge that occurs as conductive particles are carried out along the side of the electrode by the dielectric fluid.

*Spark in:* A method of locating an electrode with respect to the workpiece, using high frequency, low amperage settings so that there is no cutting action. The electrode is advanced toward the workpiece until contact is indicated and this point is used as the basis for setting up the job.

*Spark out:* A technique used in orbiting, which moves the electrode in the same path until sparking ceases.

*Square wave:* An electrical wave shape generated by a solid state power supply.

*Stroke:* The distance the ram travels under servo control.

*UV-axis:* A mechanism that provides for movement of the upper head of a wire EDM machine to allow inclined surfaces to be generated.

*White layer:* The surface layer of an EDM cut that is affected by the heat generated during the process. The characteristics of the layer depend on the material, and may be extremely hard martensite or an annealed layer.

*Wire EDM:* An EDM machine or process in which the electrode is a continuously unspooling, conducting wire that moves in preset patterns in relation to the workpiece.

*Wire guide:* A replaceable precision round diamond insert, sized to match the wire, that guides the wire at the entrance and exit points of a wire cut.

*Wire speed:* The rate at which the wire is fed axially through the workpiece (not the rate at which cutting takes place), adjusted so that clean wire is maintained in the cut slowly enough to minimize waste.

**The EDM Process.—**During the EDM process, energy from the sparks created between the electrode and the workpiece is dissipated by the melting and vaporizing of the workpiece material preferentially, only small amounts of material being lost from the electrode. When current starts to flow between the electrode and the work, the dielectric fluid in the small area in which the gap is smallest, and in which the spark will occur, is transformed into a plasma of hydrogen, carbon, and various oxides. This plasma forms a conducting passageway, consisting of ionized or electrically charged particles, through which the spark can form between the electrode and the workpiece. After current starts to flow, to heat and vaporize a tiny area, the striking voltage is reached, the voltage drops, and the field of ionized particles loses its energy, so that the spark can no longer be sustained. As the voltage then begins to rise again with the increase in resistance, the electrical supply is cut off by the control, causing the plasma to implode and creating a low-pressure pulse that draws in dielectric fluid to flush away metallic debris and cool the impinged area. Such a cycle typically lasts a few microseconds (millionths of a second, or μs), and is repeated continuously in various places on the workpiece as the electrode is moved into the work by the control system.

*Flushing:* An insulating dielectric fluid is made to flow in the space between the workpiece and the electrode to prevent premature spark discharge, cool the workpiece and the electrode, and flush away the debris. For sinker machines, this fluid is paraffin, kerosene, or a silicon-based dielectric fluid, and for wire machines, the dielectric fluid is usually deionized water. The dielectric fluid can be cooled in a heat exchanger to prevent it from rising above about 100° F (38° C), at which cooling efficiency may be reduced. The fluid must also be filtered to remove workpiece particles that would prevent efficient flushing of the spark gaps. Care must be taken to avoid the possibility of entrapment of gases generated by sparking. These gases may explode, endangering life, breaking a valuable electrode or workpiece, or causing a fire.

Flushing away of particles generated during the process is vital to successful EDM operations. A secondary consideration is the heat transferred to the side walls of a cavity, which may cause the workpiece material to expand and close in around the electrode, leading to formation of dc arcs where conductive particles are trapped. Flushing can be done by forcing the fluid to pass through the spark gap under pressure, by sucking it through the gap, or by directing a side nozzle to move the fluid in the tank surrounding the workpiece. In pressure flushing, fluid is usually pumped through strategically placed holes in the electrode or in the workpiece. Vacuum flushing is used when side walls must be accurately formed and straight, and is seldom needed on numerically controlled machines because the table can be programmed to move the workpiece sideways.

Flushing needs careful consideration because of the forces involved, especially where fluid is pumped or sucked through narrow passageways, and large hydraulic forces can easily be generated. Excessively high pressures can lead to displacement of the electrode, the workpiece, or both, causing inaccuracy in the finished product. Many low-pressure flushing holes are preferable to a few high-pressure holes. Pressure-relief valves in the system are recommended.

*Electronic Controls:* The electrical circuit that produces the sparks between the electrode and the workpiece is controlled electronically, the length of the extremely short on and off periods being matched by the operator or the programmer to the materials of the electrode and the workpiece, the dielectric, the rate of flushing, the speed of metal removal, and the quality of surface finish required. The average current flowing between the electrode and the workpiece is shown on an ammeter on the power source, and is the determining factor in machining time for a specific operation. The average spark gap voltage is shown on a voltmeter.

EDM machines can incorporate provision for orbiting the electrode so that flushing is easier, and cutting is faster and increased on one side. Numerical control can also be used to move the workpiece in relation to the electrode with the same results. Numerical control can also be used for checking dimensions and changing electrodes when necessary. The clearance on all sides between the electrode and the workpiece, after the machining operation, is called the overcut or overburn. The overcut becomes greater with increases in the on time, the spark energy, or the amperage applied, but its size is little affected by voltage changes. Allowances must be made for overcut in the dimensioning of electrodes. Sidewall encroachment and secondary discharge can take up parts of these allowances, and electrodes must always be made smaller to avoid making a cavity or hole too large.

*Polarity:* Polarity can affect processing speed, finish, wear, and stability of the EDM operation. On sinker machines, the electrode is generally made positive to protect the electrode from excessive wear and preserve its dimensional accuracy. This arrangement removes metal at a slower rate than electrode negative, which is mostly used for high-speed metal removal with graphite electrodes. Negative polarity is also used for machining carbides, titanium, and refractory alloys using metallic electrodes. Metal removal with graphite electrodes can be as much as 50 percent faster with electrode negative polarity than with electrode positive, but negative polarity results in much faster electrode wear, so it is generally restricted to electrode shapes that can be redressed easily.

Newer generators can provide less than 1 percent wear with either copper or graphite electrodes during roughing operations. Roughing is typically done with a positive-polarity electrode using elevated on times. Some electrodes, particularly micrograin graphites, have a high resistance to wear. Fine-grain, high-density graphites provide better wear characteristics than coarser, less dense grades, and copper-tungsten resists wear better than pure copper electrodes.

*Machine Settings:* For vertical machines, a rule of thumb for power selection on graphite and copper electrodes is 50 to 65 amps per square inch of electrode engagement. For example, an electrode that is ½ $in^2$ (3.2 $cm^2$) might use $0.5 \times 0.5 \times 50 = 12.5$ amps. Although each square inch of electrode surface may be able to withstand higher currents, lower settings should be used with very large jobs or the workpiece may become overheated and it may be difficult to clean up the recast layer. Lower amperage settings are required for electrodes that are thin or have sharp details. The voltage applied across the arc gap between the electrode and the workpiece is ideally about 35 volts, but should be as small as possible to maintain stability of the process.

*Spark Frequency:* Spark frequency is the number of times per second that the current is switched on and off. Higher frequencies are used for finishing operations and for work on cemented carbide, titanium, and copper alloys. The frequency of sparking affects the surface finish produced, low frequencies being used with large spark gaps for rapid metal removal with a rough finish, and higher frequencies with small gaps for finer finishes. High frequency usually increases, and low frequency reduces electrode wear.

*The Duty Cycle:* Electronic units on modern EDM machines provide extremely close control of each stage in the sparking cycle, down to millionths of a second (μs). A typical EDM cycle might last 100 μs. Of this time, the current might be on for 40 μs and off for 60 μs. The relationship between the lengths of the on and off times is called the duty cycle, and it indicates the degree of efficiency of the operation. The duty cycle states the on time as a percentage of the total cycle time, and in the previous example it is 40 percent. Although reducing the off time will increase the duty cycle, factors such as flushing efficiency, electrode and workpiece material, and dielectric condition control the minimum off time. Some EDM units incorporate sensors and fuzzy logic circuits that provide for adaptive control of cutting conditions for unattended operation. Efficiency is also reported as the amount of metal removed, expressed as $in^3$/hr.

In the EDM process, work is done only during the on time, and the longer the on time, the more material is removed in each sparking cycle. Roughing operations use extended on time for high metal-removal rates, resulting in fewer cycles per second, or lower frequency. The resulting craters are broader and deeper so that the surface is rougher and the heat-affected zone (HAZ) on the workpiece is deeper. With positively charged electrodes, the spark moves from the electrode toward the workpiece and the maximum material is removed from the workpiece. However, every spark takes a minute particle from the electrode so that the electrode also is worn away. Finishing electrodes tend to wear much faster than roughing electrodes because more sparks are generated in unit time.

The part of the cycle needed for reionizing the dielectric (the off time) greatly affects the operating speed. Although increasing the off time slows the process, longer off times can increase stability by providing more time for the ejected material to be swept away by the flow of the dielectric fluid, and for deionization of the fluid, so that erratic cycling of the servo-mechanisms that advance and retract the electrode is avoided. In any vertical

EDM operation, if the overcut, wear, and finish are satisfactory, machining speed can best be adjusted by slowly decreasing the off time setting in small increments of 1 to 5 μs until machining becomes erratic, then returning to the previous stable setting. As the off time is decreased, the machining gap or gap voltage will slowly fall and the working current will rise. The gap voltage should not be allowed to drop below 35 to 40 volts.

*Metal Removal Rates (MRR):* The amount of metal removed in any EDM process depends largely on the length of the on time, the energy/spark, and the number of sparks/second. The following data were provided by Poco Graphite, Inc., in their *EDM Technical Manual*. For a typical roughing operation using electrode positive polarity on high-carbon steel, a 67 percent duty cycle removed 0.28 $in^3$/hr (4.59 $cm^3$/hr). For the same material, a 50 percent duty cycle removed 0.15 $in^3$/hr (2.46 $cm^3$/hr), and a 33 percent duty cycle for finishing removed 0.075 $in^3$/hr (1.23 $cm^3$/hr).

In another example, shown in the top data row in Table 1, a 40 percent duty cycle with a frequency of 10 kHz and peak current of 50 amps was run for 5 minutes of cutting time. Metal was removed at the rate of 0.8 $in^3$/hr (13.11 $cm^3$/hr) with electrode wear of 2.5 percent and a surface finish of 400 μinch Ra. When the on and off times in this cycle were halved, as shown in the second data row in Table 1, the duty cycle remained at 40 percent, but the frequency doubled to 20 kHz. The result was that the peak current remained unaltered, but with only half the on time the MRR was reduced to 0.7 $in^3$/hr (11.47 $cm^3$/hr), the electrode wear increased to 6.3 percent, and the surface finish improved to 300 μinch Ra. The third and fourth rows in Table 1 show other variations in the basic cycle and the results.

**Table 1. Effect of Electrical Control Adjustments on EDM Operations**

| On Time (μs) | Off Time (μs) | Frequency (kHz) | Peak Current (Amps) | Metal Removal Rate ($in^3$/hr) | Metal Removal Rate ($cm^3$/hr) | Electrode Wear (%) | Surface Finish (μin Ra) |
|---|---|---|---|---|---|---|---|
| 40 | 60 | 10 | 50 | 0.8 | 13.1 | 2.5 | 400 |
| 20 | 30 | 20 | 50 | 0.7 | 11.47 | 6.3 | 300 |
| 40 | 10 | 20 | 50 | 1.2 | 19.66 | 1.4 | 430 |
| 40 | 60 | 10 | 25 | 0.28 | 4.59 | 2.5 | 350 |

*The Recast Layer:* One drawback of the EDM process when used for steel is the recast layer, which is created wherever sparking occurs. The oil used as a dielectric fluid causes the EDM operation to become a random heat-treatment process in which the metal surface is heated to a very high temperature, then quenched in oil. The heat breaks down the oil into hydrocarbons, tars, and resins, and the molten metal draws out the carbon atoms and traps them in the resolidified metal to form the very thin, hard, brittle surface called the recast layer that covers the heat-affected zone (HAZ). This recast layer has a white appearance and consists of particles of material that have been melted by the sparks, enriched with carbon, and drawn back to the surface or retained by surface tension. The recast layer is harder than the parent metal and can be as hard as glass, and must be reduced or removed by vapor blasting with glass beads, polishing, electrochemical or abrasive flow machining, after the shaping process is completed, to avoid cracking or flaking of surface layers that may cause failure of the part in service.

Beneath the thin recast layer, the HAZ, in steel, consists of martensite that usually has been hardened by the heating and cooling sequences coupled with the heat-sink cooling effect of a thick steel workpiece. This martensite is hard and its rates of expansion and contraction are different from those of the parent metal. If the workpiece is subjected to heating and cooling cycles in use, the two layers are constantly stressed and these stresses may cause formation of surface cracks. The HAZ is usually much deeper in a workpiece cut on a sinker than on a wire machine, especially after roughing, because of the increased heating effect caused by the higher amounts of energy applied.

The depth of the HAZ depends on the amperage and the length of the on time, increasing as these values increase, to about 0.012–0.015 inch (0.30–0.38 mm) deep. Residual stress in the HAZ can range up to 650 N/mm$^2$. As the HAZ cannot be removed easily, it is best avoided by programming the series of cuts taken on the machine so that most of the HAZ produced by one cut is removed by the following cut. If time is available, cut depth can be reduced gradually until the finishing cuts produce an HAZ having a thickness of less than 0.0001 inch (2.54 mm).

**Workpiece Materials.—**Most homogeneous materials used in metalworking can be shaped by the EDM process. Some data on typical workpiece materials are given in Table 2. Sintered materials present some difficulties caused by the use of a cobalt or other binder used to hold the carbide or other particles in the matrix. The binder usually melts at a lower temperature than the tungsten, molybdenum, titanium, or other carbides, so it is preferentially removed by the sparking sequence, and the carbide particles are thus loosened and freed from the matrix. The structures of sintered materials based on tungsten, cobalt, and molybdenum require higher EDM frequencies with very short on times, so that there is less danger of excessive heat buildup, leading to melting. Copper-tungsten electrodes are recommended for EDM of tungsten carbides. When used with high frequencies for powdered metals, graphite electrodes often suffer from excessive wear.

Workpieces of aluminum, brass, and copper should be processed with metallic electrodes of low melting points such as copper or copper-tungsten. Workpieces of carbon and stainless steel that have high melting points should be processed with graphite electrodes. The melting points and specific gravities of the electrode material and of the workpiece should preferably be similar.

**Table 2. Characteristics of Common Workpiece Materials for EDM**

| Material | Specific Gravity | Melting Point | | Vaporization Temperature | | Conductivity (Silver = 100) |
|---|---|---|---|---|---|---|
| | | °F | °C | °F | °C | |
| Aluminum | 2.70 | 1220 | 660 | 4442 | 2450 | 63.00 |
| Brass | 8.40 | 1710 | 930 | … | | … |
| Cobalt | 8.71 | 2696 | 1480 | 5520 | 2900 | 16.93 |
| Copper | 8.89 | 1980 | 1082 | 4710 | 2595 | 97.61 |
| Graphite | 2.07 | N/A | | 6330 | 3500 | 70.00 |
| Inconel | … | 2350 | 1285 | … | | … |
| Magnesium | 1.83 | 1202 | 650 | 2025 | 1110 | 39.40 |
| Manganese | 7.30 | 2300 | 1260 | 3870 | 2150 | 15.75 |
| Molybdenum | 10.20 | 4748 | 2620 | 10,040 | 5560 | 17.60 |
| Nickel | 8.80 | 2651 | 1455 | 4900 | 2730 | 12.89 |
| Carbon Steel | 7.80 | 2500 | 1371 | … | | 12.00 |
| Tool Steel | … | 2730 | 1500 | … | | … |
| Stainless Steel | … | 2750 | 1510 | … | | … |
| Titanium | 4.50 | 3200 | 1700 | 5900 | 3260 | 13.73 |
| Tungsten | 18.85 | 6098 | 3370 | 10,670 | 5930 | 14.00 |
| Zinc | 6.40 | 790 | 420 | 1663 | 906 | 26.00 |

**Electrode Materials.—**Most EDM electrodes are made from graphite, which provides a much superior rate of metal removal than copper because of the ability of graphite to resist thermal damage. Graphite has a density of 1.55 to 1.85 g/cm$^3$, lower than most metals. Instead of melting when heated, graphite sublimates, that is, it changes directly from a solid to a gas without passing through the liquid stage. Sublimation of graphite occurs at a temperature of 3350° C (6062° F). EDM graphite is made by sintering a compressed mixture of fine graphite powder (1 to 100 micron particle size) and coal tar pitch in a furnace. The open structure of graphite means that it is eroded more rapidly than metal in the EDM process. The electrode surface is also reproduced on the surface of the workpiece. The sizes of individual surface recesses may be reduced during sparking when the work is moved under numerical control of workpiece table movements.

The fine grain sizes and high densities of graphite materials that are specially made for high-quality EDM finishing provide high wear resistance, better finish, and good reproduction of fine details, but these fine grades cost more than graphite of larger grain sizes and lower densities. Premium grades of graphite cost up to five times as much as the least expensive and about three times as much as copper, but the extra cost often can be justified by savings during machining or shaping of the electrode.

Graphite has a high resistance to heat and wear at lower frequencies, but will wear more rapidly when used with high frequencies or with negative polarity. Infiltrated graphites for EDM electrodes are also available as a mixture of copper particles in a graphite matrix, for applications where good machinability of the electrode is required. This material presents a trade-off between lower arcing and greater wear with a slower metal-removal rate, but costs more than plain graphite.

EDM electrodes are also made from copper, tungsten, silver-tungsten, brass, and zinc, all of which have good electrical and thermal conductivity. However, because these metals have melting points below those encountered in the spark gap, they wear rapidly. Copper with 5 percent tellurium, added for better machining properties, is the most commonly used metal alloy. Tungsten resists wear better than brass or copper and is more rigid when used for thin electrodes but is expensive and difficult to machine. Metal electrodes, with their more even surfaces and slower wear rates, are often preferred for finishing operations on work that requires a smooth finish. In fine-finishing operations, the arc gap between the surfaces of the electrode and the workpiece is very small, and there is a danger of dc arcs being struck, causing pitting of the surface when particles dislodged from a graphite electrode during fine-finishing cuts are not flushed from the gap. If struck by a spark, such a particle may provide a path for a continuous discharge of current that will mar the almost completed work surface.

Some combinations of electrode and workpiece material, electrode polarity, and likely amounts of corner wear are listed in Table 3. Corner wear rates indicate the ability of the electrode to maintain its shape and reproduce fine detail. The column headed Capacitance refers to the use of capacitors in the control circuits to increase the impact of the spark without increasing the amperage. Such circuits can accomplish more work in a given time, at the expense of surface-finish quality and increased electrode wear.

**Table 3. Types of Electrodes Used for Various Workpiece Materials**

| Electrode | Electrode Polarity | Workpiece Material | Corner Wear (%) | Capacitance |
|---|---|---|---|---|
| Copper | + | Steel | 2–10 | No |
| Copper | + | Inconel | 2–10 | No |
| Copper | + | Aluminum | <3 | No |
| Copper | – | Titanium | 20–40 | Yes |
| Copper | – | Carbide | 35–60 | Yes |
| Copper | – | Copper | 34–45 | Yes |
| Copper | – | Copper-tungsten | 40–60 | Yes |
| Copper-tungsten | + | Steel | 1–10 | No |
| Copper-tungsten | – | Copper | 20–40 | Yes |
| Copper-tungsten | – | Copper-tungsten | 30–50 | Yes |
| Copper-tungsten | – | Titanium | 15–25 | Yes |
| Copper-tungsten | – | Carbide | 35–50 | Yes |
| Graphite | + | Steel | <1 | No |
| Graphite | – | Steel | 30–40 | No |
| Graphite | + | Inconel | <1 | No |
| Graphite | – | Inconel | 30–40 | No |
| Graphite | + | Aluminum | <1 | No |
| Graphite | – | Aluminum | 10–20 | No |
| Graphite | – | Titanium | 40–70 | No |
| Graphite | – | Copper | N/A | Yes |

*Electrode Wear:* Wear of electrodes can be reduced by leaving the smallest amounts of finishing stock possible on the workpiece and using no-wear or low-wear settings to remove most of the remaining material so that only a thin layer remains for finishing with the redressed electrode. The material left for removal in the finishing step should be only slightly more than the maximum depth of the craters left by the previous cut. Finishing operations should be regarded as only changing the quality of the finish, not removing metal or sizing. Low power with very high frequencies and minimal amounts of offset for each finishing cut are recommended.

On manually adjusted machines, fine finishing is usually carried out by several passes of a full-size finishing electrode. Removal of a few thousandths of an inch from a cavity with such an arrangement requires the leading edge of the electrode to recut the cavity over the entire vertical depth. By the time the electrode has been sunk to full depth, it is so worn that precision is lost. This problem sometimes can be avoided on a manual machine by use of an orbiting attachment that will cause the electrode to traverse the cavity walls, providing improved speed, finish, and flushing, and reducing corner wear on the electrode.

*Selection of Electrode Material:* Factors that affect selection of electrode material include metal-removal rate, wear resistance (including volumetric, corner, end, and side, with corner wear being the greatest concern), desired surface finish, costs of electrode manufacture and material, and characteristics of the material to be machined. A major factor is the ability of the electrode material to resist thermal damage, but the electrode's density, the polarity, and the frequencies used are all important factors in wear rates. Copper melts at about 1085° C (1985° F) and spark-gap temperatures must generally exceed 3800° C (6872° F), so use of copper may be made unacceptable because of its rapid wear rates. Graphites have good resistance to heat and wear at low frequencies, but will wear more with high frequency, negative polarity, or a combination of these.

**Making Electrodes.—**Electrodes made from copper and its alloys can be machined conventionally by lathes, and milling and grinding machines, but copper acquires a burr on run-off edges during turning and milling operations. For grinding copper, the wheel must often be charged with beeswax or similar material to prevent loading of the surface. Flat grinding of copper is done with wheels having open grain structures (46-J, for instance) to contain the wax and to allow room for the soft, gummy, copper chips. For finish grinding, wheels of at least 60 and up to 80 grit should be used for electrodes requiring sharp corners and fine detail. These wheels will cut hot and load up much faster, but are necessary to avoid rapid breakdown of sharp corners.

Factors to be considered in selection of electrode materials are:the electrode material cost/in$^3$; the time to manufacture electrodes; difficulty of flushing; the number of electrodes needed to complete the job; speed of the EDM; amount of electrode wear during EDM; and workpiece surface-finish requirements.

Copper electrodes have the advantage over graphite in their ability to be discharge-dressed in the EDM, usually under computer numerical control (CNC). The worn electrode is engaged with a premachined dressing block made from copper-tungsten or carbide. The process renews the original electrode shape, and can provide sharp, burr-free edges. Because of its higher vaporization temperature and wear resistance, discharge dressing of graphite is slow, but graphite has the advantage that it can be machined conventionally with ease.

*Machining Graphite:* Graphites used for EDM are very abrasive, so carbide tools are required for machining them. Graphite does not shear away and flow across the face of the tool as metal does, but fractures or is crushed by the tool pressure and floats away as a fine powder or dust. Graphite particles have sharp edges and, if allowed to mix with the machine lubricant, will form an abrasive slurry that causes rapid wear of machine guiding surfaces. Dust may also cause respiratory problems and allergic reactions, especially if the graphite is infiltrated with copper, so an efficient exhaust system is needed for machining.

Compressed air can be used to flush out the graphite dust from blind holes, for instance, but provision must be made for vacuum removal of the dust to avoid hazards to health and problems with wear caused by the hard, sharp-edged particles. Air velocities of at least 500 ft/min (152 m/min) are recommended for flushing, and of 2000 ft/min (610 m/min) in collector ducts to prevent settling out. Fluids can also be used, but small-pore filters are needed to keep the fluid clean. High-strength graphite can be clamped or chucked tightly but care must be taken to avoid crushing. Collets are preferred for turning because of the uniform pressure they apply to the workpiece. Sharp corners on electrodes made from less dense graphite are liable to chip or break away during machining.

For conventional machining of graphite, tools of high-quality tungsten carbide or polycrystaline diamond are preferred and must be kept sharp. Recommended cutting speeds for high-speed steel tools are 100–300 ft/min (30–91 m/min), tungsten carbide 500–750 ft/min (152–229 m/min), and polycrystaline diamond, 500–2000 surface ft/min (152–610 m/min). Tools for turning should have positive rake angles and nose radii of $\frac{1}{64}$ to $\frac{1}{32}$ inch. Depths of cut of 0.015 to 0.020 inch produce a better finish than light cuts such as 0.005 inch because of the tendency of graphite to chip away rather than flow across the tool face. Low feed rates of 0.005 inch/rev for rough- and 0.001 to 0.003 inch/rev for finish-turning are preferred. Cutting off is best done with a tool having an angle of 20°.

For band sawing graphite, standard carbon steel blades can be run at 2100–3100 surface ft/min (640–945 m/min). Use low power feed rates to avoid overloading the teeth and the feed rate should be adjusted until the saw has a very slight speed up at the breakthrough point. Milling operations require rigid machines, short tool extensions, and firm clamping of parts. Milling cutters will chip the exit side of the cut, but chipping can be reduced by use of sharp tools, positive rake angles, and low feed rates to reduce tool pressure. Feed/tooth for two-flute end mills is 0.003 to 0.005 inch (0.076 to 0.13 mm) for roughing and 0.001 to 0.003 inch (0.025 to 0.076 mm) for finishing.

Standard high-speed steel drills can be used for drilling holes but will wear rapidly, causing holes that are tapered or undersized, or both. High-spiral, tungsten carbide drills should be used for large numbers of holes over $\frac{1}{16}$ inch (1.59 mm) diameter, but diamond-tipped drills will last longer. Pecking cycles should be used to clear dust from the holes. Compressed air can be passed through drills with through coolant holes to clear dust. Feed rates for drilling are 0.0015 to 0.002 inch/rev for drills up to $\frac{1}{32}$, 0.001 to 0.003 inch/rev for $\frac{1}{32}$- to $\frac{1}{8}$-inch drills, and 0.002 to 0.005 inch/rev for larger drills. Standard taps without fluid are best used for through holes, and for blind holes, tapping should be completed as far as possible with a taper tap before the bottoming tap is used.

For surface grinding of graphite, a medium (60) grade, medium-open structure, vitreous-bond, green-grit, silicon-carbide wheel is most commonly used. The wheel speed should be 5300–6000 surface ft/min (1615–1829 m/min), with traversing feed rates at about 56 ft/min (17.1 m/min). Roughing cuts are taken at 0.005 to 0.010 inch/pass (0.13–0.25 mm/pass), and finishing cuts at 0.001 to 0.003 inch/pass (0.025–0.076 mm/pass). Surface finishes in the range of 18 to 32 μinch (0.457–0.813 micron) Ra are normal, and can be improved by longer spark-out times and finer grit wheels, or by lapping. Graphite can be centerless ground using a silicon-carbide, resinoid-bond work wheel and a regulating wheel speed of 195 ft/min (59.4 m/min).

Wire EDM, orbital abrading, and ultrasonic machining are also used to shape graphite electrodes. Orbital abrading uses a die containing hard particles to remove graphite, and can produce a fine surface finish. In ultrasonic machining, a water-based abrasive slurry is pumped between the die attached to the ultrasonic transducer and the graphite workpiece on the machine table. Ultrasonic machining is rapid and can reproduce small details down to 0.002 inch (0.05 mm) in size, with surface finishes down to 8 μin (0.203 micron) Ra. If coolants are used, the graphite should be dried for 1 hour at over 400° F (204° C) (but not in a microwave oven) to remove liquids before use.

**Wire EDM.—**In the wire EDM process, with deionized water as the dielectric fluid, carbon is extracted from the recast layer, rather than added to it. When copper-base wire is used, copper atoms migrate into the recast layer, softening the surface slightly so that wire-cut surfaces are sometimes softer than the parent metal. On wire EDM machines, very high amperages are used with very short on times, so that the heat-affected zone (HAZ) is quite shallow. With proper adjustment of the on and off times, the depth of the HAZ can be held below 1 micron (0.00004 inch).

The cutting wire is used only once, so that the portion in the cut is always cylindrical and has no spark-eroded sections that might affect the cut accuracy. The power source controls the electrical supply to the wire and to the drive motors on the table to maintain the preset arc gap within 0.1 micron (0.000004 inch) of the programmed position. On wire EDM machines, the water used as a dielectric fluid is deionized by a deionizer included in the cooling system, to improve its properties as an insulator. Chemical balance of the water is also important for good dielectric properties.

*Drilling Holes for Wire EDM:* Before an aperture can be cut in a die plate, a hole must be provided in the workpiece. Such holes are often "drilled" by EDM, and the wire threaded through the workpiece before starting the cut. The "EDM drill" does not need to be rotated, but rotation will help in flushing and reduce electrode wear. The EDM process can drill a hole 0.04 inch (1 mm) in diameter through 4-inch (101.6-mm) thick steel in about 3 minutes, using an electrode made from brass or copper tubing. Holes of smaller diameter can be drilled, but the practical limit is 0.012 inch (0.3 mm) because of the overcut, the lack of rigidity of tubing in small sizes, and the excessive wear on such small electrodes. The practical upper size limit on holes is about 0.12 inch (3 mm) because of the comparatively large amounts of material that must be eroded away for larger sizes. However, EDM is commonly used for making large or deep holes in such hard materials as tungsten carbide. For instance, a 0.2-inch (5-mm) hole has been made in carbide 2.9-inch (73.66-mm) thick in 49 minutes by EDM. Blind holes are difficult to produce with accuracy, and must often be made with cut-and-try methods.

Deionized water is usually used for drilling and is directed through the axial hole in the tubular electrode to flush away the debris created by the sparking sequence. Because of the need to keep the extremely small cutting area clear of metal particles, the dielectric fluid is often not filtered but is replaced continuously by clean fluid that is pumped from a supply tank to a disposal tank on the machine.

*Wire Electrodes:* Wire for EDM generally is made from yellow brass containing copper 63 and zinc 37 percent, with a tensile strength of 50,000–145,000 $lb_f/in^2$ (345–1000 MPa), and may be from 0.002–0.012 inch (0.05–0.30 mm) diameter.

In addition to yellow brass, electrode wires are also made from brass alloyed with aluminum or titanium for tensile strengths of 140,000–160,000 $lb_f/in^2$ (965–1103 MPa). Wires with homogeneous, uniform electrolytic coatings of alloys such as brass or zinc are also used. Zinc is favored as a coating on brass wires because it gives faster cutting and reduced wire breakage due to its low melting temperature of 419° C, and vaporization temperature of 906° C. The layer of zinc can boil off while the brass core, which melts at 930° C, continues to deliver current.

Some wires for EDM are made from steel for strength, with a coating of brass, copper, or other metal. Most wire machines use wire negative polarity (the wire is negative) because the wire is constantly renewed and is used only once, so wear is not important. Important qualities of wire for EDM include smooth surfaces, free from nicks, scratches and cracks, precise diameters to ±0.00004 inch (±1 micron) for drawn and ±0.00006 inch (±2 μm) for plated, high tensile strength, consistently good ductility, uniform spooling, and good protective packaging.

## METAL CASTING, MOLDING, AND EXTRUSION

### Material Properties for Iron and Steel Casting

Cast irons and cast steels encompass a large family of ferrous alloys, which, as the name implies, are cast to shape rather than being formed by working in the solid state. In general, cast irons contain more than 2 percent carbon and from 1 to 3 percent silicon. Varying the balance between carbon and silicon, alloying with different elements, and changing melting, casting, and heat-treating practices can produce a broad range of properties. In most cases, the carbon exists in two forms: free carbon in the form of graphite and combined carbon in the form of iron carbide (cementite). Mechanical and physical properties depend strongly on the shape and distribution of the free graphite and the type of matrix surrounding the graphite particles.

The four basic types of cast iron are white iron, gray iron, malleable iron, and ductile iron. In addition to these basic types, there are other specific forms of cast iron to which special names have been applied, such as chilled iron, alloy iron, and compacted graphite cast iron.

**Gray Cast Iron.**—Gray cast iron may easily be cast into any desirable form and it may also be machined readily. It usually contains from 1.7 to 4.5 percent carbon, and from 1 to 3 percent silicon. The excess carbon is in the form of graphite flakes, which impart to the material the dark-colored fracture that gives it its name. Gray iron castings are widely used for such applications as machine tools, automotive cylinder blocks, cast-iron pipe and fittings, and agricultural implements.

The American National Standard Specifications for Gray Iron Castings, ANSI/ASTM A48-76, groups the castings into two categories. Gray iron castings in Classes 20A, 20B, 20C, 25A, 25B, 25C, 30A, 30B, 30C, 35A, 35B, and 35C are characterized by excellent machinability, high damping capacity, low modulus of elasticity, and comparative ease of manufacture. Castings in Classes 40B, 40C, 45B, 45C, 50B, 50C, 60B, and 60C are usually more difficult to machine, have lower damping capacity, higher modulus of elasticity, and are more difficult to manufacture. The prefix number indicates minimum tensile strength in pounds per square inch, e.g., 20 is 20,000 psi (138 MPa), 25 is 25,000 psi (172 MPa), 30 is 30,000 psi (207 MPa), etc.

High-strength iron castings produced by the Meehanite-controlled process may have various combinations of physical properties to meet different requirements. In addition to a number of general engineering types, there are heat-resisting, wear-resisting and corrosion-resisting Meehanite castings.

**White Cast Iron.**—When nearly all of the carbon in a casting is in the combined or cementite form, it is known as white cast iron. It is so named because it has a silvery-white fracture. White cast iron is very hard and also brittle; its ductility is practically zero. Castings of this material need particular attention with respect to design since sharp corners and thin sections result in material failures at the foundry. These castings are less resistant to impact loading than gray iron castings, but they have a compressive strength that is usually higher than 200,000 psi (1379 MPa) as compared to 65,000–160,000 psi (448–1103 MPa) for gray iron castings. Some white iron castings are used for applications that require maximum wear resistance but most of them are used in the production of malleable iron castings.

**Chilled Cast Iron.**—Many gray iron castings have wear-resisting surfaces of white cast iron. These surfaces are designated by the term "chilled cast iron" since they are produced in molds having metal chills for cooling the molten metal rapidly. This rapid cooling results in the formation of cementite and white cast iron.

**Alloy Cast Iron.**—This term designates castings containing alloying elements such as nickel, chromium, molybdenum, copper, and manganese in sufficient amounts to appreciably change the physical properties. These elements may be added either to increase the strength or to obtain special properties such as higher wear resistance, corrosion resistance, or heat resistance. Alloy cast irons are used extensively for such parts as automotive cylinders, pistons, piston rings, crankcases, and brake drums; for certain machine tool castings; for certain types of dies; for parts of crushing and grinding machinery; and for applications where the casting must resist scaling at high temperatures. Machinable alloy cast irons having tensile strengths up to 70,000 psi (483 MPa) or even higher may be produced.

**Malleable-Iron Castings.**—Malleable iron is produced by the annealing or graphitization of white iron castings. The graphitization in this case produces temper carbon which is graphite in the form of compact rounded aggregates. Malleable castings are used for many industrial applications where strength, ductility, machinability, and resistance to shock are important factors. In manufacturing these castings, the usual procedure is to first produce a hard, brittle white iron from a charge of pig iron and scrap. These hard white-iron castings are then placed in stationary batch-type furnaces or car-bottom furnaces and the graphitization (malleablizing) of the castings is accomplished by means of a suitable annealing heat treatment. During this annealing period the temperature is slowly (50 hours) increased to as much as 1650 or 1700° F (899 or 927° C), after which time it is slowly (60 hours) cooled. The American National Standard Specifications for Malleable Iron Castings, ANSI/ASTM A47-77, specifies the following grades and their properties: No. 32520, having a minimum tensile strength of 50,000 psi (345 MPa), a minimum yield strength of 32,500 psi (224 MPa), and a minimum elongation in 2 inches (50.8 mm) of 10 percent; and No. 35018, having a minimum tensile strength of 53,000 psi (365 MPa), a minimum yield strength of 35,000 psi (241 MPa), and a minimum elongation in 2 inches of 18 percent.

*Cupola Malleable Iron:* Another method of producing malleable iron involves initially the use of a cupola or a cupola in conjunction with an air furnace. This type of malleable iron, called cupola malleable iron, exhibits good fluidity and will produce sound castings. It is used in the making of pipe fittings, valves, and similar parts and possesses the useful property of being well suited to galvanizing. The American National Standard Specifications for Cupola Malleable Iron, ANSI/ASTM 197-79, calls for a minimum tensile strength of 40,000 pounds per square inch (276 MPa); a minimum yield strength of 30,000 psi (207 MPa); and a minimum elongation in 2 inches of 5 percent.

*Pearlitic Malleable Iron:* This type of malleable iron contains some combined carbon in various forms. It may be produced either by stopping the heat treatment of regular malleable iron during production before the combined carbon contained therein has all been transformed to graphite or by reheating regular malleable iron above the transformation range. Pearlitic malleable irons exhibit a wide range of properties and are used in place of steel castings or forgings or to replace malleable iron when a greater strength or wear resistance is required. Some forms are made rigid to resist deformation while others will undergo considerable deformation before breaking. This material has been used in axle housings, differential housings, camshafts, and crankshafts for automobiles; machine parts; ordnance equipment; and tools. Tension test requirements of pearlitic malleable iron castings called for in ASTM A220-99 (R2018) are given in the accompanying table.

**Tension Test Requirements of Pearlitic Malleable Iron Castings *ASTM A220-99* (*R2018*)**

| Casting Grade Numbers | | 40010 | 45008 | 45006 | 50005 | 60004 | 70003 | 80002 | 90001 |
|---|---|---|---|---|---|---|---|---|---|
| Min. Tensile Strength | psi | 60000 | 65000 | 65000 | 70000 | 80000 | 85000 | 95000 | 105000 |
| | MPa | 414 | 448 | 448 | 483 | 552 | 586 | 655 | 724 |
| Min. Yield Strength | psi | 40000 | 45000 | 45000 | 50000 | 60000 | 70000 | 80000 | 90000 |
| | MPa | 276 | 310 | 310 | 345 | 414 | 483 | 552 | 621 |
| Min. Elongation in 2 Inch, % | | 10 | 8 | 6 | 5 | 4 | 3 | 2 | 1 |

**Ductile Cast Iron.**—A distinguishing feature of this widely used type of cast iron, also known as spheroidal graphite iron or nodular iron, is that the graphite is present in ball-like form instead of in flakes as in ordinary gray cast iron. The addition of small amounts of magnesium- or cerium-bearing alloys together with special processing produces this spheroidal graphite structure and results in a casting of high strength and appreciable ductility. Its toughness is intermediate between that of cast iron and steel, and its shock resistance is comparable to ordinary grades of mild carbon steel. Melting point and fluidity are similar to those of the high-carbon cast irons. It exhibits good pressure tightness under high stress and can be welded and brazed. It can be softened by annealing or hardened by normalizing and air cooling or oil quenching and drawing.

Five grades of this iron are specified in ASTM A536-84 (R2014), Standard Specification for Ductile Iron Castings. The grades and their corresponding matrix microstructures and heat treatments are as follows: Grade 60-40-18, ferritic, may be annealed; Grade 65-45-12, mostly ferritic, as-cast or annealed; Grade 80-55-06, ferritic/pearlitic, as-cast; Grade 100-70-03, mostly pearlitic, may be normalized; Grade 120-90-02, martensitic, oil quenched and tempered. The grade nomenclature identifies the minimum tensile strength, on percent yield strength, and percent elongation in 2 inches. Thus, Grade 60-40-18 has a minimum tensile strength of 60,000 psi, a minimum 0.2 percent yield strength of 40,000 psi, and minimum elongation in 2 inches of 18 percent. Several other types are commercially available to meet specific needs. The common grades of ductile iron can also be specified by just the Brinell Hardness Number, although the appropriate microstructure for the indicated hardness is also a requirement. This method is used in SAE Specification J434C for automotive castings and similar applications. Other specifications not only specify tensile properties but also have limitations in composition. Austenitic types with high nickel content, high corrosion resistance, and good strength at elevated temperatures, are specified in ASTM A439-18.

Ductile cast iron can be cast in molds containing metal chills if wear-resisting surfaces are desired. Hard carbide areas will form in a manner similar to the forming of areas of chilled cast iron in gray iron castings. Surface hardening by flame or induction methods is also feasible. Ductile cast iron can be machined with the same ease as gray cast iron. It finds use as crankshafts, pistons, and cylinder heads in the automotive industry; forging hammer anvils, cylinders, guides, and control levers in the heavy machinery field; and wrenches, clamp frames, faceplates, chuck bodies, and dies for forming metals in the tool and die field. The production of ductile iron castings involves complex metallurgy, the use of special melting stock, and close process control. The majority of applications of ductile iron have been made to utilize its excellent mechanical properties in combination with the castability, machinability, and corrosion resistance of gray iron.

**Steel Castings.—**Steel castings are especially adapted for machine parts that must withstand shocks or heavy loads. They are stronger than wrought iron, cast iron, or malleable iron, and are very tough. The steel used for making steel castings may be produced either by the open-hearth, electric arc, side-blow converter, or electric induction methods. The raw materials used are steel scrap, pig iron, and iron ore, the materials and their proportions varying according to the process and the type of furnace used. The open-hearth method is used when large tonnages are continually required, while a small electric furnace might be used for steels of widely differing analyses, which are required in small lot production. The high frequency induction furnace is used for small quantity production of expensive steels of special composition such as high-alloy steels. Steel castings are used for such parts as hydroelectric turbine wheels, forging presses, gears, railroad car frames, valve bodies, pump casings, mining machinery, marine equipment, engine casings, etc.

Steel castings can generally be made from any of the many types of carbon and alloy steels produced in wrought form and respond similarly to heat treatment; they also do not exhibit directionality effects that are typical of wrought steel. Steel castings are classified into two general groups: carbon steel and alloy steel.

**Carbon Steel Castings.—**Carbon steel castings may be designated as low-carbon medium-carbon, and high-carbon. Low-carbon steel castings have a carbon content of less than 0.20 percent (most are produced in the 0.16 to 0.19 percent range). Other elements present are: manganese, 0.50 to 0.85 percent; silicon, 0.25 to 0.70 percent; phosphorus, 0.05 percent max.; and sulfur, 0.06 percent max. Their tensile strengths (annealed condition) range from 40,000 to 70,000 pounds per square inch (276–483 MPa). Medium-carbon steel castings have a carbon content of from 0.20 to 0.50 percent. Other elements present are manganese, 0.50 to 1.00 percent; silicon, 0.20 to 0.80 percent; phosphorus, 0.05 percent max.; and sulfur, 0.06 percent max. Their tensile strengths range from 65,000 to 105,000 pounds per square inch (448–724 MPa) depending, in part, upon heat treatment. High-carbon steel castings have a carbon content of more than 0.50 percent and also contain: manganese, 0.50 to 1.00 percent; silicon, 0.20 to 0.70 percent; and phosphorus and sulfur, 0.05 percent max. each. Fully annealed high-carbon steel castings exhibit tensile strengths of from 95,000 to 125,000 pounds per square inch (655–125 MPa). See Table 1 for grades and properties of carbon steel castings.

**Table 1. Mechanical Properties of Steel Castings**

| Tensile Strength, ksi (MPa) | Yield Point, ksi (MPa) | Elongation in 2 Inch, Percent | Brinell Hardness Number | Type of Heat Treatment | Application Indicating Properties |
|---|---|---|---|---|---|
| Structural Grades of Carbon Steel Castings | | | | | |
| 60 (414) | 30 (207) | 32 | 120 | Annealed | Low electric resistivity. Desirable magnetic properties. Carburizing and case-hardening grades. Weldability. |
| 65 (448) | 35 (241) | 30 | 130 | Normalized | Good weldability. Medium strength with good machinability and high ductility. |
| 70 (483) | 38 (262) | 28 | 140 | Normalized | |
| 80 (552) | 45 (310) | 26 | 160 | Normalized and tempered | High strength carbon steels with good machinability, toughness and good fatigue resistance. |
| 85 (586) | 50 (345) | 24 | 175 | | |
| 100 (689) | 70 (483) | 20 | 200 | Quenched and tempered | Wear resistance. Hardness. |
| Engineering Grades of Low-Alloy Steel Castings | | | | | |
| 70 (483) | 45 (310) | 26 | 150 | Normalized and tempered | Good weldability. Medium strength with high toughness and good machinability. For high temperature service. |
| 80 (552) | 50 (345) | 24 | 170 | | |
| 90 (621) | 60 (414) | 22 | 190 | Normalized and tempered[a] | Certain steels of these classes have good high temperature properties and deep hardening properties. Toughness. |
| 100 (689) | 68 (469) | 20 | 209 | | |
| 110 (758) | 85 (586) | 20 | 235 | Quenched and tempered | Impact resistance. Good low temperature properties for certain steels. Deep hardening. Good combination of strength and toughness. |
| 120 (827) | 95 (655) | 16 | 245 | | |
| 150 (1034) | 125 (862) | 12 | 300 | Quenched and tempered | Deep hardening. High strength. Wear and fatigue resistance. |
| 175 (1207) | 148 (1020) | 8 | 340 | Quenched and tempered | High strength and hardness. Wear resistance. High fatigue resistance. |
| 200 (1379) | 170 (1172) | 5 | 400 | | |

[a] Quench and temper heat treatments may also be employed for these classes.

ksi = kips per square inch = 1000s of pounds per square inch; MPa = megapascals.

The data in the table above is for general information purposes only and not for use as design or specification limit values. The values listed above have been compiled by the Steel Founders' Society of America as those normally expected in the production of steel castings. The castings are classified according to tensile strength values which are given in the first column. Specifications covering steel castings are prepared by the American Society for Testing and Materials, the Association of American Railroads, the Society of Automotive Engineers, the United States government (federal and military specifications), etc. Refer to publications issued by these organizations for these specifications.

**Alloy Steel Castings.—**Alloy cast steels are those in which special alloying elements such as manganese, chromium, nickel, molybdenum, or vanadium have been added in sufficient quantities to obtain or increase certain desirable properties. Alloy cast steels are comprised of two groups—the low-alloy steels with their alloy content totaling less than 8 percent and the high-alloy steels with their alloy content totaling 8 percent or more. The addition of these various alloying elements in conjunction with suitable heat treatments, makes it possible to secure steel castings having a wide range of properties. The three accompanying tables give information on these steels. The lower portion of Table 1 gives the engineering grades of low-alloy cast steels grouped according to tensile strengths and gives properties normally expected in the production of steel castings. Table 2 and Table 3 give the standard designations and nominal chemical composition ranges of high-alloy castings which may be classified according to heat or corrosion resistance. The grades given in these tables are recognized in whole or in part by the Alloy Casting Institute (ACI), the American Society for Testing and Materials (ASTM), and the Society of Automotive Engineers (SAE).

**Table 2. Nominal Chemical Composition and Mechanical Properties of Heat-Resistant Steel Castings** *ASTM A297-17*

| Grade | Nominal Chemical Composition, Percent[a] | Tensile Strength, min | | 0.2 Percent Yield Strength, min | | Percent Elongation in 2 inch, or 50 mm, min. |
|---|---|---|---|---|---|---|
| | | ksi | MPa | ksi | MPa | |
| HF | 19 Chromium, 9 Nickel | 70 | 485 | 35 | 240 | 25 |
| HH | 25 Chromium, 12 Nickel | 75 | 515 | 35 | 240 | 10 |
| HI | 28 Chromium, 15 Nickel | 70 | 485 | 35 | 240 | 10 |
| HK | 25 Chromium, 20 Nickel | 65 | 450 | 35 | 240 | 10 |
| HE | 29 Chromium, 9 Nickel | 85 | 585 | 40 | 275 | 9 |
| HT | 15 Chromium, 35 Nickel | 65 | 450 | … | … | 4 |
| HU | 19 Chromium, 39 Nickel | 65 | 450 | … | … | 4 |
| HW | 12 Chromium, 60 Nickel | 60 | 415 | … | … | … |
| HX | 17 Chromium, 66 Nickel | 60 | 415 | … | … | … |
| HC | 28 Chromium | 55 | 380 | … | … | … |
| HD | 28 Chromium, 5 Nickel | 75 | 515 | 35 | 240 | 8 |
| HL | 29 Chromium, 20 Nickel | 65 | 450 | 35 | 240 | 10 |
| HN | 20 Chromium, 25 Nickel | 63 | 435 | … | … | 8 |
| HP | 26 Chromium, 35 Nickel | 62.5 | 430 | 34 | 235 | 4.5 |
| HG 10 MNN | 19 Chromium, 12 Nickel, 4 Manganese | 76 | 525 | 33 | 225 | 20 |
| CT15C | 20 Chromium, 33 Nickel, 1 Niobium | 63 | 435 | 25 | 170 | 20 |

ksi = kips per square inch = 1000s of pounds per square inch; MPa = megapascals.
[a] Remainder is iron.

The specifications committee of the Steel Founders Society issues a *Steel Castings Handbook* with supplements. Supplement 1 provides design rules and data based on the fluidity and solidification of steel, mechanical principles involved in production of molds and cores, cleaning of castings, machining, and functionality and weight aspects. Data and examples are included to show how these rules are applied. Supplement 2 summarizes the standard steel castings specification issued by the American Society for Testing and Materials (ASTM), Society of Automotive Engineers (SAE), Association of American Railroads (AAR), American Bureau of Shipping (ABS), and Federal authorities, and provides guidance as to their applications. Information is included for carbon and alloy cast steels, high alloy cast steels, and centrifugally cast steel pipe. Details are also given of standard test methods for steel castings, including mechanical, non-destructive (visual, liquid penetrant, magnetic particle, radiographic, and ultrasonic), and testing of qualifications of welding procedures and personnel. Other supplements cover such subjects as tolerances, drafting practices, properties, repair and fabrication welding, of carbon, low alloy and high alloy castings, foundry terms, and hardenability and heat treatment.

*Austenitic Manganese Cast Steel:* Austenitic manganese cast steel is an important high-alloy cast steel which provides a high degree of shock and wear resistance. Its composition normally falls within the following ranges: carbon, 1.00 to 1.40 percent; manganese, 10.00 to 14.00 percent; silicon, 0.30 to 1.00 percent; sulfur, 0.06 percent max.; phosphorus, 0.10 percent, max. In the as-cast condition, austenitic manganese steel is quite brittle. In order to strengthen and toughen the steel, it is heated to between 1830 and 1940° F (999 and 1060° C) and quenched in cold water. Physical properties of quenched austenitic manganese steel that has been cast to size are as follows: tensile strength, 80,000–100,000 psi (552–689 MPa); shear strength (single shear), 84,000 psi (579 MPa); elongation in 2 inches (50.8 mm), 15 to 35 percent; reduction in area, 15 to 35 percent; and 180 to 220 BHN (Brinell Hardness Number). When cold-worked, the surface of such a casting increases to a hardness of 450 to 550 BHN. In many cases the surfaces are cold-worked to maximum hardness to ensure immediate hardness in use. Heat-treated austenitic manganese steel is machined only with great difficulty since it hardens at and slightly ahead of the point of contact of the cutting tool. Grinding wheels mounted on specially adapted machines are used for boring, planing, keyway cutting, and similar operations on this steel. Where grinding cannot be employed and machining must be resorted to, high-speed tool steel or cemented carbide tools are used with heavy, rigid equipment and slow, steady operation. In any event, this procedure tends to be both tedious and expensive. Austenitic manganese cast steel can be arc-welded with manganese-nickel steel welding rods containing from 3 to 5 percent nickel, 10 to 15 percent manganese, and, usually, 0.60 to 0.80 percent carbon.

**Table 3. Nominal Chemical Composition and Mechanical Properties of Corrosion-Resistant Steel Castings** *ASTM A743-17*

| Grade | Nominal Chemical Composition, Percent[a] | Tensile Strength, min | | 0.2% Yield Strength, min | | Percent Elongation in 2 inch, or 50 mm, min | Percent Reduction of Area, min |
|---|---|---|---|---|---|---|---|
| | | ksi | MPa | ksi | MPa | | |
| CA6N | 11 Chromium, 7 Nickel | 140 | 965 | 135 | 930 | 15 | 50 |
| CA6NM | 12 Chromium, 4 Nickel | 110 | 755 | 80 | 550 | 15 | 35 |
| CA15 and CA15M | 12 Chromium | 90 | 620 | 65 | 450 | 18 | 30 |
| CA28MWV[b] | 12 Chromium, with Molybdenum, Tungsten, and Vanadium | 140 | 965 | 110 | 760 | 10 | 24 |
| CA40 | 12 Chromium | 100 | 690 | 70 | 485 | 15 | 25 |
| CA40F | 12 Chromium, Free Machining | 100 | 690 | 70 | 485 | 12 | … |
| CB6 | 16 Chromium, 4 Nickel | 115 | 790 | 85 | 580 | 16 | 35 |
| CB6 | 20 Chromium | 65 | 450 | 30 | 205 | … | … |
| CC50 | 28 Chromium | 55 | 380 | … | … | … | … |
| CE30 | 29 Chromium, 9 Nickel | 80 | 550 | 40 | 275 | 10 | … |
| CF3 | 19 Chromium, 9 Nickel | 70 | 485 | 30 | 205 | 35 | … |
| CF3M | 19 Chromium, 10 Nickel, with Molybdenum | 70 | 485 | 30 | 205 | 30 | … |
| CF3MN | 19 Chromium, 10 Nickel, with Molybdenum and Nitrogen | 75 | 515 | 37 | 255 | 35 | … |
| CF8 | 19 Chromium, 9 Nickel | 70[c] | 485[c] | 30[c] | 205[c] | 25 | … |
| CF8C | 19 Chromium, 10 Nickel, with Columbium | 70 | 485 | 30 | 205 | 30 | … |
| CF8M | 19 Chromium, 10 Nickel, with Molybdenum | 70 | 485 | 30 | 205 | 30 | … |
| CF10SMnN | 17 Chromium, 8.5 Nickel with Nitrogen, 9 Nickel | 85 | 585 | 42 | 290 | 30 | … |
| CF16F and CF16Fa | 19 Chromium, 9 Nickel, Free Machining | 70 | 485 | 30 | 205 | 25 | … |
| CF20 | 19 Chromium, 9 Nickel | 70 | 485 | 30 | 205 | 30 | … |
| CG3M | 19 Chromium, 11 Nickel, with Molybdenum | 75 | 515 | 35 | 240 | 25 | … |
| CG6MMN | Chromium-Nickel-Manganese-Molybdenum[d] | 85 | 585 | 42 | 290 | 30 | … |
| CG8M | 19 Chromium, 11 Nickel, with Molybdenum | 75 | 520 | 35 | 240 | 25 | … |
| CG12 | 22 Chromium, 12 Nickel | 70 | 485 | 28 | 195 | 35 | … |
| CH10 and CH20 | 25 Chromium, 12 Nickel | 70 | 485 | 30 | 205 | 30 | … |
| CK3MCuN | 20 Chromium, 18 Nickel, with Copper and Molybdenum | 80 | 550 | 38 | 260 | 35 | … |
| CK35MN | 23 Chromium, 21 Nickel, with Molybdenum and Nitrogen | 83 | 570 | 41 | 280 | 35 | … |
| CK20 | 25 Chromium, 20 Nickel | 65 | 450 | 28 | 195 | 30 | … |
| CN3M[e] | --- | 63 | 435 | 25 | 170 | 30 | … |
| CN3MN | 21 Chromium, 24 Nickel, with Molybdenum and Nitrogen | 80 | 550 | 38 | 260 | 35 | … |
| CN7M | 20 Chromium, 29 Nickel, with Copper and Molybdenum | 62 | 425 | 25 | 170 | 35 | … |
| CN7MS | 19 Chromium, 24 Nickel, with Copper and Molybdenum | 70 | 485 | 30 | 205 | 35 | … |
| HG10MNN | 19 Chromium, 12 Nickel, 4 Manganese | 76 | 525 | 33 | 225 | 20 | … |

[a] Remainder is iron.

[b] These mechanical properties apply only when this material is heated to 1875–1925° F (1025–1050° C), quenched in air or oil, and tempered at 1150° F (620° C) minimum, or when annealed at 1400° F (760° C) minimum.

[c] For low ferrite or non-magnetic castings of this grade, the following values shall apply: tensile strength, min, 65 ksi (450 MPa); yield point, min, 28 ksi (195 MPa).

[d] Nominal Chemical Composition for CG6MMN as noted in the ASME Standard. Other industry data indicates 20.5–23.5% Chromium, 11.5–13.5% Nickel, with 4–6% Manganese, 1.5–3% Molybdenum.

[e] Nominal Chemical Composition for this grade is not noted in the ASME Standard. Other industry data indicates 20–22% Chromium, 23–27% Nickel, with 4.5–5.5% Molybdenum.

## Nonferrous Casting Metals

Nonferrous metals include metal elements and alloys not based on iron. Offering a wide variety of material characteristics and mechanical properties, these casting materials are commonly specified for structural applications that require reduced weight, increased strength, higher melting points, nonmagnetic properties, and/or resistance to chemical and atmospheric corrosion. They also are suitable for electrical and electronic applications. Such metals include aluminum, copper, magnesium, nickel, titanium, zinc, refractory metals (molybdenum and tungsten), and noble metals.

Important considerations in selecting material for a specific mechanical or structural application include how easily the material can be shaped into a finished part and how its properties may be altered in the process—either intentionally or inadvertently. Depending on the end use, metal can be cast into a part and finished. Or it can be cast into an intermediate form (such as ingots), worked, or wrought, by forging, extruding, rolling, or applying other deformation processes, and then finished as needed. Following are the primary nonferrous metals and alloys used in casting. For more information on specific alloys, see *NONFERROUS ALLOYS* starting on page 508.

**Aluminum (Al).—**Aluminum, also spelled aluminium, is a lightweight, silvery-white metal. It has a melting point of 1220° F (660° C) and a density of 163.55 lb/ft$^3$ (2700 kg/m$^3$). The most abundant metallic element in the Earth's crust, aluminum is the most widely used nonferrous metal. While it is thermodynamically the least stable of the main engineering metals, it has the advantage of forming a dense, highly protective alumina film only 20 μin.–60 μin. (0.5 μm–1.5 μm) thick. This film can be reinforced by anodizing and destroyed by salt.

Aluminum is one of the few metals that can be cast by all of the processes. When alloyed with other metals, numerous properties are obtained that make such alloys useful over a broad range of applications. Many organizations publish specific standards for the manufacture of aluminum alloy, including ASTM International and SAE International (specifically, its aerospace standards subgroups). Alloys composed mostly of aluminum have been important in aerospace manufacturing since the introduction of metal-skinned aircraft.

Classification of cast aluminum alloys is developed by the Aluminum Association of the United States. Each cast alloy is designated by a four-digit number with a decimal point separating the third and fourth digits:

1. The first digit indicates the alloy group, according to the major alloying element:
   1xx.x aluminum (99.0% minimum)
   2xx.x copper (4% to 4.6%)
   3xx.x silicon (5% to 17%) with added copper and/or magnesium
   4xx.x silicon (5% to 12%)
   5xx.x magnesium (4% to 10%)
   7xx.x zinc (6.2% to 7.5%)
   8xx.x tin
   9xx.x others
2. The next two digits indicate the alloy purity or identify the alloy: In the alloys of the 1xx.x series, these digits indicate the level of purity of the alloy—the same as the two digits to the right of the decimal point in the minimum concentration of aluminum (in percent): 150.0 means a minimum 99.50% of aluminum in the alloy, 120.1 means a minimum 99.20% of aluminum. In all other groups of aluminum alloys (2xx.x through 9xx.x), the second and third digits together signify different alloys in the group.
3. The last digit indicates the product form: casting (designated by 0) or ingot (designated by 1 or 2, depending on chemical composition limits). A modification of the original alloy or impurity limits is indicated by a serial letter before the numerical designation. The serial letters are assigned in alphabetical order, starting with A, but omitting I, O, Q, and X. (The letter X is reserved for experimental alloys.)

*Aluminum-Copper (Al-Cu) Cast Alloys* (2xx.x series): These alloys contain 4–4.6% copper, along with small impurities of iron, silicon, and magnesium. Characteristics are heat-treatable, high strength, low fluidity, low ductility, susceptibility to hot cracks, and low corrosion resistance. Aluminum-copper cast alloys are used for cylinder heads for automotive and aircraft engines, pistons for diesel engines, and exhausting system parts.

*Aluminum-Silicon-Copper (Al-Si-Cu) Cast Alloys* (3xx.x series): Aluminum-silicon-copper alloys contain 5–17% silicon and 2–4.5% copper. Characteristics of these alloys are heat-treatable, high strength, good fluidity, low ductility, good machinability, good wear resistance, and decreased corrosion resistance. The copper contributes to strength, and the silicon improves castability and reduces hot shortness; thus, higher silicon alloys are suitable for more complex castings and for permanent mold and die casting processes, which cannot tolerate hot-short alloys. They are used for automotive cylinder blocks and heads, car wheels, aircraft fittings, casings, and other parts of compressors and pumps.

*Aluminum-Silicon (Al-Si) Cast Alloys* (4xx.x series): These alloys contain 5–12% silicon but no copper. Characteristics are non-heat-treatable, moderate strength, moderate ductility, very good cast properties, and good wear and corrosion resistance. Rapid cooling to increase strength and ductility can refine the microstructure. These alloys are used for pump casings, thin wall castings, and cookware.

*Aluminum-Magnesium (Al-Mg) Cast Alloys* (5xx.x series): Containing 4–10% magnesium, characteristics of these alloys are non-heat-treatable, good machinability, and good appearance when anodized. The moderate to relatively poor castability of such alloys and the tendency of magnesium to oxidize increase handling difficulties and, therefore, cost. In general, these alloys are used for sand cast parts.

*Aluminum-Tin (Al-Sn) Cast Alloys* (8xx.x series): Aluminum-tin alloys usually contain about 6% tin and a small amount of copper and nickel to improve strength. Though these compounds are not heat-treatable, castability and machinability are good, and wear resistance is very good. They are used for cast bearings, due to tin's excellent lubrication characteristics.

*Aluminum-Zinc (Al-Zn) Cast Alloys* (7xx.x series): With the addition of 6.2–7.5% zinc, these alloys can be heat treated and have good dimensional stability, good machinability (if the alloy contains copper), and good corrosion resistance. They are used for abrasion-resistant parts under various operating conditions, such as axle bushes, shaft sleeves, and worm gears.

**Copper (Cu).**—Copper is a reddish-yellow material with a melting point of 1984.6° F (1084.6° C) and a density of 556.85 lb/ft$^3$ (8920 kg/m$^3$). Usually a good conductor of electricity and heat, copper is one of the most ductile metals, but it is not especially strong or hard. Pure copper is extremely difficult to cast. As copper forms alloys more freely than most metals, casting and other characteristics can be improved by adding small amounts of various elements, including beryllium, chromium, nickel, silicon, silver, tin, and zinc.

*Brass:* This generic term refers to a range of copper-zinc (Cu-Zn) alloys with differing combinations of properties, including strength, hardness, ductility, machinability, resistance to wear and corrosion, and electrical and thermal conductivity.

*Bronze:* Bronze made from copper and tin (Cu-Sn) was the first manmade alloy used thousands of years ago. In modern times, wrought bronzes have been developed with 4–8% tin. These alloys are harder, stronger, and stiffer than wrought brasses and, in strip and wire form, produce a combination of high yield strength and good corrosion resistance.

*Copper-Nickel (Cu-Ni) Alloys:* This combination offers high conductivity, excellent resistance to marine corrosion, and low susceptibility to attachment of marine macro-organisms. The addition of nickel to copper improves strength and corrosion resistance, but good ductility is retained. The two main alloys are 90/10 (90% copper, 10% nickel) and 70/30 (70% copper, 30% nickel).

*Copper-Nickel-Silver (Cu-Ni-Zn) Alloys:* Also known as "Nickel Silver," such alloys contain copper, nickel, zinc, and other trace elements, but generally not silver. Regarded as special brasses, these polish to an attractive silvery color, rather than the typical brassy tone, and are abrasion resistant. Commonly used in plating over several centuries, this material has been replaced in some uses by aluminum and stainless steel.

*Beryllium-Copper (Be-Cu) Alloys:* In the fully heat-treated and cold-worked condition, this is the hardest and strongest copper alloy. It is similar in mechanical properties to many high-strength alloy steels but, compared to steels, has better corrosion resistance.

**Magnesium (Mg).**—A shiny gray, solid, lightweight metal, magnesium has a melting point of 1202° F (650° C) and a density of 108.3 lb/ft$^3$ (1740 kg/m$^3$). As the world's lightest metal, with good strength-to-weight ratio, magnesium and its alloys are prevalent in the automotive, airplane, and missile industries. Its compounds also are used as refractory material in furnace linings for producing iron, steel, other nonferrous metals, glass, and cement.

Magnesium is the most electrochemically active metal. Care must be taken in processing to avoid fire hazards, and small particles of the metal, such as metal cutting chips, oxidize rapidly. Magnesium and its alloys are available in both wrought and cast forms.

*Magnesium Alloys:* These light alloys have received renewed interest as substitutes for some conventional structural materials to reduce weight in vehicles. Cast alloys, widely used in interior and power-train components, account for more than 99 percent of magnesium alloys used today, while only a small number of wrought products are used. This is because magnesium alloys lack formability for wrought applications; their high cost also discourages use for some automotive applications.

Rare earth (RE) elements, such as cerium (Ce), gadolinium (Gd), neodymium (Nd), and yttrium (Y), often are used as major alloying elements because of their relatively high solubility in magnesium and effectiveness in precipitation hardening and creep resistance. Commercial wrought magnesium alloys, such as ZK60 and AZ61, are based on magnesium-zinc (Mg-Zn) and magnesium-zinc-aluminum (Mg-Zn-Al) compounds, both of which are age-hardenable. ZK and AZ alloys attain good strength with hot extrusion or rolling; however, age-hardening processes, such as T6 (solution heat treatment and aging) and T8 (cold-work and subsequent aging), do not add strengthening due to softening by recrystallization at the temperature for artificial aging.

## CASTING OF METALS

Casting is an age-old manufacturing process in which a liquid material usually is poured into a mold that contains a hollow cavity of the desired shape. As a result of new technological advances, this traditional form of manufacturing has become more essential than ever.

In metal casting, the mold generally includes runners and risers that facilitate metal filling the cavity. The mold and metal are cooled until the metal solidifies, then the solidified part (*casting*) is recovered from the mold. Subsequent operations are used to remove excess material caused by the casting process (such as the runners and risers), after which the part may solidify further. Metal casting processes are divided into two broad types: expendable-mold casting processes and nonexpendable-mold casting processes.

*Expendable-mold casting* includes sand, plastic, shell, plaster, and investment (lost-wax technique) molding processes, all of which rely on gravity to move the liquid material into casting cavities. The molds in which the molten material solidifies usually are made of nonmetallic materials and are temporary—that is, they must be destroyed to remove the casting.

*Nonexpendable-mold casting* refers to casting processes involving a reusable mold. Such molds are designed with two or more sections for easy, precise closing (for molding) and opening (to remove the casting) and can be used for multiple castings. For metal casting, the mold, also called a *die*, is commonly made of metal or ceramic.

*Permanent mold casting* refers to all casting technologies in which the mold cavity is reused many times and is made of a metallic material or graphite. This is the predominant casting method for manufacturing metal shapes. Specifically, about 90 percent of all aluminum castings produced use metal molds, including gravity-fed, low-pressure, and high-pressure die castings.

Permanent metal molds are commonly made of steel or cast iron, with metal or sand cores, though it is desirable and generally more economical to use permanent steel cores to form cavities. When the casting has re-entrant surfaces or cavities from which one-piece permanent metal cores cannot be withdrawn, destructive cores made of sand, shell, plaster, or other materials may be used. This process is called *semipermanent mold casting*. Sectional steel cores also may be used in some instances.

An advantage of permanent metal molds is that they heat up and expand during the pour, so the cavity does not need to be expanded as much as in sand castings. Therefore, the cavity, with the gating system included, can be machined into halves that produce more precise parts with closer dimensional tolerances and smoother surfaces. Permanent mold castings also usually have better mechanical properties than sand castings, as solidification is more rapid and fill is more laminar.

Permanent mold casting is used mostly for aluminum, copper alloys, magnesium, and gray iron, because of their generally lower melting points. Typical parts include automobile pistons, cylinder heads, gears, and kitchenware. Parts that can be made economically generally weigh less than 55 lb (25 kg), though special castings weighing a few hundred kilograms have been made using this process. It may not be economical for small, unique production runs (due to the cost of permanent molds) and generally is not suitable for casting intricate shapes (because of difficulty of removing the part from the mold).

Disadvantages of different types of castings include poor finish; wide tolerance (sand casting); limited workpiece size (shell molds and ceramic molds); patterns with low strength (expendable-pattern casting); expensive, limited shapes (centrifugal casting); porosity (all types); and environmental problems (all types).

## Heating and Pouring the Metal

*Pouring* is the process by which molten metal is transferred to the mold for cooling and solidification to be converted into the intended shape. *Pouring temperature* ($T_p$) is the temperature to which the molten metal must be heated before being poured into molds for cooling and setting.

The heat energy required for heating metal to a pouring temperature is the sum of 1) the heat needed to raise the temperature of a unit mass solid to its melting point; 2) the heat of fusion required to convert it from a solid at its melting point to a liquid without an increase in temperature (see Fig. 1); and 3) the heat needed to raise the molten metal to the desired temperature for pouring. The pouring temperature must take into account heat lost in the transfer of metal through ladles, due both to the heat absorbed by the ladles and to the distance between the furnace and the mold.

Molten metal must be poured carefully to avoid casting defects. For example, too rigorous a stream could cause mold erosion; highly turbulent flows could result in air and inclusion entrapments; and relatively slow filling might generate cold shuts. Thus, design of the gating and venting overflow systems must take into consideration proper control of the liquid metal as it fills the mold.

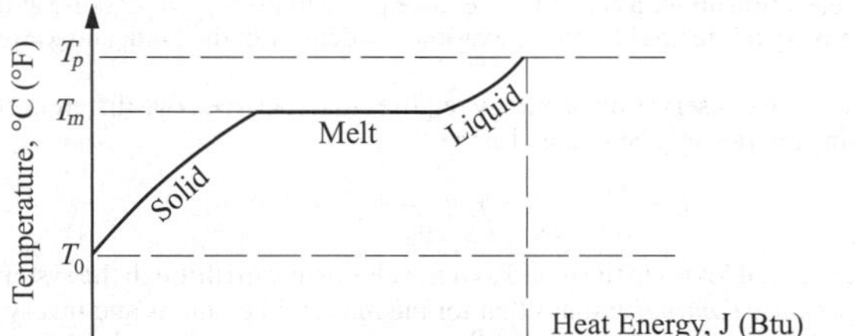


Fig. 1. Phase Change Diagram of Process of Heating Metal Casting to Pouring Temperature.

Whether the casting process takes place by expendable-mold casting (such as sand casting) or nonexpendable-mold casting (using a permanent metal mold), the basic terminology of the casting process is the same.

## Fluid Flow

Fluid flow is very important in casting. The molten metal is poured through a pouring basin. It then flows through the gating system (comprising sprue, risers, runners, and gates) to fill the mold cavity. (See the gravity casting system shown in Fig. 2.)

The *sprue* is the vertical part of the gating system that connects the *pouring cup* or *pouring basin* and runners; liquid metal enters the mold through the sprue. The *cope* is the top half of the mold. The *runners,* which are cut in the *drag* (the lower part of the mold), form the horizontal portion of the gating system that connects the sprue to the gates. The *gate* is the portion of the runners through which molten metal enters the mold cavity; there may be one or more gates. The *core* is designed to leave an unfilled space in the part. *Risers* are the part of the gating system used to identify when enough liquid material has been poured to fill the mold; it also acts as a reservoir of extra material to compensate for shrinkage during solidification. The *flask* is the molding box or outside of the mold.

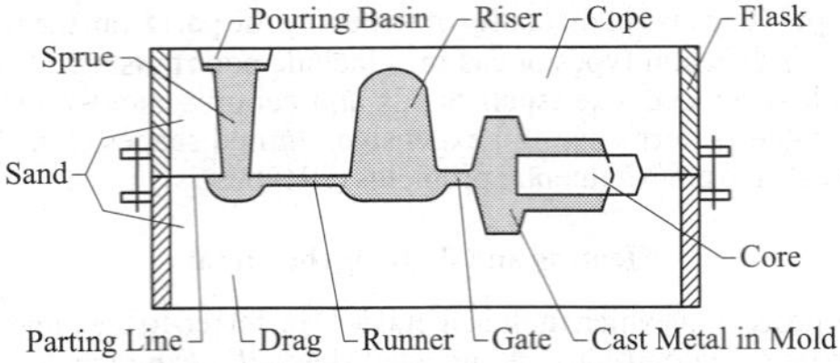


Fig. 2. Cross Section of Typical Two-Part Sand Mold

*Gating Design:* Successful casting requires proper design and control of the solidification process to ensure adequate fluid flow in the system. For example, an important function of the gating system in sand casting is to trap contaminants (oxides and other inclusions) in the molten metal by having such contaminants adhere to the walls of the gating system, thereby preventing them from reaching the mold cavity. A properly designed gating system also avoids or minimizes problems such as premature cooling, turbulence, and gas entrapment. Two basic principles of fluid flow are relevant to gating design: Bernoulli's theorem and the law of mass continuity.

*Bernoulli's theorem* states that the sum of the energies in a flowing liquid is constant at any two points. This can be written as:

$$h + \frac{p}{\rho g} + \frac{v^2}{2g} = \text{constant} \qquad (1)$$

where $h$ = elevation above a certain preference plane (datum); $p$ = pressure at that elevation; $v$ = velocity of the liquid at that elevation; $\rho$ = density of the fluid; and $g$ = gravitation constant.

This concept of conservation of energy requires that (between two different elevations) the following relationship be satisfied as:

$$h_1 + \frac{p_1}{\rho g} + \frac{v_1^2}{2g} = h_2 + \frac{p_2}{\rho g} + \frac{v_2^2}{2g} + f \qquad (2)$$

where $f$ = frictional losses in the liquid as it travels downward through the system.

The *law of mass continuity* states that for incompressible liquids and in a system with impermeable walls, the volume rate of flow remains constant throughout the liquid. So, considering two different locations in the system:

$$Q = A_1 v_1 = A_2 v_2 \tag{3}$$

where $Q$ = volume rate of flow; $A$ = cross-sectional area of the liquid stream; $v$ = the average velocity of the liquid at that cross-sectional location; and subscripts 1 and 2 indicate two different locations in the liquid flow.

*Sprue Design:* The shape of the sprue can be calculated using Equation (2) and Equation (3). Assuming that pressure at the top of the sprue is equal to pressure at the bottom and that there are no frictional losses, the relationship between height and cross-sectional area at any point in the sprue is given by the parabolic relationship:

$$\frac{A_1}{A_2} = \sqrt{\frac{h_1}{h_2}} \tag{4}$$

*Flow Characteristics:* A molten metal's flow characteristics are an important consideration in the gating system because of the possible consequences of turbulence. There are two types of real fluid flow: laminar and turbulent.

In *laminar flow,* fluid moves in layers called *laminas*. Laminar flow need not be in a straight line; the flow may follow curved surfaces. The fluid layers slide smoothly over one another, without fluid being exchanged between the layers.

In *turbulent flow*, secondary random motions are superimposed on the principal flow, and there is an exchange of fluid from one adjacent sector to another. More important, there is an exchange of momentum: slow-moving fluid particles speed up, while fast-moving particles give up their momentum to slower-moving particles and are slowed down.

The factor that determines which type of flow is present is the ratio of the inertial forces to the viscous forces within the fluid. This ratio is expressed by the dimensionless Reynolds number as:

$$R_e = \frac{\rho v L}{\mu} \tag{5}$$

where $R_e$ = Reynolds number; $v$ = mean fluid velocity; $L$ = characteristic length (equal to diameter if cross section is a circle); $\mu$ = dynamic fluid viscosity; $\rho$ = density of the fluid.

The higher the $R_e$, the greater the tendency for turbulent flow to occur in the gating system: (1) laminar flow: $0 = R_e < 2000$; (2) transition flow: $20 < R_e < 20{,}000$; (3) turbulent flow: $R_e > 20{,}000$. Turbulent flow leads to inclusions, and therefore a reduction in fluidity.

### Fluidity of Molten Metal

The phrase *fluidity of molten metal* describes the capability of molten metal to fill mold cavities, based on two factors: characteristics of the molten metal and casting parameters. The following characteristics of molten metal influence fluidity:

*Viscosity:* If viscosity (a measure of the metal's deformation in a molten state) and sensitivity to temperature (viscosity index) increase, fluidity decreases.

*Surface Tension:* High surface tension of the liquid metal reduces fluidity. Oxide films that develop on the surface of molten metal increase surface tension and thus can significantly reduce fluidity. Fluxing processes (treating metal with flux to promote melting) in the heating of metals are used to reduce or eliminate oxidation and to improve fluidity of metal surface layers.

*Inclusions:* Metallic and nonmetallic inclusions have long been recognized as one of the most important quality issues in metal casting. As insoluble particles, inclusions can have a significant adverse effect on fluidity; the presence of inclusions often is the cause of decreased fluidity.

*Solidification Pattern of the Alloy:* Fluidity is inversely proportional to the freezing temperature range. In general, pure metals and alloys of eutectic composition have the highest values of fluidity. Solid-solution and other alloys have lower fluidity.

*Superheat:* The difference between the melting temperature and the pouring temperature also influences fluidity. For given alloy compositions, if the melt temperature is higher, fluidity increases. The pouring temperature is often specified, rather than the superheat temperature, because it is easier to do so.

**Test for Fluidity.**—Although none is accepted universally, several tests have been developed to quantify fluidity. One such test is the spiral test (Fig. 3a), where the molten metal is made to flow along a channel at room temperature. The distance the metal flows before it solidifies (and stops flowing) is a measure of its fluidity. Obviously, this distance is a function of the thermal properties of both the metal and mold, as well as the design of the channel. Another test for fluidity is the vacuum test (Fig. 3b), which measures the length the metal flows inside a narrow channel when sucked from a crucible by a vacuum pump. Such tests can be useful in simulating casting situations.

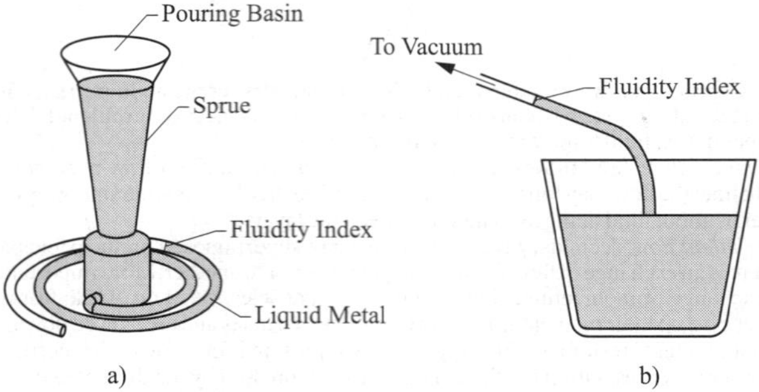


Fig. 3. Test Method for Fluidity: a) Spiral Method; b) Vacuum Method

## Heat Transfer

An important consideration in casting is heat transfer during the complete cycle from pouring to solidification and cooling to ambient temperature. Heat flow at different locations in the system is a complex phenomenon that depends on many factors relating to the casting material and mold and process parameters. For instance, when casting thin sections, metal flow rates must be high enough to avoid premature chilling and solidification. However, the flow rate must not be so fast as to cause excessive turbulence, which can have detrimental effects.

## Solidification and Cooling of Metals

At the macro level, *solidification* refers to the phase change of metal from liquid to solid. At the micro level, changes occur in the material as the disordered structure of the liquid transforms into an orderly arrangement of crystals.

Once molten metal has been poured into the mold, it cools rapidly. When the temperature of the liquid drops below the melting point of that metal or alloy, the solidification process begins. This usually takes less than a few minutes. As the temperature drops further, the molten metal loses energy, and crystals begin to form. (This process starts near the mold walls, where cooling occurs first.) These crystals eventually become grains within the final structure.

*Grain size* refers to crystals (*dendrites*) formed during the solidification process. If the metal solidifies slowly, the grains are longer. If it cools quickly, the grains are visibly shorter. Crystals continue to form and harden, until the entire melt is solidified.

Throughout the solidification process, the metal shrinks. It is important to compensate for such shrinking to ensure castings are free of voids and shrink defects. This is accomplished by using risers.

The cooling rate of a casting affects its microstructure, quality, and properties. The cooling curve illustrates how molten metals solidify. There is a fundamental difference between the cooling curve observed during solidification of a pure metal and that of an alloy.

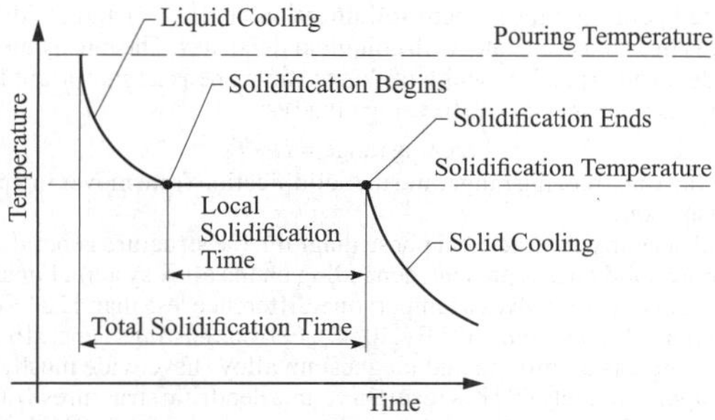


Fig. 4. Cooling Curve for a Poured Pure Metal During Casting

**Solidification of Pure Metal.—**Pure metal solidifies at constant temperature. It has a clearly defined melting (or freezing) point. Fig. 4 shows the cooling curve for a poured pure metal during casting.

After the temperature of the molten pure metal drops to its freezing point, the temperature remains constant while the latent heat of fusion is given off. The *solidification front* (solid–liquid interface) moves through the molten metal, solidifying from the mold walls toward the center.

At the mold walls, the molten metal cools rapidly and first produces a solidified shell of fine, approximately equal-dimension-in-all-direction (equiaxed) grains. Starting with these grains, the grains grow upon themselves, in the opposite direction of the heat transfer out through the mold. Those with favorable orientations—that is, away from the surface of the mold—are columnar in shape; as the driving force of the heat transfer is reduced (farther from the mold walls), the grains again become equiaxed, as well as coarse. Those grains that have substantially different orientations are blocked from further growth.

When the heat is segregated rapidly during solidification, it leads to fine structures due to a decrease in diffusion rates.

**Solidification of Alloys.—**Solidification begins when the temperature of the alloy drops below the point of liquidus, and it is complete when it reaches solidus. A phase diagram and a cooling curve for alloys during casting are shown in Fig. 5.

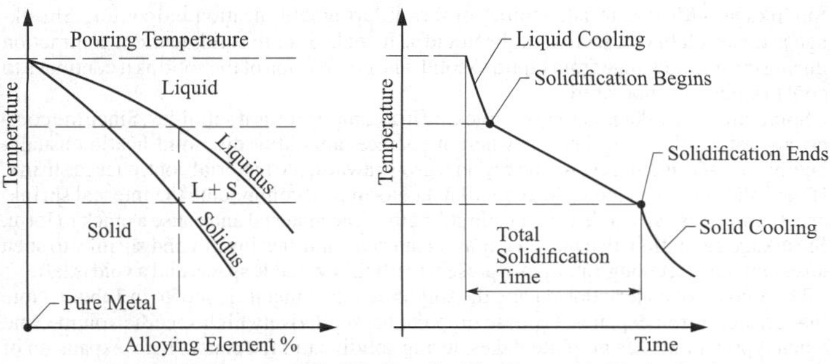


Fig. 5. Phase Diagram and Cooling Curve for Alloy Composition During Casting

Within the temperature range where solidification begins ($T_l$) and solidification ends ($T_S$), the alloy is in a "mushy" state, with columnar dendrites. The mushy metal is present between the dendrite arms. The width of this mushy zone is an important factor during solidification. It is described by the freezing range as:

$$\text{Freezing range} = T_l - T_S \tag{6}$$

Pure metals have no freezing range, and the solidification front moves in a plane without forming a mushy zone.

In alloys with a nearly symmetrical phase diagram, the structure generally is lamellar, with two or more solid phases present, depending on the alloy system. For alloys, a short freezing range generally involves a temperature difference less than 122° F (50° C) and a long freezing range higher than 230° F (110° C). Ferrous castings generally have narrow mushy zones, whereas aluminum and magnesium alloys have wide mushy zones. Slow cooling rates approximately $10^2$ K/s result in coarse dendritic structures with large spacing between the dendrite arms. For higher cooling rates, from $10^6$ to $10^8$ K/s, the structures developed are amorphous.

**Solidification Time.—**Total solidification time is the time required for the casting to solidify from molten metal after pouring. Casting geometry, material, and process determine solidification time.

Chvorinov's rule states that under the same conditions, a casting with a large surface area and small volume will cool more rapidly than a casting with a small surface area and large volume. Therefore, a large sphere solidifies and cools to ambient temperature at a much slower rate than a smaller sphere. The reason is that the volume of a sphere is proportional to the cube of its diameter, and its surface area is proportional to the square of its diameter. Similarly, molten metal in a cube-shaped mold will solidify faster than in a spherical mold of the same volume. According to this rule, solidification time is a function of the volume of a casting and its surface area:

$$\text{Solidification time} = k(V/A)^n \tag{7}$$

where $V$ is the volume of the casting; $A$ is the surface area of the casting; $k$ is the mold constant; and $n$ is the exponent ($1.5 < n \leq 2$, but usually taken as 2). The mold constant $k$ depends on the properties of the cast metal (heat of fusion, specific heat, and thermal conductivity), mold material, and pouring temperature. The value of $k$ for a given casting operation can be based on experimental data from previous operations carried out using the same mold material, metal, and pouring temperature, even though the shape of the workpiece might be complex.

During the early stages of solidification, a thin, solidified skin begins to form at the cool mold walls; as times passes, the skin thickens. With flat mold walls, this thickness is proportional to the square root of time.

**Shrinkage.—**Most materials contract or shrink during solidification and cooling. Shrinkage is the result of contraction of the liquid as it cools prior to solidification; contraction during the phase change from liquid to solid; and contraction of the solid as it continues to cool to ambient temperature.

Sometimes, shrinkage can cause cracking in a component as it solidifies. Since the coolest area of a volume of liquid is where it contacts a mold or die, solidification usually begins first at this surface. As the crystals grow inward, the material continues to shrink. If the solid surface is too rigid and will not deform to accommodate the internal shrinkage, the stresses can exceed the tensile strength of the material and cause a crack to form. Shrinkage cavitation also may occur as a material solidifies inward and shrinks to such an extent that not enough atoms are present to fill the available space, and a void is left.

The amount of contraction during the solidification of metals is shown in Table 1. Note that gray cast iron expands, because graphite has a relatively high specific volume, and when it precipitates as graphite flakes during solidification, it causes a net expansion of the metal.

**Table 1. Volumetric Solidification Contraction or Expansion for Various Metals**

| Metals | Volumetric Contraction due to: Solidification Shrinkage (%) | Volumetric Contraction due to: Cooling Shrinkage (%) | Solidus Temperature °F (°C) |
|---|---|---|---|
| Aluminum (Al) | 7.1 | 5.1 | 1220 (660) |
| Aluminum (Al), 4.5% Copper (Cu) | 6.3 | 5.3 | 1058 (570) |
| Aluminum (Al), 12% Silicone (Si) | 3.8 | 4.8 | 1076 (580) |
| Gold (Au) | 5.1 | 5.2 | 1945 (1063) |
| Bismuth (Bi) | −3.3 (expansion) | 1.0 | 520 (271) |
| Copper (Cu) | 5.1 | 6.4 | 1981 (1083) |
| Iron (Pure) (FE) | 3.1 | 6.0 to 7.5 | 2795 (1535) |
| Steel (Various) | 2.5 to 3.0 | 4.2 to 6.0 | 2732 (1500) |
| Gray Cast Iron, 2% Carbon (C) | 4.0 | 3.0 | 2111 (1155) |
| Gray Cast Iron, 5% Carbon (C) | −2.5 (expansion) | 3.0 | 2098 (1148) |
| White Cast Iron, 5% Carbon (C) | 5.0 | 3.0 | 2111 (1155) |
| Germanium (Ge) | −5.1 (expansion) | 1.6 | 1719 (937) |
| Magnesium (Mg) | 4.2 | 5.4 | 1200 (649) |
| Nickel (Ni) | 5.1 | 7.0 | 2647 (1453) |
| Lead (Pb) | 2.7 | 2.8 | 621 (327) |
| Tin (Sn) | 2.3 | 1.5 | 450 (232) |
| Zinc (Zn) | 4.7 | 4.0 | 788 (420) |

## Defects in Casting

The International Committee of Foundry Technical Associations has developed a standardized nomenclature, consisting of eight basic categories of casting defects:

*Metallic Projections:* Such defects can include fins, flash, or massive projections, such as swells and rough surfaces.

*Cavities:* Variously rounded or rough internal or exposed cavities include blowholes, pinholes, and shrinkage cavities.

*Discontinuities:* These can include cracks, cold or hot tearing, and cold shuts. If the solidifying metal is constrained from shrinking freely, cracking and tearing can occur. Coarse grain size and the presence of low-melting-point segregates along the grain boundaries (intergranular) increase the tendency for hot tearing. Cold shut is an interface in a casting that lacks complete fusion because two streams of liquid metal meet from different gates and do not completely fuse.

*Defective Surface:* Surface defects include folds, laps, scars, adhering sand layers, and oxide scale.

*Incomplete Casting:* Misruns can occur due to premature solidification, the molten metal being at too low a temperature, pouring metal too slowly, insufficient volume of metal being poured, or runout (loss of metal from a mold after pouring).

*Incorrect Dimensions or Shape:* These undesirable results may be caused by factors such as an improper shrinkage allowance, pattern-mounting error, deformed pattern, irregular contraction, or warped casting.

*Inclusions:* Inclusions form during melting, solidification, and molding and generally are nonmetallic. They increase stress and reduce strength of the casting. Inclusions may form during melting, when the molten metal reacts with the environment (usually oxygen) or with crucible or mold material. Chemical reactions among components in the molten metal itself may produce inclusions. Spalling of the mold and core surfaces also produces inclusions, indicating the importance of the quality and maintenance of molds.

*Porosity:* Porosity may be caused by shrinkage or gases or both. Thin sections in a casting solidify sooner than thicker regions. As a result, molten metal can flow into thicker regions that have not yet solidified, and porous regions may develop at their centers due to contraction as the surfaces of the thicker region begin to solidify. Micro-porosity can develop when the liquid metal solidifies and shrinks between dendrites or between dendrite branches.

Porosity caused by shrinkage can be reduced or eliminated by various means. Adequate liquid metal should be provided to avoid cavities caused by shrinkage. Internal or external chills used in sand casting also are an effective means of reducing shrinkage porosity by increasing the rate of solidification in critical regions. Internal chills usually are made of the same material as the casting and are left in the casting; external chills may be made of the same material or of iron, copper, or graphite. With alloys, porosity also may be reduced or eliminated by making the temperature gradient steep. For example, mold materials with higher thermal conductivity may be used. Subjecting the casting to hot isostatic pressing is another method of reducing porosity.

Most obvious porosity defects are caused by entrapment of gases within the molten solution. Because liquid metals have much greater solubility for gases than solid metals, when a metal begins to solidify, dissolved gases are expelled from the solution. Typically, hydrogen precipitates into melt because of contact with the atmosphere or excessive moisture in the flux. Since hydrogen is highly soluble in molten metal, it is best to avoid superheating metals beyond their melting temperature and to avoid holding the material in a molten state any longer than necessary.

To reduce absorption of gases from the atmosphere, which may leave slag or dross, cover molten metal until just prior to pouring it into the mold. Gases also may result from reactions of the molten metal with the mold materials, either accumulating in regions of existing porosity or causing micro-porosity in the casting, particularly in cast iron, aluminum, and copper. Dissolved gases may be removed from the molten metal by flushing or purging with an inert gas or by melting and pouring metal in a vacuum. If the dissolved gas is oxygen, the molten metal can be deoxidized. Steel is usually deoxidized with aluminum, copper-based alloys with phosphorus, silicon, titanium, and zirconium-bearing materials.

If the porosity is spherical and the walls are smooth, porosity usually is the result of gases. If the walls are rough and angular, porosity is likely the result of shrinkage between dendrites. Gross porosity is caused by shrinkage and usually is called a *shrinkage cavity*.

The loss in casting properties measured by a tensile test may reflect the amount of porosity in a casting; because imperfections become areas of higher stress concentration, the percentage of property loss becomes greater when the strength requirement is higher. Porosity also is detrimental to the ductility of a casting and its surface finish, making it permeable and thus affecting the pressure tightness of a cast pressure vessel. A metallographic examination can determine whether porosity exists in a casting. X-ray techniques also are used for nondestructive evaluations of porosity.

## METAL CASTING AND MOLDING PROCESSES

Metal casting processes may be classified in four different ways:

*According to the Mold Type:* (1) expendable mold (destroyed after each casting); or (2) permanent mold (reused many times).

*According to the Type of Pattern Used for Making a Sand Mold:* (1) expendable pattern (melted for each mold), using wax as the pattern material; or (2) permanent pattern (reused for many molds), using wood or metal as the pattern material.

*According to the Type of Core Used for Producing a Hole in Casting:* (1) expendable core (used in both sand and metal molds), using sand as the core material; or (2) permanent core (used with a permanent mold only), using metal as the core material.

*According to the Method by Which the Mold is Filled:* (1) gravity (sand casting, gravity die casting); (2) pressure (low- and high-pressure die casting); or (3) vacuum (vacuum investment casting).

## Sand Casting

Sand casting uses natural or synthetic sand in forming molds. Larger-sized molds use green sand (a mixture of sand, clay, and some water).

*Advantages of Sand Casting:* Ferrous, nonferrous, and even nonmetal materials can be cast in this process. Sand can be reused; excess metal poured is cut off and reused as well. Simple, inexpensive tools are required. Intricate shapes can be made by this process, as molten metal flows into small sections. Finally, this process can be used to produce many small components, with no limit on the size or weight of castings.

*Disadvantages of Sand Casting:* Accuracy and surface finish are lacking, requiring additional processing and finishing of parts. Overall, sand casting is a labor-intensive process. However, since sand casting usually is the least expensive way of making a component, its cost advantage over other methods makes it an attractive molding method.

**Sands.—**Most sand casting operations use a refractory material called silica ($SiO_2$). Sand is inexpensive and has high melting point of 3110° F (1710° C). Sand can be naturally bonded (bank sand) or synthetic (lake sand). Because its composition can be controlled more accurately, synthetic sand is preferred by most foundries.

Several factors affect the selection of sand for molds. The grains of the sand must be small enough so that it can be packed densely; sand having fine, round grains can be closely packed and form a smooth mold surface. Fine-grained sand enhances mold strength, but fine grains also lower mold permeability. The sand grains must be large enough for good permeability of molds (and cores), which allows the gases and steam that evolve during casting to escape through the pores of the mold.

For proper functioning, mold sand must be clean and preferably new. And the sand mold should have good collapsibility to allow the casting to shrink while cooling, thus avoiding defects such as hot tearing and cracking.

**Types of Sand Molds.—**Sand molds are characterized by the types of sand that compose them and by the methods used to make the molds.

*Green-Sand Molds:* Clay sand is the mixture of natural silica sand, clay, additives, and water. It is the least expensive method of making sand molds, and the sand can be recycled for the next use. Bentonite clay is used to make wet clay sand. Sand in these molds is kept moist or damp while the metal is being poured. As the wet sand has a high moisture content, good air permeability, and low strength, castings can have issues with porosity, coarse, sticky sand, and sand expansion defects.

Green-sand molds are commonly used in hand molding and machine molding. In hand molding, dimensional accuracy is low, so it is generally used only for production of small and medium-sized iron castings and nonferrous alloy castings. But in mechanical molding, the castings have much higher dimensional accuracy, so it is widely used for high-volume production of castings.

*Skin-Dried Method:* In this case, the mold surfaces need to be sprayed with a mixture of 10 percent water to one part molasses or lignin sulfonate after the surfaces are dried, either by storing the mold in air or by using torches.

*Cold-Box Mold:* Various organic and inorganic binders are blended into the sand to bond the grains chemically for greater strength. These molds are dimensionally more accurate than green-sand molds, but they are more expensive.

*No-Bake Mold:* A synthetic liquid resin is mixed with the sand; the mixture hardens at room temperature. Because bonding of the mold in this and in the cold-box process takes place without heat, these processes are called *cold-setting processes*. This type of mold has good dimensional control in high-production applications.

*Dried Molds:* Sand molds that are oven dried (baked) prior to pouring the molten metal are stronger than green-sand molds and impart better dimensional accuracy and surface finish to the casting. However, this method has drawbacks: (a) distortion of the mold is greater; (b) castings are more susceptible to hot tearing because of the lower collapsibility of the mold; and (c) production rates are slower due to the drying time required.

**Features of Sand Molds.**—Major features of sand molds (most shown in Fig. 2) are:

*Cope:* The top half of the flask, mold, or core.

*Drag:* The bottom half of the flask, mold, or core.

*Core:* A core part is inserted into the mold cavity to produce a hole within the mold.

*Core Print:* The region added to the pattern, core, or mold to locate and support the core. The prints are made of such a size and shape that it would be impossible to set the core in any position except the proper one.

*Mold Cavity:* The combined open area in the molding material and core, where the liquid metal solidifies to produce the casting.

*Gating System:* The network of connected channels that allows molten material to flow into the mold cavity; it includes the sprue, riser, runners, and gates.

*Pouring Cup or Basin:* The part of the gating system that initially receives the molten metal from the pouring vessels and controls its delivery to the rest of the mold.

*Sprue:* The vertical part of the gating system that connects the pouring cup and runners. The liquid metal enters into the mold cavity through the sprue.

*Runners:* The horizontal portion of the gating system, cut into the drab part of the mold, which connects the sprues to the gates.

*Riser:* This part of the gating system normally is used to identify the filling position of the liquid metal in the mold. It also acts as a reservoir for extra molten metal to compensate for shrinkage during solidification.

*Gates:* The controlled entrances from the runners into the mold cavities. Properly designed gates admit liquid metal into the mold cavity without turbulence.

*Vents:* These small holes, in all parts of the mold, provide passage for gases to escape during pouring and solidification of mold metals.

*Parting Line or Parting Surface:* The interface line between the cope and drag halves of the mold or flask.

*Draft:* The taper on the pattern that allows it to be easily withdrawn from the mold.

*Core Box:* This is the mold or die used to produce the cores.

**Patterns.**—Patterns are used to mold the sand mixture into the shape of casting. They may be made of wood, plastic, or metal. Because patterns are used repeatedly to make molds, the strength and durability of the materials selected must reflect the number of castings that the mold will produce. Patterns usually are coated with a parting agent to facilitate their removal from the molds. There are four types of patterns: solid one-piece patterns, split patterns, match-plate patterns, and cope-and-drag patterns.

*One-Piece Patterns:* Also called *loose* or *solid patterns*, one-piece patterns are used for simple shapes and low-quantity production. They usually are made of wood and are inexpensive.

*Split Patterns:* These two-piece patterns are made so that each part forms a portion of the cavity for the casting, enabling casting of complicated shapes.

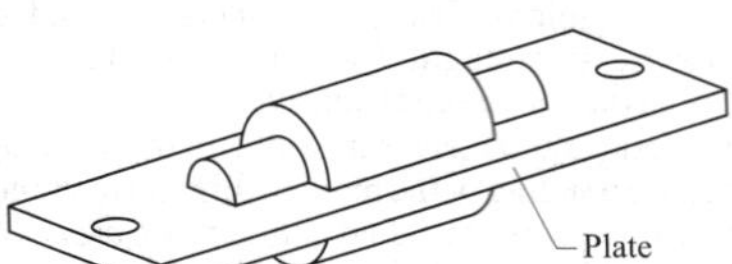


Fig. 6. Match-Plate Pattern

*Match-Plate Patterns:* These two-piece patterns are constructed by securing each half of one or more split patterns to the opposite sides of a single plate (Fig. 6). The gating system can be mounted on the drag side of the pattern.

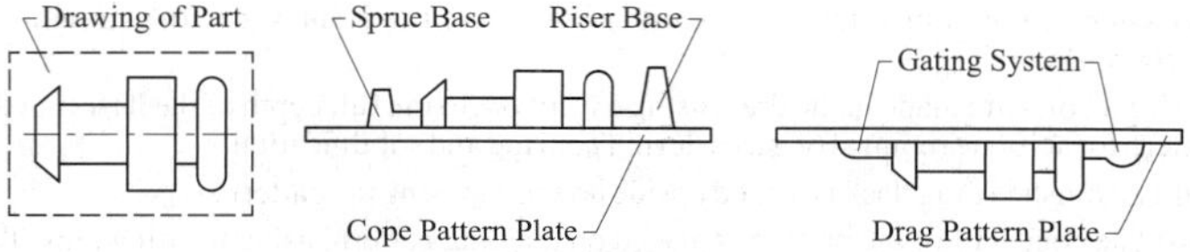


Fig. 7. Cope-and-Drag Pattern

*Cope-and-Drag Patterns:* These are similar to match-plate patterns. Except each half of the split pattern is assembled with separate plates, so separate patterns, and possibly separate machines, can be used to make the mold halves. Cope-and-drag patterns (Fig. 7) include a gating and riser system.

**Cores.—**Many cast parts have interior holes (hollow parts) or other cavities in their shape that are not directly accessible from either piece of the mold. Such interior surfaces are formed using inserts called *cores*, which are placed in the mold cavity before casting and removed from the finished part during shakeout and further processing. Like molds, cores must possess strength, permeability, collapsibility, and ability to withstand heat. Hence, cores are made of sand aggregates.

Cores are anchored and supported by *core prints*, recesses added to the pattern that also provide vents for gases to escape. To keep cores from shifting, metal supports (*chaplets*) may be used to anchor the cores.

On pouring and solidification, *chaplets* are integrated into the casting. The portions of the chaplets protruding from the castings are cut off. Fig. 8 illustrates how a core is held in the mold cavity with and without chaplets.

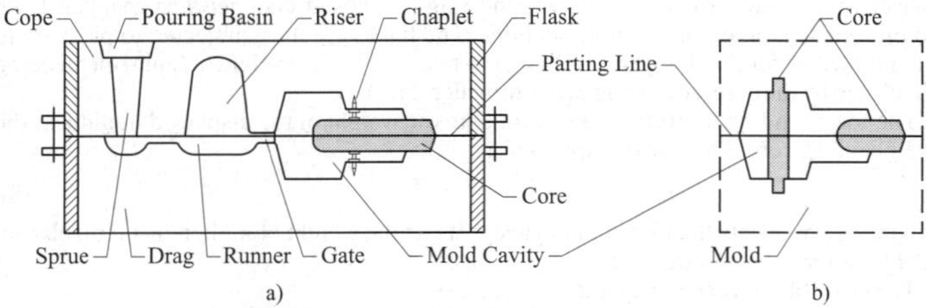


Fig. 8. Core Held in Place in the Mold Cavity: a) Core Held in with Chaplets; b) Core Held without Chaplets

**Sand Molding Techniques.—**These are the steps involved in the traditional technique of making a simple sand mold:

1) The drag flask is placed on the board.

2) Dry facing sand is sprinkled over the board to provide a nonstick layer.

3) The drag half of the pattern is located on the mold board.

4) Molding sand is poured in to cover the pattern with the fingers; the drag is filled completely.

5) Sand is tightly packed in the drag using *hand rammers*, *peen hammers* (used first close to the drag pattern), and *butt hammers* (used for surface ramming).

6) Ramming must be proper (neither too hard nor too soft). Too soft ramming will generate a weak mold and poor imprint of the pattern. Too hard ramming will not allow gases and air to escape, creating bubbles in castings, resulting in defects called *blows*. Also, the making of runners and gates will be difficult.

7) After the ramming is finished, excess sand is leveled and removed with a straight bar (*strike rod*).

8) Vent holes are made in the drag, using a *vent rod*, to the full depth of the flask and the pattern to facilitate removal of gases during pouring and solidification.

9) The finished drag flask is turned upside down, exposing the pattern.

10) The cope half of the pattern is placed on the drag pattern, using locating pins. The cope flask also is located with the help of pins. Dry parting sand is sprinkled all over the drag surface and on the pattern.

11) A sprue pin, used in making the sprue passage, is located some distance from the pattern edge. A riser pin is placed at an appropriate place.

12) Filling, ramming, and venting of the cope are done in the same manner.

13) The sprue and riser are removed. A pouring cup or basin is made at the top to pour the liquid metal.

14) The pattern is removed from the cope and drag.

15) Runners and gates are made by cutting the parting surface with a gate cutter (a piece of sheet metal bent to the desired radius).

16) The core for making a central hole is placed into the mold cavity in the drag, where it is anchored and supported by core prints.

17) The mold is now assembled and ready for pouring.

For information on using additive manufacturing as an alternate way to create sand molds, see *AM Considerations for Tooling and Parts* on page 616.

**Sand Casting Operation.—**After the mold has been shaped and the cores have been placed in position, the two halves (cope and drag) are closed, clamped, and weighed down to prevent separation of the mold sections. The flasks are then subjected to pressure to counteract the force of buoyancy. (Buoyancy results from the weight of liquid metal being displaced by the core, according to Archimedes' law.)

The force tending to lift the cope is equal to the weight of the displaced liquid less the weight of the core. This can be expressed as:

$$F_b = W_m - W_c \tag{8}$$

where $F_b$ represents the force of buoyancy; $W_m$ is the weight of molten metal displaced; and $W_c$ is the weight of the core.

The order of sand casting operation is as follows:

1) A suitable flask or molding box is selected, and space is allowed for ramming in sand.

2) The drag is placed on the molding board upside down. Then the lower part of the pattern is placed on the board inside the flask. Space is left for cutting.

3) Sand is added to fill in the flask and cover the pattern; it is firmly packed using hammers. After properly ramming in sand, the excess is leveled off with straight strike-off bar.

4) Using a vent rod, the vent holes are made in the drag to the full depth of flask to allow gases or air to escape during pouring and solidification.

5) The cope half of pattern is placed over the drag and positioned with locating pins. The cope flask is placed on the drag. The sprue pin is provided for sprue passage, located at a distance from the pattern. And the riser pin is positioned on the pattern.

6) The ramming, filling, and venting processes are repeated.

7) The cope is filled with sand and proper ramming force applied.

8) Excess sand is cut off from mold surface. The riser pin and sprue pin are removed.

9) The molding box is opened, and the pattern is removed from the cope and drag. One or more dry sand cores are mounted in position.

10) The mold is closed by placing the cope on top of the drag and securing the assembly with pins. The flasks are subjected to pressure to counteract buoyant forces in the liquid, which might lift the cope.

11) Molten metal is poured through the riser into the mold cavity.

12) After pouring and solidification, the part is removed with the required pattern shape.

13) The sprue and risers are cut off and recycled.

14) The casting is cleaned, inspected, and heat treated (when necessary).

15) The final casting is inspected using nondestructive testing and destructive methods in accordance with standards.

**Rammed Graphite Molding.—**A rammed graphite mold typically is used for large industrial casting for reactive metals such as titanium and zirconium. It uses graphite instead of sand in a process similar to sand casting. Traditionally, a mixture of properly size-fractioned graphite powder, pitch, corn syrup, and water are rammed against a wooden or fiberglass pattern to form each mold section. The mold sections are air dried, baked at 350° F (177° C), and then fired in a furnace for 24 hours at 1877° F (1025° C), causing the mold to carbonize and harden. Rammed graphite molds must be stored under controlled humidity and temperature.

Mold ramming is a labor-intensive process that cannot be easily mechanized, and the graphite mold is so hard that it must be chiseled off the cast parts. Castings made with this process usually are cleaned in an acid bath, followed, if necessary, by chemical milling, to remove any reaction zone; weld-repaired as needed; and then sandblasted for a good surface appearance.

## Shell Molding

Shell molding is a foundry process in which the molds are made in the form of thin shells. This process can produce many types of castings with close dimensional tolerance and good surface finish at relatively low cost. Shell molding applications include small mechanical parts requiring high precision, including gear housings, cylinder heads, connecting rods, and close-tolerance molding cores.

*Advantages of Shell Molding:* Rigidly bonded sand provides great reproducibility and produces castings near to net shape with intricate detail and high dimensional accuracy of ± 0.010 in. (± 0.25 mm). Castings can range from 1 oz to 25 lb (28 g to 11 kg). The resin-bond strength of the mold allows for smaller draft angles, deep draws, and built-in mold locators that prevent mold shift mismatch.

Because mold shells are thin, permeability for gas escape is increased, allowing use of finer sands. Finer sand and excellent flowability produce dense mold surfaces and contribute to producing complex casting with high-quality finishes with a roughness average of 50 μin (1.25 μm) Ra. The virtual absence of moisture eliminates moisture-related defects, and the burning resin provides a favorable anti-oxidizing atmosphere for the casting surface. In addition, heat from burning slows the casting-cooling rate, yielding a more machinable structure.

*Disadvantages of Shell Molding:* Since the tooling requires heat to cure the mold, pattern costs and pattern wear can be high. Material costs are higher than for green-sand molding, and energy costs are higher than for other processes.

Shell molds are made in the following operations:

1) A pattern made of a ferrous metal or aluminum is heated to 347–698° F (175–370° C).

2) The pattern is coated with a parting agent, such as silicone.

3) It is clamped to a box or chamber that contains fine sand, mixed with 2.5–4% thermosetting resin binder (such as phenol-formaldehyde) that coats the sand particles.

4) Either the box/chamber is rotated upside down, or the sand mixture is blown over the pattern to coat it.

5) The assembly is placed in an oven for a short period of time to complete curing of the resin.

6) The shell mold hardens around the pattern.

7) The mold is removed from the pattern using built-in ejector pins.

8) The second half of the shell mold is made, using the same steps.

9) Once complete, the two half shells are bonded or clamped together to form the mold. Support is provided by sand or metal shot in the flask when pouring the molten metal.

10) After casting and cooling, the part is cleaned and trimmed.

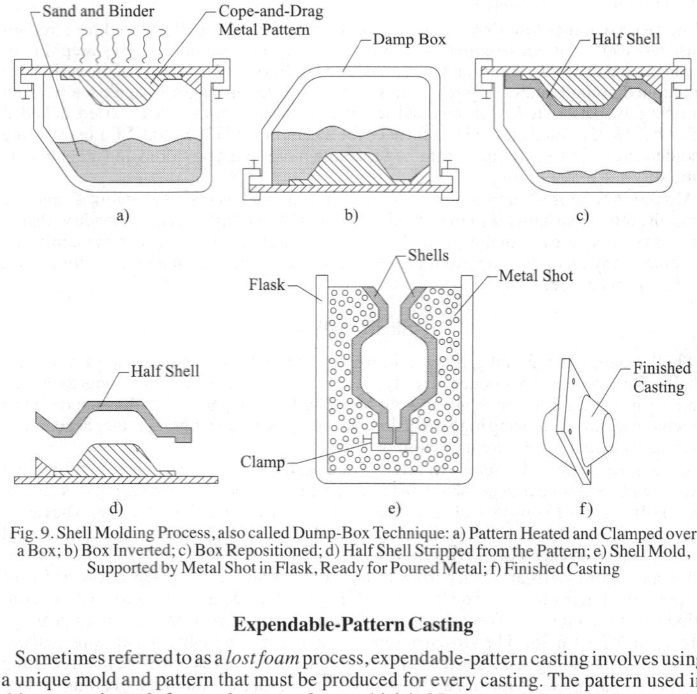


Fig. 9. Shell Molding Process, also called Dump-Box Technique: a) Pattern Heated and Clamped over a Box; b) Box Inverted; c) Box Repositioned; d) Half Shell Stripped from the Pattern; e) Shell Mold, Supported by Metal Shot in Flask, Ready for Poured Metal; f) Finished Casting

## Expendable-Pattern Casting

Sometimes referred to as a *lost foam* process, expendable-pattern casting involves using a unique mold and pattern that must be produced for every casting. The pattern used in this process is made from polystyrene foam, which is 95 percent air bubbles; the material evaporates when liquid metal is poured on it.

The pattern is made by molding. Polystyrene beads and pentane are put inside an aluminum mold and heated. The material expands to fill the mold, taking the shape of the cavity. Then the pattern is removed from the mold and used for the casting processes as follows:

1) The pattern is dipped in a slurry of water and clay (or other refractory grains). It is dried to form a hard shell around the pattern.

2) The shell-covered pattern is placed in a container with sand for support.

3) Liquid metal is poured into a hole at the top.

4) The foam evaporates as the metal fills the shell.

5) Upon cooling and solidification, the part is removed by breaking the shell.

This process is useful since it is very cheap, can produce complex geometry, and yields good surface finish. There are no runners, risers, gating, or parting lines. Thus, the design process is simplified. This method is used to manufacture crankshafts for engines, aluminum engine blocks, manifolds, and other parts.

### Plaster Mold Casting

Plaster mold casting produces parts with good surface finish and dimensional accuracy, and provide the capability to make parts with thin cross sections. The mold is made by mixing plaster of Paris or simple "plaster" ($CaSO_4$) with talc and silica flour. This produces a fine white powder that, when mixed with water, forms a clay-like consistency and can be shaped around pattern halves in a flask and allowed to set. The mold halves are extracted carefully from the pattern and then dried in an oven at a temperature range of 248–500° F (120–260° C) to remove moisture.

The mold halves are carefully assembled to form the mold cavity and used for the casting process as follows:

1) The mold is preheated to about 248° F (120° C).

2) Molten metal is poured in the mold cavity.

3) After the metal is solidified and cooled, the plaster mold is broken away from the finished casting.

Plaster casts are relatively soft and not strong enough at temperatures above 2192° F (1200° C). So this method is mainly used to make castings from nonferrous metals, such as aluminum, magnesium, zinc, and copper-based alloys. It cannot be used to cast ferrous materials, because the sulfur in gypsum slowly reacts with iron. Since plaster has lower thermal conductivity, the casting cools slowly, and therefore has more uniform grain structure, with the advantages of less warpage and less residual stresses.

### Ceramic Mold Casting

Ceramic mold casting also is called *cope-and-drag investment casting*. It is similar to the plaster mold process, and the metal is cast the same way, but it uses refractory materials suitable for high-temperature applications. Ceramic mold casting uses a permanent pattern made of plastic, wood, or metal. To make the slurries for molding, fine-grained zircon ($ZrSiO_4$), aluminum oxide, and fused silica are mixed with bonding agents and poured over the pattern, which has been placed in a flask.

With this process, parts may be cast that weigh as much as 1,540 lb (700 kg). Typically, it is used for impellers, cutters for machining operations, dies for metalworking, and molds for making plastic and rubber components.

The sequence of operations in making a ceramic mold is:

1) A pattern is placed in a flask.

2) A slurry is poured over the pattern to create the mold.

3) It is dried in a low temperature oven, and the pattern is removed.

4) The mold is backed by clay for strength and baked in a high-temperature oven to burn off any volatile substances.

5) The all-ceramic mold is firmly clamped for use in casting.

The metal is cast the same way as in plaster casting. Although this process is somewhat expensive, it is used because castings can be made that feature good dimensional accuracy and surface finish over a wide range of sizes and intricate shapes. It can be used to make very good quality castings of steel or even stainless steel and parts such as impeller blades for turbines or pumps and rotors for motorboats.

## Precision Investment Casting

Investment casting is a highly developed process that is capable of great casting accuracy and can form extremely intricate contours. The process may be utilized when metals are too hard to machine or otherwise fabricate; when it is the only practical method of producing a part; or when it is more economical than any other method of obtaining work of the quality required. Precision investment casting is especially applicable in producing either exterior or interior contours of intricate form with surfaces so located that they could not be machined readily if at all. The process provides efficient, accurate means of producing such parts as turbine blades, airplane or other parts made from alloys that have high melting points and must withstand exceptionally high temperatures, and many other products. The accuracy and finish of precision investment castings may either eliminate machining entirely or reduce it to a minimum. The quantity that may be produced economically may range from a few to thousands of duplicate parts.

Investment casting uses an expendable pattern, usually of wax or injection-molded plastics. Several wax replicas or patterns are usually joined together or to bars of wax that are shaped to form runner channels in the mold. Wax shapes that will produce pouring funnels also are fastened to the runner bars. The mold is formed by dipping the wax assembly (tree) into a thick slurry containing refractory particles. This process is known as investing. After the coating has dried, the process is repeated until a sufficient thickness of material has been built up to form a one-piece mold shell. Because the mold is in one piece, undercuts, apertures, and hollows can be produced easily. As in shell molding, this invested shell is baked to increase its strength, and the wax or plastics pattern melts and runs out or evaporates (lost-wax casting). Some molds are backed up with solid refractory material that is also dried and baked to increase the strength. Molds for lighter castings are often treated similarly to shell molds described before. Filling of the molds may take place in the atmosphere, in a chamber filled with inert gas or under vacuum, to suit the metal being cast.

**Materials That May Be Cast.—**The precision investment process may be applied to a wide range of both ferrous and nonferrous alloys. In industrial applications, these include alloys of aluminum and bronze, Stellite, Hastelloys, stainless and other alloy steels, and iron castings, especially where thick and thin sections are encountered. In producing investment castings, it is possible to control the process in various ways so as to change the porosity or density of castings, obtain hardness variations in different sections, and vary the corrosion resistance and strength by special alloying.

**General Procedure in Making Investment Castings.—**Precision investment casting is similar in principle to the "lost-wax" process that has long been used in manufacturing jewelry, ornamental pieces, and individual dentures, inlays, and other items required in dentistry, which is not discussed here. When this process is employed, both the pattern and mold used in producing the casting are destroyed after each casting operation, but they may both be replaced readily. The "dispensable patterns" (or cluster of duplicate patterns) is first formed in a permanent mold or die and is then used to form the cavity in the mold or "investment" in which the casting (or castings) is made. The investment or casting mold consists of a refractory material contained within a reinforcing steel flask. The pattern is made of wax, plastics, or a mixture of the two. The material used is evacuated from the investment to form a cavity (without parting lines) for receiving the metal to be cast. Evacuation of the pattern (by the application of sufficient heat to melt and vaporize it) and the use of a master mold or die for reproducing it quickly and accurately in making duplicate castings are distinguishing features of this casting process. Modern applications of the process include many developments such as variations in the preparation of molds, patterns, investments, etc., as well as in the casting procedure. Application of the process requires specialized knowledge and experience.

**Master Mold for Making Dispensable Patterns.—**Duplicate patterns for each casting operation are made by injecting the wax, plastics, or other pattern material into a master mold or die that usually is made either of carbon steel or of a soft metal alloy. Rubber, alloy steels, and other materials may also be used. The mold cavity commonly is designed to form a cluster of patterns for multiple castings. The mold cavity is not, as a rule, an exact duplicate of the part to be cast because it is necessary to allow for shrinkage and perhaps to compensate for distortion that might affect the accuracy of the cast product. In producing master pattern molds there is considerable variation in practice. One general method is to form the cavity by machining; another is by pouring a molten alloy around a master pattern that usually is made of monel metal or of a high-alloy stainless steel. If the cavity is not machined, a master pattern is required. Sometimes, a sample of the product itself may be used as a master pattern, when, for example, a slight reduction in size due to shrinkage is not objectionable. The dispensable pattern material, which may consist of waxes, plastics, or a combination of these materials, is injected into the mold by pressure, by gravity, or by the centrifugal method. The mold is made in sections to permit removal of the dispensable pattern. The mold while in use may be kept at the correct temperature by electrical means, by steam heating, or by a water jacket.

**Shrinkage Allowances for Patterns.—**The shrinkage allowance varies considerably for different materials. In casting accurate parts, experimental preliminary casting operations may be necessary to determine the required shrinkage allowance and possible effects of distortion. Shrinkage allowances, in inches per inch, usually average about 0.022 for steel, 0.012 for gray iron, 0.016 for brass, 0.012 to 0.022 for bronze, 0.014 for aluminum and magnesium alloys. (See also *Shrinkage* on page 1494.)

**Casting Dimensions and Tolerances.—**Generally, dimensions on investment castings can be held to ±0.005 in. and on specified dimensions to as low as ±0.002 in. Many factors, such as the grade of refractory used for the initial coating on the pattern, the alloy composition, and the pouring temperature, affect the cast surface finish. Surface discontinuities on the as-cast products therefore can range from 30–300 μin (0.76–7.6 μm) in height.

**Investment Materials.—**For investment casting of materials having low melting points, a mixture of plaster of Paris and powdered silica in water may be used to make the molds, the silica forming the refractory and the plaster acting as the binder. To cast materials having high melting points, the refractory may be changed to sillimanite, an alumina-silicate material having a low coefficient of expansion that is mixed with powdered silica as the binder. Powdered silica is then used as the binder. The interior surfaces of the mold are reproduced on the casting so, when fine finishes are needed, a first coating of fine sillimanite sand and a silicon ester such as ethyl silicate with a small amount of piperidine, is applied and built up to a thickness of about 0.06 in. (1.5 mm). This investment is covered with a coarser grade of refractory that acts to improve bonding with the main refractory coatings, before the back up coatings are applied.

With light castings, the invested material may be used as a shell, without further reinforcement. With heavy castings the shell is placed in a larger container which may be of thick waxed paper or card, and further slurry is poured around it to form a thicker mold of whatever proportions are needed to withstand the forces generated during pouring and solidification. After drying in air for several hours, the invested mold is passed through an oven where it is heated to a temperature high enough to cause the wax to run out. When pouring is to take place, the mold is pre-heated to between 700 and 1000° C (1292 and 1832° F), to get rid of any remaining wax, to harden the binder and prepare for pouring the molten alloy. Pouring metal into a hot mold helps to ensure complete filling of intricate details in the castings. Pouring may be done under gravity, under a vacuum under pressure, or with a centrifuge. When pressure is used, attention must be paid to mold permeability to ensure gases can escape as the metal enters the cavities.

**Casting Operations.—**The temperature of the flask for casting may range all the way from a chilled condition up to 2000° F (1093° C) or higher, depending upon the metal to be cast, the size and shape of the casting or cluster, and the desired metallurgical conditions. During casting, metals are nearly always subjected to centrifugal force vacuum, or other pressure. The procedure is governed by the kind of alloy, the size of the investment cavity, and its contours or shape.

**Investment Removal.—**When the casting has solidified, the investment material is removed by destroying it. Some investments are soluble in water, but those used for ferrous castings are broken by using pneumatic tools, hammers, or by shot or abrasive blasting and tumbling to remove all particles. Gates, sprues, and runners may be removed from the castings by an abrasive cutting wheel or a band saw according to the shape of the cluster and machinability of the material.

**Accuracy of Investment Castings.—**The accuracy of precision investment castings may, in general, compare favorably with that of many machined parts. The overall tolerance varies with the size of the work, the kind of metal and the skill and experience of the operators. Under normal conditions, tolerances may vary from ±0.005 or ±0.006 inch per inch (or mm/mm), down to ±0.0015 to ±0.002 inch per inch (or mm/mm), and even smaller tolerances are possible on very small dimensions. Where tolerances applying to a lengthwise dimension must be smaller than would be normal for the casting process, the casting gate may be placed at one end to permit controlling the length by a grinding operation when the gate is removed.

**Casting Weights and Sizes.—**Investment castings may vary in weight from a fractional part of an ounce up to 75 lb (34 kg) or more. Although the range of weights representing the practice of different firms specializing in investment casting may vary from about ½ pound up to 10 or 20 lb (4.5 or 9.1 kg), a practical limit of 10 or 15 lb (4.5 or 6.8 kg) is common. The length of investment castings ordinarily does not exceed 12 or 15 in. (30 or 38 cm), but much longer parts may be cast. It is possible to cast sections having a thickness of only a few thousandths of an inch, but the preferred minimum thickness, as a general rule, is about 0.020 in. (0.50 mm) for alloys of high castability and 0.040 in. (1 mm) for alloys of low castability.

**Design for Investment Casting.—**As with most casting processes, best results from investment casting are achieved when uniform wall thicknesses between 0.040 and 0.375 in. (1 and 10 mm) are used for both cast components and channels forming runners in the mold. Gradual transition from thick to thin sections is also desirable. It is important that molten metal should not have to pass through a thin section to fill a thick part of the casting. Thin edges should be avoided because of the difficulty of producing them in the wax pattern. Fillets should be used in all internal corners to avoid stress concentrations that usually accompany sharp angles. Thermal contraction usually causes distortion of the casting, and should be allowed for if machining is to be minimized. Machining allowances vary from 0.010 in. (0.25 mm) on small, to 0.04 in. (1 mm) on large parts. With proper arrangement of castings in the mold, grain size and orientation can be controlled and directional solidification can often be used to advantage to ensure desired physical properties in the finished components.

**Casting Milling Cutters by Investment Method.—**Possible applications of precision investment casting in tool manufacture and in other industrial applications are indicated by its use in producing high-speed steel milling cutters of various forms and sizes. Removal of the risers, sand blasting to improve the appearance, and grinding the cutting edges are the only machining operations required. The bore is used as cast. Numerous tests have shown that the life of these cutters compares favorably with high-speed steel cutters made in the usual way.

## Hard Mold Casting

Hard mold casting is permanent mold casting. Two halves of a mold are made from materials such as cast iron, steel, bronze, graphite, or refractory metal alloys. The mold cavity and gating system are machined into the mold and become an integral part of it.

To produce castings with internal cavities, cores made of metal or sand aggregate are placed in the mold prior to casting. Typical core materials are oil-bonded or resin-bonded sand, plaster, graphite, gray iron, low-carbon steel, and hot-work die steel.

To increase life of permanent molds, surfaces of the mold cavity usually are coated with refractory slurry (such as sodium silicate and clay) or sprayed with graphite every few castings. These coatings also serve as parting agents, as well as thermal barriers, controlling the rate of cooling of the casting.

The molds are clamped together by mechanical means and heated to about 302–392° F (150–200° C) to facilitate metal flow and reduce thermal damage to the dies due to high temperature gradients. Then molten metal is poured through the gating system. After solidification, the molds are opened, and the casting is removed. Water or fins may be used in cooling the mold, and mechanical ejectors may be needed for removal of complex castings.

Hard mold casting is used mostly for aluminum, magnesium, copper alloys, and gray iron because of their generally lower melting points. Steels also can be cast using graphite or heat-resistant metal molds. This process produces castings with good surface finish, close dimensional tolerances, and uniform and good mechanical properties, and supports high production rates. Typical parts made are automotive pistons, cylinder heads, connecting rods, gear blanks for appliances, and kitchenware.

### Vacuum Casting

Vacuum permanent mold casting (not to be confused with vacuum molding) is similar to low-pressure permanent mold casting, except for the step of filling the mold. In this case, molten metal is sucked upward into the mold by a vacuum pump (Fig. 10).

The permanent mold is enclosed in an airtight bell housing with two openings: the sprue at the bottom, through which molten metal enters the mold, and the vacuum outlet at the top. The sprue opening is submerged below the surface of the molten metal, and the vacuum is drawn within the housing, creating a pressure differential between the mold cavity and the molten metal in the crucible. This pressure differential causes the molten metal to flow up the sprue and into the mold cavity, where it solidifies. The mold is removed from the housing and opened, and the casting is ejected.

By controlling the vacuum, the pressure differential between the mold cavity and the molten metal can be varied, allowing for different fill rates necessitated by certain part designs and gating requirements. This results in tight control of the fill rate, which directly influences soundness of the casting. Because the sprue opening is submerged beneath the surface of the molten metal, only pure alloy, free from oxides and dross, can enter the die cavity. This helps to produce clean, sound castings with minimal foreign materials that detract from strength, appearance, and machinability. With proper part and mold design, this process can greatly reduce or eliminate issues such as voids, shrinks, and gas pockets in critical areas. Typical parts made by vacuum casting include vacuum pump covers, turbine wheels, intake manifolds, and train accessories.

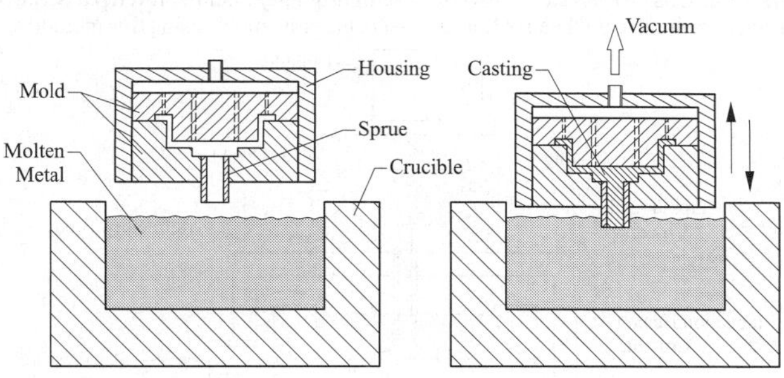


Fig. 10. Vacuum Casting Process a) Before Flow Up of Molten Metal into Cavity; b) After Flow Up of Molten Metal

## Slush Casting

Slush casting is a traditional method of permanent mold casting process in which the molten metal is not allowed to completely solidify in the mold. When the desired thickness is obtained, the remaining molten metal is poured out. This permanent mold casting process can be used to create hollow castings with thin walls, based on the principle that a solidified skin develops first in a casting and becomes thicker with time.

A disadvantage of slush casting is that the casting internal geometry cannot be effectively controlled. But good surface finish and accurate external geometry are possible. Suitable for small production runs, slush casting generally is used for making ornamental and decorative objects, including lamp bases and stems, and toys from metals with low melting temperatures, such as zinc, tin, and lead alloys.

The sequence of operations in slush casting is as follows:

1) The molten metal is poured into the metal mold.

2) After the desired thickness of the solidified skin is obtained, the mold is inverted, and the remaining molten metal is poured out.

3) Upon solidification and cooling, the mold halves are opened, and the casting is removed.

## Pressure Casting

Another variation of permanent mold casting is pressure casting, also known in industry as *low-pressure casting* or *pressure pouring* (Fig. 11). Instead of pouring molten metal into the mold and gravity distributing it throughout the mold, pressure casting uses air (gas) pressure to force the liquid metal through the gating system and the metal casting's cavity.

As in other permanent mold methods, the mold needs to be able to open and close for removal of the workpiece, so complicated casting geometry is limited. However, since the refractory tube is submerged in the molten metal and the metal drawn up into the casting comes from well below the surface, the risk of gas entrapment in the metal and oxidation are greatly reduced. The sequence of operations in pressure casting is as follows:

1) The molten metal is forced upward by gas pressure into the graphite or metal mold through the refractory tube, using an overpressure by air of about 12 to 15 psi (0.08 to 0.1 MPa).

2) Pressure is maintained until the metal has solidified completely in the mold.

Pressure casting can be used to produce metal castings with superior mechanical properties, good surface finish, and close dimensional accuracy. High setup costs make pressure casting inefficient for mass runs, but excellent productivity rates make it suitable for batch manufacture. This process can be used to cast high-quality manufactured parts. Steel castings, such as railroad car wheels, often are cast in graphite molds using this method.

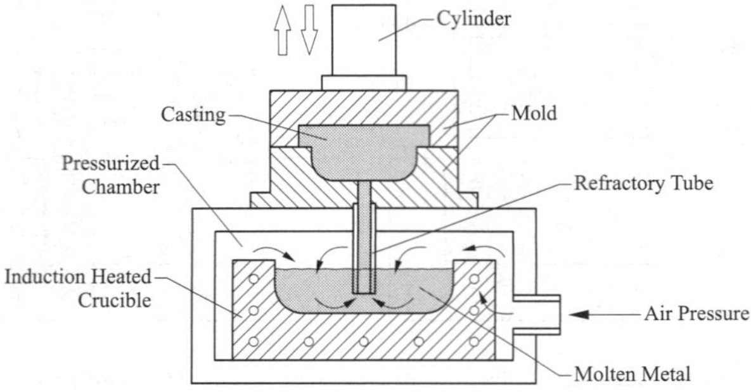


Fig. 11. Low-Pressure Casting Process

## Die Casting

Die casting is a commonly used type of permanent mold casting process used for producing many components of home appliances, such as stoves, fans, washing machines, dryers, refrigerators; various kinds of motors; toys; and hand tools. Die casting molds, commonly called *dies*, are expensive and require significant lead time to fabricate. However, surface finish and tolerance of die cast parts are so good that there is almost no post-processing required. There are two common types of die casting: hot- and cold-chamber die casting.

*Advantages of Die Casting:* Depending on the casting material, excellent dimensional accuracy can be achieved—typically within 0.005 in. for the first inch (0.127 mm for the first 25 cm) and 0.002 in. for each additional inch (0.051 mm for additional 25 cm). Compared with sand and permanent mold casting, parts can have thinner walls (approximately 0.030 in. or 0.76 mm). Inserts, such as threaded inserts, heating elements, and high-strength bearing surfaces, can be cast in. Casting tensile strength may be as high 60 ksi (414 MPa). Smooth cast surfaces of 40–100 μin (1–2.5 μm) Ra reduce or eliminate secondary machining operations. And rapid production rates can be achieved.

*Disadvantages of Die Casting:* Both the casting equipment required and the die components are very costly, so large production volume is needed to make this process economical. The process is limited to high-fluidity metals. Casting weight must be under 24 in. (600 mm) in size and weight between 1 oz (28 g) and 20 lb (9 kg). Parts that need to be hardened and tempered cannot be cast in dies.

Another consideration of the standard die casting process is that the final casting will have a small amount of porosity. This does not accommodate heat treatment or welding, because heat causes gas in the pores to expand, which causes micro-cracks inside the part and exfoliation of the surface. Thus, a related disadvantage of die casting is that it is only suitable for parts in which softness is acceptable.

**Hot-Chamber Process.—**Hot-chamber die casting is one of the two main techniques in the manufacturing process of die casting. A characteristic common to both die casting processes is the use of high pressure to force molten metal through a mold. Advantageous characteristics of manufacturing by die casting, such as a smooth casting surface, can be attributed to the use of pressure to ensure the flow of metal through the die. In hot-chamber die casting (Fig. 12), the die is attached to the casting machine and an integral part of the casting equipment.

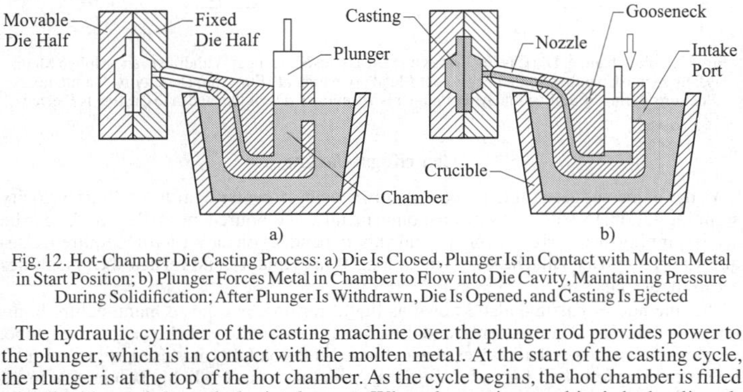


Fig. 12. Hot-Chamber Die Casting Process: a) Die Is Closed, Plunger Is in Contact with Molten Metal in Start Position; b) Plunger Forces Metal in Chamber to Flow into Die Cavity, Maintaining Pressure During Solidification; After Plunger Is Withdrawn, Die Is Opened, and Casting Is Ejected

The hydraulic cylinder of the casting machine over the plunger rod provides power to the plunger, which is in contact with the molten metal. At the start of the casting cycle, the plunger is at the top of the hot chamber. As the cycle begins, the hot chamber is filled with liquid metal through the intake port. When the casting machine's hydraulic cylinder forces the plunger downward, it travels past the intake port, cutting off the flow of molten metal to the hot chamber and ensuring the correct amount of molten metal is in the chamber to fill the cavity of the mold. In preparation for the next cycle of casting, the plunger travels back upward and opens the intake port in the hot chamber, again allowing the chamber to fill with molten metal. Hot-chamber die casting has a high rate of productivity.

Typical materials used in hot-chamber die casting are zinc, lead, and magnesium alloys and other low melting alloys. Common applications of hot-chamber die casting parts include housings with thin walls, often requiring many ribs.

**Cold-Chamber Process.—**Cold-chamber die casting is a permanent mold casting process. Large, robust machines are used to exert the great clamping force necessary to hold the mold closed against very high pressure during the manufacturing process. A metal cold chamber is located at the entrance of the mold. A piston is connected to this chamber, which, in turn, is connected to a power cylinder. At the start of the process, the correct amount of molten metal for a single part is poured into the shot chamber from an external source holding the metal for the cast.

When the power cylinder forces the piston forward in the chamber, it cuts off the intake port. The piston forward forces molten metal into the casting mold cavity with high pressure, which is maintained during solidification. After solidification, the pressure is released, the mold is opened, and the casting is removed using ejector pins.

In comparison with hot-chamber die casting, cold-chamber die casting uses higher pressure to force molten metal into the die cavity. Pressures as high as 3000–50000 psi (21 MPa–345 MPa) may be used. Fig. 13 illustrates the cold-chamber die casting process.

Cold-chamber die casting is ideal for metals with high melting points and corrosive properties, such as aluminum alloys and other alloys with high melting points. This process is used for several automobile components, including pistons, cylinder heads, and engine blocks. Other common die cast parts are propellers, gears, bushings, pumps, and valves.

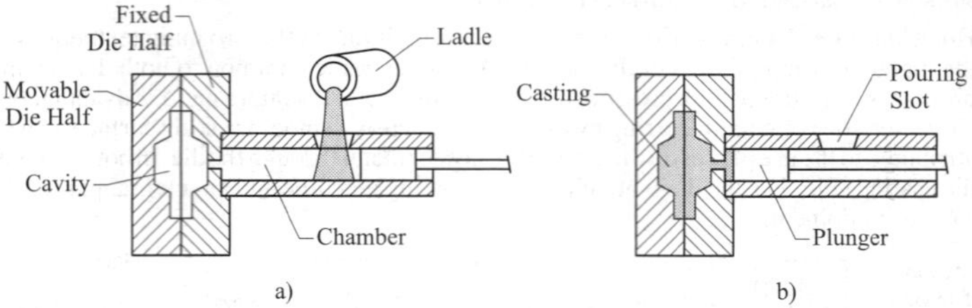


Fig. 13. Cold-Chamber Die Casting Process: a) Die is Closed, Plunger Withdrawn, and Molten Metal is Poured into Chamber; b) Plunger Forces Metal in Chamber to Flow into Die Cavity, Maintaining Pressure During Solidification; After Plunger is Withdrawn, Die is Opened, and Casting is Ejected

## Centrifugal Casting

As the name suggests, this process involves using a centrifugal force (high-velocity spinning action) to exert pressure on molten metal as it is poured into a die. The die can be set to spinning on a horizontal or vertical axis, depending on the structure required. Usually, a horizontal axis machine is used for long, thin cylinders, and vertical axis machines are used for rings.

This method of casting, also known as liquid forging, is used to manufacture high-integrity cylindrical components, and therefore is a technology of choice for pipes, tube, bearings, rolls, bushes, ball valves, and cylinder liners. Notably, it is used to cast cylinder liners and sleeve valves for piston engines, parts that cannot reliably be manufactured otherwise. Wall thickness of parts can be varied to suit specific needs; this thickness is controlled by the amount of molten metal that is poured, making the centrifugal casting system highly flexible to individual requirements.

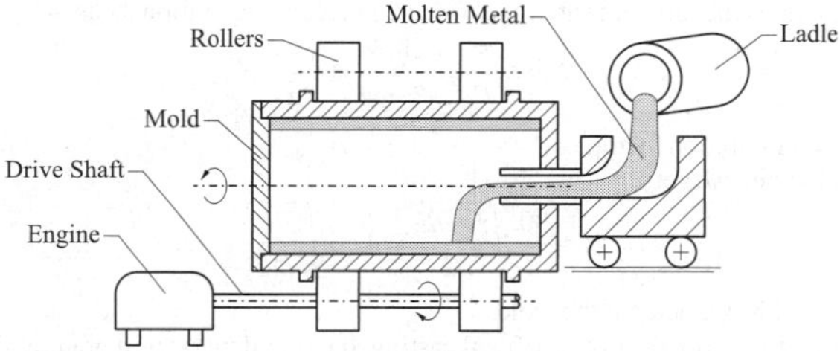


Fig. 14. True Centrifugal Casting

There are numerous advantages of centrifugal casting. The force used to distribute the molten metal as it is poured is so strong that it displaces lighter oxides and impurities to the center of the hollow cylinder; it is then easy to remove these impurities during machining. Blowholes, sand residue, hard spots, cavities, and porosity are virtually eliminated, and the resulting uniformity means machining time and material waste are significantly reduced. Due to the finer grain and denser structure of parts cast with centrifugal force, the part's usage life, endurance, and ability to withstand impact without fracturing are increased. The centrifugal process also allows economical production of a wide range of sizes, shapes, and quantities.

Thus, centrifugal casting produces quality parts, reduces manufacturing costs, and provides production flexibility. There are three types of centrifugal casting: true centrifugal casting, semicentrifugal casting, and centrifuge casting.

**True Centrifugal Casting.—**In this technique, molten metal is poured into a rotating mold. The axis of rotation usually is horizontal but can be vertical for short workpieces. (One setup is illustrated in Fig. 14.) Molds are made of steel, iron, or graphite and may be coated with a refractory lining to increase mold life. Mold surfaces can be shaped so that pipes with various external designs can be cast. Cylindrical parts ranging from 0.6 in. to 10 ft. (15 mm to 3 m) in diameter and 52 ft. (16 m) in length can be cast centrifugally, with wall thicknesses ranging from 0.24 to 5 in. (6 to 127 mm).

True centrifugal casting produces parts of good quality, dimensional accuracy, and external surface detail; better mechanical properties can be achieved than in conventional static casting. Nonmetallic impurities segregate toward the bore and can be machined off. There is less loss of metal as compared with conventional sand casting. The casting is relatively free from defects, there are no parting lines, and production rates are high. This process is used to produce hollow cylindrical parts, such as pipes, gun barrels, bushings, bearing rings, and streetlamp posts; bimetallic tubes can be made.

Consider how fast the mold must rotate in horizontal centrifugal casting for the process to work successfully. Centrifugal force acting on a rotating body is defined by the following equation:

$$F_c = \frac{mv^2}{R} \tag{9}$$

where $F_c$ = centrifugal force; $m$ = mass; $v$ = velocity; and $R$ = inside radius of the mold.

Gravitational force is its weight:

$$F_g = mg \tag{10}$$

where $F_g$ = gravitation force; $m$ = mass; and $g$ = acceleration of gravity.

Velocity $v$ can be expressed as:

$$v = \frac{2\pi RN}{60} = \frac{\pi RN}{30} \tag{11}$$

where $N$ = rotation speed.

The G-factor is the ratio of centrifugal force divided by gravitation force:

$$GF = \frac{F_c}{F_g} = \frac{mv^2}{R \cdot mg} = \frac{v^2}{Rg} \tag{12}$$

where $GF$ = gravitation factor.

Solving further, we get:

$$N = \frac{30}{\pi}\sqrt{\frac{2gGF}{D}} = 42.3\sqrt{\frac{GF}{D}} \tag{13}$$

where $D$ = inside diameter of the mold.

If the G-factor is too low in centrifugal casting, the liquid metal will not remain forced against the mold wall during the upper half of the circular path but will drop inside the cavity. Too high a speed results in excessive stresses and hot tears in the outside surface of the casting. On an empirical basis, $GF$ = 50 to 100 for a metal mold, and $GF$ = 25 to 50 for a sand cast mold.

**Semicentrifugal Casting.—**During semicentrifugal casting (Fig. 15), the mold is rotated around its axis of symmetry. The molds used can be permanent or expendable and may contain cores. The detailed shape is given by the shape cavity of the rotating mold. The mold is designed with risers in the center to supply the feed metal.

Centrifugal force is utilized for refilling of melt metal, increase of the filling power in order to cast parts with thin walls, and slag separation. In general, the rotational speed is lower than speeds used in true centrifugal casting—usually it is set so that a G-factor of around 15 is obtained. As this process produces parts in which the central zone (near the axis of rotation) usually has inclusion defects, it is suitable only for parts where these can be machined away, such as cogwheels.

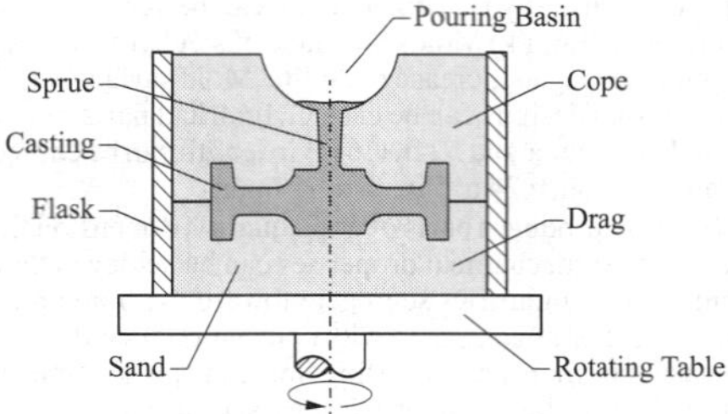


Fig. 15. Semicentrifugal Casting

**Centrifuge Casting.—**The third type of centrifugal casting, centrifuge casting (see Fig. 16) involves molds arranged about the central sprue at the same distance from the center of the axis of rotation. These molds contain all necessary geometry for the cast part, as well as the gating system. During the pouring phase, molten metal is introduced into the central sprue and then travels via runners to the entrance of each mold.

When an object is rotated, force is produced that acts directly away from the center of the axis of rotation. As the centrifuge spins, centrifugal force pushes the metal outward from the center, through the runners into the molds. Once the correct amount of molten metal for the casting is poured and distributed into the molds, the apparatuses continue to rotate while solidification occurs. After the castings have completely solidified, the apparatuses will stop rotating, and the parts can be removed.

Use of centrifugal force helps controls specific properties of the cast parts. The method is extensively used for casting smaller parts, including casting gold crowns for teeth in the dental industry.

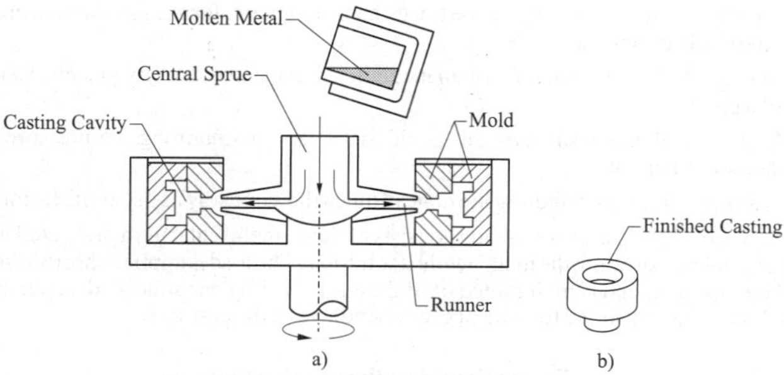


Fig. 16. Centrifuge Casting: a) Mold; b) Final Casting

## Squeeze Casting

Squeeze casting (Fig. 17) combines casting and forging technologies. In contrast to other casting techniques in which a molten metal is poured or injected into the mold cavity after the two parts of the mold are assembled, a squeeze casting mold is closed *after* a portion of molten metal has been poured into the preheated bottom die. The applied pressure and instant contact of the molten metal with the die surface produce a rapid heat transfer condition that yields a pore-free, fine-grain casting with mechanical properties approaching those of a wrought product. Squeeze casting is easily automated to produce near-net to net shape high-quality components.

Because of the high pressure applied during solidification, porosities caused by both gas and shrinkage can be prevented or eliminated. The cooling rate of the casting can be increased by applying high pressure during solidification, since that contact between the casting and the die is improved by pressurization, which results in the formation of fine-grained structures.

With the current emphasis on reducing materials consumption through virtually net shape processing and the demand for higher-strength parts for weight savings, the emergence of squeeze casting as a production process has given materials and process engineers an alternative to traditional approaches to casting and forging. By pressurizing liquid metals while they solidify, near-net shapes can be achieved in sound, dense castings. Squeeze casting is simple and economical, efficient in its use of raw material, has excellent potential for automated operation at high rates of production, and achieves the highest mechanical properties attainable in a cast product. The microstructural refinement and integrity of squeeze cast products are desirable for many critical applications.

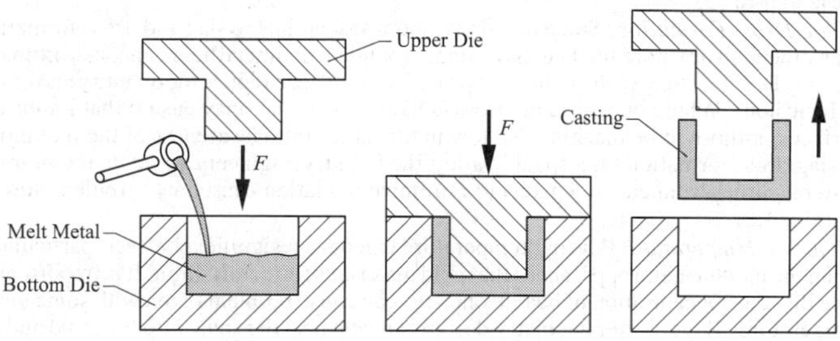


Fig. 17. Sequence of Operations in Squeeze Casting Process: a) Pouring Molten Metal in Die; b) Closing Die and Applying Pressure; c) Casting Ejecting

The origins of squeeze casting can be traced back to squeeze forming, which is a process structured in three phases:

1) Pouring a known amount of molten metal in a preheated die cavity placed at a lower plate of a press.

2) Closing the die, pressurizing the liquid metal, and maintaining the pressure until complete solidification.

3) Casting ejecting and handling are done in much the same way as in closed die forging.

As shown in Fig. 17, squeeze casting consists of liquid metal entering a preheated, lubricated die and pressurizing the metal while it solidifies. The load is applied shortly after the metal begins to set and is maintained until the entire casting has solidified. A number of variables can be controlled to influence the soundness of the castings.

## Design Considerations for Casting

When designing cast metal components, several characteristics are inherent in the pouring and solidification of the alloy and its interaction with the mold. These characteristics define the parameters within which the designer can work and affect the casting method chosen, design of casting sections and junctions between sections, surface integrity and appearance, internal integrity of the cast alloy, and dimensional accuracy.

Carefully planned casting geometry allows the foundry to work with the known pouring and solidification characteristics to produce high-quality castings that perform to required specifications, while avoiding costly and time-consuming problems. By understanding the interaction between casting geometry, the material (in both liquid and solid forms), and the casting process, design engineers can anticipate and avoid many iterations to their product design that could otherwise interfere with and delay progress of a project. The following geometry/material/process interactions dictate good casting design.

*Fluid Life:* This refers to the liquid characteristics of the material that give it the ability to flow freely throughout the mold, along narrow sections and into fine surface detail. Fluid life depends on temperature and the unique chemical and metallurgical properties of each metal. At least in part, it determines minimum wall thickness and how long a thin section can be. So, the designer's choice of alloy, with its associated fluid life, will dictate certain structural and aesthetic elements of the design.

*Solidification Shrinkage:* As molten metal cools, shrinkage occurs in three distinct stages: liquid shrinkage, liquid-to-solid shrinkage, and solid shrinkage. Solid shrinkage is the continued shrinkage that occurs as the metal casting cools to ambient temperature in its solid state. Known as "patternmaker's shrinkage," it must be compensated for within the tooling or mold design to ensure that specified final overall dimensions are achieved.

*Slag/Dross Formation:* Some metals are more susceptible to slag and dross formation than others, so are more likely to have small, rounded, nonmetallic inclusions within the casting. Design, along with quality and process control, can reduce the quantity and effect of inclusions. Where buoyant slag/dross is likely, a designer may ensure that important surfaces or those to be machined sit low in the mold. With awareness of the likelihood of slag/dross formation for a specific alloy, the foundry engineer can set up the running system (sprues, runners, and gates) to minimize oxidation caused by turbulent flow or entrained air.

*Pouring Temperature:* Pouring temperature can have a significant effect, particularly when the molten metal approaches the mold material's refractory limit. It is important to note that the concentration of heat in one area can cause problems even with some lower temperature alloys. Better geometry design can help heat dissipate into the mold and alleviate such issues.

*Fluid Flow:* Designers should take in account the flow rate of molten metal entering the mold and issues this can cause. A balance needs to be struck between the need to get the alloy into the mold quickly to avoid oxidization and dross formation, while avoiding turbulence, wall erosion, and core displacement that can occur at higher fluid speeds. Although not altogether preventable in the manufacturing process, turbulence can be reduced by using a gating system that promotes a more laminar flow of the liquid metal. Turbulence caused by sharp corners and abrupt changes in sections within the casting may be mitigated by employing radii (Fig. 18).

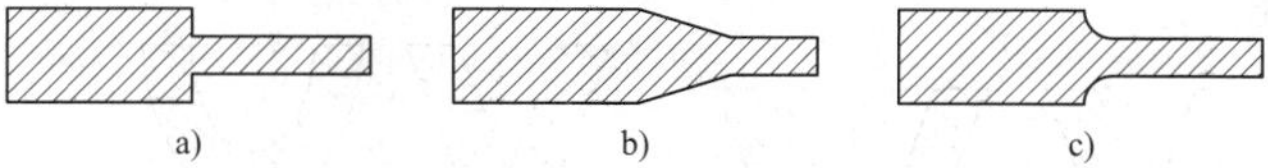


Fig. 18. Design Modification to Avoid Abrupt Changes in Sections within Metal Casting: a) Incorrect; b) and c) Correct

The designer also must be aware that fluid flow can introduce unacceptable thermal gradients, particularly if metal flows around a core and rejoins elsewhere in the mold. Designing the casting geometry and running system at the same time can help identify and eradicate these issues.

*Heat Transfer Considerations:* Management of heat within a mold is an important part of the geometry design. At high pouring temperatures, considerable heat must be transferred into the mold in a way that avoids creation of localized hot spots, where the heat cannot efficiently dissipate. This can occur on narrow peninsulas or tight corners, where molten metal surrounds thin areas of the mold. The retained heat slows solidification in that area that can lead to hot tears or pulls—particularly if the geometry design has introduced stress buildup during solid shrinkage, as the softer metal in the hot spot will have lower tensile strength.

As mentioned earlier, directional solidification is important to the manufacture of a part during the metal casting process, in order to ensure that no area of the casting is cut off from the flow of liquid material before it solidifies. To achieve directional solidification within the casting, it is important to control the flow of molten metal and the solidification rate of the different areas of the metal casting. Regulation of thermal gradients is key.

Sometimes there is an area of a metal casting that needs to solidify at a faster rate to ensure proper directional solidification. Planning sections effectively and regulating flow rates within the mold may not be sufficient. To accelerate solidification of a particular section, it may be necessary to employ the use of chills, which act as heat sinks, increasing the cooling rate in the vicinity where they are placed.

Chills are solid geometric shapes of material, manufactured for this purpose. Chills are of two basic types. Internal chills are located inside the mold cavity and usually are made of the same material as the casting. When the metal solidifies, internal chills are fused into the metal casting itself. External chills are located just outside the casting. They are made of a material that can remove heat from the metal casting faster than the surrounding mold material. Possible materials for external chills include iron, copper, and graphite. Fig. 19 demonstrates the use of the two types of chills to solve a hot-spot problem in cross and T junctions.

Designers should remember that thermal inequalities also can occur in the opposite direction if a sharp corner is surrounded by mold material. In this case, rather than a hot spot forming, the area will cool rapidly, which can cause cold cracking and other issues.

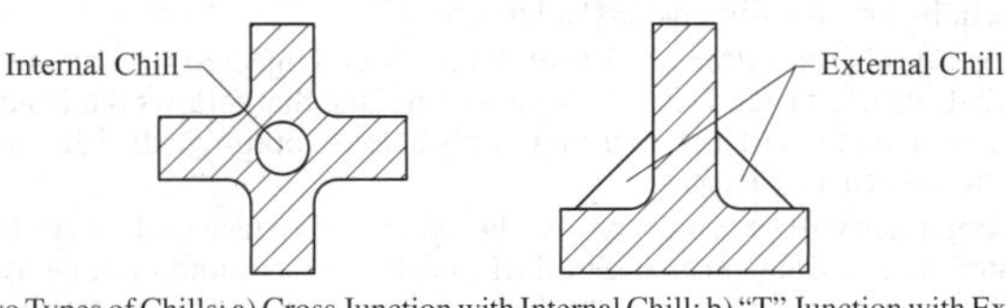


Fig. 19. Two Types of Chills: a) Cross Junction with Internal Chill; b) "T" Junction with External Chill

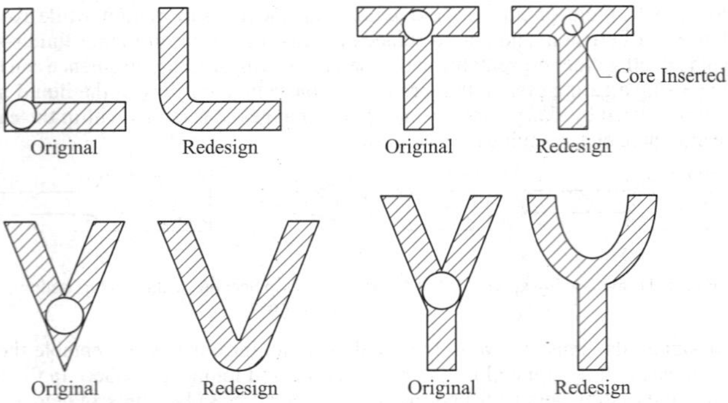


Fig. 20. Redesigned Casting Parts to Avoid Hot Spot and Fillet

**Joint Design Consideration.—**The design of a junction in a casting can cause shrinkage cavities to form. In Fig. 20, L, T, V, and Y junctions are shown. The thickness of the casting here is consistent, though at the center of the junction, there is an area where the casting's solidification is slower than the rest of the junction. These hot spots, circled in the figure, are located such that the material around them, which will undergo solidification first, will cut off the hot spots from the flow of molten metal.

The flow of casting material must be carefully considered when manufacturing such junctions. If there is flexibility in the design of the casting, the designer may want to think about redesigning the junction. Some possible design alternatives are shown in Fig. 20 that should reduce formation of hot spots and shrinkage porosity.

**Location of the Parting Line.—**A parting line in industrial casting of molds is the borderline in which draft angles change direction. That is, it is the dividing line that splits the core and cavity halves of a molded part and is sometimes a starting point for the mold parting surface. In engineering drawings, a parting line is often abbreviated as PL; ASME's Y14.8 specifies a symbol.

The parting line of a casting must be determined early in the design processes. In deciding on its location, the part designer and die cast engineer should work together to consider its influence on part functionality, draft angle, wall thickness, metal flow, casting integrity, and tolerances. Often, it is useful to have most geometry in the ejector half; this prevents the casting from sticking in the cover half. Cosmetic surfaces typically are placed on the cover side.

Gates, overflows, and vents will connect to the casting at the parting line. Also, flash will form at the parting line when the pressure from the injected metal tries to force the die halves apart. During the trimming process, the gates, overflows, vents, and flash will have to be removed by the trim die; a well-designed parting line will ease the trimming process.

When flash occurs, it increases the thickness of the parting line, altering dimensions measured across this line. For critical dimensions, it is important to have measurements in only one die half to reduce dimensional tolerance.

The parting line also can affect the life of the die. Metal injected straight onto a surface will erode that surface faster. Designating a parting line that follows the contours of the design and so that metal will flow more smoothly helps to optimize die life, casting quality, and aesthetics of a die cast part.

Since parting lines will be noticeable as a bump on the surface of the part, they should not be located on a sliding surface. Similarly, parting lines should not be located on a sealing surface, where a bump or mismatch would prevent the seal from making complete contact. The placement of the parting line and orientation of the part determine the number of cores needed, and it is preferable to avoid using cores when possible.

**Location and Design of a Gating System.—**The gating system must be designed to accomplish extremely careful transport of the liquid metal to the mold cavity. Gating system location is defined based on the geometry of the mold, method of casting, and economic aspects. The gating system includes the pouring cup, sprue, risers, runners, and gates. Fig. 21 shows a schematic illustration of a typical gating system in gravity casting.

*Pouring Cup Design:* At the top of the sprue, a *pouring cup* or *pouring basin* is often used to minimize splash and turbulence as metal flows into the sprue. The design of the pouring cup, for an optimal casting process, needs to be such that it can keep the sprue full of molten metal throughout the pour. A constant level of molten metal in the pouring cup is important for a successful casting process; also, if the level of molten metal is maintained in the pouring cup during pouring, then any dross will float and will not enter the mold cavity.

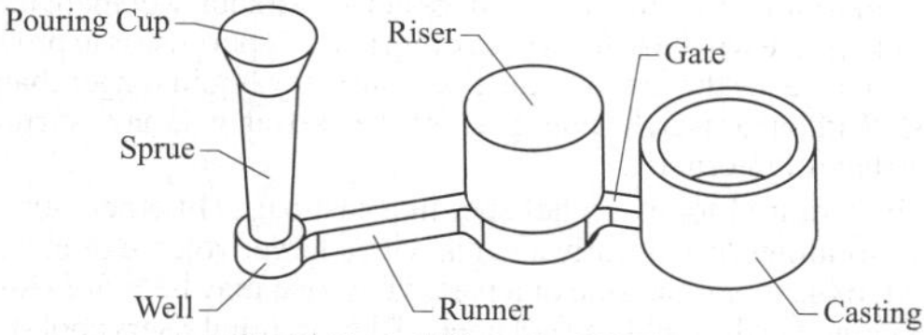

Fig. 21. Typical Riser-Gated Casting

*Sprue Design:* The design of the pouring cup and sprue can affect turbulence. For best results, these should be designed so that the sprue is kept full of molten metal throughout pour. A sprue tapered to a smaller size at its bottom will create a choke, which will help keep the sprue full of molten metal. If a non-tapered sprue is used, a choke can be added when making the runners; it is necessary to have the choke as close as possible to the bottom of the sprue.

A choke also will increase the speed of the molten metal, which is undesirable. To address this problem, an enlarged area at the bottom of the sprue, called a *sprue base,* can be added to decrease the speed of the molten metal. There are two basic types of sprue bases, *enlargement* and *well.* The general rules of thumb for enlargement bases are: the diameter is roughly 2.5 times the width of the runners; the depth is equal to the depth of the runners. The general rules for well bases are: the depth of a well base is twice that of the runners; the cross-sectional area of the base is 5 times the cross-sectional area of the sprue. The bottom of the sprue base should be flat, not rounded like a bowl, or it will cause turbulence in the metal.

*Runner Design:* One of the most important things to keep in mind when designing runners and gates is to avoid sharp edges. Any changes in direction or cross-sectional area should make use of rounded corners. Also make sure the runners and gates are well rammed and smooth; this helps avoid sand erosion and turbulence.

To ensure that metal is not flowing too fast in the runners, the cross-sectional area of the runners should be greater than the area of the choke. And the walls of the runners should be as smooth as possible to avoid causing turbulence. The runners should be filled with metal before the gates; one way to ensure this happens is to put the runners in the drag and the gates in the cope. If a choke is needed in the runner to restrict flow, it should be at least 6 in. (152 mm) from the first gate. The cross-sectional area of runners should decrease as the gates come off them to keep the same gating ratio.

For example, a good gating ratio for aluminum is 1:4:4. The 1 is the cross-sectional area of the choke; the first 4 is the total cross-sectional area of the runners (measured after the choke but before the first gate); and the final 4 is total cross-sectional area of the gates. For example, for a tapered sprue with an exit area of 0.5 in$^2$ (322 mm$^2$) and two runners with two gates from each runner, the total runner area should be 2 $in^2$ (1290 $mm^2$), so that each runner would be 1 $in^2$ (645 $mm^2$). The total gate area should be 2 $in^2$ (1290 $mm^2$), and there are four gates, so each gate would have an area of 0.5 $in^2$ (322 $mm^2$). The gate calculation only works this way if there are an equal number of gates on each runner; if not, divide the area of the runner by the number of gates on that runner to get the area of each gate.

The area of the runners should be reduced just after a gate by an amount equal to the area of that gate. This ensures that each gate in the system will have the same flow of metal, even if it is farther from the sprue. The first bit of metal poured is most likely to be contaminated by air and sand entrapment. A runner extension can be used to prevent this metal from going into the mold cavity; this extension will have the same area as that of the last gate on that runner.

*Riser Design:* Risers are important to ensure a flow of molten metal to the part being cast as it is starting to solidify. Basically, a riser is a vertical portion of the gating system (similar to a straight sprue) that stores the molten metal until it is needed by the casting.

As molten metal solidifies, it shrinks. If it does not have a source of more molten metal to feed it as it shrinks, there will be defects in casting; a riser's purpose is to provide that extra molten metal. This means the metal in the riser must stay liquid longer than metal in the part being cast. Without a riser, heavier parts of the casting will have shrinkage defects, either on the surface or internally.

A riser may be required for every hot spot in a cast part. (In other words, the part of the casting that solidifies last, usually an area with a larger volume of metal.) The risers can be attached to the top or the side of a part. They also may be *blind risers*, which are completely contained in the mold and not open to the air; blind risers cool slower and thus will stay liquid longer. No matter where it is located, any gate that connects a riser to the casting must not be too small; however, it should be as short as possible (preferably half the diameter of the riser), else the material in the gate will solidify too soon and prevent metal in the riser from reaching the casting.

Risers may be upstream from the casting in the runner/gate system. In this case, the metal must flow through the riser prior to reaching the casting. After the pour is completed, the metal in the riser will be hotter than the metal in the casting.

Risers also may be placed downstream, after the casting. Metal flows through the casting to get to the riser, so the metal in the riser will be cooler than the metal in the casting. This could cause a problem if the metal in the casting feeds the riser as it cools.

It is necessary for the metal in the riser to solidify last. Since the more surface area something has, the faster it cools, the surface area of the riser should be minimized for a given volume. The optimum shape for a riser would be a sphere, but that is not an easy shape to mold; next best is a cylinder, which is easier. Ideally, the cylinder's height should be somewhere between one-half and one-and-a-half times the diameter. If possible, the bottom (and top if it is a blind riser) should be spherical or bowl-shaped to help the metal stay molten longer.

### Using Computer Modeling to Optimize Casting Processes

In today's global manufacturing environment, time is money. Casting buyers want foundries to deliver quality cast components with short lead times, regardless of the number of steps in the component's production cycle. To this end, foundries must take a greater role in the initial design of cast components, as well as gating/riser design. The one tool that has become indispensable to foundries when working with their customers to achieve an optimized casting design with minimal lead times is computer modeling. Using solid models, finite element and fatigue analysis, and casting process modeling, foundry workers can make numerous no-cost design revisions on the computer screen.

This analysis ensures proper mold filling and solidification of the component. The accuracy of such models in predicting defects, such as solidification porosity, hot tearing, and grain coarsening, has led to their widespread use among foundries.

**Modeling Optimization at Design Conception.—**Some foundries become involved in the design of cast components at conception. In such cases, the foundry is able to optimize a cast component's structure and manufacturability from the beginning of the process, instead of reengineering it later. Two computer-aided engineering analysis methods used to optimize the structure of cast components and process of metal casting from conception are topology optimization and multidisciplinary response surface optimization. Both methods utilize casting modeling as the foundation for design iterations.

*Topology Optimization:* This method of layout optimization is most effectively employed in early stages of the design process, ideally at the start. The only input required is a layout of the package space, a definition of loads and constraints on the structure, and a mass target. The optimization process starts with a uniform topology (mass distribution), which it modifies to minimize the compliance of the structure. In other words, the mass is moved around in the package structure to achieve the stiffest design for the mass target. Multiple loading conditions can be input, as well as weighing factors.

Topology optimization can quickly provide information to the design engineer as to the optimal layout of the mass, before concept designs are even developed. This can reduce the number of iterations significantly during design, since optimal part configuration will be known in the beginning. If used on existing designs, the process may be used to reduce structural weight and optimize performance while maintaining mass.

*Multidisciplinary Response Surface Optimization:* This optimization method is used in the conceptual design stage to ensure manufacturability of design. Using solidification modeling, it can increase the quality of casting and productivity of foundries. The steps involved are:

1) Set up the optimization problem, including definition of the baseline problem and description of component geometry (finite element, material properties, etc.).

2) Define process variables and constraints, including fill rates, mold temperatures, etc.

3) Define shape variables and constraints.

4) Define the objective function (such as minimizing porosity or solidification time).

5) Set the input parameters, including a solidification modeling input deck.

6) Submit the analysis to the solver system (software) and perform an initial run.

7) Compare the results of this analysis with the objective function.

In the last step, if the convergence criteria for the objective function are met, the result will be an optimized process. If the criteria are not met, the process will repeat itself. The optimization software then will reconfigure the input and launch a new analysis. The software controls the input of the analysis each time after checking the results of the previous run. The greater the number of variables, the greater number of iterations are required.

Such computerized multiphysics process modeling simulates the flow of liquid metal in the mold, heat transfer, and solidification of the metal, and identifies when and where defects, such as porosity, are likely to occur. With guidance of skilled casting engineers, modifications to the mold and feed system can be made, and the computer model run again, until a defect-free casting design is simulated that can be used to fabricate the production mold. Limiting failures to computer simulations reduces trial-and-error efforts to obtain a sound casting.

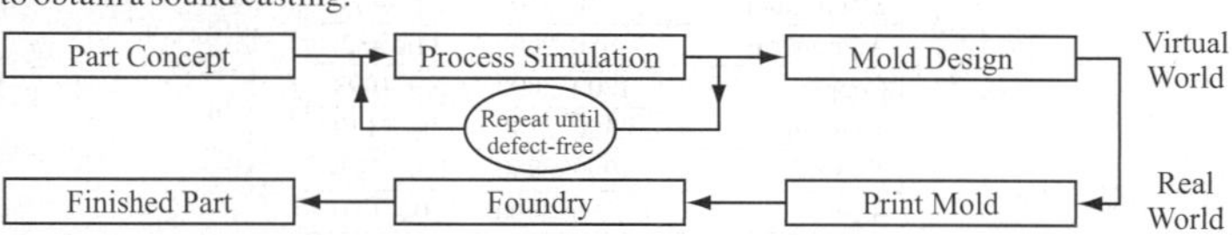


Fig. 22. Computer-Aided Optimization for Efficient Casting Production

## Extrusion of Metals

Extrusion is a metalworking process used to produce long, straight semifinished products, such as bars, tubes, solid and hollow sections, wire, and strips, by squeezing a solid slug of metal, either cast or wrought, from a closed container through a die. Depending on the materials and dies used, cross sections of varying complexity may be produced. The extruding process can be performed at room or high temperature, based on the alloy and method. During extrusion, compressive and shear forces that develop in the stock enable the material to be heavily deformed without fracturing. But if extrusion parameters are not carefully monitored and controlled, localized tensile stresses can cause cracking.

In the specially constructed presses used for extrusion, the load is transmitted by a ram through an intermediate dummy block to the stock. The press container is usually fitted with a wear-resistant liner and is constructed to withstand high radial loads. The die stack consists of the die, die holder, and die backer, all of which are supported in the press end housing or platen, which resists the axial loads.

The following are characteristics of different extrusion methods and presses:

1) The movement of the extrusion relative to the ram. In "direct extrusion," the ram is advanced toward the die stack; in "indirect extrusion," the die moves down the container bore
2) The position of the press axis, which is either horizontal or vertical
3) The type of drive, which is either hydraulic or mechanical
4) The method of load application, which is either conventional or hydrostatic

In forming a hollow extrusion, such as a tube, a mandrel integral with the ram is pushed through the previously pierced raw billet.

*Cold Extrusion:* Cold extrusion has often been considered a separate process from hot extrusion; however, the only real difference is that cold or only slightly warm billets are used as starting stock. Cold extrusion is not limited to certain materials; the only limiting factor is the stresses in the tooling. In addition to the soft metals such as lead and tin, aluminum alloys, copper, zirconium, titanium, molybdenum, beryllium, vanadium, niobium, and steel can be extruded cold or at low deformation temperatures. Cold extrusion has many advantages, such as no oxidation or gas/metal reactions; high mechanical properties due to cold-working if the heat of deformation does not initiate recrystallization; narrow tolerances; good surface finish if optimum lubrication is used; fast extrusion speeds can be used with alloys subject to hot shortness. Examples of cold extruded parts are collapsible tubes, aluminum cans, fire extinguisher cases, shock absorber cylinders, automotive pistons, and gear blanks.

*Hot Extrusion:* Most hot extrusion is performed in horizontal hydraulic presses rated in size from 250 to 12,000 tons. The extrusions are long pieces of uniform cross sections, but complex cross sections are also produced. Most types of alloys can be hot extruded.

Owing to the temperatures and pressures encountered in hot extrusion, the major problems are the construction and the preservation of the equipment. Table 2 gives approximate temperature ranges used to extrude various types of alloys.

**Table 2. Extrusion Temperatures for Common Metals**

| Alloy Type | Temperature °F | Temperature °C |
|---|---|---|
| Magnesium | 650–850 | 343–454 |
| Aluminum | 650–900 | 343–482 |
| Copper | 1200–2000 | 649–1093 |
| Titanium | 1300–2100 | 704–1149 |
| Nickel | 1900–2200 | 1038–1204 |
| Steel | 2200–2400 | 1204–1316 |
| Refractory Alloys | up to 4000 | up to 2204 |

In addition, pressures range from as low as 5000 psi (34.5 MPa) to over 100,000 psi (690 MPa). Therefore, lubrication and protection of the chamber, ram, and die are generally required. The use of oil and graphite mixtures is often sufficient at the lower temperatures; while at higher temperatures, glass powder, which becomes a molten lubricant, is used.

*Extrusion Applications:* The stress conditions in extrusion make it possible to work materials that are brittle and tend to crack when deformed by other primary metalworking processes. The most outstanding feature of the extrusion process, however, is its ability to produce a wide variety of cross-sectional configurations; shapes can be extruded that have complex, nonuniform, and nonsymmetrical sections that would be difficult or impossible to roll or forge. Use of extruded sections may reduce machining operations or even replace more costly assemblies joined with welding, bolting, or riveting. However, as extrusion temperatures increase, processing costs also increase, and the range of shapes and section sizes that can be obtained becomes narrower.

While many asymmetrical shapes are produced, symmetry is the most important factor in determining extrudability. Adjacent sections should be as nearly equal as possible to permit uniform metal flow through the die. The length of their protruding legs should not exceed 10 times their thickness.

The size and weight of extruded shapes are limited by the section configuration and properties of the material extruded. The maximum size that can be extruded on a press of a given capacity is determined by the "circumscribing circle," which is defined as the smallest diameter circle that will enclose the shape. This diameter controls the die size, which in turn is limited by the press size. For instance, the larger presses are generally capable of extruding aluminum shapes with a 25 in. (63.5 cm) diameter circumscribing circle and steel and titanium shapes with about 22 in. (55.9 cm) diameter circle.

The minimum cross-sectional area and thickness that can be extruded on a given size press are dependent on the properties of the material, the extrusion ratio (ratio of the cross-sectional area of the billet to the extruded section), and the complexity of shape. As a rule, thicker sections are required with increased section size. Table 3 gives approximate minimum cross section and thickness of some commonly extruded metals.

**Table 3. Cross Section and Thickness of Commonly Extruded Metals**

| Material | Minimum Cross Section | | Minimum Thickness | |
|---|---|---|---|---|
| | in² | mm² | in. | mm |
| Carbon and Alloy Steels | 0.40 | 258 | 0.120 | 3.0 |
| Stainless Steels | 0.45–0.70 | 290–452 | 0.120–0.187 | 3.0–4.8 |
| Titanium | 0.50 | 323 | 0.150 | 3.8 |
| Aluminum and Magnesium | <0.40 | <258 | 0.040 | 1.0 |

Extruded shapes minimize and sometimes eliminate the need for machining; however, they do not have the dimensional accuracy of machined parts. Smooth surfaces with finishes better than 30 μin. (0.762 μm) rms are attainable in magnesium and aluminum; an extruded finish of 125 μin. (3.175 μm) rms is generally obtained with most steels and titanium alloys. Minimum corner and fillet radii of $^1/_{64}$ in. (0.39 mm) are preferred for aluminum and magnesium alloys; while for steel, minimum corner radii of 0.030 in. (0.76 mm) and fillet radii of 0.125 in. (3.175 mm) are typical.

*Extrusion of Tubes:* In tube extrusion, the metal passes through a die, which determines its outer diameter, and around a central mandrel, which determines its inner diameter. Either solid or hollow billets may be used, with the solid billet being used most often. When a solid billet is extruded, the mandrel must pierce the billet by pushing axially through it before the metal can pass through the annular gap between the die and the mandrel. Special presses are used in tube extrusion to increase the output and improve the quality compared to what is obtained using ordinary extrusion presses. These special hydraulic presses independently control ram and mandrel positioning and movement.

# POWDER METALLURGY

## Introduction

Powder metallurgy (PM) is a process for forming metal parts by heating compacted metal powders to just below their melting points. The heating treatment is called sintering. Although the modern field of powder metallurgy dates to the early 19th century, over the past quarter century, it has become widely recognized as a superior way of producing high quality parts for a variety of important applications.

Powder metallurgy actually comprises several different technologies for fabricating semidense and fully dense components. The conventional PM process, referred to as press-and-sinter, has been used to produce many complex parts, such as the planetary carrier, helical gears and blades, piston rings, connecting rods, cams, brake pads, bushings, tool steels, tungsten carbides, cermets tool and die materials, magnetic materials, graphite brushes impregnated with copper for electric motors, surgical implants, and many other parts for aerospace, nuclear, and industrial applications.

PM's popularity is due to a number of attributes: a) the advantage that the process offers over other metal-forming technologies such as forging and metal casting, b) its advantages in material utilization, c) shape complexity, d) near-net-shape dimensional control, and others. PM's benefits add up to cost-effectiveness, shape and material flexibility, application versatility, and part-to-part uniformity for improved product quality.

Advantages that make powder metallurgy an important commercial technology include the following:

- eliminates or minimizes machining by producing parts at or close to final dimensions
- eliminates or minimizes scrap losses by typically using more than 97 percent of the starting raw material in the finished part
- permits a wide variety of alloy systems
- produces good surface finishes and tolerances of ±0.005 in. (±0.13 mm)
- provides materials that may be heat treated for increased strength or increased wear resistance
- relatively low processing temperatures
- provides controlled porosity for self-lubrication or filtration
- facilitates manufacture of complex or unique shapes that would be impractical or impossible with other metalworking processes
- is suited to moderate- to high-volume component production requirements
- offers long-term performance reliability in critical applications
- is cost-effective and environmentally friendly
- can be automated.

There are some disadvantages associated with PM processing. These include:

- high tooling and equipment cost
- expense of metal powder
- difficulties with storing and handing metal powders (degradation of the metal over time and fire hazards with particular metals)
- variations in material density can cause problems, especially for complex geometries
- limitations on part geometry because metal powders do not readily flow laterally in the die during pressing.

Most parts weigh less than 5.5 lb (2.5 kg), although parts weighing as much as 110 lb (50 kg) can be produced using conventional powder metallurgy equipment. While many of the early powder metallurgy parts, such as bushings and bearings, were very simple shapes, today's sophisticated powder metallurgy process produces components with complex contours and multiple levels and does so quite economically.

## Characteristics of Metal Powders

A powder is defined as a finely divided solid, smaller than 0.039 in. (1 mm) in its maximum dimension. In most cases the powders will be metal, although in many instances they are combined with other materials such as ceramics or polymers. Powders exhibit behavior that is intermediate between that of a solid and a liquid. Powders will flow under gravity to fill containers or die cavities, so in this sense they behave like liquids. They are compressible like a gas. But the compression of a metal powder is essentially irreversible, like the plastic deformation of a metal. Thus, a metal powder is easily shaped, but it has the desirable behavior of a solid after it is processed.

When one deal with powders, the properties of both the individual particles and the collective (bulk) properties of the powder must be considered. The properties of single particles include size, shape, and microstructure, which can be determined by optical or scanning electron microscopic observations. In order to characterize a bulk powder, it is necessary to be able to determine at least the following properties:

*Basic Chemical Composition:* The minimum percentage of the base metal or the percentages of main elements in case of metal alloy powders.

*Impurities:* The percentage of impurities.

*Particle Size Distribution:* (see next section).

*Apparent Density:* The weight per unit volume of a simply poured metal powder, which is always less than the density of the metal itself. It is measured by letting the powder drop freely through a funnel to fill a 1.52 in$^3$ (25 cm$^3$) cylindrical container. The ratio between mass and volume, that is, the apparent density, is provided through leveling and weighing and is expressed in kg/m$^3$. The apparent density depends on a series of factors, the more important of which are the metal's true density, powder shape and structure, particle size distribution, corrosion resistance, etc.).

*Flowability:* To assess the speed, standardized funnels with varying calibrated openings are used. A certain amount of powder is poured in the funnel and the flow time is recorded.

**Particle Size Measurement and Distribution.—**The size, distribution, and shape of the powder metal particles influence the physical properties of the parts created. A particle is defined as the smallest unit of a powder. The particles of many metal powders are 0.001 to 0.0078 in. (0.025 to 0.2 mm) in size.

Describing a three-dimensional particle is often a more complex matter than it first appears. For simplicity, it is convenient to describe particle size in terms of one single number. The sphere is the only shape that can be described by one dimension, its diameter ($D$). However, a particle is rarely a perfect sphere. Over the years, particle size analyses of metal powders have been performed using various techniques, including microscope, sieves, aerodynamic time of flight, and laser diffraction. At present, the most popular technique for measuring the size distribution of metal powders is screening and laser diffraction, typically measuring powders in their natural dry state.

*Screening Method:* The most common method used for measured particle size is the use of screens (sieves) of different mesh sizes. The term *mesh count* is used to refer to the number of openings per linear inch of screen.

The basic principle of sieving techniques is as follows. A representative sample of a known weight of particles is passed through a set of sieves of known mesh sizes. The sieves are arranged in downwardly decreasing mesh diameters. The higher the mesh size number, the smaller is the opening in the screen. For example, a mesh size No. 200 has an opening of 0.003 in. (0.074 mm), size No. 100 has an opening of 0.006 in. (0.149 mm), size No. 10 has an opening of 0.078 in. (2 mm). The sieves are mechanically vibrated for a fixed period of time. The weight of particles retained on each sieve is measured and converted into a percentage of the total sample. This method is quick and sufficiently accurate for most purposes.

*Laser Diffraction Method:* Laser diffraction, alternatively referred to as low-angle laser light scattering (LALLS), can be used for the nondestructive analysis of wet or dry samples, with particles in the size range from microscopic up to 0.080 in. (2 mm). This method has inherent advantages that make it preferable to other options for many different materials.

Laser diffraction-based particle size analysis relies on the fact that particles passing through a laser beam will scatter light at an angle that is directly related to the particles' size. Typically, an He-Ne laser (1 = 632.8nm) in the 5 mW to 10 mW range is used as the coherent light source. As particle size decreases, the observed scattering angle increases logarithmically. Scattering intensity is also dependent on particle size, diminishing with particle volume. Large particles, therefore, scatter light at narrow angles with high intensity, whereas small particles scatter light at wider angles but with low intensity.

A typical system consists of: a laser, to provide a source of coherent, intense light of fixed wavelength; a series of detectors to measure the light pattern produced over a wide range of angles; and some kind of sample presentation system to ensure that the material being tested passes through the laser beam as a homogeneous stream of particles in a known, reproducible state of dispersion. The dynamic range of the measurement is directly related to the angular range of the scattering measurement, with modern instruments making measurements from around 0.02 degree to beyond 140 degrees (Fig. 1). The wavelength of light used for the measurements is also important, with smaller wavelengths (e.g., blue light sources) providing improved sensitivity to sub-micron particles.

In laser diffraction, particle size distributions are calculated by comparing a sample's scattering pattern with an appropriate optical model. Traditionally, two different models are used: the Mie theory and Fraunhofer approximation.

*The Mie theory* describes scattering by homogeneous spheres of arbitrary size, and provides a more rigorous solution for the calculation of particle size distributions from light-scattering data. It predicts scattering intensities for all particles, small or large, transparent or opaque. The Mie theory allows for primary scattering from the surface of the particle, with the intensity predicted by the refractive index difference between the particle and the dispersion medium. It also predicts the secondary scattering caused by light refraction within the particle. This is especially important for particles below 0.002 in. (0.05 mm) in diameter, as stated in the international standard for laser diffraction measurements. The Mie approach does not work well for extremely fine particulates in the range below 100 nm, possibly because of increased sensitivity to changes in the refractive index that occur with these materials.

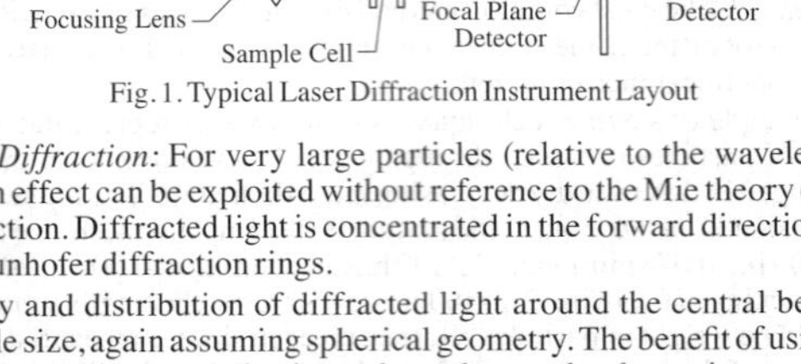

Fig. 1. Typical Laser Diffraction Instrument Layout

*Fraunhofer Diffraction:* For very large particles (relative to the wavelength of light), the diffraction effect can be exploited without reference to the Mie theory or the complex index of refraction. Diffracted light is concentrated in the forward direction, forming the so-called Fraunhofer diffraction rings.

The intensity and distribution of diffracted light around the central beam can be related to particle size, again assuming spherical geometry. The benefit of using Fraunhofer diffraction is that the interpretation is not dependent on the absorptive or refractive properties of the material. On the other hand, the particles being measured are opaque and scatter light at narrow angles. As a result, it is only applicable to large particles and will give an incorrect assessment of fine-particle fractions.

Since laser diffraction is such a popular method of measuring the size of metal powders, an ASTM Standard written to guide proper technique. ASTM B822-02, "Standard Test Method for Particle Size Distribution of Metal Powders and Related Compounds by Light Scattering," provides guidance on the use of laser diffraction for measuring metal powders. The ASTM Standard notes the requirement for a representative sample and suggests a micro-sample splitter to minimize sampling errors. When powder in a liquid suspension is measured, the Standard points out that it is important to create a stable dispersion.

Laser diffraction is a nondestructive, nonintrusive method that can be used for either dry or wet samples. As it derives particle size data using fundamental scientific principles there is no need for external calibration; well-designed instruments are easy to set up and run, and they require very little maintenance.

**Particle Shape and Internal Structure.—**There are various types of metal powder particle shapes, several of which are illustrated in Fig. 2. The choice depends on the requirements of the final product. The microstructure, bulk and surface properties, and porosity depend on the particle shape. There will be a variety of particle shapes in a collection of metal powders, just as the particle size will vary.

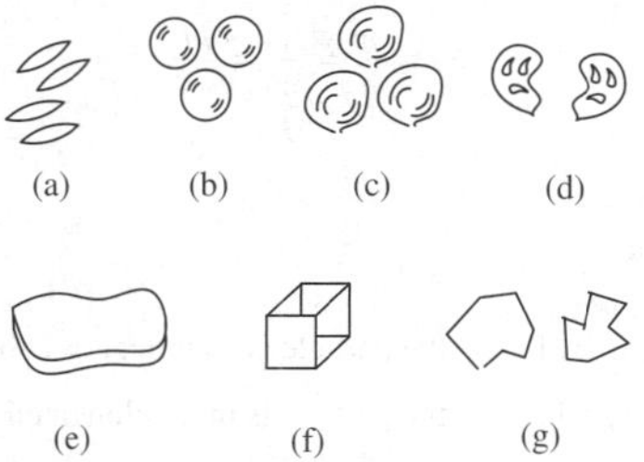


Fig. 2. Schematic Illustration of Several Particle Shapes in Metal Powders:
a) Acicular; b) Spherical; c) Rounded; d) Spongy; e) Flakey; f) Cubic; g) Angular

The particle shape is widely accepted as important if process operation is to be optimized. One of the simple and most useful measuring techniques is *digital imaging*.

By this technique, particle shape is quantified in two ways. First, the aspect ratio determination is made.

*Aspect Ratio:* In general, the aspect ratio is the ratio of the maximum dimension to the minimum dimension for a given particle and is defined more specifically in this method as the ratio of the maximum and minimum particle dimension that passes through the geometric center of the particle. This is illustrated in Fig. 3.

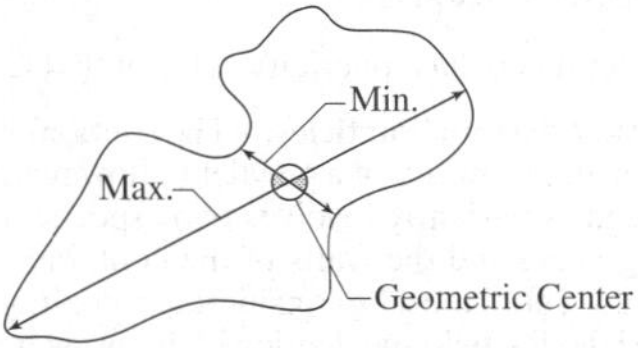


Fig. 3. Particle Aspect Ratio Definition

The aspect ratio for a spherical particle is 1.0, but for another grain, like a needle-shaped grain (acicular), the ratio might be up to 4. Any amount of loose powder contains pores between the particles. These are cooled open pores but in some cases might be voids in the structure of individual particles (closed pores) and their influence on powder characteristics is minor.

*Shape Factor:* The second determination is defined as a shape factor that attempts to quantify the particle shape as a single number. This is not easy; and there have been many attempts to employ shape factor in the past. However, combining the flatness ratio and elongation ratio allows the shape of the particle to be described by a shape factor.

Flatness ratio $p$ is the ratio of the short length (thickness $T$) to the intermediate length $w$:

$$p = \frac{T}{w} \quad (1)$$

The elongation ratio $q$ is the ratio of the intermediate length $w$ to the longest length $l$:

$$q = \frac{w}{l} \quad (2)$$

where $T$ = thickness of particle, in. (mm)

$w$ = intermediate length of particle, in. (mm)

$l$ = longest length of particle, in. (mm)

$$S_f = \frac{p}{q} = \frac{\frac{T}{w}}{\frac{w}{l}} = \frac{lT}{w^2} \quad (3)$$

where $S_f$ = shape factor

$p$ = flatness ratio

$q$ = elongation ratio

If $S_f = 1$, the particle is spherical or cubical

$S_f < 1$, the particle is more elongated and thin (4)

$S_f > 1$, the particle is blade-shaped

*Sphericity:* Sphericity is, naturally, a measure of how spherical an object is. Defined by Wadell in 1932, the sphericity $y$ of a particle is the ratio of the surface area of a sphere with the same volume as the given particle to the surface area of the particle:

$$\Psi = \frac{\pi^{\frac{1}{3}}(6V_p)^{\frac{2}{3}}}{A_p} \quad (5)$$

where $V_p$ = volume of the particle

$A_p$ = the surface area of the particle

A particle that is a perfect sphere has sphericity value of $\Psi = 1$.

**Friction and Flow Characteristics of Particles.—**The friction characteristics of particles play a very important rule in the ability of a powder to flow readily and pack tightly. The friction of particle aggregates basically involves two aspects: inter-particle friction and friction between the aggregate and the walls of the container. Both aspects constitute challenging topics, and often both must be considered in dealing with practical problems related to powder materials. Particle mechanics is the branch of knowledge that deals with the mechanics properties and flow of particles and particle aggregates. Friction is a critical aspect of many particle mechanics problems. The most common way to quantify inter-particle friction is by measuring the angle of repose.

*Angle of Repose:* Angle of repose, in general, is an engineering property of granular material. When powders are poured from a narrow funnel onto a horizontal surface, a conical pile will form (Fig. 4).

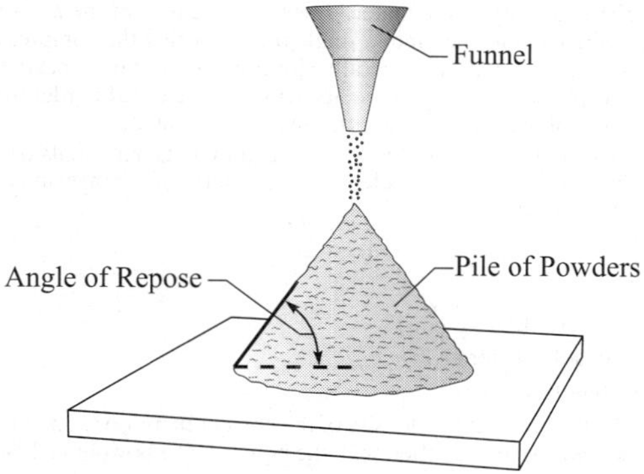


Fig. 4. Inter-Particle Friction as Indicated by the Angle of Repose

The angle between the horizontal surface and the surface of the pile is known as the angle of repose and is related to the density, surface area, and coefficient of friction of the particles. Materials with a higher angle of repose have greater friction between particles, while smaller angles suggest less friction. Smaller particle size generally indicates greater friction and steeper angles.

*Powder Flow:* A simple definition of powder flowability is the ability of a powder to flow. Flowability is the result of a number of factors and for convenience they may be grouped under the following headings:

- Physical properties of the powder particles, such as their size, shape, hardness, elasticity, porosity, mass, interactions between particles, and so on.
- Environmental factors that affect the powder's bulk properties, such as air or moisture content, external pressure, vibration, etc. These factors modify the physical distribution and arrangement of the particles in the powder mass.
- Individual particle changes caused by factors such as attrition, chemical charge, etc.

Powders will flow under gravity to fill containers or die cavities, so in this sense they behave like liquids. As particles in powders get smaller, they have a greater tendency to adhere to one another rather than to fall away under the influence of gravity. This is due to the fact that the surface area of the particles becomes large in proportion to the mass. Therefore, the surface forces of friction and cohesion begin to dominate gravitational forces as particle size decreases. Increasing interparticle friction and cohesion reduce the flowability of the powder. To reduce interparticle friction and facilitate flow during pressing, lubricants are added to the powders in small amounts.

Particle shape also greatly affects the flow characteristics of powders. Depending on the fabrication method, particles can take on many different shapes, including spherical, rounded, and angular.

The flowability of a powder can be quantified by measuring its gravity-driven flow rate through a calibrated orifice. The conventional method measures the time required for a 50-gram sample to flow through a standardized funnel.

**Packing, Density, and Porosity.—**Packing, density and porosity are based on how much the metal powder will pack down when pressed. A packing of particles is an assemblage of particles and is widely encountered in many industries. Packing characteristics depend on two density measures: true density and bulk density.

*True Density:* True density of the material would be density of the mass if the powder particles were melted and formed into a single piece. During the compression stage the particles are reshuffled, becoming generally closer together as inter-particular pores are eliminated. As the powder is compressed, its density increases. In order to process to its true density, all porosities, open and closed, must be eliminated.

True density of a solid material is defined as the ratio of its mass to its true volume; the true density value is given in many tables as one property of engineering materials. The true density value is calculated as:

$$\rho_t = \frac{M_s}{V_t} \tag{6}$$

where $\rho_t$ = true density, lb/in$^3$ (g/cm$^3$)
$M_s$ = mass of solid material, lb (g)
$V_t$ = true volume, in$^3$ (cm$^3$)

*Bulk Density:* Bulk density is the density of the powder in its loose uncompressed form. A container of known volume is filled with the powder. The powder is then weighed and the density is determined as:

$$\rho_b = \frac{M_p}{V_p} \tag{7}$$

where $\rho_b$ = bulk density, lb/in$^3$ (g/cm$^3$)
$M_p$ = powder mass, lb (g)
$V_p$ = volume of powder in uncompressed form, in$^3$ (cm$^3$)

Major factors determining bulk density of a powder are particle size, particle size distribution, particle shape, and cohesiveness. Bulk density is an important property in packaging and powder handling.

*Packing Factor:* The packing factor value is determined as:

$$PF = \frac{\rho_b}{\rho_t} \tag{8}$$

Typical packing factor values are 0.5 to 0.7, meaning that the material will compact 30 to 50 percent.

The packing factor depends on the particle shape, the distribution of the particle sizes, and external pressure applied during compacting of the particle. Vibration of the powders can also increase packing.

*Porosity:* Porosity is the ratio of volume of the pores (empty spaces) in the powder to the bulk volume. This ratio can be expressed as:

$$\varphi = \frac{V_p}{V_b} \tag{9}$$

where $\varphi$ = porosity; $V_p$ = volume of pores, in$^3$ (cm$^3$); and $V_b$ = bulk volume, in$^3$ (cm$^3$).

Pores, or spaces within the powder, are in large part determined by particle shape and size, since the shape dictates how particles will contact each other. Spaces that exist between particles of a powder and are open to the outside are called open pores. Those spaces expose the external surfaces of powder particles. During compaction of powder, these spaces are eliminated. If any amount of this space remains after pressing, it will result in porosity in the manufactured part. These open pores are permeable to the atmosphere, and they are permeable to fluids such as liquid lubricants and water. The other type of empty spaces that exist in a powder material are called closed pores. These pores are not open to the outside atmosphere. Closed pores can develop during pressing and sintering if the open-pore region becomes closed off. Another type of closed pore exists within the material of powder particles themselves.

In principle:

$$\text{Porosity} + \text{Packing factor} = 1 \tag{10}$$

The issue is complicated by the possible existence of closed pores in some of the particles. If internal pore volumes are included in the above porosity value, then the equation is exact.

## Production of Metallic Powder

A powder can be defined as a finely divided particulate solid. It can be produced in the form of sponges or atomized powders. Generally, any metal can be made into a powder. There are three basic methods for producing metallic powder: atomization, chemical, and electrolytic. Each method can be further subdivided, and powders can also be made by various combinations from each group. As a result, there is a wide range of production routes available, each producing powders with their own particular characteristics.

Which powder production process is used depends on the required production rate, the desired powder properties, and the properties desired in the final part. Chemical and electrolytic methods are used to produce high-purity powders. Mechanical milling is widely used for the production of hardmetals and oxides.

**Atomization.**—Atomization involves either spraying or smashing molten metal into smaller particles.

It is the most versatile method for metal powder production. The main operational parameters and physical conditions for a suitable atomization process for different applications in powder production are the following:

- melting (surface tension, viscosity, and temperature range–solidification temperature)
- process type (aimed throughput, energy efficiency)
- product type (particle size distribution).

Atomization is the dominant method for producing metal and pre-alloying powders from aluminum, brass, iron, low-alloy steel, stainless steel, superalloy, titanium alloy, and other alloys. Atomization is the best method because it yields high production rates and favors economies of scale, and because pre-alloying powders can only be produced by atomization.

There are many different processes employed industrially for atomization, several of which are:

- water atomization
- gas/air atomization
- centrifugal atomization.

*Water Atomization:* In water atomization (Fig. 5) the raw material is melted, and then the liquid metal is broken into individual particles. To accomplish this, the melt stock, in the form of elemental, multi-element metallic alloys, is melted in an induction, arc, or other type of furnace. After the metal is molten and homogenous, it is transferred to a tundish, which is a reservoir used to supply a constant, controlled flow of metal into the atomizing chamber. As the metal stream exits the tundish, it is struck by a high-velocity stream of the water.

The water is sprayed at pressures that range from 145 to 1450 psi (1 MPa to 10 MPa). The molten metal stream is disintegrated into fine droplets, which solidify and cool during their fall through the atomizing tank. Particles are collected at the bottom of the tank. Water atomization cools the metal very rapidly, and the process can cause the formation of irregular particles. It can also corrode some metals.

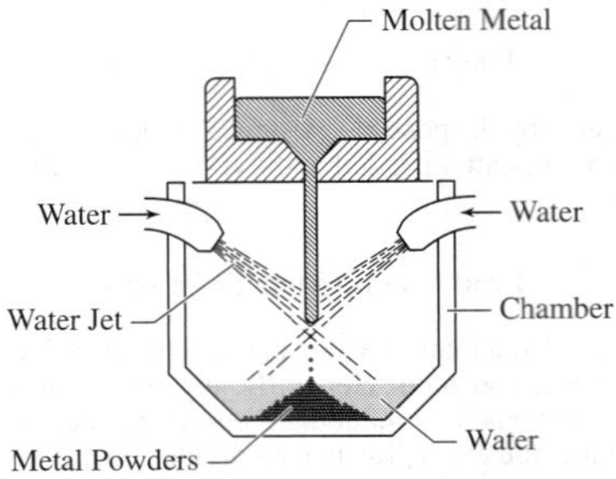


Fig. 5. Water Atomization Method for Producing Metal Powder

*Gas Atomization:* In the area of metal powder production by atomization of molten metal, twin-fluid atomization by means of inert gases is typically used. The main reasons for using this specific atomization technique are:

- the possibility of high throughputs and disintegration of high mass flow rates
- the greater amount of heat transfer between gas and particles for rapid partial cooling of the particles
- the direct delivery of kinetic energy to accelerate the particles towards the substrate/ deposit for compaction
- the minimization of oxidation risks of the atomized materials within the spray process by use of inert gases.

In this process, atomization is accomplished by forcing a molten metal stream through an orifice at moderate pressures (Fig. 6). An inert gas at pressures from 29 to 290 psi (0.2 to 2.0 MPa) is introduced into the metal stream just before it leaves the nozzle, serving to create turbulence as the entrained gas expands (due to heating) and exits into a large collection volume exterior to the orifice. The collection volume is filled with gas to promote further turbulence of the molten metal jet.

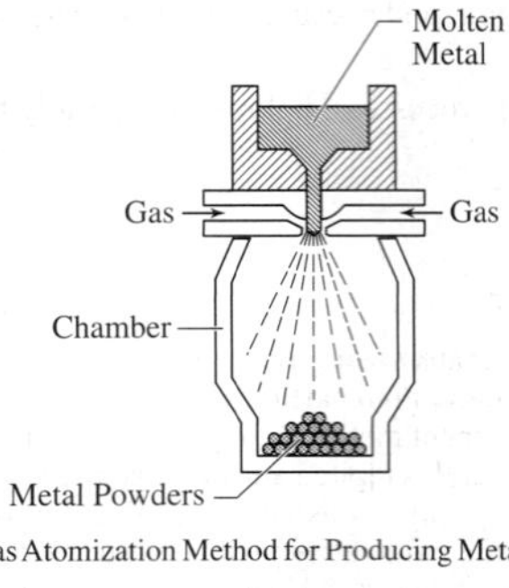


Fig. 6. Gas Atomization Method for Producing Metal Powder

This process as compared to water atomization makes a similarly broad distribution and produce very fine powders. However, the process must be operated more slowly than water atomization and the energy needed to compress the gas is far higher than that needed to pump water. Because this cools the metal at a slower rate, it creates a more circular particle. The advantage of the inert gas process is better control of oxygen levels in oxygen-sensitive materials.

*Centrifugal Atomization:* Centrifugal atomization of molten metals is a cost-effective process for powder production and spray deposition. The properties of the powder and deposit produced by this method are determined primarily by the characteristics of the atomized droplets, which in turn are largely dependent on the flow development of the melt on the atomizer.

There are basically two types of centrifugal atomization processes: the rotating disk process and the rotating electrode process.

*Rotating Disk Process:* In this process, the molten metal stream drops onto a rapidly rotating disk or cup on a vertical axis so that centrifugal forces break up the molten metal stream to form, first, a film of molten metal and then to generate particles (Fig. 7a).

Naturally, the material of which the spinning disk or cup is made takes a lot of punishment. Particle sizes are not very fine, unless very high speeds (over 10,000 rpm) are used. An advantage of this process is that it can produce much narrower distributions than gas or water atomization. Thus, excessively fine or oversized particles can be minimized.

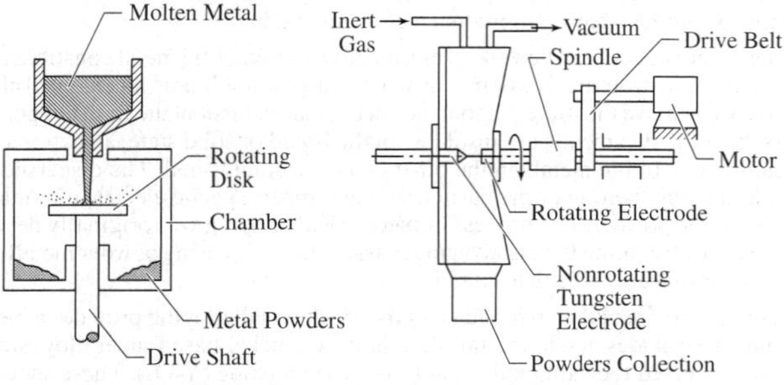


Fig. 7. Centrifugal Atomization: (left) Rotating Disk Method; (right) Rotating Electrode Method

*Rotating Electrode Process:* In the rotating electrode process (REP), an arc is struck between a horizontal tungsten cathode and a spinning electrode of the desired metal for pulverization (Fig. 7b). The electric arc (or plasma in later versions) melts the spinning workplace electrode tip, which throws off a shower of the molten metal droplets by centrifugal forces. The droplets solidify as they pass through a cooling gas or vacuum and are collected in a concentric chamber.

A salient characteristic of REP powder particles is their perfectly spherical shape and freedom from satellite particles adhering to them.

**Mechanical Processes.**—There are basically two types of mechanical processes for the production of metal powders: the mechanical comminution process and the mechanical alloying process.

*Mechanical Comminution Process:* Brittle materials such as ferro-alloys (ferro-chromium, ferro-silicon, etc.) are pulverized mechanically in ball mills (Fig. 8). A process known as the coldstream process is finding increasing application for the production of very fine powders such as are required for injection molding.

Milling is the primary method for reducing the size of large particles and particle agglomerates. Ball, hammer, vibratory, attrition, and tumbler mills are some of the commercially available comminuting devices. During milling, forces act on the feed metal to modify the resultant particles. Impact, attrition, shear, and compression all influence powder particle size and shape. Lathe turning is a technique used for materials such as magnesium for creating coarse particles from billets. These particles then are reduced in size by milling or grinding.

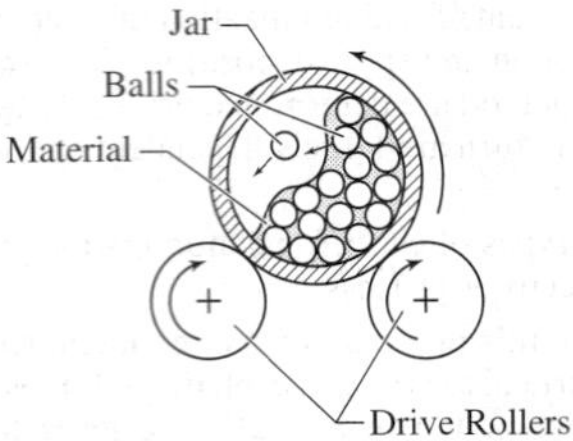


Fig. 8. Method of Mechanical Comminution to Obtain Fine Particles

*Mechanical Alloying:* Mechanical alloying is a process for the production of new materials. It takes place in a high-energy mill where two powders of different origin are mixed. By the conversion of mechanical into chemical energy, nanocrystalline structures emerge that cannot be produced by conventional melting methods.

In general, the process can be viewed as a means of assembling metal constituents with a controlled microstructure. If two metals will form a solid solution, mechanical alloying can be used to achieve this state without the need for an excursion into high temperatures. Conversely, if the two metals are insoluble in the liquid or solid state, an extremely fine dispersion of one of the metals in the other can be accomplished. The dispersed phase can result in strengthening of the particles or can impart special electrical or magnetic properties to the powder. The process of mechanical alloying was originally developed as a means of overcoming the disadvantages associated with using powder metallurgy to alloy elements that are difficult to combine.

Some oxides are insoluble in molten metals. Mechanical alloying provides a means of dispersing these oxides into the metals. Examples are nickel-based superalloys strengthened with dispersed thorium oxide ($ThO_2$) or yttrium oxide ($Y_2O_3$). These superalloys have excellent strength and corrosion resistance at elevated temperatures, making them attractive candidate materials for use in applications such as jet engine turbine blades, vanes, and combustors.

**Chemical Processes.—**There are numerous chemical methods for producing powders. Generally, chemical methods result in very fine powder particle sizes.

The most common chemical powder treatments involve oxide reduction, precipitation from solutions, and thermal decomposition. Chemical reduction involves a variety of chemical reactions that reduce the metal into elemental powders. A common process involves liberating metals from their oxides by the use of reducing agents, which attach to the oxygen in the oxide and render metal powders. The powders produced in this way have great variation in properties and yet have closely controlled particle sizes and shapes. Oxide-reduced powders are often characterized as "spongy," due to pores present within individual particles. Solution-precipitated powders can provide narrow particle size distributions and high purity. Thermal decomposition is most often used to process carbonyls. These powders, once milled and annealed, exceed 99.5 percent purity.

Mill scale and oxidized metallic products are annealed to reduce both the oxygen and carbon contents. FeO, $Fe_2O_3$, or $Fe_3O_4$, are reduced in the presence of a reducing atmosphere. In addition, the carbon within the particles is removed via the formation of CO and $CO_2$.

Hydrometallurgical manufacturing followed by thermal decomposition constitutes another chemical method. Precipitation of a metal from a solution can be accomplished by using electrolysis, cementation, or chemical reduction. This is done either from a solution containing an ore, or by means of precipitation of a metal hydroxide followed by heating which results in decomposition and reduction.

**Electrolytic Deposition.—**Electrolytic deposition is often categorized as a fourth mode of powder fabrication.

In this process, an electrolytic cell is set up in which the source of the desired metal is the anode. The process involves the precipitation of a metallic element at the cathode of an electrolytic cell (Fig. 9).

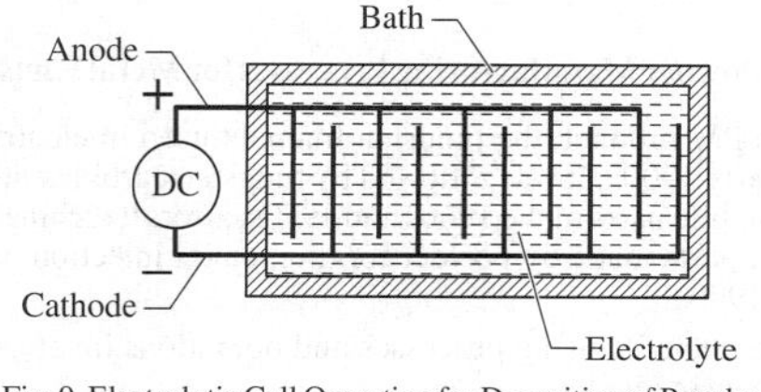


Fig. 9. Electrolytic Cell Operation for Deposition of Powder

By the choice of suitable conditions, composition and strength of the electrolyte, temperature, current density, and so on, many metals can be deposited in a spongy or powdery state. Extensive further processing—washing, drying, reducing, annealing, and crushing—may be required. Copper is the main metal produced in this way, but chromium and manganese powders are also produced by electrolysis. In these cases, however, a dense and normally brittle deposit is formed and needs to be crushed to powder. Electrolytic iron was at one time produced on a substantial scale but it has been largely superseded by powders made by less costly processes. Electrolytic powders are of high purity with a dendritic morphology.

**Hybrid Atomization.—**There is increasing demand for fine spherical powder with uniform particle size applicable to advanced powder metallurgy such as metal injection molding, solder for electronics parts, and joining and conductive inks. However, conventional technologies cannot easily produce powders that satisfy such requirements, so a new powder production technology is needed.

The National Institute for Materials Science (NIMS) developed the world's first powder production method, *hybrid atomization* (Fig. 10), which can easily produce powder having 0.0004 in. (0.01 mm) or smaller spherical particles, with uniform size and low oxygen content. Such powders cannot be produced by conventional powder production technology.

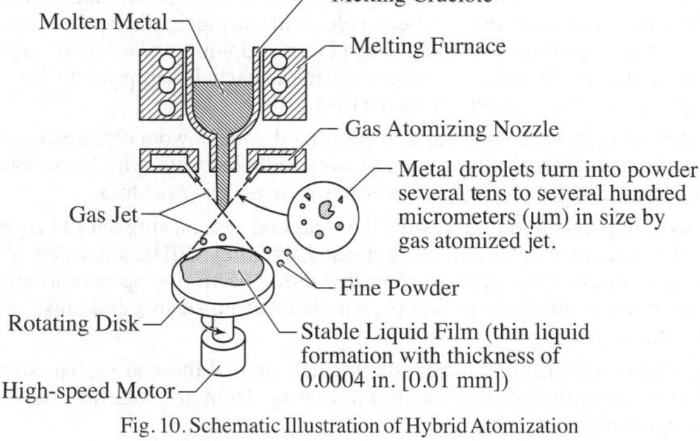


Fig. 10. Schematic Illustration of Hybrid Atomization

Hybrid atomization efficiently combines gas atomization with centrifugal atomization. The gas atomization breaks molten metal into pieces of several tens to several hundred micrometers in size using a gas jet. A rotating disk located beneath the spray is driven at high speed up to 66,000 rpm, and the molten metal is spread uniformly over the rotating disk using a gas spray flow, thereby forming a thin liquid film of 0.0004 in. (0.01 mm) or less in thickness. Then, the fine droplets are scattered from the edge of the rotating disk to result in a fine spherical powder.

### Powder Manufacturing Processes for Metal Parts

The first modern PM product, the tungsten filament used in electric light bulbs, was developed in the early 1900s. It was followed by tungsten carbide cutting tool materials in the 1930s, automobile parts in the 1960s and 1970s, aircraft turbine engine parts in the 1980s, and, finally, parts made by powder forging, metal injection molding, and warm compacting in the 1990s.

An outline of the manufacturing processes and operations involved in making metal powder parts is shown in Fig. 11.

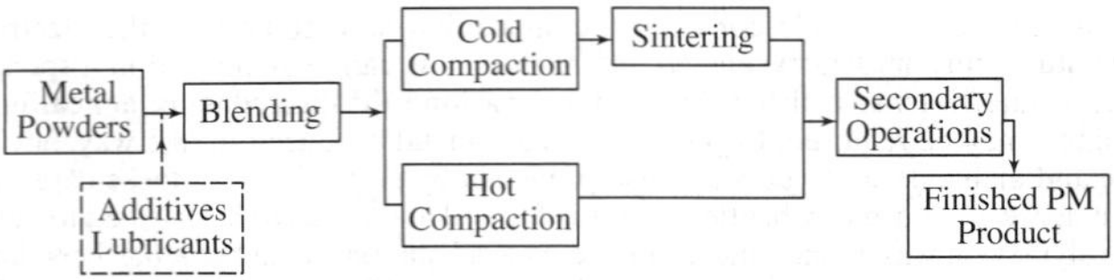


Fig. 11. Outline of Manufacturing Processes of Metal Parts.

After metallic powder is produced, it can be converted to a solid in different ways. But there are three basic steps in producing PM components that are common to most techniques. Whether the final result is sintered brass, sintered stainless steel, sintered bronze, or any other powder-based element, the three production steps are the same:

1) blending and mixing of the powders

2) compaction

3) sintering.

**Blending and Mixing of the Powders.—**Blending and mixing is carried out for the following purposes: 1) powders made by various processes have different sizes and shapes, so they must be mixed to obtain uniformity; 2) powder of different metals and other materials can be mixed in order to impart special physical and mechanical properties and characteristics to the PM parts; 3) lubricants can be mixed within powders to improve their flow characteristics (reducing friction between metal particles, improving flow of powder materials into the dies, and lengthening die life).

Many of the alloy powders are available premixed—the powder already has the correct material characteristics. Additives, such as admixed lubricants, which coat each powder particle and reduce interparticle friction, usually are added to the mix.

Mixing time depends upon the results desired, and overmixing should be prevented; otherwise the size of particles will be decreased, and they will be hardened. This makes the subsequent compacting operation more difficult. The mixing operation can be carried out in air-inert atmospheres or in liquids, which act as lubricants and make the mixture more uniform.

Blending and mixing is done in a mixer device; some of these are schematically illustrated in Fig. 12; they include the horizontal rotating drum, the rotating double cone, the screw mixer, and the blade mixer.

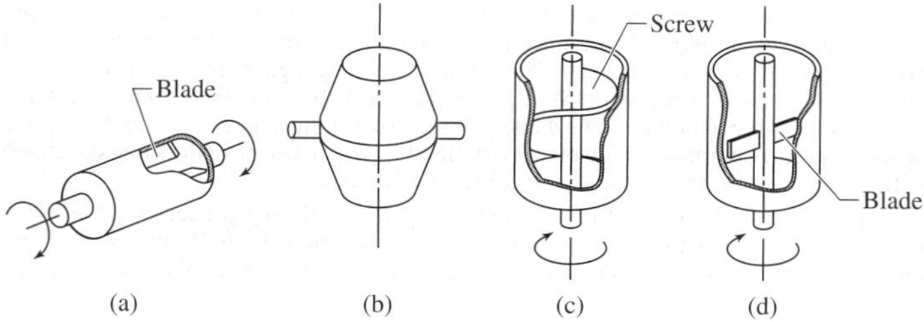


Fig. 12. Blending and Mixing Devices:
a) Rotating Drum; b) Rotating Double Cone; c) Screw Mixer; d) Blade Mixer

In the drum and double cone methods, the container has bafflers inside to prevent the mixture from the free falling, an action that separates the particles by mass. This is the point at which other materials can be added to the mixture. The other materials consist of lubricants (used to lower the friction between the powder and the die), binders (used to hold some metals together between pressing and sintering), and deflocculants (used to allow for better powder flow in the die). Lubricants typically are stearic acid or zinc stearate in a proportion of 0.25 to 5 percent by weight. Because of their high surface area-to-volume ratio, metal powders are explosive; this is particularly true of aluminum (Al), magnesium (Mg), titanium (Ti), zirconium (Zr), and thorium (Th). For that reason, great care must be exercised to protect both workers' health and the general environment during blending and handling.

**Compaction of Metallic Powder.—**After the metallic powders have been blended, compacting is the next step. The purposes of the compacting operation are to obtain the required shape, ensure particle-to-particle interlocking, and make the part sufficiently strong for further processing.

In compaction, the powder is fed into a die cavity, either automatically or by hand by a feed shoe. At the start the powder is at about 2.5 times the final part volume.

The process generally is carried out at room temperature (cold compacting), although it can be done at elevated temperatures (hot compacting). After pressing, the part, which is called a *green compact*, is ejected from the die. Green parts are very frail and can crumble or be damaged very easily. To obtain higher green strengths, the powder must be fed properly into the die cavity, and proper pressure must be developed throughout the part. Fig. 13 shows various stage metallic powder compacting process.

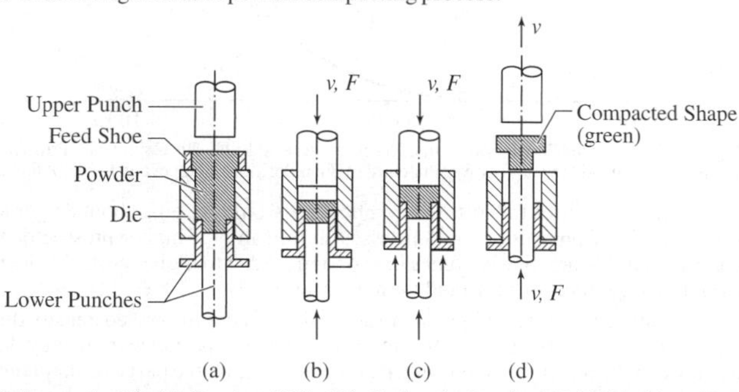


Fig. 13. Compacting Metallic Powder: a) Filled Die Cavity with Powder; b) Initial Position of Upper and Lower Punches; c) Final Position of Upper and Lower Punches; d) Ejection of Part (Green)

The amount of pressure necessary for a compaction operation is to a large degree based on the material. For example, compacting aluminum powder generally requires lower pressure, while compacting iron powder requires relatively higher pressure. The compaction pressure also depends upon the powder characteristics, additives, and the desired density of the green compact. The density of the green compact depends on the applied pressure and the size distribution of the particles. If all particles are of the same size, there will always be some porosity when they are packed together.

Friction force will oppose the movement of particles during processing; therefore, lubrication can reduce the required compaction pressure and also will cause a more uniform distribution of particles during pressing. Lubricant should be applied in the correct quantities.

As the compaction pressure increases, the density of the green compact increases, and the individual particles are pressed together to plastically deform them, increasing their contact and friction but decreasing the pore size. The higher density increases the strength and elastic modulus of a part. Fig. 14 shows density of the powders as a function of pressure.

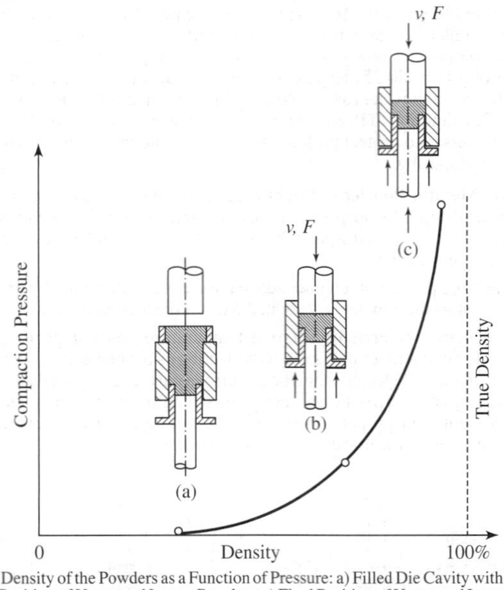


Fig. 14. Density of the Powders as a Function of Pressure: a) Filled Die Cavity with Powder; b) Initial Position of Upper and Lower Punches; c) Final Position of Upper and Lower Punches

Typical pressed densities for soft iron components would be, e.g., about 85 percent of theoretical density. Compaction pressure rises significantly if higher pressed densities are required, and this practice becomes uneconomical due to higher costs for the larger presses and stronger tools to withstand the higher pressures.

One disadvantage of this technique involves the differences in pressed density that can occur in different parts. Because of friction between metal particles in the powder and friction between the punches and die walls, the density within the part can vary considerably. Density variation can be minimized by proper punch and die design, by control of friction, and by keeping low the length-to-diameter ratio of the part.

*Pressing Force:* The required force for pressing depends on the cross-sectional area of the PM part (area in the horizontal plane for a vertical press) and the pressure needed to compact given metal powders. This force may be calculated by the following formula:

$$F = A_p p_c \quad (11)$$

where $F$ = required force, lb (N)
$A_p$ = cross-sectional area, in² (m²)
$p_c$ = compaction pressure, psi (MPa)

Compaction pressures typically range from 1007 to 116030 psi (69 MPa to 800 MPa); however, 20305 to 60191 psi (140 to 415 MPa) is the most common range. Table 1 shows compaction pressures for various metal powders.

**Table 1. Compaction Pressure for Various Metal Powders**

| Material | Compaction Pressures | |
|---|---|---|
| | psi | MPa |
| Porous metals and filters | 5802–10153 | 40–70 |
| Refractory metals and carbides | 10298–29008 | 70–200 |
| Porous bearings | 21030–50763 | 145–350 |
| Machine parts (medium-density iron and steel) | 39885–100076 | 275–690 |
| High-density copper and aluminum parts | 36259–39885 | 250–275 |
| High-density iron and steel parts | 100076–239312 | 690–1650 |

Most conventional PM compacting is done with mechanical presses, although hydraulic and hybrid (combinations of mechanical, hydraulic, or pneumatic) presses are often used. The mechanical presses may be of the eccentric or crank type. Most presses have capacities of less than 100 tons. Mechanical presses with capacities of the magnitude of a few hundred tons are usually adequate for most PM processing operations. Hydraulic presses with a capacity of several thousand tons are sometimes used for compacting parts that require more force. Double action presses are commonly used; for more complex parts, multiple action presses may be employed. Punch speed must be regulated. Faster compaction of the workpiece can result in higher productivity; however, if the punch speed is too high, air may become trapped in the parts and prevent the parts from compacting correctly.

*Die:* Punch and die surfaces are very important in the powder compacting operation. Some clearance between the punch and die must exist in order for the punch to move within the die. However, in a die with a large clearance, powder particles can become stuck in this clearance, causing problems with the proper movement of the punch. In order to prevent this problem, clearance is designed to be extremely low, typically less than the size of the powder particle.

Most punches and dies are made from hardened tool steels, and the surfaces of the die and punch in the direction of the punch movement are grounded and polished or lapped.

**Sintering.—**Sintering is the final step in the PM process. Green powder compacts are heated in a controlled-atmosphere furnace so that adjacent particles fuse together, thus resulting in a solid part with improved mechanical strength compared to the powder compact. The nature and strength of the bond between the particles and the sintered compact depend on the mechanisms of diffusion, the plastic flow, the evaporation of volatile materials in the compact, the recrystallization, the grain growth, and the pore shrinkage. The main variables in sintering are:

- temperature
- time
- furnace atmosphere.

Sintering temperature is below melting point (generally within 70 to 90 percent of the melting point of the metal or alloy), and the metal remains unmelted. The temperature in the furnace is high enough to get metal into the recrystallization zone. In this temperature zone, the metal particles begin to recrystallize into each other. Sintering times average from 1 to 1.5 hours for small parts, such as bushings, to 3 hours for average-size ferrous parts. However, for tungsten parts, average sintering time may go to 8 hours.

Sintering furnaces have three sections: first is a burn-off section for volatilizing the lubricants in the green compact in order to improve bond strength and prevent cracking; second is a high-temperature section for sintering; third is a cooling section.

Hardening sintered parts involves applying a controlled cooling rate in a separate cooling section of the sintering furnace. In the cooling section of the sintering furnace the parts are cooled in a protective atmosphere in order not to be oxidized in contact with air. Dissociated ammonia or nitrogen-based atmospheres are commonly used. Vacuum atmospheres are used for certain metals, such as stainless steel, titanium, and tungsten.

The cooling speed, especially in the temperature range of 1560 to 930° F (850 to 500° C) also affects the mechanical properties, due to phase transformations in the material. During the sintering, a moderate dimensional change takes place. Most materials shrink, but some alloying elements, such as copper, cause material growth. The design of the die must compensate for any anticipated dimensional change. In Fig. 15 is shown a typical heat-treatment cycle in a PM sintering operation.

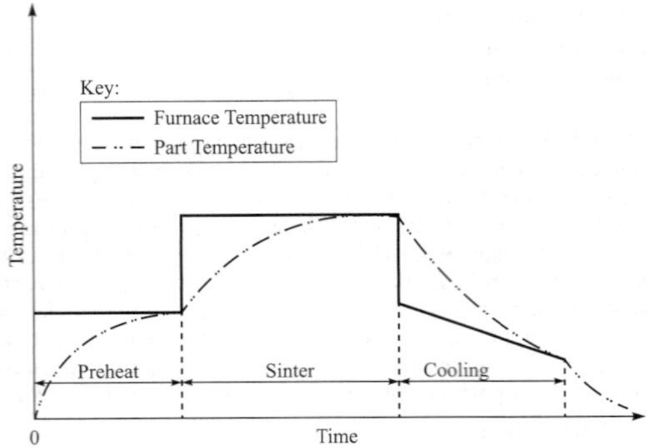


Fig. 15. Typical Heat-Treatment Cycle in Sintering Operation

*Secondary Operations:* Although PM parts can be used as finished parts, they also may need some additional operations after sintering to improve their properties or to impart special characteristics to the parts.

*Repressing:* A second pressing operation (wherein the part is squeezed in a closed die) serves to decrease porosity (i.e., improve density) for applications where density is very important to achieve the required mechanical or other physical properties. After repressing, the parts are sintered a second time.

*Sizing:* Sizing is forcing the part through a finish die to provide dimensional accuracy.

*Coining:* Coining is a pressworking operation on a sintered part to improve the surface finish by further densification or to press details into its surface.

*Finishing Operations:* Some powder metal parts require additional finishing operations that may include: machining to produce holes or threaded holes, and various other geometric features by milling or other processes; grinding to improve dimensional accordance and surface finish; plating to improve resistance to wear and corrosion and sometimes to improve appearance; and heat treating for hardness and other mechanical properties.

*Joining:* Larger parts and very complex shapes can be obtained by joining. Several techniques exist for joining, such as diffusion bonding, sinter brazing, and laser welding.

*Infiltration:* The interconnected porosity is filled with an alloy having a melting point lower than the sintering temperature of the metal of which the component is made, e.g., copper-based alloys infiltrate ferrous parts, usually during the sintering phase. Infiltration makes the components impermeable, and there is some increase in mechanical properties, but at the expense of dimensional accuracy. Infiltration simplifies some heat treatments. For instance, it is easier to obtain a defined case depth without interconnected porosity.

*Impregnation:* Oil or other fluid is permeated into the parts of a sintered PM part. Sintered parts achieve greater protection against corrosion by being impregnated by oil or another nonmetallic material such as polymer resins that seep into the pore spaces in liquid form and then solidify to create a pressure-tight part. Common products are gears and bearings. Self-lubricating bearings are manufactured by impregnating porous sintered bearings with lubricants; these bearings can only be produced by powder metallurgy.

**Liquid Phase Sintering.—**Liquid phase sintering (LPS) is a subclass of the sintering process and can be defined as sintering involving a coexisting liquid and particulate solid during some part of the thermal cycle. The most common way to obtain the liquid phase is to use a system involving a mixture of two powder metals in which there is a difference in the melting temperatures between the metals. The interaction of the two powders leads to formation of a liquid during sintering. The melted metal thoroughly wets the solid particles, leading to rapid consolidation and giving rapid compact densification without the need for an external force. It is, of course, essential to restrict the amount of liquid phase in order to avoid impairing the shape of the part. Depending on the metals involved, prolonged heating may lead to diffusion of the liquid metal into the solid or the dissolution of solid particles into the liquid melt. In either case, the resulting part is fully dense (having no pores) and strong.

**Isostatic Pressing.—**Isostatic pressing is generally used to produce large PM parts to near-net shapes of varied complexity. Unlike conventional PM, in which the powder is compacted uniaxially through direct contact with tooling, isostatic pressing confines the metal powder within a flexible membrane or hermetic metal or glass container, which acts as a pressure barrier between the powder and the pressurizing medium, whether liquid or gas, that surrounds it. The use of this pressurizing system ensures a uniform compaction pressure from all directions and the absence of die wall friction; it produces a fully homogeneous and uniform grain structure and density irrespective of the shape of the final parts. Other advantages of isostatic pressing are that parts with high length-to-diameter ratios have been produced with very uniform density, strength, toughness, and good surface detail; and that it is possible to compact much larger parts than are possible with other compacting processes. Limitations of isostatic pressing include wider dimensional tolerances than are caused by other compacting processes; greater cost and time than are required by other processes; applicability only to relatively small production quantities (fewer than 10,000). Two of the most commonly used methods are cold isostatic pressing and hot isostatic pressing.

*Cold Isostatic Pressing:* In cold isostatic pressing (CIP), metal powders are contained in a flexible mold that is typically made of rubber or another elastomer material and pressed to the green compact by a fluid. Water or oil is usually used to provide hydrostatic pressure against the mold inside the chamber. The most common pressure is 58,000 psi (400 MPa). As the pressure is isostatic, the resulting pressed component is of uniform density. In CIP processing, the part must be sintered after removal from the mold. Fig. 16 illustrates the processing sequence in cold isostatic pressing.

The advantages of CIP include more uniform density, less expensive tooling, and greater applicability to shorter production runs, typically fewer than 10,000 parts per year.

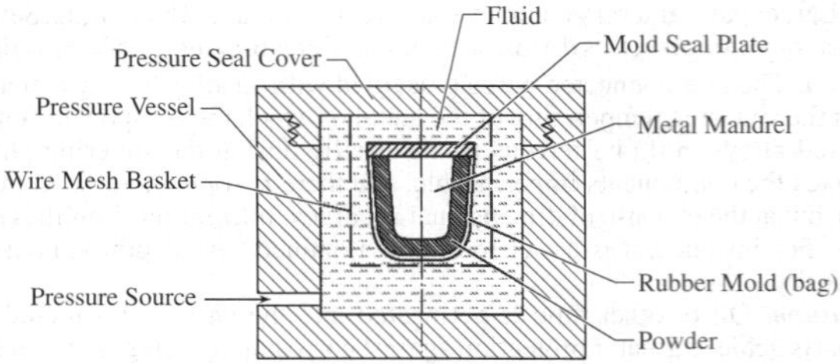

Fig. 16. Cold Isostatic Pressing

Good dimensional accuracy is difficult to achieve in isostatic pressing, due to the flexible mold. Normally, this technique is only used for roughly shaped components, all of which require considerable secondary operations to produce the final, accurately dimensioned component

*Hot Isostatic Pressing:* Hot isostatic pressing (HIP) is a manufacturing process that uniquely combines pressure and temperature to produce materials and parts with substantially better properties than other techniques. The pressure medium is an inert gas, usually argon. Powders are usually encapsulated in a sheet metal container (mold), which is assembled of steel sheets and optionally pipes and metallic inserts by TIG welding. The container is tested to make sure it is leak-free and is designed to give the part its full density and desired shape. Once the container is leak-free, the powders are filled into it via a fill tube. In order to achieve maximum and uniform packing of the powders, which is necessary to ensure a predictable and consistent shrinking, a vibration table is used. In special cases such as critical aerospace applications, the filling operation is done under inert gas or vacuum to minimize contamination of the powders. The next step is outgassing to remove adsorbed gases and water vapor. After outgassing, the fill tube is welded to seal the container. The absence of leaking is critical. Otherwise, when the HIP vessel is pressurized, argon will enter the container and become entrapped in the powder mass, creating argon-filled pores, with damaging effects on the mechanical properties.

The container is then placed in a hot isostatic vessel. The HIP vessel is evacuated to eliminate the air. Then, during heating, argon gas pressure is increased in the vessel. After the calculated pressure is reached, further increase in pressure will occur through gas thermal expansion. During the holding time, gas pressure and temperature are kept constant. After this, a rapid cooling takes place, with decreasing pressure and temperature.

The hot isostatic pressing parameters of pressure, temperature and time are predetermined to give the material full density. Chosen temperatures are below solidus (approximately 0.8 T solidus) to avoid a liquid phase.

Modern HIP systems can feature uniform rapid cooling (URC) that circulates a lower temperature gas to cool the part at a controlled rate of up to 212° F/min (100° C). The HIP quenching technique cuts cycle time dramatically by shortening the cooling stage by as much as 80 percent. It also provides the benefit of combining heat treatment with HIP in a single step. The uniform rapid cooling restricts grain growth and thermal distortion of the parts and avoids surface contamination by using high purity argon gas.

A HIP unit consists mainly of a pressure vessel, a heating system and an argon gas system. Various HIP constructions are available. Fig. 17 schematically illustrates one type of hot isostatic pressure unit.

Molybdenum furnaces are used for temperatures up to 2462° F (1350° C) and 14500 to 43511 psi (100 to 300 MPa).

The sheet metal container remains on the part after the hot isostatic pressing and heat treatment, and it is removed by machining or by chemical milling. After the container has been removed, depending on the type of material and the application, the PM HIP part will be heat treated, machined, and subjected to various types of quality control, such as ultrasonic inspection, dimensional control, and testing of mechanical properties.

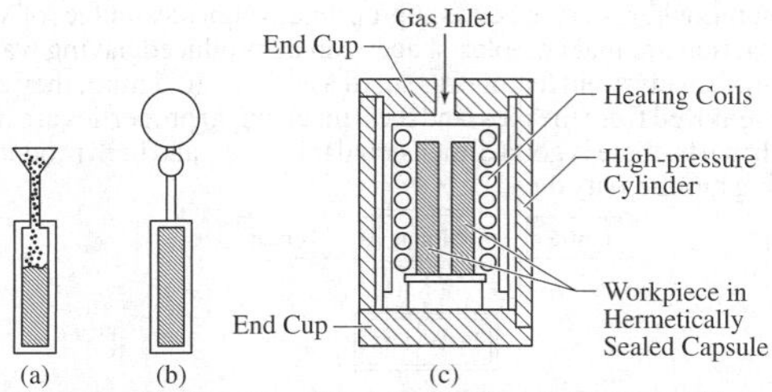


Fig. 17. Hot Isostatic Pressing: a) Container is Filled; b) Container is Evacuated and Sealed Hermetically; c) Schematic Illustration of Hot Isostatic Pressing

The main advantages of HIP are that it improves mechanical properties, produces compaction with almost 100 percent density, and increases workability, and that, when incorporated as an integral part of the manufacturing process, it reduces scrap and allows frequent replacement of wrought components.

Further, it maximizes material utilization by improving material properties, so processed parts exhibit higher reliability and longer service life, and parameters can be established to minimize subsequent heat-treatment operations.

Disadvantages of HIP include that it is a relatively expensive process and that good dimensional accuracy of parts is not easy to achieve with this method.

HIP applications include defect healing of castings, consolidation of metal powder (preform on near-net-shaped parts); improving the structural properties of premium investment castings, primarily for aerospace applications, for which the process was initially used; sintering of diamond and carbide tools; and fabrication of metal matrix composites in the aircraft industry.

**Powder Injection Molding.—**Powder injection molding (PIM) refers to the processing of both metal and ceramic powders. When one is dealing with metals or alloying powders, the term *metal injection molding* (MIM) is used. Powder injection molding is an innovative and cost-effective manufacturing process commonly used for complex, high-quality medical and dental components. It is a high-volume, high-quality, cost-effective process that helps eliminate secondary machining operations for metal part production. PIM is very efficient for manufacturing small, intricate, and complex parts with good mechanical properties and geometrical accordance.

The most common used materials for MIM are stainless steel, chrome-nickel steel, iron, and titanium.

The process is generally best suited to parts measuring less than 0.25 in. (6 mm) thick and weighing less than 100 grams. Typical parts are components for watches, precise mechanic instruments, surgical knives, and automobiles. Powder injection molding is a multi-stage process. A general outline of the process is illustrated in Fig. 18.

In MIM, very fine powders (≤ 10 μm) are processed as follows:

*Mixing:* Metallic powders are mixed with an organic binder. Regarding the maximum powder content, 60 percent is typically powder and 40 percent is binder. Powder content can be slightly increased by blending different binder particle sizes.

*Pelletizing:* Granular pellets are formed from the mixture into feedstock.

*Molding:* The feedstock then undergoes a process similar to die casting; it is injected into the mold at a temperature of 284–400° F (140–200° C) and under pressure of 445–725 psi (3–5 MPa). High pressure powder injection molding at pressures of 5,000 psi (34 MPa) allows molding of thinner features, albeit with increased mold wear.

*Debinding:* The molded green parts are heated at a low temperature to burn off the binder, or a chemical catalytic is used to remove the binder.

*Sintering:* The parts are sintered in a furnace at sintering temperature (generally within 70 to 90 percent of the melting point of the metal or alloy).

Secondary operations are performed as appropriate. Major advantages of MIM over conventional compaction are that complex shapes can be produced having wall thicknesses as small as 0.1 in. (2.5 mm) and features as small as 0.02 in. (0.5 mm); they can be molded and then easily removed from the dies, and their mechanical properties are nearly equal to those of wrought parts; there is good dimensional tolerance, and high production rates can be achieved using multi-cavity dies.

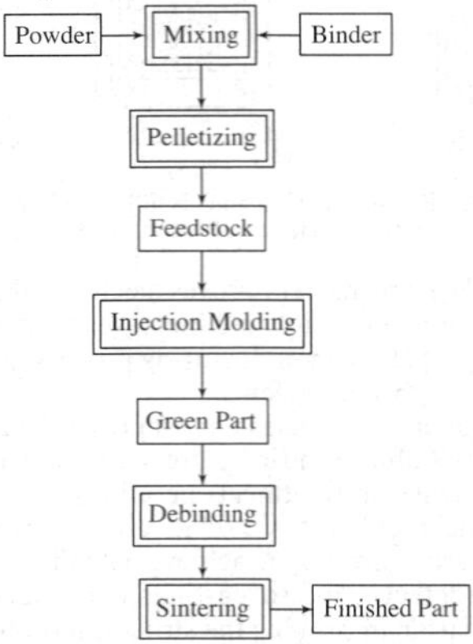


Fig. 18. Outline of the Powder Injection Molding Process

The major limitations of MIM are the limited availability of fine metal powders and the relatively high cost of production.

**Powder Rolling, Extrusion, and Forging.**—Rolling, extruding, and forging are plastic deformation metal-forming processes. In this section, these processes are described in the context of powder metallurgy.

*Powder Rolling:* This term is applied to the process, now established on an industrial scale, wherein a metal powder is fed continuously into a rolling mill (Fig. 19), which may be heated, and compacted between the rolls into strip at speeds of up to 1.6 ft/s (0.5 m/s). This strip is passed through a sintering furnace and rerolled to the finished size. In general, the product does not have any advantage over strip produced by conventional metal-rolling from bulk, although in some cases superior homogeneity can be demonstrated, as well as freedom from laminations that can arise from bulk defects.

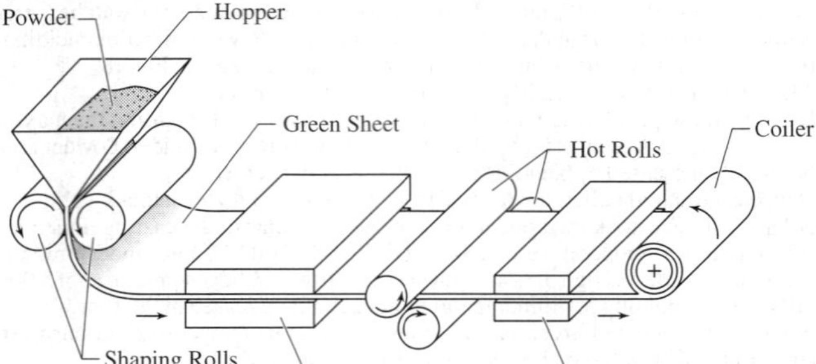


Fig. 19. Powder Rolling Illustration

The main advantages of the process are: low capital cost compared to traditional melting, forging, etc.; fewer operational steps compared with conventional rolling; availability of low-cost row materials; precision control of composition and ability to use high-purity metals; applicability to composite type mixtures or immiscible components that currently cannot be produced by the conventional methods; and adaptability and versatility, including the ability to roll multilayer structure.

Sheet metal for controlled expansion properties of products such as for alkaline batteries, fuel cell electrodes; and for coins can be made by this process.

*Powder Extrusion*: Extrusion, in general, is a process that forces metal to flow through a shape-forming die. (see *Extrusion of Metals* on page 1520). The powder is placed in a vacuum-tight sheet metal container, then heated and extruded, generally through a cylindrical die. As powder particles pass through the extrusion die, they undergo intense deformation and pressure, which bonds them together and removes porosity between the particles. Because there is a limit to the amount of reduction in cross-sectional area that can be achieved by extrusion, often the billet extruded from one die is then further extruded through a die of smaller diameter to eliminate any trace of porosity. The full-density billets produced in this way can then be cut to length, reheated, and used in a closed forging die to produce complex-shaped parts.

This method is particularly useful in production of aluminum and titanium parts for aircraft. Extrusion of powder also is used to produce billets of specialty alloys of copper and tool steels, which are then machined or formed into finished products.

*Powder Forging:* The purpose of forging, whether performed hot or cold, is to improve part performance. Densification is an essential part of the powder forging process. The powder forging (PF) process is performed in three steps, with the first two similar to normal powder metallurgy processing. A form is pressed as a conventional PM compact, called a *preform*. The mass, density, and shape of the preform are controlled closely to ensure consistency in the characteristics of the final forged part. The preform is sintered with particular attention paid to the reducing of nonmetallic inclusions. The sintered preform can be forged to full density in two ways: direct forging the preforms from the sintering temperature, or forging sintered preforms after reheating to forging temperature. The amount of deformation involved is sufficient to give a final density approaching very closely that of the solid material, and consequently, the mechanical properties are comparable with those of material forged from wrought bar.

Advantages of powder metal forging include the following: improved strength and density; high level static and dynamic properties; material flexibility, from low- to high-alloy steel; minimum weight fluctuations and reduced burr waste; high cost-effectiveness; lower tooling cost; and good dimensional accordance of forged parts.

The powder forging process is used mainly in making parts for the automotive industry. Such parts can have inside and outside spline forms, cam forms, and other forms that require extensive machining. In addition to the well-known connecting rod, other applications include bearing races, torque converter hubs, and differential gear sets.

*Additive Manufacturing with Metal Powder:* Another approach to making parts from metal powders is by additive manufacturing (AM). Additive manufacturing processes use an energy beam, such as a laser, or material jetted from an inkjet print head to bond powder particles into the desired shape, as described in a CAD file, usually layer upon layer. AM processes for metal are derived from rapid prototyping with plastics, described in the section *Additive Manufacturing Plastics* on page 612. With this approach, parts can be made into shapes that cannot be made by conventional processes. For example, AM parts may have internal channels that are not straight or round in cross section. Also, parts can be made with details that are too fine or nested to be reached by conventional machining tools. Another advantage is that multiple parts can be consolidated into one part. Also, most of the powdered material is included in the part, so there is minimal material waste. For a discussion of metal AM processes, see *METAL CASTING, MOLDING, AND EXTRUSION* on page 1480.

## Powder Metallurgy Materials

As precision-engineered materials, metal powders are available in a great variety for fabricating a wide range of products through the PM process. Generally, powder mixes for compacting are prepared from three powder types. The first type is admixed powder, in which elemental alloying powders (such as copper, nickel, graphite, and tin) are added to base element powders (such as iron or copper). The second type is partially alloyed powder, composed of two or more elements with alloying additives that are diffusion-bonded to the base powder during the powder manufacturing process. These powders produce a heterogeneous microstructure with good dimensional control and excellent as-sintered mechanical properties. The third type is prealloyed powder, which is atomized from alloyed furnace melts such that each powder particle has the same nominal composition throughout. Prealloyed powders yield homogeneous phase constituents in the microstructure.

**Conventional Materials.**—For an overview of the scope of conventional materials, refer to Table 2, Table 3, Table 4, and Table 5 that follow. Table 2 lists basic materials for ferrous structural parts. Full specifications for a variety of ferrous PM structural materials are covered in ASTM Standard B783-19 and MPIF Standard 35.

**Table 2. PM Materials for Ferrous Structural Parts**

| Material | MPIF Designation Code[a] Composition | Density[b] g/cm$^3$ | Tensile Strength MPa | Comments |
|---|---|---|---|---|
| Carbon steel | F-0008 0.8C | 6.9 | 370 | Cost-effective |
| | F-0008-HT 0.8C | 6.9 | 390 | |
| Copper steel | FC-0208 2Cu, 0.8C | 6.7 | 410 | Good sintered strength |
| | FC-0208-HT 2Cu, 0.8C | 6.7 | 590 | |
| Nickel steel | FN-0205 2Ni, 0.5C | 7.1 | 390 | Good heat-treated strength, impact energy |
| | FN-0205-HT 2Ni, 0.5C | 7.1 | 1000 | |
| Low-alloy steel | FL-4405 0.85Mo, 0.5C | 7.1 | 460 | Good hardenability, consistency in heat treatment |
| | FL-4405-HT 0.85Mo, 0.5C | 7.1 | 1100 | |
| Hybrid low-alloy steel | FNL-4405 2Ni, 0.85Mo, 0.5C | 7.05 | 550 | Best heat-treated strength, impact energy |
| | FNL-4405-HT 2Ni, 0.85Mo, 0.5C | 7.05 | 1170 | |
| Diffusion-alloyed steel | FD-0205 1.75Ni, 1.5Cu, 0.5Mo, 0.5C | 6.95 | 540 | Best sintered strength |
| | FD-0205-HT 1.75Ni, 1.5Cu, 0.5Mo, 0.5C | 6.95 | 900 | |
| Sinter hardened steel | FLC-4608-HT 1.35Ni, 1.5Cu, 0.5Mo, 0.8C | 7.0 | 690 | Hardenable in sintering |
| Stainless steel | SS-316N2 17Cr, 12Ni, 2.5Mo SS-430N2 17Cr | 6.57.1 | 410410 | Good corrosion resistance, appearance |

[a] Reference: MPIF Standard 35, "Materials Standards for PM Structural Parts."

[b] Density is given as typical values.

Table 3 shows basic materials for soft magnetic parts. Complete standard specifications for iron-phosphorous powder metallurgy parts for soft magnetic applications are covered in ASTM Standard A839-15 and MPIF Standard 35.

**Table 3. PM Materials for Soft Magnetic Parts**

| Material | MPIF Designation Code[a] Composition | Density[b] g/cm³ | Magnetic Properties | |
|---|---|---|---|---|
| | | | Induction (kG) | Maximum Permeability H/m |
| Iron | FF-0000 <0.03C | 6.6 | 9.0 | 1700 |
| Phosphorus iron | FY-4500 0.45P | 6.8 | 10.5 | 2300 |
| Silicon iron | FS-0300 3Si | 7.2 | 13.0 | 5000 |
| Nickel iron | FN-5000 | 7.5 | 12.0 | 12000 |

[a] Reference: MPIF Standard 35, "Materials Standards for PM Structural Parts."
[b] Density is given as typical values.

Some powder materials (PM) for nonferrous structural parts are given in Table 4. ASTM Standard B823-15 and MPIF Standard 35 give complete information on nonferrous powder metallurgy (PM) materials for structural parts.

**Table 4. PM Materials for Nonferrous Parts**

| Material | MPIF Designation Code[a] Composition | Density[b] g/cm³ | Usage and Characteristics |
|---|---|---|---|
| Copper | C-0000 0.2 max. other | 8.0 | Electrical parts, conductivity 85% IACS |
| Bronze | CTG-1001 10Sn, 1 Graphite | 6.2 | Self-lubricating bearings, 22% minimum oil content |
| Red brass | CZ-1000 10Zn | 7.9 | Hardware items, yield strength 76MPa. |
| Yellow brass | CZ-3000 30Zn | 7.6 | Hardware items, yield strength 110MPa |
| Nickel silver | CNZ-1818 18Ni, 18Zn | 7.9 | Appearance, corrosion resistance, yield strength 138MPa |
| Aluminum | 4Cu, 0.5Mg, 1Si | 2.6 | Light weight, corrosion resistance, good electrical, thermal conductivity, T4 condition tensile strength 220MPa. |

[a] Reference: MPIF Standard 35, "Materials Standards for PM Structural Parts."
[b] Density is given as typical values.

Table 5 shows materials used for self-lubricating bearings. More information about these materials, and other PM materials for self-lubricating bearings are given in ASTM Standard B438-17, ASTM B439-18, and MPIF Standard 35.

**Table 5. Materials for Self-Lubricating Bearings**

| Material | MPIF Designation Code[a] | Density[b] g/cm³ | Oil Content Volume% | K Strength MPa |
|---|---|---|---|---|
| Bronze | CT-1000 | 6.0–6.4 | 24 | 130 |
| | CTG1001-K | 6.4–6.8 | 17 | 160 |
| Diluted bronze | FCTG-3604-K | 6.0–6.4 | 17 | 150 |
| Iron graphite | FG-0303-K | 5.6–6.0 | 18 | 70 |

[a] Reference: MPIF Standard 35, "Materials Standards for PM Structural Parts."
[b] Density is given as typical values.

**Advanced Materials.**—Advanced materials outperform conventional materials with superior properties such as toughness, hardness, durability, and elasticity. There is always a real need for better materials; the questions are how much better and at what cost? An applied scientist, with a particular application in mind, will scour lists of known materials looking for one that meets his or her needs. The development of advanced materials can even lead to the design of completely new products, including medical implants and computers.

If existing materials are unsuitable, applied and basic scientists must work together to develop new materials. The area of advanced materials research is very broad in scope and potential applications, and synergism between what is available and what needs to be developed reflects the important and complementary roles of the basic and applied sciences in materials science.

While some advanced materials are already well known, it will take a few more years for others to appear in products. Here, we describe some advanced PM materials that are manufactured using advanced processes such as pressing, metal injection molding, and spray forming. In addition, certain alloys are produced nearly exclusively by PM techniques.

*Superalloys :* A superalloy, or high-performance alloy, is an alloy that exhibits specific desirable properties related to mechanical strength, stability at high temperatures, and surface resistance to oxidization and corrosion. Superalloys typically have a matrix with an austenitic face-centered cubic crystal structure. The use of high alloy PM materials, notably superalloys, is most common in the production of near-net shapes and the forging of preforms for aircraft turbine engines. Economic benefits have been the prime driving force in the use of PM for the manufacture of these high cost alloys.

These superalloys are processed from highly controlled metallurgically alloying elements, usually nickel, cobalt, or nickel-iron. Clean powders are produced by inert gas atomization methods that minimize surface oxidation during processing. After powder screening and cleaning, the powder is compacted using hot isostatic pressing (see *Isostatic Pressing* on page 1539). Thermochemical processing may be performed to enhance mechanical properties or microstructure. Hot extrusion of the atomized powder is an alternate consolidation method (see *Powder Extrusion* on page 1543).

PM processing offers the advantages of a homogeneous microstructure and near-net-shape configuration for a lower cost. As a result, countless pounds of PM superalloy components are now flying in both military and commercial aircraft engines.

Additional applications of superalloys include parts for space vehicles, submarines, nuclear reactors, military electric motors, racing and high-performance vehicles, chemical processing vessels, and heat exchanger tubing.

*Tool Steels:* Tool steel refers to a variety of carbon and alloy steels that are particularly well suited to be made into tools. Their suitability comes from their distinctive hardness, resistance to abrasion, their ability to hold a cutting edge, and their resistance to deformation at elevated temperatures.

It is well known that powder metallurgy provides a means of making new alloys and composites of tool materials, which cannot be made by conventional metallurgy, casting and hot-working methods. Powder metallurgy processing of high-speed steels has gained considerable significance over conventional ingot metallurgy processing.

High-performance tool steels and high-speed steels are produced in mill shapes using PM process techniques. PM tool steels offer these advantages:

- finer grain structures
- improved homogeneity and distribution of secondary and carbide phases
- elimination of secondary stringers
- less distortion in heat treatment
- improved grindability
- greater wear resistance
- improved toughness and fracture strength.

The hot isostatic process has also been used in the production of PM tool steels to near-net tool shapes, such as hobs and shaper cutters. Tool steel parts can also be made by conventional cold compacting combined with high-temperature sintering.

*Refractory Metals:* Refractory metals are a class of metals that are extraordinarily resistant to heat and wear. The expression is mostly used in the contexts of materials science, metallurgy, and engineering. The definition of the elements that belong to this group differs. By one definition, a melting point above 4000° F (2200° C) is necessary to qualify an element as refractory. Five elements, niobium, molybdenum, tantalum, tungsten and rhenium, are included in all definitions. The refractory metals, alloys, and composites are produced using powder metallurgy techniques. Powders of the pure metal are compacted into rectangular bars, sintered by heating using electric current, and further fabricated by cold-working with annealing steps. Refractory metals can be worked into wire, ingots, sheets, or foil.

Some of their applications include tools for working metals at high temperatures, casting molds, chemical-reaction vessels for corrosive environments, nuclear reaction control rods, tantalum capacitors, and tungsten-silver composite circuit breaker contacts. Refractory metals are stable against creep deformation to very high temperatures.

*Beryllium:* As a free element, beryllium is a steel-gray, strong, lightweight and brittle metal. Beryllium increases hardness and resistance to corrosion when alloyed with aluminum, cobalt, copper, iron, and nickel.

Powder metallurgy is the main processing route to produce beryllium components. The vacuum hot processing method results in a fine-grained machinable form of the material. The random orientation and fine grain size of powder metallurgy ingots enable beryllium to be used in structural applications with relatively uniform mechanical properties in all directions.

Vacuum hot pressing has been a standard consolidation technique throughout the development of beryllium technology, and hot isostatic pressing powder is the major form of beryllium in use today. Sometimes, hot isostatic pressing is substituted for the hot pressing step. Due to its light weight, beryllium is used in structural applications; its high flexural rigidity, thermal stability, thermal conductivity, and low density make beryllium a quality aerospace material for high-speed aircraft, missiles, spacecraft, and communication satellites. Because of its low density and atomic mass, beryllium is relatively transparent to X-rays and other forms of ionizing radiation; therefore, it is the most common window material for X-ray equipment and in particle physics experiments. The high thermal conductivities of beryllium and beryllium oxide have led to their use in heat transport and heat sinking applications.

Beryllium is used commercially in the manufacture of telecommunications infrastructure equipment, computers, and cellular phones, thereby helping people around the world to keep in touch. Also, medical lasers made with beryllia ceramics help ophthalmologists to restore or improve eyesight for millions. Beryllium ceramic is the only material that offers the thermal conductivity, strength, and dielectric properties required to contain and control these tiny, high-powered gas laser bores. Copper beryllium connectors transmit precise electrical signals to the delicate surgical instruments and monitoring devices used in the newest, non-invasive surgical techniques. Such techniques reduce patient trauma and infection risk, while speeding the process of healing and recovery.

*Titanium:* Titanium alloys are among the most important of the advanced materials that are key to improved performance in aerospace and conventional systems. This is because of the excellent combination of specific mechanical properties and outstanding corrosion behavior exhibited by titanium alloys. However, discouraging widespread use is the high cost of titanium alloys compared to competing materials. This has led to numerous investigations of various potentially lower cost processes, including powder metallurgy (PM). Titanium powder metallurgy can be divided into categories such as powder injection molding (use of a binder to produce complex small parts); near-net shaping (use of pre-alloyed and blended elementals in die pressing or isostatic pressing); spraying (high-speed projecting of powder onto a substrate); and laser melting (powder feed or powder layer melted with a laser.

Titanium PM alloys have a combination of low density 2.6 oz/in$^3$ (4.5 g/cm$^3$) and high strength. The strengths vary from 70,000 psi (480 MPa) for some grades of commercial titanium to about 160,000 psi (1100 MPa) for structural titanium alloy products and over 250,000 psi (1725 MPa) for special forms such as wires and springs. Another important characteristic of titanium-based materials is the reversible transformation of the crystal structure from alpha (hexagonal close-packed) structure to beta (body-centered cubic) structure when the temperature exceeds 1634° F (890° C).

The use of PM titanium metal and its alloys continues to increase since PM provides a near-net shape product as compared to more expensive cast, forged, and machined processing. Depending on the application and production quantities, titanium PM near-net products are processed either by conventional press-and-sinter techniques or by hot isostatic pressing. The latter exceeds the minimum wrought alloy specifications.

If the final product is made net-shape by application of hot isostatic pressing (HIP), a lack of texture can result, thus giving equal properties in all directions. Product applications include aerospace/aircraft components, sporting goods, chemical processing equipment, and biomedical systems.

## Design Considerations in Powder Metallurgy

The powder metallurgy manufacturer is often confronted by a drawing for a component designed with a conventional manufacturing process in mind. It is not desirable for the powder metallurgy component manufacturer or designer to attempt to follow these drawings; it is far better to redesign the component so that it can fulfill its design function, while taking full advantage of the powder metallurgy process, in particular the cost-effective manufacture to near-net shape with close dimensional tolerances.

The PM part and the complexity of tooling used to produce it are closely related. Two major factors in the compacting operation influence or control part design: the flow characteristic in the die and the degree of brittleness of the green parts.

Although powder metallurgy industry standards (MPIF Standard 35) provide useful physical and mechanical property design data for engineers familiar with the PM process, those less experienced with this manufacturing process may benefit from additional guidance. In this section, design considerations are divided into three broad areas: size and shapes, design details, and design of typical parts.

**Sizes and Shapes.—***Part Sizes:* The size limitation of PM parts is based on powder compressibility and press tonnage. Since compaction occurs in the vertical direction, using only top and bottom motions, part lengths in the pressing direction are limited. The compression ratio—the ratio of the height of the loose powder filling the die to the height of the compacted part—also tends to limit vertical part lengths. A length-to-wall thickness ratio greater than 8:1 is difficult to press, and density variations are virtually unavoidable.

*Shapes:* The shape of the part must permit ejection from the die. Part geometry must be compatible with a uniaxial compaction motion in the vertical direction. Significant variations in part length in the cross section require different tooling motions in the compaction press. Cam, gears, and sprockets are readily made. The shape of the part should permit the construction of strong tooling. The part should be designed with as few changes in section thickness as possible: the number of levels in the part should not exceed the number of pressing actions available in the compacting press. With too many levels, density varies considerably and part quality becomes a problem. Face forms on upper or lower punches can provide bosses, pads, lettering, countersinks, and other features. Parts weighing a few grams to about 26 lb (around 12 kg) or more are possible.

**Design Details.—**Successful design for PM rests on an understanding of how the unique aspects of the technology affect the countless details that make up the design of any structural part.

*Holes:* Holes in the pressing direction can be round. Tooling members, which create holes, are called core rods. Lightening holes are frequently added to large parts to reduce projected pressing area, thus making parts lighter and easier to press. Blind holes, blind steps in holes, and tapered holes are readily produced. Side holes have to be produced after a sintering operation, usually by machining.

*Wall Thickness:* Die fill is all-important. As a general rule, walls should not be made any thinner than 0.057 in. (1.45 mm). One should avoid designing long thin walls; they require tooling that is complicated, and the parts themselves have a tendency toward density variation.

*Flatness:* Total measured flatness depends on part thickness and surface area. Thin parts tend to distort more than thick parts during sintering or heat treatment. Repressing improves flatness. Projection basses are easier to flatten than entire face areas.

*Taper and Draft:* Drafts are generally not required or desired on sides of parts. While drafts on outer sections for ejection are sometimes helpful, producing one demands careful timing of the tools and slower production rates.

*Fillet and Radii:* Generous fillet radii are most desirable. Tooling with such fillets is more economical and longer lasting. Parts made with fillets have greater structural integrity (Fig. 20). Parts with generous radii are more easily and more quickly made.

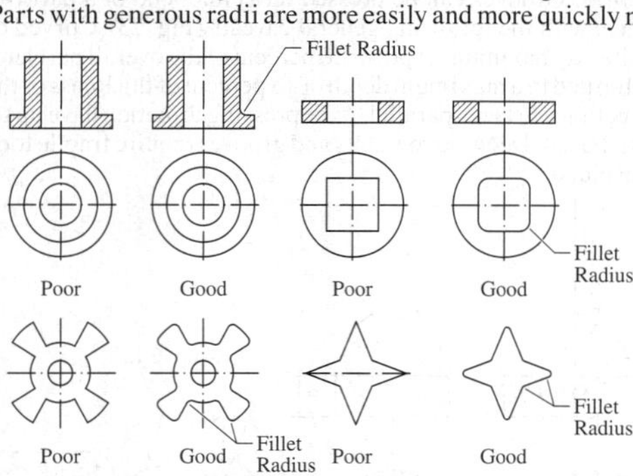


Fig. 20. Examples of PM Parts Showing Poor and Good Designs (*Source:* Metal Powder Federation)

*Chamfers and Bevels:* Chamfers are necessary on front edges to prevent burring. For example, on bushings (Fig. 21) a 30° to 45° chamber and 0.005 to 0.015 in. (0.13 to 0.38 mm) flat to eliminate feature edges are the preferred.

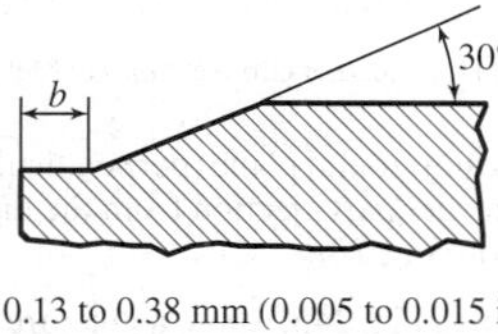


$b$ = 0.13 to 0.38 mm (0.005 to 0.015 in.)

Fig. 21. Chamfers are Preferred on Parts Edges (*Source:* Metal Powder Federation)

*Countersinks:* A countersink is a chamfer around a hole for a screw or bolt head (Fig. 22). When the countersink is formed by a punch, a 0.010 in. (0.25 mm) nominal flat is essential in order to avoid sharp, flanged edges on the punch.

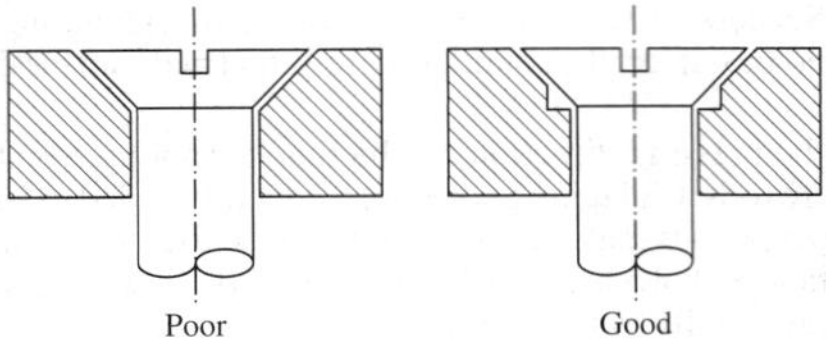


Fig. 22. Countersink in Clearance (*Source:* Metal Powder Federation)

*Flanges:* Flanges or overhands can be produced by step in the die. Too long flanges a causes ejection difficulties. A long flange should incorporate a draft around the flange, a radius at the bottom edge, and radius at the juncture of the flange and component body to reduce stress concentrations and the likelihood of fracture.

*Hubs:* The PM process can readily produce hubs, which are complementary part sections to gears, sprockets, or cams. It is important to include a generous radius between the hub and flange section and to maximize space between the hub and the root diameter of the gear or sprocket.

*Slots and Grooves:* Grooves can be pressed into either end of a part from projections on the punch face, with the following general caveats (Fig. 23). Curved or semicircular grooves are limited to maximum depth of 20 percent of the overall part length. Rectangular grooves are limited to a maximum depth of 15 percent of thickness of the workpiece in the pressing direction, surfaces parallel to the pressing direction have up to 12° draft, and all corners are radiused. Deep narrow slots and grooves require fragile tool members and should thus be avoided.

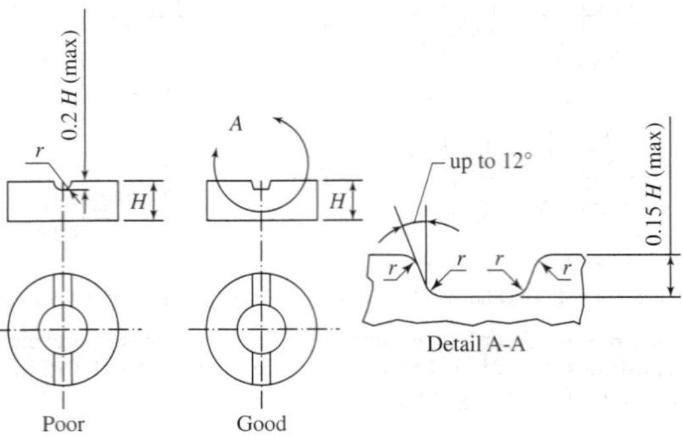


Fig. 23. Semicircular and Rectangular Groove (*Source:* Metal Powder Federation)

*Undercuts:* Undercuts on the horizontal plane (perpendicular to the pressing direction) cannot be made since they prevent part ejection from the die. Annular grooves must be machined as a second operation (Fig. 24a).

For a part such as that shown in Fig. 24b, where a juncture undercut is needed to allow fit-up to a "dead corner," an alternative approach is shown in Fig. 24c.

*Tolerances:* For reasons of economics, tolerances no closer than necessary should be specified. Table 6 illustrates tolerances characteristic of PM and competitive near-net-shape forming methods.

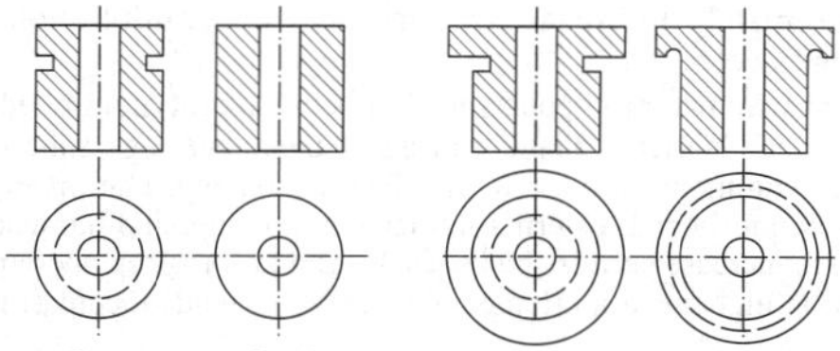


Fig. 24. Undercuts: a) Annular Groves Cannot Be Molded; b) Undercuts Cannot Be Molded; c) Undercut Can Be Molded (*Source:* Metal Powder Federation)

**Table 6. Tolerances that Can Be Produced in PM Process**

| Forming Method | Typical Dimensional Tolerances (±) (in/in) (mm/100 mm) | Critical Dimensional Tolerances (±) (in/in) (mm/100 mm) |
|---|---|---|
| PM | 0.002 (0.203) | 0.001 (0.102) |
| PM (secondary forming) | 0.001 (0.102) | 0.0005 (0.051) |
| Cold forming | 0.010 (1.016) | 0.002 (0.203) |
| Hot extrusion | 0.010 (1.016) | 0.003 (0.305) |
| Fine blanking (single level) | 0.0005 (0.013) | 0.0002 (0.020) |
| **Die casting**[a] | | |
| Magnesium and Aluminum | 0.002 (0.203) | 0.0010 (0.107) |
| Zinc | 0.0025 (0.254) | 0.0010 (0.107) |
| **Hot closed-die forging** | | |
| Conventional | 0.02 (2.032) | 0.015 (1.524) |
| PM perpendicular to press axis | 0.0032 (0.325) | 0.0025 (0.254)[b] |
| PM parallel to press axis | 0.010 (1.016) | 0.005 (0.508) |

[a] Die parting line tolerance will be two to three times greater than the data shown.
[b] Fixture-quenched forgings, i.e., forgings clamped for quenching during heat treatment immediately after hot forging, can hold tighter tolerance.

**Design of Typical Parts.—**Although PM is used for a whole host of parts in many different applications, most parts (typical categories include gears, cams, assemblies, and bearings) have their own unique design considerations.

*Gears:* Gears are well-suited to PM production. Carbide dies for production gears provide long life and good accordance. PM gears can be made with blind corners, thus eliminating the need for undercut relief. PM gears can be comminuted with other configuration such as cams, and various other components. PM helical gears are possible; copper infiltration is sometimes used to improve teeth density. Since tooth and shape is not a problem, true involve gear forms are more easily produced through PM than with other fabrication methods. One should keep in mind when designing PM gears that hubs or pinions should be located as far as possible from gear root diameters.

*Cams:* Cams are well-suited to PM production: The process provides excellent surface finishing and part-to-part consistency; the natural finish of a self-lubricating PM cam will often outwear a ground cam surface; for radial cams, the shape is formed in the die; for face cams, the shape is formed in the punch faces.

*Assemblies:* Two or more PM parts can often be joined from a unit that is difficult, if not impossible, to make as a single structure. Capitalizing on PM's flexibility, it is feasible to make assemblies of very difficult materials such as a bronze bearing in a ferrous structural part. PM parts can be joined by conventional methods and also by sintering together materials of appropriately different characteristics. Copper infiltration during sintering can be used to bond steel parts.

*Bearings:* Bearings are natural products for PM because of its controlled porosity and the resulting self-contained lubricant. Plane bearings, flanged bearings, spherical bearings, and thrust washers are commonly produced with PM technology. The operation environment should be carefully considered: external lubrication, cooling, and hardened or chromium-plated shafts tend to increase permissible loads. Repeated start-stop operation, oscillatory or reciprocating motion, high speed, and temperature extremes tend to decrease permissible loads.

## Economics of Powder Metallurgy

Powder metallurgy is a continually and rapidly evolving technology that embraces most metallic and alloy materials and a wide variety of shapes. PM is a highly developed method of manufacturing reliable ferrous and nonferrous products. The growth of the PM industry during the past few decades is largely attributable to the cost savings associated with net- or near-net-shape processing compared to other metalworking methods such as casting or forging.

The following will explain why powder metallurgy is competitive against alternative production processes. There are two principal reasons for using a powder metallurgy product:

1) cost savings
2) unique properties attainable only by the PM method

In the automotive sector, which accounts for about 80 percent of structural PM production, the reason for choosing PM is, in the majority of cases, an economic one. Why then is PM more cost-effective?

- A sintered PM part of comparable quality may be cheaper than a cast of wrought component.
- It provides better material utilization with close dimensional tolerances.
- It is suited to high-volume parts production requirements.
- It provides long-term performance reliability in critical applications.

Conventional metal forming or shaping processes, against which PM competes, generally involve significant machining operations from bar stock or from forged or cast blanks. These machining operations can be costly and are wasteful of material and energy. This fact is illustrated in Table 7, which shows that material utilization in excess of 95 percent can be achieved with close dimensional tolerances.

**Table 7. Raw Material Utilization and Energy Requirements of Various Manufacturing Processes**

| Manufacturing Process | Raw Material Utilization % | Energy Requirement per kg Finished Part MJ (BTU) |
|---|---|---|
| Casting | 90 | 30–38 (23,434–36,017) |
| Sintering | 95 | 29 (27,486) |
| Cold or warm extrusion | 85 | 41 (38,860) |
| Hot-drop forging | 75–80 | 46–49 (43,599–45,495) |
| Machining process | 40–50 | 66–82 (66,555–77,721) |

The example in Table 7 compiles data for a comparison for a production of notch segments for track transmission. The PM process (sintering) has the highest raw material utilization (95 percent) and the lowest energy requirement per kg of finished part compared with other manufacturing processes. The energy savings alone contribute significantly to the economic advantage offered by PM. Compared with forging, machining, and a number of other processes, PM consumes only around 45 to 50 percent of the energy and the number of process steps is greatly reduced.

Equipment costs for conventional PM processing are somewhat similar to those for bulk forging, but the cost increases significantly for HIP methods. Labor costs are not as high in other processes because operations are primarily performed on highly automated equipment, and the skills required are not as high.

## Powder Metallurgy Standards

Why are PM standards important? Powder metallurgy is a rather young industrial technology. The first local industries were started in the middle of the 20th century. During the next 30–40 years, business expanded primarily within the various continents, but since the last decade, the rate of expansion has accelerated and today the business is truly global.

This global business requires new modern tools to support further growth and sharpening of the weapons to compete with other manufacturing technologies. In this context, ISO, ASTM, and MPIF standards support confidence in powder metallurgy and guides end users, often with global presence, to make the right choices in using components and solutions based on metal powder.

**ISO Standards.**—International Powder Metallurgy standards are generated by the ISO technical committee, ISO/TC 119. The five sub-committees of this technical committee have created a range of standards that address sintered metals and hardmetals, as well as MIM standards. These sub-committees have created the following standards:

ISO 3252-2019, Vocabulary
ISO 3923-1 and-2, Apparent density measurement
ISO 3953, Tap Density measurement
ISO 4490, Flow Rate
ISO 18549-1 and-2, Apparent density and rate at elevated temperatures
ISO 3927 / Corrigendum 1, Compressibility and uniaxial compression
ISO 3954, Powder sampling
ISO 3995, Green strength
ISO 4491-1, -2, -3 and -4, Oxygen content
ISO 4496, Acid insoluble content in iron, copper, tin, and bronze powders
ISO 4497, Particle sizing by dry sieving
ISO 10076, Particle sizing by sedimentation analysis
ISO 10070, Specific surface area
ISO 13944, Lubricant content
ISO 13947, Nonmetallic inclusion content using powder forged specimen
ISO 14167, Testing of copper infiltration powders
ISO 4492, Dimensional changes in compaction and sintering
ISO 2738, Permeable sintered metal materials-density, oil content, open porosity
ISO 4003, Permeable sintered metal materials-bubble test pore size
ISO 4022, Permeable sintered metal materials-fluid permeability
ISO 3369, Density of impermeable sintered materials
ISO 2739, Radial crushing strength of sintered bushes
ISO 2740, Tensile test pieces
ISO 3928, Fatigue test pieces
ISO 5754, Unnotched impact test pieces
ISO 3312, Young's modulus measurement
ISO 3325 / Amendment, Transverse rupture strength
ISO 4498, Apparent hardness and microhardness
ISO 14317, Compressive Yield Strength
ISO/TR 14321, Metallographic preparation and examination
ISO 4507, Case-hardening depth measurement
ISO 7625, Sample preparation for chemical analysis of carbon content
ISO 23519, Sintered metal materials, excluding hardmetals
ISO 28279, Sintered metal materials
ISO 3326, Magnetization coercivity
ISO 3327, Transverse rupture strength

ISO 3738-1 and-2, Rockwell hardness testing
ISO 3878, Vickers hardness testing
ISO 3907, Total carbon content
ISO 3908, Insoluble carbon content
ISO 3909, Cobalt content
ISO 4489, Sampling and testing of sintered hardmetals
ISO 4499-1 and-2, Metallography
ISO 4501, Titanium content
ISO 4503 and 4883, X-ray fluorescence analysis
ISO 4505, Metallographic assessment of porosity and uncombined carbon
ISO 4506, Compression testing
ISO 4884, Sampling and testing of powders using sintered test pieces
ISO 7627-1,-2,-3,-4,-5, and -6, Chemical analysis by frame atomic spectrometry
ISO 11877 and 17352, Silicon in cobalt metal powders
ISO 28079, Palmqvist toughness testing
ISO 28080, Abrasion tests for hardmetals
ISO 26482, Lead and cadmium content
ISO 22394, Knoop hardness test
ISO 11876, Calcium, copper, iron, magnesium, manganese, sodium, nickel, and zinc in cobalt metal powder
ISO 11873, Sulfur and carbon contents in cobalt metal powders
ISO 5755-2012, Sintered metal materials specifications
ISO 22068, Sintered-metal injection-mounded material specifications

The participation of the United States in ISO/TC119 is through the member body, the American National Standards Institute (ANSI).

**ASTM Standards.—**The ASTM international committee with the responsibility for development of PM standards is Committee B09 within the general subject Metal Powders and Metal Powders Products. This committee is composed of sub-committees that address specific segments as follows:

B09.01, Terminology
B09.02, Base Metal Powders
B09.03, Refractory Metal Powders
B09.04, Bearings
B09.05, Structural Parts
B09.06, Cemented Carbides
B09.11, Near Full Density Powder Metallurgy Materials
B09.98, Long Range Planning and Awards

Standards can be found by searching the catalog of each subject area of the subcommittee.

**MPIF Standards.—**The Metal Powder Industry Federation (MPIF) Standard 35 is a family of publications that comprises four separate publications dealing with materials for conventional PM structural parts, PM self-lubricating bearings, powder forged (PF) steel, and metal injection molded parts (MIM). The same materials may appear in more than one publication or section of a standard depending upon common use. For example, some structural materials may also be used in bearing applications and vice versa, and stainless steel materials may be manufactured by more than one PM process, such as conventional PM or MIM, depending on part design and use.

Powder metallurgy standards available from MPIF include:

MPIF Standard 35, Materials Standards for PM Structural Parts
MPIF Standard 35, Material Standards for PM Self-Lubricating Bearings
MPIF Standard 35, Materials Standards for PM Steel Parts
MPIF Standard 35, Materials Standards for PM Metal Injection Molding Parts

## Metal Additive Manufacturing

Additive manufacturing (AM), also known as 3D printing, is a group of processes in which materials are joined selectively to transform information in computer-aided design (CAD) files into three-dimensional (3D) parts. The term "additive" is used to distinguish these processes from "subtractive" (machining, grinding, and chemical or electrical erosion) or forming (casting, rolling, and forging).

ASTM Committee F-42 was organized in 2009 to develop official Standards for AM. One of the committee's first actions was to define standard terminology, embodied in ASTM F2792 (Withdrawn); the current Standard is ISO/ASTM 52900-21. (For more on AM Standards, see *Standards for Additive Manufacturing* on page 1565.) This Standard groups AM processes into seven categories (listed here alphabetically):

*Binder Jetting:* A liquid bonding agent is selectively deposited to join powdered material, layer by layer, until a 3D shape is formed. The binder is deposited as fine droplets through the jets, or orifices, of an ink-jet printer onto a layer of powder.

*Directed Energy Deposition:* This process fuses material, which may be in the form of metal powder or wire, by melting as it is deposited into the focal point of a directed thermal energy source (laser, electron beam, or plasma arc).

*Material Extrusion:* In this AM process, material is selectively dispensed through a nozzle or orifice. A wide range of extruded materials may be used, from plastics to concrete to food.

*Material Jetting:* Through controlled jetting, droplets of material are selectively deposited to build a form. Materials, which may incorporate photopolymer and wax, are deposited through the orifices of an inkjet printer.

*Powder Bed Fusion:* This AM process uses thermal energy to selectively fuse areas of a powder bed. The energy source—a laser or electron beam—focuses to fuse each layer of powder that builds up the part.

*Sheet Lamination:* In this process, sheets of material are stacked and bonded across their planar interfaces to form an object.

*Vat Photopolymerization:* A specialized liquid photopolymer in a vat is selectively cured by light-activated polymerization. The light source may be a continuously moving laser or a digital array of simultaneous light projections.

Reviewing this list, it can be seen that the material feed stock may be a liquid, powder, wire or filament, or sheet. In all cases, AM machines selectively convert material in one of these simple forms into a complex, 3D part, generally by adding layer upon layer.

**Additive Manufacturing Workflow.**—Additive manufacturing processes, involving metals or other materials, follow the same basic workflow, as summarized in Table 8. A three-dimensional (3D) model is created using CAD software. (For more on this topic, see *CAD/CAM* on page 1377.) This model is converted into a surface model made up of interfacing triangles, resulting in an STL (Standard Tessellation Language) file. The build file is then sliced into two-dimensional (2D) layers. Each layer profile is used to control the AM machine's motions, which add material in that profile and build the 3D model (Table 8). Typically, the layer thickness is 0.004 in. (0.102 mm), but it can range from 0.0007 to 0.010 in. (0.0178 to 0.254 mm).

AM processes have been adapted for commercial use under various trade names and are used to produce parts in metals, polymers, ceramics, and various composites. For example, the section *Additive Manufacturing Plastics* on page 612 describes processes used for plastics: stereolithography (SLA), which is defined by ASTM as *vat polymerization*; selective laser sintering (SLS), which is an example of *powder bed fusion*; fused deposition modeling (FDM), which is an example of *extrusion*; binder jetting (BJ); and material jetting (MJ).

This section discusses AM processes used to produce metal parts: powder bed fusion (PBF), including selective laser melting (SLM) and electron beam melting (EBM), binder jetting (BJ), and powder- and wire-fed directed energy deposition (DED).

**Table 8. Additive Manufacturing Workflow**

| | |
|---|---|
| **Solid Computer Model:** A solid model of the part to be produced by an additive manufacturing (AM) process can be imported from an existing file or created using CAD (computer-aided design) software. The model may be reverse engineered, using a laser scanner or other coordinate measuring device to replicate an existing part. In biomedical applications, a CAT (computer-aided tomography) or MRI (magnetic resonance imaging) scan may be used to design a model for a custom-fit implant. |  |
| **STL File:** One of the output formats of CAD software is an STL (Standard Tessellation Language) file. This surface representation of the solid part consists of interconnected triangles generated by algorithms within the CAD software to precisely represent the surface of the solid part without gaps or overlaps. The more complex and curved the surface of the part, the more triangles are needed to represent it. | 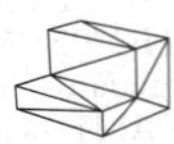 |
| **Virtual Slices:** The STL file is positioned and oriented electronically on the 3D printer's build platform. The resulting build file is separated into slices by geometric algorithms in the printer that determine the intersecting points of each horizontal plane with the sides of the triangles in the STL file. These points outline the cross-sections to be printed in each 2D layer that will build up the 3D part. | 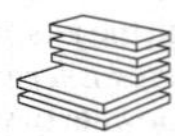 |
| **Additive Manufacturing:** CAM (computer-aided manufacturing) software in the AM machine converts the cross-section of each layer into machine instructions that control the motion of a material conversion device. In the *powder bed fusion* process, this conversion device would be a laser beam that selectively fuses metal powder into a solid layer. In *fused deposition modeling*, the device would be an extrusion head that deposits a polymer in layers corresponding to cross-sections of the part. | 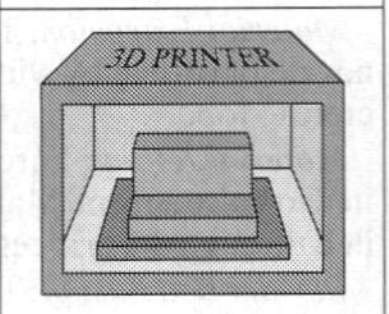  |
| **Post-Processing:** The completed 3D part is extracted from the build plate, and any supports are removed. If necessary, the part may be heat treated for material strength; in some cases, it may undergo hot isostatic pressing (HIP) to close up porosity. Surfaces requiring a precision fit can be machined or ground. The last step is finishing the part using such standard processes as sanding, shot or grit blasting, and coating. |  |

**Powder Bed Fusion (PBF).**—Powder bed fusion is ASTM's term for an additive manufacturing process in which a point heat source selectively fuses or melts a region of a powder bed. There are two types of PBF technology; each system has advantages and disadvantages compared to the other.

*Selective Laser Melting (SLM):* In the original ASTM F2792 Standard, *laser sintering* AM is defined as a "powder bed fusion process used to produce objects from powdered materials using one or more lasers to selectively fuse or melt particles at the surface, layer by layer, in an enclosed chamber." The definition goes further to note that "sintering" is a historical term and a misnomer, because the process involves full or partial melting, as opposed to traditional powdered metal sintering, in which a compacted powder part is heated below its melting point. For this reason, the term *selective laser melting* is widely used and accurately describes the process. It is similar in principle to *selective laser sintering* (SLS) described in *Additive Manufacturing Plastics* on page 612.

The SLM system is an important technical innovation, with roots dating to the 1990s. Early systems used a gas or disk laser and processed primarily low-melting-point bronze powder, mixed with stainless steel powder that had a higher melting point. Today, SLM systems use a fiber laser as the fusion heat source to completely melt the powder. Typical laser power in metal SLM systems ranges between 100 and 1000 W, and some modern systems have two or four lasers operating simultaneously to increase production rates. Of the metal AM processes, SLM is one of the most widely used.

After the heat source, the most important components of SLM systems are beam deflection optics, which provide scanning capability for selectively melting areas of the powder bed. With scan speeds of up to 23 ft/s (7 m/s), the scanning mirror must be fast, accurate, and reliable.

The final significant optical element is the correction lens, which ensures that the beam is round as it traverses the build platform at different angles. Most systems use an f-theta lens design with anti-reflection coatings to prevent damage from the beam reflecting back into the laser. A schematic illustration of the laser-beam powder bed fusion process is shown in Fig. 25.

The system consists of a powder bed build area on a build plate that moves in the Z direction. For each printed slice (see Table 8), the plate moves down by one layer thickness. The powder deposition system (sometimes called a *recoater*) then spreads a new layer of powder, and the laser traces the profile of that part slice on the layer of powder. This sequence is repeated until the full 3D part is completed.

Because the laser melts the metal powder, to prevent oxidation, the SLM mechanism is enclosed in a gas-tight chamber containing argon or other nonreactive gases. Typically, powder particle size used in SLM processes is between 0.0008 and 0.0024 in. (0.02 and 0.06 mm). Titanium, stainless steel, and nickel-base superalloys are commonly processed by SLM; aluminum, copper, and gold are more challenging, but some newer machines feature lasers with colors in a green range (~500 nm) that are more readily absorbed by these metals.

Laser spot size is typically 0.0027 in. (0.07 mm) in diameter but may be up to 0.006 in. (0.15 mm) in diameter. At any given instant, metal particles within the laser spot are melted, forming a small pool of molten metal. This pool quickly solidifies, as heat transfers to the solid metal build plate on the first layer or to the previously solidified material in the layers below the current one. Thus, the molten metal pool within the laser spot zone continuously forms and solidifies as the laser spot moves rapidly across the powder layer, much like a mini-welding process.

The high rate of solidification of the molten pool results in very fine microstructure and excellent mechanical properties. Strength of SLM-processed material is often greater than for the same alloy in conventional casting due to this fine microstructure. On the other hand, because the process involves melting and solidification, gas porosity and solidification porosity may form, leading to fine pores throughout the material. Such defects significantly reduce fatigue and impact strengths, though without reducing the final part's yield and tensile strength.

One advantage of AM, and SLM in particular, is that the processes can produce overhangs and internal channels or cavities. In SLM, when material is printed in an overhang, the molten pool sits on loose powder from previously created layers.

Even though molten metal in such an overhang cools and solidifies rapidly, some of the metal seeps down into the spaces between the loose particles, leaving a rough surface on

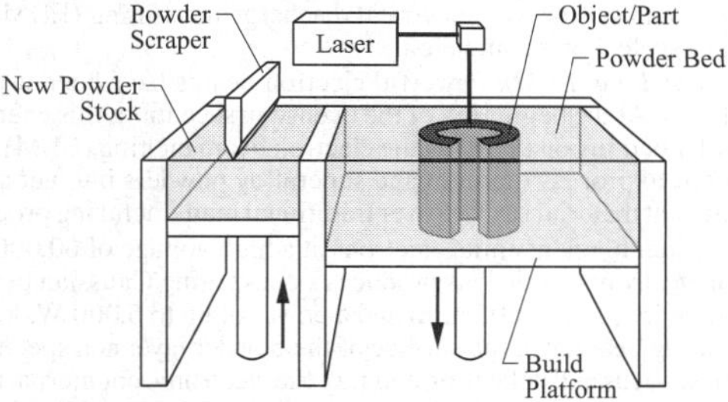


Fig. 25. Schematic Illustration of Laser Beam Powder Bed Fusion, also known as Selective Laser Melting (SLM)

the bottom of the overhang. This roughness is particularly high if the angle of the overhang, relative to horizontal, is less than 45 degrees. As a result, the as-printed finish of up-facing surfaces will be better than down-facing overhang surfaces. (Surface finish in SLM material may be improved by subsequent shot peening.) Typical surface finishes of up-facing and side-facing surfaces are equivalent to those of investment castings, so only bearing or mounting surfaces may need to be machined.

Since the rate of heat transfer into loose powder is much less than the rate into solid metal, heat transfer from overhangs is reduced and the cooling rate is lower than in layers in the body of a part. Structural supports printed beneath overhangs (Fig. 26) and connected to the base plate provide a path for increased heat transfer from overhangs. Supports also anchor the printed part to the base and reduce distortion due to differences in temperature throughout the part as it is printed. Without such supports, overhanging layers of the printed parts would deflect upward due to thermal stress. A downside of such supports is that they can interfere with printing more than one layer of parts in one build.

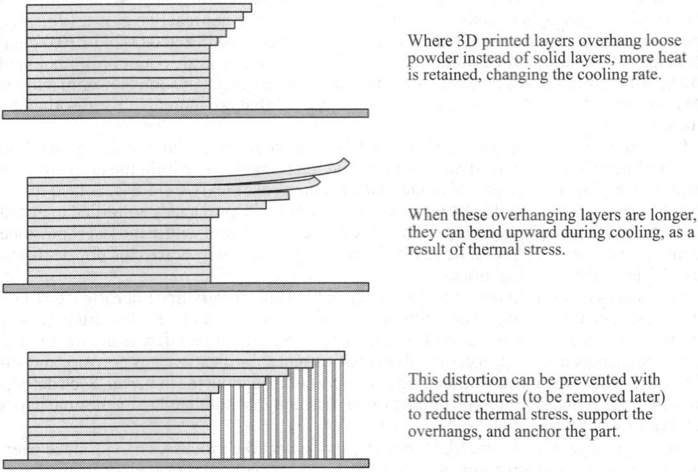


Fig. 26. Use of Supports to Anchor Parts in SLM Process

After the part is completed, the unused powder is collected for future use, and the base plate with the parts attached is removed from the SLM machine. Parts usually are cut from the plate using a band saw or wire electrical discharge machining (EDM), supports are removed, and any surfacing is completed.

*Electron Beam Melting (EBM):* Powerful electron beams have been used for welding since the late 1950s. At the beginning of the twenty-first century, this energy source was adapted to powder bed fusion (PBF) as the electron beam melting (EBM) process. EBM is commonly used to process titanium and superalloy powders into net-shaped parts—providing significant material savings over traditional manufacturing processes.

In EBM, a tungsten filament emits electrons at a high voltage of 60,000 V (60 kV). A focusing coil or electromagnetic lens produces a converging Gaussian beam with a spot size of 0.008–0.04 in. (0.203–1.02 mm) and a power of up to 6,000 W; a deflection coil directs the scanning beam over the surface of the powder layer at a speed of 26,000 ft/s (8,000 m/s). These focus and deflection coils are the electronic counterparts of the f-theta and scanning optics of the laser beam systems in SLM. Schematically, the EBM system is similar to that of SLM; however, while the laser beam is moved optically by deflection mirrors in SLM, the electron beam is moved by electromagnetic coils with no moving parts in EBM, facilitating high-speed scanning.

EBM machines require a vacuum in the build chamber of $<1 \times 10^{-4}$ mbar in order to prevent obstruction of the focused beam. Similar to SLM machines, EBM machines have a build platform that moves downward one layer thickness for each build cycle, and a powder distribution system, or *recoater*, to spread a new layer of powder. But unlike SLM systems, the electron beam repulses powder particles in the build layer. To prevent powder dispersion, each layer of powder in the build box is first partially sintered by rapidly scanning the total surface with the electron beam at low power; this causes the powder particles to adhere to the previous layer and adjoining particles. Then the electron beam power is increased, and the layer profile is scanned, melting powder particles within the beam spot. Because EBM power is much greater than the laser power in SLM, the particle size used is larger, between 0.002 and 0.004 in. (0.055 and 0.105 mm).

Because each layer is partially sintered by quick scan of the electron beam over each powder layer, it is generally not necessary to print support structures for overhangs. Therefore, unlike SLM, the EBM process allows the production of multiple layers of parts in one build. In addition, because each layer is preheated—with the temperature of powder in the build box reaching temperatures around 1300° F (700° C)—temperature variation throughout the build box is very low and distortion due to thermal stress is negligible.

Unlike the SLM process, the unused powder in the EBM process is not loose but is lightly sintered into a cake surrounding the finished part. After the build box cools down, which takes 12–18 hours, the powder cake is removed. An air gun, using the same-sized powder particles as those in the build, is used to impact the cake and break up the unused powder cake into fragments. Particles of the fragmented cake are reused in a future build.

**Binder Jetting (BJ).**—Like SLM and EBM, binder jetting begins with an STL CAD file that is sliced in preparation for layer-by-layer printing (see Table 8), the part is built in a powder bed in which a build plate is lowered one layer thickness for each slice, and a recoater spreads a new layer of metal powder. Unlike the SLM and EBM processes, a printhead with one or more jets passes over the powder surface and deposit binder droplets within the 2D slice profile defined by the CAD file of the 3D part. This sequence is repeated layer by layer, until the 3D part is completed.

In BJ, powder layer thickness ranges between 0.001 and 0.005 in. (0.0254 and 0.127 mm). A useful analogy is to think of the process as an inkjet printhead in a paper printer and each layer of metal powder as the paper sheet. Because binder jetting occurs at room temperature, there are no issues of heat transfer and supports are not needed. Therefore, as with EBM, multiple layers of parts can be made in each build box, increasing productivity.

In BJ, each binder droplet agglomerates powder particles into a building block called a *volume element* or *voxel*—the 3D equivalent of a 2D picture element or pixel. The binder serves as an adhesive, bonding particles in the new powder layer into a voxel and to material in the previous layer (Fig. 27). This bond is not enough to transmit useful loads, but it is sufficient to handle the part during removal of excess power surrounding the part, removal of the part from the build box, and setup for the next process step. Printed parts are then sent to a sintering furnace, where the binder burns off and the powder particles sinter together at a temperature below the melting point.

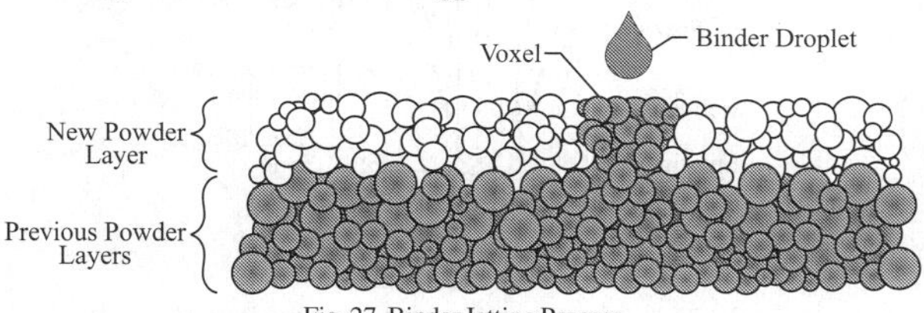


Fig. 27. Binder Jetting Process

In one version of this process, the part is lightly sintered, and the remaining volume between particles is filled by infiltration with a lower-melting-point alloy. In such cases, the printed powder is typically stainless steel and the infiltrating material bronze. Average particle size for this method is 0.001–0.002 in. (0.030–0.060 mm). The resulting material is useful for making functional prototypes, because it has properties comparable to high-strength steel.

In another variation, the printed material is sintered to full density. In this case, the materials used may be stainless steel, Inconel, or another high-strength metal. To promote sintering to full density, very small particle size is used, typically 0.0004–0.0006 in. (0.010–0.015 mm).

Metals produced by BJ processes may have residual porosity as high as 2 percent. Because these processes do not use a laser, they can be used to produce parts of copper, gold, and other metals that would reflect laser light. In addition, BJ can be used for ceramic materials. Typical print resolutions are 1200 dots per inch, corresponding to a voxel length of 0.0077 in. (0.196 mm).

**Directed Energy Deposition (DED).—**In this metal AM process, the energy source may be a laser, electron beam, or plasma arc. The material source may be metal powder or metal wire. Unlike the powder bed processes described above, DED processes the powder or feed wire into the focal point of the energy source. Typically, the substrate and part are fitted to a five-axis motion system, while the energy and powder nozzles remain stationary.

*Powder-Fed DED:* In the most common application of directed energy deposition (Fig. 28), metal powder flows through nozzles at high speed into the focal point of a laser. The powder melts and is deposited layer upon layer onto a substrate or a previously built part. As in powder bed fusion processes, liquid metal in this molten pool solidifies very quickly.

To prevent oxidation of the molten metal, an inert shield gas (argon or nitrogen) engulfs the metal deposition region. Alternately, the entire build system may be enclosed in a chamber containing inert gas. In one typical system, the five-axis motion and powder application systems are enclosed in a chamber measuring 31 by 31 by 24 in. (800 by 800 by 600 mm) in length, width, and height.

A major advantage of this process is that it does not involve spreading a layer of powder; therefore, it can be used to apply a material surface to an existing part or substrate. DED processes, as a result, are widely used for cladding and repair of existing parts. Another advantage of DED is the high build rate of up to 2.2 lb (1 kg) per hour.

Unlike powder bed processes, powder-fed DED processes have multiple nozzles, each of which can supply a different material. Therefore, material can be deposited with varying composition. For example, if one nozzle applies stainless steel and another supplies

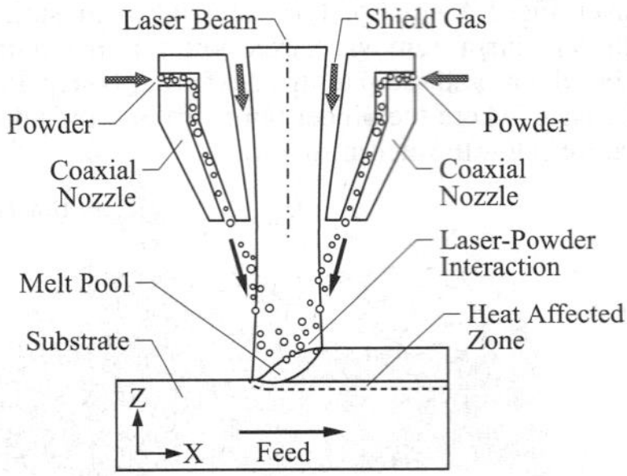


Fig. 28. Directed Energy Deposition with Powder Feed

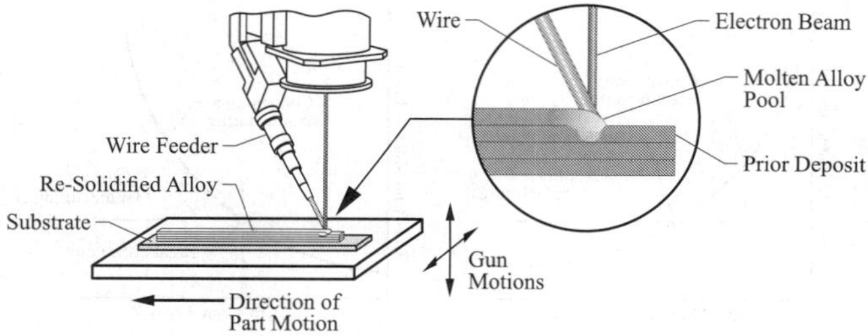

Fig. 29. Directed Energy Deposition with Wire Feed

Inconel nickel-based superalloy, a part could be built with a relatively low-cost stainless steel base that transitions to the higher cost Inconel surface where greater heat resistance is required.

A major disadvantage of this process is rough surface finish. In addition, because the powder is blown onto the surface, not all of the powder becomes melted, and material losses can range from 20 to 60 percent.

*Wire-Fed DED:* Another approach to DED involves feeding a metal wire from a spool into the focal point of an electron beam (Fig. 29). Energy from an electron beam continuously melts wire fed from a coil into the beam focal point. The motion of the base plate beneath the beam and wire feed builds up the part as directed by the CAD file.

In this case, the process must be carried out in a vacuum chamber, because any gas in the environment would interfere with the electron beam. Wire-fed electron beam DED processes are capable of very high build rates, up to 25 lb (11.3 kg) per hour, and do not result in any direct metal loss. But since such processes essentially involve weld metal build-up, the finish is rough; machining the entire surface is needed to reach desired dimensions and surface finish.

*Hybrid Processes:* Both wire-fed and powder-fed approaches to DED provide the basis for hybrid manufacturing, or combined additive/subtractive manufacturing. The basic form of the part to be produced is made by one of the DED processes, and the finished dimensions and surface finish are achieved by material removal, usually cutting. This sequence of operations can be carried out in one machine with one setup to ensure accuracy. The metal addition device (powder deposition or wire deposition) is attached to the CNC tool disk or carriage and is used to add metal where it is needed, while the usual cutting tools are then used to carry out the subsequent machining and finishing operations.

One way to use such hybrid systems is to fix a conventionally manufactured (by forging or casting) base shape in the hybrid machine, and then add features, such as nozzles or fins, that would be impossible or difficult to form in the conventional process. Then, the overall part can be finish-machined in the same setup.

Hybrid additive/subtractive manufacturing offers a convenient approach for machine shops to enter into additive manufacturing. Some hybrid equipment providers offer retrofit systems—including powder feed, laser, and software in one package—to convert a standard five-axis CNC system into a hybrid additive/subtractive manufacturing cell.

**Comparing AM with Conventional Processes.**—AM technology is just a few decades old, but great progress has been made in applying these processes to production of tailored, one-of-a-kind or limited-production parts. The majority of current applications involve specialized materials used in aerospace and medical fields. However, serial (mass) production of other manufactured items is economically feasible in some cases—increasingly so as rapid development of AM processes, materials, and applications continues.

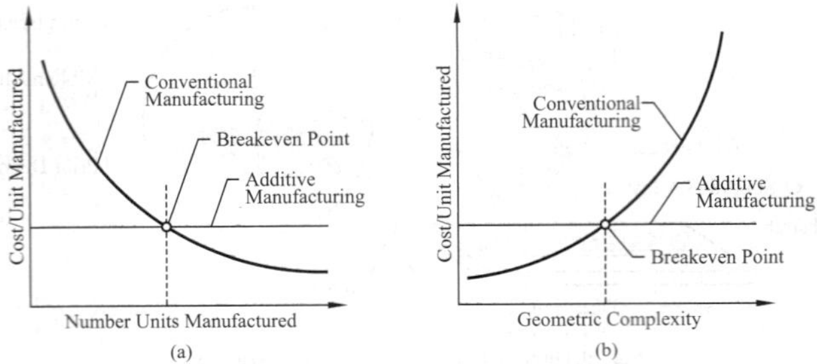


Fig. 30. a) Cost as a Function of Number of Units Manufactured; b) Cost as a Function of Geometric Complexity of a Part

Advantages of AM technology include:

- Short lead time and fast prototyping can be achieved using a 3D file.
- Design changes easily can be made, facilitating product development.
- New designs are possible that are not doable with conventional processes.
- Complex assemblies can be combined into a single part.
- Flexibility enables production of customized single- and short-run parts.
- Multiple 3D files for different parts/customers can be handled simultaneously.

Many metal parts currently being made by AM would be cost-prohibitive, difficult, or impossible to produce using conventional manufacturing methods, such as casting or machining. This gives AM an advantage, but only if the right types of parts are selected. Quality requirements play a major role. With most AM processes, good feature detail and surface finish are possible and comparable to metal castings, but surface quality is not comparable to CNC-machined parts, and additional surfacing adds processing time and expense. If a metal part can be produced by conventional processes at a reasonable cost and the number of units to be produced is relatively high, it usually is best to go that route at the present time.

Fig. 30a illustrates cost per unit as a function of number of units manufactured, comparing AM and traditional manufacturing methods. The initial cost of traditional methods is high due to the cost of tooling (molds or dies) or CNC machines and programming, but cost per unit manufactured decreases as these costs are shared over each part produced. Meanwhile, the cost of parts by AM remains the same, regardless of the number of units manufactured. Breakeven between the two production approaches occurs where these curves cross. AM is cost-efficient for small quantities, but traditional production methods yield cost advantages at higher volumes, as suggested by the declining cost curve. The breakeven point may be on the order of 100 to 1,000, with this number being higher for complex shapes.

Geometric complexity of a part has a significant influence on the cost per unit manufactured. Increasing complexity leads to exponentially increasing cost of conventional production technologies for small series production, as shown in Fig. 30b. By contrast with conventional production technologies, manufacturing costs for AM technologies do not increase with higher geometric complexity.

Additive manufacturing also is efficient in terms of energy and materials. In addition to not wasting quantities of expensive metals, the efficiency of AM processes opens the opportunity to produce lightweight components with a buy-to-fly ratio close to 1. (Buy-to-fly ratio is the ratio of the weight of raw material used for a component to the weight of the component itself.)

Fig. 31 illustrates comparative data on energy and material efficiency for production of high-performance parts by conventional CNC machining and AM. (This example is based on using electron beam powder bed fusion to produce a specialized aircraft bracket.) Material waste is extensive in CNC machining, with a buy-to-fly ratio of 8:1, compared to 1.5:1 for AM.

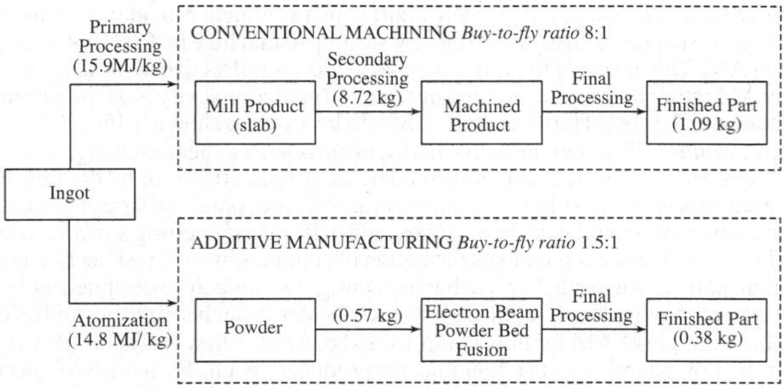


Fig. 31. Comparative Date for Conventional Machining and Additive Manufacturing of a Titanium Aircraft Bracket *(Source: MFT and LIGHTEnUp Team)*

**Unique Capabilities of Metal AM.**—For a given metal alloy, the cost of material in powder or wire form for additive manufacturing generally is much higher than the cost of bar stock used in conventional manufacturing. In addition, the cost of AM equipment is high and production rate is low, so production costs for AM are high. On the other hand, AM processes do not require tooling and associated production delays, and usually little material is lost. More important, AM can easily make much more complicated designs than can be made by conventional processes.

Conventional manufacturing processes, such as machining, sheet forming, and casting, impose constraints on the design of parts to be made by those processes, as is well known by designers. These constraints control possible configurations when designing a new component. Additive manufacturing processes remove many of those design constraints, though, despite the hype surrounding these technologies, AM also has limitations. Most limitations currently involve feature size and resolution: hole diameter and depth, wall height and thickness, and surface finish. In addition, at a practical level, for most metal AM processes, part features must be such that unused powder can be removed; that is, a completely enclosed hollow chamber must be provided with a hole for removal of internal powder. Nevertheless, AM processes provide great opportunity for expanding the designer's vision. Table 9 compares capabilities of SLM, EBM, BJ, and DED processes for producing metal parts by additive manufacturing.

Within the constraints shown, complex configurations can be conceived that would be difficult or impossible to achieve by conventional manufacturing processes. Four prominent shapes classes are discussed here, though many other complexities can be explored.

**Table 9. Comparison of the Capabilities of AM Processes for Producing Metal Parts**

| Feature | Selective Laser Melting (SLM) | Electron Beam Melting (EBM) | Directed Energy Deposition (DED) | Binder Jetting (BJ) |
|---|---|---|---|---|
| Overhang angle (°) | >45 | No limit | >45 | No limit |
| Machining stock (mm) | 0.1–0.5 | 0.5–2.0 | 1.0–2.0 | 0.05–0.1 |
| Wall thickness (mm) | 0.3–0.5 | 0.6–1.0 | 3 | 0.3 |
| Hole diameter (vertical) (mm) | 0.5 | 1–2 | — | 0.4 |
| Hole diameter (horizontal) (mm) | 8.0 | 8.0 | — | 0.4 |
| Clearance (mm) | 1 | 1 | — | 1 |
| Surface finish upfacing | 10–15 | 20–30 | 40–50 | 2 |
| Surface finish downfacing | 90–100 | 130–150 | — | 2 |
| Porosity (%) | 0–0.5 | 0.1–0.3 | — | 2–3 |
| Screw threads | Vertical axis | No | No | Yes |
| Powder recovery | Vacuum | Blast/vacuum | No | Vacuum |
| Layer thickness (μm) | 50–100 | 150–300 | 500 | 25–75 |
| Build rate (kg/hr) | 0.05 | 0.1 | 1.0 | 2 |

*Complex Fins and Blades:* Axisymmetric parts, such as impellers made up of fins or blades on a flat or cone shape, are costly and time-consuming to machine by CNC but are relatively easy using AM. This is one of the rare cases when it is cost-effective to produce an existing design by AM rather than a conventional process. As part complexity is on the extreme right side of the curve, the cost of production by AM will be lower, as shown in Fig. 30b.

*Internal Channels:* Internal fluid flow paths, in parts such as heat exchangers, fuel injectors, and sensors, normally are accomplished by assembling tubes, or by drilling and capping internal passages. These internal pathways usually are round and straight, because tubing is most easily produced with round tubes, and drilling uses cutting surfaces in circular motion. Use of AM processes enables formation of channels in any cross section and along non-straight paths. As a result, heat exchanger configurations can be designed and made by AM that allow for more effective and thus efficient heat transfer between hot and cold fluids.

A notable example of AM for flow control—and one of the first examples of AM used in serial production quantities—is a fuel injector produced by GE for the LEAP gas turbine engine; nineteen fuel injectors are required for each engine. Redesigning the component for AM reduced flow losses and, at the same time, reduced the parts count from twenty to one.

Another example of the benefits of producing internal channels by AM is injection mold tooling. Rapid removal of heat from injected polymer is key in improving production rate of injection molded parts, as this step takes about half the total time for each injection cycle. Internal cooling channels often are used to increase heat removal rates; such channels are made by drilling straight channels that intersect internally and capping holes that penetrate the outside of the mold. With conformal cooling channels (conforming to the shape of the mold), heat can be removed more quickly, decreasing injection cycle time. While savings may seem small for each cycle, consider the cumulative effect over millions of cycles typically used in injection molding of plastic consumer goods.

*Internal Structures and Topology Optimization:* One of the unique capabilities of AM processes is the ability to produce lattice structures, cellular structures, and organic geometries. Lattice structures involve a network of struts inside a part that remove a large amount of material yet maintain strength of the part. Cellular structures build up a part as an assembly of cells, usually hexagonal in shape. Organic structures are unusual shapes that appear similar to natural materials, such as bone or trees.

Such geometries can be developed using topology optimization computer programs to analyze stresses in a loaded component. By identifying elements that transmit little or no stress, the program can adjust part design to reduce or discard those elements, minimizing the material needed to transmit forces while maximizing resistance to deformation under stress. Analysis is repeated until optimum configuration is reached. (See also *Topology Optimization* on page 1519.) Used for decades in designing lightgweight structures, particularly aircraft, topology optimization is a perfect complement to AM.

*Topology Optimization:* This mathematical technique minimizes the amount of material needed to transmit forces, while maximizing resistance of the structure to distortion under those forces. By analyzing stresses in a loaded component by finite element modeling, elements that carry little or no stress can be identified and discarded; analysis is repeated until optimum configuration is reached. While topology optimization has been used for decades in design of lightweight structures, particularly aircraft, it is a perfect complement to AM, as such optimized structures typically are organic in form and best made with AM.

*Assemblies and Part Integration:* Part count reduction can be accomplished by designing assemblies as one component. This reduces the need for assembly operations, such as brazing or riveting. In addition, by producing the part in one piece, blend radii can be made at joining surfaces, reducing stress concentrations and potential sources of fatigue failure. While the cost of each part of the assembly may be small, the cumulative cost of tooling for each part, keeping an inventory of each part, plus the cost of assembly, can be greater than the cost of producing the assembly as one component by AM.

*Reasons to Choose AM:* Current AM processes cannot be expected to compete economically with traditional metal manufacturing methods, unless: (1) you absolutely need it as soon as possible without waiting for tooling fabrication or CNC programming, and cost is not an issue; (2) complexity of the part is so great that it actually is less expensive to make by

AM than by conventional processes; (3) AM processes provide unique geometric or material characteristics, such as internal conformal cooling channels or part consolidation, that improve performance, so lifecycle cost savings by AM exceed its premium production cost.

To take advantage of opportunities afforded by metal AM processes, it is necessary to consider the entire system, rather than each individual part, for part reduction possibilities and/or potential improvements in system performance. As continuous improvements are made in AM materials, machines, and software, this innovative manufacturing method will become increasingly viable for large-scale production of automobile components and other commodity items.

**Standards for Additive Manufacturing.—**Commercial additive manufacturing is less than five decades old but is expanding at an extremely rapid rate and has attracted business interest worldwide. To support this growth and provide confidence in AM technologies to compete with existing manufacturing technologies, ASTM committee F42 was initiated in 2009 for the purpose of developing standards of practice. Subsequently, a Partner Standards Developing Organization (PSDO) agreement between ASTM International and the International Organization for Standardization (ISO), signed in 2011, paved the way to create joint AM standards. The agreement's goal was to respond to this burgeoning industry's need for a set of AM standards that would be recognized worldwide. In 2016, these organizations jointly announced a common roadmap and organizational structure for AM standards. The current Standards listed below define terminology, measure the performance of different production processes, ensure the quality of the end products, and specify procedures for the calibration of additive manufacturing machines.

*Applications:*

ASTM F3456-22, Standard Guide for Powder Reuse Schema in Powder Bed Fusion Processes for Medical Applications for Additive Manufacturing Feedstock Materials

ASTM F3554-22, Standard Specification for Additive Manufacturing–Finished Part Properties–Grade 4340 (UNS G43400) via Laser Beam Powder Bed Fusion for Transportation Applications

ASTM F3572-22, Standard Practice for Additive Manufacturing–General Principles–Part Classifications for Additive Manufactured Parts Used in Aviation

ISO/ASTM 52941-20, Additive manufacturing—System performance and reliability—Acceptance tests for laser metal powder-bed fusion machines for metallic materials for aerospace application

ISO/ASTM 52942-20, Additive manufacturing—Qualification principles—Qualifying machine operators of laser metal powder bed fusion machines and equipment used in aerospace applications

*Design:*

ASTM F2915-20, Standard Specification for Additive Manufacturing File Format (AMF) Version 1.2

ASTM F3413-19, Guide for Additive Manufacturing—Design—Directed Energy Deposition

ASTM F3529-21, Guide for Additive Manufacturing—Design—Material Extrusion of Polymers

ASTM F3530-22, Standard Guide for Additive Manufacturing—Design—Post-Processing for Metal PBF-LB

ISO/ASTM 52910-18, Additive manufacturing–Design–Requirements, guidelines and recommendations

ISO/ASTM 52911-1-19, Additive manufacturing—Design—Part 1: Laser-based-powder bed fusion of metals

ISO/ASTM 52911-2-19, Additive manufacturing—Design—Part 2: Laser-based-powder bed fusion of polymers

ISO/ASTM 52915-20, Specification for additive manufacturing file format (AMF) Version 1.2

ISO/ASTM 52921-13(2019), Standard Terminology for Additive Manufacturing—Coordinate Systems and Test Methodologies

ISO/ASTM 52950-21, Additive manufacturing—General principles—Overview of data processing

*Materials and Manufacturing Processes:*

ASTM F2924-14(2021), Standard Specification for Additive Manufacturing Titanium-6 Aluminum-4 Vanadium with Powder Bed Fusion

ASTM F3001-14(2021), Standard Specification for Additive Manufacturing Titanium-6 Aluminum-4 Vanadium ELI (Extra Low Interstitial) with Powder Bed Fusion

ASTM F3049-14(2021), Standard Guide for Characterizing Properties of Metal Powders Used for Additive Manufacturing Processes

ASTM F3055-14a(2021), Standard Specification for Additive Manufacturing Nickel Alloy (UNS N07718) with Powder Bed Fusion

ASTM F3056-14(2021), Standard Specification for Additive Manufacturing Nickel Alloy (UNS N06625) with Powder Bed Fusion

ASTM F3091/F3091M-14(2021), Standard Specification for Powder Bed Fusion of Plastic Materials

ASTM F3184-16, Standard Specification for Additive Manufacturing Stainless Steel Alloy (UNS S31603) with Powder Bed Fusion

ASTM F3187-16, Standard Guide for Directed Energy Deposition of Metals

ASTM F3213-17, Standard for Additive Manufacturing–Finished Part Properties–Standard Specification for Cobalt-28 Chromium-6 Molybdenum via Powder Bed Fusion

ASTM F3301-18a, Standard for Additive Manufacturing–Post Processing Methods–Standard Specification for Thermal Post-Processing Metal Parts Made Via Powder Bed Fusion

ASTM F3302-18, Standard for Additive Manufacturing–Finished Part Properties–Standard Specification for Titanium Alloys via Powder Bed Fusion

ASTM F3318-18, Standard for Additive Manufacturing–Finished Part Properties–Specification for AlSi10Mg with Powder Bed Fusion–Laser Beam

ISO/ASTM 52901-17, Standard Guide for Additive Manufacturing–General Principles–Requirements for Purchased AM Parts

ISO/ASTM 52903-1-20, Additive manufacturing—Material extrusion-based additive manufacturing of plastic materials—Part 1: Feedstock materials

ISO/ASTM 52903-2-20, Additive manufacturing—Material extrusion-based additive manufacturing of plastic materials—Part 2: Process equipment

ISO/ASTM 52904-19, Additive Manufacturing—Process Characteristics and Performance: Practice for metal powder bed fusion process to meet critical applications

ISO/ASTM 52925-22, Additive manufacturing of polymers—Feedstock materials—Qualification of materials for laser-based powder bed fusion of parts

ISO/ASTM 52930-21, Additive manufacturing—Qualification principles—Installation, operation and performance (IQ/OQ/PQ) of PBF-LB equipment

*Terminology:*

ASTM F2792-12a (Withdrawn 2015), Standard Terminology for Additive Manufacturing Technologies

ISO/ASTM 52900-21, Additive manufacturing—General principles—Fundamentals and vocabulary

*Test Methods:*

ASTM F52921-13(2019), Standard Terminology for Additive Manufacturing—Coordinate Systems and Test Methodologies

ASTM F2971-13(2021), Standard Practice for Reporting Data for Test Specimens Prepared by Additive Manufacturing

ASTM F3122-14(2022), Standard Guide for Evaluating Mechanical Properties of Metal Materials Made Via Additive Manufacturing Processes

ASTM F3522-22, Standard Guide for Additive Manufacturing of Metals—Feedstock Materials—Assessment of Powder Spreadability

ASTM F3571-22, Standard Guide for Additive Manufacturing— Feedstock—Particle Shape Image Analysis by Optical Photography to Identify and Quantify the Agglomerates/Satellites in Metal Powder Feedstock

ASTM F3606-22, Standard Guide for Additive Manufacturing—Feedstock Materials—Testing Moisture Content in Powder Feedstock

ISO/ASTM 52902-19, Additive manufacturing—Test artifacts—Geometric capability assessment of additive manufacturing systems

ISO/ASTM52907-19, Additive manufacturing—Feedstock materials—Methods to characterize metallic powders

## SOLDERING AND BRAZING

Metals may be joined without using fasteners by employing soldering, brazing, and welding. Soldering and brazing utilize capillary action to join two or more adjoined faying (in contact) surfaces of a component; only the filler material melts and solidifies to create a strong bond. Soldering involves the use of a nonferrous metal whose melting point is below that of the base metal and in all cases below 840° F (450° C). Brazing entails the use of a nonferrous filler metal with a melting point below that of the base metal but above 840° F (450° C). In fusion welding, abutting metal surfaces are made molten, are joined in the molten state, and then allowed to cool. The use of a filler metal and the application of pressure are considered optional in the practice of fusion welding.

### Soldering

Soldering is used to provide a convenient joint that does not require any great mechanical strength. It is used in a great many instances in combination with mechanical staking, crimping, or folding, the solder being used only to seal against leakage or to assure electrical contact. Soldering employs lead- or tin-base alloys with melting points below 840° F (450° C) and is commonly referred to as soft soldering. Use of hard solders, silver solders and spelter solders which have silver, copper, or nickel bases and have melting points above 840° F (450° C) is known as brazing. The accompanying table, page 1568, gives some of the properties and uses of various solders that are generally available.

**Forms Available.—** Soft solders can be obtained in bar, cake, wire, pig, slab ingot, ribbon, segment, powder, and foil form, depending on what is needed for the specific use. In bar form they are commonly used for hand soldering. The pigs, ingots, and slabs are used in operations that employ melting kettles. Ribbon, segment, powder, and foil forms are used for special applications, and the cake form is used for wiping. Wire forms are either solid or they contain acid or rosin cores for fluxing. These wire forms, both solid and core-containing, are used in hand and automatic machine applications. Prealloyed powders, suspended in a fluxing medium, are frequently applied by brush and, upon heating, consistently wet the solderable surfaces to produce a satisfactory joint.

**Fluxes for Soldering.—** The surfaces of the metals being joined in the soldering operation must be clean in order to obtain an efficient joint. Fluxes clean the surfaces of the metal in the joint area by removing the oxide coating present, keep the area clean by preventing formation of oxide films, and lower the surface tension of the solder, thereby increasing its wetting properties. Rosin, tallow, and stearin are mild fluxes that prevent oxidation but are not too effective in removing oxides present. Rosin is used for electrical applications, since the residue is non-corrosive and non-conductive. Zinc chloride and ammonium chloride (sal ammoniac), used separately or in combination, are common fluxes that remove oxide films readily.

The residue from these fluxes may in time cause trouble, due to their corrosive effects, if they are not removed or neutralized. Two methods of inactivating and removing this residue are washing with water containing about 5 ounces (142 g) of sodium citrate (for nonferrous soldering) or 1 ounce (28.35 g) of trisodium phosphate (for ferrous and nonferrous soldering) per gallon (1 US gallon = 3.754 liters) followed by a clear water rinse, or washing with commercial water-soluble detergents.

**Methods of Application.—** Solder is applied using a soldering iron, a torch, a solder bath, electric induction or resistance heating, a stream of hot neutral gas, or by wiping. Clean surfaces that are hot enough to melt the solder being applied or accept molten solder are necessary to obtain a good clean bond. Parts being soldered should be free of oxides, dirt, oil, and scale. Scraping and the use of abrasives as well as fluxes are techniques for preparing surfaces for soldering. The procedures followed in soldering aluminum, magnesium, and stainless steel differ somewhat from conventional soldering techniques and are indicated in the material that follows.

**Properties of Soft Solder Alloys,** *Appendix, ASTM B32-08 (R2014)*

| Nominal Composition[a] Percent | | | | Specific Gravity[b] | Melting Ranges,[c] Degrees Fahrenheit | | Uses |
|---|---|---|---|---|---|---|---|
| Sn | Pb | Sb | Ag | | Solidus | Liquidus | |
| 70 | 30 | … | … | 8.32 | 361 | 377 | For soldering zinc and coating metals. |
| 63 | 37 | … | … | 8.40 | 361 | 361 | For soldering printed circuit boards where temperature limitations are critical and in applications where an extremely short melting range is required. |
| 62 | 36 | … | 2 | 8.44 | 354 | 372 | For soldering silver-coated surfaces. |
| 60 | 40 | … | … | 8.65 | 361 | 374 | "Fine solder" for soldering electrical and electronic connections and for coating metals. |
| 50 | 50 | … | … | 8.85 | 361 | 421 | For general purposes. Most popular of all. Can be used for noncritical electrical soldering and applications such as joining sheet metal, pipe, tubing, and other structural shapes. |
| 45 | 55 | … | … | 8.97 | 361 | 441 | For general purposes. Can be used for noncritical electrical soldering and applications such as joining sheet metal, pipe, tubing, and other structural shapes. |
| 40 | 60 | … | … | 9.30 | 361 | 460 | For dip soldering and as a wiping solder for joining lead pipes and cable sheaths. |
| 35 | 65 | … | … | 9.50 | 361 | 447 | A plumber's solder, but with a lower antimony content. |
| 30 | 70 | … | … | 9.70 | 361 | 491 | For automobile-body solder and for removing heat-strippable insulation during high-temperature (700 to 900° F) [370 to 480° C] tinning of wires. |
| 25 | 75 | … | … | 10.00 | 361 | 511 | For automobile-body solder. |
| 20 | 80 | … | … | 10.20 | 361 | 531 | For automobile-body solder with a lower antimony than alloy Sn20 below. |
| 15 | 85 | … | … | 10.50 | 437[d] | 554 | For coating and joining metals. |
| 10 | 90 | … | … | 10.80 | 514[d] | 576 | For coating and joining metals. |
| 5 | 95 | … | … | 11.30 | 586 | 594 | For coating and joining metals. |
| 2 | 98 | … | … | 11.22 | 601 | 611 | For soldering automobile radiator cores. |
| 40 | 58 | 2 | … | 9.23 | 365 | 448 | Similar to Sn40 above, but not recommended for use on galvanized iron. |
| 35 | 63.2 | 1.8 | … | 9.44 | 365 | 470 | For wiping solders promoting finer grain size and greater strength. |
| 30 | 68.4 | 1.6 | … | 9.65 | 364 | 482 | For automobile-body solder for filling dents and seams. |
| 25 | 73.7 | 1.3 | … | 9.96 | 365 | 504 | For automobile-body solder. |
| 20 | 79 | 1 | … | 10.17 | 363 | 517 | Used for automobile-body solder for filling dents and seams, and for general purposes such as protective coatings on steel sheet where a high tin content alloy is not required. |
| 95 | … | 5 | … | 7.25 | 452 | 464 | For joints on copper in electrical, plumbing, and heating work. |
| … | 94.5 | … | 5.5 | 11.3 | 580 | 716 | This alloy will develop a shearing strength of 1500 psi at 350° F [177° C]. When soldering hard-drawn brass or copper, the application temperature should not exceed 850° F [455° C]. Used on thermocouples for aircraft engines, where relatively high operating temperatures will not affect strength of the solder. |
| … | 97.5 | … | 2.5 | 11.35 | 580 | 580 | For use on copper, brass, and similar metals with torch heating. Not recommended in humid environments due to its known susceptibility to corrosion. |
| 1 | 97.5 | … | 1.5 | 11.28 | 588 | 588 | For use on copper, brass, and similar metals with torch heating. This alloy is used interchangeably with the above alloy Ag2.5, but it has a better shelf life and does not develop a black surface deposit when stored under humid environmental conditions. |

[a] Abbreviations of alloying elements are as follows: Sn, tin; Pb, lead; Sb, antimony; and Ag, silver.

[b] The specific gravity multiplied by 0.0361 equals the density in pounds per cubic inch.

[c] The alloys are completely solid below the lower point given, designated "solidus," and completely liquid above the higher point given, designated "liquidus." In the range of temperatures between these two points the alloys are partly solid and partly liquid.

[d] For some engineering design purposes, it is well to consider these alloys as having practically no mechanical strength above 360° F (182° C).

*Soldering Aluminum:* Two properties of aluminum which tend to make it more difficult to solder are its high thermal conductivity and the tenacity of its ever-present oxide film. Aluminum soldering is performed in a temperature range of from 550–770° F (288–410° C), compared to 375–400° F (191–204° C) temperature range for ordinary metals, because of the metal's high thermal conductivity. Two methods can be used, one using flux and one using abrasion. The flux method is most widely used and is known as flow soldering. In this method flux dissolves the aluminum oxide and keeps it from re-forming. The flux should be fluid at soldering temperatures so the solder can displace it in the joint. In the friction method the oxide film is mechanically abraded with a soldering iron, wire brush, or multi-toothed tool while being covered with molten solder. The molten solder keeps the oxygen in the atmosphere from reacting with the newly exposed aluminum surface; thus, wetting of the surface can take place.

The alloys that are used in soldering aluminum generally contain from 50 to 75 percent tin, with the remainder being zinc. The following aluminum alloys are listed in order of ease of soldering: commercial and high-purity aluminum; wrought alloys containing not more than 1 percent manganese or magnesium; and finally, the heat-treatable alloys, which are the most difficult.

Cast and forged aluminum parts are not generally soldered.

*Soldering Magnesium:* Magnesium is not ordinarily soldered to itself or other metals. Soldering is generally used for filling small surface defects, voids, or dents in castings or sheets where the soldered area is not to be subjected to a load. Two solders can be used: one with a composition of 60 percent cadmium, 30 percent zinc, and 10 percent tin has a melting point of 315° F (157° C); the other has a melting point of 500° F (260° C) and has a nominal composition of 90 percent cadmium and 10 percent zinc.

The surfaces to be soldered are cleaned to a bright metallic luster by abrasive methods before soldering. The parts are preheated with a torch to the approximate melting temperature of the solder being used. The solder is applied and the surface under the molten solder is rubbed vigorously with a sharp pointed tool or wire brush. This action results in the wetting of the magnesium surface. To completely wet the surface, the solder is kept molten and the rubbing action continued. The use of flux is not recommended.

*Soldering Stainless Steel:* Stainless steel is somewhat more difficult to solder than other common metals because of a tightly adhering oxide film on the surface of the metal and because of its low thermal conductivity. The surface of the stainless steel must be thoroughly cleaned. This can be done by abrasion or by clean white pickling with acid. Mixtures that may be used as fluxes for soldering stainless steel include: muriatic (hydrochloric) acid saturated with zinc, or this mixture combined with 25 percent additional muriatic acid, or with 10 percent additional acetic acid, or with 10 to 20 percent additional water solution of orthophosphoric acid. Tin-lead solder can be used successfully. Because of the low thermal conductivity of stainless steel, a large soldering iron is needed to bring the surfaces to the proper temperature, which is reached when the solder flows freely into the area of the joint. Removal of the corrosive flux is important in order to prevent joint failure. Soap and water or a commercial detergent may be used to remove the flux residue.

**Ultrasonic Fluxless Soldering.—**This more recently introduced method of soldering makes use of ultrasonic vibrations, which facilitates the penetration of surface films by the molten solder thus eliminating the need for flux. The equipment offered by one manufacturer consists of an ultrasonic generator and ultrasonic soldering head, which includes a transducer coupling, soldering tip, tip heater, and heating platen. Metals that can be soldered by this method include aluminum, copper, brass, silver, magnesium, germanium, and silicon.

## Brazing

Brazing is a metal joining process that uses a nonferrous filler metal with a melting point below that of the base metals but above 840° F (450° C). The filler metal wets the base metal when molten in a manner similar to that of a solder and its base metal. There is a slight diffusion of the filler metal into the hot, solid base metal or a surface alloying of the base and filler metal. The molten metal flows between the close-fitting metals because of capillary forces.

**Filler Metals for Brazing Applications.—**Brazing filler metals have melting points that are lower than those of the base metals being joined and have the ability when molten to flow readily into closely fitted surfaces by capillary action. The commonly used brazing

**Table 1a. Brazing Filler Metals,** *Based on Specifications and Appendix of American Welding Society AWS A5.8–1981*

| AWS Classification[a] | Nominal Composition,[b] Percent | | | | | | Temperature, Degrees F | | | Standard Form[c] | Uses |
|---|---|---|---|---|---|---|---|---|---|---|---|
| | Ag | Cu | Zn | Al | Ni | Other | Solidus | Liquidus | Brazing Range | | |
| BAlSi-2 | … | … | … | 92.5 | … | Si, 7.5 | 1070 | 1135 | 1110–1150 | 7 | For joining the following aluminum alloys: 1060, EC, 1100, 3003, 3004, 5005, 5050, 6053, 6061, 6062, 6063, 6951 and cast alloys A612 and C612. All of these filler metals are suitable for furnace and dip brazing. BAlSi-3, -4 and -5 are suitable for torch brazing. Used with lap and tee joints rather than butt joints. Joint clearances run from .006 to .025 inch. |
| BAlSi-3 | … | 4 | … | 86 | … | Si, 10 | 970 | 1085 | 1160–1120 | 2, 3, 5 | |
| BAlSi-4 | … | … | … | 88 | … | Si, 12 | 1070 | 1080 | 1080–1120 | 2, 3, 4, 5 | |
| BAlSi-5 | … | … | … | 90 | … | Si, 10 | 1070 | 1095 | 1090–1120 | 7 | |
| BAlSi-6 | … | … | … | 90 | … | Si, 7.5; Mg. 2.5 | 1038 | 1125 | 1110–1150 | 7 | BAlSi-6 through -11 are vacuum brazing filler metals. Magnesium is present as an $O_2$ getter. When used in vacuum, solidus & liquidus temperatures are different from those shown. |
| BAlSi-7 | … | … | … | 88.5 | … | Si, 10; Mg, 1.5 | 1038 | 1105 | 1090–1120 | 7 | |
| BAlSi-8 | … | … | … | 86.5 | … | Si, 12; Mg, 1.5 | 1038 | 1075 | 1080–1120 | 2, 7 | |
| BAlSi-9 | … | … | … | 87 | … | Si, 12; Mg, 0.3 | 1044 | 1080 | 1080–1120 | 7 | |
| BAlSi-10 | … | … | … | 86.5 | … | Si, 11; Mg, 2.5 | 1038 | 1086 | 1080–1120 | 2 | |
| BAlSi-11 | … | … | … | 88.4 | … | Si, 10; Mg, 1.5; Bi, 0.1 | 1038 | 1105 | 1090–1120 | 7 | |
| BCuP-1 | … | 95 | … | … | … | P, 5 | 1310 | 1695 | 1450–1700 | 1 | For joining copper and its alloys with some limited use on silver, tungsten and molybdenum. Not for use on ferrous or nickel-base alloys. Are used for cupro-nickels but caution should be exercised when nickel content is greater than 30 percent. Suitable for all brazing processes. Lap joints recommended but butt joints may be used. Clearances used range from .001 to .005 inch. |
| BCuP-2 | … | 93 | … | … | … | P, 7 | 1310 | 1460 | 1350–1550 | 2, 3, 4 | |
| BCuP-3 | 5 | 89 | … | … | … | P, 6 | 1190 | 1485 | 1300–1500 | 2, 3, 4 | |
| BCuP-4 | 6 | 87 | … | … | … | P, 7 | 1190 | 1335 | 1300–1450 | 2, 3, 4 | |
| BCuP-5 | 15 | 80 | … | … | … | P, 5 | 1190 | 1475 | 1300–1500 | 1, 2, 3, 4 | |
| BCuP-6 | 2 | 91 | … | … | … | P, 7 | 1190 | 1450 | 1350–1500 | 2, 3, 4 | |
| BCuP-7 | 5 | 88 | … | … | … | P, 6.8 | 1190 | 1420 | 1300–1500 | 2, 3, 4 | |
| BAg-1 | 45 | 15 | 16 | … | … | Cd, 24 | 1125 | 1145 | 1145–1400 | 1, 2, 4 | For joining most ferrous and nonferrous metals except aluminum and magnesium. These filler metals have good brazing properties and are suitable for preplacement in the joint or for manual feeding into the joint. All methods of heating may be used. Lap joints are generally used; however, butt joints may be used. Joint clearances of .002 to .005 inch are recommended. Flux is generally required. |
| BAg-1a | 50 | 15.5 | 16.5 | … | … | Cd, 18 | 1160 | 1175 | 1175–1400 | 1, 2, 4 | |
| BAg-2 | 35 | 26 | 21 | … | … | Cd, 18 | 1125 | 1295 | 1295–1550 | 1, 2, 4, 7 | |

**Table 1a.** *(Continued)* **Brazing Filler Metals,** *Based on Specifications and Appendix of American Welding Society AWS A5.8–1981*

| AWS Classification[a] | Nominal Composition,[b] Percent | | | | | | Temperature, Degrees F | | | Standard Form[c] | Uses |
|---|---|---|---|---|---|---|---|---|---|---|---|
| | Ag | Cu | Zn | Al | Ni | Other | Solidus | Liquidus | Brazing Range | | |
| BAg-2a | 30 | 27 | 23 | ... | ... | Cd, 20 | 1125 | 1310 | 1310–1550 | 1, 2, 4 | For joining most ferrous and nonferrous metals except aluminum and magnesium. These filler metals have good brazing properties and are suitable for preplacement in the joint or for manual feeding into the joint. All methods of heating may be used. Lap joints are generally used; however, butt joints may be used. Joint clearances of .002 to .005 inch are recommended. Flux is generally required. |
| BAg-3 | 50 | 15.5 | 15.5 | ... | 3 | Cd, 16 | 1170 | 1270 | 1270–1500 | 1, 2, 4, 7 | |
| BAg-4 | 40 | 30 | 28 | ... | 2 | ... | 1240 | 1435 | 1435–1650 | 1, 2 | |
| BAg-5 | 45 | 30 | 25 | ... | ... | ... | 1250 | 1370 | 1370–1550 | 1, 2 | |
| BAg-6 | 50 | 34 | 16 | ... | ... | ... | 1270 | 1425 | 1425–1600 | 1, 2 | |
| BAg-7 | 56 | 22 | 17 | ... | ... | Sn, 5 | 1145 | 1205 | 1205–1400 | 1, 2 | |
| BAg-8 | 72 | 28 | ... | ... | ... | ... | 1435 | 1435 | 1435–1650 | 1, 2, 4 | |
| BAg-8a | 72 | 27.8 | ... | ... | ... | Li, 2. | 1410 | 1410 | 1410–1600 | 1, 2 | |
| BAg-13 | 54 | 40 | 5 | ... | 1 | ... | 1325 | 1575 | 1575–1775 | 1, 2 | |
| BAg-13a | 56 | 42 | ... | ... | 2 | ... | 1420 | 1640 | 1600–1800 | 1, 2 | |
| BAg-18 | 60 | 30 | ... | ... | ... | Sn, 10 | 1115 | 1325 | 1325–1550 | 1, 2 | |
| BAg-19 | 92.5 | 7.3 | ... | ... | ... | Li, 2 | 1435 | 1635 | 1610–1800 | 1, 2 | |
| BAg-20 | 30 | 38 | 32 | ... | ... | ... | 1250 | 1410 | 1410–1600 | 1, 2, 4 | |
| BAg-21 | 63 | 28.5 | ... | ... | 2.5 | Sn, 6 | 1275 | 1475 | 1475–1650 | 1, 2, 4 | |
| BAg-22 | 49 | 16 | 23 | ... | 4.5 | Mn, 7.5 | 1260 | 1290 | 1290–1525 | 1, 2, 4, 7 | |
| BAg-23 | 85 | ... | ... | ... | ... | Mn, 15 | 1760 | 1780 | 1780–1900 | 1, 2, 4 | |
| BAg-24 | 50 | 20 | 28 | ... | 2 | ... | 1220 | 1305 | 1305–1550 | 1, 2 | |
| BAg-25 | 20 | 40 | 35 | ... | ... | Mn, 5 | 1360 | 1455 | 1455–1555 | 2, 4 | |
| BAg-26 | 25 | 38 | 33 | ... | 2 | Mn, 2 | 1305 | 1475 | 1475–1600 | 1, 2, 4, 7 | |
| BAg-27 | 25 | 35 | 26.5 | ... | ... | Cd, 13.5 | 1125 | 1375 | 1375–1575 | 1, 2, 4 | |
| BAg-28 | 40 | 30 | 28 | ... | ... | Sn, 2 | 1200 | 1310 | 1310–1550 | 1, 2, 4 | |

[a] These classifications contain chemical symbols preceded by "B" which stands for brazing filler metal.

[b] These are nominal compositions. Trace elements may be present in small amounts and are not shown. Abbreviations used are: Ag, silver; Cu, copper; Zn, zinc; Al, aluminum; Ni, nickel; Ot, other; Si, silicon; P, phosphorus; Cd, cadmium; Sn, tin; Li, lithium; Cr, chromium; B, boron; Fe, iron; O, oxygen; Mg, magnesium; W, tungsten; Pd, palladium; and Au, gold.

[c] Numbers specify standard forms as follows: 1, strip; 2, wire; 3, rod; 4, powder; 5, sheet; 6, paste; 7, clad sheet or strip; and 8, transfer tape.

**Table 1b. Brazing Filler Metals,** *Based on Specifications and Appendix of American Welding Society AWS A5.8–1981*

| AWS Classification[a] | Nominal Composition,[b] Percent | | | | | | Temperature, Degrees F | | | Standard Form[c] | Uses |
|---|---|---|---|---|---|---|---|---|---|---|---|
| | Ni | Cu | Cr | B | Si | Other | Solidus | Liquidus | Brazing Range | | |
| BNi-1 | 74 | … | 14 | 3.5 | 4 | Fe, 4.5 | 1790 | 1900 | 1950–2200 | 1, 2, 3, 4, 8 | For brazing AISI 300 and 400 series stainless steels, and nickel- and cobalt-base alloys. Particularly suited to vacuum systems and vacuum tube applications because of their very low vapor pressure. The limiting element is chromium in those alloys in which it is employed. Special brazing procedures required with filler metal containing manganese. |
| BNi-2 | 82.5 | … | 7 | 3 | 4.5 | Fe, 3 | 1780 | 1830 | 1850–2150 | 1, 2, 3, 4, 8 | |
| BNi-3 | 91 | … | … | 3 | 4.5 | Fe, 1.5 | 1800 | 1900 | 1850–2150 | 1, 2, 3, 4, 8 | |
| BNi-4 | 93.5 | … | … | 1.5 | 3.5 | Fe, 1.5 | 1800 | 1950 | 1850–2150 | 1, 2, 3, 4, 8 | |
| BNi-5 | 71 | … | 19 | … | 10 | … | 1975 | 2075 | 2100–2200 | 1, 2, 3, 4, 8 | |
| BNi-6 | 89 | … | … | … | … | P, 11 | 1610 | 1610 | 1700–1875 | 1, 2, 3, 4, 8 | |
| BNi-7 | 77 | … | 13 | … | … | P, 10 | 1630 | 1630 | 1700–1900 | 1, 2, 3, 4, 8 | |
| BNi-8 | 65.5 | 4.5 | … | … | 7 | Mn, 23 | 1800 | 1850 | 1850–2000 | 1, 2, 3, 4, 8 | |
| BCu-1 | … | 100 | … | … | … | … | 1980 | 1980 | 2000–2100 | 1, 2 | For joining various ferrous and nonferrous metals. They can also be used with various brazing processes. Avoid overheating the Cu-Zn alloys. Lap and butt joints are commonly used. |
| BCu-1a | … | 99 | … | … | … | Ot, 1 | 1980 | 1980 | 2000–2100 | 4 | |
| BCu-2 | … | 86.5 | … | … | … | O, 13.5 | 1980 | 1980 | 2000–2100 | 6 | |
| RBCuZn-A | … | 59 | … | … | … | Zn, 41 | 1630 | 1650 | 1670–1750 | 1, 2, 3 | |
| RBCuZn-C | … | 58 | … | … | 0.1 | Zn, 40; Fe, 0.7; Mn, 0.3; Sn, 1 | 1590 | 1630 | 1670–1750 | 2 | |
| RBCuZn-D | 10 | 48 | … | … | 0.2 | Zn, 42 | 1690 | 1715 | 1720–1800 | 1, 2, 3 | |
| BCuZn-E | … | 50 | … | … | … | Zn, 50 | 1595 | 1610 | 1610–1725 | 1, 2, 3, 4, 5 | |
| BCuZn-F | … | 50 | | … | … | Zn, 46.5; Sn, 3.5 | 1570 | 1580 | 1580–1700 | 1, 2, 3, 4, 5 | |
| BCuZn-G | … | 70 | … | … | … | Zn, 30 | 1680 | 1750 | 1750–1850 | 1, 2, 3, 4, 5 | |
| BCuZn-H | … | 80 | … | … | … | Zn, 20 | 1770 | 1830 | 1830–1950 | 1, 2, 3, 4, 5 | |
| BMg-1 | … | … | … | … | … | [d] | 830 | 1100 | 1120–1160 | 2, 3 | BMg-1 is used for joining AZ10A, K1A, and M1A magnesium-base metals. |
| BAu-1 | … | 63 | … | … | … | Au, 37 | 1815 | 1860 | 1860–2000 | 1, 2, 4 | For brazing of iron, nickel, and cobalt-base metals where resistance to oxidation or corrosion is required. Low rate of interaction with base metal facilitates use on thin base metals. Used with induction, furnace, or resistance heating in a reducing atmosphere or in a vacuum and with no flux. For other applications, a borax-boric acid flux is used. |
| BAu-2 | … | 20.5 | … | … | … | Au, 79.5 | 1635 | 1635 | 1635–1850 | 1, 2, 4 | |
| BAu-3 | 3 | 62.5 | … | … | … | Au, 34.5 | 1785 | 1885 | 1885–1995 | 1, 2, 4 | |
| BAu-4 | 18.5 | … | … | … | … | Au, 81.5 | 1740 | 1740 | 1740–1840 | 1, 2, 4 | |
| BAu-5 | 36 | … | … | … | … | Au, 30; Pd, 34 | 2075 | 2130 | 2130–2250 | 1, 2, 4 | |
| BAu-6 | 22 | … | … | … | … | Au, 70; Pd, 8 | 1845 | 1915 | 1915–2050 | 1, 2, 4 | |
| BCo-1 | 17 | … | … | … | 8 | Cr, 19; W, 4; B, 0.8; C, 0.4; Co, 59 | 2050 | 2100 | 2100–2250 | 1, 3, 4, 8 | Generally used for high temperature properties and compatability with cobalt-base metals. |

[a] Al, 9; Zn, 2; Mg, 89.

**Table 2. Guide to Selection of Brazing Filler Metals and Fluxes**

| Base Metals Being Brazed | Filler Metals Recommended[a] | AWS Brazing Flux Type No. | Effective Temperature Range, °F(°C) | Flux Ingredients | Flux Supplied As | Flux Method of Use[b] |
|---|---|---|---|---|---|---|
| All brazeable aluminum alloys | BA1Si | 1 | 700–1190 (371–643) | Chlorides, Fluorides | Powder | 1,2 3,4 |
| All brazeable magnesium alloys | BMg | 2 | 900–1200 (482–649) | Chloides, Fluorides | Powder | 3,4 |
| Alloys such as aluminum-bronze; aluminum- brass containing additions of aluminum of 0.5 percent or more | BCuZn, BCuP | 4[c] | 1050–1800 (566–982) | Chlorides, Fluorides, Borates, Wetting agent | Paste or Powder | 1,2,3 |
| Titanium and zirconium in base alloys | BAg | 6 | 700–1600 (371–871) | Chlorides, Fluorides, Wetting agent | Paste or Powder | 1,2,3 |
| Any other brazeable alloys not listed above | All brazing filler metals except BA1Si and BMg | 3 | 700–2000 (371–1093) | Boric acid, Borates, Fluorides, Fluorates, Wetting agent *Must contain fluorine compound* | Paste, Powder, or Liquid | 1,2,3 |
| | All brazing filler metals except BA1Si, BMg, and BAg 1 through BAg 7 | 5 | 1000–2200 (538–1204) | Borax, Boric acid Borates, Wetting agent *No fluorine in any form* | Paste, Powder, or Liquid | 1,2,3 |

[a] Abbreviations used in this column are as follows: B, brazing filler metal; Al, aluminum; Si, silicon; Mg, magnesium; Cu, copper; Zn, zinc; P, phosphorus; and Ag, silver.

[b] Explanation of numbering system used is as follows: 1—dry powder is sprinkled in joint region; 2—heated metal filler rod is dipped into powder or paste; 3—flux is mixed with alcohol, water, monochlorobenzene, etc., to form a paste or slurry; 4—flux is used molten in a bath.

[c] Types 1 and 3 fluxes, alone or in combination, may be used with some of these base metals also.

metals may be considered as grouped into the seven standard classifications shown in Table 1a and Table 1b. These are aluminum-silicon; copper-phosphorus; silver; nickel; copper and copper-zinc; magnesium; and precious metals.

To avoid confusion, Table 1a and Table 1b give the filler metal's solidus and liquidus temperatures, rather than the melting and flow points. The solidus is the highest temperature at which the metal is completely solid or, in other words, the temperature above which the melting starts. The liquidus is the lowest temperature at which the metal is completely liquid, that is, the temperature below which the solidification starts.

**Fluxes for Brazing.—**In order to obtain a sound joint the surfaces within and adjacent to the joint must be free from dirt, oil, and oxides or other foreign matter at the time of brazing. Cleaning may be achieved by chemical or mechanical means. Some of the mechanical means employed are filing, grinding, scratch brushing and machining. The chemical means include the use of trisodium phosphate, carbon tetrachloride, and trichloroethylene for removing oils and greases.

Fluxes are used mainly to prevent the formation of oxides and to remove any oxides on the base and filler metals. They also promote free flow of the filler metal during the course of the brazing operation.

They are made available in the following forms: powders; pastes or solutions; gases or vapors; and as coatings on the brazing rods.

In the powder form a flux can be sprinkled along the joint, provided that the joint has been preheated sufficiently to permit the sprinkled flux to adhere and not be blown away by the torch flame during brazing. A thin paste or solution is easily applied and when spread on evenly, with no bare spots, gives a very satisfactory flux coating. Gases or vapors are used in controlled atmosphere furnace brazing where large amounts of assemblies are mass-brazed. Coatings on the brazing rods protect the filler metal from becoming oxidized and eliminate the need for dipping rods into the flux, but it is recommended that flux be applied to the base metal since it may become oxidized in the heating operation. No matter which flux is used, it performs its task only if it is chemically active at the brazing temperature.

Chemical compounds incorporated into brazing fluxes include borates (sodium, potassium, lithium, etc.), fused borax, fluoborates (potassium, sodium, etc.), fluorides (sodium, potassium, lithium, etc.), chlorides (sodium, potassium, lithium), acids (boric, calcined boric acid), alkalies (potassium hydroxide, sodium hydroxide), wetting agents, and water (either as water of crystallization or as an addition for paste fluxes). Table 2 provides a guide which will aid in the selection of brazing fluxes that are available commercially.

**Methods of Steadying Work for Brazing.—**Pieces to be joined by brazing after being properly jointed may be held in a stable position by means of clamping devices, spot welds, or mechanical means such as crimping, staking, or spinning. When using clamping devices care must be taken to avoid the use of devices containing springs for applying pressure because springs tend to lose their properties under the influence of heat. Care must also be taken to be sure that the clamping devices are no larger than is necessary for strength considerations, because a large metal mass in contact with the base metal near the brazing area would tend to conduct heat away from the area too quickly and result in an inefficient braze. Thin sections that are to be brazed are frequently held together by spot welds. It must be remembered that these spot welds may interfere with the flow of the molten brazing alloy and appropriate steps must be taken to be sure that the alloy is placed where it can flow into all portions of the joint.

**Methods of Supplying Heat for Brazing.—**The methods of supplying heat for brazing form the basis of the classification of the different brazing methods and are as follows.

*Torch or Blowpipe Brazing:* Air-gas, oxy-acetylene, air-acetylene, and oxy-other fuel gas blowpipes are used to bring the areas of the joint and the filler material to the proper heat for brazing. The flames should generally be neutral or slightly reducing but in some instances some types of bronze welding require a slightly oxidizing flame.

*Dip Brazing:* Baths of molten alloy, covered with flux, or baths of molten salts are used for dip brazing. The parts to be brazed are first assembled, usually with the aid of jigs, and are dipped into the molten metal, then raised and allowed to drain. The molten alloy enters the joint by capillary action. When the salt bath is used, the filler metal is first inserted between the parts being joined, or, in the form of wire, is wrapped around the area of the joint. The brazing metal melts and flows into the joint, again by capillary action.

*Furnace Brazing:* Furnaces that are heated electrically or by gas or oil with auxiliary equipment that maintains a reducing or protective atmosphere and controlled temperatures therein are used for brazing large numbers of units, usually without flux.

*Resistance Brazing:* Heat is supplied by means of hot or incandescent electrodes. The heat is produced by the resistance of the electrodes to the flow of electricity and the filler metal is frequently used as an insert between the parts being joined.

*Induction Brazing:* A coil carrying an electric current heats the joining parts. Eddy current losses of the induced electric current are dissipated in the form of heat raising the temperature of the work to a point higher than the melting point of the brazing alloy. This method is both quick and clean.

*Vacuum Furnace Brazing:* Cold-wall vacuum furnaces, with electrical-resistance radiant heaters, and pumping systems capable of evacuating a conditioned chamber to moderate vacuum (about 0.01 micron) in 5 minutes are recommended for vacuum brazing. Metals commonly brazed in vacuum are the stainless steels, heat-resistant alloys, titanium, refractory metals, and aluminum. Fluxes and filler metals containing alloying elements with low boiling points or high vapor pressure are not used.

**Brazing Symbol Application.—**ANSI/AWS A2.4-2012 symbols for brazing are also used for welding with the exception of the symbol for a scarf joint (see the diagram at the top of page 1575, and the symbol for a scarf joint in the table *Basic Weld Symbols* on page 1619, for applications of brazing symbols). The second, third and fourth figures from the top of the next page show how joint clearances are indicated. If no special joint preparation is required, only the arrow is used with the brazing process indicated in the tail.

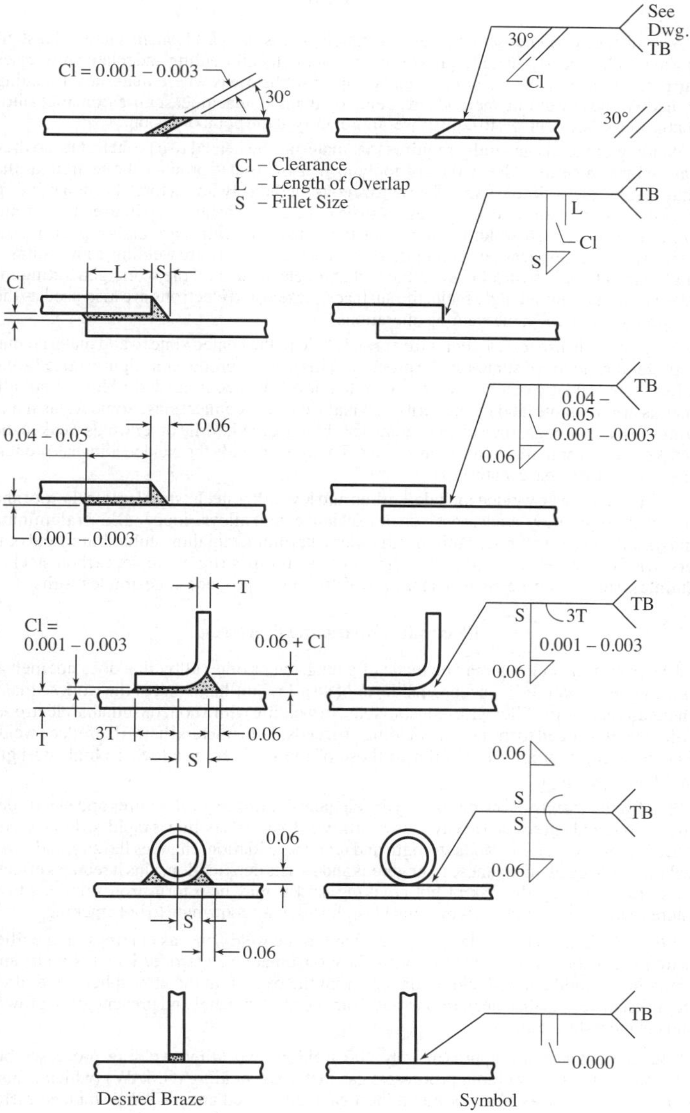


Typical Applications of Standard Brazing Symbols

## WELDING

A number of variables set welding and related processes apart from mechanical fastening methods used to join material. However, practically all welding and related processes currently in use involve heat input and/or upset in the areas where materials are being joined. Modern welding methods may employ manual, machine, semi-automatic, automatic, or robotic controls to deliver welding energy at the point of fusion.

Welding of metals generally requires that materials be heated to a molten state so they can be fused together. The method of applying heat or upset to produce the desired results may involve just a simple flame. The autogenous method—where a torch burns a mixture of (usually) acetylene and oxygen gases to heat the components—is still used for certain work. However, most welding operations today use an electric arc and/or pressure, or they involve the application of complex electric waveforms. In arc welding, a low-voltage, high-current arc is struck between the end of an electrode and the work, generating intense heat that immediately melts the surfaces. Lasers and electrons are also used as the heating medium for some welding operations.

In the welding process, a filler wire or rod is held in the heated zone to add material that replaces the metal consumed by the process. (This usually produces a slightly raised area, which subsequently can be dressed down to a level surface if needed.) Most, if not all, metals are weldable, and the majority can make capable connections. Some welds make joints that are stronger than the base materials being joined, while other welds weaken the workpiece's overall mechanical properties. Though typically the welded joint has properties similar to the base materials.

Metals that can be welded include: carbon and low-alloy steels, stainless steels, tool and die steels, high-alloy steels, cast iron, nickel and cobalt alloys, copper alloys, aluminum alloys, magnesium alloys, titanium, zirconium, hafnium, tantalum, columbium, and others. Dissimilar welds—joining two different metal types (for example, carbon steel to stainless steel)—can be made but may prove difficult or produce undesirable results.

### Electrodes, Fluxes, and Processes

Electrodes for welding may be made of a tungsten or other alloy that does not melt at welding temperatures (nonconsumable) or of an alloy similar to that of the work so that it melts and acts as the filler wire (consumable). In welding with a nonconsumable electrode, filler metal is added to the pool as welding proceeds. Filler metals that will produce welds having strength properties similar to those of the work are used where high-strength welds are specified.

Briefly, the effects of the main alloying elements in welding filler wires and electrodes are: carbon adds strength but may cause brittle weld metal if cooling is rapid, so low-carbon wire is preferred; silicon adds strength and reduces oxidation, changes fluidity, and gives a flatter weld bead; manganese strengthens and assists deoxidation, plus it reduces effects of sulfur, lowering the risk of hot cracking; sulfur may help form iron sulfide, which increases the risk of hot cracking; and phosphorus, may contribute to hot cracking.

*Fluxes* in (usually) granular form are added to the weld zone, as coatings on the filler wire or as a core in the tube that forms the (consumable) electrode. The flux melts and flows in the weld zone, shielding the arc from the oxygen in the atmosphere, and often contains materials that clean impurities from the molten metal and prevent grain growth during recrystallization.

**Processes.—**There are approximately 100 welding and allied welding processes but the four manual arc welding processes: gas metal arc welding (GMAW) (which is also commonly known as MIG for metal inert gas), flux-cored arc (FCAW), shielded metal arc (SMAW), gas tungsten arc welding (GTAW), account for over 90 percent of the arc welding used in production, fabrication, structural, and repair applications. FCAW and

SMAW use fluxes to shield the arc and FCAW uses fluxes and gases to protect the weld from oxygen and nitrogen. GMAW and GTAW use mixtures of gases to protect the weld.

There are two groups of weld types, groove and fillet, which are self-explanatory. Each type of weld may be made with the work at any angle from horizontal (flat) to inverted (overhead). In a vertical orientation, the electrode tip may move down the groove or fillet (vertical down), or up (vertical up). In any weld other than flat, skill is needed to prevent the molten metal falling from the weld area.

Because of the many variables, such as material to be welded and its thickness, equipment, fluxes, gases, electrodes, degree of skill, and strength requirements for the finished welds, it is not practicable to set up a complete list of welding recommendations that would have general validity. Instead, examples embracing a wide range of typical applications, and assuming common practices, are presented here for the most-used welding processes. The recommendations given are intended as a guide to finding the best approach to any welding job, and are to be varied by the user to fit the conditions encountered in the specific welding situation.

**Heat Input and Weld Distortion.—**When metals are joined in a welding process, a great amount of heat is generated. As this heat melts, coalesces, and joins two or more members at the point of connection, it often changes the position and dimensions of the welded structure to some extent.

When a weld joint is fit-up for welding, the gap between the parts to be welded is filled in with filler metal and/or base metal. As the metal in the gap cools, shrinkage occurs, drawing adjacent surfaces together and strengthening the connection, but also causing movement.

In jobs where part alignment and dimensional tolerances must be maintained, the effect of such weld distortion should be taken into consideration. Tactics to limit or compensate for these factors include heat input control, using welding fixtures, post-weld stress relief, and finish machining and processing operations.

## Gas Metal Arc Welding (GMAW)

The two most cost-effective manual arc welding processes are GMAW and FCAW. These two welding processes are used with more than 50 percent of the arc welding consumable electrodes purchased. Gas metal arc welding modes extend from short-circuit welding, where the consumable electrode wire is melted into the molten pool in a rapid succession of short circuits during which the arc is extinguished, to pulsed and regular spray transfer, where a stream of fine drops and vaporized weld metal is propelled across the continuous arc gap by electromagnetic forces in the arc.

GMAW is the most-used welding process and the two most common GMAW low-carbon steel electrodes used for production welding in North America are the E70S-3 and E70S-6 from the ANSI/AWS Standard A5 series of specifications for arc welding. The E70S-3 contains manganese and silicon as deoxidants and is mainly used for welding low-carbon steels, using argon mixtures as shielding gases. The wire used in E70S-6 electrodes has more silicon than wire used for E70S-3 electrodes, and is preferred where straight $CO_2$ or argon mixes are used as the shielding gas or if the metal to be welded is contaminated. The deoxidizing properties of the E70S-6 electrode also may be beneficial for high-current, deep-penetration welds, and welds in which higher than normal impact-strength properties are required.

E80S-D2 wire contains more manganese and silicon, plus 0.5 percent molybdenum for welding such steels as AISI 4130, and steels for high-temperature service. The argon + $CO_2$ mixture is preferred to exert the influence of argon's inertness over the oxidizing action of $CO_2$. E70S-2 electrodes contain aluminum, titanium, and zirconium to provide greater deoxidation action and are valuable for welding contaminated steel plate.

When the GMAW welding process is used for galvanized steels, minute welding cracks may be caused by the reaction of the zinc coating on the work with silicon in the electrode. Galvanized steel should be welded with an electrode having the lowest possible silicon content such as the E70S-3. For welding low-carbon and low-alloy steels with conventional argon mixture shielding gases, there is little difference between the E70S-3 and E70S-6.

**Electrode Diameters.—**One of the most important welding decisions is selecting the optimum GMAW electrode diameter, which should be based on the material thickness, as shown for carbon and stainless steels in Table 1, the compatibility of the electrode current requirements with the material thickness, the mode of weld metal transfer, and the deposition rate potential shown in Table 2. The two most popular GMAW electrode sizes are 0.035 inch (1.0 mm) and 0.045 inch (1.2 mm). Diameters of electrodes used for GMAW exert a strong influence on cost of welding. Table 2 also shows how the weld deposition rate varies in short-circuit and spray transfer modes in welding carbon and stainless steels.

**Table 1. GMAW Electrode Sizes for Welding Carbon and Stainless Steels**

| Material Thickness | Electrode Diameter | | | |
|---|---|---|---|---|
| | 0.030 inch (0.8 mm) | 0.035 inch (1.0 mm) | 0.045 inch (1.2 mm) | 0.062 inch (1.6 mm) |
| 25 to 21 gage (0.020 to 0.032 inch) | yes | … | … | … |
| 20 gage to ¼ inch (0.036 to 0.25 inch) | … | yes | … | … |
| ³⁄₁₆ to ⁷⁄₁₆ inch flat and horizontal | … | … | yes | … |
| ½ inch and up | … | … | … | yes |

The table is based on suitability of the electrode size to mode of weld transfer, material thickness, and cost-effectiveness. If a smaller electrode size is selected, the lower deposition rates could increase welding costs by 20 to 60 percent.

**Table 2. Typical Maximum GMAW Deposition Rates for Carbon and Stainless Steels, Constant-Voltage 450-Amp Power Source and Standard Wire Feeder**

| Weld transfer mode | Electrode Diameter | | | |
|---|---|---|---|---|
| | 0.030 inch (0.8 mm) | 0.035 inch (1.0 mm) | 0.045 inch (1.2 mm) | 0.062 inch (1.6 mm) |
| Short circuit | 5 lb/h (2.3 kg/h) | 7 lb/h (3.2 kg/h) | 9 lb/h (4 kg/h) | … |
| Spray transfer | 9 lb/h (4 kg/h) | 11 lb/h (5 kg/h) | 19 lb/h (8.6 kg/h) | 21 lb/h (9.5 kg/h) |

For the lowest-cost welds with GMAW electrodes larger than 0.030 inch in diameter, the power source should provide a minimum of 350 amps. The compatibility of the optimum current range of the 0.035-inch (1.0-mm) electrode and its deposition potential make it the first choice for welding of 20 gage to ¼ inch (0.88 to 6.4 mm) thicknesses. For welding thinner sheet metals of 25 to 21 gage, the optimum electrode diameter is 0.030 inch (0.8 mm). The 0.045-inch (1.2-mm) electrode is the most practical choice for spray transfer applications on materials over ¼ inch (6.4 mm) thick and thicker.

As an example, when welding ¼-in. (6.4-mm) thick steel, with 100 percent arc-on time and a labor cost of $15/h, the deposition rate with a 0.035-in. (0.9-mm) electrode is approximately 11 lb/h (5 kg/h). The labor cost per lb at $15/h ÷ 11 lb/h = $1.36/lb ($3.00/kg). If an electrode of 0.045-in. (1.2-mm) diameter is used for the same application, the deposition rate is 16 lb/h (7.2 kg/h) and at a $15/h labor rate, the cost of weld metal deposited = $15/h ÷ 16 lb/h = $0.93/lb ($2.00/kg). The 0.045-in. diameter electrode would also cost less per pound than a smaller wire, and the weld time with the 0.045-in. electrode would be reduced, so less shielding gas also would be consumed.

**GMAW Welding of Sheet Steel.—**In GMAW, the short-circuit transfer mode is used to weld carbon steel, low-alloy steel, and stainless steel sheet of 24 gage (0.023 in., or 0.6 mm) to 11 gage (0.12 in., or 3 mm). The most common gage sizes welded with short-circuit transfer are 20 gage to 11 gage (0.88 to 3 mm) and the best GMAW electrode for these thin, sheet metal gages is the 0.035-in. (1-mm) diameter electrode. The short-circuit current requirements for these operations are typically 50 to 200 amps with voltages in the range of 14 to 22 volts. The optimum short-circuit voltage for the majority of applications is 16 to 18 volts.

**Shielding Gases for Welding Carbon and Low-Alloy Steels.—**With more than 40 GMAW gas mixtures available for welding carbon steels, low-alloy steels, and stainless steels, selection is often confusing. Reactive oxygen and carbon dioxide ($CO_2$) are added to argon to stabilize the arc and add energy to the weld. $CO_2$ can provide more energy to the weld than oxygen. As the $CO_2$ content in a shielding gas mixture is increased to certain levels, the voltage requirements are increased. Argon + oxygen mixtures will require lower voltages than mixtures containing argon with 10 to 25 percent $CO_2$. Helium may also be added to argon if increased weld energy is required.

**Shielding Gases for Short-Circuit Welding of Carbon Steels.—**GMAW short-circuit transfer (SCT) is used mainly for welding thin metals of less than 10 gage, and gaps. With the SCT mode of weld metal transfer, the arc short circuits many times each second. The numerous short circuits switch the arc energy on and off. The short circuits and low current cause the transferred weld to freeze rapidly. Short-circuit transfer on carbon steel gage metals thicker than $\frac{1}{16}$ in. (1.6 mm) requires a shielding gas that will provide substantial weld energy. For these applications, argon with 15–25 percent $CO_2$ is recommended.

If short-circuit transfer is used on metals thinner than 18 gage (0.047 in., 1.2 mm), melt-through and distortion often occur. Melt-through and distortion can be reduced on very thin-gage carbon and low-alloy steels by using a shielding gas that provides less weld energy than argon + 15 to 25 percent $CO_2$ mixes. Argon + oxygen mixtures can utilize lower voltages to sustain the arc. Argon mixed with 2 to 5 percent oxygen is a practical mixture for thin carbon steel of less than 16 gage, where there is sensitivity to heat.

**Shielding Gases for Spray Transfer Welding of Carbon Steels.—**With GMAW spray transfer, all traditional argon gas mixtures will provide spatter-free spray weld transfer, depending on the electrode diameter and welding parameters used. The electrode diameter and the electrode current density influence the formation of the weld metal to be transferred. For example, with a 0.035-inch (0.9 mm) diameter electrode using a mixture containing argon 75 + $CO_2$ 25 percent, a small globular weld droplet is formed on the end of the electrode tip in the conventional spray transfer parameter range. With the same gas mixture, a 0.045-inch (1.14-mm) diameter electrode, and current above 330 amps, the globular formation disappears and the metal transfers in the spray mode.

Spatter potential stemming from shielding gas, with 0.035 inch (0.90 mm) and smaller diameter electrodes can be controlled by reducing the $CO_2$ content in the argon mixture to less than 21 percent. Each different shielding gas will primarily influence the open arc spray transfer mode by variations in the weld energy provided through the welding voltage requirements.

Gas selection in spray transfer must be given careful consideration. In welding of clean cold-rolled carbon steel or low-alloy steel less than $\frac{3}{8}$ inch (9.5 mm) thick, the energy potential of the arc is less important than it is for welding of steels thicker than $\frac{1}{2}$ inch (13 mm) or steels with mill scale. The energy level of the arc is also a key factor in welding steels for which higher than normal impact properties are specified.

A simple, practical multipurpose gas mixture for carbon and low-alloy steels is argon + 15 to 20 percent $CO_2$, and a mixture of argon + 17 percent $CO_2$ would be ideal. This two-part argon/$CO_2$ mixture provides higher weld energy than two-component argon + $CO_2$ mixtures having less than 10 percent $CO_2$, argon + oxygen mixtures, or argon + $CO_2$ + oxygen tri-component mixtures. The argon + 17 percent $CO_2$ mixture will provide an arc slightly less sensitive to mill scale than the other mixtures mentioned.

The argon + 17 percent $CO_2$ mixture also has practical benefits in that it provides sufficient weld energy for all GMAW short-circuit and spray transfer applications with cylinder or bulk gases. The argon + 17 percent $CO_2$ mixture may also be used for all-position FCAW electrodes in welding carbon steels, low-alloy steels, and stainless steels.

**Shielding Gases for GMAW Welding of Stainless Steels.—**The major problems encountered when using GMAW on stainless steels of thinner than 14 gage include controlling potential melt-through, controlling distortion, and black oxidation on the weld surface. These three welding problems have a common denominator, which is heat. The key to welding thin stainless steel is to minimize the potential heat when welding, by appropriate choice of gas mixture.

A popular gas mixture that is often recommended for GMAW welding of thin-gage stainless steel is the three-part helium gas mixture containing helium 90 + argon 7.5 + $CO_2$ 2.5 percent. In contrast to gas mixtures without helium, the helium tri-mixture requires the use of higher voltages to sustain the arc, which adds unnecessary heat to the heat-sensitive thin-gage welds.

A practical and lower-cost alternative for GMAW short-circuit transfer on stainless steels is an argon mixture with 2 to 4 percent $CO_2$. The argon + $CO_2$ mixture allows use of lower voltages than is practical with argon/helium mixtures, and the lower voltages resulting from the argon + $CO_2$ mixture will help to reduce distortion and oxidation, and decrease the melt-through potential. The mixture that works with short-circuit transfer is also a logical practical choice for spray transfer welding of stainless steel because it is less oxidizing than argon/oxygen mixtures. Table 3 provides practical gas mixture recommendations for specific applications.

**Table 3. Shielding Gases for Welding Carbon Steels and Stainless Steels**

| Application | Gas mixtures | | | | | |
|---|---|---|---|---|---|---|
| | Argon + Oxygen | Argon + $CO_2$ + Oxygen | Argon + 2–4% $CO_2$ | Argon + 6–10% $CO_2$ | Argon + 13–20% $CO_2$ | Argon + 25% $CO_2$ |
| Short-circuit melt-through problems; less than 20 gage | 1 | 1 | 1 | 1 | 2 | 3 |
| Short-circuit 18 to 11 gage | … | … | … | … | 1 | 1 |
| Spray if mill scale or surface problems; carbon steels | … | … | … | … | 1 | 2 |
| Spray if low energy required; carbon steel | 1 | 1 | 1 | 1 | … | … |
| Spray, best impact strengths, lowest porosity; carbon steels | … | … | … | … | 1 | … |
| Best single gas mixture for carbon steels | … | … | … | … | 1 | … |
| Short-circuit; stainless steels | … | … | 1 | … | … | … |
| Spray; stainless steels | 2 | … | 1 | … | … | … |
| Best single gas mixture for stainless and duplex steels | … | … | 1 | … | … | … |

Preferred choice of shielding gas is 1, followed by 2 and 3.

For GMAW spray transfer welding of stainless steels thicker than 11 gage, the traditional GMAW shielding gas has been argon 98 + oxygen 2 percent. The argon + oxygen mixture provides excellent, stable, spray transfer, but the oxygen promotes oxidation, leaving the weld with a black surface. To reduce the oxidation, the 2 percent oxygen can be replaced with the less oxidizing 2–4 percent $CO_2$.

**Shielding Gases for GMAW Welding of Aluminum.—**For GMAW welding of aluminum, helium is added to argon to provide additional weld energy, increasing penetration width, and reducing porosity potential. A gas mixture that has worked well in practice and can be used on the majority of aluminum applications is argon + 25 to 35 percent helium. Mixtures with higher helium content, of 50 to 90 percent, require voltages and flow rates that may be excessive for many established aluminum applications.

**Welding Controls.—**The two primary controls for welding with GMAW are the electrode wire feed control on the wire feeder and the voltage control on the power source. As shown in Fig. 1, these controls typically consist of switches and knobs but do not have the scales, seen enlarged at the upper left, that indicate combinations of wire feed rate, wire gage, volts, and amps. These scales have been added here to allow clearer explanation of the functioning of the wire feed control.

The typical wire feed unit provides maximum feed rates of 600 to 800 in/min (15.2–20.3 m/min). The scale surrounding the setting knob on a wire feed control unit usually has only 10 unnumbered graduations, somewhat like the hour markers on a clock face. On most machines, each of these graduations represents an adjustment of the feed rate of approximately 70 in/min (1.8 m/min). For each increase in the wire feed rate of 70 in/min (1.8 m/min), depending on the voltage, the welding current increases by approximately 20 to 40 amps, depending on the wire diameter and wire feed positions.

In Fig. 1, a black sector has been drawn in on the wire feed rate adjustment knob to indicate the range of wire feed rates usable with the gas mixture and the electrode diameter (gage) specified. The wire feed and voltage settings shown are for welding thin-gage carbon, low-alloy, or stainless steels with a 0.030 or 0.035 inch (0.76 or 0.89 mm) diameter electrode. The left edge of the sector on the wire feed knob is set to the eight o'clock position, corresponding to 70 in/min (1.8 m/min). The optimum voltage for this wire feed rate is 15. If a setting is too low, the knob is turned to the second (nine o'clock) or third (ten o'clock) position to increase the current. The voltage typically increases or decreases by 1 volt for each graduation of the wire feed quadrant.

The short-circuit transfer current range of 50 to 200 amps corresponds to a wire feed rate of 70 to 420 in/min (1.8–10.7 m/min), and is typically found between the eight and one o'clock positions on the scale, as indicated by the black sector on the knob in Fig. 1.

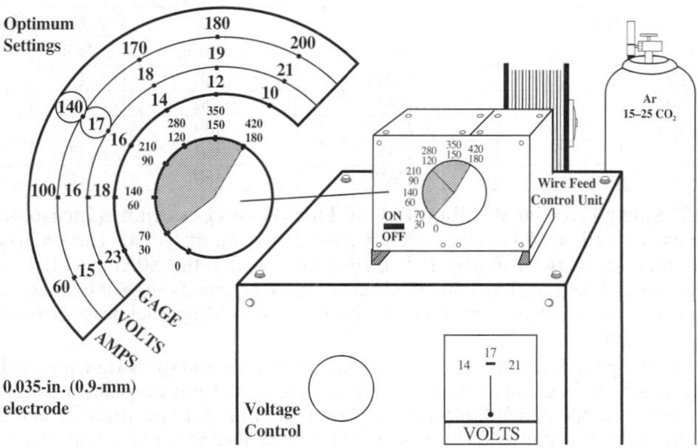


Fig. 1. Wire Feed Settings for Short-Circuit Welding of Carbon, Low-Alloy, and Stainless Steel Sheet

Diagrammatic quadrants have been added at the left in Fig. 1, to show the material thickness, voltage, and current that correspond to the setting of the wire feed rate adjustment knob. Optimum settings are easily made for short-circuit welding of sheet metals. When using a 0.030-in. (0.8-mm) or 0.035-in. (1-mm) diameter GMAW electrode, for instance, to weld 16-gage carbon or stainless steel with a conventional 200-to 450-amp constant-voltage power source and wire feeder, the wire feed control is set to the ten o'clock position for a feed rate of 210 in/min (5.3 m/min). With digital wire feed units, the short-circuit current range is typically between 100 and 400 in/min (2.54 and 10.16 m/min), so a good starting point is to set the wire feeder at 210 in/min (5.3 m/min). The welding voltage is set to 17.

Many welders set their parameters by an established mark on the equipment or by the sound of the arc as the weld is being made. The sound of the arc, influenced by the optimum current and voltage set, should be a consistent, smooth, crackling noise. If the SCT sound is harsh, the voltage should be increased slightly. If the sound is soft, the voltage should be decreased in volt increments until the sound becomes a smooth crackle. For welding metals thicker than 16 gage but less than 10 gage, the wire feed control should be moved to the eleven o'clock position (280 in/min, or 7.1 m/min), and the voltage reset to 18.

Welding of thicknesses less than 16 gage should be started with the wire feed control at the nine o'clock position (140 in/min, or 3.6 m/min) and the voltage control set to 16. The parameters discussed above apply when using argon mixtures containing 15 to 25 percent $CO_2$.

**GMAW Spray Transfer.—**In the spray transfer mode, spatter is often caused by the voltage being set so low that the electrode runs into the weld, resulting in expulsion of molten metal from the weld pool. GMAW spray transfer is normally used for welding carbon, low-alloy, and stainless steels of a minimum thickness of $\frac{1}{8}$ in. (3.2 mm).

In Table 4, typical deposition rates with a 0.045-in. (1.2-mm) carbon steel electrode are compared with rates for larger carbon steel GMAW and flux-cored electrodes. These welds are typically carried out in the flat and horizontal positions. The practical GMAW electrode diameters commonly used for spray transfer are 0.035-in. (1-mm), 0.045-in. (1.2-mm), and 0.062-in. (1.6-mm) diameter. The most cost-effective GMAW electrode that also has the greatest range of applications on metals over $\frac{3}{16}$ in. thick is the 0.045-in. (1.2-mm) diameter size.

**Table 4. Typical Deposition Rates for Carbon Steel Welding Electrodes**

| Electrode Diameter | | | | Deposit Rates | |
|---|---|---|---|---|---|
| inch | mm | Electrode Type | Amperage | lb/h | kg/h |
| 0.035 | 1.0 | GMAW | 350 | 11 | 5 |
| 0.045 | 1.2 | GMAW | 380 | 13 | 6 |
| 0.062 | 1.6 | GMAW | 400 | 14 | 6.4 |
| $\frac{1}{16}$ | 1.6 | FCAW | 350 | 15 | 7 |
| $\frac{3}{32}$ | 2.4 | FCAW | 450 | 16 | 7.3 |

**GMAW Spray Transfer Welding of Metal Thicknesses Less than $\frac{1}{4}$ Inch (6.4 mm).—** The most versatile GMAW electrode for a welding shop that welds carbon, low-alloy, and stainless steels from 20 gage to $\frac{1}{4}$ in. (6.4 mm) thick is the 0.035 in. (1.0 mm) diameter electrode. The traditional practical spray transfer current range of between 200 and 350 amps for the 0.035 in. electrode is well suited for welding thicknesses from 10 gage to $\frac{1}{4}$ in. (6.4 mm).

The correct parameters for a 0.035 in. (1 mm) electrode and spray transfer welding are found on the wire feed unit between the one and five o'clock positions, or, on a digital wire feeder, between 420 and 700 in/min (10.7 and 17.8 m/min). In the drawing at the left in Fig. 2, the spray transfer wire feed range is shaded. When the wire feed rate has been set, the voltage should be fine-tuned so that the electrode wire tip is just touching the weld and a smooth crackling sound without spatter is produced.

An optimum single spray transfer mode current setting for a 0.035-in. (1-mm) diameter electrode for most welding applications is approximately 280 amps with the wire feed set at the three o'clock position for 560 in/min. Manual or high-speed mechanized welds on material of 10 gage to $\frac{1}{4}$ in. thick can be made at the three o'clock wire feed position with only an adjustment for voltage, which should be set initially at 31 volts, when using an argon + $CO_2$ mixture.

**GMAW Spray Transfer for Metal Thicknesses $\frac{1}{4}$ Inch (6.4 mm) and Up.—**The 0.45-in. (1.2-mm) diameter is the most cost-effective GMAW electrode for spray transfer welding of carbon, low-alloy, and stainless steels $\frac{1}{4}$ in. and thicker. A $\frac{7}{16}$-in. (11.2-mm) single-pass, no-weave, fillet weld can be produced with this electrode. If larger single-pass welds are required, use of flux-cored electrodes should be considered.

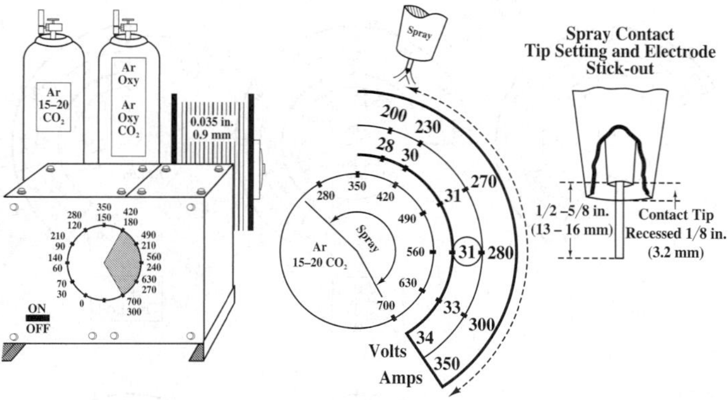


Fig. 2. GMAW Spray Transfer Parameters with 0.035-inch (0.9-mm) Diameter Electrodes

A 400-amp power source is a practical cost-effective unit to use with the 0.045-in. diameter electrode. Globular spray transfer, obtained at the ten o'clock position on the wire feed adjustment knob, starts at current levels of approximately 230 amps and requires a wire feed rate of approximately 210 in/min (90 mm/s). Most spray applications are carried out in the higher-energy, deeper-penetrating 270- to 380-amp range, or between twelve and two o'clock wire feed positions giving 350 to 490 in/min (150 to 210 mm/s). In this range, in which there is minimum weld spatter, the weld deposits are in the form of minute droplets and vaporized weld metal.

The quadrants at the top in Fig. 3 show some typical settings for feed rate, voltage, and current, with different shielding gases. An ideal starting point with a 0.045-in. (1.2-mm) diameter electrode is to set the wire feed rate knob at the one o'clock position, or 420 in/min, at which rate the current drawn, depending on the power source used, should be about 320 to 350 amps. The best starting voltage for the 0.045-in. (1.2-mm) electrode is 30 volts. The arc length should then be set as indicated in Fig. 4. With current over 400 amps at 560 in/min, the 0.045-in. diameter electrode may produce a turbulent weld puddle and a digging arc, which can lead to lack of fusion, porosity, and cracks.

**GMAW Spray Transfer with 0.062-Inch (1.6-mm) Diameter Electrodes.—**Electrode wire of 0.062-in. (1.6-mm) diameter is the largest size in normal use and is often chosen for its high deposition rates. Due to the high-current requirements for the spray transfer mode, use of these thicker electrodes is generally restricted to metal thicknesses of $\frac{1}{2}$ in. (13 mm) and thicker. The high-current requirement reduces ease of welding. This electrode size is suitable for mechanized welding in which fillet welds greater than $\frac{3}{8}$ in. (9.6 mm) are required.

As indicated at the lower left in Fig. 3, the current range for 0.062-in. (1.6-mm) electrodes is narrow and most welds are made in the range of 360 to 420 amps, or between the ten and eleven o'clock positions on the wire feed control unit for 210 to 280 in/min (90 to 120 mm/s). The quadrants at the lower center and lower right in Fig. 3 show deposition rates in lb/h and kg/h for 0.045-in. (1.2-mm) and 0.062-in. (1.6-mm) diameter electrodes.

Some optimum settings for GMAW welding with a mixture of argon + 15 to 20 percent $CO_2$ gases are given in Table 5.

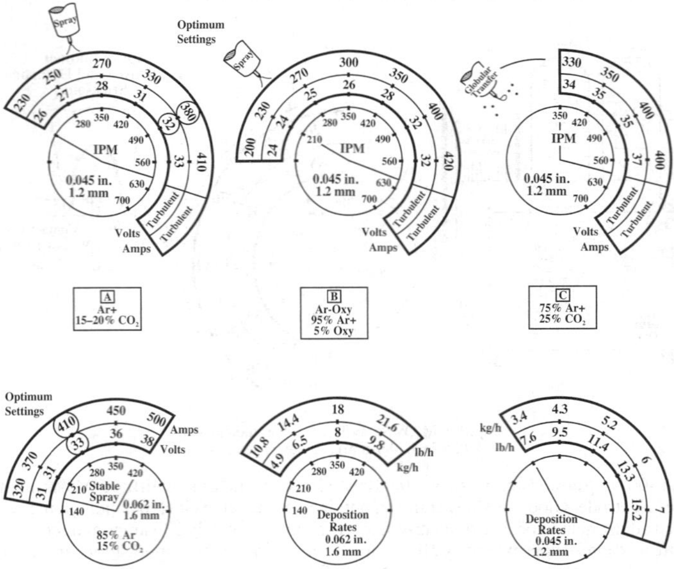


Fig. 3. GMAW Spray Transfer Parameters for Various Electrodes and Gases

## Table 5. Optimum Settings for GMAW with Argon + 15–20 Percent $CO_2$

| Diameters | | Mode | Wire Feed Rates | | Amps | Volts |
|---|---|---|---|---|---|---|
| inch | mm | | in/min | m/min | | |
| 0.035 | 1.0 | short circuit | 210 | 5.3 | 140 | 17 |
| | | spray transfer | 560 | 14.2 | 280 | 29–30 |
| 0.045 | 1.2 | short circuit | 210 | 5.3 | 190 | 18 |
| | | spray transfer | 420 | 10.7 | 380 | 30–31 |
| 0.052 | 1.4 | spray transfer | 280 | 7.1 | 370 | 31–32 |
| 0.062 | 1.6 | spray transfer | 280 | 7.1 | 410 | 31–32 |

*Note:* If argon + oxygen gas mixtures are used, voltage should be lowered by 1 to 4 volts for the spray transfer mode. The faster the weld travel speed, the lower the voltage required.

**Spray Transfer Voltage.—**The usual setting for spray transfer welding with commonly used electrode diameters is between 25 and 35 volts (see Fig. 4A). To set the optimum voltage for GMAW spray transfer, set the voltage initially so that it is too high, usually between

30 and 35 volts. With excess voltage, there should be a visible gap between the tip of the electrode and the weld, and the arc sound should be free from crackle. With the sequence shown in Fig. 4, the voltage should now be reduced until a consistent smooth crackle sound is produced. If the voltage is lowered too much, the electrode will run into the weld, making a harsh crackling sound, and the resulting weld expulsion will cause spatter.

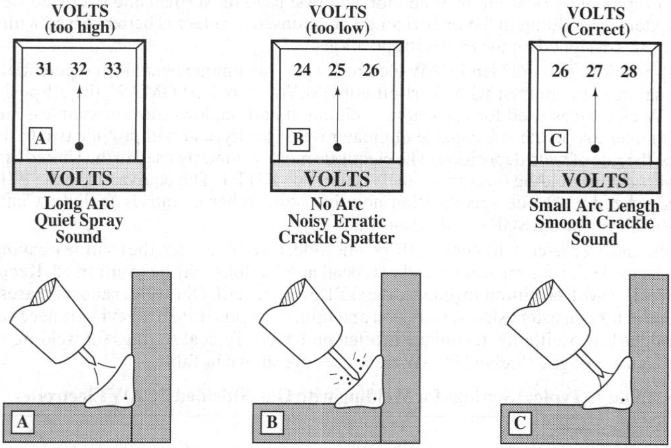


Fig. 4. Setting Optimum Voltage for GMAW Spray Transfer Welding

## Flux-Cored Arc Welding

FCAW welding offers unique benefits for specific applications, but flux-cored consumable electrodes cost more than the solid electrodes used in gas metal arc welding, so users need to be aware of FCAW benefits and disadvantages compared with those of GMAW welding. Generally, flux-cored electrodes designed for use without a shielding gas are intended for welding outdoors. Most indoor FCAW welding is done with gas-shielded FCAW welding electrode wire. Some Standards for gas-shielded FCAW electrodes for various countries are listed in Table 6.

**Table 6. Standards for Gas-Shielded, Flux-Cored Welding Electrodes**

| Steel Type | Country | Standard |
|---|---|---|
| Low-Carbon Steels | United States | AWS A5.20 |
| | Canada | CSA W48.5 |
| | Japan | JIS Z3313 |
| | Germany | DIN 8559 |
| Low-Alloy Steels | United States | AWS A5.29 |
| | Canada | CSA W48.3-M |
| | United Kingdom | BS 639-2492 |
| Stainless Steels | United States | AWS A5.22 |

**All-Position, Gas-Shielded Electrodes.**—The term "all-position" does not necessarily mean that these electrodes are the best choice for all positions. Also, flux-cored electrodes may meet all standard specifications, but there will inevitably be subtle differences in weld transfer characteristics and recommended current, voltage, and other settings between electrodes made by different manufacturers. The chemistry and slag of the electrodes developed for welding in the flat and horizontal positions (E70T-X) typically provide superior results when they are used for flat and horizontal applications where the surface conditions of the plate are suspect or large, deep-penetrating welds are required. All-position electrodes are intended for, and best used in, vertical and overhead welds. For extensive welding in flat or horizontal positions, the welder is better served with the electrodes designed for these specific positions.

All-position, gas-shielded FCAW electrodes provide unique benefits and potential for cost savings. In contrast with short-circuit GMAW, or pulsed GMAW, the all-position FCAW electrodes used for vertical up welding of carbon, low-alloy, or stainless steels are simpler to operate, are capable of greater weld quality, and will provide two to three times the rate of weld deposition. The electrode most commonly used in the United States for vertical up welding on carbon steels is the type E71T-1. The equivalent to the E71T-1 standardized electrode specification now in use in other countries include: Canada, E4801T9; Germany, SGR1; and Japan, YFW 24.

If the end user selects the correct all-position electrode diameter, the routine weaving of the electrode during vertical up and overhead applications may be minimized. Keeping the weld weave to a minimum reduces the skill level needed by the welder and increases the potential for consistent side-wall fusion and minimum porosity. If weaving is necessary, a straight-line oscillation technique is often preferred. Typical settings for welding with various sizes of gas-shielded FCAW electrodes are shown in Table 7.

**Table 7. Typical Settings for Welding with Gas-Shielded FCAW Electrodes**

| Electrode Diameter inch | (mm) | Vertical Up Welds | | Flat and Horizontal Welds | |
|---|---|---|---|---|---|
| 0.035 | (1) | Feed rate | 450 ipm (11.4 m/min) | Feed rate | 630 ipm (16.0 m/min) |
| | | Current | 165 amps | Current | 250 amps |
| | | Voltage | 28 volts | Voltage | 30 volts |
| 0.045 | (1.2) | Feed rate | 350 ipm (8.9 m/min) | Feed rate | 560 ipm (14.2 m/min) |
| | | Current | 200 amps | Current | 280 amps |
| | | Voltage | 25 volts | Voltage | 26 Volts |
| 0.052 | (1.4) | Feed rate | 240 ipm (6.1 m/min) | Feed rate | 520 ipm (13.2 m/min) |
| | | Current | 200 amps | Current | 300 amps |
| | | Voltage | 25 volts | Voltage | 30 volts |
| 0.062 | (1.6) | Feed rate | 210 ipm (5.3 m/min) | Feed rate | 350 ipm (8.9 m/min) |
| | | Current | 240 amps | Current | 340 amps |
| | | Voltage | 25 volts | Voltage | 29 volts |
| $\frac{3}{32}$ | (2.4) | … | | Feed rate | 210 ipm (5.3 m/min) |
| | | | | Current | 460 amps |
| | | | | Voltage | 32 volts |

**Material Condition and Weld Requirements.—**Practical considerations for selecting a gas-shielded, flux-cored electrode depend on the material condition and weld requirements. FCAW electrodes are beneficial if the surface of the material to be welded is contaminated with mill scale, rust, oil, or paint; the fillet weld size is to be over $\frac{3}{8}$ in. (9.6 mm) wide (a GMAW single-pass fillet weld with an electrode size of 0.045 in. is typically $\frac{3}{8}$ in. wide); the weld is vertical up, or overhead; the required impact strengths and other mechanical properties are above normal levels; crack resistance needs to be high; and increased penetration is required.

**Selecting an FCAW Electrode.—**Selection of FCAW electrodes is simplified by matching the characteristics of flux-cored types with the material and weld requirements listed above. Once the correct electrode type is selected, the next step is to choose the optimum size. In selecting an all-position, flux-cored electrode for vertical up or overhead welding, the steel thickness is the prime consideration. Selecting the optimum electrode diameter allows the high current capability of the electrode to be fully used to attain maximum deposition rates and allows use of the highest penetrating current without concern for excessive heat-related problems during welding. When used in the optimum current range, the deposited filler metal matches the required amount of filler metal for the specific size of the weld, determined by the plate thickness.

The following suggestions and recommendations are made for FCAW welding of carbon, low-alloy, and stainless steels having flat surfaces. For vertical up welds on steels of thicknesses from 1⁄8 to 3⁄16 in. (3.2 to 4.8 mm) and for vertical up welds on pipe in the thickness range of 1⁄4 to 1⁄2 in. (6.4 to 13 mm), consider the 0.035-in. (1.0-mm) diameter E71T-1 electrode. For vertical up welds on steels in the range of 1⁄4 to 3⁄8 in. (6.4 to 9.6 mm) thickness, consider the 0.045-in. (1.2-mm) diameter E71T-1 electrode or, in nonheat-sensitive applications, the 0.062-in. (1.6-mm) diameter E71T-1 electrode. For vertical up welds on steels of over 3⁄8 in. (9.6 mm) thickness, consider the 0.062-in. (1.6-mm) diameter E71T-1 electrodes for optimum deposition rates. For flat and horizontal welds on steels of 3⁄8 to 3⁄4 in. (9.6 to 19 mm) thickness, consider the 1⁄16-in. (1.6-mm) diameter E70T-X electrodes. For flat and horizontal welds on steels over 3⁄4 in. (19 mm) in thickness, consider the 3⁄32-in. (2.4-mm) E70T-X electrodes.

**FCAW Welding of Low-Carbon Steels.—**Low-carbon steel is usually called carbon steel or mild steel. The most-used FCAW electrode for welding carbon steels in the flat or horizontal welding positions is the type E70T-1, which is suited to welding of reasonably clean steel using single-pass or multi-pass welds. Type E70T-2 has added deoxidizers and is suited to surfaces with mill scale or other contamination. This type is used when no more than two layers of weld are to be applied. Type E70T-5 is used for single-pass or multi-pass welds where superior impact properties or improved crack resistance are required. The E70T-X electrodes typically range in size from 0.045 to 3⁄32 in. (1.2 to 2.4 mm) in diameter. Type E71T-1 all-position electrodes are available in diameters of 0.035 in. (1 mm) to 0.062 in. (1.6 mm). With the FCAW process, multi-pass welds are defined as a condition where three or more weld passes are placed on top of each other.

**Settings for Gas-Shielded, All-Position, FCAW Electrodes.—**The optimum setting range (volts and amps) for vertical up welding with all-position FCAW electrodes is rather narrow. The welder usually obtains the greatest degree of weld puddle control at the recommended low to medium current settings. The electrode manufacturers' recommended current range for an E71T-1 electrode of 0.045-in. (1.2-mm) diameter for vertical up welding may be approximately 130 to 250 amps. Using the 0.045-in. (1.2-mm) diameter electrode at 250 amps for a vertical up weld in 1⁄4-in. (6.4-mm) thick steel, the welder may find that after 3 to 4 inches (75–100 mm) of weld, the weld heat built up in the steel being welded is sufficient to make the weld puddle fluidity increasingly difficult to control. Reducing the current to 160–220 amps will make it possible to maintain control over the weld puddle.

A typical optimum setting for a vertical up weld with an E71T-1, 0.035-in. (1.2-mm) diameter, all-position electrode is as follows. First, set the wire feed rate. If the wire feeder maximum rate is 650 to 750 in/min (16.5–19.0 m/min), the setting mark on the adjustment knob should be set between the one and two o'clock positions on the dial to obtain a feed rate of 450 in/min (11 m/min). If the wire feeder has a digital readout, the rate setting should be the same. At the 450-in/min setting, the welding current with the 0.035-in. all-position electrode should be optimized at between 160 and 170 amps. The welding voltage should be set at 27 to 28 volts with the electrode tip just touching the weld. If there is a gap causing the weld puddle to become too fluid, the voltage should be lowered. If the electrode runs into the weld, causing spatter, the voltage needs to be increased.

With the above conditions, welding steel of 1⁄8 to 1⁄4 in. thickness will deposit 5 to 7 lb/h (2.2 to 3 kg/h). The 0.035-in. electrode is also ideal for welding steel pipe with wall thicknesses of less than 1⁄2 in. (13 mm). The thickness of the pipe after bevelling controls the size of electrode to be used. The 0.035-in. (6.4-mm) electrode can produce a 1⁄4 -in. (6.4-mm) vertical up fillet weld on such a pipe without weaving.

**Contact Tip Recess.—**The dimension labeled contact tip recess in Fig. 2, and indicated as ⅛ in.(3 mm), should be about ½ in. (13 mm) for a minimum electrode extension of ¾ in. (19 mm), for FCAW welding. This dimension is critical for obtaining high-quality welds with all-position electrodes because they have a fast-freezing slag and operate with low to medium current and voltage. If the recess dimension is less than the optimum, the voltage may be lower than the minimum recommended, and if the settings are less than the minimum, the fast-freezing slag may solidify too rapidly, causing excess porosity or "worm tracks" on the weld surface.

The recommended length of electrode extension for all-position FCAW, E71T-1 electrodes is ¾ to 1 in. (19 to 25 mm). The size of this extension not only affects the minimum required parameters, but a long electrode extension also ensures preheating of the electrode and allows lower current to be used. Preheating the electrode is further beneficial as it reduces moisture on the electrode surface, and in the electrode flux. When a change is made from the GMAW to the FCAW process, welders must be aware of the influence on weld quality of the electrode extension in the FCAW process.

**Porosity and Worm Tracks.—**As mentioned above, porosity and worm tracks typically result from a combination of incorrect electrode extension, incorrect welding settings, humidity, electrode moisture, refill scale, rust, paint, oils, or poor welding technique. Where humidity levels are high, potential for porosity and worm tracks increases. The FCAW process is less sensitive to mill scale than the GMAW spray transfer mode but mill scale will often cause excess weld porosity. The best way to avoid the effects of mill scale, rust, oil, and surface contaminants is to grind the area to be welded.

Another way to reduce porosity is to keep weaving to a minimum. If the correct size flux-cored electrode is used, weaving can be kept to a minimum for most flux-cored applications. The forehand technique produces the best weld bead surface on fillet weld beads up to ¾ in. (19 mm) steel thickness in the flat and horizontal weld positions. On larger single-pass fillet welds, the backhand technique is beneficial because the voltage directed at the weld provides additional weld puddle control to the fluid welds. The backhand technique used for flat and horizontal welds produces a more convex weld bead, reduces potential for porosity, and increases penetration.

If porosity or worm tracks occur, the prime solution is in weld practices that increase heat at the weld, but the following remedies can also be tried. Grind clean the surface to be welded; use recommended electrode extensions; increase current (wire feed rate) decrease voltage; use the backhand welding technique; slow down travel speed, consider use of a different electrode formulation containing increased deoxidizers, avoid weaving; change from argon + $CO_2$ mixture to straight $CO_2$; and provide a protective cover to keep the electrode spool clean and dry.

**Welding with 0.045-Inch (1.2-mm) Diameter All-Position Electrodes.—**Fig. 5 shows wire feed settings for welding of steel with 0.045-in. (1.2-mm) diameter, E71T-1 all-position electrodes using a mixture of argon + 15 to 25 percent $CO_2$ as the shielding gas, and an electrode extension of ¾ in. (18 mm). Parameters for vertical up welding, shown at the left in Fig. 5, include setting the wire feed rate at the twelve o'clock position, or about 350 in/min, using 200 to 190 amps, and setting the voltage between 24 and 25 volts. Optimum parameters for flat welding, shown at the right in Fig. 5 include setting the wire feed rate at three o'clock position, or 560 in/min (240 mm/s), and 270 amps at 25 to 27 volts.

**Welding with 0.052-Inch (1.3-mm) Diameter All-Position Electrodes.—**Settings for vertical up and flat welding with all-position E71T-1 electrodes of 0.052-in. (1.3-mm) diameter are seen at the left in Fig. 6. These electrodes are suited to welding steel having thicknesses of ¼ in. (6 mm) and thicker. For vertical up welding, the wire feed rate is set between the ten and eleven o'clock positions, or 250 in/min (106 mm/s), with about 200 amps at 25 volts. Flat welding with these electrodes is best done with the wire feed rate set between the two and three o'clock positions, or 490 to 560 in/min (207 to 237 mm/s), giving approximately 300 amps at 28 volts.

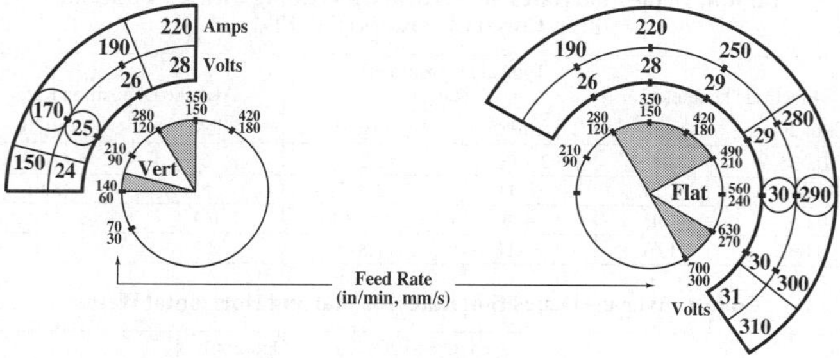


Fig. 5. Wire Feed and Voltage Settings for FCAW Welding with 0.045-in. (1.2-mm) Diameter E71T-1 Electrodes, with Optimum Settings Circled.

Settings for all-position E71T-1 electrodes of 0.062-in. (1.6-mm) diameter, shown at the right in Fig. 6, for vertical up welding are just before the ten o'clock position, or 190 in/min, giving 230 to 240 amps with voltage adjusted to 24–25 volts. For flat welding, the wire feed is set to the twelve o'clock position, giving 340–350 amps with a voltage of 29–30 volts.

**High-Deposition, All-Position Electrodes.**—Vertical up weld deposition rates of 10 to 14 lb/h can be achieved with the E71T-1, 0.062-in. (1.6-mm) and 0.045-in. (1.2-mm) flux-cored electrodes. Settings are shown in Fig. 6 for E71T-1, FCAW electrodes of 0.052- and 0.062-in. (1.4- and 1.6-mm) diameter. These electrodes are suited to applications in which the steel thickness is ¼ in. and thicker, and are the most cost-effective diameter for all-position welds on carbon and stainless steels of ¼ in. (6.35 mm) thickness and thicker. In contrast, vertical up welds, using GMAW or SMAW, may deposit an average of 2 to 4 lb/h (1 to 2 kg/h). Deposition rates are based on welding 60 minutes of each hour. Pulsed GMAW provides deposition rates of 3 to 6 lb/h (1.3–2.7 kg/h). Average rates for vertical up welding with all-position, flux-cored electrodes are shown in Table 8.

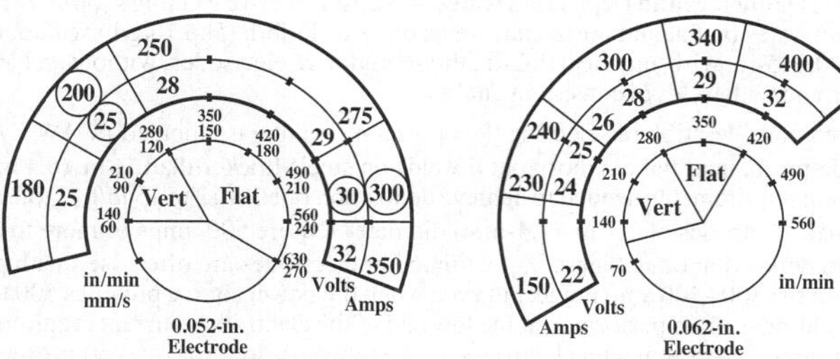


Fig. 6. Wire Feed and Voltage Settings for Vertical Up Welding with 0.052- and 0.062-in. Diameter Electrodes

The average deposition rates in Table 9 are to be expected with FCAW electrodes available today. Special electrodes are also available that are specifically designed to provide higher deposition rates. A typical manual welder, welding on steel of ¼ to ⅜ in. thickness for 30 minutes of each hour with an all-position flux-cored 0.062-in. (1.6-mm) or 0.045-in. (1.2-mm) diameter electrode would deposit about 4–5 lb/h.

**Table 8. Deposition Rates for Vertical Up Welding with All-Position, Flux-Cored Electrodes (ET71T-1)**

| Electrode Diameter | | Typical Deposition Rate Range | | Average Deposition Rate | |
|---|---|---|---|---|---|
| inch | (mm) | lb/h | (kg/h) | lb/h | (kg/h) |
| 0.035 | (1) | 2.7–6.5 | (1.2–3) | 5 | (2.3) |
| 0.045 | (1.2) | 5–11 | (2.3–5) | 8 | (3.6) |
| 0.052 | (1.4) | 4–8 | (1.8–3.6) | 6.5 | (3) |
| 0.062 | (1.6) | 4–11 | (1.8–5) | 8.5 | (4) |

**Table 9. Average Deposition Rates for Flat and Horizontal Welds**

| Process | Electrode Size | | Cost-Effective Current Range | Optimum Current | Deposition Rate | |
|---|---|---|---|---|---|---|
| | inch | (mm) | (amps) | (amps) | lb/h | (kg/h) |
| GMAW spray transfer | 0.035 | (1) | 250–350 | 285 | 9 | (4) |
| | 0.045 | (1.2) | 300–400 | 385 | 13 | (5.9) |
| | 0.052 | (1.4) | 350–470 | 410 | 11 | (5) |
| | 0.062 | (1.6) | 375–500 | 450 | 17 | (7.7) |
| FCAW | 0.045 | (1.2) | 225–310 | 300 | 14 | (6.4) |
| | 0.052 | (1.4) | 260–350 | 310 | 15 | (6.8) |
| | 0.062 | (1.6) | 300–400 | 340 | 15 | (6.8) |
| | $\frac{3}{32}$ | (2.4) | 380–560 | 460 | 17 | (7.7) |

The average deposition rates of pulsed GMAW and FCAW for vertical up welds are similar for applications where the steel thickness is $\frac{1}{8}$ in. (3.2 mm) or less. On steels thicker than $\frac{1}{8}$ in., where the current may be increased, and larger-diameter all-position FCAW electrodes may be used, deposition rates will be much greater than with pulsed GMAW. Compared with GMAW electrodes for pulsed welding, FCAW all-position electrodes require less costly equipment, less welding skill, and have potential for increased weld fusion with less porosity than with GMAW pulsed techniques.

**Electrode Diameters and Deposition Rates.—**A cost-effective welding shop can achieve deposition rates on flat and horizontal welds of 12 to 15 lb/h (5 to 7 kg/h) with both the GMAW 0.045-in. wire and the 0.062-in. flux-cored wire electrodes, without welder discomfort, and with welds of consistent quality.

The first consideration in selecting the optimum size of gas-shielded FCAW E70T-X electrode for manual flat and horizontal welds on steels thicker than $\frac{1}{4}$ in. (6.4 mm) is the current requirements needed to achieve deposition rates of 12 to 15 lb/h (5 to 7 kg/h). Large-size electrodes of $\frac{3}{32}$-in. (2.4-mm) diameter require 500 amps or more to attain optimum deposition rates. These $\frac{3}{32}$-in. diameter electrodes are often used with power sources in the 300–400 amp range, but even when the power source provides 500 to 600 amps, welding is often performed at the low end of the electrode's current requirements. With the large, $\frac{3}{32}$-in. diameter electrodes, welder appeal is low, smoke is often excessive, and deposition rates are often only comparable with smaller, easier-to-operate FCAW electrodes.

Typical deposition rates for flat and horizontal welds with various electrode sizes and weld settings are shown in Table 9. In connection with this table, it may be noted that high deposition rates in welding steel plate thicker than $\frac{1}{4}$ in. require use of currents above the minimum shown for the various sizes of electrodes. The optimum current requirements for the most popular electrode sizes indicate that a 450-amp power source is the most suitable for welding steel of more than $\frac{1}{4}$ in. thickness. The two most cost-effective and versatile consumables for thin and thick steel sections are the 0.045 in. for GMAW and the 0.062 in. for FCAW electrodes.

The approach to a welding application is critical to achievement of optimum weld quality at minimum cost. In many applications, minimal consideration is given to weld costs. Half of every man-hour of welding in many shops could be saved with selection of the correct electrode diameter used with optimum parameter settings. A practical point that is often overlooked in selection of FCAW electrodes is that the larger the electrode diameter, the more restricted is the application thickness range. Large FCAW electrodes such as the $^3/_{32}$ in. (2.4 mm) are neither suitable nor cost-effective for the common steel thickness range of $^1/_4$ to $^1/_2$ in. (6.4 to 13 mm). Smaller FCAW electrodes such as the $^1/_{16}$-in. (1.6-mm) diameter, are suitable for both thin and thick applications. A $^1/_6$-in. diameter FCAW electrode used in the 300- to 350-amp range provides excellent deposition rates with superior welder appeal and negligible smoke.

Large-diameter $^3/_{32}$-in. (2.4-mm) electrodes are popular for manual applications. However, from a practical point of view, this electrode size is often better suited to mechanized high-current welding in which the high currents required for optimum deposition rates may be safely used without health risks. Use of an electrode at 60 to 80 percent of its welding current capability indicates that the correct diameter electrode has been selected for the application. When an electrode is used at its maximum-current capability, it shows that the next size larger electrode should be preferred, and when the low end of the current capability is in use, the electrode selected is typically too large.

**The 0.062-Inch (1.6-mm) Diameter, E70T-X Electrode.**—The 0.062-in. (1.6-mm) diameter FCAW electrode is the most practical size of its type and will provide excellent deposition rate potential with a practical current range and the broadest application range. Settings for the common $^1/_{16}$-in. diameter E70T-1 electrode are shown in Fig. 7. With the GMAW process, a $^3/_8$- to $^7/_{16}$-in. (9.6- to 11-mm) minimum-weave fillet weld is typically the maximum size that can be made in a single pass. The 0.062-in. ($^1/_{16}$-in, 1.6-mm) FCAW electrode can easily produce a $^3/_4$-in. (19-mm), nonweave, single-pass fillet weld. This size of electrode is also a practical choice for welding steel of $^1/_4$ in. (6.4 mm) or greater thickness. From a cost perspective, FCAW consumable electrodes should be used whenever the GMAW process is not suitable.

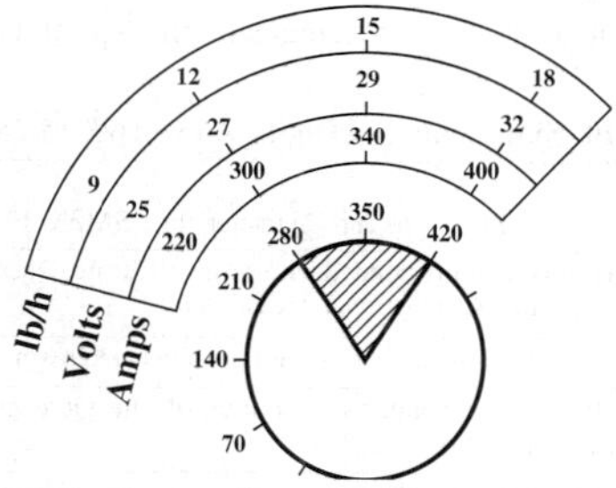


Fig. 7. Settings for $^1/_{16}$-in. (1.6-mm) FCAW, E70T-X Electrodes

The average deposition efficiency of a flux-cored electrode is 85 percent, which means that for every 100 lb or kg of electrode material used, 85 lb or kg ends up as weld material. In contrast, the average deposition efficiency of a GMAW electrode used with argon mixtures and correct equipment settings should be a minimum of 99 percent.

**Shielding Gases and FCAW Electrodes.—**The E70T-X flux-cored electrodes that are recommended for flat and horizontal welds use $CO_2$ gas shielding. Because of new OSHA welding smoke restrictions, manufacturers of FCAW electrodes now provide E70T-X consumable electrodes that can be used with less reactive argon + $CO_2$ mixtures to reduce smoke levels. The fast-freezing slag, all-position, E71T-1 flux-cored electrodes can use either $CO_2$ or argon + 15 to 25 percent $CO_2$ mixtures for welding carbon, low-alloy, or stainless steels. The argon + $CO_2$ mixture is often selected because it provides the highest energy from a reactive gas mixture with a compatible voltage range.

Instead of $CO_2$, welders often prefer the arc characteristics, lower smoke levels, and lower voltage requirements of the argon + $CO_2$ mixtures for all-position welding. However, if lower reactive argon mixtures such as argon + oxygen, or argon with less than 13 percent $CO_2$, are used, the weld voltage requirements and the arc plasma energy are reduced, adding to the possibility of changing the mechanical properties significantly, increasing the porosity, and raising the potential for forming worm tracks.

## Shielded Metal Arc Welding

With the shielded metal arc welding (SMAW) process, commonly known as stick welding, it is most important to select an electrode that is suited to the application. For welding austenitic stainless or high-alloy steels, the electrode is first selected to match the mechanical and chemical requirements of the metal to be welded. Secondary requirements such as the welding position, penetration potential, deposition capabilities, and ease of slag removal are then considered. Many electrodes for SMAW welding of low- to medium-carbon steels have unique characteristics making them the most suitable and cost-effective for a specific welding application.

In interpreting the ANSI/AWS Standard specification code for SMAW electrodes shown in Table 10, for example, E60XX, the E stands for a low-carbon steel, metal arc welding electrode. The next two digits, such as 60 or 70, indicate the approximate tensile strength of the weld deposit in thousands of psi.

Of the last two digits, the next (second to last) number indicates the positions in which the electrode can be used: 1 = indicates the electrode is suitable for use in flat, horizontal, vertical, and overhead positions; 2 = the electrode can be used in the flat position for all welding types, and in the horizontal position for fillet welds only; 4 = the electrode is suitable for use in the flat, horizontal, vertical-down, and overhead positions. Note that number "3" is no longer used as a position designation.

The final digit, combined with the above, indicates the type of flux coating, as shown in Table 10.

**Table 10. Significance of Digits,** *ANSI/AWS A5.18-1979*

| Third and Fourth Digits | Flux Type and Characteristics, SMAW Electrodes |
|---|---|
| 10 | High-cellulose coating bonded with sodium silicate. Deep penetration, energetic spray-type arc. All-positional, DCEP[a] only |
| 11 | Similar to 10 but bonded with potassium silicate to permit use with AC or DCEP |
| 12 | High-rutile coating, bonded with sodium silicate. Quiet arc, medium penetration, all-positional, AC or DCEN |
| 13 | Similar to 12 but bonded with sodium silicate and with easily ionized materials added. Gives steady arc on low voltage. All-positional, AC or DCEN |
| 14 | Similar to 12 with addition of medium amount of iron powder. All-positional, AC or DC |
| 15 | Lime-fluoride coating (basic low-hydrogen) bonded with sodium silicate. All-positional. For welding high-tensile steels. DCEP only |

**Table 10.** *(Continued)* **Significance of Digits,** *ANSI/AWS A5.18-1979*

| Third and Fourth Digits | Flux Type and Characteristics, SMAW Electrodes |
|---|---|
| 16 | Similar to 15 but bonded with potassium silicate. AC or DCEP |
| 18 | Similar to 15 but with addition of iron powder. All-positional, AC or DC |
| 20 | High iron-oxide coating bonded with sodium silicate. Flat or HV positions. Good X-ray quality. AC or DC |
| 24 | Heavy coating containing high percentage of iron powder for fast deposition rates. Flat and horizontal positions only. AC or DC |
| 27 | Very heavy coating with ingredients similar to 20 and high percentage of iron powder. Flat or horizontal positions. High X-ray quality. AC or DC |
| 28 | Similar to 18 but heavier coating and suited for use in flat and HV positions only. AC or DC |
| 30 | High-iron-oxide-type coating but produces less fluid slag than 20. For use in flat position only (primarily narrow-groove butt welds). Good X-ray quality. AC or DC |

[a] DC = direct current, AC = alternating current, EP = electrode positive, EN = electrode negative.

British Standard BS 639:1986 (Withdrawn) defines requirements for covered carbon- and carbon-manganese-steel electrodes for manual metal arc welding, depositing weld metal having a tensile strength of not more than 650 N/mm$^2$. Appendix A of this standard lists minimum mandatory and optional characteristics of these electrodes. The extensive classifications provide for electrodes to be rated for strength, toughness, and covering (STC), with codes such as E 51 5 4 BB [160 3 0 H]. In this series, E indicates that the electrode is covered and is for manual metal arc welding. The next two digits (51) indicate the strength (tensile, yield, and elongation) properties. The next digits (5 and 4) give the temperatures at which minimum average impact strengths of 28J (at −40° C) and 47J (at −30° C), using Charpy V-notch test specimens, are required. The next group is for the covering and the BB stands for basic, high efficiency. Other letters are B for basic; C for cellulosic; R for rutile, RR for rutile, heavy coated; and S for other.

The letters in brackets are optional, and the first group indicates the efficiency, which is the ratio of the mass of weld metal to the mass of nominal diameter core wire consumed with the largest diameter electrode, rounded up to the nearest multiple of 10. The next digit (3) is the maker's advice for the position(s) to be used. Codes for this category include 1, all positions; 2, all positions except vertical down; 3, flat, and for fillet welds, horizontal/vertical; 4, flat; 5, flat, vertical/down; and for fillet welds, horizontal/vertical; and 9, other. The digit at (0), which may have numbers from 0 to 9, shows the polarity, and the minimum open-circuit voltage to be used for that electrode. A 0 here indicates that the electrode is not suited for use with AC. The (H) is included only for hydrogen-controlled electrodes that will deposit not more than 15 ml of diffusible hydrogen for each 100 g of deposited weld metal. The ISO Standard is combined into BS EN ISO 2560:2020. Low-alloy steel electrodes and chromium and chromium nickel steel electrodes are covered in BS EN ISO 3580:2017 and BS EN ISO 3581:2016.

The most common electrodes used for the SMAW process are the AWS types E60XX and E70XX. SMAW welding electrode Standards are issued by the American Welding Society (AWS), the British Standards Institute (BS), Canada (CSA), Germany (DIN), and Japan (JIS) and are shown in Table 11.

**AWS E60XX Electrodes.—**Characteristics of the E60XX electrodes influence the weld position capability, ease of slag removal, penetration potential, weld travel speed capability, and weld deposition rates. These electrodes are designed for welding low-carbon steels and they provide welds with typical tensile strength in the range of 58,000–65,000 lb$_f$/in$^2$ (400–448 MPa), depending on the specific electrode utilized, the base metal condition and chemistry, and the amount of weld dilution. In selecting an electrode for SMAW welding, knowing that the mechanical and chemical requirements have been matched, it is necessary to choose electrodes with characteristics that influence the features required, as shown in Table 11.

**Table 11. Characteristics of SMAW Welding Electrodes Made to Standards of Various Countries**

| Standard | Description |
|---|---|
| AWS E6010<br>CSA E41010<br>BS E4343C10<br>DIN E4343C4<br>JIS D3410 | Designed for welding pipe and general structures. Excellent for all-position and vertical down welding. Slag is light and easy to remove. Deep, penetrating arc. Low deposition rates. Polarity DC + (electrode positive). |
| AWS E6011<br>CSA E41011<br>BS E4343C13<br>DIN E4343C4<br>JIS D4311 | Similar to E6010 but modified to allow use of AC. Excellent for welding sheet metal corner joints vertical down. Polarity AC or DC + (electrode positive). |
| AWS 6012<br>CSA E41012<br>BS E4332R12<br>DIN E4332R(C)<br>JIS D4313 | Designed for welding sheet metal and light structural steels. Medium penetration suitable for gaps or where minimum weld dilution is needed. Ideal for flat, horizontal, or vertical down welding. Will weld faster than the E6010-11 electrode. Polarity AC or DC—(electrode negative). |
| AWS E6013<br>CSA E41013<br>BS E4332R21<br>DIN E4332R3<br>JIS D4313 | Excellent AC or DC—performance. All-position. Shallow penetration. Good choice for low open-circuit welding machines. AC or DC both excellent on thin structural applications. Polarity AC or DC (DC both polarities). |
| AWS 6027<br>CSA 41027<br>BSE4343A13035<br>DIN 4343AR11<br>JIS D4327 | Iron powder is added to the flux to provide higher deposition rates. Ideal for multipass groove and fillet welding in flat and horizontal positions. Polarity AC or DC (both polarities). |

Table 12 shows approximate current requirements for AWS E60XX electrodes for welding sheet metal carbon steels. The current ranges specified vary slightly with different electrode manufacturers. For welding sheet metal start at the low end of the given current requirements with electrodes of $3/16$-in (5-mm) diameter or smaller. For metals thicker than 10 gage (0.134 in.), start in the center of the current range, then adjust to suit. A high DC current may result in arc blow, and improved results may then be obtained with AC.

**Table 12. Diameters of AWS E6010/E6011 SMAW Electrodes for Welding Low-Carbon Steel Sheet Metal**

| SWG of Sheet Metal to be Welded | Electrode Diameter | | Current Starting Level |
|---|---|---|---|
| | in. | (mm) | (amps) |
| 18 | $3/32$ | (2.5) | 45–60 |
| 16–14 | $1/8$ | (3.2) | 80–110 |
| 12 | $5/32$ | (4) | 125–135 |
| 10 | $3/16$ | (5) | 135–150 |

For welding thicker materials, a good starting setting is in the middle of the current range shown in Table 12. In welding material less than $1/4$-in. (6.4-mm) thick, vertically, with an E6010 electrode, try a $1/8$-in. (3.2-mm) electrode at 90 to 100 amps. For welding thicknesses between $3/16$ and $5/16$ in. (5 and 8 mm) with the E6010 electrode, vertically, try the $5/16$-in. (8-mm) diameter electrode at 100 to 125 amps. For thicknesses of $3/8$ to 1 in. (9.5 to 25 mm), try a $3/16$-in. (5-mm) diameter electrode at 155 to 165 amps.

Recommended current ranges, shown in Table 13 for the various sizes of AWS E60XX electrodes most commonly used for welding carbon steel, will give optimum results with SMAW electrodes. An ideal starting point for the current setting for any SMAW electrode diameter is in the middle of the range. The current ranges shown are average values taken from literature of electrode manufacturers in three different countries.

**Table 13. Current Ranges for AWS E60XX SMAW Electrodes**

| Electrode Diameter in. | (mm) | E6010/E6011 (amps) | E6012 (amps) | E6013 (amps) | E6027 (amps) |
|---|---|---|---|---|---|
| 1⁄16 | (1.6) | … | 25–50 | 20–40 | … |
| 3⁄32 | (2.5) | 40–75 | 40–100 | 50–100 | … |
| 1⁄8 | (3.2) | 75–130 | 85–140 | 75–135 | 120–180 |
| 5⁄32 | (4) | 90–170 | 115–185 | 110–185 | 155–245 |
| 3⁄16 | (5) | 135–220 | 145–240 | 150–235 | 200–300 |
| 1⁄4 | (6.4) | 205–325 | 250–390 | 240–340 | 300–410 |
| 5⁄16 | (8) | 260–420 | 290–480 | 310–425 | 370–480 |

**AWS E70XX Electrodes.—**Information on the most commonly used AWS E70XX electrodes is given in Table 15. For critical welding applications, low-hydrogen electrodes are typically used. It is most important that manufacturers' instructions regarding storage requirements for keeping low-hydrogen electrodes free from moisture are followed. Current ranges for welding low-carbon steel sheet metal with E70XX electrodes of diameters from 3⁄32 to 3⁄16 in. (2.5 to 5 mm) are shown in Table 14. The optimum starting point is in the middle of the current range indicated.

**Table 14. Current Ranges for SMAW E70XX Welding Electrodes**

| Electrode Diameter in. | (mm) | E7014 (amps) | E7018 (amps) | E7024 (amps) |
|---|---|---|---|---|
| 3⁄32 | (2.5) | 75–120 | 70–105 | 85–135 |
| 1⁄8 | (3.2) | 110–155 | 110–160 | 130–180 |
| 5⁄32 | (4) | 145–210 | 150–215 | 175–240 |
| 3⁄16 | (5) | 190–280 | 180–275 | 230–315 |
| 7⁄32 | (5.5) | 255–335 | 255–350 | 280–370 |
| 1⁄4 | (6.4) | 330–415 | 295–360 | 325–450 |
| 5⁄16 | (8) | 380–490 | 370–480 | 390–530 |

In using AWS E7018 electrodes for vertical up welding of plate thicknesses of 3⁄16 to 5⁄16 in. (5 to 8 mm), try a 1⁄8-in. (3.2-mm) diameter electrode. For vertical up welding of thicknesses greater than 5⁄16 in. (8 mm), try a 5⁄32-in. (4-mm) electrode. With AWS E7018 electrodes, to make horizontal fillet welds in plate thicknesses of 10 swg (0.135 in., 3.4 mm), try a 3⁄16-in. (5-mm) electrode, for 1⁄4-in. (6.4-mm) plate, try the 7⁄32-in. (5.5-mm) electrode, and for steel plate thicker than 1⁄4 in., try the 1⁄4-in. (6.4-mm) diameter electrode.

**Table 15. Characteristics of AWS Electrodes for SMAW Welding**

| Standard | Description |
|---|---|
| AWS E7014<br>CSA E48014<br>BS E5121RR11011<br>DIN E5121RR8<br>JIS D4313 | An iron-powder, all-position electrode for shallow penetration. Excellent for vertical down and applications with poor fit. Similar to AWS E6012-E6013 with added iron powder. For welding mild and low-alloy steels. Polarity AC or DC, + or –. |

**Table 15.** *(Continued)* **Characteristics of AWS Electrodes for SMAW Welding**

| Standard | Description |
|---|---|
| AWS E7018<br>CSA E48018<br>BS E5154B11026(H)<br>DIN E5154B(R)10<br>JIS D5016 | An iron-powder, low-hydrogen, all-position electrode. Excellent for rigid, highly stressed structures of low- to medium-carbon steel. Can also be used for welding mild and high-strength steels, high-carbon steels, and alloy steels. Polarity AC or DC + reverse polarity. |
| AWS 7024<br>CSA E48024<br>BS E5122RR13034<br>DIN E5122RR11<br>JIS D4324 | An iron-powder electrode with low hydrogen, usable in all positions. Excellent for high-amperage, large, fillet welds in flat and horizontal positions. Polarity AC or DC, + or −. |
| AWS E7028<br>CSA E48028<br>BS E514B12036(H)<br>DIN E5143B(R)12<br>JIS D5026 | An iron-powder, low-hydrogen electrode suitable for horizontal fillets and grooved flat position welding. Higher deposition rates. More cost-effective than the AWS E7018 electrode. Polarity AC or DC + reverse polarity. |

The E7024 electrode is suggested for horizontal fillet welds. For 10-gage (0.135-in, 3.4-mm) material, try the $\frac{1}{8}$-in. (3.2-mm) diameter electrode; for above 10-gage to $\frac{3}{16}$-in. (5-mm) material, try the $\frac{5}{32}$-in. (4-mm) diameter electrode. For plate of $\frac{3}{16}$- to $\frac{1}{4}$-in. thickness, try the $\frac{3}{16}$-in. size, and for plate thicker than $\frac{1}{4}$ in., try the $\frac{1}{4}$-in. (6.4-mm) electrode.

## Gas Tungsten Arc Welding

Often called TIG (for tungsten inert gas) welding, gas tungsten arc welding (GTAW) uses a nonconsumable tungsten electrode with a gas shield, and was, until the development of plasma arc welding (PAW), the most versatile of all common manual welding processes. Plasma arc welding is a modified GTAW process. In contrast to GTAW, plasma arc welding has less sensitivity to arc length variations, superior low-current arc stability, greater potential tungsten life, and the capability for single-pass, full-penetration welds on thick sections.

In examining a potential welding application, the three primary considerations are: achieving a quality weld, ease of welding, and cost. Selecting the optimum weld process becomes more complex as sophisticated electronic technology is applied to conventional welding equipment and consumable electrodes. Rapid advances in gas metal arc and PAW welding power source technology, and the development of many new flux-cored electrodes, have made selection of the optimum welding process or weld consumable more difficult. When several manual welding processes are available, the logical approach in considering GTAW for production welding is to first examine whether the job can be welded by gas metal arc or flux-cored methods.

**GTAW Welding Current.—**A major benefit offered by GTAW, compared with GMAW, FCAW, or SMAW, is the highly concentrated, spatter-free, inert heat from the tungsten arc, which is beneficial for many applications. The GTAW process can use any of three types of welding current, including direct-current straight polarity, electrode negative (DC−), direct-current reverse polarity, electrode positive (DC+), and alternating current with high frequency for arc stabilization (ACHF). Each of the different current types provides benefits that can be used for a specific application.

*GTAW Direct-Current Straight Polarity (DC−):* The most common GTAW current is straight polarity, where the electrode is connected to the negative terminal on the power source and the ground is connected to the positive terminal. Gas tungsten arc welding is used with inert gases such as argon, and argon + helium to weld most metals. During a DC− straight-polarity weld, electrons flow from the negative tungsten electrode tip and pass through the electric field in the arc plasma to the positive workpiece, as shown in Fig. 8.

Plasma is a high-temperature, ionized, gaseous column that is formed when electrons in the arc collide with the shielding gas molecules. The gas atoms lose one or more electrons, leaving them positively charged. The electrons and the resulting plasma are concentrated at the electrode tip, where they cause the plasma pressure to be at its greatest. The electron density thins out as the electrons travel from the straight-polarity, negatively charged, tungsten electrode across the open arc. As the electrons traverse the arc to the work, the resulting arc column width increases slightly, controlled in part by the electromagnetic forces generated by the current. With the increase in the arc column width, the density and pressure of the plasma decrease. The electrons collide with the work, liberating much heat. The downward pressure of the plasma is exerted against the surface of the weld pool. The gas ions in the plasma are positively charged and greater in mass than the electrons.

In DC–, straight-polarity welding, the positive gas ions are drawn to the negative electrode. The electron flow to the weld ensures that most of the arc heat is generated at the positive work side of the arc. This current setup provides maximum penetration potential, as indicated in Fig. 8. With DC–, straight polarity, the tungsten electrode can carry a higher current and operate at lower temperatures than with the other current arrangements.

*Direct-Current Reverse Polarity (DC+):* With direct-current positive polarity (DC+), the tungsten electrode is connected to the power-source positive terminal so that the electrons flow from the negative work to the positive electrode. As illustrated in Fig. 9, the electrons impinging on the electrode tip reverse the direction of the heat concentration that occurs with straight polarity, as described above. Approximately two-thirds of the heat generated with DC+ reverse polarity is at the electrode tip, and the electrode becomes very hot, even with low current levels. DC+ reverse polarity requires large-diameter electrodes.

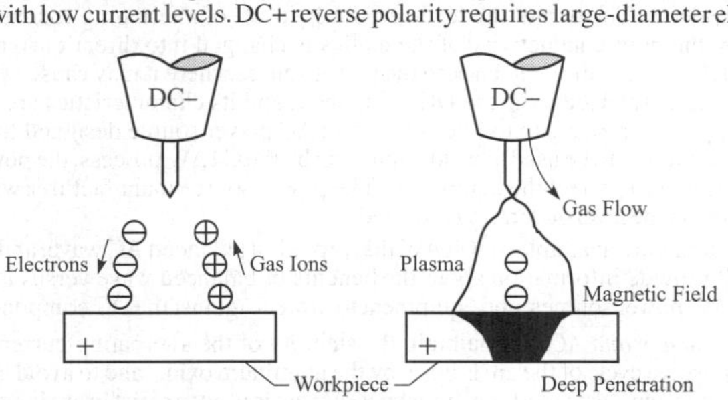


Fig. 8. Straight Polarity (DC–) Provides Highest Electrode Current Capacity and Deepest Penetration Potential

In the current range of 100 to 150 amps, DC+ reverse polarity requires a ¼-in. (6.4-mm) diameter electrode. This larger electrode produces a weld puddle almost twice as wide as that produced by a 120-amp, $\frac{1}{16}$-in. (1.6-mm) diameter, DC– straight polarity electrode. Most of the heat is generated at the electrode tip with DC+ reverse polarity, so penetration is much less than with DC– straight polarity. With DC+ reverse polarity, the positive gas ions in the arc plasma are drawn to the negative workpiece where they bombard and break up the surface oxides that form on metals such as aluminum and magnesium. However, the best welding method for aluminum and magnesium is to use alternating current (AC), which combines the benefits of DC– straight and DC+ reverse polarity.

*Alternating Current (AC):* The surface oxides formed on metals such as aluminum and magnesium disturb the arc and reduce the weld quality. Welding of these metals requires DC+ reverse or AC polarity to break up the surface oxides. An alternating current (AC) cycle consists of one-half cycle of straight polarity and one-half cycle of reverse polarity. With alternating current, the cleaning action benefits of the reverse-polarity arc can be combined with the electrode current-carrying capacity of the straight-polarity arc. In welding aluminum and magnesium, the half cycles of AC polarity may become unbalanced. During the AC cycle, the reverse electrode-positive portion of the cycle is restricted by the oxides on the surfaces of these materials. The surface oxides are poor conductors and make it difficult for the electrons generated by the reverse-polarity part of the cycle to flow from the work to the electrode tip, but they do not upset the straight polarity in which the electrons flow from the electrode to the work.

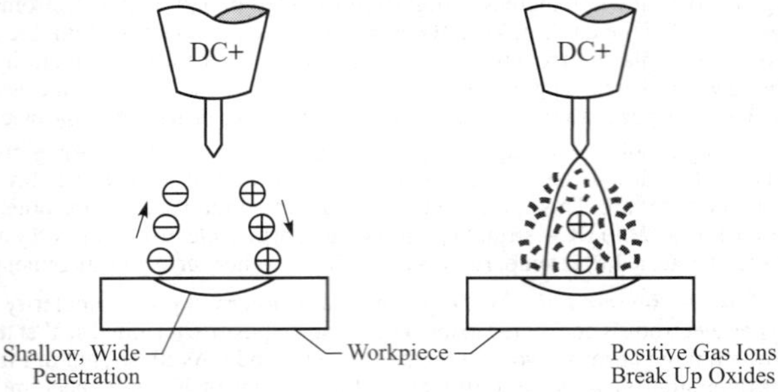


Fig. 9. Direct-Current (DC+) Reverse Polarity Provides a Shallow, Wide, Weld Pool

*DC Component:* The part of the reverse-polarity cycle of alternating current (AC) that is upset by the poor conductivity of the oxides is changed into direct-current, straight polarity (DC–) and is directed back to the power source where it may cause overheating. The feedback is referred to as the DC component and its characteristics are important in deciding which process to use because, if an AC power source designed for shielded metal arc welding is to be used to weld aluminum by the GTAW process, the power source must be derated to protect the equipment. The power-source manufacturer will provide information on the level of derating required.

Power sources are available for GTAW that provide a balanced AC wave, and manufacturers will provide information about the benefits of balanced wave versus unbalanced wave, GTAW power sources, and equipment to protect against the DC component.

*High Frequency and AC:* To maintain the stability of the alternating-current (AC) arc when the positive cycle of the arc is upset by the aluminum oxide, and to avoid contamination of the tungsten electrode, high-frequency current is used to assist in arc ignition during each AC cycle. In direct-current, straight-polarity (DC–) welding of carbon and stainless steels, the high-frequency current is typically selected by the HF arc start-only switch. During AC welding of steels without oxide problems, the HF switch may be left on the arc start-only setting. When AC is used on aluminum, magnesium, or other metals with poor electron-conductive oxides, the HF switch should be moved to the continuous setting.

High-frequency current is also beneficial in that it promotes gas ionization. The more positively charged molecules produced, the more cleaning action takes place in the direct-current, reverse-polarity (DCRP) cycle.

**Selecting the Tungsten Electrode-Type.**—Use of the correct tungsten electrode composition is vital to producing good-quality GTAW welds. Tungsten has the highest melting temperature of all metals. Pure tungsten provides a low-current capacity and requires addition of such alloying elements as thorium or zirconium to increase the current-carrying capability. The electrode diameter and the electrode tip configuration also require consideration as both have a great influence on the performance and application potential of GTAW welding.

Table 17 shows typical compositions of commonly used GTAW tungsten electrode materials from the American Welding Society AWS A5.12 Standard. New electrode compositions have been designed that utilize other alloys and rare-earth metals. These electrodes are designed for longer lives in both GTAW and plasma welding.

*Pure Tungsten:* Pure tungsten electrode material provides good arc stability with alternating current (AC). Tungsten has low current capacity and low resistance to electrode contamination. Pure tungsten is good for low-amperage welding of aluminum and magnesium alloys. On medium- to high-current ferrous applications, there is a potential for tungsten inclusions in the weld.

With DC, the current capacity of pure tungsten is lower than with the alloyed tungsten electrodes. During AC welding, a molten ball shape forms at the pure tungsten electrode tip, and this formation is desirable for welding aluminum.

**Table 16. Selection of Gas Tungsten Arc Welding (GTAW) Electrodes**

| Base Metal | Electrode | Current | Recommendations |
|---|---|---|---|
| Carbon, low-alloy, stainless, and nickel steels | Thoriated | DCEN | Use EWZr electrodes with AC on thin materials |
| Aluminum | Zirconium or pure tungsten | AC | Use EWZr on critical applications |
| Aluminum | Thoriated zirconium | DCEP | Use EWZr or EWP electrodes with DCEP on thin sections |
| Copper and copper alloys | Thoriated | DCEN | Use EWZr or EWP with AC on thin sections |
| Magnesium | Zirconium | AC | Use DCEP with same electrode on thin sections |
| Titanium | Thoriated | DCEN | … |

**Table 17. Common Tungsten Electrode Compositions**

| Classification | Color | Tungsten (%) | Thorium Oxide (%) | Zirconium Oxide (%) |
|---|---|---|---|---|
| EWP | Green | 99.50 | … | … |
| EWTh-1 | Yellow | 98.50 | 0.8–1.2 | … |
| EWTh-2 | Red | 97.50 | 1.7–2.2 | … |
| EWTh-3 | Blue | 98.95 | 0.35–0.55 | … |
| EWZr | Brown | 99.20 | … | 0.15–0.4 |

In the classification column, E = electrode; W = tungsten; P = pure; Th = thoriated (thorium oxide); Zr = zirconiated (zirconium oxide). The colors are codes used by manufacturers to identify the material. Tungsten percentages are minimum requirements. The EWTh-3 is also called striped tungsten because it is made with a strip of thoriated material along the length. This electrode needs to be preheated by striking an arc to melt the tip, providing for the thorium and the tungsten to combine before welding is started.

The electrode recommendations in Table 16 are a guide to attaining good-quality GTAW welds from the venous types of polarities available.

**Electrode and Current Selection.—**Table 18 and Table 19 show approximate current recommendations for common electrode types and diameters. The GTAW electrode size should be selected so that its midrange current provides the energy required for the intended application. If the electrode is too thin, excess current may be required, causing the electrode to wear too quickly or melt and contaminate the weld. If the electrodes used are found to be constantly at the top end of the current range, a change should be made to the next larger size. Table 20 and Table 21 show recommended sizes of electrodes and filler metal rods or wires for welding various thicknesses of carbon, low-alloy, and stainless steels and aluminum.

**Table 18. Recommended Current Ranges for Thoriated GTAW Electrodes**

| Electrode | | Current Range (amps) |
|---|---|---|
| 1⁄16 in. | (1.6 mm) | 60–150 |
| 3⁄32 in. | (2.4 mm) | 150–250 |
| 1⁄8 in. | (3.2 mm) | 250–400 |
| 5⁄32 in. | (4 mm) | 400–500 |

The electrode selected must suit the application and the current capacity of the power source.

**Table 19. Current Ranges for EWP and EWZr GTAW Electrodes**

| Electrode | | Ampere Range AC Balanced | | Ampere Range AC Unbalanced | |
|---|---|---|---|---|---|
| in. | (mm) | EWP | EWZr | EWP | EWZr |
| 1⁄16 | (1.6) | 30–80 | 60–120 | 50–100 | 70–150 |
| 3⁄32 | (2.4) | 60–130 | 100–180 | 100–160 | 140–235 |
| 1⁄8 | (3.2) | 100–180 | 160–250 | 150–210 | 225–325 |
| 5⁄32 | (4) | 160–240 | 200–320 | 200–275 | 300–400 |

**Table 20. Electrode and Current Recommendations for Carbon, Low-Alloy, and Stainless Steels**

| Material Thickness | | Electrode Diameter | | Filler Rod Diameter | | Current Range (amps) |
|---|---|---|---|---|---|---|
| in. | (mm) | in. | (mm) | in. | (mm) | DCEN EWTh |
| 1⁄16 | (1.6) | 1⁄16 | (1.6) | 1⁄16 | (1.6) | 60–100 |
| 1⁄8 | (3.2) | 3⁄32 | (2.4) | 3⁄32 | (2.4) | 150–170 |
| 3⁄16 | (4.8) | 3⁄32 | (2.4) | 1⁄8 | (3.2) | 180–220 |
| 1⁄4 | (6.4) | 1⁄8 | (3.2) | 5⁄32 | (7.2) | 260–300 |

Note: The shielding gas is argon at 15 to 20 cu ft/h (CFH). For stainless steel, reduce the current by approximately 10 percent.

**Table 21. Recommendations for GTAW Welding of Aluminum with EWP Electrodes Using AC and High-Frequency Current**

| Material Thickness | | Electrode Diameter | | Filler Rod Diameter | | AC Current Range |
|---|---|---|---|---|---|---|
| in. | (mm) | in. | (mm) | in. | (mm) | (amps) |
| 1⁄16 | (1.6) | 1⁄16 | (1.6) | 1⁄16 | (1.6) | 40–70 |
| 1⁄8 | (3.2) | 3⁄32 | (2.4) | 3⁄32 | (2.4) | 70–125 |
| 3⁄16 | (4.8) | 1⁄8 | (3.2) | 1⁄8 | (3.2) | 110–170 |
| 1⁄4 | (6.4) | 5⁄32 | (4) | 3⁄16 | (4.8) | 170–220 |

*Thoriated Electrodes:* In contrast with the pure EWP electrodes, thoriated electrodes have a higher melting temperature and up to about 50 percent more current-carrying capacity, with superior arc starting and arc stability. These electrodes are typically the first choice for critical DC welding applications, but do not have the potential to maintain a rounded ball shape at the tip. The best welding mode for these electrodes is with the tip ground to a tapered or fine point.

*Zirconiated Electrodes:* Tungsten electrodes with zirconium are practical for critical applications and have less sensitivity to contamination and superior current capacity than pure tungsten electrodes.

*Protecting and Prolonging Electrode Life:* To improve tungsten electrode life, the tip should be tapered in accordance with the manufacturer's recommendations. There must also be preflow, postflow shielding gas coverage to protect the electrode before and after the weld. When possible, high frequency should be used to avoid scratch starts, which contaminate the electrode. The shortest possible electrode extension should be employed, to avoid the possibility of the electrode touching the filler or weld metal. The grinding wheel used to sharpen the tungsten must not be contaminated from grinding other metals or with dirt.

**Filler Metals.—**Specifications covering composition and mechanical properties for GTAW filler metal are published by the American Welding Society under the following classifications: A5.7, copper and copper alloys; A5.9, chromium and chromium nickel;

A5.10, aluminum; A5.14, nickel; A5.16, titanium; A5.18, carbon steels; A5.19, magnesium; and A5.28, low-alloy steels.

Filler metals must be kept dry and clean if they are to be used satisfactorily.

**Shielding Gases.**—Inert gases such as argon, and argon + helium mixtures are most commonly used for GTAW. Helium provides greater thermal conductivity and additional arc voltage potential than argon, and is normally added to argon when more weld energy is required for improved penetration and increased mechanized welding travel speeds. Argon gas mixtures containing 30 to 75 percent helium provide benefits for manual welding of aluminum over ⅜ in. (9.6 mm) thick; mechanized welding of aluminum where high speeds are required; mechanized welding of carbon and stainless steels where good penetration is needed; mechanized welding of stainless steel where good penetration and faster speeds are required; and for copper of ¼ in. (6.4 mm) thickness and thicker.

Shielding gas purity for GTAW welding is important. Welding-grade argon is supplied at a purity of at least 99.996 percent and helium is produced to a minimum purity of 99.995 percent. However, shielding gases may be contaminated due to poor cylinder filling practices. If impure gas is suspected, the following test is suggested. With the HF and power on, create an arc without welding and hold the arc for about 30 seconds. Examine the electrode tip for signs of unusual coloration, oxidation, or contamination, which result from impurities in the shielding gas.

## PLASMA ARC WELDING (PAW)

When an electric current passes between two electrodes through certain gases, the energy of the gas molecules is increased so that they accelerate and collide with each other more often. With increases in energy, the binding forces between the nuclei and the electrons are exceeded, and electrons are released from the nuclei. The gas now consists of neutral molecules, positively charged atoms, and negatively charged electrons. The plasma gas is said to be ionized, so that it is capable of conducting electric current. Plasma forms in all welding arcs but in plasma arc welding it is generated by a series of events that begins with inert gas passing through the welding torch nozzle. High-frequency current is then generated between the tungsten electrode (cathode) and the torch nozzle (anode), forming a low-current pilot arc. The ionized path of this nontransferred arc is then transferred from the tungsten electrode to the work, and a preset plasma current is generated.

The above sequence of events provides the ionized path for the plasma current between the electrode and the work so that arcing between the electrode and the nozzle ceases. (Nontransferred arcs may be used for metal spraying or nonmetallic welds.) Forcing the ionized gas through the small orifice in the nozzle increases both the level of ionization and the arc velocity, and arc temperatures between 30,000 and 50,000° F (16,650 and 27,760° C) are generated.

**Gases for Welding.**—Argon is the preferred gas for plasma arc welding (PAW) as it is easily ionized and the plasma column formed by argon can be sustained by a low voltage. The low thermal conductivity of argon produces a plasma column with a narrow, concentrated hot core surrounded by a cooler outer zone. Argon plasmas are suited to welding steel up to ⅛ in. (3.2 mm). For thicker materials, requiring a hotter arc and using higher current melt-in technique, a mixture of argon 25 + helium 75 percent may be used. Additions of helium and hydrogen to the gas mixture improve heat transfer, reduce porosity, and increase weld travel speed. For welding materials thinner than ⅛ in. thick by the plasma gas keyhole method (full penetration welds), gases may contain up to 15 percent hydrogen with the remainder argon. Good results are obtained with argon + 5 percent hydrogen in welding stainless and nickel steels over ⅛ in. thick.

**Shielding Gases.**—A shielding gas is needed to protect the narrow plasma arc column and the weld pool, and generally is provided by mixtures of argon, argon + hydrogen, argon + helium, or argon + $O_2$ + $CO_2$, depending on compatibility with the material being welded.

Shielding gas flow rates vary from 5 to 35 cu ft/h (2.4 to 17 l/min). However, if argon is used for both plasma and shielding, the plasma gas will become less concentrated. The normally tight plasma arc column will expand in contact with the colder shielding gas, reducing ionization and thus concentration and intensity of the plasma column. With no shielding gas, the tight column is unaffected by the surrounding oxygen and nitrogen of the atmosphere, which are not easily ionized.

*Hydrogen* is added to the shielding gas when welding low-alloy steels of less than $\frac{1}{16}$ in. (1.6 mm) thickness, or stainless and nickel steels, with many benefits. The hydrogen molecules dissociate in contact with the arc at temperatures of about 7,000° F (3,870° C) and the energy thus created is released when the hydrogen molecules recombine on contact with the work surface. The diatomic molecular action creates a barrier around the plasma, maintaining column stiffness. Hydrogen in the shielding gas combines with oxygen in the weld zone, releasing it into the atmosphere and keeping the weld clean. Hydrogen reduces the surface tension of the weld pool, increasing fluidity, and the added energy increases penetration.

*Helium* mixed with the argon shielding gas is beneficial for all metals as it increases the ionization potential, allowing use of higher voltages that give increased welding temperatures. Flow rates are in the range of 15 to 50 cu ft/h (7 to 24 l/min). Arc-starting efficiency is reduced with pure helium, but adding 25 percent of argon helps both arc starting and stability. Helium additions of 25 to 75 percent are made to obtain increased thermal benefits.

Argon + $CO_2$ shielding gas mixtures are beneficial in fusion welding of carbon steels. A mixture of argon with 20 to 30 percent $CO_2$ improves weld fluidity. Shielding gas mixtures of argon + $CO_2$ with an argon + 5 percent hydrogen plasma should be considered for welding carbon steel of $\frac{1}{16}$ to $\frac{1}{4}$ in. thickness. Steels with higher amounts of carbon have higher heat conductivity and need application of more heat than is needed with stainless steels. Manufacturers usually make recommendations on types of gas mixtures to use with their equipment.

**PAW Welding Equipment.—**The PAW process uses electrode negative (DCEN) polarity in a current range from 25 to 400 amps, and equipment is offered by many manufacturers. Solid-state inverter units are available with nonmechanical contactors. Most PAW units contain a high-frequency generator, a small DC power supply, controls for welding and shielding gas mixtures, and a torch coolant control. A weld sequencer is recommended, especially for keyhole mode welding, but it is also useful in automated fusion welding. The sequencer provides control of up-slope and down-slope conditions for gas mixtures and current, so that it is possible to make welds without run-on and run-off tabs, as is necessary with circumferential welds.

Generally, plasma arc torches are liquid-cooled using deionized water in the coolant lines to the torch to avoid effects of electrolysis. Electrodes are usually tungsten with 2 percent thorium. If the welding shop already has a constant-current power supply and a coolant recirculator, plasma arc welding may be used by addition of a pilot arc welding console and a torch.

**Applications.—**Fusion welding is the main use for plasma arc welding. The process is used for high-volume, repetitive, high-duty cycle, manual and automated operations on lap, flange, butt, and corner fusion welds, in all positions. Joint design for materials less than 0.01 in. (0.254 mm) thick may require a flange type joint for rigidity and to allow use of extra, weld metal reinforcement. Filler metal may be added during fusion welding, and automated hot or cold wire feeders can be used. Fusion welding uses a soft, less-restricted arc with low gas flows, and the current level may vary from approximately 25 to 200 amps. The soft arc is obtained by setting the end of the tungsten electrode level with the face of the torch nozzle, in which position lower currents and gas flows are required. With these conditions, the weld bead is slightly wider than a bead produced with a recessed electrode.

*Low-Current Plasma Fusion Welding:* With the reduced consumption of gas and electric current, the low-current plasma fusion welding method is ideal for welding metals down to 0.001 in. (0.025 mm) in thickness, as the low-current plasma pilot arc allows arcs to be started consistently with currents of less than 1 amp. With currents below 1 amp, the pilot arc is usually left in the continuous mode to maintain the arc. In the conditions described, arc stability is improved and the process is much less sensitive to variations in the distance of the torch from the workpiece. Given this height tolerance, setting up is simplified, and with the smaller torches required, it is often easier to see the weld pool than with the GTAW process. Some plasma welding units incorporate gas flow meters that are designed for low flow rates, and currents in the range of 0.1 to 15 amps can be selected.

Low-current plasma arc welding is more economical than other gas tungsten arc welding methods, especially with solid-state inverter systems and smaller torches. The process is useful for sealing type welds where joint access is good, and for welding components of office furniture, household items, electronic and aerospace parts, metallic screening, and thin-wall tubing.

*Keyhole mode welding* describes a method whereby abutting edges of two plates are melted simultaneously, forming a vapor capillary (or keyhole) and the resulting molten-walled hole moves along the joint line. This method requires the end of the tungsten electrode to be positioned well back inside the torch nozzle to produce a high-velocity, restricted arc column with sufficient energy to pierce the workpiece. This mode is also used for the plasma cutting process, but the major difference is that welding uses very low plasma flow rates of the order of 1 to 3 cu ft/h (0.5 to 1.4 l/min) for work thicknesses of $\frac{1}{16}$ to $\frac{5}{32}$ in. (1.6 to 4 mm). These low rates avoid unwanted displacement of the weld metal. After the arc pierces the workpiece, the torch moves along the weld line and the thin layer of molten metal is supported by surface tension as it flows to the rear of the line of movement, where it solidifies and forms the weld.

As it passes through the keyhole, the high-velocity plasma gas column flushes the molten weld pool and carries away trapped gases and contaminants that otherwise would be trapped in the weld. Plasma arc keyhole welding is affected less by surface and internal defects in the work material than is the GTAW process. Most metals that can be welded by the gas tungsten arc method can be plasma arc welded with the conventional DC electrode, negative keyhole method, except aluminum, which requires a variable polarity keyhole method.

Plasma keyhole welding is usually automated because it requires consistent travel speed and torch height above the work. A typical operation is welding steel with square abutting edges (no bevels) in thicknesses of 0.09 to 0.375 in. (2.3 to 9 mm), where 100 percent penetration in a single pass is required. Producing square-groove butt welds in materials thicker than $\frac{1}{2}$ in. by the plasma arc keyhole process requires some edge preparation and several filler passes. The finished weld is uniformly narrow and the even distribution of heat means that distortion is minimized.

**Welding Aluminum.**—The variable polarity plasma arc (VPPA) process was developed for welding metals that form an oxide skin, such as aluminum. Electrode negative (straight) polarity is necessary for the plasma arc to provide sufficient heat to the workpiece and minimize heat buildup in the tungsten electrode. With electrode negative polarity, electrons move rapidly from the negative cathode tungsten electrode to the positive anode workpiece, generating most of the heat in the workpiece. Because of the oxide skin on aluminum, however, straight polarity produces an erratic arc, poor weld fluidity, and an irregularly shaped weld bead. The oxide skin must be broken up if the metal flow is to be controlled, and this breakup is effected by a power supply that constantly switches from negative to positive polarity.

A typical cycle uses a 20-ms pulse of electrode negative polarity and a 3-ms pulse of electrode positive polarity. The pulses are generated as square waves and the positive (cleaning) pulse is set at 30 to 80 amps higher than the negative pulse for greater oxide-breaking action. The tenacious oxide skin is thus broken constantly and the rapid cycle changes result in optimum cathode cleaning with minimum deterioration of the tungsten electrode and consistent arc stability. Varying polarity has advantages in both gas metal arc and plasma arc welding, but with the keyhole process it allows single-pass, square-groove, full-penetration welds in materials up to ½ in. (12.7 mm) thick.

The VPPA process ensures extremely low levels of porosity in weld areas in aluminum. VPPA welding is often used in the vertical up position for aluminum because it provides superior control of root reinforcement, which tends to be excessive when welding is done in the flat position. Pulsing in the VPPA process when welding aluminum of ⅛ to ¼ in. thickness in the flat position gives satisfactory root profiles. Pulsing gives improved arc control in keyhole welds in both ferrous and nonferrous metals and is beneficial with melt-in fusion welding of thin materials as it provides better control of heat input to the workpiece.

### Plasma Arc Surface Coating

*Plasma Arc Surfacing* uses an arc struck between the electrode and the workpiece, or transferred arc, to apply coatings of other metals or alloys to the workpiece surface. This high-temperature process produces homogeneous welds in which the ionized plasma gas stream melts both the work surface and a stream of powdered alloy or filler wire fed into the arc. Dilution of the base metal can be held below 5 percent if required. With arc temperatures between 25,000 and 50,000° F (14,000 and 28,000° C), deposition occurs rapidly, and a rate of 15 lb/h (6.8 kg/h) of powdered alloy is not unusual. Deposition from wire can be performed at rates up to 28 lb/h (12.7 kg/h), much higher than with oxygen/fuel or gas metal/arc methods.

In the nontransferred arc process used for coating of surfaces, the arc is struck between the electrode and the torch nozzle so that it does not attach to the work surface. This process is sometimes called metal spraying, and is used for building up surfaces for hard facing and for application of anticorrosion and barrier layers. Argon is frequently used as the plasma gas. As the coating material in the form of powder or wire enters the plasma, it is melted thoroughly by the plasma column and is propelled toward the work at high velocity to form a mechanical bond with the work surface. Some 500 different powder combinations are available for this process so that a variety of requirements can be fulfilled and deposition rates up to 100 lb/h (45 kg/h) can be achieved.

The plasma arc process allows parts to be modified or recovered if worn, and surfaces with unique properties can be provided on new or existing components. Low levels of porosity in the deposited metal can be achieved. Metal spraying can be performed manually or automatically, and its use depends primarily on whether a mechanical bond is acceptable. Other factors include the volume of parts to be treated, the time needed for the process and for subsequent finishing, the quality requirements for the finished parts, rejection rates, and costs of consumable materials and energy.

Some systems are available that can use either metal powder or wire as the spray material, and can be operated at higher voltage settings that result in longer plasma arc lengths at temperatures over 10,000° F (5,537° C). With these systems, the plasma velocity is increased to about 12,000 ft/s (3,658 m/s), giving an extremely dense coating with less than 1 percent porosity. Current ranges of 30 to 500 amps are available, and nitrogen is frequently used as the plasma gas, coupled with $CO_2$, nitrogen, or compressed air as the shielding gas. Gas flow rates are between 50 and 350 cubic feet per hour (24 and 165 l/min). Large or small surface areas can be coated at low cost, with minimum heat input, if other aspects of the process are compatible with the product being made.

### Plasma Arc Cutting of Metals

Higher current and gas flow rates than for plasma arc welding are used for the plasma arc cutting (PAC) process, which operates on DC straight polarity, and uses a transferred arc to melt through the material to be cut. The nozzle is positioned close to the work surface and the velocity of the plasma jet is greatly increased by a restricting nozzle orifice so that it blows away the metal as it is melted to make the cut.

The higher energy level makes the process much faster than cutting with an oxygen/fuel torch on cutting steel of less than ½ in. (12.7 mm) thick, but the process produces kerfs with some variation in the width and in the bevel angle, affecting the precision of the part. Some of the molten metal may recast itself on the edges of the cut and may be difficult to remove.

For more information on plasma arc cutting, refer to *NONTRADITIONAL MACHINING AND CUTTING* starting on page 1330.

## Cutting Metals with an Oxidizing Flame

The oxyhydrogen and oxyacetylene flames are especially adapted to cutting metals. When iron or steel is heated to a high temperature, it has a great affinity for oxygen and readily combines with it to form various oxides, and causing the metal to be disintegrated and burned with great rapidity. The metal-cutting or burning torch operates on this principle. A torch tip is designed to preheat the metal, which is then burned or oxidized by a jet of pure oxygen. The kerf or path left by the flame is suggestive of a saw cut when the cutting torch has been properly adjusted and used. The traversing motion of the torch along the work may be controlled either by hand or mechanically.

**Arc Cutting.—**According to the *Procedure Handbook of Arc-Welding Design & Practice,* published by The Lincoln Electric Co., a steel may be cut easily, and with great accuracy by means of the oxyacetylene torch. All metals, however, do not cut as easily as steel. Cast iron, stainless steels, manganese steels, and nonferrous materials are not as readily cut and shaped with the oxyacetylene cutting process because of their reluctance to oxidize. For these materials, arc cutting is often used to good advantage.

The cutting of steel is a chemical action. The oxygen combines readily with the iron to form iron oxide. In cast iron, this action is hindered by the presence of carbon in graphite form. Thus, cast iron cannot be cut as readily as steel; higher temperatures are necessary and cutting is slower. In steel, the action starts at bright red heat, whereas in cast iron, the temperature must be nearer to the melting point to obtain a sufficient reaction.

**The Cutting Torch.—**The ordinary cutting torch consists of a heating jet using oxygen and acetylene, oxygen and hydrogen, or, in fact, any other gas that, when combined with oxygen, will produce sufficient heat. By the use of this heating jet, the metal is first brought to a sufficiently high temperature, and an auxiliary jet of pure oxygen is then turned onto the red-hot metal, and the action just referred to takes place. Some cutting torches have a number of preheating flame ports surrounding the central oxygen port, so that a preheating flame will precede the oxygen regardless of the direction in which the torch is moved. This arrangement has been used to advantage in mechanically guided torches. The rate of cutting varies with the thickness of the steel, the size of the tip, and the oxygen pressure.

**Adjustment and Use of Cutting Torch.—**When using the cutting torch for the cutting of steel plate, the preheating flame first comes into contact with the edge of the plate and quickly raises it to a white-hot temperature. The oxygen valve is then opened, and as the pure oxygen comes into contact with the heated metal, the latter is burned or oxidized.

**Metals That Can Be Cut.—**Metals such as wrought iron and steels of comparatively low-carbon content can be cut readily with the cutting torch. High-carbon steels may be cut successfully if preheated to a temperature that depends somewhat on the carbon content. The higher the carbon content, the greater the degree of preheating required. A black heat is sufficient for ordinary tool steel, but a low red heat may be required for some alloy tool steels. Brass and bronze plates have been cut by interposing them between steel plates.

**Cutting Stainless Steel.—**Stainless steel can be cut readily by the flux-injection method. The elements that give stainless steels their desirable properties produce oxides that reduce the flame cutting operation to a slow melting-away process when the conventional oxyacetylene cutting equipment is used. By injecting a suitable flux directly into the stream of cutting oxygen before it enters the torch, the obstructing oxides can be removed.

Portable flux feeding units are designed to inject a predetermined amount of the flux powder. The rate of flux flow is accurately regulated by a vibrator type of dispenser with rheostat control. The flux-injection method is applicable either to machine cutting or to a hand-controlled torch. The operating procedure and speed of cutting are practically the same as in cutting mild steel.

**Cutting Cast Iron.—**The cutting of cast iron with the oxyacetylene torch is practicable, although it cannot be cut as readily as steel. The ease of cutting seems to depend largely on the physical character of the cast iron, very soft cast iron being more difficult to cut than harder varieties. The cost is much higher than that for cutting the same thickness of steel, because of the larger preheating flame necessary and the larger oxygen consumption. In spite of this extra cost, however, this method is often economical. The slag from a cast-iron cut contains considerable melted cast iron, whereas in steel, the slag is practically free from particles of the metal, indicating that cast-iron cutting is partly a melting operation. Increased speed and decreased cost often can be obtained by feeding a steel rod, about ¼ inch (6.35 mm) in diameter, into the top of the cut, beneath the torch tip. This rod furnishes a large amount of slag that flows over the cut and increases the temperature of the cast iron. Special tips are used because of the larger amounts of heat and oxygen required.

**Mechanically Guided Torches.—**Cutting torches used for cutting openings in plates or blocks or for cutting parts to some definite outline are often guided mechanically or by numerical control. Torches guided by pantograph mechanisms are especially adapted for tracing the outline to be cut from a pattern or drawing. Other designs are preferable for straight-line cutting and one type is designed for circular cutting.

**Cutting Steel Castings.—**When cutting steel castings, care should be taken to prevent burning pockets in the metal when the flame strikes a blowhole. If a blowhole is penetrated, the molten oxide will splash into the cavity and the flame will be diverted. The presence of the blowhole is generally indicated by excessive sparks. The operator should immediately move the torch back along the cut and direct it at an angle so as to strike the metal beneath the blowhole and burn it away if possible beyond the cavity. Cutting in the normal position then may be resumed.

**Thickness of Metal That Can Be Cut.—**The maximum thickness of metal that can be cut by these high-temperature flames depends largely upon the gases used and the pressure of the oxygen, which may be as high as 150 lb/in$^2$ (1034 kPa). The thicker the metal, the higher the pressure required. When using an oxyacetylene flame, it might be practicable to cut iron or steel up to 12 or 14 inches (30.5 or 35.6 cm) in thickness, whereas an oxyhydrogen flame has been used to cut steel plates 24 inches (61 cm) thick. The oxyhydrogen flame will cut thicker material principally because it is longer than the oxyacetylene flame and can penetrate to the full depth of the cut, thus keeping all the oxide in a molten condition so that it can be easily blown out by the oxygen cutting jet. A mechanically guided torch will cut thick material more satisfactorily than a hand-guided torch, because the flame is directed straight into the cut and does not wobble, as it tends to do when the torch is held by hand. With any flame, the cut is less accurate and the kerf wider, as the thickness of the metal increases. When cutting light material, the kerf might be $\frac{1}{16}$ inch (1.59 mm) wide, whereas for heavy stock, it might be ¼ or ⅜ inch (6.35 or 9.5 mm) wide.

## Hard Facing

Hard facing is a method of adding a coating, edge, or point, of a metal or alloy capable of resisting abrasion, corrosion, heat, or impact, to a metal component. The process can be applied equally well to new parts or old worn parts. The most common welding methods used to apply hard-facing materials include the oxyacetylene gas, shielded-metal arc, submerged arc, plasma arc, and inert-gas-shielded arc (consuming and nonconsuming electrode). Such coatings can also be applied by a spraying process, using equipment designed to handle the coating material in the form of a wire or a powder.

**Hard-Facing Materials.—**The first thing to be considered in the selection of a hard-facing material is the type of service the part in question is to undergo. Other considerations include machinability, cost of hard-facing material, porosity of the deposit, appearance in use, and ease of application. Only generalized information can be given here to guide the selection of a material as the choice is dependent upon experience with a particular type of service. Generally, the greater the hardness of the facing material, the greater is its resistance to abrasion and shock or impact wear. Many hardenable materials may be used for hard facing such as carbon steels, low-alloy steels, medium-alloy steels, and medium-high alloys but none of these is outstanding. Some of the materials that might be considered to be preferable are high-speed steel, austenitic manganese steel, austenitic high-chromium iron, cobalt-chromium alloy, copper-base alloy, and nickel-chromium-boron alloy.

**High-Speed Steels.—**These steels are available in the form of welding rods (RFe5) and electrodes (EFe5) for hard facing where hardness is required at service temperatures up to 1100° F (593° C) and where wear resistance and toughness are also required. Typical surfacing operations are done on cutting tools, shear blades, reamers, forming dies, shearing dies, guides, ingot tongs, and broaches using these metals.

*Hardness:* These steels have a hardness of 55 to 60 on the Rockwell C scale in the as-welded condition and a hardness of 30 Rockwell C in the annealed condition. At a temperature of 1100° F (593° C), the as-deposited hardness of 60 Rockwell C falls off very slowly to 47 Rockwell C. At about 1200° F (649° C), the maximum Rockwell C hardness is 30.

*Resistance Properties:* As deposited, the alloys can withstand only medium impact, but when tempered, the impact resistance is increased appreciably. Deposits of these alloys will oxidize readily because of their high molybdenum content but can withstand atmospheric corrosion. They do not withstand liquid corrosives.

*Other Properties or Characteristics:* The metals are well suited for metal-to-metal wear especially at elevated temperatures. They retain their hardness at elevated temperatures and can take a high polish. For machining, these alloys must first be annealed. Full hardness may be regained by a subsequent heat treatment of the metal.

**Austenitic Manganese Steels.—**These metals are available in the form of electrodes (EFeMn) for hard facing when dealing with metal-to-metal wear and impact. Uses include facing rock-crushing equipment and railway frogs and crossings.

*Hardness:* Hardness of the as-deposited metals are 170 to 230 BHN, but they can be work-hardened to 450 to 550 BHN very readily. For all practical purposes, these metals have no hot hardness as they become brittle when reheated above 500–600° F (260–316 °C).

*Resistance Properties:* These metals have high impact resistance. Their corrosion and oxidation resistance are similar to those of ordinary carbon steels. Their resistance to abrasion is only mediocre compared with hard abrasives like quartz.

*Other Properties or Characteristics:* The yield strength of the deposited metal in compression is low, but any compressive deformation rapidly raises it until plastic flow ceases. This property is an asset in impact wear situations. Machining is difficult with ordinary tools and equipment; finished surfaces are usually ground.

**Austenitic High-Chromium Irons.—**These metals are available in rod (RFeCr-A) and electrode (EFeCr-A) form and are used for facing agricultural machinery parts, coke chutes, steel mill guides, sand-blasting equipment, and brick-making machinery.

*Hardness:* The as-welded deposit ranges in hardness from 51 to 62 Rockwell C. Under impact, the deposit work hardens somewhat, but the resulting deformation also leads to cracking and impact service is therefore avoided. Hot hardness decreases slowly at temperatures up to 800 and 900° F (427 and 482° C). At 900° F (482° C), the instantaneous hardness is 43 Rockwell C. In 3 minutes under load, the hardness drops to 37 Rockwell C. At 1200° F (649° C), the instantaneous hardness is 5 Rockwell C. The decrease in hardness during hot testing is practically recovered on cooling to ambient temperatures.

*Resistance Properties:* Deposits will withstand only light impact without cracking. Dynamic compression stresses above 60,000 psi (414 MPa) should be avoided. These metals exhibit good oxidation resistance up to 1800° F (982° C) and can be considered for hot wear applications where hot plasticity is not objectionable. They are not very resistant to corrosion from liquids and will rust in moist air, but are more stable than ordinary iron and steel. Resistance to low-stress scratching is outstanding and is related to the amount of hard carbides present. However, under high-stress grinding abrasion, performance is only mediocre and they are not deemed suitable for such service.

*Other Properties or Characteristics:* The deposited metals have a yield strength (0.1 percent offset) of between 80,000 and 140,000 pounds per square inch (552 and 965 MPa) in compression and an ultimate strength of from 150,000 to 280,000 psi (1034–1930 MPa). Their tensile strength is low and therefore tension uses are avoided in design. These deposits are considered to be commercially unmachinable and are also very difficult to grind. When ground, a grinding wheel of aluminum oxide abrasive with a 24-grit size and a hard (Q) and medium-spaced resinoid bond is recommended for off-hand high-speed work and a slightly softer (P) vitrified bond for off-hand low-speed work.

**Cobalt-Base Alloys.—**These metals are available in both rod (RCoCr) and electrode (ECoCr) form and are frequently used to surface the contact surfaces of exhaust valves in aircraft, truck, and bus engines. Other uses include parts such as valve trim in steam engines, and on pump shafts, where conditions of corrosion and erosion are encountered. Several metals with a greater carbon content are available (CoCr-B, CoCr-C) and are used in applications requiring greater hardness and abrasion resistance but where impact resistance is not mandatory or expected to be a factor.

*Hardness:* Hardness ranges on the Rockwell C scale for gas-welded deposits are as follows: CoCr-A, 38 to 47; CoCr-B, 45 to 49; and CoCr-C, 48 to 58. For arc-welded deposits, hardness ranges (Rockwell C) as follows: CoCr-A, 23 to 47; CoCr-B, 34 to 47; and CoCr-C, 43 to 58. The values for arc-weld deposits depend for the most part on the base metal dilution. The greater the dilution, the lower the hardness. Many surfacing alloys are softened permanently by heating to elevated temperatures, however, these metals are exceptional. They do exhibit lower hardness values when hot but return to their approximate original hardness values upon cooling. Elevated-temperature strength and hardness are outstanding properties of this group. Their use at 1200 °F (649 °C) and above is considered advantageous but between 1000 and 1200 °F (538–649 °C), their advantages are not definitely established, and at temperatures below 1000 °F (538 °C), other surfacing metals may prove better.

*Resistance Properties:* In the temperature range from 1000 to 1200 degrees F (538–649 °C), weld deposits of these metals have a great resistance to creep. Tough martensitic steel deposits are considered superior to cobalt-base deposits in both flow resistance and toughness. The chromium in the deposited metal promotes the formation of a thin, tightly adherent scale that provides a scaling resistance to combustion products of internal combustion engines, including deposits from leaded fuels. These metals are corrosion-resistant in such media as air, food, and certain acids. It is advisable to conduct field tests to determine specific corrosion resistance for the application being considered.

*Other Properties or Characteristics:* Deposits are able to take a high polish and have a low coefficient of friction and therefore are well suited for metal-to-metal wear resistance. Machining of these deposits is difficult; the difficulty increases in proportion to the increase in carbon content. CoCr-A alloys are preferably machined with sintered carbide tools. CoCr-C deposits are finished by grinding.

**Copper-Base Alloys.—**These metals are available in rod (RCuA1-A2, RCuA1-B, RCuA1-C, RCuA1-D, RCuA1-E, RCuSi-A, RCuSn, RCuSn-D, RCuSn-E, and RCuZn-E) and electrode (ECuA1-A2, ECuA1-B, ECuA1-C, ECuA1-D, ECuA1-E, ECuSi, ECuSn-A, ECuSn-C, ECuSn-E, and ECuZn-E) forms and are used in depositing overlays and inlays for bearing, corrosion-resistant, and wear-resistant surfaces. The CuA1-A2 rods and electrodes are used for surfacing bearing surfaces between the hardness ranges of 130 to 190 BHN as well as for corrosion-resistant surfaces. The CuA1-B and CuA1-C rods and electrodes are used for surfacing bearing surfaces of hardness ranges 140 to 290 BHN. The CuA1-D and CuA1-E rods and electrodes are used on bearing and wear-resistant surfaces requiring the higher hardnesses of 230 to 390 BHN such as are found on gears, cams, wear plates, and dies. The copper-tin (CuSn) metals are used where a lower hardness is required for surfacing, for corrosion-resistant surfaces, and sometimes for wear-resistant applications.

*Hardness:* Hardness of a deposit depends upon the welding process employed and the manner of depositing the metal. Deposits made by the inert-gas metal-arc process (both consumable and nonconsumable electrode) will be higher in hardness than deposits made with the gas, metal-arc, and carbon-arc processes because lower losses of aluminum, tin, silicon, and zinc are achieved due to the better shielding from oxidation. Copper-base alloys are not recommended for use at elevated temperatures because their hardness and mechanical properties decrease consistently as the temperature goes above 400° F (204° C).

*Resistance Properties:* The highest impact resistance of the copper-base alloy metals is exhibited by CuA1-A2 deposits. As the aluminum content increases, the impact resistance decreases markedly. CuSi weld deposits have good impact properties. CuSn metals as deposited have low impact resistance and CuZn-E deposits have a very low impact resistance. Deposits of the CuA1 filler metals form a protective oxide coating upon exposure to the atmosphere. Oxidation resistance of CuSi deposits is fair and that of CuSn deposits are comparable to pure copper. With the exception of the CuSn-E and CuZn-E alloys, these metals are widely used to resist many acids, mild alkalies, and salt water. Copper-base alloy deposits are not recommended for use where severe abrasion is encountered in service. CuA1 filler metals are used to overlay surfaces subjected to excessive wear from metal-to-metal contact such as gears, cams, sheaves, wear plates, and dies.

*Other Properties or Characteristics:* All copper-base alloy metals are used for overlays and inlays for bearing surfaces with the exception of the CuSi metals. Metals selected for bearing surfaces should have a Brinell hardness of 50 to 75 units below that of the mating metal surface. Slight porosity is generally acceptable in bearing service as a porous deposit is able to retain oil for lubricating purposes. CuA1 deposits in compression have elastic limits ranging from 25,000 to 65,000 (172 to 448 MPa) and ultimate strengths of 120,000 to 171,000 (827 to 1179 MPa). The elastic limit and ultimate strength of CuSi deposits in compression are 22,000 and 60,000 (152 to 414 MPa), respectively. CuZn-E deposits in compression have an elastic limit of only about 5000 (34 MPa) and an ultimate strength of 20,000 (138 MPa). All copper-base alloy deposits can be machined.

**Nickel-Chromium-Boron Alloys.**—These metals are available in both rod (RNiCr) and electrode (ENiCr) form and their deposits have good metal-to-metal wear resistance, good low-stress, scratch-abrasion resistance, corrosion resistance, and retention of hardness at elevated temperatures. These properties make the alloys suitable for use on seal rings, cement pump screws, valves, screw conveyors, and cams. Three different formulations of these metals are recognized (NiCr-A, NiCr-B, and NiCr-C).

*Hardness:* Hardness of the deposited NiCr-A from rods range from 35 to 40 Rockwell C; of NiCr-B rods, 45 to 50 Rockwell C; of NiCr-C rods, 56 to 62 Rockwell C. Hardness of the deposited NiCr-A from electrodes ranges from 24 to 35 Rockwell C; of NiCr-B from electrodes, 30 to 45 Rockwell C; and of NiCr-C electrodes, 35 to 56. The lower hardness values and greater ranges of hardness values of the electrode deposits are attributed to the dilution of deposit and base metals. Hot Rockwell C hardness values of NiCr-A electrode deposits range from 30 to 19 in the temperature range from 600–1000° F (316–538° C) from instantaneous loading to a 3-minute loading interval. NiCr-A rod deposits range from 34 to 24 in the same temperature range and under the same load conditions. Hot Rockwell C hardness values of NiCr-B electrode deposits range from 41 to 26 in the temperature range from 600 to 1000 °F (315–538° C) from instantaneous loading to a 3-minute loading interval. NiCr-B rod deposits range from 46 to 37 in the same temperature range and under the same load conditions. Hot Rockwell C hardness values of NiCr-C electrode deposits range from 49 to 31 in the temperature range from 600 to 1000 degrees F from instantaneous loading to a 3-minute loading interval. NiCr-C rod deposits range from 55 to 40 in the same temperature range and under the same load conditions.

*Resistance Properties:* Deposits of these metal alloys will withstand light impact fairly well. When plastic deformation occurs, cracks are more likely to appear in the NiCr-C deposit than in the NiCr-A and NiCr-B deposits. NiCr deposits are oxidation-resistant up to 1800° F (982° C). Their use above 1750° F (954° C) is not recommended because fusion may begin near this temperature. NiCr deposits are completely resistant to atmospheric, steam, salt water, and salt spray corrosion and to the milder acids and many common corrosive chemicals. It is advisable to conduct field tests when a corrosion application is contemplated. These metals are not recommended for high-stress grinding abrasion. NiCr deposits have good metal-to-metal wear resistance, take a high polish under wearing conditions, and are particularly resistant to galling. These properties are especially evident in the NiCr-C alloy.

*Other Properties or Characteristics:* In compression, these alloys have an elastic limit of 42,000 psi (290 MPa). Their yield strength in compression is 92,000 psi (634 MPa) (0.01 percent offset), 150,000 psi (1034 MPa) (0.10 percent offset), and 210,000 psi (1448 MPa) (0.20 percent offset). Deposits of NiCr filler metals may be machined with tungsten carbide tools using slow speeds, light feeds, and heavy tool shanks. They are also finished by grinding using a soft-to-medium vitrified silicon carbide wheel.

**Chromium Plating.—**Chromium plating is an electrolytic process of depositing chromium on metals either as a protection against corrosion or to increase the surface-wearing qualities. The value of chromium-plating plug and ring gages has probably been more thoroughly demonstrated than any other single application of this treatment. Chromium-plated gages not only wear longer, but when worn, the chromium may be removed and the gage replated and reground to size.

In general, chromium-plated tools have operated well, giving greatly improved performance on nearly all classes of materials such as brass, bronze, copper, nickel, aluminum, cast iron, steel, plastics, asbestos compositions, and similar materials. Increased cutting life has been obtained with chromium-plated drills, taps, reamers, files, broaches, tool tips, saws, thread chasers, and the like. Dies for stamping, drawing, hot forging, die casting, and for molding plastics materials have shown greatly increased life after being plated with hard chromium.

Special care is essential in grinding and lapping tools preparatory to plating the cutting edges, because the chromium deposit is influenced materially by the grain structure and hardness of the base metal. The thickness of the plating may vary from 0.0001 to 0.001 or 0.002 inch, the thicker platings being used to build up undersize tools such as taps and reamers. A common procedure in the hard chromium plating of tools, as well as for parts to be salvaged by depositing chromium to increase diameters, is as follows:

1) Degrease with solvent; 2) Mount the tools on racks; 3) Clean in an anodic alkali bath held at a temperature of 82° C (180° F) for 3 to 5 minutes; 4) Rinse in boiling water; 5) Immerse in a 20 percent hydrochloric acid solution for 2 to 3 seconds; 6) Rinse in cold water; 7) Rinse in hot water; 8) Etch in a reverse-current chromic acid bath for 2 to 5 minutes; 9) Place work immediately in the chromium plating bath; and 10) Remove hydrogen embrittlement, if necessary, by immersing the plated tools for 2 hours in an oil bath maintained at 177° C (350° F).

Chromium has a very low coefficient of friction. The static coefficient of friction for steel on chromium-plated steel is 0.17, and the sliding coefficient of friction is 0.16. This value may be compared with the static coefficient of friction for steel on steel of 0.30 and a sliding coefficient of friction of 0.20. The static coefficient of friction for steel on babbitt is 0.25, and the sliding coefficient of friction 0.20, whereas for chromium-plated steel on babbitt, the static coefficient of friction is 0.15, and the sliding coefficient of friction is 0.13. These figures apply to highly polished bearing surfaces. Articles that are to be chromium plated in order to resist frictional wear should be highly polished before plating so that full advantage can be taken of the low coefficient of friction that is characteristic of chromium. Chromium resists attack by almost all organic and inorganic compounds, except muriatic and sulfuric acids. The melting point of chromium is 2930° F (1610° C), and it remains bright up to 1200° F (649° C). Above 1200 degrees F, a light adherent oxide forms and does not readily become detached. For this reason, chromium has been used successfully for protecting articles that must resist high temperatures, even above 2000° F (1093° C).

## Electron-Beam (EB) Welding

Heat for melting of metals in electron-beam welding is obtained by generating electrons, concentrating them into a beam, and accelerating them to between 30 and 70 percent of the speed of light, using voltages between 25 and 200 kV. The apparatus used is called an electron-beam gun, and it is provided with electrical coils to focus and deflect the beam as needed for the welding operation. Energy input depends on the number of electrons impinging on the work in unit time, their velocity, the degree of concentration of the beam, and the traveling speed of the workpiece being welded. Some $6.3 \times 10^{15}$ electrons/s are generated in a 1-mA current stream. With beam diameters of 0.01 to 0.03 in. (0.25 to 0.76 mm), beam power can reach 100 kW and power density can be as high as $10^7$ W/in$^2$ ($1.55 \times 10^4$ W/mm$^2$), higher than most arc welding levels.

At these power densities, an electron beam can penetrate steel up to 4 in thick and form a vapor capillary or keyhole, as described earlier. Although patterns can be traced by deflecting the beam, the method used in welding is to move the electron gun or the workpiece. A numerical control, or computer numerical control, program is used because of the accuracy required to position the narrow beam in relation to the weld line.

Equipment is available for electron-beam welding under atmospheric pressure or at various degrees of vacuum. The process is most efficient (produces the narrowest width and deepest penetration welds) at high levels of vacuum, of the order of $10^{-6}$ to $10^{-3}$ torr or lower (standard atmospheric pressure is about 760 torr, or 760 mm of mercury), so that a vacuum chamber large enough to enclose the work is needed. Operation in a vacuum minimizes contamination of the molten weld material by oxygen and nitrogen. Gases produced during welding are also extracted rapidly by the vacuum pump so that welding of reactive metals is eased. However, the pumping time and the size of many workpieces restrict the use of high-vacuum enclosures.

At atmospheric pressures, scattering of the beam electrons by gas molecules is increased in relation to the number of stray molecules and the distance traveled, so that penetration depth is less and the beam spread is greater. In the atmosphere, the gun-to-work distance must be less than about 1.5 in. (38 mm). Electron-beam welding at atmospheric pressure requires beam-accelerating voltages above 150 kV, but lower values can be used with a protective gas. Helium is preferred because it is lighter than air and permits greater penetration. Argon, which is heavier than air and allows less penetration, can also be used to prevent contamination.

Required safety precautions, such as radiation shields to guard workers against the effects of X-rays when the electron beam strikes the work, are essential when electron-beam welding is done at atmospheric pressure. Such barriers are usually built into enclosures that are designed specifically for electron-beam welding in a partial vacuum. Adequate ventilation is also required to remove ozone and other gases generated when the process is used in the atmosphere.

Carbon, low-alloy, and stainless steels; high-temperature and refractory alloys; copper and aluminum alloys can be electron-beam welded, and single-pass, reasonably square, butt welds can be made in materials up to 1 in. (25.4 mm) thick at good speeds with nonvacuum equipment rated at 60 kW. Edges of thick material to be electron-beam welded require precision machining to provide good joint alignment and minimize the joint gap. Dissimilar metals usually may be welded without problems.

Because of the heat-sink effect, electron-beam welds solidify and cool very rapidly, causing cracking in certain materials such as low-ferrite stainless steel. Although capital costs for electron-beam welding are generally higher than for other methods, welding of large numbers of parts and the high welding travel rates make the process competitive.

## Pipe Welding

**Pipe Welding.**—Welding of (usually steel) pipe is commonly performed manually, with the pipe joint stationary, or held in a fixture whereby rotation can be used to keep

**Positioning of Joint Components in Pipe Welding**

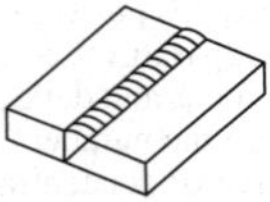

Fig. 1a. Flat Position 1G

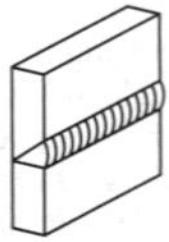

Fig. 1b. Horizontal Position 2G

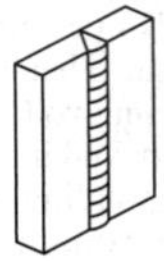

Fig. 1c. Vertical Position 3G

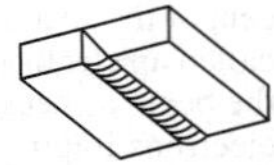

Fig. 1d. Overhead Position 4G

Fig. 1e. Horizontal Position 2G Pipe Axis Vertical

Fig. 1f. Horizontal Position 5G Pipe Fixed, Axis Horizontal

Fig. 1g. Position 6G Pipe Fixed, Axis Inclined 45°

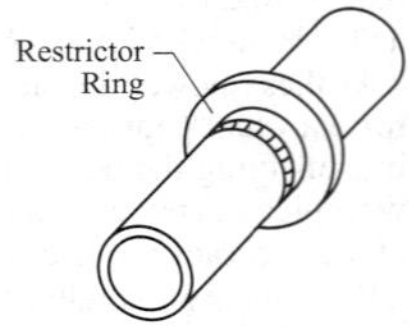


Fig. 1h. Position 6GR Pipe Fixed, Axis Inclined 45°

the weld location in a fixed, downhand (flat) position. Alternatively, pipe may need to be welded on site, without rotation, and the welder then has to exert considerable skill to produce a satisfactory, pressure-tight joint. Before welding stationary pipe, a welder must be proficient in welding in the four basic positions: 1G flat, 2G horizontal, 3G vertical, and 4G overhead, depicted in Fig. 1a, Fig. 1b, Fig. 1c, and Fig. 1d, respectively.

Figs 1e to 1h show four examples of pipe weld test positions. Fig. 1e corresponds to the 2G horizontal position, where the vertical pipe is fixed or free to rotate along its vertical axis, and the weld axis is horizontal. Fig. 1f shows a pipe in the fixed 5G position, with its axis horizontal; the weld is deposited in the flat, vertical, and overhead welding positions. Similarly, Fig. 1g typifies pipe welded in the fixed 6G position, with its axis approximately 45° from the vertical plane. Fig. 1h illustrates the 6GR position, which is again 45° from vertical, with the addition of a restriction ring (represented by the "R") that limits access to the weld joint from one side. Another difference between 6G and 6GR is the bevel preparation. 6G is prepared as a single V-groove at 45° from the horizontal plane. 6GR is prepared as a single bevel (one pipe without bevel preparation), with pipe coupons of varying thicknesses approximating $^3/_{16}$ inch (4.76 mm); the restriction ring is placed at a distance not greater than ½ inch (12.7 mm) from the weld circumferential centerline. This simulates actual restrictions observed during the welding of T, K, and Y tubular connections and is also the reason 6GR is considered the most difficult of all welding positions.

For satisfactory pipe welding, consideration must be given to the chemical composition and thickness of the metal to be welded; selection of a suitable electrode material composition and size; determination of the current, voltage and wire feed rate to be used; preparation of the joint or edges of the pipes; and ways of holding the pipes in the positions needed while welding is carried out. High-quality tack welds, each about 1.5 inches (38 mm) long, and projecting about $^1/_{16}$ inch (1.6 mm) beyond the inner wall of the pipe, are usually made to hold the parts of the assembly in position during welding.

SMAW (stick) welding was used almost exclusively for pipe welding until the advent of MIG welding with its potential for much greater rates of deposition. It cannot be emphasized too strongly that practices suitable for SMAW cannot be transferred to MIG welding, for which greater expertise is required if satisfactory welds are to be produced. MIG short-circuit, globular, and spray transfer, and pulsed MIG, with flux or metal-cored consumables (electrodes), can now all be used for pipe welding. Use of all-position, flux-cored, MIG consumables in particular, can reduce skill requirements, improve weld quality, and hold down costs in pipe welding.

Among the important items involved in the change to the MIG process is the automatic wire feeder. With today's wire feeding equipment, an increase of one increment on the dial, say from the 9 to the 10 o'clock position, can increase the wire feed rate by 70 in/min (1.8 m/min). As an example, such an increase could raise the weld current from 110 to 145 amps and the weld voltage from 16 to 17, resulting in an increase of 40 percent in the energy supplied to the weld. Another vital parameter is the amount that the wire sticks out from the contact tip. In low-parameter, short-circuit welding, a small change in the wire stick-out can alter the energy supplied to the weld by 20 to 30 percent.

*Root passes:* Whatever welding process is selected, the most important step in pipe welding, as in other types of welding, is the root pass, which helps to determine the degree of penetration of the weld metal, and affects the amount of lack of fusion in the finished weld. During the root pass, the action of the arc in the weld area should reshape the gap between the adjacent sides of the joint into a pear-shaped opening, often called a "keyhole." As the work proceeds, this keyhole opening is continuously being filled, on the trailing side of the weld, by the metal being deposited from the electrode. The keyhole travels along with the weld so that the root pass produces a weld that penetrates slightly through the inner wall of the pipe.

MIG short-circuit root welding of carbon steel pipe requires a gap of $\frac{5}{32} \pm \frac{1}{32}$ inch ($4 \pm 0.8$ mm), between the ends of the pipe, and the width of the root faces (at the base of the bevels) should be $\frac{1}{16}$ to $\frac{3}{32}$ inch (1.6 to 2.4 mm). The recommended bevel angle for MIG pipe welding is 40° (80° included angle) and the maximum root gap is $\frac{3}{16}$ inch (4.8 mm). The root pass in 1G welds should be made in the vertical-down direction with the electrode held between the 2 and 3 o'clock positions. When an 0.035-inch (0.9-mm) diameter E70S-3 MIG wire is used with the above root dimensions, weaving is not needed for the root pass except when welding over tack welds.

*Fill Passes:* In welding carbon steel pipe in the 1G position with an 0.035-inch (0.9-mm) diameter electrode wire, MIG short circuit fill passes should use a minimum of 135 amps and be done in the vertical-up position. Fill passes should deposit a maximum thickness of no more than $\frac{1}{8}$ inch (3.2 mm). Inclusion of $CO_2$ gas in the mixture will improve weld fusion. With flux-cored electrodes, the minimum amount of wire stick-out is $\frac{3}{4}$ inch (19 mm). Weld fusion can be improved in welding pipe of 0.4-inch (10-mm) wall thickness and thicker by preheating the work to a temperature between 400 and 500° F (204 and 260 °C).

*Horizontal Pipe Welding:* In 1G welds (see Fig. 1a), the pipe should be rotated in the direction that moves the solidifying area away from the wire tip, to minimize penetration and resulting breakthrough. Welding of pipe in the 2G horizontal position is made more difficult by the tendency for the molten metal to drip from the weld pool. Such dripping may cause an excessively large keyhole to form during the root-welding pass, and in subsequent passes electrode metal may be lost. Metal may also be lost from the edge of the upper pipe, causing an undercut at that side of the weld.

*Vertical-Down Welding:* With the pipe axis horizontal (as in the 5G position in Fig. 1f), vertical-down welding is usually started at the top or 12 o'clock location, and proceeds until the 6 o'clock location is reached. Welding then starts again at the 12 o'clock location and continues in the opposite direction until the 6 o'clock location is reached. Vertical-down welding is mainly used for thin-walled, low-carbon steel pipe of $\frac{1}{8}$ to $\frac{5}{16}$ inch (3.2 to 7.9 mm) wall thickness, which has low heat-retaining capacity so that the weld metal cools slowly, producing a soft and ductile structure. The slow rate of cooling also permits faster weld deposition, and, when several beads are deposited, causes an annealing effect that may refine the entire weld structure.

*Vertical-Up Welding:* In the 5G position, vertical-up welding normally begins at the 6 o'clock location and continues up to the 12 o'clock location, the weld then being completed by starting at the 6 o'clock location on the other side of the pipe and traversing up to the 12 o'clock location again. Vertical-up welding is more suited to pipe with thick walls and to alloy steels. However, the greater heat sink effect of the heavy-walled pipe may result in a faster cooling rate and embrittlement of the material, especially in alloy steels. The cooling rate can be reduced by slowing the rate of traverse and depositing a heavier bead of metal, both facilitated by welding in the vertical-up direction.

Using a thicker electrode and higher current for thicker-walled pipe to reduce the number of beads required may result in dripping from the molten puddle of metal. Defects, such as pin holes, lack of fusion, and cold lap, may then appear in the weld. Vertical-up welding of pipe in the 5G, fixed, horizontal position, Fig. 1f, used for thick-walled pipe, is probably the most difficult for a welder, but once mastered will form the basis for other methods of pipe welding. Starting at the 6 o'clock location, the arc for the root pass is struck overhead, with the electrode at an angle of 5 to 10° from the vertical, on the joint, not on the tack weld. A long arc should be maintained for a short period while weaving the electrode to pre-heat the area ahead of the weld. Only small amounts of filler metal will be transferred while this long arc is maintained in the overhead position. The electrode tip is then advanced to establish the correct arc length and held in position long enough for the keyhole to form before starting to lay down the root bead, moving up toward the 12 o'clock location.

*Thin-Wall Pipe:* The optimum globular/spray parameters for welding rotated (1G position) thin-wall pipe of less than 12-inch (305 mm) diameter are 0.035-inch (0.9-mm) electrode wire fed at 380–420 in/min (9.7–10.7 m/min) with a protective gas mixture of argon 80 to 85, $CO_2$ 15 to 20 percent, and current of 190 to 210 amps. These conditions will provide deposition rates of about 6 lb/hr (3 kg/hr).

**Use of Flux-Cored Electrodes.—**Small diameter, flux-cored electrodes developed in the 1980s are still a rarity in many pipe welding shops, but flux-cored welding can produce consistent, high-quality, low-cost welds on carbon steel or stainless pipe. Flux-cored E71T-1, 0.035-inch (0.9-mm) diameter wire provides a continuous, medium energy, open arc, with a practical current range of 135 to 165 amps for welding pipe. This current range is similar to the optimum MIG short-circuit current range, and is 25 to 30 percent less current than the minimum open arc spray transfer current for an 0.035-inch (0.9-mm) diameter MIG wire.

In contrast to MIG short circuit welding, FCAW works with an open arc and no short circuits. The FCAW arc energy is continuous, and, in contrast to short-circuit transfer, provides increased weld fusion potential. The weld metal from the flux-cored tubular wire is transferred from the periphery and the center of the wire, resulting in broad coverage of the weld. The plasma in the flux-cored arc is wider than MIG plasma, and the flux-cored arc is less focused and easier to control than the MIG spray arc.

Open arc, gas shielded, flux-cored welding can produce spray type transfer at lower currents than open arc MIG spray transfer. With FCAW, the current density is high because the electrode wire cross-sectional area is less than that of the same size MIG solid wire due to the central core of flux. This higher density provides for improved weld penetration potential. The FCAW process produces slag, which serves as a mold to hold the fluid molten metal in place, an ideal arrangement for vertical-up and overhead welds.

All position, flux-cored wires require less operator skill for vertical-up and overhead welds than MIG, SMAW, and TIG processes. Fill passes can also be completed in 30 to 50 percent less time with all-position, flux-cored wires than with MIG short circuit and SMAW wires.

For good quality FCAW, welders need to know the best root and bevel dimensions, and the importance of maintaining those dimensions for continuous weld fusion; the preferred direction of pipe rotation; the diameter of flux-cored electrode best suited for welding thin wall pipe; the optimum parameter range for that electrode on 1G and 5G welds; the preferred amount of wire stick-out (typically 0.7 inch or 18 mm); and how to fine tune the voltage. When flux-cored welding is to be used for the fill passes, MIG short circuit welding is recommended for the root welds to reduce the possibility of slag from the flux being trapped in the weld. Higher weld deposit rates are provided with flux-cored, vertical-up welding, and there is the temptation to weld faster with a process that is easy to use. Conservative wire feed settings are recommended unless the high deposition rates are shown to provide consistent weld fusion. Wire feed settings should allow the welder time to control and direct the weave into the critical groove locations.

**Complete Weld Fusion.—**It is essential that new weld metal deposits be completely fused with the pipe components, and with metal laid down in successive passes. Factors that can prevent complete fusing are too numerous to list here. Some basic rules that, if followed, will improve weld fusion and quality in MIG welding in the 1G and 5G positions are:

1) The maximum gap at the root should be $^3/_{16}$ inch (5 mm).
2) The root land should be $^1/_{16}$ to $^3/_{32}$ inch (1.6 to 2.4 mm) wide.
3) A bevel angle of 80° inclusive should be used for MIG and flux-cored welding of pipe to provide width for weaving and improve fusion.
4) A 0.035-in MIG electrode should have a minimum short circuit current of 135 amps for fill passes.
5) Tack and root welds should be made in the vertical-down position.
6) Tack welds should be about 1.5 in (38 mm) long by $^1/_{16}$ to $^3/_{32}$ inch (1.6 to 2.4 mm) thick.
7) Short circuit fill passes should be made in the vertical-up position.
8) With flux-cored electrodes, a minimum of 0.7 inch (18 mm) wire stick-out from the contact tip must be maintained.
9) Current and voltage must be related to the pipe wall thickness.
10) Argon + 25 percent $CO_2$ is recommended for short circuit welding of pipe roots.
11) Use of undiluted $CO_2$ gas will improve MIG weld fusion in fill passes because of the "digging" action of the arc and the increased weld energy.
12) With pipe wall thicknesses of 0.4 inch (10 mm) or greater, preheating to between 400 and 500° F (205 and 260° C) will help make fusion complete.

**Other Methods.—**Pulsed MIG is a viable alternative to flux-cored for all-position welds on 5G pipe, but requires more costly equipment. The pulsed MIG process, however, has few advantages over conventional MIG and flux-cored when the latter are used correctly. Pulsed MIG may have some advantage on mechanized 5G welds and on welding of stainless steel pipe in the 5G position.

Metal-cored electrode wire also has few advantages for pipe welds because they work best with low-energy gas welds, which cancels out the increased current density claimed for them.

On most manual pipe welds, the welder needs time to control and direct the weave to ensure even heating and avoid lack of fusion. Satisfactory welds are often performed at travel speeds of 4 to 12 in/min (0.10–0.31 m/min), giving deposit rates of 3–5 lb/hr (1.36–2.27 kg/hr).

## Pipe Welding Procedure

Because of the variety of parameter combinations that can be used in pipe welding, it is suggested that charts be prepared and displayed in welding booths to remind welders of the basic settings to be used. Examples of such charts for tack, root, fill, and cover passes are included in what follows:

**FCAW 5G (Non-Rotated) MIG Welding of Thick-Walled, Carbon Steel Pipes, Procedure for Root Welding.—**This procedure can be applied to most pipe sizes, and should be given special consideration for 5G (non-rotated) welds on carbon steel pipe with $^3/_8$-inch (10-mm) wall thickness and thicker.

### Pipe and Weld Data

*Pipe bevel included angle* = 80
*Root face land* = $^3/_{32} \pm ^1/_{32}$ inch (2.4 ± 0.8 mm)
*Root gap between faces* = $^5/_{32} \pm ^1/_{32}$ inch (4 ± 0.8 mm)
*Electrode for root weld* = 0.035-inch (0.9-mm) diameter, E70S-3 flux-cored
*Gas* = argon with 15–25% $CO_2$
*Gas flow rates* = 30 to 40 cubic ft/hr (0.85–1.13 $m^3$/hr)

Set wire feeder to 210–280 in/min (10 to 11 o'clock position on many feeders) for current of 140–170 amps, 17–18 volts.

*Wire extension:* For MIG root weld, set contact tip to stick outside the nozzle, $\frac{1}{16}$ to $\frac{1}{8}$ inch (1.6 to 3 mm). Maintain $\frac{3}{8}$- to $\frac{5}{8}$-inch (10- to 16-mm) maximum wire stick-out from contact tip.

*Tack Welding Procedures for FCAW 5G Pipe Welds:* Make tack welds 1.5 to 2 inches (38 to 50 mm) long. After welding, grind full length of tack to thickness of approximately $\frac{1}{16}$ inch (1.6 mm). Feather tack ends back $\frac{3}{8}$ to $\frac{1}{2}$ inches (9.5 to 13 mm).

On pipes of less than 6 inches (15 cm) outside diameter, use three tack welds, equally spaced, starting at 12 o'clock.

On pipes over 6 inches outside diameter, use 4 tack welds. Locate tack welds at 12, 3, 6, and 9 o'clock.

*Root Welding Procedures for FCAW 5G Pipe Welds:* Root weld MIG vertical-down. Weld sequence: 12 to 3, 9 to 6, 3 to 6, and 12 to 9 o'clock positions.

Start and finish MIG root welds at tack centers. Use slight weave oscillation over tacks. No weave necessary if $\frac{1}{8}$- to $\frac{5}{32}$-inch root gap is maintained. Weaving may be required if root gap is less than $\frac{1}{8}$ inch (3 mm). Weaving is also beneficial for root welds between 7 and 6 o'clock, and between 5 and 6 o'clock. After each root pass, blend the starts and stops back to the original tack thickness.

To complete the root, ensure that the weld stops and starts on the last tack, and that the root weld center is ground flat or slightly concave. Remove any slag islands.

**FCAW 5G (Non-rotated) MIG Welding of Thick-Walled Carbon-Steel Pipes, Procedure for Fill and Cover Welds.—**This procedure can be applied to most common pipe sizes, and should be given special consideration for 5G (non-rotated) welds on carbon steel pipe with $\frac{3}{8}$-inch (10-mm) wall thickness and thicker.

**Pipe and Weld Data**

*Electrode for fill and cover passes* = 0.035 inch (0.9 mm) diameter, 71T-1 flux-cored

*Gas* = argon with 15–25% $CO_2$

*Gas flow rates* = 30 to 40 cubic ft/hr

Set an initial wire feed rate of 350 to 450 in/min (12 to 1 o'clock position on typical wire feed unit), 135–165 amps, 25–28 volts. Alternatively, use a wire feed setting of 350 in/min (12 o'clock on wire feed unit), which should result in about 135–145 amps, 25–26 volts. If the weld pool and weld heat buildup permit, increase the wire feed rate to 380 in/min (between the 12 and 1 o'clock positions), 150 amps, 27 volts. Try also a wire feed setting of 420 in/min (1 o'clock on the wire feeder), 165 amps, 28 volts. Determine the low and maximum wire feed rates to be used by examination of the weld fusion obtained in sectioned test samples.

*Wire extension:* Adjust contact tip so it is recessed $\frac{1}{2}$ inch within the nozzle to provide a total wire stick-out from the contact tip of 0.7 to 1 inch (18 to 25 mm).

*Fill and Cover Pass Procedures for FCAW 5G Pipe Welds:* Weld vertical-up. If the pipe diameter allows the fill pass to be made in two passes, start at the 7 o'clock position and weld to the 1 o'clock position. This approach is preferable to starting and finishing on the root tacks. Starting at the 7 o'clock position will ensure that optimum weld energy is achieved as the first pass welds over the initial 6 o'clock root tack location. Use the grinder to feather the first 1 inch (25 mm) of the weld start and stop of the first pass, before applying the second vertical-up weld pass. Use a slight weave action for the fill pass.

Remove all flux-cored slag between weld passes. Make sure no fill pass is greater in depth than $\frac{1}{8}$ inch (3 mm). Use a straight weave across the root face. At the bevel edge use a slight upward motion with the gun. The motion should be no greater than the wire diameter. Then use a slight back step for added bevel fusion and to avoid undercuts.

Leave $\frac{1}{32}$ to $\frac{1}{16}$ inch (0.8 to 1.6 mm) of the groove depth to provide for the optimum cover pass profile. The bevel edge will act as a guide for the cover pass weld. If more weld fusion is required for pipe thicker than 3/8 inch (10 mm), after the root weld is complete, preheat the pipe to between 400 and 600° F (200–300° C) before welding. Preheating is typically not necessary for a cover pass.

For pipe diameters on which the welder needs more than two passes for the vertical-up welds, the recommended sequence for vertical-up welding is:

1) First pass, weld from the 7 to the 4 o'clock position. Start with a slight forehand nozzle angle. At the 4 o'clock position, the gun should be at the same angle as the pipe.
2) Second pass, weld from the 10 to the 1 o'clock position, then grind all stops and start again at the 1-inch (25-mm) position.
3) Third pass, weld from the 4 to the 1 o'clock position.
4) Fourth pass, weld from the 7 to the 10 o'clock position.

**FCAW 5G (Non-rotated) Welding of Thin-Walled Carbon Steel Pipes, Procedure for Root, Fill and Cover Pass Welding.—**This procedure can be applied to most common pipe sizes, and should be given special consideration for 5G (non-rotated) welding of carbon steel pipe with wall thicknesses up to 3/8 inch (10 mm).

**Pipe and Weld Data**

*Electrode for root weld* = 0.035-inch (0.89-mm) diameter, E70S-6 flux cored

*Gas* = argon with 15–25% $CO_2$

*Gas flow rates* = 30 to 40 cubic ft/hr

*Root Welding Procedure for 5G Welds:* Use root welding data from *Root Welding Procedures for FCAW 5G Pipe Welds*, above.

*Fill and Cover Pass Procedures for 5G Welds:* Use MIG short-circuit, vertical-up for fill and cover passes. Electrode wire and gas, same as for root weld.

Weld vertical-up. If the vertical-up fill pass can be made in two passes, weld from the 7 to the 1 o'clock position, to avoid starting and finishing on the root tacks. Starting just past 6 o'clock ensures that optimum weld energy is achieved as the first pass welds over the initial 6 o'clock root tack location. Feather 1 inch (25 mm) of the weld start and stop on the first pass with the grinder before applying the second vertical-up weld pass. Use a slight weave action.

Use MIG short-circuit wire feed, 200–230 in/min 125–135 amps, 19–22 volts. Start at optimum 210 in/min (10 o'clock on the wire feeder) for 130 amps, 21–22 volts. Fine tune voltage by listening to arc sound to obtain a consistent rapid crackle sound.

Electrode sticks out 1/2 to 5/8 inch; contact tip flush with nozzle end.

Remove MIG surface slag islands between weld passes. No fill pass should be thicker than 1/8 inch (3 mm). Use straight weave across the root face. At the bevel, use a slight upward motion with the gun. The motion should be no greater than the wire diameter. Then use a slight back step for added bevel fusion and to avoid possibility of undercut.

For the cover pass, leave 1/32 to 1/16 inch of the groove depth for the optimum cover pass profile. The bevel edge will act as a guide for the cover pass weld.

If more weld fusion is required after the root is complete and between fill passes, preheat pipe to 200–400° F (93–204° C).

For pipe diameters on which more than two passes are required for the circumference the weld sequence is:

1) First pass, weld from the 7 to the 4 o'clock position. Start with a slight forehand nozzle angle. At the 4 o'clock position the gun should point straight at the joint; 2) Second pass, weld from the 10 to the 1 o'clock position, then grind all stops and starts for at least 1 inch (25 mm); 3) Third pass, weld from the 4 to the 1 o'clock position; and 4) Fourth pass, weld from the 7 to the 10 o'clock position.

## Weld and Welding Symbols

**American National Standard Weld and Welding Symbols.—**Graphical symbols for welding provide a means of conveying complete welding information from the designer to the welder by means of drawings. The symbols and their method of use (examples of which are given in the table following this section) are part of the American National Standard ANSI/AWS A2.4-2012 sponsored by the American Welding Society.

In the Standard, a distinction is made between the terms *weld symbol* and *welding symbol*. Weld symbols, shown in the table *Basic Weld Symbols*, are ideographs used to indicate the type of weld desired, whereas welding symbol denotes a symbol made up of as many as eight elements conveying explicit welding instructions.

The eight elements which may appear in a welding symbol are: reference line; arrow; basic weld symbols; dimensions and other data; supplementary symbols; finish symbols; tail and specification; and process or other reference.

The standard location of elements of a welding symbol are shown in Fig. 1.

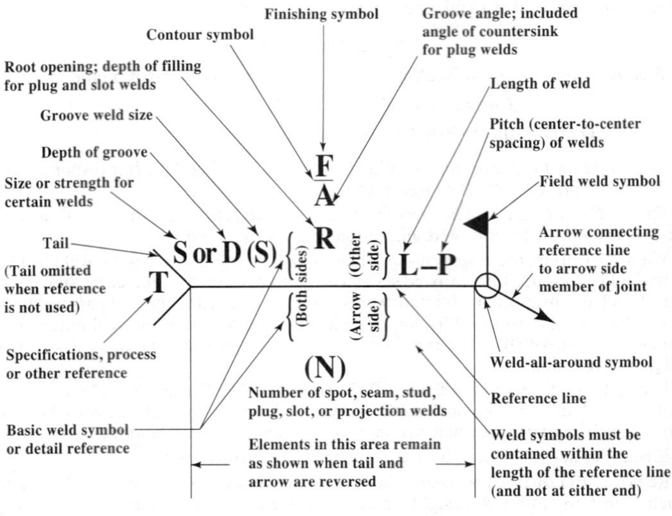


Fig. 1. Standard Location of Elements of a Welding Symbol

*Reference Line:* This is the basis of the welding symbol. All other elements are oriented with respect to this line. The arrow is affixed to one end and a tail, when necessary, is affixed to the other. Weld symbols should be contained within the length of this line.

*Arrow:* This connects the reference line to one side of the joint in the case of groove, fillet, flange, and flash or upset welding symbols. This side of the joint is known as the *arrow side* of the joint. The opposite side is known as the *other side* of the joint. In the case of plug, slot, projection, and seam welding symbols, the arrow connects the reference line to the outer surface of one of the members of the joint at the center line of the weld. In this case the member to which the arrow points is the *arrow side* member: the other member is the *other side* member. In the case of bevel and J-groove weld symbols, a two-directional arrow pointing toward a member indicates that the member is to be chamfered.

*Basic Weld Symbols:* These designate the type of welding to be performed. The basic symbols which are shown in the table *Basic Weld Symbols* are placed approximately in the center of the reference line, either above or below it or on both sides of it as shown in Fig. 1. Welds on the arrow side of the joint are shown by placing the weld symbols on the side of the reference line towards the reader (lower side). Welds on the other side of the joint are shown by placing the weld symbols on the side of the reference line away from the reader (upper side).

*Supplementary Symbols:* These convey additional information relative to the extent of the welding, where the welding is to be performed, and the contour of the weld bead. The "weld-all-around" and "field" symbols are placed at the end of the reference line at the base of the arrow as shown in Fig. 1 and the table *Supplementary Weld Symbols*.

*Dimensions:* These include the size, length, spacing, etc., of the weld or welds. The size of the weld is given to the left of the basic weld symbol and the length to the right. If the length is followed by a dash and another number, this number indicates the center-to-center spacing of intermittent welds. Other pertinent information such as groove angles, included angle of countersink for plug welds and the designation of the number of spot or projection welds are also located above or below the weld symbol. The number designating the number of spot or projection welds is always enclosed in parentheses.

*Contour and Finish Symbols:* The contour symbol is placed above or below the weld symbol. The finish symbol always appears above or below the contour symbol (see Fig. 1).

The following finish symbols indicate the method, not the degrees of finish: C—chipping; G—grinding; M—machining; R—rolling; and H—hammering.

For indication of surface finish refer to the section *SURFACE TEXTURE* starting on page 797.

*Tail:* The tail which appears on the end of the reference line opposite to the arrow end is used when a specification, process, or other reference is made in the welding symbol. When no specification, process, or other reference is used with a welding symbol, the tail may be omitted.

*Melt-Thru Symbol:* The melt-thru symbol is used only where 100 percent joint or member penetration plus reinforcement are required.

**Table 1. Basic Weld Symbols**

| Groove Symbols | | | | | | | |
|---|---|---|---|---|---|---|---|
| Square | Scarf | V | Bevel | U | J | Flare-V | Flare-bevel |

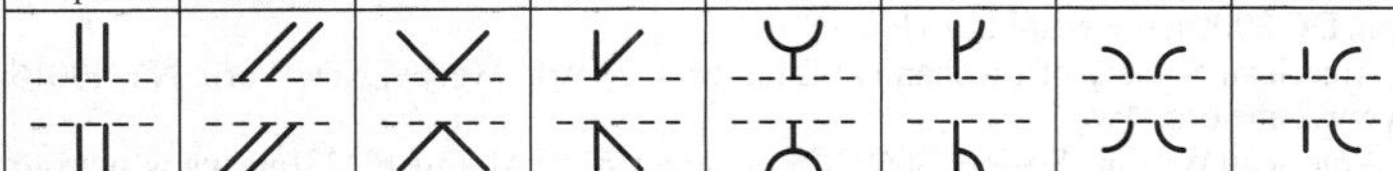

| Other Weld Symbols | | | | | | | | |
|---|---|---|---|---|---|---|---|---|
| Fillet | Plug[a] | Slot | Stud | Spot or projection | Seam | Back or backing | Surfacing | Edge |

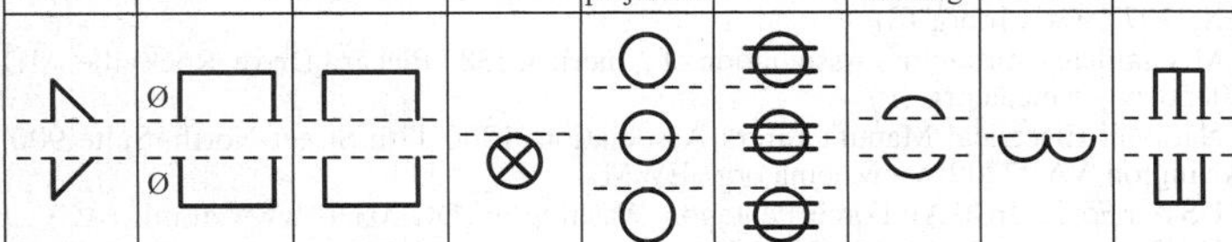

[a] Ø designates a round hole for a plug.

For examples of basic weld symbol applications see *Application of American National Standard Welding Symbols* starting on page 1622.

**Table 2. Supplementary Weld Symbols**

| Weld all around | Field weld | Melt-thru | Backing or spacer material | Contour | | |
|---|---|---|---|---|---|---|
| | | | | Flush | Convex | Concave |
|  | 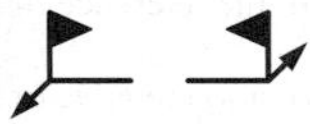 |  | 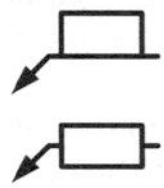 |  |  |  |

*Specification, Process, or Other Designation:* These are placed in the tail of the welding symbol and are in accordance with the American National Standard. They do not have to be used if a note is placed on the drawing indicating that the welding is to be done to some specification or that instructions are given elsewhere as to the welding procedure to be used.

*Letter Designations:* American National Standard letter designations for welding and allied processes are shown in the table on page 1621.

*Further Information:* For complete information concerning welding specification by the use of standard symbols, reference should be made to American National Standard ANSI/AWS A2.4-2012, which may be obtained from either the American National Standards Institute or the American Welding Society listed below.

**Welding Codes, Rules, Regulations, and Specifications.—**Codes recommending procedures for obtaining specified results in the welding of various structures have been established by societies, institutes, bureaus, and associations, as well as state and federal departments.

The latest codes, rules, etc., may be obtained from these agencies, whose names and addresses are listed as follows: PV = Pressure Vessels; P = Piping; T = Tanks; SB = Structural and Bridges; S = Ships; AC = Aircraft Construction; and EWM = Electrical Welding Machinery.

American Bureau of Shipping, ABS World Headquarters, 1701 City Plaza Drive, Spring, TX 77389, ww2.eagle.org. (S)

American Institute of Steel Construction, 130 East Randolph, Suite 2000, Chicago, IL, 60601, www.aisc.org. (SB)

American National Standards Institute, 25 West 43rd Street, 4th Floor, New York, NY 10036, www.ansi.org. (PV, P, EWM)

American Petroleum Institute, 200 Massachusetts Avenue NW, Suite 1100, Washington, DC 20001, www.api.org. (PV)

American Society of Mechanical Engineers, 2 Park Avenue, New York, NY 10016, www.asme.org. (PV)

American Welding Society, 8669 NW 36 Street, #130, Miami, FL 33166, www.aws.org. (T, S, SB, AC)

Federal Aviation Administration, 800 Independence Avenue SW, Room 300 East, Washington DC 20591, www.faa.gov. (AC)

Lloyd's Register of North America, 1330 Enclave Parkway, Suite 200, Houston, TX 77077, www.lr.org. (S)

Mechanical Contractors Association of America, 1385 Piccard Drive, Rockville, MD 20850, www.mcaa.org. (P)

National Electrical Manufacturers Association, 1300 17th Street North, Suite 900, Arlington, VA 22209, www.nema.org. (EWM)

US Air Force, 1690 Air Force, Pentagon, Washington, DC 20330, www.af.mil. (AC)

US Government Printing Office, 732 North Capitol Street NW, Washington, DC 20401, www.gpo.gov. (PV)

US Naval Facilities Engineering Systems Command, 1322 Patterson Avenue SE, Suite 1000, Washington Navy Yard, Washington, DC 20374, www.navfac.navy.mil. (SB)

## American National Standard Letter Designations for Welding and Allied Processes

*AWS A3.0M/A3.0-2010* **and** *AWS A2.4-2012*

**adhesive bonding (AB)**
**arc welding (AW)**
- arc stud welding (SW)
- atomic hydrogen welding (AHW)
- bare metal arc welding (BMAW)
- carbon arc welding (CAW)
  - gas carbon arc welding (CAW-G)
  - shielded carbon arc welding (CAW-S)
  - twin carbon arc welding (CAW-T)
- electrogas welding (EGW)
- flux cored arc welding (FCAW)
  - gas-shielded flux cored arc welding (FCAW-G)
  - self-shielded flux cored arc welding (FCAW-S)
- gas metal arc welding (GMAW)
  - pulsed gas metal arc welding (GMAW-P)
  - short circuit gas metal arc welding (GMAW-S)
- gas tungsten arc welding (GTAW)
  - pulsed gas tungsten arc welding (GTAW-P)
- magnetically impelled arc welding (MIAW)
- plasma arc welding (PAW)
- shielded metal arc welding (SMAW)
- submerged arc welding (SAW)
  - series submerged arc welding (SAW-S)

**brazing (B)**
- block brazing (BB)
- carbon arc brazing (CAB)
  - twin carbon arc brazing (TCAB)
- diffusion brazing (DFB)
- dip brazing (DB)
- electron beam brazing (EBB)
- exothermic brazing (EXB)
- furnace brazing (FB)
- induction brazing (IB)
- infrared brazing (IRB)
- laser beam brazing (LBB)
- resistance brazing (RB)
- torch brazing (TB)

**braze welding (BW)**
- arc braze welding (ABW)
- carbon arc braze welding (CABW)
- electron beam braze welding (EBBW)
- exothermic braze welding (EXBW)
- flow brazing (FLB)
- flow welding (FLOW)
- laser beam braze welding (LBBW)

**consumable guide electroslag welding (ESW-CG)**
**electroslag welding (ESW)**
**high energy beam welding (HEBW)**
- electron beam welding (EBW)
  - high vacuum electron beam welding (EBW-HV)
  - medium vacuum electron beam welding (EBW-MV)
  - nonvacuum electron beam welding (EBW-NV)
- laser beam welding (LBW)

**induction welding (IW)**
**oxyfuel gas welding (OFW)**
- air acetylene welding (AAW)
- oxyacetylene welding (OAW)
- oxyhydrogen welding (OHW)
- pressure gas welding (PGW)

**percussion welding (PEW)**
**resistance welding (RW)**
- flash welding (FW)
- pressure-controlled resistance welding (RW-PC)
- projection welding (PW)
- resistance seam welding (RSEW)
  - high-frequency seam welding (RSEW-HF)
  - induction seam welding (RSEW-I)
  - mash seam welding (RSEW-MS)
- resistance spot welding (RSW)
- upset welding (UW)
  - high-frequency upset welding (UW-HF)
  - induction upset welding (UW-I)

**soldering (S)**
- dip soldering (DS)
- furnace soldering (FS)
- induction soldering (IS)
- infrared soldering (IRS)
- iron soldering (INS)
- resistance soldering (RS)
- torch soldering (TS)
- ultrasonic soldering (USS)
- wave soldering (WS)

**solid-state welding (SSW)**
- coextrusion welding (CEW)
- cold welding (CW)
- diffusion welding (DFW)
  - hot isostatic pressure welding (HIPW)
- explosion welding (EXW)
- forge welding (FOW)
- friction welding (FRW)
  - direct drive friction welding (FRW-DD)
  - friction stir welding (FSW)
  - inertia friction welding (FRW-I)
- hot pressure welding (HPW)
- roll welding (ROW)
- ultrasonic welding (USW)

**thermal cutting (TC)**
- arc cutting (AC)
  - carbon arc cutting (CAC)
    - air carbon arc cutting (CAC-A)
  - gas metal arc cutting (GMAC)
  - gas tungsten arc cutting (GTAC)
  - plasma arc cutting (PAC)
  - shielded metal arc cutting (SMAC)
- high energy beam cutting (HEBC)
  - electron beam cutting (EBC)
  - laser beam cutting (LBC)
    - laser beam air cutting (LBC-A)
    - laser beam evaporative cutting (LBC-EV)
    - laser beam inert gas cutting (LBC-IG)
    - laser beam oxygen cutting (LBC-0)
- oxygen cutting (OC)
  - flux cutting (OC-F)
  - metal powder cutting (OC-P)
  - oxyfuel gas cutting (OFC)
    - oxyacetylene cutting (OFC-A)
    - oxyhydrogen gas cutting (OFC-H)
    - oxynatural gas cutting (OFC-N)
    - oxypropane cutting (OFC-P)
  - oxygen arc cutting (OAC)
  - oxygen lance cutting (OLC)

**thermal gouging (TG)**
- carbon arc gouging (CAG)
- oxygen gouging (OG)
- plasma arc gouging (PAG)

**thermal spraying (THSP)**
- arc spraying (ASP)
  - flame spraying (FLSP)
    - wire flame spraying (FLSP-W)
- high velocity oxyfuel spraying (HVOF)
  - plasma spraying (PSP)
  - vacuum plasma spraying (VPSP)

**thermite welding (TW)**

*Sources:* ANSI/AWS A3.0M/ A3.0-2010, *Standard Welding Terms and Definitions: Including Terms for Adhesive Bonding, Brazing, Soldering, Thermal Cutting, and Thermal Spraying*; ANSI/AWS A2.4-2012, *Standard Symbols for Welding, Brazing, and Nondestructive Examination*. Courtesy of the American Welding Society.

**Application of American National Standard Welding Symbols**

| Desired Weld | Symbol | Symbol Meaning |
|---|---|---|
| 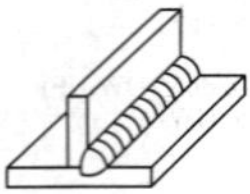 | 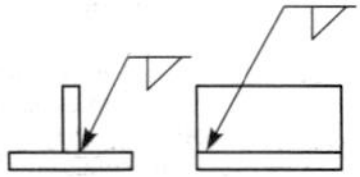 | Symbol indicates fillet weld on *arrow side* of the joint. |
| 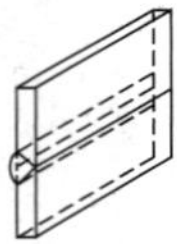 | 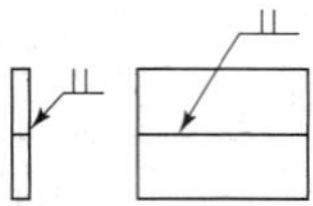 | Symbol indicates square-groove weld on *other side* of the joint. |
| 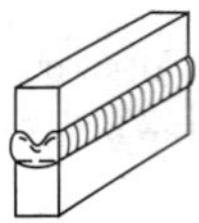 | 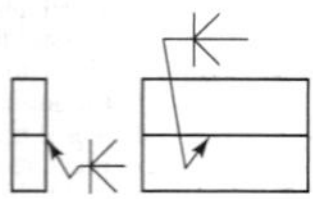 | Symbol indicates bevel-groove weld on both sides of joint. Breaks in arrow indicate bevels on upper member of joint. Breaks in arrows are used on symbols designating bevel and J-groove welds. |
| 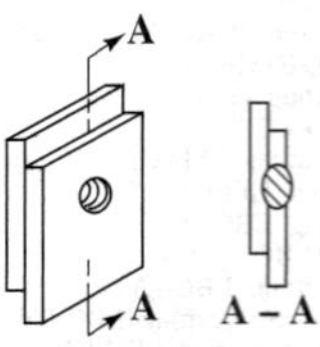 | 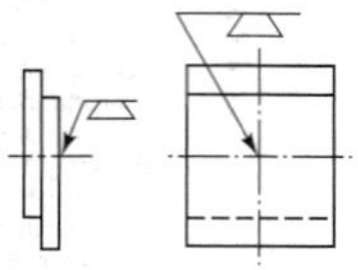 | Symbol indicates plug weld on *arrow side* of joint. |
| 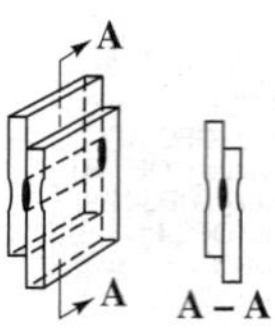 | 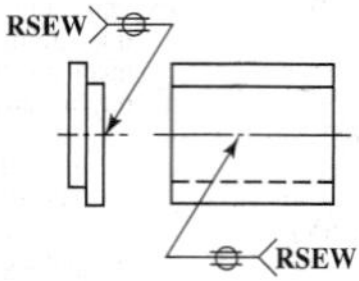 | Symbol indicates resistance-seam weld. Weld symbol appears on both sides of reference line, indicating that *arrow* and *other side* of joint references have no significance. |
| 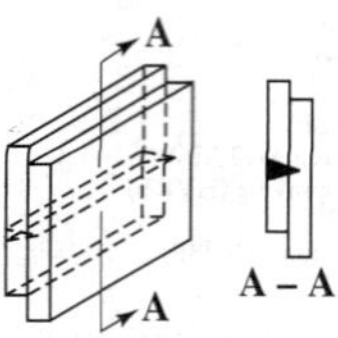 | 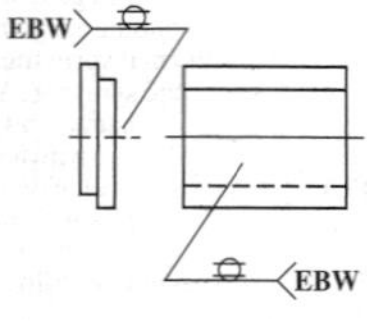 | Symbol indicates electron beam seam weld on *other side* of joint. |

**Application of American National Standard Welding Symbols** *(Continued)*

| Desired Weld | Symbol | Symbol Meaning |
|---|---|---|
| 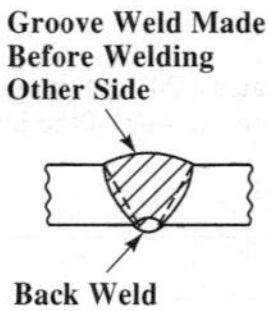  | 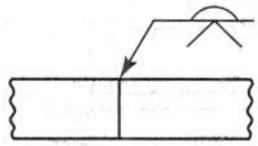 | Symbol indicates single-pass back weld. |
| 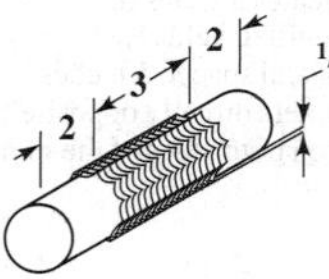  | 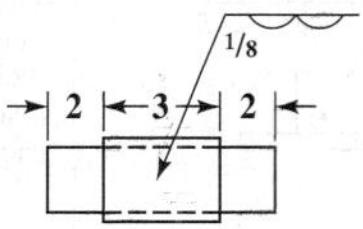  | Symbol indicates a built-up surface ⅛ inch thick. |
| 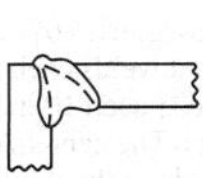 | 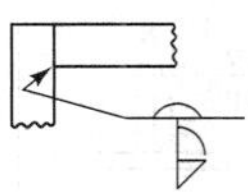 | Symbol indicates a bead-type back weld on the *other side* of joint, and a J-groove grooved horizontal member (shown by break in arrow) and fillet weld on *arrow side* of the joint. |
| 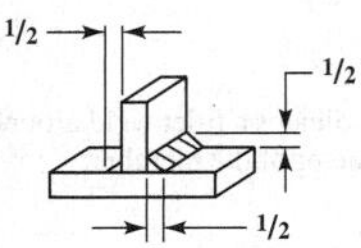  | 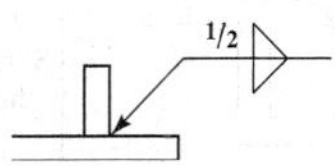  | Symbol indicates two fillet welds, both with ½-inch leg dimensions. |
| 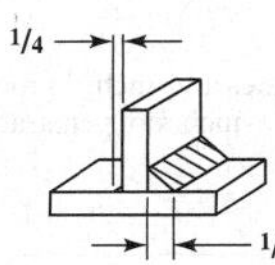  | 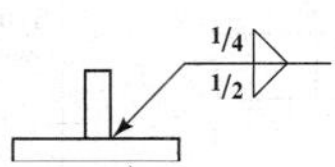  | Symbol indicates a ½-inch fillet weld on *arrow side* of the joint and a ¼-inch fillet weld on *far side* of the joint. |
| 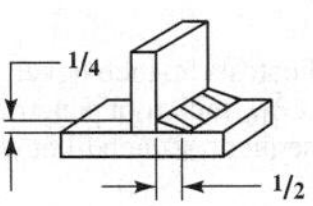  | 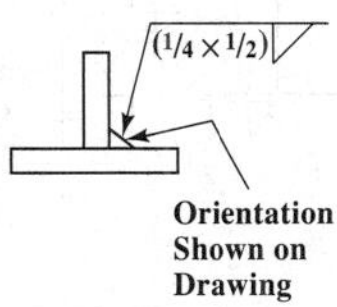  | Symbol indicates a fillet weld on *arrow side* of joint with ¼- and ½-inch legs. Orientation of legs must be shown on drawing. |

**Application of American National Standard Welding Symbols** *(Continued)*

| Desired Weld | Symbol | Symbol Meaning |
|---|---|---|
| 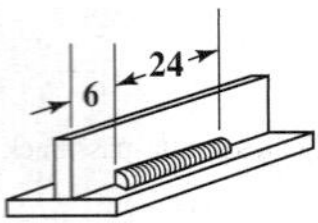  | 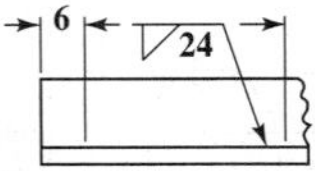  | Symbol indicates a 24-inch long fillet weld on the *arrow side* of the joint. |
| **Locate Welds at Ends of Joint** 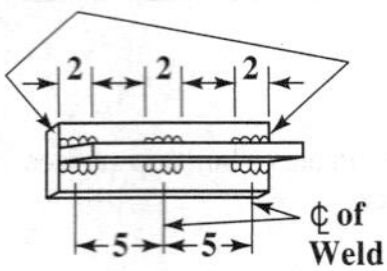  | 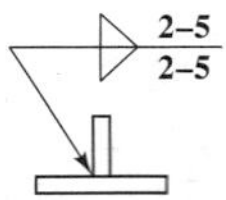  | Symbol indicates a series of intermittent fillet welds each 2 inches long and spaced 5 inches apart on centers directly opposite each other on both sides of the joint. |
| **Locate Welds at Ends of Joint** 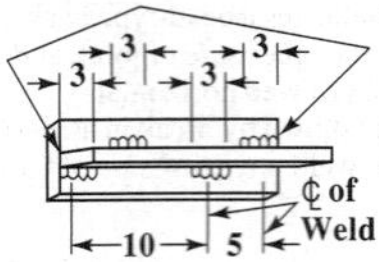  | 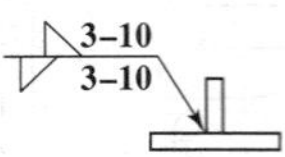  | Symbol indicates a series of intermittent fillet welds each 3 inches long and spaced 10 inches apart on centers. The centers of the welds on one side of the joint are displaced from those on the other. |
| 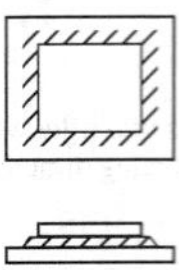 | 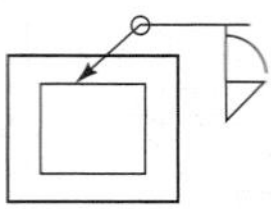 | Symbol indicates a fillet weld around the perimeter of the member. |
| 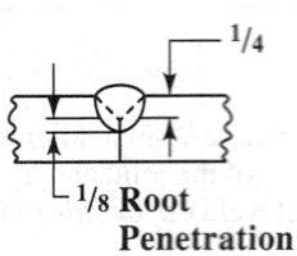  | 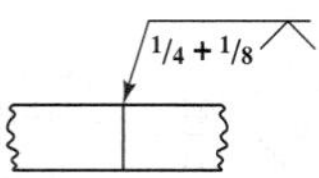  | Symbol indicates a ¼-inch V-groove weld with a ⅛-inch root penetration. |
| 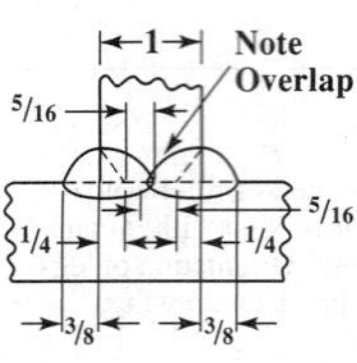  | 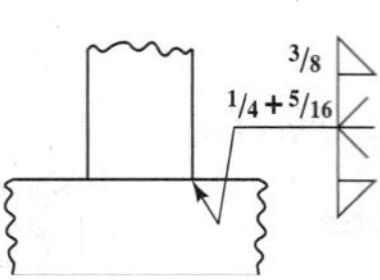  | Symbol indicates a ¼-inch bevel weld with a 5⁄16-inch root penetration plus a subsequent ⅜-inch fillet weld. |

**Application of American National Standard Welding Symbols** *(Continued)*

<table>
<tr><th>Desired Weld</th><th>Symbol</th><th>Symbol Meaning</th></tr>
<tr><td>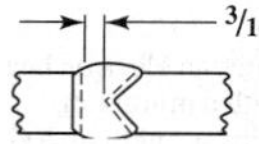</td><td>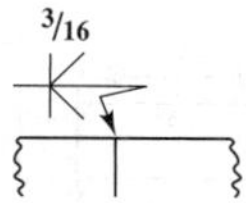</td><td>Symbol indicates a bevel weld with a root opening of 3⁄16 inch.</td></tr>
<tr><td>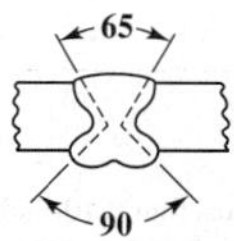</td><td>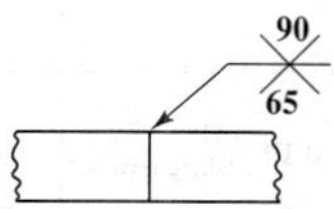</td><td>Symbol indicates a V-groove weld with a groove angle of 65 degrees on the arrow side and 90 degrees on the other side.</td></tr>
<tr><td></td><td>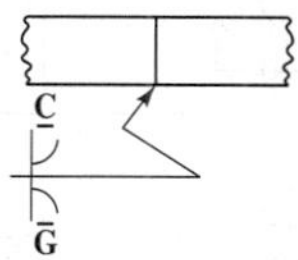</td><td>Symbol indicates a flush surface with the reinforcement removed by chipping on the other side of the joint and a smooth grind on the arrow side. The symbols C and G should be the user's standard finish symbols.</td></tr>
<tr><td>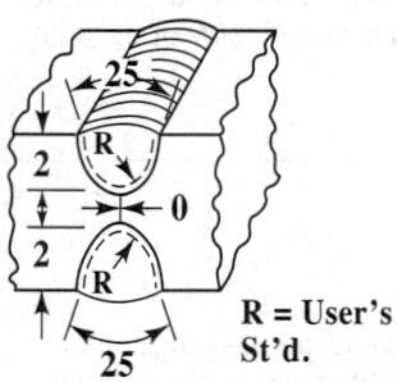</td><td>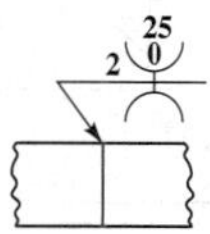</td><td>Symbol indicates a 2-inch U-groove weld with a 25-degree groove angle and no root opening for both sides of the joint.</td></tr>
<tr><td>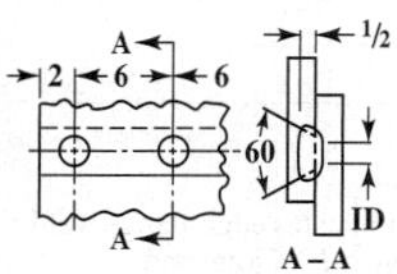</td><td>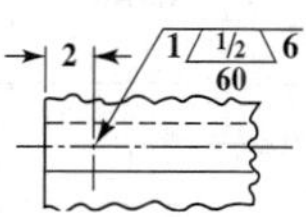</td><td>Symbol indicates plug welds of 1-inch diameter, a depth of filling of ½ inch and a 60-degree angle of countersink spaced 6 inches apart on centers.</td></tr>
<tr><td>Preparation<br>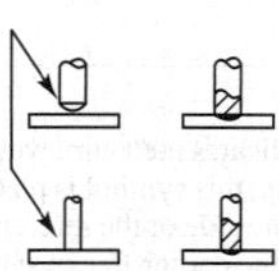</td><td>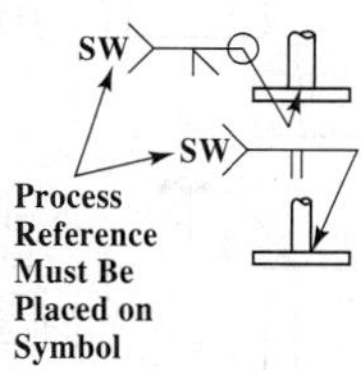</td><td>Symbol indicates all-around bevel and square-groove weld of these studs.</td></tr>
</table>

**Application of American National Standard Welding Symbols** *(Continued)*

| Desired Weld | Symbol | Symbol Meaning |
|---|---|---|
| **Min. Acceptable Shear Strength 200 lb/lin. in.** 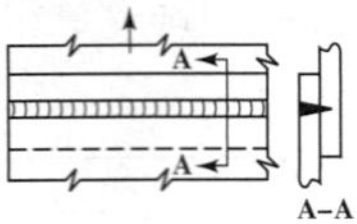  | 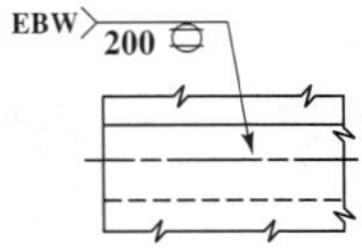  | Symbol indicates an electron beam seam weld with a minimum acceptable joint strength of 200 pounds per linear inch. |
| 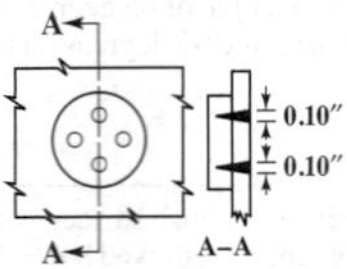  | 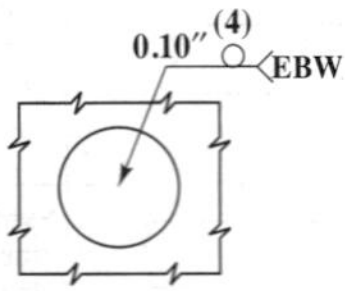  | Symbol indicates four 0.10-inch diameter electron beam spot welds located at random. |
| 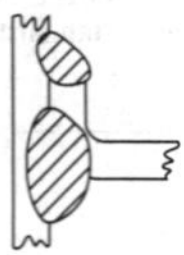 | 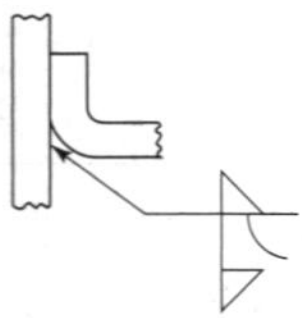 | Symbol indicates a fillet weld on the *other side* of joint and a flare-bevel-groove weld and a fillet weld on the *arrow side* of the joint. |
| 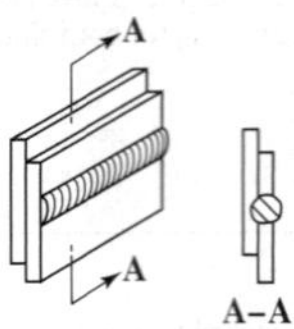  | 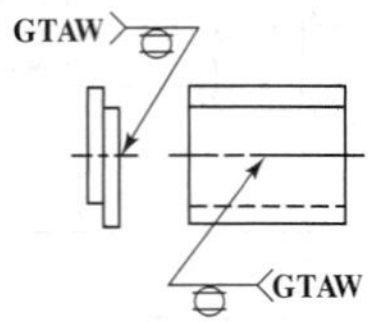  | Symbol indicates gas tungsten-arc seam weld on *arrow side* of joint. |
| 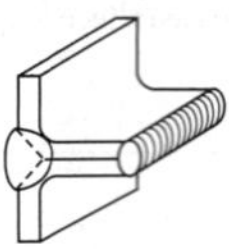 | 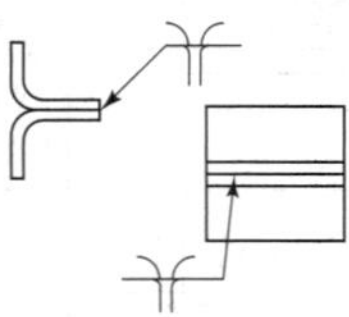 | Symbol indicates edge-flange weld on *arrow side* of joint and flare-V-groove weld on *other side* of joint. |
| 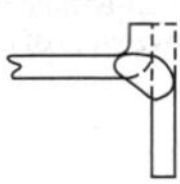 | 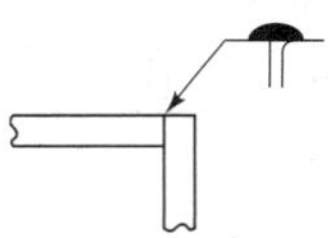 | Symbol indicates melt-thru weld. By convention, this symbol is placed on the opposite side of the reference line from the corner-flange symbol. |

## Nondestructive Testing

Nondestructive testing (NDT) methods are used to examine a material, component, or assembly to measure thickness, determine structure, and/or search for surface or internal cracks, material flaws, or other nonhomogeneities. "Nondestructive" refers to a method of examination that does not affect or impair the use of the subject for its intended purpose. NDT is particularly useful in the welding process.

**Nondestructive Testing Methods.—**The primary NDT methods, introduced below, include ultrasonic, radiography, magnetic particle, penetrant, and eddy current. Excellent references, including books and standards, as well as certification and qualification programs, provide in-depth information on the principles and applications of each approach.

*Ultrasonic Testing (UT):* The workhorse of NDT, ultrasonic testing utilizes high-frequency sound waves. An electronic pulser/receiver drives a transducer, which sends ultrasonic sound waves through a test subject. These waves travel at the same distinct speed for any given material or medium that has matter (see Table 1). The time and distance that sound waves travel through a substance and echo off a reflecting surface can be analyzed to determine medium characteristics, measure thickness, and detect and evaluate flaws.

**Table 1. Material Velocities**

| Medium | Velocity | | Time/Distance Traveled | | Distance/Time Traveled | |
|---|---|---|---|---|---|---|
| | in/second | cm/second | second/mile | second/km | mile/second | km/second |
| Air (68°F/20°C) | 14,016 | 35,601 | 4.5 | 2.80 | | |
| Water (68°F/20°C) | 57,992 | 147,300 | 1.09 | 0.68 | | |
| Steel 1020 | 232,008 | 589,300 | | | 3.7 | 5.95 |
| Aluminum 6061-T6 | 247,992 | 629,900 | | | 3.9 | 6.27 |
| Beryllium | 507,000 | 1,287,780 | | | 8.0 | 12.87 |

The calibration of the instrumentation is critical and must be carried out with reference blocks of the same material and velocity. This allows accurate measurement of the time and distance that a sound wave has traveled to and from a reflecting surface. Whether measuring a backwall for thickness or locating the depth and size of discontinuities, the accuracy of test results is only as reliable as the calibration. For the plate shown in Fig. 1, calculation of the angle, thickness, and sound path distance are required; these are obtained with formulas frequently used for ultrasonic testing.

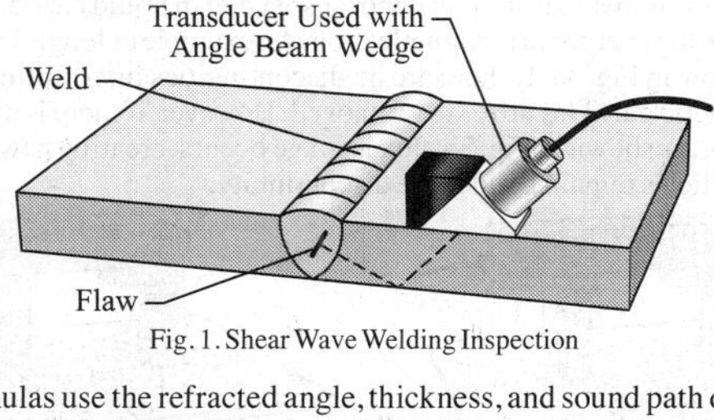


Fig. 1. Shear Wave Welding Inspection

Standard formulas use the refracted angle, thickness, and sound path distance to calculate the flaw depth and location. Other formulas are used to calculate acoustic impedance, nearfield, wavelength, and angular relationships. For instance, a formula developed from Snell's law is used to calculate incident and refracted angles.

*Radiographic Testing (RT):* Favored by many industries, due to the insight provided, radiographic testing is another method of detecting subsurface discontinuities and other material variations. It involves transmission of radiation through the material being tested, with the results recorded on photographic film or the digital equivalent. In general, areas of reduced thickness or lower-density material absorb less radiation, and more is transmitted to the film or digital sensors. As shown in Fig. 2, this produces relatively darker regions in the resulting imagery, as compared to thicker and/or denser areas.

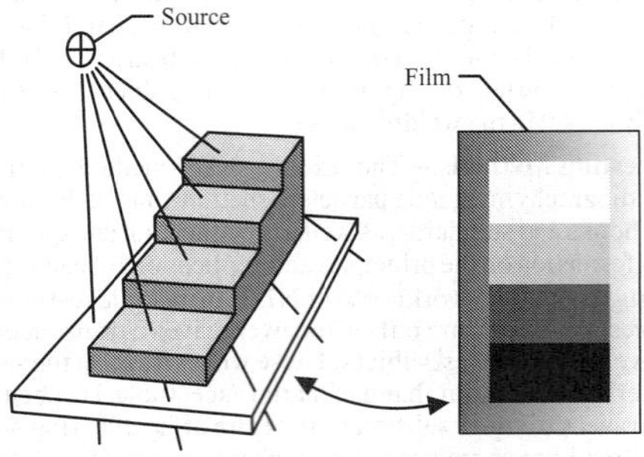


Fig. 2. Radiographic Absorption Effect on a Steel Step Wedge

RT can detect voids, such as porosity, slag, and tungsten inclusions, that indicate weld discontinuities. Critical flaws, including cracks and incomplete fusion, are more difficult to see unless the radiation source is oriented at a suitable direction with respect to the flaw. A key advantage of using radiographic film or digital media is that a permanent record of the testing can be stored.

A disadvantage of RT (unlike UT) is that it requires access to both sides of the material: one for the transmission source and the other for the film. Another issue is the danger posed to humans by exposure to radiation, particularly when using portable X-ray equipment involving isotopes such as iridium-192 and cobalt-60. Training in radiation safety is mandatory, along with safeguards to protect personnel in the test vicinity.

*Magnetic Particle Testing (MT):* Used to detect surface and near-surface discontinuities that are primarily linear in nature, magnetic particle testing is relatively easy to use, is low in cost, and provides immediate results. However, it is only effective on ferromagnetic materials.

Magnetization is achieved by passing an electric current through the material, or by using an external magnetizing coil; permanent magnets also may be used. Finely divided magnetic particles—either as a dry powder or suspended in liquid carrier—are applied to and removed from the part during magnetization. An example of longitudinal magnetization is shown below in Fig. 3a. If there are no discontinuities interrupting the flow of the current, the magnetic particles are easily removed. However, if there is an interruption in the flow of current, as shown in Fig. 3b, flux leakage occurs, creating new north and south poles, and the particles remain fixed at the discontinuity.

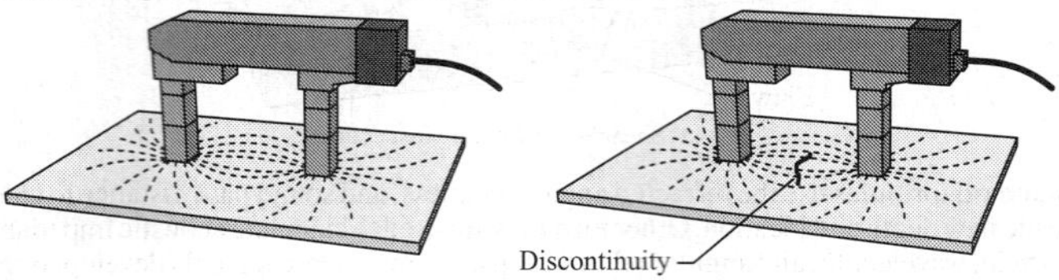


Fig. 3a. Magnetic Flux Lines

Fig. 3b. Interruption of Flow

Evaluation of test results is arguably the most critical stage of MT and requires distinguishing among false, nonrelevant, and relevant indications. Experience and familiarity with codes and specifications, application of proper techniques, stringent equipment performance checks, and adherence to the detailed and qualified procedure are essential.

*Penetrant Testing (PT):* Widely used to reveal surface discontinuities on numerous materials—all metals (including aluminum, titanium, magnesium, and other nonferrous compounds), glass, ceramics, and various composites—penetrant testing works on the principle of capillary attraction. This surface tension phenomenon permits liquids to be drawn into tight crevices, regardless of orientation or gravitational forces.

After the test area is cleaned, a dye concentrate is applied to and left on the material for a specific (dwell) time. When the penetrant is removed, the part usually is dried, and a developer is applied. The developer acts as an absorbent, drawing the remaining penetrant out of discontinuities open to the surface. Observing the test surface under appropriate lighting *immediately* after application of the developer assists proper interpretation of indications, though the final evaluation cannot be performed before the minimum developer dwell time has elapsed.

Utilizing the appropriate method as shown in Table 2, very large parts can be tested, or numerous small parts, such as turbine and compressor blades, can be processed simultaneously.

**Table 2. Classification of Penetrant Materials and Processes**

| | | |
|---|---|---|
| Type I | Fluorescent penetrants | May be used with all developers |
| Type II | Visible (color contrast) penetrants | Nonaqueous and aqueous developers |
| Method A | Water-washable visible dye penetrants | Water-washable fluorescent penetrants |
| Method B | Post-emulsified visible or fluorescent | Lipophilic emulsifier (oil-based) |
| Method C | Solvent removable | Visible or fluorescent penetrants |
| Method D | Post-emulsified fluorescent penetrants | Hydrophilic emulsifier (water-based) |
| Developers | Dry powder, nonaqueous | Water suspended, water soluble |

*Note:* Sensitivity levels available are 1, 2, 3, and 4 (4 is the highest).

PT is versatile enough to be used in the field with or without electricity, is relatively low cost, and is easy to use. However, it is time-consuming and usually requires additional cleaning of the part after testing is complete.

*Eddy Current Testing (ET):* This sophisticated NDT method can be used to detect conductivity variations; discontinuities; material thickness, including plating or cladding on a base material; and spacing between conductive layers. Based on the principle of electromagnetism, ET uses the inductive properties of alternating current flowing through a primary coil to develop an alternating magnetic field. Interruption of this magnetic field causes an impedance change in the coil, recorded as a phase change by the instrumentation.

In theory, ET is much more complex; it was not until the early 1980s that explanatory material was produced that made this process understandable to laypersons. While ET is a high-speed process, it has a limited depth of penetration, depending on such factors as the test frequency, type of coil, and conductivity of materials being tested.

**Other NDT Methods.—**In addition to the five primary NDT methods discussed above, there are six other methods recognized by the American Society of Nondestructive Testing (ASNT), including visual testing. Each has unique advantages and limitations.

*Visual Testing (VT):* Being the first inspection, visual acuity examination generally is required for NDT certification. Many factors can affect VT, including cleanliness, brightness, glare, surface condition, shape, size, and temperature of an object. While the most important instrument in visual testing is the human eye, direct visual aids, such as light sources, magnifiers, cameras, and measuring devices are commonly used to verify conformance to industry codes, standards, and specification requirements.

*Thermal/Infrared Testing (IR):* The laws of thermodynamics can be applied to the detection of many anomalies in metals, cements, and composites, as well as in the chemical industries. When there is a difference of temperature between two objects, or when an object changes temperature, the net flow energy is transferred from warmer areas to cooler areas or vice versa, until thermal equilibrium is reached. For example, IR can detect moisture that has intruded into the insulation of buildings and liquid $CO_2$ storage tanks.

*Acoustic Emission Testing (AE):* This NDT method is commonly used to detect and locate faults in mechanically loaded components or structures. When a material is subjected to mechanical loading or stress, such discontinuities release energy in the form of stress waves that can be detected by strategically placing sensors, which pick up these signals and convert the energy into voltage. The voltage is then electronically amplified and timing circuits are used to process the AE signal data and identify the fault.

*Leak Testing (LT):* Leak testing detects a flow of gas or liquid through the wall of a vessel, via a discontinuity (such as a hole, crack, or bad seal). Leaks require a pressure difference and travel from higher to lower pressure, whether from inside an object with greater pressure than the surrounding atmospheric pressure, or from the atmosphere to inside a vessel under a vacuum. Techniques for LT include mass spectrometer, halogen diode, pressure change, and bubble testing.

*Neutron Radiography (NRT):* This technique is often used with conventional radiographic testing. But neutrons interact with the nucleus of the atom, rather than with the electrons, and are absorbed by matter very differently than X-rays and gamma rays. NRT is ideal for imaging light elements with low atomic numbers (such as hydrogen, water, carbon, and plastics) and penetrating heavy elements with high atomic numbers (such as aluminum, lead, steel, and titanium). Thus, neutron radiographs can show organic materials or water, with many structural materials being nearly transparent.

*Magnetic Flux Leakage (MFL):* This method uses a direct current (DC) magnetizing field to create flux density to bring a material to near saturation. Surface or internal discontinuities, such as cracks, pits, and other defects, interrupt the flux field, causing it to leak beyond the test surface. The longitudinal and transverse magnetizing fields enable identification of defects in both directions. Sensors and configured inspection devices gather and transmit data for processing and analysis.

**Nondestructive Testing Symbol Application.—**The application of NDT symbols is covered in American National Standard ANSI/AWS A2.4-2012. This Standard also includes the letters used to indicate each basic testing method: Acoustic Emission (AET), Eddy Current (ET), Leak (LT), Magnetic Particle (MT), Neutron Radiographic (NRT), Penetrant (PT), Proof (PRT), Radiographic (RT), Ultrasonic (UT), and Visual (VT).

*Testing Symbol Elements:* The testing symbol consists of the following elements: number of tests (N), length of section to be tested (L), reference line, test in field, test-all-around symbol, tail, arrow, and specification or other reference. The standard locations of the testing symbol elements are shown in Fig. 4.

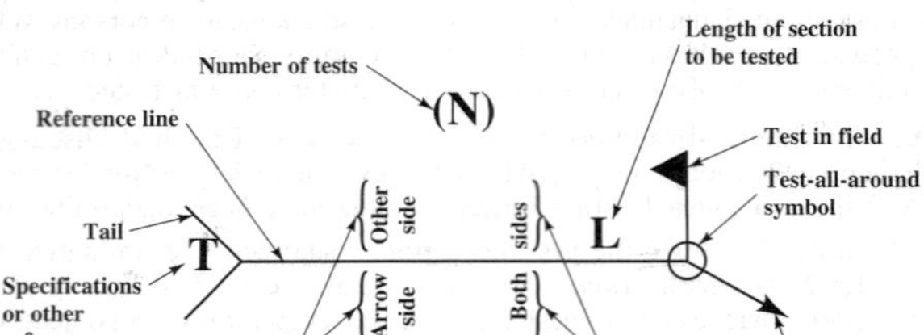


Fig. 4. Locations of Testing Symbol Elements

The arrow connects the reference line to the part to be tested. The side of the part that the arrow points to is considered the *arrow side*. The side opposite the arrow side is considered the *other side*.

*Location of Testing Symbol:* Tests to be made on the arrow side of the part are indicated by positioning the testing symbol on the side of the reference line toward the reader (Fig. 5a).

Tests to be made on the other side of the part are indicated by the testing symbol being on the side of the reference line away from the reader (Fig. 5b).

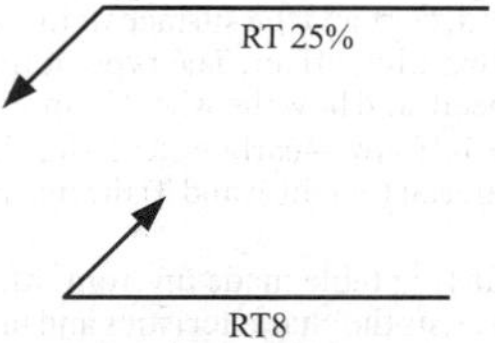


Fig. 5a. Position of Testing Symbols Indicates Tests to Be Made on the Arrow Side

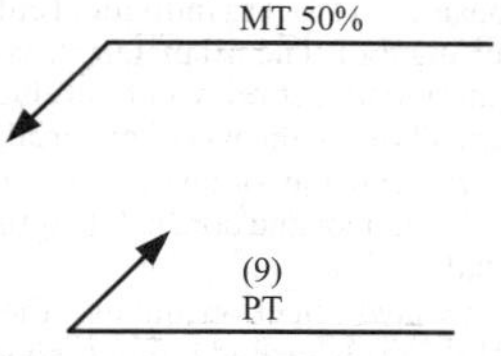


Fig. 5b. Position of Testing Symbols Indicates Tests to Be Made on the Other Side

To specify where only a certain length of a section is to be considered, the actual length or percentage of length to be tested is shown to the right of the test symbol. To specify the number of tests to be taken on a joint or part, the number of tests is shown in parentheses.

Tests to be made on both sides of the part are indicated by test symbols on both sides of the reference line. Where nondestructive symbols have no arrow or other significance, the testing symbols are centered in the reference line.

*Direction of Radiation:* When specified, the direction of radiation may be shown, in conjunction with the radiographic or neutron radiographic basic testing symbols, by means of a radiation symbol located on the drawing at the desired angle.

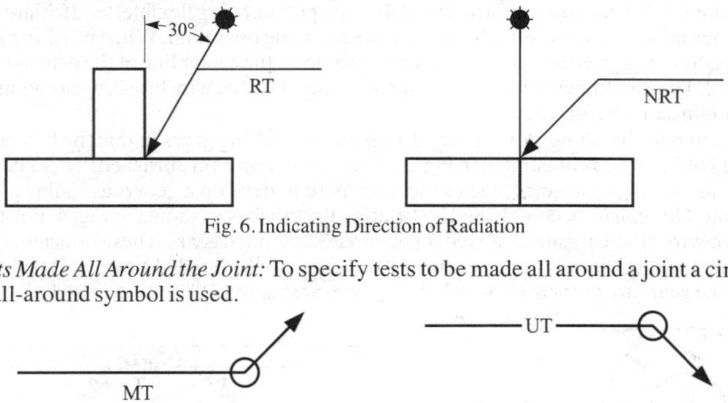


Fig. 6. Indicating Direction of Radiation

*Tests Made All Around the Joint:* To specify tests to be made all around a joint a circular test-all-around symbol is used.

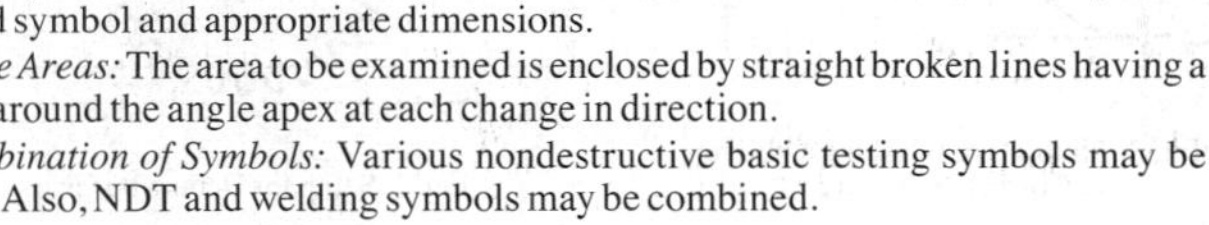


Fig. 7. Specifying Circular Testing

*Areas of Revolution:* For NDT of areas of revolution, the area is indicated by the test-all-around symbol and appropriate dimensions.

*Plane Areas:* The area to be examined is enclosed by straight broken lines having a small circle around the angle apex at each change in direction.

*Combination of Symbols:* Various nondestructive basic testing symbols may be combined. Also, NDT and welding symbols may be combined.

## FINISHING OPERATIONS

### Power Brush Finishing

Power brush finishing is a production method of metal finishing that employs wire, elastomer bonded wire, or nonmetallic (cord, natural fiber or synthetic) brushing wheels in automatic machines, semi-automatic machines and portable air tools to smooth or roughen surfaces, remove surface oxidation and weld scale or remove burrs.

**Description of Brushes.—**Brushes work in the following ways: the wire points of a brush can be considered to act as individual culling tools so that the brush, in effect, is a multiple-tipped cutting tool. The fill material, as it is rotated, contacts the surface of the work and imparts an impact action which produces a coldworking effect. The type of finish produced depends upon the wheel material, wheel speed, and how the wheel is applied.

Brushes differ in the following ways: 1) fill material (wire—carbon steel, stainless steel; synthetic; Tampico; and cord); 2) length of fill material (or trim); and 3) the density of the fill material.

To aid in wheel selection and use, the accompanying table made up from information supplied by *The Osborn Manufacturing Company* lists the characteristics and major uses of brushing wheels.

**Use of Brushes.—**The brushes should be located so as to bring the full face of the brush in contact with the work. Full face contact is necessary to avoid grooving the brush. Operations that are set up with the brush face not in full contact with the work require some provision for dressing the brush face. When the tips of a brush, used with full face contact, become dull during use with subsequent loss of working clearance, reconditioning and resharpening is necessary. This is accomplished simply and efficiently by alternately reversing the direction of rotation during use.

**Deburring and Producing a Radius on the Tooth Profile of Gears.—**The brush employed for deburring and producing a radius on the tooth profile of gears is a short trim, dense, wire-fill radial brush. The brush should be set up so as to brush across the edge as shown in Fig. 1a. Line contact brushing, as shown in Fig. 1b should be avoided because the Crisis face will wear non-uniformly; and the wire points, being flexible, tend to flare to the side, thus minimizing the effectiveness of the brushing operation. When brushing gears, the brushes are spaced and contact the tooth profile on the center line of the gear as shown in Fig. 2. This facilitates using brush reversal to maintain the wire brushing points at their maximum cutting efficiency.

The setup for brushing spline bores differs from brushing gears in that the brushes are located off-center, as illustrated in Fig. 3. When helical gears are brushed, it is sometimes necessary to favor the acute side of the gear tooth to develop a generous radius prior to shaving. This can be accomplished by locating the brushes as shown in Fig. 4. Elastomer bonded wire-filled brushes are used for deburring fine pitch gears. These brushes remove the burrs without leaving any secondary roll. The use of bonded brushes is necessary when the gears are not shaved after hobbing or gear shaping.

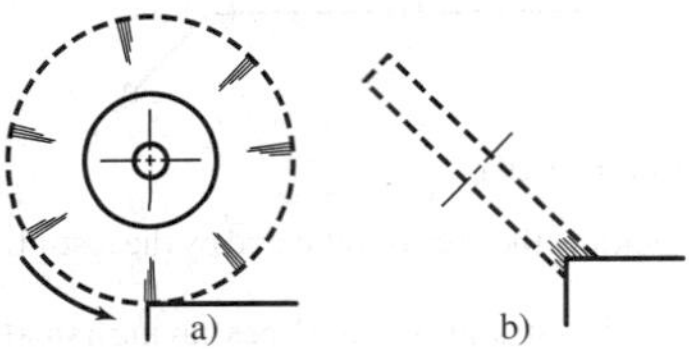


Fig. 1. Methods of Brushing an Edge:
a) Correct, b) Incorrect

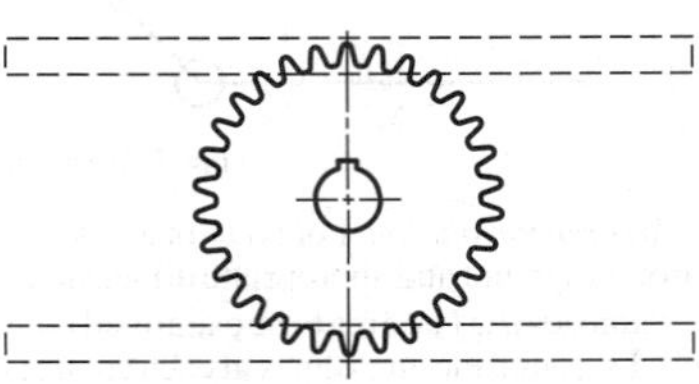

Fig. 2. Setup for Deburring Gears

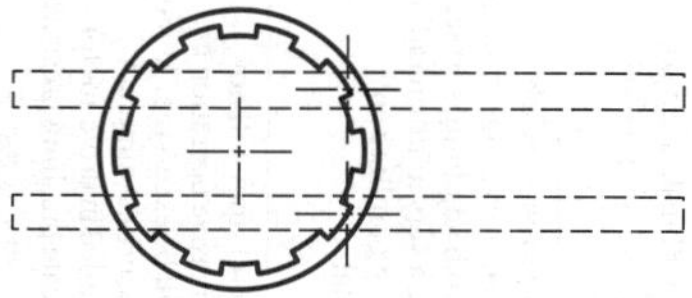

Fig. 3. Setup for Brushing Broached Splines

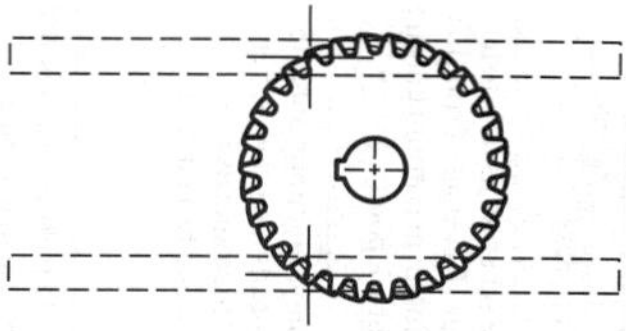

Fig. 4. Setup for Finishing Helical Gears

**Adjustments for Eliminating Undesirable Conditions in Power Brush Finishing**

| Undesirable Condition | Possible Adjustments for Eliminating Condition |
|---|---|
| Brush works too slowly | (1) Decrease trim length and increase fill density.<br>(2) Increase filament diameter.<br>(3) Increase surface speed by increasing R.P.M. or outside diameter. |
| Brush works too fast | (1) Reduce filament diameter.<br>(2) Reduce surface speed by reducing R.P.M. or outside diameter.<br>(3) Reduce fill density.<br>(4) Increase trim length. |
| Action of brush peens burr to adjacent surface | (1) Decrease trim length and increase fill density.<br>(2) If wire brush tests indicate metal too ductile (burr is peened rather than removed), change to nonmetallic brush such as a treated Tampico brush used with a burring compound. |
| Finer or smoother finish required | (1) Decrease trim length and increase fill density.<br>(2) Decrease filament diameter.<br>(3) Try treated Tampico or cord brushes with suitable compounds at recommended speeds.<br>(4) Use auxiliary buffing compound with brush. |
| Finish too smooth and lustrous | (1) Increase trim length.<br>(2) Reduce brush fill density.<br>(3) Reduce surface speed.<br>(4) Increase filament diameter. |
| Brushing action not sufficiently uniform | (1) Devise handheld or mechanical fixture or machine which will avoid irregular off-hand manipulation.<br>(2) Increase trim length and decrease fill density. |

## Polishing and Buffing

The terms "polishing" and "buffing" are sometimes applied to similar classes of work in different plants, but according to approved usage of the terms, there is the following distinction: Polishing is any operation performed with wheels having abrasive glued to the working surfaces, whereas buffing is done with wheels having the abrasive applied loosely instead of imbedding it into glue; moreover, buffing is not so harsh an operation as ordinary polishing, and it is commonly utilized to obtain very fine surfaces having a "grainless finish."

**Polishing Wheels.**—The principal materials from which polishing wheels are made are wood, leather, canvas, cotton cloth, plastics, felt, paper, sheepskin, impregnated rubber, canvas composition, and wool. Leather and canvas are the materials most commonly used in polishing wheel construction. Wooden wheels covered with material to which emery or some other abrasive is glued are employed extensively for polishing flat surfaces,

**Characteristics and Applications of Brushes Used in Power Finishing**

| Brush Type | Description | Operating Speed Range, sfpm | Uses | Remarks |
|---|---|---|---|---|
| Radial, short trim dense wire fill | Develops very little impact action but maximum cutting action. | 6500 | Removal of burrs from gear teeth and sprockets. Produces blends and radii at juncture of intersecting surfaces. | Brush should be set up so as to brush across any edge. Reversal of rotation needed to maintain maximum cutting efficiency of brush points. |
| Radial, medium to long trim twisted knot wire fill | Normally used singly and on portable tools. Brush is versatile and provides high impact action. | 7500-9500 for high speeds. 1200 for slow speeds. | For cleaning welds in the automotive and pipeline industries. Also for cleaning surfaces prior to painting, stripping rubber flash from molded products and cleaning mesh-wire conveyor belts. | Surface speed plays an important role since at low speeds the brush is very flexible and at high speeds it is extremely hard and fast cutting. |
| Radial, medium to long trim crimped wire fill | With the 4- to 8-inch diameter brush, part is handheld. With the 10- to 15-inch diameter brush, part is held by machine. | 4500-6000 | Serves as utility tool on bench grinder for removing feather grinding burrs, machining burrs, and for cleaning and producing a satin or matte finish. | Good for handheld parts as brush is soft enough to conform to irregular surfaces and hard-to-reach areas. Smaller diameter brushes are not recommended for high-production operations. |
| Radial, sectional, nonmetallic fill (treated and untreated Tampico or cord) | Provides means for improving finish or improving surface for plating. Works best with grease base deburring or buffing compound. | 5500-6500<br>7500 for polishing | For producing radii and improving surface finish. Removes the sharp peaks that fixed abrasives leave on a surface so that surface will accept a uniform plating. Polishing marks and draw marks can be successfully blended. | Brush is selective to an edge which means that it removes metal from an edge but not from adjoining surfaces. It will produce a very uniform radius without peening or rolling any secondary metal. |
| Radial, wide-face, nonmetallic fill (natural fibers or synthetics) | Can be used with flow-through mounting which facilitates feeding of cold water and hot alkaline solutions through brush face to prevent buildup. | 750-1200 for cleaning steel. 600 when used with slurries | For cleaning steel. Used in electrolytic tinplate lines, continuous galvanizing and annealing lines, and cold reduction lines. Used to produce dull or matte-type finishes on stainless steel and synthetics. | Speeds above 3600 sfpm will not appreciably improve operation as brush wear will be excessive. Avoid excessive pressures. Ammeters should be installed in drive-motor circuit to indicate brushing pressure. |

**Characteristics and Applications of Brushes Used in Power Finishing** *(Continued)*

| Brush Type | Description | Operating Speed Range, sfpm | Uses | Remarks |
|---|---|---|---|---|
| Radial, wide face, metallic fill | This brush is made to customer's specifications. It is dynamically balanced at the speed at which it will operate. | 2000-4000 | Removes buildup of aluminum oxide from work rolls in aluminum mill. Removes lime or magnesium coatings from certain types of steel. Burnishes hot-dipped galvanized steel to produce a minimum spangled surface. | Each brush should have its own drive. An ammeter should be present in drive-motor circuit to measure brushing pressure. If strip is being brushed, a steel backup roll should be opposite the brush roll. |
| Radial, wide face, strip (interrupted brush face) | Performs cleaning operations that would cause a solid face brush to become loaded and unusable. | When cleaning conveyor belts, brush speed is 2 to 3 times that of conveyer belt. | Need for cleaning rubber and fabric conveyor belts of carry-back material which would normally foul snubber pulley and return idlers. | Designed for medium- to light-duty work. Brush face does not load. |
| Radial, Cup, Flared End, and Straight End, wire fill elastomer bonded | Extremely fast cutting with maximum operator safety. No loss of wire through fatigue. Always has uniform face. | 3600-9000 | For removing oxide weld scale, burrs, and insulation from wire. | Periodic reversing of brush direction will result in a brush life ten times greater than non-bonded wheels. Fast cutting action necessitates precise holding of part with respect to brush. |
| Cup, twisted knot wire fill | Fast cutting wheel used on portable tools to clean welds, scale, rust, and other oxides. | 8000-10,000<br><br>4500-6500 for deburring and producing a radius around periphery of holes. | Used in shipyards and in structural steel industry. For cleaning outside diameter of pipe and removing burrs and producing radii on heat exchanger tube sheets and laminations for stator cores. | Fast acting brush cleans large areas economically. Setup time is short. |
| Radial, wire or treated Tampico or cord | For use with standard centerless grinders. Brush will not remove metal from a cylindrical surface. Parts must be ground to size before brushing. | … | For removing feather grinding burrs and improving surface finish. Parts of 24 microinches can be finished down to 15 to 10 microinches. Parts of 10 to 12 microinches can be finished down to 7 to 4 microinches. | Follows centerless grinding principles, except that accuracy in pressure and adjustment is not critical. A machine no longer acceptable for grinding can be used for brushing. |

especially when good edges must be maintained. Cloth wheels are made in various ways; wheels having disks that are cemented together are very hard and used for rough, coarse work, whereas those having sewn disks are made of varying densities by sewing together a larger or smaller number of disks into sections and gluing them.

Wheels in which the disks are held together by thread or metal stitches and which are not stiffened by the use of glue usually require metal side plates to support the canvas disks. Muslin wheels are made from sewed or stapled buffs glued together, but the outer edges of a wheel frequently are left open or free from glue to provide an open face of any desired depth. Wool felt wheels are flexible and resilient, and the density may be varied by sewing two or more disks together and then cementing to form a wheel. Solid felt wheels are quite popular for fine finishing but have little value as general utility wheels. Paper wheels are made from strawboard paper disks and are cemented together under pressure to form a very hard wheel for rough work. Softer wheels are similarly made from felt paper. The "compress" canvas wheel has a cushion of polishing material formed by pieces of leather, canvas, or felt, that are held in a crosswise radial position by two side plates attached to the wheel hub. This cushion of polishing material may be varied in density to suit the requirements; it may be readily shaped to conform to the curvature of the work and this shape can be maintained. Sheepskin polishing wheels and paper wheels are little used.

**Polishing Operations and Abrasives.—**Polishing operations on such parts as chisels, hammers, screwdrivers, wrenches, and similar parts that are given a fine finish but are not plated, usually require four operations, which are "roughing," "dry fining," "greasing," and "coloring." Roughing is frequently regarded as a solid grinding wheel job. Sometimes there are two steps to the greasing operation—rough and fine greasing. For some hardware, such as the cheaper screwdrivers, wrenches, etc., the operations of roughing and dry fining are considered sufficient. For knife blades and cutlery, the roughing operation is performed with solid grinding wheels and the polishing is known as fine or blue glazing, but these terms are never used when referring to the polishing of hardware parts, plumbers' supplies, etc. A term used in finishing German silver, white metal, and similar materials is "sand-buffing," which, in distinction from the ordinary buffing operation that is used only to produce a very high finish, actually removes considerable metal, as in rough polishing or flexible grinding. For sand-buffing, pumice and other abrasive powders are loosely applied.

Aluminum oxide abrasives are widely used for polishing high-tensile-strength metals such as carbon and alloy steels, tough iron, and nonferrous alloys. Silicon carbide abrasives are recommended for hard, brittle substances such as gray iron, cemented carbide tools, and materials of low tensile strength such as brass, aluminum, and copper.

**Buffing Wheels.—**Buffing wheels are manufactured from disks (either whole or pieced) of bleached or unbleached cotton or woolen cloth, and they are used as the agent for carrying abrasive powders, such as tripoli, crocus, rouge, lime, etc., which are mixed with waxes or greases as a bond. There are two main classes of buffs, one of which is known as the "pieced-sewed" buffs, and is made from various weaves and weights of cloth. The other is the "full-disk" buffs, which are made from specially woven material. Bleached cloth is harder and stiffer than unbleached cloth, and is used for the faster cutting buffs. Coarsely woven unbleached cloth is recommended for highly colored work on soft metals, and the finer woven unbleached cloths are better adapted for harder metals. When working at the usual speed, a stiff buff is not suitable for "cutting down" soft metal or for light plated ware, but is used on harder metals and for heavy nickel-plated articles.

**Speed of Polishing Wheels.—**The proper speed for polishing is governed to some extent by the nature of the work, but for ordinary operations, the polishing wheel should have a peripheral speed of about 7500 ft/min (2286 m/min). If run at a lower speed, the work tends to tear the polishing material from the wheel too readily, and the work is not as good in quality. Muslin, felt, or leather polishing wheels having wood or iron centers should be run at peripheral speeds varying from 300 to 7000 ft/min (91–2133 m/min). It is rarely necessary to exceed 6000 ft/min (1829 m/min), and for most purposes, 4000 ft/min (1219 m/min) is sufficient. If the wheels are kept in good condition, in perfect balance, and are suitably mounted on substantial buffing lathes, they can be used safely at speeds within the limits given. However, manufacturers' recommendations concerning wheel speeds should be followed, where they apply.

**Grain Numbers of Emery.—**The numbers commonly used in designating the different grains of emery, corundum, and other abrasives are 10, 12, 14, 16, 18, 20, 24, 30, 36, 40, 46, 54, 60, 70, 80, 90, 100, 120, 150, 180, and 200, ranging from coarse to fine, respectively. These numbers represent the number of meshes per linear inch in the grading sieve. An abrasive finer than No. 200 is known as "flour" and the degree of fineness is designated by the letters CF, F, FF, FFF, FFFF, and PCF or SF, ranging from coarse to fine. The methods of grading flour-emery adopted by different manufacturers do not exactly agree, the letters differing somewhat for the finer grades. Again, manufacturers' recommendations should be followed.

**Grades of Emery Cloth.—**The coarseness of emery cloth is indicated by letters and numbers corresponding to the grain number of the loose emery used in the manufacture of the cloth. The letters and numbers for grits ranging from fine to coarse are as follows: FF, F, 120, 100, 90, 80, 70, 60, 54, 46, and 40. For large work roughly filed, use coarse cloth such as numbers 46 or 54, and then finer grades to obtain the required polish. If the work has been carefully filed, a good polish can be obtained with numbers 60 and 90 cloth, and a brilliant polish can be achieved by finishing with number 120 and flour-emery.

**Mixture for Cementing Emery Cloth to a Lapping Wheel.—**Many proprietary adhesives are available for application of emery cloth to the periphery of a buffing or lapping wheel, and generally are supplied with application instructions. In the absence of such instructions, clean the wheel thoroughly before applying the adhesive, and then rub the emery cloth down so as to exclude all air from between the surface of the wheel and the cloth.

## Etching and Etching Fluids

**Etching Fluids for Different Metals.—**A common method of etching names or simple designs upon steel is to apply a thin, even coating of beeswax or some similar substance which will resist acid; then mark the required lines or letters in the wax with a sharp-pointed scriber, thus exposing the steel (where the wax has been removed by the scriber point) to the action of an acid, which is finally applied. To apply a very thin coating of beeswax, place the latter in a silk cloth, warm the piece to be etched, and tub the pad over it. Regular coach varnish is also used instead of wax, as a "resist."

An etching fluid ordinarily used for carbon steel consists of nitric acid, 1 part; water, 4 parts. It may be necessary to vary the amount of water, as the exact proportion depends upon the carbon content and whether the steel is hard or soft. For hard steel, use nitric acid, 2 parts; acetic acid, 1 part. For high-speed steel, nickel or brass, use nitro-hydrochloric acid (nitric, 1 part; hydrochloric, 4 parts). For high-speed steel it is sometimes better to add a little more nitric acid. For etching bronze, use nitric acid, 100 parts; muriatic acid, 5 parts. For brass, nitric acid, 16 parts; water, 160 parts. Dissolve 6 parts potassium chlorate in 100 parts of water; then mix the two solutions and apply.

A fluid which may be used either for producing a frosted effect or for deep etching (depending upon the time it is allowed to act) is composed of 1 ounce sulphate of copper (blue vitriol); $\frac{1}{4}$ ounce alum; $\frac{1}{2}$ teaspoonful of salt; 1 gill of vinegar, and 20 drops of nitric acid. For aluminum, use a solution composed of alcohol, 4 ounces; acetic acid, 6 ounces; antimony chloride, 4 ounces; water, 40 ounces (1 ounce = 0.02957 liter).

Various acid-resisting materials are used for covering the surfaces of steel rules etc., prior to marking off the lines on a graduating machine. When the graduation lines are fine and very closely spaced, as on machinists' scales which are divided into hundredths or sixty-fourths, it is very important to use a thin resist that will cling to the metal and prevent any under-cutting of the acid: the resist should also enable fine lines to be drawn without tearing or crumbling as the tool passes through it. One resist that has been extensively used is composed of about 50 percent of asphaltum, 25 percent of beeswax, and, in addition, a small percentage of Burgundy pitch, black pitch, and turpentine. A thin covering of this resisting material is applied to the clean polished surface to be graduated and, after it is dry, the work is ready for the graduating machine. For some classes of work, paraffin is used for protecting the surface surrounding the graduation lines which are to be etched. The method of application consists in melting the paraffin and raising its temperature high enough so that it will flow freely; then the work is held at a slight angle and the paraffin is poured on its upper surface. The melted paraffin forms a thin protective coating.

## Conversion Coatings and the Coloring of Metals

**Conversion Coatings.—**Conversion coatings are thin, adherent chemical compounds that are produced on metallic surfaces by chemical or electrochemical treatment. These coatings are insoluble, passive, and protective, and are divided into two basic systems: oxides or mixtures of oxides with other compounds, usually chromates or phosphates. Conversion coatings are used for corrosion protection, as an adherent paint base; and for decorative purposes because of their inherent color and because they can absorb dyes and colored sealants.

Conversion coatings are produced in three or four steps. Pretreatment often involves mechanical surface preparation followed by decreasing and/or chemical or electrochemical cleaning or etching. Then thermal, chemical, or electrochemical surface conversion processes take place in acid or alkaline solutions applied by immersion spraying, or brushing. A post treatment follows, which includes rinsing and drying, and may also include sealing or dyeing. If coloring is the main purpose of the coating, then oiling, waxing, or lacquering may be required.

**Passivation of Copper.—**The blue-green patina that forms on copper alloys during atmospheric exposure is a passivated film; i.e., it prevents corrosion. This patina may be produced artificially or its growth may be accelerated by a solution of ammonium sulfate, 6 pounds (2.7 kg); copper sulfate, 3 ounces (85 g); ammonia (technical grade, 0.90 specific gravity), 1.34 fluid ounces (39.6 cc); and water, 6.5 gallons (24.6 liters). This solution is applied as a fine spray to a chemically cleaned surface and is allowed to dry between each of five or six applications. In about 6 hours a patina somewhat bluer than natural begins to develop and continues after exposure to weathering.

Small copper parts can be coated with a passivated film by immersion in or brushing with a solution consisting of the following weight proportions: copper, 30; nitric acid, concentrated, 60; acetic acid (6%), 600; ammonium chloride, 11; and ammonium hydroxide (technical grade, 0.90 specific gravity), 20. To prepare the solution, the copper is dissolved in the nitric acid before the remaining chemicals are added, and the solution is allowed to stand for several days before use. A coating of linseed oil is applied to the treated parts.

**Coloring of Copper Alloys.—**Metals are colored to enhance their appearance, to produce an undercoat for an organic finish, or to reduce light reflection. Copper alloys can be treated to produce a variety of colors, with the final color depending on the base metal composition, the coloring solution's composition, the immersion time, and the operator's skill. Cleaning is an important part of the pretreatment; nitric and sulfuric acid solutions are used to remove oxides and to activate the surface.

The following solutions are used to color alloys that contain 85 percent or more of copper. A dark red color is produced by immersing the parts in molten potassium nitrate, at 1200–1300° F (649–704° C), for up to 20 seconds, followed by a hot water quench. The parts must then be lacquered. A steel black color can be obtained by immersing the parts in a 180° F (82° C) solution of arsenious oxide (white arsenic), 4 ounces (113 g); hydrochloric acid

(1.16 specific gravity), 8 fluid ounces (23.6 cc); and water, 1 gallon (3.9 liters). The parts are immersed until a uniform color is obtained; they are scratch brushed while wet, and then dried and lacquered. A light brown color is obtained using a room-temperature solution of barium sufate, 0.5 ounce (14g); ammonium carbonate, 0.25 ounce (7g); and water, 1 gallon (3.8 liters).

The following solutions are used to color alloys that contain less than 85 percent copper. To color brass black, parts are placed in an oblique tumbling barrel made of stainless steel and covered with 3 to 5 gallons (11.2 to 19.8 liters) of water. Three ounces (85 g) of copper sulfate and 6 ounces (170 g) of sodium thiosulfate are dissolved in warm water and added to the barrel's contents. After tumbling for 15 to 30 minutes to obtain the finish, the solution is drained from the barrel, and the parts are washed thoroughly in clean water, dried in sawdust or air-blasted and, if necessary, lacquered. To produce a blue-black color, the parts are immersed in a 130–175° F (54–79° C) solution of copper carbonate 1 pound (450 g); ammonium hydroxide (0.89 specific gravity), 1 quart (940 cc); and water, 3 quarts. Excess copper carbonate should be present. The proper color is obtained in 1 minute. To color brass a hardware green, immerse the parts in a 160° F (71° C) solution of ferric nitrate, 1 ounce (28 g); sodium thiosulfate, 6 ounces; and water, 1 gallon. To color brass a light brown, immerse the parts in a 195–212° F (91–100° C) solution of potassium chlorate, 5.5 ounces; nickel sulfate, 2.75 ounces (170 g); copper sulfate, 24 ounces (680 g); and water, 1 gallon (3.8 liters).

*Post treatment:* The treated parts should be scratch brushed to remove any excess or loose deposits. A contrast of colors may be obtained by brushing with a slurry of fine pumice, hand nabbing with an abrasive paste, mass finishing, or buffing to remove the color from the highlights. In order to prolong the life of parts used for outdoor decorative purposes, a clear lacquer should be applied. Parts intended for indoor purposes are often used without additional protection.

**Coloring of Iron and Steel.—**Thin black oxide coatings are applied to steel by immersing the parts to be coated in a boiling solution of sodium hydroxide and mixtures of nitrates and nitrites. These coatings serve as paint bases and, in some cases, as final finishes. When the coatings are impregnated with oil or wax, they furnish fairly good corrosion resistance. These finishes are relatively inexpensive compared to other coatings.

*Phosphate Coatings* are applied to iron and steel parts by reacting them with a dilute solution of phosphoric acid and other chemicals. The surface of the metal is converted into an integral, mildly protective layer of insoluble crystalline phosphate. Small items are coated in tumbling barrels; large items are spray coated on conveyors.

The three types of phosphate coatings in general use are zinc, iron, and manganese. Zinc phosphate coatings vary from light to dark gray. The color depends on the carbon content and pretreatment of the steel's surface, as well as the composition of the solution. Zinc phosphate coatings are generally used as a base for paint or oil, as an aid in cold-working, for increased wear resistance, or for rustproofing. Iron phosphate coatings were the first type to be used; they produce dark gray coatings and their chief application is as a paint base. Manganese phosphate coatings are usually dark gray; however, since they are used almost exclusively as an oil base, for break in and to prevent galling, they become black in appearance. In general, stainless steels and certain alloy steels cannot be phosphated. Most cast irons and alloy steels accept coating with various degrees of difficulty depending on alloy content.

**Anodizing Aluminum Alloys.—**In the anodizing process, the aluminum object to be treated is immersed as the anode in an acid electrolyte, and a direct current is applied. Oxidation of the surface occurs, producing a greatly thickened, hard, porous film of aluminum oxide. The object is then immersed in boiling water to seal the porosity and render the film impermeable. Before sealing, the film can be colored by impregnation with dyes or pigments. Special electrolytes may also be used to produce colored anodic films directly in the anodizing bath. The anodic coatings are used primarily for corrosion protection and abrasion resistance, and as a paint base.

The three principal types of anodizing processes are: chromic, in which the active agent is chromic acid; sulfuric, in which the active agent is sulfuric acid, and hard anodizing, in which sulfuric acid is used by itself or with additives in a low-temperature electrolyte bath. Most of the anodic coatings range in thickness from 0.2 to 0.7 mil. The hard anodizing process can produce coatings up to 2 mils. The chromic acid coating is less brittle than the sulfuric, and, since the chromic electrolyte does not attack aluminum, it does not present a corrosion problem when it is trapped in crevices. The chromic coating is less resistant to abrasion than the sulfuric, but it cannot be used with alloys containing more than 5 percent copper due to corrosion of the base metal.

*Chemical Conversion Coatings for Aluminum:* Chemical conversion coatings for aluminum alloys are adherent surface layers of low volubility oxide, phosphate, or chromate compounds produced by the reaction of the metal surface with suitable reagents. The conversion coatings are much thinner and softer than anodic coatings but they are less expensive and serve as an excellent paint base.

**Magnesium Alloys.—**Chemical treatment of magnesium alloys is used to provide a paint base and to improve corrosion resistance. The popular conversion "dip" coatings are chrome pickle and dichromate treatments, and they are very thin. Anodic coatings are thicker and harder, and, after sealing, give the same protection against corrosion, although painting is still desirable.

**Titanium Alloys.—**Chemical conversion coatings are used on titanium alloys to improve lubricity by acting as a base for the retention of lubricants. The coatings are applied by immersion, spraying, or brushing. A popular coating bath is an aqueous solution of phosphates, fluorides, and hydrofluoric acid. The coating is composed primarily of titanium and potassium fluorides and phosphates.

## Plating

**Surface Coatings.—**The following is a list of military plating and coating specifications.

*Anodize (Chromic and Sulfuric), MIL-A-8625F:* Conventional Types I, IB, and II anodic coatings are intended to improve surface corrosion protection under severe conditions or as a base for paint systems. Coatings can be colored with a large variety of dyes and pigments. Class 1 is non-dyed; Class 2 dyed. Color is to be specified on the contract. Prior to dying or sealing, coatings shall meet the weight requirements.

Type I and IB coatings should be used on fatigue critical components (due to thinness of coating). Type I unless otherwise specified shall not be applied to aluminum alloys with over 5% copper or 7% silicon or total alloying constituents over 7.5%. Type IC is a mineral or mixed mineral/organic acid that anodizes. It provides a non-chromate alternative for Type I and IB coatings where corrosion resistance, paint adhesion, and fatigue resistance are required. Type IIB is a thin sulfuric anodizing coating for use as non-chromate alternatives for Type I and IB coatings where corrosion resistance, paint adhesion, and fatigue resistance are required. Be sure to specify the class of anodic coating and any special sealing requirements.

Types I, IB, IC, and IIB shall have a thickness between 0.00002 and 0.0007 inch (0.5 and 17.78 μm). Type II shall be between 0.0007 and 0.0010 inch (17.8 and 25.4 μm).

*Black Chrome, MIL-C-14538C:* A hard, non-reflective, abrasion, heat and corrosion resistant coating approximately 0.0002 in. (5.08 μm) thick. Provides limited corrosion protection, but added protection can be obtained by specifying underplate such as nickel. Color is a dull dark gray, approaching black and may be waxed or oiled to darken.

Black chromium has poor throwing power, and conforming anodes are necessary for intricate shapes. Apply coating after heat treating and all mechanical operations are performed. Steel parts with hardness in excess of 40 RC (Rockwell C scale) shall be stress relieved prior to plating by baking one hour or more, 300–500° F (149–260° C) and baked after plating, 375 ± 25° F (190 ± 14° C) for 3 hours.

*Black Oxide Coating, MIL-C-13924C:* A uniform, mostly decorative black coating for ferrous metals used to decrease light reflection. Only very limited corrosion protection under mild corrosion conditions. Black oxide coatings should normally be given a supplementary treatment.

Used for moving parts that cannot tolerate the dimensional change of a more corrosion resistant finish. Use alkaline oxidizing for wrought iron, cast and malleable irons, plain carbon, low alloy steel and corrosion resistant steel alloys. Alkaline-chromite oxidizing may be used on certain corrosion resistant steel alloys tempered at less than 900° F (482° C). Salt oxidizing is suitable for corrosion resistant steel alloys that are tempered at 900° F or higher.

*Cadmium, QQ-P-416F:* Cadmium plating is required to be smooth, adherent, uniform in appearance, free from blisters, pits, nodules, burning, and other defects when examined visually without magnification. Unless otherwise specified in the engineering or procurement documentation, the use of brightening agents in the plating solution to modify luster is prohibited on components with a specified heat treatment of 180 ksi (1241 MPa) minimum tensile strength (or 40 RC) and higher. Either a bright (not caused by brightening agents) or dull luster shall be acceptable. Baking on Types II and III shall be done prior to application of supplementary coatings. For Classes 1, 2, and 3 the minimum thicknesses shall be 0.0005, 0.0003, and 0.0002 inch (12.7, 7.62, and 5.08 μm) respectively.

Type I is to be used as plated. Types II and III require supplementary chromate and phosphate treatment respectively. Chromate treatment required for type II may be colored iridescent bronze to brown including olive drab, yellow and forest green. Type II is recommended for corrosion resistance. Type III is used as a paint base and is excellent for plating stainless steels that are to be used in conjunction with aluminum to prevent galvanic corrosion. For Types II and III the minimum cadmium thickness requirement shall be met after the supplementary treatment.

*Chemical Films, MIL-C-5541E:* The materials that qualify produce coatings that range in color from clear to iridescent yellow or brown. Inspection difficulties may arise with clear coatings because of their invisibility.

Class 1A chemical conversion coatings are intended to provide corrosion prevention when left unpainted as well as to improve adhesion of paint finish systems on aluminum and aluminum alloys. May be used on tanks, tubings, and component structures where paint finishes are not required for the exterior surfaces but are required for the interior surfaces.

Class 3 chemical conversion coatings are intended for use as a corrosive film for electrical and electronic applications where lower resistant contacts are required. The primary difference between Class 1A and Class 3 coating is thickness.

*Chemical Finish: Black, MIL-F-495E:* A uniform black corrosion retardant for copper. Coating has no abrasion resistance. Used to blacken color and reduce gloss on copper-alloy surfaces other than food service and water supply items. Also used as a base for subsequent coatings such as lacquer, varnish, oil, and wax.

*Chrome, QQ-C-320B:* Has excellent hardness, wear resistance, and erosion resistance. In addition chrome has a low coefficient of friction, is resistant to heat, and can be rendered porous for lubrication purposes.

Types I and II have bright and satin appearances respectively.

Class 1 is used as plating for corrosion protection and Class 2, for engineering plating. Class 1 and 2 both shall have a minimum thickness of 0.00001 in. (0.25 μm) on all visible surfaces. If thickness is not specified use 0.002 in (50.8 μm).

Class 2a will be plated to specified dimensions or processed to specified dimensions after plating. Class 2b will be used on parts below 40 RC and subject to static loads or designed for limited life under dynamic loads. Class 2c will be used on parts below 40 RC and designed for unlimited life under dynamic loads. Class 2d parts have hardness of 40 RC or above, which are subject to static loads or designed for unlimited life under dynamic loads. Class 2e parts have hardness of 40 RC or above, which are designed for unlimited life under dynamic loads.

All coated steel parts having a hardness of 36 RC and higher shall be baked at a minimum of 375 ± 25° F (190 ± 14° C) per the following conditions. With a tensile strength of 160–180 ksi (1103–1241 MPa), the time at temperature will be 3 hr.; at 181–220 ksi (1248–1517 MPa), the time will be 8 hr.; and at 221 ksi (1524 MPa) and above, the time will be 12 hr.

*Copper, MIL-C-14550B:* Has good corrosion resistance when used as an undercoat. A number of copper processes are available, each designed for a specific purpose such as, to improve brightness (to eliminate the need for buffing), high speed (for electroforming), and fine grain (to prevent case hardening).

All steel parts having a hardness of 35 RC and higher shall be baked at 375 ± 25° F (190 ± 14° C) for 24 hours, within 4 hours after plating to provide hydrogen embrittlement relief. Plated springs and other parts subject to flexure shall not be flexed prior to baking operations.

Class 0 will have a thickness 0.001–0.005 in. (25.4–127 μm) and is used for heat-treatment stop-off; Class 1 is 0.001 in. (25.4 μm) and is used to provide carburizing shield, also for plated through printed circuit boards. Class 2 is 0.0005 in. (12.7 μm) thick and is used as an undercoat for nickel and other platings. Class 3 is 0.0002 in. (5.08 μm) thick and is used to prevent basis metal migration into tin (prevents poisoning solderability). Class 4 is 0.0001 in. (2.54 μm) thick.

*Tin Lead, MIL-P-81728A:* It has excellent solderability. Either a matte or bright luster is acceptable. For electronics components, use only parts with a matte or flow brightened finish.

For brightened electronic components, the maximum thickness will be 0.0003 in. (7.62 μm) Tin 50 to 70% by weight and with a lead remainder, 0.0003–0.0005 in (7.62–12.7 μm).

*Magnesium Process, MIL-M-3171C:* Process #1-A chrome pickle treatment for magnesium. Color varies from matte gray to yellow-red. Has only fair corrosion resistance (< 24 hours, 20 percent salt spray resistance).

#7-A dichromate treatment for magnesium. Color varies from light brown to gray depending on alloy. Only fair corrosion resistance (< 24 hours, 20 percent salt spray resistance).

#9-A galvanic anodize treatment for magnesium. Produces a dark brown to black coating. Designed to give a protective film on alloys which do not react to Dow No. 7 treatment. Only fair corrosion resistance (< 24 hours, 20 percent salt spray resistance).

| Type/Class | Thickness (inch) | Comments |
|---|---|---|
| Type 1 | Removes metal. (approx. 0.0006 for wrought, less for die castings.) No dimensional change | Used for protecting magnesium during shipment, storage and machining. Can be used as a paint base. NOTE: Must remove Type I coating before applying Type III and Type IV treatments. |
| Type III | … | *Note:* precleaning and pickling may result in dimensional changes due to metal loss. |
| Type IV | No dimensional change | Can be used as a paint base, and is applicable to all magnesium alloys. Used where optical properties (black) are required on close tolerance parts. NOTE: Precleaning and pickling may result in dimensional changes due to metal loss. |

*Magnesium Anodic Treatment, MIL-M-45202C:* The HAE anodic finish is probably the hardest coating currently available for magnesium. It exhibits stability at high temperatures and has good dielectric strength. It serves as an excellent paint base. It requires resin seal or paint for maximum corrosion protection.

Coatings range from thin clear to light gray-green, to thick dark-green coatings. The clear coatings are used as a base for subsequent clear lacquers or paints to produce a final appearance similar to clear anodizing on aluminum. The light gray-green coatings are used in most applications which are to be painted. The thick, dark-green coating offers the best combination of abrasion resistance, protective value and paint base characteristics.

| Type/Class | Typical Thickness | Comments |
|---|---|---|
| Type I, Light coating. | | |
| Class A | 0.2 mil | Tan coating (HAE) |
| Grade 1 | … | Without post treatment (dyed) |
| Grade 2 | … | With biflouride-dichromate post treatment |
| Class C | 0.3 mil | Light green coating (Dow #17) |
| Type II, Heavy coating | | |
| Class A | 1.5 mil | Hard brown coating (HAE) |
| Grade 1 | … | Without post treatment |
| Grade 3 | … | With biflouride-dichromate post treatment |
| Grade 4 | … | With biflouride-dichromate post treatment including moist heat aging |
| Grade 5 | | With double application of biflouride-dichromate post treatment including moist heat aging. |
| Class D | 1.2 mil | Dark green coating (Dow #17) |

*Electroless Nickel, AMS 2404C, AMS 2405B, AMS 2433B:* Is typically used as a coating to provide a hard-ductile, wear-resistant, and corrosion-resistant surface for operation in service up to 1000F, to provide uniform build-up on complex shapes.

AMS 2404C, is deposited directly on the basis metal without a flash coating of other metal, unless otherwise specified. AMS 2405B, is deposited directly on the basis metal except where parts fabricated from corrosion resistant steels or alloys where a "strike" coating of nickel or other suitable metal is required, unless otherwise specified. AMS 2433B, is a type of electroless nickel typically used to enhance the solderability of surfaces, but usage is not limited to such applications. Generally, the plate shall be placed directly on the basis metal. However, aluminum alloys shall be zinc immersion coated per ASTM B253 followed by copper flash; corrosion resistant steels and nickel and cobalt alloys or other basis metals may use a nickel or copper flash undercoat when the purchaser permits.

Electroless Nickel Preparation: Parts having a hardness higher than 40 RC and have been machined or ground after heat treatment shall be suitably stress-relieved before cleaning and plating.

After treatment, parts having a hardness of 33 RC and over shall be heated to 375 ± 15° F (190.5 ± 8.5° C) for 3 hours. If such treatment is injurious to the parts, bake at 275 ±15° F (135 ± 8.5° C) for 4 hours.

*Electroless Nickel, Low-Phosphorous, Note:* If permitted by drawing, the maximum hardness and wear resistance are obtained by heating parts for 30–60 minutes, preferably in an inert atmosphere, at 750° F ± 15° F (399 ± 8° C) except aluminum parts shall be baked at 450 ± 15° F (231.5 ± 8.5° C) for 4 hours. If such heating is not specified, bake at 375 ± 15° F (190.5 ± 8.5° C) for 3 hours. If this treatment is injurious to parts or assemblies, bake at 275° F (135° C) for 5 hours.

Plating: nickel-thallium-boron (Electroless Deposition) and nickel-boron (Electroless Deposition)

Preparation: All fabrication-type operations shall be completed.

Post-treatment: Cold-worked or heat-treated parts and aluminum alloys and other parts requiring special thermal treatment shall be post treated as agreed upon by purchaser and vendor. Other plated parts within 4 hours after plating shall be heat treated for 90 ± 10 minutes at 675° F ± 15° F (357.5 ± 8.5° C).

*Electropolishing, ASTM B912-02 (2018):* This process electrolytically removes or diminishes scratches, burrs, and unwanted sharp edges from most metals. Electropolishing can be expected to reduce roughness average (Ra) by approximately 10 to 35 percent, with 50 percent improvement possible in some cases. By eliminating microscopic defects and impurities, electropolishing can improve both corrosion resistance and fatigue life of parts.

In addition, this process passivates exposed surfaces, with results superior to chemical passivation methods. Finishes from satin to mirror-bright are produced by controlling time, temperature, or both. Stainless steels are the most commonly electropolished materials, but many metals are good candidates for the process.

Material thickness loss during electropolishing typically is 0.0002–0.0007 in. (5–18 μm) but may be as much as 0.003 inch (76 μm). The amount removed depends on desired surface improvement and usually can be controlled to within 0.0001 in. (2.5 μm). Parts with tight tolerances can be electropolished if process parameters are well planned and controlled.

Parts to be electropolished must be placed on racks. Features such as deep bores may require additional cathodes, and complex parts may need custom process tooling.

*Gold, MIL-G-45204C:* Has a yellow to orange color depending on the proprietary process used. Will range from matte to bright finish depending on basis metal. It has good corrosive resistance and a high tarnish resistance. It provides a low contact resistance, is a good conductor of electricity, and has excellent solderability. If the hardness grade for the gold coating is not specified, Type I shall be furnished at a hardness of Grade A, and Type II furnished at a hardness of Grade C.

For soldering, a thin pure soft gold coating is preferred. A minimum and maximum thickness 0.00005 and 0.00010 inch (1.27 and 2.54 μm), respectively, shall be plated.

Unless otherwise specified, gold over silver underplate combinations shall be excluded from electronics hardware. Silver or copper plus silver may not be used as an underplate unless required by the item specification. When gold is applied to brass bronze or beryllium copper, or a copper plate or strike, an antidiffusion underplate such as nickel shall be applied.

Type I is 99.7% gold minimum (Grades A, B, or C); Type II is 99.0% (Grades B, C, or D); and Type III is 99.9% (Grade A only).

Grade A is 90 Knoop maximum; Grade B is 91-129 Knoop; Grade C is 130–200 Knoop; and Grade D is 201 Knoop and over.

Class 00 has a thickness of 0.00002 inch (0.5 μm) minimum; Class 0, 0.00003 inch (0.76 μm); Class 1, 0.00005 inch (1.3 μm); Class 2, 0.0001 inch (2.5 μm); Class 3, 0.0002 inch (5 μm); Class 4, 0.0003 inch (7.6 μm); Class 5, 0.0005 inch (12.7 μm); and Class 6, 0.0015 inch (38.1 μm).

*Hard Anodize, MIL-A-8625F:* The color will vary from light tan to black depending on alloy and thickness. Can be dyed in darker colors depending on the thickness. Coating penetrates base metal as much as builds up on the surface. The term thickness includes both the buildup and penetration. It provides a very hard ceramic type coating. Abrasion resistance will vary with alloy and thickness of coating. Has good dielectric properties.

Do not seal coatings where the main function is to obtain maximum abrasion or wear resistance. When used for exterior applications requiring corrosion resistance but permitting reduced abrasion, the coating shall be sealed (boiling deionized water or hot 5% sodium dichromate solution, or other suitable chemical solutions).

Type III will have a thickness specified on the contract or applicable drawing. If not specified use a nominal thickness of 0.002 inch (50.8 μm). Hard coatings may vary in thickness by 0.0005–0.0045 inch (12.7–114.3 μm).

Class 1 shall not be dyed or pigmented. Class 2 shall be dyed and the color specified on the contract. The process can be controlled to very close thickness tolerances. Where maximum serviceability or special properties are required, consult metal finisher for best alloy choice. Thick coatings, those over 0.004 inch (101 μm), will tend to break down sharp edges. Can be used as an electrical insulation coating. "Flash" hard anodize may be used instead of conventional anodize for corrosion resistance and may be more economical in conjunction with other hard anodized areas.

*Lubrication, Solid Film MIL-L-46010D:* The Military Plating Specification establishes the requirements for three types of heat cured solid film lubricants that are intended to reduce wear and prevent galling, corrosion, and seizure of metals. For use on aluminum, copper, steel, stainless steel, titanium, and chromium, and nickel bearing surfaces.

Types I, II, and III have a thicknesses of 0.008 - 0.013 mm. No single reading less than 0.005 mm or greater than 0.018 mm.

Type I has a curing temperature of 150 ± 15° C and an endurance life of 250 minutes; Type II, 204 ± 15° C and 450 minutes; and Type III is a low volatile organic compound (VOC) content lubricant with cure cycles of 150 ± 15° C for 2 hours, or 204 ± 15° C for 1 hour with an endurance life of 450 minutes. Color 1 has a natural product color and Color 2 has a black color.

*Nickel, QQ-N-290A:* There is a nickel finish for almost any need. Nickel can be deposited soft, hard-dull, or bright, depending on process used and conditions employed in plating. Thus, hardness can range from 150–500 HV (Vickers). Nickel can be similar to stainless steel in color, or can be a dull gray (almost white) color. Corrosion resistance is a function of thickness. Nickel has a low coefficient of thermal expansion. All steel parts having a tensile strength of 220,000 or greater shall not be a nickel plate without specific approval of procuring agency.

Class 1 is used for corrosion protection. Plating shall be applied over an underplating of copper or yellow brass on zinc and zinc based alloys. In no case, shall the copper underplate be substituted for any part of the specified nickel thickness. Class 2 is used in engineering applications.

Grade A has a thickness of 0.0016 inch (41 μm); Grade B, 0.0012 in. (30.48 μm); Grade C, 0.001 in. (25.4 μm); Grade D, 0.0008 in. (20.32 μm); Grade E, 0.0006 in. (15.24 μm); Grade F, 0.0004 in. (10.16 μm); and Grade G, 0.0002 in. (5.08 μm).

*Palladium, MIL-P-45209B:* A gray, dense deposit good for undercoats. Has good wear characteristics, corrosion resistance, catalytic properties, and good conductivity. The thickness shall be 0.00005 in. (1.27 μm) unless otherwise specified.

Steel springs and other steel parts subject to flexure or repeated impact and of hardness greater than 40 RC are heated to 375 ± 25° F (190 ± 14° C) for 3 hours after plating.

*Chemical Passivation, ASTM A967:* This process aims to improve the corrosion resistance of parts by removing contaminants from surfaces and facilitating formation of a passive oxide layer. Commercial passivation is performed on austenitic, ferritic, and martensitic stainless steels of the 200, 300, and 400 series, and related variants such as precipitation-hardening stainless steels.

Passivation methods included in the Standard are nitric acid immersion, citric acid immersion, and electrochemical treatment (See *Electropolishing, ASTM B912-02 (2018)* on page 1644). Nitric acid has long been used for chemical passivation, but safety and environmental concerns have led to the increasing use of citric acid, when possible. Various grades of stainless steel respond to passivation and related chemistry differently, so care must be taken when specifying a process.

The efficacy of chemical passivation depends on the amount of dynamic contact between the fluid and critical part surfaces. For parts with complex geometry, deep bores, or blind holes, the process may require agitation, repositioning, and use of fixtures.

*Phosphate Coating: Light, TT-C-490D:* This specification covers cleaning methods and pretreatment processes.

| Methods / Types | Typical Thickness (in.) | Comments |
|---|---|---|
| Cleaning Methods | | |
| | … | Light coating for use as a paint base. |
| Method I | … | Mechanical or abrasive cleaning (for ferrous surfaces only). |
| Method II | … | Used for solvent cleaning. |
| Method III | … | Used for hot alkalines (for ferrous surfaces only). |
| Method IV | … | Emulsion. |
| Method V | … | Used for alkaline derusting (for ferrous surfaces only). |
| Method VI | … | Phosphoric acid. |

| Methods / Types | Typical Thickness (in.) | Comments |
|---|---|---|
| Pretreatment Coatings | | |
| Type I | … | Zinc phosphate. Class 1-spray application: Class 2A and 2B-Immersion or Dip application |
| Type II | … | Aqueous Iron Phosphate |
| Type III | 0.0003 – 0.0005 | Organic pretreatment coating |
| Type IV | … | Non-aqueous iron phosphate |
| Type V | … | Zinc phosphate |

Type I is intended as a general all-purpose pretreatment prior to painting. Type II and IV are intended primarily for use where metal parts are to be formed after painting. Type III is intended for use where size and shape preclude using Type I, II, or IV and where items containing mixed metal components are assembled prior to treatment.

*Phosphate Coating: Heavy, DOD-P-16232-F:* The primary differences are that Type M is used as a heavy manganese phosphate coating for corrosion and wear resistance and Type Z is used as a zinc phosphate coating.

Type M has a thickness from 0.0002-0.0004 in. (5–10 μm) and Type Z, 0.0002-0.0006 in. (5–15 μm) Class 1, for both types has a supplementary preservative treatment or coating as specified; Class 2, has a supplementary treatment with lubricating oil; and Class 3, no supplementary treatment is required. For Type M, Class 4 is chemically converted (may be dyed to color as specified) with no supplementary coating or supplementary coating as specified. For Type Z, Class 4 is the same as Class 3.

This coating is for medium and low alloy steels. The coatings range from gray to black in color. The "heavy" phosphate coatings covered by this specification are intended as a base for holding/retaining supplemental coatings which provide the major portion of the corrosion resistance. "Light" phosphate coatings used for a paint base are covered by other specifications. Heavy zinc phosphate coatings may be used when paint and supplemental oil coatings are required on various parts or assemblies.

*Rhodium, MIL-R-46085B:* Rhodium is metallic and similar to stainless steel in color, has excellent corrosion and abrasion resistance, is almost as hard as chromium, and has a high reflectivity. Thicker coatings of Rhodium are very brittle.

| Class/Types | Thickness (in.) | Comments |
|---|---|---|
| Type I | … | Over nickel, silver, gold, or platinum. |
| Type II | … | Over other metals, requires nickel undercoat. |
| Class 1 | 0.000002 | Used on silver for tarnish resistance. |
| Class 2 | 0.00001 | Applications range from electronic to nose cones — wherever wear, corrosion resist solderability and reflectivity are important. |
| Class 3 | 0.00002 | |
| Class 4 | 0.00010 | |
| Class 5 | 0.00025 | |

Parts having a hardness of 33 RC or above shall be baked at 375° F (191° C) for 3 hours prior to cleaning. Parts having hardness of 40 RC and above shall be baked within 4 hours after plating at 375° F (191° C) for 3 hours.

*Silver, QQ-S-365D:* Silver has an increasing use in both decorative and engineering fields, including electrical and electronic fields.

Silver is white matte to very bright in appearance. Has good corrosion resistance, depending on base metal and will tarnish easily. Its hardness varies from about 90–135 BHN (Brinell Hardness Number) depending on process and plating conditions. Solderability is excellent, but decreases with age. Silver is the best conductor of electricity. Has excellent lubricity and smear characteristics for antigalling uses on static seals, bushing, etc. Stress relief steel parts at a minimum 375 ± 25° F (190 ± 14° C) or more prior to cleaning and plating if they contain or are suspected of having damaging residual tensile stresses.

All types and grades will have a minimum thickness of 0.0005 in. (12.7 μm) unless otherwise specified. Type I is matte, Type II is semi-bright, and Type III is bright. Grade A has a chromate post-treatment to improve tarnish resistance. In contrast Grade B has no supplementary treatment.

*Tin, MIL-T-10727C:* There are two different types of coating methods used, electrodeposited (based on ASTM B545 specification for electrodeposited coatings of tin) and hot dipped.

Thickness as specified on drawing (thickness is not part of the specification) is 0.0001–0.0025 inch (2.5–63.5 μm), flash for soldering; 0.0002–0.0004 inch (5.08–10.16 μm), to prevent galling and seizing; 0.0003 inch (7.62 μm) minimum, where corrosion resistance is important; and 0.0002–0.0006 inch (5.08–15.24 μm) to prevent formation of case during nitriding.

Color is a gray-white color in plated condition. Tin is soft, but very ductile. It has good corrosion resistance, and has excellent solderability. Tin is not good for low temperature applications.

If a bright finish is desired to be used in lieu of fused tin, specify Bright Tin plate. Thickness can exceed that of fused tin and deposit shows excellent corrosion resistance and solderability.

*Vacuum Cadmium, MIL-C-8837B:* Is used primarily to provide corrosion resistance to ferrous parts free from hydrogen contamination and possible embrittlement. Recommended on steels with a strength of $2.2 \times 10^5$ psi (1517 MPa) or above.

Coating is applied after all machining, brazing, welding, and forming has been completed. Prior to coating, all steel parts shall be stress relieved by baking at 375 ± 25° F (190 ± 14° C) for 3 hours if suspected of having residual tensile stresses. Immediately prior to coating, lightly dry abrasive blast areas are to be coated.

Type I shall be as plated; and Types II and III require supplementary chromate and phosphate treatments respectively.

Classes 1, 2, and 3 have thicknesses of 0.0005, 0.0003, and 0.0002 inch (12.7, 7.62, and 5.08 μm) respectively.

Cadmium coating shall not be used, if in service, temperature reaches 450° F (232° C).

A salt spray test is required for type II and is 96 hours.

*Zinc, ASTM B633:* This specification covers requirements for electrodeposited zinc coatings applied to iron or steel articles to protect them from corrosion. It does not cover zinc-coated wire or sheets.

Type I will be as plated; Type II will have colored chromate conversion coatings; Type III will have colorless chromate conversion coatings; and Type IV will have phosphate conversion coatings.

High strength steels (tensile strength over 1700 MPa or 246,500 psi) shall not be electroplated.

Stress relief: All parts with ultimate tensile strength 1000 MPa (145,000 psi) and above at minimum 190° C (374° F) for 3 hours or more before cleaning and plating.

Hydrogen embrittlement relief: All electroplated parts 1200 MPa (174,000 psi) or higher shall be baked at 190° C (374° F) for 3 hours or more within 4 hours after electroplating.

| Corrosion Resistance Requirements | |
|---|---|
| Types | Test Period Hr. |
| II | 96 |
| III | 12 |

## Flame Spraying Process

In this process, the forerunner of which was called the metal spraying process, metals, alloys, ceramics, and cermets are deposited on metallic or other surfaces. The object may be to build up worn or undersize parts, provide wear-resisting or corrosion-resisting surfaces, correct defective castings, etc.

Different types of equipment are available that provide the means of depositing the coatings on the surfaces. In one, wire is fed automatically through the nozzle of the spray gun; then a combustible gas, oxygen and compressed air serve to melt and blow the atomized metal against the surface to be coated. The gas usually used is acetylene but other gases may be used. Any desired thickness of metal may be deposited and the metals include steels, ranging from low to high carbon content, various brass and bronze compositions, babbitt metal, tin, zinc, lead, nickel, copper, and aluminum. The movement of the spray gun, in covering a given surface, is controlled either mechanically or by hand. In enlarging worn or undersize shafts, spindles, etc., it is common practice to clamp the gun in a lathe toolholder and use the feed mechanism to traverse the gun at a uniform rate while the metal is being deposited upon the rotating workpiece. The spraying operation may be followed by machining or grinding to obtain a more precise dimension.

Some typical production applications using the wire process are the coating of automotive exhaust valves, refinishing of transfer ink rollers for the printing industry and the rebuilding of worn truck clutch plates. Other production applications include the metallizing of glass meter box windows, the spraying of aluminum onto cloth gauze to produce electrolytic condenser plates, and the spraying of zinc or copper for coating ceramic insulators.

With another type of equipment, metal, refractory, and ceramic powder are used instead of wire. Ordinarily this equipment employs the use of two gases, oxygen and a fuel gas. The fuel gas is usually acetylene but in some instances hydrogen may be used. When handheld, a small reservoir supplies the powder to the equipment but a larger reservoir is used for lathe-mounted equipment or for large-scale production work. The four basic types of coating powders used with this equipment are ceramics, oxidation-resistant metals and alloys, self-bonding alloys, and alloys for fused coatings. These powders are used to produce wear-resistant, corrosion-resistant, heat-resistant, and electrically conductive coatings.

Still other equipment employs the use of plasma flame with which vapors of materials are raised to a higher energy level than the ordinary gaseous state. Its use raises the temperature ceiling and provides a controlled atmosphere by permitting employment of an inert or chemically inactive gas so that chemical action, such as oxidation, during the heating and application of the spray material can be controlled. The temperatures that can be obtained with commercially available plasma equipment often exceed 30,000° F (16,650° C) but for most plasma flame spray processes the temperature range of from 12,000 to 20,000° F (6650–11,093° C) is optimum. Plasma flame spray materials include alumina, zirconia, tungsten, molybdenum, tantalum, copper, aluminum, carbides, and nickel-base alloys.

Regardless of the equipment used, what is important is the proper preparation of the surface that will receive the sprayed coating. Preparation activities include the degreasing or solvent cleaning of the surface, undercutting of the surface to provide room for the proper coating thickness, abrasive or grit blasting the substrate to provide a roughened surface, grooving (in the case of flat surfaces) or rough threading (in the case of cylindrical work) the surface to be coated, preheating the base metal. Methods of obtaining a bond between the sprayed material and the substrate are: heating the base, roughening the base, or spraying a "self-bonding" material onto a smooth surface; however, heating alone is seldom used in machine element work as the elevated temperatures required to obtain the proper bond causes problems of warpage and surface corrosion.

FASTENERS

# TABLE OF CONTENTS

## FASTENERS

### TORQUE AND TENSION IN FASTENERS



### INCH THREADED FASTENERS



### INCH THREADED FASTENERS

*(Continued)*



### METRIC THREADED FASTENERS

## TABLE OF CONTENTS

## FASTENERS

TABLE OF CONTENTS

FASTENERS

## TABLE OF CONTENTS

## FASTENERS

### RIVETS AND RIVETED JOINTS

*(Continued)*



### PINS AND STUDS



### RETAINING RINGS



### WING NUTS, WING SCREWS, AND THUMB SCREWS



### NAILS, SPIKES, AND WOOD SCREWS

## TORQUE AND TENSION IN FASTENERS

**Tightening Bolts.—**Bolts are often tightened by applying torque to the head or nut, which causes the bolt to stretch. The stretching results in bolt tension or preload, which is the force that holds a joint together. Torque is relatively easy to measure with a torque wrench, so it is the most frequently used indicator of bolt tension. Unfortunately, a torque wrench does not measure bolt tension accurately, mainly because it does not take friction into account. The friction depends on bolt, nut, and washer material, surface smoothness, machining accuracy, degree of lubrication, and the number of times a bolt has been installed. Fastener manufacturers often provide information for determining torque requirements for tightening various bolts, accounting for friction and other effects. If this information is not available, the methods described in what follows give general guidelines for determining how much tension should be present in a bolt, and how much torque may need to be applied to arrive at that tension.

High preload tension helps keep bolts tight, increases joint strength, creates friction between parts to resist shear, and improves the fatigue resistance of bolted connections. The recommended preload $F_i$, which can be used for either static (stationary) or fatigue (alternating) applications, can be determined from: $F_i = 0.75 \times A_t \times S_p$ for reusable connections, and $F_i = 0.9 \times A_t \times S_p$ for permanent connections. In these formulas, $F_i$ is the bolt preload, $A_t$ is the tensile stress area of the bolt, and $S_p$ is the proof strength of the bolt. Determine $A_t$ from screw-thread tables or by means of formulas in this section. Proof strength $S_p$ of commonly used ASTM and SAE steel fasteners is given in this section and in the section on metric screws and bolts for those fasteners. For other materials, an approximate value of proof strength can be obtained from: $S_p = 0.85 \times S_y$, where $S_y$ is the yield strength of the material. Soft materials should not be used for threaded fasteners.

Once the required preload has been determined, one of the best ways to be sure that a bolt is properly tensioned is to measure its tension directly with a strain gage. Next best is to measure the change in length (elongation) of the bolt during tightening, using a micrometer or dial indicator. Each of the following two formulas calculates the required change in length of a bolt needed to make the bolt tension equal to the recommended preload. The change in length δ of the bolt is given by:

$$\delta = F_i \times \frac{A_d \times l_t + A_t \times l_d}{A_d \times A_t \times E} \quad (1) \qquad \text{or} \qquad \delta = \frac{F_i \times l}{A \times E} \quad (2)$$

In Equation (1), $F_i$ is the bolt preload; $A_d$ is the major-diameter area of the bolt; $A_t$ is the tensile-stress area of the bolt; $E$ is the bolt modulus of elasticity; $l_t$ is the length of the threaded portion of the fastener within the grip; and $l_d$ is the length of the unthreaded portion of the grip. Here, the grip is defined as the total thickness of the clamped material. Equation (2) is a simplified formula for use when the area of the fastener is constant, and gives approximately the same results as Equation (1). In Equation (2), $l$ is the bolt length; $A$ is the bolt area; and $\delta$, $F_i$, and $E$ are as described before.

If measuring bolt elongation is not possible, the torque necessary to tighten the bolt must be estimated. If the recommended preload is known, use the following general relation for the torque: $T = K \times F_i \times d$, where $T$ is the wrench torque, $K$ is a constant that depends on the bolt material and size, $F_i$ is the preload, and $d$ is the nominal bolt diameter. A value of $K = 0.2$ may be used in this equation for mild-steel bolts in the size range of ¼ to 1 inch (6.35–25.4 mm). For other steel bolts, use the following values of $K$: nonplated black finish, 0.3; zinc-plated, 0.2; lubricated, 0.18; cadmium-plated, 0.16. Check with bolt manufacturers and suppliers for values of $K$ to use with bolts of other sizes and materials.

The proper torque to use for tightening bolts in sizes up to about ½ inch (12.7 mm) may also be determined by trial. Test a bolt by measuring the amount of torque required to fracture it (use bolt, nut, and washers equivalent to those chosen for the real application). Then, use a tightening torque of about 50 to 60 percent of the fracture torque determined by the test. The tension in a bolt tightened using this procedure will be about 60 to 70 percent of the elastic limit (yield strength) of the bolt material.

The table that follows can be used to get a rough idea of the torque necessary to properly tension a bolt by using the bolt diameter *d* and the coefficients *b* and *m* from the table; the approximate tightening torque *T* in ft-lb for the listed fasteners is obtained by solving the equation $T = 10^{b+m \log d}$. This equation is approximate, for use with unlubricated fasteners as supplied by the mill. See the notes at the end of the table for more details on using the equation.

**Wrench Torque $T = 10^{b+m \log d}$ for Steel Bolts, Studs, and Cap Screws (*see notes*)**

| Fastener Grade(s) | Bolt Diameter *d* (in.) | *m* | *b* |
|---|---|---|---|
| SAE 2, ASTM A307 | ¼ to 3 | 2.940 | 2.533 |
| SAE 3 | ¼ to 3 | 3.060 | 2.775 |
| ASTM A449, A354-BB, SAE 5 | ¼ to 3 | 2.965 | 2.759 |
| ASTM A325[a] | ½ to 1½ | 2.922 | 2.893 |
| ASTM A354-BC | ¼ to ⅝ | 3.046 | 2.837 |
| SAE 6, SAE 7 | ¼ to 3 | 3.095 | 2.948 |
| SAE 8 | ¼ to 3 | 3.095 | 2.983 |
| ASTM A354-BD, ASTM A490[a] | ⅜ to 1¾ | 3.092 | 3.057 |
| Socket Head Cap Screws | ¼ to 3 | 3.096 | 3.014 |

[a] Values for permanent fastenings on steel structures.

Usage: Values calculated using the preceding equation are for standard, unplated industrial fasteners as received from the manufacturer; for cadmium-plated cap screws, multiply the torque by 0.9; for cadmium-plated nuts and bolts, multiply the torque by 0.8; for fasteners used with special lubricants, multiply the torque by 0.9; for studs, use cap screw values for equivalent grade.

**Preload for Bolts in Loaded Joints.—**The following recommendations are based on MIL-HDBK-60, a subsection of FED-STD-H28, Screw Thread Standards for Federal Service. Generally, bolt preload in joints should be high enough to maintain joint members in contact and in compression. Loss of compression in a joint may result in leakage of pressurized fluids past compression gaskets, loosening of fasteners under conditions of cyclic loading, and reduction of fastener fatigue life.

The relationship between fastener fatigue life and fastener preload is illustrated by Fig. 1. An axially loaded bolted joint in which there is no bolt preload is represented by line OAB, that is, the bolt load is equal to the joint load. When joint load varies between $P_a$ and $P_b$, the bolt load varies accordingly between $P_{Ba}$ and $P_{Bb}$. However, if preload $P_{B1'}$ is applied to the bolt, the joint is compressed and bolt load changes more slowly than the joint load (indicated by line $P_{B1'}$A, whose slope is less than line OAB) because some of the load is absorbed as a reduction of compression in the joint. Thus, the axial load applied to the joint varies between $P_{Ba'}$ and $P_{Bb'}$ as joint load varies between $P_a$ and $P_b$. This condition results in a considerable reduction in cyclic bolt-load variation and thereby increases the fatigue life of the fastener.

**Preload for Bolts in Shear.—**In shear-loaded joints, with members that slide, the joint members transmit shear loads to the fasteners in the joint and the preload must be sufficient to hold the joint members in contact. In joints that do not slide (i.e., there is no relative motion between joint members), shear loads are transmitted within the joint by frictional forces that mainly result from the preload. Therefore, preload must be great enough for the resulting friction forces to be greater than the applied shear force. With high applied shear loads, the shear stress induced in the fastener during application of the preload must also be considered in the bolted-joint design. Joints with combined axial and shear loads must be analyzed to ensure that the bolts will not fail in either tension or shear.

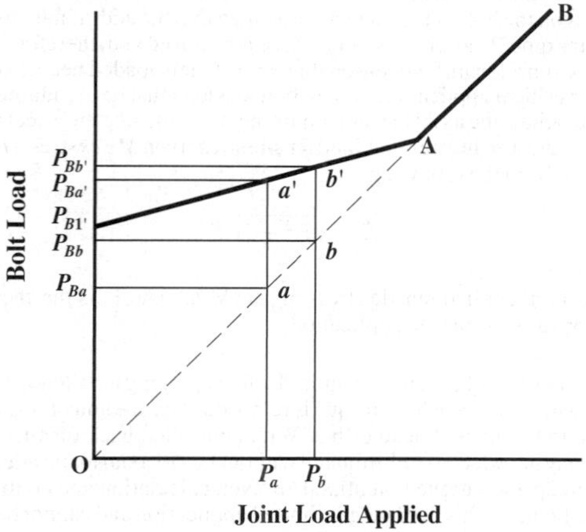


Fig. 1. Bolt Load in a Joint with Applied Axial Load

**General Application of Preload.—**Preload values should be based on joint requirements, as outlined before. Fastener applications are generally designed for maximum utilization of the fastener material; that is to say, the fastener size is the minimum required to perform its function and a maximum safe preload is generally applied to it. However, if a low-strength fastener is replaced by one of higher strength, for the sake of convenience or standardization, the preload in the replacement should not be increased beyond that required in the original fastener.

To utilize the maximum amount of bolt strength, bolts are sometimes tightened to or beyond the yield point of the material. This practice is generally limited to ductile materials, where there is considerable difference between the yield strength and the ultimate (breaking) strength, because low-ductility materials are more likely to fail due to unexpected overloads when preloaded to yield. Joints designed for primarily static load conditions that use ductile bolts, with a yield strain that is relatively far from the strain at fracture, are often preloaded above the yield point of the bolt material. Methods for tightening up to and beyond the yield point include tightening by feel without special tools, and the use of electronic equipment designed to compare the applied torque with the angular rotation of the fastener and detect changes that occur in the elastic properties of fasteners at yield.

Bolt loads are maintained below the yield point in joints subjected to cyclic loading and in joints using bolts of high-strength material where the yield strain is close to the strain at fracture. For these conditions, the maximum preloads generally fall within the following ranges: 50 to 80 percent of the minimum tensile ultimate strength; 75 to 90 percent of the minimum tensile yield strength or proof load; or 100 percent of the observed proportional limit or onset of yield.

Bolt heads, driving recesses (in socket screws, for example), and the juncture of head and shank must be sufficiently strong to withstand the preload and any additional stress encountered during tightening. There must also be sufficient thread to prevent stripping (generally, at least three fully engaged threads). Materials susceptible to stress-corrosion cracking may require further preload limitations.

**Preload Adjustments.—**Preloads may be applied directly by axial loading or indirectly by turning of the nut or bolt. When preload is applied by turning of nuts or bolts, a torsion load component is added to the desired axial bolt load. This combined loading increases the tensile stress on the bolt. It is frequently assumed that the additional torsion load component dissipates quickly after the driving force is removed and, therefore, can be largely ignored. This assumption may be reasonable for fasteners loaded near to or beyond yield strength, but for critical applications where bolt tension must be maintained below yield, it is important to adjust the axial tension requirements to include the effects of the preload torsion. For this adjustment, the combined tensile stress (*von Mises* stress) $F_{tc}$ in psi (MPa) can be calculated from the following:

$$F_{tc} = \sqrt{F_t^2 + 3F_s^2} \quad (3)$$

where $F_t$ is the axial applied tensile stress in psi (MPa), and $F_s$ is the shear stress in psi (MPa) caused by the torsion load application.

Some of the torsion load on a bolt, acquired when applying a preload, may be released by springback when the wrenching torque is removed. The amount of relaxation depends on the friction under the bolt head or nut. With controlled back turning of the nut, the torsional load may be reduced or eliminated without loss of axial load, reducing bolt stress and lowering creep and fatigue potential. However, calculation and control of the back-turn angle is difficult, so this method has limited application and cannot be used for short bolts because of the small angles involved.

For relatively soft work-hardenable materials, tightening bolts in a joint slightly beyond yield will work-harden the bolt to some degree. Back turning of the bolt to the desired tension will reduce embedment and metal flow and improve resistance to preload loss.

The following formula for use with single-start Unified inch screw threads calculates the combined tensile stress, $F_{tc}$:

$$F_{tc} = F_t\sqrt{1 + 3\left(\frac{1.96 + 2.31\mu}{1 - 0.325P/d_2} - 1.96\right)^2} \quad (4)$$

Single-start UNJ screw threads in accordance with MIL-S-8879 have a thread stress diameter equal to the bolt pitch diameter. For these threads, $F_{tc}$ can be calculated from:

$$F_{tc} = F_t\sqrt{1 + 3\left(\frac{0.637P}{d_2} + 2.31\mu\right)^2} \quad (5)$$

where $\mu$ is the coefficient of friction between threads, $P$ is the thread pitch ($P = 1/n$, and $n$ is the number of threads per inch), and $d_2$ is the bolt-thread pitch diameter in inches. Both Equations (2) and (3) are derived from Equation (1); thus, the quantity within the radical ($\sqrt{\ }$) represents the proportion of increase in axial bolt tension resulting from preload torsion. In these equations, tensile stress due to torsion load application becomes most significant when the thread friction, $\mu$, is high.

**Coefficients of Friction for Bolts and Nuts.—**Table 1 gives examples of coefficients of friction that are frequently used in determining torque requirements. Dry threads, indicated by the words "None added" in the Lubricant column, are assumed to have some residual machine oil lubrication. Table 1 values are not valid for threads that have been cleaned to remove all traces of lubrication because the coefficient of friction of these threads may be very much higher unless a plating or other film is acting as a lubricant.

**Table 1. Coefficients of Friction of Bolts and Nuts**

| Bolt/Nut Materials | Lubricant | Coefficient of Friction, $\mu \pm 20\%$ |
|---|---|---|
| Steel[a] | Graphite in petrolatum or oil | 0.07 |
| | Molybdenum disulfide grease | 0.11 |
| | Machine oil | 0.15 |
| Steel,[a] cadmium-plated | None added | 0.12 |
| Steel,[a] zinc-plated | None added | 0.17 |
| Steel[a]/bronze | None added | 0.15 |
| Corrosion-resistant steel or nickel-base alloys/silver-plated materials | None added | 0.14 |
| Titanium/steel[a] | Graphite in petrolatum | 0.08 |
| Titanium | Molybdenum disulfide grease | 0.10 |

[a] "Steel" includes carbon and low-alloy steels but not corrosion-resistant steels.

Where two materials are separated by a slash (/), either may be the bolt material; the other is the nut material.

**Preload Relaxation.—**Local yielding, due to excess bearing stress under nuts and bolt heads (caused by high local spots, rough surface finish, and lack of perfect squareness of bolt and nut bearing surfaces), may result in preload relaxation after preloads are first applied to a bolt. Bolt tension also may be unevenly distributed over the threads in a joint, so thread deformation may occur, causing the load to be redistributed more evenly over the threaded length. Preload relaxation occurs over a period of minutes to hours after the application of the preload, so retightening after several minutes to several days may be required. As a general rule, an allowance for loss of preload of about 10 percent may be made when designing a joint.

Increasing the resilience of a joint will make it more resistant to local yielding, that is, there will be less loss of preload due to yielding. When practical, a joint-length to bolt-diameter ratio of 4 or more is recommended; for example, a $^1/_4$-inch (6.35-mm) bolt and a 1-inch (25.4-mm) or greater joint length. Through bolts, far-side tapped holes, spacers, and washers can be used in the joint design to improve the joint-length to bolt-diameter ratio.

Over an extended period of time, preload may be reduced or completely lost due to vibration; temperature cycling, including changes in ambient temperature; creep; joint load; and other factors. An increase in the initial bolt preload or the use of thread-locking methods that prevent relative motion of the joint may reduce the problem of preload relaxation due to vibration and temperature cycling. Creep is generally a high-temperature effect, although some loss of bolt tension can be expected even at normal temperatures. Harder materials and creep-resistant materials should be considered if creep is a problem or high-temperature service of the joint is expected.

Mechanical properties of fastener materials vary significantly with temperature, and allowance must be made for these changes when ambient temperatures range beyond 30 to 200° F (−1 to 93° C). Mechanical properties that may change include tensile strength, yield strength, and modulus of elasticity. Where bolts and flange materials are generically dissimilar, such as carbon steel and corrosion-resistant steel, or steel and brass, differences in thermal expansion that might cause preload to increase or decrease must be taken into consideration.

**Methods of Applying and Measuring Preload.—**Depending on the tightening method, the accuracy of preload application may vary up to 25 percent or more. Care must be taken to maintain the calibration of torque and load indicators. Allowance should be made for uncertainties in bolt load to prevent overstressing the bolts or failing to obtain sufficient preload. The method of tensioning should be based on required accuracy and relative costs.

The most common methods of bolt tension control are indirect because it is usually difficult or impractical to measure the tension produced in each fastener during assembly. Table 2 lists the most frequently used methods of applying bolt preload and the approximate accuracy of each method. For many applications, fastener tension can be satisfactorily controlled within certain limits by applying a known torque to the fastener. Laboratory tests have shown that whereas a satisfactory torque tension relationship can be established for a given set of conditions, a change of any of the variables, such as fastener material, surface finish, and the presence or absence of lubrication, may severely alter the relationship. Because most of the applied torque is absorbed in intermediate friction, a change in the surface roughness of the bearing surfaces or a change in the lubrication will drastically affect the friction and thus the torque tension relationship. Regardless of the method or accuracy of applying the preload, tension will decrease in time if the bolt, nut, or washer seating faces deform under load, if the bolt stretches or creeps under tensile load, or if cyclic loading causes relative motion between joint members.

**Table 2. Accuracy of Bolt Preload Application Methods**

| Method | Accuracy | Method | Accuracy |
|---|---|---|---|
| By feel | ±35% | Computer-controlled wrench | |
| Torque wrench | ±25% | below yield (turn-of-nut) | ±15% |
| Turn-of-nut | ±15% | yield-point sensing | ±8% |
| Preload indicating washer | ±10% | Bolt elongation | ±3–5% |
| Strain gages | ±1% | Ultrasonic sensing | ±1% |

Tightening methods using power drivers are similar in accuracy to equivalent manual methods.

**Elongation Measurement.—**Bolt elongation is directly proportional to axial stress when the applied stress is within the elastic range of the material. If both ends of a bolt are accessible, a micrometer measurement of bolt length made before and after the application of tension will ensure the required axial stress is applied. The elongation δ in inches (mm) can be determined from the formula $\delta = F_t \times L_B \div E$, given the required axial stress $F_t$ in psi (MPa), the bolt modulus of elasticity $E$ in psi (MPa), and the effective bolt length $L_B$ in inches (mm). $L_B$, as indicated in Fig. 2, includes the contribution of bolt area and ends (head and nut) and is calculated from:

$$L_B = \left(\frac{d_{ts}}{d}\right)^2 \times \left(L_s + \frac{H_B}{2}\right) + L_J - L_S + \frac{H_N}{2} \tag{6}$$

where $d_{ts}$ is the thread stress diameter, $d$ is the bolt diameter, $L_s$ is the unthreaded length of the bolt shank, $L_j$ is the overall joint length, $H_B$ is the height of the bolt head, and $H_N$ is the height of the nut.

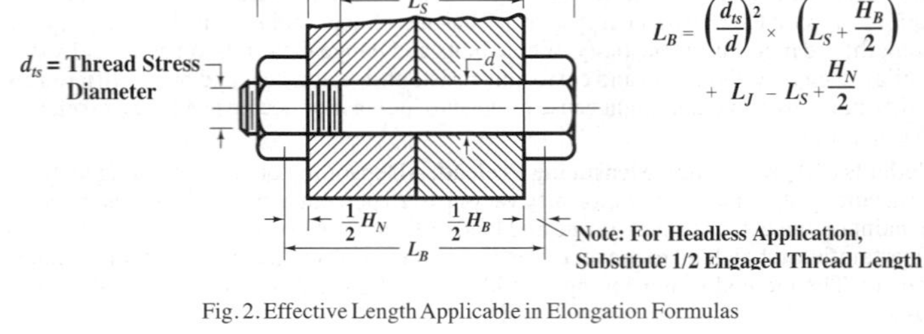


Fig. 2. Effective Length Applicable in Elongation Formulas

The micrometer method is most easily and accurately applied to bolts that are essentially uniform throughout the bolt length, that is, threaded along the entire length or that have only a few threads in the bolt grip area. If the bolt geometry is complex, such as tapered or stepped, the elongation is equal to the sum of the elongations of each section with allowances made for transitional stresses in bolt head height and nut engagement length.

The direct method of measuring elongation is practical only if both ends of a bolt are accessible. Otherwise, if the diameter of the bolt or stud is sufficiently large, an axial hole can be drilled, as shown in Fig. 3, and a micrometer depth gage or other means used to determine the change in length of the hole as the fastener is tightened. A similar method uses a special indicating bolt that has a blind axial hole containing a pin fixed at the bottom. The pin is usually made flush with the bolt head surface before load application. As the bolt is loaded, the elongation causes the end of the pin to move below the reference surface. The displacement of the pin can be converted directly into unit stress by means of a calibrated gage. In some bolts of this type, the pin is set a distance above the bolt so that the pin is flush with the bolt head when the required axial load is reached.

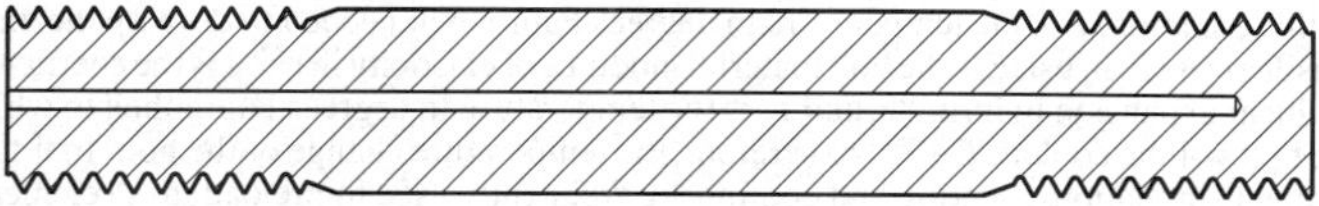

Fig. 3. Hole Drilled to Measure Elongation When One End of Stud or Bolt Is Not Accessible

The *ultrasonic method* of measuring elongation uses a sound pulse, generated at one end of a bolt, that travels the length of a bolt, bounces off the far end, and returns to the sound generator in a measured period of time. The time required for the sound pulse to return depends on the length of the bolt and the speed of sound in the bolt material. The speed of sound in the bolt depends on the material, the temperature, and the stress level. The ultrasonic measurement system can compute the stress, load, or elongation of the bolt at any time by comparing the pulse travel time in the loaded and unstressed conditions. In a similar method, measuring round-trip transit times of longitudinal and shear wave sonic pulses allows calculation of tensile stress in a bolt without consideration of bolt length. This method permits checking bolt tension at any time and does not require a record of the ultrasonic characteristics of each bolt at zero load.

To ensure consistent results, the ultrasonic method requires that both ends of the bolt be finished square to the bolt axis. The accuracy of ultrasonic measurement compares favorably with strain gage methods, but is limited by sonic velocity variations between bolts of the same material and by corrections that must be made for unstressed portions of the bolt heads and threads.

The *turn-of-nut method* applies preload by turning a nut through an angle that corresponds to a given elongation. The elongation of the bolt is related to the angle turned by the formula: $\delta_B = \theta \times l \div 360$, where $\delta_B$ is the elongation in inches (mm), $\theta$ is the turn angle of the nut in degrees, and $l$ is the lead of the thread helix in inches (mm). Substituting $F_t \times L_B \div E$ for elongation $\delta_B$ in this equation gives the turn-of-nut angle required to attain preload $F_t$:

$$\theta = 360\frac{F_t\,L_B}{El} \qquad (7)$$

where $L_B$ is given by Equation (6), and $E$ is the modulus of elasticity.

Accuracy of the turn-of-nut method is affected by elastic deformation of the threads, by roughness of the bearing surfaces, and by the difficulty of determining the starting point for measuring the angle. The starting point is usually found by tightening the nut enough to seat the contact surfaces firmly, and then loosening it just enough to release any tension and twisting in the bolt. The nut-turn angle will be different for each bolt size, length, material, and thread lead. The preceding method of calculating the nut-turn angle also requires elongation of the bolt without a corresponding compression of the joint material. The turn-of-nut method, as just outlined, is not valid for joints with compressible gaskets or other soft material, or if there is a significant deformation of the nut and joint material relative to that of the bolt. The nut-turn angle would then have to be determined empirically using a simulated joint and a tension-measuring device.

The Japanese Industrial Standards (JIS) handbook, *Fasteners and Screw Threads,* indicates that the turn-of-nut tightening method is applicable in both elastic and plastic region tightening. Refer to JIS B 1083 for more detail on this subject.

*Heating* causes a bolt to expand at a rate proportional to its coefficient of expansion. When a hot bolt and nut are fastened in a joint and cooled, the bolt shrinks and tension is developed. The temperature necessary to develop an axial stress, $F_t$ (when the stress is below the elastic limit) can be found as follows:

$$T = \frac{F_t}{Ee} + T_o \tag{8}$$

In this equation, $T$ is the temperature in degrees Fahrenheit needed to develop the axial tensile stress $F_t$ in psi, $E$ is the bolt material modulus of elasticity in psi, $e$ is the coefficient of linear expansion in in/in-°F, and $T_o$ is the temperature in degrees Fahrenheit to which the bolt will be cooled. $T - T_o$ is, therefore, the temperature change of the bolt. In finite-element simulations, heating and cooling are frequently used to preload mesh elements in tension or compression. Equation (8) can be used to determine required temperature changes in such problems.

*Example:* A tensile stress of 40,000 psi is required for a steel bolt in a joint operating at 70° F. If $E$ is $30 \times 10^6$ psi and $e$ is $6.2 \times 10^{-6}$ in/in-°F, determine the temperature of the bolt needed to develop the required stress on cooling.

$$T = \frac{40{,}000}{(30 \times 10^6)(6.2 \times 10^{-6})} + 70 = 285°\text{F}$$

In practice, the bolt is heated slightly above the required temperature (to allow for some cooling while the nut is screwed down) and the nut is tightened snugly. Tension develops as the bolt cools. In another method, the nut is tightened snugly on the bolt, and the bolt is heated in place. When the bolt has elongated sufficiently, as indicated by inserting a thickness gage between the nut and the bearing surface of the joint, the nut is tightened. The bolt develops the required tension as it cools; however, preload may be lost if the joint temperature increases appreciably while the bolt is being heated.

**Calculating Thread Tensile-Stress Area.—**The tensile-stress area for Unified threads is based on a diameter equivalent to the mean of the pitch and minor diameters. The pitch and the minor diameters for Unified screw threads can be found from the major (nominal) diameter, $d$, and the screw pitch, $P = 1/n$, where $n$ is the number of threads per inch, by use of the following formulas: the pitch diameter $d_p = d - 0.649519 \times P$; the minor diameter $d_m = d - 1.299038 \times P$. The tensile stress area, $A_s$, for Unified threads can then be found as follows:

$$A_s = \frac{\pi}{4}\left(\frac{d_m + d_p}{2}\right)^2 \tag{9}$$

UNJ threads in accordance with MIL-S-8879 have a tensile thread area that is usually considered to be at the basic bolt pitch diameter; for these threads, $A_s = (\pi d_p^2)/4$. The tensile stress area for Unified screw threads is smaller than this area, so the required tightening torque for UNJ threaded bolts is greater than for an equally stressed Unified threaded bolt in an equivalent joint. To convert tightening torque for a Unified fastener to the equivalent torque required with a UNJ fastener, use the following relationship:

$$\text{UNJ}_{\text{torque}} = \left(\frac{d \times n - 0.6495}{d \times n - 0.9743}\right)^2 \times \text{Unified}_{\text{torque}} \quad (10)$$

where $d$ is the basic thread major diameter, and $n$ is the number of threads per inch.

The tensile stress area for metric threads is based on a diameter equivalent to the mean of the pitch diameter and a diameter obtained by subtracting ⅙ the height of the fundamental thread triangle from the external-thread minor diameter. The Japanese Industrial Standard JIS B 1082 (see also ISO 898/1) defines the stress area of metric screw threads as follows:

$$A_s = \frac{\pi}{4}\left(\frac{d_2 + d_3}{2}\right)^2 \quad (11)$$

In Equation (11), $A_s$ is the stress area of the metric screw thread in mm²; $d_2$ is the pitch diameter of the external thread in mm, given by $d_2 = d - 0.649515 \times P$; and $d_3$ is defined by $d_3 = d_1 - H/6$. Here, $d$ is the nominal bolt diameter; $P$ is the thread pitch; $d_1 = d - 1.082532 \times P$ is the minor diameter of the external thread in mm; and $H = 0.866025 \times P$ is the height of the fundamental thread triangle. Substituting the formulas for $d_2$ and $d_3$ into Equation (11) results in $A_s = 0.7854(d - 0.9382P)^2$.

The stress area, $A_s$, of Unified threads in mm² is given in JIS B 1082 as:

$$A_s = 0.7854\left(d - \frac{0.9743}{n} \times 25.4\right)^2 \quad (12)$$

**Relation between Torque and Clamping Force.**—The Japanese Industrial Standard JIS B 1803 defines fastener tightening torque $T_f$ as the sum of the bearing surface torque $T_w$ and the shank (threaded) portion torque $T_s$. The relationship between the applied tightening torque and bolt preload $F_{ft}$ is as follows: $T_f = T_s + T_w = K \times F_f \times d$. In the preceding, $d$ is the nominal diameter of the screw thread, and $K$ is the torque coefficient defined as follows:

$$K = \frac{1}{2d}\left(\frac{P}{\pi} + \mu_s d_2 \sec\alpha' + \mu_w D_w\right) \quad (13)$$

where $P$ is the screw thread pitch; $\mu_s$ is the coefficient of friction between threads; $d_2$ is the pitch diameter of the thread; $\mu_w$ is the coefficient of friction between bearing surfaces; $D_w$ is the equivalent diameter of the friction torque bearing surfaces; and $\alpha'$ is the flank angle at the ridge perpendicular section of the thread ridge, defined by $\tan\alpha' = \tan\alpha\cos\beta$, where $\alpha$ is the thread half angle (30°, for example), and $\beta$ is the thread helix, or lead, angle. $\beta$ can be found from $\tan\beta = 1 \div 2\pi r$, where $l$ is the thread lead, and $r$ is the thread radius (i.e., one-half the nominal diameter $d$). When the bearing surface contact area is circular, $D_w$ can be obtained as follows:

$$D_w = \frac{2}{3} \times \frac{D_o^3 - D_i^3}{D_o^2 - D_i^2} \quad (14)$$

where $D_o$ and $D_i$ are the outside and inside diameters, respectively, of the bearing surface contact area.

The torques attributable to the threaded portion of a fastener, $T_s$, and bearing surfaces of a joint, $T_w$, are as follows:

$$T_s = \frac{F_f}{2}\left(\frac{P}{\pi} + \mu_s d_2 \sec\alpha'\right) \quad (15) \qquad T_w = \frac{F_f}{2}\mu_w D_w \quad (16)$$

where $F_f$, $P$, $\mu$, $d_2$, $\alpha'$, $\mu_w$, and $D_w$ are as previously defined.

Table 3 and Table 4 give values of torque coefficient $K$ for coarse- and fine-pitch metric screw threads corresponding to various values of $\mu_s$ and $\mu_w$. When a fastener material yields according to the shearing-strain energy theory, the torque corresponding to the yield clamping force (see Fig. 4) is $T_{fy} = K \times F_{fy} \times d$, where the yield clamping force $F_{fy}$ is given by:

$$F_{fy} = \frac{\sigma_y A_s}{\sqrt{1 + 3\left[\frac{2}{d_A}\left(\frac{P}{\pi} + \mu_s d_2 \sec\alpha'\right)\right]^2}} \qquad (17)$$

**Table 3. Torque Coefficients $K$ for Metric Hexagon Head Bolt and Nut Coarse Screw Threads**

| Between Threads, $\mu_s$ | Coefficient of Friction: Between Bearing Surfaces, $\mu_w$ | | | | | | | | | |
|---|---|---|---|---|---|---|---|---|---|---|
| | 0.08 | 0.10 | 0.12 | 0.15 | 0.20 | 0.25 | 0.30 | 0.35 | 0.40 | 0.45 |
| 0.08 | 0.117 | 0.130 | 0.143 | 0.163 | 0.195 | 0.228 | 0.261 | 0.293 | 0.326 | 0.359 |
| 0.10 | 0.127 | 0.140 | 0.153 | 0.173 | 0.206 | 0.239 | 0.271 | 0.304 | 0.337 | 0.369 |
| 0.12 | 0.138 | 0.151 | 0.164 | 0.184 | 0.216 | 0.249 | 0.282 | 0.314 | 0.347 | 0.380 |
| 0.15 | 0.153 | 0.167 | 0.180 | 0.199 | 0.232 | 0.265 | 0.297 | 0.330 | 0.363 | 0.396 |
| 0.20 | 0.180 | 0.193 | 0.206 | 0.226 | 0.258 | 0.291 | 0.324 | 0.356 | 0.389 | 0.422 |
| 0.25 | 0.206 | 0.219 | 0.232 | 0.252 | 0.284 | 0.317 | 0.350 | 0.383 | 0.415 | 0.448 |
| 0.30 | 0.232 | 0.245 | 0.258 | 0.278 | 0.311 | 0.343 | 0.376 | 0.409 | 0.442 | 0.474 |
| 0.35 | 0.258 | 0.271 | 0.284 | 0.304 | 0.337 | 0.370 | 0.402 | 0.435 | 0.468 | 0.500 |
| 0.40 | 0.285 | 0.298 | 0.311 | 0.330 | 0.363 | 0.396 | 0.428 | 0.461 | 0.494 | 0.527 |
| 0.45 | 0.311 | 0.324 | 0.337 | 0.357 | 0.389 | 0.422 | 0.455 | 0.487 | 0.520 | 0.553 |

Values in the table are average values of torque coefficient calculated using: Equations (13) and (14) for $K$ and $D_w$; diameters $d$ of 4, 5, 6, 8, 10, 12, 16, 20, 24, 30, and 36 mm; and selected corresponding pitches $P$ and pitch diameters $d_2$ according to JIS B 0205 (ISO 724) thread standard. Dimension $D_i$ was obtained for a Class 2 fit without chamfer from JIS B 1001, Diameters of Clearance Holes and Counterbores for Bolts and Screws (equivalent to ISO 273-1979). The value of $D_o$ was obtained by multiplying the reference dimension from JIS B 1002, width across the flats of the hexagon head, by 0.95.

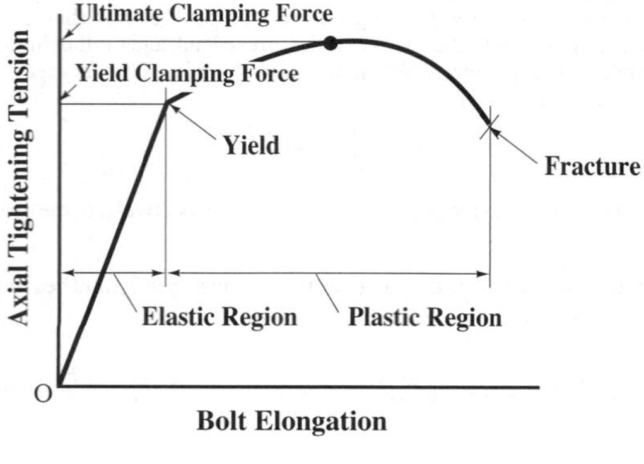


Fig. 4. Relationship between Bolt Elongation and Axial Tightening Tension

**Table 4. Torque Coefficients *K* for Metric Hexagon Head Bolt and Nut Fine Screw Threads**

| Between Threads, $\mu_s$ | Coefficient of Friction: Between Bearing Surfaces, $\mu_w$ 0.08 | 0.10 | 0.12 | 0.15 | 0.20 | 0.25 | 0.30 | 0.35 | 0.40 | 0.45 |
|---|---|---|---|---|---|---|---|---|---|---|
| 0.08 | 0.106 | 0.118 | 0.130 | 0.148 | 0.177 | 0.207 | 0.237 | 0.267 | 0.296 | 0.326 |
| 0.10 | 0.117 | 0.129 | 0.141 | 0.158 | 0.188 | 0.218 | 0.248 | 0.278 | 0.307 | 0.337 |
| 0.12 | 0.128 | 0.140 | 0.151 | 0.169 | 0.199 | 0.229 | 0.259 | 0.288 | 0.318 | 0.348 |
| 0.15 | 0.144 | 0.156 | 0.168 | 0.186 | 0.215 | 0.245 | 0.275 | 0.305 | 0.334 | 0.364 |
| 0.20 | 0.171 | 0.183 | 0.195 | 0.213 | 0.242 | 0.272 | 0.302 | 0.332 | 0.361 | 0.391 |
| 0.25 | 0.198 | 0.210 | 0.222 | 0.240 | 0.270 | 0.299 | 0.329 | 0.359 | 0.389 | 0.418 |
| 0.30 | 0.225 | 0.237 | 0.249 | 0.267 | 0.297 | 0.326 | 0.356 | 0.386 | 0.416 | 0.445 |
| 0.35 | 0.252 | 0.264 | 0.276 | 0.294 | 0.324 | 0.353 | 0.383 | 0.413 | 0.443 | 0.472 |
| 0.40 | 0.279 | 0.291 | 0.303 | 0.321 | 0.351 | 0.381 | 0.410 | 0.440 | 0.470 | 0.500 |
| 0.45 | 0.306 | 0.318 | 0.330 | 0.348 | 0.378 | 0.408 | 0.437 | 0.467 | 0.497 | 0.527 |

Values in the table are average values of torque coefficient calculated using Equations (13) and (14) for $K$ and $D_w$; diameters $d$ of 8, 10, 12, 16, 20, 24, 30, and 36 mm; and selected respective pitches $P$ and pitch diameters $d_2$ according to JIS B 0207 thread standard (ISO 724). Dimension $D_i$ was obtained for a Class 1 fit without chamfer from JIS B 1001, Diameters of Clearance Holes and Counterbores for Bolts and Screws (equivalent to ISO 273-1979). The value of $D_o$ was obtained by multiplying the reference dimension from JIS B 1002 (small type series), width across the flats of the hexagon head, by 0.95.

In Equation (17), $\sigma_y$ is the yield point or proof stress of the bolt, $A_s$ is the stress area of the thread, and $d_A = (4A_s/\pi)^{1/2}$ is the diameter of a circle having an area equal to the stress area of the thread. The other variables have been identified previously.

*Example:* Find the torque required to tighten a 10-mm coarse-threaded ($P = 1.5$) grade 8.8 bolt to yield assuming that both the thread- and bearing-friction coefficients are 0.12.

*Solution:* From Equation (17), calculate $F_{fy}$ and then solve $T_{fy} = KF_{fy}d$ to obtain the torque required to stress the bolt to the yield point.

$\sigma_y = 640$ N/mm² (MPa) (minimum, based on 8.8 grade rating)

$A_s = 0.7854(10 - 0.9382 \times 1.5)^2 = 57.99\text{ mm}^2$

$d_A = (4A_s/\pi)^{1/2} = 8.6\text{ mm}$

$d_2 = 9.026$ mm (see JIS B 0205 or ISO 724)

Find $\alpha'$ from $\tan\alpha' = \tan\alpha\cos\beta$ using:

$\alpha = 30°$; $\tan\beta = l \div 2\pi r$; $l = P = 1.5$; and $r = d \div 2 = 5$ mm

$\tan\beta = 1.5 \div 10\pi = 0.0477$, therefore $\beta = 2.73°$

$\tan\alpha' = \tan\alpha\cos\beta = \tan 30° \times \cos 2.73° = 0.577$, and $\alpha' = 29.97°$

Solving Equation (17) gives the yield clamping force as follows:

$$F_{fy} = \frac{640 \times 57.99}{\sqrt{1 + 3\left[\frac{2}{8.6}\left(\frac{1.5}{\pi} + 0.12 \times 9.026 \times \sec 29.97°\right)\right]^2}} = 30{,}463\text{N}$$

$K$ can be determined from Table 3 (coarse thread) and Table 4 (fine thread) or from Equations (13) and (14). From Table 3, for $\mu_s$ and $\mu_w$ equal to 0.12, $K = 0.164$. The yield-point tightening torque can then be found from $T_{fy} = K \times F_{fy} \times d = 0.164 \times 30{,}463 \times 10 = 49.9 \times 10^3$ N-mm = 49.9 N-m.

**Obtaining Torque and Friction Coefficients.—**Given suitable test equipment, the torque coefficient $K$ and friction coefficients between threads $\mu_s$ or between bearing surfaces $\mu_w$ can be determined experimentally as follows: Measure the value of the axial tightening tension and the corresponding tightening torque at an arbitrary point in the 50 to 80 percent range of the bolt yield point or proof stress (for steel bolts, use the minimum value of the yield point or proof stress multiplied by the stress area of the bolt). Repeat this test several times and average the results. The tightening torque may be considered as the sum of the torque on the threads plus the torque on the bolt head- or nut-to-joint bearing surface. The torque coefficient can be found from $K = T_f \div F_f \times d$, where $F_f$ is the measured axial tension, and $T_f$ is the measured tightening torque.

To measure the coefficient of friction between threads or bearing surfaces, obtain the total tightening torque and that portion of the torque due to the thread or bearing surface friction. If only tightening torque and the torque on the bearing surfaces can be measured, then the difference between these two measurements can be taken as the thread-tightening torque. Likewise, if only the tightening torque and threaded-portion torque are known, the torque due to bearing can be taken as the difference between the known torques. The coefficients of friction between threads and bearing surfaces, respectively, can be obtained from the following:

$$\mu_s = \frac{2T_s \cos\alpha'}{d_2 F_f} - \cos\alpha' \tan\beta \qquad (18) \qquad\qquad \mu_w = \frac{2T_w}{D_w F_f} \qquad (19)$$

As before, $T_s$ is the torque attributable to the threaded portion of the screw, $T_w$ is the torque due to bearing, $D_w$ is the equivalent diameter of friction torque on bearing surfaces according to Equation (14), and $F_f$ is the measured axial tension.

**Torque-Tension Relationships.—**Torque is usually applied to develop an axial load in a bolt. To achieve the desired axial load in a bolt, the torque must overcome friction in the threads and friction under the nut or bolt head. In Fig. 5, the axial load $P_B$ is a component of the normal force developed between threads. The normal-force component perpendicular to the thread helix is $P_{N\beta}$ and the other component of this force is the torque load $P_B \tan\beta$ that is applied in tightening the fastener. Assuming the turning force is applied at the pitch diameter of the thread, the torque $T_1$ needed to develop the axial load is $T_1 = P_B \times \tan\beta \times d_2/2$. Substituting $\tan\beta = l \div \pi d_2$ into the previous expression gives $T_1 = P_B \times l \div 2\pi$.

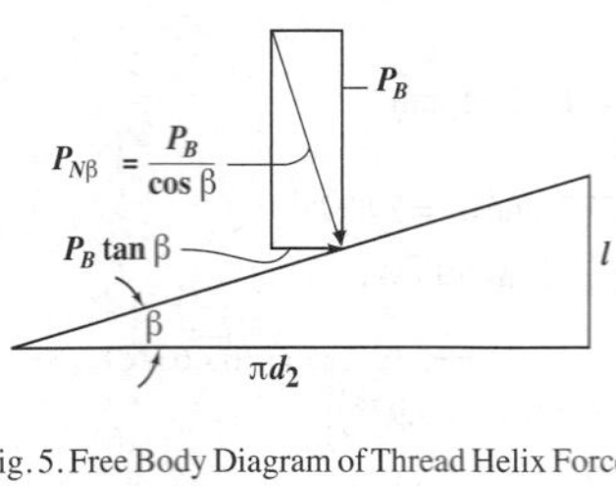


Fig. 5. Free Body Diagram of Thread Helix Forces

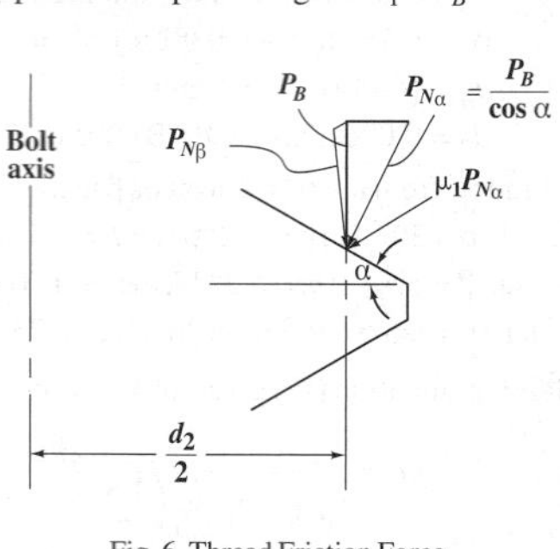


Fig. 6. Thread Friction Force

In Fig. 6, the normal-force component perpendicular to the thread flanks is $P_{N\alpha}$. With a coefficient of friction $\mu_1$ between the threads, the friction load is equal to $\mu_1 P_{N\alpha}$, or $\mu_1 P_B \div \cos\alpha$. Assuming the force is applied at the pitch diameter of the thread, the torque $T_2$ to overcome thread friction is given by:

$$T_2 = \frac{d_2 \mu_1 P_B}{2\cos\alpha} \qquad (20)$$

With the coefficient of friction $\mu_2$ between a nut or bolt-head pressure face and a component face, as in Fig. 7, the friction load is equal to $\mu_2 P_B$. Assuming the force is applied midway between the nominal (bolt) diameter $d$ and the pressure-face diameter $b$, the torque $T_3$ to overcome the nut or bolt underhead friction is:

$$T_3 = \frac{d+b}{4}\mu_2 P_B \tag{21}$$

The total torque, $T$, required to develop axial bolt load, $P_B$, is equal to the sum of the torques $T_1$, $T_2$, and $T_3$ as follows:

$$T = P_B\left(\frac{l}{2\pi} + \frac{d_2\mu_1}{2\cos\alpha} + \frac{(d+b)\mu_2}{4}\right) \tag{22}$$

For a fastener system with 60° threads, $\alpha = 30°$ and $d_2$ is approximately $0.92d$. If no loose washer is used under the rotated nut or bolt head, $b$ is approximately $1.5d$ and Equation (22) reduces to:

$$T = P_B\,[0.159 \times l + d(0.531\mu_1 + 0.625\mu_2)] \tag{23}$$

In addition to the conditions of Equation (23), if the thread and bearing friction coefficients, $\mu_1$ and $\mu_2$, are equal (which is not necessarily so), then $\mu_1 = \mu_2 = \mu$, and the previous equation reduces to:

$$T = P_B\,(0.159l + 1.156\mu d) \tag{24}$$

*Example:* Estimate the torque required to tighten a UNC ½-13 grade 8 steel bolt to a preload equivalent to 55 percent of the minimum tensile bolt strength. Assume that the bolt is unplated and both the thread and bearing friction coefficients equal 0.15.

*Solution:* The minimum tensile strength for SAE grade 8 bolt material is 150,000 psi (from page 1666). To use Equation (24), find the stress area of the bolt using Equation (9) with $P = 1/13$, $d_m = d - 1.2990P$, and $d_p = d - 0.6495P$, and then calculate the necessary preload, $P_B$, and the applied torque, $T$.

$$A_s = \frac{\pi}{4}\left(\frac{0.4500 + 0.4001}{2}\right)^2 = 0.1419\text{ in.}^2$$

$$P_B = \sigma_{allow} \times A_s = 0.55 \times 150000 \times 0.1419 = 11707\text{ lb}_f$$

$$T = 11707\left(\frac{0.159}{13} + 1.156 \times 0.15 \times 0.500\right) = 1158\text{ lb-in.} = 96.5\text{1b-ft}$$

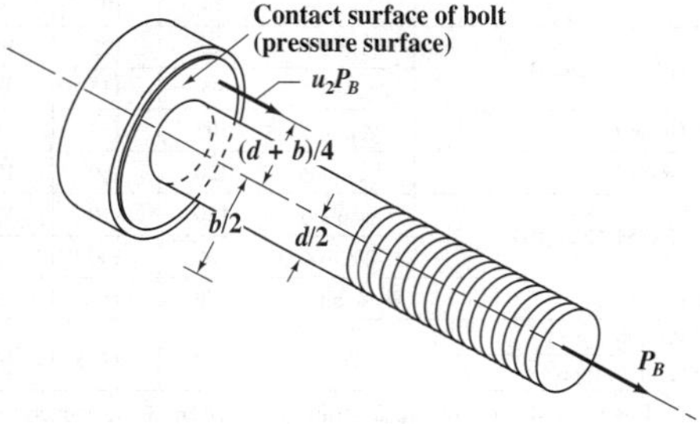


Fig. 7. Nut or Bolt Head Friction Force

**Grade Marks and Material Properties for Bolts and Screws.—**Bolts, screws, and other fasteners are marked on the head with a symbol that identifies the grade of the fastener. The grade specification establishes the minimum mechanical properties that the fastener must meet. Additionally, industrial fasteners must be stamped with a registered head mark that identifies the manufacturer. The grade identification table identifies the grade markings and gives mechanical properties for some commonly used ASTM and SAE steel fasteners. Metric fasteners are identified by property grade marks, which are specified in ISO and SAE standards. These marks are discussed with metric fasteners.

### Grade Identification Marks and Mechanical Properties of Bolts and Screws

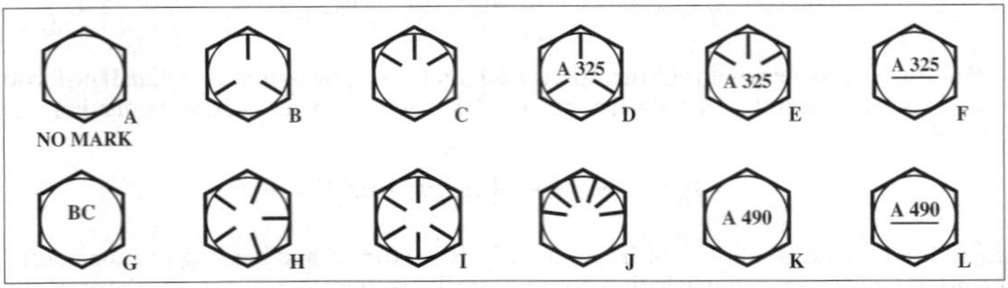


| Identifier | Grade | Size (in.) | Min. Strength ($10^3$ psi) | | | Material & Treatment |
|---|---|---|---|---|---|---|
| | | | Proof | Tensile | Yield | |
| A | SAE Grade 1 | ¼ to 1½ | 33 | 60 | 36 | 1 |
| | ASTM A307 | ¼ to 1½ | 33 | 60 | 36 | 3 |
| | SAE Grade 2 | ¼ to ¾ | 55 | 74 | 57 | 1 |
| | | ⅞ to 1½ | 33 | 60 | 36 | |
| | SAE Grade 4 | ¼ to 1½ | 65 | 115 | 100 | 2, a |
| B | SAE Grade 5, ASTM A449 | ¼ to 1 | 85 | 120 | 92 | 2,b |
| | ASTM A449 | 1⅛ to 1½ | 74 | 105 | 81 | |
| | ASTM A449 | 1¾ to 3 | 55 | 90 | 58 | |
| C | SAE Grade 5.2 | ¼ to 1 | 85 | 120 | 92 | 4, b |
| D | ASTM A325, Type 1 | ½ to 1 | 85 | 120 | 92 | 2,b |
| | | 1⅛ to 1½ | 74 | 105 | 81 | |
| E | ASTM A325, Type 2 | ½ to 1 | 85 | 120 | 92 | 4, b |
| | | 1⅛ to 1½ | 74 | 105 | 81 | |
| F | ASTM A325, Type 3 | ½ to 1 | 85 | 120 | 92 | 5, b |
| | | 1⅛ to 1½ | 74 | 105 | 81 | |
| G | ASTM A354, Grade BC | ¼ to 2½ | 105 | 125 | 109 | 5, b |
| | | 2⅝ to 4 | 95 | 115 | 99 | |
| H | SAE Grade 7 | ¼ to 1½ | 105 | 133 | 115 | 7, b |
| I | SAE Grade 8 | ¼ to 1½ | 120 | 150 | 130 | 7, b |
| | ASTM A354, Grade BD | ¼ to 2½ | 120 | 150 | 130 | 6, b |
| | | 2⅝ to 4 | 105 | 140 | 115 | |
| J | SAE Grade 8.2 | ¼ to 1 | 120 | 150 | 130 | 4, b |
| K | ASTM A490, Type 1 | ½ to 1½ | 120 | 150 | 130 | 6, b |
| L | ASTM A490, Type 3 | | | | | 5, b |

Material Steel: 1—low or medium carbon; 2—medium carbon; 3—low carbon; 4—low-carbon martensite; 5—weathering steel; 6—alloy steel; 7—medium-carbon alloy. Treatment: a—cold drawn; b—quench and temper.

**Detecting Counterfeit Fasteners.**—Fasteners that have markings identifying them as belonging to a specific grade or property class are counterfeit if they do not meet the standards established for that class. Counterfeit fasteners may break unexpectedly at smaller loads than expected. Generally, these fasteners are made from the wrong material or they are not properly strengthened during manufacture. Either way, counterfeit fasteners can lead to dangerous failures in assemblies. The law now requires testing of fasteners used in some critical applications. Detection of counterfeit fasteners is difficult because the counterfeits look genuine. The only sure way to determine if a fastener meets its specification is to test it. However, reputable distributors will assist in verifying the authenticity of the fasteners they sell. For important applications, fasteners can be checked to determine whether they perform according to the standard. Typical laboratory checks used to detect fakes include testing hardness, elongation, and ultimate loading, and a variety of chemical tests.

**Mechanical Properties and Grade Markings of Nuts.**—Three grades of hex and square nuts designated Grades 2, 5, and 8 are specified by the SAE J995 Standard covering nuts in the $\frac{1}{4}$- to $1\frac{1}{2}$-inch diameter range. Grades 2, 5, and 8 nuts roughly correspond to the SAE specified bolts of the same grade. Additional specifications are given for miscellaneous nuts such as hex jam nuts, hex slotted nuts, heavy hex nuts, etc. Generally speaking, use nuts of a grade equal to or greater than the grade of the bolt being used. Grade 2 nuts are not required to be marked, however, all Grades 5 and 8 nuts in the $\frac{1}{4}$- to $1\frac{1}{2}$-inch range must be marked in one of three ways: Grade 5 nuts may be marked with a dot on the face of the nut and a radial or circumferential mark at 120° counterclockwise from the dot; or a dot at one corner of the nut and a radial line at 120° clockwise from the nut, or one notch at each of the six corners of the nut. Grade 8 nuts may be identified by a dot on the face of the nut with a radial or circumferential mark at 60° counterclockwise from the dot; or a dot at one corner of the nut and a radial line at 60° clockwise from the nut, or two notches at each of the six corners of the nut.

**Working Strength of Bolts.**—When the nut on a bolt is tightened, an initial tensile load is placed on the bolt that must be taken into account in determining its safe working strength or external load-carrying capacity. The total load on the bolt theoretically varies from a maximum equal to the sum of the initial and external loads (when the bolt is absolutely rigid and the parts held together are elastic) to a minimum equal to either the initial or external loads, whichever is the greater (where the bolt is elastic and the parts held together are absolutely rigid). No material is absolutely rigid, so in practice the total load values fall somewhere between these maximum and minimum limits, depending upon the relative elasticity of the bolt and joint members.

Some experiments made at Cornell University to determine the initial stress due to tightening nuts on bolts sufficiently to make a packed joint steam-tight showed that experienced mechanics tighten nuts with a pull roughly proportional to the bolt diameter. It was also found that the stress due to nut tightening was often sufficient to break a $\frac{1}{2}$-inch (12.7-mm) bolt, but not larger sizes, assuming that the nut is tightened by an experienced mechanic. It may be concluded, therefore, that bolts smaller than $\frac{5}{8}$ inch (15.9 mm) should not be used for holding cylinder heads or other parts requiring a tight joint.

As a result of these tests, the following empirical formula was established for the working strength of bolts used for packed joints or joints where the elasticity of a gasket is greater than the elasticity of the studs or bolts.

$$W = S_t\,(0.55d^2 - 0.25d)$$

In this formula, $W$ = working strength of bolt or permissible load, in pounds, after allowance is made for initial load due to tightening; $S_t$ = allowable working stress in tension, pounds per square inch; and $d$ = nominal outside diameter of stud or bolt, inches.

A somewhat more convenient formula, and one that gives approximately the same results, is

$$W = S_t(A - 0.25d)$$

In this formula, $W$, $S_t$, and $d$ are as previously given, and $A$ = area at the root of the thread, square inches.

*Example:* What is the working strength of a 1-inch bolt that is screwed tightly in a packed joint when the allowable working stress is 10,000 psi?

$$W = 10{,}000(0.55 \times 1^2 - 0.25 \times 1) = 3000 \text{ pounds approx.}$$

**Formulas for Stress Areas and Lengths of Engagement of Screw Threads.**—The critical areas of stress of mating screw threads are: 1) The effective cross-sectional area, or tensile-stress area, of the external thread; 2) the shear area of the external thread, which depends principally on the minor diameter of the tapped hole; and 3) the shear area of the internal thread, which depends principally on the major diameter of the external thread. The relation of these three stress areas to each other is an important factor in determining how a threaded connection will fail, whether by breakage in the threaded section of the screw (or bolt) or by stripping of either the external or internal thread.

If failure of a threaded assembly should occur, it is preferable for the screw to break rather than have either the external or internal thread strip. In other words, the length of engagement of mating threads should be sufficient to carry the full load necessary to break the screw without the threads stripping.

If mating internal and external threads are manufactured of materials having equal tensile strengths, to prevent stripping of the external thread, the length of engagement should be not less than that given by Formula (1):

$$L_e = \frac{2 \times A_t}{3.1416 \times K_n \max\left[\frac{1}{2} + 0.57735n(E_s \min - K_n \max)\right]} \qquad (1)$$

In this formula, the factor of 2 means that it is assumed the area of the screw in shear must be twice the tensile-stress area to attain the full strength of the screw (this value is slightly larger than required, and thus provides a small factor of safety against stripping); $L_e$ = length of engagement, in inches; $n$ = number of threads per inch; $K_n$ max = maximum minor diameter of internal thread; $E_s$ min = minimum pitch diameter of external thread for the class of thread specified; and $A_t$ = tensile-stress area of screw thread (in pounds-force per square inch) given by Formula (2).

$$A_t = 0.7854\left(D - \frac{0.9743}{n}\right)^2 \qquad (2)$$

In this formula, $D$ = basic major diameter of the thread. The other symbols have the same meanings as before, as referenced in ASTM F606-21. For basic dimensions of Unified screw threads, also see Table 4a through Table 5h, starting on page 1981, which are based on Formula (2).

*Stripping of Internal Thread:* If the internal thread is made of material of lower strength than the external thread, stripping of the internal thread may take place before the screw breaks. To determine whether this condition exists, it is necessary to calculate the factor $J$ for the relative strength of the external and internal threads given by Formula (3):

$$J = \frac{A_s \times \text{tensile strength of external thread material}}{A_n \times \text{tensile strength of internal thread material}} \qquad (3)$$

If $J$ is less than or equal to 1, the length of engagement determined by Formula (1) is adequate to prevent stripping of the internal thread; if $J$ is greater than 1, the required length of engagement $Q$ to prevent stripping of the internal thread is obtained by multiplying the length of engagement $L_e$, Formula (1), by $J$:

$$Q = JL_e \qquad (4)$$

In Formula (3), $A_s$ and $A_n$ are the shear areas of the external and internal threads, respectively, given by Formulas (5) and (6):

$$A_s = 3.1416nL_eK_n\max\left[\frac{1}{2n} + 0.57735(E_s\min - K_n\max)\right] \qquad (5)$$

$$A_n = 3.1416nL_eD_s\min\left[\frac{1}{2n} + 0.57735(D_s\min - E_n\max)\right] \qquad (6)$$

In these formulas, $n$ = threads per inch; $L_e$ = length of engagement from Formula (1); $K_n$ max = maximum minor diameter of internal thread; $E_s$ min = minimum pitch diameter of the external thread for the class of thread specified; $D_s$ min = minimum major diameter of the external thread; and $E_n$ max = maximum pitch diameter of internal thread.

**Load to Break Threaded Portion of Screws and Bolts.**—The direct tensile load $P$ to break the threaded portion of a screw or bolt (assuming that no shearing or torsional stresses are acting) can be determined from the following formula:

$$P = SA_t$$

where $P$ = load in pounds to break screw; $S$ = ultimate tensile strength of material of screw or bolt in pounds per square inch; and $A_t$ = tensile-stress area in square inches from Formula (2) or from the screw thread tables.

**Lock Wire Procedure Detail.**—Wire ties are frequently used as a locking device for bolted connections to prevent loosening due to vibration and loading conditions, or tampering. The use of safety wire ties is illustrated in Fig. 1 and Fig. 2 below. The illustrations assume the use of right-hand threaded fasteners and the following additional rules apply:

1) No more that three bolts may be tied together; 2) bolt heads may be tied as shown only when the female thread receiver is captive; 3) pre-drilled nuts may be tied in a fashion similar to that illustrated with the following conditions: a) Nuts must be heat treated; and b) nuts are factory drilled for use with lock wire; 4) lock wire must fill a minimum of 75 percent of the drilled hole provided for the use of lock wire; and 5) lock wire must be aircraft quality stainless steel of 0.508 mm (0.020 inch) diameter, 0.8128 mm (0.032 inch) diameter, or 1.067 mm (0.042 inch) diameter. Diameter of lock wire is determined by the thread size of the fastener to be safe-tied: a) Thread sizes of 6 mm (0.25 inch) and smaller use 0.508 mm (0.020 inch) wire; b) thread sizes of 6 mm (0.25 inch) to 12 mm (0.5 inch) use 0.8128 mm (0.032 inch) wire; c) thread sizes > 12 mm (0.5 inch) use 1.067 mm (0.042 inch) wire; and d) the larger wire may be used in smaller bolts in cases of convenience, but smaller wire must not be used in larger fasteners sizes.

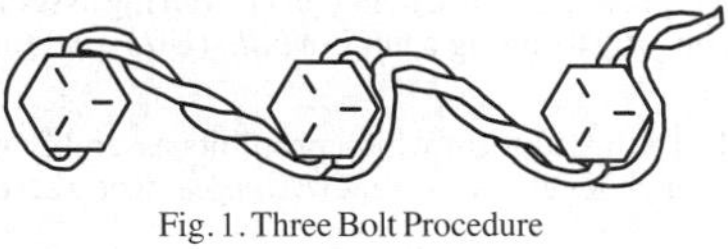

Fig. 1. Three Bolt Procedure

Fig. 2. Two Bolt Procedure

## INCH THREADED FASTENERS

Dimensions of bolts, screws, nuts, and washers used in machine construction are given here. For data on thread forms, see the section *SCREW THREAD SYSTEMS* starting on page 1943.

**American Square and Hexagon Bolts, Screws, and Nuts.—**The 1941 American Standard ASA B18.2 covered head dimensions only. In 1952 and 1955, the Standard was revised to cover the entire product. Some bolt and nut classifications were simplified by elimination or consolidation in agreements reached with the British and Canadians. In 1965, ASA B18.2 was redesignated into two standards: B18.2.1 covering square and hexagon bolts and screws including hexagon cap screws and lag screws and B18.2.2 covering square and hexagon nuts. In B18.2.1-1965, hexagon head cap screws and finished hexagon bolts were consolidated into a single product; heavy semifinished hexagon bolts and heavy finished hexagon bolts were consolidated into a single product; regular semifinished hexagon bolts were eliminated; a new tolerance pattern for all bolts and screws and a positive identification procedure for determining whether an externally threaded product should be designated as a bolt or screw were established. Also included in this Standard are heavy hexagon bolts and heavy hexagon structural bolts. In B18.2.2-1965, regular semifinished nuts were discontinued; regular hexagon and heavy hexagon nuts in sizes $\frac{1}{4}$ through 1 inch, finished hexagon nuts in sizes larger than $1\frac{1}{2}$ inches, washer-faced semifinished style of finished nuts in sizes $\frac{5}{8}$-inch and smaller and heavy series nuts in sizes $\frac{7}{16}$-inch and smaller were eliminated.

Further revisions and refinements include the addition of a skew head bolts and hex head lag screws and the specifying of countersunk diameters for the various hex nuts. Heavy hex structural bolts and heavy hex nuts were moved to a new structural applications Standard, ANSI/ASME B18.2.6-2019, "Fasteners for Use in Structural Applications." Additionally, B18.2.1 has been revised to allow easier conformance to Public Law 101-592. All these changes are reflected in ANSI/ASME B18.2.1-2012 (R2021), and ANSI/ASME B18.2.2-2022.

**Unified Square and Hexagon Bolts, Screws, and Nuts.—**Items that are recognized in the Standard as "unified" dimensionally with British and Canadian Standards are shown in bold-face in certain tables.

The other items in the same tables are based on formulas accepted and published by the British for sizes outside the ranges listed in their Standards which, as a matter of information, are BS 1768:1963 for Precision (Normal Series) Unified Hexagon Bolts, Screws, Nuts (UNC and UNF Threads) and BS 1769 and amendments for Black (Heavy Series) Unified Hexagon Bolts, etc. Tolerances applied to comparable dimensions of American and British Unified bolts and nuts may differ because of rounding off practices and other factors.

**Differentiation between Bolt and Screw.—**A bolt is an externally threaded fastener designed for insertion through holes in assembled parts, and is normally intended to be tightened or released by torquing a nut.

A screw is an externally threaded fastener capable of being inserted into holes in assembled parts, of mating with a preformed internal thread or forming its own thread and of being tightened or released by torquing the head.

An externally threaded fastener which is prevented from being turned during assembly, and which can be tightened or released only by torquing a nut is a *bolt*. (*Example:* round head bolts, track bolts, plow bolts.)

An externally threaded fastener that has a thread form which prohibits assembly with a nut having a straight thread of multiple pitch length is a *screw*. (*Example:* wood screws, tapping screws.)

An externally threaded fastener that must be assembled with a nut to perform its intended service is a *bolt*. (*Example:* heavy hex structural bolt.)

An externally threaded fastener that must be torqued by its head into a tapped or other preformed hole to perform its intended service is a *screw*. (*Example:* square head set screw.)

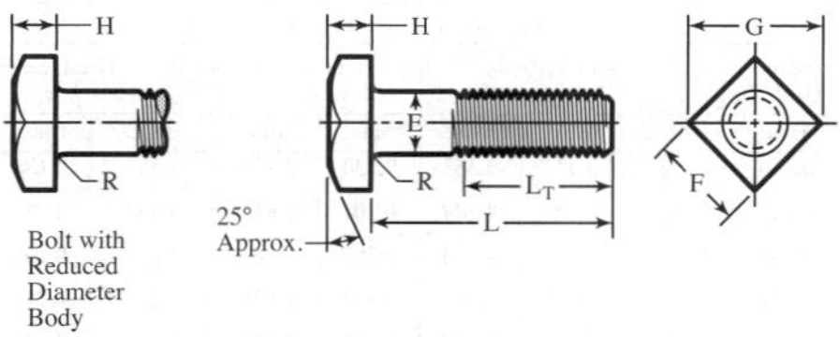


Fig. 1. Square Bolts ( Table 1 )

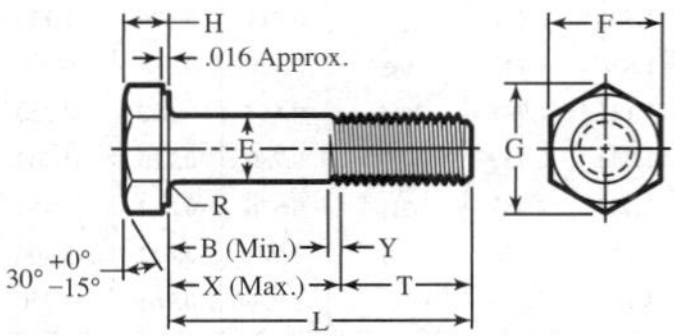


Fig. 2. Heavy Hex Structural Bolts ( Table 2 )

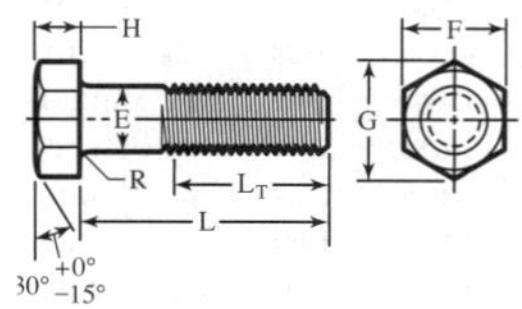


Fig. 3. Hex Bolts, Heavy Hex Bolts (Table 3 )

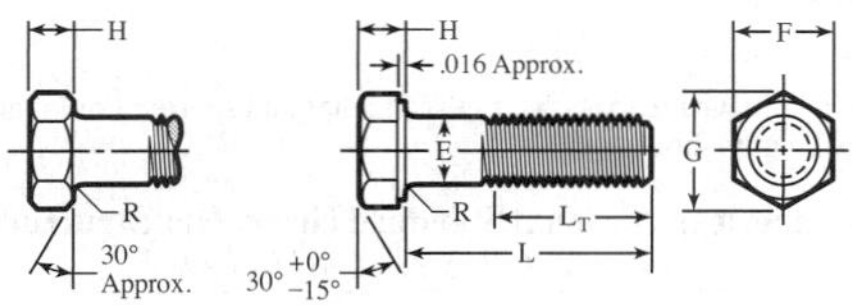


Fig. 4. Hex Cap Screws, Heavy Hex Screws (Table 4)

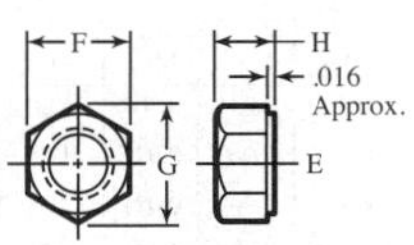


Fig. 5. Hex Nuts, Heavy Hex Nuts (Table 7)

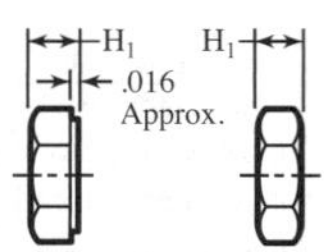


Fig. 6. Hex Jam Nuts, Heavy Hex Jam Nuts (Table 7)

**Square and Hex Bolts, Screws, and Nuts.—**The dimensions for square and hex bolts and screws given in the following tables have been taken from American National Standard ANSI/ASME B18.2.1-2012 (R2021) and for nuts from American National Standard ANSI/ASME B18.2.2-2022. Reference should be made to these Standards for information or data not found in the following text and tables:

*Designation:* Bolts and screws should be designated by the following data in the sequence shown: nominal size (fractional and decimal equivalent); threads per inch (omit for lag screws); product length for bolts and screws (fractional or two-place decimal equivalent); product name; material, including specification, where necessary; and protective finish, if required. Examples: (1) $^3/_8$-16 × 1$^1/_2$ Square Bolt, Steel, Zinc Plated; (2) $^1/_2$-13 × 3 Hex Cap

Screw, SAE Grade 8 Steel; and (3) .75 × 5.00 Hex Lag Screw, Steel. (4) 1/2-13 Square Nut, Steel, Zinc Plated; (5) 3/4-16 Heavy Hex Nut, SAE J995 Grade 5 Steel; and (6) 1000-8 Hex Thick Slotted Nut, ASTM F594 (Alloy Group 1) Corrosion-Resistant Steel.

### Table 1. American National Standard and Unified Standard Square Bolts *ANSI/ASME B18.2.1-2012 (R2021)*

| SQUARE BOLTS (Fig. 1) | | | | | | | | | | | |
|---|---|---|---|---|---|---|---|---|---|---|---|
| Nominal Size[a] or Basic Product Dia. | | Body Dia.[b] *E* | Width Across Flats *F* | | | Width Across Corners *G* | | Head Height *H* | | | Thread Length[c] $L_T$ |
| | | Max. | Basic | Max. | Min. | Max. | Min. | Basic | Max. | Min. | Nom. |
| 1/4 | **0.2500** | **0.260** | 3/8 | **0.375** | **0.362** | **0.530** | **0.498** | 11/64 | **0.188** | **0.156** | 0.750 |
| 5/16 | **0.3125** | **0.324** | 1/2 | **0.500** | **0.484** | **0.707** | **0.665** | 13/64 | **0.220** | **0.186** | 0.875 |
| 3/8 | **0.3750** | **0.388** | 9/16 | **0.562** | **0.544** | **0.795** | **0.747** | 1/4 | **0.268** | **0.232** | 1.000 |
| 7/16 | **0.4375** | **0.452** | 5/8 | **0.625** | **0.603** | **0.884** | **0.828** | 19/64 | **0.316** | **0.278** | 1.125 |
| 1/2 | **0.5000** | **0.515** | 3/4 | **0.750** | **0.725** | **1.061** | **0.995** | 21/64 | **0.348** | **0.308** | 1.250 |
| 5/8 | **0.6250** | **0.642** | 15/16 | **0.938** | **0.906** | **1.326** | **1.244** | 27/64 | **0.444** | **0.400** | 1.500 |
| 3/4 | **0.7500** | **0.768** | 1 1/8 | **1.125** | **1.088** | **1.591** | **1.494** | 1/2 | **0.524** | **0.476** | 1.750 |
| 7/8 | **0.8750** | **0.895** | 1 5/16 | **1.312** | **1.269** | **1.856** | **1.742** | 19/32 | **0.620** | **0.568** | 2.000 |
| 1 | **1.0000** | **1.022** | 1 1/2 | **1.500** | **1.450** | **2.121** | **1.991** | 21/32 | **0.684** | **0.628** | 2.250 |
| 1 1/8 | **1.1250** | **1.149** | 1 11/16 | **1.688** | **1.631** | **2.386** | **2.239** | 3/4 | **0.780** | **0.720** | 2.500 |
| 1 1/4 | **1.2500** | **1.277** | 1 7/8 | **1.875** | **1.812** | **2.652** | **2.489** | 27/32 | **0.876** | **0.812** | 2.750 |
| 1 3/8 | **1.3750** | **1.404** | 2 1/16 | **2.062** | **1.994** | **2.917** | **2.738** | 29/32 | **0.940** | **0.872** | 3.000 |
| 1 1/2 | **1.5000** | **1.531** | 2 1/4 | **2.250** | **2.175** | **3.182** | **2.986** | 1 | **1.036** | **0.964** | 3.250 |

[a] Where specifying nominal size in decimals, zeros before the decimal point and in the fourth decimal place are omitted.

[b] See *Body Diameter* footnote in Table 3.

[c] Thread lengths, $L_T$, shown are for bolt lengths 6 inches and shorter. For longer bolt lengths add 0.250 inch to thread lengths shown.

### Table 2. American National Standard Heavy Hex Structural Bolts *ANSI/ASME B18.2.1-1981 (R1992)*[a]

| HEAVY HEX STRUCTURAL BOLTS (Fig. 2) | | | | | | | | | | | | | |
|---|---|---|---|---|---|---|---|---|---|---|---|---|---|
| Nominal Size[a] or Basic Product Dia. | | Body Dia. *E* | | Width Across Flats *F* | | Width Across Corners *G* | | Height *H* | | Radius of Fillet *R* | | Thread Length $L_T$ | Trans. Thread *Y* |
| | | Max. | Min. | Max. | Min. | Max. | Min. | Max. | Min. | Max. | Min. | Basic | Max. |
| 1/2 | 0.5000 | 0.515 | 0.482 | 0.875 | 0.850 | 1.010 | 0.969 | 0.323 | 0.302 | 0.031 | 0.009 | 1.00 | 0.19 |
| 5/8 | 0.6250 | 0.642 | 0.605 | 1.062 | 1.031 | 1.227 | 1.175 | 0.403 | 0.378 | 0.062 | 0.021 | 1.25 | 0.22 |
| 3/4 | 0.7500 | 0.768 | 0.729 | 1.250 | 1.212 | 1.443 | 1.383 | 0.483 | 0.455 | 0.062 | 0.021 | 1.38 | 0.25 |
| 7/8 | 0.8750 | 0.895 | 0.852 | 1.438 | 1.394 | 1.660 | 1.589 | 0.563 | 0.531 | 0.062 | 0.031 | 1.50 | 0.28 |
| 1 | 1.0000 | 1.022 | 0.976 | 1.625 | 1.575 | 1.876 | 1.796 | 0.627 | 0.591 | 0.093 | 0.062 | 1.75 | 0.31 |
| 1 1/8 | 1.1250 | 1.149 | 1.098 | 1.812 | 1.756 | 2.093 | 2.002 | 0.718 | 0.658 | 0.093 | 0.062 | 2.00 | 0.34 |
| 1 1/4 | 1.2500 | 1.277 | 1.223 | 2.000 | 1.938 | 2.309 | 2.209 | 0.813 | 0.749 | 0.093 | 0.062 | 2.00 | 0.38 |
| 1 3/8 | 1.3750 | 1.404 | 1.345 | 2.188 | 2.119 | 2.526 | 2.416 | 0.878 | 0.810 | 0.093 | 0.062 | 2.25 | 0.44 |
| 1 1/2 | 1.5000 | 1.531 | 1.470 | 2.375 | 2.300 | 2.742 | 2.622 | 0.974 | 0.902 | 0.093 | 0.062 | 2.25 | 0.44 |

[a] The table has been included for reference only. Heavy hex structural bolts have been removed from ANSI/ASME B18.2.1 and are now included in ANSI/ASME B18.2.6.

All dimensions are in inches. **Bold type shows bolts unified dimensionally with British and Canadian Standards.** Threads, when rolled, shall be Unified Coarse, Fine, or 8-thread series (UNRC, UNRF, or 8 UNR Series), Class 2A. Threads produced by other methods may be Unified Coarse, Fine, or 8-thread series (UNC, UNF, or 8 UN Series), Class 2A.

## Table 3. American National Standard and Unified Standard Hex and Heavy Hex Bolts *ANSI/ASME B18.2.1-2012 (R2021)*

| Nominal Size[a] or Basic Dia. | | Full Size Body Dia. *E* | Width Across Flats *F* | | | Width Across Corners *G* | | Head Height *H* | | | Thread Length[b] $L_T$ |
|---|---|---|---|---|---|---|---|---|---|---|---|
| | | Max. | Basic | Max. | Min. | Max. | Min. | Basic | Max. | Min. | Nom. |
| HEX BOLTS (Fig. 3) | | | | | | | | | | | |
| **1/4** | **0.2500** | **0.260** | **7/16** | **0.438** | **0.425** | **0.505** | **0.484** | **11/64** | **0.188** | **0.150** | 0.750 |
| **5/16** | **0.3125** | **0.324** | **1/2** | **0.500** | **0.484** | **0.577** | **0.552** | **7/32** | **0.235** | **0.195** | 0.875 |
| **3/8** | **0.3750** | **0.388** | **9/16** | **0.562** | **0.544** | **0.650** | **0.620** | **1/4** | **0.268** | **0.226** | 1.000 |
| **7/16** | **0.4375** | **0.452** | **5/8** | **0.625** | **0.603** | **0.722** | **0.687** | **19/64** | **0.316** | **0.272** | 1.125 |
| **1/2** | **0.5000** | **0.515** | **3/4** | **0.750** | **0.725** | **0.866** | **0.826** | **11/32** | **0.364** | **0.302** | 1.250 |
| **5/8** | **0.6250** | **0.642** | **15/16** | **0.938** | **0.906** | **1.083** | **1.033** | **27/64** | **0.444** | **0.378** | 1.500 |
| **3/4** | **0.7500** | **0.768** | **1 1/8** | **1.125** | **1.088** | **1.299** | **1.240** | **1/2** | **0.524** | **0.455** | 1.750 |
| **7/8** | **0.8750** | **0.895** | **1 5/16** | **1.312** | **1.269** | **1.516** | **1.447** | **37/64** | **0.604** | **0.531** | 2.000 |
| **1** | **1.0000** | **1.022** | **1 1/2** | **1.500** | **1.450** | **1.732** | **1.653** | **43/64** | **0.700** | **0.591** | 2.250 |
| **1 1/8** | **1.1250** | **1.149** | **1 11/16** | **1.688** | **1.631** | **1.949** | **1.859** | **3/4** | **0.780** | **0.658** | 2.500 |
| **1 1/4** | **1.2500** | **1.277** | **1 7/8** | **1.875** | **1.812** | **2.165** | **2.066** | **27/32** | **0.876** | **0.749** | 2.750 |
| **1 3/8** | **1.3750** | **1.404** | **2 1/16** | **2.062** | **1.994** | **2.382** | **2.273** | **29/32** | **0.940** | **0.810** | 3.000 |
| **1 1/2** | **1.5000** | **1.531** | **2 1/4** | **2.250** | **2.175** | **2.598** | **2.480** | **1** | **1.036** | **0.902** | 3.250 |
| **1 5/8** | **1.6250** | **1.658** | **2 7/16** | **2.438** | **2.356** | **2.815** | **2.616** | **1 3/32** | **1.116** | **0.978** | 3.500 |
| **1 3/4** | **1.7500** | **1.785** | **2 5/8** | **2.625** | **2.538** | **3.031** | **2.893** | **1 5/32** | **1.196** | **1.054** | 3.750 |
| **1 3/4** | **1.7500** | **1.912** | **2 13/16** | **2.812** | **2.719** | **3.248** | **3.099** | **1 1/4** | **1.276** | **1.130** | 4.000 |
| **2** | **2.0000** | **2.039** | **3** | **3.000** | **2.900** | **3.464** | **3.306** | **1 11/32** | **1.388** | **1.175** | 4.250 |
| 2 1/4 | 2.2500 | 2.305 | 3 3/8 | 3.375 | 3.262 | 3.897 | 3.719 | 1 1/2 | 1.548 | 1.327 | 4.750 |
| 2 1/2 | 2.5000 | 2.559 | 3 3/4 | 3.750 | 3.625 | 4.330 | 4.133 | 1 21/32 | 1.708 | 1.479 | 5.250 |
| 2 3/4 | 2.7500 | 2.827 | 4 1/8 | 4.125 | 3.988 | 4.763 | 4.546 | 1 13/16 | 1.869 | 1.632 | 5.750 |
| 3 | 3.0000 | 3.081 | 4 1/2 | 4.500 | 4.350 | 5.196 | 4.959 | 2 | 2.060 | 1.815 | 6.250 |
| 3 1/4 | 3.2500 | 3.335 | 4 7/8 | 4.875 | 4.712 | 5.629 | 5.372 | 2 3/16 | 2.251 | 1.936 | 6.750 |
| 3 1/2 | 3.5000 | 3.589 | 5 1/4 | 5.250 | 5.075 | 6.062 | 5.786 | 2 5/16 | 2.380 | 2.057 | 7.250 |
| 3 3/4 | 3.7500 | 3.858 | 5 5/8 | 5.625 | 5.437 | 6.495 | 6.198 | 2 1/2 | 2.572 | 2.241 | 7.750 |
| 4 | 4.0000 | 4.111 | 6 | 6.000 | 5.800 | 6.928 | 6.612 | 2 11/16 | 2.764 | 2.424 | 8.250 |
| HEAVY HEX BOLTS (Fig. 3) | | | | | | | | | | | |
| **3/8** | **0.3750** | **0.388** | **11/16** | **0.688** | **0.669** | **0.794** | **0.763** | 1/4 | 0.268 | 0.226 | 1.000 |
| **1/2** | **0.5000** | **0.515** | **7/8** | **0.875** | **0.850** | **1.010** | **0.969** | 11/32 | 0.364 | 0.302 | 1.250 |
| **5/8** | **0.6250** | **0.642** | **1 1/16** | **1.062** | **1.031** | **1.227** | **1.175** | 27/64 | 0.444 | 0.378 | 1.500 |
| **3/4** | **0.7500** | **0.768** | **1 1/4** | **1.250** | **1.212** | **1.443** | **1.383** | 1/2 | 0.524 | 0.455 | 1.750 |
| **7/8** | **0.8750** | **0.895** | **1 7/16** | **1.438** | **1.394** | **1.660** | **1.589** | 37/64 | 0.604 | 0.531 | 2.000 |
| **1** | **1.0000** | **1.022** | **1 5/8** | **1.625** | **1.575** | **1.876** | **1.796** | 43/64 | 0.700 | 0.591 | 2.250 |
| **1 1/8** | **1.1250** | **1.149** | **1 13/16** | **1.812** | **1.756** | **2.093** | **2.002** | 3/4 | 0.780 | 0.658 | 2.500 |
| **1 1/4** | **1.2500** | **1.277** | **2** | **2.000** | **1.938** | **2.309** | **2.209** | 27/32 | 0.876 | 0.749 | 2.750 |
| **1 3/8** | **1.3750** | **1.404** | **2 3/16** | **2.188** | **2.119** | **2.526** | **2.416** | 29/32 | 0.940 | 0.810 | 3.000 |
| **1 1/2** | **1.5000** | **1.531** | **2 3/8** | **2.375** | **2.300** | **2.742** | **2.622** | 1 | 1.036 | 0.902 | 3.250 |
| **1 5/8** | **1.6250** | **1.658** | **2 9/16** | **2.562** | **2.481** | **3.959** | **2.829** | 1 3/32 | 1.116 | 0.978 | 3.500 |
| **1 3/4** | **1.7500** | **1.785** | **2 3/4** | **2.750** | **2.662** | **3.175** | **3.035** | 1 5/32 | 1.196 | 1.054 | 3.750 |
| **1 7/8** | **1.8750** | **1.912** | **2 15/16** | **2.938** | **2.844** | **3.392** | **3.242** | 1 1/4 | 1.276 | 1.130 | 4.000 |
| **2** | **2.0000** | **2.039** | **3 1/8** | **3.125** | **3.025** | **3.608** | **3.449** | 1 11/32 | 1.388 | 1.175 | 4.250 |
| 2 1/4 | 2.2500 | 2.305 | 3 1/2 | 3.500 | 3.388 | 4.041 | 3.862 | 1 1/2 | 1.548 | 1.327 | 4.750 |
| 2 1/2 | 2.5000 | 2.559 | 3 7/8 | 3.875 | 3.750 | 4.474 | 4.275 | 1 21/32 | 1.708 | 1.479 | 5.250 |
| 2 3/4 | 2.7500 | 2.827 | 4 1/4 | 4.250 | 4.112 | 4.907 | 4.688 | 1 13/16 | 1.869 | 1.632 | 5.750 |
| 3 | 3.0000 | 3.081 | 4 5/8 | 4.625 | 4.475 | 5.340 | 5.102 | 2 | 2.060 | 1.815 | 6.250 |

[a] *Nominal Size:* Where specifying nominal size in decimals, zeros preceding the decimal point and in the fourth decimal place are omitted.

[b] Thread lengths, $L_T$, shown are for bolt lengths 6 inches and shorter. For longer bolt lengths add 0.250 inch to thread lengths shown.

All dimensions are in inches.

**Bold type shows bolts unified dimensionally with British and Canadian Standards.**

*Threads:* Threads, when rolled, are Unified Coarse, Fine, or 8-thread series (UNRC, UNRF, or 8 UNR Series), Class 2A. Threads produced by other methods may be Unified Coarse, Fine or 8-thread series (UNC, UNF, or 8 UN Series), Class 2A.

*Body Diameter:* Bolts may be obtained in "reduced diameter body." Where "reduced diameter body" is specified, the body diameter may be reduced to approximately the pitch diameter of the thread. A shoulder of full body diameter under the head may be supplied at the option of the manufacturer.

*Material:* Unless otherwise specified, chemical and mechanical properties of steel bolts conform to ASTM A307, Grade A. Other materials are as agreed upon by manufacturer and purchaser.

## Table 4. American National Standard and Unified Standard Heavy Hex Screws and Hex Cap Screws *ANSI/ASME B18.2.1-2012 (R2021)*

| Nominal Size[a] or Basic Product Dia. | | Body Dia. *E* | | Width Across Flats *F* | | | Width Across Corners *G* | | Height *H* | | | Thread Length[b] $L_T$ |
|---|---|---|---|---|---|---|---|---|---|---|---|---|
| | | Max. | Min. | Basic | Max. | Min. | Max. | Min. | Basic | Max. | Min. | Basic |
| HEAVY HEX SCREWS (Fig. 4) | | | | | | | | | | | | |
| **3/8** | **0.3750** | **0.3750** | **0.360** | **11/16** | **0.688** | **0.669** | **0.794** | **0.763** | 15/64 | 0.243 | 0.226 | 1.000 |
| **1/2** | **0.5000** | **0.5000** | **0.482** | **7/8** | **0.875** | **0.850** | **1.010** | **0.969** | 5/16 | 0.323 | 0.302 | 1.250 |
| **5/8** | **0.6250** | **0.6250** | **0.605** | **1 1/16** | **1.062** | **1.031** | **1.227** | **1.175** | 25/64 | 0.403 | 0.378 | 1.500 |
| **3/4** | **0.7500** | **0.7500** | **0.729** | **1 1/4** | **1.250** | **1.212** | **1.443** | **1.383** | 15/32 | 0.483 | 0.455 | 1.750 |
| **7/8** | **0.8750** | **0.8750** | **0.852** | **1 7/16** | **1.438** | **1.394** | **1.660** | **1.589** | 35/64 | 0.563 | 0.531 | 2.000 |
| **1** | **1.0000** | **1.0000** | **0.976** | **1 5/8** | **1.625** | **1.575** | **1.876** | **1.796** | 39/64 | 0.627 | 0.591 | 2.250 |
| **1 1/8** | **1.1250** | **1.1250** | **1.098** | **1 13/16** | **1.812** | **1.756** | **2.093** | **2.002** | 11/16 | 0.718 | 0.658 | 2.500 |
| **1 1/4** | **1.2500** | **1.2500** | **1.223** | **2** | **2.000** | **1.938** | **2.309** | **2.209** | 25/32 | 0.813 | 0.749 | 2.750 |
| **1 3/8** | **1.3750** | **1.3750** | **1.345** | **2 3/16** | **2.188** | **2.119** | **2.526** | **2.416** | 27/32 | 0.878 | 0.810 | 3.000 |
| **1 1/2** | **1.5000** | **1.5000** | **1.470** | **2 5/8** | **2.375** | **2.300** | **2.742** | **2.622** | 15/16 | 0.974 | 0.902 | 3.250 |
| **1 5/8** | **1.6250** | **1.6250** | **1.591** | **2 9/16** | **2.562** | **2.481** | **2.959** | **2.829** | 1 | 1.038 | 0.962 | 3.500 |
| **1 3/4** | **1.7500** | **1.7500** | **1.716** | **2 3/4** | **2.750** | **2.662** | **3.175** | **3.035** | 1 3/32 | 1.134 | 1.054 | 3.750 |
| **1 7/8** | **1.8750** | **1.8750** | **1.839** | **2 15/16** | **2.938** | **2.844** | **3.392** | **3.242** | 1 5/32 | 1.198 | 1.114 | 4.000 |
| **2** | **2.0000** | **2.0000** | **1.964** | **3 1/8** | **3.125** | **3.025** | **3.608** | **3.449** | 1 7/32 | 1.263 | 1.175 | 4.250 |
| 2 1/4 | 2.2500 | 2.2500 | 2.214 | 3 1/2 | 3.500 | 3.388 | 4.041 | 3.862 | 1 3/8 | 1.423 | 1.327 | 5.000[c] |
| 2 1/2 | 2.5000 | 2.5000 | 2.461 | 3 7/8 | 3.875 | 3.750 | 4.474 | 4.275 | 1 17/32 | 1.583 | 1.479 | 5.500[c] |
| 2 3/4 | 2.7500 | 2.7500 | 2.711 | 4 1/4 | 4.250 | 41.112 | 4.907 | 4.688 | 1 11/16 | 1.744 | 1.632 | 6.000[c] |
| 3 | 3.0000 | 3.0000 | 2.961 | 4 5/8 | 4.625 | 4.475 | 5.340 | 5.102 | 1 7/8 | 1.935 | 1.815 | 6.500[c] |
| 3 1/4 | 3.2500 | 3.2500 | 3.210 | 5 | 5.000 | 4.838 | 5.774 | 5.515 | 2 | 2.126 | 1.998 | 7.000[c] |
| 3 1/2 | 3.5000 | 3.5000 | 3.461 | 5 3/8 | 5.375 | 5.200 | 6.207 | 5.928 | 2 1/4 | 2.256 | 2.120 | 7.500[c] |
| 3 3/4 | 3.7500 | 3.7500 | 3.711 | 5 3/4 | 5.750 | 5.562 | 6.640 | 6.341 | 2 3/8 | 2.447 | 2.303 | 8.000[c] |
| 4 | 4.0000 | 4.0000 | 3.961 | 6 1/8 | 6.125 | 5.925 | 7.073 | 6.755 | 2 1/2 | 2.576 | 2.424 | 8.500[c] |
| 4 1/4 | 4.2500 | 4.2500 | 4.223 | 6 1/2 | 6.500 | 6.288 | 7.506 | 7.168 | 2 3/4 | 2.768 | 2.608 | 9.000[c] |
| 4 1/2 | 4.5000 | 4.5000 | 4.473 | 6 7/8 | 6.875 | 6.650 | 7.939 | 7.581 | 2 7/8 | 2.896 | 2.728 | 9.500[c] |
| 4 3/4 | 4.7500 | 4.7500 | 4.723 | 7 1/4 | 7.250 | 7.012 | 8.372 | 7.994 | 3 | 3.088 | 2.912 | 10.000[c] |
| 5 | 5.0000 | 5.0000 | 4.973 | 7 5/8 | 7.625 | 7.375 | 8.805 | 8.408 | 3 1/8 | 3.217 | 3.033 | 10.500[c] |
| 5 1/4 | 5.2500 | 5.2500 | 5.223 | 8 | 8.000 | 7.738 | 9.238 | 8.821 | 3 3/8 | 3.408 | 3.216 | 11.000[c] |
| 5 1/2 | 5.5000 | 5.5000 | 5.473 | 8 3/8 | 8.375 | 8.100 | 9.671 | 9.234 | 3 1/2 | 3.538 | 3.338 | 11.500[c] |
| 5 3/4 | 5.7500 | 5.7500 | 5.723 | 8 3/4 | 8.750 | 8.462 | 10.104 | 9.647 | 3 5/8 | 3.729 | 3.521 | 12.000[c] |
| 6 | 6.0000 | 6.0000 | 5.973 | 9 1/8 | 9.125 | 8.825 | 10.537 | 10.060 | 3 3/4 | 3.858 | 3.642 | 12.500[c] |
| HEX CAP SCREWS (Finished Hex Bolts) (Fig. 4) | | | | | | | | | | | | |
| **1/4** | **0.2500** | **0.2500** | **0.2450** | **7/16** | **0.438** | **0.428** | **0.505** | **0.488** | **5/32** | **0.163** | **0.150** | 0.750 |
| **5/16** | **0.3125** | **0.3125** | **0.3065** | **1/2** | **0.500** | **0.489** | **0.577** | **0.557** | **13/64** | **0.211** | **0.195** | 0.875 |
| **3/8** | **0.3750** | **0.3750** | **0.3690** | **9/16** | **0.562** | **0.551** | **0.650** | **0.628** | **15/64** | **0.243** | **0.226** | 1.000 |
| **7/16** | **0.4375** | **0.4375** | **0.4305** | **5/8** | **0.625** | **0.612** | **0.722** | **0.698** | **9/32** | **0.291** | **0.272** | 1.125 |
| **1/2** | **0.5000** | **0.5000** | **0.4930** | **3/4** | **0.750** | **0.736** | **0.866** | **0.840** | **5/16** | **0.323** | **0.302** | 1.250 |
| **9/16** | **0.5625** | **0.5625** | **0.5545** | **1 3/16** | **0.812** | **0.798** | **0.938** | **0.910** | **23/64** | **0.371** | **0.348** | 1.375 |
| **5/8** | **0.6250** | **0.6250** | **0.6170** | **1 5/16** | **0.938** | **0.922** | **1.083** | **1.051** | **25/64** | **0.403** | **0.378** | 1.500 |
| **3/4** | **0.7500** | **0.7500** | **0.7410** | **1 1/8** | **1.125** | **1.100** | **1.299** | **1.254** | **15/32** | **0.483** | **0.455** | 1.750 |
| **7/8** | **0.8750** | **0.8750** | **0.8660** | **1 5/16** | **1.312** | **1.285** | **1.516** | **1.465** | **35/64** | **0.563** | **0.531** | 2.000 |
| **1** | **1.0000** | **1.0000** | **0.9900** | **1 1/2** | **1.500** | **1.469** | **1.732** | **1.675** | **39/64** | **0.627** | **0.591** | 2.250 |
| **1 1/8** | **1.1250** | **1.1250** | **1.1140** | **1 11/16** | **1.688** | **1.631** | **1.949** | **1.859** | **11/16** | **0.718** | **0.658** | 2.500 |
| **1 1/4** | **1.2500** | **1.2500** | **1.2390** | **1 7/8** | **1.875** | **1.812** | **2.165** | **2.066** | **25/32** | **0.813** | **0.749** | 2.750 |
| **1 3/8** | **1.3750** | **1.3750** | **1.3630** | **2 1/16** | **2.062** | **1.994** | **2.382** | **2.273** | **27/32** | **0.878** | **0.810** | 3.000 |
| **1 1/2** | **1.5000** | **1.5000** | **1.4880** | **2 1/4** | **2.250** | **2.175** | **2.598** | **2.480** | **15/16** | **0.974** | **0.902** | 3.250 |
| **1 5/8** | **1.6250** | **1.6250** | **1.6130** | **2 7/16** | **2.438** | **2.356** | **2.815** | **2.686** | **1** | **1.038** | **0.962** | 3.500 |
| **1 3/4** | **1.7500** | **1.7500** | **1.7380** | **2 5/8** | **2.625** | **2.538** | **3.031** | **2.893** | **1 3/32** | **1.134** | **1.054** | 3.750 |
| **1 7/8** | **1.8750** | **1.8750** | **1.8630** | **2 13/16** | **2.812** | **2.719** | **3.248** | **3.099** | **1 5/32** | **1.198** | **1.114** | 4.000 |
| **2** | **2.0000** | **2.0000** | **1.9880** | **3** | **3.000** | **2.900** | **3.464** | **3.306** | **1 7/32** | **1.263** | **1.175** | 4.250 |
| 2 1/4 | 2.2500 | 2.2500 | 2.2380 | 3 3/8 | 3.375 | 3.262 | 3.897 | 3.719 | 1 3/8 | 1.423 | 1.327 | 5.000[c] |
| 2 1/2 | 2.5000 | 2.5000 | 2.4880 | 3 3/4 | 3.750 | 3.625 | 4.330 | 4.133 | 1 17/32 | 1.583 | 1.479 | 5.500[c] |
| 2 3/4 | 2.7500 | 2.7500 | 2.7380 | 4 1/8 | 4.125 | 3.988 | 4.763 | 4.546 | 1 11/16 | 1.744 | 1.632 | 6.000[c] |
| 3 | 3.0000 | 3.0000 | 2.9880 | 4 1/2 | 4.500 | 4.350 | 5.196 | 4.959 | 1 7/8 | 1.935 | 1.815 | 6.500[c] |

[a] *Nominal Size:* Where specifying nominal size in decimals, zeros preceding the decimal and in the fourth decimal place are omitted.

[b] Thread lengths, $L_T$, shown are for bolt lengths 6 inches and shorter. For longer bolt lengths add 0.250 inch to thread lengths shown.

[c] Thread lengths, $L_T$, shown are for bolt lengths over 6 inches.

All dimensions are in inches.

*Unification:* **Bold type indicates product features unified dimensionally with British and Canadian Standards.** Unification of fine thread products is limited to sizes 1 inch and smaller.

*Bearing Surface:* Bearing surface is flat and washer faced. Diameter of bearing surface is equal to the maximum width across flats within a tolerance of minus 10 percent.

*Threads Series:* Threads, when rolled, are Unified Coarse, Fine, or 8-thread series (UNRC, UNRF, or 8 UNR Series), Class 2A. Threads produced by other methods shall preferably be UNRC, UNRF or 8 UNR but, at manufacturer's option, may be Unified Coarse, Fine or 8-thread series (UNC, UNF, or 8 UN Series), Class 2A.

*Material:* Chemical and mechanical properties of steel screws normally conform to Grades 2, 5, or 8 of SAE J429, ASTM A449 or ASTM A354 Grade BD. Where specified, screws may also be made from brass, bronze, corrosion-resisting steel, aluminum alloy or other materials.

**Table 5. American National Standard Square Lag Screws** *ANSI/ASME B18.2.1-2012 (R2021)*

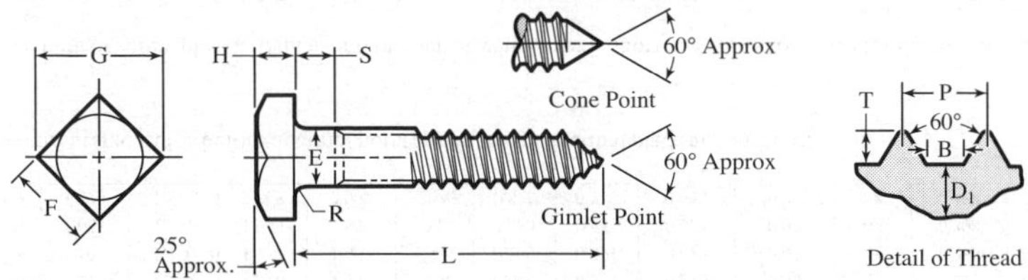


| Nominal Size[a] or Basic Product Dia. | | Body or Shoulder Dia. *E* | | Width Across Flats *F* | | | Width Across Corners *G* | | Height *H* | | | Shoulder Length *S* | Radius of Fillet *R* | Thds. per Inch | Thread Dimensions | | | |
|---|---|---|---|---|---|---|---|---|---|---|---|---|---|---|---|---|---|---|
| | | Max. | Min. | Basic | Max. | Min. | Max. | Min. | Basic | Max. | Min. | Min. | Max. | | Pitch *P* | Flat at Root *B* | Depth of Thd. *T* | Root Dia. $D_1$ |
| No. 10 | 0.1900 | 0.199 | 0.178 | 9/32 | 0.281 | 0.271 | 0.398 | 0.372 | ⅛ | 0.140 | 0.110 | 0.094 | 0.03 | 11 | 0.091 | 0.039 | 0.035 | 0.120 |
| ¼ | 0.2500 | 0.260 | 0.237 | ⅜ | 0.375 | 0.362 | 0.530 | 0.498 | 11/64 | 0.188 | 0.156 | 0.094 | 0.03 | 10 | 0.100 | 0.043 | 0.039 | 0.173 |
| 5/16 | 0.3125 | 0.324 | 0.298 | ½ | 0.500 | 0.484 | 0.707 | 0.665 | 13/64 | 0.220 | 0.186 | 0.125 | 0.03 | 9 | 0.111 | 0.048 | 0.043 | 0.227 |
| ⅜ | 0.3750 | 0.388 | 0.360 | 9/16 | 0.562 | 0.544 | 0.795 | 0.747 | ¼ | 0.268 | 0.232 | 0.125 | 0.03 | 7 | 0.143 | 0.062 | 0.055 | 0.265 |
| 7/16 | 0.4375 | 0.452 | 0.421 | ⅝ | 0.625 | 0.603 | 0.884 | 0.828 | 19/64 | 0.316 | 0.278 | 0.156 | 0.03 | 7 | 0.143 | 0.062 | 0.055 | 0.328 |
| ½ | 0.5000 | 0.515 | 0.482 | ¾ | 0.750 | 0.725 | 1.061 | 0.995 | 21/64 | 0.348 | 0.308 | 0.156 | 0.03 | 6 | 0.167 | 0.072 | 0.064 | 0.371 |
| ⅝ | 0.6250 | 0.642 | 0.605 | 15/16 | 0.938 | 0.906 | 1.326 | 1.244 | 27/64 | 0.444 | 0.400 | 0.312 | 0.06 | 5 | 0.200 | 0.086 | 0.077 | 0.471 |
| ¾ | 0.7500 | 0.768 | 0.729 | 1⅛ | 1.125 | 1.088 | 1.591 | 1.494 | ½ | 0.524 | 0.476 | 0.375 | 0.06 | 4½ | 0.222 | 0.096 | 0.085 | 0.579 |
| ⅞ | 0.8750 | 0.895 | 0.852 | 1 5/16 | 1.312 | 1.269 | 1.856 | 1.742 | 19/32 | 0.620 | 0.568 | 0.375 | 0.06 | 4 | 0.250 | 0.108 | 0.096 | 0.683 |
| 1 | 1.0000 | 1.022 | 0.976 | 1½ | 1.500 | 1.450 | 2.121 | 1.991 | 21/32 | 0.684 | 0.628 | 0.625 | 0.09 | 3½ | 0.286 | 0.123 | 0.110 | 0.780 |
| 1⅛ | 1.1250 | 1.149 | 1.098 | 1 11/16 | 1.688 | 1.631 | 2.386 | 2.239 | ¾ | 0.780 | 0.720 | 0.625 | 0.09 | 3¼ | 0.308 | 0.133 | 0.119 | 0.887 |
| 1¼ | 1.2500 | 1.277 | 1.223 | 1⅞ | 1.875 | 1.812 | 2.652 | 2.489 | 27/32 | 0.876 | 0.812 | 0.625 | 0.09 | 3¼ | 0.308 | 0.133 | 0.119 | 1.012 |

[a] When specifying decimal nominal size, zeros before decimal point and in fourth decimal place are omitted.

All dimensions in inches.

Minimum thread length is ½ length of screw plus 0.50 inch, or 6.00 inches, whichever is shorter. Screws too short for the formula thread length shall be threaded as close to the head as practicable.

Thread formulas: Pitch = 1 ÷ threads per inch. Flat at root = 0.4305 × pitch. Depth of single thread = 0.385 × pitch.

**Table 6. American National Standard Hex Lag Screws** *ANSI/ASME B18.2.1-2012 (R2021)*

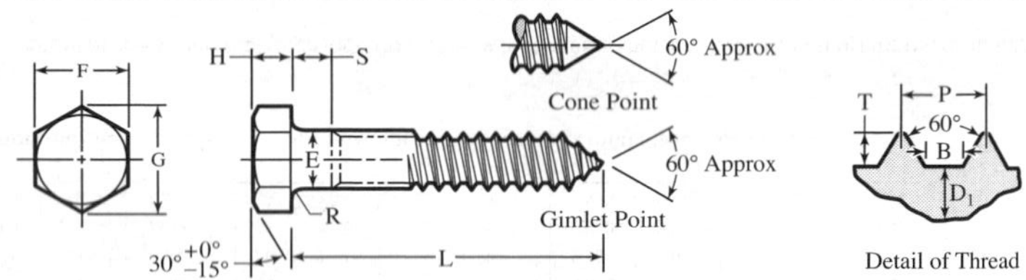


| Nominal Size[a] or Basic Product Dia. | | Body or Shoulder Dia. *E* | | Width Across Flats *F* | | | Width Across Corners *G* | | Height *H* | | | Shoulder Length *S* | Radius of Fillet *R* | Thds. per Inch | Thread Dimensions | | | |
|---|---|---|---|---|---|---|---|---|---|---|---|---|---|---|---|---|---|---|
| | | Max. | Min. | Basic | Max. | Min. | Max. | Min. | Basic | Max. | Min. | Min. | Max. | | Pitch *P* | Flat at Root *B* | Depth of Thd. *T* | Root Dia. $D_1$ |
| No. 10 | 0.1900 | 0.199 | 0.178 | 9⁄32 | 0.281 | 0.271 | 0.323 | 0.309 | 1⁄8 | 0.140 | 0.110 | 0.094 | 0.03 | 11 | 0.091 | 0.039 | 0.035 | 0.120 |
| 1⁄4 | 0.2500 | 0.260 | 0.237 | 3⁄8 | 0.438 | 0.425 | 0.505 | 0.484 | 11⁄64 | 0.188 | 0.150 | 0.094 | 0.03 | 10 | 0.100 | 0.043 | 0.039 | 0.173 |
| 5⁄16 | 0.3125 | 0.324 | 0.298 | 1⁄2 | 0.500 | 0.484 | 0.577 | 0.552 | 7⁄32 | 0.235 | 0.195 | 0.125 | 0.03 | 9 | 0.111 | 0.048 | 0.043 | 0.227 |
| 3⁄8 | 0.3750 | 0.388 | 0.360 | 9⁄16 | 0.562 | 0.544 | 0.650 | 0.620 | 1⁄4 | 0.268 | 0.226 | 0.125 | 0.03 | 7 | 0.143 | 0.062 | 0.055 | 0.265 |
| 7⁄16 | 0.4375 | 0.452 | 0.421 | 5⁄8 | 0.625 | 0.603 | 0.722 | 0.687 | 19⁄64 | 0.316 | 0.272 | 0.156 | 0.03 | 7 | 0.143 | 0.062 | 0.055 | 0.328 |
| 1⁄2 | 0.5000 | 0.515 | 0.482 | 3⁄4 | 0.750 | 0.725 | 0.866 | 0.826 | 11⁄32 | 0.364 | 0.302 | 0.156 | 0.03 | 6 | 0.167 | 0.072 | 0.064 | 0.371 |
| 5⁄8 | 0.6250 | 0.642 | 0.605 | 15⁄16 | 0.938 | 0.906 | 1.083 | 1.033 | 27⁄64 | 0.444 | 0.378 | 0.312 | 0.06 | 5 | 0.200 | 0.086 | 0.077 | 0.471 |
| 3⁄4 | 0.7500 | 0.768 | 0.729 | 1 1⁄8 | 1.125 | 1.088 | 1.299 | 1.240 | 1⁄2 | 0.524 | 0.455 | 0.375 | 0.06 | 4 1⁄2 | 0.222 | 0.096 | 0.085 | 0.579 |
| 7⁄8 | 0.8750 | 0.895 | 0.852 | 1 5⁄16 | 1.312 | 1.269 | 1.516 | 1.447 | 37⁄64 | 0.604 | 0.531 | 0.375 | 0.06 | 4 | 0.250 | 0.108 | 0.096 | 0.683 |
| 1 | 1.0000 | 1.022 | 0.976 | 1 1⁄2 | 1.500 | 1.450 | 1.732 | 1.653 | 43⁄64 | 0.700 | 0.591 | 0.625 | 0.09 | 3 1⁄2 | 0.286 | 0.123 | 0.110 | 0.780 |
| 1 1⁄8 | 1.1250 | 1.149 | 1.098 | 1 11⁄16 | 1.688 | 1.631 | 1.949 | 1.859 | 3⁄4 | 0.780 | 0.658 | 0.625 | 0.09 | 3 1⁄4 | 0.308 | 0.133 | 0.119 | 0.887 |
| 1 1⁄4 | 1.2500 | 1.277 | 1.223 | 1 7⁄8 | 1.875 | 1.812 | 2.165 | 2.066 | 27⁄32 | 0.876 | 0.749 | 0.625 | 0.09 | 3 1⁄4 | 0.308 | 0.133 | 0.119 | 1.012 |

[a] When specifying decimal nominal size, zeros before decimal point and in fourth decimal place are omitted.

All dimensions in inches.

Minimum thread length is 1⁄2 length of screw plus 0.50 inch, or 6.00 inches, whichever is shorter. Screws too short for the formula thread length shall be threaded as close to the head as practicable.

Thread formulas: Pitch = 1 ÷ thds. per inch. Flat at root = 0.4305 × pitch. Depth of single thread = 0.385 × pitch.

## Table 7. American National Standard and Unified Standard Hex Nuts and Jam Nuts and Heavy Hex Nuts and Jam Nuts *ANSI/ASME B18.2.2-2022*

| Nominal Size or Basic Major Dia. of Thread | | Width Across Flats *F* | | | Width Across Corners *G* | | Thickness, Nuts *H* | | | Thickness, Jam Nuts $H_1$ | | |
|---|---|---|---|---|---|---|---|---|---|---|---|---|
| | | Basic | Max. | Min. | Max. | Min. | Basic | Max. | Min. | Basic | Max. | Min. |
| Hex Nuts (Fig. 5) and Hex Jam Nuts (Fig. 6) | | | | | | | | | | | | |
| **1/4** | **0.2500** | **7/16** | **0.438** | **0.428** | **0.505** | **0.488** | **7/32** | **0.226** | **0.212** | **5/32** | **0.163** | **0.150** |
| **5/16** | **0.3125** | **1/2** | **0.500** | **0.489** | **0.577** | **0.557** | **17/64** | **0.273** | **0.258** | **3/16** | **0.195** | **0.180** |
| **3/8** | **0.3750** | **9/16** | **0.562** | **0.551** | **0.650** | **0.628** | **21/64** | **0.337** | **0.320** | **7/32** | **0.227** | **0.210** |
| **7/16** | **0.4375** | **11/16** | **0.688** | **0.675** | **0.794** | **0.768** | **3/8** | **0.385** | **0.365** | **1/4** | **0.260** | **0.240** |
| **1/2** | **0.5000** | **3/4** | **0.750** | **0.736** | **0.866** | **0.840** | **7/16** | **0.448** | **0.427** | **5/16** | **0.323** | **0.302** |
| **9/16** | **0.5625** | **7/8** | **0.875** | **0.861** | **1.010** | **0.982** | **31/64** | **0.496** | **0.473** | **5/16** | **0.324** | **0.301** |
| **5/8** | **0.6250** | **15/16** | **0.938** | **0.922** | **1.083** | **1.051** | **35/64** | **0.559** | **0.535** | **3/8** | **0.387** | **0.363** |
| **3/4** | **0.7500** | **1 1/8** | **1.125** | **1.088** | **1.299** | **1.240** | **41/64** | **0.665** | **0.617** | **27/64** | **0.446** | **0.398** |
| **7/8** | **0.8750** | **1 5/16** | **1.312** | **1.269** | **1.516** | **1.447** | **3/4** | **0.776** | **0.724** | **31/64** | **0.510** | **0.458** |
| **1** | **1.0000** | **1 1/2** | **1.500** | **1.450** | **1.732** | **1.653** | **55/64** | **0.887** | **0.831** | **35/64** | **0.575** | **0.519** |
| **1 1/8** | **1.1250** | **1 11/16** | **1.688** | **1.631** | **1.949** | **1.859** | **31/32** | **0.999** | **0.939** | **39/64** | **0.639** | **0.579** |
| **1 1/4** | **1.2500** | **1 7/8** | **1.875** | **1.812** | **2.165** | **2.066** | **1 1/16** | **1.094** | **1.030** | **23/32** | **0.751** | **0.687** |
| **1 3/8** | **1.3750** | **2 1/16** | **2.062** | **1.994** | **2.382** | **2.273** | **1 11/64** | **1.206** | **1.138** | **25/32** | **0.815** | **0.747** |
| **1 1/2** | **1.5000** | **2 1/4** | **2.250** | **2.175** | **2.598** | **2.480** | **1 9/32** | **1.317** | **1.245** | **27/32** | **0.880** | **0.808** |
| Heavy Hex Nuts (Fig. 5) and Heavy Hex Jam Nuts (Fig. 6) | | | | | | | | | | | | |
| 1/4 | 0.2500 | 1/2 | 0.500 | 0.488 | 0.577 | 0.556 | 15/64 | 0.250 | 0.218 | 11/64 | 0.188 | 0.156 |
| 5/16 | 0.3125 | 9/16 | 0.562 | 0.546 | 0.650 | 0.622 | 19/64 | 0.314 | 0.280 | 13/64 | 0.220 | 0.186 |
| 3/8 | 0.3750 | 11/16 | 0.688 | 0.669 | 0.794 | 0.763 | 23/64 | 0.377 | 0.341 | 15/64 | 0.252 | 0.216 |
| 7/16 | 0.4375 | 3/4 | 0.750 | 0.728 | 0.866 | 0.830 | 27/64 | 0.441 | 0.403 | 17/64 | 0.285 | 0.247 |
| **1/2** | **0.5000** | **7/8** | **0.875** | **0.850** | **1.010** | **0.969** | **31/64** | **0.504** | **0.464** | **19/64** | **0.317** | **0.277** |
| 9/16 | 0.5625 | 15/16 | 0.938 | 0.909 | 1.083 | 1.037 | 35/64 | 0.568 | 0.526 | 21/64 | 0.349 | 0.307 |
| **5/8** | **0.6250** | **1 1/16** | **1.062** | **1.031** | **1.227** | **1.1175** | **39/64** | **0.631** | **0.587** | **23/64** | **0.381** | **0.337** |
| **3/4** | **0.7500** | **1 1/4** | **1.250** | **1.212** | **1.443** | **1.382** | **47/64** | **0.758** | **0.710** | **27/64** | **0.446** | **0.398** |
| **7/8** | **0.8750** | **1 7/16** | **1.438** | **1.394** | **1660** | **1.589** | **55/64** | **0.885** | **0.833** | **31/64** | **0.510** | **0.458** |
| **1** | **1.0000** | **1 5/8** | **1.625** | **1.575** | **1.876** | **1.796** | **63/64** | **1.012** | **0.956** | **35/64** | **0.575** | **0.519** |
| **1 1/8** | **1.1250** | **1 13/16** | **1.812** | **1.756** | **2.093** | **2.002** | **1 7/64** | **1.139** | **1.079** | **39/64** | **0.639** | **0.579** |
| **1 1/4** | **1.2500** | **2** | **2.000** | **1.938** | **2.309** | **2.209** | **1 7/32** | **1.251** | **1.187** | **23/32** | **0.751** | **0.687** |
| **1 3/8** | **1.3750** | **2 3/16** | **2.188** | **2.119** | **2.526** | **2.416** | **1 11/32** | **1.378** | **1.310** | **25/32** | **0.815** | **0.747** |
| **1 1/2** | **1.5000** | **2 3/8** | **2.375** | **2.300** | **2.742** | **2.622** | **1 15/32** | **1.505** | **1.433** | **27/32** | **0.880** | **0.808** |
| 1 5/8 | 1.6250 | 2 9/16 | 2.562 | 2.481 | 2.959 | 2.828 | 1 19/32 | 1.632 | 1.556 | 29/32 | 0.944 | 0.868 |
| **1 3/4** | **1.7500** | **2 3/4** | **2.750** | **2.662** | **3.175** | **3.035** | **1 23/32** | **1.759** | **1.679** | **31/32** | **1.009** | **0.929** |
| 1 7/8 | 1.8750 | 2 15/16 | 2.938 | 2.844 | 3.392 | 3.242 | 1 27/32 | 1.886 | 1.802 | 1 1/32 | 1.073 | 0.989 |
| **2** | **2.0000** | **3 1/8** | **3.125** | **3.025** | **3.608** | **3.449** | **1 31/32** | **2.013** | **1.925** | **1 3/32** | **1.138** | **1.050** |
| 2 1/4 | 2.2500 | 3 1/2 | 3.500 | 3.388 | 4.041 | 3.862 | 2 13/64 | 2.251 | 2.155 | 1 13/64 | 1.251 | 1.155 |
| 2 1/2 | 2.5000 | 3 7/8 | 3.875 | 3.750 | 4.474 | 4.275 | 2 29/64 | 2.505 | 2.401 | 1 29/64 | 1.505 | 1.401 |
| 2 3/4 | 2.7500 | 4 1/4 | 4.250 | 4.112 | 4.907 | 4.688 | 2 45/64 | 2.759 | 2.647 | 1 37/64 | 1.634 | 1.522 |
| 3 | 3.0000 | 4 5/8 | 4.625 | 4.475 | 5.340 | 5.102 | 2 61/64 | 3.013 | 2.893 | 1 45/64 | 1.763 | 1.643 |
| 3 1/4 | 3.2500 | 5 | 5.000 | 4.838 | 5.774 | 5.515 | 3 3/16 | 3.252 | 3.124 | 1 13/16 | 1.876 | 1.748 |
| 3 1/2 | 3.5000 | 5 3/8 | 5.375 | 5.200 | 6.207 | 5.928 | 3 7/16 | 3.506 | 3.370 | 1 15/16 | 2.006 | 1.870 |
| 3 3/4 | 3.7500 | 5 3/4 | 5.750 | 5.562 | 6.640 | 6.341 | 3 11/16 | 3.760 | 3.616 | 2 1/16 | 2.134 | 1.990 |
| 4 | 4.0000 | 6 1/8 | 6.125 | 5.925 | 7.073 | 6.755 | 3 15/16 | 4.014 | 3.862 | 2 3/16 | 2.264 | 2.112 |

All dimensions are in inches.

**Bold type shows nuts unified dimensionally with British and Canadian Standards.**

Threads are Unified Coarse-, Fine-, or 8-thread series (UNC, UNF or 8UN), Class 2B. Unification of fine-thread nuts is limited to sizes 1 inch and under.

## Table 8. American National Standard and Unified Standard Hex Flat Nuts and Flat Jam Nuts and Heavy Hex Flat Nuts and Flat Jam Nuts

*ANSI/ASME B18.2.2-2022*

| Nominal Size or Basic Major Dia. of Thread | | Width Across Flats $F$ | | | Width Across Corners $G$ | | Thickness, Flat Nuts $H$ | | | Thickness, Flat Jam Nuts $H_1$ | | |
|---|---|---|---|---|---|---|---|---|---|---|---|---|
| | | Basic | Max. | Min. | Max. | Min. | Basic | Max. | Min. | Basic | Max. | Min. |
| Hex Flat Nuts and Hex Flat Jam Nuts (Fig. 7) | | | | | | | | | | | | |
| **1⅛** | **1.1250** | **1 11/16** | **1.688** | **1.631** | **1.949** | **1.859** | **1** | **1.030** | **0.970** | **5/8** | **0.655** | **0.595** |
| **1¼** | **1.2500** | **1⅞** | **1.875** | **1.812** | **2.165** | **2.066** | **1 3/32** | **1.126** | **1.062** | **3/4** | **0.782** | **0.718** |
| **1⅜** | **1.3750** | **2 1/16** | **2.062** | **1.994** | **2.382** | **2.273** | **1 13/64** | **1.237** | **1.169** | **13/16** | **0.846** | **0.778** |
| **1½** | **1.5000** | **2¼** | **2.250** | **2.175** | **2.598** | **2.480** | **1 5/16** | **1.348** | **1.276** | **7/8** | **0.911** | **0.839** |
| Heavy Hex Flat Nuts and Heavy Hex Flat Jam Nuts (Fig. 7) | | | | | | | | | | | | |
| **1⅛** | **1.1250** | **1 13/16** | **1.812** | **1.756** | **2.093** | **2.002** | **1⅛** | **1.155** | **1.079** | **5/8** | **0.655** | **0.579** |
| **1¼** | **1.2500** | **2** | **2.000** | **1.938** | **2.309** | **2.209** | **1¼** | **1.282** | **1.187** | **3/4** | **0.782** | **0.687** |
| **1⅜** | **1.3750** | **2 3/16** | **2.188** | **2.119** | **2.526** | **2.416** | **1⅜** | **1.409** | **1.310** | **13/16** | **0.846** | **0.747** |
| **1½** | **1.5000** | **2⅜** | **2.375** | **2.300** | **2.742** | **2.622** | **1½** | **1.536** | **1.433** | **7/8** | **0.911** | **0.808** |
| **1¾** | **1.7500** | **2¾** | **2.750** | **2.662** | **3.175** | **3.035** | **1¾** | **1.790** | **1.679** | **1** | **1.040** | **0.929** |
| **2** | **2.0000** | **3⅛** | **3.125** | **3.025** | **3.608** | **3.449** | **2** | **2.044** | **1.925** | **1⅛** | **1.169** | **1.050** |
| 2¼ | 2.2500 | 3½ | 3.500 | 3.388 | 4.041 | 3.862 | 2¼ | 2.298 | 2.155 | 1¼ | 1.298 | 1.155 |
| 2½ | 2.5000 | 3⅞ | 3.875 | 3.750 | 4.474 | 4.275 | 2½ | 2.552 | 2.401 | 1½ | 1.552 | 1.401 |
| 2¾ | 2.7500 | 4¼ | 4.250 | 4.112 | 4.907 | 4.688 | 2¾ | 2.806 | 2.647 | 1⅝ | 1.681 | 1.522 |
| 3 | 3.0000 | 4⅝ | 4.625 | 4.475 | 5.340 | 5.102 | 3 | 3.060 | 2.893 | 1¾ | 1.810 | 1.643 |
| 3¼ | 3.2500 | 5 | 5.000 | 4.838 | 5.774 | 5.515 | 3¼ | 3.314 | 3.124 | 1⅞ | 1.939 | 1.748 |
| 3½ | 3.5000 | 5⅜ | 5.375 | 5.200 | 6.207 | 5.928 | 3½ | 3.568 | 3.370 | 2 | 2.068 | 1.870 |
| 3¾ | 3.7500 | 5¾ | 5.750 | 5.562 | 6.640 | 6.341 | 3¾ | 3.822 | 3.616 | 2⅛ | 2.197 | 1.990 |
| 4 | 4.0000 | 6⅛ | 6.125 | 5.925 | 7.073 | 6.755 | 4 | 4.076 | 3.862 | 2¼ | 2.326 | 2.112 |

All dimensions are in inches.

**Bold type indicates nuts unified dimensionally with British and Canadian Standards.**

Threads are Unified Coarse-thread series (UNC), Class 2B.

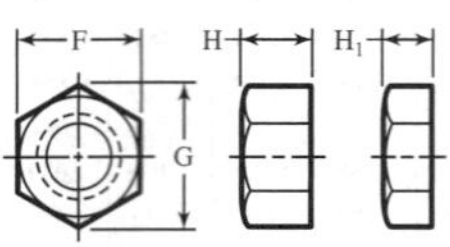


Fig. 7. Hex Flat Nuts, Heavy Hex Flat Nuts, Hex Flat Jam Nuts, and Heavy Hex Flat Jam Nuts (Table 8)

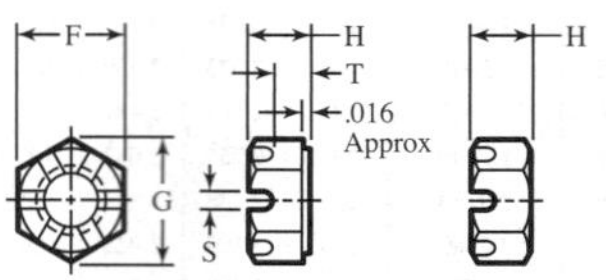


Fig. 8. Hex Slotted Nuts, Heavy Hex Slotted Nuts, and Hex Thick Slotted Nuts (Table 9)

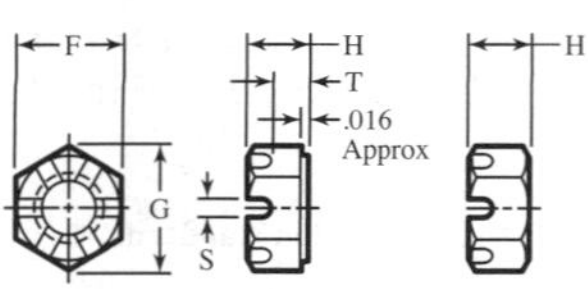


Fig. 9. Hex Thick Nuts (Table 10)

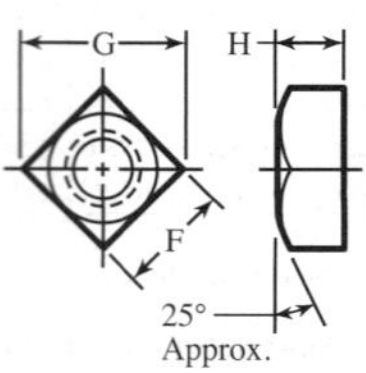


Fig. 10. Square Nuts, Heavy Square Nuts (Table 10)

### Table 9. American National and Unified Standard Hex Slotted Nuts, Heavy Hex Slotted Nuts, and Hex Thick Slotted Nuts *ANSI/ASME B18.2.2-2022*

| Nominal Size or Basic Major Dia. of Thread | | Width Across Flats *F* | | | Width Across Corners *G* | | Thickness *H* | | | Unslotted Thickness *T* | | Width of Slot *S* | |
|---|---|---|---|---|---|---|---|---|---|---|---|---|---|
| | | Basic | Max. | Min. | Max. | Min. | Basic | Max. | Min. | Max. | Min. | Max. | Min. |
| Hex Slotted Nuts (Fig. 8) | | | | | | | | | | | | | |
| **1/4** | **0.2500** | **7/16** | **0.438** | **0.428** | **0.505** | **0.488** | **7/32** | **0.226** | **0.212** | 0.14 | 0.12 | 0.10 | 0.07 |
| **5/16** | **0.3125** | **1/2** | **0.500** | **0.489** | **0.577** | **0.577** | **17/64** | **0.273** | **0.258** | 0.18 | 0.16 | 0.12 | 0.09 |
| **3/8** | **0.3750** | **9/16** | **0.562** | **0.551** | **0.650** | **0.628** | **21/64** | **0.337** | **0.320** | 0.21 | 0.19 | 0.15 | 0.12 |
| **7/16** | **0.4375** | **11/16** | **0.688** | **0.675** | **0.794** | **0.768** | **3/8** | **0.385** | **0.365** | 0.23 | 0.21 | 0.15 | 0.12 |
| **1/2** | **0.5000** | **3/4** | **0.750** | **0.736** | **0.866** | **0.840** | **7/16** | **0.448** | **0.427** | 0.29 | 0.27 | 0.18 | 0.15 |
| **9/16** | **0.5625** | **7/8** | **0.875** | **0.861** | **1.010** | **0.982** | **31/64** | **0.496** | **0.473** | 0.31 | 0.29 | 0.18 | 0.15 |
| **5/8** | **0.6250** | **15/16** | **0.938** | **0.922** | **1.083** | **1.051** | **35/64** | **0.559** | **0.535** | 0.34 | 0.32 | 0.24 | 0.18 |
| **3/4** | **0.7500** | **1 1/8** | **1.125** | **1.088** | **1.299** | **1.240** | **41/64** | **0.665** | **0.617** | 0.40 | 0.38 | 0.24 | 0.18 |
| **7/8** | **0.8750** | **1 5/16** | **1.312** | **1.269** | **1.516** | **1.447** | **3/4** | **0.776** | **0.724** | 0.52 | 0.49 | 0.24 | 0.18 |
| **1** | **1.0000** | **1 1/2** | **1.500** | **1.450** | **1.732** | **1.653** | **55/64** | **0.887** | **0.831** | 0.59 | 0.56 | 0.30 | 0.24 |
| **1 1/8** | **1.1250** | **1 11/16** | **1.688** | **1.631** | **1.949** | **1.859** | **31/32** | **0.999** | **0.939** | 0.64 | 0.61 | 0.33 | 0.24 |
| **1 1/4** | **1.2500** | **1 7/8** | **1.875** | **1.812** | **2.165** | **2.066** | **1 1/16** | **1.094** | **1.030** | 0.70 | 0.67 | 0.40 | 0.31 |
| **1 3/8** | **1.3750** | **2 1/16** | **2.062** | **1.994** | **2.382** | **2.273** | **1 11/64** | **1.206** | **1.138** | 0.82 | 0.78 | 0.40 | 0.31 |
| **1 1/2** | **1.5000** | **2 1/4** | **2.250** | **2.175** | **2.598** | **2.480** | **1 9/32** | **1.317** | **1.245** | 0.86 | 0.82 | 0.46 | 0.37 |
| Heavy Hex Slotted Nuts (Fig. 8) | | | | | | | | | | | | | |
| 1/4 | 0.2500 | 1/2 | 0.500 | 0.488 | 0.577 | 0.556 | 15/64 | 0.250 | 0.218 | 0.15 | 0.13 | 0.10 | 0.07 |
| 5/16 | 0.3125 | 9/16 | 0.562 | 0.546 | 0.650 | 0.622 | 19/64 | 0.314 | 0.280 | 0.21 | 0.19 | 0.12 | 0.09 |
| 3/8 | 0.3750 | 11/16 | 0.688 | 0.669 | 0.794 | 0.763 | 23/64 | 0.377 | 0.341 | 0.24 | 0.22 | 0.15 | 0.12 |
| 7/16 | 0.4375 | 3/4 | 0.750 | 0.728 | 0.866 | 0.830 | 27/64 | 0.441 | 0.403 | 0.28 | 0.26 | 0.15 | 0.12 |
| **1/2** | **0.5000** | **7/8** | **0.875** | **0.850** | **1.010** | **0.969** | **31/64** | **0.504** | **0.464** | 0.34 | 0.32 | 0.18 | 0.15 |
| 9/16 | 0.5625 | 15/16 | 0.938 | 0.909 | 1.083 | 1.037 | 35/64 | 0.568 | 0.526 | 0.37 | 0.35 | 0.18 | 0.15 |
| **5/8** | **0.6250** | **1 1/16** | **1.062** | **1.031** | **1.227** | **1.175** | **39/64** | **0.631** | **0.587** | 0.40 | 0.38 | 0.24 | 0.18 |
| **3/4** | **0.7500** | **1 1/4** | **1.250** | **1.212** | **1.443** | **1.382** | **47/64** | **0.758** | **0.710** | 0.49 | 0.47 | 0.24 | 0.18 |
| **7/8** | **0.8750** | **1 7/16** | **1.438** | **1.394** | **1.660** | **1.589** | **55/64** | **0.885** | **0.833** | 0.62 | 0.59 | 0.24 | 0.18 |
| **1** | **1.0000** | **1 5/8** | **1.625** | **1.575** | **1.876** | **1.796** | **63/64** | **1.012** | **0.956** | 0.72 | 0.69 | 0.30 | 0.24 |
| **1 1/8** | **1.1250** | **1 13/16** | **1.812** | **1.756** | **2.093** | **2.002** | **1 7/64** | **1.139** | **1.079** | 0.78 | 0.75 | 0.33 | 0.24 |
| **1 1/4** | **1.2500** | **2** | **2.000** | **1.938** | **2.309** | **2.209** | **1 7/32** | **1.251** | **1.187** | 0.86 | 0.83 | 0.40 | 0.31 |
| **1 3/8** | **1.3750** | **2 3/16** | **2.188** | **2.119** | **2.526** | **2.416** | **1 11/32** | **1.378** | **1.310** | 0.99 | 0.95 | 0.40 | 0.31 |
| **1 1/2** | **1.5000** | **2 3/8** | **2.375** | **2.300** | **2.742** | **2.622** | **1 15/32** | **1.505** | **1.433** | 1.05 | 1.01 | 0.46 | 0.37 |
| **1 3/4** | **1.7500** | **2 3/4** | **2.750** | **2.662** | **3.175** | **3.035** | **1 23/32** | **1.759** | **1.679** | 1.24 | 1.20 | 0.52 | 0.43 |
| **2** | **2.0000** | **3 1/8** | **3.125** | **3.025** | **3.608** | **3.449** | **1 31/32** | **2.013** | **1.925** | 1.43 | 1.38 | 0.52 | 0.43 |
| 2 1/4 | 2.2500 | 3 1/2 | 3.500 | 3.388 | 4.041 | 3.862 | 2 13/64 | 2.251 | 2.155 | 1.67 | 1.62 | 0.52 | 0.43 |
| 2 1/2 | 2.5000 | 3 7/8 | 3.875 | 3.750 | 4.474 | 4.275 | 2 29/64 | 2.505 | 2.401 | 1.79 | 1.74 | 0.64 | 0.55 |
| 2 3/4 | 2.7500 | 4 1/4 | 4.250 | 4.112 | 4.907 | 4.688 | 2 45/64 | 2.759 | 2.647 | 2.05 | 1.99 | 0.64 | 0.55 |
| 3 | 3.0000 | 4 5/8 | 4.625 | 4.475 | 5.340 | 5.102 | 2 61/64 | 3.013 | 2.893 | 2.23 | 2.17 | 0.71 | 0.62 |
| 3 1/4 | 3.2500 | 5 | 5.000 | 4.838 | 5.774 | 5.515 | 3 3/16 | 3.252 | 3.124 | 2.47 | 2.41 | 0.71 | 0.62 |
| 3 1/2 | 3.5000 | 5 3/8 | 5.375 | 5.200 | 6.207 | 5.928 | 3 7/16 | 3.506 | 3.370 | 2.72 | 2.65 | 0.71 | 0.62 |
| 3 3/4 | 3.7500 | 5 3/4 | 5.750 | 5.562 | 6.640 | 6.341 | 3 11/16 | 3.760 | 3.616 | 2.97 | 2.90 | 0.71 | 0.62 |
| 4 | 4.0000 | 6 1/8 | 6.125 | 5.925 | 7.073 | 6.755 | 3 15/16 | 4.014 | 3.862 | 3.22 | 3.15 | 0.71 | 0.62 |
| Hex Thick Slotted Nuts (Fig. 8) | | | | | | | | | | | | | |
| **1/4** | **0.2500** | **7/16** | **0.438** | **0.428** | **0.505** | **0.488** | **9/32** | **0.288** | **0.274** | 0.20 | 0.18 | 0.10 | 0.07 |
| **5/16** | **0.3125** | **1/2** | **0.500** | **0.489** | **0.577** | **0.557** | **21/64** | **0.336** | **0.320** | 0.24 | 0.22 | 0.12 | 0.09 |
| **3/8** | **0.3750** | **9/16** | **0.562** | **0.551** | **0.650** | **0.628** | **13/32** | **0.415** | **0.398** | 0.29 | 0.27 | 0.15 | 0.12 |
| **7/16** | **0.4375** | **11/16** | **0.688** | **0.675** | **0.794** | **0.768** | **29/24** | **0.463** | **0.444** | 0.31 | 0.29 | 0.15 | 0.12 |
| **1/2** | **0.5000** | **3/4** | **0.750** | **0.736** | **0.866** | **0.840** | **9/16** | **0.573** | **0.552** | 0.42 | 0.40 | 0.18 | 0.15 |
| **9/16** | **0.5625** | **7/8** | **0.875** | **0.861** | **1.010** | **0.982** | **39/64** | **0.621** | **0.598** | 0.43 | 0.41 | 0.18 | 0.15 |
| **5/8** | **0.6250** | **15/16** | **0.938** | **0.922** | **1.083** | **1.051** | **23/32** | **0.731** | **0.706** | 0.51 | 0.49 | 0.24 | 0.18 |
| **3/4** | **0.7500** | **1 1/8** | **1.125** | **1.088** | **1.299** | **1.240** | **13/16** | **0.827** | **0.798** | 0.57 | 0.55 | 0.24 | 0.18 |
| **7/8** | **0.8750** | **1 5/16** | **1.312** | **1.269** | **1.516** | **1.447** | **29/32** | **0.922** | **0.890** | 0.67 | 0.64 | 0.24 | 0.18 |
| **1** | **1.0000** | **1 1/2** | **1.500** | **1.450** | **1.732** | **1.653** | **1** | **1.018** | **0.982** | 0.73 | 0.70 | 0.30 | 0.24 |
| **1 1/8** | **1.1250** | **1 11/16** | **1.688** | **1.631** | **1.949** | **1.859** | **1 5/32** | **1.176** | **1.136** | 0.83 | 0.80 | 0.33 | 0.24 |
| **1 1/4** | **1.2500** | **1 7/8** | **1.875** | **1.812** | **2.165** | **2.066** | **1 1/4** | **1.272** | **1.228** | 0.89 | 0.86 | 0.40 | 0.31 |
| **1 3/8** | **1.3750** | **2 1/16** | **2.062** | **1.994** | **2.382** | **2.273** | **1 3/8** | **1.399** | **1.351** | 1.02 | 0.98 | 0.40 | 0.31 |
| **1 1/2** | **1.5000** | **2 1/4** | **2.250** | **2.175** | **2.598** | **2.480** | **1 1/2** | **1.526** | **1.474** | 1.08 | 1.04 | 0.46 | 0.37 |

All dimensions are in inches.

**Bold type indicates nuts unified dimensionally with British and Canadian Standards.**

Threads are Unified Coarse-, Fine-, or 8-thread series (UNC, UNF, or 8UN), Class 2B.

Unification of fine-thread nuts is limited to sizes 1 inch and under.

## Table 10. American National and Unified Standard Square Nuts and Heavy Square Nuts and American National Standard Hex Thick Nuts

*ANSI/ASME B18.2.2-2022*

| Nominal Size or Basic Major Dia. of Thread | | Width Across Flats *F* | | | Width Across Corners *G* | | Thickness *H* | | |
|---|---|---|---|---|---|---|---|---|---|
| | | Basic | Max. | Min. | Max. | Min. | Basic | Max. | Min. |
| Square Nuts[a] (Fig. 10) | | | | | | | | | |
| **1/4** | **0.2500** | **7/16** | **0.438** | **0.425** | **0.619** | **0.554** | **7/32** | **0.235** | **0.203** |
| **5/16** | **0.3125** | **9/16** | **0.562** | **0.547** | **0.795** | **0.721** | **17/64** | **0.283** | **0.249** |
| **3/8** | **0.3750** | **5/8** | **0.625** | **0.606** | **0.884** | **0.802** | **21/64** | **0.346** | **0.310** |
| **7/16** | **0.4375** | **3/4** | **0.750** | **0.728** | **1.061** | **0.970** | **3/8** | **0.394** | **0.356** |
| **1/2** | **0.5000** | **13/16** | **0.812** | **0.788** | **1.149** | **1.052** | **7/16** | **0.458** | **0.418** |
| **5/8** | **0.6250** | **1** | **1.000** | **0.969** | **1.414** | **1.300** | **35/64** | **0.569** | **0.525** |
| **3/4** | **0.7500** | **1 1/8** | **1.125** | **1.088** | **1.591** | **1.464** | **21/32** | **0.680** | **0.632** |
| **7/8** | **0.8750** | **1 5/16** | **1.312** | **1.269** | **1.856** | **1.712** | **49/64** | **0.792** | **0.740** |
| **1** | **1.0000** | **1 1/2** | **1.500** | **1.450** | **2.121** | **1.961** | **7/8** | **0.903** | **0.847** |
| **1 1/8** | **1.1250** | **1 11/16** | **1.688** | **1.631** | **2.386** | **2.209** | **1** | **1.030** | **0.970** |
| **1 1/4** | **1.2500** | **1 7/8** | **1.875** | **1.812** | **2.652** | **2.458** | **1 3/32** | **1.126** | **1.062** |
| **1 3/8** | **1.3750** | **2 1/16** | **2.062** | **1.994** | **2.917** | **2.708** | **1 13/64** | **1.237** | **1.169** |
| **1 1/2** | **1.5000** | **2 1/4** | **2.250** | **2.175** | **3.182** | **2.956** | **1 5/16** | **1.348** | **1.276** |
| Heavy Square Nuts[a] (Fig. 10) | | | | | | | | | |
| 1/4 | 0.2500 | 1/2 | 0.500 | 0.488 | 0.707 | 0.640 | 1/4 | 0.266 | 0.218 |
| 5/16 | 0.3125 | 9/16 | 0.562 | 0.546 | 0.795 | 0.720 | 5/16 | 0.330 | 0.280 |
| 3/8 | 0.3750 | 11/16 | 0.688 | 0.669 | 0.973 | 0.889 | 3/8 | 0.393 | 0.341 |
| 7/16 | 0.4375 | 3/4 | 0.750 | 0.728 | 1.060 | 0.970 | 7/16 | 0.456 | 0.403 |
| 1/2 | 0.5000 | 7/8 | 0.875 | 0.850 | 1.237 | 1.137 | 1/2 | 0.520 | 0.464 |
| 5/8 | 0.6250 | 1 1/16 | 1.062 | 1.031 | 1.503 | 1.386 | 5/8 | 0.647 | 0.587 |
| 3/4 | 0.7500 | 1 1/4 | 1.250 | 1.212 | 1.768 | 1.635 | 3/4 | 0.774 | 0.710 |
| 7/8 | 0.8750 | 1 7/16 | 1.438 | 1.394 | 2.033 | 1.884 | 7/8 | 0.901 | 0.833 |
| 1 | 1.0000 | 1 5/8 | 1.625 | 1.575 | 2.298 | 2.132 | 1 | 1.028 | 0.956 |
| 1 1/8 | 1.1250 | 1 13/16 | 1.812 | 1.756 | 2.563 | 2.381 | 1 1/8 | 1.155 | 1.079 |
| 1 1/4 | 1.2500 | 2 | 2.000 | 1.938 | 2.828 | 2.631 | 1 1/4 | 1.282 | 1.187 |
| 1 3/8 | 1.3750 | 2 3/16 | 2.188 | 2.119 | 3.094 | 2.879 | 1 3/8 | 1.409 | 1.310 |
| 1 1/2 | 1.5000 | 2 3/8 | 2.375 | 2.300 | 3.359 | 3.128 | 1 1/2 | 1.536 | 1.433 |
| Hex Thick Nuts[b] (Fig. 10) | | | | | | | | | |
| 1/4 | 0.2500 | 7/16 | 0.438 | 0.428 | 0.505 | 0.488 | 9/32 | 0.288 | 0.274 |
| 5/16 | 0.3125 | 1/2 | 0.500 | 0.489 | 0.577 | 0.557 | 21/64 | 0.336 | 0.320 |
| 3/8 | 0.3750 | 9/16 | 0.562 | 0.551 | 0.650 | 0.628 | 13/32 | 0.415 | 0.398 |
| 7/16 | 0.4375 | 11/16 | 0.688 | 0.675 | 0.794 | 0.768 | 29/64 | 0.463 | 0.444 |
| 1/2 | 0.5000 | 3/4 | 0.750 | 0.736 | 0.866 | 0.840 | 9/16 | 0.573 | 0.552 |
| 9/16 | 0.5625 | 7/8 | 0.875 | 0.861 | 1.010 | 0.982 | 39/64 | 0.621 | 0.598 |
| 5/8 | 0.6250 | 15/16 | 0.938 | 0.922 | 1.083 | 1.051 | 23/32 | 0.731 | 0.706 |
| 3/4 | 0.7500 | 1 1/8 | 1.125 | 1.088 | 1.299 | 1.240 | 13/16 | 0.827 | 0.798 |
| 7/8 | 0.8750 | 1 5/16 | 1.312 | 1.269 | 1.516 | 1.447 | 29/32 | 0.922 | 0.890 |
| 1 | 1.0000 | 1 1/2 | 1.500 | 1.450 | 1.732 | 1.653 | 1 | 1.018 | 0.982 |
| 1 1/8 | 1.1250 | 1 11/16 | 1.688 | 1.631 | 1.949 | 1.859 | 1 5/32 | 1.176 | 1.136 |
| 1 1/4 | 1.2500 | 1 7/8 | 1.875 | 1.812 | 2.165 | 2.066 | 1 1/4 | 1.272 | 1.228 |
| 1 3/8 | 1.3750 | 2 1/16 | 2.062 | 1.994 | 2.382 | 2.273 | 1 3/8 | 1.399 | 1.351 |
| 1 1/2 | 1.5000 | 2 1/4 | 2.250 | 2.175 | 2.598 | 2.480 | 1 1/2 | 1.526 | 1.474 |

[a] Coarse-thread series, Class 2B.

[b] Unified Coarse-, Fine-, or 8-thread series (8 UN), Class 2B.

All dimensions are in inches.

**Bold type indicates nuts unified dimensionally with British and Canadian Standards.**

## Low and High Crown (Blind, Acorn) Nuts *SAE Recommended Practice J483a*

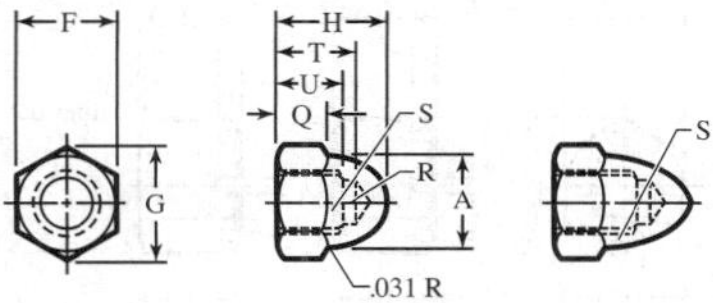


| Nom. Size[a] or Basic Major Dia. of Thread | | Width Across Flats, *F* | | | Width Across Corners, *G* | | Body Dia., *A* | Overall Height, *H* | Hexagon Height, *Q* | Nose Radius, *R* | Body Radius, *S* | Drill Dep., *T* | Full Thd., *U* |
|---|---|---|---|---|---|---|---|---|---|---|---|---|---|
| | | Max. | (Basic) | Min. | Max. | Min. | | | | | | Max. | Min. |
| **Low Crown** | | | | | | | | | | | | | |
| 6 | 0.1380 | 5/16 | 0.3125 | 0.302 | 0.361 | 0.344 | 0.30 | 0.34 | 0.16 | 0.08 | 0.17 | 0.25 | 0.16 |
| 8 | 0.1640 | 5/16 | 0.3125 | 0.302 | 0.361 | 0.344 | 0.30 | 0.34 | 0.16 | 0.08 | 0.17 | 0.25 | 0.16 |
| 10 | 0.1900 | 3/8 | 0.3750 | 0.362 | 0.433 | 0.413 | 0.36 | 0.41 | 0.19 | 0.09 | 0.22 | 0.28 | 0.19 |
| 12 | 0.2160 | 3/8 | 0.3750 | 0.362 | 0.433 | 0.413 | 0.36 | 0.41 | 0.19 | 0.09 | 0.22 | 0.31 | 0.22 |
| 1/4 | 0.2500 | 7/16 | 0.4375 | 0.428 | 0.505 | 0.488 | 0.41 | 0.47 | 0.22 | 0.11 | 0.25 | 0.34 | 0.25 |
| 5/16 | 0.3125 | 1/2 | 0.5000 | 0.489 | 0.577 | 0.557 | 0.47 | 0.53 | 0.25 | 0.12 | 0.28 | 0.41 | 0.31 |
| 3/8 | 0.3750 | 9/16 | 0.5625 | 0.551 | 0.650 | 0.628 | 0.53 | 0.62 | 0.28 | 0.14 | 0.33 | 0.45 | 0.38 |
| 7/16 | 0.4375 | 5/8 | 0.6250 | 0.612 | 0.722 | 0.698 | 0.59 | 0.69 | 0.31 | 0.16 | 0.36 | 0.52 | 0.44 |
| 1/2 | 0.5000 | 3/4 | 0.7500 | 0.736 | 0.866 | 0.840 | 0.72 | 0.81 | 0.38 | 0.19 | 0.42 | 0.59 | 0.50 |
| 9/16 | 0.5625 | 7/8 | 0.8750 | 0.861 | 1.010 | 0.982 | 0.84 | 0.94 | 0.44 | 0.22 | 0.50 | 0.69 | 0.56 |
| 5/8 | 0.6250 | 15/16 | 0.9375 | 0.922 | 1.083 | 1.051 | 0.91 | 1.00 | 0.47 | 0.23 | 0.53 | 0.75 | 0.62 |
| 3/4 | 0.7500 | 1 1/16 | 1.0625 | 1.045 | 1.227 | 1.191 | 1.03 | 1.16 | 0.53 | 0.27 | 0.59 | 0.88 | 0.75 |
| 7/8 | 0.8750 | 1 1/4 | 1.2500 | 1.231 | 1.443 | 1.403 | 1.22 | 1.36 | 0.62 | 0.31 | 0.70 | 1.00 | 0.88 |
| 1 | 1.0000 | 1 7/16 | 1.4375 | 1.417 | 1.660 | 1.615 | 1.41 | 1.55 | 0.72 | 0.36 | 0.81 | 1.12 | 1.00 |
| 1 1/8 | 1.1250 | 1 5/8 | 1.6250 | 1.602 | 1.876 | 1.826 | 1.59 | 1.75 | 0.81 | 0.41 | 0.92 | 1.31 | 1.12 |
| 1 1/4 | 1.2500 | 1 13/16 | 1.8125 | 1.788 | 2.093 | 2.038 | 1.78 | 1.95 | 0.91 | 0.45 | 1.03 | 1.44 | 1.25 |
| **High Crown** | | | | | | | | | | | | | |
| Nom. Size[a] or Basic Major Dia. of Thread | | Width Across Flats, *F* | | | Width Across Corners, *G* | | Body Dia., *A* | Overall Height, *H* | Hexagon Height, *Q* | Nose Radius, *R* | Body Radius, *S* | Drill Dep., *T* | Full Thd., *U* |
| | | Max. | (Basic) | Min. | Max. | Min. | | | | | | Max. | Min. |
| 6 | 0.1380 | 5/16 | 0.3125 | 0.302 | 0.361 | 0.344 | 0.30 | 0.42 | 0.17 | 0.05 | 0.25 | 0.28 | 0.19 |
| 8 | 0.1640 | 5/16 | 0.3125 | 0.302 | 0.361 | 0.344 | 0.30 | 0.42 | 0.17 | 0.05 | 0.25 | 0.28 | 0.19 |
| 10 | 0.1900 | 3/8 | 0.3750 | 0.362 | 0.433 | 0.413 | 0.36 | 0.52 | 0.20 | 0.06 | 0.30 | 0.34 | 0.25 |
| 12 | 0.2160 | 3/8 | 0.3750 | 0.362 | 0.433 | 0.413 | 0.36 | 0.52 | 0.20 | 0.06 | 0.30 | 0.38 | 0.28 |
| 1/4 | 0.2500 | 7/16 | 0.4375 | 0.428 | 0.505 | 0.488 | 0.41 | 0.59 | 0.23 | 0.06 | 0.34 | 0.41 | 0.31 |
| 5/16 | 0.3125 | 1/2 | 0.5000 | 0.489 | 0.577 | 0.557 | 0.47 | 0.69 | 0.28 | 0.08 | 0.41 | 0.47 | 0.38 |
| 3/8 | 0.3750 | 9/16 | 0.5625 | 0.551 | 0.650 | 0.628 | 0.53 | 0.78 | 0.31 | 0.09 | 0.44 | 0.56 | 0.47 |
| 7/16 | 0.4375 | 5/8 | 0.6250 | 0.612 | 0.722 | 0.698 | 0.59 | 0.88 | 0.34 | 0.09 | 0.50 | 0.62 | 0.53 |
| 1/2 | 0.5000 | 3/4 | 0.7500 | 0.736 | 0.866 | 0.840 | 0.72 | 1.03 | 0.42 | 0.12 | 0.59 | 0.75 | 0.62 |
| 9/16 | 0.5625 | 7/8 | 0.8750 | 0.861 | 1.010 | 0.982 | 0.84 | 1.19 | 0.48 | 0.16 | 0.69 | 0.81 | 0.69 |
| 5/8 | 0.6250 | 15/16 | 0.9375 | 0.922 | 1.083 | 1.051 | 0.91 | 1.28 | 0.53 | 0.16 | 0.75 | 0.91 | 0.78 |
| 3/4 | 0.7500 | 1 1/16 | 1.0625 | 1.045 | 1.227 | 1.191 | 1.03 | 1.45 | 0.59 | 0.17 | 0.84 | 1.06 | 0.94 |
| 7/8 | 0.8750 | 1 1/4 | 1.12500 | 1.231 | 1.443 | 1.403 | 1.22 | 1.72 | 0.70 | 0.20 | 0.98 | 1.22 | 1.09 |
| 1 | 1.0000 | 1 7/16 | 1.4375 | 1.417 | 1.660 | 1.615 | 1.41 | 1.97 | 0.81 | 0.23 | 1.14 | 1.38 | 1.25 |
| 1 1/8 | 1.1250 | 1 5/8 | 1.6250 | 1.602 | 1.876 | 1.826 | 1.59 | 2.22 | 0.92 | 0.27 | 1.28 | 1.59 | 1.41 |
| 1 1/4 | 1.2500 | 1 13/16 | 1.8125 | 1.788 | 2.093 | 2.038 | 1.78 | 2.47 | 1.03 | 0.28 | 1.44 | 1.75 | 1.56 |

[a] When specifying a nominal size in decimals, any zero in the fourth decimal place is omitted.



All dimensions are in inches. Threads are Unified Standard Class 2B, UNC or UNF Series.

## Hex High and Hex Slotted High Nuts *SAE J482a*

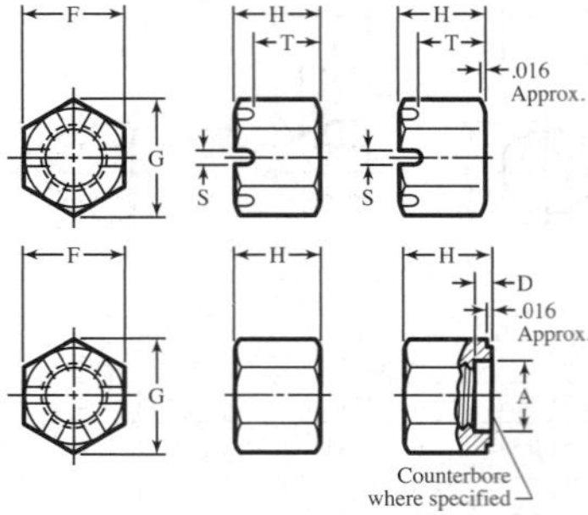


| Nominal Size[a] or Basic Major Diameter of Thread | | Width Across Flats, *F* | | | Width Across Corners, *G* | | Slot Width, *S* | |
|---|---|---|---|---|---|---|---|---|
| | | Basic | Max. | Min. | Max. | Min. | Min. | Max. |
| 1⁄4 | 0.2500 | 7⁄16 | 0.4375 | 0.428 | 0.505 | 0.488 | 0.07 | 0.10 |
| 5⁄16 | 0.3125 | 1⁄2 | 0.5000 | 0.489 | 0.577 | 0.557 | 0.09 | 0.12 |
| 3⁄8 | 0.3750 | 9⁄16 | 0.5625 | 0.551 | 0.650 | 0.628 | 0.12 | 0.15 |
| 7⁄16 | 0.4375 | 11⁄16 | 0.6875 | 0.675 | 0.794 | 0.768 | 0.12 | 0.15 |
| 1⁄2 | 0.5000 | 3⁄4 | 0.7500 | 0.736 | 0.866 | 0.840 | 0.15 | 0.18 |
| 9⁄16 | 0.5625 | 7⁄8 | 0.8750 | 0.861 | 1.010 | 0.982 | 0.15 | 0.18 |
| 5⁄8 | 0.6250 | 15⁄16 | 0.9375 | 0.922 | 1.083 | 1.051 | 0.18 | 0.24 |
| 3⁄4 | 0.7500 | 1 1⁄8 | 1.1250 | 1.088 | 1.299 | 1.240 | 0.18 | 0.24 |
| 7⁄8 | 0.8750 | 1 5⁄16 | 1.3125 | 1.269 | 1.516 | 1.447 | 0.18 | 0.24 |
| 1 | 1.0000 | 1 1⁄2 | 1.5000 | 1.450 | 1.732 | 1.653 | 0.24 | 0.30 |
| 1 1⁄8 | 1.1250 | 1 11⁄16 | 1.6875 | 1.631 | 1.949 | 1.859 | 0.24 | 0.33 |
| 1 1⁄4 | 1.2500 | 1 7⁄8 | 1.8750 | 1.812 | 2.165 | 2.066 | 0.31 | 0.40 |
| Nominal Size[a] or Basic Major Diameter of Thread | | Thickness, *H* | | | Unslotted Thickness, *T* | | Counterbore (Optional) | |
| | | Basic | Max. | Min. | Max. | Min. | Dia., *A* | Depth, *D* |
| 1⁄4 | 0.2500 | 3⁄8 | 0.382 | 0.368 | 0.29 | 0.27 | 0.266 | 0.062 |
| 5⁄16 | 0.3125 | 29⁄64 | 0.461 | 0.445 | 0.37 | 0.35 | 0.328 | 0.078 |
| 3⁄8 | 0.3750 | 1⁄2 | 0.509 | 0.491 | 0.38 | 0.36 | 0.391 | 0.094 |
| 7⁄16 | 0.4375 | 39⁄64 | 0.619 | 0.599 | 0.46 | 0.44 | 0.453 | 0.109 |
| 1⁄2 | 0.5000 | 21⁄32 | 0.667 | 0.645 | 0.51 | 0.49 | 0.516 | 0.125 |
| 9⁄16 | 0.5625 | 49⁄64 | 0.778 | 0.754 | 0.59 | 0.57 | 0.594 | 0.141 |
| 5⁄8 | 0.6250 | 27⁄32 | 0.857 | 0.831 | 0.63 | 0.61 | 0.656 | 0.156 |
| 3⁄4 | 0.7500 | 1 | 1.015 | 0.985 | 0.76 | 0.73 | 0.781 | 0.188 |
| 7⁄8 | 0.8750 | 1 5⁄32 | 1.172 | 1.140 | 0.92 | 0.89 | 0.906 | 0.219 |
| 1 | 1.0000 | 1 5⁄16 | 1.330 | 1.292 | 1.05 | 1.01 | 1.031 | 0.250 |
| 1 1⁄8 | 1.1250 | 1 1⁄2 | 1.520 | 1.480 | 1.18 | 1.14 | 1.156 | 0.281 |
| 1 1⁄4 | 1.2500 | 1 11⁄16 | 1.710 | 1.666 | 1.34 | 1.29 | 1.281 | 0.312 |

[a] When specifying a nominal size in decimals, any zero in the fourth decimal place is omitted.

All dimensions are in inches. Threads are Unified Standard Class 2B, UNC or UNF Series.

## American National Standard Round Head and Round Head Square Neck Bolts
*ANSI/ASME B18.5-2012 (R2023)*

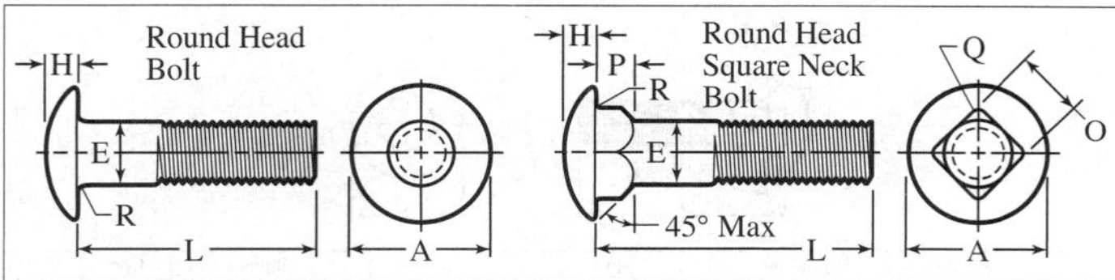


| Nominal Size | Body Dia., *E* | | Dia. of Head, *A* | | Height of Head, *H* | | Fillet Rad., *R* | Width of Square, *O* | | Depth of Square, *P* | | Corner Rad. on Square, *Q* |
|---|---|---|---|---|---|---|---|---|---|---|---|---|
| | Max. | Min. | Max. | Min. | Max. | Min. | Max. | Max. | Min. | Max. | Min. | Max. |
| No. 10 | 0.199 | .182 | .469 | .438 | .114 | .094 | .031 | .199 | .185 | .125 | .094 | .031 |
| ¼ | 0.260 | .237 | .594 | .563 | .145 | .125 | .031 | .260 | .245 | .156 | .125 | .031 |
| 5⁄16 | 0.324 | .298 | .719 | .688 | .176 | .156 | .031 | .324 | .307 | .187 | .156 | .031 |
| ⅜ | 0.388 | .360 | .844 | .782 | .208 | .188 | .031 | .388 | .368 | .219 | .188 | .047 |
| 7⁄16 | 0.452 | .421 | .969 | .907 | .239 | .219 | .031 | .452 | .431 | .250 | .219 | .047 |
| ½ | 0.515 | .483 | 1.094 | 1.032 | .270 | .250 | .031 | .515 | .492 | .281 | .250 | .047 |
| ⅝ | 0.642 | .605 | 1.344 | 1.219 | .344 | .313 | .062 | .642 | .616 | .344 | .313 | .078 |
| ¾ | 0.768 | .729 | 1.594 | 1.469 | .406 | .375 | .062 | .768 | .741 | .406 | .375 | .078 |
| ⅞ | 0.895 | .852 | 1.844 | 1.719 | .469 | .438 | .062 | .895 | .865 | .469 | .438 | .094 |
| 1 | 1.022 | .976 | 2.094 | 1.969 | .531 | .500 | .062 | 1.022 | .990 | .531 | .500 | .094 |

All dimensions are in inches unless otherwise specified.

Threads are Unified Standard, Class 2A, UNC Series, in accordance with ANSI/ASME B1.1. For threads with additive finish, the maximum diameters of Class 2A shall apply before plating or coating, whereas the basic diameters (Class 2A maximum diameters plus the allowance) shall apply to a bolt after plating or coating.

Bolts are designated in the sequence shown: nominal size (number, fraction or decimal equivalent); threads per inch; nominal length (fraction or decimal equivalent); product name; material; and protective finish, if required.

i.e.: ½-13 × 3 Round Head Square Neck Bolt, Steel .375-16 × 2.50 Step Bolt, Steel, Zinc Plated

## American National Standard T-Head Bolts *ANSI/ASME B18.5-2012 (R2023)*

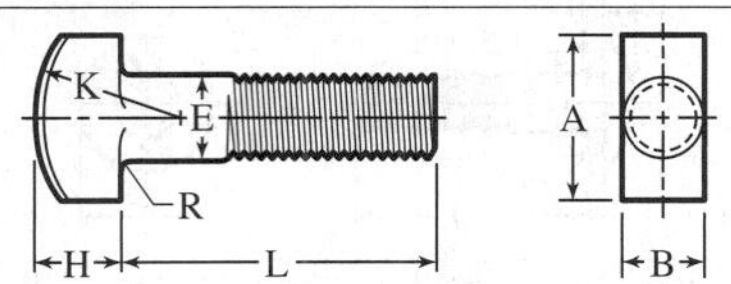


| Nom. Size[a] or Basic Bolt Dia. | | Body Dia., *E* | | Head Length, *A* | | Head Width, *B* | | Head Height, *H* | | Head Rad., *K* | Fillet Rad., *R* |
|---|---|---|---|---|---|---|---|---|---|---|---|
| | | Max. | Min. | Max. | Min. | Max. | Min. | Max. | Min. | Basic | Max. |
| ¼ | 0.2500 | .260 | .237 | .500 | .488 | .280 | .245 | .204 | .172 | .438 | .031 |
| 5⁄16 | 0.3125 | .324 | .298 | .625 | .609 | .342 | .307 | .267 | .233 | .500 | .031 |
| ⅜ | 0.3750 | .388 | .360 | .750 | .731 | .405 | .368 | .331 | .295 | .625 | .031 |
| 7⁄16 | 0.4375 | .452 | .421 | .875 | .853 | .468 | .431 | .394 | .356 | .875 | .031 |
| ½ | 0.5000 | .515 | .483 | 1.000 | .975 | .530 | .492 | .458 | .418 | .875 | .031 |
| ⅝ | 0.6250 | .642 | .605 | 1.250 | 1.218 | .675 | .616 | .585 | .541 | 1.062 | .062 |
| ¾ | 0.7500 | .768 | .729 | 1.500 | 1.462 | .800 | .741 | .649 | .601 | 1.250 | .062 |
| ⅞ | 0.8750 | .895 | .852 | 1.750 | 1.706 | .938 | .865 | .776 | .724 | 1.375 | .062 |
| 1 | 1.0000 | 1.022 | .976 | 2.000 | 1.950 | 1.063 | .990 | .903 | .847 | 1.500 | .062 |

[a] Where specifying nominal size in decimals, zeros preceding the decimal point and in the fourth decimal place are omitted. For information as to threads and method of bolt designation, see footnotes to preceding table.

All dimensions are given in inches.

## American National Standard Round Head Short Square Neck Bolts

*ANSI/ASME B18.5-2012 (R2023)*

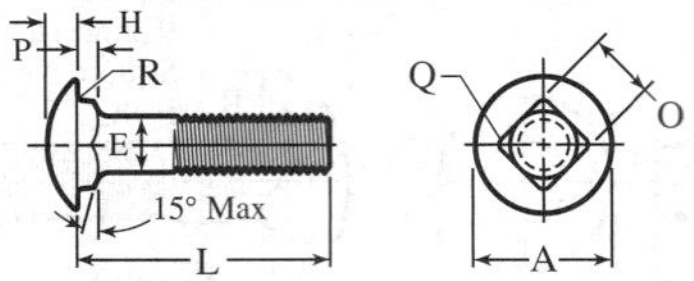


| Nominal Size | Body Dia., *E* | | Head Dia., *A* | | Head Height, *H* | | Square Width, *O* | | Square Depth, *P* | | Cor. Rad. on Sq., *Q* | Fillet Rad., *R* |
|---|---|---|---|---|---|---|---|---|---|---|---|---|
| | Max. | Min. | Max. | Min. | Max. | Min. | Max. | Min. | Max. | Min. | Max. | Max. |
| ¼ | 0.260 | 0.213 | 0.594 | 0.563 | 0.145 | 0.125 | 0.260 | 0.245 | 0.124 | 0.093 | 0.031 | 0.031 |
| 5⁄16 | 0.324 | 0.272 | 0.719 | 0.688 | 0.176 | 0.156 | 0.324 | 0.307 | 0.124 | 0.093 | 0.031 | 0.031 |
| 3⁄8 | 0.388 | 0.329 | 0.844 | 0.782 | 0.208 | 0.188 | 0.388 | 0.368 | 0.156 | 0.125 | 0.047 | 0.031 |
| 7⁄16 | 0.452 | 0.385 | 0.969 | 0.907 | 0.239 | 0.219 | 0.452 | 0.431 | 0.156 | 0.125 | 0.047 | 0.031 |
| ½ | 0.515 | 0.444 | 1.094 | 1.032 | 0.270 | 0.250 | 0.515 | 0.492 | 0.156 | 0.125 | 0.047 | 0.031 |
| 5⁄8 | 0.642 | 0.559 | 1.344 | 1.219 | 0.344 | 0.313 | 0.642 | 0.616 | 0.218 | 0.187 | 0.078 | 0.062 |
| ¾ | 0.768 | 0.678 | 1.594 | 1.469 | 0.406 | 0.375 | 0.768 | 0.741 | 0.218 | 0.187 | 0.078 | 0.062 |

All dimensions are given in inches.

Threads are Unified Standard, Class 2A, UNC Series, in accordance with ANSI/ASME B1.1. For threads with additive finish, the maximum diameters of Class 2A apply before plating or coating, whereas the basic diameters (Class 2A maximum diameters plus the allowance) apply to a bolt after plating or coating.

Bolts are designated in the sequence shown: nominal size (number, fraction or decimal equivalent); threads per inch; nominal length (fraction or decimal equivalent); product name; material; and protective finish, if required. For example,

½-13 × 3 Round Head Short Square Neck Bolt, Steel

.375-16 × 2.50 Round Head Short Square Neck Bolt, Steel, Zinc Plated

## American National Standard Round Head Fin Neck Bolts

*ANSI/ASME B18.5-2012 (R2023)*

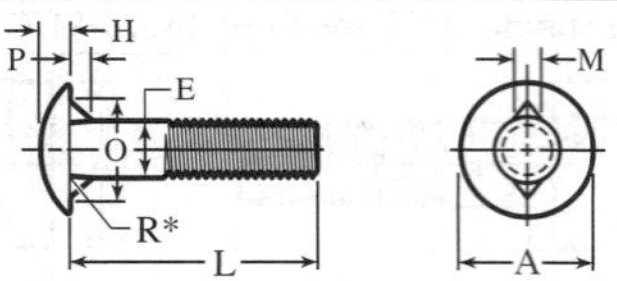


| Nominal Size | Body Dia., *E* | | Head Dia., *A* | | Head Height, *H* | | Fin Thickness, *M* | | Dist. Across Fins, *O* | | Fin Depth, *P* | |
|---|---|---|---|---|---|---|---|---|---|---|---|---|
| | Max. | Min. | Max. | Min. | Max. | Min. | Max. | Min. | Max. | Min. | Max. | Min. |
| No. 10 | 0.199 | 0.182 | 0.469 | 0.438 | 0.114 | 0.094 | 0.098 | 0.078 | 0.395 | 0.375 | 0.088 | 0.078 |
| ¼ | 0.260 | 0.237 | 0.594 | 0.563 | 0.145 | 0.125 | 0.114 | 0.094 | 0.458 | 0.438 | 0.104 | 0.094 |
| 5⁄16 | 0.324 | 0.298 | 0.719 | 0.688 | 0.176 | 0.156 | 0.145 | 0.125 | 0.551 | 0.531 | 0.135 | 0.125 |
| 3⁄8 | 0.388 | 0.360 | 0.844 | 0.782 | 0.208 | 0.188 | 0.161 | 0.141 | 0.645 | 0.625 | 0.151 | 0.141 |
| 7⁄16 | 0.452 | 0.421 | 0.969 | 0.907 | 0.239 | 0.219 | 0.192 | 0.172 | 0.739 | 0.719 | 0.182 | 0.172 |
| ½ | 0.515 | 0.483 | 1.094 | 1.032 | 0.270 | 0.250 | 0.208 | 0.188 | 0.833 | 0.813 | 0.198 | 0.188 |

All dimensions are given in inches unless otherwise specified.

*Maximum fillet radius *R* is 0.031 inch for all sizes.

For information as to threads and method of bolt designation, see footnotes to the preceding table.

## American National Standard Round Head Ribbed Neck Bolts *ANSI/ASME B18.5-2012 (R2023)*

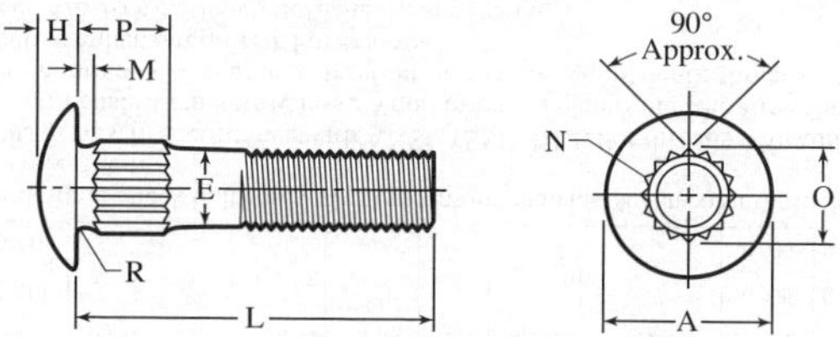


| Nominal Size[a] or Basic Bolt Diameter | | Body Diameter, *E* | | Head Diameter, *A* | | Head Height, *H* | | Head to Ribs, *M* For Lengths of | | Number of Ribs, *N* | Dia. Over Ribs, *O* | Depth Over Ribs, *P* For Lengths of | | | Fillet Radius, *R* |
|---|---|---|---|---|---|---|---|---|---|---|---|---|---|---|---|
| | | | | | | | | ⅞ in. and Shorter | 1 in. and Longer | | | ⅞ in. and Shorter | 1 in. and 1⅛ in. | ¼ in. and Longer | |
| | | Max. | Min. | Max. | Min. | Max. | Min. | ±0.031[b] | | Approx. | Min. | ±0.031 | | | Max.[c] |
| No. 10 | 0.1900 | 0.199 | 0.182 | 0.469 | 0.438 | 0.114 | 0.094 | 0.031† | 0.063 | 9 | 0.210 | 0.250 | 0.407 | 0.594 | 0.031 |
| ¼ | 0.2500 | 0.260 | 0.237 | 0.594 | 0.563 | 0.145 | 0.125 | 0.031† | 0.063 | 10 | 0.274 | 0.250 | 0.407 | 0.594 | 0.031 |
| 5⁄16 | 0.3125 | 0.324 | 0.298 | 0.719 | 0.688 | 0.176 | 0.156 | 0.031† | 0.063 | 12 | 0.340 | 0.250 | 0.407 | 0.594 | 0.031 |
| ⅜ | 0.3750 | 0.388 | 0.360 | 0.844 | 0.782 | 0.208 | 0.188 | 0.031† | 0.063 | 12 | 0.405 | 0.250 | 0.407 | 0.594 | 0.031 |
| 7⁄16 | 0.4375 | 0.452 | 0.421 | 0.969 | 0.907 | 0.239 | 0.219 | 0.031† | 0.063 | 14 | 0.470 | 0.250 | 0.407 | 0.594 | 0.031 |
| ½ | 0.5000 | 0.515 | 0.483 | 1.094 | 1.032 | 0.270 | 0.250 | 0.031† | 0.063 | 16 | 0.534 | 0.250 | 0.407 | 0.594 | 0.031 |
| ⅝ | 0.6250 | 0.642 | 0.605 | 1.344 | 1.219 | 0.344 | 0.313 | 0.094 | 0.094 | 19 | 0.660 | 0.313 | 0.438 | 0.625 | 0.062 |
| ¾ | 0.7500 | 0.768 | 0.729 | 1.594 | 1.469 | 0.406 | 0.375 | 0.094 | 0.094 | 22 | 0.785 | 0.313 | 0.438 | 0.625 | 0.062 |

[a] Where specifying nominal size in decimals, zeros preceding decimal and in the fourth decimal place shall be omitted.
[b] Tolerance on the No. 10 through ½ in. sizes for nominal lengths ⅞ in. and shorter shall be +0.031 and −0.000.
[c] The minimum radius is one half of the value shown.
All dimensions are given in inches unless otherwise specified.
For information as to threads and method of designating bolts, see following table.

## American National Standard Step and 114-Degree Countersunk Square Neck Bolts
*ANSI/ASME B18.5-2012 (R2023)*

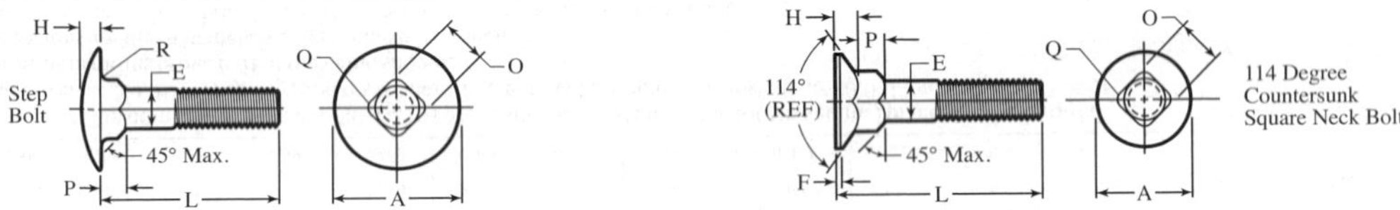


| Nominal Size | Step & 114° Countersunk Bolts | | | | | Step Bolts | | | | | | | 114° Countersunk Square Neck Bolts | | | | | | |
|---|---|---|---|---|---|---|---|---|---|---|---|---|---|---|---|---|---|---|---|
| | Body Dia., *E* | | Corner Rad. on Square, *Q* | Width of Square, *O* | | Depth of Square, *P* | | Dia. of Head, *A* | | Height of Head, *H* | | Fillet Radius, *R* | Depth of Square, *P* | | Dia. of Head, *A* | | Flat on Head, *F* | Height of Head, *H* | |
| | Max. | Min. | Max. | Max. | Min. | Max. | Min. | Max. | Min. | Max. | Min. | Max. | Max. | Min. | Max. | Min. | Min. | Max. | Min. |
| No. 10 | 0.199 | 0.182 | 0.031 | 0.199 | 0.185 | 0.125 | 0.094 | 0.656 | 0.625 | 0.114 | 0.094 | 0.031 | 0.125 | 0.094 | 0.548 | 0.500 | 0.015 | 0.131 | 0.112 |
| ¼ | 0.260 | 0.237 | 0.031 | 0.260 | 0.245 | 0.156 | 0.125 | 0.844 | 0.813 | 0.145 | 0.125 | 0.031 | 0.156 | 0.125 | 0.682 | 0.625 | 0.018 | 0.154 | 0.135 |
| 5⁄16 | 0.324 | 0.298 | 0.031 | 0.324 | 0.307 | 0.187 | 0.156 | 1.031 | 1.000 | 0.176 | 0.156 | 0.031 | 0.219 | 0.188 | 0.821 | 0.750 | 0.023 | 0.184 | 0.159 |
| ⅜ | 0.388 | 0.360 | 0.047 | 0.388 | 0.368 | 0.219 | 0.188 | 1.219 | 1.188 | 0.208 | 0.188 | 0.031 | 0.250 | 0.219 | 0.960 | 0.875 | 0.027 | 0.212 | 0.183 |
| 7⁄16 | 0.452 | 0.421 | 0.047 | 0.452 | 0.431 | 0.250 | 0.219 | 1.406 | 1.375 | 0.239 | 0.219 | 0.031 | 0.281 | 0.250 | 1.093 | 1.000 | 0.030 | 0.235 | 0.205 |
| ½ | 0.515 | 0.483 | 0.047 | 0.515 | 0.492 | 0.281 | 0.250 | 1.594 | 1.563 | 0.270 | 0.250 | 0.031 | 0.312 | 0.281 | 1.233 | 1.125 | 0.035 | 0.265 | 0.229 |
| ⅝[a] | 0.642 | 0.605 | 0.078 | 0.642 | 0.616 | … | … | … | … | … | … | … | 0.406 | 0.375 | 1.495 | 1.375 | 0.038 | 0.316 | 0.272 |
| ¾[a] | 0.768 | 0.729 | 0.078 | 0.768 | 0.741 | … | … | … | … | … | … | … | 0.500 | 0.469 | 10.754 | 1.625 | 0.041 | 0.368 | 0.314 |

[a] These sizes pertain to 114 degree countersunk square neck bolts only. Dimensions given in last seven columns to the right are for these bolts only.

All dimensions are in inches unless otherwise specified.

Threads are Unified Standard, Class 2A, UNC Series, in accordance with ANSI/ASME B1.1. For threads with additive finish, the maximum diameters of Class 2A shall apply before plating or coating, whereas the basic diameters (Class 2A maximum diameters plus the allowance) shall apply to a bolt after plating or coating.

Bolts are designated in the sequence shown: nominal size (number, fraction or decimal equivalent); threads per inch; nominal length (fraction or decimal equivalent); product name; material; and protective finish, if required. For example

½-13 × 3 Round Head Square Neck Bolt, Steel .375-16 × 2.50 Step Bolt, Steel, Zinc Plated

**American National Standard Countersunk Bolts and Slotted Countersunk Bolts**
*ANSI/ASME B18.5-2012 (R2023)*

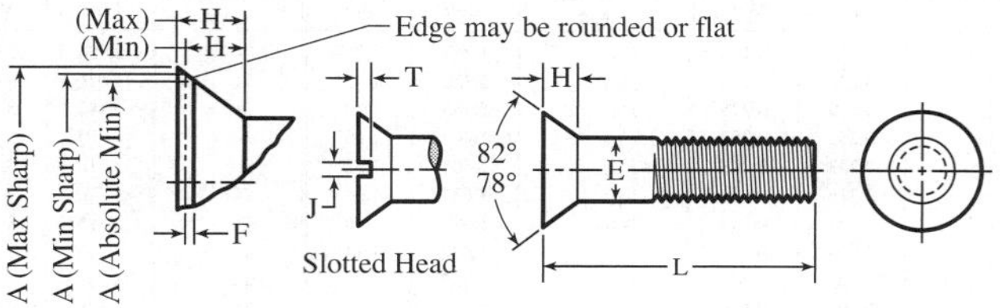


| Nominal Size[a] or Basic Bolt Diameter | | Body Diameter, *E* | | Head Diameter, *A* | | | Flat on Min. Dia., Head, *F*[b] |
|---|---|---|---|---|---|---|---|
| | | Max. | Min. | Max. Edge Sharp | Min. Edge Sharp | Absolute Min. Edge Rounded or Flat | Max. |
| 1/4 | 0.2500 | 0.260 | 0.237 | 0.493 | 0.477 | 0.445 | 0.018 |
| 5/16 | 0.3125 | 0.324 | 0.298 | 0.618 | 0.598 | 0.558 | 0.023 |
| 3/8 | 0.3750 | 0.388 | 0.360 | 0.740 | 0.715 | 0.668 | 0.027 |
| 7/16 | 0.4375 | 0.452 | 0.421 | 0.803 | 0.778 | 0.726 | 0.030 |
| 1/2 | 0.5000 | 0.515 | 0.483 | 0.935 | 0.905 | 0.845 | 0.035 |
| 5/8 | 0.6250 | 0.642 | 0.605 | 1.169 | 1.132 | 1.066 | 0.038 |
| 3/4 | 0.7500 | 0.768 | 0.729 | 1.402 | 1.357 | 1.285 | 0.041 |
| 7/8 | 0.8750 | 0.895 | 0.852 | 1.637 | 1.584 | 1.511 | 0.042 |
| 1 | 1.0000 | 1.022 | 0.976 | 1.869 | 1.810 | 1.735 | 0.043 |
| 1 1/8 | 1.1250 | 1.149 | 1.098 | 2.104 | 2.037 | 1.962 | 0.043 |
| 1 1/4 | 1.2500 | 1.277 | 1.223 | 2.337 | 2.262 | 2.187 | 0.043 |
| 1 3/8 | 1.3750 | 1.404 | 1.345 | 2.571 | 2.489 | 2.414 | 0.043 |
| 1 1/2 | 1.5000 | 1.531 | 1.470 | 2.804 | 2.715 | 2.640 | 0.043 |
| Nom. Size or Basic Bolt Dia. | | Head Height, *H* | | Slot Width, *J* | | Slot Depth, *T* | |
| | | Max.[c] | Min.[d] | Max. | Min. | Max. | Min. |
| 1/4 | 0.2500 | 0.150 | 0.131 | 0.075 | 0.064 | 0.068 | 0.045 |
| 5/16 | 0.3125 | 0.189 | 0.164 | 0.084 | 0.072 | 0.086 | 0.057 |
| 3/8 | 0.3750 | 0.225 | 0.196 | 0.094 | 0.081 | 0.103 | 0.068 |
| 7/16 | 0.4375 | 0.226 | 0.196 | 0.094 | 0.081 | 0.103 | 0.068 |
| 1/2 | 0.5000 | 0.269 | 0.233 | 0.106 | 0.091 | 0.103 | 0.068 |
| 5/8 | 0.6250 | 0.336 | 0.292 | 0.133 | 0.116 | 0.137 | 0.091 |
| 3/4 | 0.7500 | 0.403 | 0.349 | 0.149 | 0.131 | 0.171 | 0.115 |
| 7/8 | 0.8750 | 0.470 | 0.408 | 0.167 | 0.147 | 0.206 | 0.138 |
| 1 | 1.0000 | 0.537 | 0.466 | 0.188 | 0.166 | 0.240 | 0.162 |
| 1 1/8 | 1.1250 | 0.604 | 0.525 | 0.196 | 0.178 | 0.257 | 0.173 |
| 1 1/4 | 1.2500 | 0.671 | 0.582 | 0.211 | 0.193 | 0.291 | 0.197 |
| 1 3/8 | 1.3750 | 0.738 | 0.641 | 0.226 | 0.208 | 0.326 | 0.220 |
| 1 1/2 | 1.5000 | 0.805 | 0.698 | 0.258 | 0.240 | 0.360 | 0.244 |

[a] Where specifying size in decimals, zeros preceding decimal and in fourth decimal place omitted.

[b] Flat on minimum diameter head calculated on minimum sharp and absolute minimum head diameters and 82° head angle.

[c] Maximum head height calculated on maximum sharp head diameter, basic bolt diameter, and 78° head angle.

[d] Minimum head height calculated on minimum sharp head diameter, basic bolt diameter, and 82° head angle.

All dimensions are given in inches. For thread information and method of bolt designation see footnotes to previous table. Heads are unslotted unless otherwise specified. For slot dimensions see Table 1 in *Slotted Head Cap Screws* on page 1815.

**Wrench Openings and Clearances.—**Wrench openings for nuts are given in Table 1, clearances for open end engineers wrenches in Table 2, clearances for single and double hexagon socket wrenches in Table 3a (inch) and Table 3b (metric), clearances for 12-point box wrenches (inch and metric) are given in Table 4. They are based on the dimensions across the flats of the fastener.

**Table 1. Wrench Openings for Nuts** *ANSI/ASME B18.2.2-2022, Appendix*

| Max.[a] Width Across Flats of Nut | Wrench Opening[b] | | Max.[a] Width Across Flats of Nut | Wrench Opening[b] | | Max.[a] Width Across Flats of Nut | Wrench Opening[b] | |
|---|---|---|---|---|---|---|---|---|
| | Min. | Max. | | Min. | Max. | | Min. | Max. |
| 5/32 | 0.158 | 0.163 | 1 1/4 | 1.257 | 1.267 | 2 15/16 | 2.954 | 2.973 |
| 3/16 | 0.190 | 0.195 | 1 5/16 | 1.320 | 1.331 | *3* | *3.016* | *3.035* |
| 7/32 | 0.220 | 0.225 | *1 3/8* | *1.383* | *1.394* | 3 1/8 | 3.142 | 3.162 |
| 1/4 | 0.252 | 0.257 | 1 7/16 | 1.446 | 1.457 | *3 3/8* | *3.393* | *3.414* |
| 9/32 | 0.283 | 0.288 | 1 1/2 | 1.508 | 1.520 | 3 1/2 | 3.518 | 3.540 |
| 5/16 | 0.316 | 0.322 | 1 5/8 | 1.634 | 1.646 | *3 3/4* | *3.770* | *3.793* |
| 11/32 | 0.347 | 0.353 | 1 11/16 | 1.696 | 1.708 | 3 7/8 | 3.895 | 3.918 |
| 3/8 | 0.378 | 0.384 | 1 13/16 | 1.822 | 1.835 | *4 1/8* | *4.147* | *4.172* |
| 7/16 | 0.440 | 0.446 | 1 7/8 | 1.885 | 1.898 | 4 1/4 | 4.272 | 4.297 |
| 1/2 | 0.504 | 0.510 | 2 | 2.011 | 2.025 | 4 1/2 | 4.524 | 4.550 |
| 9/16 | 0.566 | 0.573 | 2 1/16 | 2.074 | 2.088 | 4 5/8 | 4.649 | 4.676 |
| 5/8 | 0.629 | 0.636 | 2 3/16 | 2.200 | 2.215 | *4 7/8* | *4.900* | *4.928* |
| 11/16 | 0.692 | 0.699 | 2 1/4 | 2.262 | 2.277 | 5 | 5.026 | 5.055 |
| 3/4 | 0.755 | 0.763 | 2 3/8 | 2.388 | 2.404 | *5 1/4* | *5.277* | *5.307* |
| 13/16 | 0.818 | 0.826 | *2 7/16* | *2.450* | *2.466* | 5 3/8 | 5.403 | 5.434 |
| 7/8 | 0.880 | 0.888 | 2 9/16 | 2.576 | 2.593 | *5 5/8* | *5.654* | *5.686* |
| 15/16 | 0.944 | 0.953 | *2 5/8* | *2.639* | *2.656* | 5 3/4 | 5.780 | 5.813 |
| 1 | 1.006 | 1.015 | 2 3/4 | 2.766 | 2.783 | *6* | *6.031* | 6.157 |
| 1 1/16 | 1.068 | 1.077 | *2 13/16* | *2.827* | *2.845* | 6 1/8 | 6.065 | 6.192 |
| 1 1/8 | 1.132 | 1.142 | | | | | | |

[a] Wrenches are marked with the "Nominal Size of Wrench," which is equal to the basic or maximum width across flats of the corresponding nut. Minimum wrench opening is (1.005W + 0.001). Tolerance on wrench opening is (0.005W + 0.004) from minimum, where W equals nominal size of wrench.

[b] Openings for 5/32 to 3/8 widths from old ASA B18.2-1960 and italic values are from former ANSI B18.2.2-1972.

All dimensions given in inches.

**Table 2. Clearances for Open End Engineers Wrench (15°)**

| Nominal Wrench Size | | *A* Min. (in.) | *B*[a] Max. (in.) | *C* Min. (in.) | *D* Min. (in.) | *E* Min. (in.) | *F*[b] Max. (in.) | *G* Ref. (in.) | *H*[c] Max. (in.) | *J* Min.[d] in.-lbf |
|---|---|---|---|---|---|---|---|---|---|---|
| 5/32 | 0.156 | 0.220 | 0.250 | 0.390 | 0.160 | 0.250 | 0.200 | 0.030 | 0.094 | 35 |
| 3/16 | 0.188 | 0.250 | 0.280 | 0.430 | 0.190 | 0.270 | 0.230 | 0.030 | 0.172 | 45 |
| 1/4 | 0.250 | 0.280 | 0.340 | 0.530 | 0.270 | 0.310 | 0.310 | 0.030 | 0.172 | 67 |
| 5/16 | 0.313 | 0.380 | 0.470 | 0.660 | 0.280 | 0.390 | 0.390 | 0.050 | 0.203 | 138 |
| 11/32 | 0.344 | 0.420 | 0.500 | 0.750 | 0.340 | 0.450 | 0.450 | 0.050 | 0.203 | 193 |
| 3/8 | 0.375 | 0.420 | 0.500 | 0.780 | 0.360 | 0.450 | 0.520 | 0.050 | 0.219 | 275 |
| 7/16 | 0.438 | 0.470 | 0.590 | 0.890 | 0.420 | 0.520 | 0.640 | 0.050 | 0.250 | 413 |
| 1/2 | 0.500 | 0.520 | 0.640 | 1.000 | 0.470 | 0.580 | 0.660 | 0.050 | 0.266 | 550 |
| 9/16 | 0.563 | 0.590 | 0.770 | 1.130 | 0.520 | 0.660 | 0.700 | 0.050 | 0.297 | 770 |
| 5/8 | 0.625 | 0.640 | 0.830 | 1.230 | 0.550 | 0.700 | 0.700 | 0.050 | 0.344 | 1100 |
| 11/16 | 0.688 | 0.770 | 0.920 | 1.470 | 0.660 | 0.880 | 0.800 | 0.060 | 0.375 | 1375 |
| 3/4 | 0.750 | 0.770 | 0.920 | 1.510 | 0.670 | 0.880 | 0.800 | 0.060 | 0.375 | 1650 |
| 13/16 | 0.813 | 0.910 | 1.120 | 1.660 | 0.720 | 0.970 | 0.860 | 0.060 | 0.406 | 2200 |
| 7/8 | 0.875 | 0.970 | 1.150 | 1.810 | 0.800 | 1.060 | 0.910 | 0.060 | 0.438 | 2475 |
| 15/16 | 0.938 | 0.970 | 1.150 | 1.850 | 0.810 | 1.060 | 0.950 | 0.060 | 0.438 | 3025 |
| 1 | 1.000 | 1.050 | 1.230 | 2.000 | 0.880 | 1.160 | 1.060 | 0.060 | 0.500 | 3575 |
| 1 1/16 | 1.063 | 1.090 | 1.250 | 2.100 | 0.970 | 1.200 | 1.200 | 0.080 | 0.500 | 3850 |
| 1 1/8 | 1.125 | 1.140 | 1.370 | 2.210 | 1.000 | 1.270 | 1.230 | 0.080 | 0.500 | 4400 |
| 1 1/4 | 1.250 | 1.270 | 1.420 | 2.440 | 1.080 | 1.390 | 1.310 | 0.080 | 0.562 | 5775 |
| 1 5/16 | 1.313 | 1.390 | 1.690 | 2.630 | 1.170 | 1.520 | 1.340 | 0.080 | 0.562 | 6600 |
| 1 7/16 | 1.438 | 1.470 | 1.720 | 2.800 | 1.250 | 1.590 | 1.340 | 0.090 | 0.641 | 8250 |
| 1 1/2 | 1.500 | 1.470 | 1.720 | 2.840 | 1.270 | 1.590 | 1.450 | 0.090 | 0.641 | 8500 |
| 1 5/8 | 1.625 | 1.560 | 1.880 | 3.100 | 1.380 | 1.750 | 1.560 | 0.090 | 0.641 | 9000 |

[a] *B* = arc radius created by the swing of the wrench.

[b] *F* = inside arc radius of part.

[c] *H* = thickness of wrench head. (Dimension line not shown.)

[d] *J* = torque that wrench will withstand in inch-pounds. Values updated from ANSI/ASME B107, 100-2020, Flat Wrenches.

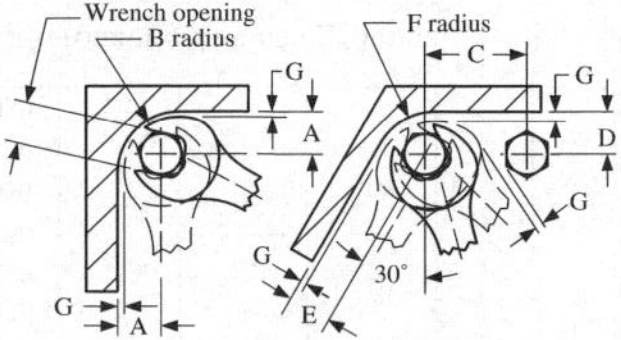


Fig. 1. Clearances for Open End Engineers Wrench (See Table 2)

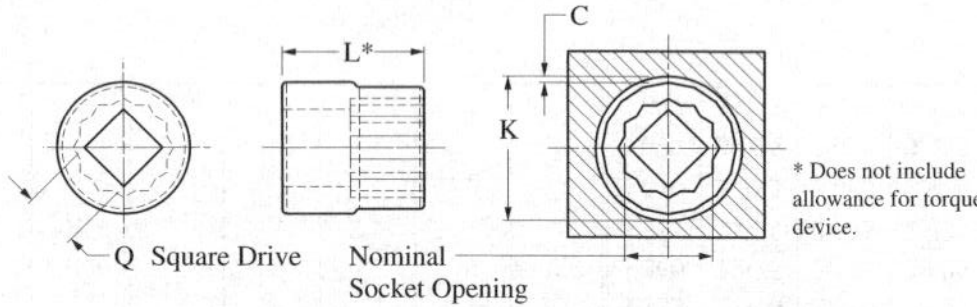


Fig. 2. Clearance Dimensions for Single and Double Hexagon Socket Wrenches (See Table 3a and Table 3b)

**Table 3a. Clearances for Single and Double Hexagon Socket Wrenches, Regular Length—Inch Series**

See Fig. 2, page 1689 for Dimensions

| Nominal Opening | | Radial Clearance C Ref.[a] | 1/4 in. Square Drive, Q | | | | | 3/8 in. Square Drive, Q | | | | | 1/2 in. Square Drive, Q | | | | | 3/4 in. Square Drive, Q | | | | |
|---|---|---|---|---|---|---|---|---|---|---|---|---|---|---|---|---|---|---|---|---|---|---|
| | | | Length L Max. | Nut End Dia. D1 Max. | Drive End Dia. D2 Max. | C-Bore Dia. K Min. | Proof Torque P (lbf-in.) Min. | Length L Max. | Nut End Dia. D1 Max. | Drive End Dia. D2 Max. | Counter-Bore Dia. K Min. | Proof Torque P (lbf-in.) Min. | Length L Max. | Nut End Dia. D1 Max. | Drive End Dia. D2 Max. | Counter-Bore Dia. K Min. | Proof Torque P (lbf-in.) Min. | Length L Max. | Nut End Dia. D1 Max. | Drive End Dia. D2 Max. | Counter-Bore Dia. K Min. | Proof Torque P (lbf-in.) Min. |
| 1/8 | (0.125) | 0.030 | 1.010 | 0.250 | 0.510 | 0.540 | 35 | … | … | … | … | … | … | … | … | … | … | … | … | … | … | … |
| 5/32 | (0.156) | 0.030 | 1.010 | 0.281 | 0.510 | 0.540 | 60 | … | … | … | … | … | … | … | … | … | … | … | … | … | … | … |
| 3/16 | (0.188) | 0.030 | 1.010 | 0.338 | 0.510 | 0.540 | 95 | … | … | … | … | … | … | … | … | … | … | … | … | … | … | … |
| 7/32 | (0.219) | 0.030 | 1.010 | 0.382 | 0.510 | 0.540 | 135 | … | … | … | … | … | … | … | … | … | … | … | … | … | … | … |
| 1/4 | (0.250) | 0.030 | 1.010 | 0.425 | 0.510 | 0.540 | 190 | 1.260 | 0.472 | 0.690 | 0.720 | 270 | … | … | … | … | … | … | … | … | … | … |
| 9/32 | (0.281) | 0.030 | 1.010 | 0.457 | 0.510 | 0.540 | 250 | 1.260 | 0.496 | 0.690 | 0.720 | 350 | … | … | … | … | … | … | … | … | … | … |
| 5/16 | (0.313) | 0.030 | 1.010 | 0.510 | 0.510 | 0.540 | 320 | 1.260 | 0.521 | 0.690 | 0.720 | 440 | … | … | … | … | … | … | … | … | … | … |
| 11/32 | (0.344) | 0.030 | 1.010 | 0.547 | 0.547 | 0.577 | 400 | 1.260 | 0.567 | 0.690 | 0.720 | 550 | … | … | … | … | … | … | … | … | … | … |
| 3/8 | (0.375) | 0.030 | 1.010 | 0.597 | 0.597 | 0.627 | 500 | 1.260 | 0.613 | 0.690 | 0.720 | 660 | 1.525 | 0.655 | 0.940 | 0.970 | 1100 | … | … | … | … | … |
| 7/16 | (0.438) | 0.030 | 1.010 | 0.683 | 0.683 | 0.713 | 500 | 1.260 | 0.683 | 0.690 | 0.720 | 930 | 1.525 | 0.730 | 0.940 | 0.970 | 1500 | … | … | … | … | … |
| 1/2 | (0.500) | 0.030 | 1.010 | 0.697 | 0.697 | 0.727 | 500 | 1.260 | 0.751 | 0.880 | 0.910 | 1240 | 1.525 | 0.775 | 0.940 | 0.970 | 2000 | … | … | … | … | … |
| 9/16 | (0.563) | 0.030 | 1.010 | 0.778 | 0.778 | 0.808 | 500 | 1.260 | 0.814 | 0.880 | 0.910 | 1610 | 1.572 | 0.845 | 0.940 | 0.970 | 2600 | … | … | … | … | … |
| 5/8 | (0.625) | 0.030 | … | … | … | … | … | 1.260 | 0.890 | 0.890 | 0.920 | 200 | 1.572 | 0.942 | 0.970 | 1.000 | 3300 | … | … | … | … | … |

**Table 3a.** *(Continued)* **Clearances for Single and Double Hexagon Socket Wrenches, Regular Length — Inch Series**

| See Fig. 2, page 1689 for Dimensions | | | | | | | | | | | | | | | | | | | | | | | | | | |
|---|---|---|---|---|---|---|---|---|---|---|---|---|---|---|---|---|---|---|---|---|---|---|---|---|---|---|
| Nominal Opening | | Radial Clearance C Ref.[a] | 1/4 in. Square Drive, Q | | | | | 3/8 in. Square Drive, Q | | | | | 1/2 in. Square Drive, Q | | | | | 3/4 in. Square Drive, Q | | | | |
| | | | Length L Max. | Nut End Dia. D1 Max. | Drive End Dia. D2 Max. | C-Bore Dia. K Min. | Proof Torque P (lbf-in.) Min. | Length L Max. | Nut End Dia. D1 Max. | Drive End Dia. D2 Max. | Counter-Bore Dia. K Min. | Proof Torque P (lbf-in.) Min. | Length L Max. | Nut End Dia. D1 Max. | Drive End Dia. D2 Max. | Counter-Bore Dia. K Min. | Proof Torque P (lbf-in.) Min. | Length L Max. | Nut End Dia. D1 Max. | Drive End Dia. D2 Max. | Counter-Bore Dia. K Min. | Proof Torque P (lbf-in.) Min. |
| 11/16 | (0.688) | 0.030 | … | … | … | … | … | 1.260 | 0.968 | 0.968 | 0.998 | 2200 | 1.572 | 1.010 | 1.010 | 1.040 | 4100 | … | … | … | … | … |
| 3/4 | (0.750) | 0.030 | … | … | … | … | … | 1.260 | 1.110 | 1.110 | 1.140 | 2200 | 1.572 | 1.080 | 1.080 | 1.110 | 5000 | 2.000 | 1.285 | 1.450 | 1.480 | 6000 |
| 13/16 | (0.813) | 0.030 | … | … | … | … | … | 1.406 | 1.141 | 1.141 | 1.171 | 2200 | 1.635 | 1.145 | 1.145 | 1.175 | 5000 | 2.000 | 1.300 | 1.450 | 1.480 | 6800 |
| 7/8 | (0.875) | 0.030 | … | … | … | … | … | 1.406 | 1.250 | 1.250 | 1.280 | 2200 | 1.760 | 1.218 | 1.218 | 1.248 | 5000 | 2.010 | 1.385 | 1.575 | 1.605 | 7700 |
| 15/16 | (0.938) | 0.030 | … | … | … | … | … | 1.650 | 1.310 | 1.310 | 1.340 | 2200 | 1.760 | 1.300 | 1.300 | 1.330 | 5000 | 2.010 | 1.450 | 1.575 | 1.605 | 8700 |
| 1 | (1.000) | 0.030 | … | … | … | … | … | 1.650 | 1.380 | 1.380 | 1.410 | 2200 | 1.760 | 1.375 | 1.375 | 1.405 | 5000 | 2.072 | 1.520 | 1.575 | 1.605 | 9700 |
| 1 1/16 | (1.063) | 0.030 | … | … | … | … | … | … | … | … | … | … | 1.853 | 1.480 | 1.480 | 1.510 | 5000 | 2.200 | 1.595 | 1.595 | 1.625 | 10,800 |
| 1 1/8 | (1.125) | 0.030 | … | … | … | … | … | … | … | … | … | … | 1.947 | 1.540 | 1.540 | 1.570 | 5000 | 2.322 | 1.600 | 1.680 | 1.710 | 11,900 |
| 1 3/16 | (1.188) | 0.030 | … | … | … | … | … | … | … | … | … | … | 1.947 | 1.675 | 1.675 | 1.705 | 5000 | 2.322 | 1.735 | 1.735 | 1.765 | 13,000 |
| 1 1/4 | (1.250) | 0.030 | … | … | … | … | … | … | … | … | … | … | 2.015 | 1.750 | 1.750 | 1.780 | 5000 | 2.385 | 1.870 | 1.870 | 1.900 | 14,200 |
| 1 5/16 | (1.313) | 0.030 | … | … | … | … | … | … | … | … | … | … | 2.015 | 1.820 | 1.820 | 1.850 | 5000 | 2.510 | 1.920 | 1.920 | 1.950 | 15,400 |
| 1 3/8 | (1.375) | 0.030 | … | … | … | … | … | … | … | … | … | … | 2.155 | 1.885 | 1.885 | 1.915 | 5000 | 2.635 | 1.980 | 1.980 | 2.010 | 16,700 |
| 1 7/16 | (1.438) | 0.030 | … | … | … | … | … | … | … | … | … | … | 2.295 | 1.955 | 1.955 | 1.985 | 5000 | 2.635 | 2.075 | 2.075 | 2.105 | 18,000 |
| 1 1/2 | (1.500) | 0.030 | … | … | … | … | … | … | … | … | … | … | 2.295 | 2.025 | 2.025 | 2.055 | 5000 | 2.635 | 2.145 | 2.145 | 2.175 | 18,000 |
| 1 5/8 | (1.625) | 0.030 | … | … | … | … | … | … | … | … | … | … | … | … | … | … | … | 2.760 | 2.260 | 2.260 | 2.290 | 18,000 |
| 1 3/4 | (1.750) | 0.030 | … | … | … | … | … | … | … | … | … | … | … | … | … | … | … | 2.760 | 2.325 | 2.325 | 2.355 | 18,000 |
| 1 13/16 | (1.813) | 0.030 | … | … | … | … | … | … | … | … | … | … | … | … | … | … | … | 3.135 | 2.400 | 2.400 | 2.430 | 18,000 |
| 1 7/8 | (1.875) | 0.030 | … | … | … | … | … | … | … | … | … | … | … | … | … | … | … | 3.135 | 2.510 | 2.510 | 2.540 | 18,000 |
| 2 | (2.000) | 0.030 | … | … | … | … | … | … | … | … | … | … | … | … | … | … | … | 3.260 | 2.575 | 2.575 | 2.605 | 18,000 |
| 2 1/16 | (2.063) | 0.030 | … | … | … | … | … | … | … | … | … | … | … | … | … | … | … | 3.385 | 2.695 | 2.695 | 2.725 | 18,000 |
| 2 1/8 | (2.125) | 0.030 | … | … | … | … | … | … | … | … | … | … | … | … | … | … | … | 3.510 | 2.885 | 2.885 | 2.915 | 18,000 |
| 2 3/16 | (2.188) | 0.030 | … | … | … | … | … | … | … | … | … | … | … | … | … | … | … | 3.697 | 3.025 | 3.025 | 3.055 | 18,000 |
| 2 1/4 | (2.250) | 0.030 | … | … | … | … | … | … | … | … | … | … | … | … | … | … | … | 3.697 | 3.075 | 3.075 | 3.105 | 18,000 |

[a] From the SAE Aeronautical Drafting Manual

All dimensions are in inches. For details not shown and additional socket sizes, see ANSI/ASME B107.1-2002, Socket Wrenches, Hand (Inch Series).

## Table 3b. Single and Double Hexagon Socket, Regular Length — Metric Series

(See Fig. 2, page 1689 for Dimensions)

| | | 6.3 mm Square Drive Q | | | | | 10 mm Square Drive Q | | | | | 12.5 mm Square Drive Q | | | | | 20 mm Square Drive Q | | | | |
|---|---|---|---|---|---|---|---|---|---|---|---|---|---|---|---|---|---|---|---|---|---|
| Nominal Opening | Radial Clearance C Ref[a] | Length L Max. | Nut End Dia. D1 Max. | Drive End Dia D2 Max. | C-Bore Dia. K Min. | Proof Torque P (N-m) Min. | Length L Max. | Nut End Dia. D1 Max. | Drive End Dia. D2 Max. | Counter-Bore Dia. K Min. | Proof Torque P (N-m) Min. | Length L Max. | Nut End Dia. D1 Max. | Drive End Dia. D2 Max. | Counter-Bore Dia. K Min. | Proof Torque P (N-m) Min. | Length L Max. | Nut End Dia. D1 Max. | Drive End Dia. D2 Max. | Counter-Bore Dia. K Min. | Proof Torque P (N-m) Min. |
| 3.2 | 0.762 | 26 | 6.10 | 12.95 | 14.47 | 7 | … | … | … | … | … | … | … | … | … | … | … | … | … | … | … |
| 4 | 0.762 | 26 | 7.10 | 12.95 | 14.47 | 8 | … | … | … | … | … | … | … | … | … | … | … | … | … | … | … |
| 4.5 | 0.762 | 26 | 7.60 | 12.95 | 14.47 | 9 | … | … | … | … | … | … | … | … | … | … | … | … | … | … | … |
| 5 | 0.762 | 26 | 8.15 | 12.95 | 14.47 | 10 | … | … | … | … | … | … | … | … | … | … | … | … | … | … | … |
| 5.5 | 0.762 | 26 | 8.90 | 12.95 | 14.47 | 14 | 32 | 10.10 | 17.60 | 19.124 | 270 | … | … | … | … | … | … | … | … | … | … |
| 6 | 0.762 | 26 | 9.90 | 12.95 | 14.47 | 16 | 32 | 10.10 | 17.60 | 19.124 | 350 | … | … | … | … | … | … | … | … | … | … |
| 6.3 | 0.762 | 26 | 9.90 | 12.95 | 14.47 | 21 | 32 | 10.10 | 17.60 | 19.124 | 440 | … | … | … | … | … | … | … | … | … | … |
| 7 | 0.762 | 26 | 10.90 | 12.95 | 14.47 | 27 | 32 | 11.05 | 17.60 | 19.124 | 550 | … | … | … | … | … | … | … | … | … | … |
| 8 | 0.762 | 26 | 12.20 | 12.95 | 14.47 | 38 | 32 | 12.20 | 17.60 | 19.124 | 660 | 39 | 14.00 | 23.87 | 25.39 | 80 | … | … | … | … | … |
| 9 | 0.762 | 26 | 13.45 | 13.45 | 14.97 | 49 | 32 | 13.60 | 17.60 | 19.124 | 930 | 39 | 15.10 | 23.87 | 25.39 | 110 | … | … | … | … | … |
| 10 | 0.762 | 26 | 14.75 | 14.75 | 16.27 | 63 | 32 | 15.00 | 17.60 | 19.124 | 1240 | 39 | 16.80 | 23.87 | 25.39 | 153 | … | … | … | … | … |
| 11 | 0.762 | 26 | 16.00 | 16.00 | 17.52 | 68 | 32 | 16.75 | 17.60 | 19.124 | 1610 | 39 | 18.20 | 23.87 | 25.39 | 170 | … | … | … | … | … |
| 12 | 0.762 | 26 | 17.30 | 17.30 | 18.82 | 68 | 32 | 17.80 | 22.40 | 23.924 | 200 | 39 | 18.70 | 23.87 | 25.39 | 203 | … | … | … | … | … |
| 13 | 0.762 | 26 | 18.55 | 18.55 | 20.07 | 68 | 32 | 18.80 | 22.40 | 23.924 | 2200 | 39 | 20.25 | 23.87 | 25.39 | 249 | … | … | … | … | … |
| 14 | 0.762 | 26 | 19.80 | 19.80 | 21.32 | 68 | 32 | 20.00 | 22.40 | 23.924 | 2200 | 39 | 21.80 | 23.87 | 25.39 | 282 | … | … | … | … | … |
| 15 | 0.762 | 26 | 21.50 | 21.50 | 23.02 | 68 | 32 | 22.40 | 22.40 | 23.924 | 2200 | 40 | 22.40 | 23.87 | 25.39 | 339 | … | … | … | … | … |
| 16 | 0.762 | 26 | 22.00 | 22.00 | 23.52 | 68 | 32 | 22.50 | 22.50 | 24.024 | 2200 | 40 | 23.87 | 23.87 | 25.39 | 407 | … | … | … | … | … |
| 17 | 0.762 | … | … | … | … | … | 32 | 23.80 | 23.80 | 25.324 | 2200 | 40 | 24.75 | 24.75 | 26.27 | 475 | … | … | … | … | … |
| 18 | 0.762 | … | … | … | … | … | 32 | 24.60 | 24.60 | 26.124 | 2200 | 40 | 26.14 | 26.14 | 27.66 | 542 | … | … | … | … | … |
| 19 | 0.762 | … | … | … | … | … | 32 | 25.70 | 25.70 | 27.224 | … | 40 | 27.20 | 27.20 | 28.72 | 575 | 51 | 30.50 | 33.00 | 33.76 | 780 |
| 20 | 0.762 | … | … | … | … | … | 32 | 27.76 | 27.76 | 29.284 | … | 42 | 27.95 | 27.95 | 29.47 | 570 | … | … | … | … | … |
| 21 | 0.762 | … | … | … | … | … | 34 | 28.80 | 28.80 | 30.324 | … | 42 | 28.95 | 28.95 | 30.47 | 570 | 51 | 33.00 | 33.00 | 33.76 | 930 |
| 22 | 0.762 | … | … | … | … | … | 34 | 30.00 | 30.00 | 31.524 | … | 45 | 30.20 | 30.20 | 31.72 | 570 | 51 | 35.05 | 38.10 | 38.86 | 972 |
| 23 | 0.762 | … | … | … | … | … | 35 | 31.30 | 31.30 | 32.824 | … | 45 | 31.25 | 31.25 | 32.77 | 570 | 51 | 36.10 | 39.10 | 39.86 | 1015 |

## Table 3b. *(Continued)* Single and Double Hexagon Socket, Regular Length — Metric Series

(See Fig. 2, page 1689 for Dimensions)

| | | 6.3 mm Square Drive Q | | | | | 10 mm Square Drive Q | | | | | 12.5 mm Square Drive Q | | | | | 20 mm Square Drive Q | | | | |
|---|---|---|---|---|---|---|---|---|---|---|---|---|---|---|---|---|---|---|---|---|---|
| Nominal Opening | Radial Clearance C Ref[a] | Length L Max. | Nut End Dia. D1 Max. | Drive End Dia D2 Max. | C-Bore Dia. K Min. | Proof Torque P (N-m) Min. | Length L Max. | Nut End Dia. D1 Max. | Drive End Dia. D2 Max. | Counter-Bore Dia. K Min. | Proof Torque P (N-m) Min. | Length L Max. | Nut End Dia. D1 Max. | Drive End Dia. D2 Max. | Counter-Bore Dia. K Min. | Proof Torque P (N-m) Min. | Length L Max. | Nut End Dia. D1 Max. | Drive End Dia. D2 Max. | Counter-Bore Dia. K Min. | Proof Torque P (N-m) Min. |
| 24 | 0.762 | … | … | … | … | … | 36 | 32.50 | 32.50 | 34.024 | … | 45 | 32.15 | 32.15 | 33.67 | 570 | 51 | 37.00 | 40.00 | 40.76 | 1085 |
| 25 | 0.762 | … | … | … | … | … | 38 | 33.00 | 33.00 | 34.524 | … | 45 | 33.40 | 33.40 | 34.92 | 570 | 52 | 37.85 | 40.00 | 40.76 | 1160 |
| 26 | 0.762 | … | … | … | … | … | 38 | 35.00 | 35.00 | 36.524 | … | 48 | 35.05 | 35.05 | 36.57 | 570 | 53 | 38.85 | 40.00 | 40.76 | 1240 |
| 27 | 0.762 | … | … | … | … | … | … | … | … | … | … | 48 | 36.75 | 36.75 | 38.27 | 570 | 54 | 41.00 | 41.00 | 41.76 | 1330 |
| 28 | 0.762 | … | … | … | … | … | … | … | … | … | … | 50 | 37.80 | 37.80 | 39.32 | 570 | 57 | 41.00 | 41.00 | 41.76 | 1420 |
| 29 | 0.762 | … | … | … | … | … | … | … | … | … | … | 50 | 39.50 | 39.50 | 41.02 | 570 | 59 | 42.10 | 42.10 | 42.86 | 1520 |
| 30 | 0.762 | … | … | … | … | … | … | … | … | … | … | 50 | 42.40 | 42.40 | 43.92 | 570 | 59 | 43.00 | 43.00 | 43.76 | 1640 |
| 31 | 0.762 | … | … | … | … | … | … | … | … | … | … | 50 | 43.20 | 43.20 | 44.72 | 570 | 60 | 45.10 | 45.10 | 45.86 | 1730 |
| 32 | 0.762 | … | … | … | … | … | … | … | … | … | … | 51 | 44.05 | 44.05 | 45.57 | 570 | 60 | 47.05 | 47.05 | 47.81 | 1820 |
| 34 | 0.762 | … | … | … | … | … | … | … | … | … | … | … | … | … | … | … | 64 | 49.00 | 49.00 | 49.76 | 2000 |
| 35 | 0.762 | … | … | … | … | … | … | … | … | … | … | … | … | … | … | … | 67 | 50.40 | 50.40 | 51.16 | 2030 |
| 36 | 0.762 | … | … | … | … | … | … | … | … | … | … | … | … | … | … | … | 67 | 51.80 | 51.80 | 52.56 | 2030 |
| 38 | 0.762 | … | … | … | … | … | … | … | … | … | … | … | … | … | … | … | 67 | 54.10 | 54.10 | 54.86 | 2030 |
| 40 | 0.762 | … | … | … | … | … | … | … | … | … | … | … | … | … | … | … | 70 | 57.65 | 57.65 | 58.41 | 2030 |
| 41 | 0.762 | … | … | … | … | … | … | … | … | … | … | … | … | … | … | … | 70 | 58.80 | 58.80 | 59.56 | 2030 |
| 42 | 0.762 | … | … | … | … | … | … | … | … | … | … | … | … | … | … | … | 70 | 58.80 | 58.80 | 59.56 | 2030 |
| 46 | 0.762 | … | … | … | … | … | … | … | … | … | … | … | … | … | … | … | 83 | 65.40 | 65.40 | 66.16 | 2030 |
| 50 | 0.762 | … | … | … | … | … | … | … | … | … | … | … | … | … | … | … | 89 | 72.15 | 72.15 | 72.91 | 2030 |
| 54 | 0.762 | … | … | … | … | … | … | … | … | … | … | … | … | … | … | … | 94 | 78.10 | 78.10 | 78.86 | 2030 |
| 55 | 0.762 | … | … | … | … | … | … | … | … | … | … | … | … | … | … | … | 95 | 79.10 | 79.10 | 79.86 | 2030 |
| 58 | 0.762 | … | … | … | … | … | … | … | … | … | … | … | … | … | … | … | 97 | 80.00 | 80.00 | 80.76 | 2030 |
| 60 | 0.762 | … | … | … | … | … | … | … | … | … | … | … | … | … | … | … | 100 | 84.45 | 84.45 | 85.21 | 2030 |

[a] Converted from inch dimensions given in the SAE Aeronautical Drafting Manual.
All dimensions are in mm. For details not shown and additional socket sizes, see ANSI/ASME B107.5M-2002, Socket Wrenches, Hand (Metric Series).

**Table 4. Clearances for Box Wrenches — 12-Point Inch and Metric Series**

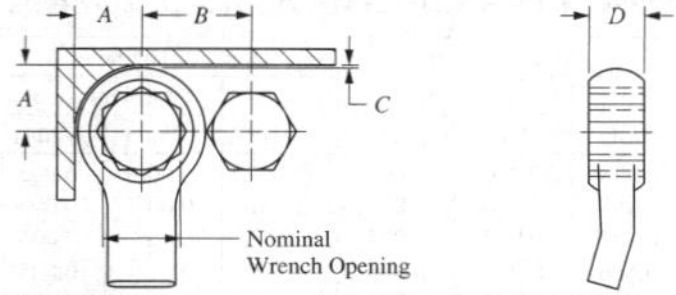


| US Customary (inch) | | | | | | | Metric (mm) | | | | | |
|---|---|---|---|---|---|---|---|---|---|---|---|---|
| Nominal Wrench Opening (in.) | | A Min. (in.) | B Min. (in.) | C Ref.[a] (in.) | Head Thickness D Max., (in.) | Proof Torque (lbf-in.) | Nominal Wrench Opening (mm) | A Min. (mm) | B Min. (mm) | C Ref.[b] (mm) | Head Thickness D Max., (mm) | Proof Torque (N-m) |
| 1/8 | (0.125) | 0.179 | 0.219 | 0.030 | 0.172 | 60 | 4 | 4.56 | 6.03 | 0.762 | 4.0 | 12 |
| 5/32 | (0.156) | 0.187 | 0.244 | 0.030 | 0.172 | 90 | 5 | 5.26 | 7.29 | 0.762 | 4.6 | 17 |
| 3/16 | (0.188) | 0.218 | 0.301 | 0.030 | 0.203 | 150 | 5.5 | 6.66 | 8.97 | 0.762 | 6.0 | 18 |
| 7/32 | (0.219) | 0.233 | 0.325 | 0.030 | 0.234 | 165 | 6 | 7.11 | 9.69 | 0.762 | 7.4 | 20 |
| 1/4 | (0.250) | 0.269 | 0.378 | 0.030 | 0.295 | 220 | 7 | 7.91 | 11.05 | 0.762 | 7.7 | 27 |
| 9/32 | (0.281) | 0.280 | 0.407 | 0.030 | 0.280 | 248 | 8 | 8.26 | 11.98 | 0.762 | 8.2 | 30 |
| 5/16 | (0.313) | 0.316 | 0.461 | 0.030 | 0.330 | 275 | 9 | 9.46 | 13.76 | 0.762 | 9.0 | 40 |
| 11/32 | (0.344) | 0.336 | 0.499 | 0.030 | 0.335 | 275 | 10 | 10.16 | 15.04 | 0.762 | 9.0 | 71 |
| 3/8 | (0.375) | 0.362 | 0.543 | 0.030 | 0.344 | 605 | 11 | 10.71 | 16.15 | 0.762 | 10.0 | 80 |
| 7/16 | (0.438) | 0.395 | 0.612 | 0.030 | 0.391 | 715 | 12 | 11.46 | 17.47 | 0.762 | 10.0 | 91 |
| 1/2 | (0.500) | 0.442 | 0.694 | 0.030 | 0.394 | 1020 | 13 | 12.31 | 18.89 | 0.762 | 10.5 | 115 |
| 9/16 | (0.563) | 0.492 | 0.779 | 0.030 | 0.425 | 1500 | 14 | 12.96 | 20.10 | 0.762 | 11.5 | 158 |
| 5/8 | (0.625) | 0.530 | 0.853 | 0.030 | 0.500 | 2200 | 15 | 13.76 | 21.46 | 0.762 | 11.5 | 200 |
| 11/16 | (0.688) | 0.577 | 0.935 | 0.030 | 0.535 | 2640 | 16 | 14.26 | 22.53 | 0.762 | 12.1 | 248 |
| 3/4 | (0.750) | 0.618 | 1.012 | 0.030 | 0.594 | 2860 | 17 | 15.41 | 24.25 | 0.762 | 12.7 | 267 |
| 13/16 | (0.813) | 0.702 | 1.132 | 0.030 | 0.609 | 3300 | 18 | 15.41 | 24.83 | 0.762 | 12.7 | 304 |
| 7/8 | (0.875) | 0.718 | 1.183 | 0.030 | 0.688 | 3630 | 19 | 16.36 | 26.35 | 0.762 | 14.8 | 323 |
| 15/16 | (0.938) | 0.765 | 1.266 | 0.030 | 0.701 | 4510 | 20 | 17.21 | 27.77 | 0.762 | 14.8 | 347 |
| 1 | (1.000) | 0.796 | 1.330 | 0.030 | 0.719 | 5390 | 21 | 17.66 | 28.79 | 0.762 | 16.3 | 372 |
| 1 1/16 | (1.063) | 0.874 | 1.445 | 0.030 | 0.790 | 5940 | 22 | 18.56 | 30.27 | 0.762 | 16.3 | 408 |
| 1 1/8 | (1.125) | 0.892 | 1.498 | 0.030 | 0.860 | 6430 | 23 | 19.41 | 31.69 | 0.762 | 16.5 | 455 |
| 1 3/16 | (1.188) | 0.937 | 1.579 | 0.030 | 0.890 | 7200 | 24 | 19.81 | 32.65 | 0.762 | 17.8 | 509 |
| 1 1/4 | (1.250) | 0.983 | 1.661 | 0.030 | 0.940 | 7920 | 25 | 20.86 | 34.24 | 0.762 | 17.9 | 559 |
| 1 5/16 | (1.313) | 1.062 | 1.775 | 0.030 | 0.940 | 8400 | 26 | 12.86 | 26.79 | 0.762 | 18.0 | 608 |
| 1 3/8 | (1.375) | 1.087 | 1.836 | 0.030 | 0.940 | 8970 | 27 | 22.86 | 37.37 | 0.762 | 19.8 | 671 |
| 1 7/16 | (1.438) | 1.144 | 1.929 | 0.030 | 0.953 | 9240 | 28 | 23.41 | 38.49 | 0.762 | 19.8 | 710 |
| 1 1/2 | (1.500) | 1.228 | 2.049 | 0.030 | 1.008 | 10,365 | 29 | 23.41 | 39.06 | 0.762 | 19.8 | 750 |
| 1 9/16 | (1.563) | 1.249 | 2.104 | 0.030 | 1.031 | 11,495 | 30 | 24.51 | 40.73 | 0.762 | 20.0 | 795 |
| 1 5/8 | (1.625) | 1.351 | 2.241 | 0.030 | 1.063 | 12,800 | 31 | 25.06 | 41.85 | 0.762 | 20.5 | 850 |
| 1 11/16 | (1.688) | 1.425 | 2.351 | 0.030 | 1.063 | 13,570 | 32 | 25.66 | 43.03 | 0.762 | 22.0 | 905 |
| 1 3/4 | (1.750) | 1.499 | 2.461 | 0.030 | 1.125 | 14,300 | 33 | 25.91 | 43.84 | 0.762 | 22.3 | 950 |
| 1 13/16 | (1.813) | 1.499 | 2.496 | 0.030 | 1.125 | 15,100 | 34 | 26.76 | 45.26 | 0.762 | 23.2 | 994 |
| 1 7/8 | (1.875) | 1.593 | 2.625 | 0.030 | 1.125 | 15,900 | 36 | 28.81 | 48.47 | 0.762 | 25.1 | 1165 |
| 2 | (2.000) | 1.593 | 2.696 | 0.030 | 1.125 | 17,400 | 41 | 32.21 | 54.68 | 0.762 | 25.3 | 1579 |
| 2 1/16 | (2.063) | 1.687 | 2.825 | 0.030 | 1.234 | 18,200 | 46 | 34.76 | 60.06 | 0.762 | 25.8 | 2067 |
| 2 1/8 | (2.125) | 1.687 | 2.861 | 0.030 | 1.234 | 19,000 | 50 | 38.76 | 66.33 | 0.762 | 27.6 | 2512 |
| 2 3/16 | (2.188) | 1.687 | 2.896 | 0.030 | 1.234 | 19,700 | … | … | … | … | … | … |
| 2 1/4 | (2.250) | 1.687 | 2.931 | 0.030 | 1.234 | 20,500 | … | … | … | … | … | … |

[a] From SAE Aeronautical Drafting Manual

[b] Converted from SAE Aeronautical Drafting Manual. For details not shown, including material, see ANSI/ASME B107, 100-2020 Flat Wrenches

## Table 1a. American National Standard Type A Plain Washers — Preferred Sizes *ANSI/ASME B18.21.1-2009 (R2016)*

| Nominal Washer Size[a] | | Series | Inside Diameter | | | Outside Diameter | | | Thickness | | |
|---|---|---|---|---|---|---|---|---|---|---|---|
| | | | Basic | Tolerance | | Basic | Tolerance | | Basic | Max. | Min. |
| | | | | Plus | Minus | | Plus | Minus | | | |
| No. 0 | — | | 0.078 | 0.000 | 0.005 | 0.188 | 0.000 | 0.005 | 0.020 | 0.025 | 0.016 |
| No. 2 | — | | 0.094 | 0.000 | 0.005 | 0.250 | 0.000 | 0.005 | 0.020 | 0.025 | 0.016 |
| No. 4 | — | | 0.125 | 0.008 | 0.005 | 0.312 | 0.008 | 0.005 | 0.032 | 0.040 | 0.025 |
| No. 6 | 0.138 | | 0.156 | 0.008 | 0.005 | 0.375 | 0.015 | 0.005 | 0.049 | 0.065 | 0.036 |
| No. 8 | 0.164 | | 0.188 | 0.008 | 0.005 | 0.438 | 0.015 | 0.005 | 0.049 | 0.065 | 0.036 |
| No. 10 | 0.190 | | 0.219 | 0.008 | 0.005 | 0.500 | 0.015 | 0.005 | 0.049 | 0.065 | 0.036 |
| 3/16 | 0.188 | | 0.250 | 0.015 | 0.005 | 0.562 | 0.015 | 0.005 | 0.049 | 0.065 | 0.036 |
| No. 12 | 0.216 | | 0.250 | 0.015 | 0.005 | 0.562 | 0.015 | 0.005 | 0.065 | 0.080 | 0.051 |
| 1/4 | 0.250 | N | 0.281 | 0.015 | 0.005 | 0.625 | 0.015 | 0.005 | 0.065 | 0.080 | 0.051 |
| 1/4 | 0.250 | W | 0.312 | 0.015 | 0.005 | 0.734[b] | 0.015 | 0.007 | 0.065 | 0.080 | 0.051 |
| 5/16 | 0.312 | N | 0.344 | 0.015 | 0.005 | 0.688 | 0.015 | 0.007 | 0.065 | 0.080 | 0.051 |
| 5/16 | 0.312 | W | 0.375 | 0.015 | 0.005 | 0.875 | 0.030 | 0.007 | 0.083 | 0.104 | 0.064 |
| 3/8 | 0.375 | N | 0.406 | 0.015 | 0.005 | 0.812 | 0.015 | 0.007 | 0.065 | 0.080 | 0.051 |
| 3/8 | 0.375 | W | 0.438 | 0.015 | 0.005 | 1.000 | 0.030 | 0.007 | 0.083 | 0.104 | 0.064 |
| 7/16 | 0.438 | N | 0.469 | 0.015 | 0.005 | 0.922 | 0.015 | 0.007 | 0.065 | 0.080 | 0.051 |
| 7/16 | 0.438 | W | 0.500 | 0.015 | 0.005 | 1.250 | 0.030 | 0.007 | 0.083 | 0.104 | 0.064 |
| 1/2 | 0.500 | N | 0.531 | 0.015 | 0.005 | 1.062 | 0.030 | 0.007 | 0.095 | 0.121 | 0.074 |
| 1/2 | 0.500 | W | 0.562 | 0.015 | 0.005 | 1.375 | 0.030 | 0.007 | 0.109 | 0.132 | 0.086 |
| 9/16 | 0.562 | N | 0.594 | 0.015 | 0.005 | 1.156[b] | 0.030 | 0.007 | 0.095 | 0.121 | 0.074 |
| 9/16 | 0.562 | W | 0.625 | 0.015 | 0.005 | 1.469[b] | 0.030 | 0.007 | 0.109 | 0.132 | 0.086 |
| 5/8 | 0.625 | N | 0.656 | 0.030 | 0.007 | 1.312 | 0.030 | 0.007 | 0.095 | 0.121 | 0.074 |
| 5/8 | 0.625 | W | 0.688 | 0.030 | 0.007 | 1.750 | 0.030 | 0.007 | 0.134 | 0.160 | 0.108 |
| 3/4 | 0.750 | N | 0.812 | 0.030 | 0.007 | 1.469 | 0.030 | 0.007 | 0.134 | 0.160 | 0.108 |
| 3/4 | 0.750 | W | 0.812 | 0.030 | 0.007 | 2.000 | 0.030 | 0.007 | 0.148 | 0.177 | 0.122 |
| 7/8 | 0.875 | N | 0.938 | 0.030 | 0.007 | 1.750 | 0.030 | 0.007 | 0.134 | 0.160 | 0.108 |
| 7/8 | 0.875 | W | 0.938 | 0.030 | 0.007 | 2.250 | 0.030 | 0.007 | 0.165 | 0.192 | 0.136 |
| 1 | 1.000 | N | 1.062 | 0.030 | 0.007 | 2.000 | 0.030 | 0.007 | 0.134 | 0.160 | 0.108 |
| 1 | 1.000 | W | 1.062 | 0.030 | 0.007 | 2.500 | 0.030 | 0.007 | 0.165 | 0.192 | 0.136 |
| 1 1/8 | 1.125 | N | 1.250 | 0.030 | 0.007 | 2.250 | 0.030 | 0.007 | 0.134 | 0.160 | 0.108 |
| 1 1/8 | 1.125 | W | 1.250 | 0.030 | 0.007 | 2.750 | 0.030 | 0.007 | 0.165 | 0.192 | 0.136 |
| 1 1/4 | 1.250 | N | 1.375 | 0.030 | 0.007 | 2.500 | 0.030 | 0.007 | 0.165 | 0.192 | 0.136 |
| 1 1/4 | 1.250 | W | 1.375 | 0.030 | 0.007 | 3.000 | 0.030 | 0.007 | 0.165 | 0.192 | 0.136 |
| 1 3/8 | 1.375 | N | 1.500 | 0.030 | 0.007 | 2.750 | 0.030 | 0.007 | 0.165 | 0.192 | 0.136 |
| 1 3/8 | 1.375 | W | 1.500 | 0.045 | 0.010 | 3.250 | 0.045 | 0.010 | 0.180 | 0.213 | 0.153 |
| 1 1/2 | 1.500 | N | 1.625 | 0.030 | 0.007 | 3.000 | 0.030 | 0.007 | 0.165 | 0.192 | 0.136 |
| 1 1/2 | 1.500 | W | 1.625 | 0.045 | 0.010 | 3.500 | 0.045 | 0.010 | 0.180 | 0.213 | 0.153 |
| 1 5/8 | 1.625 | | 1.750 | 0.045 | 0.010 | 3.750 | 0.045 | 0.010 | 0.180 | 0.213 | 0.153 |
| 1 3/4 | 1.750 | | 1.875 | 0.045 | 0.010 | 4.000 | 0.045 | 0.010 | 0.180 | 0.213 | 0.153 |
| 1 7/8 | 1.875 | | 2.000 | 0.045 | 0.010 | 4.250 | 0.045 | 0.010 | 0.180 | 0.213 | 0.153 |
| 2 | 2.000 | | 2.125 | 0.045 | 0.010 | 4.500 | 0.045 | 0.010 | 0.180 | 0.213 | 0.153 |
| 2 1/4 | 2.250 | | 2.375 | 0.045 | 0.010 | 4.750 | 0.045 | 0.010 | 0.220 | 0.248 | 0.193 |
| 2 1/2 | 2.500 | | 2.625 | 0.045 | 0.010 | 5.000 | 0.045 | 0.010 | 0.238 | 0.280 | 0.210 |
| 2 3/4 | 2.750 | | 2.875 | 0.065 | 0.010 | 5.250 | 0.065 | 0.010 | 0.259 | 0.310 | 0.228 |
| 3 | 3.000 | | 3.125 | 0.065 | 0.010 | 5.500 | 0.065 | 0.010 | 0.284 | 0.327 | 0.249 |

[a] Nominal washer sizes are intended for use with comparable nominal screw or bolt sizes.

[b] The 0.734-inch, 1.156-inch, and 1.469-inch outside diameters avoid washers which could be used in coin operated devices.

All dimensions are in inches.

Preferred sizes are for the most part from series previously designated "Standard Plate" and "SAE." Where common sizes existed in the two series, the SAE size is designated "N" (narrow) and the Standard Plate "W" (wide). These sizes as well as all other sizes of Type A Plain Washers are to be ordered by inside diameter, outside diameter, and thickness dimensions.

Additional selected sizes of Type A Plain Washers are shown in Table 1b.

## Table 1b. American National Standard Type A Plain Washers—Additional Selected Sizes *ANSI/ASME B18.21.1-2009 (R2016)*

| Inside Diameter | | | Outside Diameter | | | Thickness | | |
|---|---|---|---|---|---|---|---|---|
| | Tolerance | | | Tolerance | | | | |
| Basic | Plus | Minus | Basic | Plus | Minus | Basic | Max. | Min. |
| 0.094 | 0.000 | 0.005 | 0.219 | 0.000 | 0.005 | 0.020 | 0.025 | 0.016 |
| 0.125 | 0.000 | 0.005 | 0.250 | 0.000 | 0.005 | 0.022 | 0.028 | 0.017 |
| 0.156 | 0.008 | 0.005 | 0.312 | 0.008 | 0.005 | 0.035 | 0.048 | 0.027 |
| 0.172 | 0.008 | 0.005 | 0.406 | 0.015 | 0.005 | 0.049 | 0.065 | 0.036 |
| 0.188 | 0.008 | 0.005 | 0.375 | 0.015 | 0.005 | 0.049 | 0.065 | 0.036 |
| 0.203 | 0.008 | 0.005 | 0.469 | 0.015 | 0.005 | 0.049 | 0.065 | 0.036 |
| 0.219 | 0.008 | 0.005 | 0.438 | 0.015 | 0.005 | 0.049 | 0.065 | 0.036 |
| 0.234 | 0.008 | 0.005 | 0.531 | 0.015 | 0.005 | 0.049 | 0.065 | 0.036 |
| 0.250 | 0.015 | 0.005 | 0.500 | 0.015 | 0.005 | 0.049 | 0.065 | 0.036 |
| 0.266 | 0.015 | 0.005 | 0.625 | 0.015 | 0.005 | 0.049 | 0.065 | 0.036 |
| 0.312 | 0.015 | 0.005 | 0.875 | 0.015 | 0.007 | 0.065 | 0.080 | 0.051 |
| 0.375 | 0.015 | 0.005 | 0.734[a] | 0.015 | 0.007 | 0.065 | 0.080 | 0.051 |
| 0.375 | 0.015 | 0.005 | 1.125 | 0.015 | 0.007 | 0.065 | 0.080 | 0.051 |
| 0.438 | 0.015 | 0.005 | 0.875 | 0.030 | 0.007 | 0.083 | 0.104 | 0.064 |
| 0.438 | 0.015 | 0.005 | 1.375 | 0.030 | 0.007 | 0.083 | 0.104 | 0.064 |
| 0.500 | 0.015 | 0.005 | 1.125 | 0.030 | 0.007 | 0.083 | 0.104 | 0.064 |
| 0.500 | 0.015 | 0.005 | 1.625 | 0.030 | 0.007 | 0.083 | 0.104 | 0.064 |
| 0.562 | 0.015 | 0.005 | 1.250 | 0.030 | 0.007 | 0.109 | 0.132 | 0.086 |
| 0.562 | 0.015 | 0.005 | 1.875 | 0.030 | 0.007 | 0.109 | 0.132 | 0.086 |
| 0.625 | 0.015 | 0.005 | 1.375 | 0.030 | 0.007 | 0.109 | 0.132 | 0.086 |
| 0.625 | 0.015 | 0.005 | 2.125 | 0.030 | 0.007 | 0.134 | 0.160 | 0.108 |
| 0.688 | 0.030 | 0.007 | 1.469[a] | 0.030 | 0.007 | 0.134 | 0.160 | 0.108 |
| 0.688 | 0.030 | 0.007 | 2.375 | 0.030 | 0.007 | 0.165 | 0.192 | 0.136 |
| 0.812 | 0.030 | 0.007 | 1.750 | 0.030 | 0.007 | 0.148 | 0.177 | 0.122 |
| 0.812 | 0.030 | 0.007 | 2.875 | 0.030 | 0.007 | 0.165 | 0.192 | 0.136 |
| 0.938 | 0.030 | 0.007 | 2.000 | 0.030 | 0.007 | 0.165 | 0.192 | 0.136 |
| 0.938 | 0.030 | 0.007 | 3.375 | 0.045 | 0.010 | 0.180 | 0.213 | 0.153 |
| 1.062 | 0.030 | 0.007 | 2.250 | 0.030 | 0.007 | 0.165 | 0.192 | 0.136 |
| 1.062 | 0.045 | 0.010 | 3.875 | 0.045 | 0.010 | 0.238 | 0.280 | 0.210 |
| 1.250 | 0.030 | 0.007 | 2.500 | 0.030 | 0.007 | 0.165 | 0.192 | 0.136 |
| 1.375 | 0.030 | 0.007 | 2.750 | 0.030 | 0.007 | 0.165 | 0.192 | 0.136 |
| 1.500 | 0.045 | 0.010 | 3.000 | 0.045 | 0.010 | 0.180 | 0.213 | 0.153 |
| 1.625 | 0.045 | 0.010 | 3.250 | 0.045 | 0.010 | 0.180 | 0.213 | 0.153 |
| 1.688 | 0.045 | 0.010 | 3.500 | 0.045 | 0.010 | 0.180 | 0.213 | 0.153 |
| 1.812 | 0.045 | 0.010 | 3.750 | 0.045 | 0.010 | 0.180 | 0.213 | 0.153 |
| 1.938 | 0.045 | 0.010 | 4.000 | 0.045 | 0.010 | 0.180 | 0.213 | 0.153 |
| 2.062 | 0.045 | 0.010 | 4.250 | 0.045 | 0.010 | 0.180 | 0.213 | 0.153 |

[a] The 0.734-inch and 1.469-inch outside diameters avoid washers which could be used in coin operated devices.

All dimensions are in inches.

The above sizes are to be ordered by inside diameter, outside diameter, and thickness dimensions.

Preferred Sizes of Type A Plain Washers are shown in Table 1a.

**ANSI Standard Plain Washers.—**The Type A plain washers were originally developed in a light, medium, heavy and extra heavy series. These series have been discontinued and the washers are now designated by their nominal dimensions.

The Type B plain washers are available in a narrow, regular and wide series with proportions designed to distribute the load over larger areas of lower strength materials.

Plain washers are made of ferrous or nonferrous metal, plastic or other material as specified. The tolerances indicated in the tables are intended for metal washers only.

## Table 2. American National Standard Type B Plain Washers

| Nominal Washer Size[a] | | Series[b] | Inside Diameter | | | Outside Diameter | | | Thickness | | |
|---|---|---|---|---|---|---|---|---|---|---|---|
| | | | | Tolerance | | | Tolerance | | | | |
| | | | Basic | Plus | Minus | Basic | Plus | Minus | Basic | Max. | Min. |
| No. 0 | 0.060 | N | 0.068 | 0.000 | 0.005 | 0.125 | 0.000 | 0.005 | 0.025 | 0.028 | 0.022 |
| | | R | 0.068 | 0.000 | 0.005 | 0.188 | 0.000 | 0.005 | 0.025 | 0.028 | 0.022 |
| | | W | 0.068 | 0.000 | 0.005 | 0.250 | 0.000 | 0.005 | 0.025 | 0.028 | 0.022 |
| No. 1 | 0.073 | N | 0.084 | 0.000 | 0.005 | 0.156 | 0.000 | 0.005 | 0.025 | 0.028 | 0.022 |
| | | R | 0.084 | 0.000 | 0.005 | 0.219 | 0.000 | 0.005 | 0.025 | 0.028 | 0.022 |
| | | W | 0.084 | 0.000 | 0.005 | 0.281 | 0.000 | 0.005 | 0.032 | 0.036 | 0.028 |
| No. 2 | 0.086 | N | 0.094 | 0.000 | 0.005 | 0.188 | 0.000 | 0.005 | 0.025 | 0.028 | 0.022 |
| | | R | 0.094 | 0.000 | 0.005 | 0.250 | 0.000 | 0.005 | 0.032 | 0.036 | 0.028 |
| | | W | 0.094 | 0.000 | 0.005 | 0.344 | 0.000 | 0.005 | 0.032 | 0.036 | 0.028 |
| No. 3 | 0.099 | N | 0.109 | 0.000 | 0.005 | 0.219 | 0.000 | 0.005 | 0.025 | 0.028 | 0.022 |
| | | R | 0.109 | 0.000 | 0.005 | 0.312 | 0.000 | 0.005 | 0.032 | 0.036 | 0.028 |
| | | W | 0.109 | 0.008 | 0.005 | 0.406 | 0.008 | 0.005 | 0.040 | 0.045 | 0.036 |
| No. 4 | 0.112 | N | 0.125 | 0.000 | 0.005 | 0.250 | 0.000 | 0.005 | 0.032 | 0.036 | 0.028 |
| | | R | 0.125 | 0.008 | 0.005 | 0.375 | 0.008 | 0.005 | 0.040 | 0.045 | 0.036 |
| | | W | 0.125 | 0.008 | 0.005 | 0.438 | 0.008 | 0.005 | 0.040 | 0.045 | 0.036 |
| No. 5 | 0.125 | N | 0.141 | 0.000 | 0.005 | 0.281 | 0.000 | 0.005 | 0.032 | 0.036 | 0.028 |
| | | R | 0.141 | 0.008 | 0.005 | 0.406 | 0.008 | 0.005 | 0.040 | 0.045 | 0.036 |
| | | W | 0.141 | 0.008 | 0.005 | 0.500 | 0.008 | 0.005 | 0.040 | 0.045 | 0.036 |
| No. 6 | 0.138 | N | 0.156 | 0.000 | 0.005 | 0.312 | 0.000 | 0.005 | 0.032 | 0.036 | 0.028 |
| | | R | 0.156 | 0.008 | 0.005 | 0.438 | 0.008 | 0.005 | 0.040 | 0.045 | 0.036 |
| | | W | 0.156 | 0.008 | 0.005 | 0.562 | 0.008 | 0.005 | 0.040 | 0.045 | 0.036 |
| No. 8 | 0.164 | N | 0.188 | 0.008 | 0.005 | 0.375 | 0.008 | 0.005 | 0.040 | 0.045 | 0.036 |
| | | R | 0.188 | 0.008 | 0.005 | 0.500 | 0.008 | 0.005 | 0.040 | 0.045 | 0.036 |
| | | W | 0.188 | 0.008 | 0.005 | 0.625 | 0.015 | 0.005 | 0.063 | 0.071 | 0.056 |
| No. 10 | 0.190 | N | 0.203 | 0.008 | 0.005 | 0.406 | 0.008 | 0.005 | 0.040 | 0.045 | 0.036 |
| | | R | 0.203 | 0.008 | 0.005 | 0.562 | 0.008 | 0.005 | 0.040 | 0.045 | 0.036 |
| | | W | 0.203 | 0.008 | 0.005 | 0.734[c] | 0.015 | 0.007 | 0.063 | 0.071 | 0.056 |
| No. 12 | 0.216 | N | 0.234 | 0.008 | 0.005 | 0.438 | 0.008 | 0.005 | 0.040 | 0.045 | 0.036 |
| | | R | 0.234 | 0.008 | 0.005 | 0.625 | 0.015 | 0.005 | 0.063 | 0.071 | 0.056 |
| | | W | 0.234 | 0.008 | 0.005 | 0.875 | 0.015 | 0.007 | 0.063 | 0.071 | 0.056 |
| 1/4 | 0.250 | N | 0.281 | 0.015 | 0.005 | 0.500 | 0.015 | 0.005 | 0.063 | 0.071 | 0.056 |
| | | R | 0.281 | 0.015 | 0.005 | 0.734[c] | 0.015 | 0.007 | 0.063 | 0.071 | 0.056 |
| | | W | 0.281 | 0.015 | 0.005 | 1.000 | 0.015 | 0.007 | 0.063 | 0.071 | 0.056 |
| 5/16 | 0.312 | N | 0.344 | 0.015 | 0.005 | 0.625 | 0.015 | 0.005 | 0.063 | 0.071 | 0.056 |
| | | R | 0.344 | 0.015 | 0.005 | 0.875 | 0.015 | 0.007 | 0.063 | 0.071 | 0.056 |
| | | W | 0.344 | 0.015 | 0.005 | 1.125 | 0.015 | 0.007 | 0.063 | 0.071 | 0.056 |
| 3/8 | 0.375 | N | 0.406 | 0.015 | 0.005 | 0.734[c] | 0.015 | 0.007 | 0.063 | 0.071 | 0.056 |
| | | R | 0.406 | 0.015 | 0.005 | 1.000 | 0.015 | 0.007 | 0.063 | 0.071 | 0.056 |
| | | W | 0.406 | 0.015 | 0.005 | 1.250 | 0.030 | 0.007 | 0.100 | 0.112 | 0.090 |
| 7/16 | 0.438 | N | 0.469 | 0.015 | 0.005 | 0.875 | 0.015 | 0.007 | 0.063 | 0.071 | 0.056 |
| | | R | 0.469 | 0.015 | 0.005 | 1.125 | 0.015 | 0.007 | 0.063 | 0.071 | 0.056 |
| | | W | 0.469 | 0.015 | 0.005 | 1.469[c] | 0.030 | 0.007 | 0.100 | 0.112 | 0.090 |
| 1/2 | 0.500 | N | 0.531 | 0.015 | 0.005 | 1.000 | 0.015 | 0.007 | 0.063 | 0.071 | 0.056 |
| | | R | 0.531 | 0.015 | 0.005 | 1.250 | 0.030 | 0.007 | 0.100 | 0.112 | 0.090 |
| | | W | 0.531 | 0.015 | 0.005 | 1.750 | 0.030 | 0.007 | 0.100 | 0.112 | 0.090 |
| 9/16 | 0.562 | N | 0.594 | 0.015 | 0.005 | 1.125 | 0.015 | 0.007 | 0.063 | 0.071 | 0.056 |
| | | R | 0.594 | 0.015 | 0.005 | 1.469[c] | 0.030 | 0.007 | 0.100 | 0.112 | 0.090 |
| | | W | 0.594 | 0.015 | 0.005 | 2.000 | 0.030 | 0.007 | 0.100 | 0.112 | 0.090 |
| 5/8 | 0.625 | N | 0.656 | 0.030 | 0.007 | 1.250 | 0.030 | 0.007 | 0.100 | 0.112 | 0.090 |
| | | R | 0.656 | 0.030 | 0.007 | 1.750 | 0.030 | 0.007 | 0.100 | 0.112 | 0.090 |
| | | W | 0.656 | 0.030 | 0.007 | 2.250 | 0.030 | 0.007 | 0.160 | 0.174 | 0.146 |

**Table 2.** *(Continued)* **American National Standard Type B Plain Washers**

| Nominal Washer Size[a] | | Series[b] | Inside Diameter Basic | Inside Diameter Tolerance Plus | Inside Diameter Tolerance Minus | Outside Diameter Basic | Outside Diameter Tolerance Plus | Outside Diameter Tolerance Minus | Thickness Basic | Thickness Max. | Thickness Min. |
|---|---|---|---|---|---|---|---|---|---|---|---|
| ¾ | 0.750 | N | 0.812 | 0.030 | 0.007 | 1.375 | 0.030 | 0.007 | 0.100 | 0.112 | 0.090 |
| | | R | 0.812 | 0.030 | 0.007 | 2.000 | 0.030 | 0.007 | 0.100 | 0.112 | 0.090 |
| | | W | 0.812 | 0.030 | 0.007 | 2.500 | 0.030 | 0.007 | 0.160 | 0.174 | 0.146 |
| ⅞ | 0.875 | N | 0.938 | 0.030 | 0.007 | 1.469[c] | 0.030 | 0.007 | 0.100 | 0.112 | 0.090 |
| | | R | 0.938 | 0.030 | 0.007 | 2.250 | 0.030 | 0.007 | 0.160 | 0.174 | 0.146 |
| | | W | 0.938 | 0.030 | 0.007 | 2.750 | 0.030 | 0.007 | 0.160 | 0.174 | 0.146 |
| 1 | 1.000 | N | 1.062 | 0.030 | 0.007 | 1.750 | 0.030 | 0.007 | 0.100 | 0.112 | 0.090 |
| | | R | 1.062 | 0.030 | 0.007 | 2.500 | 0.030 | 0.007 | 0.160 | 0.174 | 0.146 |
| | | W | 1.062 | 0.030 | 0.007 | 3.000 | 0.030 | 0.007 | 0.160 | 0.174 | 0.146 |
| 1⅛ | 1.125 | N | 1.188 | 0.030 | 0.007 | 2.000 | 0.030 | 0.007 | 0.100 | 0.112 | 0.090 |
| | | R | 1.188 | 0.030 | 0.007 | 2.750 | 0.030 | 0.007 | 0.160 | 0.174 | 0.146 |
| | | W | 1.188 | 0.030 | 0.007 | 3.250 | 0.030 | 0.007 | 0.160 | 0.174 | 0.146 |
| 1¼ | 1.250 | N | 1.312 | 0.030 | 0.007 | 2.250 | 0.030 | 0.007 | 0.160 | 0.174 | 0.146 |
| | | R | 1.312 | 0.030 | 0.007 | 3.000 | 0.030 | 0.007 | 0.160 | 0.174 | 0.146 |
| | | W | 1.312 | 0.045 | 0.010 | 3.500 | 0.045 | 0.010 | 0.250 | 0.266 | 0.234 |
| 1⅜ | 1.375 | N | 1.438 | 0.030 | 0.007 | 2.500 | 0.030 | 0.007 | 0.160 | 0.174 | 0.146 |
| | | R | 1.438 | 0.030 | 0.007 | 3.250 | 0.030 | 0.007 | 0.160 | 0.174 | 0.146 |
| | | W | 1.438 | 0.045 | 0.010 | 3.750 | 0.045 | 0.010 | 0.250 | 0.266 | 0.234 |
| 1½ | 1.500 | N | 1.562 | 0.030 | 0.007 | 2.750 | 0.030 | 0.007 | 0.160 | 0.174 | 0.146 |
| | | R | 1.562 | 0.045 | 0.010 | 3.500 | 0.045 | 0.010 | 0.250 | 0.266 | 0.234 |
| | | W | 1.562 | 0.045 | 0.010 | 4.000 | 0.045 | 0.010 | 0.250 | 0.266 | 0.234 |
| 1⅝ | 1.625 | N | 1.750 | 0.030 | 0.007 | 3.000 | 0.030 | 0.007 | 0.160 | 0.174 | 0.146 |
| | | R | 1.750 | 0.045 | 0.010 | 3.750 | 0.045 | 0.010 | 0.250 | 0.266 | 0.234 |
| | | W | 1.750 | 0.045 | 0.010 | 4.250 | 0.045 | 0.010 | 0.250 | 0.266 | 0.234 |
| 1¾ | 1.750 | N | 1.875 | 0.030 | 0.007 | 3.250 | 0.030 | 0.007 | 0.160 | 0.174 | 0.146 |
| | | R | 1.875 | 0.045 | 0.010 | 4.000 | 0.045 | 0.010 | 0.250 | 0.266 | 0.234 |
| | | W | 1.875 | 0.045 | 0.010 | 4.500 | 0.045 | 0.010 | 0.250 | 0.266 | 0.234 |
| 1⅞ | 1.875 | N | 2.000 | 0.045 | 0.010 | 3.500 | 0.045 | 0.010 | 0.250 | 0.266 | 0.234 |
| | | R | 2.000 | 0.045 | 0.010 | 4.250 | 0.045 | 0.010 | 0.250 | 0.266 | 0.234 |
| | | W | 2.000 | 0.045 | 0.010 | 4.750 | 0.045 | 0.010 | 0.250 | 0.266 | 0.234 |
| 2 | 2.000 | N | 2.125 | 0.045 | 0.010 | 3.750 | 0.045 | 0.010 | 0.250 | 0.266 | 0.234 |
| | | R | 2.125 | 0.045 | 0.010 | 4.500 | 0.045 | 0.010 | 0.250 | 0.266 | 0.234 |
| | | W | 2.125 | 0.045 | 0.010 | 5.000 | 0.045 | 0.010 | 0.250 | 0.266 | 0.234 |

[a] Nominal washer sizes are intended for use with comparable nominal screw or bolt sizes.

[b] N indicates Narrow; R, Regular; and W, Wide Series.

[c] The 0.734-inch and 1.469-inch outside diameter avoids washers which could be used in coin operated devices.

All dimensions are in inches.

Inside and outside diameters shall be concentric within at least the inside diameter tolerance.

Washers shall be flat within 0.005-inch for basic outside diameters up through 0.875-inch and within 0.010 inch for larger outside diameters.

For 2¼-, 2½-, 2¾-, and 3-inch sizes see ANSI/ASME B18.21.1-2009 (R2016).

**American National Standard Helical Spring and Tooth Lock Washers ANSI/ASME B18.21.1-2009 (R2016).—**This Standard covers helical spring lock washers of carbon steel; boron steel; corrosion resistant steel, Types 302 and 305; aluminum-zinc alloy; phosphor-bronze; silicon-bronze; and K-Monel; in various series. Tooth lock washers of carbon steel having internal teeth, external teeth, and both internal and external teeth, of two constructions, designated as Type A and Type B. Washers intended for general industrial application are also covered. American National Standard Lock Washers (Metric Series) ANSI/ASME B18.21.2M-1999 (R2014) covers metric sizes for helical spring and tooth lock washers.

*Helical Spring Lock Washers:* These washers are used to provide: 1) good bolt tension per unit of applied torque for tight assemblies; 2) hardened bearing surfaces to create uniform torque control; 3) uniform load distribution through controlled radii—section—cut-off; and 4) protection against looseness resulting from vibration and corrosion.

Nominal washer sizes are intended for use with comparable nominal screw or bolt sizes. These washers are designated by the following data in the sequence shown: Product name; nominal size (number, fraction or decimal equivalent); series; material; and protective finish, if required. For example: Helical Spring Lock Washer, 0.375 Extra Duty, Steel, Phosphate Coated.

Helical spring lock washers are available in four series: Regular, heavy, extra duty and hi-collar as given in Table 2 and Table 1. Helical spring lock washers made of materials other than carbon steel are available in the regular series as given in Table 2.

**Table 1. American National Standard High Collar Helical Spring Lock Washers**
*ANSI/ASME B18.21.1-2009 (R2016)*

| Nominal Washer Size | | Inside Diameter | | Outside Diameter | Washer Section | |
|---|---|---|---|---|---|---|
| | | | | | Width | Thickness[a] |
| | | Min. | Max. | Max. | Min. | Min. |
| No. 4 | 0.112 | 0.114 | 0.120 | 0.173 | 0.022 | 0.022 |
| No. 5 | 0.125 | 0.127 | 0.133 | 0.202 | 0.030 | 0.030 |
| No. 6 | 0.138 | 0.141 | 0.148 | 0.216 | 0.030 | 0.030 |
| No. 8 | 0.164 | 0.167 | 0.174 | 0.267 | 0.042 | 0.047 |
| No. 10 | 0.190 | 0.193 | 0.200 | 0.294 | 0.042 | 0.047 |
| 1/4 | 0.250 | 0.252 | 0.260 | 0.363 | 0.047 | 0.078 |
| 5/16 | 0.3125 | 0.314 | 0.322 | 0.457 | 0.062 | 0.093 |
| 3/8 | 0.375 | 0.377 | 0.385 | 0.550 | 0.076 | 0.125 |
| 7/16 | 0.4375 | 0.440 | 0.450 | 0.644 | 0.090 | 0.140 |
| 1/2 | 0.500 | 0.502 | 0.512 | 0.733 | 0.103 | 0.172 |
| 5/8 | 0.625 | 0.628 | 0.641 | 0.917 | 0.125 | 0.203 |
| 3/4 | 0.750 | 0.753 | 0.766 | 1.105 | 0.154 | 0.218 |
| 7/8 | 0.875 | 0.878 | 0.894 | 1.291 | 0.182 | 0.234 |
| 1 | 1.000 | 1.003 | 1.024 | 1.478 | 0.208 | 0.250 |
| 1 1/8 | 1.125 | 1.129 | 1.153 | 1.663 | 0.236 | 0.313 |
| 1 1/4 | 1.250 | 1.254 | 1.280 | 1.790 | 0.236 | 0.313 |
| 1 3/8 | 1.375 | 1.379 | 1.408 | 2.031 | 0.292 | 0.375 |
| 1 1/2 | 1.500 | 1.504 | 1.534 | 2.159 | 0.292 | 0.375 |
| 1 3/4 | 1.750 | 1.758 | 1.789 | 2.596 | 0.383 | 0.469 |
| 2 | 2.000 | 2.008 | 2.039 | 2.846 | 0.383 | 0.469 |
| 2 1/4 | 2.250 | 2.262 | 2.293 | 3.345 | 0.508 | 0.508 |
| 2 1/2 | 2.500 | 2.512 | 2.543 | 3.595 | 0.508 | 0.508 |
| 2 3/4 | 2.750 | 2.762 | 2.793 | 4.095 | 0.633 | 0.633 |
| 3 | 3.000 | 3.012 | 3.043 | 4.345 | 0.633 | 0.633 |

[a] Mean section thickness = (inside thickness + outside thickness) ÷ 2.

**Table 2. American National Standard Helical Spring Lock Washers** *ANSI/ASME B18.21.1-2009 (R2016)*

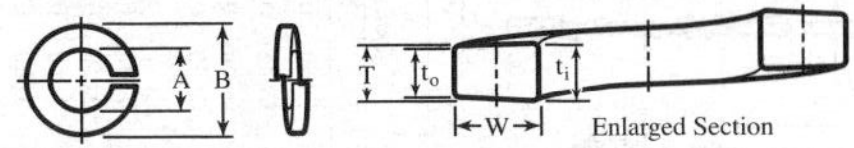


| Nominal Washer Size | | Inside Diameter, A | | Regular | | | Heavy | | | Extra Duty | | |
|---|---|---|---|---|---|---|---|---|---|---|---|---|
| | | Max. | Min. | O.D., B Max. | Section Width, W | Section Thickness, T[a] | O.D., B Max. | Section Width, W | Section Thickness, T[a] | O.D., B Max. | Section Width, W | Section Thickness, T[a] |
| No. 2 | 0.086 | 0.094 | 0.088 | 0.172 | 0.035 | 0.020 | 0.182 | 0.040 | 0.025 | 0.208 | 0.053 | 0.027 |
| No. 3 | 0.099 | 0.107 | 0.101 | 0.195 | 0.040 | 0.025 | 0.209 | 0.047 | 0.031 | 0.239 | 0.062 | 0.034 |
| No. 4 | 0.112 | 0.120 | 0.114 | 0.209 | 0.040 | 0.025 | 0.223 | 0.047 | 0.031 | 0.253 | 0.062 | 0.034 |
| No. 5 | 0.125 | 0.133 | 0.127 | 0.236 | 0.047 | 0.031 | 0.252 | 0.055 | 0.040 | 0.300 | 0.079 | 0.045 |
| No. 6 | 0.138 | 0.148 | 0.141 | 0.250 | 0.047 | 0.031 | 0.266 | 0.055 | 0.040 | 0.314 | 0.079 | 0.045 |
| No. 8 | 0.164 | 0.174 | 0.167 | 0.293 | 0.055 | 0.040 | 0.307 | 0.062 | 0.047 | 0.375 | 0.096 | 0.057 |
| No. 10 | 0.190 | 0.200 | 0.193 | 0.334 | 0.062 | 0.047 | 0.350 | 0.070 | 0.056 | 0.434 | 0.112 | 0.068 |
| No. 12 | 0.216 | 0.227 | 0.220 | 0.377 | 0.070 | 0.056 | 0.391 | 0.077 | 0.063 | 0.497 | 0.130 | 0.080 |
| 1⁄4 | 0.250 | 0.260 | 0.252 | 0.487 | 0.109 | 0.062 | 0.489 | 0.110 | 0.077 | 0.533 | 0.132 | 0.084 |
| 5⁄16 | 0.3125 | 0.322 | 0.314 | 0.583 | 0.125 | 0.078 | 0.593 | 0.130 | 0.097 | 0.619 | 0.143 | 0.108 |
| 3⁄8 | 0.375 | 0.385 | 0.377 | 0.680 | 0.141 | 0.094 | 0.688 | 0.145 | 0.115 | 0.738 | 0.170 | 0.123 |
| 7⁄16 | 0.4375 | 0.450 | 0.440 | 0.776 | 0.156 | 0.109 | 0.784 | 0.160 | 0.133 | 0.836 | 0.186 | 0.143 |
| 1⁄2 | 0.500 | 0.512 | 0.502 | 0.869 | 0.171 | 0.125 | 0.879 | 0.176 | 0.151 | 0.935 | 0.204 | 0.162 |
| 9⁄16 | 0.5625 | 0.574 | 0.564 | 0.965 | 0.188 | 0.141 | 0.975 | 0.193 | 0.170 | 1.035 | 0.223 | 0.182 |
| 5⁄8 | 0.625 | 0.641 | 0.628 | 1.073 | 0.203 | 0.156 | 1.087 | 0.210 | 0.189 | 1.151 | 0.242 | 0.202 |
| 11⁄16 | 0.6875 | 0.704 | 0.691 | 1.170 | 0.219 | 0.172 | 1.186 | 0.227 | 0.207 | 1.252 | 0.260 | 0.221 |
| 3⁄4 | 0.750 | 0.766 | 0.753 | 1.265 | 0.234 | 0.188 | 1.285 | 0.244 | 0.226 | 1.355 | 0.279 | 0.241 |
| 13⁄16 | 0.8125 | 0.832 | 0.816 | 1.363 | 0.250 | 0.203 | 1.387 | 0.262 | 0.246 | 1.458 | 0.298 | 0.261 |
| 7⁄8 | 0.875 | 0.894 | 0.878 | 1.459 | 0.266 | 0.219 | 1.489 | 0.281 | 0.266 | 1.571 | 0.322 | 0.285 |
| 15⁄16 | 0.9375 | 0.958 | 0.941 | 1.556 | 0.281 | 0.234 | 1.590 | 0.298 | 0.284 | 1.684 | 0.345 | 0.308 |
| 1 | 1.000 | 1.024 | 1.003 | 1.656 | 0.297 | 0.250 | 1.700 | 0.319 | 0.306 | 1.794 | 0.366 | 0.330 |
| 1 1⁄16 | 1.0625 | 1.087 | 1.066 | 1.751 | 0.312 | 0.266 | 1.803 | 0.338 | 0.326 | 1.905 | 0.389 | 0.352 |
| 1 1⁄8 | 1.125 | 1.153 | 1.129 | 1.847 | 0.328 | 0.281 | 1.903 | 0.356 | 0.345 | 2.013 | 0.411 | 0.375 |
| 1 3⁄16 | 1.1875 | 1.217 | 1.192 | 1.943 | 0.344 | 0.297 | 2.001 | 0.373 | 0.364 | 2.107 | 0.431 | 0.396 |
| 1 1⁄4 | 1.250 | 1.280 | 1.254 | 2.036 | 0.359 | 0.312 | 2.104 | 0.393 | 0.384 | 2.222 | 0.452 | 0.417 |
| 1 5⁄16 | 1.3125 | 1.344 | 1.317 | 2.133 | 0.375 | 0.328 | 2.203 | 0.410 | 0.403 | 2.327 | 0.472 | 0.438 |
| 1 3⁄8 | 1.375 | 1.408 | 1.379 | 2.219 | 0.391 | 0.344 | 2.301 | 0.427 | 0.422 | 2.429 | 0.491 | 0.458 |
| 1 7⁄16 | 1.4375 | 1.472 | 1.442 | 2.324 | 0.406 | 0.359 | 2.396 | 0.442 | 0.440 | 2.530 | 0.509 | 0.478 |
| 1 1⁄2 | 1.500 | 1.534 | 1.504 | 2.419 | 0.422 | 0.375 | 2.491 | 0.458 | 0.458 | 2.627 | 0.526 | 0.496 |

[a] $T$ = mean section thickness = $(t_i + t_o) \div 2$.

All dimensions are given in inches.*See ANSI/ASME B18.21.1-2009 (R2016) Standard for sizes over 1½ to 3, inclusive, for regular and heavy helical spring lock washers and over 1½ to 2, inclusive, for extra-duty helical spring lock washers.

When carbon steel helical spring lock washers are to be hot-dipped galvanized for use with hot-dipped galvanized bolts or screws, they are to be coiled to limits onto inch in excess of those specified in Table 2 and Table 1 for minimum inside diameter and maximum outside diameter. Galvanizing washers under ¼ inch nominal size are not recommended.

*Tooth Lock Washers:* These washers serve to lock fasteners, such as bolts and nuts, to the component parts of an assembly, or increase the friction between the fasteners and the assembly. They are designated in a manner similar to helical spring lock washers, and are available in carbon steel. Dimensions are given in Table 3 and Table 4.

**Table 3. American National Standard Internal-External Tooth Lock Washers**
*ANSI/ASME B18.21.1-2009 (R2016)*

All dimensions are given in inches except whole numbers under "Size"

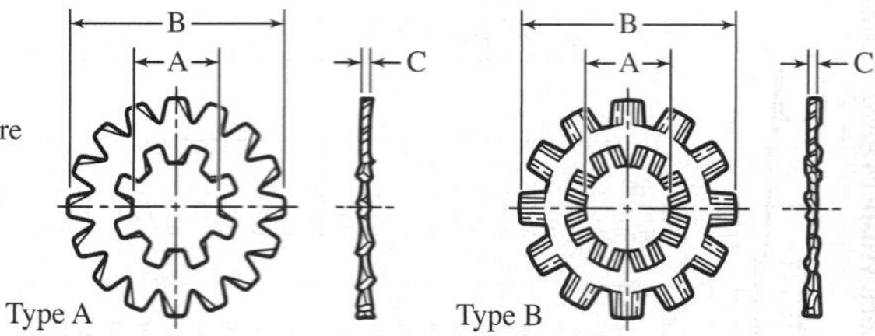


| Size | A Inside Diameter Max. | A Inside Diameter Min. | B Outside Diameter Max. | B Outside Diameter Min. | C Thickness Max. | C Thickness Min. |
|---|---|---|---|---|---|---|
| No. 4 (0.112) | 0.123 | 0.115 | 0.475 | 0.460 | 0.021 | 0.016 |
| | | | 0.510 | 0.495 | 0.021 | 0.017 |
| | | | 0.610 | 0.580 | 0.021 | 0.017 |
| No. 6 (0.138) | .150 | .0141 | 0.510 | 0.495 | 0.028 | 0.023 |
| | | | 0.610 | 0.580 | 0.028 | 0.023 |
| | | | 0.690 | 0.670 | 0.028 | 0.023 |
| No. 8 (0.164) | 0.176 | 0.168 | 0.610 | 0.580 | 0.034 | 0.028 |
| | | | 0.690 | 0.670 | 0.034 | 0.028 |
| | | | 0.760 | 0.740 | 0.034 | 0.028 |
| No. 10 (0.190) | 0.204 | 0.195 | 0.610 | 0.580 | 0.034 | 0.028 |
| | | | 0.690 | 0.670 | 0.040 | 0.032 |
| | | | 0.760 | 0.740 | 0.040 | 0.032 |
| | | | 0.900 | 0.880 | 0.040 | 0.032 |
| No. 12 (0.216) | 0.231 | 0.221 | 0.690 | 0.670 | 0.040 | 0.032 |
| | | | 0.760 | 0.725 | 0.040 | 0.032 |
| | | | 0.900 | 0.880 | 0.040 | 0.032 |
| | | | 0.985 | 0.965 | 0.045 | 0.037 |
| ¼ (0.250) | 0.267 | 0.256 | 0.760 | 0.725 | 0.040 | 0.032 |
| | | | 0.900 | 0.880 | 0.040 | 0.032 |
| | | | 0.985 | 0.965 | 0.045 | 0.037 |
| | | | 1.070 | 1.045 | 0.045 | 0.037 |

| Size | A Inside Diameter Max. | A Inside Diameter Min. | B Outside Diameter Max. | B Outside Diameter Min. | C Thickness Max. | C Thickness Min. |
|---|---|---|---|---|---|---|
| 5⁄16 (0.312) | 0.332 | 0.320 | 0.900 | 0.865 | 0.040 | 0.032 |
| | | | 0.985 | 0.965 | 0.045 | 0.037 |
| | | | 1.070 | 1.045 | 0.050 | 0.042 |
| | | | 1.155 | 1.130 | 0.050 | 0.042 |
| ⅜ (0.375) | 0.398 | 0.384 | .985 | .965 | 0.045 | 0.037 |
| | | | 1.070 | 1.045 | 0.050 | 0.042 |
| | | | 1.155 | 1.130 | 0.050 | 0.042 |
| | | | 1.260 | 1.220 | 0.050 | 0.042 |
| 7⁄16 (0.438) | 0.464 | 0.448 | 1.070 | 1.045 | 0.050 | 0.042 |
| | | | 1.155 | 1.130 | 0.050 | 0.042 |
| | | | 1.260 | 1.220 | 0.055 | 0.047 |
| | | | 1.315 | 1.290 | 0.055 | 0.047 |
| ½ (0.500) | 0.530 | 0.512 | 1.260 | 1.220 | 0.055 | 0.047 |
| | | | 1.315 | 1.290 | 0.055 | 0.047 |
| | | | 1.410 | 1.380 | 0.060 | 0.052 |
| | | | 1.620 | 1.590 | 0.067 | 0.059 |
| 9⁄16 (0.562) | .596 | .576 | 1.315 | 1.290 | 0.055 | 0.047 |
| | | | 1.430 | 1.380 | 0.060 | 0.052 |
| | | | 1.620 | 1.590 | 0.067 | 0.059 |
| | | | 1.830 | 1.797 | 0.067 | 0.059 |
| ⅝ (0.625) | .663 | .640 | 1.410 | 1.380 | 0.060 | 0.052 |
| | | | 1.620 | 1.590 | 0.067 | 0.059 |
| | | | 1.830 | 1.797 | 0.067 | 0.059 |
| | | | 1.975 | 1.935 | 0.067 | 0.059 |

All dimensions are given in inches.

**Table 4. American National Standard Internal and External Tooth Lock Washers** *ANSI/ASME B18.21.1-2009 (R2016)*

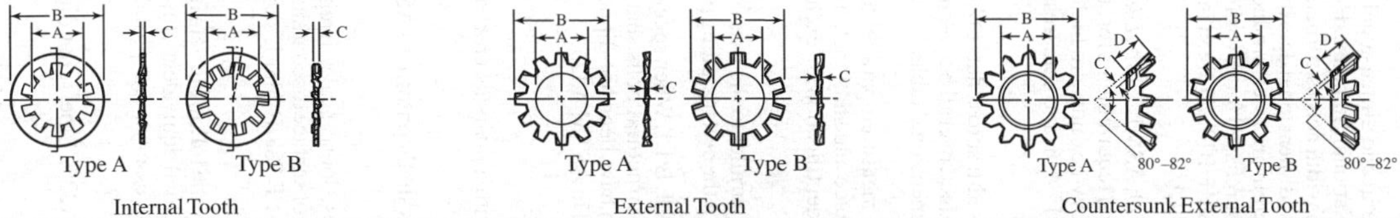


Internal Tooth Lock Washers

| | Size | #2 | #3 | #4 | #5 | #6 | #8 | #10 | #12 | 1⁄4 | 5⁄16 | 3⁄8 | 7⁄16 | 1⁄2 | 9⁄16 | 5⁄8 | 11⁄16 | 3⁄4 | 13⁄16 | 7⁄8 | 1 | 1 1⁄8 | 1 1⁄4 |
|---|---|---|---|---|---|---|---|---|---|---|---|---|---|---|---|---|---|---|---|---|---|---|---|
| *A* | Max. | 0.095 | 0.109 | 0.123 | 0.136 | 0.150 | 0.176 | 0.204 | 0.231 | 0.267 | 0.332 | 0.398 | 0.464 | 0.530 | 0.596 | 0.663 | 0.728 | 0.795 | 0.861 | 0.927 | 1.060 | 1.192 | 1.325 |
| | Min. | 0.089 | 0.102 | 0.115 | 0.129 | 0.141 | 0.168 | 0.195 | 0.221 | 0.256 | 0.320 | 0.384 | 0.448 | 0.512 | 0.576 | 0.640 | 0.704 | 0.769 | 0.832 | 0.894 | 1.019 | 1.144 | 1.275 |
| *B* | Max. | 0.200 | 0.232 | 0.270 | 0.280 | 0.295 | 0.340 | 0.381 | 0.410 | 0.478 | 0.610 | 0.692 | 0.789 | 0.900 | 0.985 | 1.071 | 1.166 | 1.245 | 1.315 | 1.410 | 1.637 | 1.830 | 1.975 |
| | Min. | 0.175 | 0.215 | 0.245 | 0.255 | 0.275 | 0.325 | 0.365 | 0.394 | 0.460 | 0.594 | 0.670 | 0.740 | 0.867 | 0.957 | 1.045 | 1.130 | 1.220 | 1.290 | 1.364 | 1.590 | 1.799 | 1.921 |
| *C* | Max. | 0.016 | 0.016 | 0.018 | 0.020 | 0.022 | 0.023 | 0.024 | 0.027 | 0.028 | 0.034 | 0.040 | 0.040 | 0.045 | 0.045 | 0.050 | 0.050 | 0.055 | 0.055 | 0.060 | 0.067 | 0.067 | 0.067 |
| | Min. | 0.010 | 0.010 | 0.012 | 0.014 | 0.016 | 0.018 | 0.018 | 0.020 | 0.023 | 0.028 | 0.032 | 0.032 | 0.037 | 0.037 | 0.042 | 0.042 | 0.047 | 0.047 | 0.052 | 0.059 | 0.059 | 0.059 |
| External Tooth Lock Washers | | | | | | | | | | | | | | | | | | | | | | | |
| *A* | Max. | … | 0.109 | 0.123 | 0.136 | 0.150 | 0.176 | 0.204 | 0.231 | 0.267 | 0.332 | 0.398 | 0.464 | 0.530 | 0.596 | 0.663 | 0.728 | 0.795 | 0.861 | 0.927 | 1.060 | … | … |
| | Min. | … | 0.102 | 0.115 | 0.129 | 0.141 | 0.168 | 0.195 | 0.221 | 0.256 | 0.320 | 0.384 | 0.448 | 0.513 | 0.576 | 0.641 | 0.704 | 0.768 | 0.833 | 0.897 | 1.025 | … | … |
| *B* | Max. | … | 0.235 | 0.260 | 0.285 | 0.320 | 0.381 | 0.410 | 0.475 | 0.510 | 0.610 | 0.694 | 0.760 | 0.900 | 0.985 | 1.070 | 1.155 | 1.260 | 1.315 | 1.410 | 1.620 | … | … |
| | Min. | … | 0.220 | 0.245 | 0.270 | 0.305 | 0.365 | 0.395 | 0.460 | 0.494 | 0.588 | 0.670 | 0.740 | 0.880 | 0.960 | 1.045 | 1.130 | 1.220 | 1.290 | 1.380 | 1.590 | … | … |
| *C* | Max. | … | 0.016 | 0.018 | 0.020 | 0.022 | 0.023 | 0.024 | 0.027 | 0.028 | 0.034 | 0.040 | 0.040 | 0.045 | 0.045 | 0.050 | 0.050 | 0.055 | 0.055 | 0.060 | 0.067 | … | … |
| | Min. | … | 0.010 | 0.012 | 0.014 | 0.016 | 0.018 | 0.018 | 0.020 | 0.023 | 0.028 | 0.032 | 0.032 | 0.037 | 0.037 | 0.042 | 0.042 | 0.047 | 0.047 | 0.052 | 0.059 | … | … |

Heavy Internal Tooth Lock Washers

| | Size | 1⁄4 | 5⁄16 | 3⁄8 | 7⁄16 | 1⁄2 | 9⁄16 | 5⁄8 | 3⁄4 | 7⁄8 |
|---|---|---|---|---|---|---|---|---|---|---|
| *A* | Max. | 0.267 | 0.332 | 0.398 | 0.464 | 0.530 | 0.596 | 0.663 | 0.795 | 0.927 |
| | Min. | 0.256 | 0.320 | 0.384 | 0.448 | 0.512 | 0.576 | 0.640 | 0.768 | 0.894 |
| *B* | Max. | 0.536 | 0.607 | 0.748 | 0.858 | 0.924 | 1.034 | 1.135 | 1.265 | 1.447 |
| | Min. | 0.500 | 0.590 | 0.700 | 0.800 | 0.880 | 0.990 | 1.100 | 1.240 | 1.400 |
| *C* | Max. | 0.045 | 0.050 | 0.050 | 0.067 | 0.067 | 0.067 | 0.067 | 0.084 | 0.084 |
| | Min. | 0.035 | 0.040 | 0.042 | 0.050 | 0.055 | 0.055 | 0.059 | 0.070 | 0.075 |

Countersunk External Tooth Lock Washers[a]

| | Size | #4 | #6 | #8 | #10 | #12 | 1⁄4 | #16 | 5⁄16 | 3⁄8 | 7⁄16 | 1⁄2 |
|---|---|---|---|---|---|---|---|---|---|---|---|---|
| *A* | Max. | 0.123 | 0.150 | 0.177 | 0.205 | 0.231 | 0.267 | 0.287 | 0.333 | 0.398 | 0.463 | 0.529 |
| | Min. | 0.113 | 0.140 | 0.167 | 0.195 | 0.220 | 0.255 | 0.273 | 0.318 | 0.383 | 0.448 | 0.512 |
| *C* | Max. | 0.019 | 0.021 | 0.021 | 0.025 | 0.025 | 0.025 | 0.028 | 0.028 | 0.034 | 0.045 | 0.045 |
| | Min. | 0.015 | 0.017 | 0.017 | 0.020 | 0.020 | 0.020 | 0.023 | 0.023 | 0.028 | 0.037 | 0.037 |
| *D* | Max. | 0.065 | 0.092 | 0.099 | 0.105 | 0.128 | 0.128 | 0.147 | 0.192 | 0.255 | 0.270 | 0.304 |
| | Min. | 0.050 | 0.082 | 0.083 | 0.088 | 0.118 | 0.113 | 0.137 | 0.165 | 0.242 | 0.260 | 0.294 |

[a] Starting with #4, approx. O.D.'s are: 0.213, 0.289, 0.322, 0.354, 0.421 0.454, 0.505, 0.599, 0.765, 0.867, and 0.976.

All dimensions are given in inches.

## Fasteners for Use in Structural Applications

The ANSI/ASME B18.2.6 Standard covers the complete general and dimensional data for four products in the inch series recognized as American National Standard. Heavy hex nut thickness formulas and those for width across flats and width across corners are found in the Appendix of ANSI/ASME B18.2.2. The inclusion of dimensional data in this Standard is not intended to imply that all products described herein are stock production sizes. Fasteners intended for use in structural applications and purchased for government use shall conform to this Standard. All dimensions in ANSI/ASME B18.2.6 are in inches unless stated otherwise, and apply to unplated or uncoated product.

Symbols specifying geometric characteristics are in accord with ASME Y14.5, *Dimensioning and Tolerancing*. Standards for chemical and mechanical requirements for structural bolts are included in ASTM A325 and ASTM A490. Heavy hex nuts are included in ASTM A563. Hardened steel washers are included in ASTM F436, and compressible washer type direct tension indicators in ASTM F 959.

**Heavy Hex Structural Bolts.—***Head Width Across Flats:* The width across flats of heads shall be the distance measured perpendicular to the axis of product, overall between two opposite sides of the head in accordance with the notes in the dimensional table.

*Head Height:* The head height shall be that overall distance measured parallel to the axis of the product from the top of the head to the bearing surface and shall include the thickness of the washer face. Raised grade and manufacturer's identification are excluded from head height.

*Bolt Length:* The bolt length shall be the distance measured parallel to the axis of the product from the bearing surface of the head to the extreme end of the bolt including point.

*Threads:* Threads shall be cut or rolled in accordance with ASME B 1.1. When specified, 8 thread series may be used on bolts over 1 inch in diameter. Structural bolts shall not be undersized to accommodate heavy coatings. Threads which have been hot-dipped or mechanically zinc coated shall meet the requirements specified in ASTM A325.

*Body Diameter:* The body diameter limits are shown in Table 5.

*Finish:* Unless otherwise specified, bolts shall be supplied with a plain (as processed) finish. Bolts to ASTM A490 shall not be metallic coated.

*Materials:* Chemical and mechanical properties of steel bolts shall conform to ASTM A325 or ASTM A490.

*Workmanship:* Bolts shall be free from burrs, seams, laps, loose scale, irregular surfaces, and any defects affecting serviceability. When control of surface discontinuities is required, the purchaser shall specify conformance to ASTM F 788/F 788M, surface discontinuities of bolts, screws, and studs, inch and metric series.

*Designation:* Heavy hex structural bolts shall be designated by the following data in the sequence shown: product name, nominal size (fractional or decimal equivalent), threads per inch, material (including specification and type where necessary), and protective finish (if required). See example below:

*Example:* Heavy Hex Structural Bolt, $^3/_4$-10 × 2 $^1/_4$, ASTM A325, Type 1, hot dipped zinc coated.

*Identification Grade Symbols:* Each bolt shall be marked in accordance with the requirements of the applicable specification; ASTM A325 or ASTM A490 and requirements of Table 5, Note (13).

## Table 5. Dimensions of Heavy Hex Structural Bolts *ASME B18.2.6-2019*

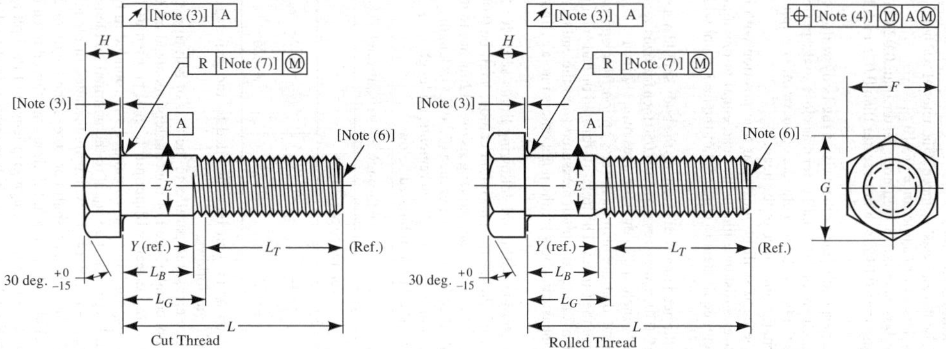


| Nominal Size or Basic Product Diameter Note (15) | | Body Diameter, *E* Note (5) | | Width Across Flats *F* Note (2) | | | Width Across Corners *G* | | Head Height *H* | | | Radius of Fillet *R* | | Thread Length $L_T$ Note (10) | Transition Thread Length *Y* Note (10) | Total Runout of Bearing Surface FIM Note (3) |
|---|---|---|---|---|---|---|---|---|---|---|---|---|---|---|---|---|
| [No a (15)] | | Max. | Min. | Basic | Max. | Min. | Max. | Min. | Basic | Max. | Min. | Max. | Min. | Ref. | Max., Ref. | Max. |
| ½ | 0.5000 | 0.515 | 0.482 | 7⁄8 | 0.875 | 0.850 | 1.010 | 0.969 | 5⁄16 | 0.323 | 0.302 | 0.031 | 0.009 | 1.00 | 0.19 | 0.016 |
| 5⁄8 | 0.6250 | 0.642 | 0.605 | 1 1⁄16 | 1.062 | 1.031 | 1.227 | 1.175 | 25⁄64 | 0.403 | 0.378 | 0.062 | 0.021 | 1.25 | 0.22 | 0.019 |
| ¾ | 0.7500 | 0.768 | 0.729 | 1¼ | 1.250 | 1.212 | 1.443 | 1.383 | 15⁄32 | 0.483 | 0.455 | 0.062 | 0.021 | 1.38 | 0.25 | 0.022 |
| 7⁄8 | 0.8750 | 0.895 | 0.852 | 1 7⁄16 | 1.438 | 1.394 | 1.660 | 1.589 | 35⁄64 | 0.563 | 0.531 | 0.062 | 0.031 | 1.50 | 0.28 | 0.025 |
| 1 | 1.0000 | 1.022 | 0.976 | 1 5⁄8 | 1.625 | 1.575 | 1.876 | 1.796 | 39⁄64 | 0.627 | 0.591 | 0.093 | 0.062 | 1.75 | 0.31 | 0.028 |
| 1 1⁄8 | 1.1250 | 1.149 | 1.098 | 1 13⁄16 | 1.812 | 1.756 | 2.093 | 2.002 | 11⁄16 | 0.718 | 0.658 | 0.093 | 0.062 | 2.00 | 0.34 | 0.032 |
| 1¼ | 1.2500 | 1.277 | 1.223 | 2 | 2.000 | 1.938 | 2.309 | 2.209 | 25⁄32 | 0.813 | 0.749 | 0.093 | 0.062 | 2.00 | 0.38 | 0.035 |
| 1 3⁄8 | 1.3750 | 1.404 | 1.345 | 2 3⁄16 | 2.188 | 2.119 | 2.526 | 2.416 | 27⁄32 | 0.878 | 0.810 | 0.093 | 0.062 | 2.25 | 0.44 | 0.038 |
| 1½ | 1.5000 | 1.531 | 1.470 | 2 3⁄8 | 2.375 | 2.300 | 2.742 | 2.622 | 15⁄16 | 0.974 | 0.902 | 0.093 | 0.062 | 2.25 | 0.44 | 0.041 |

(1) *Top of Head*: Top of head shall be full form and chamfered or rounded with the diameter of chamfer circle or start of rounding being equal to the maximum width across flats within a tolerance of −15%.

(2) *Head Taper*: Maximum width across flats shall not be exceeded. No transverse section through the head between 25 and 75 percent of actual head height, as measured from the bearing surface, shall be less than the minimum width across flats.

(3) *Bearing Surface*: Bearing surface shall be flat and washer faced. Diameter of washer face shall be equal to the maximum width across flats within a tolerance of −10%.

Thickness of the washer face shall be not less than 0.015 inch nor greater than 0.025 inch for bolt sizes ¾ inch and smaller; and not less than 0.015 inch nor greater than 0.035 inch for sizes larger than ¾ inch.

The plane of the bearing surface shall be perpendicular to the axis of the body within the FIM limits specified for total runout. Measurement of FIM shall extend as close to the periphery of the bearing surface as possible while the bolt is being held in a collet or other gripping device at a distance of one bolt diameter from the underside of the head.

(4) *True Position of Head*: The axis of the head shall be located at true position with respect to the axis of the body (determined over a distance under the head equal to one diameter) within a tolerance zone having a diameter equivalent to 6 percent of the maximum width across flats at maximum material condition.

(5) *Body Diameter*: Any swell or fin under the head or any die seam on the body should not exceed the basic bolt diameter by the following: 0.030 in, for sizes ½ inch; 0.050 inch for sizes ⅝ and ¾ inch; 0.060 inch for sizes over ¾ inch to 1¼ inch; 0.090 inch for sizes over 1¼ inch.

(6) *Point*: Point shall be chamfered or rounded at the manufacturer's option from approximately 0.016 inch below the minor diameter of the thread. The first full formed thread at major diameter is located a distance no greater than 2 times the pitch measured from the end of the screw. This distance is to be determined by measuring how far the point enters into a cylindrical NOT GO major diameter ring gage.

(7) *Straightness*: Shanks of bolts shall be straight within the following limits at MMC; for bolts with nominal lengths to and including 12 inches the maximum camber shall be 0.006 inch per inch (0.006L) of bolt length, and for bolts with nominal lengths over 12 inches to and including 24 inches the maximum camber shall be 0.008 inch per inch (0.008L) of length.

(8) *Bolt Length*: Bolts are normally supplied in ¼ inch length increments, all lengths.

(9) *Length Tolerance*: Bolt length tolerances shall be as tabulated below:

| Nominal Bolt Size | ½ | ⅝ | ¾ through 1 | 1⅛ through 1½ |
|---|---|---|---|---|
| Nominal Bolt Length | Tolerance of Length | | | |
| Through 6 inches | −0.12 | −0.12 | −0.19 | −0.25 |
| Over 6 inches | −0.19 | −0.25 | −0.25 | −0.25 |

(10) *Thread Length*: The length of thread on bolts shall be controlled by the grip gaging length, LG, max., and the body length, $L_B$, min., as follows.

Grip gaging length, $L_G$, max., is the distance measured parallel to the axis of bolt from the underhead bearing surface to the face of a noncounterbored or noncountersunk standard G0 thread ring gage assembled by hand as far as the thread will permit. It shall be used as the criterion for inspection. The maximum grip gaging length, as calculated and rounded to two decimal places for any bolt not threaded full length, shall be equal to the nominal bolt length minus the basic thread length ($L_G$, max. = $L_{nom} - L_T$). For bolts which are threaded full length, $L_G$, max. defines the unthreaded length under the head and shall not exceed the length of 2.5 times the thread pitch for sizes up to and including 1 inch, and 3.5 times the thread pitch for sizes larger than 1 inch, $L_G$, max. represents the minimum design grip length of the bolt and may be used for determining thread availability when selecting bolt lengths even though usable threads may extend beyond this point.

Basic thread length, $L_T$, is a reference dimension, intended for calculation purposes only, which represents the distance from the extreme end of the bolt to the last complete (full form) thread.

Body length, $L_B$, min., is the distance measured parallel to the axis of bolt from the underhead bearing surface to the last scratch of thread or to the top of the extrusion angle. It shall be used as a criterion for inspection. The minimum body length, as calculated and rounded to two decimal places, shall be equal to the maximum grip gaging length minus the maximum transition thread length ($L_{B,\,min} = L_{G,\,max} - Y_{max}$). Bolts of nominal lengths which have a calculated $L_{B,\,min}$ length equal to or shorter than 2.5 times the thread pitch for sizes 1 inch and smaller, and 3.5 times the thread pitch for sizes larger than 1 inch shall be threaded for full length.

Transition thread length $Y$ is a reference dimension, intended for calculation purposes only, which represents the length of incomplete threads and tolerance on grip gaging length.

(11) *Incomplete Thread Diameter*: The major diameter of incomplete thread shall not exceed the actual major diameter of the full form thread.

(12) *Threads*: Threads, when rolled, shall be in the Unified inch coarse or 8 thread series (UNRC or 8 UNR Series), Class 2A. Threads produced by other methods may be Unified inch coarse or 8 thread series (UNC or 8 UN Series), Class 2A. Acceptability of screw threads shall be determined based on System 21, ANSI/ASME B1.3, screw thread gaging systems for dimensional acceptability, unless otherwise specified.

(13) *Identification Symbols*: Identification marking symbols on the tops of heads for bolt sizes ⅝ inch and smaller shall project not less than 0.005 inch above the surface nor more than 0.015 inch over the specified maximum head height. Bolt sizes larger than ⅝ inch shall project not less than the equivalent in inches of 0.0075 times the basic bolt diameter above the surface nor more than 0.030 inch over the specified maximum head height.

(14) *Material*. Chemical and mechanical properties of steel bolts shall conform to ASTM A325 or ASTM A490.

(15) *Nominal Size*. Where specifying nominal size in decimals, zeros preceding the decimal and in the fourth decimal place shall be omitted.

(16) *Dimensional Conformance*. Heavy hex structural bolts shall have the following characteristics inspected to ANSI/ASME B18.18-2017 (R2021) to the inspection level C: threads, width across comers, head height, grip length, visual.

If verifiable in-process inspection is used, see *Dimensional Characteristics* on page 1705.

*Identification Source Symbols:* Each bolt shall be marked to identify its source (manufacturer, or private label distributor) accepting the responsibility for conformance to this and other applicable specifications.

*Quality Assurance:* Unless otherwise specified, products shall be furnished in accordance with ANSI/ASME B18.18-2017 (R2021) as noted as below.

*Dimensional Characteristics:* Bolts shall conform to the dimensions indicated in Table 5. The designated characteristics defined in Table 5, Note (16) shall be inspected in accordance with ANSI/ASME B18.18. For nondesignated characteristics, the provisions of ANSI/ASME B18.18 shall apply. Should a nondesignated dimension be determined to have a variance, it shall be deemed conforming to this Standard if the user, who is the installer, accepts the variance based on fit, form, and function considerations. Where verifiable in-process inspection is used in accordance with ANSI/ASME B18.18-2017 (R2021), the final inspection level sample sizes of those respective standards shall apply.

**Heavy Hex Nuts.—** *Width Across Flats:* The width across flats of heavy hex nuts shall be the overall distance measured, perpendicular to the axis of the nut, between two opposite sides of the nut in accordance with the notes of Table 6. For milled-from-bar hex nuts, the nominal bar size used shall be the closest commercially available size to the specified basic width across flats of the nut.

**Table 6. Dimensions of Heavy Hex Nuts for Use with Structural Bolts** *ANSI/ASME B18.2.6-2019*

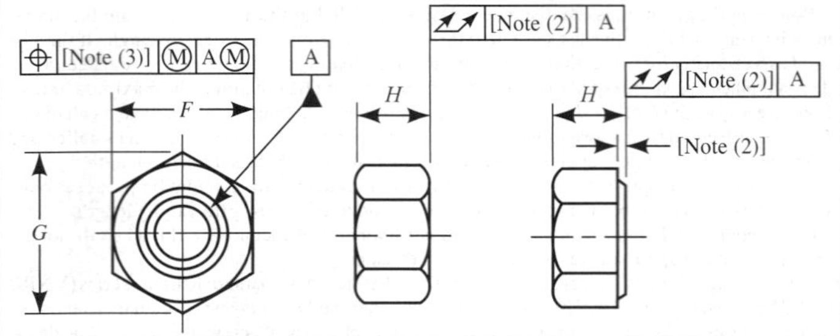


| Nominal Size or Basic Product Diameter | | F Width Across Flats | | | G Width Across Corners | | H Thickness | | | Runout of Bearing Face FIM Heavy Hex Nuts Specified Proof Load | |
|---|---|---|---|---|---|---|---|---|---|---|---|
| | | Basic | Max. | Min. | Max. | Min. | Basic | Max. | Min. | Up to 150,000 psi | 150,000 psi and Greater |
| [Note (1)] | | [Note (6)] | | | [Note (5)] | | | | | [Note (2)] | |
| 1/2 | 0.5000 | 7/8 | 0.875 | 0.850 | 1.010 | 0.969 | 31/64 | 0.504 | 0.464 | 0.023 | 0.016 |
| 5/8 | 0.6250 | 11/16 | 1.062 | 1.031 | 1.227 | 1.175 | 39/64 | 0.631 | 0.587 | 0.025 | 0.018 |
| 3/4 | 0.7500 | 1 1/4 | 1.250 | 1.212 | 1.443 | 1.382 | 47/64 | 0.758 | 0.710 | 0.027 | 0.020 |
| 7/8 | 0.8750 | 1 7/16 | 1.438 | 1.394 | 1.660 | 1.589 | 55/64 | 0.885 | 0.833 | 0.029 | 0.022 |
| 1 | 1.0000 | 1 5/8 | 1.625 | 1.575 | 1.876 | 1.796 | 63/64 | 1.012 | 0.956 | 0.031 | 0.024 |
| 1 1/8 | 1.1250 | 1 13/16 | 1.812 | 1.756 | 2.093 | 2.002 | 1 7/64 | 1.139 | 1.079 | 0.033 | 0.027 |
| 1 1/4 | 1.2500 | 2 | 2.000 | 1.938 | 2.309 | 2.209 | 1 7/32 | 1.215 | 1.187 | 0.035 | 0.030 |
| 1 3/8 | 1.3750 | 2 3/16 | 2.188 | 2.119 | 2.526 | 2.416 | 1 11/32 | 1.378 | 1.310 | 0.038 | 0.033 |
| 1 1/2 | 1.5000 | 2 3/8 | 2.375 | 2.300 | 2.742 | 2.622 | 1 15/32 | 1.505 | 1.433 | 0.041 | 0.036 |

Note: Complete table included in B18.2.2 Square and Hex Nuts (Inch Series)

(1) *Unification*: Only the 9/16 size is not unified dimensionally with British and Canadian standards. Unification of fine thread products is limited to sizes 1 inch and under.

(2) *Tops and Bearing Surfaces of Nuts*: Nuts may be double chamfered or have washer faced bearing surface and chamfered top.

The diameter of chamfer circle on double chamfered nuts and diameter of washer face shall be within the limits of the maximum width across flats and 95 percent of the minimum width across flats.

The tops of washer faced nuts shall be flat and the diameter of chamfer circle shall be equal to the maximum width across flats within a tolerance of −15%. The length of chamfer at hex corners shall be 5 to 15 percent of the basic thread diameter.The surface of chamfer may be slightly convex or rounded.

Bearing surfaces shall be flat and, unless otherwise specified, shall be perpendicular to the axis of the threaded hole within the total runout (FIM) tabulated for the respective nut size, type, and strength level.

(3) *True Position of Tapped Hole*: At maximum material condition, the axis of nut body shall be located at true position with respect to the axis of the tapped hole within a tolerance zone having a diameter equivalent to 4 percent of the maximum width across flats for 1 1/2 inch nominal size nuts or smaller.

(4) *Countersink*: Tapped hole shall be countersunk on the bearing face or faces. The maximum countersink diameter shall be 1.08 times the thread basic (nominal) major diameter. No part of the threaded portion shall project beyond the bearing surface.

(5) *Corner Fill*: A rounding or lack of fill at junction of hex corners with chamfer shall be permissible provided the width across corners is within specified limits at and beyond a distance equal to 17.5% of the basic thread diameter from the chamfered faces.

(6) *Width Across Flats*: Maximum width across flats shall not be exceeded (see exception in *Heavy Hex Nuts.—Width Across Flats*). No transverse section through the nut between 25 and 75 percent of the actual nut thickness, as measured from the bearing surface, shall be less than the minimum width across flats. For milled-from-bar nuts, *Heavy Hex Nuts.—Width Across Flats*, pertaining to the nominal bar size to be used.

(7) *Threads*: Threads shall be UNC or 8 UN Class 2B in accordance with ANSI/ASME B1.1, Unified Inch Screw Threads. When specified, 8 thread series may be used on nuts over 1 inch in diameter.

(8) *Dimensional Conformance*: Heavy hex nuts shall have the following characteristics inspected to ANSI/ASME B18.18-2017 (R2021) to the inspection levels as follows: Width across corners, inspection level C; Thickness, inspection level B; Visual, inspection level C.

If verifiable in-process inspection is used, see *Dimensional Characteristics* on page 1707.

*Nut Thickness:* The nut thickness shall be the overall distance measured parallel to the axis of nut, from the top of the nut to the bearing surface, and shall include the thickness of the washer face where provided.

*Threads:* Threads shall be in accordance with Table 6, Note (7).

*Thread Gaging:* Unless otherwise specified by the purchaser, gaging for screw thread dimensional acceptability shall be in accordance with Gaging System 21 as specified in ANSI/ASME B1.3 Screw Thread Gaging Systems for Acceptability: Inch and Metric Screw Threads (UN, UNR, UNJ, M, and MJ).

*Overtapping:* When nuts are zinc coated, they shall be overtapped after coating in accordance with the provisions of ASTM A563.

*Finish:* Unless otherwise specified, nuts shall be supplied with a plain (as-processed) finish, unplated or uncoated.

*Materials:* Chemical and mechanical properties of heavy hex nuts shall conform to ASTM A563.

*Workmanship:* Nuts shall be free from burrs, seams, laps, loose scale, irregular surfaces, and any defects affecting their serviceability. When control of surface discontinuities is required, the purchaser shall specify conformance to ASTM F 812/F 812M, Surface Discontinuities of Nuts—Inch and Metric Series.

*Designation:* Nuts shall be designated by the following data in sequence shown: product name, nominal size (fraction or decimal), threads per inch, material (including specification where necessary), protective finish (if required). See example below:

*Example:* Heavy Hex Nut, $^1/_2$-13, ASTM A563 Grade C, Plain Finish

*Identification Grade Symbols:* Each nut shall be marked in accordance with the requirements of the applicable specification: ASTM A563 or ASTM A194.

*Identification Source Symbols:* Each nut shall be marked to identify its source (manufacturer, or private label distributor) accepting the responsibility for conformance to this and other applicable specifications.

*Quality Assurance:* Unless otherwise specified, products shall be furnished in accordance with ANSI/ASME B18.18 as noted in *Dimensional Characteristics*.

*Dimensional Characteristics:* Products shall conform to the dimensions indicated for the heavy hex nut in Table 6. The designated characteristics defined in Table 6, Note (8) shall be inspected in accordance with ANSI/ASME B18.18. For nondesignated characteristics, the provision of ANSI/ASME B18.18 shall apply. Should a nondesignated dimension be determined to have a variance, it shall be deemed conforming to this Standard if the user, who is the installer, accepts the variance based on fit, form, and function considerations. Where verifiable in-process inspection is used in accordance with ANSI/ASME B18.18-2017 (R2021), the final inspection level sample sizes of those respective standards shall apply.

## Hardened Steel Washers

**Flat Washers.—***Flat Washers Dimensions:* All circular and circular clipped washers shall conform to the dimensions shown in Table 7.

**Table 7. Dimensions for Hardened Steel Circular and Circular Clipped Washers** *ANSI/ASME B18.2.6-2019*

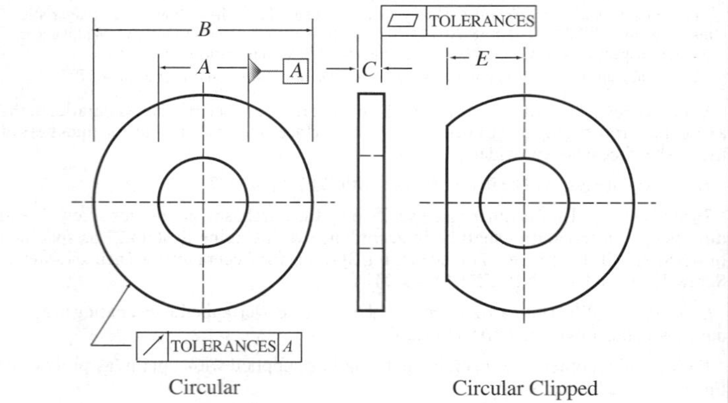


| Bolt Size or Nominal[a] Washer Size, Inch | Inside Diameter A | | | Outside Diameter B | | | Thickness C | | Minimum Edge Distance E |
|---|---|---|---|---|---|---|---|---|---|
| | Basic Washer | Tolerance | | Basic | Tolerance | | Min. | Max. | |
| | | Plus | Minus | | Plus | Minus | | | |
| ½ | 0.531 | 0.0313 | 0 | 1.063 | 0.0313 | 0.0313 | 0.097 | 0.177 | 0.438 |
| ⅝ | 0.688 | 0.0313 | 0 | 1.313 | 0.0313 | 0.0313 | 0.122 | 0.177 | 0.547 |
| ¾ | 0.813 | 0.0313 | 0 | 1.469 | 0.0313 | 0.0313 | 0.122 | 0.177 | 0.656 |
| ⅞ | 0.938 | 0.0313 | 0 | 1.750 | 0.0313 | 0.0313 | 0.136 | 0.177 | 0.766 |
| 1 | 1.125 | 0.0313 | 0 | 2.000 | 0.0313 | 0.0313 | 0.136 | 0.177 | 0.875 |
| 1⅛ | 1.250 | 0.0313 | 0 | 2.250 | 0.0313 | 0.0313 | 0.136 | 0.177 | 0.984 |
| 1¼ | 1.375 | 0.0313 | 0 | 2.500 | 0.0313 | 0.0313 | 0.136 | 0.177 | 1.094 |
| 1⅜ | 1.500 | 0.0313 | 0 | 2.750 | 0.0313 | 0.0313 | 0.136 | 0.177 | 1.203 |
| 1½ | 1.625 | 0.0313 | 0 | 3.000 | 0.0313 | 0.0313 | 0.136 | 0.177 | 1.313 |

[a] Nominal washer sizes are intended for use with comparable nominal bolt diameters.

General Notes:

(1) *Dimensional Conformance*: Circular and circular clipped washers shall have the following characteristics inspected to ANSI/ASME B18.18-2017 (R2021) to the inspection levels as follows: Width across corners, inspection level B; Visual, inspection level C.

If verifiable in-process inspection is used, see *Dimensional Characteristics* on page 1709.

(2) Nominal washer sizes are intended for use with comparable nominal bolt diameters.

(3) Additional requirements are in *Beveled Washers* on page 1709.

*Flat Washers Tolerances:* Washer inside diameter, outside diameter, thickness, and edge distance shall be in accordance with Table 7. The deviation from flatness shall not exceed 0.010 inch as the maximum deviation from a straight edge placed on the cut side. Circular runout of the outside diameter with respect to the hole shall not exceed 0.030 FIM. Burrs shall not project above immediately adjacent washer surface more than 0.010 inch.

*Finish:* Unless otherwise specified, washers shall be supplied with a plain (as processed) finish. If zinc coatings are required, they shall be in accordance with ASTM F 436.

*Materials:* Materials shall conform to the requirements established by ASTM F 436.

*Workmanship:* Washers shall be free from burrs, seams, laps, loose scale, irregular surfaces, and any defects affecting serviceability.

*Designation:* Washers shall be designated by the following data in the sequence shown: product name, nominal size (fraction or decimal), material specification, protective finish. See example below:

*Example:* Hardened Steel Circular Washer, $1\frac{1}{8}$ ASTM F 436, Hot-Dip Galvanized in Accordance with ASTM A153 Class C.

*Identification Symbols:* Grade and source marking and symbols shall conform to the requirements of ASTM F436. The source marking is intended to identify the source accepting the responsibility for the conformance to this and other applicable specifications.

*Quality Assurance:* Unless otherwise specified, products shall be furnished in accordance with ANSI/ASME B18.18-2017 (R2021) as noted in *Dimensional Characteristics*.

*Dimensional Characteristics:* Washers shall conform to the dimensions indicated in Table 7. The designated characteristics defined in Table 7, Note (1) shall be inspected in accordance with ANSI/ASME B18.18-2017 (R2021). For nondesignated characteristics, the provisions of ANSI/ASME B18.18-2017 (R2021) shall also apply. Should a nondesignated dimension be determined to have a variance, it shall be deemed conforming to this Standard if the user, who is the installer, accepts the variance based on fit, form, and function considerations. Where verifiable in-process inspection is used in accordance with ANSI/ASME B18.18-2017 (R2021), the final inspection level sample sizes of those respective standards shall apply.

**Beveled Washers.—***Dimensions:* All square beveled and clipped square beveled washers shall conform to the dimensions shown in Table 8.

*Tolerances:* Tolerances for inside diameter for beveled washers shall be in accordance with Table 8. The flatness shall not exceed 0.010 inch as the maximum deviation from a straight edge placed on the cut side. Burrs shall not project above immediately adjacent washer surface more than 0.010 inch.

*Finish:* Unless otherwise specified, washers shall be supplied with a plain (as-processed) finish. If zinc coatings are required, they shall be in accordance with ASTM F 436.

*Materials and Mechanical Properties:* Materials and properties shall conform to the requirements established by ASTM F 436.

*Workmanship:* Washers shall be free from burrs, seams, laps, loose scale, irregular surfaces, and any defects affecting serviceability.

*Designation:* Washers shall be designated by the following data in the sequence shown: product name, nominal washer size (fraction or decimal), material specification, protective finish. See example below:

*Example:* Square Beveled Washer, $1\frac{1}{4}$ ASTM F 436, hot-dip galvanized in accordance with ASTM A153 Class C.

**Table 8. Dimensions for Hardened Beveled Washers with Slope or Taper in Thickness 1:6** *ANSI/ASME B18.2.6-2019*

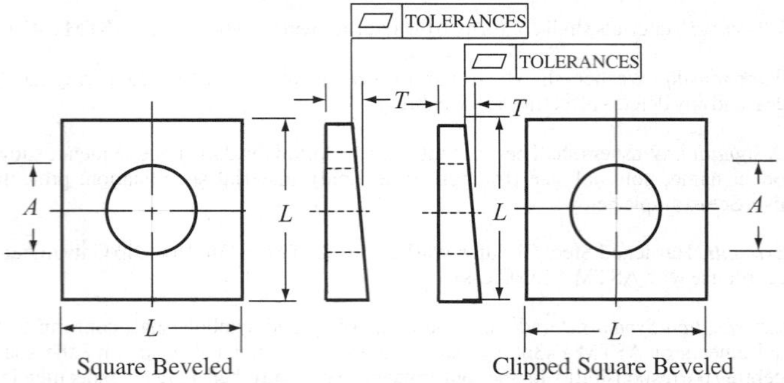


| Nominal Washer Size [Note (3)] | | Inside Diameter *A* | | | Minimum Side Length *L* | Thickness *T* | Minimum Edge Distance *E* |
|---|---|---|---|---|---|---|---|
| | | Basic | Tolerance Plus | Tolerance Minus | | | |
| ½ | 0.500 | 0.531 | 0.0313 | 0 | 1.750 | 0.313 | 0.438 |
| ⅝ | 0.625 | 0.688 | 0.0313 | 0 | 1.750 | 0.313 | 0.547 |
| ¾ | 0.750 | 0.813 | 0.0313 | 0 | 1.750 | 0.313 | 0.656 |
| ⅞ | 0.875 | 0.938 | 0.0313 | 0 | 1.750 | 0.313 | 0.766 |
| 1 | 1.000 | 1.125 | 0.0313 | 0 | 1.750 | 0.313 | 0.875 |
| 1⅛ | 1.125 | 1.250 | 0.0313 | 0 | 2.250 | 0.313 | 0.984 |
| 1¼ | 1.250 | 1.375 | 0.0313 | 0 | 2.250 | 0.313 | 1.094 |
| 1⅜ | 1.375 | 1.500 | 0.0313 | 0 | 2.250 | 0.313 | 1.203 |
| 1½ | 1.500 | 1.625 | 0.0313 | 0 | 2.250 | 0.313 | 1.313 |

*General notes for Table 8*:

(1) *Dimensional Conformance*: Beveled washers shall have the following characteristics inspected to ANSI/ASME B18.18-2017 (R2021) to the inspection levels as follows: Width across corners, inspection level B; Visual, inspection level C.

If verifiable in-process inspection is used, see *Dimensional Characteristics* on page 1709.

(2) Nonclipped washers may be rectangular providing neither side dimension is less than *L*.

(3) Nominal washer sizes are intended for use with comparable nominal bolt diameters.

(4) Additional requirements are in *Beveled Washers* on page 1709.

*Identification Symbols:* Grade and source marking and symbols shall conform to the requirements of ASTM F 436. The source marking is intended to identify the source accepting the responsibility for conformance to this and other applicable specifications.

*Quality Assurance:* Unless otherwise specified, products shall be furnished in accordance with ANSI/ASME B18.18 as noted in *Dimensional Characteristics* on page 1710.

*Dimensional Characteristics:* Washers shall conform to the dimensions specified in Table 8. The designated characteristics defined in Table 8, Note (1) shall be inspected in accordance with ANSI/ASME B18.18. For nondesignated characteristics, the provisions of ANSI/ASME B18.18 shall apply. Should a nondesignated dimension be determined to have a variance, it shall be deemed conforming to this Standard if the user, who is the installer, accepts the variance based on fit, form, and function considerations. Where verifiable in-process inspection is used in accordance with ANSI/ASME B18.18-2017 (R2021), the final inspection level sample sizes of these respective standards shall apply.

**Compressible Washer-type Direct Tension Indicators.—***Dimensions:* The dimensions for the two washer types of direct tension indicators, Type A 325 and A 490, shall be in accordance with Table 9.

*Finish:* Unless otherwise specified, direct tension indicators shall be supplied with a plain (as-processed) finish, tinplated, or uncoated. If zinc coatings are required, they shall be in accordance with ASTM F 959.

*Materials and Performance:* Direct tension indicators shall conform to the requirements of ASTM F 959.

*Workmanship:* The workmanship shall be smooth and free of burrs, laps, seams, excess mill scale, and foreign material on bearing surfaces or in protrusion welds, or other defects which would make them unsuitable for intended application.

*Designation:* Compressible washer-type direct tension indicators shall be designated by the following data in the sequence shown: product name, nominal size (fractional or decimal equivalent), type (325 or 490), finish (plain, zinc, or epoxy). See example below:

*Example:* DTI, ½ Type 325, Plain Finish

*Identification Symbols:* Grade and source marking and symbols shall conform to the requirements of ASTM F 959.

*Lot Number:* Each direct tension indicator shall be marked with a lot number in accordance with ASTM F 959.

*Quality Assurance:* Unless otherwise specified, products shall be furnished in accordance with ANSI/ASME B18.18 as noted in *Dimensional Characteristics*.

*Dimensional Characteristics:* Direct tension indicators shall conform to the dimensions indicated in Table 9. The designated characteristics defined in Table 9, Note (1) shall be inspected in accordance with ANSI/ASME B18.18. For nondesignated characteristics, the provisions of ANSI/ASME B18.18 shall apply. Should a nondesignated dimension be determined to have a variance, it shall be deemed conforming to this Standard if the user, who is the installer, accepts the variance based on fit, form, and function considerations. Where verifiable in-process inspection is used in accordance with ANSI/ASME B18.18-2017 (R2021), the final inspection level sample sizes of these respective standards shall apply.

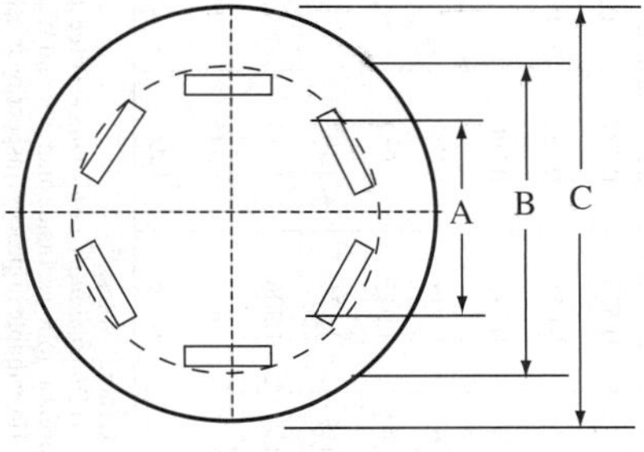


Fig. 3. Compressible Washer Dimensions

**Table 9. Dimensions for Compressible Washer-Type Direct Tension Indicators** *ANSI/ASME B18.2.6-2019*

| Bolt Size, inch | All Types | | | | | Type A 325 | | | Type A 490 | | |
|---|---|---|---|---|---|---|---|---|---|---|---|
| | Inside Diameter, inch | | Protrusion Tangential Diameter, inch | Thickness, inch | | Outside Diameter, inch *C* | | Number of Protrusions (Equally Spaced) | Outside Diameter, inch | | Number of Protrusions (Equally Spaced) |
| | | | | Without Protrusion | With Protrusion | | | | | | |
| | Min. | Max. | *B*, Max. | Min. | Max. | Min. | Max. | | Min. | Max. | |
| ½ | 0.520 | 0.527 | 0.788 | 0.104 | 0.180 | 1.031 | 1.187 | 4 | 1.031 | 1.375 | 5 |
| ⅝ | 0.651 | 0.658 | 0.956 | 0.126 | 0.220 | 1.281 | 1.375 | 4 | 1.281 | 1.625 | 5 |
| ¾ | 0.783 | 0.790 | 1.125 | 0.126 | 0.240 | 1.437 | 1.625 | 5 | 1.437 | 1.750 | 6 |
| ⅞ | 0.914 | 0.921 | 1.294 | 0.142 | 0.260 | 1.718 | 1.875 | 5 | 1.718 | 2.000 | 6 |
| 1 | 1.043 | 1.052 | 1.463 | 0.158 | 0.270 | 1.937 | 2.000 | 6 | 1.937 | 2.250 | 7 |
| 1⅛ | 1.174 | 1.183 | 1.631 | 0.158 | 0.280 | 2.187 | 2.250 | 6 | 2.187 | 2.500 | 7 |
| 1¼ | 1.306 | 1.315 | 1.800 | 0.158 | 0.280 | 2.437 | 2.500 | 7 | 2.437 | 2.750 | 8 |
| 1⅜ | 1.437 | 1.446 | 1.969 | 0.158 | 0.280 | 2.687 | 2.750 | 7 | 2.687 | 3.000 | 8 |
| 1½ | 1.568 | 1.577 | 2.138 | 0.158 | 0.280 | 2.937 | 3.000 | 8 | 2.937 | 3.250 | 9 |

General Notes:

(1) *Dimensional Conformance*: Direct tension indicators shall have the following characteristics inspected to ANSI/ASME B18.18-2017 (R2021) to the inspection levels as follows: Inside diameter and Width across corners, inspection level B; Visual, inspection level C.

If verifiable in-process inspection is used, *Dimensional Characteristics* on page 1709.

(2) Nonclipped washers may be rectangular providing neither side dimension is less than *L*.

(3) Nominal washer sizes are intended for use with comparable nominal bolt diameters.

(4) Additional requirements are in *Compressible Washer-type Direct Tension Indicators* on page 1711.

## METRIC THREADED FASTENERS

A number of American National Standards covering metric bolts, screws, nuts, and washers have been established in cooperation with the Department of Defense in such a way that they could be used by the government for procurement purposes. Extensive information concerning these metric fasteners is given in the following text and tables, but for additional manufacturing and acceptance specifications reference should be made to the respective Standards which may be obtained by nongovernmental agencies from the American National Standards Institute, 25 West 43rd Street, New York, NY 10036, www.ansi.org. These Standards are:

| | |
|---|---|
| ANSI/ASME B18.2.3.1M-1999 (Withdrawn), Metric Hex Cap Screws | Table 1 |
| ANSI/ASME B18.2.3.2M-2005 (Withdrawn), Metric Formed Hex Screws | Table 2 |
| ANSI/ASME B18.2.3.3M-2007 (R2014), Metric Heavy Hex Screws | Table 3 |
| ANSI/ASME B18.2.3.8M-1981 (Withdrawn), Metric Hex Lag Screws | Table 5 |
| ANSI/ASME B18.3.3M-1986 (Withdrawn), Metric Socket Hd. Shldr. Screws | Table 6a |
| ASME B18.2.3.9M-2001 (R2020), Metric Heavy Hex Flange Screws | Table 7 |
| ASME B18.2.3.4M-2001 (Withdrawn), Metric Hex Flange Screws | Table 8 |
| ANSI/ASME B18.5.2.2M-1982, Metric Round Head Square Neck Bolts | Table 10 |
| ANSI/ASME B18.2.3.6M-1979 (Withdrawn), Metric Heavy Hex Bolts | Table 11 |
| ANSI/ASME B18.2.6M-2012 (R2021), Metric Fasteners for Use in Structural Applications | Table 12 |
| ASME B18.2.3.5M-1979 (R2016), Metric Hex Bolts | Table 13 |
| ANSI/ASME B18.3.1M-1986 (Withdrawn), Socket Head Cap Screws (Metric Series) | Table 21 |
| ANSI/ASME B18.2.6M-2012(R2021), Metric Fasteners for Use in Structural Applications (For section on Metric Hex Nuts, Style) | Table 26 |
| ANSI/ASME B18.2.6-2019, Metric Hex Nuts, Style 2 | Table 26 |
| ANSI/ASME B18.2.4.3M-1979 (R2017), Metric Slotted Hex Nuts | Table 25 |
| ANSI/ASME B18.2.4.4M-1982 (Withdrawn), Metric Hex Flange Nuts | Table 27 |
| ANSI/ASME B18.16M-2004 (R2021), Prevailing-Torque Type Steel Metric Hex Nuts and Hex Flange Nuts | Table 30 |
| ANSI/ASME B18.16M-2004 (R2021), Prevailing-Torque Type Steel Metric Hex Nuts and Hex Flange Nuts | Table 28 |
| ANSI/ASME B18.2.6-2019, Metric Hex Jam Nuts | Table 29 |
| ANSI/ASME B18.2.6M-2012 (R2021), Metric Fasteners for Use in Structural Applications (For section on Metric Heavy Hex Nuts) | Table 29 |
| ANSI/ASME B18.22M-1981 (R2017), Metric Plain Washers | Table 33 |

Manufacturers should be consulted concerning items and sizes in stock production.

**Comparison with ISO Standards.—**American National Standards for metric bolts, screws, and nuts have been coordinated to the extent possible with the comparable ISO Standards or proposed Standards. The dimensional differences between the ANSI and the comparable ISO Standards or proposed Standards are few, relatively minor, and none will affect the functional interchangeability of bolts, screws, and nuts manufactured to the requirements of either.

Where no comparable ISO Standard had been developed, as was the case when the ANSI Standards for Metric Heavy Hex Screws, Metric Heavy Hex Bolts, and Metric Hex Lag Screws were adopted, nominal diameters, thread pitches, body diameters, widths across flats, head heights, thread lengths, thread dimensions, and nominal lengths are in accord with ISO Standards for related hex head screws and bolts. At the time of ANSI adoption (1982) there was no ISO Standard for round head square neck bolts.

The following functional characteristics of hex head screws and bolts are in agreement between the respective ANSI Standard and the comparable ISO Standard or proposed Standard: diameters and thread pitches, body diameters, widths across flats (see exception below), bearing surface diameters (except for metric hex bolts), flange diameters (for metric hex flange screws), head heights, thread lengths, thread dimensions, and nominal lengths.

**Table 1. American National Standard Metric Hex Cap Screws**
*ANSI/ASME B18.2.3.1M-1999 (Withdrawn)*

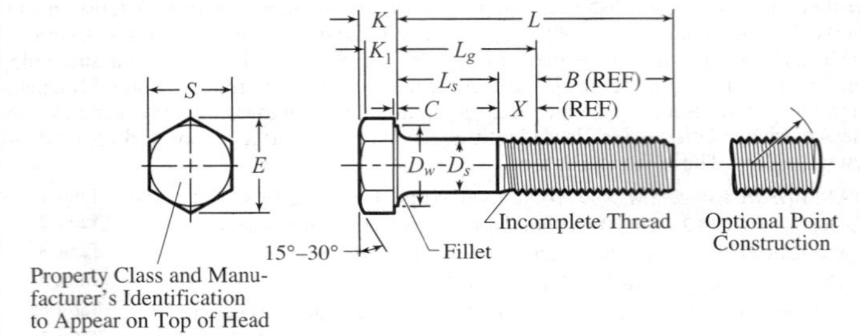


| Nominal Screw Diameter, D and Thread Pitch | Body Diameter $D_s$ | | Width Across Flats, S | | Width Across Corners, E | | Head Height, K | | Wrenching Height, $K_1$ | Washer Face Thickness, C | | Washer Face Dia., $D_w$ |
|---|---|---|---|---|---|---|---|---|---|---|---|---|
| | Max. | Min. | Max. | Min. | Max. | Min. | Max. | Min. | Min. | Max. | Min. | Min. |
| M5×0.8 | 5.00 | 4.82 | 8.00 | 7.78 | 9.24 | 8.79 | 3.65 | 3.35 | 2.4 | 0.5 | 0.2 | 7.0 |
| M6×1 | 6.00 | 5.82 | 10.00 | 9.78 | 11.55 | 11.05 | 4.15 | 3.85 | 2.8 | 0.5 | 0.2 | 8.9 |
| M8×1.25 | 8.00 | 7.78 | 13.00 | 12.73 | 15.01 | 14.38 | 5.50 | 5.10 | 3.7 | 0.6 | 0.3 | 11.6 |
| [a]M10×1.5 | 10.00 | 9.78 | 15.00 | 14.73 | 17.32 | 16.64 | 6.63 | 6.17 | 4.5 | 0.6 | 0.3 | 13.6 |
| M10×1.5 | 10.00 | 9.78 | 16.00 | 15.73 | 18.48 | 17.77 | 6.63 | 6.17 | 4.5 | 0.6 | 0.3 | 14.6 |
| M12×1.75 | 12.00 | 11.73 | 18.00 | 17.73 | 20.78 | 20.03 | 7.76 | 7.24 | 5.2 | 0.6 | 0.3 | 16.6 |
| M14×2 | 14.00 | 13.73 | 21.00 | 20.67 | 24.25 | 23.35 | 9.09 | 8.51 | 6.2 | 0.6 | 0.3 | 19.6 |
| M16×2 | 16.00 | 15.73 | 24.00 | 23.67 | 27.71 | 26.75 | 10.32 | 9.68 | 7.0 | 0.8 | 0.4 | 22.49 |
| M20×2.5 | 20.00 | 19.67 | 30.00 | 29.16 | 34.64 | 32.95 | 12.88 | 12.12 | 8.8 | 0.8 | 0.4 | 27.7 |
| M24×3 | 24.00 | 23.67 | 36.00 | 35.00 | 41.57 | 39.55 | 15.44 | 14.56 | 10.5 | 0.8 | 0.4 | 33.2 |
| M30×3.5 | 30.00 | 29.67 | 46.00 | 45.00 | 53.12 | 50.85 | 19.48 | 17.92 | 13.1 | 0.8 | 0.4 | 42.7 |
| M36×4 | 36.00 | 35.61 | 55.00 | 53.80 | 63.51 | 60.79 | 23.38 | 21.62 | 15.8 | 0.8 | 0.4 | 51.1 |
| M42×4.5 | 42.00 | 41.38 | 65.00 | 62.90 | 75.06 | 71.71 | 26.97 | 25.03 | 18.2 | 1.0 | 0.5 | 59.8 |
| M48×5 | 48.00 | 47.38 | 75.00 | 72.60 | 86.60 | 82.76 | 31.07 | 28.93 | 21.0 | 1.0 | 0.5 | 69.0 |
| M56×5.5 | 56.00 | 55.26 | 85.00 | 82.20 | 98.15 | 93.71 | 36.20 | 33.80 | 24.5 | 1.0 | 0.5 | 78.1 |
| M64×6 | 64.00 | 63.26 | 95.00 | 91.80 | 109.70 | 104.65 | 41.32 | 38.68 | 28.0 | 1.0 | 0.5 | 87.2 |
| M72×6 | 72.00 | 71.26 | 105.00 | 101.40 | 121.24 | 115.60 | 46.45 | 43.55 | 31.5 | 1.2 | 0.6 | 96.3 |
| M80×6 | 80.00 | 79.26 | 115.00 | 111.00 | 132.72 | 126.54 | 51.58 | 48.42 | 35.0 | 1.2 | 0.6 | 105.4 |
| M90×6 | 90.00 | 89.13 | 130.00 | 125.50 | 150.11 | 143.07 | 57.75 | 54.26 | 39.2 | 1.2 | 0.6 | 119.2 |
| M100×6 | 100.00 | 99.13 | 145.00 | 140.00 | 167.43 | 159.60 | 63.90 | 60.10 | 43.4 | 1.2 | 0.6 | 133.0 |

[a] This size with width across flats of 15 mm is not standard. M10 screws with 15 mm width across flats are commonly produced in United States and other countries. The width across flats for all M10's should be specified. All sizes except the following are included in ISO 4014 and ISO 4017:M10×1.5 with 15 mm width across flats, and M72 through M100.

All dimensions are in millimeters.

Basic thread lengths, *B*, are the same as given in Table 13.

Transition thread length, *X*, includes the length of incomplete threads and tolerances on grip gaging length and body length. It is intended for calculation purposes.

For additional manufacturing and acceptance specifications, reference should be made to the ANSI/ASME B18.2.3.1M-1999 (Withdrawn).

**Table 2. American National Standard Metric Formed Hex Screws**
*ANSI/ASME B18.2.3.2M-2005 (Withdrawn)*

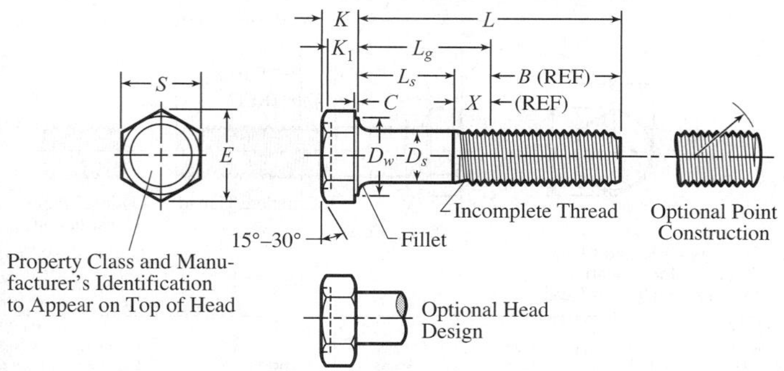


| Nominal Screw Dia., *D*, and Thread Pitch | Body Dia., $D_s$ | | Width Across Flats, *S* | | Width Across Corners, *E* | | Head Height, *K* | | Wrenching Height, $K_1$ | Washer Face Thickness, *C* | | Washer Face Dia., $D_w$ |
|---|---|---|---|---|---|---|---|---|---|---|---|---|
| | Max. | Min. | Max. | Min. | Max. | Min. | Max. | Min. | Min. | Max. | Min. | Max. |
| M5 × 0.8 | 5.00 | 4.82 | 8.00 | 7.64 | 9.24 | 8.56 | 3.65 | 3.35 | 2.4 | 0.5 | 0.2 | 6.9 |
| M6 × 1 | 6.00 | 5.82 | 10.00 | 9.64 | 11.55 | 10.80 | 4.15 | 3.85 | 2.0 | 0.5 | 0.2 | 8.9 |
| M8 × 1.25 | 8.00 | 7.78 | 13.00 | 12.57 | 15.01 | 14.08 | 5.50 | 5.10 | 3.7 | 0.6 | 0.3 | 11.6 |
| [a]M10 × 1.5 | 10.00 | 9.78 | 15.00 | 14.57 | 17.32 | 16.32 | 6.63 | 6.17 | 4.5 | 0.6 | 0.3 | 13.6 |
| M10 × 1.5 | 10.00 | 9.78 | 16.00 | 15.57 | 18.48 | 17.43 | 6.63 | 6.17 | 4.5 | 0.6 | 0.3 | 14.6 |
| M12 × 1.75 | 12.00 | 11.73 | 18.00 | 17.57 | 20.78 | 19.68 | 7.76 | 7.24 | 5.2 | 0.6 | 0.3 | 16.6 |
| M14 × 2 | 14.00 | 13.73 | 21.00 | 20.16 | 24.25 | 22.58 | 9.09 | 8.51 | 6.2 | 0.6 | 0.3 | 19.6 |
| M16 × 2 | 16.00 | 15.73 | 24.00 | 23.16 | 27.71 | 25.94 | 10.32 | 9.68 | 7.0 | 0.8 | 0.4 | 22.5 |
| M20 × 2.5 | 20.00 | 19.67 | 30.00 | 29.16 | 34.64 | 32.66 | 12.88 | 12.12 | 8.8 | 0.8 | 0.4 | 27.7 |
| M24 × 3 | 24.00 | 23.67 | 36.00 | 35.00 | 41.57 | 39.20 | 15.44 | 14.46 | 10.5 | 0.8 | 0.4 | 33.2 |

[a] This size with width across flats of 15 mm is not standard. Unless specifically ordered, M10 formed hex screws with 16 mm width across flats will be furnished.

All dimensions are in millimeters.

†Basic thread lengths, *B*, are the same as given in Table 13.

‡Transition thread length, *X*, includes the length of incomplete threads and tolerances on the grip gaging length and body length. It is intended for calculation purposes.

For additional manufacturing and acceptance specifications, reference should be made to the Standard.

Socket head cap screws ANSI/ASME B18.3.1M-1986 (Withdrawn) are functionally interchangeable with screws which conform to ISO R861-1968 (Withdrawn) or ISO 4762-2004. However, the thread lengths specified in the ANSI Standard are equal to or longer than required by either ISO Standard. Consequently the grip lengths also vary on screws where the North American thread length practice differs. Minor variations in head diameter, head height, key engagement and wall thickness are due to diverse tolerancing practice and will be found documented in the ANSI Standard.

One exception with respect to width across flats for metric hex cap screws, formed hex screws, and hex bolts is the M10 size. These are currently being produced in the United States with a width across flats of 15 mm. This size, however, is not an ISO Standard. Unless these M10 screws and bolts with 15 mm width across flats are specifically ordered, the M10 size with 16 mm across flats will be furnished.

**Table 3. American National Standard Metric Heavy Hex Screws**
*ANSI/ASME B18.2.3.3M-2007 (R2014)*

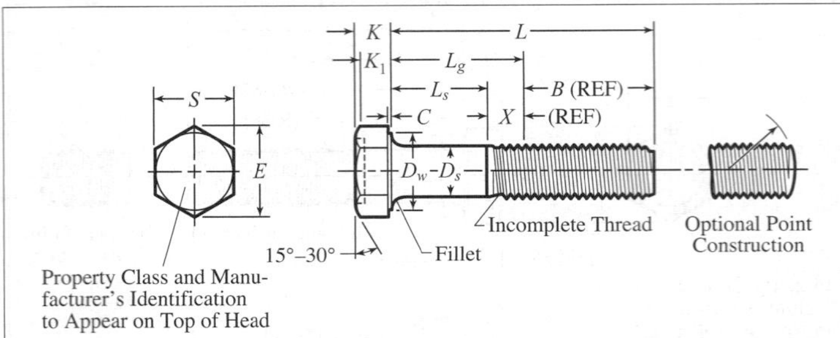


| Nominal Screw Dia., *D*, and Thread Pitch | Body Diameter, $D_s$ | | Width Across Flats, *S* | | Width Across Corners, *E* | | Head Height, *K* | | Wrenching Height, $K_1$ | Washer Face Thickness, *C* | | Washer Face Dia., $D_w$ |
|---|---|---|---|---|---|---|---|---|---|---|---|---|
| | Max. | Min. | Max. | Min. | Max. | Min. | Max. | Min. | Min. | Max. | Min. | Min. |
| M12 × 1.75 | 12.00 | 11.73 | 21.00 | 20.67 | 24.25 | 23.35 | 7.76 | 7.24 | 5.2 | 0.6 | 0.3 | 19.6 |
| M14 × 2 | 14.00 | 13.73 | 24.00 | 23.67 | 27.71 | 26.75 | 9.09 | 8.51 | 6.2 | 0.6 | 0.3 | 22.5 |
| M16 × 2 | 16.00 | 15.73 | 27.00 | 26.67 | 31.18 | 30.14 | 10.32 | 9.68 | 7.0 | 0.8 | 0.4 | 25.3 |
| M20 × 2.5 | 20.00 | 19.67 | 34.00 | 33.00 | 39.26 | 37.29 | 12.88 | 12.12 | 8.8 | 0.8 | 0.4 | 31.4 |
| M24 × 3 | 24.00 | 23.67 | 41.00 | 40.00 | 47.34 | 45.20 | 15.44 | 14.56 | 10.5 | 0.8 | 0.4 | 38.0 |
| M30 × 3.5 | 30.00 | 29.67 | 50.00 | 49.00 | 57.74 | 55.37 | 19.48 | 17.92 | 13.1 | 0.8 | 0.4 | 46.6 |
| M36 × 4 | 36.00 | 35.61 | 60.00 | 58.80 | 69.28 | 66.44 | 23.38 | 21.72 | 15.8 | 0.8 | 0.4 | 55.9 |

All dimensions are in millimeters.

Basic thread lengths, *B*, are the same as given in Table 13.

Transition thread length, *X*, includes the length of incomplete threads and tolerances on grip gaging length and body length. It is intended for calculation purposes.

For additional manufacturing and acceptance specifications, reference should be made to the Standard.

ANSI letter symbols designating dimensional characteristics are in accord with those used in ISO Standards except capitals have been used for data processing convenience instead of the lower case letters used in the ISO Standards.

**Metric Screw and Bolt Diameters.—**Metric screws and bolts are furnished with full diameter body within the limits shown in the respective dimensional tables, or are threaded to the head (see *Metric Screw and Bolt Thread Lengths* on page 1725) unless the purchaser specifies "reduced body diameter." Metric formed hex screws (Table 4), hex flange screws (Table 4), hex bolts (Table 4), heavy hex bolts (Table 4), hex lag screws (Table 5), heavy hex flange screws (Table 7), and round head square neck bolts (Table 9) may be obtained with reduced diameter body, if so specified; however, formed hex screws, hex flange screws, heavy hex flange screws, hex bolts, or heavy hex bolts with nominal lengths shorter than 4*D*, where *D* is the nominal diameter, are not recommended. Metric formed hex screws, hex flange screws, heavy hex flange screws, and hex lag screws with reduced body diameter will be furnished with a shoulder under the head. For metric hex bolts and heavy hex bolts this is optional with the manufacturer.

For bolts and lag screws there may be a reasonable swell, fin, or die seam on the body adjacent to the head not exceeding the nominal bolt diameter by: 0.50 mm for M5, 0.65 mm for M6, 0.75 mm for M8 through M14, 1.25 mm for M16, 1.50 mm for M20 through M30, 2.30 mm for M36 through M48, 3.00 mm for M56 through M72, and 4.80 mm for M80 through M100.

## Table 4. American National Standard Metric Hex Screws and Bolts — Reduced Body Diameters

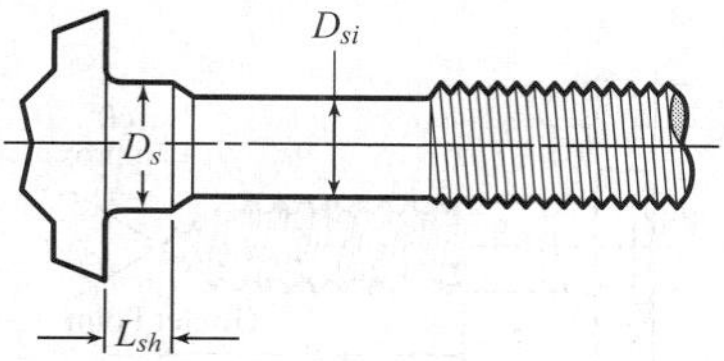


| Nominal Dia., *D*, and Thread Pitch | Shoulder Diameter,[a] $D_s$ | | Body Diameter, $D_{si}$ | | Shoulder Length,[a] $L_{sh}$ | | Nominal Dia., *D*, and Thread Pitch | Shoulder Diameter,[a] $D_s$ | | Body Diameter, $D_{si}$ | | Shoulder Length,[a] $L_{sh}$ | |
|---|---|---|---|---|---|---|---|---|---|---|---|---|---|
| | Max. | Min. | Max. | Min. | Max. | Min. | | Max. | Min. | Max. | Min. | Max. | Min. |
| Metric Formed Hex Screws (ANSI/ASME B18.2.3.2M-2005 [Withdrawn]) | | | | | | | | | | | | | |
| M5 × 0.8 | 5.00 | 4.82 | 4.46 | 4.36 | 3.5 | 2.5 | M14 × 2 | 14.00 | 13.73 | 12.77 | 12.50 | 8.0 | 7.0 |
| M6 × 1 | 6.00 | 5.82 | 5.39 | 5.21 | 4.0 | 3.0 | M16 × 2 | 16.00 | 15.73 | 14.77 | 14.50 | 9.0 | 8.0 |
| M8 × 1.25 | 8.00 | 7.78 | 7.26 | 7.04 | 5.0 | 4.0 | M20 × 2.5 | 20.00 | 19.67 | 18.49 | 18.16 | 11.0 | 10.0 |
| M10 × 1.5 | 10.00 | 9.78 | 9.08 | 8.86 | 6.0 | 5.0 | M24 × 3 | 24.00 | 23.67 | 22.13 | 21.80 | 13.0 | 12.0 |
| M12 × 1.75 | 12.00 | 11.73 | 10.95 | 10.68 | 7.0 | 6.0 | … | … | … | … | … | … | … |
| Metric Hex Flange Screws (ANSI/ASME B18.2.3.4M-2001 [Withdrawn]) | | | | | | | | | | | | | |
| M5 × 0.8 | 5.00 | 4.82 | 4.54 | 4.36 | 3.5 | 2.5 | M12 × 1.75 | 12.00 | 11.73 | 10.95 | 10.68 | 7.0 | 6.0 |
| M6 × 1 | 6.00 | 5.82 | 5.39 | 5.21 | 4.0 | 3.0 | M14 × 2 | 14.00 | 13.73 | 12.77 | 12.50 | 8.0 | 7.0 |
| M8 × 1.25 | 8.00 | 7.78 | 7.26 | 7.04 | 5.0 | 4.0 | M16 × 2 | 16.00 | 15.73 | 14.77 | 14.50 | 9.0 | 8.0 |
| M10 × 1.5 | 10.00 | 9.78 | 9.08 | 8.86 | 6.0 | 5.0 | … | … | … | … | … | … | … |
| Metric Hex Bolts (ANSI/ASME B18.2.3.5M-1979 [R2016]) | | | | | | | | | | | | | |
| M5 × 0.8 | 5.48 | 4.52 | 4.46 | 4.36 | 3.5 | 2.5 | M14 × 2 | 14.70 | 13.30 | 12.77 | 12.50 | 8.0 | 7.0 |
| M6 × 1 | 6.48 | 5.52 | 5.39 | 5.21 | 4.0 | 3.0 | M16 × 2 | 16.70 | 15.30 | 14.77 | 14.50 | 9.0 | 8.0 |
| M8 × 1.25 | 8.58 | 7.42 | 7.26 | 7.04 | 5.0 | 4.0 | M20 × 2.5 | 20.84 | 19.16 | 18.49 | 18.16 | 11.0 | 10.0 |
| M10 × 1.5 | 10.58 | 9.42 | 9.08 | 8.86 | 6.0 | 5.0 | M24 × 3 | 24.84 | 23.16 | 22.13 | 21.80 | 13.0 | 12.0 |
| M12 × 1.75 | 12.70 | 11.30 | 10.95 | 10.68 | 7.0 | 6.0 | … | … | … | … | … | … | … |
| Metric Heavy Hex Bolts (ANSI/ASME B18.2.3.6M-1979 [Withdrawn]) | | | | | | | | | | | | | |
| M12 × 1.75 | 12.70 | 11.30 | 10.95 | 10.68 | 7.0 | 6.0 | M20 × 2.5 | 20.84 | 19.16 | 18.49 | 18.16 | 11.0 | 10.0 |
| M14 × 2 | 14.70 | 13.30 | 12.77 | 12.50 | 8.0 | 7.0 | M24 × 3 | 24.84 | 23.16 | 22.13 | 21.80 | 13.0 | 12.0 |
| M16 × 2 | 16.70 | 15.30 | 14.77 | 14.50 | 9.0 | 8.0 | … | … | … | … | … | … | … |
| Metric Heavy Hex Flange Screws (ANSI/ASME B18.2.3.9M-2001 [R2020]) | | | | | | | | | | | | | |
| M10 × 1.5 | 10.00 | 9.78 | 9.08 | 8.86 | 6.0 | 5.0 | M16 × 2 | 16.00 | 15.73 | 14.77 | 14.50 | 9.0 | 8.0 |
| M12 × 1.75 | 12.00 | 11.73 | 10.95 | 10.68 | 7.0 | 6.0 | M20 × 2.5 | 20.00 | 19.67 | 18.49 | 18.16 | 11.0 | 10.0 |
| M14 × 2 | 14.00 | 13.73 | 12.77 | 12.50 | 8.0 | 7.0 | … | … | … | … | … | … | … |

[a] Shoulder is mandatory for formed hex screws, hex flange screws, and heavy hex flange screws. Shoulder is optional for hex bolts and heavy hex bolts.

All dimensions are in millimeters.

**Table 5. American National Standard Metric Hex Lag Screws**
*ANSI/ASME B18.2.3.8M-1981 (Withdrawn)*

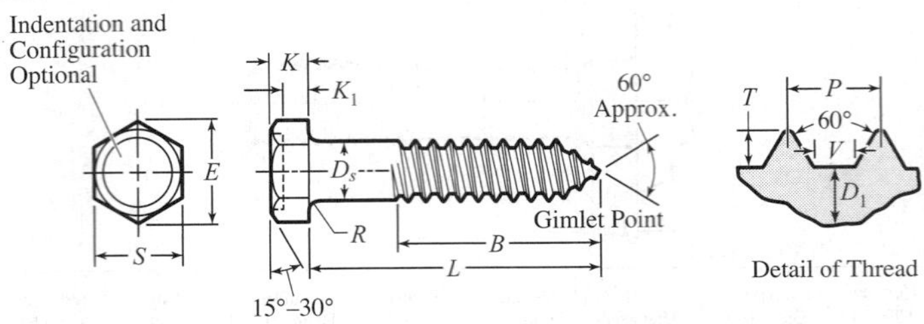


| Nominal Screw Dia., $D$ | Body Diameter, $D_s$ | | Width Across Flats, $S$ | | Width Across Corners, $E$ | | Head Height, $K$ | | Wrenching Height, $K_1$ |
|---|---|---|---|---|---|---|---|---|---|
| | Max. | Min. | Max. | Min. | Max. | Min. | Max. | Min. | Min. |
| 5 | 5.48 | 4.52 | 8.00 | 7.64 | 9.24 | 8.63 | 3.9 | 3.1 | 2.4 |
| 6 | 6.48 | 5.52 | 10.00 | 9.64 | 11.55 | 10.89 | 4.4 | 3.6 | 2.8 |
| 8 | 8.58 | 7.42 | 13.00 | 12.57 | 15.01 | 14.20 | 5.7 | 4.9 | 3.7 |
| 10 | 10.58 | 9.42 | 16.00 | 15.57 | 18.48 | 17.59 | 6.9 | 5.9 | 4.5 |
| 12 | 12.70 | 11.30 | 18.00 | 17.57 | 20.78 | 19.85 | 8.0 | 7.0 | 5.2 |
| 16 | 16.70 | 15.30 | 24.00 | 23.16 | 27.71 | 26.17 | 10.8 | 9.3 | 7.0 |
| 20 | 20.84 | 19.16 | 30.00 | 29.16 | 34.64 | 32.95 | 13.4 | 11.6 | 8.8 |
| 24 | 24.84 | 23.16 | 36.00 | 35.00 | 41.57 | 39.55 | 15.9 | 14.1 | 10.5 |

| Nominal Screw Dia., $D$ | Thread Dimensions | | | | Nominal Screw Dia., $D$ | Thread Dimensions | | | |
|---|---|---|---|---|---|---|---|---|---|
| | Thread Pitch, $P$ | Flat at Root, $V$ | Depth of Thread, $T$ | Root Dia., $D_1$ | | Thread Pitch, $P$ | Flat at Root, $V$ | Depth of Thread, $T$ | Root Dia., $D_1$ |
| 5 | 2.3 | 1.0 | 0.9 | 3.2 | 12 | 4.2 | 1.8 | 1.6 | 8.7 |
| 6 | 2.5 | 1.1 | 1.0 | 4.0 | 16 | 5.1 | 2.2 | 2.0 | 12.0 |
| 8 | 2.8 | 1.2 | 1.1 | 5.8 | 20 | 5.6 | 2.4 | 2.2 | 15.6 |
| 10 | 3.6 | 1.6 | 1.4 | 7.2 | 24 | 7.3 | 3.1 | 2.8 | 18.1 |

REDUCED BODY DIAMETER

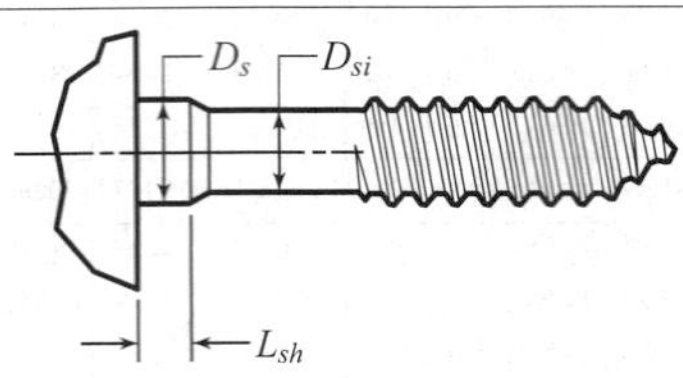


| Nominal Screw Dia., $D$ | Shoulder Diameter, $D_s$ | | Shoulder Length, $L_{sh}$ | | Nominal Screw Dia., $D$ | Shoulder Diameter, $D_s$ | | Shoulder Length, $L_{sh}$ | |
|---|---|---|---|---|---|---|---|---|---|
| | Max. | Min. | Max. | Min. | | Max. | Min. | Max. | Min. |
| 5 | 5.48 | 4.52 | 3.5 | 2.5 | 12 | 12.70 | 11.30 | 7.0 | 6.0 |
| 6 | 6.48 | 5.52 | 4.0 | 3.0 | 16 | 16.70 | 15.30 | 9.0 | 8.0 |
| 8 | 8.58 | 7.42 | 5.0 | 4.0 | 20 | 20.84 | 19.16 | 11.0 | 10.0 |
| 10 | 10.58 | 9.42 | 6.0 | 5.0 | 24 | 24.84 | 23.16 | 13.0 | 12.0 |

All dimensions are in millimeters. Reduced body diameter, $D_{si}$, is the blank diameter before rolling. Shoulder is mandatory when body diameter is reduced.

## Table 6a. Hexagon Socket Head Shoulder Screws — Metric Series

*ANSI/ASME B18.3.3M-1986 (Withdrawn)*

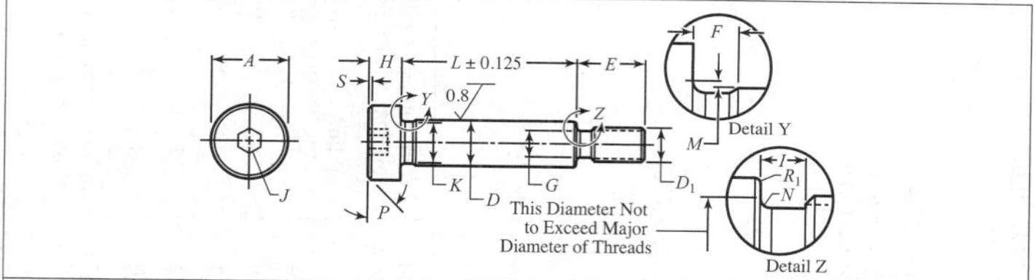


| Nominal Shoulder, Dia. | Shoulder Diameter[a], D Max. | Shoulder Diameter[a], D Min. | Head Diameter, A Max. | Head Diameter, A Min. | Head Height, H Max. | Head Height, H Min. | Chamfer or Radius, S Max. | Nominal Thread Size, $D_1$ | Thread Length, E Max. |
|---|---|---|---|---|---|---|---|---|---|
| 6.5 | 6.487 | 6.451 | 10.00 | 9.78 | 4.50 | 4.32 | 0.6 | M5 × 0.8 | 9.75 |
| 8.0 | 7.987 | 7.951 | 13.00 | 12.73 | 5.50 | 5.32 | 0.8 | M6 × 1 | 11.25 |
| 10.0 | 9.987 | 9.951 | 16.00 | 15.73 | 7.00 | 6.78 | 1.0 | M8 × 1.25 | 13.25 |
| 13.0 | 12.984 | 12.941 | 18.00 | 17.73 | 9.00 | 8.78 | 1.2 | M10 × 1.5 | 16.40 |
| 16.0 | 15.984 | 15.941 | 24.00 | 23.67 | 11.00 | 10.73 | 1.6 | M12 × 1.75 | 18.40 |
| 20.0 | 19.980 | 19.928 | 30.00 | 29.67 | 14.00 | 13.73 | 2.0 | M16 × 2 | 22.40 |
| 25.0 | 24.980 | 24.928 | 36.00 | 35.61 | 16.00 | 15.73 | 2.4 | M20 × 2.5 | 27.40 |

| Nominal Shoulder Diameter | Thread Neck Dia., G Max. | Thread Neck Dia., G Min. | Thread Neck Width, I Max. | Shoulder Neck Dia, K Min. | Shoulder Neck Width, F Max. | Thread Neck Fillet, N Max. | Thread Neck Fillet, N Min. | Head Fillet Extension above D, M Max. | Hexagon Socket Size, J |
|---|---|---|---|---|---|---|---|---|---|
| 6.5 | 3.86 | 3.68 | 2.4 | 5.92 | 2.5 | 0.66 | 0.50 | 7.5 | 3 |
| 8.0 | 4.58 | 4.40 | 2.6 | 7.42 | 2.5 | 0.69 | 0.53 | 9.2 | 4 |
| 10.0 | 6.25 | 6.03 | 2.8 | 9.42 | 2.5 | 0.80 | 0.64 | 11.2 | 5 |
| 13.0 | 7.91 | 7.69 | 3.0 | 12.42 | 2.5 | 0.93 | 0.77 | 15.2 | 6 |
| 16.0 | 9.57 | 9.35 | 4.0 | 15.42 | 2.5 | 1.03 | 0.87 | 18.2 | 8 |
| 20.0 | 13.23 | 12.96 | 4.8 | 19.42 | 2.5 | 1.30 | 1.14 | 22.4 | 10 |
| 25.0 | 16.57 | 16.30 | 5.6 | 14.42 | 3.0 | 1.46 | 1.30 | 27.4 | 12 |

[a] The shoulder is the enlarged, unthreaded portion of the screw.

## Table 6b. Standard Sizes and Socket Dimensions

*ANSI/ASME B18.3.3M-1986 (Withdrawn)*

**Standard Sizes for Government Use**

| Nominal Shoulder Length | Nominal Shoulder Diameter 6.5 | 8 | 10 | 13 | 16 | 20 | 25 |
|---|---|---|---|---|---|---|---|
| 10.0 | 065010 | 080010 | 100010 | | | | |
| 12.0 | | | | 130012 | | | |
| 16.0 | | | | | | | |
| 20.0 | | | | | | | |
| 25.0 | | | | | | | |
| 30.0 | | | | | 160030 | | |
| 40.0 | 065040 | | | | | 200040 | |
| 50.0 | | 080050 | | | | | 250050 |
| 60.0 | | | | | | | |
| 70.0 | | | | | | | |
| 80.0 | | | | | | | |
| 90.0 | | | | | | | |
| 100.0 | | | 100100 | | | | |
| 110.0 | | | | | | | |
| 120.0 | | | | 130120 | 160120 | 200120 | 250120 |

Range of standard diameter and length combinations and their part numbers. (PIN)

**Dimensions Hexagon Sockets - Metric**

| Nominal Socket Size | Socket Width Across Flats J Max. | Socket Width Across Flats J Min. | Socket Width Across Corners, C Min. |
|---|---|---|---|
| 3 | 3.071 | 3.020 | 3.44 |
| 4 | 4.084 | 4.020 | 4.58 |
| 5 | 5.084 | 5.020 | 5.72 |
| 6 | 6.095 | 6.020 | 6.86 |
| 8 | 8.115 | 8.025 | 9.15 |
| 10 | 10.127 | 10.025 | 11.50 |
| 12 | 12.146 | 12.032 | 13.80 |

All dimensions are in millimeters.

$R_1$ rounded or chamfered: $R_1 \leq 0.15$ mm for M5, M6, M10; $R_1 \leq 0.20$ mm for M12, M16, M20 Unless specified otherwise, threads are metric coarse series in accordance with ANSI/ASME B1.13M, Metric Screw Threads - M Profile. Tolerances for threads are to ISO Tolerance Class 4g6g. Standard length and diameter combinations are shown in the lower chart. The government encourages the general use of the part number system (PIN) to achieve maximum parts standardization. For details not shown, including material and complete PIN system, see ANSI/ASME B18.3.3M-1986 (Withdrawn).

**Table 7. American National Standard Metric Heavy Hex Flange Screws** *ANSI/ASME B18.2.3.9M-2001 (R2020)*

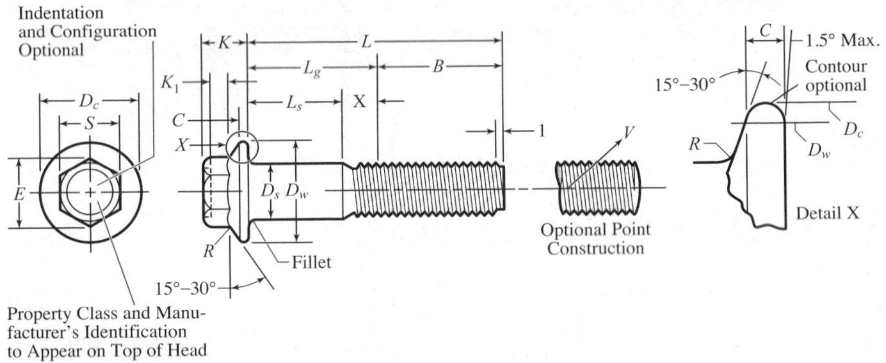


| Nominal Screw Dia., $D$, and Thread Pitch | Body Dia., $D_s$ | | Width Across Flats, $S$ | | Width Across Corners, $E$ | | Flange Dia., $D_c$ | Bearing Circle Dia., $D_w$ | Flange Edge Thickness, $C$ | Head Height, $K$ | Wrenching Height $K_1$ | Fillet Radius, $R$ |
|---|---|---|---|---|---|---|---|---|---|---|---|---|
| | Max. | Min. | Max. | Min. | Max. | Min. | Max. | Min. | Min. | Max. | Min. | Max. |
| M10 × 1.5 | 10.00 | 9.78 | 15.00 | 14.57 | 17.32 | 16.32 | 22.3 | 19.6 | 1.5 | 10.4 | 4.6 | 0.6 |
| M12 × 1.75 | 12.00 | 11.73 | 18.00 | 17.57 | 20.78 | 19.68 | 26.6 | 23.8 | 1.8 | 11.8 | 5.2 | 0.7 |
| M14 × 2 | 14.00 | 13.73 | 21.00 | 20.16 | 24.25 | 22.58 | 30.5 | 27.6 | 2.1 | 13.7 | 6.3 | 0.9 |
| M16 × 2 | 16.00 | 15.73 | 24.00 | 23.16 | 27.71 | 25.94 | 35.0 | 31.9 | 2.4 | 15.4 | 7.1 | 1.0 |
| M20 × 2.5 | 20.00 | 19.67 | 30.00 | 29.16 | 34.64 | 32.66 | 43.0 | 39.9 | 3.0 | 18.9 | 8.8 | 1.2 |

All dimensions are in millimeters. Basic thread lengths, $B$, are as given in Table 13. Transition thread length, X, includes the length of incomplete threads and tolerances on grip gaging length and body length. It is intended for calculation purposes. For additional manufacturing and acceptance specifications, reference should be made to ANSI/ASME B18.2.3.9M-2001 (R2020).

## Table 8. American National Standard Metric Hex Flange Screws
*ANSI/ASME B18.2.3.4M-2001 (Withdrawn)*

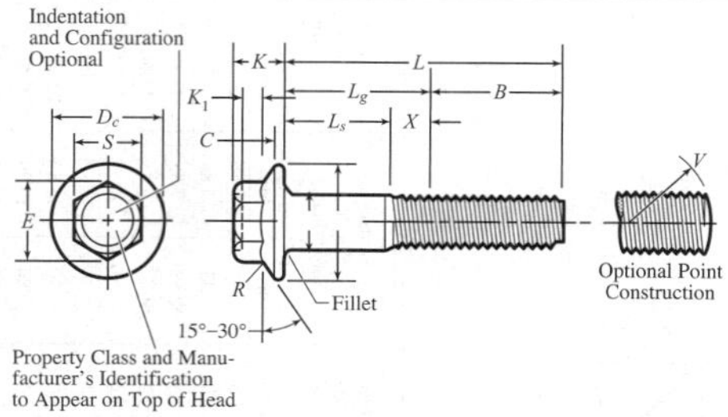


| Nominal Screw Dia., D, and Thread Pitch | Body Dia., $D_s$ | | Width Across Flats, S | | Width Across Corners, E | | Flange Dia., $D_c$ | Bearing Circle Dia., $D_w$ | Flange Edge Thickness, C | Head Height, K | Wrenching Height, $K_1$ | Fillet Radius, R |
|---|---|---|---|---|---|---|---|---|---|---|---|---|
| | Max. | Min. | Max. | Min. | Max. | Min. | Max. | Min. | Min. | Max. | Min. | Max. |
| M5 × 0.8 | 5.00 | 4.82 | 7.00 | 6.64 | 8.08 | 7.44 | 11.4 | 9.4 | 1.0 | 5.6 | 2.30 | 0.3 |
| M6 × 1 | 6.00 | 5.82 | 8.00 | 7.64 | 9.24 | 8.56 | 13.6 | 11.6 | 1.1 | 6.8 | 2.90 | 0.4 |
| M8 × 1.25 | 8.00 | 7.78 | 10.00 | 9.64 | 11.55 | 10.80 | 17.0 | 14.9 | 1.2 | 8.5 | 3.80 | 0.5 |
| M10 × 1.5 | 10.00 | 9.78 | 13.00 | 12.57 | 15.01 | 14.08 | 20.8 | 18.7 | 1.5 | 9.7 | 4.30 | 0.6 |
| M12 × 1.75 | 12.00 | 11.73 | 15.00 | 14.57 | 17.32 | 16.32 | 24.7 | 22.0 | 1.8 | 11.9 | 5.40 | 0.7 |
| M14 × 2 | 14.00 | 13.73 | 18.00 | 17.57 | 20.78 | 19.68 | 28.6 | 25.9 | 2.1 | 12.9 | 5.60 | 0.8 |
| M16 × 2 | 16.00 | 15.73 | 21.00 | 20.48 | 24.25 | 22.94 | 32.8 | 30.1 | 2.4 | 15.1 | 6.70 | 1.0 |

All dimensions are in millimeters. Basic thread lengths, B, are the same as given in Table 13. Transition thread length, X, includes the length of incomplete threads and tolerances on grip gaging length and body length. This dimension is intended for calculation purposes only. For additional manufacturing and acceptance specifications, reference should be made to ANSI/ ASME B18.2.3.4M-2001.

## Table 9. American National Standard Metric Round Head Square Neck Bolts Reduced Body Diameters *ANSI/ASME B18.5.2.2M-1982*

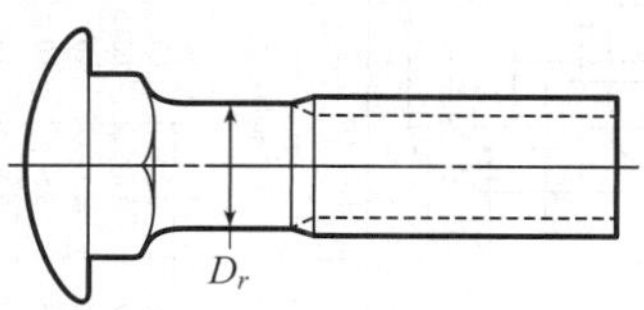


| Nominal Bolt Dia., D and Thread Pitch | Diameter of Reduced Body $D_r$ | | Nominal Bolt Dia., D and Thread Pitch | Diameter of Reduced Body $D_r$ | |
|---|---|---|---|---|---|
| | Max. | Min. | | Max. | Min. |
| M5 × 0.8 | 5.00 | 4.36 | M14 × 2 | 14.00 | 12.50 |
| M6 × 1 | 6.00 | 5.21 | M16 × 2 | 16.00 | 14.50 |
| M8 × 1.25 | 8.00 | 7.04 | M20 × 2.5 | 20.00 | 18.16 |
| M10 × 1.5 | 10.00 | 8.86 | M24 × 3 | 24.00 | 21.80 |
| M12 × 1.75 | 12.00 | 10.68 | … | … | … |

All dimensions are in millimeters.

**Table 10. American National Standard Metric Round Head Square Neck Bolts** *ANSI/ASME B18.5.2.2M-1982 (Withdrawn)*

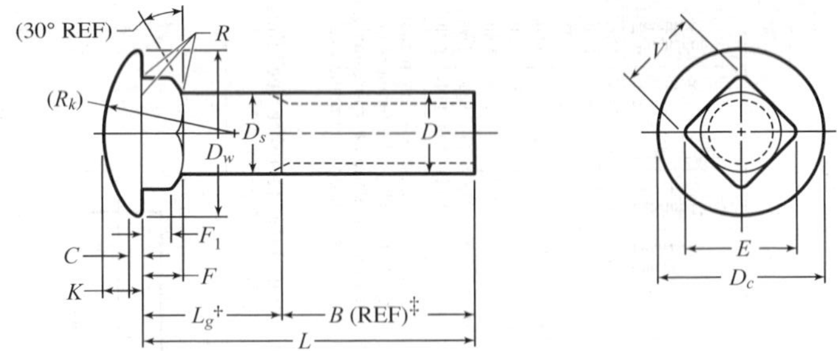


| Nominal Bolt Dia., *D* and Thread Pitch | Diameter of Full Body, $D_s$ | | Head Radius, ($R_k$) | Head height, *K* | | Head Edge Thickness, *C* | | Head Dia., $D_c$ | Bearing Surface Dia., $D_w$ | Square Depth, *F* | | Square Corner Depth, $F_1$ | Square Width Across Flats, *V* | | Square Width Across Corners, *E* | |
|---|---|---|---|---|---|---|---|---|---|---|---|---|---|---|---|---|
| | Max. | Min. | Ref. | Max. | Min. | Max. | Min. | Max. | Min. | Max. | Min. | Min. | Max. | Min. | Max. | Min. |
| M5×0.8 | 5.48 | 4.52 | 8.8 | 3.1 | 2.5 | 1.8 | 1.0 | 11.8 | 9.8 | 3.1 | 2.5 | 1.6 | 5.48 | 4.88 | 7.75 | 6.34 |
| M6×1 | 6.48 | 5.52 | 10.7 | 3.6 | 3.0 | 1.9 | 1.1 | 14.2 | 12.2 | 3.6 | 3.0 | 1.9 | 6.48 | 5.88 | 9.16 | 7.64 |
| M8×1.25 | 8.58 | 7.42 | 12.5 | 4.8 | 4.0 | 2.2 | 1.2 | 18.0 | 15.8 | 4.8 | 4.0 | 2.5 | 8.58 | 7.85 | 12.13 | 10.20 |
| M10×1.5 | 10.58 | 9.42 | 15.5 | 5.8 | 5.0 | 2.5 | 1.5 | 22.3 | 19.6 | 5.8 | 5.0 | 3.2 | 10.58 | 9.85 | 14.96 | 12.80 |
| M12×1.75 | 12.70 | 11.30 | 19.0 | 6.8 | 6.0 | 2.8 | 1.8 | 26.6 | 23.8 | 6.8 | 6.0 | 3.8 | 12.70 | 11.82 | 17.96 | 15.37 |
| M14×2 | 14.70 | 13.30 | 21.9 | 7.9 | 7.0 | 3.3 | 2.1 | 30.5 | 27.6 | 7.9 | 7.0 | 4.4 | 14.70 | 13.82 | 20.79 | 17.97 |
| M16×2 | 16.70 | 15.30 | 25.5 | 8.9 | 8.0 | 3.6 | 2.4 | 35.0 | 31.9 | 8.9 | 8.0 | 5.0 | 16.70 | 15.82 | 23.62 | 20.57 |
| M20×2.5 | 20.84 | 19.16 | 31.9 | 10.9 | 10.0 | 4.2 | 3.0 | 43.0 | 39.9 | 10.9 | 10.0 | 6.3 | 20.84 | 19.79 | 29.47 | 25.73 |
| M24×3 | 24.84 | 23.16 | 37.9 | 13.1 | 12.0 | 5.1 | 3.6 | 51.0 | 47.6 | 13.1 | 12.0 | 7.6 | 24.84 | 23.79 | 35.13 | 30.93 |

All dimensions are in millimeters.

†$L_g$ is the grip gaging length which controls the length of thread *B*.

‡*B* is the basic thread length and is a reference dimension (see Table 14).

For additional manufacturing and acceptance specifications, see ANSI/ASME B18.5.2.2M-1982.

**Table 11. ANSI Heavy Hex Bolts** *ANSI/ASME B18.2.3.6M-1979 (Withdrawn)*

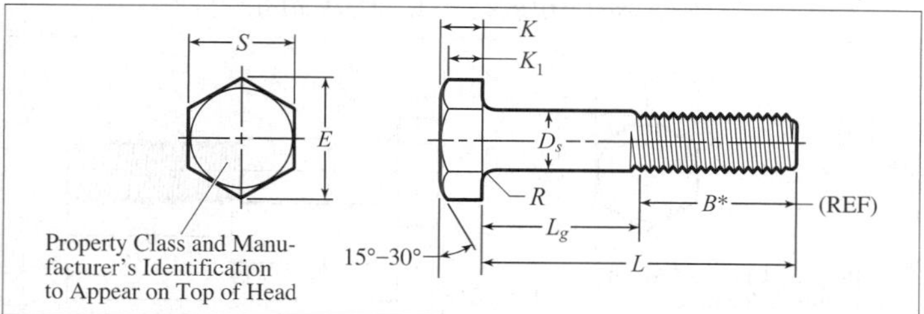


| Nominal Dia., $D$ and Thread Pitch | Body Diameter, $D_s$ | | Width Across Flats, $S$ | | Width Across Corners, $E$ | | Head Height, $K$ | | Wrenching Height, $K_1$ |
|---|---|---|---|---|---|---|---|---|---|
| | Max. | Min. | Max. | Min. | Max. | Min. | Max. | Min. | Min. |
| M12 × 1.75 | 12.70 | 11.30 | 21.00 | 20.16 | 24.25 | 22.78 | 7.95 | 7.24 | 5.2 |
| M14 × 2 | 14.70 | 13.30 | 24.00 | 23.16 | 27.71 | 26.17 | 9.25 | 8.51 | 6.2 |
| M16 × 2 | 16.70 | 15.30 | 27.00 | 26.16 | 31.18 | 29.56 | 10.75 | 9.68 | 7.0 |
| M20 × 2.5 | 20.84 | 19.16 | 34.00 | 33.00 | 39.26 | 37.29 | 13.40 | 12.12 | 8.8 |
| M24 × 3 | 24.84 | 23.16 | 41.00 | 40.00 | 47.34 | 45.20 | 15.90 | 14.56 | 10.5 |
| M30 × 3.5 | 30.84 | 29.16 | 50.00 | 49.00 | 57.74 | 55.37 | 19.75 | 17.92 | 13.1 |
| M36 × 4 | 37.00 | 35.00 | 60.00 | 58.80 | 69.28 | 66.44 | 23.55 | 21.72 | 15.8 |

All dimensions are in millimeters.*Basic thread lengths, $B$, are the same as given in Table 13. For additional manufacturing and acceptance specifications, reference should be made to the B18.2.3.6M-1979 (Withdrawn) Standard.

**Table 12. Metric Heavy Hex Structural Bolts** *ANSI/ASME B18.2.6M-2012 (R2021)*

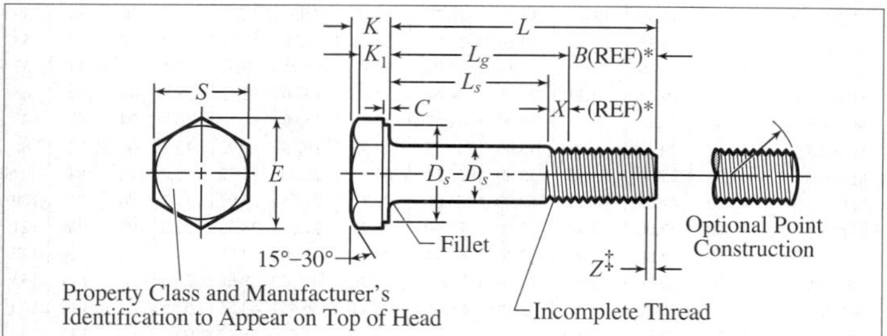


| Dia. $D$, and Thread Pitch | Body Diameter, $D_s$ | | Width Across Flats, $S$ | | Width Across Corners, $E$ | | Head Height, $K$ | | Wrenching Height, $K_1$ | Washer Face Dia., $D_w$ | Washer Face Thickness, $C$ | | Thread Length, $B$[a] Lengths ≤ 100 | Thread Length, $B$[a] Lengths > 100 | Transition Thread Length, $X$[b] |
|---|---|---|---|---|---|---|---|---|---|---|---|---|---|---|---|
| | Max. | Min. | Max. | Min. | Max. | Min. | Max. | Min. | Min. | Min. | Max. | Min. | Basic | | Max. |
| M16 × 2 | 16.70 | 15.30 | 27.00 | 26.16 | 31.18 | 29.56 | 10.75 | 9.25 | 6.5 | 24.9 | 0.8 | 0.4 | 31 | 38 | 6.0 |
| M20 × 2.5 | 20.84 | 19.16 | 34.00 | 33.00 | 39.26 | 37.29 | 13.40 | 11.60 | 8.1 | 31.4 | 0.8 | 0.4 | 36 | 43 | 7.5 |
| M22 × 2.5 | 22.84 | 21.16 | 36.00 | 35.00 | 41.57 | 39.55 | 14.90 | 13.10 | 9.2 | 33.3 | 0.8 | 0.4 | 38 | 45 | 7.5 |
| M24 × 3 | 24.84 | 23.16 | 41.00 | 40.00 | 47.34 | 45.20 | 15.90 | 14.10 | 9.9 | 38.0 | 0.8 | 0.4 | 41 | 48 | 9.0 |
| M27 × 3 | 27.84 | 26.16 | 46.00 | 45.00 | 53.12 | 50.85 | 17.90 | 16.10 | 11.3 | 42.8 | 0.8 | 0.4 | 44 | 51 | 9.0 |
| M30 × 3.5 | 30.84 | 29.16 | 50.00 | 49.00 | 57.74 | 55.37 | 19.75 | 17.65 | 12.4 | 46.5 | 0.8 | 0.4 | 49 | 56 | 10.5 |
| M36 × 4 | 37.00 | 35.00 | 60.00 | 58.80 | 69.28 | 66.44 | 23.55 | 21.45 | 15.0 | 55.9 | 0.8 | 0.4 | 56 | 63 | 12.0 |

[a] Basic thread length, $B$, is a reference dimension.

[b] Transition thread length, $X$, includes the length of incomplete threads and tolerances on grip gaging length and body length. It is intended for calculation purposes.

All dimensions are in millimeters.

For additional manufacturing and acceptance specifications, reference should be made to the ANSI/ASME B18.2.6M-2012 (R2021) Standard.

**Table 13. American National Standard Metric Hex Bolts**
*ANSI/ASME B18.2.3.5M-1979 (R2016)*

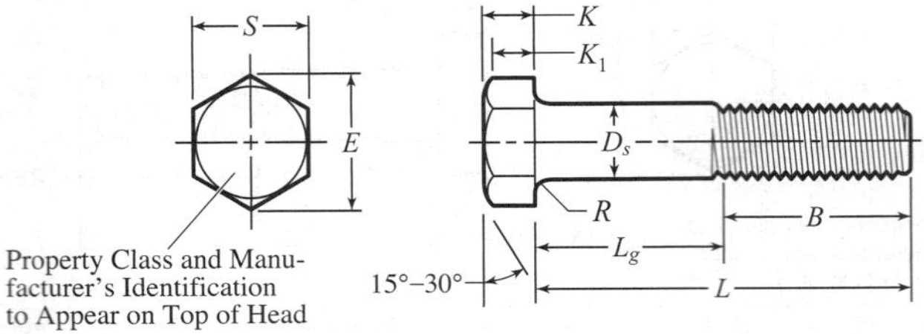


| Nominal Bolt Dia., *D* and Thread Pitch | Body Diameter, $D_s$ | | Width Across Flats, *S* | | Width Across Corners, *E* | | Head Height, *K* | | Wrenching Height, $K_1$ | For Bolt Lengths ≤ 125 mm | > 125 and ≤ 200 mm | > 200 mm |
|---|---|---|---|---|---|---|---|---|---|---|---|---|
| | Max. | Min. | Max. | Min. | Max. | Min. | Max. | Min. | Min. | Basic Thread Length,[a] *B* | | |
| M5 × 0.8 | 5.48 | 4.52 | 8.00 | 7.64 | 9.24 | 8.63 | 3.58 | 3.35 | 2.4 | 16 | 22 | 35 |
| M6 × 1 | 6.19 | 5.52 | 10.00 | 9.64 | 11.55 | 10.89 | 4.38 | 3.55 | 2.8 | 18 | 24 | 37 |
| M8 × 1.25 | 8.58 | 7.42 | 13.00 | 12.57 | 15.01 | 14.20 | 5.68 | 5.10 | 3.7 | 22 | 28 | 41 |
| [b]M10 × 1.5 | 10.58 | 9.42 | 15.00 | 14.57 | 17.32 | 16.46 | 6.85 | 6.17 | 4.5 | 26 | 32 | 45 |
| M10 × 1.5 | 10.58 | 9.42 | 16.00 | 15.57 | 18.48 | 17.59 | 6.85 | 6.17 | 4.5 | 26 | 32 | 45 |
| M12 × 1.75 | 12.70 | 11.30 | 18.00 | 17.57 | 20.78 | 19.85 | 7.95 | 7.24 | 5.2 | 30 | 36 | 49 |
| M14 × 2 | 14.70 | 13.30 | 21.00 | 20.16 | 24.25 | 22.78 | 9.25 | 8.51 | 6.2 | 34 | 40 | 53 |
| M16 × 2 | 16.70 | 15.30 | 24.00 | 23.16 | 27.71 | 26.17 | 10.75 | 9.68 | 7.0 | 38 | 44 | 57 |
| M20 × 2.5 | 20.84 | 19.16 | 30.00 | 29.16 | 34.64 | 32.95 | 13.40 | 12.12 | 8.8 | 46 | 52 | 65 |
| M24 × 3 | 24.84 | 23.16 | 36.00 | 35.00 | 41.57 | 39.55 | 15.90 | 14.56 | 10.5 | 54 | 60 | 73 |
| M30 × 3.5 | 30.84 | 29.16 | 46.00 | 45.00 | 53.12 | 50.55 | 19.75 | 17.92 | 13.1 | 66 | 72 | 85 |
| M36 × 4 | 37.00 | 35.00 | 55.00 | 53.80 | 63.51 | 60.79 | 23.55 | 21.72 | 15.8 | 78 | 84 | 97 |
| M42 × 4.5 | 43.00 | 41.00 | 65.00 | 62.90 | 75.06 | 71.71 | 27.05 | 25.03 | 18.2 | 90 | 96 | 109 |
| M48 × 5 | 49.00 | 47.00 | 75.00 | 72.60 | 86.60 | 82.76 | 31.07 | 28.93 | 21.0 | 102 | 108 | 121 |
| M56 × 5.5 | 57.20 | 54.80 | 85.00 | 82.20 | 98.15 | 93.71 | 36.20 | 33.80 | 24.5 | … | 124 | 137 |
| M64 × 6 | 65.52 | 62.80 | 95.00 | 91.80 | 109.70 | 104.65 | 41.32 | 38.68 | 28.0 | … | 140 | 153 |
| M72 × 6 | 73.84 | 70.80 | 105.00 | 101.40 | 121.24 | 115.60 | 46.45 | 43.55 | 31.5 | … | 156 | 169 |
| M80 × 6 | 82.16 | 78.80 | 115.00 | 111.00 | 132.79 | 126.54 | 51.58 | 48.42 | 35.0 | … | 172 | 185 |
| M90 × 6 | 92.48 | 88.60 | 130.00 | 125.50 | 150.11 | 143.07 | 57.74 | 54.26 | 39.2 | … | 192 | 205 |
| M100 × 6 | 102.80 | 98.60 | 145.00 | 140.00 | 167.43 | 159.60 | 63.90 | 60.10 | 43.4 | … | 212 | 225 |

[a] Basic thread length, *B*, is a reference dimension.

[b] This size with width across flats of 15 mm is not standard. Unless specifically ordered, M10 hex bolts with 16 mm width across flats will be furnished.

All dimensions are in millimeters.

For additional manufacturing and acceptance specifications, reference should be made to the ASME B18.2.3.5M-1979 (R2016) Standard.

**Materials and Mechanical Properties.—**Unless otherwise specified, steel metric screws and bolts, with the exception of heavy hex structural bolts, hex lag screws, and socket head cap screws, conform to the requirements specified in SAE J1199 or ASTM F568. Steel heavy hex structural bolts conform to ASTM A325M or ASTM A490M. Alloy steel socket head cap screws conform to ASTM A574M, property class 12.9, where the numeral 12 represents approximately one-hundredth of the minimum tensile strength in megapascals and the decimal .9 approximates the ratio of the minimum yield stress to the minimum tensile stress. This is in accord with ISO designation practice. Screws and bolts of other materials, and all materials for hex lag bolts, have properties as agreed upon by the purchaser and the manufacturer.

Except for socket head cap screws, metric screws and bolts are furnished with a natural (as processed) finish, unplated or uncoated unless otherwise specified.

Alloy steel socket head cap screws are furnished with an oiled black oxide coating (thermal or chemical) unless a protective plating or coating is specified by the purchaser.

**Metric Screw and Bolt Identification Symbols.—**Screws and bolts are identified on the top of the head by property class symbols and manufacturer's identification symbol.

**Metric Screw and Bolt Designation.—**Metric screws and bolts with the exception of socket head cap screws are designated by the following data, preferably in the sequence shown: product name, nominal diameter and thread pitch (except for hex lag screws), nominal length, steel property class or material identification, and protective coating, if required.

*Example:* Hex cap screw, M10 × 1.5 × 50, class 9.8, zinc plated

Heavy hex structural bolt, M24 × 3 × 80, ASTM A490M

Hex lag screw, 6 × 35, silicon bronze.

Socket head cap screws (metric series) are designated by the following data in the order shown: ANSI Standard number, nominal size, thread pitch, nominal screw length, name of product (may be abbreviated SHCS), material and property class (alloy steel screws are supplied to property class 12.9 as specified in ASTM A574M: corrosion-resistant steel screws are specified to the property class and material requirements in ASTM F837M), and protective finish, if required.

*Example:* B18.3.1M—6 × 1 × 20 Hexagon Socket Head Cap Screw, Alloy Steel

B18.3.1M—10 × 1.5 × 40 SHCS, Alloy Steel Zinc Plated.

**Metric Screw and Bolt Thread Lengths.—**The length of thread on metric screws and bolts (except for metric lag screws) is controlled by the grip gaging length, $L_g$ max. This is the distance measured parallel to the axis of the screw or bolt, from under the head bearing surface to the face of a noncounterbored or noncountersunk standard GO thread ring gage assembled by hand as far as the thread will permit. The maximum grip gaging length, as calculated and rounded to one decimal place, is equal to the nominal screw length, $L$, minus the basic thread length, $B$, or in the case of socket head cap screws, minus the minimum thread length $L_T$. $B$ and $L_T$ are reference dimensions intended for calculation purposes only and will be found in Table 13 and Table 15, respectively.

**Table 14. Basic Thread Lengths for Metric Round Head Square Neck Bolts**
*ANSI/ASME B18.5.2.2M-1982 (R2010)*

| Nom. Bolt Dia., D and Thread Pitch | Bolt Length, L | | | Nom. Bolt Dia., D and Thread Pitch | Bolt Length, L | | |
|---|---|---|---|---|---|---|---|
| | ≤ 125 | > 125 and ≤ 200 | > 200 | | ≤ 125 | > 125 and ≤ 200 | > 200 |
| | Basic Thread Length, B | | | | Basic Thread length, B | | |
| M5 × 0.8 | 16 | 22 | 35 | M14 × 2 | 34 | 40 | 53 |
| M6 × 1 | 18 | 24 | 37 | M16 × 2 | 38 | 44 | 57 |
| M8 × 1.25 | 22 | 28 | 41 | M20 × 2.5 | 46 | 52 | 65 |
| M10 × 1.5 | 26 | 32 | 45 | M24 × 3 | 54 | 60 | 73 |
| M12 × 1.75 | 30 | 36 | 49 | … | … | … | … |

All dimensions are in millimeters

Basic thread length $B$ is a reference dimension intended for calculation purposes only.

**Table 15. Socket Head Cap Screws (Metric Series)—Length of Complete Thread**
*ANSI/ASME B18.3.1M-1986 (Withdrawn)*

| Nominal Size | Length of Complete Thread, $L_T$ | Nominal Size | Length of Complete Thread, $L_T$ | Nominal Size | Length of Complete Thread, $L_T$ |
|---|---|---|---|---|---|
| M1.6 | 15.2 | M6 | 24.0 | M20 | 52.0 |
| M2 | 16.0 | M8 | 28.0 | M24 | 60.0 |
| M2.5 | 17.0 | M10 | 32.0 | M30 | 72.0 |
| M3 | 18.0 | M12 | 36.0 | M36 | 84.0 |
| M4 | 20.0 | M14 | 40.0 | M42 | 96.0 |
| M5 | 22.0 | M16 | 44.0 | M48 | 108.0 |

Grip length, $L_G$ equals screw length, $L$, minus $L_T$. Total length of thread $L_{TT}$ equals $L_T$ plus 5 times the pitch of the coarse thread for the respective screw size. Body length $L_B$ equals $L$ minus $L_{TT}$.

The minimum thread length for hex lag screws is equal to one-half the nominal screw length plus 12 mm, or 150 mm, whichever is shorter. Screws too short for this formula to apply are threaded as close to the head as practicable.

**Metric Screw and Bolt Diameter-Length Combinations.—**For a given diameter, the recommended range of lengths of metric cap screws, formed hex screws, heavy hex screws, hex flange screws, and heavy hex flange screws can be found in Table 17, for heavy hex structural bolts in Table 18, for hex lag screws in Table 16, for round head square neck bolts in Table 19, and for socket head cap screws in Table 20. No recommendations for diameter-length combinations are given in the Standards for hex bolts and heavy hex bolts.

Hex bolts in sizes M5 through M24 and heavy hex bolts in sizes M12 through M24 are standard only in lengths longer than 150 mm or 10*D*, whichever is shorter. When shorter lengths of these sizes are ordered, hex cap screws are normally supplied in place of hex bolts and heavy hex screws in place of heavy hex bolts. Hex bolts in sizes M30 and larger and heavy hex bolts in sizes M30 and M36 are standard in all lengths; however, at manufacturer's option, hex cap screws may be substituted for hex bolts and heavy hex screws for heavy hex bolts for any diameter-length combination.

**Table 16. Recommended Diameter-Length Combinations for Metric Hex Lag Screws** *ANSI/ASME B18.2.3.8M-1981 (Withdrawn)*

| Nominal Length, *L* | Nominal Screw Diameter | | | | | | | | Nominal Length, *L* | Nominal Screw Diameter | | | | |
|---|---|---|---|---|---|---|---|---|---|---|---|---|---|---|
| | 5 | 6 | 8 | 10 | 12 | 16 | 20 | 24 | | 10 | 12 | 16 | 20 | 24 |
| 8 | ● | … | … | … | … | … | … | … | 90 | ● | ● | ● | ● | ● |
| 10 | ● | ● | … | … | … | … | … | … | 100 | ● | ● | ● | ● | ● |
| 12 | ● | ● | ● | … | … | … | … | … | 110 | … | ● | ● | ● | ● |
| 14 | ● | ● | ● | … | … | … | … | … | 120 | … | ● | ● | ● | ● |
| 16 | ● | ● | ● | ● | … | … | … | … | 130 | … | … | ● | ● | ● |
| 20 | ● | ● | ● | ● | ● | … | … | … | 140 | … | … | ● | ● | ● |
| 25 | ● | ● | ● | ● | ● | ● | … | … | 150 | … | … | ● | ● | ● |
| 30 | ● | ● | ● | ● | ● | ● | ● | … | 160 | … | … | ● | ● | ● |
| 35 | ● | ● | ● | ● | ● | ● | ● | ● | 180 | … | … | … | ● | ● |
| 40 | ● | ● | ● | ● | ● | ● | ● | ● | 200 | … | … | … | ● | ● |
| 45 | ● | ● | ● | ● | ● | ● | ● | ● | 220 | … | … | … | … | ● |
| 50 | ● | ● | ● | ● | ● | ● | ● | ● | 240 | … | … | … | … | ● |
| 60 | … | ● | ● | ● | ● | ● | ● | ● | 260 | … | … | … | … | ● |
| 70 | … | … | ● | ● | ● | ● | ● | ● | 280 | … | … | … | … | ● |
| 80 | … | … | ● | ● | ● | ● | ● | ● | 300 | … | … | … | … | ● |

All dimensions are in millimeters.

Recommended diameter-length combinations are indicated by the symbol ●.

**Table 17. Rec'd Diameter-Length Combinations for Metric Hex Cap Screws, Formed Hex and Heavy Hex Screws, Hex Flange and Heavy Hex Flange Screws**

| Nominal Length[a] | Diameter—Pitch | | | | | | | | | | |
|---|---|---|---|---|---|---|---|---|---|---|---|
| | M5 ×0.8 | M6 ×1 | M8 ×1.25 | M10 ×1.5 | M12 ×1.75 | M14 ×2 | M16 ×2 | M20 ×2.5 | M24 ×3 | M30 ×3.5 | M36 ×4 |
| 8 | ● | … | … | … | … | … | … | … | … | … | … |
| 10 | ● | ● | … | … | … | … | … | … | … | … | … |
| 12 | ● | ● | ● | … | … | … | … | … | … | … | … |
| 14 | ● | ● | ● | ●[b] | … | … | … | … | … | … | … |
| 16 | ● | ● | ● | ● | ●[b] | ●[b] | … | … | … | … | … |
| 20 | ● | ● | ● | ● | ● | ● | … | … | … | … | … |
| 25 | ● | ● | ● | ● | ● | ● | ● | … | … | … | … |
| 30 | ● | ● | ● | ● | ● | ● | ● | ● | … | … | … |
| 35 | ● | ● | ● | ● | ● | ● | ● | ● | ● | … | … |
| 40 | ● | ● | ● | ● | ● | ● | ● | ● | ● | ● | … |
| 45 | ● | ● | ● | ● | ● | ● | ● | ● | ● | ● | … |
| 50 | ● | ● | ● | ● | ● | ● | ● | ● | ● | ● | ● |
| (55) | … | ● | ● | ● | ● | ● | ● | ● | ● | ● | ● |
| 60 | … | ● | ● | ● | ● | ● | ● | ● | ● | ● | ● |
| (65) | … | … | ● | ● | ● | ● | ● | ● | ● | ● | ● |
| 70 | … | … | ● | ● | ● | ● | ● | ● | ● | ● | ● |
| (75) | … | … | ● | ● | ● | ● | ● | ● | ● | ● | ● |
| 80 | … | … | ● | ● | ● | ● | ● | ● | ● | ● | ● |
| (85) | … | … | … | ● | ● | ● | ● | ● | ● | ● | ● |
| 90 | … | … | … | ● | ● | ● | ● | ● | ● | ● | ● |
| 100 | … | … | … | ● | ● | ● | ● | ● | ● | ● | ● |
| 110 | … | … | … | … | ● | ● | ● | ● | ● | ● | ● |
| 120 | … | … | … | … | ● | ● | ● | ● | ● | ● | ● |
| 130 | … | … | … | … | … | ● | ● | ● | ● | ● | ● |
| 140 | … | … | … | … | … | ● | ● | ● | ● | ● | ● |
| 150 | … | … | … | … | … | … | ● | ● | ● | ● | ● |
| 160 | … | … | … | … | … | … | ● | ● | ● | ● | ● |
| (170) | … | … | … | … | … | … | … | ● | ● | ● | ● |
| 180 | … | … | … | … | … | … | … | ● | ● | ● | ● |
| (190) | … | … | … | … | … | … | … | ● | ● | ● | ● |
| 200 | … | … | … | … | … | … | … | ● | ● | ● | ● |
| 220 | … | … | … | … | … | … | … | … | ● | ● | ● |
| 240 | … | … | … | … | … | … | … | … | ● | ● | ● |
| 260 | … | … | … | … | … | … | … | … | … | ● | ● |
| 280 | … | … | … | … | … | … | … | … | … | ● | ● |
| 300 | … | … | … | … | … | … | … | … | … | ● | ● |

[a] Lengths in parentheses are not recommended. Recommended lengths of formed hex screws, hex flange screws, and heavy hex flange screws do not extend above 150 mm. Recommended lengths of heavy hex screws do not extend below 20 mm. Standard sizes for government use. Recommended diameter-length combinations are indicated by the symbol ●. Screws with lengths above heavy cross lines are threaded full length.

[b] Does not apply to hex flange screws and heavy hex flange screws.

All dimensions are in millimeters.

For available diameters of each type of screw, see respective dimensional table.

## Table 18. Recommended Diameter-Length Combinations for Metric Heavy Hex Structural Bolts

| Nominal Length, *L* | Nominal Diameter and Thread Pitch | | | | | | |
|---|---|---|---|---|---|---|---|
| | M16 × 2 | M20 × 2.5 | M22 × 2.5 | M24 × 3 | M27 × 3 | M30 × 3.5 | M36 × 4 |
| 45 | ● | … | … | … | … | … | … |
| 50 | ● | ● | … | … | … | … | … |
| 55 | ● | ● | ● | … | … | … | … |
| 60 | ● | ● | ● | ● | … | … | … |
| 65 | ● | ● | ● | ● | ● | … | … |
| 70 | ● | ● | ● | ● | ● | ● | … |
| 75 | ● | ● | ● | ● | ● | ● | … |
| 80 | ● | ● | ● | ● | ● | ● | ● |
| 85 | ● | ● | ● | ● | ● | ● | ● |
| 90 | ● | ● | ● | ● | ● | ● | ● |
| 95 | ● | ● | ● | ● | ● | ● | ● |
| 100 | ● | ● | ● | ● | ● | ● | ● |
| 110 | ● | ● | ● | ● | ● | ● | ● |
| 120 | ● | ● | ● | ● | ● | ● | ● |
| 130 | ● | ● | ● | ● | ● | ● | ● |
| 140 | ● | ● | ● | ● | ● | ● | ● |
| 150 | ● | ● | ● | ● | ● | ● | ● |
| 160 | ● | ● | ● | ● | ● | ● | ● |
| 170 | ● | ● | ● | ● | ● | ● | ● |
| 180 | ● | ● | ● | ● | ● | ● | ● |
| 190 | ● | ● | ● | ● | ● | ● | ● |
| 200 | ● | ● | ● | ● | ● | ● | ● |
| 210 | ● | ● | ● | ● | ● | ● | ● |
| 220 | ● | ● | ● | ● | ● | ● | ● |
| 230 | ● | ● | ● | ● | ● | ● | ● |
| 240 | ● | ● | ● | ● | ● | ● | ● |
| 250 | ● | ● | ● | ● | ● | ● | ● |
| 260 | ● | ● | ● | ● | ● | ● | ● |
| 270 | ● | ● | ● | ● | ● | ● | ● |
| 280 | ● | ● | ● | ● | ● | ● | ● |
| 290 | ● | ● | ● | ● | ● | ● | ● |
| 300 | ● | ● | ● | ● | ● | ● | ● |

All dimensions are in millimeters.

Recommended diameter-length combinations are indicated by the symbol ●.

Bolts with lengths above the heavy cross lines are threaded full length.

## Table 19. Recommended Diameter-Length Combinations for Metric Round Head Square Neck Bolts

| Nominal Length,[a] *L* | Nominal Diameter and Thread Pitch | | | | | | | | |
|---|---|---|---|---|---|---|---|---|---|
| | M5 ×0.8 | M6 ×1 | M8 ×1.25 | M10 ×1.5 | M12 ×1.75 | M14 ×2 | M16 ×2 | M20 ×2.5 | M24 ×3 |
| 10 | ● | … | … | … | … | … | … | … | … |
| 12 | ● | ● | … | … | … | … | … | … | … |
| (14) | ● | ● | … | … | … | … | … | … | … |
| 16 | ● | ● | ● | … | … | … | … | … | … |
| 20 | ● | ● | ● | ● | … | … | … | … | … |
| 25 | ● | ● | ● | ● | ● | … | … | … | … |
| 30 | ● | ● | ● | ● | ● | ● | ● | … | … |
| 35 | ● | ● | ● | ● | ● | ● | ● | … | … |
| 40 | ● | ● | ● | ● | ● | ● | ● | ● | … |
| 45 | ● | ● | ● | ● | ● | ● | ● | ● | ● |
| 50 | ● | ● | ● | ● | ● | ● | ● | ● | ● |
| (55) | … | ● | ● | ● | ● | ● | ● | ● | ● |
| 60 | … | ● | ● | ● | ● | ● | ● | ● | ● |
| (65) | … | … | ● | ● | ● | ● | ● | ● | ● |
| 70 | … | … | ● | ● | ● | ● | ● | ● | ● |
| (75) | … | … | ● | ● | ● | ● | ● | ● | ● |
| 80 | … | … | ● | ● | ● | ● | ● | ● | ● |

**Table 19.** *(Continued)* **Recommended Diameter-Length Combinations for Metric Round Head Square Neck Bolts**

| Nominal Length,[a] L | Nominal Diameter and Thread Pitch | | | | | | | | |
|---|---|---|---|---|---|---|---|---|---|
| | M5 ×0.8 | M6 ×1 | M8 ×1.25 | M10 ×1.5 | M12 ×1.75 | M14 ×2 | M16 ×2 | M20 ×2.5 | M24 ×3 |
| (85) | … | … | … | ● | ● | ● | ● | ● | ● |
| 90 | … | … | … | ● | ● | ● | ● | ● | ● |
| 100 | … | … | … | ● | ● | ● | ● | ● | ● |
| 110 | … | … | … | … | ● | ● | ● | ● | ● |
| 120 | … | … | … | … | ● | ● | ● | ● | ● |
| 130 | … | … | … | … | … | ● | ● | ● | ● |
| 140 | … | … | … | … | … | ● | ● | ● | ● |
| 150 | … | … | … | … | … | … | ● | ● | ● |
| 160 | … | … | … | … | … | … | ● | ● | ● |
| (170) | … | … | … | … | … | … | … | ● | ● |
| 180 | … | … | … | … | … | … | … | ● | ● |
| (190) | … | … | … | … | … | … | … | ● | ● |
| 200 | … | … | … | … | … | … | … | ● | ● |
| 220 | … | … | … | … | … | … | … | … | ● |
| 240 | … | … | … | … | … | … | … | … | ● |

[a] Bolts with lengths above the heavy cross lines are threaded full length. Lengths in ( ) are not recommended.

All dimensions are in millimeters. Recommended diameter-length combinations are indicated by the symbol ●. Standard sizes for government use.

**Table 20. Diameter-Length Combinations for Socket Head Cap Screws (Metric Series)**

| Nominal Length, L | Nominal Size | | | | | | | | | | | | | |
|---|---|---|---|---|---|---|---|---|---|---|---|---|---|---|
| | M1.6 | M2 | M2.5 | M3 | M4 | M5 | M6 | M8 | M10 | M12 | M14 | M16 | M20 | M24 |
| 20 | ● | ● | | | | | | | | | | | | |
| 25 | ● | ● | ● | ● | | | | | | | | | | |
| 30 | ● | ● | ● | ● | ● | | | | | | | | | |
| 35 | … | ● | ● | ● | ● | ● | ● | | | | | | | |
| 40 | … | ● | ● | ● | ● | ● | ● | | | | | | | |
| 45 | … | … | ● | ● | ● | ● | ● | ● | | | | | | |
| 50 | … | … | ● | ● | ● | ● | ● | ● | ● | | | | | |
| 55 | … | … | … | ● | ● | ● | ● | ● | ● | | | | | |
| 60 | … | … | … | ● | ● | ● | ● | ● | ● | ● | | | | |
| 65 | … | … | … | ● | ● | ● | ● | ● | ● | ● | ● | | | |
| 70 | … | … | … | … | ● | ● | ● | ● | ● | ● | ● | ● | | |
| 80 | … | … | … | … | ● | ● | ● | ● | ● | ● | ● | ● | | |
| 90 | … | … | … | … | … | ● | ● | ● | ● | ● | ● | ● | ● | |
| 100 | … | … | … | … | … | ● | ● | ● | ● | ● | ● | ● | ● | ● |
| 110 | … | … | … | … | … | … | ● | ● | ● | ● | ● | ● | ● | ● |
| 120 | … | … | … | … | … | … | ● | ● | ● | ● | ● | ● | ● | ● |
| 130 | … | … | … | … | … | … | … | ● | ● | ● | ● | ● | ● | ● |
| 140 | … | … | … | … | … | … | … | ● | ● | ● | ● | ● | ● | ● |
| 150 | … | … | … | … | … | … | … | ● | ● | ● | ● | ● | ● | ● |
| 160 | … | … | … | … | … | … | … | ● | ● | ● | ● | ● | ● | ● |
| 180 | … | … | … | … | … | … | … | … | ● | ● | ● | ● | ● | ● |
| 200 | … | … | … | … | … | … | … | … | ● | ● | ● | ● | ● | ● |
| 220 | … | … | … | … | … | … | … | … | … | ● | ● | ● | ● | ● |
| 240 | … | … | … | … | … | … | … | … | … | ● | ● | ● | ● | ● |
| 260 | … | … | … | … | … | … | … | … | … | … | ● | ● | ● | ● |
| 300 | … | … | … | … | … | … | … | … | … | … | … | ● | ● | ● |

All dimensions are in millimeters. Screws with lengths above heavy cross lines are threaded full length. Diameter-length combinations are indicated by the symbol ●. Standard sizes for government use. In addition to the lengths shown, the following lengths are standard: 3, 4, 5, 6, 8, 10, 12, and 16 mm. No diameter-length combinations are given in the Standard for these lengths. Screws larger than M24 with lengths equal to or shorter than $L_{TT}$ (see Table 15 footnote) are threaded full length.

**Table 21. American National Standard Socket Head Cap Screws— Metric Series** *ANSI/ASME B18.3.1M-1986 (Withdrawn)*

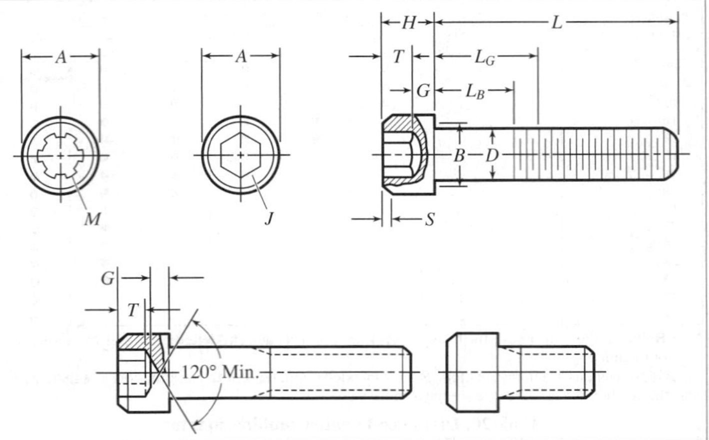


| Nom. Size and Thread Pitch | Body Diameter, D | | Head Diameter A | | Head Height H | | Chamfer or Radius S | Hexagon Socket Size[a] J | Spline Socket Size[a] M | Key Engagement T | Transition Dia. B[a] |
|---|---|---|---|---|---|---|---|---|---|---|---|
| | Max. | Min. | Max. | Min. | Max. | Min. | Max. | Nom. | Nom. | Min. | Max. |
| M1.6 × 0.35 | 1.60 | 1.46 | 3.00 | 2.87 | 1.60 | 1.52 | 0.16 | 1.5 | 1.829 | 0.80 | 2.0 |
| M2 × 0.4 | 2.00 | 1.86 | 3.80 | 3.65 | 2.00 | 1.91 | 0.20 | 1.5 | 1.829 | 1.00 | 2.6 |
| M2.5 × 0.45 | 2.50 | 2.36 | 4.50 | 4.33 | 2.50 | 2.40 | 0.25 | 2.0 | 2.438 | 1.25 | 3.1 |
| M3 × 0.5 | 3.00 | 2.86 | 5.50 | 5.32 | 3.00 | 2.89 | 0.30 | 2.5 | 2.819 | 1.50 | 3.6 |
| M4 × 0.7 | 4.00 | 3.82 | 7.00 | 6.80 | 4.00 | 3.88 | 0.40 | 3.0 | 3.378 | 2.00 | 4.7 |
| M5 × 0.8 | 5.00 | 4.82 | 8.50 | 8.27 | 5.00 | 4.86 | 0.50 | 4.0 | 4.648 | 2.50 | 5.7 |
| M6 × 1 | 6.00 | 5.82 | 10.00 | 9.74 | 6.00 | 5.85 | 0.60 | 5.0 | 5.486 | 3.00 | 6.8 |
| M8 × 1.25 | 8.00 | 7.78 | 13.00 | 12.70 | 8.00 | 7.83 | 0.80 | 6.0 | 7.391 | 4.00 | 9.2 |
| M10 × 1.5 | 10.00 | 9.78 | 16.00 | 15.67 | 10.00 | 9.81 | 1.00 | 8.0 | … | 5.00 | 11.2 |
| M12 × 1.75 | 12.00 | 11.73 | 18.00 | 17.63 | 12.00 | 11.79 | 1.20 | 10.0 | … | 6.00 | 14.2 |
| M14 × 2[b] | 14.00 | 13.73 | 21.00 | 20.60 | 14.00 | 13.77 | 1.40 | 12.0 | … | 7.00 | 16.2 |
| M16 × 2 | 16.00 | 15.73 | 24.00 | 23.58 | 16.00 | 15.76 | 1.60 | 14.0 | … | 8.00 | 18.2 |
| M20 × 2.5 | 20.00 | 19.67 | 30.00 | 29.53 | 20.00 | 19.73 | 2.00 | 17.0 | … | 10.00 | 22.4 |
| M24 × 3 | 24.00 | 23.67 | 36.00 | 35.48 | 24.00 | 23.70 | 2.40 | 19.0 | … | 12.00 | 26.4 |
| M30 × 3.5 | 30.00 | 29.67 | 45.00 | 44.42 | 30.00 | 29.67 | 3.00 | 22.0 | … | 15.00 | 33.4 |
| M36 × 4 | 36.00 | 35.61 | 54.00 | 53.37 | 36.00 | 35.64 | 3.60 | 27.0 | … | 18.00 | 39.4 |
| M42 × 4.5 | 42.00 | 41.61 | 63.00 | 62.31 | 42.00 | 41.61 | 4.20 | 32.0 | … | 21.00 | 45.6 |
| M48 × 5 | 48.00 | 47.61 | 72.00 | 71.27 | 48.00 | 47.58 | 4.80 | 36.0 | … | 24.00 | 52.6 |

[a] See also Table 23.

[b] The M14 × 2 size is not recommended for use in new designs.

All dimensions are in millimeters

$L_G$ is grip length and $L_B$ is body length (see Table 15). For length of complete thread, see Table 15.

For additional manufacturing and acceptance specifications, see ANSI/ASME B18.3.1M (Withdrawn).

## Table 22. Drilled Head Dimensions for Metric Hex Socket Head Cap Screws

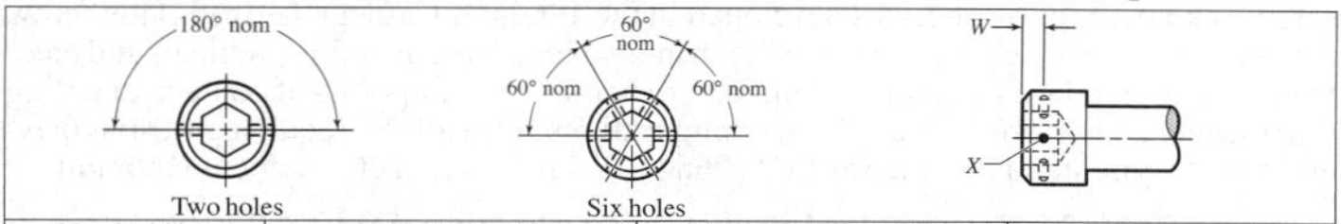


| Nominal Size or Basic Screw Diameter | Hole Center Location, *W* | | Drilled Hole Diameter, *X* | | Hole Alignment Check Plug Diameter |
|---|---|---|---|---|---|
| | Max. | Min. | Max. | Min. | Basic |
| M3 | 1.20 | 0.80 | 0.95 | 0.80 | 0.75 |
| M4 | 1.60 | 1.20 | 1.35 | 1.20 | 0.90 |
| M5 | 2.00 | 1.50 | 1.35 | 1.20 | 0.90 |
| M6 | 2.30 | 1.80 | 1.35 | 1.20 | 0.90 |
| M8 | 2.70 | 2.20 | 1.35 | 1.20 | 0.90 |
| M10 | 3.30 | 2.80 | 1.65 | 1.50 | 1.40 |
| M12 | 4.00 | 3.50 | 1.65 | 1.50 | 1.40 |
| M16 | 5.00 | 4.50 | 1.65 | 1.50 | 1.40 |
| M20 | 6.30 | 5.80 | 2.15 | 2.00 | 1.80 |
| M24 | 7.30 | 6.80 | 2.15 | 2.00 | 1.80 |
| M30 | 9.00 | 8.50 | 2.15 | 2.00 | 1.80 |
| M36 | 10.50 | 10.00 | 2.15 | 2.00 | 1.80 |

All dimensions are in millimeters.

Drilled head metric hexagon socket head cap screws normally are not available in screw sizes smaller than M3 nor larger than M36. The M3 and M4 nominal screw sizes have two drilled holes spaced 180 degrees apart. Nominal screw sizes M5 and larger have six drilled holes spaced 60 degrees apart unless the purchaser specifies two drilled holes. The positioning of holes on opposite sides of the socket should be such that the hole alignment check plug will pass completely through the head without any deflection. When so specified by the purchaser, the edges of holes on the outside surface of the head will be chamfered 45 degrees to a depth of 0.30 to 0.50 mm.

## Table 23. American National Standard Hexagon and Spline Sockets for Socket Head Cap Screws—Metric Series *ANSI/ASME B18.3.1M-1986 (Withdrawn)*

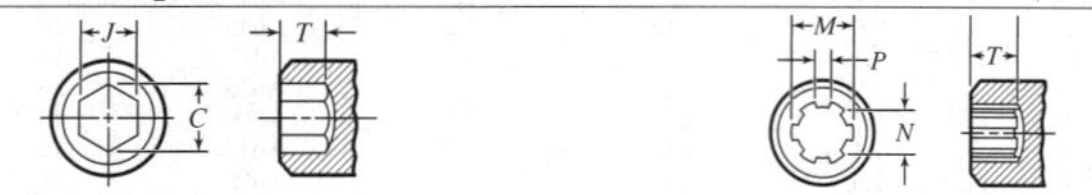


METRIC HEXAGON SOCKETS — See Table 21

METRIC SPLINE SOCKET — See Table 21

| Nominal Hexagon Socket Size | Socket Width Across Flats, *J* | | Socket Width Across Corners, *C* | Nominal Hexagon Socket Size | Socket Width Across Flats, *J* | | Socket Width Across Corners, *C* |
|---|---|---|---|---|---|---|---|
| Metric Hexagon Sockets | | | | | | | |
| | Max. | Min. | Min. | | Max. | Min. | Min. |
| 1.5 | 1.545 | 1.520 | 1.73 | 12 | 12.146 | 12.032 | 13.80 |
| 2 | 2.045 | 2.020 | 2.30 | 14 | 14.159 | 14.032 | 16.09 |
| 2.5 | 2.560 | 2.520 | 2.87 | 17 | 17.216 | 17.050 | 19.56 |
| 3 | 3.071 | 3.020 | 3.44 | 19 | 19.243 | 19.065 | 21.87 |
| 4 | 4.084 | 4.020 | 4.58 | 22 | 22.319 | 22.065 | 25.31 |
| 5 | 5.084 | 5.020 | 5.72 | 24 | 24.319 | 24.065 | 27.60 |
| 6 | 6.095 | 6.020 | 6.86 | 27 | 27.319 | 27.065 | 31.04 |
| 8 | 8.115 | 8.025 | 9.15 | 32 | 32.461 | 32.080 | 36.80 |
| 10 | 10.127 | 10.025 | 11.50 | 36 | 36.461 | 36.080 | 41.38 |

| Metric Spline Sockets[a] | | | | | | |
|---|---|---|---|---|---|---|
| Nominal Spline Socket Size | Socket Major Diameter, *M* | | Socket Minor Diameter, *N* | | Width of Tooth, *P* | |
| | Max. | Min. | Max. | Min. | Max. | Min. |
| 1.829 | 1.8796 | 1.8542 | 1.6256 | 1.6002 | 0.4064 | 0.3810 |
| 2.438 | 2.4892 | 2.4638 | 2.0828 | 2.0320 | 0.5588 | 0.5334 |
| 2.819 | 2.9210 | 2.8702 | 2.4892 | 2.4384 | 0.6350 | 0.5842 |
| 3.378 | 3.4798 | 3.4290 | 2.9972 | 2.9464 | 0.7620 | 0.7112 |
| 4.648 | 4.7752 | 4.7244 | 4.1402 | 4.0894 | 0.9906 | 0.9398 |
| 5.486 | 5.6134 | 5.5626 | 4.8260 | 4.7752 | 1.2700 | 1.2192 |
| 7.391 | 7.5692 | 7.5184 | 6.4516 | 6.4008 | 1.7272 | 2.6764 |

[a] The tabulated dimensions represent direct metric conversions of the equivalent inch size spline sockets shown in American National Standard Socket Cap, Shoulder and Set Screws.—Inch Series ANSI/ASME B18.3. Therefore, the spline keys and bits shown therein are applicable for wrenching the corresponding size metric spline sockets.

**Torx Screws.—**The hexalobular internal driving feature, commonly known as the Torx® screw or sometimes 6-lobe, was developed in the 1960s by Camcar Textron. This screw type has become widely used worldwide in many industries, including vehicles and electronics. The standard internal driving feature has six lobular points of contact, enabling significant turning torque. Many head configurations are available requiring various drivers; refer to type standards such as ISO 10664-2014 and manufacturer documentation.

**Table 24. Hexalobular Internal Driving Feature** *ISO 10664-2014*

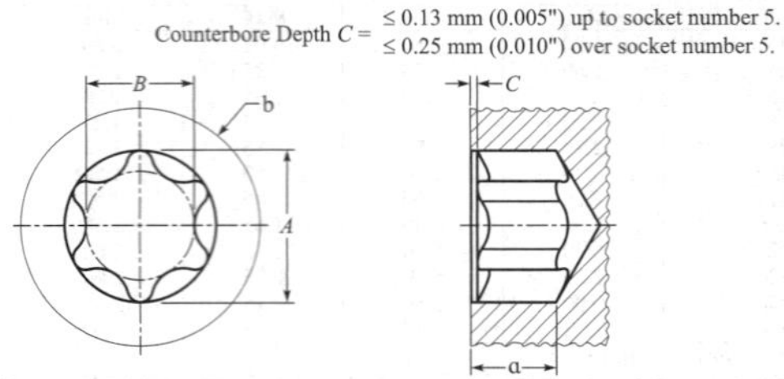


[a] Refer to the relevant product standard for the penetration depth.
[b] Refer to the relevant product standard for the head configuration.

| Socket Number | External Diameter *A* | | Internal Diameter *B* | | Torx Driver Number |
|---|---|---|---|---|---|
| | mm | inch | mm | inch | |
| 1 | 0.9 | 0.035 | 0.6 | 0.024 | T1 |
| 2 | 1.0 | 0.039 | 0.7 | 0.028 | T2 |
| 3 | 1.2 | 0.047 | 0.85 | 0.033 | T3 |
| 4 | 1.35 | 0.053 | 1.0 | 0.039 | T4 |
| 5 | 1.5 | 0.059 | 1.1 | 0.043 | T5 |
| 6 | 1.75 | 0.069 | 1.27 | 0.050 | T6 |
| 7 | 2.1 | 0.083 | 1.5 | 0.059 | T7 |
| 8 | 2.4 | 0.094 | 1.75 | 0.069 | T8 |
| 9 | 2.6 | 0.102 | 1.9 | 0.075 | T9 |
| 10 | 2.8 | 0.110 | 2.05 | 0.081 | T10 |
| 15 | 3.35 | 0.132 | 2.4 | 0.094 | T15 |
| 20 | 3.95 | 0.156 | 2.85 | 0.112 | T20 |
| 25 | 4.5 | 0.177 | 3.25 | 0.128 | T25 |
| 27 | 5.1 | 0.201 | 3.68 | 0.145 | T27 |
| 30 | 5.6 | 0.220 | 4.05 | 0.159 | T30 |
| 40 | 6.75 | 0.266 | 4.85 | 0.191 | T40 |
| 45 | 7.93 | 0.312 | 5.64 | 0.222 | T45 |
| 50 | 8.95 | 0.352 | 6.45 | 0.254 | T50 |
| 55 | 11.35 | 0.447 | 8.05 | 0.317 | T55 |
| 60 | 13.45 | 0.530 | 9.6 | 0.378 | T60 |
| 70 | 15.7 | 0.618 | 11.2 | 0.441 | T70 |
| 80 | 17.75 | 0.699 | 12.8 | 0.504 | T80 |
| 90 | 20.2 | 0.795 | 14.4 | 0.567 | T90 |
| 100 | 22.4 | 0.882 | 16 | 0.630 | T100 |

## Metric Nuts

The American National Standards covering metric were established in cooperation with the Department of Defense so that they could be used by the government for procurement purposes. Extensive information concerning these nuts is given in the following text and tables, but for more complete manufacturing and acceptance specifications, reference should be made to the respective Standards, which may be obtained by non-governmental agencies from the American National Standards Institute, 25 West 43rd Street, 4th Floor, New York, NY 10036, www.ansi.org. Manufacturers should be consulted about items and sizes in stock production.

**Comparison with ISO Standards.—**American National Standards for metric nuts have been coordinated to the extent possible with comparable ISO Standards or proposed Standards, thus: ANSI B18.2.4.1M (Withdrawn) Metric Hex Nuts, Style 1 with ISO 4032; B18.2.4.2M (Withdrawn) Metric Hex Nuts, Style 2 with ISO 4033; B18.2.4.4M

(Withdrawn) Metric Hex Flange Nuts with ISO 4161; B18.2.4.5M (Withdrawn) Metric Hex Jam Nuts with ISO 4035; and B18.2.4.3M Metric Slotted Hex Nuts, B18.2.4.6M Metric Heavy Hex Nuts in sizes M12 through M36, and B18.16M Prevailing-Torque Type Steel Metric Hex Nuts and Hex Flange Nuts with comparable draft ISO Standards. The dimensional differences between each ANSI Standard and the comparable ISO Standard or draft Standard are very few, relatively minor, and none will affect the interchangeability of nuts manufactured to the requirements of either.

At its meeting in Varna, May 1977, ISO/TC2 studied several technical reports analyzing design considerations influencing determination of the best series of widths across flats for hex bolts, screws, and nuts. A primary technical objective was to achieve a logical ratio between under head (nut) bearing surface area (which determines the magnitude of compressive stress on the bolted members) and the tensile stress area of the screw thread (which governs the clamping force that can be developed by tightening the fastener). The series of widths across flats in the ANSI Standards agree with those which were selected by ISO/TC2 to be ISO Standards.

One exception for width across flats of metric hex nuts, styles 1 and 2, metric slotted hex nuts, metric hex jam nuts, and prevailing-torque metric hex nuts is the M10 size. These nuts in M10 size are currently being produced in the United States with a width across flats of 15 mm. This width, however, is not an ISO Standard. Unless these M10 nuts with width across flats of 15 mm are specifically ordered, the M10 size with 16 mm width across flats will be furnished.

In ANSI Standards for metric nuts, letter symbols designating dimensional characteristics are in accord with those used in ISO Standards, except capitals have been used for data processing convenience instead of lower case letters used in ISO Standards.

**Metric Nut Tops and Bearing Surfaces.—**Metric hex nuts, styles 1 and 2, slotted hex nuts, and hex jam nuts are double chamfered in sizes M16 and smaller and in sizes M20 and larger may either be double chamfered or have a washer-faced bearing surface and a chamfered top at the option of the manufacturer. Metric heavy hex nuts are optional either way in all sizes. Metric hex flange nuts have a flange bearing surface and a chamfered top and prevailing-torque type metric hex nuts have a chamfered bearing surface. Prevailing-torque type metrix hex flange nuts have a flange bearing surface. All types of metric nuts have the tapped hole countersunk on the bearing face and metric slotted hex nuts, hex flange nuts, and prevailing-torque type hex nuts and hex flange nuts may be countersunk on the top face.

**Table 25. American National Standard Metric Slotted Hex Nuts**

*ANSI/ASME B18.2.4.3M-1979 (R2017)*

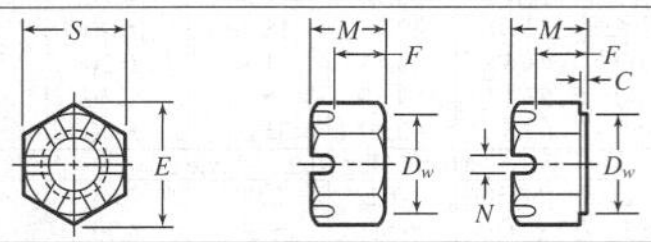


| Nominal Nut Dia. and Thread Pitch | Width Across Flats, S | | Width Across Corners, E | | Thickness, M | | Bearing Face Dia., $D_w$ | Unslotted Thickness, F | | Width of Slot, N | | Washer Face Thickness C | |
|---|---|---|---|---|---|---|---|---|---|---|---|---|---|
| | Max. | Min. | Max. | Min. | Max. | Min. | Min. | Max. | Min. | Max. | Min. | Max. | Min. |
| M5 × 0.8 | 8.00 | 7.78 | 9.24 | 8.79 | 5.10 | 4.80 | 6.9 | 3.2 | 2.9 | 2.0 | 1.4 | … | … |
| M6 × 1 | 10.00 | 9.78 | 11.55 | 11.05 | 5.70 | 5.40 | 8.9 | 3.5 | 3.2 | 2.4 | 1.8 | … | … |
| M8 × 1.25 | 13.00 | 12.73 | 15.01 | 14.38 | 7.50 | 7.14 | 11.6 | 4.4 | 4.1 | 2.9 | 2.3 | … | … |
| [a]M10 × 1.5 | 15.00 | 14.73 | 17.32 | 16.64 | 10.0 | 9.6 | 13.6 | 5.7 | 5.4 | 3.4 | 2.8 | 0.6 | 0.3 |
| M10 × 1.5 | 16.00 | 15.73 | 18.48 | 17.77 | 9.30 | 8.94 | 14.6 | 5.2 | 4.9 | 3.4 | 2.8 | … | … |
| M12 × 1.75 | 18.00 | 17.73 | 20.78 | 20.03 | 12.00 | 11.57 | 16.6 | 7.3 | 6.9 | 4.0 | 3.2 | … | … |
| M14 × 2 | 21.00 | 20.67 | 24.25 | 23.35 | 14.10 | 13.40 | 19.6 | 8.6 | 8.0 | 4.3 | 3.5 | … | … |
| M16 × 2 | 24.00 | 23.67 | 27.71 | 26.75 | 16.40 | 15.70 | 22.5 | 9.9 | 9.3 | 5.3 | 4.5 | … | … |
| M20 × 2.5 | 30.00 | 29.16 | 34.64 | 32.95 | 20.30 | 19.00 | 27.7 | 13.3 | 12.2 | 5.7 | 4.5 | 0.8 | 0.4 |
| M24 × 3 | 36.00 | 35.00 | 41.57 | 39.55 | 23.90 | 22.60 | 33.2 | 15.4 | 14.3 | 6.7 | 5.5 | 0.8 | 0.4 |
| M30 × 3.5 | 46.00 | 45.00 | 53.12 | 50.85 | 28.60 | 27.30 | 42.7 | 18.1 | 16.8 | 8.5 | 7.0 | 0.8 | 0.4 |
| M36 × 4 | 55.00 | 53.80 | 63.51 | 60.79 | 34.70 | 33.10 | 51.1 | 23.7 | 22.4 | 8.5 | 7.0 | 0.8 | 0.4 |

[a] This size with width across flats of 15 mm is not standard. Unless specifically ordered, M10 slotted hex nuts with 16 mm width across flats will be furnished.

All dimensions are in millimeters.

**Materials and Mechanical Properties.—**Non-heat-treated carbon steel metric hex nuts, style 1 and slotted hex nuts conform to material and property class requirements specified for property class 5 nuts; hex nuts, style 2 and hex flange nuts to property class 9 nuts; hex jam nuts to property class 4 nuts, and non-heat-treated carbon and alloy steel heavy hex nuts to property classes 5, 9, 8S, or 8S3 nuts; all as covered in ASTM A563M. Carbon steel metric hex nuts, style 1 and slotted hex nuts that have specified heat treatment conform to material and property class requirements specified for property class 10 nuts; hex nuts, style 2 to property class 12 nuts; hex jam nuts to property class 05 nuts; hex flange nuts to property classes 10 and 12 nuts; and carbon or alloy steel heavy hex nuts to property classes 10S, 10S3, or 12 nuts, all as covered in ASTM A563M. Carbon steel prevailing-torque type hex nuts and hex flange nuts conform to mechanical and property class requirements as given in ANSI/ASME B18.16M-2004 (R2021).

**Table 26. American National Standard Metric Hex Nuts, Styles 1 and 2**
*ANSI/ASME B18.2.6M-2012 (R2021) and B18.2.6-2019*

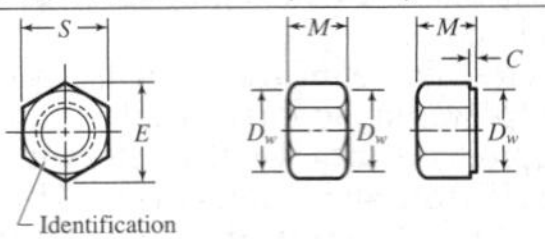


| Nominal Nut Dia. and Thread Pitch | Width Across Flats,[a] $S$ | | Width Across Corners,[b] $E$ | | Thickness,[c] $M$ | | Bearing Face Dia.,[d] $D_w$ | Washer Face Thickness,[d] $C$ | |
|---|---|---|---|---|---|---|---|---|---|
| | Max. | Min. | Max. | Min. | Max. | Min. | Min. | Max. | Min. |
| Metric Hex Nuts.—Style 1 | | | | | | | | | |
| M1.6 × 0.35 | 3.20 | 3.02 | 3.70 | 3.41 | 1.30 | 1.05 | 2.3 | … | … |
| M2 × 0.4 | 4.00 | 3.82 | 4.62 | 4.32 | 1.60 | 1.35 | 3.1 | … | … |
| M2.5 × 0.45 | 5.00 | 4.82 | 5.77 | 5.45 | 2.00 | 1.75 | 4.1 | … | … |
| M3 × 0.5 | 5.50 | 5.32 | 6.35 | 6.01 | 2.40 | 2.15 | 4.6 | … | … |
| M3.5 × 0.6 | 6.00 | 5.82 | 6.93 | 6.58 | 2.80 | 2.55 | 5.1 | … | … |
| M4 × 0.7 | 7.00 | 6.78 | 8.08 | 7.66 | 3.20 | 2.90 | 6.0 | … | … |
| M5 × 0.8 | 8.00 | 7.78 | 9.24 | 8.79 | 4.70 | 4.40 | 7.0 | … | … |
| M6 × 1 | 10.00 | 9.78 | 11.55 | 11.05 | 5.20 | 4.90 | 8.9 | … | … |
| M8 × 1.25 | 13.00 | 12.73 | 15.01 | 14.38 | 6.80 | 6.44 | 11.6 | … | … |
| [e]**M10 × 1.5** | **15.00** | **14.73** | **17.32** | **16.64** | **9.1** | **8.7** | **13.6** | **…** | **…** |
| [f]M10 × 1.5 | 16.00 | 15.73 | 18.48 | 17.77 | 8.40 | 8.04 | 14.6 | … | … |
| M12 × 1.75 | 18.00 | 17.73 | 20.78 | 20.03 | 10.80 | 10.37 | 16.6 | … | … |
| M14 × 2 | 21.00 | 20.67 | 24.25 | 23.36 | 12.80 | 12.10 | 19.4 | … | … |
| M16 × 2 | 24.00 | 23.67 | 27.71 | 26.75 | 14.80 | 14.10 | 22.4 | … | … |
| M20 × 2.5 | 30.00 | 29.16 | 34.64 | 32.95 | 18.00 | 16.90 | 27.9 | 0.8 | 0.4 |
| M24 × 3 | 36.00 | 35.00 | 41.57 | 39.55 | 21.50 | 20.20 | 32.5 | 0.8 | 0.4 |
| M30 × 3.5 | 46.00 | 45.00 | 53.12 | 50.85 | 25.60 | 24.30 | 42.5 | 0.8 | 0.4 |
| M36 × 4 | 55.00 | 53.80 | 63.51 | 60.79 | 31.00 | 29.40 | 50.8 | 0.8 | 0.4 |
| Metric Hex Nuts.—Style 2 | | | | | | | | | |
| M3 × 0.5 | 5.50 | 5.32 | 6.35 | 6.01 | 2.90 | 2.65 | 4.6 | … | … |
| M3.5 × 0.6 | 6.00 | 5.82 | 6.93 | 6.58 | 3.30 | 3.00 | 5.1 | … | … |
| M4 × 0.7 | 7.00 | 6.78 | 8.08 | 7.66 | 3.80 | 3.50 | 5.9 | … | … |
| M5 × 0.8 | 8.00 | 7.78 | 9.24 | 8.79 | 5.10 | 4.80 | 6.9 | … | … |
| M6 × 1 | 10.00 | 9.78 | 11.55 | 11.05 | 5.70 | 5.40 | 8.9 | … | … |
| M8 × 1.25 | 13.00 | 12.73 | 15.01 | 14.38 | 7.50 | 7.14 | 11.6 | … | … |
| [e]**M10 × 1.5** | **15.00** | **14.73** | **17.32** | **16.64** | **10.0** | **9.6** | **13.6** | **…** | **…** |
| [f]M10 × 1.5 | 16.00 | 15.73 | 18.48 | 17.77 | 9.30 | 8.94 | 14.6 | … | … |
| M12 × 1.75 | 18.00 | 17.73 | 20.78 | 20.03 | 12.00 | 11.57 | 16.6 | … | … |
| M14 × 2 | 21.00 | 20.67 | 24.25 | 23.35 | 14.10 | 13.40 | 19.6 | … | … |
| M16 × 2 | 24.00 | 23.67 | 27.71 | 26.75 | 16.40 | 15.70 | 22.5 | … | … |
| M20 × 2.5 | 30.00 | 29.16 | 34.64 | 32.95 | 20.30 | 19.00 | 27.7 | 0.8 | 0.4 |
| M24 × 3 | 36.00 | 35.00 | 41.57 | 39.55 | 23.90 | 22.60 | 33.2 | 0.8 | 0.4 |
| M30 × 3.5 | 46.00 | 45.00 | 53.12 | 50.85 | 28.60 | 27.30 | 42.7 | 0.8 | 0.4 |
| M36 × 4 | 55.00 | 53.80 | 63.51 | 60.79 | 34.70 | 33.10 | 51.1 | 0.8 | 0.4 |

[a] The width across flats shall be the distance, measured perpendicular to the axis of the nut, between two opposite wrenching flats.
[b] A rounding or lack of fill at the junction of hex corners with the chamfer shall be permissible.
[c] The nut thickness shall be the overall distance, measured parallel to the axis of the nut, from the top of the nut to the bearing surface, and shall include the thickness of the washer face where provided.
[d] M16 and smaller nuts shall be double chamfered. M20 and larger nuts shall be either double chamfered or have a washer faced bearing surface and a chamfered top.
[e] Dimensional requirements shown in bold type are in addition to or differ from ISO 4032.
[f] When M10 hex nuts are ordered, nuts with 16 mm width across flats shall be furnished unless 15 mm width across flats is specified.

**Table 27. American National Standard Metric Hex Flange Nuts**
*ANSI/ASME B18.2.4.4M-1982 (Withdrawn)*

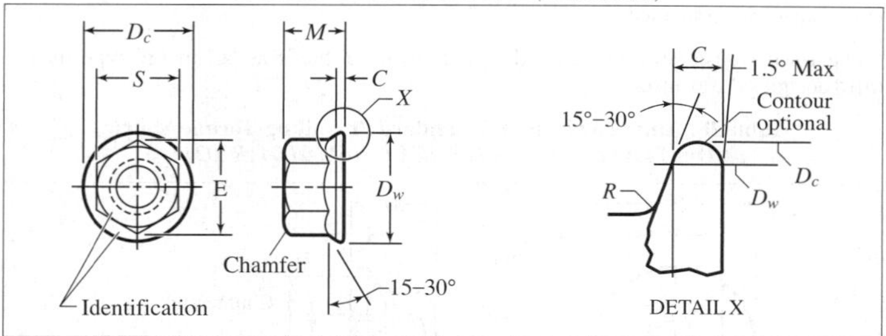


| Nominal Nut Dia. and Thread Pitch | Width Across Flats, $S$ | | Width Across Corners, $E$ | | Flange Dia., $D_c$ | Bearing Circle Dia., $D_w$ | Flange Edge Thick-ness, $C$ | Thickness, $M$ | | Flange Top Fillet Radius, $R$ |
|---|---|---|---|---|---|---|---|---|---|---|
| | Max. | Min. | Max. | Min. | Max. | Min. | Min. | Max. | Min. | Max. |
| M5 × 0.8 | 8.00 | 7.78 | 9.24 | 8.79 | 11.8 | 9.8 | 1.0 | 5.00 | 4.70 | 0.3 |
| M6 × 1 | 10.00 | 9.78 | 11.55 | 11.05 | 14.2 | 12.2 | 1.1 | 6.00 | 5.70 | 0.4 |
| M8 × 1.25 | 13.00 | 12.73 | 15.01 | 14.38 | 17.9 | 15.8 | 1.2 | 8.00 | 7.60 | 0.5 |
| M10 × 1.5 | 15.00 | 14.73 | 17.32 | 16.64 | 21.8 | 19.6 | 1.5 | 10.00 | 9.60 | 0.6 |
| M12 × 1.75 | 18.00 | 17.73 | 20.78 | 20.03 | 26.0 | 23.8 | 1.8 | 12.00 | 11.60 | 0.7 |
| M14 × 2 | 21.00 | 20.67 | 24.25 | 23.35 | 29.9 | 27.6 | 2.1 | 14.00 | 13.30 | 0.9 |
| M16 × 2 | 24.00 | 23.67 | 27.71 | 26.75 | 34.5 | 31.9 | 2.4 | 16.00 | 15.30 | 1.0 |
| M20 × 2.5 | 30.00 | 29.16 | 34.64 | 32.95 | 42.8 | 39.9 | 3.0 | 20.00 | 18.90 | 1.2 |

All dimensions are in millimeters.

Metric nuts of other materials, such as stainless steel, brass, bronze, and aluminum alloys, have properties as agreed upon by the manufacturer and purchaser. Properties of nuts of several grades of nonferrous materials are covered in ASTM F467M.

Unless otherwise specified, metric nuts are furnished with a natural (unprocessed) finish, unplated or uncoated.

**Metric Nut Thread Series.—**Metric nuts have metric coarse threads with class 6H tolerances in accordance with ANSI/ASME B1.13M (see *Metric Screw and Bolt Diameter-Length Combinations* on page 1726 ). For prevailing-torque type metric nuts this condition applies before introduction of the prevailing torque feature. Nuts intended for use with externally threaded fasteners which are plated or coated with a plating or coating thickness (e.g., hot dip galvanized) requiring overtapping of the nut thread to permit assembly, have over-tapped threads in conformance with requirements specified in ASTM A563M.

**Types of Metric Prevailing-Torque Type Nuts.—**There are three basic designs for prevailing-torque type nuts:

1) All-metal, one-piece construction nuts which derive their prevailing-torque characteristics from controlled distortion of the nut thread and/or body.

2) Metal nuts which derive their prevailing-torque characteristics from addition or fusion of a nonmetallic insert, plug or patch in their threads.

3) Top insert, two-piece construction nuts which derive their prevailing-torque characteristics from an insert, usually a full ring of nonmetallic material, located and retained in the nut at its top surface.

The first two designs are designated in Table 30 and Table 28 as "all-metal" type and the third design as "top-insert" type.

**Table 28. American National Standard Prevailing-Torque Metric Hex Flange Nuts** *ANSI/ASME B18.16M-2004 (R2021)*

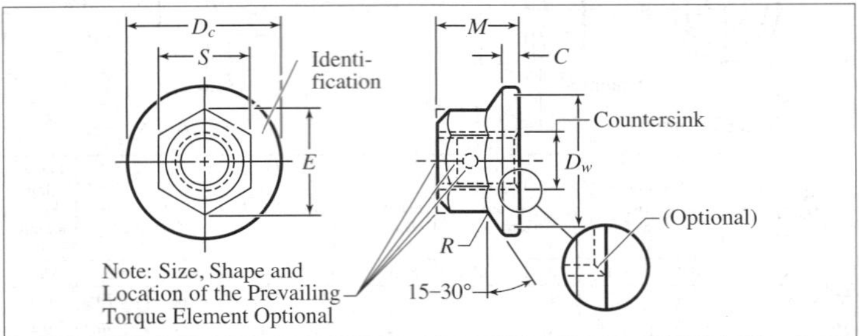


| Nominal Dia. and Thread Pitch | Width Across Flats, $S$ | | Width Across Corners, $E$ | | Thickness, $M$ (All Nut Property Classes): All Metal Type[a] | | Thickness, $M$ (All Nut Property Classes): Top Insert Type | | Flange Dia., $D_c$ | Bearing Circle Dia., $D_w$ | Flange Edge Thickness, $C$ | Flange Top Fillet Radius, $R$ |
|---|---|---|---|---|---|---|---|---|---|---|---|---|
| | Max. | Min. | Max. | Min. | Max. | Min. | Max. | Min. | Max. | Min. | Min. | Max. |
| M6 × 1 | 10.00 | 9.78 | 11.55 | 11.05 | 7.30 | 5.70 | 8.80 | 8.00 | 14.2 | 12.2 | 1.1 | 0.4 |
| M8 × 1.25 | 13.00 | 12.73 | 15.01 | 14.38 | 9.40 | 7.60 | 10.70 | 9.70 | 17.9 | 15.8 | 1.2 | 0.5 |
| M10 × 1.5 | 15.00 | 14.73 | 17.32 | 16.64 | 11.40 | 9.60 | 13.50 | 12.50 | 21.8 | 19.6 | 1.5 | 0.6 |
| M12 × 1.75 | 18.00 | 17.73 | 20.78 | 20.03 | 13.80 | 11.60 | 16.10 | 15.10 | 26.0 | 23.8 | 1.8 | 0.7 |
| M14 × 2 | 21.00 | 20.67 | 24.25 | 23.35 | 15.90 | 13.30 | 18.20 | 17.00 | 29.9 | 27.6 | 2.1 | 0.9 |
| M16 × 2 | 24.00 | 23.67 | 27.71 | 26.75 | 18.30 | 15.30 | 20.30 | 19.10 | 34.5 | 31.9 | 2.4 | 1.0 |
| M20 × 2.5 | 30.00 | 29.16 | 34.64 | 32.95 | 22.40 | 18.90 | 24.80 | 23.50 | 42.8 | 39.9 | 3.0 | 1.2 |

[a] Also includes metal nuts with nonmetallic inserts, plugs, or patches in their threads.

All dimensions are in millimeters.

**Metric Nut Identification Symbols.—**Carbon steel hex nuts, styles 1 and 2, hex flange nuts, and carbon and alloy steel heavy hex nuts are marked to identify the property class and manufacturer in accordance with requirements specified in ASTM A563M. The aforementioned nuts when made of other materials, as well as slotted hex nuts and hex jam nuts, are marked to identify the property class and manufacturer as agreed upon by manufacturer and purchaser. Carbon steel prevailing-torque type hex nuts and hex flange nuts are marked to identify property class and manufacturer as specified in ANSI/ASME B18.16M. Prevailing-torque type nuts of other materials are identified as agreed upon by the manufacturer and purchaser.

**Metric Nut Designation.—**Metric nuts are designated by the following data, preferably in the sequence shown: product name, nominal diameter and thread pitch, steel property class or material identification, and protective coating, if required. (Note: It is common practice in ISO Standards to omit thread pitch from the product designation when the nut threads are the metric coarse thread series, e.g., M10 stands for M10 × 1.5.)

*Example:* Hex nut, style 1, M10 × 1.5, ASTM A563M class 10, zinc plated

Heavy hex nut, M20 × 2.5, silicon bronze, ASTM F467, grade 651

Slotted hex nut, M20, ASTM A563M class 10.

**Table 29. American National Standard Metric Hex Jam Nuts and Heavy Hex Nuts**
*ASME B18.2.6-2019 and ANSI/ASME B18.2.6M-2012 (R2021)*

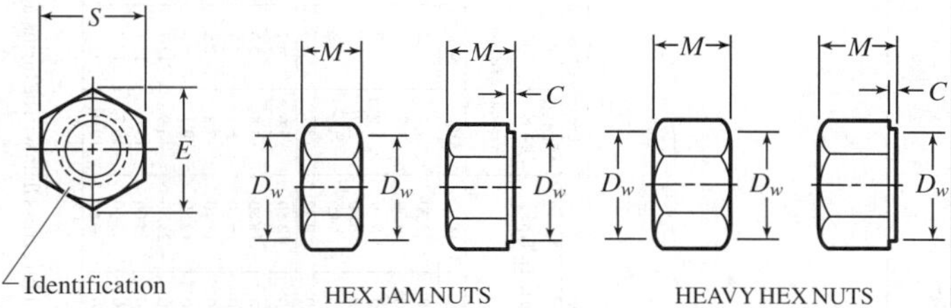


| Nominal Nut Dia. and Thread Pitch | Width Across Flats, S | | Width Across Corners, E | | Thickness, M | | Bearing Face Dia., $D_w$ | Washer Face Thickness, C | |
|---|---|---|---|---|---|---|---|---|---|
| | Max. | Min. | Max. | Min. | Max. | Min. | Min. | Max. | Min. |
| **Metric Hex Jam Nuts** | | | | | | | | | |
| M5 × 0.8 | 8.00 | 7.78 | 9.24 | 8.79 | 2.70 | 2.45 | 6.9 | … | … |
| M6 × 1 | 10.00 | 9.78 | 11.55 | 11.05 | 3.20 | 2.90 | 8.9 | … | … |
| M8 × 1.25 | 13.00 | 12.73 | 15.01 | 14.38 | 4.00 | 3.70 | 11.6 | … | … |
| [a]M10 × 1.5 | 15.00 | 14.73 | 17.32 | 16.64 | 5.00 | 4.70 | 13.6 | … | … |
| M10 × 1.5 | 16.00 | 15.73 | 18.48 | 17.77 | 5.00 | 4.70 | 14.6 | … | … |
| M12 × 1.75 | 18.00 | 17.73 | 20.78 | 20.03 | 6.00 | 5.70 | 16.6 | … | … |
| M14 × 2 | 21.00 | 20.67 | 24.25 | 23.35 | 7.00 | 6.42 | 19.6 | … | … |
| M16 × 2 | 24.00 | 23.67 | 27.71 | 26.75 | 8.00 | 7.42 | 22.5 | … | … |
| M20 × 2.5 | 30.00 | 29.16 | 34.64 | 32.95 | 10.00 | 9.10 | 27.7 | 0.8 | 0.4 |
| M24 × 3 | 36.00 | 35.00 | 41.57 | 39.55 | 12.00 | 10.90 | 33.2 | 0.8 | 0.4 |
| M30 × 3.5 | 46.00 | 45.00 | 53.12 | 50.85 | 15.00 | 13.90 | 42.7 | 0.8 | 0.4 |
| M36 × 4 | 55.00 | 53.80 | 63.51 | 60.79 | 18.00 | 16.90 | 51.1 | 0.8 | 0.4 |
| **Metric Heavy Hex Nuts** | | | | | | | | | |
| M12 × 1.75 | 21.00 | 20.16 | 24.25 | 22.78 | 12.3 | 11.9 | 19.2 | 0.8 | 0.4 |
| M14 × 2 | 24.00 | 23.16 | 27.71 | 26.17 | 14.3 | 13.6 | 22.0 | 0.8 | 0.4 |
| M16 × 2 | 27.00 | 26.16 | 31.18 | 29.56 | 17.1 | 16.4 | 24.9 | 0.8 | 0.4 |
| M20 × 2.5 | 34.00 | 33.00 | 39.26 | 37.29 | 20.7 | 19.4 | 31.4 | 0.8 | 0.4 |
| M22 × 2.5 | 36.00 | 35.00 | 41.57 | 39.55 | 23.6 | 22.3 | 33.3 | 0.8 | 0.4 |
| M24 × 3 | 41.00 | 40.00 | 47.34 | 45.20 | 24.2 | 22.9 | 38.0 | 0.8 | 0.4 |
| M27 × 3 | 46.00 | 45.00 | 53.12 | 50.85 | 27.6 | 26.3 | 42.8 | 0.8 | 0.4 |
| M30 × 3.5 | 50.00 | 49.00 | 57.74 | 55.37 | 30.7 | 29.1 | 46.6 | 0.8 | 0.4 |
| M36 × 4 | 60.00 | 58.80 | 69.28 | 66.44 | 36.6 | 35.0 | 55.9 | 0.8 | 0.4 |
| M42 × 4.5 | 70.00 | 67.90 | 80.83 | 77.41 | 42.0 | 40.4 | 64.5 | 1.0 | 0.5 |
| M48 × 5 | 80.00 | 77.60 | 92.38 | 88.46 | 48.0 | 46.4 | 73.7 | 1.0 | 0.5 |
| M56 × 5.5 | 90.00 | 87.20 | 103.92 | 99.41 | 56.0 | 54.1 | 82.8 | 1.0 | 0.5 |
| M64 × 6 | 100.00 | 96.80 | 115.47 | 110.35 | 64.0 | 62.1 | 92.0 | 1.0 | 0.5 |
| M72 × 6 | 110.00 | 106.40 | 127.02 | 121.30 | 72.0 | 70.1 | 101.1 | 1.2 | 0.6 |
| M80 × 6 | 120.00 | 116.00 | 138.56 | 132.24 | 80.0 | 78.1 | 110.2 | 1.2 | 0.6 |
| M90 × 6 | 135.00 | 130.50 | 155.88 | 148.77 | 90.0 | 87.8 | 124.0 | 1.2 | 0.6 |
| M100 × 6 | 150.00 | 145.00 | 173.21 | 165.30 | 100.0 | 97.8 | 137.8 | 1.2 | 0.6 |

[a] This size with width across flats of 15 mm is not standard. Unless specifically ordered, M10 hex jam nuts with 16 mm width across flats will be furnished.

All dimensions are in millimeters.

**Table 30. American National Standard Prevailing-Torque Metric Hex Nuts— Property Classes 5, 9, and 10** *ANSI/ASME B18.16M-2004 (R2021)*

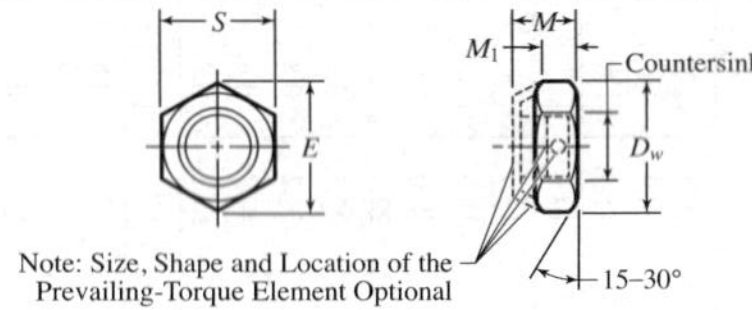


| Nominal Nut Dia. and Thread Pitch | Width Across Flats, *S* | | Width Across Corners, *E* | | Property Classes 5 and 10 Nuts: All Metal[a] Type — Thickness, *M* | | Property Classes 5 and 10 Nuts: Top Insert Type — Thickness, *M* | | Property Class 9 Nuts: All Metal Type — Thickness, *M* | | Property Class 9 Nuts: Top Insert Type — Thickness, *M* | | Property Class 5 and 10 Nuts — Wrenching Height, $M_1$ | Property Class 9 Nuts — Wrenching Height, $M_1$ | Bearing Face Dia., $D_w$ |
|---|---|---|---|---|---|---|---|---|---|---|---|---|---|---|---|
| | Max. | Min. | Max. | Min. | Max. | Min. | Max. | Min. | Max. | Min. | Max. | Min. | Min. | Min. | Min. |
| M3 × 0.5 | 5.50 | 5.32 | 6.35 | 6.01 | 3.10 | 2.65 | 4.50 | 3.90 | 3.10 | 2.65 | 4.50 | 3.90 | 1.4 | 1.4 | 4.6 |
| M3.5 × 0.6 | 6.00 | 5.82 | 6.93 | 6.58 | 3.50 | 3.00 | 5.00 | 4.30 | 3.50 | 3.00 | 5.00 | 4.30 | 1.7 | 1.7 | 5.1 |
| M4 × 0.7 | 7.00 | 6.78 | 8.08 | 7.66 | 4.00 | 3.50 | 6.00 | 5.30 | 4.00 | 3.50 | 6.00 | 5.30 | 1.9 | 1.9 | 5.9 |
| M5 × 0.8 | 8.00 | 7.78 | 9.24 | 8.79 | 5.30 | 4.80 | 6.80 | 6.00 | 5.30 | 4.80 | 7.20 | 6.40 | 2.7 | 2.7 | 6.9 |
| M6 × 1 | 10.00 | 9.78 | 11.55 | 11.05 | 5.90 | 5.40 | 8.00 | 7.20 | 6.70 | 5.40 | 8.50 | 7.70 | 3.0 | 3.0 | 8.9 |
| M8 × 1.25 | 13.00 | 12.73 | 15.01 | 14.38 | 7.10 | 6.44 | 9.50 | 8.50 | 8.00 | 7.14 | 10.20 | 9.20 | 3.7 | 4.3 | 11.6 |
| [b]M10 × 1.5 | 15.00 | 14.73 | 17.32 | 16.64 | 9.70 | 8.70 | 12.50 | 11.50 | 11.20 | 9.60 | 13.50 | 12.50 | 5.6 | 6.2 | 13.6 |
| M10 × 1.5 | 16.00 | 15.73 | 18.48 | 17.77 | 9.00 | 8.04 | 11.90 | 10.90 | 10.50 | 8.94 | 12.80 | 11.80 | 4.8 | 5.6 | 14.6 |
| M12 × 1.75 | 18.00 | 17.73 | 20.78 | 20.03 | 11.60 | 10.37 | 14.90 | 13.90 | 13.30 | 11.57 | 16.10 | 15.10 | 6.7 | 7.7 | 16.6 |
| M14 × 2 | 21.00 | 20.67 | 24.25 | 23.35 | 13.20 | 12.10 | 17.00 | 15.80 | 15.40 | 13.40 | 18.30 | 17.10 | 7.8 | 8.9 | 19.6 |
| M16 × 2 | 24.00 | 23.67 | 27.71 | 26.75 | 15.20 | 14.10 | 19.10 | 17.90 | 17.90 | 15.70 | 20.70 | 19.50 | 9.1 | 10.5 | 22.5 |
| M20 × 2.5 | 30.00 | 29.16 | 34.64 | 32.95 | 19.00 | 16.90 | 22.80 | 21.50 | 21.80 | 19.00 | 25.10 | 23.80 | 10.9 | 12.7 | 27.7 |
| M24 × 3 | 36.00 | 35.00 | 41.57 | 39.55 | 23.00 | 20.20 | 27.10 | 25.60 | 26.40 | 22.60 | 29.50 | 28.00 | 13.0 | 15.1 | 33.2 |
| M30 × 3.5 | 46.00 | 45.00 | 53.12 | 50.85 | 26.90 | 24.30 | 32.60 | 30.60 | 31.80 | 27.30 | 35.60 | 33.60 | 15.7 | 18.2 | 42.7 |
| M36 × 4 | 55.00 | 53.80 | 63.51 | 60.79 | 32.50 | 29.40 | 38.90 | 36.90 | 38.50 | 33.10 | 42.60 | 40.60 | 19.0 | 22.1 | 51.1 |

[a] Also includes metal nuts with nonmetallic inserts, plugs, or patches in their threads.
[b] This size with width across flats of 15 mm is not standard. Unless specifically ordered, M10 slotted hex nuts with 16 mm width across flats will be furnished.
All dimensions are in millimeters.

**Table 31. Metric Hexagon Regular Nuts (Style 1), Grades A and B** *ISO 4032-2012*

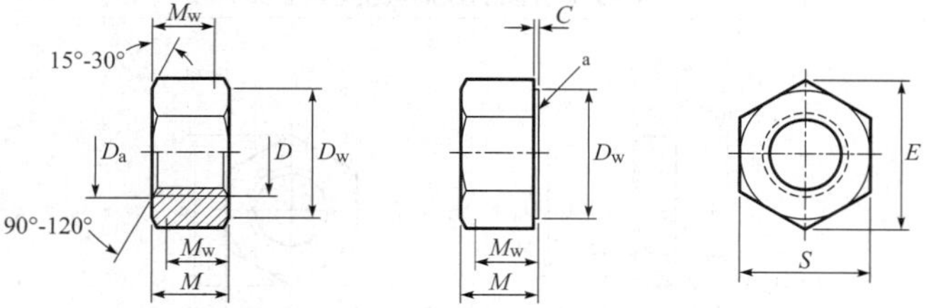


[a] The nuts are delivered without the washer face unless specified at the time of order.

| Nom. Thread Diameter and Thread Pitch *D* | Washer Face Thickness *C* | | Countersink Diameter *Da* | | Bearing Face Diameter *Dw* | Width Across Corners *E* | Thickness *M* | | Wrenching Height *Mw* | Width Across Flats *S* | |
|---|---|---|---|---|---|---|---|---|---|---|---|
| | Max. | Min. | Max. | Min. | Min. | Min. | Max. | Min. | Min. | Nom. = Max. | Min. |
| **Preferred Threads** | | | | | | | | | | | |
| M1.6 × 0.35 | 0.20 | 0.10 | 1.84 | 1.60 | 2.40 | 3.41 | 1.30 | 1.05 | 0.80 | 3.20 | 3.02 |
| M2 × 0.4 | 0.20 | 0.10 | 2.30 | 2.00 | 3.10 | 4.32 | 1.60 | 1.35 | 1.10 | 4.00 | 3.82 |
| M2.5 × 0.45 | 0.30 | 0.10 | 2.90 | 2.50 | 4.10 | 5.45 | 2.00 | 1.75 | 1.40 | 5.00 | 4.82 |
| M3 × 0.5 | 0.40 | 0.15 | 3.45 | 3.00 | 4.60 | 6.01 | 2.40 | 2.15 | 1.70 | 5.50 | 5.32 |
| M4 × 0.7 | 0.40 | 0.15 | 4.60 | 4.00 | 5.90 | 7.66 | 3.20 | 2.90 | 2.30 | 7.00 | 6.78 |
| M5 × 0.8 | 0.50 | 0.15 | 5.75 | 5.00 | 6.90 | 8.79 | 4.70 | 4.40 | 3.50 | 8.00 | 7.78 |
| M6 × 1 | 0.50 | 0.15 | 6.75 | 6.00 | 8.90 | 11.05 | 5.20 | 4.90 | 3.90 | 10.00 | 9.78 |
| M8 × 1.25 | 0.60 | 0.15 | 8.75 | 8.00 | 11.60 | 14.38 | 6.80 | 6.44 | 5.20 | 13.00 | 12.73 |
| M10 × 1.5 | 0.60 | 0.15 | 10.80 | 10.00 | 14.60 | 17.77 | 8.40 | 8.04 | 6.40 | 16.00 | 15.73 |
| M12 × 1.75 | 0.60 | 0.15 | 13.00 | 12.00 | 16.60 | 20.03 | 10.80 | 10.37 | 8.30 | 18.00 | 17.73 |
| M16 × 2 | 0.80 | 0.20 | 17.30 | 16.00 | 22.50 | 26.75 | 14.80 | 14.10 | 11.30 | 24.00 | 23.67 |
| M20 × 2.5 | 0.80 | 0.20 | 21.60 | 20.00 | 27.70 | 32.95 | 18.00 | 16.90 | 13.50 | 30.00 | 29.16 |
| M24 × 3 | 0.80 | 0.20 | 25.90 | 24.00 | 33.30 | 39.55 | 21.50 | 20.20 | 16.20 | 36.00 | 35.00 |
| M30 × 3.5 | 0.80 | 0.20 | 32.40 | 30.00 | 42.80 | 50.85 | 25.60 | 24.30 | 19.40 | 46.00 | 45.00 |
| M36 × 4 | 0.80 | 0.20 | 38.90 | 36.00 | 51.10 | 60.79 | 31.00 | 29.40 | 23.50 | 55.00 | 53.80 |
| M42 × 4.5 | 1.00 | 0.30 | 45.40 | 42.00 | 60.00 | 71.30 | 34.00 | 32.40 | 25.90 | 65.00 | 63.10 |
| M48 × 5 | 1.00 | 0.30 | 51.80 | 48.00 | 69.50 | 82.60 | 38.00 | 36.40 | 29.10 | 75.00 | 73.10 |
| M56 × 5.5 | 1.00 | 0.30 | 60.50 | 56.00 | 78.70 | 93.56 | 45.00 | 43.40 | 34.70 | 85.00 | 82.80 |
| M64 × 6 | 1.00 | 0.30 | 69.10 | 64.00 | 88.20 | 104.86 | 51.00 | 49.10 | 39.30 | 95.00 | 92.80 |
| **Non-Preferred Threads** | | | | | | | | | | | |
| M3.5 × 0.6 | 0.40 | 0.15 | 4.00 | 3.50 | 5.00 | 6.58 | 2.80 | 2.55 | 2.00 | 6.00 | 5.82 |
| M14 × 2 | 0.60 | 0.15 | 15.10 | 14.00 | 19.60 | 23.36 | 12.80 | 12.10 | 9.70 | 21.00 | 20.67 |
| M18 × 2.5 | 0.80 | 0.20 | 19.50 | 18.00 | 24.90 | 29.56 | 15.80 | 15.10 | 12.10 | 27.00 | 26.16 |
| M22 × 2.5 | 0.80 | 0.20 | 23.70 | 22.00 | 31.40 | 37.29 | 19.40 | 18.10 | 14.50 | 34.00 | 33.00 |
| M27 × 3 | 0.80 | 0.20 | 29.10 | 27.00 | 38.00 | 45.20 | 23.80 | 22.50 | 18.00 | 41.00 | 40.00 |
| M33 × 3.5 | 0.80 | 0.20 | 35.60 | 33.00 | 46.60 | 55.37 | 28.70 | 27.40 | 21.90 | 50.00 | 49.00 |
| M39 × 4 | 1.00 | 0.30 | 42.10 | 39.00 | 55.90 | 66.44 | 33.40 | 31.80 | 25.40 | 60.00 | 58.80 |
| M45 × 4.5 | 1.00 | 0.30 | 48.60 | 45.00 | 64.70 | 76.95 | 36.00 | 34.40 | 27.50 | 70.00 | 68.10 |
| M52 × 5 | 1.00 | 0.30 | 56.20 | 52.00 | 74.20 | 88.25 | 42.00 | 40.40 | 32.30 | 80.00 | 78.10 |
| M60 × 5.5 | 1.00 | 0.30 | 64.80 | 60.00 | 83.40 | 99.21 | 48.00 | 46.40 | 37.10 | 90.00 | 87.80 |

All dimensions are in millimeters.

**Table 32. Metric Hexagon Thin Nuts Chamfered (Style 0), Grades A and B *ISO 4035-2012***

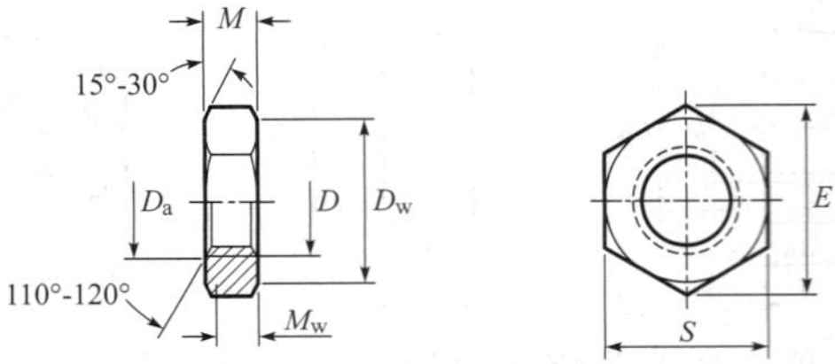


| Nom. Thread Diameter and Thread Pitch D | Countersink Diameter Da | | Bearing Face Diameter Dw | Width Across Corners E | Thickness M | | Wrenching Height Mw | Width Across Flats S | |
|---|---|---|---|---|---|---|---|---|---|
| | Max. | Min. | Min. | Min. | Max. | Min. | Min. | Nom. = Max. | Min. |
| **Preferred Threads** | | | | | | | | | |
| M1.6 × 0.35 | 1.84 | 1.60 | 2.40 | 3.41 | 1.00 | 0.75 | 0.60 | 3.20 | 3.02 |
| M2 × 0.4 | 2.30 | 2.00 | 3.10 | 4.32 | 1.20 | 0.95 | 0.80 | 4.00 | 3.82 |
| M2.5 × 0.45 | 2.90 | 2.50 | 4.10 | 5.45 | 1.60 | 1.35 | 1.10 | 5.00 | 4.82 |
| M3 × 0.5 | 3.45 | 3.00 | 4.60 | 6.01 | 1.80 | 1.55 | 1.20 | 5.50 | 5.32 |
| M4 × 0.7 | 4.60 | 4.00 | 5.90 | 7.66 | 2.20 | 1.95 | 1.60 | 7.00 | 6.78 |
| M5 × 0.8 | 5.75 | 5.00 | 6.90 | 8.79 | 2.70 | 2.45 | 2.00 | 8.00 | 7.78 |
| M6 × 1 | 6.75 | 6.00 | 8.90 | 11.05 | 3.20 | 2.90 | 2.30 | 10.00 | 9.78 |
| M8 × 1.25 | 8.75 | 8.00 | 11.60 | 14.38 | 4.00 | 3.70 | 3.00 | 13.00 | 12.73 |
| M10 × 1.5 | 10.80 | 10.00 | 14.60 | 17.77 | 5.00 | 4.70 | 3.80 | 16.00 | 15.73 |
| M12 × 1.75 | 13.00 | 12.00 | 16.60 | 20.03 | 6.00 | 5.70 | 4.60 | 18.00 | 17.73 |
| M16 × 2 | 17.30 | 16.00 | 22.50 | 26.75 | 8.00 | 7.42 | 5.90 | 24.00 | 23.67 |
| M20 × 2.5 | 21.60 | 20.00 | 27.70 | 32.95 | 10.00 | 9.10 | 7.30 | 30.00 | 29.16 |
| M24 × 3 | 25.90 | 24.00 | 33.20 | 39.55 | 12.00 | 10.90 | 8.70 | 36.00 | 35.00 |
| M30 × 3.5 | 32.40 | 30.00 | 42.80 | 50.85 | 15.00 | 13.90 | 11.10 | 46.00 | 45.00 |
| M36 × 4 | 38.90 | 36.00 | 51.10 | 60.79 | 18.00 | 16.90 | 13.50 | 55.00 | 53.80 |
| M42 × 4.5 | 45.40 | 42.00 | 60.00 | 71.30 | 21.00 | 19.70 | 15.80 | 65.00 | 63.10 |
| M48 × 5 | 51.80 | 48.00 | 69.50 | 82.60 | 24.00 | 22.70 | 18.20 | 75.00 | 73.10 |
| M56 × 5.5 | 60.50 | 56.00 | 78.70 | 93.56 | 28.00 | 26.70 | 21.40 | 85.00 | 82.80 |
| M64 × 6 | 69.10 | 64.00 | 88.20 | 104.86 | 32.00 | 30.40 | 24.30 | 95.00 | 92.80 |
| **Non-Preferred Threads** | | | | | | | | | |
| M3.5 × 0.6 | 4.00 | 3.50 | 5.10 | 6.58 | 2.00 | 1.75 | 1.40 | 6.00 | 5.82 |
| M14 × 2 | 15.10 | 14.00 | 19.60 | 23.36 | 7.00 | 6.42 | 5.10 | 21.00 | 20.67 |
| M18 × 2.5 | 19.50 | 18.00 | 24.90 | 29.56 | 9.00 | 8.42 | 6.70 | 27.00 | 26.16 |
| M22 × 2.5 | 23.70 | 22.00 | 31.40 | 37.29 | 11.00 | 9.90 | 7.90 | 34.00 | 33.00 |
| M27 × 3 | 29.10 | 27.00 | 38.00 | 45.20 | 13.50 | 12.40 | 9.90 | 41.00 | 40.00 |
| M33 × 3.5 | 35.60 | 33.00 | 46.60 | 55.37 | 16.50 | 15.40 | 12.30 | 50.00 | 49.00 |
| M39 × 4 | 42.10 | 39.00 | 55.90 | 66.44 | 19.50 | 18.20 | 14.60 | 60.00 | 58.80 |
| M45 × 4.5 | 48.60 | 45.00 | 64.70 | 76.95 | 22.50 | 21.20 | 17.00 | 70.00 | 68.10 |
| M52 × 5 | 56.20 | 52.00 | 74.20 | 88.25 | 26.00 | 24.70 | 19.80 | 80.00 | 78.10 |
| M60 × 5.5 | 64.80 | 60.00 | 83.40 | 99.21 | 30.00 | 28.70 | 23.00 | 90.00 | 87.80 |

All dimensions are in millimeters.

## Metric Washers

**Metric Plain Washers.—**American National Standard ANSI/ASME B18.22M-1981 (R2017) covers general specifications and dimensions for flat, round-hole washers, both soft (as fabricated) and hardened, intended for use in general-purpose applications. Dimensions are given in the following table. Manufacturers should be consulted for current information on stock sizes.

**Comparison with ISO Standards.—**The washers covered by this ASME Standard are nominally similar to those covered in various ISO documents. Outside diameters were selected, where possible, from ISO/TC2/WG6/N47 "General Plan for Plain Washers for Metric Bolts, Screws, and Nuts." The thicknesses given in the ASME Standard are similar to the nominal ISO thicknesses, however the tolerances differ. Inside diameters also differ.

ISO metric washers are currently covered in ISO 887, "Plain Washers for Metric Bolts, Screws, and Nuts - General Plan."

**Types of Metric Plain Washers.—**Soft (as fabricated) washers are generally available in nominal sizes 1.6 mm through 36 mm in a variety of materials. They are normally used in low-strength applications to distribute bearing load, to provide a uniform bearing surface, and to prevent marring of the work surface.

Hardened steel washers are normally available in sizes 6 mm through 36 mm in the narrow and regular series. They are intended primarily for use in high-strength joints to minimize embedment, to provide a uniform bearing surface, and to bridge large clearance holes and slots.

**Metric Plain Washer Materials and Finish.—**Soft (as fabricated) washers are made of nonhardened steel unless otherwise specified by the purchaser. Hardened washers are made of through-hardened steel tempered to a hardness of 38 to 45 RC (Rockwell C scale).

Unless otherwise specified, washers are furnished with a natural (as fabricated) finish, unplated or uncoated with a light film of oil or rust inhibitor.

**Metric Plain Washer Designation.—**When specifying metric plain washers, the designation should include the following data in the sequence shown: description, nominal size, series, material type, and finish, if required.

*Example:* Plain washer, 6 mm, narrow, soft, steel, zinc plated

Plain washer, 10 mm, regular, hardened steel.

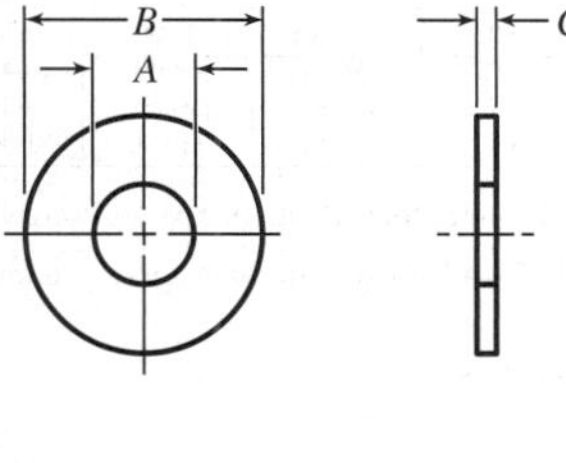

**Table 33. American National Standard Metric Plain Washers**
*ANSI/ASME B18.22M-1981 (R2017)*

| Nominal Washer Size[a] | Washer Series | Inside Diameter, *A* | | Outside Diameter, *B* | | Thickness, *C* | |
|---|---|---|---|---|---|---|---|
| | | Max. | Min. | Max. | Min. | Max. | Min. |
| 1.6 | Narrow | 2.09 | 1.95 | 4.00 | 3.70 | 0.70 | 0.50 |
| | Regular | 2.09 | 1.95 | 5.00 | 4.70 | 0.70 | 0.50 |
| | Wide | 2.09 | 1.95 | 6.00 | 5.70 | 0.90 | 0.60 |
| 2 | Narrow | 2.64 | 2.50 | 5.00 | 4.70 | 0.90 | 0.60 |
| | Regular | 2.64 | 2.50 | 6.00 | 5.70 | 0.90 | 0.60 |
| | Wide | 2.64 | 2.50 | 8.00 | 7.64 | 0.90 | 0.60 |
| 2.5 | Narrow | 3.14 | 3.00 | 6.00 | 5.70 | 0.90 | 0.60 |
| | Regular | 3.14 | 3.00 | 8.00 | 7.64 | 0.90 | 0.60 |
| | Wide | 3.14 | 3.00 | 10.00 | 9.64 | 1.20 | 0.80 |
| 3 | Narrow | 3.68 | 3.50 | 7.00 | 6.64 | 0.90 | 0.60 |
| | Regular | 3.68 | 3.50 | 10.00 | 9.64 | 1.20 | 0.80 |
| | Wide | 3.68 | 3.50 | 12.00 | 11.57 | 1.40 | 1.00 |
| 3.5 | Narrow | 4.18 | 4.00 | 9.00 | 8.64 | 1.20 | 0.80 |
| | Regular | 4.18 | 4.00 | 10.00 | 9.64 | 1.40 | 1.00 |
| | Wide | 4.18 | 4.00 | 15.00 | 14.57 | 1.75 | 1.20 |
| 4 | Narrow | 4.88 | 4.70 | 10.00 | 9.64 | 1.20 | 0.80 |
| | Regular | 4.88 | 4.70 | 12.00 | 11.57 | 1.40 | 1.00 |
| | Wide | 4.88 | 4.70 | 16.00 | 15.57 | 2.30 | 1.60 |
| 5 | Narrow | 5.78 | 5.50 | 11.00 | 10.57 | 1.40 | 1.00 |
| | Regular | 5.78 | 5.50 | 15.00 | 14.57 | 1.75 | 1.20 |
| | Wide | 5.78 | 5.50 | 20.00 | 19.48 | 2.30 | 1.60 |
| 6 | Narrow | 6.87 | 6.65 | 13.00 | 12.57 | 1.75 | 1.20 |
| | Regular | 6.87 | 6.65 | 18.80 | 18.37 | 1.75 | 1.20 |
| | Wide | 6.87 | 6.65 | 25.40 | 24.88 | 2.30 | 1.60 |
| 8 | Narrow | 9.12 | 8.90 | 18.80[b] | 18.37[b] | 2.30 | 1.60 |
| | Regular | 9.12 | 8.90 | 25.40[b] | 24.48[b] | 2.30 | 1.60 |
| | Wide | 9.12 | 8.90 | 32.00 | 31.38 | 2.80 | 2.00 |
| 10 | Narrow | 11.12 | 10.85 | 20.00 | 19.48 | 2.30 | 1.60 |
| | Regular | 11.12 | 10.85 | 28.00 | 27.48 | 2.80 | 2.00 |
| | Wide | 11.12 | 10.85 | 39.00 | 38.38 | 3.50 | 2.50 |
| 12 | Narrow | 13.57 | 13.30 | 25.40 | 24.88 | 2.80 | 2.00 |
| | Regular | 13.57 | 13.30 | 34.00 | 33.38 | 3.50 | 2.50 |
| | Wide | 13.57 | 13.30 | 44.00 | 43.38 | 3.50 | 2.50 |
| 14 | Narrow | 15.52 | 15.25 | 28.00 | 27.48 | 2.80 | 2.00 |
| | Regular | 15.52 | 15.25 | 39.00 | 38.38 | 3.50 | 2.50 |
| | Wide | 15.52 | 15.25 | 50.00 | 49.38 | 4.00 | 3.00 |
| 16 | Narrow | 17.52 | 17.25 | 32.00 | 31.38 | 3.50 | 2.50 |
| | Regular | 17.52 | 17.25 | 44.00 | 43.38 | 4.00 | 3.00 |
| | Wide | 17.52 | 17.25 | 56.00 | 54.80 | 4.60 | 3.50 |
| 20 | Narrow | 22.32 | 21.80 | 39.00 | 38.38 | 4.00 | 3.00 |
| | Regular | 22.32 | 21.80 | 50.00 | 49.38 | 4.60 | 3.50 |
| | Wide | 22.32 | 21.80 | 66.00 | 64.80 | 5.10 | 4.00 |
| 24 | Narrow | 26.12 | 25.60 | 44.00 | 43.38 | 4.60 | 3.50 |
| | Regular | 26.12 | 25.60 | 56.00 | 54.80 | 5.10 | 4.00 |
| | Wide | 26.12 | 25.60 | 72.00 | 70.80 | 5.60 | 4.50 |
| 30 | Narrow | 33.02 | 32.40 | 56.00 | 54.80 | 5.10 | 4.00 |
| | Regular | 33.02 | 32.40 | 72.00 | 70.80 | 5.60 | 4.50 |
| | Wide | 33.02 | 32.40 | 90.00 | 88.60 | 6.40 | 5.00 |
| 36 | Narrow | 38.92 | 38.30 | 66.00 | 64.80 | 5.60 | 4.50 |
| | Regular | 38.92 | 38.30 | 90.00 | 88.60 | 6.40 | 5.00 |
| | Wide | 38.92 | 38.30 | 110.00 | 108.60 | 8.50 | 7.00 |

[a] Nominal washer sizes are intended for use with comparable screw and bolt sizes.

[b] The 18.80/18.37 and 25.40/24.48 mm outside diameters avoid washers which could be used in coin-operated devices.

All dimensions are in millimeters.

**Table 34. Metric Plain Washers Normal Series, Grade A *ISO 7089-2000* and Large Series, Grade A *ISO 7093-1-2000***

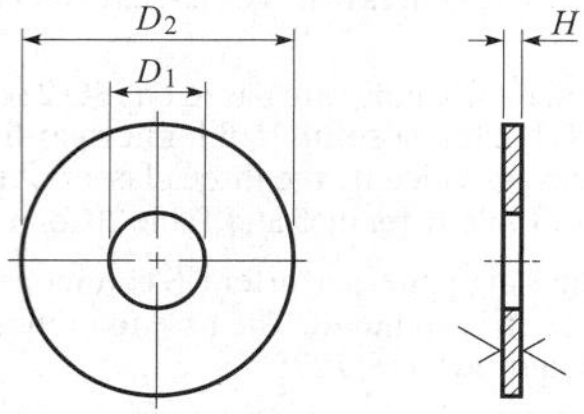


Surface Roughness Values in Micrometers

$$\sqrt{} = \begin{cases} \sqrt{Ra\ 1.6} & \text{for } H \leq 3 \\ \sqrt{Ra\ 3.2} & \text{for } 3 < H \leq 6 \\ \sqrt{Ra\ 6.3} & \text{for } H > 6 \end{cases}$$

| Nominal Thread Diameter | Washer Series | Clearance Hole $D1$ | | Outside Diameter $D2$ | | Thickness $H$ | | |
|---|---|---|---|---|---|---|---|---|
| | | Nom. = Min. | Max. | Nom. = Max. | Min. | Nom. | Max. | Min. |
| M1.6 | Normal | 1.70 | 1.84 | 4.00 | 3.70 | 0.30 | 0.35 | 0.25 |
| M2 | Normal | 2.20 | 2.34 | 5.00 | 4.70 | 0.30 | 0.35 | 0.25 |
| M2.5 | Normal | 2.70 | 2.84 | 6.00 | 5.70 | 0.50 | 0.55 | 0.45 |
| M3 | Normal | 3.20 | 3.38 | 7.00 | 6.64 | 0.50 | 0.55 | 0.45 |
| | Large | 3.20 | 3.38 | 9.00 | 8.64 | 0.80 | 0.90 | 0.70 |
| M4 | Normal | 4.30 | 4.48 | 9.00 | 8.64 | 0.80 | 0.90 | 0.70 |
| | Large | 4.30 | 4.48 | 12.00 | 11.57 | 1.00 | 1.10 | 0.90 |
| M5 | Normal | 5.30 | 5.48 | 10.00 | 9.64 | 1.00 | 1.10 | 0.90 |
| | Large | 5.30 | 5.48 | 15.00 | 14.57 | 1.00 | 1.10 | 0.90 |
| M6 | Normal | 6.40 | 6.62 | 12.00 | 11.57 | 1.60 | 1.80 | 1.40 |
| | Large | 6.40 | 6.62 | 18.00 | 17.57 | 1.60 | 1.80 | 1.40 |
| M8 | Normal | 8.40 | 8.62 | 16.00 | 15.57 | 1.60 | 1.80 | 1.40 |
| | Large | 8.40 | 8.62 | 24.00 | 23.48 | 2.00 | 2.20 | 1.80 |
| M10 | Normal | 10.50 | 10.77 | 20.00 | 19.48 | 2.00 | 2.20 | 1.80 |
| | Large | 10.50 | 10.77 | 30.00 | 29.48 | 2.50 | 2.70 | 2.30 |
| M12 | Normal | 13.00 | 13.27 | 24.00 | 23.48 | 2.50 | 2.70 | 2.30 |
| | Large | 13.00 | 13.27 | 37.00 | 36.38 | 3.00 | 3.30 | 2.70 |
| M16 | Normal | 17.00 | 17.27 | 30.00 | 29.48 | 3.00 | 3.30 | 2.70 |
| | Large | 17.00 | 17.27 | 50.00 | 49.38 | 3.00 | 3.30 | 2.70 |
| M20 | Normal | 21.00 | 21.33 | 37.00 | 36.38 | 3.00 | 3.30 | 2.70 |
| | Large | 21.00 | 21.33 | 60.00 | 59.26 | 4.00 | 4.30 | 3.70 |
| M24 | Normal | 25.00 | 25.33 | 44.00 | 43.38 | 4.00 | 4.30 | 3.70 |
| | Large | 25.00 | 25.52 | 72.00 | 70.80 | 5.00 | 5.60 | 4.40 |
| M30 | Normal | 31.00 | 31.39 | 56.00 | 55.26 | 4.00 | 4.30 | 3.70 |
| | Large | 33.00 | 33.62 | 92.00 | 90.60 | 6.00 | 6.60 | 5.40 |
| M36 | Normal | 37.00 | 37.62 | 66.00 | 64.80 | 5.00 | 5.60 | 4.40 |
| | Large | 39.00 | 39.62 | 110.00 | 108.60 | 8.00 | 9.00 | 7.00 |
| M42 | Normal | 45.00 | 45.62 | 78.00 | 76.80 | 8.00 | 9.00 | 7.00 |
| M48 | Normal | 52.00 | 52.74 | 92.00 | 90.60 | 8.00 | 9.00 | 7.00 |
| M56 | Normal | 62.00 | 62.74 | 105.00 | 103.60 | 10.00 | 11.00 | 9.00 |
| M64 | Normal | 70.00 | 70.74 | 115.00 | 113.60 | 10.00 | 11.00 | 9.00 |

All dimensions are in millimeters.

## Clearance Holes for Bolts, Screws, and Studs

ANSI/ASME B18.2.8-1999 (R2017) covers the recommended clearance hole sizes for #0 through 1.5 inch and M1.6 through M100 metric fasteners in three classes of clearance using a close-, normal-, and loose-fit category.

The clearance hole tolerances for both inch and metric holes are based on ISO 286, *ISO System of Limits and Fits*, using tolerance class H12 for close-fit, H13 for normal-fit, and H14 for loose-fit clearance holes. The clearances provided by the three classes of fit are based on regularly stepped clearances as listed in Table 1a for inch and Table 2b for metric.

**Inch Fasteners.—**The hole sizes for inch fasteners are patterned after US common usage and the general clearances translated from the metric standard. The hole tolerances are based on the *ISO System of Limits and Fits*, as required by ISO 273.

The tabulated drill and hole sizes, Table 1a, lists the inch fastener clearance hole recommendations. The recommended drill sizes for inch fasteners are tabulated by nominal drill designation as letter, numbers, or fractional sizes. The drill sizes were selected to provide as nearly as practical a step-patterned clearance size for the minimum recommended hole (Table 1b). The maximum recommended hole size is based on standard hole tolerances.

**Table 1a. Clearance Holes for Inch Fasteners** *ANSI/ASME B18.2.8-1999 (R2017)*

| Nominal Screw Size | Normal | | | Close | | | Loose | | |
|---|---|---|---|---|---|---|---|---|---|
| | Nominal Drill Size | Hole Diameter | | Nominal Drill Size | Hole Diameter | | Nominal Drill Size | Hole Diameter | |
| | | Min. | Max. | | Min. | Max. | | Min. | Max. |
| #0 | #48 | 0.076 | 0.082 | #51 | 0.067 | 0.071 | 3⁄32 | 0.094 | 0.104 |
| #1 | #43 | 0.089 | 0.095 | #46 | 0.081 | 0.085 | #37 | 0.104 | 0.114 |
| #2 | #38 | 0.102 | 0.108 | 3⁄32 | 0.094 | 0.098 | #32 | 0.116 | 0.126 |
| #3 | #32 | 0.116 | 0.122 | #36 | 0.106 | 0.110 | #30 | 0.128 | 0.140 |
| #4 | #30 | 0.128 | 0.135 | #31 | 0.120 | 0.124 | #27 | 0.144 | 0.156 |
| #5 | 5⁄32 | 0.156 | 0.163 | 9⁄64 | 0.141 | 0.146 | 11⁄64 | 0.172 | 0.184 |
| #6 | #18 | 0.170 | 0.177 | #23 | 0.154 | 0.159 | #13 | 0.185 | 0.197 |
| #8 | #9 | 0.196 | 0.203 | #15 | 0.180 | 0.185 | #3 | 0.213 | 0.225 |
| #10 | #2 | 0.221 | 0.228 | #5 | 0.206 | 0.211 | B | 0.238 | 0.250 |
| 1⁄4 | 9⁄32 | 0.281 | 0.290 | 17⁄64 | 0.266 | 0.272 | 19⁄64 | 0.297 | 0.311 |
| 5⁄16 | 11⁄32 | 0.344 | 0.354 | 21⁄64 | 0.328 | 0.334 | 23⁄64 | 0.359 | 0.373 |
| 3⁄8 | 13⁄32 | 0.406 | 0.416 | 25⁄64 | 0.391 | 0.397 | 27⁄64 | 0.422 | 0.438 |
| 7⁄16 | 15⁄32 | 0.469 | 0.479 | 29⁄64 | 0.453 | 0.460 | 31⁄64 | 0.484 | 0.500 |
| 1⁄2 | 9⁄16 | 0.562 | 0.572 | 17⁄32 | 0.531 | 0.538 | 39⁄64 | 0.609 | 0.625 |
| 5⁄8 | 11⁄16 | 0.688 | 0.698 | 21⁄32 | 0.656 | 0.663 | 47⁄64 | 0.734 | 0.754 |
| 3⁄4 | 13⁄16 | 0.812 | 0.824 | 25⁄32 | 0.781 | 0.789 | 29⁄32 | 0.906 | 0.926 |
| 7⁄8 | 15⁄16 | 0.938 | 0.950 | 29⁄32 | 0.906 | 0.914 | 1 1⁄32 | 1.031 | 1.051 |
| 1 | 1 3⁄32 | 1.094 | 1.106 | 1 1⁄32 | 1.031 | 1.039 | 1 5⁄32 | 1.156 | 1.181 |
| 1 1⁄8 | 1 7⁄32 | 1.219 | 1.235 | 1 5⁄32 | 1.156 | 1.164 | 1 5⁄16 | 1.312 | 1.337 |
| 1 1⁄4 | 1 11⁄32 | 1.344 | 1.360 | 1 9⁄32 | 1.281 | 1.291 | 1 7⁄16 | 1.438 | 1.463 |
| 1 3⁄8 | 1 1⁄2 | 1.500 | 1.516 | 1 7⁄16 | 1.438 | 1.448 | 1 39⁄64 | 1.609 | 1.634 |
| 1 1⁄2 | 1 5⁄8 | 1.625 | 1.641 | 1 9⁄16 | 1.562 | 1.572 | 1 47⁄64 | 1.734 | 1.759 |

**Table 1b. Inch Clearance Hole Allowances**

| Nominal Screw Size | Fit Classes | | | Nominal Screw Size | Fit Classes | | |
|---|---|---|---|---|---|---|---|
| | Normal | Close | Loose | | Normal | Close | Loose |
| #0 – #4 | 1⁄64 | 0.008 | 1⁄32 | 1 | 3⁄32 | 1⁄32 | 5⁄32 |
| #5 – 7⁄16 | 1⁄32 | 1⁄64 | 3⁄64 | 1 1⁄8 , 1 1⁄4 | 3⁄32 | 1⁄32 | 3⁄16 |
| 1⁄2 , 5⁄8 | 1⁄16 | 1⁄32 | 7⁄64 | 1 3⁄8 , 1 1⁄2 | 1⁄8 | 1⁄16 | 15⁄64 |
| 3⁄4 , 7⁄8 | 1⁄16 | 1⁄32 | 5⁄32 | … | … | … | … |

Dimensions are in inches.

**Metric Fasteners.**—The recommended drill and hole sizes for metric fasteners are tabulated in Table 2a. The minimum recommended hole is the drill size and the maximum recommended hole size is based on standard tolerances. The hole sizes for metric fasteners are in agreement with ISO 273, *Fasteners-Clearance Holes for Bolts and Screws*, except that ISO 273 covers fastener sizes M1 through M150.

**Table 2a. Clearance Holes for Metric Fasteners** *ANSI/ASME B18.2.8-1999 (R2017)*

| Nominal Screw Size | Normal | | | Close | | | Loose | | |
|---|---|---|---|---|---|---|---|---|---|
| | Nominal Drill Size | Hole Diameter | | Nominal Drill Size | Hole Diameter | | Nominal Drill Size | Hole Diameter | |
| | | Min. | Max. | | Min. | Max. | | Min. | Max. |
| M1.6 | 1.8 | 1.8 | 1.94 | 1.7 | 1.7 | 1.8 | 2 | 2 | 2.25 |
| M2 | 2.4 | 2.4 | 2.54 | 2.2 | 2.2 | 2.3 | 2.6 | 2.6 | 2.85 |
| M2.5 | 2.9 | 2.9 | 3.04 | 2.7 | 2.7 | 2.8 | 3.1 | 3.1 | 3.4 |
| M3 | 3.4 | 3.4 | 3.58 | 3.2 | 3.2 | 3.32 | 3.6 | 3.6 | 3.9 |
| M4 | 4.5 | 4.5 | 4.68 | 4.3 | 4.3 | 4.42 | 4.8 | 4.8 | 5.1 |
| M5 | 5.5 | 5.5 | 5.68 | 5.3 | 5.3 | 5.42 | 5.8 | 5.8 | 6.1 |
| M6 | 6.6 | 6.6 | 6.82 | 6.4 | 6.4 | 6.55 | 7 | 7 | 7.36 |
| M8 | 9 | 9 | 9.22 | 8.4 | 8.4 | 8.55 | 10 | 10 | 10.36 |
| M10 | 11 | 11 | 11.27 | 10.5 | 10.5 | 10.68 | 12 | 12 | 12.43 |
| M12 | 13.5 | 13.5 | 13.77 | 13 | 13 | 13.18 | 14.5 | 14.5 | 14.93 |
| M14 | 15.5 | 15.5 | 15.77 | 15 | 15 | 15.18 | 16.5 | 16.5 | 16.93 |
| M16 | 17.5 | 17.5 | 17.77 | 17 | 17 | 17.18 | 18.5 | 18.5 | 19.02 |
| M20 | 22 | 22 | 22.33 | 21 | 21 | 21.21 | 24 | 24 | 24.52 |
| M24 | 26 | 26 | 26.33 | 25 | 25 | 25.21 | 28 | 28 | 28.52 |
| M30 | 33 | 33 | 33.39 | 31 | 31 | 31.25 | 35 | 35 | 35.62 |
| M36 | 39 | 39 | 39.39 | 37 | 37 | 37.25 | 42 | 42 | 42.62 |
| M42 | 45 | 45 | 45.39 | 43 | 43 | 43.25 | 48 | 48 | 48.62 |
| M48 | 52 | 52 | 52.46 | 50 | 50 | 50.25 | 56 | 56 | 56.74 |
| M56 | 62 | 62 | 62.46 | 58 | 58 | 58.3 | 66 | 66 | 66.74 |
| M64 | 70 | 70 | 70.46 | 66 | 66 | 66.3 | 74 | 74 | 74.74 |
| M72 | 78 | 78 | 78.46 | 74 | 74 | 74.3 | 82 | 82 | 82.87 |
| M80 | 86 | 86 | 86.54 | 82 | 82 | 82.35 | 91 | 91 | 91.87 |
| M90 | 96 | 96 | 96.54 | 93 | 93 | 93.35 | 101 | 101 | 101.87 |
| M100 | 107 | 107 | 107.54 | 104 | 104 | 104.35 | 112 | 112 | 112.87 |

**Table 2b. Metric Clearance Hole Allowances**

| Nominal Screw Size | Fit Classes | | | Nominal Screw Size | Fit Classes | | |
|---|---|---|---|---|---|---|---|
| | Normal | Close | Loose | | Normal | Close | Loose |
| M1.6 | 0.2 | 0.1 | 0.25 | M20, M24 | 2 | 1 | 4 |
| M2 | 0.4 | 0.1 | 0.3 | M30 | 3 | 1 | 5 |
| M2.5 | 0.4 | 0.1 | 0.3 | M36, M42 | 3 | 1 | 6 |
| M3 | 0.4 | 0.2 | 0.6 | M48 | 4 | 2 | 8 |
| M4, M5 | 0.5 | 0.3 | 0.8 | M56-M72 | 6 | 2 | 10 |
| M6 | 0.6 | 0.4 | 1 | M80 | 6 | 2 | 11 |
| M8 | 1 | 0.4 | 2 | M90 | 6 | 3 | 11 |
| M10 | 1 | 0.5 | 2 | M100 | 7 | 4 | 12 |
| M12-M16 | 1.5 | 1 | 2.5 | … | … | … | … |

Dimensions are in millimeters.

**Recommended Substitute Drills.**—If the clearance hole application is dimensioned in metric drill sizes for inch fasteners, or inch drill sizes for metric fasteners, Table 3a and Table 3b list the nearest standard drill size translations for the designated drills of Table 1a and Table 2a.

## Table 3a. Standard Metric Drills For Inch Fasteners

*ANSI/ASME B18.2.8-1999 (R2017)*

| Nominal Screw Size, inch | Nominal Drill Size, mm — Fit Classes: Normal | Close | Loose | Nominal Screw Size, inch | Nominal Drill Size, mm — Fit Classes: Normal | Close | Loose |
|---|---|---|---|---|---|---|---|
| #0 | 1.9 | 1.7 | 2.4 | 3/8 | 10.2 | 9.9 | 10.5 |
| #1 | 2.25 | 2.05 | 2.6 | 7/16 | 11.8 | 11.5 | 12.2 |
| #2 | 2.6 | 2.4 | 2.9 | 1/2 | 14.25 | 13.5 | 15.5 |
| #3 | 2.9 | 2.7 | 3.3 | 5/8 | 17.5 | 16.75 | 19 |
| #4 | 3.3 | 3 | 3.7 | 3/4 | 20.5 | 20 | 23 |
| #5 | 4 | 3.6 | 4.4 | 7/8 | 24 | 23 | 26 |
| #6 | 4.3 | 3.9 | 4.7 | 1 | 27.5 | 26 | 29.5 |
| #8 | 5 | 4.6 | 5.4 | 1 1/8 | 31 | 29.5 | 33.5 |
| #10 | 5.6 | 5.2 | 6 | 1 1/4 | 34 | 32.5 | 36.5 |
| 1/4 | 7.1 | 6.7 | 7.5 | 1 3/8 | 38 | 36.5 | 41 |
| 5/16 | 8.7 | 8.3 | 9.1 | 1 1/2 | 41 | 39.5 | 44 |

## Table 3b. Standard Inch Drills For Metric Fasteners

*ANSI/ASME B18.2.8-1999 (R2017)*

| Nominal Screw Size, mm | Nominal Drill Size, inch — Fit Classes: Normal | Close | Loose | Nominal Screw Size, mm | Nominal Drill Size, inch — Fit Classes: Normal | Close | Loose |
|---|---|---|---|---|---|---|---|
| M1.6 | #50 | #51 | #47 | M16 | 11/16 | 43/64 | 47/64 |
| M2 | 3/32 | #44 | #38 | M20 | 55/64 | 53/64 | 15/16 |
| M2.5 | #33 | #36 | #31 | M24 | 1 1/32 | 63/64 | 1 7/64 |
| M3 | #29 | 1/8 | 9/64 | M30 | 1 9/32 | 1 7/32 | 1 3/8 |
| M4 | #16 | #19 | #12 | M36 | 1 17/32 | 1 15/32 | 1 21/32 |
| M5 | 7/32 | #4 | #1 | M42 | 1 25/32 | 1 11/16 | 1 29/32 |
| M6 | G | 1/4 | J | M48 | 2 1/32 | 1 31/32 | 2 3/16 |
| M8 | T | Q | 25/64 | M56 | 2 7/16 | 2 5/16 | 2 5/8 |
| M10 | 7/16 | Z | 31/64 | M64 | 2 3/4 | 2 5/8 | 2 15/16 |
| M12 | 17/32 | 33/64 | 37/64 | M72 | 3 1/8 | 2 15/16 | 3 1/4 |
| M14 | 39/64 | 19/32 | 21/32 | … | … | … | … |

## Table 4. Recommended Clearance Holes for Metric Round Head Square Neck Bolts

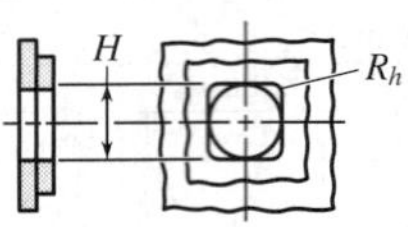


*Close Clearance:* Close clearance should be specified only for square holes in very thin and/or soft material, or for slots, or where conditions such as critical alignment of assembled parts, wall thickness, or other limitations necessitate use of a minimal hole. Allowable swell or fins on the bolt body and/or fins on the corners of the square neck may interfere with close clearance round or square holes.

*Normal Clearance:* Normal clearance hole sizes are preferred for general purpose applications and should be specified unless special design considerations dictate the need for either a close or loose clearance hole.

| Nom. Bolt Dia., $D$ and Thd. Pitch | Clearance: Close | Normal | Loose | Corner Radius $R_h$ | Nom. Bolt Dia., $D$ and Thd. Pitch | Clearance: Close | Normal | Loose | Corner Radius $R_h$ |
|---|---|---|---|---|---|---|---|---|---|
| | Min. Hole Diameter or Square Width, $H$ | | | | | Min. Hole Diameter or Square Width, $H$ | | | |
| M5 × 0.8 | 5.5 | … | 5.8 | 0.2 | M14 × 2 | 15.0 | 15.5 | 16.5 | 0.6 |
| M6 × 1 | 6.6 | … | 7.0 | 0.3 | M16 × 2 | 17.0 | 17.5 | 18.5 | 0.6 |
| M8 × 1.25 | … | 9.0 | 10.0 | 0.4 | M20 × 2.5 | 21.0 | 22.0 | 24.0 | 0.8 |
| M10 × 1.5 | … | 11.0 | 12.0 | 0.4 | M24 × 3 | 25.0 | 26.0 | 28.0 | 1.0 |
| M12 × 1.75 | 13.0 | 13.5 | 14.5 | 0.6 | … | … | … | … | … |

*Loose Clearance:* Loose clearance hole sizes should be specified only for applications where maximum adjustment capability between components being assembled is necessary. Loose clearance square hole or slots may not prevent bolt turning during wrenching.

All dimensions are in millimeters. Source: ANSI/ASME B18.5.2.2M-1982, Appendix II

**Table 5. Drill and Counterbore Sizes for Metric Socket Head Cap Screws**

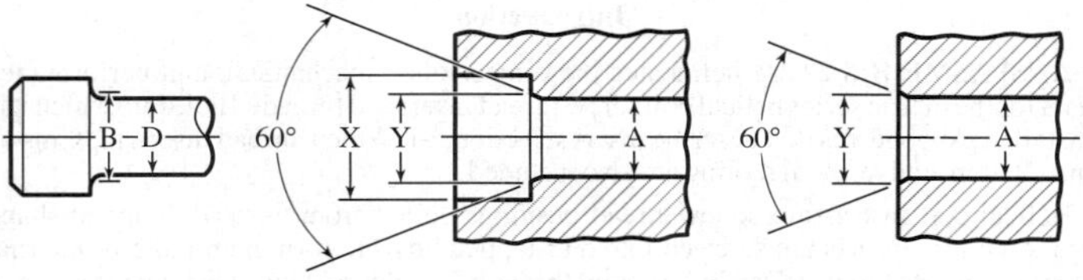


| Nominal Size or Basic Screw Diameter | Nominal Drill Size, A | | Counterbore Diameter, X | Countersink Diameter,[a] Y |
|---|---|---|---|---|
| | Close Fit[b] | Normal Fit[c] | | |
| M1.6 | 1.80 | 1.95 | 3.50 | 2.0 |
| M2 | 2.20 | 2.40 | 4.40 | 2.6 |
| M2.5 | 2.70 | 3.00 | 5.40 | 3.1 |
| M3 | 3.40 | 3.70 | 6.50 | 3.6 |
| M4 | 4.40 | 4.80 | 8.25 | 4.7 |
| M5 | 5.40 | 5.80 | 9.75 | 5.7 |
| M6 | 6.40 | 6.80 | 11.25 | 6.8 |
| M8 | 8.40 | 8.80 | 14.25 | 9.2 |
| M10 | 10.50 | 10.80 | 17.25 | 11.2 |
| M12 | 12.50 | 12.80 | 19.25 | 14.2 |
| M14 | 14.50 | 14.75 | 22.25 | 16.2 |
| M16 | 16.50 | 16.75 | 25.50 | 18.2 |
| M20 | 20.50 | 20.75 | 31.50 | 22.4 |
| M24 | 24.50 | 24.75 | 37.50 | 26.4 |
| M30 | 30.75 | 31.75 | 47.50 | 33.4 |
| M36 | 37.00 | 37.50 | 56.50 | 39.4 |
| M42 | 43.00 | 44.00 | 66.00 | 45.6 |
| M48 | 49.00 | 50.00 | 75.00 | 52.6 |

[a] *Countersink:* It is considered good practice to countersink or break the edges of holes which are smaller than *B* Max. (see Table 21, page 1730) in parts having a hardness which approaches, equals, or exceeds the screw hardness. If such holes are not countersunk, the heads of screws may not seat properly or the sharp edges on holes may deform the fillets on screws, thereby making them susceptible to fatigue in applications involving dynamic loading. The countersink or corner relief, however, should not be larger than is necessary to ensure that the fillet on the screw is cleared. Normally, the diameter of countersink does not have to exceed *B* Max. Countersinks or corner reliefs in excess of this diameter reduce the effective bearing area and introduce the possibility of embedment where the parts to be fastened are softer than the screws or of brinnelling or flaring the heads of the screws where the parts to be fastened are harder than the screws.

[b] *Close Fit:* The close fit is normally limited to holes for those lengths of screws which are threaded to the head in assemblies where only one screw is to be used or where two or more screws are to be used and the mating holes are to be produced either at assembly or by matched and coordinated tooling.

[c] *Normal Fit:* The normal fit is intended for screws of relatively long length or for assemblies involving two or more screws where the mating holes are to be produced by conventional tolerancing methods. It provides for the maximum allowable eccentricity of the longest standard screws and for certain variations in the parts to be fastened, such as: deviations in hole straightness, angularity between the axis of the tapped hole and that of the hole for shank, differences in center distances of the mating holes, etc.

All dimensions are in millimeters.

## HELICAL COIL SCREW THREAD INSERTS

### Introduction

ANSI/ASME B18.29.2M delineates the dimensional, mechanical, and performance data for the metric series helical coil screw thread insert and threaded hole into which it is installed. Appendices that describe insert selection, STI (screw thread insert) taps, insert installation, and removal tooling are also included.

Helical coil inserts are screw thread bushings coiled from wire of diamond-shape cross section. Inserts are screwed into STI-tapped holes to form nominal size internal threads. Inserts are installed by torquing through a diametral tang. This tang is notched for removal after installation. In the free state, they are larger in diameter than the tapped hole into which they are installed. In the assembly operation, the torque applied to the tang reduces the diameter of the leading coil and permits it to enter the tapped thread. The remaining coils are reduced in diameter as they, in turn, are screwed into the tapped hole. When the torque or rotation is stopped, the coils expand with a spring-like action anchoring the insert in place against the tapped hole.

**Dimensions.—**Dimensions in this Standard are in millimeters and apply before any coating. Symbols specifying geometric characteristics are in accordance with ASME Y14.5.

**Tolerance Classes 4H5H and 5H.—**Because helical coil inserts are flexible, the class of fit of the final assembly is a function of the size of the tapped hole. Helical coil STI taps are available for both tolerance class 4H5H (or class 4H6H) and class 5H tapped holes. Tolerance class 5H tapped holes provide maximum production tolerances but result in lower locking torques when screw-locking inserts are used. The higher and more consistent torques given in Table 5 are met by the screw-locking inserts when assembled and tested in tolerance class 4H5H (or class 4H6H) tapped holes.

**Compatibility.—**Assembled helical coil inserts will mate properly with items that have M Profile external threads in accordance with ANSI/ASME B1.13M. Also, due to the radius on the crest of the insert at the minor diameter, the assembled insert will mate with MJ Profile externally threaded parts with controlled radius root threads per ANSI/ASME B1.21M.

**Types of Inserts.—***Free-running* inserts provides a smooth, hard, and free-running thread. *Screw-locking* inserts provides a resilient locking thread produced by a series of chords on one or more of the insert coils.

**STI-Tapped Hole.—**The tapped hole into which the insert is installed shall be in accordance with ANSI/ASME B1.13M, except that diameters are larger to accommodate the wire cross section of the insert (See Fig. 1). Dimensions of the STI-tapped hole are shown in Table 1 and are calculated per General Note (c) to Table 1.

*Screw Thread Designation for Tapped Hole:* The drawing note for the STI-threaded hole per Table 1 to accept the helical coil insert shall be in accordance with the following:

*Example 1:* MS × 1.25-5H STI; 23.5 T per ANSI/ASME B18.29.2M.

*Designation for a Helical Coil Insert:* Helical coil inserts shall be designated by the following data, in the sequence shown:

a) product name; b) designation of the Standard; c) nominal diameter and thread pitch (4) nominal length; and d) insert type (free-running or screw-locking).

*Example 2:* Helical Coil insert, ANSI/ASME B18.29.2M, M8 × 1.25 × 12.0 free-running.

Helical Coil insert, ANSI/ASME B18.29.2M, M5 × 0.8 × 7.5 screw-locking.

The recommended B18 part number (PIN) code system for helical coil inserts is included in ANSI/ASME B18.24. This system may be used by user needing definitive part-numbering.

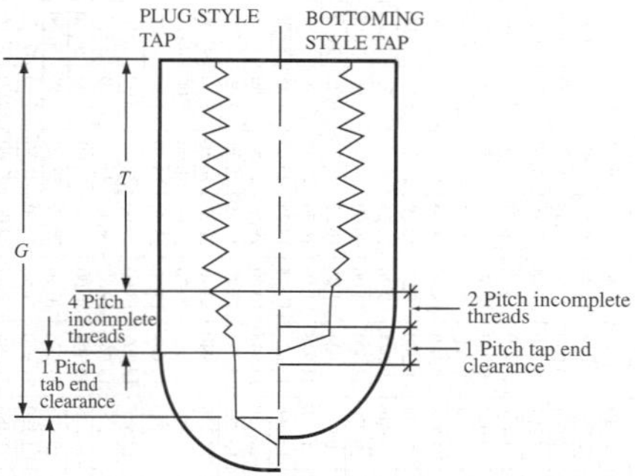


Fig. 1. Tapping Depth

*Designation for STI-Threaded Hole Including Installed Helical Coil Insert:* The drawing note for the STI-threaded hole per Table 1 having a helical coil insert installed shall be in accordance with this example.

*Example 3:* M8 × 1.25 STI 23.5 deep;

Helical Coil insert, ANSI/ASME B18.29.2M, M8 × 1.25 × 12.0, free running

*Gages and Gaging:* Acceptance of the threaded hole is determined by gaging with STI GO, NOT GO (HI), and plain cylindrical gages designed and applied in accordance with System 21 of ANSI/ASME B1.3 and with ANSI/ASME B1.16M.

## Helical Coil Insert

**Material.—**Chemical composition of the inserts is austenitic corrosion-resistant (stainless) steel material within the limits of Table 2.

**Properties.—**Wire, before coiling into inserts, shall have tensile strength not lower than 1035 MPa, determined in accordance with ASTM A370. Wire shall withstand, without cracking, bending in accordance with ASTM E 290 at room temperature through an angle of 180° around a diameter equal to twice the cross-sectional dimension of the wire in the plane of the bend. The formed wire shall be of uniform quality and temper; it shall be smooth, clean, and free from kinks, waviness, splits, cracks, laps, seams, scale, segregation, and other defects that may impair the serviceability of the insert.

**Coatings.—**At the option of the user, dry film lubricant coating can be applied to helical coil inserts.The color of dry film-lubricated inserts is dark gray to black. Lubricant shall meet requirements of Aerospace Standard SAE AS5272, type I, lubricant, solid film heat cured, and corrosion inhibiting. Coating shall be uniformly deposited on the insert with the minimum thickness being complete coverage. Maximum thickness shall be the avoidance of bridging between coils. Slight fill in between closely wound coils, which immediately separates as the coils are axially pulled apart by hand, shall not be considered bridging.

**Configuration and Dimensions.—**Insert configurations shall be in accordance with Fig. 2, and dimensions shall be in accordance with Table 3 and Table 4. Each nominal insert size is standardized in five lengths, which are multiples of the insert's nominal diameter. These are 1, 1.5, 2, 2.5, and 3 times nominal diameter. Each nominal length is the minimum through-hole length (material thickness), without countersink, into which that insert can be installed. The nominal insert length is a reference value and cannot be measured. Actual assembled length of the insert equals nominal length minus 0.5 pitch to minus 0.75 pitch, with insert installed in a basic STI threaded hole. Assembled length cannot be measured in the insert's free state.

## Table 1. Screw Thread Insert Threaded Hole Data *ANSI/ASME B18.29.2M-2005 (R2017)*

| Nominal Thread Size | Minimum Drilling Depth for Each Insert Length, *G* | | | | | | | | | | Countersink Diameter, *M* (120°±5° included angle) | | Minor Diameter | | Pitch Diameter | | | | Minor Major Dia. | Minimum Tapping Depth, *T* | | | | |
|---|---|---|---|---|---|---|---|---|---|---|---|---|---|---|---|---|---|---|---|---|---|---|---|---|
| | Plug Taps | | | | | Bottoming Taps | | | | | | | | | | | | | | Insert Length | | | | |
| | 1*D* | 1.5*D* | 2*D* | 2.5*D* | 3*D* | 1*D* | 1.5*D* | 2*D* | 2.5*D* | 3*D* | Min. | Max. | Min. | Max. | Min. | 4H Max. | 5H Max. | 6H Max. | All Classes | 1*D* | 1.5*D* | 2*D* | 2.5*D* | 3*D* |
| M2 × 0.4 | 5.40 | 6.40 | 7.40 | 8.40 | 9.40 | 3.60 | 4.60 | 5.60 | 6.60 | 7.60 | 2.30 | 2.70 | 2.087 | 2.199 | 2.260 | 2.295 | 2.310 | 2.329 | 2.520 | 2.40 | 3.40 | 4.40 | 5.40 | 6.40 |
| M2.5 ×0.45 | 6.45 | 7.70 | 8.95 | 10.20 | 11.45 | 4.30 | 5.55 | 6.80 | 8.05 | 9.30 | 2.90 | 3.40 | 2.597 | 2.722 | 2.792 | 2.832 | 2.847 | 2.867 | 3.084 | 2.95 | 4.20 | 5.45 | 6.70 | 7.95 |
| M3 × 0.5 | 7.50 | 9.00 | 10.50 | 12.00 | 13.50 | 5.00 | 5.00 | 8.00 | 9.50 | 11.00 | 3.40 | 4.00 | 3.108 | 3.248 | 3.326 | 3.367 | 3.384 | 3.404 | 3.650 | 3.50 | 5.00 | 6.50 | 8.00 | 9.50 |
| M3.5 × 0.6 | 8.86 | 10.60 | 12.35 | 14.10 | 15.85 | 5.90 | 7.65 | 9.40 | 11.15 | 12.90 | 4.10 | 4.70 | 3.630 | 3.790 | 3.890 | 3.940 | 3.959 | 3.981 | 4.280 | 4.10 | 5.85 | 7.60 | 9.35 | 11.10 |
| M4 × 0.7 | 10.20 | 12.20 | 14.20 | 16.20 | 18.20 | 6.80 | 8.80 | 10.80 | 12.80 | 14.80 | 4.70 | 5.30 | 4.162 | 4.332 | 4.455 | 4.508 | 4.529 | 4.552 | 4.910 | 4.70 | 6.70 | 8.70 | 10.70 | 12.70 |
| M5 × 0.8 | 12.30 | 14.80 | 17.30 | 19.80 | 22.30 | 8.20 | 10.70 | 13.20 | 15.70 | 18.20 | 5.80 | 6.40 | 5.174 | 5.374 | 5.520 | 5.577 | 5.597 | 5.622 | 6.040 | 5.80 | 8.30 | 10.80 | 13.30 | 15.80 |
| M6 × 1 | 15.00 | 18.00 | 21.00 | 24.00 | 27.00 | 10.00 | 13.00 | 16.00 | 19.00 | 22.00 | 7.10 | 7.70 | 6.217 | 6.407 | 6.650 | 6.719 | 6.742 | 6.774 | 7.300 | 7.00 | 10.00 | 13.00 | 16.00 | 19.00 |
| M7 × 1 | 16.50 | 20.00 | 23.50 | 27.00 | 30.50 | 11.00 | 14.50 | 18.00 | 21.50 | 25.00 | 8.10 | 8.70 | 7.217 | 7.407 | 7.650 | 7.719 | 7.742 | 7.774 | 8.300 | 8.00 | 11.50 | 15.00 | 18.50 | 22.00 |
| M8 × 1 | 18.00 | 22.00 | 26.00 | 30.00 | 34.00 | 12.00 | 16.00 | 20.00[a] | 24.00[a] | 28.00[a] | 0.10 | 9.70 | 8.217 | 8.407 | 8.650 | 8.719 | 8.742 | 8.774 | 9.300 | 9.00 | 13.00 | 17.00 | 21.00 | 25.00 |
| M8 × 1.25 | 19.50 | 23.60 | 27.50 | 31.50 | 35.50 | 13.00 | 17.00 | 21.00[a] | 25.00[a] | 29.00[a] | 9.50 | 10.10 | 8.271 | 8.483 | 8.812 | 8.886 | 8.911 | 8.946 | 9.624 | 9.26 | 13.25 | 17.26 | 21.25 | 25.25 |
| M10 × 1 | 16.00 | 21.00 | 26.00 | 31.00 | 36.00 | 14.00 | 19.00 | 24.00[a] | 29.00[a] | 34.00[a] | 11.10 | 11.70 | 10.217 | 10.407 | 10.650 | 10.719 | 10.742 | 10.774 | 11.300 | 11.00 | 16.00 | 21.00 | 26.00 | 31.00 |
| M10 × 1.25 | 17.50 | 22.60 | 27.50 | 32.50 | 37.50 | 15.00 | 20.00 | 25.00[a] | 30.00[a] | 35.00[a] | 11.50 | 12.10 | 10.271 | 10.483 | 10.812 | 10.886 | 10.911 | 10.946 | 11.624 | 11.26 | 16.25 | 21.26 | 26.25 | 31.25 |
| M10 × 1.5 | 19.00 | 24.00 | 29.00 | 34.00 | 39.00 | 16.00 | 21.00 | 26.00[a] | 31.00[a] | 38.00[a] | 11.80 | 12.40 | 10.324 | 10.580 | 10.974 | 11.061 | 11.089 | 11.129 | 11.948 | 11.50 | 16.50 | 21.50 | 26.50 | 31.50 |
| M12 × 1.25 | 19.50 | 25.50 | 31.50 | 37.50 | 43.50 | 17.00 | 23.00 | 29.00 | 35.00 | 41.00 | 13.50 | 14.10 | 12.271 | 12.483 | 12.812 | 12.896 | 12.926 | 12.966 | 13.624 | 13.25 | 19.25 | 25.25 | 31.25 | 37.25 |
| M12 × 1.5 | 21.00 | 27.00 | 33.00 | 39.00 | 45.00 | 18.00 | 24.00 | 30.00 | 36.00 | 42.00 | 13.00 | 14.40 | 12.324 | 12.560 | 12.974 | 13.067 | 13.099 | 13.139 | 13.948 | 13.50 | 19.50 | 25.50 | 31.50 | 37.60 |
| M12 × 1.75 | 22.50 | 28.50 | 34.50 | 40.50 | 48.50 | 19.00 | 25.00 | 31.00 | 37.00 | 43.00 | 14.20 | 14.80 | 12.379 | 12.644 | 13.137 | 13.236 | 13.271 | 13.311 | 14.274 | 13.75 | 19.75 | 25.75 | 31.75 | 37.75 |
| M14 × 1.5 | 23.00 | 30.00 | 37.00 | 44.00 | 51.00 | 20.00 | 27.00 | 34.00 | 41.00 | 48.00 | 15.80 | 16.40 | 14.324 | 14.560 | 14.974 | 15.067 | 15.099 | 15.139 | 15.940 | 15.50 | 22.50 | 29.50 | 38.50 | 43.50 |
| M14 × 2 | 26.00 | 33.00 | 40.00 | 47.00 | 54.00 | 22.00 | 29.00 | 36.00 | 43.00 | 50.00 | 16.50 | 17.10 | 14.433 | 14.733 | 15.299 | 15.406 | 15.444 | 15.486 | 16.958 | 16.00 | 23.00 | 30.00 | 37.00 | 44.00 |
| M16 × 1.5 | 25.00 | 33.00 | 41.00 | 49.00 | 57.00 | 22.00 | 30.00 | 38.00 | 46.00 | 50.00 | 17.80 | 18.40 | 16.324 | 16.560 | 16.974 | 17.067 | 17.099 | 17.139 | 17.948 | 17.50 | 25.50 | 33.50 | 41.50 | 49.50 |
| M16 × 2 | 28.00 | 36.00 | 44.00 | 52.00 | 60.00 | 24.00 | 32.00 | 40.00 | 48.00 | 56.00 | 18.50 | 19.10 | 16.433 | 16.733 | 17.299 | 17.406 | 17.444 | 17.486 | 18.598 | 18.00 | 26.00 | 34.00 | 42.00 | 50.00 |
| M18 × 1.5 | 27.00 | 36.00 | 45.00 | 54.00 | 63.00 | 24.00 | 33.00 | 42.00 | 51.00 | 60.00 | 19.80 | 20.40 | 18.324 | 18.560 | 18.974 | 19.067 | 19.099 | 19.139 | 19.948 | 19.50 | 28.50 | 37.50 | 46.50 | 55.50 |
| M18 × 2 | 30.00 | 39.00 | 48.00 | 57.00 | 66.00 | 26.00 | 35.00 | 44.00 | 53.00 | 62.00 | 20.50 | 21.10 | 18.433 | 18.733 | 19.299 | 19.406 | 19.444 | 19.486 | 20.598 | 20.00 | 29.00 | 38.00 | 47.00 | 56.00 |
| M18 × 2.5 | 33.00 | 42.00 | 51.00 | 60.00 | 69.00 | 28.00 | 37.00 | 46.00 | 55.00 | 64.00 | 21.20 | 21.80 | 18.541 | 18.896 | 19.624 | 19.738 | 19.778 | 19.822 | 21.248 | 20.50 | 29.50 | 38.50 | 47.50 | 56.50 |
| M20 × 1.5 | 29.00 | 39.00 | 49.00 | 59.00 | 69.00 | 26.00 | 36.00 | 46.00 | 56.00 | 66.00 | 21.80 | 22.40 | 20.324 | 20.560 | 20.974 | 21.067 | 21.099 | 21.139 | 21.940 | 21.50 | 31.50 | 41.50 | 51.50 | 61.50 |
| M20 × 2 | 32.00 | 42.00 | 52.00 | 62.00 | 72.00 | 28.00 | 38.00 | 48.00 | 58.00 | 68.00 | 22.50 | 23.10 | 20.433 | 20.733 | 21.299 | 21.406 | 21.444 | 21.486 | 22.598 | 22.00 | 32.00 | 42.00 | 52.00 | 62.00 |
| M20 × 2.5 | 35.00 | 45.00 | 55.00 | 65.00 | 75.00 | 30.00 | 40.00 | 50.00 | 60.00 | 70.00 | 23.20 | 23.80 | 20.541 | 20.896 | 21.624 | 21.738 | 21.778 | 21.822 | 23.248 | 22.50 | 32.50 | 42.50 | 52.50 | 62.50 |
| M22 × 1.5 | 31.00 | 42.00 | 53.00 | 64.00 | 75.00 | 28.00 | 39.00 | 50.00 | 61.00 | 72.00 | 23.80 | 24.40 | 22.324 | 22.560 | 22.974 | 23.067 | 23.099 | 23.139 | 23.948 | 23.50 | 34.50 | 45.50 | 56.50 | 67.50 |
| M22 × 2 | 34.00 | 45.00 | 56.00 | 67.00 | 78.00 | 30.00 | 41.00 | 52.00 | 63.00 | 74.00 | 24.50 | 25.10 | 22.433 | 22.733 | 23.299 | 23.406 | 23.444 | 23.486 | 24.598 | 24.00 | 35.00 | 46.00 | 57.00 | 68.00 |
| M22 × 2.5 | 37.00 | 48.00 | 59.00 | 70.00 | 81.00 | 32.00 | 43.00 | 54.00 | 65.00 | 76.00 | 25.20 | 25.80 | 22.541 | 22.896 | 23.624 | 23.738 | 23.778 | 23.822 | 25.248 | 24.50 | 35.50 | 46.50 | 57.50 | 68.50 |
| M24 × 2 | 38.00 | 48.00 | 60.00 | 72.00 | 84.00 | 32.00 | 44.00 | 56.00 | 68.00 | 80.00 | 26.50 | 27.10 | 24.433 | 24.733 | 25.299 | 25.414 | 15.454 | 25.498 | 26.598 | 26.00 | 38.00 | 50.00 | 62.00 | 74.00 |
| M24 × 3 | 42.00 | 54.00 | 66.00 | 78.00 | 90.00 | 36.00 | 48.00 | 60.00 | 72.00 | 84.00 | 27.90 | 28.50 | 24.649 | 25.049 | 25.948 | 26.093 | 26.135 | 26.188 | 27.897 | 27.00 | 39.00 | 51.00 | 63.00 | 75.00 |
| M27 × 2 | 39.00 | 52.50 | 66.00 | 79.50 | 93.00 | 35.00 | 48.50 | 62.00 | 75.50 | 89.00 | 29.50 | 30.10 | 27.433 | 27.733 | 28.299 | 28.414 | 28.454 | 28.498 | 29.598 | 29.00 | 42.50 | 58.00 | 69.50 | 83.00 |

[a] Note: These table values do not match the referenced table in ANSI/ASME B18.29.2M-2005 (R2017).

**Table 1.** *(Continued)* **Screw Thread Insert Threaded Hole Data** *ANSI/ASME B18.29.2M-2005 (R2017)*

| Nominal Thread Size | Minimum Drilling Depth for Each Insert Length, $G$ | | | | | | | | | | Countersink Diameter, $M$ (120°±5° included angle) | | Minor Diameter | | Pitch Diameter | | | | Minor Major Dia. | Minimum Tapping Depth, $T$ | | | | |
|---|---|---|---|---|---|---|---|---|---|---|---|---|---|---|---|---|---|---|---|---|---|---|---|---|
| | Plug Taps | | | | | Bottoming Taps | | | | | | | | | | | | | | Insert Length | | | | |
| | 1*D* | 1.5*D* | 2*D* | 2.5*D* | 3*D* | 1*D* | 1.5*D* | 2*D* | 2.5*D* | 3*D* | Min. | Max. | Min. | Max. | Min. | 4H Max. | 5H Max. | 6H Max. | All Classes | 1*D* | 1.5*D* | 2*D* | 2.5*D* | 3*D* |
| M27 × 3 | 45.00 | 68.50 | 72.00 | 85.50 | 99.00 | 39.00 | 52.50 | 66.00 | 79.50 | 93.00 | 30.90 | 31.50 | 27.649 | 28.049 | 28.948 | 29.093 | 29.135 | 29.188 | 30.897 | 30.00 | 43.50 | 57.00 | 70.50 | 84.00 |
| M30 × 2 | 42.00 | 67.00 | 72.00 | 87.00 | 102.00 | 38.00 | 53.00 | 68.00 | 83.00 | 98.00 | 32.50 | 33.10 | 30.433 | 30.733 | 31.299 | 31.414 | 31.454 | 31.489 | 32.598 | 32.00 | 47.00 | 62.00 | 77.00 | 92.00 |
| M30 × 3 | 48.00 | 63.00 | 78.00 | 93.00 | 108.00 | 42.00 | 57.00 | 72.00 | 87.00 | 102.00 | 33.90 | 34.50 | 30.649 | 31.049 | 31.948 | 32.093 | 32.136 | 32.188 | 33.897 | 33.00 | 48.00 | 63.00 | 78.00 | 93.00 |
| M30 × 3.5 | 51.00 | 66.00 | 81.00 | 96.00 | 111.00 | 44.00 | 59.00 | 74.00 | 89.00 | 104.00 | 34.60 | 35.20 | 30.767 | 31.207 | 32.273 | 32.428 | 32.472 | 32.628 | 34.546 | 33.50 | 48.50 | 63.50 | 78.50 | 93.50 |
| M33 × 2 | 45.00 | 61.60 | 78.00 | 94.50 | 111.00 | 41.00 | 57.50 | 74.00 | 90.50 | 107.00 | 35.50 | 36.10 | 33.433 | 33.733 | 34.299 | 34.414 | 34.454 | 34.498 | 35.598 | 35.00 | 51.50 | 68.00 | 84.50 | 101.00 |
| M33 × 3 | 51.00 | 67.60 | 84.00 | 104.50 | 117.00 | 45.00 | 61.50 | 78.00 | 94.50 | 111.00 | 36.90 | 37.50 | 33.649 | 34.049 | 34.948 | 35.093 | 35.135 | 36.188 | 36.897 | 36.00 | 52.50 | 69.00 | 85.50 | 102.00 |
| M36 × 2 | 48.00 | 66.00 | 84.00 | 102.00 | 120.00 | 44.00 | 62.00 | 80.00 | 98.00 | 116.00 | 38.50 | 39.10 | 36.433 | 36.733 | 37.299 | 37.414 | 37.464 | 37.498 | 38.598 | 38.00 | 58.00 | 74.00 | 92.00 | 110.00 |
| M36 × 3 | 54.00 | 72.00 | 90.00 | 108.00 | 126.00 | 48.00 | 66.00 | 84.00 | 102.00 | 120.00 | 39.90 | 40.50 | 36.649 | 37.049 | 37.948 | 38.093 | 38.135 | 38.188 | 39.897 | 39.00 | 57.00 | 75.00 | 93.00 | 111.00 |
| M36 × 4 | 60.00 | 78.00 | 96.00 | 114.00 | 132.00 | 52.00 | 70.00 | 88.00 | 106.00 | 124.00 | 41.30 | 41.90 | 36.866 | 37.341 | 28.598 | 38.763 | 38.809 | 38.873 | 41.196 | 40.00 | 58.00 | 76.00 | 94.00 | 112.00 |
| M39 × 2 | 51.00 | 70.50 | 90.00 | 109.50 | 129.00 | 47.00 | 66.50 | 88.00 | 105.50 | 125.00 | 41.50 | 42.10 | 39.433 | 39.733 | 40.299 | 40.414 | 40.454 | 40.498 | 41.598 | 41.00 | 60.50 | 80.00 | 99.50 | 119.00 |
| M39 × 3 | 57.00 | 76.50 | 96.00 | 115.50 | 135.00 | 51.00 | 70.50 | 90.00 | 109.50 | 129.00 | 42.90 | 43.50 | 39.649 | 40.049 | 40.948 | 41.093 | 41.136 | 41.188 | 42.897 | 42.00 | 61.50 | 81.00 | 100.50 | 120.00 |

*Notes:*

(1) The minimum drilling depths allow for

a) countersinking the drilled hole to prevent a feather edge at the start of the tapped hole.

b) 0.75 to 1.5 pitch of insert set-down to allow for maximum production tolerance.

c) Dimensions are shown for both plug and bottoming taps. Plug taps 8 mm and smaller have a male center, and the drilled hole depth dimensions allow for this length (one-half of the diameter of the bolt). Calculation of minimum drilling depth dimension $G$ is as follows:

Plug taps 8 mm and smaller, $G$ = insert nominal length + 0.5 × nominal bolt diameter + 4 pitchs for tap chamfer + 1 pitch for tap end clearance + 1 pitch allowance for countersink and maximum insert set-down.

Plug taps larger than 8 mm, $G$ = insert nominal length + 4 pitchs for tap chamfer + 1 pitch for tap end clearance + 1 pitch allowance for countersink and maximum insert set-down.

Bottoming taps, $G$ = insert nominal length + 2 pitchs for tap chamfer + 1 pitch for tap end clearance + 1 pitch allowance for countersink and maximum insert set-down.

(2) The minimum tapping depth (dimension $T$) is the minimum for countersink holes with insert set-down of 1.5 pitch maximum (See Fig. 1). The dimension $T$ = insert nominal length + 1 pitch.

(3) Thread diameters are calculated as follows:

Pitch diameter, min. = Pitch diameter, min. of nominal thread + $2 \times H_{max}$

Pitch diameter, max. = Pitch diameter, max. of nominal thread + $2 \times H_{min}$

Major diameter, min. = Pitch diameter, min. + $0.649519 \times P$

Minor diameter, min. = Pitch diameter min. − $0.433013 \times P$

Minor diameter, max. = Minor diameter min. + tolerance

where $H_{max}$ and $H_{min}$ are from Table 14, and tolerance is selected from the appropriate table in ANSI/ASME B1.13M with basic major diameter equal to the minimum major diameter of the STI thread.

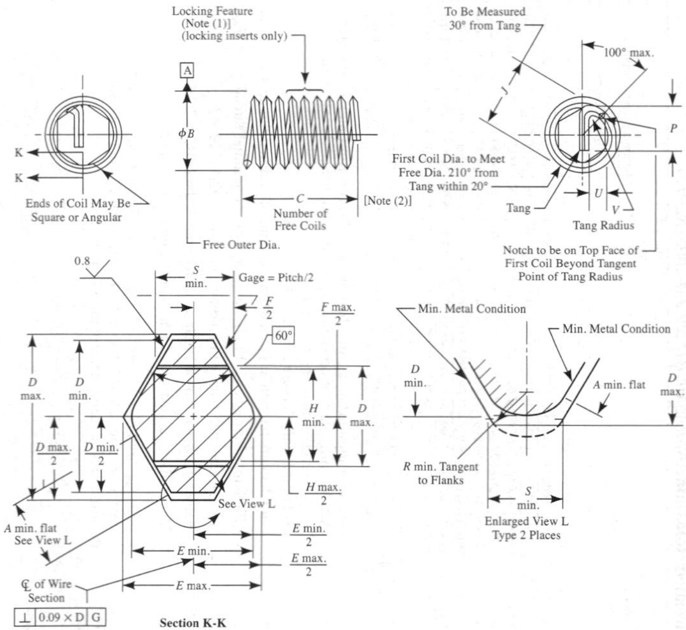


Fig. 2. Insert Configuration

*General Notes for* Fig. 2*:*

(a) Assembled length of insert to be measured from notch.

(b) Dimensions apply before supplementary coating (see Table 3 and Table 4).

(c) Surface texture; symbols per ASME Y14.35, requirements per ANSI/ASME B46.1.

(d) Dimensions and tolerancing; ASME Y14.5.

Notes:

(1) Number of locking coils, spacing of locking coils, number of locking deformations, shape and orientation optional locking feature for 1, 1.5, and 2 dia. length inserts symmetrically positioned about the center of insert, and for 2.5 and 3 dia. length inserts at 1 dia. from tang end of insert.

(2) Number of free coils to be counted from notch.

## Inspection and Quality Assurance

The inspection of inserts shall be in accordance with ANSI/ASME B18.18, with inspection level 3 for the 15 cycle torque test.

**Inspection (Nondestructive).—**Inserts shall be visually examined for conformance with drawings and workmanship requirements in accordance with ANSI/ASME B18.18.

*Threads:* The inserts, when assembled in STI threaded holes conforming to Table 1, shall form threads conforming to ANSI/ASME B1.13M tolerance class 4H5H or 5H except for the locking feature of screw-locking inserts. The assembled insert, both types, shall accept and function with parts having external MJ threads per ANSI/ASME B1.21M.

**Table 2. Screw Thread Insert Chemical Composition**
*ANSI/ASME B18.29.2M-2005 (R2017)*

| Element | Analysis, % | Check Analysis | |
|---|---|---|---|
| | | Under, Min. | Over, Max. |
| Carbon | 0.15 max. | … | 0.01 |
| Manganese | 2.00 max. | … | 0.04 |
| Silicon | 1.00 max. | … | 0.05 |
| Phosphorous | 0.045 max. | … | 0.01 |
| Sulphur | 0.035 max. | … | 0.005 |
| Chromium | 17.00 to 20.00 | 0.20 | 0.20 |
| Nickel | 8.00 to 10.50 | 0.15 | 0.15 |
| Molybdenum | 0.75 max. | … | 0.05 |
| Copper | 0.75 max. | … | 0.05 |
| Iron | Remainder | … | … |

The accuracy of the finished thread when the insert is installed depends on the accuracy of the tapped hole. If the finished tapped hole gages satisfactorily, the installed insert will be within the thread tolerance when the insert meets the requirements of the Standard. It is, therefore, not necessary to gage the installed insert. After the insert is installed, the GO thread plug gage may not enter freely because the insert may not have been fully seated in the tapped hole. However, the insert should become seated after a bolt or screw is installed and tightened.

*Tang Removal Notch:* The tang removal notch shall be located as shown in Fig. 2 and of such depth that the part may be installed without failure of the tang and that the tang may be removed, after assembly, without affecting the function of the installed insert.

*Torque Test Bolts:* Assembled screw-locking inserts shall be torque tested with bolts in accordance with ANSI/ASME B1.13M or ANSI/ASME B1.21M, cadmium plated, or having other coating with a similar coefficient of friction and hardness of 36 to 44 RC (Rockwell C scale). The bolts selected for this test shall be of sufficient length so the thread runout does not enter the insert and that a minimum of one full thread extends past the end of the insert when the bolt is fully seated. Acceptability of bolt threads shall be determined based on System 22 of ANSI/ASME B1.3.

Until a replacement for cadmium plating on the torque test bolts is found, and test data completed, an alternate coating/lubricant can be used to perform the torque test.

**Self-Locking Torque (Destructive).—**The screw-locking insert, when assembled in threaded holes conforming to Table 1 and tested in accordance with the following paragraphs, shall provide a frictional lock to retain the bolt threads within the torque limits specified in Table 5.

*Torque Test Block and Spacer:* The insert to be tested shall be installed in a tolerance class 4H5H or 4H6H threaded hole conforming to Table 1 in a test block made from 2024-T4 (SAE AMS4120 or ASTM B209M) aluminum alloy. After installation, the tang shall be removed. The surface of the test block from which the insert is assembled shall be marked "TOP" and shall be marked to indicate the radial location where the assembled insert begins. A steel spacer meeting the requirements of Fig. 3 and Table 6 shall be used for developing the bolt load.

**Table 3. Screw Thread Insert Length Data** *ANSI/ASME B18.29.2M-2005 (R2017)*

| Nominal Thread Size | 1 × Dia. | | | | 1½ × Dia. | | | | 2 × Dia. | | | | 2 ½ × Dia. | | | | 3 × Dia. | | | |
|---|---|---|---|---|---|---|---|---|---|---|---|---|---|---|---|---|---|---|---|---|
| | Nominal | Assembled | | *C* | Nominal | Assembled | | *C* | Nominal | Assembled | | *C* | Nominal | Assembled | | *C* | Nominal | Assembled | | *C* |
| | | Max. | Min. | (Ref.) | | Max. | Min. | (Ref.) | | Max. | Min. | (Ref.) | | Max. | Min. | (Ref.) | | Max. | Min. | (Ref.) |
| M2 × 0.4 | 2.00 | 1.80 | 1.70 | 3.250 | 3.00 | 2.80 | 2.70 | 5.500 | 4.00 | 3.80 | 3.70 | 7.750 | 5.00 | 4.80 | 4.70 | 10.125 | 6.00 | 5.80 | 5.70 | 12.375 |
| M2.5 × 0.45 | 2.50 | 2.28 | 2.16 | 3.575 | 3.80 | 3.52 | 3.41 | 5.750 | 5.00 | 4.78 | 4.66 | 8.125 | 6.30 | 6.02 | 5.91 | 10.500 | 7.50 | 7.28 | 7.16 | 12.750 |
| M3 × 0.5 | 3.00 | 2.75 | 2.62 | 3.750 | 4.50 | 4.25 | 4.12 | 6.375 | 6.00 | 5.75 | 5.62 | 8.875 | 7.50 | 7.25 | 7.12 | 11.375 | 9.00 | 8.75 | 8.62 | 13.875 |
| M3.5 × 0.6 | 3.50 | 3.20 | 3.05 | 3.750 | 5.30 | 5.00 | 4.80 | 6.375 | 7.00 | 6.70 | 6.55 | 8.750 | 8.80 | 8.50 | 8.30 | 11.375 | 10.50 | 10.20 | 10.05 | 13.750 |
| M4 × 0.7 | 4.00 | 3.65 | 3.47 | 3.625 | 6.00 | 5.65 | 5.47 | 6.125 | 8.00 | 7.65 | 7.47 | 8.625 | 10.00 | 9.65 | 9.47 | 11.125 | 12.00 | 11.65 | 11.47 | 13.625 |
| M5 × 0.8 | 5.00 | 4.60 | 4.40 | 4.125 | 7.50 | 7.10 | 6.90 | 6.875 | 10.00 | 9.60 | 9.40 | 9.625 | 12.50 | 12.10 | 11.90 | 12.375 | 15.00 | 14.60 | 14.40 | 15.125 |
| M6 × 1 | 6.00 | 5.50 | 5.25 | 4.000 | 9.00 | 8.50 | 8.25 | 6.750 | 12.00 | 11.50 | 11.25 | 9.500 | 15.00 | 14.50 | 14.25 | 12.125 | 18.00 | 17.50 | 17.25 | 14.875 |
| M7 × 1 | 7.00 | 6.50 | 6.25 | 4.875 | 10.50 | 10.00 | 9.75 | 8.000 | 14.00 | 13.50 | 13.25 | 11.125 | 17.50 | 17.00 | 16.75 | 14.125 | 21.00 | 20.50 | 20.25 | 17.250 |
| M8 × 1 | 8.00 | 7.50 | 7.25 | 5.875 | 12.00 | 11.50 | 11.25 | 9.375 | 16.00 | 15.50 | 15.25 | 13.000 | 20.00 | 19.50 | 19.25 | 16.500 | 24.00 | 23.50 | 23.25 | 20.125 |
| M8 × 1.25 | 8.00 | 7.38 | 7.06 | 4.500 | 12.00 | 11.38 | 11.06 | 7.375 | 16.00 | 15.38 | 15.06 | 10.250 | 20.00 | 19.38 | 19.06 | 13.250 | 24.00 | 23.38 | 23.06 | 16.125 |
| M10 × 1 | 10.00 | 9.50 | 9.25 | 7.625 | 15.00 | 14.50 | 14.25 | 12.000 | 20.00 | 19.50 | 19.25 | 16.500 | 25.00 | 24.50 | 24.25 | 21.000 | 30.00 | 29.50 | 29.25 | 25.500 |
| M10 × 1.25 | 10.00 | 9.38 | 9.06 | 5.875 | 15.00 | 14.38 | 14.06 | 9.500 | 20.00 | 19.38 | 19.06 | 13.125 | 25.00 | 24.38 | 24.06 | 16.750 | 30.00 | 29.38 | 29.06 | 20.375 |
| M10 × 1.5 | 10.00 | 9.25 | 8.87 | 4.875 | 15.00 | 14.25 | 13.87 | 8.000 | 20.00 | 19.25 | 18.87 | 11.125 | 25.00 | 24.25 | 23.87 | 14.250 | 30.00 | 29.25 | 28.87 | 17.375 |
| M12 × 1.25 | 12.00 | 11.38 | 11.06 | 7.250 | 18.00 | 17.38 | 17.06 | 11.625 | 24.00 | 23.38 | 23.06 | 15.875 | 30.00 | 29.38 | 29.06 | 20.250 | 36.00 | 35.38 | 35.06 | 24.500 |
| M12 × 1.5 | 12.00 | 11.25 | 10.87 | 6.000 | 18.00 | 17.25 | 16.87 | 9.625 | 24.00 | 23.25 | 22.87 | 13.375 | 30.00 | 29.25 | 28.87 | 17.000 | 36.00 | 35.25 | 34.87 | 20.750 |
| M12 × 1.75 | 12.00 | 11.12 | 10.68 | 5.000 | 18.00 | 17.12 | 16.68 | 8.250 | 24.00 | 23.12 | 22.68 | 11.500 | 30.00 | 29.12 | 28.68 | 14.625 | 36.00 | 35.12 | 34.68 | 17.875 |
| M14 × 1.5 | 14.00 | 13.25 | 12.87 | 7.125 | 21.00 | 20.25 | 19.87 | 11.375 | 28.00 | 27.25 | 26.87 | 15.625 | 35.00 | 4.25 | 33.87 | 20.000 | 42.00 | 41.25 | 40.87 | 24.250 |
| M14 × 2 | 14.00 | 13.00 | 12.50 | 5.125 | 21.00 | 20.00 | 19.50 | 8.500 | 28.00 | 27.00 | 26.50 | 11.750 | 35.00 | 34.00 | 33.50 | 15.000 | 42.00 | 41.00 | 40.50 | 18.375 |
| M16 × 1.5 | 16.00 | 15.25 | 14.87 | 8.250 | 24.00 | 23.25 | 22.87 | 13.125 | 32.00 | 31.25 | 30.87 | 18.000 | 40.00 | 39.25 | 38.87 | 22.750 | 48.00 | 47.25 | 46.87 | 27.625 |
| M16 × 2 | 16.00 | 15.00 | 14.50 | 6.125 | 24.00 | 23.00 | 22.50 | 9.750 | 32.00 | 31.00 | 30.50 | 13.500 | 40.00 | 39.00 | 38.50 | 17.250 | 48.00 | 47.00 | 46.50 | 21.000 |
| M18 × 1.5 | 18.00 | 17.25 | 16.87 | 9.500 | 27.00 | 26.25 | 25.87 | 15.000 | 36.00 | 35.25 | 34.87 | 20.375 | 45.00 | 44.25 | 43.87 | 25.875 | 54.00 | 53.25 | 52.87 | 31.375 |
| M18 × 2 | 18.00 | 17.00 | 16.50 | 7.000 | 27.00 | 26.00 | 25.50 | 11.125 | 36.00 | 35.00 | 34.50 | 15.375 | 45.00 | 44.00 | 43.50 | 19.500 | 54.00 | 53.00 | 52.50 | 23.625 |

**Table 3.** *(Continued)* **Screw Thread Insert Length Data** *ANSI/ASME B18.29.2M-2005 (R2017)*

| Nominal Thread Size | 1 × Dia. | | | | 1½ × Dia. | | | | 2 × Dia. | | | | 2 ½ × Dia. | | | | 3 × Dia. | | | |
|---|---|---|---|---|---|---|---|---|---|---|---|---|---|---|---|---|---|---|---|---|
| | Nominal | Assembled | | *C* | Nominal | Assembled | | *C* | Nominal | Assembled | | *C* | Nominal | Assembled | | *C* | Nominal | Assembled | | *C* |
| | | Max. | Min. | (Ref.) | | Max. | Min. | (Ref.) | | Max. | Min. | (Ref.) | | Max. | Min. | (Ref.) | | Max. | Min. | (Ref.) |
| M18 × 2.5 | 18.00 | 16.75 | 16.12 | 5.375 | 27.00 | 25.75 | 25.12 | 8.875 | 36.00 | 34.75 | 34.12 | 12.250 | 45.00 | 43.75 | 43.12 | 15.625 | 54.00 | 52.75 | 52.12 | 19.000 |
| M20 × 1.5 | 20.00 | 19.25 | 18.87 | 10.750 | 30.00 | 29.25 | 28.87 | 16.875 | 40.00 | 39.25 | 38.87 | 22.875 | 50.00 | 49.25 | 48.87 | 28.875 | 60.00 | 59.25 | 58.87 | 35.000 |
| M20 × 2 | 20.00 | 19.00 | 18.50 | 7.875 | 30.00 | 29.00 | 28.50 | 12.500 | 40.00 | 39.00 | 38.50 | 17.250 | 50.00 | 49.00 | 48.50 | 21.875 | 60.00 | 59.00 | 58.50 | 26.500 |
| M20 × 2.5 | 20.00 | 18.75 | 18.12 | 6.125 | 30.00 | 28.75 | 28.12 | 9.875 | 40.00 | 38.75 | 38.12 | 13.625 | 50.00 | 48.75 | 48.12 | 17.375 | 60.00 | 58.75 | 58.12 | 21.125 |
| M22 × 1.5 | 22.00 | 21.25 | 20.87 | 11.875 | 33.00 | 32.25 | 31.87 | 18.500 | 44.00 | 43.25 | 42.87 | 25.125 | 55.00 | 54.25 | 53.87 | 31.625 | 66.00 | 65.25 | 64.87 | 38.250 |
| M22 × 2 | 22.00 | 21.00 | 20.50 | 8.750 | 33.00 | 32.00 | 31.50 | 13.750 | 44.00 | 43.00 | 42.50 | 18.875 | 55.00 | 54.00 | 53.50 | 23.875 | 66.00 | 65.00 | 64.50 | 29.000 |
| M22 × 2.5 | 22.00 | 20.75 | 20.12 | 6.750 | 33.00 | 31.75 | 31.12 | 10.875 | 44.00 | 42.75 | 42.12 | 14.875 | 55.00 | 53.75 | 53.12 | 19.000 | 66.00 | 64.75 | 64.12 | 23.125 |
| M24 × 2 | 24.00 | 23.00 | 22.50 | 9.500 | 36.00 | 35.00 | 34.50 | 15.000 | 48.00 | 47.00 | 16.50 | 20.375 | 60.00 | 59.00 | 58.50 | 25.875 | 72.00 | 71.00 | 70.50 | 31.250 |
| M24 × 3 | 24.00 | 22.50 | 21.75 | 6.125 | 36.00 | 34.50 | 33.75 | 10.000 | 48.00 | 46.50 | 45.75 | 13.750 | 60.00 | 58.50 | 57.75 | 17.500 | 72.00 | 70.50 | 69.75 | 21.375 |
| M27 × 2 | 27.00 | 26.00 | 25.50 | 10.875 | 40.50 | 39.50 | 39.00 | 17.000 | 54.00 | 53.00 | 52.50 | 23.250 | 67.50 | 66.50 | 66.00 | 29.375 | 81.00 | 80.00 | 79.50 | 35.500 |
| M27 × 3 | 27.00 | 25.50 | 24.75 | 7.000 | 40.50 | 39.00 | 38.25 | 11.250 | 54.00 | 52.50 | 51.75 | 15.500 | 67.50 | 66.50 | 65.25 | 19.750 | 81.00 | 79.50 | 78.75 | 24.000 |
| M30 × 2 | 30.00 | 29.00 | 28.50 | 12.250 | 45.00 | 44.00 | 43.50 | 19.125 | 60.00 | 59.00 | 58.50 | 25.875 | 75.00 | 74.00 | 73.50 | 32.750 | 90.00 | 89.00 | 88.50 | 39.500 |
| M30 × 3 | 30.00 | 28.50 | 27.75 | 7.875 | 45.00 | 43.50 | 42.75 | 12.500 | 60.00 | 58.50 | 57.75 | 17.125 | 75.00 | 73.50 | 72.75 | 21.875 | 90.00 | 88.50 | 87.75 | 26.500 |
| M30 × 3.5 | 30.00 | 28.25 | 27.37 | 6.750 | 45.00 | 43.25 | 42.37 | 10.750 | 60.00 | 58.25 | 57.37 | 14.875 | 75.00 | 73.25 | 72.37 | 18.875 | 90.00 | 88.25 | 87.37 | 23.000 |
| M33 × 2 | 33.00 | 32.00 | 31.50 | 13.625 | 49.50 | 48.50 | 48.00 | 21.125 | 66.00 | 65.00 | 64.50 | 28.625 | 82.50 | 81.50 | 81.00 | 35.000 | 99.00 | 98.00 | 97.50 | 43.500 |
| M33 × 3 | 33.00 | 32.50 | 30.75 | 8.750 | 49.50 | 48.00 | 47.25 | 13.875 | 66.00 | 64.50 | 63.75 | 19.000 | 82.50 | 81.00 | 80.25 | 24.125 | 99.00 | 97.50 | 96.75 | 29.250 |
| M36 × 2 | 36.00 | 35.00 | 34.50 | 15.000 | 54.00 | 53.00 | 52.50 | 23.250 | 72.00 | 71.00 | 70.50 | 31.375 | 90.00 | 89.00 | 88.50 | 39.500 | 108.00 | 107.00 | 106.50 | 47.750 |
| M36 × 3 | 36.00 | 34.50 | 33.75 | 9.750 | 54.00 | 52.50 | 51.75 | 15.250 | 72.00 | 70.50 | 69.75 | 20.875 | 90.00 | 88.50 | 87.75 | 26.500 | 108.00 | 106.50 | 105.75 | 32.000 |
| M36 × 4 | 36.00 | 34.00 | 33.00 | 7.125 | 54.00 | 52.00 | 51.00 | 11.375 | 72.00 | 70.00 | 69.00 | 15.625 | 90.00 | 88.00 | 87.00 | 19.875 | 108.00 | 106.00 | 105.00 | 24.250 |
| M39 × 2 | 39.00 | 38.00 | 37.50 | 16.375 | 58.50 | 57.50 | 57.00 | 25.250 | 78.00 | 77.00 | 76.50 | 34.125 | 97.50 | 96.50 | 96.00 | 43.000 | 117.00 | 116.00 | 115.50 | 51.875 |
| M39 × 3 | 39.00 | 37.50 | 36.75 | 10.750 | 58.50 | 57.00 | 56.25 | 15.750 | 78.00 | 76.50 | 75.75 | 22.750 | 97.50 | 96.00 | 95.25 | 28.875 | 117.00 | 115.50 | 114.75 | 34.875 |

**Table 4. Screw Thread Insert Dimensions** *ANSI/ASME B18.29.2M-2005 (R2017)*

| Nominal Thread Size | *A*, Min. | *B* | | *D* | | *E* | | Gage, *F* | *H* | | *J* | | *P* | | *R*, Min. | *S*, Min. | *U* | | *V*, Max. |
|---|---|---|---|---|---|---|---|---|---|---|---|---|---|---|---|---|---|---|---|
| | | Min. | Max. | Min. | Max. | Min. | Max. | | Min. | Max. | Min. | Max. | Min. | Max. | | | Min. | Max. | |
| M2 × 0.4 | 0.074 | 2.50 | 2.70 | 0.389 | 0.433 | 0.274 | 0.350 | 0.200 | 0.2495 | 0.2600 | 2.50 | 2.70 | 1.30 | 1.90 | 0.072 | 0.125 | 0.66 | 0.37 | 0.22 |
| M2.5 × 0.45 | 0.082 | 3.20 | 3.70 | 0.437 | 0.487 | 0.318 | 0.394 | 0.225 | 0.2820 | 0.2920 | 3.05 | 3.65 | 1.60 | 2.25 | 0.081 | 0.141 | 1.22 | 0.81 | 0.30 |
| M3 × 0.5 | 0.105 | 3.80 | 4.35 | 0.482 | 0.541 | 0.352 | 0.438 | 0.250 | 0.3145 | 0.3250 | 3.60 | 4.30 | 1.95 | 2.80 | 0.090 | 0.156 | 1.33 | 0'.56 | 0.30 |
| M3.5 × 0.6 | 1.160 | 4.40 | 4.95 | 0.586 | 0.650 | 0.449 | 0.525 | 0.300 | 0.3795 | 0.3900 | 4.25 | 4.90 | 2.20 | 3.00 | 0.108 | 0.158 | 1.47 | 0.92 | 0.30 |
| M4 × 0.7 | 0.163 | 5.05 | 5.60 | 0.683 | 0.758 | 0.510 | 0.612 | 0.350 | 0.4445 | 0.4550 | 4.90 | 5.55 | 2.50 | 3.55 | 0.126 | 0.219 | 1.67 | 1.02 | 0.45 |
| M5 × 0.8 | 0.209 | 6.25 | 6.80 | 0.775 | 0.866 | 0.598 | 0.700 | 0.400 | 0.5085 | 0.5200 | 6.10 | 6.75 | 3.15 | 4.55 | 0.144 | 0.250 | 2.09 | 1.41 | 0.60 |
| M6 × 1 | 0.267 | 7.40 | 7.95 | 0.975 | 1.083 | 0.748 | 0.875 | 0.500 | 0.6370 | 0.6500 | 7.25 | 7.90 | 3.70 | 4.85 | 0.180 | 0.312 | 2.55 | 1.65 | 0.60 |
| M7 × 1 | 0.267 | 8.65 | 9.20 | 0.975 | 1.083 | 0.748 | 0.875 | 0.500 | 0.6370 | 0.6500 | 8.40 | 9.15 | 4.30 | 5.50 | 0.180 | 0.312 | 3.10 | 2.09 | 0.75 |
| M8 × 1 | 0.267 | 9.70 | 10.25 | 0.975 | 1.083 | 0.748 | 0.875 | 0.500 | 0.6370 | 0.6500 | 9.20 | 9.65 | 4.75 | 6.50 | 0.180 | 0.312 | 3.58 | 2.27 | 0.75 |
| M8 × 1.25 | 0.415 | 9.80 | 10.35 | 1.251 | 1.353 | 0.967 | 1.094 | 0.625 | 0.7990 | 0.8120 | 9.50 | 9.90 | 4.75 | 6.50 | 0.226 | 0.391 | 3.60 | 2.02 | 0.75 |
| M10 × 1 | 0.267 | 11.95 | 12.50 | 0.975 | 1.083 | 0.748 | 0.875 | 0.500 | 0.6370 | 0.6500 | 11.10 | 11.55 | 5.50 | 8.00 | 0.180 | 0.312 | 4.90 | 2.95 | 0.75 |
| M10 × 1.25 | 0.415 | 12.10 | 12.65 | 1.251 | 1.353 | 0.967 | 1.094 | 0.625 | 0.7990 | 0.8120 | 11.50 | 11.95 | 5.50 | 8.00 | 0.226 | 0.391 | 4.77 | 2.56 | 0.75 |
| M10 × 1.5 | 0.511 | 11.95 | 12.50 | 1.522 | 1.624 | 1.160 | 1.312 | 0.750 | 0.9615 | 0.9740 | 11.80 | 12.25 | 5.50 | 8.00 | 0.271 | 0.469 | 4.54 | 2.56 | 0.75 |
| M12 × 1.25 | 0.415 | 14.30 | 15.00 | 1.251 | 1.353 | 0.967 | 1.094 | 0.625 | 0.7990 | 0.8120 | 13.50 | 14.00 | 6.70 | 9.75 | 0.226 | 0.391 | 5.84 | 3.77 | 1.00 |
| M12 × 1.5 | 0.511 | 14.25 | 14.95 | 1.522 | 1.624 | 1.160 | 1.312 | 0.750 | 0.9615 | 0.9740 | 13.80 | 14.30 | 6.70 | 9.75 | 0.271 | 0.469 | 5.58 | 3.50 | 1.20 |
| M12 × 1.75 | 0.654 | 14.30 | 15.00 | 1.792 | 1.894 | 1.379 | 1.531 | 0.875 | 1.1240 | 1.1370 | 14.10 | 14.60 | 6.70 | 9.75 | 0.316 | 0.547 | 5.36 | 3.23 | 1.40 |
| M14 × 1.5 | 0.511 | 16.55 | 17.25 | 1.522 | 1.624 | 1.160 | 1.312 | 0.750 | 0.9615 | 0.9740 | 15.80 | 16.30 | 7.20 | 11.25 | 0.271 | 0.469 | 6.76 | 4.34 | 1.15 |
| M14 × 2 | 0.799 | 16.65 | 17.35 | 2.063 | 2.165 | 1.598 | 1.750 | 1.000 | 1.2865 | 1.2990 | 16.40 | 16.90 | 7.20 | 11.25 | 0.361 | 0.625 | 6.26 | 3.79 | 1.40 |
| M16 × 1.5 | 0.511 | 18.90 | 19.60 | 1.522 | 1.624 | 1.160 | 1.312 | 0.750 | 0.9615 | 0.9740 | 17.80 | 18.30 | 8.30 | 12.75 | 0.271 | 0.469 | 7.78 | 5.32 | 1.45 |
| M16 × 2 | 0.799 | 18.90 | 19.60 | 2.063 | 2.165 | 1.598 | 1.750 | 1.000 | 1.2865 | 1.2990 | 18.40 | 18.90 | 8.30 | 12.75 | 0.361 | 0.625 | 7.30 | 4.76 | 2.70 |
| M18 × 1.5 | 0.511 | 21.05 | 21.75 | 1.522 | 1.624 | 1.160 | 1.312 | 0.750 | 0.9615 | 0.9740 | 19.80 | 20.35 | 9.30 | 14.00 | 0.271 | 0.469 | 8.83 | 6.26 | 1.75 |
| M18 × 2 | 0.799 | 21.15 | 21.85 | 2.063 | 2.165 | 1.598 | 1.750 | 1.000 | 1.2865 | 1.2990 | 20.40 | 20.95 | 9.30 | 14.00 | 0.361 | 0.625 | 8.30 | 5.74 | 2.70 |

**Table 4.** *(Continued)* **Screw Thread Insert Dimensions** *ANSI/ASME B18.29.2M-2005 (R2017)*

| Nominal Thread Size | *A*, Min. | *B* | | *D* | | *E* | | Gage, *F* | *H* | | *J* | | *P* | | *R*, Min. | *S*, Min. | *U* | | *V*, Max. |
|---|---|---|---|---|---|---|---|---|---|---|---|---|---|---|---|---|---|---|---|
| | | Min. | Max. | Min. | Max. | Min. | Max. | | Min. | Max. | Min. | Max. | Min. | Max. | | | Min. | Max. | |
| M18 × 2.5 | 1.017 | 21.30 | 22.00 | 2.604 | 2.706 | 1.998 | 2.188 | 1.250 | 1.6110 | 1.6240 | 20.90 | 21.45 | 9.30 | 14.00 | 0.451 | 0.781 | 7.79 | 5.20 | 2.85 |
| M20 × 1.5 | 0.511 | 23.15 | 24.00 | 1.522 | 1.624 | 1.160 | 1.312 | 0.750 | 0.9615 | 0.9740 | 21.80 | 22.50 | 10.40 | 14.50 | 0.271 | 0.469 | 9.77 | 7.19 | 2.85 |
| M20 × 2 | 0.799 | 23.20 | 24.05 | 2.063 | 2.165 | 1.598 | 1.750 | 1.000 | 1.2865 | 1.2990 | 22.40 | 23.10 | 10.40 | 14.50 | 0.361 | 0.625 | 9.40 | 6.65 | 2.85 |
| M20 × 2.5 | 1.017 | 23.55 | 24.40 | 2.604 | 2.706 | 1.998 | 2.188 | 1.250 | 1.6110 | 1.6240 | 22.90 | 23.60 | 10.40 | 14.50 | 0.451 | 0.781 | 8.89 | 6.11 | 2.85 |
| M20 × 1.5 | 0.511 | 23.15 | 24.00 | 1.522 | 1.624 | 1.160 | 1.312 | 0.750 | 0.9615 | 0.9740 | 24.10 | 24.80 | 11.40 | 16.00 | 0.271 | 0.469 | 11.10 | 8.01 | 2.85 |
| M22 × 2 | 0.799 | 25.60 | 26.50 | 2.063 | 2.165 | 1.598 | 1.750 | 1.000 | 1.2865 | 1.2990 | 24.40 | 25.10 | 11.40 | 16.00 | 0.361 | 0.625 | 10.45 | 7.61 | 2.85 |
| M22 × 2.5 | 1.017 | 25.90 | 26.90 | 2.604 | 2.706 | 1.998 | 2.188 | 1.250 | 1.6110 | 1.6240 | 24.90 | 25.60 | 11.40 | 16.00 | 0.451 | 0.781 | 9.94 | 7.07 | 2.85 |
| M24 × 2 | 0.799 | 28.10 | 29.10 | 2.063 | 2.165 | 1.598 | 1.750 | 1.000 | 1.2865 | 1.2990 | 26.40 | 27.10 | 12.50 | 16.50 | 0.361 | 0.625 | 11.48 | 8.60 | 2.85 |
| M24 × 3 | 1.234 | 28.00 | 29.00 | 3.146 | 3.248 | 2.396 | 2.625 | 1.500 | 1.9360 | 1.9485 | 27.50 | 28.20 | 12.50 | 16.50 | 0.541 | 0.938 | 10.45 | 7.51 | 2.85 |
| M27 × 2 | 0.799 | 31.30 | 32.30 | 2.063 | 2.165 | 1.598 | 1.750 | 1.000 | 1.2865 | 1.2990 | 29.40 | 30.10 | 14.00 | 17.50 | 0.361 | 0.625 | 13.14 | 9.93 | 2.85 |
| M27 × 3 | 1.234 | 31.40 | 32.40 | 3.146 | 3.248 | 2.396 | 2.625 | 1.500 | 1.9360 | 1.9485 | 30.50 | 31.20 | 14.00 | 17.50 | 0.541 | 0.938 | 12.13 | 8.85 | 2.85 |
| M30 × 2 | 0.799 | 34.50 | 35.70 | 2.063 | 2.165 | 1.598 | 1.750 | 1.000 | 1.2865 | 1.2990 | 32.50 | 33.20 | 15.00 | 19.00 | 0.361 | 0.625 | 14.81 | 11.26 | 2.85 |
| M30 × 3 | 1.234 | 34.90 | 36.10 | 3.146 | 3.248 | 2.396 | 2.625 | 1.500 | 1.9360 | 1.9485 | 33.50 | 34.20 | 15.00 | 19.00 | 0.541 | 0.938 | 13.65 | 10.32 | 2.85 |
| M30 × 3.5 | 1.451 | 34.90 | 36.10 | 3.687 | 3.789 | 2.833 | 3.062 | 1.750 | 2.2605 | 2.2750 | 34.10 | 34.60 | 15.00 | 19.00 | 0.631 | 1.094 | 13.13 | 9.65 | 2.85 |
| M33 × 2 | 0.799 | 37.80 | 39.20 | 2.063 | 2.165 | 1.598 | 1.750 | 1.000 | 1.2865 | 1.2990 | 35.80 | 36.50 | 17.00 | 21.00 | 0.361 | 0.625 | 16.35 | 12.74 | 2.85 |
| M33 × 3 | 1.234 | 38.10 | 39.50 | 3.146 | 3.248 | 2.396 | 2.625 | 1.500 | 1.9360 | 1.9485 | 36.50 | 37.20 | 17.00 | 21.00 | 0.541 | 0.938 | 15.19 | 11.78 | 2.85 |
| M36 × 2 | 0.799 | 41.00 | 42.40 | 2.063 | 2.165 | 1.598 | 1.750 | 1.000 | 1.2865 | 1.2990 | 39.00 | 39.70 | 18.50 | 22.50 | 0.361 | 0.625 | 17.77 | 14.29 | 2.85 |
| M36 × 3 | 1.234 | 41.30 | 42.70 | 3.146 | 3.248 | 2.396 | 2.625 | 1.500 | 1.9360 | 1.9485 | 39.50 | 40.20 | 18.50 | 22.50 | 0.541 | 0.938 | 16.73 | 13.23 | 2.85 |
| M36 × 4 | 1.688 | 41.50 | 42.90 | 4.228 | 4.330 | 3.271 | 3.500 | 2.000 | 2.5855 | 2.5980 | 40.60 | 41.10 | 18.50 | 22.50 | 0.722 | 1.250 | 15.57 | 12.12 | 2.85 |
| M39 × 2 | 0.799 | 44.30 | 45.70 | 2.063 | 2.165 | 1.598 | 1.750 | 1.000 | 1.2865 | 1.2990 | 42.30 | 43.00 | 20.00 | 24.00 | 0.361 | 0.625 | 19.28 | 15.77 | 2.85 |
| M39 × 3 | 1.234 | 44.40 | 45.80 | 3.146 | 3.248 | 2.396 | 2.625 | 1.500 | 1.9360 | 1.9485 | 42.50 | 43.20 | 20.00 | 24.00 | 0.541 | 0.938 | 18.28 | 14.68 | 2.85 |

*Torque Test Method:* The torque test shall consist of a 15-cycle, room temperature test. A new bolt or screw and new tapped hole shall be used for each complete 15-cycle test For each of the 15 cycles, bolts shall be assembled and seated to the assembly torque specified in Table 5. Bolts shall be completely disengaged from the locking coils of the insert at the end of each cycle. The test shall be run at less than 40 rpm to yield a dependable measure of torque and avoid heating of the bolt.

*Maximum Locking Torque:* Maximum locking torque shall be the highest torque value encountered on any installation or removal cycle and shall not exceed the values specified in Table 5. Maximum locking torque readings shall be taken on the first and seventh installation cycles before the assembly torque is applied and on the 15th removal cycle.

**Table 5. Self-Locking Torque** *ANSI/ASME B18.29.2M-2005 (R2017)*

| Nominal Thread Size | Maximum Locking Torque Installation or Removal, N-m | Minimum Breakaway Torque, N-m | Nominal Thread Size | Maximum Locking Torque Installation or Removal, N-m | Minimum Breakaway Torque, N-m |
|---|---|---|---|---|---|
| M2 × 0.4 | 0.12 | 0.03 | M18 × 1.5 | 42 | 5.5 |
| M2.5 × 0.45 | 0.22 | 0.06 | M18 × 2 | 42 | 5.5 |
| M3 × 0.5 | 0.44 | 0.1 | M18 × 2.5 | 42 | 5.5 |
| M3.5 × 0.6 | 0.68 | 0.12 | M20 × 1.5 | 54 | 7 |
| M4 × 0.7 | 0.9 | 0.16 | M20 × 2 | 54 | 7 |
| M5 × 0.8 | 1.6 | 0.3 | M20 × 2.5 | 54 | 7 |
| M6 × 1 | 3 | 0.4 | M22 × 1.5 | 70 | 9 |
| M7 × 1 | 4.4 | 0.6 | M22 × 2 | 70 | 9 |
| M8 × 1 | 6 | 0.8 | M22 × 2.5 | 70 | 9 |
| M8 × 1.25 | 6 | 0.8 | M24 × 2 | 80 | 11 |
| M10 × 1 | 10 | 1.4 | M24 × 3 | 80 | 11 |
| M10 × 1.25 | 10 | 1.4 | M27 × 2 | 95 | 12 |
| M10 × 1.5 | 10 | 1.4 | M27 × 3 | 95 | 12 |
| M12 × 1.25 | 15 | 2.2 | M30 × 2 | 110 | 14 |
| M12 × 1.5 | 15 | 2.2 | M30 × 3 | 110 | 14 |
| M12 × 1.75 | 15 | 2.2 | M30 × 3.5 | 110 | 14 |
| M14 × 1.5 | 23 | 3 | M33 × 2 | 125 | 16 |
| M14 × 2 | 23 | 3 | M33 × 3 | 125 | 16 |
| M16 × 1.5 | 32 | 4.2 | M36 × 2 | 140 | 18 |
| M16 × 2 | 32 | 4.2 | M36 × 3 | 140 | 18 |
| … | … | … | M36 × 4 | 140 | 18 |
| … | … | … | M39 × 2 | 150 | 20 |
| … | … | … | M39 × 3 | 150 | 20 |

*Minimum Breakaway Torque:* Minimum breakaway torque shall be the torque required to overcome static friction when 100 percent of the locking feature is engaged and the bolt or screw is not seated (no axial load). It shall be recorded at the start of the 15th removal cycle. The torque value for any cycle shall be not less than the applicable value shown in Table 5.

*Acceptance:* The inserts shall be considered to have failed if, at the completion of any of the tests and inspection, any of the following conditions exist:

a) any break or crack in the insert

b) installation or removal torque exceeds the maximum locking torque value in Table 5

c) breakaway torque less than the values in Table 5

d) movement of the insert beyond 90° relative to the top surface when installing or removing the test bolt

e) seizure or galling of the insert or test bolt

f) tang not broken off, which interferes with the test bolt at installation

g) tang breaks off during insert installation

**Table 6. Torque Test Spacer Dimensions** *ANSI/ASME B18.29.2M-2005 (R2017)*

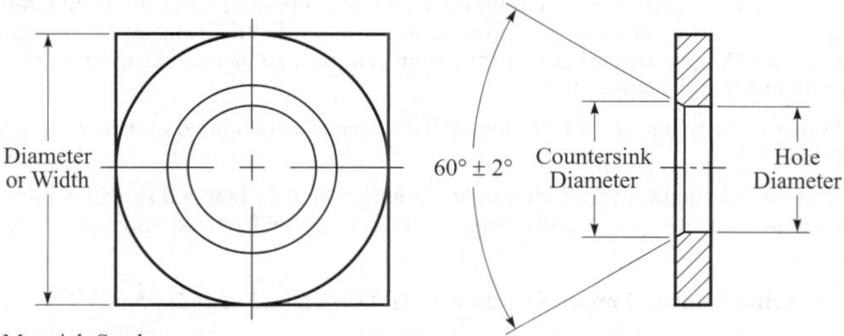


Material: Steel
Hardness: 45-50 RC

Fig. 3. Torque Test Spacer

| Nominal Insert Size | Minimum Diameter or Width | Hole Diameter | | Countersink Diameter | | Minimum Thickness |
|---|---|---|---|---|---|---|
| | | Max. | Min. | Max. | Min. | |
| 2 | 7.0 | 2.3 | 2.1 | 2.7 | 2.5 | 1.5 |
| 2.5 | 8.0 | 2.8 | 2.6 | 3.3 | 3.1 | 1.5 |
| 3 | 9.0 | 3.5 | 3.3 | 3.8 | 3.6 | 2.0 |
| 3.5 | 10.0 | 4.0 | 3.8 | 4.3 | 4.1 | 2.0 |
| 4 | 11.0 | 4.5 | 4.3 | 4.9 | 4.7 | 3.0 |
| 5 | 12.0 | 5.5 | 5.3 | 5.9 | 5.7 | 3.0 |
| 6 | 14.0 | 6.5 | 6.3 | 7.0 | 6.8 | 3.5 |
| 7 | 17.0 | 7.6 | 7.3 | 8.4 | 8.2 | 3.5 |
| 8 | 19.0 | 8.6 | 8.3 | 9.5 | 9.2 | 4.0 |
| 10 | 23.0 | 10.7 | 10.4 | 11.5 | 11.2 | 4.0 |
| 12 | 27.0 | 12.7 | 12.4 | 14.5 | 14.2 | 4.5 |
| 14 | 31.0 | 14.8 | 14.4 | 16.5 | 16.2 | 4.5 |
| 16 | 35.0 | 16.8 | 16.4 | 18.5 | 18.2 | 4.5 |
| 18 | 39.0 | 18.8 | 18.4 | 20.7 | 20.4 | 4.5 |
| 20 | 43.0 | 20.8 | 20.4 | 22.7 | 22.4 | 5.0 |
| 22 | 47.0 | 22.8 | 22.4 | 24.7 | 24.4 | 5.0 |
| 24 | 51.0 | 24.8 | 24.4 | 26.7 | 26.4 | 5.0 |
| 27 | 56.0 | 28.3 | 27.9 | 29.8 | 29.4 | 5.0 |
| 30 | 62.0 | 31.3 | 30.9 | 33.8 | 33.4 | 6.0 |
| 33 | 67.0 | 34.3 | 33.9 | 36.8 | 36.4 | 6.0 |
| 36 | 72.0 | 37.3 | 36.9 | 39.8 | 39.4 | 6.0 |
| 39 | 77.0 | 40.3 | 39.9 | 42.8 | 42.4 | 6.0 |

## Insert Length Selection

**Engaged Length of Bolt.—**Normally, the engaged length of bolt in an insert is determined by strength considerations.

**Material Strengths.—**The standard engineering practice of balancing the tensile strength of the bolt material against the shear strength of the parent or boss material also applies to helical coil inserts. Table 7 and Table 8 will aid in developing the full load value of the bolt rather than stripping the parent or tapped material.

In using this table, the following factors must be considered:

a) The parent material shear strengths are for room temperature. Elevated temperatures call for significant shear value reductions; compensation should be made when required. Shear values are appropriate because the parent material is subject to shearing stress at the major diameter of the tapped threads.

b) When parent material shear strength falls between two tabulated values, use the lower of the two.

c) Bolt thread length; overall length, insert length, and full tapped thread depth must be adequate to ensure full-thread engagement when assembled to comply with its design function.

**Table 7. Insert Length Selection** *ANSI/ASME B18.29.2M-2005 (R2017)*

| Parent Material Shear Strength, MPa | Bolt Property Class | | | | | | |
|---|---|---|---|---|---|---|---|
| | 4.6 | 4.8 | 5.8 | 8.8 | 9.8 | 10.9 | 12.9 |
| | Insert Length in Terms of Diameters | | | | | | |
| 70 | 3 | 3 | 3 | … | … | … | … |
| 100 | 2 | 2 | 2 | 3 | … | … | … |
| 150 | 1.5 | 1.5 | 1.5 | 2 | 2.5 | 2.5 | 3 |
| 200 | 1.5 | 1.5 | 1.5 | 2 | 2 | 2 | 2 |
| 250 | 1 | 1 | 1 | 1.5 | 1.5 | 1.5 | 1.5 |
| 300 | 1 | 1 | 1 | 1.5 | 1.5 | 1.5 | 1.5 |
| 350 | 1 | 1 | 1 | 1 | 1 | 1.5 | 1.5 |

**Table 8. Hardness Number Conversion** *ANSI/ASME B18.29.2M-2005 (R2017)*

| Bolt Property Class | Max. Rockwell Hardness | Max. Tensile Strength, MPa | Bolt Property Class | Max. Rockwell Hardness | Max. Tensile Strength, MPa |
|---|---|---|---|---|---|
| 4.6 | 95 RB | 705 | 9.8 | 36 RC | 1115 |
| 4.8 | 95 RB | 705 | 10.9 | 39 RC | 1215 |
| 5.8 | 95 RB | 705 | 12.9 | 44 RC | 1435 |
| 8.8 | 34 RC | 1055 | | | |

Bolt strength upon which insert length recommendations are based is developed by taking the maximum hardness per ASTM F568M *Carbon and Alloy Steel Externally Threaded Metric Fasteners* and the equivalent tensile strength from SAE J417 *Hardness Tests and Hardness Number Conversions*.

**Screw Thread Insert Taps.—**ANSI/ASME B94.9 covers design and dimensions for taps for producing Metric Series STI-threaded holes required for the installation of helical coil screw thread inserts. Threaded hole dimensions are shown in Table 1 of this Standard. Helical coil screw thread insert taps are identified by the designation STI. Various types and styles of STI taps are available. General dimensions and tolerances are in accordance with ANSI/ASME B94.9.

*Tap Thread Limits:* Ground thread taps are recommended for screw thread inserts. Tap thread limits are in accordance with ANSI/ASME B94.9. Basic pitch diameter used for determining values is the "Pitch Diameter, min." from Table 1.

*Marking:* Taps are marked in accordance with ANSI/ASME B94.9.

*Example:* M6 × 1 STI HS G H2.

## BRITISH FASTENERS

**British Standard Square and Hexagon Bolts, Screws, and Nuts.—**Important dimensions of precision hexagon bolts, screws, and nuts (BSW and BSF threads) as covered by British Standard BS 1083:1965 are given in Table 1 and Table 2. The use of fasteners in this standard will decrease as fasteners having Unified inch and ISO metric threads come into increasing use.

Dimensions of Unified precision hexagon bolts, screws, and nuts (UNC and UNF threads) are given in BS 1768:1963; of Unified black hexagon bolts, screws, and nuts (UNC and UNF threads) in BS 1769:1951; and of Unified black square and hexagon bolts, screws, and nuts (UNC and UNF threads) in BS 2708:1956 (Withdrawn). Unified nominal and basic dimensions in these British Standards are the same as the comparable dimensions in the American Standards, but the tolerances applied to these basic dimensions may differ because of rounding-off practices and other factors. Unified dimensions of square and hexagon bolts and nuts are given in ANSI/ASME B18.2.1-2012 (R2021) and ANSI/ASME B18.2.2-2022; see Table 1 through Table 4 starting on page 1672, and Table 7 to Table 10 starting on page 1677.

ISO metric precision hexagon bolts, screws, and nuts are specified in the British Standard BS 3692:2014 (see *British Standard ISO Metric Precision Hexagon Bolts, Screws, and Nuts* starting on page 1769), and ISO metric black hexagon bolts, screws, and nuts are covered by British Standard BS 4190:2014.

See the section *MACHINE SCREWS AND NUTS* starting on page 1778 for information on British Standard metric, Unified, Whitworth, and BSF machine screws and nuts.

**British Standard Screwed Studs.—**General purpose screwed studs are covered in British Standard BS 2693: Part 1:1956. The aim in this Standard is to provide for a stud having tolerances which would not render it expensive to manufacture and which could be used in association with standard tapped holes for most purposes. Provision has been made for the use of both Unified Fine threads, Unified Coarse threads, British Standard Fine threads, and British Standard Whitworth threads as shown in the table on page 1764.

*Designations:* The *metal end* of the stud is the end which is screwed into the component. The *nut end* is the end of the screw of the stud which is not screwed into the component. The *plain portion* of the stud is the unthreaded length.

*Recommended Fitting Practices for Metal End of Stud:* It is recommended that holes tapped to Class 3B limits (see Table 3, page 1954) in accordance with BS 1580:2007 "Unified Screw Threads" or to Close Class limits in accordance with BS 84 "Parallel screw threads of Whitworth form-Requirements" as appropriate, be used in association with the metal end of the stud specified in this Standard. Where fits are not critical, however, holes may be tapped to Class 2B limits (see table on page 1954) in accordance with BS 1580:2007 or Normal Class limits in accordance with BS 84.

It is recommended that the B.A. stud specified in this Standard be associated with holes tapped to the limits specified for nuts in BS 93, 1919 edition. Where fits for these studs are not critical, holes may be tapped to limits specified for nuts in BS 93:2008.

In general, it will be found that the amount of oversize specified for the studs will produce a satisfactory fit in conjunction with the standard tapping as above. Even when interference is not present, locking will take place on the thread runout which has been carefully controlled for this purpose. Where it is considered essential to assure a true interference fit, higher grade studs should be used. Standard studs should be used even under special conditions where selective assembly may be necessary.

**British Standard Whitworth (BSW) and Fine (BSF) Precision Hexagon Bolts, Screws, and Nuts**

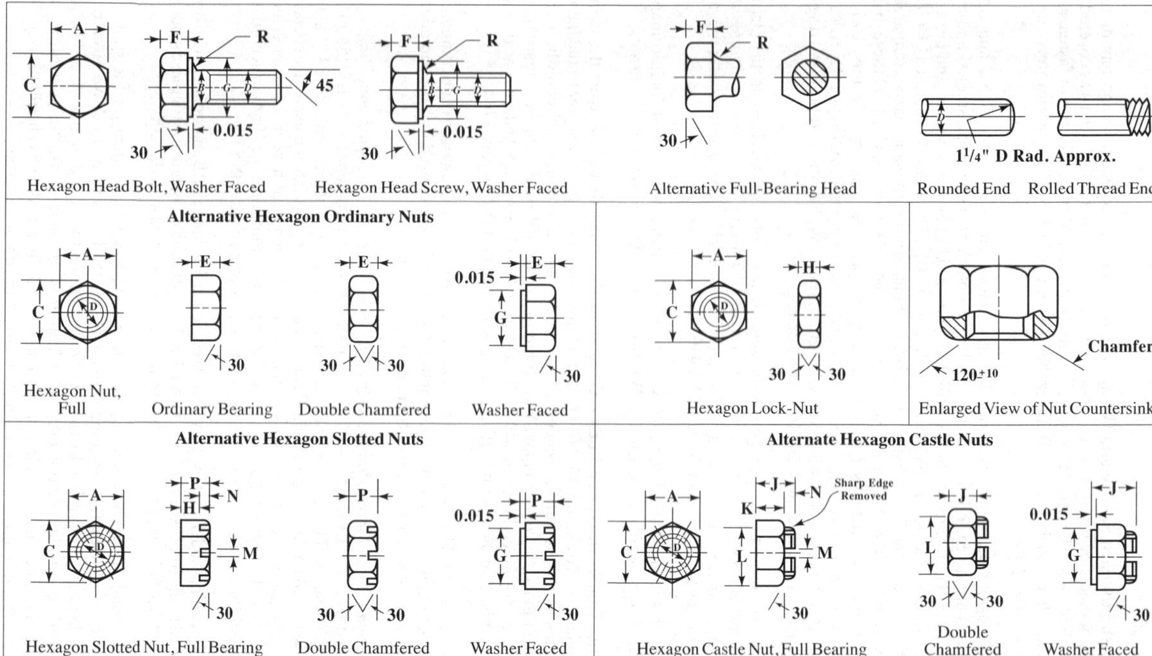


For dimensions, see Table 1 and Table 2.

**Table 1. British Standard Whitworth (BSW) and Fine (BSF) Precision Hexagon Slotted and Castle Nuts** *BS 1083:1965*

| Nominal Size *D* | Number of Threads per Inch | | Bolts, Screws, and Nuts: Width, Across Flats *A* | | Width, Across Corners *C* | Diameter of Washer Face *G* | | Bolts and Screws: Radius Under Head *R* | | Diameter of Unthreaded Portion of Shank *B* | | Thickness, Head *F* | | Nuts: Thickness, Ordinary *E* | | Thickness, Lock *H* | |
|---|---|---|---|---|---|---|---|---|---|---|---|---|---|---|---|---|---|
| | BSW | BSF | Max. | Min.[a] | Max. | Max. | Min. | Max. | Min. | Max. | Min. | Max. | Min. | Max. | Min. | Max. | Min. |
| 1/4 | 20 | 26 | 0.445 | 0.438 | 0.51 | 0.428 | 0.418 | 0.025 | 0.015 | 0.2500 | 0.2465 | 0.176 | 0.166 | 0.200 | 0.190 | 0.185 | 0.180 |
| 5/16 | 18 | 22 | 0.525 | 0.518 | 0.61 | 0.508 | 0.498 | 0.025 | 0.015 | 0.3125 | 0.3090 | 0.218 | 0.208 | 0.250 | 0.240 | 0.210 | 0.200 |
| 3/8 | 16 | 20 | 0.600 | 0.592 | 0.69 | 0.582 | 0.572 | 0.025 | 0.015 | 0.3750 | 0.3715 | 0.260 | 0.250 | 0.312 | 0.302 | 0.260 | 0.250 |
| 7/16 | 14 | 18 | 0.710 | 0.702 | 0.82 | 0.690 | 0.680 | 0.025 | 0.015 | 0.4375 | 0.4335 | 0.302 | 0.292 | 0.375 | 0.365 | 0.275 | 0.265 |
| 1/2 | 12 | 16 | 0.820 | 0.812 | 0.95 | 0.800 | 0.790 | 0.025 | 0.015 | 0.5000 | 0.4960 | 0.343 | 0.333 | 0.437 | 0.427 | 0.300 | 0.290 |
| 9/16 | 12 | 16 | 0.920 | 0.912 | 1.06 | 0.900 | 0.890 | 0.045 | 0.020 | 0.5625 | 0.5585 | 0.375 | 0.365 | 0.500 | 0.490 | 0.333 | 0.323 |
| 5/8 | 11 | 14 | 1.010 | 1.000 | 1.17 | 0.985 | 0.975 | 0.045 | 0.020 | 0.6250 | 0.6190 | 0.417 | 0.407 | 0.562 | 0.552 | 0.375 | 0.365 |
| 3/4 | 10 | 12 | 1.200 | 1.190 | 1.39 | 1.175 | 1.165 | 0.045 | 0.020 | 0.7500 | 0.7440 | 0.500 | 0.480 | 0.687 | 0.677 | 0.458 | 0.448 |
| 7/8 | 9 | 11 | 1.300 | 1.288 | 1.50 | 1.273 | 1.263 | 0.065 | 0.040 | 0.8750 | 0.8670 | 0.583 | 0.563 | 0.750 | 0.740 | 0.500 | 0.490 |
| 1 | 8 | 10 | 1.480 | 1.468 | 1.71 | 1.453 | 1.443 | 0.095 | 0.060 | 1.0000 | 0.9920 | 0.666 | 0.636 | 0.875 | 0.865 | 0.583 | 0.573 |
| 1 1/8 | 7 | 9 | 1.670 | 1.640 | 1.93 | 1.620 | 1.610 | 0.095 | 0.060 | 1.1250 | 1.1170 | 0.750 | 0.710 | 1.000 | 0.990 | 0.666 | 0.656 |
| 1 1/4 | 7 | 9 | 1.860 | 1.815 | 2.15 | 1.795 | 1.785 | 0.095 | 0.060 | 1.2500 | 1.2420 | 0.830 | 0.790 | 1.125 | 1.105 | 0.750 | 0.730 |
| 1 3/8[b] | … | 8 | 2.050 | 2.005 | 2.37 | 1.985 | 1.975 | 0.095 | 0.060 | 1.3750 | 1.3650 | 0.920 | 0.880 | 1.250 | 1.230 | 0.833 | 0.813 |
| 1 1/2 | 6 | 8 | 2.220 | 2.175 | 2.56 | 2.155 | 2.145 | 0.095 | 0.060 | 1.5000 | 1.4900 | 1.000 | 0.960 | 1.375 | 1.355 | 0.916 | 0.896 |
| 1 3/4 | 5 | 7 | 2.580 | 2.520 | 2.98 | 2.495 | 2.485 | 0.095 | 0.060 | 1.7500 | 1.7400 | 1.170 | 1.110 | 1.625 | 1.605 | 1.083 | 1.063 |
| 2 | 4.5 | 7 | 2.760 | 2.700 | 3.19 | 2.675 | 2.665 | 0.095 | 0.060 | 2.0000 | 1.9900 | 1.330 | 1.270 | 1.750 | 1.730 | 1.166 | 1.146 |

[a] When bolts from 1/4 to 1 inch are hot forged, the tolerance on the width across flats shall be two and a half times the tolerance shown in the table and shall be unilaterally minus from maximum size. For dimensional notation, see diagram on page 1762.

[b] Noted standard with BSW thread.

All dimensions in inches except where otherwise noted.

**Table 2. British Standard Whitworth (BSW) and Fine (BSF) Precision Hexagon Slotted and Castle Nuts** *BS 1083:1965*

| Nominal Size D | Number of Threads per Inch | | Slotted Nuts | | | | Castle Nuts | | | | | | Slotted and Castle Nuts | | |
|---|---|---|---|---|---|---|---|---|---|---|---|---|---|---|---|
| | | | Thickness P | | Lower Face to Bottom of Slot H | | Total Thickness J | | Lower Face to Bottom of Slot K | | Castellated Portion Diameter L | | Slots Width M | | Slots Depth N |
| | BSW | BSF | Max. | Min. | Max. | Min. | Max. | Min. | Max. | Min. | Max. | Min. | Max. | Min. | Approx. |
| 1/4 | 20 | 26 | 0.200 | 0.190 | 0.170 | 0.160 | 0.290 | 0.280 | 0.200 | 0.190 | 0.430 | 0.425 | 0.100 | 0.090 | 0.090 |
| 5/16 | 18 | 22 | 0.250 | 0.240 | 0.190 | 0.180 | 0.340 | 0.330 | 0.250 | 0.240 | 0.510 | 0.500 | 0.100 | 0.090 | 0.090 |
| 3/8 | 16 | 20 | 0.312 | 0.302 | 0.222 | 0.212 | 0.402 | 0.392 | 0.312 | 0.302 | 0.585 | 0.575 | 0.100 | 0.090 | 0.090 |
| 7/16 | 14 | 18 | 0.375 | 0.365 | 0.235 | 0.225 | 0.515 | 0.505 | 0.375 | 0.365 | 0.695 | 0.685 | 0.135 | 0.125 | 0.140 |
| 1/2 | 12 | 16 | 0.437 | 0.427 | 0.297 | 0.287 | 0.577 | 0.567 | 0.437 | 0.427 | 0.805 | 0.795 | 0.135 | 0.125 | 0.140 |
| 9/16 | 12 | 16 | 0.500 | 0.490 | 0.313 | 0.303 | 0.687 | 0.677 | 0.500 | 0.490 | 0.905 | 0.895 | 0.175 | 0.165 | 0.187 |
| 5/8 | 11 | 14 | 0.562 | 0.552 | 0.375 | 0.365 | 0.749 | 0.739 | 0.562 | 0.552 | 0.995 | 0.985 | 0.175 | 0.165 | 0.187 |
| 3/4 | 10 | 12 | 0.687 | 0.677 | 0.453 | 0.443 | 0.921 | 0.911 | 0.687 | 0.677 | 1.185 | 1.165 | 0.218 | 0.208 | 0.234 |
| 7/8 | 9 | 11 | 0.750 | 0.740 | 0.516 | 0.506 | 0.984 | 0.974 | 0.750 | 0.740 | 1.285 | 1.265 | 0.218 | 0.208 | 0.234 |
| 1 | 8 | 10 | 0.875 | 0.865 | 0.595 | 0.585 | 1.155 | 1.145 | 0.875 | 0.865 | 1.465 | 1.445 | 0.260 | 0.250 | 0.280 |
| 1 1/8 | 7 | 9 | 1.000 | 0.990 | 0.720 | 0.710 | 1.280 | 1.270 | 1.000 | 0.990 | 1.655 | 1.635 | 0.260 | 0.250 | 0.280 |
| 1 1/4 | 7 | 9 | 1.125 | 1.105 | 0.797 | 0.777 | 1.453 | 1.433 | 1.125 | 1.105 | 1.845 | 1.825 | 0.300 | 0.290 | 0.328 |
| 1 3/8[a] | … | 8 | 1.250 | 1.230 | 0.922 | 0.902 | 1.578 | 1.558 | 1.250 | 1.230 | 2.035 | 2.015 | 0.300 | 0.290 | 0.328 |
| 1 1/2 | 6 | 8 | 1.375 | 1.355 | 1.047 | 1.027 | 1.703 | 1.683 | 1.375 | 1.355 | 2.200 | 2.180 | 0.300 | 0.290 | 0.328 |
| 1 3/4 | 5 | 7 | 1.625 | 1.605 | 1.250 | 1.230 | 2.000 | 1.980 | 1.625 | 1.605 | 2.555 | 2.535 | 0.343 | 0.333 | 0.375 |
| 2 | 4.5 | 7 | 1.750 | 1.730 | 1.282 | 1.262 | 2.218 | 2.198 | 1.750 | 1.730 | 2.735 | 2.715 | 0.426 | 0.416 | 0.468 |

[a] Not standard with BSW thread. For widths across flats, widths across corners, and diameter of washer face see Table 1. For dimensional notation, see diagram on page 1762.

All dimensions in inches except where otherwise noted.

**Table 3. British Standard ISO Metric Precision Hexagon Bolts, Screws, and Nuts**
*BS 3692:2014*

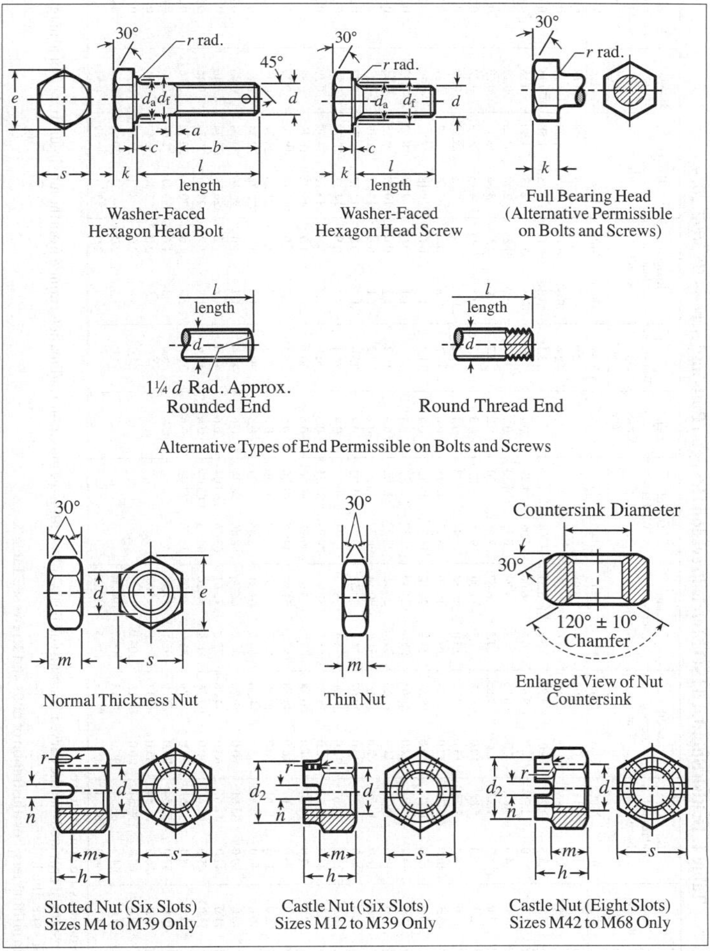

**Table 4. British Standard ISO Metric Precision Hexagon Bolts and Screws** *BS 3692:2014*

| Nom. Size and Thread Dia.[a] *d* | Pitch of Thread (Coarse Pitch Series) | Thread Runout *a* | Dia. of Unthreaded Shank *d* | | Width Across Flats *s* | | Width Across Corners *e* | | Dia. of Washer Face $d_t$ | | Depth of Washer Face *c* | Transition Dia.[b] $d_a$ | Radius Under Head[b] *r* | | Height of Head *k* | | Eccentricity of Head | Eccentricity of Shank and Split Pin Hole to the Thread |
|---|---|---|---|---|---|---|---|---|---|---|---|---|---|---|---|---|---|---|
| | | Max. | Max. | Min. | Max. | Min. | Max. | Min. | Max. | Min. | | Max. | Max. | Min. | Max. | Min. | Max. | Max. |
| M1.6 | 0.35 | 0.8 | 1.6 | 1.46 | 3.2 | 3.08 | 3.7 | 3.48 | … | … | … | 2.0 | 0.2 | 0.1 | 1.23 | 0.98 | 0.18 | 0.14 |
| M2 | 0.4 | 1.0 | 2.0 | 1.86 | 4.0 | 3.88 | 4.6 | 4.38 | … | … | … | 2.6 | 0.3 | 0.1 | 1.53 | 1.28 | 0.18 | 0.14 |
| M2.5 | 0.45 | 1.0 | 2.5 | 2.23 | 5.0 | 4.88 | 5.8 | 5.51 | … | … | … | 3.1 | 0.3 | 0.1 | 1.83 | 1.58 | 0.18 | 0.14 |
| M3 | 0.5 | 1.2 | 3.0 | 2.86 | 5.5 | 5.38 | 6.4 | 6.08 | 5.08 | 4.83 | 0.1 | 3.6 | 0.3 | 0.1 | 2.13 | 1.90 | 0.18 | 0.14 |
| M4 | 0.7 | 1.6 | 4.0 | 3.82 | 7.0 | 6.85 | 8.1 | 7.74 | 6.55 | 6.30 | 0.1 | 4.7 | 0.35 | 0.2 | 2.93 | 2.68 | 0.22 | 0.18 |
| M5 | 0.8 | 2.0 | 5.0 | 4.82 | 8.0 | 7.85 | 9.2 | 8.87 | 7.55 | 7.30 | 0.2 | 5.7 | 0.35 | 0.2 | 3.66 | 3.35 | 0.22 | 0.18 |
| M6 | 1 | 2.5 | 6.0 | 5.82 | 10.0 | 9.78 | 11.5 | 11.05 | 9.48 | 9.23 | 0.3 | 6.8 | 0.4 | 0.25 | 4.15 | 3.85 | 0.22 | 0.18 |
| M8 | 1.25 | 3.0 | 8.0 | 7.78 | 13.0 | 12.73 | 15.0 | 14.38 | 12.43 | 12.18 | 0.4 | 9.2 | 0.6 | 0.4 | 5.65 | 5.35 | 0.27 | 0.22 |
| M10 | 1.5 | 3.5 | 10.0 | 9.78 | 17.0 | 16.73 | 19.6 | 18.90 | 16.43 | 16.18 | 0.4 | 11.2 | 0.6 | 0.4 | 7.18 | 6.82 | 0.27 | 0.22 |
| M12 | 1.75 | 4.0 | 12.0 | 11.73 | 19.0 | 18.67 | 21.9 | 21.10 | 18.37 | 18.12 | 0.4 | 14.2 | 1.1 | 0.6 | 8.18 | 7.82 | 0.33 | 0.27 |
| (M14) | 2 | 5.0 | 14.0 | 13.73 | 22.0 | 21.67 | 25.4 | 24.49 | 21.37 | 21.12 | 0.4 | 16.2 | 1.1 | 0.6 | 9.18 | 8.82 | 0.33 | 0.27 |
| M16 | 2 | 5.0 | 16.0 | 15.73 | 24.0 | 23.67 | 27.7 | 26.75 | 23.27 | 23.02 | 0.4 | 18.2 | 1.1 | 0.6 | 10.18 | 9.82 | 0.33 | 0.27 |
| (M18) | 2.5 | 6.0 | 18.0 | 17.73 | 27.0 | 26.67 | 31.2 | 30.14 | 26.27 | 26.02 | 0.4 | 20.2 | 1.1 | 0.6 | 12.22 | 11.79 | 0.33 | 0.27 |
| M20 | 2.5 | 6.0 | 20.0 | 19.67 | 30.0 | 29.67 | 34.6 | 33.53 | 29.27 | 28.80 | 0.4 | 22.4 | 1.2 | 0.8 | 13.22 | 12.79 | 0.33 | 0.33 |
| (M22) | 2.5 | 6.0 | 22.0 | 21.67 | 32.0 | 31.61 | 36.9 | 35.72 | 31.21 | 30.74 | 0.4 | 24.4 | 1.2 | 0.8 | 14.22 | 13.79 | 0.39 | 0.33 |
| M24 | 3 | 7.0 | 24.0 | 23.67 | 36.0 | 35.38 | 41.6 | 39.98 | 34.98 | 34.51 | 0.5 | 26.4 | 1.2 | 0.8 | 15.22 | 14.79 | 0.39 | 0.33 |
| (M27) | 3 | 7.0 | 27.0 | 26.67 | 41.0 | 40.38 | 47.3 | 45.63 | 39.98 | 39.36 | 0.5 | 30.4 | 1.7 | 1.0 | 17.22 | 16.79 | 0.39 | 0.33 |
| M30 | 3.5 | 8.0 | 30.0 | 29.67 | 46.0 | 45.38 | 53.1 | 51.28 | 44.98 | 44.36 | 0.5 | 33.4 | 1.7 | 1.0 | 19.26 | 18.74 | 0.39 | 0.33 |
| (M33) | 3.5 | 8.0 | 33.0 | 32.61 | 50.0 | 49.38 | 57.7 | 55.80 | 48.98 | 48.36 | 0.5 | 36.4 | 1.7 | 1.0 | 21.26 | 20.74 | 0.39 | 0.39 |
| M36 | 4 | 10.0 | 36.0 | 35.61 | 55.0 | 54.26 | 63.5 | 61.31 | 53.86 | 53.24 | 0.5 | 39.4 | 1.7 | 1.0 | 23.26 | 22.74 | 0.46 | 0.39 |
| (M39) | 4 | 10.0 | 39.0 | 38.61 | 60.0 | 59.26 | 69.3 | 66.96 | 58.86 | 58.24 | 0.6 | 42.4 | 1.7 | 1.0 | 25.26 | 24.74 | 0.46 | 0.39 |
| M42 | 4.5 | 11.0 | 42.0 | 41.61 | 65.0 | 64.26 | 75.1 | 72.61 | 63.76 | 63.04 | 0.6 | 45.6 | 1.8 | 1.2 | 26.26 | 25.74 | 0.46 | 0.39 |
| (M45) | 4.5 | 11.0 | 45.0 | 44.61 | 70.0 | 69.26 | 80.8 | 78.26 | 68.76 | 68.04 | 0.6 | 48.6 | 1.8 | 1.2 | 28.26 | 27.74 | 0.46 | 0.39 |
| M48 | 5 | 12.0 | 48.0 | 47.61 | 75.0 | 74.26 | 86.6 | 83.91 | 73.76 | 73.04 | 0.6 | 52.6 | 2.3 | 1.6 | 30.26 | 29.74 | 0.46 | 0.39 |
| (M52) | 5 | 12.0 | 52.0 | 51.54 | 80.0 | 79.26 | 92.4 | 89.56 | … | … | … | 56.6 | 2.3 | 1.6 | 33.21 | 32.69 | 0.46 | 0.46 |
| M56 | 5.5 | 19.0 | 56.0 | 55.54 | 85.0 | 84.13 | 98.1 | 95.07 | … | … | … | 63.0 | 3.5 | 2.0 | 35.31 | 34.69 | 0.54 | 0.46 |
| (M60) | 5.5 | 19.0 | 60.0 | 59.54 | 90.0 | 89.13 | 103.9 | 100.72 | … | … | … | 67.0 | 3.5 | 2.0 | 38.31 | 37.69 | 0.54 | 0.46 |
| M64 | 6 | 21.0 | 64.0 | 63.54 | 95.0 | 94.13 | 109.7 | 106.37 | … | … | … | 71.0 | 3.5 | 2.0 | 40.31 | 39.69 | 0.54 | 0.46 |
| (M68) | 6 | 21.0 | 68.0 | 67.54 | 100.0 | 99.13 | 115.5 | 112.02 | … | … | … | 75.0 | 3.5 | 2.0 | 43.31 | 42.69 | 0.54 | 0.46 |

[a] Sizes shown in parentheses are non-preferred.

[b] A true radius is not essential provided that the curve is smooth and lies wholly within the maximum radius, determined from the maximum transitional diameter, and the minimum radius specified.

All dimensions are in millimeters. For illustration of bolts and screws see Table 3.

**Table 5. British Standard ISO Metric Precision Hexagon Nuts and Thin Nuts** *BS 3692:2014*

| Nominal Size and Thread Diameter[a] *d* | Pitch of Thread (Coarse Pitch Series) | Width Across Flats *s* | | Width Across Corners *e* | | Thickness of Normal Nut *m* | | Tolerance on Squareness of Thread to Face of Nut[b] | Eccentricity of Hexagon | Thickness of Thin Nut *t* | |
|---|---|---|---|---|---|---|---|---|---|---|---|
| | | Max. | Min. | Max. | Min. | Max. | Min. | Max. | Max. | Max. | Min. |
| M1.6 | 0.35 | 3.20 | 3.08 | 3.70 | 3.48 | 1.30 | 1.05 | 0.05 | 0.14 | … | … |
| M2 | 0.4 | 4.00 | 3.88 | 4.60 | 4.38 | 1.60 | 1.35 | 0.06 | 0.14 | … | … |
| M2.5 | 0.45 | 5.00 | 4.88 | 5.80 | 5.51 | 2.00 | 1.75 | 0.08 | 0.14 | … | … |
| M3 | 0.5 | 5.50 | 5.38 | 6.40 | 6.08 | 2.40 | 2.15 | 0.09 | 0.14 | … | … |
| M4 | 0.7 | 7.00 | 6.85 | 8.10 | 7.74 | 3.20 | 2.90 | 0.11 | 0.18 | … | … |
| M5 | 0.8 | 8.00 | 7.85 | 9.20 | 8.87 | 4.00 | 3.70 | 0.13 | 0.18 | … | … |
| M6 | 1 | 10.00 | 9.78 | 11.50 | 11.05 | 5.00 | 4.70 | 0.17 | 0.18 | … | … |
| M8 | 1.25 | 13.00 | 12.73 | 15.00 | 14.38 | 6.50 | 6.14 | 0.22 | 0.22 | 5.0 | 4.70 |
| M10 | 1.5 | 17.00 | 16.73 | 19.60 | 18.90 | 8.00 | 7.64 | 0.29 | 0.22 | 6.0 | 5.70 |
| M12 | 1.75 | 19.00 | 18.67 | 21.90 | 21.10 | 10.00 | 9.64 | 0.32 | 0.27 | 7.0 | 6.64 |
| (M14) | 2 | 22.00 | 21.67 | 25.4 | 24.49 | 11.00 | 10.57 | 0.37 | 0.27 | 8.0 | 7.64 |
| M16 | 2 | 24.00 | 23.67 | 27.7 | 6.75 | 13.00 | 12.57 | 0.41 | 0.27 | 8.0 | 7.64 |
| (M18) | 2.5 | 27.00 | 26.67 | 31.20 | 30.14 | 15.00 | 14.57 | 0.46 | 0.27 | 9.0 | 8.64 |
| M20 | 2.5 | 30.00 | 29.67 | 34.60 | 33.53 | 16.00 | 15.57 | 0.51 | 0.33 | 9.0 | 8.64 |
| (M22) | 2.5 | 32.00 | 31.61 | 36.90 | 35.72 | 18.00 | 17.57 | 0.54 | 0.33 | 10.0 | 9.64 |
| M24 | 3 | 36.00 | 35.38 | 41.60 | 39.98 | 19.00 | 18.48 | 0.61 | 0.33 | 10.0 | 9.64 |
| (M27) | 3 | 41.00 | 40.38 | 47.3 | 45.63 | 22.00 | 21.48 | 0.70 | 0.33 | 12.0 | 11.57 |
| M30 | 3.5 | 46.00 | 45.38 | 53.1 | 51.28 | 24.00 | 23.48 | 0.78 | 0.33 | 12.0 | 11.57 |
| (M33) | 3.5 | 50.00 | 49.38 | 57.70 | 55.80 | 26.00 | 25.48 | 0.85 | 0.39 | 14.0 | 13.57 |
| M36 | 4 | 55.00 | 54.26 | 63.50 | 61.31 | 29.00 | 28.48 | 0.94 | 0.39 | 14.0 | 13.57 |
| (M39) | 4 | 60.00 | 59.26 | 69.30 | 66.96 | 31.00 | 30.38 | 1.03 | 0.39 | 16.0 | 15.57 |
| M42 | 4.5 | 65.00 | 64.26 | 75.10 | 72.61 | 34.00 | 33.38 | 1.11 | 0.39 | 16.0 | 15.57 |
| (M45) | 4.5 | 70.00 | 69.26 | 80.80 | 78.26 | 36.00 | 35.38 | 1.20 | 0.39 | 18.0 | 17.57 |
| M48 | 5 | 75.00 | 74.26 | 86.60 | 83.91 | 38.00 | 37.38 | 1.29 | 0.39 | 18.0 | 17.57 |
| (M52) | 5 | 80.00 | 79.26 | 92.40 | 89.56 | 42.00 | 41.38 | 1.37 | 0.46 | 20.0 | 19.48 |
| M56 | 5.5 | 85.00 | 84.13 | 98.10 | 95.07 | 45.00 | 44.38 | 1.46 | 0.46 | … | … |
| (M60) | 5.5 | 90.00 | 89.13 | 103.90 | 100.72 | 48.00 | 47.38 | 1.55 | 0.46 | … | … |
| M64 | 6 | 95.00 | 94.13 | 109.70 | 106.37 | 51.00 | 50.26 | 1.63 | 0.46 | … | … |
| (M68) | 6 | 100.00 | 99.13 | 115.50 | 112.02 | 54.00 | 53.26 | 1.72 | 0.46 | … | … |

[a] Sizes shown in parentheses are non-preferred.
[b] As measured with the nut squareness gage described in the text and illustrated in Appendix A of the Standard and a feeler gage.
All dimensions are in millimeters. For illustration of hexagon nuts and thin nuts see Table 3.

**Table 6. British Standard ISO Metric Precision Hexagon Slotted Nuts and Castle Nuts** *BS 3692:2014*

| Nominal Size and Thread Diameter[a] $d$ | Width Across Flats $s$ | | Width Across Corners $e$ | | Diameter $d_2$ | | Thickness $h$ | | Lower Face of Nut to Bottom of Slot $m$ | | Width of Slot $n$ | | Radius (0.25 $n$) $r$ | Eccentricity of the Slots |
|---|---|---|---|---|---|---|---|---|---|---|---|---|---|---|
| | Max. | Min. | Max. | Min. | Max. | Min. | Max. | Min. | Max. | Min. | Max. | Min. | Min. | Max. |
| M4 | 7.00 | 6.85 | 8.10 | 7.74 | … | … | 5 | 4.70 | 3.2 | 2.90 | 1.45 | 1.2 | 0.3 | 0.18 |
| M5 | 8.00 | 7.85 | 9.20 | 8.87 | … | … | 6 | 5.70 | 4.0 | 3.70 | 1.65 | 1.4 | 0.35 | 0.18 |
| M6 | 10.00 | 9.78 | 11.50 | 11.05 | … | … | 7.5 | 7.14 | 5 | 4.70 | 2.25 | 2 | 0.5 | 0.18 |
| M8 | 13.00 | 12.73 | 15.00 | 14.38 | … | … | 9.5 | 9.14 | 6.5 | 6.14 | 2.75 | 2.5 | 0.625 | 0.22 |
| M10 | 17.00 | 16.73 | 19.60 | 18.90 | … | … | 12 | 11.57 | 8 | 7.64 | 3.05 | 2.8 | 0.70 | 0.22 |
| M12 | 19.00 | 18.67 | 21.90 | 21.10 | 17 | 16.57 | 15 | 14.57 | 10 | 9.64 | 3.80 | 3.5 | 0.875 | 0.27 |
| (M14) | 22.00 | 21.67 | 25.4 | 24.49 | 19 | 18.48 | 16 | 15.57 | 11 | 10.57 | 3.80 | 3.5 | 0.875 | 0.27 |
| M16 | 24.00 | 23.67 | 27.7 | 26.75 | 22 | 21.48 | 19 | 18.48 | 13 | 12.57 | 4.80 | 4.5 | 1.125 | 0.27 |
| (M18) | 27.00 | 26.67 | 31.20 | 30.14 | 25 | 24.48 | 21 | 20.48 | 15 | 14.57 | 4.80 | 4.5 | 1.125 | 0.27 |
| M20 | 30.00 | 29.67 | 34.60 | 33.53 | 28 | 27.48 | 22 | 21.48 | 16 | 15.57 | 4.80 | 4.5 | 1.125 | 0.33 |
| (M22) | 32.00 | 31.61 | 36.90 | 35.72 | 30 | 29.48 | 26 | 25.48 | 18 | 17.57 | 5.80 | 5.5 | 1.375 | 0.33 |
| M24 | 36.00 | 35.38 | 41.60 | 39.98 | 34 | 33.38 | 27 | 26.48 | 19 | 18.48 | 5.80 | 5.5 | 1.375 | 0.33 |
| (M27) | 41.00 | 40.38 | 47.3 | 45.63 | 38 | 37.38 | 30 | 29.48 | 22 | 21.48 | 5.80 | 5.5 | 1.375 | 0.33 |
| M30 | 46.00 | 45.38 | 53.1 | 51.28 | 42 | 41.38 | 33 | 32.38 | 24 | 23.48 | 7.36 | 7 | 1.75 | 0.33 |
| (M33) | 50.00 | 49.38 | 57.70 | 55.80 | 46 | 45.38 | 35 | 34.38 | 26 | 25.48 | 7.36 | 7 | 1.75 | 0.39 |
| M36 | 55.00 | 54.26 | 63.50 | 61.31 | 50 | 49.38 | 38 | 37.38 | 29 | 28.48 | 7.36 | 7 | 1.75 | 0.39 |
| (M39) | 60.00 | 59.26 | 69.30 | 66.96 | 55 | 54.26 | 40 | 39.38 | 31 | 30.38 | 7.36 | 7 | 1.75 | 0.39 |
| M42 | 65.00 | 64.26 | 75.10 | 72.61 | 58 | 57.26 | 46 | 45.38 | 34 | 33.38 | 9.36 | 9 | 2.25 | 0.39 |
| (M45) | 70.00 | 69.26 | 80.80 | 78.26 | 62 | 61.26 | 48 | 47.38 | 36 | 35.38 | 9.36 | 9 | 2.25 | 0.39 |
| M48 | 75.00 | 74.26 | 86.60 | 83.91 | 65 | 64.26 | 50 | 49.38 | 38 | 37.38 | 9.36 | 9 | 2.25 | 0.39 |
| (M52) | 80.00 | 79.26 | 92.40 | 89.56 | 70 | 69.26 | 54 | 53.26 | 42 | 41.38 | 9.36 | 9 | 2.25 | 0.46 |
| M56 | 85.00 | 84.13 | 98.10 | 95.07 | 75 | 74.26 | 57 | 56.26 | 45 | 44.38 | 9.36 | 9 | 2.25 | 0.46 |
| (M60) | 90.00 | 89.13 | 103.90 | 100.72 | 80 | 79.26 | 63 | 62.26 | 48 | 47.38 | 11.43 | 11 | 2.75 | 0.46 |
| M64 | 95.00 | 94.13 | 109.70 | 106.37 | 85 | 84.13 | 66 | 65.26 | 51 | 50.26 | 11.43 | 11 | 2.75 | 0.46 |
| (M68) | 100.00 | 99.13 | 115.50 | 112.02 | 90 | 89.13 | 69 | 68.26 | 54 | 53.26 | 11.43 | 11 | 2.75 | 0.46 |

[a] Sizes shown in parentheses are non-preferred.

All dimensions are in millimeters. For illustration of hexagon slotted nuts and castle nuts see Table 3.

After several years of use of BS 2693:Part 1:1956, it was recognized that it would not meet the requirements of all stud users. The thread tolerances specified could result in clearance of interference fits because locking depended on the run-out threads. Thus, some users felt that true interference fits were essential for their needs. As a result, the British Standards Committee has incorporated the Class 5 interference fit threads specified in American Standard ASA B1.12 into the BS 2693:Part 2:1964 (Withdrawn), "Recommendations for High Grade Studs."

**British Standard ISO Metric Precision Hexagon Bolts, Screws, and Nuts.—**This British Standard BS 3692:2014 gives the general dimensions and tolerances of precision hexagon bolts, screws, and nuts with ISO metric threads in diameters from 1.6 to 68 mm. It is based on the following ISO recommendations and draft recommendations: R 272, R 288, DR 911, DR 947, DR 950, DR 952 and DR 987. Mechanical properties are given only with respect to carbon or alloy steel bolts, screws, and nuts, which are not to be used for special applications such as those requiring weldability, corrosion resistance or ability to withstand temperatures above 300° C or below –50° C. The dimensional requirements of this Standard also apply to nonferrous and stainless steel bolts, screws, and nuts.

*Finish:* Finishes may be dull black which results from the heat-treating operation or may be bright finish, the result of bright drawing. Other finishes are possible by mutual agreement between purchaser and producer. It is recommended that reference be made to BS 3382 (Withdrawn) "Electroplated Coatings on Threaded Components" in this respect.

*General Dimensions:* The bolts, screws, and nuts conform to the general dimensions given in Table 3, Table 4, Table 5 and Table 6.

*Nominal Lengths of Bolts and Screws:* The nominal length of a bolt or screw is the distance from the underside of the head to the extreme end of the shank including any chamfer or radius. Standard nominal lengths and tolerances thereon are given in Table 7.

**Table 7. British Standard ISO Metric Bolt and Screw Nominal Lengths**
*BS 3692:2014*

| Nominal Length[a] $l$ | Tolerance | Nominal Length[a] $l$ | Tolerance | Nominal Length[a] $l$ | Tolerance | Nominal Length[a] $l$ | Tolerance |
|---|---|---|---|---|---|---|---|
| 5 | ±0.24 | 30 | ±0.42 | 90 | ±0.70 | 200 | ±0.925 |
| 6 | ±0.24 | (32) | ±0.50 | (95) | ±0.70 | 220 | ±0.925 |
| (7) | ±0.29 | 35 | ±0.50 | 100 | ±0.70 | 240 | ±0.925 |
| 8 | ±0.29 | (38) | ±0.50 | (105) | ±0.70 | 260 | ±1.05 |
| (9) | ±0.29 | 40 | ±0.50 | 110 | ±0.70 | 280 | ±1.05 |
| 10 | ±0.29 | 45 | ±0.50 | (115) | ±0.70 | 300 | ±1.05 |
| (11) | ±0.35 | 50 | ±0.50 | 120 | ±0.70 | 325 | ±1.15 |
| 12 | ±0.35 | 55 | ±0.60 | (125) | ±0.80 | 350 | ±1.15 |
| 14 | ±0.35 | 60 | ±0.60 | 130 | ±0.80 | 375 | ±1.15 |
| 16 | ±0.35 | 65 | ±0.60 | 140 | ±0.80 | 400 | ±1.15 |
| (18) | ±0.35 | 70 | ±0.60 | 150 | ±0.80 | 425 | ±1.25 |
| 20 | ±0.42 | 75 | ±0.60 | 160 | ±0.80 | 450 | ±1.25 |
| (22) | ±0.42 | 80 | ±0.60 | 170 | ±0.80 | 475 | ±1.25 |
| 25 | ±0.42 | 85 | ±0.70 | 180 | ±0.80 | 500 | ±1.25 |
| (28) | ±0.42 | … | … | 190 | ±0.925 | … | … |

[a] Nominal lengths shown in parentheses are non-preferred.
All dimensions are in millimeters.

*Bolt and Screw Ends:* The ends of bolts and screws may be finished with either a 45-degree chamfer to a depth slightly exceeding the depth of thread or a radius approximately equal to $1\frac{1}{4}$ times the nominal diameter of the shank. With rolled threads, the lead formed at the end of the bolt by the thread rolling operation may be regarded as providing the necesssary chamfer to the end; the end being reasonably square with the center line of the shank.

*Screw Thread Form:* The form of thread and diameters and associated pitches of standard ISO metric bolts, screws, and nuts are in accordance with BS 3643: Part 1:2007, "Principles and Basic Data." The screw threads are made to the tolerances for the medium class of fit (*6H*/*6g*) as specified in BS 3643: Part 2:2007, "Specification for Selected Limits of Size."

*Length of Thread on Bolts:* The length of thread on bolts is the distance from the end of the bolt (including any chamfer or radius) to the leading face of a screw ring gage which has been screwed as far as possible onto the bolt by hand. Standard thread lengths of bolts are $2d + 6$ mm for a nominal length of bolt up to and including 125 mm, $2d + 12$ mm for a nominal bolt length over 125 mm up to and including 200 mm, and $2d + 25$ mm for a nominal bolt length over 200 mm. Bolts that are too short for minimum thread lengths are threaded as screws and designated as screws. The tolerance on bolt thread lengths are plus two pitches for all diameters.

*Length of Thread on Screws:* Screws are threaded to permit a screw ring gage being screwed by hand to within a distance from the underside of the head not exceeding two and a half times the pitch for diameters up to and including 52 mm and three and a half times the pitch for diameters over 52 mm.

*Angularity and Eccentricity of Bolts, Screws, and Nuts:* The axis of the thread of the nut is square to the face of the nut subject to the "squareness tolerance" given in Table 5.

In gaging, the nut is screwed by hand onto a gage, having a truncated taper thread, until the thread of the nut is tight on the thread of the gage. A sleeve sliding on a parallel extension of the gage, which has a face of diameter equal to the minimum distance across the flats of the nut and exactly at 90 degrees to the axis of the gage, is brought into contact with the leading face of the nut. With the sleeve in this position, it should not be possible for a feeler gage of thickness equal to the "squareness tolerance" to enter anywhere between the leading nut face and sleeve face.

The hexagon flats of bolts, screws, and nuts are square to the bearing face, and the angularity of the head is within the limits of 90 degrees, plus or minus 1 degree. The eccentricity of the hexagon flats of nuts relative to the thread diameter should not exceed the values given in Table 5 and the eccentricity of the head relative to the width across flats and eccentricity between the shank and thread of bolts and screws should not exceed the values given in Table 4.

*Chamfering, Washer Facing and Countersinking:* Bolt and screw heads have a chamfer of approximately 30 degrees on their upper faces and, at the option of the manufacturer, a washer face or full bearing face on the underside. Nuts are countersunk at an included angle of 120 degrees plus or minus 10 degrees at both ends of the thread. The diameter of the countersink should not exceed the nominal major diameter of the thread plus 0.13 mm up to and including 12 mm diameter, and plus 0.25 mm above 12 mm diameter. This stipulation does not apply to slotted, castle or thin nuts.

*Strength Grade Designation System for Steel Bolts and Screws:* This Standard includes a strength grade designation system consisting of two figures. The first figure is one tenth of the minimum tensile strength in $kgf/mm^2$, and the second figure is one tenth of the ratio between the minimum yield stress (or stress at permanent set limit, $R_{0.2}$) and the minimum tensile strength, expressed as a percentage. For example with the strength designation grade 8.8, the first figure 8 represents $\frac{1}{10}$ the minimum tensile strength of 80 $kgf/mm^2$ and the second figure 8 represents $\frac{1}{10}$ the ratio

$$\frac{\text{stress at permanent set limit } R_{0.2}\%}{\text{minimum tensile strength}} = \frac{1}{10} \times \frac{64}{80} \times \frac{100}{1}$$

the numerical values of stress and strength being obtained from the accompanying table.

**Strength Grade Designations of Steel Bolts and Screws**

| Strength Grade Designation | 4.6 | 4.8 | 5.6 | 5.8 | 6.6 | 6.8 | 8.8 | 10.9 | 12.9 | 14.9 |
|---|---|---|---|---|---|---|---|---|---|---|
| Tensile Strength ($R_m$), Min. | 40 | 40 | 50 | 50 | 60 | 60 | 80 | 100 | 120 | 140 |
| Yield Stress ($R_e$), Min. | 24 | 32 | 30 | 40 | 36 | 48 | … | … | … | … |
| Stress at Permanent Set Limit ($R_{0.2}$), Min. | … | … | … | … | … | … | 64 | 90 | 108 | 126 |
| All stress and strength values are in kgf/mm$^2$ units. | | | | | | | | | | |

*Strength Grade Designation System for Steel Nuts:* The strength grade designation system for steel nuts is a number which is one-tenth of the specified proof load stress in kgf/mm$^2$. The proof load stress corresponds to the minimum tensile strength of the highest grade of bolt or screw with which the nut can be used.

**Strength Grade Designations of Steel Nuts**

| Strength Grade Designation | 4 | 5 | 6 | 8 | 12 | 14 |
|---|---|---|---|---|---|---|
| Proof Load Stress (kgf/mm$^2$) | 40 | 50 | 60 | 80 | 120 | 140 |

**Recommended Bolt and Nut Combinations**

| Grade of Bolt | 4.6 | 4.8 | 5.6 | 5.8 | 6.6 | 6.8 | 8.8 | 10.9 | 12.9 | 14.9 |
|---|---|---|---|---|---|---|---|---|---|---|
| Recommended Grade of Nut | 4 | 4 | 5 | 5 | 6 | 6 | 8 | 12 | 12 | 14 |
| *Note:* Nuts of a higher strength grade may be substituted for nuts of a lower strength grade. | | | | | | | | | | |

*Marking:* The marking and identification requirements of this Standard are only mandatory for steel bolts, screws, and nuts of 6 mm diameter and larger; manufactured to strength grade designations 8.8 (for bolts or screws) and 8 (for nuts) or higher. Bolts and screws are identified as ISO metric by either of the symbols "ISO M" or "M", embossed or indented on top of the head. Nuts may be indented or embossed by alternative methods depending on their method of manufacture.

*Designation:* Bolts 10 mm diameter, 50 mm long manufactured from steel of strength grade 8.8, would be designated:

"Bolts M10 × 50 to BS 3692.—8.8."

Brass screws 8 mm diameter, 20 mm long would be designated:

"Brass screws M8 × 20 to BS 3692."

Nuts 12 mm diameter, manufactured from steel of strength grade 6, cadmium plated could be designated:

"Nuts M12 to BS 3692.—6, plated to BS 3382: Part 1 (Withdrawn)."

*Miscellaneous Information:* The Standard also gives mechanical properties of steel bolts, screws, and nuts [i.e., tensile strengths; hardnesses (Brinell, Rockwell, Vickers); stresses (yield, proof load); etc.], material and manufacture of steel bolts, screws, and nuts; and information on inspection and testing. Appendices to the Standard give information on gaging; chemical composition; testing of mechanical properties; examples of marking of bolts, screws, and nuts; and a table of preferred standard sizes of bolts and screws, to name some.

## British Standard General Purpose Studs *BS 2693:Part 1:1956*

| Limits for End Screwed into Component (All threads except BA) | | | | | | | | | | | | | |
|---|---|---|---|---|---|---|---|---|---|---|---|---|---|
| Nom. Dia. *D* | Major Dia. | Thds. per In. | Major Dia. | Effective Diameter | | Minor Diameter | | Thds. per In. | Major Dia. | Effective Diameter | | Minor Dia. | |
| | Max. | | Min. | Max. | Min. | Max. | Min. | | Min. | Max. | Min. | Max. | Min. |
| UN THREADS | | UNF THREADS | | | | | | UNC THREADS | | | | | |
| ¼ | 0.2500 | 28 | 0.2435 | 0.2294 | 0.2265 | 0.2088 | 0.2037 | 20 | 0.2419 | 0.2201 | 0.2172 | 0.1913 | 0.1849 |
| 5⁄16 | 0.3125 | 24 | 0.3053 | 0.2883 | 0.2852 | 0.2643 | 0.2586 | 18 | 0.3038 | 0.2793 | 0.2762 | 0.2472 | 0.2402 |
| ⅜ | 0.3750 | 24 | 0.3678 | 0.3510 | 0.3478 | 0.3270 | 0.3211 | 16 | 0.3656 | 0.3375 | 0.3343 | 0.3014 | 0.2936 |
| 7⁄16 | 0.4375 | 20 | 0.4294 | 0.4084 | 0.4050 | 0.3796 | 0.3729 | 14 | 0.4272 | 0.3945 | 0.3911 | 0.3533 | 0.3447 |
| ½ | 0.5000 | 20 | 0.4919 | 0.4712 | 0.4675 | 0.4424 | 0.4356 | 13 | 0.4891 | 0.4537 | 0.4500 | 0.4093 | 0.4000 |
| 9⁄16 | 0.5625 | 18 | 0.5538 | 0.5302 | 0.5264 | 0.4981 | 0.4907 | 12 | 0.5511 | 0.5122 | 0.5084 | 0.4641 | 0.4542 |
| ⅝ | 0.6250 | 18 | 0.6163 | 0.5929 | 0.5889 | 0.5608 | 0.5533 | 11 | 0.6129 | 0.5700 | 0.5660 | 0.5175 | 0.5069 |
| ¾ | 0.7500 | 16 | 0.7406 | 0.7137 | 0.7094 | 0.6776 | 0.6693 | 10 | 0.7371 | 0.6893 | 0.6850 | 0.6316 | 0.6200 |
| ⅞ | 0.8750 | 14 | 0.8647 | 0.8332 | 0.8286 | 0.7920 | 0.7828 | 9 | 0.8611 | 0.8074 | 0.8028 | 0.7433 | 0.7306 |
| 1 | 1.0000 | 12 | 0.9886 | 0.9510 | 0.9459 | 0.9029 | 0.8925 | 8 | 0.9850 | 0.9239 | 0.9188 | 0.8517 | 0.8376 |
| 1⅛ | 1.1250 | 12 | 1.1136 | 1.0762 | 1.0709 | 1.0281 | 1.0176 | 7 | 1.1086 | 1.0375 | 1.0322 | 0.9550 | 0.9393 |
| 1¼ | 1.2500 | 12 | 1.2386 | 1.2014 | 1.1959 | 1.1533 | 1.1427 | 7 | 1.2336 | 1.1627 | 1.1572 | 1.0802 | 1.0644 |
| 1⅜ | 1.3750 | 12 | 1.3636 | 1.3265 | 1.3209 | 1.2784 | 1.2677 | 6 | 1.3568 | 1.2723 | 1.2667 | 1.1761 | 1.1581 |
| 1½ | 1.5000 | 12 | 1.4886 | 1.4517 | 1.4459 | 1.4036 | 1.3928 | 6 | 1.4818 | 1.3975 | 1.3917 | 1.3013 | 1.2832 |
| BS THREADS | | BSF THREADS | | | | | | BSW THREADS | | | | | |
| ¼ | 0.2500 | 26 | 0.2455 | 0.2280 | 0.2251 | 0.2034 | 0.1984 | 20 | 0.2452 | 0.2206 | 0.2177 | 0.1886 | 0.1831 |
| 5⁄16 | 0.3125 | 22 | 0.3077 | 0.2863 | 0.2832 | 0.2572 | 0.2517 | 18 | 0.3073 | 0.2798 | 0.2767 | 0.2442 | 0.2383 |
| ⅜ | 0.3750 | 20 | 0.3699 | 0.3461 | 0.3429 | 0.3141 | 0.3083 | 16 | 0.3695 | 0.3381 | 0.3349 | 0.0981 | 0.2919 |
| 7⁄16 | 0.4375 | 18 | 0.4320 | 0.4053 | 0.4019 | 0.3697 | 0.3635 | 14 | 0.4316 | 0.3952 | 0.3918 | 0.3495 | 0.3428 |
| ½ | 0.5000 | 16 | 0.4942 | 0.4637 | 0.4600 | 0.4237 | 0.4172 | 12 | 0.4937 | 0.4503 | 0.4466 | 0.3969 | 0.3897 |
| 9⁄16 | 0.5625 | 16 | 0.5566 | 0.5263 | 0.5225 | 0.4863 | 0.4797 | 12 | 0.5560 | 0.5129 | 0.5091 | 0.4595 | 0.4521 |
| ⅝ | 0.6250 | 14 | 0.6187 | 0.5833 | 0.5793 | 0.5376 | 0.5305 | 11 | 0.6183 | 0.5708 | 0.5668 | 0.5126 | 0.5050 |
| ¾ | 0.7500 | 12 | 0.7432 | 0.7009 | 0.6966 | 0.6475 | 0.6398 | 10 | 0.7428 | 0.6903 | 0.6860 | 0.6263 | 0.6182 |
| ⅞ | 0.8750 | 11 | 0.8678 | 0.8214 | 0.8168 | 0.7632 | 0.7551 | 9 | 0.8674 | 0.8085 | 0.8039 | 0.7374 | 0.7288 |
| 1 | 1.0000 | 10 | 0.9924 | 0.9411 | 0.9360 | 0.8771 | 0.8686 | 8 | 0.9920 | 0.9251 | 0.9200 | 0.8451 | 0.8360 |
| 1⅛ | 1.1250 | 9 | 1.1171 | 1.0592 | 1.0539 | 0.9881 | 0.9792 | 7 | 1.1164 | 1.0388 | 1.0335 | 0.9473 | 0.9376 |
| 1¼ | 1.2500 | 9 | 1.2419 | 1.1844 | 1.1789 | 1.1133 | 1.1042 | 7 | 1.2413 | 1.1640 | 1.1585 | 1.0725 | 1.0627 |
| 1⅜ | 1.3750 | 8 | 1.3665 | 1.3006 | 1.2950 | 1.2206 | 1.2110 | 6 | 1.4906 | 1.3991 | 1.3933 | 1.2924 | 1.2818 |
| 1½ | 1.5000 | 8 | 1.4913 | 1.4258 | 1.4200 | 1.3458 | 1.3360 | … | … | … | … | … | … |

| Limits for End Screwed into Component (BA Threads)[a] | | | | | | | |
|---|---|---|---|---|---|---|---|
| Designation No. | Pitch | Major Diameter | | Effective Diameter | | Minor Diameter | |
| | | Max. | Min. | Max. | Min. | Max. | Min. |
| 2 | 0.8100 mm<br>0.03189 in. | 4.700 mm<br>0.1850 in. | 4.580 mm<br>0.1803 in. | 4.275 mm<br>0.1683 in. | 4.200 mm<br>0.1654 in. | 3.790 mm<br>0.1492 in. | 3.620 mm<br>0.1425 in. |
| 4 | 0.6600 mm<br>0.2598 in. | 3.600 mm<br>0.1417 in. | 3.500 mm<br>0.1378 in. | 3.260 mm<br>0.1283 in. | 3.190 mm<br>0.1256 in. | 2.865 mm<br>0.1128 in. | 2.720 mm<br>0.1071 in. |

[a] Approximate inch equivalents are shown below the dimensions given in mm.

| Minimum Nominal Lengths of Studs[a] | | | | | | | | |
|---|---|---|---|---|---|---|---|---|
| Nom. Stud. Dia. | For Thread Length (Component End) of | | Nom. Stud. Dia. | For Thread Length (Component End) of | | Nom. Stud Dia. | For Thread Length (Component End) of | |
| | 1*D* | 1.5*D* | | 1*D* | 1.5*D* | | 1*D* | 1.5*D* |
| ¼ | ⅞ | 1 | 9⁄16 | 2 | 2⅜ | 1⅛ | 4 | 4⅝ |
| 5⁄16 | 1⅛ | 1⅜ | ⅝ | 2¼ | 2⅝ | 1¼ | 4¾ | 5½ |
| ⅜ | 1⅜ | 1⅝ | ¾ | 2⅝ | 3 | 1⅜ | 5 | 5¾ |
| 7⁄16 | 1⅝ | 1⅞ | ⅞ | 3⅛ | 3⅝ | 1½ | 5¼ | 6 |
| ½ | 1¾ | 2 | 1 | 3½ | 4 | … | … | … |

[a] The Standard also gives preferred and standard lengths of studs: *Preferred* lengths of studs: ⅞, 1, 1⅛, 1¼, 1⅜, 1½, 1¾, 2, 2¼, 2½, 2¾, 3, 3¼, 3½ and for lengths above 3½ the preferred increment is ½.

*Standard* lengths of studs: ⅞, 1, 1⅛, 1¼, 1⅜, 1½, 1⅝, 1¾, 1⅞, 2, 2⅛, 2¼, 2⅜, 2½, 2⅝, 2¾, 2⅞, 3, 3⅛, 3¼, 3⅜, 3½ and for lengths above 3½ the standard increment is ¼.

All dimensions are in inches except where otherwise noted.

See page 2109 for interference-fit threads.

**British Standard Single Coil Rectangular Section Spring Washers— Metric Series—Types B and BP** *BS 4464:1969*

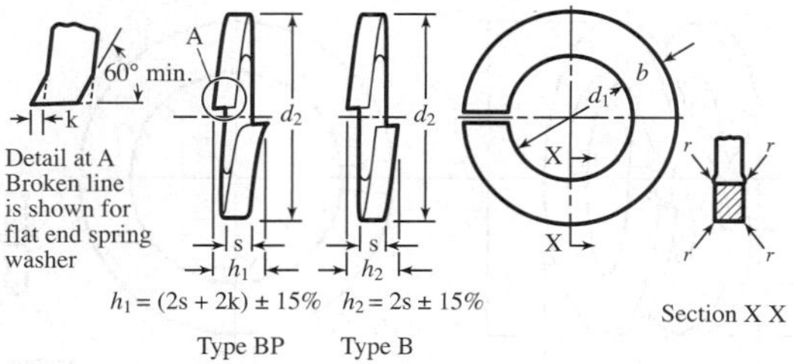


| Nom. Size & Thread Dia., $d$ | Inside Dia., $d_1$ | | Width, $b$ | Thickness, $s$ | Outside Dia., $d_2$ Max. | Radius, $r$ Max. | $k$ (Type BP Only) |
|---|---|---|---|---|---|---|---|
| | Max. | Min. | | | | | |
| M1.6 | 1.9 | 1.7 | 0.7 ± 0.1 | 0.4 ± 0.1 | 3.5 | 0.15 | … |
| M2 | 2.3 | 2.1 | 0.9 ± 0.1 | 0.5 ± 0.1 | 4.3 | 0.15 | … |
| (M2.2) | 2.5 | 2.3 | 1.0 ± 0.1 | 0.6 ± 0.1 | 4.7 | 0.2 | … |
| M2.5 | 2.8 | 2.6 | 1.0 ± 0.1 | 0.6 ± 0.1 | 5.0 | 0.2 | … |
| M3 | 3.3 | 3.1 | 1.3 ± 0.1 | 0.8 ± 0.1 | 6.1 | 0.25 | … |
| (M3.5) | 3.8 | 3.6 | 1.3 ± 0.1 | 0.8 ± 0.1 | 6.6 | 0.25 | 0.15 |
| M4 | 4.35 | 4.1 | 1.5 ± 0.1 | 0.9 ± 0.1 | 7.55 | 0.3 | 0.15 |
| M5 | 5.35 | 5.1 | 1.8 ± 0.1 | 1.2 ± 0.1 | 9.15 | 0.4 | 0.15 |
| M6 | 6.4 | 6.1 | 2.5 ± 0.15 | 1.6 ± 0.1 | 11.7 | 0.5 | 0.2 |
| M8 | 8.55 | 8.2 | 3 ± 0.15 | 2 ± 0.1 | 14.85 | 0.65 | 0.3 |
| M10 | 10.6 | 10.2 | 3.5 ± 0.2 | 2.2 ± 0.15 | 18.0 | 0.7 | 0.3 |
| M12 | 12.6 | 12.2 | 4 ± 0.2 | 2.5 ± 0.15 | 21.0 | 0.8 | 0.4 |
| (M14) | 14.7 | 14.2 | 4.5 ± 0.2 | 3 ± 0.15 | 24.1 | 1.0 | 0.4 |
| M16 | 16.9 | 16.3 | 5 ± 0.2 | 3.5 ± 0.2 | 27.3 | 1.15 | 0.4 |
| (M18) | 19.0 | 18.3 | 5 ± 0.2 | 3.5 ± 0.2 | 29.4 | 1.15 | 0.4 |
| M20 | 21.1 | 20.3 | 6 ± 0.2 | 4 ± 0.2 | 33.5 | 1.3 | 0.4 |
| (M22) | 23.3 | 22.4 | 6 ± 0.2 | 4 ± 0.2 | 35.7 | 1.3 | 0.4 |
| M24 | 25.3 | 24.4 | 7 ± 0.25 | 5 ± 0.2 | 39.8 | 1.65 | 0.5 |
| (M27) | 28.5 | 27.5 | 7 ± 0.25 | 5 ± 0.2 | 43.0 | 1.65 | 0.5 |
| M30 | 31.5 | 30.5 | 8 ± 0.25 | 6 ± 0.25 | 48.0 | 2.0 | 0.8 |
| (M33) | 34.6 | 33.5 | 10 ± 0.25 | 6 ± 0.25 | 55.1 | 2.0 | 0.8 |
| M36 | 37.6 | 36.5 | 10 ± 0.25 | 6 ± 0.25 | 58.1 | 2.0 | 0.8 |
| (M39) | 40.8 | 39.6 | 10 ± 0.25 | 6 ± 0.25 | 61.3 | 2.0 | 0.8 |
| M42 | 43.8 | 42.6 | 12 ± 0.25 | 7 ± 0.25 | 68.3 | 2.3 | 0.8 |
| (M45) | 46.8 | 45.6 | 12 ± 0.25 | 7 ± 0.25 | 71.3 | 2.3 | 0.8 |
| M48 | 50.0 | 48.8 | 12 ± 0.25 | 7 ± 0.25 | 74.5 | 2.3 | 0.8 |
| (M52) | 54.1 | 52.8 | 14 ± 0.25 | 8 ± 0.25 | 82.6 | 2.65 | 1.0 |
| M56 | 58.1 | 56.8 | 14 ± 0.25 | 8 ± 0.25 | 86.6 | 2.65 | 1.0 |
| (M60) | 62.3 | 60.9 | 14 ± 0.25 | 8 ± 0.25 | 90.8 | 2.65 | 1.0 |
| M64 | 66.3 | 64.9 | 14 ± 0.25 | 8 ± 0.25 | 93.8 | 2.65 | 1.0 |
| (M68) | 70.5 | 69.0 | 14 ± 0.25 | 8 ± 0.25 | 99.0 | 2.65 | 1.0 |

All dimensions are given in millimeters. Sizes shown in parentheses are non-preferred, and are not usually stock sizes.

**British Standard Double Coil Rectangular Section Spring Washers— Metric Series—Type D** *BS 4464:1969*

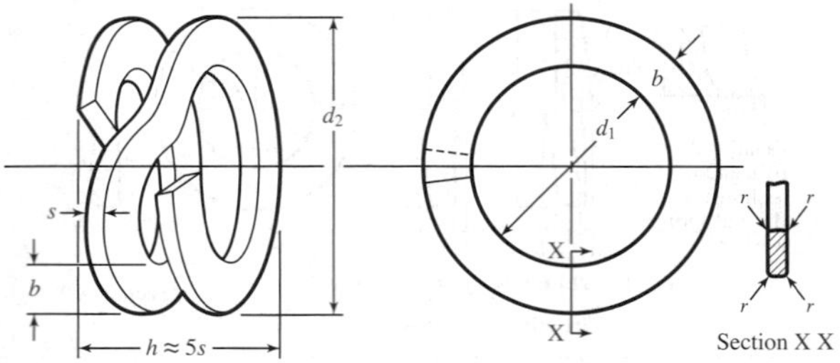


| Nom. Size, $d$ | Inside Dia., $d_1$ | | Width, $b$ | Thickness, $s$ | O.D., $d_2$ Max. | Radius, $r$ Max. |
|---|---|---|---|---|---|---|
| | Max. | Min. | | | | |
| M2 | 2.4 | 2.1 | 0.9±0.1 | 0.5±0.05 | 4.4 | 0.15 |
| (M2.2) | 2.6 | 2.3 | 1.0±0.1 | 0.6±0.05 | 4.8 | 0.2 |
| M2.5 | 2.9 | 2.6 | 1.2±0.1 | 0.7±0.1 | 5.5 | 0.23 |
| M3.0 | 3.6 | 3.3 | 1.2±0.1 | 0.8±0.1 | 6.2 | 0.25 |
| (M3.5) | 4.1 | 3.8 | 1.6±0.1 | 0.8±0.1 | 7.5 | 0.25 |
| M4 | 4.6 | 4.3 | 1.6±0.1 | 0.8±0.1 | 8.0 | 0.25 |
| M5 | 5.6 | 5.3 | 2±0.1 | 0.9±0.1 | 9.8 | 0.3 |
| M6 | 6.6 | 6.3 | 3±0.15 | 1±0.1 | 12.9 | 0.33 |
| M8 | 8.8 | 8.4 | 3±0.15 | 1.2±0.1 | 15.1 | 0.4 |
| M10 | 10.8 | 10.4 | 3.5±0.20 | 1.2±0.1 | 18.2 | 0.4 |
| M12 | 12.8 | 12.4 | 3.5±0.2 | 1.6±0.1 | 20.2 | 0.5 |
| (M14) | 15.0 | 14.5 | 5±0.2 | 1.6±0.1 | 25.4 | 0.5 |
| M16 | 17.0 | 16.5 | 5±0.2 | 2±0.1 | 27.4 | 0.65 |
| (M18) | 19.0 | 18.5 | 5±0.2 | 2±0.1 | 29.4 | 0.65 |
| M20 | 21.5 | 20.8 | 5±0.2 | 2±0.1 | 31.9 | 0.65 |
| (M22) | 23.5 | 22.8 | 6±0.2 | 2.5±0.15 | 35.9 | 0.8 |
| M24 | 26.0 | 25.0 | 6.5±0.2 | 3.25±0.15 | 39.4 | 1.1 |
| (M27) | 29.5 | 28.0 | 7±0.25 | 3.25±0.15 | 44.0 | 1.1 |
| M30 | 33.0 | 31.5 | 8±0.25 | 3.25±0.15 | 49.5 | 1.1 |
| (M33) | 36.0 | 34.5 | 8±0.25 | 3.25±0.15 | 52.5 | 1.1 |
| M36 | 40.0 | 38.0 | 10±0.25 | 3.25±0.15 | 60.5 | 1.1 |
| (M39) | 43.0 | 41.0 | 10±0.25 | 3.25±0.15 | 63.5 | 1.1 |
| M42 | 46.0 | 44.0 | 10±0.25 | 4.5±0.2 | 66.5 | 1.5 |
| M48 | 52.0 | 50.0 | 10±0.25 | 4.5±0.2 | 72.5 | 1.5 |
| M56 | 60.0 | 58.0 | 12±0.25 | 4.5±0.2 | 84.5 | 1.5 |
| M64 | 70.0 | 67.0 | 12±0.25 | 4.5±0.2 | 94.5 | 1.5 |

All dimensions are given in millimeters. Sizes shown in parentheses are non-preferred, and are not usually stock sizes. The free height of double coil washers before compression is normally approximately five times the thickness but, if required, washers with other free heights may be obtained by arrangement with manufacturer.

## British Standard Single Coil Square Section Spring Washers—Metric Series—Type A-1 *BS 4464:1969*

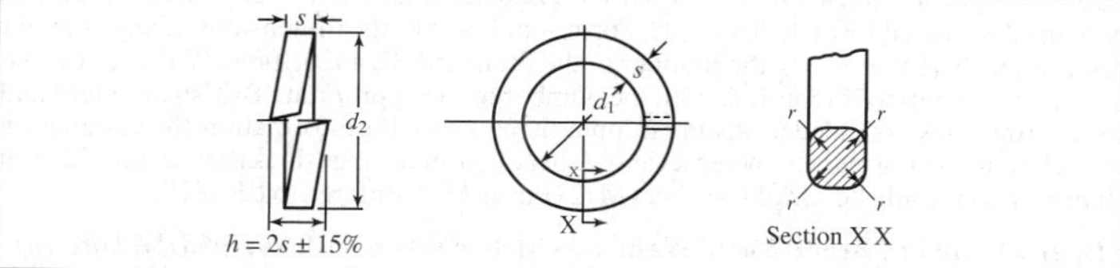


## British Standard Single Coil Square Section Spring Washers—Metric Series—Type A-2 *BS 4464:1969*

| Nom. Size, $d$ | Inside Dia., $d_1$ | | Thickness & Width, $s$ | O.D., $d_2$ Max. | Radius, $r$ Max. |
|---|---|---|---|---|---|
| | Max. | Min. | | | |
| M3 | 3.3 | 3.1 | 1 ± 0.1 | 5.5 | 0.3 |
| (M3.5) | 3.8 | 3.6 | 1 ± 0.1 | 6.0 | 0.3 |
| M4 | 4.35 | 4.1 | 1.2 ± 0.1 | 6.95 | 0.4 |
| M5 | 5.35 | 5.1 | 1.5 ± 0.1 | 8.55 | 0.5 |
| M6 | 6.4 | 6.1 | 1.5 ± 0.1 | 9.6 | 0.5 |
| M8 | 8.55 | 8.2 | 2 ± 0.1 | 12.75 | 0.65 |
| M10 | 10.6 | 10.2 | 2.5 ± 0.15 | 15.9 | 0.8 |
| M12 | 12.6 | 12.2 | 2.5 ± 0.15 | 17.9 | 0.8 |
| (M14) | 14.7 | 14.2 | 3 ± 0.2 | 21.1 | 1.0 |
| M16 | 16.9 | 16.3 | 3.5 ± 0.2 | 24.3 | 1.15 |
| (M18) | 19.0 | 18.3 | 3.5 ± 0.2 | 26.4 | 1.15 |
| M20 | 21.1 | 20.3 | 4.5 ± 0.2 | 30.5 | 1.5 |
| (M22) | 23.3 | 22.4 | 4.5 ± 0.2 | 32.7 | 1.5 |
| M24 | 25.3 | 24.4 | 5 ± 0.2 | 35.7 | 1.65 |
| (M27) | 28.5 | 27.5 | 5 ± 0.2 | 38.9 | 1.65 |
| M30 | 31.5 | 30.5 | 6 ± 0.2 | 43.9 | 2.0 |
| (M33) | 34.6 | 33.5 | 6 ± 0.2 | 47.0 | 2.0 |
| M36 | 37.6 | 36.5 | 7 ± 0.25 | 52.1 | 2.3 |
| (M39) | 40.8 | 39.6 | 7 ± 0.25 | 55.3 | 2.3 |
| M42 | 43.8 | 42.6 | 8 ± 0.25 | 60.3 | 2.65 |
| (M45) | 46.8 | 45.6 | 8 ± 0.25 | 63.3 | 2.65 |
| M48 | 50.0 | 48.8 | 8 ± 0.25 | 66.5 | 2.65 |

All dimensions are in millimeters. Sizes shown in parentheses are nonpreferred and are not usually stock sizes.

**British Standard for Metric Series Metal Washers.**—BS 4320:1968 (Withdrawn) specifies bright and black metal washers for general engineering purposes.

*Bright Metal Washers:* These washers are made from either CS4 cold-rolled strip steel BS 1449: Part 3B:1964 or from CZ 108 brass strip BS 2870:1980 (Withdrawn), both in the hard condition. However, by mutual agreement between purchaser and supplier, washers may be made available with the material in any other condition, or they may be made from another material, or may be coated with a protective or decorative finish to some appropriate British Standard. Washers are reasonably flat and free from burrs and are normally supplied unchamfered. They may, however, have a 30-degree chamfer on one edge of the external diameter. These washers are made available in two size categories, normal and large diameter, and in two thicknesses, normal (Form A or C) and light (Form B or D). The thickness of a light-range washer is from ½ to ⅔ the thickness of a normal range washer.

*Black Metal Washers:* These washers are made from mild steel, and can be supplied in three size categories designated normal, large, and extra large diameters. The normal-diameter series is intended for bolts ranging from M5 to M68 (Form E washers), the large-diameter series for bolts ranging from M8 to M39 (Form F washers), and the extra large series for bolts from M5 to M39 (Form G washers). A protective finish can be specified by the purchaser in accordance with any appropriate British Standard.

*Washer Designations:* The Standard specifies the details that should be given when ordering or placing an inquiry for washers. These details are the general description, namely, bright or black washers; the nominal size of the bolt or screw involved, for example, M5; the designated form, for example, Form A or Form E; the dimensions of any chamfer required on bright washers; the number of the Standard BS 4320:1968 (Withdrawn), and coating information if required, with the number of the appropriate British Standard and the coating thickness needed. As an example, in the use of this information, the designation for a chamfered, normal-diameter series washer of normal-range thickness to suit a 12-mm diameter bolt would be: Bright washers M12 (Form A) chamfered to BS 4320.

**British Standard Bright Metal Washers—Metric Series** *BS 4320:1968 (Withdrawn)*

| NORMAL DIAMETER SIZES | | | | | | | | | | | | |
|---|---|---|---|---|---|---|---|---|---|---|---|---|
| Nominal Size of Bolt or Screw | Inside Diameter | | | Outside Diameter | | | Thickness | | | | | |
| | | | | | | | Form A (Normal Range) | | | Form B (Light Range) | | |
| | Nom. | Max. | Min. | Nom. | Max. | Min. | Nom. | Max. | Min. | Nom. | Max. | Min. |
| M 1.0 | 1.1 | 1.25 | 1.1 | 2.5 | 2.5 | 2.3 | 0.3 | 0.4 | 0.2 | … | … | … |
| M 1.2 | 1.3 | 1.45 | 1.3 | 3.0 | 3.0 | 2.8 | 0.3 | 0.4 | 0.2 | … | … | … |
| (M 1.4) | 1.5 | 1.65 | 1.5 | 3.0 | 3.0 | 2.8 | 0.3 | 0.4 | 0.2 | … | … | … |
| M 1.6 | 1.7 | 1.85 | 1.7 | 4.0 | 4.0 | 3.7 | 0.3 | 0.4 | 0.2 | … | … | … |
| M 2.0 | 2.2 | 2.35 | 2.2 | 5.0 | 5.0 | 4.7 | 0.3 | 0.4 | 0.2 | … | … | … |
| (M 2.2) | 2.4 | 2.55 | 2.4 | 5.0 | 5.0 | 4.7 | 0.5 | 0.6 | 0.4 | … | … | … |
| M 2.5 | 2.7 | 2.85 | 2.7 | 6.5 | 6.5 | 6.2 | 0.5 | 0.6 | 0.4 | … | … | … |
| M3 | 3.2 | 3.4 | 3.2 | 7 | 7 | 6.7 | 0.5 | 0.6 | 0.4 | … | … | … |
| (M 3.5) | 3.7 | 3.9 | 3.7 | 7 | 7 | 6.7 | 0.5 | 0.6 | 0.4 | … | … | … |
| M4 | 4.3 | 4.5 | 4.3 | 9 | 9 | 8.7 | 0.8 | 0.9 | 0.7 | … | … | … |
| (M 4.5) | 4.8 | 5.0 | 4.8 | 9 | 9 | 8.7 | 0.8 | 0.9 | 0.7 | … | … | … |
| M 5 | 5.3 | 5.5 | 5.3 | 10 | 10 | 9.7 | 1.0 | 1.1 | 0.9 | … | … | … |
| M 6 | 6.4 | 6.7 | 6.4 | 12.5 | 12.5 | 12.1 | 1.6 | 1.8 | 1.4 | 0.8 | 0.9 | 0.7 |
| (M 7) | 7.4 | 7.7 | 7.4 | 14 | 14 | 13.6 | 1.6 | 1.8 | 1.4 | 0.8 | 0.9 | 0.7 |
| M 8 | 8.4 | 8.7 | 8.4 | 17 | 17 | 16.6 | 1.6 | 1.8 | 1.4 | 1.0 | 1.1 | 0.9 |
| M 10 | 10.5 | 10.9 | 10.5 | 21 | 21 | 20.5 | 2.0 | 2.2 | 1.8 | 1.25 | 1.45 | 1.05 |
| M 12 | 13.0 | 13.4 | 13.0 | 24 | 24 | 23.5 | 2.5 | 2.7 | 2.3 | 1.6 | 1.80 | 1.40 |
| (M 14) | 15.0 | 15.4 | 15.0 | 28 | 28 | 27.5 | 2.5 | 2.7 | 2.3 | 1.6 | 1.8 | 1.4 |
| M 16 | 17.0 | 17.4 | 17.0 | 30 | 30 | 29.5 | 3.0 | 3.3 | 2.7 | 2.0 | 2.2 | 1.8 |
| (M 18) | 19.0 | 19.5 | 19.0 | 34 | 34 | 33.2 | 3.0 | 3.3 | 2.7 | 2.0 | 2.2 | 1.8 |
| M 20 | 21 | 21.5 | 21 | 37 | 37 | 36.2 | 3.0 | 3.3 | 2.7 | 2.0 | 2.2 | 1.8 |
| (M 22) | 23 | 23.5 | 23 | 39 | 39 | 38.2 | 3.0 | 3.3 | 2.7 | 2.0 | 2.2 | 1.8 |
| M24 | 25 | 25.5 | 25 | 44 | 44 | 43.2 | 4.0 | 4.3 | 3.7 | 2.5 | 2.7 | 2.3 |
| (M 27) | 28 | 28.5 | 28 | 50 | 50 | 49.2 | 4.0 | 4.3 | 3.7 | 2.5 | 2.7 | 2.3 |
| M30 | 31 | 31.6 | 31 | 56 | 56 | 55.0 | 4.0 | 4.3 | 3.7 | 2.5 | 2.7 | 2.3 |
| (M 33) | 34 | 34.6 | 34 | 60 | 60 | 59.0 | 5.0 | 5.6 | 4.4 | 3.0 | 3.3 | 2.7 |
| M 36 | 37 | 37.6 | 37 | 66 | 66 | 65.0 | 5.0 | 5.6 | 4.4 | 3.0 | 3.3 | 2.7 |
| (M 39) | 40 | 40.6 | 40 | 72 | 72 | 71.0 | 6.0 | 6.6 | 5.4 | 3.0 | 3.3 | 2.7 |
| LARGE DIAMETER SIZES | | | | | | | | | | | | |
| Nominal Size of Bolt or Screw | Inside Diameter | | | Outside Diameter | | | Thickness | | | | | |
| | | | | | | | Form C (Normal Range) | | | Form D (Light Range) | | |
| | Nom. | Max. | Min. | Nom. | Max. | Min. | Nom. | Max. | Min. | Nom. | Max. | Min. |
| M 4 | 4.3 | 4.5 | 4.3 | 10.0 | 10.0 | 9.7 | 0.8 | 0.9 | 0.7 | … | … | … |
| M 5 | 5.3 | 5.5 | 5.3 | 12.5 | 12.5 | 12.1 | 1.0 | 1.1 | 0.9 | … | … | … |
| M 6 | 6.4 | 6.7 | 6.4 | 14 | 14 | 13.6 | 1.6 | 1.8 | 1.4 | 0.8 | 0.9 | 0.7 |
| M 8 | 8.4 | 8.7 | 8.4 | 21 | 21 | 20.5 | 1.6 | 1.8 | 1.4 | 1.0 | 1.1 | 0.9 |
| M 10 | 10.5 | 10.9 | 10.5 | 24 | 24 | 23.5 | 2.0 | 2.2 | 1.8 | 1.25 | 1.45 | 1.05 |
| M 12 | 13.0 | 13.4 | 13.0 | 28 | 28 | 27.5 | 2.5 | 2.7 | 2.3 | 1.6 | 1.8 | 1.4 |
| (M 14) | 15.0 | 15.4 | 15 | 30 | 30 | 29.5 | 2.5 | 2.7 | 2.3 | 1.6 | 1.8 | 1.4 |
| M 16 | 17.0 | 17.4 | 17 | 34 | 34 | 33.2 | 3.0 | 3.3 | 2.7 | 2.0 | 2.2 | 1.8 |
| (M 18) | 19.0 | 19.5 | 19 | 37 | 37 | 36.2 | 3.0 | 3.3 | 2.7 | 2.0 | 2.2 | 1.8 |
| M 20 | 21 | 21.5 | 21 | 39 | 39 | 38.2 | 3.0 | 3.3 | 2.7 | 2.0 | 2.2 | 1.8 |
| (M 22) | 23 | 23.5 | 23 | 44 | 44 | 43.2 | 3.0 | 3.3 | 2.7 | 2.0 | 2.2 | 1.8 |
| M 24 | 25 | 25.5 | 25 | 50 | 50 | 49.2 | 4.0 | 4.3 | 3.7 | 2.5 | 2.7 | 2.3 |
| (M 27) | 28 | 28.5 | 28 | 56 | 56 | 55 | 4.0 | 4.3 | 3.7 | 2.5 | 2.7 | 2.3 |
| M 30 | 31 | 31.6 | 31 | 60 | 60 | 59 | 4.0 | 4.3 | 3.7 | 2.5 | 2.7 | 2.3 |
| (M 33) | 34 | 34.6 | 34 | 66 | 66 | 65 | 5.0 | 5.6 | 4.4 | 3.0 | 3.3 | 2.7 |
| M 36 | 37 | 37.6 | 37 | 72 | 72 | 71 | 5.0 | 5.6 | 4.4 | 3.0 | 3.3 | 2.7 |
| (M 39) | 40 | 40.6 | 40 | 77 | 77 | 76 | 6.0 | 6.6 | 5.4 | 3.0 | 3.3 | 2.7 |

All dimensions are in millimeters.

Nominal bolt or screw sizes shown in parentheses are nonpreferred.

## British Standard Black Metal Washers—Metric Series *BS 4320:1968 (Withdrawn)*

| Nom Bolt or Screw Size | Inside Diameter | | | Outside Diameter | | | Thickness | | |
|---|---|---|---|---|---|---|---|---|---|
| | Nom. | Max. | Min. | Nom. | Max. | Min. | Nom. | Max. | Min. |
| NORMAL DIAMETER SIZES (Form E) | | | | | | | | | |
| M 5 | 5.5 | 5.8 | 5.5 | 10.0 | 10.0 | 9.2 | 1.0 | 1.2 | 0.8 |
| M 6 | 6.6 | 7.0 | 6.6 | 12.5 | 12.5 | 11.7 | 1.6 | 1.9 | 1.3 |
| (M 7) | 7.6 | 8.0 | 7.6 | 14.0 | 14.0 | 13.2 | 1.6 | 1.9 | 1.3 |
| M 8 | 9.0 | 9.4 | 9.0 | 17 | 17 | 16.2 | 1.6 | 1.9 | 1.3 |
| M 10 | 11.0 | 11.5 | 11.0 | 21 | 21 | 20.2 | 2.0 | 2.3 | 1.7 |
| M 12 | 14 | 14.5 | 14 | 24 | 24 | 23.2 | 2.5 | 2.8 | 2.2 |
| (M 14) | 16 | 16.5 | 16 | 28 | 28 | 27.2 | 2.5 | 2.8 | 2.2 |
| M 16 | 18 | 18.5 | 18 | 30 | 30 | 29.2 | 3.0 | 3.6 | 2.4 |
| (M 18) | 20 | 20.6 | 20 | 34 | 34 | 32.8 | 3.0 | 3.6 | 2.4 |
| M 20 | 22 | 22.6 | 22 | 37 | 37 | 35.8 | 3.0 | 3.6 | 2.4 |
| (M 22) | 24 | 24.6 | 24 | 39 | 39 | 37.8 | 3.0 | 3.6 | 2.4 |
| M 24 | 26 | 26.6 | 26 | 44 | 44 | 42.8 | 4 | 4.6 | 3.4 |
| (M 27) | 30 | 30.6 | 30 | 50 | 50 | 48.8 | 4 | 4.6 | 3.4 |
| M 30 | 33 | 33.8 | 33 | 56 | 56 | 54.5 | 4 | 4.6 | 3.4 |
| (M 33) | 36 | 36.8 | 36 | 60 | 60 | 58.5 | 5 | 6.0 | 4.0 |
| M 36 | 39 | 39.8 | 39 | 66 | 66 | 64.5 | 5 | 6.0 | 4.0 |
| (M 39) | 42 | 42.8 | 42 | 72 | 72 | 70.5 | 6 | 7.0 | 5.0 |
| M 42 | 45 | 45.8 | 45 | 78 | 78 | 76.5 | 7 | 8.2 | 5.8 |
| (M 45) | 48 | 48.8 | 48 | 85 | 85 | 83 | 7 | 8.2 | 5.8 |
| M 48 | 52 | 53 | 52 | 92 | 92 | 90 | 8 | 9.2 | 6.8 |
| (M 52) | 56 | 57 | 56 | 98 | 98 | 96 | 8 | 9.2 | 6.8 |
| M 56 | 62 | 63 | 62 | 105 | 105 | 103 | 9 | 10.2 | 7.8 |
| (M 60) | 66 | 67 | 66 | 110 | 110 | 108 | 9 | 10.2 | 7.8 |
| M 64 | 70 | 71 | 70 | 115 | 115 | 113 | 9 | 10.2 | 7.8 |
| (M 68) | 74 | 75 | 74 | 120 | 120 | 118 | 10 | 11.2 | 8.8 |
| LARGE DIAMETER SIZES (Form F) | | | | | | | | | |
| M 8 | 9 | 9.4 | 9.0 | 21 | 21 | 20.2 | 1.6 | 1.9 | 1.3 |
| M 10 | 11 | 11.5 | 11 | 24 | 24 | 23.2 | 2 | 2.3 | 1.7 |
| M 12 | 14 | 14.5 | 14 | 28 | 28 | 27.2 | 2.5 | 2.8 | 2.2 |
| (M 14) | 16 | 16.5 | 16 | 30 | 30 | 29.2 | 2.5 | 2.8 | 2.2 |
| M 16 | 18 | 18.5 | 18 | 34 | 34 | 32.8 | 3 | 3.6 | 2.4 |
| (M 18) | 20 | 20.6 | 20 | 37 | 37 | 35.8 | 3 | 3.6 | 2.4 |
| M 20 | 22 | 22.6 | 22 | 39 | 39 | 37.8 | 3 | 3.6 | 2.4 |
| (M 22) | 24 | 24.6 | 24 | 44 | 44 | 42.8 | 3 | 3.6 | 2.4 |
| M 24 | 26 | 26.6 | 26 | 50 | 50 | 48.8 | 4 | 4.6 | 3.4 |
| (M 27) | 30 | 30.6 | 30 | 56 | 56 | 54.5 | 4 | 4.6 | 3.4 |
| M 30 | 33 | 33.8 | 33 | 60 | 60 | 58.5 | 4 | 4.6 | 3.4 |
| (M 33) | 36 | 36.8 | 36 | 66 | 66 | 64.5 | 5 | 6.0 | 4 |
| M 36 | 39 | 39.8 | 39 | 72 | 72 | 70.5 | 5 | 6.0 | 4 |
| (M 39) | 42 | 42.8 | 42 | 77 | 77 | 75.5 | 6 | 7 | 5 |
| EXTRA LARGE DIAMETER SIZES (Form G) | | | | | | | | | |
| M 5 | 5.5 | 5.8 | 5.5 | 15 | 15 | 14.2 | 1.6 | 1.9 | 1.3 |
| M 6 | 6.6 | 7.0 | 6.6 | 18 | 18 | 17.2 | 2 | 2.3 | 1.7 |
| (M 7) | 7.6 | 8.0 | 7.6 | 21 | 21 | 20.2 | 2 | 2.3 | 1.7 |
| M 8 | 9 | 9.4 | 9.0 | 24 | 24 | 23.2 | 2 | 2.3 | 1.7 |
| M 10 | 11 | 11.5 | 11.0 | 30 | 30 | 29.2 | 2.5 | 2.8 | 2.2 |
| M 12 | 14 | 14.5 | 14.0 | 36 | 36 | 34.8 | 3 | 3.6 | 2.4 |
| (M 14) | 16 | 16.5 | 16.0 | 42 | 42 | 40.8 | 3 | 3.6 | 2.4 |
| M 16 | 18 | 18.5 | 18 | 48 | 48 | 46.8 | 4 | 4.6 | 3.4 |
| (M 18) | 20 | 20.6 | 20 | 54 | 54 | 52.5 | 4 | 4.6 | 3.4 |
| M 20 | 22 | 22.6 | 22 | 60 | 60 | 58.5 | 5 | 6.0 | 4 |
| (M 22) | 24 | 24.6 | 24 | 66 | 66 | 64.5 | 5 | 6.0 | 4 |
| M 24 | 26 | 26.6 | 26 | 72 | 72 | 70.5 | 6 | 7 | 5 |
| (M 27) | 30 | 30.6 | 30 | 81 | 81 | 79 | 6 | 7 | 5 |
| M 30 | 33 | 33.8 | 33 | 90 | 90 | 88 | 8 | 9.2 | 6.8 |
| (M 33) | 36 | 36.8 | 36 | 99 | 99 | 97 | 8 | 9.2 | 6.8 |
| M 36 | 39 | 39.8 | 39 | 108 | 108 | 106 | 10 | 11.2 | 8.8 |
| (M39) | 42 | 42.8 | 42 | 117 | 117 | 115 | 10 | 11.2 | 8.8 |

All dimensions are in millimeters.
Nominal bolt or screw sizes shown in parentheses are nonpreferred.

## MACHINE SCREWS AND NUTS

### American National Standard Machine Screws and Machine Screw Nuts

ANSI/ASME B18.6.3 covers both slotted and recessed head machine screws. Dimensions of various types of slotted machine screws, machine screw nuts, and header points are given in Table 1 through Table 12. The Standard also covers flat trim head, oval trim head and drilled fillister head machine screws and gives cross recess dimensions and gaging dimensions for all types of machine screw heads. Information on metric machine screws in ANSI/ASME B18.6.7M (Withdrawn) is given beginning on page 1793.

**Threads.—**Except for sizes 0000, 000, and 00, machine screw threads may be either Unified Coarse (UNC) and Fine thread (UNF) Class 2A (see *American Standard for Unified Screw Threads* starting on page 1950) or UNRC and UNRF Series, at option of manufacturer. Thread dimensions for sizes 0000, 000, and 00 are given in Table 7 on page 1783.

Threads for hexagon machine screw nuts may be either UNC or UNF, Class 2B, and for square machine screw nuts are UNC Class 2B.

**Length of Thread.—**Machine screws of sizes No. 5 and smaller with nominal lengths equal to 3 diameters and shorter have full form threads extending to within 1 pitch (thread) of the bearing surface of the head, or closer, if practicable. Nominal lengths greater than 3 diameters, up to and including $1\frac{1}{8}$ inch, have full form threads extending to within two pitches (threads) of the bearing surface of the head, or closer, if practicable. Unless otherwise specified, screws of longer nominal length have a minimum length of full form thread of 1.00 inch. Machine screws of sizes No. 6 and larger with nominal length equal to 3 diameters and shorter have full form threads extending to within 1 pitch (thread) of the bearing surface of the head, or closer, if practicable. Nominal lengths greater than 3 diameters, up to and including 2 inches, have full form threads extending to within 2 pitches (threads) of the bearing surface of the head, or closer, if practicable. Screws of longer nominal length, unless otherwise specified, have a minimum length of full form thread of 1.50 inches.

**Table 1. Square and Hexagon Machine Screw Nuts** *ANSI/ASME B18.6.3-1972 (R1991)*

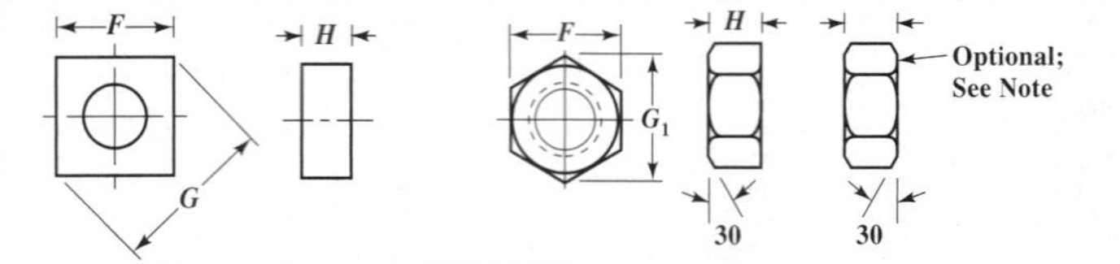


| Nom. Size | Basic Dia. | Basic *F* | Max. *F* | Min. *F* | Max. *G* | Min. *G* | Max. $G_1$ | Min. $G_1$ | Max. *H* | Min. *H* |
|---|---|---|---|---|---|---|---|---|---|---|
| 0 | 0.0600 | 5⁄32 | 0.156 | 0.150 | 0.221 | 0.206 | 0.180 | 0.171 | 0.050 | 0.043 |
| 1 | 0.0730 | 5⁄32 | 0.156 | 0.150 | 0.221 | 0.206 | 0.180 | 0.171 | 0.050 | 0.043 |
| 2 | 0.0860 | 3⁄16 | 0.188 | 0.180 | 0.265 | 0.247 | 0.217 | 0.205 | 0.066 | 0.057 |
| 3 | 0.0990 | 3⁄16 | 0.188 | 0.180 | 0.265 | 0.247 | 0.217 | 0.205 | 0.066 | 0.057 |
| 4 | 0.1120 | 1⁄4 | 0.250 | 0.241 | 0.354 | 0.331 | 0.289 | 0.275 | 0.098 | 0.087 |
| 5 | 0.1250 | 5⁄16 | 0.312 | 0.302 | 0.442 | 0.415 | 0.361 | 0.344 | 0.114 | 0.102 |
| 6 | 0.1380 | 5⁄16 | 0.312 | 0.302 | 0.442 | 0.415 | 0.361 | 0.344 | 0.114 | 0.102 |
| 8 | 0.1640 | 11⁄32 | 0.344 | 0.332 | 0.486 | 0.456 | 0.397 | 0.378 | 0.130 | 0.117 |
| 10 | 0.1900 | 3⁄8 | 0.375 | 0.362 | 0.530 | 0.497 | 0.433 | 0.413 | 0.130 | 0.117 |
| 12 | 0.2160 | 7⁄16 | 0.438 | 0.423 | 0.619 | 0.581 | 0.505 | 0.482 | 0.161 | 0.148 |
| 1⁄4 | 0.2500 | 7⁄16 | 0.438 | 0.423 | 0.619 | 0.581 | 0.505 | 0.482 | 0.193 | 0.178 |
| 5⁄16 | 0.3125 | 9⁄16 | 0.562 | 0.545 | 0.795 | 0.748 | 0.650 | 0.621 | 0.225 | 0.208 |
| 3⁄8 | 0.3750 | 5⁄8 | 0.625 | 0.607 | 0.884 | 0.833 | 0.722 | 0.692 | 0.257 | 0.239 |

All dimensions in inches. Hexagon machine screw nuts have tops flat and chamfered. Diameter of top circle should be the maximum width across flats within a tolerance of minus 15 percent. Bottoms are flat but may be chamfered if so specified. Square machine screw nuts have tops and bottoms flat without chamfer.

**Diameter of Body.—**The diameter of machine screw bodies is not less than Class 2A thread minimum pitch diameter nor greater than the basic major diameter of the thread. Cross-recessed trim head machine screws not threaded to the head have an 0.062 in. minimum length shoulder under the head with diameter limits as specified in the dimensional tables in the Standard.

**Designation.—**Machine screws are designated by the following data in the sequence shown: Nominal size (number, fraction, or decimal equivalent); threads per inch; nominal length (fraction or decimal equivalent); product name, including head type and driving provision; header point, if desired; material; and protective finish, if required. For example:

¼ – 20 × 1¼ Slotted Pan Head Machine Screw, Steel, Zinc Plated

6 – 32 × ¾ Type IA Cross Recessed Fillister Head Machine Screw, Brass

Machine screw nuts are designated by the following data in the sequence shown: Nominal size (number, fraction, or decimal equivalent); threads per inch; product name; material; and protective finish, if required. For example:

10 – 24 Hexagon Machine Screw Nut, Steel, Zinc Plated

0.138 – 32 Square Machine Screw Nut, Brass

**Table 2. American National Standard Slotted 100-Degree Flat Countersunk Head Machine Screws** *ANSI/ASME B18.6.3-2013*

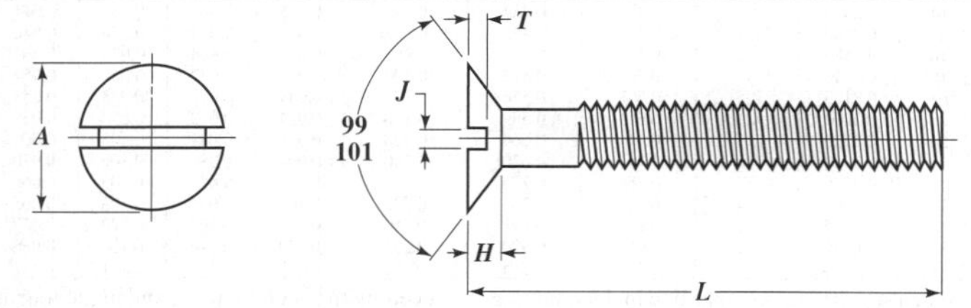


| Nominal Size[a] or Basic Screw Dia. | | Head Dia., A | | Head Height, H | Slot Width, J | | Slot Depth, T | |
|---|---|---|---|---|---|---|---|---|
| | | Max. | Min. | Ref. | Max. | Min. | Max. | Min. |
| 0000 | 0.0210 | 0.040 | 0.035 | 0.009 | 0.008 | 0.005 | 0.008 | 0.004 |
| 000 | 0.0340 | 0.060 | 0.055 | 0.014 | 0.012 | 0.008 | 0.011 | 0.007 |
| 00 | 0.0470 | 0.087 | 0.080 | 0.020 | 0.017 | 0.010 | 0.013 | 0.008 |
| 0 | 0.0600 | 0.112 | 0.095 | 0.026 | 0.023 | 0.016 | 0.013 | 0.008 |
| 1 | 0.0730 | 0.137 | 0.118 | 0.031 | 0.026 | 0.019 | 0.016 | 0.010 |
| 2 | 0.0860 | 0.162 | 0.142 | 0.037 | 0.031 | 0.023 | 0.019 | 0.012 |
| 3 | 0.0990 | 0.187 | 0.165 | 0.043 | 0.035 | 0.027 | 0.022 | 0.014 |
| 4 | 0.1120 | 0.212 | 0.188 | 0.049 | 0.039 | 0.031 | 0.024 | 0.017 |
| 6 | 0.1380 | 0.262 | 0.235 | 0.060 | 0.048 | 0.039 | 0.030 | 0.022 |
| 8 | 0.1640 | 0.312 | 0.282 | 0.072 | 0.054 | 0.045 | 0.036 | 0.027 |
| 10 | 0.1900 | 0.362 | 0.329 | 0.083 | 0.060 | 0.050 | 0.042 | 0.031 |
| ¼ | 0.2500 | 0.477 | 0.437 | 0.110 | 0.075 | 0.064 | 0.055 | 0.042 |
| 5⁄16 | 0.3125 | 0.597 | 0.550 | 0.138 | 0.084 | 0.072 | 0.069 | 0.053 |
| ⅜ | 0.3750 | 0.717 | 0.662 | 0.165 | 0.094 | 0.081 | 0.083 | 0.065 |

[a] When specifying nominal size in decimals, zeros preceding the decimal point and in the fourth decimal place are omitted.

All dimensions are in inches.

**Table 3. American National Standard Slotted Flat Countersunk Head and Close Tolerance 100-Degree Flat Countersunk Head Machine Screws** *ANSI/ASME B18.6.3-2013*

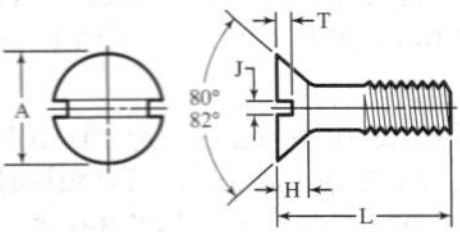


SLOTTED FLAT COUNTERSUNK HEAD TYPE

| Nominal Size[a] or Basic Screw Dia. | | Max., $L$[b] | Head Dia., $A$ | | Head Height, $H$ | Slot Width, $J$ | | Slot Depth, $T$ | |
|---|---|---|---|---|---|---|---|---|---|
| | | | Max. | Min.[c] | Ref. | Max. | Min. | Max. | Min. |
| 0000 | 0.0210 | … | 0.040 | 0.035 | 0.011 | 0.008 | 0.004 | 0.007 | 0.003 |
| 000 | 0.0340 | … | 0.060 | 0.055 | 0.016 | 0.011 | 0.007 | 0.009 | 0.005 |
| 00 | 0.0470 | … | 0.087 | 0.080 | 0.028 | 0.017 | 0.010 | 0.014 | 0.009 |
| 0 | 0.0600 | 1/8 | 0.112 | 0.096 | 0.035 | 0.023 | 0.016 | 0.015 | 0.010 |
| 1 | 0.0730 | 1/8 | 0.137 | 0.120 | 0.043 | 0.026 | 0.019 | 0.019 | 0.012 |
| 2 | 0.0860 | 1/8 | 0.162 | 0.144 | 0.051 | 0.031 | 0.023 | 0.023 | 0.015 |
| 3 | 0.0990 | 1/8 | 0.187 | 0.167 | 0.059 | 0.035 | 0.027 | 0.027 | 0.017 |
| 4 | 0.1120 | 3/16 | 0.212 | 0.191 | 0.067 | 0.039 | 0.031 | 0.030 | 0.020 |
| 5 | 0.1250 | 3/16 | 0.237 | 0.215 | 0.075 | 0.043 | 0.035 | 0.034 | 0.022 |
| 6 | 0.1380 | 3/16 | 0.262 | 0.238 | 0.083 | 0.048 | 0.039 | 0.038 | 0.024 |
| 7 | 0.1510 | … | 0.287 | 0.262 | 0.091 | 0.048 | 0.039 | 0.041 | 0.027 |
| 8 | 0.1640 | 1/4 | 0.312 | 0.285 | 0.100 | 0.054 | 0.045 | 0.045 | 0.029 |
| 10 | 0.1900 | 5/16 | 0.362 | 0.333 | 0.116 | 0.060 | 0.050 | 0.053 | 0.034 |
| 12 | 0.2160 | 3/8 | 0.412 | 0.380 | 0.132 | 0.067 | 0.056 | 0.060 | 0.039 |
| 14 | 0.2420 | … | 0.462 | 0.427 | 0.148 | 0.075 | 0.064 | 0.068 | 0.044 |
| 1/4 | 0.2500 | 7/16 | 0.477 | 0.442 | 0.153 | 0.075 | 0.064 | 0.070 | 0.046 |
| 16 | 0.2680 | … | 0.512 | 0.475 | 0.164 | 0.075 | 0.064 | 0.075 | 0.049 |
| 18 | 0.2940 | … | 0.561 | 0.522 | 0.180 | 0.084 | 0.072 | 0.083 | 0.054 |
| 5/16 | 0.3125 | 1/2 | 0.597 | 0.556 | 0.191 | 0.084 | 0.072 | 0.088 | 0.058 |
| 20 | 0.3200 | … | 0.611 | 0.569 | 0.196 | 0.084 | 0.072 | 0.090 | 0.059 |
| 24 | 0.3720 | … | 0.711 | 0.664 | 0.228 | 0.094 | 0.081 | 0.105 | 0.069 |
| 3/8 | 0.3750 | 9/16 | 0.717 | 0.670 | 0.230 | 0.094 | 0.081 | 0.106 | 0.070 |
| 7/16 | 0.4375 | 5/8 | 0.760 | 0.715 | 0.223 | 0.094 | 0.081 | 0.103 | 0.066 |
| 1/2 | 0.5000 | 3/4 | 0.815 | 0.765 | 0.223 | 0.106 | 0.091 | 0.103 | 0.065 |
| 9/16 | 0.5625 | … | 0.932 | 0.878 | 0.260 | 0.118 | 0.102 | 0.120 | 0.077 |
| 5/8 | 0.6250 | … | 1.050 | 0.990 | 0.298 | 0.133 | 0.116 | 0.137 | 0.088 |
| 3/4 | 0.7500 | … | 1.285 | 1.215 | 0.372 | 0.149 | 0.131 | 0.171 | 0.111 |

[a] When specifying nominal size in decimals, zeros preceding the decimal point and in the fourth decimal place are omitted.
[b] These lengths or shorter are undercut.
[c] May be rounded or flat.

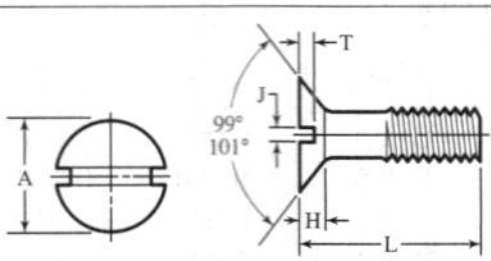


CLOSE TOLERANCE 100-DEGREE FLAT COUNTERSUNK HEAD TYPE

| Nominal Size[a] or Basic Screw Dia. | | Head Diameter, $A$ | | Head Height, $H$ | Slot Width, $J$ | | Slot Depth, $T$ | |
|---|---|---|---|---|---|---|---|---|
| | | Max. | Min.[c] | Ref. | Max. | Min. | Max. | Min. |
| 4 | 0.1120 | 0.212 | 0.188 | 0.049 | 0.039 | 0.031 | 0.024 | 0.017 |
| 6 | 0.1380 | 0.262 | 0.235 | 0.060 | 0.048 | 0.039 | 0.030 | 0.022 |
| 8 | 0.1640 | 0.312 | 0.282 | 0.072 | 0.054 | 0.045 | 0.036 | 0.027 |
| 10 | 0.1900 | 0.362 | 0.329 | 0.083 | 0.060 | 0.050 | 0.042 | 0.031 |
| 1/4 | 0.2500 | 0.477 | 0.437 | 0.110 | 0.075 | 0.064 | 0.055 | 0.042 |
| 5/16 | 0.3125 | 0.597 | 0.550 | 0.138 | 0.084 | 0.072 | 0.069 | 0.053 |
| 3/8 | 0.3750 | 0.717 | 0.662 | 0.165 | 0.094 | 0.081 | 0.083 | 0.065 |
| 7/16 | 0.4375 | 0.837 | 0.774 | 0.193 | 0.094 | 0.081 | 0.097 | 0.076 |
| 1/2 | 0.5000 | 0.957 | 0.887 | 0.221 | 0.106 | 0.091 | 0.111 | 0.088 |
| 9/16 | 0.5625 | 1.077 | 1.000 | 0.249 | 0.118 | 0.102 | 0.125 | 0.099 |
| 5/8 | 0.6250 | 1.197 | 1.112 | 0.276 | 0.133 | 0.116 | 0.139 | 0.111 |

All dimensions are in inches.

## Table 4. American National Standard Slotted Undercut Flat Countersunk Head and Plain and Slotted Hex Washer Head Machine Screws *ANSI/ASME B18.6.3-2013*

SLOTTED UNDERCUT FLAT COUNTERSUNK HEAD TYPE

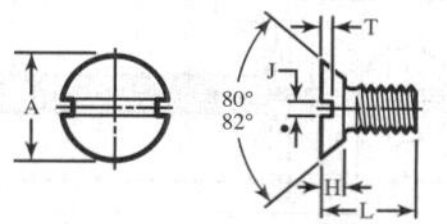


| Nominal Size[a] or Basic Screw Dia. | | Max., L[b] | Head Dia., A | | Head Height, H | | Slot Width, J | | Slot Depth, T | |
|---|---|---|---|---|---|---|---|---|---|---|
| | | | Max., Edge Sharp | Min., Edge Rounded or Flat | Max. | Min. | Max. | Min. | Max. | Min. |
| 0 | 0.0600 | 1/8 | 0.112 | 0.096 | 0.025 | 0.018 | 0.023 | 0.016 | 0.011 | 0.007 |
| 1 | 0.0730 | 1/8 | 0.137 | 0.120 | 0.031 | 0.023 | 0.026 | 0.019 | 0.014 | 0.009 |
| 2 | 0.0860 | 1/8 | 0.162 | 0.144 | 0.036 | 0.028 | 0.031 | 0.023 | 0.016 | 0.011 |
| 3 | 0.0990 | 1/8 | 0.187 | 0.167 | 0.042 | 0.033 | 0.035 | 0.027 | 0.019 | 0.012 |
| 4 | 0.1120 | 3/16 | 0.212 | 0.191 | 0.047 | 0.038 | 0.039 | 0.031 | 0.022 | 0.014 |
| 5 | 0.1250 | 3/16 | 0.237 | 0.215 | 0.053 | 0.043 | 0.043 | 0.035 | 0.024 | 0.016 |
| 6 | 0.1380 | 3/16 | 0.262 | 0.238 | 0.059 | 0.048 | 0.048 | 0.039 | 0.027 | 0.017 |
| 7 | 0.1510 | … | 0.287 | 0.262 | 0.064 | 0.053 | 0.048 | 0.039 | 0.030 | 0.019 |
| 8 | 0.1640 | 1/4 | 0.312 | 0.285 | 0.070 | 0.058 | 0.054 | 0.045 | 0.032 | 0.021 |
| 10 | 0.1900 | 5/16 | 0.362 | 0.333 | 0.081 | 0.068 | 0.060 | 0.050 | 0.037 | 0.024 |
| 12 | 0.2160 | 3/8 | 0.412 | 0.380 | 0.092 | 0.078 | 0.067 | 0.056 | 0.043 | 0.028 |
| 14 | 0.2420 | … | 0.462 | 0.427 | 0.103 | 0.089 | 0.086 | 0.064 | 0.047 | 0.032 |
| 1/4 | 0.2500 | 7/16 | 0.477 | 0.442 | 0.107 | 0.092 | 0.075 | 0.064 | 0.050 | 0.032 |
| 16 | 0.2680 | … | 0.512 | 0.475 | 0.114 | 0.099 | 0.075 | 0.064 | 0.053 | 0.035 |
| 18 | 0.2940 | … | 0.581 | 0.522 | 0.126 | 0.109 | 0.084 | 0.072 | 0.058 | 0.039 |
| 5/16 | 0.3125 | 1/2 | 0.597 | 0.556 | 0.134 | 0.116 | 0.084 | 0.072 | 0.062 | 0.041 |
| 20 | 0.3200 | … | 0.611 | 0.569 | 0.137 | 0.119 | 0.084 | 0.072 | 0.063 | 0.042 |
| 24 | 0.3720 | … | 0.711 | 0.664 | 0.159 | 0.139 | 0.094 | 0.081 | 0.074 | 0.049 |
| 3/8 | 0.3750 | 9/16 | 0.717 | 0.670 | 0.161 | 0.140 | 0.094 | 0.081 | 0.075 | 0.049 |
| 7/16 | 0.4375 | 5/8 | 0.760 | 0.715 | 0.156 | 0.133 | 0.094 | 0.081 | 0.072 | 0.045 |
| 1/2 | 0.5000 | 3/4 | 0.815 | 0.765 | 0.156 | 0.130 | 0.106 | 0.091 | 0.072 | 0.046 |

[a] When specifying nominal size in decimals, zeros preceding the decimal point and in the fourth decimal place are omitted.

[b] These lengths or shorter are undercut.

PLAIN AND SLOTTED HEX WASHER HEAD TYPES

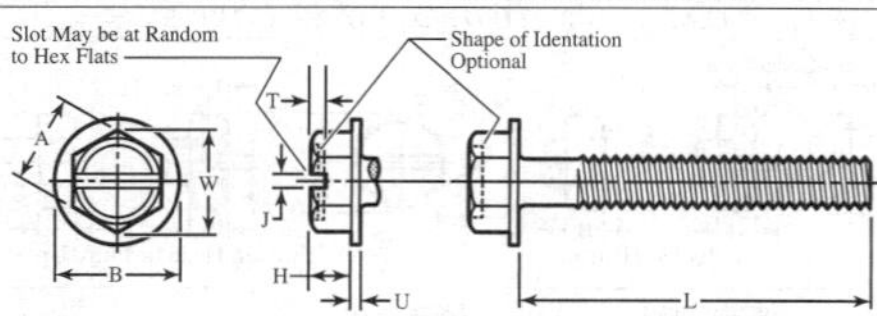


| Nominal Size[a] or Basic Screw Dia. | | Width Across Flats, A | | Width Across-Corners, W | Head Height, H | | Washer Dia., B | | Washer Thickness, U | | Slot Width, J | | Slot[a] Depth, T | |
|---|---|---|---|---|---|---|---|---|---|---|---|---|---|---|
| | | Max. | Min. | Min. | Max. | Min. | Max. | Min. | Max. | Min. | Max. | Min. | Max. | Min. |
| 2 | 0.0860 | 0.125 | 0.120 | 0.134 | 0.050 | 0.040 | 0.166 | 0.154 | 0.016 | 0.010 | …. | …. | …. | …. |
| 3 | 0.0990 | 0.125 | 0.120 | 0.134 | 0.055 | 0.044 | 0.177 | 0.163 | 0.016 | 0.010 | …. | …. | …. | …. |
| 4 | 0.1120 | 0.188 | 0.181 | 0.202 | 0.060 | 0.049 | 0.243 | 0.225 | 0.019 | 0.011 | 0.039 | 0.031 | 0.042 | 0.025 |
| 5 | 0.1250 | 0.188 | 0.181 | 0.202 | 0.070 | 0.058 | 0.260 | 0.240 | 0.025 | 0.015 | 0.043 | 0.035 | 0.049 | 0.030 |
| 6 | 0.1380 | 0.250 | 0.244 | 0.272 | 0.093 | 0.080 | 0.328 | 0.302 | 0.025 | 0.015 | 0.048 | 0.039 | 0.053 | 0.033 |
| 8 | 0.1640 | 0.250 | 0.244 | 0.272 | 0.110 | 0.096 | 0.348 | 0.322 | 0.031 | 0.019 | 0.054 | 0.045 | 0.074 | 0.052 |
| 10 | 0.1900 | 0.312 | 0.305 | 0.340 | 0.120 | 0.105 | 0.414 | 0.384 | 0.031 | 0.019 | 0.060 | 0.050 | 0.080 | 0.057 |
| 12 | 0.2160 | 0.312 | 0.305 | 0.340 | 0.155 | 0.139 | 0.432 | 0.398 | 0.039 | 0.022 | 0.067 | 0.056 | 0.103 | 0.077 |
| 1/4 | 0.2500 | 0.375 | 0.367 | 0.409 | 0.190 | 0.172 | 0.520 | 0.480 | 0.050 | 0.030 | 0.075 | 0.064 | 0.111 | 0.083 |
| 5/16 | 0.3125 | 0.500 | 0.489 | 0.545 | 0.230 | 0.208 | 0.676 | 0.624 | 0.055 | 0.035 | 0.084 | 0.072 | 0.134 | 0.100 |
| 3/8 | 0.3750 | 0.562 | 0.551 | 0.614 | 0.295 | 0.270 | 0.780 | 0.720 | 0.063 | 0.037 | 0.094 | 0.081 | 0.168 | 0.131 |

[a] Unless otherwise specified, hexagon washer head machine screws are not slotted.

All dimensions are in inches.

**Table 5. American National Standard Slotted Truss Head and Plain and Slotted Hexagon Head Machine Screws** *ANSI/ASME B18.6.3-2013*

SLOTTED TRUSS HEAD TYPE

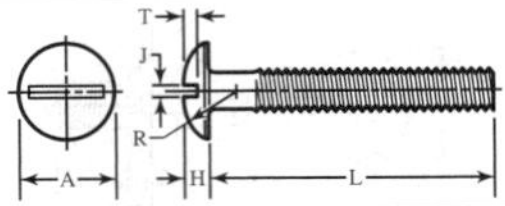


| Nominal Size[a] or Basic Screw Dia. | | Head Dia., *A* | | Head Height, *H* | | Head Radius, *R* | Slot Width, *J* | | Slot Depth, *T* | |
|---|---|---|---|---|---|---|---|---|---|---|
| | | Max. | Min. | Max. | Min. | Max. | Max. | Min. | Max. | Min. |
| 0000 | 0.0210 | 0.049 | 0.043 | 0.014 | 0.010 | 0.032 | 0.009 | 0.005 | 0.009 | 0.005 |
| 000 | 0.0340 | 0.077 | 0.071 | 0.022 | 0.018 | 0.051 | 0.013 | 0.009 | 0.013 | 0.009 |
| 00 | 0.0470 | 0.106 | 0.098 | 0.030 | 0.024 | 0.070 | 0.017 | 0.010 | 0.018 | 0.012 |
| 0 | 0.0600 | 0.131 | 0.119 | 0.037 | 0.029 | 0.087 | 0.023 | 0.016 | 0.022 | 0.014 |
| 1 | 0.0730 | 0.164 | 0.149 | 0.045 | 0.037 | 0.107 | 0.026 | 0.019 | 0.027 | 0.018 |
| 2 | 0.0860 | 0.194 | 0.180 | 0.053 | 0.044 | 0.129 | 0.031 | 0.023 | 0.031 | 0.022 |
| 3 | 0.0990 | 0.226 | 0.211 | 0.061 | 0.051 | 0.151 | 0.035 | 0.027 | 0.036 | 0.026 |
| 4 | 0.1120 | 0.257 | 0.241 | 0.069 | 0.059 | 0.169 | 0.039 | 0.031 | 0.040 | 0.030 |
| 5 | 0.1250 | 0.289 | 0.272 | 0.078 | 0.066 | 0.191 | 0.043 | 0.035 | 0.045 | 0.034 |
| 6 | 0.1380 | 0.321 | 0.303 | 0.086 | 0.074 | 0.211 | 0.048 | 0.039 | 0.050 | 0.037 |
| 7 | 0.1510 | 0.352 | 0.333 | 0.094 | 0.081 | 0.231 | 0.048 | 0.039 | 0.054 | 0.041 |
| 8 | 0.1640 | 0.384 | 0.364 | 0.102 | 0.088 | 0.254 | 0.054 | 0.045 | 0.058 | 0.045 |
| 10 | 0.1900 | 0.448 | 0.425 | 0.118 | 0.103 | 0.283 | 0.060 | 0.050 | 0.068 | 0.053 |
| 12 | 0.2160 | 0.511 | 0.487 | 0.134 | 0.118 | 0.336 | 0.067 | 0.056 | 0.077 | 0.061 |
| 14 | 0.2420 | 0.557 | 0.530 | 0.146 | 0.129 | 0.375 | 0.075 | 0.064 | 0.085 | 0.068 |
| 1/4 | 0.2500 | 0.573 | 0.546 | 0.150 | 0.133 | 0.375 | 0.075 | 0.064 | 0.087 | 0.070 |
| 16 | 0.2680 | 0.609 | 0.580 | 0.159 | 0.141 | 0.410 | 0.075 | 0.064 | 0.093 | 0.074 |
| 18 | 0.2940 | 0.661 | 0.630 | 0.173 | 0.153 | 0.446 | 0.084 | 0.072 | 0.100 | 0.080 |
| 5/16 | 0.3125 | 0.698 | 0.666 | 0.183 | 0.162 | 0.457 | 0.084 | 0.072 | 0.106 | 0.085 |
| 20 | 0.3200 | 0.713 | 0.680 | 0.186 | 0.165 | 0.484 | 0.084 | 0.072 | 0.108 | 0.087 |
| 24 | 0.3720 | 0.817 | 0.780 | 0.213 | 0.190 | 0.557 | 0.081 | 0.081 | 0.123 | 0.100 |
| 3/8 | 0.3750 | 0.823 | 0.787 | 0.215 | 0.191 | 0.538 | 0.094 | 0.081 | 0.124 | 0.100 |
| 7/16 | 0.4375 | 0.948 | 0.907 | 0.248 | 0.221 | 0.619 | 0.094 | 0.081 | 0.142 | 0.116 |
| 1/2 | 0.5000 | 1.073 | 1.028 | 0.280 | 0.250 | 0.701 | 0.106 | 0.091 | 0.161 | 0.131 |
| 9/16 | 0.5625 | 1.198 | 1.149 | 0.312 | 0.279 | 0.783 | 0.118 | 0.102 | 0.179 | 0.146 |
| 5/8 | 0.6250 | 1.323 | 1.269 | 0.345 | 0.309 | 0.863 | 0.133 | 0.116 | 0.196 | 0.162 |
| 3/4 | 0.7500 | 1.573 | 1.511 | 0.410 | 0.368 | 1.024 | 0.149 | 0.131 | 0.234 | 0.182 |

[a] Where specifying nominal size in decimals, zeros preceding decimal points and in the fourth decimal place are omitted.

PLAIN AND SLOTTED HEXAGON HEAD TYPES

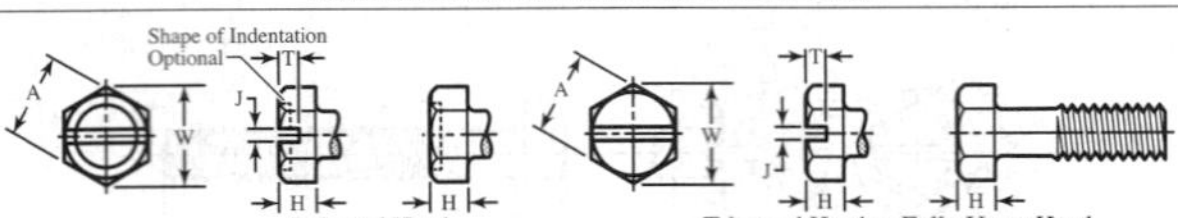


| Nominal Size[a] or Basic Screw Dia. | | Regular Head: Width Across Flats, *A* | | Regular Head: Across Corners, *W* | Large Head: Width Across Flats, *A* | | Large Head: Across Corners, *W* | Head Height, *H* | | Slot Width, *J* | | Slot[a] Depth, *T* | |
|---|---|---|---|---|---|---|---|---|---|---|---|---|---|
| | | Max. | Min. | Min. | Max. | Min. | Min. | Max. | Min. | Max. | Min. | Max. | Min. |
| 1 | 0.0730 | .125 | .120 | .134 | … | … | … | .044 | .036 | … | … | … | … |
| 2 | 0.0860 | .125 | .120 | .134 | … | … | … | .050 | .040 | … | … | … | … |
| 3 | 0.0990 | .188 | .181 | .202 | … | … | … | .055 | .044 | … | … | … | … |
| 4 | 0.1120 | .188 | .181 | .202 | .219 | .213 | .238 | .060 | .049 | .039 | .031 | .036 | .02 |
| 5 | 0.1250 | .188 | .181 | .202 | .250 | .244 | .272 | .070 | .058 | .043 | .035 | .042 | .03 |
| 6 | 0.1380 | .250 | .244 | .272 | … | … | … | .093 | .080 | .048 | .039 | .046 | .03 |
| 8 | 0.1640 | .250 | .244 | .272 | .312 | .305 | .340 | .110 | .096 | .054 | .045 | .066 | .05 |
| 10 | 0.1900 | .312 | .305 | .340 | … | … | … | .120 | .105 | .060 | .050 | .072 | .057 |
| 12 | 0.2160 | .312 | .305 | .340 | .375 | .367 | .409 | .155 | .139 | .067 | .056 | .093 | .07 |
| 1/4 | 0.2500 | .375 | .367 | .409 | .438 | .428 | .477 | .190 | .172 | .075 | .064 | .101 | .08 |
| 5/16 | 0.3125 | .500 | .489 | .545 | … | … | … | .230 | .208 | .084 | .072 | .122 | .10 |
| 3/8 | 0.3750 | .562 | .551 | .614 | … | … | … | .295 | .270 | .094 | .081 | .156 | .13 |

[a] Unless otherwise specified, hexagon head machine screws are not slotted.

All dimensions are in inches.

## Table 6. American National Standard Slotted Pan Head Machine Screws *ANSI/ASME B18.6.3-2013*

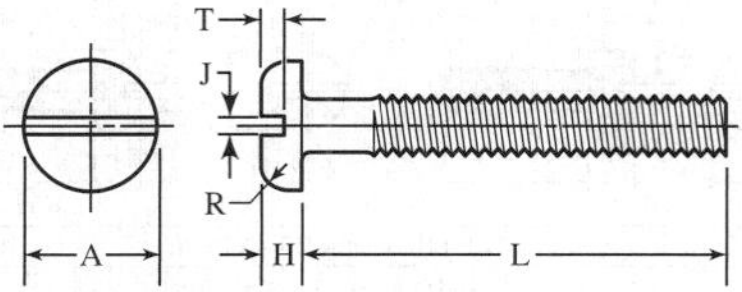


| Nominal Size[a] or Basic Screw Dia. | | Head Dia., *A* | | Head Height, *H* | | Head Radius, *R* | Slot Width, *J* | | Slot Depth, *T* | |
|---|---|---|---|---|---|---|---|---|---|---|
| | | Max. | Min. | Max. | Min. | Max. | Max. | Min. | Max. | Min. |
| 0000 | 0.0210 | .042 | .036 | .016 | .010 | .007 | .008 | .004 | .008 | .004 |
| 000 | 0.0340 | .066 | .060 | .023 | .017 | .010 | .012 | .008 | .012 | .008 |
| 00 | 0.0470 | .090 | .082 | .032 | .025 | .015 | .017 | .010 | .016 | .010 |
| 0 | 0.0600 | .116 | .104 | .039 | .031 | .020 | .023 | .016 | .022 | .014 |
| 1 | 0.0730 | .142 | .130 | .046 | .038 | .025 | .026 | .019 | .027 | .018 |
| 2 | 0.0860 | .167 | .155 | .053 | .045 | .035 | .031 | .023 | .031 | .022 |
| 3 | 0.0990 | .193 | .180 | .060 | .051 | .037 | .035 | .027 | .036 | .026 |
| 4 | 0.1120 | .219 | .205 | .068 | .058 | .042 | .039 | .031 | .040 | .030 |
| 5 | 0.1250 | .245 | .231 | .075 | .065 | .044 | .043 | .035 | .045 | .034 |
| 6 | 0.1380 | .270 | .256 | .082 | .072 | .046 | .048 | .039 | .050 | .037 |
| 7 | 0.1510 | .296 | .281 | .089 | .079 | .049 | .048 | .039 | .054 | .041 |
| 8 | 0.1640 | .322 | .306 | .096 | .085 | .052 | .054 | .045 | .058 | .045 |
| 10 | 0.1900 | .373 | .357 | .110 | .099 | .061 | .060 | .050 | .068 | .053 |
| 12 | 0.2160 | .425 | .407 | .125 | .112 | .078 | .067 | .056 | .077 | .061 |
| 14 | 0.2420 | .476 | .457 | .139 | .126 | .087 | .075 | .064 | .085 | .068 |
| 1/4 | 0.2500 | .492 | .473 | .144 | .130 | .087 | .075 | .064 | .087 | .070 |
| 16 | 0.2680 | .528 | .508 | .153 | .139 | .094 | .075 | .064 | .093 | .074 |
| 18 | 0.2940 | .579 | .558 | .168 | .153 | .099 | .084 | .072 | .100 | .080 |
| 5/16 | 0.3125 | .615 | .594 | .178 | .162 | .099 | .084 | .072 | .106 | .085 |
| 20 | 0.3200 | .631 | .608 | .182 | .166 | .121 | .084 | .072 | .108 | .087 |
| 24 | 0.3720 | .734 | .709 | .211 | .193 | .143 | .094 | .081 | .123 | .100 |
| 3/8 | 0.3750 | .740 | .716 | .212 | .195 | .143 | .094 | .081 | .124 | .100 |
| 7/16 | 0.4375 | .863 | .837 | .247 | .228 | .153 | .094 | .081 | .142 | .116 |
| 1/2 | 0.5000 | .987 | .958 | .281 | .260 | .175 | .106 | .091 | .161 | .131 |
| 9/16 | 0.5625 | 1.041 | 1.000 | .315 | .293 | .197 | .118 | .102 | .179 | .146 |
| 5/8 | 0.6250 | 1.172 | 1.125 | .350 | .325 | .219 | .133 | .116 | .197 | .162 |
| 3/4 | 0.7500 | 1.435 | 1.375 | .419 | .390 | .263 | .149 | .131 | .234 | .192 |

[a] Where specifying nominal size in decimals, zeros preceding decimal and in the fourth decimal place are omitted.

All dimensions are in inches.

## Table 7. Nos. 0000, 000 and 00 Threads *ANSI/ASME B18.6.3-2013 Appendix*

| Nominal Size[a] and Threads Per Inch | Series Designat. | External[b] | | | | | | | Internal[c] | | | | |
|---|---|---|---|---|---|---|---|---|---|---|---|---|---|
| | | Class | Major Diameter | | Pitch Diameter | | | Minor Dia. | Class | Pitch Diameter | | | Major Dia. |
| | | | Max. | Min. | Max. | Min. | Tol. | | | Min. | Max. | Tol. | Min. |
| 0000-160 or 0.0210-160 | NS | 2 | .0210 | .0195 | .0169 | .0158 | .0011 | .0128 | 2 | .0169 | .0181 | .0012 | .0210 |
| 000-120 or 0.0340-120 | NS | 2 | .0340 | .0325 | .0286 | 0.272 | .0014 | .0232 | 2 | .0286 | .0300 | .0014 | .034 |
| 00-90 or 0.0470-90 | NS | 2 | .0470 | .0450 | .0398 | .0382 | .0016 | .0326 | 2 | .0398 | .0414 | .0016 | .047 |
| 00-96 or 0.0470-96 | NS | 2 | .0470 | .0450 | .0402 | .0386 | .0016 | .0334 | 2 | .0402 | .0418 | .0016 | .047 |

[a] Where specifying nominal size in decimals, zeros preceding decimal and in the fourth decimal place are omitted.

[b] There is no allowance provided on the external threads.

[c] The minor diameter limits for internal threads are not specified, they being determined by the amount of thread engagement necessary to satisfy the strength requirements and tapping performance in the intended application.

All dimensions are in inches.

## Table 8. American National Standard Slotted Fillister and Slotted Drilled Fillister Head Machine Screws *ANSI/ASME B18.6.3-2013*

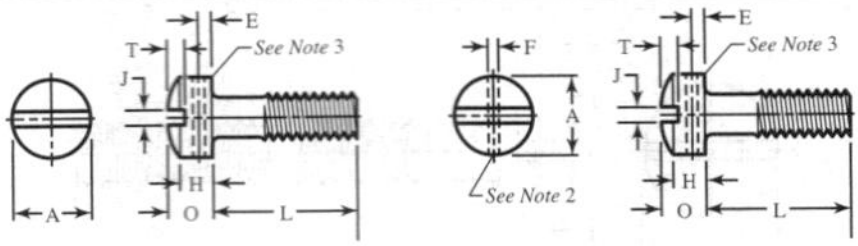


SLOTTED FILLISTER HEAD TYPE

| Nominal Size[1] or Basic Screw Dia. | | Head Dia., *A* | | Head Side Height, *H* | | Total Head Height, *O* | | Slot Width, *J* | | Slot Depth, *T* | |
|---|---|---|---|---|---|---|---|---|---|---|---|
| | | Max. | Min. | Max. | Min. | Max. | Min. | Max | Min. | Max. | Min. |
| 0000 | 0.0210 | .038 | .032 | .019 | .011 | .025 | .015 | .008 | .004 | .012 | .006 |
| 000 | 0.0340 | .059 | .053 | .029 | .021 | .035 | .027 | .012 | .006 | .017 | .011 |
| 00 | 0.0470 | .082 | .072 | .037 | .028 | .047 | .039 | .017 | .010 | .022 | .015 |
| 0 | 0.0600 | .096 | .083 | .043 | .038 | .055 | .047 | .023 | .016 | .025 | .015 |
| 1 | 0.0730 | .118 | .104 | .053 | .045 | .066 | .058 | .026 | .019 | .031 | .020 |
| 2 | 0.0860 | .140 | .124 | .062 | .053 | .083 | .066 | .031 | .023 | .037 | .025 |
| 3 | 0.0990 | .161 | .145 | .070 | .061 | .095 | .077 | .035 | .027 | .043 | .030 |
| 4 | 0.1120 | .183 | .166 | .079 | .069 | .107 | .088 | .039 | .031 | .048 | .035 |
| 5 | 0.1250 | .205 | .187 | .088 | .078 | .120 | .100 | .043 | .035 | .054 | .040 |
| 6 | 0.1380 | .226 | .208 | .096 | .086 | .132 | .111 | .048 | .039 | .060 | .045 |
| 8 | 0.1640 | .270 | .250 | .113 | .102 | .156 | .133 | .054 | .045 | .071 | .054 |
| 10 | 0.1900 | .313 | .292 | .130 | .118 | .180 | .156 | .060 | .050 | .083 | .064 |
| 12 | 0.2160 | .357 | .334 | .148 | .134 | .205 | .178 | .067 | .056 | .094 | .074 |
| 1/4 | 0.2500 | .414 | .389 | .170 | .155 | .237 | .207 | .075 | .064 | .109 | .087 |
| 5/16 | 0.3125 | .518 | .490 | .211 | .194 | .295 | .262 | .084 | .072 | .137 | .110 |
| 3/8 | 0.3750 | .622 | .590 | .253 | .233 | .355 | .315 | .094 | .081 | .164 | .133 |
| 7/16 | 0.4375 | .625 | .589 | .265 | .242 | .368 | .321 | .094 | .081 | .170 | .135 |
| 1/2 | 0.5000 | .750 | .710 | .297 | .273 | .412 | .362 | .106 | .091 | .190 | .151 |
| 9/16 | 0.5625 | .812 | .768 | .336 | .308 | .466 | .410 | .118 | .102 | .214 | .172 |
| 5/8 | 0.6250 | .875 | .827 | .375 | .345 | .521 | .461 | .133 | .116 | .240 | .193 |
| 3/4 | 0.7500 | 1.000 | .945 | .441 | .406 | .612 | .542 | .149 | .131 | .281 | .226 |

SLOTTED DRILLED FILLISTER HEAD TYPE

| Nominal Size[1] or Basic Screw Dia. | | Head Dia., *A* | | Head Side Height, *H* | | Total Head Height, *O* | | Slot Width, *J* | | Slot Depth, *T* | | Drilled Hole Locat., *E* | Drilled Hole. Dia., *F* |
|---|---|---|---|---|---|---|---|---|---|---|---|---|---|
| | | Max. | Min. | Max. | Min. | Max. | Min. | Max. | Min. | Max. | Min. | Basic | Basic |
| 2 | 0.0860 | .140 | .124 | .062 | .055 | .083 | .070 | .031 | .023 | .030 | .022 | .026 | .031 |
| 3 | 0.0990 | .161 | .145 | .070 | .064 | .095 | .082 | .035 | .027 | .034 | .026 | .030 | .037 |
| 4 | 0.1120 | .183 | .166 | .079 | .072 | .107 | .094 | .039 | .031 | .038 | .030 | .035 | .037 |
| 5 | 0.1250 | .205 | .187 | .088 | .081 | .120 | .106 | .043 | .035 | .042 | .033 | .038 | .046 |
| 6 | 0.1380 | .226 | .208 | .096 | .089 | .132 | .118 | .048 | .039 | .045 | .035 | .043 | .046 |
| 8 | 0.1640 | .270 | .250 | .113 | .106 | .156 | .141 | .054 | .045 | .065 | .054 | .043 | .046 |
| 10 | 0.1900 | .313 | .292 | .130 | .123 | .180 | .165 | .060 | .050 | .075 | .064 | .043 | .046 |
| 12 | 0.2160 | .357 | .334 | .148 | .139 | .205 | .188 | .067 | .056 | .087 | .074 | .053 | .046 |
| 1/4 | 0.2500 | .414 | .389 | .170 | .161 | .237 | .219 | .075 | .064 | .102 | .087 | .062 | .062 |
| 5/16 | 0.3125 | .518 | .490 | .211 | .201 | .295 | .276 | .084 | .072 | .130 | .110 | .078 | .070 |
| 3/8 | 0.3750 | .622 | .590 | .253 | .242 | .355 | .333 | .094 | .081 | .154 | .134 | .094 | .070 |

All dimensions are in inches.

[1] Where specifying nominal size in decimals, zeros preceding decimal points and in the fourth decimal place are omitted.

[2] Drilled hole shall be approximately perpendicular to the axis of slot and may be permitted to break through bottom of the slot. Edges of the hole shall be free from burrs.

[3] A slight rounding of the edges at periphery of head is permissible provided the diameter of the bearing circle is equal to no less than 90 percent of the specified minimum head diameter.

## Table 9. American National Standard Slotted Oval Countersunk Head Machine Screws *ANSI/ASME B18.6.3-2013*

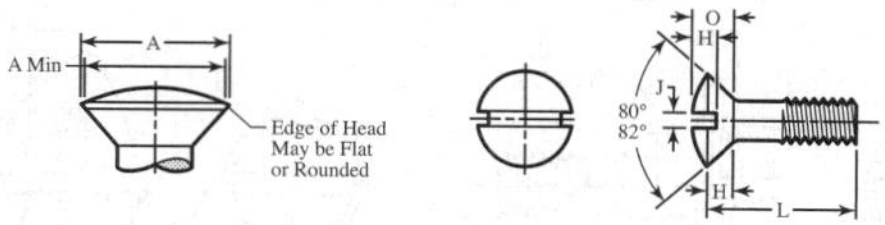


| Nominal Size[a] or Basic Screw Dia. | | Max $L^b$ | Head Dia., *A* Max., Edge Sharp | Head Dia., *A* Min., Edge Rounded or Flat | Head Side Height, *H* Ref. | Total Head Height, *O* Max. | Total Head Height, *O* Min. | Slot Width, *J* Max. | Slot Width, *J* Min. | Slot Depth, *T* Max. | Slot Depth, *T* Min. |
|---|---|---|---|---|---|---|---|---|---|---|---|
| 00 | 0.0470 | … | .087 | .080 | .028 | .042 | .034 | .017 | .010 | .023 | .016 |
| 0 | 0.0600 | 1/8 | .112 | .096 | .035 | .056 | .041 | .023 | .016 | .030 | .025 |
| 1 | 0.0730 | 1/8 | .137 | .120 | .043 | .068 | .052 | .026 | .019 | .038 | .031 |
| 2 | 0.0860 | 1/8 | .162 | .144 | .051 | .080 | .063 | .031 | .023 | .045 | .037 |
| 3 | 0.0990 | 1/8 | .187 | .167 | .059 | .092 | .073 | .035 | .027 | .052 | .043 |
| 4 | 0.1120 | 3/16 | .212 | .191 | .067 | .104 | .084 | .039 | .031 | .059 | .049 |
| 5 | 0.1250 | 3/16 | .237 | .215 | .075 | .116 | .095 | .043 | .035 | .067 | .055 |
| 6 | 0.1380 | 3/16 | .262 | .238 | .083 | .128 | .105 | .048 | .039 | .074 | .060 |
| 7 | 0.1510 | … | .287 | .262 | .091 | .140 | .115 | .048 | .039 | .081 | .066 |
| 8 | 0.1640 | 1/4 | .312 | .285 | .100 | .152 | .126 | .054 | .045 | .088 | .072 |
| 10 | 0.1900 | 5/16 | .362 | .333 | .116 | .176 | .148 | .060 | .050 | .103 | .084 |
| 12 | 0.2160 | 3/8 | .412 | .380 | .132 | .200 | .169 | .067 | .056 | .117 | .096 |
| 14 | 0.2420 | … | .462 | .427 | .148 | .224 | .190 | .075 | .064 | .132 | .108 |
| 1/4 | 0.2500 | 7/16 | .477 | .442 | .153 | .232 | .197 | .075 | .064 | .136 | .112 |
| 16 | 0.2680 | … | .512 | .475 | .164 | .248 | .212 | .075 | .064 | .146 | .120 |
| 18 | 0.2940 | … | .561 | .522 | .180 | .272 | .233 | .084 | .072 | .160 | .132 |
| 5/16 | 0.3125 | 1/2 | .597 | .556 | .191 | .290 | .248 | .084 | .072 | .171 | .141 |
| 20 | 0.3200 | … | .611 | .569 | .196 | .296 | .254 | .094 | .072 | .175 | .144 |
| 24 | 0.3720 | … | .711 | .664 | .228 | .344 | .297 | .094 | .081 | .204 | .168 |
| 3/8 | 0.3750 | 9/16 | .717 | .670 | .230 | .347 | .300 | .094 | .081 | .206 | .170 |
| 7/16 | 0.4375 | 5/8 | .760 | .715 | .223 | .345 | .295 | .094 | .081 | .210 | .174 |
| 1/2 | 0.5000 | 3/4 | .815 | .765 | .223 | .354 | .299 | .106 | .091 | .216 | .176 |
| 9/16 | 0.5625 | … | .932 | .878 | .260 | .410 | .350 | .118 | .102 | .250 | .207 |
| 5/8 | 0.6250 | … | 1.050 | .990 | .298 | .467 | .399 | .133 | .116 | .285 | .235 |
| 3/4 | 0.7500 | … | 1.285 | 1.215 | .372 | .578 | .497 | .149 | .131 | .353 | .293 |

[a] When specifying nominal size in decimals, zeros preceding decimal points and in the fourth decimal place are omitted.

[b] These lengths or shorter are undercut.

All dimensions are in inches.

## Table 10. American National Standard Header Points for Machine Screws before Threading *ANSI/ASME B18.6.3-2013*

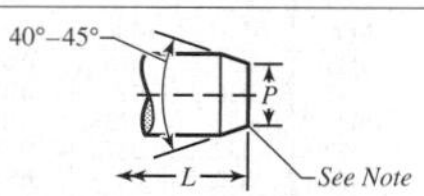


| Nom. Size. | Threads per Inch | Max. *P* | Min. *P* | Max. *L* |
|---|---|---|---|---|
| 2 | 56 | 0.057 | 0.050 | ½ |
| | 64 | 0.060 | 0.053 | |
| 4 | 40 | 0.074 | 0.065 | ½ |
| | 48 | 0.079 | 0.070 | |
| 5 | 40 | 0.086 | 0.076 | ½ |
| | 44 | 0.088 | 0.079 | |
| 6 | 32 | 0.090 | 0.080 | ¾ |
| | 40 | 0.098 | 0.087 | |
| 8 | 32 | 0.114 | 0.102 | 1 |
| | 36 | 0.118 | 0.106 | |
| 10 | 24 | 0.125 | 0.112 | 1¼ |
| | 32 | 0.138 | 0.124 | |
| 12 | 24 | 0.149 | 0.134 | 1⅜ |
| | 28 | 0.156 | 0.141 | |
| ¼ | 20 | 0.170 | 0.153 | 1½ |
| | 28 | 0.187 | 0.169 | |
| 5/16 | 18 | 0.221 | 0.200 | 1½ |
| | 24 | 0.237 | 0.215 | |
| ⅜ | 16 | 0.270 | 0.244 | 1½ |
| | 24 | 0.295 | 0.267 | |
| 7/16 | 14 | 0.316 | 0.287 | 1½ |
| | 20 | 0.342 | 0.310 | |
| ½ | 13 | 0.367 | 0.333 | 1½ |
| | 20 | 0.399 | 0.362 | |

All dimensions in inches. Edges of point may be rounded and end of point need not be flat nor perpendicular to shank. Machine screws normally have plain sheared ends but when specified may have header points, as shown above.

**Table 11. American National Standard Slotted Binding Head and Slotted Undercut Oval Countersunk Head Machine Screws** *ANSI/ASME B18.6.3-2013*

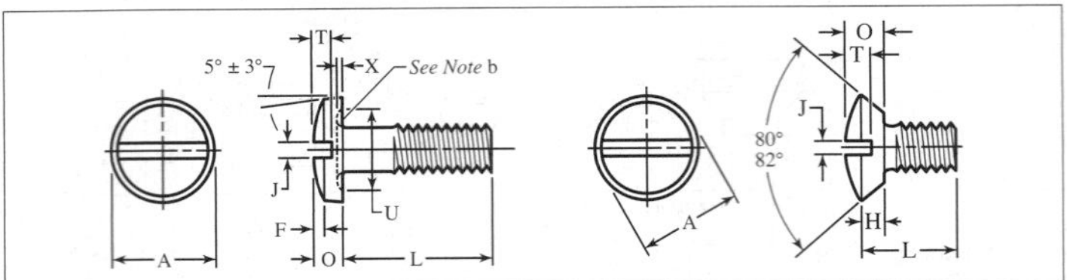


SLOTTED BINDING HEAD TYPE

| Nominal Size[a] or Basic Screw Dia. | | Head Dia., *A* | | Total Head Height, *O* | | Head Oval Height, *F* | | Slot Width, *J* | | Slot Depth, *T* | | Undercut[b] Dia., *U* | | Undercut[b] Depth, *X* | |
|---|---|---|---|---|---|---|---|---|---|---|---|---|---|---|---|
| | | Max. | Min. | Max. | Min. | Max. | Min. | Max. | Min. | Max. | Min. | Max. | Min. | Max. | Min. |
| 0000 | 0.0210 | .046 | .040 | .014 | .009 | .006 | .003 | .008 | .004 | .009 | .005 | … | … | … | … |
| 000 | 0.0340 | .073 | .067 | .021 | .015 | .008 | .005 | .012 | .006 | .013 | .009 | … | … | … | … |
| 00 | 0.0470 | .098 | .090 | .028 | .023 | .011 | .007 | .017 | .010 | .018 | .012 | … | … | … | … |
| 0 | 0.0600 | .126 | .119 | .032 | .026 | .012 | .008 | .023 | .016 | .018 | .009 | .098 | .086 | .007 | .002 |
| 1 | 0.0730 | .153 | .145 | .041 | .035 | .015 | .011 | .026 | .019 | .024 | .014 | .120 | .105 | .008 | .003 |
| 2 | 0.0860 | .181 | .171 | .050 | .043 | .018 | .013 | .031 | .023 | .030 | .020 | .141 | .124 | .010 | .005 |
| 3 | 0.0990 | .208 | .197 | .059 | .052 | .022 | .016 | .035 | .027 | .036 | .025 | .162 | .143 | .011 | .006 |
| 4 | 0.1120 | .235 | .223 | .068 | .061 | .025 | .018 | .039 | .031 | .042 | .030 | .184 | .161 | .012 | .007 |
| 5 | 0.1250 | .263 | .249 | .078 | .069 | .029 | .021 | .043 | .035 | .048 | .035 | .205 | .180 | .014 | .009 |
| 6 | 0.1380 | .290 | .275 | .087 | .078 | .032 | .024 | .048 | .039 | .053 | .040 | .226 | .199 | .015 | .010 |
| 8 | 0.1640 | .344 | .326 | .105 | .095 | .039 | .029 | .054 | .045 | .065 | .050 | .269 | .236 | .017 | .012 |
| 10 | 0.1900 | .399 | .378 | .123 | .112 | .045 | .034 | .060 | .050 | .077 | .060 | .312 | .274 | .020 | .015 |
| 12 | 0.2160 | .454 | .430 | .141 | .130 | .052 | .039 | .067 | .056 | .089 | .070 | .354 | .311 | .023 | .018 |
| 1⁄4 | 0.2500 | .525 | .498 | .165 | .152 | .061 | .046 | .075 | .064 | .105 | .084 | .410 | .360 | .026 | .021 |
| 5⁄16 | 0.3125 | .656 | .622 | .209 | .194 | .077 | .059 | .084 | .072 | .134 | .108 | .513 | .450 | .032 | .027 |
| 3⁄8 | 0.3750 | .788 | .746 | .253 | .235 | .094 | .071 | .094 | .081 | .163 | .132 | .615 | .540 | .039 | .034 |

[a] Where specifying nominal size in decimals, zeros preceding decimal points and in the fourth decimal place are omitted.

[b] Unless otherwise specified, slotted binding head machine screws are not undercut.

SLOTTED UNDERCUT OVAL COUNTERSUNK HEAD TYPES

| Nominal Size[a] or Basic Screw Dia. | | Max. *L*[a] | Head Dia., *A* | | Head Side Height, *H* | Total Head Height, *O* | | Slot Width, *J* | | Slot Depth, *T* | |
|---|---|---|---|---|---|---|---|---|---|---|---|
| | | | Max. | Min. | Ref. | Max. | Min. | Max. | Min. | Max. | Min. |
| 0 | 0.0600 | 1/8 | .112 | .096 | .025 | .046 | .033 | .023 | .016 | .028 | .022 |
| 1 | 0.0730 | 1/8 | .137 | .120 | .031 | .056 | .042 | .026 | .019 | .034 | .027 |
| 2 | 0.0860 | 1/8 | .162 | .144 | .036 | .065 | .050 | .031 | .023 | .040 | .033 |
| 3 | 0.0990 | 1/8 | .187 | .167 | .042 | .075 | .059 | .035 | .027 | .047 | .038 |
| 4 | 0.1120 | 3/16 | .212 | .191 | .047 | .084 | .067 | .039 | .031 | .053 | .043 |
| 5 | 0.1250 | 3/16 | .237 | .215 | .053 | .094 | .076 | .043 | .035 | .059 | .048 |
| 6 | 0.1380 | 3/16 | .262 | .238 | .059 | .104 | .084 | .048 | .039 | .065 | .053 |
| 7 | 0.1510 | … | .287 | .262 | .064 | .113 | .093 | .048 | .039 | .071 | .059 |
| 8 | 0.1640 | 1/4 | .312 | .285 | .070 | .123 | .101 | .054 | .045 | .078 | .064 |
| 10 | 0.1900 | 5/16 | .362 | .333 | .081 | .142 | .118 | .060 | .050 | .090 | .074 |
| 12 | 0.2160 | 3/8 | .412 | .380 | .092 | .161 | .135 | .067 | .056 | .103 | .085 |
| 14 | 0.2420 | … | .462 | .427 | .104 | .180 | .153 | .075 | .064 | .115 | .096 |
| 1/4 | 0.2500 | 7/16 | .477 | .442 | .107 | .186 | .158 | .075 | .064 | .119 | .098 |
| 16 | 0.2680 | … | .512 | .475 | .115 | .199 | .170 | .075 | .064 | .127 | .106 |
| 18 | 0.2940 | … | .561 | .522 | .126 | .218 | .187 | .084 | .072 | .140 | .116 |
| 5/16 | 0.3125 | 1/2 | .597 | .556 | .134 | .232 | .198 | .084 | .072 | .149 | .124 |
| 20 | 0.3200 | … | .611 | .569 | .137 | .237 | .204 | .084 | .072 | .152 | .127 |
| 24 | 0.3720 | … | .711 | .664 | .160 | .275 | .238 | .094 | .081 | .177 | .148 |
| 3/8 | 0.3750 | 9/16 | .717 | .670 | .161 | .278 | .239 | .094 | .081 | .179 | .149 |
| 7/16 | 0.4375 | 5/8 | .760 | .715 | .156 | .279 | .239 | .094 | .081 | .184 | .154 |
| 1/2 | 0.5000 | 3/4 | .815 | .765 | .156 | .288 | .244 | .106 | .091 | .204 | .169 |

[a] These lengths or shorter are undercut.

All dimensions are in inches.

**Table 12. Slotted Round Head Machine Screws**
*ANSI/ASME B18.6.3-2013 Appendix*

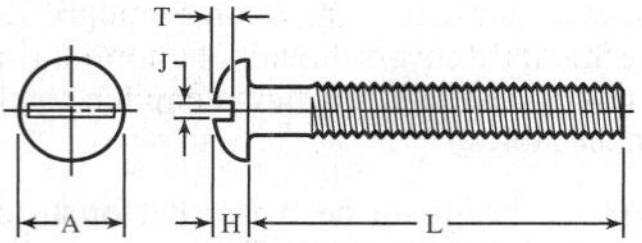


| Nominal Size[a] or Basic Screw Dia. | | Head Diameter, A | | Head Height, H | | Slot Width, J | | Slot Depth, T | |
|---|---|---|---|---|---|---|---|---|---|
| | | Max. | Min. | Max. | Min. | Max. | Min. | Max. | Min. |
| 0000 | 0.0210 | .041 | .035 | .022 | .016 | .008 | .004 | .017 | .013 |
| 000 | 0.0340 | .062 | .056 | .031 | .025 | .012 | .008 | .018 | .012 |
| 00 | 0.0470 | .089 | .080 | .045 | .036 | .017 | .010 | .026 | .018 |
| 0 | 0.0600 | .113 | .099 | .053 | .043 | .023 | .016 | .039 | .029 |
| 1 | 0.0730 | .138 | .122 | .061 | .051 | .026 | .019 | .044 | .033 |
| 2 | 0.0860 | .162 | .146 | .069 | .059 | .031 | .023 | .048 | .037 |
| 3 | 0.0990 | .187 | .169 | .078 | .067 | .035 | .027 | .053 | .040 |
| 4 | 0.1120 | .211 | .193 | .086 | .075 | .039 | .031 | .058 | .044 |
| 5 | 0.1250 | .236 | .217 | .095 | .083 | .043 | .035 | .063 | .047 |
| 6 | 0.1380 | .260 | .240 | .103 | .091 | .048 | .039 | .068 | .051 |
| 7 | 0.1510 | .285 | .264 | .111 | .099 | .048 | .039 | .072 | .055 |
| 8 | 0.1640 | .309 | .287 | .120 | .107 | .054 | .045 | .077 | .058 |
| 10 | 0.1900 | .359 | .334 | .137 | .123 | .060 | .050 | .087 | .065 |
| 12 | 0.2160 | .408 | .382 | .153 | .139 | .067 | .056 | .096 | .073 |
| 14 | 0.2420 | .457 | .429 | .170 | .155 | .075 | .064 | .106 | .080 |
| 1/4 | 0.2500 | .472 | .443 | .175 | .160 | .075 | .064 | .109 | .082 |
| 16 | 0.2680 | .506 | .476 | .187 | .171 | .075 | .064 | .115 | .087 |
| 18 | 0.2940 | .555 | .523 | .204 | .187 | .084 | .072 | .125 | .094 |
| 5/16 | 0.3125 | .590 | .557 | .216 | .198 | .084 | .072 | .132 | .099 |
| 20 | 0.3200 | .604 | .570 | .220 | .203 | .084 | .072 | .134 | .101 |
| 24 | 0.3720 | .702 | .664 | .254 | .235 | .094 | .081 | .154 | .116 |
| 3/8 | 0.3750 | .708 | .670 | .256 | .237 | .094 | .081 | .155 | .117 |
| 7/16 | 0.4375 | .750 | .707 | .328 | .307 | .094 | .081 | .196 | .148 |
| 1/2 | 0.5000 | .813 | .766 | .355 | .332 | .106 | .091 | .211 | .159 |
| 9/16 | 0.5625 | .938 | .887 | .410 | .385 | .118 | .102 | .242 | .183 |
| 5/8 | 0.6250 | 1.000 | .944 | .438 | .411 | .133 | .116 | .258 | .195 |
| 3/4 | 0.7500 | 1.250 | 1.185 | .547 | .516 | .149 | .131 | .320 | .242 |

[a] When specifying nominal size in decimals, zeros preceding decimal point and in the fourth decimal place are omitted.

All dimensions are in inches.

Not recommended, use Pan Head machine screws.

**Machine Screw Cross Recesses.—**Four cross recesses, Types I, IA, II, and III, may be used in lieu of slots in machine screw heads. Dimensions for recess diameter *M*, width *N*, and depth *T* (not shown above) together with recess penetration gaging depths are given in American National Standard ANSI/ASME B18.6.3-2013 for machine screws, and in ANSI/ASME B18.6.7M-1999 (Withdrawn) for metric machine screws.

**ANSI/ASME Cross Recesses for Machine Screws and Metric Machine Screw**

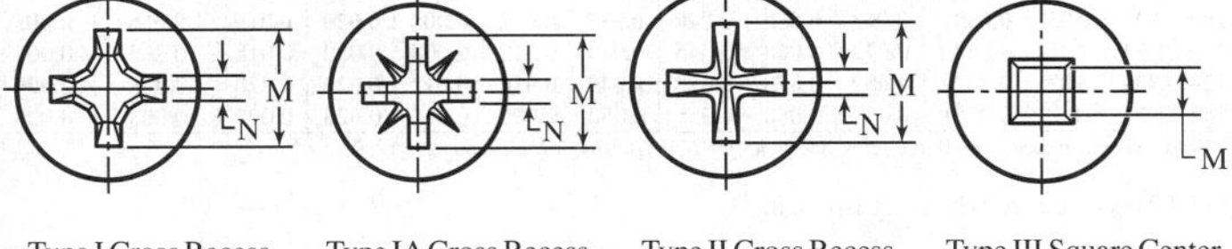


Type I Cross Recess Type IA Cross Recess Type II Cross Recess Type III Square Center

## Slotted Head Miniature Screws

ASA B18.11 establishes head types, dimensions, and lengths of slotted head miniature screws, threaded in conformance with American Standard Unified Miniature Screw Threads, ASA B1.10. The Standard covers threads of a nominal diameter from 0.0118 inch (0.3 mm) to 0.0551 inch (1.4 mm). Preferred diameter pitch combinations for general use are shown in bold type in the tables.

**Head Types.—***Fillister Head:* The fillister head has a flat top surface (oval crown optional) with cylindrical sides and a flat bearing surface. The head proportions are given in Table 1.

*Pan Head:* The pan head has a flat top surface, cylindrical sides, and a flat bearing surface. The head height is less than the fillister but the head diameter is slightly larger. Head proportions are given in Table 2.

*Flat Head:* The flat head has a flat top surface and a conical bearing surface with an included angle of approximately 100°. Head proportions are given in Table 3.

*Binding Head:* The head height is less than the pan head but the head diameter is greater, and is intended for applications which would otherwise require washers. Head proportions are given in Table 4.

**Table 1. Miniature Screws - Fillister Head** *ASA B18.11-1961, R2010*

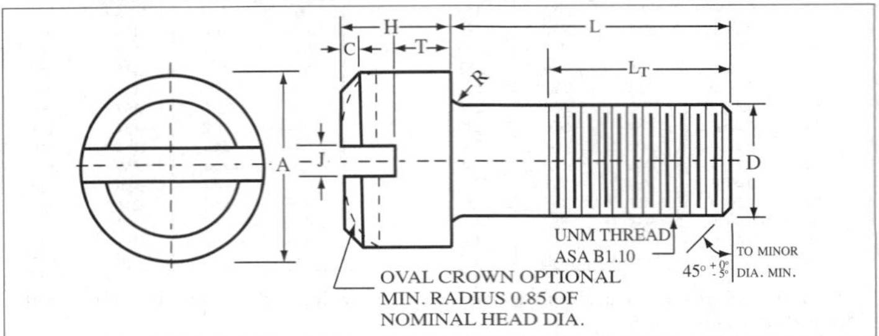


| Size Desig-nation | Thds per Inch | *D* Basic Major Dia. | Fillister Head Dimensions | | | | | | | | | |
|---|---|---|---|---|---|---|---|---|---|---|---|---|
| | | | *A* Head Dia. | | *H* Head Height | | *J* Slot Width | | *T* Slot Depth [a] | | *C* Chamfer | *R* Radius[b] |
| | | Max. | Max. | Min. | Max. | Min. | Max. | Min. | Max. | Min. | Max. | Min. |
| **30 UNM** | **318** | **0.0118** | **0.021** | **0.019** | **0.012** | **0.010** | **0.004** | **0.003** | **0.006** | **0.004** | **0.002** | **0.002** |
| 35 UNM | 282 | 0.0138 | 0.023 | 0.021 | 0.014 | 0.012 | 0.004 | 0.003 | 0.007 | 0.005 | 0.002 | 0.002 |
| **40 UNM** | **254** | **0.0157** | **0.025** | **0.023** | **0.016** | **0.013** | **0.005** | **0.003** | **0.008** | **0.006** | **0.002** | **0.002** |
| 45 UNM | 254 | 0.0177 | 0.029 | 0.027 | 0.018 | 0.015 | 0.005 | 0.003 | 0.009 | 0.007 | 0.002 | 0.002 |
| **50 UNM** | **203** | **0.0197** | **0.033** | **0.031** | **0.020** | **0.017** | **0.006** | **0.004** | **0.010** | **0.007** | **0.003** | **0.002** |
| 55 UNM | 203 | 0.0217 | 0.037 | 0.035 | 0.022 | 0.019 | 0.006 | 0.004 | 0.011 | 0.008 | 0.003 | 0.002 |
| **60 UNM** | **169** | **0.0236** | **0.041** | **0.039** | **0.025** | **0.021** | **0.008** | **0.005** | **0.012** | **0.009** | **0.004** | **0.003** |
| 70 UNM | 145 | 0.0276 | 0.045 | 0.043 | 0.028 | 0.024 | 0.008 | 0.005 | 0.014 | 0.011 | 0.004 | 0.003 |
| **80 UNM** | **127** | **0.0315** | **0.051** | **0.049** | **0.032** | **0.028** | **0.010** | **0.007** | **0.016** | **0.012** | **0.005** | **0.004** |
| 90 UNM | 113 | 0.0354 | 0.056 | 0.054 | 0.036 | 0.032 | 0.010 | 0.007 | 0.018 | 0.014 | 0.005 | 0.004 |
| **100 UNM** | **102** | **0.0394** | **0.062** | **0.058** | **0.040** | **0.035** | **0.012** | **0.008** | **0.020** | **0.016** | **0.006** | **0.005** |
| 110 UNM | 102 | 0.0433 | 0.072 | 0.068 | 0.045 | 0.040 | 0.012 | 0.008 | 0.022 | 0.018 | 0.006 | 0.005 |
| **120 UNM** | **102** | **0.0472** | **0.082** | **0.078** | **0.050** | **0.045** | **0.016** | **0.012** | **0.025** | **0.020** | **0.008** | **0.006** |
| 140 UNM | 85 | 0.0551 | 0.092 | 0.088 | 0.055 | 0.050 | 0.016 | 0.012 | 0.028 | 0.023 | 0.008 | 0.006 |

Bold face type indicates preferred sizes. See *Notes for Table 1 through Table 4* on page 1789.

[a] *T* measured from bearing surface.

[b] Relative to maximum major diameter.

## Notes for Table 1 through Table 4

*Material:*

Corrosion resistant steels: ASTM Designation A276

CLASS 303, COND A

CLASS 416, COND A, heat treat to approx 120,000–150,000 PSI (28–34 on Rockwell C scale)

CLASS 420, COND A, heat treat to approx 220,000–240,000 PSI (50–53 on Rockwell C scale)

Brass: Temper half hard ASTM Designation B16

Nickel Silver: Temper hard ASTM Designation B151, Alloy C

*Machine Finish:*

Machined surface roughness of heads shall be approximately 63 μin. arithmetical average determined by visual comparison.

*Applied coatings:*

Corrosion resistant steel: passivate; Brass: bare, black oxide, or nickel flash.

Nickel silver: none

*Notes:*

1) The diameter of the unthreaded body shall not be more than the maximum major diameter nor less than the minimum pitch diameter of the thread.
2) For screw lengths four times the major diameter or less, thread length ($L_T$) shall extend to within two threads of the head bearing surface. Screws of greater length shall have complete threads for a minimum of four major diameters.
3) Screws shall be free of all projecting burrs, observed at 3× magnification.
4) All dimensions are in inches.

## Table 2. Miniature Screws - Pan Head *ASA B18.11-1961, R2010*

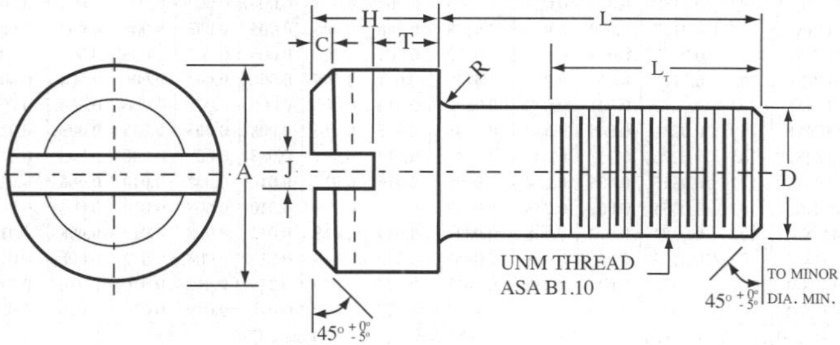


| Size Designation | Thds per Inch | *D* Basic Major Dia., Max. | *A* Head Dia. Max. | *A* Head Dia. Min. | *H* Head Height Max. | *H* Head Height Min. | *J* Slot Width Max. | *J* Slot Width Min. | *T* Slot Depth[a] Max. | *T* Slot Depth[a] Min. | *C* Chamfer Max. | *R* Radius[b] Min. |
|---|---|---|---|---|---|---|---|---|---|---|---|---|
| **30 UNM** | **318** | **0.0118** | **0.025** | **0.023** | **0.010** | **0.008** | **0.005** | **0.003** | **0.005** | **0.003** | **0.002** | **0.002** |
| 35 UNM | 282 | 0.0138 | 0.029 | 0.027 | 0.011 | 0.009 | 0.005 | 0.003 | 0.006 | 0.004 | 0.002 | 0.002 |
| **40 UNM** | **254** | **0.0157** | **0.033** | **0.031** | **0.012** | **0.010** | **0.006** | **0.004** | **0.006** | **0.004** | **0.002** | **0.002** |
| 45 UNM | 254 | 0.0177 | 0.037 | 0.035 | 0.014 | 0.012 | 0.006 | 0.004 | 0.007 | 0.005 | 0.002 | 0.002 |
| **50 UNM** | **203** | **0.0197** | **0.041** | **0.039** | **0.016** | **0.013** | **0.008** | **0.005** | **0.008** | **0.006** | **0.003** | **0.002** |
| 55 UNM | 203 | 0.0217 | 0.045 | 0.043 | 0.018 | 0.015 | 0.008 | 0.005 | 0.009 | 0.007 | 0.003 | 0.002 |
| **60 UNM** | **169** | **0.0236** | **0.051** | **0.049** | **0.020** | **0.017** | **0.010** | **0.007** | **0.010** | **0.007** | **0.004** | **0.003** |
| 70 UNM | 145 | 0.0276 | 0.056 | 0.054 | 0.022 | 0.019 | 0.010 | 0.007 | 0.011 | 0.008 | 0.004 | 0.003 |
| **80 UNM** | **127** | **0.0315** | **0.062** | **0.058** | **0.025** | **0.021** | **0.012** | **0.008** | **0.012** | **0.009** | **0.005** | **0.004** |
| 90 UNM | 113 | 0.0354 | 0.072 | 0.068 | 0.028 | 0.024 | 0.012 | 0.008 | 0.014 | 0.011 | 0.005 | 0.004 |
| **100 UNM** | **102** | **0.0394** | **0.082** | **0.078** | **0.032** | **0.028** | **0.016** | **0.012** | **0.018** | **0.014** | **0.006** | **0.005** |
| 110 UNM | 102 | 0.0433 | 0.092 | 0.088 | 0.036 | 0.032 | 0.016 | 0.012 | 0.018 | 0.014 | 0.006 | 0.005 |
| **120 UNM** | **102** | **0.0472** | **0.103** | **0.097** | **0.040** | **0.035** | **0.020** | **0.015** | **0.020** | **0.016** | **0.008** | **0.006** |
| 140 UNM | 85 | 0.0551 | 0.113 | 0.107 | 0.045 | 0.040 | 0.020 | 0.015 | 0.022 | 0.018 | 0.008 | 0.006 |

Bold face type indicates preferred sizes. See *Notes for Table 1 through Table 4*.

[a] *T* measured from bearing surface.

[b] Relative to maximum major diameter.

**Table 3. Miniature Screws - 100° Flat Head** *ASA B18.11-1961, R2010*

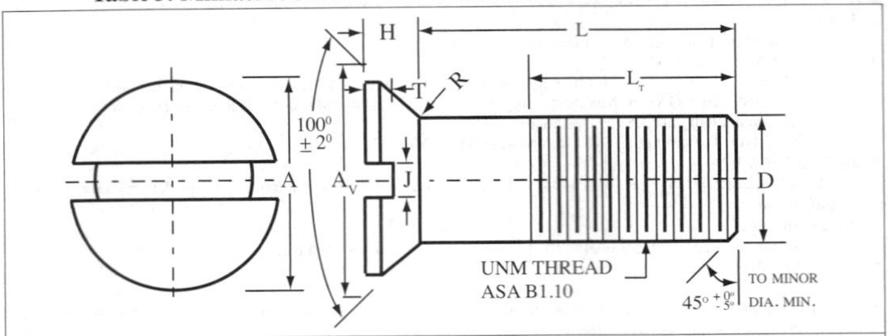


| Size Desig-nation | Thds per Inch | *D* Basic Major Dia. | *A* Head Dia. | | *Av* at Full Cone[a] at max. *H* | Head Dimensions *H* Head Height | | *J* Slot Width | | *T* Slot Depth | | *R* Radius[b] |
|---|---|---|---|---|---|---|---|---|---|---|---|---|
| | | Max. | Max. | Min. | | Max. | Min. | Max. | Min. | Max. | Min. | Max. |
| **30 UNM** | **318** | **0.0118** | **0.023** | **0.021** | **0.0285** | **0.007** | **0.005** | **0.004** | **0.003** | **0.004** | **0.002** | **0.005** |
| 35 UNM | 282 | 0.0138 | 0.025 | 0.023 | 0.0305 | 0.007 | 0.005 | 0.004 | 0.003 | 0.004 | 0.002 | 0.005 |
| **40 UNM** | **254** | **0.0157** | **0.029** | **0.027** | **0.0348** | **0.008** | **0.006** | **0.005** | **0.003** | **0.005** | **0.003** | **0.006** |
| 45 UNM | 254 | 0.0177 | 0.033 | 0.031 | 0.0392 | 0.009 | 0.007 | 0.005 | 0.003 | 0.005 | 0.003 | 0.006 |
| **50 UNM** | **203** | **0.0197** | **0.037** | **0.035** | **0.0459** | **0.011** | **0.008** | **0.006** | **0.004** | **0.006** | **0.004** | **0.008** |
| 55 UNM | 203 | 0.0217 | 0.041 | 0.039 | 0.0503 | 0.012 | 0.009 | 0.006 | 0.004 | 0.006 | 0.004 | 0.008 |
| **60 UNM** | **169** | **0.0236** | **0.045** | **0.043** | **0.0546** | **0.013** | **0.010** | **0.008** | **0.005** | **0.008** | **0.005** | **0.010** |
| 70 UNM | 145 | 0.0276 | 0.051 | 0.049 | 0.0610 | 0.014 | 0.011 | 0.008 | 0.005 | 0.008 | 0.005 | 0.010 |
| **80 UNM** | **127** | **0.0315** | **0.056** | **0.054** | **0.0696** | **0.016** | **0.012** | **0.010** | **0.007** | **0.010** | **0.006** | **0.012** |
| 90 UNM | 113 | 0.0354 | 0.062 | 0.058 | 0.0759 | 0.017 | 0.013 | 0.010 | 0.007 | 0.010 | 0.006 | 0.012 |
| **100 UNM** | **102** | **0.0394** | **0.072** | **0.068** | **0.0847** | **0.019** | **0.015** | **0.012** | **0.008** | **0.012** | **0.008** | **0.016** |
| 110 UNM | 102 | 0.0433 | 0.082 | 0.078 | 0.0957 | 0.022 | 0.018 | 0.012 | 0.008 | 0.012 | 0.008 | 0.016 |
| **120 UNM** | **102** | **0.0472** | **0.092** | **0.088** | **0.1068** | **0.025** | **0.020** | **0.016** | **0.012** | **0.016** | **0.010** | **0.020** |
| 140 UNM | 85 | 0.0551 | 0.103 | 0.097 | 0.1197 | 0.027 | 0.022 | 0.016 | 0.012 | 0.016 | 0.010 | 0.020 |

Bold face type indicates preferred sizes. See *Notes for Table 1 through Table 4* on page 1789.

[a] *Av* derived from maximum *D*, maximum *H*, and mean angle.

[b] Relative to maximum major diameter.

**Specifications.—***Head Height:* The head heights given in the dimensional tables represent the metal measurement (after slotting).

*Depth of Slots:* The depth of slots on fillister, pan and binding head screws is measured from the bearing surface to the intersection of the bottom of the slot with the head diameter. On heads with a conical bearing surface, the depth of slots is measured parallel to the axis of the screw from the flat top surface to the intersection of the bottom of the slot with the bearing surface. The maximum permissible concavity of the slot shall not exceed 3 percent of the mean head diameter.

*Bearing Surface:* The bearing surface of fillister, pan and binding head screws shall be at right angles to the axis of the body within 2°.

*Eccentricity:* Eccentricity is defined as one half of the total indicator reading.

*Head Eccentricity:* The heads of miniature fastening screws shall not be eccentric with the screw bodies by more than 2 percent of the maximum head diameter or 0.001 inch, whichever is the greater.

*Eccentricity of Slots:* Slots in miniature fastening screw heads shall not be eccentric with screw bodies by more than 5 percent of the nominal body diameter.

**Table 4. Miniature Screws - Binding Head** *ASA B18.11-1961, R2010*

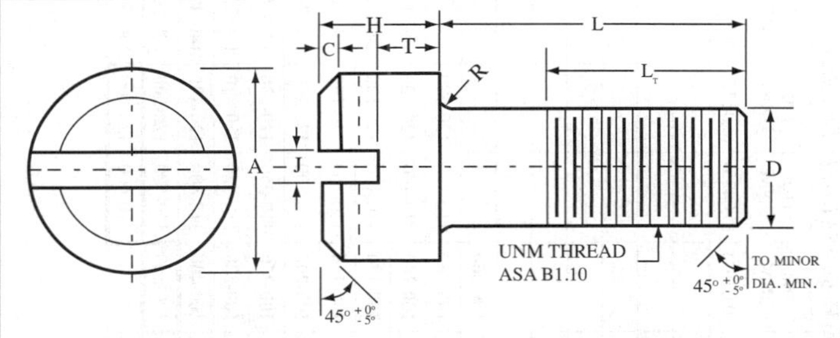


| Size Desig-nation | Thds per Inch | *D* Basic Major Dia | Binding Head Dimensions | | | | | | | | | | |
|---|---|---|---|---|---|---|---|---|---|---|---|---|---|
| | | | *A* Head Dia. | | *H* Head Height | | *J* Slot Width | | *T* Slot Depth[a] | | *C* Chamfer | *R* Radius | |
| | | Max. | Max. | Min. | Max. | Min. | Max. | Min. | Max. | Min. | Max. | Max. | Min. |
| **40 UNM** | **254** | **0.0157** | **0.041** | **0.039** | **0.010** | **0.008** | **0.006** | **0.004** | **0.005** | **0.003** | **0.002** | **0.004** | **0.002** |
| 45 UNM | 254 | 0.0177 | 0.045 | 0.043 | 0.011 | 0.009 | 0.006 | 0.004 | 0.006 | 0.004 | 0.002 | 0.004 | 0.002 |
| **50 UNM** | **203** | **0.0197** | **0.051** | **0.049** | **0.012** | **0.010** | **0.008** | **0.005** | **0.006** | **0.004** | **0.003** | **0.004** | **0.002** |
| 55 UNM | 203 | 0.0217 | 0.056 | 0.054 | 0.014 | 0.012 | 0.008 | 0.005 | 0.007 | 0.005 | 0.003 | 0.004 | 0.002 |
| **60 UNM** | **169** | **0.0236** | **0.062** | **0.058** | **0.016** | **0.013** | **0.010** | **0.007** | **0.008** | **0.006** | **0.004** | **0.006** | **0.003** |
| 70 UNM | 145 | 0.0276 | 0.072 | 0.068 | 0.018 | 0.015 | 0.010 | 0.007 | 0.009 | 0.007 | 0.004 | 0.006 | 0.003 |
| **80 UNM** | **127** | **0.0315** | **0.082** | **0.078** | **0.020** | **0.017** | **0.012** | **0.008** | **0.010** | **0.007** | **0.005** | **0.008** | **0.004** |
| 90 UNM | 113 | 0.0354 | 0.092 | 0.088 | 0.022 | 0.019 | 0.012 | 0.008 | 0.011 | 0.008 | 0.005 | 0.008 | 0.004 |
| **100 UNM** | **102** | **0.0394** | **0.103** | **0.097** | **0.025** | **0.021** | **0.016** | **10.012** | **0.012** | **0.009** | **0.006** | **0.010** | **0.005** |
| 110 UNM | 102 | 0.0433 | 0.113 | 0.107 | 0.028 | 0.024 | 0.016 | 0.012 | 0.014 | 0.011 | 0.006 | 0.010 | 0.005 |
| **120 UNM** | **102** | **0.0472** | **0.124** | **0.116** | **0.032** | **0.028** | **0.020** | **0.015** | **0.016** | **0.012** | **0.008** | **0.012** | **0.006** |
| 140 UNM | 85 | 0.0551 | 0.144 | 0.136 | 0.036 | 0.032 | 0.020 | 0.015 | 0.018 | 0.014 | 0.008 | 0.012 | 0.006 |

Bold face type indicates preferred sizes. See *Notes for Table 1 through Table 4* on page 1789.

[a] *T* measured from bearing surface.

*Underhead Fillets:* The radius of the fillet under perpendicular bearing surface type heads shall not exceed ½ times the pitch of the thread. The radius of the fillet under conical bearing surface type heads shall not exceed 2 times the pitch of the thread. The radius of the fillet under the binding head is given in Table 4.

*Unthreaded Diameter:* On miniature fastening screws not threaded to the head, the diameter of the unthreaded body shall not be more than the maximum major diameter of the thread nor less than the minimum pitch diameter of the thread.

*Length:* The length of miniature screws having perpendicular bearing surface type heads shall be measured from the bearing surface to the extreme end in a line parallel to the axis of the screw. The length of screws with conical bearing surface type heads shall be measured from the top of the head to the extreme end in a line parallel to the axis of the screw. Preferred lengths are those listed in Table 5.

*Tolerance on Length:* The length tolerance of miniature screws shall conform to the limits given in Table 5.

*Length of Thread:* On all miniature screws having a length four times the nominal body diameter or less the threaded length shall extend to within two threads of the bearing surface of the head. Screws of greater length shall possess complete threads for a minimum of four diameters.

**Table 5. Miniature Screw Standard Lengths - Fillister Head, Pan Head, Binding Head, and 100° Flat Head** *ASA B18.11-1961, R21010*

| Length (In.) | | **30 UNM** [a] | 35 UNM [a] | **40 UNM** | 45 UNM | **50 UNM** | 55 UNM | **60 UNM** | 70 UNM | **80 UNM** | 90 UNM | **100 UNM** | 110 UNM | **120 UNM** | 140 UNM |
|---|---|---|---|---|---|---|---|---|---|---|---|---|---|---|---|
| Min. | Max. | (0.0118) | (0.0138) | (0.0157) | (0.0177 | (0.0197) | (0.0217) | (0.0236) | (0.0276) | (0.0315) | (0.0354) | (0.0394) | (0.0433) | (0.0472) | (0.0551) |
| 0.016 | 0.020 | **30-020** [b] | | | | | | | | | | | | | |
| 0.020 | 0.025 | **30-025** | 35-025 | **40-025** | | | | | | | | | | | |
| 0.021 | 0.025 | **30-025** | 35-025 | **40-025** | | | | | | | | | | | |
| 0.027 | 0.032 | **30-032** | 35-032 | **40-032** | 45-032 | **50-032** | | | | | | | | | |
| 0.035 | 0.040 | **30-040** | 35-040 | **40-040** | 45-040 | **50-040** | 55-040 | **60-040** | | | | | | | |
| 0.044 | 0.050 | **30-050** | 35-050 | **40-050** | 45-050 | **50-050** | 50-050 | **60-050** | 70-050 | **80-050** | | | | | |
| 0.054 | 0.060 | **30-060** | 35-060 | **40-060** | 45-060 | **50-060** | 55-060 | **60-060** | 70-060 | **80-060** | 90-060 | **100-060** | | | |
| 0.072 | 0.080 | **30-080** | 35-080 | **40-080** | 45-080 | **50-080** | 55-080 | **60-080** | 70-080 | **80-080** | 90-080 | **100-080** | 110-080 | **120-080** | |
| 0.092 | 0.100 | **30-100** | 35-100 | **40-100** | 45-100 | **50-100** | 55-100 | **60-100** | 70-100 | **80-100** | 90-100 | **100-100** | 110-100 | **120-100** | 140-100 |
| 0.110 | 0.120 | **30-120** | 35-120 | **40-120** | 45-120 | **50-120** | 55-120 | **60-120** | 70-120 | **80-120** | 90-120 | **100-120** | 110-120 | **120-120** | 140-120 |
| 0.150 | 0.160 | 30-160 | 35-160 | **40-160** | 45-160 | **50-160** | 55-160 | **60-160** | 70-160 | **80-160** | 90-160 | **100-160** | 110-160 | **120-160** | 140-160 |
| 0.188 | 0.200 | | 35-200 | **40-200** | 45-200 | **50-200** | 55-200 | **60-200** | 70-200 | **80-200** | 90-200 | **100-200** | 110-200 | **120-200** | 140-200 |
| 0.238 | 0.250 | | | | 45-250 | 50-250 | 55-250 | **60-250** | 70-250 | **80-250** | 90-250 | **100-250** | 110-250 | **120-250** | 140-250 |
| 0.304 | 0.320 | | | | | | 55-320 | 60-320 | 70-320 | **80-320** | 90-320 | **100-320** | 110-320 | **120-320** | 140-320 |
| 0.384 | 0.400 | | | | | | | | 70-400 | 80-400 | 90-400 | **100-400** | 110-400 | **120-400** | 140-400 |
| 0.480 | 0.500 | | | | | | | | | | 90-500 | 100-500 | 110-500 | **120-500** | 140-500 |
| 0.580 | 0.600 | | | | | | | | | | | | 110-600 | 120-600 | 140-600 |

[a] Sizes 30 UMN and 35 UMN are not specified for Binding Head.
[b] Does not apply to 100° Flat Head.
Bold face type indicates preferred sizes. Sizes surrounded by heavy line apply to 100° Flat Head only.

*End of Body:* Miniature fastening screws shall be regularly supplied with flat ends having a chamfer of approximately 45° extending to the minor diameter of the thread as a minimum depth.

*Thread Series and Tolerances:* The screw threads of miniature screws shall be in conformance with American Standard Unified Miniature Screw Threads, ASA B1.10-1958, R1988.

*Material and Finish:* Miniature screws are generally supplied in ferrous and nonferrous materials, coatings, and heat treatments which must be specified by the user. Coatings, when required, are limited to those of electroplating or chemical oxidation.

*Designation:* Screws in conformance with this Standard shall be identified by the designation for thread size in conformance with American Standard ASA B1.10 followed by the nominal length in units of $^1/_{1000}$ inch (omitting the decimal point) and the head type. Typical examples are:

60 UNM × 040 FIL HD

100 UNM × 080 PAN HD

120 UNM × 120 FLAT HD

140 UNM × 250 BIND HD

*Machined Finish:* Roughness of the machined surfaces of heads shall not exceed 63 micro-inches arithmetical average (per ASA B46.1, Surface Texture) determined by visual comparison with roughness comparison specimens.

## American National Standard Metric Machine Screws

ANSI/ASME B18.6.7M (Withdrawn) covers metric flat and oval countersunk and slotted and recessed pan head machine screws and metric hex head and hex flange head machine screws. Dimensions are given in Table 1 through Table 4 and Table 5.

*Threads:* Threads for metric machine screws are coarse M profile threads, as given in ANSI/ASME B1.13M (see page 2015), unless otherwise specified.

*Length of Thread:* The lengths of threads on metric machine screws are given in Table 1 for the applicable screw type, size, and length. Also see Table 6.

*Diameter of Body:* The body diameters of metric machine screws are within the limits specified in the dimensional tables (Table 3 through Table 4 and Table 5).

*Designation:* Metric machine screws are designated by the following data in the sequence shown: Nominal size and thread pitch; nominal length; product name, including head type and driving provision; header point if desired; material (including property class, if steel); and protective finish, if required. For example:

M8 × 1.25 × 30 Slotted Pan Head Machine Screw, Class 4.8 Steel, Zinc Plated

M3.5 × 0.6 × 20 Type IA Cross Recessed Oval Countersunk Head Machine Screw, Header Point, Brass

It is common ISO practice to omit the thread pitch from the product size designation when screw threads are the metric coarse thread series, e.g., M10 stands for M10 × 1.5.

**Table 1. American National Standard Thread Lengths for Metric Machine Screws**
*ANSI/ASME B18.6.7M-1999 (Withdrawn)*

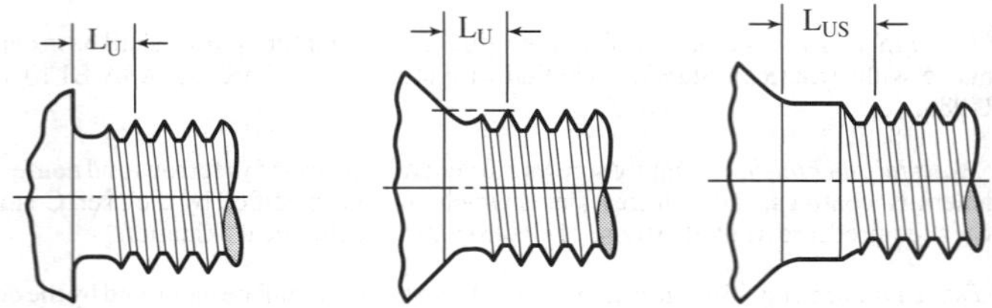


Pan, Hex, and Hex Flange Head Screws — Flat and Oval Countersunk Head Screws — Heat-Treated Recessed Flat Countersunk Head Screws

| Nominal Screw Size and Thread Pitch | L | $L_{US}$ | $L_U$ | L | | $L_{US}$ | $L_UL$ | L | B |
|---|---|---|---|---|---|---|---|---|---|
| | Nominal Screw Length Equal to or Shorter than[a] | Unthreaded Length[b] | | Nominal Screw Length[a] | | Unthreaded Length[b] | | Nominal Screw Length Longer than[a] | Full Form Thread Length[c] |
| | | Max[d] | Max[e] | Over | To and Including | Max[d] | Max[e] | | Min |
| M2 × 0.4 | 6 | 1.0 | 0.4 | 6 | 30 | 1.0 | 0.8 | 30 | 25.0 |
| M2.5 × 0.45 | 8 | 1.1 | 0.5 | 8 | 30 | 1.1 | 0.9 | 30 | 25.0 |
| M3 × 0.5 | 9 | 1.2 | 0.5 | 9 | 30 | 1.2 | 1.0 | 30 | 25.0 |
| M3.5 × 0.6 | 10 | 1.5 | 0.6 | 10 | 50 | 1.5 | 1.2 | 50 | 38.0 |
| M4 × 0.7 | 12 | 1.8 | 0.7 | 12 | 50 | 1.8 | 1.4 | 50 | 38.0 |
| M5 × 0.8 | 15 | 2.0 | 0.8 | 15 | 50 | 2.0 | 1.6 | 50 | 38.0 |
| M6 × 1 | 18 | 2.5 | 1.0 | 18 | 50 | 2.5 | 2.0 | 50 | 38.0 |
| M8 × 1.25 | 24 | 3.1 | 1.2 | 24 | 50 | 3.1 | 2.5 | 50 | 38.0 |
| M10 × 1.5 | 30 | 3.8 | 1.5 | 30 | 50 | 3.8 | 3.0 | 50 | 38.0 |
| M12 × 1.75 | 36 | 4.4 | 1.8 | 36 | 50 | 4.4 | 3.5 | 50 | 38.0 |

[a] The length tolerances for metric machine screws are: up to 3 mm, incl., ± 0.2 mm; over 3 to 10 mm, incl., ± 0.3 mm; over 10 to 16 mm, incl., ± 0.4 mm; over 16 to 50 mm, incl., ± 0.5 mm; over 50 mm, ± 1.0 mm.

[b] Unthreaded lengths $L_U$ and $L_{US}$ represent the distance, measured parallel to the axis of screw, from the underside of the head to the face of a nonchamfered or noncounterbored standard GO thread ring gage assembled by hand as far as the thread will permit.

[c] Refer to the illustrations for respective screw head styles.

[d] The $L_{US}$ values apply only to heat-treated recessed flat countersunk head screws.

[e] The $L_U$ values apply to all screws except heat-treated recessed flat countersunk head screws.

All dimensions in millimeters.

**Table 2. American National Standard Slotted, Cross and Square Recessed Flat Countersunk Head Metric Machine Screws**
*ANSI/ASME B18.6.7M-1999 (Withdrawn)*

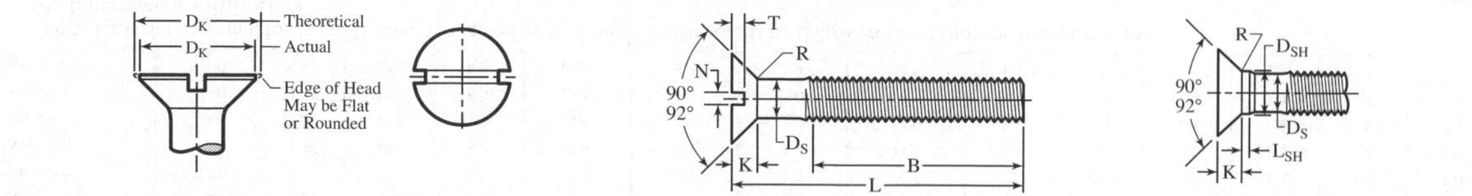


| Nominal Screw Size and Thread Pitch | Slotted and Style A: $D_S$ Body Diameter Max. | Min. | Style B: $D_{SH}$[a] Body and Shoulder Diameter Max. | Shoulder Diameter Min. | $D_S$ Body Diameter Min. | $L_{SH}$[a] Shoulder Length Max. | Min. | $D_K$ Head Diameter, Theoretical Sharp Max. | Min. | Actual Min. | K Head Height Max. Ref. | R Underhead Fillet Radius Max. | Min. | N Slot Width Max. | Min. | T Slot Depth Max. | Min. |
|---|---|---|---|---|---|---|---|---|---|---|---|---|---|---|---|---|---|
| M2×0.4[b] | 2.00 | 1.65 | 2.00 | 1.86 | 1.65 | 0.50 | 0.30 | 4.4 | 4.1 | 3.5 | 1.2 | 0.8 | 0.4 | 0.7 | 0.5 | 0.6 | 0.4 |
| M2.5×0.45 | 2.50 | 2.12 | 2.50 | 2.36 | 2.12 | 0.55 | 0.35 | 5.5 | 5.1 | 4.4 | 1.5 | 1.0 | 0.5 | 0.8 | 0.6 | 0.7 | 0.5 |
| M3×0.5 | 3.00 | 2.58 | 3.00 | 2.86 | 2.58 | 0.60 | 0.40 | 6.3 | 5.9 | 5.2 | 1.7 | 1.2 | 0.6 | 1.0 | 0.8 | 0.9 | 0.6 |
| M3.5×0.6 | 3.50 | 3.00 | 3.50 | 3.32 | 3.00 | 0.70 | 0.50 | 8.2 | 7.7 | 6.9 | 2.3 | 1.4 | 0.7 | 1.2 | 1.0 | 1.2 | 0.9 |
| M4×0.7 | 4.00 | 3.43 | 4.00 | 3.82 | 3.43 | 0.80 | 0.60 | 9.4 | 8.9 | 8.0 | 2.7 | 1.6 | 0.8 | 1.5 | 1.2 | 1.3 | 1.0 |
| M5×0.8 | 5.00 | 4.36 | 5.00 | 4.82 | 4.36 | 0.90 | 0.70 | 10.4 | 9.8 | 8.9 | 2.7 | 2.0 | 1.0 | 1.5 | 1.2 | 1.4 | 1.1 |
| M6×1 | 6.00 | 5.21 | 6.00 | 5.82 | 5.21 | 1.10 | 0.90 | 12.6 | 11.9 | 10.9 | 3.3 | 2.4 | 1.2 | 1.9 | 1.6 | 1.6 | 1.2 |
| M8×1.25 | 8.00 | 7.04 | 8.00 | 7.78 | 7.04 | 1.40 | 1.10 | 17.3 | 16.5 | 15.4 | 4.6 | 3.2 | 1.6 | 2.3 | 2.0 | 2.3 | 1.8 |
| M10×1.5 | 10.00 | 8.86 | 10.00 | 9.78 | 8.86 | 1.70 | 1.30 | 20.0 | 19.2 | 17.8 | 5.0 | 4.0 | 2.0 | 2.8 | 2.5 | 2.6 | 2.0 |

[a] All recessed head heat-treated steel screws of property class 9.8 or higher strength have the Style B head form. Recessed head screws other than those specifically designated to be Style B have the Style A head form. The underhead shoulder on the Style B head form is mandatory and all other head dimensions are common to both the Style A and Style B head forms.

[b] This size is not specified for Type III square recessed flat countersunk heads; Type II cross recess is not specified for any size.

All dimensions in millimeters.

For dimension *B*, see Table 1.

For dimension *L*, see Table 6.

**Table 3. American National Standard Slotted, Cross and Square Recessed Oval Countersunk Head Metric Machine Screws**
*ANSI/ASME B18.6.7M-1999 (Withdrawn)*

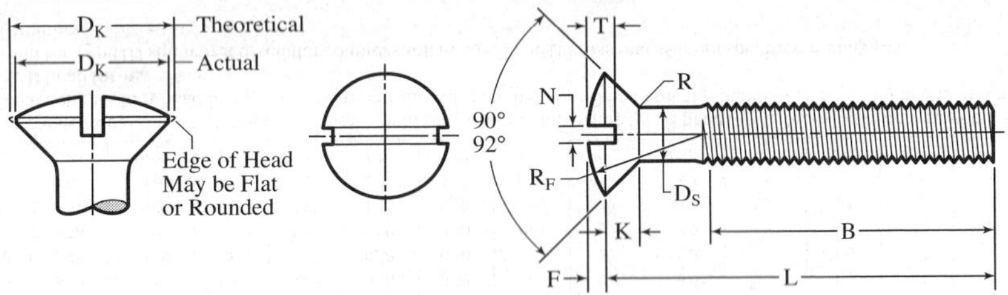


| Nominal Screw Size and Thread Pitch | $D_S$ | | $D_K$ | | | K | F | $R_F$ | R | | N | | T | |
|---|---|---|---|---|---|---|---|---|---|---|---|---|---|---|
| | Body Diameter | | Head Diameter | | | Head Side Height | Raised Head Height | Head Top Radius | Underhead Fillet Radius | | Slot Width | | Slot Depth | |
| | | | Theoretical Sharp | | Actual | | | | | | | | | |
| | Max. | Min. | Max. | Min. | Min. | Max. Ref. | Max. | Approx. | Max. | Min. | Max. | Min. | Max. | Min. |
| M2 × 0.4[a] | 2.00 | 1.65 | 4.4 | 4.1 | 3.5 | 1.2 | 0.5 | 5.0 | 0.8 | 0.4 | 0.7 | 0.5 | 1.0 | 0.8 |
| M2.5 × 0.45 | 2.50 | 2.12 | 5.5 | 5.1 | 4.4 | 1.5 | 0.6 | 6.6 | 1.0 | 0.5 | 0.8 | 0.6 | 1.2 | 1.0 |
| M3 × 0.5 | 3.00 | 2.58 | 6.3 | 5.9 | 5.2 | 1.7 | 0.7 | 7.4 | 1.2 | 0.6 | 1.0 | 0.8 | 1.5 | 1.2 |
| M3.5 × 0.6 | 3.50 | 3.00 | 8.2 | 7.7 | 6.9 | 2.3 | 0.8 | 10.9 | 1.4 | 0.7 | 1.2 | 1.0 | 1.7 | 1.4 |
| M4 × 0.7 | 4.00 | 3.43 | 9.4 | 8.9 | 8.0 | 2.7 | 1.0 | 11.6 | 1.6 | 0.8 | 1.5 | 1.2 | 1.9 | 1.6 |
| M5 × 0.8 | 5.00 | 4.36 | 10.4 | 9.8 | 8.9 | 2.7 | 1.2 | 11.9 | 2.0 | 1.0 | 1.5 | 1.2 | 2.4 | 2.0 |
| M6 × 1 | 6.00 | 5.21 | 12.6 | 11.9 | 10.9 | 3.3 | 1.4 | 14.9 | 2.4 | 1.2 | 1.9 | 1.6 | 2.8 | 2.4 |
| M8 × 1.25 | 8.00 | 7.04 | 17.3 | 16.5 | 15.4 | 4.6 | 2.0 | 19.7 | 3.2 | 1.6 | 2.3 | 2.0 | 3.7 | 3.2 |
| M10 × 1.5 | 10.00 | 8.86 | 20.0 | 19.2 | 17.8 | 5.0 | 2.3 | 22.9 | 4.0 | 2.0 | 2.8 | 2.5 | 4.4 | 3.8 |

[a] This size is not specified for Type III square recessed oval countersunk heads; Type II cross recess is not specified for any size.
All dimensions in millimeters.
For dimension *B*, see Table 1.
For dimension *L*, see Table 6.

**Table 4. American National Standard Slotted and Cross and Square Recessed Pan Head Metric Machine Screws**
*ANSI/ASME B18.6.7M-1999 (Withdrawn)*

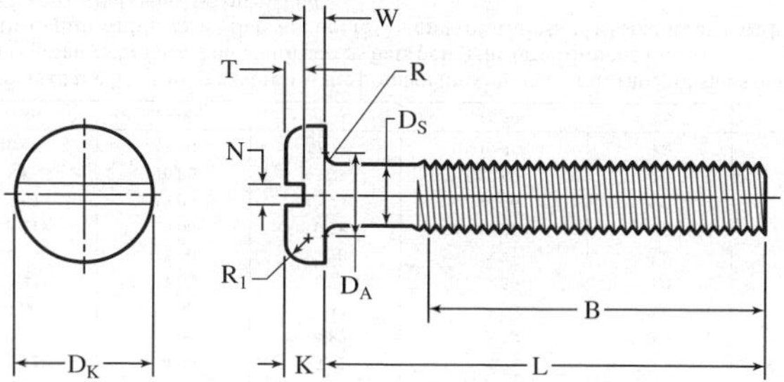


| Nominal Screw Size and Thread Pitch | $D_s$ | | $D_K$ | | Slotted | | | Cross and Square Recess | | | $D_A$ | R | N | | T | W |
|---|---|---|---|---|---|---|---|---|---|---|---|---|---|---|---|---|
| | | | | | K | | $R_1$ | K | | $R_1$ | Underhead Fillet | | | | | |
| | Body Diameter | | Head Diameter | | Head Height | | Head Radius | Head Height | | Head Radius | Transition Dia. | Radius | Slot Width | | Slot Depth | Unslotted Head Thickness |
| | Max. | Min. | Max. | Min. | Max. | Min. | Max. | Max. | Min. | Ref. | Max. | Min. | Max. | Min. | Min. | Min. |
| M2 × 0.4[a] | 2.00 | 1.65 | 4.0 | 3.7 | 1.3 | 1.1 | 0.8 | 1.6 | 1.4 | 3.2 | 2.6 | 0.1 | 0.7 | 0.5 | 0.5 | 0.4 |
| M2.5 × 0.45 | 2.50 | 2.12 | 5.0 | 4.7 | 1.5 | 1.3 | 1.0 | 2.1 | 1.9 | 4.0 | 3.1 | 0.1 | 0.8 | 0.6 | 0.6 | 0.5 |
| M3 × 0.5 | 3.00 | 2.58 | 5.6 | 5.3 | 1.8 | 1.6 | 1.2 | 2.4 | 2.2 | 5.0 | 3.6 | 0.1 | 1.0 | 0.8 | 0.7 | 0.7 |
| M3.5 × 0.6 | 3.50 | 3.00 | 7.0 | 6.6 | 2.1 | 1.9 | 1.4 | 2.6 | 2.3 | 6.0 | 4.1 | 0.1 | 1.2 | 1.0 | 0.8 | 0.8 |
| M4 × 0.7 | 4.00 | 3.43 | 8.0 | 7.6 | 2.4 | 2.2 | 1.6 | 3.1 | 2.8 | 6.5 | 4.7 | 0.2 | 1.5 | 1.2 | 1.0 | 0.9 |
| M5 × 0.8 | 5.00 | 4.36 | 9.5 | 9.1 | 3.0 | 2.7 | 2.0 | 3.7 | 3.4 | 8.0 | 5.7 | 0.2 | 1.5 | 1.2 | 1.2 | 1.2 |
| M6 × 1 | 6.00 | 5.21 | 12.0 | 11.5 | 3.6 | 3.3 | 2.5 | 4.6 | 4.3 | 10.0 | 6.8 | 0.3 | 1.9 | 1.6 | 1.4 | 1.4 |
| M8 × 1.25 | 8.00 | 7.04 | 16.0 | 15.5 | 4.8 | 4.5 | 3.2 | 6.0 | 5.6 | 13.0 | 9.2 | 0.4 | 2.3 | 2.0 | 1.9 | 1.9 |
| M10 × 1.5 | 10.00 | 8.86 | 20.0 | 19.4 | 6.0 | 5.7 | 4.0 | 7.5 | 7.1 | 16.0 | 11.2 | 0.4 | 2.8 | 2.5 | 2.4 | 2.4 |

[a] This size not specified for Type III square recessed pan heads; Type II cross recess is not specified for any size.
All dimensions in millimeters.
For dimension *B*, see Table 1.
For dimension *L*, see Table 6.

**Table 5. American National Standard Hex and Hex Flange Head Metric Machine Screws** *ANSI/ASME B18.6.7M-1999 (Withdrawn)*

Hex Head

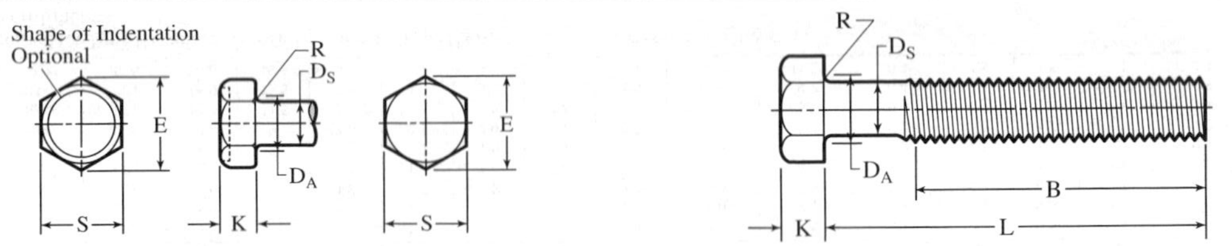


| Nominal Screw Size and Thread Pitch | $D_S$ Body Diameter | | $S^a$ Hex Width Across Flats | | $E^a$ Hex Width Across Corners | K Head Height | | $D_A$ Underhead Fillet Transition Dia. | R Underhead Fillet Radius |
|---|---|---|---|---|---|---|---|---|---|
| | Max. | Min. | Max. | Min. | Min. | Max. | Min. | Max. | Min. |
| M2×0.4 | 2.00 | 1.65 | 3.20 | 3.02 | 3.38 | 1.6 | 1.3 | 2.6 | 0.1 |
| M2.5×0.45 | 2.50 | 2.12 | 4.00 | 3.82 | 4.28 | 2.1 | 1.8 | 3.1 | 0.1 |
| M3×0.5 | 3.00 | 2.58 | 5.00 | 4.82 | 5.40 | 2.3 | 2.0 | 3.6 | 0.1 |
| M3.5×0.6 | 3.50 | 3.00 | 5.50 | 5.32 | 5.96 | 2.6 | 2.3 | 4.1 | 0.1 |
| M4×0.7 | 4.00 | 3.43 | 7.00 | 6.78 | 7.59 | 3.0 | 2.6 | 4.7 | 0.2 |
| M5×0.8 | 5.00 | 4.36 | 8.00 | 7.78 | 8.71 | 3.8 | 3.3 | 5.7 | 0.2 |
| M6×1 | 6.00 | 5.21 | 10.00 | 9.78 | 10.95 | 4.7 | 4.1 | 6.8 | 0.3 |
| M8×1.25 | 8.00 | 7.04 | 13.00 | 12.73 | 14.26 | 6.0 | 5.2 | 9.2 | 0.4 |
| M10×1.5 | 10.00 | 8.86 | 16.00 | 15.73 | 17.62 | 7.5 | 6.5 | 11.2 | 0.4 |
| M12×1.75 | 12.00 | 10.68 | 18.00 | 17.73 | 19.86 | 9.0 | 7.8 | 13.2 | 0.4 |
| M10×1.5[b] | 10.00 | 8.86 | 15.00 | 14.73 | 16.50 | 7.5 | 6.5 | 11.2 | 0.4 |

[a] Dimensions across flats and across corners of the head are measured at the point of maximum metal. Taper of sides of head (angle between one side and the axis) shall not exceed 2° or 0.10 mm, whichever is greater, the specified width across flats being the large dimension.

[b] The M10 size screws having heads with 15 mm width across flats are not ISO standard. Unless M10 size screws with 15 mm width across flats are specifically ordered, M10 size screws with 16 mm width across flats shall be furnished.

**Table 5.** *(Continued)* **American National Standard Hex and Hex Flange Head Metric Machine Screws** *ANSI/ASME B18.6.7M-1999 (Withdrawn)*

Hex Flange Head

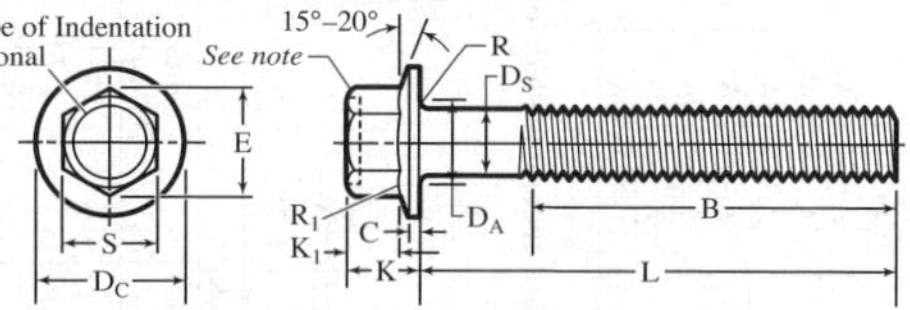


| Nominal Screw Size and Thread Pitch | Body Diameter, $D_S$ | | Hex Width Across Flats, S[a] | | Hex Width Across Corners, E[a] | Flange Diameter, $D_C$ | | Overall Head Height, K | Hex Height, $K_1$ | Flange Edge Thickness, C[b] | Flange Top Fillet Radius, $R_1$ | Underhead Fillet | |
|---|---|---|---|---|---|---|---|---|---|---|---|---|---|
| | Max. | Min. | Max. | Min. | Min. | Max. | Min. | | Min. | Min. | Max. | Max. Transition Dia., $D_A$ | Min. Radius, R |
| M2 × 0.4 | 2.00 | 1.65 | 3.00 | 2.84 | 3.16 | 4.5 | 4.1 | 2.2 | 1.3 | 0.3 | 0.1 | 2.6 | 0.1 |
| M2.5 × 0.45 | 2.50 | 2.12 | 3.20 | 3.04 | 3.39 | 5.4 | 5.0 | 2.7 | 1.6 | 0.3 | 0.2 | 3.1 | 0.1 |
| M3 × 0.5 | 3.00 | 2.58 | 4.00 | 3.84 | 4.27 | 6.4 | 5.9 | 3.2 | 1.9 | 0.4 | 0.2 | 3.6 | 0.1 |
| M3.5 × 0.6 | 3.50 | 3.00 | 5.00 | 4.82 | 5.36 | 7.5 | 6.9 | 3.8 | 2.4 | 0.5 | 0.2 | 4.1 | 0.1 |
| M4 × 0.7 | 4.00 | 3.43 | 5.50 | 5.32 | 5.92 | 8.5 | 7.8 | 4.3 | 2.8 | 0.6 | 0.2 | 4.7 | 0.2 |
| M5 × 0.8 | 5.00 | 4.36 | 7.00 | 6.78 | 7.55 | 10.6 | 9.8 | 5.4 | 3.5 | 0.7 | 0.3 | 5.7 | 0.2 |
| M6 × 1 | 6.00 | 5.21 | 8.00 | 7.78 | 8.66 | 12.8 | 11.8 | 6.7 | 4.2 | 1.0 | 0.4 | 6.8 | 0.3 |
| M8 × 1.25 | 8.00 | 7.04 | 10.00 | 9.78 | 10.89 | 16.8 | 15.5 | 8.6 | 5.6 | 1.2 | 0.5 | 9.2 | 0.4 |
| M10 × 1.5 | 10.00 | 8.86 | 13.00 | 12.72 | 14.16 | 21.0 | 19.3 | 10.7 | 7.0 | 1.4 | 0.6 | 11.2 | 0.4 |
| M12 × 1.75 | 12.00 | 10.68 | 15.00 | 14.72 | 16.38 | 24.8 | 23.3 | 13.7 | 8.4 | 1.8 | 0.7 | 13.2 | 0.4 |

[a] Dimensions across flats and across corners of the head are measured at the point of maximum metal. Taper of sides of head (angle between one side and the axis) shall not exceed 2° or 0.10 mm, whichever is greater, the specified width across flats being the large dimension.

[b] The contour of the edge at periphery of flange is optional provided the minimum flange thickness is maintained at the minimum flange diameter. The top surface of flange may be straight or slightly rounded (convex) upward.

All dimensions in millimeters.

A slight rounding of all edges of the hexagon surfaces of indented hex heads is permissible provided the diameter of the bearing circle is not less than the equivalent of 90 percent of the specified minimum width across flats dimension.

Heads may be indented, trimmed, or fully upset at the option of the manufacturer.

For dimension *B*, see Table 1.

For dimension *L*, see Table 6.

**Table 6. Recommended Nominal Screw Lengths for Metric Machine Screws**

| Nominal Screw Length | Nominal Screw Size | | | | | | | | | |
|---|---|---|---|---|---|---|---|---|---|---|
| | M2 | M2.5 | M3 | M3.5 | M4 | M5 | M6 | M8 | M10 | M12 |
| 2.5 | PH | | | | | | | | | |
| 3 | A | PH | | | | | | | | |
| 4 | A | A | PH | | | | | | | |
| 5 | A | A | A | PH | PH | | | | | |
| 6 | A | A | A | A | A | PH | | | | |
| 8 | A | A | A | A | A | A | A | | | |
| 10 | A | A | A | A | A | A | A | A | | |
| 13 | A | A | A | A | A | A | A | A | A | |
| 16 | A | A | A | A | A | A | A | A | A | H |
| 20 | A | A | A | A | A | A | A | A | A | H |
| 25 | | A | A | A | A | A | A | A | A | H |
| 30 | | | A | A | A | A | A | A | A | H |
| 35 | | | | A | A | A | A | A | A | H |
| 40 | | | | | A | A | A | A | A | H |
| 45 | | | | | | A | A | A | A | H |
| 50 | | | | | | A | A | A | A | H |
| 55 | | | | | | | A | A | A | H |
| 60 | | | | | | | A | A | A | H |
| 65 | | | | | | | | A | A | H |
| 70 | | | | | | | | A | A | H |
| 80 | | | | | | | | A | A | H |
| 90 | | | | | | | | | A | H |

All dimensions in millimeters.

[1]The nominal screw lengths included between the heavy lines are recommended for the respective screw sizes and screw head styles as designated by the symbols.

A—Signifies screws of all head styles covered in this Standard.

P—Signifies pan head screws.

H—Signifies hex and hex flange head screws.

**Table 7. Clearance Holes for Metric Machine Screws**

*ANSI/ASME B18.6.7M-1999 (Withdrawn), Appendix*

| Nominal Screw Size | Basic Clearance Hole Diameter[a] | | |
|---|---|---|---|
| | Close Clearance[b] | Normal Clearance (Preferred)[b] | Loose Clearance[b] |
| M2 | 2.20 | 2.40 | 2.60 |
| M2.5 | 2.70 | 2.90 | 3.10 |
| M3 | 3.20 | 3.40 | 3.60 |
| M3.5 | 3.70 | 3.90 | 4.20 |
| M4 | 4.30 | 4.50 | 4.80 |
| M5 | 5.30 | 5.50 | 5.80 |
| M6 | 6.40 | 6.60 | 7.00 |
| M8 | 8.40 | 9.00 | 10.00 |
| M10 | 10.50 | 11.00 | 12.00 |
| M12 | 13.00 | 13.50 | 14.50 |

[a] The values given in this table are minimum limits. The recommended plus tolerances are as follows: for clearance hole diameters over 1.70 to and including 5.80 mm, plus 0.12, 0.20, and 0.30 mm for close, normal, and loose clearances, respectively; for clearance hole diameters over 5.80 to 14.50 mm, plus 0.18, 0.30, and 0.45 mm for close, normal, and loose clearances, respectively.

[b] Normal clearance hole sizes are preferred. Close clearance hole sizes are for situations such as critical alignment of assembled components, wall thickness, or other limitations which necessitate the use of a minimal hole. Countersinking or counterboring at the fastener entry side may be necessary for the proper seating of the head. Loose clearance hole sizes are for applications where maximum adjustment capability between the components being assembled is necessary.

All dimensions in millimeters.

**Table 8. American National Standard Header Points for Metric Machine Screws Before Threading** *ANSI/ASME B18.6.7M-1999 (Withdrawn)*

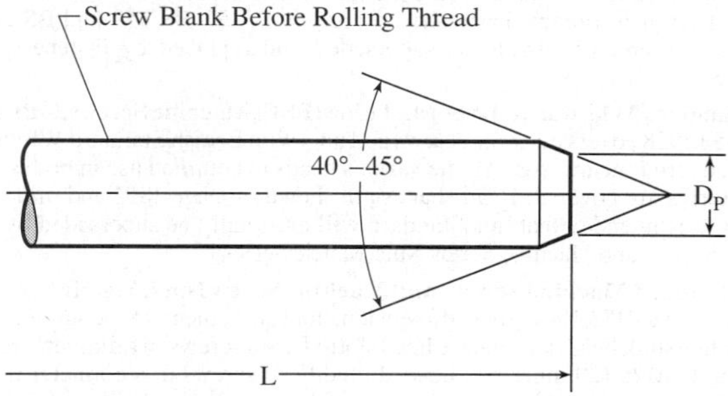


| Nominal Screw Size and Thread Pitch | $D_P$ Point Diameter | | $L^a$ Nominal Screw Length |
|---|---|---|---|
| | Max. | Min. | Max. |
| M2 × 0.4 | 1.33 | 1.21 | 13 |
| M2.5 × 0.45 | 1.73 | 1.57 | 13 |
| M3 × 0.5 | 2.12 | 1.93 | 16 |
| M3.5 × 0.6 | 2.46 | 2.24 | 20 |
| M4 × 0.7 | 2.80 | 2.55 | 25 |
| M5 × 0.8 | 3.60 | 3.28 | 30 |
| M6 × 1 | 4.25 | 3.85 | 40 |
| M8 × 1.25 | 5.82 | 5.30 | 40 |
| M10 × 1.5 | 7.36 | 6.71 | 40 |
| M12 × 1.75 | 8.90 | 8.11 | 45 |

[a] Header points apply to these nominal lengths or shorter. The pointing of longer lengths may require machining to the dimensions specified.

All dimensions in millimeters.

The edge of the point may be rounded and the end of point need not be flat nor perpendicular to the axis of screw shank.

## British Machine Screws

Many of these classifications of fasteners are covered in British Standards BS 57:1951, "B.A. Screws, Bolts and Nuts"; BS 450:1958, "Machine Screws and Machine Screw Nuts (BSW and BSF Threads)"; BS 1981:1991, "Unified Machine Screws and Machine Screw Nuts"; BS 2827:1957, "Machine Screw Nuts, Pressed Type (B.A. and Whitworth Form Threads)"; BS 3155:1960, "American Machine Screws and Nuts in Sizes Below ¼ inch Diameter"; and BS 4183:1967, "Machine Screws and Machine Screw Nuts, Metric Series." At a conference organized by the British Standards Institution in 1965 at which the major sectors of British industry were represented, a policy statement was approved that urged British firms to regard the traditional screw thread systems—Whitworth, B.A. and BSF—as obsolescent, and to make the internationally-agreed ISO metric thread their first choice (with ISO Unified thread as second choice) for all future designs. It is recognized that some sections of British industry already using ISO inch (Unified) screw threads may find it necessary, for various reasons, to continue with their use for some time: Whitworth and B.A. threads should, however, be superseded by ISO metric threads in preference to an intermediate change to ISO inch threads. Fasteners covered by BS 57, BS 450 and BS 2827:1957 (obsolescent) eventually would be superseded and replaced by fasteners specified by BS 4183.

**British Standard Whitworth (BSW) and Fine (BSF) Machine Screws.—**British Standard BS 450:1958 covers machine screws and nuts with British Standard Whitworth and British Standard Fine threads. All the various heads in common use in both slotted and recessed forms are covered. Head shapes are shown on page 1812 and dimensions on page 1814. It is intended that this Standard will eventually be superseded by BS 4183, "Machine Screws and Machine Screw Nuts, Metric Series."

**British Standard Machine Screws and Machine Screw Nuts, Metric Series.—**British Standard BS 4183:1967 gives dimensions and tolerances for: countersunk head, raised countersunk head, and cheese head slotted head screws in a diameter range from M1 (1 mm) to M20 (20 mm); pan head slotted head screws in a diameter range from M2.5 (2.5 mm) to M10 (10 mm); countersunk head and raised countersunk head recessed head screws in a diameter range from M2.5 (2.5 mm) to M12 (12 mm); pan head recessed head screws in a diameter range from M2.5 (2.5 mm) to M10 (10 mm); and square and hexagon machine screw nuts in a diameter range from M1.6 (1.6 mm) to M10 (10 mm). Mechanical properties are also specified for steel, brass and aluminum alloy machine screws and machine screw nuts in this Standard.

*Material:* The materials from which the screws and nuts are manufactured have a tensile strength not less than the following: steel, 40 kgf/mm$^2$ (392 N/mm$^2$); brass, 32 kgf/mm$^2$ (314 N/mm$^2$); and aluminum alloy, 32 kgf/mm$^2$ (314 N/mm$^2$). The unit, kgf/mm$^2$ is in accordance with ISO DR 911 and the unit in parentheses has the relationship, 1 kgf = 9.80665 newtons. These minimum strengths are applicable to the finished products. Steel machine screws conform to the requirements for strength grade designation 4.8. The strength grade designation system for machine screws consists of two figures, the first is $^1/_{10}$ of the minimum tensile strength in kgf/mm$^2$, the second is $^1/_{10}$ of the ratio between the yield stress and the minimum tensile strength expressed as a percentage: $^1/_{10}$ minimum tensile strength of 40 kgf/mm$^2$ gives the symbol "4"; $^1/_{10}$ ratio $\frac{\text{yield stress}}{\text{minimum tensile strength}}\% = {}^1/_{10} \times {}^{32}/_{40} \times 100/1$ = "8"; giving the strength grade designation "4.8." Multiplication of these two figures gives the minimum yield stress in kgf/mm$^2$.

*Coating of Screws and Nuts:* It is recommended that the coating comply with the appropriate part of BS 3382. "Electroplated Coatings on Threaded Components."

*Screw Threads:* Screw threads are ISO metric coarse pitch series threads in accordance with BS 3643. "ISO Metric Screw Threads," Part 1, "Thread Data and Standard Thread Series." The external threads used for screws conform to tolerance Class 6g limits (medium fit) as given in BS 3643, "ISO Metric Screw Threads," Part 2, "Limits and Tolerances for Coarse Pitch Series Threads." The internal threads used for nuts conform to tolerance Class 6H limits (medium fit) as given in BS 3643: Part 2.

*Nominal Lengths of Screws:* For countersunk head screws the nominal length is the distance from the upper surface of the head to the extreme end of the shank, including any chamfer, radius, or cone point. For raised countersunk head screws the nominal length is the distance from the upper surface of the head (excluding the raised portion) to the extreme end of the shank, including any chamfer, radius, or cone point. For pan and cheese head screws the nominal length is the distance from the underside of the head to the extreme end of the shank, including any chamfer, radius, or cone point. Standard nominal lengths and tolerances are given in Table 5.

*Lengths of Thread on Screws:* The length of thread is the distance from the end of the screw (including any chamfer, radius, or cone point) to the leading face of a nut without countersink which has been screwed as far as possible onto the screw by hand. The minimum thread length is shown in the following table:

| Nominal Thread Dia., $d$[a] | M1 | M1.2 | (M1.4) | M1.6 | M2 | (M2.2) | M2.5 | M3 | (M3.5) | M4 |
|---|---|---|---|---|---|---|---|---|---|---|
| Thread Length $b$ (Min.) | b | b | b | 15 | 16 | 17 | 18 | 19 | 20 | 22 |
| Nominal Thread Dia., $d$[a] | (M4.5) | M5 | M6 | M8 | M10 | M12 | (M14) | M16 | (M18) | M20 |
| Thread Length $b$ (Min.) | 24 | 25 | 28 | 34 | 40 | 46 | 52 | 58 | 64 | 70 |

[a] Items shown in parentheses are non-preferred.
[b] Threaded up to the head.
All dimensions are in millimeters.

Screws of nominal thread diameter M1, M1.2 and M1.4 and screws of larger diameters that are too short for the above thread lengths are threaded as far as possible up to the head.

In these screws the length of unthreaded shank under the head does not exceed 1½ pitches for lengths up to twice the diameter and 2 pitches for longer lengths, and is defined as the distance from the leading face of a nut that has been screwed as far as possible onto the screw by hand to: 1) the junction of the basic major diameter and the countersunk portion of the head on countersunk and raised countersunk heads; and 2) the underside of the head on other types of heads.

*Diameter of Unthreaded Shank on Screws:* The diameter of the unthreaded portion of the shank on screws is not greater than the basic major diameter of the screw thread and not less than the minimum effective diameter of the screw thread. The diameter of the unthreaded portion of shank is closely associated with the method of manufacture; it will generally be nearer the major diameter of the thread for turned screws and nearer the effective diameter for those produced by cold heading.

*Radius Under the Head of Screws:* The radius under the head of pan and cheese head screws runs smoothly into the face of the head and shank without any step or discontinuity. A true radius is not essential providing that the curve is smooth and lies wholly within the maximum radius. Any radius under the head of countersunk head screws runs smoothly into the conical bearing surface of the head and the shank without any step or discontinuity. The radius values given in Table 1 and Table 2 are regarded as the maximum where the shank diameter is equal to the major diameter of the thread and minimum where the shank diameter is approximately equal to the effective diameter of the thread.

## Table 1. British Standard Slotted Countersunk Head Machine Screws—Metric Series *BS 4183:1967*

| Nominal Size $d$[a] | Head Diameter $D$ | | Head Height $k$ | | Radius $r$[b] | Thread Length $b$ | Thread Run-out $a$ | Flushness Tolerance[c] | Slot Width $n$ | | Slot Depth $t$ | |
|---|---|---|---|---|---|---|---|---|---|---|---|---|
| | Max. (Theor. Sharp) $2d$ | Min. $1.75d$ | Max. $0.5d$ | Min. $0.45d$ | | Min. | Max. $2p$[d] | Max. | Max. | Min. | Max. $0.3d$ | Min. $0.2d$ |
| M1 | 2.00 | 1.75 | 0.50 | 0.45 | 0.1 | [e] | 0.50 | … | 0.45 | 0.31 | 0.30 | 0.20 |
| M1.2 | 2.40 | 2.10 | 0.60 | 0.54 | 0.1 | [e] | 0.50 | … | 0.50 | 0.36 | 0.36 | 0.24 |
| (M1.4) | 2.80 | 2.45 | 0.70 | 0.63 | 0.1 | [e] | 0.60 | … | 0.50 | 0.36 | 0.42 | 0.28 |
| M1.6 | 3.20 | 2.80 | 0.80 | 0.72 | 0.1 | 15.0 | 0.70 | … | 0.60 | 0.46 | 0.48 | 0.32 |
| M2.0 | 4.00 | 3.50 | 1.00 | 0.90 | 0.1 | 16.0 | 0.80 | … | 0.70 | 0.56 | 0.60 | 0.40 |
| (M2.2) | 4.40 | 3.85 | 1.10 | 0.99 | 0.1 | 17.0 | 0.90 | … | 0.80 | 0.66 | 0.66 | 0.44 |
| M2.5 | 5.00 | 4.38 | 1.25 | 1.12 | 0.1 | 18.0 | 0.90 | 0.10 | 0.80 | 0.66 | 0.75 | 0.50 |
| M3 | 6.00 | 5.25 | 1.50 | 1.35 | 0.1 | 19.0 | 1.00 | 0.12 | 1.00 | 0.86 | 0.90 | 0.60 |
| (M3.5) | 7.00 | 6.10 | 1.75 | 1.57 | 0.2 | 20.0 | 1.20 | 0.13 | 1.00 | 0.86 | 1.05 | 0.70 |
| M4 | 8.00 | 7.00 | 2.00 | 1.80 | 0.2 | 22.0 | 1.40 | 0.15 | 1.20 | 1.06 | 1.20 | 0.80 |
| (M4.5) | 9.00 | 7.85 | 2.25 | 2.03 | 0.2 | 24.0 | 1.50 | 0.17 | 1.20 | 1.06 | 1.35 | 0.90 |
| M5 | 10.00 | 8.75 | 2.50 | 2.25 | 0.2 | 25.0 | 1.60 | 0.19 | 1.51 | 1.26 | 1.50 | 1.00 |
| M6 | 12.00 | 10.50 | 3.00 | 2.70 | 0.25 | 28.0 | 2.00 | 0.23 | 1.91 | 1.66 | 1.80 | 1.20 |
| M8 | 16.00 | 14.00 | 4.00 | 3.60 | 0.4 | 34.0 | 2.50 | 0.29 | 2.31 | 2.06 | 2.40 | 1.60 |
| M10 | 20.00 | 17.50 | 5.00 | 4.50 | 0.4 | 40.0 | 3.00 | 0.37 | 2.81 | 2.56 | 3.00 | 2.00 |
| M12 | 24.00 | 21.00 | 6.00 | 5.40 | 0.6 | 46.0 | 3.50 | 0.44 | 3.31 | 3.06 | 3.60 | 2.40 |
| (M14) | 28.00 | 24.50 | 7.00 | 6.30 | 0.6 | 52.0 | 4.00 | 0.52 | 3.31 | 3.06 | 4.20 | 2.80 |
| M16 | 32.00 | 28.00 | 8.00 | 7.20 | 0.6 | 58.0 | 4.00 | 0.60 | 4.37 | 4.07 | 4.80 | 3.20 |
| (M18) | 36.00 | 31.50 | 9.00 | 8.10 | 0.6 | 64.0 | 5.00 | 0.67 | 4.37 | 4.07 | 5.40 | 3.60 |
| M20 | 40.00 | 35.00 | 10.00 | 9.00 | 0.8 | 70.0 | 5.00 | 0.75 | 5.37 | 5.07 | 6.00 | 4.00 |

[a] Nominal sizes shown in parentheses are non-preferred.
[b] See *Radius Under the Head of Screws* description in text.
[c] See *Dimensions of 90-Degree Countersunk Head Screws* description in text.
[d] See text following table in *Lengths of Thread on Screws* description in text.
[e] Threaded up to head.

All dimensions are given in millimeters. For dimensional notation, see diagram on page 1807. Recessed head screws are also standard and are available. For dimensions see British Standard.

## Table 2. British Standard Slotted Raised Countersunk Head Machine Screws—Metric Series *BS 4183:1967*

| Nominal Size $d$[a] | Head Diameter $D$ | | Head Height $k$ | | Radius Under Head $r$[b] | Thread Length $b$ | Thread Run-out $a$ | Height of Raised Portion $f$ | Head Radius $R$ | Slot Width $n$ | | Slot Depth $t$ | |
|---|---|---|---|---|---|---|---|---|---|---|---|---|---|
| | Max. (Theor. Sharp) $2d$ | Min. $1.75d$ | Max. $0.5d$ | Min. $0.45d$ | | Min. | Max. $2p$[c] | Nom. $0.25d$ | Nom. | Max. | Min. | Max. $0.5d$ | Min. $0.4d$ |
| M1 | 2.00 | 1.75 | 0.50 | 0.45 | 0.1 | [d] | 0.50 | 0.25 | 2.0 | 0.45 | 0.31 | 0.50 | 0.40 |
| M1.2 | 2.40 | 2.10 | 0.60 | 0.54 | 0.1 | [d] | 0.50 | 0.30 | 2.5 | 0.50 | 0.36 | 0.60 | 0.48 |
| (M1.4) | 2.80 | 2.45 | 0.70 | 0.63 | 0.1 | [d] | 0.60 | 0.35 | 2.5 | 0.50 | 0.36 | 0.70 | 0.56 |
| M1.6 | 3.20 | 2.80 | 0.80 | 0.72 | 0.1 | 15.0 | 0.70 | 0.40 | 3.0 | 0.60 | 0.46 | 0.80 | 0.64 |
| M2.0 | 4.00 | 3.50 | 1.00 | 0.90 | 0.1 | 16.0 | 0.80 | 0.50 | 4.0 | 0.70 | 0.56 | 1.00 | 0.80 |
| (M2.2) | 4.40 | 3.85 | 1.10 | 0.99 | 0.1 | 17.0 | 0.90 | 0.55 | 4.0 | 0.80 | 0.66 | 1.10 | 0.88 |
| M2.5 | 5.00 | 4.38 | 1.25 | 1.12 | 0.1 | 18.0 | 0.90 | 0.60 | 5.0 | 0.80 | 0.66 | 1.25 | 1.00 |
| M3 | 6.00 | 5.25 | 1.50 | 1.35 | 0.1 | 19.0 | 1.00 | 0.75 | 6.0 | 1.00 | 0.86 | 1.50 | 1.20 |
| (M3.5) | 7.00 | 6.10 | 1.75 | 1.57 | 0.2 | 20.0 | 1.20 | 0.90 | 6.0 | 1.00 | 0.86 | 1.75 | 1.40 |
| M4 | 8.00 | 7.00 | 2.00 | 1.80 | 0.2 | 22.0 | 1.40 | 1.00 | 8.0 | 1.20 | 1.06 | 2.00 | 1.60 |
| (M4.5) | 9.00 | 7.85 | 2.25 | 2.03 | 0.2 | 24.0 | 1.50 | 1.10 | 8.0 | 1.20 | 1.06 | 2.25 | 1.80 |
| M5 | 10.00 | 8.75 | 2.50 | 2.25 | 0.2 | 25.0 | 1.60 | 1.25 | 10.0 | 1.51 | 1.26 | 2.50 | 2.00 |
| M6 | 12.00 | 10.50 | 3.00 | 2.70 | 0.25 | 28.0 | 2.00 | 1.50 | 12.0 | 1.91 | 1.66 | 3.00 | 2.40 |
| M8 | 16.00 | 14.00 | 4.00 | 3.60 | 0.4 | 34.0 | 2.50 | 2.00 | 16.0 | 2.31 | 2.06 | 4.00 | 3.20 |
| M10 | 20.00 | 17.50 | 5.00 | 4.50 | 0.4 | 40.0 | 3.00 | 2.50 | 20.0 | 2.81 | 2.56 | 5.00 | 4.00 |
| M12 | 24.00 | 21.00 | 6.00 | 5.40 | 0.6 | 46.0 | 3.50 | 3.00 | 25.0 | 3.31 | 3.06 | 6.00 | 4.80 |
| (M14) | 28.00 | 24.50 | 7.00 | 6.30 | 0.6 | 52.0 | 4.00 | 3.50 | 25.0 | 3.31 | 3.06 | 7.00 | 5.60 |
| M16 | 32.00 | 28.00 | 8.00 | 7.20 | 0.6 | 58.0 | 4.00 | 4.00 | 32.0 | 4.37 | 4.07 | 8.00 | 6.40 |
| (M18) | 36.00 | 31.50 | 9.00 | 8.10 | 0.6 | 64.0 | 5.00 | 4.50 | 32.0 | 4.37 | 4.07 | 9.00 | 7.20 |
| M20 | 40.00 | 35.00 | 10.00 | 9.00 | 0.8 | 70.0 | 5.00 | 5.00 | 40.0 | 5.37 | 5.07 | 10.00 | 8.00 |

[a] Nominal sizes shown in parentheses are non-preferred.
[b] See *Radius Under the Head of Screws* description in text.
[c] See text following table in *Lengths of Thread on Screws* description in text.
[d] Threaded up to head.

All dimensions are given in millimeters. For dimensional notation see diagram on page 1807. Recessed head screws are also standard and available. For dimensions see British Standard.

*Ends of Screws:* When screws are made with rolled threads, the "lead" formed by the thread rolling operation is normally regarded as providing the necessary chamfer and no other machining is necessary. The ends of screws with cut threads are normally finished with a chamfer conforming to the dimension in Fig. 1a through Fig. 1d. At the option of the manufacturer, the ends of screws smaller than M6 (6-mm diameter) may be finished with a radius approximately equal to $1\frac{1}{4}$ times the nominal diameter of the shank. When cone point ends are required, they should have the dimensions given in Fig. 1a through Fig. 1d.

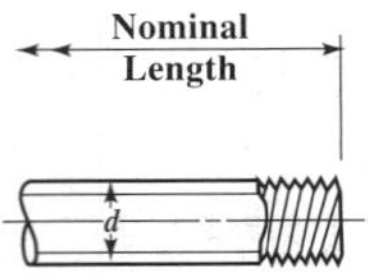


Fig. 1a. Rolled Thread End (Approximate Form as Rolled)

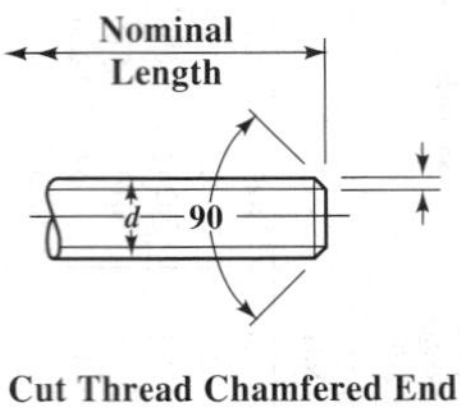


Fig. 1b. Chamfer to Extend to Slightly Below the Minor Diameter

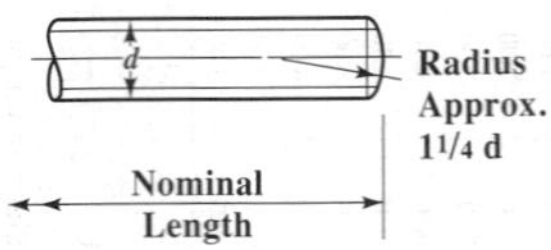


Fig. 1c. Cut Thread Radiused End (Permissible on Sizes Below M6 Diameter)

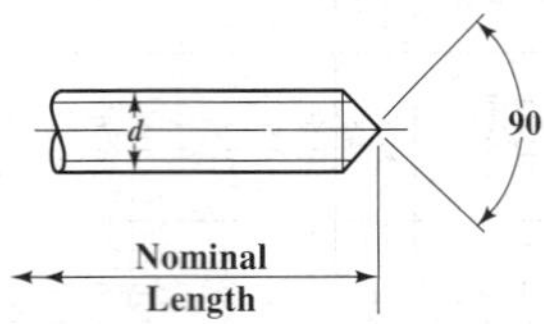


Fig. 1d. Cone Pointed End (Permissible on Cut or Rolled Thread Screws, but Regarded as "Special")

*Dimensions of 90-Degree Countersunk Head Screws:* One of the appendices to this British Standard states that countersunk head screws should fit into the countersunk hole with as great a degree of flushness as possible. To achieve this condition, it is necessary for the dimensions of both the head of the screw and the countersunk hole to be controlled within prescribed limits. The maximum or design size of the head is controlled by a theoretical diameter to a sharp corner and the minimum head angle of 90 degrees. The minimum head size is controlled by a minimum head diameter, the maximum head angle of 92 degrees and a flushness tolerance (see Fig. 2). The edge of the head may be flat or rounded, as shown in Fig. 3.

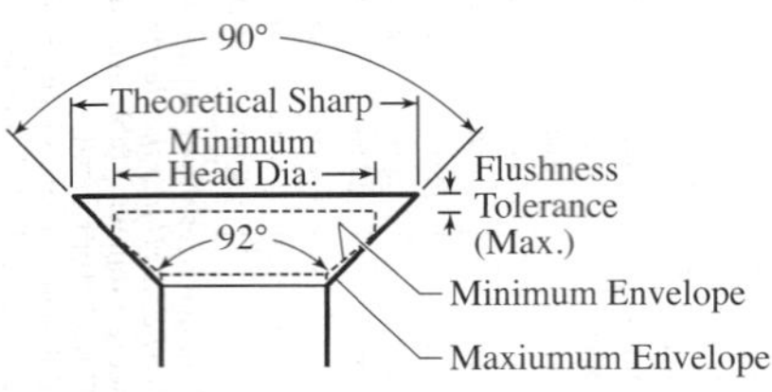


Fig. 2. Head Configuration

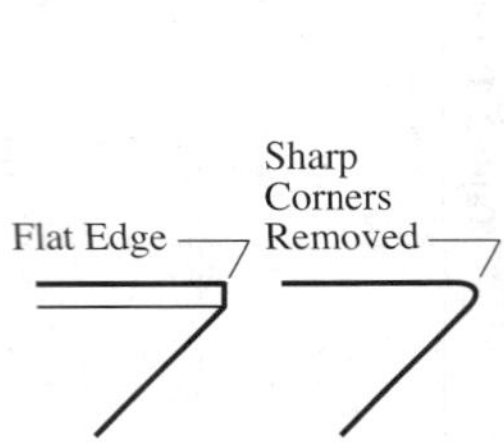


Fig. 3. Edge Configuration

## British Standard Machine Screws and Machine Screw Nuts—Metric Series

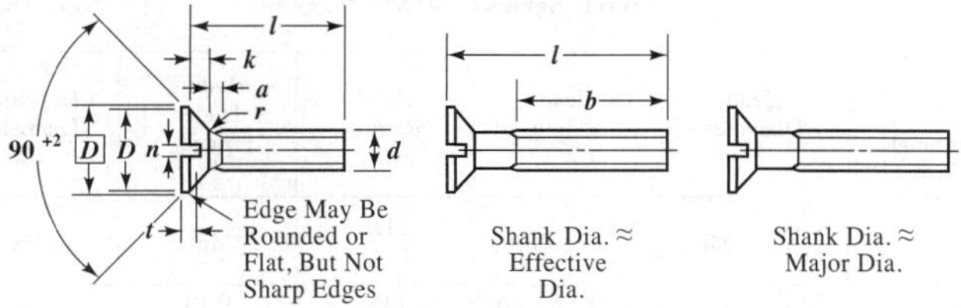


**Slotted Countersunk Head Machine Screws**

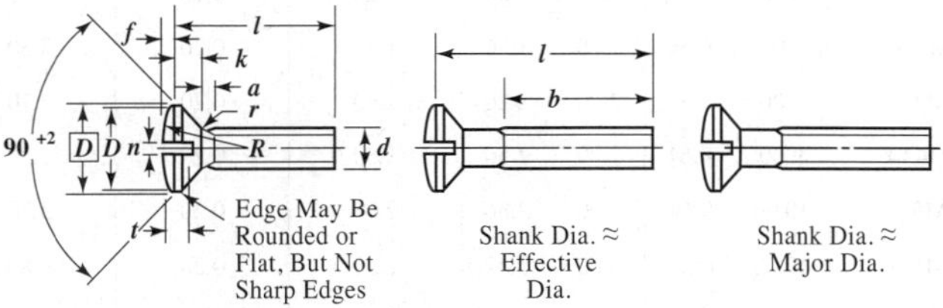


**Slotted Raised Countersunk Head Machine Screws**

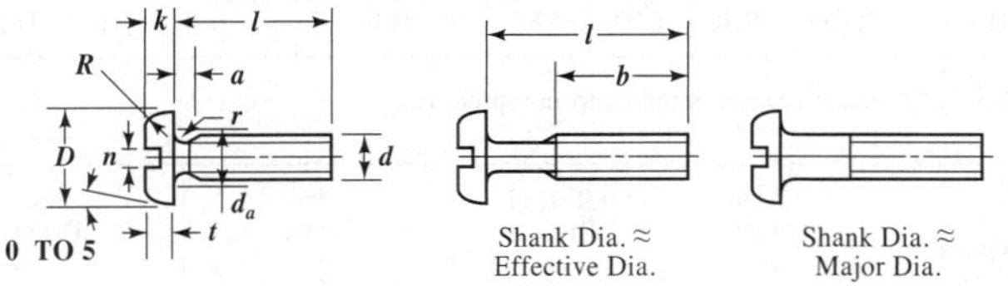


**Slotted Pan Head Machine Screws**

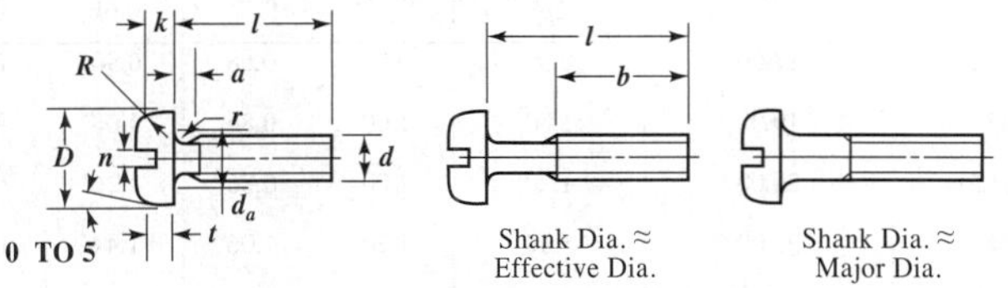


**Slotted Cheese Head Machine Screws**

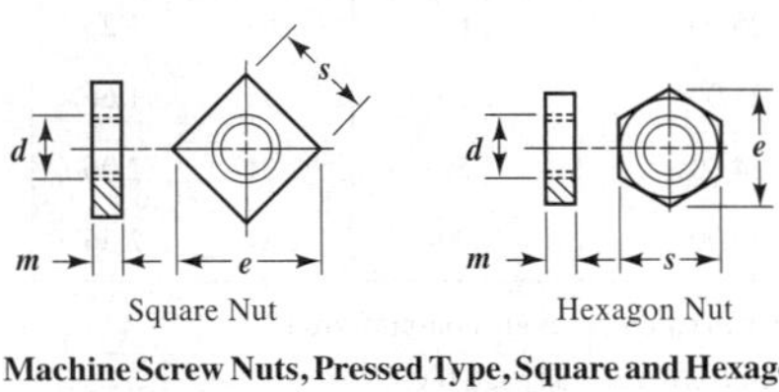


**Machine Screw Nuts, Pressed Type, Square and Hexagon**

For dimensions, see Table 1 through Table 5.

**Table 3. British Standard Slotted Pan Head Machine Screws— Metric Series** *BS 4183:1967*

| Nominal Size $d$[a] | Head Diameter $D$ | | Head Height $k$ | | Head Radius $R$ | Radius Under Head $r$ | Transition Diameter $d_a$ |
|---|---|---|---|---|---|---|---|
| | Max. $2d$ | Min. | Max. $0.6d$ | Min. | Max $0.4d$ | Min. | Max. |
| M2.5 | 5.00 | 4.70 | 1.50 | 1.36 | 1.00 | 0.10 | 3.10 |
| M3 | 6.00 | 5.70 | 1.80 | 1.66 | 1.20 | 0.10 | 3.60 |
| (M3.5) | 7.00 | 6.64 | 2.10 | 1.96 | 1.40 | 0.20 | 4.30 |
| M4 | 8.00 | 7.64 | 2.40 | 2.26 | 1.60 | 0.20 | 4.70 |
| (M4.5) | 9.00 | 8.64 | 2.70 | 2.56 | 1.80 | 0.20 | 5.20 |
| M5 | 10.00 | 9.64 | 3.00 | 2.86 | 2.00 | 0.20 | 5.70 |
| M6 | 12.00 | 11.57 | 3.60 | 3.42 | 2.50 | 0.25 | 6.80 |
| M8 | 16.00 | 15.57 | 4.80 | 4.62 | 3.20 | 0.40 | 9.20 |
| M10 | 20.00 | 19.48 | 6.00 | 5.82 | 4.00 | 0.40 | 11.20 |

[a] Nominal sizes shown in parentheses are non-preferred.

| Nominal Size $d$[a] | Thread Length $b$ | Thread Run-out $a$ | Slot Width $n$ | | Slot Depth $t$ | |
|---|---|---|---|---|---|---|
| | Min. | Max. $2p$[b] | Max. | Min. | Max. $0.6k$ | Min. $0.4k$ |
| M2.5 | 18.00 | 0.90 | 0.80 | 0.66 | 0.90 | 0.60 |
| M3 | 19.00 | 1.00 | 1.00 | 0.86 | 1.08 | 0.72 |
| (M3.5) | 20.00 | 1.20 | 1.00 | 0.86 | 1.26 | 0.84 |
| M4 | 22.00 | 1.40 | 1.20 | 1.06 | 1.44 | 0.96 |
| (M4.5) | 24.00 | 1.50 | 1.20 | 1.06 | 1.62 | 1.08 |
| M5 | 25.00 | 1.60 | 1.51 | 1.26 | 1.80 | 1.20 |
| M6 | 28.00 | 2.00 | 1.91 | 1.66 | 2.16 | 1.44 |
| M8 | 34.00 | 2.50 | 2.31 | 2.06 | 2.88 | 1.92 |
| M10 | 40.00 | 3.00 | 2.81 | 2.56 | 3.60 | 2.40 |

[a] Nominal sizes shown in parentheses are non-preferred.

[b] See *Lengths of Thread on Screws* on page 1803.

All dimensions are in millimeters. For dimensional notation, see diagram on page 1807. Recessed head screws are also standard and available. For dimensions, see British Standard.

**Table 4. British Standard Slotted Cheese Head Machine Screws—Metric Series** *BS 4183:1967*

| Nominal Size $d$[a] | Head Diameter $D$ | | Head Height $k$ | | Radius $r$[b] | Transition Diameter $d_a$ | Thread Length $b$ | Thread Run-out $a$ | Slot Width $n$ | | Slot Depth $t$ | |
|---|---|---|---|---|---|---|---|---|---|---|---|---|
| | Max. | Min. | Max. | Min. | Min. | Max. | Min. | Max.[c] | Max. | Min. | Max. | Min. |
| M1 | 2.00 | 1.75 | 0.70 | 0.56 | 0.10 | 1.30 | [b] | 0.50 | 0.45 | 0.31 | 0.44 | 0.30 |
| M1.2 | 2.30 | 2.05 | 0.80 | 0.66 | 0.10 | 1.50 | [b] | 0.50 | 0.50 | 0.36 | 0.49 | 0.35 |
| (M1.4) | 2.60 | 2.35 | 0.90 | 0.76 | 0.10 | 1.70 | [b] | 0.60 | 0.50 | 0.36 | 0.60 | 0.40 |
| M1.6 | 3.00 | 2.75 | 1.00 | 0.86 | 0.10 | 2.00 | 15.00 | 0.70 | 0.60 | 0.46 | 0.65 | 0.45 |
| M2 | 3.80 | 3.50 | 1.30 | 1.16 | 0.10 | 2.60 | 16.00 | 0.80 | 0.70 | 0.56 | 0.85 | 0.60 |
| (M2.2) | 4.00 | 3.70 | 1.50 | 1.36 | 0.10 | 2.80 | 17.00 | 0.90 | 0.80 | 0.66 | 1.00 | 0.70 |
| M2.5 | 4.50 | 4.20 | 1.60 | 1.46 | 0.10 | 3.10 | 18.00 | 0.90 | 0.80 | 0.66 | 1.00 | 0.70 |
| M3 | 5.50 | 5.20 | 2.00 | 1.86 | 0.10 | 3.60 | 19.00 | 1.00 | 1.00 | 0.86 | 1.30 | 0.90 |
| (M3.5) | 6.00 | 5.70 | 2.40 | 2.26 | 0.10 | 4.10 | 20.00 | 1.20 | 1.00 | 0.86 | 1.40 | 1.00 |
| M4 | 7.00 | 6.64 | 2.60 | 2.46 | 0.20 | 4.70 | 22.00 | 1.40 | 1.20 | 1.06 | 1.60 | 1.20 |
| (M4.5) | 8.00 | 7.64 | 3.10 | 2.92 | 0.20 | 5.20 | 24.00 | 1.50 | 1.20 | 1.06 | 1.80 | 1.40 |
| M5 | 8.50 | 8.14 | 3.30 | 3.12 | 0.20 | 5.70 | 25.00 | 1.60 | 1.51 | 1.26 | 2.00 | 1.50 |
| M6 | 10.00 | 9.64 | 3.90 | 3.72 | 0.25 | 6.80 | 28.00 | 2.00 | 1.91 | 1.66 | 2.30 | 1.80 |
| M8 | 13.00 | 12.57 | 5.00 | 4.82 | 0.40 | 9.20 | 34.00 | 2.50 | 2.31 | 2.06 | 2.80 | 2.30 |
| M10 | 16.00 | 15.57 | 6.00 | 5.82 | 0.40 | 11.20 | 40.00 | 3.00 | 2.81 | 2.56 | 3.20 | 2.70 |
| M12 | 18.00 | 17.57 | 7.00 | 6.78 | 0.60 | 14.20 | 46.00 | 3.50 | 3.31 | 3.06 | 3.80 | 3.20 |
| (M14) | 21.00 | 20.48 | 8.00 | 7.78 | 0.60 | 16.20 | 52.00 | 4.00 | 3.31 | 3.06 | 4.20 | 3.60 |
| M16 | 24.00 | 23.48 | 9.00 | 8.78 | 0.60 | 18.20 | 58.00 | 4.00 | 4.37 | 4.07 | 4.60 | 4.00 |
| (M18) | 27.00 | 26.48 | 10.00 | 9.78 | 0.60 | 20.20 | 64.00 | 5.00 | 4.37 | 4.07 | 5.10 | 4.50 |
| M20 | 30.00 | 29.48 | 11.00 | 10.73 | 0.80 | 22.40 | 70.00 | 5.00 | 5.27 | 5.07 | 5.60 | 5.00 |

[a] Nominal sizes shown in parentheses are non-preferred.
[b] Threaded up to head.
[c] See text following table in *Lengths of Thread on Screws* description in text.
All dimensions are given in millimeters. For dimensional notation, see diagram on page 1807.

**Table 5. British Standard Machine Screws and Nuts—Metric Series** *BS 4183:1967*

Concentricity Tolerances

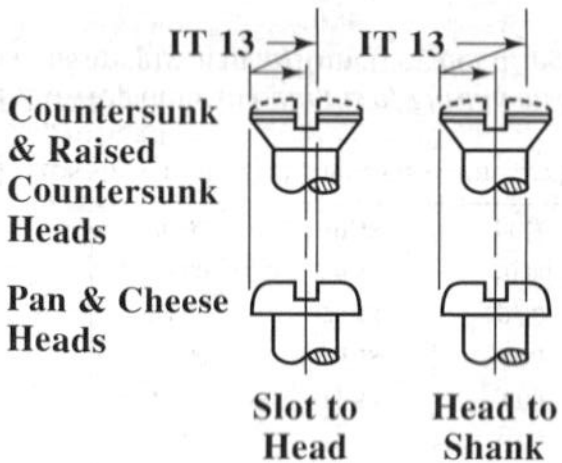


| Nominal Size $d^a$ | Head to Shank and Slot to Head (IT 13): Countersunk, Raised Csk., and Pan Heads | Head to Shank and Slot to Head (IT 13): Cheese Heads |
|---|---|---|
| M1, M1.2, (M1.4) | 0.14 | 0.14 |
| M1.6 | 0.18 | 0.14 |
| M2, (M2.2), M2.5, M3 | 0.18 | 0.18 |
| (M3.5) | 0.22 | 0.18 |
| M4, (M4.5), M5 | 0.22 | 0.22 |
| M6 | 0.27 | 0.22 |
| M8 | 0.27 | 0.27 |
| M10, M12 | 0.33 | 0.27 |
| (M14) | 0.33 | 0.33 |
| M16, (M18), M20 | 0.39 | 0.33 |

Nominal Lengths and Tolerances on Length for Machine Screws

| Nominal Length[a] | Tolerance | Nominal Length[a] | Tolerance |
|---|---|---|---|
| 1.5 | ±0.12 | 45 | ±0.50 |
| 2 | ±0.12 | 50 | ±0.060 |
| 2.5 | ±0.20 | 55 | ±0.60 |
| 3 | ±0.20 | 60 | ±0.60 |
| 4 | ±0.24 | 65 | ±0.60 |
| 5 | ±0.24 | 70 | ±0.60 |
| 6 | ±0.24 | 75 | ±0.60 |
| (7) | ±0.29 | 80 | ±0.60 |
| 8 | ±0.29 | 85 | ±0.70 |
| (9) | ±0.29 | 90 | ±0.70 |
| 10 | ±0.29 | (95) | ±0.70 |
| (11) | ±0.35 | 100 | ±0.70 |
| 12 | ±0.35 | (105) | ±0.70 |
| 14 | ±0.35 | 110 | ±0.70 |
| 16 | ±0.35 | (115) | ±0.70 |
| (18) | ±0.35 | 120 | ±0.70 |
| 20 | ±0.42 | (125) | ±0.70 |
| (22) | ±0.42 | 130 | ±0.80 |
| 25 | ±0.42 | 140 | ±0.80 |
| (28) | ±0.42 | 150 | ±0.80 |
| 30 | ±0.42 | 160 | ±0.80 |
| (38) | ±0.50 | 190 | ±0.925 |
| 40 | ±0.50 | 200 | ±0.925 |

Dimensions of Machine Screw Nuts, Pressed Type, Square and Hexagon

| Nominal Size $d^a$ | Width Across Flats $s$ Max. | Width Across Flats $s$ Min. | Width Across Corners $e$ Square |
|---|---|---|---|
| M1.6 | 3.2 | 3.02 | 4.5 |
| M2 | 4.0 | 3.82 | 5.7 |
| (M2.2) | 4.5 | 4.32 | 6.4 |
| (M2.5) | 5.0 | 4.82 | 7.1 |
| M3 | 5.5 | 5.32 | 7.8 |
| (M3.5) | 6.0 | 5.82 | 8.5 |
| M4 | 7.0 | 6.78 | 9.9 |
| M5 | 8.0 | 7.78 | 11.3 |
| M6 | 10.0 | 9.78 | 14.1 |
| M8 | 13.0 | 12.73 | 18.4 |
| M10 | 17.0 | 16.73 | 24.0 |

| Nominal Size $d^a$ | Width Across Corners $e$ Hexagon | Thickness $m$ Max. | Thickness $m$ Min |
|---|---|---|---|
| M1.6 | 3.7 | 1.0 | 0.75 |
| M2 | 4.6 | 1.2 | 0.95 |
| (M2.2) | 5.2 | 1.2 | 0.95 |
| M2.5 | 5.8 | 1.6 | 1.35 |
| M3 | 6.4 | 1.6 | 1.35 |
| (M3.5) | 6.9 | 2.0 | 1.75 |
| M4 | 8.1 | 2.0 | 1.75 |
| M5 | 9.2 | 2.5 | 2.25 |
| M6 | 11.5 | 3.0 | 2.75 |
| M8 | 15.0 | 4.0 | 3.70 |
| M10 | 19.6 | 5.0 | 4.70 |
| M8 | 13.0 | 12.73 | 18.4 |
| M10 | 17.0 | 16.73 | 24.0 |

[a] Nominal sizes and lengths shown in parentheses are non-preferred.

All dimensions are given in millimeters. For dimensional notation, see diagram on page 1807.

*General Dimensions:* The general dimensions and tolerances for screws and nuts are given in the accompanying tables. Although slotted screw dimensions are given, recessed head screws are also standard and available. Dimensions of recessed head screws are given in BS 4183:1967.

**British Unified Machine Screws and Nuts.**—British Standard BS 1981:1991 covers certain types of machine screws and machine screw nuts for which agreement has been reached with the United States and Canada as to general dimensions for interchangeability. These types are: countersunk, raised-countersunk, pan, and raised-cheese head screws with slotted or recessed heads; small hexagon head screws; and precision and pressed nuts. All have Unified threads. Head shapes are shown on page 1812 and dimensions are given on page 1813.

*Identification:* The above-mentioned screws and nuts that conform to this Standard should have a distinguishing feature applied to identify them as Unified. All recessed head screws are to be identified as Unified by a groove in the form of four arcs of a circle in the upper surface of the head. All hexagon head screws are to be identified as Unified by: 1) a circular recess in the upper surface of the head; 2) a continuous line of circles indented on one or more of the flats of the hexagon and parallel to the screw axis; and 3) at least two contiguous circles indented on the upper surface of the head. All *machine screw* nuts of the pressed type shall be identified as Unified by means of the application of a groove indented in one face of the nut approximately midway between the major diameter of the thread and flats of the square or hexagon. *Slotted head screws* shall be identified as Unified either by a circular recess or by a circular platform or raised portion on the upper surface of the head. *Machine screw nuts* of the *precision type* shall be identified as Unified by either a groove indented on one face of the front approximately midway between the major diameter of the thread and the flats of the hexagon or a continuous line of circles indented on one or more of the flats of the hexagon and parallel to the nut axis.

**Identification Markings for British Standard Unified Machine Screws**

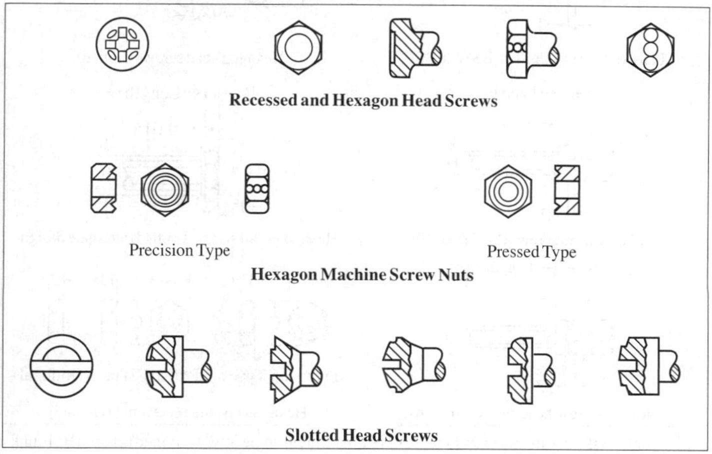

**British Standard Machine Screws and Nuts**

*BS 450:1958 and BS 1981:1991*

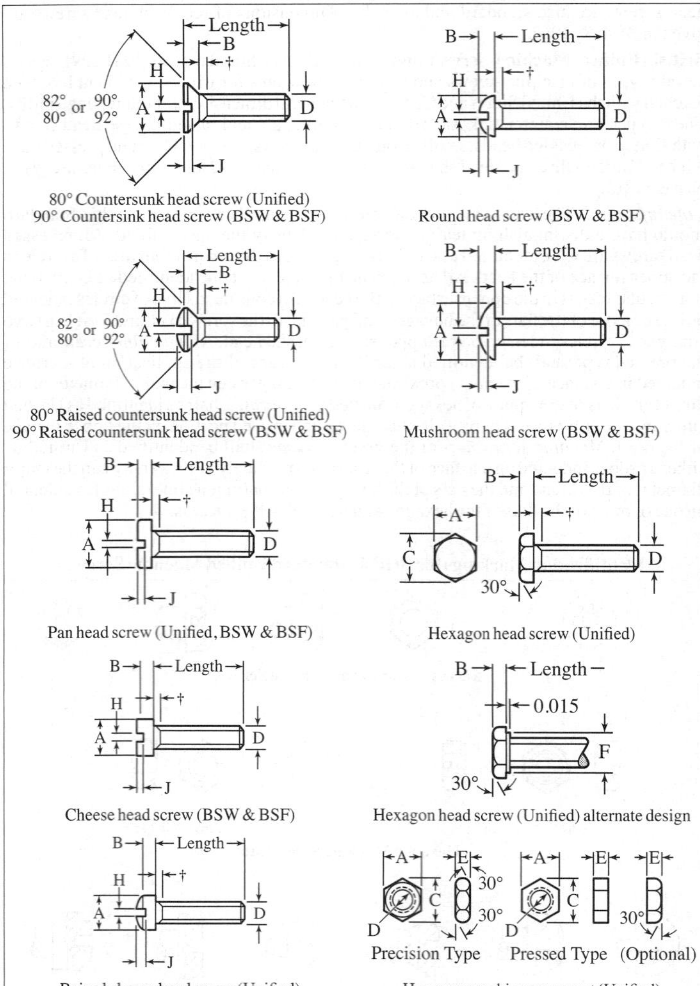

80° Countersunk head screw (Unified)
90° Countersink head screw (BSW & BSF)

Round head screw (BSW & BSF)

80° Raised countersunk head screw (Unified)
90° Raised countersunk head screw (BSW & BSF)

Mushroom head screw (BSW & BSF)

Pan head screw (Unified, BSW & BSF)

Hexagon head screw (Unified)

Cheese head screw (BSW & BSF)

Hexagon head screw (Unified) alternate design

Raised cheese head screw (Unified)

Hexagon machine screw nut (Unified)

*Countersinks to suit the screws should have a maximum angle of 80° (Unified) or 90° (BSF and BSW) with a negative tolerance.

†Unified countersunk and raised countersunk head screws 2 inches long and under are threaded right up to the head. Other Unified, BSW and BSF machine screws 2 inches long and under have an unthread shank equal to twice the pitch. All Unified, BSW and BSF machine screws longer than 2 inches have a minimum thread length of 1¾ inches.

## British Standard Unified Machine Screws and Nuts *BS 1981:1991*

| Nom. Size of Screw | Basic Dia. *D* | Threads per Inch | | Dia. of Head *A* | | Depth of Head *B* | | Width of Slot *H* | | Depth of Slot *J*[d] |
|---|---|---|---|---|---|---|---|---|---|---|
| | | UNC | UNF | Max. | Min. | Max. | Min. | Max. | Min. | |
| 80° Countersunk Head Screws[a,b] | | | | | | | | | | |
| 4 | 0.112 | 40 | … | 0.225 | 0.195 | 0.067 | … | 0.039 | 0.031 | 0.025 |
| 6 | 0.138 | 32 | … | 0.279 | 0.244 | 0.083 | … | 0.048 | 0.039 | 0.031 |
| 8 | 0.164 | 32 | … | 0.332 | 0.292 | 0.100 | … | 0.054 | 0.045 | 0.037 |
| 10 | 0.190 | 24[c] | 32 | 0.385 | 0.340 | 0.116 | … | 0.060 | 0.050 | 0.044 |
| 1/4 | 0.250 | 20 | 28 | 0.507 | 0.452 | 0.153 | … | 0.075 | 0.064 | 0.058 |
| 5/16 | 0.3125 | 18 | 24 | 0.635 | 0.568 | 0.191 | … | 0.084 | 0.072 | 0.073 |
| 3/8 | 0.375 | 16 | 24 | 0.762 | 0.685 | 0.230 | … | 0.094 | 0.081 | 0.088 |
| 7/16 | 0.4375 | 14 | 20 | 0.812 | 0.723 | 0.223 | … | 0.094 | 0.081 | 0.085 |
| 1/2 | 0.500 | 13 | 20 | 0.875 | 0.775 | 0.223 | … | 0.106 | 0.091 | 0.084 |
| 5/8 | 0.625 | 11 | 18 | 1.125 | 1.002 | 0.298 | … | 0.133 | 0.116 | 0.113 |
| 3/4 | 0.750 | 10 | 16 | 1.375 | 1.230 | 0.372 | … | 0.149 | 0.131 | 0.141 |
| Pan Head Screws[b] | | | | | | | | | | |
| 4 | 0.112 | 40 | … | 0.219 | 0.205 | 0.068 | 0.058 | 0.039 | 0.031 | 0.035 |
| 6 | 0.138 | 32 | … | 0.270 | 0.256 | 0.082 | 0.072 | 0.048 | 0.039 | 0.044 |
| 8 | 0.164 | 32 | … | 0.322 | 0.306 | 0.096 | 0.085 | 0.054 | 0.045 | 0.051 |
| 10 | 0.190 | 24[c] | 32 | 0.373 | 0.357 | 0.110 | 0.099 | 0.060 | 0.050 | 0.061 |
| 1/4 | 0.250 | 20 | 28 | 0.492 | 0.473 | 0.144 | 0.130 | 0.075 | 0.064 | 0.079 |
| 5/16 | 0.3125 | 18 | 24 | 0.615 | 0.594 | 0.178 | 0.162 | 0.084 | 0.072 | 0.096 |
| 3/8 | 0.375 | 16 | 24 | 0.740 | 0.716 | 0.212 | 0.195 | 0.094 | 0.081 | 0.112 |
| 7/16 | 0.4375 | 14 | 20 | 0.863 | 0.838 | 0.247 | 0.227 | 0.094 | 0.081 | 0.134 |
| 1/2 | 0.500 | 13 | 20 | 0.987 | 0.958 | 0.281 | 0.260 | 0.106 | 0.091 | 0.153 |
| 5/8 | 0.625 | 11 | 18 | 1.125 | 1.090 | 0.350 | 0.325 | 0.133 | 0.116 | 0.191 |
| 3/4 | 0.750 | 10 | 16 | 1.250 | 1.209 | 0.419 | 0.390 | 0.149 | 0.131 | 0.229 |
| Raised Cheese-Head Screws[b] | | | | | | | | | | |
| 4 | 0.112 | 40 | … | 0.183 | 0.166 | 0.107 | 0.088 | 0.039 | 0.031 | 0.042 |
| 6 | 0.138 | 32 | … | 0.226 | 0.208 | 0.132 | 0.111 | 0.048 | 0.039 | 0.053 |
| 8 | 0.164 | 32 | … | 0.270 | 0.250 | 0.156 | 0.133 | 0.054 | 0.045 | 0.063 |
| 10 | 0.190 | 24[c] | 32 | 0.313 | 0.292 | 0.180 | 0.156 | 0.060 | 0.050 | 0.074 |
| 1/4 | 0.250 | 20 | 28 | 0.414 | 0.389 | 0.237 | 0.207 | 0.075 | 0.064 | 0.098 |
| 5/16 | 0.3125 | 18 | 24 | 0.518 | 0.490 | 0.295 | 0.262 | 0.084 | 0.072 | 0.124 |
| 3/8 | 0.375 | 16 | 24 | 0.622 | 0.590 | 0.355 | 0.315 | 0.094 | 0.081 | 0.149 |
| 7/16 | 0.4375 | 14 | 20 | 0.625 | 0.589 | 0.368 | 0.321 | 0.094 | 0.081 | 0.153 |
| 1/2 | 0.500 | 13 | 20 | 0.750 | 0.710 | 0.412 | 0.362 | 0.106 | 0.091 | 0.171 |
| 5/8 | 0.625 | 11 | 18 | 0.875 | 0.827 | 0.521 | 0.461 | 0.133 | 0.116 | 0.217 |
| 3/4 | 0.750 | 10 | 16 | 1.000 | 0.945 | 0.612 | 0.542 | 0.149 | 0.131 | 0.254 |

[a] All dimensions, except *J*, given for the No. 4 to 3/8-inch sizes, incl., also apply to all the 80° Raised Countersunk Head Screws given in the Standard.

[b] Also available with recessed heads.

[c] Non-preferred.

[d] The depth of slot is the average of the maximum and minimum range of depths as mentioned in BS 1981:1991.

| Nom. Size | Basic Dia. *D* | Threads per Inch | | Width Across Flats *A* | | Width Across Corners *C* | Head Depth *B* Nut Thickness *E* | | Washer Face Dia. *F* | |
|---|---|---|---|---|---|---|---|---|---|---|
| | | UNC | UNF | Max. | Min. | Max. | Max. | Min. | Max. | Min. |
| Hexagon Head Screws | | | | | | | | | | |
| 4 | 0.112 | 40 | … | 0.187 | 0.181 | 0.202 | 0.060 | 0.049 | 0.171 | 0.161 |
| 6 | 0.138 | 32 | … | 0.250 | 0.244 | 0.272 | 0.093 | 0.080 | 0.234 | 0.224 |
| 8 | 0.164 | 32 | … | 0.250 | 0.244 | 0.272 | 0.110 | 0.096 | 0.234 | 0.224 |
| 10 | 0.190 | 24[c] | 32 | 0.312 | 0.305 | 0.340 | 0.120 | 0.105 | 0.295 | 0.285 |
| Hexagon Machine Screw Nuts—Precision Type | | | | | | | | | | |
| 4 | 0.112 | 40 | … | 0.2500 | 0.241 | 0.289 | 0.098 | 0.087 | … | … |
| 6 | 0.138 | 32 | … | 0.3125 | 0.302 | 0.361 | 0.114 | 0.102 | … | … |
| 8 | 0.164 | 32 | … | 0.3438 | 0.332 | 0.397 | 0.130 | 0.117 | … | … |
| 10 | 0.190 | 24[c] | … | 0.3750 | 0.362 | 0.433 | 0.130 | 0.117 | … | … |
| Hexagon Machine Screw Nuts—Pressed Type | | | | | | | | | | |
| 4 | 0.112 | 40 | … | 0.250 | 0.241 | 0.289 | 0.098 | 0.087 | … | … |
| 6 | 0.138 | 32 | … | 0.312 | 0.302 | 0.361 | 0.114 | 0.102 | … | … |
| 8 | 0.164 | 32 | … | 0.344 | 0.332 | 0.397 | 0.130 | 0.117 | … | … |
| 10 | 0.190 | 24[c] | 32 | 0.375 | 0.362 | 0.433 | 0.130 | 0.117 | … | … |
| 1/4 | 0.250 | 20 | 28 | 0.437 | 0.423 | 0.505 | 0.193 | 0.178 | … | … |
| 5/16 | 0.3125 | 18 | 24 | 0.562 | 0.545 | 0.649 | 0.225 | 0.208 | … | … |
| 3/8 | 0.375 | 16 | 24 | 0.625 | 0.607 | 0.722 | 0.257 | 0.239 | … | … |

All dimensions in inches. See page 1812 for a pictorial representation and letter dimensions.

## British Standard Whitworth (BSW) and Fine (BSF) Machine Screws

*BS 450:1958*

| | Nom. Size of Screw | Basic Dia. *D* | Threads per Inch | | Dia. of Head *A* | | Depth of Head *B* | | Width of Slot *H* | | Depth of Slot *J* |
|---|---|---|---|---|---|---|---|---|---|---|---|
| | | | BSW | BSF | Max. | Min. | Max. | Min. | Max. | Min. | |
| 90° Countersunk Head Screws[a,b] | 1/8 | 0.1250 | 40 | … | 0.219 | 0.201 | 0.056 | … | 0.039 | 0.032 | 0.027 |
| | 3/16 | 0.1875 | 24 | 32[c] | 0.328 | 0.307 | 0.084 | … | 0.050 | 0.042 | 0.041 |
| | 7/32 | 0.2188 | … | 28[c] | 0.383 | 0.360 | 0.098 | … | 0.055 | 0.046 | 0.048 |
| | 1/4 | 0.2500 | 20 | 26 | 0.438 | 0.412 | 0.113 | … | 0.061 | 0.051 | 0.055 |
| | 5/16 | 0.3125 | 18 | 22 | 0.547 | 0.518 | 0.141 | … | 0.071 | 0.061 | 0.069 |
| | 3/8 | 0.3750 | 16 | 20 | 0.656 | 0.624 | 0.169 | … | 0.082 | 0.072 | 0.083 |
| | 7/16 | 0.4375 | 14 | 18 | 0.766 | 0.729 | 0.197 | … | 0.093 | 0.082 | 0.097 |
| | 1/2 | 0.5000 | 12 | 16 | 0.875 | 0.835 | 0.225 | … | 0.104 | 0.092 | 0.111 |
| | 9/16 | 0.5625 | 12[c] | 16[c] | 0.984 | 0.941 | 0.253 | … | 0.115 | 0.103 | 0.125 |
| | 5/8 | 0.6250 | 11 | 14 | 1.094 | 1.046 | 0.281 | … | 0.126 | 0.113 | 0.138 |
| | 3/4 | 0.7500 | 10 | 12 | 1.312 | 1.257 | 0.338 | … | 0.148 | 0.134 | 0.166 |
| Round Head Screws[b] | 1/8 | 0.1250 | 40 | … | 0.219 | 0.206 | 0.087 | 0.082 | 0.039 | 0.032 | 0.048 |
| | 3/16 | 0.1875 | 24 | 32[c] | 0.328 | 0.312[d] | 0.131 | 0.124 | 0.050 | 0.042 | 0.072 |
| | 7/32 | 0.2188 | … | 28[c] | 0.383 | 0.365 | 0.153 | 0.145 | 0.055 | 0.046 | 0.084 |
| | 1/4 | 0.2500 | 20 | 26 | 0.438 | 0.417 | 0.175 | 0.165 | 0.061 | 0.051 | 0.096 |
| | 5/16 | 0.3125 | 18 | 22 | 0.547 | 0.524 | 0.219 | 0.207 | 0.071 | 0.061 | 0.120 |
| | 3/8 | 0.3750 | 16 | 20 | 0.656 | 0.629 | 0.262 | 0.249 | 0.082 | 0.072 | 0.144 |
| | 7/16 | 0.4375 | 14 | 18 | 0.766 | 0.735 | 0.306 | 0.291 | 0.093 | 0.082 | 0.168 |
| | 1/2 | 0.5000 | 12 | 16 | 0.875 | 0.840 | 0.350 | 0.333 | 0.104 | 0.092 | 0.192 |
| | 9/16 | 0.5625 | 12[c] | 16[c] | 0.984 | 0.946 | 0.394 | 0.375 | 0.115 | 0.103 | 0.217 |
| | 5/8 | 0.6250 | 11 | 14 | 1.094 | 1.051 | 0.437 | 0.417 | 0.126 | 0.113 | 0.240 |
| | 3/4 | 0.7500 | 10 | 12 | 1.312 | 1.262 | 0.525 | 0.500 | 0.148 | 0.134 | 0.288 |
| Pan Head Screws[b] | 1/8 | 0.1250 | 40 | … | 0.245 | 0.231 | 0.075 | 0.065 | 0.039 | 0.032 | 0.040 |
| | 3/16 | 0.1875 | 24 | 32[c] | 0.373 | 0.375 | 0.110 | 0.099 | 0.050 | 0.042 | 0.061 |
| | 7/32 | 0.2188 | … | 28[c] | 0.425 | 0.407 | 0.125 | 0.112 | 0.055 | 0.046 | 0.069 |
| | 1/4 | 0.2500 | 20 | 26 | 0.492 | 0.473[e] | 0.144 | 0.130 | 0.061 | 0.051 | 0.078 |
| | 5/16 | 0.3125 | 18 | 22 | 0.615 | 0.594 | 0.178 | 0.162 | 0.071 | 0.061 | 0.095 |
| | 3/8 | 0.3750 | 16 | 20 | 0.740 | 0.716 | 0.212 | 0.195 | 0.082 | 0.072 | 0.112 |
| | 7/16 | 0.4375 | 14 | 18 | 0.863 | 0.838 | 0.247 | 0.227 | 0.093 | 0.082 | 0.129 |
| | 1/2 | 0.5000 | 12 | 16 | 0.987 | 0.958 | 0.281 | 0.260 | 0.104 | 0.092 | 0.145 |
| | 9/16 | 0.5625 | 12[c] | 16[c] | 1.031 | 0.999 | 0.315 | 0.293 | 0.115 | 0.103 | 0.162 |
| | 5/8 | 0.6250 | 11 | 14 | 1.125 | 1.090 | 0.350 | 0.325 | 0.126 | 0.113 | 0.179 |
| | 3/4 | 0.7500 | 10 | 12 | 1.250 | 1.209 | 0.419 | 0.390 | 0.148 | 0.134 | 0.213 |
| Cheese Head Screws[b] | 1/8 | 0.1250 | 40 | … | 0.188 | 0.180 | 0.087 | 0.082 | 0.039 | 0.032 | 0.039 |
| | 3/16 | 0.1875 | 24 | 32[c] | 0.281 | 0.270 | 0.131 | 0.124 | 0.050 | 0.042 | 0.059 |
| | 7/32 | 0.2188 | … | 28[c] | 0.328 | 0.315 | 0.153 | 0.145 | 0.055 | 0.046 | 0.069 |
| | 1/4 | 0.2500 | 20 | 26 | 0.375 | 0.360 | 0.175 | 0.165 | 0.061 | 0.051 | 0.079 |
| | 5/16 | 0.3125 | 18 | 22 | 0.469 | 0.450 | 0.219 | 0.207 | 0.071 | 0.061 | 0.098 |
| | 3/8 | 0.3750 | 16 | 20 | 0.562 | 0.540 | 0.262 | 0.249 | 0.082 | 0.072 | 0.118 |
| | 7/16 | 0.4375 | 14 | 18 | 0.656 | 0.630 | 0.306 | 0.291 | 0.093 | 0.082 | 0.138 |
| | 1/2 | 0.5000 | 12 | 16 | 0.750 | 0.720 | 0.350 | 0.333 | 0.104 | 0.092 | 0.157 |
| | 9/16 | 0.5625 | 12[c] | 16[c] | 0.844 | 0.810 | 0.394 | 0.375 | 0.115 | 0.103 | 0.177 |
| | 5/8 | 0.6250 | 11 | 14 | 0.938 | 0.900 | 0.437 | 0.417 | 0.126 | 0.113 | 0.197 |
| | 3/4 | 0.7500 | 10 | 12 | 1.125 | 1.080 | 0.525 | 0.500 | 0.148 | 0.134 | 0.236 |
| Mushroom Head Screws[b] | 1/8 | 0.1250 | 40 | … | 0.289 | 0.272 | 0.078 | 0.066 | 0.043 | 0.035 | 0.040 |
| | 3/16 | 0.1875 | 24 | 32[c] | 0.448 | 0.425 | 0.118 | 0.103 | 0.060 | 0.050 | 0.061 |
| | 1/4 | 0.2500 | 20 | 26 | 0.573 | 0.546 | 0.150 | 0.133 | 0.075 | 0.064 | 0.079 |
| | 5/16 | 0.3125 | 18 | 22 | 0.698 | 0.666 | 0.183 | 0.162 | 0.084 | 0.072 | 0.096 |
| | 3/8 | 0.3750 | 16 | 20 | 0.823 | 0.787 | 0.215 | 0.191 | 0.094 | 0.081 | 0.112 |

[a] All dimensions, except *J*, given for the 1/8- through 3/8-inch sizes also apply to all the 90° Raised Countersunk Head Screw dimensions given in the Standard.

[b] These screws are also available with recessed heads; dimensions of recess are not given here but may be found in the Standard.

[c] *Non-preferred size; avoid use whenever possible.*

[d] By arrangement may also be 0.309.

[e] By arrangement may also be 0.468.

All dimensions in inches.

See diagram on page 1812 for a pictorial representation of screws and letter dimensions.

## CAP AND SET SCREWS

**Slotted Head Cap Screws.—**ANSI/ASME B18.6.2 is intended to cover the complete general and dimensional data for the various styles of slotted head cap screws as well as square head and slotted headless set screws (see page 1822). Reference should be made to this Standard for information or data not found in the following text or tables.

*Length of Thread:* The length of complete (full form) thread on cap screws is equal to twice the basic screw diameter plus 0.250 in. with a plus tolerance of 0.188 in. or an amount equal to $2\frac{1}{2}$ times the pitch of the thread, whichever is greater. Cap screws of lengths too short to accommodate the minimum thread length have full form threads extending to within a distance equal to $2\frac{1}{2}$ pitches (threads) of the head.

*Designation:* Slotted head cap screws are designated by the following data in the sequence shown: Nominal size (fraction or decimal equivalent); threads per inch; screw length (fraction or decimal equivalent); product name; material; and protective finish, if required. Examples: ½-13 × 3 Slotted Round Head Cap Screw, SAE Grade 2 Steel, Zinc Plated; .750-16 × 2.25 Slotted Flat Countersunk Head Cap Screw, Corrosion Resistant Steel.

**Table 1. American National Standard Slotted Flat Countersunk Head Cap Screws**
*ANSI/ASME B18.6.2-1998 (Withdrawn)*[b]

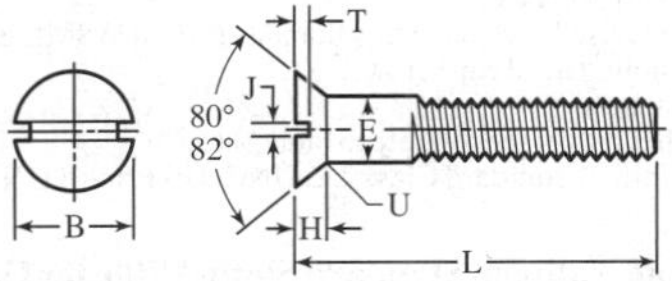


| Nominal Size[a] or Basic Screw Dia. | | Body Dia., *E* | | Head Dia., *A* Edge Sharp | Head Dia., *A* Edge Rnd'd. or Flat | Head Height, *H* | Slot Width, *J* | | Slot Depth, *T* | | Fillet Rad., *U* |
|---|---|---|---|---|---|---|---|---|---|---|---|
| | | Max. | Min. | Max. | Min. | Ref. | Max. | Min. | Max. | Min. | Max. |
| ¼ | 0.2500 | .2500 | .2450 | .500 | .452 | .140 | .075 | .064 | .068 | .045 | .100 |
| 5⁄16 | 0.3125 | .3125 | .3070 | .625 | .567 | .177 | .084 | .072 | .086 | .057 | .125 |
| ⅜ | 0.3750 | .3750 | .3690 | .750 | .682 | .210 | .094 | .081 | .103 | .068 | .150 |
| 7⁄16 | 0.4375 | .4375 | .4310 | .812 | .736 | .210 | .094 | .081 | .103 | .068 | .175 |
| ½ | 0.5000 | .5000 | .4930 | .875 | .791 | .210 | .106 | .091 | .103 | .068 | .200 |
| 9⁄16 | 0.5625 | .5625 | .5550 | 1.000 | .906 | .244 | .118 | .102 | .120 | .080 | .225 |
| ⅝ | 0.6250 | .6250 | .6170 | 1.125 | 1.020 | .281 | .133 | .116 | .137 | .091 | .250 |
| ¾ | 0.7500 | .7500 | .7420 | 1.375 | 1.251 | .352 | .149 | .131 | .171 | .115 | .300 |
| ⅞ | 0.8750 | .8750 | .8660 | 1.625 | 1.480 | .423 | .167 | .147 | .206 | .138 | .350 |
| 1 | 1.0000 | 1.0000 | .9900 | 1.875 | 1.711 | .494 | .188 | .166 | .240 | .162 | .400 |
| 1⅛ | 1.1250 | 1.1250 | 1.1140 | 2.062 | 1.880 | .529 | .196 | .178 | .257 | .173 | .450 |
| 1¼ | 1.2500 | 1.2500 | 1.2390 | 2.312 | 2.110 | .600 | .211 | .193 | .291 | .197 | .500 |
| 1⅜ | 1.3750 | 1.3750 | 1.3630 | 2.562 | 2.340 | .665 | .226 | .208 | .326 | .220 | .550 |
| 1½ | 1.5000 | 1.5000 | 1.4880 | 2.812 | 2.570 | .742 | .258 | .240 | .360 | .244 | .600 |

[a] When specifying a nominal size in decimals, the zero preceding the decimal point is omitted as is any zero in the fourth decimal place.

[b] Standard Update: The revised version of this Standard ANSI/ASME B18.6.2, released in 2020, no longer includes these slotted head cap screws.

All dimensions are in inches.

*Threads:* Threads are Unified Standard Class 2A; UNC, UNF and 8 UN Series or UNRC, UNRF, and 8 UNR Series.

**Table 2. American National Standard Slotted Round Head Cap Screws**
*ANSI/ASME B18.6.2-1998 (Withdrawn)*[b]

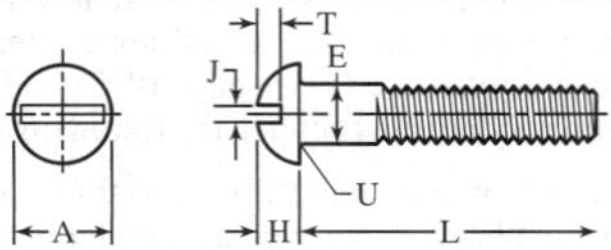


| Nom. Size[a] or Basic Screw Diameter | | Body Diameter, *E* | | Head Diameter, *A* | | Head Height, *H* | | Slot Width, *J* | | Slot Depth, *T* | |
|---|---|---|---|---|---|---|---|---|---|---|---|
| | | Max. | Min. | Max. | Min. | Max. | Min. | Max. | Min. | Max. | Min. |
| ¼ | 0.2500 | .2500 | .2450 | .437 | .418 | .191 | .175 | .075 | .064 | .117 | .097 |
| 5⁄16 | 0.3125 | .3125 | .3070 | .562 | .540 | .245 | .226 | .084 | .072 | .151 | .126 |
| ⅜ | 0.3750 | .3750 | .3690 | .625 | .603 | .273 | .252 | .094 | .081 | .168 | .138 |
| 7⁄16 | 0.4375 | .4375 | .4310 | .750 | .725 | .328 | .302 | .094 | .081 | .202 | .167 |
| ½ | 0.5000 | .5000 | .4930 | .812 | .786 | .354 | .327 | .106 | .091 | .218 | .178 |
| 9⁄16 | 0.5625 | .5625 | .5550 | .937 | .909 | .409 | .378 | .118 | .102 | .252 | .207 |
| ⅝ | 0.6250 | .6250 | .6170 | 1.000 | .970 | .437 | .405 | .133 | .116 | .270 | .220 |
| ¾ | 0.7500 | .7500 | .7420 | 1.250 | 1.215 | .546 | .507 | .149 | .131 | .338 | .278 |

[a] When specifying a nominal size in decimals, the zero preceding the decimal point is omitted as is any zero in the fourth decimal place.
[b] Standard Update: The revised version of this Standard ANSI/ASME B18.6.2, released in 2020, no longer includes these slotted head cap screws.
All dimensions are in inches.
*Fillet Radius, U*: For fillet radius see foonote to table below.
*Threads:* Threads are Unified Standard Class 2A; UNC, UNF and 8 UN Series or UNRC, UNRF and 8 UNR Series.

**Table 3. American National Standard Slotted Fillister Head Cap Screws**
*ANSI/ASME B18.6.2-1998 (Withdrawn)*[b]

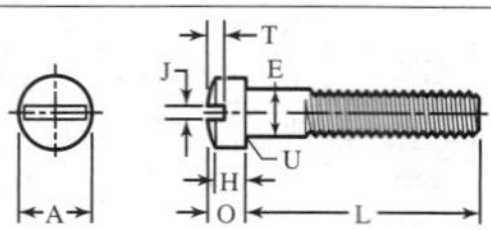


| Nom. Size[a] or Basic Screw Dia. | | Body Dia., *E* | | Head Dia., *A* | | Head Side Height, *H* | | Total Head Height, *O* | | Slot Width, *J* | | Slot Depth, *T* | |
|---|---|---|---|---|---|---|---|---|---|---|---|---|---|
| | | Max. | Min. | Max. | Min. | Max. | Min. | Max. | Min. | Max. | Min. | Max. | Min. |
| ¼ | 0.2500 | .2500 | .2450 | .375 | .363 | .172 | .157 | .216 | .194 | .075 | .064 | .097 | .077 |
| 5⁄16 | 0.3125 | .3125 | .3070 | .437 | .424 | .203 | .186 | .253 | .230 | .084 | .072 | .115 | .090 |
| ⅜ | 0.3750 | .3750 | .3690 | .562 | .547 | .250 | .229 | .314 | .284 | .094 | .081 | .142 | .112 |
| 7⁄16 | 0.4375 | .4375 | .4310 | .625 | .608 | .297 | .274 | .368 | .336 | .094 | .081 | .168 | .133 |
| ½ | 0.5000 | .5000 | .4930 | .750 | .731 | .328 | .301 | .413 | .376 | .106 | .091 | .193 | .153 |
| 9⁄16 | 0.5625 | .5625 | .5550 | .812 | .792 | .375 | .346 | .467 | .427 | .118 | .102 | .213 | .168 |
| ⅝ | 0.6250 | .6250 | .6170 | .875 | .853 | .422 | .391 | .521 | .478 | .133 | .116 | .239 | .189 |
| ¾ | 0.7500 | .7500 | .7420 | 1.000 | .976 | .500 | .466 | .612 | .566 | .149 | .131 | .283 | .223 |
| ⅞ | 0.8750 | .8750 | .8660 | 1.125 | 1.098 | .594 | .556 | .720 | .668 | .167 | .147 | .334 | .264 |
| 1 | 1.0000 | 1.0000 | .9900 | 1.312 | 1.282 | .656 | .612 | .803 | .743 | .188 | .166 | .371 | .291 |

[a] When specifying nominal size in decimals, the zero preceding the decimal point is omitted as is any zero in the fourth decimal place.
[b] Standard Update: The revised version of this Standard ANSI/ASME B18.6.2, released in 2020, no longer includes these slotted head cap screws.
All dimensions are in inches.
*Fillet Radius, U*: The fillet radius is as follows: For screw sizes ¼ to ⅜ incl., .031 max. and .016 min.; 7⁄16 to 9⁄16, incl., .047 max., .016 min.; and for ⅝ to 1, incl., .062 max., .031 min.
*Threads:* Threads are Unified Standard Class 2A; UNC, UNF and 8 UN Series or UNRC, UNRF and 8 UNR Series.

**Table 4. American National Standard Hexagon and Spline Socket Head Cap Screws** *ANSI/ASME B18.3-2012*

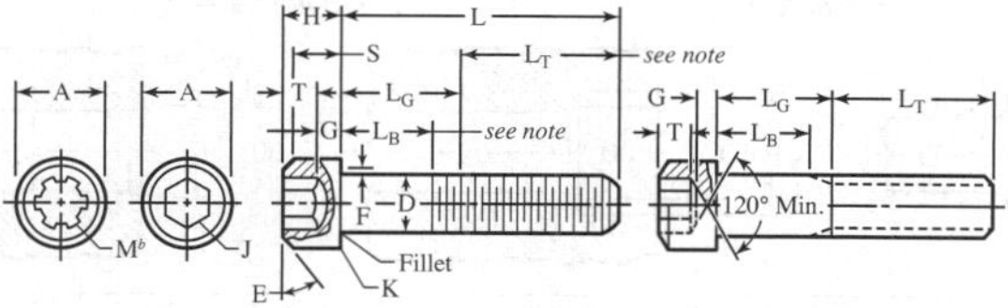


| Nominal Size | Body Diameter, *D* | | Head Diameter, *A* | | Head Height, *H* | | Spline Socket[a] Size, *M*[b] | Nom. Hex. Socket Size, *J* | Fillet Ext., *F* | Key Engagement,[a] *T* |
|---|---|---|---|---|---|---|---|---|---|---|
| | Max. | Min. | Max. | Min. | Max. | Min. | | | Max. | |
| 0 | 0.0600 | 0.0568 | 0.096 | 0.091 | 0.060 | 0.057 | 0.060 | 0.050 | 0.007 | 0.025 |
| 1 | 0.0730 | 0.0695 | 0.118 | 0.112 | 0.073 | 0.070 | 0.072 | 1/16 0.062 | 0.007 | 0.031 |
| 2 | 0.0860 | 0.0822 | 0.140 | 0.134 | 0.086 | 0.083 | 0.096 | 5/64 0.078 | 0.008 | 0.038 |
| 3 | 0.0990 | 0.0949 | 0.161 | 0.154 | 0.099 | 0.095 | 0.096 | 5/64 0.078 | 0.008 | 0.044 |
| 4 | 0.1120 | 0.1075 | 0.183 | 0.176 | 0.112 | 0.108 | 0.111 | 3/32 0.094 | 0.009 | 0.051 |
| 5 | 0.1250 | 0.1202 | 0.205 | 0.198 | 0.125 | 0.121 | 0.111 | 3/32 0.094 | 0.010 | 0.057 |
| 6 | 0.1380 | 0.1329 | 0.226 | 0.218 | 0.138 | 0.134 | 0.133 | 7/64 0.109 | 0.010 | 0.064 |
| 8 | 0.1640 | 0.1585 | 0.270 | 0.262 | 0.164 | 0.159 | 0.168 | 9/64 0.141 | 0.012 | 0.077 |
| 10 | 0.1900 | 0.1840 | 0.312 | 0.303 | 0.190 | 0.185 | 0.183 | 5/32 0.156 | 0.014 | 0.090 |
| 1/4 | 0.2500 | 0.2435 | 0.375 | 0.365 | 0.250 | 0.244 | 0.216 | 3/16 0.188 | 0.014 | 0.120 |
| 5/16 | 0.3125 | 0.3053 | 0.469 | 0.457 | 0.312 | 0.306 | 0.291 | 1/4 0.250 | 0.017 | 0.151 |
| 3/8 | 0.3750 | 0.3678 | 0.562 | 0.550 | 0.375 | 0.368 | 0.372 | 5/16 0.312 | 0.020 | 0.182 |
| 7/16 | 0.4375 | 0.4294 | 0.656 | 0.642 | 0.438 | 0.430 | 0.454 | 3/8 0.375 | 0.023 | 0.213 |
| 1/2 | 0.5000 | 0.4919 | 0.750 | 0.735 | 0.500 | 0.492 | 0.454 | 3/8 0.375 | 0.026 | 0.245 |
| 5/8 | 0.6250 | 0.6163 | 0.938 | 0.921 | 0.625 | 0.616 | 0.595 | 1/2 0.500 | 0.032 | 0.307 |
| 3/4 | 0.7500 | 0.7406 | 1.125 | 1.107 | 0.750 | 0.740 | 0.620 | 5/8 0.625 | 0.039 | 0.370 |
| 7/8 | 0.8750 | 0.8647 | 1.312 | 1.293 | 0.875 | 0.864 | 0.698 | 3/4 0.750 | 0.044 | 0.432 |
| 1 | 1.0000 | 0.9886 | 1.500 | 1.479 | 1.000 | 0.988 | 0.790 | 3/4 0.750 | 0.050 | 0.495 |
| 1 1/8 | 1.1250 | 1.1086 | 1.688 | 1.665 | 1.125 | 1.111 | … | 7/8 0.875 | 0.055 | 0.557 |
| 1 1/4 | 1.2500 | 1.2336 | 1.875 | 1.852 | 1.250 | 1.236 | … | 7/8 0.875 | 0.060 | 0.620 |
| 1 3/8 | 1.3750 | 1.3568 | 2.062 | 2.038 | 1.375 | 1.360 | … | 1 1.000 | 0.065 | 0.682 |
| 1 1/2 | 1.5000 | 1.4818 | 2.250 | 2.224 | 1.500 | 1.485 | … | 1 1.000 | 0.070 | 0.745 |
| 1 3/4 | 1.7500 | 1.7295 | 2.625 | 2.597 | 1.750 | 1.734 | … | 1 1/4 1.250 | 0.080 | 0.870 |
| 2 | 2.0000 | 1.9780 | 3.000 | 2.970 | 2.000 | 1.983 | … | 1 1/2 1.500 | 0.090 | 0.995 |
| 2 1/4 | 2.2500 | 2.2280 | 3.375 | 3.344 | 2.250 | 2.232 | … | 1 3/4 1.750 | 0.100 | 1.120 |
| 2 1/2 | 2.5000 | 2.4762 | 3.750 | 3.717 | 2.500 | 2.481 | … | 1 3/4 1.750 | 0.110 | 1.245 |
| 2 3/4 | 2.7500 | 2.7262 | 4.125 | 4.090 | 2.750 | 2.730 | … | 2 2.000 | 0.120 | 1.370 |
| 3 | 3.0000 | 2.9762 | 4.500 | 4.464 | 3.000 | 2.979 | … | 2 1/4 2.250 | 0.130 | 1.495 |
| 3 1/4 | 3.2500 | 3.2262 | 4.875 | 4.837 | 3.250 | 3.228 | … | 2 1/4 2.250 | 0.140 | 1.620 |
| 3 1/2 | 3.5000 | 3.4762 | 5.250 | 5.211 | 3.500 | 3.478 | … | 2 3/4 2.750 | 0.150 | 1.745 |
| 3 3/4 | 3.7500 | 3.7262 | 5.625 | 5.584 | 3.750 | 3.727 | … | 2 3/4 2.750 | 0.160 | 1.870 |
| 4 | 4.0000 | 3.9762 | 6.000 | 5.958 | 4.000 | 3.976 | … | 3 3.000 | 0.170 | 1.995 |

[a] Key engagement depths are minimum. Spline socket sizes are nominal.

[b] Spline driven fasteners are removed from the latest revision. For additional information on these fasteners, see ANSI/ASME B18.3-1998.

All dimensions in inches. The body length $L_B$ of the screw is the length of the unthreaded cylindrical portion of the shank. The length of thread, $L_T$, is the distance from the extreme point to the last complete (full form) thread. Standard length increments for screw diameters up to 1 inch are $^1/_{16}$ inch for lengths $^1/_8$ through $^1/_4$ inch, $^1/_8$ inch for lengths $^1/_4$ through 1 inch, $^1/_4$ inch for lengths 1 through $3^1/_2$ inches, $^1/_2$ inch for lengths $3^1/_2$ through 7 inches, 1 inch for lengths 7 through 10 inches and for diameters over 1 inch are $^1/_2$ inch for lengths 1 through 7 inches, 1 inch for lengths 7 through 10 inches, and 2 inches for lengths over 10 inches.

Heads may be plain or knurled, and chamfered to an angle *E* of 30 to 45 degrees with the surface of the flat. The thread conforms to the Unified Standard with radius root, Class 3A UNRC and UNRF for screw sizes No. 0 through 1 inch inclusive, Class 2A UNRC and UNRF for over 1 inch through $1^1/_2$ inches inclusive, and Class 2A UNRC for larger sizes. Socket dimensions are given in Table 11. For details not shown, including materials, see ANSI/ASME B18.3-2012 and ANSI/ASME B18.3-1998 (for Spline driven fasteners).

**Table 5. Drill and Counterbore Sizes For Socket Head Cap Screws (1960 Series)**

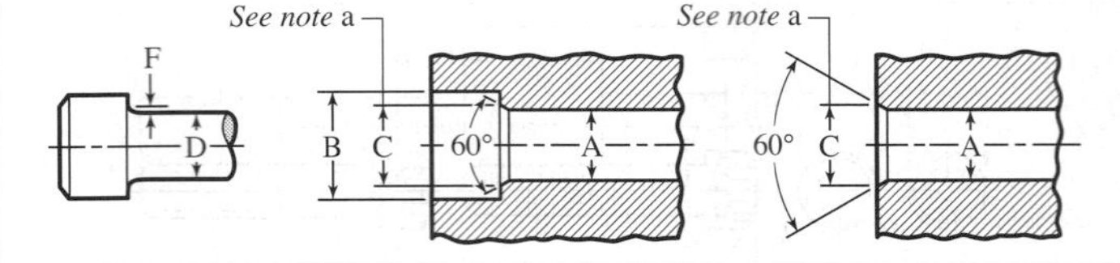


| Nominal Size or Basic Screw Diameter | | Nominal Drill Size: Close Fit[b] Number or Fractional Size | Close Fit[b] Decimal Size | Normal Fit[c] Number or Fractional Size | Normal Fit[c] Decimal Size | Counterbore Diameter | Countersink Diameter[a] |
|---|---|---|---|---|---|---|---|
| | | A | | | | B | C |
| 0 | 0.0600 | 51 | 0.067 | 49 | 0.073 | 1/8 | 0.074 |
| 1 | 0.0730 | 46 | 0.081 | 43 | 0.089 | 5/32 | 0.087 |
| 2 | 0.0860 | 3/32 | 0.094 | 36 | 0.106 | 3/16 | 0.102 |
| 3 | 0.0990 | 36 | 0.106 | 31 | 0.120 | 7/32 | 0.115 |
| 4 | 0.1120 | 1/8 | 0.125 | 29 | 0.136 | 7/32 | 0.130 |
| 5 | 0.1250 | 9/64 | 0.141 | 23 | 0.154 | 1/4 | 0.145 |
| 6 | 0.1380 | 23 | 0.154 | 18 | 0.170 | 9/32 | 0.158 |
| 8 | 0.1640 | 15 | 0.180 | 10 | 0.194 | 5/16 | 0.188 |
| 10 | 0.1900 | 5 | 0.206 | 2 | 0.221 | 3/8 | 0.218 |
| 1/4 | 0.2500 | 17/64 | 0.266 | 9/32 | 0.281 | 7/16 | 0.278 |
| 5/16 | 0.3125 | 21/64 | 0.328 | 11/32 | 0.344 | 17/32 | 0.346 |
| 3/8 | 0.3750 | 25/64 | 0.391 | 13/32 | 0.406 | 5/8 | 0.415 |
| 7/16 | 0.4375 | 29/64 | 0.453 | 15/32 | 0.469 | 23/32 | 0.483 |
| 1/2 | 0.5000 | 33/64 | 0.516 | 17/32 | 0.531 | 13/16 | 0.552 |
| 5/8 | 0.6250 | 41/64 | 0.641 | 21/32 | 0.656 | 1 | 0.689 |
| 3/4 | 0.7500 | 49/64 | 0.766 | 25/32 | 0.781 | 1 3/16 | 0.828 |
| 7/8 | 0.8750 | 57/64 | 0.891 | 29/32 | 0.906 | 1 3/8 | 0.963 |
| 1 | 1.0000 | 1 1/64 | 1.016 | 1 1/32 | 1.031 | 1 5/8 | 1.100 |
| 1 1/4 | 1.2500 | 1 9/32 | 1.281 | 1 5/16 | 1.312 | 2 | 1.370 |
| 1 1/2 | 1.5000 | 1 17/32 | 1.531 | 1 9/16 | 1.562 | 2 3/8 | 1.640 |
| 1 3/4 | 1.7500 | 1 25/32 | 1.781 | 1 13/16 | 1.812 | 2 3/4 | 1.910 |
| 2 | 2.0000 | 2 1/32 | 2.031 | 2 1/16 | 2.062 | 3 1/8 | 2.180 |

[a] *Countersink:* It is considered good practice to countersink or break the edges of holes which are smaller than ($D$ Max + $2F$ Max) in parts having a hardness which approaches, equals or exceeds the screw hardness. If such holes are not countersunk, the heads of screws may not seat properly or the sharp edges on holes may deform the fillets on screws thereby making them susceptible to fatigue in applications involving dynamic loading. The countersink or corner relief, however, should not be larger than is necessary to ensure that the fillet on the screw is cleared.

[b] *Close Fit:* The close fit is normally limited to holes for those lengths of screws which are threaded to the head in assemblies where only one screw is to be used or where two or more screws are to be used and the mating holes are to be produced either at assembly or by matched and coordinated tooling.

[c] *Normal Fit:* The normal fit is intended for screws of relatively long length or for assemblies involving two or more screws where the mating holes are to be produced by conventional tolerancing methods. It provides for the maximum allowable eccentricity of the longest standard screws and for certain variations in the parts to be fastened, such as: deviations in hole straightness, angularity between the axis of the tapped hole and that of the hole for the shank, differences in center distances of the mating holes, etc.

All dimensions in inches.

*Source:* Appendix to American National Standard ANSI/ASME B18.3-2012 and ANSI/ASME B18.3-1998 (for Spline driven fasteners).

**Table 6. American National Standard Hexagon and Spline Socket Flat Countersunk Head Cap Screws** *ANSI/ASME B18.3-2012*

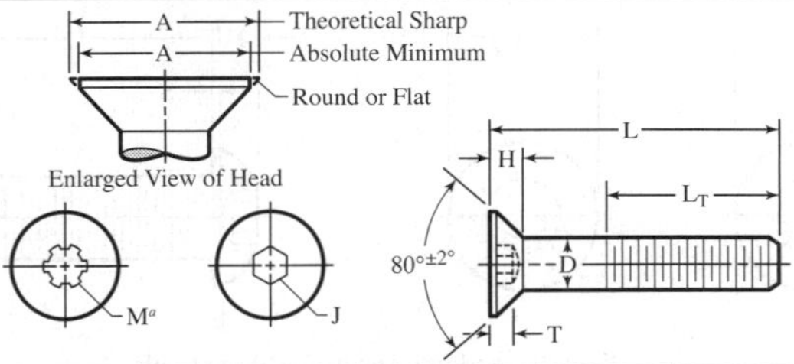


| Nominal Size | Body Diameter Max. | Body Diameter Min. | Head Diameter Theoretical Sharp Max. | Head Diameter Abs. Min. | Head Height Reference | Spline Socket Size | Hexagon Socket Size Nom. | Key Engagement Min. |
|---|---|---|---|---|---|---|---|---|
| | *D* | | *A* | | *H* | *M*[a] | *J* | *T* |
| 0 | 0.0600 | 0.0568 | 0.138 | 0.117 | 0.044 | 0.048 | 0.035 | 0.025 |
| 1 | 0.0730 | 0.0695 | 0.168 | 0.143 | 0.054 | 0.060 | 0.050 | 0.031 |
| 2 | 0.0860 | 0.0822 | 0.197 | 0.168 | 0.064 | 0.060 | 0.050 | 0.038 |
| 3 | 0.0990 | 0.0949 | 0.226 | 0.193 | 0.073 | 0.072 | 1⁄16 | 0.044 |
| 4 | 0.1120 | 0.1075 | 0.255 | 0.218 | 0.083 | 0.072 | 1⁄16 | 0.055 |
| 5 | 0.1250 | 0.1202 | 0.281 | 0.240 | 0.090 | 0.096 | 5⁄64 | 0.061 |
| 6 | 0.1380 | 0.1329 | 0.307 | 0.263 | 0.097 | 0.096 | 5⁄64 | 0.066 |
| 8 | 0.1640 | 0.1585 | 0.359 | 0.311 | 0.112 | 0.111 | 3⁄32 | 0.076 |
| 10 | 0.1900 | 0.1840 | 0.411 | 0.359 | 0.127 | 0.145 | 1⁄8 | 0.087 |
| 1⁄4 | 0.2500 | 0.2435 | 0.531 | 0.480 | 0.161 | 0.183 | 5⁄32 | 0.111 |
| 5⁄16 | 0.3125 | 0.3053 | 0.656 | 0.600 | 0.198 | 0.216 | 3⁄16 | 0.135 |
| 3⁄8 | 0.3750 | 0.3678 | 0.781 | 0.720 | 0.234 | 0.251 | 7⁄32 | 0.159 |
| 7⁄16 | 0.4375 | 0.4294 | 0.844 | 0.781 | 0.234 | 0.291 | 1⁄4 | 0.159 |
| 1⁄2 | 0.5000 | 0.4919 | 0.938 | 0.872 | 0.251 | 0.372 | 5⁄16 | 0.172 |
| 5⁄8 | 0.6250 | 0.6163 | 1.188 | 1.112 | 0.324 | 0.454 | 3⁄8 | 0.220 |
| 3⁄4 | 0.7500 | 0.7406 | 1.438 | 1.355 | 0.396 | 0.454 | 1⁄2 | 0.220 |
| 7⁄8 | 0.8750 | 0.8647 | 1.688 | 1.604 | 0.468 | … | 9⁄16 | 0.248 |
| 1 | 1.0000 | 0.9886 | 1.938 | 1.841 | 0.540 | … | 5⁄8 | 0.297 |
| 1 1⁄8 | 1.1250 | 1.1086 | 2.188 | 2.079 | 0.611 | … | 3⁄4 | 0.325 |
| 1 1⁄4 | 1.2500 | 1.2336 | 2.438 | 2.316 | 0.683 | … | 7⁄8 | 0.358 |
| 1 3⁄8 | 1.3750 | 1.3568 | 2.688 | 2.553 | 0.755 | … | 7⁄8 | 0.402 |
| 1 1⁄2 | 1.5000 | 1.4818 | 2.938 | 2.791 | 0.827 | … | 1 | 0.435 |

[a] Spline driven fasteners are removed from the latest revision. For additional information on these fasteners, see ANSI/ASME B18.3-1998.

All dimensions in inches.

The body of the screw is the unthreaded cylindrical portion of the shank where not threaded to the head; the shank being the portion of the screw from the point of juncture of the conical bearing surface and the body to the flat of the point. The length of thread $L_T$ is the distance measured from the extreme point to the last complete (full form) thread.

Standard length increments of No. 0 through 1-inch sizes are as follows: 1⁄16 inch for nominal screw lengths of 1⁄8 through 1⁄4 inch; 1⁄8 inch for lengths of 1⁄4 through 1 inch; 1⁄4 inch for lengths of 1 inch through 3 1⁄2 inches; 1⁄2 inch for lengths of 3 1⁄2 through 7 inches; and 1 inch for lengths of 7 through 10 inches, incl. For screw sizes over 1 inch, length increments are: 1⁄2 inch for nominal screw lengths of 1 inch through 7 inches; 1 inch for lengths of 7 through 10 inches; and 2 inches for lengths over 10 inches.

Threads shall be Unified external threads with radius root; Class 3A UNRC and UNRF series for sizes No. 0 through 1 inch and Class 2A UNRC and UNRF series for sizes over 1 inch to 1 1⁄2 inches, incl.

For manufacturing details not shown, including materials, see American National Standard ANSI/ASME B18.3-2012 and ANSI/ASME B18.3-1998 (for Spline driven fasteners). Socket dimensions are given in Table 11.

**Table 7. American National Standard Hexagon Socket and Spline Socket Button Head Cap Screws** *ANSI/ASME B18.3-2012*

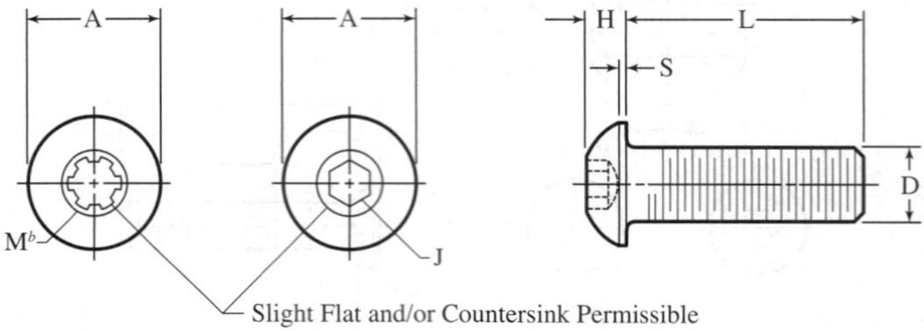


| Nominal Size | Screw Diameter | Head Diameter | | Head Height | | Head Side Height | Spline Socket Size[a] | Hexagon Socket Size[a] | Standard Length |
|---|---|---|---|---|---|---|---|---|---|
| | Basic | Max. | Min. | Max. | Min. | Ref. | Nom. | Nom. | Max. |
| | *D* | *A* | | *H* | | *S* | *M*[b] | *J* | *L* |
| 0 | 0.0600 | 0.114 | 0.104 | 0.032 | 0.026 | 0.010 | 0.048 | 0.035 | ½ |
| 1 | 0.0730 | 0.139 | 0.129 | 0.039 | 0.033 | 0.010 | 0.060 | 0.050 | ½ |
| 2 | 0.0860 | 0.164 | 0.154 | 0.046 | 0.038 | 0.010 | 0.060 | 0.050 | ½ |
| 3 | 0.0990 | 0.188 | 0.176 | 0.052 | 0.044 | 0.010 | 0.072 | 1⁄16 | ½ |
| 4 | 0.1120 | 0.213 | 0.201 | 0.059 | 0.051 | 0.015 | 0.072 | 1⁄16 | ½ |
| 5 | 0.1250 | 0.238 | 0.226 | 0.066 | 0.058 | 0.015 | 0.096 | 5⁄64 | ½ |
| 6 | 0.1380 | 0.262 | 0.250 | 0.073 | 0.063 | 0.015 | 0.096 | 5⁄64 | ⅝ |
| 8 | 0.1640 | 0.312 | 0.298 | 0.087 | 0.077 | 0.015 | 0.111 | 3⁄32 | ¾ |
| 10 | 0.1900 | 0.361 | 0.347 | 0.101 | 0.091 | 0.020 | 0.145 | ⅛ | 1 |
| ¼ | 0.2500 | 0.437 | 0.419 | 0.132 | 0.122 | 0.031 | 0.183 | 5⁄32 | 1 |
| 5⁄16 | 0.3125 | 0.547 | 0.527 | 0.166 | 0.152 | 0.031 | 0.216 | 3⁄16 | 1 |
| ⅜ | 0.3750 | 0.656 | 0.636 | 0.199 | 0.185 | 0.031 | 0.251 | 7⁄32 | 1¼ |
| ½ | 0.5000 | 0.875 | 0.851 | 0.265 | 0.245 | 0.046 | 0.372 | 5⁄16 | 2 |
| ⅝ | 0.6250 | 1.000 | 0.970 | 0.331 | 0.311 | 0.062 | 0.454 | ⅜ | 2 |

[a] Socket dimensions are given in Table 11.

[b] Spline driven fasteners are removed from the latest revision. For additional information on these fasteners, see ANSI/ASME B18.3-1998.

All dimensions in inches.

These cap screws have been designed and recommended for light fastening applications. They are not suggested for use in critical high-strength applications where socket head cap screws should normally be used.

Standard length increments for socket button head cap screws are as follows: 1⁄16 inch for nominal screw lengths of ⅛ through ¼ inch, ⅛ inch for nominal screw lengths of ¼ through 1 inch, and ¼ inch for nominal screw lengths of 1 inch through 2 inches. Tolerances on lengths are −0.03 inch for lengths up to 1 inch inclusive. For lengths from 1 through 2 inches, inclusive, length tolerances are −0.04 inch.

The thread conforms to the Unified Standard, Class 3A, with radius root, UNRC and UNRF.

To prevent interference, American National Standard ANSI/ASME B18.3.4M-1986 gives metric dimensional and general requirements for a lower head profile hexagon socket button head cap screw. Because of its design, wrenchability and other design factors are reduced; therefore, B18.3.4M should be reviewed carefully. Available only in metric sizes and with metric threads.

For manufacturing details, including materials, not shown, see American National Standard ANSI/ASME B18.3-2012 and ANSI/ASME B18.3-1998 (for Spline driven fasteners).

**Table 8. American National Standard Hexagon Socket Head Shoulder Screws**
*ANSI/ASME B18.3-2012*

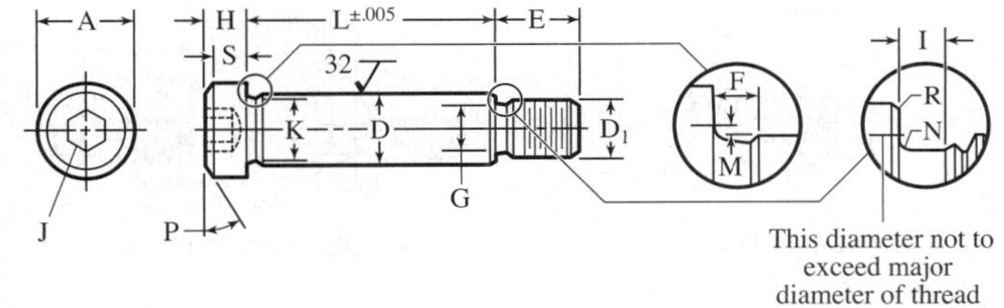


| Nominal Size | Shoulder Diameter | | Head Diameter | | Head Height | | Head Side Height | Nominal Thread Size | Thread Length |
|---|---|---|---|---|---|---|---|---|---|
| | Max. | Min. | Max. | Min. | Max. | Min. | Min. | | |
| | *D* | | *A* | | *H* | | *S* | $D_1$ | *E* |
| ¼ | 0.2480 | 0.2460 | 0.375 | 0.357 | 0.188 | 0.177 | 0.157 | 10-24 | 0.375 |
| 5/16 | 0.3105 | 0.3085 | 0.438 | 0.419 | 0.219 | 0.209 | 0.183 | ¼-20 | 0.438 |
| ⅜ | 0.3730 | 0.3710 | 0.562 | 0.543 | 0.250 | 0.240 | 0.209 | 5/16-18 | 0.500 |
| ½ | 0.4980 | 0.4960 | 0.750 | 0.729 | 0.312 | 0.302 | 0.262 | ⅜-16 | 0.625 |
| ⅝ | 0.6230 | 0.6210 | 0.875 | 0.853 | 0.375 | 0.365 | 0.315 | ½-13 | 0.750 |
| ¾ | 0.7480 | 0.7460 | 1.000 | 0.977 | 0.500 | 0.490 | 0.421 | ⅝-11 | 0.875 |
| 1 | 0.9980 | 0.9960 | 1.312 | 1.287 | 0.625 | 0.610 | 0.527 | ¾-10 | 1.000 |
| 1¼ | 1.2480 | 1.2460 | 1.750 | 1.723 | 0.750 | 0.735 | 0.633 | ⅞-9 | 1.125 |
| 1½ | 1.4980 | 1.4960 | 2.125 | 2.095 | 1.000 | 0.980 | 0.842 | 1⅛-7 | 1.500 |
| 1¾ | 1.7480 | 1.7460 | 2.375 | 2.345 | 1.125 | 1.105 | 0.948 | 1¼-7 | 1.750 |
| 2 | 1.9980 | 1.9960 | 2.750 | 2.720 | 1.250 | 1.230 | 1.054 | 1½-6 | 2.000 |

| Nominal Size | Thread Neck Diameter | | Thread Neck Width | Shoulder Neck Dia. | Shoulder Neck Width | Thread Neck Fillet | | Head Fillet Extension Above *D* | Hexagon Socket Size |
|---|---|---|---|---|---|---|---|---|---|
| | Max. | Min. | Max. | Min. | Max. | Max. | Min. | Max. | Nom. |
| | *G* | | *I* | *K* | *F* | *N* | | *M* | *J* |
| ¼ | 0.142 | 0.133 | 0.083 | 0.227 | 0.093 | 0.023 | 0.017 | 0.014 | ⅛ |
| 5/16 | 0.193 | 0.182 | 0.100 | 0.289 | 0.093 | 0.028 | 0.022 | 0.017 | 5/32 |
| ⅜ | 0.249 | 0.237 | 0.111 | 0.352 | 0.093 | 0.031 | 0.025 | 0.020 | 3/16 |
| ½ | 0.304 | 0.291 | 0.125 | 0.477 | 0.093 | 0.035 | 0.029 | 0.026 | ¼ |
| ⅝ | 0.414 | 0.397 | 0.154 | 0.602 | 0.093 | 0.042 | 0.036 | 0.032 | 5/16 |
| ¾ | 0.521 | 0.502 | 0.182 | 0.727 | 0.093 | 0.051 | 0.045 | 0.039 | ⅜ |
| 1 | 0.638 | 0.616 | 0.200 | 0.977 | 0.125 | 0.055 | 0.049 | 0.050 | ½ |
| 1¼ | 0.750 | 0.726 | 0.222 | 1.227 | 0.125 | 0.062 | 0.056 | 0.060 | ⅝ |
| 1½ | 0.964 | 0.934 | 0.286 | 1.478 | 0.125 | 0.072 | 0.066 | 0.070 | ⅞ |
| 1¾ | 1.089 | 1.059 | 0.286 | 1.728 | 0.125 | 0.072 | 0.066 | 0.080 | 1 |
| 2 | 1.307 | 1.277 | 0.333 | 1.978 | 0.125 | 0.102 | 0.096 | 0.090 | 1¼ |

All dimensions are in inches. The shoulder is the enlarged, unthreaded portion of the screw. Standard length increments for shoulder screws are: ⅛ inch for nominal screw lengths of ¼ through ¾ inch; ¼ inch for lengths above ¾ through 5 inches; and ½ inch for lengths over 5 inches. The thread conforms to the Unified Standard Class 3A, UNC. Hexagon socket sizes for the respective shoulder screw sizes are the same as for set screws of the same nominal size (see Table 7) except for shoulder screw size 1 inch, socket size is ½ inch, for screw size 1 ½ inches, socket size is ⅞ inch, and for screw size 2 inches, socket size is 1 ¼ inches. For details not shown, including materials, see ANSI/ASME B18.3-2012.

**Table 9. American National Standard Slotted Headless Set Screws**
*ANSI/ASME B18.6.2-2020*

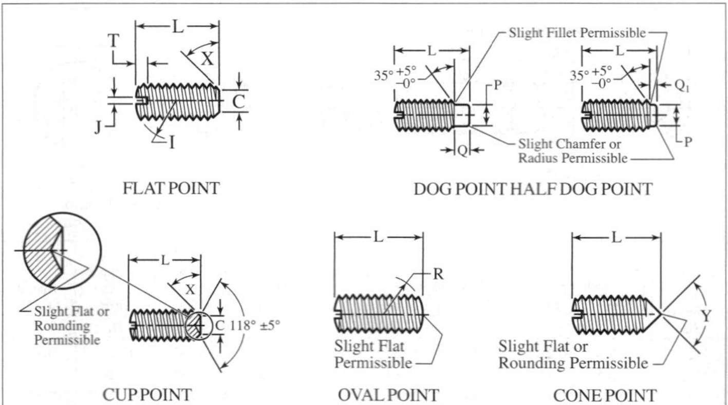


| Nominal Size[a] or Basic Screw Diameter | | Slot Width, *J* | | Slot Depth, *T* | | Cup and Flat Point Dia., *C* | | Dog Point Dia., *P* | | Point Length | | | |
|---|---|---|---|---|---|---|---|---|---|---|---|---|---|
| | | | | | | | | | | Dog, *Q* | | Half Dog, *Q*1 | |
| | | Max. | Min. | Max. | Min. | Max. | Min. | Max. | Min. | Max. | Min. | Max. | Min. |
| 0 | 0.0600 | .014 | .010 | .020 | .016 | .033 | .027 | .040 | .037 | .032 | .028 | .017 | .013 |
| 1 | 0.0730 | .016 | .012 | .020 | .016 | .040 | .033 | .049 | .045 | .040 | .036 | .021 | .017 |
| 2 | 0.0860 | .018 | .014 | .025 | .019 | .047 | .039 | .057 | .053 | .046 | .042 | .024 | .020 |
| 3 | 0.0990 | .020 | .016 | .028 | .022 | .054 | .045 | .066 | .062 | .052 | .048 | .027 | .023 |
| 4 | 0.1120 | .024 | .018 | .031 | .025 | .061 | .051 | .075 | .070 | .058 | .054 | .030 | .026 |
| 5 | 0.1250 | .026 | .020 | .036 | .026 | .067 | .057 | .083 | .078 | .063 | .057 | .033 | .027 |
| 6 | 0.1380 | .028 | .022 | .040 | .030 | .074 | .064 | .092 | .087 | .073 | .067 | .038 | .032 |
| 8 | 0.1640 | .032 | .026 | .046 | .036 | .087 | .076 | .109 | .103 | .083 | .077 | .043 | .037 |
| 10 | 0.1900 | .035 | .029 | .053 | .043 | .102 | .088 | .127 | .120 | .095 | .085 | .050 | .040 |
| 12 | 0.2160 | .042 | .035 | .061 | .051 | .115 | .101 | .144 | .137 | .115 | .105 | .060 | .050 |
| 1⁄4 | 0.2500 | .049 | .041 | .068 | .058 | .132 | .118 | .156 | .149 | .130 | .120 | .068 | .058 |
| 5⁄16 | 0.3125 | .055 | .047 | .083 | .073 | .172 | .156 | .203 | .195 | .161 | .151 | .083 | .073 |
| 3⁄8 | 0.3750 | .068 | .060 | .099 | .089 | .212 | .194 | .250 | .241 | .193 | .183 | .099 | .089 |
| 7⁄16 | 0.4375 | .076 | .068 | .114 | .104 | .252 | .232 | .297 | .287 | .224 | .214 | .114 | .104 |
| 1⁄2 | 0.5000 | .086 | .076 | .130 | .120 | .291 | .270 | .344 | .334 | .255 | .245 | .130 | .120 |
| 9⁄16 | 0.5625 | .096 | .086 | .146 | .136 | .332 | .309 | .391 | .379 | .287 | .275 | .146 | .134 |
| 5⁄8 | 0.6250 | .107 | .097 | .161 | .151 | .371 | .347 | .469 | .456 | .321 | .305 | .164 | .148 |
| 3⁄4 | 0.7500 | .134 | .124 | .193 | .183 | .450 | .425 | .562 | .549 | .383 | .367 | .196 | .180 |

[a] When specifying a nominal size in decimals a zero preceding the decimal point or any zero in the fourth decimal place is omitted.

All dimensions are in inches.

*Crown Radius, I*: The crown radius has the same value as the basic screw diameter to three decimal places.

*Oval Point Radius, R*: Values of the oval point radius according to nominal screw size are: For a screw size of 0, a radius of .045; 1, .055; 2, .064; 3, .074; 4, .084; 5, .094; 6, .104; 8, .123; 10, .142; 12, .162; 1⁄4, .188; 5⁄16, .234; 3⁄8, .281; 7⁄16, .328; 1⁄2, .375; 9⁄16, .422; 5⁄8, .469; and for 3⁄4, .562.

*Cone Point Angle, Y*: The cone point angle is 90° ± 2° for the following nominal lengths, or longer, shown according to screw size: For nominal size 0, a length of 5⁄64; 1, 3⁄32; 2, 7⁄64; 3, 1⁄8; 4, 5⁄32; 5, 3⁄16; 6, 3⁄16; 8, 1⁄4; 10, 1⁄4; 12, 5⁄16; 1⁄4, 5⁄16; 5⁄16, 3⁄8; 3⁄8, 7⁄16; 7⁄16, 1⁄2; 1⁄2, 9⁄16; 9⁄16, 5⁄8; 5⁄8, 3⁄4; and for 3⁄4, 7⁄8. For shorter screws, the cone point angle is 118° ± 2°.

*Point Angle X:* The point angle is 45°, + 5°, − 0°, for screws of nominal lengths, or longer, as given just above for cone point angle, and 30°, min. for shorter screws.

*Threads:* are Unified Standard Class 2A; UNC and UNF Series or UNRC and UNRF Series.

**Table 10. American National Standard Hexagon and Spline Socket Set Screw Optional Cup Points** *ANSI/ASME B18.3-2012* and *ANSI/ASME B18.3-1998 (for Spline driven fasteners).*

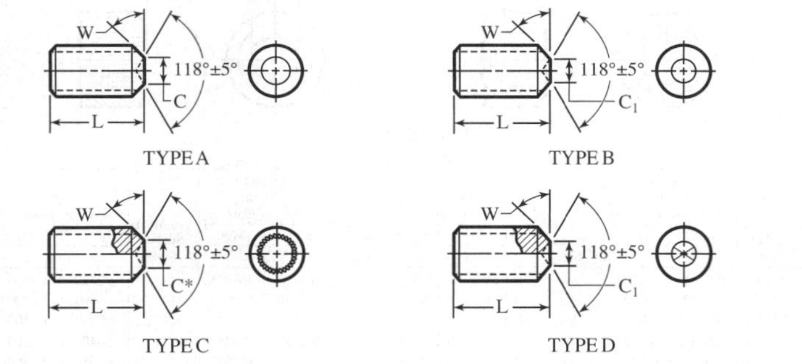


* This diameter may be counterbored.

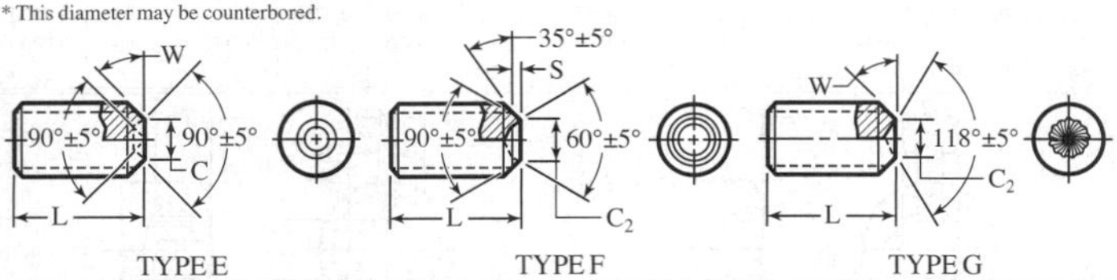


| Nom. Size | Point Dia. | | Point Dia. | | Point Dia. | | Point Length | |
|---|---|---|---|---|---|---|---|---|
| | Max. | Min. | Max. | Min. | Max. | Min. | Max. | Min. |
| | *C* | | *C*1 | | *C*2 | | *S* | |
| 0 | 0.033 | 0.027 | 0.032 | 0.027 | 0.027 | 0.022 | 0.007 | 0.004 |
| 1 | 0.040 | 0.033 | 0.038 | 0.033 | 0.035 | 0.030 | 0.008 | 0.005 |
| 2 | 0.047 | 0.039 | 0.043 | 0.038 | 0.043 | 0.038 | 0.010 | 0.007 |
| 3 | 0.054 | 0.045 | 0.050 | 0.045 | 0.051 | 0.046 | 0.011 | 0.007 |
| 4 | 0.061 | 0.051 | 0.056 | 0.051 | 0.059 | 0.054 | 0.013 | 0.008 |
| 5 | 0.067 | 0.057 | 0.062 | 0.056 | 0.068 | 0.063 | 0.014 | 0.009 |
| 6 | 0.074 | 0.064 | 0.069 | 0.062 | 0.074 | 0.068 | 0.017 | 0.012 |
| 8 | 0.087 | 0.076 | 0.082 | 0.074 | 0.090 | 0.084 | 0.021 | 0.016 |
| 10 | 0.102 | 0.088 | 0.095 | 0.086 | 0.101 | 0.095 | 0.024 | 0.019 |
| ¼ | 0.132 | 0.118 | 0.125 | 0.114 | 0.156 | 0.150 | 0.027 | 0.022 |
| 5⁄16 | 0.172 | 0.156 | 0.156 | 0.144 | 0.190 | 0.185 | 0.038 | 0.033 |
| ⅜ | 0.212 | 0.194 | 0.187 | 0.174 | 0.241 | 0.236 | 0.041 | 0.036 |
| 7⁄16 | 0.252 | 0.232 | 0.218 | 0.204 | 0.286 | 0.281 | 0.047 | 0.042 |
| ½ | 0.291 | 0.270 | 0.250 | 0.235 | 0.333 | 0.328 | 0.054 | 0.049 |
| ⅝ | 0.371 | 0.347 | 0.312 | 0.295 | 0.425 | 0.420 | 0.067 | 0.062 |
| ¾ | 0.450 | 0.425 | 0.375 | 0.357 | 0.523 | 0.518 | 0.081 | 0.076 |
| ⅞ | 0.530 | 0.502 | 0.437 | 0.418 | … | … | … | … |
| 1 | 0.609 | 0.579 | 0.500 | 0.480 | … | … | … | … |
| 1⅛ | 0.689 | 0.655 | 0.562 | 0.542 | … | … | … | … |
| 1¼ | 0.767 | 0.733 | 0.625 | 0.605 | … | … | … | … |
| 1⅜ | 0.848 | 0.808 | 0.687 | 0.667 | … | … | … | … |
| 1½ | 0.926 | 0.886 | 0.750 | 0.730 | … | … | … | … |
| 1¾ | 1.086 | 1.039 | 0.875 | 0.855 | … | … | … | … |
| 2 | 1.244 | 1.193 | 1.000 | 0.980 | … | … | … | … |

All dimensions are in inches.

The cup point types shown are those available from various manufacturers.

## Table 11. American National Standard Hexagon and Spline Sockets

*ANSI/ASME B18.3-2012*

BROACHED SOCKET

HEXAGON SOCKETS

| Nominal Socket Size | Socket Width Across Flats Max. *J* | Socket Width Across Flats Min. *J* | Nominal Socket Size | Socket Width Across Flats Max. *J* | Socket Width Across Flats Min. *J* | Nominal Socket Size | Socket Width Across Flats Max. *J* | Socket Width Across Flats Min. *J* | Nominal Socket Size | Socket Width Across Flats Max. *J* | Socket Width Across Flats Min. *J* |
|---|---|---|---|---|---|---|---|---|---|---|---|
| 0.028 | 0.0285 | 0.0280 | 9/64 | 0.1426 | 0.1406 | 7/16 | 0.4420 | 0.4375 | 1 1/4 | 1.2750 | 1.2500 |
| 0.035 | 0.0355 | 0.0350 | 5/32 | 0.1587 | 0.1562 | 1/2 | 0.5050 | 0.5000 | 1 1/2 | 1.5300 | 1.5000 |
| 0.050 | 0.0510 | 0.0500 | 3/16 | 0.1900 | 0.1875 | 9/16 | 0.5680 | 0.5625 | 1 3/4 | 1.7850 | 1.7500 |
| 1/16 | 0.0635 | 0.0625 | 7/32 | 0.2217 | 0.2187 | 5/8 | 0.6310 | 0.6250 | 2 | 2.0400 | 2.0000 |
| 5/64 | 0.0791 | 0.0781 | 1/4 | 0.2530 | 0.2500 | 3/4 | 0.7570 | 0.7500 | 2 1/4 | 2.2950 | 2.2500 |
| 3/32 | 0.0952 | 0.0937 | 5/16 | 0.3160 | 0.3125 | 7/8 | 0.8850 | 0.8750 | 2 3/4 | 2.8050 | 2.7500 |
| 7/64 | 0.1111 | 0.1094 | 3/8 | 0.3790 | 0.3750 | 1 | 1.0200 | 1.0000 | 3 | 3.0600 | 3.0000 |
| 1/8 | 0.1270 | 0.1250 | … | … | … | … | … | … | … | … | … |

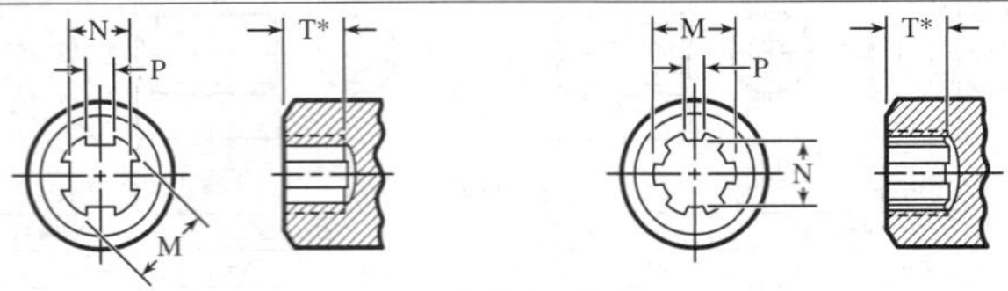


SPLINE SOCKETS[a]

| Nominal Socket Size | Number of Teeth | Socket Major Diameter Max. *M* | Socket Major Diameter Min. *M* | Socket Minor Diameter Max. *N* | Socket Minor Diameter Min. *N* | Width of Tooth Max. *P* | Width of Tooth Min. *P* |
|---|---|---|---|---|---|---|---|
| 0.033 | 4 | 0.0350 | 0.0340 | 0.0260 | 0.0255 | 0.0120 | 0.0115 |
| 0.048 | 6 | 0.050 | 0.049 | 0.041 | 0.040 | 0.011 | 0.010 |
| 0.060 | 6 | 0.062 | 0.061 | 0.051 | 0.050 | 0.014 | 0.013 |
| 0.072 | 6 | 0.074 | 0.073 | 0.064 | 0.063 | 0.016 | 0.015 |
| 0.096 | 6 | 0.098 | 0.097 | 0.082 | 0.080 | 0.022 | 0.021 |
| 0.111 | 6 | 0.115 | 0.113 | 0.098 | 0.096 | 0.025 | 0.023 |
| 0.133 | 6 | 0.137 | 0.135 | 0.118 | 0.116 | 0.030 | 0.028 |
| 0.145 | 6 | 0.149 | 0.147 | 0.128 | 0.126 | 0.032 | 0.030 |
| 0.168 | 6 | 0.173 | 0.171 | 0.150 | 0.147 | 0.036 | 0.033 |
| 0.183 | 6 | 0.188 | 0.186 | 0.163 | 0.161 | 0.039 | 0.037 |
| 0.216 | 6 | 0.221 | 0.219 | 0.190 | 0.188 | 0.050 | 0.048 |
| 0.251 | 6 | 0.256 | 0.254 | 0.221 | 0.219 | 0.060 | 0.058 |
| 0.291 | 6 | 0.298 | 0.296 | 0.254 | 0.252 | 0.068 | 0.066 |
| 0.372 | 6 | 0.380 | 0.377 | 0.319 | 0.316 | 0.092 | 0.089 |
| 0.454 | 6 | 0.463 | 0.460 | 0.386 | 0.383 | 0.112 | 0.109 |
| 0.595 | 6 | 0.604 | 0.601 | 0.509 | 0.506 | 0.138 | 0.134 |
| 0.620 | 6 | 0.631 | 0.627 | 0.535 | 0.531 | 0.149 | 0.145 |
| 0.698 | 6 | 0.709 | 0.705 | 0.604 | 0.600 | 0.168 | 0.164 |
| 0.790 | 6 | 0.801 | 0.797 | 0.685 | 0.681 | 0.189 | 0.185 |

[a] Spline driven fasteners are removed from the latest revision. For additional information on these fasteners, see ANSI/ASME B18.3-1998.

All dimensions are in inches.

* Socket depths, *T*, for various screw types are given in the Standard but are not shown here.

Where sockets are chamfered, the depth of chamfer shall not exceed 10 percent of the nominal socket size for sizes up to and including 1/16 inch for hexagon sockets and 0.060 for spline sockets, and 7.5 percent for larger sizes.

**Table 12. American National Standard Square Head Set Screws**
*ANSI/ASME B18.6.2-2020*

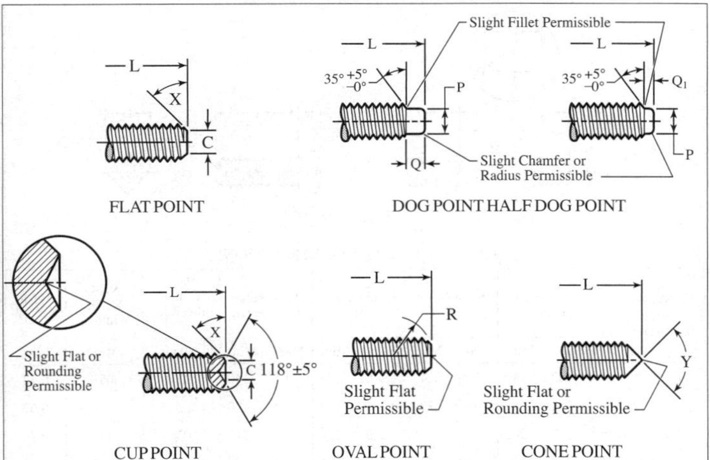


| Nominal Size* or Basic Screw Diameter | | Cup and Flat Point Diams., C | | Dog and Half Dog Point Diams., P | | Point Length | | | | Oval Point Rad., R |
|---|---|---|---|---|---|---|---|---|---|---|
| | | | | | | Dog, Q | | Half Dog, Q1 | | +.031 −.000 |
| | | Max. | Min | Max. | Min. | Max. | Min. | Max. | Min. | |
| 10 | 0.1900 | .102 | .088 | .127 | .120 | .095 | .085 | .050 | .040 | .142 |
| 1/4 | 0.2500 | .132 | .118 | .156 | .149 | .130 | .120 | .068 | .058 | .188 |
| 5/16 | 0.3125 | .172 | .156 | .203 | .195 | .161 | .151 | .083 | .073 | .234 |
| 3/8 | 0.3750 | .212 | .194 | .250 | .241 | .193 | .183 | .099 | .089 | .281 |
| 7/16 | 0.4375 | .252 | .232 | .297 | .287 | .224 | .214 | .114 | .104 | .328 |
| 1/2 | 0.500 | .291 | .270 | .344 | .334 | .255 | .245 | .130 | .120 | .375 |
| 9/16 | 0.5625 | .332 | .309 | .391 | .379 | .287 | .275 | .146 | .134 | .422 |
| 5/8 | 0.6250 | .371 | .347 | .469 | .456 | .321 | .305 | .164 | .148 | .469 |
| 3/4 | 0.7500 | .450 | .425 | .562 | .549 | .383 | .367 | .196 | .180 | .562 |
| 7/8 | 0.8750 | .530 | .502 | .656 | .642 | .446 | .430 | .227 | .211 | .656 |
| 1 | 1.0000 | .609 | .579 | .750 | .734 | .510 | .490 | .260 | .240 | .750 |
| 1 1/8 | 1.1250 | .689 | .655 | .844 | .826 | .572 | .552 | .291 | .271 | .844 |
| 1 1/4 | 1.2500 | .767 | .733 | .938 | .920 | .635 | .615 | .323 | .303 | .938 |
| 1 3/8 | 1.3750 | .848 | .808 | 1.031 | 1.011 | .698 | .678 | .354 | .334 | 1.031 |
| 1 1/2 | 1.5000 | .926 | .886 | 1.125 | 1.105 | .760 | .740 | .385 | .365 | 1.125 |

All dimensions are in inches.

*Threads:* Threads are Unified Standard Class 2A; UNC, UNF and 8 UN Series or UNRC, UNRF and 8 UNR Series.

*Length of Thread:* Square head set screws have complete (full form) threads extending over that portion of the screw length which is not affected by the point. For the respective constructions, threads extend into the neck relief, to the conical underside of head, or to within one thread (as measured with a thread ring gage) from the flat underside of the head. Threads through angular or crowned portions of points have fully formed roots with partial crests.

* When specifying a nominal size in decimals, the zero preceding the decimal point is omitted as is any zero in the fourth decimal place.

**Table 13. American National Standard Square Head Set Screws**
*ANSI/ASME B18.6.2-2020*

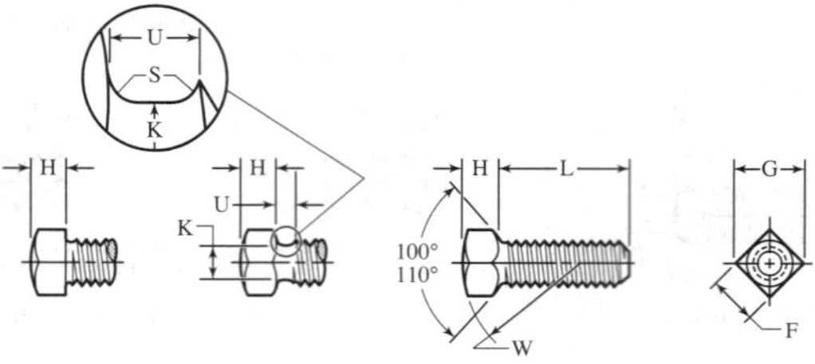


OPTIONAL HEAD CONSTRUCTIONS

| Nominal Size[a] or Basic Screw Diameter | | Width Across Flats, *F* | | Width Across Corners, *G* | | Head Height, *H* | | Neck Relief Diameter, *K* | | Neck Relief Fillet Rad., *S* | Neck Relief Width, *U* | Head Rad., *W* |
|---|---|---|---|---|---|---|---|---|---|---|---|---|
| | | Max. | Min. | Max. | Min. | Max. | Min. | Max. | Min. | Max. | Min. | Min. |
| 10 | 0.1900 | .188 | .180 | .265 | .247 | .148 | .134 | .145 | .140 | .027 | .083 | 0.48 |
| ¼ | 0.2500 | .250 | .241 | .354 | .331 | .196 | .178 | .185 | .170 | .032 | .100 | 0.62 |
| 5⁄16 | 0.3125 | .312 | .302 | .442 | .415 | .245 | .224 | .240 | .225 | .036 | .111 | 0.78 |
| ⅜ | 0.3750 | .375 | .362 | .530 | .497 | .293 | .270 | .294 | .279 | .041 | .125 | 0.94 |
| 7⁄16 | 0.4375 | .438 | .423 | .619 | .581 | .341 | .315 | .345 | .330 | .046 | .143 | 1.09 |
| ½ | 0.5000 | .500 | .484 | .707 | .665 | .389 | .361 | .400 | .385 | .050 | .154 | 1.25 |
| 9⁄16 | 0.5625 | .562 | .545 | .795 | .748 | .437 | .407 | .454 | .439 | .054 | .167 | 1.41 |
| ⅝ | 0.6250 | .625 | .606 | .884 | .833 | .485 | .452 | .507 | .492 | .059 | .182 | 1.56 |
| ¾ | 0.7500 | .750 | .729 | 1.060 | 1.001 | .582 | .544 | .620 | .605 | .065 | .200 | 1.88 |
| ⅞ | 0.8750 | .875 | .852 | 1.237 | 1.170 | .678 | .635 | .731 | .716 | .072 | .222 | 2.19 |
| 1 | 1.0000 | 1.000 | .974 | 1.414 | 1.337 | .774 | .726 | .838 | .823 | .081 | .250 | 2.50 |
| 1⅛ | 1.1250 | 1.125 | 1.096 | 1.591 | 1.505 | .870 | .817 | .939 | .914 | .092 | .283 | 2.81 |
| 1¼ | 1.2500 | 1.250 | 1.219 | 1.768 | 1.674 | .966 | .908 | 1.064 | 1.039 | .092 | .283 | 3.12 |
| 1⅜ | 1.3750 | 1.375 | 1.342 | 1.945 | 1.843 | 1.063 | 1.000 | 1.159 | 1.134 | .109 | .333 | 3.44 |
| 1½ | 1.5000 | 1.500 | 1.464 | 2.121 | 2.010 | 1.159 | 1.091 | 1.284 | 1.259 | .109 | .333 | 3.75 |

*Designation:* Square head set screws are designated by the following data in the sequence shown: Nominal size (number, fraction or decimal equivalent); threads per inch; screw length (fraction or decimal equivalent); product name; point style; material; and protective finish, if required. Examples: ¼ - 20 × ¾ Square Head Set Screw, Flat Point, Steel, Cadmium Plated; .500 – 13 × 1.25 Square Head Set Screw, Cone Point, Corrosion Resistant Steel.

*Cone Point Angle, Y:* For the following nominal lengths, or longer, shown according to nominal size, the cone point angle is 90° ± 2°: For size No. 10, ¼; ¼, 5⁄16; 5⁄16, ⅜; ⅜, 7⁄16; 7⁄16, ½; ½, 9⁄16; 9⁄16, ⅝; ⅝, ¾; ¾, ⅞; ⅞, 1; 1, 1⅛; 1⅛, 1¼; 1¼, 1½; 1⅜, 1⅝; and for 1½, 1¾. For shorter screws the cone point angle is 118° ± 2°.

*Point Types:* Unless otherwise specified, square head set screws are supplied with cup points. Cup points as furnished by some manufacturers may be externally or internally knurled. Where so specified by the purchaser, screws have cone, dog, half-dog, flat or oval points as given on the following page.

*Point Angle, X:* The point angle is 45°, + 5°, – 0° for screws of the nominal lengths, or longer, given just above for cone point angle, and 30° min. for shorter lengths.

## Table 14. Applicability of Hexagon and Spline Keys and Bits

| Nominal Key or Bit Size | | Cap Screws 1960 Series | Flat Countersunk Head Cap Screws | Button Head Cap Screws | Shoulder Screws | Set Screws |
|---|---|---|---|---|---|---|
| | | Nominal Screw Sizes | | | | |
| HEXAGON KEYS AND BITS | | | | | | |
| | 0.028 | … | … | … | … | 0 |
| | 0.035 | … | 0 | 0 | … | 1 & 2 |
| | 0.050 | 0 | 1 & 2 | 1 & 2 | … | 3 & 4 |
| 1⁄16 | 0.062 | 1 | 3 & 4 | 3 & 4 | … | 5 & 6 |
| 5⁄64 | 0.078 | 2 & 3 | 5 & 6 | 5 & 6 | … | 8 |
| 3⁄32 | 0.094 | 4 & 5 | 8 | 8 | … | 10 |
| 7⁄64 | 0.109 | 6 | … | … | … | … |
| 1⁄8 | 0.125 | … | 10 | 10 | 1⁄4 | 1⁄4 |
| 9⁄64 | 0.141 | 8 | … | … | … | … |
| 5⁄32 | 0.156 | 10 | 1⁄4 | 1⁄4 | 5⁄16 | 5⁄16 |
| 3⁄16 | 0.188 | 1⁄4 | 5⁄16 | 5⁄16 | 3⁄8 | 3⁄8 |
| 7⁄32 | 0.219 | … | 3⁄8 | 3⁄8 | … | 7⁄16 |
| 1⁄4 | 0.250 | 5⁄16 | 7⁄16 | … | 1⁄2 | 1⁄2 |
| 5⁄16 | 0.312 | 3⁄8 | 1⁄2 | 1⁄2 | 5⁄8 | 5⁄8 |
| 3⁄8 | 0.375 | 7⁄16 & 1⁄2 | 5⁄8 | 5⁄8 | 3⁄4 | 3⁄4 |
| 7⁄16 | 0.438 | … | … | … | … | … |
| 1⁄2 | 0.500 | 5⁄8 | 3⁄4 | … | 1 | 7⁄8 |
| 9⁄16 | 0.562 | … | 7⁄8 | … | … | 1 & 1 1⁄8 |
| 5⁄8 | 0.625 | 3⁄4 | 1 | … | 1 1⁄4 | 1 1⁄4 & 1 3⁄8 |
| 3⁄4 | 0.750 | 7⁄8 & 1 | 1 1⁄8 | … | … | 1 1⁄2 |
| 7⁄8 | 0.875 | 1 1⁄8 & 1 1⁄4 | 1 1⁄4 & 1 3⁄8 | … | 1 1⁄2 | … |
| 1 | 1.000 | 1 3⁄8 & 1 1⁄2 | 1 1⁄2 | … | 1 3⁄4 | 1 3⁄4 & 2 |
| 1 1⁄4 | 1.250 | 1 3⁄4 | … | … | 2 | … |
| 1 1⁄2 | 1.500 | 2 | … | … | … | … |
| 1 3⁄4 | 1.750 | 2 1⁄4 & 2 1⁄2 | … | … | … | … |
| 2 | 2.000 | 2 3⁄4 | … | … | … | … |
| 2 1⁄4 | 2.250 | 3 & 3 1⁄4 | … | … | … | … |
| 2 3⁄4 | 2.750 | 3 1⁄2 & 3 3⁄4 | … | … | … | … |
| 3 | 3.000 | 4 | … | … | … | … |
| SPLINE KEYS AND BITS | | | | | | |
| | 0.033 | … | … | … | … | 0 & 1 |
| | 0.048 | … | 0 | 0 | … | 2 & 3 |
| | 0.060 | 0 | 1 & 2 | 1 & 2 | … | 4 |
| | 0.072 | 1 | 3 & 4 | 3 & 4 | … | 5 & 6 |
| | 0.096 | 2 & 3 | 5 & 6 | 5 & 6 | … | 8 |
| | 0.111 | 4 & 5 | 8 | 8 | … | 10 |
| | 0.133 | 6 | … | … | … | … |
| | 0.145 | … | 10 | 10 | … | 1⁄4 |
| | 0.168 | 8 | … | … | … | … |
| | 0.183 | 10 | 1⁄4 | 1⁄4 | … | 5⁄16 |
| | 0.216 | 1⁄4 | 5⁄16 | 5⁄16 | … | 3⁄8 |
| | 0.251 | … | 3⁄8 | 3⁄8 | … | 7⁄16 |
| | 0.291 | 5⁄16 | 7⁄16 | … | … | 1⁄2 |
| | 0.372 | 3⁄8 | 1⁄2 | 1⁄2 | … | 5⁄8 |
| | 0.454 | 7⁄16 & 1⁄2 | 5⁄8 & 3⁄4 | 5⁄8 | … | 3⁄4 |
| | 0.595 | 5⁄8 | … | … | … | 7⁄8 |
| | 0.620 | 3⁄4 | … | … | … | … |
| | 0.698 | 7⁄8 | … | … | … | … |
| | 0.790 | 1 | … | … | … | … |

*Source:* Appendix to American National Standard ANSI/ASME B18.3-2012 and ANSI/ASME B18.3-1998 (for spline driven fasteners).

**Table 15. ANSI Hexagon and Spline Socket Set Screws** *ANSI/ASME B18.3-2012* and *ANSI/ASME B18.3-1998 (for Spline driven fasteners).*

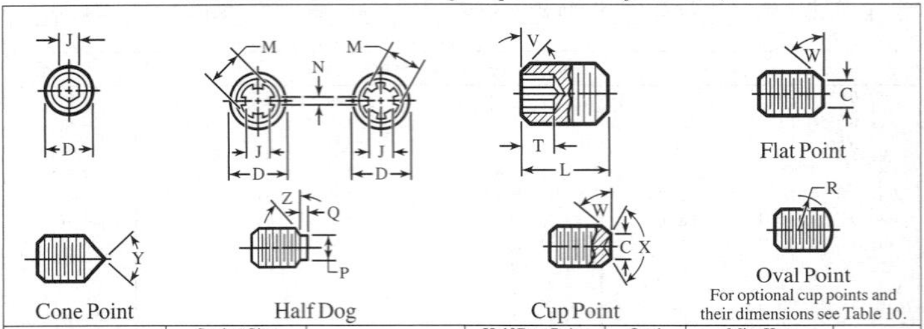


| Nominal Size or Basic Screw Diameter | | Socket Size Hex. Nom. *J* | Socket Size Spl. Nom. *M* | Cup and Flat Point Diameters *C* Max. | Cup and Flat Point Diameters *C* Min. | Half Dog Point Dia. Max. *P* | Half Dog Point Lgth. Max. *Q* | Oval Point Radius Basic *R* | Min. Key Engagement Depth Hex. $T_H$[a] | Min. Key Engagement Depth Spl. $T_S$[a] | Lgth. Limit for Angle *Y*[b] |
|---|---|---|---|---|---|---|---|---|---|---|---|
| 0 | 0.0600 | 0.028 | 0.033 | 0.033 | 0.027 | 0.040 | 0.017 | 0.045 | 0.050 | 0.026 | 0.09 |
| 1 | 0.0730 | 0.035 | 0.033 | 0.040 | 0.033 | 0.049 | 0.021 | 0.055 | 0.060 | 0.035 | 0.09 |
| 2 | 0.0860 | 0.035 | 0.048 | 0.047 | 0.039 | 0.057 | 0.024 | 0.064 | 0.060 | 0.040 | 0.13 |
| 3 | 0.0990 | 0.050 | 0.048 | 0.054 | 0.045 | 0.066 | 0.027 | 0.074 | 0.070 | 0.040 | 0.13 |
| 4 | 0.1120 | 0.050 | 0.060 | 0.061 | 0.051 | 0.075 | 0.030 | 0.084 | 0.070 | 0.045 | 0.19 |
| 5 | 0.1250 | 1/16 | 0.072 | 0.067 | 0.057 | 0.083 | 0.033 | 0.094 | 0.080 | 0.055 | 0.19 |
| 6 | 0.1380 | 1/16 | 0.072 | 0.074 | 0.064 | 0.092 | 0.038 | 0.104 | 0.080 | 0.055 | 0.19 |
| 8 | 0.1640 | 5/64 | 0.096 | 0.087 | 0.076 | 0.109 | 0.043 | 0.123 | 0.090 | 0.080 | 0.25 |
| 10 | 0.1900 | 3/32 | 0.111 | 0.102 | 0.088 | 0.127 | 0.049 | 0.142 | 0.100 | 0.080 | 0.25 |
| 1/4 | 0.2500 | 1/8 | 0.145 | 0.132 | 0.118 | 0.156 | 0.067 | 0.188 | 0.125 | 0.125 | 0.31 |
| 5/16 | 0.3125 | 5/32 | 0.183 | 0.172 | 0.156 | 0.203 | 0.082 | 0.234 | 0.156 | 0.156 | 0.38 |
| 3/8 | 0.3750 | 3/16 | 0.216 | 0.212 | 0.194 | 0.250 | 0.099 | 0.281 | 0.188 | 0.188 | 0.44 |
| 7/16 | 0.4375 | 7/32 | 0.251 | 0.252 | 0.232 | 0.297 | 0.114 | 0.328 | 0.219 | 0.219 | 0.50 |
| 1/2 | 0.5000 | 1/4 | 0.291 | 0.291 | 0.270 | 0.344 | 0.130 | 0.375 | 0.250 | 0.250 | 0.57 |
| 5/8 | 0.6250 | 5/16 | 0.372 | 0.371 | 0.347 | 0.469 | 0.164 | 0.469 | 0.312 | 0.312 | 0.75 |
| 3/4 | 0.7500 | 3/8 | 0.454 | 0.450 | 0.425 | 0.562 | 0.196 | 0.562 | 0.375 | 0.375 | 0.88 |
| 7/8 | 0.8750 | 1/2 | 0.595 | 0.530 | 0.502 | 0.656 | 0.227 | 0.656 | 0.500 | 0.500 | 1.00 |
| 1 | 1.0000 | 9/16 | … | 0.609 | 0.579 | 0.750 | 0.260 | 0.750 | 0.562 | … | 1.13 |
| 1 1/8 | 1.1250 | 9/16 | … | 0.689 | 0.655 | 0.844 | 0.291 | 0.844 | 0.562 | … | 1.25 |
| 1 1/4 | 1.2500 | 5/8 | … | 0.767 | 0.733 | 0.938 | 0.323 | 0.938 | 0.625 | … | 1.50 |
| 1 3/8 | 1.3750 | 5/8 | … | 0.848 | 0.808 | 1.031 | 0.354 | 1.031 | 0.625 | … | 1.63 |
| 1 1/2 | 1.5000 | 3/4 | … | 0.926 | 0.886 | 1.125 | 0.385 | 1.125 | 0.750 | … | 1.75 |
| 1 3/4 | 1.7500 | 1 | … | 1.086 | 1.039 | 1.312 | 0.448 | 1.321 | 1.000 | … | 2.00 |
| 2 | 2.0000 | 1 | … | 1.244 | 1.193 | 1.500 | 0.510 | 1.500 | 1.000 | … | 2.25 |

[a] Reference should be made to the Standard for shortest optimum nominal lengths to which the minimum key engagement depths $T_H$ and $T_S$ apply.

[b] Cone point angle *Y* is 90 degrees plus or minus 2 degrees for these nominal lengths or longer and 118 degrees plus or minus 2 degrees for shorter nominal lengths.

All dimensions are in inches. The thread conforms to the Unified Standard, Class 3A, UNC and UNF series. The socket depth *T* is included in the Standard and some are shown here. The nominal length *L* of all socket type set screws is the total or overall length. For nominal screw lengths of 1/16 through 3/16 inch (0 through 3 sizes incl.) the standard length increment is 0.06 inch; for lengths 1/8 through 1 inch the increment is 1/8 inch; for lengths 1 through 2 inches the increment is 1/4 inch; for lengths 2 through 6 inches the increment is 1/2 inch; for lengths 6 inches and longer the increment is 1 inch. Socket dimensions are given in Table 11.

*Length Tolerance:* The allowable tolerance on length *L* for all set screws of the socket type is ± 0.01 inch for set screws up to 5/8 inch long; ± 0.02 inch for screws over 5/8 to 2 inches long; ± 0.03 inch for screws over 2 to 6 inches long and ± 0.06 inch for screws over 6 inches long. Socket dimensions are given in Table 11.

For manufacturing details, including materials, not shown, see American National Standard ANSI/ASME B18.3-2012 and ANSI/ASME B18.3-1998 (for Spline driven fasteners).

**British Standard Hexagon Socket Screws—Metric Series.**—The first five parts of British Standard BS 4168:1981 provide specifications for hexagon socket head cap screws and hexagon socket set screws.

*Hexagon Socket Head Cap Screws:* The dimensional data in Table 1 are based upon BS 4168: Part 1:1981. These screws are available in stainless steel and alloy steel, the latter having class 12.9 properties as specified in BS 6104: Part 1:1981 (Withdrawn). When ordering these screws, the designation "Hexagon socket head cap screw BS 4168 M5 × 20-12.9" would mean, as an example, a cap screw having a thread size of $d$ = M5, nominal length $l$ = 20 mm, and property class 12.9. Alloy steel cap screws are furnished with a black oxide finish (thermal or chemical); stainless steel cap screws with a plain finish. Combinations of thread size, nominal length, and length of thread are shown in Table 2; the screw threads in these combinations are in the ISO metric coarse pitch series specified in BS 3643 with tolerances in the 5g6g class. (See Metric Screw Threads in Index.)

*Hexagon Socket Set Screws:* Part 2 of BS 4168:1994 (Withdrawn) specifies requirements for hexagon socket set screws with flat point having ISO metric threads, and diameters from 1.6 mm up to and including 24 mm. The dimensions of these set screws along with those of cone-point, dog-point, and cup-point set screws in accord, respectively, with Parts 3, 4, and 5 of the Standard are given in Table 3 and the accompanying illustration.

All of these set screws are available in either steel processed to mechanical properties class 45H BS 6104: Part 3:1981 (Withdrawn); or stainless steel processed to mechanical properties described in BS 6105 (Withdrawn). Steel set screws are furnished with black oxide (thermal or chemical) finish; stainless steel set screws are furnished plain. The tolerances applied to the threads of these set screws are for ISO product grade A, based on ISO 4759/1-1978 "Tolerances for fasteners.—Part 1: Bolts, screws, and nuts with thread diameters greater than or equal to 1.6 mm and less than or equal to 150 mm and product grades A, B, and C."

Hexagon socket set screws are designated by the type, the thread size, nominal length, and property class. As an example, for a flat-point set screw of thread size $d$ = M6, nominal length $l$ = 12 mm, and property class 45H:

Hexagon socket set screw flat point BS 4168 M6 × 12-45H

*British Standard Hexagon Socket Countersunk and Button Head Screws.—Metric Series:* British Standard BS 4168:1982 (Withdrawn) provides a metric series of hexagon socket countersunk and button head screws. The dimensions of these screws are given in Table 4. The revision of this Standard will constitute Parts 6 and 8 of BS 4168.

*British Standards for Mechanical Properties of Fasteners:* BS 6104: Part 1:1981 specifies mechanical properties for bolts, screws, and studs with nominal diameters up to and including 39 mm of any triangular ISO thread and made of carbon or alloy steel. It does not apply to set screws and similar threaded fasteners. Part 2 of this Standard specifies the mechanical properties of set screws and similar fasteners, not under tensile stress, in the range from M1.6 up to and including M39 and made of carbon or alloy steel.

BS 6105:1981 (Withdrawn) provides specifications for bolts, screws, studs, and nuts made from austenitic, ferritic, and martensitic grades of corrosion-resistant steels. This Standard applies only to fastener components after completion of manufacture with nominal diameters from M1.6 up to and including M39.

These Standards are not described further here. Copies may be obtained from the British Standards Institution, 389 Chiswick High Road, London, W4 4AL, www.bsigroup.com, and also from the American National Standards Institute, 25 West 43rd Street, 4th Floor, New York, NY 10036, www.ansi.org.

**Table 1. British Standard Hexagon Socket Head Cap Screws—Metric Series**
*BS 4168: Part 1:1981 (Withdrawn)*

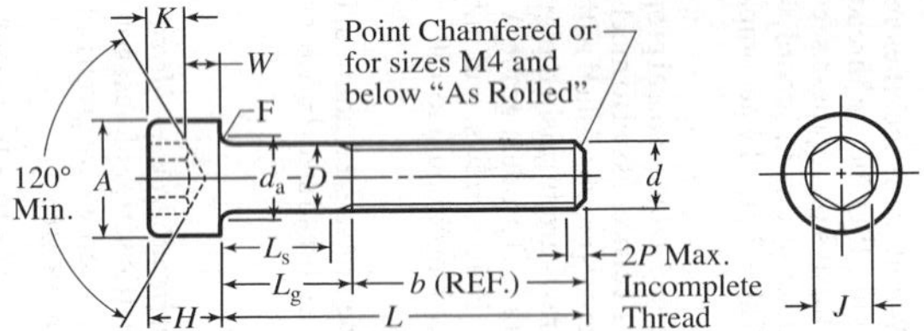


| Nominal Size,[a] $d$ | Body Diameter, $D$ | | Head Diameter, $A$ | | | Head Height, $H$ | | Hexagon Socket Size, $J$[b] | Key Engagement, $K$ | Wall Thickness, $W$ | Fillet Rad., $F$ | Fillet Dia., $d_a$ |
|---|---|---|---|---|---|---|---|---|---|---|---|---|
| | Max. | Min. | Max.[c] | Max.[d] | Min. | Max. | Min. | Nom. | Min. | Min. | Min. | Max. |
| M1.6 | 1.6 | 1.46 | 3 | 3.14 | 2.86 | 1.6 | 1.46 | 1.5 | 0.7 | 0.55 | 0.1 | 2 |
| M2 | 2 | 1.86 | 3.8 | 3.98 | 3.62 | 2 | 1.86 | 1.5 | 1 | 0.55 | 0.1 | 2.6 |
| M2.5 | 2.5 | 2.36 | 4.5 | 4.68 | 4.32 | 2.5 | 2.36 | 2 | 1.1 | 0.85 | 0.1 | 3.1 |
| M3 | 3 | 2.86 | 5.5 | 5.68 | 5.32 | 3 | 2.86 | 2.5 | 1.3 | 1.15 | 0.1 | 3.6 |
| M4 | 4 | 3.82 | 7 | 7.22 | 6.78 | 4 | 3.82 | 3 | 2 | 1.4 | 0.2 | 4.7 |
| M5 | 5 | 4.82 | 8.5 | 8.72 | 8.28 | 5 | 4.82 | 4 | 2.5 | 1.9 | 0.2 | 5.7 |
| M6 | 6 | 5.82 | 10 | 10.22 | 9.78 | 6 | 5.70 | 5 | 3 | 2.3 | 0.25 | 6.8 |
| M8 | 8 | 7.78 | 13 | 13.27 | 12.73 | 8 | 7.64 | 6 | 4 | 3.3 | 0.4 | 9.2 |
| M10 | 10 | 9.78 | 16 | 16.27 | 15.73 | 10 | 9.64 | 8 | 5 | 4 | 0.4 | 11.2 |
| M12 | 12 | 11.73 | 18 | 18.27 | 17.73 | 12 | 11.57 | 10 | 6 | 4.8 | 0.6 | 14.2 |
| (M14) | 14 | 13.73 | 21 | 21.33 | 20.67 | 14 | 13.57 | 12 | 7 | 5.8 | 0.6 | 16.2 |
| M16 | 16 | 15.73 | 24 | 24.33 | 23.67 | 16 | 15.57 | 14 | 8 | 6.8 | 0.6 | 18.2 |
| M20 | 20 | 19.67 | 30 | 30.33 | 29.67 | 20 | 19.48 | 17. | 10 | 8.6 | 0.8 | 22.4 |
| M24 | 24 | 23.67 | 36 | 36.39 | 35.61 | 24 | 23.48 | 19 | 12 | 10.4 | 0.8 | 26.4 |
| M30 | 30 | 29.67 | 45 | 45.39 | 44.61 | 30 | 29.48 | 22 | 15.5 | 13.1 | 1 | 33.4 |
| M36 | 36 | 35.61 | 54 | 54.46 | 53.54 | 36 | 35.38 | 27 | 19 | 15.3 | 1 | 39.4 |

[a] The size shown in ( ) is non-preferred.

[b] See Table 2 for min/max.

[c] For plain heads.

[d] For knurled heads.

All dimensions are given in millimeters.

## Table 2. British Standard Hexagon Socket Screws—Metric Series

*BS 4168: Part 1:1981 (Withdrawn)*

**Dimensions of Hexagon Sockets**

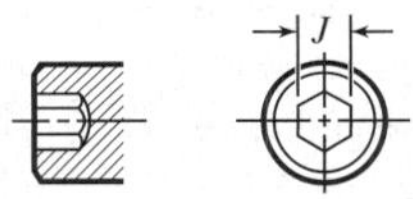


| Nominal Socket Size | Socket Width Across Flats, J | | Nominal Socket Size | Socket Width Across Flats, J | |
|---|---|---|---|---|---|
| | Max. | Min. | | Max. | Min. |
| 1.5 | 1.545 | 1.52 | 6 | 6.095 | 6.02 |
| 2.0 | 2.045 | 2.02 | 8 | 8.115 | 8.025 |
| 2.5 | 2.56 | 2.52 | 10 | 10.115 | 10.025 |
| 3 | 3.08 | 3.02 | 12 | 12.142 | 12.032 |
| 4 | 4.095 | 4.02 | 14 | 14.142 | 14.032 |
| 5 | 5.095 | 5.02 | 17 | 17.23 | 17.05 |
| … | … | … | 19 | 19.275 | 19.065 |

**Association of Nominal and Thread Lengths for Each Thread Size**

| Nominal Length, L | Nominal Thread Size, D | | | | | | | |
|---|---|---|---|---|---|---|---|---|
| | M1.6 | M2 | M2.5 | M3 | M4 | M5 | M6 | M8 |
| 2.5 | | | | | | | | |
| 3 | | | | | | | | |
| 4 | | | | | | | | |
| 5 | | | | | | | | |
| 6 | | | | | | | | |
| 8 | | | | | | | | |
| 10 | | | | | | | | |
| 12 | | | | | | | | |
| 16 | | | | | | | | |
| 20 | | | | | | | | |
| 25 | | | | | | | | |
| 30 | | | | | | | | |
| 35 | | | | | | | | |
| 40 | | | | | | | | |
| 45 | | | | | | | | |
| 50 | | | | | | | | |
| 55 | | | | | | | | |
| 60 | | | | | | | | |
| 65 | | | | | | | | |
| 70 | | | | | | | | |
| 80 | | | | | | | | |
| … | … | … | … | … | … | … | … | … |
| … | … | … | … | … | … | … | … | … |
| *b* (ref) | 15 | 16 | 17 | 18 | 20 | 22 | 24 | 28 |

| Nominal Length, L | Nominal Thread Size, D | | | | | | | |
|---|---|---|---|---|---|---|---|---|
| | M10 | M12 | (M14) | M16 | M20 | M24 | M30 | M36 |
| 16 | | | | | | | | |
| 20 | | | | | | | | |
| 25 | | | | | | | | |
| 30 | | | | | | | | |
| 35 | | | | | | | | |
| 40 | | | | | | | | |
| 45 | | | | | | | | |
| 50 | | | | | | | | |
| 55 | | | | | | | | |
| 60 | | | | | | | | |
| 65 | | | | | | | | |
| 70 | | | | | | | | |
| 80 | | | | | | | | |
| 90 | | | | | | | | |
| 100 | | | | | | | | |
| 110 | | | | | | | | |
| 120 | | | | | | | | |
| 130 | | | | | | | | |
| 140 | | | | | | | | |
| 150 | | | | | | | | |
| 160 | | | | | | | | |
| 180 | | | | | | | | |
| 200 | | | | | | | | |
| *b* (ref) | 32 | 36 | 40 | 44 | 52 | 60 | 72 | 84 |

All dimensions are in millimeters.

The popular lengths are those between the stepped solid lines. Lengths above the shaded areas are threaded to the head within 3 pitch lengths (3*P*). Lengths within and below the shaded areas have values of $L_g$ and $L_s$ (see Table 1) given by the formulas: $L_g\text{max} = L\text{ nom} - b\text{ ref}$, and $L_s\text{min} = L_g\text{max} - 5P$.

**Table 3. British Standard Hexagon Socket Set Screws—Metric Series**
*BS 4168: Parts 2, 3, 4, and 5:1994 (Withdrawn)*

| Nom. Size, $d$ | Pitch, $P$ | Socket Size, $s$ | Depth of Key Engagement, $t$[a] | | Range of Popular Lengths | | | | Length of Dog on Dog Point Screws[b] | | | | | End Diameters | | | |
|---|---|---|---|---|---|---|---|---|---|---|---|---|---|---|---|---|---|
| | | | | | Flat Point | Cone Point | Dog Point | Cup Point | Short Dog, $z$ | | Long Dog, $z$ | | | Flat Point, $d_z$ | Cone Point $d_t$ | Dog Point, $d_p$ | Cup Point, $d_z$ |
| | | Nom. | Min. | Min. | $l$ | $l$ | $l$ | $l$ | Min. | Max. | Min. | Max. | *[b] | Max. | Max. | Max. | Max. |
| M1.6 | 0.35 | 0.7 | 0.7 | 1.5 | 2-8 | 2-8 | 2-8 | 2-8 | 0.4 | 0.65 | 0.8 | 1.05 | 2.5 | 0.8 | 0 | 0.8 | 0.8 |
| M2 | 0.4 | 0.9 | 0.8 | 1.7 | 2-10 | 2-10 | 2.5-10 | 2-10 | 0.5 | 0.75 | 1.0 | 1.25 | 3.0 | 1.0 | 0 | 1.0 | 1.0 |
| M2.5 | 0.45 | 1.3 | 1.2 | 2.0 | 2-12 | 2.5-12 | 3-12 | 2-12 | 0.63 | 0.88 | 1.25 | 1.5 | 4 | 1.5 | 0 | 1.5 | 1.2 |
| M3 | 0.5 | 1.5 | 1.2 | 2.0 | 2-16 | 2.5-16 | 4-16 | 2.5-16 | 0.75 | 1.0 | 1.5 | 1.75 | 5 | 2.0 | 0 | 2.0 | 1.4 |
| M4 | 0.7 | 2.0 | 1.5 | 2.5 | 2.5-20 | 3-20 | 5-20 | 3-20 | 1.0 | 1.25 | 2.0 | 2.25 | 6 | 2.5 | 0 | 2.5 | 2.0 |
| M5 | 0.8 | 2.5 | 2.0 | 3.0 | 3-25 | 4-25 | 6-25 | 4-25 | 1.25 | 1.5 | 2.5 | 2.75 | 6 | 3.5 | 0 | 3.5 | 2.5 |
| M6 | 1.0 | 3.0 | 2.0 | 3.5 | 4-30 | 5-30 | 8-30 | 5-30 | 1.5 | 1.75 | 3.0 | 3.25 | 8 | 4.0 | 1.5 | 4.0 | 3.0 |
| M8 | 1.25 | 4.0 | 3.0 | 5.0 | 5-40 | 6-40 | 8-40 | 6-40 | 2.0 | 2.25 | 4.0 | 4.3 | 10 | 5.5 | 2.0 | 5.5 | 5.0 |
| M10 | 1.5 | 5.0 | 4.0 | 6.0 | 6-50 | 8-50 | 10-50 | 8-50 | 2.5 | 2.75 | 5.0 | 5.3 | 12 | 7.0 | 2.5 | 7.0 | 6.0 |
| M12 | 1.75 | 6.0 | 4.8 | 8.0 | 8-60 | 10-60 | 12-60 | 10-60 | 3.0 | 3.25 | 6.0 | 6.3 | 16 | 8.5 | 3.0 | 8.5 | 8.0 |
| M16 | 2.0 | 8.0 | 6.4 | 10.0 | 10-60 | 12-60 | 16-60 | 12-60 | 4.0 | 4.3 | 8.0 | 8.36 | 20 | 12.0 | 4.0 | 12.0 | 10.0 |
| M20 | 2.5 | 10.0 | 8.0 | 12.0 | 12-60 | 16-60 | 20-60 | 16-60 | 5.0 | 5.3 | 10.0 | 10.36 | 25 | 15.0 | 5.0 | 15.0 | 14.0 |
| M24 | 3.0 | 12.0 | 10.0 | 15.0 | 16-60 | 20-60 | 25-60 | 20-60 | 6.0 | 6.3 | 12.0 | 12.43 | 30 | 18.0 | 6.0 | 18.0 | 16.0 |

[a] The smaller of the two $t$ min. values applies to certain short-length set screws. These short-length screws are those whose length is approximately equal to the diameter of the screw. The larger $t$ min. values apply to longer-length screws.

[b] A dog point set screw having a nominal length equal to or less than the length shown in the (*) column of the table is supplied with length $z$ shown in the short dog column. For set screws of lengths greater than shown in the (*) column, $z$ for long dogs applies.

All dimensions are in millimeters. For dimensional notation, see diagram, page 1834.

**Table 4. British Standard Hexagon Socket Countersunk and Button Head Screws—Metric Series** *BS 4168: Parts 6 and 8:1982 (Withdrawn)*

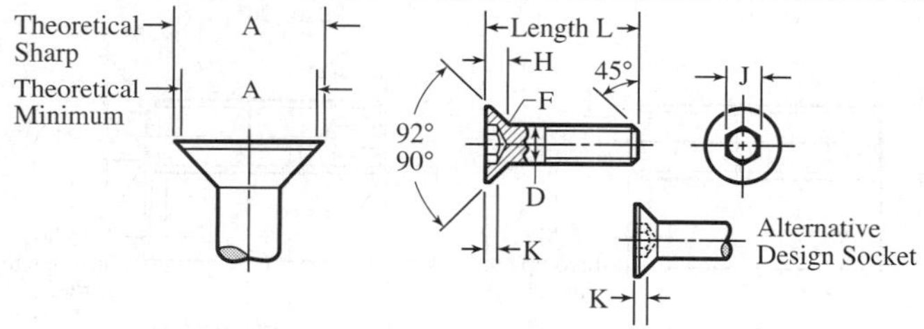


COUNTERSUNK HEAD SCREWS

| Nom. Size | Body Diameter, *D* | | Head Diameter, *A* | | Head Height, *H* | | Hexagon Socket Size, *J* | Key Engagement, *K* | Fillet Radius, *F* |
|---|---|---|---|---|---|---|---|---|---|
| | Max. | Min. | Theor. Sharp Max. | Absolute Min. | Ref. | Flushness Tolerance | Nom. | Min. | Max. |
| M3 | 3.00 | 2.86 | 6.72 | 5.82 | 1.86 | 0.20 | 2.00 | 1.05 | 0.40 |
| M4 | 4.00 | 3.82 | 8.96 | 7.78 | 2.48 | 0.20 | 2.50 | 1.49 | 0.40 |
| M5 | 5.00 | 4.82 | 11.20 | 9.78 | 3.10 | 0.20 | 3.00 | 1.86 | 0.40 |
| M6 | 6.00 | 5.82 | 13.44 | 11.73 | 3.72 | 0.20 | 4.00 | 2.16 | 0.60 |
| M8 | 8.00 | 7.78 | 17.92 | 15.73 | 4.96 | 0.24 | 5.00 | 2.85 | 0.70 |
| M10 | 10.00 | 9.78 | 22.40 | 19.67 | 6.20 | 0.30 | 6.00 | 3.60 | 0.80 |
| M12 | 12.00 | 11.73 | 26.88 | 23.67 | 7.44 | 0.36 | 8.00 | 4.35 | 1.10 |
| (M14) | 14.00 | 13.73 | 30.24 | 26.67 | 8.12 | 0.40 | 10.00 | 4.65 | 1.10 |
| M16 | 16.00 | 15.73 | 33.60 | 29.67 | 8.80 | 0.45 | 10.00 | 4.89 | 1.10 |
| (M18) | 18.00 | 17.73 | 36.96 | 32.61 | 9.48 | 0.50 | 12.00 | 5.25 | 1.10 |
| M20 | 20.00 | 19.67 | 40.32 | 35.61 | 10.16 | 0.54 | 12.00 | 5.45 | 1.10 |

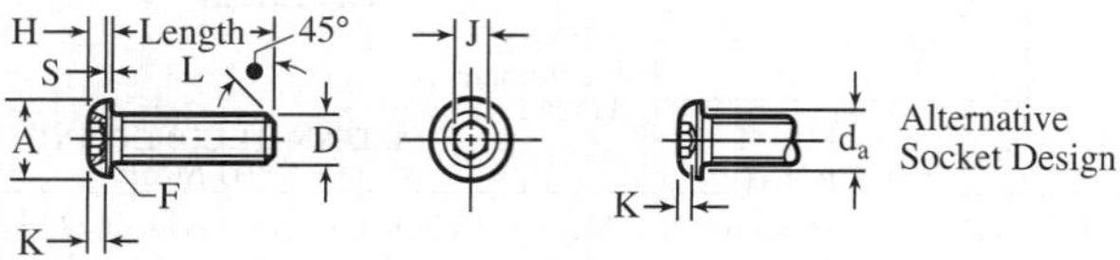


BUTTON HEAD SCREWS

| Nom. Size, *D* | Head Diameter, *A* | | Head Height, *H* | | Head Side Height, *S* | Hexagon Socket Size, *J* | Key Engagement, *K* | Fillet Radius | |
|---|---|---|---|---|---|---|---|---|---|
| | | | | | | | | *F* | $d_a$ |
| | Max. | Min. | Max. | Min. | Ref. | Nom. | Min. | Min. | Max. |
| M3 | 5.50 | 5.32 | 1.60 | 1.40 | 0.38 | 2.00 | 1.04 | 0.10 | 3.60 |
| M4 | 7.50 | 7.28 | 2.10 | 1.85 | 0.38 | 2.50 | 1.30 | 0.20 | 4.70 |
| M5 | 9.50 | 9.28 | 2.70 | 2.45 | 0.50 | 3.00 | 1.56 | 0.20 | 5.70 |
| M6 | 10.50 | 10.23 | 3.20 | 2.95 | 0.80 | 4.00 | 2.08 | 0.25 | 6.80 |
| M8 | 14.00 | 13.73 | 4.30 | 3.95 | 0.80 | 5.00 | 2.60 | 0.40 | 9.20 |
| M10 | 18.00 | 17.73 | 5.30 | 4.95 | 0.80 | 6.00 | 3.12 | 0.40 | 11.20 |
| M12 | 21.00 | 20.67 | 6.40 | 5.90 | 0.80 | 8.00 | 4.16 | 0.60 | 14.20 |

All dimensions are given in millimeters.

**British Standard Hexagon Socket Set Screws—Metric Series**

*BS 4168: Parts 2, 3, 4, and 5:1994 (Withdrawn)*

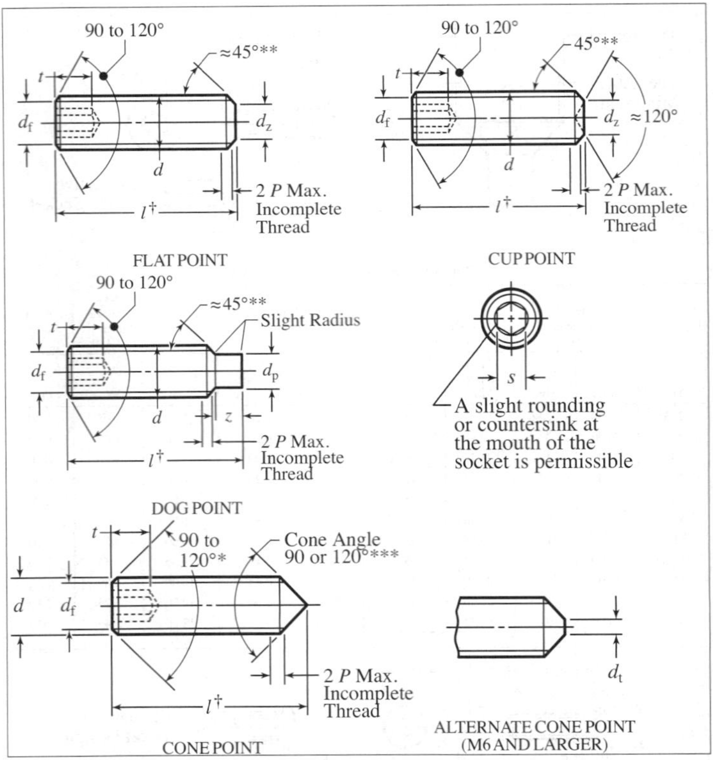


*The 120° angle is mandatory for short-length screws shown in the Standard. Short-length screws are those whose length is, approximately, equal to the diameter of the screw.

**The 45° angle applies only to that portion of the point below the root diameter, *df*, of the thread.

***The cone angle applies only to the portion of the point below the root diameter, *df*, of the thread and shall be 120° for certain short lengths listed in the Standard. All other lengths have a 90° cone angle.

†The popular length ranges of these set screws are listed in Table 3. These lengths have been selected from the following nominal lengths: 2, 2.5, 3, 4, 6, 8, 10, 12, 16, 20, 25, 30, 35, 40, 45, 50, 55, and 60 millimeters.

**Holding Power of Set Screws.—**While the amount of power a set screw of given size will transmit without slipping (when used for holding a pulley, gear, or other part from turning relative to a shaft) varies somewhat according to the physical properties of both set screw and shaft and other variable factors, experiments have shown that the safe holding force in pounds for different diameters of set screws should be approximately as follows: For $\frac{1}{4}$-inch (6.35-mm) diameter set screws the safe holding force is 100 pounds (445 N), for $\frac{3}{8}$-inch (9.5-mm) diameter set screws the safe holding force is 250 pounds (1112 N), for

½-inch (12.7-mm) diameter set screws the safe holding force is 500 pounds (2224 N), for ¾-inch (19-mm) diameter set screws the safe holding force is 1300 pounds (5783 N), and for 1-inch (25.4-mm) diameter set screws the safe holding force is 2500 pounds (11121 N).

The power or torque that can be safely transmitted by a set screw may be determined from the formulas, $P = (DNd^{2.3}) \div 50$; or $T = 1250Dd^{2.3}$ in which $P$ is the horsepower transmitted; $T$ is the torque in inch-pounds transmitted; $D$ is the shaft diameter in inches; $N$ is the speed of the shaft in revolutions per minute; and $d$ is the diameter of the set screw in inches.

*Example:* How many ½-inch diameter set screws would be required to transmit 3 horsepower at a shaft speed of 1000 rpm if the shaft diameter is 1 inch?

Using the first formula given above, the power transmitted by a single ½-inch diameter set screw is determined: $P = [1 \times 1000 \times (1/2)^{2.3}] \div 50 = 4.1$ hp. Therefore a single ½-inch diameter set screw is sufficient.

*Example:* In the previous example, how many ⅜-inch diameter set screws would be required? $P = [1 \times 1000 \times (3/8)^{2.3}] \div 50 = 2.1$ hp. Therefore two ⅜-inch diameter set screws are required.

**Table 5. British Standard Whitworth (BSW) and British Standard Fine (BSF) Bright Square Head Set Screws (With Flat Chamfered Ends)**

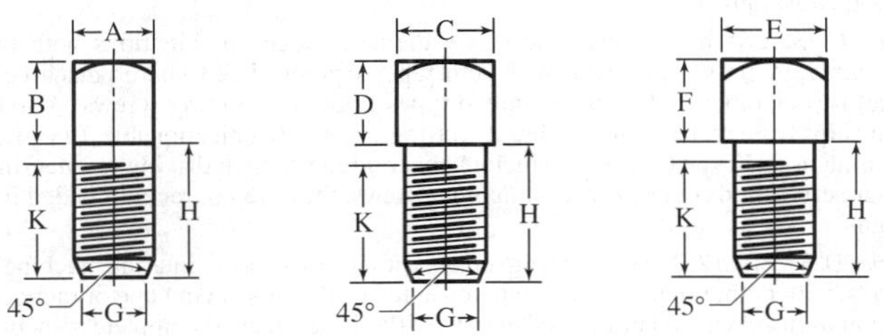


| Nominal Size and Max. Dia., Inches | Number of Threads per Inch | | No. 1 Standard | | No. 2 Standard | | No. 3 Standard | |
|---|---|---|---|---|---|---|---|---|
| | BSW | BSF | Width Across Flats *A* | Depth of Head *B* | Width Across Flats *C* | Depth of Head *D* | Width Across Flats *E* | Depth of Head *F* |
| ¼ | 20 | 26 | 0.250 | 0.250 | 0.313 | 0.250 | 0.375 | 0.250 |
| 5⁄16 | 18 | 22 | 0.313 | 0.313 | 0.375 | 0.313 | 0.438 | 0.313 |
| ⅜ | 16 | 20 | 0.375 | 0.375 | 0.438 | 0.375 | 0.500 | 0.375 |
| 7⁄16 | 14 | 18 | 0.438 | 0.438 | 0.500 | 0.438 | 0.625 | 0.438 |
| ½ | 12 | 16 | 0.500 | 0.500 | 0.563 | 0.500 | 0.750 | 0.500 |
| ⅝ | 11 | 14 | 0.625 | 0.625 | 0.750 | 0.625 | 0.875 | 0.625 |
| ¾ | 10 | 12 | 0.750 | 0.750 | 0.875 | 0.750 | 1.000 | 0.750 |
| ⅞ | 9 | 11 | 0.875 | 0.875 | 1.000 | 0.875 | 1.125 | 0.875 |
| 1 | 8 | 10 | 1.000 | 1.000 | 1.125 | 1.000 | 1.250 | 1.000 |

* Depth of Head *B*, *D* and *F* same as for Width Across Flats, No. 1 Standard.

Dimensions *A*, *B*, *C*, *D*, *E*, and *F* are in inches.

## SELF-THREADING SCREWS

**ANSI Standard Sheet Metal, Self-Tapping, and Metallic Drive Screws.—**Table 1 shows the various types of "self-tapping" screw threads covered by the ANSI/ASME B18.6.3-2013 Standard. (Metric thread forming and thread cutting tapping screws are discussed beginning on page 1851). ANSI designations are also shown. Types A, AB, B, BP and C when turned into a hole of proper size form a thread by a displacing action. Types D, F, G, T, BF and BT when turned into a hole of proper size form a thread by a cutting action. Type U when driven into a hole of proper size forms a series of multiple threads by a displacing action. These screws have the following descriptions and applications:

*Type A:* Spaced-thread screw with gimlet point primarily for use in light sheet metal, resin-impregnated plywood, and asbestos compositions. This type is no longer recommended. Use Type AB in new designs and whenever possible substitute for Type A in existing designs.

*Type AB:* Spaced-thread screw with same pitches as Type B but with gimlet point, primarily for similar uses as for Type A.

*Type B:* Spaced-thread screw with a blunt point with pitches generally somewhat finer than Type A. Used for thin metal, nonferrous castings, plastics, resin-impregnated plywood, and asbestos compositions.

*Type BP:* Spaced-thread screw, the same as Type B but having a conical point extending beyond incomplete entering threads. Used for piercing fabrics or in assemblies where holes are misaligned.

*Type C:* Screws having machine screw diameter-pitch combinations with threads approximately Unified Form and with blunt tapered points. Used where a machine screw thread is preferable to the spaced-thread types of thread forming screws. Also useful when chips from machine screw thread-cutting screws are objectionable. In view of the declining use of Type C screws, which in general require high driving torques, in favor of more efficient designs of thread tapping screws, they are not recommended for new designs.

*Types D, F, G, and T:* Thread-cutting screws with threads approximating machine screw threads, with blunt point, and with tapered entering threads having one or more cutting edges and chip cavities. The tapered threads of the Type F may be complete or incomplete at the producer's option; all other types have incomplete tapered threads. These screws can be used in materials such as aluminum, zinc, and lead die castings; steel sheets and shapes; cast iron; brass; and plastics.

*Types BF and BT:* Thread-cutting screws with spaced threads as in Type B, with blunt points, and one or more cutting grooves. Used in plastics, asbestos, and other similar compositions.

*Type U:* Multiple-threaded drive screw with large helix angle, having a pilot point, for use in metal and plastics. This screw is forced into the work by pressure and is intended for making permanent fastenings.

*ANSI Standard Head Types for Tapping and Metallic Drive Screws:* Many of the head types used with "self-tapping" screw threads are similar to the head types of American National Standard machine screws shown in the section with that heading.

*Round Head:* The round head has a semi-elliptical top surface and a flat bearing surface. Because of the superior slot driving characteristics of pan head screws over round head screws, and the overlap in dimensions of cross recessed pan heads and round heads, it is recommended that pan head screws be used in new designs and wherever possible substituted in existing designs.

*Undercut Flat and Oval Countersunk Heads:* For short lengths, 82-degree and oval countersunk head tapping screws have heads undercut to 70 percent of normal side height to afford greater length of thread on the screws.

*Flat Countersunk Head:* The flat countersunk head has a flat top surface and a conical bearing surface with a head angle for one design of approximately 82 degrees and for another design of approximately 100 degrees. Because of its limited usage and in the interest of curtailing product varieties, the 100-degree flat countersunk head is considered non-preferred.

*Oval Countersunk Head:* The oval countersunk head has a rounded top surface and a conical bearing surface with a head angle of approximately 82 degrees.

*Flat and Oval Countersunk Trim Heads:* Flat and oval countersunk trim heads are similar to the 82-degree flat and oval countersunk heads except that the size of head for a given size screw is one (large trim head) or two (small trim head) sizes smaller than the regular flat and oval countersunk head size. Oval countersunk trim heads have a definite radius where the curved top surface meets the conical bearing surface. Trim heads are furnished only in cross recessed types.

*Pan Head:* The slotted pan head has a flat top surface rounded into cylindrical sides and a flat bearing surface. The recessed pan head has a rounded top and a flat bearing surface. This head type is now preferred to the round head.

*Fillister Head:* The fillister head has a rounded top surface, cylindrical sides, and a flat bearing surface.

*Hex Head:* The hex head has a flat or indented top surface, six flat sides, and a flat bearing surface. Because the slotted hex head requires a secondary operation in manufacture which often results in burrs at the extremity of the slot that interfere with socket wrench engagement and the wrenching capability of the hex far exceeds that of the slot, it is not recommended for new designs.

*Hex Washer Head:* The hex washer head has an indented top surface and six flat sides formed integrally with a flat washer that projects beyond the sides and provides a flat bearing surface. Because the slotted hex washer head requires a secondary operation in manufacture, which often results in burrs at the extremity of the slot that often interferes with socket wrench engagement, and because the wrenching capability of the hex far exceeds that of the slot in the indented head, it is not recommended for new designs.

*Truss Head:* The truss head has a low rounded top surface with a flat bearing surface, the diameter of which for a given screw size is larger than the diameter of the corresponding round head. In the interest of product simplification and recognizing that the truss head is an inherently weak design, it is not recommended for new designs.

**Method of Designation.—**Tapping screws are designated by the following data in the sequence shown: Nominal size (number, fraction or decimal equivalent); threads per inch; nominal length (fraction or decimal equivalent); point type; product name, including head type and driving provision; material; and protective finish, if required. Examples:

$^1/_4$-14 × 1$^1/_2$ Type AB Slotted Pan Head Tapping Screw, Steel, Nickel Plated

6-32 × $^3/_4$ Type T, Type 1A Cross Recessed Pan Head Tapping Screw, Corrosion Resistant Steel

0.375-16 × 1.50 Type D, Washer Head Tapping Screw, Steel

*Metallic Drive Screws:* Type U metallic drive screws are designated by the following data in the sequence shown: Nominal size (number, fraction, or decimal equivalent); nominal length (fraction or decimal equivalent); product name, including head type; material; and protective finish, if required. Examples:

10 × $^5/_{16}$ Round Head Metallic Drive Screw, Steel

0.312 × 0.50 Round Head Metallic Drive Screw, Steel, Zinc Plated

**Table 1. ANSI Standard Threads and Points for Thread Forming Self-Tapping Screws** *ANSI/ASME B18.6.3-2013*

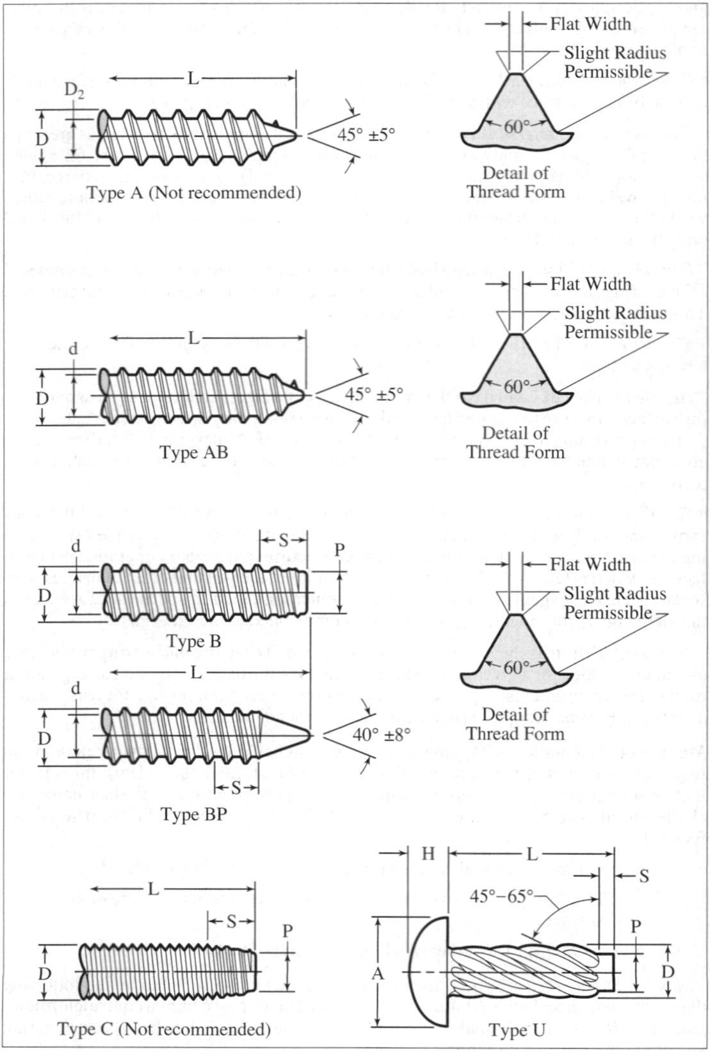


See Table 3, Table 5, and Table 6 for thread data.

**Table 2. ANSI Standard Threads and Points for Thread Cutting Self-Tapping Screws** *ANSI/ASME B18.6.3-2013*

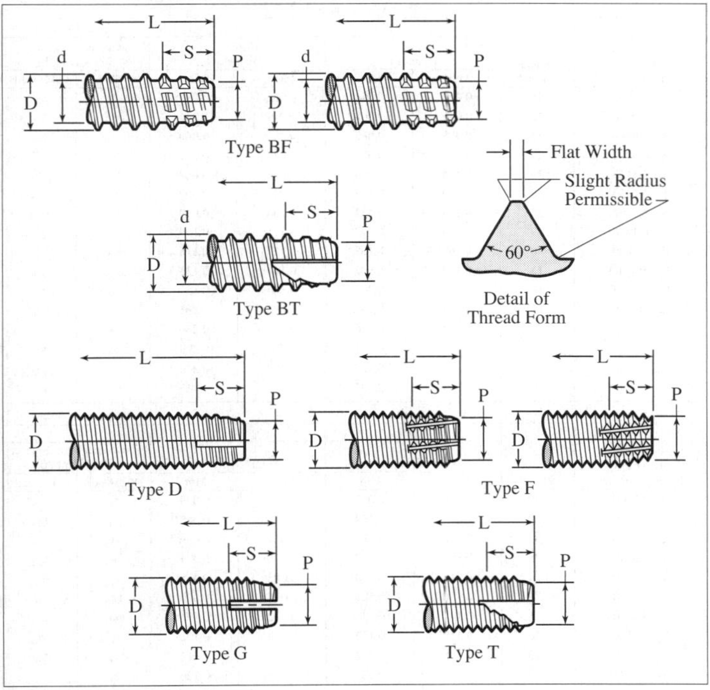


See Table 5 and Table 7 for thread data.

**Cross Recesses.—**Type I cross recess has a large center opening, tapered wings, and blunt bottom, with all edges relieved or rounded. Type IA cross recess has a large center opening, wide straight wings, and blunt bottom, with all edges relieved or rounded. Type II consists of two intersecting slots with parallel sides converging to a slightly truncated apex at the bottom of the recess. Type III has a square center opening, slightly tapered side walls, and a conical bottom, with top edges relieved or rounded.

**Table 3. ANSI Standard Cross Recesses for Self-Tapping Screws** *ANSI/ASME B18.6.3-2013* **and Metric Thread Forming and Thread Cutting Tapping Screws** *ANSI/ASME B18.6.5M-2000 (Withdrawn)*

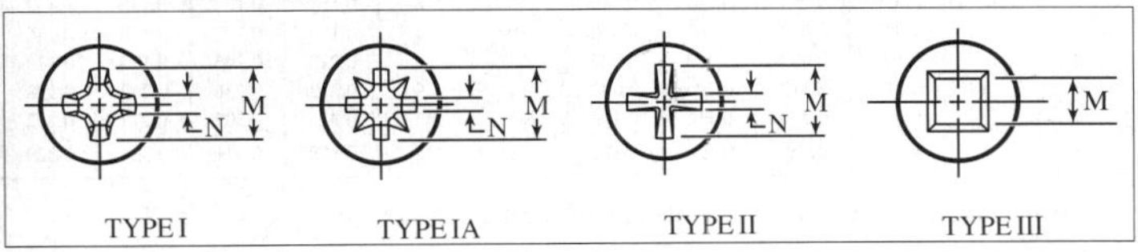

**Table 4. ANSI Standard Thread and Point Dimensions for Types AB, A and U Thread Forming Tapping Screws** *ANSI/ASME B18.6.3-2013*

| Type AB (Formerly BA) | | | | | | | |
|---|---|---|---|---|---|---|---|
| | | D | | d | | L | |
| Nominal Size or Basic Screw Diameter | Threads per inch | Major Diameter | | Minor Diameter | | Minimum Practical Screw Lengths | |
| | | Max. | Min. | Max. | Min. | 90° Heads | Csk. Heads |
| 0 0.0600 | 48 | 0.060 | 0.054 | 0.036 | 0.033 | 1/8 | 5/32 |
| 1 0.0730 | 42 | 0.075 | 0.069 | 0.049 | 0.046 | 5/32 | 3/16 |
| **2 0.0860** | **32** | **0.088** | **0.082** | **0.064** | **0.060** | 3/16 | 7/32 |
| 3 0.0990 | 28 | 0.101 | 0.095 | 0.075 | 0.071 | 3/16 | 1/4 |
| **4 0.1120** | **24** | **0.114** | **0.108** | **0.086** | **0.082** | 7/32 | 9/32 |
| 5 0.1250 | 20 | 0.130 | 0.123 | 0.094 | 0.090 | 1/4 | 5/16 |
| **6 0.1380** | **20** | **0.139** | **0.132** | **0.104** | **0.099** | 9/32 | 11/32 |
| 7 0.1510 | 19 | 0.154 | 0.147 | 0.115 | 0.109 | 5/16 | 3/8 |
| **8 0.1640** | **18** | **0.166** | **0.159** | **0.122** | **0.116** | 5/16 | 3/8 |
| **10 0.1900** | **16** | **0.189** | **0.182** | **0.141** | **0.135** | 3/8 | 7/16 |
| 12 0.2160 | 14 | 0.215 | 0.208 | 0.164 | 0.157 | 7/16 | 21/32 |
| **1/4 0.2500** | **14** | **0.246** | **0.237** | **0.192** | **0.185** | 1/2 | 19/32 |
| 5/16 0.3125 | 12 | 0.315 | 0.306 | 0.244 | 0.236 | 5/8 | 3/4 |
| 3/8 0.3750 | 12 | 0.380 | 0.371 | 0.309 | 0.299 | 3/4 | 29/32 |
| 7/16 0.4375 | 10 | 0.440 | 0.429 | 0.359 | 0.349 | 7/8 | 1 1/32 |
| 1/2 0.5000 | 10 | 0.504 | 0.493 | 0.423 | 0.413 | 1 | 1 5/32 |
| Type A | | | | | | | |
| | | D | | d | | L | |
| Nominal Size[a] Basic Screw Diameter | Threads per inch | Major Diameter | | Minor Diameter | | These Lengths or Shorter — Use Type AB | |
| | | Max. | Min. | Max. | Min. | 90° Heads | Csk. Heads |
| 0 0.0600 | 40 | 0.060 | 0.057 | 0.042 | 0.039 | 1/8 | 3/16 |
| 1 0.0730 | 32 | 0.075 | 0.072 | 0.051 | 0.048 | 1/8 | 3/16 |
| 2 0.0860 | 32 | 0.088 | 0.084 | 0.061 | 0.056 | 5/32 | 3/16 |
| 3 0.0990 | 28 | 0.101 | 0.097 | 0.076 | 0.071 | 3/16 | 7/32 |
| 4 0.1120 | 24 | 0.114 | 0.110 | 0.083 | 0.078 | 3/16 | 1/4 |
| 5 0.1250 | 20 | 0.130 | 0.126 | 0.095 | 0.090 | 3/16 | 1/4 |
| 6 0.1380 | 18 | 0.141 | 0.136 | 0.102 | 0.096 | 1/4 | 5/16 |
| 7 0.1510 | 16 | 0.158 | 0.152 | 0.114 | 0.108 | 5/16 | 3/8 |
| 8 0.1640 | 15 | 0.168 | 0.162 | 0.123 | 0.116 | 3/8 | 7/16 |
| 10 0.1900 | 12 | 0.194 | 0.188 | 0.133 | 0.126 | 3/8 | 1/2 |
| 12 0.2160 | 11 | 0.221 | 0.215 | 0.162 | 0.155 | 7/16 | 9/16 |
| 14 0.2420 | 10 | 0.254 | 0.248 | 0.185 | 0.178 | 1/2 | 5/8 |
| 16 0.2680 | 10 | 0.280 | 0.274 | 0.197 | 0.189 | 9/16 | 3/4 |
| 18 0.2940 | 9 | 0.306 | 0.300 | 0.217 | 0.209 | 5/8 | 13/16 |
| 20 0.3200 | 9 | 0.333 | 0.327 | 0.234 | 0.226 | 11/16 | 13/16 |
| 24 0.3720 | 9 | 0.390 | 0.383 | 0.291 | 0.282 | 3/4 | 1 |

[a] Where specifying nominal size in decimals, zeros preceding decimal and in fourth place are omitted.

| Type U Metallic Drive Screws | | | | | | | | | | | |
|---|---|---|---|---|---|---|---|---|---|---|---|
| Nom. Size | No. of Starts | Out. Dia. | | Pilot Dia. | | Nom. Size | No. of Starts | Out. Dia. | | Pilot Dia. | |
| | | Max. | Min. | Max. | Min. | | | Max. | Min. | Max. | Min. |
| 00 | 6 | 0.060 | 0.057 | 0.049 | 0.046 | 8 | 8 | 0.167 | 0.162 | 0.136 | 0.132 |
| 0 | 6 | 0.075 | 0.072 | 0.063 | 0.060 | 10 | 8 | 0.182 | 0.177 | 0.150 | 0.146 |
| 2 | 8 | 0.100 | 0.097 | 0.083 | 0.080 | 12 | 8 | 0.212 | 0.206 | 0.177 | 0.173 |
| 4 | 7 | 0.116 | 0.112 | 0.096 | 0.092 | 14 | 9 | 0.242 | 0.236 | 0.202 | 0.198 |
| 6 | 7 | 0.140 | 0.136 | 0.116 | 0.112 | 5/16 | 11 | 0.315 | 0.309 | 0.272 | 0.267 |
| 7 | 8 | 0.154 | 0.150 | 0.126 | 0.122 | 3/8 | 12 | 0.378 | 0.371 | 0.334 | 0.329 |

All dimensions are in inches. See Table 1 for thread diagrams.

Sizes shown in bold face type are preferred. Type A screws are no longer recommended.

## Table 5. ANSI Standard Thread and Point Dimensions for B and BP Thread Forming and BF and BT Thread Cutting Tapping Screws

*ANSI/ASME B18.6.3-2013*

THREAD FORMING TYPES B AND BP

| Nominal Size[a] or Basic Screw Diameter | Thds per Inch[b] | D Major Diameter | | d Minor Diameter | | P Point Diameter[c] | | S Point Taper Length[d] | | L Minimum Practical Nominal Screw Lengths | | | |
|---|---|---|---|---|---|---|---|---|---|---|---|---|---|
| | | | | | | | | | | Type B | | Type BP | |
| | | Max. | Min. | Max. | Min. | Max. | Min. | Max. | Min. | 90° Heads | Csk Heads | 90° Heads | Csk Heads |
| 0 0.0600 | 48 | 0.060 | 0.054 | 0.036 | 0.033 | 0.031 | 0.027 | 0.042 | 0.031 | 1/8 | 1/8 | 5/32 | 3/16 |
| 1 0.0730 | 42 | 0.075 | 0.069 | 0.049 | 0.046 | 0.044 | 0.040 | 0.048 | 0.036 | 1/8 | 5/32 | 3/16 | 7/32 |
| 2 0.0860 | 32 | 0.088 | 0.082 | 0.064 | 0.060 | 0.058 | 0.054 | 0.062 | 0.047 | 5/32 | 3/16 | 1/4 | 9/32 |
| 3 0.0990 | 28 | 0.101 | 0.095 | 0.075 | 0.071 | 0.068 | 0.063 | 0.071 | 0.054 | 3/16 | 7/32 | 9/32 | 5/16 |
| 4 0.1120 | 24 | 0.114 | 0.108 | 0.086 | 0.082 | 0.079 | 0.074 | 0.083 | 0.063 | 3/16 | 1/4 | 5/16 | 11/32 |
| 5 0.1250 | 20 | 0.130 | 0.123 | 0.094 | 0.090 | 0.087 | 0.082 | 0.100 | 0.075 | 7/32 | 9/32 | 11/32 | 13/32 |
| 6 0.1380 | 20 | 0.139 | 0.132 | 0.104 | 0.099 | 0.095 | 0.089 | 0.100 | 0.075 | 1/4 | 9/32 | 3/8 | 7/16 |
| 7 0.1510 | 19 | 0.154 | 0.147 | 0.115 | 0.109 | 0.105 | 0.099 | 0.105 | 0.079 | 1/4 | 5/16 | 13/32 | 15/32 |
| 8 0.1640 | 18 | 0.166 | 0.159 | 0.122 | 0.116 | 0.112 | 0.106 | 0.111 | 0.083 | 9/32 | 11/32 | 7/16 | 1/2 |
| 10 0.1900 | 16 | 0.189 | 0.182 | 0.141 | 0.135 | 0.130 | 0.123 | 0.125 | 0.094 | 5/16 | 3/8 | 1/2 | 19/32 |
| 12 0.2160 | 14 | 0.215 | 0.208 | 0.164 | 0.157 | 0.152 | 0.145 | 0.143 | 0.107 | 11/32 | 7/16 | 9/16 | 21/32 |
| 1/4 0.2500 | 14 | 0.246 | 0.237 | 0.192 | 0.185 | 0.179 | 0.171 | 0.143 | 0.107 | 3/8 | 1/2 | 21/32 | 3/4 |
| 5/16 0.3125 | 12 | 0.315 | 0.306 | 0.244 | 0.236 | 0.230 | 0.222 | 0.167 | 0.125 | 15/32 | 19/32 | 27/32 | 31/32 |
| 3/8 0.3750 | 12 | 0.380 | 0.371 | 0.309 | 0.299 | 0.293 | 0.285 | 0.167 | 0.125 | 17/32 | 11/16 | 15/16 | 1 1/8 |
| 7/16 0.4375 | 10 | 0.440 | 0.429 | 0.359 | 0.349 | 0.343 | 0.335 | 0.200 | 0.150 | 5/8 | 25/32 | 1 1/8 | 1 1/4 |
| 1/2 0.5000 | 10 | 0.504 | 0.493 | 0.423 | 0.413 | 0.407 | 0.399 | 0.200 | 0.150 | 11/16 | 27/32 | 1 1/4 | 1 13/32 |

[a] Where specifying nominal size in decimals, zeros preceding decimal and in the fourth decimal place shall be omitted.

[b] The width of flat at crest of thread shall not exceed 0.004 inch for sizes up to No. 8, inclusive, and 0.006 inch for larger sizes.

[c] Point diameters specified apply to screw threads before roll threading.

[d] Points of screws are tapered and fluted or slotted. The flute on Type BT screws has an included angle of 90 to 95 degrees and the thread cutting edge is located above the axis of the screw. Flutes and slots extend through first full form thread beyond taper except for Type BF screw on which tapered threads may be complete at manufacturer's option and flutes may be one pitch short of first full form thread.

THREAD CUTTING TYPES BF AND BT[d]

| Nominal Size[a] or Basic Screw Diameter | Thds per Inch[b] | D Major Diameter | | d Minor Diameter | | P Point Diameter[c] | | S Point Taper Length[d] | | L Minimum Practical Nominal Screw Lengths | |
|---|---|---|---|---|---|---|---|---|---|---|---|
| | | Max. | Min. | Max. | Min. | Max. | Min. | Max. | Min. | 90° Heads | Csk Heads |
| 0 0.0600 | 48 | 0.060 | 0.054 | 0.036 | 0.033 | 0.031 | 0.027 | 0.042 | 0.031 | 1/8 | 1/8 |
| 1 0.0730 | 42 | 0.075 | 0.069 | 0.049 | 0.046 | 0.044 | 0.040 | 0.048 | 0.036 | 1/8 | 5/32 |
| 2 0.0860 | 32 | 0.088 | 0.082 | 0.064 | 0.060 | 0.058 | 0.054 | 0.062 | 0.047 | 5/32 | 3/16 |
| 3 0.0990 | 28 | 0.101 | 0.095 | 0.075 | 0.071 | 0.068 | 0.063 | 0.071 | 0.054 | 3/16 | 7/32 |
| 4 0.1120 | 24 | 0.114 | 0.108 | 0.086 | 0.082 | 0.079 | 0.074 | 0.083 | 0.063 | 3/16 | 1/4 |
| 5 0.1250 | 20 | 0.130 | 0.123 | 0.094 | 0.090 | 0.087 | 0.082 | 0.100 | 0.075 | 7/32 | 9/32 |
| 6 0.1380 | 20 | 0.139 | 0.132 | 0.104 | 0.099 | 0.095 | 0.089 | 0.100 | 0.075 | 1/4 | 9/32 |
| 7 0.1510 | 19 | 0.154 | 0.147 | 0.115 | 0.109 | 0.105 | 0.099 | 0.105 | 0.079 | 1/4 | 5/16 |
| 8 0.1640 | 18 | 0.166 | 0.159 | 0.122 | 0.116 | 0.112 | 0.106 | 0.111 | 0.083 | 9/32 | 11/32 |
| 10 0.1900 | 16 | 0.189 | 0.182 | 0.141 | 0.135 | 0.130 | 0.123 | 0.125 | 0.094 | 5/16 | 3/8 |
| 12 0.2160 | 14 | 0.215 | 0.208 | 0.164 | 0.157 | 0.152 | 0.145 | 0.143 | 0.107 | 11/32 | 7/16 |
| 1/4 0.2500 | 14 | 0.246 | 0.237 | 0.192 | 0.185 | 0.179 | 0.171 | 0.143 | 0.107 | 3/8 | 1/2 |
| 5/16 0.3125 | 12 | 0.315 | 0.306 | 0.244 | 0.236 | 0.230 | 0.222 | 0.167 | 0.125 | 15/32 | 19/32 |
| 3/8 0.3750 | 12 | 0.380 | 0.371 | 0.309 | 0.299 | 0.293 | 0.285 | 0.167 | 0.125 | 17/32 | 11/16 |
| 7/16 0.4375 | 10 | 0.440 | 0.429 | 0.359 | 0.349 | 0.343 | 0.335 | 0.200 | 0.150 | 5/8 | 25/32 |
| 1/2 0.5000 | 10 | 0.504 | 0.493 | 0.423 | 0.413 | 0.407 | 0.399 | 0.200 | 0.150 | 11/16 | 27/32 |

All dimensions are in inches. See Table 1 and Table 2 for thread diagrams.

**Table 6. Thread and Point Dimensions for Type C Thread Forming Tapping Screws** *ANSI/ASME B18.6.3-2013, Appendix*

| Nominal Size[a] or Basic Screw Diameter | Threads per inch | D | | P | | S | | | | L | | | |
|---|---|---|---|---|---|---|---|---|---|---|---|---|---|
| | | Major Diameter | | Point Diameter[b] | | Point Taper Length[c] | | | | Determinant Lengths for Point Taper[c] | | Minimum Practical Nominal Screw Lengths | |
| | | | | | | For Short Screws | | For Long Screws | | | | | |
| | | Max. | Min. | Max. | Min. | Max. | Min. | Max. | Min. | 90° Heads | Csk Heads | 90° Heads | Csk Heads |
| 2 0.0860 | 56 | 0.0860 | 0.0813 | 0.068 | 0.061 | 0.062 | 0.045 | 0.080 | 0.062 | 5/32 | 3/16 | 5/32 | 3/16 |
| 2 0.0860 | 64 | 0.0860 | 0.0816 | 0.070 | 0.064 | 0.055 | 0.039 | 0.070 | 0.055 | 1/8 | 3/16 | 1/8 | 5/32 |
| 3 0.0990 | 48 | 0.0990 | 0.0938 | 0.078 | 0.070 | 0.073 | 0.052 | 0.094 | 0.073 | 3/16 | 7/32 | 5/32 | 7/32 |
| 3 0.0990 | 56 | 0.0990 | 0.0942 | 0.081 | 0.074 | 0.062 | 0.045 | 0.080 | 0.062 | 5/32 | 3/16 | 5/32 | 3/16 |
| 4 0.1120 | 40 | 0.1120 | 0.1061 | 0.087 | 0.078 | 0.088 | 0.062 | 0.112 | 0.088 | 7/32 | 1/4 | 3/16 | 1/4 |
| 4 0.1120 | 48 | 0.1120 | 0.1068 | 0.091 | 0.083 | 0.073 | 0.052 | 0.094 | 0.073 | 3/16 | 7/32 | 5/32 | 7/32 |
| 5 0.1250 | 40 | 0.1250 | 0.1191 | 0.100 | 0.091 | 0.088 | 0.062 | 0.112 | 0.088 | 7/32 | 9/32 | 3/16 | 1/4 |
| 5 0.1250 | 44 | 0.1250 | 0.1195 | 0.102 | 0.094 | 0.080 | 0.057 | 0.102 | 0.080 | 3/16 | 1/4 | 3/16 | 1/4 |
| 6 0.1380 | 32 | 0.1380 | 0.1312 | 0.107 | 0.096 | 0.109 | 0.078 | 0.141 | 0.109 | 1/4 | 5/16 | 1/4 | 5/16 |
| 6 0.1380 | 40 | 0.1380 | 0.1321 | 0.113 | 0.104 | 0.088 | 0.062 | 0.112 | 0.088 | 7/32 | 9/32 | 3/16 | 1/4 |
| 8 0.1640 | 32 | 0.1640 | 0.1571 | 0.132 | 0.122 | 0.109 | 0.078 | 0.141 | 0.109 | 1/4 | 11/32 | 1/4 | 5/16 |
| 8 0.1640 | 36 | 0.1640 | 0.1577 | 0.136 | 0.126 | 0.097 | 0.069 | 0.125 | 0.097 | 7/32 | 5/16 | 7/32 | 9/32 |
| 10 0.1900 | 24 | 0.1900 | 0.1818 | 0.148 | 0.135 | 0.146 | 0.104 | 0.188 | 0.146 | 11/32 | 7/16 | 5/16 | 13/32 |
| 10 0.1900 | 32 | 0.1900 | 0.1831 | 0.158 | 0.148 | 0.109 | 0.078 | 0.141 | 0.109 | 1/4 | 11/32 | 1/4 | 5/16 |
| 12 0.2160 | 24 | 0.2160 | 0.2078 | 0.174 | 0.161 | 0.146 | 0.104 | 0.188 | 0.146 | 11/32 | 7/16 | 5/16 | 13/32 |
| 12 0.2160 | 28 | 0.2160 | 0.2085 | 0.180 | 0.168 | 0.125 | 0.089 | 0.161 | 0.125 | 5/16 | 13/32 | 9/32 | 3/8 |
| 1/4 0.2500 | 20 | 0.2500 | 0.2408 | 0.200 | 0.184 | 0.175 | 0.125 | 0.225 | 0.175 | 13/32 | 17/32 | 3/8 | 1/2 |
| 1/4 0.2500 | 28 | 0.2500 | 0.2425 | 0.214 | 0.202 | 0.125 | 0.089 | 0.161 | 0.125 | 5/16 | 13/32 | 9/32 | 3/8 |
| 5/16 0.3125 | 18 | 0.3125 | 0.3026 | 0.257 | 0.239 | 0.194 | 0.139 | 0.250 | 0.194 | 15/32 | 19/32 | 7/16 | 9/16 |
| 5/16 0.3125 | 24 | 0.3125 | 0.3042 | 0.271 | 0.257 | 0.146 | 0.104 | 0.188 | 0.146 | 11/32 | 15/32 | 5/16 | 15/32 |
| 3/8 0.3750 | 16 | 0.3750 | 0.3643 | 0.312 | 0.293 | 0.219 | 0.156 | 0.281 | 0.219 | 1/2 | 11/16 | 15/32 | 5/8 |
| 3/8 0.3750 | 24 | 0.3750 | 0.3667 | 0.333 | 0.319 | 0.146 | 0.104 | 0.188 | 0.146 | 11/32 | 1/2 | 5/16 | 1/2 |
| 7/16 0.4375 | 14 | 0.4375 | 0.4258 | 0.366 | 0.344 | 0.250 | 0.179 | 0.321 | 0.250 | 19/32 | 3/4 | 9/16 | 23/32 |
| 7/16 0.4375 | 20 | 0.4375 | 0.4281 | 0.387 | 0.371 | 0.175 | 0.125 | 0.225 | 0.175 | 13/32 | 9/16 | 3/8 | 17/32 |
| 1/2 0.5000 | 13 | 0.5000 | 0.4876 | 0.423 | 0.399 | 0.269 | 0.192 | 0.346 | 0.269 | 5/8 | 25/32 | 19/32 | 3/4 |
| 1/2 0.5000 | 20 | 0.5000 | 0.4906 | 0.450 | 0.433 | 0.175 | 0.125 | 0.225 | 0.175 | 13/32 | 9/16 | 3/8 | 17/32 |

[a] Where specifying nominal size in decimals, zeros preceding decimal and in the fourth decimal place shall be omitted.

[b] The tabulated values apply to screw blanks before roll threading.

[c] Screws of these nominal lengths and shorter shall have point taper length specified above for short screws. Longer lengths shall have point taper length specified for long screws.

All dimensions are in inches. See Table 1 for thread diagrams. Type C is not recommended for new designs.

Tapered threads shall have unfinished crests.

## Table 7. ANSI Standard Thread and Point Dimensions for Types D, F, G, and T Thread Cutting Tapping Screws *ANSI/ASME B18.6.3-2013*

| Nominal Size[a] or Basic Screw Diameter | Threads per inch | D | | P | | S | | | | L | | | |
|---|---|---|---|---|---|---|---|---|---|---|---|---|---|
| | | Major Diameter | | Point Diameter[b] | | Point Taper Length[c] | | | | Determinant Lengths for Point Taper[c] | | Minimum Practical Nominal Screw Lengths | |
| | | | | | | For Short Screws | | For Long Screws | | | | | |
| | | Max. | Min. | Max. | Min. | Max. | Min. | Max. | Min. | 90° Heads | Csk Heads | 90° Heads | Csk Heads |
| 2 0.0860 | 56 | 0.0860 | 0.0813 | 0.068 | 0.061 | 0.062 | 0.045 | 0.080 | 0.062 | 5/32 | 3/16 | 5/32 | 3/16 |
| 2 0.0860 | 64 | 0.0860 | 0.0816 | 0.070 | 0.064 | 0.055 | 0.039 | 0.070 | 0.055 | 1/8 | 3/16 | 1/8 | 5/32 |
| 3 0.0990 | 48 | 0.0990 | 0.0938 | 0.078 | 0.070 | 0.073 | 0.052 | 0.094 | 0.073 | 3/16 | 7/32 | 5/32 | 7/32 |
| 3 0.0990 | 56 | 0.0990 | 0.0942 | 0.081 | 0.074 | 0.062 | 0.045 | 0.080 | 0.062 | 5/32 | 3/16 | 5/32 | 3/16 |
| 4 0.1120 | 40 | 0.1120 | 0.1061 | 0.087 | 0.078 | 0.088 | 0.062 | 0.112 | 0.088 | 7/32 | 1/4 | 3/16 | 1/4 |
| 4 0.1120 | 48 | 0.1120 | 0.1068 | 0.091 | 0.083 | 0.073 | 0.052 | 0.094 | 0.073 | 3/16 | 7/32 | 5/32 | 7/32 |
| 5 0.1250 | 40 | 0.1250 | 0.1191 | 0.100 | 0.091 | 0.088 | 0.062 | 0.112 | 0.088 | 7/32 | 9/32 | 3/16 | 1/4 |
| 5 0.1250 | 44 | 0.1250 | 0.1195 | 0.102 | 0.094 | 0.080 | 0.057 | 0.102 | 0.080 | 3/16 | 1/4 | 3/16 | 1/4 |
| 6 0.1380 | 32 | 0.1380 | 0.1312 | 0.107 | 0.096 | 0.109 | 0.078 | 0.141 | 0.109 | 1/4 | 5/16 | 1/4 | 5/16 |
| 6 0.1380 | 40 | 0.1380 | 0.1321 | 0.113 | 0.104 | 0.088 | 0.062 | 0.112 | 0.088 | 7/32 | 9/32 | 3/16 | 1/4 |
| 8 0.1640 | 32 | 0.1640 | 0.1571 | 0.132 | 0.122 | 0.109 | 0.078 | 0.141 | 0.109 | 1/4 | 11/32 | 1/4 | 5/16 |
| 8 0.1640 | 36 | 0.1640 | 0.1577 | 0.136 | 0.126 | 0.097 | 0.069 | 0.125 | 0.097 | 7/32 | 5/16 | 7/32 | 9/32 |
| 10 0.1900 | 24 | 0.1900 | 0.1818 | 0.148 | 0.135 | 0.146 | 0.104 | 0.188 | 0.146 | 11/32 | 7/16 | 5/16 | 13/32 |
| 10 0.1900 | 32 | 0.1900 | 0.1831 | 0.158 | 0.148 | 0.109 | 0.078 | 0.141 | 0.109 | 1/4 | 11/32 | 1/4 | 5/16 |
| 12 0.2160 | 24 | 0.2160 | 0.2078 | 0.174 | 0.161 | 0.146 | 0.104 | 0.188 | 0.146 | 11/32 | 7/16 | 5/16 | 13/32 |
| 12 0.2160 | 28 | 0.2160 | 0.2085 | 0.180 | 0.168 | 0.125 | 0.089 | 0.161 | 0.125 | 5/16 | 13/32 | 9/32 | 3/8 |
| 1/4 0.2500 | 20 | 0.2500 | 0.2408 | 0.200 | 0.184 | 0.175 | 0.125 | 0.225 | 0.175 | 13/32 | 17/32 | 3/8 | 1/2 |
| 1/4 0.2500 | 28 | 0.2500 | 0.2425 | 0.214 | 0.202 | 0.125 | 0.089 | 0.161 | 0.125 | 5/16 | 13/32 | 9/32 | 3/8 |
| 5/16 0.3125 | 18 | 0.3125 | 0.3026 | 0.257 | 0.239 | 0.194 | 0.139 | 0.250 | 0.194 | 15/32 | 19/32 | 7/16 | 9/16 |
| 5/16 0.3125 | 24 | 0.3125 | 0.3042 | 0.271 | 0.257 | 0.146 | 0.104 | 0.188 | 0.146 | 11/32 | 15/32 | 5/16 | 15/32 |
| 3/8 0.3750 | 16 | 0.3750 | 0.3643 | 0.312 | 0.293 | 0.219 | 0.156 | 0.281 | 0.219 | 1/2 | 11/16 | 15/32 | 5/8 |
| 3/8 0.3750 | 24 | 0.3750 | 0.3667 | 0.333 | 0.319 | 0.146 | 0.104 | 0.188 | 0.146 | 11/32 | 1/2 | 5/16 | 1/2 |
| 7/16 0.4375 | 14 | 0.4375 | 0.4258 | 0.366 | 0.344 | 0.250 | 0.179 | 0.321 | 0.250 | 19/32 | 3/4 | 9/16 | 23/32 |
| 7/16 0.4375 | 20 | 0.4375 | 0.4281 | 0.387 | 0.371 | 0.175 | 0.125 | 0.225 | 0.175 | 13/32 | 9/16 | 3/8 | 17/32 |
| 1/2 0.5000 | 13 | 0.5000 | 0.4876 | 0.423 | 0.399 | 0.269 | 0.192 | 0.346 | 0.269 | 5/8 | 25/32 | 19/32 | 3/4 |
| 1/2 0.5000 | 20 | 0.5000 | 0.4906 | 0.450 | 0.433 | 0.175 | 0.125 | 0.225 | 0.175 | 13/32 | 9/16 | 3/8 | 17/32 |

[a] Where specifying nominal size in decimals, zeros preceding decimal and in the fourth decimal place shall be omitted.

[b] The tabulated values apply to screw blanks before roll threading.

[c] Screws of these nominal lengths and shorter shall have point taper length specified above for short screws. Longer lengths shall have point taper length specified for long screws.

All dimensions are in inches. See Table 2 for thread diagrams.

"Type D" otherwise designated "Type 1."

"Type T" otherwise designated "Type 23."

## Table 8. Approximate Hole Sizes for Type A Steel Thread Forming Screws

| In Steel, Stainless Steel, Monel Metal, Brass, and Aluminum Sheet Metal | | | | | | | | | |
|---|---|---|---|---|---|---|---|---|---|
| | | Hole Size | | | | | Hole Size | | |
| Screw Size | Metal Thick-ness | Pierced or Extruded | Drilled or Clean Punched | Drill Size | Screw Size | Metal Thick-ness | Pierced or Extruded | Drilled or Clean Punched | Drill Size |
| 4 | 0.015 | … | 0.086 | 44 | 8 | 0.024 | 0.136 | 0.125 | 1/8 |
| | 0.018 | … | 0.086 | 44 | | 0.030 | 0.136 | 0.125 | 1/8 |
| | 0.024 | 0.098 | 0.094 | 42 | | 0.036 | 0.136 | 0.125 | 1/8 |
| | 0.030 | 0.098 | 0.094 | 42 | | 0.048 | 0.136 | 0.128 | 30 |
| | 0.036 | 0.098 | 0.098 | 40 | 10 | 0.018 | … | 0.136 | 29 |
| 6 | 0.015 | … | 0.104 | 37 | | 0.024 | 0.157 | 0.136 | 29 |
| | 0.018 | … | 0.104 | 37 | | 0.030 | 0.157 | 0.136 | 29 |
| | 0.024 | 0.111 | 0.104 | 37 | | 0.036 | 0.157 | 0.136 | 29 |
| | 0.030 | 0.111 | 0.104 | 37 | | 0.048 | 0.157 | 0.149 | 25 |
| | 0.036 | 0.111 | 0.106 | 36 | 12 | 0.024 | … | 0.161 | 20 |
| 7 | 0.015 | … | 0.116 | 32 | | 0.030 | 0.185 | 0.161 | 20 |
| | 0.018 | … | 0.116 | 32 | | 0.036 | 0.185 | 0.161 | 20 |
| | 0.024 | 0.120 | 0.116 | 32 | | 0.048 | 0.185 | 0.161 | 20 |
| | 0.030 | 0.120 | 0.116 | 32 | 14 | 0.024 | … | 0.185 | 13 |
| | 0.036 | 0.120 | 0.116 | 32 | | 0.030 | 0.209 | 0.189 | 12 |
| | 0.048 | 0.120 | 0.120 | 31 | | 0.036 | 0.209 | 0.191 | 11 |
| 8 | 0.018 | … | 0.125 | 1/8 | | 0.048 | 0.209 | 0.196 | 9 |

| In Plywood (Resin Impregnated) | | | | | | In Asbestos Compositions | | | | | |
|---|---|---|---|---|---|---|---|---|---|---|---|
| | | | | Penetration in Blind Holes | | | | | | Penetration in Blind Holes | |
| Screw Size | Hole Size | Drill Size | Min. Mat'l Thickness | Min. | Max. | Screw Size | Hole Size | Drill Size | Min. Mat'l Thickness | Min. | Max. |
| 4 | 0.098 | 40 | 0.188 | 0.250 | 0.750 | 4 | 0.094 | 42 | 0.188 | 0.250 | 0.750 |
| 6 | 0.110 | 35 | 0.188 | 0.250 | 0.750 | 6 | 0.106 | 36 | 0.188 | 0.250 | 0.750 |
| 7 | 0.128 | 30 | 0.250 | 0.312 | 0.750 | 7 | 0.125 | 1/8 | 0.250 | 0.312 | 0.750 |
| 8 | 0.140 | 28 | 0.250 | 0.312 | 0.750 | 8 | 0.136 | 29 | 0.250 | 0.312 | 0.750 |
| 10 | 0.170 | 18 | 0.312 | 0.375 | 1.000 | 10 | 0.161 | 20 | 0.312 | 0.375 | 1.000 |
| 12 | 0.189 | 12 | 0.312 | 0.375 | 1.000 | 12 | 0.185 | 13 | 0.312 | 0.375 | 1.000 |
| 14 | 0.228 | 1 | 0.438 | 0.500 | 1.000 | 14 | 0.213 | 3 | 0.438 | 0.500 | 1.000 |

Type A is not recommended, use Type AB.
See footnote at bottom of Table 9.

## Table 9. Approximate Hole Sizes for Type C Steel Thread Forming Screws

| In Sheet Steel | | | | | | | | | | | |
|---|---|---|---|---|---|---|---|---|---|---|---|
| Screw Size | Metal Thick-ness | Hole Size | Drill Size | Screw Size | Metal Thick-ness | Hole Size | Drill Size | Screw Size | Metal Thick-ness | Hole Size | Drill Size |
| 4-40 | 0.037 | 0.094 | 42 | 10-24 | 0.037 | 0.154 | 23 | 1/4-20 | 0.037 | 0.221 | 2 |
| | 0.048 | 0.094 | 42 | | 0.048 | 0.161 | 20 | | 0.048 | 0.221 | 2 |
| | 0.062 | 0.096 | 41 | | 0.062 | 0.166 | 19 | | 0.062 | 0.228 | 1 |
| | 0.075 | 0.100 | 39 | | 0.075 | 0.170 | 18 | | 0.075 | 0.234 | A |
| | 0.105 | 0.102 | 38 | | 0.105 | 0.173 | 17 | | 0.105 | 0.234 | A |
| | 0.134 | 0.102 | 38 | | 0.134 | 0.177 | 16 | | 0.134 | 0.236 | 6 mm |
| 6-32 | 0.037 | 0.113 | 33 | 10-32 | 0.037 | 0.170 | 18 | 1/4-28 | 0.037 | 0.224 | 5.7 mm |
| | 0.048 | 0.116 | 32 | | 0.048 | 0.170 | 18 | | 0.048 | 0.228 | 1 |
| | 0.062 | 0.116 | 32 | | 0.062 | 0.170 | 18 | | 0.062 | 0.232 | 5.9 mm |
| | 0.075 | 0.122 | 3.1 mm | | 0.075 | 0.173 | 17 | | 0.075 | 0.234 | A |
| | 0.105 | 0.125 | 1/8 | | 0.105 | 0.177 | 16 | | 0.105 | 0.238 | B |
| | 0.134 | 0.125 | 1/8 | | 0.134 | 0.177 | 16 | | 0.134 | 0.238 | B |
| 8-32 | 0.037 | 0.136 | 29 | 12-24 | 0.037 | 0.189 | 12 | 5/16-18 | 0.037 | 0.290 | L |
| | 0.048 | 0.144 | 27 | | 0.048 | 0.194 | 10 | | 0.048 | 0.290 | L |
| | 0.062 | 0.144 | 27 | | 0.062 | 0.194 | 10 | | 0.062 | 0.290 | L |
| | 0.075 | 0.147 | 26 | | 0.075 | 0.199 | 8 | | 0.075 | 0.295 | M |
| | 0.105 | 0.150 | 25 | | 0.105 | 0.199 | 8 | | 0.105 | 0.295 | M |
| | 0.134 | 0.150 | 25 | | 0.134 | 0.199 | 8 | | 0.134 | 0.295 | M |

All dimensions are in inches except drill sizes. It may be necessary to vary the hole size to suit a particular application.

Type C is not recommended for new designs.

**Table 10. Approximate Pierced or Extruded Hole Sizes for Types AB, B, and BP Steel Thread Forming Screws**

| Screw Size | Metal Thickness | Pierced or Extruded Hole Size | Screw Size | Metal Thickness | Pierced or Extruded Hole Size | Screw Size | Metal Thickness | Pierced or Extruded Hole Size |
|---|---|---|---|---|---|---|---|---|
| In Steel, Stainless Steel, Monel Metal, and Brass Sheet Metal | | | | | | | | |
| 4 | 0.015 | 0.086 | 7 | 0.024 | 0.120 | 10 | 0.030 | 0.157 |
| | 0.018 | 0.086 | | 0.030 | 0.120 | | 0.036 | 0.157 |
| | 0.024 | 0.098 | | 0.036 | 0.120 | | 0.048 | 0.157 |
| | 0.030 | 0.098 | | 0.048 | 0.120 | 12 | 0.024 | 0.185 |
| | 0.036 | 0.098 | 8 | 0.018 | 0.136 | | 0.030 | 0.185 |
| 6 | 0.015 | 0.111 | | 0.024 | 0.136 | | 0.036 | 0.185 |
| | 0.018 | 0.111 | | 0.030 | 0.136 | | 0.048 | 0.185 |
| | 0.024 | 0.111 | | 0.036 | 0.136 | ¼ | 0.030 | 0.209 |
| | 0.030 | 0.111 | | 0.048 | 0.136 | | 0.036 | 0.209 |
| | 0.036 | 0.111 | 10 | 0.018 | 0.157 | | 0.048 | 0.209 |
| 7 | 0.018 | 0.120 | | 0.024 | 0.157 | … | … | … |
| In Aluminum Alloy Sheet Metal | | | | | | | | |
| 4 | 0.024 | 0.086 | 6 | 0.048 | 0.111 | 8 | 0.036 | 0.136 |
| | 0.030 | 0.086 | 7 | 0.024 | 0.120 | | 0.048 | 0.136 |
| | 0.036 | 0.086 | | 0.030 | 0.120 | 10 | 0.024 | 0.157 |
| | 0.048 | 0.086 | | 0.036 | 0.120 | | 0.030 | 0.157 |
| 6 | 0.024 | 0.111 | | 0.048 | 0.120 | | 0.036 | 0.157 |
| | 0.030 | 0.111 | 8 | 0.024 | 0.136 | | 0.048 | 0.157 |
| | 0.036 | 0.111 | | 0.030 | 0.136 | … | … | … |

All dimensions are in inches except whole number screw and drill sizes.

Since conditions differ widely, it may be necessary to vary the hole size to suit a particular application.

**Table 11. Drilled Hole Sizes for Types AB, B, and BP Steel Thread Forming Screws**

| Screw Size | Hole Size | Drill Size | Min. Mat'l Thickness | Penetration in Blind Holes Min. | Penetration in Blind Holes Max. | Screw Size | Hole Size | Drill Size | Min. Mat'l Thickness | Penetration in Blind Holes Min. | Penetration in Blind Holes Max. |
|---|---|---|---|---|---|---|---|---|---|---|---|
| In Plywood (Resin Impregnated) | | | | | | In Asbestos Compositions | | | | | |
| 2 | 0.073 | 49 | 0.125 | 0.188 | 0.500 | 2 | 0.076 | 48 | 0.125 | 0.188 | 0.500 |
| 4 | 0.100 | 39 | 0.188 | 0.250 | 0.625 | 4 | 0.101 | 38 | 0.188 | 0.250 | 0.625 |
| 6 | 0.125 | ⅛ | 0.188 | 0.250 | 0.625 | 6 | 0.120 | 31 | 0.188 | 0.250 | 0.625 |
| 7 | 0.136 | 29 | 0.188 | 0.250 | 0.750 | 7 | 0.136 | 29 | 0.250 | 0.312 | 0.750 |
| 8 | 0.144 | 27 | 0.188 | 0.250 | 0.750 | 8 | 0.147 | 26 | 0.312 | 0.375 | 0.750 |
| 10 | 0.173 | 17 | 0.250 | 0.312 | 1.000 | 10 | 0.166 | 19 | 0.312 | 0.375 | 1.000 |
| 12 | 0.194 | 10 | 0.312 | 0.375 | 1.000 | 12 | 0.196 | 9 | 0.312 | 0.375 | 1.000 |
| ¼ | 0.228 | 1 | 0.312 | 0.375 | 1.000 | ¼ | 0.228 | 1 | 0.438 | 0.500 | 1.000 |
| In Aluminum, Magnesium, Zinc, Brass, and Bronze Castings[a] | | | | | | In Phenol Formaldehyde Plastics[a] | | | | | |
| 2 | 0.078 | 47 | … | 0.125 | … | 2 | 0.078 | 47 | … | 0.188 | … |
| 4 | 0.104 | 37 | … | 0.188 | … | 4 | 0.100 | 39 | … | 0.250 | … |
| 6 | 0.128 | 30 | … | 0.250 | … | 6 | 0.128 | 30 | … | 0.250 | … |
| 7 | 0.144 | 27 | … | 0.250 | … | 7 | 0.136 | 29 | … | 0.250 | … |
| 8 | 0.152 | 24 | … | 0.250 | … | 8 | 0.150 | 25 | … | 0.312 | … |
| 10 | 0.177 | 16 | … | 0.250 | … | 10 | 0.177 | 16 | … | 0.312 | … |
| 12 | 0.199 | 8 | … | 0.281 | … | 12 | 0.199 | 8 | … | 0.375 | … |
| ¼ | 0.234 | 15⁄64 | … | 0.312 | … | ¼ | 0.234 | 15⁄64 | … | 0.375 | … |
| In Cellulose Acetate and Nitrate, and Acrylic and Styrene Resins[a] | | | | | | | | | | | |
| 2 | 0.078 | 47 | … | 0.188 | … | 8 | 0.144 | 27 | … | 0.312 | … |
| 4 | 0.094 | 42 | … | 0.250 | … | 10 | 0.170 | 18 | … | 0.312 | … |
| 6 | 0.120 | 31 | … | 0.250 | … | 12 | 0.191 | 11 | … | 0.375 | … |
| 7 | 0.128 | 30 | … | 0.250 | … | ¼ | 0.221 | 2 | … | 0.375 | … |

[a] Data below apply to Types B and BP only.

All dimensions are in inches except whole number screw and drill sizes.

Since conditions differ widely, it may be necessary to vary the hole size to suit a particular application.

## Table 12a. Approximate Drilled or Clean-Punched Hole Sizes for Types AB, B, and BP Steel Thread Forming Screws

| Screw Size | Metal Thick-ness | Hole Size | Drill Size | Screw Size | Metal Thick-ness | Hole Size | Drill Size | Screw Size | Metal Thick-ness | Hole Size | Drill Size |
|---|---|---|---|---|---|---|---|---|---|---|---|
| In Steel, Stainless Steel, Monel Metal, and Brass Sheet Metal | | | | | | | | | | | |
| 2 | 0.015 | 0.064 | 52 | 7 | 0.018 | 0.116 | 32 | 10 | 0.125 | 0.170 | 18 |
| | 0.018 | 0.064 | 52 | | 0.024 | 0.116 | 32 | | 0.135 | 0.170 | 18 |
| | 0.024 | 0.067 | 51 | | 0.030 | 0.116 | 32 | | 0.164 | 0.173 | 17 |
| | 0.030 | 0.070 | 50 | | 0.036 | 0.116 | 32 | 12 | 0.024 | 0.166 | 19 |
| | 0.036 | 0.073 | 49 | | 0.048 | 0.120 | 31 | | 0.030 | 0.166 | 19 |
| | 0.048 | 0.073 | 49 | | 0.060 | 0.128 | 30 | | 0.036 | 0.166 | 19 |
| | 0.060 | 0.076 | 48 | | 0.075 | 0.136 | 29 | | 0.048 | 0.170 | 18 |
| 4 | 0.015 | 0.086 | 44 | | 0.105 | 0.140 | 28 | | 0.060 | 0.177 | 16 |
| | 0.018 | 0.086 | 44 | 8 | 0.024 | 0.125 | 1/8 | | 0.075 | 0.182 | 14 |
| | 0.024 | 0.089 | 43 | | 0.030 | 0.125 | 1/8 | | 0.105 | 0.185 | 13 |
| | 0.030 | 0.094 | 42 | | 0.036 | 0.125 | 1/8 | | 0.125 | 0.196 | 9 |
| | 0.036 | 0.094 | 42 | | 0.048 | 0.128 | 30 | | 0.135 | 0.196 | 9 |
| | 0.048 | 0.096 | 41 | | 0.060 | 0.136 | 29 | | 0.164 | 0.201 | 7 |
| | 0.060 | 0.100 | 39 | | 0.075 | 0.140 | 28 | 1/4 | 0.030 | 0.194[a] | 10[a] |
| | 0.075 | 0.102 | 38 | | 0.105 | 0.150 | 25 | | 0.036 | 0.194[a] | 10[a] |
| 6 | 0.015 | 0.104 | 37 | | 0.125 | 0.150 | 25 | | 0.048 | 0.194[a] | 10[a] |
| | 0.018 | 0.104 | 37 | | 0.135 | 0.152 | 24 | | 0.060 | 0.199[a] | 8[a] |
| | 0.024 | 0.106 | 36 | 10 | 0.024 | 0.144 | 27 | | 0.075 | 0.204[a] | 6[a] |
| | 0.030 | 0.106 | 36 | | 0.030 | 0.144 | 27 | | 0.105 | 0.209 | 4 |
| | 0.036 | 0.110 | 35 | | 0.036 | 0.147 | 26 | | 0.125 | 0.228 | 1 |
| | 0.048 | 0.111 | 34 | | 0.048 | 0.152[a] | 24[a] | | 0.135 | 0.228 | 1 |
| | 0.060 | 0.116 | 32 | | 0.060 | 0.152[a] | 24[a] | | 0.164 | 0.234 | 15/64 |
| | 0.075 | 0.120 | 31 | | 0.075 | 0.157 | 22 | | 0.187 | 0.234 | 15/64 |
| | 0.105 | 0.128 | 30 | | 0.105 | 0.161 | 20 | | 0.194 | 0.234 | 15/64 |
| In Aluminum Alloy Sheet Metal | | | | | | | | | | | |
| 2 | 0.024 | 0.064 | 52 | 7 | 0.060 | 0.120 | 31 | 10 | 0.164 | 0.159 | 21 |
| | 0.030 | 0.064 | 52 | | 0.075 | 0.128 | 30 | | 0.200 | | |
| | 0.036 | 0.064 | 52 | | 0.105 | 0.136 | 29 | | to | 0.166 | 19 |
| | 0.048 | 0.067 | 51 | | 0.128 | | | | 0.375 | | |
| | 0.060 | 0.070 | 50 | | to | 0.136 | 29 | 12 | 0.048 | 0.161 | 20 |
| 4 | 0.030 | 0.086 | 44 | | 0.250 | | | | 0.060 | 0.166 | 19 |
| | 0.036 | 0.086 | 44 | 8 | 0.030 | 0.116 | 32 | | 0.075 | 0.173 | 17 |
| | 0.048 | 0.086 | 44 | | 0.036 | 0.120 | 31 | | 0.105 | 0.180 | 15 |
| | 0.060 | 0.089 | 43 | | 0.048 | 0.128 | 30 | | 0.125 | 0.182 | 14 |
| | 0.075 | 0.089 | 43 | | 0.060 | 0.136 | 29 | | 0.135 | 0.182 | 14 |
| | 0.105 | 0.094 | 42 | | 0.075 | 0.140 | 28 | | 0.164 | 0.189 | 12 |
| 6 | 0.030 | 0.104 | 37 | | 0.105 | 0.147 | 26 | | 0.200 | | |
| | 0.036 | 0.104 | 37 | | 0.125 | 0.147 | 26 | | to | 0.196 | 9 |
| | 0.048 | 0.104 | 37 | | 0.135 | 0.149 | 25 | | 0.375 | | |
| | 0.060 | 0.106 | 36 | | 0.162 | | | 1/4 | 0.060 | 0.199 | 8 |
| | 0.75 | 0.110 | 35 | | to | 0.152 | 24 | | 0.075 | 0.201 | 7 |
| | 0.105 | 0.111 | 34 | | 0.375 | | | | 0.105 | 0.204 | 6 |
| | 0.128 | | | 10 | 0.036 | 0.144 | 27 | | 0.125 | 0.209 | 4 |
| | to | 120 | 31 | | 0.048 | 0.144 | 27 | | 0.135 | 0.209 | 4 |
| | 250 | | | | 0.060 | 0.144 | 27 | | 0.164 | 0.213 | 3 |
| 7 | 0.030 | 0.113 | 33 | | 0.075 | 0.147 | 26 | | 0.187 | 0.213 | 3 |
| | 0.036 | 0.113 | 33 | | 0.105 | 0.147 | 26 | | 0.194 | 0.221 | 2 |
| | 0.048 | 0.116 | 32 | | 0.125 | 0.154 | 23 | | 0.200 | | |
| | | | | | 0.135 | 0.154 | 23 | | to | 228 | 1 |
| | | | | | | | | | 0.375 | | |

[a] For Types B and BP only; for Type AB see concluded Table 12b following.

Since conditions differ widely, it may be necessary to vary the hole size to suit a particular application. Hole sizes for metal thicknesses above 0.075 inch are for Types B and BP only.

## Table 12b. Supplementary Data for Types AB Thread Forming Screws in Steel, Stainless Steel, Monel Metal, and Brass Sheet Metal

| Screw Size | Metal Thick-ness | Hole Size | Drill Size | Screw Size | Metal Thick-ness | Hole Size | Drill Size | Screw Size | Metal Thick-ness | Hole Size | Drill Size |
|---|---|---|---|---|---|---|---|---|---|---|---|
| In Steel, Stainless Steel, Monel Metal, and Brass Sheet Metal | | | | | | | | | | | |
| 10 | 0.018 | 0.144 | 27 | ¼ | 0.018 | 0.196 | 9 | ¼ | 0.048 | 0.205 | 5 |
| 10 | 0.048 | 0.149 | 25 | ¼ | 0.024 | 0.196 | 9 | ¼ | 0.060 | 0.228 | 1 |
| 10 | 0.060 | 0.154 | 23 | ¼ | 0.030 | 0.196 | 9 | ¼ | 0.075 | 0.232 | 5.9 mm |
| … | … | … | … | ¼ | 0.036 | 0.196 | 9 | … | … | … | … |

All dimensions are in inches except numbered screw and drill sizes.

## Table 13. Approximate Hole Sizes for Types D, F, G, and T Steel Thread Cutting Screws in Sheet Metals

| Screw Size | Thick-ness | Steel Hole Size | Steel Drill Size | Aluminum Alloy Hole Size | Aluminum Alloy Drill Size |
|---|---|---|---|---|---|
| 2-56 | 0.050 | 0.073 | 49 | 0.070 | 50 |
| | 0.060 | 0.073 | 49 | 0.073 | 49 |
| | 0.083 | 0.073 | 49 | 0.073 | 49 |
| | 0.109 | 0.073 | 49 | 0.073 | 49 |
| | 0.125 | 0.076 | 48 | 0.073 | 49 |
| | 0.140 | 0.076 | 48 | 0.073 | 49 |
| 3-48 | 0.050 | 0.081 | 46 | 0.078 | 5/64 |
| | 0.060 | 0.081 | 46 | 0.081 | 46 |
| | 0.083 | 0.082 | 45 | 0.082 | 45 |
| | 0.109 | 0.086 | 44 | 0.082 | 45 |
| | 0.125 | 0.086 | 44 | 0.082 | 45 |
| | 0.140 | 0.086 | 44 | 0.086 | 44 |
| | 0.187 | 0.089 | 43 | 0.086 | 44 |
| 4-40 | 0.050 | 0.089 | 43 | 0.089 | 43 |
| | 0.060 | 0.089 | 43 | 0.089 | 43 |
| | 0.083 | 0.094 | 42 | 0.089 | 43 |
| | 0.109 | 0.096 | 41 | 0.094 | 42 |
| | 0.125 | 0.098 | 40 | 0.094 | 42 |
| | 0.140 | 0.098 | 40 | 0.094 | 3/32 |
| | 0.187 | 0.102 | 38 | 0.098 | 40 |
| 5-40 | 0.050 | 0.106 | 36 | 0.102 | 38 |
| | 0.060 | 0.106 | 36 | 0.102 | 38 |
| | 0.083 | 0.106 | 36 | 0.104 | 37 |
| | 0.109 | 0.106 | 36 | 0.104 | 37 |
| | 0.125 | 0.109 | 7/64 | 0.106 | 36 |
| | 0.140 | 0.110 | 35 | 0.106 | 36 |
| | 0.187 | 0.116 | 32 | 0.110 | 35 |
| | 0.250 | 0.116 | 32 | 0.113 | 33 |
| 6-32 | 0.050 | 0.110 | 35 | 0.109 | 7/64 |
| | 0.060 | 0.113 | 33 | 0.109 | 7/64 |
| | 0.083 | 0.116 | 32 | 0.111 | 34 |
| | 0.109 | 0.116 | 32 | 0.113 | 33 |
| | 0.125 | 0.116 | 32 | 0.116 | 32 |
| | 0.140 | 0.120 | 31 | 0.116 | 32 |
| | 0.187 | 0.125 | 1/8 | 0.120 | 31 |
| | 0.250 | 0.125 | 1/8 | 0.125 | 1/8 |
| 8-32 | 0.050 | 0.136 | 29 | 0.136 | 29 |
| | 0.060 | 0.140 | 28 | 0.136 | 29 |
| | 0.083 | 0.140 | 28 | 0.136 | 29 |
| | 0.109 | 0.144 | 27 | 0.140 | 28 |
| | 0.125 | 0.144 | 27 | 0.140 | 28 |
| | 0.140 | 0.147 | 26 | 0.144 | 27 |

| Screw Size | Thick-ness | Steel Hole Size | Steel Drill Size | Aluminum Alloy Hole Size | Aluminum Alloy Drill Size |
|---|---|---|---|---|---|
| 8-32 | 0.187 | 0.150 | 25 | 0.147 | 26 |
| | 0.250 | 0.150 | 25 | 0.150 | 25 |
| | 0.312 | 0.150 | 25 | 0.150 | 25 |
| 10-24 | 0.050 | 0.152 | 24 | 0.150 | 25 |
| | 0.060 | 0.154 | 23 | 0.152 | 24 |
| | 0.083 | 0.161 | 20 | 0.154 | 23 |
| | 0.109 | 0.161 | 20 | 0.157 | 22 |
| | 0.125 | 0.166 | 19 | 0.159 | 21 |
| | 0.140 | 0.170 | 18 | 0.161 | 20 |
| | 0.187 | 0.173 | 17 | 0.166 | 19 |
| | 0.250 | 0.173 | 17 | 0.172 | 11/64 |
| | 0.312 | 0.173 | 17 | 0.173 | 17 |
| | 0.375 | 0.173 | 17 | 0.173 | 17 |
| 10-32 | 0.050 | 0.159 | 21 | 0.161 | 20 |
| | 0.060 | 0.166 | 19 | 0.161 | 20 |
| | 0.083 | 0.166 | 19 | 0.161 | 20 |
| | 0.109 | 0.170 | 18 | 0.166 | 19 |
| | 0.125 | 0.170 | 18 | 0.166 | 19 |
| | 0.140 | 0.170 | 18 | 0.166 | 19 |
| | 0.187 | 0.177 | 16 | 0.172 | 11/64 |
| | 0.250 | 0.177 | 16 | 0.177 | 16 |
| | 0.312 | 0.177 | 16 | 0.177 | 16 |
| | 0.375 | 0.177 | 16 | 0.177 | 16 |
| 12-24 | 0.060 | 0.180 | 15 | 0.177 | 16 |
| | 0.083 | 0.182 | 14 | 0.180 | 15 |
| | 0.109 | 0.188 | 3/16 | 0.182 | 14 |
| | 0.125 | 0.191 | 11 | 0.185 | 13 |
| | 0.140 | 0.191 | 11 | 0.188 | 3/16 |
| | 0.187 | 0.199 | 8 | 0.191 | 11 |
| | 0.250 | 0.199 | 8 | 0.199 | 8 |
| | 0.312 | 0.199 | 8 | 0.199 | 8 |
| | 0.375 | 0.199 | 8 | 0.199 | 8 |
| | 0.500 | 0.199 | 8 | 0.199 | 8 |
| ¼-20 | 0.083 | 0.213 | 3 | 0.206 | 5 |
| | 0.109 | 0.219 | 7/32 | 0.209 | 4 |
| | 0.125 | 0.221 | 2 | 0.213 | 3 |
| | 0.140 | 0.221 | 2 | 0.213 | 3 |
| | 0.187 | 0.228 | 1 | 0.221 | 2 |
| | 0.250 | 0.228 | 1 | 0.228 | 1 |
| | 0.312 | 0.228 | 1 | 0.228 | 1 |
| | 0.375 | 0.228 | 1 | 0.228 | 1 |
| | 0.500 | 0.228 | 1 | 0.228 | 1 |

**Table 13.** *(Continued)* **Approximate Hole Sizes for Types D, F, G, and T Steel Thread Cutting Screws in Sheet Metals**

| Screw Size | Thick-ness | Steel Hole Size | Steel Drill Size | Aluminum Alloy Hole Size | Aluminum Alloy Drill Size | Screw Size | Thick-ness | Steel Hole Size | Steel Drill Size | Aluminum Alloy Hole Size | Aluminum Alloy Drill Size |
|---|---|---|---|---|---|---|---|---|---|---|---|
| 1/4-28 | 0.083 | 0.221 | 2 | 0.219 | 7/32 | 5/16-24 | 0.187 | 0.295 | M | 0.290 | L |
| | 0.109 | 0.228 | 1 | 0.221 | 2 | | 0.250 | 0.295 | M | 0.295 | M |
| | 0.125 | 0.228 | 1 | 0.221 | 2 | | 0.312 | 0.295 | M | 0.295 | M |
| | 0.140 | 0.234 | A | 0.221 | 2 | | 0.375 | 0.295 | M | 0.295 | M |
| | 0.187 | 0.234 | 15/64 | 0.228 | 1 | | 0.500 | 0.295 | M | 0.295 | M |
| | 0.250 | 0.234 | 15/64 | 0.234 | 15/64 | 3/8-16 | 0.125 | 0.339 | R | 0.328 | 21/64 |
| | 0.312 | 0.234 | 15/64 | 0.234 | 15/64 | | 0.140 | 0.339 | R | 0.332 | Q |
| | 0.375 | 0.234 | 15/64 | 0.234 | 15/64 | | 0.187 | 0.348 | S | 0.339 | R |
| | 0.500 | 0.234 | 15/64 | 0.234 | 15/64 | | 0.250 | 0.358 | T | 0.348 | S |
| 5/16-18 | 0.109 | 0.277 | J | 0.266 | H | | 0.312 | 0.358 | T | 0.348 | S |
| | 0.125 | 0.277 | J | 0.272 | I | | 0.375 | 0.358 | T | 0.348 | S |
| | 0.140 | 0.281 | 9/32 | 0.272 | I | | 0.500 | 0.358 | T | 0.348 | S |
| | 0.187 | 0.290 | L | 0.281 | K | 3/8-24 | 0.125 | 0.348 | S | 0.344 | 11/32 |
| | 0.250 | 0.290 | L | 0.290 | L | | 0.140 | 0.348 | S | 0.344 | 11/32 |
| | 0.312 | 0.290 | L | 0.290 | L | | 0.187 | 0.358 | T | 0.348 | S |
| | 0.375 | 0.290 | L | 0.290 | L | | 0.250 | 0.358 | T | 0.358 | T |
| | 0.500 | 0.290 | L | 0.290 | L | | 0.312 | 0.358 | T | 0.358 | T |
| 5/16-24 | 0.109 | 0.290 | L | 0.281 | K | | 0.375 | 0.358 | T | 0.358 | T |
| | 0.125 | 0.290 | L | 0.281 | 9/32 | | 0.500 | 0.358 | T | 0.358 | T |
| | 0.140 | 0.290 | L | 0.281 | 9/32 | | … | … | … | … | … |

All dimensions are in inches except numbered drill and screw sizes. It may be necessary to vary the hole size to suit a particular application.

**Table 14. Approximate Hole Sizes for Types D, F, G, and T Steel Thread Cutting Screws in Cast Metals and Plastics**

| Screw Size | Thick-ness | Cast Iron Hole Size | Cast Iron Drill Size | Zinc and Aluminum[a] Hole Size | Zinc and Aluminum[a] Drill Size | Screw Size | Thick-ness | Cast Iron Hole Size | Cast Iron Drill Size | Zinc and Aluminum[a] Hole Size | Zinc and Aluminum[a] Drill Size |
|---|---|---|---|---|---|---|---|---|---|---|---|
| 2-56 | 0.050 | 0.076 | 48 | 0.073 | 49 | 5-40 | 0.083 | 0.113 | 33 | 0.106 | 36 |
| | 0.060 | 0.076 | 48 | 0.073 | 49 | | 0.109 | 0.113 | 33 | 0.110 | 35 |
| | 0.083 | 0.076 | 48 | 0.076 | 48 | | 0.125 | 0.116 | 32 | 0.110 | 35 |
| | 0.109 | 0.078 | 5/64 | 0.076 | 48 | | 0.140 | 0.116 | 32 | 0.110 | 35 |
| | 0.125 | 0.078 | 5/64 | 0.076 | 48 | | 0.187 | 0.116 | 32 | 0.111 | 34 |
| | 0.140 | 0.078 | 5/64 | 0.076 | 48 | | 0.250 | 0.116 | 32 | 0.113 | 33 |
| 3-48 | 0.050 | 0.089 | 43 | 0.082 | 45 | 6-32 | 0.050 | 0.120 | 31 | 0.116 | 32 |
| | 0.060 | 0.089 | 43 | 0.082 | 45 | | 0.060 | 0.120 | 31 | 0.120 | 31 |
| | 0.083 | 0.089 | 43 | 0.082 | 45 | | 0.083 | 0.125 | 1/8 | 0.120 | 31 |
| | 0.109 | 0.089 | 43 | 0.086 | 44 | | 0.109 | 0.125 | 1/8 | 0.120 | 31 |
| | 0.125 | 0.089 | 43 | 0.089 | 43 | | 0.125 | 0.125 | 1/8 | 0.120 | 31 |
| | 0.140 | 0.094 | 42 | 0.089 | 43 | | 0.140 | 0.125 | 1/8 | 0.120 | 31 |
| | 0.187 | 0.094 | 42 | 0.089 | 43 | | 0.187 | 0.128 | 30 | 0.120 | 31 |
| 4-40 | 0.050 | 0.100 | 39 | 0.090 | 41 | | 0.250 | 0.128 | 30 | 0.120 | 31 |
| | 0.060 | 0.100 | 39 | 0.096 | 41 | 8-32 | 0.050 | 0.147 | 26 | 0.144 | 27 |
| | 0.083 | 0.102 | 38 | 0.096 | 41 | | 0.060 | 0.150 | 25 | 0.144 | 27 |
| | 0.109 | 0.102 | 38 | 0.096 | 41 | | 0.083 | 0.150 | 25 | 0.144 | 27 |
| | 0.125 | 0.102 | 38 | 0.100 | 39 | | 0.109 | 0.150 | 25 | 0.144 | 27 |
| | 0.140 | 0.102 | 38 | 0.100 | 39 | | 0.125 | 0.150 | 25 | 0.147 | 26 |
| | 0.187 | 0.104 | 37 | 0.100 | 39 | | 0.140 | 0.150 | 25 | 0.147 | 26 |
| 5-40 | 0.050 | 0.111 | 34 | 0.106 | 36 | | 0.187 | 0.154 | 23 | 0.147 | 26 |
| | 0.060 | 0.111 | 34 | 0.106 | 36 | | 0.250 | 0.154 | 23 | 0.150 | 25 |
| | | | | | | | 0.312 | 0.154 | 23 | 0.150 | 25 |

**Table 14.** *(Continued)* **Approximate Hole Sizes for Types D, F, G, and T Steel Thread Cutting Screws in Cast Metals and Plastics**

| Screw Size | Thick- ness | Cast Iron Hole Size | Cast Iron Drill Size | Zinc and Aluminum[a] Hole Size | Zinc and Aluminum[a] Drill Size |
|---|---|---|---|---|---|
| 10-24 | 0.050 | 0.170 | 18 | 0.161 | 20 |
| | 0.060 | 0.170 | 18 | 0.166 | 19 |
| | 0.083 | 0.172 | 11/64 | 0.166 | 19 |
| | 0.109 | 0.173 | 17 | 0.166 | 19 |
| | 0.125 | 0.173 | 17 | 0.166 | 19 |
| | 0.140 | 0.173 | 17 | 0.166 | 19 |
| | 0.187 | 0.177 | 16 | 0.170 | 18 |
| | 0.250 | 0.177 | 16 | 0.170 | 18 |
| | 0.312 | 0.177 | 16 | 0.172 | 11/64 |
| | 0.375 | 0.177 | 16 | 0.172 | 11/64 |
| 10-32 | 0.050 | 0.173 | 17 | 0.170 | 18 |
| | 0.060 | 0.173 | 17 | 0.170 | 18 |
| | 0.083 | 0.177 | 16 | 0.172 | 11/64 |
| | 0.109 | 0.177 | 16 | 0.172 | 11/64 |
| | 0.125 | 0.177 | 16 | 0.172 | 11/64 |
| | 0.140 | 0.177 | 16 | 0.172 | 11/64 |
| | 0.187 | 0.180 | 15 | 0.172 | 11/64 |
| | 0.250 | 0.180 | 15 | 0.173 | 17 |
| | 0.312 | 0.180 | 15 | 0.173 | 17 |
| | 0.375 | 0.180 | 15 | 0.177 | 16 |
| 12-24 | 0.060 | 0.196 | 9 | 0.189 | 12 |
| | 0.083 | 0.199 | 8 | 0.191 | 11 |
| | 0.109 | 0.199 | 8 | 0.191 | 11 |
| | 0.125 | 0.199 | 8 | 0.191 | 11 |
| | 0.140 | 0.199 | 8 | 0.194 | 10 |
| | 0.187 | 0.203 | 13/64 | 0.194 | 10 |
| | 0.250 | 0.204 | 6 | 0.196 | 9 |
| | 0.312 | 0.204 | 6 | 0.196 | 9 |
| | 0.375 | 0.204 | 6 | 0.199 | 8 |
| | 0.500 | 0.204 | 6 | 0.199 | 8 |
| ¼-20 | 0.083 | 0.228 | 1 | 0.219 | 7/32 |
| | 0.109 | 0.228 | 1 | 0.219 | 7/32 |
| | 0.125 | 0.228 | 1 | 0.221 | 2 |
| | 0.140 | 0.228 | 1 | 0.221 | 2 |
| | 0.187 | 0.234 | 15/64 | 0.221 | 2 |
| | 0.250 | 0.234 | 15/64 | 0.228 | 1 |
| | 0.312 | 0.234 | 15/64 | 0.228 | 1 |
| | 0.375 | 0.234 | 15/64 | 0.228 | 1 |
| | 0.500 | 0.234 | 15/64 | 0.228 | 1 |

| Screw Size | Thick- ness | Cast Iron Hole Size | Cast Iron Drill Size | Zinc and Aluminum[a] Hole Size | Zinc and Aluminum[a] Drill Size |
|---|---|---|---|---|---|
| ¼-28 | 0.083 | 0.234 | A | 0.228 | 1 |
| | 0.109 | 0.234 | 15/64 | 0.228 | 1 |
| | 0.125 | 0.234 | 15/64 | 0.228 | 1 |
| | 0.140 | 0.234 | 15/64 | 0.228 | 1 |
| | 0.187 | 0.238 | B | 0.228 | 1 |
| | 0.250 | 0.238 | B | 0.234 | A |
| | 0.312 | 0.238 | B | 0.234 | A |
| | 0.375 | 0.238 | B | 0.234 | 15/64 |
| | 0.500 | 0.238 | B | 0.234 | 15/64 |
| 5/16-18 | 0.109 | 0.290 | L | 0.277 | J |
| | 0.125 | 0.290 | L | 0.281 | K |
| | 0.140 | 0.290 | L | 0.281 | K |
| | 0.187 | 0.295 | M | 0.281 | 9/32 |
| | 0.250 | 0.295 | M | 0.281 | 9/32 |
| | 0.312 | 0.295 | M | 0.290 | L |
| | 0.375 | 0.295 | M | 0.290 | L |
| | 0.500 | 0.295 | M | 0.290 | L |
| 5/16-24 | 0.109 | 0.295 | M | 0.290 | L |
| | 0.125 | 0.295 | M | 0.290 | L |
| | 0.140 | 0.295 | M | 0.290 | L |
| | 0.187 | 0.302 | N | 0.290 | L |
| | 0.250 | 0.302 | N | 0.290 | L |
| | 0.312 | 0.302 | N | 0.295 | M |
| | 0.375 | 0.302 | N | 0.295 | M |
| | 0.500 | 0.302 | N | 0.295 | M |
| ⅜-16 | 0.125 | 0.348 | S | 0.339 | R |
| | 0.140 | 0.348 | S | 0.339 | R |
| | 0.187 | 0.348 | S | 0.339 | R |
| | 0.250 | 0.348 | S | 0.344 | 11/32 |
| | 0.312 | 0.348 | S | 0.344 | 11/32 |
| | 0.375 | 0.348 | S | 0.348 | S |
| | 0.500 | 0.348 | S | 0.348 | S |
| ⅜-24 | 0.125 | 0.358 | T | 0.348 | S |
| | 0.140 | 0.358 | T | 0.348 | S |
| | 0.187 | 0.358 | T | 0.348 | S |
| | 0.250 | 0.358 | T | 0.358 | T |
| | 0.312 | 0.358 | T | 0.358 | T |
| | 0.375 | 0.358 | T | 0.358 | T |
| | 0.500 | 0.358 | T | 0.358 | T |

[a] Die Castings

| Screw Size | Phenol Formaldehyde[a] Hole Size | Phenol Formaldehyde[a] Drill Size | Depth of Penetration Min. | Depth of Penetration Max. | Cellulose Acetate, Cellulose Nitrate, Acrylic Resin, and Styrene Resin[a] Hole Size | Drill Size | Depth of Penetration Min. | Depth of Penetration Max. |
|---|---|---|---|---|---|---|---|---|
| 2-56 | 0.078 | 5/64 | 0.219 | 0.375 | 0.076 | 48 | 0.219 | 0.375 |
| 3-48 | 0.089 | 43 | 0.219 | 0.375 | 0.086 | 44 | 0.219 | 0.375 |
| 4-40 | 0.098 | 40 | 0.250 | 0.312 | 0.093 | 42 | 0.250 | 0.312 |
| 5-40 | 0.113 | 33 | 0.250 | 0.438 | 0.110 | 35 | 0.250 | 0.438 |
| 6-32 | 0.116 | 32 | 0.250 | 0.312 | 0.116 | 32 | 0.250 | 0.312 |
| 8-32 | 0.144 | 27 | 0.312 | 0.500 | 0.144 | 27 | 0.312 | 0.500 |
| 10-24 | 0.161 | 20 | 0.375 | 0.500 | 0.161 | 20 | 0.375 | 0.500 |
| 10-32 | 0.166 | 19 | 0.375 | 0.500 | 0.166 | 19 | 0.375 | 0.500 |
| ¼-20 | 0.228 | 1 | 0.375 | 0.625 | 0.228 | 1 | 0.375 | 1.000 |

[a] Plastics

For footnotes see Table 13.

## Table 15. Approximate Hole Sizes for Types BF and BT Steel Thread Cutting Screws in Cast Metals

| In Die-Cast Zinc and Aluminum | | | | | | | |
|---|---|---|---|---|---|---|---|
| Screw Size | Thickness | Hole Size | Drill Size | Screw Size | Thickness | Hole Size | Drill Size |
| 2 | 0.060 | 0.073 | 49 | 10 | 0.125 | 0.166 | 19 |
| | 0.083 | 0.073 | 49 | | 0.140 | 0.166 | 19 |
| | 0.109 | 0.076 | 48 | | 0.188 | 0.166 | 19 |
| | 0.125 | 0.076 | 48 | | 0.250 | 0.170 | 18 |
| | 0.140 | 0.076 | 48 | | 0.312 | 0.172 | 11/64 |
| 3 | 0.060 | 0.086 | 44 | | 0.375 | 0.172 | 11/64 |
| | 0.083 | 0.086 | 44 | 12 | 0.125 | 0.191 | 11 |
| | 0.109 | 0.086 | 44 | | 0.140 | 0.191 | 11 |
| | 0.125 | 0.086 | 44 | | 0.188 | 0.191 | 11 |
| | 0.140 | 0.089 | 43 | | 0.250 | 0.196 | 9 |
| | 0.188 | 0.089 | 43 | | 0.312 | 0.196 | 9 |
| 4 | 0.109 | 0.098 | 40 | | 0.375 | 0.196 | 9 |
| | 0.125 | 0.100 | 39 | 1/4 | 0.125 | 0.221 | 2 |
| | 0.140 | 0.100 | 39 | | 0.140 | 0.221 | 2 |
| | 0.188 | 0.100 | 39 | | 0.188 | 0.221 | 2 |
| | 0.250 | 0.102 | 38 | | 0.250 | 0.228 | 1 |
| 5 | 0.109 | 0.111 | 34 | | 0.312 | 0.228 | 1 |
| | 0.125 | 0.111 | 34 | | 0.375 | 0.228 | 1 |
| | 0.140 | 0.113 | 33 | 5/16 | 0.125 | 0.281 | K |
| | 0.188 | 0.113 | 33 | | 0.140 | 0.281 | K |
| | 0.250 | 0.116 | 32 | | 0.188 | 0.281 | K |
| 6 | 0.125 | 0.120 | 31 | | 0.250 | 0.281 | K |
| | 0.140 | 0.120 | 31 | | 0.312 | 0.290 | L |
| | 0.188 | 0.120 | 31 | | 0.375 | 0.290 | L |
| | 0.250 | 0.125 | 1/8 | 3/8 | 0.125 | 0.344 | 11/32 |
| | 0.312 | 0.125 | 1/8 | | 0.140 | 0.344 | 11/32 |
| 8 | 0.125 | 0.149 | 25 | | 0.188 | 0.344 | 11/32 |
| | 0.140 | 0.149 | 25 | | 0.250 | 0.344 | 11/32 |
| | 0.188 | 0.149 | 25 | | 0.312 | 0.348 | S |
| | 0.250 | 0.152 | 24 | | 0.375 | 0.348 | S |
| | 0.312 | 0.152 | 24 | | … | … | … |

All dimensions are in inches except numbered drill and screw sizes. It may be necessary to vary the hole size to suit a particular application.

## Table 16. Approximate Hole Size for Types BF and BT Steel Thread Cutting Screws in Plastics

| | Phenol Formaldehyde | | | | Cellulose Acetate, Cellulose Nitrate, Acrylic Resin and Styrene Resin | | | |
|---|---|---|---|---|---|---|---|---|
| | | | Depth of Penetration | | | | Depth of Penetration | |
| Screw Size | Hole Size | Drill Size | Min. | Max. | Hole Size | Drill Size | Min. | Max. |
| 2 | 0.078 | 5/64 | 0.094 | 0.250 | 0.076 | 48 | 0.094 | 0.250 |
| 3 | 0.089 | 43 | 0.125 | 0.312 | 0.089 | 43 | 0.125 | 0.312 |
| 4 | 0.104 | 37 | 0.125 | 0.312 | 0.100 | 39 | 0.125 | 0.312 |
| 5 | 0.116 | 32 | 0.188 | 0.375 | 0.113 | 33 | 0.188 | 0.375 |
| 6 | 0.125 | 1/8 | 0.188 | 0.375 | 0.120 | 31 | 0.188 | 0.375 |
| 8 | 0.147 | 26 | 0.250 | 0.500 | 0.144 | 27 | 0.250 | 0.500 |
| 10 | 0.170 | 18 | 0.312 | 0.625 | 0.166 | 19 | 0.312 | 0.625 |
| 12 | 0.194 | 10 | 0.375 | 0.625 | 0.189 | 12 | 0.375 | 0.625 |
| 1/4 | 0.228 | 1 | 0.375 | 0.750 | 0.221 | 2 | 0.375 | 0.750 |

For footnotes see above table.

**Table 17. Approximate Hole Sizes for Type U Hardened Steel Metallic Drive Screws**

| In Ferrous and Nonferrous Castings, Sheet Metals, Plastics, Plywood (Resin-Impregnated) and Fiber | | | | | | | | |
|---|---|---|---|---|---|---|---|---|
| Screw Size | Hole Size | Drill Size | Screw Size | Hole Size | Drill Size | Screw Size | Hole Size | Drill Size |
| 00 | .052 | 55 | 6 | .120 | 31 | 12 | .191 | 11 |
| 0 | .067 | 51 | 7 | .136 | 29 | 14 | .221 | 2 |
| 2 | .086 | 44 | 8 | .144 | 27 | 5⁄16 | .295 | M |
| 4 | .104 | 37 | 10 | .161 | 20 | 3⁄8 | .358 | T |

All dimensions are in inches except whole number screw and drill sizes and letter drill sizes.

**Table 18. ANSI Standard Torsional Strength Requirements for Tapping Screws**
*ANSI/ASME B18.6.3-2013*

| Nom. Screw Size | Type A | Types AB, B, BF, BP, and BT | Types C, D, F, G, and T | | Nom-Screw Size | Type A | Types AB, B, BF, BP, and BT | Types C, D, F, G, and T | |
|---|---|---|---|---|---|---|---|---|---|
| | | | Coarse Thread | Fine Thread | | | | Coarse Thread | Fine Thread |
| 2 | 4 | 4 | 5 | 6 | 1⁄4 | … | 142 | 140 | 179 |
| 3 | 9 | 9 | 9 | 10 | 16 | 152 | … | … | … |
| 4 | 12 | 13 | 13 | 15 | 18 | 196 | … | … | … |
| 5 | 18 | 18 | 18 | 20 | 5⁄16 | … | 290 | 306 | 370 |
| 6 | 24 | 24 | 23 | 27 | 20 | 250 | … | … | … |
| 7 | 30 | 30 | … | … | 24 | 492 | … | … | … |
| 8 | 39 | 39 | 42 | 47 | 3⁄8 | … | 590 | 560 | 710 |
| 10 | 48 | 56 | 56 | 74 | 7⁄16 | … | 620 | 700 | 820 |
| 12 | 83 | 88 | 93 | 108 | 1⁄2 | … | 1020 | 1075 | 1285 |
| 14 | 125 | … | … | … | … | … | … | … | … |

Torsional strength data are in pound-inches.

**Self-Tapping Thread Inserts.—**Self-tapping screw thread inserts are essentially hard bushings with internal and external threads. The internal threads conform to Unified and American Standard classes 2B and 3B, depending on the type of insert used. The external thread has cutting edges on the end that provide the self-tapping feature. These inserts may be used in magnesium, aluminum, cast iron, zinc, plastics, and other materials. Self-tapping inserts are made of case-hardened carbon steel, stainless steel, and brass, the brass type being designed specifically for installation in wood.

**Screw Thread Inserts.—**Screw thread inserts are helically formed coils of diamond-shaped stainless steel or phosphor bronze wire that screw into a threaded hole to form a mating internal thread for a screw or stud. These inserts provide a convenient means of repairing stripped-out threads and are also used to provide stronger threads in soft materials such as aluminum, zinc die castings, wood, magnesium, etc. than can be obtained by direct tapping of the base metal involved.

According to the Heli-Coil Corp., conventional design practice in specifying boss diameters or edge distances can usually be applied since the major diameter of a hole tapped to receive a thread insert is not much larger than the major diameter of thread the insert provides.

Screw thread inserts are available in thread sizes from 4-40 to $1\frac{1}{2}$-6 inch National and Unified Coarse Thread Series and in 6-40 to $1\frac{1}{2}$-12 sizes in the fine-thread series. When used in conjunction with appropriate taps and gages, screw thread inserts will meet requirements of 2, 2B, 3, and 3B thread classes.

**ANSI Standard Metric Thread Forming and Thread Cutting Tapping Screws.—**Table 1 shows the various types of metric thread forming and thread cutting screw threads covered by ANSI/ASME B18.6.5M-2000 (Withdrawn). The designations of the American National Standards Institute are shown.

**Table 1. ANSI Standard Threads and Points for Metric Thread Forming and Thread Cutting Tapping Screws** *ANSI/ASME B18.6.5M-2000 (Withdrawn)*

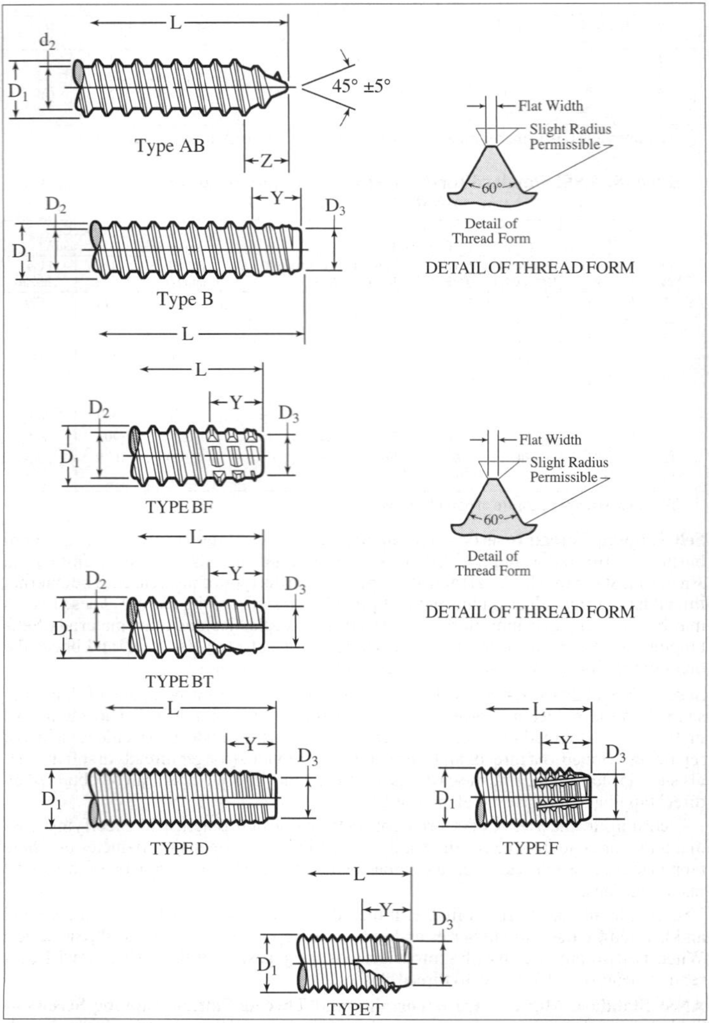


See Table 3 and Table 4 for thread data.

*Thread Forming Tapping Screws:* These types are generally for application in materials where large internal stresses are permissible or desirable, to increase resistance to loosening. These screws have the following descriptions and applications:

*Type AB:* Spaced thread screw with gimlet point primarily intended for use in thin metal, resin impregnated plywood, and asbestos compositions.

*Type B:* Spaced thread screw with a blunt point that has tapered entering threads with unfinished crests and same pitches as Type AB. Used for thin metal, nonferrous castings, resin impregnated plywood, certain resilient plastics, and asbestos compositions.

*Thread Cutting Tapping Screws:* These screws are generally for application in materials where disruptive internal stresses are undesirable or where excessive driving torques are encountered with thread forming tapping screws. These screws have the following descriptions and applications:

*Types BF and BT:* Spaced threads with blunt point and tapered entering threads having unfinished crests, as on Type B, with one or more cutting edges or chip cavities, intended for use in plastics, asbestos compositions, and other similar materials.

*Types D, F, and T:* Tapping screws with threads of machine screw diameter-pitch combinations (metric coarse thread series) approximating a 60 degree basic thread form (not necessarily conforming to any standard thread profile) with a blunt point and tapered entering threads with unfinished crests and having one or more cutting edges and chip cavities, intended for use in materials such as aluminum, zinc, and lead die castings; steel sheets and shapes; cast iron; brass; and plastics.

**ANSI Standard Head Types for Metric Thread Forming and Cutting Tapping Screws.—**The head types covered by ANSI/ASME B18.6.5M-2000 (Withdrawn) include those commonly applicable to metric tapping screws and are described as follows:

*Flat Countersunk Head:* The flat countersunk head has a flat top surface and a conical bearing surface with a head angle of 90 to 92 degrees.

*Oval Countersunk Head:* The oval countersunk head has a rounded top surface and a conical bearing surface with a head angle of 90 to 92 degrees.

*Pan Head:* The slotted pan head has a flat top surface rounding into cylindrical sides and a flat bearing surface. The recessed pan head has a rounded top surface blending into cylindrical sides and a flat bearing surface.

*Hex Head:* The hex head has a flat or indented top surface, six flat sides, and a flat bearing surface.

*Hex Flange Head:* The hex flange head has a flat or indented top surface and six flat sides formed integrally with a frustroconical or slightly rounded (convex) flange that projects beyond the sides and provides a flat bearing surface.

**Method of Designation.—**Metric tapping screws are designated with the following data, preferably in the sequence shown: Nominal size; thread pitch; nominal length; thread and point type; product name, including head style and driving provision; material; and protective finish, if required.

*Examples:*

6.3 × 1.8 × 30 Type AB, Slotted Pan Head Tapping Screw, Steel, Zinc Plated

6 × 1 × 20 Type T, Type 1A Cross Recessed Pan Head Tapping Screw, Corrosion Resistant Steel

4.2 × 1.4 × 13 Type BF, Type 1 Cross Recessed Oval Countersunk Head Tapping Screw, Steel, Chromium Plated

10 × 1.5 × 40 Type D, Hex Flange Head Tapping Screw, Steel

## Table 2. Recommended Nominal Screw Lengths for Metric Tapping Screws *ANSI/ASME B18.6.5M-2000 (Withdrawn)*

| Nominal Screw Length | Nominal Screw Size for Types AB, B, BF, and BT | | | | | | | | | |
|---|---|---|---|---|---|---|---|---|---|---|
| | 2.2 | - | 2.9 | 3.5 | 4.2 | 4.8 | 5.5 | 6.3 | 8 | 9.5 |
| | Nominal Screw Size for Types D, F, and T | | | | | | | | | |
| | 2 | 2.5 | 3 | 3.5 | 4 | 5 | - | 6 | 8 | 10 |
| 4 | PH | PH | | | | | | | | |
| 5 | PH | PH | | | | | | | | |
| 6 | A | A | PH | | | | | | | |
| 8 | A | A | A | PH | PH | | | | | |
| 10 | A | A | A | A | A | PH | | | | |
| 13 | A | A | A | A | A | A | A | PH | | |
| 16 | | A | A | A | A | A | A | A | PH | |
| 20 | | | | A | A | A | A | A | A | PH |
| 25 | | | | A | A | A | A | A | A | A |
| 30 | | | | | | A | A | A | A | A |
| 35 | | | | | | A | A | A | A | A |
| 40 | | | | | | | A | A | A | A |
| 45 | | | | | | | | | A | A |
| 50 | | | | | | | | | A | A |
| 55 | | | | | | | | | | A |
| 60 | | | | | | | | | | A |

## Table 3. ANSI Standard Thread and Point Dimensions for Types AB and B Metric Thread Forming Tapping Screws *ANSI/ASME B18.6.5M-2000 (Withdrawn)*

| Nominal Screw Size and Thread Pitch[a] | Basic Screw Diameter | Basic Thread Pitch | $D_1$ | | $D_2$ | | $D_3$ | | Y | | Z | L | | | |
|---|---|---|---|---|---|---|---|---|---|---|---|---|---|---|---|
| | | | Thread Major Diameter | | Thread Minor Diameter | | Point Diameter[b] | | Point Taper Length Type B[c] | | Point Length Factor Type AB | Min. Practical Nominal Screw Length[d] | | | |
| | | | | | | | | | | | | Type AB | | Type B | |
| | Ref.[e] | Ref.[e] | Max. | Min. | Max. | Min. | Max. | Min. | Max. | Min. | Ref.[f] | Note 1 | Note 2 | Note 1 | Note 2 |
| 2.2×0.8 | 2.184 | 0.79 | 2.24 | 2.10 | 1.63 | 1.52 | 1.47 | 1.37 | 1.6 | 1.2 | 2.0 | 4 | 6 | 4 | 5 |
| 2.9×1 | 2.845 | 1.06 | 2.90 | 2.76 | 2.18 | 2.08 | 2.01 | 1.88 | 2.1 | 1.6 | 2.6 | 6 | 7 | 5 | 7 |
| 3.5×1.3 | 3.505 | 1.27 | 3.53 | 3.35 | 2.64 | 2.51 | 2.41 | 2.26 | 2.5 | 1.9 | 3.2 | 7 | 9 | 6 | 8 |
| 4.2×1.4 | 4.166 | 1.41 | 4.22 | 4.04 | 3.10 | 2.95 | 2.84 | 2.69 | 2.8 | 2.1 | 3.7 | 8 | 10 | 7 | 10 |
| 4.8×1.6 | 4.826 | 1.59 | 4.80 | 4.62 | 3.58 | 3.43 | 3.30 | 3.12 | 3.2 | 2.4 | 4.3 | 9 | 12 | 8 | 11 |
| 5.5×1.8 | 5.486 | 1.81 | 5.46 | 5.28 | 4.17 | 3.99 | 3.86 | 3.68 | 3.6 | 2.7 | 5.0 | 11 | 14 | 9 | 12 |
| 6.3×1.8 | 6.350 | 1.81 | 6.25 | 6.03 | 4.88 | 4.70 | 4.55 | 4.34 | 3.6 | 2.7 | 6.0 | 12 | 16 | 10 | 13 |
| 8×2.1 | 7.938 | 2.12 | 8.00 | 7.78 | 6.20 | 5.99 | 5.84 | 5.64 | 4.2 | 3.2 | 7.5 | 16 | 20 | 12 | 17 |
| 9.5×2.1 | 9.525 | 2.12 | 9.65 | 9.43 | 7.85 | 7.59 | 7.44 | 7.24 | 4.2 | 3.2 | 8.0 | 19 | 24 | 14 | 19 |

[a] The body diameter (unthreaded portion) is not less than the minimum minor diameter nor greater than the maximum major diameter of the thread.

[b] The tabulated values shall apply to screw blanks prior to roll threading.

[c] The tabulated maximum limits are equal to approximately two times the thread pitch.

[d] Lengths shown are theoretical minimums and are intended to assist the user in the selection of appropriate short screw lengths. Refer to Table 2 for recommended diameter-length combinations.

[e] Basic screw diameter and basic thread pitch shall be used for calculation purposes wherever these factors appear in formulations for dimensions.

[f] The minimum effective grip length on Type AB tapping screws shall be determined by subtracting the point length factor from the minimum screw length.

All dimensions are in millimeters. See Table 1 for thread diagrams.

Note 1: Pan, hex, and hex flange heads.

Note 2: Flat and oval countersunk heads.

**Table 4. ANSI Standard Thread and Point Dimensions for Types BF, BT, D, F, and T Metric Thread Cutting Tapping Screws** *ANSI/ASME B18.6.5M-2000 (Withdrawn)*

| Types BF and BT | | | | | | | | | | | | |
|---|---|---|---|---|---|---|---|---|---|---|---|---|
| | | | $D_1$ | | $D_2$ | | $D_3$ | | Y | | L | |
| Nominal Screw Size and Thread Pitch | Basic Screw Diameter | Basic Thread Pitch | Thread Major Diameter | | Thread Minor Diameter | | Point Diameter[a] | | Point Taper Length Type B[b] | | Minimal Practical Nominal Screw Length[c] | |
| | Ref.[d] | Ref.[d] | Max. | Min. | Max. | Min. | Max. | Min. | Max. | Min. | Pan, Hex and Hex Flange Heads | Flat and Oval Countersunk Heads |
| 2.2×0.8 | 2.184 | 0.79 | 2.24 | 2.10 | 1.63 | 1.52 | 1.47 | 1.37 | 1.6 | 1.2 | 4 | 5 |
| 2.9×1 | 2.845 | 1.06 | 2.90 | 2.76 | 2.18 | 2.08 | 2.01 | 1.88 | 2.1 | 1.6 | 5 | 7 |
| 3.5×1.3 | 3.505 | 1.27 | 3.53 | 3.35 | 2.64 | 2.51 | 2.41 | 2.26 | 2.5 | 1.9 | 6 | 8 |
| 4.2×1.4 | 4.166 | 1.41 | 4.22 | 4.04 | 3.10 | 2.95 | 2.84 | 2.69 | 2.8 | 2.1 | 7 | 10 |
| 4.8×1.6 | 4.826 | 1.59 | 4.80 | 4.62 | 3.58 | 3.43 | 3.30 | 3.12 | 3.2 | 2.4 | 8 | 11 |
| 5.5×1.8 | 5.486 | 1.81 | 5.46 | 5.28 | 4.17 | 3.99 | 3.86 | 3.68 | 3.6 | 2.7 | 9 | 12 |
| 6.3×1.8 | 6.350 | 1.81 | 6.25 | 6.03 | 4.88 | 4.70 | 4.55 | 4.34 | 3.6 | 2.7 | 10 | 13 |
| 8×2.1 | 7.938 | 2.12 | 8.00 | 7.78 | 6.20 | 5.99 | 5.84 | 5.64 | 4.2 | 3.2 | 12 | 17 |
| 9.5×2.1 | 9.525 | 2.12 | 9.65 | 9.43 | 7.85 | 7.59 | 7.44 | 7.24 | 4.2 | 3.2 | 14 | 19 |

[a] The tabulated values apply to screw blanks prior to roll threading.

[b] The tabulated maximum limits are equal to approximately two times the thread pitch.

[c] Lengths shown are theoretical minimums and are intended to assist in the selection of appropriate short screw lengths. See Table 2 for recommended length-diameter combinations. For Types D, F, and T, shorter screws are available with the point length reduced to the limits tabulated for short screws.

[d] Basic screw diameter and basic thread pitch are used for calculation purposes whenever these factors appear in formulations for dimensions.

| Types D, F, T | | | | | | | | | | | |
|---|---|---|---|---|---|---|---|---|---|---|---|
| | $D_1$ | | $D_3$ | | $D_S$ | Y | | | | L | |
| | | | | | | Point Taper Length | | | | Minimum Practical Nominal Screw Length[c] | |
| Nominal Screw Size and Thread Pitch | Thread Major Diameter | | Point Diameter[a] | | Body Diameter[a] | For Short Screws | | For Long Screws[b] | | Pan, Hex and Hex Flange Heads | Flat and Oval Countersunk Heads |
| | Max. | Min. | Max. | Min. | Min. | Max. | Min. | Max. | Min. | | |
| 2×0.4 | 2.00 | 1.88 | 1.45 | 1.39 | 1.65 | 1.4 | 1.0 | 1.8 | 1.4 | 4 | 5 |
| 2.5×0.45 | 2.50 | 2.37 | 1.88 | 1.82 | 2.12 | 1.6 | 1.1 | 2.0 | 1.6 | 4 | 6 |
| 3×0.5 | 3.00 | 2.87 | 2.32 | 2.26 | 2.58 | 1.8 | 1.3 | 2.3 | 1.8 | 5 | 6 |
| 3.5×0.6 | 3.50 | 3.35 | 2.68 | 2.60 | 3.00 | 2.1 | 1.5 | 2.7 | 2.1 | 5 | 8 |
| 4×0.7 | 4.00 | 3.83 | 3.07 | 2.97 | 3.43 | 2.5 | 1.8 | 3.2 | 2.5 | 6 | 9 |
| 5×0.8 | 5.00 | 4.82 | 3.94 | 3.84 | 4.36 | 2.8 | 2.0 | 3.6 | 2.8 | 7 | 10 |
| 6×1 | 6.00 | 5.79 | 4.69 | 4.55 | 5.21 | 3.5 | 2.5 | 4.5 | 3.5 | 9 | 12 |
| 8×1.25 | 8.00 | 7.76 | 6.40 | 6.24 | 7.04 | 4.4 | 3.1 | 5.6 | 4.4 | 11 | 16 |
| 10×1.5 | 10.00 | 9.73 | 8.08 | 7.88 | 8.86 | 5.3 | 3.8 | 6.8 | 5.3 | 13 | 18 |

[a] Minimum limits for body diameter (unthreaded portion) are tabulated for convenient reference. For Types BF and BT, the body diameter is not less than the minimum minor diameter nor greater than the maximum major diameter of the thread.

[b] Long screws are screws of nominal lengths equal to or longer than those listed under L.

All dimensions are in millimeters. See Table 1 for thread diagrams.

**Material and Heat Treatment.**—Tapping screws are normally fabricated from carbon steel and are suitably processed to meet the performance and test requirements outlined in ANSI/ASME B18.6.5M-2000 (Withdrawn). Tapping screws may also be made from corrosion resistant steel, Monel, brass, and aluminum alloys. The materials, properties, and performance characteristics applicable to such screws should be mutually agreed upon between the manufacturer and the purchaser.

**Table 5. Clearance Holes for Metric Tapping Screws**
*ANSI/ASME B18.6.5M-2000 (Withdrawn), Appendix*

| Nominal Screw Size and Thread Pitch | Basic Clearance Hole Diameter[a] | | | Nominal Screw Size and Thread Pitch | Basic Clearance Hole Diameter[a] | | |
|---|---|---|---|---|---|---|---|
| | Close Clearance[b] | Normal Clearance (Preferred)[b] | Loose Clearance[b] | | Close Clearance[b] | Normal Clearance (Preferred)[b] | Loose Clearance[b] |
| Types AB, B, BF, and BT | | | | Types D, F, and T | | | |
| 2.2 × 0.8 | 2.40 | 2.60 | 2.80 | 2 × 0.4 | 2.20 | 2.40 | 2.60 |
| 2.9 × 1 | 3.10 | 3.30 | 3.50 | 2.5 × 0.45 | 2.70 | 2.90 | 3.10 |
| 3.5 × 1.3 | 3.70 | 3.90 | 4.20 | 3 × 0.5 | 3.20 | 3.40 | 3.60 |
| 4.2 × 1.4 | 4.50 | 4.70 | 5.00 | 3.5 × 0.6 | 3.70 | 3.90 | 4.20 |
| 4.8 × 1.6 | 5.10 | 5.30 | 5.60 | 4 × 0.7 | 4.30 | 4.50 | 4.80 |
| 5.5 × 1.8 | 5.90 | 6.10 | 6.50 | 5 × 0.8 | 5.30 | 5.50 | 5.80 |
| 6.3 × 1.8 | 6.70 | 6.90 | 7.30 | 6 × 1 | 6.40 | 6.60 | 7.00 |
| 8 × 2.1 | 8.40 | 9.00 | 10.00 | 8 × 1.25 | 8.40 | 9.00 | 10.00 |
| 9.5 × 2.1 | 10.00 | 10.50 | 11.50 | 10 × 1.5 | 10.50 | 11.00 | 12.00 |

[a] The values given in this table are minimum limits. The recommended plus tolerances are as follows: for clearance hole diameters over 1.70 to and including 5.80 mm, plus 0.12, 0.20, and 0.30 mm for close, normal, and loose clearances, respectively; over 5.80 to and including 14.50 mm, plus 0.18, 0.30, and 0.45 mm for close, normal, and loose clearances, respectively.

[b] Normal clearance hole sizes are preferred. Close clearance hole sizes are for situations such as critical alignment of assembled components, wall thickness, or other limitations that necessitate the use of a minimal hole. Countersinking or counterboring at the fastener entry side may be necessary for the proper seating of the head. Loose clearance hole sizes are for applications where maximum adjustment capability between the components being assembled is necessary.

All dimensions are in millimeters.

**Approximate Installation Hole Sizes for Metric Tapping Screws.—**The approximate hole sizes given in Table 7 through Table 9 provide general guidance in selecting holes for installing the respective types of metric thread forming and thread cutting tapping screws in various commonly used materials. Types AB, B, BF, and BT metric tapping screws are covered in these tables; hole sizes for Types D, F, and T metric thread cutting tapping screws are still under development.

**Table 6. Approximate Pierced or Extruded Hole Sizes for Steel Types AB and B Metric Thread Forming Tapping Screws**

| Nominal Screw Size and Thread Pitch | Metal Thickness | Hole Size | Nominal Screw Size and Thread Pitch | Metal Thickness | Hole Size | Nominal Screw Size and Thread Pitch | Metal Thickness | Hole Size |
|---|---|---|---|---|---|---|---|---|
| In Steel, Stainless Steel, Monel, and Brass Sheet Metal | | | | | | | | |
| 2.9 × 1 | 0.38 | 2.18 | 4.2 × 1.4 | 0.46 | 3.45 | 5.5 × 1.8 | 0.61 | 4.70 |
| | 0.46 | 2.18 | | 0.61 | 3.45 | | 0.76 | 4.70 |
| | 0.61 | 2.49 | | 0.76 | 3.45 | | 0.91 | 4.70 |
| | 0.76 | 2.49 | | 0.91 | 3.45 | | 1.22 | 4.70 |
| | 0.91 | 2.49 | | 1.22 | 3.45 | | … | … |
| 3.5 × 1.3 | 0.38 | 2.82 | 4.8 × 1.6 | 0.46 | 3.99 | 6.3 × 1.8 | 0.76 | 5.31 |
| | 0.46 | 2.82 | | 0.61 | 3.99 | | 0.91 | 5.31 |
| | 0.61 | 2.82 | | 0.76 | 3.99 | | 1.22 | 5.31 |
| | 0.76 | 2.82 | | 0.91 | 3.99 | | … | … |
| | 0.91 | 2.82 | | 1.22 | 3.99 | | … | … |
| In Aluminum Alloy | | | | | | | | |
| 2.9 × 1 | 0.61 | 2.18 | 3.5 × 1.3 | 0.91 | 2.82 | 4.8 × 1.6 | 0.61 | 3.99 |
| | 0.76 | 2.18 | | 1.22 | 2.82 | | 0.76 | 3.99 |
| | 0.91 | 2.18 | 4.2 × 1.4 | 0.61 | 3.45 | | 0.91 | 3.99 |
| | 1.22 | 2.18 | | 0.76 | 3.45 | | 1.22 | 3.99 |
| 3.5 × 1.3 | 0.61 | 2.82 | | 0.91 | 3.45 | | … | … |
| | 0.76 | 2.82 | | 1.22 | 3.45 | | … | … |

All dimensions are in millimeters.

**Table 7. Approximate Drilled or Clean-Punched Hole Sizes for Steel Type AB Metric Thread Forming Tapping Screws in Sheet Metal**

| Nominal Screw Size and Thread Pitch | Metal Thick-ness | Hole Size | Drill Size[a] | Nominal Screw Size and Thread Pitch | Metal Thick-ness | Hole Size | Drill Size[a] | Nominal Screw Size and Thread Pitch | Metal Thick-ness | Hole Size | Drill Size[a] |
|---|---|---|---|---|---|---|---|---|---|---|---|
| In Steel, Stainless Steel, Monel, and Brass Sheet Metal | | | | | | | | | | | |
| 2.2 × 0.8 | 0.38 | 1.63 | 52 | 3.5 × 1.3 | 0.61 | 2.69 | 36 | 4.8 × 1.6 | 1.22 | 3.78 | 25 |
| | 0.46 | 1.63 | 52 | | 0.76 | 2.69 | 36 | | 1.52 | 3.91 | 23 |
| | 0.61 | 1.70 | 51 | | 0.91 | 2.79 | 35 | | 1.90 | 3.99 | 22 |
| | 0.76 | 1.78 | 50 | | 1.22 | 2.82 | 34 | 5.5 × 1.8 | 0.46 | … | … |
| | 0.91 | 1.85 | 49 | | 1.52 | 2.95 | 32 | | 0.61 | 4.22 | 19 |
| | 1.22 | 1.85 | 49 | | 1.90 | 3.05 | 31 | | 0.76 | 4.22 | 19 |
| | 1.52 | 1.93 | 48 | 4.2 × 1.4 | 0.46 | … | … | | 0.91 | 4.22 | 19 |
| 2.9 × 1 | 0.38 | 2.18 | 44 | | 0.61 | 3.18 | … | | 1.22 | 4.32 | 18 |
| | 0.46 | 2.18 | 44 | | 0.76 | 3.18 | … | | 1.52 | 4.50 | 16 |
| | 0.61 | 2.26 | 43 | | 0.91 | 3.18 | … | | 1.90 | 4.62 | 14 |
| | 0.76 | 2.39 | 42 | | 1.22 | 3.25 | 30 | 6.3 × 1.8 | 0.46 | 4.98 | 9 |
| | 0.91 | 2.39 | 42 | | 1.52 | 3.45 | 29 | | 0.61 | 4.98 | 9 |
| | 1.22 | 2.44 | 41 | | 1.90 | 3.56 | 28 | | 0.76 | 4.98 | 9 |
| | 1.52 | 2.54 | 39 | 4.8 × 1.6 | 0.46 | 3.66 | 27 | | 0.91 | 4.98 | 9 |
| | 1.90 | 2.59 | 38 | | 0.61 | 3.66 | 27 | | 1.22 | 5.21 | W |
| 3.5 × 1.3 | 0.38 | 2.64 | 37 | | 0.76 | 3.66 | 27 | | 1.52 | 5.79 | 1 |
| | 0.46 | 2.64 | 37 | | 0.91 | 3.73 | 26 | | 1.90 | 5.89 | … |
| In Aluminum Alloy Sheet Metal | | | | | | | | | | | |
| 2.2 × 0.8 | 0.38 | … | … | 3.5 × 1.3 | 0.61 | … | … | 4.8 × 1.6 | 1.22 | 3.66 | 27 |
| | 0.46 | … | … | | 0.76 | 2.64 | 37 | | 1.52 | 3.66 | 27 |
| | 0.61 | 1.63 | 52 | | 0.91 | 2.64 | 37 | | 1.90 | 3.73 | 26 |
| | 0.76 | 1.63 | 52 | | 1.22 | 2.64 | 37 | 5.5 × 1.8 | 0.46 | … | … |
| | 0.91 | 1.63 | 52 | | 1.52 | 2.69 | 36 | | 0.61 | … | … |
| | 1.22 | 1.70 | 51 | | 1.90 | 2.79 | 35 | | 0.76 | … | … |
| | 1.52 | 1.78 | 50 | 4.2 × 1.4 | 0.46 | … | … | | 0.91 | … | … |
| 2.9 × 1 | 0.38 | … | … | | 0.61 | … | … | | 1.22 | 4.09 | 20 |
| | 0.46 | … | … | | 0.76 | 2.95 | 32 | | 1.52 | 4.22 | 19 |
| | 0.61 | … | … | | 0.91 | 3.05 | 31 | | 1.90 | 4.39 | 17 |
| | 0.76 | 2.18 | 44 | | 1.22 | 3.25 | 30 | 6.3 × 1.8 | 0.46 | … | … |
| | 0.91 | 2.18 | 44 | | 1.52 | 3.45 | 29 | | 0.61 | … | … |
| | 1.22 | 2.18 | 44 | | 1.90 | 3.56 | 28 | | 0.76 | … | … |
| | 1.52 | 2.26 | 43 | 4.8 × 1.6 | 0.46 | … | … | | 0.91 | … | … |
| | 1.90 | 2.26 | 43 | | 0.61 | … | … | | 1.22 | … | … |
| 3.5 × 1.3 | 0.38 | … | … | | 0.76 | … | … | | 1.52 | 5.05 | 8 |
| | 0.46 | … | … | | 0.91 | 3.66 | 27 | | 1.90 | 5.11 | 7 |

[a] Customary drill size references have been retained where the metric hole diameters are direct conversions of their decimal inch equivalents.

All dimensions are in millimeters except drill sizes.

**Table 8. Approximate Hole Sizes for Steel Type AB Metric Thread Forming Tapping Screws in Plywoods and Asbestos**

| Nominal Screw Size and Thread Pitch | Hole Size | Drill Size[a] | Min. Mat. Thickness | Penetration in Blind Holes | | Hole Size | Drill Size[a] | Min. Mat. Thickness | Penetration in Blind Holes | |
|---|---|---|---|---|---|---|---|---|---|---|
| | | | | Min. | Max. | | | | Min. | Max. |
| | In Plywood (Resin Impregnated) | | | | | In Asbestos Compositions | | | | |
| 2.2 × 0.8 | 1.85 | 49 | 3.18 | 4.78 | 12.70 | 1.93 | 48 | 3.18 | 4.78 | 12.70 |
| 2.9 × 1 | 2.54 | 39 | 4.78 | 6.35 | 15.88 | 2.57 | 38 | 4.78 | 6.35 | 15.88 |
| 3.5 × 1.3 | 3.18 | … | 4.78 | 6.35 | 15.88 | 3.05 | 31 | 4.78 | 6.35 | 15.88 |
| 4.2 × 1.4 | 3.66 | 27 | 4.78 | 6.35 | 19.05 | 3.73 | 26 | 7.92 | 9.52 | 19.05 |
| 4.8 × 1.6 | 4.39 | 17 | 6.35 | 7.92 | 25.40 | 4.22 | 19 | 7.92 | 9.52 | 25.40 |
| 5.5 × 1.8 | 4.93 | 10 | 7.92 | 9.52 | 25.40 | 4.98 | 9 | 7.92 | 9.52 | 25.40 |
| 6.3 × 1.8 | 5.79 | 1 | 7.92 | 9.52 | 25.40 | 5.79 | 1 | 11.13 | 12.70 | 25.40 |

[a] Customary drill size references have been retained where the metric hole diameters are direct conversions of their decimal inch equivalents.

All dimensions are in millimeters except drill sizes.

**Table 9. Approximate Hole Sizes for Steel Type B Metric Thread Forming Tapping Screws in Plywoods, Asbestos, and Plastics**

| Nominal Screw Size and Thread Pitch | Hole Size | Drill Size[a] | Min. Mat. Thick-ness | Penetration in Blind Holes | | Nominal Screw Size and Thread Pitch | Hole Size | Drill Size[a] | Min. Mat. Thick-ness | Penetration in Blind Holes | |
|---|---|---|---|---|---|---|---|---|---|---|---|
| | | | | Min. | Max. | | | | | Min. | Max. |
| In Plywood (Resin Impregnated) | | | | | | | | | | | |
| 2.2 × 0.8 | 1.85 | 49 | 3.18 | 4.78 | 12.70 | 4.8 × 1.6 | 4.39 | 17 | 6.35 | 7.92 | 25.40 |
| 2.9 × 1 | 2.54 | 39 | 4.78 | 6.35 | 15.88 | 5.5 × 1.8 | 4.93 | 10 | 7.92 | 9.52 | 25.40 |
| 3.5 × 1.3 | 3.18 | … | 4.78 | 6.35 | 15.88 | 6.3 × 1.8 | 5.79 | 1 | 7.92 | 9.52 | 25.40 |
| 4.2 × 1.4 | 3.66 | 27 | 4.78 | 6.35 | 19.05 | … | … | … | … | … | … |

[a] Customary drill size references have been retained where the metric hole diameters are direct conversions of their decimal inch equivalents.

| Nominal Screw Size and Thread Pitch | Hole Size | Drill Size[a] | Min. Mat. Thickness | Penetration in Blind Holes | |
|---|---|---|---|---|---|
| | | | | Min. | Max. |
| In Asbestos Compositions | | | | | |
| 2.2 × 0.8 | 1.93 | 48 | 3.18 | 4.78 | 12.70 |
| 2.9 × 1 | 2.57 | 38 | 4.78 | 6.35 | 15.88 |
| 3.5 × 1.3 | 3.05 | 31 | 4.78 | 6.35 | 15.88 |
| 4.2 × 1.4 | 3.73 | 26 | 7.92 | 9.52 | 19.05 |
| 4.8 × 1.6 | 4.22 | 19 | 7.92 | 9.52 | 25.40 |
| 5.5 × 1.8 | 4.98 | 9 | 7.92 | 9.52 | 25.40 |
| 6.3 × 1.8 | 5.79 | 1 | 11.13 | 12.70 | 25.40 |

| Nominal Screw Size and Thread Pitch | Hole Size | Drill Size[a] | Min. Penetration in Blind Holes | Hole Size | Drill Size[a] | Min. Penetration in Blind Holes |
|---|---|---|---|---|---|---|
| | In Phenol Formaldehyde | | | In Cellulose Acetate and Nitrate, Acrylic and Styrene Resins | | |
| 2.2 × 0.8 | 1.98 | 47 | 4.78 | 1.98 | 47 | 4.78 |
| 2.9 × 1 | 2.54 | 39 | 6.35 | 2.39 | 42 | 6.35 |
| 3.5 × 1.3 | 3.25 | 30 | 6.35 | 3.05 | 32 | 6.35 |
| 4.2 × 1.4 | 3.81 | 25 | 7.92 | 3.66 | 27 | 7.92 |
| 4.8 × 1.6 | 4.50 | 16 | 7.92 | 4.32 | 18 | 7.92 |
| 5.5 × 1.8 | 5.05 | 8 | 9.52 | 4.85 | 11 | 9.52 |
| 6.3 × 1.8 | 5.94 | … | 9.52 | 5.61 | 2 | 9.52 |

All dimensions are in millimeters except drill sizes.

**Table 10. Approximate Drilled or Clean-Punched Hole Sizes for Steel Type B Metric Thread Forming Tapping Screws in Sheet Metal and Cast Metals**

| Nominal Screw Size and Thread Pitch | Metal Thick-ness | Hole Size | Drill Size[a] | Nominal Screw Size and Thread Pitch | Metal Thick-ness | Hole Size | Drill Size[a] | Nominal Screw Size and Thread Pitch | Metal Thick-ness | Hole Size | Drill Size[a] |
|---|---|---|---|---|---|---|---|---|---|---|---|
| In Steel, Stainless Steel, Monel, and Brass Sheet Metal | | | | | | | | | | | |
| 2.2 × 0.8 | 0.38 | 1.63 | 52 | 3.5 × 1.3 | 1.90 | 3.05 | 31 | 5.5 × 1.8 | 0.61 | 4.22 | 19 |
| | 0.46 | 1.63 | 52 | | 2.67 | 3.25 | 30 | | 0.76 | 4.22 | 19 |
| | 0.61 | 1.70 | 51 | 4.2 × 1.4 | 0.61 | 3.18 | … | | 0.91 | 4.22 | 19 |
| | 0.76 | 1.78 | 50 | | 0.76 | 3.18 | … | | 1.22 | 4.32 | 18 |
| | 0.91 | 1.85 | 49 | | 0.91 | 3.18 | … | | 1.52 | 4.50 | 16 |
| | 1.22 | 1.85 | 49 | | 1.22 | 3.25 | 30 | | 1.90 | 4.62 | 14 |
| | 1.52 | 1.93 | 48 | | 1.52 | 3.45 | 29 | | 2.67 | 4.70 | 13 |
| 2.9 × 1 | 0.38 | 2.18 | 44 | | 1.90 | 3.56 | 28 | | 3.18 | 4.98 | 9 |
| | 0.46 | 2.18 | 44 | | 2.67 | 3.81 | 25 | | 3.43 | 4.98 | 9 |
| | 0.61 | 2.26 | 43 | | 3.18 | 3.81 | 25 | | 4.17 | 5.11 | 7 |
| | 0.76 | 2.39 | 42 | | 3.43 | 3.86 | 24 | 6.3 × 1.8 | 0.76 | 4.93 | 10 |
| | 0.91 | 2.39 | 42 | 4.8 × 1.6 | 0.61 | 3.66 | 27 | | 0.91 | 4.93 | 10 |
| | 1.22 | 2.44 | 41 | | 0.76 | 3.66 | 27 | | 1.22 | 4.93 | 10 |
| | 1.52 | 2.54 | 39 | | 0.91 | 3.73 | 26 | | 1.52 | 5.05 | 8 |
| | 1.90 | 2.59 | 38 | | 1.22 | 3.86 | 24 | | 1.90 | 5.18 | 6 |
| 3.5 × 1.3 | 0.38 | 2.64 | 37 | | 1.52 | 3.86 | 24 | | 2.67 | 5.31 | 4 |
| | 0.46 | 2.64 | 37 | | 1.90 | 3.99 | 22 | | 3.18 | 5.79 | 1 |
| | 0.61 | 2.69 | 36 | | 2.67 | 4.09 | 20 | | 3.43 | 5.79 | 1 |
| | 0.76 | 2.69 | 36 | | 3.18 | 4.32 | 18 | | 4.17 | 5.94 | … |
| | 0.91 | 2.79 | 35 | | 3.43 | 4.32 | 18 | | 4.75 | 5.94 | … |
| | 1.22 | 2.82 | 34 | | 4.17 | 4.39 | 17 | | 4.93 | 5.94 | … |
| | 1.52 | 2.95 | 32 | | | | | | | | |
| In Aluminum Alloy Sheet Metal | | | | | | | | | | | |
| 2.2 × 0.8 | 0.61 | 1.63 | 52 | 4.2 × 1.4 | 0.76 | 2.95 | 32 | 5.5 × 1.8 | 1.22 | 4.09 | 20 |
| | 0.76 | 1.63 | 52 | | 0.91 | 3.05 | 31 | | 1.52 | 4.22 | 19 |
| | 0.91 | 1.63 | 52 | | 1.22 | 3.25 | 30 | | 1.90 | 4.39 | 17 |
| | 1.22 | 1.70 | 51 | | 1.52 | 3.45 | 29 | | 2.67 | 4.57 | 15 |
| | 1.52 | 1.78 | 50 | | 1.90 | 3.56 | 28 | | 3.18 | 4.62 | 14 |
| 2.9 × 1 | 0.76 | 2.18 | 44 | | 2.67 | 3.73 | 26 | | 3.43 | 4.62 | 14 |
| | 0.91 | 2.18 | 44 | | 3.18 | 3.73 | 26 | | 4.17 | 4.80 | 12 |
| | 1.22 | 2.18 | 44 | | 3.43 | 3.78 | 25 | | 5.08 to 9.52 | 4.98 | 9 |
| | 1.52 | 2.26 | 43 | | 4.11 to 9.52 | 3.86 | 24 | 6.3 × 1.8 | 1.52 | 5.05 | 8 |
| | 1.90 | 2.26 | 43 | 4.8 × 1.6 | 0.91 | 3.66 | 27 | | 1.90 | 5.11 | 7 |
| | 2.67 | 2.39 | 42 | | 1.22 | 3.66 | 27 | | 2.67 | 5.18 | 6 |
| 3.5 × 1.3 | 0.76 | 2.64 | 37 | | 1.52 | 3.66 | 27 | | 3.18 | 5.31 | 4 |
| | 0.91 | 2.64 | 37 | | 1.90 | 3.73 | 26 | | 3.43 | 5.31 | 4 |
| | 1.22 | 2.64 | 37 | | 2.67 | 3.73 | 26 | | 4.17 | 5.41 | 3 |
| | 1.52 | 2.69 | 36 | | 3.18 | 3.91 | 23 | | 4.75 | 5.41 | 3 |
| | 1.90 | 2.79 | 35 | | 3.43 | 3.91 | 23 | | 4.93 | 5.61 | 2 |
| | 2.67 | 2.82 | 34 | | 4.17 | 4.04 | 21 | | 5.08 to 9.52 | 5.79 | 1 |
| | 3.25 to 6.25 | 3.05 | 31 | | 5.08 to 9.52 | 4.22 | 19 | | | | |

[a] Customary drill size references have been retained where the metric hole diameters are direct conversions of their decimal inch equivalents.

| Nominal Screw Size and Thread Pitch | Hole Size | Drill Size[a] | Min. Penetration in Blind Holes | Nominal Screw Size and Thread Pitch | Hole Size | Drill Size[a] | Min. Penetration in Blind Holes |
|---|---|---|---|---|---|---|---|
| In Aluminum, Magnesium, Zinc, Brass, and Bronze Cast Metals | | | | | | | |
| 2.2 × 0.8 | 1.98 | 47 | 3.18 | 4.8 × 1.6 | 4.50 | 16 | 6.35 |
| 2.9 × 1 | 2.64 | 37 | 4.78 | 5.5 × 1.8 | 5.05 | 8 | 7.14 |
| 3.5 × 1.3 | 3.25 | 30 | 6.35 | 6.3 × 1.8 | 5.94 | 4 | 7.92 |
| 4.2 × 1.4 | 3.86 | 24 | 6.35 | … | … | … | … |

All dimensions are in millimeters, except drill sizes.

**Table 11. Approximate Hole Sizes for Steel Types BF and BT Metric Thread Cutting Tapping Screws for Cast Metals and Plastics**

| Nominal Screw Size and Thread Pitch | Material Thickness | Hole Size | Drill Size[a] | Nominal Screw Size and Thread Pitch | Material Thickness | Hole Size | Drill Size[a] |
|---|---|---|---|---|---|---|---|
| In Die-Cast Zinc and Aluminum | | | | | | | |
| 2.2 × 0.8 | 1.52 | 1.85 | 49 | 3.5 × 1.3 | 3.18 | 3.05 | 31 |
| | 2.11 | 1.85 | 49 | | 3.56 | 3.05 | 31 |
| | 2.77 | 1.93 | 48 | | 4.78 | 3.05 | 31 |
| | 3.18 | 1.93 | 48 | | 6.35 | 3.18 | … |
| | 3.56 | 1.93 | 48 | | 7.92 | 3.18 | … |
| 2.9 × 1 | 2.77 | 2.49 | 40 | 4.2 × 1.4 | 3.18 | 3.78 | 25 |
| | 3.18 | 2.54 | 39 | | 3.56 | 3.78 | 25 |
| | 3.56 | 2.54 | 39 | | 4.78 | 3.78 | 25 |
| | 4.78 | 2.54 | 39 | | 6.35 | 3.86 | 24 |
| | 6.35 | 2.59 | 38 | | 7.92 | 3.86 | 24 |
| 4.8 × 1.6 | 3.18 | 4.22 | 19 | 6.3 × 1.8 | 6.35 | 5.79 | 1 |
| | 3.56 | 4.22 | 19 | | 7.92 | 5.79 | 1 |
| | 4.78 | 4.22 | 19 | | 9.52 | 5.79 | 1 |
| | 6.35 | 4.32 | 18 | 8 × 2.1 | 3.18 | 7.14 | K |
| | 7.92 | 4.37 | … | | 3.56 | 7.14 | K |
| | 9.52 | 4.37 | … | | 4.78 | 7.14 | K |
| 5.5 × 1.8 | 3.18 | 4.85 | 11 | | 6.35 | 7.14 | K |
| | 3.56 | 4.85 | 11 | | 7.92 | 7.37 | L |
| | 4.78 | 4.85 | 11 | | 9.52 | 7.37 | L |
| | 6.35 | 4.98 | 9 | 9.5 × 2.1 | 3.18 | 8.74 | … |
| | 7.92 | 4.98 | 9 | | 3.56 | 8.74 | … |
| | 9.52 | 4.98 | 9 | | 4.78 | 8.74 | … |
| 6.3 × 1.8 | 3.18 | 5.61 | 2 | | 6.35 | 8.74 | … |
| | 3.56 | 5.61 | 2 | | 7.92 | 8.84 | S |
| | 4.78 | 5.61 | 2 | | 9.52 | 8.84 | S |

| Nominal Screw Size and Thread Pitch | Hole Size | Drill Size[a] | Depth of Penetration | |
|---|---|---|---|---|
| | | | Min. | Max. |
| In Phenol Formaldehyde | | | | |
| 2.2 × 0.8 | 1.98 | … | 2.39 | 6.35 |
| 2.9 × 1.0 | 2.64 | 37 | 3.18 | 7.92 |
| 3.5 × 1.3 | 3.18 | … | 4.78 | 9.52 |
| 4.2 × 1.4 | 3.73 | 26 | 6.35 | 12.70 |
| 4.8 × 1.6 | 4.32 | 18 | 7.92 | 15.88 |
| 5.5 × 1.8 | 4.93 | 10 | 9.52 | 15.88 |
| 6.3 × 1.8 | 5.79 | 1 | 9.52 | 19.05 |
| In Cellulose Acetate and Nitrate, Acrylic and Styrene Resins | | | | |
| 2.2 × 0.8 | 1.93 | 48 | 2.39 | 6.35 |
| 2.9 × 1.0 | 2.54 | 39 | 3.18 | 7.92 |
| 3.5 × 1.3 | 3.05 | 31 | 4.78 | 9.52 |
| 4.2 × 1.4 | 3.66 | 27 | 6.35 | 12.70 |
| 4.8 × 1.6 | 4.22 | 19 | 7.92 | 15.88 |
| 5.5 × 1.8 | 4.80 | 12 | 9.52 | 15.88 |
| 6.3 × 1.8 | 5.61 | 2 | 9.52 | 19.05 |

[a] Customary drill size references have been retained where the metric hole sizes are direct conversions of their decimal inch equivalents.

All dimensions are in millimeters except drill sizes.

The finish (plating or coating) on metric tapping screws and the material composition and hardness of the mating component are factors that affect assembly torques in individual applications. Although the recommended installation hole sizes given in Table 7 through Table 9 were based on the use of plain unfinished carbon steel metric tapping screws, experience has shown that the specified holes are also suitable for screws having most types of commercial finishes. However, owing to various finishes providing different degrees of lubricity, some adjustment of installation torques may be necessary to suit individual applications. Also, where exceptionally heavy finishes are involved or screws are to be assembled into materials of higher hardness, some deviation from the specified hole sizes may be required to provide optimum assembly. The necessity and extent of such deviations can best be determined by experiment in the particular assembly environment.

# T-SLOTS, T-BOLTS, AND T-NUTS

## Table 1. American National Standard T-Slots *ANSI/ASME B5.1M-1985 (R2014)*

T-SLOTS

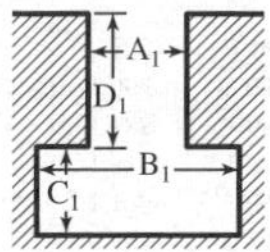


Basic Dimensions

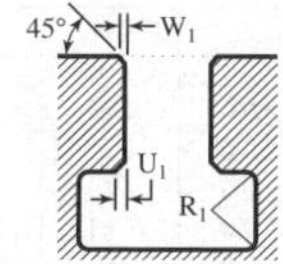


Suggested Approximate Dimensions for Rounding or Breaking of Corners

| Nominal T-Bolt Size[a] | | Width of Throat $A_1$[b] | | Width of Headspace $B_1$ | | | | Depth of Headspace $C_1$ | | | | Depth of Throat $D_1$ | | | | Rounding or Breaking of Corners[c] | | | | | |
|---|---|---|---|---|---|---|---|---|---|---|---|---|---|---|---|---|---|---|---|---|---|
| | | | | inch | | mm | | inch | | mm | | inch | | mm | | inch | | | mm | | |
| inch | mm | inch | mm | Min. | Max. | Man. | Max. | Min. | Max. | Min. | Max. | Min. | Max. | Min. | Max. | $R_1$ Max. | $W_1$ Max. | $U_1$ Max. | $R_1$ Max. | $W_1$ Max. | $U_1$ Max. |
| | 4 | | 5 | | | 10 | 11 | | | 3 | 3.5 | | | 4.5 | 7 | | | | 0.5 | 0.8 | 0.8 |
| | 5 | | 6 | | | 11 | 12.5 | | | 5 | 6 | | | 5 | 8 | | | | 0.5 | 0.8 | 0.8 |
| 0.250 | 6 | 0.282 | 8 | 0.500 | 0.562 | 14.5 | 16 | 0.203 | 0.234 | 7 | 8 | 0.125 | 0.375 | 7 | 11 | 0.02 | 0.02 | 0.03 | 0.5 | 0.8 | 0.8 |
| 0.312 | 8 | 0.344 | 10 | 0.594 | 0.656 | 16 | 18 | 0.234 | 0.266 | 7 | 8 | 0.156 | 0.438 | 9 | 14 | 0.02 | 0.03 | 0.03 | 0.5 | 0.8 | 0.8 |
| 0.375 | 10 | 0.438 | 12 | 0.719 | 0.781 | 19 | 21 | 0.297 | 0.328 | 8 | 9 | 0.219 | 0.562 | 11 | 17 | 0.02 | 0.03 | 0.03 | 0.5 | 0.8 | 0.8 |
| 0.500 | 12 | 0.562 | 14 | 0.906 | 0.969 | 23 | 25 | 0.359 | 0.391 | 9 | 11 | 0.312 | 0.688 | 12 | 19 | 0.02 | 0.03 | 0.03 | 0.5 | 0.8 | 0.8 |
| 0.625 | 16 | 0.688 | 18 | 1.188 | 1.250 | 30 | 32 | 0.453 | 0.484 | 12 | 14 | 0.438 | 0.875 | 16 | 24 | 0.03 | 0.03 | 0.05 | 0.8 | 0.8 | 1.3 |
| 0.750 | 20 | 0.812 | 22 | 1.375 | 1.469 | 37 | 40 | 0.594 | 0.625 | 16 | 18 | 0.562 | 1.062 | 20 | 29 | 0.03 | 0.03 | 0.05 | 0.8 | 0.8 | 1.3 |
| 1.000 | 24 | 1.062 | 28 | 1.750 | 1.844 | 46 | 50 | 0.781 | 0.828 | 20 | 22 | 0.750 | 1.250 | 26 | 36 | 0.03 | 0.06 | 0.05 | 0.8 | 1.5 | 1.3 |
| 1.250 | 30 | 1.312 | 36 | 2.125 | 2.219 | 56 | 60 | 1.031 | 1.094 | 25 | 28 | 1.000 | 1.562 | 33 | 46 | 0.03 | 0.06 | 0.05 | 0.8 | 1.5 | 1.3 |
| 1.500 | 36 | 1.562 | 42 | 2.562 | 2.656 | 68 | 72 | 1.281 | 1.344 | 32 | 35 | 1.250 | 1.938 | 39 | 53 | 0.03 | 0.06 | 0.05 | 0.8 | 1.5 | 1.3 |
| | 42 | | 48 | | | 80 | 85 | | | 36 | 40 | | | 44 | 59 | | | | 1.5 | 2.5 | 2 |
| | 48 | | 54 | | | 90 | 95 | | | 40 | 44 | | | 50 | 66 | | | | 1.5 | 2.5 | 2 |

[a] Width of tongue (tenon) to be used with the above t-slots will be found in the complete Standard B5.1M.

[b] Throat dimensions are basic. When slots are intended to be used for holding only, tolerances can be 0.0 + 0.010 inch or H12 Metric (ISO/R286); when intended for location, tolerance can be 0.0 + 0.001 inch or H8 Metric (see page 666).

[c] Corners of t-slots may be square or may be rounded or broken to the indicated maximum dimensions at the manufacturer's option.

For the dimensions of tongue seats, inserted tongues, and solid tongues refer to the complete Standard B5.1M.

**Table 2. American National Standard T-Bolts** *ANSI/ASME B5.1M-1985 (R2014)*

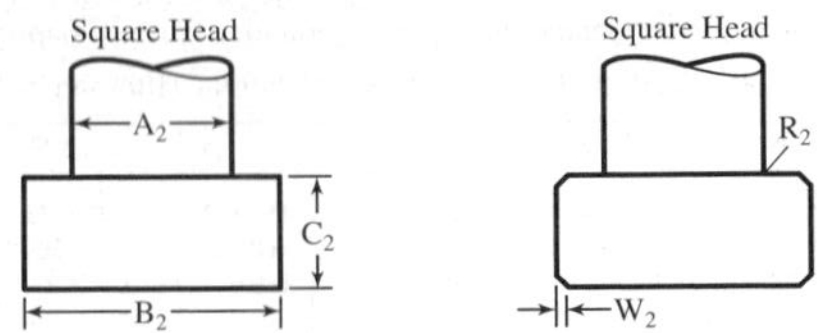


| Nominal T-Bolt Size and Thread $A_2$[a,b] | | Bolt Head Dimensions | | | | | | | | | | Rounding of Corners[c] | | | |
|---|---|---|---|---|---|---|---|---|---|---|---|---|---|---|---|
| | | Width Across Flats $B_2$ | | | | Width Across Corners | | Height $C_2$ | | | | $R_2$ | | $W_2$ | |
| inch | metric | inch | | mm | | inch | mm | inch | | mm | | inch | mm | inch | mm |
| UNC-2A | ISO[d] | Max. | Min. | Max. | Min. | Max. | Max. | Max. | Min. | Max. | Min. | Max. | Max. | Max. | Max. |
| | M4 | | | 9 | 8.5 | | 12.7 | | | 2.5 | 2.1 | | 0.3 | | 0.5 |
| | M5 | | | 10 | 9.5 | | 14.1 | | | 4 | 3.6 | | 0.3 | | 0.5 |
| 0.250-20 | M6 | 0.469 | 0.438 | 13 | 12 | 0.663 | 18.4 | 0.156 | 0.141 | 6 | 5.6 | 0.02 | 0.5 | 0.03 | 0.8 |
| 0.312-18 | M8 | 0.562 | 0.531 | 15 | 14 | 0.796 | 21.2 | 0.188 | 0.172 | 6 | 5.6 | 0.02 | 0.5 | 0.03 | 0.8 |
| 0.375-16 | M10 | 0.688 | 0.656 | 18 | 17 | 0.972 | 25.5 | 0.250 | 0.234 | 7 | 6.6 | 0.02 | 0.5 | 0.03 | 0.8 |
| 0.500-13 | M12 | 0.875 | 0.844 | 22 | 21 | 1.238 | 31.1 | 0.312 | 0.297 | 8 | 7.6 | 0.02 | 0.5 | 0.06 | 1.5 |
| 0.625-11 | M16 | 1.125 | 1.094 | 28 | 27 | 1.591 | 39.6 | 0.406 | 0.391 | 10 | 9.6 | 0.03 | 0.8 | 0.06 | 1.5 |
| 0.750-10 | M20 | 1.312 | 1.281 | 34 | 33 | 1.856 | 48.1 | 0.531 | 0.500 | 14 | 13.2 | 0.03 | 0.8 | 0.06 | 1.5 |
| 1.000-8 | M24 | 1.688 | 1.656 | 43 | 42 | 2.387 | 60.8 | 0.688 | 0.656 | 18 | 17.2 | 0.03 | 0.8 | 0.06 | 1.5 |
| 1.250-7 | M30 | 2.062 | 2.031 | 53 | 52 | 2.917 | 75 | 0.938 | 0.906 | 23 | 22.2 | 0.03 | 0.8 | 0.06 | 1.5 |
| 1.500-6 | M36 | 2.500 | 2.469 | 64 | 63 | 3.536 | 90.5 | 1.188 | 1.156 | 28 | 27.2 | 0.03 | 0.8 | 0.06 | 1.5 |
| | M42 | | | 75 | 74 | | 106.1 | | | 32 | 30.5 | | 1 | | 2 |
| | M48 | | | 85 | 84 | | 120.2 | | | 36 | 34.5 | | 1 | | 2 |

[a] For inch tolerances for thread diameters of bolts or studs and for threads see page 1954.
[b] T-slots to be used with these bolts will be found in Table 1.
[c] Corners of t-bolts may be square or may be rounded or broken to the indicated maximum dimensions at the manufacturer's option.
[d] Metric thread grade and tolerance position is 5g 6g (see page 2022).

**Table 3. American National Standard T-Nuts** *ANSI/ASME B5.1M-1985 (R2014)*

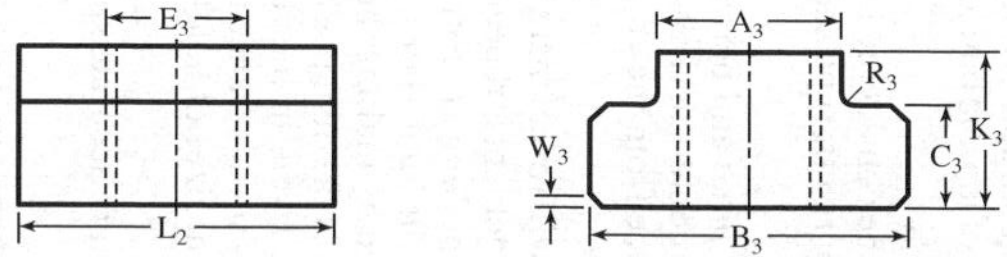


| Nominal T-Bolt Size[a] | | Width of Tongue $A_3$ | | | | Tap for Stud[b] $E_3$ | | Width of Nut $B_3$ | | | | Height of Nut $C_3$ | | | | Total Thickness Including Tongue[c] $K_3$ | | Length of Nut[c] $L_3$ | | Rounding of Corners $R_3$ | | Rounding of Corners $W_3$ | |
|---|---|---|---|---|---|---|---|---|---|---|---|---|---|---|---|---|---|---|---|---|---|---|---|---|---|
| | | inch | | mm | | inch | mm | inch | | mm | | inch | | mm | | | | | | inch | mm | inch | mm |
| inch | mm | Max. | Min. | Max. | Min. | UNC-3B | ISO[d] | Max. | Min. | Max. | Min. | Max. | Min. | Max. | Min. | inch | mm | inch | mm | Max. | Max. | Max. | Max. |
| | 4 | | … | | … | | … | | … | … | … | | | … | … | | … | | … | | … | | … |
| | 5 | | … | | … | | … | | … | … | … | | | … | … | | … | | … | | … | | … |
| 0.250 | 6 | … | … | … | … | … | … | … | … | … | … | … | … | … | … | … | … | … | … | … | … | … | … |
| 0.312 | 8 | 0.330 | 0.320 | 8.7 | 8.5 | 0.250-20 | M6 | 0.562 | 0.531 | 15 | 14 | 0.188 | 0.172 | 6 | 5.6 | 0.281 | 9 | 0.562 | 18 | 0.02 | 0.5 | 0.03 | 0.8 |
| 0.375 | 10 | 0.418 | 0.408 | 11 | 10.75 | 0.312-18 | M8 | 0.688 | 0.656 | 18 | 17 | 0.250 | 0.234 | 7 | 6.6 | 0.375 | 10.5 | 0.688 | 20 | 0.02 | 0.5 | 0.03 | 0.8 |
| 0.500 | 12 | 0.543 | 0.533 | 13.5 | 13.25 | 0.375-16 | 6M10 | 0.875 | 0.844 | 22 | 21 | 0.312 | 0.297 | 8 | 7.6 | 0.531 | 12 | 0.875 | 23 | 0.02 | 0.5 | 0.06 | 1.5 |
| 0.625 | 16 | 0.668 | 0.658 | 17.25 | 17 | 0.500-13 | M12 | 1.125 | 1.094 | 28 | 27 | 0.406 | 0.391 | 10 | 9.6 | 0.625 | 15 | 1.125 | 27 | 0.03 | 0.8 | 0.06 | 1.5 |
| 0.750 | 20 | 0.783 | 0.773 | 20.5 | 20.25 | 0.625-11 | M16 | 1.312 | 1.281 | 34 | 33 | 0.531 | 0.500 | 14 | 13.2 | 0.781 | 21 | 1.312 | 35 | 0.03 | 0.8 | 0.06 | 1.5 |
| 1.000 | 24 | 1.033 | 1.018 | 26.5 | 26 | 0.750-10 | M20 | 1.688 | 1.656 | 43 | 42 | 0.688 | 0.656 | 18 | 17.2 | 1.000 | 27 | 1.688 | 46 | 0.03 | 0.8 | 0.06 | 1.5 |
| 1.250 | 30 | 1.273 | 1.258 | 33 | 32.5 | 1.000-8 | M24 | 2.062 | 2.031 | 53 | 52 | 0.938 | 0.906 | 23 | 22.2 | 1.312 | 34 | 2.062 | 53 | 0.03 | 0.8 | 0.06 | 1.5 |
| 1.500 | 36 | 1.523 | 1.508 | 39.25 | 38.75 | 1.250-7 | M30 | 2.500 | 2.469 | 64 | 63 | 1.188 | 1.156 | 28 | 27.2 | 1.625 | 42 | 2.500 | 65 | 0.03 | 0.8 | 0.06 | 1.5 |
| | 42 | | | 46.75 | 46.25 | | M36 | | | 75 | 74 | | | 32 | 30.5 | | 48 | | 75 | | 1 | | 2 |
| | 48 | | | 52.5 | 51.75 | | M42 | | | 85 | 84 | | | 36 | 34.5 | | 54 | | 85 | | 1 | | 2 |

[a] T-slot dimensions to fit the above nuts will be found in Table 1.

[b] For tolerances of inch threads see page 1954.

[c] No tolerances are given for "Total Thickness" or "Nut Length" as they need not be held to close limits.

[d] Metric tapped thread grade and tolerance position is 5H (see page 2022).

# RIVETS AND RIVETED JOINTS

## Riveted Joint Design

**Classes and Types of Riveted Joints.—**Riveted joints may be classified by application as: 1) pressure vessel; 2) structural; and 3) machine member.

For information and data concerning joints for pressure vessels such as boilers, reference should be made to standard sources such as the ASME Boiler Code. The following sections will cover only structural and machine-member riveted joints.

Basically there are two kinds of riveted joints, the *lap-joint* and the *butt-joint*. In the ordinary *lap-joint*, the plates overlap each other and are held together by one or more rows of rivets. In the *butt-joint*, the plates being joined are in the same plane and are joined by means of a cover plate or butt strap, which is riveted to both plates by one or more rows of rivets. The term *single riveting* means one row of rivets in a lap-joint or one row on each side of a butt-joint; *double riveting* means two rows of rivets in a lap-joint or two rows on each side of the joint in butt riveting. Joints are also triple and quadruple riveted. Lap-joints may also be made with inside or outside cover plates. Types of lap and butt joints are illustrated in the tables on starting at page 1868.

**General Design Considerations for Riveted Joints.—**Factors to be considered in the design or specification of a riveted joint are: type of joint; spacing of rivets; type and size of rivet; type and size of hole; and rivet material.

*Spacing of Rivets:* The spacing between rivet centers is called *pitch* and between row center lines, *back pitch* or *transverse pitch*. The distance between centers of rivets nearest each other in adjacent rows is called *diagonal pitch*. The distance from the edge of the plate to the center line of the nearest row of rivets is called *margin*.

Examination of a riveted joint made up of several rows of rivets will reveal that after progressing along the joint a given distance, the rivet pattern or arrangement is repeated. (For a butt joint, the length of a *repeating section* is usually equal to the *long pitch* or pitch of the rivets in the outer row, that is the row farthest from the edge of the joint.) For structural and machine-member joints, the proper pitch may be determined by making the tensile strength of the plate over the length of the repeating section, that is the distance between rivets in the outer row, equal to the total shear strength of the rivets in the repeating section. Minimum pitch and diagonal pitch are also governed by the clearance required for the hold-on (Dolly bar) and rivet set. Dimensions for different sizes of hold-ons and rivet sets are given in the table on page 1873.

When fastening thin plate, it is particularly important to maintain accurate spacing to avoid buckling.

*Size and Type of Rivets:* The rivet diameter $d$ commonly falls between $d = 1.2\sqrt{t}$ and $d = 1.4\sqrt{t}$, where $t$ is the thickness of the plate. Dimensions for various types of American National Standard large ($^1\!/_2$-inch diameter and up) rivets and small solid rivets are shown in tables that follow. It may be noted that countersunk heads are not as strong as other types.

*Size and Type of Hole:* Rivet holes may be punched, punched and reamed, or drilled. Rivet holes are usually made $^1\!/_{16}$ inch (1.6 mm) larger in diameter than the nominal diameter of the rivet although in some classes of work in which the rivet is driven cold, as in automatic machine riveting, the holes are reamed to provide minimum clearance so that the rivet fills the hole completely.

When holes are punched in heavy steel plate, there may be considerable loss of strength unless the holes are reamed to remove the inferior metal immediately surrounding them. This results in the diameter of the punched hole being increased by from $^1\!/_{16}$ to $^1\!/_8$ inch (1.6 to 3.2 mm). Annealing after punching tends to restore the strength of the plate in the vicinity of the holes.

*Rivet Material:* Rivets for structural and machine-member purposes are usually made of wrought iron or soft steel, but for aircraft and other applications where light weight or resistance to corrosion is important, copper, aluminum alloy, Monel, Inconel, etc., may be used as rivet material.

*Simplified Design Assumptions:* In the design of riveted joints, a simplified treatment is frequently used in which the following assumptions are made:

1) The load is carried equally by the rivets.

2) No combined stresses act on a rivet to cause failure.

3) The shearing stress in a rivet is uniform across the cross section under question.

4) The load that would cause failure in single shear would have to be doubled to cause failure in double shear.

5) The bearing stress of rivet and plate is distributed equally over the projected area of the rivet.

6) The tensile stress is uniform in the section of metal between the rivets.

**Failure of Riveted Joints.**—Rivets may fail by:

1) Shearing through one cross section (single shear)

2) Shearing through two cross sections (double shear)

3) Crushing

Plates may fail by:

4) Shearing along two parallel lines extending from opposite sides of the rivet hole to the edge of the plate

5) Tearing along a single line from middle of rivet hole to edge of plate

6) Crushing

7) Tearing between adjacent rivets (tensile failure) in the same row or in adjacent rows

Types 4 and 5 failures are caused by rivets being placed too close to the edge of the plate. These types of failure are avoided by placing the center of the rivet at a minimum of one and one-half times the rivet diameter away from the edge.

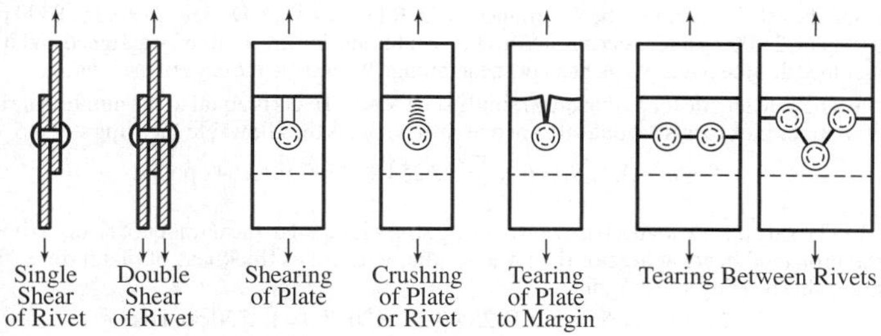


Types of Rivet and Plate Failure

Failure due to tearing on a diagonal between rivets in adjacent rows when the pitch is four times the rivet diameter or less is avoided by making the transverse pitch one and three-quarters times the rivet diameter.

*Theoretical versus Actual Riveted Joint Failure:* If it is assumed that the rivets are placed the suggested distance from the edge of the plate and each row the suggested distance from another row, then the failure of a joint is most likely to occur as a result of shear failure of the rivets, bearing failure (crushing) of the plate or rivets, or tensile failure of the plate, alone or in combination depending on the makeup of the joints.

Joint failure in actuality is more complex than this. Rivets do not undergo pure shear especially in lap-joints where rivets are subjected to single shear. The rivet, in this instance, would be subject to a combination of tensile and shearing stresses and it would fail because of combined stresses, not a single stress. Furthermore, the shearing stress is usually considered to be distributed evenly over the cross section, which is also not the case.

Rivets that are usually driven hot contract on cooling. This contraction in the length of the rivet draws the plates together and sets up a stress in the rivet estimated to be equal in magnitude to the yield point of the rivet steel. The contraction in the diameter of the rivet results in a little clearance between the rivet and the hole in the plate. The tightness in the plates caused by the contraction in length of the rivet gives rise to a condition in which quite a sizeable frictional force would have to be overcome before the plates would slip over one another and subject the rivets to a shearing force. It is European practice to design joints for resistance to this slipping. It has been found, however, that the strength-basis designs obtained in American and English practice are not very different from European designs.

**Allowable Stresses.—**The design stresses for riveted joints are usually set by codes, practices, or specifications. The American Institute of Steel Construction issues specifications for the design, fabrication, and erection of structural steel for buildings in which the allowable stress permitted in tension for structural steel and rivets is specified at 20,000 pounds per square inch, the allowable bearing stress for rivets is 40,000 psi in double shear and 32,000 psi in single shear, and the allowable shearing stress for rivets is 15,000 psi. The American Society of Mechanical Engineers in its Boiler Code lists the following ultimate stresses: tensile, 55,000 psi; shearing, 44,000 psi; compressive or bearing, 95,000 psi. The design stresses usually are one-fifth of these, that is tensile, 11,000 psi; shearing, 8800 psi; compressive or bearing, 19,000 psi. In machine design work, values close to these or somewhat lower are commonly used.

**Analysis of Joint Strength.—**The following examples and strength analyses of riveted joints are based on the six previously outlined *Simplified Design Assumptions*.

*Example 1:* Consider a 12-inch section of single-riveted lap-joint made up with plates of $\frac{1}{4}$-inch thickness and six rivets, $\frac{5}{8}$ inch in diameter. Assume that rivet holes are $\frac{1}{16}$ inch larger in diameter than the rivets. In this joint, the entire load is transmitted from one plate to the other by means of the rivets. Each plate and the six rivets carry the entire load. The safe tensile load $L$ and the efficiency $\eta$ may be determined in the following way: Design stresses of 8500 psi for shear, 20,000 psi for bearing, and 10,000 psi for tension are arbitrarily assigned and it is assumed that the rivets will not tear or shear through the plate to the edge of the joint.

a) The safe tensile load $L$ based on single shear of the rivets is equal to the number of rivets $n$ times the cross-sectional area of one rivet $A_r$ times the allowable shearing stress $S_s$ or

$$L = n \times A_r \times S_s = 6 \times \frac{\pi}{4}(0.625)^2 \times 8500 = 15{,}647 \text{ pounds}$$

b) The safe tensile load $L$ based on bearing stress is equal to the number of rivets $n$ times the projected bearing area of the rivet $A_b$ (diameter times thickness of plate) times the allowable bearing stress $S_c$ or

$$L = n \times A_b \times S_c = 6 \times (0.625 \times 0.25) \times 20{,}000 = 18{,}750 \text{ pounds}$$

c) The safe load $L$ based on the tensile stress is equal to the net cross-sectional area of the plate between rivet holes $A_p$ times the allowable tensile stress $S_t$ or

$$L = A_p \times S_t = 0.25[12 - 6(0.625 + 0.0625)] \times 10{,}000 = 19{,}688 \text{ pounds}$$

The safe tensile load for the joint would be the least of the three loads just computed or 15,647 pounds, and the efficiency $\eta$ would be equal to this load divided by the tensile strength of the section of plate under consideration if it were unperforated, or

$$\eta = \frac{15{,}647}{12 \times 0.25 \times 10{,}000} \times 100 = 52.2 \text{ percent}$$

*Example 2:* Under consideration is a 12-inch section of double-riveted butt-joint with main plates $\frac{1}{2}$ inch thick and two cover plates each $\frac{5}{16}$ inch thick. There are 3 rivets in the inner row and 2 on the outer and their diameters are $\frac{7}{8}$ inch. Assume that the diameter of the rivet holes is $\frac{1}{16}$ inch larger than that of the rivets. The rivets are so placed that the main plates will not tear diagonally from one rivet row to the others nor will they tear or fail in shear out to their edges. The safe tensile load $L$ and the efficiency $\eta$ may be determined in the following way: Design stresses for 8500 psi for shear, 20,000 psi for bearing, and 10,000 psi for tension are arbitrarily assigned.

a) The safe tensile load $L$ based on double shearing of the rivets is equal to the number of rivets $n$ times the number of shearing planes per rivet times the cross-sectional area of one rivet $A_r$ times the allowable shearing stress $S_s$ or

$$L = n \times 2 \times A_r \times S_s = 5 \times 2 \times \frac{\pi}{4}(0.875)^2 \times 8500 = 51{,}112 \text{ pounds}$$

b) The safe tensile load $L$ based on bearing stress is equal to the number of rivets $n$ times the projected bearing area of the rivet $A_b$ (diameter times thickness of plate) times the allowable bearing stress $S_c$ or $L = n \times A_b \times S_c = 5 \times (0.875 \times 0.5) \times 20{,}000 = 43{,}750$ pounds. (Cover plates are not considered since their combined thickness is $\frac{1}{4}$ inch greater than the main plate thickness.)

c) The safe tensile load $L$ based on the tensile stress is equal to the net cross-sectional area of the plate between the two rivets in the outer row $A_p$ times the allowable tensile stress $S_t$ or $L = A_p \times S_t = 0.5[12 - 2(0.875 + 0.0625)] \times 10{,}000 = 50{,}625$ pounds.

In completing the analysis, the sum of the load that would cause tearing between rivets in the three-hole section plus the load carried by the two rivets in the two-hole section is also investigated. The sum is necessary because if the joint is to fail, it must fail at both sections simultaneously. The least safe load that can be carried by the two rivets of the two-hole section is based on the bearing stress (see the foregoing calculations).

1) The safe tensile load $L$ based on the bearing strength of two rivets of the two-hole section is $L = n \times A_b \times S_c = 2 \times (0.875 \times 0.5) \times 20{,}000 = 17{,}500$ pounds.

2) The safe tensile load $L$ based on the tensile strength of the main plate between holes in the three-hole section is $L \times A_p \times S_t = 0.5[12 - 3(0.875 + 0.0625)] \times 10{,}000 =$ 45,938 pounds.

The total safe tensile load based on this combination is 17,500 + 45,938 = 63,438 pounds, which is greater than any of the other results obtained.

The safe tensile load for the joint would be the least of the loads just computed or 43,750 pounds and the efficiency $\eta$ would be equal to this load divided by the tensile strength of the section of plate under consideration if it were unperforated or

$$\eta = \frac{43{,}750}{0.5 \times 12 \times 10{,}000} \times 100 = 72.9 \text{ percent}$$

**Formulas for Riveted Joint Design.—**A riveted joint may fail by shearing through the rivets (single or double shear), crushing the rivets, tearing the plate between the rivets, crushing the plate or by a combination of two or more of the foregoing causes. Rivets placed too close to the edge of the plate may tear or shear the plate out to the edge but this type of failure is avoided by placing the center of the rivet 1.5 times the rivet diameter away from the edge.

The efficiency of a riveted joint is equal to the strength of the joint divided by the strength of the unriveted plate, expressed as a percentage.

In the following formulas, let

$d$ = diameter of holes  $t$ = thickness of plate  $t_c$ = thickness of cover plates

$p$ = pitch of inner row of rivets  $P$ = pitch of outer row of rivets

$S_s$ = shear stress for rivets  $S_t$ = tensile stress for plates

$S_c$ = compressive or bearing stress for rivets or plates

In the joint examples that follow, dimensions are usually specified in inches and stresses in pounds per square inch. See page 1866 for a discussion of allowable stresses that may be used in calculating the strengths given by the formulas. The design stresses are usually set by codes, practices, or specifications.

*Single-Riveted Lap-Joint*

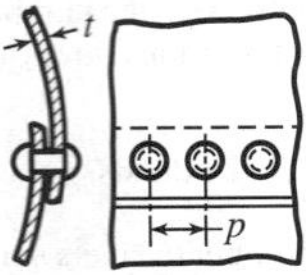


(1) Resistance to shearing one rivet = $\frac{\pi d^2}{4}S_s$

(2) Resistance to tearing plate between rivets = $(p - D)tS_t$

(3) Resistance to crushing rivet or plate = $dtS_c$

*Double-Riveted Lap-Joint*

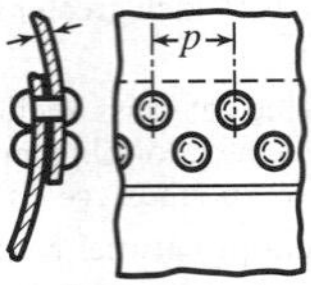


(1) Resistance to shearing two rivets = $\frac{2\pi d^2}{4}S_s$

(2) Resistance to tearing between two rivets = $(p - D)tS_t$

(3) Resistance to crushing in front of two rivets = $2dtS_c$

*Single-Riveted Lap-Joint with Inside Cover Plate*

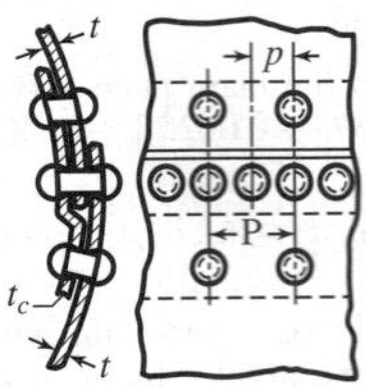


(1) Resistance to tearing between outer row of rivets = $(P - D)tS_t$

(2) Resistance to tearing between inner row of rivets, and shearing outer row of rivets = $(P - 2D)tS_t + \frac{\pi d^2}{4}S_s$

(3) Resistance to shearing three rivets = $\frac{3\pi d^2}{4}S_s$

(4) Resistance to crushing in front of three rivets = $3tdS_c$

(5) Resistance to tearing at inner row of rivets, and crushing in front of one rivet in outer row = $(P - 2D)tS_t + tdS_c$

*Double-Riveted Lap-Joint with Inside Cover Plate*

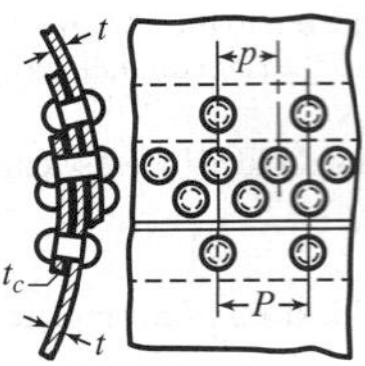


(1) Resistance to tearing at outer row of rivets = $(P - D)tS_t$

(2) Resistance to shearing four rivets = $\frac{4\pi d^2}{4}S_s$

(3) Resistance to tearing at inner row and shearing outer row of rivets = $\left(P - 1\frac{1}{2}D\right)tS_t + \frac{\pi d^2}{4}S_s$

(4) Resistance to crushing in front of four rivets = $4tdS_c$

(5) Resistance to tearing at inner row of rivets, and crushing in front of one rivet = $\left(P - 1\frac{1}{2}D\right)tS_t + tdS_c$

*Double-Riveted Butt-Joint*

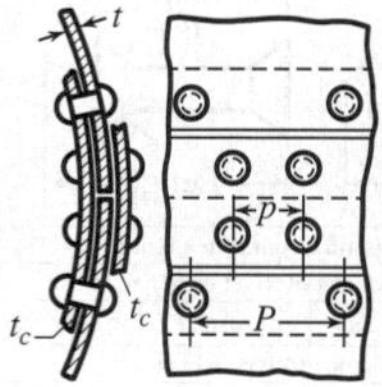


(1) Resistance to tearing at outer row of rivets = $(P-D)tS_t$

(2) Resistance to shearing two rivets in double shear and one in single shear = $\frac{5\pi d^2}{4}S_s$

(3) Resistance to tearing at inner row of rivets and shearing one rivet of the outer row = $(P-2D)tS_t+\frac{\pi d^2}{4}S_s$

(4) Resistance to crushing in front of three rivets = $3tdS_c$

(5) Resistance to tearing at inner row of rivets, and crushing in front of one rivet in outer row = $(P-2D)\,tS_t+tdS_c$

*Triple-Riveted Butt-Joint*

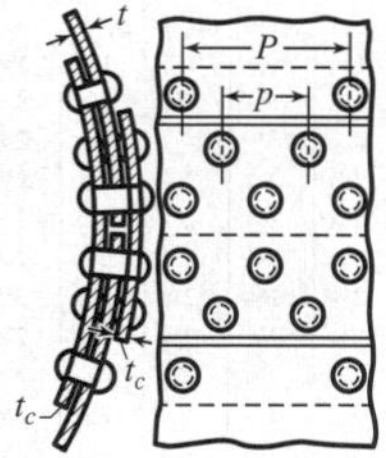


(1) Resistance to tearing at outer row of rivets = $(P-D)tS_t$

(2) Resistance to shearing four rivets in double shear and one in single shear = $\frac{9\pi d^2}{4}S_s$

(3) Resistance to tearing at middle row of rivets and shearing one rivet = $(P-2D)tS_t+\frac{\pi d^2}{4}S_s$

(4) Resistance to crushing in front of four rivets and shearing one rivet = $4dtS_c+\frac{\pi d^2}{4}S_s$

(5) Resistance to crushing in front of five rivets = $4dtS_c+dt_cS_c$

## American National Standard Rivets

Standards for rivets published by the American National Standards Institute and the British Standards Institution are as follows:

**American National Standard Large Rivets.—**The types of rivets covered by this Standard, ANSI/ASME B18.1.2-1972 (R2021), are shown on pages 1871, 1872, and 1873. It may be noted, however, that when specified, the swell neck included in this Standard is applicable to all standard large rivets except the flat countersunk head and oval countersunk head types. Also shown are the hold-on (dolly bar) and rivet set impression dimensions (see page 1873). All standard large rivets have fillets under the head not exceeding a 0.062-inch radius. The length tolerances for these rivets are given as follows: through 6 inches in length, $\frac{1}{2}$- and $\frac{5}{8}$-inch diameters, ±0.03 inch; $\frac{3}{4}$- and $\frac{7}{8}$-inch diameters, ±0.06-inch; and 1- through $1\frac{3}{4}$-inch diameters, ±0.09 inch. For rivets over 6 inches in length, $\frac{1}{2}$- and $\frac{5}{8}$-inch diameters, ±0.06 inch; $\frac{3}{4}$- and $\frac{7}{8}$-inch diameters, ±0.12 inch; and 1- through $1\frac{3}{4}$-inch diameters, ±0.19 inch. Steel and wrought iron rivet materials appear in ASTM Specifications A31, A131, A152, and A502.

**American National Standard Small Solid Rivets.—**The types of rivets covered by ANSI/ASME B18.1.1-1972 (R2021) are shown on pages 1874 through 1876. In addition, the standard gives the dimensions of 60-degree flat countersunk head rivets used to assemble ledger plates and guards for mower cutter bars, but these are not shown. As the heads of standard rivets are not machined or trimmed, the circumference may be somewhat irregular and edges may be rounded or flat. Rivets other than countersunk types are furnished with a definite fillet under the head, whose radius should not exceed 10 percent of the maximum shank diameter or 0.03 inch, whichever is the smaller. With regard to head dimensions, tolerances shown in the dimensional tables are applicable to rivets produced by the normal cold heading process. Unless otherwise specified, rivets should have plain sheared ends that should be at right angles within 2 degrees to the axis of the rivet and be reasonably flat. When so specified by the user, rivets may have the standard header points shown on page 1871. Rivets may be made of ASTM Specification A31, Grade A steel; or may adhere to SAE Recommended Practice, Mechanical and Chemical Requirements for Nonthreaded Fasteners—SAE J430, Grade 0. When specified, rivets may be made of other materials.

## Rivet Lengths for Forming Round and Countersunk Heads[a]

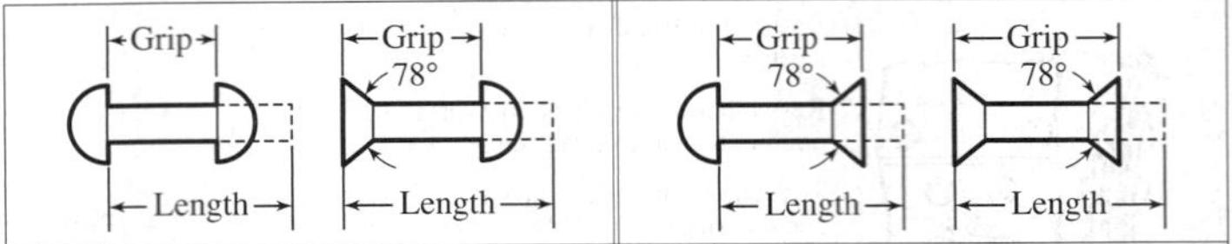


| Grip in Inches | To Form Round Head: Diameter of Rivet in Inches — Length of Rivet in Inches | | | | | | |
|---|---|---|---|---|---|---|---|
| | ½ | ⅝ | ¾ | ⅞ | 1 | 1⅛ | 1¼ |
| ½ | 1⅝ | 1⅞ | 1⅞ | 2 | 2⅛ | … | … |
| ⅝ | 1¾ | 2 | 2 | 2⅛ | 2¼ | … | … |
| ¾ | 1⅞ | 2⅛ | 2⅛ | 2¼ | 2⅜ | … | … |
| ⅞ | 2 | 2¼ | 2¼ | 2⅜ | 2½ | … | … |
| 1 | 2¼ | 2⅜ | 2⅜ | 2½ | 2⅝ | 2¾ | 2⅞ |
| 1⅛ | 2⅜ | 2½ | 2½ | 2⅝ | 2¾ | 2⅞ | 3 |
| 1¼ | 2½ | 2⅝ | 2⅝ | 2¾ | 2⅞ | 3 | 3⅛ |
| 1⅜ | 2⅝ | 2¾ | 2¾ | 2⅞ | 3 | 3⅛ | 3¼ |
| 1½ | 2⅞ | 3 | 3 | 3⅛ | 3¼ | 3⅜ | 3½ |
| 1⅝ | 3 | 3⅛ | 3⅛ | 3¼ | 3⅜ | 3½ | 3½ |
| 1¾ | 3⅛ | 3¼ | 3¼ | 3½ | 3⅝ | 3¾ | 3¾ |
| 1⅞ | 3¼ | 3⅜ | 3⅜ | 3⅝ | 3¾ | 3⅞ | 3⅞ |
| 2 | 3½ | 3½ | 3⅝ | 3¾ | 3⅞ | 4 | 4 |
| 2⅛ | 3⅝ | 3⅝ | 3¾ | 3⅞ | 4 | 4⅛ | 4⅛ |
| 2¼ | 3¾ | 3⅞ | 3⅞ | 4 | 4⅛ | 4¼ | 4¼ |
| 2⅜ | 4 | 4 | 4 | 4⅛ | 4¼ | 4⅜ | 4⅜ |
| 2½ | 4⅛ | 4⅛ | 4⅛ | 4¼ | 4⅜ | 4½ | 4½ |
| 2⅝ | 4¼ | 4¼ | 4¼ | 4⅜ | 4½ | 4⅝ | 4⅝ |
| 2¾ | 4⅜ | 4⅜ | 4⅜ | 4½ | 4⅝ | 4¾ | 4¾ |
| 2⅞ | 4⅝ | 4⅝ | 4⅝ | 4⅝ | 4¾ | 4⅞ | 5 |
| 3 | … | 4¾ | 4¾ | 4⅞ | 5 | 5 | 5⅛ |
| 3⅛ | … | 4⅞ | 4⅞ | 5 | 5⅛ | 5¼ | 5¼ |
| 3¼ | … | 5 | 5 | 5⅛ | 5¼ | 5⅜ | 5⅜ |
| 3⅜ | … | 5⅛ | 5⅛ | 5¼ | 5⅜ | 5½ | 5½ |
| 3½ | … | 5⅜ | 5⅜ | 5⅜ | 5½ | 5⅝ | 5⅝ |
| 3⅝ | … | 5½ | 5½ | 5½ | 5⅝ | 5¾ | 5¾ |
| 3¾ | … | 5⅝ | 5⅝ | 5⅝ | 5¾ | 5⅞ | 5⅞ |
| 3⅞ | … | 5¾ | 5¾ | 5¾ | 5⅞ | 6 | 6 |
| 4 | … | … | 5⅞ | 6 | 6 | 6⅛ | 6¼ |
| 4⅛ | … | … | 6 | 6⅛ | 6¼ | 6⅜ | 6⅜ |
| 4¼ | … | … | 6⅛ | 6¼ | 6½ | 6½ | 6½ |
| 4⅜ | … | … | 6⅜ | 6½ | 6½ | 6⅝ | 6⅝ |
| 4½ | … | … | 6½ | 6⅝ | 6⅝ | 6¾ | 6¾ |
| 4⅝ | … | … | 6⅝ | 6¾ | 6¾ | 6¾ | 6⅞ |
| 4¾ | … | … | 6¾ | 6⅞ | 6⅞ | 7 | 7 |
| 4⅞ | … | … | 6⅞ | 7 | 7 | 7⅛ | 7⅛ |
| 5 | … | … | … | 7⅛ | 7⅛ | 7¼ | 7¼ |
| 5⅛ | … | … | … | 7¼ | 7¼ | 7⅜ | 7⅜ |
| 5¼ | … | … | … | 7⅜ | 7⅜ | 7½ | 7½ |
| 5⅜ | … | … | … | 7⅝ | 7⅝ | 7¾ | 7⅝ |
| 5½ | … | … | … | 7¾ | 7¾ | 7⅞ | 7⅞ |
| 5⅝ | … | … | … | 7⅞ | 7⅞ | 8 | 8 |
| 5¾ | … | … | … | 8 | 8 | 8⅛ | 8⅛ |
| 5⅞ | … | … | … | 8⅛ | 8⅛ | 8¼ | 8¼ |

| Grip in Inches | To Form Countersunk Head: Diameter of Rivet in Inches — Length of Rivet in Inches | | | | | | |
|---|---|---|---|---|---|---|---|
| | ½ | ⅝ | ¾ | ⅞ | 1 | 1⅛ | 1¼ |
| ½ | 1 | 1 | 1⅛ | 1¼ | 1¼ | … | … |
| ⅝ | 1⅛ | 1¼ | 1¼ | 1⅜ | 1⅜ | … | … |
| ¾ | 1⅜ | 1⅜ | 1⅜ | 1½ | 1½ | … | … |
| ⅞ | 1½ | 1½ | 1½ | 1⅝ | 1⅝ | … | … |
| 1 | 1⅝ | 1⅝ | 1⅝ | 1¾ | 1¾ | 1⅞ | 1⅞ |
| 1⅛ | 1¾ | 1¾ | 1⅞ | 1⅞ | 1⅞ | 2 | 2 |
| 1¼ | 2 | 2 | 2 | 2 | 2 | 2⅛ | 2⅛ |
| 1⅜ | 2⅛ | 2⅛ | 2⅛ | 2¼ | 2¼ | 2⅜ | 2⅜ |
| 1½ | 2¼ | 2¼ | 2¼ | 2⅜ | 2⅜ | 2½ | 2½ |
| 1⅝ | 2⅜ | 2⅜ | 2⅜ | 2½ | 2½ | 2⅝ | 2⅝ |
| 1¾ | 2⅝ | 2⅝ | 2⅝ | 2⅝ | 2⅝ | 2¾ | 2¾ |
| 1⅞ | 2¾ | 2¾ | 2¾ | 2¾ | 2¾ | 2⅞ | 2⅞ |
| 2 | 2⅞ | 2⅞ | 2⅞ | 2⅞ | 2⅞ | 3 | 3 |
| 2⅛ | 3⅛ | 3 | 3 | 3 | 3 | 3⅛ | 3⅛ |
| 2¼ | 3¼ | 3⅛ | 3⅛ | 3⅛ | 3¼ | 3¼ | 3¼ |
| 2⅜ | 3⅜ | 3⅜ | 3⅜ | 3⅜ | 3⅜ | 3⅜ | 3⅜ |
| 2½ | 3½ | 3½ | 3½ | 3½ | 3½ | 3⅝ | 3⅝ |
| 2⅝ | 3¾ | 3⅝ | 3⅝ | 3⅝ | 3⅝ | 3¾ | 3¾ |
| 2¾ | 3⅞ | 3¾ | 3¾ | 3¾ | 3¾ | 3⅞ | 3⅞ |
| 2⅞ | 4 | 3⅞ | 3⅞ | 3⅞ | 3⅞ | 4 | 4 |
| 3 | … | 4⅛ | 4⅛ | 4⅛ | 4⅛ | 4⅛ | 4⅛ |
| 3⅛ | … | 4¼ | 4¼ | 4¼ | 4¼ | 4¼ | 4¼ |
| 3¼ | … | 4⅜ | 4⅜ | 4⅜ | 4⅜ | 4⅜ | 4⅜ |
| 3⅜ | … | 4½ | 4½ | 4½ | 4½ | 4½ | 4½ |
| 3½ | … | 4⅝ | 4⅝ | 4⅝ | 4⅝ | 4⅝ | 4⅝ |
| 3⅝ | … | 4¾ | 4¾ | 4¾ | 4¾ | 4⅞ | 4⅞ |
| 3¾ | … | 5 | 5 | 5 | 5 | 5 | 5 |
| 3⅞ | … | 5⅛ | 5⅛ | 5⅛ | 5⅛ | 5⅛ | 5⅛ |
| 4 | … | … | 5¼ | 5¼ | 5¼ | 5¼ | 5¼ |
| 4⅛ | … | … | 5⅜ | 5⅜ | 5⅜ | 5⅜ | 5⅜ |
| 4¼ | … | … | 5½ | 5½ | 5½ | 5½ | 5½ |
| 4⅜ | … | … | 5⅝ | 5⅝ | 5⅝ | 5⅝ | 5⅝ |
| 4½ | … | … | 5¾ | 5¾ | 5¾ | 5¾ | 5¾ |
| 4⅝ | … | … | 6 | 6 | 6 | 6 | 6 |
| 4¾ | … | … | 6⅛ | 6⅛ | 6⅛ | 6⅛ | 6⅛ |
| 4⅞ | … | … | 6¼ | 6¼ | 6¼ | 6¼ | 6¼ |
| 5 | … | … | … | 6⅜ | 6⅜ | 6⅜ | 6⅜ |
| 5⅛ | … | … | … | 6½ | 6½ | 6½ | 6½ |
| 5¼ | … | … | … | 6⅝ | 6⅝ | 6⅝ | 6⅝ |
| 5⅜ | … | … | … | 6¾ | 6¾ | 6¾ | 6¾ |
| 5½ | … | … | … | 6⅞ | 6⅞ | 6⅞ | 6⅞ |
| 5⅝ | … | … | … | 7 | 7 | 7 | 7 |
| 5¾ | … | … | … | 7¼ | 7¼ | 7¼ | 7¼ |
| 5⅞ | … | … | … | 7⅜ | 7⅜ | 7⅜ | 7⅜ |

[a] As given by the American Institute of Steel Construction. Values may vary from standard practice of individual fabricators and should be checked against the fabricator's standard.

### Table 1a. American National Standard Large Rivets
*ANSI/ASME B18.1.2-1972 (R2021)*

Flat Countersunk Head    Oval Countersunk Head

**Flat and Oval Countersunk Head**

| Body Diameter[a] *D* | | | | Head Dia. *A* | | Head Depth *H* | Oval Crown Height[a] *C* | Oval Crown Radius[a] *G* |
|---|---|---|---|---|---|---|---|---|
| Nominal[a] | | Max. | Min. | Max.[b] | Min.[c] | Ref. | | |
| ½ | 0.500 | 0.520 | 0.478 | 0.936 | 0.872 | 0.260 | 0.095 | 1.125 |
| ⅝ | 0.625 | 0.655 | 0.600 | 1.194 | 1.112 | 0.339 | 0.119 | 1.406 |
| ¾ | 0.750 | 0.780 | 0.725 | 1.421 | 1.322 | 0.400 | 0.142 | 1.688 |
| ⅞ | 0.875 | 0.905 | 0.850 | 1.647 | 1.532 | 0.460 | 0.166 | 1.969 |
| 1 | 1.000 | 1.030 | 0.975 | 1.873 | 1.745 | 0.520 | 0.190 | 2.250 |
| 1⅛ | 1.125 | 1.160 | 1.098 | 2.114 | 1.973 | 0.589 | 0.214 | 2.531 |
| 1¼ | 1.250 | 1.285 | 1.223 | 2.340 | 2.199 | 0.650 | 0.238 | 2.812 |
| 1⅜ | 1.375 | 1.415 | 1.345 | 2.567 | 2.426 | 0.710 | 0.261 | 3.094 |
| 1½ | 1.500 | 1.540 | 1.470 | 2.793 | 2.652 | 0.771 | 0.285 | 3.375 |
| 1⅝ | 1.625 | 1.665 | 1.588 | 3.019 | 2.878 | 0.831 | 0.309 | 3.656 |
| 1¾ | 1.750 | 1.790 | 1.713 | 3.262 | 3.121 | 0.901 | 0.332 | 3.938 |

[a] All dimensions are given in inches. Basic dimension as manufactured. For tolerances see table footnote on page 1869. The following formulas give basic dimensions for manufactured shapes: *Flat Countersunk Head,* $A = 1.810D$; $H = 1.192(\text{Max } A - D)/2$; included angle $Q$ of head = 78 degrees. *Oval Countersunk Head,* $A = 1.810D$; $H = 1.192(\text{Max } A - D)/2$; included angle of head = 78 degrees. Length $L$ is measured parallel to the rivet axis, from the extreme end to the intersection of the head top surface with the head diameter for countersunk head-type rivets.

[b] Sharp edged head.

[c] Rounded or flat edged irregularly shaped head (heads are not machined or trimmed).

ANSI/ASME B18.1.3M-1983 (R2021), Metric Small Solid Rivets, provides data for small, solid rivets with flat, round, and flat countersunk heads in metric dimensions. The main series of rivets has body diameters, in millimeters, of 1.6, 2, 2.5, 3, 4, 5, 6, 8, 10, and 12. A secondary series (nonpreferred) consists of sizes, 1, 1.2, 1.4, 3.5, 7, 9, and 11 millimeters.

## British Standard Rivets

**British Standard Rivets for General Engineering.—**Dimensions in metric units of rivets for general engineering purposes are given in this British Standard, BS4620:1970 (Withdrawn), which is based on ISO Recommendation ISO/R 1051. The snap head rivet dimensions of 14 millimeters and above are taken from the German Standard DIN 124, Round Head Rivets for Steel Structures. The shapes of heads have been restricted to those in common use in the United Kingdom. Table 3b shows the rivet dimensions. Table 3a shows a tentative range of preferred nominal lengths as given in an appendix to the Standard. It is stated that these lengths will be reviewed in the light of usage. The rivets are made by cold or hot forging methods from mild steel, copper, brass, pure aluminum, aluminum alloys, or other suitable metal. It is stated that the radius under the head of a rivet shall run smoothly into the face of the head and shank without step or discontinuity.

**Table 1b. American National Standard Large Rivet**
*ANSI/ASME B18.1.2-1972 (R2021)*

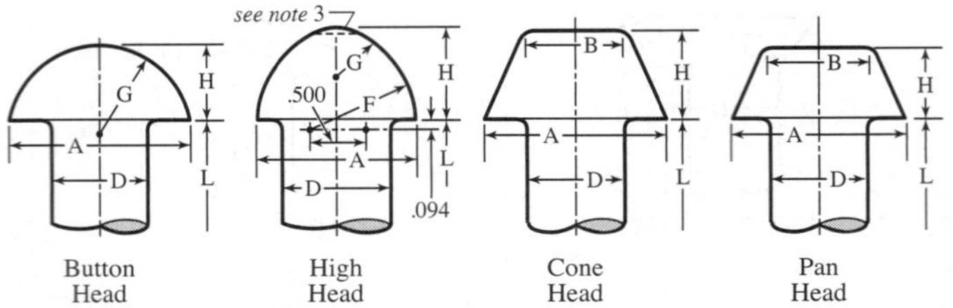


| Nom. Body Dia. $D$[a] | Head Dia. $A$ | | Height $H$ | | Head Dia. $A$ | | Height $H$ | |
|---|---|---|---|---|---|---|---|---|
| | Mfd.[b] | Driven[c] | Mfd.[b] | Driven[c] | Mfd.[b] | Driven[c] | Mfd.[b,d] | Driven[c,d] |
| | **Button Head** | | | | **High Button Head (Acorn)** | | | |
| ½ | 0.875 | 0.922 | 0.375 | 0.344 | 0.781 | 0.875 | 0.500 | 0.375 |
| ⅝ | 1.094 | 1.141 | 0.469 | 0.438 | 0.969 | 1.062 | 0.594 | 0.453 |
| ¾ | 1.312 | 1.375 | 0.562 | 0.516 | 1.156 | 1.250 | 0.688 | 0.531 |
| ⅞ | 1.531 | 1.594 | 0.656 | 0.609 | 1.344 | 1.438 | 0.781 | 0.609 |
| 1 | 1.750 | 1.828 | 0.750 | 0.688 | 1.531 | 1.625 | 0.875 | 0.688 |
| 1⅛ | 1.969 | 2.062 | 0.844 | 0.781 | 1.719 | 1.812 | 0.969 | 0.766 |
| 1¼ | 2.188 | 2.281 | 0.938 | 0.859 | 1.906 | 2.000 | 1.062 | 0.844 |
| 1⅜ | 2.406 | 2.516 | 1.031 | 0.953 | 2.094 | 2.188 | 1.156 | 0.938 |
| 1½ | 2.625 | 2.734 | 1.125 | 1.031 | 2.281 | 2.375 | 1.250 | 1.000 |
| 1⅝ | 2.844 | 2.969 | 1.219 | 1.125 | 2.469 | 2.562 | 1.344 | 1.094 |
| 1¾ | 3.062 | 3.203 | 1.312 | 1.203 | 2.656 | 2.750 | 1.438 | 1.172 |
| | **Cone Head** | | | | **Pan Head** | | | |
| ½ | 0.875 | 0.922 | 0.438 | 0.406 | 0.800 | 0.844 | 0.350 | 0.328 |
| ⅝ | 1.094 | 1.141 | 0.547 | 0.516 | 1.000 | 1.047 | 0.438 | 0.406 |
| ¾ | 1.312 | 1.375 | 0.656 | 0.625 | 1.200 | 1.266 | 0.525 | 0.484 |
| ⅞ | 1.531 | 1.594 | 0.766 | 0.719 | 1.400 | 1.469 | 0.612 | 0.578 |
| 1 | 1.750 | 1.828 | 0.875 | 0.828 | 1.600 | 1.687 | 0.700 | 0.656 |
| 1⅛ | 1.969 | 2.063 | 0.984 | 0.938 | 1.800 | 1.891 | 0.788 | 0.734 |
| 1¼ | 2.188 | 2.281 | 1.094 | 1.031 | 2.000 | 2.094 | 0.875 | 0.812 |
| 1⅜ | 2.406 | 2.516 | 1.203 | 1.141 | 2.200 | 2.312 | 0.962 | 0.906 |
| 1½ | 2.625 | 2.734 | 1.312 | 1.250 | 2.400 | 2.516 | 1.050 | 0.984 |
| 1⅝ | 2.844 | 2.969 | 1.422 | 1.344 | 2.600 | 2.734 | 1.138 | 1.062 |
| 1¾ | 3.062 | 3.203 | 1.531 | 1.453 | 2.800 | 2.938 | 1.225 | 1.141 |

[a] Tolerance for diameter of body is plus and minus from nominal and for ½-in. size equals +0.020, −0.022; for sizes ⅝ to 1-in., incl., equals +0.030, −.025; for sizes 1⅛ and 1¼-in. equals +0.035; −0.027; for sizes 1⅜ and 1½-in. equals +0.040, −0.030; for sizes 1⅝ and 1¾-in. equals +0.040, −0.037.

[b] Note 1. Basic dimensions of head as manufactured. All dimensions are given in inches. The following formulas give the basic dimensions for manufactured shapes: *Button Head,* $A = 1.750D$; $H = 0.750D$; $G = 0.885\,D$. *High Button Head,* $A = 1.500D + 0.031$; $H = 0.750D + 0.125$; $F = 0.750D + 0.281$; $G = 0.750D - 0.281$. *Cone Head,* $A = 1.750D$; $B = 0.938D$; $H = 0.875D$. *Pan Head,* $A = 1.600D$; $B = 1.000D$; $H = 0.700D$. Length L is measured parallel to the rivet axis, from the extreme end to the bearing surface plane for flat bearing surface head type rivets, or to the intersection of the head top surface with the head diameter for countersunk head-type rivets.

[c] Note 2. Dimensions of manufactured head after driving and also of driven head.

[d] Note 3. Slight flat permissible within the specified head-height tolerance.

## Table 1c. American National Standard Large Rivets
*ANSI/ASME B18.1.2-1972 (R2021)*

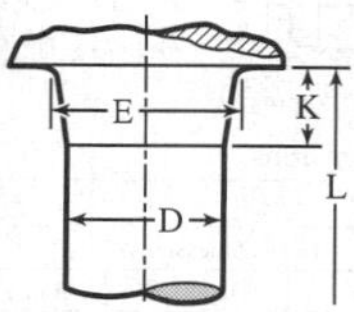

Swell Neck

| Swell Neck[a] | | | | | | |
|---|---|---|---|---|---|---|
| Body Diameter *D* | | | | Diameter Under Head *E* | | Neck Length *K*[b] |
| Nominal[b] | | Max. | Min. | Max. (Basic) | Min. | |
| ½ | 0.500 | 0.520 | 0.478 | 0.563 | 0.543 | 0.250 |
| ⅝ | 0.625 | 0.655 | 0.600 | 0.688 | 0.658 | 0.312 |
| ¾ | 0.750 | 0.780 | 0.725 | 0.813 | 0.783 | 0.375 |
| ⅞ | 0.875 | 0.905 | 0.850 | 0.938 | 0.908 | 0.438 |
| 1 | 1.000 | 1.030 | 0.975 | 1.063 | 1.033 | 0.500 |
| 1⅛ | 1.125 | 1.160 | 1.098 | 1.188 | 1.153 | 0.562 |
| 1¼ | 1.250 | 1.285 | 1.223 | 1.313 | 1.278 | 0.625 |
| 1⅜ | 1.375 | 1.415 | 1.345 | 1.438 | 1.398 | 0.688 |
| 1½ | 1.500 | 1.540 | 1.470 | 1.563 | 1.523 | 0.750 |
| 1⅝ | 1.625 | 1.665 | 1.588 | 1.688 | 1.648 | 0.812 |
| 1¾ | 1.750 | 1.790 | 1.713 | 1.813 | 1.773 | 0.875 |

[a] The swell neck is applicable to all standard forms of large rivets except the flat countersunk and oval countersunk head types.

[b] All dimensions are given in inches. The following formulas give basic dimensions for manufactured shapes: *Swell Neck*, $E = D + 0.063$; $K = 0.500D$. Length $L$ is measured parallel to the rivet axis, from the extreme end to the bearing surface plane for flat bearing surface head-type rivets. Basic dimension as manufactured. For tolerances see table footnote on page 1869.

## American National Standard Dimensions for Hold-On (Dolly Bar) and Rivet Set Impression *ANSI/ASME B18.1.2-1972 (R2021)*

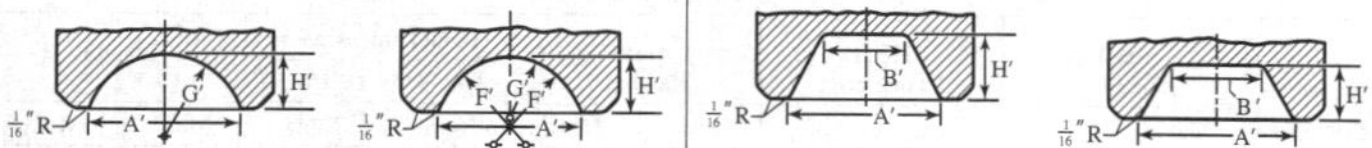

Button Head High Button Head Cone Head Pan Head

| Rivet Body Dia.[a] | Button Head | | | High Button Head | | | | Cone Head | | | Pan Head | | |
|---|---|---|---|---|---|---|---|---|---|---|---|---|---|
| | *A′* | *H′* | *G′* | *A′* | *H′* | *F′* | *G′* | *A′* | *B′* | *H′* | *A′* | *B′* | *H′* |
| ½ | 0.906 | 0.312 | 0.484 | 0.859 | 0.344 | 0.562 | 0.375 | 0.891 | 0.469 | 0.391 | 0.812 | 0.500 | 0.297 |
| ⅝ | 1.125 | 0.406 | 0.594 | 1.047 | 0.422 | 0.672 | 0.453 | 1.109 | 0.594 | 0.484 | 1.031 | 0.625 | 0.375 |
| ¾ | 1.344 | 0.484 | 0.719 | 1.234 | 0.500 | 0.797 | 0.531 | 1.328 | 0.703 | 0.578 | 1.234 | 0.750 | 0.453 |
| ⅞ | 1.578 | 0.562 | 0.844 | 1.422 | 0.578 | 0.922 | 0.609 | 1.562 | 0.828 | 0.688 | 1.438 | 0.875 | 0.531 |
| 1 | 1.812 | 0.641 | 0.953 | 1.609 | 0.656 | 1.031 | 0.688 | 1.781 | 0.938 | 0.781 | 1.641 | 1.000 | 0.609 |
| 1⅛ | 2.031 | 0.719 | 1.078 | 1.797 | 0.719 | 1.156 | 0.766 | 2.000 | 1.063 | 0.875 | 1.844 | 1.125 | 0.688 |
| 1¼ | 2.250 | 0.797 | 1.188 | 1.984 | 0.797 | 1.266 | 0.844 | 2.219 | 1.172 | 0.969 | 2.047 | 1.250 | 0.766 |
| 1⅜ | 2.469 | 0.875 | 1.312 | 2.172 | 0.875 | 1.406 | 0.938 | 2.453 | 1.297 | 1.078 | 2.250 | 1.375 | 0.844 |
| 1½ | 2.703 | 0.953 | 1.438 | 2.344 | 0.953 | 1.500 | 1.000 | 2.672 | 1.406 | 1.172 | 2.453 | 1.500 | 0.906 |
| 1⅝ | 2.922 | 1.047 | 1.547 | 2.531 | 1.031 | 1.641 | 1.094 | 2.891 | 1.531 | 1.266 | 2.656 | 1.625 | 0.984 |
| 1¾ | 3.156 | 1.125 | 1.672 | 2.719 | 1.109 | 1.750 | 1.172 | 3.109 | 1.641 | 1.375 | 2.875 | 1.750 | 1.063 |

[a] All dimensions are given in inches.

## Table 2a. American National Standard Small Solid Rivets

*ANSI/ASME B18.1.1-1972 (R2021) and Appendix*

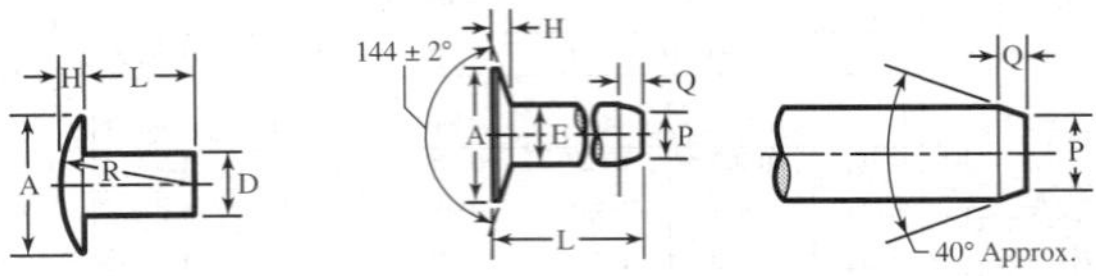


$P = D \times 0.818$

$Q = D \times 0.25$

Truss Head Rivets    Coopers Rivets    Point Dimensions

**Truss Head Rivets**[a]

| Shank Dia.,[b] D Nominal | | Head Dimensions Dia., A Max. | Min. | Height, H Max. | Min. | Rad. R Approx. | Shank Dia.,[b] D Nominal | | Head Dimensions Dia., A Max. | Min. | Height, H Max. | Min. | Rad. R Approx. |
|---|---|---|---|---|---|---|---|---|---|---|---|---|---|
| 3/32 | 0.094 | 0.226 | 0.206 | 0.038 | 0.026 | 0.239 | 9/32 | 0.281 | 0.661 | 0.631 | 0.103 | 0.085 | 0.706 |
| 1/8 | 0.125 | 0.297 | 0.277 | 0.048 | 0.036 | 0.314 | 5/16 | 0.312 | 0.732 | 0.702 | 0.113 | 0.095 | 0.784 |
| 5/32 | 0.156 | 0.368 | 0.348 | 0.059 | 0.045 | 0.392 | 11/32 | 0.344 | 0.806 | 0.776 | 0.124 | 0.104 | 0.862 |
| 3/16 | 0.188 | 0.442 | 0.422 | 0.069 | 0.055 | 0.470 | 3/8 | 0.375 | 0.878 | 0.848 | 0.135 | 0.115 | 0.942 |
| 7/32 | 0.219 | 0.515 | 0.495 | 0.080 | 0.066 | 0.555 | 13/32 | 0.406 | 0.949 | 0.919 | 0.145 | 0.123 | 1.028 |
| 1/4 | 0.250 | 0.590 | 0.560 | 0.091 | 0.075 | 0.628 | 7/16 | 0.438 | 1.020 | 0.990 | 0.157 | 0.135 | 1.098 |

[a] All dimensions in inches except where otherwise noted. Length tolerance of rivets is + or − .016 inch. Approximate proportions of rivets: $A = 2.300 \times D, H = 0.330 \times D, R = 2.512 \times D$.

[b] Tolerances on the nominal shank diameter in inches are given for the following body diameter ranges: $^3\!/_{32}$. to $^5\!/_{32}$, plus 0.002, minus 0.004; $^3\!/_{16}$ to $^1\!/_4$, plus 0.003, minus 0.006; $^9\!/_{32}$ to $^{11}\!/_{32}$, plus 0.004, minus 0.008; and $^3\!/_8$ to $^7\!/_{16}$, plus 0.005, minus 0.010.

**Coopers Rivets**

| Size No.[a] | Shank Diameter, E Max. | Min. | Head Diameter, A Max. | Min. | Head Height, H Max. | Min. | Point Dimensions[b] Dia., P Nom. | Length, Q Nom. | Length, L Max. | Min. |
|---|---|---|---|---|---|---|---|---|---|---|
| 1 lb | 0.111 | 0.105 | 0.291 | 0.271 | 0.045 | 0.031 | Not Pointed | | 0.249 | 0.219 |
| 1¼ lb | 0.122 | 0.116 | 0.324 | 0.302 | 0.050 | 0.036 | Not Pointed | | 0.285 | 0.255 |
| 1½ lb | 0.132 | 0.126 | 0.324 | 0.302 | 0.050 | 0.036 | Not Pointed | | 0.285 | 0.255 |
| 1¾ lb | 0.136 | 0.130 | 0.324 | 0.302 | 0.052 | 0.034 | Not Pointed | | 0.318 | 0.284 |
| 2 lb | 0.142 | 0.136 | 0.355 | 0.333 | 0.056 | 0.038 | Not Pointed | | 0.322 | 0.288 |
| 3 lb | 0.158 | 0.152 | 0.386 | 0.364 | 0.058 | 0.040 | 0.123 | 0.062 | 0.387 | 0.353 |
| 4 lb | 0.168 | 0.159 | 0.388 | 0.362 | 0.058 | 0.040 | 0.130 | 0.062 | 0.418 | 0.388 |
| 5 lb | 0.183 | 0.174 | 0.419 | 0.393 | 0.063 | 0.045 | 0.144 | 0.062 | 0.454 | 0.420 |
| 6 lb | 0.206 | 0.197 | 0.482 | 0.456 | 0.073 | 0.051 | 0.160 | 0.094 | 0.498 | 0.457 |
| 7 lb | 0.223 | 0.214 | 0.513 | 0.487 | 0.076 | 0.054 | 0.175 | 0.094 | 0.561 | 0.523 |
| 8 lb | 0.241 | 0.232 | 0.546 | 0.516 | 0.081 | 0.059 | 0.182 | 0.094 | 0.597 | 0.559 |
| 9 lb | 0.248 | 0.239 | 0.578 | 0.548 | 0.085 | 0.063 | 0.197 | 0.094 | 0.601 | 0.563 |
| 10 lb | 0.253 | 0.244 | 0.578 | 0.548 | 0.085 | 0.063 | 0.197 | 0.094 | 0.632 | 0.594 |
| 12 lb | 0.263 | 0.251 | 0.580 | 0.546 | 0.086 | 0.060 | 0.214 | 0.094 | 0.633 | 0.575 |
| 14 lb | 0.275 | 0.263 | 0.611 | 0.577 | 0.091 | 0.065 | 0.223 | 0.094 | 0.670 | 0.612 |
| 16 lb | 0.285 | 0.273 | 0.611 | 0.577 | 0.089 | 0.063 | 0.223 | 0.094 | 0.699 | 0.641 |
| 18 lb | 0.285 | 0.273 | 0.642 | 0.608 | 0.108 | 0.082 | 0.230 | 0.125 | 0.749 | 0.691 |
| 20 lb | 0.316 | 0.304 | 0.705 | 0.671 | 0.128 | 0.102 | 0.250 | 0.125 | 0.769 | 0.711 |
| 3/8 in. | 0.380 | 0.365 | 0.800 | 0.762 | 0.136 | 0.106 | 0.312 | 0.125 | 0.840 | 0.778 |

[a] All dimensions in inches except where otherwise noted. Size numbers in pounds refer to the approximate weight of 1000 rivets.

[b] When specified American National Standard Small Solid Rivets may be obtained with points. Point dimensions for belt and coopers rivets are given in the accompanying tables. Formulas for calculating point dimensions of other rivets are given alongside the right diagram in Table 2a.

## Table 2b. American National Standard Small Solid Rivets

*ANSI/ASME B18.1.1-1972 (R2021)*

Tinners Rivets

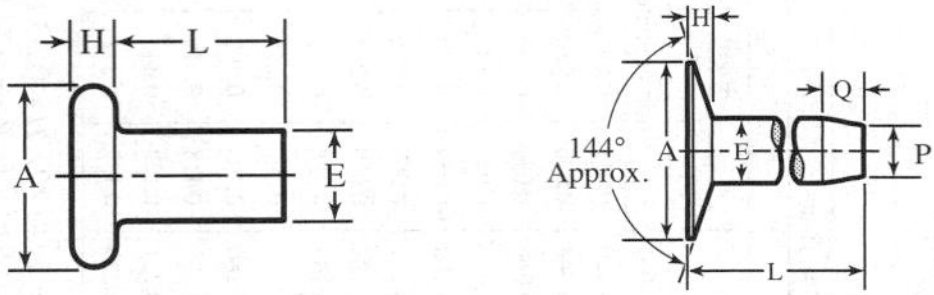


Belt Rivets

| Tinners Rivets | | | | | | | | | |
|---|---|---|---|---|---|---|---|---|---|
| Size No.[a] | Shank Diameter, *E* | | Head Dia., *A* | | Head Height, *H* | | Length, *L* | | |
| | Max. | Min. | Max. | Min. | Max. | Min. | Nom. | Max. | Min. |
| 6 oz. | 0.081 | 0.075 | 0.213 | 0.193 | 0.028 | 0.016 | 1/8 | 0.135 | 0.115 |
| 8 oz. | 0.091 | 0.085 | 0.225 | 0.205 | 0.036 | 0.024 | 5/32 | 0.166 | 0.146 |
| 10 oz. | 0.097 | 0.091 | 0.250 | 0.230 | 0.037 | 0.025 | 11/64 | 0.182 | 0.162 |
| 12 oz. | 0.107 | 0.101 | 0.265 | 0.245 | 0.037 | 0.025 | 3/16 | 0.198 | 0.178 |
| 14 oz. | 0.111 | 0.105 | 0.275 | 0.255 | 0.038 | 0.026 | 3/16 | 0.198 | 0.178 |
| 1 lb | 0.113 | 0.107 | 0.285 | 0.265 | 0.040 | 0.028 | 13/64 | 0.213 | 0.193 |
| 1 1/4 lb | 0.122 | 0.116 | 0.295 | 0.275 | 0.045 | 0.033 | 7/32 | 0.229 | 0.209 |
| 1 1/2 lb | 0.132 | 0.126 | 0.316 | 0.294 | 0.046 | 0.034 | 15/64 | 0.244 | 0.224 |
| 1 3/4 lb | 0.136 | 0.130 | 0.331 | 0.309 | 0.049 | 0.035 | 1/4 | 0.260 | 0.240 |
| 2 lb | 0.146 | 0.140 | 0.341 | 0.319 | 0.050 | 0.036 | 17/64 | 0.276 | 0.256 |
| 2 1/2 lb | 0.150 | 0.144 | 0.311 | 0.289 | 0.069 | 0.055 | 9/32 | 0.291 | 0.271 |
| 3 lb | 0.163 | 0.154 | 0.329 | 0.303 | 0.073 | 0.059 | 5/16 | 0.323 | 0.303 |
| 3 1/2 lb | 0.168 | 0.159 | 0.348 | 0.322 | 0.074 | 0.060 | 21/64 | 0.338 | 0.318 |
| 4 lb | 0.179 | 0.170 | 0.368 | 0.342 | 0.076 | 0.062 | 11/32 | 0.354 | 0.334 |
| 5 lb | 0.190 | 0.181 | 0.388 | 0.362 | 0.084 | 0.070 | 3/8 | 0.385 | 0.365 |
| 6 lb | 0.206 | 0.197 | 0.419 | 0.393 | 0.090 | 0.076 | 25/64 | 0.401 | 0.381 |
| 7 lb | 0.223 | 0.214 | 0.431 | 0.405 | 0.094 | 0.080 | 13/32 | 0.416 | 0.396 |
| 8 lb | 0.227 | 0.218 | 0.475 | 0.445 | 0.101 | 0.085 | 7/16 | 0.448 | 0.428 |
| 9 lb | 0.241 | 0.232 | 0.490 | 0.460 | 0.103 | 0.087 | 29/64 | 0.463 | 0.443 |
| 10 lb | 0.241 | 0.232 | 0.505 | 0.475 | 0.104 | 0.088 | 15/32 | 0.479 | 0.459 |
| 12 lb | 0.263 | 0.251 | 0.532 | 0.498 | 0.108 | 0.090 | 1/2 | 0.510 | 0.490 |
| 14 lb | 0.288 | 0.276 | 0.577 | 0.543 | 0.113 | 0.095 | 33/64 | 0.525 | 0.505 |
| 16 lb | 0.304 | 0.292 | 0.597 | 0.563 | 0.128 | 0.110 | 17/32 | 0.541 | 0.521 |
| 18 lb | 0.347 | 0.335 | 0.706 | 0.668 | 0.156 | 0.136 | 19/32 | 0.603 | 0.583 |

[a] All dimensions in inches. Size numbers refer to the approximate weight of 1000 rivets.

| Belt Rivets[a] | | | | | | | | |
|---|---|---|---|---|---|---|---|---|
| Size No.[b] | Shank Diameter, *E* | | Head Dia., *A* | | Head Height, *H* | | Point Dimensions[c] | |
| | | | | | | | Dia., *P* | Length, *Q* |
| | Max. | Min. | Max. | Min. | Max. | Min. | Nominal | Nominal |
| 14 | 0.085 | 0.079 | 0.260 | 0.240 | 0.042 | 0.030 | 0.065 | 0.078 |
| 13 | 0.097 | 0.091 | 0.322 | 0.302 | 0.051 | 0.039 | 0.073 | 0.078 |
| 12 | 0.111 | 0.105 | 0.353 | 0.333 | 0.054 | 0.040 | 0.083 | 0.078 |
| 11 | 0.122 | 0.116 | 0.383 | 0.363 | 0.059 | 0.045 | 0.097 | 0.078 |
| 10 | 0.136 | 0.130 | 0.417 | 0.395 | 0.065 | 0.047 | 0.109 | 0.094 |
| 9 | 0.150 | 0.144 | 0.448 | 0.426 | 0.069 | 0.051 | 0.122 | 0.094 |
| 8 | 0.167 | 0.161 | 0.481 | 0.455 | 0.072 | 0.054 | 0.135 | 0.094 |
| 7 | 0.183 | 0.174 | 0.513 | 0.487 | 0.075 | 0.056 | 0.151 | 0.125 |
| 6 | 0.206 | 0.197 | 0.606 | 0.580 | 0.090 | 0.068 | 0.165 | 0.125 |
| 5 | 0.223 | 0.214 | 0.700 | 0.674 | 0.105 | 0.083 | 0.185 | 0.125 |
| 4 | 0.241 | 0.232 | 0.921 | 0.893 | 0.138 | 0.116 | 0.204 | 0.141 |

[a] All dimensions in inches. Length tolerance on belt rivets is plus 0.031 inch, minus 0 inch.

[b] Size number refers to the Stub's iron wire gage number of the stock used in the shank of the rivet.

[c] *Note:* American National Standard Small Solid Rivets may be obtained with or without points. Point proportions are given in the diagram in Table 2a.

## Table 2c. American National Standard Small Solid Rivets *ANSI/ASME B18.1.1-1972 (R2021) and Appendix*

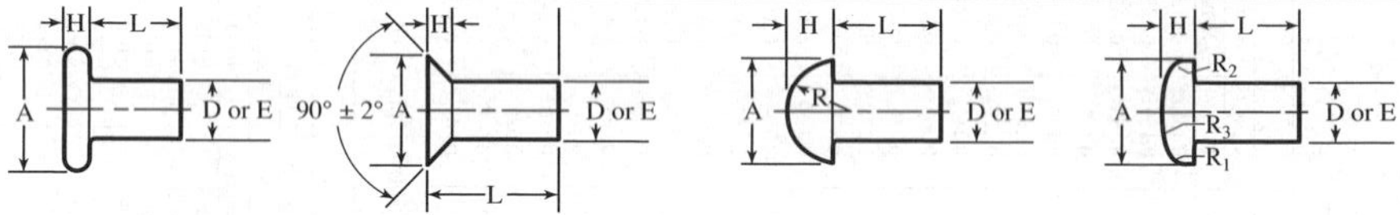


| Shank Diameter | | | | Flat Head[a] | | | | Flat Countersunk Head[a] | | | Button Head[a] | | | | | Pan Head[a] | | | | | | |
|---|---|---|---|---|---|---|---|---|---|---|---|---|---|---|---|---|---|---|---|---|---|---|
| | | | | Head Dimensions | | | | | | | | | | | | | | | | | | |
| *D* | | *E* | | Dia., *A* | | Height, *H* | | Dia., *A* Sharp | | Height[b] *H* | Dia., *A* | | Height, *H* | | Radius, *R* | Dia., *A* | | Height, *H* | | Radii *R1* | *R2* | *R3* |
| Nominal | | Max. | Min. | Max. | Min. | Max. | Min. | Max.[c] | Min.[d] | Ref. | Max. | Min. | Max. | Min. | Approx. | Max. | Min. | Max. | Min. | Approximate | | |
| 1/16 | 0.062 | 0.064 | 0.059 | 0.140 | 0.120 | 0.027 | 0.017 | 0.118 | 0.110 | 0.027 | 0.122 | 0.102 | 0.052 | 0.042 | 0.055 | 0.118 | 0.098 | 0.040 | 0.030 | 0.019 | 0.052 | 0.217 |
| 3/32 | 0.094 | 0.096 | 0.090 | 0.200 | 0.180 | 0.038 | 0.026 | 0.176 | 0.163 | 0.040 | 0.182 | 0.162 | 0.077 | 0.065 | 0.084 | 0.173 | 0.153 | 0.060 | 0.048 | 0.030 | 0.080 | 0.326 |
| 1/8 | 0.125 | 0.127 | 0.121 | 0.260 | 0.240 | 0.048 | 0.036 | 0.235 | 0.217 | 0.053 | 0.235 | 0.215 | 0.100 | 0.088 | 0.111 | 0.225 | 0.205 | 0.078 | 0.066 | 0.039 | 0.106 | 0.429 |
| 5/32 | 0.156 | 0.158 | 0.152 | 0.323 | 0.301 | 0.059 | 0.045 | 0.293 | 0.272 | 0.066 | 0.290 | 0.268 | 0.124 | 0.110 | 0.138 | 0.279 | 0.257 | 0.096 | 0.082 | 0.049 | 0.133 | 0.535 |
| 3/16 | 0.188 | 0.191 | 0.182 | 0.387 | 0.361 | 0.069 | 0.055 | 0.351 | 0.326 | 0.079 | 0.348 | 0.322 | 0.147 | 0.133 | 0.166 | 0.334 | 0.308 | 0.114 | 0.100 | 0.059 | 0.159 | 0.641 |
| 7/32 | 0.219 | 0.222 | 0.213 | 0.453 | 0.427 | 0.080 | 0.065 | 0.413 | 0.384 | 0.094 | 0.405 | 0.379 | 0.172 | 0.158 | 0.195 | 0.391 | 0.365 | 0.133 | 0.119 | 0.069 | 0.186 | 0.754 |
| 1/4 | 0.250 | 0.253 | 0.244 | 0.515 | 0.485 | 0.091 | 0.075 | 0.469 | 0.437 | 0.106 | 0.460 | 0.430 | 0.196 | 0.180 | 0.221 | 0.444 | 0.414 | 0.151 | 0.135 | 0.079 | 0.213 | 0.858 |
| 9/32 | 0.281 | 0.285 | 0.273 | 0.579 | 0.545 | 0.103 | 0.085 | 0.528 | 0.491 | 0.119 | 0.518 | 0.484 | 0.220 | 0.202 | 0.249 | 0.499 | 0.465 | 0.170 | 0.152 | 0.088 | 0.239 | 0.963 |
| 5/16 | 0.312 | 0.316 | 0.304 | 0.641 | 0.607 | 0.113 | 0.095 | 0.588 | 0.547 | 0.133 | 0.572 | 0.538 | 0.243 | 0.225 | 0.276 | 0.552 | 0.518 | 0.187 | 0.169 | 0.098 | 0.266 | 1.070 |
| 11/32 | 0.344 | 0.348 | 0.336 | 0.705 | 0.667 | 0.124 | 0.104 | 0.646 | 0.602 | 0.146 | 0.630 | 0.592 | 0.267 | 0.247 | 0.304 | 0.608 | 0.570 | 0.206 | 0.186 | 0.108 | 0.292 | 1.176 |
| 3/8 | 0.375 | 0.380 | 0.365 | 0.769 | 0.731 | 0.135 | 0.115 | 0.704 | 0.656 | 0.159 | 0.684 | 0.646 | 0.291 | 0.271 | 0.332 | 0.663 | 0.625 | 0.225 | 0.205 | 0.118 | 0.319 | 1.286 |
| 13/32 | 0.406 | 0.411 | 0.396 | 0.834 | 0.790 | 0.146 | 0.124 | 0.763 | 0.710 | 0.172 | 0.743 | 0.699 | 0.316 | 0.294 | 0.358 | 0.719 | 0.675 | 0.243 | 0.221 | 0.127 | 0.345 | 1.392 |
| 7/16 | 0.438 | 0.443 | 0.428 | 0.896 | 0.852 | 0.157 | 0.135 | 0.823 | 0.765 | 0.186 | 0.798 | 0.754 | 0.339 | 0.317 | 0.387 | 0.772 | 0.728 | 0.261 | 0.239 | 0.137 | 0.372 | 1.500 |

[a] All dimensions in inches. Length tolerance of all rivets is plus or minus 0.016 inch. Approximate proportions of rivets: flat head, $A = 2.00 \times D$, $H = 0.33\ D$; flat countersunk head, $A = 1.850 \times D$, $H = 0.425 \times D$; button head, $A = 1.750 \times D$, $H = 0.750 \times D$, $R = 0.885 \times D$; pan head, $A = 1.720 \times D$, $H = 0.570 \times D$, $R1 = 0.314 \times D$, $R2 = 0.850 \times D$, $R3 = 3.430 \times D$. *Note:* ANSI Small Solid Rivets may be obtained with or without points. Point proportions are given in the diagram in Table 2a.

[b] Given for reference purposes only. Variations in this dimension are controlled by the head and shank diameters and the included angle of the head.

[c] Tabulated maximum values calculated on basic diameter of rivet and 92° included angle extended to a sharp edge.

[d] Minimum of rounded or flat-edged irregularly shaped head. Rivet heads are not machined or trimmed and the circumference may be irregular and edges rounded or flat.

In this Standard, Table 3a and Table 3b, the following definitions apply: 1) *Nominal diameter:* The diameter of the shank; 2) *Nominal length of rivets other than countersunk or raised countersunk rivets:* The length from the underside of the head to the end of the shank; 3) *Nominal length of countersunk and raised countersunk rivets:* The distance from the periphery of the head to the end of the rivet measured parallel to the rivet axis; and 4) *Manufactured head:* The head on the rivet as received from the manufacturer.

**Table 3a. Tentative Range of Lengths for Rivets**

*Appendix to BS 4620:1970 (Withdrawn)*

| Nom. Shank Dia. | Nominal Length | | | | | | | | | | | | | | | | | | | | |
|---|---|---|---|---|---|---|---|---|---|---|---|---|---|---|---|---|---|---|---|---|---|
| | 3 | 4 | 5 | 6 | 8 | 10 | 12 | 14 | 16 | (18) | 20 | (22) | 25 | (28) | 30 | (32) | 35 | (38) | 40 | 45 | … |
| 1 | ● | ● | ● | ● | ● | ● | ● | ● | ● | … | ● | … | … | … | … | … | … | … | … | … | … |
| 1.2 | ● | ● | ● | ● | ● | ● | ● | ● | ● | … | ● | … | … | … | … | … | … | … | … | … | … |
| 1.6 | ● | ● | ● | ● | ● | ● | ● | ● | ● | ● | … | ● | ● | … | … | … | … | … | … | … | … |
| 2 | ● | ● | ● | ● | ● | ● | ● | ● | ● | … | ● | ● | ● | … | … | … | … | … | … | … | … |
| 2.5 | ● | ● | ● | ● | ● | ● | ● | ● | ● | ● | … | ● | ● | … | … | … | … | … | … | … | … |
| 3 | … | ● | ● | ● | ● | ● | ● | ● | ● | ● | ● | ● | ● | … | … | … | … | … | … | … | … |
| (3.5) | … | … | … | … | … | … | … | … | … | … | … | … | … | … | … | … | … | … | … | … | … |
| 4 | … | … | … | ● | ● | ● | ● | ● | ● | ● | … | ● | ● | … | … | … | … | … | … | … | … |
| 5 | … | … | … | ● | ● | ● | ● | ● | ● | ● | ● | ● | ● | ● | ● | … | ● | ● | … | … | … |
| 6 | … | … | … | ● | ● | ● | ● | ● | ● | ● | ● | ● | ● | ● | ● | … | ● | ● | … | ● | … |

| Nom. Shank Dia. | Nominal Length | | | | | | | | | | | | | | | | | | | | |
|---|---|---|---|---|---|---|---|---|---|---|---|---|---|---|---|---|---|---|---|---|---|
| | 10 | 12 | 14 | 16 | (18) | 20 | (22) | 25 | (28) | 30 | (32) | 35 | (38) | 40 | 45 | 50 | 55 | 60 | 65 | 70 | 75 |
| (7) | … | … | … | … | … | … | … | … | … | … | … | … | … | … | … | … | … | … | … | … | … |
| 8 | ● | ● | ● | ● | ● | ● | ● | ● | ● | ● | … | ● | ● | … | ● | ● | … | … | … | … | … |
| 10 | … | … | ● | ● | ● | ● | ● | ● | ● | ● | … | ● | ● | ● | ● | ● | … | … | … | … | … |
| 12 | … | … | … | … | … | … | … | ● | … | ● | … | ● | … | ● | … | ● | ● | … | … | … | ● |
| (14) | … | … | … | … | … | … | … | … | … | … | … | … | … | … | … | … | … | … | … | … | … |
| 16 | … | … | … | … | … | … | … | … | … | … | … | … | … | ● | ● | ● | ● | ● | ● | … | ● |

| Nom. Shank Dia. | Nominal Length | | | | | | | | | | | | | | | | | | | | |
|---|---|---|---|---|---|---|---|---|---|---|---|---|---|---|---|---|---|---|---|---|---|
| | 45 | 50 | 55 | 60 | 65 | 70 | 75 | 80 | 85 | 90 | (95) | 100 | (105) | 110 | (115) | 120 | (125) | 130 | 140 | 150 | 160 |
| (18) | … | … | … | … | … | … | … | … | … | … | … | … | … | … | … | … | … | … | … | … | … |
| 20 | ● | … | … | ● | … | ● | … | ● | … | … | … | … | … | … | … | … | … | … | … | … | … |
| (22) | … | … | … | … | … | … | … | … | … | … | … | … | … | … | … | … | … | … | … | … | … |
| 24 | … | … | … | … | ● | … | ● | … | ● | ● | … | ● | … | … | … | … | … | … | … | … | … |
| (27) | … | … | … | … | … | … | … | … | … | … | … | … | … | … | … | … | … | … | … | … | … |
| 30 | … | … | … | … | … | … | … | … | ● | ● | … | ● | … | ● | … | ● | … | … | … | … | … |
| (33) | … | … | … | … | … | … | … | … | … | … | … | … | … | … | … | … | … | … | … | … | … |
| 36 | … | … | … | … | … | … | … | … | … | … | … | ● | … | ● | … | ● | … | ● | … | … | … |
| (39) | … | … | … | … | … | … | … | … | … | … | … | … | … | … | … | ● | … | ● | ● | ● | ● |

All dimensions are in millimeters.

*Note:* Sizes and lengths shown in parenthesis are nonpreferred and should be avoided if possible.

**British Standard Small Rivets for General Purposes.**—Dimensions of small rivets for general purposes are given in British Standard 641:1951 and are shown in Table 4 on page 1879. In addition, the Standard lists the standard lengths of these rivets, gives the dimensions of washers to be used with countersunk head rivets (140°), indicates that the rivets may be made from mild steel, copper, brass, and a range of aluminum alloys and pure aluminum specified in BS 1473, and gives the dimensions of Coopers' flat head rivets $\frac{1}{2}$ inch in diameter and below, in an appendix. In all types of rivets, except those with countersunk heads, there is a small radius or chamfer at the junction of the head and the shank.

**British Standard Dimensions of Rivets ($\frac{1}{2}$ to $1\frac{3}{4}$ Inch Diameter).**—The dimensions of rivets covered in BS 275:1927 (Withdrawn) are given on page 1880 and do not apply to boiler rivets. With regard to this Standard the terms "nominal diameter" and "standard diameter" are synonymous. The term "tolerance" refers to the variation from the nominal diameter of the rivet and not to the difference between the diameter under the head and the diameter near the point.

**Table 3b. British Standard Rivets for General Engineering Purposes**

*BS 4620:1970 (Withdrawn)*

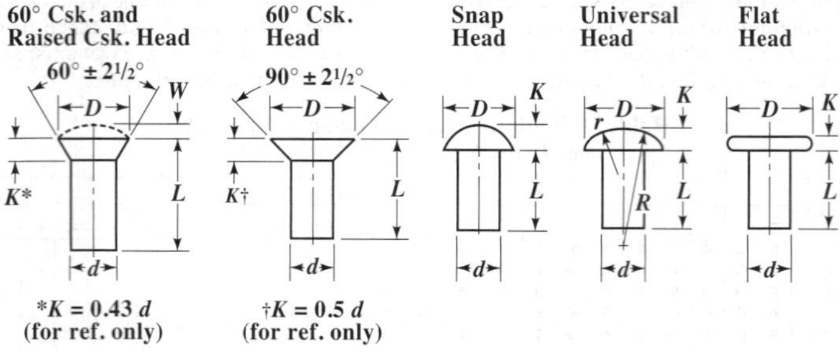


| Hot Forged Rivets | | | | | | | | | |
|---|---|---|---|---|---|---|---|---|---|
| | | 60° Csk. and Raised Csk. Head | | Snap Head | | Universal Head | | | |
| | | Head Dimensions | | | | | | | |
| Nom. Shank Dia.[a] *d* | Tol. on Dia. *d* | Nom. Dia. *D* | Height of Raise *W* | Nom. Dia. *D* | Nom. Depth *K* | Nom. Dia. *D* | Nom. Depth *K* | Rad. *R* | Rad. *r* |
| (14) | | 21 | 2.8 | 22 | 9 | 28 | 5.6 | 42 | 8.4 |
| 16 | ±0.43 | 24 | 3.2 | 25 | 10 | 32 | 6.4 | 48 | 9.6 |
| (18) | | 27 | 3.6 | 28 | 11.5 | 36 | 7.2 | 54 | 11 |
| 20 | | 30 | 4.0 | 32 | 13 | 40 | 8.0 | 60 | 12 |
| (22) | ±0.52 | 33 | 4.4 | 36 | 14 | 44 | 8.8 | 66 | 13 |
| 24 | | 36 | 4.8 | 40 | 16 | 48 | 9.6 | 72 | 14 |
| (27) | | 40 | 5.4 | 43 | 17 | 54 | 10.8 | 81 | 16 |
| 30 | | 45 | 6.0 | 48 | 19 | 60 | 12.0 | 90 | 18 |
| (33) | ±0.62 | 50 | 6.6 | 53 | 21 | 66 | 13.2 | 99 | 20 |
| 36 | | 55 | 7.2 | 58 | 23 | 72 | 14.4 | 108 | 22 |
| (39) | | 59 | 7.8 | 62 | 25 | 78 | 15.6 | 117 | 23 |

| Cold Forged Rivets | | | | | | | | | | |
|---|---|---|---|---|---|---|---|---|---|---|
| | | 90° Csk. Head | Snap Head | | Universal Head | | | | Flat Head | |
| | | Head Dimensions | | | | | | | | |
| Nom. Shank Dia.[a] *d* | Tol. on Dia. *d* | Nom. Dia. *D* | Nom. Dia. *D* | Nom. Dia. *K* | Nom. Dia. *D* | Nom. Depth *K* | Rad. *R* | Rad. *r* | Nom. Dia. *D* | Nom. Depth *K* |
| 1 | | 2 | 1.8 | 0.6 | 2 | 0.4 | 3.0 | 0.6 | 2 | 0.25 |
| 1.2 | | 2.4 | 2.1 | 0.7 | 2.4 | 0.5 | 3.6 | 0.7 | 2.4 | 0.3 |
| 1.6 | ±0.07 | 3.2 | 2.8 | 1.0 | 3.2 | 0.6 | 4.8 | 1.0 | 3.2 | 0.4 |
| 2 | | 4 | 3.5 | 1.2 | 4 | 0.8 | 6.0 | 1.2 | 4 | 0.5 |
| 2.5 | | 5 | 4.4 | 1.5 | 5 | 1.0 | 7.5 | 1.5 | 5 | 0.6 |
| 3 | | 6 | 5.3 | 1.8 | 6 | 1.2 | 9.0 | 1.8 | 6 | 0.8 |
| (3.5) | | 7 | 6.1 | 2.1 | 7 | 1.4 | 10.5 | 2.1 | 7 | 0.9 |
| 4 | ±0.09 | 8 | 7 | 2.4 | 8 | 1.6 | 12 | 2.4 | 8 | 1.0 |
| 5 | | 10 | 8.8 | 3.0 | 10 | 2.0 | 15 | 3.0 | 10 | 1.3 |
| 6 | | 12 | 10.5 | 3.6 | 12 | 2.4 | 18 | 3.6 | 12 | 1.5 |
| (7) | | 14 | 12.3 | 4.2 | 14 | 2.8 | 21 | 4.2 | 14 | 1.8 |
| 8 | ±0.11 | 16 | 14 | 4.8 | 16 | 3.2 | 24 | 4.8 | 16 | 2 |
| 10 | | 20 | 18 | 6.0 | 20 | 4.0 | 30 | 6 | 20 | 2.5 |
| 12 | | 24 | 21 | 7.2 | 24 | 4.8 | 36 | 7.2 | … | … |
| (14) | ±0.14 | … | 25 | 8.4 | 28 | 5.6 | 42 | 8.4 | … | … |
| 16 | | … | 28 | 9.6 | 32 | 6.4 | 48 | 9.6 | … | … |

[a] All dimensions are in millimeters. Sizes shown in parentheses are nonpreferred.

## Table 4. British Standard Small Rivets for General Purposes

*BS 641:1951 (Withdrawn)*

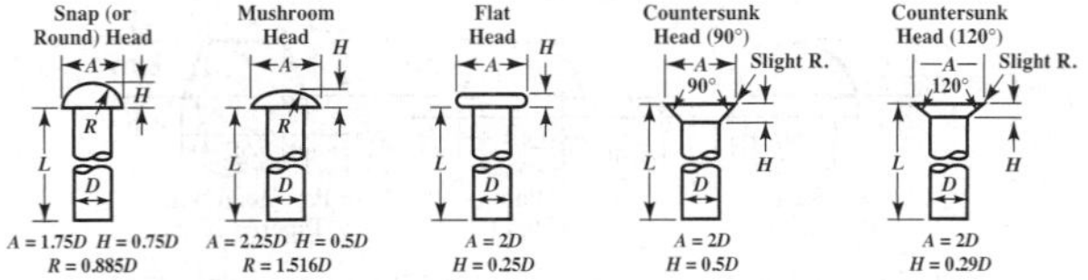


| Nom. Dia.[a] D | Snap (or Round Head) | | | Mushroom Head | | | Flat Head | | Countersunk Head (90°) | | Countersunk Head (120°) | |
|---|---|---|---|---|---|---|---|---|---|---|---|---|
| | Head Dimensions | | | | | | | | | | | |
| | Dia. *A* | Ht. *H* | Rad. *R* | Dia. *A* | Ht. *H* | Rad. *R* | Dia. *A* | Ht. *H* | Dia. *A* | Ht. *H* | Dia. *A* | Ht. *H* |
| 1/16 | 0.109 | 0.047 | 0.055 | 0.141 | 0.031 | 0.095 | 0.125 | 0.016 | 0.125 | 0.031 | … | … |
| 3/32 | 0.164 | 0.070 | 0.083 | 0.211 | 0.047 | 0.142 | 0.188 | 0.023 | 0.188 | 0.047 | … | … |
| 1/8 | 0.219 | 0.094 | 0.111 | 0.281 | 0.063 | 0.189 | 0.250 | 0.031 | 0.250 | 0.063 | 0.250 | 0.036 |
| 5/32 | 0.273 | 0.117 | 0.138 | 0.352 | 0.078 | 0.237 | 0.313 | 0.039 | 0.313 | 0.078 | … | … |
| 3/16 | 0.328 | 0.141 | 0.166 | 0.422 | 0.094 | 0.284 | 0.375 | 0.047 | 0.375 | 0.094 | 0.375 | 0.054 |
| 1/4 | 0.438 | 0.188 | 0.221 | 0.563 | 0.125 | 0.379 | 0.500 | 0.063 | 0.500 | 0.125 | 0.500 | 0.073 |
| 5/16 | 0.547 | 0.234 | 0.277 | 0.703 | 0.156 | 0.474 | 0.625 | 0.078 | 0.625 | 0.156 | 0.625 | 0.091 |
| 3/8 | 0.656 | 0.281 | 0.332 | 0.844 | 0.188 | 0.568 | 0.750 | 0.094 | 0.750 | 0.188 | 0.750 | 0.109 |
| 7/16 | 0.766 | 0.328 | 0.387 | 0.984 | 0.219 | 0.663 | 0.875 | 0.109 | 0.875 | 0.219 | … | … |

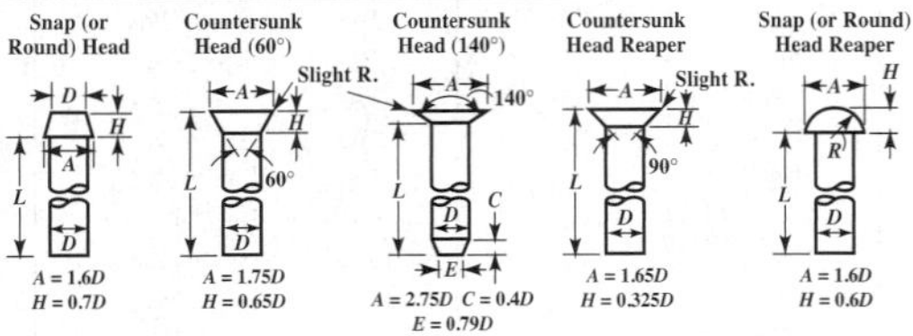


| Nominal Diameter *D* | | Pan Head | | Countersunk Head (60°) | | Countersunk Head (140°) | | | Countersunk Head Reaper | | Snap (or Round) Head Reaper | |
|---|---|---|---|---|---|---|---|---|---|---|---|---|
| | | Head Dimensions | | | | | | | | | | |
| Inch | Gage No.[b] | Dia. *A* | Ht. *H* | Dia. *A* | Ht. *H* | Dia. *A* | Ht. *C* | Dia. *E* | Dia. *A* | Ht. *H* | Dia. *A* | Ht. *H* |
| 0.104 | 12 | … | … | … | … | 0.286 | 0.042 | 0.082 | … | … | … | … |
| 0.116 | 11 | … | … | … | … | 0.319 | 0.046 | 0.092 | … | … | … | … |
| 0.128 | 10 | … | … | … | … | 0.352 | 0.051 | 0.101 | … | … | … | … |
| 0.144 | 9 | … | … | … | … | 0.396 | 0.058 | 0.114 | … | … | … | … |
| 0.160 | 8 | … | … | … | … | 0.440 | 0.064 | 0.126 | … | … | … | … |
| 0.176 | 7 | … | … | … | … | 0.484 | 0.070 | 0.139 | … | … | … | … |
| 3/16 | … | 0.300 | 0.131 | 0.328 | 0.122 | … | … | … | … | … | … | … |
| 0.192 | 6 | … | … | … | … | 0.528 | 0.077 | 0.152 | 0.317 | 0.062 | 0.307 | 0.115 |
| 0.202 | … | … | … | … | … | … | … | … | 0.333 | 0.066 | 0.323 | 0.121 |
| 0.212 | 5 | … | … | … | … | 0.583 | 0.085 | 0.167 | 0.350 | 0.069 | 0.339 | 0.127 |
| 0.232 | 4 | … | … | … | … | 0.638 | 0.093 | 0.183 | 0.383 | 0.075 | 0.371 | 0.139 |
| 1/4 | … | 0.400 | 0.175 | 0.438 | 0.162 | 0.688 | 0.100 | 0.198 | … | … | … | … |
| 0.252 | 3 | … | … | … | … | … | … | … | 0.416 | 0.082 | 0.403 | 0.151 |
| 5/16 | … | 0.500 | 0.219 | 0.547 | 0.203 | 0.859 | 0.125 | 0.247 | … | … | … | … |
| 3/8 | … | 0.600 | 0.263 | 0.656 | 0.244 | 10.031 | 0.150 | 0.296 | … | … | … | … |
| 7/16 | … | 0.700 | 0.306 | 0.766 | 0.284 | … | … | … | … | … | … | … |

[a] All dimensions in inches unless specified otherwise.

[b] Gage numbers are British Standard Wire Gage (S.W.G.) numbers.

## Head Dimensions and Diameters of British Standard Rivets

*BS 275:1927 (Withdrawn)*

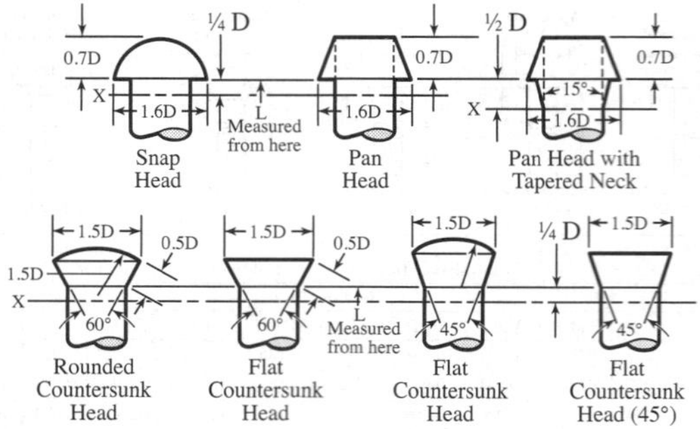


| Nominal Rivet Diameter,[a] D | Shank Diameter[b] At Position X[c] | | At Position Y[c] | | At Position Z[c] |
|---|---|---|---|---|---|
| | Min. | Max. | Min. | Max. | Min. |
| 1/2 | 1/2 | 17/32 | 31/64 | 1/2 | 31/64 |
| 9/16[d] | 9/16 | 19/32 | 35/64 | 9/16 | 35/64 |
| 5/8 | 5/8 | 21/32 | 39/64 | 5/8 | 39/64 |
| 11/16[d] | 11/16 | 23/32 | 43/64 | 11/16 | 43/64 |
| 3/4 | 3/4 | 25/32 | 47/64 | 3/4 | 47/64 |
| 13/16[d] | 13/16 | 27/32 | 51/64 | 13/16 | 51/64 |
| 7/8 | 7/8 | 29/32 | 55/64 | 7/8 | 55/64 |
| 15/16[d] | 15/16 | 31/32 | 59/64 | 15/16 | 59/64 |
| 1 | 1 | 1 1/32 | 63/64 | 1 | 63/64 |
| 1 1/16[d] | 1 1/16 | 1 3/32 | 1 3/64 | 1 1/16 | 1 3/64 |
| 1 1/8 | 1 1/8 | 1 5/32 | 1 7/64 | 1 1/8 | 1 7/64 |
| 1 3/16[d] | 1 3/16 | 1 7/32 | 1 11/64 | 1 3/16 | 1 11/64 |
| 1 1/4 | 1 1/4 | 1 9/32 | 1 15/64 | 1 1/4 | 1 15/64 |
| 1 5/16[d] | 1 5/16 | 1 11/32 | 1 19/64 | 1 5/16 | 1 19/64 |
| 1 3/8 | 1 3/8 | 1 13/32 | 1 23/64 | 1 3/8 | 1 23/64 |
| 1 7/16[d] | 1 7/16 | 1 15/32 | 1 27/64 | 1 7/16 | 1 27/64 |
| 1 1/2 | 1 1/2 | 1 17/32 | 1 31/64 | 1 1/2 | 1 31/64 |
| 1 9/16[d] | 1 9/16 | 1 19/32 | 1 35/64 | 1 9/16 | 1 35/64 |
| 1 5/8 | 1 5/8 | 1 21/32 | 1 39/64 | 1 5/8 | 1 39/64 |
| 1 11/16[d] | 1 11/16 | 1 23/32 | 1 43/64 | 1 11/16 | 1 43/64 |
| 1 3/4 | 1 3/4 | 1 25/32 | 1 47/64 | 1 3/4 | 1 47/64 |

[a] All dimensions that are tabulated are given in inches. This Standard does not apply to boiler rivets.

[b] Tolerances of the rivet diameter are as follows: at position *X*, plus 1/32 inch, minus zero; at position *Y*, plus zero, minus 1/64 inch; at position *Z*, minus 1/64 inch but in no case shall the difference between the diameters at positions *X* and *Y* exceed 1/32 inch, nor shall the diameter of the shank between positions *X* and *Y* be less than the minimum diameter specified at position *Y*.

[c] The location of positions *Y* and *Z* are as follows: Position *Y* is located 1/2*D* from the end of the rivet for rivet lengths 5 diameters long and under. For longer rivets, position *Y* is located 4 1/2 *D* from the head of the rivet. Position *Z* (found only on rivets longer than 5*D*) is located 1/2*D* from the end of the rivet.

[d] At the recommendation of the British Standards Institution, these sizes are to be dispensed with wherever possible.

## PINS AND STUDS

### American National Standard Cotter Pins *ANSI/ASME B18.8.1-2014 (R2019)*

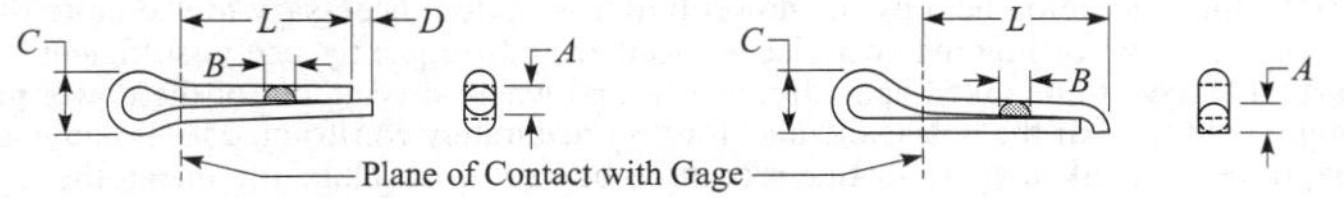


| Nom. Size | Dia. A[a] & Width B Max. | Wire Width B Min. | Head Dia. C Min. | Prong Length D Min. | Hole Size | Nom. Size | Dia. A[a] & Width B Max. | Wire Width B Min. | Head Dia. C Min. | Prong Length D Min. | Hole Size |
|---|---|---|---|---|---|---|---|---|---|---|---|
| 1⁄32 | 0.032 | 0.022 | 0.06 | 0.01 | 0.047 | 3⁄16 | 0.176 | 0.137 | 0.38 | 0.09 | 0.207 |
| 3⁄64 | 0.048 | 0.035 | 0.09 | 0.02 | 0.062 | 7⁄32 | 0.207 | 0.161 | 0.44 | 0.10 | 0.241 |
| 1⁄16 | 0.060 | 0.044 | 0.12 | 0.03 | 0.078 | 1⁄4 | 0.225 | 0.176 | 0.50 | 0.11 | 0.275 |
| 5⁄64 | 0.076 | 0.057 | 0.16 | 0.04 | 0.094 | 5⁄16 | 0.280 | 0.220 | 0.62 | 0.14 | 0.343 |
| 3⁄32 | 0.090 | 0.069 | 0.19 | 0.04 | 0.109 | 3⁄8 | 0.335 | 0.263 | 0.75 | 0.16 | 0.413 |
| 7⁄64 | 0.104 | 0.080 | 0.22 | 0.05 | 0.125 | 7⁄16 | 0.406 | 0.320 | 0.88 | 0.20 | 0.482 |
| 1⁄8 | 0.120 | 0.093 | 0.25 | 0.06 | 0.141 | 1⁄2 | 0.473 | 0.373 | 1.00 | 0.23 | 0.550 |
| 9⁄64 | 0.134 | 0.104 | 0.28 | 0.06 | 0.156 | 5⁄8 | 0.598 | 0.472 | 1.25 | 0.30 | 0.688 |
| 5⁄32 | 0.150 | 0.116 | 0.31 | 0.07 | 0.172 | 3⁄4 | 0.723 | 0.572 | 1.50 | 0.36 | 0.825 |

[a] Tolerances are: −0.004 inch for the 1⁄32- to 3⁄16-inch sizes, incl.; −0.005 inch for the 7⁄32- to 5⁄16-inch sizes, incl.; −0.006 inch for the 3⁄8- to 1⁄2-inch sizes, incl.; and −0.008 inch for the 5⁄8- and 3⁄4-inch sizes. *Note:* Tolerances for length are: up to 1 inch ± 0.030 inch, over 1 inch ±0.060 inch. All dimensions are in inches.

### American National Standard Clevis Pins *ANSI/ASME B18.8.1-2014 (R2019)*

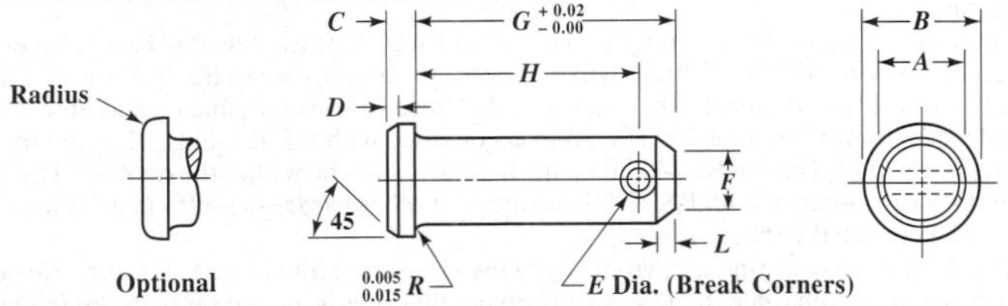


| Nom.Size (Basic Pin Dia.) | Shank Dia. A Max | Head Dia. B Max. | Head Height C Max. | Head Chamfer D Nom. | Hole Dia. E Max. | Point Dia. F Max. | Pin Lgth. G Basic[a] | Head to Hole Center H Max. | Point Length L Max. | Point Length L Min. | Cotter Pin Size for Hole |
|---|---|---|---|---|---|---|---|---|---|---|---|
| 3⁄16 | 0.186 | 0.32 | 0.07 | 0.02 | 0.088 | 0.15 | 0.58 | 0.504 | 0.055 | 0.035 | 1⁄16 |
| 1⁄4 | 0.248 | 0.38 | 0.10 | 0.03 | 0.088 | 0.21 | 0.77 | 0.692 | 0.055 | 0.035 | 1⁄16 |
| 5⁄16 | 0.311 | 0.44 | 0.10 | 0.03 | 0.119 | 0.26 | 0.94 | 0.832 | 0.071 | 0.049 | 3⁄32 |
| 3⁄8 | 0.373 | 0.51 | 0.13 | 0.03 | 0.119 | 0.33 | 1.06 | 0.958 | 0.071 | 0.049 | 3⁄32 |
| 7⁄16 | 0.436 | 0.57 | 0.16 | 0.04 | 0.119 | 0.39 | 1.19 | 1.082 | 0.071 | 0.049 | 3⁄32 |
| 1⁄2 | 0.496 | 0.63 | 0.16 | 0.04 | 0.151 | 0.44 | 1.36 | 1.223 | 0.089 | 0.063 | 1⁄8 |
| 5⁄8 | 0.621 | 0.82 | 0.21 | 0.06 | 0.151 | 0.56 | 1.61 | 1.473 | 0.089 | 0.063 | 1⁄8 |
| 3⁄4 | 0.746 | 0.94 | 0.26 | 0.07 | 0.182 | 0.68 | 1.91 | 1.739 | 0.110 | 0.076 | 5⁄32 |
| 7⁄8 | 0.871 | 1.04 | 0.32 | 0.09 | 0.182 | 0.80 | 2.16 | 1.989 | 0.110 | 0.076 | 5⁄32 |
| 1 | 0.996 | 1.19 | 0.35 | 0.10 | 0.182 | 0.93 | 2.41 | 2.239 | 0.110 | 0.076 | 5⁄32 |
| Tolerance | … | −0.05 in. | −0.02 in. | ±0.01 in. | −0.015 in. | −0.01 in. | … | −0.020 in. | … | … | … |

[a] Lengths tabulated are intended for use with standard clevises, without spacers. When other lengths are required, it is recommended that they be limited wherever possible to nominal lengths in 0.06-inch increments.

All dimensions are in inches. For sizes, greater than 1-inch, refer to ANSI/ASME B18.8.1-2014 (R2019).

**Dowel Pins.—**Dowel pins are used either to retain parts in a fixed position or to preserve alignment. Under normal conditions a properly fitted dowel pin is subjected solely to shearing strain, and this strain occurs only at the junction of the surfaces of the two parts which are being held by the dowel pin. It is seldom necessary to use more than two dowel pins for holding two pieces together and frequently one is sufficient. For parts that have to be taken apart frequently, and where driving out of the dowel pins would tend to wear the holes, and also for very accurately constructed tools and gages that have to be taken apart, or that require to be kept in absolute alignment, the taper dowel pin is preferable. The taper dowel pin is most commonly used for average machine work, but the straight type is given the preference on tool and gage work, except where extreme accuracy is required, or where the tool or gage is to be subjected to rough handling.

The size of the dowel pin is governed by its application. For locating nests, gage plates, etc., pins from $\frac{1}{8}$ to $\frac{3}{16}$ inch (3.2–4.8 mm) in diameter are satisfactory. For locating dies, the diameter of the dowel pin should never be less than $\frac{1}{4}$ inch (6.35 mm); the general rule is to use dowel pins of the same size as the screws used in fastening the work. The length of the dowel pin should be about $1\frac{1}{2}$ to 2 times its diameter in each plate or part to be doweled.

When hardened cylindrical dowel pins are inserted in soft parts, ream the hole about 0.001 inch (0.025 mm) smaller than the dowel pin. If the doweled parts are hardened, grind (or lap) the hole 0.0002 to 0.0003 inch (0.005–0.0076 mm) under size. The hole should be ground or lapped straight, that is, without taper or "bell-mouth."

See *Assembly with Dowel Pins* starting on page 1888.

**British Standard for Metric Series Dowel Pins.—**Steel parallel dowel pins specified in BS 1804: Part 2:1968 are divided into three grades which provide different degrees of pin accuracy.

*Grade 1* is a precision ground pin made from En 32A or En 32B low carbon steel in accordance with BS 970 (Withdrawn) or from high carbon steel to BS 1407 or BS 1423. Pins below 4 mm diameter are unhardened. Those of 4 mm diameter and above are hardened to a minimum of 750 HV 30 in accordance with BS 427, but if they are made from steels to BS 1407 or BS 1423 then the hardness shall be within the range 600 to 700 HV 30, in accordance with BS 427. The values of other hardness scales may be used in accordance with BS 860.

*Grade 2* is a ground pin made from any of the steels used for Grade 1. The pins are normally supplied unhardened, unless a different condition is agreed on between the purchaser and supplier.

*Grade 3* pins are made from En 1A free cutting steel in accordance with BS 970 (Withdrawn) and are supplied with a machined, bright rolled or drawn finish. They are normally supplied unhardened unless a different condition is agreed on between the purchaser and supplier.

Pins of any grade may be made from different steels in accordance with BS 970 (Withdrawn), by mutual agreement between the purchaser and manufacturer. If steels other than those in the standard range are used, the hardness of the pins shall also be decided on by mutual agreement between purchaser and supplier. As shown in the illustration at the head of the accompanying table, one end of each pin is chamfered to provide a lead. The other end may be similarly chamfered, or domed.

If a dowel pin is driven into a blind hole where no provision is made for releasing air, the worker assembling the pin may be endangered, and damage may be caused to the associated component, or stresses may be set up. The appendix of the Standard describes one method of overcoming this problem by providing a small flat surface along the length of a pin to permit the release of air.

For purposes of marking, the Standard states that each package or lot of dowel pins shall bear the manufacturer's name or trademark, the BS number, and the grade of pin.

**British Standard Parallel Steel Dowel Pins—Metric Series** *BS 1804: Part 2: 1968*

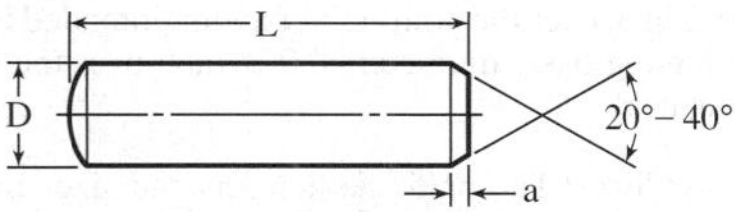


| Nom. Length L, mm | Nominal Diameter D, mm | | | | | | | | | | | | | |
|---|---|---|---|---|---|---|---|---|---|---|---|---|---|---|
| | 1 | 1.5 | 2 | 2.5 | 3 | 4 | 5 | 6 | 8 | 10 | 12 | 16 | 20 | 25 |
| | Chamfer *a* Max., mm | | | | | | | | | | | | | |
| | 0.3 | 0.3 | 0.3 | 0.4 | 0.45 | 0.6 | 0.75 | 0.9 | 1.2 | 1.5 | 1.8 | 2.5 | 3 | 4 |
| | Standard Sizes | | | | | | | | | | | | | |
| 4 | ● | ● | | | | | | | | | | | | |
| 6 | ● | ● | ● | ● | | | | | | | | | | |
| 8 | ● | ● | ● | ● | ● | | | | | | | | | |
| 10 | | ● | ● | ● | ● | ● | | | | | | | | |
| 12 | | ● | ● | ● | ● | ● | ● | | | | | | | |
| 16 | | | ● | ● | ● | ● | ● | ● | ● | | | | | |
| 20 | | | | ● | ● | ● | ● | ● | ● | ● | | | | |
| 25 | | | | | ● | ● | ● | ● | ● | ● | ● | | | |
| 30 | | | | | | ● | ● | ● | ● | ● | ● | ● | | |
| 35 | | | | | | | ● | ● | ● | ● | ● | ● | | |
| 40 | | | | | | | ● | ● | ● | ● | ● | ● | ● | |
| 45 | | | | | | | | ● | ● | ● | ● | ● | ● | |
| 50 | | | | | | | | | ● | ● | ● | ● | ● | ● |
| 60 | | | | | | | | | ● | ● | ● | ● | ● | ● |
| 70 | | | | | | | | | | ● | ● | ● | ● | ● |
| 80 | | | | | | | | | | ● | ● | ● | ● | ● |
| 90 | | | | | | | | | | | ● | ● | ● | ● |
| 100 | | | | | | | | | | | | ● | ● | ● |
| 110 | | | | | | | | | | | | ● | ● | ● |
| 120 | | | | | | | | | | | | | ● | ● |

| Limits of Tolerance on Diameter | | | | | | | |
|---|---|---|---|---|---|---|---|
| Grade[a] | | 1 | | 2 | | 3 | |
| Tolerance Zone | | m5 | | h7 | | h11 | |
| Nom. Dia., mm | | | | | | | |
| Over | To & Incl. | Limits of Tolerance, 0.001 mm | | | | | |
| | 3 | +7 | +2 | 0 | −12[b] | 0 | −60 |
| 3 | 6 | +9 | +4 | 0 | −12 | 0 | −75 |
| 6 | 10 | +12 | +6 | 0 | −15 | 0 | −90 |
| 10 | 14 | +15 | +7 | 0 | −18 | 0 | −110 |
| 14 | 18 | +15 | +7 | 0 | −18 | 0 | −110 |
| 18 | 24 | +17 | +8 | 0 | −21 | 0 | −130 |
| 24 | 30 | +17 | +8 | 0 | −21 | 0 | −130 |

[a] The limits of tolerance for grades 1 and 2 dowel pins have been chosen to provide satisfactory assembly when used in standard reamed holes (H7 and H8 tolerance zones). If the assembly is not satisfactory, refer to BS 1916: Part 1, Limits and Fits for Engineering, and select a different class of fit.

[b] This tolerance is larger than that given in BS 1916, and has been included because the use of a closer tolerance would involve precision grinding by the manufacturer, which is uneconomic for a grade 2 dowel pin.

The tolerance limits on the overall length of all grades of dowel pin up to and including 50 mm long are +0.5, −0.0 mm, and for pins over 50 mm long are +0.8, −0.0 mm. The Standard specifies that the roughness of the cylindrical surface of grades 1 and 2 dowel pins, when assessed in accordance with BS 1134, shall not be greater than 0.4 μm CLA (16 CLA).

**American National Standard Hardened Ground Machine Dowel Pins.—**Hardened ground machine dowel pins are furnished in two diameter series: Standard Series having basic diameters 0.0002 inch over the nominal diameter, intended for initial installations; and Oversize Series having basic diameters 0.001 inch over the nominal diameter, intended for replacement use.

*Preferred Lengths and Sizes:* The preferred lengths and sizes in which these pins are normally available are given in Table 1. Other sizes and lengths are produced as required by the purchaser.

*Effective Length:* The effective length, $L_e$, must not be less than 75 percent of the overall length of the pin.

*Shear Strength:* Single shear strength values are listed in Table 1. Prior versions of ANSI/ASME B18.8.2-2020 had listed double shear load minimum values and had specified a minimum single shear strength of 130,000 psi. See ANSI/ASME B18.8.2-2020, Appendix B, for a description of the double shear test.

*Designation:* These pins are designated by the following data in the sequence shown: Product name (noun first), including pin series, nominal pin diameter (fraction or decimal equivalent), length (fraction or decimal equivalent), material, and protective finish, if required.

*Examples:* Pins, Hardened Ground Machine Dowel.—Standard Series, $\frac{3}{8} \times 1\frac{1}{2}$, Steel, Phosphate Coated.

Pins, Hardened Ground Machine Dowel.—Oversize Series, 0.625 × 2.500, Steel

*Installation Precaution:* Pins should not be installed by striking or hammering and when installing with a press, a shield should be used and safety glasses worn.

**American National Standard Hardened Ground Production Dowel Pins.—**Hardened ground production dowel pins have basic diameters that are 0.0002 inch over the nominal pin diameter.

*Preferred Lengths and Sizes:* The preferred lengths and sizes in which these pins are available are given in Table 2. Other sizes and lengths are produced as required by the purchaser.

*Shear Strength:* Single shear strength values are listed in Table 2. Prior versions of ANSI/ASME B18.8.2-2020 had listed double shear load minimum values and had specified a minimum single shear strength of 102,000 psi. See ANSI/ASME B18.8.2-2020, Appendix B, for a description of the double shear test.

*Ductility:* These standard pins are sufficiently ductile to withstand being pressed into holes 0.0005 inch smaller than the nominal pin diameter in hardened steel without cracking or shattering.

*Designation:* These pins are designated by the following data in the sequence shown: Product name (noun first), nominal pin diameter (fraction or decimal equivalent), length (fraction or decimal equivalent), material, and protective finish, if required.

*Examples:* Pins, Hardened Ground Production Dowel, $\frac{1}{8} \times \frac{3}{4}$, Steel, Phosphate Coated

Pins, Hardened Ground Production Dowel, 0.375 × 1.500, Steel

## Table 1. American National Standard Hardened Ground Machine Dowel Pins *ANSI/ASME B18.8.2-2020*

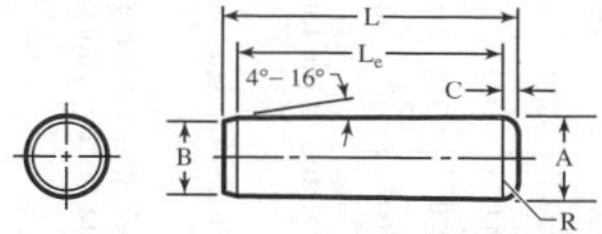


| Nominal Size[a] or Nominal Pin Diameter | | Pin Diameter, *A* | | | | | | Point Diameter, *B* | | Crown Height, *C* | Crown Radius, *R* | Range of Preferred Lengths,[b] *L* | Single Shear Load, for Carbon or Alloy Steel, Calculated, lb | Suggested Hole Diameter[c] | |
|---|---|---|---|---|---|---|---|---|---|---|---|---|---|---|---|
| | | Standard Series Pins | | | Oversize Series Pins | | | | | | | | | | |
| | | Basic | Max. | Min. | Basic | Max. | Min. | Max. | Min. | Max. | Min. | | | Max. | Min. |
| 1/16 | 0.0625 | 0.0627 | 0.0628 | 0.0626 | 0.0635 | 0.0636 | 0.0634 | 0.058 | 0.048 | 0.020 | 0.008 | 3/16-3/4 | 400 | 0.0625 | 0.0620 |
| 5/64[d] | 0.0781 | 0.0783 | 0.0784 | 0.0782 | 0.0791 | 0.0792 | 0.0790 | 0.074 | 0.064 | 0.026 | 0.010 | … | 620 | 0.0781 | 0.0776 |
| 3/32 | 0.0938 | 0.0940 | 0.0941 | 0.0939 | 0.0948 | 0.0949 | 0.0947 | 0.089 | 0.079 | 0.031 | 0.012 | 5/16-1 | 900 | 0.0937 | 0.0932 |
| 1/8 | 0.1250 | 0.1252 | 0.1253 | 0.1251 | 0.1260 | 0.1261 | 0.1259 | 0.120 | 0.110 | 0.041 | 0.016 | 3/8-2 | 1,600 | 0.1250 | 0.1245 |
| 5/32[d] | 0.1562 | 0.1564 | 0.1565 | 0.1563 | 0.1572 | 0.1573 | 0.1571 | 0.150 | 0.140 | 0.052 | 0.020 | … | 2,500 | 0.1562 | 0.1557 |
| 3/16 | 0.1875 | 0.1877 | 0.1878 | 0.1876 | 0.1885 | 0.1886 | 0.1884 | 0.180 | 0.170 | 0.062 | 0.023 | 1/2-2 | 3,600 | 0.1875 | 0.1870 |
| 1/4 | 0.2500 | 0.2502 | 0.2503 | 0.2501 | 0.2510 | 0.2511 | 0.2509 | 0.240 | 0.230 | 0.083 | 0.031 | 1/2-2 1/2 | 6,400 | 0.2500 | 0.2495 |
| 5/16 | 0.3125 | 0.3127 | 0.3128 | 0.3126 | 0.3135 | 0.3136 | 0.3134 | 0.302 | 0.290 | 0.104 | 0.039 | 1/2-2 1/2 | 10,000 | 0.3125 | 0.3120 |
| 3/8 | 0.3750 | 0.3752 | 0.3753 | 0.3751 | 0.3760 | 0.3761 | 0.3759 | 0.365 | 0.350 | 0.125 | 0.047 | 1/2-3 | 14,350 | 0.3750 | 0.3745 |
| 7/16 | 0.4375 | 0.4377 | 0.4378 | 0.4376 | 0.4385 | 0.4386 | 0.4384 | 0.424 | 0.409 | 0.146 | 0.055 | 7/8-3 | 19,550 | 0.4375 | 0.4370 |
| 1/2 | 0.5000 | 0.5002 | 0.5003 | 0.5001 | 0.5010 | 0.5011 | 0.5009 | 0.486 | 0.471 | 0.167 | 0.063 | 3/4, 1-4 | 25,500 | 0.5000 | 0.4995 |
| 5/8 | 0.6250 | 0.6252 | 0.6253 | 0.6251 | 0.6260 | 0.6261 | 0.6259 | 0.611 | 0.595 | 0.208 | 0.078 | 1 1/4-5 | 39,900 | 0.6250 | 0.6245 |
| 3/4 | 0.7500 | 0.7502 | 0.7503 | 0.7501 | 0.7510 | 0.7511 | 0.7509 | 0.735 | 0.715 | 0.250 | 0.094 | 1 1/2-6 | 57,000 | 0.7500 | 0.7495 |
| 7/8 | 0.8750 | 0.8752 | 0.8753 | 0.8751 | 0.8760 | 0.8761 | 0.8759 | 0.860 | 0.840 | 0.293 | 0.109 | 2, 2 1/2-6 | 78,000 | 0.8750 | 0.8745 |
| 1 | 1.0000 | 1.0002 | 1.0003 | 1.0001 | 1.0010 | 1.0011 | 1.0009 | 0.980 | 0.960 | 0.333 | 0.125 | 2, 2 1/2-5, 6 | 102,000 | 1.0000 | 0.9995 |

[a] Where specifying nominal size as basic diameter, zeros preceding decimal and in the fourth decimal place are omitted.

[b] Lengths increase in 1/16-inch steps up to 3/8 inch, in 1/8-inch steps from 3/8 inch to 1 inch, in 1/4-inch steps from 1 inch to 2 1/2 inches, and in 1/2-inch steps above 2 1/2 inches. Tolerance on length is ±0.010 inch.

[c] These hole sizes have been commonly used for press fitting Standard Series machine dowel pins into materials such as mild steels and cast iron. In soft materials such as aluminum or zinc die castings, hole size limits are usually decreased by 0.0005 inch to increase the press fit.

[d] Nonpreferred sizes, not recommended for use in new designs.

All dimensions are in inches.

**Table 2. American National Standard Hardened Ground Production Dowel Pins**
*ANSI/ASME B18.8.2-2020*

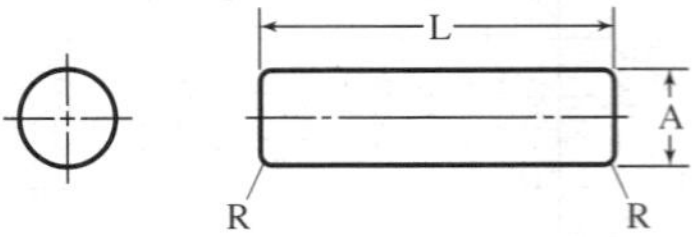


| Nominal Size[a] or Nominal Pin Diameter | Pin Diameter, *A* | | | Corner Radius, *R* | | Range of Preferred Lengths,[b] *L* | Single Shear Load, Calculated, lb | Suggested Hole Diameter[c] | |
|---|---|---|---|---|---|---|---|---|---|
| | Basic | Max. | Min. | Max. | Min. | | | Max. | Min. |
| 1/16 0.0625 | 0.0627 | 0.0628 | 0.0626 | 0.020 | 0.010 | 3/16-1 | 395 | 0.0625 | 0.0620 |
| 3/32 0.0938 | 0.0939 | 0.0940 | 0.0938 | 0.020 | 0.010 | 3/16-2 | 700 | 0.0937 | 0.0932 |
| 7/64 0.1094 | 0.1095 | 0.1096 | 0.1094 | 0.020 | 0.010 | 3/16-2 | 950 | 0.1094 | 0.1089 |
| 1/8 0.1250 | 0.1252 | 0.1253 | 0.1251 | 0.020 | 0.010 | 3/16-2 | 1,300 | 0.1250 | 0.1245 |
| 5/32 0.1562 | 0.1564 | 0.1565 | 0.1563 | 0.020 | 0.010 | 3/16-2 | 2,050 | 0.1562 | 0.1557 |
| 3/16 0.1875 | 0.1877 | 0.1878 | 0.1876 | 0.020 | 0.010 | 3/16-2 | 2,950 | 0.1875 | 0.1870 |
| 7/32 0.2188 | 0.2189 | 0.2190 | 0.2188 | 0.020 | 0.010 | 1/4-2 | 3,800 | 0.2188 | 0.2183 |
| 1/4 0.2500 | 0.2502 | 0.2503 | 0.2501 | 0.020 | 0.010 | 1/4-1 1/2, 1 3/4, 2-2 1/2 | 5,000 | 0.2500 | 0.2495 |
| 5/16 0.3125 | 0.3127 | 0.3128 | 0.3126 | 0.020 | 0.010 | 5/16-1 1/2, 1 3/4, 2-2 1/2 | 8,000 | 0.3125 | 0.3120 |
| 3/8 0.3750 | 0.3752 | 0.3753 | 0.3751 | 0.020 | 0.010 | 3/8-1 1/2, 1 3/4, 2-3 | 11,500 | 0.3750 | 0.3745 |

[a] Where specifying nominal pin size in decimals, zeros preceding decimal and in the fourth decimal place are omitted.

[b] Lengths increase in 1/16-inch steps up to 1 inch, in 1/8-inch steps from 1 inch to 2 inches; and then are 2 1/4, 2 1/2, and 3 inches.

[c] These hole sizes have been commonly used for press fitting production dowel pins into materials such as mild steels and cast iron. In soft materials such as aluminum or zinc die castings, hole size limits are usually decreased by 0.0005 inch to increase the press fit.

All dimensions are in inches.

**American National Standard Unhardened Ground Dowel Pins.—**Unhardened ground dowel pins are normally produced by grinding the outside diameter of commercial wire or rod material to size. Consequently, the maximum diameters of the pins, as specified in Table 3, are below the minimum commercial stock sizes by graduated amounts from 0.0005 inch on the 1/16-inch nominal pin size to 0.0028 inch on the 1-inch nominal pin size.

*Preferred Lengths and Sizes:* The preferred lengths and sizes in which unhardened ground pins are normally available are given in Table 3. Other sizes and lengths are produced as required by the purchaser.

*Shear Strength:* These pins must have a single shear strength of 64,000 psi minimum for pins made from steel and 40,000 psi minimum for pins made from brass and must be capable of withstanding the minimum double shear loads given in Table 3 when tested in accordance with the procedure outlined in ANSI/ASME B18.8.2-2020, Appendix B.

*Designation:* These pins are designated by the following data in the order shown: Product name (noun first), nominal pin diameter (fraction or decimal equivalent), length (fraction or decimal equivalent), material, and protective finish, if required.

*Examples:* Pins, Unhardened Ground Dowel, 1/8 × 3/4, Steel

Pins, Unhardened Ground Dowel, 0.250 × 2.500, Steel, Zinc Plated

## Table 3. American National Standard Unhardened Ground Dowel Pins

*ANSI/ASME B18.8.2-2020*

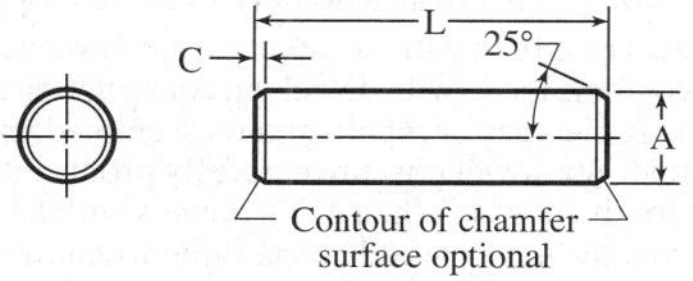


| Nominal Size[a] or Basic Pin Diameter | | Pin Diameter, A | | Chamfer Length, C | | Range of Preferred Lengths,[b] L | Suggested Hole Diameter[c] | | Double Shear Load Min., lb. | |
|---|---|---|---|---|---|---|---|---|---|---|
| | | Max. | Min. | Max. | Min. | | Max. | Min. | Carbon-Steel | Brass |
| 1/16 | 0.0625 | 0.0600 | 0.0595 | 0.025 | 0.005 | 1/4-1 | 0.0595 | 0.0580 | 350 | 220 |
| 3/32 | 0.0938 | 0.0912 | 0.0907 | 0.025 | 0.005 | 1/4-1 1/2 | 0.0907 | 0.0892 | 820 | 510 |
| 7/64[d] | 0.1094 | 0.1068 | 0.1063 | 0.025 | 0.005 | … | 0.1062 | 0.1047 | 1,130 | 710 |
| 1/8 | 0.1250 | 0.1223 | 0.1218 | 0.025 | 0.005 | 1/4-2 | 0.1217 | 0.1202 | 1,490 | 930 |
| 5/32 | 0.1562 | 0.1535 | 0.1530 | 0.025 | 0.005 | 1/4-2 | 0.1528 | 0.1513 | 2,350 | 1,470 |
| 3/16 | 0.1875 | 0.1847 | 0.1842 | 0.025 | 0.005 | 1/4-2 | 0.1840 | 0.1825 | 3,410 | 2,130 |
| 7/32 | 0.2188 | 0.2159 | 0.2154 | 0.025 | 0.005 | 1/4-2 | 0.2151 | 0.2136 | 4,660 | 2,910 |
| 1/4 | 0.2500 | 0.2470 | 0.2465 | 0.025 | 0.005 | 1/4-1 1/2, 1 3/4, 2-2 1/2 | 0.2462 | 0.2447 | 6,120 | 3,810 |
| 5/16 | 0.3125 | 0.3094 | 0.3089 | 0.040 | 0.020 | 5/16-1 1/2, 1 3/4, 2-2 1/2 | 0.3085 | 0.3070 | 9,590 | 5,990 |
| 3/8 | 0.3750 | 0.3717 | 0.3712 | 0.040 | 0.020 | 3/8-1 1/2, 1 3/4, 2-2 1/2 | 0.3708 | 0.3693 | 13,850 | 8,650 |
| 7/16 | 0.4375 | 0.4341 | 0.4336 | 0.040 | 0.020 | 7/16-5/8, 3/4, 7/8-1 1/2, 1 3/4, 2-2 1/2 | 0.4331 | 0.4316 | 18,900 | 11,810 |
| 1/2 | 0.5000 | 0.4964 | 0.4959 | 0.040 | 0.020 | 1/2, 5/8, 3/4, 7/8, 1-1 1/2, 1 3/4, 2-3 | 0.4954 | 0.4939 | 24,720 | 15,450 |
| 5/8 | 0.6250 | 0.6211 | 0.6206 | 0.055 | 0.035 | 5/8, 3/4, 7/8, 1-1 1/2, 1 3/4, 2, 2 1/2-4 | 0.6200 | 0.6185 | 38,710 | 24,190 |
| 3/4 | 0.7500 | 0.7458 | 0.7453 | 0.055 | 0.035 | 3/4, 7/8, 1, 1 1/4, 1 1/2, 1 3/4, 2, 2 1/2-4 | 0.7446 | 0.7431 | 55,840 | 34,900 |
| 7/8 | 0.8750 | 0.8705 | 0.8700 | 0.070 | 0.050 | 7/8, 1, 1 1/4, 1 1/2, 1 3/4, 2, 2 1/2-4 | 0.8692 | 0.8677 | 76,090 | 47,550 |
| 1 | 1.0000 | 0.9952 | 0.9947 | 0.070 | 0.050 | 1, 1 1/4, 1 1/2, 1 3/4, 2, 2 1/2-4 | 0.9938 | 0.9923 | 99,460 | 62,160 |

[a] Where specifying pin size in decimals, zeros preceding decimal and in the fourth decimal place are omitted.

[b] Lengths increase in 1/16-inch increments from 1/4 to 1 inch, in 1/8-inch increments from 1 inch to 2 inches, and in 1/4-inch increments from 2 to 2 1/2 inches, and in 1/2-inch increments from 2 1/2 to 4 inches.

[c] These hole sizes have been found to be satisfactory for press fitting pins into mild steel and cast and malleable irons. In soft materials such as aluminum alloys or zinc die castings, hole size limits are usually decreased by 0.0005 inch to increase the press fit.

[d] Nonpreferred size, not recommended for use in new designs.

All dimensions are in inches.

**Assembly with Dowel Pins.—** Hardened and ground dowel pins are made to a high degree of accuracy. Dowel pins are used to position and align machine components, provide a positive stop, add structural rigidity to an assembly, or to create a part nest.

When components are assembled with dowel pins, the dowel is a very close fit with its mating part. Machines or tools made with dowel pin construction can be disassembled for maintenance, engineering changes or repair, and reassembled, maintaining the original positional accuracy. Holes for dowel pins are carefully produced to obtain either a close sliding fit or a press fit for the dowel. When using dowels with fasteners, proper machine shop practice dictates that the sizes should be the same whenever possible. For example, if $\frac{3}{8}$-inch socket head cap screws are used, $\frac{3}{8}$-inch diameter dowels should also be used.

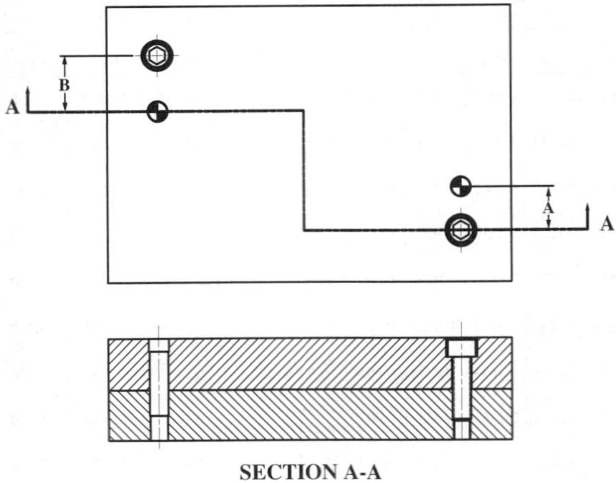


Fig. 1.

Fig. 1 is a typical example of an assembly using socket head cap screws and dowels. The fasteners and dowels are positioned as far apart as possible for rigidity and locational stability. Engineering drawings often show dowel locations as a circle with a partly filled-in cross. The dowel holes on this assembly are relieved in the top plate and have a smaller knockout hole in the bottom plate. The hole for the socket head cap screw is counterbored deep enough so that the screw is recessed 0.03 inch below the surface of the plate. Making dimension "A" different from dimension "B" prevents the plates from being assembled backwards.

When dowels are used for positioning:

- Position dowel pins as far apart as possible.
- Length of engagement for the dowel should be twice the dowel diameter in each component.
- Avoid pressing a dowel into a blind hole. If options are limited, pull dowels have a threaded hole that can be used to remove the dowel from a blind hole.
- Avoid pressing a dowel into a hardened part.
- Provide a knock out hole.
- Relieve deep holes.
- Dowels should be a press fit in one component and a slip fit in the other.

Fig. 2 shows a cross section of two plates with four different assembly possibilities for dowels.

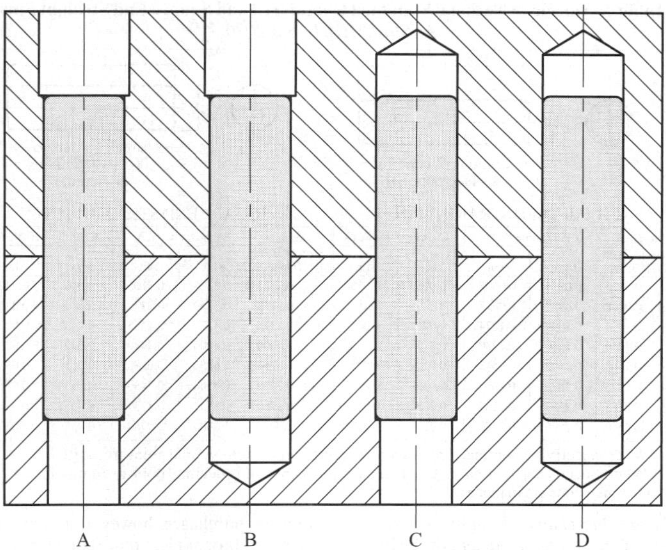


Fig. 2. Four Options for Dowel Pin Installation

*Option A:* In the lower plate the dowel is a light press fit which means that there is 0.0002 to 0.0005 inch interference between the dowel and the hole. The hole in the bottom part of the lower plate is made 0.03 inch smaller than the dowel hole to prevent the dowel from falling out while providing a knockout hole for a punch. The top plate also has a stepped hole that is 0.0002 to 0.0005 inch larger than the dowel for a slip fit, and a larger diameter for a relief. The relief is 0.03 inch larger than the dowel. It is a good practice to relieve portions of the hole that are not used for locating. Dowels should be installed with two times their diameter as a bearing surface. In other words, a $^3/_8$-inch diameter dowel should contact the hole for $^3/_4$ of its length in each component. A $^1/_4$-inch dowel should contact each plate for $^1/_2$ inch of the dowel's length in each component.

*Option B:* In this example the design does not permit the dowel hole to go through the lower plate. Since pressing dowels into blind (dead end) holes is not recommended, the dowel is a slip fit in the lower plate and a press fit in the top plate. Again, the top plate is relieved allowing the dowel the proper amount of effective engagement.

*Option C:* This example is the opposite of Option B. A knockout hole which is 0.03 inch smaller than the dowel is provided in the lower plate. The press fit is in the lower plate, the slip fit is in the top plate because it is very difficult to remove a dowel that is pressed in a blind hole.

*Option D:* In the event that through holes are not permitted in either plate, both sides are made to a slip fit for the dowel.

**American National Standard Straight Pins.—**The diameter of both chamfered and square end straight pins is that of the commercial wire or rod from which the pins are made. The tolerances shown in Table 4 are applicable to carbon steel and some deviations in the diameter limits may be necessary for pins made from other materials.

**Table 4. American National Standard Chamfered and Square End Straight Pins**
*ANSI/ASME B18.8.2-2020*

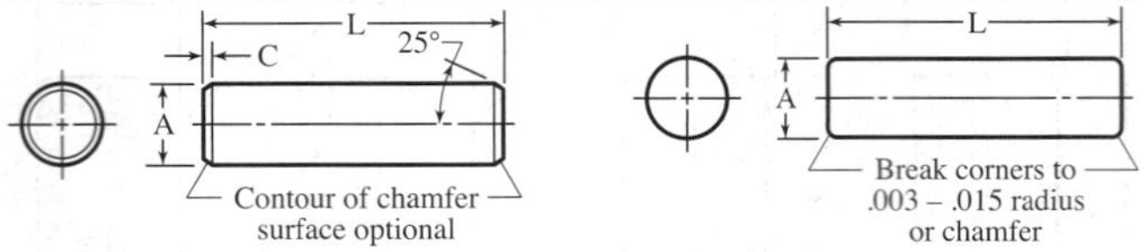


CHAMFERED STRAIGHT PIN SQUARE END STRAIGHT PIN

| Nominal Size[a] or Basic Pin Diameter | Pin Diameter, *A* | | Chamfer Length, *C* | | Nominal Size[b] or Basic Pin Diameter | Pin Diameter, *A* | | Chamfer Length, *C* | |
|---|---|---|---|---|---|---|---|---|---|
| | Max. | Min. | Max. | Min. | | Max. | Min. | Max. | Min. |
| 1/16 0.0625 | 0.0625 | 0.0605 | 0.025 | 0.005 | 5/16 0.3125 | 0.3125 | 0.3105 | 0.040 | 0.020 |
| 3/32 0.0938 | 0.0937 | 0.0917 | 0.025 | 0.00 | 3/8 0.3750 | 0.3750 | 0.3730 | 0.040 | 0.020 |
| 7/64 0.1094 | 0.1094 | 0.1074 | 0.025 | 0.005 | 7/16 0.4375 | 0.4375 | 0.4355 | 0.040 | 0.020 |
| 1/8 0.1250 | 0.1250 | 0.1230 | 0.025 | 0.005 | 1/2 0.5000 | 0.5000 | 0.4980 | 0.040 | 0.020 |
| 5/32 0.1563 | 0.1562 | 0.1542 | 0.025 | 0.005 | 5/8 0.6250 | 0.6250 | 0.6230 | 0.055 | 0.035 |
| 3/16 0.1875 | 0.1875 | 0.1855 | 0.025 | 0.005 | 3/4 0.7500 | 0.7500 | 0.7480 | 0.055 | 0.035 |
| 7/32 0.2188 | 0.2187 | 0.2167 | 0.025 | 0.005 | 7/8 0.8750 | 0.8750 | 0.8730 | 0.055 | 0.035 |
| 1/4 0.2500 | 0.2500 | 0.2480 | 0.025 | 0.005 | 1 1.0000 | 1.0000 | 0.9980 | 0.055 | 0.035 |

[a] Where specifying nominal size in decimals, zeros preceding decimal point are omitted.
[b] Where specifying nominal size in decimals, zeros preceding decimal point are omitted.
All dimensions are in inches.

*Length Increments:* Lengths are as specified by the purchaser; however, it is recommended that nominal pin lengths be limited to increments of not less than 0.0625 inch.

*Material:* Straight pins are normally made from cold drawn steel wire or rod having a maximum carbon content of 0.28 percent. Where required, pins may also be made from corrosion resistant steel, brass, or other metals.

*Designation:* Straight pins are designated by the following data, in the sequence shown: Product name (noun first), nominal size (fraction or decimal equivalent), material, and protective finish, if required.

*Examples:* Pin, Chamfered Straight, $^1/_8 \times 1.500$, Steel

Pin, Square End Straight, $0.2500 \times 2.250$, Steel, Zinc Plated

**American National Standard Taper Pins.—**Taper pins have a uniform taper over the pin length with both ends crowned. Most sizes are supplied in commercial and precision classes, the latter having generally tighter tolerances and being more closely controlled in manufacture.

*Diameters:* The major diameter of both commercial and precision classes of pins is the diameter of the large end and is the basis for pin size. The diameter at the small end is computed by multiplying the nominal length of the pin by the factor 0.02083 and subtracting the result from the basic pin diameter. See also Table 5.

*Taper:* The taper on commercial class pins is $0.250 \pm 0.006$ inch per foot and on the precision class pins is $0.250 \pm 0.004$ inch per foot of length.

*Materials:* Unless otherwise specified, taper pins are made from SAE 1211 steel or cold drawn SAE 1212 or 1213 steel or equivalents, and no mechanical property requirements apply.

*Hole Sizes:* Under most circumstances, holes for taper pins require taper reaming. Sizes and lengths of taper pins for which standard reamers are available are given in Table 6. Drilling specifications for taper pins are given below.

*Designation:* Taper pins are designated by the following data in the sequence shown: Product name (noun first), class, size number (or decimal equivalent), length (fraction or three-place decimal equivalent), material, and protective finish, if required.

*Examples:* Pin, Taper (Commercial Class) No. 0 × ¾, Steel

Pin, Taper (Precision Class) 0.219 × 1.750, Steel, Zinc Plated

**Table 5. Nominal Diameter at Small Ends of Standard Taper Pins**

| Pin Length in inches | Pin Number and Small End Diameter for Given Length | | | | | | | | | | |
|---|---|---|---|---|---|---|---|---|---|---|---|
| | 0 | 1 | 2 | 3 | 4 | 5 | 6 | 7 | 8 | 9 | 10 |
| ¾ | 0.140 | 0.156 | 0.177 | 0.203 | 0.235 | 0.273 | 0.325 | 0.393 | 0.476 | 0.575 | 0.690 |
| 1 | 0.135 | 0.151 | 0.172 | 0.198 | 0.230 | 0.268 | 0.320 | 0.388 | 0.471 | 0.570 | 0.685 |
| 1¼ | 0.130 | 0.146 | 0.167 | 0.192 | 0.224 | 0.263 | 0.315 | 0.382 | 0.466 | 0.565 | 0.680 |
| 1½ | 0.125 | 0.141 | 0.162 | 0.187 | 0.219 | 0.258 | 0.310 | 0.377 | 0.460 | 0.560 | 0.675 |
| 1¾ | 0.120 | 0.136 | 0.157 | 0.182 | 0.214 | 0.252 | 0.305 | 0.372 | 0.455 | 0.554 | 0.669 |
| 2 | 0.114 | 0.130 | 0.151 | 0.177 | 0.209 | 0.247 | 0.299 | 0.367 | 0.450 | 0.549 | 0.664 |
| 2¼ | 0.109 | 0.125 | 0.146 | 0.172 | 0.204 | 0.242 | 0.294 | 0.362 | 0.445 | 0.544 | 0.659 |
| 2½ | 0.104 | 0.120 | 0.141 | 0.166 | 0.198 | 0.237 | 0.289 | 0.356 | 0.440 | 0.539 | 0.654 |
| 2¾ | 0.099 | 0.115 | 0.136 | 0.161 | 0.193 | 0.232 | 0.284 | 0.351 | 0.434 | 0.534 | 0.649 |
| 3 | 0.094 | 0.110 | 0.131 | 0.156 | 0.188 | 0.227 | 0.279 | 0.346 | 0.429 | 0.528 | 0.643 |
| 3¼ | … | … | … | 0.151 | 0.182 | 0.221 | 0.273 | 0.340 | 0.424 | 0.523 | 0.638 |
| 3½ | … | … | … | 0.146 | 0.177 | 0.216 | 0.268 | 0.335 | 0.419 | 0.518 | 0.633 |
| 3¾ | … | … | … | 0.141 | 0.172 | 0.211 | 0.263 | 0.330 | 0.414 | 0.513 | 0.628 |
| 4 | … | … | … | 0.136 | 0.167 | 0.206 | 0.258 | 0.326 | 0.409 | 0.508 | 0.623 |
| 4¼ | … | … | … | 0.131 | 0.162 | 0.201 | 0.253 | 0.321 | 0.403 | 0.502 | 0.617 |
| 4½ | … | … | … | 0.125 | 0.156 | 0.195 | 0.247 | 0.315 | 0.398 | 0.497 | 0.612 |
| 5 | … | … | … | … | 0.146 | 0.185 | 0.237 | 0.305 | 0.389 | 0.487 | 0.602 |
| 5½ | … | … | … | … | … | … | … | 0.294 | 0.377 | 0.476 | 0.591 |
| 6 | … | … | … | … | … | … | … | 0.284 | 0.367 | 0.466 | 0.581 |

**Drilling Specifications for Taper Pins.**—When helically fluted taper pin reamers are used, the diameter of the through hole drilled prior to reaming is equal to the diameter at the small end of the taper pin. (See Table 5.) However, when straight fluted taper reamers are to be used, it may be necessary, for long pins, to step drill the hole before reaming, the number and sizes of the drills to be used depending on the depth of the hole (pin length).

To determine the number and sizes of step drills required: Find the length of pin to be used at the top of the chart on page 1892 and follow this length down to the intersection with that heavy line which represents the size of taper pin (see taper pin numbers at the right-hand end of each heavy line). If the length of pin falls between the first and second dots, counting from the left, only one drill is required. Its size is indicated by following the nearest horizontal line from the point of intersection (of the pin length) on the heavy line over to the drill diameter values at the left. If the intersection of pin length comes between the second and third dots, then two drills are required. The size of the smaller drill then corresponds to the intersection of the pin length and the heavy line and the larger is the corresponding drill diameter for the intersection of one-half this length with the heavy line. Should the pin length fall between the third and fourth dots, three drills are required. The smallest drill will have a diameter corresponding to the intersection of the total pin length with the heavy line, the next in size will have a diameter corresponding to the intersection of two-thirds of this length with the heavy line, and the largest will have a diameter corresponding to the intersection of one-third of this length with the heavy line. Where the intersection falls between two drill sizes, use the smaller.

**Chart to Facilitate Selection of Number and Sizes of Drills for Step-Drilling Prior to Taper Reaming**

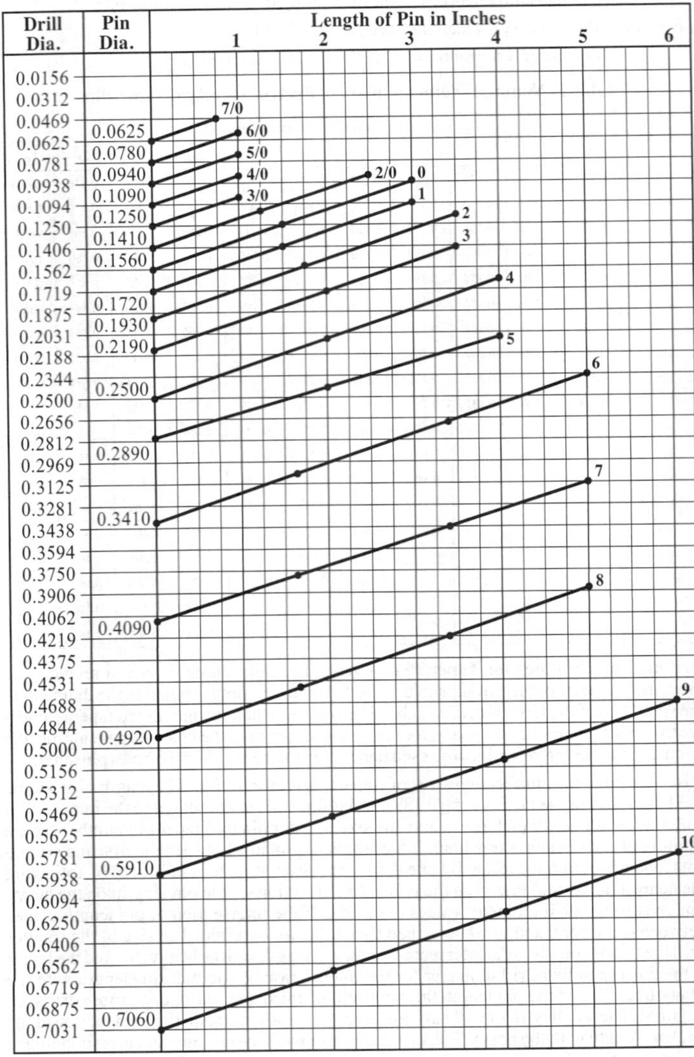

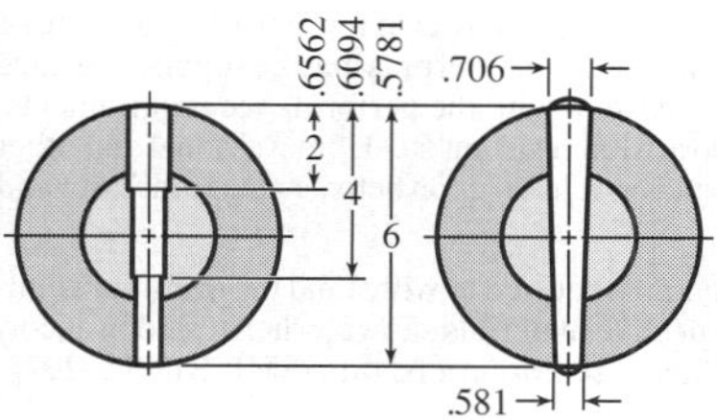


*Examples:* For a No. 10 taper pin 6 inches long, three drills would be used, of the sizes and for the depths shown in the accompanying diagram.

For a No. 10 taper pin 3 inches long, two drills would be used because the 3-inch length falls between the second and third dots. The first or through drill will be 0.6406 inch and the second drill, 0.6719 inch for a depth of 1½ inches.

**Table 6. American National Standard Taper Pins** *ANSI/ASME B18.8.2-2020*

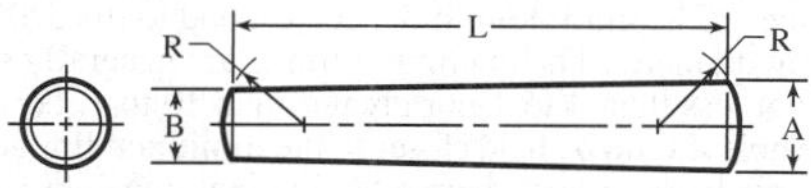


| Pin Size Number and Basic Pin Dia.[a] | | Major Diameter (Large End), *A* | | | | End Crown Radius, *R* | | Range of Lengths,[b] *L* | |
|---|---|---|---|---|---|---|---|---|---|
| | | Commercial Class | | Precision Class | | | | | |
| | | Max. | Min. | Max. | Min. | Max. | Min. | Stand. Reamer Avail.[c] | Other |
| 7/0 | 0.0625 | 0.0638 | 0.0618 | 0.0635 | 0.0625 | 0.072 | 0.052 | … | ¼-1 |
| 6/0 | 0.0780 | 0.0793 | 0.0773 | 0.0790 | 0.0780 | 0.088 | 0.068 | … | ¼-½ |
| 5/0 | 0.0940 | 0.0953 | 0.0933 | 0.0950 | 0.0940 | 0.104 | 0.084 | ¼-1 | 1¼, 1½ |
| 4/0 | 0.1090 | 0.1103 | 0.1083 | 0.1100 | 0.1090 | 0.119 | 0.099 | ¼-1 | 1¼-2 |
| 3/0 | 0.1250 | 0.1263 | 0.1243 | 0.1260 | 0.1250 | 0.135 | 0.115 | ¼-1 | 1¼-2 |
| 2/0 | 0.1410 | 0.1423 | 0.1403 | 0.1420 | 0.1410 | 0.151 | 0.131 | ½-1¼ | 1½-2½ |
| 0 | 0.1560 | 0.1573 | 0.1553 | 0.1570 | 0.1560 | 0.166 | 0.146 | ½-1¼ | 1½-3 |
| 1 | 0.1720 | 0.1733 | 0.1713 | 0.1730 | 0.1720 | 0.182 | 0.162 | ¾-1¼ | 1½-3 |
| 2 | 0.1930 | 0.1943 | 0.1923 | 0.1940 | 0.1930 | 0.203 | 0.183 | ¾-1½ | 1¾-3 |
| 3 | 0.2190 | 0.2203 | 0.2183 | 0.2200 | 0.2190 | 0.229 | 0.209 | ¾-1¾ | 2-4 |
| 4 | 0.2500 | 0.2513 | 0.2493 | 0.2510 | 0.2500 | 0.260 | 0.240 | ¾-2 | 2¼-4 |
| 5 | 0.2890 | 0.2903 | 0.2883 | 0.2900 | 0.2890 | 0.299 | 0.279 | 1-2½ | 2¾-6 |
| 6 | 0.3410 | 0.3423 | 0.3403 | 0.3420 | 0.3410 | 0.351 | 0.331 | 1¼-3 | 3¼-6 |
| 7 | 0.4090 | 0.4103 | 0.4083 | 0.4100 | 0.4090 | 0.419 | 0.399 | 1¼-3¾ | 4-8 |
| 8 | 0.4920 | 0.4933 | 0.4913 | 0.4930 | 0.4920 | 0.502 | 0.482 | 1¼-4½ | 4¾-8 |
| 9 | 0.5910 | 0.5923 | 0.5903 | 0.5920 | 0.5910 | 0.601 | 0.581 | 1¼-5¼ | 5½-8 |
| 10 | 0.7060 | 0.7073 | 0.7053 | 0.7070 | 0.7060 | 0.716 | 0.696 | 1½-6 | 6¼-8 |
| 11 | 0.8600 | 0.8613 | 0.8593 | … | … | 0.870 | 0.850 | … | 2-8 |
| 12 | 1.0320 | 1.0333 | 1.0313 | … | … | 1.042 | 1.022 | … | 2-9 |
| 13 | 1.2410 | 1.2423 | 1.2403 | … | … | 1.251 | 1.231 | … | 3-11 |
| 14 | 1.5230 | 1.5243 | 1.5223 | … | … | 1.533 | 1.513 | … | 3-13 |

[a] When specifying nominal pin size in decimals, zeros preceding the decimal and in the fourth decimal place are omitted.

[b] Lengths increase in ⅛-inch steps up to 1 inch and in ¼-inch steps above 1 inch.

[c] Standard reamers are available for pin lengths in this column.

All dimensions are in inches.

For nominal diameters, B, see Table 5.

**American National Standard Grooved Pins.—**These pins have three equally spaced longitudinal grooves and an expanded diameter over the crests of the ridges formed by the material displaced when the grooves are produced. The grooves are aligned with the axes of the pins. There are seven types of grooved pins as shown in the illustration on page 1895.

*Standard Sizes and Lengths:* The standard sizes and lengths in which grooved pins are normally available are given in Table 7.

*Materials:* Grooved pins are normally made from cold drawn low carbon steel wire or rod. Where additional performance is required, carbon steel pins may be supplied surface hardened and heat treated to a hardness consistent with the performance requirements. Pins may also be made from alloy steel, corrosion resistant steel, brass, Monel and other nonferrous metals having chemical properties as agreed upon between manufacturer and purchaser.

*Performance Requirements:* Grooved pins are required to withstand the minimum double shear loads given in Table 7 for the respective materials shown, when tested in accordance with the Double Shear Testing of Pins as set forth in ANSI/ASME B18.8.2-2020, Appendix B.

*Hole Sizes:* To obtain maximum product retention under average conditions, it is recommended that holes for the installation of grooved pins be held as close as possible to the limits shown in Table 7. The minimum limits correspond to the drill size, which is the same as the basic pin diameter. The maximum limits are generally suitable for length-diameter ratios of not less than 4 to 1 nor greater than 10 to 1. For smaller length-to-diameter ratios, the hole should be held closer to the minimum limits where retention is critical. Conversely for larger ratios where retention requirements are less important, it may be desirable to increase the hole diameters beyond the maximum limits shown.

*Designation:* Grooved pins are designated by the following data in the sequence shown: Product name (noun first) including type designation, nominal size (number, fraction or decimal equivalent), length (fraction or decimal equivalent), material, including specification or heat treatment where necessary, protective finish, if required.

*Examples:* Pin, Type A Grooved, $\frac{3}{32} \times \frac{3}{4}$, Steel, Zinc Plated

Pin, Type F Grooved, $0.250 \times 1.500$, Corrosion Resistant Steel

**American National Standard Grooved T-Head Cotter Pins and Round Head Grooved Drive Studs.—**The cotter pins have a T-head and the studs a round head. Both pins and studs have three equally spaced longitudinal grooves and an expanded diameter over the crests of the raised ridges formed by the material displaced when the grooves are formed.

*Standard Sizes and Lengths:* The standard sizes and range of standard lengths are given in Table 8 and Table 9.

*Material:* Unless otherwise specified these pins are made from low carbon steel. Where so indicated by the purchaser they may be made from corrosion resistant steel, brass or other nonferrous alloys.

*Hole Sizes:* To obtain optimum product retention under average conditions, it is recommended that holes for the installation of grooved T-head cotter pins and grooved drive studs be held as close as possible to the limits tabulated. The minimum limits given correspond to the drill size, which is equivalent to the basic shank diameter. The maximum limits shown are generally suitable for length-diameter ratios of not less than 4 to 1 and not greater than 10 to 1. For smaller length-to-diameter ratios, the holes should be held closer to minimum limits where retention is critical. Conversely, for larger length-to-diameter ratios or where retention requirements are not essential, it may be desirable to increase the hole diameter beyond the maximum limits shown.

*Designation:* Grooved T-head cotter pins and round head grooved drive studs are designated by the following data, in the order shown: Product name (noun first), nominal size (number, fraction or decimal equivalent), length (fraction or decimal equivalent), material including specification or heat treatment where necessary, and protective finish, if required.

*Examples:* Pin, Grooved T-Head Cotter, $\frac{1}{4} \times 1\frac{1}{4}$, Steel, Zinc Plated

Drive Stud, Round Head Grooved, No. $10 \times \frac{1}{2}$, Corrosion Resistant Steel

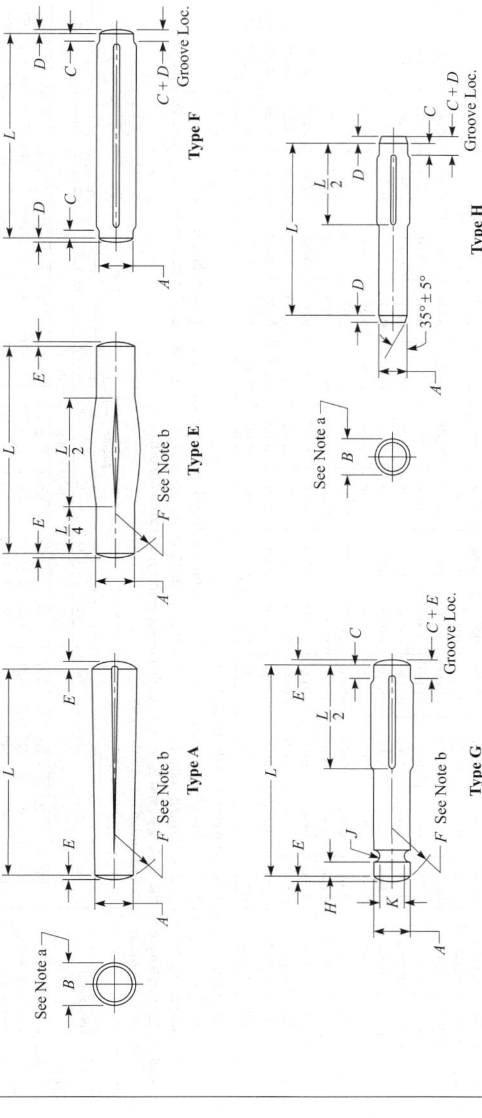


Types of American National Standard Grooved Pins, ANSI/ASME B18.8.2-2020 (For notes see bottom of Table 7.)

## Table 7. American National Standard Grooved Pins *ANSI/ASME B18.8.2-2020*

| Nominal Size or Basic Pin Diameter | | Pin Diameter,[a] *A* | | Pilot Length, *C* | Chamfer Length,[b] *D* | Crown Height,[b] *E* | | Crown Radius,[b] *F* | | Neck Width, *G* | | Shoulder Length, *H* | | Neck Radius, *J* | Neck Diameter, *K* | | Range of Standard Lengths[c] |
|---|---|---|---|---|---|---|---|---|---|---|---|---|---|---|---|---|---|
| | | Max. | Min. | Ref. | Min. | Max. | Min. | Max. | Min. | Max. | Min. | Max. | Min. | Ref. | Max. | Min. | |
| 1/32[d] | 0.0312 | 0.0312 | 0.0297 | 0.015 | … | … | … | … | … | … | … | … | … | … | … | … | 1/8-1/2 |
| 3/64[d] | 0.0469 | 0.0469 | 0.0454 | 0.031 | … | … | … | … | … | … | … | … | … | … | … | … | 1/8-5/8 |
| 1/16 | 0.0625 | 0.0625 | 0.0610 | 0.031 | 0.005 | 0.0115 | 0.0015 | 0.088 | 0.068 | … | … | … | … | … | … | … | 1/8-1 |
| 5/64[d] | 0.0781 | 0.0781 | 0.0766 | 0.031 | 0.005 | 0.0137 | 0.0037 | 0.104 | 0.084 | … | … | … | … | … | … | … | 1/4-1 |
| 3/32 | 0.0938 | 0.0938 | 0.0923 | 0.031 | 0.005 | 0.0141 | 0.0041 | 0.135 | 0.115 | 0.038 | 0.028 | 0.041 | 0.031 | 0.016 | 0.067 | 0.057 | 1/4-1 1/4 |
| 7/64[d] | 0.1094 | 0.1094 | 0.1074 | 0.031 | 0.005 | 0.0160 | 0.0060 | 0.150 | 0.130 | 0.038 | 0.028 | 0.041 | 0.031 | 0.016 | 0.082 | 0.072 | 1/4-1 1/4 |
| 1/8 | 0.1250 | 0.1250 | 0.1230 | 0.031 | 0.005 | 0.0180 | 0.0080 | 0.166 | 0.146 | 0.069 | 0.059 | 0.041 | 0.031 | 0.031 | 0.088 | 0.078 | 1/4-1 1/2 |
| 5/32 | 0.1563 | 0.1563 | 0.1543 | 0.062 | 0.005 | 0.0220 | 0.0120 | 0.198 | 0.178 | 0.069 | 0.059 | 0.057 | 0.047 | 0.031 | 0.109 | 0.099 | 3/8-2 |
| 3/16 | 0.1875 | 0.1875 | 0.1855 | 0.062 | 0.016 | 0.0230 | 0.0130 | 0.260 | 0.240 | 0.069 | 0.059 | 0.057 | 0.047 | 0.031 | 0.130 | 0.120 | 3/8-2 1/4 |
| 7/32 | 0.2188 | 0.2188 | 0.2168 | 0.062 | 0.016 | 0.0270 | 0.0170 | 0.291 | 0.271 | 0.101 | 0.091 | 0.072 | 0.062 | 0.047 | 0.151 | 0.141 | 1/2-3 |
| 1/4 | 0.2500 | 0.2500 | 0.2480 | 0.062 | 0.016 | 0.0310 | 0.0210 | 0.322 | 0.302 | 0.101 | 0.091 | 0.072 | 0.062 | 0.047 | 0.172 | 0.162 | 1/2-3 1/4 |
| 5/16 | 0.3125 | 0.3125 | 0.3105 | 0.094 | 0.031 | 0.0390 | 0.0290 | 0.385 | 0.365 | 0.132 | 0.122 | 0.104 | 0.094 | 0.062 | 0.214 | 0.204 | 5/8-3 1/2 |
| 3/8 | 0.3750 | 0.3750 | 0.3730 | 0.094 | 0.031 | 0.0440 | 0.0340 | 0.479 | 0.459 | 0.132 | 0.122 | 0.135 | 0.125 | 0.062 | 0.255 | 0.245 | 3/4-4 1/4 |
| 7/16 | 0.4375 | 0.4375 | 0.4355 | 0.094 | 0.031 | 0.0520 | 0.0420 | 0.541 | 0.521 | 0.195 | 0.185 | 0.135 | 0.125 | 0.094 | 0.298 | 0.288 | 7/8-4 1/2 |
| 1/2 | 0.5000 | 0.5000 | 0.4980 | 0.094 | 0.031 | 0.0570 | 0.0470 | 0.635 | 0.615 | 0.195 | 0.185 | 0.135 | 0.125 | 0.094 | 0.317 | 0.307 | 1-4 1/2 |

[a] For expanded diameters, *B*, see ANSI/ASME B18.8.2-2020.

[b] Pins in 1/32- and 3/64-inch sizes of any length and all sizes of 1/4-inch nominal length or shorter are not crowned or chamfered.

[c] Standard lengths increase in 1/8-inch steps from 1/8 to 1 inch, and in 1/4-inch steps above 1 inch. Standard lengths for the 1/32-, 3/64-, 1/16-, and 5/64-inch sizes and the 1/4-inch length for the 3/32-, 7/64-, and 1/8-inch sizes do not apply to Type G grooved pins.

[d] Non-stock items, not recommended for new designs.

| Pin Material | Nominal Pin Size | | | | | | | | | | | | | | |
|---|---|---|---|---|---|---|---|---|---|---|---|---|---|---|---|
| | 1/32 | 3/64 | 1/16 | 5/64 | 3/32 | 7/64 | 1/8 | 5/32 | 3/16 | 7/32 | 1/4 | 5/16 | 3/8 | 7/16 | 1/2 |
| Steels | Double Shear Load, Min, lb | | | | | | | | | | | | | | |
| Low Carbon | 100 | 220 | 410 | 620 | 890 | 1,220 | 1,600 | 2,300 | 3,310 | 4,510 | 5,880 | 7,660 | 11,000 | 15,000 | 19,600 |
| Alloy (hardness 40 - 48 RC) | 180 | 400 | 720 | 1,120 | 1,600 | 2,180 | 2,820 | 4,520 | 6,440 | 8,770 | 11,500 | 17,900 | 26,000 | 35,200 | 46,000 |
| Corrosion Resistant | 140 | 300 | 540 | 860 | 1,240 | 1,680 | 2,200 | 3,310 | 4,760 | 6,480 | 8,460 | 12,700 | 18,200 | 24,800 | 32,400 |
| Brass | 60 | 140 | 250 | 390 | 560 | 760 | 990 | 1,540 | 2,220 | 3,020 | 3,950 | 6,170 | 9,050 | 12,100 | 15,800 |
| | Recommended Hole Sizes for Unplated Pins (The minimum drill size is the same as the pin size. See also text on page 1894.) | | | | | | | | | | | | | | |
| Maximum Diameter | 0.0324 | 0.0482 | 0.0640 | 0.0798 | 0.0956 | 0.1113 | 0.1271 | 0.1587 | 0.1903 | 0.2219 | 0.2534 | 0.3166 | 0.3797 | 0.4428 | 0.5060 |
| Minimum Diameter | 0.0312 | 0.0469 | 0.0625 | 0.0781 | 0.0938 | 0.1094 | 0.1250 | 0.1563 | 0.1875 | 0.2188 | 0.2500 | 0.3125 | 0.3750 | 0.4375 | 0.5000 |

All dimensions are in inches.

**Table 8. American National Standard Grooved T-Head Cotter Pins**
*ANSI/ASME B18.8.2-2020*

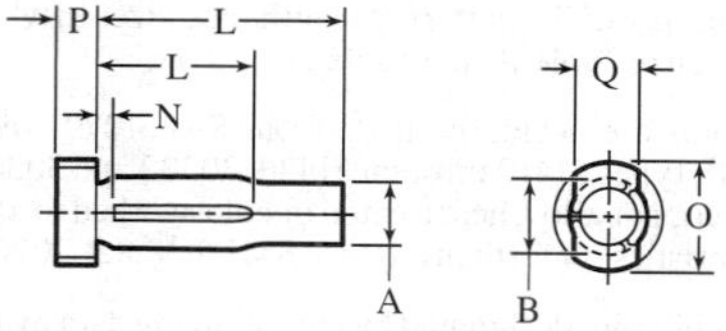


| Nominal Size[a] or Basic Shank Dia. | | Shank Diameter, A | | Length, N | Head Dia., O | | Head Height, P | | Head Width, Q | | Range of Standard Lengths,[b] L | Recommended Hole Size | |
|---|---|---|---|---|---|---|---|---|---|---|---|---|---|
| | | Max. | Min. | Max. | Max. | Min. | Max. | Min. | Max. | Min. | | Max. | Min. |
| 5⁄32 | 0.156 | 0.154 | 0.150 | 0.08 | 0.26 | 0.24 | 0.11 | 0.09 | 0.18 | 0.15 | ¾-1⅛ | 0.161 | 0.156 |
| 3⁄16 | 0.187 | 0.186 | 0.182 | 0.09 | 0.30 | 0.28 | 0.13 | 0.11 | 0.22 | 0.18 | ¾-1¼ | 0.193 | 0.187 |
| ¼ | 0.250 | 0.248 | 0.244 | 0.12 | 0.40 | 0.38 | 0.17 | 0.15 | 0.28 | 0.24 | 1-1½ | 0.257 | 0.250 |
| 5⁄16 | 0.312 | 0.310 | 0.305 | 0.16 | 0.51 | 0.48 | 0.21 | 0.19 | 0.34 | 0.30 | 1⅛-2 | 0.319 | 0.312 |
| 23⁄64 | 0.359 | 0.358 | 0.353 | 0.18 | 0.57 | 0.54 | 0.24 | 0.22 | 0.38 | 0.35 | 1¼-2 | 0.366 | 0.359 |
| ½ | 0.500 | 0.498 | 0.493 | 0.25 | 0.79 | 0.76 | 0.32 | 0.30 | 0.54 | 0.49 | 2-3 | 0.508 | 0.500 |

[a] When specifying nominal size in decimals, zeros preceding decimal point and in the fourth decimal place are omitted.

[b] Lengths increase in ⅛-inch steps from ¾ to 1¼ inch and in ¼-inch steps above 1¼ inches. For groove length, *M*, dimensions see ANSI/ASME B18.8.2-2020.

All dimensions are in inches.

For expanded diameter, *B*, dimensions, see ANSI/ASME B18.8.2-2020.

**Table 9. American National Standard Round Head Grooved Drive Studs**
*ANSI/ASME B18.8.2-2020*

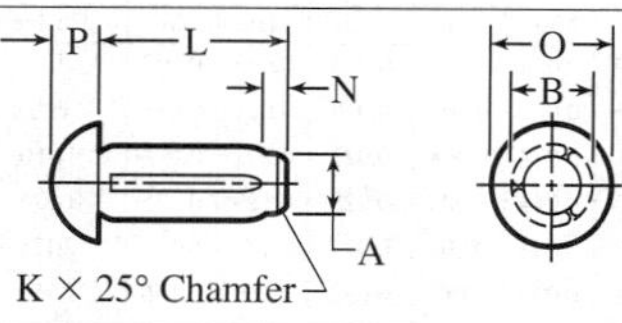


| Stud Size Number and Basic Shank Diameter[a] | | Shank Diameter, A | | Head Diameter, O | | Head Height, P | | Range of Standard Lengths,[b] L | Recommended Hole Size | | Drill Size |
|---|---|---|---|---|---|---|---|---|---|---|---|
| | | Max. | Min. | Max. | Min. | Max. | Min. | | Max. | Min. | |
| 0 | 0.067 | 0.067 | 0.065 | 0.130 | 0.120 | 0.050 | 0.040 | ⅛-¼ | 0.0686 | 0.0670 | 51 |
| 2 | 0.086 | 0.086 | 0.084 | 0.162 | 0.146 | 0.070 | 0.059 | ⅛-¼ | 0.0877 | 0.0860 | 44 |
| 4 | 0.104 | 0.104 | 0.102 | 0.211 | 0.193 | 0.086 | 0.075 | 3⁄16-5⁄16 | 0.1059 | 0.1040 | 37 |
| 6 | 0.120 | 0.120 | 0.118 | 0.260 | 0.240 | 0.103 | 0.091 | ¼-⅜ | 0.1220 | 0.1200 | 31 |
| 7 | 0.136 | 0.136 | 0.134 | 0.309 | 0.287 | 0.119 | 0.107 | 5⁄16-½ | 0.1382 | 0.1360 | 29 |
| 8 | 0.144 | 0.144 | 0.142 | 0.309 | 0.287 | 0.119 | 0.107 | ⅜-⅝ | 0.1463 | 0.1440 | 27 |
| 10 | 0.161 | 0.161 | 0.159 | 0.359 | 0.334 | 0.136 | 0.124 | ⅜-⅝ | 0.1636 | 0.1610 | 20 |
| 12 | 0.196 | 0.196 | 0.194 | 0.408 | 0.382 | 0.152 | 0.140 | ½¾ | 0.1990 | 0.1960 | 9 |
| 14 | 0.221 | 0.221 | 0.219 | 0.457 | 0.429 | 0.169 | 0.156 | ½-¾ | 0.2240 | 0.2210 | 2 |
| 16 | 0.250 | 0.250 | 0.248 | 0.472 | 0.443 | 0.174 | 0.161 | ½ | 0.2534 | 0.2500 | ¼ |

[a] Where specifying nominal size in decimals, zeros preceding decimal point and in the fourth decimal place are omitted.

[b] Lengths increase in 1⁄16-inch steps from ⅛ to ⅜ inch and in ⅛-inch steps above ⅜ inch.

All dimensions are in inches.

For pilot length, *M*, and expanded diameter, *B*, dimensions see ANSI/ASME B18.8.2-2020.

**American National Standard Spring Pins.**—These pins are made in two types: one type has a slot throughout its length; the other is shaped into a coil.

*Preferred Lengths and Sizes:* The preferred lengths and sizes in which these pins are normally available are given in Table 10 and Table 11.

*Materials:* Spring pins are normally made from SAE 1070-1095 carbon steel, SAE 6150H alloy steel, SAE types 51410 through 51420, 30302 and 30304 corrosion resistant steels, and beryllium copper alloy, heat treated or cold-worked to attain the hardness and performance characteristics set forth in ANSI/ASME B18.8.2-2020.

*Designation:* Spring pins are designated by the following data in the sequence shown:

*Examples:* Pin, Coiled Spring, 1/4 × 1 1/4, Standard Duty, Steel, Zinc Plated

Pin, Slotted Spring, 1/2 × 3, Steel, Phosphate Coated

**Table 10. American National Standard Slotted-Type Spring Pins**
*ANSI/ASME B18.8.2-2020*

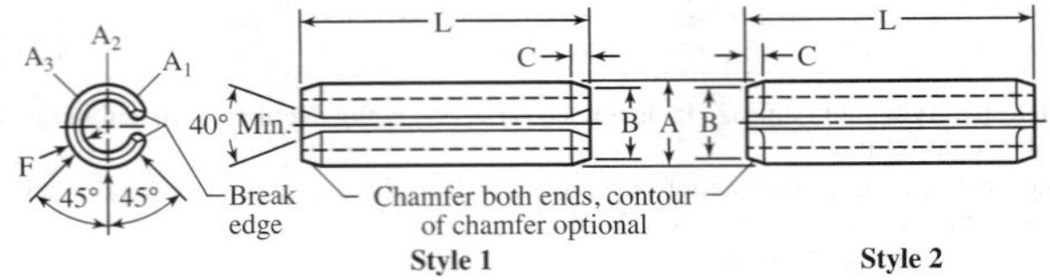


| Nominal Size[a] or Basic Pin Diameter | Average Pin Diameter, *A* | | Chamfer Dia., *B* | Chamfer Length, *C* | | Stock Thickness, *F* | Recommended Hole Size | | Material: SAE 1070, 1095, and SAE 51420 | Material: SAE 30302 and 30304 | Material: Beryllium Copper | Range of Practical Lengths[b] |
|---|---|---|---|---|---|---|---|---|---|---|---|---|
| | Max. | Min. | Max. | Max. | Min. | Basic | Max. | Min. | Double Shear Load, Min., lb | | | |
| 1/16 0.062 | 0.069 | 0.066 | 0.059 | 0.028 | 0.007 | 0.012 | 0.065 | 0.062 | 430 | 250 | 270 | 3/16-1 |
| 5/64 0.078 | 0.086 | 0.083 | 0.075 | 0.032 | 0.008 | 0.018 | 0.081 | 0.078 | 800 | 460 | 500 | 3/16-1 1/2 |
| 3/32 0.094 | 0.103 | 0.099 | 0.091 | 0.038 | 0.008 | 0.022 | 0.097 | 0.094 | 1,150 | 670 | 710 | 3/16-1 1/2 |
| 1/8 0.125 | 0.135 | 0.131 | 0.122 | 0.044 | 0.008 | 0.028 | 0.129 | 0.125 | 1,875 | 1,090 | 1,170 | 5/16-2 |
| 9/64 0.141 | 0.149 | 0.145 | 0.137 | 0.044 | 0.008 | 0.028 | 0.144 | 0.140 | 2,175 | 1,260 | 1,350 | 3/8-2 |
| 5/32 0.156 | 0.167 | 0.162 | 0.151 | 0.048 | 0.010 | 0.032 | 0.160 | 0.156 | 2,750 | 1,600 | 1,725 | 7/16-2 1/2 |
| 3/16 0.188 | 0.199 | 0.194 | 0.182 | 0.055 | 0.011 | 0.040 | 0.192 | 0.187 | 4,150 | 2,425 | 2,600 | 1/2-2 1/2 |
| 7/32 0.219 | 0.232 | 0.226 | 0.214 | 0.065 | 0.011 | 0.048 | 0.224 | 0.219 | 5,850 | 3,400 | 3,650 | 1/2-3 |
| 1/4 0.250 | 0.264 | 0.258 | 0.245 | 0.065 | 0.012 | 0.048 | 0.256 | 0.250 | 7,050 | 4,100 | 4,400 | 1/2-3 1/2 |
| 5/16 0.312 | 0.330 | 0.321 | 0.306 | 0.080 | 0.014 | 0.062 | 0.318 | 0.312 | 10,800 | 6,300 | 6,750 | 3/4-4 |
| 3/8 0.375 | 0.395 | 0.385 | 0.368 | 0.095 | 0.016 | 0.077 | 0.382 | 0.375 | 16,300 | 9,500 | 10,200 | 3/4, 7/8, 1, 1 1/4, 1 1/2, 1 3/4, 2-4 |
| 7/16 0.438 | 0.459 | 0.448 | 0.430 | 0.095 | 0.017 | 0.077 | 0.445 | 0.437 | 19,800 | 11,500 | 12,300 | 1, 1 1/4, 1 1/2, 1 3/4, 2-4 |
| 1/2 0.500 | 0.524 | 0.513 | 0.485 | 0.110 | 0.025 | 0.094 | 0.510 | 0.500 | 27,100 | 15,800 | 17,000 | 1 1/4, 1 1/2, 1 3/4, 2-4 |
| 5/8 0.625 | 0.653 | 0.640 | 0.608 | 0.125 | 0.030 | 0.125 | 0.636 | 0.625 | 46,000 | 18,800 | … | 2-6 |
| 3/4 0.750 | 0.784 | 0.769 | 0.730 | 0.150 | 0.030 | 0.150 | 0.764 | 0.750 | 66,000 | 23,200 | … | 2-6 |

[a] Where specifying nominal size in decimals, zeros preceding decimal point are omitted.

[b] Length increments are 1/16 inch from 1/8 to 1 inch; 1/8 from 1 inch to 2 inches; and 1/4 inch from 2 inches to 6 inches.

All dimensions are in inches.

**Table 11. American National Standard Coiled-Type Spring Pins** *ANSI/ASME B18.8.2-2020*

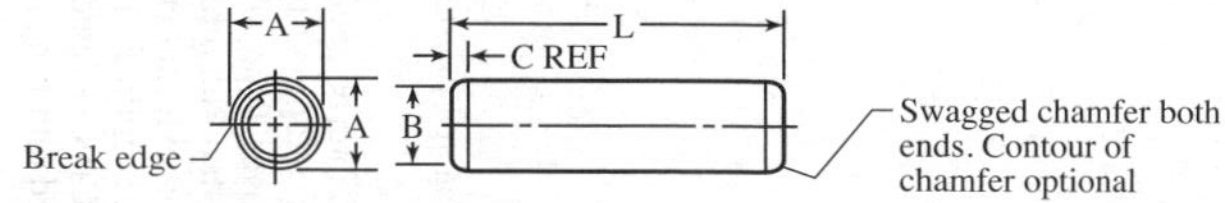


| Nominal Size or Basic Pin Diameter | Pin Diameter, *A*: Standard Duty | | Pin Diameter, *A*: Heavy Duty | | Pin Diameter, *A*: Light Duty | | Chamfer: Dia., *B* | Chamfer: Length, *C* | Recommended Hole Size | | SAE Material Number (Double Shear Load, Min., lb): Standard Duty, 1070-1095 and 51420 | Standard Duty, 30302 and 30304 | Heavy Duty, 1070-1095 and 51420 | Heavy Duty, 30302 and 30304 | Light Duty, 1070-1095 and 51420 | Light Duty, 30302 and 30304 |
|---|---|---|---|---|---|---|---|---|---|---|---|---|---|---|---|---|
| | Max. | Min. | Max. | Min. | Max. | Min. | Max. | Ref. | Max. | Min. | | | | | | |
| 1/32 0.031 | 0.035 | 0.033 | … | … | … | … | 0.029 | 0.024 | 0.032 | 0.031 | 90[a] | 65 | … | … | … | … |
| 0.039 | 0.044 | 0.041 | … | … | … | … | 0.037 | 0.024 | 0.040 | 0.039 | 135[a] | 100 | … | … | … | … |
| 3/64 0.047 | 0.052 | 0.049 | … | … | … | … | 0.045 | 0.024 | 0.048 | 0.047 | 190[a] | 145 | … | … | … | … |
| 0.052 | 0.057 | 0.054 | … | … | … | … | 0.050 | 0.024 | 0.053 | 0.051 | 250[a] | 190 | … | … | … | … |
| 1/16 0.062 | 0.072 | 0.067 | 0.070 | 0.066 | 0.073 | 0.067 | 0.059 | 0.028 | 0.065 | 0.061 | 330 | 265 | 475 | 360 | 205 | 160 |
| 5/64 0.078 | 0.088 | 0.083 | 0.086 | 0.082 | 0.089 | 0.083 | 0.075 | 0.032 | 0.081 | 0.077 | 550 | 425 | 800 | 575 | 325 | 250 |
| 3/32 0.094 | 0.105 | 0.099 | 0.103 | 0.098 | 0.106 | 0.099 | 0.091 | 0.038 | 0.097 | 0.093 | 775 | 600 | 1,150 | 825 | 475 | 360 |
| 7/64 0.109 | 0.120 | 0.114 | 0.118 | 0.113 | 0.121 | 0.114 | 0.106 | 0.038 | 0.112 | 0.108 | 1,050 | 825 | 1,500 | 1,150 | 650 | 500 |
| 1/8 0.125 | 0.138 | 0.131 | 0.136 | 0.130 | 0.139 | 0.131 | 0.121 | 0.044 | 0.129 | 0.124 | 1,400 | 1,100 | 2,000 | 1,700 | 825 | 650 |
| 5/32 0.156 | 0.171 | 0.163 | 0.168 | 0.161 | 0.172 | 0.163 | 0.152 | 0.048 | 0.160 | 0.155 | 2,200 | 1,700 | 3,100 | 2,400 | 1,300 | 1,000 |
| 3/16 0.188 | 0.205 | 0.196 | 0.202 | 0.194 | 0.207 | 0.196 | 0.182 | 0.055 | 0.192 | 0.185 | 3,150 | 2,400 | 4,500 | 3,500 | 1,900 | 1,450 |
| 7/32 0.219 | 0.238 | 0.228 | 0.235 | 0.226 | 0.240 | 0.228 | 0.214 | 0.065 | 0.224 | 0.217 | 4,200 | 3,300 | 5,900 | 4,600 | 2,600 | 2,000 |
| 1/4 0.250 | 0.271 | 0.260 | 0.268 | 0.258 | 0.273 | 0.260 | 0.243 | 0.065 | 0.256 | 0.247 | 5,500 | 4,300 | 7,800 | 6,200 | 3,300 | 2,600 |
| 5/16 0.312 | 0.337 | 0.324 | 0.334 | 0.322 | 0.339 | 0.324 | 0.304 | 0.080 | 0.319 | 0.308 | 8,700 | 6,700 | 12,000 | 9,300 | 5,200 | 4,000 |
| 3/8 0.375 | 0.403 | 0.388 | 0.400 | 0.386 | 0.405 | 0.388 | 0.366 | 0.095 | 0.383 | 0.370 | 12,600 | 9,600 | 18,000 | 14,000 | … | … |
| 7/16 0.438 | 0.469 | 0.452 | 0.466 | 0.450 | 0.471 | 0.452 | 0.427 | 0.095 | 0.446 | 0.431 | 17,000 | 13,300 | 23,500 | 18,000 | … | … |
| 1/2 0.500 | 0.535 | 0.516 | 0.532 | 0.514 | 0.537 | 0.516 | 0.488 | 0.110 | 0.510 | 0.493 | 22,500 | 17,500 | 32,000 | 25,000 | … | … |
| 5/8 0.625 | 0.661 | 0.642 | 0.658 | 0.640 | … | … | 0.613 | 0.125 | 0.635 | 0.618 | 35,000[b] | … | 48,000[b] | … | … | … |
| 3/4 0.750 | 0.787 | 0.768 | 0.784 | 0.766 | … | … | 0.738 | 0.150 | 0.760 | 0.743 | 50,000[b] | … | 70,000[b] | … | … | … |

[a] Sizes 1/32 inch through 0.052 inch are not available in SAE 1070-1095 carbon steel.

[b] Sizes 5/8 inch and larger are produced from SAE 6150H alloy steel, not SAE 1070-1095 carbon steel. Practical lengths, *L*, for sizes 1/32 through 0.052 inch are 1/8 through 5/8 inch and for the 7/64-inch size, 1/4 through 1 3/4 inches. For lengths of other sizes see Table 10.

All dimensions are in inches.

## RETAINING RINGS

**Retaining Ring Types, Uses, and Calculations.—**The purpose of a retaining ring is to act as an artificial shoulder that will retain an object in a housing (internal ring), as shown in Fig. 1, or on a shaft (external ring). Two types of retaining ring are common, the stamped ring and the spiral-wound ring. The stamped-type of retaining ring, or snap ring, is stamped from tempered sheet metal and has a nonuniform cross section. The typical spiral-wound retaining ring has a uniform cross section and is made up of two or more turns of coiled, spring-tempered steel, although one-turn spiral-wound rings are common. Spiral-wound retaining rings provide a continuous gapless shoulder to a housing or shaft. Most stamped rings can only be installed at or near the end of a shaft or housing. The spiral-wound design generally requires installation from the end of a shaft or housing. Both types, stamped and spiral, are usually installed into grooves on the shaft or housing.

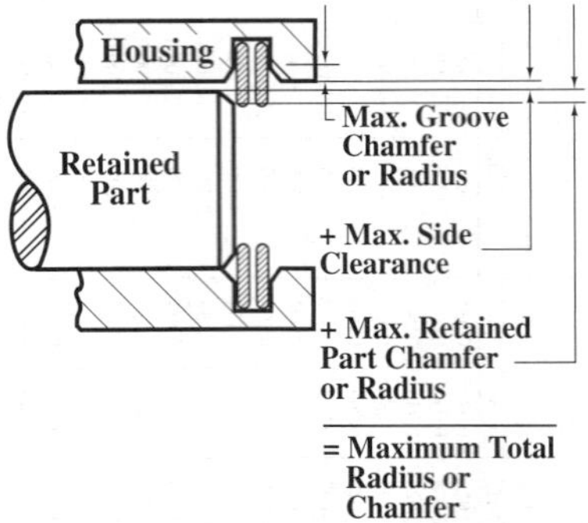


Fig. 1. Typical Retaining Ring Installation Showing Maximum Total Radius or Chamfer (*Courtesy Spirolox Retaining Rings*)

In the section that follows, Table 1 through Table 6 give dimensions and data on general-purpose tapered and reduced-cross-section metric retaining rings (stamped-type) covered by ANSI/ASME B27.7-1977 (R2017). Table 1 and Table 4 cover Type 3AM1 tapered external retaining rings, Table 2 and Table 5 cover Type 3BM1 tapered internal rings, and Table 3 and Table 6 cover Type 3CM1 reduced-cross-section external rings. Table 7 through Table 10 cover inch sizes of internal and external spiral retaining rings corresponding to MIL-R-27426 Types A (external) and B (internal), Class 1 (medium duty) and Class 2 (heavy duty). Table 11 through Table 17 cover stamped retaining rings in inch sizes.

**Table 1. American National Standard Metric Tapered Retaining Rings— Basic External Series—3AM1** *ANSI/ASME B27.7-1977 (R2017)*

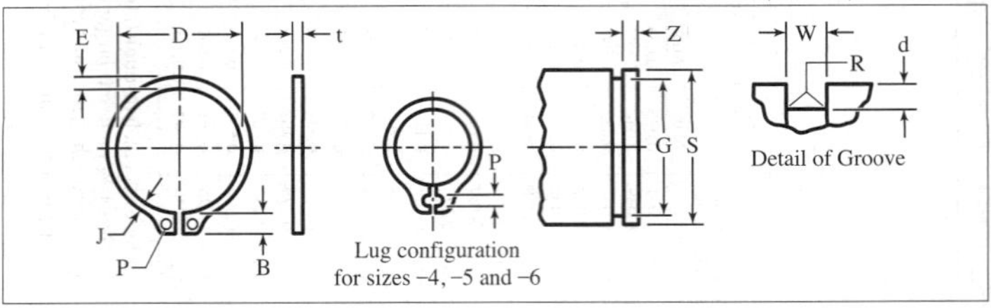

**Table 1.** *(Continued)* **American National Standard Metric Tapered Retaining Rings — Basic External Series — 3AM1** *ANSI/ASME B27.7-1977 (R2017)*

| Shaft Dia. | Ring | | Groove | | | | Shaft Diam | Ring | | Groove | | | |
|---|---|---|---|---|---|---|---|---|---|---|---|---|---|
| | Free Dia. | Thick-ness | Dia. | Width | Depth | Edge Margin | | Free Dia. | Thick-ness | Dia. | Width | Depth | Edge Margin |
| *S* | *D* | *t* | *G* | *W* | *d* ref. | *Z* min. | *S* | *D* | *t* | *G* | *W* | *d* ref. | *Z* min. |
| 4 | 3.60 | 0.25 | 3.80 | 0.32 | 0.1 | 0.3 | 36 | 33.25 | 1.3 | 33.85 | 1.4 | 1.06 | 3.2 |
| 5 | 4.55 | 0.4 | 4.75 | 0.5 | 0.13 | 0.4 | 38 | 35.20 | 1.3 | 35.8 | 1.4 | 1.10 | 3.3 |
| 6 | 5.45 | 0.4 | 5.70 | 0.5 | 0.15 | 0.5 | 40 | 36.75 | 1.6 | 37.7 | 1.75 | 1.15 | 3.4 |
| 7 | 6.35 | 0.6 | 6.60 | 0.7 | 0.20 | 0.6 | 42 | 38.80 | 1.6 | 39.6 | 1.75 | 1.20 | 3.6 |
| 8 | 7.15 | 0.6 | 7.50 | 0.7 | 0.25 | 0.8 | 43 | 39.65 | 1.6 | 40.5 | 1.75 | 1.25 | 3.8 |
| 9 | 8.15 | 0.6 | 8.45 | 0.7 | 0.28 | 0.8 | 45 | 41.60 | 1.6 | 42.4 | 1.75 | 1.30 | 3.9 |
| 10 | 9.00 | 0.6 | 9.40 | 0.7 | 0.30 | 0.9 | 46 | 42.55 | 1.6 | 43.3 | 1.75 | 1.35 | 4.0 |
| 11 | 10.00 | 0.6 | 10.35 | 0.7 | 0.33 | 1.0 | 48 | 44.40 | 1.6 | 45.2 | 1.75 | 1.40 | 4.2 |
| 12 | 10.85 | 0.6 | 11.35 | 0.7 | 0.33 | 1.0 | 50 | 46.20 | 1.6 | 47.2 | 1.75 | 1.40 | 4.2 |
| 13 | 11.90 | 0.9 | 12.30 | 1.0 | 0.35 | 1.0 | 52 | 48.40 | 2.0 | 49.1 | 2.15 | 1.45 | 4.3 |
| 14 | 12.90 | 0.9 | 13.25 | 1.0 | 0.38 | 1.2 | 54 | 49.9 | 2.0 | 51.0 | 2.15 | 1.50 | 4.5 |
| 15 | 13.80 | 0.9 | 14.15 | 1.0 | 0.43 | 1.3 | 55 | 50.6 | 2.0 | 51.8 | 2.15 | 1.60 | 4.8 |
| 16 | 14.70 | 0.9 | 15.10 | 1.0 | 0.45 | 1.4 | 57 | 52.9 | 2.0 | 53.8 | 2.15 | 1.60 | 4.8 |
| 17 | 15.75 | 0.9 | 16.10 | 1.0 | 0.45 | 1.4 | 58 | 53.6 | 2.0 | 54.7 | 2.15 | 1.65 | 4.9 |
| 18 | 16.65 | 1.1 | 17.00 | 1.2 | 0.50 | 1.5 | 60 | 55.8 | 2.0 | 56.7 | 2.15 | 1.65 | 4.9 |
| 19 | 17.60 | 1.1 | 17.95 | 1.2 | 0.53 | 1.6 | 62 | 57.3 | 2.0 | 58.6 | 2.15 | 1.70 | 5.1 |
| 20 | 18.35 | 1.1 | 18.85 | 1.2 | 0.58 | 1.7 | 65 | 60.4 | 2.0 | 61.6 | 2.15 | 1.70 | 5.1 |
| 21 | 19.40 | 1.1 | 19.80 | 1.2 | 0.60 | 1.8 | 68 | 63.1 | 2.0 | 64.5 | 2.15 | 1.75 | 5.3 |
| 22 | 20.30 | 1.1 | 20.70 | 1.2 | 0.65 | 1.9 | 70 | 64.6 | 2.4 | 66.4 | 2.55 | 1.80 | 5.4 |
| 23 | 21.25 | 1.1 | 21.65 | 1.2 | 0.67 | 2.0 | 72 | 66.6 | 2.4 | 68.3 | 2.55 | 1.85 | 5.5 |
| 24 | 22.20 | 1.1 | 22.60 | 1.2 | 0.70 | 2.1 | 75 | 69.0 | 2.4 | 71.2 | 2.55 | 1.90 | 5.7 |
| 25 | 23.10 | 1.1 | 23.50 | 1.2 | 0.75 | 2.3 | 78 | 72.0 | 2.4 | 74.0 | 2.55 | 2.00 | 6.0 |
| 26 | 24.05 | 1.1 | 24.50 | 1.2 | 0.75 | 2.3 | 80 | 74.2 | 2.4 | 75.9 | 2.55 | 2.05 | 6.1 |
| 27 | 24.95 | 1.3 | 25.45 | 1.4 | 0.78 | 2.3 | 82 | 76.4 | 2.4 | 77.8 | 2.55 | 2.10 | 6.3 |
| 28 | 25.80 | 1.3 | 26.40 | 1.4 | 0.80 | 2.4 | 85 | 78.6 | 2.4 | 80.6 | 2.55 | 2.20 | 6.6 |
| 30 | 27.90 | 1.3 | 28.35 | 1.4 | 0.83 | 2.5 | 88 | 81.4 | 2.8 | 83.5 | 2.95 | 2.25 | 6.7 |
| 32 | 29.60 | 1.3 | 30.20 | 1.4 | 0.90 | 2.7 | 90 | 83.2 | 2.8 | 85.4 | 2.95 | 2.30 | 6.9 |
| 34 | 31.40 | 1.3 | 32.00 | 1.4 | 1.00 | 3.0 | 95 | 88.1 | 2.8 | 90.2 | 2.95 | 2.40 | 7.2 |
| 35 | 32.30 | 1.3 | 32.90 | 1.4 | 1.05 | 3.1 | 100 | 92.5 | 2.8 | 95.0 | 2.95 | 2.50 | 7.5 |

All dimensions are in millimeters. Sizes −4, −5, and −6 are available in beryllium copper only.

These rings are designated by series symbol and shaft diameter, thus: for a 4 mm diameter shaft, 3AM1-4; for a 20 mm diameter shaft, 3AM1-20; etc.

*Ring Free Diameter Tolerances:* For ring sizes −4 through −6, +0.05, −0.10 mm; for sizes −7 through −12, +0.05, −0.15 mm; for sizes −13 through − 26, +0.15, −0.25 mm; for sizes −27 through −38, +0.25, −0.40 mm; for sizes −40 through −50, +0.35, −0.50 mm; for sizes −52 through −62, +0.35, −0.65 mm; and for sizes −65 through −100, +0.50, −0.75 mm.

*Groove Diameter Tolerances:* For ring sizes −4 through −6, −0.08 mm; for sizes −7 through −10, −0.10 mm; for sizes −11 through −15, −0.12 mm; for sizes −16 through −26, −0.15 mm; for sizes −27 through − 36, −0.20 mm; for sizes −38 through −55, −0.30 mm; and for sizes −57 through −100, −0.40 mm.

*Groove Diameter F.I.M. (full indicator movement)* or maximum allowable deviation of concentricity between groove and shaft: For ring sizes −4 through −6, 0.03 mm; for ring sizes −7 through −12, 0.05 mm; for sizes −13 through −28, 0.10 mm; for sizes −30 through −55, 0.15 mm; and for sizes −57 through − 00, 0.20 mm.

*Groove Width Tolerances:* For ring size −4, +0.05 mm; for sizes −5 and −6, +0.10 mm, for sizes −7 through −38, +0.15 mm; and for sizes −40 through − 100, +0.20 mm.

*Groove Maximum Bottom Radii, R*: For ring sizes −4 through −6, none; for sizes −7 through −18, 0.1 mm; for sizes −19 through −30, 0.2 mm; for sizes −32 through −50, 0.3 mm; and for sizes −52 through −100, 0.4 mm. For manufacturing details not shown, including materials, see ANSI/ASME B27.7-1977 (R2017).

**Table 2. American National Standard Metric Tapered Retaining Rings— Basic Internal Series—3BM1** *ANSI/ASME B27.7-1977 (R2017)*

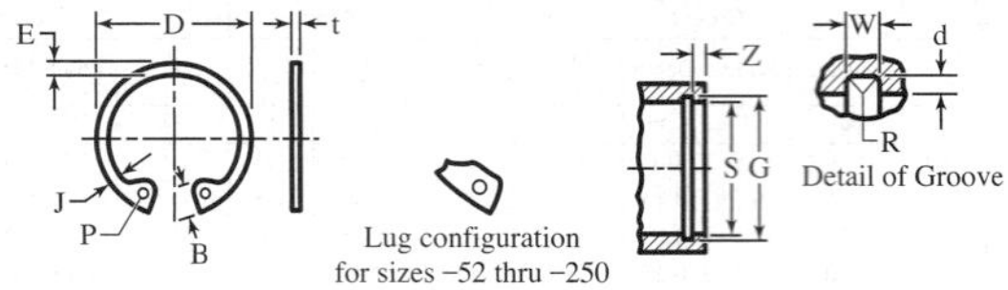


| Shaft Dia. | Ring Free Dia. | Ring Thickness | Groove Dia. | Groove Width | Groove Depth | Groove Edge Margin | Shaft Dia. | Ring Free Dia. | Ring Thickness | Groove Dia. | Groove Width | Groove Depth | Groove Edge Margin |
|---|---|---|---|---|---|---|---|---|---|---|---|---|---|
| *S* | *D* | *t* | *G* | *W* | *d* ref. | *Z* min. | *S* | *D* | *t* | *G* | *W* | *d* ref. | *Z* min. |
| 8 | 8.80 | 0.4 | 8.40 | 0.5 | 0.2 | 0.6 | 65 | 72.2 | 2.4 | 69.0 | 2.55 | 2.00 | 6.0 |
| 9 | 10.00 | 0.6 | 9.45 | 0.7 | 0.23 | 0.7 | 68 | 75.7 | 2.4 | 72.2 | 2.55 | 2.10 | 6.3 |
| 10 | 11.10 | 0.6 | 10.50 | 0.7 | 0.25 | 0.8 | 70 | 77.5 | 2.4 | 74.4 | 2.55 | 2.20 | 6.6 |
| 11 | 12.20 | 0.6 | 11.60 | 0.7 | 0.3 | 0.9 | 72 | 79.6 | 2.4 | 76.5 | 2.55 | 2.25 | 6.7 |
| 12 | 13.30 | 0.6 | 12.65 | 0.7 | 0.33 | 1.0 | 75 | 83.3 | 2.4 | 79.7 | 2.55 | 2.35 | 7.1 |
| 13 | 14.25 | 0.9 | 13.70 | 1.0 | 0.35 | 1.1 | 78 | 86.8 | 2.8 | 82.8 | 2.95 | 2.40 | 7.2 |
| 14 | 15.45 | 0.9 | 14.80 | 1.0 | 0.40 | 1.2 | 80 | 89.1 | 2.8 | 85.0 | 2.95 | 2.50 | 7.5 |
| 15 | 16.60 | 0.9 | 15.85 | 1.0 | 0.43 | 1.3 | 82 | 91.1 | 2.8 | 87.2 | 2.95 | 2.60 | 7.8 |
| 16 | 17.70 | 0.9 | 16.90 | 1.0 | 0.45 | 1.4 | 85 | 94.4 | 2.8 | 90.4 | 2.95 | 2.70 | 8.1 |
| 17 | 18.90 | 0.9 | 18.00 | 1.0 | 0.50 | 1.5 | 88 | 97.9 | 2.8 | 93.6 | 2.95 | 2.80 | 8.4 |
| 18 | 20.05 | 0.9 | 19.05 | 1.0 | 0.53 | 1.6 | 90 | 100.0 | 2.80 | 95.7 | 2.95 | 2.85 | 8.6 |
| 19 | 21.10 | 0.9 | 20.10 | 1.0 | 0.55 | 1.7 | 92 | 102.2 | 2.8 | 97.8 | 2.95 | 2.90 | 8.7 |
| 20 | 22.25 | 0.9 | 21.15 | 1.0 | 0.57 | 1.7 | 95 | 105.6 | 2.8 | 101.0 | 2.95 | 3.00 | 9.0 |
| 21 | 23.30 | 0.9 | 22.20 | 1.0 | 0.60 | 1.8 | 98 | 109.0 | 2.8 | 104.2 | 2.95 | 3.10 | 9.3 |
| 22 | 24.40 | 1.1 | 23.30 | 1.2 | 0.65 | 1.9 | 100 | 110.7 | 2.8 | 106.3 | 2.95 | 3.15 | 9.5 |
| 23 | 25.45 | 1.1 | 24.35 | 1.2 | 0.67 | 2.0 | 102 | 112.4 | 2.8 | 108.4 | 2.95 | 3.20 | 9.6 |
| 24 | 26.55 | 1.1 | 25.4 | 1.2 | 0.70 | 2.1 | 105 | 115.8 | 2.8 | 111.5 | 2.95 | 3.25 | 9.8 |
| 25 | 27.75 | 1.1 | 26.6 | 1.2 | 0.80 | 2.4 | 108 | 119.2 | 2.8 | 114.6 | 2.95 | 3.30 | 9.9 |
| 26 | 28.85 | 1.1 | 27.7 | 1.2 | 0.85 | 2.6 | 110 | 120.8 | 2.8 | 116.7 | 2.95 | 3.35 | 10.1 |
| 27 | 29.95 | 1.3 | 28.8 | 1.4 | 0.90 | 2.7 | 115 | 126.0 | 2.8 | 121.9 | 2.95 | 3.45 | 10.4 |
| 28 | 31.10 | 1.3 | 29.8 | 1.4 | 0.90 | 2.7 | 120 | 132.4 | 2.8 | 127.0 | 2.95 | 3.50 | 10.5 |
| 30 | 33.40 | 1.3 | 31.9 | 1.4 | 0.95 | 2.9 | 125 | 137.1 | 2.8 | 132.1 | 2.95 | 3.55 | 10.7 |
| 32 | 35.35 | 1.3 | 33.9 | 1.4 | 0.95 | 2.9 | 130 | 142.5 | 2.8 | 137.2 | 2.95 | 3.60 | 10.8 |
| 34 | 37.75 | 1.3 | 36.1 | 1.4 | 1.05 | 3.2 | 135 | 148.5 | 3.2 | 142.3 | 3.40 | 3.65 | 11.0 |
| 35 | 38.75 | 1.3 | 37.2 | 1.4 | 1.10 | 3.3 | 140 | 154.1 | 3.2 | 147.4 | 3.40 | 3.70 | 11.1 |
| 36 | 40.00 | 1.3 | 38.3 | 1.4 | 1.15 | 3.5 | 145 | 159.5 | 3.2 | 152.5 | 3.40 | 3.75 | 11.3 |
| 37 | 41.05 | 1.3 | 39.3 | 1.4 | 1.15 | 3.5 | 150 | 164.5 | 3.2 | 157.6 | 3.40 | 3.80 | 11.4 |
| 38 | 42.15 | 1.3 | 40.4 | 1.4 | 1.20 | 3.6 | 155 | 168.8 | 3.2 | 162.7 | 3.40 | 3.85 | 11.6 |
| 40 | 44.25 | 1.6 | 42.4 | 1.75 | 1.20 | 3.6 | 160 | 175.1 | 4.0 | 167.8 | 4.25 | 3.90 | 11.7 |
| 42 | 46.60 | 1.6 | 44.5 | 1.75 | 1.25 | 3.7 | 165 | 180.3 | 4.0 | 172.9 | 4.25 | 3.95 | 11.9 |
| 45 | 49.95 | 1.6 | 47.6 | 1.75 | 1.30 | 3.9 | 170 | 185.6 | 4.0 | 178.0 | 4.25 | 4.00 | 12.0 |
| 46 | 51.05 | 1.6 | 48.7 | 1.75 | 1.35 | 4.0 | 175 | 191.3 | 4.0 | 183.2 | 4.25 | 4.10 | 12.3 |
| 47 | 52.15 | 1.6 | 49.8 | 1.75 | 1.40 | 4.2 | 180 | 196.6 | 4.0 | 188.4 | 4.25 | 4.20 | 12.6 |
| 48 | 53.30 | 1.6 | 50.9 | 1.75 | 1.45 | 4.3 | 185 | 202.7 | 4.8 | 193.6 | 5.10 | 4.30 | 12.9 |
| 50 | 55.35 | 1.6 | 53.1 | 1.75 | 1.55 | 4.6 | 190 | 207.7 | 4.8 | 198.8 | 5.10 | 4.40 | 13.2 |
| 52 | 57.90 | 2.0 | 55.3 | 2.15 | 1.65 | 5.0 | 200 | 217.8 | 4.8 | 209.0 | 5.10 | 4.50 | 13.5 |
| 55 | 61.10 | 2.0 | 58.4 | 2.15 | 1.70 | 5.1 | 210 | 230.3 | 4.8 | 219.4 | 5.10 | 4.70 | 14.1 |
| 57 | 63.25 | 2.0 | 60.5 | 2.15 | 1.75 | 5.3 | 220 | 240.5 | 4.8 | 230.0 | 5.10 | 5.00 | 15.0 |
| 58 | 64.4 | 2.0 | 61.6 | 2.15 | 1.80 | 5.4 | 230 | 251.4 | 4.8 | 240.6 | 5.10 | 5.30 | 15.9 |
| 60 | 66.8 | 2.0 | 63.8 | 2.15 | 1.90 | 5.7 | 240 | 262.3 | 4.8 | 251.0 | 5.10 | 5.50 | 16.5 |
| 62 | 68.6 | 2.0 | 65.8 | 2.15 | 1.90 | 5.7 | 250 | 273.3 | 4.8 | 261.4 | 5.10 | 5.70 | 17.1 |
| 63 | 69.9 | 2.0 | 66.9 | 2.15 | 1.95 | 5.9 | … | … | … | … | … | … | … |

All dimensions are in millimeters.

These rings are designated by series symbol and shaft diameter, thus: for a 9 mm diameter shaft, 3BM1-9; for a 22 mm diameter shaft, 3BM1-22; etc.

*Ring Free Diameter Tolerances:* For ring sizes −8 through −20, +0.25, −0.13 mm; for sizes −21 through −26, +0.40, −0.25 mm; for sizes −27 through −38, +0.65, −0.50 mm; for sizes −40 through

−50, +0.90, −0.65 mm; for sizes −52 through −75, +1.00, −0.75 mm; for sizes −78 through −92, +1.40, −1.40 mm; for sizes −95 through −155, +1.65, −1.65 mm; for sizes −160 through −180, +2.05, −2.05 mm; and for sizes −185 through −250, +2.30, −2.30 mm.

*Groove Diameter Tolerances:* For ring sizes −8 and −9, +0.06 mm; for sizes −10 through −18, +0.10 mm; for sizes −19 through −28, +0.15 mm; for sizes −30 through −50, +0.20 mm; for sizes −52 through −98, +0.30; for sizes −100 through −160, +0.40 mm; and for sizes −165 through −250, +0.50 mm.

*Groove Diameter F.I.M. (Full Indicator Movement)* or maximum allowable deviation of concentricity between groove and shaft: For ring sizes −8 through −10, 0.03 mm; for sizes −11 through −15, 0.05 mm; for sizes −16 through −25, 0.10 mm; for sizes −26 through −45, 0.15 mm; for sizes −46 through −80, 0.20 mm; for sizes −82 through −150, 0.25 mm; and for sizes −155 through −250, 0.30 mm.

*Groove Width Tolerances:* For ring size −8, +0.10 mm; for sizes −9 through −38, +0.15 mm; for sizes −40 through −130, +0.20 mm; and for sizes −135 through −250, +0.25 mm.

*Groove Maximum Bottom Radii:* For ring sizes −8 through −17, 0.1 mm; for sizes −18 through −30, 0.2 mm; for sizes −32 through −55, 0.3 mm; and for sizes −56 through −250, 0.4 mm.

For manufacturing details not shown, including materials, see ANSI/ASME B27.7-1977 (R2017).

## Table 3. American National Standard Metric Reduced Cross Section Retaining Rings—E Ring External Series—3CM1 *ANSI/ASME B27.7-1977 (R2017)*

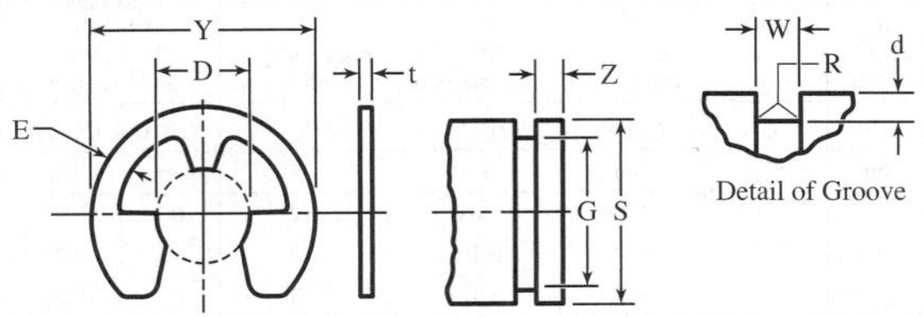


| Shaft Dia. | Ring | | | Groove | | | | Shaft Dia. | Ring | | | Groove | | | |
|---|---|---|---|---|---|---|---|---|---|---|---|---|---|---|---|
| | Free Dia. | Thickness | Outer Dia. | Dia. | Width | Depth | Edge Margin | | Free Dia. | Thickness | Outer Dia. | Dia. | Width | Depth | Edge Margin |
| *S* | *D* | *t* | *Y* nom | *G* | *W* | *d* ref. | *Z* min. | *S* | *D* | *t* | *Y* nom | *G* | *W* | *d* ref. | *Z* min. |
| 1 | 0.64 | 0.25 | 2.0 | 0.72 | 0.32 | 0.14 | 0.3 | 11 | 8.55 | 0.9 | 17.4 | 8.90 | 1.0 | 1.05 | 2.1 |
| 2 | 1.30 | 0.25 | 4.0 | 1.45 | 0.32 | 0.28 | 0.6 | 12 | 9.20 | 1.1 | 18.6 | 9.60 | 1.2 | 1.20 | 2.4 |
| 3 | 2.10 | 0.4 | 5.6 | 2.30 | 0.5 | 0.35 | 0.7 | 13 | 9.95 | 1.1 | 20.3 | 10.30 | 1.2 | 1.35 | 2.7 |
| 4 | 2.90 | 0.6 | 7.2 | 3.10 | 0.7 | 0.45 | 0.9 | 15 | 11.40 | 1.1 | 22.8 | 11.80 | 1.2 | 1.60 | 3.2 |
| 5 | 3.70 | 0.6 | 8.5 | 3.90 | 0.7 | 0.55 | 1.1 | 16 | 12.15 | 1.1 | 23.8 | 12.50 | 1.2 | 1.75 | 3.5 |
| 6 | 4.70 | 0.6 | 11.1 | 4.85 | 0.7 | 0.58 | 1.2 | 18 | 13.90 | 1.3 | 27.2 | 14.30 | 1.4 | 1.85 | 3.7 |
| 7 | 5.25 | 0.6 | 13.4 | 5.55 | 0.7 | 0.73 | 1.5 | 20 | 15.60 | 1.3 | 30.0 | 16.00 | 1.4 | 2.00 | 4.0 |
| 8 | 6.15 | 0.6 | 14.6 | 6.40 | 0.7 | 0.80 | 1.6 | 22 | 17.00 | 1.3 | 33.0 | 17.40 | 1.4 | 2.30 | 4.6 |
| 9 | 6.80 | 0.9 | 15.8 | 7.20 | 1.0 | 0.90 | 1.8 | 25 | 19.50 | 1.3 | 37.1 | 20.00 | 1.4 | 2.50 | 5.0 |
| 10 | 7.60 | 0.9 | 16.8 | 8.00 | 1.0 | 1.00 | 2.0 | … | … | … | … | … | … | … | … |

All dimensions are in millimeters. Size −1 is available in beryllium copper only.

These rings are designated by series symbol and shaft diameter, thus: for a 2 mm diameter shaft, 3CM1-2; for a 13 mm shaft, 3CMI -13; etc.

*Ring Free Diameter Tolerances:* For ring sizes −1 through −7, +0.03, −0.08 mm; for sizes −8 through −13, +0.05, −0.10 mm; and for sizes −15 through −25, +0.10, −0.15 mm.

*Groove Diameter Tolerances:* For ring sizes −1 and −2, −0.05 mm; for sizes −3 through −6, −0.08; for sizes −7 through −11, −0.10 mm; for sizes −12 through −18, − 0.15 mm; and for sizes −20 through −25, − 0.20 mm.

*Groove Diameter F.I.M. (Full Indicator Movement)* or maximum allowable deviation of concentricity between groove and shaft: For ring sizes −1 through −3, 0.04 mm; for −4 through −6, 0.05 mm; for −7 through −10, 0.08 mm; for −11 through −25, 0.10 mm.

*Groove Width Tolerances:* For ring sizes − 1 and −2, +0.05 mm; for size −3, +0.10 mm; and for sizes −4 through − 25, +0.15 mm.

*Groove Maximum Bottom Radii:* For ring sizes −1 and −2, 0.05 mm; for −3 through −7, 0.15 mm; for −8 through −13, 0.25 mm; and for −15 through −25, 0.4 mm.

For manufacturing details not shown, including materials, see ANSI/ASME B27.7-1977 (R2017).

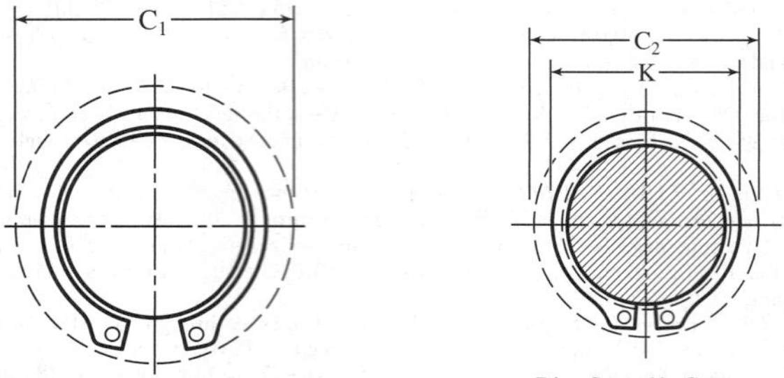


Ring Expanded over Shaft

Ring Seated in Groove

**Table 4. American National Standard Metric Basic External Series—3AM1 Retaining Rings—Checking and Performance Data** *ANSI/ASME B27.7-1977 (R2017)*

| Ring Series and Size No. | Clearance Dia. Ring Over Shaft | Clearance Dia. Ring in Groove | Gaging Diameter[a] | Allowable Thrust Loads Sharp Corner Abutment | | Maximum Allowable Corner Radii and Chamfers | | Allowable Assembly Speed[b] |
|---|---|---|---|---|---|---|---|---|
| 3AM1 | $C_1$ | $C_2$ | *K* max. | $P_r$[c] | $P_g$[d] | *R* max. | *Ch* max. | … |
| No. | mm | mm | mm | kN | kN | mm | mm | rpm |
| –4[a] | 7.0 | 6.8 | 4.90 | 0.6 | 0.2 | 0.35 | 0.25 | 70 000 |
| –5[a] | 8.2 | 7.9 | 5.85 | 1.1 | 0.3 | 0.35 | 0.25 | 70 000 |
| –6[a] | 9.1 | 8.8 | 6.95 | 1.4 | 0.4 | 0.35 | 0.25 | 70 000 |
| –7 | 12.3 | 11.8 | 8.05 | 2.6 | 0.7 | 0.45 | 0.3 | 60 000 |
| –8 | 13.6 | 13.0 | 9.15 | 3.1 | 1.0 | 0.5 | 0.35 | 55 000 |
| –9 | 14.5 | 13.8 | 10.35 | 3.5 | 1.2 | 0.6 | 0.35 | 48 000 |
| –10 | 15.5 | 14.7 | 11.50 | 3.9 | 1.5 | 0.7 | 0.4 | 42 000 |
| –11 | 16.4 | 15.6 | 12.60 | 4.3 | 1.8 | 0.75 | 0.45 | 38 000 |
| –12 | 17.4 | 16.6 | 13.80 | 4.7 | 2.0 | 0.8 | 0.45 | 34 000 |
| –13 | 19.7 | 18.8 | 15.05 | 7.5 | 2.2 | 0.8 | 0.5 | 31 000 |
| –14 | 20.7 | 19.7 | 15.60 | 8.1 | 2.6 | 0.9 | 0.5 | 28 000 |
| –15 | 21.7 | 20.6 | 17.20 | 8.7 | 3.2 | 1.0 | 0.6 | 27 000 |
| –16 | 22.7 | 21.6 | 18.35 | 9.3 | 3.5 | 1.1 | 0.6 | 25 000 |
| –17 | 23.7 | 22.6 | 19.35 | 9.9 | 4.0 | 1.1 | 0.6 | 24 000 |
| –18 | 26.2 | 25.0 | 20.60 | 16.0 | 4.4 | 1.2 | 0.7 | 23 000 |
| –19 | 27.2 | 25.9 | 21.70 | 16.9 | 4.9 | 1.2 | 0.7 | 21 500 |
| –20 | 28.2 | 26.8 | 22.65 | 17.8 | 5.7 | 1.2 | 0.7 | 20 000 |
| –21 | 29.2 | 27.7 | 23.80 | 18.6 | 6.2 | 1.3 | 0.7 | 19 000 |
| –22 | 30.3 | 28.7 | 24.90 | 19.6 | 7.0 | 1.3 | 0.8 | 18 500 |
| –23 | 31.3 | 29.6 | 26.00 | 20.5 | 7.6 | 1.3 | 0.8 | 18 000 |
| –24 | 34.1 | 32.4 | 27.15 | 21.4 | 8.2 | 1.4 | 0.8 | 17 500 |
| –25 | 35.1 | 33.3 | 28.10 | 22.3 | 9.2 | 1.4 | 0.8 | 17 000 |
| –26 | 36.0 | 34.2 | 29.25 | 23.2 | 9.6 | 1.5 | 0.9 | 16 500 |
| –27 | 37.8 | 35.9 | 30.35 | 28.4 | 10.3 | 1.5 | 0.9 | 16 300 |
| –28 | 38.8 | 36.9 | 31.45 | 28.4 | 11.0 | 1.6 | 1.0 | 15 800 |
| –30 | 40.8 | 38.8 | 33.6 | 31.6 | 12.3 | 1.6 | 1.0 | 15 000 |
| –32 | 42.8 | 40.7 | 35.9 | 33.6 | 14.1 | 1.7 | 1.0 | 14 800 |

**Table 4.** *(Continued)* **American National Standard Metric Basic External Series—3AM1 Retaining Rings—Checking and Performance Data** *ANSI/ASME B27.7-1977 (R2017)*

| Ring Series and Size No. | Clearance Dia. Ring Over Shaft | Ring in Groove | Gaging Diameter[a] | Allowable Thrust Loads Sharp Corner Abutment | | Maximum Allowable Corner Radii and Chamfers | | Allowable Assembly Speed[b] |
|---|---|---|---|---|---|---|---|---|
| 3AM1 | $C_1$ | $C_2$ | *K* max. | $P_r$[c] | $P_g$[d] | *R* max. | *Ch* max. | … |
| No. | mm | mm | mm | kN | kN | mm | mm | rpm |
| –34 | 44.9 | 42.5 | 37.9 | 36 | 16.7 | 1.7 | 1.1 | 14 000 |
| –35 | 45.9 | 43.4 | 39.0 | 37 | 18.1 | 1.8 | 1.1 | 13 500 |
| –36 | 48.6 | 46.1 | 40.2 | 38 | 18.9 | 1.9 | 1.2 | 13 300 |
| –38 | 50.6 | 48.0 | 42.5 | 40 | 20.5 | 2.0 | 1.2 | 12 700 |
| –40 | 54.0 | 51.3 | 44.5 | 52 | 22.6 | 2.1 | 1.2 | 12 000 |
| –42 | 56.0 | 53.2 | 46.9 | 54 | 24.8 | 2.2 | 1.3 | 11 000 |
| –43 | 57.0 | 54.0 | 47.9 | 55 | 26.4 | 2.3 | 1.4 | 10 800 |
| –45 | 59.0 | 55.9 | 50.0 | 58 | 28.8 | 2.3 | 1.4 | 10 000 |
| –46 | 60.0 | 56.8 | 50.9 | 59 | 30.4 | 2.4 | 1.4 | 9 500 |
| –48 | 62.4 | 59.1 | 53.0 | 62 | 33 | 2.4 | 1.4 | 8 800 |
| –50 | 64.4 | 61.1 | 55.2 | 64 | 35 | 2.4 | 1.4 | 8 000 |
| –52 | 67.6 | 64.1 | 57.4 | 84 | 37 | 2.5 | 1.5 | 7 700 |
| –54 | 69.6 | 66.1 | 59.5 | 87 | 40 | 2.5 | 1.5 | 7 500 |
| –55 | 70.6 | 66.9 | 60.4 | 89 | 44 | 2.5 | 1.5 | 7 400 |
| –57 | 72.6 | 68.9 | 62.7 | 91 | 45 | 2.6 | 1.5 | 7 200 |
| –58 | 73.6 | 69.8 | 63.6 | 93 | 46 | 2.6 | 1.6 | 7 100 |
| –60 | 75.6 | 71.8 | 65.8 | 97 | 49 | 2.6 | 1.6 | 7 000 |
| –62 | 77.6 | 73.6 | 67.9 | 100 | 52 | 2.7 | 1.6 | 6 900 |
| –65 | 80.6 | 76.6 | 71.2 | 105 | 54 | 2.8 | 1.7 | 6 700 |
| –68 | 83.6 | 79.5 | 74.5 | 110 | 58 | 2.9 | 1.7 | 6 500 |
| –70 | 88.1 | 83.9 | 76.4 | 136 | 62 | 2.9 | 1.7 | 6 400 |
| –72 | 90.1 | 85.8 | 78.5 | 140 | 65 | 2.9 | 1.7 | 6 200 |
| –75 | 93.1 | 88.7 | 81.7 | 147 | 69 | 3.0 | 1.8 | 5 900 |
| –78 | 95.4 | 92.1 | 84.6 | 151 | 76 | 3.0 | 1.8 | 5 600 |
| –80 | 97.9 | 93.1 | 87.0 | 155 | 80 | 3.1 | 1.9 | 5 400 |
| –82 | 100.0 | 95.1 | 89.0 | 159 | 84 | 3.2 | 1.9 | 5 200 |
| –85 | 103.0 | 97.9 | 92.1 | 165 | 91 | 3.2 | 1.9 | 5 000 |
| –88 | 107.0 | 100.8 | 95.1 | 199 | 97 | 3.2 | 1.9 | 4 800 |
| –90 | 109.0 | 103.6 | 97.1 | 204 | 101 | 3.2 | 1.9 | 4 500 |
| –95 | 114.0 | 108.6 | 102.7 | 215 | 112 | 3.4 | 2.1 | 4 350 |
| –100 | 119.5 | 113.7 | 108.0 | 227 | 123 | 3.5 | 2.1 | 4 150 |

[a] For checking when ring is seated in groove.

[b] These values have been calculated for steel rings.

[c] These values apply to rings made from SAE 1060-1090 steels and PH 15-7 Mo stainless steel used on shafts hardened to 50 RC minimum, with the exception of sizes –4, –5, and –6 which are supplied in beryllium copper only. Values for other sizes made from beryllium copper can be calculated by multiplying the listed values by 0.75. The values listed include a safety factor of 4.

[d] These values are for all standard rings used on low carbon steel shafts. They include a safety factor of 2.

Maximum allowable assembly loads with *R* max or *Ch* max are: For rings sizes –4, 0.2 kN; for sizes –5 and –6, 0.5 kN; for sizes –7 through –12, 2.1 kN; for sizes –13 through –17, 4.0 kN; for sizes –18 through –26, 6.0 kN; for sizes –27 through –38, 8.6 kN; for sizes –40 through – 50, 13.2 kN; for sizes –52 through –68, 22.0 kN; for sizes –70 through –85, 32 kN; and for sizes –88 through –100, 47 kN.

*Source:* Appendix to American National Standard ANSI/ASME B27.7-1977 (R2017).

**Table 5. American National Standard Metric Basic Internal Series—3BMI Retaining Rings—Checking and Performance Data** *ANSI/ASME B27.7-1977 (R2017)*

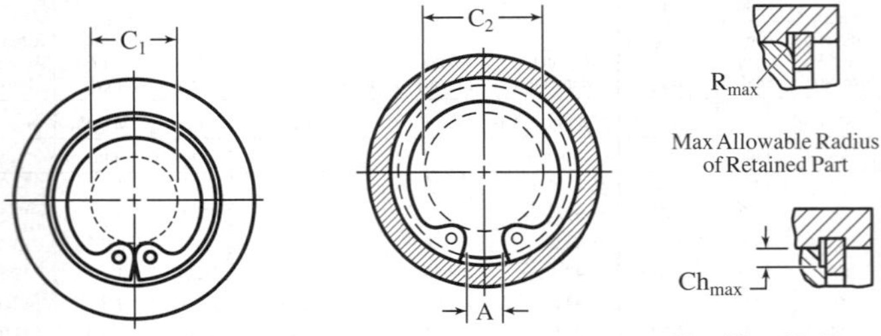


Ring Compressed in Bore — Ring Seated in Groove — Max Allowable Chamfer of Retained Part

| Ring Series and Size No. | Clearance Dia. Ring in Bore | Clearance Dia. Ring in Groove | Gaging Diameter[a] | Allowable Thrust Loads Sharp Corner Abutment | | Maximum Allowable Corner Radii and Chamfers | |
|---|---|---|---|---|---|---|---|
| 3BMI | $C_1$ | $C_2$ | *A* min. | $P_r$[b] | $P_g$[c] | *R* max. | *Ch* max. |
| No. | mm | mm | mm | kN | kN | mm | mm |
| –8 | 4.4 | 4.8 | 1.40 | 2.4 | 1.0 | 0.4 | 0.3 |
| –9 | 4.6 | 5.0 | 1.50 | 4.4 | 1.2 | 0.5 | 0.35 |
| –10 | 5.5 | 6.0 | 1.85 | 4.9 | 1.5 | 0.5 | 0.35 |
| –11 | 5.7 | 6.3 | 1.95 | 5.4 | 2.0 | 0.6 | 0.4 |
| –12 | 6.7 | 7.3 | 2.25 | 5.8 | 2.4 | 0.6 | 0.4 |
| –13 | 6.8 | 7.5 | 2.35 | 8.9 | 2.6 | 0.7 | 0.5 |
| –14 | 6.9 | 7.7 | 2.65 | 9.7 | 3.2 | 0.7 | 0.5 |
| –15 | 7.9 | 8.7 | 2.80 | 10.4 | 3.7 | 0.7 | 0.5 |
| –16 | 8.8 | 9.7 | 2.80 | 11.0 | 4.2 | 0.7 | 0.5 |
| –17 | 9.8 | 10.8 | 3.35 | 11.7 | 4.9 | 0.75 | 0.6 |
| –18 | 10.3 | 11.3 | 3.40 | 12.3 | 5.5 | 0.75 | 0.6 |
| –19 | 11.4 | 12.5 | 3.40 | 13.1 | 6.0 | 0.8 | 0.65 |
| –20 | 11.6 | 12.7 | 3.8 | 13.7 | 6.6 | 0.9 | 0.7 |
| –21 | 12.6 | 13.8 | 4.2 | 14.5 | 7.3 | 0.9 | 0.7 |
| –22 | 13.5 | 14.8 | 4.3 | 22.5 | 8.3 | 0.9 | 0.7 |
| –23 | 14.5 | 15.9 | 4.9 | 23.5 | 8.9 | 1.0 | 0.8 |
| –24 | 15.5 | 16.9 | 5.2 | 24.8 | 9.7 | 1.0 | 0.8 |
| –25 | 16.5 | 18.1 | 6.0 | 25.7 | 11.6 | 1.0 | 0.8 |
| –26 | 17.5 | 19.2 | 5.7 | 26.8 | 12.7 | 1.2 | 1.0 |
| –27 | 17.4 | 19.2 | 5.9 | 33 | 14.0 | 1.2 | 1.0 |
| –28 | 18.2 | 20.0 | 6.0 | 34 | 14.6 | 1.2 | 1.0 |
| –30 | 20.0 | 21.9 | 6.0 | 37 | 16.5 | 1.2 | 1.0 |
| –32 | 22.0 | 23.9 | 7.3 | 39 | 17.6 | 1.2 | 1.0 |
| –34 | 24.0 | 26.1 | 7.6 | 42 | 20.6 | 1.2 | 1.0 |
| –35 | 25.0 | 27.2 | 8.0 | 43 | 22.3 | 1.2 | 1.0 |
| –36 | 26.0 | 28.3 | 8.3 | 44 | 23.9 | 1.2 | 1.0 |
| –37 | 27.0 | 29.3 | 8.4 | 45 | 24.6 | 1.2 | 1.0 |
| –38 | 28.0 | 30.4 | 8.6 | 46 | 26.4 | 1.2 | 1.0 |
| –40 | 29.2 | 31.6 | 9.7 | 62 | 27.7 | 1.7 | 1.3 |
| –42 | 29.7 | 32.2 | 9.0 | 65 | 30.2 | 1.7 | 1.3 |
| –45 | 32.3 | 34.9 | 9.6 | 69 | 33.8 | 1.7 | 1.3 |
| –46 | 33.3 | 36.0 | 9.7 | 71 | 36 | 1.7 | 1.3 |
| –47 | 34.3 | 37.1 | 10.0 | 72 | 38 | 1.7 | 1.3 |
| –48 | 35.0 | 37.9 | 10.5 | 74 | 40 | 1.7 | 1.3 |
| –50 | 36.9 | 40.0 | 12.1 | 77 | 45 | 1.7 | 1.3 |
| –52 | 38.6 | 41.9 | 11.7 | 99 | 50 | 2.0 | 1.6 |
| –55 | 40.8 | 44.2 | 11.9 | 105 | 54 | 2.0 | 1.6 |
| –57 | 42.2 | 45.7 | 12.5 | 109 | 58 | 2.0 | 1.6 |

**Table 5.** *(Continued)* **American National Standard Metric Basic Internal Series—3BMI Retaining Rings—Checking and Performance Data** *ANSI/ASME B27.7-1977 (R2017)*

| Ring Series and Size No. | Clearance Dia. | | Gaging Diameter[a] | Allowable Thrust Loads Sharp Corner Abutment | | Maximum Allowable Corner Radii and Chamfers | |
|---|---|---|---|---|---|---|---|
| | Ring in Bore | Ring in Groove | | | | | |
| 3BMI | $C_1$ | $C_2$ | *A* min. | $P_r$[b] | $P_g$[c] | *R* max. | *Ch* max. |
| No. | mm | mm | mm | kN | kN | mm | mm |
| −58 | 43.2 | 46.8 | 13.0 | 111 | 60 | 2.0 | 1.6 |
| −60 | 45.5 | 49.3 | 12.7 | 115 | 66 | 2.0 | 1.6 |
| −62 | 47.0 | 50.8 | 14.0 | 119 | 68 | 2.0 | 1.6 |
| −63 | 47.8 | 51.7 | 14.2 | 120 | 71 | 2.0 | 1.6 |
| −65 | 49.4 | 53.4 | 14.2 | 149 | 75 | 2.0 | 1.6 |
| −68 | 52.0 | 56.2 | 14.4 | 156 | 82 | 2.3 | 1.8 |
| −70 | 53.8 | 58.2 | 16.1 | 161 | 88 | 2.3 | 1.8 |
| −72 | 55.9 | 60.4 | 17.4 | 166 | 93 | 2.3 | 1.8 |
| −75 | 58.2 | 62.9 | 16.8 | 172 | 101 | 2.3 | 1.8 |
| −78 | 61.2 | 66.0 | 17.6 | 209 | 108 | 2.5 | 2.0 |
| −80 | 63.0 | 68.0 | 17.2 | 215 | 115 | 2.5 | 2.0 |
| −82 | 63.5 | 68.7 | 18.8 | 220 | 122 | 2.6 | 2.1 |
| −85 | 66.8 | 72.2 | 19.1 | 228 | 131 | 2.6 | 2.1 |
| −88 | 69.6 | 75.2 | 20.4 | 236 | 141 | 2.8 | 2.2 |
| −90 | 71.6 | 77.3 | 21.4 | 241 | 147 | 2.8 | 2.2 |
| −92 | 73.6 | 79.4 | 22.2 | 247 | 153 | 2.9 | 2.4 |
| −95 | 76.7 | 82.7 | 22.6 | 255 | 164 | 3.0 | 2.5 |
| −98 | 78.3 | 84.5 | 22.6 | 263 | 174 | 3.0 | 2.5 |
| −100 | 80.3 | 86.6 | 24.1 | 269 | 181 | 3.1 | 2.5 |
| −102 | 82.2 | 88.6 | 25.5 | 273 | 187 | 3.2 | 2.6 |
| −105 | 85.1 | 91.6 | 26.0 | 281 | 196 | 3.3 | 2.6 |
| −108 | 88.1 | 94.7 | 26.4 | 290 | 205 | 3.5 | 2.7 |
| −110 | 88.4 | 95.1 | 27.5 | 295 | 212 | 3.6 | 2.8 |
| −115 | 93.2 | 100.1 | 29.4 | 309 | 227 | 3.7 | 2.9 |
| −120 | 98.2 | 105.2 | 27.2 | 321 | 241 | 3.9 | 3.1 |
| −125 | 103.1 | 110.2 | 30.3 | 335 | 255 | 4.0 | 3.2 |
| −130 | 108.0 | 115.2 | 31.0 | 349 | 269 | 4.0 | 3.2 |
| −135 | 110.4 | 117.7 | 30.4 | 415 | 283 | 4.3 | 3.4 |
| −140 | 115.3 | 122.7 | 30.4 | 429 | 298 | 4.3 | 3.4 |
| −145 | 120.4 | 127.9 | 31.6 | 444 | 313 | 4.3 | 3.4 |
| −150 | 125.3 | 132.9 | 33.5 | 460 | 327 | 4.3 | 3.4 |
| −155 | 130.4 | 138.1 | 37.0 | 475 | 343 | 4.3 | 3.4 |
| −160 | 133.8 | 141.6 | 35.0 | 613 | 359 | 4.5 | 3.6 |
| −165 | 138.7 | 146.6 | 33.1 | 632 | 374 | 4.6 | 3.7 |
| −170 | 143.6 | 151.6 | 38.2 | 651 | 390 | 4.6 | 3.7 |
| −175 | 146.0 | 154.2 | 37.7 | 670 | 403 | 4.8 | 3.8 |
| −180 | 151.4 | 159.8 | 39.0 | 690 | 434 | 5.0 | 4.0 |
| −185 | 154.7 | 163.3 | 37.3 | 851 | 457 | 5.1 | 4.1 |
| −190 | 159.5 | 168.3 | 35.0 | 873 | 480 | 5.3 | 4.3 |
| −200 | 169.2 | 178.2 | 43.9 | 919 | 517 | 5.4 | 4.3 |
| −210 | 177.5 | 186.9 | 40.6 | 965 | 566 | 5.8 | 4.6 |
| −220 | 184.1 | 194.1 | 38.3 | 1000 | 608 | 6.1 | 4.9 |
| −230 | 194.0 | 204.6 | 49.0 | 1060 | 686 | 6.3 | 5.1 |
| −240 | 200.4 | 211.4 | 45.4 | 1090 | 725 | 6.6 | 5.3 |
| −250 | 210.0 | 221.4 | 53.0 | 1150 | 808 | 6.7 | 5.4 |

[a] For checking when ring is seated in groove.

[b] These values apply to rings made from SAE 1060-1090 steels and PH 15-7 Mo stainless steel used in bores hardened to 50 RC minimum. Values for rings made from beryllium copper can be calculated by multiplying the listed values by 0.75. The values listed include a safety factor of 4.

[c] These values are for standard rings used in low carbon steel bores. They include a safety factor of 2.

Maximum allowable assembly loads for *R* max or *Ch* max are: For ring size −8, 0.8 kN; for sizes −9 through −12, 2.0 kN; for sizes −13 through −21, 4.0 kN; for sizes −22 through −26, 7.4 kN; for sizes −27 through −38, 10.8 kN; for sizes −40 through −50, 17.4 kN; for sizes −52 through −63, 27.4 kN; for size −65, 42.0 kN; for sizes −68 through −72, 39 kN; for sizes −75 through −130, 54 kN; for sizes −135 through −155, 67 kN; for sizes −160 through −180, 102 kN; and for sizes −185 through −250, 151 kN.

*Source:* Appendix to American National Standard ANSI/ASME B27.7-1977 (R2017).

**Table 6. American National Standard Metric E-Type External Series—3CM1 Retaining Rings—Checking and Performance Data** *ANSI/ASME B27.7-1977 (R2017)*

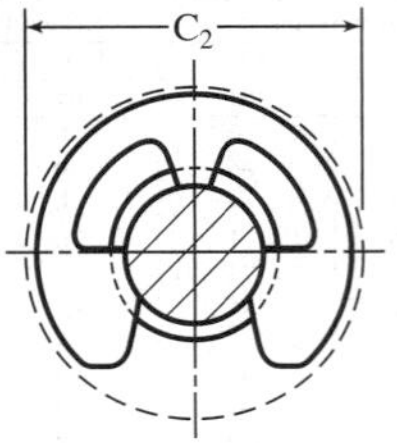


Ring Seated in Groove

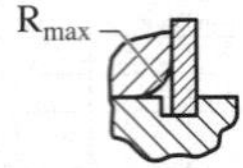


Max. Allowable Radius of Retained Part

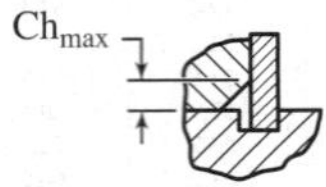


Max. Allowable Chamfer of Retained Part

| Ring Series and Size No. | Clearance Diameter: Ring in Groove | Allowable Thrust Loads Sharp Corner Abutment | | Maximum Allowable Corner Radii and Chamfers | | Allowable Assembly Speed[a] |
|---|---|---|---|---|---|---|
| 3CM1 | $C_2$ | $P_r$[b] | $P_g$[c] | *R* max | *Ch* max | … |
| No. | mm | kN | kN | mm | mm | rpm |
| –1 | 2.2 | 0.06 | 0.02 | 0.4 | 0.25 | 40 000 |
| –2 | 4.3 | 0.13 | 0.09 | 0.8 | 0.5 | 40 000 |
| –3 | 6.0 | 0.3 | 0.17 | 1.1 | 0.7 | 34 000 |
| –4 | 7.6 | 0.7 | 0.3 | 1.6 | 1.2 | 31 000 |
| –5 | 8.9 | 0.9 | 0.4 | 1.6 | 1.2 | 27 000 |
| –6 | 11.5 | 1.1 | 0.6 | 1.6 | 1.2 | 25 000 |
| –7 | 14.0 | 1.2 | 0.8 | 1.6 | 1.2 | 23 000 |
| –8 | 15.1 | 1.4 | 1.0 | 1.7 | 1.3 | 21 500 |
| –9 | 16.5 | 3.0 | 1.3 | 1.7 | 1.3 | 19 500 |
| –10 | 17.5 | 3.4 | 1.6 | 1.7 | 1.3 | 18 000 |
| –11 | 18.0 | 3.7 | 1.9 | 1.7 | 1.3 | 16 500 |
| –12 | 19.3 | 4.9 | 2.3 | 1.9 | 1.4 | 15 000 |
| –13 | 21.0 | 5.4 | 2.9 | 2.0 | 1.5 | 13 000 |
| –15 | 23.5 | 6.2 | 4.0 | 2.0 | 1.5 | 11 500 |
| –16 | 24.5 | 6.6 | 4.5 | 2.0 | 1.5 | 10 000 |
| –18 | 27.9 | 8.7 | 5.4 | 2.1 | 1.6 | 9 000 |
| –20 | 30.7 | 9.8 | 6.5 | 2.2 | 1.7 | 8 000 |
| –22 | 33.7 | 10.8 | 8.1 | 2.2 | 1.7 | 7 000 |
| –25 | 37.9 | 12.2 | 10.1 | 2.4 | 1.9 | 5 000 |

[a] These values have been calculated for steel rings.

[b] These values apply to rings made from SAE 1060-1090 steels and PH 15-7 Mo stainless steel used on shafts hardened to 50 RC minimum, with the exception of size –1 which is supplied in beryllium copper only. Values for other sizes made from beryllium copper can be calculated by multiplying the listed values by 0.75. The values listed include a safety factor of 4.

[c] These values apply to all standard rings used on low carbon steel shafts. They include a safety factor of 2.

Maximum allowable assembly loads with *R* max or *Ch* max are as follows:

| Ring Size No. | Maximum Allowable Load, kN | Ring Size No. | Maximum Allowable Load, kN | Ring Size No. | Maximum Allowable Load, kN |
|---|---|---|---|---|---|
| –1 | 0.06 | –8 | 1.4 | –16 | 6.6 |
| –2 | 0.13 | –9 | 3.0 | –18 | 8.7 |
| –3 | 0.3 | –10 | 3.4 | –20 | 9.8 |
| –4 | 0.7 | –11 | 3.7 | –22 | 10.8 |
| –5 | 0.9 | –12 | 4.9 | –25 | 12.2 |
| –6 | 1.1 | –13 | 5.4 | … | … |
| –7 | 1.2 | –15 | 6.2 | … | … |

*Source:* Appendix to American National Standard ANSI/ASME B27.7-1977 (R2017).

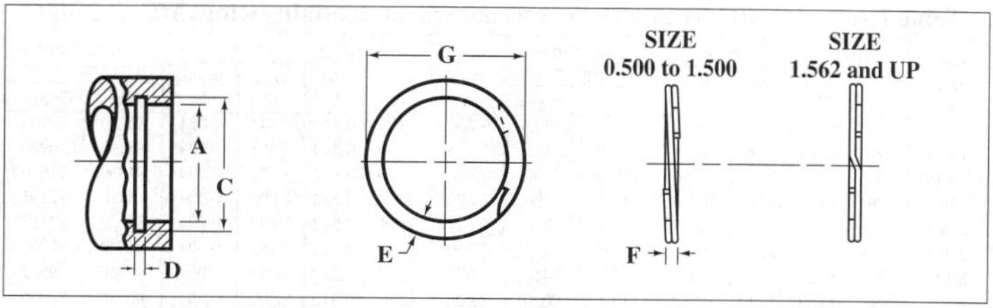


**Table 7. Medium Duty Internal Spiral Retaining Rings** *MIL-R-27426*

| Bore Dia. A | Ring Dia. G | Ring Wall E | Groove Dia. C | Groove Width D | Static Thrust Load (lb) Ring | Static Thrust Load (lb) Groove | Bore Dia. A | Ring Dia. G | Ring Wall E | Groove Dia. C | Groove Width D | Static Thrust Load (lb) Ring | Static Thrust Load (lb) Groove |
|---|---|---|---|---|---|---|---|---|---|---|---|---|---|
| 0.500 | 0.532 | 0.045 | 0.526 | 0.030 | 2000 | 405 | 3.437 | 3.574 | 0.188 | 3.543 | 0.068 | 27660 | 18240 |
| 0.512 | 0.544 | 0.045 | 0.538 | 0.030 | 2050 | 420 | 3.500 | 3.636 | 0.188 | 3.606 | 0.068 | 28170 | 18575 |
| 0.531 | 0.564 | 0.045 | 0.557 | 0.030 | 2130 | 455 | 3.543 | 3.684 | 0.198 | 3.653 | 0.068 | 28520 | 19515 |
| 0.562 | 0.594 | 0.045 | 0.588 | 0.030 | 2250 | 495 | 3.562 | 3.703 | 0.198 | 3.672 | 0.068 | 28670 | 19620 |
| 0.594 | 0.626 | 0.045 | 0.619 | 0.030 | 2380 | 535 | 3.625 | 3.769 | 0.198 | 3.737 | 0.068 | 29180 | 20330 |
| 0.625 | 0.658 | 0.045 | 0.651 | 0.030 | 2500 | 610 | 3.687 | 3.832 | 0.198 | 3.799 | 0.068 | 29680 | 20675 |
| 0.656 | 0.689 | 0.045 | 0.682 | 0.030 | 2630 | 670 | 3.740 | 3.885 | 0.198 | 3.852 | 0.068 | 30100 | 20975 |
| 0.687 | 0.720 | 0.045 | 0.713 | 0.030 | 2750 | 725 | 3.750 | 3.894 | 0.198 | 3.862 | 0.068 | 30180 | 21030 |
| 0.718 | 0.751 | 0.045 | 0.744 | 0.030 | 2870 | 790 | 3.812 | 3.963 | 0.208 | 3.930 | 0.068 | 30680 | 22525 |
| 0.750 | 0.790 | 0.065 | 0.782 | 0.036 | 3360 | 800 | 4.437 | 4.611 | 0.238 | 4.573 | 0.068 | 35710 | 30215 |
| 0.777 | 0.817 | 0.065 | 0.808 | 0.036 | 3480 | 835 | 4.500 | 4.674 | 0.238 | 4.636 | 0.068 | 36220 | 30645 |
| 0.781 | 0.821 | 0.065 | 0.812 | 0.036 | 3500 | 840 | 4.527 | 4.701 | 0.238 | 4.663 | 0.068 | 36440 | 30830 |
| 0.812 | 0.853 | 0.065 | 0.843 | 0.036 | 3640 | 915 | 4.562 | 4.737 | 0.238 | 4.698 | 0.079 | 36720 | 31065 |
| 0.843 | 0.889 | 0.065 | 0.880 | 0.036 | 3780 | 1155 | 4.625 | 4.803 | 0.250 | 4.765 | 0.079 | 43940 | 32420 |
| 0.866 | 0.913 | 0.065 | 0.903 | 0.036 | 3880 | 1250 | 4.687 | 4.867 | 0.250 | 4.827 | 0.079 | 44530 | 32855 |
| 0.875 | 0.922 | 0.065 | 0.912 | 0.036 | 3920 | 1250 | 4.724 | 4.903 | 0.250 | 4.864 | 0.079 | 44880 | 33115 |
| 0.906 | 0.949 | 0.065 | 0.939 | 0.036 | 4060 | 1335 | 4.750 | 4.930 | 0.250 | 4.890 | 0.079 | 45130 | 33300 |
| 0.938 | 0.986 | 0.065 | 0.975 | 0.036 | 4200 | 1430 | 4.812 | 4.993 | 0.250 | 4.952 | 0.079 | 45710 | 33735 |
| 0.968 | 1.025 | 0.075 | 1.015 | 0.042 | 4340 | 1950 | 4.875 | 5.055 | 0.250 | 5.015 | 0.079 | 46310 | 34175 |
| 0.987 | 1.041 | 0.075 | 1.030 | 0.042 | 4420 | 1865 | 4.921 | 5.102 | 0.250 | 5.061 | 0.079 | 46750 | 34495 |
| 1.000 | 1.054 | 0.075 | 1.043 | 0.042 | 4480 | 1910 | 4.937 | 5.122 | 0.250 | 5.081 | 0.079 | 46900 | 35595 |
| 1.023 | 1.078 | 0.075 | 1.066 | 0.042 | 5470 | 1660 | 5.000 | 5.185 | 0.250 | 5.144 | 0.079 | 47500 | 36050 |
| 1.031 | 1.084 | 0.075 | 1.074 | 0.042 | 5510 | 1650 | 5.118 | 5.304 | 0.250 | 5.262 | 0.079 | 48620 | 36905 |
| 1.062 | 1.117 | 0.075 | 1.104 | 0.042 | 5680 | 1745 | 5.125 | 5.311 | 0.250 | 5.269 | 0.079 | 48690 | 36955 |
| 1.093 | 1.147 | 0.075 | 1.135 | 0.042 | 5840 | 1820 | 5.250 | 5.436 | 0.250 | 5.393 | 0.079 | 49880 | 37590 |
| 1.125 | 1.180 | 0.075 | 1.167 | 0.042 | 6010 | 1935 | 5.375 | 5.566 | 0.250 | 5.522 | 0.079 | 51050 | 39565 |
| 1.156 | 1.210 | 0.075 | 1.198 | 0.042 | 6180 | 2020 | 5.500 | 5.693 | 0.250 | 5.647 | 0.079 | 52250 | 40485 |
| 1.188 | 1.249 | 0.085 | 1.236 | 0.048 | 7380 | 2115 | 5.511 | 5.703 | 0.250 | 5.658 | 0.079 | 52350 | 40565 |
| 1.218 | 1.278 | 0.085 | 1.266 | 0.048 | 7570 | 2195 | 5.625 | 5.818 | 0.250 | 5.772 | 0.079 | 53440 | 41405 |
| 1.250 | 1.312 | 0.085 | 1.298 | 0.048 | 7770 | 2510 | 5.708 | 5.909 | 0.250 | 5.861 | 0.079 | 54230 | 43730 |
| 1.281 | 1.342 | 0.085 | 1.329 | 0.048 | 7960 | 2425 | 5.750 | 5.950 | 0.250 | 5.903 | 0.079 | 54630 | 44050 |
| 1.312 | 1.374 | 0.085 | 1.360 | 0.048 | 8150 | 2532 | 5.875 | 6.077 | 0.250 | 6.028 | 0.079 | 55810 | 45010 |
| 1.343 | 1.408 | 0.085 | 1.395 | 0.048 | 8340 | 2875 | 5.905 | 6.106 | 0.250 | 6.058 | 0.079 | 56100 | 45240 |
| 1.375 | 1.442 | 0.095 | 1.427 | 0.048 | 8540 | 3070 | 6.000 | 6.202 | 0.312 | 6.153 | 0.079 | 57000 | 45965 |
| 1.406 | 1.472 | 0.095 | 1.458 | 0.048 | 8740 | 3180 | 6.125 | 6.349 | 0.312 | 6.297 | 0.094 | 69500 | 52750 |
| 1.437 | 1.504 | 0.095 | 1.489 | 0.048 | 8930 | 3330 | 6.250 | 6.474 | 0.312 | 6.422 | 0.094 | 70920 | 53825 |
| 1.456 | 1.523 | 0.095 | 1.508 | 0.048 | 9050 | 3410 | 6.299 | 6.524 | 0.312 | 6.471 | 0.094 | 71480 | 54250 |
| 1.468 | 1.535 | 0.095 | 1.520 | 0.048 | 9120 | 3460 | 6.375 | 6.601 | 0.312 | 6.547 | 0.094 | 72340 | 54905 |
| 1.500 | 1.567 | 0.095 | 1.552 | 0.048 | 9320 | 3605 | 6.500 | 6.726 | 0.312 | 6.672 | 0.094 | 73760 | 55980 |
| 1.562 | 1.634 | 0.108 | 1.617 | 0.056 | 10100 | 3590 | 6.625 | 6.863 | 0.312 | 6.807 | 0.094 | 75180 | 60375 |
| 1.574 | 1.649 | 0.108 | 1.633 | 0.056 | 10180 | 3640 | 6.692 | 6.931 | 0.312 | 6.874 | 0.094 | 75940 | 60985 |
| 1.625 | 1.701 | 0.108 | 1.684 | 0.056 | 10510 | 3875 | 6.750 | 6.987 | 0.312 | 6.932 | 0.094 | 76590 | 61515 |
| 1.653 | 1.730 | 0.108 | 1.712 | 0.056 | 10690 | 4020 | 6.875 | 7.114 | 0.312 | 7.057 | 0.094 | 78010 | 62655 |
| 1.687 | 1.768 | 0.118 | 1.750 | 0.056 | 10910 | 4510 | 7.000 | 7.239 | 0.312 | 7.182 | 0.094 | 79430 | 63790 |
| 1.750 | 1.834 | 0.118 | 1.813 | 0.056 | 11310 | 4895 | 7.086 | 7.337 | 0.312 | 7.278 | 0.094 | 80410 | 68125 |
| 1.813 | 1.894 | 0.118 | 1.875 | 0.056 | 11720 | 5080 | 7.125 | 7.376 | 0.312 | 7.317 | 0.094 | 80850 | 68500 |

**Table 7.** *(Continued)* **Medium Duty Internal Spiral Retaining Rings** *MIL-R-27426*

| Bore Dia. A | Ring | | Groove | | Static Thrust Load (lb) | | Bore Dia. A | Ring | | Groove | | Static Thrust Load (lb) | |
|---|---|---|---|---|---|---|---|---|---|---|---|---|---|
| | Dia. G | Wall E | Dia. C | Width D | Ring | Groove | | Dia. G | Wall E | Dia. C | Width D | Ring | Groove |
| 1.850 | 1.937 | 0.118 | 1.917 | 0.056 | 11960 | 5735 | 7.250 | 7.501 | 0.312 | 7.442 | 0.094 | 82270 | 69700 |
| 1.875 | 1.960 | 0.118 | 1.942 | 0.056 | 12120 | 5825 | 7.375 | 7.628 | 0.312 | 7.567 | 0.094 | 83690 | 70900 |
| 1.938 | 2.025 | 0.118 | 2.005 | 0.056 | 12530 | 6250 | 7.480 | 7.734 | 0.312 | 7.672 | 0.094 | 84880 | 71910 |
| 2.000 | 2.091 | 0.128 | 2.071 | 0.056 | 12930 | 7090 | 7.500 | 7.754 | 0.312 | 7.692 | 0.094 | 85110 | 72105 |
| 2.047 | 2.138 | 0.128 | 2.118 | 0.056 | 13230 | 7275 | 7.625 | 7.890 | 0.312 | 7.827 | 0.094 | 86520 | 77125 |
| 2.062 | 2.154 | 0.128 | 2.132 | 0.056 | 13330 | 7225 | 7.750 | 8.014 | 0.312 | 7.952 | 0.094 | 87940 | 78390 |
| 2.125 | 2.217 | 0.128 | 2.195 | 0.056 | 13740 | 7450 | 7.875 | 8.131 | 0.312 | 8.077 | 0.094 | 89360 | 79655 |
| 2.165 | 2.260 | 0.138 | 2.239 | 0.056 | 14000 | 8020 | 8.000 | 8.266 | 0.312 | 8.202 | 0.094 | 90780 | 80920 |
| 2.188 | 2.284 | 0.138 | 2.262 | 0.056 | 14150 | 8105 | 8.250 | 8.528 | 0.375 | 8.462 | 0.094 | 93620 | 87575 |
| 2.250 | 2.347 | 0.138 | 2.324 | 0.056 | 14550 | 8335 | 8.267 | 8.546 | 0.375 | 8.479 | 0.094 | 93810 | 87755 |
| 2.312 | 2.413 | 0.138 | 2.390 | 0.056 | 14950 | 9030 | 8.464 | 8.744 | 0.375 | 8.676 | 0.094 | 96040 | 89850 |
| 2.375 | 2.476 | 0.138 | 2.453 | 0.056 | 15350 | 9275 | 8.500 | 8.780 | 0.375 | 8.712 | 0.094 | 96450 | 90230 |
| 2.437 | 2.543 | 0.148 | 2.519 | 0.056 | 15760 | 10005 | 8.750 | 9.041 | 0.375 | 8.972 | 0.094 | 99290 | 97265 |
| 2.440 | 2.546 | 0.148 | 2.522 | 0.056 | 15780 | 10015 | 8.858 | 9.151 | 0.375 | 9.080 | 0.094 | 100520 | 98465 |
| 2.500 | 2.606 | 0.148 | 2.582 | 0.056 | 16160 | 10625 | 9.000 | 9.293 | 0.375 | 9.222 | 0.094 | 102130 | 100045 |
| 2.531 | 2.641 | 0.148 | 2.617 | 0.056 | 16360 | 10900 | 9.055 | 9.359 | 0.375 | 9.287 | 0.094 | 102750 | 105190 |
| 2.562 | 2.673 | 0.148 | 2.648 | 0.056 | 16560 | 11030 | 9.250 | 9.555 | 0.375 | 9.482 | 0.094 | 104960 | 107455 |
| 2.625 | 2.736 | 0.148 | 2.711 | 0.056 | 16970 | 11305 | 9.448 | 9.755 | 0.375 | 9.680 | 0.094 | 107210 | 109755 |
| 2.677 | 2.789 | 0.158 | 2.767 | 0.056 | 17310 | 12065 | 9.500 | 9.806 | 0.375 | 9.732 | 0.094 | 107800 | 110360 |
| 2.688 | 2.803 | 0.158 | 2.778 | 0.056 | 17380 | 12115 | 9.750 | 10.068 | 0.375 | 9.992 | 0.094 | 110640 | 118145 |
| 2.750 | 2.865 | 0.158 | 2.841 | 0.056 | 17780 | 12530 | 10.000 | 10.320 | 0.375 | 10.242 | 0.094 | 113470 | 121175 |
| 2.813 | 2.929 | 0.158 | 2.903 | 0.056 | 18190 | 12675 | 10.250 | 10.582 | 0.375 | 10.502 | 0.094 | 116310 | 129340 |
| 2.834 | 2.954 | 0.168 | 2.928 | 0.056 | 18320 | 13340 | 10.500 | 10.834 | 0.375 | 10.752 | 0.094 | 119150 | 132490 |
| 2.875 | 2.995 | 0.168 | 2.969 | 0.056 | 18590 | 13530 | 10.750 | 11.095 | 0.375 | 11.012 | 0.094 | 121980 | 141030 |
| 2.937 | 3.058 | 0.168 | 3.031 | 0.056 | 18990 | 13825 | 11.000 | 11.347 | 0.375 | 11.262 | 0.094 | 124820 | 144310 |
| 2.952 | 3.073 | 0.168 | 3.046 | 0.056 | 19090 | 13890 | 3.875 | 4.025 | 0.208 | 3.993 | 0.068 | 30680 | 22525 |
| 3.000 | 3.122 | 0.168 | 3.096 | 0.068 | 24150 | 14420 | 3.938 | 4.089 | 0.208 | 4.056 | 0.068 | 31700 | 23265 |
| 3.062 | 3.186 | 0.168 | 3.158 | 0.068 | 24640 | 14720 | 4.000 | 4.157 | 0.218 | 5.124 | 0.068 | 32190 | 24835 |
| 3.125 | 3.251 | 0.178 | 3.223 | 0.068 | 25150 | 15335 | 4.063 | 4.222 | 0.218 | 4.187 | 0.068 | 32700 | 25225 |
| 3.149 | 3.276 | 0.178 | 3.247 | 0.068 | 25340 | 15450 | 4.125 | 4.284 | 0.218 | 4.249 | 0.068 | 33200 | 25610 |
| 3.187 | 3.311 | 0.178 | 3.283 | 0.068 | 25650 | 15640 | 4.188 | 4.347 | 0.218 | 4.311 | 0.068 | 33710 | 25795 |
| 3.250 | 3.379 | 0.178 | 3.350 | 0.068 | 26160 | 16270 | 4.250 | 4.416 | 0.228 | 4.380 | 0.068 | 34210 | 27665 |
| 3.312 | 3.446 | 0.188 | 3.416 | 0.068 | 26660 | 17245 | 4.312 | 4.479 | 0.228 | 4.442 | 0.068 | 34710 | 28065 |
| 3.346 | 3.479 | 0.188 | 3.450 | 0.068 | 26930 | 17425 | 4.330 | 4.497 | 0.228 | 4.460 | 0.068 | 34850 | 28185 |
| 3.375 | 3.509 | 0.188 | 3.479 | 0.068 | 27160 | 17575 | 4.375 | 4.543 | 0.228 | 4.505 | 0.068 | 32210 | 28475 |

*Source:* Spirolox Retaining Rings, RR Series. All dimensions are in inches. Depth of groove $d = (C - A)/2$. Standard material: carbon spring steel (SAE 1070-1090).

*Ring Thickness, F:* For shaft sizes 0.500 through 0.718, 0.025; for sizes 0.750 through 0.938, 0.031; for sizes 0.968 through 1.156, 0.037; for sizes 1.188 through 1.500, 0.043; for sizes 1.562 through 2.952, 0.049; for sizes 3.000 through 4.562, 0.061; for sizes 4.625 through 6.000, 0.072; for sizes 6.125 through 11.000, 0.086.

*Ring Free Diameter Tolerances:* For housing sizes 0.500 through 1.031, +0.013, −0.000; for sizes 1.062 through 1.500, +0.015, −0.000; for sizes 1.562 through 2.047, +0.020, −0.000; for sizes 2.062 through 3.000, +0.025, −0.000; for sizes 3.062 through 4.063, +0.030, −0.000; for sizes 4.125 through 5.125, +0.035, −0.000; for sizes 5.250 through 6.125, +0.045, −0.000; for sizes 6.250 through 7.125, +0.055, −0.000; for sizes 7.250 through 11.000, +0.065, −0.000.

*Ring Thickness Tolerances:* Thickness indicated is for unplated rings; add 0.002 to upper thickness tolerance for plated rings. For housing sizes 0.500 through 1.500, ±0.002; for sizes 1.562 through 4.562, ±0.003; for sizes 4.625 through 11.000, ±0.004.

*Groove Diameter Tolerances:* For housing sizes 0.500 through 0.750, ±0.002; for sizes 0.777 through 1.031, ±0.003; for sizes 1.062 through 1.500, ±0.004; for sizes 1.562 through 2.047, ±0.005; for sizes 2.062 through 5.125, ±0.006; for sizes 5.250 through 6.000, ±0.007; for sizes 6.125 through 11.000, ±0.008.

*Groove Width Tolerances:* For housing sizes 0.500 through 1.156, +0.003, −0.000; for sizes 1.188 through 2.952, +0.004, −0.000; for sizes 3.000 through 6.000, +0.005, −0.000; for sizes 6.125 through 11.000, +0.006, −0.000.

## Table 8. Medium Duty External Spiral Retaining Rings *MIL-R-27426*

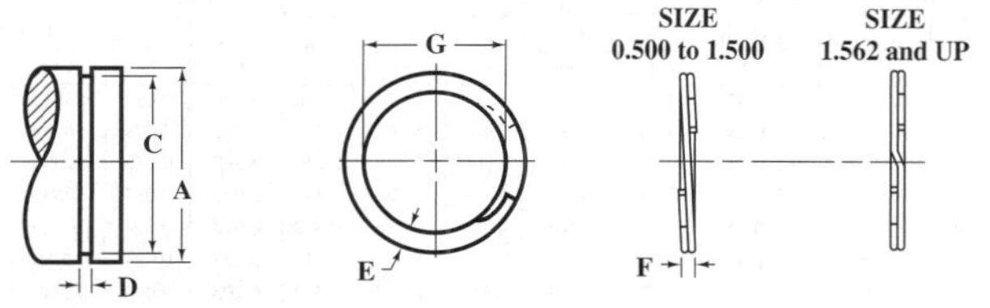


| Shaft Dia. A | Ring Dia. G | Ring Wall E | Groove Dia. C | Groove Width D | Static Thrust Load (lb) Ring | Static Thrust Load (lb) Groove | Shaft Dia. A | Ring Dia. G | Ring Wall E | Groove Dia. C | Groove Width D | Static Thrust Load (lb) Ring | Static Thrust Load (lb) Groove |
|---|---|---|---|---|---|---|---|---|---|---|---|---|---|
| 0.500 | 0.467 | 0.045 | 0.474 | 0.030 | 2000 | 550 | 3.343 | 3.210 | 0.188 | 3.239 | 0.068 | 26910 | 17410 |
| 0.531 | 0.498 | 0.045 | 0.505 | 0.030 | 2130 | 640 | 3.375 | 3.242 | 0.188 | 3.271 | 0.068 | 27160 | 17570 |
| 0.551 | 0.518 | 0.045 | 0.525 | 0.030 | 2210 | 700 | 3.437 | 3.301 | 0.188 | 3.331 | 0.068 | 27660 | 18240 |
| 0.562 | 0.529 | 0.045 | 0.536 | 0.030 | 2250 | 730 | 3.500 | 3.363 | 0.188 | 3.394 | 0.068 | 28170 | 18580 |
| 0.594 | 0.561 | 0.045 | 0.569 | 0.030 | 2380 | 740 | 3.543 | 3.402 | 0.198 | 3.433 | 0.068 | 28520 | 19510 |
| 0.625 | 0.585 | 0.055 | 0.594 | 0.030 | 2500 | 970 | 3.562 | 3.422 | 0.198 | 3.452 | 0.068 | 28670 | 19620 |
| 0.656 | 0.617 | 0.055 | 0.625 | 0.030 | 2630 | 1020 | 3.625 | 3.483 | 0.198 | 3.515 | 0.068 | 29180 | 19970 |
| 0.669 | 0.629 | 0.055 | 0.638 | 0.030 | 2680 | 1040 | 3.687 | 3.543 | 0.198 | 3.575 | 0.068 | 29680 | 20680 |
| 0.687 | 0.647 | 0.055 | 0.656 | 0.030 | 2750 | 1060 | 3.740 | 3.597 | 0.198 | 3.628 | 0.068 | 30100 | 20970 |
| 0.718 | 0.679 | 0.055 | 0.687 | 0.030 | 2870 | 1110 | 3.750 | 3.606 | 0.198 | 3.638 | 0.068 | 30180 | 21030 |
| 0.750 | 0.710 | 0.065 | 0.719 | 0.036 | 3360 | 1100 | 3.812 | 3.668 | 0.198 | 3.700 | 0.068 | 30680 | 21380 |
| 0.781 | 0.741 | 0.065 | 0.750 | 0.036 | 3500 | 1210 | 3.875 | 3.724 | 0.208 | 3.757 | 0.068 | 31190 | 22890 |
| 0.812 | 0.771 | 0.065 | 0.781 | 0.036 | 3640 | 1260 | 3.938 | 3.784 | 0.208 | 3.820 | 0.068 | 31700 | 23270 |
| 0.843 | 0.803 | 0.065 | 0.812 | 0.036 | 3780 | 1310 | 4.000 | 3.842 | 0.218 | 3.876 | 0.068 | 32190 | 24840 |
| 0.875 | 0.828 | 0.065 | 0.838 | 0.036 | 3920 | 1620 | 4.063 | 3.906 | 0.218 | 3.939 | 0.068 | 32700 | 25230 |
| 0.906 | 0.860 | 0.065 | 0.869 | 0.036 | 4060 | 1680 | 4.125 | 3.967 | 0.218 | 4.000 | 0.068 | 33200 | 25820 |
| 0.937 | 0.889 | 0.065 | 0.900 | 0.036 | 4200 | 1740 | 4.134 | 3.975 | 0.218 | 4.010 | 0.068 | 33270 | 25670 |
| 0.968 | 0.916 | 0.075 | 0.925 | 0.042 | 5180 | 2080 | 4.188 | 4.030 | 0.218 | 4.058 | 0.068 | 33710 | 27260 |
| 0.984 | 0.930 | 0.075 | 0.941 | 0.042 | 5260 | 2120 | 4.250 | 4.084 | 0.228 | 4.120 | 0.068 | 34210 | 27660 |
| 1.000 | 0.946 | 0.075 | 0.957 | 0.042 | 5350 | 2150 | 4.312 | 4.147 | 0.218 | 4.182 | 0.068 | 34710 | 28070 |
| 1.023 | 0.968 | 0.075 | 0.980 | 0.042 | 5470 | 2200 | 4.331 | 4.164 | 0.218 | 4.200 | 0.068 | 34860 | 28410 |
| 1.031 | 0.978 | 0.075 | 0.988 | 0.042 | 5510 | 2220 | 4.375 | 4.208 | 0.218 | 4.245 | 0.068 | 35210 | 28480 |
| 1.062 | 1.007 | 0.075 | 1.020 | 0.042 | 5680 | 2230 | 4.437 | 4.271 | 0.218 | 4.307 | 0.068 | 35710 | 28880 |
| 1.093 | 1.040 | 0.075 | 1.051 | 0.042 | 5840 | 2300 | 4.500 | 4.326 | 0.238 | 4.364 | 0.068 | 36220 | 30640 |
| 1.125 | 1.070 | 0.075 | 1.083 | 0.042 | 6010 | 2370 | 4.562 | 4.384 | 0.250 | 4.422 | 0.079 | 43340 | 31980 |
| 1.156 | 1.102 | 0.075 | 1.114 | 0.042 | 6180 | 2430 | 4.625 | 4.447 | 0.250 | 4.485 | 0.079 | 43940 | 32420 |
| 1.188 | 1.127 | 0.085 | 1.140 | 0.048 | 7380 | 2850 | 4.687 | 4.508 | 0.250 | 4.457 | 0.079 | 44530 | 32860 |
| 1.218 | 1.159 | 0.085 | 1.170 | 0.048 | 7570 | 2930 | 4.724 | 4.546 | 0.250 | 4.584 | 0.079 | 44880 | 33120 |
| 1.250 | 1.188 | 0.085 | 1.202 | 0.048 | 7770 | 3000 | 4.750 | 4.571 | 0.250 | 4.610 | 0.079 | 45130 | 33300 |
| 1.281 | 1.221 | 0.085 | 1.233 | 0.048 | 7960 | 3080 | 4.812 | 4.633 | 0.250 | 4.672 | 0.079 | 45710 | 33730 |
| 1.312 | 1.251 | 0.095 | 1.264 | 0.048 | 8150 | 3150 | 4.875 | 4.695 | 0.250 | 4.735 | 0.079 | 46310 | 34170 |
| 1.343 | 1.282 | 0.095 | 1.295 | 0.048 | 8340 | 3230 | 4.937 | 4.757 | 0.250 | 4.797 | 0.079 | 46900 | 34610 |
| 1.375 | 1.308 | 0.095 | 1.323 | 0.048 | 8540 | 3580 | 5.000 | 4.820 | 0.250 | 4.856 | 0.079 | 47500 | 36050 |
| 1.406 | 1.340 | 0.095 | 1.354 | 0.048 | 8740 | 3660 | 5.118 | 4.934 | 0.250 | 4.974 | 0.079 | 48620 | 36900 |
| 1.437 | 1.370 | 0.095 | 1.385 | 0.048 | 8930 | 3740 | 5.125 | 4.939 | 0.250 | 4.981 | 0.079 | 48690 | 36950 |
| 1.468 | 1.402 | 0.095 | 1.416 | 0.048 | 9120 | 3820 | 5.250 | 5.064 | 0.250 | 5.107 | 0.079 | 49880 | 37590 |
| 1.500 | 1.433 | 0.095 | 1.448 | 0.048 | 9320 | 3910 | 5.375 | 5.187 | 0.250 | 5.228 | 0.079 | 51060 | 39560 |
| 1.562 | 1.490 | 0.108 | 1.507 | 0.056 | 10100 | 4300 | 5.500 | 5.308 | 0.250 | 5.353 | 0.079 | 52250 | 40480 |
| 1.575 | 1.503 | 0.108 | 1.520 | 0.056 | 10190 | 4340 | 5.511 | 5.320 | 0.250 | 5.364 | 0.079 | 52350 | 40560 |
| 1.625 | 1.549 | 0.108 | 1.566 | 0.056 | 10510 | 4800 | 5.625 | 5.433 | 0.250 | 5.478 | 0.079 | 53440 | 41400 |
| 1.687 | 1.610 | 0.118 | 1.628 | 0.056 | 10910 | 4980 | 5.750 | 5.550 | 0.250 | 5.597 | 0.079 | 54630 | 44050 |

**Table 8.** *(Continued)* **Medium Duty External Spiral Retaining Rings** *MIL-R-27426*

| Shaft Dia. | Ring | | Groove | | Static Thrust Load (lb) | | Shaft Dia. | Ring | | Groove | | Static Thrust Load (lb) | |
|---|---|---|---|---|---|---|---|---|---|---|---|---|---|
| A | Dia. G | Wall E | Dia. C | Width D | Ring | Groove | A | Dia. G | Wall E | Dia. C | Width D | Ring | Groove |
| 1.750 | 1.673 | 0.118 | 1.691 | 0.056 | 11310 | 5170 | 5.875 | 5.674 | 0.250 | 5.722 | 0.079 | 55810 | 45010 |
| 1.771 | 1.690 | 0.118 | 1.708 | 0.056 | 11450 | 5590 | 5.905 | 5.705 | 0.250 | 5.752 | 0.079 | 56100 | 45240 |
| 1.813 | 1.730 | 0.118 | 1.749 | 0.056 | 11720 | 5810 | 6.000 | 5.798 | 0.250 | 5.847 | 0.079 | 57000 | 45970 |
| 1.875 | 1.789 | 0.128 | 1.808 | 0.056 | 12120 | 6290 | 6.125 | 5.903 | 0.312 | 5.953 | 0.094 | 69500 | 52750 |
| 1.938 | 1.844 | 0.128 | 1.861 | 0.056 | 12530 | 7470 | 6.250 | 6.026 | 0.312 | 6.078 | 0.094 | 70920 | 53830 |
| 1.969 | 1.882 | 0.128 | 1.902 | 0.056 | 12730 | 6610 | 6.299 | 6.076 | 0.312 | 6.127 | 0.094 | 71480 | 54250 |
| 2.000 | 1.909 | 0.128 | 1.992 | 0.056 | 12930 | 7110 | 6.375 | 6.152 | 0.312 | 6.203 | 0.094 | 72340 | 54900 |
| 2.062 | 1.971 | 0.128 | 2.051 | 0.056 | 13330 | 7870 | 6.500 | 6.274 | 0.312 | 6.328 | 0.094 | 73760 | 55980 |
| 2.125 | 2.029 | 0.128 | 2.082 | 0.056 | 13740 | 7990 | 6.625 | 6.390 | 0.312 | 6.443 | 0.094 | 75180 | 60380 |
| 2.156 | 2.060 | 0.138 | 2.091 | 0.056 | 13940 | 8020 | 6.750 | 6.513 | 0.312 | 6.568 | 0.094 | 76590 | 61515 |
| 2.188 | 2.070 | 0.138 | 2.113 | 0.056 | 14150 | 8220 | 6.875 | 6.638 | 0.312 | 6.693 | 0.094 | 78010 | 62650 |
| 2.250 | 2.092 | 0.138 | 2.176 | 0.056 | 14550 | 8340 | 7.000 | 6.761 | 0.312 | 6.818 | 0.094 | 79430 | 63790 |
| 2.312 | 2.153 | 0.138 | 2.234 | 0.056 | 14950 | 9030 | 7.125 | 6.877 | 0.312 | 6.933 | 0.094 | 80850 | 68500 |
| 2.362 | 2.211 | 0.138 | 2.284 | 0.056 | 15270 | 9230 | 7.250 | 6.999 | 0.312 | 7.058 | 0.094 | 82270 | 69700 |
| 2.375 | 2.273 | 0.138 | 2.297 | 0.056 | 15350 | 9280 | 7.375 | 7.125 | 0.312 | 7.183 | 0.094 | 83690 | 70900 |
| 2.437 | 2.331 | 0.148 | 2.355 | 0.056 | 15760 | 10000 | 7.500 | 7.250 | 0.312 | 7.308 | 0.094 | 85110 | 72100 |
| 2.500 | 2.394 | 0.148 | 2.418 | 0.056 | 16160 | 10260 | 7.625 | 7.363 | 0.312 | 7.423 | 0.094 | 86520 | 77120 |
| 2.559 | 2.449 | 0.148 | 2.473 | 0.056 | 16540 | 11020 | 7.750 | 7.486 | 0.312 | 7.548 | 0.094 | 87940 | 78390 |
| 2.562 | 2.452 | 0.148 | 2.476 | 0.056 | 16560 | 11030 | 7.875 | 7.611 | 0.312 | 7.673 | 0.094 | 89360 | 79650 |
| 2.625 | 2.514 | 0.148 | 2.539 | 0.056 | 16970 | 11300 | 8.000 | 7.734 | 0.312 | 7.798 | 0.094 | 90780 | 80920 |
| 2.688 | 2.572 | 0.158 | 2.597 | 0.056 | 17380 | 12250 | 8.250 | 7.972 | 0.375 | 8.038 | 0.094 | 93620 | 87580 |
| 2.750 | 2.635 | 0.158 | 2.660 | 0.056 | 17780 | 12390 | 8.500 | 8.220 | 0.375 | 8.288 | 0.094 | 96450 | 90230 |
| 2.813 | 2.696 | 0.168 | 2.722 | 0.056 | 18190 | 12820 | 8.750 | 8.459 | 0.375 | 8.528 | 0.094 | 99290 | 97270 |
| 2.875 | 2.755 | 0.168 | 2.781 | 0.056 | 18590 | 13530 | 9.000 | 8.707 | 0.375 | 8.778 | 0.094 | 102130 | 100050 |
| 2.937 | 2.817 | 0.168 | 2.843 | 0.056 | 18990 | 13820 | 9.250 | 8.945 | 0.375 | 9.018 | 0.094 | 104960 | 107560 |
| 2.952 | 2.831 | 0.168 | 2.858 | 0.056 | 19090 | 13890 | 9.500 | 9.194 | 0.375 | 9.268 | 0.094 | 107800 | 110360 |
| 3.000 | 2.877 | 0.168 | 2.904 | 0.068 | 24150 | 14420 | 9.750 | 9.432 | 0.375 | 9.508 | 0.094 | 110640 | 118150 |
| 3.062 | 2.938 | 0.168 | 2.966 | 0.068 | 24640 | 14720 | 10.000 | 9.680 | 0.375 | 9.758 | 0.094 | 113470 | 121180 |
| 3.125 | 3.000 | 0.178 | 3.027 | 0.068 | 25150 | 15335 | 10.250 | 9.918 | 0.375 | 9.998 | 0.094 | 116310 | 129340 |
| 3.149 | 3.023 | 0.178 | 3.051 | 0.068 | 25340 | 15450 | 10.500 | 10.166 | 0.375 | 10.248 | 0.094 | 119150 | 132490 |
| 3.187 | 3.061 | 0.178 | 3.089 | 0.068 | 25650 | 15640 | 10.750 | 10.405 | 0.375 | 10.488 | 0.094 | 121980 | 141030 |
| 3.250 | 3.121 | 0.178 | 3.150 | 0.068 | 26160 | 16270 | 11.000 | 10.653 | 0.375 | 10.738 | 0.094 | 124820 | 144310 |
| 3.312 | 3.180 | 0.188 | 3.208 | 0.068 | 26660 | 17250 | | | | | | | |

*Source*: Spirolox Retaining Rings, RS Series. All dimensions are in inches.

Depth of groove $d = (A - C)/2$. Standard material: carbon spring steel (SAE 1070-1090).

*Ring Thickness, F*: For shaft sizes 0.500 through 0.718, 0.025; for sizes 0.750 through 0.937, 0.031; for sizes 0.968 through 1.156, 0.037; for sizes 1.188 through 1.500, 0.043; for sizes 1.562 through 2.952, 0.049; for sizes 3.000 through 4.500, 0.061; for sizes 4.562 through 6.000, 0.072; for sizes 6.125 through 11.000, 0.086.

*Ring Free Diameter Tolerances*: For shaft sizes 0.500 through 1.031, +0.000, + 0.000, −0.013; for sizes 1.062 through 1.500, +0.000, −0.015; for sizes 1.562 through 2.125, +0.000, −0.020; for sizes 2.156 through 2.688, +0.000, −0.025; for sizes 2.750 through 3.437, +0.000, −0.030; for sizes 3.500 through 5.125, +0.000, −0.040; for sizes 5.250 through 6.125, +0.000, −0.050; for sizes 6.250 through 7.375, +0.000, −0.060; for sizes 7.500 through 11.000, +0.000, −0.070.

*Ring Thickness Tolerances*: Thickness indicated is for unplated rings; add 0.002 to upper tolerance for plated rings. For shaft sizes 0.500 through 1.500, ± 0.002; for sizes 1.562 through 4.500, ±0.003; for sizes 4.562 through 11.000, ±0.004.

*Groove Diameter Tolerances*: For shaft sizes 0.500 through 0.562, ±0.002; for sizes 0.594 through 1.031, ± 0.003; for sizes 1.062 through 1.500, ± 0.004; for sizes 1.562 through 2.000, ± 0.005; for sizes 2.062 through 5.125, ±0.006; for sizes 5.250 through 6.000, ± 0.007; for sizes 6.125 through 11.000, ±0.008.

*Groove Width Tolerances*: For shaft sizes 0.500 through 1.156, +0.003, −0.000; for sizes 1.188 through 2.952, +0.004, −0.000; for sizes 3.000 through 6.000, +0.005, −0.000; for sizes 6.125 through 11.000, +0.006, −0.000.

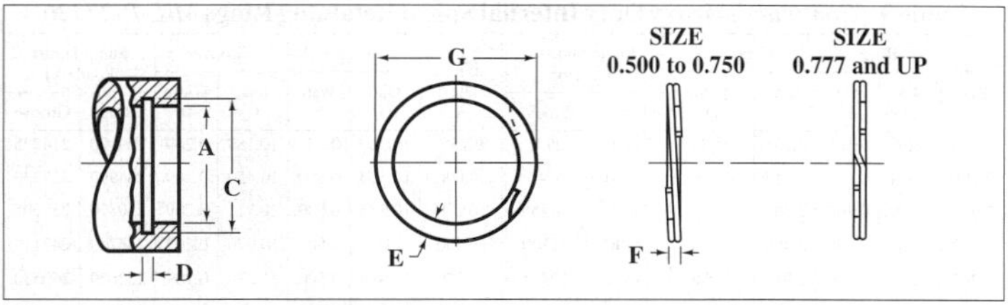


## Table 9. Heavy Duty Internal Spiral Retaining Rings *MIL-R-27426*

| Bore Dia. A | Ring Dia. G | Ring Wall E | Groove Dia. C | Groove Width D | Static Thrust Load (lb) Ring | Static Thrust Load (lb) Groove | Bore Dia. A | Ring Dia. G | Ring Wall E | Groove Dia. C | Groove Width D | Static Thrust Load (lb) Ring | Static Thrust Load (lb) Groove |
|---|---|---|---|---|---|---|---|---|---|---|---|---|---|
| 0.500 | 0.538 | 0.045 | 0.530 | 0.039 | 2530 | 310 | 3.543 | 3.781 | 0.281 | 3.755 | 0.120 | 49420 | 28250 |
| 0.512 | 0.550 | 0.045 | 0.542 | 0.039 | 2590 | 325 | 3.562 | 3.802 | 0.281 | 3.776 | 0.120 | 49680 | 28815 |
| 0.562 | 0.605 | 0.055 | 0.596 | 0.039 | 2840 | 455 | 3.625 | 3.868 | 0.281 | 3.841 | 0.120 | 50560 | 30160 |
| 0.625 | 0.675 | 0.055 | 0.655 | 0.039 | 3160 | 655 | 3.750 | 4.002 | 0.312 | 3.974 | 0.120 | 52310 | 33720 |
| 0.688 | 0.743 | 0.065 | 0.732 | 0.039 | 3480 | 965 | 3.875 | 4.136 | 0.312 | 4.107 | 0.120 | 54050 | 37250 |
| 0.750 | 0.807 | 0.065 | 0.796 | 0.039 | 3790 | 1065 | 3.938 | 4.203 | 0.312 | 4.174 | 0.120 | 54930 | 39045 |
| 0.777 | 0.836 | 0.075 | 0.825 | 0.046 | 4720 | 1026 | 4.000 | 4.270 | 0.312 | 4.240 | 0.120 | 55790 | 41025 |
| 0.812 | 0.873 | 0.075 | 0.862 | 0.046 | 4930 | 1150 | 4.125 | 4.369 | 0.312 | 4.339 | 0.120 | 57540 | 38495 |
| 0.866 | 0.931 | 0.075 | 0.920 | 0.046 | 5260 | 1395 | 4.250 | 4.501 | 0.312 | 4.470 | 0.120 | 59280 | 41955 |
| 0.875 | 0.943 | 0.085 | 0.931 | 0.046 | 5310 | 1520 | 4.330 | 4.588 | 0.312 | 4.556 | 0.120 | 60400 | 44815 |
| 0.901 | 0.972 | 0.085 | 0.959 | 0.046 | 5470 | 1675 | 4.500 | 4.768 | 0.312 | 4.735 | 0.120 | 62770 | 50290 |
| 0.938 | 1.013 | 0.085 | 1.000 | 0.046 | 5690 | 1925 | 4.625 | 4.899 | 0.312 | 4.865 | 0.120 | 64510 | 54155 |
| 1.000 | 1.080 | 0.085 | 1.066 | 0.046 | 6070 | 2310 | 4.750 | 5.030 | 0.312 | 4.995 | 0.120 | 66260 | 58270 |
| 1.023 | 1.105 | 0.085 | 1.091 | 0.046 | 6210 | 2480 | 5.000 | 5.297 | 0.312 | 5.260 | 0.120 | 69740 | 65095 |
| 1.062 | 1.138 | 0.103 | 1.130 | 0.056 | 7010 | 1940 | 5.250 | 5.559 | 0.350 | 5.520 | 0.139 | 83790 | 68315 |
| 1.125 | 1.205 | 0.103 | 1.197 | 0.056 | 7420 | 2280 | 5.375 | 5.690 | 0.350 | 5.650 | 0.139 | 85780 | 72840 |
| 1.188 | 1.271 | 0.103 | 1.262 | 0.056 | 7840 | 2615 | 5.500 | 5.810 | 0.350 | 5.770 | 0.139 | 87780 | 74355 |
| 1.250 | 1.339 | 0.103 | 1.330 | 0.056 | 8250 | 3110 | 5.750 | 6.062 | 0.350 | 6.020 | 0.139 | 91770 | 77735 |
| 1.312 | 1.406 | 0.118 | 1.396 | 0.056 | 8650 | 3650 | 6.000 | 6.314 | 0.350 | 6.270 | 0.139 | 95760 | 81120 |
| 1.375 | 1.471 | 0.118 | 1.461 | 0.056 | 9070 | 4075 | 6.250 | 6.576 | 0.380 | 6.530 | 0.174 | 122520 | 80655 |
| 1.439 | 1.539 | 0.118 | 1.528 | 0.056 | 9490 | 4670 | 6.500 | 6.838 | 0.380 | 6.790 | 0.174 | 127420 | 90295 |
| 1.456 | 1.559 | 0.118 | 1.548 | 0.056 | 9600 | 4890 | 6.625 | 6.974 | 0.380 | 6.925 | 0.174 | 129870 | 92060 |
| 1.500 | 1.605 | 0.118 | 1.594 | 0.056 | 9900 | 5275 | 6.750 | 7.105 | 0.380 | 7.055 | 0.174 | 132320 | 102475 |
| 1.562 | 1.675 | 0.128 | 1.658 | 0.068 | 12780 | 4840 | 7.000 | 7.366 | 0.380 | 7.315 | 0.174 | 137220 | 110410 |
| 1.625 | 1.742 | 0.128 | 1.725 | 0.068 | 13290 | 5415 | 7.250 | 7.628 | 0.418 | 7.575 | 0.209 | 170370 | 103440 |
| 1.653 | 1.772 | 0.128 | 1.755 | 0.068 | 13520 | 5695 | 7.500 | 7.895 | 0.418 | 7.840 | 0.209 | 176240 | 115780 |
| 1.688 | 1.810 | 0.128 | 1.792 | 0.068 | 13810 | 6070 | 7.750 | 8.157 | 0.418 | 8.100 | 0.209 | 182120 | 127270 |
| 1.750 | 1.876 | 0.128 | 1.858 | 0.068 | 14320 | 7635 | 8.000 | 8.419 | 0.418 | 8.360 | 0.209 | 187990 | 139370 |
| 1.812 | 1.940 | 0.128 | 1.922 | 0.068 | 14820 | 7305 | 8.250 | 8.680 | 0.437 | 8.620 | 0.209 | 193870 | 152695 |
| 1.850 | 1.981 | 0.158 | 1.962 | 0.068 | 15130 | 7960 | 8.500 | 8.942 | 0.437 | 8.880 | 0.209 | 199740 | 161735 |
| 1.875 | 2.008 | 0.158 | 1.989 | 0.068 | 15340 | 8305 | 8.750 | 9.209 | 0.437 | 9.145 | 0.209 | 205620 | 173065 |
| 1.938 | 2.075 | 0.158 | 2.056 | 0.068 | 15850 | 9125 | 9.000 | 9.471 | 0.437 | 9.405 | 0.209 | 211490 | 182515 |
| 2.000 | 2.142 | 0.158 | 2.122 | 0.068 | 16360 | 10040 | 9.250 | 9.737 | 0.437 | 9.669 | 0.209 | 217370 | 194070 |
| 2.062 | 2.201 | 0.168 | 2.186 | 0.086 | 21220 | 8280 | 9.500 | 10.000 | 0.500 | 9.930 | 0.209 | 223240 | 204550 |

**Table 9.** *(Continued)* **Heavy Duty Internal Spiral Retaining Rings** *MIL-R-27426*

| Bore Dia. A | Ring Dia. G | Ring Wall E | Groove Dia. C | Groove Width D | Static Thrust Load (lb) Ring | Static Thrust Load (lb) Groove | Bore Dia. A | Ring Dia. G | Ring Wall E | Groove Dia. C | Groove Width D | Static Thrust Load (lb) Ring | Static Thrust Load (lb) Groove |
|---|---|---|---|---|---|---|---|---|---|---|---|---|---|
| 2.125 | 2.267 | 0.168 | 2.251 | 0.086 | 21870 | 8935 | 9.750 | 10.260 | 0.500 | 10.189 | 0.209 | 229120 | 214325 |
| 2.188 | 2.334 | 0.168 | 2.318 | 0.086 | 22520 | 9745 | 10.000 | 10.523 | 0.500 | 10.450 | 0.209 | 234990 | 225330 |
| 2.250 | 2.399 | 0.168 | 2.382 | 0.086 | 23160 | 10455 | 10.250 | 10.786 | 0.500 | 10.711 | 0.209 | 240870 | 236605 |
| 2.312 | 2.467 | 0.200 | 2.450 | 0.086 | 23790 | 11700 | 10.500 | 11.047 | 0.500 | 10.970 | 0.209 | 246740 | 247110 |
| 2.357 | 2.535 | 0.200 | 2.517 | 0.086 | 24440 | 12715 | 10.750 | 11.313 | 0.500 | 11.234 | 0.209 | 252620 | 260530 |
| 2.440 | 2.602 | 0.200 | 2.584 | 0.086 | 25110 | 13550 | 11.000 | 11.575 | 0.500 | 11.495 | 0.209 | 258490 | 272645 |
| 2.500 | 2.667 | 0.200 | 2.648 | 0.086 | 25730 | 14640 | 11.250 | 11.838 | 0.500 | 11.756 | 0.209 | 264360 | 285040 |
| 2.531 | 2.700 | 0.200 | 2.681 | 0.086 | 26050 | 15185 | 11.500 | 12.102 | 0.562 | 12.018 | 0.209 | 270240 | 298285 |
| 2.562 | 2.733 | 0.225 | 2.714 | 0.103 | 29940 | 12775 | 11.750 | 12.365 | 0.562 | 12.279 | 0.209 | 276120 | 311240 |
| 2.625 | 2.801 | 0.225 | 2.781 | 0.103 | 30680 | 13780 | 12.000 | 12.628 | 0.562 | 12.540 | 0.209 | 281990 | 324475 |
| 2.688 | 2.868 | 0.225 | 2.848 | 0.103 | 31410 | 14775 | 12.250 | 12.891 | 0.562 | 12.801 | 0.209 | 287860 | 337980 |
| 2.750 | 2.934 | 0.225 | 2.914 | 0.103 | 32140 | 15790 | 12.500 | 13.154 | 0.562 | 13.063 | 0.209 | 293740 | 352390 |
| 2.813 | 3.001 | 0.225 | 2.980 | 0.103 | 32870 | 16845 | 12.750 | 13.417 | 0.562 | 13.324 | 0.209 | 299610 | 366460 |
| 2.834 | 3.027 | 0.225 | 3.006 | 0.103 | 33120 | 17595 | 13.000 | 13.680 | 0.662 | 13.585 | 0.209 | 305490 | 380805 |
| 2.875 | 3.072 | 0.225 | 3.051 | 0.103 | 33600 | 18505 | 13.250 | 13.943 | 0.662 | 13.846 | 0.209 | 311360 | 395430 |
| 3.000 | 3.204 | 0.225 | 3.182 | 0.103 | 35060 | 20795 | 13.500 | 14.207 | 0.662 | 14.108 | 0.209 | 317240 | 411000 |
| 3.062 | 3.271 | 0.281 | 3.248 | 0.120 | 42710 | 18735 | 13.750 | 14.470 | 0.662 | 14.369 | 0.209 | 323110 | 426185 |
| 3.125 | 3.338 | 0.281 | 3.315 | 0.120 | 43590 | 19865 | 14.000 | 14.732 | 0.662 | 14.630 | 0.209 | 328990 | 441645 |
| 3.157 | 3.371 | 0.281 | 3.348 | 0.120 | 44020 | 20345 | 14.250 | 14.995 | 0.662 | 14.891 | 0.209 | 334860 | 457380 |
| 3.250 | 3.470 | 0.281 | 3.446 | 0.120 | 45330 | 22120 | 14.500 | 15.259 | 0.750 | 15.153 | 0.209 | 340740 | 474120 |
| 3.346 | 3.571 | 0.281 | 3.546 | 0.120 | 46670 | 23905 | 14.750 | 15.522 | 0.750 | 15.414 | 0.209 | 346610 | 490415 |
| 3.469 | 3.701 | 0.281 | 3.675 | 0.120 | 48390 | 26405 | 15.000 | 15.785 | 0.750 | 15.675 | 0.209 | 352490 | 506990 |
| 3.500 | 3.736 | 0.281 | 3.710 | 0.120 | 48820 | 27370 | | | | | | | |

*Source:* Spirolox Retaining Rings, RRN Series. All dimensions are in inches. Depth of groove $d = (C - A)/2$. Thickness indicated is for unplated rings; add 0.002 to upper thickness tolerance for plated rings. Standard material: carbon spring steel (SAE 1070-1090).

*Ring Thickness, F*: For housing sizes 0.500 through 0.750, 0.035; for sizes 0.777 through 1.023, 0.042; for sizes 1.062 through 1.500, 0.050; for sizes 1.562 through 2.000, 0.062; for sizes 2.062 through 2.531, 0.078; for sizes 2.562 through 3.000, 0.093; for sizes 3.062 through 5.000, 0.111; for sizes 5.250 through 7.000, 0.156; for sizes 7.250 through 15.000, 0.187.

*Ring Free Diameter Tolerances:* For housing sizes 0.500 through 1.500, +0.013, −0.000; for sizes 1.562 through 2.000, +0.020, −0.000; for sizes 2.062 through 2.531, + 0.025, −0.000; for sizes 2.562 through 3.000, +0.030, −0.000; for sizes 3.062 through 5.000, +0.035, −0.000; for sizes 5.250 through 6.000, +0.050, −0.000; for sizes 6.250 through 7.000, +0.055. −0.000; for sizes 7.250 through 10.500, +0.070, −0.000; for sizes 10.750 through 12.750, +0.120, −0.000; for sizes 13.000 through 15.000, +0.140, −0.000.

*Ring Thickness Tolerances:* For housing sizes 0.500 through 1.500, ± 0.002; for sizes 1.562 through 5.000, ± 0.003; for sizes 5.250 through 6.000, ± 0.004; for sizes 6.250 through 15.000, ±0.005.

*Groove Diameter Tolerances:* For housing sizes 0.500 through 0.750, ± 0.002; for sizes 0.777 through 1.023, ± 0.003; for sizes 1.062 through 1.500, ± 0.004; for sizes 1.562 through 2.000, ± 0.005; for sizes 2.062 through 5.000, ± 0.006; for sizes 5.250 through 6.000, ± 0.007; for sizes 6.250 through 10.500, ± 0.008; for sizes 10.750 through 12.500, ± 0.010; for sizes 12.750 through 15.000, ± 0.012.

*Groove Width Tolerances:* For housing sizes 0.500 through 1.023, +0.003, −0.000; for sizes 1.062 through 2.000, +0.004, −0.000; for sizes 2.062 through 5.000, +0.005, −0.000; for sizes 5.250 through 6.000, +0.006, −0.000; for sizes 6.250 through 7.000, +0.008, −0.000; for sizes 7.250 through 15.000, +0.008, −0.000.

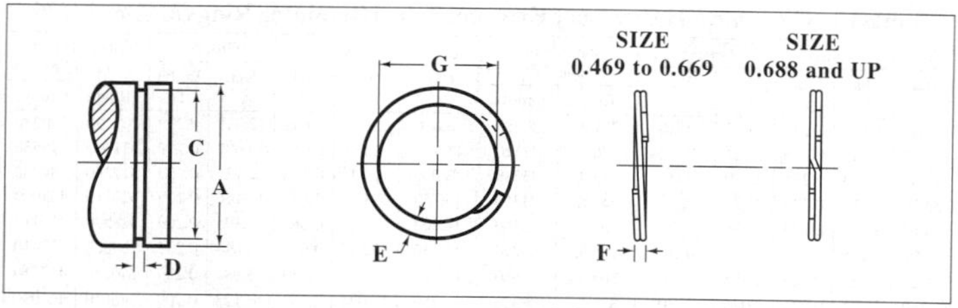


## Table 10. Heavy Duty External Spiral Retaining Rings *MIL-R-27426*

| Shaft Dia. | Ring | | Groove | | Static Thrust Load (lb) | | Shaft Dia. | Ring | | Groove | | Static Thrust Load (lb) | |
|---|---|---|---|---|---|---|---|---|---|---|---|---|---|
| A | Dia. G | Wall E | Dia. C | Width D | Ring | Groove | A | Dia. G | Wall E | Dia. C | Width D | Ring | Groove |
| 0.469 | 0.439 | 0.045 | 0.443 | 0.029 | 1880 | 510 | 3.500 | 3.293 | 0.270 | 3.316 | 0.120 | 48820 | 32250 |
| 0.500 | 0.464 | 0.050 | 0.468 | 0.039 | 2530 | 440 | 3.543 | 3.333 | 0.270 | 3.357 | 0.120 | 49420 | 33000 |
| 0.551 | 0.514 | 0.050 | 0.519 | 0.039 | 2790 | 540 | 3.625 | 3.411 | 0.270 | 3.435 | 0.120 | 50560 | 34490 |
| 0.562 | 0.525 | 0.050 | 0.530 | 0.039 | 2840 | 560 | 3.687 | 3.469 | 0.270 | 3.493 | 0.120 | 51430 | 35820 |
| 0.594 | 0.554 | 0.050 | 0.559 | 0.039 | 3000 | 700 | 3.750 | 3.527 | 0.270 | 3.552 | 0.120 | 52310 | 37180 |
| 0.625 | 0.583 | 0.055 | 0.588 | 0.039 | 3160 | 820 | 3.875 | 3.647 | 0.270 | 3.673 | 0.120 | 54050 | 39190 |
| 0.669 | 0.623 | 0.055 | 0.629 | 0.039 | 3380 | 1070 | 3.938 | 3.708 | 0.270 | 3.734 | 0.120 | 54930 | 40230 |
| 0.688 | 0.641 | 0.065 | 0.646 | 0.046 | 4170 | 960 | 4.000 | 3.765 | 0.270 | 3.792 | 0.120 | 55790 | 41660 |
| 0.750 | 0.698 | 0.065 | 0.704 | 0.046 | 4550 | 1250 | 4.250 | 4.037 | 0.270 | 4.065 | 0.120 | 59280 | 39370 |
| 0.781 | 0.727 | 0.065 | 0.733 | 0.046 | 4740 | 1430 | 4.375 | 4.161 | 0.270 | 4.190 | 0.120 | 61020 | 40530 |
| 0.812 | 0.756 | 0.065 | 0.762 | 0.046 | 4930 | 1620 | 4.500 | 4.280 | 0.270 | 4.310 | 0.120 | 62770 | 42810 |
| 0.875 | 0.814 | 0.075 | 0.821 | 0.046 | 5310 | 2000 | 4.750 | 4.518 | 0.270 | 4.550 | 0.120 | 66260 | 47570 |
| 0.938 | 0.875 | 0.075 | 0.882 | 0.046 | 5690 | 2440 | 5.000 | 4.756 | 0.270 | 4.790 | 0.120 | 69740 | 52580 |
| 0.984 | 0.919 | 0.085 | 0.926 | 0.046 | 5970 | 2790 | 5.250 | 4.995 | 0.350 | 5.030 | 0.139 | 83790 | 57830 |
| 1.000 | 0.932 | 0.085 | 0.940 | 0.046 | 6070 | 2950 | 5.500 | 5.228 | 0.350 | 5.265 | 0.139 | 87780 | 64720 |
| 1.023 | 0.953 | 0.085 | 0.961 | 0.046 | 6210 | 3170 | 5.750 | 5.466 | 0.350 | 5.505 | 0.139 | 91770 | 70540 |
| 1.062 | 0.986 | 0.103 | 0.998 | 0.056 | 7010 | 2810 | 6.000 | 5.705 | 0.350 | 5.745 | 0.139 | 95760 | 76610 |
| 1.125 | 1.047 | 0.103 | 1.059 | 0.056 | 7420 | 2890 | 6.250 | 5.938 | 0.418 | 5.985 | 0.174 | 122520 | 82930 |
| 1.188 | 1.105 | 0.103 | 1.118 | 0.056 | 7840 | 3450 | 6.500 | 6.181 | 0.418 | 6.225 | 0.174 | 127420 | 89510 |
| 1.250 | 1.163 | 0.103 | 1.176 | 0.056 | 8250 | 4110 | 6.750 | 6.410 | 0.418 | 6.465 | 0.174 | 132320 | 96330 |
| 1.312 | 1.218 | 0.118 | 1.232 | 0.056 | 8650 | 4810 | 7.000 | 6.648 | 0.418 | 6.705 | 0.174 | 137220 | 103400 |
| 1.375 | 1.277 | 0.118 | 1.291 | 0.056 | 9070 | 5650 | 7.250 | 6.891 | 0.418 | 6.942 | 0.174 | 142130 | 111810 |
| 1.438 | 1.336 | 0.118 | 1.350 | 0.056 | 9490 | 6340 | 7.500 | 7.130 | 0.437 | 7.180 | 0.209 | 176240 | 120170 |
| 1.500 | 1.385 | 0.118 | 1.406 | 0.056 | 9900 | 7060 | 7.750 | 7.368 | 0.437 | 7.420 | 0.209 | 182120 | 128060 |
| 1.562 | 1.453 | 0.128 | 1.468 | 0.068 | 12780 | 6600 | 8.000 | 7.606 | 0.437 | 7.660 | 0.209 | 187990 | 136200 |
| 1.625 | 1.513 | 0.128 | 1.529 | 0.068 | 13290 | 7330 | 8.250 | 7.845 | 0.437 | 7.900 | 0.209 | 193870 | 144590 |
| 1.687 | 1.573 | 0.128 | 1.589 | 0.068 | 13800 | 8190 | 8.500 | 8.083 | 0.437 | 8.140 | 0.209 | 199740 | 153220 |
| 1.750 | 1.633 | 0.128 | 1.650 | 0.068 | 14320 | 8760 | 8.750 | 8.324 | 0.437 | 8.383 | 0.209 | 205620 | 160800 |
| 1.771 | 1.651 | 0.128 | 1.669 | 0.068 | 14490 | 9040 | 9.000 | 8.560 | 0.500 | 8.620 | 0.209 | 211490 | 171250 |
| 1.812 | 1.690 | 0.128 | 1.708 | 0.068 | 14820 | 9440 | 9.250 | 8.798 | 0.500 | 8.860 | 0.209 | 217370 | 180640 |
| 1.875 | 1.751 | 0.158 | 1.769 | 0.068 | 15340 | 9950 | 9.500 | 9.036 | 0.500 | 9.100 | 0.209 | 223240 | 190280 |
| 1.969 | 1.838 | 0.158 | 1.857 | 0.068 | 16110 | 11040 | 9.750 | 9.275 | 0.500 | 9.338 | 0.209 | 229120 | 201140 |
| 2.000 | 1.867 | 0.158 | 1.886 | 0.068 | 16360 | 11420 | 10.000 | 9.508 | 0.500 | 9.575 | 0.209 | 234990 | 212810 |
| 2.062 | 1.932 | 0.168 | 1.946 | 0.086 | 21220 | 11820 | 10.250 | 9.745 | 0.500 | 9.814 | 0.209 | 240870 | 223780 |
| 2.125 | 1.989 | 0.168 | 2.003 | 0.086 | 21870 | 12980 | 10.500 | 9.984 | 0.500 | 10.054 | 0.209 | 246740 | 234490 |
| 2.156 | 2.018 | 0.168 | 2.032 | 0.086 | 22190 | 13390 | 10.750 | 10.221 | 0.500 | 10.293 | 0.209 | 252620 | 246000 |
| 2.250 | 2.105 | 0.168 | 2.120 | 0.086 | 23160 | 14650 | 11.000 | 10.459 | 0.500 | 10.533 | 0.209 | 258490 | 257230 |
| 2.312 | 2.163 | 0.168 | 2.178 | 0.086 | 23790 | 15510 | 11.250 | 10.692 | 0.500 | 10.772 | 0.209 | 264360 | 269270 |
| 2.375 | 2.223 | 0.200 | 2.239 | 0.086 | 24440 | 16170 | 11.500 | 10.934 | 0.562 | 11.011 | 0.209 | 270240 | 281590 |
| 2.437 | 2.283 | 0.200 | 2.299 | 0.086 | 25080 | 16840 | 11.750 | 11.171 | 0.562 | 11.250 | 0.209 | 276120 | 294180 |
| 2.500 | 2.343 | 0.200 | 2.360 | 0.086 | 25730 | 17530 | 12.000 | 11.410 | 0.562 | 11.490 | 0.209 | 281990 | 306450 |
| 2.559 | 2.402 | 0.200 | 2.419 | 0.086 | 26340 | 17940 | 12.250 | 11.647 | 0.562 | 11.729 | 0.209 | 287860 | 319580 |
| 2.625 | 2.464 | 0.200 | 2.481 | 0.086 | 27020 | 18930 | 12.500 | 11.885 | 0.562 | 11.969 | 0.209 | 293740 | 332360 |
| 2.687 | 2.523 | 0.200 | 2.541 | 0.086 | 27650 | 19640 | 12.750 | 12.124 | 0.562 | 12.208 | 0.209 | 299610 | 346030 |

**Table 10.** *(Continued)* **Heavy Duty External Spiral Retaining Rings** *MIL-R-27426*

| Shaft Dia. | Ring | | Groove | | Static Thrust Load (lb) | | Shaft Dia. | Ring | | Groove | | Static Thrust Load (lb) | |
|---|---|---|---|---|---|---|---|---|---|---|---|---|---|
| A | Dia. G | Wall E | Dia. C | Width D | Ring | Groove | A | Dia. G | Wall E | Dia. C | Width D | Ring | Groove |
| 2.750 | 2.584 | 0.225 | 2.602 | 0.103 | 32140 | 20380 | 13.000 | 12.361 | 0.662 | 12.448 | 0.209 | 305490 | 359330 |
| 2.875 | 2.702 | 0.225 | 2.721 | 0.103 | 33600 | 22170 | 13.250 | 12.598 | 0.662 | 12.687 | 0.209 | 311360 | 373530 |
| 2.937 | 2.760 | 0.225 | 2.779 | 0.103 | 34320 | 23240 | 13.500 | 12.837 | 0.662 | 12.927 | 0.209 | 317240 | 387340 |
| 3.000 | 2.818 | 0.225 | 2.838 | 0.103 | 35060 | 24340 | 13.750 | 13.074 | 0.662 | 13.166 | 0.209 | 323110 | 402090 |
| 3.062 | 2.878 | 0.225 | 2.898 | 0.103 | 35780 | 25140 | 14.000 | 13.311 | 0.662 | 13.405 | 0.209 | 328990 | 417110 |
| 3.125 | 2.936 | 0.225 | 2.957 | 0.103 | 36520 | 26290 | 14.250 | 13.548 | 0.662 | 13.644 | 0.209 | 334860 | 432410 |
| 3.156 | 2.965 | 0.225 | 2.986 | 0.103 | 36880 | 26860 | 14.500 | 13.787 | 0.750 | 13.884 | 0.209 | 340740 | 447250 |
| 3.250 | 3.054 | 0.225 | 3.076 | 0.103 | 37980 | 28320 | 14.750 | 14.024 | 0.750 | 14.123 | 0.209 | 346610 | 463090 |
| 3.344 | 3.144 | 0.225 | 3.166 | 0.103 | 39080 | 29800 | 15.000 | 14.262 | 0.750 | 14.363 | 0.209 | 352490 | 478450 |
| 3.437 | 3.234 | 0.225 | 3.257 | 0.103 | 40170 | 30980 | | | | | | | |

*Source:* Spirolox Retaining Rings, RSN Series. All dimensions are in inches. Depth of groove $d = (A - C)/2$. Thickness indicated is for unplated rings; add 0.002 to upper tolerance for plated rings. Standard material: carbon spring steel (SAE 1070-1090).

*Ring Thickness, F*: For shaft size 0.469, 0.025; for sizes 0.500 through 0.669, 0.035; for sizes 0.688 through 1.023, 0.042; for sizes 1.062 through 1.500, 0.050; for sizes 1.562 through 2.000, 0.062; for sizes 2.062 through 2.687, 0.078; for sizes 2.750 through 3.437, 0.093; for sizes 3.500 through 5.000, 0.111; for sizes 5.250 through 6.000, 0.127; for sizes 6.250 through 7.250, 0.156; for sizes 7.500 through 15.000, 0.187.

*Ring Free Diameter Tolerances:* For shaft sizes 0.469 through 1.500, +0.000, −0.013; for sizes 1.562 through 2.000, +0.000, −0.020; for sizes 2.062 through 2.687, + 0.000, −0.025; for sizes 2.750 through 3.437, +0.000, −0.030; for sizes 3.500 through 5.000, +0.000, −0.035; for sizes 5.250 through 6.000, +0.000, −0.050; for sizes 6.250 through 7.000, +0.000, −0.060; for sizes 7.250 through 10.000, +0.000, −0.070; for sizes 10.250 through 12.500, +0.000, −0.090; for sizes 12.750 through 15.000, +0.000, −0.110.

*Ring Thickness Tolerances:* For shaft sizes 0.469 through 1.500, ±0.002; for sizes 1.562 through 5.000, ±0.003; for sizes 5.250 through 6.000, ±0.004; for sizes 6.250 through 15.000, ±0.005.

*Groove Diameter Tolerances:* For shaft sizes 0.469 through 0.562, ±0.002; for sizes 0.594 through 1.023, ±0.003; for sizes 1.062 through 1.500, ±0.004; for sizes 1.562 through 2.000, ±0.005; for sizes 2.062 through 5.000, ±0.006; for sizes 5.250 through 6.000, ±0.007; for sizes 6.250 through 10.000, ±0.008; for sizes 10.250 through 12.500, ±0.010; for sizes 12.750 through 15.000, ±0.012.

*Groove Width Tolerances:* For shaft sizes 0.469 through 1.023, +0.003, −0.000; for sizes 1.062 through 2.000, +0.004, −0.000; for sizes 2.062 through 5.000, +0.005, −0.000; for sizes 5.250 through 6.000, +0.006; −0.000; for sizes 6.250 through 7.250, + 0.008, −0.000; for sizes 7.500 through 15.000, +0.008, −0.000.

*Thrust Load Capacity:* The most important criterion in determining which ring is best suited for a specific application is thrust load capacity. The strength of the retaining ring and groove must both be considered when analyzing the thrust load capacity of an application to determine whether the groove or the retaining ring is likely to fail first. When a retaining ring application fails, the fault will usually be with the groove, unless the groove material is of very high strength.

*Ring Material:* The standard materials for spiral-wound retaining rings are SAE 1070 to 1090 carbon spring steels and 18-8 type 302 stainless steels. The 1070 to 1090 carbon spring steels provide high-strength retaining rings at low cost. Type 302 stainless steel withstands ordinary rusting. Other materials are used for specialized applications, such as the type 316 stainless frequently used in the food industry. For high-temperature use, superalloy A286 rings can be used at up to 900° F (482° C) and Inconel X-750 at up to 1200° F (649° C). Other materials, such as 316 stainless steel, 17-7PH and Inconel stainless steels are sometimes used for special-purpose and custom-made rings. Standard ring are typically supplied uncoated, however, special finishes such as cadmium, phosphate, zinc, or black oxide coatings for carbon spring steel rings and passivation of stainless steel rings are available.

## Table 11. Important Dimensions of Inch Series External Retaining Rings

*MS 16624*

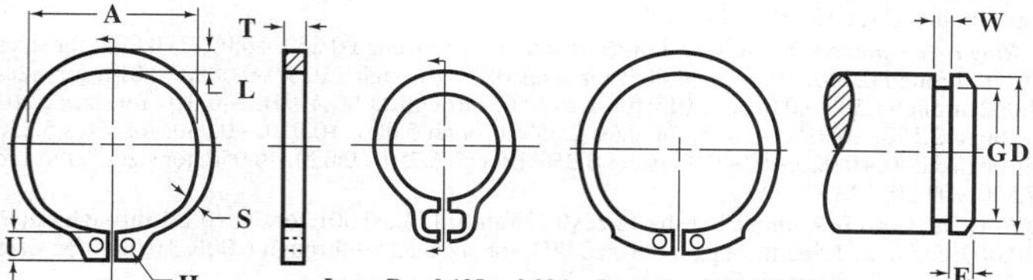


**Lugs: D = 0.125 to 0.236 Lugs: D = 4.25 to 8.00**

| Shaft | Ring | | Groove | | | Shaft | Ring | | Groove | | |
|---|---|---|---|---|---|---|---|---|---|---|---|
| Dia. D | Dia. A | Thickness T | Dia. G | Width W | Margin E | Dia. D | Dia. A | Thickness T | Dia. G | Width W | Margin E |
| 0.125 | 0.112 | 0.010 | 0.117 | 0.012 | 0.012 | 1.812 | 1.675 | 0.062 | 1.708 | 0.068 | 0.156 |
| 0.156 | 0.142 | 0.010 | 0.146 | 0.012 | 0.015 | 1.875 | 1.735 | 0.062 | 1.769 | 0.068 | 0.159 |
| 0.188 | 0.168 | 0.015 | 0.175 | 0.018 | 0.018 | 1.969 | 1.819 | 0.062 | 1.857 | 0.068 | 0.168 |
| 0.197 | 0.179 | 0.015 | 0.185 | 0.018 | 0.018 | 2.000 | 1.850 | 0.062 | 1.886 | 0.068 | 0.171 |
| 0.219 | 0.196 | 0.015 | 0.205 | 0.018 | 0.021 | 2.062 | 1.906 | 0.078 | 1.946 | 0.086 | 0.174 |
| 0.236 | 0.215 | 0.015 | 0.222 | 0.018 | 0.021 | 2.125 | 1.964 | 0.078 | 2.003 | 0.086 | 0.183 |
| 0.250 | 0.225 | 0.025 | 0.230 | 0.029 | 0.030 | 2.156 | 1.993 | 0.078 | 2.032 | 0.086 | 0.186 |
| 0.276 | 0.250 | 0.025 | 0.255 | 0.029 | 0.030 | 2.250 | 2.081 | 0.078 | 2.120 | 0.086 | 0.195 |
| 0.281 | 0.256 | 0.025 | 0.261 | 0.029 | 0.030 | 2.312 | 2.139 | 0.078 | 2.178 | 0.086 | 0.201 |
| 0.312 | 0.281 | 0.025 | 0.290 | 0.029 | 0.033 | 2.375 | 2.197 | 0.078 | 2.239 | 0.086 | 0.204 |
| 0.344 | 0.309 | 0.025 | 0.321 | 0.029 | 0.033 | 2.438 | 2.255 | 0.078 | 2.299 | 0.086 | 0.207 |
| 0.354 | 0.320 | 0.025 | 0.330 | 0.029 | 0.036 | 2.500 | 2.313 | 0.078 | 2.360 | 0.086 | 0.210 |
| 0.375 | 0.338 | 0.025 | 0.352 | 0.029 | 0.036 | 2.559 | 2.377 | 0.078 | 2.419 | 0.086 | 0.210 |
| 0.394 | 0.354 | 0.025 | 0.369 | 0.029 | 0.036 | 2.625 | 2.428 | 0.078 | 2.481 | 0.086 | 0.216 |
| 0.406 | 0.366 | 0.025 | 0.382 | 0.029 | 0.036 | 2.688 | 2.485 | 0.078 | 2.541 | 0.086 | 0.219 |
| 0.438 | 0.395 | 0.025 | 0.412 | 0.029 | 0.039 | 2.750 | 2.543 | 0.093 | 2.602 | 0.103 | 0.222 |
| 0.469 | 0.428 | 0.025 | 0.443 | 0.029 | 0.039 | 2.875 | 2.659 | 0.093 | 2.721 | 0.103 | 0.231 |
| 0.500 | 0.461 | 0.035 | 0.468 | 0.039 | 0.048 | 2.938 | 2.717 | 0.093 | 2.779 | 0.103 | 0.237 |
| 0.551 | 0.509 | 0.035 | 0.519 | 0.039 | 0.048 | 3.000 | 2.775 | 0.093 | 2.838 | 0.103 | 0.243 |
| 0.562 | 0.521 | 0.035 | 0.530 | 0.039 | 0.048 | 3.062 | 2.832 | 0.093 | 2.898 | 0.103 | 0.246 |
| 0.594 | 0.550 | 0.035 | 0.559 | 0.039 | 0.051 | 3.125 | 2.892 | 0.093 | 2.957 | 0.103 | 0.252 |
| 0.625 | 0.579 | 0.035 | 0.588 | 0.039 | 0.054 | 3.156 | 2.920 | 0.093 | 2.986 | 0.103 | 0.255 |
| 0.669 | 0.621 | 0.035 | 0.629 | 0.039 | 0.060 | 3.250 | 3.006 | 0.093 | 3.076 | 0.103 | 0.261 |
| 0.672 | 0.621 | 0.035 | 0.631 | 0.039 | 0.060 | 3.346 | 3.092 | 0.093 | 3.166 | 0.103 | 0.270 |
| 0.688 | 0.635 | 0.042 | 0.646 | 0.046 | 0.063 | 3.438 | 3.179 | 0.093 | 3.257 | 0.103 | 0.270 |
| 0.750 | 0.693 | 0.042 | 0.704 | 0.046 | 0.069 | 3.500 | 3.237 | 0.109 | 3.316 | 0.120 | 0.276 |
| 0.781 | 0.722 | 0.042 | 0.733 | 0.046 | 0.072 | 3.543 | 3.277 | 0.109 | 3.357 | 0.120 | 0.279 |
| 0.812 | 0.751 | 0.042 | 0.762 | 0.046 | 0.075 | 3.625 | 3.352 | 0.109 | 3.435 | 0.120 | 0.285 |
| 0.844 | 0.780 | 0.042 | 0.791 | 0.046 | 0.080 | 3.688 | 3.410 | 0.109 | 3.493 | 0.120 | 0.291 |
| 0.875 | 0.810 | 0.042 | 0.821 | 0.046 | 0.081 | 3.750 | 3.468 | 0.109 | 3.552 | 0.120 | 0.297 |
| 0.938 | 0.867 | 0.042 | 0.882 | 0.046 | 0.084 | 3.875 | 3.584 | 0.109 | 3.673 | 0.120 | 0.303 |
| 0.984 | 0.910 | 0.042 | 0.926 | 0.046 | 0.087 | 3.938 | 3.642 | 0.109 | 3.734 | 0.120 | 0.306 |
| 1.000 | 0.925 | 0.042 | 0.940 | 0.046 | 0.090 | 4.000 | 3.700 | 0.109 | 3.792 | 0.120 | 0.312 |
| 1.023 | 0.946 | 0.042 | 0.961 | 0.046 | 0.093 | 4.250 | 3.989 | 0.109 | 4.065 | 0.120 | 0.276 |
| 1.062 | 0.982 | 0.050 | 0.998 | 0.056 | 0.096 | 4.375 | 4.106 | 0.109 | 4.190 | 0.120 | 0.276 |
| 1.125 | 1.041 | 0.050 | 1.059 | 0.056 | 0.099 | 4.500 | 4.223 | 0.109 | 4.310 | 0.120 | 0.285 |
| 1.188 | 1.098 | 0.050 | 1.118 | 0.056 | 0.105 | 4.750 | 4.458 | 0.109 | 4.550 | 0.120 | 0.300 |
| 1.250 | 1.156 | 0.050 | 1.176 | 0.056 | 0.111 | 5.000 | 4.692 | 0.109 | 4.790 | 0.120 | 0.315 |
| 1.312 | 1.214 | 0.050 | 1.232 | 0.056 | 0.120 | 5.250 | 4.927 | 0.125 | 5.030 | 0.139 | 0.330 |
| 1.375 | 1.272 | 0.050 | 1.291 | 0.056 | 0.126 | 5.500 | 5.162 | 0.125 | 5.265 | 0.139 | 0.351 |
| 1.438 | 1.333 | 0.050 | 1.350 | 0.056 | 0.132 | 5.750 | 5.396 | 0.125 | 5.505 | 0.139 | 0.366 |
| 1.500 | 1.387 | 0.050 | 1.406 | 0.056 | 0.141 | 6.000 | 5.631 | 0.125 | 5.745 | 0.139 | 0.381 |
| 1.562 | 1.446 | 0.062 | 1.468 | 0.068 | 0.141 | 6.250 | 5.866 | 0.156 | 5.985 | 0.174 | 0.396 |
| 1.625 | 1.503 | 0.062 | 1.529 | 0.068 | 0.144 | 6.500 | 6.100 | 0.156 | 6.225 | 0.174 | 0.411 |
| 1.687 | 1.560 | 0.062 | 1.589 | 0.068 | 0.147 | 6.750 | 6.335 | 0.156 | 6.465 | 0.174 | 0.426 |
| 1.750 | 1.618 | 0.062 | 1.650 | 0.068 | 0.150 | 7.000 | 6.570 | 0.156 | 6.705 | 0.174 | 0.441 |
| 1.772 | 1.637 | 0.062 | 1.669 | 0.068 | 0.153 | 7.500 | 7.009 | 0.187 | 7.180 | 0.209 | 0.480 |

*Source:* Industrial Retaining Rings, 3100 Series. All dimensions are in inches. Depth of groove $d = (D - G)/2$. Thickness indicated is for unplated rings; for most plated rings, the maximum ring thickness will not exceed the minimum groove width (W) minus 0.0002 inch. Standard material: carbon spring steel (SAE 1060-1090).

*Ring Free Diameter Tolerances:* For shaft sizes 0.125 through 0.250, +0.002, −0.004; for sizes 0.276 through 0.500, +0.002, −0.005; for sizes 0.551 through 1.023, +0.005, −0.010; for sizes 1.062 through 1.500, +0.010, −0.015; for sizes 1.562 through 2.000, +0.013, −0.020; for sizes 2.062 through 2.500, +0.015, −0.025; for sizes 2.559 through 5.000, +0.020, −0.030; for sizes 5.250 through 6.000, +0.020, −0.040; for sizes 6.250 through 6.750, +0.020, −0.050; for sizes 7.000 and 7.500, +0.050, −0.130.

*Ring Thickness Tolerances:* For shaft sizes 0.125 and 0.156, ±0.001; for sizes 0.188 through 1.500, ±0.002; for sizes 1.562 through 5.000, ±0.003; for sizes 5.250 through 6.000, ±0.004; for sizes 6.250 through 7.500, ±0.005.

*Groove Diameter Tolerances:* For shaft sizes 0.125 through 0.250, ±0.0015; for sizes 0.276 through 0.562, ±0.002; for sizes 0.594 through 1.023, ±0.003; for sizes 1.062 though 1.500, ±0.004; for sizes 1.562 through 2.000, ±0.005; for sizes 2.062 through 5.000, ±0.006; for sizes 5.250 through 6.000, ±0.007; for sizes 6.250 through 7.500, ±0.008.

*Groove Width Tolerances:* For shaft sizes 0.125 through 0.236, +0.002, −0.000; for sizes 0.250 through 1.023, +0.003, −0.000; for sizes 1.062 through 2.000, +0.004, −0.000; for sizes 2.062 through 5.000, +0.005, −0.000; for sizes 5.250 through 6.000, +0.006, −0.000; for sizes 6.250 through 7.500, +0.008, −0.000.

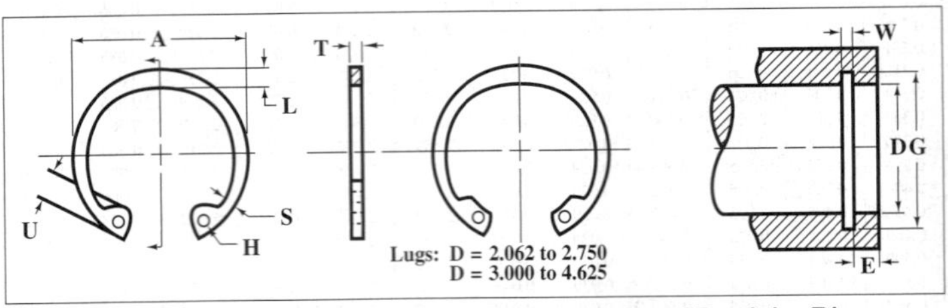


**Table 12. Important Dimensions of Inch Series Internal Retaining Rings**

| Housing Dia. D | Ring Dia. A | Ring Thickness T | Groove Dia. G | Groove Width W | Groove Margin E | Housing Dia. D | Ring Dia. A | Ring Thickness T | Groove Dia. G | Groove Width W | Groove Margin E |
|---|---|---|---|---|---|---|---|---|---|---|---|
| 0.250 | 0.280 | 0.015 | 0.268 | 0.018 | 0.027 | 2.500 | 2.775 | 0.078 | 2.648 | 0.086 | 0.222 |
| 0.312 | 0.346 | 0.015 | 0.330 | 0.018 | 0.027 | 2.531 | 2.775 | 0.078 | 2.681 | 0.086 | 0.225 |
| 0.375 | 0.415 | 0.025 | 0.397 | 0.029 | 0.033 | 2.562 | 2.844 | 0.093 | 2.714 | 0.103 | 0.228 |
| 0.438 | 0.482 | 0.025 | 0.461 | 0.029 | 0.036 | 2.625 | 2.910 | 0.093 | 2.781 | 0.103 | 0.234 |
| 0.453 | 0.498 | 0.025 | 0.477 | 0.029 | 0.036 | 2.677 | 2.980 | 0.093 | 2.837 | 0.103 | 0.240 |
| 0.500 | 0.548 | 0.035 | 0.530 | 0.039 | 0.045 | 2.688 | 2.980 | 0.093 | 2.848 | 0.103 | 0.240 |
| 0.512 | 0.560 | 0.035 | 0.542 | 0.039 | 0.045 | 2.750 | 3.050 | 0.093 | 2.914 | 0.103 | 0.246 |
| 0.562 | 0.620 | 0.035 | 0.596 | 0.039 | 0.051 | 2.812 | 3.121 | 0.093 | 2.980 | 0.103 | 0.252 |
| 0.625 | 0.694 | 0.035 | 0.665 | 0.039 | 0.060 | 2.835 | 3.121 | 0.093 | 3.006 | 0.103 | 0.255 |
| 0.688 | 0.763 | 0.035 | 0.732 | 0.039 | 0.066 | 2.875 | 3.191 | 0.093 | 3.051 | 0.103 | 0.264 |
| 0.750 | 0.831 | 0.035 | 0.796 | 0.039 | 0.069 | 2.953 | 3.325 | 0.093 | 3.135 | 0.103 | 0.273 |
| 0.777 | 0.859 | 0.042 | 0.825 | 0.046 | 0.072 | 3.000 | 3.325 | 0.093 | 3.182 | 0.103 | 0.273 |
| 0.812 | 0.901 | 0.042 | 0.862 | 0.046 | 0.075 | 3.062 | 3.418 | 0.109 | 3.248 | 0.120 | 0.279 |
| 0.866 | 0.961 | 0.042 | 0.920 | 0.046 | 0.081 | 3.125 | 3.488 | 0.109 | 3.315 | 0.120 | 0.285 |
| 0.875 | 0.971 | 0.042 | 0.931 | 0.046 | 0.084 | 3.149 | 3.523 | 0.109 | 3.341 | 0.120 | 0.288 |
| 0.901 | 1.000 | 0.042 | 0.959 | 0.046 | 0.087 | 3.156 | 3.523 | 0.109 | 3.348 | 0.120 | 0.288 |
| 0.938 | 1.041 | 0.042 | 1.000 | 0.046 | 0.093 | 3.250 | 3.623 | 0.109 | 3.446 | 0.120 | 0.294 |
| 1.000 | 1.111 | 0.042 | 1.066 | 0.046 | 0.099 | 3.346 | 3.734 | 0.109 | 3.546 | 0.120 | 0.300 |
| 1.023 | 1.136 | 0.042 | 1.091 | 0.046 | 0.102 | 3.469 | 3.857 | 0.109 | 3.675 | 0.120 | 0.309 |
| 1.062 | 1.180 | 0.050 | 1.130 | 0.056 | 0.102 | 3.500 | 3.890 | 0.109 | 3.710 | 0.120 | 0.315 |

**Table 12.** *(Continued)* **Important Dimensions of Inch Series Internal Retaining Rings**

| Housing Dia. D | Ring Dia. A | Ring Thickness T | Groove Dia. G | Groove Width W | Groove Margin E | Housing Dia. D | Ring Dia. A | Ring Thickness T | Groove Dia. G | Groove Width W | Groove Margin E |
|---|---|---|---|---|---|---|---|---|---|---|---|
| 1.125 | 1.249 | 0.050 | 1.197 | 0.056 | 0.108 | 3.543 | 3.936 | 0.109 | 3.755 | 0.120 | 0.318 |
| 1.181 | 1.319 | 0.050 | 1.255 | 0.056 | 0.111 | 3.562 | 3.936 | 0.109 | 3.776 | 0.120 | 0.321 |
| 1.188 | 1.319 | 0.050 | 1.262 | 0.056 | 0.111 | 3.625 | 4.024 | 0.109 | 3.841 | 0.120 | 0.324 |
| 1.250 | 1.388 | 0.050 | 1.330 | 0.056 | 0.120 | 3.740 | 4.157 | 0.109 | 3.964 | 0.120 | 0.336 |
| 1.259 | 1.388 | 0.050 | 1.339 | 0.056 | 0.120 | 3.750 | 4.157 | 0.109 | 3.974 | 0.120 | 0.336 |
| 1.312 | 1.456 | 0.050 | 1.396 | 0.056 | 0.126 | 3.875 | 4.291 | 0.109 | 4.107 | 0.120 | 0.348 |
| 1.375 | 1.526 | 0.050 | 1.461 | 0.056 | 0.129 | 3.938 | 4.358 | 0.109 | 4.174 | 0.120 | 0.354 |
| 1.378 | 1.526 | 0.050 | 1.464 | 0.056 | 0.129 | 4.000 | 4.424 | 0.109 | 4.240 | 0.120 | 0.360 |
| 1.438 | 1.596 | 0.050 | 1.528 | 0.056 | 0.135 | 4.125 | 4.558 | 0.109 | 4.365 | 0.120 | 0.360 |
| 1.456 | 1.616 | 0.050 | 1.548 | 0.056 | 0.138 | 4.250 | 4.691 | 0.109 | 4.490 | 0.120 | 0.360 |
| 1.500 | 1.660 | 0.050 | 1.594 | 0.056 | 0.141 | 4.331 | 4.756 | 0.109 | 4.571 | 0.120 | 0.360 |
| 1.562 | 1.734 | 0.062 | 1.658 | 0.068 | 0.144 | 4.500 | 4.940 | 0.109 | 4.740 | 0.120 | 0.360 |
| 1.575 | 1.734 | 0.062 | 1.671 | 0.068 | 0.144 | 4.625 | 5.076 | 0.109 | 4.865 | 0.120 | 0.360 |
| 1.625 | 1.804 | 0.062 | 1.725 | 0.068 | 0.150 | 4.724 | 5.213 | 0.109 | 4.969 | 0.120 | 0.366 |
| 1.653 | 1.835 | 0.062 | 1.755 | 0.068 | 0.153 | 4.750 | 5.213 | 0.109 | 4.995 | 0.120 | 0.366 |
| 1.688 | 1.874 | 0.062 | 1.792 | 0.068 | 0.156 | 5.000 | 5.485 | 0.109 | 5.260 | 0.120 | 0.390 |
| 1.750 | 1.942 | 0.062 | 1.858 | 0.068 | 0.162 | 5.250 | 5.770 | 0.125 | 5.520 | 0.139 | 0.405 |
| 1.812 | 2.012 | 0.062 | 1.922 | 0.068 | 0.165 | 5.375 | 5.910 | 0.125 | 5.650 | 0.139 | 0.405 |
| 1.850 | 2.054 | 0.062 | 1.962 | 0.068 | 0.168 | 5.500 | 6.066 | 0.125 | 5.770 | 0.139 | 0.405 |
| 1.875 | 2.054 | 0.062 | 1.989 | 0.068 | 0.171 | 5.750 | 6.336 | 0.125 | 6.020 | 0.139 | 0.405 |
| 1.938 | 2.141 | 0.062 | 2.056 | 0.068 | 0.177 | 6.000 | 6.620 | 0.125 | 6.270 | 0.139 | 0.405 |
| 2.000 | 2.210 | 0.062 | 2.122 | 0.068 | 0.183 | 6.250 | 6.895 | 0.156 | 6.530 | 0.174 | 0.420 |
| 2.047 | 2.280 | 0.078 | 2.171 | 0.086 | 0.186 | 6.500 | 7.170 | 0.156 | 6.790 | 0.174 | 0.435 |
| 2.062 | 2.280 | 0.078 | 2.186 | 0.086 | 0.186 | 6.625 | 7.308 | 0.156 | 6.925 | 0.174 | 0.450 |
| 2.125 | 2.350 | 0.078 | 2.251 | 0.086 | 0.189 | 6.750 | 7.445 | 0.156 | 7.055 | 0.174 | 0.456 |
| 2.165 | 2.415 | 0.078 | 2.295 | 0.086 | 0.195 | 7.000 | 7.720 | 0.156 | 7.315 | 0.174 | 0.471 |
| 2.188 | 2.415 | 0.078 | 2.318 | 0.086 | 0.195 | 7.250 | 7.995 | 0.187 | 7.575 | 0.209 | 0.486 |
| 2.250 | 2.490 | 0.078 | 2.382 | 0.086 | 0.198 | 7.500 | 8.270 | 0.187 | 7.840 | 0.209 | 0.510 |
| 2.312 | 2.560 | 0.078 | 2.450 | 0.086 | 0.207 | 7.750 | 8.545 | 0.187 | 8.100 | 0.209 | 0.525 |
| 2.375 | 2.630 | 0.078 | 2.517 | 0.086 | 0.213 | 8.000 | 8.820 | 0.187 | 8.360 | 0.209 | 0.540 |
| 2.440 | 2.702 | 0.078 | 2.584 | 0.086 | 0.216 | 8.250 | 9.095 | 0.187 | 8.620 | 0.209 | 0.555 |

*Source:* Industrial Retaining Rings, 3000 Series. All dimensions are in inches. Depth of groove $d = (G - D)/2$. Thickness indicated is for unplated rings. Standard material: carbon spring steel (SAE 1060-1090).

*Ring Free Diameter Tolerances:* For housing sizes 0.250 through 0.777, +0.010, −0.005; for sizes 0.812 through 1.023, +0.015, − 0.010; for sizes 1.062 through 1.500, + 0.025, − 0.020; for sizes 1.562 through 2.000, +0.035, −0.025; for sizes 2.047 through 3.000, +0.040, −0.030; for sizes 3.062 through 3.625, ±0.055; for sizes 3.740 through 6.000, ±0.065; for sizes 6.250 through 7.000, ±0.080; for sizes 7.250 through 8.250, ±0.090.

*Ring Thickness Tolerances:* For housing sizes 0.250 through 1.500, ±0.002; for sizes 1.562 through 5.000, ±0.003; for sizes 5.250 through 6.000, ±0.004; for sizes 6.250 through 8.250, ±0.005.

*Groove Diameter Tolerances:* For housing sizes 0.250 and 0.312, ±0.001; for sizes 0.375 through 0.750, ±0.002; for sizes 0.777 through 1.023, ±0.003; for sizes 1.062 through 1.500, ±0.004; for sizes 1.562 through 2.000, ±0.005; for sizes 2.047 through 5.000, ±0.006; for sizes 5.250 through 6.000, ±0.007; for sizes 6.250 through 8.250, ±0.008.

*Groove Width Tolerances:* For housing sizes 0.250 and 0.312, +0.002, − 0.000; for sizes 0.375 through 1.023, +0.003, −0.000; for sizes 1.062 through 2.000, +0.004, −0.000; for sizes 2.047 through 5.000, +0.005; −0.000; for sizes 5.250 through 6.000, +0.006, −0.000; for sizes 6.250 through 8.250, +0.008, −0.000.

**Table 13. Important Dimensions of Inch Series External Retaining Rings** *MS16632*

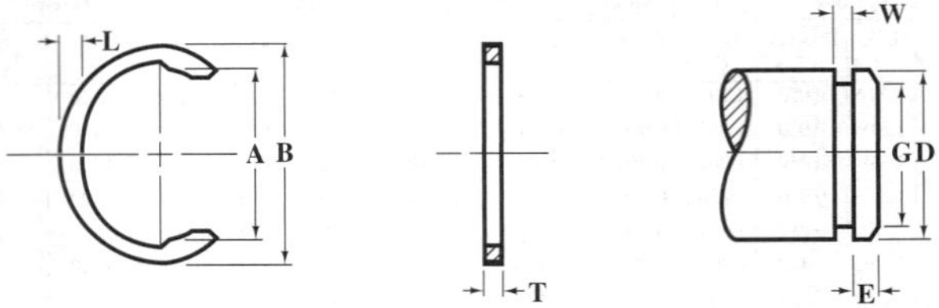


| Shaft Diameter D | Ring | | | Groove | | | [a]Static Thrust Load (lb) | |
|---|---|---|---|---|---|---|---|---|
| | Free Dia. A | Thickness T | Diameter B | Diameter G | Width W | Margin E | Ring | Groove |
| 0.125 | 0.102 | 0.015 | 0.164 | 0.106 | 0.018 | 0.020 | 85 | 40 |
| 0.156 | 0.131 | 0.015 | 0.205 | 0.135 | 0.018 | 0.020 | 110 | 55 |
| 0.188 | 0.161 | 0.015 | 0.245 | 0.165 | 0.018 | 0.022 | 130 | 70 |
| 0.219 | 0.187 | 0.025 | 0.275 | 0.193 | 0.029 | 0.026 | 260 | 100 |
| 0.236 | 0.203 | 0.025 | 0.295 | 0.208 | 0.029 | 0.028 | 280 | 115 |
| 0.250 | 0.211 | 0.025 | 0.311 | 0.220 | 0.029 | 0.030 | 295 | 130 |
| 0.281 | 0.242 | 0.025 | 0.344 | 0.247 | 0.029 | 0.034 | 330 | 170 |
| 0.312 | 0.270 | 0.025 | 0.376 | 0.276 | 0.029 | 0.036 | 370 | 200 |
| 0.375 | 0.328 | 0.025 | 0.448 | 0.335 | 0.029 | 0.040 | 440 | 265 |
| 0.406 | 0.359 | 0.025 | 0.485 | 0.364 | 0.029 | 0.042 | 480 | 300 |
| 0.437 | 0.386 | 0.025 | 0.516 | 0.393 | 0.029 | 0.044 | 515 | 340 |
| 0.500 | 0.441 | 0.035 | 0.581 | 0.450 | 0.039 | 0.050 | 825 | 440 |
| 0.562 | 0.497 | 0.035 | 0.653 | 0.507 | 0.039 | 0.056 | 930 | 550 |
| 0.625 | 0.553 | 0.035 | 0.715 | 0.563 | 0.039 | 0.062 | 1030 | 690 |
| 0.687 | 0.608 | 0.042 | 0.780 | 0.619 | 0.046 | 0.068 | 1700 | 820 |
| 0.750 | 0.665 | 0.042 | 0.845 | 0.676 | 0.046 | 0.074 | 1850 | 985 |
| 0.812 | 0.721 | 0.042 | 0.915 | 0.732 | 0.046 | 0.080 | 2010 | 1150 |
| 0.875 | 0.777 | 0.042 | 0.987 | 0.789 | 0.046 | 0.086 | 2165 | 1320 |
| 0.937 | 0.830 | 0.042 | 1.054 | 0.843 | 0.046 | 0.094 | 2320 | 1550 |
| 1.000 | 0.887 | 0.042 | 1.127 | 0.900 | 0.046 | 0.100 | 2480 | 1770 |
| 1.125 | 0.997 | 0.050 | 1.267 | 1.013 | 0.056 | 0.112 | 3300 | 2200 |
| 1.188 | 1.031 | 0.050 | 1.321 | 1.047 | 0.056 | 0.140 | 3500 | 2900 |
| 1.250 | 1.110 | 0.050 | 1.410 | 1.126 | 0.056 | 0.124 | 3600 | 2700 |
| 1.375 | 1.220 | 0.050 | 1.550 | 1.237 | 0.056 | 0.138 | 4000 | 3300 |
| 1.500 | 1.331 | 0.050 | 1.691 | 1.350 | 0.056 | 0.150 | 4400 | 4000 |
| 1.750 | 1.555 | 0.062 | 1.975 | 1.576 | 0.068 | 0.174 | 6400 | 5300 |
| 2.000 | 1.777 | 0.062 | 2.257 | 1.800 | 0.068 | 0.200 | 7300 | 7000 |

[a] *Thrust Load Safety Factors:* Ring, 4; groove, 2. Groove wall thrust loads are for grooves machined in cold-rolled steel with a tensile yield strength of 45,000 psi; for other shaft materials, the thrust load varies proportionally with the yield strength.

*Source:* Industrial Retaining Rings, 2000 Series. All dimensions are in inches. Depth of groove $d = (D - G)/2$. Standard material: carbon spring steel (SAE 1060-1090). Thickness indicated is for unplated rings; for most plated rings with shaft sizes less than 1.000 inch, the maximum thickness will not exceed the minimum groove width (W) minus 0.0002 inch; for larger rings, the ring thickness may increase by 0.002 inch.

*Groove Maximum Bottom Radii:* For shaft diameters less than 0.500 inch, 0.005 inch; for shaft sizes 0.500 through 1.000 inch, 0.010 inch; all larger sizes, 0.015 inch.

*Ring Free Diameter Tolerances:* For shaft sizes 0.125 through 0.188, +0.002, −0.004; for sizes 0.219 through 0.437, +0.003, −0.005; for sizes 0.500 through 0.625, ±0.006; for sizes 0.687 through 1.000, ±0.007; for sizes 1.125 through 1.500, ±0.008; for sizes 1.750 and 2.000, ±0.010.

*Ring Thickness Tolerances:* For shaft sizes 0.125 through 1.500, ±0.002; for sizes 1.750 and 2.000, ±0.003.

*Groove Diameter Tolerances:* For shaft sizes 0.125 through 0.188, ±0.0015; for sizes 0.219 through 0.437, ±0.002; for sizes 0.500 through 1.000, ±0.003; for sizes 1.125 through 1.500, ±0.004; for sizes 1.750 and 2.000, ±0.005.

*Groove Width Tolerances:* For shaft sizes 0.125 through 0.188, +0.002, −0.000; for sizes 0.219 through 1.000, +0.003, −0.000; for sizes 1.125 through 2.000, +0.004, −0.000.

**Table 14. Important Dimensions of Inch Series External Retaining Rings** *MS16633*

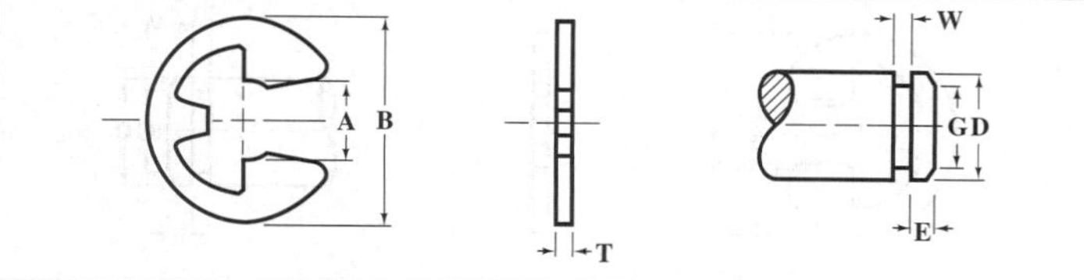


| Shaft Diameter | Ring | | | Groove | | | [a]Static Thrust Load (lb) | |
|---|---|---|---|---|---|---|---|---|
| | Free Dia. | Thickness | Diameter | Diameter | Width | Margin | | |
| D | A | T | B | G | W | E | Ring | Groove |
| 0.040 | 0.025 | 0.010 | 0.079 | 0.026 | 0.012 | 0.014 | 13 | 7 |
| 0.062 | 0.051 | 0.010 | 0.140 | 0.052 | 0.012 | 0.010 | 20 | 7 |
| 0.062[a] | 0.051 | 0.010 | 0.156 | 0.052 | 0.012 | 0.010 | 20 | 7 |
| 0.062[b] | 0.051 | 0.020 | 0.187 | 0.052 | 0.023 | 0.010 | 40 | 7 |
| 0.094 | 0.073 | 0.015 | 0.187 | 0.074 | 0.018 | 0.020 | 45 | 20 |
| 0.094 | 0.069 | 0.015 | 0.230 | 0.074 | 0.018 | 0.020 | 45 | 20 |
| 0.110 | 0.076 | 0.015 | 0.375 | 0.079 | 0.018 | 0.030 | 55 | 40 |
| 0.125 | 0.094 | 0.015 | 0.230 | 0.095 | 0.018 | 0.030 | 65 | 45 |
| 0.140 | 0.100 | 0.015 | 0.203 | 0.102 | 0.018 | 0.038 | 70 | 60 |
| 0.140[c] | 0.108 | 0.015 | 0.250 | 0.110 | 0.018 | 0.030 | 70 | 45 |
| 0.140[d] | 0.102 | 0.025 | 0.270 | 0.105 | 0.029 | 0.034 | 150 | 55 |
| 0.156 | 0.114 | 0.025 | 0.282 | 0.116 | 0.029 | 0.040 | 165 | 70 |
| 0.172 | 0.125 | 0.025 | 0.312 | 0.127 | 0.029 | 0.044 | 180 | 90 |
| 0.188 | 0.145 | 0.025 | 0.335 | 0.147 | 0.029 | 0.040 | 195 | 90 |
| 0.188 | 0.122 | 0.025 | 0.375 | 0.125 | 0.029 | 0.062 | 195 | 135 |
| 0.218 | 0.185 | 0.025 | 0.437 | 0.188 | 0.029 | 0.030 | 225 | 75 |
| 0.250 | 0.207 | 0.025 | 0.527 | 0.210 | 0.029 | 0.040 | 260 | 115 |
| 0.312 | 0.243 | 0.025 | 0.500 | 0.250 | 0.029 | 0.062 | 325 | 225 |
| 0.375 | 0.300 | 0.035 | 0.660 | 0.303 | 0.039 | 0.072 | 685 | 315 |
| 0.437 | 0.337 | 0.035 | 0.687 | 0.343 | 0.039 | 0.094 | 800 | 485 |
| 0.437 | 0.375 | 0.035 | 0.600 | 0.380 | 0.039 | 0.058 | 800 | 290 |
| 0.500 | 0.392 | 0.042 | 0.800 | 0.396 | 0.046 | 0.104 | 1100 | 600 |
| 0.625 | 0.480 | 0.042 | 0.940 | 0.485 | 0.046 | 0.140 | 1370 | 1040 |
| 0.744 | 0.616 | 0.050 | 1.000 | 0.625 | 0.056 | 0.118 | 1940 | 1050 |
| 0.750 | 0.616 | 0.050 | 1.000 | 0.625 | 0.056 | 0.124 | 1960 | 1100 |
| 0.750 | 0.574 | 0.050 | 1.120 | 0.580 | 0.056 | 0.170 | 1960 | 1500 |
| 0.875 | 0.668 | 0.050 | 1.300 | 0.675 | 0.056 | 0.200 | 2200 | 2050 |
| 0.985 | 0.822 | 0.050 | 1.500 | 0.835 | 0.056 | 0.148 | 2570 | 1710 |
| 1.000 | 0.822 | 0.050 | 1.500 | 0.835 | 0.056 | 0.164 | 2620 | 1900 |
| 1.188 | 1.066 | 0.062 | 1.626 | 1.079 | 0.068 | 0.108 | 3400 | 1500 |
| 1.375 | 1.213 | 0.062 | 1.875 | 1.230 | 0.068 | 0.144 | 4100 | 2300 |

[a] *Thrust Load Safety Factors:* Ring 3; groove, 2.

*Source:* Industrial Retaining Rings, 1000 Series. All dimensions are in inches. Depth of groove $d = (D - G)/2$. Standard material: carbon spring steel (SAE 1060 -1090). Thickness indicated is for unplated rings; for most plated rings with shaft sizes less than 0.625, the maximum ring thickness will not exceed the minimum groove width (W) minus 0.0002 inch; for larger rings, the thickness may increase by 0.002 inch.

*Groove Maximum Bottom Radii:* For shaft sizes 0.040 and 0.062, 0.003 inch; for sizes 0.094 through 0.250, 0.005 inch; for sizes 0.312 through 0.437, 0.010 inch; for sizes 0.500 through 1.375, 0.015 inch.

*Ring Free Diameter Tolerances:* For shaft sizes 0.040 through 0.250, +0.001, −0.003; for sizes 0.312 through 0.500, +0.002, −0.004; for sizes 0.625 through 1.000, +0.003, −0.005; for sizes 1.188 and 1.375, +0.006, −0.010.

*Ring Thickness Tolerances:* For shaft sizes 0.040 and 0.062, ±0.001; for sizes 0.062 through 1.000, ±0.002; for sizes 1.188 and 1.375, ±0.003.

*Groove Diameter Tolerances:* For shaft sizes 0.040 through 0.218, +0.002, −0.000; for sizes 0.250 through 1.000, +0.003, −0.000; for sizes 1.188 and 1.375, +0.005, −0.000.

*Grove Width Tolerances:* For shaft sizes 0.040 through 0.140, +0.002, −0.000; for sizes 0.140 through 1.000, +0.003, −0.000; for sizes 1.188 and 1.375, +0.004, −0.000.

**Table 15. Dimensions of Inch Series External Retaining Rings** *MS3215*

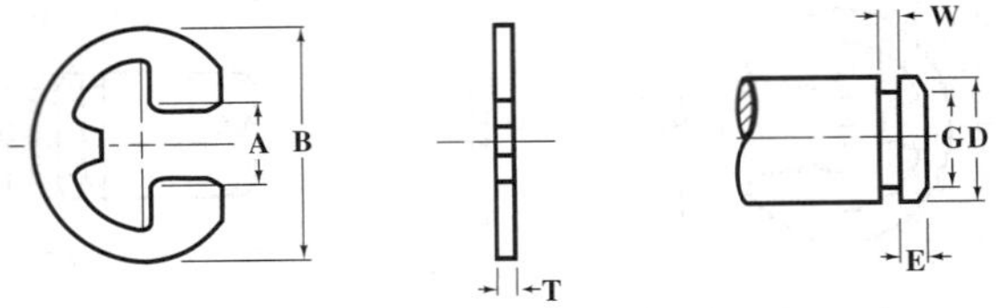


| Shaft Diameter D | Ring Free Dia. A | Ring Thickness T | Ring Diameter B | Groove Diameter G | Groove Width W | Groove Margin E | [a]Static Thrust Load (lb) Ring | [a]Static Thrust Load (lb) Groove |
|---|---|---|---|---|---|---|---|---|
| 0.094 | 0.072 | 0.015 | 0.206 | 0.074 | 0.018 | 0.020 | 55 | 13 |
| 0.125 | 0.093 | 0.015 | 0.270 | 0.095 | 0.018 | 0.030 | 75 | 25 |
| 0.156 | 0.113 | 0.025 | 0.335 | 0.116 | 0.029 | 0.040 | 150 | 40 |
| 0.188 | 0.143 | 0.025 | 0.375 | 0.147 | 0.029 | 0.040 | 180 | 50 |
| 0.219 | 0.182 | 0.025 | 0.446 | 0.188 | 0.029 | 0.031 | 215 | 50 |
| 0.250 | 0.204 | 0.025 | 0.516 | 0.210 | 0.029 | 0.040 | 250 | 75 |
| 0.312 | 0.242 | 0.025 | 0.588 | 0.250 | 0.029 | 0.062 | 300 | 135 |
| 0.312 | 0.242 | 0.035 | 0.588 | 0.250 | 0.039 | 0.062 | 420 | 135 |
| 0.375 | 0.292 | 0.035 | 0.660 | 0.303 | 0.039 | 0.072 | 520 | 190 |
| 0.438 | 0.332 | 0.035 | 0.746 | 0.343 | 0.039 | 0.096 | 600 | 285 |
| 0.500 | 0.385 | 0.042 | 0.810 | 0.396 | 0.046 | 0.104 | 820 | 360 |
| 0.562 | 0.430 | 0.042 | 0.870 | 0.437 | 0.046 | 0.124 | 930 | 480 |

[a] *Thrust Load Safety Factors:* Ring, 3; groove, 2.

*Source:* Industrial Retaining Rings, 1200 Series. All dimensions are in inches. Depth of groove $d = (D - G)/2$. Standard material: carbon spring steel (SAE 1060-1090). Thickness indicated is for unplated rings; for most plated rings the maximum thickness will not exceed the minimum groove width (W) minus 0.0002 inch.

*Groove Maximum Bottom Radii:* For shaft sizes 0.250 and smaller, 0.005 inch; for sizes 0.312 through 0.438, 0.010 inch; for sizes 0.500 and 0.562, 0.015 inch. *Ring Free Diameter Tolerances:* For shaft sizes 0.094 through 0.156, +0.001, −0.003; for sizes 0.188 through 0.312, ±0.003; for sizes 0.375 through 0.562, ±0.004. *Ring Thickness Tolerances:* For all shaft sizes, ±0.002. *Groove Diameter Tolerances:* For shaft sizes 0.094 through 0.188, +0.002, −0.000; for sizes 0.219 and 0.250, ±0.002; for sizes 0.312 through 0.562, ±0.003. *Groove Width Tolerances:* For shaft sizes 0.094 and 0.125, +0.002, −0.000; for sizes 0.156 through 0.562, +0.003, −0.000.

*Thrust Load Capacity:* The thrust load capacities shown in the tables of this section include safety factors. Usually, a safety factor of 2 is used for groove thrust load calculations when the load is applied through a retained part and groove with both having sharp corners and where the minimum side clearance exists between the retained part and the shaft or bore. Groove thrust load values in the tables of this section are based on these conditions. A safety factor of 3 is usual for calculations of thrust load capacity based on ring shear.

Ideally, the corner of a retained part in contact with a retaining ring should have square corners and contact the ring as closely as possible to the shaft or housing. The tabulated thrust capacities assume that minimum clearances exist between the retained part and shaft or housing, that the groove and retained part have square corners, and that contact between the retained part and the ring occurs close to the shaft or housing. If these conditions apply, the tabulated thrust loads apply. If the application does not meet the previous conditions but the side clearances, radii, and chamfers are less than the maximum total radius or chamfer of Fig. 1, then the thrust load capacity must be reduced by dividing the tabulated value by 2. The maximum total radius is given by $0.5(b - d)$ and the maximum total chamfer by $0.375(b - d)$, where $b$ is the radial wall thickness, and $d$ is the groove depth. The recommended maximum total radius or chamfer specifications are intended to be used as guidelines by the designer, and to ensure the ring application will withstand published and calculated values of static thrust loads.

In analyzing the retaining ring loading conditions, a static, uniformly applied load is usually assumed. Dynamic and eccentric loads, however, are frequently encountered. Eccentric loading occurs when the load is concentrated on a small portion of the ring, such as may be caused by incorrectly machined surfaces, cocking of the retained part, and axial misalignment of parts. Conditions leading to eccentric loading on the ring should be avoided. In addition to the factors that affect the static thrust capacity, applications in which shock or impact loading occurs must be evaluated very carefully and tested in service to assess the effect of the mass and velocity of the retained part striking the ring. Vibration caused by impact loading can also cause the ring to fail if the resonant frequency of the system (retaining ring application) coincides with the resonant frequency of the retaining ring.

**Table 16. Dimensions of Inch Series Self-Locking External Retaining Rings**

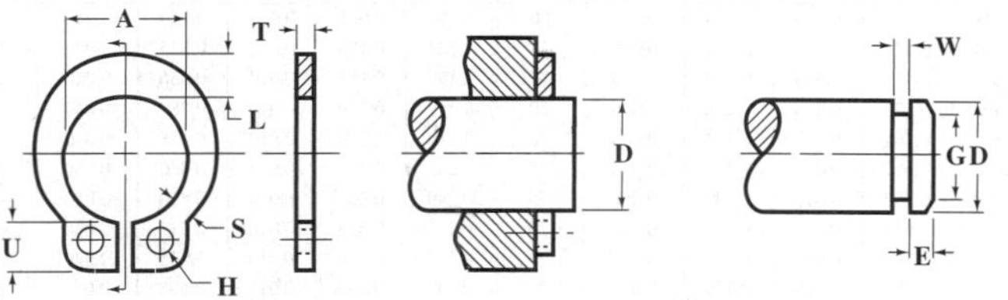


| Shaft Diameter | | Ring | | Optical Groove | | | [a]Static Thrust Load (lb) | |
|---|---|---|---|---|---|---|---|---|
| Min. D | Max. D | Free Dia. A | Thickness T | Diameter G | Width W | Margin E | Ring | Groove |
| 0.078 | 0.080 | 0.074 | 0.025 | The use of grooves with these shaft sizes is not suggested. | | | 10 | 0 |
| 0.092 | 0.096 | 0.089 | 0.025 | | | | 10 | 0 |
| 0.123 | 0.127 | 0.120 | 0.025 | | | | 20 | 0 |
| 0.134 | 0.138 | 0.130 | 0.025 | | | | 20 | 0 |
| 0.154 | 0.158 | 0.150 | 0.025 | | | | 22 | 0 |
| 0.185 | 0.189 | 0.181 | 0.035 | | | | 25 | 0 |
| 0.248 | 0.252 | 0.238 | 0.035 | 0.240 | 0.041 | 0.030 | 35 | 90 |
| 0.310 | 0.316 | 0.298 | 0.042 | 0.303 | 0.048 | 0.030 | 50 | 110 |
| 0.373 | 0.379 | 0.354 | 0.042 | 0.361 | 0.048 | 0.030 | 55 | 185 |
| 0.434 | 0.440 | 0.412 | 0.050 | 0.419 | 0.056 | 0.030 | 60 | 280 |
| 0.497 | 0.503 | 0.470 | 0.050 | 0.478 | 0.056 | 0.040 | 65 | 390 |
| 0.622 | 0.628 | 0.593 | 0.062 | 0.599 | 0.069 | 0.045 | 85 | 570 |
| 0.745 | 0.755 | 0.706 | 0.062 | 0.718 | 0.069 | 0.050 | 90 | 845 |

[a] *Thrust Load Safety Factors:* Ring, 1; groove, 2.

*Source:* Industrial Retaining Rings, 7100 Series. All dimensions are in inches. Depth of groove $d = (D - G)/2$. Standard material: carbon spring steel (SAE 1060-1090). Thickness indicated is for unplated rings; for plated, phosphate coated, and stainless steel rings, the maximum ring thickness may be exceeded by 0.002 inch.

*Ring Free Diameter Tolerances:* For shaft sizes 0.078 through 0.138, +0.002, −0.003; for sizes 0.154 through 0.252, +0.002, −0.004; for sizes 0.310 through 0.440, +0.003, −0.005; for sizes 0.497 through 0.755, +0.004, −0.006. *Ring Thickness Tolerances:* For shaft sizes 0.078 through 0.158, ±0.002; for sizes 0.185 through 0.503, ±0.003; for sizes 0.622 through 0.755, ±0.004. *Groove Diameter Tolerances:* For shaft sizes less than 0.248, grooves are not recommended; for other sizes, grooves are optional. For shaft sizes 0.248 through 0.316, +0.005, −0.0015; for sizes 0.373 through 0.628, +0.001, −0.002; for sizes 0.745 and 0.755, +0.002, −0.003. *Groove Width Tolerances:* For shaft sizes 0.248 through 0.379, +0.003, −0.000; for sizes 0.434 through 0.755, +0.004, −0.000.

**Table 17. Inch Series Internal and External Self-Locking Retaining Rings**

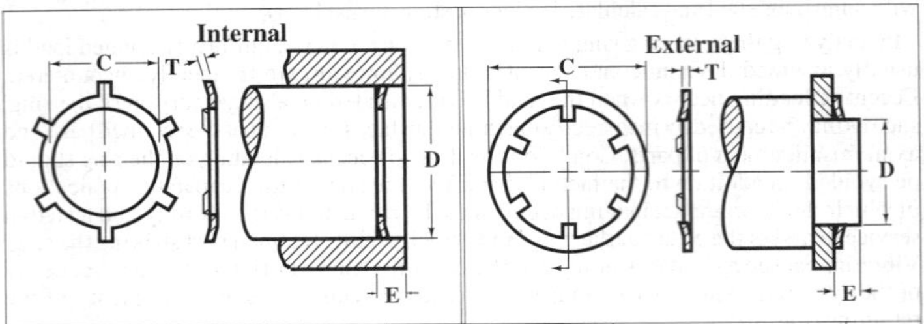


| Housing | | Ring Dimensions | | | Static Thrust Load | Shaft | | Ring Dimensions | | | Static Thrust Load |
|---|---|---|---|---|---|---|---|---|---|---|---|
| Min. D | Max. D | Thickness T | Dia. D | Margin E | | Min. D | Max. D | Thickness T | Dia. D | Margin E | |
| 0.311 | 0.313 | 0.010 | 0.136 | 0.040 | 80 | 0.093 | 0.095 | 0.010 | 0.250 | 0.040 | 15 |
| 0.374 | 0.376 | 0.010 | 0.175 | 0.040 | 75 | 0.124 | 0.126 | 0.010 | 0.325 | 0.040 | 20 |
| 0.437 | 0.439 | 0.010 | 0.237 | 0.040 | 70 | 0.155 | 0.157 | 0.010 | 0.356 | 0.040 | 25 |
| 0.498 | 0.502 | 0.010 | 0.258 | 0.040 | 60 | 0.187 | 0.189 | 0.010 | 0.387 | 0.040 | 35 |
| 0.560 | 0.564 | 0.010 | 0.312 | 0.040 | 50 | 0.218 | 0.220 | 0.010 | 0.418 | 0.040 | 35 |
| 0.623 | 0.627 | 0.010 | 0.390 | 0.040 | 45 | 0.239 | 0.241 | 0.015 | 0.460 | 0.060 | 35 |
| 0.748 | 0.752 | 0.015 | 0.500 | 0.060 | 75 | 0.249 | 0.251 | 0.010 | 0.450 | 0.040 | 40 |
| 0.873 | 0.877 | 0.015 | 0.625 | 0.060 | 70 | 0.311 | 0.313 | 0.010 | 0.512 | 0.040 | 40 |
| 0.936 | 0.940 | 0.015 | 0.687 | 0.060 | 70 | 0.374 | 0.376 | 0.010 | 0.575 | 0.040 | 40 |
| 0.998 | 1.002 | 0.015 | 0.750 | 0.060 | 70 | 0.437 | 0.440 | 0.015 | 0.638 | 0.060 | 50 |
| 1.248 | 1.252 | 0.015 | 0.938 | 0.060 | 60 | 0.498 | 0.502 | 0.015 | 0.750 | 0.060 | 50 |
| 1.436 | 1.440 | 0.015 | 1.117 | 0.060 | 60 | 0.560 | 0.564 | 0.015 | 0.812 | 0.060 | 50 |
| 1.498 | 1.502 | 0.015 | 1.188 | 0.060 | 60 | 0.623 | 0.627 | 0.015 | 0.875 | 0.060 | 50 |
| | | | | | | 0.748 | 0.752 | 0.015 | 1.000 | 0.060 | 50 |
| | | | | | | 0.873 | 0.877 | 0.015 | 1.125 | 0.060 | 55 |
| | | | | | | 0.998 | 1.002 | 0.015 | 1.250 | 0.060 | 60 |

*Source:* Industrial Retaining Rings, 6000 Series (internal) and 6100 Series (external). All dimensions are in inches, thrust loads are in pounds. Thickness indicated is for unplated rings. Standard material: carbon spring steel (SAE 1060-1090).

*Internal Rings:* Thrust loads are for rings made of standard material inserted into cold-rolled, low-carbon housing. *Ring Thickness Tolerances:* For housing sizes 0.311 through 0.627, ±0.001; for sizes 0.748 through 1.502, ±0.002. *Ring Diameter Tolerances:* For housing sizes 0.311 through 0.439, ±0.005; for sizes 0.498 through 1.502, ±0.010.

*External Rings:* Thrust loads are for rings made of standard material installed onto cold-rolled, low-carbon shafts. *Ring Thickness Tolerances:* For shaft sizes 0.093 through 0.220, ±0.001; for size 0.239, ±0.002; for sizes 0.249 through 0.376, ±0.001; for sizes 0.437 through 1.002, ±0.002. *Ring Diameter Tolerances:* For shaft sizes 0.093 through 0.502, ±0.005; for sizes 0.560 through 1.002, ±0.010.

*Centrifugal Capacity:* Proper functioning of a retaining ring depends on the ring remaining seated on the groove bottom. External rings "cling" to the groove bottom because the ring ID is slightly smaller than the diameter at the bottom of the groove. Ring speed should be kept below the allowable steady-state speed of the ring, or self-locking rings specially designed for high-speed applications should be used, otherwise an external ring can lose its grip on the groove. Applications of large retaining rings that tend to spin in their grooves when subjected to sudden acceleration or deceleration of the retained part can benefit from a ring with more "cling" (i.e., a smaller interior diameter) as long as the stress of installation is within permissible limits. Special rings are also available that lock into a hole in the bottom of the groove, thereby preventing rotation. The following equation can be used to determine the allowable steady-state speed $N$ of an external spiral retaining ring:

$$N = \sqrt{\frac{0.466 C_1 E^3 \times 10^{12}}{R_n^3 (1 + C_1)(R_o^3 - R_i^3)}} \tag{1}$$

where the speed $N$ is in revolutions per minute, $C_1$ is the minimum ring cling to groove bottom, $E$ is the ring radial wall, $R_n$ is the free neutral ring radius, $R_o$ is the free outside ring radius, and $R_i$ is the free inside ring radius, all in inches. For external spiral rings, the minimum ring cling is given by: $C_1 = (C - G)/G$, where $C$ is the mean groove diameter in inches, and $G$ is the maximum ring free ID in inches.

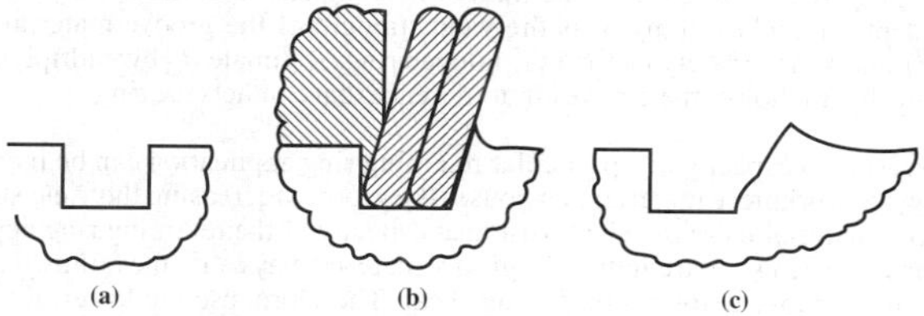


Fig. 2. Localized Groove Yielding under Load. (a) Groove Profile before Loading; (b) Localized Yielding of Retained Part and Groove under Load; (c) Groove Profile after Loading beyond Thrust Capacity (*Courtesy Spirolox Retaining Rings*)

*Rotation Between Parts:* The use of spiral-wound rings to retain a rotating part should be limited to applications with rotation in only one direction. The ring should be matched so that the rotation tends to wind the spring into the groove. External rings should be wound in the direction of rotation of the retained part but internal rings should be wound against the direction of rotation of the rotating part. Failure to observe these precautions will cause the ring to wind out of the groove. Spiral-wound rings can be obtained with either right-hand (normal rotation) or left-hand (reverse rotation) wound configurations. Stamped retaining rings do not have these limitations, and may be used for applications that require rotation of the retained part, regardless of the direction of rotation.

**Retaining Ring Failure and Prevention.**—Failure of a retaining ring application can result from failure of the ring itself, failure of the groove, or both. If a ring fails, the cause is likely to be from shearing of the ring. Shear failure occurs when a ring is installed in a groove and loaded by a retained part with both the groove and the retained part having a compressive yield strength greater than 45,000 psi (310 MPa); or when the load is applied through a retained part and groove, both having sharp corners and line-to-line contact; or when the ring is too thin in section compared with its diameter. To examine the possibility of ring shear, the allowable thrust $P_s$, based on the shear strength of the ring material, is given by

$$P_s = \frac{\pi D t S_s}{K} \tag{2}$$

where $P_s$ is in $lb_f$ (N), $D$ is the shaft or housing diameter in inches (mm), $t$ is the ring thickness in inches (mm), $S_s$ is the shear strength of the ring material in $lb/in^2$ ($N/mm^2$), and $K$ is the factor of safety.

*Groove Failure:* The most common type of groove failure is yielding of the groove material that occurs when the thrust load, applied through the retaining ring against the corner of the groove, exceeds the compressive yield strength of the groove. This yielding of the groove results from a low compressive yield strength of the groove material, and allows the ring to tilt and come out of the groove, as illustrated in Fig. 2(b).

When dishing of a ring occurs as a result of yielding in the groove material, a bending moment across the cross section of the ring generates a tensile stress that is highest at the interior diameter of the ring. If the maximum stress is greater than the yield strength of the ring material, the ring ID will grow and the ring will become permanently dished in shape. To determine the thrust load capacity of a ring based on groove deformation, the allowable angle of ring deflection must be calculated, then the thrust load based on groove yield can be determined. However, for spiral-wound rings, the thrust load $P_G$ that initiates the onset of groove deformation can be estimated from the following:

$$P_G = \frac{\pi D d S_y}{K} \qquad (3)$$

where $P_G$ is given in $lb_f$ (N), $D$ is the shaft or housing diameter in inches (mm), $d$ is the groove depth in inches (mm), $S_y$ is the yield strength of the groove material in lb/in$^2$ (N/mm$^2$), and $K$ is the safety factor. For stamped rings, estimate $P_G$ by multiplying Equation (3) by the fraction of the groove circumference that contacts the ring.

The thrust load capacity of a particular retaining ring application can be increased by changing the workpiece material that houses the groove. Increasing the yield strength of the groove material increases the thrust load capacity of the retaining ring application. However, increasing the strength of the groove material may cause the failure mechanism to shift from groove deformation to ring shear. Therefore, use the lower of the values obtained from Equations (2) and (3) for the allowable thrust load.

*Groove Design and Machining:* In most applications, grooves are located near the end of a shaft or housing bore to facilitate installation and removal of the rings. The groove is normally located a distance at least two to three times the groove depth from the end of the shaft or bore. If the groove is too close to the end of the shaft or bore, the groove may shear or yield. The following equation can be used to determine the minimum safe distance $Y$ of a groove from the end of a shaft or housing:

$$Y = \frac{K P_t}{\pi D S_c} \qquad (4)$$

where $K$ is the factor of safety, $P_t$ is the thrust load on the groove in pounds (N), $S_c$ is the shear strength of the groove material in lb/in$^2$ (N/mm$^2$), and $D$ is the shaft or housing diameter in inches (mm).

A properly designed and machined groove is just as important in a retaining ring application as the ring itself. The walls of grooves should be perpendicular to the shaft or bore diameter; the grooves should have square corners on the top edges, and radii at the bottom, within the tolerances specified by the manufacturers, as shown in Fig. 1 (page 1900). Test data indicate that the ultimate thrust capacity for both static and dynamic loading conditions is greatly affected if these groove requirements are not met. For spiral-wound rings, the maximum bottom groove radius is 0.005 inch (0.127 mm) for rings up to 1.000 inch (25.4 mm) free diameter, and 0.010 inch (0.254 mm) for larger rings, internal or external. For stamped rings, the maximum bottom groove radius varies with ring size and style.

**Table 18. Retaining Ring Standards**

| Military | |
|---|---|
| MIL-R-21248B | MS-16633 Open-type external uniform cross-section |
| | MS-16634 Open-type external uniform cross-section cylindrically |
| | MS-3215 Open-type external tapered cross-section |
| | MS-16632 Crescent-type external |
| | MS-16625 Internal |
| | MS-16629 Internal cylindrically bowed |
| | MS-16624 Closed-type external tapered cross-section |

**Table 18.** *(Continued)* **Retaining Ring Standards**

| | |
|---|---|
| Military | |
| MIL-R-21248B | MS-16628 Closed-type external tapered cross-section cylindrically bowed |
| | MS-16627 Internal inverted |
| | MS-16626 Closed-type external tapered cross-section |
| | MS-90707 Self-locking external tapered cross-section |
| | MS-3217 External heavy-duty tapered cross-section |
| MIL-R-27426 | Uniform cross-section spiral retaining rings, Type 1-External, Type 2-Internal |
| Aerospace Standard | |
| AS 3215 | Ring, Retaining—Spiral, Internal, Heavy Duty, Stainless Steel |
| AS 3216 | Ring, Retaining—Spiral, External, Heavy Duty, Stainless Steel |
| AS 3217 | Ring, Retaining—Spiral, Internal, Light Duty, Stainless Steel |
| AS 3218 | Ring, Retaining—Spiral, External, Light Duty, Stainless Steel |
| AS 3219 | Ring, Wound, Dimensional and Acceptance Standard for Spiral Wound Retaining Rings |
| ANSI/ASME | |
| B27.6-1972 (R2017) | General Purpose Uniform Cross-Section Spiral Retaining Rings |
| B27.7-1977 (R2017) | General Purpose Tapered and Reduced Cross-Section Retaining Rings (Metric) |
| B27.8M-1977 (Withdrawn) | General Purpose Metric Tapered and Reduced Cross-Section Retaining Rings |
| | Type 3DM1—Heavy Duty External Rings |
| | Type 3EM1—Reinforced "E" Rings |
| | Type 3FM1—"C" Type Rings |
| ANSI/SAE | |
| MA4016 | Ring, Retaining—External Spiral Wound, Heavy and Medium Duty, Crescent, Metric |
| MA4017 | Ring, Retaining—External Spiral Wound, Heavy and Medium Duty, Crescent, Metric |
| MA4020 | Ring, Retaining—External Tapered, Type 1, Class 2, AMS 5520, Metric |
| MA4021 | Ring, Retaining—Internal Tapered, Type 1, Class 1, AMS 5520, Metric |
| MA4029 | Ring, Retaining—Internal, Beveled, Tapered, Type 2, Class 1, AMS 5520, Metric |
| MA4030 | Ring, Retaining—External, Reinforced E-Ring, Type 1, Class 3, AMS 5520, Metric |
| MA4035 | Rings, Retaining—Spiral Wound, Uniform Section, Corrosion Resistant, Procurement Specification for, Metric |
| MA4036 | Ring, Retaining—Tapered Width, Uniform Thickness, Corrosion Resistant, Procurement Specification for, Metric |
| DIN | |
| DIN 471, 472, 6799, 984, 5417, 7993 | Standards for normal and heavy type, internal and external retaining rings and retaining washers |
| LN 471, 472, 6799 | Aerospace standards for internal and external retaining rings |

## WING NUTS, WING SCREWS, AND THUMB SCREWS

**Wing Nuts.—**A wing nut is a nut having wings designed for manual turning without driver or wrench. As covered by ANSI/ASME B18.6.9-2010 (R2017), wing nuts are classified first, by type on the basis of the method of manufacture; and second, by style on the basis of design characteristics. They consist of:

*Type A:* Type A wing nuts are cold forged or cold formed solid nuts having wings of moderate height. In some sizes they are produced in regular, light, and heavy series to best suit the requirements of specific applications. Dimensions are given in Table 1.

**Table 1. American National Standard Type A Wing Nuts**
*ANSI/ASME B18.6.9-2010 (R2017)*

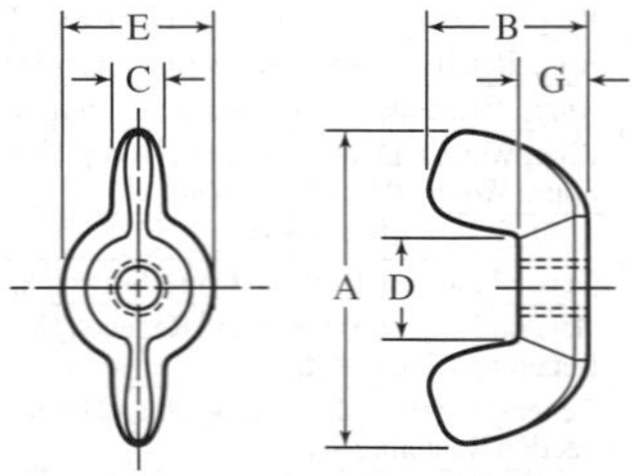


| Nominal Size or Basic Major Diameter of Thread[a] | Thds. per Inch | Series[b] | Nut Blank Size (Ref) | *A* Wing Spread | | *B* Wing Height | | *C* Wing Thickness | | *D* Between Wings | | *E* Boss Dia. | | *G* Boss Height | |
|---|---|---|---|---|---|---|---|---|---|---|---|---|---|---|---|
| | | | | Max. | Min. | Max. | Min. | Max. | Min. | Max. | Min. | Max. | Min. | Max. | Min. |
| 3 (0.0990) | 48, 56 | Hvy. | AA | 0.72 | 0.59 | 0.41 | 0.28 | 0.11 | 0.07 | 0.21 | 0.17 | 0.33 | 0.29 | 0.14 | 0.10 |
| 4 (0.1120) | 40, 48 | Hvy. | AA | 0.72 | 0.59 | 0.41 | 0.28 | 0.11 | 0.07 | 0.21 | 0.17 | 0.33 | 0.29 | 0.14 | 0.10 |
| 5 (0.1250) | 40, 44 | **Lgt.** | **AA** | **0.72** | **0.59** | **0.41** | **0.28** | **0.11** | **0.07** | **0.21** | **0.17** | **0.33** | **0.29** | **0.14** | **0.10** |
| | | Hvy. | A | 0.91 | 0.78 | 0.47 | 0.34 | 0.14 | 0.10 | 0.27 | 0.22 | 0.43 | 0.39 | 0.18 | 0.14 |
| 6 (0.1380) | 32, 40 | Lgt. | AA | 0.72 | 0.59 | 0.41 | 0.28 | 0.11 | 0.07 | 0.21 | 0.17 | 0.33 | 0.29 | 0.14 | 0.10 |
| | | **Hvy.** | **A** | **0.91** | **0.78** | **0.47** | **0.34** | **0.14** | **0.10** | **0.27** | **0.22** | **0.43** | **0.39** | **0.18** | **0.14** |
| 8 (0.1640) | 32, 36 | **Lgt.** | **A** | **0.91** | **0.78** | **0.47** | **0.34** | **0.14** | **0.10** | **0.27** | **0.22** | **0.43** | **0.39** | **0.18** | **0.14** |
| | | Hvy. | B | 1.10 | 0.97 | 0.57 | 0.43 | 0.18 | 0.14 | 0.33 | 0.26 | 0.50 | 0.45 | 0.22 | 0.17 |
| 10 (0.1900) | 24, 32 | **Lgt.** | **A** | **0.91** | **0.78** | **0.47** | **0.34** | **0.14** | **0.10** | **0.27** | **0.22** | **0.43** | **0.39** | **0.18** | **0.14** |
| | | Hvy. | B | 1.10 | 0.97 | 0.57 | 0.43 | 0.18 | 0.14 | 0.33 | 0.26 | 0.50 | 0.45 | 0.22 | 0.17 |
| 12 (0.2160) | 24, 28 | **Lgt.** | **B** | **1.10** | **0.97** | **0.57** | **0.43** | **0.18** | **0.14** | **0.33** | **0.26** | **0.50** | **0.45** | **0.22** | **0.17** |
| | | Hvy. | C | 1.25 | 1.12 | 0.66 | 0.53 | 0.21 | 0.17 | 0.39 | 0.32 | 0.58 | 0.51 | 0.25 | 0.20 |
| ¼ (0.2500) | 20, 28 | **Lgt.** | **B** | **1.10** | **0.97** | **0.57** | **0.43** | **0.18** | **0.14** | **0.33** | **0.26** | **0.50** | **0.45** | **0.22** | **0.17** |
| | | Reg. | C | 1.25 | 1.12 | 0.66 | 0.53 | 0.21 | 0.17 | 0.39 | 0.32 | 0.58 | 0.51 | 0.25 | 0.20 |
| | | Hvy. | D | 1.44 | 1.31 | 0.79 | 0.65 | 0.24 | 0.20 | 0.48 | 0.42 | 0.70 | 0.64 | 0.30 | 0.26 |
| 5⁄16 (0.3125) | 18, 24 | **Lgt.** | **C** | **1.25** | **1.12** | **0.66** | **0.53** | **0.21** | **0.17** | **0.39** | **0.32** | **0.58** | **0.51** | **0.25** | **0.20** |
| | | Reg. | D | 1.44 | 1.31 | 0.79 | 0.65 | 0.24 | 0.20 | 0.48 | 0.42 | 0.70 | 0.64 | 0.30 | 0.26 |
| | | Hvy. | E | 1.94 | 1.81 | 1.00 | 0.87 | 0.33 | 0.26 | 0.65 | 0.54 | 0.93 | 0.86 | 0.39 | 0.35 |
| ⅜ (0.3750) | 16, 24 | **Lgt.** | **D** | **1.44** | **1.31** | **0.79** | **0.65** | **0.24** | **0.20** | **0.48** | **0.42** | **0.70** | **0.64** | **0.30** | **0.26** |
| | | Reg. | E | 1.94 | 1.81 | 1.00 | 0.87 | 0.33 | 0.26 | 0.65 | 0.54 | 0.93 | 0.86 | 0.39 | 0.35 |
| 7⁄16 (0.4375) | 14, 20 | **Lgt.** | **E** | **1.94** | **1.81** | **1.00** | **0.87** | **0.33** | **0.26** | **0.65** | **0.54** | **0.93** | **0.86** | **0.39** | **0.35** |
| | | Hvy. | F | 2.76 | 2.62 | 1.44 | 1.31 | 0.40 | 0.34 | 0.90 | 0.80 | 1.19 | 1.13 | 0.55 | 0.51 |
| ½ (0.5000) | 13, 20 | **Lgt.** | **E** | **1.94** | **1.81** | **1.00** | **0.87** | **0.33** | **0.26** | **0.65** | **0.54** | **0.93** | **0.86** | **0.39** | **0.35** |
| | | Hvy. | F | 2.76 | 2.62 | 1.44 | 1.31 | 0.40 | 0.34 | 0.90 | 0.80 | 1.19 | 1.13 | 0.55 | 0.51 |
| 9⁄16 (0.5625) | 12, 18 | Hvy. | F | 2.76 | 2.62 | 1.44 | 1.31 | 0.40 | 0.34 | 0.90 | 0.80 | 1.19 | 1.13 | 0.55 | 0.51 |
| ⅝ (0.6250) | 11, 18 | Hvy. | F | 2.76 | 2.62 | 1.44 | 1.31 | 0.40 | 0.34 | 0.90 | 0.80 | 1.19 | 1.13 | 0.55 | 0.51 |
| ¾ (0.7500) | 10, 16 | Hvy. | F | 2.76 | 2.62 | 1.44 | 1.31 | 0.40 | 0.34 | 0.90 | 0.80 | 1.19 | 1.13 | 0.55 | 0.51 |

[a] Where specifying nominal size in decimals, zeros in the fourth decimal place are omitted.
[b] Lgt. = Light; Hvy. = Heavy; Reg. = Regular. Sizes shown in **bold face** are preferred.

All dimensions in inches.

*Type B:* Type B wing nuts are hot forged solid nuts available in two wing styles: Style 1, having wings of moderate height; and Style 2, having high wings. Dimensions are given in Table 2.

**Table 2. American National Standard Type B Wing Nuts**
*ANSI/ASME B18.6.9-2010 (R2017)*

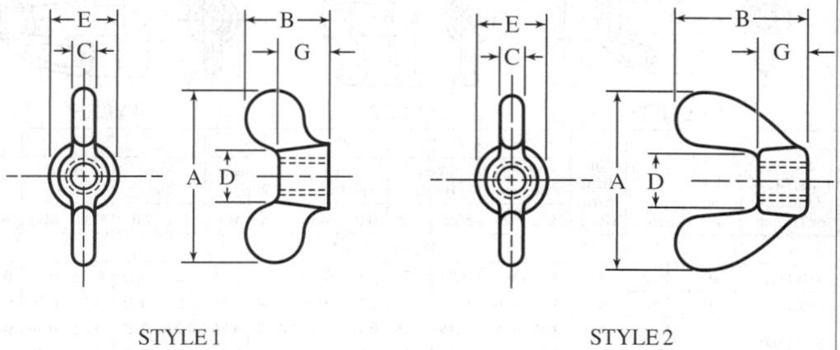


| Nominal Size or Basic Major Diameter of Thread[a] | Thds. per Inch | *A* Wing Spread | | *B* Wing Height | | *C* Wing Thickness | | *D* Between Wings | | *E* Boss Dia. | | *G* Boss Height | |
|---|---|---|---|---|---|---|---|---|---|---|---|---|---|
| | | Max. | Min. | Max. | Min. | Max. | Min. | Max. | Min. | Max. | Min. | Max. | Min. |
| **Type B, Style 1** | | | | | | | | | | | | | |
| 5 (0.1250) | 40 | 0.78 | 0.72 | 0.36 | 0.30 | 0.13 | 0.10 | 0.28 | 0.22 | 0.31 | 0.28 | 0.22 | 0.16 |
| 10 (0.1900) | 24 | 0.97 | 0.91 | 0.45 | 0.39 | 0.15 | 0.12 | 0.34 | 0.28 | 0.39 | 0.36 | 0.28 | 0.22 |
| ¼ (0.2500) | 20 | 1.16 | 1.09 | 0.56 | 0.50 | 0.17 | 0.14 | 0.41 | 0.34 | 0.47 | 0.44 | 0.34 | 0.28 |
| 5⁄16 (0.3125) | 18 | 1.44 | 1.38 | 0.67 | 0.61 | 0.18 | 0.15 | 0.50 | 0.44 | 0.55 | 0.52 | 0.41 | 0.34 |
| ⅜ (0.3750) | 16 | 1.72 | 1.66 | 0.80 | 0.73 | 0.20 | 0.17 | 0.59 | 0.53 | 0.63 | 0.60 | 0.47 | 0.41 |
| 7⁄16 (0.4375) | 14 | 2.00 | 1.94 | 0.91 | 0.84 | 0.21 | 0.18 | 0.69 | 0.62 | 0.71 | 0.68 | 0.53 | 0.47 |
| ½ (0.5000) | 13 | 2.31 | 2.22 | 1.06 | 0.94 | 0.23 | 0.20 | 0.78 | 0.69 | 0.79 | 0.76 | 0.62 | 0.50 |
| 9⁄16 (0.5625) | 12 | 2.59 | 2.47 | 1.17 | 1.05 | 0.25 | 0.21 | 0.88 | 0.78 | 0.88 | 0.84 | 0.69 | 0.56 |
| ⅝ (0.6250) | 11 | 2.84 | 2.72 | 1.31 | 1.19 | 0.27 | 0.23 | 0.94 | 0.84 | 0.96 | 0.92 | 0.75 | 0.62 |
| ¾ (0.7500) | 10 | 3.31 | 3.19 | 1.52 | 1.39 | 0.29 | 0.25 | 1.10 | 1.00 | 1.12 | 1.08 | 0.88 | 0.75 |
| **Type B, Style 2** | | | | | | | | | | | | | |
| 5 (0.1250) | 40 | 0.81 | 0.75 | 0.62 | 0.56 | 0.12 | 0.09 | 0.28 | 0.22 | 0.31 | 0.28 | 0.22 | 0.16 |
| 10 (0.1900) | 24 | 1.01 | 0.95 | 0.78 | 0.72 | 0.14 | 0.11 | 0.35 | 0.29 | 0.39 | 0.36 | 0.28 | 0.22 |
| ¼ (0.2500) | 20 | 1.22 | 1.16 | 0.94 | 0.88 | 0.16 | 0.13 | 0.41 | 0.35 | 0.47 | 0.44 | 0.34 | 0.28 |
| 5⁄16 (0.3125) | 18 | 1.43 | 1.37 | 1.09 | 1.03 | 0.17 | 0.14 | 0.48 | 0.42 | 0.55 | 0.52 | 0.41 | 0.34 |
| ⅜ (0.3750) | 16 | 1.63 | 1.57 | 1.25 | 1.19 | 0.18 | 0.15 | 0.55 | 0.49 | 0.63 | 0.60 | 0.47 | 0.41 |
| 7⁄16 (0.4375) | 14 | 1.90 | 1.84 | 1.42. | 1.36 | 0.19 | 0.16 | 0.62 | 0.56 | 0.71 | 0.68 | 0.53 | 0.47 |
| ½ (0.5000) | 13 | 2.13 | 2.04 | 1.58 | 1.45 | 0.20 | 0.17 | 0.69 | 0.60 | 0.79 | 0.76 | 0.62 | 0.50 |
| 9⁄16 (0.5625) | 12 | 2.40 | 2.28 | 1.75 | 1.62 | 0.22 | 0.18 | 0.76 | 0.67 | 0.88 | 0.84 | 0.69 | 0.56 |
| ⅝ (0.6250) | 11 | 2.60 | 2.48 | 1.91 | 1.78 | 0.23 | 0.19 | 0.83 | 0.74 | 0.96 | 0.92 | 0.75 | 0.62 |
| ¾ (0.7500) | 10 | 3.02 | 2.90 | 2.22 | 2.09 | 0.24 | 0.20 | 0.97 | 0.88 | 1.12 | 1.08 | 0.88 | 0.75 |

[a] Where specifying nominal size in decimals, zeros in the fourth decimal place are omitted.

All dimensions in inches.

**Table 3. American National Standard Type C Wing Nuts** *ANSI/ASME B18.6.9-2010 (R2017)*

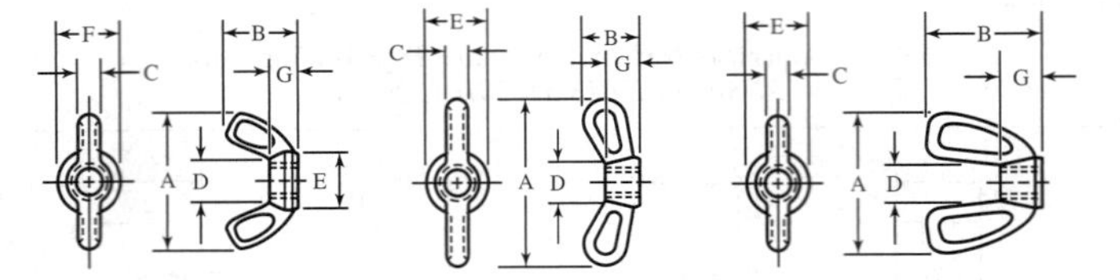


STYLE 1 STYLE 2 STYLE 3

| Nominal Size or Basic Major Diameter of Thread[a] | Thds. per Inch | Series | Nut Blank Size (Ref) | *A* Wing Spread | | *B* Wing Height | | *C* Wing Thickness | | *D* Between Wings | | *E* Boss Dia. | | *F* Boss Dia. | | *G* Boss Height | |
|---|---|---|---|---|---|---|---|---|---|---|---|---|---|---|---|---|---|
| | | | | Max. | Min. | Max. | Min. | Max. | Min. | Max. | Min. | Max. | Min. | Max. | Min. | Max. | Min. |
| **Type C, Style 1** | | | | | | | | | | | | | | | | | |
| 4 (0.1120) | 40 | Reg. | AA | 0.66 | 0.64 | 0.36 | 0.35 | 0.11 | 0.09 | 0.18 | 0.16 | 0.27 | 0.25 | 0.32 | 0.30 | 0.16 | 0.14 |
| 5 (0.1250) | 40 | Reg. | AA | 0.66 | 0.64 | 0.36 | 0.35 | 0.11 | 0.09 | 0.18 | 0.16 | 0.27 | 0.25 | 0.32 | 0.30 | 0.16 | 0.14 |
| 6 (0.1380) | 32 | **Reg.** | **AA** | **0.66** | **0.64** | **0.36** | **0.35** | **0.11** | **0.09** | **0.18** | **0.16** | **0.27** | **0.25** | **0.32** | **0.30** | **0.16** | **0.14** |
| | | Hvy. | A | 0.85 | 0.83 | 0.43 | 0.42 | 0.14 | 0.12 | 0.29 | 0.27 | 0.38 | 0.36 | 0.41 | 0.40 | 0.20 | 0.18 |
| 8 (0.1640) | 32 | Reg. | A | 0.85 | 0.83 | 0.43 | 0.42 | 0.14 | 0.12 | 0.29 | 0.27 | 0.38 | 0.36 | 0.41 | 0.40 | 0.20 | 0.18 |
| 10 (0.1900) | 24, 32 | Reg. | A | 0.85 | 0.83 | 0.43 | 0.42 | 0.14 | 0.12 | 0.29 | 0.27 | 0.38 | 0.36 | 0.41 | 0.40 | 0.20 | 0.18 |
| 12 (0.2160) | 24 | **Reg.** | A | **0.85** | **0.83** | **0.43** | **0.42** | **0.14** | **0.12** | **0.29** | **0.27** | **0.38** | **0.36** | **0.41** | **0.40** | **0.20** | **0.18** |
| | | Hvy. | B | 1.08 | 1.05 | 0.57 | 0.53 | 0.16 | 0.14 | 0.32 | 0.30 | 0.44 | 0.42 | 0.48 | 0.46 | 0.23 | 0.21 |
| ¼ (0.2500) | 20, 28 | Reg. | B | 1.08 | 1.05 | 0.57 | 0.53 | 0.16 | 0.14 | 0.32 | 0.30 | 0.44 | 0.42 | 0.48 | 0.46 | 0.23 | 0.21 |
| 5⁄16 (0.3125) | 18, 24 | Reg. | C | 1.23 | 1.20 | 0.64 | 0.62 | 0.20 | 0.18 | 0.39 | 0.35 | 0.50 | 0.49 | 0.57 | 0.55 | 0.26 | 0.24 |
| ⅜ (0.3750) | 16, 24 | Reg. | D | 1.45 | 1.42 | 0.74 | 0.72 | 0.23 | 0.21 | 0.46 | 0.42 | 0.62 | 0.60 | 0.69 | 0.67 | 0.29 | 0.27 |
| 7⁄16 (0.4375) | 14, 20 | **Reg.** | E | **1.89** | **1.86** | **0.91** | **0.90** | **0.29** | **0.28** | **0.67** | **0.65** | **0.75** | **0.73** | **0.83** | **0.82** | **0.38** | **0.37** |
| | | Hvy. | EH | 1.89 | 1.86 | 0.93 | 0.91 | 0.34 | 0.33 | 0.63 | 0.62 | 0.81 | 0.79 | 0.89 | 0.87 | 0.42 | 0.40 |
| ½ (0.5000) | 13, 20 | **Reg.** | E | **1.89** | **1.86** | **0.91** | **0.90** | **0.29** | **0.28** | **0.67** | **0.65** | **0.75** | **0.73** | **0.83** | **0.82** | **0.38** | **0.37** |
| | | Hvy. | EH | 1.89 | 1.86 | 0.93 | 0.91 | 0.34 | 0.33 | 0.63 | 0.62 | 0.81 | 0.79 | 0.89 | 0.87 | 0.42 | 0.40 |
| **Type C, Style 2** | | | | | | | | | | | | | | | | | |
| 5 (0.1250) | 40 | … | … | 0.82 | 0.80 | 0.25 | 0.23 | 0.09 | 0.08 | 0.21 | 0.19 | 0.26 | 0.24 | … | … | 0.17 | 0.15 |
| 6 (0.1380) | 32 | … | … | 0.82 | 0.80 | 0.25 | 0.23 | 0.09 | 0.08 | 0.21 | 0.19 | 0.26 | 0.24 | … | … | 0.17 | 0.15 |
| 8 (0.1640) | 32 | … | … | 1.01 | 0.99 | 0.28 | 0.27 | 0.11 | 0.09 | 0.29 | 0.28 | 0.36 | 0.34 | … | … | 0.19 | 0.18 |
| 10 (0.1900) | 24, 32 | … | … | 1.01 | 0.99 | 0.28 | 0.27 | 0.11 | 0.09 | 0.29 | 0.28 | 0.36 | 0.34 | … | … | 0.19 | 0.18 |
| 12 (0.2160) | 24 | … | … | 1.20 | 1.18 | 0.32 | 0.31 | 0.12 | 0.11 | 0.38 | 0.37 | 0.44 | 0.43 | … | … | 0.22 | 0.20 |
| ¼ (0.2500) | 20 | … | … | 1.20 | 1.18 | 0.32 | 0.31 | 0.12 | 0.11 | 0.38 | 0.37 | 0.44 | 0.43 | … | … | 0.22 | 0.20 |
| 5⁄16 (0.3125) | 18 | … | … | 1.51 | 1.49 | 0.36 | 0.35 | 0.14 | 0.12 | 0.44 | 0.43 | 0.51 | 0.49 | … | … | 0.24 | 0.23 |
| ⅜ (0.3750) | 16 | … | … | 1.89 | 1.86 | 0.58 | 0.55 | 0.20 | 0.17 | 0.44 | 0.43 | 0.63 | 0.62 | … | … | 0.37 | 0.35 |
| **Type C, Style 3** | | | | | | | | | | | | | | | | | |
| 5 (0.1250) | 40 | … | … | 0.92 | 0.89 | 0.70 | 0.67 | 0.16 | 0.15 | 0.26 | 0.24 | 0.38 | 0.36 | … | … | 0.25 | 0.24 |
| 6 (0.1380) | 32 | … | … | 0.92 | 0.89 | 0.70 | 0.67 | 0.16 | 0.15 | 0.26 | 0.24 | 0.38 | 0.36 | … | … | 0.25 | 0.24 |
| 8 (0.1640) | 32 | … | … | 0.92 | 0.89 | 0.70 | 0.67 | 0.16 | 0.15 | 0.26 | 0.24 | 0.38 | 0.36 | … | … | 0.25 | 0.24 |
| 10 (0.1900) | 24, 32 | … | … | 1.14 | 1.12 | 0.85 | 0.83 | 0.19 | 0.17 | 0.32 | 0.30 | 0.44 | 0.42 | … | … | 0.29 | 0.27 |
| 12 (0.2160) | 24 | … | … | 1.14 | 1.12 | 0.85 | 0.83 | 0.19 | 0.17 | 0.32 | 0.30 | 0.44 | 0.42 | … | … | 0.29 | 0.27 |
| ¼ (0.2500) | 20 | … | … | 1.14 | 1.12 | 0.85 | 0.83 | 0.19 | 0.17 | 0.32 | 0.30 | 0.44 | 0.42 | … | … | 0.29 | 0.27 |
| 5⁄16 (0.3125) | 18 | … | … | 1.29 | 1.27 | 1.04 | 1.02 | 0.23 | 0.22 | 0.39 | 0.36 | 0.50 | 0.49 | … | … | 0.35 | 0.34 |
| ⅜ (0.3750) | 16 | … | … | 1.51 | 1.49 | 1.20 | 1.18 | 0.27 | 0.25 | 0.45 | 0.42 | 0.62 | 0.60 | … | … | 0.43 | 0.42 |

[a] Where specifying nominal size in decimals, zeros in the fourth decimal place are omitted.

All dimensions in inches. Sizes shown in **bold face** are preferred.

*Type C:* Type C wing nuts are die-cast solid nuts and are available in three wing styles: Style 1, having wings of moderate height; Style 2, having low wings; and Style 3, having high wings. In some sizes, the Style 1 nuts are produced in regular, light, and heavy series to best suit the requirements of specific applications. Dimensions are given in Table 3.

**Table 4. American National Standard Type D Wing Nuts**
*ANSI/ASME B18.6.9-2010 (R2017)*

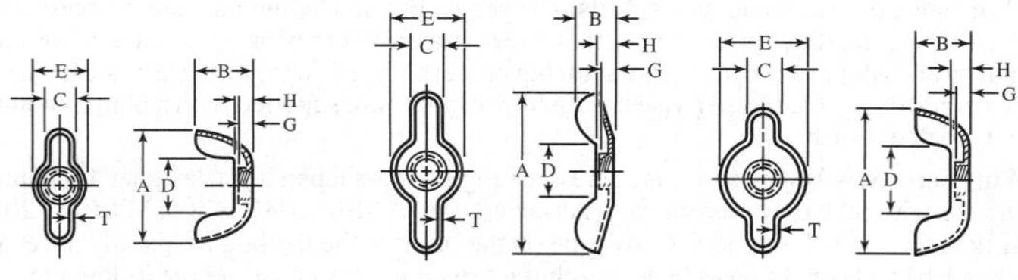


STYLE 1 STYLE 2 (LOW WING) STYLE 3 (LARGE BASE)

| Nominal Size or Basic Major Diameter of Thread[a] | Thds. per Inch | Series[b] | *A* Wing Spread Max. | *A* Wing Spread Min. | *B* Wing Height Max. | *B* Wing Height Min. | *C* Wing Thickness Max. | *C* Wing Thickness Min. | *D* Between Wings Min. | *E* Boss Dia. Max. | *E* Boss Dia. Min. | *G* Boss Height Min. | *H* Wall Height Min. | *T* Stock Thickness Max. | *T* Stock Thickness Min. |
|---|---|---|---|---|---|---|---|---|---|---|---|---|---|---|---|
| **Type D, Style 1** | | | | | | | | | | | | | | | |
| 8 (0.1640) | 32, 36 | … | 0.78 | 0.72 | 0.40 | 0.34 | 0.18 | 0.14 | 0.25 | 0.41 | 0.35 | 0.08 | 0.12 | 0.04 | 0.03 |
| 10 (0.1900) | 24, 32 | … | 0.91 | 0.85 | 0.47 | 0.41 | 0.21 | 0.17 | 0.34 | 0.53 | 0.47 | 0.10 | 0.12 | 0.04 | 0.03 |
| 12 (0.2160) | 24, 28 | … | 1.09 | 1.03 | 0.47 | 0.41 | 0.21 | 0.17 | 0.34 | 0.53 | 0.47 | 0.10 | 0.12 | 0.05 | 0.04 |
| 1⁄4 (0.2500) | 20, 28 | … | 1.11 | 1.05 | 0.50 | 0.44 | 0.25 | 0.21 | 0.34 | 0.62 | 0.56 | 0.11 | 0.12 | 0.05 | 0.04 |
| 5⁄16 (0.3125) | 18, 24 | … | 1.30 | 1.24 | 0.59 | 0.53 | 0.30 | 0.26 | 0.46 | 0.73 | 0.67 | 0.14 | 0.18 | 0.06 | 0.05 |
| 3⁄8 (0.3750) | 16, 24 | … | 1.41 | 1.34 | 0.67 | 0.61 | 0.34 | 0.30 | 0.69 | 0.83 | 0.77 | 0.16 | 0.18 | 0.06 | 0.05 |
| **Type D, Style 2** | | | | | | | | | | | | | | | |
| 5 (0.1250) | 40 | Reg. | 1.03 | 0.97 | 0.25 | 0.19 | 0.19 | 0.13 | 0.30 | 0.40 | 0.34 | 0.07 | 0.09 | 0.04 | 0.03 |
| 6 (0.1380) | 32 | Reg. | 1.03 | 0.97 | 0.25 | 0.19 | 0.19 | 0.13 | 0.30 | 0.40 | 0.34 | 0.08 | 0.09 | 0.04 | 0.03 |
| 8 (0.1640) | 32 | Reg. | 1.03 | 0.97 | 0.25 | 0.19 | 0.19 | 0.13 | 0.30 | 0.40 | 0.34 | 0.08 | 0.09 | 0.04 | 0.03 |
| 10 (0.1900) | 24, 32 | Reg. | 1.40 | 1.34 | 0.34 | 0.28 | 0.25 | 0.18 | 0.32 | 0.53 | 0.47 | 0.09 | 0.16 | 0.05 | 0.04 |
| | | Hvy. | 1.21 | 1.16 | 0.28 | 0.26 | 0.31 | 0.25 | 0.60 | 0.61 | 0.55 | 0.09 | 0.13 | 0.05 | 0.04 |
| 12 (0.2160) | 24 | Reg. | 1.21 | 1.16 | 0.28 | 0.26 | 0.31 | 0.25 | 0.60 | 0.61 | 0.55 | 0.11 | 0.13 | 0.05 | 0.04 |
| 1⁄4 (0.2500) | 20 | Reg. | 1.21 | 1.16 | 0.28 | 0.26 | 0.31 | 0.25 | 0.60 | 0.61 | 0.55 | 0.11 | 0.13 | 0.05 | 0.04 |
| **Type D, Style 3** | | | | | | | | | | | | | | | |
| 10 (0.1900) | 24, 32 | Lgt. | 1.31 | 1.25 | 0.48 | 0.42 | 0.29 | 0.23 | 0.47 | 0.65 | 0.59 | 0.08 | 0.12 | 0.04 | 0.03 |
| | | Reg. | 1.40 | 1.34 | 0.53 | 0.47 | 0.25 | 0.19 | 0.50 | 0.75 | 0.69 | 0.08 | 0.14 | 0.04 | 0.03 |
| 12 (0.2160) | 24 | Reg. | 1.28 | 1.22 | 0.40 | 0.34 | 0.23 | 0.17 | 0.59 | 0.73 | 0.67 | 0.11 | 0.12 | 0.04 | 0.03 |
| 1⁄4 (0.2500) | 20 | Lgt. | 1.28 | 1.22 | 0.40 | 0.34 | 0.23 | 0.17 | 0.59 | 0.73 | 0.67 | 0.11 | 0.12 | 0.04 | 0.03 |
| | | Reg. | 1.78 | 1.72 | 0.66 | 0.60 | 0.31 | 0.25 | 0.70 | 1.03 | 0.97 | 0.14 | 0.17 | 0.06 | 0.04 |
| | | Hvy. | 1.47 | 1.40 | 0.50 | 0.44 | 0.37 | 0.31 | 0.66 | 1.03 | 0.97 | 0.14 | 0.14 | 0.08 | 0.06 |
| 5⁄16 (0.3125) | 18 | Reg. | 1.78 | 1.72 | 0.66 | 0.60 | 0.31 | 0.25 | 0.70 | 1.03 | 0.97 | 0.14 | 0.17 | 0.06 | 0.04 |
| | | Hvy. | 1.47 | 1.40 | 0.50 | 0.44 | 0.37 | 0.31 | 0.66 | 1.03 | 0.97 | 0.14 | 0.14 | 0.08 | 0.06 |

[a] Where specifying nominal size in decimals, zeros in the fourth decimal place are omitted.
[b] Lgt. = Light; Hvy. = Heavy; Reg. = Regular.
All dimensions in inches.

*Type D:* Type D wing nuts are stamped sheet metal nuts and are available in three styles: Style 1, having wings of moderate height; Style 2, having low wings; and Style 3, having wings of moderate height and a larger bearing surface. In some sizes, Styles 2 and 3 are produced in regular, light, and heavy series to best suit the requirements of specific applications. Dimensions are given in Table 4.

**Specification of Wing Nuts.—**When specifying wing nuts, the following data should be included in the designation and should appear in the following sequence: nominal size (number, fraction or decimal equivalent), threads per inch, type, style and/or series, material, and finish.

*Examples:* 10–32 Type A Wing Nut, Regular Series, Steel, Zinc Plated.
0.250–20 Type C Wing Nut, Style 1, Zinc Alloy, Plain.

**Threads for Wing Nuts.—**Threads are in conformance with the American National Standard Unified Thread, Class 2B for all types of wing nuts except type D which have a modified Class 2B thread. Because of the method of manufacture, the minor diameter of the thread in type D nuts may be somewhat larger than the Unified Thread Class 2B maximum but shall in no case exceed the minimum pitch diameter.

**Materials and Finish for Wing Nuts.—**Types A, B, and D wing nuts are normally supplied as specified by the user in carbon steel, brass or corrosion resistant steel of good quality and adaptable to the manufacturing process. Type C wing nuts are made from die-cast zinc alloy. Unless otherwise specified, wing nuts are supplied with a plain (unplated or uncoated) finish.

**Wing Screws.—**A wing screw is a screw having a wing-shaped head designed for manual turning without a driver or wrench. As covered by ANSI/ASME B18.6.9-2010 (R2017) wing screws are classified first, by type on the basis of the method of manufacture, and second, by style on the basis of design characteristics. They consist of the following:

*Type A:* Type A wing screws are of two-piece construction having cold formed or cold forged wing portions of moderate height. In some sizes they are produced in regular, light, and heavy series to best suit the requirements of specific applications. Dimensions are given in Table 5.

*Type B:* Type B wing screws are of hot forged one-piece construction available in two wing styles: Style 1, having wings of moderate height; and Style 2, having high wings. Dimensions are given in Table 5.

*Type C:* Type C wing screws are available in two styles: Style 1, of a one-piece die-cast construction having wings of moderate height; and Style 2, of a two-piece construction having a die-cast wing portion of moderate height. Dimensions are given in Table 6.

*Type D:* Type D wing screws are of two-piece welded construction having stamped sheet metal wing portions of moderate height. Dimensions are given in Table 6.

*Materials for Wing Screws and Thumb Screws:* Type A wing screws are normally supplied in carbon steel with the shank portion case hardened. When so specified, they also may be made from corrosion resistant steel, brass or other materials as agreed upon by the manufacturer and user.

Type B wing screws are normally made from carbon steel but also may be made from corrosion resistant steel, brass or other materials.

Type C, Style 1, wing screws are supplied only in die-cast zinc alloy. Type C, Style 2, wing screws have the wing portion made from die-cast zinc alloy with the shank portion normally made from carbon steel. Where so specified, the shank portion may be made from corrosion resistant steel, brass or other materials as agreed upon by the manufacturer and user.

Type D wing screws are normally supplied in carbon steel but also may be made from corrosion resistant steel, brass or other materials.

Thumb screws of all types are normally made from a good commercial quality of carbon steel having a maximum ultimate tensile strength of 48,000 psi (331 MPa). Where so specified, carbon steel thumb screws are case hardened. They are also made from corrosion resistant steel, brass, and other materials as agreed upon by the manufacturer and user.

Unless otherwise specified, wing screws and thumb screws are supplied with a plain (unplated or uncoated) finish.

**Thumb Screws.—**A thumb screw is a screw having a flattened head designed for manual turning without a driver or wrench. As covered by ANSI/ASME B18.6.9-2010 (R2017) thumb screws are classified by type on the basis of design characteristics. They consist of the following:

*Type A:* Type A thumb screws are forged one-piece screws having a shoulder under the head and are available in two series: regular and heavy. Dimensions are given in Table 7.

*Type B:* Type B thumb screws are forged one-piece screws without a shoulder and are available in two series: regular and heavy. Dimensions are given in Table 7.

## Table 5. American National Standard Types A and B Wing Screws
*ANSI/ASME B18.6.9-2010 (R2017)*

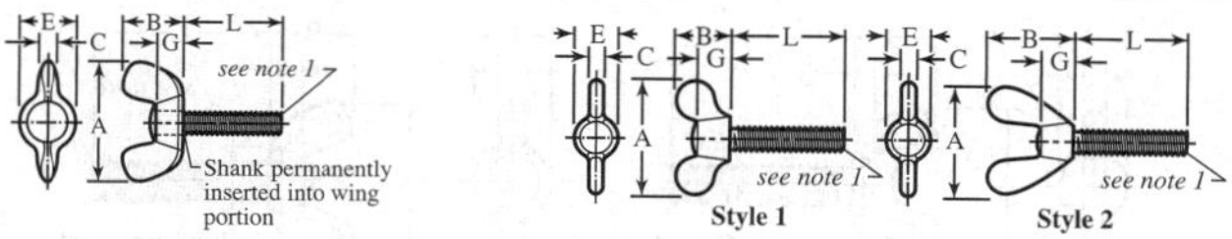


TYPE A TYPE B

| Nominal Size or Basic Major Diameter[a] | Thds. per Inch | Series[b] | Head Blank size (Ref) | A Wing Spread | | B Wing Height | | C Wing Thickness | | E Boss Dia. | | G Boss Height | | L Practical Screw Lengths | |
|---|---|---|---|---|---|---|---|---|---|---|---|---|---|---|---|
| | | | | Max. | Min. | Max. | Min. | Max. | Min. | Max. | Min. | Max. | Min. | Max. | Min. |
| **Type A** | | | | | | | | | | | | | | | |
| 4 (0.1120) | 40 | Hvy. | AA | 0.72 | 0.59 | 0.41 | 0.28 | 0.11 | 0.07 | 0.33 | 0.29 | 0.14 | 0.10 | 0.75 | 0.25 |
| 6 (0.1380) | 32 | Lgt. | AA | 0.72 | 0.59 | 0.41 | 0.28 | 0.11 | 0.07 | 0.33 | 0.29 | 0.14 | 0.10 | } 0.75 | 0.25 |
| | | **Hvy.** | **A** | **0.91** | **0.78** | **0.47** | **0.34** | **0.14** | **0.10** | **0.43** | **0.39** | **0.18** | **0.14** | | |
| 8 (0.1640) | 32 | **Lgt.** | **A** | **0.91** | **0.78** | **0.47** | **0.34** | **0.14** | **0.10** | **0.43** | **0.39** | **0.18** | **0.14** | } 0.75 | 0.38 |
| | | Hvy. | B | 1.10 | 0.97 | 0.57 | 0.43 | 0.18 | 0.14 | 0.50 | 0.45 | 0.22 | 0.17 | | |
| 10 (0.1900) | 24, 32 | **Lgt.** | **A** | **0.91** | **0.78** | **0.47** | **0.34** | **0.14** | **0.10** | **0.43** | **0.39** | **0.18** | **0.14** | } 1.00 | 0.38 |
| | | Hvy. | B | 1.10 | 0.97 | 0.57 | 0.43 | 0.18 | 0.14 | 0.50 | 0.45 | 0.22 | 0.17 | | |
| 12 (0.2160) | 24 | **Lgt.** | **B** | **1.10** | **0.97** | **0.57** | **0.43** | **0.18** | **0.14** | **0.50** | **0.45** | **0.22** | **0.17** | } 1.00 | 0.38 |
| | | Hvy. | C | 1.25 | 1.12 | 0.66 | 0.53 | 0.21 | 0.17 | 0.58 | 0.51 | 0.25 | 0.20 | | |
| ¼ (0.2500) | 20 | **Lgt.** | **B** | **1.10** | **0.97** | **0.57** | **0.43** | **0.18** | **0.14** | **0.50** | **0.45** | **0.22** | **0.17** | } 1.50 | 0.50 |
| | | Reg. | C | 1.25 | 1.12 | 0.66 | 0.53 | 0.21 | 0.17 | 0.58 | 0.51 | 0.25 | 0.20 | | |
| | | Hvy. | D | 1.44 | 1.31 | 0.79 | 0.65 | 0.24 | 0.20 | 0.70 | 0.64 | 0.30 | 0.26 | | |
| 5⁄16 (0.3125) | 18 | **Lgt.** | **C** | **1.25** | **1.12** | **0.66** | **0.53** | **0.21** | **0.17** | **0.58** | **0.51** | **0.25** | **0.20** | } 1.50 | 0.50 |
| | | Reg. | D | 1.44 | 1.31 | 0.79 | 0.65 | 0.24 | 0.20 | 0.70 | 0.64 | 0.30 | 0.26 | | |
| | | Hvy. | E | 1.94 | 1.81 | 1.00 | 0.87 | 0.33 | 0.26 | 0.93 | 0.86 | 0.39 | 0.35 | | |
| ⅜ (0.3750) | 16 | **Lgt.** | **D** | **1.44** | **1.31** | **0.79** | **0.65** | **0.24** | **0.20** | **0.70** | **0.64** | **0.30** | **0.26** | } 2.00 | 0.75 |
| | | Reg. | E | 1.94 | 1.81 | 1.00 | 0.87 | 0.33 | 0.26 | 0.93 | 0.86 | 0.39 | 0.35 | | |
| | | Hvy. | F | 2.76 | 2.62 | 1.44 | 1.31 | 0.40 | 0.34 | 1.19 | 1.13 | 0.55 | 0.51 | | |
| 7⁄16 (0.4375) | 14 | **Lgt.** | **E** | **1.94** | **1.81** | **1.00** | 0.87 | **0.33** | **0.26** | **0.93** | **0.86** | **0.39** | **0.35** | } 4.00 | 1.00 |
| | | Hvy. | F | 2.76 | 2.62 | 1.44 | 1.31 | 0.40 | 0.34 | 1.19 | 1.13 | 0.55 | 0.51 | | |
| ½ (0.5000) | 13 | **Lgt.** | **E** | **1.94** | **1.81** | **1.00** | **0.87** | **0.33** | **0.26** | **0.93** | **0.86** | **0.39** | **0.35** | } 4.00 | 1.00 |
| | | Hvy. | F | 2.76 | 2.62 | 1.44 | 1.31 | 0.40 | 0.34 | 1.19 | 1.13 | 0.55 | 0.51 | | |
| ⅝ (0.6250) | 11 | Hvy. | F | 2.76 | 2.62 | 1.44 | 1.31 | 0.40 | 0.34 | 1.19 | 1.13 | 0.55 | 0.51 | 4.00 | 1.25 |
| **Type B, Style 1** | | | | | | | | | | | | | | | |
| 10 (0.1900) | 24 | … | … | 0.97 | 0.91 | 0.45 | 0.39 | 0.15 | 0.12 | 0.39 | 0.36 | 0.28 | 0.22 | 2.00 | 0.50 |
| ¼ (0.2500) | 20 | … | … | 1.16 | 1.09 | 0.56 | 0.50 | 0.17 | 0.14 | 0.47 | 0.44 | 0.34 | 0.28 | 3.00 | 0.50 |
| 5⁄16 (0.3125) | 18 | … | … | 1.44 | 1.38 | 0.67 | 0.61 | 0.18 | 0.15 | 0.55 | 0.52 | 0.41 | 0.34 | 3.00 | 0.50 |
| ⅜ (0.3750) | 16 | … | … | 1.72 | 1.66 | 0.80 | 0.73 | 0.20 | 0.17 | 0.63 | 0.60 | 0.47 | 0.41 | 4.00 | 0.50 |
| 7⁄16 (0.4375) | 14 | … | … | 2.00 | 1.94 | 0.91 | 0.84 | 0.21 | 0.18 | 0.71 | 0.68 | 0.53 | 0.47 | 3.00 | 1.00 |
| ½ (0.5000) | 13 | … | … | 2.31 | 2.22 | 1.06 | 0.94 | 0.23 | 0.20 | 0.79 | 0.76 | 0.62 | 0.50 | 3.00 | 1.00 |
| ⅝ (0.6250) | 11 | … | … | 2.84 | 2.72 | 1.31 | 1.19 | 0.27 | 0.23 | 0.96 | 0.92 | 0.75 | 0.62 | 2.50 | 1.00 |
| **Type B, Style 2** | | | | | | | | | | | | | | | |
| 10 (0.1900) | 24 | … | … | 1.01 | 0.95 | 0.78 | 0.72 | 0.14 | 0.11 | 0.39 | 0.36 | 0.28 | 0.22 | 1.25 | 0.50 |
| ¼ (0.2500) | 20 | … | … | 1.22 | 1.16 | 0.94 | 0.88 | 0.16 | 0.13 | 0.47 | 0.44 | 0.34 | 0.28 | 2.00 | 0.50 |
| 5⁄16 (0.3125) | 18 | … | … | 1.43 | 1.37 | 1.09 | 1.03 | 0.17 | 0.14 | 0.55 | 0.52 | 0.41 | 0.34 | 2.00 | 0.50 |
| ⅜ (0.3750) | 16 | … | … | 1.63 | 1.57 | 1.25 | 1.19 | 0.18 | 0.15 | 0.63 | 0.60 | 0.47 | 0.41 | 2.00 | 0.50 |

All dimensions in inches. Sizes shown in **bold face** are preferred.

[1]Plain point, unless alternate point from styles shown in Table 8 is specified by user.

[a] Where specifying nominal size in decimals, zeros in the fourth decimal place are omitted.

[b] Hvy. = Heavy; Lgt. = Light; Reg. = Regular.

**Table 6. American National Standard Types C and D Wing Screws**
*ANSI/ASME B18.6.9-2010 (R2017)*

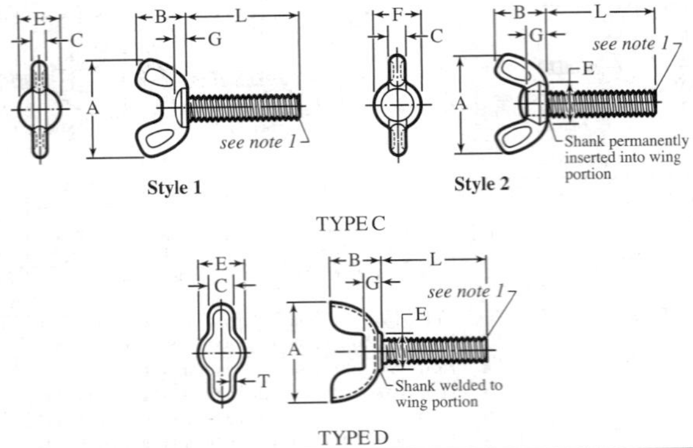


| Nominal Size or Basic Screw Diameter[a] | Thds. per Inch | *A* Wing Spread Max. | *A* Wing Spread Min. | *B* Wing Height Max. | *B* Wing Height Min. | *C* Wing Thickness Max. | *C* Wing Thickness Min. | *E* Boss Dia. Max. | *E* Boss Dia. Min. | *F* Boss Dia. Max. | *F* Boss Dia. Min. | *G* Height Max. | *G* Height Min. | *L* Practical Screw Lengths Max. | *L* Practical Screw Lengths Min. |
|---|---|---|---|---|---|---|---|---|---|---|---|---|---|---|---|
| **Type C, Style 1** | | | | | | | | | | | | | | | |
| 6 (0.1380) | 32 | 0.85 | 0.83 | 0.45 | 0.43 | 0.15 | 0.12 | … | … | 0.41 | 0.39 | 0.12 | 0.07 | 0.75 | 0.25 |
| 8 (0.1640) | 32 | 0.85 | 0.83 | 0.45 | 0.43 | 0.15 | 0.12 | … | … | 0.41 | 0.39 | 0.12 | 0.07 | 1.00 | 0.38 |
| 10 (0.1900) | 24, 32 | 0.85 | 0.83 | 0.45 | 0.43 | 0.15 | 0.12 | … | … | 0.41 | 0.39 | 0.12 | 0.07 | 1.25 | 0.38 |
| ¼ (0.2500) | 20 | 1.08 | 1.05 | 0.56 | 0.53 | 0.17 | 0.14 | … | … | 0.46 | 0.44 | 0.12 | 0.07 | 1.50 | 0.50 |
| 5⁄16 (0.3125) | 18 | 1.23 | 1.20 | 0.64 | 0.62 | 0.22 | 0.19 | … | … | 0.51 | 0.49 | 0.14 | 0.10 | 1.50 | 0.50 |
| 3⁄8 (0.3750) | 16 | 1.45 | 1.42 | 0.74 | 0.72 | 0.24 | 0.21 | … | … | 0.63 | 0.62 | 0.15 | 0.12 | 1.50 | 0.50 |
| **Type C, Style 2** | | | | | | | | | | | | | | | |
| 6 (0.1380) | 32 | 0.85 | 0.83 | 0.43 | 0.42 | 0.14 | 0.12 | 0.38 | 0.36 | 0.41 | 0.40 | 0.20 | 0.18 | 1.00 | 0.25 |
| 8 (0.1640) | 32 | 0.85 | 0.83 | 0.43 | 0.42 | 0.14 | 0.12 | 0.38 | 0.36 | 0.41 | 0.40 | 0.20 | 0.18 | 1.00 | 0.38 |
| 10 (0.1900) | 24, 32 | 0.85 | 0.83 | 0.43 | 0.42 | 0.14 | 0.12 | 0.38 | 0.36 | 0.41 | 0.40 | 0.20 | 0.18 | 2.00 | 0.38 |
| ¼ (0.2500) | 20 | 1.08 | 1.05 | 0.57 | 0.53 | 0.16 | 0.14 | 0.44 | 0.42 | 0.48 | 0.46 | 0.23 | 0.21 | 2.50 | 0.50 |
| 5⁄16 (0.3125) | 18 | 1.23 | 1.20 | 0.64 | 0.62 | 0.20 | 0.18 | 0.50 | 0.49 | 0.57 | 0.55 | 0.26 | 0.24 | 3.00 | 0.50 |
| 3⁄8 (0.3750) | 16 | 1.45 | 1.42 | 0.74 | 0.72 | 0.23 | 0.21 | 0.62 | 0.60 | 0.69 | 0.67 | 0.29 | 0.27 | 3.00 | 0.75 |
| 7⁄16 (0.4375) | 14 | 1.89 | 1.86 | 0.91 | 0.90 | 0.29 | 0.28 | 0.75 | 0.73 | 0.83 | 0.82 | 0.38 | 0.37 | 4.00 | 1.00 |
| ½ (0.5000) | 13 | 1.89 | 1.86 | 0.91 | 0.90 | 0.29 | 0.28 | 0.75 | 0.73 | 0.83 | 0.82 | 0.38 | 0.37 | 4.00 | 1.00 |
| **Type D** | | | | | | | | | | | | | | | |
| 6 (0.1380) | 32 | 0.78 | 0.72 | 0.40 | 0.34 | 0.18 | 0.12 | 0.35 | 0.31 | 0.40 | 0.34 | 0.21 | 0.14 | 0.75 | 0.25 |
| 8 (0.1640) | 32 | 0.78 | 0.72 | 0.40 | 0.34 | 0.18 | 0.12 | 0.35 | 0.31 | 0.40 | 0.34 | 0.21 | 0.14 | 0.75 | 0.38 |
| 10 (0.1900) | 24 | 0.90 | 0.84 | 0.46 | 0.40 | 0.21 | 0.15 | 0.35 | 0.31 | 0.53 | 0.47 | 0.22 | 0.16 | 1.00 | 0.38 |
| 12 (0.2160) | 24 | 1.09 | 1.03 | 0.46 | 0.40 | 0.26 | 0.20 | 0.44 | 0.39 | 0.61 | 0.55 | 0.24 | 0.18 | 1.00 | 0.38 |
| ¼ (0.2500) | 20 | 1.09 | 1.03 | 0.46 | 0.40 | 0.26 | 0.20 | 0.47 | 0.43 | 0.61 | 0.55 | 0.24 | 0.18 | 1.50 | 0.50 |
| 5⁄16 (0.3125) | 18 | 1.31 | 1.25 | 0.62 | 0.56 | 0.29 | 0.23 | 0.57 | 0.53 | 0.68 | 0.62 | 0.29 | 0.23 | 1.50 | 0.50 |
| 3⁄8 (0.3750) | 16 | 1.31 | 1.25 | 0.62 | 0.56 | 0.29 | 0.23 | 0.63 | 0.59 | 0.68 | 0.62 | 0.29 | 0.23 | 2.00 | 0.75 |

All dimensions in inches.

[1]Plain point, unless alternate point from styles shown in Table 8 is specified by user.

[a] Where specifying nominal size in decimals, zeros in the fourth decimal place are omitted.

**Wing Screw and Thumb Screw Designation.—**When specifying wing and thumb screws, the following data should be included in the designation and should appear in the following sequence: nominal size (number, fraction or decimal equivalent), threads per inch, length (fractions or decimal equivalents), type, style and/or series, point (if other than plain point), materials, and finish.

*Examples:* 10–32 × 1¼, Thumb Screw, Type A, Regular, Steel, Zinc Plated.
0.375–16 × 2.00, Wing Screw, Type B, Style 2, Steel, Cadmium Plated.
0.250–20 × 1.50, Wing Screw, Type C, Style 2, Zinc Alloy Wings, Steel Shank, Brass Plated.

**Table 7. American National Standard Types A and B Thumb Screws**
*ANSI/ASME B18.6.9-2010 (R2017)*

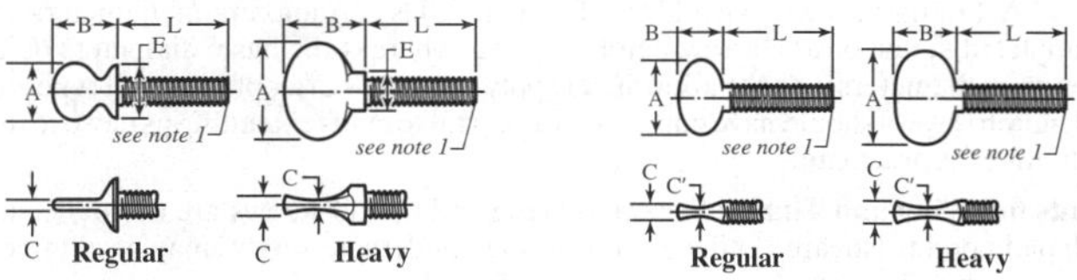


TYPE A TYPE B

| Nominal Size or Basic Screw Diameter[a] | Thds. per Inch | *A* Head Width Max. | *A* Head Width Min. | *B* Head Height Max. | *B* Head Height Min. | *C* Head Thickness Max. | *C* Head Thickness Min. | *C′* Head Thickness Max. | *C′* Head Thickness Min. | *E* Shoulder Diameter Max. | *E* Shoulder Diameter Min. | *L* Practical Screw Lengths Max. | *L* Practical Screw Lengths Min. |
|---|---|---|---|---|---|---|---|---|---|---|---|---|---|
| **Type A, Regular** | | | | | | | | | | | | | |
| 6 (0.1380) | 32 | 0.31 | 0.29 | 0.33 | 0.31 | 0.05 | 0.04 | … | … | 0.25 | 0.23 | 0.75 | 0.25 |
| 8 (0.1640) | 32 | 0.36 | 0.34 | 0.38 | 0.36 | 0.06 | 0.05 | … | … | 0.31 | 0.29 | 0.75 | 0.38 |
| 10 (0.1900) | 24, 32 | 0.42 | 0.40 | 0.48 | 0.46 | 0.06 | 0.05 | … | … | 0.35 | 0.32 | 1.00 | 0.38 |
| 12 (0.2160) | 24 | 0.48 | 0.46 | 0.54 | 0.52 | 0.06 | 0.05 | … | … | 0.40 | 0.38 | 1.00 | 0.38 |
| ¼ (0.2500) | 20 | 0.55 | 0.52 | 0.64 | 0.61 | 0.07 | 0.05 | … | … | 0.47 | 0.44 | 1.50 | 0.50 |
| 5⁄16 (0.3125) | 18 | 0.70 | 0.67 | 0.78 | 0.75 | 0.09 | 0.07 | … | … | 0.59 | 0.56 | 1.50 | 0.50 |
| ⅜ (0.3750) | 16 | 0.83 | 0.80 | 0.95 | 0.92 | 0.11 | 0.09 | … | … | 0.76 | 0.71 | 2.00 | 0.75 |
| **Type A, Heavy** | | | | | | | | | | | | | |
| 10 (0.1900) | 24 | 0.89 | 0.83 | 0.84 | 0.72 | 0.18 | 0.16 | 0.10 | 0.08 | 0.33 | 0.31 | 2.00 | 0.50 |
| ¼ (0.2500) | 20 | 1.05 | 0.99 | 0.94 | 0.81 | 0.24 | 0.22 | 0.10 | 0.08 | 0.40 | 0.38 | 3.00 | 0.50 |
| 5⁄16 (0.3125) | 18 | 1.21 | 1.15 | 1.00 | 0.88 | 0.27 | 0.25 | 0.11 | 0.09 | 0.46 | 0.44 | 4.00 | 0.50 |
| ⅜ (0.3750) | 16 | 1.41 | 1.34 | 1.16 | 1.03 | 0.30 | 0.28 | 0.11 | 0.09 | 0.55 | 0.53 | 4.00 | 0.50 |
| 7⁄16 (0.4375) | 14 | 1.59 | 1.53 | 1.22 | 1.09 | 0.36 | 0.34 | 0.13 | 0.11 | 0.71 | 0.69 | 2.50 | 1.00 |
| ½ (0.5000) | 13 | 1.81 | 1.72 | 1.28 | 1.16 | 0.40 | 0.38 | 0.14 | 0.12 | 0.83 | 0.81 | 3.00 | 1.00 |
| **Type B, Regular** | | | | | | | | | | | | | |
| 6 (0.1380) | 32 | 0.45 | 0.43 | 0.28 | 0.26 | 0.08 | 0.06 | 0.03 | 0.02 | … | … | 1.00 | 0.25 |
| 8 (0.1640) | 32 | 0.51 | 0.49 | 0.32 | 0.30 | 0.09 | 0.07 | 0.04 | 0.02 | … | … | 1.00 | 0.38 |
| 10 (0.1900) | 24, 32 | 0.58 | 0.54 | 0.39 | 0.36 | 0.10 | 0.08 | 0.05 | 0.03 | … | … | 2.00 | 0.38 |
| 12 (0.2160) | 24 | 0.71 | 0.67 | 0.45 | 0.43 | 0.11 | 0.09 | 0.05 | 0.03 | … | … | 2.00 | 0.38 |
| ¼ (0.2500) | 20 | 0.83 | 0.80 | 0.52 | 0.48 | 0.16 | 0.14 | 0.06 | 0.03 | … | … | 2.50 | 0.50 |
| 5⁄16 (0.3125) | 18 | 0.96 | 0.91 | 0.64 | 0.60 | 0.17 | 0.14 | 0.09 | 0.06 | … | … | 3.00 | 0.50 |
| ⅜ (0.3750) | 16 | 1.09 | 1.03 | 0.71 | 0.67 | 0.22 | 0.18 | 0.11 | 0.08 | … | … | 3.00 | 0.75 |
| 7⁄16 (0.4375) | 14 | 1.40 | 1.35 | 0.96 | 0.91 | 0.27 | 0.24 | 0.14 | 0.11 | … | … | 4.00 | 1.00 |
| ½ (0.5000) | 13 | 1.54 | 1.46 | 1.09 | 1.03 | 0.33 | 0.29 | 0.15 | 0.11 | … | … | 4.00 | 1.00 |
| **Type B, Heavy** | | | | | | | | | | | | | |
| 10 (0.1900) | 24 | 0.89 | 0.83 | 0.78 | 0.66 | 0.18 | 0.16 | 0.08 | 0.06 | … | … | 2.00 | 0.50 |
| ¼ (0.2500) | 20 | 1.05 | 0.99 | 0.81 | 0.72 | 0.24 | 0.22 | 0.11 | 0.09 | … | … | 3.00 | 0.50 |
| 5⁄16 (0.3125) | 18 | 1.21 | 1.15 | 0.88 | 0.78 | 0.27 | 0.25 | 0.11 | 0.09 | … | … | 4.00 | 0.50 |
| ⅜ (0.3750) | 16 | 1.41 | 1.34 | 0.94 | 0.84 | 0.30 | 0.28 | 0.14 | 0.12 | … | … | 4.00 | 0.50 |
| 7⁄16 (0.4375) | 14 | 1.59 | 1.53 | 1.00 | 0.91 | 0.36 | 0.34 | 0.14 | 0.12 | … | … | 3.00 | 1.00 |
| ½ (0.5000) | 13 | 1.81 | 1.72 | 1.09 | 0.97 | 0.40 | 0.38 | 0.18 | 0.16 | … | … | 3.00 | 1.00 |

[a] Where specifying nominal size in decimals, zeroes in fourth decimal place are omitted.
All dimensions in inches.
[1] Plain point, unless alternate point from styles shown in Table 8 is specified by user.

**Lengths of Wing and Thumb Screws.—**The length of wing or thumb screws is measured parallel to the axis of the screw from the intersection of the head or shoulder with the shank to the extreme point of the screw. Standard length increments are as follows: For sizes No. 4 through $\frac{1}{4}$ inch and for nominal lengths of 0.25 to 0.75 inch, 0.12-inch increments; from 0.75- to 1.50-inch lengths, 0.25-inch increments; and for 1.50- to 3.00-inch lengths, 0.50-inch increments. For sizes $\frac{5}{16}$ through $\frac{1}{2}$ inch and for 0.50- to 1.50-inch lengths, 0.25-inch increments; for 1.50- to 3.00-inch lengths, 0.50-inch increments; and for 3.00- to 4.00-inch lengths, 1.00-inch increments.

**Threads for Wing Screws and Thumb Screws.—**Threads for all types of wing screws and thumb screws are in conformance with American National Standard Unified Thread, Class 2A. For threads with an additive finish the Class 2A maximum diameters apply to an unplated screw or to a screw before plating, whereas the basic diameters (Class 2A maximum diameters plus the allowance) apply to a screw after plating. All types of wing and thumb screws should have complete (full form) threads extending as close to the head or shoulder as practicable.

**Points for Wing and Thumb Screws.—**Wing and thumb screws are normally supplied with plain points (sheared ends). Where so specified, these screws may be obtained with cone, cup, dog, flat or oval points as shown in Table 8.

**Table 8. American National Standard Alternate Points for Wing and Thumb Screws**
*ANSI/ASME B18.6.9-2010 (R2017)*

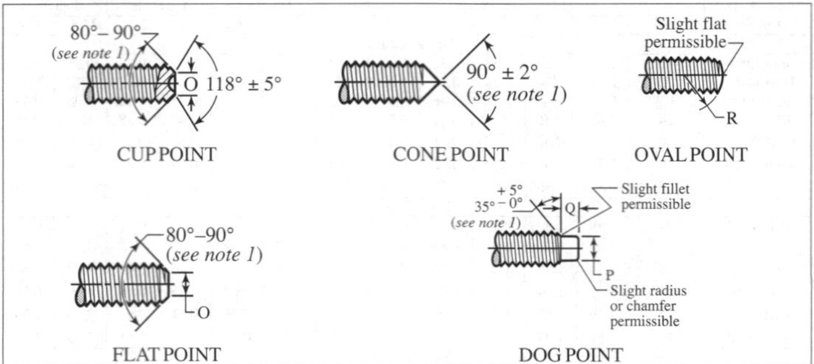


| Nominal Size or Basic Screw Diamter[a] | *O* | | *P* | | *Q* | | *R* | |
|---|---|---|---|---|---|---|---|---|
| | Cup and Flat Point Diameter | | Dog Point[b] Diameter | | Dog Point[b] Length | | Oval Point Radius | |
| | Max. | Min. | Max. | Min. | Max. | Min. | Max. | Min. |
| 4 (0.1120) | 0.061 | 0.051 | 0.075 | 0.070 | 0.061 | 0.051 | 0.099 | 0.084 |
| 6 (0.1380) | 0.074 | 0.064 | 0.092 | 0.087 | 0.075 | 0.065 | 0.140 | 0.109 |
| 8 (0.1640) | 0.087 | 0.076 | 0.109 | 0.103 | 0.085 | 0.075 | 0.156 | 0.125 |
| 10 (0.1900) | 0.102 | 0.088 | 0.127 | 0.120 | 0.095 | 0.085 | 0.172 | 0.141 |
| 12 (0.2160) | 0.115 | 0.101 | 0.144 | 0.137 | 0.115 | 0.105 | 0.188 | 0.156 |
| ¼ (0.2500) | 0.132 | 0.118 | 0.156 | 0.149 | 0.130 | 0.120 | 0.219 | 0.188 |
| 5⁄16 (0.3125) | 0.172 | 0.156 | 0.203 | 0.195 | 0.161 | 0.151 | 0.256 | 0.231 |
| ⅜ (0.3750) | 0.212 | 0.194 | 0.250 | 0.241 | 0.193 | 0.183 | 0.312 | 0.281 |
| 7⁄16 (0.4375) | 0.252 | 0.232 | 0.297 | 0.287 | 0.224 | 0.214 | 0.359 | 0.328 |
| ½ (0.5000) | 0.291 | 0.270 | 0.344 | 0.334 | 0.255 | 0.245 | 0.406 | 0.375 |
| ⅝ (0.6250) | 0.371 | 0.347 | 0.469 | 0.456 | 0.321 | 0.305 | 0.500 | 0.469 |

[a] Where specifying nominal size in decimals, zeros in the fourth decimal place are omitted.

[b] The axis of dog points shall not be eccentric with the axis of the screw by more than 3 percent of the basic screw diameter or 0.005 in., whichever is the smaller.

All dimensions in inches.

[1] The external point angles specified shall apply to those portions of the angles which lie below the thread root diameter, it being recognized the angle within the thread profile may be varied due to the manufacturing processes.

# NAILS, SPIKES, AND WOOD SCREWS

## Standard Wire Nails and Spikes (Size, Length, and Approximate Number to Pound)

| Size of Nail | Length, Inches | Gage | Num/lb | Gage | Num/lb | Gage | Num/lb | Gage | Num/lb | Gage | Num/lb |
|---|---|---|---|---|---|---|---|---|---|---|---|
| | | Common Wire Nails and Brads | | Flooring Brads | | Fence Nails | | Casing, Smooth and Barbed Box | | Finishing Nails | |
| 2 *d* | 1 | 15 | 876 | … | … | … | … | 15 ½ | 1010 | 16 ½ | 1351 |
| 3 *d* | 1 ¼ | 14 | 568 | … | … | … | … | 14 ½ | 635 | 15 ½ | 807 |
| 4 *d* | 1 ½ | 12 ½ | 316 | … | … | … | … | 14 | 473 | 15 | 584 |
| 5 *d* | 1 ¾ | 12 ½ | 271 | … | … | 10 | 142 | 14 | 406 | 15 | 500 |
| 6 *d* | 2 | 11 ½ | 181 | 11 | 157 | 10 | 124 | 12 ½ | 236 | 13 ½ | 309 |
| 7 *d* | 2 ¼ | 11 ½ | 161 | 11 | 139 | 9 | 92 | 12 ½ | 210 | 13 | 238 |
| 8 *d* | 2 ½ | 10 ¼ | 106 | 10 | 99 | 9 | 82 | 11 ½ | 145 | 12 ½ | 189 |
| 9 *d* | 2 ¾ | 10 ¼ | 96 | 10 | 90 | 8 | 62 | 11 ½ | 132 | 12 ½ | 172 |
| 10 *d* | 3 | 9 | 69 | 9 | 69 | 7 | 50 | 10 ½ | 94 | 11 ½ | 121 |
| 12 *d* | 3 ¼ | 9 | 64 | 8 | 54 | 6 | 40 | 10 ½ | 87 | 11 ½ | 113 |
| 16 *d* | 3 ½ | 8 | 49 | 7 | 43 | 5 | 30 | 10 | 71 | 11 | 90 |
| 20 *d* | 4 | 6 | 31 | 6 | 31 | 4 | 23 | 9 | 52 | 10 | 62 |
| 30 *d* | 4 ½ | 5 | 24 | … | … | … | … | 9 | 46 | … | … |
| 40 *d* | 5 | 4 | 18 | … | … | … | … | 8 | 35 | … | … |
| 50 *d* | 5 ½ | 3 | 16 | … | … | … | … | … | … | … | … |
| 60 *d* | 6 | 2 | 11 | … | … | … | … | … | … | … | … |
| Size and Length | | Hinge Nails, Heavy | | Hinge Nails, Light | | Clinch Nails | | Barbed Car Nails, Heavy | | Barbed Car Nails, Light | |
| 2 *d* | 1 | … | … | … | … | 14 | 710 | … | … | … | … |
| 3 *d* | 1 ¼ | … | … | … | … | 13 | 429 | … | … | … | … |
| 4 *d* | 1 ½ | 3 | 50 | 6 | 82 | 12 | 274 | 10 | 165 | 12 | 274 |
| 5 *d* | 1 ¾ | 3 | 38 | 6 | 62 | 12 | 235 | 9 | 118 | 10 | 142 |
| 6 *d* | 2 | 3 | 30 | 6 | 50 | 11 | 157 | 9 | 103 | 10 | 124 |
| 7 *d* | 2 ¼ | 00 | 12 | 3 | 25 | 11 | 139 | 8 | 76 | 9 | 92 |
| 8 *d* | 2 ½ | 00 | 11 | 3 | 23 | 10 | 99 | 8 | 69 | 9 | 82 |
| 9 *d* | 2 ¾ | 00 | 10 | 3 | 22 | 10 | 90 | 7 | 54 | 8 | 62 |
| 10 *d* | 3 | 00 | 9 | 3 | 19 | 9 | 69 | 7 | 50 | 8 | 57 |
| 12 *d* | 3 ¼ | … | … | … | … | 9 | 62 | 6 | 42 | 7 | 50 |
| 16 *d* | 3 ½ | … | … | … | … | 8 | 49 | 6 | 35 | 7 | 43 |
| 20 *d* | 4 | … | … | … | … | 7 | 37 | 5 | 26 | 6 | 31 |
| 30 *d* | 4 ½ | … | … | … | … | … | … | 5 | 24 | 6 | 28 |
| 40 *d* | 5 | … | … | … | … | … | … | 4 | 18 | 5 | 21 |
| 50 *d* | 5 ½ | … | … | … | … | … | … | 3 | 15 | 4 | 17 |
| 60 *d* | 6 | … | … | … | … | … | … | 3 | 13 | 4 | 15 |
| Size and Length | | Boat Nails, Heavy | | Boat Nails, Light | | Slating Nails | | Spikes | | | |
| 2 *d* | 1 | … | … | … | … | 12 | 411 | | | | |
| 3 *d* | 1 ¼ | … | … | … | … | 10 ½ | 225 | | | | |
| 4 *d* | 1 ½ | ¼ | 44 | 3/16 | 82 | 10 ½ | 187 | Size and Length | | Gage | No. to Lb |
| 5 *d* | 1 ¾ | … | … | … | … | 10 | 142 | 10 *d* | 3 | 6 | 41 |
| 6*d* | 2 | ¼ | 32 | 3/16 | 62 | 9 | 103 | 12 *d* | 3 ¼ | 6 | 38 |
| 7 *d* | 2 ¼ | … | … | … | … | … | … | 16 *d* | 3 ½ | 5 | 30 |
| 8 *d* | 2 ½ | ¼ | 26 | 3/16 | 50 | … | … | 20 *d* | 4 | 4 | 23 |
| 9 *d* | 2 ¾ | … | … | … | … | … | … | 30 *d* | 4 ½ | 3 | 17 |
| 10 *d* | 3 | ⅜ | 14 | ¼ | 22 | … | … | 40 *d* | 5 | 2 | 13 |
| 12 *d* | 3 ¼ | ⅜ | 13 | ¼ | 20 | … | … | 50 *d* | 5 ½ | 1 | 10 |
| 16 *d* | 3 ½ | ⅜ | 12 | ¼ | 18 | … | … | 60 *d* | 6 | 1 | 8 |
| 20 *d* | 4 | ⅜ | 10 | ¼ | 16 | … | … | … | 7 | 0 | 7 |
| 30 *d* | 4 ½ | … | … | … | … | … | … | … | 8 | 00 | 6 |
| 40 *d* | 5 | … | … | … | … | … | … | … | 9 | 00 | 5 |
| 50 *d* | 5 ½ | … | … | … | … | … | … | … | 10 | ⅜ | 4 |
| 60 *d* | 6 | … | … | … | … | … | … | … | 12 | ⅜ | 3 |

## ANSI Flat, Pan, and Oval Head Wood Screws *ANSI/ASME B18.6.1-1981 (R2016)*

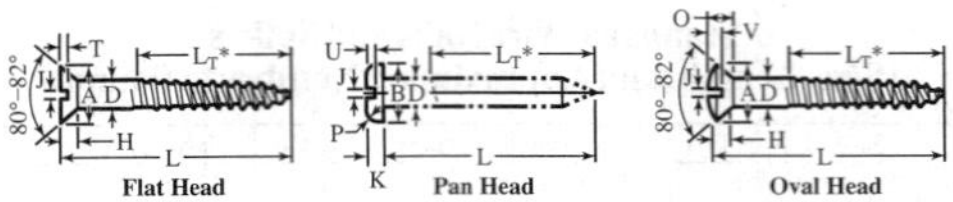


| Nominal Size | Threads per inch | *D* Basic Dia. of Screw | *J* Width of Slot Max. | *J* Width of Slot Min. | *A* Head Diameter Max., Sharp Edge | *A* Head Diameter Min., Edge Rounded or Flat | *B* Head Diameter Max. | *B* Head Diameter Min. | *P* Head Radius Max. | *H* Height of Head Ref. |
|---|---|---|---|---|---|---|---|---|---|---|
| 0 | 32 | .060 | .023 | .016 | .119 | .099 | .116 | .104 | .020 | .035 |
| 1 | 28 | .073 | .026 | .019 | .146 | .123 | .142 | .130 | .025 | .043 |
| 2 | 26 | .086 | .031 | .023 | .172 | .147 | .167 | .155 | .035 | .051 |
| 3 | 24 | .099 | .035 | .027 | .199 | .171 | .193 | .180 | .037 | .059 |
| 4 | 22 | .112 | .039 | .031 | .225 | .195 | .219 | .205 | .042 | .067 |
| 5 | 20 | .125 | .043 | .035 | .252 | .220 | .245 | .231 | .044 | .075 |
| 6 | 18 | .138 | .048 | .039 | .279 | .244 | .270 | .256 | .046 | .083 |
| 7 | 16 | .151 | .048 | .039 | .305 | .268 | .296 | .281 | .049 | .091 |
| 8 | 15 | .164 | .054 | .045 | .332 | .292 | .322 | .306 | .052 | .100 |
| 9 | 14 | .177 | .054 | .045 | .358 | .316 | .348 | .331 | .056 | .108 |
| 10 | 13 | .190 | .060 | .050 | .385 | .340 | .373 | .357 | .061 | .116 |
| 12 | 11 | .216 | .067 | .056 | .438 | .389 | .425 | .407 | .078 | .132 |
| 14 | 10 | .242 | .075 | .064 | .507 | .452 | .492 | .473 | .087 | .153 |
| 16 | 9 | .268 | .075 | .064 | .544 | .485 | .528 | .508 | .094 | .164 |
| 18 | 8 | .294 | .084 | .072 | .635 | .568 | .615 | .594 | .099 | .191 |
| 20 | 8 | .320 | .084 | .072 | .650 | .582 | .631 | .608 | .121 | .196 |
| 24 | 7 | .372 | .094 | .081 | .762 | .685 | .740 | .716 | .143 | .230 |

| Nominal Size | Threads per Inch | *O* Total Height of Head Max. | *O* Total Height of Head Min. | *K* Height of Head Max. | *K* Height of Head Min. | *T* Depth of Slot Max. | *T* Depth of Slot Min. | *U* Depth of Slot Max. | *U* Depth of Slot Min. | *V* Depth of Slot Max. | *V* Depth of Slot Min. |
|---|---|---|---|---|---|---|---|---|---|---|---|
| 0 | 32 | .056 | .041 | .039 | .031 | .015 | .010 | .022 | .014 | .030 | .025 |
| 1 | 28 | .068 | .052 | .046 | .038 | .019 | .012 | .027 | .018 | .038 | .031 |
| 2 | 26 | .080 | .063 | .053 | .045 | .023 | .015 | .031 | .022 | .045 | .037 |
| 3 | 24 | .092 | .073 | .060 | .051 | .027 | .017 | .035 | .027 | .052 | .043 |
| 4 | 22 | .104 | .084 | .068 | .058 | .030 | .020 | .040 | .030 | .059 | .049 |
| 5 | 20 | .116 | .095 | .075 | .065 | .034 | .022 | .045 | .034 | .067 | .055 |
| 6 | 18 | .128 | .105 | .082 | .072 | .038 | .024 | .050 | .037 | .074 | .060 |
| 7 | 16 | .140 | .116 | .089 | .079 | .041 | .027 | .054 | .041 | .081 | .066 |
| 8 | 15 | .152 | .126 | .096 | .085 | .045 | .029 | .058 | .045 | .088 | .072 |
| 9 | 14 | .164 | .137 | .103 | .092 | .049 | .032 | .063 | .049 | .095 | .078 |
| 10 | 13 | .176 | .148 | .110 | .099 | .053 | .034 | .068 | .053 | .103 | .084 |
| 12 | 11 | .200 | .169 | .125 | .112 | .060 | .039 | .077 | .061 | .117 | .096 |
| 14 | 10 | .232 | .197 | .144 | .130 | .070 | .046 | .087 | .070 | .136 | .112 |
| 16 | 9 | .248 | .212 | .153 | .139 | .075 | .049 | .093 | .074 | .146 | .120 |
| 18 | 8 | .290 | .249 | .178 | .162 | .083 | .054 | .106 | .085 | .171 | .141 |
| 20 | 8 | .296 | .254 | .182 | .166 | .090 | .059 | .108 | .087 | .175 | .144 |
| 24 | 7 | .347 | .300 | .212 | .195 | .106 | .070 | .124 | .100 | .204 | .168 |

All dimensions in inches. The edge of flat and oval head screws may be flat or rounded. Wood screws are also available with Types I, IA, and II recessed heads. Consult the Standard for recessed head dimensions. *The length of the thread, $L_T$, on wood screws having cut threads shall be equivalent to approximately two-thirds of the nominal length of the screw. For rolled threads, $L_T$ shall be at least four times the basic screw diameter or two-thirds of the nominal screw length, whichever is greater. Screws of nominal lengths that are too short to accommodate the minimum thread length shall have threads extending as close to the underside of the head as practicable.

### Pilot Hole Drill Sizes for Wood Screws

| Work Material | Wood Screw Size 2 | 4 | 6 | 8 | 10 | 12 | 14 |
|---|---|---|---|---|---|---|---|
| Hardwood | 3/64 | 1/16 | 5/64 | 3/32 | 7/64 | 1/8 | 9/64 |
| Softwood | 1/32 | 3/64 | 1/16 | 5/64 | 3/32 | 7/64 | 1/8 |

## TABLE OF CONTENTS

## THREADS AND THREADING

## TABLE OF CONTENTS

## THREADS AND THREADING

## TABLE OF CONTENTS

## THREADS AND THREADING

### PIPE AND HOSE THREADS

*(Continued)*



### OTHER THREADS



### MEASURING SCREW THREADS



### TAPPING AND THREAD CUTTING

## TABLE OF CONTENTS

## THREADS AND THREADING

### TAPPING AND THREAD CUTTING

*(Continued)*



### THREAD ROLLING



### THREAD ROLLING

*(Continued)*



### THREAD GRINDING



### THREAD MILLING

## SCREW THREAD SYSTEMS

### Screw Thread Forms

Of the various screw thread forms which have been developed, the most used are those having symmetrical sides inclined at equal angles with a vertical center line through the thread apex. Present-day examples of such threads would include the Unified, the Whitworth and the Acme forms. One of the early forms was the Sharp V which is now used only occasionally. Symmetrical threads are relatively easy to manufacture and inspect and hence are widely used on mass-produced general-purpose threaded fasteners of all types. In addition to general-purpose fastener applications, certain threads are used to repeatedly move or translate machine parts against heavy loads. For these so-called translation threads a stronger form is required. The most widely used translation thread forms are the square, the Acme, and the buttress. Of these, the square thread is the most efficient, but it is also the most difficult to cut owing to its parallel sides and it cannot be adjusted to compensate for wear. Although less efficient, the Acme form of thread has none of the disadvantages of the square form and has the advantage of being somewhat stronger. The buttress form is used for translation of loads in one direction only because of its non-symmetrical form and combines the high efficiency and strength of the square thread with the ease of cutting and adjustment of the Acme thread.

**V-Thread, Sharp V-Thread.**—The sides of the thread form an angle of 60 degrees with each other. The top and bottom or root of this thread form are theoretically sharp, but in actual practice the thread is made with a slight flat, owing to the difficulty of producing a perfectly sharp edge and because of the tendency of such an edge to wear away or become battered. This flat is usually equal to about one twenty-fifth of the pitch, although there is no generally recognized standard.

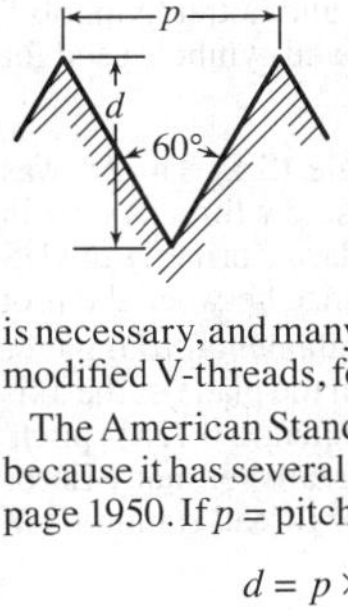


Owing to the difficulties connected with the V-thread, the tap manufacturers agreed in 1909 to discontinue the making of sharp V-thread taps, except when ordered. One advantage of the V-thread is that the same cutting tool may be used for all pitches, whereas, with the American Standard form, the width of the point or the flat varies according to the pitch.

The V-thread is regarded as a good form where a steam-tight joint is necessary, and many of the taps used on locomotive work have this form of thread. Some modified V-threads, for locomotive boiler taps particularly, have a depth of 0.8 × pitch.

The American Standard screw thread is used largely in preference to the sharp V-thread because it has several advantages; see *American Standard for Unified Screw Threads* on page 1950. If $p$ = pitch of thread, and $d$ depth of thread, then

$$d = p \times \cos 30\text{ deg.} = 0.866 \times p = \frac{0.866}{\text{No. of threads per inch}}$$

**United States Standard Screw Thread.**—William Sellers of Philadelphia, in a paper read before the Franklin Institute in 1864, originally proposed the screw thread system that later became known as the US Standard system for screw threads. A report was made to the United States Navy in May, 1868, in which the Sellers system was recommended as a standard for the Navy Department, which accounts for the name of US Standard. The American Standard Screw Thread system is a further development of the US Standard. The thread form which is known as the American (National) form is the same as the US Standard form. See *American Standard for Unified Screw Threads* on page 1950.

**American National and Unified Screw Thread Forms.**—The American National form (formerly known as the US Standard) was used for many years for most screws, bolts, and miscellaneous threaded products produced in the United States. The American National Standard for Unified Screw Threads now in use includes certain modifications of the former Standard as is explained below and on page 1950. The basic profile is shown in Fig. 1 and is identical for both UN and UNR screw threads. In this figure $H$ is the height of a sharp V-thread, $P$ is the pitch, $D$ and $d$ are the basic major diameters, $D_2$ and $d_2$ are the basic pitch diameters, and $D_1$ and $d_1$ are the basic minor diameters. Capital letters are used to designate the internal thread dimensions ($D, D_2, D_1$), and lowercase letters to designate the external thread dimensions ($d, d_2, d_1$). Definitions of *Basic Size* and *Basic Profile of Thread* are given on page 1945.

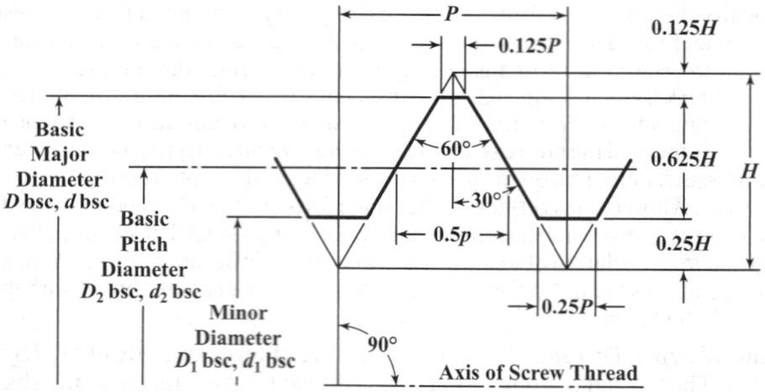


Fig. 1. Basic Profile of UN and UNF Screw Threads

In the past, other symbols were used for some of the thread dimensions illustrated above. These symbols were changed to conform with current practice in nomenclature as defined in ANSI/ASME B1.7, "Screw Threads: Nomenclature, Definitions, and Letter Symbols." The symbols used above are also in accordance with terminology and symbols used for threads of the ISO metric thread system.

**International Metric Thread System.**—The Système Internationale (S.I.) Thread was adopted at the International Congress for the standardization of screw threads held in Zurich in 1898. The thread form is similar to the American Standard (formerly the US Standard), excepting the depth which is greater. There is a clearance between the root and mating crest fixed at a maximum of $\frac{1}{16}$ the height of the fundamental triangle or 0.054 × pitch. A rounded root profile is recommended. The angle in the plane of the axis is 60 degrees and the crest has a flat like the American Standard equal to 0.125 × pitch. This system formed the basis of the normal metric series (ISO threads) of many European countries, Japan, and many other countries, including metric thread standards of the United States.

*Depth* $d = 0.7035\,P$ max.; $0.6855\,P$ min.

*Flat* $f = 0.125\,P$

*Radius* $r = 0.0633\,P$ max.; $0.054\,P$ min.

*Tap drill dia* = major diameter − pitch

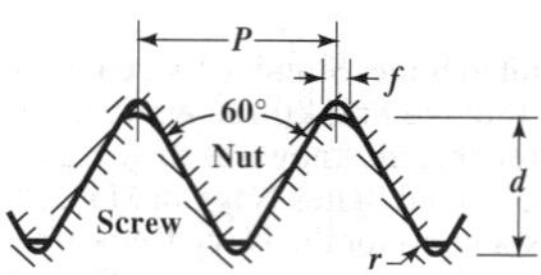


*International Metric Fine Thread:* The International Metric Fine Thread form of thread is the same as the International system but the pitch for a given diameter is smaller.

*German Metric Thread Form:* The German metric thread form is like the International Standard but the thread depth = 0.6945 *P*. The root radius is the same as the maximum for the International Standard or 0.0633 *P*.

**ISO Metric Thread System.—**ISO refers to the International Organization for Standardization, a worldwide federation of national standards bodies (for example, the American National Standards Institute is the ISO national body representing the United States) that develops standards on a very wide variety of subjects.

The basic profile of ISO metric threads is specified in ISO 68 and shown in Fig. 2. The basic profile of this thread is very similar to that of the Unified thread, and as previously discussed, $H$ is the height of a sharp V-thread, $P$ is the pitch, $D$ and $d$ are the basic major diameters, $D_2$ and $d_2$ are the basic pitch diameters, and $D_1$ and $d_1$ are the basic minor diameters. Here also, capital letters designate the internal thread dimensions ($D, D_2, D_1$), and lowercase letters designate the external thread dimensions ($d, d_2, d_1$). This metric thread is discussed in detail in the section *METRIC SCREW THREADS* starting on page 2015.

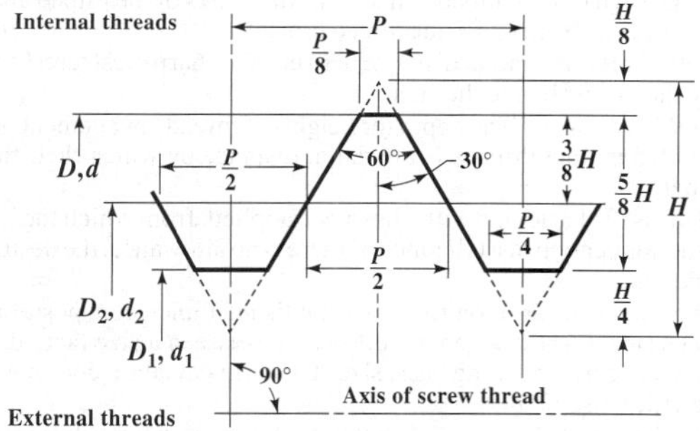


$$H = \frac{\sqrt{3}}{2} \times P = 0.866025404P$$

$$0.125H = 0.108253175P \quad 0.250H = 0.216506351P \quad 0.375H = 0.324759526P \quad 0.625H = 0.541265877P$$

Fig. 2. ISO 68 Basic Profile

## Definitions of Screw Threads

The following definitions are based on American National Standard ANSI/ASME B1.7-2006 (R2021) "Screw Threads: Nomenclature, Definitions, and Letter Symbols," and refer to both straight and taper threads.

*Actual Size:* An actual size is a measured size.

*Allowance:* An allowance is the prescribed difference between the design (maximum material) size and the basic size. It is numerically equal to the absolute value of the ISO term *fundamental deviation*.

*Axis of Thread:* Thread axis is coincident with the axis of its pitch cylinder or cone.

*Basic Profile of Thread:* The basic profile of a thread is the cyclical outline, in an axial plane, of the permanently established boundary between the provinces of the external and internal threads. All deviations are with respect to this boundary.

*Basic Size:* The basic size is that size from which the limits of size are derived by the application of allowances and tolerances.

*Bilateral Tolerance:* This is a tolerance in which variation is permitted in both directions from the specified dimension.

*Black Crest Thread:* This is a thread whose crest displays an unfinished cast, rolled, or forged surface.

*Blunt Start Thread:* "Blunt start" designates the removal of the incomplete thread at the starting end of the thread. This is a feature of threaded parts that are repeatedly assembled by hand, such as hose couplings and thread plug gages, to prevent cutting of hands and crossing of threads. It was formerly known as a Higbee cut.

*Chamfer:* This is a conical surface at the starting end of a thread.

*Class of Thread:* The class of a thread is an alphanumerical designation to indicate the standard grade of tolerance and allowance specified for a thread.

*Clearance Fit:* This is a fit having limits of size so prescribed that a clearance always results when mating parts are assembled at their maximum material condition.

*Complete Thread:* The complete thread is that thread whose profile lies within the size limits. (See also *Effective Thread* and *Length of Complete Thread.*) *Note:* Formerly in pipe thread terminology this was referred to as "the perfect thread" but that term is no longer considered desirable.

*Crest:* This is that surface of a thread which joins the flanks of the thread and is farthest from the cylinder or cone from which the thread projects.

*Crest Truncation:* This is the radial distance between the sharp crest (crest apex) and the cylinder or cone that would bound the crest.

*Depth of Thread Engagement:* The depth (or height) of thread engagement between two coaxially assembled mating threads is the radial distance by which their thread forms overlap each other.

*Design Size:* This is the basic size with allowance applied, from which the limits of size are derived by the application of a tolerance. If there is no allowance, the design size is the same as the basic size.

*Deviation:* Deviation is a variation from an established dimension, position, standard, or value. In ISO usage, it is the algebraic difference between a size (actual, maximum, or minimum) and the corresponding basic size. The term deviation does not necessarily indicate an error. (See also *Error.*)

*Deviation, Fundamental (ISO term):* For standard threads, the fundamental deviation is the upper or lower deviation closer to the basic size. It is the upper deviation *es* for an external thread and the lower deviation *EI* for an internal thread. (See also *Allowance* and *Tolerance Position.*)

*Deviation, Lower (ISO term):* The algebraic difference between the minimum limit of size and the basic size. It is designated *EI* for internal and *ei* for external thread diameters.

*Deviation, Upper (ISO term):* The algebraic difference between the maximum limit of size and the basic size. It is designated *ES* for internal and *es* for external thread diameters.

*Dimension:* A numerical value expressed in appropriate units of measure and indicated on drawings along with lines, symbols, and notes to define the geometrical characteristic of an object.

*Effective Size:* See *Pitch Diameter, Functional Diameter.*

*Effective Thread:* The effective (or useful) thread includes the complete thread, and those portions of the incomplete thread which are fully formed at the root but not at the crest (in taper pipe threads it includes the so-called black crest threads); thus excluding the vanish thread.

*Error:* The algebraic difference between an observed or measured value beyond tolerance limits, and the specified value.

*External Thread:* A thread on a cylindrical or conical external surface.

*Fit:* Fit is the relationship resulting from the designed difference, before assembly, between the sizes of two mating parts which are to be assembled.

*Flank:* The flank of a thread is either surface connecting the crest with the root. The flank surface intersection with an axial plane is theoretically a straight line.

*Flank Angle:* The flank angles are the angles between the individual flanks and the perpendicular to the axis of the thread, measured in an axial plane. A flank angle of a symmetrical thread is commonly termed the half-angle of thread.

*Flank Diametral Displacement:* In a boundary profile defined system, flank diametral displacement is twice the radial distance between the straight thread flank segments of the maximum and minimum boundary profiles. The value of flank diametral displacement is equal to pitch diameter tolerance in a pitch line reference thread system.

*Height of Thread:* The height (or depth) of thread is the distance, measured radially, between the major and minor cylinders or cones, respectively.

*Helix Angle:* On a straight thread, the helix angle is the angle made by the helix of the thread and its relation to the thread axis. On a taper thread, the helix angle at a given axial position is the angle made by the conical spiral of the thread with the axis of the thread. The helix angle is the complement of the lead angle. (See also page 2202 for diagram.)

*Higbee Cut:* See *Blunt Start Thread*.

*Imperfect Thread:* See *Incomplete Thread*.

*Included Angle:* This is the angle between the flanks of the thread measured in an axial plane.

*Incomplete Thread:* A threaded profile having either crests or roots or both, not fully formed, resulting from their intersection with the cylindrical or end surface of the work or the vanish cone. It may occur at either end of the thread.

*Interference Fit:* A fit having limits of size so prescribed that an interference always results when mating parts are assembled.

*Internal Thread:* A thread on a cylindrical or conical internal surface.

*Lead:* Lead is the axial distance between two consecutive points of intersection of a helix by a line parallel to the axis of the cylinder on which it lies, i.e., the axial movement of a threaded part rotated one turn in its mating thread.

*Lead Angle:* On a straight thread, the lead angle is the angle made by the helix of the thread at the pitch line with a plane perpendicular to the axis. On a taper thread, the lead angle at a given axial position is the angle made by the conical spiral of the thread with the perpendicular to the axis at the pitch line.

*Lead Thread:* That portion of the incomplete thread that is fully formed at the root but not fully formed at the crest that occurs at the entering end of either an external or internal thread.

*Left-Hand Thread:* A thread is a left-hand thread if, when viewed axially, it winds in a counterclockwise and receding direction. Left-hand threads are designated LH.

*Length of Complete Thread:* The axial length of a thread section having full form at both crest and root but also including a maximum of two pitches at the start of the thread which may have a chamfer or incomplete crests.

*Length of Thread Engagement:* The length of thread engagement of two mating threads is the axial distance over which the two threads, each having full form at both crest and root, are designed to contact. (See also *Length of Complete Thread*.)

*Limits of Size:* The applicable maximum and minimum sizes.

*Major Clearance:* The radial distance between the root of the internal thread and the crest of the external thread of the coaxially assembled designed forms of mating threads.

*Major Cone:* The imaginary cone that would bound the crests of an external taper thread or the roots of an internal taper thread.

*Major Cylinder:* The imaginary cylinder that would bound the crests of an external straight thread or the roots of an internal straight thread.

*Major Diameter:* On a straight thread the major diameter is that of the major cylinder. On a taper thread the major diameter at a given position on the thread axis is that of the major cone at that position. (See also *Major Cylinder* and *Major Cone*.)

*Maximum Material Condition:* (*MMC*): The condition where a feature of size contains the maximum amount of material within the stated limits of size. For example, minimum internal thread size or maximum external thread size.

*Minimum Material Condition:* (*Least Material Condition* (*LMC*)): The condition where a feature of size contains the least amount of material within the stated limits of size. For example, maximum internal thread size or minimum external thread size.

*Minor Clearance:* The radial distance between the crest of the internal thread and the root of the external thread of the coaxially assembled design forms of mating threads.

*Minor Cone:* The imaginary cone that would bound the roots of an external taper thread or the crests of an internal taper thread.

*Minor Cylinder:* The imaginary cylinder that would bound the roots of an external straight thread or the crests of an internal straight thread.

*Minor Diameter:* On a straight thread the minor diameter is that of the minor cylinder. On a taper thread the minor diameter at a given position on the thread axis is that of the minor cone at that position. (See also *Minor Cylinder* and *Minor Cone*.)

*Multiple-Start Thread:* A thread in which the lead is an integral multiple, other than one, of the pitch.

*Nominal Size:* Designation used for general identification.

*Parallel Thread:* See *Screw Thread*.

*Partial Thread:* See *Vanish Thread*.

*Pitch:* The pitch of a thread having uniform spacing is the distance measured parallel with its axis between corresponding points on adjacent thread forms in the same axial plane and on the same side of the axis. Pitch is equal to the lead divided by the number of thread starts.

*Pitch Cone:* The pitch cone is an imaginary cone of such apex angle and location of its vertex and axis that its surface would pass through a taper thread in such a manner as to make the widths of the thread ridge and the thread groove equal. It is, therefore, located equidistantly between the sharp major and minor cones of a given thread form. On a theoretically perfect taper thread, these widths are equal to one-half the basic pitch. (See also *Axis of Thread* and *Pitch Diameter*.)

*Pitch Cylinder:* The pitch cylinder is an imaginary cylinder of such diameter and location of its axis that its surface would pass through a straight thread in such a manner as to make the widths of the thread ridge and groove equal. It is, therefore, located equidistantly between the sharp major and minor cylinders of a given thread form. On a theoretically perfect thread these widths are equal to one-half the basic pitch. (See also *Axis of Thread* and *Pitch Diameter*.)

*Pitch Diameter:* On a straight thread the pitch diameter is the diameter of the pitch cylinder. On a taper thread the pitch diameter at a given position on the thread axis is the diameter of the pitch cone at that position. *Note:* When the crest of a thread is truncated beyond the pitch line, the pitch diameter and pitch cylinder or pitch cone would be based on a theoretical extension of the thread flanks.

*Pitch Diameter, Functional Diameter:* The functional diameter is the pitch diameter of an enveloping thread with perfect pitch, lead, and flank angles and having a specified length of engagement. It includes the cumulative effect of variations in lead (pitch), flank angle, taper, straightness, and roundness. Variations at the thread crest and root are excluded. Other, nonpreferred terms are *virtual diameter, effective size, virtual effective diameter,* and *thread assembly diameter.*

*Pitch Line:* The generator of the cylinder or cone specified in *Pitch Cylinder* and *Pitch Cone*.

*Right-Hand Thread:* A thread is a fight-hand thread if, when viewed axially, it winds in a clockwise and receding direction. A thread is considered to be right-hand unless specifically indicated otherwise.

*Root:* That surface of the thread which joins the flanks of adjacent thread forms and is immediately adjacent to the cylinder or cone from which the thread projects.

*Root Truncation:* The radial distance between the sharp root (root apex) and the cylinder or cone that would bound the root. See also *Sharp Root (Root Apex)*.

*Runout:* As applied to screw threads, unless otherwise specified, runout refers to circular runout of major and minor cylinders with respect to the pitch cylinder. Circular runout, in accordance with ASME Y14.5, controls cumulative variations of circularity and coaxiality. Runout includes variations due to eccentricity and out-of-roundness. The amount of runout is usually expressed in terms of full indicator movement (FIM).

*Screw Thread:* A screw thread is a continuous and projecting helical ridge usually of uniform section on a cylindrical or conical surface.

*Sharp Crest (Crest Apex):* The apex formed by the intersection of the flanks of a thread when extended, if necessary, beyond the crest.

*Sharp Root (Root Apex):* The apex formed by the intersection of the adjacent flanks of adjacent threads when extended, if necessary, beyond the root.

*Standoff:* The axial distance between specified reference points on external and internal taper thread members or gages, when assembled with a specified torque or under other specified conditions.

*Straight Thread:* A straight thread is a screw thread projecting from a cylindrical surface.

*Taper Thread:* A taper thread is a screw thread projecting from a conical surface.

*Tensile Stress Area:* The tensile stress area is an arbitrarily selected area for computing the tensile strength of an externally threaded fastener so that the fastener strength is consistent with the basic material strength of the fastener. It is typically defined as a function of pitch diameter and/or minor diameter to calculate a circular cross section of the fastener correcting for the notch and helix effects of the threads.

*Thread:* A thread is a portion of a screw thread encompassed by one pitch. On a single-start thread it is equal to one turn. (See also *Threads per Inch* and *Turns per Inch.*)

*Thread Angle:* See *Included Angle.*

*Thread Runout:* See *Vanish Thread.*

*Thread Series:* Thread Series are groups of diameter/pitch combinations distinguished from each other by the number of threads per inch applied to specific diameters.

*Thread Shear Area:* The thread shear area is the total ridge cross-sectional area intersected by a specified cylinder with diameter and length equal to the mating thread engagement. Usually the cylinder diameter for external thread shearing is the minor diameter of the internal thread and for internal thread shearing it is the major diameter of the external thread.

*Threads per Inch:* The number of threads per inch is the reciprocal of the axial pitch in inches.

*Tolerance:* The total amount by which a specific dimension is permitted to vary. The tolerance is the difference between the maximum and minimum limits.

*Tolerance Class:* (*metric*): The tolerance class (metric) is the combination of a tolerance position with a tolerance grade. It specifies the allowance (fundamental deviation), pitch diameter tolerance (flank diametral displacement), and the crest diameter tolerance.

*Tolerance Grade:* (*metric*): The tolerance grade (metric) is a numerical symbol that designates the tolerances of crest diameters and pitch diameters applied to the design profiles.

*Tolerance Limit:* The variation, positive or negative, by which a size is permitted to depart from the design size.

*Tolerance Position:* (*metric*): The tolerance position (metric) is a letter symbol that designates the position of the tolerance zone in relation to the basic size. This position provides the allowance (fundamental deviation).

*Total Thread:* Includes the complete and all the incomplete thread, thus including the vanish thread and the lead thread.

*Transition Fit:* A fit having limits of size so prescribed that either a clearance or an interference may result when mating parts are assembled.

*Turns per Inch:* The number of turns per inch is the reciprocal of the lead in inches.

*Unilateral Tolerance:* A tolerance in which variation is permitted in one direction from the specified dimension.

*Vanish Thread:* (*Partial Thread, Washout Thread,* or *Thread Runout*): That portion of the incomplete thread which is not fully formed at the root or at crest and root. It is produced by the chamfer at the starting end of the thread forming tool.

*Virtual Diameter:* See *Pitch Diameter, Functional Diameter.*

*Washout Thread:* See *Vanish Thread.*

# UNIFIED SCREW THREADS

## American Standard for Unified Screw Threads

American Standard B1.1-1949 was the first American standard to cover those Unified Thread Series agreed upon by the United Kingdom, Canada, and the United States to obtain screw thread interchangeability among these three nations. These Unified threads are now the basic American standard for fastening types of screw threads. In relation to previous American practice, Unified threads have substantially the same thread form and are mechanically interchangeable with the former American National threads of the same diameter and pitch.

The principal differences between the two systems lie in: 1) application of allowances; 2) variation of tolerances with size; 3) difference in amount of pitch diameter tolerance on external and internal threads; and 4) differences in thread designation.

In the Unified system an allowance is provided on both the Classes 1A and 2A external threads whereas in the American National system only the Class I external thread has an allowance. Also, in the Unified system, the pitch diameter tolerance of an internal thread is 30 percent greater than that of the external thread, whereas they are equal in the American National system.

The Unified standard is designed to correct certain production difficulties resulting from the former standard. Often, under the old system, the tolerances of the product were practically absorbed by the combined tool and gage tolerances, leaving little for a working tolerance in manufacture. Somewhat greater tolerances are now provided for nut threads. As contrasted with the old "classes of fit" 1, 2, and 3, for each of which the pitch diameter tolerance on the external and internal threads were equal, the Classes 1B, 2B, and 3B (internal) threads in the new standard have, respectively, a 30 percent larger pitch diameter tolerance than the 1A, 2A, and 3A (external) threads. Relatively more tolerance is provided for fine threads than for coarse threads of the same pitch. Where previous tolerances were more liberal than required, they were reduced.

**Revised Standard.—**The revised screw thread Standard ANSI/ASME B1.1-2019 is much the same as that of ANSI B1.1-1982. The latest symbols in accordance with ANSI/ASME B1.7-2006 (R2021) Nomenclature, are used. Acceptability criteria are described in ANSI/ASME B1.3-2007 (R2022) Screw Thread Gaging Systems for Acceptability: Inch and Metric Screw Threads (UN, UNR, UNJ, M, and MJ).

Where the letters U, A or B do not appear in the thread designations, the threads conform to the outdated American National screw threads.

**Thread Form.—**The Design Profiles for Unified screw threads, shown on page 1951, define the maximum material condition for external and internal threads with no allowance and are derived from the Basic Profile, shown on page 1944.

*UN External Screw Threads:* A flat root contour is specified, but it is necessary to provide for some threading tool crest wear, hence a rounded root contour cleared beyond the $0.25P$ flat width of the Basic Profile is optional.

*UNR External Screw Threads:* To reduce the rate of threading tool crest wear and to improve fatigue strength of a flat root thread, the Design Profile of the UNR thread has a smooth, continuous, non-reversing contour with a radius of curvature not less than $0.108P$ at any point and blends tangentially into the flanks and any straight segment. At the maximum material condition, the point of tangency is specified to be at a distance not less than $0.625H$ (where $H$ is the height of a sharp V-thread) below the basic major diameter.

*UN and UNR External Screw Threads:* The Design Profiles of both UN and UNR external screw threads have flat crests. However, in practice, product threads are produced with partially or completely rounded crests. A rounded crest tangent at $0.125P$ flat is shown as an option on page 1951.

*UN Internal Screw Thread:* In practice it is necessary to provide for some threading tool crest wear, therefore the root of the Design Profile is rounded and cleared beyond the 0.125*P* flat width of the Basic Profile. There is no internal UNR screw thread.

**American National Standard Unified Internal and External Screw Thread Design Profiles (Maximum Material Condition)**

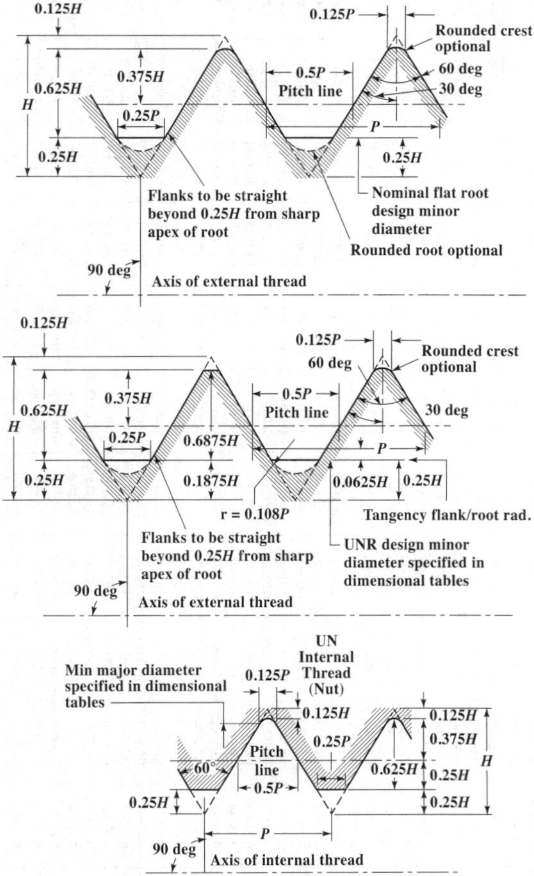

(*H* = height of sharp V-thread = 0.86603 × pitch)

**Thread Series.—**Thread series are groups of diameter-pitch combinations distinguished from each other by the numbers of threads per inch applied to a specific diameter. The various diameter-pitch combinations of eleven standard series are shown in Table 2. The limits of size of threads in the eleven standard series together with certain selected combinations of diameter and pitch, as well as the symbols for designating the various threads, are given in Table 3. *(Text continues on page 1981.)*

**Table 1. American Standard Unified Inch Screw Thread Form Data**

| Threads per Inch $n$ | Pitch $P$ | Depth of Sharp V-Thread $0.86603P$ | Depth of Int. Thd. and UN Ext. Thd.[a] $0.54127P$ | Depth of UNR Ext. Thd. $0.59539P$ | Truncation of Ext. Thd. Root $0.21651P$ | Truncation of UNR Ext. Thd. Root[b] $0.16238P$ | Truncation of Ext. Thd. Crest $0.10825P$ | Truncation of Int. Thd. Root $0.10825P$ | Truncation of Int. Thd. Crest $0.2165P$ | Flat at Ext. Thd. Crest and Int. Thd. Root $0.125P$ | Basic Flat at Int. Thd. Crest[c] $0.25P$ | Maximum Ext. Thd. Root Radius $0.14434P$ | Addendum of Ext. Thd. $0.32476P$ |
|---|---|---|---|---|---|---|---|---|---|---|---|---|---|
| 80 | 0.01250 | 0.01083 | 0.00677 | 0.00744 | 0.00271 | 0.00203 | 0.00135 | 0.00135 | 0.00271 | 0.00156 | 0.00312 | 0.00180 | 0.00406 |
| 72 | 0.01389 | 0.01203 | 0.00752 | 0.00827 | 0.00301 | 0.00226 | 0.00150 | 0.00150 | 0.00301 | 0.00174 | 0.00347 | 0.00200 | 0.00451 |
| 64 | 0.01563 | 0.01353 | 0.00846 | 0.00930 | 0.00338 | 0.00254 | 0.00169 | 0.00169 | 0.00338 | 0.00195 | 0.00391 | 0.00226 | 0.00507 |
| 56 | 0.01786 | 0.01546 | 0.00967 | 0.01063 | 0.00387 | 0.00290 | 0.00193 | 0.00193 | 0.00387 | 0.00223 | 0.00446 | 0.00258 | 0.00580 |
| 48 | 0.02083 | 0.01804 | 0.01128 | 0.01240 | 0.00451 | 0.00338 | 0.00226 | 0.00226 | 0.00451 | 0.00260 | 0.00521 | 0.00301 | 0.00677 |
| 44 | 0.02273 | 0.01968 | 0.01230 | 0.01353 | 0.00492 | 0.00369 | 0.00246 | 0.00246 | 0.00492 | 0.00284 | 0.00568 | 0.00328 | 0.00738 |
| 40 | 0.02500 | 0.02165 | 0.01353 | 0.01488 | 0.00541 | 0.00406 | 0.00271 | 0.00271 | 0.00541 | 0.00312 | 0.00625 | 0.00361 | 0.00812 |
| 36 | 0.02778 | 0.02406 | 0.01504 | 0.01654 | 0.00601 | 0.00451 | 0.00301 | 0.00301 | 0.00601 | 0.00347 | 0.00694 | 0.00401 | 0.00902 |
| 32 | 0.03125 | 0.02706 | 0.01691 | 0.01861 | 0.00677 | 0.00507 | 0.00338 | 0.00338 | 0.00677 | 0.00391 | 0.00781 | 0.00451 | 0.01015 |
| 28 | 0.03571 | 0.03093 | 0.01933 | 0.02126 | 0.00773 | 0.00580 | 0.00387 | 0.00387 | 0.00773 | 0.00446 | 0.00893 | 0.00515 | 0.01160 |
| 27 | 0.03704 | 0.03208 | 0.02005 | 0.02205 | 0.00802 | 0.00601 | 0.00401 | 0.00401 | 0.00802 | 0.00463 | 0.00926 | 0.00535 | 0.01203 |
| 24 | 0.04167 | 0.03608 | 0.02255 | 0.02481 | 0.00902 | 0.00677 | 0.00451 | 0.00451 | 0.00902 | 0.00521 | 0.01042 | 0.00601 | 0.01353 |
| 20 | 0.05000 | 0.04330 | 0.02706 | 0.02977 | 0.01083 | 0.00812 | 0.00541 | 0.00541 | 0.01083 | 0.00625 | 0.01250 | 0.00722 | 0.01624 |
| 18 | 0.05556 | 0.04811 | 0.03007 | 0.03308 | 0.01203 | 0.00902 | 0.00601 | 0.00601 | 0.01203 | 0.00694 | 0.01389 | 0.00802 | 0.01804 |
| 16 | 0.06250 | 0.05413 | 0.03383 | 0.03721 | 0.01353 | 0.01015 | 0.00677 | 0.00677 | 0.01353 | 0.00781 | 0.01562 | 0.00902 | 0.02030 |
| 14 | 0.07143 | 0.06186 | 0.03866 | 0.04253 | 0.01546 | 0.01160 | 0.00773 | 0.00773 | 0.01546 | 0.00893 | 0.01786 | 0.01031 | 0.02320 |
| 13 | 0.07692 | 0.06662 | 0.04164 | 0.04580 | 0.01655 | 0.01249 | 0.00833 | 0.00833 | 0.01665 | 0.00962 | 0.01923 | 0.01110 | 0.02498 |
| 12 | 0.08333 | 0.07217 | 0.04511 | 0.04962 | 0.01804 | 0.01353 | 0.00902 | 0.00902 | 0.01804 | 0.01042 | 0.02083 | 0.01203 | 0.02706 |
| 11½ | 0.08696 | 0.07531 | 0.04707 | 0.05177 | 0.01883 | 0.01412 | 0.00941 | 0.00941 | 0.01883 | 0.01087 | 0.02174 | 0.01255 | 0.02824 |
| 11 | 0.09091 | 0.07873 | 0.04921 | 0.05413 | 0.01968 | 0.01476 | 0.00984 | 0.00984 | 0.01968 | 0.01136 | 0.02273 | 0.01312 | 0.02952 |
| 10 | 0.10000 | 0.08660 | 0.05413 | 0.05954 | 0.02165 | 0.01624 | 0.01083 | 0.01083 | 0.02165 | 0.01250 | 0.02500 | 0.01443 | 0.03248 |
| 9 | 0.11111 | 0.09623 | 0.06014 | 0.06615 | 0.02406 | 0.01804 | 0.01203 | 0.01203 | 0.02406 | 0.01389 | 0.02778 | 0.01604 | 0.03608 |
| 8 | 0.12500 | 0.10825 | 0.06766 | 0.07442 | 0.02706 | 0.02030 | 0.01353 | 0.01353 | 0.02706 | 0.01562 | 0.03125 | 0.01804 | 0.04059 |
| 7 | 0.14286 | 0.12372 | 0.07732 | 0.08506 | 0.03093 | 0.02320 | 0.01546 | 0.01546 | 0.03093 | 0.01786 | 0.03571 | 0.02062 | 0.04639 |
| 6 | 0.16667 | 0.14434 | 0.09021 | 0.09923 | 0.03608 | 0.02706 | 0.01804 | 0.01804 | 0.03608 | 0.02083 | 0.04167 | 0.02406 | 0.05413 |
| 5 | 0.20000 | 0.17321 | 0.10825 | 0.11908 | 0.04330 | 0.03248 | 0.02165 | 0.02165 | 0.04330 | 0.02500 | 0.05000 | 0.02887 | 0.06495 |
| 4½ | 0.22222 | 0.19245 | 0.12028 | 0.13231 | 0.04811 | 0.03608 | 0.02406 | 0.02406 | 0.04811 | 0.02778 | 0.05556 | 0.03208 | 0.07217 |
| 4 | 0.25000 | 0.21651 | 0.13532 | 0.14885 | 0.05413 | 0.04059 | 0.02706 | 0.02706 | 0.05413 | 0.03125 | 0.06250 | 0.03608 | 0.08119 |

[a] Also depth of thread engagement.
[b] Design profile.
[c] Also basic flat at external UN thread root.
All dimensions are in inches.

**Table 2. Diameter-Pitch Combinations for Standard Series of Threads (UN/UNR)**

| Sizes[a] No. or Inches | Basic Major Dia. Inches | Threads per Inch | | | | | | | | | | |
|---|---|---|---|---|---|---|---|---|---|---|---|---|
| | | Series with Graded Pitches | | | Series with Uniform (Constant) Pitches | | | | | | | |
| | | Coarse UNC | Fine[b] UNF | Extra fine[c] UNEF | 4-UN | 6-UN | 8-UN | 12-UN | 16-UN | 20-UN | 28-UN | 32-UN |
| 0 | 0.0600 | … | 80 | Series designation shown indicates the UN thread form; however, the UNR thread form may be specified by substituting UNR in place of UN in all designations for external threads. | | | | | | | | |
| (1) | 0.0730 | 64 | 72 | | | | | | | | | |
| 2 | 0.0860 | 56 | 64 | | | | | | | | | |
| (3) | 0.0990 | 48 | 56 | | | | | | | | | |
| 4 | 0.1120 | 40 | 48 | | | | | | | | | |
| 5 | 0.1250 | 40 | 44 | … | … | … | … | … | … | … | … | … |
| 6 | 0.1380 | 32 | 40 | … | … | … | … | … | … | … | … | UNC |
| 8 | 0.1640 | 32 | 36 | … | … | … | … | … | … | … | … | UNC |
| 10 | 0.1900 | 24 | 32 | … | … | … | … | … | … | … | … | UNF |
| (12) | 0.2160 | 24 | 28 | 32 | … | … | … | … | … | … | UNF | UNEF |
| 1/4 | 0.2500 | 20 | 28 | 32 | … | … | … | … | … | UNC | UNF | UNEF |
| 5/16 | 0.3125 | 18 | 24 | 32 | … | … | … | … | … | 20 | 28 | UNEF |
| 3/8 | 0.3750 | 16 | 24 | 32 | … | … | … | … | UNC | 20 | 28 | UNEF |
| 7/16 | 0.4375 | 14 | 20 | 28 | … | … | … | … | 16 | UNF | UNEF | 32 |
| 1/2 | 0.5000 | 13 | 20 | 28 | … | … | … | … | 16 | UNF | UNEF | 32 |
| 9/16 | 0.5625 | 12 | 18 | 24 | … | … | … | UNC | 16 | 20 | 28 | 32 |
| 5/8 | 0.6250 | 11 | 18 | 24 | … | … | … | 12 | 16 | 20 | 28 | 32 |
| (11/16) | 0.6875 | … | … | 24 | … | … | … | 12 | 16 | 20 | 28 | 32 |
| 3/4 | 0.7500 | 10 | 16 | 20 | … | … | … | 12 | UNF | UNEF | 28 | 32 |
| (13/16) | 0.8125 | … | … | 20 | … | … | … | 12 | 16 | UNEF | 28 | 32 |
| 7/8 | 0.8750 | 9 | 14 | 20 | … | … | … | 12 | 16 | UNEF | 28 | 32 |
| (15/16) | 0.9375 | … | … | 20 | … | … | … | 12 | 16 | UNEF | 28 | 32 |
| 1 | 1.0000 | 8 | 12 | 20 | … | … | UNC | UNF | 16 | UNEF | 28 | 32 |
| (1 1/16) | 1.0625 | … | … | 18 | … | … | 8 | 12 | 16 | 20 | 28 | … |
| 1 1/8 | 1.1250 | 7 | 12 | 18 | … | … | 8 | UNF | 16 | 20 | 28 | … |
| (1 3/16) | 1.1875 | … | … | 18 | … | … | 8 | 12 | 16 | 20 | 28 | … |
| 1 1/4 | 1.2500 | 7 | 12 | 18 | … | … | 8 | UNF | 16 | 20 | 28 | … |
| 1 5/16 | 1.3125 | … | … | 18 | … | … | 8 | 12 | 16 | 20 | 28 | … |
| 1 3/8 | 1.3750 | 6 | 12 | 18 | … | UNC | 8 | UNF | 16 | 20 | 28 | … |
| (1 7/16) | 1.4375 | … | … | 18 | … | 6 | 8 | 12 | 16 | 20 | 28 | … |
| 1 1/2 | 1.5000 | 6 | 12 | 18 | … | UNC | 8 | UNF | 16 | 20 | 28 | … |
| (1 9/16) | 1.5625 | … | … | 18 | … | 6 | 8 | 12 | 16 | 20 | … | … |
| 1 5/8 | 1.6250 | … | … | 18 | … | 6 | 8 | 12 | 16 | 20 | … | … |
| (1 11/16) | 1.6875 | … | … | 18 | … | 6 | 8 | 12 | 16 | 20 | … | … |
| 1 3/4 | 1.7500 | 5 | … | … | … | 6 | 8 | 12 | 16 | 20 | … | … |
| (1 13/16) | 1.8125 | … | … | … | … | 6 | 8 | 12 | 16 | 20 | … | … |
| 1 7/8 | 1.8750 | … | … | … | … | 6 | 8 | 12 | 16 | 20 | … | … |
| (1 15/16) | 1.9375 | … | … | … | … | 6 | 8 | 12 | 16 | 20 | … | … |
| 2 | 2.0000 | 4 1/2 | … | … | … | 6 | 8 | 12 | 16 | 20 | … | … |
| (2 1/8) | 2.1250 | … | … | … | … | 6 | 8 | 12 | 16 | 20 | … | … |
| 2 1/4 | 2.2500 | 4 1/2 | … | … | … | 6 | 8 | 12 | 16 | 20 | … | … |
| (2 3/8) | 2.3750 | … | … | … | … | 6 | 8 | 12 | 16 | 20 | … | … |
| 2 1/2 | 2.5000 | 4 | … | … | UNC | 6 | 8 | 12 | 16 | 20 | … | … |
| (2 5/8) | 2.6250 | … | … | … | 4 | 6 | 8 | 12 | 16 | 20 | … | … |
| 2 3/4 | 2.7500 | 4 | … | … | UNC | 6 | 8 | 12 | 16 | 20 | … | … |
| (2 7/8) | 2.8750 | … | … | … | 4 | 6 | 8 | 12 | 16 | 20 | … | … |
| 3 | 3.0000 | 4 | … | … | UNC | 6 | 8 | 12 | 16 | 20 | … | … |
| (3 1/8) | 3.1250 | … | … | … | 4 | 6 | 8 | 12 | 16 | … | … | … |
| 3 1/4 | 3.2500 | 4 | … | … | UNC | 6 | 8 | 12 | 16 | … | … | … |
| (3 3/8) | 3.3750 | … | … | … | 4 | 6 | 8 | 12 | 16 | … | … | … |
| 3 1/2 | 3.5000 | 4 | … | … | UNC | 6 | 8 | 12 | 16 | … | … | … |
| (3 5/8) | 3.6250 | … | … | … | 4 | 6 | 8 | 12 | 16 | … | … | … |
| 3 3/4 | 3.7500 | 4 | … | … | UNC | 6 | 8 | 12 | 16 | … | … | … |
| (3 7/8) | 3.8750 | … | … | … | 4 | 6 | 8 | 12 | 16 | … | … | … |
| 4 | 4.0000 | 4 | … | … | UNC | 6 | 8 | 12 | 16 | … | … | … |

[a] Sizes shown in parentheses are secondary sizes. Primary sizes of $4\frac{1}{4}$, $4\frac{1}{2}$, $4\frac{3}{4}$, 5, $5\frac{1}{4}$, $5\frac{1}{2}$, $5\frac{3}{4}$ and 6 inches also are in the 4, 6, 8, 12, and 16 thread series; secondary sizes of $4\frac{1}{8}$, $4\frac{3}{8}$, $4\frac{5}{8}$, $4\frac{7}{8}$, $5\frac{1}{8}$, $5\frac{3}{8}$, $5\frac{5}{8}$, and $5\frac{7}{8}$ also are in the 4, 6, 8, 12, and 16 thread series.

[b] For diameters over $1\frac{1}{2}$ inches, use 12-thread series.

[c] For diameters over $1\frac{11}{16}$ inches, use 16-thread series.

For UNR thread form substitute UNR for UN for external threads only.

## Table 3. Standard Series and Selected Combinations — Unified Screw Threads

| Nominal Size, Threads per Inch, and Series Designation[a] | External[b] | | | | | | | | Internal[b] | | | | | |
|---|---|---|---|---|---|---|---|---|---|---|---|---|---|---|
| | Class | Allowance | Major Diameter | | | Pitch Diameter | | UNR Minor Dia.,[c] Max (Ref.) | Class | Minor Diameter | | Pitch Diameter | | Major Diameter |
| | | | Max[d] | Min | Min[e] | Max[d] | Min | | | Min | Max | Min | Max | Min |
| 0-80 UNF | 2A | 0.0005 | 0.0595 | 0.0563 | — | 0.0514 | 0.0496 | 0.0446 | 2B | 0.0465 | 0.0514 | 0.0519 | 0.0542 | 0.0600 |
| | 3A | 0.0000 | 0.0600 | 0.0568 | — | 0.0519 | 0.0506 | 0.0451 | 3B | 0.0465 | 0.0514 | 0.0519 | 0.0536 | 0.0600 |
| 1-64 UNC | 2A | 0.0006 | 0.0724 | 0.0686 | — | 0.0623 | 0.0603 | 0.0538 | 2B | 0.0561 | 0.0622 | 0.0629 | 0.0655 | 0.0730 |
| | 3A | 0.0000 | 0.0730 | 0.0692 | — | 0.0629 | 0.0614 | 0.0544 | 3B | 0.0561 | 0.0622 | 0.0629 | 0.0648 | 0.0730 |
| 1-72 UNF | 2A | 0.0006 | 0.0724 | 0.0689 | — | 0.0634 | 0.0615 | 0.0559 | 2B | 0.0580 | 0.0634 | 0.0640 | 0.0665 | 0.0730 |
| | 3A | 0.0000 | 0.0730 | 0.0695 | — | 0.0640 | 0.0626 | 0.0565 | 3B | 0.0580 | 0.0634 | 0.0640 | 0.0659 | 0.0730 |
| 2-56 UNC | 2A | 0.0006 | 0.0854 | 0.0813 | — | 0.0738 | 0.0717 | 0.0641 | 2B | 0.0667 | 0.0737 | 0.0744 | 0.0772 | 0.0860 |
| | 3A | 0.0000 | 0.0860 | 0.0819 | — | 0.0744 | 0.0728 | 0.0647 | 3B | 0.0667 | 0.0737 | 0.0744 | 0.0765 | 0.0860 |
| 2-64 UNF | 2A | 0.0006 | 0.0854 | 0.0816 | — | 0.0753 | 0.0733 | 0.0668 | 2B | 0.0691 | 0.0752 | 0.0759 | 0.0786 | 0.0860 |
| | 3A | 0.0000 | 0.0860 | 0.0822 | — | 0.0759 | 0.0744 | 0.0674 | 3B | 0.0691 | 0.0752 | 0.0759 | 0.0779 | 0.0860 |
| 3-48 UNC | 2A | 0.0007 | 0.0983 | 0.0938 | — | 0.0848 | 0.0825 | 0.0735 | 2B | 0.0764 | 0.0845 | 0.0855 | 0.0885 | 0.0990 |
| | 3A | 0.0000 | 0.0990 | 0.0945 | — | 0.0855 | 0.0838 | 0.0742 | 3B | 0.0764 | 0.0845 | 0.0855 | 0.0877 | 0.0990 |
| 3-56 UNF | 2A | 0.0007 | 0.0983 | 0.0942 | — | 0.0867 | 0.0845 | 0.0770 | 2B | 0.0797 | 0.0865 | 0.0874 | 0.0902 | 0.0990 |
| | 3A | 0.0000 | 0.0990 | 0.0949 | — | 0.0874 | 0.0858 | 0.0777 | 3B | 0.0797 | 0.0865 | 0.0874 | 0.0895 | 0.0990 |
| 4-40 UNC | 2A | 0.0008 | 0.1112 | 0.1061 | — | 0.0950 | 0.0925 | 0.0814 | 2B | 0.0849 | 0.0939 | 0.0958 | 0.0991 | 0.1120 |
| | 3A | 0.0000 | 0.1120 | 0.1069 | — | 0.0958 | 0.0939 | 0.0822 | 3B | 0.0849 | 0.0939 | 0.0958 | 0.0982 | 0.1120 |
| 4-48 UNF | 2A | 0.0007 | 0.1113 | 0.1068 | — | 0.0978 | 0.0954 | 0.0865 | 2B | 0.0894 | 0.0968 | 0.0985 | 0.1016 | 0.1120 |
| | 3A | 0.0000 | 0.1120 | 0.1075 | — | 0.0985 | 0.0967 | 0.0872 | 3B | 0.0894 | 0.0968 | 0.0985 | 0.1008 | 0.1120 |
| 5-40 UNC | 2A | 0.0008 | 0.1242 | 0.1191 | — | 0.1080 | 0.1054 | 0.0944 | 2B | 0.0979 | 0.1062 | 0.1088 | 0.1121 | 0.1250 |
| | 3A | 0.0000 | 0.1250 | 0.1199 | — | 0.1088 | 0.1069 | 0.0952 | 3B | 0.0979 | 0.1062 | 0.1088 | 0.1113 | 0.1250 |
| 5-44 UNF | 2A | 0.0007 | 0.1243 | 0.1195 | — | 0.1095 | 0.1070 | 0.0972 | 2B | 0.1004 | 0.1079 | 0.1102 | 0.1134 | 0.1250 |
| | 3A | 0.0000 | 0.1250 | 0.1202 | — | 0.1102 | 0.1083 | 0.0979 | 3B | 0.1004 | 0.1079 | 0.1102 | 0.1126 | 0.1250 |
| 6-32 UNC | 2A | 0.0008 | 0.1372 | 0.1312 | — | 0.1169 | 0.1141 | 0.1000 | 2B | 0.104 | 0.114 | 0.1177 | 0.1214 | 0.1380 |
| | 3A | 0.0000 | 0.1380 | 0.1320 | — | 0.1177 | 0.1156 | 0.1008 | 3B | 0.1040 | 0.1139 | 0.1177 | 0.1204 | 0.1380 |
| 6-40 UNF | 2A | 0.0008 | 0.1372 | 0.1321 | — | 0.1210 | 0.1184 | 0.1074 | 2B | 0.111 | 0.119 | 0.1218 | 0.1252 | 0.1380 |
| | 3A | 0.0000 | 0.1380 | 0.1329 | — | 0.1218 | 0.1198 | 0.1082 | 3B | 0.1110 | 0.1186 | 0.1218 | 0.1243 | 0.1380 |
| 8-32 UNC | 2A | 0.0009 | 0.1631 | 0.1571 | — | 0.1428 | 0.1399 | 0.1259 | 2B | 0.130 | 0.139 | 0.1437 | 0.1475 | 0.1640 |
| | 3A | 0.0000 | 0.1640 | 0.1580 | — | 0.1437 | 0.1415 | 0.1268 | 3B | 0.1300 | 0.1388 | 0.1437 | 0.1465 | 0.1640 |
| 8-36 UNF | 2A | 0.0008 | 0.1632 | 0.1577 | — | 0.1452 | 0.1424 | 0.1301 | 2B | 0.134 | 0.142 | 0.1460 | 0.1496 | 0.1640 |
| | 3A | 0.0000 | 0.1640 | 0.1585 | — | 0.1460 | 0.1439 | 0.1309 | 3B | 0.1340 | 0.1416 | 0.1460 | 0.1487 | 0.1640 |

**Table 3.** *(Continued)* **Standard Series and Selected Combinations — Unified Screw Threads**

| Nominal Size, Threads per Inch, and Series Designation[a] | External[b] | | | | | | | | Internal[b] | | | | | |
|---|---|---|---|---|---|---|---|---|---|---|---|---|---|---|
| | Class | Allowance | Major Diameter | | | Pitch Diameter | | UNR Minor Dia.,[c] Max (Ref.) | Class | Minor Diameter | | Pitch Diameter | | Major Diameter |
| | | | Max[d] | Min | Min[e] | Max[d] | Min | | | Min | Max | Min | Max | Min |
| 10-24 UNC | 2A | 0.0010 | 0.1890 | 0.1818 | — | 0.1619 | 0.1586 | 0.1394 | 2B | 0.145 | 0.155 | 0.1629 | 0.1672 | 0.1900 |
| | 3A | 0.0000 | 0.1900 | 0.1828 | — | 0.1629 | 0.1604 | 0.1404 | 3B | 0.1450 | 0.1555 | 0.1629 | 0.1661 | 0.1900 |
| 10-28 UNS | 2A | 0.0010 | 0.1890 | 0.1825 | — | 0.1658 | 0.1625 | 0.1465 | 2B | 0.151 | 0.160 | 0.1668 | 0.1711 | 0.1900 |
| 10-32 UNF | 2A | 0.0009 | 0.1891 | 0.1831 | — | 0.1688 | 0.1658 | 0.1519 | 2B | 0.156 | 0.164 | 0.1697 | 0.1736 | 0.1900 |
| | 3A | 0.0000 | 0.1900 | 0.1840 | — | 0.1697 | 0.1674 | 0.1528 | 3B | 0.1560 | 0.1641 | 0.1697 | 0.1726 | 0.1900 |
| 10-36 UNS | 2A | 0.0009 | 0.1891 | 0.1836 | — | 0.1711 | 0.1681 | 0.1560 | 2B | 0.160 | 0.166 | 0.1720 | 0.1759 | 0.1900 |
| 10-40 UNS | 2A | 0.0009 | 0.1891 | 0.1840 | — | 0.1729 | 0.1700 | 0.1593 | 2B | 0.163 | 0.169 | 0.1738 | 0.1775 | 0.1900 |
| 10-48 UNS | 2A | 0.0008 | 0.1892 | 0.1847 | — | 0.1757 | 0.1731 | 0.1644 | 2B | 0.167 | 0.172 | 0.1765 | 0.1799 | 0.1900 |
| 10-56 UNS | 2A | 0.0007 | 0.1893 | 0.1852 | — | 0.1777 | 0.1752 | 0.1680 | 2B | 0.171 | 0.175 | 0.1784 | 0.1816 | 0.1900 |
| 12-24 UNC | 2A | 0.0010 | 0.2150 | 0.2078 | — | 0.1879 | 0.1845 | 0.1654 | 2B | 0.171 | 0.181 | 0.1889 | 0.1933 | 0.2160 |
| | 3A | 0.0000 | 0.2160 | 0.2088 | — | 0.1889 | 0.1863 | 0.1664 | 3B | 0.1710 | 0.1807 | 0.1889 | 0.1922 | 0.2160 |
| 12-28 UNF | 2A | 0.0010 | 0.2150 | 0.2085 | — | 0.1918 | 0.1886 | 0.1725 | 2B | 0.177 | 0.186 | 0.1928 | 0.1970 | 0.2160 |
| | 3A | 0.0000 | 0.2160 | 0.2095 | — | 0.1928 | 0.1904 | 0.1735 | 3B | 0.1770 | 0.1857 | 0.1928 | 0.1959 | 0.2160 |
| 12-32 UNEF | 2A | 0.0010 | 0.2150 | 0.2090 | — | 0.1947 | 0.1915 | 0.1778 | 2B | 0.182 | 0.190 | 0.1957 | 0.1998 | 0.2160 |
| | 3A | 0.0000 | 0.2160 | 0.2100 | — | 0.1957 | 0.1933 | 0.1788 | 3B | 0.1820 | 0.1895 | 0.1957 | 0.1988 | 0.2160 |
| 12-36 UNS | 2A | 0.0009 | 0.2151 | 0.2096 | — | 0.1971 | 0.1941 | 0.1820 | 2B | 0.186 | 0.193 | 0.1980 | 0.2019 | 0.2160 |
| 12-40 UNS | 2A | 0.0009 | 0.2151 | 0.2100 | — | 0.1989 | 0.1960 | 0.1853 | 2B | 0.189 | 0.195 | 0.1998 | 0.2036 | 0.2160 |
| 12-48 UNS | 2A | 0.0008 | 0.2152 | 0.2107 | — | 0.2017 | 0.1990 | 0.1904 | 2B | 0.193 | 0.198 | 0.2025 | 0.2060 | 0.2160 |
| 12-56 UNS | 2A | 0.0008 | 0.2152 | 0.2111 | — | 0.2036 | 0.2011 | 0.1939 | 2B | 0.197 | 0.201 | 0.2044 | 0.2077 | 0.2160 |
| 1⁄4-20 UNC | 1A | 0.0011 | 0.2489 | 0.2367 | — | 0.2164 | 0.2108 | 0.1894 | 1B | 0.196 | 0.207 | 0.2175 | 0.2248 | 0.2500 |
| | 2A | 0.0011 | 0.2489 | 0.2408 | 0.2367 | 0.2164 | 0.2127 | 0.1894 | 2B | 0.196 | 0.207 | 0.2175 | 0.2224 | 0.2500 |
| | 3A | 0.0000 | 0.2500 | 0.2419 | — | 0.2175 | 0.2147 | 0.1905 | 3B | 0.1960 | 0.2067 | 0.2175 | 0.2211 | 0.2500 |
| 1⁄4-24 UNS | 2A | 0.0011 | 0.2489 | 0.2417 | — | 0.2218 | 0.2181 | 0.1993 | 2B | 0.205 | 0.215 | 0.2229 | 0.2277 | 0.2500 |
| 1⁄4-27 UNS | 2A | 0.0010 | 0.2490 | 0.2423 | — | 0.2249 | 0.2214 | 0.2049 | 2B | 0.210 | 0.219 | 0.2259 | 0.2304 | 0.2500 |
| 1⁄4-28 UNF | 1A | 0.0010 | 0.2490 | 0.2392 | — | 0.2258 | 0.2208 | 0.2065 | 1B | 0.211 | 0.220 | 0.2268 | 0.2333 | 0.2500 |
| | 2A | 0.0010 | 0.2490 | 0.2425 | — | 0.2258 | 0.2225 | 0.2065 | 2B | 0.211 | 0.220 | 0.2268 | 0.2311 | 0.2500 |
| | 3A | 0.0000 | 0.2500 | 0.2435 | — | 0.2268 | 0.2243 | 0.2075 | 3B | 0.2110 | 0.2190 | 0.2268 | 0.2300 | 0.2500 |
| 1⁄4-32 UNEF | 2A | 0.0010 | 0.2490 | 0.2430 | — | 0.2287 | 0.2255 | 0.2118 | 2B | 0.216 | 0.224 | 0.2297 | 0.2339 | 0.2500 |
| | 3A | 0.0000 | 0.2500 | 0.2440 | — | 0.2297 | 0.2273 | 0.2128 | 3B | 0.2160 | 0.2229 | 0.2297 | 0.2328 | 0.2500 |

**Table 3.** *(Continued)* **Standard Series and Selected Combinations — Unified Screw Threads**

| Nominal Size, Threads per Inch, and Series Designation[a] | External[b] | | | | | | | | Internal[b] | | | | | |
|---|---|---|---|---|---|---|---|---|---|---|---|---|---|---|
| | Class | Allowance | Major Diameter | | | Pitch Diameter | | UNR Minor Dia.,[c] Max (Ref.) | Class | Minor Diameter | | Pitch Diameter | | Major Diameter |
| | | | Max[d] | Min | Min[e] | Max[d] | Min | | | Min | Max | Min | Max | Min |
| 1/4-36 UNS | 2A | 0.0009 | 0.2491 | 0.2436 | — | 0.2311 | 0.2280 | 0.2160 | 2B | 0.220 | 0.227 | 0.2320 | 0.2360 | 0.2500 |
| 1/4-40 UNS | 2A | 0.0009 | 0.2491 | 0.2440 | — | 0.2329 | 0.2300 | 0.2193 | 2B | 0.223 | 0.229 | 0.2338 | 0.2376 | 0.2500 |
| 1/4-48 UNS | 2A | 0.0008 | 0.2492 | 0.2447 | — | 0.2357 | 0.2330 | 0.2244 | 2B | 0.227 | 0.232 | 0.2365 | 0.2401 | 0.2500 |
| 1/4-56 UNS | 2A | 0.0008 | 0.2492 | 0.2451 | — | 0.2376 | 0.2350 | 0.2279 | 2B | 0.231 | 0.235 | 0.2384 | 0.2417 | 0.2500 |
| 5/16-18 UNC | 1A | 0.0012 | 0.3113 | 0.2982 | — | 0.2752 | 0.2691 | 0.2451 | 1B | 0.252 | 0.265 | 0.2764 | 0.2843 | 0.3125 |
| | 2A | 0.0012 | 0.3113 | 0.3026 | 0.2982 | 0.2752 | 0.2712 | 0.2451 | 2B | 0.252 | 0.265 | 0.2764 | 0.2817 | 0.3125 |
| | 3A | 0.0000 | 0.3125 | 0.3038 | — | 0.2764 | 0.2734 | 0.2463 | 3B | 0.2520 | 0.2630 | 0.2764 | 0.2803 | 0.3125 |
| 5/16-20 UN | 2A | 0.0012 | 0.3113 | 0.3032 | — | 0.2788 | 0.2747 | 0.2518 | 2B | 0.258 | 0.270 | 0.2800 | 0.2853 | 0.3125 |
| | 3A | 0.0000 | 0.3125 | 0.3044 | — | 0.2800 | 0.2770 | 0.2530 | 3B | 0.2580 | 0.2680 | 0.2800 | 0.2840 | 0.3125 |
| 5/16-24 UNF | 1A | 0.0011 | 0.3114 | 0.3006 | — | 0.2843 | 0.2788 | 0.2618 | 1B | 0.267 | 0.277 | 0.2854 | 0.2925 | 0.3125 |
| | 2A | 0.0011 | 0.3114 | 0.3042 | — | 0.2843 | 0.2806 | 0.2618 | 2B | 0.267 | 0.277 | 0.2854 | 0.2902 | 0.3125 |
| | 3A | 0.0000 | 0.3125 | 0.3053 | — | 0.2854 | 0.2827 | 0.2629 | 3B | 0.2670 | 0.2754 | 0.2854 | 0.2890 | 0.3125 |
| 5/16-27 UNS | 2A | 0.0011 | 0.3114 | 0.3047 | — | 0.2873 | 0.2837 | 0.2673 | 2B | 0.272 | 0.281 | 0.2884 | 0.2930 | 0.3125 |
| 5/16-28 UN | 2A | 0.0010 | 0.3115 | 0.3050 | — | 0.2883 | 0.2848 | 0.2690 | 2B | 0.274 | 0.282 | 0.2893 | 0.2938 | 0.3125 |
| | 3A | 0.0000 | 0.3125 | 0.3060 | — | 0.2893 | 0.2867 | 0.2700 | 3B | 0.2740 | 0.2807 | 0.2893 | 0.2927 | 0.3125 |
| 5/16-32 UNEF | 2A | 0.0010 | 0.3115 | 0.3055 | — | 0.2912 | 0.2879 | 0.2743 | 2B | 0.279 | 0.286 | 0.2922 | 0.2965 | 0.3125 |
| | 3A | 0.0000 | 0.3125 | 0.3065 | — | 0.2922 | 0.2897 | 0.2753 | 3B | 0.2790 | 0.2846 | 0.2922 | 0.2954 | 0.3125 |
| 5/16-36 UNS | 2A | 0.0009 | 0.3116 | 0.3061 | — | 0.2936 | 0.2905 | 0.2785 | 2B | 0.282 | 0.289 | 0.2945 | 0.2986 | 0.3125 |
| 5/16-40 UNS | 2A | 0.0009 | 0.3116 | 0.3065 | — | 0.2954 | 0.2924 | 0.2818 | 2B | 0.285 | 0.291 | 0.2963 | 0.3002 | 0.3125 |
| 5/16-48 UNS | 2A | 0.0008 | 0.3117 | 0.3072 | — | 0.2982 | 0.2954 | 0.2869 | 2B | 0.290 | 0.295 | 0.2990 | 0.3026 | 0.3125 |
| 3/8-16 UNC | 1A | 0.0013 | 0.3737 | 0.3595 | — | 0.3331 | 0.3266 | 0.2993 | 1B | 0.307 | 0.321 | 0.3344 | 0.3429 | 0.3750 |
| | 2A | 0.0013 | 0.3737 | 0.3643 | 0.3595 | 0.3331 | 0.3287 | 0.2993 | 2B | 0.307 | 0.321 | 0.3344 | 0.3401 | 0.3750 |
| | 3A | 0.0000 | 0.3750 | 0.3656 | — | 0.3344 | 0.3311 | 0.3006 | 3B | 0.3070 | 0.3182 | 0.3344 | 0.3387 | 0.3750 |
| 3/8-18 UNS | 2A | 0.0013 | 0.3737 | 0.3650 | — | 0.3376 | 0.3333 | 0.3075 | 2B | 0.315 | 0.328 | 0.3389 | 0.3445 | 0.3750 |
| 3/8-20 UN | 2A | 0.0012 | 0.3738 | 0.3657 | — | 0.3413 | 0.3372 | 0.3143 | 2B | 0.321 | 0.332 | 0.3425 | 0.3479 | 0.3750 |
| | 3A | 0.0000 | 0.3750 | 0.3669 | — | 0.3425 | 0.3394 | 0.3155 | 3B | 0.3210 | 0.3297 | 0.3425 | 0.3465 | 0.3750 |

**Table 3.** *(Continued)* **Standard Series and Selected Combinations — Unified Screw Threads**

| Nominal Size, Threads per Inch, and Series Designation[a] | External[b] | | | | | | | | Internal[b] | | | | | |
|---|---|---|---|---|---|---|---|---|---|---|---|---|---|---|
| | Class | Allowance | Major Diameter | | | Pitch Diameter | | UNR Minor Dia.,[c] Max (Ref.) | Class | Minor Diameter | | Pitch Diameter | | Major Diameter |
| | | | Max[d] | Min | Min[e] | Max[d] | Min | | | Min | Max | Min | Max | Min |
| 3/8-24 UNF | 1A | 0.0011 | 0.3739 | 0.3631 | — | 0.3468 | 0.3411 | 0.3243 | 1B | 0.330 | 0.340 | 0.3479 | 0.3553 | 0.3750 |
| | 2A | 0.0011 | 0.3739 | 0.3667 | — | 0.3468 | 0.3430 | 0.3243 | 2B | 0.330 | 0.340 | 0.3479 | 0.3528 | 0.3750 |
| | 3A | 0.0000 | 0.3750 | 0.3678 | — | 0.3479 | 0.3450 | 0.3254 | 3B | 0.3300 | 0.3372 | 0.3479 | 0.3516 | 0.3750 |
| 3/8-27 UNS | 2A | 0.0011 | 0.3739 | 0.3672 | — | 0.3498 | 0.3462 | 0.3298 | 2B | 0.335 | 0.344 | 0.3509 | 0.3556 | 0.3750 |
| 3/8-28 UN | 2A | 0.0011 | 0.3739 | 0.3674 | — | 0.3507 | 0.3471 | 0.3314 | 2B | 0.336 | 0.345 | 0.3518 | 0.3564 | 0.3750 |
| | 3A | 0.0000 | 0.3750 | 0.3685 | — | 0.3518 | 0.3491 | 0.3325 | 3B | 0.3360 | 0.3426 | 0.3518 | 0.3553 | 0.3750 |
| 3/8-32 UNEF | 2A | 0.0010 | 0.3740 | 0.3680 | — | 0.3537 | 0.3503 | 0.3368 | 2B | 0.341 | 0.349 | 0.3547 | 0.3591 | 0.3750 |
| | 3A | 0.0000 | 0.3750 | 0.3690 | — | 0.3547 | 0.3522 | 0.3378 | 3B | 0.3410 | 0.3469 | 0.3547 | 0.3580 | 0.3750 |
| 3/8-36 UNS | 2A | 0.0010 | 0.3740 | 0.3685 | — | 0.3560 | 0.3528 | 0.3409 | 2B | 0.345 | 0.352 | 0.3570 | 0.3612 | 0.3750 |
| 3/8-40 UNS | 2A | 0.0009 | 0.3741 | 0.3690 | — | 0.3579 | 0.3548 | 0.3443 | 2B | 0.348 | 0.354 | 0.3588 | 0.3628 | 0.3750 |
| 0.390-27 UNS | 2A | 0.0011 | 0.3889 | 0.3822 | — | 0.3648 | 0.3612 | 0.3448 | 2B | 0.350 | 0.359 | 0.3659 | 0.3706 | 0.3900 |
| 7/16-14 UNC | 1A | 0.0014 | 0.4361 | 0.4206 | — | 0.3897 | 0.3826 | 0.3510 | 1B | 0.360 | 0.376 | 0.3911 | 0.4003 | 0.4375 |
| | 2A | 0.0014 | 0.4361 | 0.4258 | 0.4206 | 0.3897 | 0.3850 | 0.3510 | 2B | 0.360 | 0.376 | 0.3911 | 0.3972 | 0.4375 |
| | 3A | 0.0000 | 0.4375 | 0.4272 | — | 0.3911 | 0.3876 | 0.3524 | 3B | 0.3600 | 0.3717 | 0.3911 | 0.3957 | 0.4375 |
| 7/16-16 UN | 2A | 0.0014 | 0.4361 | 0.4267 | — | 0.3955 | 0.3909 | 0.3617 | 2B | 0.370 | 0.384 | 0.3969 | 0.4029 | 0.4375 |
| | 3A | 0.0000 | 0.4375 | 0.4281 | — | 0.3969 | 0.3934 | 0.3631 | 3B | 0.3700 | 0.3800 | 0.3969 | 0.4014 | 0.4375 |
| 7/16-18 UNS | 2A | 0.0013 | 0.4362 | 0.4275 | — | 0.4001 | 0.3957 | 0.3700 | 2B | 0.377 | 0.390 | 0.4014 | 0.4071 | 0.4375 |
| 7/16-20 UNF | 1A | 0.0013 | 0.4362 | 0.4240 | — | 0.4037 | 0.3974 | 0.3767 | 1B | 0.383 | 0.395 | 0.4050 | 0.4131 | 0.4375 |
| | 2A | 0.0013 | 0.4362 | 0.4281 | — | 0.4037 | 0.3995 | 0.3767 | 2B | 0.383 | 0.395 | 0.4050 | 0.4104 | 0.4375 |
| | 3A | 0.0000 | 0.4375 | 0.4294 | — | 0.4050 | 0.4019 | 0.3780 | 3B | 0.3830 | 0.3916 | 0.4050 | 0.4091 | 0.4375 |
| 7/16-24 UNS | 2A | 0.0012 | 0.4363 | 0.4291 | — | 0.4092 | 0.4053 | 0.3867 | 2B | 0.392 | 0.402 | 0.4104 | 0.4154 | 0.4375 |
| 7/16-27 UNS | 2A | 0.0011 | 0.4364 | 0.4297 | — | 0.4123 | 0.4086 | 0.3923 | 2B | 0.397 | 0.406 | 0.4134 | 0.4182 | 0.4375 |
| 7/16-28 UNEF | 2A | 0.0011 | 0.4364 | 0.4299 | — | 0.4132 | 0.4096 | 0.3939 | 2B | 0.399 | 0.407 | 0.4143 | 0.4190 | 0.4375 |
| | 3A | 0.0000 | 0.4375 | 0.4310 | — | 0.4143 | 0.4116 | 0.3950 | 3B | 0.3990 | 0.4051 | 0.4143 | 0.4178 | 0.4375 |
| 7/16-32 UN | 2A | 0.0010 | 0.4365 | 0.4305 | — | 0.4162 | 0.4128 | 0.3993 | 2B | 0.404 | 0.411 | 0.4172 | 0.4216 | 0.4375 |
| | 3A | 0.0000 | 0.4375 | 0.4315 | — | 0.4172 | 0.4146 | 0.4003 | 3B | 0.4040 | 0.4094 | 0.4172 | 0.4205 | 0.4375 |

**Table 3.** *(Continued)* **Standard Series and Selected Combinations — Unified Screw Threads**

| Nominal Size, Threads per Inch, and Series Designation[a] | External[b] | | | | | | | | Internal[b] | | | | | |
|---|---|---|---|---|---|---|---|---|---|---|---|---|---|---|
| | Class | Allowance | Major Diameter | | | Pitch Diameter | | UNR Minor Dia.,[c] Max (Ref.) | Class | Minor Diameter | | Pitch Diameter | | Major Diameter |
| | | | Max[d] | Min | Min[e] | Max[d] | Min | | | Min | Max | Min | Max | Min |
| ½-12 UNS | 2A | 0.0016 | 0.4984 | 0.4870 | — | 0.4443 | 0.4389 | 0.3992 | 2B | 0.410 | 0.428 | 0.4459 | 0.4529 | 0.5000 |
| | 3A | 0.0000 | 0.5000 | 0.4886 | — | 0.4459 | 0.4419 | 0.4008 | 3B | 0.4100 | 0.4185 | 0.4459 | 0.4511 | 0.5000 |
| ½-13 UNC | 1A | 0.0015 | 0.4985 | 0.4822 | — | 0.4485 | 0.4411 | 0.4069 | 1B | 0.417 | 0.434 | 0.4500 | 0.4597 | 0.5000 |
| | 2A | 0.0015 | 0.4985 | 0.4876 | 0.4822 | 0.4485 | 0.4435 | 0.4069 | 2B | 0.417 | 0.434 | 0.4500 | 0.4565 | 0.5000 |
| | 3A | 0.0000 | 0.5000 | 0.4891 | — | 0.4500 | 0.4463 | 0.4084 | 3B | 0.4170 | 0.4284 | 0.4500 | 0.4548 | 0.5000 |
| ½-14 UNS | 2A | 0.0015 | 0.4985 | 0.4882 | — | 0.4521 | 0.4471 | 0.4134 | 2B | 0.423 | 0.438 | 0.4536 | 0.4601 | 0.5000 |
| ½-16 UN | 2A | 0.0014 | 0.4986 | 0.4892 | — | 0.4580 | 0.4533 | 0.4242 | 2B | 0.432 | 0.446 | 0.4594 | 0.4655 | 0.5000 |
| | 3A | 0.0000 | 0.5000 | 0.4906 | — | 0.4594 | 0.4559 | 0.4256 | 3B | 0.4320 | 0.4420 | 0.4594 | 0.4640 | 0.5000 |
| ½-18 UNS | 2A | 0.0013 | 0.4987 | 0.4900 | — | 0.4626 | 0.4582 | 0.4325 | 2B | 0.440 | 0.453 | 0.4639 | 0.4697 | 0.5000 |
| ½-20 UNF | 1A | 0.0013 | 0.4987 | 0.4865 | — | 0.4662 | 0.4598 | 0.4392 | 1B | 0.446 | 0.457 | 0.4675 | 0.4759 | 0.5000 |
| | 2A | 0.0013 | 0.4987 | 0.4906 | — | 0.4662 | 0.4619 | 0.4392 | 2B | 0.446 | 0.457 | 0.4675 | 0.4731 | 0.5000 |
| | 3A | 0.0000 | 0.5000 | 0.4919 | — | 0.4675 | 0.4643 | 0.4405 | 3B | 0.4460 | 0.4537 | 0.4675 | 0.4717 | 0.5000 |
| ½-24 UNS | 2A | 0.0012 | 0.4988 | 0.4916 | — | 0.4717 | 0.4678 | 0.4492 | 2B | 0.455 | 0.465 | 0.4729 | 0.4780 | 0.5000 |
| ½-27 UNS | 2A | 0.0011 | 0.4989 | 0.4922 | — | 0.4748 | 0.4711 | 0.4548 | 2B | 0.460 | 0.469 | 0.4759 | 0.4807 | 0.5000 |
| ½-28 UNEF | 2A | 0.0011 | 0.4989 | 0.4924 | — | 0.4757 | 0.4720 | 0.4564 | 2B | 0.461 | 0.470 | 0.4768 | 0.4816 | 0.5000 |
| | 3A | 0.0000 | 0.5000 | 0.4935 | — | 0.4768 | 0.4740 | 0.4575 | 3B | 0.4610 | 0.4676 | 0.4768 | 0.4804 | 0.5000 |
| ½-32 UN | 2A | 0.0010 | 0.4990 | 0.4930 | — | 0.4787 | 0.4752 | 0.4618 | 2B | 0.466 | 0.474 | 0.4797 | 0.4842 | 0.5000 |
| | 3A | 0.0000 | 0.5000 | 0.4940 | — | 0.4797 | 0.4771 | 0.4628 | 3B | 0.4660 | 0.4719 | 0.4797 | 0.4831 | 0.5000 |
| 9⁄16-12 UNC | 1A | 0.0016 | 0.5609 | 0.5437 | — | 0.5068 | 0.4990 | 0.4617 | 1B | 0.472 | 0.490 | 0.5084 | 0.5186 | 0.5625 |
| | 2A | 0.0016 | 0.5609 | 0.5495 | 0.5437 | 0.5068 | 0.5016 | 0.4617 | 2B | 0.472 | 0.490 | 0.5084 | 0.5152 | 0.5625 |
| | 3A | 0.0000 | 0.5625 | 0.5511 | — | 0.5084 | 0.5045 | 0.4633 | 3B | 0.4720 | 0.4843 | 0.5084 | 0.5135 | 0.5625 |
| 9⁄16-14 UNS | 2A | 0.0015 | 0.5610 | 0.5507 | — | 0.5146 | 0.5096 | 0.4759 | 2B | 0.485 | 0.501 | 0.5161 | 0.5226 | 0.5625 |
| 9⁄16-16 UN | 2A | 0.0014 | 0.5611 | 0.5517 | — | 0.5205 | 0.5158 | 0.4867 | 2B | 0.495 | 0.509 | 0.5219 | 0.5280 | 0.5625 |
| | 3A | 0.0000 | 0.5625 | 0.5531 | — | 0.5219 | 0.5184 | 0.4881 | 3B | 0.4950 | 0.5041 | 0.5219 | 0.5265 | 0.5625 |
| 9⁄16-18 UNF | 1A | 0.0014 | 0.5611 | 0.5480 | — | 0.5250 | 0.5182 | 0.4949 | 1B | 0.502 | 0.515 | 0.5264 | 0.5353 | 0.5625 |
| | 2A | 0.0014 | 0.5611 | 0.5524 | — | 0.5250 | 0.5205 | 0.4949 | 2B | 0.502 | 0.515 | 0.5264 | 0.5323 | 0.5625 |
| | 3A | 0.0000 | 0.5625 | 0.5538 | — | 0.5264 | 0.5230 | 0.4963 | 3B | 0.5020 | 0.5106 | 0.5264 | 0.5308 | 0.5625 |

**Table 3.** *(Continued)* **Standard Series and Selected Combinations — Unified Screw Threads**

| Nominal Size, Threads per Inch, and Series Designation[a] | External[b] | | | | | | | | Internal[b] | | | | | |
|---|---|---|---|---|---|---|---|---|---|---|---|---|---|---|
| | Class | Allowance | Major Diameter | | | Pitch Diameter | | UNR Minor Dia.,[c] Max (Ref.) | Class | Minor Diameter | | Pitch Diameter | | Major Diameter |
| | | | Max[d] | Min | Min[e] | Max[d] | Min | | | Min | Max | Min | Max | Min |
| 9/16-20 UN | 2A | 0.0013 | 0.5612 | 0.5531 | — | 0.5287 | 0.5244 | 0.5017 | 2B | 0.508 | 0.520 | 0.5300 | 0.5356 | 0.5625 |
| | 3A | 0.0000 | 0.5625 | 0.5544 | — | 0.5300 | 0.5268 | 0.5030 | 3B | 0.5080 | 0.5161 | 0.5300 | 0.5342 | 0.5625 |
| 9/16-24 UNEF | 2A | 0.0012 | 0.5613 | 0.5541 | — | 0.5342 | 0.5302 | 0.5117 | 2B | 0.517 | 0.527 | 0.5354 | 0.5405 | 0.5625 |
| | 3A | 0.0000 | 0.5625 | 0.5553 | — | 0.5354 | 0.5324 | 0.5129 | 3B | 0.5170 | 0.5244 | 0.5354 | 0.5393 | 0.5625 |
| 9/16-27 UNS | 2A | 0.0011 | 0.5614 | 0.5547 | — | 0.5373 | 0.5335 | 0.5173 | 2B | 0.522 | 0.531 | 0.5384 | 0.5433 | 0.5625 |
| 9/16-28 UN | 2A | 0.0011 | 0.5614 | 0.5549 | — | 0.5382 | 0.5345 | 0.5189 | 2B | 0.524 | 0.532 | 0.5393 | 0.5441 | 0.5625 |
| | 3A | 0.0000 | 0.5625 | 0.5560 | — | 0.5393 | 0.5365 | 0.5200 | 3B | 0.5240 | 0.5301 | 0.5393 | 0.5429 | 0.5625 |
| 9/16-32 UN | 2A | 0.0011 | 0.5614 | 0.5554 | — | 0.5411 | 0.5376 | 0.5242 | 2B | 0.529 | 0.536 | 0.5422 | 0.5468 | 0.5625 |
| | 3A | 0.0000 | 0.5625 | 0.5565 | — | 0.5422 | 0.5396 | 0.5253 | 3B | 0.5290 | 0.5344 | 0.5422 | 0.5456 | 0.5625 |
| 5/8-11 UNC | 1A | 0.0017 | 0.6233 | 0.6051 | — | 0.5643 | 0.5560 | 0.5150 | 1B | 0.527 | 0.546 | 0.5660 | 0.5767 | 0.6250 |
| | 2A | 0.0017 | 0.6233 | 0.6112 | 0.6052 | 0.5643 | 0.5588 | 0.5150 | 2B | 0.527 | 0.546 | 0.5660 | 0.5732 | 0.6250 |
| | 3A | 0.0000 | 0.6250 | 0.6129 | — | 0.5660 | 0.5619 | 0.5167 | 3B | 0.5270 | 0.5391 | 0.5660 | 0.5714 | 0.6250 |
| 5/8-12 UN | 2A | 0.0016 | 0.6234 | 0.6120 | — | 0.5693 | 0.5639 | 0.5242 | 2B | 0.535 | 0.553 | 0.5709 | 0.5780 | 0.6250 |
| | 3A | 0.0000 | 0.6250 | 0.6136 | — | 0.5709 | 0.5668 | 0.5258 | 3B | 0.5350 | 0.5463 | 0.5709 | 0.5762 | 0.6250 |
| 5/8-14 UNS | 2A | 0.0015 | 0.6235 | 0.6132 | — | 0.5771 | 0.5720 | 0.5384 | 2B | 0.548 | 0.563 | 0.5786 | 0.5852 | 0.6250 |
| 5/8-16 UN | 2A | 0.0014 | 0.6236 | 0.6142 | — | 0.5830 | 0.5782 | 0.5492 | 2B | 0.557 | 0.571 | 0.5844 | 0.5906 | 0.6250 |
| | 3A | 0.0000 | 0.6250 | 0.6156 | — | 0.5844 | 0.5808 | 0.5506 | 3B | 0.5570 | 0.5662 | 0.5844 | 0.5890 | 0.6250 |
| 5/8-18 UNF | 1A | 0.0014 | 0.6236 | 0.6105 | — | 0.5875 | 0.5805 | 0.5574 | 1B | 0.565 | 0.578 | 0.5889 | 0.5980 | 0.6250 |
| | 2A | 0.0014 | 0.6236 | 0.6149 | — | 0.5875 | 0.5828 | 0.5574 | 2B | 0.565 | 0.578 | 0.5889 | 0.5949 | 0.6250 |
| | 3A | 0.0000 | 0.6250 | 0.6163 | — | 0.5889 | 0.5854 | 0.5588 | 3B | 0.5650 | 0.5730 | 0.5889 | 0.5934 | 0.6250 |
| 5/8-20 UN | 2A | 0.0013 | 0.6237 | 0.6156 | — | 0.5912 | 0.5869 | 0.5642 | 2B | 0.571 | 0.582 | 0.5925 | 0.5981 | 0.6250 |
| | 3A | 0.0000 | 0.6250 | 0.6169 | — | 0.5925 | 0.5893 | 0.5655 | 3B | 0.5710 | 0.5786 | 0.5925 | 0.5967 | 0.6250 |
| 5/8-24 UNEF | 2A | 0.0012 | 0.6238 | 0.6166 | — | 0.5967 | 0.5927 | 0.5742 | 2B | 0.580 | 0.590 | 0.5979 | 0.6031 | 0.6250 |
| | 3A | 0.0000 | 0.6250 | 0.6178 | — | 0.5979 | 0.5949 | 0.5754 | 3B | 0.5800 | 0.5869 | 0.5979 | 0.6018 | 0.6250 |
| 5/8-27 UNS | 2A | 0.0011 | 0.6239 | 0.6172 | — | 0.5998 | 0.5960 | 0.5798 | 2B | 0.585 | 0.594 | 0.6009 | 0.6059 | 0.6250 |
| 5/8-28 UN | 2A | 0.0011 | 0.6239 | 0.6174 | — | 0.6007 | 0.5969 | 0.5814 | 2B | 0.586 | 0.595 | 0.6018 | 0.6067 | 0.6250 |
| | 3A | 0.0000 | 0.6250 | 0.6185 | — | 0.6018 | 0.5990 | 0.5825 | 3B | 0.5860 | 0.5926 | 0.6018 | 0.6055 | 0.6250 |

**Table 3.** *(Continued)* **Standard Series and Selected Combinations — Unified Screw Threads**

| Nominal Size, Threads per Inch, and Series Designation[a] | External[b] | | | | | | | | Internal[b] | | | | | |
|---|---|---|---|---|---|---|---|---|---|---|---|---|---|---|
| | Class | Allow-ance | Major Diameter | | | Pitch Diameter | | UNR Minor Dia.,[c] Max (Ref.) | Class | Minor Diameter | | Pitch Diameter | | Major Diameter |
| | | | Max[d] | Min | Min[e] | Max[d] | Min | | | Min | Max | Min | Max | Min |
| 5⁄8-32 UN | 2A | 0.0011 | 0.6239 | 0.6179 | — | 0.6036 | 0.6000 | 0.5867 | 2B | 0.591 | 0.599 | 0.6047 | 0.6093 | 0.6250 |
| | 3A | 0.0000 | 0.6250 | 0.6190 | — | 0.6047 | 0.6020 | 0.5878 | 3B | 0.5910 | 0.5969 | 0.6047 | 0.6082 | 0.6250 |
| 11⁄16-12 UN | 2A | 0.0016 | 0.6859 | 0.6745 | — | 0.6318 | 0.6263 | 0.5867 | 2B | 0.597 | 0.615 | 0.6334 | 0.6405 | 0.6875 |
| | 3A | 0.0000 | 0.6875 | 0.6761 | — | 0.6334 | 0.6293 | 0.5883 | 3B | 0.5970 | 0.6085 | 0.6334 | 0.6387 | 0.6875 |
| 11⁄16-16 UN | 2A | 0.0014 | 0.6861 | 0.6767 | — | 0.6455 | 0.6407 | 0.6117 | 2B | 0.620 | 0.634 | 0.6469 | 0.6532 | 0.6875 |
| | 3A | 0.0000 | 0.6875 | 0.6781 | — | 0.6469 | 0.6433 | 0.6131 | 3B | 0.6200 | 0.6284 | 0.6469 | 0.6516 | 0.6875 |
| 11⁄16-20 UN | 2A | 0.0013 | 0.6862 | 0.6781 | — | 0.6537 | 0.6493 | 0.6267 | 2B | 0.633 | 0.645 | 0.6550 | 0.6607 | 0.6875 |
| | 3A | 0.0000 | 0.6875 | 0.6794 | — | 0.6550 | 0.6517 | 0.6280 | 3B | 0.6330 | 0.6411 | 0.6550 | 0.6593 | 0.6875 |
| 11⁄16-24 UNEF | 2A | 0.0012 | 0.6863 | 0.6791 | — | 0.6592 | 0.6552 | 0.6367 | 2B | 0.642 | 0.652 | 0.6604 | 0.6657 | 0.6875 |
| | 3A | 0.0000 | 0.6875 | 0.6803 | — | 0.6604 | 0.6574 | 0.6379 | 3B | 0.6420 | 0.6494 | 0.6604 | 0.6643 | 0.6875 |
| 11⁄16-28 UN | 2A | 0.0011 | 0.6864 | 0.6799 | — | 0.6632 | 0.6594 | 0.6439 | 2B | 0.649 | 0.657 | 0.6643 | 0.6692 | 0.6875 |
| | 3A | 0.0000 | 0.6875 | 0.6810 | — | 0.6643 | 0.6614 | 0.6450 | 3B | 0.6490 | 0.6551 | 0.6643 | 0.6680 | 0.6875 |
| 11⁄16-32 UN | 2A | 0.0011 | 0.6864 | 0.6804 | — | 0.6661 | 0.6625 | 0.6492 | 2B | 0.654 | 0.661 | 0.6672 | 0.6719 | 0.6875 |
| | 3A | 0.0000 | 0.6875 | 0.6815 | — | 0.6672 | 0.6645 | 0.6503 | 3B | 0.6540 | 0.6594 | 0.6672 | 0.6707 | 0.6875 |
| 3⁄4-10 UNC | 1A | 0.0018 | 0.7482 | 0.7288 | — | 0.6832 | 0.6744 | 0.6291 | 1B | 0.642 | 0.663 | 0.6850 | 0.6965 | 0.7500 |
| | 2A | 0.0018 | 0.7482 | 0.7353 | 0.7288 | 0.6832 | 0.6773 | 0.6291 | 2B | 0.642 | 0.663 | 0.6850 | 0.6927 | 0.7500 |
| | 3A | 0.0000 | 0.7500 | 0.7371 | — | 0.6850 | 0.6806 | 0.6309 | 3B | 0.6420 | 0.6545 | 0.6850 | 0.6907 | 0.7500 |
| 3⁄4-12 UN | 2A | 0.0017 | 0.7483 | 0.7369 | — | 0.6942 | 0.6887 | 0.6491 | 2B | 0.660 | 0.678 | 0.6959 | 0.7031 | 0.7500 |
| | 3A | 0.0000 | 0.7500 | 0.7386 | — | 0.6959 | 0.6918 | 0.6508 | 3B | 0.6600 | 0.6707 | 0.6959 | 0.7013 | 0.7500 |
| 3⁄4-14 UNS | 2A | 0.0015 | 0.7485 | 0.7382 | — | 0.7021 | 0.6970 | 0.6634 | 2B | 0.673 | 0.688 | 0.7036 | 0.7103 | 0.7500 |
| 3⁄4-16 UNF | 1A | 0.0015 | 0.7485 | 0.7343 | — | 0.7079 | 0.7004 | 0.6741 | 1B | 0.682 | 0.696 | 0.7094 | 0.7192 | 0.7500 |
| | 2A | 0.0015 | 0.7485 | 0.7391 | — | 0.7079 | 0.7029 | 0.6741 | 2B | 0.682 | 0.696 | 0.7094 | 0.7159 | 0.7500 |
| | 3A | 0.0000 | 0.7500 | 0.7406 | — | 0.7094 | 0.7056 | 0.6756 | 3B | 0.6820 | 0.6909 | 0.7094 | 0.7143 | 0.7500 |
| 3⁄4-18 UNS | 2A | 0.0014 | 0.7486 | 0.7399 | — | 0.7125 | 0.7079 | 0.6824 | 2B | 0.690 | 0.703 | 0.7139 | 0.7199 | 0.7500 |
| 3⁄4-20 UNEF | 2A | 0.0013 | 0.7487 | 0.7406 | — | 0.7162 | 0.7118 | 0.6892 | 2B | 0.696 | 0.707 | 0.7175 | 0.7232 | 0.7500 |
| | 3A | 0.0000 | 0.7500 | 0.7419 | — | 0.7175 | 0.7142 | 0.6905 | 3B | 0.6960 | 0.7036 | 0.7175 | 0.7218 | 0.7500 |
| 3⁄4-24 UNS | 2A | 0.0012 | 0.7488 | 0.7416 | — | 0.7217 | 0.7176 | 0.6992 | 2B | 0.705 | 0.715 | 0.7229 | 0.7282 | 0.7500 |
| 3⁄4-27 UNS | 2A | 0.0012 | 0.7488 | 0.7421 | — | 0.7247 | 0.7208 | 0.7047 | 2B | 0.710 | 0.719 | 0.7259 | 0.7310 | 0.7500 |

**Table 3.** *(Continued)* **Standard Series and Selected Combinations — Unified Screw Threads**

| Nominal Size, Threads per Inch, and Series Designation[a] | External[b] | | | | | | | | Internal[b] | | | | | |
|---|---|---|---|---|---|---|---|---|---|---|---|---|---|---|
| | Class | Allowance | Major Diameter | | | Pitch Diameter | | UNR Minor Dia.,[c] Max (Ref.) | Class | Minor Diameter | | Pitch Diameter | | Major Diameter |
| | | | Max[d] | Min | Min[e] | Max[d] | Min | | | Min | Max | Min | Max | Min |
| ¾-28 UN | 2A | 0.0012 | 0.7488 | 0.7423 | — | 0.7256 | 0.7218 | 0.7063 | 2B | 0.711 | 0.720 | 0.7268 | 0.7318 | 0.7500 |
| | 3A | 0.0000 | 0.7500 | 0.7435 | — | 0.7268 | 0.7239 | 0.7075 | 3B | 0.7110 | 0.7176 | 0.7268 | 0.7305 | 0.7500 |
| ¾-32 UN | 2A | 0.0011 | 0.7489 | 0.7429 | — | 0.7286 | 0.7250 | 0.7117 | 2B | 0.716 | 0.724 | 0.7297 | 0.7344 | 0.7500 |
| | 3A | 0.0000 | 0.7500 | 0.7440 | — | 0.7297 | 0.7270 | 0.7128 | 3B | 0.7160 | 0.7219 | 0.7297 | 0.7333 | 0.7500 |
| 13⁄16-12 UN | 2A | 0.0017 | 0.8108 | 0.7994 | — | 0.7567 | 0.7511 | 0.7116 | 2B | 0.722 | 0.740 | 0.7584 | 0.7656 | 0.8125 |
| | 3A | 0.0000 | 0.8125 | 0.8011 | — | 0.7584 | 0.7542 | 0.7133 | 3B | 0.7220 | 0.7329 | 0.7584 | 0.7638 | 0.8125 |
| 13⁄16-16 UN | 2A | 0.0015 | 0.8110 | 0.8016 | — | 0.7704 | 0.7655 | 0.7366 | 2B | 0.745 | 0.759 | 0.7719 | 0.7783 | 0.8125 |
| | 3A | 0.0000 | 0.8125 | 0.8031 | — | 0.7719 | 0.7682 | 0.7381 | 3B | 0.7450 | 0.7534 | 0.7719 | 0.7767 | 0.8125 |
| 13⁄16-20 UNEF | 2A | 0.0013 | 0.8112 | 0.8031 | — | 0.7787 | 0.7743 | 0.7517 | 2B | 0.758 | 0.770 | 0.7800 | 0.7858 | 0.8125 |
| | 3A | 0.0000 | 0.8125 | 0.8044 | — | 0.7800 | 0.7767 | 0.7530 | 3B | 0.7580 | 0.7661 | 0.7800 | 0.7843 | 0.8125 |
| 13⁄16-28 UN | 2A | 0.0012 | 0.8113 | 0.8048 | — | 0.7881 | 0.7842 | 0.7688 | 2B | 0.774 | 0.782 | 0.7893 | 0.7943 | 0.8125 |
| | 3A | 0.0000 | 0.8125 | 0.8060 | — | 0.7893 | 0.7864 | 0.7700 | 3B | 0.7740 | 0.7801 | 0.7893 | 0.7931 | 0.8125 |
| 13⁄16-32 UN | 2A | 0.0011 | 0.8114 | 0.8054 | — | 0.7911 | 0.7874 | 0.7742 | 2B | 0.779 | 0.786 | 0.7922 | 0.7970 | 0.8125 |
| | 3A | 0.0000 | 0.8125 | 0.8065 | — | 0.7922 | 0.7894 | 0.7753 | 3B | 0.7790 | 0.7844 | 0.7922 | 0.7958 | 0.8125 |
| ⅞-9 UNC | 1A | 0.0019 | 0.8731 | 0.8523 | — | 0.8009 | 0.7914 | 0.7408 | 1B | 0.755 | 0.778 | 0.8028 | 0.8151 | 0.8750 |
| | 2A | 0.0019 | 0.8731 | 0.8592 | 0.8523 | 0.8009 | 0.7946 | 0.7408 | 2B | 0.755 | 0.778 | 0.8028 | 0.8110 | 0.8750 |
| | 3A | 0.0000 | 0.8750 | 0.8611 | — | 0.8028 | 0.7981 | 0.7427 | 3B | 0.7550 | 0.7681 | 0.8028 | 0.8089 | 0.8750 |
| ⅞-10 UNS | 2A | 0.0018 | 0.8732 | 0.8603 | — | 0.8082 | 0.8021 | 0.7541 | 2B | 0.767 | 0.788 | 0.8100 | 0.8179 | 0.8750 |
| ⅞-12 UN | 2A | 0.0017 | 0.8733 | 0.8619 | — | 0.8192 | 0.8136 | 0.7741 | 2B | 0.785 | 0.803 | 0.8209 | 0.8282 | 0.8750 |
| | 3A | 0.0000 | 0.8750 | 0.8636 | — | 0.8209 | 0.8167 | 0.7758 | 3B | 0.7850 | 0.7952 | 0.8209 | 0.8264 | 0.8750 |
| ⅞-14 UNF | 1A | 0.0016 | 0.8734 | 0.8579 | — | 0.8270 | 0.8189 | 0.7883 | 1B | 0.798 | 0.813 | 0.8286 | 0.8392 | 0.8750 |
| | 2A | 0.0016 | 0.8734 | 0.8631 | — | 0.8270 | 0.8216 | 0.7883 | 2B | 0.798 | 0.813 | 0.8286 | 0.8356 | 0.8750 |
| | 3A | 0.0000 | 0.8750 | 0.8647 | — | 0.8286 | 0.8245 | 0.7899 | 3B | 0.7980 | 0.8067 | 0.8286 | 0.8339 | 0.8750 |
| ⅞-16 UN | 2A | 0.0015 | 0.8735 | 0.8641 | — | 0.8329 | 0.8280 | 0.7991 | 2B | 0.807 | 0.821 | 0.8344 | 0.8408 | 0.8750 |
| | 3A | 0.0000 | 0.8750 | 0.8656 | — | 0.8344 | 0.8307 | 0.8006 | 3B | 0.8070 | 0.8159 | 0.8344 | 0.8392 | 0.8750 |
| ⅞-18 UNS | 2A | 0.0014 | 0.8736 | 0.8649 | — | 0.8375 | 0.8328 | 0.8074 | 2B | 0.815 | 0.828 | 0.8389 | 0.8450 | 0.8750 |
| ⅞-20 UNEF | 2A | 0.0013 | 0.8737 | 0.8656 | — | 0.8412 | 0.8367 | 0.8142 | 2B | 0.821 | 0.832 | 0.8425 | 0.8483 | 0.8750 |
| | 3A | 0.0000 | 0.8750 | 0.8669 | — | 0.8425 | 0.8391 | 0.8155 | 3B | 0.8210 | 0.8286 | 0.8425 | 0.8469 | 0.8750 |

**Table 3.** *(Continued)* **Standard Series and Selected Combinations—Unified Screw Threads**

| Nominal Size, Threads per Inch, and Series Designation[a] | External[b] | | | | | | | | Internal[b] | | | | | |
|---|---|---|---|---|---|---|---|---|---|---|---|---|---|---|
| | | | Major Diameter | | | Pitch Diameter | | UNR Minor Dia.,[c] Max | | Minor Diameter | | Pitch Diameter | | Major Diameter |
| | Class | Allow-ance | Max[d] | Min | Min[e] | Max[d] | Min | (Ref.) | Class | Min | Max | Min | Max | Min |
| 7/8-24 UNS | 2A | 0.0012 | 0.8738 | 0.8666 | — | 0.8467 | 0.8425 | 0.8242 | 2B | 0.830 | 0.840 | 0.8479 | 0.8533 | 0.8750 |
| 7/8-27 UNS | 2A | 0.0012 | 0.8738 | 0.8671 | — | 0.8497 | 0.8457 | 0.8297 | 2B | 0.835 | 0.844 | 0.8509 | 0.8561 | 0.8750 |
| 7/8-28 UN | 2A | 0.0012 | 0.8738 | 0.8673 | — | 0.8506 | 0.8467 | 0.8313 | 2B | 0.836 | 0.845 | 0.8518 | 0.8569 | 0.8750 |
| | 3A | 0.0000 | 0.8750 | 0.8685 | — | 0.8518 | 0.8489 | 0.8325 | 3B | 0.8360 | 0.8426 | 0.8518 | 0.8556 | 0.8750 |
| 7/8-32 UN | 2A | 0.0011 | 0.8739 | 0.8679 | — | 0.8536 | 0.8499 | 0.8367 | 2B | 0.841 | 0.849 | 0.8547 | 0.8595 | 0.8750 |
| | 3A | 0.0000 | 0.8750 | 0.8690 | — | 0.8547 | 0.8519 | 0.8378 | 3B | 0.8410 | 0.8469 | 0.8547 | 0.8583 | 0.8750 |
| 15/16-12 UN | 2A | 0.0017 | 0.9358 | 0.9244 | — | 0.8817 | 0.8761 | 0.8366 | 2B | 0.847 | 0.865 | 0.8834 | 0.8907 | 0.9375 |
| | 3A | 0.0000 | 0.9375 | 0.9261 | — | 0.8834 | 0.8792 | 0.8383 | 3B | 0.8470 | 0.8575 | 0.8834 | 0.8889 | 0.9375 |
| 15/16-16 UN | 2A | 0.0015 | 0.9360 | 0.9266 | — | 0.8954 | 0.8904 | 0.8616 | 2B | 0.870 | 0.884 | 0.8969 | 0.9033 | 0.9375 |
| | 3A | 0.0000 | 0.9375 | 0.9281 | — | 0.8969 | 0.8932 | 0.8631 | 3B | 0.8700 | 0.8784 | 0.8969 | 0.9017 | 0.9375 |
| 15/16-20 UNEF | 2A | 0.0014 | 0.9361 | 0.9280 | — | 0.9036 | 0.8991 | 0.8766 | 2B | 0.883 | 0.895 | 0.9050 | 0.9109 | 0.9375 |
| | 3A | 0.0000 | 0.9375 | 0.9294 | — | 0.9050 | 0.9016 | 0.8780 | 3B | 0.8830 | 0.8911 | 0.9050 | 0.9094 | 0.9375 |
| 15/16-28 UN | 2A | 0.0012 | 0.9363 | 0.9298 | — | 0.9131 | 0.9092 | 0.8938 | 2B | 0.899 | 0.907 | 0.9143 | 0.9194 | 0.9375 |
| | 3A | 0.0000 | 0.9375 | 0.9310 | — | 0.9143 | 0.9113 | 0.8950 | 3B | 0.8990 | 0.9051 | 0.9143 | 0.9181 | 0.9375 |
| 15/16-32 UN | 2A | 0.0011 | 0.9364 | 0.9304 | — | 0.9161 | 0.9123 | 0.8992 | 2B | 0.904 | 0.911 | 0.9172 | 0.9221 | 0.9375 |
| | 3A | 0.0000 | 0.9375 | 0.9315 | — | 0.9172 | 0.9144 | 0.9003 | 3B | 0.9040 | 0.9094 | 0.9172 | 0.9209 | 0.9375 |
| 1-8 UNC | 1A | 0.0020 | 0.9980 | 0.9755 | — | 0.9168 | 0.9067 | 0.8492 | 1B | 0.865 | 0.890 | 0.9188 | 0.9320 | 1.0000 |
| | 2A | 0.0020 | 0.9980 | 0.9830 | 0.9755 | 0.9168 | 0.9101 | 0.8492 | 2B | 0.865 | 0.890 | 0.9188 | 0.9276 | 1.0000 |
| | 3A | 0.0000 | 1.0000 | 0.9850 | — | 0.9188 | 0.9137 | 0.8512 | 3B | 0.8650 | 0.8797 | 0.9188 | 0.9254 | 1.0000 |
| 1-10 UNS | 2A | 0.0018 | 0.9982 | 0.9853 | — | 0.9332 | 0.9270 | 0.8791 | 2B | 0.892 | 0.913 | 0.9350 | 0.9430 | 1.0000 |
| 1-12 UNF | 1A | 0.0018 | 0.9982 | 0.9810 | — | 0.9441 | 0.9353 | 0.8990 | 1B | 0.910 | 0.928 | 0.9459 | 0.9573 | 1.0000 |
| | 2A | 0.0018 | 0.9982 | 0.9868 | — | 0.9441 | 0.9382 | 0.8990 | 2B | 0.910 | 0.928 | 0.9459 | 0.9535 | 1.0000 |
| | 3A | 0.0000 | 1.0000 | 0.9886 | — | 0.9459 | 0.9415 | 0.9008 | 3B | 0.9100 | 0.9198 | 0.9459 | 0.9516 | 1.0000 |
| 1-14 UNS[f] | 2A | 0.0016 | 0.9984 | 0.9881 | — | 0.9520 | 0.9467 | 0.9133 | 2B | 0.923 | 0.938 | 0.9536 | 0.9605 | 1.0000 |
| | 3A | 0.0000 | 1.0000 | 0.9897 | — | 0.9536 | 0.9496 | 0.9149 | 3B | 0.9230 | 0.9315 | 0.9536 | 0.9588 | 1.0000 |
| 1-16 UN | 2A | 0.0015 | 0.9985 | 0.9891 | — | 0.9579 | 0.9529 | 0.9241 | 2B | 0.932 | 0.946 | 0.9594 | 0.9659 | 1.0000 |
| | 3A | 0.0000 | 1.0000 | 0.9906 | — | 0.9594 | 0.9557 | 0.9256 | 3B | 0.9320 | 0.9409 | 0.9594 | 0.9643 | 1.0000 |
| 1-18 UNS | 2A | 0.0014 | 0.9986 | 0.9899 | — | 0.9625 | 0.9578 | 0.9324 | 2B | 0.940 | 0.953 | 0.9639 | 0.9701 | 1.0000 |

**Table 3.** *(Continued)* **Standard Series and Selected Combinations — Unified Screw Threads**

| Nominal Size, Threads per Inch, and Series Designation[a] | External[b] | | | | | | | | Internal[b] | | | | | |
|---|---|---|---|---|---|---|---|---|---|---|---|---|---|---|
| | Class | Allowance | Major Diameter | | | Pitch Diameter | | UNR Minor Dia.,[c] Max (Ref.) | Class | Minor Diameter | | Pitch Diameter | | Major Diameter |
| | | | Max[d] | Min | Min[e] | Max[d] | Min | | | Min | Max | Min | Max | Min |
| 1-20 UNEF | 2A | 0.0014 | 0.9986 | 0.9905 | — | 0.9661 | 0.9616 | 0.9391 | 2B | 0.946 | 0.957 | 0.9675 | 0.9734 | 1.0000 |
| | 3A | 0.0000 | 1.0000 | 0.9919 | — | 0.9675 | 0.9641 | 0.9405 | 3B | 0.9460 | 0.9536 | 0.9675 | 0.9719 | 1.0000 |
| 1-24 UNS | 2A | 0.0013 | 0.9987 | 0.9915 | — | 0.9716 | 0.9674 | 0.9491 | 2B | 0.955 | 0.965 | 0.9729 | 0.9784 | 1.0000 |
| 1-27 UNS | 2A | 0.0012 | 0.9988 | 0.9921 | — | 0.9747 | 0.9707 | 0.9547 | 2B | 0.960 | 0.969 | 0.9759 | 0.9811 | 1.0000 |
| 1-28 UN | 2A | 0.0012 | 0.9988 | 0.9923 | — | 0.9756 | 0.9716 | 0.9562 | 2B | 0.961 | 0.970 | 0.9768 | 0.9820 | 1.0000 |
| | 3A | 0.0000 | 1.0000 | 0.9935 | — | 0.9768 | 0.9738 | 0.9575 | 3B | 0.9610 | 0.9676 | 0.9768 | 0.9807 | 1.0000 |
| 1-32 UN | 2A | 0.0011 | 0.9989 | 0.9929 | — | 0.9786 | 0.9748 | 0.9617 | 2B | 0.966 | 0.974 | 0.9797 | 0.9846 | 1.0000 |
| | 3A | 0.0000 | 1.0000 | 0.9940 | — | 0.9797 | 0.9769 | 0.9628 | 3B | 0.9660 | 0.9719 | 0.9797 | 0.9834 | 1.0000 |
| 1 1/16-8 UN | 2A | 0.0020 | 1.0605 | 1.0455 | — | 0.9793 | 0.9725 | 0.9117 | 2B | 0.927 | 0.952 | 0.9813 | 0.9902 | 1.0625 |
| | 3A | 0.0000 | 1.0625 | 1.0475 | — | 0.9813 | 0.9762 | 0.9137 | 3B | 0.9270 | 0.9422 | 0.9813 | 0.9880 | 1.0625 |
| 1 1/16-12 UN | 2A | 0.0017 | 1.0608 | 1.0494 | — | 1.0067 | 1.0010 | 0.9616 | 2B | 0.972 | 0.990 | 1.0084 | 1.0158 | 1.0625 |
| | 3A | 0.0000 | 1.0625 | 1.0511 | — | 1.0084 | 1.0041 | 0.9633 | 3B | 0.9720 | 0.9823 | 1.0084 | 1.0139 | 1.0625 |
| 1 1/16-16 UN | 2A | 0.0015 | 1.0610 | 1.0516 | — | 1.0204 | 1.0154 | 0.9866 | 2B | 0.995 | 1.009 | 1.0219 | 1.0284 | 1.0625 |
| | 3A | 0.0000 | 1.0625 | 1.0531 | — | 1.0219 | 1.0181 | 0.9881 | 3B | 0.9950 | 1.0034 | 1.0219 | 1.0268 | 1.0625 |
| 1 1/16-18 UNEF | 2A | 0.0014 | 1.0611 | 1.0524 | — | 1.0250 | 1.0202 | 0.9949 | 2B | 1.002 | 1.015 | 1.0264 | 1.0326 | 1.0625 |
| | 3A | 0.0000 | 1.0625 | 1.0538 | — | 1.0264 | 1.0228 | 0.9963 | 3B | 1.0020 | 1.0105 | 1.0264 | 1.0311 | 1.0625 |
| 1 1/16-20 UN | 2A | 0.0014 | 1.0611 | 1.0530 | — | 1.0286 | 1.0240 | 1.0016 | 2B | 1.008 | 1.020 | 1.0300 | 1.0359 | 1.0625 |
| | 3A | 0.0000 | 1.0625 | 1.0544 | — | 1.0300 | 1.0266 | 1.0030 | 3B | 1.0080 | 1.0161 | 1.0300 | 1.0345 | 1.0625 |
| 1 1/16-28 UN | 2A | 0.0012 | 1.0613 | 1.0548 | — | 1.0381 | 1.0341 | 1.0188 | 2B | 1.024 | 1.032 | 1.0393 | 1.0445 | 1.0625 |
| | 3A | 0.0000 | 1.0625 | 1.0560 | — | 1.0393 | 1.0363 | 1.0200 | 3B | 1.0240 | 1.0301 | 1.0393 | 1.0432 | 1.0625 |
| 1 1/8-7 UNC | 1A | 0.0022 | 1.1228 | 1.0982 | — | 1.0300 | 1.0191 | 0.9527 | 1B | 0.970 | 0.998 | 1.0322 | 1.0463 | 1.1250 |
| | 2A | 0.0022 | 1.1228 | 1.1064 | 1.0982 | 1.0300 | 1.0228 | 0.9527 | 2B | 0.970 | 0.998 | 1.0322 | 1.0416 | 1.1250 |
| | 3A | 0.0000 | 1.1250 | 1.1086 | — | 1.0322 | 1.0268 | 0.9549 | 3B | 0.9700 | 0.9875 | 1.0322 | 1.0393 | 1.1250 |
| 1 1/8-8 UN | 2A | 0.0021 | 1.1229 | 1.1079 | 1.1004 | 1.0417 | 1.0348 | 0.9741 | 2B | 0.990 | 1.015 | 1.0438 | 1.0528 | 1.1250 |
| | 3A | 0.0000 | 1.1250 | 1.1100 | — | 1.0438 | 1.0386 | 0.9762 | 3B | 0.9900 | 1.0047 | 1.0438 | 1.0505 | 1.1250 |
| 1 1/8-10 UNS | 2A | 0.0019 | 1.1231 | 1.1102 | — | 1.0581 | 1.0519 | 1.0040 | 2B | 1.017 | 1.038 | 1.0600 | 1.0681 | 1.1250 |
| 1 1/8-12 UNF | 1A | 0.0018 | 1.1232 | 1.1060 | — | 1.0691 | 1.0601 | 1.0240 | 1B | 1.035 | 1.053 | 1.0709 | 1.0826 | 1.1250 |
| | 2A | 0.0018 | 1.1232 | 1.1118 | — | 1.0691 | 1.0631 | 1.0240 | 2B | 1.035 | 1.053 | 1.0709 | 1.0787 | 1.1250 |
| | 3A | 0.0000 | 1.1250 | 1.1136 | — | 1.0709 | 1.0664 | 1.0258 | 3B | 1.0350 | 1.0448 | 1.0709 | 1.0768 | 1.1250 |

**Table 3.** *(Continued)* **Standard Series and Selected Combinations — Unified Screw Threads**

| Nominal Size, Threads per Inch, and Series Designation[a] | External[b] | | | | | | | | Internal[b] | | | | | |
|---|---|---|---|---|---|---|---|---|---|---|---|---|---|---|
| | Class | Allowance | Major Diameter | | | Pitch Diameter | | UNR Minor Dia.,[c] Max (Ref.) | Class | Minor Diameter | | Pitch Diameter | | Major Diameter |
| | | | Max[d] | Min | Min[e] | Max[d] | Min | | | Min | Max | Min | Max | Min |
| 1⅛-14 UNS | 2A | 0.0016 | 1.1234 | 1.1131 | — | 1.0770 | 1.0717 | 1.0383 | 2B | 1.048 | 1.063 | 1.0786 | 1.0855 | 1.1250 |
| 1⅛-16 UN | 2A | 0.0015 | 1.1235 | 1.1141 | — | 1.0829 | 1.0779 | 1.0491 | 2B | 1.057 | 1.071 | 1.0844 | 1.0910 | 1.1250 |
| | 3A | 0.0000 | 1.1250 | 1.1156 | — | 1.0844 | 1.0806 | 1.0506 | 3B | 1.0570 | 1.0659 | 1.0844 | 1.0893 | 1.1250 |
| 1⅛-18 UNEF | 2A | 0.0014 | 1.1236 | 1.1149 | — | 1.0875 | 1.0827 | 1.0574 | 2B | 1.065 | 1.078 | 1.0889 | 1.0951 | 1.1250 |
| | 3A | 0.0000 | 1.1250 | 1.1163 | — | 1.0889 | 1.0853 | 1.0588 | 3B | 1.0650 | 1.0730 | 1.0889 | 1.0936 | 1.1250 |
| 1⅛-20 UN | 2A | 0.0014 | 1.1236 | 1.1155 | — | 1.0911 | 1.0865 | 1.0641 | 2B | 1.071 | 1.082 | 1.0925 | 1.0985 | 1.1250 |
| | 3A | 0.0000 | 1.1250 | 1.1169 | — | 1.0925 | 1.0890 | 1.0655 | 3B | 1.0710 | 1.0786 | 1.0925 | 1.0970 | 1.1250 |
| 1⅛-24 UNS | 2A | 0.0013 | 1.1237 | 1.1165 | — | 1.0966 | 1.0923 | 1.0741 | 2B | 1.080 | 1.090 | 1.0979 | 1.1035 | 1.1250 |
| 1⅛-28 UN | 2A | 0.0012 | 1.1238 | 1.1173 | — | 1.1006 | 1.0966 | 1.0813 | 2B | 1.086 | 1.095 | 1.1018 | 1.1070 | 1.1250 |
| | 3A | 0.0000 | 1.1250 | 1.1185 | — | 1.1018 | 1.0988 | 1.0825 | 3B | 1.0860 | 1.0926 | 1.1018 | 1.1057 | 1.1250 |
| 1 3/16-8 UN | 2A | 0.0021 | 1.1854 | 1.1704 | — | 1.1042 | 1.0972 | 1.0366 | 2B | 1.052 | 1.077 | 1.1063 | 1.1154 | 1.1875 |
| | 3A | 0.0000 | 1.1875 | 1.1725 | — | 1.1063 | 1.1011 | 1.0387 | 3B | 1.0520 | 1.0672 | 1.1063 | 1.1131 | 1.1875 |
| 1 3/16-12 UN | 2A | 0.0017 | 1.1858 | 1.1744 | — | 1.1317 | 1.1260 | 1.0866 | 2B | 1.097 | 1.115 | 1.1334 | 1.1409 | 1.1875 |
| | 3A | 0.0000 | 1.1875 | 1.1761 | — | 1.1334 | 1.1291 | 1.0883 | 3B | 1.0970 | 1.1073 | 1.1334 | 1.1390 | 1.1875 |
| 1 3/16-16 UN | 2A | 0.0015 | 1.1860 | 1.1766 | — | 1.1454 | 1.1403 | 1.1116 | 2B | 1.120 | 1.134 | 1.1469 | 1.1535 | 1.1875 |
| | 3A | 0.0000 | 1.1875 | 1.1781 | — | 1.1469 | 1.1431 | 1.1131 | 3B | 1.1200 | 1.1284 | 1.1469 | 1.1518 | 1.1875 |
| 1 3/16-18 UNEF | 2A | 0.0014 | 1.1861 | 1.1774 | — | 1.1500 | 1.1452 | 1.1199 | 2B | 1.127 | 1.140 | 1.1514 | 1.1577 | 1.1875 |
| | 3A | 0.0000 | 1.1875 | 1.1788 | — | 1.1514 | 1.1478 | 1.1213 | 3B | 1.1270 | 1.1355 | 1.1514 | 1.1561 | 1.1875 |
| 1 3/16-20 UN | 2A | 0.0014 | 1.1861 | 1.1780 | — | 1.1536 | 1.1490 | 1.1266 | 2B | 1.133 | 1.145 | 1.1550 | 1.1610 | 1.1875 |
| | 3A | 0.0000 | 1.1875 | 1.1794 | — | 1.1550 | 1.1515 | 1.1280 | 3B | 1.1330 | 1.1411 | 1.1550 | 1.1595 | 1.1875 |
| 1 3/16-28 UN | 2A | 0.0012 | 1.1863 | 1.1798 | — | 1.1631 | 1.1590 | 1.1438 | 2B | 1.149 | 1.157 | 1.1643 | 1.1696 | 1.1875 |
| | 3A | 0.0000 | 1.1875 | 1.1810 | — | 1.1643 | 1.1613 | 1.1450 | 3B | 1.1490 | 1.1551 | 1.1643 | 1.1683 | 1.1875 |
| 1¼-7 UNC | 1A | 0.0022 | 1.2478 | 1.2232 | — | 1.1550 | 1.1439 | 1.0777 | 1B | 1.095 | 1.123 | 1.1572 | 1.1716 | 1.2500 |
| | 2A | 0.0022 | 1.2478 | 1.2314 | 1.2232 | 1.1550 | 1.1476 | 1.0777 | 2B | 1.095 | 1.123 | 1.1572 | 1.1668 | 1.2500 |
| | 3A | 0.0000 | 1.2500 | 1.2336 | — | 1.1572 | 1.1517 | 1.0799 | 3B | 1.0950 | 1.1125 | 1.1572 | 1.1644 | 1.2500 |
| 1¼-8 UN | 2A | 0.0021 | 1.2479 | 1.2329 | 1.2254 | 1.1667 | 1.1597 | 1.0991 | 2B | 1.115 | 1.140 | 1.1688 | 1.1780 | 1.2500 |
| | 3A | 0.0000 | 1.2500 | 1.2350 | — | 1.1688 | 1.1635 | 1.1012 | 3B | 1.1150 | 1.1297 | 1.1688 | 1.1757 | 1.2500 |
| 1¼-10 UNS | 2A | 0.0019 | 1.2481 | 1.2352 | — | 1.1831 | 1.1768 | 1.1290 | 2B | 1.142 | 1.163 | 1.1850 | 1.1932 | 1.2500 |

**Table 3.** *(Continued)* **Standard Series and Selected Combinations — Unified Screw Threads**

| Nominal Size, Threads per Inch, and Series Designation[a] | External[b] | | | | | | | | Internal[b] | | | | | |
|---|---|---|---|---|---|---|---|---|---|---|---|---|---|---|
| | Class | Allowance | Major Diameter | | | Pitch Diameter | | UNR Minor Dia.,[c] Max (Ref.) | Class | Minor Diameter | | Pitch Diameter | | Major Diameter |
| | | | Max[d] | Min | Min[e] | Max[d] | Min | | | Min | Max | Min | Max | Min |
| 1¼-12 UNF | 1A | 0.0018 | 1.2482 | 1.2310 | — | 1.1941 | 1.1849 | 1.1490 | 1B | 1.160 | 1.178 | 1.1959 | 1.2079 | 1.2500 |
| | 2A | 0.0018 | 1.2482 | 1.2368 | — | 1.1941 | 1.1879 | 1.1490 | 2B | 1.160 | 1.178 | 1.1959 | 1.2039 | 1.2500 |
| | 3A | 0.0000 | 1.2500 | 1.2386 | — | 1.1959 | 1.1913 | 1.1508 | 3B | 1.1600 | 1.1698 | 1.1959 | 1.2019 | 1.2500 |
| 1¼-14 UNS | 2A | 0.0016 | 1.2484 | 1.2381 | — | 1.2020 | 1.1966 | 1.1633 | 2B | 1.173 | 1.188 | 1.2036 | 1.2106 | 1.2500 |
| 1¼-16 UN | 2A | 0.0015 | 1.2485 | 1.2391 | — | 1.2079 | 1.2028 | 1.1741 | 2B | 1.182 | 1.196 | 1.2094 | 1.2160 | 1.2500 |
| | 3A | 0.0000 | 1.2500 | 1.2406 | — | 1.2094 | 1.2056 | 1.1756 | 3B | 1.1820 | 1.1909 | 1.2094 | 1.2144 | 1.2500 |
| 1¼-18 UNEF | 2A | 0.0015 | 1.2485 | 1.2398 | — | 1.2124 | 1.2075 | 1.1823 | 2B | 1.190 | 1.203 | 1.2139 | 1.2202 | 1.2500 |
| | 3A | 0.0000 | 1.2500 | 1.2413 | — | 1.2139 | 1.2103 | 1.1838 | 3B | 1.1900 | 1.1980 | 1.2139 | 1.2186 | 1.2500 |
| 1¼-20 UN | 2A | 0.0014 | 1.2486 | 1.2405 | — | 1.2161 | 1.2114 | 1.1891 | 2B | 1.196 | 1.207 | 1.2175 | 1.2236 | 1.2500 |
| | 3A | 0.0000 | 1.2500 | 1.2419 | — | 1.2175 | 1.2140 | 1.1905 | 3B | 1.1960 | 1.2036 | 1.2175 | 1.2220 | 1.2500 |
| 1¼-24 UNS | 2A | 0.0013 | 1.2487 | 1.2415 | — | 1.2216 | 1.2173 | 1.1991 | 2B | 1.205 | 1.215 | 1.2229 | 1.2285 | 1.2500 |
| 1¼-28 UN | 2A | 0.0012 | 1.2488 | 1.2423 | — | 1.2256 | 1.2215 | 1.2063 | 2B | 1.211 | 1.220 | 1.2268 | 1.2321 | 1.2500 |
| | 3A | 0.0000 | 1.2500 | 1.2435 | — | 1.2268 | 1.2237 | 1.2075 | 3B | 1.2110 | 1.2176 | 1.2268 | 1.2308 | 1.2500 |
| 1⁵⁄₁₆-8 UN | 2A | 0.0021 | 1.3104 | 1.2954 | — | 1.2292 | 1.2221 | 1.1616 | 2B | 1.177 | 1.202 | 1.2313 | 1.2405 | 1.3125 |
| | 3A | 0.0000 | 1.3125 | 1.2975 | — | 1.2313 | 1.2260 | 1.1637 | 3B | 1.1770 | 1.1922 | 1.2313 | 1.2382 | 1.3125 |
| 1⁵⁄₁₆-12 UN | 2A | 0.0017 | 1.3108 | 1.2994 | — | 1.2567 | 1.2509 | 1.2116 | 2B | 1.222 | 1.240 | 1.2584 | 1.2659 | 1.3125 |
| | 3A | 0.0000 | 1.3125 | 1.3011 | — | 1.2584 | 1.2540 | 1.2133 | 3B | 1.2220 | 1.2323 | 1.2584 | 1.2641 | 1.3125 |
| 1⁵⁄₁₆-16 UN | 2A | 0.0015 | 1.3110 | 1.3016 | — | 1.2704 | 1.2653 | 1.2366 | 2B | 1.245 | 1.259 | 1.2719 | 1.2786 | 1.3125 |
| | 3A | 0.0000 | 1.3125 | 1.3031 | — | 1.2719 | 1.2681 | 1.2381 | 3B | 1.2450 | 1.2534 | 1.2719 | 1.2769 | 1.3125 |
| 1⁵⁄₁₆-18 UNEF | 2A | 0.0015 | 1.3110 | 1.3023 | — | 1.2749 | 1.2700 | 1.2448 | 2B | 1.252 | 1.265 | 1.2764 | 1.2828 | 1.3125 |
| | 3A | 0.0000 | 1.3125 | 1.3038 | — | 1.2764 | 1.2727 | 1.2463 | 3B | 1.2520 | 1.2605 | 1.2764 | 1.2812 | 1.3125 |
| 1⁵⁄₁₆-20 UN | 2A | 0.0014 | 1.3111 | 1.3030 | — | 1.2786 | 1.2739 | 1.2516 | 2B | 1.258 | 1.270 | 1.2800 | 1.2861 | 1.3125 |
| | 3A | 0.0000 | 1.3125 | 1.3044 | — | 1.2800 | 1.2765 | 1.2530 | 3B | 1.2580 | 1.2661 | 1.2800 | 1.2846 | 1.3125 |
| 1⁵⁄₁₆-28 UN | 2A | 0.0012 | 1.3113 | 1.3048 | — | 1.2881 | 1.2840 | 1.2688 | 2B | 1.274 | 1.282 | 1.2893 | 1.2947 | 1.3125 |
| | 3A | 0.0000 | 1.3125 | 1.3060 | — | 1.2893 | 1.2862 | 1.2700 | 3B | 1.2740 | 1.2801 | 1.2893 | 1.2933 | 1.3125 |
| 1⅜-6 UNC | 1A | 0.0024 | 1.3726 | 1.3453 | — | 1.2643 | 1.2523 | 1.1741 | 1B | 1.195 | 1.225 | 1.2667 | 1.2822 | 1.3750 |
| | 2A | 0.0024 | 1.3726 | 1.3544 | 1.3453 | 1.2643 | 1.2563 | 1.1741 | 2B | 1.195 | 1.225 | 1.2667 | 1.2771 | 1.3750 |
| | 3A | 0.0000 | 1.3750 | 1.3568 | — | 1.2667 | 1.2607 | 1.1765 | 3B | 1.1950 | 1.2146 | 1.2667 | 1.2745 | 1.3750 |

**Table 3.** *(Continued)* **Standard Series and Selected Combinations — Unified Screw Threads**

| Nominal Size, Threads per Inch, and Series Designation[a] | External[b] | | | | | | | | Internal[b] | | | | | |
|---|---|---|---|---|---|---|---|---|---|---|---|---|---|---|
| | Class | Allow-ance | Major Diameter | | | Pitch Diameter | | UNR Minor Dia.,[c] Max (Ref.) | Class | Minor Diameter | | Pitch Diameter | | Major Diameter |
| | | | Max[d] | Min | Min[e] | Max[d] | Min | | | Min | Max | Min | Max | Min |
| 1$\frac{3}{8}$-8 UN | 2A | 0.0022 | 1.3728 | 1.3578 | 1.3503 | 1.2916 | 1.2844 | 1.2240 | 2B | 1.240 | 1.265 | 1.2938 | 1.3031 | 1.3750 |
| | 3A | 0.0000 | 1.3750 | 1.3600 | — | 1.2938 | 1.2884 | 1.2262 | 3B | 1.2400 | 1.2547 | 1.2938 | 1.3008 | 1.3750 |
| 1$\frac{3}{8}$-10 UNS | 2A | 0.0019 | 1.3731 | 1.3602 | — | 1.3081 | 1.3018 | 1.2540 | 2B | 1.267 | 1.288 | 1.3100 | 1.3182 | 1.3750 |
| 1$\frac{3}{8}$-12 UNF | 1A | 0.0019 | 1.3731 | 1.3559 | — | 1.3190 | 1.3096 | 1.2739 | 1B | 1.285 | 1.303 | 1.3209 | 1.3332 | 1.3750 |
| | 2A | 0.0019 | 1.3731 | 1.3617 | — | 1.3190 | 1.3127 | 1.2739 | 2B | 1.285 | 1.303 | 1.3209 | 1.3291 | 1.3750 |
| | 3A | 0.0000 | 1.3750 | 1.3636 | — | 1.3209 | 1.3162 | 1.2758 | 3B | 1.2850 | 1.2948 | 1.3209 | 1.3270 | 1.3750 |
| 1$\frac{3}{8}$-14 UNS | 2A | 0.0016 | 1.3734 | 1.3631 | — | 1.3270 | 1.3215 | 1.2883 | 2B | 1.298 | 1.313 | 1.3286 | 1.3357 | 1.3750 |
| 1$\frac{3}{8}$-16 UN | 2A | 0.0015 | 1.3735 | 1.3641 | — | 1.3329 | 1.3277 | 1.2991 | 2B | 1.307 | 1.321 | 1.3344 | 1.3411 | 1.3750 |
| | 3A | 0.0000 | 1.3750 | 1.3656 | — | 1.3344 | 1.3305 | 1.3006 | 3B | 1.3070 | 1.3159 | 1.3344 | 1.3394 | 1.3750 |
| 1$\frac{3}{8}$-18 UNEF | 2A | 0.0015 | 1.3735 | 1.3648 | — | 1.3374 | 1.3325 | 1.3073 | 2B | 1.315 | 1.328 | 1.3389 | 1.3453 | 1.3750 |
| | 3A | 0.0000 | 1.3750 | 1.3663 | — | 1.3389 | 1.3352 | 1.3088 | 3B | 1.3150 | 1.3230 | 1.3389 | 1.3437 | 1.3750 |
| 1$\frac{3}{8}$-20 UN | 2A | 0.0014 | 1.3736 | 1.3655 | — | 1.3411 | 1.3364 | 1.3141 | 2B | 1.321 | 1.332 | 1.3425 | 1.3486 | 1.3750 |
| | 3A | 0.0000 | 1.3750 | 1.3669 | — | 1.3425 | 1.3390 | 1.3155 | 3B | 1.3210 | 1.3286 | 1.3425 | 1.3471 | 1.3750 |
| 1$\frac{3}{8}$-24 UNS | 2A | 0.0013 | 1.3737 | 1.3665 | — | 1.3466 | 1.3422 | 1.3241 | 2B | 1.330 | 1.340 | 1.3479 | 1.3536 | 1.3750 |
| 1$\frac{3}{8}$-28 UN | 2A | 0.0012 | 1.3738 | 1.3673 | — | 1.3506 | 1.3465 | 1.3313 | 2B | 1.336 | 1.345 | 1.3518 | 1.3572 | 1.3750 |
| | 3A | 0.0000 | 1.3750 | 1.3685 | — | 1.3518 | 1.3487 | 1.3325 | 3B | 1.3360 | 1.3426 | 1.3518 | 1.3558 | 1.3750 |
| 1$\frac{7}{16}$-6 UN | 2A | 0.0024 | 1.4351 | 1.4169 | — | 1.3268 | 1.3188 | 1.2366 | 2B | 1.257 | 1.288 | 1.3292 | 1.3396 | 1.4375 |
| | 3A | 0.0000 | 1.4375 | 1.4193 | — | 1.3292 | 1.3232 | 1.2390 | 3B | 1.2570 | 1.2771 | 1.3292 | 1.3370 | 1.4375 |
| 1$\frac{7}{16}$-8 UN | 2A | 0.0022 | 1.4353 | 1.4203 | — | 1.3541 | 1.3469 | 1.2865 | 2B | 1.302 | 1.327 | 1.3563 | 1.3657 | 1.4375 |
| | 3A | 0.0000 | 1.4375 | 1.4225 | — | 1.3563 | 1.3509 | 1.2887 | 3B | 1.3020 | 1.3172 | 1.3563 | 1.3634 | 1.4375 |
| 1$\frac{7}{16}$-12 UN | 2A | 0.0018 | 1.4357 | 1.4243 | — | 1.3816 | 1.3757 | 1.3365 | 2B | 1.347 | 1.365 | 1.3834 | 1.3910 | 1.4375 |
| | 3A | 0.0000 | 1.4375 | 1.4261 | — | 1.3834 | 1.3790 | 1.3383 | 3B | 1.3470 | 1.3573 | 1.3834 | 1.3891 | 1.4375 |
| 1$\frac{7}{16}$-16 UN | 2A | 0.0016 | 1.4359 | 1.4265 | — | 1.3953 | 1.3901 | 1.3615 | 2B | 1.370 | 1.384 | 1.3969 | 1.4036 | 1.4375 |
| | 3A | 0.0000 | 1.4375 | 1.4281 | — | 1.3969 | 1.3930 | 1.3631 | 3B | 1.3700 | 1.3784 | 1.3969 | 1.4020 | 1.4375 |
| 1$\frac{7}{16}$-18 UNEF | 2A | 0.0015 | 1.4360 | 1.4273 | — | 1.3999 | 1.3950 | 1.3698 | 2B | 1.377 | 1.390 | 1.4014 | 1.4078 | 1.4375 |
| | 3A | 0.0000 | 1.4375 | 1.4288 | — | 1.4014 | 1.3977 | 1.3713 | 3B | 1.3770 | 1.3855 | 1.4014 | 1.4062 | 1.4375 |
| 1$\frac{7}{16}$-20 UN | 2A | 0.0014 | 1.4361 | 1.4280 | — | 1.4036 | 1.3989 | 1.3766 | 2B | 1.383 | 1.395 | 1.4050 | 1.4112 | 1.4375 |
| | 3A | 0.0000 | 1.4375 | 1.4294 | — | 1.4050 | 1.4014 | 1.3780 | 3B | 1.3830 | 1.3911 | 1.4050 | 1.4096 | 1.4375 |

**Table 3.** *(Continued)* **Standard Series and Selected Combinations — Unified Screw Threads**

| Nominal Size, Threads per Inch, and Series Designation[a] | External[b] | | | | | | | | Internal[b] | | | | | |
|---|---|---|---|---|---|---|---|---|---|---|---|---|---|---|
| | Class | Allowance | Major Diameter | | | Pitch Diameter | | UNR Minor Dia.,[c] Max (Ref.) | Class | Minor Diameter | | Pitch Diameter | | Major Diameter |
| | | | Max[d] | Min | Min[e] | Max[d] | Min | | | Min | Max | Min | Max | Min |
| 1 7/16-28 UN | 2A | 0.0013 | 1.4362 | 1.4297 | — | 1.4130 | 1.4088 | 1.3937 | 2B | 1.399 | 1.407 | 1.4143 | 1.4197 | 1.4375 |
| | 3A | 0.0000 | 1.4375 | 1.4310 | — | 1.4143 | 1.4112 | 1.3950 | 3B | 1.3990 | 1.4051 | 1.4143 | 1.4184 | 1.4375 |
| 1 1/2-6 UNC | 1A | 0.0024 | 1.4976 | 1.4703 | — | 1.3893 | 1.3772 | 1.2991 | 1B | 1.320 | 1.350 | 1.3917 | 1.4075 | 1.5000 |
| | 2A | 0.0024 | 1.4976 | 1.4794 | 1.4703 | 1.3893 | 1.3812 | 1.2991 | 2B | 1.320 | 1.350 | 1.3917 | 1.4022 | 1.5000 |
| | 3A | 0.0000 | 1.5000 | 1.4818 | — | 1.3917 | 1.3856 | 1.3015 | 3B | 1.3200 | 1.3396 | 1.3917 | 1.3996 | 1.5000 |
| 1 1/2-8 UN | 2A | 0.0022 | 1.4978 | 1.4828 | 1.4753 | 1.4166 | 1.4093 | 1.3490 | 2B | 1.365 | 1.390 | 1.4188 | 1.4283 | 1.5000 |
| | 3A | 0.0000 | 1.5000 | 1.4850 | — | 1.4188 | 1.4133 | 1.3512 | 3B | 1.3650 | 1.3797 | 1.4188 | 1.4259 | 1.5000 |
| 1 1/2-10 UNS | 2A | 0.0019 | 1.4981 | 1.4852 | — | 1.4331 | 1.4267 | 1.3790 | 2B | 1.392 | 1.413 | 1.4350 | 1.4433 | 1.5000 |
| 1 1/2-12 UNF | 1A | 0.0019 | 1.4981 | 1.4809 | — | 1.4440 | 1.4344 | 1.3989 | 1B | 1.410 | 1.428 | 1.4459 | 1.4584 | 1.5000 |
| | 2A | 0.0019 | 1.4981 | 1.4867 | — | 1.4440 | 1.4376 | 1.3989 | 2B | 1.410 | 1.428 | 1.4459 | 1.4542 | 1.5000 |
| | 3A | 0.0000 | 1.5000 | 1.4886 | — | 1.4459 | 1.4411 | 1.4008 | 3B | 1.4100 | 1.4198 | 1.4459 | 1.4522 | 1.5000 |
| 1 1/2-14 UNS | 2A | 0.0017 | 1.4983 | 1.4880 | — | 1.4519 | 1.4464 | 1.4132 | 2B | 1.423 | 1.438 | 1.4536 | 1.4608 | 1.5000 |
| 1 1/2-16 UN | 2A | 0.0016 | 1.4984 | 1.4890 | — | 1.4578 | 1.4526 | 1.4240 | 2B | 1.432 | 1.446 | 1.4594 | 1.4662 | 1.5000 |
| | 3A | 0.0000 | 1.5000 | 1.4906 | — | 1.4594 | 1.4555 | 1.4256 | 3B | 1.4320 | 1.4409 | 1.4594 | 1.4645 | 1.5000 |
| 1 1/2-18 UNEF | 2A | 0.0015 | 1.4985 | 1.4898 | — | 1.4624 | 1.4574 | 1.4323 | 2B | 1.440 | 1.453 | 1.4639 | 1.4704 | 1.5000 |
| | 3A | 0.0000 | 1.5000 | 1.4913 | — | 1.4639 | 1.4602 | 1.4338 | 3B | 1.4400 | 1.4480 | 1.4639 | 1.4687 | 1.5000 |
| 1 1/2-20 UN | 2A | 0.0014 | 1.4986 | 1.4905 | — | 1.4661 | 1.4613 | 1.4391 | 2B | 1.446 | 1.457 | 1.4675 | 1.4737 | 1.5000 |
| | 3A | 0.0000 | 1.5000 | 1.4919 | — | 1.4675 | 1.4639 | 1.4405 | 3B | 1.4460 | 1.4536 | 1.4675 | 1.4721 | 1.5000 |
| 1 1/2-24 UNS | 2A | 0.0013 | 1.4987 | 1.4915 | — | 1.4716 | 1.4672 | 1.4491 | 2B | 1.455 | 1.465 | 1.4729 | 1.4787 | 1.5000 |
| 1 1/2-28 UN | 2A | 0.0013 | 1.4987 | 1.4922 | — | 1.4755 | 1.4713 | 1.4562 | 2B | 1.461 | 1.470 | 1.4768 | 1.4823 | 1.5000 |
| | 3A | 0.0000 | 1.5000 | 1.4935 | — | 1.4768 | 1.4737 | 1.4575 | 3B | 1.4610 | 1.4676 | 1.4768 | 1.4809 | 1.5000 |
| 1 9/16-6 UN | 2A | 0.0024 | 1.5601 | 1.5419 | — | 1.4518 | 1.4436 | 1.3616 | 2B | 1.382 | 1.413 | 1.4542 | 1.4648 | 1.5625 |
| | 3A | 0.0000 | 1.5625 | 1.5443 | — | 1.4542 | 1.4481 | 1.3640 | 3B | 1.3820 | 1.4021 | 1.4542 | 1.4622 | 1.5625 |
| 1 9/16-8 UN | 2A | 0.0022 | 1.5603 | 1.5453 | — | 1.4791 | 1.4717 | 1.4115 | 2B | 1.427 | 1.452 | 1.4813 | 1.4909 | 1.5625 |
| | 3A | 0.0000 | 1.5625 | 1.5475 | — | 1.4813 | 1.4758 | 1.4137 | 3B | 1.4270 | 1.4422 | 1.4813 | 1.4885 | 1.5625 |
| 1 9/16-12 UN | 2A | 0.0018 | 1.5607 | 1.5493 | — | 1.5066 | 1.5007 | 1.4615 | 2B | 1.472 | 1.490 | 1.5084 | 1.5161 | 1.5625 |
| | 3A | 0.0000 | 1.5625 | 1.5511 | — | 1.5084 | 1.5040 | 1.4633 | 3B | 1.4720 | 1.4823 | 1.5084 | 1.5142 | 1.5625 |

**Table 3.** *(Continued)* **Standard Series and Selected Combinations — Unified Screw Threads**

| Nominal Size, Threads per Inch, and Series Designation[a] | External[b] | | | | | | | | Internal[b] | | | | | |
|---|---|---|---|---|---|---|---|---|---|---|---|---|---|---|
| | Class | Allowance | Major Diameter | | | Pitch Diameter | | UNR Minor Dia.,[c] Max (Ref.) | Class | Minor Diameter | | Pitch Diameter | | Major Diameter |
| | | | Max[d] | Min | Min[e] | Max[d] | Min | | | Min | Max | Min | Max | Min |
| 1 9/16-16 UN | 2A | 0.0016 | 1.5609 | 1.5515 | — | 1.5203 | 1.5151 | 1.4865 | 2B | 1.495 | 1.509 | 1.5219 | 1.5287 | 1.5625 |
| | 3A | 0.0000 | 1.5625 | 1.5531 | — | 1.5219 | 1.5180 | 1.4881 | 3B | 1.4950 | 1.5034 | 1.5219 | 1.5270 | 1.5625 |
| 1 9/16-18 UNEF | 2A | 0.0015 | 1.5610 | 1.5523 | — | 1.5249 | 1.5199 | 1.4948 | 2B | 1.502 | 1.515 | 1.5264 | 1.5329 | 1.5625 |
| | 3A | 0.0000 | 1.5625 | 1.5538 | — | 1.5264 | 1.5227 | 1.4963 | 3B | 1.5020 | 1.5105 | 1.5264 | 1.5313 | 1.5625 |
| 1 9/16-20 UN | 2A | 0.0014 | 1.5611 | 1.5530 | — | 1.5286 | 1.5238 | 1.5016 | 2B | 1.508 | 1.520 | 1.5300 | 1.5362 | 1.5625 |
| | 3A | 0.0000 | 1.5625 | 1.5544 | — | 1.5300 | 1.5264 | 1.5030 | 3B | 1.5080 | 1.5161 | 1.5300 | 1.5347 | 1.5625 |
| 1 5/8-6 UN | 2A | 0.0025 | 1.6225 | 1.6043 | — | 1.5142 | 1.5060 | 1.4240 | 2B | 1.445 | 1.475 | 1.5167 | 1.5274 | 1.6250 |
| | 3A | 0.0000 | 1.6250 | 1.6068 | — | 1.5167 | 1.5105 | 1.4265 | 3B | 1.4450 | 1.4646 | 1.5167 | 1.5247 | 1.6250 |
| 1 5/8-8 UN | 2A | 0.0022 | 1.6228 | 1.6078 | 1.6003 | 1.5416 | 1.5342 | 1.4740 | 2B | 1.490 | 1.515 | 1.5438 | 1.5535 | 1.6250 |
| | 3A | 0.0000 | 1.6250 | 1.6100 | — | 1.5438 | 1.5382 | 1.4762 | 3B | 1.4900 | 1.5047 | 1.5438 | 1.5510 | 1.6250 |
| 1 5/8-10 UNS | 2A | 0.0019 | 1.6231 | 1.6102 | — | 1.5581 | 1.5517 | 1.5040 | 2B | 1.517 | 1.538 | 1.5600 | 1.5683 | 1.6250 |
| 1 5/8-12 UN | 2A | 0.0018 | 1.6232 | 1.6118 | — | 1.5691 | 1.5632 | 1.5240 | 2B | 1.535 | 1.553 | 1.5709 | 1.5786 | 1.6250 |
| | 3A | 0.0000 | 1.6250 | 1.6136 | — | 1.5709 | 1.5665 | 1.5258 | 3B | 1.5350 | 1.5448 | 1.5709 | 1.5767 | 1.6250 |
| 1 5/8-14 UNS | 2A | 0.0017 | 1.6233 | 1.6130 | — | 1.5769 | 1.5714 | 1.5382 | 2B | 1.548 | 1.563 | 1.5786 | 1.5858 | 1.6250 |
| 1 5/8-16 UN | 2A | 0.0016 | 1.6234 | 1.6140 | — | 1.5828 | 1.5775 | 1.5490 | 2B | 1.557 | 1.571 | 1.5844 | 1.5912 | 1.6250 |
| | 3A | 0.0000 | 1.6250 | 1.6156 | — | 1.5844 | 1.5805 | 1.5506 | 3B | 1.5570 | 1.5659 | 1.5844 | 1.5895 | 1.6250 |
| 1 5/8-18 UNEF | 2A | 0.0015 | 1.6235 | 1.6148 | — | 1.5874 | 1.5824 | 1.5573 | 2B | 1.565 | 1.578 | 1.5889 | 1.5954 | 1.6250 |
| | 3A | 0.0000 | 1.6250 | 1.6163 | — | 1.5889 | 1.5851 | 1.5588 | 3B | 1.5650 | 1.5730 | 1.5889 | 1.5938 | 1.6250 |
| 1 5/8-20 UN | 2A | 0.0014 | 1.6236 | 1.6155 | — | 1.5911 | 1.5863 | 1.5641 | 2B | 1.571 | 1.582 | 1.5925 | 1.5987 | 1.6250 |
| | 3A | 0.0000 | 1.6250 | 1.6169 | — | 1.5925 | 1.5889 | 1.5655 | 3B | 1.5710 | 1.5786 | 1.5925 | 1.5972 | 1.6250 |
| 1 5/8-24 UNS | 2A | 0.0013 | 1.6237 | 1.6165 | — | 1.5966 | 1.5921 | 1.5741 | 2B | 1.580 | 1.590 | 1.5979 | 1.6037 | 1.6250 |
| 1 11/16-6 UN | 2A | 0.0025 | 1.6850 | 1.6668 | — | 1.5767 | 1.5684 | 1.4865 | 2B | 1.507 | 1.538 | 1.5792 | 1.5900 | 1.6875 |
| | 3A | 0.0000 | 1.6875 | 1.6693 | — | 1.5792 | 1.5730 | 1.4890 | 3B | 1.5070 | 1.5271 | 1.5792 | 1.5873 | 1.6875 |
| 1 11/16-8 UN | 2A | 0.0022 | 1.6853 | 1.6703 | — | 1.6041 | 1.5966 | 1.5365 | 2B | 1.552 | 1.577 | 1.6063 | 1.6160 | 1.6875 |
| | 3A | 0.0000 | 1.6875 | 1.6725 | — | 1.6063 | 1.6007 | 1.5387 | 3B | 1.5520 | 1.5672 | 1.6063 | 1.6136 | 1.6875 |
| 1 11/16-12 UN | 2A | 0.0018 | 1.6857 | 1.6743 | — | 1.6316 | 1.6257 | 1.5865 | 2B | 1.597 | 1.615 | 1.6334 | 1.6411 | 1.6875 |
| | 3A | 0.0000 | 1.6875 | 1.6761 | — | 1.6334 | 1.6289 | 1.5883 | 3B | 1.5970 | 1.6073 | 1.6334 | 1.6392 | 1.6875 |

**Table 3.** *(Continued)* **Standard Series and Selected Combinations — Unified Screw Threads**

| Nominal Size, Threads per Inch, and Series Designation[a] | External[b] | | | | | | | | Internal[b] | | | | | |
|---|---|---|---|---|---|---|---|---|---|---|---|---|---|---|
| | Class | Allow-ance | Major Diameter | | | Pitch Diameter | | UNR Minor Dia.,[c] Max (Ref.) | Class | Minor Diameter | | Pitch Diameter | | Major Diameter |
| | | | Max[d] | Min | Min[e] | Max[d] | Min | | | Min | Max | Min | Max | Min |
| 1 11/16-16 UN | 2A | 0.0016 | 1.6859 | 1.6765 | — | 1.6453 | 1.6400 | 1.6115 | 2B | 1.620 | 1.634 | 1.6469 | 1.6538 | 1.6875 |
| | 3A | 0.0000 | 1.6875 | 1.6781 | — | 1.6469 | 1.6429 | 1.6131 | 3B | 1.6200 | 1.6284 | 1.6469 | 1.6520 | 1.6875 |
| 1 11/16-18 UNEF | 2A | 0.0015 | 1.6860 | 1.6773 | — | 1.6499 | 1.6449 | 1.6198 | 2B | 1.627 | 1.640 | 1.6514 | 1.6579 | 1.6875 |
| | 3A | 0.0000 | 1.6875 | 1.6788 | — | 1.6514 | 1.6476 | 1.6213 | 3B | 1.6270 | 1.6355 | 1.6514 | 1.6563 | 1.6875 |
| 1 11/16-20 UN | 2A | 0.0014 | 1.6861 | 1.6780 | — | 1.6536 | 1.6488 | 1.6266 | 2B | 1.633 | 1.645 | 1.6550 | 1.6613 | 1.6875 |
| | 3A | 0.0000 | 1.6875 | 1.6794 | — | 1.6550 | 1.6514 | 1.6280 | 3B | 1.6330 | 1.6411 | 1.6550 | 1.6597 | 1.6875 |
| 1 3/4-5 UNC | 1A | 0.0027 | 1.7473 | 1.7165 | — | 1.6174 | 1.6040 | 1.5091 | 1B | 1.533 | 1.567 | 1.6201 | 1.6375 | 1.7500 |
| | 2A | 0.0027 | 1.7473 | 1.7268 | 1.7165 | 1.6174 | 1.6085 | 1.5091 | 2B | 1.533 | 1.567 | 1.6201 | 1.6317 | 1.7500 |
| | 3A | 0.0000 | 1.7500 | 1.7295 | — | 1.6201 | 1.6134 | 1.5118 | 3B | 1.5330 | 1.5575 | 1.6201 | 1.6288 | 1.7500 |
| 1 3/4-6 UN | 2A | 0.0025 | 1.7475 | 1.7293 | — | 1.6392 | 1.6309 | 1.5490 | 2B | 1.570 | 1.600 | 1.6417 | 1.6525 | 1.7500 |
| | 3A | 0.0000 | 1.7500 | 1.7318 | — | 1.6417 | 1.6354 | 1.5515 | 3B | 1.5700 | 1.5896 | 1.6417 | 1.6498 | 1.7500 |
| 1 3/4-8 UN | 2A | 0.0023 | 1.7477 | 1.7327 | 1.7252 | 1.6665 | 1.6590 | 1.5989 | 2B | 1.615 | 1.640 | 1.6688 | 1.6786 | 1.7500 |
| | 3A | 0.0000 | 1.7500 | 1.7350 | — | 1.6688 | 1.6631 | 1.6012 | 3B | 1.6150 | 1.6297 | 1.6688 | 1.6762 | 1.7500 |
| 1 3/4-10 UNS | 2A | 0.0019 | 1.7481 | 1.7352 | — | 1.6831 | 1.6766 | 1.6290 | 2B | 1.642 | 1.663 | 1.6850 | 1.6934 | 1.7500 |
| 1 3/4-12 UN | 2A | 0.0018 | 1.7482 | 1.7368 | — | 1.6941 | 1.6881 | 1.6490 | 2B | 1.660 | 1.678 | 1.6959 | 1.7037 | 1.7500 |
| | 3A | 0.0000 | 1.7500 | 1.7386 | — | 1.6959 | 1.6914 | 1.6508 | 3B | 1.6600 | 1.6698 | 1.6959 | 1.7017 | 1.7500 |
| 1 3/4-14 UNS | 2A | 0.0017 | 1.7483 | 1.7380 | — | 1.7019 | 1.6963 | 1.6632 | 2B | 1.673 | 1.688 | 1.7036 | 1.7109 | 1.7500 |
| 1 3/4-16 UN | 2A | 0.0016 | 1.7484 | 1.7390 | — | 1.7078 | 1.7025 | 1.6740 | 2B | 1.682 | 1.696 | 1.7094 | 1.7163 | 1.7500 |
| | 3A | 0.0000 | 1.7500 | 1.7406 | — | 1.7094 | 1.7054 | 1.6756 | 3B | 1.6820 | 1.6909 | 1.7094 | 1.7146 | 1.7500 |
| 1 3/4-18 UNS | 2A | 0.0015 | 1.7485 | 1.7398 | — | 1.7124 | 1.7073 | 1.6823 | 2B | 1.690 | 1.703 | 1.7139 | 1.7205 | 1.7500 |
| 1 3/4-20 UN | 2A | 0.0015 | 1.7485 | 1.7404 | — | 1.7160 | 1.7112 | 1.6890 | 2B | 1.696 | 1.707 | 1.7175 | 1.7238 | 1.7500 |
| | 3A | 0.0000 | 1.7500 | 1.7419 | — | 1.7175 | 1.7139 | 1.6905 | 3B | 1.6960 | 1.7036 | 1.7175 | 1.7222 | 1.7500 |
| 1 13/16-6 UN | 2A | 0.0025 | 1.8100 | 1.7918 | — | 1.7017 | 1.6933 | 1.6115 | 2B | 1.632 | 1.663 | 1.7042 | 1.7151 | 1.8125 |
| | 3A | 0.0000 | 1.8125 | 1.7943 | — | 1.7042 | 1.6979 | 1.6140 | 3B | 1.6320 | 1.6521 | 1.7042 | 1.7124 | 1.8125 |
| 1 13/16-8 UN | 2A | 0.0023 | 1.8102 | 1.7952 | — | 1.7290 | 1.7214 | 1.6614 | 2B | 1.677 | 1.702 | 1.7313 | 1.7412 | 1.8125 |
| | 3A | 0.0000 | 1.8125 | 1.7975 | — | 1.7313 | 1.7256 | 1.6637 | 3B | 1.6770 | 1.6922 | 1.7313 | 1.7387 | 1.8125 |
| 1 13/16-12 UN | 2A | 0.0018 | 1.8107 | 1.7993 | — | 1.7566 | 1.7506 | 1.7115 | 2B | 1.722 | 1.740 | 1.7584 | 1.7662 | 1.8125 |
| | 3A | 0.0000 | 1.8125 | 1.8011 | — | 1.7584 | 1.7539 | 1.7133 | 3B | 1.7220 | 1.7323 | 1.7584 | 1.7642 | 1.8125 |

**Table 3.** *(Continued)* **Standard Series and Selected Combinations — Unified Screw Threads**

| Nominal Size, Threads per Inch, and Series Designation[a] | External[b] | | | | | | | | Internal[b] | | | | | |
|---|---|---|---|---|---|---|---|---|---|---|---|---|---|---|
| | Class | Allow-ance | Major Diameter | | | Pitch Diameter | | UNR Minor Dia.,[c] Max (Ref.) | Class | Minor Diameter | | Pitch Diameter | | Major Diameter |
| | | | Max[d] | Min | Min[e] | Max[d] | Min | | | Min | Max | Min | Max | Min |
| 1 13/16-16 UN | 2A | 0.0016 | 1.8109 | 1.8015 | — | 1.7703 | 1.7650 | 1.7365 | 2B | 1.745 | 1.759 | 1.7719 | 1.7788 | 1.8125 |
| | 3A | 0.0000 | 1.8125 | 1.8031 | — | 1.7719 | 1.7679 | 1.7381 | 3B | 1.7450 | 1.7534 | 1.7719 | 1.7771 | 1.8125 |
| 1 13/16-20 UN | 2A | 0.0015 | 1.8110 | 1.8029 | — | 1.7785 | 1.7736 | 1.7515 | 2B | 1.758 | 1.770 | 1.7800 | 1.7863 | 1.8125 |
| | 3A | 0.0000 | 1.8125 | 1.8044 | — | 1.7800 | 1.7763 | 1.7530 | 3B | 1.7580 | 1.7661 | 1.7800 | 1.7847 | 1.8125 |
| 1 7/8-6 UN | 2A | 0.0025 | 1.8725 | 1.8543 | — | 1.7642 | 1.7558 | 1.6740 | 2B | 1.695 | 1.725 | 1.7667 | 1.7777 | 1.8750 |
| | 3A | 0.0000 | 1.8750 | 1.8568 | — | 1.7667 | 1.7604 | 1.6765 | 3B | 1.6950 | 1.7146 | 1.7667 | 1.7749 | 1.8750 |
| 1 7/8-8 UN | 2A | 0.0023 | 1.8727 | 1.8577 | 1.8502 | 1.7915 | 1.7838 | 1.7239 | 2B | 1.740 | 1.765 | 1.7938 | 1.8038 | 1.8750 |
| | 3A | 0.0000 | 1.8750 | 1.8600 | — | 1.7938 | 1.7881 | 1.7262 | 3B | 1.7400 | 1.7547 | 1.7938 | 1.8013 | 1.8750 |
| 1 7/8-10 UNS | 2A | 0.0020 | 1.8730 | 1.8601 | — | 1.8080 | 1.8015 | 1.7539 | 2B | 1.767 | 1.788 | 1.8100 | 1.8185 | 1.8750 |
| 1 7/8-12 UN | 2A | 0.0018 | 1.8732 | 1.8618 | — | 1.8191 | 1.8131 | 1.7740 | 2B | 1.785 | 1.803 | 1.8209 | 1.8287 | 1.8750 |
| | 3A | 0.0000 | 1.8750 | 1.8636 | — | 1.8209 | 1.8164 | 1.7758 | 3B | 1.7850 | 1.7948 | 1.8209 | 1.8268 | 1.8750 |
| 1 7/8-14 UNS | 2A | 0.0017 | 1.8733 | 1.8630 | — | 1.8269 | 1.8213 | 1.7882 | 2B | 1.798 | 1.813 | 1.8286 | 1.8359 | 1.8750 |
| 1 7/8-16 UN | 2A | 0.0016 | 1.8734 | 1.8640 | — | 1.8328 | 1.8275 | 1.7990 | 2B | 1.807 | 1.821 | 1.8344 | 1.8413 | 1.8750 |
| | 3A | 0.0000 | 1.8750 | 1.8656 | — | 1.8344 | 1.8304 | 1.8006 | 3B | 1.8070 | 1.8159 | 1.8344 | 1.8396 | 1.8750 |
| 1 7/8-18 UNS | 2A | 0.0015 | 1.8735 | 1.8648 | — | 1.8374 | 1.8323 | 1.8073 | 2B | 1.815 | 1.828 | 1.8389 | 1.8455 | 1.8750 |
| 1 7/8-20 UN | 2A | 0.0015 | 1.8735 | 1.8654 | — | 1.8410 | 1.8361 | 1.8140 | 2B | 1.821 | 1.832 | 1.8425 | 1.8489 | 1.8750 |
| | 3A | 0.0000 | 1.8750 | 1.8669 | — | 1.8425 | 1.8388 | 1.8155 | 3B | 1.8210 | 1.8286 | 1.8425 | 1.8473 | 1.8750 |
| 1 15/16-6 UN | 2A | 0.0026 | 1.9349 | 1.9167 | — | 1.8266 | 1.8181 | 1.7364 | 2B | 1.757 | 1.788 | 1.8292 | 1.8403 | 1.9375 |
| | 3A | 0.0000 | 1.9375 | 1.9193 | — | 1.8292 | 1.8228 | 1.7390 | 3B | 1.7570 | 1.7771 | 1.8292 | 1.8375 | 1.9375 |
| 1 15/16-8 UN | 2A | 0.0023 | 1.9352 | 1.9202 | — | 1.8540 | 1.8463 | 1.7864 | 2B | 1.802 | 1.827 | 1.8563 | 1.8663 | 1.9375 |
| | 3A | 0.0000 | 1.9375 | 1.9225 | — | 1.8563 | 1.8505 | 1.7887 | 3B | 1.8020 | 1.8172 | 1.8563 | 1.8638 | 1.9375 |
| 1 15/16-12 UN | 2A | 0.0018 | 1.9357 | 1.9243 | — | 1.8816 | 1.8756 | 1.8365 | 2B | 1.847 | 1.865 | 1.8834 | 1.8912 | 1.9375 |
| | 3A | 0.0000 | 1.9375 | 1.9261 | — | 1.8834 | 1.8789 | 1.8383 | 3B | 1.8470 | 1.8573 | 1.8834 | 1.8893 | 1.9375 |
| 1 15/16-16 UN | 2A | 0.0016 | 1.9359 | 1.9265 | — | 1.8953 | 1.8899 | 1.8615 | 2B | 1.870 | 1.884 | 1.8969 | 1.9039 | 1.9375 |
| | 3A | 0.0000 | 1.9375 | 1.9281 | — | 1.8969 | 1.8929 | 1.8631 | 3B | 1.8700 | 1.8784 | 1.8969 | 1.9021 | 1.9375 |
| 1 15/16-20 UN | 2A | 0.0015 | 1.9360 | 1.9279 | — | 1.9035 | 1.8986 | 1.8765 | 2B | 1.883 | 1.895 | 1.9050 | 1.9114 | 1.9375 |
| | 3A | 0.0000 | 1.9375 | 1.9294 | — | 1.9050 | 1.9013 | 1.8780 | 3B | 1.8830 | 1.8911 | 1.9050 | 1.9098 | 1.9375 |

**Table 3.** *(Continued)* **Standard Series and Selected Combinations — Unified Screw Threads**

| Nominal Size, Threads per Inch, and Series Designation[a] | External[b] | | | | | | | | Internal[b] | | | | | |
|---|---|---|---|---|---|---|---|---|---|---|---|---|---|---|
| | Class | Allowance | Major Diameter | | | Pitch Diameter | | UNR Minor Dia.,[c] Max (Ref.) | Class | Minor Diameter | | Pitch Diameter | | Major Diameter |
| | | | Max[d] | Min | Min[e] | Max[d] | Min | | | Min | Max | Min | Max | Min |
| 2-4½ UNC | 1A | 0.0029 | 1.9971 | 1.9641 | — | 1.8528 | 1.8385 | 1.7325 | 1B | 1.759 | 1.795 | 1.8557 | 1.8743 | 2.0000 |
| | 2A | 0.0029 | 1.9971 | 1.9751 | 1.9641 | 1.8528 | 1.8433 | 1.7325 | 2B | 1.759 | 1.795 | 1.8557 | 1.8681 | 2.0000 |
| | 3A | 0.0000 | 2.0000 | 1.9780 | — | 1.8557 | 1.8486 | 1.7354 | 3B | 1.7590 | 1.7861 | 1.8557 | 1.8650 | 2.0000 |
| 2-6 UN | 2A | 0.0026 | 1.9974 | 1.9792 | — | 1.8891 | 1.8805 | 1.7989 | 2B | 1.820 | 1.850 | 1.8917 | 1.9028 | 2.0000 |
| | 3A | 0.0000 | 2.0000 | 1.9818 | — | 1.8917 | 1.8853 | 1.8015 | 3B | 1.8200 | 1.8396 | 1.8917 | 1.9000 | 2.0000 |
| 2-8 UN | 2A | 0.0023 | 1.9977 | 1.9827 | 1.9752 | 1.9165 | 1.9087 | 1.8489 | 2B | 1.865 | 1.890 | 1.9188 | 1.9289 | 2.0000 |
| | 3A | 0.0000 | 2.0000 | 1.9850 | — | 1.9188 | 1.9130 | 1.8512 | 3B | 1.8650 | 1.8797 | 1.9188 | 1.9264 | 2.0000 |
| 2-10 UNS | 2A | 0.0020 | 1.9980 | 1.9851 | — | 1.9330 | 1.9265 | 1.8789 | 2B | 1.892 | 1.913 | 1.9350 | 1.9435 | 2.0000 |
| 2-12 UN | 2A | 0.0018 | 1.9982 | 1.9868 | — | 1.9441 | 1.9380 | 1.8990 | 2B | 1.910 | 1.928 | 1.9459 | 1.9538 | 2.0000 |
| | 3A | 0.0000 | 2.0000 | 1.9886 | — | 1.9459 | 1.9414 | 1.9008 | 3B | 1.9100 | 1.9198 | 1.9459 | 1.9518 | 2.0000 |
| 2-14 UNS | 2A | 0.0017 | 1.9983 | 1.9880 | — | 1.9519 | 1.9462 | 1.9132 | 2B | 1.923 | 1.938 | 1.9536 | 1.9610 | 2.0000 |
| 2-16 UN | 2A | 0.0016 | 1.9984 | 1.9890 | — | 1.9578 | 1.9524 | 1.9240 | 2B | 1.932 | 1.946 | 1.9594 | 1.9664 | 2.0000 |
| | 3A | 0.0000 | 2.0000 | 1.9906 | — | 1.9594 | 1.9554 | 1.9256 | 3B | 1.9320 | 1.9409 | 1.9594 | 1.9646 | 2.0000 |
| 2-18 UNS | 2A | 0.0015 | 1.9985 | 1.9898 | — | 1.9624 | 1.9573 | 1.9323 | 2B | 1.940 | 1.953 | 1.9639 | 1.9706 | 2.0000 |
| 2-20 UN | 2A | 0.0015 | 1.9985 | 1.9904 | — | 1.9660 | 1.9611 | 1.9390 | 2B | 1.946 | 1.957 | 1.9675 | 1.9739 | 2.0000 |
| | 3A | 0.0000 | 2.0000 | 1.9919 | — | 1.9675 | 1.9638 | 1.9405 | 3B | 1.9460 | 1.9536 | 1.9675 | 1.9723 | 2.0000 |
| 2 1/16-16 UNS | 2A | 0.0016 | 2.0609 | 2.0515 | — | 2.0203 | 2.0149 | 1.9865 | 2B | 1.995 | 2.009 | 2.0219 | 2.0289 | 2.0625 |
| | 3A | 0.0000 | 2.0625 | 2.0531 | — | 2.0219 | 2.0179 | 1.9881 | 3B | 1.9950 | 2.0034 | 2.0219 | 2.0272 | 2.0625 |
| 2 1/8-6 UN | 2A | 0.0026 | 2.1224 | 2.1042 | — | 2.0141 | 2.0054 | 1.9239 | 2B | 1.945 | 1.975 | 2.0167 | 2.0280 | 2.1250 |
| | 3A | 0.0000 | 2.1250 | 2.1068 | — | 2.0167 | 2.0102 | 1.9265 | 3B | 1.9450 | 1.9646 | 2.0167 | 2.0251 | 2.1250 |
| 2 1/8-8 UN | 2A | 0.0024 | 2.1226 | 2.1076 | 2.1001 | 2.0414 | 2.0335 | 1.9738 | 2B | 1.990 | 2.015 | 2.0438 | 2.0540 | 2.1250 |
| | 3A | 0.0000 | 2.1250 | 2.1100 | — | 2.0438 | 2.0379 | 1.9762 | 3B | 1.9900 | 2.0047 | 2.0438 | 2.0515 | 2.1250 |
| 2 1/8-12 UN | 2A | 0.0018 | 2.1232 | 2.1118 | — | 2.0691 | 2.0630 | 2.0240 | 2B | 2.035 | 2.053 | 2.0709 | 2.0788 | 2.1250 |
| | 3A | 0.0000 | 2.1250 | 2.1136 | — | 2.0709 | 2.0663 | 2.0258 | 3B | 2.0350 | 2.0448 | 2.0709 | 2.0768 | 2.1250 |
| 2 1/8-16 UN | 2A | 0.0016 | 2.1234 | 2.1140 | — | 2.0828 | 2.0774 | 2.0490 | 2B | 2.057 | 2.071 | 2.0844 | 2.0914 | 2.1250 |
| | 3A | 0.0000 | 2.1250 | 2.1156 | — | 2.0844 | 2.0803 | 2.0506 | 3B | 2.0570 | 2.0659 | 2.0844 | 2.0897 | 2.1250 |
| 2 1/8-20 UN | 2A | 0.0015 | 2.1235 | 2.1154 | — | 2.0910 | 2.0860 | 2.0640 | 2B | 2.071 | 2.082 | 2.0925 | 2.0990 | 2.1250 |
| | 3A | 0.0000 | 2.1250 | 2.1169 | — | 2.0925 | 2.0888 | 2.0655 | 3B | 2.0710 | 2.0786 | 2.0925 | 2.0973 | 2.1250 |

**Table 3.** *(Continued)* **Standard Series and Selected Combinations — Unified Screw Threads**

| Nominal Size, Threads per Inch, and Series Designation[a] | External[b] | | | | | | | | Internal[b] | | | | | |
|---|---|---|---|---|---|---|---|---|---|---|---|---|---|---|
| | Class | Allowance | Major Diameter | | | Pitch Diameter | | UNR Minor Dia.,[c] Max (Ref.) | Class | Minor Diameter | | Pitch Diameter | | Major Diameter |
| | | | Max[d] | Min | Min[e] | Max[d] | Min | | | Min | Max | Min | Max | Min |
| 2 3/16-16 UNS | 2A | 0.0016 | 2.1859 | 2.1765 | — | 2.1453 | 2.1399 | 2.1115 | 2B | 2.120 | 2.134 | 2.1469 | 2.1540 | 2.1875 |
| | 3A | 0.0000 | 2.1875 | 2.1781 | — | 2.1469 | 2.1428 | 2.1131 | 3B | 2.1200 | 2.1284 | 2.1469 | 2.1522 | 2.1875 |
| 2 1/4-4 1/2 UNC | 1A | 0.0029 | 2.2471 | 2.2141 | — | 2.1028 | 2.0882 | 1.9825 | 1B | 2.009 | 2.045 | 2.1057 | 2.1247 | 2.2500 |
| | 2A | 0.0029 | 2.2471 | 2.2251 | 2.2141 | 2.1028 | 2.0931 | 1.9825 | 2B | 2.009 | 2.045 | 2.1057 | 2.1183 | 2.2500 |
| | 3A | 0.0000 | 2.2500 | 2.2280 | — | 2.1057 | 2.0984 | 1.9854 | 3B | 2.0090 | 2.0361 | 2.1057 | 2.1152 | 2.2500 |
| 2 1/4-6 UN | 2A | 0.0026 | 2.2474 | 2.2292 | — | 2.1391 | 2.1303 | 2.0489 | 2B | 2.070 | 2.100 | 2.1417 | 2.1531 | 2.2500 |
| | 3A | 0.0000 | 2.2500 | 2.2318 | — | 2.1417 | 2.1351 | 2.0515 | 3B | 2.0700 | 2.0896 | 2.1417 | 2.1502 | 2.2500 |
| 2 1/4-8 UN | 2A | 0.0024 | 2.2476 | 2.2326 | 2.2251 | 2.1664 | 2.1584 | 2.0988 | 2B | 2.115 | 2.140 | 2.1688 | 2.1792 | 2.2500 |
| | 3A | 0.0000 | 2.2500 | 2.2350 | — | 2.1688 | 2.1628 | 2.1012 | 3B | 2.1150 | 2.1297 | 2.1688 | 2.1766 | 2.2500 |
| 2 1/4-10 UNS | 2A | 0.0020 | 2.2480 | 2.2351 | — | 2.1830 | 2.1764 | 2.1289 | 2B | 2.142 | 2.163 | 2.1850 | 2.1936 | 2.2500 |
| 2 1/4-12 UN | 2A | 0.0018 | 2.2482 | 2.2368 | — | 2.1941 | 2.1880 | 2.1490 | 2B | 2.160 | 2.178 | 2.1959 | 2.2039 | 2.2500 |
| | 3A | 0.0000 | 2.2500 | 2.2386 | — | 2.1959 | 2.1913 | 2.1508 | 3B | 2.1600 | 2.1698 | 2.1959 | 2.2019 | 2.2500 |
| 2 1/4-14 UNS | 2A | 0.0017 | 2.2483 | 2.2380 | — | 2.2019 | 2.1961 | 2.1632 | 2B | 2.173 | 2.188 | 2.2036 | 2.2111 | 2.2500 |
| 2 1/4-16 UN | 2A | 0.0016 | 2.2484 | 2.2390 | — | 2.2078 | 2.2023 | 2.1740 | 2B | 2.182 | 2.196 | 2.2094 | 2.2165 | 2.2500 |
| | 3A | 0.0000 | 2.2500 | 2.2406 | — | 2.2094 | 2.2053 | 2.1756 | 3B | 2.1820 | 2.1909 | 2.2094 | 2.2147 | 2.2500 |
| 2 1/4-18 UNS | 2A | 0.0016 | 2.2484 | 2.2397 | — | 2.2123 | 2.2071 | 2.1822 | 2B | 2.190 | 2.203 | 2.2139 | 2.2207 | 2.2500 |
| 2 1/4-20 UN | 2A | 0.0015 | 2.2485 | 2.2404 | — | 2.2160 | 2.2110 | 2.1890 | 2B | 2.196 | 2.207 | 2.2175 | 2.2240 | 2.2500 |
| | 3A | 0.0000 | 2.2500 | 2.2419 | — | 2.2175 | 2.2137 | 2.1905 | 3B | 2.1960 | 2.2036 | 2.2175 | 2.2223 | 2.2500 |
| 2 5/16-16 UNS | 2A | 0.0016 | 2.3109 | 2.3015 | — | 2.2703 | 2.2648 | 2.2365 | 2B | 2.245 | 2.259 | 2.2719 | 2.2790 | 2.3125 |
| | 3A | 0.0000 | 2.3125 | 2.3031 | — | 2.2719 | 2.2678 | 2.2381 | 3B | 2.2450 | 2.2534 | 2.2719 | 2.2772 | 2.3125 |
| 2 3/8-6 UN | 2A | 0.0027 | 2.3723 | 2.3541 | — | 2.2640 | 2.2551 | 2.1738 | 2B | 2.195 | 2.225 | 2.2667 | 2.2782 | 2.3750 |
| | 3A | 0.0000 | 2.3750 | 2.3568 | — | 2.2667 | 2.2601 | 2.1765 | 3B | 2.1950 | 2.2146 | 2.2667 | 2.2753 | 2.3750 |
| 2 3/8-8 UN | 2A | 0.0024 | 2.3726 | 2.3576 | — | 2.2914 | 2.2833 | 2.2238 | 2B | 2.240 | 2.265 | 2.2938 | 2.3043 | 2.3750 |
| | 3A | 0.0000 | 2.3750 | 2.3600 | — | 2.2938 | 2.2878 | 2.2262 | 3B | 2.2400 | 2.2547 | 2.2938 | 2.3017 | 2.3750 |
| 2 3/8-12 UN | 2A | 0.0018 | 2.3732 | 2.3618 | — | 2.3191 | 2.3129 | 2.2740 | 2B | 2.285 | 2.303 | 2.3209 | 2.3289 | 2.3750 |
| | 3A | 0.0000 | 2.3750 | 2.3636 | — | 2.3209 | 2.3163 | 2.2758 | 3B | 2.2850 | 2.2948 | 2.3209 | 2.3269 | 2.3750 |
| 2 3/8-16 UN | 2A | 0.0016 | 2.3734 | 2.3640 | — | 2.3328 | 2.3273 | 2.2990 | 2B | 2.307 | 2.321 | 2.3344 | 2.3415 | 2.3750 |
| | 3A | 0.0000 | 2.3750 | 2.3656 | — | 2.3344 | 2.3303 | 2.3006 | 3B | 2.3070 | 2.3159 | 2.3344 | 2.3398 | 2.3750 |

**Table 3.** *(Continued)* **Standard Series and Selected Combinations — Unified Screw Threads**

| Nominal Size, Threads per Inch, and Series Designation[a] | External[b] | | | | | | | | Internal[b] | | | | | |
|---|---|---|---|---|---|---|---|---|---|---|---|---|---|---|
| | Class | Allowance | Major Diameter | | | Pitch Diameter | | UNR Minor Dia.,[c] Max (Ref.) | Class | Minor Diameter | | Pitch Diameter | | Major Diameter |
| | | | Max[d] | Min | Min[e] | Max[d] | Min | | | Min | Max | Min | Max | Min |
| 2⅜-20 UN | 2A | 0.0015 | 2.3735 | 2.3654 | — | 2.3410 | 2.3360 | 2.3140 | 2B | 2.321 | 2.332 | 2.3425 | 2.3491 | 2.3750 |
| | 3A | 0.0000 | 2.3750 | 2.3669 | — | 2.3425 | 2.3387 | 2.3155 | 3B | 2.3210 | 2.3286 | 2.3425 | 2.3474 | 2.3750 |
| 2 7/16-16 UNS | 2A | 0.0017 | 2.4358 | 2.4264 | — | 2.3952 | 2.3897 | 2.3614 | 2B | 2.370 | 2.384 | 2.3969 | 2.4041 | 2.4375 |
| | 3A | 0.0000 | 2.4375 | 2.4281 | — | 2.3969 | 2.3928 | 2.3631 | 3B | 2.3700 | 2.3784 | 2.3969 | 2.4023 | 2.4375 |
| 2½-4 UNC | 1A | 0.0031 | 2.4969 | 2.4612 | — | 2.3345 | 2.3190 | 2.1992 | 1B | 2.229 | 2.267 | 2.3376 | 2.3578 | 2.5000 |
| | 2A | 0.0031 | 2.4969 | 2.4731 | 2.4612 | 2.3345 | 2.3241 | 2.1992 | 2B | 2.229 | 2.267 | 2.3376 | 2.3511 | 2.5000 |
| | 3A | 0.0000 | 2.5000 | 2.4762 | — | 2.3376 | 2.3298 | 2.2023 | 3B | 2.2290 | 2.2594 | 2.3376 | 2.3477 | 2.5000 |
| 2½-6 UN | 2A | 0.0027 | 2.4973 | 2.4791 | — | 2.3890 | 2.3800 | 2.2988 | 2B | 2.320 | 2.350 | 2.3917 | 2.4033 | 2.5000 |
| | 3A | 0.0000 | 2.5000 | 2.4818 | — | 2.3917 | 2.3850 | 2.3015 | 3B | 2.3200 | 2.3396 | 2.3917 | 2.4004 | 2.5000 |
| 2½-8 UN | 2A | 0.0024 | 2.4976 | 2.4826 | 2.4751 | 2.4164 | 2.4082 | 2.3488 | 2B | 2.365 | 2.390 | 2.4188 | 2.4294 | 2.5000 |
| | 3A | 0.0000 | 2.5000 | 2.4850 | — | 2.4188 | 2.4127 | 2.3512 | 3B | 2.3650 | 2.3797 | 2.4188 | 2.4268 | 2.5000 |
| 2½-10 UNS | 2A | 0.0020 | 2.4980 | 2.4851 | — | 2.4330 | 2.4263 | 2.3789 | 2B | 2.392 | 2.413 | 2.4350 | 2.4437 | 2.5000 |
| 2½-12 UN | 2A | 0.0019 | 2.4981 | 2.4867 | — | 2.4440 | 2.4378 | 2.3989 | 2B | 2.410 | 2.428 | 2.4459 | 2.4540 | 2.5000 |
| | 3A | 0.0000 | 2.5000 | 2.4886 | — | 2.4459 | 2.4413 | 2.4008 | 3B | 2.4100 | 2.4198 | 2.4459 | 2.4519 | 2.5000 |
| 2½-14 UNS | 2A | 0.0017 | 2.4983 | 2.4880 | — | 2.4519 | 2.4461 | 2.4132 | 2B | 2.423 | 2.438 | 2.4536 | 2.4612 | 2.5000 |
| 2½-16 UN | 2A | 0.0017 | 2.4983 | 2.4889 | — | 2.4577 | 2.4522 | 2.4239 | 2B | 2.432 | 2.446 | 2.4594 | 2.4666 | 2.5000 |
| | 3A | 0.0000 | 2.5000 | 2.4906 | — | 2.4594 | 2.4553 | 2.4256 | 3B | 2.4320 | 2.4409 | 2.4594 | 2.4648 | 2.5000 |
| 2½-18 UNS | 2A | 0.0016 | 2.4984 | 2.4897 | — | 2.4623 | 2.4570 | 2.4322 | 2B | 2.440 | 2.453 | 2.4639 | 2.4708 | 2.5000 |
| 2½-20 UN | 2A | 0.0015 | 2.4985 | 2.4904 | — | 2.4660 | 2.4609 | 2.4390 | 2B | 2.446 | 2.457 | 2.4675 | 2.4741 | 2.5000 |
| | 3A | 0.0000 | 2.5000 | 2.4919 | — | 2.4675 | 2.4637 | 2.4405 | 3B | 2.4460 | 2.4536 | 2.4675 | 2.4725 | 2.5000 |
| 2⅝-6 UN | 2A | 0.0027 | 2.6223 | 2.6041 | — | 2.5140 | 2.5050 | 2.4238 | 2B | 2.445 | 2.475 | 2.5167 | 2.5285 | 2.6250 |
| | 3A | 0.0000 | 2.6250 | 2.6068 | — | 2.5167 | 2.5099 | 2.4265 | 3B | 2.4450 | 2.4646 | 2.5167 | 2.5255 | 2.6250 |
| 2⅝-8 UN | 2A | 0.0025 | 2.6225 | 2.6075 | — | 2.5413 | 2.5331 | 2.4737 | 2B | 2.490 | 2.515 | 2.5438 | 2.5545 | 2.6250 |
| | 3A | 0.0000 | 2.6250 | 2.6100 | — | 2.5438 | 2.5376 | 2.4762 | 3B | 2.4900 | 2.5047 | 2.5438 | 2.5518 | 2.6250 |
| 2⅝-12 UN | 2A | 0.0019 | 2.6231 | 2.6117 | — | 2.5690 | 2.5628 | 2.5239 | 2B | 2.535 | 2.553 | 2.5709 | 2.5790 | 2.6250 |
| | 3A | 0.0000 | 2.6250 | 2.6136 | — | 2.5709 | 2.5662 | 2.5258 | 3B | 2.5350 | 2.5448 | 2.5709 | 2.5770 | 2.6250 |
| 2⅝-16 UN | 2A | 0.0017 | 2.6233 | 2.6139 | — | 2.5827 | 2.5771 | 2.5489 | 2B | 2.557 | 2.571 | 2.5844 | 2.5916 | 2.6250 |
| | 3A | 0.0000 | 2.6250 | 2.6156 | — | 2.5844 | 2.5802 | 2.5506 | 3B | 2.5570 | 2.5659 | 2.5844 | 2.5898 | 2.6250 |

**Table 3.** *(Continued)* **Standard Series and Selected Combinations—Unified Screw Threads**

| Nominal Size, Threads per Inch, and Series Designation[a] | External[b] | | | | | | | | Internal[b] | | | | | |
|---|---|---|---|---|---|---|---|---|---|---|---|---|---|---|
| | Class | Allow-ance | Major Diameter | | | Pitch Diameter | | UNR Minor Dia.,[c] Max (Ref.) | Class | Minor Diameter | | Pitch Diameter | | Major Diameter |
| | | | Max[d] | Min | Min[e] | Max[d] | Min | | | Min | Max | Min | Max | Min |
| 2⅝-20 UN | 2A | 0.0015 | 2.6235 | 2.6154 | — | 2.5910 | 2.5859 | 2.5640 | 2B | 2.571 | 2.582 | 2.5925 | 2.5991 | 2.6250 |
| | 3A | 0.0000 | 2.6250 | 2.6169 | — | 2.5925 | 2.5887 | 2.5655 | 3B | 2.5710 | 2.5786 | 2.5925 | 2.5975 | 2.6250 |
| 2¾-4 UNC | 1A | 0.0032 | 2.7468 | 2.7111 | — | 2.5844 | 2.5686 | 2.4491 | 1B | 2.479 | 2.517 | 2.5876 | 2.6082 | 2.7500 |
| | 2A | 0.0032 | 2.7468 | 2.7230 | 2.7111 | 2.5844 | 2.5739 | 2.4491 | 2B | 2.479 | 2.517 | 2.5876 | 2.6013 | 2.7500 |
| | 3A | 0.0000 | 2.7500 | 2.7262 | — | 2.5876 | 2.5797 | 2.4523 | 3B | 2.4790 | 2.5094 | 2.5876 | 2.5979 | 2.7500 |
| 2¾-6 UN | 2A | 0.0027 | 2.7473 | 2.7291 | — | 2.6390 | 2.6299 | 2.5488 | 2B | 2.570 | 2.600 | 2.6417 | 2.6536 | 2.7500 |
| | 3A | 0.0000 | 2.7500 | 2.7318 | — | 2.6417 | 2.6349 | 2.5515 | 3B | 2.5700 | 2.5896 | 2.6417 | 2.6506 | 2.7500 |
| 2¾-8 UN | 2A | 0.0025 | 2.7475 | 2.7325 | 2.7250 | 2.6663 | 2.6580 | 2.5987 | 2B | 2.615 | 2.640 | 2.6688 | 2.6796 | 2.7500 |
| | 3A | 0.0000 | 2.7500 | 2.7350 | — | 2.6688 | 2.6625 | 2.6012 | 3B | 2.6150 | 2.6297 | 2.6688 | 2.6769 | 2.7500 |
| 2¾-10 UNS | 2A | 0.0020 | 2.7480 | 2.7351 | — | 2.6830 | 2.6762 | 2.6289 | 2B | 2.642 | 2.663 | 2.6850 | 2.6938 | 2.7500 |
| 2¾-12 UN | 2A | 0.0019 | 2.7481 | 2.7367 | — | 2.6940 | 2.6877 | 2.6489 | 2B | 2.660 | 2.678 | 2.6959 | 2.7040 | 2.7500 |
| | 3A | 0.0000 | 2.7500 | 2.7386 | — | 2.6959 | 2.6912 | 2.6508 | 3B | 2.6600 | 2.6698 | 2.6959 | 2.7020 | 2.7500 |
| 2¾-14 UNS | 2A | 0.0018 | 2.7482 | 2.7379 | — | 2.7018 | 2.6959 | 2.6631 | 2B | 2.673 | 2.688 | 2.7036 | 2.7113 | 2.7500 |
| 2¾-16 UN | 2A | 0.0017 | 2.7483 | 2.7389 | — | 2.7077 | 2.7021 | 2.6739 | 2B | 2.682 | 2.696 | 2.7094 | 2.7167 | 2.7500 |
| | 3A | 0.0000 | 2.7500 | 2.7406 | — | 2.7094 | 2.7052 | 2.6756 | 3B | 2.6820 | 2.6909 | 2.7094 | 2.7148 | 2.7500 |
| 2¾-18 UNS | 2A | 0.0016 | 2.7484 | 2.7397 | — | 2.7123 | 2.7070 | 2.6822 | 2B | 2.690 | 2.703 | 2.7139 | 2.7209 | 2.7500 |
| 2¾-20 UN | 2A | 0.0015 | 2.7485 | 2.7404 | — | 2.7160 | 2.7109 | 2.6890 | 2B | 2.696 | 2.707 | 2.7175 | 2.7242 | 2.7500 |
| | 3A | 0.0000 | 2.7500 | 2.7419 | — | 2.7175 | 2.7136 | 2.6905 | 3B | 2.6960 | 2.7036 | 2.7175 | 2.7225 | 2.7500 |
| 2⅞-6 UN | 2A | 0.0028 | 2.8722 | 2.8540 | — | 2.7639 | 2.7547 | 2.6737 | 2B | 2.695 | 2.725 | 2.7667 | 2.7787 | 2.8750 |
| | 3A | 0.0000 | 2.8750 | 2.8568 | — | 2.7667 | 2.7598 | 2.6765 | 3B | 2.6950 | 2.7146 | 2.7667 | 2.7757 | 2.8750 |
| 2⅞-8 UN | 2A | 0.0025 | 2.8725 | 2.8575 | — | 2.7913 | 2.7829 | 2.7237 | 2B | 2.740 | 2.765 | 2.7938 | 2.8048 | 2.8750 |
| | 3A | 0.0000 | 2.8750 | 2.8600 | — | 2.7938 | 2.7875 | 2.7262 | 3B | 2.7400 | 2.7547 | 2.7938 | 2.8020 | 2.8750 |
| 2⅞-12 UN | 2A | 0.0019 | 2.8731 | 2.8617 | — | 2.8190 | 2.8127 | 2.7739 | 2B | 2.785 | 2.803 | 2.8209 | 2.8291 | 2.8750 |
| | 3A | 0.0000 | 2.8750 | 2.8636 | — | 2.8209 | 2.8162 | 2.7758 | 3B | 2.7850 | 2.7948 | 2.8209 | 2.8270 | 2.8750 |
| 2⅞-16 UN | 2A | 0.0017 | 2.8733 | 2.8639 | — | 2.8327 | 2.8271 | 2.7989 | 2B | 2.807 | 2.821 | 2.8344 | 2.8417 | 2.8750 |
| | 3A | 0.0000 | 2.8750 | 2.8656 | — | 2.8344 | 2.8302 | 2.8006 | 3B | 2.8070 | 2.8159 | 2.8344 | 2.8399 | 2.8750 |
| 2⅞-20 UN | 2A | 0.0016 | 2.8734 | 2.8653 | — | 2.8409 | 2.8357 | 2.8139 | 2B | 2.821 | 2.832 | 2.8425 | 2.8492 | 2.8750 |
| | 3A | 0.0000 | 2.8750 | 2.8669 | — | 2.8425 | 2.8386 | 2.8155 | 3B | 2.8210 | 2.8286 | 2.8425 | 2.8475 | 2.8750 |

**Table 3.** *(Continued)* **Standard Series and Selected Combinations — Unified Screw Threads**

| Nominal Size, Threads per Inch, and Series Designation[a] | External[b] | | | | | | | | Internal[b] | | | | | |
|---|---|---|---|---|---|---|---|---|---|---|---|---|---|---|
| | Class | Allow-ance | Major Diameter | | | Pitch Diameter | | UNR Minor Dia.,[c] Max (Ref.) | Class | Minor Diameter | | Pitch Diameter | | Major Diameter |
| | | | Max[d] | Min | Min[e] | Max[d] | Min | | | Min | Max | Min | Max | Min |
| 3-4 UNC | 1A | 0.0032 | 2.9968 | 2.9611 | — | 2.8344 | 2.8183 | 2.6991 | 1B | 2.729 | 2.767 | 2.8376 | 2.8585 | 3.0000 |
| | 2A | 0.0032 | 2.9968 | 2.9730 | 2.9611 | 2.8344 | 2.8237 | 2.6991 | 2B | 2.729 | 2.767 | 2.8376 | 2.8515 | 3.0000 |
| | 3A | 0.0000 | 3.0000 | 2.9762 | — | 2.8376 | 2.8296 | 2.7023 | 3B | 2.7290 | 2.7594 | 2.8376 | 2.8480 | 3.0000 |
| 3-6 UN | 2A | 0.0028 | 2.9972 | 2.9790 | — | 2.8889 | 2.8796 | 2.7987 | 2B | 2.820 | 2.850 | 2.8917 | 2.9038 | 3.0000 |
| | 3A | 0.0000 | 3.0000 | 2.9818 | — | 2.8917 | 2.8847 | 2.8015 | 3B | 2.8200 | 2.8396 | 2.8917 | 2.9008 | 3.0000 |
| 3-8 UN | 2A | 0.0026 | 2.9974 | 2.9824 | 2.9749 | 2.9162 | 2.9077 | 2.8486 | 2B | 2.865 | 2.890 | 2.9188 | 2.9299 | 3.0000 |
| | 3A | 0.0000 | 3.0000 | 2.9850 | — | 2.9188 | 2.9124 | 2.8512 | 3B | 2.8650 | 2.8797 | 2.9188 | 2.9271 | 3.0000 |
| 3-10 UNS | 2A | 0.0020 | 2.9980 | 2.9851 | — | 2.9330 | 2.9262 | 2.8789 | 2B | 2.892 | 2.913 | 2.9350 | 2.9439 | 3.0000 |
| 3-12 UN | 2A | 0.0019 | 2.9981 | 2.9867 | — | 2.9440 | 2.9377 | 2.8989 | 2B | 2.910 | 2.928 | 2.9459 | 2.9541 | 3.0000 |
| | 3A | 0.0000 | 3.0000 | 2.9886 | — | 2.9459 | 2.9412 | 2.9008 | 3B | 2.9100 | 2.9198 | 2.9459 | 2.9521 | 3.0000 |
| 3-14 UNS | 2A | 0.0018 | 2.9982 | 2.9879 | — | 2.9518 | 2.9459 | 2.9131 | 2B | 2.923 | 2.938 | 2.9536 | 2.9613 | 3.0000 |
| 3-16 UN | 2A | 0.0017 | 2.9983 | 2.9889 | — | 2.9577 | 2.9521 | 2.9239 | 2B | 2.932 | 2.946 | 2.9594 | 2.9667 | 3.0000 |
| | 3A | 0.0000 | 3.0000 | 2.9906 | — | 2.9594 | 2.9552 | 2.9256 | 3B | 2.9320 | 2.9409 | 2.9594 | 2.9649 | 3.0000 |
| 3-18 UNS | 2A | 0.0016 | 2.9984 | 2.9897 | — | 2.9623 | 2.9569 | 2.9322 | 2B | 2.940 | 2.953 | 2.9639 | 2.9709 | 3.0000 |
| 3-20 UN | 2A | 0.0016 | 2.9984 | 2.9903 | — | 2.9659 | 2.9607 | 2.9389 | 2B | 2.946 | 2.957 | 2.9675 | 2.9743 | 3.0000 |
| | 3A | 0.0000 | 3.0000 | 2.9919 | — | 2.9675 | 2.9636 | 2.9405 | 3B | 2.9460 | 2.9536 | 2.9675 | 2.9726 | 3.0000 |
| 3⅛-6 UN | 2A | 0.0028 | 3.1222 | 3.1040 | — | 3.0139 | 3.0045 | 2.9237 | 2B | 2.945 | 2.975 | 3.0167 | 3.0289 | 3.1250 |
| | 3A | 0.0000 | 3.1250 | 3.1068 | — | 3.0167 | 3.0097 | 2.9265 | 3B | 2.9450 | 2.9646 | 3.0167 | 3.0259 | 3.1250 |
| 3⅛-8 UN | 2A | 0.0026 | 3.1224 | 3.1074 | — | 3.0412 | 3.0326 | 2.9736 | 2B | 2.990 | 3.015 | 3.0438 | 3.0550 | 3.1250 |
| | 3A | 0.0000 | 3.1250 | 3.1100 | — | 3.0438 | 3.0374 | 2.9762 | 3B | 2.9900 | 3.0047 | 3.0438 | 3.0522 | 3.1250 |
| 3⅛-12 UN | 2A | 0.0019 | 3.1231 | 3.1117 | — | 3.0690 | 3.0626 | 3.0239 | 2B | 3.035 | 3.053 | 3.0709 | 3.0792 | 3.1250 |
| | 3A | 0.0000 | 3.1250 | 3.1136 | — | 3.0709 | 3.0661 | 3.0258 | 3B | 3.0350 | 3.0448 | 3.0709 | 3.0771 | 3.1250 |
| 3⅛-16 UN | 2A | 0.0017 | 3.1233 | 3.1139 | — | 3.0827 | 3.0770 | 3.0849 | 2B | 3.057 | 3.071 | 3.0844 | 3.0918 | 3.1250 |
| | 3A | 0.0000 | 3.1250 | 3.1156 | — | 3.0844 | 3.0801 | 3.0506 | 3B | 3.0570 | 3.0659 | 3.0844 | 3.0899 | 3.1250 |
| 3¼-4 UNC | 1A | 0.0033 | 3.2467 | 3.2110 | — | 3.0843 | 3.0680 | 2.9490 | 1B | 2.979 | 3.017 | 3.0876 | 3.1088 | 3.2500 |
| | 2A | 0.0033 | 3.2467 | 3.2229 | 3.2110 | 3.0843 | 3.0734 | 2.9490 | 2B | 2.979 | 3.017 | 3.0876 | 3.1017 | 3.2500 |
| | 3A | 0.0000 | 3.2500 | 3.2262 | — | 3.0876 | 3.0794 | 2.9523 | 3B | 2.9790 | 3.0094 | 3.0876 | 3.0982 | 3.2500 |
| 3¼-6 UN | 2A | 0.0028 | 3.2472 | 3.2290 | — | 3.1389 | 3.1294 | 3.0487 | 2B | 3.070 | 3.100 | 3.1417 | 3.1540 | 3.2500 |
| | 3A | 0.0000 | 3.2500 | 3.2318 | — | 3.1417 | 3.1346 | 3.0515 | 3B | 3.0700 | 3.0896 | 3.1417 | 3.1509 | 3.2500 |

**Table 3.** *(Continued)* **Standard Series and Selected Combinations — Unified Screw Threads**

| Nominal Size, Threads per Inch, and Series Designation[a] | External[b] | | | | | | | | Internal[b] | | | | | |
|---|---|---|---|---|---|---|---|---|---|---|---|---|---|---|
| | Class | Allowance | Major Diameter | | | Pitch Diameter | | UNR Minor Dia.,[c] Max (Ref.) | Class | Minor Diameter | | Pitch Diameter | | Major Diameter |
| | | | Max[d] | Min | Min[e] | Max[d] | Min | | | Min | Max | Min | Max | Min |
| 3¼-8 UN | 2A | 0.0026 | 3.2474 | 3.2324 | 3.2249 | 3.1662 | 3.1575 | 3.0986 | 2B | 3.115 | 3.140 | 3.1688 | 3.1801 | 3.2500 |
| | 3A | 0.0000 | 3.2500 | 3.2350 | — | 3.1688 | 3.1623 | 3.1012 | 3B | 3.1150 | 3.1297 | 3.1688 | 3.1773 | 3.2500 |
| 3¼-10 UNS | 2A | 0.0021 | 3.2479 | 3.2350 | — | 3.1829 | 3.1760 | 3.1288 | 2B | 3.142 | 3.163 | 3.1850 | 3.1939 | 3.2500 |
| 3¼-12 UN | 2A | 0.0019 | 3.2481 | 3.2367 | — | 3.1940 | 3.1876 | 3.1489 | 2B | 3.160 | 3.178 | 3.1959 | 3.2042 | 3.2500 |
| | 3A | 0.0000 | 3.2500 | 3.2386 | — | 3.1959 | 3.1911 | 3.1508 | 3B | 3.1600 | 3.1698 | 3.1959 | 3.2021 | 3.2500 |
| 3¼-14 UNS | 2A | 0.0018 | 3.2482 | 3.2379 | — | 3.2018 | 3.1958 | 3.1631 | 2B | 3.173 | 3.188 | 3.2036 | 3.2114 | 3.2500 |
| 3¼-16 UN | 2A | 0.0017 | 3.2483 | 3.2389 | — | 3.2077 | 3.2020 | 3.1739 | 2B | 3.182 | 3.196 | 3.2094 | 3.2168 | 3.2500 |
| | 3A | 0.0000 | 3.2500 | 3.2406 | — | 3.2094 | 3.2051 | 3.1756 | 3B | 3.1820 | 3.1909 | 3.2094 | 3.2150 | 3.2500 |
| 3¼-18 UNS | 2A | 0.0016 | 3.2484 | 3.2397 | — | 3.2123 | 3.2068 | 3.1823 | 2B | 3.190 | 3.203 | 3.2139 | 3.2210 | 3.2500 |
| 3⅜-6 UN | 2A | 0.0029 | 3.3721 | 3.3539 | — | 3.2638 | 3.2543 | 3.1736 | 2B | 3.195 | 3.225 | 3.2667 | 3.2791 | 3.3750 |
| | 3A | 0.0000 | 3.3750 | 3.3568 | — | 3.2667 | 3.2595 | 3.1765 | 3B | 3.1950 | 3.2146 | 3.2667 | 3.2760 | 3.3750 |
| 3⅜-8 UN | 2A | 0.0026 | 3.3724 | 3.3574 | — | 3.2912 | 3.2824 | 3.2236 | 2B | 3.240 | 3.265 | 3.2938 | 3.3052 | 3.3750 |
| | 3A | 0.0000 | 3.3750 | 3.3600 | — | 3.2938 | 3.2872 | 3.2262 | 3B | 3.2400 | 3.2547 | 3.2938 | 3.3023 | 3.3750 |
| 3⅜-12 UN | 2A | 0.0019 | 3.3731 | 3.3617 | — | 3.3190 | 3.3126 | 3.2739 | 2B | 3.285 | 3.303 | 3.3209 | 3.3292 | 3.3750 |
| | 3A | 0.0000 | 3.3750 | 3.3636 | — | 3.3209 | 3.3161 | 3.2758 | 3B | 3.2850 | 3.2948 | 3.3209 | 3.3272 | 3.3750 |
| 3⅜-16 UN | 2A | 0.0017 | 3.3733 | 3.3639 | — | 3.3327 | 3.3270 | 3.2989 | 2B | 3.307 | 3.321 | 3.3344 | 3.3419 | 3.3750 |
| | 3A | 0.0000 | 3.3750 | 3.3656 | — | 3.3344 | 3.3301 | 3.3006 | 3B | 3.3070 | 3.3159 | 3.3344 | 3.3400 | 3.3750 |
| 3½-4 UNC | 1A | 0.0033 | 3.4967 | 3.4610 | — | 3.3343 | 3.3177 | 3.1990 | 1B | 3.229 | 3.267 | 3.3376 | 3.3591 | 3.5000 |
| | 2A | 0.0033 | 3.4967 | 3.4729 | 3.4610 | 3.3343 | 3.3233 | 3.1990 | 2B | 3.229 | 3.267 | 3.3376 | 3.3519 | 3.5000 |
| | 3A | 0.0000 | 3.5000 | 3.4762 | — | 3.3376 | 3.3293 | 3.2023 | 3B | 3.2290 | 3.2594 | 3.3376 | 3.3484 | 3.5000 |
| 3½-6 UN | 2A | 0.0029 | 3.4971 | 3.4789 | — | 3.3888 | 3.3792 | 3.2986 | 2B | 3.320 | 3.350 | 3.3917 | 3.4042 | 3.5000 |
| | 3A | 0.0000 | 3.5000 | 3.4818 | — | 3.3917 | 3.3845 | 3.3015 | 3B | 3.3200 | 3.3396 | 3.3917 | 3.4011 | 3.5000 |
| 3½-8 UN | 2A | 0.0026 | 3.4974 | 3.4824 | 3.4749 | 3.4162 | 3.4074 | 3.3486 | 2B | 3.365 | 3.390 | 3.4188 | 3.4303 | 3.5000 |
| | 3A | 0.0000 | 3.5000 | 3.4850 | — | 3.4188 | 3.4122 | 3.3512 | 3B | 3.3650 | 3.3797 | 3.4188 | 3.4274 | 3.5000 |
| 3½-10 UNS | 2A | 0.0021 | 3.4979 | 3.4850 | — | 3.4329 | 3.4260 | 3.3788 | 2B | 3.392 | 3.413 | 3.4350 | 3.4440 | 3.5000 |
| 3½-12 UN | 2A | 0.0019 | 3.4981 | 3.4867 | — | 3.4440 | 3.4376 | 3.3989 | 2B | 3.410 | 3.428 | 3.4459 | 3.4543 | 3.5000 |
| | 3A | 0.0000 | 3.5000 | 3.4886 | — | 3.4459 | 3.4411 | 3.4008 | 3B | 3.4100 | 3.4198 | 3.4459 | 3.4522 | 3.5000 |

**Table 3.** *(Continued)* **Standard Series and Selected Combinations — Unified Screw Threads**

| Nominal Size, Threads per Inch, and Series Designation[a] | External[b] | | | | | | | | Internal[b] | | | | | |
|---|---|---|---|---|---|---|---|---|---|---|---|---|---|---|
| | Class | Allowance | Major Diameter | | | Pitch Diameter | | UNR Minor Dia.,[c] Max (Ref.) | Class | Minor Diameter | | Pitch Diameter | | Major Diameter |
| | | | Max[d] | Min | Min[e] | Max[d] | Min | | | Min | Max | Min | Max | Min |
| 3½-14 UNS | 2A | 0.0018 | 3.4982 | 3.4879 | — | 3.4518 | 3.4457 | 3.4131 | 2B | 3.423 | 3.438 | 3.4536 | 3.4615 | 3.5000 |
| 3½-16 UN | 2A | 0.0017 | 3.4983 | 3.4889 | — | 3.4577 | 3.4519 | 3.4239 | 2B | 3.432 | 3.446 | 3.4594 | 3.4669 | 3.5000 |
| | 3A | 0.0000 | 3.5000 | 3.4906 | — | 3.4594 | 3.4551 | 3.4256 | 3B | 3.4320 | 3.4409 | 3.4594 | 3.4650 | 3.5000 |
| 3½-18 UNS | 2A | 0.0017 | 3.4983 | 3.4896 | — | 3.4622 | 3.4567 | 3.4321 | 2B | 3.440 | 3.453 | 3.4639 | 3.4711 | 3.5000 |
| 3⅝-6 UN | 2A | 0.0029 | 3.6221 | 3.6039 | — | 3.5138 | 3.5041 | 3.4236 | 2B | 3.445 | 3.475 | 3.5167 | 3.5293 | 3.6250 |
| | 3A | 0.0000 | 3.6250 | 3.6068 | — | 3.5167 | 3.5094 | 3.4265 | 3B | 3.4450 | 3.4646 | 3.5167 | 3.5262 | 3.6250 |
| 3⅝-8 UN | 2A | 0.0027 | 3.6223 | 3.6073 | — | 3.5411 | 3.5322 | 3.4735 | 2B | 3.490 | 3.515 | 3.5438 | 3.5554 | 3.6250 |
| | 3A | 0.0000 | 3.6250 | 3.6100 | — | 3.5438 | 3.5371 | 3.4762 | 3B | 3.4900 | 3.5047 | 3.5438 | 3.5525 | 3.6250 |
| 3⅝-12 UN | 2A | 0.0019 | 3.6231 | 3.6117 | — | 3.5690 | 3.5625 | 3.5239 | 2B | 3.535 | 3.553 | 3.5709 | 3.5793 | 3.6250 |
| | 3A | 0.0000 | 3.6250 | 3.6136 | — | 3.5709 | 3.5661 | 3.5258 | 3B | 3.5350 | 3.5448 | 3.5709 | 3.5772 | 3.6250 |
| 3⅝-16 UN | 2A | 0.0017 | 3.6233 | 3.6139 | — | 3.5827 | 3.5769 | 3.5489 | 2B | 3.557 | 3.571 | 3.5844 | 3.5919 | 3.6250 |
| | 3A | 0.0000 | 3.6250 | 3.6156 | — | 3.5844 | 3.5801 | 3.5506 | 3B | 3.5570 | 3.5659 | 3.5844 | 3.5900 | 3.6250 |
| 3¾-4 UNC | 1A | 0.0034 | 3.7466 | 3.7109 | — | 3.5842 | 3.5674 | 3.4489 | 1B | 3.479 | 3.517 | 3.5876 | 3.6094 | 3.7500 |
| | 2A | 0.0034 | 3.7466 | 3.7228 | 3.7109 | 3.5842 | 3.5730 | 3.4489 | 2B | 3.479 | 3.517 | 3.5876 | 3.6021 | 3.7500 |
| | 3A | 0.0000 | 3.7500 | 3.7262 | — | 3.5876 | 3.5792 | 3.4523 | 3B | 3.4790 | 3.5094 | 3.5876 | 3.5985 | 3.7500 |
| 3¾-6 UN | 2A | 0.0029 | 3.7471 | 3.7289 | — | 3.6388 | 3.6290 | 3.5486 | 2B | 3.570 | 3.600 | 3.6417 | 3.6544 | 3.7500 |
| | 3A | 0.0000 | 3.7500 | 3.7318 | — | 3.6417 | 3.6344 | 3.5515 | 3B | 3.5700 | 3.5896 | 3.6417 | 3.6512 | 3.7500 |
| 3¾-8 UN | 2A | 0.0027 | 3.7473 | 3.7323 | 3.7248 | 3.6661 | 3.6571 | 3.5985 | 2B | 3.615 | 3.640 | 3.6688 | 3.6805 | 3.7500 |
| | 3A | 0.0000 | 3.7500 | 3.7350 | — | 3.6688 | 3.6621 | 3.6012 | 3B | 3.6150 | 3.6297 | 3.6688 | 3.6776 | 3.7500 |
| 3¾-10 UNS | 2A | 0.0021 | 3.7479 | 3.7350 | — | 3.6829 | 3.6759 | 3.6288 | 2B | 3.642 | 3.663 | 3.6850 | 3.6941 | 3.7500 |
| 3¾-12 UN | 2A | 0.0019 | 3.7481 | 3.7367 | — | 3.6940 | 3.6875 | 3.6489 | 2B | 3.660 | 3.678 | 3.6959 | 3.7043 | 3.7500 |
| | 3A | 0.0000 | 3.7500 | 3.7386 | — | 3.6959 | 3.6910 | 3.6508 | 3B | 3.6600 | 3.6698 | 3.6959 | 3.7022 | 3.7500 |
| 3¾-14 UNS | 2A | 0.0018 | 3.7482 | 3.7379 | — | 3.7018 | 3.6957 | 3.6631 | 2B | 3.673 | 3.688 | 3.7036 | 3.7115 | 3.7500 |
| 3¾-16 UN | 2A | 0.0017 | 3.7483 | 3.7389 | — | 3.7077 | 3.7019 | 3.6739 | 2B | 3.682 | 3.696 | 3.7094 | 3.7170 | 3.7500 |
| | 3A | 0.0000 | 3.7500 | 3.7406 | — | 3.7094 | 3.7050 | 3.6756 | 3B | 3.6820 | 3.6909 | 3.7094 | 3.7151 | 3.7500 |
| 3¾-18 UNS | 2A | 0.0017 | 3.7483 | 3.7396 | — | 3.7122 | 3.7066 | 3.6821 | 2B | 3.690 | 3.703 | 3.7139 | 3.7211 | 3.7500 |

**Table 3.** *(Continued)* **Standard Series and Selected Combinations — Unified Screw Threads**

| Nominal Size, Threads per Inch, and Series Designation[a] | External[b] | | | | | | | | Internal[b] | | | | | |
|---|---|---|---|---|---|---|---|---|---|---|---|---|---|---|
| | Class | Allowance | Major Diameter | | | Pitch Diameter | | UNR Minor Dia.,[c] Max (Ref.) | Class | Minor Diameter | | Pitch Diameter | | Major Diameter |
| | | | Max[d] | Min | Min[e] | Max[d] | Min | | | Min | Max | Min | Max | Min |
| 3⅞-6 UN | 2A | 0.0030 | 3.8720 | 3.8538 | — | 3.7637 | 3.7538 | 3.6735 | 2B | 3.695 | 3.725 | 3.7667 | 3.7795 | 3.8750 |
| | 3A | 0.0000 | 3.8750 | 3.8568 | — | 3.7667 | 3.7593 | 3.6765 | 3B | 3.6950 | 3.7146 | 3.7667 | 3.7763 | 3.8750 |
| 3⅞-8 UN | 2A | 0.0027 | 3.8723 | 3.8573 | — | 3.7911 | 3.7820 | 3.7235 | 2B | 3.740 | 3.765 | 3.7938 | 3.8056 | 3.8750 |
| | 3A | 0.0000 | 3.8750 | 3.8600 | — | 3.7938 | 3.7870 | 3.7262 | 3B | 3.7400 | 3.7547 | 3.7938 | 3.8026 | 3.8750 |
| 3⅞-12 UN | 2A | 0.0020 | 3.8730 | 3.8616 | — | 3.8189 | 3.8124 | 3.7738 | 2B | 3.785 | 3.803 | 3.8209 | 3.8294 | 3.8750 |
| | 3A | 0.0000 | 3.8750 | 3.8636 | — | 3.8209 | 3.8160 | 3.7758 | 3B | 3.7850 | 3.7948 | 3.8209 | 3.8273 | 3.8750 |
| 3⅞-16 UN | 2A | 0.0018 | 3.8732 | 3.8638 | — | 3.8326 | 3.8268 | 3.7988 | 2B | 3.807 | 3.821 | 3.8344 | 3.8420 | 3.8750 |
| | 3A | 0.0000 | 3.8750 | 3.8656 | — | 3.8344 | 3.8300 | 3.8006 | 3B | 3.8070 | 3.8159 | 3.8344 | 3.8401 | 3.8750 |
| 4-4 UNC | 1A | 0.0034 | 3.9966 | 3.9609 | — | 3.8342 | 3.8172 | 3.6989 | 1B | 3.729 | 3.767 | 3.8376 | 3.8597 | 4.0000 |
| | 2A | 0.0034 | 3.9966 | 3.9728 | 3.9609 | 3.8342 | 3.8229 | 3.6989 | 2B | 3.729 | 3.767 | 3.8376 | 3.8523 | 4.0000 |
| | 3A | 0.0000 | 4.0000 | 3.9762 | — | 3.8376 | 3.8291 | 3.7023 | 3B | 3.7290 | 3.7594 | 3.8376 | 3.8487 | 4.0000 |
| 4-6 UN | 2A | 0.0030 | 3.9970 | 3.9788 | — | 3.8887 | 3.8788 | 3.7985 | 2B | 3.820 | 3.850 | 3.8917 | 3.9046 | 4.0000 |
| | 3A | 0.0000 | 4.0000 | 3.9818 | — | 3.8917 | 3.8843 | 3.8015 | 3B | 3.8200 | 3.8396 | 3.8917 | 3.9014 | 4.0000 |
| 4-8 UN | 2A | 0.0027 | 3.9973 | 3.9823 | 3.9748 | 3.9161 | 3.9070 | 3.8485 | 2B | 3.865 | 3.890 | 3.9188 | 3.9307 | 4.0000 |
| | 3A | 0.0000 | 4.0000 | 3.9850 | — | 3.9188 | 3.9120 | 3.8512 | 3B | 3.8650 | 3.8797 | 3.9188 | 3.9277 | 4.0000 |
| 4-10 UNS | 2A | 0.0021 | 3.9979 | 3.9850 | — | 3.9329 | 3.9259 | 3.8788 | 2B | 3.892 | 3.913 | 3.9350 | 3.9441 | 4.0000 |
| 4-12 UN | 2A | 0.0020 | 3.9980 | 3.9866 | — | 3.9439 | 3.9374 | 3.8988 | 2B | 3.910 | 3.928 | 3.9459 | 3.9544 | 4.0000 |
| | 3A | 0.0000 | 4.0000 | 3.9886 | — | 3.9459 | 3.9410 | 3.9008 | 3B | 3.9100 | 3.9198 | 3.9459 | 3.9523 | 4.0000 |
| 4-14 UNS | 2A | 0.0018 | 3.9982 | 3.9879 | — | 3.9518 | 3.9456 | 3.9131 | 2B | 3.923 | 3.938 | 3.9536 | 3.9616 | 4.0000 |
| 4-16 UN | 2A | 0.0018 | 3.9982 | 3.9888 | — | 3.9576 | 3.9517 | 3.9238 | 2B | 3.932 | 3.946 | 3.9594 | 3.9670 | 4.0000 |
| | 3A | 0.0000 | 4.0000 | 3.9906 | — | 3.9594 | 3.9550 | 3.9256 | 3B | 3.9320 | 3.9409 | 3.9594 | 3.9651 | 4.0000 |
| 4¼-10 UNS | 2A | 0.0021 | 4.2479 | 4.2350 | — | 4.1829 | 4.1758 | 4.1288 | 2B | 4.142 | 4.163 | 4.1850 | 4.1942 | 4.2500 |
| 4¼-12 UN | 2A | 0.0020 | 4.2480 | 4.2366 | — | 4.1939 | 4.1873 | 4.1488 | 2B | 4.160 | 4.178 | 4.1959 | 4.2045 | 4.2500 |
| | 3A | 0.0000 | 4.2500 | 4.2386 | — | 4.1959 | 4.1910 | 4.1508 | 3B | 4.1600 | 4.1698 | 4.1959 | 4.2023 | 4.2500 |
| 4¼-14 UNS | 2A | 0.0019 | 4.2481 | 4.2378 | — | 4.2017 | 4.1955 | 4.1630 | 2B | 4.173 | 4.188 | 4.2036 | 4.2117 | 4.2500 |
| 4¼-16 UN | 2A | 0.0018 | 4.2482 | 4.2388 | — | 4.2076 | 4.2017 | 4.1738 | 2B | 4.182 | 4.196 | 4.2094 | 4.2171 | 4.2500 |
| | 3A | 0.0000 | 4.2500 | 4.2406 | — | 4.2094 | 4.2050 | 4.1756 | 3B | 4.1820 | 4.1909 | 4.2094 | 4.2152 | 4.2500 |
| 4½-10 UNS | 2A | 0.0021 | 4.4979 | 4.4850 | — | 4.4329 | 4.4258 | 4.3788 | 2B | 4.392 | 4.413 | 4.4350 | 4.4443 | 4.5000 |

**Table 3.** *(Continued)* **Standard Series and Selected Combinations — Unified Screw Threads**

| Nominal Size, Threads per Inch, and Series Designation[a] | External[b] | | | | | | | | Internal[b] | | | | | |
|---|---|---|---|---|---|---|---|---|---|---|---|---|---|---|
| | Class | Allowance | Major Diameter | | | Pitch Diameter | | UNR Minor Dia.,[c] Max (Ref.) | Class | Minor Diameter | | Pitch Diameter | | Major Diameter |
| | | | Max[d] | Min | Min[e] | Max[d] | Min | | | Min | Max | Min | Max | Min |
| 4½-12 UN | 2A | 0.0020 | 4.4980 | 4.4866 | — | 4.4439 | 4.4373 | 4.3988 | 2B | 4.410 | 4.428 | 4.4459 | 4.4545 | 4.5000 |
| | 3A | 0.0000 | 4.5000 | 4.4886 | — | 4.4459 | 4.4409 | 4.4008 | 3B | 4.4100 | 4.4198 | 4.4459 | 4.4524 | 4.5000 |
| 4½-14 UNS | 2A | 0.0019 | 4.4981 | 4.4878 | — | 4.4517 | 4.4454 | 4.4130 | 2B | 4.423 | 4.438 | 4.4536 | 4.4617 | 4.5000 |
| 4½-16 UN | 2A | 0.0018 | 4.4982 | 4.4888 | — | 4.4576 | 4.4516 | 4.4238 | 2B | 4.432 | 4.446 | 4.4594 | 4.4672 | 4.5000 |
| | 3A | 0.0000 | 4.5000 | 4.4906 | — | 4.4594 | 4.4549 | 4.4256 | 3B | 4.4320 | 4.4409 | 4.4594 | 4.4652 | 4.5000 |
| 4¾-10 UNS | 2A | 0.0022 | 4.7478 | 4.7349 | — | 4.6828 | 4.6756 | 4.6287 | 2B | 4.642 | 4.663 | 4.6850 | 4.6943 | 4.7500 |
| 4¾-12 UN | 2A | 0.0020 | 4.7480 | 4.7366 | — | 4.6939 | 4.6872 | 4.6488 | 2B | 4.660 | 4.678 | 4.6959 | 4.7046 | 4.7500 |
| | 3A | 0.0000 | 4.7500 | 4.7386 | — | 4.6959 | 4.6909 | 4.6508 | 3B | 4.6600 | 4.6698 | 4.6959 | 4.7024 | 4.7500 |
| 4¾-14 UNS | 2A | 0.0019 | 4.7481 | 4.7378 | — | 4.7017 | 4.6954 | 4.6630 | 2B | 4.673 | 4.688 | 4.7036 | 4.7118 | 4.7500 |
| 4¾-16 UN | 2A | 0.0018 | 4.7482 | 4.7388 | — | 4.7076 | 4.7016 | 4.6738 | 2B | 4.682 | 4.696 | 4.7094 | 4.7172 | 4.7500 |
| | 3A | 0.0000 | 4.7500 | 4.7406 | — | 4.7094 | 4.7049 | 4.6756 | 3B | 4.6820 | 4.6909 | 4.7094 | 4.7153 | 4.7500 |
| 5.00-10 UNS | 2A | 0.0022 | 4.9978 | 4.9849 | — | 4.9328 | 4.9256 | 4.8787 | 2B | 4.892 | 4.913 | 4.9350 | 4.9444 | 5.0000 |
| 5.00-12 UN | 2A | 0.0020 | 4.9980 | 4.9866 | — | 4.9439 | 4.9372 | 4.8988 | 2B | 4.910 | 4.928 | 4.9459 | 4.9546 | 5.0000 |
| | 3A | 0.0000 | 5.0000 | 4.9886 | — | 4.9459 | 4.9409 | 4.9008 | 3B | 4.9100 | 4.9198 | 4.9459 | 4.9525 | 5.0000 |
| 5.00-14 UNS | 2A | 0.0019 | 4.9981 | 4.9878 | — | 4.9517 | 4.9454 | 4.9130 | 2B | 4.923 | 4.938 | 4.9536 | 4.9619 | 5.0000 |
| 5.00-16 UN | 2A | 0.0018 | 4.9982 | 4.9888 | — | 4.9576 | 4.9515 | 4.9238 | 2B | 4.932 | 4.946 | 4.9594 | 4.9673 | 5.0000 |
| | 3A | 0.0000 | 5.0000 | 4.9906 | — | 4.9594 | 4.9549 | 4.9256 | 3B | 4.9320 | 4.9409 | 4.9594 | 4.9653 | 5.0000 |
| 5¼-10 UNS | 2A | 0.0022 | 5.2478 | 5.2349 | — | 5.1828 | 5.1755 | 5.1287 | 2B | 5.142 | 5.163 | 5.1850 | 5.1944 | 5.2500 |
| 5¼-12 UN | 2A | 0.0020 | 5.2480 | 5.2366 | — | 5.1939 | 5.1871 | 5.1488 | 2B | 5.160 | 5.178 | 5.1959 | 5.2047 | 5.2500 |
| | 3A | 0.0000 | 5.2500 | 5.2386 | — | 5.1959 | 5.1908 | 5.1508 | 3B | 5.1600 | 5.1698 | 5.1959 | 5.2025 | 5.2500 |
| 5¼-14 UNS | 2A | 0.0019 | 5.2481 | 5.2378 | — | 5.2017 | 5.1953 | 5.1630 | 2B | 5.173 | 5.188 | 5.2036 | 5.2119 | 5.2500 |
| 5¼-16 UN | 2A | 0.0018 | 5.2482 | 5.2388 | — | 5.2076 | 5.2015 | 5.1738 | 2B | 5.182 | 5.196 | 5.2094 | 5.2173 | 5.2500 |
| | 3A | 0.0000 | 5.2500 | 5.2406 | — | 5.2094 | 5.2048 | 5.1756 | 3B | 5.1820 | 5.1909 | 5.2094 | 5.2153 | 5.2500 |
| 5½-10 UNS | 2A | 0.0022 | 5.4978 | 5.4849 | — | 5.4328 | 5.4255 | 5.3787 | 2B | 5.392 | 5.413 | 5.4350 | 5.4445 | 5.5000 |
| 5½-12 UN | 2A | 0.0020 | 5.4980 | 5.4866 | — | 5.4439 | 5.4371 | 5.3988 | 2B | 5.410 | 5.428 | 5.4459 | 5.4548 | 5.5000 |
| | 3A | 0.0000 | 5.5000 | 5.4886 | — | 5.4459 | 5.4408 | 5.4008 | 3B | 5.4100 | 5.4198 | 5.4459 | 5.4525 | 5.5000 |
| 5½-14 UNS | 2A | 0.0019 | 5.4981 | 5.4878 | — | 5.4517 | 5.4453 | 5.4130 | 2B | 5.423 | 5.438 | 5.4536 | 5.4620 | 5.5000 |

**Table 3.** *(Continued)* **Standard Series and Selected Combinations — Unified Screw Threads**

| Nominal Size, Threads per Inch, and Series Designation[a] | External[b] | | | | | | | | Internal[b] | | | | | |
|---|---|---|---|---|---|---|---|---|---|---|---|---|---|---|
| | Class | Allow-ance | Major Diameter | | | Pitch Diameter | | UNR Minor Dia.,[c] Max (Ref.) | Class | Minor Diameter | | Pitch Diameter | | Major Diameter |
| | | | Max[d] | Min | Min[e] | Max[d] | Min | | | Min | Max | Min | Max | Min |
| 5½-16 UN | 2A | 0.0018 | 5.4982 | 5.4888 | — | 5.4576 | 5.4515 | 5.4238 | 2B | 5.432 | 5.446 | 5.4594 | 5.4674 | 5.5000 |
| | 3A | 0.0000 | 5.5000 | 5.4906 | — | 5.4594 | 5.4548 | 5.4256 | 3B | 5.4320 | 5.4409 | 5.4594 | 5.4654 | 5.5000 |
| 5¾-10 UNS | 2A | 0.0022 | 5.7478 | 5.7349 | — | 5.6828 | 5.6755 | 5.6287 | 2B | 5.642 | 5.663 | 5.6850 | 5.6945 | 5.7500 |
| 5¾-12 UN | 2A | 0.0021 | 5.7479 | 5.7365 | — | 5.6938 | 5.6870 | 5.6487 | 2B | 5.660 | 5.678 | 5.6959 | 5.7048 | 5.7500 |
| | 3A | 0.0000 | 5.7500 | 5.7386 | — | 5.6959 | 5.6908 | 5.6508 | 3B | 5.6600 | 5.6698 | 5.6959 | 5.7026 | 5.7500 |
| 5¾-14 UNS | 2A | 0.0019 | 5.7481 | 5.7378 | — | 5.7017 | 5.6952 | 5.6630 | 2B | 5.673 | 5.688 | 5.7036 | 5.7120 | 5.7500 |
| 5¾-16 UN | 2A | 0.0019 | 5.7481 | 5.7387 | — | 5.7075 | 5.7013 | 5.6737 | 2B | 5.682 | 5.696 | 5.7094 | 5.7174 | 5.7500 |
| | 3A | 0.0000 | 5.7500 | 5.7406 | — | 5.7094 | 5.7048 | 5.6756 | 3B | 5.6820 | 5.6909 | 5.7094 | 5.7154 | 5.7500 |
| 6-10 UNS | 2A | 0.0022 | 5.9978 | 5.9849 | — | 5.9328 | 5.9254 | 5.8787 | 2B | 5.892 | 5.913 | 5.9350 | 5.9446 | 6.0000 |
| 6-14 UNS | 2A | 0.0020 | 5.9980 | 5.9877 | — | 5.9516 | 5.9451 | 5.9129 | 2B | 5.923 | 5.938 | 5.9536 | 5.9621 | 6.0000 |
| 6-12 UN | 2A | 0.0021 | 5.9979 | 5.9865 | — | 5.9438 | 5.9369 | 5.8987 | 2B | 5.910 | 5.928 | 5.9459 | 5.9549 | 6.0000 |
| | 3A | 0.0000 | 6.0000 | 5.9886 | — | 5.9459 | 5.9407 | 5.9008 | 3B | 5.9100 | 5.9198 | 5.9459 | 5.9526 | 6.0000 |
| 6-16 UN | 2A | 0.0019 | 5.9981 | 5.9887 | — | 5.9575 | 5.9513 | 5.9237 | 2B | 5.932 | 5.946 | 5.9594 | 5.9675 | 6.0000 |
| | 3A | 0.0000 | 6.0000 | 5.9906 | — | 5.9594 | 5.9547 | 5.9256 | 3B | 5.9320 | 5.9409 | 5.9594 | 5.9655 | 6.0000 |

[a] Use UNR designation instead of UN wherever UNR thread form is desired for external use.
[b] Regarding combinations of thread classes, see text on page 1991.
[c] UN series external thread maximum minor diameter is basic for Class 3A and basic minus allowance for Classes 1A and 2A.
[d] For Class 2A threads having an additive finish the maximum is increased, by the allowance, to the basic size, the value being the same as for Class 3A.
[e] For unfinished hot-rolled material not including standard fasteners with rolled threads.
[f] Formerly NF, tolerances and allowances are based on one diameter length of engagement.
All dimensions in inches.
Use UNS threads only if the standard series does not meet requirements (see pages 1951, 1983, and 1994). For additional sizes above 4 inches see ANSI/ASME B1.1-2019.

*Coarse-Thread Series:* This series, UNC/UNRC, is the one most commonly used in the bulk production of bolts, screws, nuts and other general engineering applications. It is also used for threading into lower tensile strength materials such as cast iron, mild steel and softer materials (bronze, brass, aluminum, magnesium and plastics) to obtain the optimum resistance to stripping of the internal thread. It is applicable for rapid assembly or disassembly, or if corrosion or slight damage is possible.

**Table 4a. Coarse-Thread Series, UNC and UNRC—Basic Dimensions**

| Sizes No. or Inches | Basic Major Dia., $D$ | Thds. per Inch, | Basic Pitch Dia.,[a] $D_2$ | Minor Diameter Ext. Thds.,[c] $d_3$ (Ref.) | Minor Diameter Int. Thds.,[d] $D_1$ | Lead Angle λ at Basic P.D. | | Area of Minor Dia. at $D\text{-}2h_b$ | Tensile Stress Area[b] |
|---|---|---|---|---|---|---|---|---|---|
| | Inches | $n$ | Inches | Inches | Inches | Deg. | Min | Sq. In. | Sq. In. |
| **1 (0.073)**[e] | **0.0730** | **64** | **0.0629** | **0.0544** | **0.0561** | **4** | **31** | 0.00218 | 0.00263 |
| **2 (0.086)** | **0.0860** | **56** | **0.0744** | **0.0648** | **0.0667** | **4** | **22** | 0.00310 | 0.00370 |
| **3 (0.099)**[e] | **0.0990** | **48** | **0.0855** | **0.0741** | **0.0764** | **4** | **26** | 0.00406 | 0.00487 |
| **4 (0.112)** | **0.1120** | **40** | **0.0958** | **0.0822** | **0.0849** | **4** | **45** | 0.00496 | 0.00604 |
| **5 (0.125)** | **0.1250** | **40** | **0.1088** | **0.0952** | **0.0979** | **4** | **11** | 0.00672 | 0.00796 |
| **6 (0.138)** | **0.1380** | **32** | **0.1177** | **0.1008** | **0.1042** | **4** | **50** | 0.00745 | 0.00909 |
| **8 (0.164)** | **0.1640** | **32** | **0.1437** | **0.1268** | **0.1302** | **3** | **58** | 0.01196 | 0.0140 |
| **10 (0.190)** | **0.1900** | **24** | **0.1629** | **0.1404** | **0.1449** | **4** | **39** | 0.01450 | 0.0175 |
| **12 (0.216)**[e] | **0.2160** | **24** | **0.1889** | **0.1664** | **0.1709** | **4** | **1** | 0.0206 | 0.0242 |
| **1/4** | **0.2500** | **20** | **0.2175** | **0.1905** | **0.1959** | **4** | **11** | 0.0269 | 0.0318 |
| **5/16** | **0.3125** | **18** | **0.2764** | **0.2464** | **0.2524** | **3** | **40** | 0.0454 | 0.0524 |
| **3/8** | **0.3750** | **16** | **0.3344** | **0.3005** | **0.3073** | **3** | **24** | 0.0678 | 0.0775 |
| **7/16** | **0.4375** | **14** | **0.3911** | **0.3525** | **0.3602** | **3** | **20** | 0.0933 | 0.1063 |
| **1/2** | **0.5000** | **13** | **0.4500** | **0.4084** | **0.4167** | **3** | **7** | 0.1257 | 0.1419 |
| **9/16** | **0.5625** | **12** | **0.5084** | **0.4633** | **0.4723** | **2** | **59** | 0.162 | 0.182 |
| **5/8** | **0.6250** | **11** | **0.5660** | **0.5168** | **0.5266** | **2** | **56** | 0.202 | 0.226 |
| **3/4** | **0.7500** | **10** | **0.6850** | **0.6309** | **0.6417** | **2** | **40** | 0.302 | 0.334 |
| **7/8** | **0.8750** | **9** | **0.8028** | **0.7427** | **0.7547** | **2** | **31** | 0.419 | 0.462 |
| **1** | **1.0000** | **8** | **0.9188** | **0.8512** | **0.8647** | **2** | **29** | 0.551 | 0.606 |
| **1 1/8** | **1.1250** | **7** | **1.0322** | **0.9549** | **0.9704** | **2** | **31** | 0.693 | 0.763 |
| **1 1/4** | **1.2500** | **7** | **1.1572** | **1.0799** | **1.0954** | **2** | **15** | 0.890 | 0.969 |
| **1 3/8** | **1.3750** | **6** | **1.2667** | **1.1766** | **1.1946** | **2** | **24** | 1.054 | 1.155 |
| **1 1/2** | **1.5000** | **6** | **1.3917** | **1.3016** | **1.3196** | **2** | **11** | 1.294 | 1.405 |
| **1 3/4** | **1.7500** | **5** | **1.6201** | **1.5119** | **1.5335** | **2** | **15** | 1.74 | 1.90 |
| **2** | **2.0000** | **4 1/2** | **1.8557** | **1.7353** | **1.7594** | **2** | **11** | 2.30 | 2.50 |
| **2 1/4** | **2.2500** | **4 1/2** | **2.1057** | **1.9853** | **2.0094** | **1** | **55** | 3.02 | 3.25 |
| **2 1/2** | **2.5000** | **4** | **2.3376** | **2.2023** | **2.2294** | **1** | **57** | 3.72 | 4.00 |
| **2 3/4** | **2.7500** | **4** | **2.5876** | **2.4523** | **2.4794** | **1** | **46** | 4.62 | 4.93 |
| **3** | **3.0000** | **4** | **2.8376** | **2.7023** | **2.7294** | **1** | **36** | 5.62 | 5.97 |
| **3 1/4** | **3.2500** | **4** | **3.0876** | **2.9523** | **2.9794** | **1** | **29** | 6.72 | 7.10 |
| **3 1/2** | **3.500** | **4** | **3.3376** | **3.2023** | **3.2294** | **1** | **22** | 7.92 | 8.33 |
| **3 3/4** | **3.7500** | **4** | **3.5876** | **3.4523** | **3.4794** | **1** | **16** | 9.21 | 9.66 |
| **4** | **4.0000** | **4** | **3.8376** | **3.7023** | **3.7294** | **1** | **11** | 10.61 | 11.08 |

[a] British: Effective Diameter.

[b] See formula, pages 1660 and 1668.

[c] Design form for UNR threads. (See figure on page 1951.)

[d] Basic minor diameter.

[e] Secondary sizes.

*Fine-Thread Series:* This series, UNF/UNRF, is suitable for the production of bolts, screws, and nuts and for other applications where the Coarse series is not applicable. External threads of this series have greater tensile stress area than comparable sizes of the Coarse series. The Fine series is suitable when the resistance to stripping of both external and mating internal threads equals or exceeds the tensile load-carrying capacity of the externally threaded member (see page 1668). It is also used where the length of engagement is short, where a smaller lead angle is desired, where the wall thickness demands a fine pitch, or where finer adjustment is needed.

**Table 4b. Fine-Thread Series, UNF and UNRF—Basic Dimensions**

| Sizes No. or Inches | Basic Major Dia., $D$ Inches | Thds. per Inch, $n$ | Basic Pitch Dia.,[a] $D_2$ Inches | Minor Diameter Ext. Thds.,[c] $d_3$ (Ref.) Inches | Minor Diameter Int. Thds.,[d] $D_1$ Inches | Lead Angle λ at Basic P.D. Deg. | Lead Angle λ at Basic P.D. Min | Area of Minor Dia. at $D\text{-}2h_b$ Sq. In. | Tensile Stress Area[b] Sq. In. |
|---|---|---|---|---|---|---|---|---|---|
| 0 (0.060) | 0.0600 | 80 | 0.0519 | 0.0451 | 0.0465 | 4 | 23 | 0.00151 | 0.00180 |
| 1 (0.073)[e] | 0.0730 | 72 | 0.0640 | 0.0565 | 0.0580 | 3 | 57 | 0.00237 | 0.00278 |
| 2 (0.086) | 0.0860 | 64 | 0.0759 | 0.0674 | 0.0691 | 3 | 45 | 0.00339 | 0.00394 |
| 3 (0.099)[e] | 0.0990 | 56 | 0.0874 | 0.0778 | 0.0797 | 3 | 43 | 0.00451 | 0.00523 |
| 4 (0.112) | 0.1120 | 48 | 0.0985 | 0.0871 | 0.0894 | 3 | 51 | 0.00566 | 0.00661 |
| 5 (0.125) | 0.1250 | 44 | 0.1102 | 0.0979 | 0.1004 | 3 | 45 | 0.00716 | 0.00830 |
| 6 (0.138) | 0.1380 | 40 | 0.1218 | 0.1082 | 0.1109 | 3 | 44 | 0.00874 | 0.01015 |
| 8 (0.164) | 0.1640 | 36 | 0.1460 | 0.1309 | 0.1339 | 3 | 28 | 0.01285 | 0.01474 |
| 10 (0.190) | 0.1900 | 32 | 0.1697 | 0.1528 | 0.1562 | 3 | 21 | 0.0175 | 0.0200 |
| 12 (0.216)[e] | 0.2160 | 28 | 0.1928 | 0.1734 | 0.1773 | 3 | 22 | 0.0226 | 0.0258 |
| 1/4 | 0.2500 | 28 | 0.2268 | 0.2074 | 0.2113 | 2 | 52 | 0.0326 | 0.0364 |
| 5/16 | 0.3125 | 24 | 0.2854 | 0.2629 | 0.2674 | 2 | 40 | 0.0524 | 0.0580 |
| 3/8 | 0.3750 | 24 | 0.3479 | 0.3254 | 0.3299 | 2 | 11 | 0.0809 | 0.0878 |
| 7/16 | 0.4375 | 20 | 0.4050 | 0.3780 | 0.3834 | 2 | 15 | 0.1090 | 0.1187 |
| 1/2 | 0.5000 | 20 | 0.4675 | 0.4405 | 0.4459 | 1 | 57 | 0.1486 | 0.1599 |
| 9/16 | 0.5625 | 18 | 0.5264 | 0.4964 | 0.5024 | 1 | 55 | 0.189 | 0.203 |
| 5/8 | 0.6250 | 18 | 0.5889 | 0.5589 | 0.5649 | 1 | 43 | 0.240 | 0.256 |
| 3/4 | 0.7500 | 16 | 0.7094 | 0.6763 | 0.6823 | 1 | 36 | 0.351 | 0.373 |
| 7/8 | 0.8750 | 14 | 0.8286 | 0.7900 | 0.7977 | 1 | 34 | 0.480 | 0.509 |
| 1 | 1.0000 | 12 | 0.9459 | 0.9001 | 0.9098 | 1 | 36 | 0.625 | 0.663 |
| 1 1/8 | 1.1250 | 12 | 1.0709 | 1.0258 | 1.0348 | 1 | 25 | 0.812 | 0.856 |
| 1 1/4 | 1.2500 | 12 | 1.1959 | 1.1508 | 1.1598 | 1 | 16 | 1.024 | 1.073 |
| 1 3/8 | 1.3750 | 12 | 1.3209 | 1.2758 | 1.2848 | 1 | 9 | 1.260 | 1.315 |
| 1 1/2 | 1.5000 | 12 | 1.4459 | 1.4008 | 1.4098 | 1 | 3 | 1.521 | 1.581 |

[a] British: Effective Diameter.

[b] See formula, pages 1660 and 1668.

[c] Design form for UNR threads. (See figure on page 1951.)

[d] Basic minor diameter.

[e] Secondary sizes.

*Extra-Fine-Thread Series:* This series, UNEF/UNREF, is applicable where even finer pitches of threads are desirable, as for short lengths of engagement and for thin-walled tubes, nuts, ferrules, or couplings. It is also generally applicable under the conditions stated above for the fine threads. See Table 4c.

*Fine Threads for Thin-Wall Tubing:* Dimensions for a 27-thread series, ranging from 1/4- to 1-inch nominal size, also are included in Table 3. These threads are recommended for general use on thin-wall tubing. The minimum length of complete thread is one-third of the basic major diameter plus 5 threads (+0.185 in.).

*Selected Combinations:* Thread data are tabulated in Table 3 for certain additional selected special combinations of diameter and pitch, with pitch diameter tolerances based on a length of thread engagement of 9 times the pitch. The pitch diameter limits are applicable to a length of engagement of from 5 to 15 times the pitch. (This provision should not be confused with the lengths of thread on mating parts, as they may exceed the length of engagement by a considerable amount.) Thread symbols are UNS and UNRS.

**Table 4c. Extra-Fine-Thread Series, UNEF and UNREF — Basic Dimensions**

| Sizes No. or Inches | Basic Major Dia., $D$ | Thds. per Inch, $n$ | Basic Pitch Dia.,[a] $D_2$ | Minor Diameter | | Lead Angle λ at Basic P.D. | | Area of Minor Dia. at $D-2h_b$ | Tensile Stress Area[b] |
|---|---|---|---|---|---|---|---|---|---|
| | | | | Ext. Thds.,[c] $d_3$ (Ref.) | Int. Thds.,[d] $D_1$ | | | | |
| | Inches | | Inches | Inches | Inches | Deg. | Min | Sq. In. | Sq. In. |
| **12 (0.216)[e]** | **0.2160** | **32** | **0.1957** | **0.1788** | **0.1822** | **2** | **55** | 0.0242 | 0.0270 |
| **¼** | **0.2500** | **32** | **0.2297** | **0.2128** | **0.2162** | **2** | **29** | 0.0344 | 0.0379 |
| **5⁄16** | **0.3125** | **32** | **0.2922** | **0.2753** | **0.2787** | **1** | **57** | 0.0581 | 0.0625 |
| **⅜** | **0.3750** | **32** | **0.3547** | **0.3378** | **0.3412** | **1** | **36** | 0.0878 | 0.0932 |
| **7⁄16** | **0.4375** | **28** | **0.4143** | **0.3949** | **0.3988** | **1** | **34** | 0.1201 | 0.1274 |
| **½** | **0.5000** | **28** | **0.4768** | **0.4574** | **0.4613** | **1** | **22** | 0.162 | 0.170 |
| **9⁄16** | **0.5625** | **24** | **0.5354** | **0.5129** | **0.5174** | **1** | **25** | 0.203 | 0.214 |
| **⅝** | **0.6250** | **24** | **0.5979** | **0.5754** | **0.5799** | **1** | **16** | 0.256 | 0.268 |
| **11⁄16[e]** | **0.6875** | **24** | **0.6604** | **0.6379** | **0.6424** | **1** | **9** | 0.315 | 0.329 |
| **¾** | **0.7500** | **20** | **0.7175** | **0.6905** | **0.6959** | **1** | **16** | 0.369 | 0.386 |
| **13⁄16[e]** | **0.8125** | **20** | **0.7800** | **0.7530** | **0.7584** | **1** | **10** | 0.439 | 0.458 |
| **⅞** | **0.8750** | **20** | **0.8425** | **0.8155** | **0.8209** | **1** | **5** | 0.515 | 0.536 |
| **15⁄16[e]** | **0.9375** | **20** | **0.9050** | **0.8780** | **0.8834** | **1** | **0** | 0.598 | 0.620 |
| **1** | **1.0000** | **20** | **0.9675** | **0.9405** | **0.9459** | **0** | **57** | 0.687 | 0.711 |
| **1 1⁄16[e]** | **1.0625** | **18** | **1.0264** | **0.9964** | **1.0024** | **0** | **59** | 0.770 | 0.799 |
| **1⅛** | **1.1250** | **18** | **1.0889** | **1.0589** | **1.0649** | **0** | **56** | 0.871 | 0.901 |
| **1 3⁄16[e]** | **1.1875** | **18** | **1.1514** | **1.1214** | **1.1274** | **0** | **53** | 0.977 | 1.009 |
| **1¼** | **1.2500** | **18** | **1.2139** | **1.1839** | **1.1899** | **0** | **50** | 1.090 | 1.123 |
| **1 5⁄16[e]** | **1.3125** | **18** | **1.2764** | **1.2464** | **1.2524** | **0** | **48** | 1.208 | 1.244 |
| **1⅜** | **1.3750** | **18** | **1.3389** | **1.3089** | **1.3149** | **0** | **45** | 1.333 | 1.370 |
| **1 7⁄16[e]** | **1.4375** | **18** | **1.4014** | **1.3714** | **1.3774** | **0** | **43** | 1.464 | 1.503 |
| **1½** | **1.5000** | **18** | **1.4639** | **1.4339** | **1.4399** | **0** | **42** | 1.60 | 1.64 |
| **1 9⁄16[e]** | **1.5625** | **18** | **1.5264** | **1.4964** | **1.5024** | **0** | **40** | 1.74 | 1.79 |
| **1⅝** | **1.6250** | **18** | **1.5889** | **1.5589** | **1.5649** | **0** | **38** | 1.89 | 1.94 |
| **1 11⁄16[e]** | **1.6875** | **18** | **1.6514** | **1.6214** | **1.6274** | **0** | **37** | 2.05 | 2.10 |

[a] British: Effective Diameter.

[b] See formula, pages 1660 and 1668.

[c] Design form for UNR threads. (See figure on page 1951.)

[d] Basic minor diameter.

[e] Secondary sizes.

*Other Threads of Special Diameters, Pitches, and Lengths of Engagement:* Thread data for special combinations of diameter, pitch, and length of engagement not included in selected combinations are also given in the Standard but are not given here. Also, when design considerations require nonstandard pitches or extreme conditions of engagement not covered by the tables, the allowance and tolerances should be derived from the formulas in the Standard. The thread symbol for such special threads is UNS.

**Constant Pitch Series.**—The various constant-pitch series, UN, with 4, 6, 8, 12, 16, 20, 28 and 32 threads per inch, given in Table 3, offer a comprehensive range of diameter-pitch combinations for those purposes where the threads in the Coarse, Fine, and Extra-Fine series do not meet the particular requirements of the design.

When selecting threads from these constant-pitch series, preference should be given wherever possible to those tabulated in the 8-, 12-, or 16-thread series.

*8-Thread Series:* The 8-thread series (8-UN) is a uniform-pitch series for large diameters. Although originally intended for high-pressure-joint bolts and nuts, it is now widely used as a substitute for the Coarse-Thread Series for diameters larger than 1 inch.

*12-Thread Series:* The 12-thread series (12-UN) is a uniform pitch series for large diameters requiring threads of medium-fine pitch. Although originally intended for boiler practice, it is now used as a continuation of the Fine-Thread Series for diameters larger than 1½ inches.

*16-Thread Series:* The 16-thread series (16-UN) is a uniform pitch series for large diameters requiring fine-pitch threads. It is suitable for adjusting collars and retaining nuts, and also serves as a continuation of the Extra-fine Thread Series for diameters larger than 1¹¹⁄₁₆ inches.

*4-, 6-, 20-, 28-, and 32-Thread Series:* These thread series have been used more or less widely in industry for various applications where the Standard Coarse, Fine or Extra-fine Series were not as applicable. They are now recognized as Standard Unified Thread Series in a specified selection of diameters for each pitch (see Table 2).

Whenever a thread in a constant-pitch series also appears in the UNC, UNF, or UNEF series, the symbols and tolerances for limits of size of UNC, UNF, or UNEF series are applicable, as will be seen in Table 2 and Table 3. *(Text continues on page 1991)*

**Table 5a. 4-Thread Series, 4-UN and 4-UNR—Basic Dimensions**

| Sizes | | Basic Major Dia., $D$ | Basic Pitch Dia.,[a] $D_2$ | Minor Diameter | | Lead Angle λ at Basic P.D. | | Area of Minor Dia. at $D-2h_b$ | Tensile Stress Area[b] |
|---|---|---|---|---|---|---|---|---|---|
| Primary | Secondary | | | Ext. Thds.[c] $d_{3s}$ (Ref.) | Int. Thds.,[d] $D_1$ | | | | |
| Inches | Inches | Inches | Inches | Inches | Inches | Deg. | Min. | Sq. In. | Sq. In. |
| 2½[e] | | 2.5000 | 2.3376 | 2.2023 | 2.2294 | 1 | 57 | 3.72 | 4.00 |
| | 2⅝ | 2.6250 | 2.4626 | 2.3273 | 2.3544 | 1 | 51 | 4.16 | 4.45 |
| 2¾[e] | | 2.7500 | 2.5876 | 2.4523 | 2.4794 | 1 | 46 | 4.62 | 4.93 |
| | 2⅞ | 2.8750 | 2.7126 | 2.5773 | 2.6044 | 1 | 41 | 5.11 | 5.44 |
| 3[e] | | 3.0000 | 2.8376 | 2.7023 | 2.7294 | 1 | 36 | 5.62 | 5.97 |
| | 3⅛ | 3.1250 | 2.9626 | 2.8273 | 2.8544 | 1 | 32 | 6.16 | 6.52 |
| 3¼[e] | | 3.2500 | 3.0876 | 2.9523 | 2.9794 | 1 | 29 | 6.72 | 7.10 |
| | 3⅜ | 3.3750 | 3.2126 | 3.0773 | 3.1044 | 1 | 25 | 7.31 | 7.70 |
| 3½[e] | | 3.5000 | 3.3376 | 3.2023 | 3.2294 | 1 | 22 | 7.92 | 8.33 |
| | 3⅝ | 3.6250 | 3.4626 | 3.3273 | 3.3544 | 1 | 19 | 8.55 | 9.00 |
| 3¼[e] | | 3.7500 | 3.5876 | 3.4523 | 3.4794 | 1 | 16 | 9.21 | 9.66 |
| | 3⅞ | 3.8750 | 3.7126 | 3.5773 | 3.6044 | 1 | 14 | 9.90 | 10.36 |
| 4[e] | | 4.0000 | 3.8376 | 3.7023 | 3.7294 | 1 | 11 | 10.61 | 11.08 |
| | 4⅛ | 4.1250 | 3.9626 | 3.8273 | 3.8544 | 1 | 9 | 11.34 | 11.83 |
| 4¼ | | 4.2500 | 4.0876 | 3.9523 | 3.9794 | 1 | 7 | 12.10 | 12.61 |
| | 4⅜ | 4.3750 | 4.2126 | 4.0773 | 4.1044 | 1 | 5 | 12.88 | 13.41 |
| 4½ | | 4.5000 | 4.3376 | 4.2023 | 4.2294 | 1 | 3 | 13.69 | 14.23 |
| | 4⅝ | 4.6250 | 4.4626 | 4.3273 | 4.3544 | 1 | 1 | 14.52 | 15.1 |
| 4¾ | | 4.7500 | 4.5876 | 4.4523 | 4.4794 | 1 | 0 | 15.4 | 15.9 |
| | 4⅞ | 4.8750 | 4.7126 | 4.5773 | 4.6044 | 0 | 58 | 16.3 | 16.8 |
| 5 | | 5.0000 | 4.8376 | 4.7023 | 4.7294 | 0 | 57 | 17.2 | 17.8 |
| | 5⅛ | 5.1250 | 4.9626 | 4.8273 | 4.8544 | 0 | 55 | 18.1 | 18.7 |
| 5¼ | | 5.2500 | 5.0876 | 4.9523 | 4.9794 | 0 | 54 | 19.1 | 19.7 |
| | 5⅜ | 5.3750 | 5.2126 | 5.0773 | 5.1044 | 0 | 52 | 20.0 | 20.7 |
| 5½ | | 5.5000 | 5.3376 | 5.2023 | 5.2294 | 0 | 51 | 21.0 | 21.7 |
| | 5⅝ | 5.6250 | 5.4626 | 5.3273 | 5.3544 | 0 | 50 | 22.1 | 22.7 |
| 5¾ | | 5.7500 | 5.5876 | 5.4523 | 5.4794 | 0 | 49 | 23.1 | 23.8 |
| | 5⅞ | 5.8750 | 5.7126 | 5.5773 | 5.6044 | 0 | 48 | 24.2 | 24.9 |
| 6 | | 6.0000 | 5.8376 | 5.7023 | 5.7294 | 0 | 47 | 25.3 | 26.0 |

[a] British: Effective Diameter.

[b] See formula, pages 1660 and 1668.

[c] Design form for UNR threads. (See figure on page 1951).

[d] Basic minor diameter.

[e] These are standard sizes of the UNC series.

## Table 5b. 6-Thread Series, 6-UN and 6-UNR—Basic Dimensions

| Sizes | | Basic Major Dia., $D$ | Basic Pitch Dia.,[a] $D_2$ | Minor Diameter | | Lead Angle λ at Basic P.D. | | Area of Minor Dia. at $D - 2h_b$ | Tensile Stress Area[b] |
|---|---|---|---|---|---|---|---|---|---|
| Primary | Secondary | | | Ext. Thds.,[c] $d_3$ (Ref.) | Int. Thds.,[d] $D_1$ | | | | |
| Inches | Inches | Inches | Inches | Inches | Inches | Deg. | Min. | Sq. In. | Sq. In. |
| 1⅜[e] | | 1.3750 | 1.2667 | 1.1766 | 1.1946 | 2 | 24 | 1.054 | 1.155 |
| | 1⁷⁄₁₆ | 1.4375 | 1.3292 | 1.2391 | 1.2571 | 2 | 17 | 1.171 | 1.277 |
| 1½[e] | | 1.5000 | 1.3917 | 1.3016 | 1.3196 | 2 | 11 | 1.294 | 1.405 |
| | 1⁹⁄₁₆ | 1.5625 | 1.4542 | 1.3641 | 1.3821 | 2 | 5 | 1.423 | 1.54 |
| 1⅝ | | 1.6250 | 1.5167 | 1.4271 | 1.4446 | 2 | 0 | 1.56 | 1.68 |
| | 1¹¹⁄₁₆ | 1.6875 | 1.5792 | 1.4891 | 1.5071 | 1 | 55 | 1.70 | 1.83 |
| 1¾ | | 1.7500 | 1.6417 | 1.5516 | 1.5696 | 1 | 51 | 1.85 | 1.98 |
| | 1¹³⁄₁₆ | 1.8125 | 1.7042 | 1.6141 | 1.6321 | 1 | 47 | 2.00 | 2.14 |
| 1⅞ | | 1.8750 | 1.7667 | 1.6766 | 1.6946 | 1 | 43 | 2.16 | 2.30 |
| | 1¹⁵⁄₁₆ | 1.9375 | 1.8292 | 1.7391 | 1.7571 | 1 | 40 | 2.33 | 2.47 |
| 2 | | 2.0000 | 1.8917 | 1.8016 | 1.8196 | 1 | 36 | 2.50 | 2.65 |
| | 2⅛ | 2.1250 | 2.0167 | 1.9266 | 1.9446 | 1 | 30 | 2.86 | 3.03 |
| 2¼ | | 2.2500 | 2.1417 | 2.0516 | 2.0696 | 1 | 25 | 3.25 | 3.42 |
| | 2⅜ | 2.3750 | 2.2667 | 2.1766 | 2.1946 | 1 | 20 | 3.66 | 3.85 |
| 2½ | | 2.5000 | 2.3917 | 2.3016 | 2.3196 | 1 | 16 | 4.10 | 4.29 |
| | 2⅝ | 2.6250 | 2.5167 | 2.4266 | 2.4446 | 1 | 12 | 4.56 | 4.76 |
| 2¾ | | 2.7500 | 2.6417 | 2.5516 | 2.5696 | 1 | 9 | 5.04 | 5.26 |
| | 2⅞ | 2.8750 | 2.7667 | 2.6766 | 2.6946 | 1 | 6 | 5.55 | 5.78 |
| 3 | | 3.0000 | 2.8917 | 2.8016 | 2.8196 | 1 | 3 | 6.09 | 6.33 |
| | 3⅛ | 3.1250 | 3.0167 | 2.9266 | 2.9446 | 1 | 0 | 6.64 | 6.89 |
| 3¼ | | 3.2500 | 3.1417 | 3.0516 | 3.0696 | 0 | 58 | 7.23 | 7.49 |
| | 3⅜ | 3.3750 | 3.2667 | 3.1766 | 3.1946 | 0 | 56 | 7.84 | 8.11 |
| 3½ | | 3.5000 | 3.3917 | 3.3016 | 3.3196 | 0 | 54 | 8.47 | 8.75 |
| | 3⅝ | 3.6250 | 3.5167 | 3.4266 | 3.4446 | 0 | 52 | 9.12 | 9.42 |
| 3¾ | | 3.7500 | 3.6417 | 3.5516 | 3.5696 | 0 | 50 | 9.81 | 10.11 |
| | 3⅞ | 3.8750 | 3.7667 | 3.6766 | 3.6946 | 0 | 48 | 10.51 | 10.83 |
| 4 | | 4.0000 | 3.8917 | 3.8016 | 3.8196 | 0 | 47 | 11.24 | 11.57 |
| | 4⅛ | 4.1250 | 4.0167 | 3.9266 | 3.9446 | 0 | 45 | 12.00 | 12.33 |
| 4¼ | | 4.2500 | 4.1417 | 4.0516 | 4.0696 | 0 | 44 | 12.78 | 13.12 |
| | 4⅜ | 4.3750 | 4.2667 | 4.1766 | 4.1946 | 0 | 43 | 13.58 | 13.94 |
| 4½ | | 4.5000 | 4.3917 | 4.3016 | 4.3196 | 0 | 42 | 14.41 | 14.78 |
| | 4⅝ | 4.6250 | 4.5167 | 4.4266 | 4.4446 | 0 | 40 | 15.3 | 15.6 |
| 4¾ | | 4.7500 | 4.6417 | 4.5516 | 4.5696 | 0 | 39 | 16.1 | 16.5 |
| | 4⅞ | 4.8750 | 4.7667 | 4.6766 | 4.6946 | 0 | 38 | 17.0 | 17.5 |
| 5 | | 5.0000 | 4.8917 | 4.8016 | 4.8196 | 0 | 37 | 18.0 | 18.4 |
| | 5⅛ | 5.1250 | 5.0167 | 4.9266 | 4.9446 | 0 | 36 | 18.9 | 19.3 |
| 5¼ | | 5.2500 | 5.1417 | 5.0516 | 5.0696 | 0 | 35 | 19.9 | 20.3 |
| | 5⅜ | 5.3750 | 5.2667 | 5.1766 | 5.1946 | 0 | 35 | 20.9 | 21.3 |
| 5½ | | 5.5000 | 5.3917 | 5.3016 | 5.3196 | 0 | 34 | 21.9 | 22.4 |
| | 5⅝ | 5.6250 | 5.5167 | 5.4266 | 5.4446 | 0 | 33 | 23.0 | 23.4 |
| 5¾ | | 5.7500 | 5.6417 | 5.5516 | 5.5696 | 0 | 32 | 24.0 | 24.5 |
| | 5⅞ | 5.8750 | 5.7667 | 5.6766 | 5.6946 | 0 | 32 | 25.1 | 25.6 |
| 6 | | 6.0000 | 5.8917 | 5.8016 | 5.8196 | 0 | 31 | 26.3 | 26.8 |

[a] British: Effective Diameter.

[b] See formula, pages 1660 and 1668.

[c] Design form for UNR threads. (See figure on page 1951).

[d] Basic minor diameter.

[e] These are standard sizes of the UNC series.

## Table 5c. 8-Thread Series, 8-UN and 8-UNR — Basic Dimensions

| Sizes | | Basic Major Dia., $D$ | Basic Pitch Dia.,[a] $D_2$ | Minor Diameter | | Lead Angle λ at Basic P.D. | | Area of Minor Dia. at $D - 2h_b$ | Tensile Stress Area[b] |
|---|---|---|---|---|---|---|---|---|---|
| Primary | Secondary | | | Ext.Thds.,[c] $d_3$ (Ref.) | Int.Thds.,[d] $D_1$ | | | | |
| Inches | Inches | Inches | Inches | Inches | Inches | Deg. | Min. | Sq. In. | Sq. In. |
| 1[e] | | **1.0000** | **0.9188** | **0.8512** | **0.8647** | **2** | **29** | 0.551 | 0.606 |
| | 1 1/16 | **1.0625** | **0.9813** | **0.9137** | **0.9272** | **2** | **19** | 0.636 | 0.695 |
| 1 1/8 | | **1.1250** | **1.0438** | **0.9792** | **0.9897** | **2** | **11** | 0.728 | 0.790 |
| | 1 3/16 | **1.1875** | **1.1063** | **1.0387** | **1.0522** | **2** | **4** | 0.825 | 0.892 |
| 1 1/4 | | **1.2500** | **1.1688** | **1.1012** | **1.1147** | **1** | **57** | 0.929 | 1.000 |
| | 1 5/16 | **1.3125** | **1.2313** | **1.1637** | **1.1772** | **1** | **51** | 1.039 | 1.114 |
| 1 3/8 | | **1.3750** | **1.2938** | **1.2262** | **1.2397** | **1** | **46** | 1.155 | 1.233 |
| | 1 7/16 | **1.4375** | **1.3563** | **1.2887** | **1.3022** | **1** | **41** | 1.277 | 1.360 |
| 1 1/2 | | **1.5000** | **1.4188** | **1.3512** | **1.3647** | **1** | **36** | 1.405 | 1.492 |
| | 1 9/16 | **1.5625** | **1.4813** | **1.4137** | **1.4272** | **1** | **32** | 1.54 | 1.63 |
| 1 5/8 | | **1.6250** | **1.5438** | **1.4806** | **1.4897** | **1** | **29** | 1.68 | 1.78 |
| | 1 11/16 | **1.6875** | **1.6063** | **1.5387** | **1.5522** | **1** | **25** | 1.83 | 1.93 |
| 1 3/4 | | **1.7500** | **1.6688** | **1.6012** | **1.6147** | **1** | **22** | 1.98 | 2.08 |
| | 1 13/16 | **1.8125** | **1.7313** | **1.6637** | **1.6772** | **1** | **19** | 2.14 | 2.25 |
| 1 7/8 | | **1.8750** | **1.7938** | **1.7262** | **1.7397** | **1** | **16** | 2.30 | 2.41 |
| | 1 15/16 | **1.9375** | **1.8563** | **1.7887** | **1.8022** | **1** | **14** | 2.47 | 2.59 |
| 2 | | **2.0000** | **1.9188** | **1.8512** | **1.8647** | **1** | **11** | 2.65 | 2.77 |
| | 2 1/8 | **2.1250** | **2.0438** | **1.9762** | **1.9897** | **1** | **7** | 3.03 | 3.15 |
| 2 1/4 | | **2.2500** | **2.1688** | **2.1012** | **2.1147** | **1** | **3** | 3.42 | 3.56 |
| | 2 3/8 | **2.3750** | **2.2938** | **2.2262** | **2.2397** | **1** | **0** | 3.85 | 3.99 |
| 2 1/2 | | **2.5000** | **2.4188** | **2.3512** | **2.3647** | **0** | **57** | 4.29 | 4.44 |
| | 2 5/8 | **2.6250** | **2.5438** | **2.4762** | **2.4897** | **0** | **54** | 4.76 | 4.92 |
| 2 3/4 | | **2.7500** | **2.6688** | **2.6012** | **2.6147** | **0** | **51** | 5.26 | 5.43 |
| | 2 7/8 | **2.8750** | **2.7938** | **2.7262** | **2.7397** | **0** | **49** | 5.78 | 5.95 |
| 3 | | **3.0000** | **2.9188** | **2.8512** | **2.8647** | **0** | **47** | 6.32 | 6.51 |
| | 3 1/8 | **3.1250** | **3.0438** | **2.9762** | **2.9897** | **0** | **45** | 6.89 | 7.08 |
| 3 1/4 | | **3.2500** | **3.1688** | **3.1012** | **3.1147** | **0** | **43** | 7.49 | 7.69 |
| | 3 3/8 | **3.3750** | **3.2938** | **3.2262** | **3.2397** | **0** | **42** | 8.11 | 8.31 |
| 3 1/2 | | **3.5000** | **3.4188** | **3.3512** | **3.3647** | **0** | **40** | 8.75 | 8.96 |
| | 3 5/8 | **3.6250** | **3.5438** | **3.4762** | **3.4897** | **0** | **39** | 9.42 | 9.64 |
| 3 3/4 | | **3.7500** | **3.6688** | **3.6012** | **3.6147** | **0** | **37** | 10.11 | 10.34 |
| | 3 7/8 | **3.8750** | **3.7938** | **3.7262** | **3.7397** | **0** | **36** | 10.83 | 11.06 |
| 4 | | **4.0000** | **3.9188** | **3.8512** | **3.8647** | **0** | **35** | 11.57 | 11.81 |
| | 4 1/8 | **4.1250** | **4.0438** | **3.9762** | **3.9897** | **0** | **34** | 12.34 | 12.59 |
| 4 1/4 | | **4.2500** | **4.1688** | **4.1012** | **4.1147** | **0** | **33** | 13.12 | 13.38 |
| | 4 3/8 | **4.3750** | **4.2938** | **4.2262** | **4.2397** | **0** | **32** | 13.94 | 14.21 |
| 4 1/2 | | **4.5000** | **4.4188** | **4.3512** | **4.3647** | **0** | **31** | 14.78 | 15.1 |
| | 4 5/8 | **4.6250** | **4.5438** | **4.4762** | **4.4897** | **0** | **30** | 15.6 | 15.9 |
| 4 3/4 | | **4.7500** | **4.6688** | **4.6012** | **4.6147** | **0** | **29** | 16.5 | 16.8 |
| | 4 7/8 | **4.8750** | **4.7938** | **4.7262** | **4.7397** | **0** | **29** | 17.4 | 17.7 |
| 5 | | **5.0000** | **4.9188** | **4.8512** | **4.8647** | **0** | **28** | 18.4 | 18.7 |
| | 5 1/8 | **5.1250** | **5.0438** | **4.9762** | **4.9897** | **0** | **27** | 19.3 | 19.7 |
| 5 1/4 | | **5.2500** | **5.1688** | **5.1012** | **5.1147** | **0** | **26** | 20.3 | 20.7 |
| | 5 3/8 | **5.3750** | **5.2938** | **5.2262** | **5.2397** | **0** | **26** | 21.3 | 21.7 |
| 5 1/2 | | **5.5000** | **5.4188** | **5.3512** | **5.3647** | **0** | **25** | 22.4 | 22.7 |
| | 5 5/8 | **5.6250** | **5.5438** | **5.4762** | **5.4897** | **0** | **25** | 23.4 | 23.8 |
| 5 3/4 | | **5.7500** | **5.6688** | **5.6012** | **5.6147** | **0** | **24** | 24.5 | 24.9 |
| | 5 7/8 | **5.8750** | **5.7938** | **5.7262** | **5.7397** | **0** | **24** | 25.6 | 26.0 |
| 6 | | **6.0000** | **5.9188** | **5.8512** | **5.8647** | **0** | **23** | 26.8 | 27.1 |

[a] British: Effective Diameter.

[b] See formula, pages 1660 and 1668.

[c] Design form for UNR threads. (See figure on page 1951).

[d] Basic minor diameter.

[e] This is a standard size of the UNC series.

**Table 5d. 12-Thread series, 12-UN and 12-UNR — Basic Dimensions**

| Sizes | | Basic Major Dia., $D$ | Basic Pitch Dia.,[a] $D_2$ | Minor Diameter | | Lead Angle λ at Basic P.D. | | Area of Minor Dia. at $D - 2h_b$ | Tensile Stress Area[b] |
|---|---|---|---|---|---|---|---|---|---|
| Primary | Secondary | | | Ext. Thds.,[c] $d_3$ (Ref.) | Int. Thds.,[d] $D_1$ | | | | |
| Inches | Inches | Inches | Inches | Inches | Inches | Deg. | Min. | Sq. In. | Sq. In. |
| 9/16[e] | | 0.5625 | 0.5084 | 0.4633 | 0.4723 | 2 | 59 | 0.162 | 0.182 |
| 5/8 | | 0.6250 | 0.5709 | 0.5258 | 0.5348 | 2 | 40 | 0.210 | 0.232 |
| | 11/16 | 0.6875 | 0.6334 | 0.5883 | 0.5973 | 2 | 24 | 0.264 | 0.289 |
| 3/4 | | 0.7500 | 0.6959 | 0.6508 | 0.6598 | 2 | 11 | 0.323 | 0.351 |
| | 13/16 | 0.8125 | 0.7584 | 0.7133 | 0.7223 | 2 | 0 | 0.390 | 0.420 |
| 7/8 | | 0.8750 | 0.8209 | 0.7758 | 0.7848 | 1 | 51 | 0.462 | 0.495 |
| | 15/16 | 0.9375 | 0.8834 | 0.8383 | 0.8473 | 1 | 43 | 0.540 | 0.576 |
| 1[e] | | 1.0000 | 0.9459 | 0.9008 | 0.9098 | 1 | 36 | 0.625 | 0.663 |
| | 1 1/16 | 1.0625 | 1.0084 | 0.9633 | 0.9723 | 1 | 30 | 0.715 | 0.756 |
| 1 1/8[e] | | 1.1250 | 1.0709 | 1.0258 | 1.0348 | 1 | 25 | 0.812 | 0.856 |
| | 1 3/16 | 1.1875 | 1.1334 | 1.0883 | 1.0973 | 1 | 20 | 0.915 | 0.961 |
| 1 1/4[e] | | 1.2500 | 1.1959 | 1.1508 | 1.1598 | 1 | 16 | 1.024 | 1.073 |
| | 1 5/16 | 1.3125 | 1.2584 | 1.2133 | 1.2223 | 1 | 12 | 1.139 | 1.191 |
| 1 3/8 | | 1.3750 | 1.3209 | 1.2758 | 1.2848 | 1 | 9 | 1.260 | 1.315 |
| | 1 7/16 | 1.4375 | 1.3834 | 1.3383 | 1.3473 | 1 | 6 | 1.388 | 1.445 |
| 1 1/2[e] | | 1.5000 | 1.4459 | 1.4008 | 1.4098 | 1 | 3 | 1.52 | 1.58 |
| | 1 9/16 | 1.5625 | 1.5084 | 1.4633 | 1.4723 | 1 | 0 | 1.66 | 1.72 |
| 1 5/8 | | 1.6250 | 1.5709 | 1.5258 | 1.5348 | 0 | 58 | 1.81 | 1.87 |
| | 1 11/16 | 1.6875 | 1.6334 | 1.5883 | 1.5973 | 0 | 56 | 1.96 | 2.03 |
| 1 3/4 | | 1.7500 | 1.6959 | 1.6508 | 1.6598 | 0 | 54 | 2.12 | 2.19 |
| | 1 13/16 | 1.8125 | 1.7584 | 1.7133 | 1.7223 | 0 | 52 | 2.28 | 2.35 |
| 1 7/8 | | 1.8750 | 1.8209 | 1.7758 | 1.7848 | 0 | 50 | 2.45 | 2.53 |
| | 1 15/16 | 1.9375 | 1.8834 | 1.8383 | 1.8473 | 0 | 48 | 2.63 | 2.71 |
| 2 | | 2.0000 | 1.9459 | 1.9008 | 1.9098 | 0 | 47 | 2.81 | 2.89 |
| | 2 1/8 | 2.1250 | 2.0709 | 2.0258 | 2.0348 | 0 | 44 | 3.19 | 3.28 |
| 2 1/4 | | 2.2500 | 2.1959 | 2.1508 | 2.1598 | 0 | 42 | 3.60 | 3.69 |
| | 2 3/8 | 2.3750 | 2.3209 | 2.2758 | 2.2848 | 0 | 39 | 4.04 | 4.13 |
| 2 1/2 | | 2.5000 | 2.4459 | 2.4008 | 2.4098 | 0 | 37 | 4.49 | 4.60 |
| | 2 5/8 | 2.6250 | 2.5709 | 2.5258 | 2.5348 | 0 | 35 | 4.97 | 5.08 |
| 2 3/4 | | 2.7500 | 2.6959 | 2.6508 | 2.6598 | 0 | 34 | 5.48 | 5.59 |
| | 2 7/8 | 2.8750 | 2.8209 | 2.7758 | 2.7848 | 0 | 32 | 6.01 | 6.13 |
| 3 | | 3.0000 | 2.9459 | 2.9008 | 2.9098 | 0 | 31 | 6.57 | 6.69 |
| | 3 1/8 | 3.1250 | 3.0709 | 3.0258 | 3.0348 | 0 | 30 | 7.15 | 7.28 |
| 3 1/4 | | 3.2500 | 3.1959 | 3.1508 | 3.1598 | 0 | 29 | 7.75 | 7.89 |
| | 3 3/8 | 3.3750 | 3.3209 | 3.2758 | 3.2848 | 0 | 27 | 8.38 | 8.52 |
| 3 1/2 | | 3.5000 | 3.4459 | 3.4008 | 3.4098 | 0 | 26 | 9.03 | 9.18 |
| | 3 5/8 | 3.6250 | 3.5709 | 3.5258 | 3.5348 | 0 | 26 | 9.71 | 9.86 |
| 3 3/4 | | 3.7500 | 3.6959 | 3.6508 | 3.6598 | 0 | 25 | 10.42 | 10.57 |
| | 3 7/8 | 3.8750 | 3.8209 | 3.7758 | 3.7848 | 0 | 24 | 11.14 | 11.30 |
| 4 | | 4.0000 | 3.9459 | 3.9008 | 3.9098 | 0 | 23 | 11.90 | 12.06 |
| | 4 1/8 | 4.1250 | 4.0709 | 4.0258 | 4.0348 | 0 | 22 | 12.67 | 12.84 |
| 4 1/4 | | 4.2500 | 4.1959 | 4.1508 | 4.1598 | 0 | 22 | 13.47 | 13.65 |
| | 4 3/8 | 4.3750 | 4.3209 | 4.2758 | 4.2848 | 0 | 21 | 14.30 | 14.48 |
| 4 1/2 | | 4.5000 | 4.4459 | 4.4008 | 4.4098 | 0 | 21 | 15.1 | 15.3 |
| | 4 5/8 | 4.6250 | 4.5709 | 4.5258 | 4.5348 | 0 | 20 | 16.0 | 16.2 |
| 4 3/4 | | 4.7500 | 4.6959 | 4.6508 | 4.6598 | 0 | 19 | 16.9 | 17.1 |
| | 4 7/8 | 4.8750 | 4.8209 | 4.7758 | 4.7848 | 0 | 19 | 17.8 | 18.0 |
| 5 | | 5.0000 | 4.9459 | 4.9008 | 4.9098 | 0 | 18 | 18.8 | 19.0 |
| | 5 1/8 | 5.1250 | 5.0709 | 5.0258 | 5.0348 | 0 | 18 | 19.8 | 20.0 |
| 5 1/4 | | 5.2500 | 5.1959 | 5.1508 | 5.1598 | 0 | 18 | 20.8 | 21.0 |
| | 5 3/8 | 5.3750 | 5.3209 | 5.2758 | 5.2848 | 0 | 17 | 21.8 | 22.0 |
| 5 1/2 | | 5.5000 | 5.4459 | 5.4008 | 5.4098 | 0 | 17 | 22.8 | 23.1 |
| | 5 5/8 | 5.6250 | 5.5709 | 5.5258 | 5.5348 | 0 | 16 | 23.9 | 24.1 |
| 5 3/4 | | 5.7500 | 5.6959 | 5.6508 | 5.6598 | 0 | 16 | 25.0 | 25.2 |
| | 5 7/8 | 5.8750 | 5.8209 | 5.7758 | 5.7848 | 0 | 16 | 26.1 | 26.4 |
| 6 | | 6.0000 | 5.9459 | 5.9008 | 5.9098 | 0 | 15 | 27.3 | 27.5 |

[a] British: Effective Diameter.

[b] See formula, pages 1660 and 1668.

[c] Design form for UNR threads. (See figure on page 1951).

[d] Basic minor diameter.

[e] These are standard sizes of the UNC or UNF series.

## Table 5e. 16-Thread Series, 16-UN and 16-UNR — Basic Dimensions

| Sizes | | Basic Major Dia., $D$ | Basic Pitch Dia.,[a] $D_2$ | Minor Diameter | | Lead Angle λ at Basic P.D. | | Area of Minor Dia. at $D - 2h_b$ | Tensile Stress Area[b] |
|---|---|---|---|---|---|---|---|---|---|
| Primary | Secondary | | | Ext. Thds.,[c] $d_3$ (Ref.) | Int. Thds.,[d] $D_1$ | | | | |
| Inches | Inches | Inches | Inches | Inches | Inches | Deg. | Min. | Sq. In. | Sq. In. |
| 3/8[e] | | **0.3750** | **0.3344** | **0.3005** | **0.3073** | **3** | **24** | 0.0678 | 0.0775 |
| 7/16 | | **0.4375** | **0.3969** | **0.3630** | **0.3698** | **2** | **52** | 0.0997 | 0.1114 |
| 1/2 | | **0.5000** | **0.4594** | **0.4255** | **0.4323** | **2** | **29** | 0.1378 | 0.151 |
| 9/16 | | **0.5625** | **0.5219** | **0.4880** | **0.4948** | **2** | **11** | 0.182 | 0.198 |
| 5/8 | | **0.6250** | **0.5844** | **0.5505** | **0.5573** | **1** | **57** | 0.232 | 0.250 |
| | 11/16 | **0.6875** | **0.6469** | **0.6130** | **0.6198** | **1** | **46** | 0.289 | 0.308 |
| 3/4[e] | | **0.7500** | **0.7094** | **0.6755** | **0.6823** | **1** | **36** | 0.351 | 0.373 |
| | 13/16 | **0.8125** | **0.7719** | **0.7380** | **0.7448** | **1** | **29** | 0.420 | 0.444 |
| 7/8 | | **0.8750** | **0.8344** | **0.8005** | **0.8073** | **1** | **22** | 0.495 | 0.521 |
| | 15/16 | **0.9375** | **0.8969** | **0.8630** | **0.8698** | **1** | **16** | 0.576 | 0.604 |
| 1 | | **1.0000** | **0.9594** | **0.9255** | **0.9323** | **1** | **11** | 0.663 | 0.693 |
| | 1 1/16 | **1.0625** | **1.0219** | **0.9880** | **0.9948** | **1** | **7** | 0.756 | 0.788 |
| 1 1/8 | | **1.1250** | **1.0844** | **1.0505** | **1.0573** | **1** | **3** | 0.856 | 0.889 |
| | 1 3/16 | **1.1875** | **1.1469** | **1.1130** | **1.1198** | **1** | **0** | 0.961 | 0.997 |
| 1 1/4 | | **1.2500** | **1.2094** | **1.1755** | **1.1823** | **0** | **57** | 1.073 | 1.111 |
| | 1 5/16 | **1.3125** | **1.2719** | **1.2380** | **1.2448** | **0** | **54** | 1.191 | 1.230 |
| 1 3/8 | | **1.3750** | **1.3344** | **1.3005** | **1.3073** | **0** | **51** | 1.315 | 1.356 |
| | 1 7/16 | **1.4375** | **1.3969** | **1.3630** | **1.3698** | **0** | **49** | 1.445 | 1.488 |
| 1 1/2 | | **1.5000** | **1.4594** | **1.4255** | **1.4323** | **0** | **47** | 1.58 | 1.63 |
| | 1 9/16 | **1.5625** | **1.5219** | **1.4880** | **1.4948** | **0** | **45** | 1.72 | 1.77 |
| 1 5/8 | | **1.6250** | **1.5844** | **1.5505** | **1.5573** | **0** | **43** | 1.87 | 1.92 |
| | 1 11/16 | **1.6875** | **1.6469** | **1.6130** | **1.6198** | **0** | **42** | 2.03 | 2.08 |
| 1 3/4 | | **1.7500** | **1.7094** | **1.6755** | **1.6823** | **0** | **40** | 2.19 | 2.24 |
| | 1 13/16 | **1.8125** | **1.7719** | **1.7380** | **1.7448** | **0** | **39** | 2.35 | 2.41 |
| 1 7/8 | | **1.8750** | **1.8344** | **1.8005** | **1.8073** | **0** | **37** | 2.53 | 2.58 |
| | 1 15/16 | **1.9375** | **1.8969** | **1.8630** | **1.8698** | **0** | **36** | 2.71 | 2.77 |
| 2 | | **2.0000** | **1.9594** | **1.9255** | **1.9323** | **0** | **35** | 2.89 | 2.95 |
| | 2 1/8 | **2.1250** | **2.0844** | **2.0505** | **2.0573** | **0** | **33** | 3.28 | 3.35 |
| 2 1/4 | | **2.2500** | **2.2094** | **2.1755** | **2.1823** | **0** | **31** | 3.69 | 3.76 |
| | 2 3/8 | **2.3750** | **2.3344** | **2.3005** | **2.3073** | **0** | **29** | 4.13 | 4.21 |
| 2 1/2 | | **2.5000** | **2.4594** | **2.4255** | **2.4323** | **0** | **28** | 4.60 | 4.67 |
| | 2 5/8 | **2.6250** | **2.5844** | **2.5505** | **2.5573** | **0** | **26** | 5.08 | 5.16 |
| 2 3/4 | | **2.7500** | **2.7094** | **2.6755** | **2.6823** | **0** | **25** | 5.59 | 5.68 |
| | 2 7/8 | **2.8750** | **2.8344** | **2.8005** | **2.8073** | **0** | **24** | 6.13 | 6.22 |
| 3 | | **3.0000** | **2.9594** | **2.9255** | **2.9323** | **0** | **23** | 6.69 | 6.78 |
| | 3 1/8 | **3.1250** | **3.0844** | **3.0505** | **3.0573** | **0** | **22** | 7.28 | 7.37 |
| 3 1/4 | | **3.2500** | **3.2094** | **3.1755** | **3.1823** | **0** | **21** | 7.89 | 7.99 |
| | 3 3/8 | **3.3750** | **3.3344** | **3.3005** | **3.3073** | **0** | **21** | 8.52 | 8.63 |
| 3 1/2 | | **3.5000** | **3.4594** | **3.4255** | **3.4323** | **0** | **20** | 9.18 | 9.29 |
| | 3 5/8 | **3.6250** | **3.5844** | **3.5505** | **3.5573** | **0** | **19** | 9.86 | 9.98 |
| 3 3/4 | | **3.7500** | **3.7094** | **3.6755** | **3.6823** | **0** | **18** | 10.57 | 10.69 |
| | 3 7/8 | **3.8750** | **3.8344** | **3.8005** | **3.8073** | **0** | **18** | 11.30 | 11.43 |
| 4 | | **4.0000** | **3.9594** | **3.9255** | **3.9323** | **0** | **17** | 12.06 | 12.19 |
| | 4 1/8 | **4.1250** | **4.0844** | **4.0505** | **4.0573** | **0** | **17** | 12.84 | 12.97 |
| 4 1/4 | | **4.2500** | **4.2094** | **4.1755** | **4.1823** | **0** | **16** | 13.65 | 13.78 |
| | 4 3/8 | **4.3750** | **4.3344** | **4.3005** | **4.3073** | **0** | **16** | 14.48 | 14.62 |
| 4 1/2 | | **4.5000** | **4.4594** | **4.4255** | **4.4323** | **0** | **15** | 15.34 | 15.5 |
| | 4 5/8 | **4.6250** | **4.5844** | **4.5505** | **4.5573** | **0** | **15** | 16.2 | 16.4 |
| 4 3/4 | | **4.7500** | **4.7094** | **4.6755** | **4.6823** | **0** | **15** | 17.1 | 17.3 |
| | 4 7/8 | **4.8750** | **4.8344** | **4.8005** | **4.8073** | **0** | **14** | 18.0 | 18.2 |
| 5 | | **5.0000** | **4.9594** | **4.9255** | **4.9323** | **0** | **14** | 19.0 | 19.2 |
| | 5 1/8 | **5.1250** | **5.0844** | **5.0505** | **5.0573** | **0** | **13** | 20.0 | 20.1 |
| 5 1/4 | | **5.2500** | **5.2094** | **5.1755** | **5.1823** | **0** | **13** | 21.0 | 21.1 |
| | 5 3/8 | **5.3750** | **5.3344** | **5.3005** | **5.3073** | **0** | **13** | 22.0 | 22.2 |

## Table 5e. *(Continued)* 16-Thread Series, 16-UN and 16-UNR — Basic Dimensions

| Sizes | | Basic Major Dia., $D$ | Basic Pitch Dia.,[a] $D_2$ | Minor Diameter | | Lead Angle λ at Basic P.D. | | Area of Minor Dia. at $D - 2h_b$ | Tensile Stress Area[b] |
|---|---|---|---|---|---|---|---|---|---|
| Primary | Secondary | | | Ext. Thds.,[c] $d_3$ (Ref.) | Int. Thds.,[d] $D_1$ | | | | |
| Inches | Inches | Inches | Inches | Inches | Inches | Deg. | Min. | Sq. In. | Sq. In. |
| 5½ | | **5.5000** | **5.4594** | **5.4255** | **5.4323** | **0** | **13** | 23.1 | 23.2 |
| | 5⅝ | **5.6250** | **5.5844** | **5.5505** | **5.5573** | **0** | **12** | 24.1 | 24.3 |
| 5¾ | | **5.7500** | **5.7094** | **5.6755** | **5.6823** | **0** | **12** | 25.2 | 25.4 |
| | 5⅞ | **5.8750** | **5.8344** | **5.8005** | **5.8073** | **0** | **12** | 26.4 | 26.5 |
| 6 | | **6.0000** | **5.9594** | **5.9255** | **5.9323** | **0** | **11** | 27.5 | 27.7 |

[a] British: Effective Diameter.

[b] See formula, pages 1660 and 1668.

[c] Design form for UNR threads. (See figure on page 1951).

[d] Basic minor diameter.

[e] These are standard sizes of the UNC or UNF series.

## Table 5f. 20-Thread Series, 20-UN and 20-UNR — Basic Dimensions

| Sizes | | Basic Major Dia., $D$ | Basic Pitch Dia.,[a] $D_2$ | Minor Diameter | | Lead Angle λ at Basic P.D. | | Area of Minor Dia. at $D - 2h_b$ | Tensile Stress Area[b] |
|---|---|---|---|---|---|---|---|---|---|
| Primary | Secondary | | | Ext. Thds.,[c] $d_3$ (Ref.) | Int. Thds.,[d] $D_1$ | | | | |
| Inches | Inches | Inches | Inches | Inches | Inches | Deg. | Min. | Sq. In. | Sq. In. |
| ¼[e] | | **0.2500** | **0.2175** | **0.1905** | **0.1959** | **4** | **11** | 0.0269 | 0.0318 |
| 5/16 | | **0.3125** | **0.2800** | **0.2530** | **0.2584** | **3** | **15** | 0.0481 | 0.0547 |
| ⅜ | | **0.3750** | **0.3425** | **0.3155** | **0.3209** | **2** | **40** | 0.0755 | 0.0836 |
| 7/16[e] | | **0.4375** | **0.4050** | **0.3780** | **0.3834** | **2** | **15** | 0.1090 | 0.1187 |
| ½[e] | | **0.5000** | **0.4675** | **0.4405** | **0.4459** | **1** | **57** | 0.1486 | 0.160 |
| 9/16 | | **0.5625** | **0.5300** | **0.5030** | **0.5084** | **1** | **43** | 0.194 | 0.207 |
| ⅝ | | **0.6250** | **0.5925** | **0.5655** | **0.5709** | **1** | **32** | 0.246 | 0.261 |
| | 11/16 | **0.6875** | **0.6550** | **0.6280** | **0.6334** | **1** | **24** | 0.304 | 0.320 |
| ¾[e] | | **0.7500** | **0.7175** | **0.6905** | **0.6959** | **1** | **16** | 0.369 | 0.386 |
| | 13/16[e] | **0.8125** | **0.7800** | **0.7530** | **0.7584** | **1** | **10** | 0.439 | 0.458 |
| ⅞[e] | | **0.8750** | **0.8425** | **0.8155** | **0.8209** | **1** | **5** | 0.515 | 0.536 |
| | 15/16[e] | **0.9375** | **0.9050** | **0.8780** | **0.8834** | **1** | **0** | 0.598 | 0.620 |
| 1[e] | | **1.0000** | **0.9675** | **0.9405** | **0.9459** | **0** | **57** | 0.687 | 0.711 |
| | 1 1/16 | **1.0625** | **1.0300** | **1.0030** | **1.0084** | **0** | **53** | 0.782 | 0.807 |
| 1⅛ | | **1.1250** | **1.0925** | **1.0655** | **1.0709** | **0** | **50** | 0.882 | 0.910 |
| | 1 3/16 | **1.1875** | **1.1550** | **1.1280** | **1.1334** | **0** | **47** | 0.990 | 1.018 |
| 1¼ | | **1.2500** | **1.2175** | **1.1905** | **1.1959** | **0** | **45** | 1.103 | 1.133 |
| | 1 5/16 | **1.3125** | **1.2800** | **1.2530** | **1.2584** | **0** | **43** | 1.222 | 1.254 |
| 1⅜ | | **1.3750** | **1.3425** | **1.3155** | **1.3209** | **0** | **41** | 1.348 | 1.382 |
| | 1 7/16 | **1.4375** | **1.4050** | **1.3780** | **1.3834** | **0** | **39** | 1.479 | 1.51 |
| 1½ | | **1.5000** | **1.4675** | **1.4405** | **1.4459** | **0** | **37** | 1.62 | 1.65 |
| | 1 9/16 | **1.5625** | **1.5300** | **1.5030** | **1.5084** | **0** | **36** | 1.76 | 1.80 |
| 1⅝ | | **1.6250** | **1.5925** | **1.5655** | **1.5709** | **0** | **34** | 1.91 | 1.95 |
| | 1 11/16 | **1.6875** | **1.6550** | **1.6280** | **1.6334** | **0** | **33** | 2.07 | 2.11 |
| 1¾ | | **1.7500** | **1.7175** | **1.6905** | **1.6959** | **0** | **32** | 2.23 | 2.27 |
| | 1 13/16 | **1.8125** | **1.7800** | **1.7530** | **1.7584** | **0** | **31** | 2.40 | 2.44 |
| 1⅞ | | **1.8750** | **1.8425** | **1.8155** | **1.8209** | **0** | **30** | 2.57 | 2.62 |
| | 1 15/16 | **1.9375** | **1.9050** | **1.8780** | **1.8834** | **0** | **29** | 2.75 | 2.80 |
| 2 | | **2.0000** | **1.9675** | **1.9405** | **1.9459** | **0** | **28** | 2.94 | 2.99 |
| | 2⅛ | **2.1250** | **2.0925** | **2.0655** | **2.0709** | **0** | **26** | 3.33 | 3.39 |
| 2¼ | | **2.2500** | **2.2175** | **2.1905** | **2.1959** | **0** | **25** | 3.75 | 3.81 |
| | 2⅜ | **2.3750** | **2.3425** | **2.3155** | **2.3209** | **0** | **23** | 4.19 | 4.25 |
| 2½ | | **2.5000** | **2.4675** | **2.4405** | **2.4459** | **0** | **22** | 4.66 | 4.72 |
| | 2⅝ | **2.6250** | **2.5925** | **2.5655** | **2.5709** | **0** | **21** | 5.15 | 5.21 |
| 2¾ | | **2.7500** | **2.7175** | **2.6905** | **2.6959** | **0** | **20** | 5.66 | 5.73 |
| | 2⅞ | **2.8750** | **2.8425** | **2.8155** | **2.8209** | **0** | **19** | 6.20 | 6.27 |
| 3 | | **3.0000** | **2.9675** | **2.9405** | **2.9459** | **0** | **18** | 6.77 | 6.84 |

[a] British: Effective Diameter.

[b] See formula, pages 1660 and 1668.

[c] Design form for UNR threads. (See figure on page 1951).

[d] Basic minor diameter.

[e] These are standard sizes of the UNC, UNF, or UNEF Series.

## Table 5g. 28-Thread Series, 28-UN and 28-UNR — Basic Dimensions

| Sizes | | Basic Major Dia., $D$ | Basic Pitch Dia.,[a] $D_2$ | Minor Diameter | | Lead Angel λ at Basic P.D. | | Area of Minor Dia. at $D\text{-}2h_b$ | Tensile Stress Area[b] |
|---|---|---|---|---|---|---|---|---|---|
| Primary | Secondary | | | Ext. Thds.,[c] $d_3$ (Ref.) | Int. Thds.,[d] $D_1$ | | | | |
| Inches | Inches | Inches | Inches | Inches | Inches | Deg. | Min. | Sq. In. | Sq. In. |
| | **12 (0.216)[e]** | **0.2160** | **0.1928** | **0.1734** | **0.1773** | **3** | **22** | 0.0226 | 0.0258 |
| ¼[e] | | **0.2500** | **0.2268** | **0.2074** | **0.2113** | **2** | **52** | 0.0326 | 0,0364 |
| 5/16 | | **0.3125** | **0.2893** | **0.2699** | **0.2738** | **2** | **15** | 0.0556 | 0.0606 |
| 3/8 | | **0.3750** | **0.3518** | **0.3324** | **0.3363** | **1** | **51** | 0.0848 | 0.0909 |
| 7/16[e] | | **0.4375** | **0.4143** | **0.3949** | **0.3988** | **1** | **34** | 0.1201 | 0.1274 |
| ½[e] | | **0.5000** | **0.4768** | **0.4574** | **0.4613** | **1** | **22** | 0.162 | 0.170 |
| 9/16 | | **0.5625** | **0.5393** | **0.5199** | **0.5238** | **1** | **12** | 0.209 | 0.219 |
| 5/8 | | **0.6250** | **0.6018** | **0.5824** | **0.5863** | **1** | **5** | 0.263 | 0.274 |
| | 11/16 | **0.6875** | **0.6643** | **0.6449** | **0.6488** | **0** | **59** | 0.323 | 0.335 |
| ¾ | | **0.7500** | **0.7268** | **0.7074** | **0.7113** | **0** | **54** | 0.389 | 0.402 |
| | 13/16 | **0.8125** | **0.7893** | **0.7699** | **0.7738** | **0** | **50** | 0.461 | 0.475 |
| 7/8 | | **0.8750** | **0.8518** | **0.8324** | **0.8363** | **0** | **46** | 0.539 | 0.554 |
| | 15/16 | **0.9375** | **0.9143** | **0.8949** | **0.8988** | **0** | **43** | 0.624 | 0.640 |
| 1 | | **1.0000** | **0.9768** | **0.9574** | **0.9613** | **0** | **40** | 0.714 | 0.732 |
| | 1 1/16 | **1.0625** | **1.0393** | **1.0199** | **1.0238** | **0** | **38** | 0.811 | 0.830 |
| 1 1/8 | | **1.1250** | **1.1018** | **1.0824** | **1.0863** | **0** | **35** | 0.914 | 0.933 |
| | 1 3/16 | **1.1875** | **1.1643** | **1.1449** | **1.1488** | **0** | **34** | 1.023 | 1.044 |
| 1¼ | | **1.2500** | **1.2268** | **1.2074** | **1.2113** | **0** | **32** | 1.138 | 1.160 |
| | 1 5/16 | **1.3125** | **1.2893** | **1.2699** | **1.2738** | **0** | **30** | 1.259 | 1.282 |
| 1 3/8 | | **1.3750** | **1.3518** | **1.3324** | **1.3363** | **0** | **29** | 1.386 | 1.411 |
| | 1 7/16 | **1.4375** | **1.4143** | **1.3949** | **1.3988** | **0** | **28** | 1.52 | 1.55 |
| 1½ | | **1.5000** | **1.4768** | **1.4574** | **1.4613** | **0** | **26** | 1.66 | 1.69 |

[a] British: Effective Diameter.

[b] See formula, pages 1660 and 1668.

[c] Design form for UNR threads. (See figure on page 1951).

[d] Basic minor diameter.

[e] These are standard sizes of the UNC or UNF Series.

## Table 5h. 32-Thread Series, 32-UN and 32-UNR — Basic Dimensions

| Sizes | | Basic Major Dia., $D$ | Basic Pitch Dia.,[a] $D_2$ | Minor Diameter | | Lead Angel λ at Basic P.D. | | Area of Minor Dia. at $D - 2h_b$ | Tensile Stress Area[b] |
|---|---|---|---|---|---|---|---|---|---|
| Primary | Secondary | | | Ext.Thds.,[c] $d_3$ (Ref.) | Int.Thds.,[d] $D_1$ | | | | |
| Inches | Inches | Inches | Inches | Inches | Inches | Deg. | Min. | Sq. In. | Sq. In. |
| **6 (0.138)[e]** | | **0.1380** | **0.1177** | **0.1008** | **0.1042** | **4** | **50** | 0.00745 | 0.00909 |
| **8 (0.164)[e]** | | **0.1640** | **0.1437** | **0.1268** | **0.1302** | **3** | **58** | 0.01196 | 0.0140 |
| **10 (0.190)[e]** | | **0.1900** | **0.1697** | **0.1528** | **0.1562** | **3** | **21** | 0.01750 | 0.0200 |
| | **12 (0.216)[e]** | **0.2160** | **0.1957** | **0.1788** | **0.1822** | **2** | **55** | 0.0242 | 0.0270 |
| ¼[e] | | **0.2500** | **0.2297** | **0.2128** | **0.2162** | **2** | **29** | 0.0344 | 0.0379 |
| 5/16[e] | | **0.3125** | **0.2922** | **0.2753** | **0.2787** | **1** | **57** | 0.0581 | 0.0625 |
| 3/8[e] | | **0.3750** | **0.3547** | **0.3378** | **0.3412** | **1** | **36** | 0.0878 | 0.0932 |
| 7/16 | | **0.4375** | **0.4172** | **0.4003** | **0.4037** | **1** | **22** | 0.1237 | 0.1301 |
| ½ | | **0.5000** | **0.4797** | **0.4628** | **0.4662** | **1** | **11** | 0.166 | 0.173 |
| 9/16 | | **0.5625** | **0.5422** | **0.5253** | **0.5287** | **1** | **3** | 0.214 | 0.222 |
| 5/8 | | **0.6250** | **0.6047** | **0.5878** | **0.5912** | **0** | **57** | 0.268 | 0.278 |
| | 11/16 | **0.6875** | **0.6672** | **0.6503** | **0.6537** | **0** | **51** | 0.329 | 0.339 |
| ¾ | | **0.7500** | **0.7297** | **0.7128** | **0.7162** | **0** | **47** | 0.395 | 0.407 |
| | 13/16 | **0.8125** | **0.7922** | **0.7753** | **0.7787** | **0** | **43** | 0.468 | 0.480 |
| 7/8 | | **0.8750** | **0.8547** | **0.8378** | **0.8412** | **0** | **40** | 0.547 | 0.560 |
| | 15/16 | **0.9375** | **0.9172** | **0.9003** | **0.9037** | **0** | **37** | 0.632 | 0.646 |
| 1 | | **1.0000** | **0.9797** | **0.9628** | **0.9662** | **0** | **35** | 0.723 | 0.738 |

[a] British: Effective Diameter.

[b] See formula, pages 1660 and 1668.

[c] Design form for UNR threads. (See figure on page 1951).

[d] Basic minor diameter.

[e] These are standard sizes of the UNC,UNF, or UNEF Series.

**Thread Classes.**—Thread classes are distinguished from each other by the amounts of tolerance and allowance. Classes identified by a numeral followed by the letters A and B are derived from certain Unified formulas (not shown here) in which the pitch diameter tolerances are based on increments of the basic major (nominal) diameter, the pitch, and the length of engagement. These formulas and the class identification or symbols apply to all of the Unified threads.

Classes 1A, 2A, and 3A apply to external threads only, and Classes 1B, 2B, and 3B apply to internal threads only. The disposition of the tolerances, allowances, and crest clearances for the various classes is illustrated on page 1992.

*Classes 2A and 2B:* Classes 2A and 2B are the most commonly used for general applications, including production of bolts, screws, nuts, and similar fasteners.

The maximum diameters of Class 2A (external) uncoated threads are less than basic by the amount of the allowance. The allowance minimizes galling and seizing in high-cycle wrench assembly, or it can be used to accommodate plated finishes or other coating. However, for threads with additive finish, the maximum diameters of Class 2A may be exceeded by the amount of the allowance; for example, the 2A maximum diameters apply to an unplated part or to a part before plating whereas the basic diameters (the 2A maximum diameter plus allowance) apply to a part after plating. The minimum diameters of Class 2B (internal) threads, whether or not plated or coated, are basic, affording no allowance or clearance in assembly at maximum metal limits.

*Class 2AG:* Certain applications require an allowance for rapid assembly to permit application of the proper lubricant or for residual growth due to high-temperature expansion. In these applications, when the thread is coated and the 2A allowance is not permitted to be consumed by such coating, the thread class symbol is qualified by G following the class symbol.

*Classes 3A and 3B:* Classes 3A and 3B may be used if closer tolerances are desired than those provided by Classes 2A and 2B. The maximum diameters of Class 3A (external) threads and the minimum diameters of Class 3B (internal) threads, whether or not plated or coated, are basic, affording no allowance or clearance for assembly of maximum metal components.

*Classes 1A and 1B:* Classes 1A and 1B threads replaced American National Class 1. These classes are intended for ordnance and other special uses. They are used on threaded components where quick and easy assembly is necessary and where a liberal allowance is required to permit ready assembly, even with slightly bruised or dirty threads.

Maximum diameters of Class 1A (external) threads are less than basic by the amount of the same allowance as applied to Class 2A. For the intended applications in American practice the allowance is not available for plating or coating. Where the thread is plated or coated, special provisions are necessary. The minimum diameters of Class 1B (internal) threads, whether or not plated or coated, are basic, affording no allowance or clearance for assembly with maximum metal external thread components having maximum diameters which are basic.

**Coated 60-Degree Threads.**—Although the Standard does not make recommendations for thicknesses of, or specify limits for coatings, it does outline certain principles that will aid mechanical interchangeability if followed whenever conditions permit.

To keep finished threads within the limits of size established in the Standard, external threads should not exceed basic size after plating and internal threads should not be below basic size after plating. This recommendation does not apply to threads coated by certain commonly used processes such as hot-dip galvanizing where it may not be required to maintain these limits.

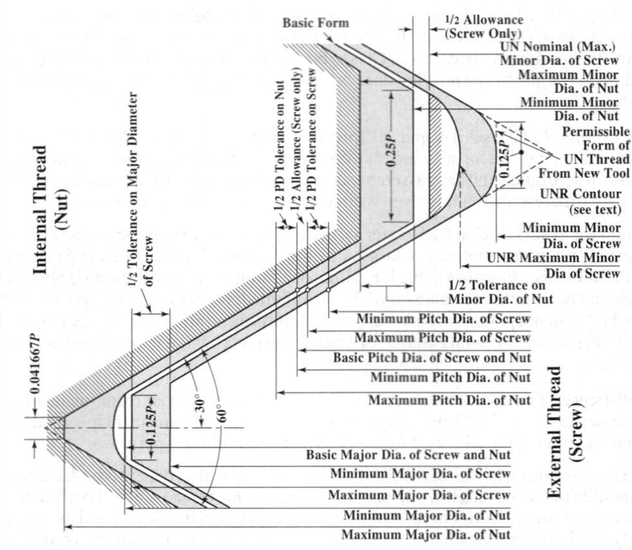


Limits of Size Showing Tolerances, Allowances (Neutral Space), and Crest Clearances for Unified Classes 1A, 2A, 1B, and 2B

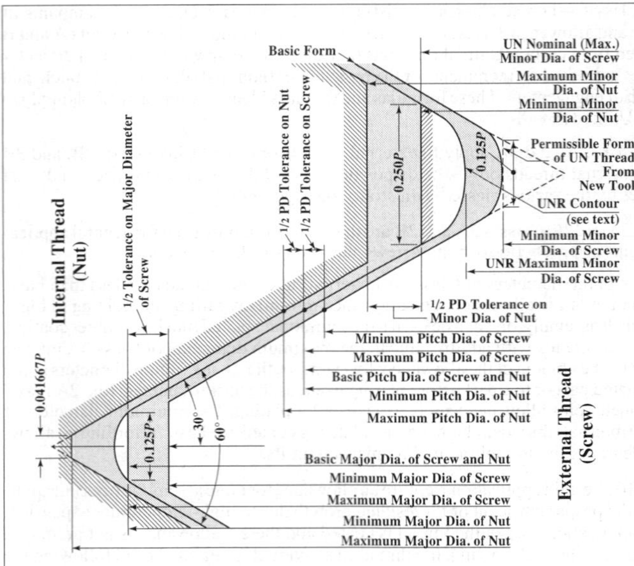


Limits of Size Showing Tolerances and Crest Clearances for Unified Classes 3A and 3B and American National Classes 2 and 3

Class 2A provides both a tolerance and an allowance. Many thread requirements call for coatings such as those deposited by electroplating processes and, in general, the 2A allowance provides adequate undercut for such coatings. There may be variations in thickness and symmetry of coating resulting from commercial processes but after plating the threads should be accepted by a basic Class 3A size GO gage and a Class 2A gage as a NOT-GO gage. Class 1A provides an allowance which is maintained for both coated and uncoated product, i.e., it is not available for coating.

Class 3A does not include an allowance so it is suggested that the limits of size before plating be reduced by the amount of the 2A allowance whenever that allowance is adequate.

No provision is made for overcutting internal threads as coatings on such threads are not generally required. Further, it is very difficult to deposit a significant thickness of coating on the flanks of internal threads. Where a specific thickness of coating is required on an internal thread, it is suggested that the thread be overcut so that the thread as coated will be accepted by a GO thread plug gage of basic size.

ANSI/ASME B1.1-2019 specifies limits of size that pertain whether threads are coated or uncoated. Only in Class 2A threads is an allowance available to accommodate coatings. Thus, in all classes of internal threads and in all Class 1A, 2AG, and 3A external threads, limits of size must be adjusted to provide suitable provision for the desired coating.

For further information concerning dimensional accommodation of coating or plating for 60-degree threads, see Section 7, ANSI/ASME B1.1-2019, Section 7.

**Screw Thread Selection—Combination of Classes.—**Whenever possible, selection should be made from Table 2, Standard Series Unified Screw Threads, preference being given to the Coarse- and Fine- thread Series. If threads in the standard series do not meet the requirements of design, reference should be made to the selected combinations in Table 3. The third expedient is to compute the limits of size from the tolerance tables or tolerance increment tables given in the Standard. The fourth and last resort is calculation by the formulas given in the Standard.

The requirements for screw thread fits for specific applications depend on end use and can be met by specifying the proper combinations of thread classes for the components. For example, a Class 2A external thread may be used with a Class 1B, 2B, or 3B internal thread.

**Pitch Diameter Tolerances, All Classes.—**The pitch diameter tolerances in Table 3 for all classes of the UNC, UNF, 4-UN, 6-UN, and 8-UN series are based on a length of engagement equal to the basic major (nominal) diameter and are applicable for lengths of engagement up to $1\frac{1}{2}$ diameters.

The pitch diameter tolerances used in Table 3 for all classes of the UNEF, 12-UN, 16-UN, 20-UN, 28-UN, and 32-UN series and the UNS series, are based on a length of engagement of 9 pitches and are applicable for lengths of engagement of from 5 to 15 pitches.

**Screw Thread Designation.—**The basic method of designating a screw thread is used where the standard tolerances or limits of size based on the standard length of engagement are applicable. The designation specifies in sequence the nominal size, number of threads per inch, thread series symbol, thread class symbol, and the gaging system number per ANSI/ASME B1.3. The nominal size is the basic major diameter and is specified as the fractional diameter, screw number, or their decimal equivalent. Where decimal equivalents are used for size callout, they shall be interpreted as being nominal size designations only and shall have no dimensional significance beyond the fractional size or number designation. The symbol LH is placed after the thread class symbol to indicate a left-hand thread:

*Examples:*

$\frac{1}{4}$-20 UNC-2A (21) or 0.250-20 UNC-2A (21)

10-32 UNF-2A (22) or 0.190-32 UNF-2A (22)

$\frac{7}{16}$-20 UNRF-2A (23) or 0.4375-20 UNRF-2A (23)

2-12 UN-2A(21) or 2.000-12 UN-2A(21)

¼-20 UNC-3A-LH (21) or 0.250-20 UNC-3A-LH (21)

For uncoated standard series threads these designations may optionally be supplemented by the addition of the pitch diameter limits of size.

*Examples:*

¼-20 UNC-2A(21)

PD 0.2164-0.2127 (Optional for uncoated threads)

**Designating Coated Threads.—**For coated (or plated) Class 2A external threads, the basic (max) major and basic (max) pitch diameters are given followed by the words AFTER COATING. The major and pitch diameter limits of size before coating are also given followed by the words BEFORE COATING.

| | | |
|---|---|---|
| *Example:* | ¾-10 UNC-2A (21) | |
| | [a]Major dia 0.7500 max<br>PD 0.6850 max | } AFTER COATING |
| | [b]Major dia 0.7482-0.7353<br>PD 0.6832-0.6773 | } BEFORE COATING |

[a] Major and PD values are equal to basic and correspond to those in Table 3 for Class 3A.

[b] Major and PD limits are those in Table 3 for Class 2A.

Certain applications require an allowance for rapid assembly, to permit application of a proper lubricant, or for residual growth due to high-temperature expansion. In such applications where the thread is to be coated and the 2A allowance is not permitted to be consumed by such coating, the thread class symbol is qualified by the addition of the letter G (symbol for allowance) following the class symbol, and the maximum major and maximum pitch diameters are reduced below basic size by the amount of the 2A allowance and followed by the words AFTER COATING. This arrangement ensures that the allowance is maintained. The major and pitch diameter limits of size before coating are also given followed by SPL and BEFORE COATING. For information concerning the designating of this and other special coating conditions reference should be made to ANSI/ASME B1.1-2019.

**Designating UNS Threads.—**UNS screw threads that have special combinations of diameter and pitch with tolerance to Unified formulation have the basic form designation set out first followed always by the limits of size.

**Designating Multiple Start Threads.—**If a screw thread is of multiple start, it is designated by specifying in sequence the nominal size, pitch (in decimals or threads per inch) and lead (in decimals or fractions).

**Other Special Designations.—**For other special designations including threads with modified limits of size or with special lengths of engagement, reference should be made to ANSI/ASME B1.1-2019.

**Hole Sizes for Tapping.—**Hole size limits for tapping Classes 1B, 2B, and 3B threads of various lengths of engagement are given in Table 3 on page 2162.

**Internal Thread Minor Diameter Tolerances.—**Internal thread minor diameter tolerances in Table 3 are based on a length of engagement equal to the nominal diameter. For general applications these tolerances are suitable for lengths of engagement up to 1½ diameters. However, some thread applications have lengths of engagement which are greater than 1½ diameters or less than the nominal diameter. For such applications it may be advantageous to increase or decrease the tolerance, respectively, as explained in the Tapping Section.

## American National Standard for Unified Miniature Screw Threads

American National Standard ASA B1.10-1958, R1988 introduces a new series to be known as Unified Miniature Screw Threads and intended for general purpose fastening screws and similar uses in watches, instruments, and miniature mechanisms. Use of this series is recommended on all new products in place of the many improvised and unsystematized sizes now in existence which have never achieved broad acceptance nor recognition by standardization bodies. The series covers a diameter range from 0.30 to 1.40 millimeters (0.0118 to 0.0551 inch) and thus supplements the Unified and American thread series which begins at 0.060 inch (number 0 of the machine screw series). It comprises a total of fourteen sizes which, together with their respective pitches, are those endorsed by the American-British-Canadian Conference of April 1955 as the basis for a Unified standard among the inch-using countries, and coincide with the corresponding range of sizes in ISO (International Organization for Standardization) Recommendation No. 68. Additionally, it utilizes thread forms which are compatible in all significant respects with both the Unified and ISO basic thread profiles. Thus, threads in this series are interchangeable with the corresponding sizes in both the American-British-Canadian and ISO Standards.

**Basic Form of Thread.—**The basic profile by which the design forms of the threads covered by this Standard are governed is shown in Table 1. The thread angle is 60 degrees and except for basic height and depth of engagement which are 0.52*p*, instead of 0.54127*p*, the basic profile for this thread standard is identical with the Unified and American basic thread form. The selection of 0.52 as the exact value of the coefficient for the height of this basic form is based on practical manufacturing considerations and a plan to simplify calculations and achieve more precise agreement between metric and inch dimensional tables.

Products made to this Standard will be interchangeable with products made to other standards that allow a maximum depth of engagement (or combined addendum height) of 0.54127*p*. The resulting difference is negligible (only 0.00025 inch for the coarsest pitch) and is completely offset by practical considerations in tapping, since internal thread heights exceeding 0.52*p* are avoided in these (Unified Miniature) small thread sizes in order to reduce excessive tap breakage.

**Design Forms of Threads.—**The design (maximum material) forms of the external and internal threads are shown in Table 2. These forms are derived from the basic profile shown in Table 1 by the application of clearances for the crests of the addenda at the roots of the mating dedendum forms. Basic and design form dimensions are given in Table 3.

*Nominal Sizes:* The thread sizes comprising this series and their respective pitches are shown in the first two columns of Table 5. The fourteen sizes shown in Table 5 have been systematically distributed to provide a uniformly proportioned selection over the entire range. They are separated alternately into two categories: The sizes shown in bold type are selections made in the interest of simplification and are those to which it is recommended that usage be confined wherever the circumstances of design permit. Where these sizes do not meet requirements the intermediate sizes shown in light type are available.

**Table 1. Unified Miniature Screw Threads—Basic Thread Form**

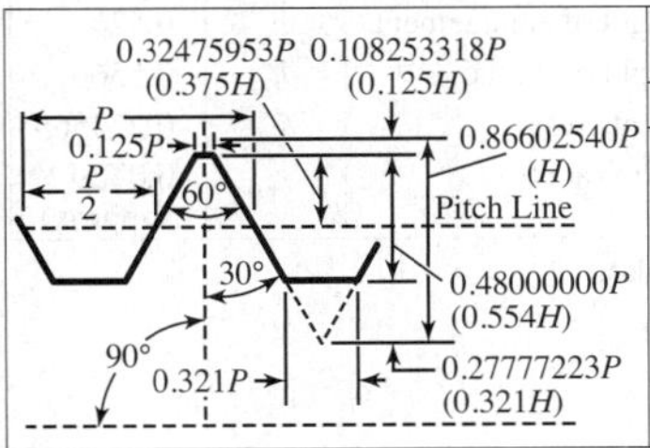


| Formulas for Basic Thread Form<br>Metric units (millimeters) used in formulas | | |
|---|---|---|
| Thread Element | Symbol | Formula |
| Angle of thread | 2α | 60° |
| Half angle of thread | α | 30° |
| Pitch of thread | *p* | |
| No. of threads per inch | *n* | 25.4/*p* |
| Height of sharp V thread | *H* | 0.86603*p* |
| Addendum of basic thread | $h_{ab}$ | 0.32476*p* |
| Height of basic thread | $h_b$ | 0.52*p* |

**Table 2. Unified Miniature Screw Threads—Design Thread Form**

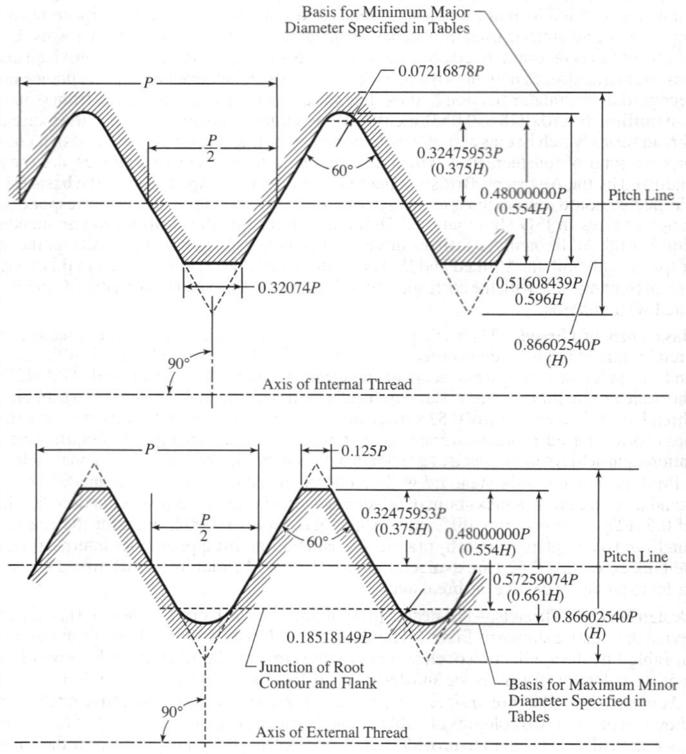


Formulas for Design Thread Form (maximum material)[a]

| External Thread | | | Internal Thread | | |
|---|---|---|---|---|---|
| Thread Element | Symbol | Formula | Thread Element | Symbol | Formula |
| Addendum | $h_{as}$ | 0.32476*p* | Height of engagement | $h_e$ | 0.52*p* |
| Height | $h_s$ | 0.60*p* | Height of thread | $h_n$ | 0.556*p* |
| Flat at crest | $F_{cs}$ | 0.125*p* | Flat at crest | $F_{cn}$ | 0.27456*p* |
| Radius at root | $r_{rs}$ | 0.158*p* (approx) | Radius at root | $r_{rn}$ | 0.072*p* (approx) |

[a] Metric units (millimeters) are used in all formulas.

## Table 3. Unified Miniature Screw Threads—Basic and Design Form Dimensions

| Basic Thread Form | | | | | External Thread Design Form | | | Internal Thread Design Form | | |
|---|---|---|---|---|---|---|---|---|---|---|
| Threads per inch $n^a$ | Pitch $p$ | Height of Sharp V $H = 0.86603p$ | Height $h_b = 0.52p$ | Addendum $h_{ab} = h_{as} = 0.32476p$ | Height $h_s = 0.60p$ | Flat at Crest $F_{cs} = 0.125p$ | Radius at Root $r_{rs} = 0.158p$ | Height $h_n = 0.556p$ | Flat at Crest $F_{cn} = 0.27456p$ | Radius at Root $r_{rn} = 0.072p$ |
| **Millimeter Dimensions** | | | | | | | | | | |
| … | .080 | .0693 | .0416 | .0260 | .048 | .0100 | .0126 | .0445 | .0220 | .0058 |
| … | .090 | .0779 | .0468 | .0292 | .054 | .0112 | .0142 | .0500 | .0247 | .0065 |
| … | .100 | .0866 | .0520 | .0325 | .060 | .0125 | .0158 | .0556 | .0275 | .0072 |
| … | .125 | .1083 | .0650 | .0406 | .075 | .0156 | .0198 | .0695 | .0343 | .0090 |
| … | .150 | .1299 | .0780 | .0487 | .090 | .0188 | .0237 | .0834 | .0412 | .0108 |
| … | .175 | .1516 | .0910 | .0568 | .105 | .0219 | .0277 | .0973 | .0480 | .0126 |
| … | .200 | .1732 | .1040 | .0650 | .120 | .0250 | .0316 | .1112 | .0549 | .0144 |
| … | .225 | .1949 | .1170 | .0731 | .135 | .0281 | .0356 | .1251 | .0618 | .0162 |
| … | .250 | .2165 | .1300 | .0812 | .150 | .0312 | .0395 | .1390 | .0686 | .0180 |
| … | .300 | .2598 | .1560 | .0974 | .180 | .0375 | .0474 | .1668 | .0824 | .0216 |
| **Inch Dimensions** | | | | | | | | | | |
| $317\frac{1}{2}$ | .003150 | .00273 | .00164 | .00102 | .00189 | .00039 | .00050 | .00175 | .00086 | .00023 |
| $282\frac{2}{9}$ | .003543 | .00307 | .00184 | .00115 | .00213 | .00044 | .00056 | .00197 | .00097 | .00026 |
| 254 | .003937 | .00341 | .00205 | .00128 | .00236 | .00049 | .00062 | .00219 | .00108 | .00028 |
| $203\frac{1}{5}$ | .004921 | .00426 | .00256 | .00160 | .00295 | .00062 | .00078 | .00274 | .00135 | .00035 |
| $169\frac{1}{3}$ | .005906 | .00511 | .00307 | .00192 | .00354 | .00074 | .00093 | .00328 | .00162 | .00043 |
| $145\frac{1}{7}$ | .006890 | .00597 | .00358 | .00224 | .00413 | .00086 | .00109 | .00383 | .00189 | .00050 |
| 127 | .007874 | .00682 | .00409 | .00256 | .00472 | .00098 | .00124 | .00438 | .00216 | .00057 |
| $112\frac{8}{9}$ | .008858 | .00767 | .00461 | .00288 | .00531 | .00111 | .00140 | .00493 | .00243 | .00064 |
| $101\frac{3}{5}$ | .009843 | .00852 | .00512 | .00320 | .00591 | .00123 | .00156 | .00547 | .00270 | .00071 |
| $84\frac{2}{3}$ | .011811 | .01023 | .00614 | .00384 | .00709 | .00148 | .00187 | .00657 | .00324 | .00085 |

[a] In Table 5 and Table 6 these values are shown rounded to the nearest whole number.

## Table 4. Unified Miniature Screw Threads—Formulas for Basic and Design Dimensions and Tolerances

| Formulas for Basic Dimensions | |
|---|---|
| D = Basic Major Diameter and Nominal Size in millimeters; $p$ = Pitch in millimeters; $E$ = Basic Pitch Diameter in millimeters = $D - 0.64952p$; and $K$ = Basic Minor Diameter in millimeters = $D - 1.04p$ | |
| Formulas for Design Dimensions (Maximum Material) | |
| **External Thread** | **Internal Thread** |
| $D_s$ = Major Diameter = $D$ | $D_n$ = Major Diameter = $D + 0.072p$ |
| $E_s$ = Pitch Diameter = $E$ | $E_n$ = Pitch Diameter = $E$ |
| $K_s$ = Minor Diameter = $D - 1.20p$ | $K_n$ = Minor Diameter = $K$ |
| Formulas for Tolerances on Design Dimensions[a] | |
| **External Thread (−)** | **Internal Thread (+)** |
| Major Diameter Tol., $0.12p + 0.006$ | [b]Major Diameter Tol., $0.168p + 0.008$ |
| Pitch Diameter Tol., $0.08p + 0.008$ | Pitch Diameter Tol., $0.08p + 0.008$ |
| [c]Minor Diameter Tol., $0.16p + 0.008$ | Minor Diameter Tol., $0.32p + 0.012$ |

[a] These tolerances are based on lengths of engagement of $\frac{2}{3}D$ to $1\frac{1}{2}D$.

[b] This tolerance establishes the maximum limit of the major diameter of the internal thread. In practice, this limit is applied to the threading tool (tap) and not gaged on the product. Values for this tolerance are, therefore, not given in Table 5.

[c] This tolerance establishes the minimum limit of the minor diameter of the external thread. In practice, this limit is applied to the threading tool and only gaged on the product in confirming new tools. Values for this tolerance are, therefore, not given in Table 5.

Metric units (millimeters) apply in all formulas. Inch tolerances are not derived by direct conversion of the metric values. They are the differences between the rounded off limits of size in inch units.

## Table 5. Unified Miniature Screw Threads—Limits of Size and Tolerances

| Size Designation[a] | Pitch | External Threads | | | | | | Internal Threads | | | | | | Lead Angle at Basic Pitch Dia. | | Sectional Area at Minor Dia. at D−1.28p |
|---|---|---|---|---|---|---|---|---|---|---|---|---|---|---|---|---|
| | | Major Dia. | | Pitch Dia. | | Minor Dia. | | Minor Dia. | | Pitch Dia. | | Major Dia. | | | | |
| | | Max[b] | Min | Max[b] | Min | Max[c] | Min[d] | Min[b] | Max | Min[b] | Max | Min[e] | Max[d] | | | |
| | mm | mm | mm | mm | mm | mm | mm | mm | mm | mm | mm | mm | mm | deg | min | sq mm |
| **0.30 UNM** | **0.080** | **0.300** | **0.284** | **0.248** | **0.234** | **0.204** | **0.183** | **0.217** | **0.254** | **0.248** | **0.262** | **0.306** | **0.327** | **5** | **52** | **0.0307** |
| 0.35 UNM | 0.090 | 0.350 | 0.333 | 0.292 | 0.277 | 0.242 | 0.220 | 0.256 | 0.297 | 0.292 | 0.307 | 0.356 | 0.380 | 5 | 37 | 0.0433 |
| **0.40 UNM** | **0.100** | **0.400** | **0.382** | **0.335** | **0.319** | **0.280** | **0.256** | **0.296** | **0.340** | **0.335** | **0.351** | **0.407** | **0.432** | **5** | **26** | **0.0581** |
| 0.45 UNM | 0.100 | 0.450 | 0.432 | 0.385 | 0.369 | 0.330 | 0.306 | 0.346 | 0.390 | 0.385 | 0.401 | 0.457 | 0.482 | 4 | 44 | 0.0814 |
| **0.50 UNM** | **0.125** | **0.500** | **0.479** | **0.419** | **0.401** | **0.350** | **0.322** | **0.370** | **0.422** | **0.419** | **0.437** | **0.509** | **0.538** | **5** | **26** | **0.0908** |
| 0.55 UNM | 0.125 | 0.550 | 0.529 | 0.469 | 0.451 | 0.400 | 0.372 | 0.420 | 0.472 | 0.469 | 0.487 | 0.559 | 0.588 | 4 | 51 | 0.1195 |
| **0.60 UNM** | **0.150** | **0.600** | **0.576** | **0.503** | **0.483** | **0.420** | **0.388** | **0.444** | **0.504** | **0.503** | **0.523** | **0.611** | **0.644** | **5** | **26** | **0.1307** |
| 0.70 UNM | 0.175 | 0.700 | 0.673 | 0.586 | 0.564 | 0.490 | 0.454 | 0.518 | 0.586 | 0.586 | 0.608 | 0.713 | 0.750 | 5 | 26 | 0.1780 |
| **0.80 UNM** | **0.200** | **0.800** | **0.770** | **0.670** | **0.646** | **0.560** | **0.520** | **0.592** | **0.668** | **0.670** | **0.694** | **0.814** | **0.856** | **5** | **26** | **0.232** |
| 0.90 UNM | 0.225 | 0.900 | 0.867 | 0.754 | 0.728 | 0.630 | 0.586 | 0.666 | 0.750 | 0.754 | 0.780 | 0.916 | 0.962 | 5 | 26 | 0.294 |
| **1.00 UNM** | **0.250** | **1.000** | **0.964** | **0.838** | **0.810** | **0.700** | **0.652** | **0.740** | **0.832** | **0.838** | **0.866** | **1.018** | **1.068** | **5** | **26** | **0.363** |
| 1.10 UNM | 0.250 | 1.100 | 1.064 | 0.938 | 0.910 | 0.800 | 0.752 | 0.840 | 0.932 | 0.938 | 0.966 | 1.118 | 1.168 | 4 | 51 | 0.478 |
| **1.20 UNM** | **0.250** | **1.200** | **1.164** | **1.038** | **1.010** | **0.900** | **0.852** | **0.940** | **1.032** | **1.038** | **1.066** | **1.218** | **1.268** | **4** | **23** | **0.608** |
| 1.40 UNM | 0.300 | 1.400 | 1.358 | 1.205 | 1.173 | 1.040 | 0.984 | 1.088 | 1.196 | 1.205 | 1.237 | 1.422 | 1.480 | 4 | 32 | 0.811 |
| | Thds. per in. | inch | inch | inch | inch | inch | inch | inch | inch | inch | inch | inch | inch | deg | min | sq in |
| **0.30 UNM** | **318** | **0.0118** | **0.0112** | **0.0098** | **0.0092** | **0.0080** | **0.0072** | **0.0085** | **0.0100** | **0.0098** | **0.0104** | **0.0120** | **0.0129** | **5** | **52** | **0.0000475** |
| 0.35 UNM | 282 | 0.0138 | 0.0131 | 0.0115 | 0.0109 | 0.0095 | 0.0086 | 0.0101 | 0.0117 | 0.0115 | 0.0121 | 0.0140 | 0.0149 | 5 | 37 | 0.0000671 |
| **0.40 UNM** | **254** | **0.0157** | **0.0150** | **0.0132** | **0.0126** | **0.0110** | **0.0101** | **0.0117** | **0.0134** | **0.0132** | **0.0138** | **0.0160** | **0.0170** | **5** | **26** | **0.0000901** |
| 0.45 UNM | 254 | 0.0177 | 0.0170 | 0.0152 | 0.0145 | 0.0130 | 0.0120 | 0.0136 | 0.0154 | 0.0152 | 0.0158 | 0.0180 | 0.0190 | 4 | 44 | 0.0001262 |
| **0.50 UNM** | **203** | **0.0197** | **0.0189** | **0.0165** | **0.0158** | **0.0138** | **0.0127** | **0.0146** | **0.0166** | **0.0165** | **0.0172** | **0.0200** | **0.0212** | **5** | **26** | **0.0001407** |
| 0.55 UNM | 203 | 0.0217 | 0.0208 | 0.0185 | 0.0177 | 0.0157 | 0.0146 | 0.0165 | 0.0186 | 0.0185 | 0.0192 | 0.0220 | 0.0231 | 4 | 51 | 0.0001852 |
| **0.60 UNM** | **169** | **0.0236** | **0.0227** | **0.0198** | **0.0190** | **0.0165** | **0.0153** | **0.0175** | **0.0198** | **0.0198** | **0.0206** | **0.0240** | **0.0254** | **5** | **26** | **0.000203** |
| 0.70 UNM | 145 | 0.0276 | 0.0265 | 0.0231 | 0.0222 | 0.0193 | 0.0179 | 0.0204 | 0.0231 | 0.0231 | 0.0240 | 0.0281 | 0.0295 | 5 | 26 | 0.000276 |
| **0.80 UNM** | **127** | **0.0315** | **0.0303** | **0.0264** | **0.0254** | **0.0220** | **0.0205** | **0.0233** | **0.0263** | **0.0264** | **0.0273** | **0.0321** | **0.0337** | **5** | **26** | **0.000360** |
| 0.90 UNM | 113 | 0.0354 | 0.0341 | 0.0297 | 0.0287 | 0.0248 | 0.0231 | 0.0262 | 0.0295 | 0.0297 | 0.0307 | 0.0361 | 0.0379 | 5 | 26 | 0.000456 |
| **1.00 UNM** | **102** | **0.0394** | **0.0380** | **0.0330** | **0.0319** | **0.0276** | **0.0257** | **0.0291** | **0.0327** | **0.0330** | **0.0341** | **0.0401** | **0.0420** | **5** | **26** | **0.000563** |
| 1.10 UNM | 102 | 0.0433 | 0.0419 | 0.0369 | 0.0358 | 0.0315 | 0.0296 | 0.0331 | 0.0367 | 0.0369 | 0.0380 | 0.0440 | 0.0460 | 4 | 51 | 0.000741 |
| **1.20 UNM** | **102** | **0.0472** | **0.0458** | **0.0409** | **0.0397** | **0.0354** | **0.0335** | **0.0370** | **0.0406** | **0.0409** | **0.0420** | **0.0480** | **0.0499** | **4** | **23** | **0.000943** |
| 1.40 UNM | 85 | 0.0551 | 0.0535 | 0.0474 | 0.0462 | 0.0409 | 0.0387 | 0.0428 | 0.0471 | 0.0474 | 0.0487 | 0.0560 | 0.0583 | 4 | 32 | 0.001257 |

[a] Sizes shown in bold type are preferred.

[b] This is also the basic dimension.

[c] This limit, in conjunction with root form shown in Table 2, is advocated for use when optical projection methods of gaging are employed. For mechanical gaging the minimum minor diameter of the internal thread is applied.

[d] This limit is provided for reference only. In practice, the form of the threading tool is relied upon for this limit.

[e] This limit is provided for reference only, and is not gaged. For gaging, the maximum major diameter of the external thread is applied.

**Table 6. Unified Miniature Screw Threads— Minimum Root Flats for External Threads**

| Pitch | No. of Threads Per Inch | Thread Height for Min. Flat at Root 0.64$p$ | | Minimum Flat at Root $F_{rs} = 0.136p$ | |
|---|---|---|---|---|---|
| mm | | mm | Inch | mm | Inch |
| 0.080 | 318 | 0.0512 | 0.00202 | 0.0109 | 0.00043 |
| 0.090 | 282 | 0.0576 | 0.00227 | 0.0122 | 0.00048 |
| 0.100 | 254 | 0.0640 | 0.00252 | 0.0136 | 0.00054 |
| 0.125 | 203 | 0.0800 | 0.00315 | 0.0170 | 0.00067 |
| 0.150 | 169 | 0.0960 | 0.00378 | 0.0204 | 0.00080 |
| 0.175 | 145 | 0.1120 | 0.00441 | 0.0238 | 0.00094 |
| 0.200 | 127 | 0.1280 | 0.00504 | 0.0272 | 0.00107 |
| 0.225 | 113 | 0.1440 | 0.00567 | 0.0306 | 0.00120 |
| 0.250 | 102 | 0.1600 | 0.00630 | 0.0340 | 0.00134 |
| 0.300 | 85 | 0.1920 | 0.00756 | 0.0408 | 0.00161 |

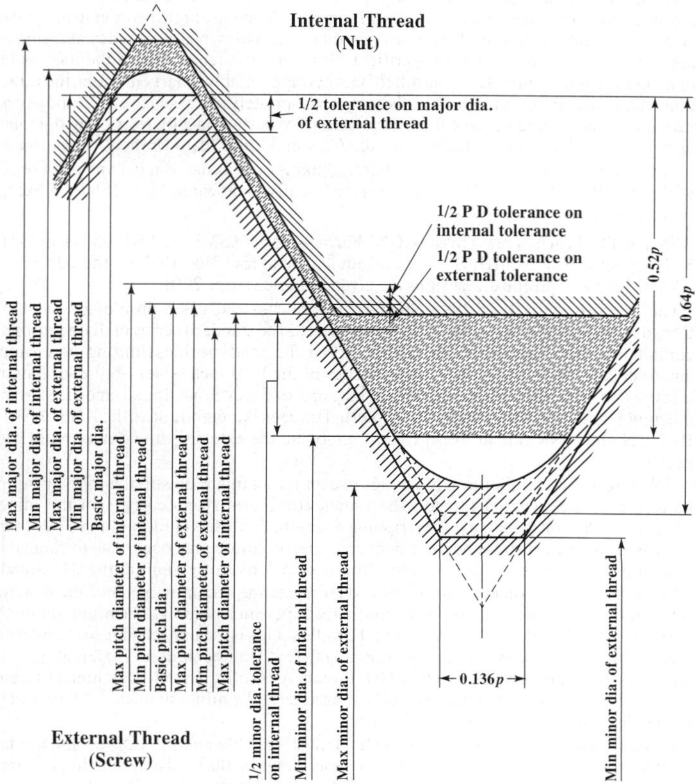


Limits of Size Showing Tolerances and Crest Clearances for UNM Threads

*Limits of Size:* Formulas used to determine limits of size are given in Table 4; the limits of size are given in Table 5. The diagram on page 1999 illustrates the limits of size and Table 6 gives values for the minimum flat at the root of the external thread shown on the diagram.

*Classes of Threads:* The Standard establishes one class of thread with zero allowance on all diameters. When coatings of a measurable thickness are required, they should be included within the maximum material limits of the threads since these limits apply to both coated and uncoated threads.

*Hole Sizes for Tapping:* Suggested hole sizes are given in the Tapping Section.

### Unified Screw Threads of UNJ Basic Profile

**British Standard UNJ Threads.**—This British Standard BS 4084:1996 (Withdrawn) arises from a request originating from within the British aircraft industry and is based upon specifications for Unified screw threads and American military Standard MIL-S-8879.

These UNJ threads, having an enlarged root radius, were introduced for applications requiring high fatigue strength where working stress levels are high, in order to minimize size and weight, as in aircraft engines, airframes, missiles, space vehicles and similar designs where size and weight are critical. To meet these requirements the root radius of external Unified threads is controlled between appreciably enlarged limits, the minor diameter of the mating internal threads being appropriately increased to ensure the necessary clearance. The requirement for high strength is further met by restricting the tolerances for UNJ threads to the highest classes, Classes 3A and 3B, of Unified screw threads.

The Standard, not described further here, contains both a coarse and a fine pitch series of threads. BS 4084:1996 (Withdrawn) is technically identical to ISO 3161-1977 except for Appendix A.

**ASME Unified Inch Screw Threads, UNJ Form.**—ANSI/ASME B1.15-1995 is similar to Military Specification MIL-S-8879, and equivalent to ISO 3161-1977 for thread Classes 3A and 3B. Basic profile dimensions are given in Table 1, page 2001.

The ANSI/ASME B1.15-1995 Standard establishes the basic profile for the UNJ thread form, specifies a system of designation, lists the standard series of diameter-pitch combinations for diameters from 0.060 to 6.00 inches, and specifies limiting dimensions and tolerances. It specifies the characteristics of the UNJ inch series of threads having $0.15011P$ to $0.18042P$ designated radius at the root of the external thread, and also having the minor diameter of the external and internal threads increased above the ANSI/ASME B1.1 UN and UNR thread forms to accommodate the external thread maximum root radius.

UNJ threads are similar to UN threads except for a large radius in the root, or minor diameter, of the external thread. The radius eliminates sharp corners in the minor diameter of the bolt to increase the stripping strength. The fillets or radius in sharp corners increases strength at stress points where cracking or failure may occur due to change in temperature, heavy loads, or vibration. Other dimensions are the same as the UN thread.

Because the radius on the external thread increases the minor diameter of the bolt, the internal thread, or nut, is modified accordingly to permit assembly. The minor diameter of the internal thread is enlarged to clear the radius. This is the only change to the internal thread. All other dimensions are the same as standard Unified threads. Different types of tap drill sizes are required to produce UNJ thread. All tooling for external threads, thread rolls, and chasers must be made to produce a radius at the minor diameter. All runout or incomplete threads shall have a radius also.

Thread conforming to the ANSI/ASME B1.1 UN profile and the UNJ profile are not interchangeable because of possible interference between the UNJ external thread minor diameter and the UN internal thread minor diameter. However, the UNJ internal thread will assemble with the UN external thread.

**Table 1. Basic Profile Dimensions UNJ Threads** *ANSI/ASME B1.15-1995*

| Threads per inch $n$ | Pitch $P = 1/n$ | Pitch Line $0.5P$ | Flat at Internal Thread Crest $0.3125P$ | Flat at Internal Thread Root and External Thread Crest $0.125P$ | Height of Sharp V Thread $H = 0.866025P$ | Height of Internal Thread and Depth of Thread Engagement $0.5625H = 0.487139P$ | Addendum of External Thread $0.375H = 0.324760P$ | Truncation of Internal Thread Crest $0.3125H = 0.270633P$ | Truncation of Internal Thread Root and External Thread Crest $0.125H = 0.108253P$ | Half Addendum of External Thread (REF. ONLY) $0.1875H = 0.16238P$ |
|---|---|---|---|---|---|---|---|---|---|---|
| 80 | 0.012500 | 0.006250 | 0.00391 | 0.00156 | 0.010825 | 0.00609 | 0.00406 | 0.00338 | 0.00135 | 0.00203 |
| 72 | 0.013889 | 0.006944 | 0.00434 | 0.00174 | 0.012028 | 0.00677 | 0.00451 | 0.00376 | 0.00150 | 0.00226 |
| 64 | 0.015625 | 0.007813 | 0.00488 | 0.00195 | 0.013532 | 0.00761 | 0.00507 | 0.00423 | 0.00169 | 0.00254 |
| 56 | 0.017857 | 0.008929 | 0.00558 | 0.00223 | 0.015465 | 0.00870 | 0.00580 | 0.00483 | 0.00193 | 0.00290 |
| 48 | 0.020833 | 0.010417 | 0.00651 | 0.00260 | 0.018042 | 0.01015 | 0.00677 | 0.00564 | 0.00226 | 0.00338 |
| 44 | 0.022727 | 0.011364 | 0.00710 | 0.00284 | 0.019682 | 0.01107 | 0.00738 | 0.00615 | 0.00246 | 0.00369 |
| 40 | 0.025000 | 0.012500 | 0.00781 | 0.00313 | 0.021651 | 0.01218 | 0.00812 | 0.00677 | 0.00271 | 0.00406 |
| 36 | 0.027778 | 0.013889 | 0.00868 | 0.00347 | 0.024056 | 0.01353 | 0.00902 | 0.00752 | 0.00301 | 0.00451 |
| 32 | 0.031250 | 0.015625 | 0.00977 | 0.00391 | 0.027063 | 0.01522 | 0.01015 | 0.00846 | 0.00338 | 0.00507 |
| 28 | 0.035714 | 0.017857 | 0.01116 | 0.00446 | 0.030929 | 0.01740 | 0.01160 | 0.00967 | 0.00387 | 0.00580 |
| 24 | 0.041667 | 0.020833 | 0.01302 | 0.00521 | 0.036084 | 0.02030 | 0.01353 | 0.01128 | 0.00451 | 0.00677 |
| 20 | 0.050000 | 0.025000 | 0.01563 | 0.00625 | 0.043301 | 0.02436 | 0.01624 | 0.01353 | 0.00541 | 0.00812 |
| 18 | 0.055556 | 0.027778 | 0.01736 | 0.00694 | 0.048113 | 0.02706 | 0.01804 | 0.01504 | 0.00601 | 0.00902 |
| 16 | 0.062500 | 0.031250 | 0.01953 | 0.00781 | 0.054127 | 0.03045 | 0.02030 | 0.01691 | 0.00677 | 0.01015 |
| 14 | 0.071429 | 0.035714 | 0.02232 | 0.00893 | 0.061859 | 0.03480 | 0.02320 | 0.01933 | 0.00773 | 0.01160 |
| 12 | 0.083333 | 0.041667 | 0.02604 | 0.01042 | 0.072169 | 0.04059 | 0.02706 | 0.02255 | 0.00902 | 0.01353 |
| 11 | 0.090909 | 0.045455 | 0.02841 | 0.01136 | 0.078730 | 0.04429 | 0.02952 | 0.02460 | 0.00984 | 0.01476 |
| 10 | 0.100000 | 0.050000 | 0.03125 | 0.01250 | 0.086603 | 0.04871 | 0.03248 | 0.02706 | 0.01083 | 0.01624 |
| 9 | 0.111111 | 0.055556 | 0.03472 | 0.01389 | 0.096225 | 0.05413 | 0.03608 | 0.03007 | 0.01203 | 0.01804 |
| 8 | 0.125000 | 0.062500 | 0.03906 | 0.01563 | 0.108253 | 0.06089 | 0.04060 | 0.03383 | 0.01353 | 0.02030 |
| 7 | 0.142857 | 0.071429 | 0.04464 | 0.01786 | 0.123718 | 0.06959 | 0.04639 | 0.03866 | 0.01546 | 0.02320 |
| 6 | 0.166667 | 0.083333 | 0.05208 | 0.02083 | 0.144338 | 0.08119 | 0.05413 | 0.04511 | 0.01804 | 0.02706 |
| 5 | 0.200000 | 0.100000 | 0.06250 | 0.02500 | 0.173205 | 0.09743 | 0.06495 | 0.05413 | 0.02165 | 0.03248 |
| 4.5 | 0.222222 | 0.111111 | 0.06944 | 0.02778 | 0.192450 | 0.10825 | 0.07217 | 0.06014 | 0.02406 | 0.03608 |
| 4 | 0.250000 | 0.125000 | 0.07813 | 0.03125 | 0.216506 | 0.12178 | 0.08119 | 0.06766 | 0.02706 | 0.04060 |

## CALCULATING THREAD DIMENSIONS

### Introduction

The purpose of the ANSI/ASME B1.30 Standard is to establish uniform and specific practices for calculating and rounding the numeric values used for inch and metric screw thread design data dimensions only. No attempt has been made to establish a policy of rounding actual thread characteristics measured by the manufacturer or user of thread gages. Covered is the Standard Rounding Policy* regarding the last figure or decimal place to be retained by a numeric value and the number of decimal places to be retained by values used in intermediate calculations of thread design data dimensions. Values calculated to ANSI/ASME B1.30 for inch and metric screw thread design data dimensions may vary slightly from values shown in previously issued ASME B1 screw thread standards and are to take precedence in all new or future revisions of ASME B1 standards as applicable except as noted in following paragraph.

**Metric Application.—**Allowances (fundamental deviations) and tolerances for metric M and MJ screw threads are based upon formulas which appear in applicable standards. Values of allowances for standard tolerance positions and values of tolerances for standard tolerance grades are tabulated in these standards for a selection of pitches. Rounding rules specified in ANSI/ASME B1.30 have not been applied to these values but have followed practices of the International Organization for Standardization (ISO). For pitches which are not included in the tables, standard formulas and the rounding rules specified herein are applicable.

ISO rounding practices, for screw thread tolerances and allowances, use rounding to the nearest values in the R40 series of numbers in accordance with ISO 3 (see page 690). In some cases, the rounded values have been adjusted to produce a smooth progression. Since the ISO rounded values have been standardized internationally for metric screw threads, it would lead to confusion if tolerances and allowances were recalculated using B1.30 rules for use in the United States. The B1.30 rounding rules are, therefore, only applicable to special threads where tabulated values do not exist in ISO standards. Values calculated using the ISO R40 series values may differ from those calculated using B1.30. In such a case the special thread values generated using B1.30 take precedence.

**Purpose.—**Thread dimensions calculated from published formulas frequently may not yield the exact values published in the standards. The difference in most cases are due to rounding policy.

ANSI/ASME B1.30 specifies that pitch, *P,* values shall be rounded to eight decimal places. In Example 1 that follows on page 2004, the pitch of 28 threads per inch, 0.03571429, is correct; using $\frac{1}{28}$ or 0.0357 or 0.0357142856 instead of 0.03571429 will not produce values that conform to values calculated according this Standard.

The rounding rules specified by the Standard are not uniform, and vary by feature. Pitch is held to eight decimal places, maximum major diameter to four decimal places, and tolerances to six decimal places. In order to maintain the same screw dimensions, everybody has to follow the same rounding practice.

The basic profile of UN and UNF screw threads are shown in Fig. 1. Examples given on the following pages, including the detailed calculations of UNEF and UNS external and internal threads, are intended to provide a basis for better understanding of the formulas, the calculations, and the rounding process.

* It is recognized that ANSI/ASME B1.30 is not in agreement with other published documents, e.g., ASME SI-9, "Guide for Metrication of Codes and Standards SI (Metric) Units," and IEEE/ASTM SI-10, "Standard for Metric Practice." The rounding practices used in the forenamed documents are designed to produce even distribution of numerical values. The purpose of this document is to define the most practical and commonly used method of rounding numerical thread form values. Application of this method is far more practical in the rounding of thread form values.

## Calculating and Rounding Dimensions

**Rounding of Decimal Values.—**The following rounding practice represents the method to be used in new or future revisions of ASME B1 thread standards.

*Rounding Policy:* When the figure next beyond the last figure or place retained is less than 5, the figure in the last place retained is kept unchanged.

*Example:*

| | |
|---|---|
| 1.012342 | 1.01234 |
| 1.012342 | 1.0123 |
| 1.012342 | 1.012 |

When the figure next beyond the last figure or place retained is greater than 5, the figure in the last place retained is increased by 1.

*Example:*

| | |
|---|---|
| 1.56789 | 1.5679 |
| 1.56789 | 1.568 |
| 1.56789 | 1.57 |

When the figure next beyond the last figure or place retained is 5, and:

1) There are no figures, or only zeros, beyond the 5, the last figure should be increased by 1.

*Example:*

| | |
|---|---|
| 1.01235 | 1.0124 |
| 1.0123500 | 1.0124 |
| 1.012345 | 1.01235 |
| 1.01234500 | 1.01235 |

2) If the 5 next beyond the figure in the last place to be retained is followed by any figures other than zero, the figure in the last place retained should be increased by 1.

*Example:*

| | |
|---|---|
| 1.0123501 | 1.0124 |
| 1.0123599 | 1.0124 |
| 1.01234501 | 1.01235 |
| 1.01234599 | 1.01235 |

The final rounded value is obtained from the most precise value available and not from a series of successive rounding. For example, 0.5499 should be rounded to 0.550, 0.55 and 0.5 (not 0.6), since the most precise value available is less than 0.55. Similarly, 0.5501 should be rounded as 0.550, 0.55 and 0.6, since the most precise value available is more than 0.55. In the case of 0.5500 rounding should be 0.550, 0.55 and 0.6, since the most precise value available is 0.5500.

**Calculations from Formulas, General Rules.—**1) Values for pitch and constants derived from a function of pitch are used out to eight decimal places for inch series. The eight place values are obtained by rounding their truncated ten place values.

Seven decimal place values for metric series constants are derived by rounding their truncated nine place values.

Values used in intermediate calculations are rounded to two places beyond the number of decimal places retained for the final value, see Table 1 and Table 7.

2) Rounding to the final value is the last step in a calculation.

*Example 1, Rounding Inch Series:*

$$n = 28 \text{ threads per inch} \qquad P = \frac{1}{n} = \frac{1}{28}$$

$$P = 0.0357142857 \quad \text{(calculated and truncated to 10 places)}$$

$$P = 0.03571429 \quad \text{(rounded to 8 places)}$$

**Table 1. Number of Decimal Places Used in Calculations**

| Units | Pitch | Constants | Intermediate | Final |
|---|---|---|---|---|
| Inch | 8 | 8 | 6 | 4 |
| Metric | as designated | 7 | 5 | 3 |

3) For inch screw thread dimensions, four decimal places are required for the final values of pitch diameter, major diameter, and minor diameter with the exception of Class 1B and 2B internal thread minor diameters for thread sizes 0.138 and larger.

The final values for the allowances and tolerances applied to thread elements are expressed to four decimal places except for external thread pitch diameter tolerance, $Td_2$, which is expressed to six decimal places.

*Minor Diameter Exceptions for Internal Threads*:

*Minimum Minor Diameter:* All classes are calculated and then rounded off to the nearest 0.001 inch and expressed in three decimal places for sizes 0.138 inch and larger. For Class 3B, a zero is added to yield four decimal places.

*Maximum Minor Diameter:* All classes are calculated before rounding, then rounded for Classes 1B and 2B to the nearest 0.001 in. for sizes 0.138 in. and larger. Class 3B values are rounded to four decimal places.

4) Metric screw threads are dimensioned in millimeters. The final values of pitch diameter, major diameter, minor diameter, allowance and thread element tolerances are expressed to three decimal places.

5) Values containing multiple trailing zeros out to the required number of decimal places can be expressed by displaying only two of them beyond the last significant digit.

*Example:* 20 threads per inch has a pitch equal to 0.05000000 and can be expressed as 0.0500.

## Examples

**Inch Screw Threads.—**The formulas in the examples for inch screw threads are based on those listed in ANSI/ASME B1.1, *Unified Inch Screw Threads*. Table 3 and Table 4 are based on a size that when converted from a fraction to a decimal will result in a number that has only four decimal places. Table 5 and Table 6 are based on a size that when converted will result in a number with infinite numbers of digits after the decimal point. Fig. 1 is provided for reference.

**Metric Screw Threads.—**The formulas for metric screw threads are based on those listed in ANSI/ASME B1.13M, *Metric Screw Threads*. The calculation of size limits for standard diameter/pitch combinations listed in both ISO 261 and ANSI/ASME B1.13M use the tabulated values for allowances and tolerances (in accordance with ISO 965-1). The constant values differ from those used for inch screw threads, in accordance with the policy of rounding of this Standard, because metric limits of size are expressed to only three decimal places rather than four.

**Thread Form Constants.—**For thread form data see Table 2. The number of decimal places and the manner in which they are listed should be consistent. Thread form constants printed in older thread standards are based on a function of thread height ($H$) or pitch ($P$). The equivalent of the corresponding function is also listed. There are some constants that would require these values to 8 or 7 decimal places before they would round to equivalent values. For standardization the tabulated listing of thread values based on a function of pitch has been established, with thread height used as a reference only All thread calculations are to be performed using a function of pitch (*P*), rounded to 8 decimal places for inch series and as designated for metric series, not a function of thread height (*H*). Thread height is to be used for reference only. See Table 7.

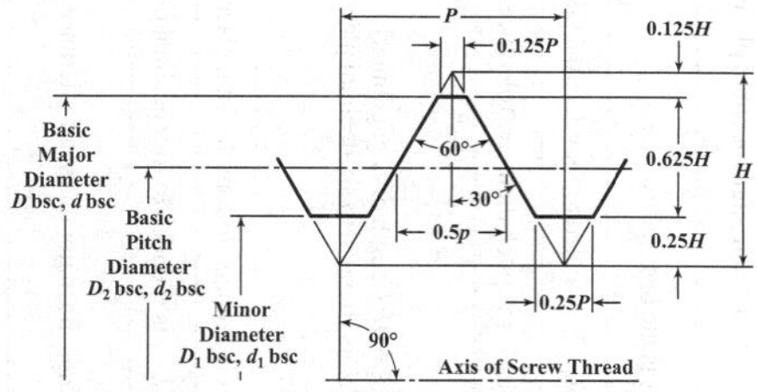


Fig. 1. Basic Profile of UN and UNF Screw Threads

**Table 2. Thread Form Data**

| Constant for Inch Series (8-place) | Reference Values | | Constant for Metric Series (7-place) |
|---|---|---|---|
| 0.04811252*P* | 1/18 *H* | 0.0556*H* | 0.0481125*P* |
| 0.05412659*P* | 1/16 *H* | 0.0625*H* | 0.0541266*P* |
| 0.08660254*P* | 1/10 *H* | 0.1000*H* | 0.0866025*P* |
| 0.09622504*P* | 1/9 *H* | 0.1111*H* | 0.0962250*P* |
| 0.10825318*P* | 1/8 *H* | 0.1250*H* | 0.1082532*P* |
| 0.12990381*P* | 3/20 *H* | 0.1500*H* | 0.1299038*P* |
| 0.14433757*P* | 1/6 *H* | 0.1667*H* | 0.1443376*P* |
| 0.16237976*P* | 3/16 *H* | 0.1875*H* | 0.1623798*P* |
| 0.21650635*P* | 1/4 *H* | 0.2500*H* | 0.2165064*P* |
| 0.28867513*P* | 1/3 *H* | 0.3333*H* | 0.2886751*P* |
| 0.32475953*P* | 3/8 *H* | 0.3750*H* | 0.3247595*P* |
| 0.36084392*P* | 5/12 *H* | 0.4167*H* | 0.3608439*P* |
| 0.39692831*P* | 11/24 *H* | 0.4583*H* | 0.3969283*P* |
| 0.43301270*P* | 1/2 *H* | 0.5000*H* | 0.4330127*P* |
| 0.48713929*P* | 9/16 *H* | 0.5625*H* | 0.4871393*P* |
| 0.54126588*P* | 5/8 *H* | 0.6250*H* | 0.5412659*P* |
| 0.57735027*P* | 2/3 *H* | 0.6667*H* | 0.5773503*P* |
| 0.59539246*P* | 11/16 *H* | 0.6875*H* | 0.5953925*P* |
| 0.61343466*P* | 17/24 *H* | 0.7083*H* | 0.6134347*P* |
| 0.61602540*P* | … | 0.7113*H* | 0.6160254*P* |
| 0.64951905*P* | 3/4 *H* | 0.7500*H* | 0.6495191*P* |
| 0.72168783*P* | 5/6 *H* | 0.8333*H* | 0.7216878*P* |
| 0.79385662*P* | 11/12 *H* | 0.9167*H* | 0.7938566*P* |
| 0.86602540*P* | *H* | 1.0000*H* | 0.8660254*P* |
| 1.08253175*P* | 5/4 *H* | 1.2500*H* | 1.0825318*P* |
| 1.19078493*P* | 11/8 *H* | 1.3750*H* | 1.1907849*P* |
| 1.22686932*P* | 17/12 *H* | 1.4167*H* | 1.2268693*P* |

**Table 3. External Inch Screw Thread Calculations for ½ -28 UNEF-2A**

| Characteristic Description | Calculation | Notes |
|---|---|---|
| Basic major diameter, $d_{bsc}$ | $d_{bsc} = \frac{1}{2} = 0.5 = 0.5000$ | $d_{bsc}$ is rounded to four decimal places |
| Pitch, $P$ | $P = \frac{1}{28} = 0.035714285714 = 0.03571429$ | $P$ is rounded to eight decimal places |
| **Maximum external major diameter** ($d_{max}$) = basic major diameter ($d_{bsc}$) – allowance ($es$) | $d_{max} = d_{bsc} - es$ | $es$ is the basic allowance |
| Basic major diameter ($d_{bsc}$) | $d_{bsc} = 0.5000$ | $d_{bsc}$ is rounded to four decimal places |
| Allowance ($es$) | $es = 0.300 \times Td_2$ for Class 2A | $Td_2$ is the pitch diameter tolerance for Class 2A |
| External pitch diameter tolerance $Td_2$ | $Td_2 = 0.0015D^{\frac{1}{3}} + 0.0015\sqrt{LE} + 0.015P^{\frac{2}{3}}$<br>$= 0.0015 \times 0.5^{\frac{1}{3}} + 0.0015\sqrt{9 \times 0.03571429} + 0.015(0.03571429)^{\frac{2}{3}}$<br>$= 0.001191 + 0.000850 + 0.001627 = 0.003668$ | $LE = 9P$ (length of engagement)<br>$Td_2$ is rounded to six decimal places |
| Allowance ($es$) | $es = 0.300 \times 0.003668 = 0.0011004 = 0.0011$ | $es$ is rounded to four decimal places |
| **Maximum external major diameter** ($d_{max}$) | $d_{max} = d_{base} - es = 0.5000 - 0.0011 = 0.4989$ | $d_{max}$ is rounded to four decimal places |
| **Minimum external major diameter** ($d_{min}$) = maximum external major diameter ($d_{max}$) – major diameter tolerance ($Td$) | $d_{min} = d_{max} - Td$ | $Td$ is the major diameter tolerance |
| Major diameter tolerance ($Td$) | $Td = 0.060\sqrt[3]{P^2} = 0.060 \times \sqrt[3]{0.03571429^2}$<br>$= 0.060 \times \sqrt[3]{0.001276} = 0.060 \times 0.108463$<br>$= 0.00650778 = 0.0065$ | $Td$ is rounded to four decimal places |

**Table 3.** *(Continued)* **External Inch Screw Thread Calculations for $^1/_2$-28 UNEF-2A**

| Characteristic Description | Calculation | Notes |
|---|---|---|
| **Minimum external major diameter** ($d_{min}$) | $d_{min} = d_{max} - Td = 0.4989 - 0.006508$<br>$= 0.492392 = 0.4924$ | $d_{min}$ is rounded to four decimal places |
| **Maximum external pitch diameter** ($d_{2max}$) = maximum external major diameter ($d_{max}$) – twice the external thread addendum ($h_{as}$) | $d_{2max} = d_{max} - 2 \times h_{as}$ | $h_{as}$ = external thread addendum |
| External thread addendum | $h_{as} = \frac{0.64951905P}{2}$ $\quad 2h_{as} = 0.64951905P$<br>$2h_{as} = 0.64951905 \times 0.03571429 = 0.02319711$<br>$= 0.023197$ | $2h_{as}$ is rounded to six decimal places |
| **Maximum external pitch diameter** ($d_{2\,max}$) | $d_{2max} = d_{max} - 2 \times h_{as} = 0.4989 - 0.23197$<br>$= 0.475703 = 0.4757$ | $d_{2\,max}$ is rounded to four decimal places |
| **Minimum external pitch diameter** ($d_{2min}$) = maximum external pitch diameter ($d_{2max}$) – external pitch diameter tolerance ($Td_2$) | $d_{2min} = d_{2max} - Td_2$ | $Td_2$ = external pitch diameter tolerance (see previous $Td_2$ calculation in this table) |
| **Minimum external pitch diameter** ($d_{2\,min}$) | $d_{2min} = d_{2max} - Td_2 = 0.4757 - 0.003668$<br>$= 0.472032 = 0.4720$ | $d_{2\,min}$ is rounded to four decimal places |
| **Maximum external UNR minor diameter** ($d_{3max}$) = maximum external major diameter ($d_{max}$) – double height of external UNR thread $2h_s$ | $d_{3max} = d_{max} - 2 \times h_s$ | $h_s$ = external UNR thread height, |
| External UNR thread height ($2h_s$) | $2h_s = 1.19078493P = 1.19078493 \times 0.03571429$<br>$= 0.042528$ | $2h_s$ rounded to six decimal places |
| **Maximum external UNR minor diameter** ($d_{3max}$) | $d_{3max} = d_{max} - 2 \times h_s = 0.4989 - 0.042528$<br>$= 0.456372 = 0.4564$ | $d_{3\,max}$ is rounded to four decimal places |

**Table 3.** *(Continued)* **External Inch Screw Thread Calculations for $\frac{1}{2}$-28 UNEF-2A**

| Characteristic Description | Calculation | Notes |
|---|---|---|
| **Maximum external UN minor diameter** ($d_{1max}$) = maximum external major diameter ($d_{max}$) – double height of external UN thread $2h_s$ | $d_{1max} = d_{max} - 2 \times h_s$ | For UN threads, $2h_s = 2h_n$ |
| Double height of external UN thread $2h_s$ | $2h_s = 1.08253175P$<br>$= 1.08253175 \times 0.03571429 = 0.03866185$<br>$= 0.038662$ | $2h_s$ is rounded to six decimal places |
| **Maximum external UN minor diameter** ($d_{1max}$) | $d_{1max} = d_{max} - 2 \times h_s$<br>$= 0.4989 - 0.038662 = 0.460238 = 0.4602$ | $d_{1max}$ is rounded to four decimal places |

**Table 4. Internal Inch Screw Thread Calculations for $\frac{1}{2}$-28 UNEF-2B**

| Characteristic Description | Calculation | Notes |
|---|---|---|
| Basic major diameter, $d_{bsc}$ | $d_{bsc} = \frac{1}{2} = 0.5 = 0.5000$ | $d_{bsc}$ is rounded to four decimal places |
| Pitch, $P$ | $P = \frac{1}{28} = 0.035714285714 = 0.03571429$ | $P$ is rounded to eight decimal places |
| **Minimum internal minor diameter** ($D_{1min}$) = basic major diameter ($D_{bsc}$) – double height of external UN thread $2h_n$ | $D_{1min} = D_{bsc} - 2h_n$ | $2h_n$ is the double height of external UN thread |
| Double height of external UN thread $2h_s$ | $2h_n = 1.08253175P = 1.08253175 \times 0.03571429$<br>$= 0.03866185 = 0.038662$ | $2h_n$ is rounded to six decimal places |
| **Minimum internal major diameter** ($D_{1min}$) | $D_{1min} = D_{bsc} - 2 \times h_n = 0.5000 - 0.038662$<br>$= 0.461338 = 0.461$ | For class 2B the value is rounded to three decimal places to obtain the final values |

**Table 4.** *(Continued)* **Internal Inch Screw Thread Calculations for ½-28 UNEF-2B**

| Characteristic Description | Calculation | Notes |
|---|---|---|
| **Maximum internal minor diameter** ($D_{1max}$) = minimum internal minor diameter ($D_{1min}$) + internal minor diameter tolerance $TD_1$ | $D_{1max} = D_{1min} + TD_1$ | $D_{1min}$ is rounded to six decimal places |
| Internal minor diameter tolerance $TD_1$ | $TD_1 = 0.25P - 0.40P^2$<br>$= 0.25 \times 0.03571429 - 0.40 \times 0.03571429^2$<br>$= 0.008929 - 0.000510 = 0.008419 = 0.003127$ | $TD_1$ is rounded to four decimal places |
| **Maximum internal minor diameter** ($D_{1max}$) | $D_{1max} = D_{1min} + TD_1 = 0.461338 + 0.008419$<br>$= 0.469757 = 0.470$ | For the Class 2B thread $D_{1max}$ is rounded to three decimal places to obtain final values. Other sizes and classes are expressed in four decimal places |
| **Minimum internal pitch diameter** ($D_{2min}$) = basic major diameter ($D_{bsc}$) − twice the external thread addendum ($h_b$) | $D_{2min} = D_{bsc} - h_b$ | $h_b$ = external thread addendum |
| External thread addendum ($h_b$) | $h_b = 0.64951905P = 0.64951905 \times 0.03571429$<br>$= 0.02319711 = 0.023197$ | $h_b$ is rounded to six decimal places |
| **Minimum internal pitch diameter** ($D_{2min}$) | $D_{2min} = D_{bsc} - h_b = 0.5000 - 0.023197$<br>$= 0.476803 = 0.4768$ | $D_{2min}$ is rounded to four decimal places |
| **Maximum internal pitch diameter** ($D_{2max}$) = minimum internal pitch diameter ($D_{2min}$) + internal pitch diameter tolerance ($TD_2$) | $D_{2max} = D_{2min} + TD_2$ | $TD_2$ = external pitch diameter tolerance |
| External pitch diameter tolerance $TD_2$ | $TD_2 = 1.30 \times (Td_2 \text{ for Class 2A}) = 1.30 \times 0.003668$<br>$= 0.0047684 = 0.0048$ | Constant 1.30 is for this Class 2B example, and will be different for Classes 1B and 3B. $Td_2$ for Class 2A (see Table 3) is rounded to six decimal places. $TD_2$ is rounded to four places |
| **Maximum internal pitch diameter** ($D_{2max}$) | $D_{2max} = D_{2min} + TD_2 = 0.4768 + 0.0048 = 0.4816$ | $D_{2max}$ is rounded to four decimal places |

**Table 4.** *(Continued)* **Internal Inch Screw Thread Calculations for ½ -28 UNEF-2B**

| Characteristic Description | Calculation | Notes |
|---|---|---|
| **Minimum internal major diameter** ($D_{min}$) = basic major diameter ($D_{bsc}$) | $D_{min} = D_{bsc} = 0.5000$ | $D_{min}$ is rounded to four decimal places |

**Table 5. External Inch Screw Thread Calculations for 19⁄64 -36 UNS-2A**

| Characteristic Description | Calculation | Notes |
|---|---|---|
| Basic major diameter, $d_{bsc}$ | $d_{bsc} = \frac{19}{64} = 0.296875 = 0.2969$ | $d_{bsc}$ is rounded to four decimal places |
| Pitch, $P$ | $P = \frac{1}{36} = 0.0277777777778 = 0.02777778$ | $P$ is rounded to eight decimal places |
| **Maximum external major diameter** ($d_{max}$) = basic major diameter ($d_{bsc}$) – allowance ($es$) | $d_{max} = d_{bsc} - es$ | |
| Allowance ($es$) | $es = 0.300 \times Td_2$ for Class 2A | $Td_2$ is Pitch diameter tolerance for Class 2A |
| External pitch diameter tolerance, $Td_2$ | $Td_2 = 0.0015D^{\frac{1}{3}} + 0.0015\sqrt{LE} + 0.015P^{\frac{2}{3}}$ $= 0.0015 \times 0.2969^{\frac{1}{3}} + 0.0015\sqrt{9 \times 0.02777778} + 0.015(0.02777778)^{\frac{2}{3}}$ $= 0.001000679 + 0.00075 + 0.001375803 = 0.003126482$ $= 0.003127$ | $LE = 9P$ (length of engagement)<br>$Td_2$ is rounded to six decimal places |
| Allowance ($es$) | $es = 0.300 \times 0.003127 = 0.0009381 = 0.0009$ | $es$ is rounded to four decimal places |
| **Maximum external major diameter** ($d_{max}$) | $d_{max} = d_{bsc} - es = 0.2969 - 0.0009 = 0.2960$ | $d_{max}$ is rounded to four decimal places |

**Table 5.** *(Continued)* **External Inch Screw Thread Calculations for $^{19}/_{64}$-36 UNS-2A**

| Characteristic Description | Calculation | Notes |
|---|---|---|
| **Minimum external major diameter** ($d_{min}$) = maximum external major diameter ($d_{max}$) – major diameter tolerance ($Td$) | $d_{min} = d_{max} - Td$ | $Td$ is the major diameter tolerance |
| Major diameter tolerance ($Td$) | $Td = 0.060\sqrt[3]{P^2} = 0.060 \times \sqrt[3]{0.02777778^2}$ $= 0.060 \times \sqrt[3]{0.000772} = 0.060 \times 0.091736$ $= 0.00550416 = 0.0055$ | $Td$ is rounded to four decimal places |
| **Minimum external major diameter** ($d_{min}$) | $d_{min} = d_{max} - Td = 0.2960 - 0.0055 = 0.2905$ | $d_{min}$ is rounded to four decimal places |
| **Maximum external pitch diameter** ($d_{2max}$) = maximum external major diameter ($d_{max}$) – twice the external thread addendum | $d_{2max} = d_{max} - 2 \times h_{as}$ | $h_{as}$ = external thread addendum |
| External thread addendum | $h_{as} = \frac{0.64951905P}{2}$  $2h_{as} = 0.64951905P$ $2h_{as} = 0.64951905 \times 0.02777778 = 0.0180421972$ $= 0.018042$ | $h_{as}$ is rounded to six decimal places |
| **Maximum external pitch diameter** ($d_{2max}$) | $d_{2max} = d_{max} - 2h_{as} = 0.2960 - 0.018042$ $= 0.277958 = 0.2780$ | $d_{2max}$ is rounded to four decimal places |
| **Minimum external pitch diameter** ($d_{2min}$) = maximum external pitch diameter ($d_{2max}$) – external pitch diameter tolerance ($Td_2$) | $d_{2min} = d_{2max} - Td_2$ | $Td_2$ = external pitch diameter tolerance (see previous $Td_2$ calculation in this table) |
| **Minimum external pitch diameter** ($d_{2min}$) | $d_{2min} = d_{2max} - Td_2 = 0.2780 - 0.003127$ $= 0.274873 = 0.2749$ | $d_{2min}$ is rounded to four decimal places |

**Table 5.** *(Continued)* **External Inch Screw Thread Calculations for $^{19}/_{64}$-36 UNS-2A**

| Characteristic Description | Calculation | Notes |
|---|---|---|
| **Maximum external UNR minor diameter** ($d_{3max}$) = maximum external major diameter ($d_{max}$) − double height of external UNR thread $2h_s$ | $d_{3max} = d_{max} - 2h_s$ | $h_s$ = external UNR thread height, |
| External UNR thread height | $2h_s = 1.19078493P = 1.19078493 \times 0.02777778$ $= 0.033077362 = 0.033077$ | $2h_s$ is rounded to six decimal places |
| **Maximum external UNR minor diameter** ($d_{3max}$) | $d_{3max} = d_{max} - 2h_s = 0.2960 - 0.033077$ $= 0.262923 = 0.2629$ | $d_{3max}$ is rounded to four decimal places |
| **Maximum external UN minor diameter** ($d_{1max}$) = maximum external major diameter ($d_{max}$) − double height of external UN thread $2h_s$ | $d_{1max} = d_{max} - 2 \times h_s$ | For UN threads, $2h_s = 2h_n$ |
| Double height of external UN thread $2h_s$ | $2h_s = 1.08253175P = 1.08253175 \times 0.02777778$ $= 0.030070329 = 0.030070$ | For UN threads, $2h_s = 2h_n$<br>$2h_s$ is rounded to six decimal places |
| **Maximum external UN minor diameter** ($d_{1max}$) | $d_{1max} = d_{max} - 2h_s = 0.2960 - 0.030070$ $= 0.265930 = 0.2659$ | Maximum external UN minor diameter is rounded to four decimal places |

**Table 6. Internal Inch Screw Thread Calculations for $^{19}/_{64}$-28 UNS-2B**

| Characteristic Description | Calculation | Notes |
|---|---|---|
| **Minimum internal minor diameter** ($D_{1min}$) = basic major diameter ($D_{bsc}$) − double height of external UN thread $2h_n$ | $D_{1min} = D_{bsc} - 2h_n$ | $2h_n$ is the double height of external UN threads |
| Basic major diameter ($D_{bsc}$) | $D_{bsc} = \frac{19}{64} = 0.296875 = 0.2969$ | This is the final value of basic major diameter (given) and rounded to four decimal places |

**Table 6.** *(Continued)* **Internal Inch Screw Thread Calculations for $^{19}/_{64}$-28 UNS-2B**

| Characteristic Description | Calculation | Notes |
|---|---|---|
| Double height of external UN thread $2h_s$ | $2h_n = 1.08253175P = 1.08253175 \times 0.02777778$<br>$= 0.030070329 = 0.030070$ | $P$ is rounded to eight decimal places |
| **Minimum internal major diameter** ($D_{1min}$) | $D_{1min} = D_{bsc} - 2h_n = 0.2969 - 0.030070$<br>$= 0.266830 = 0.267$ | For class 2B the value is rounded to three decimal places to obtain the final value; other sizes and classes are expressed in a four place decimal. |
| **Maximum internal minor diameter** ($D_{1max}$) = minimum internal minor diameter ($D_{1min}$) + internal minor diameter tolerance $TD_1$ | $D_{1max} = D_{1min} + TD_1$ | $D_{1min}$ is rounded to six decimal places |
| Internal minor diameter tolerance $TD_1$ | $TD_1 = 0.25P - 0.40P^2$<br>$= 0.25 \times 0.02777778 - 0.40 \times 0.02777778^2$<br>$= 0.006944 - 0.000309 = 0.006635 = 0.0066$ | $TD_1$ is rounded to four decimal places. |
| **Maximum internal minor diameter** ($D_{1max}$) | $D_{1max} = D_{1min} + TD_1 = 0.266830 + 0.006635$<br>$= 0.273465 = 0.273$ | For Class 2B thread the value is rounded to three decimal places to obtain the final values. Other sizes and classes are expressed to four decimal places |
| **Minimum internal pitch diameter** ($D_{2min}$) = basic major diameter ($D_{bsc}$) – twice the external thread addendum ($h_b$) | $D_{2min} = D_{1max} - h_b$ | $h_b$ = external thread addendum |
| External thread addendum | $h_b = 0.64951905P = 0.64951905 \times 0.02777778$<br>$= 0.018042197 = 0.018042$ | $h_b$ is rounded to six decimal places |
| **Minimum internal pitch diameter** ($D_{2min}$) | $D_{2min} = D_{bsc} - h_b = 0.2969 - 0.018042$<br>$= 0.278858 = 0.2789$ | $D_{2\,min}$ is rounded to four decimal places |

**Table 6.** *(Continued)* **Internal Inch Screw Thread Calculations for $^{19}/_{64}$ -28 UNS-2B**

| Characteristic Description | Calculation | Notes |
|---|---|---|
| **Maximum internal pitch diameter** ($D_{2max}$) = minimum internal pitch diameter ($D_{2min}$) + internal pitch diameter tolerance ($TD_2$) | $D_{2max} = D_{2min} + TD_2$ | $TD_2$ = external pitch diameter tolerance |
| External pitch diameter tolerance $TD_2$ | $TD_2 = 1.30 \times (Td_2 \text{ for Class } 2A)$<br>$= 1.30 \times 0.003127 = 0.0040651 = 0.0041$ | The constant 1.30 is for this Class 2B example, and will be different for Classes 1B and 3B. $Td_2$ for Class 2A (see calculation, Table 5) is rounded to six decimal places |
| **Maximum internal pitch diameter** ($D_{2max}$) | $D_{2max} = D_{2min} + TD_2 = 0.2789 + 0.0041 = 0.2830$ | $D_{2max}$ is rounded to four decimal places |
| **Minimum internal major diameter** ($D_{min}$) = basic major diameter ($D_{bsc}$) | $D_{min} = D_{bsc} = 0.2969$ | $D_{min}$ is rounded to four decimal places |

**Table 7. Number of Decimal Places for Intermediate and Final Calculations of Thread Characteristics**

| Symbol | Dimensions | Final | | Symbol | Dimensions | Intermediate | | Final | |
|---|---|---|---|---|---|---|---|---|---|
| | | Inch | Metric | | | Inch | Metric | Inch | Metric |
| $d$ | Major diameter, external thread | 4 | 3 | $LE$ | Length of thread engagement | 6 | N/A | … | … |
| $D$ | Major diameter, internal thread | 4 | 3 | $P$ | Pitch | … | … | 8 | Note [a] |
| $d_2$ | Pitch diameter, external thread | 4 | 3 | $Td$ | Major diameter tolerance | … | … | 4 | 3 |
| $D_2$ | Pitch diameter, internal thread | 4 | 3 | $Td_2$ | Pitch diameter tolerance, external thread | … | … | 6 | 3 |
| $d_1$ | Minor diameter, external thread | 4 | 3 | $TD_2$ | Pitch diameter tolerance, internal thread | … | … | 4 | 3 |
| $d_3$ | Minor diameter, rounded root external thread | 4 | 3 | $TD_1$ | Minor diameter tolerance, internal thread | … | … | 4 | 3 |
| $D_1$ | Minor diameter, internal threads for sizes 0.138 and larger for Classes 1B and 2B only | 3 | N/A | $h_b = 2h_{as}$ | Twice the external thread addendum | 6 | N/A | … | … |
| $D_1$ | Minor diameter, internal threads for sizes smaller than 0.138 for Classes 1B and 2B, and all sizes for Class 3B | 4 | N/A | $2h_s$ | Double height of UNR external thread | 6 | N/A | … | … |
| $D_1$ | Minor diameter, internal metric thread | N/A | 3 | $2h_n$ | Double height of internal thread and UN external thread | 6 | N/A | … | … |
| $es$ | Allowance at major pitch and minor diameters of external thread | | 3 | | Twice the external thread addendum | 6 | N/A | … | … |

[a] Metric pitches are not calculated. They are stated in the screw thread designation and are to be used out to the number of decimal places as stated.

*Note:* Constants based on a function of $P$ are rounded to an 8-place decimal for inch threads and a 7-place decimal for metric threads.

# METRIC SCREW THREADS

## American National Standard Metric Screw Threads M Profile

American National Standard ANSI/ASME B1.13M-2005 (R2020) describes a system of metric threads for general fastening purposes in mechanisms and structures. The Standard is in basic agreement with ISO screw standards and resolutions, as of the date of publication, and features detailed information for diameter-pitch combinations selected as to preferred standard sizes. This Standard contains general metric standards for a 60-degree symmetrical screw thread with a basic ISO 68 designated profile.

**Application Comparison with Inch Threads.**—The metric M profile threads of tolerance class 6H/6g (see page 2022) are intended for metric applications where the inch class 2A/2B have been used. At the minimum material limits, the 6H/6g results in a looser fit than the 2A/2B. Tabular data are also provided for a tighter tolerance fit external thread of class 4g6g which is approximately equivalent to the inch class 3A but with an allowance applied. It may be noted that a 4H5H/4h6h fit is approximately equivalent to class 3A/3B fit in the inch system.

**Interchangeability with Other System Threads.**—Threads produced to ANSI/ASME B1.13M are fully interchangeable with threads conforming to other Standards based on ISO 68 basic profile and ISO 965/1 tolerance practices.

Threads produced to this Standard should be mechanically interchangeable with those produced to ANSI/ASME B1.18M-1982 (Withdrawn), Metric Screw Threads for Commercial Mechanical Fasteners—Boundary Profile Defined, of the same size and tolerance class. However, there is a possibility that some parts may be accepted by conventional gages used for threads made to ANSI/ASME B1.13M and rejected by the Double-NOT-GO gages required for threads made to ANSI/ASME B1.18M (Withdrawn).

Threads produced in accordance with M profile and MJ profile ANSI/ASME B1.21M-1997 (R2018) design data will assemble with each other. However, external MJ threads will encounter interference on the root radii with internal M thread crests when both threads are at maximum material condition.

**Definitions.**—The following definitions apply to metric screw threads—M profile.

*Allowance:* The minimum nominal clearance between a prescribed dimension and its basic dimension. Allowance is not an ISO metric screw thread term but it is numerically equal to the absolute value of the ISO term *fundamental deviation*.

*Basic Thread Profile:* The cyclical outline in an axial plane of the permanently established boundary between the provinces of the external and internal threads. All deviations are with respect to this boundary. (See Fig. 1 and Fig. 5.)

*Bolt Thread (External Thread):* The term used in ISO metric thread standards to describe all external threads. All symbols associated with external threads are designated with lower case letters. This Standard uses the term external threads in accordance with United States practice.

*Clearance:* The difference between the size of the internal thread and the size of the external thread when the latter is smaller.

*Crest Diameter:* The major diameter of an external thread and the minor diameter of an internal thread.

*Design Profiles:* The maximum material profiles permitted for external and internal threads for a specified tolerance class. (See Fig. 2 and Fig. 3.)

*Deviation:* An ISO term for the algebraic difference between a given size (actual, measured, maximum, minimum, etc.) and the corresponding basic size. The term deviation does not necessarily indicate an error.

*Fit:* The relationship existing between two corresponding external and internal threads with respect to the amount of clearance or interference which is present when they are assembled.

*Fundamental Deviation:* For standard threads, the deviation (upper or lower) closer to the basic size. It is the upper deviation, *es*, for an external thread and the lower deviation, *EI*, for an internal thread. (See Fig. 5.)

*Limiting Profiles:* The limiting M profile for internal threads is shown in Fig. 6. The limiting M profile for external threads is shown in Fig. 7.

*Lower Deviation:* The algebraic difference between the minimum limit of size and the corresponding basic size.

*Nut Thread (Internal Thread):* A term used in ISO metric thread standards to describe all internal threads. All symbols associated with internal threads are designated with upper case letters. This Standard uses the term *internal thread* in accordance with United States practice.

*Tolerance:* The total amount of variation permitted for the size of a dimension. It is the difference between the maximum limit of size and the minimum limit of size (i.e., the algebraic difference between the upper deviation and the lower deviation). The tolerance is an absolute value without sign. Tolerance for threads is applied to the design size in the direction of the minimum material. On external threads the tolerance is applied negatively. On internal threads the tolerance is applied positively.

*Tolerance Class:* The combination of a tolerance position with a tolerance grade. It specifies the allowance (fundamental deviation) and tolerance for the pitch and major diameters of external threads and pitch and minor diameters of internal threads.

*Tolerance Grade:* A numerical symbol that designates the tolerances of crest diameters and pitch diameters applied to the design profiles.

*Tolerance Position:* A letter symbol that designates the position of the tolerance zone in relation to the basic size. This position provides the allowance (fundamental deviation).

*Upper Deviation:* The algebraic difference between the maximum limit of size and the corresponding basic size.

**Basic M Profile.**—The basic M thread profile also known as ISO 68 basic profile for metric screw threads is shown in Fig. 1 with associated dimensions listed in Table 3.

**Design M Profile for Internal Thread.**—The design M profile for the internal thread at maximum material condition is the basic ISO 68 profile. It is shown in Fig. 2 with associated thread data listed in Table 3.

**Design M Profile for External Thread.**—The design M profile for the external thread at the no allowance maximum material condition is the basic ISO 68 profile except where a rounded root is required. For the standard $0.125P$ minimum radius, the ISO 68 profile is modified at the root with a $0.17783H$ truncation blending into two arcs with radii of $0.125P$ tangent to the thread flanks as shown in Fig. 3 with associated thread data in Table 3.

**M Crest and Root Form.**—The form of crest at the major diameter of the external thread is flat, permitting corner rounding. The external thread is truncated $0.125H$ from a sharp crest. The form of the crest at the minor diameter of the internal thread is flat. It is truncated $0.25H$ from a sharp crest.

The crest and root tolerance zones at the major and minor diameters will permit rounded crest and root forms in both external and internal threads.

The root profile of the external thread must lie within the "section lined" tolerance zone shown in Fig. 4. For the rounded root thread, the root profile must lie within the "section lined" rounded root tolerance zone shown in Fig. 4. The profile must be a continuous, smoothly blended non-reversing curve, no part of which has a radius of less than $0.125P$, and which is tangential to the thread flank. The profile may comprise tangent flank arcs that are joined by a tangential flat at the root.

The root profile of the internal thread must not be smaller than the basic profile. The maximum major diameter must not be sharp.

**General Symbols.—**The general symbols used to describe the metric screw thread forms are shown in Table 1.

**Table 1. American National Standard Symbols for Metric Threads**
*ANSI/ASME B1.13M-2005 (R2020)*

| Symbol | Explanation |
|---|---|
| $D$ | Major Diameter Internal Thread |
| $D_1$ | Minor Diameter Internal Thread |
| $D_2$ | Pitch Diameter Internal Thread |
| $d$ | Major Diameter External Thread |
| $d_1$ | Minor Diameter External Thread |
| $d_2$ | Pitch Diameter External Thread |
| $d_3$ | Rounded Form Minor Diameter External Thread |
| $P$ | Pitch |
| $r$ | External Thread Root Radius |
| $T$ | Tolerance |
| $T_{D1}, T_{D2}$ | Tolerances for $D_1$, $D_2$ |
| $T_d, T_{d2}$ | Tolerances for $d$, $d_2$ |
| $ES$ | Upper Deviation, Internal Thread [Equals the Allowance (Fundamental Deviation) Plus the Tolerance]. See Fig. 5. |
| $EI$ | Lower Deviation, Internal Thread Allowance (Fundamental Deviation). See Fig. 5. |
| $G, H$ | Letter Designations for Tolerance Positions for Lower Deviation, Internal Thread |
| $g, h$ | Letter Designations for Tolerance Positions for Upper Deviation, External Thread |
| $es$ | Upper Deviation, External Thread Allowance (Fundamental Deviation). See Fig. 5. In the ISO system *es* is always negative for an allowance fit or zero for no allowance. |
| $ei$ | Lower Deviation, External Thread [Equals the Allowance (Fundamental Deviation) Plus the Tolerance]. See Fig. 5. In the ISO system *ei* is always negative for an allowance fit. |
| $H$ | Height of Fundamental Triangle |
| $LE$ | Length of Engagement |
| $LH$ | Left-Hand Thread |

**Standard M Profile Screw Thread Series.—**The standard metric screw thread series for general purpose equipment's threaded components design and mechanical fasteners is a *coarse thread* series. Their diameter/pitch combinations are shown in Table 2. These diameter/pitch combinations are the preferred sizes and should be the first choice as applicable. Additional *fine pitch* diameter/pitch combinations are shown in Table 5.

**Table 2. American National Standard General Purpose and Mechanical Fastener Coarse Pitch Metric Thread—M Profile Series** *ANSI/ASME B1.13M-2005 (R2020)*

| Nom.Size | Pitch | Nom.Size | Pitch | Nom.Size | Pitch | Nom.Size | Pitch |
|---|---|---|---|---|---|---|---|
| 1.6 | 0.35 | 6 | 1 | 22 | 2.5[a] | 56 | 5.5 |
| 2 | 0.4 | 8 | 1.25 | 24 | 3 | 64 | 6 |
| 2.5 | 0.45 | 10 | 1.5 | 27 | 3[a] | 72 | 6[b] |
| 3 | 0.5 | 12 | 1.75 | 30 | 3.5 | 80 | 6[b] |
| 3.5 | 0.6 | 14 | 2 | 36 | 4 | 90 | 6[b] |
| 4 | 0.7 | 16 | 2 | 42 | 4.5 | 100 | 6[b] |
| 5 | 0.8 | 20 | 2.5 | 48 | 5 | … | … |

[a] For high strength structural steel fasteners only.

[b] Designated as part of 6 mm fine pitch series in ISO 261.

All dimensions are in millimeters.

**Table 3. American National Standard Metric Thread—M Profile Data** *ANSI/ASME B1.13M-2005 (R2020)*

| Pitch $P$ | Truncation of Internal Thread Root and External Thread Crest $\frac{H}{8}$ 0.1082532$P$ | Addendum of Internal Thread and Truncation of Internal Thread $\frac{H}{4}$ 0.2165064$P$ | Dedendum of Internal Thread and Addendum External Thread $\frac{3}{8}H$ 0.3247595$P$ | Difference[a] $\frac{H}{2}$ 0.4330127$P$ | Height of Internal Thread and Depth of Thread Engagement $\frac{5}{8}H$ 0.5412659$P$ | Difference[b] 0.711325$H$ 0.6160254$P$ | Twice the External Thread Addendum $\frac{3}{4}H$ 0.6495191$P$ | Difference[c] $\frac{11}{12}H$ 0.7938566$P$ | Height of Sharp V-Thread $H$ 0.8660254$P$ | Double Height of Internal Thread $\frac{5}{4}H$ 1.0825318$P$ |
|---|---|---|---|---|---|---|---|---|---|---|
| 0.2 | 0.02165 | 0.04330 | 0.06495 | 0.08660 | 0.10825 | 0.12321 | 0.12990 | 0.15877 | 0.17321 | 0.21651 |
| 0.25 | 0.02706 | 0.05413 | 0.08119 | 0.10825 | 0.13532 | 0.15401 | 0.16238 | 0.19846 | 0.21651 | 0.27063 |
| 0.3 | 0.03248 | 0.06495 | 0.09743 | 0.12990 | 0.16238 | 0.18481 | 0.19486 | 0.23816 | 0.25981 | 0.32476 |
| 0.35 | 0.03789 | 0.07578 | 0.11367 | 0.15155 | 0.18944 | 0.21561 | 0.22733 | 0.27785 | 0.30311 | 0.37889 |
| 0.4 | 0.04330 | 0.08660 | 0.12990 | 0.17321 | 0.21651 | 0.24541 | 0.25981 | 0.31754 | 0.34641 | 0.43301 |
| 0.45 | 0.04871 | 0.09743 | 0.14614 | 0.19486 | 0.24357 | 0.27721 | 0.29228 | 0.35724 | 0.38971 | 0.48714 |
| 0.5 | 0.05413 | 0.10825 | 0.16238 | 0.21651 | 0.27063 | 0.30801 | 0.32476 | 0.39693 | 0.43301 | 0.54127 |
| 0.6 | 0.06495 | 0.12990 | 0.19486 | 0.25981 | 0.32476 | 0.36962 | 0.38971 | 0.47631 | 0.51962 | 0.64952 |
| 0.7 | 0.07578 | 0.15155 | 0.22733 | 0.30311 | 0.37889 | 0.43122 | 0.45466 | 0.55570 | 0.60622 | 0.75777 |
| 0.75 | 0.08119 | 0.16238 | 0.24357 | 0.32476 | 0.40595 | 0.46202 | 0.48714 | 0.59539 | 0.64952 | 0.81190 |
| 0.8 | 0.08660 | 0.17321 | 0.25981 | 0.34641 | 0.43301 | 0.49282 | 0.51962 | 0.63509 | 0.69282 | 0.86603 |
| 1 | 0.10825 | 0.21651 | 0.32476 | 0.43301 | 0.54127 | 0.61603 | 0.64952 | 0.79386 | 0.86603 | 1.08253 |
| 1.25 | 0.13532 | 0.27063 | 0.40595 | 0.54127 | 0.67658 | 0.77003 | 0.81190 | 0.99232 | 1.08253 | 1.35316 |
| 1.5 | 0.16238 | 0.32476 | 0.48714 | 0.64952 | 0.81190 | 0.92404 | 0.97428 | 1.19078 | 1.29904 | 1.62380 |
| 1.75 | 0.18944 | 0.37889 | 0.56833 | 0.75777 | 0.94722 | 1.07804 | 1.13666 | 1.38925 | 1.51554 | 1.89443 |
| 2 | 0.21651 | 0.43301 | 0.64952 | 0.86603 | 1.08253 | 1.23205 | 1.29904 | 1.58771 | 1.73205 | 2.16506 |
| 2.5 | 0.27063 | 0.54127 | 0.81190 | 1.08253 | 1.35316 | 1.54006 | 1.62380 | 1.98464 | 2.16506 | 2.70633 |
| 3 | 0.32476 | 0.64652 | 0.97428 | 1.29904 | 1.62380 | 1.84808 | 1.94856 | 2.38157 | 2.59808 | 3.24760 |
| 3.5 | 0.37889 | 0.75777 | 1.13666 | 1.51554 | 1.89443 | 2.15609 | 2.27332 | 2.77850 | 3.03109 | 3.78886 |
| 4 | 0.43301 | 0.86603 | 1.29904 | 1.73205 | 2.16506 | 2.46410 | 2.59808 | 3.17543 | 3.46410 | 4.33013 |
| 4.5 | 0.48714 | 0.97428 | 1.46142 | 1.94856 | 2.43570 | 2.77211 | 2.92284 | 3.57235 | 3.89711 | 4.87139 |
| 5 | 0.54127 | 1.08253 | 1.62380 | 2.16506 | 2.70633 | 3.08013 | 3.24760 | 3.96928 | 4.33013 | 5.41266 |
| 5.5 | 0.59539 | 1.19079 | 1.78618 | 2.38157 | 2.97696 | 3.38814 | 3.57236 | 4.36621 | 4.76314 | 5.95392 |
| 6 | 0.64952 | 1.29904 | 1.94856 | 2.59808 | 3.24760 | 3.69615 | 3.89711 | 4.76314 | 5.19615 | 6.49519 |
| 8 | 0.86603 | 1.73205 | 2.59808 | 3.46410 | 4.33013 | 4.92820 | 5.19615 | 6.35085 | 6.92820 | 8.66025 |

[a] Difference between max theoretical pitch diameter and max minor diameter of external thread and between min theoretical pitch diameter and min minor diameter of internal thread.

[b] Difference between min theoretical pitch diameter and min design minor diameter of external thread for 0.125$P$ root radius.

[c] Difference between max major diameter and max theoretical pitch diameter of internal thread.

All dimensions are in millimeters.

**Table 4. American National Standard Minimum Rounded Root Radius— M Profile Series** *ANSI/ASME B1.13M-2005 (R2020)*

| Pitch *P* | Min. Root Radius, 0.125*P* | Pitch *P* | Min. Root Radius, 0.125*P* | Pitch *P* | Min. Root Radius, 0.125*P* | Pitch *P* | Min. Root Radius, 0.125*P* |
|---|---|---|---|---|---|---|---|
| 0.2 | 0.025 | 0.6 | 0.075 | 1.5 | 0.188 | 4 | 0.500 |
| 0.25 | 0.031 | 0.7 | 0.088 | 1.75 | 0.219 | 4.5 | 0.563 |
| 0.3 | 0.038 | 0.75 | 0.094 | 2 | 0.250 | 5 | 0.625 |
| 0.35 | 0.044 | 0.8 | 0.100 | 2.5 | 0.313 | 5.5 | 0.688 |
| 0.4 | 0.050 | 1 | 0.125 | 3 | 0.375 | 6 | 0.750 |
| 0.45 | 0.056 | 1.25 | 0.156 | 3.5 | 0.438 | 8 | 1.000 |
| 0.5 | 0.063 | … | … | … | … | … | … |

All dimensions are in millimeters.

**Table 5. American National Standard Fine Pitch Metric Thread—M Profile Series** *ANSI/ASME B1.13M-2005 (R2020)*

| Nom. Size | Pitch | | | Nom. Size | Pitch | | Nom. Size | Pitch | | Nom. Size | Pitch |
|---|---|---|---|---|---|---|---|---|---|---|---|
| 8 | 1 | | … | 27 | … | 2 | 56 | … | 2 | 105 | 2 |
| 10 | 0.75 | 1.0 | 1.25 | 30 | 1.5 | 2 | 60 | 1.5 | … | 110 | 2 |
| 12 | 1 | 1.5 | 1.25 | 33 | … | 2 | 64 | … | 2 | 120 | 2 |
| 14 | … | | 1.5 | 35 | 1.5 | … | 65 | 1.5 | … | 130 | 2 |
| 15 | 1 | | … | 36 | … | 2 | 70 | 1.5 | … | 140 | 2 |
| 16 | … | | 1.5 | 39 | … | 2 | 72 | … | 2 | 150 | 2 |
| 17 | 1 | | … | 40 | 1.5 | … | 75 | 1.5 | … | 160 | 3 |
| 18 | … | | 1.5 | 42 | … | 2 | 80 | 1.5 | 2 | 170 | 3 |
| 20 | 1 | | 1.5 | 45 | 1.5 | … | 85 | … | 2 | 180 | 3 |
| 22 | … | | 1.5 | 48 | … | 2 | 90 | … | 2 | 190 | 3 |
| 24 | … | | 2 | 50 | 1.5 | … | 95 | … | 2 | 200 | 3 |
| 25 | 1.5 | | … | 55 | 1.5 | … | 100 | … | 2 | | |

All dimensions are in millimeters.

**Limits and Fits for Metric Screw Threads—M Profile.**—The International (ISO) metric tolerance system is based on a system of limits and fits. The limits of the tolerances on the mating parts together with their allowances (fundamental deviations) determine the fit of the assembly. For simplicity the system is described for cylindrical parts (see *British and ISO Metric Limits and Fits* starting on page 675) but in this Standard it is applied to screw threads. Holes are equivalent to internal threads and shafts to external threads.

*Basic Size:* This is the zero line or surface at assembly where the interface of the two mating parts have a common reference.*

*Upper Deviation:* This is the algebraic difference between the maximum limit of size and the basic size. It is designated by the French term "écart supérieur" (*ES* for internal and *es* for external threads).

*Lower Deviation:* This is the algebraic difference between the minimum limit of size and the basic size. It is designated by the French term "écart inférieur" (*EI* for internal and *ei* for external threads).

*Fundamental Deviations (Allowances):* These are the deviations which are closest to the basic size. In the accompanying figure they would be *EI* and *es*.

* "Basic," when used to identify a particular dimension in this Standard, such as basic major diameter, refers to the h/H tolerance position (zero fundamental deviation) value.

*Tolerance:* The tolerance is defined by a series of numerical grades. Each grade provides numerical values for the various nominal sizes corresponding to the standard tolerance for that grade.

In the schematic diagram the tolerance for the external thread is shown as negative. Thus the tolerance plus the fit define the lower deviation (*ei*). The tolerance for the mating internal thread is shown as positive. Thus the tolerance plus the fit defines the upper deviation (*ES*).

*Fits:* Fits are determined by the fundamental deviations assigned to the mating parts and may be positive or negative. The selected fits can be clearance, transition, or interference. To illustrate the fits schematically, a zero line is drawn to represent the basic size as shown in Fig. 5. By convention, the external thread lies below the zero line and the internal thread lies above it (except for interference fits). This makes the fundamental deviation negative for the external thread and equal to its upper deviation (*es*). The fundamental deviation is positive for the internal thread and equal to its lower deviation (*EI*).

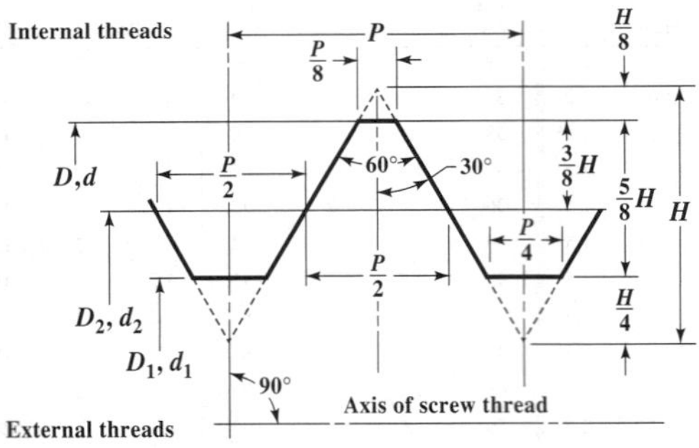


$$H = \frac{\sqrt{3}}{2} \times P = 0.866025P$$

$$0.125H = 0.108253P \quad 0.250H = 0.216506P \quad 0.375H = 0.324760P \quad 0.625H = 0.541266P$$

Fig. 1. Basic M Thread Profile (ISO 68 Basic Profile)

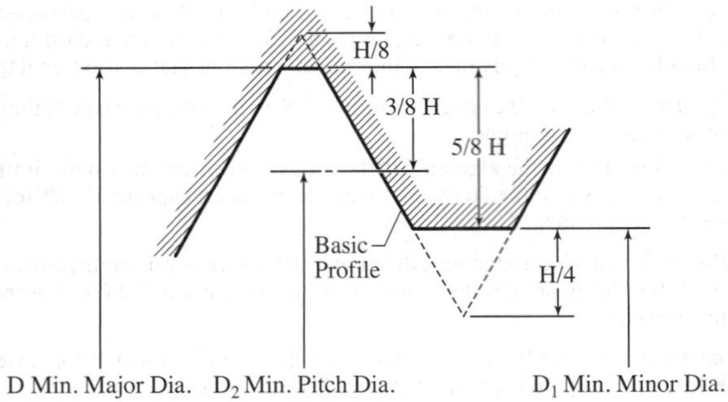


Fig. 2. Internal Thread Design M Profile with No Allowance (Fundamental Deviation) (Maximum Material Condition). For Dimensions see Table 3

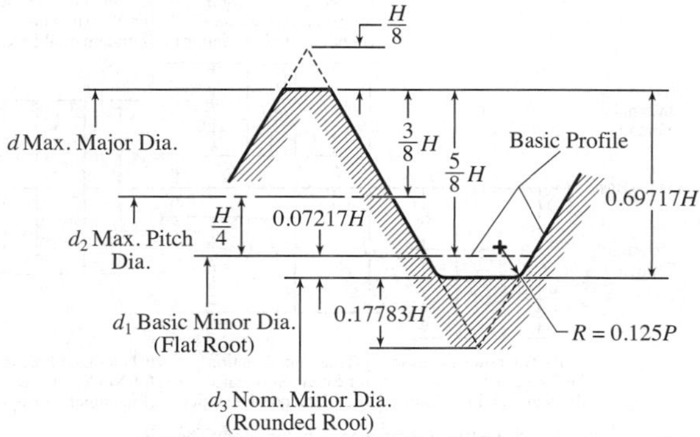


Fig. 3. External Thread Design M Profile with No Allowance (Fundamental Deviation) (Flanks at Maximum Material Condition). For Dimensions see Table 3

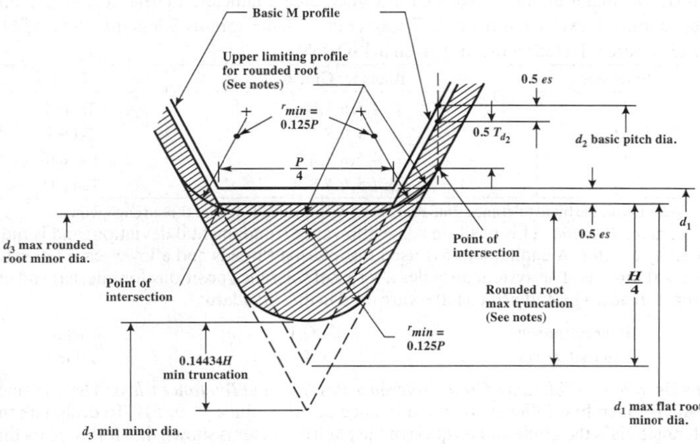


Fig. 4. M Profile, External Thread Root, Upper and Lower Limiting Profiles for $r_{min} = 0.125\,P$ and for Flat Root (Shown for Tolerance Position g)

*Notes:*

1) "Section lined" portions identify tolerance zone and unshaded portions identify allowance (fundamental deviation).
2) The upper limiting profile for rounded root is not a design profile; rather it indicates the limiting acceptable condition for the rounded root which will pass a GO thread gage.
3) Max truncation $= \dfrac{H}{4} - r_{min}\left(1 - \cos\left[60° - \text{arc}\ \cos\left(1 - \dfrac{T_{d2}}{4r_{min}}\right)\right]\right)$

where $H$ = Height of fundamental triangle

$r_{min}$ = Minimum external thread root radius

$T_{d2}$ = Tolerance on pitch diameter of external threasd

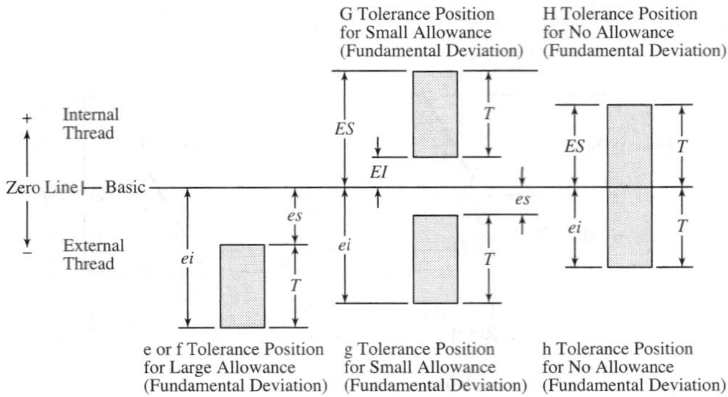


Fig. 5. Metric Tolerance System for Screw Threads

*Tolerance Grade:* This is indicated by a number. The system provides for a series of tolerance grades for each of the four screw thread parameters: minor diameter, internal thread, $D_1$; major diameter, external thread, $d$; pitch diameter, internal thread, $D_2$; and pitch diameter, external thread, $d_2$. The tolerance grades for this Standard ANSI/ASME B1.13M were selected from those given in ISO 965/1.

| Dimension | Tolerance Grades | Table |
|---|---|---|
| $D_1$ | 4, 5, 6, 7, 8 | Table 8 |
| $d$ | 4, 6, 8 | Table 9 |
| $D_2$ | 4, 5, 6, 7, 8 | Table 10 |
| $d_2$ | 3, 4, 5, 6, 7, 8, 9 | Table 11 |

*Note:* The underlined tolerance grades are used with normal length of thread engagement.

*Tolerance Position:* This position is the allowance (fundamental deviation) and is indicated by a letter. A capital letter is used for internal threads and a lower case letter for external threads. The system provides a series of tolerance positions for internal and external threads. The underlined letters are used in this Standard:

| | | |
|---|---|---|
| Internal threads | G, H | Table 6 |
| External threads | e, f, g, h | Table 6 |

*Designations of Tolerance Grade, Tolerance Position, and Tolerance Class:* The tolerance grade is given first followed by the tolerance position, thus: 4g or 5H. To designate the tolerance class the grade and position of the pitch diameter is shown first followed by that for the major diameter in the case of the external thread or that for the minor diameter in the case of the internal thread, thus 4g6g for an external thread and 5H6H for an internal thread. If the two grades and positions are identical, it is not necessary to repeat the symbols, thus 4g, alone, stands for 4g4g and 5H, alone, stands for 5H5H.

*Lead and Flank Angle Tolerances:* For acceptance of lead and flank angles of product screw threads, see Section 10 of ANSI/ASME B1.13M-2005 (R2020).

*Short and Long Lengths of Thread Engagement when Gaged with Normal Length Contacts:* For short lengths of thread engagement, LE, reduce the pitch diameter tolerance of the external thread by one tolerance grade number. For long lengths of thread engagement, LE, increase the allowance (fundamental deviation) at the pitch diameter of the external thread. Examples of tolerance classes required for normal, short, and long gage length contacts are given in the following table.

For lengths of thread engagement classified as normal, short, and long, see Table 7.

**Table 6. Allowance (Fundamental Deviation) for Internal and External Metric Threads** *ISO 965/1 ANSI/ASME B1.13M-2005 (R2020)*

| Pitch | Allowance (Fundamental Deviation)[a] | | | | | |
|---|---|---|---|---|---|---|
| | Internal Thread $D_2, D_1$ | | External Thread $d, d_2$ | | | |
| | G | H[b] | e | f | g[c] | h |
| $P$ | *EI* | *EI* | *es* | *es* | *es* | *es* |
| 0.2 | +0.017 | 0 | … | … | −0.017 | 0 |
| 0.25 | +0.018 | 0 | … | … | −0.018 | 0 |
| 0.3 | +0.018 | 0 | … | … | −0.018 | 0 |
| 0.35 | +0.019 | 0 | … | −0.034 | −0.019 | 0 |
| 0.4 | +0.019 | 0 | … | −0.034 | −0.019 | 0 |
| 0.45 | +0.020 | 0 | … | −0.035 | −0.020 | 0 |
| 0.5 | +0.020 | 0 | −0.050 | −0.036 | −0.020 | 0 |
| 0.6 | +0.021 | 0 | −0.053 | −0.036 | −0.021 | 0 |
| 0.7 | +0.022 | 0 | −0.056 | −0.038 | −0.022 | 0 |
| 0.75 | +0.022 | 0 | −0.056 | −0.038 | −0.022 | 0 |
| 0.8 | +0.024 | 0 | −0.060 | −0.038 | −0.024 | 0 |
| 1 | +0.026 | 0 | −0.060 | −0.040 | −0.026 | 0 |
| 1.25 | +0.028 | 0 | −0.063 | −0.042 | −0.028 | 0 |
| 1.5 | +0.032 | 0 | −0.067 | −0.045 | −0.032 | 0 |
| 1.75 | +0.034 | 0 | −0.071 | −0.048 | −0.034 | 0 |
| 2 | +0.038 | 0 | −0.071 | −0.052 | −0.038 | 0 |
| 2.5 | +0.042 | 0 | −0.080 | −0.058 | −0.042 | 0 |
| 3 | +0.048 | 0 | −0.085 | −0.063 | −0.048 | 0 |
| 3.5 | +0.053 | 0 | −0.090 | −0.070 | −0.053 | 0 |
| 4 | +0.060 | 0 | −0.095 | −0.075 | −0.060 | 0 |
| 4.5 | +0.063 | 0 | −0.100 | −0.080 | −0.063 | 0 |
| 5 | +0.071 | 0 | −0.106 | −0.085 | −0.071 | 0 |
| 5.5 | +0.075 | 0 | −0.112 | −0.090 | −0.075 | 0 |
| 6 | +0.080 | 0 | −0.118 | −0.095 | −0.080 | 0 |
| 8 | +0.100 | 0 | −0.140 | −0.118 | −0.100 | 0 |

All dimensions are in millimeters.

[a] Allowance is the absolute value of fundamental deviation.
[b] Tabulated in this standard for M internal threads.
[c] Tabulated in this standard for M external threads.

| Normal LE | Short LE | Long LE |
|---|---|---|
| 6g | 5g6g | 6e6g |
| 4g6g | 3g6g | 4e6g |
| 6h[a] | 5h6h | 6g6h |
| 4h6h[a] | 3h6h | 4g6h |
| 6H | 5H | 6G |
| 4H6H | 3H6H | 4G6G |

[a] Applies to maximum material functional size (GO thread gage) for plated 6g and 4g6g class threads, respectively.

**Material Limits for Coated Threads.—**Unless otherwise specified, size limits for standard external tolerance classes 6g and 4g6g apply prior to coating. The external thread allowance may thus be used to accommodate the coating thickness on coated parts, provided that the maximum coating thickness is no more than $\frac{1}{4}$ of the allowance. Thus, a 6g thread after coating is subject to acceptance using a basic size 6*h* GO thread gage and a 4*g*6*g* thread, a 4*h*6*h* or 6*h* GO thread gage. Minimum material, LO, or NOT-GO gages would be 6g and 4*g*6*g*, respectively. Where the external thread has no allowance or the allowance must be maintained after coating, and for standard internal threads, sufficient allowance must be provided prior to coating to ensure that finished product threads do not exceed the maximum material limits specified. For thread classes with tolerance position *H* or *h*, coating allowances in accordance with Table 6 for position *G* or *g*, respectively, should be applied wherever possible.

**Table 7. Length of Metric Thread Engagement**
*ISO 965/1 ANSI/ASME B1.13M-2005 (R2020)*

| Basic Major Diameter $d_{bsc}$ | | Pitch | Length of Thread Engagement | | | |
|---|---|---|---|---|---|---|
| | | | Short LE | Normal LE | | Long LE |
| Over | Up to and incl. | $P$ | Up to and incl. | Over | Up to and incl. | Over |
| 1.5 | 2.8 | 0.2 | 0.5 | 0.5 | 1.5 | 1.5 |
| | | 0.25 | 0.6 | 0.6 | 1.9 | 1.9 |
| | | 0.35 | 0.8 | 0.8 | 2.6 | 2.6 |
| | | 0.4 | 1 | 1 | 3 | 3 |
| | | 0.45 | 1.3 | 1.3 | 3.8 | 3.8 |
| 2.8 | 5.6 | 0.35 | 1 | 1 | 3 | 3 |
| | | 0.5 | 1.5 | 1.5 | 4.5 | 4.5 |
| | | 0.6 | 1.7 | 1.7 | 5 | 5 |
| | | 0.7 | 2 | 2 | 6 | 6 |
| | | 0.75 | 2.2 | 2.2 | 6.7 | 6.7 |
| | | 0.8 | 2.5 | 2.5 | 7.5 | 7.5 |
| 5.6 | 11.2 | 0.75 | 2.4 | 2.4 | 7.1 | 7.1 |
| | | 1 | 3 | 3 | 9 | 9 |
| | | 1.25 | 4 | 4 | 12 | 12 |
| | | 1.5 | 5 | 5 | 15 | 15 |
| 11.2 | 22.4 | 1 | 3.8 | 3.8 | 11 | 11 |
| | | 1.25 | 4.5 | 4.5 | 13 | 13 |
| | | 1.5 | 5.6 | 5.6 | 16 | 16 |
| | | 1.75 | 6 | 6 | 18 | 18 |
| | | 2 | 8 | 8 | 24 | 24 |
| | | 2.5 | 10 | 10 | 30 | 30 |
| 22.4 | 45 | 1 | 4 | 4 | 12 | 12 |
| | | 1.5 | 6.3 | 6.3 | 19 | 19 |
| | | 2 | 8.5 | 8.5 | 25 | 25 |
| | | 3 | 12 | 12 | 36 | 36 |
| | | 3.5 | 15 | 15 | 45 | 45 |
| | | 4 | 18 | 18 | 53 | 53 |
| | | 4.5 | 21 | 21 | 63 | 63 |
| 45 | 90 | 1.5 | 7.5 | 7.5 | 22 | 22 |
| | | 2 | 9.5 | 9.5 | 28 | 28 |
| | | 3 | 15 | 15 | 45 | 45 |
| | | 4 | 19 | 19 | 56 | 56 |
| | | 5 | 24 | 24 | 71 | 71 |
| | | 5.5 | 28 | 28 | 85 | 85 |
| | | 6 | 32 | 32 | 95 | 95 |
| 90 | 180 | 2 | 12 | 12 | 36 | 36 |
| | | 3 | 18 | 18 | 53 | 53 |
| | | 4 | 24 | 24 | 71 | 71 |
| | | 6 | 36 | 36 | 106 | 106 |
| | | 8 | 45 | 45 | 132 | 132 |
| 180 | 355 | 3 | 20 | 20 | 60 | 60 |
| | | 4 | 26 | 26 | 80 | 80 |
| | | 6 | 40 | 40 | 118 | 118 |
| | | 8 | 50 | 50 | 150 | 150 |

All dimensions are in millimeters.

**Dimensional Effect of Coating.—**On a cylindrical surface, the effect of coating is to change the diameter by twice the coating thickness. On a 60-degree thread, however, since the coating thickness is measured perpendicular to the thread surface while the pitch diameter is measured perpendicular to the thread axis, the effect of a uniformly coated flank on the pitch diameter is to change it by four times the thickness of the coating on the flank.

*External Thread with No Allowance for Coating:* To determine gaging limits before coating for a uniformly coated thread, decrease: 1) maximum pitch diameter by four times maximum coating thickness; 2) minimum pitch diameter by four times minimum coating thickness; 3) maximum major diameter by two times maximum coating thickness; and 4) minimum major diameter by two times minimum coating thickness.

*External Thread with Only Nominal or Minimum Thickness Coating:* If no coating thickness tolerance is given, it is recommended that a tolerance of plus 50 percent of the nominal or minimum thickness be assumed.

Then, to determine before coating gaging limits for a uniformly coated thread, decrease: 1) maximum pitch diameter by six times coating thickness; 2) minimum pitch diameter by four times coating thickness; 3) maximum major diameter by three times coating thickness; and 4) minimum major diameter by two times coating thickness.

*Adjusted Size Limits:* It should be noted that the before coating material limit tolerances are less than the tolerance after coating. This is because the coating tolerance consumes some of the product tolerance. In cases there may be insufficient pitch diameter tolerance available in the before coating condition so that additional adjustments and controls will be necessary.

*Strength:* On small threads (5 mm and smaller) there is a possibility that coating thickness adjustments will cause base material minimum material conditions which may significantly affect strength of externally threaded parts. Limitations on coating thickness or part redesign may then be necessary.

*Internal Threads:* Standard internal threads provide no allowance for coating thickness.

To determine before coating, gaging limits for a uniformly coated thread, increase: 1) minimum pitch diameter by four times maximum coating thickness, if specified, or by six times minimum or nominal coating thickness when a tolerance is not specified; 2) maximum pitch diameter by four times minimum or nominal coating thickness; 3) minimum minor diameter by two times maximum coating thickness, if specified, or by three times minimum or nominal coating thickness; and 4) maximum minor diameter by two times minimum or nominal coating thickness.

*Other Considerations:* It is essential to review all possibilities adequately and consider limitations in the threading and coating production processes before finally deciding on the coating process and the allowance required to accommodate the coating. A no-allowance thread after coating must not transgress the basic profile and is, therefore, subject to acceptance using a basic (tolerance position H/h) size GO thread gage.

**Formulas for M Profile Screw Thread Limiting Dimensions.—**The limiting dimensions for M profile screw threads are calculated from the following formulas.

**Internal Threads:**

*Min major dia.* = basic major dia. + *EI* (Table 6)

*Min pitch dia.* = basic major dia. − 0.6495191*P* (Table 3) + *EI* for $D_2$ (Table 6)

*Max pitch dia.* = min pitch dia. + $TD_2$ (Table 10)

*Max major dia.* = max pitch dia. + 0.7938566*P* (Table 3)

*Min minor dia.* = min major dia. − 1.0825318*P* (Table 3)

*Max minor dia.* = min minor dia. + $TD_1$ (Table 8)

**External Threads:**

*Max major dia.* = basic major dia. − *es* (Table 6) (Note that *es* is an absolute value.)

*Min major dia.* = max major dia. − *Td* (Table 9)

*Max pitch dia.* = basic major dia. − 0.6495191*P* (Table 3) − *es* for $d_2$ (Table 6)

*Min pitch dia.* = max pitch dia. − $Td_2$ (Table 11)

*Max flat form minor dia.* = max pitch dia. − 0.433013*P* (Table 3)

*Max rounded root minor dia.* = max pitch dia. − H + 2 × max trunc. (See Fig. 4)

*Min rounded root minor dia.* = min pitch dia. − 0.616025*P* (Table 3)

*Min root radius* = 0.125*P*

**Table 8. Minor Diameter Tolerances of Internal Metric Threads $TD_1$**
*ISO 965/1 ANSI/ASME B1.13M-2005 (R2020)*

| Pitch *P* | Tolerance Grade 4 | 5 | 6[a] | 7 | 8 |
|---|---|---|---|---|---|
| 0.2 | 0.038 | … | … | … | … |
| 0.25 | 0.045 | 0.056 | … | … | … |
| 0.3 | 0.053 | 0.067 | 0.085 | … | … |
| 0.35 | 0.063 | 0.080 | 0.100 | … | … |
| 0.4 | 0.071 | 0.090 | 0.112 | … | … |
| 0.45 | 0.080 | 0.100 | 0.125 | … | … |
| 0.5 | 0.090 | 0.112 | 0.140 | 0.180 | … |
| 0.6 | 0.100 | 0.125 | 0.160 | 0.200 | … |
| 0.7 | 0.112 | 0.140 | 0.180 | 0.224 | … |
| 0.75 | 0.118 | 0.150 | 0.190 | 0.236 | … |
| 0.8 | 0.125 | 0.160 | 0.200 | 0.250 | 0.315 |
| 1 | 0.150 | 0.190 | 0.236 | 0.300 | 0.375 |
| 1.25 | 0.170 | 0.212 | 0.265 | 0.335 | 0.425 |
| 1.5 | 0.190 | 0.236 | 0.300 | 0.375 | 0.475 |
| 1.75 | 0.212 | 0.265 | 0.335 | 0.425 | 0.530 |
| 2 | 0.236 | 0.300 | 0.375 | 0.475 | 0.600 |
| 2.5 | 0.280 | 0.355 | 0.450 | 0.560 | 0.710 |
| 3 | 0.315 | 0.400 | 0.500 | 0.630 | 0.800 |
| 3.5 | 0.355 | 0.450 | 0.560 | 0.710 | 0.900 |
| 4 | 0.375 | 0.475 | 0.600 | 0.750 | 0.950 |
| 4.5 | 0.425 | 0.530 | 0.670 | 0.850 | 1.060 |
| 5 | 0.450 | 0.560 | 0.710 | 0.900 | 1.120 |
| 5.5 | 0.475 | 0.600 | 0.750 | 0.950 | 1.180 |
| 6 | 0.500 | 0.630 | 0.800 | 1.000 | 1.250 |
| 8 | 0.630 | 0.800 | 1.000 | 1.250 | 1.600 |

[a] Tabulated in this standard for M internal threads.

All dimensions are in millimeters.

## Table 9. Major Diameter Tolerances of External Metric Threads, *Td*
*ISO 965/1 ANSI/ASME B1.13M-2005 (R2020)*

| Pitch P | Tolerance Grade 4 | 6[a] | 8 | Pitch P | Tolerance Grade 4 | 6[a] | 8 |
|---|---|---|---|---|---|---|---|
| 0.2 | 0.036 | 0.056 | … | 1.5 | 0.150 | 0.236 | 0.375 |
| 0.25 | 0.042 | 0.067 | … | 1.75 | 0.170 | 0.265 | 0.425 |
| 0.3 | 0.048 | 0.075 | … | 2 | 0.180 | 0.280 | 0.450 |
| 0.35 | 0.053 | 0.085 | … | 2.5 | 0.212 | 0.335 | 0.530 |
| 0.4 | 0.060 | 0.095 | … | 3 | 0.236 | 0.375 | 0.600 |
| 0.45 | 0.063 | 0.100 | … | 3.5 | 0.265 | 0.425 | 0.670 |
| 0.5 | 0.067 | 0.106 | … | 4 | 0.300 | 0.475 | 0.750 |
| 0.6 | 0.080 | 0.125 | … | 4.5 | 0.315 | 0.500 | 0.800 |
| 0.7 | 0.090 | 0.140 | … | 5 | 0.335 | 0.530 | 0.850 |
| 0.75 | 0.090 | 0.140 | … | 5.5 | 0.355 | 0.560 | 0.900 |
| 0.8 | 0.095 | 0.150 | 0.236 | 6 | 0.375 | 0.600 | 0.950 |
| 1 | 0.112 | 0.180 | 0.280 | 8 | 0.450 | 0.710 | 1.180 |
| 1.25 | 0.132 | 0.212 | 0.335 | … | … | … | … |

[a] Tabulated in this standard for M internal threads.

All dimensions are in millimeters.

## Table 10. Pitch-Diameter Tolerances of Internal Metric Thread, $TD_2$
*ISO 965/1 ANSI/ASME B1.13M-2005 (R2020)*

| Basic Major Diameter, *D* Over | Up to and incl. | Pitch P | Tolerance Grade 4 | 5 | 6[a] | 7 | 8 |
|---|---|---|---|---|---|---|---|
| 1.5 | 2.8 | 0.2 | 0.042 | … | … | … | … |
| | | 0.25 | 0.048 | 0.060 | … | … | … |
| | | 0.35 | 0.053 | 0.067 | 0.085 | … | … |
| | | 0.4 | 0.056 | 0.071 | 0.090 | … | … |
| | | 0.45 | 0.060 | 0.075 | 0.095 | … | … |
| 2.8 | 5.6 | 0.35 | 0.056 | 0.071 | 0.090 | … | … |
| | | 0.5 | 0.063 | 0.080 | 0.100 | 0.125 | … |
| | | 0.6 | 0.071 | 0.090 | 0.112 | 0.140 | … |
| | | 0.7 | 0.075 | 0.095 | 0.118 | 0.150 | … |
| | | 0.75 | 0.075 | 0.095 | 0.118 | 0.150 | … |
| | | 0.8 | 0.080 | 0.100 | 0.125 | 0.160 | 0.200 |
| 5.6 | 11.2 | 0.75 | 0.085 | 0.106 | 0.132 | 0.170 | … |
| | | 1 | 0.095 | 0.118 | 0.150 | 0.190 | 0.236 |
| | | 1.25 | 0.100 | 0.125 | 0.160 | 0.200 | 0.250 |
| | | 1.5 | 0.112 | 0.140 | 0.180 | 0.224 | 0.280 |
| 11.2 | 22.4 | 1 | 0.100 | 0.125 | 0.160 | 0.200 | 0.250 |
| | | 1.25 | 0.112 | 0.140 | 0.180 | 0.224 | 0.280 |
| | | 1.5 | 0.118 | 0.150 | 0.190 | 0.236 | 0.300 |
| | | 1.75 | 0.125 | 0.160 | 0.200 | 0.250 | 0.315 |
| | | 2 | 0.132 | 0.170 | 0.212 | 0.265 | 0.335 |
| | | 2.5 | 0.140 | 0.180 | 0.224 | 0.280 | 0.355 |
| 22.4 | 45 | 1 | 0.106 | 0.132 | 0.170 | 0.212 | … |
| | | 1.5 | 0.125 | 0.160 | 0.200 | 0.250 | 0.315 |
| | | 2 | 0.140 | 0.180 | 0.224 | 0.280 | 0.355 |
| | | 3 | 0.170 | 0.212 | 0.265 | 0.335 | 0.425 |
| | | 3.5 | 0.180 | 0.224 | 0.280 | 0.355 | 0.450 |
| | | 4 | 0.190 | 0.236 | 0.300 | 0.375 | 0.475 |
| | | 4.5 | 0.200 | 0.250 | 0.315 | 0.400 | 0.500 |
| 45 | 90 | 1.5 | 0.132 | 0.170 | 0.212 | 0.265 | 0.335 |
| | | 2 | 0.150 | 0.190 | 0.236 | 0.300 | 0.375 |
| | | 3 | 0.180 | 0.224 | 0.280 | 0.355 | 0.450 |
| | | 4 | 0.200 | 0.250 | 0.315 | 0.400 | 0.500 |
| | | 5 | 0.212 | 0.265 | 0.335 | 0.425 | 0.530 |
| | | 5.5 | 0.224 | 0.280 | 0.355 | 0.450 | 0.560 |
| | | 6 | 0.236 | 0.300 | 0.375 | 0.475 | 0.600 |

**Table 10.** *(Continued)* **Pitch-Diameter Tolerances of Internal Metric Thread,** $TD_2$
*ISO 965/1 ANSI/ASME B1.13M-2005 (R2020)*

| Basic Major Diameter, *D* | | Pitch | Tolerance Grade | | | | |
|---|---|---|---|---|---|---|---|
| Over | Up to and incl. | *P* | 4 | 5 | 6[a] | 7 | 8 |
| 90 | 180 | 2 | 0.160 | 0.200 | 0.250 | 0.315 | 0.400 |
| | | 3 | 0.190 | 0.236 | 0.300 | 0.375 | 0.475 |
| | | 4 | 0.212 | 0.265 | 0.335 | 0.425 | 0.530 |
| | | 6 | 0.250 | 0.315 | 0.400 | 0.500 | 0.630 |
| | | 8 | 0.280 | 0.355 | 0.450 | 0.560 | 0.710 |
| 180 | 355 | 3 | 0.212 | 0.265 | 0.335 | 0.425 | 0.530 |
| | | 4 | 0.236 | 0.300 | 0.375 | 0.475 | 0.600 |
| | | 6 | 0.265 | 0.335 | 0.425 | 0.530 | 0.670 |
| | | 8 | 0.300 | 0.375 | 0.475 | 0.600 | 0.750 |

[a] Tabulated in this standard for M threads.
All dimensions are in millimeters.

**Table 11. Pitch-Diameter Tolerances of External Metric Threads,** $Td_2$
*ISO 965/1 ANSI/ASME B1.13M-2005 (R2020)*

| Basic Major Diameter, *d* | | Pitch | Tolerance Grade | | | | | | |
|---|---|---|---|---|---|---|---|---|---|
| Over | Up to and incl. | *P* | 3 | 4[a] | 5 | 6[a] | 7 | 8 | 9 |
| 1.5 | 2.8 | 0.2 | 0.025 | 0.032 | 0.040 | 0.050 | … | … | … |
| | | 0.25 | 0.028 | 0.036 | 0.045 | 0.056 | … | … | … |
| | | 0.35 | 0.032 | 0.040 | 0.050 | 0.063 | 0.080 | … | … |
| | | 0.4 | 0.034 | 0.042 | 0.053 | 0.067 | 0.085 | … | … |
| | | 0.45 | 0.036 | 0.045 | 0.056 | 0.071 | 0.090 | … | … |
| 2.8 | 5.6 | 0.35 | 0.034 | 0.042 | 0.053 | 0.067 | 0.085 | … | … |
| | | 0.5 | 0.038 | 0.048 | 0.060 | 0.075 | 0.095 | … | … |
| | | 0.6 | 0.042 | 0.053 | 0.067 | 0.085 | 0.106 | … | … |
| | | 0.7 | 0.045 | 0.056 | 0.071 | 0.090 | 0.112 | … | … |
| | | 0.75 | 0.045 | 0.056 | 0.071 | 0.090 | 0.112 | … | … |
| | | 0.8 | 0.048 | 0.060 | 0.075 | 0.095 | 0.118 | 0.150 | 0.190 |
| 5.6 | 11.2 | 0.75 | 0.050 | 0.063 | 0.080 | 0.100 | 0.125 | … | … |
| | | 1 | 0.056 | 0.071 | 0.090 | 0.112 | 0.140 | 0.180 | 0.224 |
| | | 1.25 | 0.060 | 0.075 | 0.095 | 0.118 | 0.150 | 0.190 | 0.236 |
| | | 1.5 | 0.067 | 0.085 | 0.106 | 0.132 | 0.170 | 0.212 | 0.265 |
| 11.2 | 22.4 | 1 | 0.060 | 0.075 | 0.095 | 0.118 | 0.150 | 0.190 | 0.236 |
| | | 1.25 | 0.067 | 0.085 | 0.106 | 0.132 | 0.170 | 0.212 | 0.265 |
| | | 1.5 | 0.071 | 0.090 | 0.112 | 0.140 | 0.180 | 0.224 | 0.280 |
| | | 1.75 | 0.075 | 0.095 | 0.118 | 0.150 | 0.190 | 0.236 | 0.300 |
| | | 2 | 0.080 | 0.100 | 0.125 | 0.160 | 0.200 | 0.250 | 0.315 |
| | | 2.5 | 0.085 | 0.106 | 0.132 | 0.170 | 0.212 | 0.265 | 0.335 |
| 22.4 | 45 | 1 | 0.063 | 0.080 | 0.100 | 0.125 | 0.160 | 0.200 | 0.250 |
| | | 1.5 | 0.075 | 0.095 | 0.118 | 0.150 | 0.190 | 0.236 | 0.300 |
| | | 2 | 0.085 | 0.106 | 0.132 | 0.170 | 0.212 | 0.265 | 0.335 |
| | | 3 | 0.100 | 0.125 | 0.160 | 0.200 | 0.250 | 0.315 | 0.400 |
| | | 3.5 | 0.106 | 0.132 | 0.170 | 0.212 | 0.265 | 0.335 | 0.425 |
| | | 4 | 0.112 | 0.140 | 0.180 | 0.224 | 0.280 | 0.355 | 0.450 |
| | | 4.5 | 0.118 | 0.150 | 0.190 | 0.236 | 0.300 | 0.375 | 0.475 |
| 45 | 90 | 1.5 | 0.080 | 0.100 | 0.125 | 0.160 | 0.200 | 0.250 | 0.315 |
| | | 2 | 0.090 | 0.112 | 0.140 | 0.180 | 0.224 | 0.280 | 0.355 |
| | | 3 | 0.106 | 0.132 | 0.170 | 0.212 | 0.265 | 0.335 | 0.425 |
| | | 4 | 0.118 | 0.150 | 0.190 | 0.236 | 0.300 | 0.375 | 0.475 |
| | | 5 | 0.125 | 0.160 | 0.200 | 0.250 | 0.315 | 0.400 | 0.500 |
| | | 5.5 | 0.132 | 0.170 | 0.212 | 0.265 | 0.335 | 0.425 | 0.530 |
| | | 6 | 0.140 | 0.180 | 0.224 | 0.280 | 0.355 | 0.450 | 0.560 |
| 90 | 180 | 2 | 0.095 | 0.118 | 0.150 | 0.190 | 0.236 | 0.300 | 0.375 |
| | | 3 | 0.112 | 0.140 | 0.180 | 0.224 | 0.280 | 0.355 | 0.450 |
| | | 4 | 0.125 | 0.160 | 0.200 | 0.250 | 0.315 | 0.400 | 0.500 |
| | | 6 | 0.150 | 0.190 | 0.236 | 0.300 | 0.375 | 0.475 | 0.600 |
| | | 8 | 0.170 | 0.212 | 0.265 | 0.335 | 0.425 | 0.530 | 0.670 |

**Table 11.** *(Continued)* **Pitch-Diameter Tolerances of External Metric Threads, $Td_2$**
*ISO 965/1 ANSI/ASME B1.13M-2005 (R2020)*

| Basic Major Diameter, *d* | | Pitch | Tolerance Grade | | | | | | |
|---|---|---|---|---|---|---|---|---|---|
| Over | Up to and incl. | *P* | 3 | 4[a] | 5 | 6[a] | 7 | 8 | 9 |
| 180 | 355 | 3 | 0.125 | 0.160 | 0.200 | 0.250 | 0.315 | 0.400 | 0.500 |
| | | 4 | 0.140 | 0.180 | 0.224 | 0.280 | 0.355 | 0.450 | 0.560 |
| | | 6 | 0.160 | 0.200 | 0.250 | 0.315 | 0.400 | 0.500 | 0.630 |
| | | 8 | 0.180 | 0.224 | 0.280 | 0.355 | 0.450 | 0.560 | 0.710 |

[a] Tabulated in this Standard for M threads.
All dimensions are in millimeters.

**Tolerance Grade Comparisons.—**The approximate ratios of the tolerance grades shown in Table 8, Table 9, Table 10, and Table 11 in terms of Grade 6 are as follows:

*Minor Diameter Tolerance of Internal Thread:* Grade 6 is $TD_1$ (Table 8)*:* Grade 4 is 0.63 $TD_1$ (6); Grade 5 is 0.8 $TD_1$ (6); Grade 7 is 1.25 $TD_1$ (6); and Grade 8 is 1.6 $TD_1$ (6).

*Pitch Diameter Tolerance of Internal Thread: $Td_2$* (Table 10)*:* Grade 4 is 0.85 $Td_2$ (6); Grade 5 is 1.06 $Td_2$ (6); Grade 6 is 1.32 $Td_2$ (6); Grade 7 is 1.7 $Td_2$ (6); and Grade 8 is 2.12 $Td_2$ (6). It should be noted that these ratios are in terms of the Grade 6 pitch diameter tolerance for the external thread.

*Major Diameter Tolerance of External Thread: Td*(6) (Table 9)*:* Grade 4 is 0.63 *Td* (6); and Grade 8 is 1.6 *Td* (6).

*Pitch Diameter Tolerance of External Thread: $Td_2$* (Table 11)*:* Grade 3 is 0.5 $Td_2$ (6); Grade 4 is 0.63 $Td_2$ (6); Grade 5 is 0.8 $Td_2$ (6); Grade 7 is 1.25 $Td_2$ (6); Grade 8 is 1.6 $Td_2$ (6); and Grade 9 is 2 $Td_2$ (6).

**Standard M Profile Screw Threads, Limits of Size.—**The limiting M profile for internal threads is shown in Fig. 6 with associated dimensions for standard sizes in Table 12. The limiting M profiles for external threads are shown in Fig. 7 with associated dimensions for standard sizes in Table 13.

If the required values are not listed in these tables, they may be calculated using the data in Table 3, Table 6, Table 7, Table 8, Table 9, Table 10, and Table 11 together with the preceding formulas. If the required data are not included in any of the tables listed above, reference should be made to Sections 6 and 9.3 of ANSI/ASME B1.13M, which gives design formulas.

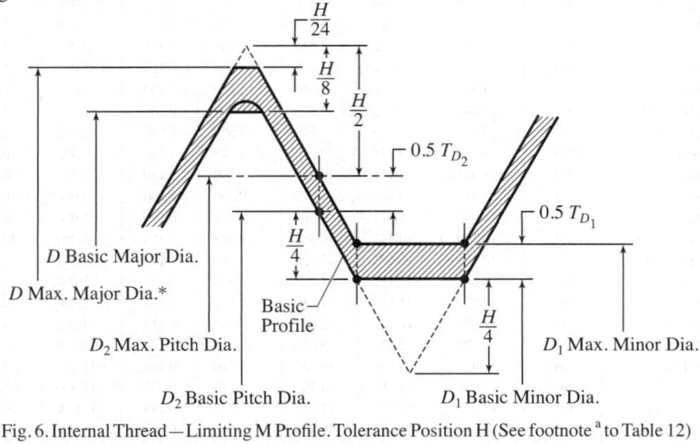


Fig. 6. Internal Thread—Limiting M Profile. Tolerance Position H (See footnote [a] to Table 12)

*Note:* "Section Lined" portions identify tolerance zone.

*Dimension *D* in Fig. 6 is used in the design of tools, etc. For internal threads it is not normally specified. Generally, major diameter acceptance is based on maximum material condition gaging.

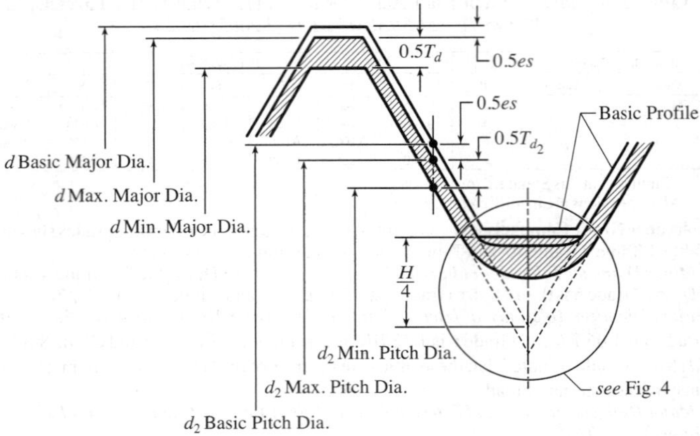


Fig. 7. External Thread—Limiting M Profile. Tolerance Position g

*Note:* "Section Lined" portions identify tolerance zone and unshaded portions identify allowance (fundamental deviation.)

**Table 12. Internal Metric Thread—M Profile Limiting Dimensions**
*ANSI/ASME B1.13M-2005 (R2020)*

| Basic Thread Designation | Toler. Class | Minor Diameter $D_1$ | | Pitch Diameter $D_2$ | | | Major Diameter $D$ | |
|---|---|---|---|---|---|---|---|---|
| | | Min. | Max. | Min. | Max. | Tol. | Min. | Max.[a] |
| M1.6 × 0.35 | 6H | 1.221 | 1.321 | 1.373 | 1.458 | 0.085 | 1.600 | 1.736 |
| M2 × 0.4 | 6H | 1.567 | 1.679 | 1.740 | 1.830 | 0.090 | 2.000 | 2.148 |
| M2.5 × 0.45 | 6H | 2.013 | 2.138 | 2.208 | 2.303 | 0.095 | 2.500 | 2.660 |
| M3 × 0.5 | 6H | 2.459 | 2.599 | 2.675 | 2.775 | 0.100 | 3.000 | 3.172 |
| M3.5 × 0.6 | 6H | 2.850 | 3.010 | 3.110 | 3.222 | 0.112 | 3.500 | 3.698 |
| M4 × 0.7 | 6H | 3.242 | 3.422 | 3.545 | 3.663 | 0.118 | 4.000 | 4.219 |
| M5 × 0.8 | 6H | 4.134 | 4.334 | 4.480 | 4.605 | 0.125 | 5.000 | 5.240 |
| M6 × 1 | 6H | 4.917 | 5.153 | 5.350 | 5.500 | 0.150 | 6.000 | 6.294 |
| M8 × 1.25 | 6H | 6.647 | 6.912 | 7.188 | 7.348 | 0.160 | 8.000 | 8.340 |
| M8 × 1 | 6H | 6.917 | 7.153 | 7.350 | 7.500 | 0.150 | 8.000 | 8.294 |
| M10 × 0.75 | 6H | 9.188 | 9.378 | 9.513 | 9.645 | 0.132 | 10.000 | 10.240 |
| M10 × 1 | 6H | 8.917 | 9.153 | 9.350 | 9.500 | 0.150 | 10.000 | 10.294 |
| M10 × 1.5 | 6H | 8.376 | 8.676 | 9.026 | 9.206 | 0.180 | 10.000 | 10.397 |
| M10 × 1.25 | 6H | 8.647 | 8.912 | 9.188 | 9.348 | 0.160 | 10.000 | 10.340 |
| M12 × 1.75 | 6H | 10.106 | 10.441 | 10.863 | 11.063 | 0.200 | 12.000 | 12.452 |
| M12 × 1.5 | 6H | 10.376 | 10.676 | 11.026 | 11.216 | 0.190 | 12.000 | 12.407 |
| M12 × 1.25 | 6H | 10.647 | 10.912 | 11.188 | 11.368 | 0.180 | 12.000 | 12.360 |
| M12 × 1 | 6H | 10.917 | 11.153 | 11.350 | 11.510 | 0.160 | 12.000 | 12.304 |
| M14 × 2 | 6H | 11.835 | 12.210 | 12.701 | 12.913 | 0.212 | 14.000 | 14.501 |
| M14 × 1.5 | 6H | 12.376 | 12.676 | 13.026 | 13.216 | 0.190 | 14.000 | 14.407 |
| M15 × 1 | 6H | 13.917 | 14.153 | 14.350 | 14.510 | 0.160 | 15.000 | 15.304 |
| M16 × 2 | 6H | 13.835 | 14.210 | 14.701 | 14.913 | 0.212 | 16.000 | 16.501 |
| M16 × 1.5 | 6H | 14.376 | 14.676 | 15.026 | 15.216 | 0.190 | 16.000 | 16.407 |
| M17 × 1 | 6H | 15.917 | 16.153 | 16.350 | 16.510 | 0.160 | 17.000 | 17.304 |
| M18 × 1.5 | 6H | 16.376 | 16.676 | 17.026 | 17.216 | 0.190 | 18.000 | 18.407 |
| M20 × 2.5 | 6H | 17.294 | 17.744 | 18.376 | 18.600 | 0.224 | 20.000 | 20.585 |
| M20 × 1.5 | 6H | 18.376 | 18.676 | 19.026 | 19.216 | 0.190 | 20.000 | 20.407 |

**Table 12.** *(Continued)* **Internal Metric Thread — M Profile Limiting Dimensions**
*ANSI/ASME B1.13M-2005 (R2020)*

| Basic Thread Designation | Toler. Class | Minor Diameter $D_1$ | | Pitch Diameter $D_2$ | | | Major Diameter $D$ | |
|---|---|---|---|---|---|---|---|---|
| | | Min. | Max. | Min. | Max. | Tol. | Min. | Max.[a] |
| M20 × 1 | 6H | 18.917 | 19.153 | 19.350 | 19.510 | 0.160 | 20.000 | 20.304 |
| M22 × 2.5 | 6H | 19.294 | 19.744 | 20.376 | 20.600 | 0.224 | 22.000 | 22.585 |
| M22 × 1.5 | 6H | 20.376 | 20.676 | 21.026 | 21.216 | 0.190 | 22.000 | 22.407 |
| M24 × 3 | 6H | 20.752 | 21.252 | 22.051 | 22.316 | 0.265 | 24.000 | 24.698 |
| M24 × 2 | 6H | 21.835 | 22.210 | 22.701 | 22.925 | 0.224 | 24.000 | 24.513 |
| M25 × 1.5 | 6H | 23.376 | 23.676 | 24.026 | 24.226 | 0.200 | 25.000 | 25.417 |
| M27 × 3 | 6H | 23.752 | 24.252 | 25.051 | 25.316 | 0.265 | 27.000 | 27.698 |
| M27 × 2 | 6H | 24.835 | 25.210 | 25.701 | 25.925 | 0.224 | 27.000 | 27.513 |
| M30 × 3.5 | 6H | 26.211 | 26.771 | 27.727 | 28.007 | 0.280 | 30.000 | 30.786 |
| M30 × 2 | 6H | 27.835 | 28.210 | 28.701 | 28.925 | 0.224 | 30.000 | 30.513 |
| M30 × 1.5 | 6H | 28.376 | 28.676 | 29.026 | 29.226 | 0.200 | 30.000 | 30.417 |
| M33 × 2 | 6H | 30.835 | 31.210 | 31.701 | 31.925 | 0.224 | 33.000 | 33.513 |
| M35 × 1.5 | 6H | 33.376 | 33.676 | 34.026 | 34.226 | 0.200 | 35.000 | 35.417 |
| M36 × 4 | 6H | 31.670 | 32.270 | 33.402 | 33.702 | 0.300 | 36.000 | 36.877 |
| M36 × 2 | 6H | 33.835 | 34.210 | 34.701 | 34.925 | 0.224 | 36.000 | 36.513 |
| M39 × 2 | 6H | 36.835 | 37.210 | 37.701 | 37.925 | 0.224 | 39.000 | 39.513 |
| M40 × 1.5 | 6H | 38.376 | 38.676 | 39.026 | 39.226 | 0.200 | 40.000 | 40.417 |
| M42 × 4.5 | 6H | 37.129 | 37.799 | 39.077 | 39.392 | 0.315 | 42.000 | 42.964 |
| M42 × 2 | 6H | 39.835 | 40.210 | 40.701 | 40.925 | 0.224 | 42.000 | 42.513 |
| M45 × 1.5 | 6H | 43.376 | 43.676 | 44.026 | 44.226 | 0.200 | 45.000 | 45.417 |
| M48 × 5 | 6H | 42.587 | 43.297 | 44.752 | 45.087 | 0.335 | 48.000 | 49.056 |
| M48 × 2 | 6H | 45.835 | 46.210 | 46.701 | 46.937 | 0.236 | 48.000 | 48.525 |
| M50 × 1.5 | 6H | 48.376 | 48.676 | 49.026 | 49.238 | 0.212 | 50.000 | 50.429 |
| M55 × 1.5 | 6H | 53.376 | 53.676 | 54.026 | 54.238 | 0.212 | 55.000 | 55.429 |
| M56 × 5.5 | 6H | 50.046 | 50.796 | 52.428 | 52.783 | 0.355 | 56.000 | 57.149 |
| M56 × 2 | 6H | 53.835 | 54.210 | 54.701 | 54.937 | 0.236 | 56.000 | 56.525 |
| M60 × 1.5 | 6H | 58.376 | 58.676 | 59.026 | 59.238 | 0.212 | 60.000 | 60.429 |
| M64 × 6 | 6H | 57.505 | 58.305 | 60.103 | 60.478 | 0.375 | 64.000 | 65.241 |
| M64 × 2 | 6H | 61.835 | 62.210 | 62.701 | 62.937 | 0.236 | 64.000 | 64.525 |
| M65 × 1.5 | 6H | 63.376 | 63.676 | 64.026 | 64.238 | 0.212 | 65.000 | 65.429 |
| M70 × 1.5 | 6H | 68.376 | 68.676 | 69.026 | 69.238 | 0.212 | 70.000 | 70.429 |
| M72 × 6 | 6H | 65.505 | 66.305 | 68.103 | 68.478 | 0.375 | 72.000 | 73.241 |
| M72 × 2 | 6H | 69.835 | 70.210 | 70.701 | 70.937 | 0.236 | 72.000 | 72.525 |
| M75 × 1.5 | 6H | 73.376 | 73.676 | 74.026 | 74.238 | 0.212 | 75.000 | 75.429 |
| M80 × 6 | 6H | 73.505 | 74.305 | 76.103 | 76.478 | 0.375 | 80.000 | 81.241 |
| M80 × 2 | 6H | 77.835 | 78.210 | 78.701 | 78.937 | 0.236 | 80.000 | 80.525 |
| M80 × 1.5 | 6H | 78.376 | 78.676 | 79.026 | 79.238 | 0.212 | 80.000 | 80.429 |
| M85 × 2 | 6H | 82.835 | 83.210 | 83.701 | 83.937 | 0.236 | 85.000 | 85.525 |
| M90 × 6 | 6H | 83.505 | 84.305 | 86.103 | 86.478 | 0.375 | 90.000 | 91.241 |
| M90 × 2 | 6H | 87.835 | 88.210 | 88.701 | 88.937 | 0.236 | 90.000 | 90.525 |
| M95 × 2 | 6H | 92.835 | 93.210 | 93.701 | 93.951 | 0.250 | 95.000 | 95.539 |
| M100 × 6 | 6H | 93.505 | 94.305 | 96.103 | 96.503 | 0.400 | 100.000 | 101.266 |
| M100 × 2 | 6H | 97.835 | 98.210 | 98.701 | 98.951 | 0.250 | 100.000 | 100.539 |
| M105 × 2 | 6H | 102.835 | 103.210 | 103.701 | 103.951 | 0.250 | 105.000 | 105.539 |
| M110 × 2 | 6H | 107.835 | 108.210 | 108.701 | 108.951 | 0.250 | 110.000 | 110.539 |
| M120 × 2 | 6H | 117.835 | 118.210 | 118.701 | 118.951 | 0.250 | 120.000 | 120.539 |
| M130 × 2 | 6H | 127.835 | 128.210 | 128.701 | 128.951 | 0.250 | 130.000 | 130.539 |
| M140 × 2 | 6H | 137.835 | 138.210 | 138.701 | 138.951 | 0.250 | 140.000 | 140.539 |
| M150 × 2 | 6H | 147.835 | 148.210 | 148.701 | 148.951 | 0.250 | 150.000 | 150.539 |
| M160 × 3 | 6H | 156.752 | 157.252 | 158.051 | 158.351 | 0.300 | 160.000 | 160.733 |
| M170 × 3 | 6H | 166.752 | 167.252 | 168.051 | 168.351 | 0.300 | 170.000 | 170.733 |
| M180 × 3 | 6H | 176.752 | 177.252 | 178.051 | 178.351 | 0.300 | 180.000 | 180.733 |
| M190 × 3 | 6H | 186.752 | 187.252 | 188.051 | 188.386 | 0.335 | 190.000 | 190.768 |
| M200 × 3 | 6H | 196.752 | 197.252 | 198.051 | 198.386 | 0.335 | 200.000 | 200.768 |

[a] This reference dimension is used in design of tools, etc., and is not normally specified. Generally, major diameter acceptance is based upon maximum material condition gaging.

All dimensions are in millimeters.

## Table 13. External Metric Thread — M Profile Limiting Dimensions *ANSI/ASME B1.13M-2005 (R2020)*

| Basic Thread Designation | Tol. Class | Allowance[a] $es$ | Major Diameter[b] $d$ | | Pitch Diameter[b,c] $d_2$ | | | Minor Dia.[b] $d_1$ | Minor Dia.[d] $d_3$ |
|---|---|---|---|---|---|---|---|---|---|
| | | | Max. | Min. | Max. | Min. | Tol. | Max. | Min. |
| M1.6×0.35 | 6g | 0.019 | 1.581 | 1.496 | 1.354 | 1.291 | 0.063 | 1.202 | 1.075 |
| M1.6×0.35 | 6h | 0.000 | 1.600 | 1.515 | 1.373 | 1.310 | 0.063 | 1.221 | 1.094 |
| M1.6×0.35 | 4g6g | 0.019 | 1.581 | 1.496 | 1.354 | 1.314 | 0.040 | 1.202 | 1.098 |
| M2×0.4 | 6g | 0.019 | 1.981 | 1.886 | 1.721 | 1.654 | 0.067 | 1.548 | 1.408 |
| M2×0.4 | 6h | 0.000 | 2.000 | 1.905 | 1.740 | 1.673 | 0.067 | 1.567 | 1.427 |
| M2×0.4 | 4g6g | 0.019 | 1.981 | 1.886 | 1.721 | 1.679 | 0.042 | 1.548 | 1.433 |
| M2.5×0.45 | 6g | 0.020 | 2.480 | 2.380 | 2.188 | 2.117 | 0.071 | 1.993 | 1.840 |
| M2.5×0.45 | 6h | 0.000 | 2.500 | 2.400 | 2.208 | 2.137 | 0.071 | 2.013 | 1.860 |
| M2.5×0.45 | 4g6g | 0.020 | 2.480 | 2.380 | 2.188 | 2.143 | 0.045 | 1.993 | 1.866 |
| M3×0.5 | 6g | 0.020 | 2.980 | 2.874 | 2.655 | 2.580 | 0.075 | 2.438 | 2.272 |
| M3×0.5 | 6h | 0.000 | 3.000 | 2.894 | 2.675 | 2.600 | 0.075 | 2.458 | 2.292 |
| M3×0.5 | 4g6g | 0.020 | 2.980 | 2.874 | 2.655 | 2.607 | 0.048 | 2.438 | 2.299 |
| M3.5×0.6 | 6g | 0.021 | 3.479 | 3.354 | 3.089 | 3.004 | 0.085 | 2.829 | 2.634 |
| M3.5×0.6 | 6h | 0.000 | 3.500 | 3.375 | 3.110 | 3.025 | 0.085 | 2.850 | 2.655 |
| M3.5×0.6 | 4g6g | 0.021 | 3.479 | 3.354 | 3.089 | 3.036 | 0.053 | 2.829 | 2.666 |
| M4×0.7 | 6g | 0.022 | 3.978 | 3.838 | 3.523 | 3.433 | 0.090 | 3.220 | 3.002 |
| M4×0.7 | 6h | 0.000 | 4.000 | 3.860 | 3.545 | 3.455 | 0.090 | 3.242 | 3.024 |
| M4×0.7 | 4g6g | 0.022 | 3.978 | 3.838 | 3.523 | 3.467 | 0.056 | 3.220 | 3.036 |
| M5×0.8 | 6g | 0.024 | 4.976 | 4.826 | 4.456 | 4.361 | 0.095 | 4.110 | 3.868 |
| M5×0.8 | 6h | 0.000 | 5.000 | 4.850 | 4.480 | 4.385 | 0.095 | 4.134 | 3.892 |
| M5×0.8 | 4g6g | 0.024 | 4.976 | 4.826 | 4.456 | 4.396 | 0.060 | 4.110 | 3.903 |
| M6×1 | 6g | 0.026 | 5.974 | 5.794 | 5.324 | 5.212 | 0.112 | 4.891 | 4.596 |
| M6×1 | 6h | 0.000 | 6.000 | 5.820 | 5.350 | 5.238 | 0.112 | 4.917 | 4.622 |
| M6×1 | 4g6g | 0.026 | 5.974 | 5.794 | 5.324 | 5.253 | 0.071 | 4.891 | 4.637 |
| M8×1.25 | 6g | 0.028 | 7.972 | 7.760 | 7.160 | 7.042 | 0.118 | 6.619 | 6.272 |
| M8×1.25 | 6h | 0.000 | 8.000 | 7.788 | 7.188 | 7.070 | 0.118 | 6.647 | 6.300 |
| M8×1.25 | 4g6g | 0.028 | 7.972 | 7.760 | 7.160 | 7.085 | 0.075 | 6.619 | 6.315 |
| M8×1 | 6g | 0.026 | 7.974 | 7.794 | 7.324 | 7.212 | 0.112 | 6.891 | 6.596 |
| M8×1 | 6h | 0.000 | 8.000 | 7.820 | 7.350 | 7.238 | 0.112 | 6.917 | 6.622 |
| M8×1 | 4g6g | 0.026 | 7.974 | 7.794 | 7.324 | 7.253 | 0.071 | 6.891 | 6.637 |
| M10×1.5 | 6g | 0.032 | 9.968 | 9.732 | 8.994 | 8.862 | 0.132 | 8.344 | 7.938 |
| M10×1.5 | 6h | 0.000 | 10.000 | 9.764 | 9.026 | 8.894 | 0.132 | 8.376 | 7.970 |
| M10×1.5 | 4g6g | 0.032 | 9.968 | 9.732 | 8.994 | 8.909 | 0.085 | 8.344 | 7.985 |
| M10×1.25 | 6g | 0.028 | 9.972 | 9.760 | 9.160 | 9.042 | 0.118 | 8.619 | 8.272 |
| M10×1.25 | 6h | 0.000 | 10.000 | 9.788 | 9.188 | 9.070 | 0.118 | 8.647 | 8.300 |
| M10×1.25 | 4g6g | 0.028 | 9.972 | 9.760 | 9.160 | 9.085 | 0.075 | 8.619 | 8.315 |
| M10×1 | 6g | 0.026 | 9.974 | 9.794 | 9.324 | 9.212 | 0.112 | 8.891 | 8.596 |
| M10×1 | 6h | 0.000 | 10.000 | 9.820 | 9.350 | 9.238 | 0.112 | 8.917 | 8.622 |
| M10×1 | 4g6g | 0.026 | 9.974 | 9.794 | 9.324 | 9.253 | 0.071 | 8.891 | 8.637 |
| M10×0.75 | 6g | 0.022 | 9.978 | 9.838 | 9.491 | 9.391 | 0.100 | 9.166 | 8.929 |
| M10×0.75 | 6h | 0.000 | 10.000 | 9.860 | 9.513 | 9.413 | 0.100 | 9.188 | 8.951 |
| M10×0.75 | 4g6g | 0.022 | 9.978 | 9.838 | 9.491 | 9.428 | 0.063 | 9.166 | 8.966 |
| M12×1.75 | 6g | 0.034 | 11.966 | 11.701 | 10.829 | 10.679 | 0.150 | 10.071 | 9.601 |
| M12×1.75 | 6h | 0.000 | 12.000 | 11.735 | 10.863 | 10.713 | 0.150 | 10.105 | 9.635 |
| M12×1.75 | 4g6g | 0.034 | 11.966 | 11.701 | 10.829 | 10.734 | 0.095 | 10.071 | 9.656 |
| M12×1.5 | 6g | 0.032 | 11.968 | 11.732 | 10.994 | 10.854 | 0.140 | 10.344 | 9.930 |
| M12×1.5 | 6h | 0.000 | 12.000 | 11.764 | 11.026 | 10.886 | 0.140 | 10.376 | 9.962 |
| M12×1.5 | 4g6g | 0.032 | 11.968 | 11.732 | 10.994 | 10.904 | 0.090 | 10.344 | 9.980 |
| M12×1.25 | 6g | 0.028 | 11.972 | 11.760 | 11.160 | 11.028 | 0.132 | 10.619 | 10.258 |
| M12×1.25 | 6h | 0.000 | 12.000 | 11.788 | 11.188 | 11.056 | 0.132 | 10.647 | 10.286 |
| M12×1.25 | 4g6g | 0.028 | 11.972 | 11.760 | 11.160 | 11.075 | 0.085 | 10.619 | 10.305 |
| M12×1 | 6g | 0.026 | 11.974 | 11.794 | 11.324 | 11.206 | 0.118 | 10.891 | 10.590 |
| M12×1 | 6h | 0.000 | 12.000 | 11.820 | 11.350 | 11.232 | 0.118 | 10.917 | 10.616 |
| M12×1 | 4g6g | 0.026 | 11.974 | 11.794 | 11.324 | 11.249 | 0.075 | 10.891 | 10.633 |
| M14×2 | 6g | 0.038 | 13.962 | 13.682 | 12.663 | 12.503 | 0.160 | 11.797 | 11.271 |
| M14×2 | 6h | 0.000 | 14.000 | 13.720 | 12.701 | 12.541 | 0.160 | 11.835 | 11.309 |
| M14×2 | 4g6g | 0.038 | 13.962 | 13.682 | 12.663 | 12.563 | 0.100 | 11.797 | 11.331 |
| M14×1.5 | 6g | 0.032 | 13.968 | 13.732 | 12.994 | 12.854 | 0.140 | 12.344 | 11.930 |
| M14×1.5 | 6h | 0.000 | 14.000 | 13.764 | 13.026 | 12.886 | 0.140 | 12.376 | 11.962 |
| M14×1.5 | 4g6g | 0.032 | 13.968 | 13.732 | 12.994 | 12.904 | 0.090 | 12.344 | 11.980 |
| M15×1 | 6g | 0.026 | 14.974 | 14.794 | 14.324 | 14.206 | 0.118 | 13.891 | 13.590 |
| M15×1 | 6h | 0.000 | 15.000 | 14.820 | 14.350 | 14.232 | 0.118 | 13.917 | 13.616 |
| M15×1 | 4g6g | 0.026 | 14.974 | 14.794 | 14.324 | 14.249 | 0.075 | 13.891 | 13.633 |

**Table 13.** *(Continued)* **External Metric Thread—M Profile Limiting Dimensions**
*ANSI/ASME B1.13M-2005 (R2020)*

| Basic Thread Designation | Tol. Class | Allowance[a] *es* | Major Diameter[b] *d* Max. | Major Diameter[b] *d* Min. | Pitch Diameter[b,c] $d_2$ Max. | Pitch Diameter[b,c] $d_2$ Min. | Pitch Diameter[b,c] $d_2$ Tol. | Minor Dia.[b] $d_1$ Max. | Minor Dia.[d] $d_3$ Min. |
|---|---|---|---|---|---|---|---|---|---|
| M16 × 2 | 6g | 0.038 | 15.962 | 15.682 | 14.663 | 14.503 | 0.160 | 13.797 | 13.271 |
| M16 × 2 | 6h | 0.000 | 16.000 | 15.720 | 14.701 | 14.541 | 0.160 | 13.835 | 13.309 |
| M16 × 2 | 4g6g | 0.038 | 15.962 | 15.682 | 14.663 | 14.563 | 0.100 | 13.797 | 13.331 |
| M16 × 1.5 | 6g | 0.032 | 15.968 | 15.732 | 14.994 | 14.854 | 0.140 | 14.344 | 13.930 |
| M16 × 1.5 | 6h | 0.000 | 16.000 | 15.764 | 15.026 | 14.886 | 0.140 | 14.376 | 13.962 |
| M16 × 1.5 | 4g6g | 0.032 | 15.968 | 15.732 | 14.994 | 14.904 | 0.090 | 14.344 | 13.980 |
| M17 × 1 | 6g | 0.026 | 16.974 | 16.794 | 16.324 | 16.206 | 0.118 | 15.891 | 15.590 |
| M17 × 1 | 6h | 0.000 | 17.000 | 16.820 | 16.350 | 16.232 | 0.118 | 15.917 | 15.616 |
| M17 × 1 | 4g6g | 0.026 | 16.974 | 16.794 | 16.324 | 16.249 | 0.075 | 15.891 | 15.633 |
| M18 × 1.5 | 6g | 0.032 | 17.968 | 17.732 | 16.994 | 16.854 | 0.140 | 16.344 | 15.930 |
| M18 × 1.5 | 6h | 0.000 | 18.000 | 17.764 | 17.026 | 16.886 | 0.140 | 16.376 | 15.962 |
| M18 × 1.5 | 4g6g | 0.032 | 17.968 | 17.732 | 16.994 | 16.904 | 0.090 | 16.344 | 15.980 |
| M20 × 2.5 | 6g | 0.042 | 19.958 | 19.623 | 18.334 | 18.164 | 0.170 | 17.251 | 16.624 |
| M20 × 2.5 | 6h | 0.000 | 20.000 | 19.665 | 18.376 | 18.206 | 0.170 | 17.293 | 16.666 |
| M20 × 2.5 | 4g6g | 0.042 | 19.958 | 19.623 | 18.334 | 18.228 | 0.106 | 17.251 | 16.688 |
| M20 × 1.5 | 6g | 0.032 | 19.968 | 19.732 | 18.994 | 18.854 | 0.140 | 18.344 | 17.930 |
| M20 × 1.5 | 6h | 0.000 | 20.000 | 19.764 | 19.026 | 18.886 | 0.140 | 18.376 | 17.962 |
| M20 × 1.5 | 4g6g | 0.032 | 19.968 | 19.732 | 18.994 | 18.904 | 0.090 | 18.344 | 17.980 |
| M20 × 1 | 6g | 0.026 | 19.974 | 19.794 | 19.324 | 19.206 | 0.118 | 18.891 | 18.590 |
| M20 × 1 | 6h | 0.000 | 20.000 | 19.820 | 19.350 | 19.232 | 0.118 | 18.917 | 18.616 |
| M20 × 1 | 4g6g | 0.026 | 19.974 | 19.794 | 19.324 | 19.249 | 0.075 | 18.891 | 18.633 |
| M22 × 2.5 | 6g | 0.042 | 21.958 | 21.623 | 20.334 | 20.164 | 0.170 | 19.251 | 18.624 |
| M22 × 2.5 | 6h | 0.000 | 22.000 | 21.665 | 20.376 | 20.206 | 0.170 | 19.293 | 18.666 |
| M22 × 1.5 | 6g | 0.032 | 21.968 | 21.732 | 20.994 | 20.854 | 0.140 | 20.344 | 19.930 |
| M22 × 1.5 | 6h | 0.000 | 22.000 | 21.764 | 21.026 | 20.886 | 0.140 | 20.376 | 19.962 |
| M22 × 1.5 | 4g6g | 0.032 | 21.968 | 21.732 | 20.994 | 20.904 | 0.090 | 20.344 | 19.980 |
| M24 × 3 | 6g | 0.048 | 23.952 | 23.577 | 22.003 | 21.803 | 0.200 | 20.704 | 19.955 |
| M24 × 3 | 6h | 0.000 | 24.000 | 23.625 | 22.051 | 21.851 | 0.200 | 20.752 | 20.003 |
| M24 × 3 | 4g6g | 0.048 | 23.952 | 23.577 | 22.003 | 21.878 | 0.125 | 20.704 | 20.030 |
| M24 × 2 | 6g | 0.038 | 23.962 | 23.682 | 22.663 | 22.493 | 0.170 | 21.797 | 21.261 |
| M24 × 2 | 6h | 0.000 | 24.000 | 23.720 | 22.701 | 22.531 | 0.170 | 21.835 | 21.299 |
| M24 × 2 | 4g6g | 0.038 | 23.962 | 23.682 | 22.663 | 22.557 | 0.106 | 21.797 | 21.325 |
| M25 × 1.5 | 6g | 0.032 | 24.968 | 24.732 | 23.994 | 23.844 | 0.150 | 23.344 | 22.920 |
| M25 × 1.5 | 6h | 0.000 | 25.000 | 24.764 | 24.026 | 23.876 | 0.150 | 23.376 | 22.952 |
| M25 × 1.5 | 4g6g | 0.032 | 24.968 | 24.732 | 23.994 | 23.899 | 0.095 | 23.344 | 22.975 |
| M27 × 3 | 6g | 0.048 | 26.952 | 26.577 | 25.003 | 24.803 | 0.200 | 23.704 | 22.955 |
| M27 × 3 | 6h | 0.000 | 27.000 | 26.625 | 25.051 | 24.851 | 0.200 | 23.752 | 23.003 |
| M27 × 2 | 6g | 0.038 | 26.962 | 26.682 | 25.663 | 25.493 | 0.170 | 24.797 | 24.261 |
| M27 × 2 | 6h | 0.000 | 27.000 | 26.720 | 25.701 | 25.531 | 0.170 | 24.835 | 24.299 |
| M27 × 2 | 4g6g | 0.038 | 26.962 | 26.682 | 25.663 | 25.557 | 0.106 | 24.797 | 24.325 |
| M30 × 3.5 | 6g | 0.053 | 29.947 | 29.522 | 27.674 | 27.462 | 0.212 | 26.158 | 25.306 |
| M30 × 3.5 | 6h | 0.000 | 30.000 | 29.575 | 27.727 | 27.515 | 0.212 | 26.211 | 25.359 |
| M30 × 3.5 | 4g6g | 0.053 | 29.947 | 29.522 | 27.674 | 27.542 | 0.132 | 26.158 | 25.386 |
| M30 × 2 | 6g | 0.038 | 29.962 | 29.682 | 28.663 | 28.493 | 0.170 | 27.797 | 27.261 |
| M30 × 2 | 6h | 0.000 | 30.000 | 29.720 | 28.701 | 28.531 | 0.170 | 27.835 | 27.299 |
| M30 × 2 | 4g6g | 0.038 | 29.962 | 29.682 | 28.663 | 28.557 | 0.106 | 27.797 | 27.325 |
| M30 × 1.5 | 6g | 0.032 | 29.968 | 29.732 | 28.994 | 28.844 | 0.150 | 28.344 | 27.920 |
| M30 × 1.5 | 6h | 0.000 | 30.000 | 29.764 | 29.026 | 28.876 | 0.150 | 28.376 | 27.952 |
| M30 × 1.5 | 4g6g | 0.032 | 29.968 | 29.732 | 28.994 | 28.899 | 0.095 | 28.344 | 27.975 |
| M33 × 2 | 6g | 0.038 | 32.962 | 32.682 | 31.663 | 31.493 | 0.170 | 30.797 | 30.261 |
| M33 × 2 | 6h | 0.000 | 33.000 | 32.720 | 31.701 | 31.531 | 0.170 | 30.835 | 30.299 |
| M33 × 2 | 4g6g | 0.038 | 32.962 | 32.682 | 31.663 | 31.557 | 0.106 | 30.797 | 30.325 |
| M35 × 1.5 | 6g | 0.032 | 34.968 | 34.732 | 33.994 | 33.844 | 0.150 | 33.344 | 32.920 |
| M35 × 1.5 | 6h | 0.000 | 35.000 | 34.764 | 34.026 | 33.876 | 0.150 | 33.376 | 32.952 |
| M36 × 4 | 6g | 0.060 | 35.940 | 35.465 | 33.342 | 33.118 | 0.224 | 31.610 | 30.654 |
| M36 × 4 | 6h | 0.000 | 36.000 | 35.525 | 33.402 | 33.178 | 0.224 | 31.670 | 30.714 |
| M36 × 4 | 4g6g | 0.060 | 35.940 | 35.465 | 33.342 | 33.202 | 0.140 | 31.610 | 30.738 |
| M36 × 2 | 6g | 0.038 | 35.962 | 35.682 | 34.663 | 34.493 | 0.170 | 33.797 | 33.261 |
| M36 × 2 | 6h | 0.000 | 36.000 | 35.720 | 34.701 | 34.531 | 0.170 | 33.835 | 33.299 |
| M36 × 2 | 4g6g | 0.038 | 35.962 | 35.682 | 34.663 | 34.557 | 0.106 | 33.797 | 33.325 |
| M39 × 2 | 6g | 0.038 | 38.962 | 38.682 | 37.663 | 37.493 | 0.170 | 36.797 | 36.261 |
| M39 × 2 | 6h | 0.000 | 39.000 | 38.720 | 37.701 | 37.531 | 0.170 | 36.835 | 36.299 |
| M39 × 2 | 4g6g | 0.038 | 38.962 | 38.682 | 37.663 | 37.557 | 0.106 | 36.797 | 36.325 |
| M40 × 1.5 | 6g | 0.032 | 39.968 | 39.732 | 38.994 | 38.844 | 0.150 | 38.344 | 37.920 |

**Table 13.** *(Continued)* **External Metric Thread—M Profile Limiting Dimensions**
*ANSI/ASME B1.13M-2005 (R2020)*

| Basic Thread Designation | Tol. Class | Allowance[a] *es* | Major Diameter[b] *d* | | Pitch Diameter[b, c] $d_2$ | | | Minor Dia.[b] $d_1$ | Minor Dia.[d] $d_3$ |
|---|---|---|---|---|---|---|---|---|---|
| | | | Max. | Min. | Max. | Min. | Tol. | Max. | Min. |
| M40 × 1.5 | 6h | 0.000 | 40.000 | 39.764 | 39.026 | 38.876 | 0.150 | 38.376 | 37.952 |
| M40 × 1.5 | 4g6g | 0.032 | 39.968 | 39.732 | 38.994 | 38.899 | 0.095 | 38.344 | 37.975 |
| M42 × 4.5 | 6g | 0.063 | 41.937 | 41.437 | 39.014 | 38.778 | 0.236 | 37.065 | 36.006 |
| M42 × 4.5 | 6h | 0.000 | 42.000 | 41.500 | 39.077 | 38.841 | 0.236 | 37.128 | 36.069 |
| M42 × 4.5 | 4g6g | 0.063 | 41.937 | 41.437 | 39.014 | 38.864 | 0.150 | 37.065 | 36.092 |
| M42 × 2 | 6g | 0.038 | 41.962 | 41.682 | 40.663 | 40.493 | 0.170 | 39.797 | 39.261 |
| M42 × 2 | 6h | 0.000 | 42.000 | 41.720 | 40.701 | 40.531 | 0.170 | 39.835 | 39.299 |
| M42 × 2 | 4g6g | 0.038 | 41.962 | 41.682 | 40.663 | 40.557 | 0.106 | 39.797 | 39.325 |
| M45 × 1.5 | 6g | 0.032 | 44.968 | 44.732 | 43.994 | 43.844 | 0.150 | 43.344 | 42.920 |
| M45 × 1.5 | 6h | 0.000 | 45.000 | 44.764 | 44.026 | 43.876 | 0.150 | 43.376 | 42.952 |
| M45 × 1.5 | 4g6g | 0.032 | 44.968 | 44.732 | 43.994 | 43.899 | 0.095 | 43.344 | 42.975 |
| M48 × 5 | 6g | 0.071 | 47.929 | 47.399 | 44.681 | 44.431 | 0.250 | 42.516 | 41.351 |
| M48 × 5 | 6h | 0.000 | 48.000 | 47.470 | 44.752 | 44.502 | 0.250 | 42.587 | 41.422 |
| M48 × 5 | 4g6g | 0.071 | 47.929 | 47.399 | 44.681 | 44.521 | 0.160 | 42.516 | 41.441 |
| M48 × 2 | 6g | 0.038 | 47.962 | 47.682 | 46.663 | 46.483 | 0.180 | 45.797 | 45.251 |
| M48 × 2 | 6h | 0.000 | 48.000 | 47.720 | 46.701 | 46.521 | 0.180 | 45.835 | 45.289 |
| M48 × 2 | 4g6g | 0.038 | 47.962 | 47.682 | 46.663 | 46.551 | 0.112 | 45.797 | 45.319 |
| M50 × 1.5 | 6g | 0.032 | 49.968 | 49.732 | 48.994 | 48.834 | 0.160 | 48.344 | 47.910 |
| M50 × 1.5 | 6h | 0.000 | 50.000 | 49.764 | 49.026 | 48.866 | 0.160 | 48.376 | 47.942 |
| M50 × 1.5 | 4g6g | 0.032 | 49.968 | 49.732 | 48.994 | 48.894 | 0.100 | 48.344 | 47.970 |
| M55 × 1.5 | 6g | 0.032 | 54.968 | 54.732 | 53.994 | 53.834 | 0.160 | 53.344 | 52.910 |
| M55 × 1.5 | 6h | 0.000 | 55.000 | 54.764 | 54.026 | 53.866 | 0.160 | 53.376 | 52.942 |
| M55 × 1.5 | 4g6g | 0.032 | 54.968 | 54.732 | 53.994 | 53.894 | 0.100 | 53.344 | 52.970 |
| M56 × 5.5 | 6g | 0.075 | 55.925 | 55.365 | 52.353 | 52.088 | 0.265 | 49.971 | 48.700 |
| M56 × 5.5 | 6h | 0.000 | 56.000 | 55.440 | 52.428 | 52.163 | 0.265 | 50.046 | 48.775 |
| M56 × 5.5 | 4g6g | 0.075 | 55.925 | 55.365 | 52.353 | 52.183 | 0.170 | 49.971 | 48.795 |
| M56 × 2 | 6g | 0.038 | 55.962 | 55.682 | 54.663 | 54.483 | 0.180 | 53.797 | 53.251 |
| M56 × 2 | 6h | 0.000 | 56.000 | 55.720 | 54.701 | 54.521 | 0.180 | 53.835 | 53.289 |
| M56 × 2 | 4g6g | 0.038 | 55.962 | 55.682 | 54.663 | 54.551 | 0.112 | 53.797 | 53.319 |
| M60 × 1.5 | 6g | 0.032 | 59.968 | 59.732 | 58.994 | 58.834 | 0.160 | 58.344 | 57.910 |
| M60 × 1.5 | 6h | 0.000 | 60.000 | 59.764 | 59.026 | 58.866 | 0.160 | 58.376 | 57.942 |
| M60 × 1.5 | 4g6g | 0.032 | 59.968 | 59.732 | 58.994 | 58.894 | 0.100 | 58.344 | 57.970 |
| M64 × 6 | 6g | 0.080 | 63.920 | 63.320 | 60.023 | 59.743 | 0.280 | 57.425 | 56.047 |
| M64 × 6 | 6h | 0.000 | 64.000 | 63.400 | 60.103 | 59.823 | 0.280 | 57.505 | 56.127 |
| M64 × 6 | 4g6g | 0.080 | 63.920 | 63.320 | 60.023 | 59.843 | 0.180 | 57.425 | 56.147 |
| M64 × 2 | 6g | 0.038 | 63.962 | 63.682 | 62.663 | 62.483 | 0.180 | 61.797 | 61.251 |
| M64 × 2 | 6h | 0.000 | 64.000 | 63.720 | 62.701 | 62.521 | 0.180 | 61.835 | 61.289 |
| M64 × 2 | 4g6g | 0.038 | 63.962 | 63.682 | 62.663 | 62.551 | 0.112 | 61.797 | 61.319 |
| M65 × 1.5 | 6g | 0.032 | 64.968 | 64.732 | 63.994 | 63.834 | 0.160 | 63.344 | 62.910 |
| M65 × 1.5 | 6h | 0.000 | 65.000 | 64.764 | 64.026 | 63.866 | 0.160 | 63.376 | 62.942 |
| M65 × 1.5 | 4g6g | 0.032 | 64.968 | 64.732 | 63.994 | 63.894 | 0.100 | 63.344 | 62.970 |
| M70 × 1.5 | 6g | 0.032 | 69.968 | 69.732 | 68.994 | 68.834 | 0.160 | 68.344 | 67.910 |
| M70 × 1.5 | 6h | 0.000 | 70.000 | 69.764 | 69.026 | 68.866 | 0.160 | 68.376 | 67.942 |
| M70 × 1.5 | 4g6g | 0.032 | 69.968 | 69.732 | 68.994 | 68.894 | 0.100 | 68.344 | 67.970 |
| M72 × 6 | 6g | 0.080 | 71.920 | 71.320 | 68.023 | 67.743 | 0.280 | 65.425 | 64.047 |
| M72 × 6 | 6h | 0.000 | 72.000 | 71.400 | 68.103 | 67.823 | 0.280 | 65.505 | 64.127 |
| M72 × 6 | 4g6g | 0.080 | 71.920 | 71.320 | 68.023 | 67.843 | 0.180 | 65.425 | 64.147 |
| M72 × 2 | 6g | 0.038 | 71.962 | 71.682 | 70.663 | 70.483 | 0.180 | 69.797 | 69.251 |
| M72 × 2 | 6h | 0.000 | 72.000 | 71.720 | 70.701 | 70.521 | 0.180 | 69.835 | 69.289 |
| M72 × 2 | 4g6g | 0.038 | 71.962 | 71.682 | 70.663 | 70.551 | 0.112 | 69.797 | 69.319 |
| M75 × 1.5 | 6g | 0.032 | 74.968 | 74.732 | 73.994 | 73.834 | 0.160 | 73.344 | 72.910 |
| M75 × 1.5 | 6h | 0.000 | 75.000 | 74.764 | 74.026 | 73.866 | 0.160 | 73.376 | 72.942 |
| M75 × 1.5 | 4g6g | 0.032 | 74.968 | 74.732 | 73.994 | 73.894 | 0.100 | 73.344 | 72.970 |
| M80 × 6 | 6g | 0.080 | 79.920 | 79.320 | 76.023 | 75.743 | 0.280 | 73.425 | 72.047 |
| M80 × 6 | 6h | 0.000 | 80.000 | 79.400 | 76.103 | 75.823 | 0.280 | 73.505 | 72.127 |
| M80 × 6 | 4g6g | 0.080 | 79.920 | 79.320 | 76.023 | 75.843 | 0.180 | 73.425 | 72.147 |
| M80 × 2 | 6g | 0.038 | 79.962 | 79.682 | 78.663 | 78.483 | 0.180 | 77.797 | 77.251 |
| M80 × 2 | 6h | 0.000 | 80.000 | 79.720 | 78.701 | 78.521 | 0.180 | 77.835 | 77.289 |
| M80 × 2 | 4g6g | 0.038 | 79.962 | 79.682 | 78.663 | 78.551 | 0.112 | 77.797 | 77.319 |
| M80 × 1.5 | 6g | 0.032 | 79.968 | 79.732 | 78.994 | 78.834 | 0.160 | 78.344 | 77.910 |
| M80 × 1.5 | 6h | 0.000 | 80.000 | 79.764 | 79.026 | 78.866 | 0.160 | 78.376 | 77.942 |
| M80 × 1.5 | 4g6g | 0.032 | 79.968 | 79.732 | 78.994 | 78.894 | 0.100 | 78.344 | 77.970 |
| M85 × 2 | 6g | 0.038 | 84.962 | 84.682 | 83.663 | 83.483 | 0.180 | 82.797 | 82.251 |
| M85 × 2 | 6h | 0.000 | 85.000 | 84.720 | 83.701 | 83.521 | 0.180 | 82.835 | 82.289 |

**Table 13.** *(Continued)* **External Metric Thread—M Profile Limiting Dimensions**
*ANSI/ASME B1.13M-2005 (R2020)*

| Basic Thread Designation | Tol. Class | Allowance [a] *es* | Major Diameter [b] *d* | | Pitch Diameter [b,c] $d_2$ | | | Minor Dia. [b] $d_1$ | Minor Dia. [d] $d_3$ |
|---|---|---|---|---|---|---|---|---|---|
| | | | Max. | Min. | Max. | Min. | Tol. | Max. | Min. |
| M85 × 2 | 4g6g | 0.038 | 84.962 | 84.682 | 83.663 | 83.551 | 0.112 | 82.797 | 82.319 |
| M90 × 6 | 6g | 0.080 | 89.920 | 89.320 | 86.023 | 85.743 | 0.280 | 83.425 | 82.047 |
| M90 × 6 | 6h | 0.000 | 90.000 | 89.400 | 86.103 | 85.823 | 0.280 | 83.505 | 82.127 |
| M90 × 6 | 4g6g | 0.080 | 89.920 | 89.320 | 86.023 | 85.843 | 0.180 | 83.425 | 82.147 |
| M90 × 2 | 6g | 0.038 | 89.962 | 89.682 | 88.663 | 88.483 | 0.180 | 87.797 | 87.251 |
| M90 × 2 | 6h | 0.000 | 90.000 | 89.720 | 88.701 | 88.521 | 0.180 | 87.835 | 87.289 |
| M90 × 2 | 4g6g | 0.038 | 89.962 | 89.682 | 88.663 | 88.551 | 0.112 | 87.797 | 87.319 |
| M95 × 2 | 6g | 0.038 | 94.962 | 94.682 | 93.663 | 93.473 | 0.190 | 92.797 | 92.241 |
| M95 × 2 | 6h | 0.000 | 95.000 | 94.720 | 93.701 | 93.511 | 0.190 | 92.835 | 92.279 |
| M95 × 2 | 4g6g | 0.038 | 94.962 | 94.682 | 93.663 | 93.545 | 0.118 | 92.797 | 92.313 |
| M100 × 6 | 6g | 0.080 | 99.920 | 99.320 | 96.023 | 95.723 | 0.300 | 93.425 | 92.027 |
| M100 × 6 | 6h | 0.000 | 100.000 | 99.400 | 96.103 | 95.803 | 0.300 | 93.505 | 92.107 |
| M100 × 6 | 4g6g | 0.080 | 99.920 | 99.320 | 96.023 | 95.833 | 0.190 | 93.425 | 92.137 |
| M100 × 2 | 6g | 0.038 | 99.962 | 99.682 | 98.663 | 98.473 | 0.190 | 97.797 | 97.241 |
| M100 × 2 | 6h | 0.000 | 100.000 | 99.720 | 98.701 | 98.511 | 0.190 | 97.835 | 97.279 |
| M100 × 2 | 4g6g | 0.038 | 99.962 | 99.682 | 98.663 | 98.545 | 0.118 | 97.797 | 97.313 |
| M105 × 2 | 6g | 0.038 | 104.962 | 104.682 | 103.663 | 103.473 | 0.190 | 102.797 | 102.241 |
| M105 × 2 | 6h | 0.000 | 105.000 | 104.720 | 103.701 | 103.511 | 0.190 | 102.835 | 102.279 |
| M105 × 2 | 4g6g | 0.038 | 104.962 | 104.682 | 103.663 | 103.545 | 0.118 | 102.797 | 102.313 |
| M110 × 2 | 6g | 0.038 | 109.962 | 109.682 | 108.663 | 108.473 | 0.190 | 107.797 | 107.241 |
| M110 × 2 | 6h | 0.000 | 110.000 | 109.720 | 108.701 | 108.511 | 0.190 | 107.835 | 107.279 |
| M110 × 2 | 4g6g | 0.038 | 109.962 | 109.682 | 108.663 | 108.545 | 0.118 | 107.797 | 107.313 |
| M120 × 2 | 6g | 0.038 | 119.962 | 119.682 | 118.663 | 118.473 | 0.190 | 117.797 | 117.241 |
| M120 × 2 | 6h | 0.000 | 120.000 | 119.720 | 118.701 | 118.511 | 0.190 | 117.835 | 117.279 |
| M120 × 2 | 4g6g | 0.038 | 119.962 | 119.682 | 118.663 | 118.545 | 0.118 | 117.797 | 117.313 |
| M130 × 2 | 6g | 0.038 | 129.962 | 129.682 | 128.663 | 128.473 | 0.190 | 127.797 | 127.241 |
| M130 × 2 | 6h | 0.000 | 130.000 | 129.720 | 128.701 | 128.511 | 0.190 | 127.835 | 127.279 |
| M130 × 2 | 4g6g | 0.038 | 129.962 | 129.682 | 128.663 | 128.545 | 0.118 | 127.797 | 127.313 |
| M140 × 2 | 6g | 0.038 | 139.962 | 139.682 | 138.663 | 138.473 | 0.190 | 137.797 | 137.241 |
| M140 × 2 | 6h | 0.000 | 140.000 | 139.720 | 138.701 | 138.511 | 0.190 | 137.835 | 137.279 |
| M140 × 2 | 4g6g | 0.038 | 139.962 | 139.682 | 138.663 | 138.545 | 0.118 | 137.797 | 137.313 |
| M150 × 2 | 6g | 0.038 | 149.962 | 149.682 | 148.663 | 148.473 | 0.190 | 147.797 | 147.241 |
| M150 × 2 | 6h | 0.000 | 150.000 | 149.720 | 148.701 | 148.511 | 0.190 | 147.835 | 147.279 |
| M150 × 2 | 4g6g | 0.038 | 149.962 | 149.682 | 148.663 | 148.545 | 0.118 | 147.797 | 147.313 |
| M160 × 3 | 6g | 0.048 | 159.952 | 159.577 | 158.003 | 157.779 | 0.224 | 156.704 | 155.931 |
| M160 × 3 | 6h | 0.000 | 160.000 | 159.625 | 158.051 | 157.827 | 0.224 | 156.752 | 155.979 |
| M160 × 3 | 4g6g | 0.048 | 159.952 | 159.577 | 158.003 | 157.863 | 0.140 | 156.704 | 156.015 |
| M170 × 3 | 6g | 0.048 | 169.952 | 169.577 | 168.003 | 167.779 | 0.224 | 166.704 | 165.931 |
| M170 × 3 | 6h | 0.000 | 170.000 | 169.625 | 168.051 | 167.827 | 0.224 | 166.752 | 165.979 |
| M170 × 3 | 4g6g | 0.048 | 169.952 | 169.577 | 168.003 | 167.863 | 0.140 | 166.704 | 166.015 |
| M180 × 3 | 6g | 0.048 | 179.952 | 179.577 | 178.003 | 177.779 | 0.224 | 176.704 | 175.931 |
| M180 × 3 | 6h | 0.000 | 180.000 | 179.625 | 178.051 | 177.827 | 0.224 | 176.752 | 175.979 |
| M180 × 3 | 4g6g | 0.048 | 179.952 | 179.577 | 178.003 | 177.863 | 0.140 | 176.704 | 176.015 |
| M190 × 3 | 6g | 0.048 | 189.952 | 189.577 | 188.003 | 187.753 | 0.250 | 186.704 | 185.905 |
| M190 × 3 | 6h | 0.000 | 190.000 | 189.625 | 188.051 | 187.801 | 0.250 | 186.752 | 185.953 |
| M190 × 3 | 4g6g | 0.048 | 189.952 | 189.577 | 188.003 | 187.843 | 0.160 | 186.704 | 185.995 |
| M200 × 3 | 6g | 0.048 | 199.952 | 199.577 | 198.003 | 197.753 | 0.250 | 196.704 | 195.905 |
| M200 × 3 | 6h | 0.000 | 200.000 | 199.625 | 198.051 | 197.801 | 0.250 | 196.752 | 195.953 |
| M200 × 3 | 4g6g | 0.048 | 199.952 | 199.577 | 198.003 | 197.843 | 0.160 | 196.704 | 195.995 |

[a] *es* is an absolute value.

[b] Coated threads with tolerance classes 6g or 4g6g, see *Material Limits for Coated Threads,* 2023.

[c] Functional diameter size includes the effects of all variations in pitch diameter, thread form, and profile. The variations in the individual thread characteristics such as flank angle, lead, taper, and roundness on a given thread, cause the measurements of the pitch diameter and functional diameter to vary from one another on most threads. The pitch diameter and the functional diameter on a given thread are equal to one another only when the thread form is perfect. When required to inspect either the pitch diameter, the functional diameter, or both, for thread acceptance, use the same limits of size for the appropriate thread size and class.

[d] Dimension used in the design of tools, etc.; in dimensioning external threads it is not normally specified. Generally, minor diameter acceptance is based on maximum material condition gaging.

All dimensions are in millimeters.

**Metric Screw Thread Designations.**—Metric screw threads are identified by the letter (M) for the thread form profile, followed by the nominal diameter size and the pitch expressed in millimeters, separated by the sign (×) and followed by the tolerance class separated by a dash (–) from the pitch.

The simplified international practice for designating coarse pitch M profile metric screw threads is to leave off the pitch. Thus a M14 × 2 thread is designated just M14. However, to prevent misunderstanding, it is mandatory to use the value for pitch in all designations.

Thread acceptability gaging system requirements of ANSI/ASME B1.3 may be added to the thread size designation as noted in the examples (numbers in parentheses) or as specified in pertinent documentation, such as the drawing or procurement document.

Unless otherwise specified in the designation, the screw thread is right-hand.

*Examples:* External thread of M profile, right-hand: M6 × 1 – 4g6g (22)

Internal thread of M profile, right-hand: M6 × 1 – 5H6H (21)

*Designation of Left-Hand Thread:* When a left-hand thread is specified, the tolerance class designation is followed by a dash and LH.

M6 × 1 – 5H6H – LH (23)

*Designation for Identical Tolerance Classes:* If the two tolerance class designations for a thread are identical, it is not necessary to repeat the symbols.

M6 × 1 – 6H (21)

*Designation Using All Capital Letters:* When computer and teletype thread designations use all capital letters, the external or internal thread may need further identification. Thus the tolerance class is followed by the abbreviations EXT or INT in capital letters.

*Example:* M6 × 1 – 4G6G EXT; M6 × 1 – 6H INT

*Designation for Thread Fit:* A fit between mating threads is indicated by the internal thread tolerance class followed by the external thread tolerance class and separated by a slash.

*Examples:* M6 × 1 – 6H/6g; M6 × 1 – 6H/4g6g

*Designation for Rounded Root External Thread:* The M profile with a minimum root radius of 0.125P on the external thread is desirable for all threads but is mandatory for threaded mechanical fasteners of ISO 898/I property class 8.8 (minimum tensile strength 800 MPa) and stronger. No special designation is required for these threads. Other parts requiring a 0.125P root radius must have that radius specified.

When a special rounded root is required, its external thread designation is suffixed by the minimum root radius value in millimeters and the letter R.

*Example:* M42 × 4.5 – 6g – 0.63R

*Designation of Threads Having Modified Crests:* Where the limits of size of the major diameter of an external thread or the minor diameter of an internal thread are modified, the thread designation is suffixed by the letters MOD followed by the modified diameter limits.

*Examples:*

| | |
|---|---|
| External thread M profile, major diameter reduced 0.075 mm.<br>M6 × 1 – 4h6h MOD<br>Major dia = 5.745 – 5.925 MOD | Internal thread M profile, minor diameter increased 0.075 mm.<br>M6 × 1 – 4H5H MOD<br>Minor dia = 5.101 – 5.291 MOD |

*Designation of Special Threads:* Special diameter-pitch threads developed in accordance with this Standard ANSI/ASME B1.13M are identified by the letters SPL following the tolerance class. The limits of size for the major diameter, pitch diameter, and minor diameter are specified below this designation.

*Examples:*

| External thread | Internal thread |
|---|---|
| M6.5 × 1 – 4h6h – SPL (22)<br>Major dia = 6.320 – 6.500<br>Pitch dia = 5.779 – 5.850<br>Minor dia = 5.163 – 5.386 | M6.5 × 1 – 4H5H – SPL (23)<br>Major dia = 6.500 min<br>Pitch dia = 5.850 – 5.945<br>Minor dia = 5.417 – 5.607 |

*Designation of Multiple Start Threads:* When a thread is required with a multiple start, it is designated by specifying sequentially: M for metric thread, nominal diameter size, × L for lead, lead value, dash, P for pitch, pitch value, dash, tolerance class, parenthesis, script number of starts, and the word starts, close parenthesis.

*Examples:* M16 × L4 – P2 – 4h6h (TWO STARTS)
M14 × L6 – P2 – 6H (THREE STARTS)

*Designation of Coated or Plated Threads:* In designating coated or plated M threads the tolerance class should be specified as after coating or after plating. If no designation of after coating or after plating is specified, the tolerance class applies before coating or plating in accordance with ISO practice. After plating, the thread must not transgress the maximum material limits for the tolerance position H/h.

*Examples:* M6 × 1 – 6h AFTER COATING or AFTER PLATING
M6 × 1 – 6g AFTER COATING or AFTER PLATING

Where the tolerance position G/g is insufficient relief for the application to hold the threads within product limits, the coating or plating allowance may be specified as the maximum and minimum limits of size for minor and pitch diameters of internal threads or major and pitch diameters for external threads before coating or plating.

*Example:* Allowance on external thread M profile based on 0.010 mm minimum coating thickness.

M6 × 1 – 4h6h – AFTER COATING

BEFORE COATING

Major dia = 5.780 – 5.940

Pitch dia = 5.239 – 5.290

## Metric Screw Threads — MJ Profile

The MJ screw thread is intended for aerospace metric threaded parts and for other highly stressed applications requiring high temperature or high fatigue strength, or for "no allowance" applications. The MJ profile thread is a hard metric version similar to the UNJ inch standards, ANSI/ASME B1.15 and MIL-S-8879. The MJ profile thread has a 0.15011$P$ to 0.180424$P$ controlled root radius in the external thread and the internal thread minor diameter truncated to accommodate the external thread maximum root radius.

First issued in 1978, the American National Standard ANSI/ASME B1.21M-1997 (R2018) establishes the basic triangular profile for the MJ form of thread; gives a system of designations; lists the standard series of diameter-pitch combinations for diameters from 1.6 to 200 mm; and specifies limiting dimensions and tolerances. Changes included in the 1997 revision are the addition of tolerance class 4G6G and 4G5G/4g6g comparable to ANSI/ASME B1.15 (UNJ thread); the addition of tolerance class 6H/6g comparable to ANSI/ASME B1.13M; and changes in the rounding proceedure as set forth in ANSI/ASME B1.30.

**Diameter-Pitch Combinations.—**This Standard includes a selected series of diameter-pitch combinations of threads taken from International Standard ISO 261 plus some additional sizes in the constant pitch series. These are given in Table 1. It also includes the standard series of diameter-pitch combinations for aerospace screws, bolts, nuts, and fluid system fittings as shown in Table 2.

**Table 1. American National Standard Metric Screw Threads MJ Profile Diameter-Pitch Combinations** *ANSI/ASME B1.21M-1997 (R2018)*

| Nominal Diameter | | Pitch | | Nominal Diameter | | Pitch | |
|---|---|---|---|---|---|---|---|
| Choices | | | | Choices | | | |
| 1st | 2nd | Coarse | Fine | 1st | 2nd | Coarse | Fine |
| 1.6 | … | 0.35 | … | … | 52 | … | 3, 2, 1.5 |
| … | 1.8 | 0.35 | … | 55 | … | … | 3, 2, 1.5 |
| 2.0 | … | 0.4 | … | | 56 | 5.5 | 3, 2, 1.5 |
| … | 2.2 | 0.45 | … | … | 58 | … | 3, 2, 1.5 |
| 2.5 | … | 0.45 | … | 60 | | … | 3, 2, 1.5 |
| 3 | … | 0.5 | … | … | 62 | … | 3, 2, 1.5 |
| 3.5 | … | 0.6 | … | … | 64 | 6 | 3, 2, 1.5 |
| 4 | … | 0.7 | … | 65 | … | … | 3, 2, 1.5 |
| … | 4.5 | 0.75 | … | … | 68 | … | 3, 2, 1.5 |
| 5 | … | 0.8 | … | 70 | … | … | 3, 2, 1.5 |
| 6 | … | 1 | 0.75 | … | 72 | 6 | 3, 2, 1.5 |
| 7 | … | 1 | 0.75 | 75 | … | … | 3, 2, 1.5 |
| 8 | … | 1.25 | 1, 0.75 | … | 76 | … | 3, 2, 1.5 |
| … | 9 | 1.25 | 1, 0.75 | … | 78 | … | 3[a], 2, 1.5[a] |
| 10 | … | 1.5 | 1.25, 1, 0.75 | 80 | … | 6 | 3, 2, 1.5 |
| … | 11 | 1.5 | 1.25[b], 1, 0.75 | … | 82 | … | 3[a], 2, 1.5[a] |
| 12 | … | 1.75 | 1.5, 1.25, 1 | 85 | … | … | 3, 2, 1.5[a] |
| 14 | … | 2 | 1.5, 1.25[c], 1 | 90 | … | 6 | 3, 2, 1.5[a] |
| … | 15 | … | 1.5, 1 | 95 | … | … | 3, 2, 1.5[a] |
| 16 | … | 2 | 1.5, 1 | 100 | … | 6 | 3, 2, 1.5[a] |
| … | 17 | … | 1.5, 1 | 105 | … | … | 3, 2, 1.5[a] |
| 18 | … | 2.5 | 2, 1.5, 1 | 110 | … | … | 3, 2, 1.5[a] |
| 20 | … | 2.5 | 2, 1.5, 1 | … | 115 | … | 3, 2, 1.5[a] |
| 22 | … | 2.5 | 2, 1.5, 1 | 120 | … | … | 3, 2, 1.5[a] |
| 24 | … | 3 | 2, 1.5, 1 | … | 125 | … | 3, 2, 1.5[a] |
| … | 25 | … | 2, 1.5, 1 | 130 | … | … | 3, 2, 1.5[a] |
| … | 26 | … | 1.5 | … | 135 | … | 3, 2, 1.5[a] |
| 27 | … | 3 | 2, 1.5, 1 | 140 | … | … | 3, 2, 1.5[a] |
| … | 28 | … | 2, 1.5, 1 | … | 145 | … | 3, 2, 1.5[a] |
| 30 | … | 3.5 | 3, 2, 1.5, 1 | 150 | … | … | 3, 2, 1.5[a] |
| … | 32 | … | 2, 1.5 | … | 155 | … | 3 |
| 33 | … | … | 3, 2, 1.5 | 160 | … | … | 3 |
| … | 35 | … | 1.5 | … | 165 | … | 3 |
| 36 | … | 4 | 3, 2, 1.5 | 170 | … | … | 3 |
| … | 38 | … | 1.5 | … | 175 | … | 3 |
| 39 | … | … | 3, 2, 1.5 | 180 | … | … | 3 |
| … | 40 | … | 3, 2, 1.5 | … | 185 | … | 3 |
| . | 42 | 4.5 | 3, 2, 1.5 | 190 | … | … | 3 |
| 45 | … | … | 3, 2, 1.5 | … | 195 | … | 3 |
| … | 48 | 5 | 3, 2, 1.5 | 200 | … | … | 3 |
| 50 | … | … | 3, 2, 1.5 | … | … | … | … |

[a] Not included in ISO 261.

[b] Only for aircraft control cable fittings.

[c] Only for spark plugs for engines.

All dimensions are in millimeters. Pitches in parentheses ( ) are to be avoided as far as possible.

**Table 2. American National Standard Metric Screw Threads MJ Profile, Diameter-Pitch Combinations for Aerospace** *ANSI/ASME B1.21M-1997 (R2018)*

| Aerospace Screws, Bolts and Nuts | | | | | | | | Aerospace Fluid System Fittings | | | | | |
|---|---|---|---|---|---|---|---|---|---|---|---|---|---|
| Nom. Size[a] | Pitch | Nom. Size | Pitch | Nom. Size | Pitch | Nom. Size | Pitch | Nom. Size | Pitch | Nom. Size | Pitch | Nom. Size | Pitch |
| 1.6 | 0.35 | 5 | 0.8 | 14 | 1.5 | 27 | 2 | 8 | 1 | 20 | 1.5 | 36 | 1.5 |
| 2 | 0.4 | 6 | 1 | 16 | 1.5 | 30 | 2 | 10 | 1 | 22 | 1.5 | 39 | 1.5 |
| 2.5 | 0.45 | 7 | 1 | 18 | 1.5 | 33 | 2 | 12 | 1.25 | 24 | 1.5 | 42 | 2 |
| 3 | 0.5 | 8 | 1 | 20 | 1.5 | 36 | 2 | 14 | 1.5 | 27 | 1.5 | 48 | 2 |
| 3.5 | 0.6 | 10 | 1.25 | 22 | 1.5 | 39 | 2 | 16 | 1.5 | 30 | 1.5 | 50 | 2 |
| 4 | 0.7 | 12 | 1.25 | 24 | 2 | … | … | 18 | 1.5 | 33 | 1.5 | … | … |

All dimensions are in millimeters.

[a] For threads smaller than 1.6 mm nominal size, use miniature screw threads (ANSI B1.10M).

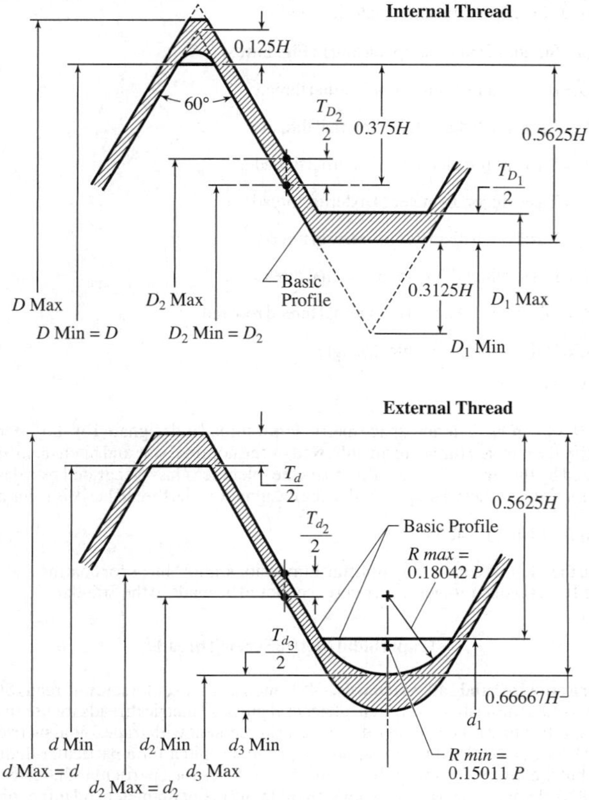


Fig. 1. Internal MJ Thread Basic and Design Profiles (Top) and External MJ Thread Basic and Design Profiles (Bottom) Showing Tolerance Zones

*Tolerances:* The thread tolerance system is based on ISO 965/1, *Metric Screw thread System of Tolerance Positions and Grades*. Tolerances are positive for internal threads and negative for external threads, that is, in the direction of minimum material.

For aerospace applications, except for fluid fittings, tolerance classes 4H5H or 4G6G and 4g6g should be used. These classes approximate classes 3B/3A in the inch system. Aerospace fluid fittings use classes 4H5H or 4H6H and 4g6g.

Tolerance classes 4G5G or 4G6G and 4g6g are provided for use when thread allowances are required. These classes provide a slightly tighter fit than the inch classes 2B/2A at minimum material condition.

Additional tolerance classes 6H/6g are included in this Standard to provide appropriate product selection based on general applications. These classes and the selection of standard diameter/pitch combinations are the same as those provided for the M profile metric screw threads in ANSI/ASME B1.13M. Classes 6H/6g result in a slightly looser fit than inch classes 2B/2A at minimum material condition.

*Symbols:* Standard symbols appearing in Fig. 1 are:

$D$ = Basic major diameter of internal thread

$D_2$ = Basic pitch diameter of internal thread

$D_1$ = Basic minor diameter of internal thread

$d$ = Basic major diameter of external thread

$d_2$ = Basic pitch diameter of external thread

$d_1$ = Basic minor diameter of internal thread

$d_3$ = Diameter to bottom of external thread root radius

$H$ = Height of fundamental triangle

$P$ = Pitch

*Basic Designations:* The aerospace metric screw thread is designated by the letters "MJ" to identify the metric J thread form, followed by the nominal size and pitch in millimeters (separated by the sign "×") and followed by the tolerance class (separated by a dash from the pitch). Unless otherwise specified in the designation, the thread helix is right-hand.

*Example:* MJ6 × 1 – 4h6h

For further details concerning limiting dimensions, allowances for coating and plating, modified and special threads, etc., reference should be made to the Standard.

## Trapezoidal Metric Screw Threads

**Comparison of ISO and DIN Standards.—**ISO metric trapezoidal screw threads Standard ISO 2904-2020 describes the system of general purpose metric threads for use in mechanisms and structures. The Standard is in basic agreement with trapezoidal metric thread DIN 103. The DIN 103 Standard applies a particular pitch for a particular diameter of thread, but the ISO Standard applies a variety of pitches for a particular diameter. In ISO 2904-2020, the same clearance is applied to both the major diameter and minor diameter, but in DIN 103 the clearance in the minor diameter is two or three times greater than clearance in the major diameter. A comparison of DIN 103 is given in Table 1.

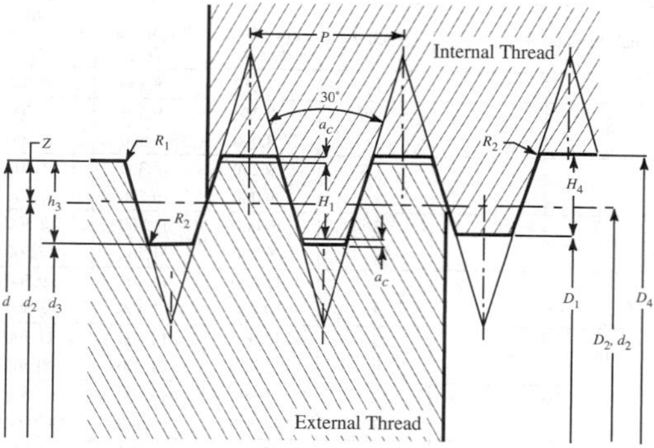


Metric Trapezoidal Thread, ISO 2904

*Terminology:* The term "bolt threads" is used for external screw threads, the term "nut threads" for internal screw threads.

*Calculation:* The values given in the International standards have been calculated by using the following formulas:

$$H_1 = 0.5P \qquad H_4 = H_1 + a_c = 0.5P + a_c \qquad H_3 = H_1 + a_c = 0.5P + a_c$$

$$D_4 = d + 2a_c \qquad Z = 0.25P = H_1/2 \qquad D_1 = d - 2H_1 = d - p$$

$$D_3 = D - 2h_3 \qquad d_2 = D_2 = d - 2Z = d - 0.5P \qquad R_{1max.} = 0.5a_c \quad R_{2max.} = a_c$$

where $a_c$ = clearance on the crest; $D$ = major diameter for nut threads; $D_2$ = pitch diameter for nut threads; $D_1$ = minor diameter for nut threads; $d$ = major diameter for bolt threads = nominal diameter; $d_2$ = pitch diameter for bolt threads; $d_3$ = minor diameter for bolt threads; $h_1$ = Height of overlapping; $h_4$ = height of nut threads; $h_3$ = height of bolt threads; and, $P$ = pitch.

**Table 1. Comparison of ISO Metric Trapezoidal Screw Thread** *ISO 2904-2020* **and Trapezoidal Metric Screw Thread** *DIN 103*

| | ISO 2904 | DIN 103 | Comment |
|---|---|---|---|
| Nominal Diameter | $D$ | $D_S$ | |
| Pitch | $p$ | $p$ | Same |
| Clearances (Bolt Circle) | $a_c$ | $b$ | Same |
| Clearances (Nut Circle) | $a_c$ | $a$ | Not same |
| Height of Overlapping | $h_1$ | $h_e$ | Same |
| Bolt Circle | | | |
| | $h_3 = 0.50P + a_c$ | $h_s = 0.50P + a$ | Same |
| | $h_{as} = 0.25p$ | $z = 0.25p$ | Same |
| Minor diameter for external thread | $D_3 = d - 2h_3$ | $k_s = d - 2h_s$ | Same |
| Pitch diameter for external thread | $D_2 = d - 2h_{as}$ | $d_2 = d - 2z$ | Same |
| Nut Circle | | | |
| Basic major diameter for nut thread | $D_4 = d + 2a_c$ | $d_n = d + a + b$ | Not same |
| Height of internal thread | $h_4 = h_3$ | $h_n = h_3 + a$ | Not same |
| Minor diameter of internal thread | $D_1 = D - 2h_1$ | $K_n = D_n - 2h_n$ | Not same |

**Table 2. ISO Metric Trapezoidal Screw Thread** *ISO 2904-2020*

| Nominal Diameter, $d$ | | | Pitch, $P$ | Pitch Dia. $d_2 = D_2$ | Major Dia. $D_4$ | Minor Diameter | |
|---|---|---|---|---|---|---|---|
| | | | | | | $d_3$ | $D_1$ |
| 8 | | | 1.5 | 7.250 | 8.300 | 6.200 | 6.500 |
| | 9 | | 1.5 | 8.250 | 9.300 | 7.200 | 7.500 |
| | | | 2 | 8.000 | 9.500 | 6.500 | 7.000 |
| 10 | | | 1.5 | 9.250 | 10.300 | 8.200 | 8.500 |
| | | | 2 | 9.000 | 10.500 | 7.500 | 8.000 |
| | 11 | | 2 | 10.000 | 11.500 | 8.500 | 9.000 |
| | | | 3 | 9.500 | 11.500 | 7.500 | 8.000 |
| 12 | | | 2 | 11.000 | 12.500 | 9.500 | 10.000 |
| | | | 3 | 10.500 | 12.500 | 8.500 | 9.000 |
| | 14 | | 2 | 13.000 | 14.500 | 11.500 | 12.000 |
| | | | 3 | 12.500 | 14.500 | 10.500 | 11.000 |
| 16 | | | 2 | 15.000 | 16.500 | 13.500 | 14.000 |
| | | | 3 | 14.500 | 16.500 | 12.500 | 13.000 |
| | 18 | | 2 | 17.000 | 18.500 | 15.500 | 16.000 |
| | | | 4 | 16.000 | 18.500 | 13.500 | 14.000 |
| 20 | | | 2 | 19.000 | 20.500 | 17.500 | 18.000 |
| | | | 4 | 18.000 | 20.500 | 15.500 | 16.000 |
| | 22 | | 3 | 20.500 | 22.500 | 18.500 | 19.000 |
| | | | 5 | 19.500 | 22.500 | 16.500 | 17.000 |
| | | | 8 | 18.000 | 23.000 | 13.000 | 14.000 |
| 24 | | | 3 | 22.500 | 24.500 | 20.500 | 21.000 |
| | | | 5 | 21.500 | 24.500 | 18.500 | 19.000 |
| | | | 8 | 20.000 | 25.000 | 15.000 | 16.000 |
| | 26 | | 3 | 24.500 | 26.500 | 22.500 | 23.000 |
| | | | 5 | 23.500 | 26.500 | 20.500 | 21.000 |
| | | | 8 | 22.000 | 27.000 | 17.000 | 18.000 |
| 28 | | | 3 | 26.500 | 28.500 | 24.500 | 25.000 |
| | | | 5 | 25.500 | 28.500 | 22.500 | 23.000 |
| | | | 8 | 24.000 | 29.000 | 19.000 | 20.000 |
| | 30 | | 3 | 28.500 | 30.500 | 26.500 | 27.000 |
| | | | 6 | 27.000 | 31.000 | 23.000 | 24.000 |
| | | | 10 | 25.000 | 31.000 | 19.000 | 20.000 |
| 32 | | | 3 | 30.500 | 32.500 | 28.500 | 29.000 |
| | | | 6 | 29.000 | 33.000 | 25.000 | 26.000 |
| | | | 10 | 27.000 | 33.000 | 21.000 | 22.000 |
| | 34 | | 3 | 32.500 | 34.500 | 30.500 | 31.000 |
| | | | 6 | 31.000 | 35.000 | 27.000 | 28.000 |
| | | | 10 | 29.000 | 35.000 | 23.000 | 24.000 |
| 36 | | | 3 | 34.500 | 36.500 | 32.500 | 33.000 |
| | | | 6 | 33.000 | 37.000 | 29.000 | 30.000 |
| | | | 10 | 31.000 | 37.000 | 25.000 | 26.000 |
| | 38 | | 3 | 36.500 | 38.500 | 34.500 | 35.000 |
| | | | 7 | 34.500 | 39.000 | 30.000 | 31.000 |
| | | | 10 | 33.000 | 39.000 | 27.000 | 28.000 |
| 40 | | | 3 | 38.500 | 40.500 | 36.500 | 37.000 |
| | | | 7 | 36.500 | 41.000 | 32.000 | 33.000 |
| | | | 10 | 35.000 | 41.000 | 29.000 | 30.000 |

**Table 2.** *(Continued)* **ISO Metric Trapezoidal Screw Thread** *ISO 2904-2020*

| Nominal Diameter, $d$ | | | Pitch, $P$ | Pitch Dia. $d_2 = D_2$ | Major Dia. $D_4$ | Minor Diameter $d_3$ | Minor Diameter $D_1$ |
|---|---|---|---|---|---|---|---|
| | 42 | | 3 | 40.500 | 42.500 | 38.500 | 39.000 |
| | | | 7 | 38.500 | 43.000 | 34.000 | 35.000 |
| | | | 10 | 37.000 | 43.000 | 31.000 | 32.000 |
| 44 | | | 3 | 42.500 | 44.500 | 40.500 | 41.000 |
| | | | 7 | 40.500 | 45.000 | 36.000 | 37.000 |
| | | | 12 | 38.000 | 45.000 | 31.000 | 32.000 |
| | 46 | | 3 | 44.500 | 46.500 | 42.500 | 43.000 |
| | | | 8 | 42.000 | 47.000 | 37.000 | 38.000 |
| | | | 12 | 40.000 | 47.000 | 33.000 | 34.000 |
| 48 | | | 3 | 46.500 | 48.500 | 44.500 | 45.000 |
| | | | 8 | 44.000 | 49.000 | 39.000 | 40.000 |
| | | | 12 | 42.000 | 49.000 | 35.000 | 36.000 |
| | 50 | | 3 | 48.500 | 50.500 | 46.500 | 47.000 |
| | | | 8 | 46.000 | 51.000 | 41.000 | 42.000 |
| | | | 12 | 44.000 | 51.000 | 37.000 | 38.000 |
| 52 | | | 3 | 50.500 | 52.500 | 48.500 | 49.000 |
| | | | 8 | 48.000 | 53.000 | 43.000 | 44.000 |
| | | | 12 | 46.000 | 53.000 | 39.000 | 40.000 |
| | 55 | | 3 | 53.500 | 55.500 | 51.500 | 52.000 |
| | | | 9 | 50.500 | 56.000 | 45.000 | 46.000 |
| | | | 14 | 48.000 | 57.000 | 39.000 | 41.000 |
| 60 | | | 3 | 58.500 | 60.500 | 56.500 | 57.000 |
| | | | 9 | 55.500 | 61.000 | 50.000 | 51.000 |
| | | | 14 | 53.000 | 62.000 | 44.000 | 46.000 |
| | 65 | | 4 | 63.000 | 65.500 | 60.500 | 61.000 |
| | | | 10 | 60.000 | 66.000 | 54.000 | 55.000 |
| | | | 16 | 57.000 | 67.000 | 47.000 | 49.000 |
| 70 | | | 4 | 68.000 | 70.500 | 65.500 | 66.000 |
| | | | 10 | 65.000 | 71.000 | 59.000 | 60.000 |
| | | | 16 | 62.000 | 72.000 | 52.000 | 54.000 |
| | 75 | | 4 | 73.000 | 75.500 | 70.500 | 71.000 |
| | | | 10 | 70.000 | 76.000 | 64.000 | 65.000 |
| | | | 16 | 67.000 | 77.000 | 57.000 | 59.000 |
| 80 | | | 4 | 78.000 | 80.500 | 75.500 | 76.000 |
| | | | 10 | 75.000 | 81.000 | 69.000 | 70.000 |
| | | | 16 | 72.000 | 82.000 | 62.000 | 64.000 |
| | 85 | | 4 | 83.000 | 85.500 | 80.500 | 81.000 |
| | | | 12 | 79.000 | 86.000 | 72.000 | 73.000 |
| | | | 18 | 76.000 | 87.000 | 65.000 | 67.000 |
| 90 | | | 4 | 88.000 | 90.500 | 85.500 | 86.000 |
| | | | 12 | 84.000 | 91.000 | 77.000 | 78.000 |
| | | | 18 | 81.000 | 92.000 | 70.000 | 72.000 |
| | 95 | | 4 | 93.000 | 95.500 | 90.500 | 91.000 |
| | 95 | | 12 | 89.000 | 96.000 | 82.000 | 83.000 |
| | | | 18 | 86.000 | 97.000 | 75.000 | 77.000 |
| 100 | | | 4 | 98.000 | 100.500 | 95.500 | 96.000 |
| | | | 12 | 94.000 | 101.000 | 87.000 | 88.000 |
| | | | 20 | 90.000 | 102.000 | 78.000 | 80.000 |

**Table 2.** *(Continued)* **ISO Metric Trapezoidal Screw Thread** *ISO 2904-2020*

| Nominal Diameter, $d$ | | | Pitch, $P$ | Pitch Dia. $d_2 = D_2$ | Major Dia. $D_4$ | Minor Diameter | |
|---|---|---|---|---|---|---|---|
| | | | | | | $d_3$ | $D_1$ |
| | | 105 | 4 | 103.000 | 105.500 | 100.500 | 101.000 |
| | | | 12 | 103.000 | 106.000 | 92.000 | 93.000 |
| | | | 20 | 95.000 | 107.000 | 83.000 | 85.000 |
| | 110 | | 4 | 108.000 | 110.500 | 105.500 | 106.000 |
| | | | 12 | 104.000 | 111.000 | 97.000 | 98.000 |
| | | | 20 | 100.000 | 112.000 | 88.000 | 90.000 |
| | | 115 | 6 | 112.000 | 116.000 | 108.000 | 109.000 |
| | | | 14 | 112.000 | 117.000 | 99.000 | 101.000 |
| | | | 22 | 104.000 | 117.000 | 91.000 | 93.000 |
| 120 | | | 6 | 117.000 | 121.000 | 113.000 | 114.000 |
| | | | 14 | 113.000 | 122.000 | 104.000 | 106.000 |
| | | | 22 | 109.000 | 122.000 | 96.000 | 98.000 |
| | | 125 | 6 | 122.000 | 126.000 | 118.000 | 119.000 |
| | | | 14 | 122.000 | 127.000 | 109.000 | 111.000 |
| | | | 22 | 114.000 | 127.000 | 101.000 | 103.000 |
| | 130 | | 6 | 127.000 | 131.000 | 123.000 | 124.000 |
| | | | 14 | 123.000 | 132.000 | 114.000 | 116.000 |
| | | | 22 | 119.000 | 132.000 | 106.000 | 108.000 |
| | | 135 | 6 | 132.000 | 136.000 | 128.000 | 129.000 |
| | | | 14 | 132.000 | 137.000 | 119.000 | 121.000 |
| | | | 24 | 123.000 | 137.000 | 109.000 | 111.000 |
| 140 | | | 6 | 137.000 | 141.000 | 133.000 | 134.000 |
| | | | 14 | 133.000 | 142.000 | 124.000 | 126.000 |
| | | | 24 | 128.000 | 142.000 | 114.000 | 116.000 |
| | | 145 | 6 | 142.000 | 146.000 | 138.000 | 139.000 |
| | | | 14 | 142.000 | 147.000 | 129.000 | 131.000 |
| | | | 24 | 133.000 | 147.000 | 119.000 | 121.000 |
| | 150 | | 6 | 147.000 | 151.000 | 143.000 | 144.000 |
| | | | 16 | 142.000 | 152.000 | 132.000 | 134.000 |
| | | | 24 | 138.000 | 152.000 | 124.000 | 126.000 |
| | | 155 | 6 | 152.000 | 156.000 | 148.000 | 149.000 |
| | | | 16 | 152.000 | 157.000 | 137.000 | 139.000 |
| | | | 24 | 143.000 | 157.000 | 129.000 | 131.000 |
| 160 | | | 6 | 157.000 | 161.000 | 153.000 | 154.000 |
| | | | 16 | 152.000 | 162.000 | 142.000 | 144.000 |
| | | | 28 | 146.000 | 162.000 | 130.000 | 132.000 |
| | | 165 | 6 | 162.000 | 166.000 | 158.000 | 159.000 |
| | | | 16 | 162.000 | 167.000 | 147.000 | 149.000 |
| | | | 28 | 151.000 | 167.000 | 135.000 | 137.000 |
| | 170 | | 6 | 167.000 | 171.000 | 163.000 | 164.000 |
| | | | 16 | 162.000 | 172.000 | 152.000 | 154.000 |
| | | | 28 | 156.000 | 172.000 | 140.000 | 142.000 |
| | | 175 | 8 | 171.000 | 176.000 | 166.000 | 167.000 |
| | | | 16 | 171.000 | 177.000 | 157.000 | 159.000 |
| | | | 28 | 161.000 | 177.000 | 145.000 | 147.000 |
| 180 | | | 8 | 176.000 | 181.000 | 171.000 | 172.000 |
| | | | 18 | 171.000 | 182.000 | 160.000 | 162.000 |
| | | | 28 | 166.000 | 182.000 | 150.000 | 152.000 |

**Table 2.** *(Continued)* **ISO Metric Trapezoidal Screw Thread** *ISO 2904-2020*

| Nominal Diameter, $d$ | | | Pitch, $P$ | Pitch Dia. $d_2 = D_2$ | Major Dia. $D_4$ | Minor Diameter | |
|---|---|---|---|---|---|---|---|
| | | | | | | $d_3$ | $D_1$ |
| | | 185 | 8 | 181.000 | 186.000 | 176.000 | 177.000 |
| | | | 18 | 181.000 | 187.000 | 165.000 | 167.000 |
| | | | 32 | 169.000 | 187.000 | 151.000 | 153.000 |
| | 190 | | 8 | 186.000 | 191.000 | 181.000 | 182.000 |
| | | | 18 | 181.000 | 192.000 | 170.000 | 172.000 |
| | | | 32 | 174.000 | 192.000 | 156.000 | 158.000 |
| | | 195 | 8 | 191.000 | 196.000 | 186.000 | 187.000 |
| | | | 18 | 191.000 | 197.000 | 175.000 | 177.000 |
| | | | 32 | 179.000 | 197.000 | 161.000 | 163.000 |
| 200 | | | 8 | 196.000 | 201.000 | 191.000 | 192.000 |
| | | | 18 | 191.000 | 202.000 | 180.000 | 182.000 |
| | | | 32 | 184.000 | 202.000 | 166.000 | 168.000 |
| | 210 | | 8 | 206.000 | 211.000 | 201.000 | 202.000 |
| | | | 20 | 200.000 | 212.000 | 188.000 | 190.000 |
| | | | 36 | 192.000 | 212.000 | 172.000 | 174.000 |
| 220 | | | 8 | 216.000 | 221.000 | 211.000 | 212.000 |
| | | | 20 | 210.000 | 222.000 | 198.000 | 200.000 |
| | | | 36 | 202.000 | 222.000 | 182.000 | 184.000 |
| | 230 | | 8 | 226.000 | 231.000 | 221.000 | 222.000 |
| | | | 20 | 220.000 | 232.000 | 208.000 | 210.000 |
| | | | 36 | 212.000 | 232.000 | 192.000 | 194.000 |
| 240 | | | 8 | 236.000 | 241.000 | 231.000 | 232.000 |
| | | | 22 | 229.000 | 242.000 | 216.000 | 218.000 |
| | | | 36 | 222.000 | 242.000 | 202.000 | 204.000 |
| | 250 | | 12 | 244.000 | 251.000 | 237.000 | 238.000 |
| | | | 22 | 239.000 | 252.000 | 226.000 | 228.000 |
| | | | 40 | 230.000 | 252.000 | 208.000 | 210.000 |
| 260 | | | 12 | 254.000 | 261.000 | 247.000 | 248.000 |
| | | | 22 | 249.000 | 262.000 | 236.000 | 238.000 |
| | | | 40 | 240.000 | 262.000 | 218.000 | 220.000 |
| | 270 | | 12 | 264.000 | 271.000 | 257.000 | 258.000 |
| | | | 24 | 258.000 | 272.000 | 244.000 | 246.000 |
| | | | 40 | 250.000 | 272.000 | 228.000 | 230.000 |
| 280 | | | 12 | 274.000 | 281.000 | 267.000 | 268.000 |
| | | | 24 | 268.000 | 282.000 | 254.000 | 256.000 |
| | | | 40 | 260.000 | 282.000 | 238.000 | 240.000 |
| | 290 | | 12 | 284.000 | 291.000 | 277.000 | 278.000 |
| | | | 24 | 278.000 | 292.000 | 264.000 | 266.000 |
| | | | 44 | 268.000 | 292.000 | 244.000 | 246.000 |
| 300 | | | 12 | 294.000 | 301.000 | 287.000 | 288.000 |
| | | | 24 | 288.000 | 302.000 | 274.000 | 276.000 |
| | | | 44 | 278.000 | 302.000 | 254.000 | 256.000 |

All dimensions in millimeters.

## Trapezoidal Metric Screw Thread—Preferred Basic Sizes *DIN 103*

$H = 1.866P$

$h_s = 0.5P + a$

$h_e = 0.5P + a - b$

$h_n = 0.5P + 2a - b$

$h_{as} = 0.25P$

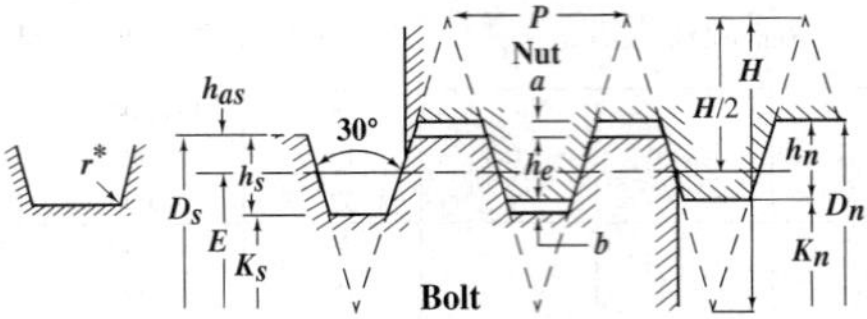


| Nom. & Major Dia. of Bolt, $D_s$ | Pitch, $P$ | Pitch Dia., $E$ | Depth of Engagement, $h_e$ | Clearance | | Bolt | | Nut | | |
|---|---|---|---|---|---|---|---|---|---|---|
| | | | | $a$ | $b$ | Minor Dia., $K_s$ | Depth of Thread, $h_s$ | Major Dia., $D_n$ | Minor Dia., $K_n$ | Depth of Thread, $h_n$ |
| 10 | 3 | 8.5 | 1.25 | 0.25 | 0.5 | 6.5 | 1.75 | 10.5 | 7.5 | 1.50 |
| 12 | 3 | 10.5 | 1.25 | 0.25 | 0.5 | 8.5 | 1.75 | 12.5 | 9.5 | 1.50 |
| 14 | 4 | 12 | 1.75 | 0.25 | 0.5 | 9.5 | 2.25 | 14.5 | 10.5 | 2.00 |
| 16 | 4 | 14 | 1.75 | 0.25 | 0.5 | 11.5 | 2.25 | 16.5 | 12.5 | 2.00 |
| 18 | 4 | 16 | 1.75 | 0.25 | 0.5 | 13.5 | 2.25 | 18.5 | 14.5 | 2.00 |
| 20 | 4 | 18 | 1.75 | 0.25 | 0.5 | 15.5 | 2.25 | 20.5 | 16.5 | 2.00 |
| 22 | 5 | 19.5 | 2 | 0.25 | 0.75 | 16.5 | 2.75 | 22.5 | 18 | 2.00 |
| 24 | 5 | 21.5 | 2 | 0.25 | 0.75 | 18.5 | 2.75 | 24.5 | 20 | 2.25 |
| 26 | 5 | 23.5 | 2 | 0.25 | 0.75 | 20.5 | 2.75 | 26.5 | 22 | 2.25 |
| 28 | 5 | 25.5 | 2 | 0.25 | 0.75 | 22.5 | 2.75 | 28.5 | 24 | 2.25 |
| 30 | 6 | 27 | 2.5 | 0.25 | 0.75 | 23.5 | 3.25 | 30.5 | 25 | 2.75 |
| 32 | 6 | 29 | 2.5 | 0.25 | 0.75 | 25.5 | 3.25 | 32.5 | 27 | 2.75 |
| 36 | 6 | 33 | 2.5 | 0.25 | 0.75 | 29.5 | 3.25 | 36.5 | 31 | 2.75 |
| 40 | 7 | 36.5 | 3 | 0.25 | 0.75 | 32.5 | 3.75 | 40.5 | 34 | 3.25 |
| 44 | 7 | 40.5 | 3 | 0.25 | 0.75 | 36.5 | 3.75 | 44.5 | 38 | 3.25 |
| 48 | 8 | 44 | 3.5 | 0.25 | 0.75 | 39.5 | 4.25 | 48.5 | 41 | 3.75 |
| 50 | 8 | 46 | 3.5 | 0.25 | 0.75 | 41.5 | 4.25 | 50.5 | 43 | 3.75 |
| 52 | 8 | 48 | 3.5 | 0.25 | 0.75 | 43.5 | 4.25 | 52.5 | 45 | 3.75 |
| 55 | 9 | 50.5 | 4 | 0.25 | 0.75 | 45.5 | 4.75 | 55.5 | 47 | 4.25 |
| 60 | 9 | 55.5 | 4 | 0.25 | 0.75 | 50.5 | 4.75 | 60.5 | 52 | 4.25 |
| 65 | 10 | 60 | 4.5 | 0.25 | 0.75 | 54.5 | 5.25 | 65.5 | 56 | 4.75 |
| 70 | 10 | 65 | 4.5 | 0.25 | 0.75 | 59.5 | 5.25 | 70.5 | 61 | 4.75 |
| 75 | 10 | 70 | 4.5 | 0.25 | 0.75 | 64.5 | 5.25 | 75.5 | 66 | 4.75 |
| 80 | 10 | 75 | 4.5 | 0.25 | 0.75 | 69.5 | 5.25 | 80.5 | 71 | 4.75 |
| 85 | 12 | 79 | 5.5 | 0.25 | 0.75 | 72.5 | 6.25 | 85.5 | 74 | 5.75 |
| 90 | 12 | 84 | 5.5 | 0.25 | 0.75 | 77.5 | 6.25 | 90.5 | 79 | 5.75 |
| 95 | 12 | 89 | 5.5 | 0.25 | 0.75 | 82.5 | 6.25 | 95.5 | 84 | 5.75 |
| 100 | 12 | 94 | 5.5 | 0.25 | 0.75 | 87.5 | 6.25 | 100.5 | 89 | 5.75 |
| 110 | 12 | 104 | 5.5 | 0.25 | 0.75 | 97.5 | 6.25 | 110.5 | 99 | 5.75 |
| 120 | 14 | 113 | 6 | 0.5 | 1.5 | 105 | 7.5 | 121 | 108 | 6.5 |
| 130 | 14 | 123 | 6 | 0.5 | 1.5 | 115 | 7.5 | 131 | 118 | 6.5 |
| 140 | 14 | 133 | 6 | 0.5 | 1.5 | 125 | 7.5 | 141 | 128 | 6.5 |
| 150 | 16 | 142 | 7 | 0.5 | 1.5 | 133 | 8.5 | 151 | 136 | 7.5 |
| 160 | 16 | 152 | 7 | 0.5 | 1.5 | 143 | 8.5 | 161 | 146 | 7.5 |
| 170 | 16 | 162 | 7 | 0.5 | 1.5 | 153 | 8.5 | 171 | 156 | 7.5 |
| 180 | 18 | 171 | 8 | 0.5 | 1.5 | 161 | 9.5 | 181 | 164 | 8.5 |
| 190 | 18 | 181 | 8 | 0.5 | 1.5 | 171 | 9.5 | 191 | 174 | 8.5 |
| 200 | 18 | 191 | 8 | 0.5 | 1.5 | 181 | 9.5 | 201 | 184 | 8.5 |
| 210 | 20 | 200 | 9 | 0.5 | 1.5 | 189 | 10.5 | 211 | 192 | 9.5 |
| 220 | 20 | 210 | 9 | 0.5 | 1.5 | 199 | 10.5 | 221 | 202 | 9.5 |
| 230 | 20 | 220 | 9 | 0.5 | 1.5 | 209 | 10.5 | 231 | 212 | 9.5 |
| 240 | 22 | 229 | 10 | 0.5 | 1.5 | 217 | 11.5 | 241 | 220 | 10.5 |
| 250 | 22 | 239 | 10 | 0.5 | 1.5 | 227 | 11.5 | 251 | 230 | 10.5 |
| 260 | 22 | 249 | 10 | 0.5 | 1.5 | 237 | 11.5 | 261 | 240 | 10.5 |
| 270 | 24 | 258 | 11 | 0.5 | 1.5 | 245 | 12.5 | 271 | 248 | 11.5 |
| 280 | 24 | 268 | 11 | 0.5 | 1.5 | 255 | 12.5 | 281 | 258 | 11.5 |
| 290 | 24 | 278 | 11 | 0.5 | 1.5 | 265 | 12.5 | 291 | 268 | 11.5 |
| 300 | 26 | 287 | 12 | 0.5 | 1.5 | 273 | 13.5 | 301 | 276 | 12.5 |

All dimensions are in millimeters.

*Roots are rounded to a radius, *r*, equal to 0.25 mm for pitches of from 3 to 12 mm inclusive and 0.5 mm for pitches of from 14 to 26 mm inclusive for power transmission.

## ISO Miniature Screw Threads

### ISO Miniature Screw Threads, Basic Form *ISO/R1501:2009*

| Pitch *P* | *H* = 0.866025*P* | 0.554256*H* = 0.48*P* | 0.375*H* = 0.324760*P* | 0.320744*H* = 0.320744*P* | 0.125*H* = 0.108253*P* |
|---|---|---|---|---|---|
| 0.08 | 0.069282 | 0.038400 | 0.025981 | 0.022222 | 0.008660 |
| 0.09 | 0.077942 | 0.043200 | 0.029228 | 0.024999 | 0.009743 |
| 0.1 | 0.086603 | 0.048000 | 0.032476 | 0.027777 | 0.010825 |
| 0.125 | 0.108253 | 0.060000 | 0.040595 | 0.034722 | 0.013532 |
| 0.15 | 0.129904 | 0.072000 | 0.048714 | 0.041666 | 0.016238 |
| 0.175 | 0.151554 | 0.084000 | 0.056833 | 0.048610 | 0.018944 |
| 0.2 | 0.173205 | 0.096000 | 0.064952 | 0.055554 | 0.021651 |
| 0.225 | 0.194856 | 0.108000 | 0.073071 | 0.062499 | 0.024357 |
| 0.25 | 0.216506 | 0.120000 | 0.081190 | 0.069443 | 0.027063 |
| 0.3 | 0.259808 | 0.144000 | 0.097428 | 0.083332 | 0.032476 |

### ISO Miniature Screw Threads, Basic Dimensions *ISO/R1501:2009*

| Nominal Diameter | Pitch *P* | Major Diameter $D, d$ | Pitch Diameter $D_2, d_2$ | Minor Diameter $D_1, d_1$ |
|---|---|---|---|---|
| 0.30 | 0.080 | 0.300000 | 0.248039 | 0.223200 |
| 0.35 | 0.090 | 0.350000 | 0.291543 | 0.263600 |
| 0.40 | 0.100 | 0.400000 | 0.335048 | 0.304000 |
| 0.45 | 0.100 | 0.450000 | 0.385048 | 0.354000 |
| 0.50 | 0.125 | 0.500000 | 0.418810 | 0.380000 |
| 0.55 | 0.125 | 0.550000 | 0.468810 | 0.430000 |
| 0.60 | 0.150 | 0.600000 | 0.502572 | 0.456000 |
| 0.70 | 0.175 | 0.700000 | 0.586334 | 0.532000 |
| 0.80 | 0.200 | 0.800000 | 0.670096 | 0.608000 |
| 0.90 | 0.225 | 0.900000 | 0.753858 | 0.684000 |
| 1.00 | 0.250 | 1.000000 | 0.837620 | 0.760000 |
| 1.10 | 0.250 | 1.100000 | 0.937620 | 0.860000 |
| 1.20 | 0.250 | 1.200000 | 1.037620 | 0.960000 |
| 1.40 | 0.300 | 1.400000 | 1.205144 | 1.112000 |

*D* and *d* dimensions refer to the nut (internal) and screw (external) threads, respectively.

## British Standard ISO Metric Screw Threads

BS 3643: Part 1:2007 provides principles and basic data for ISO metric screw threads. It covers single-start, parallel screw threads of from 1 to 300 millimeters in diameter. Part 2 of the Standard gives the specifications for selected limits of size.

**Basic Profile.**—The ISO basic profile for triangular screw threads is shown in Fig. 1. and basic dimensions of this profile are given in Table 1.

### Table 1. British Standard ISO Metric Screw Threads Basic Profile Dimensions *BS 3643: Part 1:2007*

| Pitch *P* | *H* = 0.86603*P* | ⅝*H* = 0.54127*P* | ⅜*H* = 0.32476*P* | *H*/4 = 0.21651*P* | *H*/8 = 0.10825*P* |
|---|---|---|---|---|---|
| 0.2 | 0.173 205 | 0.108 253 | 0.064 952 | 0.043 301 | 0.021 651 |
| 0.25 | 0.216 506 | 0.135 316 | 0.081 190 | 0.054 127 | 0.027 063 |
| 0.3 | 0.259 808 | 0.162 380 | 0.097 428 | 0.064 952 | 0.032 476 |
| 0.35 | 0.303 109 | 0.189 443 | 0.113 666 | 0.075 777 | 0.037 889 |
| 0.4 | 0.346 410 | 0.216 506 | 0.129 904 | 0.086 603 | 0.043 301 |
| 0.45 | 0.389 711 | 0.243 570 | 0.146 142 | 0.097 428 | 0.048 714 |
| 0.5 | 0.433 013 | 0.270 633 | 0.162 380 | 0.108 253 | 0.054 127 |
| 0.6 | 0.519 615 | 0.324 760 | 0.194 856 | 0.129 904 | 0.064 952 |
| 0.7 | 0.606 218 | 0.378 886 | 0.227332 | 0.151 554 | 0.075 777 |

**Table 1.** *(Continued)* **British Standard ISO Metric Screw Threads Basic Profile Dimensions** *BS 3643: Part 1:2007*

| Pitch $P$ | $H =$ $0.86603P$ | $\frac{5}{8}H =$ $0.54127P$ | $\frac{3}{8}H =$ $0.32476P$ | $H/4 =$ $0.21651P$ | $H/8 =$ $0.10825P$ |
|---|---|---|---|---|---|
| 0.75 | 0.649 519 | 0.405 949 | 0.243 570 | 0.162 380 | 0.081 190 |
| 0.8 | 0.692 820 | 0.433 013 | 0.259 808 | 0.173 205 | 0.086 603 |
| 1 | 0.866 025 | 0.541 266 | 0.324 760 | 0.216 506 | 0.108 253 |
| 1.25 | 1.082 532 | 0.676 582 | 0.405 949 | 0.270 633 | 0.135 316 |
| 1.5 | 1.299 038 | 0.811 899 | 0.487 139 | 0.324 760 | 0.162 380 |
| 1.75 | 1.515 544 | 0.947 215 | 0.568 329 | 0.378 886 | 0.189 443 |
| 2 | 1.732 051 | 1.082 532 | 0.649 519 | 0.433 013 | 0.216 506 |
| 2.5 | 2.165 063 | 1.353 165 | 0.811 899 | 0.541 266 | 0.270 633 |
| 3 | 2.598 076 | 1.623 798 | 0.974 279 | 0.649 519 | 0.324 760 |
| 3.5 | 3.031 089 | 1.894 431 | 1.136 658 | 0.757 772 | 0.378 886 |
| 4 | 3.464 102 | 2.165 063 | 1.299 038 | 0.866 025 | 0.433 013 |
| 4.5 | 3.897 114 | 2.435 696 | 1.461 418 | 0.974 279 | 0.487 139 |
| 5 | 4.330 127 | 2.706 329 | 1.623 798 | 1.082 532 | 0.541 266 |
| 5.5 | 4.763 140 | 2.976 962 | 1.786 177 | 1.190 785 | 0.595 392 |
| 6 | 5.196 152 | 3.247 595 | 1.948 557 | 1.299 038 | 0.649 519 |
| 8[a] | 6.928 203 | 4.330 127 | 2.598 076 | 1.732 051 | 0.866 025 |

[a] This pitch is not used in any of the ISO metric standard series.
All dimensions are given in millimeters.

**Tolerance System.—**The tolerance system defines *tolerance classes* in terms of a combination of a *tolerance grade* (figure) and a *tolerance position* (letter). The tolerance position is defined by the distance between the basic size and the nearest end of the tolerance zone, this distance being known as the *fundamental deviation,* EI, in the case of internal threads, and es in the case of external threads. These tolerance positions with respect to the basic size (zero line) are shown in Fig. 2 and fundamental deviations for nut and bolt threads are given in Table 2.

**Table 2. Fundamental Deviations for Nut Threads and Bolt Threads**

| | Nut Thread $D_2, D_1$ | | Bolt Thread $d, d_2$ | | | | | Nut Thread $D_2, D_1$ | | Bolt Thread $d, d_2$ | | | |
|---|---|---|---|---|---|---|---|---|---|---|---|---|---|
| | Tolerance Position | | | | | | | Tolerance Position | | | | | |
| | G | H | e | f | g | h | | G | H | e | f | g | h |
| | Fundamental Deviation | | | | | | | Fundamental Deviation | | | | | |
| Pitch $P$ | EI | EI | es | es | es | es | Pitch $P$ | EI | EI | es | es | es | es |
| mm | μm | μm | μm | μm | μm | μm | mm | μm | μm | μm | μm | μm | μm |
| 0.2 | +17 | 0 | … | … | −17 | 0 | 1.5 | +32 | 0 | −67 | −45 | −32 | 0 |
| 0.25 | +18 | 0 | … | … | −18 | 0 | 1.75 | +34 | 0 | −71 | −48 | −34 | 0 |
| 0.3 | +18 | 0 | … | … | −18 | 0 | 2 | +38 | 0 | −71 | −52 | −38 | 0 |
| 0.35 | +19 | 0 | … | −34 | −19 | 0 | 2.5 | +42 | 0 | −80 | −58 | −42 | 0 |
| 0.4 | +19 | 0 | … | −34 | −19 | 0 | 3 | +48 | 0 | −85 | −63 | −48 | 0 |
| 0.45 | +20 | 0 | … | −35 | −20 | 0 | 3.5 | +53 | 0 | −90 | −70 | −53 | 0 |
| 0.5 | +20 | 0 | −50 | −36 | −20 | 0 | 4 | +60 | 0 | −95 | −75 | −60 | 0 |
| 0.6 | +21 | 0 | −53 | −36 | −21 | 0 | 4.5 | +63 | 0 | −100 | −80 | −63 | 0 |
| 0.7 | +22 | 0 | −56 | −38 | −22 | 0 | 5 | +71 | 0 | −106 | −85 | −71 | 0 |
| 0.75 | +22 | 0 | −56 | −38 | −22 | 0 | 5.5 | +75 | 0 | −112 | −90 | −75 | 0 |
| 0.8 | +24 | 0 | −60 | −38 | −24 | 0 | 6 | +80 | 0 | −118 | −95 | −80 | 0 |
| 1 | +26 | 0 | −60 | −40 | −26 | 0 | 8 | +100 | 0 | −140 | −118 | −100 | 0 |
| 1.25 | +28 | 0 | −63 | −42 | −28 | 0 | | | | | | | |

See Fig. 1 and Fig. 2 for meaning of symbols.

**Tolerance Grades.—**Tolerance grades specified in the Standard for each of the four main screw thread diameters are as follows:

Minor diameter of nut threads ($D_1$): tolerance grades 4, 5, 6, 7, and 8.

Major diameter of bolt threads ($d$): tolerance grades 4, 6, and 8.

Pitch diameter of nut threads ($D_2$): tolerance grades 4, 5, 6, 7, and 8.

Pitch diameter of bolt threads ($d_2$): tolerance grades 3, 4, 5, 6, 7, 8, and 9.

**Tolerance Positions.—**Tolerance positions are G and H for nut threads and e, f, g, and h for bolt threads. The relationship of these tolerance position identifying letters to the amount of fundamental deviation is shown in Table 2.

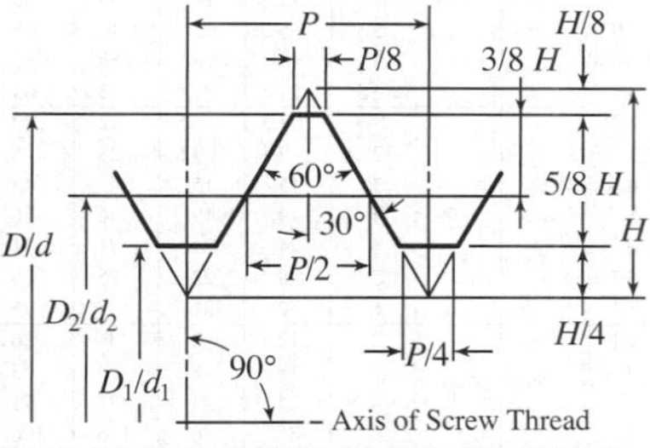


$D$ = major dia. of internal thread
$d$ = major dia. of external thread
$D_2$ = pitch dia. of internal thread
$d_2$ = pitch dia. of external thread
$D_1$ = minor dia. of internal thread
$d_1$ = minor dia. of external thread
$P$ = pitch
$H$ = height of fundamental angle

Fig. 1. Basic Profile of ISO Metric Thread

**Tolerance Classes.—**To reduce the number of gages and tools, the Standard specifies that the tolerance positions and classes shall be chosen from those listed in Table 3 for short, normal, and long lengths of thread engagement. The following rules apply for the choice of tolerance quality: *Fine:* for precision threads when little variation of fit character is needed; *Medium:* for general use; and *Coarse:* for cases where manufacturing difficulties can arise as, for example, when threading hot-rolled bars and long blind holes. If the actual length of thread engagement is unknown, as in the manufacturing of standard bolts, normal is recommended.

**Table 3. Tolerance Classes for Nuts and Bolts[a,b,c]**

**Tolerance Classes for Nuts**

| Tolerance Quality | Tolerance Position G | | | Tolerance Position H | | |
|---|---|---|---|---|---|---|
| | Short | Normal | Long | Short | Normal | Long |
| Fine | … | … | … | 4H[b] | 5H[b] | 6H[b] |
| Medium | 5G[a] | 6G[c] | 7G[c] | 5H[a] | 6H[a,d] | 7H[a] |
| Coarse | … | 7G[c] | 8G[c] | … | 7H[b] | 8H[b] |

**Tolerance Classes for Bolts**

| Tolelance Quality | Tolerance Position e | | | Tolerance Position f | | | Tolerance Position g | | | Tolerance Position h | | |
|---|---|---|---|---|---|---|---|---|---|---|---|---|
| | Short | Normal | Long | Short | Normal | Long | Short | Normal | Long | Short | Normal | Long |
| Fine | … | … | … | … | … | … | … | … | … | 3h4h[c] | 4h[a] | 5h4h[c] |
| Medium | … | 6e[a] | 7e6e[c] | … | 6f[a] | … | 5g6g[c] | 6g[a,d] | 7g6g[c] | 5h6h[c] | 6h[b] | 7h6h[c] |
| Coarse | … | … | … | … | … | … | … | 8g[b] | 9g8g[c] | … | … | … |

[a] First choice.

[b] Second choice.

[c] Third choice; these are to be avoided.

[d] For commercial nut and bolt threads.

*Note:* See Table 4 for short, normal, and long categories. Any of the recommended tolerance classes for nuts can be combined with any of the recommended tolerance classes for bolts with the exception of sizes M1.4 and smaller for which the combination 5H/6h or finer shall be chosen. However, to guarantee a sufficient overlap, the finished components should preferably be made to form the fits H/g, H/h, or G/h.

## Table 4. Lengths of Thread Engagements for Short, Normal, and Long Categories

| Basic Major Diameter *d* Over | Basic Major Diameter *d* Up to and Incl. | Pitch *P* | Short (Length of Thread Engagement) Up to and Incl. | Normal Over | Normal Up to and Incl. | Long Over |
|---|---|---|---|---|---|---|
| 0.99 | 1.4 | 0.2 | 0.5 | 0.5 | 1.4 | 1.4 |
| | | 0.25 | 0.6 | 0.6 | 1.7 | 1.7 |
| | | 0.3 | 0.7 | 0.7 | 2 | 2 |
| 1.4 | 2.8 | 0.2 | 0.5 | 0.5 | 1.5 | 1.5 |
| | | 0.25 | 0.6 | 0.6 | 1.9 | 1.9 |
| | | 0.35 | 0.8 | 0.8 | 2.6 | 2.6 |
| | | 0.4 | 1 | 1 | 3 | 3 |
| | | 0.45 | 1.3 | 1.3 | 3.8 | 3.8 |
| 2.8 | 5.6 | 0.35 | 1 | 1 | 3 | 3 |
| | | 0.5 | 1.5 | 1.5 | 4.5 | 4.5 |
| | | 0.6 | 1.7 | 1.7 | 5 | 5 |
| | | 0.7 | 2 | 2 | 6 | 6 |
| | | 0.75 | 2.2 | 2.2 | 6.7 | 6.7 |
| | | 0.8 | 2.5 | 2.5 | 7.5 | 7.5 |
| 5.6 | 11.2 | 0.75 | 2.4 | 2.4 | 7.1 | 7.1 |
| | | 1 | 3 | 3 | 9 | 9 |
| | | 1.25 | 4 | 4 | 12 | 12 |
| | | 1.5 | 5 | 5 | 15 | 15 |
| 11.2 | 22.4 | 1 | 3.8 | 3.8 | 11 | 11 |
| | | 1.25 | 4.5 | 4.5 | 13 | 13 |
| | | 1.5 | 5.6 | 5.6 | 16 | 16 |
| | | 1.75 | 6 | 6 | 18 | 18 |
| | | 2 | 8 | 8 | 24 | 24 |
| | | 2.5 | 10 | 10 | 30 | 30 |

| Basic Major Diameter *d* Over | Basic Major Diameter *d* Up to and Incl. | Pitch *P* | Short (Length of Thread Engagement) Up to and Incl. | Normal Over | Normal Up to and Incl. | Long Over |
|---|---|---|---|---|---|---|
| 22.4 | 45 | 1 | 4 | 4 | 12 | 12 |
| | | 1.5 | 6.3 | 6.3 | 19 | 19 |
| | | 2 | 8.5 | 8.5 | 25 | 25 |
| | | 3 | 12 | 12 | 36 | 36 |
| | | 3.5 | 15 | 15 | 45 | 45 |
| | | 4 | 18 | 18 | 53 | 53 |
| | | 4.5 | 21 | 21 | 63 | 63 |
| 45 | 90 | 1.5 | 7.5 | 7.5 | 22 | 22 |
| | | 2 | 9.5 | 9.5 | 28 | 28 |
| | | 3 | 15 | 15 | 45 | 45 |
| | | 4 | 19 | 19 | 56 | 56 |
| | | 5 | 24 | 24 | 71 | 71 |
| | | 5.5 | 28 | 28 | 85 | 85 |
| | | 6 | 32 | 32 | 95 | 95 |
| 90 | 180 | 2 | 12 | 12 | 36 | 36 |
| | | 3 | 18 | 18 | 53 | 53 |
| | | 4 | 24 | 24 | 71 | 71 |
| | | 6 | 36 | 36 | 106 | 106 |
| 180 | 300 | 3 | 20 | 20 | 60 | 60 |
| | | 4 | 26 | 26 | 80 | 80 |
| | | 6 | 40 | 40 | 118 | 118 |

All dimensions are given in millimeters

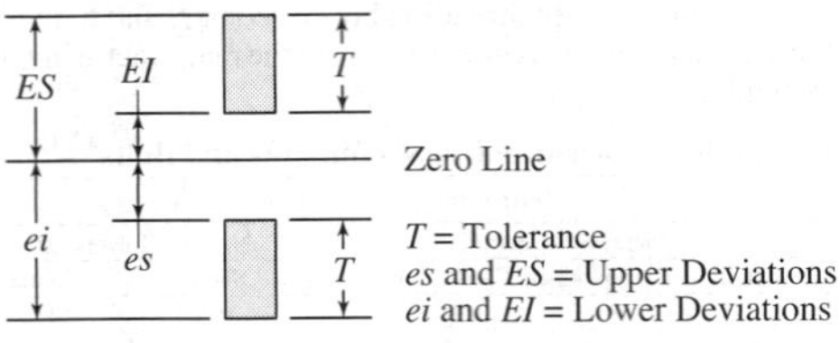


Fig. 2. Tolerance Positions with Respect to Zero Line (Basic Size)

**Design Profiles.—**The design profiles for ISO metric internal and external screw threads are shown in Fig. 3. These represent the profiles of the threads at their maximum metal condition. It may be noted that the root of each thread is deepened so as to clear the basic flat crest of the other thread. The contact between the thread is thus confined to their sloping flanks. However, for nut threads as well as bolt threads, the actual root contours shall not at any point violate the basic profile.

**Designation.—**Screw threads complying with the requirements of the Standard shall be designated by the letter M followed by values of the nominal diameter and of the pitch, expressed in millimeters, and separated by the sign ×. *Example:* M6×0.75. The absence of the indication of pitch means that a coarse pitch is specified.

The complete designation of a screw thread consists of a designation for the thread system and size, and a designation for the crest diameter tolerance. Each class designation consists of: a figure indicating the tolerance grade; and a letter indicating the tolerance position, capital for nuts, lower case for bolts. If the two class designations for a thread are the same (one for the pitch diameter and one for the crest diameter), it is not necessary to repeat the symbols. As examples, a bolt thread designated M10-6g signifies a thread of 10 mm nominal diameter in the Coarse Thread Series having a tolerance class 6g for both pitch and major diameters. A designation M10 × 1-5g6g signifies a bolt thread of 10 mm nominal diameter having a pitch of 1 mm, a tolerance class 5g for pitch diameter, and a tolerance class 6g for major diameter. A designation M10-6H signifies a nut thread of 10 mm diameter in the Coarse Thread Series having a tolerance class 6H for both pitch and minor diameters.

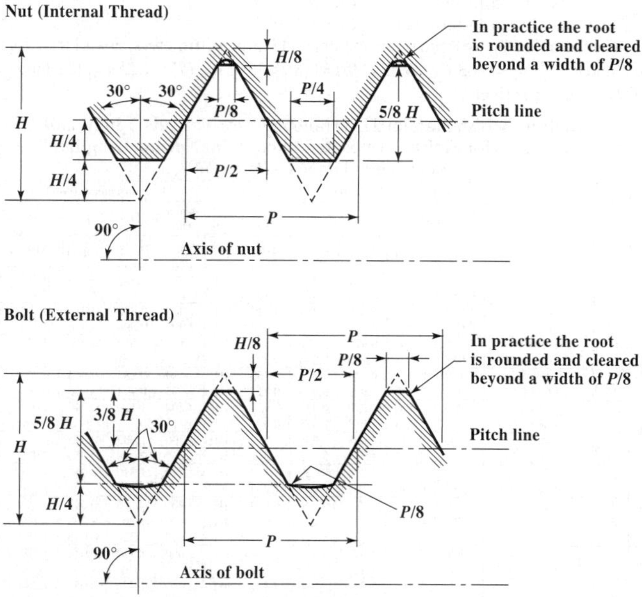


Fig. 3. Maximum Material Profiles for Internal and External Threads

A fit between mating parts is indicated by the nut thread tolerance class followed by the bolt thread tolerance class separated by an oblique stroke. *Examples:* M6-6H/6g and M20 × 2-6H/5g6g. For coated threads, the tolerances apply to the parts before coating, unless otherwise specified. After coating, the actual thread profile shall not at any point exceed the maximum material limits for either tolerance position H or h.

**Fundamental Deviation Formulas.—**The formulas used to calculate the fundamental deviations in Table 2 are:

$EI_G = +(15 + 11P)$

$EI_H = 0$

$es_e = -(50 + 11P)$ except for threads with $P \leq 0.45$ mm

$es_f = -(30 + 11P)$

$es_g = -(15 + 11P)$

$es_h = 0$

In these formulas, EI and es are expressed in micrometers and $P$ is in millimeters.

**Crest Diameter Tolerance Formulas.—**The tolerances for the major diameter of bolt threads ($T_d$), grade 6, in Table 5, were calculated from the formula:

$$T_d(6) = 180\sqrt[3]{P^2} - \frac{3.15}{\sqrt{P}}$$

In this formula, $T_d$ (6) is in micrometers and $P$ is in millimeters. For tolerance grades 4 and 8: $T_d(4) = 0.63\ T_d(6)$ and $T_d(8) = 1.6\ T_d(6)$, respectively.

The tolerances for the minor diameter of nut threads ($T_{D1}$), grade 6, in Table 5, were calculated as follows:

For pitches 0.2 to 0.8 mm, $T_{D1}(6) = 433P - 190P^{1.22}$.

For pitches 1 mm and coarser, $T_{D1}(6) = 230P^{0.7}$.

In these formulas, $T_{D1}$ (6) is in micrometers and $P$ is in millimeters. For tolerance grades 4, 5, 7, and 8: $T_{D1}(4) = 0.63\ T_{D1}(6)$; $T_{D1}(5) = 0.8\ T_{D1}(6)$; $T_{D1}(7) = 1.25\ T_{D1}(6)$; and $T_{D1}(8) = 1.6\ T_{D1}(6)$, respectively.

**Table 5. British Standard ISO Metric Screw Threads: Limits and Tolerances for Finished Uncoated Threads for Normal Lengths of Engagement** *BS 3643: Part 2:2007*

| Nominal Diameter[a] | Pitch | | External Threads (Bolts) | | | | | | | Internal Threads (Nuts)[b] | | | | | |
|---|---|---|---|---|---|---|---|---|---|---|---|---|---|---|---|
| | Coarse | Fine | Tol. Class | Fund dev. | Major Dia. | | Pitch Dia. | | Minor Dia | Tol. Class | Major Dia. | Pitch Dia. | | Minor Dia | |
| | | | | | Max | Tol(−) | Max | Tol(−) | Min | | Min | Max | Tol(−) | Max | Tol(−) |
| 1 | | 0.2 | 4h | 0 | 1.000 | 0.036 | 0.870 | 0.030 | 0.717 | 4H | 1.000 | 0.910 | 0.040 | 0.821 | 0.038 |
| | | | 6g | 0.017 | 0.983 | 0.056 | 0.853 | 0.048 | 0.682 | | | | | | |
| | 0.25 | | 4h | 0 | 1.000 | 0.042 | 0.838 | 0.034 | 0.649 | 4H | 1.000 | 0.883 | 0.045 | 0.774 | 0.045 |
| | | | 6g | 0.018 | 0.982 | 0.067 | 0.820 | 0.053 | 0.613 | 5H | 1.000 | 0.894 | 0.056 | 0.785 | 0.056 |
| 1.1 | | 0.2 | 4h | 0 | 1.100 | 0.036 | 0.970 | 0.030 | 0.817 | 4H | 1.100 | 1.010 | 0.040 | 0.921 | 0.038 |
| | | | 6g | 0.017 | 1.083 | 0.056 | 0.953 | 0.048 | 0.782 | | | | | | |
| | 0.25 | | 4h | 0 | 1.100 | 0.042 | 0.938 | 0.034 | 0.750 | 4H | 1.100 | 0.983 | 0.045 | 0.874 | 0.045 |
| | | | 6g | 0.018 | 1.082 | 0.067 | 0.920 | 0.053 | 0.713 | 5H | 1.100 | 0.994 | 0.056 | 0.885 | 0.056 |
| 1.2 | | 0.2 | 4h | 0 | 1.200 | 0.036 | 1.070 | 0.030 | 0.917 | 4H | 1.200 | 1.110 | 0.040 | 1.021 | 0.038 |
| | | | 6g | 0.017 | 1.183 | 0.056 | 1.053 | 0.048 | 0.882 | | | | | | |
| | 0.25 | | 4h | 0 | 1.200 | 0.042 | 1.038 | 0.034 | 0.850 | 4H | 1.200 | 1.083 | 0.045 | 0.974 | 0.045 |
| | | | 6g | 0.018 | 1.182 | 0.067 | 1.020 | 0.053 | 0.813 | 5H | 1.200 | 1.094 | 0.056 | 0.985 | 0.056 |
| 1.4 | | 0.2 | 4h | 0 | 1.400 | 0.036 | 1.270 | 0.030 | 1.117 | 4H | 1.400 | 1.310 | 0.040 | 1.221 | 0.038 |
| | | | 6g | 0.017 | 1.383 | 0.056 | 1.253 | 0.048 | 1.082 | | | | | | |
| | 0.3 | | 4h | 0 | 1.400 | 0.048 | 1.205 | 0.036 | 0.984 | 4H | 1.400 | 1.253 | 0.048 | 1.128 | 0.053 |
| | | | 6g | 0.018 | 1.382 | 0.075 | 1.187 | 0.056 | 0.946 | 5H | 1.400 | 1.265 | 0.060 | 1.142 | 0.067 |
| | | | | | | | | | | 6H | 1.400 | 1.280 | 0.075 | 1.160 | 0.085 |
| 1.6 | | 0.2 | 4h | 0 | 1.600 | 0.036 | 1.470 | 0.032 | 1.315 | 4H | 1.600 | 1.512 | 0.042 | 1.421 | 0.038 |
| | | | 6g | 0.017 | 1.583 | 0.056 | 1.453 | 0.050 | 1.280 | | | | | | |
| | 0.35 | | 4h | 0 | 1.600 | 0.053 | 1.373 | 0.040 | 1.117 | 4H | 1.600 | 1.426 | 0.053 | 1.284 | 0.063 |
| | | | 6g | 0.019 | 1.581 | 0.085 | 1.354 | 0.063 | 1.075 | 5H | 1.600 | 1.440 | 0.067 | 1.301 | 0.080 |
| | | | | | | | | | | 6H | 1.600 | 1.458 | 0.085 | 1.321 | 0.100 |
| 1.8 | | 0.2 | 4h | 0 | 1.800 | 0.036 | 1.670 | 0.032 | 1.515 | 4H | 1.800 | 1.712 | 0.042 | 1.621 | 0.038 |
| | | | 6g | 0.017 | 1.783 | 0.056 | 1.653 | 0.050 | 1.480 | | | | | | |
| | 0.35 | | 4h | 0 | 1.800 | 0.053 | 1.573 | 0.040 | 1.317 | 4H | 1.800 | 1.626 | 0.053 | 1.484 | 0.063 |
| | | | 6g | 0.019 | 1.781 | 0.085 | 1.554 | 0.063 | 1.275 | 5H | 1.800 | 1.640 | 0.067 | 1.501 | 0.080 |
| | | | | | | | | | | 6H | 1.800 | 1.658 | 0.085 | 1.521 | 0.100 |
| 2 | | 0.25 | 4h | 0 | 2.000 | 0.042 | 1.838 | 0.036 | 1.648 | 4H | 2.000 | 1.886 | 0.048 | 1.774 | 0.045 |
| | | | 6g | 0.018 | 1.982 | 0.067 | 1.820 | 0.056 | 1.610 | 5H | 2.000 | 1.898 | 0.060 | 1.785 | 0.056 |
| | 0.4 | | 4h | 0 | 2.000 | 0.060 | 1.740 | 0.042 | 1.452 | 4H | 2.000 | 1.796 | 0.056 | 1.638 | 0.071 |
| | | | 6g | 0.019 | 1.981 | 0.095 | 1.721 | 0.067 | 1.408 | 5H | 2.000 | 1.811 | 0.071 | 1.657 | 0.090 |
| | | | | | | | | | | 6H | 2.000 | 1.830 | 0.090 | 1.679 | 0.112 |
| 2.2 | | 0.25 | 4h | 0 | 2.200 | 0.042 | 2.038 | 0.036 | 1.848 | 4H | 2.200 | 2.086 | 0.048 | 1.974 | 0.045 |
| | | | 6g | 0.018 | 2.182 | 0.067 | 2.020 | 0.056 | 1.810 | 5H | 2.200 | 2.098 | 0.060 | 1.985 | 0.056 |
| | 0.45 | | 4h | 0 | 2.200 | 0.063 | 1.908 | 0.045 | 1.585 | 4H | 2.200 | 1.968 | 0.060 | 1.793 | 0.080 |
| | | | 6g | 0.020 | 2.180 | 0.100 | 1.888 | 0.071 | 1.539 | 5H | 2.200 | 1.983 | 0.075 | 1.813 | 0.100 |
| | | | | | | | | | | 6H | 2.000 | 2.003 | 0.095 | 1.838 | 0.125 |
| 2.5 | | 0.35 | 4h | 0 | 2.500 | 0.053 | 2.273 | 0.040 | 2.017 | 4H | 2.500 | 2.326 | 0.053 | 2.184 | 0.063 |
| | | | 6g | 0.019 | 2.481 | 0.085 | 2.254 | 0.063 | 1.975 | 5H | 2.500 | 2.340 | 0.067 | 2.201 | 0.080 |
| | | | | | | | | | | 6H | 2.500 | 2.358 | 0.085 | 2.221 | 0.100 |
| | 0.45 | | 4h | 0 | 2.500 | 0.063 | 2.208 | 0.045 | 1.885 | 4H | 2.500 | 2.268 | 0.060 | 2.093 | 0.080 |
| | | | 6g | 0.020 | 2.480 | 0.100 | 2.188 | 0.071 | 1.839 | 5H | 2.500 | 2.283 | 0.075 | 2.113 | 0.100 |
| | | | | | | | | | | 6H | 2.500 | 2.303 | 0.095 | 2.138 | 0.125 |

**Table 5.** *(Continued)* **British Standard ISO Metric Screw Threads: Limits and Tolerances for Finished Uncoated Threads for Normal Lengths of Engagement** *BS 3643: Part 2:2007*

| Nominal Diameter[a] | Pitch | | External Threads (Bolts) | | | | | | | Internal Threads (Nuts)[b] | | | | | |
|---|---|---|---|---|---|---|---|---|---|---|---|---|---|---|---|
| | Coarse | Fine | Tol. Class | Fund dev. | Major Dia. | | Pitch Dia. | | Minor Dia | Tol. Class | Major Dia. | Pitch Dia. | | Minor Dia | |
| | | | | | Max | Tol(−) | Max | Tol(−) | Min | | Min | Max | Tol(−) | Max | Tol(−) |
| 3 | | 0.35 | 4h | 0 | 3.000 | 0.053 | 2.773 | 0.042 | 2.515 | 4H | 3.000 | 2.829 | 0.056 | 2.684 | 0.063 |
| | | | 6g | 0.019 | 2.981 | 0.085 | 2.754 | 0.067 | 2.471 | 5H | 3.000 | 2.844 | 0.071 | 2.701 | 0.080 |
| | | | | | | | | | | 6H | 3.000 | 2.863 | 0.090 | 2.721 | 0.100 |
| | 0.5 | | 4h | 0 | 3.000 | 0.067 | 2.675 | 0.048 | 2.319 | 5H | 3.000 | 2.755 | 0.080 | 2.571 | 0.112 |
| | | | 6g | 0.020 | 2.980 | 0.106 | 2.655 | 0.075 | 2.272 | 6H | 3.000 | 2.775 | 0.100 | 2.599 | 0.140 |
| | | | | | | | | | | 7H | 3.000 | 2.800 | 0.125 | 2.639 | 0.180 |
| 3.5 | | 0.35 | 4h | 0 | 3.500 | 0.053 | 3.273 | 0.042 | 3.015 | 4H | 3.500 | 3.329 | 0.056 | 3.184 | 0.063 |
| | | | 6g | 0.019 | 3.481 | 0.085 | 3.254 | 0.067 | 2.971 | 5H | 3.500 | 3.344 | 0.071 | 3.201 | 0.080 |
| | | | | | | | | | | 6H | 3.500 | 3.363 | 0.090 | 3.221 | 0.100 |
| | 0.6 | | 4h | 0 | 3.500 | 0.080 | 3.110 | 0.053 | 2.688 | 5H | 3.500 | 3.200 | 0.090 | 2.975 | 0.125 |
| | | | 6g | 0.021 | 3.479 | 0.125 | 3.089 | 0.085 | 2.635 | 6H | 3.500 | 3.222 | 0.112 | 3.010 | 0.160 |
| | | | | | | | | | | 7H | 3.500 | 3.250 | 0.140 | 3.050 | 0.200 |
| 4 | | 0.5 | 4h | 0 | 4.000 | 0.067 | 3.675 | 0.048 | 3.319 | 5H | 4.000 | 3.755 | 0.080 | 3.571 | 0.112 |
| | | | 6g | 0.020 | 3.980 | 0.106 | 3.655 | 0.075 | 3.272 | 6H | 4.000 | 3.775 | 0.100 | 3.599 | 0.140 |
| | | | | | | | | | | 7H | 4.000 | 3.800 | 0.125 | 3.639 | 0.180 |
| | 0.7 | | 4h | 0 | 4.000 | 0.090 | 3.545 | 0.056 | 3.058 | 5H | 4.000 | 3.640 | 0.095 | 3.382 | 0.140 |
| | | | 6g | 0.022 | 3.978 | 0.140 | 3.523 | 0.090 | 3.002 | 6H | 4.000 | 3.663 | 0.118 | 3.422 | 0.180 |
| | | | | | | | | | | 7H | 4.000 | 3.695 | 0.150 | 3.466 | 0.224 |
| 4.5 | | 0.5 | 4h | 0 | 4.500 | 0.067 | 4.175 | 0.048 | 3.819 | 5H | 4.500 | 4.255 | 0.080 | 4.071 | 0.112 |
| | | | 6g | 0.020 | 4.480 | 0.106 | 4.155 | 0.075 | 3.772 | 6H | 4.500 | 4.275 | 0.100 | 4.099 | 0.140 |
| | | | | | | | | | | 7H | 4.500 | 4.300 | 0.125 | 4.139 | 0.180 |
| | 0.75 | | 4h | 0 | 4.500 | 0.090 | 4.013 | 0.056 | 3.495 | 5H | 4.500 | 4.108 | 0.095 | 3.838 | 0.150 |
| | | | 6g | 0.022 | 4.478 | 0.140 | 3.991 | 0.090 | 3.439 | 6H | 4.500 | 4.131 | 0.118 | 3.878 | 0.190 |
| | | | | | | | | | | 7H | 4.500 | 4.163 | 0.150 | 3.924 | 0.236 |
| 5 | | 0.5 | 4h | 0 | 5.000 | 0.067 | 4.675 | 0.048 | 4.319 | 5H | 5.000 | 4.755 | 0.080 | 4.571 | 0.112 |
| | | | 6g | 0.020 | 4.980 | 0.106 | 4.655 | 0.075 | 4.272 | 6H | 5.000 | 4.775 | 0.100 | 4.599 | 0.140 |
| | | | | | | | | | | 7H | 5.000 | 4.800 | 0.125 | 4.639 | 0.180 |
| | 0.8 | | 4h | 0 | 5.000 | 0.095 | 4.480 | 0.060 | 3.927 | 5H | 5.000 | 4.580 | 0.100 | 4.294 | 0.160 |
| | | | 6g | 0.024 | 4.976 | 0.150 | 4.456 | 0.095 | 3.868 | 6H | 5.000 | 4.605 | 0.125 | 4.334 | 0.200 |
| | | | | | | | | | | 7H | 5.000 | 4.640 | 0.160 | 4.384 | 0.250 |
| 5.5 | | 0.5 | 4h | 0 | 5.500 | 0.067 | 5.175 | 0.048 | 4.819 | 5H | 5.500 | 5.255 | 0.080 | 5.071 | 0.112 |
| | | | 6g | 0.020 | 5.480 | 0.106 | 5.155 | 0.075 | 4.772 | 6H | 5.500 | 5.275 | 0.100 | 5.099 | 0.140 |
| | | | | | | | | | | 7H | 5.500 | 5.300 | 0.125 | 5.139 | 0.180 |
| 6 | | 0.75 | 4h | 0 | 6.000 | 0.090 | 5.513 | 0.063 | 4.988 | 5H | 6.000 | 5.619 | 0.106 | 5.338 | 0.150 |
| | | | 6g | 0.022 | 5.978 | 0.140 | 5.491 | 0.100 | 4.929 | 6H | 6.000 | 5.645 | 0.132 | 5.378 | 0.190 |
| | | | | | | | | | | 7H | 6.000 | 5.683 | 0.170 | 5.424 | 0.236 |
| | 1 | | 4h | 0 | 6.000 | 0.112 | 5.350 | 0.071 | 4.663 | 5H | 6.000 | 5.468 | 0.118 | 5.107 | 0.190 |
| | | | 6g | 0.026 | 5.974 | 0.180 | 5.324 | 0.112 | 4.597 | 6H | 6.000 | 5.500 | 0.150 | 5.153 | 0.236 |
| | | | 8g | 0.026 | 5.974 | 0.280 | 5.324 | 0.180 | 4.528 | 7H | 6.000 | 5.540 | 0.190 | 5.217 | 0.300 |
| 7 | | 0.75 | 4h | 0 | 7.000 | 0.090 | 6.513 | 0.063 | 5.988 | 5H | 7.000 | 6.619 | 0.106 | 6.338 | 0.150 |
| | | | 6g | 0.022 | 6.978 | 0.140 | 6.491 | 0.100 | 5.929 | 6H | 7.000 | 6.645 | 0.132 | 6.378 | 0.190 |
| | | | | | | | | | | 7H | 7.000 | 6.683 | 0.170 | 6.424 | 0.236 |
| | 1 | | 4h | 0 | 7.000 | 0.112 | 6.350 | 0.071 | 5.663 | 5H | 7.000 | 6.468 | 0.118 | 6.107 | 0.190 |
| | | | 6g | 0.026 | 6.974 | 0.180 | 6.324 | 0.112 | 5.596 | 6H | 7.000 | 6.500 | 0.150 | 6.153 | 0.236 |
| | | | 8g | 0.026 | 6.974 | 0.280 | 6.324 | 0.180 | 5.528 | 7H | 7.000 | 6.540 | 0.190 | 6.217 | 0.300 |
| 8 | | 1 | 4h | 0 | 8.000 | 0.112 | 7.350 | 0.071 | 6.663 | 5H | 8.000 | 7.468 | 0.118 | 7.107 | 0.190 |
| | | | 6g | 0.026 | 7.974 | 0.180 | 7.324 | 0.112 | 6.596 | 6H | 8.000 | 7.500 | 0.150 | 7.153 | 0.236 |
| | | | 8g | 0.026 | 7.974 | 0.280 | 7.324 | 0.180 | 6.528 | 7H | 8.000 | 7.540 | 0.190 | 7.217 | 0.300 |
| | 1.25 | | 4h | 0 | 8.000 | 0.132 | 7.188 | 0.075 | 6.343 | 5H | 8.000 | 7.313 | 0.125 | 6.859 | 0.212 |
| | | | 6g | 0.028 | 7.972 | 0.212 | 7.160 | 0.118 | 6.272 | 6H | 8.000 | 7.348 | 0.160 | 6.912 | 0.265 |
| | | | 8g | 0.028 | 7.972 | 0.335 | 7.160 | 0.190 | 6.200 | 7H | 8,000 | 7.388 | 0.200 | 6.982 | 0.335 |

**Table 5.** *(Continued)* **British Standard ISO Metric Screw Threads: Limits and Tolerances for Finished Uncoated Threads for Normal Lengths of Engagement** *BS 3643: Part 2:2007*

| Nominal Diameter[a] | Pitch | | External Threads (Bolts) | | | | | | | Internal Threads (Nuts)[b] | | | | | |
|---|---|---|---|---|---|---|---|---|---|---|---|---|---|---|---|
| | Coarse | Fine | Tol. Class | Fund dev. | Major Dia. Max | Major Dia. Tol(−) | Pitch Dia. Max | Pitch Dia. Tol(−) | Minor Dia Min | Tol. Class | Major Dia. Min | Pitch Dia. Max | Pitch Dia. Tol(−) | Minor Dia Max | Minor Dia Tol(−) |
| 9 | 1.25 | | 4h | 0 | 9.000 | 0.132 | 8.188 | 0.075 | 7.343 | 5H | 9.000 | 8.313 | 0.125 | 7.859 | 0.212 |
| | | | 6g | 0.028 | 8.972 | 0.212 | 8.160 | 0.008 | 7.272 | 6H | 9.000 | 8.348 | 0.160 | 7.912 | 0.265 |
| | | | 8g | 0.028 | 8.972 | 0.335 | 8.160 | 0.190 | 7.200 | 7H | 9.000 | 8.388 | 0.200 | 7.982 | 0.335 |
| 10 | | 1.25 | 4h | 0 | 10.000 | 0.132 | 9.188 | 0.075 | 8.343 | 5H | 10.000 | 9.313 | 0.125 | 8.859 | 0.212 |
| | | | 6g | 0.028 | 9.972 | 0.212 | 9.160 | 0.118 | 8.272 | 6H | 10.000 | 9.348 | 0.160 | 8.912 | 0.265 |
| | | | 8g | 0.028 | 9.972 | 0.335 | 9.160 | 0.190 | 8.200 | 7H | 10.000 | 9.388 | 0.200 | 8.982 | 0.335 |
| | 1.5 | | 4h | 0 | 10.000 | 0.150 | 9.026 | 0.085 | 8.018 | 5H | 10.000 | 9.166 | 0.140 | 8.612 | 0.236 |
| | | | 6g | 0.032 | 9.968 | 0.236 | 8.994 | 0.132 | 7.938 | 6H | 10.000 | 9.206 | 0.180 | 8.676 | 0.300 |
| | | | 8g | 0.032 | 9.968 | 0.375 | 8.994 | 0.212 | 7.858 | 7H | 10.000 | 9.250 | 0.224 | 8.751 | 0.375 |
| 11 | 1.5 | | 4h | 0 | 11.000 | 0.150 | 10.026 | 0.085 | 9.018 | 5H | 11.000 | 10.166 | 0.140 | 9.612 | 0.236 |
| | | | 6g | 0.032 | 10.968 | 0.236 | 9.994 | 0.132 | 8.938 | 6H | 11.000 | 10.206 | 0.180 | 9.676 | 0.300 |
| | | | 8g | 0.032 | 10.968 | 0.375 | 9.994 | 0.212 | 8.858 | 7H | 11.000 | 10.250 | 0.224 | 9.751 | 0.375 |
| 12 | | 1.25 | 4h | 0 | 12.000 | 0.132 | 11.188 | 0.085 | 10.333 | 5H | 12.000 | 11.328 | 0.140 | 10.859 | 0.212 |
| | | | 6g | 0.028 | 11.972 | 0.212 | 11.160 | 0.132 | 10.257 | 6H | 12.000 | 11.398 | 0.180 | 10.912 | 0.265 |
| | | | 8g | 0.028 | 11.972 | 0.335 | 11.160 | 0.212 | 10.177 | 7H | 12.000 | 11.412 | 0.224 | 10.985 | 0.335 |
| | 1.75 | | 4h | 0 | 12.000 | 0.170 | 10.863 | 0.095 | 9.692 | 5H | 12.000 | 11.023 | 0.160 | 10.371 | 0.265 |
| | | | 6g | 0.034 | 11.966 | 0.265 | 10.829 | 0.150 | 9.602 | 6H | 12.000 | 11.063 | 0.200 | 10.441 | 0.335 |
| | | | 8g | 0.034 | 11.966 | 0.425 | 10.829 | 0.236 | 9.516 | 7H | 12.000 | 11.113 | 0.250 | 10.531 | 0.425 |
| 14 | | 1.5 | 4h | 0 | 14.000 | 0.150 | 13.026 | 0.090 | 12.012 | 5H | 14.000 | 13.176 | 0.150 | 12.612 | 0.236 |
| | | | 6g | 0.032 | 13.968 | 0.236 | 12.994 | 0.140 | 11.930 | 6H | 14.000 | 13.216 | 0.190 | 12.676 | 0.300 |
| | | | 8g | 0.032 | 13.968 | 0.375 | 12.994 | 0.224 | 11.846 | 7H | 14.000 | 13.262 | 0.236 | 12.751 | 0.375 |
| | 2 | | 4h | 0 | 14.000 | 0.180 | 12.701 | 0.100 | 11.369 | 5H | 14.000 | 12.871 | 0.170 | 12.135 | 0.300 |
| | | | 6g | 0.038 | 13.962 | 0.280 | 12.663 | 0.160 | 11.271 | 6H | 14.000 | 12.913 | 0.212 | 12.210 | 0.375 |
| | | | 8g | 0.038 | 13.962 | 0.450 | 12.663 | 0.250 | 11.181 | 7H | 14.000 | 12.966 | 0.265 | 12.310 | 0.475 |
| 16 | | 1.5 | 4h | 0 | 16.000 | 0.150 | 15.026 | 0.090 | 14.012 | 5H | 16.000 | 15.176 | 0.150 | 14.612 | 0.236 |
| | | | 6g | 0.032 | 15.968 | 0.236 | 14.994 | 0.140 | 13.930 | 6H | 16.000 | 15.216 | 0.190 | 14.676 | 0.300 |
| | | | 8g | 0.032 | 15.968 | 0.375 | 14.994 | 0.224 | 13.846 | 7H | 16.000 | 15.262 | 0.236 | 14.751 | 0.375 |
| | 2 | | 4h | 0 | 16.000 | 0.180 | 14.701 | 0.100 | 13.369 | 5H | 16.000 | 14.871 | 0.170 | 14.135 | 0.300 |
| | | | 6g | 0.038 | 15.962 | 0.280 | 14.663 | 0.160 | 13.271 | 6H | 16.000 | 14.913 | 0.212 | 14.210 | 0.375 |
| | | | 8g | 0.038 | 15.962 | 0.450 | 14.663 | 0.250 | 13.181 | 7H | 16.000 | 14.966 | 0.265 | 14.310 | 0.475 |
| 18 | | 1.5 | 4h | 0 | 18.000 | 0.150 | 17.026 | 0.090 | 16.012 | 5H | 18.000 | 17.176 | 0.150 | 16.612 | 0.236 |
| | | | 6g | 0.032 | 17.968 | 0.236 | 16.994 | 0.140 | 15.930 | 6H | 18.000 | 17.216 | 0.190 | 16.676 | 0.300 |
| | | | 8g | 0.032 | 17.968 | 0.375 | 16.994 | 0.224 | 15.846 | 7H | 18.000 | 17.262 | 0.236 | 16.751 | 0.375 |
| | 2.5 | | 4h | 0 | 18.000 | 0.212 | 16.376 | 0.106 | 14.730 | 5H | 18.000 | 16.556 | 0.180 | 15.649 | 0.355 |
| | | | 6g | 0.042 | 17.958 | 0.335 | 16.334 | 0.170 | 14.624 | 6H | 18.000 | 16.600 | 0.224 | 15.774 | 0.450 |
| | | | 8g | 0.042 | 17.958 | 0.530 | 16.334 | 0.265 | 14.529 | 7H | 18.000 | 16.656 | 0.280 | 15.854 | 0.560 |
| 20 | | 1.5 | 4h | 0 | 20.000 | 0.150 | 19.026 | 0.090 | 18.012 | 5H | 20.000 | 19.176 | 0.150 | 18.612 | 0.236 |
| | | | 6g | 0.032 | 19.968 | 0.236 | 18.994 | 0.140 | 17.930 | 6H | 20.000 | 19.216 | 0.190 | 18.676 | 0.300 |
| | | | 8g | 0.032 | 19.968 | 0.375 | 18.994 | 0.224 | 17.846 | 7H | 20.000 | 19.262 | 0.236 | 18.751 | 0.375 |
| | 2.5 | | 4h | 0 | 20.000 | 0.212 | 18.376 | 0.106 | 16.730 | 5H | 20.000 | 18.556 | 0.180 | 17.649 | 0.355 |
| | | | 6g | 0.042 | 19.958 | 0.335 | 18.334 | 0.170 | 16.624 | 6H | 20.000 | 18.600 | 0.224 | 17.744 | 0.450 |
| | | | 8g | 0.042 | 19.958 | 0.530 | 18.334 | 0.265 | 16.529 | 7H | 20.000 | 18.650 | 0.280 | 17.854 | 0.560 |
| 22 | | 1.5 | 4h | 0 | 22.000 | 0.150 | 21.026 | 0.090 | 20.012 | 5H | 22.000 | 21.176 | 0.150 | 20.612 | 0.236 |
| | | | 6g | 0.032 | 21.968 | 0.236 | 20.994 | 0.140 | 19.930 | 6H | 22.000 | 21.216 | 0.190 | 20.676 | 0.300 |
| | | | 8g | 0.032 | 21.968 | 0.375 | 20.994 | 0.224 | 19.846 | 7H | 22.000 | 21.262 | 0.236 | 20.751 | 0.375 |
| | 2.5 | | 4h | 0 | 22.000 | 0.212 | 20.376 | 0.106 | 18.730 | 5H | 22.000 | 20.556 | 0.180 | 19.649 | 0.335 |
| | | | 6g | 0.042 | 21.958 | 0.335 | 20.334 | 0.170 | 18.624 | 6H | 22.000 | 20.600 | 0.224 | 19.744 | 0.450 |
| | | | 8g | 0.042 | 21.958 | 0.530 | 20.334 | 0.265 | 18.529 | 7H | 22.000 | 20.656 | 0.280 | 19.854 | 0.560 |
| 24 | | 2 | 4h | 0 | 24.000 | 0.180 | 22.701 | 0.106 | 21.363 | 5H | 24.000 | 22.881 | 0.180 | 22.135 | 0.300 |
| | | | 6g | 0.038 | 23.962 | 0.280 | 22.663 | 0.170 | 21.261 | 6H | 24.000 | 22.925 | 0.224 | 22.210 | 0.375 |
| | | | 8g | 0.038 | 23.962 | 0.450 | 22.663 | 0.265 | 21.166 | 7H | 24.000 | 22.981 | 0.280 | 22.310 | 0.475 |
| | 3 | | 4h | 0 | 24.000 | 0.236 | 22.051 | 0.125 | 20.078 | 5H | 24.000 | 22.263 | 0.212 | 21.152 | 0.400 |
| | | | 6g | 0.048 | 23.952 | 0.375 | 22.003 | 0.200 | 19.955 | 6H | 24.000 | 22.316 | 0.265 | 21.252 | 0.500 |
| | | | 8g | 0.048 | 23.952 | 0.600 | 22.003 | 0.315 | 19.840 | 7H | 24.000 | 22.386 | 0.335 | 21.382 | 0.630 |
| 27 | | 2 | 4h | 0 | 27.000 | 0.180 | 25.701 | 0.106 | 24.363 | 5H | 27.000 | 25.881 | 0.180 | 25.135 | 0.300 |
| | | | 6g | 0.038 | 26.962 | 0.280 | 25.663 | 0.170 | 24.261 | 6H | 27.000 | 25.925 | 0.224 | 25.210 | 0.375 |
| | | | 8g | 0.038 | 26.962 | 0.450 | 25.663 | 0.265 | 24.166 | 7H | 27.000 | 25.981 | 0.280 | 25.310 | 0.475 |
| | 3 | | 4h | 0 | 27.000 | 0.236 | 25.051 | 0.125 | 23.078 | 5H | 27.000 | 25.263 | 0.212 | 24.152 | 0.400 |
| | | | 6g | 0.048 | 26.952 | 0.375 | 25.003 | 0.200 | 22.955 | 6H | 27.000 | 25.316 | 0.265 | 24.252 | 0.500 |
| | | | 8g | 0.048 | 26.952 | 0.600 | 25.003 | 0.315 | 22.840 | 7H | 27.000 | 25.386 | 0.335 | 24.382 | 0.630 |

**Table 5.** *(Continued)* **British Standard ISO Metric Screw Threads: Limits and Tolerances for Finished Uncoated Threads for Normal Lengths of Engagement** *BS 3643: Part 2:2007*

| Nominal Diameter[a] | Pitch | | External Threads (Bolts) | | | | | | | Internal Threads (Nuts)[b] | | | | | |
|---|---|---|---|---|---|---|---|---|---|---|---|---|---|---|---|
| | Coarse | Fine | Tol. Class | Fund dev. | Major Dia. | | Pitch Dia. | | Minor Dia | Tol. Class | Major Dia. | Pitch Dia. | | Minor Dia | |
| | | | | | Max | Tol(−) | Max | Tol(−) | Min | | Min | Max | Tol(−) | Max | Tol(−) |
| 30 | | 2 | 4h | 0 | 30.000 | 0.180 | 28.701 | 0.106 | 27.363 | 5H | 30.000 | 28.881 | 0.180 | 28.135 | 0.300 |
| | | | 6g | 0.038 | 29.962 | 0.280 | 28.663 | 0.170 | 27.261 | 6H | 30.000 | 27.925 | 0.224 | 28.210 | 0.375 |
| | | | 8g | 0.038 | 29.962 | 0.450 | 28.663 | 0.265 | 27.166 | 7H | 30.000 | 28.981 | 0.280 | 28.310 | 0.475 |
| | 3.5 | | 4h | 0 | 30.000 | 0.265 | 27.727 | 0.132 | 25.439 | 5H | 30.000 | 27.951 | 0.224 | 26.661 | 0.450 |
| | | | 6g | 0.053 | 29.947 | 0.425 | 27.674 | 0.212 | 25.305 | 6H | 30.000 | 28.007 | 0.280 | 26.771 | 0.560 |
| | | | 8g | 0.053 | 29.947 | 0.670 | 27.674 | 0.335 | 25.183 | 7H | 30.000 | 28.082 | 0.355 | 26.921 | 0.710 |
| 33 | | 2 | 4h | 0 | 33.000 | 0.180 | 31.701 | 0.106 | 30.363 | 5H | 33.000 | 31.881 | 0.180 | 31.135 | 0.300 |
| | | | 6g | 0.038 | 32.962 | 0.280 | 31.663 | 0.170 | 30.261 | 6H | 33.000 | 31.925 | 0.224 | 31.210 | 0.375 |
| | | | 8g | 0.038 | 32.962 | 0.450 | 30.663 | 0.265 | 30.166 | 7H | 33.000 | 31.981 | 0.280 | 31.310 | 0.475 |
| | 3.5 | | 4h | 0 | 33.000 | 0.265 | 30.727 | 0.132 | 28.438 | 5H | 33.000 | 30.951 | 0.224 | 29.661 | 0.450 |
| | | | 6g | 0.053 | 32.947 | 0.425 | 30.674 | 0.212 | 28.305 | 6H | 33.000 | 31.007 | 0.280 | 29.771 | 0.560 |
| | | | 8g | 0.053 | 32.947 | 0.670 | 30.674 | 0.335 | 28.182 | 7H | 33.000 | 31.082 | 0.355 | 29.921 | 0.710 |
| 36 | 4 | | 4h | 0 | 36.000 | 0.300 | 33.402 | 0.140 | 30.798 | 5H | 36.000 | 33.638 | 0.236 | 32.145 | 0.475 |
| | | | 6g | 0.060 | 35.940 | 0.475 | 33.342 | 0.224 | 30.654 | 6H | 36.000 | 33.702 | 0.300 | 32.270 | 0.600 |
| | | | 8g | 0.060 | 35.940 | 0.750 | 33.342 | 0.355 | 30.523 | 7H | 36.000 | 33.777 | 0.375 | 32.420 | 0.750 |
| 39 | 4 | | 4h | 0 | 39.000 | 0.300 | 36.402 | 0.140 | 33.798 | 5H | 39.000 | 36.638 | 0.236 | 35.145 | 0.475 |
| | | | 6g | 0.060 | 38.940 | 0.475 | 36.342 | 0.224 | 33.654 | 6H | 39.000 | 36.702 | 0.300 | 35.270 | 0.600 |
| | | | 8g | 0.060 | 38.940 | 0.750 | 36.342 | 0.355 | 33.523 | 7H | 39.000 | 36.777 | 0.375 | 35.420 | 0.750 |

[a] This table provides coarse- and fine-pitch series data for threads listed in Table 6 for first, second, and third choices. For constant-pitch series and for larger sizes than are shown, refer to the Standard.

[b] The fundamental deviation for internal threads (nuts) is zero for threads in this table.

All dimensions are in millimeters.

**Diameter/Pitch Combinations.—**Part 1 of BS 3643 provides a choice of diameter/pitch combinations shown here in Table 6. The use of first-choice items is preferred but if necessary, second, then third choice combinations may be selected. If pitches finer than those given in Table 6 are necessary, only the following pitches should be used: 3, 2, 1.5, 1, 0.75, 0.5, 0.35, 0.25, and 0.2 mm. When selecting such pitches it should be noted that there is increasing difficulty in meeting tolerance requirements as the diameter is increased for a given pitch. It is suggested that diameters greater than the following should not be used with the pitches indicated:

| Pitch, mm | 0.5 | 0.75 | 1 | 1.5 | 2 | 3 |
|---|---|---|---|---|---|---|
| Maximum Diameter, mm | 22 | 33 | 80 | 150 | 200 | 300 |

In cases where it is necessary to use a thread with a pitch larger than 6 mm, in the diameter range of 150 to 300 mm, the 8 mm pitch should be used.

**Limits and Tolerances for Finished Uncoated Threads.—**Part 2 of BS 3643 specifies the fundamental deviations, tolerances, and limits of size for the tolerance classes 4H, 5H, 6H, and 7H for internal threads (nuts) and 4h, 6g, and 8g for external threads (bolts) for coarse-pitch series within the range of 1 to 68 mm; fine-pitch series within the range of 1 to 33 mm; and constant pitch series within the range of 8 to 300 mm diameter.

The data in Table 5 provide the first, second, and third choice combinations shown in Table 6 except that constant-pitch series threads are omitted. For diameters larger than shown in Table 5, and for constant-pitch series data, refer to the Standard.

**Table 6. British Standard ISO Metric Screw Threads— Diameter/Pitch Combinations** *BS 3643: Part 1:2007*

| Nominal Diameter | | | | | | Nominal Diameter | | | |
|---|---|---|---|---|---|---|---|---|---|
| Choices | | | | | | Choices | | | |
| 1st | 2nd | 3rd | Coarse Pitch | Fine Pitch | Constant Pitch | 1st | 2nd | 3rd | Constant Pitch |
| 1 | … | … | 0.25 | 0.2 | … | … | … | 70 | 6, 4, 3, 2, 1.5 |
| … | 1.1 | … | 0.25 | 0.2 | … | 72 | … | … | 6, 4, 3, 2, 1.5 |
| 1.2 | … | … | 0.25 | 0.2 | … | … | … | 75 | 4, 3, 2, 1.5 |
| … | 1.4 | … | 0.3 | 0.2 | … | … | 76 | … | 6, 4, 3, 2, 1.5 |
| 1.6 | … | … | 0.35 | 0.2 | … | … | … | 78 | 2 |
| … | 1.8 | … | 0.35 | 0.2 | … | 80 | … | … | 6, 4, 3, 2, 1.5 |
| 2.0 | … | … | 0.4 | 0.25 | … | … | … | 82 | 2 |
| … | 2.2 | … | 0.45 | 0.25 | … | … | 85 | … | 6, 4, 3, 2 |
| 2.5 | … | … | 0.45 | 0.35 | … | 90 | … | … | 6, 4, 3, 2 |
| 3 | … | … | 0.5 | 0.35 | … | … | 95 | … | 6, 4, 3, 2 |
| … | 3.5 | … | 0.6 | 0.35 | … | 100 | … | … | 6, 4, 3, 2 |
| 4 | … | … | 0.7 | 0.5 | … | … | 105 | … | 6, 4, 3, 2 |
| … | 4.5 | … | 0.75 | 0.5 | … | 110 | … | … | 6, 4, 3, 2 |
| 5 | … | … | 0.8 | 0.5 | … | … | 115 | … | 6, 4, 3, 2 |
| … | … | 5.5 | … | (0.5) | … | … | 120 | … | 6, 4, 3, 2 |
| 6 | … | … | 1 | 0.75 | … | 125 | … | … | 6, 4, 3, 2 |
| … | 7 | … | 1 | 0.75 | … | … | 130 | … | 6, 4, 3, 2 |
| 8 | … | … | 1.25 | 1 | 0.75 | … | … | 135 | 6, 4, 3, 2 |
| … | … | 9 | 1.25 | … | 1, 0.75 | 140 | … | … | 6, 4, 3, 2 |
| 10 | … | … | 1.5 | 1.25 | 1, 0.75 | … | … | 145 | 6, 4, 3, 2 |
| … | … | 11 | 1.5 | … | 1, 0.75 | … | 150 | … | 6, 4, 3, 2 |
| 12 | … | … | 1.75 | 1.25 | 1.5, 1 | … | … | 155 | 6, 4, 3 |
| … | 14 | … | 2 | 1.5 | 1.25[a], 1 | 160 | … | … | 6, 4, 3 |
| … | … | 15 | … | … | 1.5, 1 | … | … | 165 | 6, 4, 3 |
| 16 | … | … | 2 | 1.5 | 1 | … | 170 | … | 6, 4, 3 |
| … | … | 17 | … | … | 1.5, 1 | … | … | 175 | 6, 4, 3 |
| … | 18 | … | 2.5 | 1.5 | 2, 1 | 180 | … | … | 6, 4, 3 |
| 20 | … | … | 2.5 | 1.5 | 2, 1 | … | … | 185 | 6, 4, 3 |
| … | 22 | … | 2.5 | 1.5 | 2, 1 | … | 190 | … | 6, 4, 3 |
| 24 | … | … | 3 | 2 | 1.5, 1 | … | … | 195 | 6, 4, 3 |
| … | … | 25 | … | … | 2, 1.5, 1 | 200 | … | … | 6, 4, 3 |
| … | … | 26 | … | … | 1.5 | … | … | 205 | 6, 4, 3 |
| … | 27 | … | 3 | 2 | 1.5, 1 | … | 210 | … | 6, 4, 3 |
| … | … | 28 | … | … | 2, 1.5, 1 | … | … | 215 | 6, 4, 3 |
| 30 | … | … | 3.5 | 2 | (3), 1.5, 1 | 220 | … | … | 6, 4, 3 |
| … | … | 32 | … | … | 2, 1.5 | … | … | 225 | 6, 4, 3 |
| … | 33 | … | 3.5 | 2 | (3), 1.5 | … | … | 230 | 6, 4, 3 |
| … | … | 35[b] | … | … | 1.5 | … | … | 235 | 6, 4, 3 |
| 36 | … | … | 4 | … | 3, 2, 1.5 | … | 240 | … | 6, 4, 3 |
| … | … | 38 | … | … | 1.5 | … | … | 245 | 6, 4, 3 |
| … | 39 | … | 4 | … | 3, 2, 1.5 | 250 | … | … | 6, 4, 3 |
| … | … | 40 | … | … | 3, 2, 1.5 | … | … | 255 | 6, 4 |
| 42 | 45 | … | 4.5 | … | 4, 3, 2, 1.5 | … | 260 | … | 6, 4 |
| 48 | … | … | 5 | … | 4, 3, 2, 1.5 | … | … | 265 | 6, 4 |
| … | … | 50 | … | … | 3, 2, 1.5 | … | … | 270 | 6, 4 |
| … | 52 | … | 5 | … | 4, 3, 2, 1.5 | … | … | 275 | 6, 4 |
| … | … | 55 | … | … | 4, 3, 2, 1.5 | 280 | … | … | 6, 4 |
| 56 | … | … | 5.5 | … | 4, 3, 2, 1.5 | … | … | 285 | 6, 4 |
| … | … | 58 | … | … | 4, 3, 2, 1.5 | … | … | 290 | 6, 4 |
| … | 60 | … | 5.5 | … | 4, 3, 2, 1.5 | … | … | 295 | 6, 4 |
| … | … | 62 | … | … | 4, 3, 2, 1.5 | … | 300 | … | 6, 4 |
| 64 | … | … | 6 | … | 4, 3, 2, 1.5 | … | … | … | … |
| … | … | 65 | … | … | 4, 3, 2, 1.5 | … | … | … | … |
| … | 68 | … | 6 | … | 4, 3, 2, 1.5 | … | … | … | … |

[a] Only for spark plugs for engines.
[b] Only for locking nuts for bearings.

All dimensions are in millimeters. Pitches in parentheses ( ) are to be avoided as far as possible.

## ACME SCREW THREADS

### American National Standard Acme Screw Threads

American National Standard ANSI/ASME B1.5-1997 is a revision of ANSI/ASME B1.5-1988 and provides for two general applications of Acme screw threads, namely, General Purpose and Centralizing.

The limits and tolerances in this Standard relate to single-start Acme threads, and may be used, if considered suitable, for multi-start Acme threads, which provide fast relative traversing motion when this is necessary. For information on additional allowances for multi-start Acme threads, see later section on page 2059.

**General Purpose Acme Threads.—**Three classes of General Purpose threads, 2G, 3G, and 4G, are provided in the Standard, each having clearance on all diameters for free movement, and may be used in assemblies with the internal thread rigidly fixed and movement of the external thread in a direction perpendicular to its axis limited by its bearing or bearings. It is suggested that external and internal threads of the same class be used together for general purpose assemblies, Class 2G being the preferred choice. If less backlash or end play is desired, Classes 3G and 4G are provided. Class 5G is not recommended for new designs.

*Thread Form:* The accompanying Fig. 1 shows the thread form of these General Purpose threads, and the formulas accompanying the figure determine their basic dimensions. Table 1 gives the basic dimensions for the most generally used pitches.

*Angle of Thread:* The angle between the sides of the thread, measured in an axial plane, is 29 degrees. The line bisecting this 29-degree angle shall be perpendicular to the axis of the screw thread.

*Thread Series:* A series of diameters and associated pitches is recommended in the Standard as preferred. These diameters and pitches have been chosen to meet present needs with the fewest number of items in order to reduce to a minimum the inventory of both tools and gages. This series of diameters and associated pitches is given in Table 3.

*Chamfers and Fillets:* General Purpose external threads may have the crest corner chamfered to an angle of 45 degrees with the axis to a maximum width of $P/15$, where $P$ is the pitch. This corresponds to a maximum depth of chamfer flat of $0.0945P$.

*Basic Diameters:* The max major diameter of the external thread is basic and is the nominal major diameter for all classes. The min pitch diameter of the internal thread is basic and is equal to the basic major diameter minus the basic height of the thread, $h$. The basic minor diameter is the min minor diameter of the internal thread. It is equal to the basic major diameter minus twice the basic thread height, $2h$.

*Length of Engagement:* The tolerances specified in this Standard are applicable to lengths of engagement not exceeding twice the nominal major diameter.

*Major and Minor Diameter Allowances:* A minimum diametral clearance is provided at the minor diameter of all external threads by establishing the maximum minor diameter 0.020 inch below the basic minor diameter of the nut for pitches of 10 threads per inch and coarser, and 0.010 inch for finer pitches. A minimum diametral clearance at the major diameter is obtained by establishing the minimum major diameter of the internal thread 0.020 inch above the basic major diameter of the screw for pitches of 10 threads per inch and coarser, and 0.010 inch for finer pitches.

*Major and Minor Diameter Tolerances:* The tolerance on the external thread major diameter is $0.05P$, where $P$ is the pitch, with a minimum of 0.005 inch. The tolerance on the internal thread major diameter is 0.020 inch for 10 threads per inch and coarser and 0.010 for finer pitches. The tolerance on the external thread minor diameter is $1.5 \times$ pitch diameter tolerance. The tolerance on the internal thread minor diameter is $0.05P$ with a minimum of 0.005 inch.

**ANSI General Purpose Acme Thread Form** *ANSI/ASME B1.5-1997 (R2014)*, **and Stub Acme Screw Thread Form** *ANSI/ASME B1.8-1988 (R2021)*

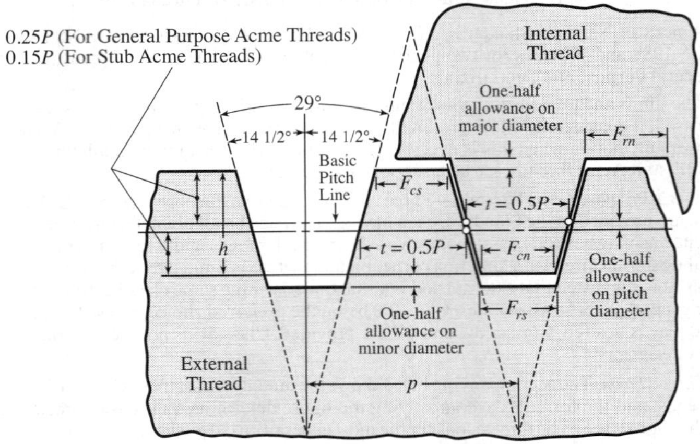


Fig. 1. General Purpose and Stub Acme Thread Forms

**Formulas for Basic Dimensions of General Purpose and Stub Acme Screw Threads**

| General Purpose | Stub Acme Threads |
|---|---|
| Pitch = $P$ = 1 ÷ No. threads per inch, $n$ | Pitch = $P$ = 1 ÷ No. threads per inch, $n$ |
| Basic thread height $h = 0.5P$ | Basic thread height $h = 0.3P$ |
| Basic thread thickness $t = 0.5P$ | Basic thread thickness $t = 0.5P$ |
| Basic flat at crest $F_{cn} = 0.3707P$ (internal thread) | Basic flat at crest $F_{cn} = 0.4224P$ (internal thread) |
| Basic flat at crest $F_{cs} = 0.3707P - 0.259 \times$ (pitch dia. allowance on ext. thd.) | Basic flat at crest $F_{cs} = 0.4224P - 0.259 \times$ (pitch dia. allowance on ext. thread) |
| $F_{rn} = 0.3707P - 0.259 \times$ (major dia. allowance on internal thread) | $F_{rn} = 0.4224P - 0.259 \times$ (major dia. allowance on internal thread) |
| $F_{rs} = 0.3707P - 0.259 \times$ (minor dia. allowance on ext. thread – pitch dia. allowance on ext. thread) | $F_{rs} = 0.4224P - 0.259 \times$ (minor dia. allowance on ext. thread – pitch dia. allowance on ext. thread) |

*Pitch Diameter Allowances and Tolerances:* Allowances on the pitch diameter of General Purpose Acme threads are given in Table 4. Pitch diameter tolerances are given in Table 5. The ratios of the pitch diameter tolerances of Classes 2G, 3G, and 4G, General Purpose threads are 3.0, 1.4, and 1, respectively.

An increase of 10 percent in the allowance is recommended for each inch, or fraction thereof, that the length of engagement exceeds two diameters.

*Application of Tolerances:* The tolerances specified are designed to ensure interchangeability and maintain a high grade of product. The tolerances on diameters of the internal thread are plus, being applied from minimum sizes to above the minimum sizes. The tolerances on diameters of the external thread are minus, being applied from the maximum sizes to below the maximum sizes. The pitch diameter (or thread thickness) tolerances for an external or internal thread of a given class are the same. The thread thickness tolerance is 0.259 times the pitch diameter tolerance.

*Limiting Dimensions:* Limiting dimensions of General Purpose Acme screw threads in the recommended series are given in Table 2b and Table 2c. These limits are based on the formulas in Table 2a.

For combinations of pitch and diameter other than those in the recommended series, the formulas in Table 2a and the data in Table 4 and Table 5 make it possible to readily determine the limiting dimensions required.

A diagram showing the disposition of allowances, tolerances, and crest clearances for General Purpose Acme threads appears on page 2058.

*Stress Area of General Purpose Acme Threads:* For computing the tensile strength of the thread section, the minimum stress area based on the mean of the minimum pitch diameter $d_2$ min. and the minimum minor diameter $d_1$ min. of the external thread is used:

$$\text{Stress Area} = 3.1416\left(\frac{d_2\text{min.} + d_1\text{min.}}{4}\right)^2$$

where $d_2$ min. and $d_1$ min. may be computed by Formulas 4 and 6, Table 2a or taken from Table 2b or Table 2c.

*Shear Area of General Purpose Acme Threads:* For computing the shear area per inch length of engagement of the external thread, the maximum minor diameter of the internal thread $D_1$ max., and the minimum pitch diameter of the external thread $d_2$ min., Table 2b or Table 2c, or Formulas 12 and 4, Table 2a, are used:

$$\text{Shear Area} = 3.1416 D_1\,\text{max.}\left[0.5 + n\tan 14\tfrac{1}{2}^\circ(d_2\text{min.} - D_1\text{max.})\right]$$

**Acme Thread Abbreviations.—**The following abbreviations are recommended for use on drawings and in specifications, and on tools and gages:

*ACME* = Acme threads

*G* = General Purpose

*C* = Centralizing

*P* = pitch

*L* = lead

*LH* = left-hand

**Designation of General Purpose Acme Threads.—**The examples listed below are given here to show how General Purpose Acme threads are designated on drawings and tools:

1.750-4 ACME-2G indicates a General Purpose Class 2G Acme thread of 1.750-inch major diameter, 4 threads per inch, single thread, right-hand. The same thread, but left-hand, is designated 1.750-4 ACME-2G-LH.

2.875-0.4*P*-0.8*L*-ACME-3G indicates a General Purpose Class 3G Acme thread of 2.875-inch major diameter, pitch 0.4 inch, lead 0.8 inch, double thread, right-hand.

**Multiple Start Acme Threads.—**The tabulated diameter-pitch data with allowances and tolerances relate to single-start threads. These data, as tabulated, may be and often are used for two-start Class 2G threads but this usage generally requires reduction of the full working tolerances to provide a greater allowance or clearance zone between the mating threads to assure satisfactory assembly.

When the class of thread requires smaller working tolerances than the 2G class or when threads with 3, 4, or more starts are required, some additional allowances or increased tolerances or both may be needed to ensure adequate working tolerances and satisfactory assembly of mating parts.

It is suggested that the allowances shown in Table 4 be used for all external threads and that allowances be applied to internal threads in the following ratios: for two-start threads, 50 percent of the allowances shown in the Class 2G, 3G and 4G columns of Table 4; for three-start threads, 75 percent of these allowances; and for four-start threads, 100 percent of these same values.

**Table 1. American National Standard General Purpose Acme Screw Thread Form—Basic Dimensions** *ANSI/ASME B1.5-1997 (R2014)*

| Thds. per Inch $n$ | Pitch, $P = 1/n$ | Height of Thread (Basic), $h = P/2$ | Total Height of Thread, $h_s = P/2 + 1/2$ allowance[a] | Thread Thickness (Basic), $t = P/2$ | Width of Flat: Crest of Internal Thread (Basic), $F_{cn} = 0.3707P$ | Width of Flat: Root of Internal Thread, $F_{rn}$ $0.3707P - 0.259 \times$ allowance[a] |
|---|---|---|---|---|---|---|
| 16 | 0.06250 | 0.03125 | 0.0362 | 0.03125 | 0.0232 | 0.0206 |
| 14 | 0.07143 | 0.03571 | 0.0407 | 0.03571 | 0.0265 | 0.0239 |
| 12 | 0.08333 | 0.04167 | 0.0467 | 0.04167 | 0.0309 | 0.0283 |
| 10 | 0.10000 | 0.05000 | 0.0600 | 0.05000 | 0.0371 | 0.0319 |
| 8 | 0.12500 | 0.06250 | 0.0725 | 0.06250 | 0.0463 | 0.0411 |
| 6 | 0.16667 | 0.08333 | 0.0933 | 0.08333 | 0.0618 | 0.0566 |
| 5 | 0.20000 | 0.10000 | 0.1100 | 0.10000 | 0.0741 | 0.0689 |
| 4 | 0.25000 | 0.12500 | 0.1350 | 0.12500 | 0.0927 | 0.0875 |
| 3 | 0.33333 | 0.16667 | 0.1767 | 0.16667 | 0.1236 | 0.1184 |
| 2½ | 0.40000 | 0.20000 | 0.2100 | 0.20000 | 0.1483 | 0.1431 |
| 2 | 0.50000 | 0.25000 | 0.2600 | 0.25000 | 0.1853 | 0.1802 |
| 1½ | 0.66667 | 0.33333 | 0.3433 | 0.33333 | 0.2471 | 0.2419 |
| 1⅓ | 0.75000 | 0.37500 | 0.3850 | 0.37500 | 0.2780 | 0.2728 |
| 1 | 1.00000 | 0.50000 | 0.5100 | 0.50000 | 0.3707 | 0.3655 |

All dimensions are in inches.

[a] Allowance is 0.020 inch for 10 threads per inch and coarser, and 0.010 inch for finer threads.

**Table 2a. American National Standard General Purpose Acme Single-Start Screw Threads—Formulas for Determining Diameters** *ANSI/ASME B1.5-1997 (R2014)*

$D$ = Basic Major Diameter and Nominal Size, in Inches.

$P$ = Pitch = 1 ÷ Number of Threads per Inch.

$E$ = Basic Pitch Diameter = $D - 0.5P$

$K$ = Basic Minor Diameter = $D - P$

| No. | External Threads (Screws) |
|---|---|
| 1 | Major Dia., Max. = $D$ |
| 2 | Major Dia., Min. = $D$ *minus* $0.05P^a$ but not less than 0.005. |
| 3 | Pitch Dia., Max. = $E$ *minus* allowance from Table 4. |
| 4 | Pitch Dia., Min. = Pitch Dia., Max. (Formula 3) *minus* tolerance from Table 5. |
| 5 | Minor Dia., Max. = $K$ *minus* 0.020 for 10 threads per inch and coarser and 0.010 for finer pitches. |
| 6 | Minor Dia., Min. = Minor Dia., Max. (Formula 5) *minus* 1.5 × pitch diameter tolerance from Table 5. |
| | **Internal Threads (Nuts)** |
| 7 | Major Dia., Min. = $D$ *plus* 0.020 for 10 threads per inch and coarser and 0.010 for finer pitches. |
| 8 | Major Dia., Max. = Major Dia., Min. (Formula 7) *plus* 0.020 for 10 threads per inch and coarser and 0.010 for finer pitches. |
| 9 | Pitch Dia., Min. = $E$ |
| 10 | Pitch Dia., Max. = Pitch Dia., Min. (Formula 9) *plus* tolerance from Table 5. |
| 11 | Minor Dia., Min. = $K$ |
| 12 | Minor Dia., Max. = Minor Dia., Min. (Formula 11) *plus* $0.05P^a$ but not less than 0.005. |

[a] If $P$ is between two recommended pitches listed in Table 3, use the coarser of the two pitches in this formula instead of the actual value of $P$.

**Table 2b. Limiting Dimensions of ANSI General Purpose Acme Single-Start Screw Threads** *ANSI/ASME B1.5-1997 (R2014)*

| Nominal Diameter, *D* | | 1/4 | 5/16 | 3/8 | 7/16 | 1/2 | 5/8 | 3/4 | 7/8 | 1 | 1 1/8 | 1 1/4 | 1 3/8 |
|---|---|---|---|---|---|---|---|---|---|---|---|---|---|
| Threads per Inch[a] | | 16 | 14 | 12 | 12 | 10 | 8 | 6 | 6 | 5 | 5 | 5 | 4 |
| Limiting Diameters | | External Threads | | | | | | | | | | | |
| Classes 2G, 3G, and 4G Major Diameter | Max (*D*) | 0.2500 | 0.3125 | 0.3750 | 0.4375 | 0.5000 | 0.6250 | 0.7500 | 0.8750 | 1.0000 | 1.1250 | 1.2500 | 1.3750 |
| | Min | 0.2450 | 0.3075 | 0.3700 | 0.4325 | 0.4950 | 0.6188 | 0.7417 | 0.8667 | 0.9900 | 1.1150 | 1.2400 | 1.3625 |
| Classes 2G, 3G, and 4G Minor Diameter | Max | 0.1775 | 0.2311 | 0.2817 | 0.3442 | 0.3800 | 0.4800 | 0.5633 | 0.6883 | 0.7800 | 0.9050 | 1.0300 | 1.1050 |
| Class 2G, Minor Diameter | Min | 0.1618 | 0.2140 | 0.2632 | 0.3253 | 0.3594 | 0.4570 | 0.5372 | 0.6615 | 0.7509 | 0.8753 | 0.9998 | 1.0720 |
| Class 3G, Minor Diameter | Min | 0.1702 | 0.2231 | 0.2730 | 0.3354 | 0.3704 | 0.4693 | 0.5511 | 0.6758 | 0.7664 | 0.8912 | 1.0159 | 1.0896 |
| Class 4G, Minor Diameter | Min | 0.1722 | 0.2254 | 0.2755 | 0.3379 | 0.3731 | 0.4723 | 0.5546 | 0.6794 | 0.7703 | 0.8951 | 1.0199 | 1.0940 |
| Class 2G, Pitch Diameter | Max | 0.2148 | 0.2728 | 0.3284 | 0.3909 | 0.4443 | 0.5562 | 0.6598 | 0.7842 | 0.8920 | 1.0165 | 1.1411 | 1.2406 |
| | Min | 0.2043 | 0.2614 | 0.3161 | 0.3783 | 0.4306 | 0.5408 | 0.6424 | 0.7663 | 0.8726 | 0.9967 | 1.1210 | 1.2188 |
| Class 3G, Pitch Diameter | Max | 0.2158 | 0.2738 | 0.3296 | 0.3921 | 0.4458 | 0.5578 | 0.6615 | 0.7861 | 0.8940 | 1.0186 | 1.1433 | 1.2430 |
| | Min | 0.2109 | 0.2685 | 0.3238 | 0.3862 | 0.4394 | 0.5506 | 0.6534 | 0.7778 | 0.8849 | 1.0094 | 1.1339 | 1.2327 |
| Class 4G, Pitch Diameter | Max | 0.2168 | 0.2748 | 0.3309 | 0.3934 | 0.4472 | 0.5593 | 0.6632 | 0.7880 | 0.8960 | 1.0208 | 1.1455 | 1.2453 |
| | Min | 0.2133 | 0.2710 | 0.3268 | 0.3892 | 0.4426 | 0.5542 | 0.6574 | 0.7820 | 0.8895 | 1.0142 | 1.1388 | 1.2380 |
| | | Internal Threads | | | | | | | | | | | |
| Classes 2G, 3G, and 4G Major Diameter | Min | 0.2600 | 0.3225 | 0.3850 | 0.4475 | 0.5200 | 0.6450 | 0.7700 | 0.8950 | 1.0200 | 1.1450 | 1.2700 | 1.3950 |
| | Max | 0.2700 | 0.3325 | 0.3950 | 0.4575 | 0.5400 | 0.6650 | 0.7900 | 0.9150 | 1.0400 | 1.1650 | 1.2900 | 1.4150 |
| Classes 2G, 3G, and 4G Minor Diameter | Min | 0.1875 | 0.2411 | 0.2917 | 0.3542 | 0.4000 | 0.5000 | 0.5833 | 0.7083 | 0.8000 | 0.9250 | 1.0500 | 1.1250 |
| | Max | 0.1925 | 0.2461 | 0.2967 | 0.3592 | 0.4050 | 0.5062 | 0.5916 | 0.7166 | 0.8100 | 0.9350 | 1.0600 | 1.1375 |
| Class 2G, Pitch Diameter | Min | 0.2188 | 0.2768 | 0.3333 | 0.3958 | 0.4500 | 0.5625 | 0.6667 | 0.7917 | 0.9000 | 1.0250 | 1.1500 | 1.2500 |
| | Max | 0.2293 | 0.2882 | 0.3456 | 0.4084 | 0.4637 | 0.5779 | 0.6841 | 0.8096 | 0.9194 | 1.0448 | 1.1701 | 1.2720 |
| Class 3G, Pitch Diameter | Min | 0.2188 | 0.2768 | 0.3333 | 0.3958 | 0.4500 | 0.5625 | 0.6667 | 0.7917 | 0.9000 | 1.0250 | 1.1500 | 1.2500 |
| | Max | 0.2237 | 0.2821 | 0.3391 | 0.4017 | 0.4564 | 0.5697 | 0.6748 | 0.8000 | 0.9091 | 1.0342 | 1.1594 | 1.2603 |
| Class 4G, Pitch Diameter | Min | 0.2188 | 0.2768 | 0.3333 | 0.3958 | 0.4500 | 0.5625 | 0.6667 | 0.7917 | 0.9000 | 1.0250 | 1.1500 | 1.2500 |
| | Max | 0.2223 | 0.2806 | 0.3374 | 0.4000 | 0.4546 | 0.5676 | 0.6725 | 0.7977 | 0.9065 | 1.0316 | 1.1567 | 1.2573 |

**Table 2c. Limiting Dimensions of ANSI General Purpose Acme Single-Start Screw Threads** *ANSI/ASME B1.5-1997 (R2014)*

| Nominal Diameter, *D* | | 1½ | 1¾ | 2 | 2¼ | 2½ | 2¾ | 3 | 3½ | 4 | 4½ | 5 |
|---|---|---|---|---|---|---|---|---|---|---|---|---|
| Threads per Inch[a] | | 4 | 4 | 4 | 3 | 3 | 3 | 2 | 2 | 2 | 2 | 2 |
| Limiting Diameters | | External Threads | | | | | | | | | | |
| Classes 2G, 3G, and 4G Major Diameter | Max (*D*) | 1.5000 | 1.7500 | 2.0000 | 2.2500 | 2.5000 | 2.7500 | 3.0000 | 3.5000 | 4.0000 | 4.5000 | 5.0000 |
| | Min | 1.4875 | 1.7375 | 1.9875 | 2.2333 | 2.4833 | 2.7333 | 2.9750 | 3.4750 | 3.9750 | 4.4750 | 4.9750 |
| Classes 2G, 3G, and 4G Minor Diameter | Max | 1.2300 | 1.4800 | 1.7300 | 1.8967 | 2.1467 | 2.3967 | 2.4800 | 2.9800 | 3.4800 | 3.9800 | 4.4800 |
| Class 2G, Minor Diameter | Min | 1.1965 | 1.4456 | 1.6948 | 1.8572 | 2.1065 | 2.3558 | 2.4326 | 2.9314 | 3.4302 | 3.9291 | 4.4281 |
| Class 3G, Minor Diameter | Min | 1.2144 | 1.4640 | 1.7136 | 1.8783 | 2.1279 | 2.3776 | 2.4579 | 2.9574 | 3.4568 | 3.9563 | 4.4558 |
| Class 4G, Minor Diameter | Min | 1.2189 | 1.4686 | 1.7183 | 1.8835 | 2.1333 | 2.3831 | 2.4642 | 2.9638 | 3.4634 | 3.9631 | 4.4627 |
| Class 2G, Pitch Diameter | Max | 1.3652 | 1.6145 | 1.8637 | 2.0713 | 2.3207 | 2.5700 | 2.7360 | 3.2350 | 3.7340 | 4.2330 | 4.7319 |
| | Min | 1.3429 | 1.5916 | 1.8402 | 2.0450 | 2.2939 | 2.5427 | 2.7044 | 3.2026 | 3.7008 | 4.1991 | 4.6973 |
| Class 3G, Pitch Diameter | Max | 1.3677 | 1.6171 | 1.8665 | 2.0743 | 2.3238 | 2.5734 | 2.7395 | 3.2388 | 3.7380 | 4.2373 | 4.7364 |
| | Min | 1.3573 | 1.6064 | 1.8555 | 2.0620 | 2.3113 | 2.5607 | 2.7248 | 3.2237 | 3.7225 | 4.2215 | 4.7202 |
| Class 4G, Pitch Diameter | Max | 1.3701 | 1.6198 | 1.8693 | 2.0773 | 2.3270 | 2.5767 | 2.7430 | 3.2425 | 3.7420 | 4.2415 | 4.7409 |
| | Min | 1.3627 | 1.6122 | 1.8615 | 2.0685 | 2.3181 | 2.5676 | 2.7325 | 3.2317 | 3.7309 | 4.2302 | 4.7294 |
| | | Internal Threads | | | | | | | | | | |
| Classes 2G, 3G, and 4G Major Diameter | Min | 1.5200 | 1.7700 | 2.0200 | 2.2700 | 2.5200 | 2.7700 | 3.0200 | 3.5200 | 4.0200 | 4.5200 | 5.0200 |
| | Max | 1.5400 | 1.7900 | 2.0400 | 2.2900 | 2.5400 | 2.7900 | 3.0400 | 3.5400 | 4.0400 | 4.5400 | 5.0400 |
| Classes 2G, 3G, and 4G Minor Diameter | Min | 1.2500 | 1.5000 | 1.7500 | 1.9167 | 2.1667 | 2.4167 | 2.5000 | 3.0000 | 3.5000 | 4.0000 | 4.5000 |
| | Max | 1.2625 | 1.5125 | 1.7625 | 1.9334 | 2.1834 | 2.4334 | 2.5250 | 3.0250 | 3.5250 | 4.0250 | 4.5250 |
| Class 2G, Pitch Diameter | Min | 1.3750 | 1.6250 | 1.8750 | 2.0833 | 2.3333 | 2.5833 | 2.7500 | 3.2500 | 3.7500 | 4.2500 | 4.7500 |
| | Max | 1.3973 | 1.6479 | 1.8985 | 2.1096 | 2.3601 | 2.6106 | 2.7816 | 3.2824 | 3.7832 | 4.2839 | 4.7846 |
| Class 3G, Pitch Diameter | Min | 1.3750 | 1.6250 | 1.8750 | 2.0833 | 2.3333 | 2.5833 | 2.7500 | 3.2500 | 3.7500 | 4.2500 | 4.7500 |
| | Max | 1.3854 | 1.6357 | 1.8860 | 2.0956 | 2.3458 | 2.5960 | 2.7647 | 3.2651 | 3.7655 | 4.2658 | 4.7662 |
| Class 4G, Pitch Diameter | Min | 1.3750 | 1.6250 | 1.8750 | 2.0833 | 2.3333 | 2.5833 | 2.7500 | 3.2500 | 3.7500 | 4.2500 | 4.7500 |
| | Max | 1.3824 | 1.6326 | 1.8828 | 2.0921 | 2.3422 | 2.5924 | 2.7605 | 3.2608 | 3.7611 | 4.2613 | 4.7615 |

[a] All dimensions are given in inches. The selection of threads per inch is arbitrary and for the purpose of establishing a standard.

## Table 3. General Purpose Acme Single-Start Screw Thread Data *ANSI/ASME B1.5-1997 (R2014)*

| Identification | | Basic Diameters | | | Thread Data | | | | | | | |
|---|---|---|---|---|---|---|---|---|---|---|---|---|
| | | Classes 2G, 3G, and 4G | | | | | | | Lead Angle λ at Basic Pitch Diameter[a] Classes 2G, 3G, and 4G | | | |
| Nominal Sizes (All Classes) | Threads per Inch,[a] $n$ | Major Diameter, $D$ | Pitch Diameter, $D_2 = D - h$ | Minor Diameter, $D_1 = D - 2h$ | Pitch, $P$ | Thickness at Pitch Line, $t = P/2$ | Basic Height of Thread, $h = P/2$ | Basic Width of Flat, $F = 0.3707P$ | Deg | Min | Shear Area[b] Class 3G | Stress Area[c] Class 3G |
| 1/4 | 16 | 0.2500 | 0.2188 | 0.1875 | 0.06250 | 0.03125 | 0.03125 | 0.0232 | 5 | 12 | 0.350 | 0.0285 |
| 5/16 | 14 | 0.3125 | 0.2768 | 0.2411 | 0.07143 | 0.03571 | 0.03571 | 0.0265 | 4 | 42 | 0.451 | 0.0474 |
| 3/8 | 12 | 0.3750 | 0.3333 | 0.2917 | 0.08333 | 0.04167 | 0.04167 | 0.0309 | 4 | 33 | 0.545 | 0.0699 |
| 7/16 | 12 | 0.4375 | 0.3958 | 0.3542 | 0.08333 | 0.04167 | 0.04167 | 0.0309 | 3 | 50 | 0.660 | 0.1022 |
| 1/2 | 10 | 0.5000 | 0.4500 | 0.4000 | 0.10000 | 0.05000 | 0.05000 | 0.0371 | 4 | 3 | 0.749 | 0.1287 |
| 5/8 | 8 | 0.6250 | 0.5625 | 0.5000 | 0.12500 | 0.06250 | 0.06250 | 0.0463 | 4 | 3 | 0.941 | 0.2043 |
| 3/4 | 6 | 0.7500 | 0.6667 | 0.5833 | 0.16667 | 0.08333 | 0.08333 | 0.0618 | 4 | 33 | 1.108 | 0.2848 |
| 7/8 | 6 | 0.8750 | 0.7917 | 0.7083 | 0.16667 | 0.08333 | 0.08333 | 0.0618 | 3 | 50 | 1.339 | 0.4150 |
| 1 | 5 | 1.0000 | 0.9000 | 0.8000 | 0.20000 | 0.10000 | 0.10000 | 0.0741 | 4 | 3 | 1.519 | 0.5354 |
| 1 1/8 | 5 | 1.1250 | 1.0250 | 0.9250 | 0.20000 | 0.10000 | 0.10000 | 0.0741 | 3 | 33 | 1.751 | 0.709 |
| 1 1/4 | 5 | 1.2500 | 1.1500 | 1.0500 | 0.20000 | 0.10000 | 0.10000 | 0.0741 | 3 | 10 | 1.983 | 0.907 |
| 1 3/8 | 4 | 1.3750 | 1.2500 | 1.1250 | 0.25000 | 0.12500 | 0.12500 | 0.0927 | 3 | 39 | 2.139 | 1.059 |
| 1 1/2 | 4 | 1.5000 | 1.3750 | 1.2500 | 0.25000 | 0.12500 | 0.12500 | 0.0927 | 3 | 19 | 2.372 | 1.298 |
| 1 3/4 | 4 | 1.7500 | 1.6250 | 1.5000 | 0.25000 | 0.12500 | 0.12500 | 0.0927 | 2 | 48 | 2.837 | 1.851 |
| 2 | 4 | 2.0000 | 1.8750 | 1.7500 | 0.25000 | 0.12500 | 0.12500 | 0.0927 | 2 | 26 | 3.301 | 2.501 |
| 2 1/4 | 3 | 2.2500 | 2.0833 | 1.9167 | 0.33333 | 0.16667 | 0.16667 | 0.1236 | 2 | 55 | 3.643 | 3.049 |
| 2 1/2 | 3 | 2.5000 | 2.3333 | 2.1667 | 0.33333 | 0.16667 | 0.16667 | 0.1236 | 2 | 36 | 4.110 | 3.870 |
| 2 3/4 | 3 | 2.7500 | 2.5833 | 2.4167 | 0.33333 | 0.16667 | 0.16667 | 0.1236 | 2 | 21 | 4.577 | 4.788 |
| 3 | 2 | 3.0000 | 2.7500 | 2.5000 | 0.50000 | 0.25000 | 0.25000 | 0.1853 | 3 | 19 | 4.786 | 5.27 |
| 3 1/2 | 2 | 3.5000 | 3.2500 | 3.0000 | 0.50000 | 0.25000 | 0.25000 | 0.1853 | 2 | 48 | 5.73 | 7.50 |
| 4 | 2 | 4.0000 | 3.7500 | 3.5000 | 0.50000 | 0.25000 | 0.25000 | 0.1853 | 2 | 26 | 6.67 | 10.12 |
| 4 1/2 | 2 | 4.5000 | 4.2500 | 4.0000 | 0.50000 | 0.25000 | 0.25000 | 0.1853 | 2 | 9 | 7.60 | 13.13 |
| 5 | 2 | 5.0000 | 4.7500 | 4.5000 | 0.50000 | 0.25000 | 0.25000 | 0.1853 | 1 | 55 | 8.54 | 16.53 |

[a] All other dimensions are given in inches.

[b] Per inch length of engagement of the external thread in line with the minor diameter crests of the internal thread. Figures given are the minimum shear area based on max $D_1$ and min $d_2$.

[c] Figures given are the minimum stress area based on the mean of the minimum minor and pitch diameters of the external thread. See formulas for shear area and stress area on page 2059.

**Table 4. American National Standard General Purpose Acme Single-Start Screw Threads—Pitch Diameter Allowances** *ANSI/ASME B1.5-1997 (R2014)*

| Nominal Size Range[a] | | Allowances on External Threads[b] | | | Nominal Size Range[a] | | Allowances on External Threads[b] | | |
|---|---|---|---|---|---|---|---|---|---|
| Above | To and Including | Class 2G[c], $0.008\sqrt{D}$ | Class 3G, $0.006\sqrt{D}$ | Class 4G, $0.004\sqrt{D}$ | Above | To and Including | Class 2G[c], $0.008\sqrt{D}$ | Class 3G, $0.006\sqrt{D}$ | Class 4G, $0.004\sqrt{D}$ |
| 0 | 3⁄16 | 0.0024 | 0.0018 | 0.0012 | 1 7⁄16 | 1 9⁄16 | 0.0098 | 0.0073 | 0.0049 |
| 3⁄16 | 5⁄16 | 0.0040 | 0.0030 | 0.0020 | 1 9⁄16 | 1 7⁄8 | 0.0105 | 0.0079 | 0.0052 |
| 5⁄16 | 7⁄16 | 0.0049 | 0.0037 | 0.0024 | 1 7⁄8 | 2 1⁄8 | 0.0113 | 0.0085 | 0.0057 |
| 7⁄16 | 9⁄16 | 0.0057 | 0.0042 | 0.0028 | 2 1⁄8 | 2 3⁄8 | 0.0120 | 0.0090 | 0.0060 |
| 9⁄16 | 11⁄16 | 0.0063 | 0.0047 | 0.0032 | 2 3⁄8 | 2 5⁄8 | 0.0126 | 0.0095 | 0.0063 |
| 11⁄16 | 13⁄16 | 0.0069 | 0.0052 | 0.0035 | 2 5⁄8 | 2 7⁄8 | 0.0133 | 0.0099 | 0.0066 |
| 13⁄16 | 15⁄16 | 0.0075 | 0.0056 | 0.0037 | 2 7⁄8 | 3 1⁄4 | 0.0140 | 0.0105 | 0.0070 |
| 15⁄16 | 1 1⁄16 | 0.0080 | 0.0060 | 0.0040 | 3 1⁄4 | 3 3⁄4 | 0.0150 | 0.0112 | 0.0075 |
| 1 1⁄16 | 1 3⁄16 | 0.0085 | 0.0064 | 0.0042 | 3 3⁄4 | 4 1⁄4 | 0.0160 | 0.0120 | 0.0080 |
| 1 3⁄16 | 1 5⁄16 | 0.0089 | 0.0067 | 0.0045 | 4 1⁄4 | 4 3⁄4 | 0.0170 | 0.0127 | 0.0085 |
| 1 5⁄16 | 1 7⁄16 | 0.0094 | 0.0070 | 0.0047 | 4 3⁄4 | 5 1⁄2 | 0.0181 | 0.0136 | 0.0091 |

All dimensions in inches. It is recommended that the sizes given in Table 3 be used whenever possible.

[a] The values in columns for Classes 2G, 3G, and 4G are to be used for any size within the nominal size range shown. These values are calculated from the mean of the range.

[b] An increase of 10 percent in the allowance is recommended for each inch, or fraction thereof, that the length of engagement exceeds two diameters.

[c] Allowances for the 2G Class of thread in this table also apply to American National Standard Stub Acme threads ASME B 1.8-1988.

These values will provide for a 0.25-16 ACME-2G thread size, 0.002, 0.003, and 0.004 inch additional clearance for 2-, 3-, and 4-start threads, respectively. For a 5-2 ACME-3G thread size the additional clearances would be 0.0091, 0.0136, and 0.0181 inch, respectively. GO thread plug gages and taps would be increased by these same values. To maintain the same working tolerances on multi-start threads, the pitch diameter of the NOT GO thread plug gage would also be increased by these same values.

For multi-start threads with more than four starts, it is believed that the 100 percent allowance provided by the above procedures would be adequate as index spacing variables would generally be no greater than on a four-start thread.

In general, for multi-start threads of Classes 2G, 3G, and 4G the percentages would be applied, usually, to allowances for the same class, respectively. However, where exceptionally good control over lead, angle, and spacing variables would produce close to theoretical values in the product, it is conceivable that these percentages could be applied to Class 3G or Class 4G allowances used on Class 2G internally threaded product. Also, these percentages could be applied to Class 4G allowances used on Class 3G internally threaded product. It is not advocated that any change be made in externally threaded products.

Designations for gages or tools for internal threads could cover allowance requirements as follows:

GO and NOT GO thread plug gages for: 2.875-0.4*P*-0.8L-ACME-2G with 50 percent of the 4G internal thread allowance.

**Centralizing Acme Threads.—**The three classes of Centralizing Acme threads in American National Standard ANSI/ASME B1.5-1997 (R2014), designated as 2C, 3C, and 4C, have limited clearance at the major diameters of internal and external threads so that a bearing at the major diameters maintains approximate alignment of the thread axis and prevents wedging on the flanks of the thread. An alternative series having centralizing control on the *minor* diameter is described on page 2075. For any combination of the three classes of threads covered in this Standard some end play or backlash will result. Classes 5C and 6C are not recommended for new designs.

**Table 5. American National Standard General Purpose Acme Single-Start Screw Threads—Pitch Diameter Tolerances** *ANSI/ASME B1.5-1997 (R2014)*

| Nom. Dia.,[a] $D$ | Class of Thread | | | Nom. Dia.,[a] $D$ | Class of Thread | | |
|---|---|---|---|---|---|---|---|
| | 2G[b] | 3G | 4G | | 2G[b] | 3G | 4G |
| | Diameter Increment | | | | Diameter Increment | | |
| | $0.006\sqrt{D}$ | $0.0028\sqrt{D}$ | $0.002\sqrt{D}$ | | $0.006\sqrt{D}$ | $0.0028\sqrt{D}$ | $0.002\sqrt{D}$ |
| ¼ | .00300 | .00140 | .00100 | 1½ | .00735 | .00343 | .00245 |
| 5/16 | .00335 | .00157 | .00112 | 1¾ | .00794 | .00370 | .00265 |
| ⅜ | .00367 | .00171 | .00122 | 2 | .00849 | .00396 | .00283 |
| 7/16 | .00397 | .00185 | .00132 | 2¼ | .00900 | .00420 | .00300 |
| ½ | .00424 | .00198 | .00141 | 2½ | .00949 | .00443 | .00316 |
| ⅝ | .00474 | .00221 | .00158 | 2¾ | .00995 | .00464 | .00332 |
| ¾ | .00520 | .00242 | .00173 | 3 | .01039 | .00485 | .00346 |
| ⅞ | .00561 | .00262 | .00187 | 3½ | .01122 | .00524 | .00374 |
| 1 | .00600 | .00280 | .00200 | 4 | .01200 | .00560 | .00400 |
| 1⅛ | .00636 | .00297 | .00212 | 4½ | .01273 | .00594 | .00424 |
| 1¼ | .00671 | .00313 | .00224 | 5 | .01342 | .00626 | .00447 |
| 1⅜ | .00704 | .00328 | .00235 | … | … | … | … |
| Thds. per Inch,[c] $n$ | Class of Thread | | | Thds. per Inch[c], $n$ | Class of Thread | | |
| | 2G[b] | 3G | 4G | | 2G[b] | 3G | 4G |
| | Pitch Increment | | | | Pitch Increment | | |
| | $0.030\sqrt{\frac{1}{n}}$ | $0.014\sqrt{\frac{1}{n}}$ | $0.010\sqrt{\frac{1}{n}}$ | | $0.030\sqrt{\frac{1}{n}}$ | $0.014\sqrt{\frac{1}{n}}$ | $0.010\sqrt{\frac{1}{n}}$ |
| 16 | .00750 | .00350 | .00250 | 4 | .01500 | .00700 | .00500 |
| 14 | .00802 | .00374 | .00267 | 3 | .01732 | .00808 | .00577 |
| 12 | .00866 | .00404 | .00289 | 2½ | .01897 | .00885 | .00632 |
| 10 | .00949 | .00443 | .00316 | 2 | .02121 | .00990 | .00707 |
| 8 | .01061 | .00495 | .00354 | 1½ | .02449 | .01143 | .00816 |
| 6 | .01225 | .00572 | .00408 | 1⅓ | .02598 | .01212 | .00866 |
| 5 | .01342 | .00626 | .00447 | 1 | .03000 | .01400 | .01000 |

For any particular size of thread, the pitch diameter tolerance is obtained by adding the *diameter increment* from the upper half of the table to the *pitch increment* from the lower half of the table. *Example:* A ¼-16 Acme-2G thread has a pitch diameter tolerance of 0.00300 + 0.00750 = 0.0105 inch.

The equivalent tolerance on thread thickness is 0.259 times the pitch diameter tolerance.

[a] For a nominal diameter between any two tabulated nominal diameters, use the diameter increment for the larger of the two tabulated nominal diameters.

[b] Columns for the 2G Class of thread in this table also apply to American National Standard Stub Acme threads, ANSI/ASME B1.8-1988 (R2021).

[c] All other dimensions are given in inches.

*Application:* These three classes together with the accompanying specifications are for the purpose of ensuring the interchangeable manufacture of Centralizing Acme threaded parts. Users are free to select the classes best adapted to their particular needs. It is suggested that external and internal threads of the same class be used together for centralizing assemblies, Class 2C providing the maximum end play or backlash. If less backlash or end play is desired, Classes 3C and 4C are provided. The requirement for a centralizing fit is that the sum of the major diameter tolerance plus the major diameter allowance on the internal thread, and the major diameter tolerance on the external thread shall equal or be less than the pitch diameter allowance on the external thread. A Class 2C external thread, which has a larger pitch diameter allowance than either a Class 3C or 4C, can be used interchangeably with a Class 2C, 3C, or 4C internal thread and fulfill this requirement. Similarly, a Class 3C external thread can be used interchangeably with a Class 3C or 4C internal thread, but only a Class 4C internal thread can be used with a Class 4C external thread.

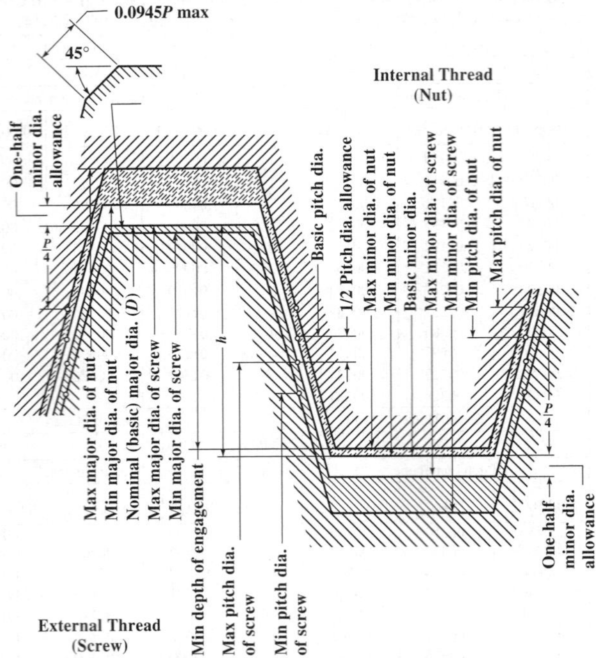


Fig. 2. Disposition of Allowances, Tolerances, and Crest Clearances for General Purpose Single-start Acme Threads (All Classes)

*Thread Form:* The thread form is the same as the General Purpose Acme Thread and is shown in Fig. 3. The formulas in Table 7 determine the basic dimensions, which are given in Table 6 for the most generally used pitches.

*Angle of Thread:* The angle between the sides of the thread measured in an axial plane is 29 degrees. The line bisecting this 29-degree angle shall be perpendicular to the axis of the thread.

*Chamfers and Fillets:* External threads have the crest corners chamfered at an angle of 45 degrees with the axis to a minimum depth of $P/20$ and a maximum depth of $P/15$. These modifications correspond to a minimum width of chamfer flat of $0.0707P$ and a maximum width of $0.0945P$ (see Table 6, columns 6 and 7).

External threads for Classes 2C, 3C, and 4C may have a fillet at the minor diameter not greater than $0.1P$

*Thread Series:* A series of diameters and pitches is recommended in the Standard as preferred. These diameters and pitches have been chosen to meet present needs with the fewest number of items in order to reduce to a minimum the inventory of both tools and gages. This series of diameters and associated pitches is given in Table 9.

**Table 6. American National Standard Centralizing Acme Screw Thread Form— Basic Dimensions** *ANSI/ASME B1.5-1997 (R2014)*

| Thds per Inch, $n$ | Pitch, $P$ | Height of Thread (Basic), $h = P/2$ | Total Height of Thread (All External Threads) $h_s = h + ½$ allowance[a] | Thread Thickness (Basic), $t = P/2$ | 45-Deg Chamfer Crest of External Threads: Min Depth, $0.05P$ | 45-Deg Chamfer Crest of External Threads: Min Width of Chamfer Flat, $0.0707P$ | Max Fillet Radius, Root of Tapped Hole, $0.06P$ | Fillet Radius at Min or Diameter of Screws Max (All) $0.10P$ |
|---|---|---|---|---|---|---|---|---|
| 16 | 0.06250 | 0.03125 | 0.0362 | 0.03125 | 0.0031 | 0.0044 | 0.0038 | 0.0062 |
| 14 | 0.07143 | 0.03571 | 0.0407 | 0.03571 | 0.0036 | 0.0050 | 0.0038 | 0.0071 |
| 12 | 0.08333 | 0.04167 | 0.0467 | 0.04167 | 0.0042 | 0.0059 | 0.0050 | 0.0083 |
| 10 | 0.10000 | 0.05000 | 0.0600 | 0.05000 | 0.0050 | 0.0071 | 0.0060 | 0.0100 |
| 8 | 0.12500 | 0.06250 | 0.0725 | 0.06250 | 0.0062 | 0.0088 | 0.0075 | 0.0125 |
| 6 | 0.16667 | 0.08333 | 0.0933 | 0.08333 | 0.0083 | 0.0119 | 0.0100 | 0.0167 |
| 5 | 0.20000 | 0.10000 | 0.1100 | 0.10000 | 0.0100 | 0.0141 | 0.0120 | 0.0200 |
| 4 | 0.25000 | 0.12500 | 0.1350 | 0.12500 | 0.0125 | 0.0177 | 0.0150 | 0.0250 |
| 3 | 0.33333 | 0.16667 | 0.1767 | 0.16667 | 0.0167 | 0.0236 | 0.0200 | 0.0333 |
| 2½ | 0.40000 | 0.20000 | 0.2100 | 0.20000 | 0.0200 | 0.0283 | 0.0240 | 0.0400 |
| 2 | 0.50000 | 0.25000 | 0.2600 | 0.25000 | 0.0250 | 0.0354 | 0.0300 | 0.0500 |
| 1½ | 0.66667 | 0.33333 | 0.3433 | 0.33333 | 0.0330 | 0.0471 | 0.0400 | 0.0667 |
| 1⅓ | 0.75000 | 0.37500 | 0.3850 | 0.37500 | 0.0380 | 0.0530 | 0.0450 | 0.0750 |
| 1 | 1.00000 | 0.50000 | 0.5100 | 0.50000 | 0.0500 | 0.0707 | 0.0600 | 0.1000 |

All dimensions in inches. See Fig. 3.

[a] Allowance is 0.020 inch for 10 or less threads per inch and 0.010 inch for more than 10 threads per inch.

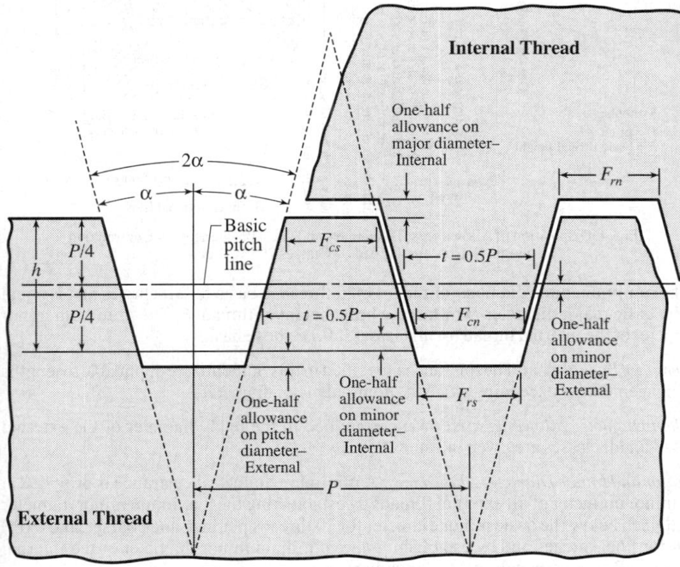


Fig. 3. Centralizing Acme Screw Thread Form

*Basic Diameters:* The maximum major diameter of the external thread is basic and is the nominal major diameter for all classes.

### Table 7. Formulas for Finding Basic Dimensions of Centralizing Acme Screw Threads

Pitch $= P = 1 \div$ No. threads per inch, $n$: Basic thread height $h = 0.5P$

Basic thread thickness $t = 0.5P$

Basic flat at crest $F_{cn} = 0.3707P + 0.259 \times$ (minor. diameter allowance on internal threads) (internal thread)

Basic flat at crest $F_{cs} = 0.3707P - 0.259 \times$ (pitch diameter allowance on external thread) (external thread)

$F_{rn} = 0.3707P - 0.259 \times$ (major dia. allowance on internal thread)

$F_{rs} = 0.3707P - 0.259 \times$ (minor dia. allowance on external thread—pitch dia. allowance on external thread)

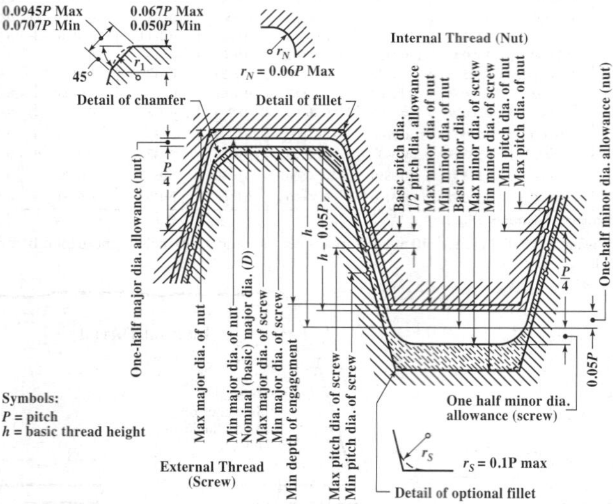


Fig. 4. Disposition of Allowances, Tolerances, and Crest Clearances for Centralizing Single-Start Acme Threads—Classes 2C, 3C, and 4C

The minimum pitch diameter of the internal thread is basic for all classes and is equal to the basic major diameter $D$ minus the basic height of thread, $h$. The minimum minor diameter of the internal thread for all classes is $0.1P$ above basic.

*Length of Engagement:* The tolerances specified in this Standard are applicable to lengths of engagement not exceeding twice the nominal major diameter.

*Pitch Diameter Allowances:* Allowances applied to the pitch diameter of the external thread for all classes are given in Table 10.

*Major and Minor Diameter Allowances:* A minimum diametral clearance is provided at the minor diameter of all external threads by establishing the maximum minor diameter 0.020 inch below the basic minor diameter for 10 threads per inch and coarser, and 0.010 inch for finer pitches and by establishing the minimum minor diameter of the internal thread $0.1P$ greater than the basic minor diameter.

A minimum diametral clearance at the major diameter is obtained by establishing the minimum major diameter of the internal thread $0.001\sqrt{D}$ above the basic major diameter. These allowances are shown in Table 12.

**Table 8a. American National Standard Centralizing Acme Single-Start Screw Threads—Formulas for Determining Diameters** *ANSI/ASME B1.5-1997 (R2014)*

| | |
|---|---|
| | $D$ = Nominal Size or Diameter in Inches<br>$P$ = Pitch = 1 ÷ Number of Threads per Inch |
| No. | Classes 2C, 3C, and 4C External Threads (Screws) |
| 1 | Major Dia., Max = $D$ (Basic). |
| 2 | Major Dia., Min = $D$ *minus* tolerance from Table 12, columns 7, 8, or 10. |
| 3 | Pitch Dia., Max = Int. Pitch Dia., Min (Formula 9) *minus* allowance from the appropriate Class 2C, 3C, or 4C column of Table 10. |
| 4 | Pitch Dia., Min = Ext. Pitch Dia., Max (Formula 3) *minus* tolerance from Table 11. |
| 5 | Minor Dia., Max = $D$ *minus* $P$ *minus* allowance from Table 12, column 3. |
| 6 | Minor Dia., Min = Ext. Minor Dia., Max (Formula 5) *minus* 1.5 × Pitch Dia. tolerance from Table 11. |
| | Classes 2C, 3C, and 4C Internal Threads (Nuts) |
| 7 | Major Dia., Min = $D$ *plus* allowance from Table 12, column 4. |
| 8 | Major Dia., Max = Int. Major Dia., Min (Formula 7) *plus* tolerance from Table 12, columns 7, 9, or 11. |
| 9 | Pitch Dia., Min = $D$ *minus* $P/2$ (Basic). |
| 10 | Pitch Dia., Max = Int. Pitch Dia., Min (Formula 9) *plus* tolerance from Table 11. |
| 11 | Minor Dia., Min = $D$ *minus* $0.9P$. |
| 12 | Minor Dia., Max = Int. Minor Dia., Min (Formula 11) *plus* tolerance from Table 12, column 6. |

*Major and Minor Diameter Tolerances:* The tolerances on the major and minor diameters of the external and internal threads are listed in Table 12 and are based upon the formulas given in the column headings.

An increase of 10 percent in the allowance is recommended for each inch or fraction thereof that the length of engagement exceeds two diameters.

For information on gages for Centralizing Acme threads, the Standard ANSI/ASME B1.5 should be consulted.

*Pitch Diameter Tolerances:* Pitch diameter tolerances for Classes 2C, 3C and 4C for various practicable combinations of diameter and pitch are given in Table 11. The ratios of the pitch diameter tolerances of Classes 2C, 3C, and 4C are 3.0, 1.4, and 1, respectively.

*Application of Tolerances:* The tolerances specified are such as to ensure interchangeability and maintain a high grade of product. The tolerances on the diameters of internal threads are plus, being applied from the minimum sizes to above the minimum sizes. The tolerances on the diameters of external threads are minus, being applied from the maximum sizes to below the maximum sizes. The pitch diameter tolerances for an external or internal thread of a given class are the same.

*Limiting Dimensions:* Limiting dimensions for Centralizing Acme threads in the preferred series of diameters and pitches are given in Table 8b and Table 8c. These limits are based on the formulas in Table 8a.

For combinations of pitch and diameter other than those in the preferred series the formulas in Table 8b and Table 8c and the data in the tables referred to therein make it possible to readily determine the limiting dimension required.

**Designation of Centralizing Acme Threads.—**The following examples are given to show how these Acme threads are designated on drawings, in specifications, and on tools and gages:

*Example, 1.750-6-ACME-4C:* Indicates a Centralizing Class 4C Acme thread of 1.750-inch major diameter, 0.1667-inch pitch, single thread, right-hand.

**Table 8b. Limiting Dimensions of American National Standard Centralizing Acme Single-Start Screw Threads, Classes 2C, 3C, and 4C** *ANSI/ASME B1.5-1997 (R2014)*

| Nominal Diameter, *D* | | ½ | ⅝ | ¾ | ⅞ | 1 | 1⅛ | 1¼ | 1⅜ | 1½ |
|---|---|---|---|---|---|---|---|---|---|---|
| Threads per Inch[a] | | 10 | 8 | 6 | 6 | 5 | 5 | 5 | 4 | 4 |
| Limiting Diameters | | **External Threads** | | | | | | | | |
| Classes 2C, 3C, and 4C, Major Diameter | Max | 0.5000 | 0.6250 | 0.7500 | 0.8750 | 1.0000 | 1.1250 | 1.2500 | 1.3750 | 1.5000 |
| Class 2C, Major Diameter | Min | 0.4975 | 0.6222 | 0.7470 | 0.8717 | 0.9965 | 1.1213 | 1.2461 | 1.3709 | 1.4957 |
| Class 3C, Major Diameter | Min | 0.4989 | 0.6238 | 0.7487 | 0.8736 | 0.9985 | 1.1234 | 1.2483 | 1.3732 | 1.4982 |
| Class 4C, Major Diameter | Min | 0.4993 | 0.6242 | 0.7491 | 0.8741 | 0.9990 | 1.1239 | 1.2489 | 1.3738 | 1.4988 |
| Classes 2C, 3C, and 4C, Minor Diameter | Max | 0.3800 | 0.4800 | 0.5633 | 0.6883 | 0.7800 | 0.9050 | 1.0300 | 1.1050 | 1.2300 |
| Class 2C, Minor Diameter | Min | 0.3594 | 0.4570 | 0.5371 | 0.6615 | 0.7509 | 0.8753 | 0.9998 | 1.0719 | 1.1965 |
| Class 3C, Minor Diameter | Min | 0.3704 | 0.4693 | 0.5511 | 0.6758 | 0.7664 | 0.8912 | 1.0159 | 1.0896 | 1.2144 |
| Class 4C, Minor Diameter | Min | 0.3731 | 0.4723 | 0.5546 | 0.6794 | 0.7703 | 0.8951 | 1.0199 | 1.0940 | 1.2188 |
| Class 2C, Pitch Diameter | Max | 0.4443 | 0.5562 | 0.6598 | 0.7842 | 0.8920 | 1.0165 | 1.1411 | 1.2406 | 1.3652 |
| | Min | 0.4306 | 0.5408 | 0.6424 | 0.7663 | 0.8726 | 0.9967 | 1.1210 | 1.2186 | 1.3429 |
| Class 3C, Pitch Diameter | Max | 0.4458 | 0.5578 | 0.6615 | 0.7861 | 0.8940 | 1.0186 | 1.1433 | 1.2430 | 1.3677 |
| | Min | 0.4394 | 0.5506 | 0.6534 | 0.7778 | 0.8849 | 1.0094 | 1.1339 | 1.2327 | 1.3573 |
| Class 4C, Pitch Diameter | Max | 0.4472 | 0.5593 | 0.6632 | 0.7880 | 0.8960 | 1.0208 | 1.1455 | 1.2453 | 1.3701 |
| | Min | 0.4426 | 0.5542 | 0.6574 | 0.7820 | 0.8895 | 1.0142 | 1.1388 | 1.2380 | 1.3627 |
| | | **Internal Threads** | | | | | | | | |
| Classes 2C, 3C, and 4C, Major Diameter | Min | 0.5007 | 0.6258 | 0.7509 | 0.8759 | 1.0010 | 1.1261 | 1.2511 | 1.3762 | 1.5012 |
| Classes 2C and 3C, Major Diameter | Max | 0.5032 | 0.6286 | 0.7539 | 0.8792 | 1.0045 | 1.1298 | 1.2550 | 1.3803 | 1.5055 |
| Class 4C, Major Diameter | Max | 0.5021 | 0.6274 | 0.7526 | 0.8778 | 1.0030 | 0.1282 | 1.2533 | 1.3785 | 1.5036 |
| Classes 2C, 3C, and 4C, Minor Diameter | Min | 0.4100 | 0.5125 | 0.6000 | 0.7250 | 0.8200 | 0.9450 | 1.0700 | 1.1500 | 1.2750 |
| | Max | 0.4150 | 0.5187 | 0.6083 | 0.7333 | 0.8300 | 0.9550 | 1.0800 | 1.1625 | 1.2875 |
| Class 2C, Pitch Diameter | Min | 0.4500 | 0.5625 | 0.6667 | 0.7917 | 0.9000 | 1.0250 | 1.1500 | 1.2500 | 1.3750 |
| | Max | 0.4637 | 0.5779 | 0.6841 | 0.8096 | 0.9194 | 1.0448 | 1.1701 | 1.2720 | 1.3973 |
| Class 3C, Pitch Diameter | Min | 0.4500 | 0.5625 | 0.6667 | 0.7917 | 0.9000 | 1.0250 | 1.1500 | 1.2500 | 1.3750 |
| | Max | 0.4564 | 0.5697 | 0.6748 | 0.8000 | 0.9091 | 1.0342 | 1.1594 | 1.2603 | 1.3854 |
| Class 4C, Pitch Diameter | Min | 0.4500 | 0.5625 | 0.6667 | 0.7917 | 0.9000 | 1.0250 | 1.1500 | 1.2500 | 1.3750 |
| | Max | 0.4546 | 0.5676 | 0.6725 | 0.7977 | 0.9065 | 1.0316 | 1.1567 | 1.2573 | 1.3824 |

**Table 8c. Limiting Dimensions of American National Standard Centralizing Acme Single-Start Screw Threads, Classes 2C, 3C, and 4C** *ANSI/ASME B1.5-1997 (R2014)*

| Nominal Diameter, $D$ | | 1¾ | 2 | 2¼ | 2½ | 2¾ | 3 | 3½ | 4 | 4½ | 5 |
|---|---|---|---|---|---|---|---|---|---|---|---|
| Threads per Inch[a] | | 4 | 4 | 3 | 3 | 3 | 2 | 2 | 2 | 2 | 2 |
| Limiting Diameters | | External Threads | | | | | | | | | |
| Classes 2C, 3C, and 4C, Major Diameter | Max | 1.7500 | 2.0000 | 2.2500 | 2.5000 | 2.7500 | 3.0000 | 3.5000 | 4.0000 | 4.5000 | 5.0000 |
| Class 2C, Major Diameter | Min | 1.7454 | 1.9951 | 2.2448 | 2.4945 | 2.7442 | 2.9939 | 3.4935 | 3.9930 | 4.4926 | 4.9922 |
| Class 3C, Major Diameter | Min | 1.7480 | 1.9979 | 2.2478 | 2.4976 | 2.7475 | 2.9974 | 3.4972 | 3.9970 | 4.4968 | 4.9966 |
| Class 4C, Major Diameter | Min | 1.7487 | 1.9986 | 2.2485 | 2.4984 | 2.7483 | 2.9983 | 3.4981 | 3.9980 | 4.4979 | 4.9978 |
| Classes 2C, 3C, and 4C, Minor Diameter | Max | 1.4800 | 1.7300 | 1.8967 | 2.1467 | 2.3967 | 2.4800 | 2.9800 | 3.4800 | 3.9800 | 4.4800 |
| Class 2C, Minor Diameter | Min | 1.4456 | 1.6948 | 1.8572 | 2.1065 | 2.3558 | 2.4326 | 2.9314 | 3.4302 | 3.9291 | 4.4281 |
| Class 3C, Minor Diameter | Min | 1.4640 | 1.7136 | 1.8783 | 2.1279 | 2.3776 | 2.4579 | 2.9574 | 3.4568 | 3.9563 | 4.4558 |
| Class 4C, Minor Diameter | Min | 1.4685 | 1.7183 | 1.8835 | 2.1333 | 2.3831 | 2.4642 | 2.9638 | 3.4634 | 3.9631 | 4.4627 |
| Class 2C, Pitch Diameter | Max | 1.6145 | 1.8637 | 2.0713 | 2.3207 | 2.5700 | 2.7360 | 3.2350 | 3.7340 | 4.2330 | 4.7319 |
| | Min | 1.5916 | 1.8402 | 2.0450 | 2.2939 | 2.5427 | 2.7044 | 3.2026 | 3.7008 | 4.1991 | 4.6973 |
| Class 3C, Pitch Diameter | Max | 1.6171 | 1.8665 | 2.0743 | 2.3238 | 2.5734 | 2.7395 | 3.2388 | 3.7380 | 4.2373 | 4.7364 |
| | Min | 1.6064 | 1.8555 | 2.0620 | 2.3113 | 2.5607 | 2.7248 | 3.2237 | 3.7225 | 4.2215 | 4.7202 |
| Class 4C, Pitch Diameter | Max | 1.6198 | 1.8693 | 2.0773 | 2.3270 | 2.5767 | 2.7430 | 3.2425 | 3.7420 | 4.2415 | 4.7409 |
| | Min | 1.6122 | 1.8615 | 2.0685 | 2.3181 | 2.5676 | 2.7325 | 3.2317 | 3.7309 | 4.2302 | 4.7294 |
| | | Internal Threads | | | | | | | | | |
| Classes 2C, 3C, and 4C, Major Diameter | Min | 1.7513 | 2.0014 | 2.2515 | 2.5016 | 2.7517 | 3.0017 | 3.5019 | 4.0020 | 4.5021 | 5.0022 |
| Classes 2C and 3C, Major Diameter | Max | 1.7559 | 2.0063 | 2.2567 | 2.5071 | 2.7575 | 3.0078 | 3.5084 | 4.0090 | 4.5095 | 5.0100 |
| Class 4C, Major Diameter | Max | 1.7539 | 2.0042 | 2.2545 | 2.5048 | 2.7550 | 3.0052 | 3.5056 | 4.0060 | 4.5063 | 5.0067 |
| Classes 2C, 3C, and 4C, Minor Diameter | Min | 1.5250 | 1.7750 | 1.9500 | 2.2000 | 2.4500 | 2.5500 | 3.0500 | 3.5500 | 4.0500 | 4.5500 |
| | Max | 1.5375 | 1.7875 | 1.9667 | 2.2167 | 2.4667 | 2.5750 | 3.0750 | 3.5750 | 4.0750 | 4.5750 |
| Class 2C, Pitch Diameter | Min | 1.6250 | 1.8750 | 2.0833 | 2.3333 | 2.5833 | 2.7500 | 3.2500 | 3.7500 | 4.2500 | 4.7500 |
| | Max | 1.6479 | 1.8985 | 2.1096 | 2.3601 | 2.6106 | 2.7816 | 3.2824 | 3.7832 | 4.2839 | 4.7846 |
| Class 3C, Pitch Diameter | Min | 1.6250 | 1.8750 | 2.0833 | 2.3333 | 2.5833 | 2.7500 | 3.2500 | 3.7500 | 4.2500 | 4.7500 |
| | Max | 1.6357 | 1.8860 | 2.0956 | 2.3458 | 2.5960 | 2.7647 | 3.2651 | 3.7655 | 4.2658 | 4.7662 |
| Class 4C Pitch Diameter | Min | 1.6250 | 1.8750 | 2.0833 | 2.3333 | 2.5833 | 2.7500 | 3.2500 | 3.7500 | 4.2500 | 4.7500 |
| | Max | 1.6326 | 1.8828 | 2.0921 | 2.3422 | 2.5924 | 2.7605 | 3.2608 | 3.7611 | 4.2613 | 4.7615 |

[a] All other dimensions are in inches. The selection of threads per inch is arbitrary and for the purpose of establishing a standard.

**Table 9. American National Standard Centralizing Acme Single-Start Screw Thread Data** *ANSI/ASME B1.5-1997 (R2014)*

| Identification | | Diameters | | | Thread Data | | | | | |
|---|---|---|---|---|---|---|---|---|---|---|
| | | Centralizing, Classes 2C, 3C, and 4C | | | | | | | Lead Angle λ at Basic Pitch Diameter[a] Centralizing Classes 2C, 3C, and 4C, | |
| Nominal Sizes (All Classes) | Threads per Inch,[a] $n$ | Basic Major Diameter, $D$ | Pitch Diameter, $D_2 = (D - h)$ | Minor Diameter, $D_1 = (D - 2h)$ | Pitch, $P$ | Thickness at Pitch Line, $t = P/2$ | Basic Height of Thread, $h = P/2$ | Basic Width of Flat, $F = 0.3707P$ | Deg | Min |
| ¼ | 16 | 0.2500 | 0.2188 | 0.1875 | 0.06250 | 0.03125 | 0.03125 | 0.0232 | 5 | 12 |
| 5⁄16 | 14 | 0.3125 | 0.2768 | 0.2411 | 0.07143 | 0.03571 | 0.03571 | 0.0265 | 4 | 42 |
| ⅜ | 12 | 0.3750 | 0.3333 | 0.2917 | 0.08333 | 0.04167 | 0.04167 | 0.0309 | 4 | 33 |
| 7⁄16 | 12 | 0.4375 | 0.3958 | 0.3542 | 0.08333 | 0.04167 | 0.04167 | 0.0309 | 3 | 50 |
| ½ | 10 | 0.5000 | 0.4500 | 0.4000 | 0.10000 | 0.05000 | 0.05000 | 0.0371 | 4 | 3 |
| ⅝ | 8 | 0.6250 | 0.5625 | 0.5000 | 0.12500 | 0.06250 | 0.06250 | 0.0463 | 4 | 3 |
| ¾ | 6 | 0.7500 | 0.6667 | 0.5833 | 0.16667 | 0.08333 | 0.08333 | 0.0618 | 4 | 33 |
| ⅞ | 6 | 0.8750 | 0.7917 | 0.7083 | 0.16667 | 0.08333 | 0.08333 | 0.0618 | 3 | 50 |
| 1 | 5 | 1.0000 | 0.9000 | 0.8000 | 0.20000 | 0.10000 | 0.10000 | 0.0741 | 4 | 3 |
| 1⅛ | 5 | 1.1250 | 1.0250 | 0.9250 | 0.20000 | 0.10000 | 0.10000 | 0.0741 | 3 | 33 |
| 1¼ | 5 | 1.2500 | 1.1500 | 1.0500 | 0.20000 | 0.10000 | 0.10000 | 0.0741 | 3 | 10 |
| 1⅜ | 4 | 1.3750 | 1.2500 | 1.1250 | 0.25000 | 0.12500 | 0.12500 | 0.0927 | 3 | 39 |
| 1½ | 4 | 1.5000 | 1.3750 | 1.2500 | 0.25000 | 0.12500 | 0.12500 | 0.0927 | 3 | 19 |
| 1¾ | 4 | 1.7500 | 1.6250 | 1.5000 | 0.25000 | 0.12500 | 0.12500 | 0.0927 | 2 | 48 |
| 2 | 4 | 2.0000 | 1.8750 | 1.7500 | 0.25000 | 0.12500 | 0.12500 | 0.0927 | 2 | 26 |
| 2¼ | 3 | 2.2500 | 2.0833 | 1.9167 | 0.33333 | 0.16667 | 0.16667 | 0.1236 | 2 | 55 |
| 2½ | 3 | 2.5000 | 2.3333 | 2.1667 | 0.33333 | 0.16667 | 0.16667 | 0.1236 | 2 | 36 |
| 2¾ | 3 | 2.7500 | 2.5833 | 2.4167 | 0.33333 | 0.16667 | 0.16667 | 0.1236 | 2 | 21 |
| 3 | 2 | 3.0000 | 2.7500 | 2.5000 | 0.50000 | 0.25000 | 0.25000 | 0.1853 | 3 | 19 |
| 3½ | 2 | 3.5000 | 3.2500 | 3.0000 | 0.50000 | 0.25000 | 0.25000 | 0.1853 | 2 | 48 |
| 4 | 2 | 4.0000 | 3.7500 | 3.5000 | 0.50000 | 0.25000 | 0.25000 | 0.1853 | 2 | 26 |
| 4½ | 2 | 4.5000 | 4.2500 | 4.0000 | 0.50000 | 0.25000 | 0.25000 | 0.1853 | 2 | 9 |
| 5 | 2 | 5.0000 | 4.7500 | 4.5000 | 0.50000 | 0.25000 | 0.25000 | 0.1853 | 1 | 55 |

[a] All other dimensions are given in inches.

**Table 10. American National Standard Centralizing Acme Single-Start Screw Threads — Pitch Diameter Allowances** *ANSI/ASME B1.5-1997 (R2014)*

| Nominal Size Range[a] | | Allowances on External Threads[b] | | | Nominal Size Range[a] | | Allowances on External Threads[b] | | |
|---|---|---|---|---|---|---|---|---|---|
| | | Centralizing | | | | | Centralizing | | |
| Above | To and Including | Class 2C, $0.008\sqrt{D}$ | Class 3C, $0.006\sqrt{D}$ | Class 4C, $0.004\sqrt{D}$ | Above | To and Including | Class 2C, $0.008\sqrt{D}$ | Class 3C, $0.006\sqrt{D}$ | Class 4C, $0.004\sqrt{D}$ |
| 0 | 3/16 | 0.0024 | 0.0018 | 0.0012 | 1 7/16 | 1 9/16 | 0.0098 | 0.0073 | 0.0049 |
| 3/16 | 5/16 | 0.0040 | 0.0030 | 0.0020 | 1 9/16 | 1 7/8 | 0.0105 | 0.0079 | 0.0052 |
| 5/16 | 7/16 | 0.0049 | 0.0037 | 0.0024 | 1 7/8 | 2 1/8 | 0.0113 | 0.0085 | 0.0057 |
| 7/16 | 9/16 | 0.0057 | 0.0042 | 0.0028 | 2 1/8 | 2 3/8 | 0.0120 | 0.0090 | 0.0060 |
| 9/16 | 11/16 | 0.0063 | 0.0047 | 0.0032 | 2 3/8 | 2 5/8 | 0.0126 | 0.0095 | 0.0063 |
| 11/16 | 13/16 | 0.0069 | 0.0052 | 0.0035 | 2 5/8 | 2 7/8 | 0.0133 | 0.0099 | 0.0066 |
| 13/16 | 15/16 | 0.0075 | 0.0056 | 0.0037 | 2 7/8 | 3 1/4 | 0.0140 | 0.0105 | 0.0070 |
| 15/16 | 1 1/16 | 0.0080 | 0.0060 | 0.0040 | 3 1/4 | 3 3/4 | 0.0150 | 0.0112 | 0.0075 |
| 1 1/16 | 1 3/16 | 0.0085 | 0.0064 | 0.0042 | 3 3/4 | 4 1/4 | 0.0160 | 0.0120 | 0.0080 |
| 1 3/16 | 1 5/16 | 0.0089 | 0.0067 | 0.0045 | 4 1/4 | 4 3/4 | 0.0170 | 0.0127 | 0.0085 |
| 1 5/16 | 1 7/16 | 0.0094 | 0.0070 | 0.0047 | 4 3/4 | 5 1/2 | 0.0181 | 0.0136 | 0.0091 |

All dimensions are given in inches.

It is recommended that the sizes given in Table 9 be used whenever possible.

[a] The values in columns for Classes 2C, 3C, and 4C are to be used for any size within the nominal size range columns. These values are calculated from the mean of the range.

[b] An increase of 10 percent in the allowance is recommended for each inch, or fraction thereof, that the length of engagement exceeds two diameters.

**Table 11. American National Standard Centralizing Acme Single-Start Screw Threads — Pitch Diameter Tolerances** *ANSI/ASME B1.5-1997 (R2014)*

| Nom. Dia.,[a] D | Class of Thread and Diameter Increment | | | Nom. Dia.,[a] D | Class of Thread and Diameter Increment | | |
|---|---|---|---|---|---|---|---|
| | 2C | 3C | 4C | | 2C | 3C | 4C |
| | $0.006\sqrt{D}$ | $0.0028\sqrt{D}$ | $0.002\sqrt{D}$ | | $0.006\sqrt{D}$ | $0.0028\sqrt{D}$ | $0.002\sqrt{D}$ |
| 1/4 | .00300 | .00140 | .00100 | 1 1/2 | .00735 | .00343 | .00245 |
| 5/16 | .00335 | .00157 | .00112 | 1 3/4 | .00794 | .00370 | .00265 |
| 3/8 | .00367 | .00171 | .00122 | 2 | .00849 | .00396 | .00283 |
| 7/16 | .00397 | .00185 | .00132 | 2 1/4 | .00900 | .00420 | .00300 |
| 1/2 | .00424 | .00198 | .00141 | 2 1/2 | .00949 | .00443 | .00316 |
| 5/8 | .00474 | .00221 | .00158 | 2 3/4 | .00995 | .00464 | .00332 |
| 3/4 | .00520 | .00242 | .00173 | 3 | .01039 | .00485 | .00346 |
| 7/8 | .00561 | .00262 | .00187 | 3 1/2 | .01122 | .00524 | .00374 |
| 1 | .00600 | .00280 | .00200 | 4 | .01200 | .00560 | .00400 |
| 1 1/8 | .00636 | .00297 | .00212 | 4 1/2 | .01273 | .00594 | .00424 |
| 1 1/4 | .00671 | .00313 | .00224 | 5 | .01342 | .00626 | .00447 |
| 1 3/8 | .00704 | .00328 | .00235 | … | … | … | … |

| Thds. per Inch, n | Class of Thread and Pitch Increment | | | Thds. per Inch, n | Class of Thread and Pitch Increment | | |
|---|---|---|---|---|---|---|---|
| | 2C | 3C | 4C | | 2C | 3C | 4C |
| | $0.030\sqrt{\frac{1}{n}}$ | $0.014\sqrt{\frac{1}{n}}$ | $0.010\sqrt{\frac{1}{n}}$ | | $0.030\sqrt{\frac{1}{n}}$ | $0.014\sqrt{\frac{1}{n}}$ | $0.010\sqrt{\frac{1}{n}}$ |
| 16 | .00750 | .00350 | .00250 | 4 | .01500 | .00700 | .00500 |
| 14 | .00802 | .00374 | .00267 | 3 | .01732 | .00808 | .00577 |
| 12 | .00866 | .00404 | .00289 | 2 1/2 | .01897 | .00885 | .00632 |
| 10 | .00949 | .00443 | .00316 | 2 | .02121 | .00990 | .00707 |
| 8 | .01061 | .00495 | .00354 | 1 1/2 | .02449 | .01143 | .00816 |
| 6 | .01225 | .00572 | .00408 | 1 1/3 | .02598 | .01212 | .00866 |
| 5 | .01342 | .00626 | .00447 | 1 | .03000 | .01400 | .01000 |

All dimensions are given in inches.

For any particular size of thread, the pitch diameter tolerance is obtained by adding the *diameter increment* from the upper half of the table to the *pitch increment* from the lower half of the table. *Example:* A 0.250-16-ACME-2C thread has a pitch diameter tolerance of 0.00300 + 0.00750 = 0.0105 inch.

The equivalent tolerance on thread thickness is 0.259 times the pitch diameter tolerance.

[a] For a nominal diameter between any two tabulated nominal diameters, use the diameter increment for the larger of the two tabulated nominal diameters.

**Table 12. American National Standard Centralizing Acme Single-Start Screw Threads— Tolerances and Allowances for Major and Minor Diameters** *ANSI/ASME B1.5-1997 (R2014)*

| Size (Nom.) | Thds[a] per Inch | Allowance From Basic Major and Minor Diameters (All Classes): Minor Diam,[d] All External Threads (Minus) | Allowance, Internal Thread: Major Diam,[e] (Plus $0.0010\sqrt{D}$) | Allowance, Internal Thread: Minor Diam,[d] (Plus $0.1P$) | Tolerance on Minor Diam,[b,c] All Internal Threads, (Plus $0.05P$) | Tolerance on Major Diameter Plus on Internal, Minus on External Threads: Class 2C External and Internal Threads, $0.0035\sqrt{D}$ | Class 3C External Thread, $0.0015\sqrt{D}$ | Class 3C Internal Thread, $0.0035\sqrt{D}$ | Class 4C External Thread, $0.0010\sqrt{D}$ | Class 4C Internal Thread, $0.0020\sqrt{D}$ |
|---|---|---|---|---|---|---|---|---|---|---|
| ¼ | 16 | 0.010 | 0.0005 | 0.0062 | 0.0050 | 0.0017 | 0.0007 | 0.0017 | 0.0005 | 0.0010 |
| 5⁄16 | 14 | 0.010 | 0.0006 | 0.0071 | 0.0050 | 0.0020 | 0.0008 | 0.0020 | 0.0006 | 0.0011 |
| ⅜ | 12 | 0.010 | 0.0006 | 0.0083 | 0.0050 | 0.0021 | 0.0009 | 0.0021 | 0.0006 | 0.0012 |
| 7⁄16 | 12 | 0.010 | 0.0007 | 0.0083 | 0.0050 | 0.0023 | 0.0010 | 0.0023 | 0.0007 | 0.0013 |
| ½ | 10 | 0.020 | 0.0007 | 0.0100 | 0.0050 | 0.0025 | 0.0011 | 0.0025 | 0.0007 | 0.0014 |
| ⅝ | 8 | 0.020 | 0.0008 | 0.0125 | 0.0062 | 0.0028 | 0.0012 | 0.0028 | 0.0008 | 0.0016 |
| ¾ | 6 | 0.020 | 0.0009 | 0.0167 | 0.0083 | 0.0030 | 0.0013 | 0.0030 | 0.0009 | 0.0017 |
| ⅞ | 6 | 0.020 | 0.0009 | 0.0167 | 0.0083 | 0.0033 | 0.0014 | 0.0033 | 0.0009 | 0.0019 |
| 1 | 5 | 0.020 | 0.0010 | 0.0200 | 0.0100 | 0.0035 | 0.0015 | 0.0035 | 0.0010 | 0.0020 |
| 1⅛ | 5 | 0.020 | 0.0011 | 0.0200 | 0.0100 | 0.0037 | 0.0016 | 0.0037 | 0.0011 | 0.0021 |
| 1¼ | 5 | 0.020 | 0.0011 | 0.0200 | 0.0100 | 0.0039 | 0.0017 | 0.0039 | 0.0011 | 0.0022 |
| 1⅜ | 4 | 0.020 | 0.0012 | 0.0250 | 0.0125 | 0.0041 | 0.0018 | 0.0041 | 0.0012 | 0.0023 |
| 1½ | 4 | 0.020 | 0.0012 | 0.0250 | 0.0125 | 0.0043 | 0.0018 | 0.0043 | 0.0012 | 0.0024 |
| 1¾ | 4 | 0.020 | 0.0013 | 0.0250 | 0.0125 | 0.0046 | 0.0020 | 0.0046 | 0.0013 | 0.0026 |
| 2 | 4 | 0.020 | 0.0014 | 0.0250 | 0.0125 | 0.0049 | 0.0021 | 0.0049 | 0.0014 | 0.0028 |
| 2¼ | 3 | 0.020 | 0.0015 | 0.0333 | 0.0167 | 0.0052 | 0.0022 | 0.0052 | 0.0015 | 0.0030 |
| 2½ | 3 | 0.020 | 0.0016 | 0.0333 | 0.0167 | 0.0055 | 0.0024 | 0.0055 | 0.0016 | 0.0032 |
| 2¾ | 3 | 0.020 | 0.0017 | 0.0333 | 0.0167 | 0.0058 | 0.0025 | 0.0058 | 0.0017 | 0.0033 |
| 3 | 2 | 0.020 | 0.0017 | 0.0500 | 0.0250 | 0.0061 | 0.0026 | 0.0061 | 0.0017 | 0.0035 |
| 3½ | 2 | 0.020 | 0.0019 | 0.0500 | 0.0250 | 0.0065 | 0.0028 | 0.0065 | 0.0019 | 0.0037 |
| 4 | 2 | 0.020 | 0.0020 | 0.0500 | 0.0250 | 0.0070 | 0.0030 | 0.0070 | 0.0020 | 0.0040 |
| 4½ | 2 | 0.020 | 0.0021 | 0.0500 | 0.0250 | 0.0074 | 0.0032 | 0.0074 | 0.0021 | 0.0042 |
| 5 | 2 | 0.020 | 0.0022 | 0.0500 | 0.0250 | 0.0078 | 0.0034 | 0.0078 | 0.0022 | 0.0045 |

[a] All other dimensions are given in inches. Intermediate pitches take the values of the next coarser pitch listed. Values for intermediate diameters should be calculated from the formulas in column headings, but ordinarily may be interpolated.

[b] To avoid a complicated formula and still provide an adequate tolerance, the pitch factor is used as a basis, with the minimum tolerance set at 0.005 in.

[c] Tolerance on minor diameter of all external threads is 1.5 × pitch diameter tolerance.

[d] The minimum clearance at the minor diameter between the internal and external thread is the sum of the values in columns 3 and 5.

[e] The minimum clearance at the major diameter between the internal and external thread is equal to column 4.

*Example, 1.750-6-ACME-4C-LH:* Indicates the same thread left-hand.

*Example, 2.875-0.4P-0.8L-ACME-3C (Two Start):* Indicates a Centralizing Class 3C Acme thread with 2.875-inch major diameter, 0.4-inch pitch, 0.8-inch lead, double thread, right-hand.

*Example, 2.500-0.3333P-0.6667L-ACME-4C (Two Start):* Indicates a Centralizing Class 4C Acme thread with 2.500-inch nominal major diameter (basic major diameter 2.500 inches), 0.3333-inch pitch, 0.6667-inch lead, double thread, right-hand. The same thread left-hand would have LH at the end of the designation.

**Acme Centralizing Threads—Alternative Series with Minor Diameter Centralizing Control.**—When Acme centralizing threads are produced in single units or in very small quantities (and principally in sizes larger than the range of commercial taps and dies) where the manufacturing process employs cutting tools (such as lathe cutting), it may be economically advantageous and therefore desirable to have the centralizing control of the mating threads located at the *minor diameters*.

Particularly under the above-mentioned type of manufacturing, the two advantages cited for minor diameter centralizing control over centralizing control at the major diameters of the mating threads are: 1) Greater ease and faster checking of machined thread dimensions. It is much easier to measure the minor diameter (root) of the external thread and the mating minor diameter (crest or bore) of the internal thread than it is to determine the major diameter (root) of the internal thread and the major diameter (crest or turn) of the external thread; and 2) better manufacturing control of the machined size due to greater ease of checking.

In the event that minor diameter centralizing is necessary, recalculate all thread dimensions, reversing major and minor diameter allowances, tolerances, radii, and chamfer.

**American National Standard Stub Acme Threads.**—American National Standard ANSI/ASME B1.8-1988 (R2021) provides a Stub Acme screw thread for those unusual applications where, due to mechanical or metallurgical considerations, a coarse-pitch thread of shallow depth is required. The fit of Stub Acme threads corresponds to the Class 2G General Purpose Acme thread in American National Standard ANSI/ASME B1.5-1997 (R2014). For a fit having less backlash, the tolerances and allowances for Classes 3G or 4G General Purpose Acme threads may be used.

*Thread Form:* The thread form and basic formulas for Stub Acme threads are given on page 2057 and the basic dimensions in Table 13.

*Allowances and Tolerances:* The major and minor diameter allowances for Stub Acme threads are the same as those given for General Purpose Acme threads on page 2057.

Pitch diameter allowances for Stub Acme threads are the same as for Class 2G General Purpose Acme threads and are given in Table 4. Pitch diameter tolerances for Stub Acme threads are the same as for Class 2G General Purpose Acme threads given in Table 5.

*Limiting Dimensions:* Limiting dimensions of Stub Acme threads may be determined by using the formulas given in Table 14a, or directly from Table 14b. The diagram below shows the limits of size for Stub Acme threads.

*Thread Series:* A preferred series of diameters and pitches for General Purpose Acme threads (Table 15) is recommended for Stub Acme threads.

**Stub Acme Thread Designations.**—The method of designation for American National Standard Stub Acme threads is illustrated in the following examples: 0.500-20 Stub Acme indicates a ½-inch major diameter, 20 threads per inch, right-hand, single thread, Standard Stub Acme thread. The designation 0.500-20 Stub Acme-LH indicates the same thread except that it is left-hand.

**Table 13. American National Standard Stub Acme Screw Thread Form— Basic Dimensions** *ANSI/ASME B1.8-1988 (R2021)*

| Thds. per Inch[a] $n$ | Pitch, $P = 1/n$ | Height of Thread (Basic), $0.3P$ | Total Height of Thread, $0.3P + ½$ allowance[b] | Thread Thickness (Basic), $P/2$ | Width of Flat: Crest of Internal Thread (Basic), $0.4224P$ | Width of Flat: Root of Internal Thread, $0.4224P - 0.259 \times$ allowance[b] |
|---|---|---|---|---|---|---|
| 16 | 0.06250 | 0.01875 | 0.0238 | 0.03125 | 0.0264 | 0.0238 |
| 14 | 0.07143 | 0.02143 | 0.0264 | 0.03571 | 0.0302 | 0.0276 |
| 12 | 0.08333 | 0.02500 | 0.0300 | 0.04167 | 0.0352 | 0.0326 |
| 10 | 0.10000 | 0.03000 | 0.0400 | 0.05000 | 0.0422 | 0.0370 |
| 9 | 0.11111 | 0.03333 | 0.0433 | 0.05556 | 0.0469 | 0.0417 |
| 8 | 0.12500 | 0.03750 | 0.0475 | 0.06250 | 0.0528 | 0.0476 |
| 7 | 0.14286 | 0.04285 | 0.0529 | 0.07143 | 0.0603 | 0.0551 |
| 6 | 0.16667 | 0.05000 | 0.0600 | 0.08333 | 0.0704 | 0.0652 |
| 5 | 0.20000 | 0.06000 | 0.0700 | 0.10000 | 0.0845 | 0.0793 |
| 4 | 0.25000 | 0.07500 | 0.0850 | 0.12500 | 0.1056 | 0.1004 |
| 3½ | 0.28571 | 0.08571 | 0.0957 | 0.14286 | 0.1207 | 0.1155 |
| 3 | 0.33333 | 0.10000 | 0.1100 | 0.16667 | 0.1408 | 0.1356 |
| 2½ | 0.40000 | 0.12000 | 0.1300 | 0.20000 | 0.1690 | 0.1638 |
| 2 | 0.50000 | 0.15000 | 0.1600 | 0.25000 | 0.2112 | 0.2060 |
| 1½ | 0.66667 | 0.20000 | 0.2100 | 0.33333 | 0.2816 | 0.2764 |
| 1⅓ | 0.75000 | 0.22500 | 0.2350 | 0.37500 | 0.3168 | 0.3116 |
| 1 | 1.00000 | 0.30000 | 0.3100 | 0.50000 | 0.4224 | 0.4172 |

[a] All other dimensions in inches. See Fig. 1, page 2058.

[b] Allowance is 0.020 inch for 10 or less threads per inch and 0.010 inch for more than 10 threads per inch.

**Table 14a. American National Standard Stub Acme Single-Start Screw Threads— Formulas for Determining Diameters** *ANSI/ASME B1.8-1988 (R2021)*

$D$ = Basic Major Diameter and Nominal Size in Inches

$D_2$ = Basic Pitch Diameter = $D - 0.3P$

$D_1$ = Basic Minor Diameter = $D - 0.6P$

| No. | External Threads (Screws) |
|---|---|
| 1 | Major Dia., Max = $D$. |
| 2 | Major Dia., Min. = $D$ *minus* $0.05P$. |
| 3 | Pitch Dia., Max. = $D_2$ *minus* allowance from the appropriate Class 2G column, Table 4. |
| 4 | Pitch Dia., Min. = Pitch Dia., Max. (Formula 3) *minus* Class 2G tolerance from Table 5. |
| 5 | Minor Dia., Max. = $D_1$ *minus* 0.020 for 10 threads per inch and coarser and 0.010 for finer pitches. |
| 6 | Minor Dia., Min. = Minor Dia., Max. (Formula 5) *minus* Class 2G pitch diameter tolerance from Table 5. |
| | **Internal Threads (Nuts)** |
| 7 | Major Dia., Min. = $D$ *plus* 0.020 for 10 threads per inch and coarser and 0.010 for finer pitches. |
| 8 | Major Dia., Max.= Major Dia., Min. (Formula 7) *plus* Class 2G pitch diameter tolerance from Table 5. |
| 9 | Pitch Dia., Min. = $D_2 = D - 0.3P$ |
| 10 | Pitch Dia., Max. = Pitch Dia., Min. (Formula 9) *plus* Class 2G tolerance from Table 5. |
| 11 | Minor Dia., Min. = $D_1 = D - 0.6P$ |
| 12 | Minor Dia., Max = Minor Dia., Min. (Formula 11) *plus* $0.05P$. |

**Table 14b. Limiting Dimensions for American National Standard Stub Acme Single-Start Screw Threads** *ANSI/ASME B1.8-1988 (R2021)*

| Nominal Diameter, D | | 1/4 | 5/16 | 3/8 | 7/16 | 1/2 | 5/8 | 3/4 | 7/8 | 1 | 1 1/8 | 1 1/4 | 1 3/8 |
|---|---|---|---|---|---|---|---|---|---|---|---|---|---|
| Threads per Inch[a] | | 16 | 14 | 12 | 12 | 10 | 8 | 6 | 6 | 5 | 5 | 5 | 4 |
| Limiting Diameters | | External Threads | | | | | | | | | | | |
| Major Dia. | Max (D) | 0.2500 | 0.3125 | 0.3750 | 0.4375 | 0.5000 | 0.6250 | 0.7500 | 0.8750 | 1.0000 | 1.1250 | 1.2500 | 1.3750 |
| | Min | 0.2469 | 0.3089 | 0.3708 | 0.4333 | 0.4950 | 0.6188 | 0.7417 | 0.8667 | 0.9900 | 1.1150 | 1.2400 | 1.3625 |
| Pitch Dia. | Max | 0.2272 | 0.2871 | 0.3451 | 0.4076 | 0.4643 | 0.5812 | 0.6931 | 0.8175 | 0.9320 | 1.0565 | 1.1811 | 1.2906 |
| | Min | 0.2167 | 0.2757 | 0.3328 | 0.3950 | 0.4506 | 0.5658 | 0.6757 | 0.7996 | 0.9126 | 1.0367 | 1.1610 | 1.2686 |
| Minor Dia. | Max | 0.2024 | 0.2597 | 0.3150 | 0.3775 | 0.4200 | 0.5300 | 0.6300 | 0.7550 | 0.8600 | 0.9850 | 1.1100 | 1.2050 |
| | Min | 0.1919 | 0.2483 | 0.3027 | 0.3649 | 0.4063 | 0.5146 | 0.6126 | 0.7371 | 0.8406 | 0.9652 | 1.0899 | 1.1830 |
| | | Internal Threads | | | | | | | | | | | |
| Major Dia. | Min | 0.2600 | 0.3225 | 0.3850 | 0.4475 | 0.5200 | 0.6450 | 0.7700 | 0.8950 | 1.0200 | 1.1450 | 1.2700 | 1.3950 |
| | Max | 0.2705 | 0.3339 | 0.3973 | 0.4601 | 0.5337 | 0.6604 | 0.7874 | 0.9129 | 1.0394 | 1.1648 | 1.2901 | 1.4170 |
| Pitch Dia. | Min | 0.2312 | 0.2911 | 0.3500 | 0.4125 | 0.4700 | 0.5875 | 0.7000 | 0.8250 | 0.9400 | 1.0650 | 1.1900 | 1.3000 |
| | Max | 0.2417 | 0.3025 | 0.3623 | 0.4251 | 0.4837 | 0.6029 | 0.7174 | 0.8429 | 0.9594 | 1.0848 | 1.2101 | 1.3220 |
| Minor Dia. | Min | 0.2125 | 0.2696 | 0.3250 | 0.3875 | 0.4400 | 0.5500 | 0.6500 | 0.7750 | 0.8800 | 1.0050 | 1.1300 | 1.2250 |
| | Max | 0.2156 | 0.2732 | 0.3292 | 0.3917 | 0.4450 | 0.5562 | 0.6583 | 0.7833 | 0.8900 | 1.0150 | 1.1400 | 1.2375 |

| Nominal Diameter, D | | 1 1/2 | 1 3/4 | 2 | 2 1/4 | 2 1/2 | 2 3/4 | 3 | 3 1/2 | 4 | 4 1/2 | 5 |
|---|---|---|---|---|---|---|---|---|---|---|---|---|
| Threads per Inch[a] | | 4 | 4 | 4 | 3 | 3 | 3 | 2 | 2 | 2 | 2 | 2 |
| Limiting Diameters | | External Threads | | | | | | | | | | |
| Major Dia. | Max (D) | 1.5000 | 1.7500 | 2.0000 | 2.2500 | 2.5000 | 2.7500 | 3.0000 | 3.5000 | 4.0000 | 4.5000 | 5.0000 |
| | Min | 1.4875 | 1.7375 | 1.9875 | 2.2333 | 2.4833 | 2.7333 | 2.9750 | 3.4750 | 3.9750 | 4.4750 | 4.9750 |
| Pitch Dia. | Max | 1.4152 | 1.6645 | 1.9137 | 2.1380 | 2.3874 | 2.6367 | 2.8360 | 3.3350 | 3.8340 | 4.3330 | 4.8319 |
| | Min | 1.3929 | 1.6416 | 1.8902 | 2.1117 | 2.3606 | 2.6094 | 2.8044 | 3.3026 | 3.8008 | 4.2991 | 4.7973 |
| Minor Dia. | Max | 1.3300 | 1.5800 | 1.8300 | 2.0300 | 2.2800 | 2.5300 | 2.6800 | 3.1800 | 3.6800 | 4.1800 | 4.6800 |
| | Min | 1.3077 | 1.5571 | 1.8065 | 2.0037 | 2.2532 | 2.5027 | 2.6484 | 3.1476 | 3.6468 | 4.1461 | 4.6454 |
| | | Internal Threads | | | | | | | | | | |
| Major Dia. | Min | 1.5200 | 1.7700 | 2.0200 | 2.2700 | 2.5200 | 2.7700 | 3.0200 | 3.5200 | 4.0200 | 4.5200 | 5.0200 |
| | Max | 1.5423 | 1.7929 | 2.0435 | 2.2963 | 2.5468 | 2.7973 | 3.0516 | 3.5524 | 4.0532 | 4.5539 | 5.0546 |
| Pitch Dia. | Min | 1.4250 | 1.6750 | 1.9250 | 2.1500 | 2.4000 | 2.6500 | 2.8500 | 3.3500 | 3.8500 | 4.3500 | 4.8500 |
| | Max | 1.4473 | 1.6979 | 1.9485 | 2.1763 | 2.4268 | 2.6773 | 2.8816 | 3.3824 | 3.8832 | 4.3839 | 4.8846 |
| Minor Dia. | Min | 1.3500 | 1.6000 | 1.8500 | 2.0500 | 2.3000 | 2.5500 | 2.7000 | 3.2000 | 3.7000 | 4.2000 | 4.7000 |
| | Max | 1.3625 | 1.6125 | 1.8625 | 2.0667 | 2.3167 | 2.5667 | 2.7250 | 3.2250 | 3.7250 | 4.2250 | 4.7250 |

[a] All other dimensions are given in inches.

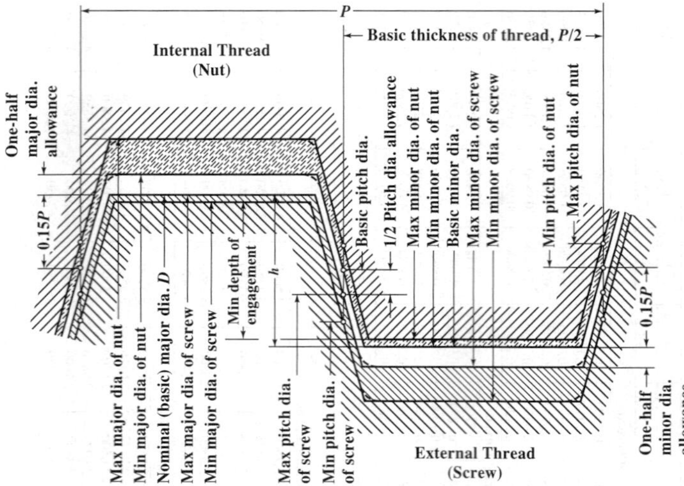


Limits of Size, Allowances, Tolerances, and Crest Clearances for American National Standard Stub Acme Threads

**Alternate Stub Acme Threads.—**Since one Stub Acme thread form may not meet the requirements of all applications, basic data for two of the other commonly used forms are included in the appendix of the American National Standard for Stub Acme Threads. These so-called Modified Form 1 and Modified Form 2 threads utilize the same tolerances and allowances as Standard Stub Acme threads and have the same major diameter and basic thread thickness at the pitchline (0.5$P$). The basic height of Form 1 threads, $h$, is 0.375$P$; for Form 2 it is 0.250$P$. The basic width of flat at the crest of the internal thread is 0.4030$P$ for Form 1 and 0.4353$P$ for Form 2.

The pitch diameter and minor diameter for Form 1 threads will be smaller than similar values for the Standard Stub Acme Form and for Form 2 they will be larger owing to the differences in basic thread height $h$. Therefore, in calculating the dimensions of Form 1 and Form 2 threads using Formulas 1 through 12 in Table 14a, it is only necessary to substitute the following values in applying the formulas: For Form 1, $D_2 = D - 0.375P$, $D_1 = D - 0.75P$; for Form 2, $D_2 = D - 0.25P$, $D_1 = D - 0.5P$.

*Thread Designation:* These threads are designated in the same manner as Standard Stub Acme threads except for the insertion of either M1 or M2 after "Acme." Thus, 0.500-20 Stub Acme M1 for a Form 1 thread; and 0.500-20 Stub Acme M2 for a Form 2 thread.

**Former 60-Degree Stub Thread.—**Former American Standard B1.3-1941 included a 60-degree stub thread for use where design or operating conditions could be better satisfied by the use of this thread, or other modified threads, than by Acme threads. Data for 60-Degree Stub thread form are given in the accompanying diagram.

**Table 15. Stub Acme Screw Thread Data** *ANSI/ASME B1.8-1988 (R2021)*

| Identification | | Basic Diameters | | | Thread Data | | | | | |
|---|---|---|---|---|---|---|---|---|---|---|
| Nominal Sizes | Threads per Inch,[a] $n$ | Major Diameter, $D$ | Pitch Diameter, $D_2 = D - h$ | Minor Diameter, $D_1 = D - 2h$ | Pitch, $P$ | Thread Thickness at Pitch Line, $t = P/2$ | Basic Thread Height, $h = 0.3P$ | Basic Width of Flat, $0.4224P$ | Lead Angle at Basic Pitch Diameter Deg | Min |
| 1/4 | 16 | 0.2500 | 0.2312 | 0.2125 | 0.06250 | 0.03125 | 0.01875 | 0.0264 | 4 | 54 |
| 5/16 | 14 | 0.3125 | 0.2911 | 0.2696 | 0.07143 | 0.03572 | 0.02143 | 0.0302 | 4 | 28 |
| 3/8 | 12 | 0.3750 | 0.3500 | 0.3250 | 0.08333 | 0.04167 | 0.02500 | 0.0352 | 4 | 20 |
| 7/16 | 12 | 0.4375 | 0.4125 | 0.3875 | 0.08333 | 0.04167 | 0.02500 | 0.0352 | 3 | 41 |
| 1/2 | 10 | 0.5000 | 0.4700 | 0.4400 | 0.10000 | 0.05000 | 0.03000 | 0.0422 | 3 | 52 |
| 5/8 | 8 | 0.6250 | 0.5875 | 0.5500 | 0.12500 | 0.06250 | 0.03750 | 0.0528 | 3 | 52 |
| 3/4 | 6 | 0.7500 | 0.7000 | 0.6500 | 0.16667 | 0.08333 | 0.05000 | 0.0704 | 4 | 20 |
| 7/8 | 6 | 0.8750 | 0.8250 | 0.7750 | 0.16667 | 0.08333 | 0.05000 | 0.0704 | 3 | 41 |
| 1 | 5 | 1.0000 | 0.9400 | 0.8800 | 0.20000 | 0.10000 | 0.06000 | 0.0845 | 3 | 52 |
| 1 1/8 | 5 | 1.1250 | 1.0650 | 1.0050 | 0.20000 | 0.10000 | 0.06000 | 0.0845 | 3 | 25 |
| 1 1/4 | 5 | 1.2500 | 1.1900 | 1.1300 | 0.20000 | 0.10000 | 0.06000 | 0.0845 | 3 | 4 |
| 1 3/8 | 4 | 1.3750 | 1.3000 | 1.2250 | 0.25000 | 0.12500 | 0.07500 | 0.1056 | 3 | 30 |
| 1 1/2 | 4 | 1.5000 | 1.4250 | 1.3500 | 0.25000 | 0.12500 | 0.07500 | 0.1056 | 3 | 12 |
| 1 3/4 | 4 | 1.7500 | 1.6750 | 1.6000 | 0.25000 | 0.12500 | 0.07500 | 0.1056 | 2 | 43 |
| 2 | 4 | 2.0000 | 1.9250 | 1.8500 | 0.25000 | 0.12500 | 0.07500 | 0.1056 | 2 | 22 |
| 2 1/4 | 3 | 2.2500 | 2.1500 | 2.0500 | 0.33333 | 0.16667 | 0.10000 | 0.1408 | 2 | 50 |
| 2 1/2 | 3 | 2.5000 | 2.4000 | 2.3000 | 0.33333 | 0.16667 | 0.10000 | 0.1408 | 2 | 32 |
| 2 3/4 | 3 | 2.7500 | 2.6500 | 2.5500 | 0.33333 | 0.16667 | 0.10000 | 0.1408 | 2 | 18 |
| 3 | 2 | 3.0000 | 2.8500 | 2.7000 | 0.50000 | 0.25000 | 0.15000 | 0.2112 | 3 | 12 |
| 3 1/2 | 2 | 3.5000 | 3.3500 | 3.2000 | 0.50000 | 0.25000 | 0.15000 | 0.2112 | 2 | 43 |
| 4 | 2 | 4.0000 | 3.8500 | 3.7000 | 0.50000 | 0.25000 | 0.15000 | 0.2112 | 2 | 22 |
| 4 1/2 | 2 | 4.5000 | 4.3500 | 4.2000 | 0.50000 | 0.25000 | 0.15000 | 0.2112 | 2 | 6 |
| 5 | 2 | 5.0000 | 4.8500 | 4.7000 | 0.50000 | 0.25000 | 0.15000 | 0.2112 | 1 | 53 |

[a] All other dimensions are given in inches.

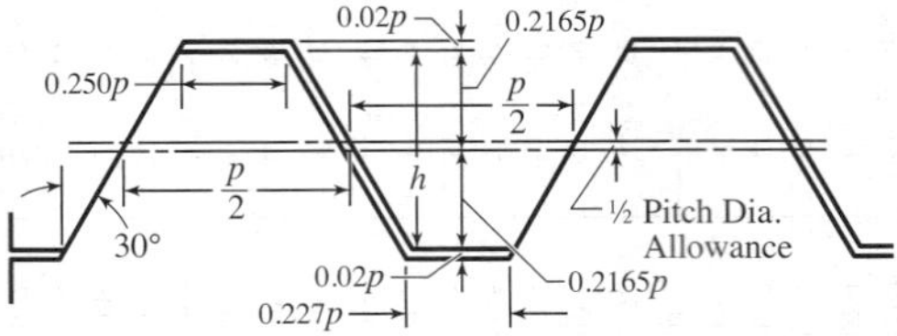


60-Degree Stub Thread

A clearance of at least 0.02 × pitch is added to depth $h$ to produce extra depth, thus avoiding interference with threads of mating part at minor or major diameters.

Basic thread thickness at pitch line = 0.5 × pitch $p$; basic depth $h$ = 0.433 × pitch; basic width of flat at crest = 0.25 × pitch; width of flat at root of screw thread = 0.227 × pitch; basic pitch diameter = basic major diameter − 0.433 × pitch; basic minor diameter = basic major diameter − 0.866 × pitch.

**Square Thread.—**The square thread is so named because the section is square, the depth, in the case of a screw, being equal to the width or one-half the pitch. The thread groove in a square-threaded nut is made a little greater than one-half the pitch in order to provide a slight clearance for the screw; hence, the tools used for threading square-threaded taps are a little less in width at the point than one-half the pitch. The pitch of a square thread is usually twice the pitch of an American Standard thread of corresponding diameter. The square thread has been superseded quite largely by the Acme form which has several advantages. See *ACME SCREW THREADS* on page 2057.

*10-Degree Modified Square Thread:* The included angle between the sides of the thread is 10 degrees (see accompanying diagram). The angle of 10 degrees results in a thread which is the practical equivalent of a "square thread," and yet is capable of economical production. Multiple thread milling cutters and ground thread taps should not be specified for modified square threads of the larger lead angles without consulting the cutting tool manufacturer.

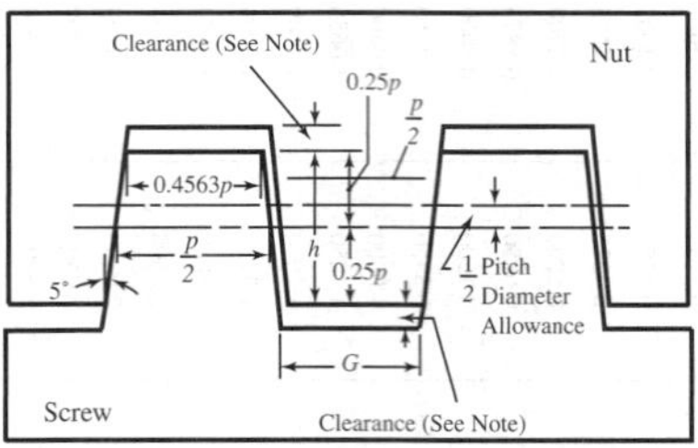


In the following formulas, $D$ = basic major diameter; $E$ = basic pitch diameter; $K$ = basic minor diameter; $p$ = pitch; $h$ = basic depth of thread on screw depth when there is no clearance between root of screw and crest of thread on nut; $t$ = basic thickness of thread at pitch line; $F$ = basic width of flat at crest of screw thread; $G$ = basic width of flat at root of screw thread; $C$ = clearance between root of screw and crest of thread on nut: $E = D - 0.5p$; $K = D - p$; $h = 0.5p$ (see Note); $t = 0.5p$; $F = 0.4563p$; $G = 0.4563p - (0.17 \times C)$.

*Note:* A clearance should be added to depth $h$ to avoid interference with threads of mating parts at minor or major diameters.

## BUTTRESS THREADS

### Threads of Buttress Form

The buttress form of thread has certain advantages in applications involving exceptionally high stresses along the thread axis in one direction only. The contacting flank of the thread, which takes the thrust, is referred to as the *pressure flank* and is so nearly perpendicular to the thread axis that the radial component of the thrust is reduced to a minimum. Because of the small radial thrust, this form of thread is particularly applicable where tubular members are screwed together, as in the case of breech mechanisms of large guns and airplane propeller hubs.

Fig. 1a shows a common form. The front or load-resisting face is perpendicular to the axis of the screw and the thread angle is 45 degrees. According to one rule, the pitch $P = 2 \times$ screw diameter $\div$ 15. The thread depth $d$ may equal $\frac{3}{4} \times$ pitch, making the flat $f = \frac{1}{8} \times$ pitch. Sometimes depth $d$ is reduced to $\frac{2}{3} \times$ pitch, making $f = \frac{1}{6} \times$ pitch.

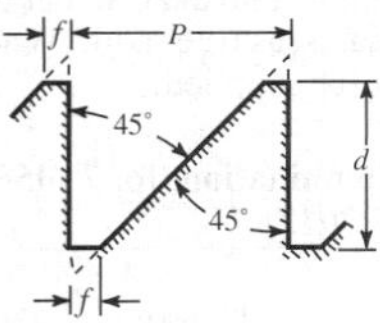


Fig. 1a.

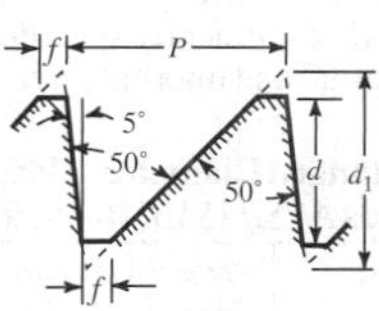


Fig. 1b.

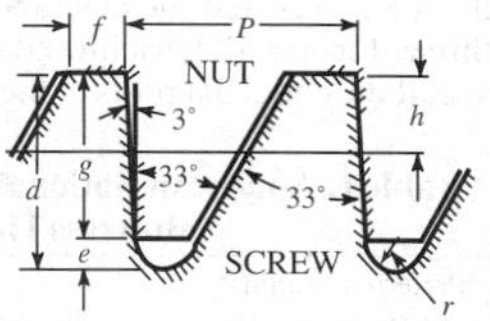


Fig. 1c.

The load-resisting side or flank may be inclined an amount (Fig. 1b) ranging usually from 1 to 5 degrees to avoid cutter interference in milling the thread. With an angle of 5 degrees and an included thread angle of 50 degrees, if the width of the flat $f$ at both crest and root equals $\frac{1}{8} \times$ pitch, then the thread depth equals 0.69 × pitch or $\frac{3}{4}\, d_1$.

The saw-tooth form of thread illustrated by Fig. 1c is known in Germany as the "Sägengewinde" and in Italy as the "Fillettatura a dente di Sega." Pitches are standardized from 2 millimeters up to 48 millimeters in the German and Italian specifications. The front face inclines 3 degrees from the perpendicular and the included angle is 33 degrees.

The thread depth $d$ for the screw = 0.86777 × pitch $P$. The thread depth $g$ for the nut = 0.75 × pitch. Dimension $h = 0.341 \times P$. The width $f$ of flat at the crest of the thread on the screw = 0.26384 × pitch. Radius $r$ at the root = 0.12427 × pitch. The clearance space $e$ = 0.11777 × pitch.

**British Standard Buttress Threads BS 1657: 1950.—**Specifications for buttress threads in this Standard are similar to those in the American National Standard (see page 2082) except: 1) A basic depth of thread of $0.4p$ is used instead of $0.6p$; 2) Sizes below 1 inch are not included; 3) Tolerances on major and minor diameters are the same as the pitch diameter tolerances, whereas in the American Standard separate tolerances are provided; however, provision is made for smaller major and minor diameter tolerances when crest surfaces of screws or nuts are used as datum surfaces, or when the resulting reduction in depth of engagement must be limited; and 4) Certain combinations of large diameters with fine pitches are provided that are not encouraged in the American Standard.

### American National Standard Buttress Inch Screw Threads

The buttress form of thread has certain advantages in applications involving exceptionally high stresses along the thread axis in one direction only. As the thrust side (load flank) of the standard buttress thread is made very nearly perpendicular to the thread axis, the radial component of the thrust is reduced to a minimum. On account of the small radial thrust, the buttress form of thread is particularly applicable when tubular members are screwed together. Examples of actual applications are the breech assemblies of large guns, airplane propeller hubs, and columns for hydraulic presses.

**7°/45° Buttress Thread Form.—**Milling, grinding, rolling, or other suitable manufacturing means have been taken into consideration in selecting the thread form, as shown in ANSI/ASME B1.9-1973 (R2017).

*Form of Thread:* The form of the buttress thread is shown in the accompanying Fig. 2a and Fig. 2b, and has the following characteristics:

a) A load flank angle, measured in an axial plane, of 7 degrees from the normal to the axis.

b) A clearance flank angle, measured in an axial plane, of 45 degrees from the normal to the axis.

c) Equal truncations at the crests of the external and internal threads such that the basic height of thread engagement (assuming no allowance) is equal to 0.6 of the pitch

b) Equal radii, at the roots of the external and internal basic thread forms tangential to the load flank and the clearance flank. (There is, in practice, almost no chance that the thread forms will be achieved strictly as basically specified, that is, as true radii.) When specified, equal flat roots of the external and internal thread may be supplied.

**Table 1. American National Standard Diameter—Pitch Combinations for 7°/45° Buttress Threads** *ANSI/ASME B1.9-1973 (R2017)*

| Preferred Nominal Major Diameters, Inches | Threads per Inch[a] | Preferred Nominal Major Diameters, Inches | Threads per Inch[a] |
|---|---|---|---|
| 0.5, 0.625, 0.75 | (20, 16, 12) | 4.5, 5, 5.5, 6 | 12, 10, 8, (6, 5, 4), 3 |
| 0.875, 1.0 | (16, 12, 10) | 7, 8, 9, 10 | 10, 8, 6, (5, 4, 3), 2.5, 2 |
| 1.25, 1.375, 1.5 | 16, (12, 10, 8), 6 | 11, 12, 14, 16 | 10, 8, 6, 5, (4, 3, 2.5), 2, 1.5, 1.25 |
| 1.75, 2, 2.25, 2.5 | 16, 12, (10, 8, 6), 5, 4 | 18, 20, 22, 24 | 8, 6, 5, 4, (3, 2.5, 2), 1.5, 1.25, 1 |
| 2.75, 3, 3.5, 4 | 16, 12, 10, (8, 6, 5), 4 | | |

[a] Preferred threads per inch are in parentheses.

**Table 2. American National Standard Inch Buttress Screw Threads—Basic Dimensions** *ANSI/ASME B1.9-1973 (R2017)*

| Thds.[a] per Inch | Pitch, $p$ | Basic Height of Thread, $h = 0.6p$ | Height of Sharp-V Thread, $H = 0.89064p$ | Crest Truncation, $f = 0.14532p$ | Height of Thread, $h_s$ or $h_n = 0.66271p$ | Max. Root Truncation,[b] $s = 0.0826p$ | Max. Root Radius,[c] $r = 0.0714p$ | Width of Flat at Crest, $F = 0.16316p$ |
|---|---|---|---|---|---|---|---|---|
| 20 | 0.0500 | 0.0300 | 0.0445 | 0.0073 | 0.0331 | 0.0041 | 0.0036 | 0.0082 |
| 16 | 0.0625 | 0.0375 | 0.0557 | 0.0091 | 0.0414 | 0.0052 | 0.0045 | 0.0102 |
| 12 | 0.0833 | 0.0500 | 0.0742 | 0.0121 | 0.0552 | 0.0069 | 0.0059 | 0.0136 |
| 10 | 0.1000 | 0.0600 | 0.0891 | 0.0145 | 0.0663 | 0.0083 | 0.0071 | 0.0163 |
| 8 | 0.1250 | 0.0750 | 0.1113 | 0.0182 | 0.0828 | 0.0103 | 0.0089 | 0.0204 |
| 6 | 0.1667 | 0.1000 | 0.1484 | 0.0242 | 0.1105 | 0.0138 | 0.0119 | 0.0271 |
| 5 | 0.2000 | 0.1200 | 0.1781 | 0.0291 | 0.1325 | 0.0165 | 0.0143 | 0.0326 |
| 4 | 0.2500 | 0.1500 | 0.2227 | 0.0363 | 0.1657 | 0.0207 | 0.0179 | 0.0408 |
| 3 | 0.3333 | 0.2000 | 0.2969 | 0.0484 | 0.2209 | 0.0275 | 0.0238 | 0.0543 |
| 2½ | 0.4000 | 0.2400 | 0.3563 | 0.0581 | 0.2651 | 0.0330 | 0.0286 | 0.0653 |
| 2 | 0.5000 | 0.3000 | 0.4453 | 0.0727 | 0.3314 | 0.0413 | 0.0357 | 0.0816 |
| 1½ | 0.6667 | 0.4000 | 0.5938 | 0.0969 | 0.4418 | 0.0551 | 0.0476 | 0.1088 |
| 1¼ | 0.8000 | 0.4800 | 0.7125 | 0.1163 | 0.5302 | 0.0661 | 0.0572 | 0.1305 |
| 1 | 1.0000 | 0.6000 | 0.8906 | 0.1453 | 0.6627 | 0.0826 | 0.0714 | 0.1632 |

[a] All other dimensions are in inches.

[b] Minimum root truncation is one-half of maximum.

[c] Minimum root radius is one-half of maximum.

**Form of American National Standard 7°/45° Buttress Thread with 0.6*p* Basic Height of Thread Engagement**

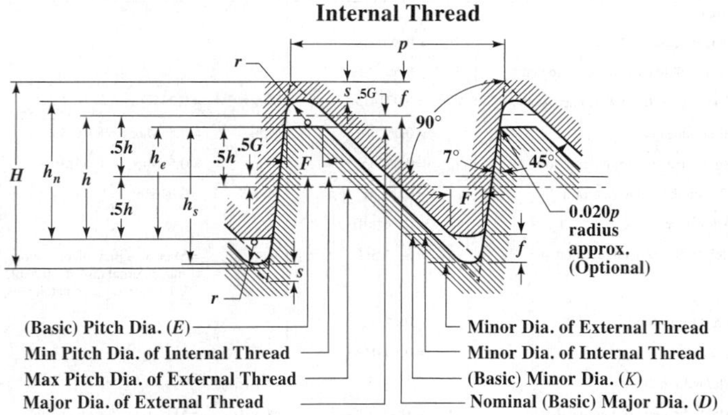


Fig. 2a. Round Root External Thread

Note: Heavy Line Indicates Basic Form

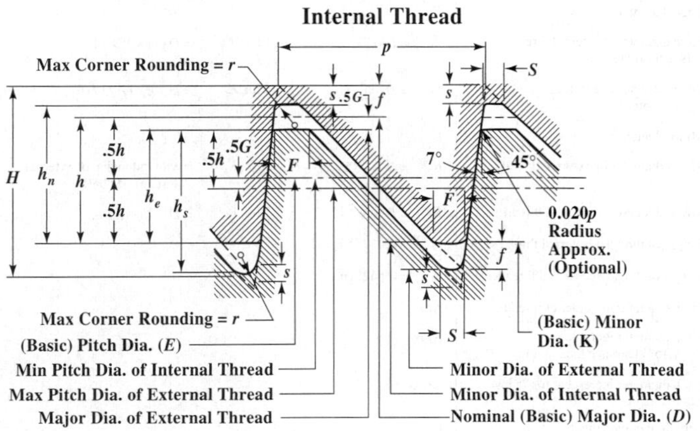


Fig. 2b. Flat Root External Thread

Note: Heavy Line Indicates Basic Form

**Table 3. American National Standard Buttress Inch Screw Thread Symbols and Form**

| Thread Element | Max. Material (Basic) | | Min. Material | |
|---|---|---|---|---|
| Pitch | $p$ | | | |
| Height of sharp-V thread | $H$ | $= 0.89064p$ | | |
| Basic height of thread engagement | $h$ | $= 0.6p$ | | |
| Root radius (theoretical)(see footnote [a]) | $r$ | $= 0.07141p$ | Min. $r$ | $= 0.0357p$ |
| Root truncation | $s$ | $= 0.0826p$ | Min. $s$ | $= 0.5$; Max. $s = 0.0413p$ |
| Root truncation for flat root form | $s$ | $= 0.0826p$ | Min. $s$ | $= 0.5$; Max. $s = 0.0413p$ |
| Flat width for flat root form | $S$ | $= 0.0928p$ | Min. $S$ | $= 0.0464p$ |
| Allowance | $G$ | (see text) | | |
| Height of thread engagement | $h_e$ | $= h - 0.5G$ | Min. $h_e$ | = Max. $h_e$ − [0.5 tol. on major dia. external thread + 0.5 tol. on minor dia. internal thread]. |
| Crest truncation | $f$ | $= 0.14532p$ | | |
| Crest width | $F$ | $= 0.16316p$ | | |
| Major diameter | $D$ | | | |
| Major diameter of internal thread | $D_n$ | $= D + 0.12542p$ | Max. $D_n$ | = Max. pitch dia.of internal thread + $0.80803p$ |
| Major diameter of external thread | $D_s$ | $= D - G$ | Min. $D_s$ | $= D - G - D$ tol. |
| Pitch diameter | $E$ | | | |
| Pitch diameter of internal thread (see footnote [b]) | $E_n$ | $= D - h$ | Max. $E_n$ | $= D - h + PD$ tol. |
| Pitch diameter of external thread (see footnote [c]) | $E_s$ | $= D - h - G$ | Min. $E_s$ | $= D - h - G - PD$ tol. |
| Minor diameter | $K$ | | | |
| Minor diameter of external thread | $K_s$ | $= D - 1.32542p - G$ | Min. $K_s$ | = Min. pitch dia. of external thread − $0.80803p$ |
| Minor diameter of internal thread | $K_n$ | $= D - 2h$ | Min. $K_n$ | $= D - 2h + K$ tol. |
| Height of thread of internal thread | $h_n$ | $= 0.66271p$ | | |
| Height of thread of external thread | $h_s$ | $= 0.66271p$ | | |
| Pitch diameter increment for lead | $\Delta El$ | | | |
| Pitch diameter increment for 45° clearance flank angle | $\Delta E\alpha_1$ | | | |
| Pitch diameter increment for 7° load flank angle | $\Delta E\alpha_2$ | | | |
| Length of engagement | $L_e$ | | | |

[a] Unless the flat root form is specified, the rounded root form of the external and internal thread shall be a continuous, smoothly blended curve within the zone defined by $0.07141p$ maximum to $0.0357p$ minimum radius. The resulting curve shall have no reversals or sudden angular variations, and shall be tangent to the flanks of the thread. There is, in practice, almost no chance that the rounded thread form will be achieved strictly as basically specified, that is, as a true radius.

[b] The pitch diameter $X$ tolerances for GO and NOT GO threaded plug gages are applied to the internal product limits for $E_n$ and Max. $E_n$.

[c] The pitch diameter $W$ tolerances for GO and NOT GO threaded setting plug gages are applied to the external product limits for $E_s$ and Min. $E_s$.

**Buttress Thread Tolerances.—**Tolerances from basic size on external threads are applied in a minus direction and on internal threads in a plus direction.

*Pitch Diameter Tolerances:* The following formula is used for determining the pitch diameter product tolerance for Class 2 (standard grade) external or internal threads:

$$PD\text{ tolerance } = 0.002\ \sqrt[3]{D} + 0.00278\sqrt{L_e} + 0.00854\sqrt{p}$$

where $D$ = basic major diameter of external thread (assuming no allowance)

$L_e$ = length of engagement

$p$ = pitch of thread

When the length of engagement is taken as $10p$, the formula reduces to

$$0.002\ \sqrt[3]{D} + 0.0173\sqrt{p}$$

It is to be noted that this formula relates specifically to Class 2 (standard grade) PD tolerances. Class 3 (precision grade) PD tolerances are two-thirds of Class 2 PD tolerances. Pitch diameter tolerances based on this latter formula, for various diameter pitch combinations, are given in Table 4.

*Functional Size:* Deviations in lead and flank angle of product threads increase the functional size of an external thread and decrease the functional size of an internal thread by the cumulative effect of the diameter equivalents of these deviations. The functional size of all buttress product threads shall not exceed the maximum-material limit.

*Tolerances on Major Diameter of External Thread and Minor Diameter of Internal Thread:* Unless otherwise specified, these tolerances should be the same as the pitch diameter tolerance for the class used.

*Tolerances on Minor Diameter of External Thread and Major Diameter of Internal Thread:* It will be sufficient in most instances to state only the maximum minor diameter of the external thread and the minimum major diameter of the internal thread without any tolerance. However, the root truncation from a sharp V should not be greater than $0.0826p$ nor less than $0.0413p$.

*Lead and Flank Angle Deviations for Class 2:* The deviations in lead and flank angles may consume the entire tolerance zone between maximum and minimum material product limits given in Table 4.

*Diameter Equivalents for Variations in Lead and Flank Angles for Class 3:* The combined diameter equivalents of variations in lead (including helix deviations), and flank angle for Class 3, shall not exceed 50 percent of the Class 2 pitch diameter tolerances given in Table 4.

*Tolerances on Taper and Roundness:* There are no requirements for taper and roundness for Class 2 buttress screw threads.

The major and minor diameters of Class 3 buttress threads shall not taper nor be out of round to the extent that specified limits for major and minor diameter are exceeded. The taper and out-of-roundness of the pitch diameter for Class 3 buttress threads shall not exceed 50 percent of the pitch-diameter tolerances.

**Allowances for Easy Assembly.—**An allowance (clearance) should be provided on all external threads to secure easy assembly of parts. The amount of the allowance is deducted from the nominal major, pitch, and minor diameters of the external thread when the maximum material condition of the external thread is to be determined.

The minimum internal thread is basic.

The amount of the allowance is the same for both classes and is equal to the Class 3 pitch-diameter tolerance as calculated by the formulas previously given. The allowances for various diameter-pitch combinations are given in Table 5.

**Table 4. American National Standard Buttress Inch Screw Threads Tolerances Class 2 (Standard Grade) and Class 3 (Precision Grade)**
*ANSI/ASME B1.9-1973 (R2017)*

| Thds. per Inch | Pitch,[a] $p$ Inch | Basic Major Diameter, Inch: From 0.5 thru 0.7 | Over 0.7 thru 1.0 | Over 1.0 thru 1.5 | Over 1.5 thru 2.5 | Over 2.5 thru 4 | Over 4 thru 6 | Over 6 thru 10 | Over 10 thru 16 | Over 16 thru 24 | Pitch[b] Increment, $0.0173\sqrt{p}$ Inch |
|---|---|---|---|---|---|---|---|---|---|---|---|
| | | Tolerance on Major Diameter of External Thread, Pitch Diameter of External and Internal Threads, and Minor Diameter of Internal Thread, Inch | | | | | | | | | |
| | | **Class 2, Standard Grade** | | | | | | | | | |
| 20 | 0.0500 | .0056 | .... | .... | .... | .... | .... | .... | .... | .... | .00387 |
| 16 | 0.0625 | .0060 | .0062 | .0065 | .0068 | .0073 | .... | .... | .... | .... | .00432 |
| 12 | 0.0833 | .0067 | .0069 | .0071 | .0075 | .0080 | .0084 | .... | .... | .... | .00499 |
| 10 | 0.1000 | .... | .0074 | .0076 | .0080 | .0084 | .0089 | .0095 | .0102 | .... | .00547 |
| 8 | 0.1250 | .... | .... | .0083 | .0086 | .0091 | .0095 | .0101 | .0108 | .0115 | .00612 |
| 6 | 0.1667 | .... | .... | .0092 | .0096 | .0100 | .0105 | .0111 | .0118 | .0125 | .00706 |
| 5 | 0.2000 | .... | .... | .... | .0103 | .0107 | .0112 | .0117 | .0124 | .0132 | .00774 |
| 4 | 0.2500 | .... | .... | .... | .0112 | .0116 | .0121 | .0127 | .0134 | .0141 | .00865 |
| 3 | 0.3333 | .... | .... | .... | .... | .... | .0134 | .0140 | .0147 | .0154 | .00999 |
| 2.5 | 0.4000 | .... | .... | .... | .... | .... | .... | .0149 | .0156 | .0164 | .01094 |
| 2.0 | 0.5000 | .... | .... | .... | .... | .... | .... | .0162 | .0169 | .0177 | .01223 |
| 1.5 | 0.6667 | .... | .... | .... | .... | .... | .... | .... | .0188 | .0196 | .01413 |
| 1.25 | 0.8000 | .... | .... | .... | .... | .... | .... | .... | .0202 | .0209 | .01547 |
| 1.0 | 1.0000 | .... | .... | .... | .... | .... | .... | .... | .... | .0227 | .01730 |
| Diameter Increment,[c] $0.002\sqrt[3]{D}$ | | .00169 | .00189 | .00215 | .00252 | .00296 | .00342 | .00400 | .00470 | .00543 | |
| | | **Class 3, Precision Grade** | | | | | | | | | |
| 20 | 0.0500 | .0037 | .... | .... | .... | .... | .... | .... | .... | .... | |
| 16 | 0.0625 | .0040 | .0042 | .0043 | .0046 | .0049 | .... | .... | .... | .... | |
| 12 | 0.0833 | .0044 | .0046 | .0048 | .0050 | .0053 | .0056 | .... | .... | .... | |
| 10 | 0.1000 | .... | .0049 | .0051 | .0053 | .0056 | .0059 | .0063 | .0068 | .... | |
| 8 | 0.1250 | .... | .... | .0055 | .0058 | .0061 | .0064 | .0067 | .0072 | .0077 | |
| 6 | 0.1667 | .... | .... | .0061 | .0064 | .0067 | .0070 | .0074 | .0078 | .0083 | |
| 5 | .02000 | .... | .... | .... | .0068 | .0071 | .0074 | .0078 | .0083 | .0088 | |
| 4 | 0.2500 | .... | .... | .... | .0074 | .0077 | .0080 | .0084 | .0089 | .0094 | |
| 3 | .03333 | .... | .... | .... | .... | .... | .0089 | .0093 | .0098 | .0103 | |
| 2.5 | 0.4000 | .... | .... | .... | .... | .... | .... | .0100 | .0104 | .0109 | |
| 2.0 | 0.5000 | .... | .... | .... | .... | .... | .... | .0108 | .0113 | .0118 | |
| 1.5 | 0.6667 | .... | .... | .... | .... | .... | .... | .... | .0126 | .0130 | |
| 1.25 | 0.8000 | .... | .... | .... | .... | .... | .... | .... | .0135 | .0139 | |
| 1.0 | 1.0000 | .... | .... | .... | .... | .... | .... | .... | .... | .0152 | |

[a] For threads with pitches not shown in this table, pitch increment to be used in tolerance formula is to be determined by use of formula PD Tolerance $= 0.002\sqrt[3]{D} + 0.00278\sqrt{L_e} + 0.00854\sqrt{p}$, where: $D$ = basic major diameter of external thread (assuming no allowance), $L_e$ = length of engagement, and $p$ = pitch of thread. This formula relates specifically to Class 2 (standard grade) PD tolerances. Class 3 (precision grade) PD tolerances are two-thirds of Class 2 PD tolerances. See text

[b] When the length of engagement is taken as 10p, the formula reduces to: $0.002\sqrt[3]{D} + 0.0173\sqrt{p}$

[c] Diameter $D$, used in diameter increment formula, is based on the average of the range.

**Table 5. American National Standard External Thread Allowances for Classes 2 and 3 Buttress Inch Screw Threads** *ANSI/ASME B1.9-1973 (R2017)*

| Threads per Inch | Pitch, $p$ Inch | Basic Major Diameter, Inch | | | | | | | | |
|---|---|---|---|---|---|---|---|---|---|---|
| | | From 0.5 thru 0.7 | Over 0.7 thru 1.0 | Over 1.0 thru 1.5 | Over 1.5 thru 2.5 | Over 2.5 thru 4 | Over 4 thru 6 | Over 6 thru 10 | Over 10 thru 16 | Over 16 thru 24 |
| | | Allowance on Major, Minor and Pitch Diameters of External Thread, Inch | | | | | | | | |
| 20 | 0.0500 | .0037 | .... | .... | .... | .... | .... | .... | .... | .... |
| 16 | 0.0625 | .0040 | .0042 | .0043 | .0046 | .0049 | .... | .... | .... | .... |
| 12 | 0.0833 | .0044 | .0046 | .0048 | .0050 | .0053 | .0056 | .... | .... | .... |
| 10 | 0.1000 | .... | .0049 | .0051 | .0053 | .0056 | .0059 | .0063 | .0068 | .... |
| 8 | 0.1250 | .... | .... | .0055 | .0058 | .0061 | .0064 | .0067 | .0072 | .0077 |
| 6 | 0.1667 | .... | .... | .0061 | .0064 | .0067 | .0070 | .0074 | .0078 | .0083 |
| 5 | 0.2000 | .... | .... | .... | .0068 | .0071 | .0074 | .0078 | .0083 | .0088 |
| 4 | 0.2500 | .... | .... | .... | .0074 | .0077 | .0080 | .0084 | .0089 | .0094 |
| 3 | 0.3333 | .... | .... | .... | .... | .... | .0089 | .0093 | .0098 | .0103 |
| 2.5 | 0.4000 | .... | .... | .... | .... | .... | .... | .0100 | .0104 | .0109 |
| 2.0 | 0.5000 | .... | .... | .... | .... | .... | .... | .0108 | .0113 | .0118 |
| 1.5 | 0.6667 | .... | .... | .... | .... | .... | .... | .... | .0126 | .0130 |
| 1.25 | 0.8000 | .... | .... | .... | .... | .... | .... | .... | .0135 | .0139 |
| 1.0 | 1.0000 | .... | .... | .... | .... | .... | .... | .... | .... | .0152 |

**Example Showing Dimensions for a Typical Buttress Thread.—**The dimensions for a 2-inch diameter, 4 threads per inch, Class 2 buttress thread with flank angles of 7 degrees and 45 degrees are

$h$ = basic thread height = 0.1500 (Table 2)

$h_s = h_n$ = height of thread in external and internal threads = 0.1657 (Table 2)

$G$ = pitch-diameter allowance on external thread = 0.0074 (Table 5)

Tolerance on PD of external and internal threads = 0.0112 (Table 4)

Tolerance on major diameter of external thread and minor diameter of internal thread = 0.0112 (Table 4)

**Internal Thread:**

*Basic Major Diameter:* $D = 2.0000$

*Min. Major Diameter:* $D - 2h + 2h_n = 2.0314$ (see Table 2)

*Min. Pitch Diameter:* $D - h = 1.8500$ (see Table 2)

*Max. Pitch Diameter:* $D - h + PD$ Tolerance = 1.8612 (see Table 4)

*Min. Minor Diameter:* $D - 2h = 1.7000$ (see Table 2)

*Max. Minor Diameter:* $D - 2h$ + Minor Diameter Tolerance = 1.7112 (see Table 4)

**External Thread:**

*Max. Major Diameter:* $D - G = 1.9926$ (see Table 5)

*Min. Major Diameter:* $D - G$ − Major Diameter Tolerance = 1.9814 (see Table 4 and Table 5)

*Max. Pitch Diameter:* $D - h - G = 1.8426$ (see Table 2 and Table 5)

*Min. Pitch Diameter:* $D - h - G - PD$ Tolerance = 1.8314 (see Table 4)

*Max. Minor Diameter:* $D - G - 2h_s = 1.6612$ (see Table 2 and Table 5)

**Buttress Thread Designations.—**When only the designation, BUTT is used, the thread is "pull" type buttress (external thread pulls) with the clearance flank leading and the 7-degree pressure flank following. When the designation, PUSH-BUTT is used, the thread is a push type buttress (external thread pushes) with the 7-degree load flank leading and the 45-degree clearance flank following. Whenever possible this description should be confirmed by a simplified view showing thread angles on the drawing of the product that has the buttress thread.

*Standard Buttress Threads:* A buttress thread is considered to be standard when: 1) opposite flank angles are 7-degrees and 45-degrees; 2) basic thread height is 0.6*p*; 3) tolerances and allowances are as shown in Table 4 and Table 5; and 4) length of engagement is 10*p* or less.

*Thread Designation Abbreviations:* In thread designations on drawings, tools, gages, and in specifications, the following abbreviations and letters are to be used:

| | | |
|---|---|---|
| BUTT | for buttress thread, pull type | |
| PUSH-BUTT | for buttress thread, push type | |
| LH | for left-hand thread | *Note:* Absence of LH indicates that the thread is a right-hand thread. |
| P | for pitch | |
| L | for lead | |
| A | for external thread | *Note:* Absence of A or B after thread class indicates that designation covers both the external and internal threads. |
| B | for internal thread | |
| Le | for length of thread engagement | |
| SPL | for special | |
| FL | for flat root thread | |
| E | for pitch diameter | |
| TPI | for threads per inch | |
| THD | for thread | |

*Designation Sequence for Buttress Inch Screw Threads:* When designating single-start standard buttress threads the nominal size is given first, the threads per inch next, then PUSH if the internal member is to push, but nothing if it is to pull, then the class of thread (2 or 3), then whether external (A) or internal (B), then LH if left-hand, but nothing if right-hand, and finally FL if a flat root thread, but nothing if a radiused root thread; thus, 2.5-8 BUTT-2A indicates a 2.5 inch, 8 threads per inch buttress thread, Class 2 external, right-hand, internal member to pull, with radiused root of thread. The designation 2.5-8 PUSH-BUTT-2A-LH-FL signifies a 2.5 inch size, 8 threads per inch buttress thread with internal member to push, Class 2 external, left-hand, and flat root.

A multiple-start standard buttress thread is similarly designated but the pitch is given instead of the threads per inch, followed by the lead and the number of starts is indicated in parentheses after the class of thread. Thus, 10-0.25P-0.5L - BUTT-3B (2 start) indicates a 10-inch thread with 4 threads per inch, 0.5 inch lead, buttress form with internal member to pull, Class 3 internal, 2 starts, with radiused root of thread.

## WHITWORTH THREADS

### British Standard Whitworth (BSW) and British Standard Fine (BSF) Threads

The BSW is the Coarse Thread series and the BSF is the Fine Thread series of British Standard 84:1956—Parallel Screw Threads of Whitworth Form. The dimensions given in the tables on the following pages for the major, effective, and minor diameters are, respectively, the maximum limits of these diameters for bolts and the minimum limits for nuts. Formulas for the tolerances on these diameters are given in the table below.

**Whitworth Standard Thread Form.—**This thread form is used for the British Standard Whitworth (BSW) and British Standard Fine (BSF) screw threads. More recently, both threads have been known as parallel screw threads of Whitworth form. With standardization of the Unified thread, the Whitworth thread form is expected to be used only for replacements or spare parts. Tables of British Standard Parallel Screw Threads of Whitworth Form will be found on the following pages; tolerance formulas are given in the table below. The form of the thread is shown by the diagram. If $p$ = pitch, $d$ = depth of thread, $r$ = radius at crest and root, and $n$ = number of threads per inch, then

$$d = \frac{1}{3}p \times \cot 27°30' = 0.640327p = 0.640327 \div n$$
$$r = 0.137329p = 0.137329 \div n$$

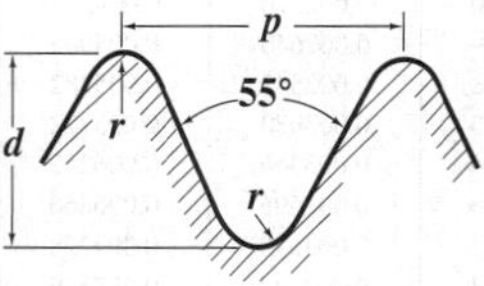


It is recommended that stainless steel bolts of nominal size ¾ inch and below should not be made to Close Class limits but rather to Medium or Free Class limits. Nominal sizes above ¾ inch should have maximum and minimum limits 0.001 inch smaller than the values obtained from the table.

*Tolerance Classes : Close Class bolts.* Applies to screw threads requiring a fine snug fit, and should be used only for special work where refined accuracy of pitch and thread form are particularly required. *Medium Class bolts and nuts.* Applies to the better class of ordinary interchangeable screw threads. *Free Class bolts.* Applies to the majority of bolts of ordinary commercial quality. *Normal Class nuts.* Applies to ordinary commercial quality nuts; this class is intended for use with Medium or Free Class bolts.

**Table 1. Tolerance Formulas for BSW and BSF Threads**

| | Class or Fit | Tolerance in inches[a] (+ for nuts, – for bolts) | | |
|---|---|---|---|---|
| | | Major Dia. | Effective Dia. | Minor Dia. |
| Bolts | Close | $\frac{2}{3}T + 0.01\sqrt{p}$ | $\frac{2}{3}T$ | $\frac{2}{3}T + 0.013\sqrt{p}$ |
| | Medium | $T + 0.01\sqrt{p}$ | $T$ | $T + 0.02\sqrt{p}$ |
| | Free | $\frac{3}{2}T + 0.01\sqrt{p}$ | $\frac{3}{2}T$ | $\frac{3}{2}T + 0.02\sqrt{p}$ |
| Nuts | Close | … | $\frac{2}{3}T$ | $0.2p + 0.004$[b] |
| | Medium | … | $T$ | $0.2p + 0.005$[c] |
| | Normal | … | $\frac{3}{2}T$ | $0.2p + 0.007$[d] |

[a] The symbol $T = 0.002\sqrt[3]{D} + 0.003\sqrt{L} + 0.005\sqrt{p}$, where $D$ = major diameter of thread in inches; $L$ = length of engagement in inches; $p$ = pitch in inches. The symbol $p$ signifies pitch.

[b] For 26 threads per inch and finer.

[c] For 24 and 22 threads per inch.

[d] For 20 threads per inch and coarser.

**Table 2. Threads of Whitworth Form—Basic Dimensions**

$p = 1 \div n$
$H = 0.960491p$
$H/6 = 0.160082p$
$h = 0.640327p$
$e = 0.0739176p$
$r = 0.137329p$

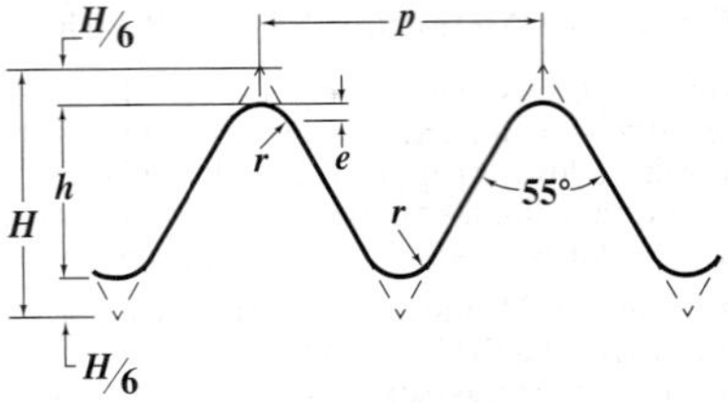


| Threads per Inch | Pitch | Triangular Height | Shortening | Depth of Thread | Depth of Rounding | Radius |
|---|---|---|---|---|---|---|
| $n$ | $p$ | $H$ | $H/6$ | $h$ | $e$ | $r$ |
| 72 | 0.013889 | 0.013340 | 0.002223 | 0.008894 | 0.001027 | 0.001907 |
| 60 | 0.016667 | 0.016009 | 0.002668 | 0.010672 | 0.001232 | 0.002289 |
| 56 | 0.017857 | 0.017151 | 0.002859 | 0.011434 | 0.001320 | 0.002452 |
| 48 | 0.020833 | 0.020010 | 0.003335 | 0.013340 | 0.001540 | 0.002861 |
| 40 | 0.025000 | 0.024012 | 0.004002 | 0.016008 | 0.0011848 | 0.003433 |
| 36 | 0.027778 | 0.026680 | 0.004447 | 0.017787 | 0.002053 | 0.003815 |
| 32 | 0.031250 | 0.030015 | 0.005003 | 0.020010 | 0.002310 | 0.004292 |
| 28 | 0.035714 | 0.034303 | 0.005717 | 0.022869 | 0.002640 | 0.004905 |
| 26 | 0.038462 | 0.036942 | 0.006157 | 0.024628 | 0.002843 | 0.005282 |
| 24 | 0.041667 | 0.040020 | 0.006670 | 0.026680 | 0.003080 | 0.005722 |
| 22 | 0.045455 | 0.043659 | 0.007276 | 0.029106 | 0.003366 | 0.006242 |
| 20 | 0.050000 | 0.048025 | 0.008004 | 0.032016 | 0.003696 | 0.006866 |
| 19 | 0.052632 | 0.050553 | 0.008425 | 0.033702 | 0.003890 | 0.007228 |
| 18 | 0.055556 | 0.053361 | 0.008893 | 0.035574 | 0.004107 | 0.007629 |
| 16 | 0.062500 | 0.060031 | 0.010005 | 0.040020 | 0.004620 | 0.008583 |
| 14 | 0.071429 | 0.068607 | 0.011434 | 0.045738 | 0.005280 | 0.009809 |
| 12 | 0.083333 | 0.080041 | 0.013340 | 0.053361 | 0.006160 | 0.011444 |
| 11 | 0.090909 | 0.087317 | 0.014553 | 0.058212 | 0.006720 | 0.012484 |
| 10 | 0.100000 | 0.096049 | 0.016008 | 0.064033 | 0.007392 | 0.013733 |
| 9 | 0.111111 | 0.106721 | 0.017787 | 0.071147 | 0.008213 | 0.015259 |
| 8 | 0.125000 | 0.120061 | 0.020010 | 0.080041 | 0.009240 | 0.017166 |
| 7 | 0.142857 | 0.137213 | 0.022869 | 0.091475 | 0.010560 | 0.019618 |
| 6 | 0.166667 | 0.160082 | 0.026680 | 0.106721 | 0.012320 | 0.022888 |
| 5 | 0.20000 | 0.192098 | 0.032016 | 0.128065 | 0.014784 | 0.027466 |
| 4.5 | 0.222222 | 0.213442 | 0.035574 | 0.142295 | 0.016426 | 0.030518 |
| 4 | 0.250000 | 0.240123 | 0.040020 | 0.160082 | 0.018479 | 0.034332 |
| 3.5 | 0.285714 | 0.274426 | 0.045738 | 0.182951 | 0.021119 | 0.039237 |
| 3.25 | 0.307692 | 0.295536 | 0.049256 | 0.197024 | 0.022744 | 0.042255 |
| 3 | 0.333333 | 0.320164 | 0.053361 | 0.213442 | 0.024639 | 0.045776 |
| 2.875 | 0.347826 | 0.334084 | 0.055681 | 0.222722 | 0.025710 | 0.047767 |
| 2.75 | 0.363636 | 0.349269 | 0.058212 | 0.232846 | 0.026879 | 0.049938 |
| 2.625 | 0.380952 | 0.365901 | 0.060984 | 0.243934 | 0.028159 | 0.052316 |
| 2.5 | 0.400000 | 0.384196 | 0.064033 | 0.256131 | 0.029567 | 0.054932 |

Dimensions are in inches.

*Allowances:* Only Free Class and Medium Class bolts have an allowance. For nominal sizes of $\frac{3}{4}$ inch down to $\frac{1}{4}$ inch, the allowance is 30 percent of the Medium Class bolt effective-diameter tolerance (0.3$T$); for sizes less than $\frac{1}{4}$ inch, the allowance for the $\frac{1}{4}$-inch size applies. Allowances are applied minus from the basic bolt dimensions; the tolerances are then applied to the reduced dimensions.

**Table 3. British Standard Whitworth (BSW) and British Standard Fine (BSF) Screw Thread Series—Basic Dimensions** *BS 84:2007*

| Nominal Size, Inches | Threads per Inch | Pitch, Inches | Depth of Thread, Inches | Major Diameter, Inches | Effective Diameter, Inches | Minor Diameter, Inches | Area at Bottom of Thread, Sq. in. | Tap Drill Dia. |
|---|---|---|---|---|---|---|---|---|
| **Coarse Thread Series (BSW)** | | | | | | | | |
| 1/8 [a] | 40 | 0.02500 | 0.0160 | 0.1250 | 0.1090 | 0.0930 | 0.0068 | 2.55 mm |
| 3/16 | 24 | 0.04167 | 0.0267 | 0.1875 | 0.1608 | 0.1341 | 0.0141 | 3.70 mm |
| 1/4 | 20 | 0.05000 | 0.0320 | 0.2500 | 0.2180 | 0.1860 | 0.0272 | 5.10 mm |
| 5/16 | 18 | 0.05556 | 0.0356 | 0.3125 | 0.2769 | 0.2413 | 0.0457 | 6.50 mm |
| 3/8 | 16 | 0.06250 | 0.0400 | 0.3750 | 0.3350 | 0.2950 | 0.0683 | 7.90 mm |
| 7/16 | 14 | 0.07143 | 0.0457 | 0.4375 | 0.3918 | 0.3461 | 0.0941 | 9.30 mm |
| 1/2 | 12 | 0.08333 | 0.0534 | 0.5000 | 0.4466 | 0.3932 | 0.1214 | 10.50 mm |
| 9/16 [a] | 12 | 0.08333 | 0.0534 | 0.5625 | 0.5091 | 0.4557 | 0.1631 | 12.10. mm |
| 5/8 | 11 | 0.09091 | 0.0582 | 0.6250 | 0.5668 | 0.5086 | 0.2032 | 13.50 mm |
| 11/16 [a] | 11 | 0.09091 | 0.0582 | 0.6875 | 0.6293 | 0.5711 | 0.2562 | 15.00 mm |
| 3/4 | 10 | 0.10000 | 0.0640 | 0.7500 | 0.6860 | 0.6220 | 0.3039 | 16.25 mm |
| 7/8 | 9 | 0.11111 | 0.0711 | 0.8750 | 0.8039 | 0.7328 | 0.4218 | 19.25 mm |
| 1 | 8 | 0.12500 | 0.0800 | 1.0000 | 0.9200 | 0.8400 | 0.5542 | 22.00 mm |
| 1 1/8 | 7 | 0.14286 | 0.0915 | 1.1250 | 1.0335 | 0.9420 | 0.6969 | 24.75 mm |
| 1 1/4 | 7 | 0.14286 | 0.0915 | 1.2500 | 1.1585 | 1.0670 | 0.8942 | 28.00 mm |
| 1 1/2 | 6 | 0.16667 | 0.1067 | 1.5000 | 1.3933 | 1.2866 | 1.3000 | 33.50 mm |
| 1 3/4 | 5 | 0.20000 | 0.1281 | 1.7500 | 1.6219 | 1.4938 | 1.7530 | 39.00 mm |
| 2 | 4.5 | 0.22222 | 0.1423 | 2.0000 | 1.8577 | 1.7154 | 2.3110 | 44.50 mm |
| 2 1/4 | 4 | 0.25000 | 0.1601 | 2.2500 | 2.0899 | 1.9298 | 2.9250 | |
| 2 1/2 | 4 | 0.25000 | 0.1601 | 2.5000 | 2.3399 | 2.1798 | 3.7320 | Tap drill diameters shown in this column are recommended sizes listed in BS 1157:1975 and provide from 77 to 87% of full thread. |
| 2 3/4 | 3.5 | 0.28571 | 0.1830 | 2.7500 | 2.5670 | 2.3840 | 4.4640 | |
| 3 | 3.5 | 0.28571 | 0.1830 | 3.0000 | 2.8170 | 2.6340 | 5.4490 | |
| 3 1/4 [a] | 3.25 | 0.30769 | 0.1970 | 3.2500 | 3.0530 | 2.8560 | 6.4060 | |
| 3 1/2 | 3.25 | 0.30769 | 0.1970 | 3.5000 | 3.3030 | 3.1060 | 7.5770 | |
| 3 3/4 [a] | 3 | 0.33333 | 0.2134 | 3.7500 | 3.5366 | 3.3232 | 8.6740 | |
| 4 | 3 | 0.33333 | 0.2134 | 4.0000 | 3.7866 | 3.5732 | 10.0300 | |
| 4 1/2 | 2.875 | 0.34783 | 0.2227 | 4.5000 | 4.2773 | 4.0546 | 12.9100 | |
| 5 | 2.75 | 0.36364 | 0.2328 | 5.0000 | 4.7672 | 4.5344 | 16.1500 | |
| 5 1/2 | 2.625 | 0.38095 | 0.2439 | 5.5000 | 5.2561 | 5.0122 | 19.7300 | |
| 6 | 2.5 | 0.40000 | 0.2561 | 6.0000 | 5.7439 | 5.4878 | 23.6500 | |
| **Fine Thread Series (BSF)** | | | | | | | | |
| 3/16 [a,b] | 32 | 0.03125 | 0.0200 | 0.1875 | 0.1675 | 0.1475 | 0.0171 | 4.00 mm |
| 7/32 [a] | 28 | 0.03571 | 0.0229 | 0.2188 | 0.1959 | 0.1730 | 0.0235 | 4.60 mm |
| 1/4 | 26 | 0.03846 | 0.0246 | 0.2500 | 0.2254 | 0.2008 | 0.0317 | 5.30 mm |
| 9/32 [a] | 26 | 0.03846 | 0.0246 | 0.2812 | 0.2566 | 0.2320 | 0.0423 | 6.10 mm |
| 5/16 | 22 | 0.04545 | 0.0291 | 0.3125 | 0.2834 | 0.2543 | 0.0508 | 6.80 mm |
| 3/8 | 20 | 0.05000 | 0.0320 | 0.3750 | 0.3430 | 0.3110 | 0.0760 | 8.30 mm |
| 7/16 | 18 | 0.05556 | 0.0356 | 0.4375 | 0.4019 | 0.3663 | 0.1054 | 9.70 mm |
| 1/2 | 16 | 0.06250 | 0.0400 | 0.5000 | 0.4600 | 0.4200 | 0.1385 | 11.10 mm |
| 9/16 | 16 | 0.06250 | 0.0400 | 0.5625 | 0.5225 | 0.4825 | 0.1828 | 12.70 mm |
| 5/8 | 14 | 0.07143 | 0.0457 | 0.6250 | 0.5793 | 0.5336 | 0.2236 | 14.00 mm |
| 11/16 [a] | 14 | 0.07143 | 0.0457 | 0.6875 | 0.6418 | 0.5961 | 0.2791 | 15.50 mm |
| 3/4 | 12 | 0.08333 | 0.0534 | 0.7500 | 0.6966 | 0.6432 | 0.3249 | 16.75 mm |
| 7/8 | 11 | 0.09091 | 0.0582 | 0.8750 | 0.8168 | 0.7586 | 0.4520 | 19.75 mm |
| 1 | 10 | 0.10000 | 0.0640 | 1.0000 | 0.9360 | 0.8720 | 0.5972 | 22.75 mm |
| 1 1/8 | 9 | 0.11111 | 0.0711 | 1.1250 | 1.0539 | 0.9828 | 0.7586 | 25.50 mm |
| 1 1/4 | 9 | 0.11111 | 0.0711 | 1.2500 | 1.1789 | 1.1078 | 0.9639 | 28.50 mm |
| 1 3/8 [a] | 8 | 0.12500 | 0.0800 | 1.3750 | 1.2950 | 1.2150 | 1.1590 | 31.50 mm |
| 1 1/2 | 8 | 0.12500 | 0.0800 | 1.5000 | 1.4200 | 1.3400 | 1.4100 | 34.50 mm |
| 1 5/8 [a] | 8 | 0.12500 | 0.0800 | 1.6250 | 1.5450 | 1.4650 | 1.6860 | |
| 1 3/4 | 7 | 0.14286 | 0.0915 | 1.7500 | 1.6585 | 1.5670 | 1.9280 | |
| 2 | 7 | 0.14286 | 0.0915 | 2.0000 | 1.9085 | 1.8170 | 2.5930 | Tap drill sizes listed in this column are recommended sizes shown in BS 1157:1975 and provide from 78 to 88% of full thread. |
| 2 1/4 | 6 | 0.16667 | 0.1067 | 2.2500 | 2.1433 | 2.0366 | 3.2580 | |
| 2 1/2 | 6 | 0.16667 | 0.1067 | 2.5000 | 2.3933 | 2.2866 | 4.1060 | |
| 2 3/4 | 6 | 0.16667 | 0.1067 | 2.7500 | 2.6433 | 2.5366 | 5.0540 | |
| 3 | 5 | 0.20000 | 0.1281 | 3.0000 | 2.8719 | 2.7438 | 5.9130 | |
| 3 1/4 | 5 | 0.20000 | 0.1281 | 3.2500 | 3.1219 | 2.9938 | 7.0390 | |
| 3 1/2 | 4.5 | 0.22222 | 0.1423 | 3.5000 | 3.3577 | 3.2154 | 8.1200 | |
| 3 3/4 | 4.5 | 0.22222 | 0.1423 | 3.7500 | 3.6077 | 3.4654 | 9.4320 | |
| 4 | 4.5 | 0.22222 | 0.1423 | 4.0000 | 3.8577 | 3.7154 | 10.8400 | |
| 4 1/4 | 4 | 0.25000 | 0.1601 | 4.2500 | 4.0899 | 3.9298 | 12.1300 | |

[a] To be dispensed with wherever possible.

[b] The use of number 2 BA threads is recommended in place of 3/16-inch BSF thread, see page 2117.

## PIPE AND HOSE THREADS

The types of threads used on pipe and pipe fittings may be classed according to their intended use: 1) threads that when assembled with a sealer will produce a pressure-tight joint; 2) threads that when assembled without a sealer will produce a pressure-tight joint; 3) threads that provide free- and loose-fitting mechanical joints without pressure tightness; and 4) threads that produce rigid mechanical joints without pressure tightness.

### American National Standard Pipe Threads

American National Standard pipe threads described in the following paragraphs provide taper and straight pipe threads for use in various combinations and with certain modifications to meet these specific needs.

**Thread Designation and Notation.—**American National Standard Pipe Threads are designated by specifying in sequence the nominal size, number of threads per inch, and the symbols for the thread series and form, as: $^3/_8$—18 NPT. The symbol designations are as follows: NPT—American National Standard Taper Pipe Thread; NPTR—American National Standard Taper Pipe Thread for Railing Joints; NPSC—American National Standard Straight Pipe Thread for Couplings; NPSM—American National Standard Straight Pipe Thread for Free-fitting Mechanical Joints; NPSL—American National Standard Straight Pipe Thread for Loose-fitting Mechanical Joints with Locknuts; and NPSH—American National Standard Straight Pipe Thread for Hose Couplings.

**American National Standard Taper Pipe Threads.—**The basic dimensions of the ANSI/ASME Standard taper pipe thread are given in Table 1a.

*Form of Thread:* The angle between the sides of the thread is 60 degrees when measured in an axial plane, and the line bisecting this angle is perpendicular to the axis. The depth of the truncated thread is based on factors entering into the manufacture of cutting tools and the making of tight joints and is given by the formulas in Table 1a or the data in Table 2 obtained from these formulas. Although the Standard shows flat surfaces at the crest and root of the thread, some rounding may occur in commercial practice, and it is intended that the pipe threads of product shall be acceptable when crest and root of the tools or chasers lie within the limits shown in Table 2.

*Pitch Diameter Formulas:* In the following formulas, which apply to the American National Standard taper pipe thread, $E_0$ = pitch diameter at end of pipe; $E_1$ = pitch diameter at the large end of the internal thread and at the gaging notch; $D$ = outside diameter of pipe; $L_1$ = length of handtight or normal engagement between external and internal threads; $L_2$ = basic length of effective external taper thread; and $p$ = pitch = 1 ÷ number of threads per inch.

$$E_0 = D - (0.05D + 1.1)p$$

$$E_1 = E_0 + 0.0625L_1$$

*Thread Length:* The formula for $L_2$ determines the length of the effective thread and includes approximately two usable threads that are slightly imperfect at the crest. The normal length of engagement, $L_1$, between external and internal taper threads, when assembled by hand, is controlled by the use of the gages.

$$L_2 = (0.80D + 6.8)p$$

*Taper:* The taper of the thread is 1 in 16, or 0.75 inch per foot, measured on the diameter and along the axis. The corresponding half-angle of taper or angle with the center line is 1 degree, 47 minutes.

**Table 1a. Basic Dimensions, American National Standard Taper Pipe Threads, NPT**
*ANSI/ASME B1.20.1-2013 (R2018)*

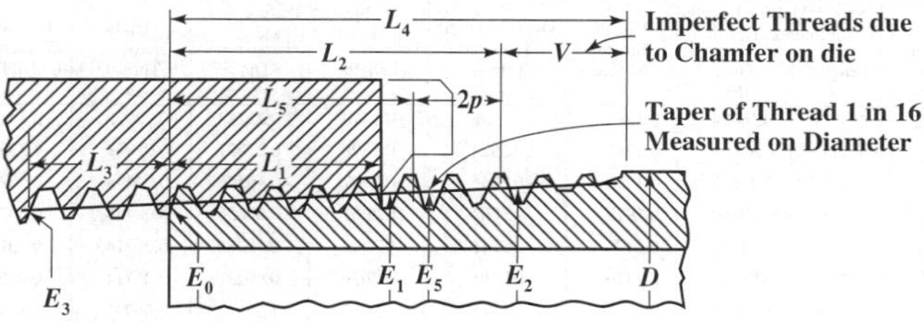


For all dimensions, see corresponding reference letter in table.

Angle between sides of thread is 60 degrees. Taper of thread, on diameter, is ¾ inch per foot. Angle of taper with center line is 1°47′.

The basic maximum thread height, $h$, of the truncated thread is 0.8 × pitch of thread. The crest and root are truncated a minimum of 0.033 × pitch for all pitches. For maximum depth of truncation, see Table 2.

| Nominal Pipe Size | Outside Dia. of Pipe, $D$ | Threads per Inch, $n$ | Pitch of Thread, $p$ | Pitch Diameter at Beginning of External Thread, $E_0$ | Handtight Engagement: Length,[a] $L_1$ (Inch) | Handtight Engagement: Dia.,[b] $E_1$ (Inch) | Effective Thread, External: Length,[c] $L_2$ (Inch) | Effective Thread, External: Dia., $E_2$ (Inch) |
|---|---|---|---|---|---|---|---|---|
| 1⁄16 | 0.3125 | 27 | 0.03704 | 0.27118 | 0.160 | 0.28118 | 0.2611 | 0.28750 |
| 1⁄8 | 0.405 | 27 | 0.03704 | 0.36351 | 0.1615 | 0.37360 | 0.2639 | 0.38000 |
| 1⁄4 | 0.540 | 18 | 0.05556 | 0.47739 | 0.2278 | 0.49163 | 0.4018 | 0.50250 |
| 3⁄8 | 0.675 | 18 | 0.05556 | 0.61201 | 0.240 | 0.62701 | 0.4078 | 0.63750 |
| 1⁄2 | 0.840 | 14 | 0.07143 | 0.75843 | 0.320 | 0.77843 | 0.5337 | 0.79179 |
| 3⁄4 | 1.050 | 14 | 0.07143 | 0.96768 | 0.339 | 0.98887 | 0.5457 | 1.00179 |
| 1 | 1.315 | 11½ | 0.08696 | 1.21363 | 0.400 | 1.23863 | 0.6828 | 1.25630 |
| 1¼ | 1.660 | 11½ | 0.08696 | 1.55713 | 0.420 | 1.58338 | 0.7068 | 1.60130 |
| 1½ | 1.900 | 11½ | 0.08696 | 1.79609 | 0.420 | 1.82234 | 0.7235 | 1.84130 |
| 2 | 2.375 | 11½ | 0.08696 | 2.26902 | 0.436 | 2.29627 | 0.7565 | 2.31630 |
| 2½ | 2.875 | 8 | 0.12500 | 2.71953 | 0.682 | 2.76216 | 1.1375 | 2.79062 |
| 3 | 3.500 | 8 | 0.12500 | 3.34062 | 0.766 | 3.38850 | 1.2000 | 3.41562 |
| 3½ | 4.000 | 8 | 0.12500 | 3.83750 | 0.821 | 3.88881 | 1.2500 | 3.91562 |
| 4 | 4.500 | 8 | 0.12500 | 4.33438 | 0.844 | 4.38712 | 1.3000 | 4.41562 |
| 5 | 5.563 | 8 | 0.12500 | 5.39073 | 0.937 | 5.44929 | 1.4063 | 5.47862 |
| 6 | 6.625 | 8 | 0.12500 | 6.44609 | 0.958 | 6.50597 | 1.5125 | 6.54062 |
| 8 | 8.625 | 8 | 0.12500 | 8.43359 | 1.063 | 8.50003 | 1.7125 | 8.54062 |
| 10 | 10.750 | 8 | 0.12500 | 10.54531 | 1.210 | 10.62094 | 1.9250 | 10.66562 |
| 12 | 12.750 | 8 | 0.12500 | 12.53281 | 1.360 | 12.61781 | 2.1250 | 12.66562 |
| 14 OD | 14.000 | 8 | 0.12500 | 13.77500 | 1.562 | 13.87262 | 2.2500 | 13.91562 |
| 16 OD | 16.000 | 8 | 0.12500 | 15.76250 | 1.812 | 15.87575 | 2.4500 | 15.91562 |
| 18 OD | 18.000 | 8 | 0.12500 | 17.75000 | 2.000 | 17.87500 | 2.6500 | 17.91562 |
| 20 OD | 20.000 | 8 | 0.12500 | 19.73750 | 2.125 | 19.87031 | 2.8500 | 19.91562 |
| 24 OD | 24.000 | 8 | 0.12500 | 23.71250 | 2.375 | 23.86094 | 3.2500 | 23.91562 |

[a] Also length of thin ring gage and length from gaging notch to small end of plug gage.
[b] Also pitch diameter at gaging notch (handtight plane).
[c] Also length of plug gage.

**Table 1b. Basic Dimensions, American National Standard Taper Pipe Threads, NPT**
*ANSI/ASME B1.20.1-2013 (R2018)*

| Nominal Pipe Size | Wrench Makeup Length for Internal Thread: Length,[c] $L_3$ | Wrench Makeup Length for Internal Thread: Dia., $E_3$ | Vanish Thread, (3.47 thds.), $V$ | Overall Length External Thread, $L_4$ | Nominal Perfect External Threads[a]: Length, $L_5$ | Nominal Perfect External Threads[a]: Dia., $E_5$ | Height of Thread, $h$ | Basic Minor Dia. at Small End of Pipe,[b] $K_0$ |
|---|---|---|---|---|---|---|---|---|
| 1⁄16 | 0.1111 | 0.26424 | 0.1285 | 0.3896 | 0.1870 | 0.28287 | 0.02963 | 0.2415 |
| 1⁄8 | 0.1111 | 0.35656 | 0.1285 | 0.3924 | 0.1898 | 0.37537 | 0.02963 | 0.3338 |
| 1⁄4 | 0.1667 | 0.46697 | 0.1928 | 0.5946 | 0.2907 | 0.49556 | 0.04444 | 0.4329 |
| 3⁄8 | 0.1667 | 0.60160 | 0.1928 | 0.6006 | 0.2967 | 0.63056 | 0.04444 | 0.5675 |
| 1⁄2 | 0.2143 | 0.74504 | 0.2479 | 0.7815 | 0.3909 | 0.78286 | 0.05714 | 0.7014 |
| 3⁄4 | 0.2143 | 0.95429 | 0.2479 | 0.7935 | 0.4029 | 0.99286 | 0.05714 | 0.9106 |
| 1 | 0.2609 | 1.19733 | 0.3017 | 0.9845 | 0.5089 | 1.24543 | 0.06957 | 1.1441 |
| 1¼ | 0.2609 | 1.54083 | 0.3017 | 1.0085 | 0.5329 | 1.59043 | 0.06957 | 1.4876 |
| 1½ | 0.2609 | 1.77978 | 0.3017 | 1.0252 | 0.5496 | 1.83043 | 0.06957 | 1.7266 |
| 2 | 0.2609 | 2.25272 | 0.3017 | 1.0582 | 0.5826 | 2.30543 | 0.06957 | 2.1995 |
| 2½ | 0.2500[d] | 2.70391 | 0.4338 | 1.5712 | 0.8875 | 2.77500 | 0.100000 | 2.6195 |
| 3 | 0.2500[d] | 3.32500 | 0.4338 | 1.6337 | 0.9500 | 3.40000 | 0.100000 | 3.2406 |
| 3½ | 0.2500 | 3.82188 | 0.4338 | 1.6837 | 1.0000 | 3.90000 | 0.100000 | 3.7374 |
| 4 | 0.2500 | 4.31875 | 0.4338 | 1.7337 | 1.0500 | 4.40000 | 0.100000 | 4.2343 |
| 5 | 0.2500 | 5.37511 | 0.4338 | 1.8400 | 1.1563 | 5.46300 | 0.100000 | 5.2907 |
| 6 | 0.2500 | 6.43047 | 0.4338 | 1.9462 | 1.2625 | 6.52500 | 0.100000 | 6.3460 |
| 8 | 0.2500 | 8.41797 | 0.4338 | 2.1462 | 1.4625 | 8.52500 | 0.100000 | 8.3335 |
| 10 | 0.2500 | 10.52969 | 0.4338 | 2.3587 | 1.6750 | 10.65000 | 0.100000 | 10.4453 |
| 12 | 0.2500 | 12.51719 | 0.4338 | 2.5587 | 1.8750 | 12.65000 | 0.100000 | 12.4328 |
| 14 OD | 0.2500 | 13.75938 | 0.4338 | 2.6837 | 2.0000 | 13.90000 | 0.100000 | 13.6749 |
| 16 OD | 0.2500 | 15.74688 | 0.4338 | 2.8837 | 2.2000 | 15.90000 | 0.100000 | 15.6624 |
| 18 OD | 0.2500 | 17.73438 | 0.4338 | 3.0837 | 2.4000 | 17.90000 | 0.100000 | 17.6499 |
| 20 OD | 0.2500 | 19.72188 | 0.4338 | 3.2837 | 2.6000 | 19.90000 | 0.100000 | 19.6374 |
| 24 OD | 0.2500 | 23.69688 | 0.4338 | 3.6837 | 3.0000 | 23.90000 | 0.100000 | 23.6124 |

[a] The length $L_5$ from the end of the pipe determines the plane beyond which the thread form is imperfect at the crest. The next two threads are perfect at the root. At this plane the cone formed by the crests of the thread intersects the cylinder forming the external surface of the pipe. $L_5 = L_2 - 2p$.

[b] Given as information for use in selecting tap drills.

[c] Three threads for 2-inch size and smaller; two threads for larger sizes.

[d] Military Specification MIL-P-7105 gives the wrench makeup as three threads for 3 in. and smaller. The $E_3$ dimensions are then as follows: Size 2½ in., 2.69609 and size 3 in., 3.31719.

All dimensions given in inches.

Increase in diameter per thread is equal to $0.0625/n$.

The basic dimensions of the American National Standard Taper Pipe Thread are given in inches to four or five decimal places. While this implies a greater degree of precision than is ordinarily attained, these dimensions are the basis of gage dimensions and are so expressed for the purpose of eliminating errors in computations.

**Engagement Between External and Internal Taper Threads.—**The normal length of engagement between external and internal taper threads when screwed together handtight is shown as $L_1$ in Table 1a. This length is controlled by the construction and use of the pipe thread gages. It is recognized that in special applications, such as flanges for high-pressure work, longer thread engagement is used, in which case the pitch diameter $E_1$ (Table 1a) is maintained and the pitch diameter $E_0$ at the end of the pipe is proportionately smaller.

**Tolerances on Thread Elements.—**The maximum allowable variation in the commercial product (manufacturing tolerance) is one turn large or small from the basic dimensions.

The permissible variations in thread elements on steel products and all pipe made of steel, wrought iron, or brass, exclusive of butt-weld pipe, are given in Table 3. This table is a guide for establishing the limits of the thread elements of taps, dies, and thread chasers. These limits may be required on product threads.

On pipe fittings and valves (not steel) for steam pressures 300 pounds and below, it is intended that plug and ring gage practice as set up in the Standard ANSI/ASME B1.20.1-2013 (R2018) will provide for a satisfactory check of accumulated variations of taper, lead, and angle in such product. Therefore, no tolerances on thread elements have been established for this class.

For service conditions where a more exact check is required, procedures have been developed by industry to supplement the regulation plug and ring method of gaging.

### Table 2. Limits on Crest and Root of American National Standard External and Internal Taper Pipe Threads, NPT *ANSI/ASME B1.20.1-2013 (R2018)*

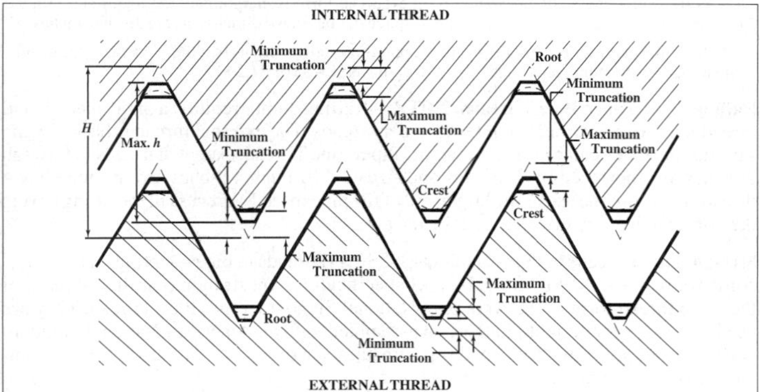


| Threads per Inch | Height of Sharp V Thread, $H$ | Height of Pipe Thread, $h$ | | Truncation, $f$ | | Width of Flat, $F$, Equivalent to Truncation | |
|---|---|---|---|---|---|---|---|
| | | Max. | Min. | Min. | Max. | Min. | Max. |
| 27 | 0.03208 | 0.02963 | 0.02496 | 0.0012 | 0.0036 | 0.0014 | 0.0041 |
| 18 | 0.04811 | 0.04444 | 0.03833 | 0.0018 | 0.0049 | 0.0021 | 0.0057 |
| 14 | 0.06186 | 0.05714 | 0.05072 | 0.0024 | 0.0056 | 0.0027 | 0.0064 |
| 11½ | 0.07531 | 0.06957 | 0.06261 | 0.0029 | 0.0063 | 0.0033 | 0.0073 |
| 8 | 0.10825 | 0.10000 | 0.09275 | 0.0041 | 0.0078 | 0.0048 | 0.0090 |

All dimensions are in inches and are given to four or five decimal places only to avoid errors in computations, not to indicate required precision.

### Table 3. Tolerances on Taper, Lead, and Angle of Pipe Threads of Steel Products and All Pipe of Steel, Wrought Iron, or Brass *ANSI/ASME B1.20.1-2013 (R2018)* (Exclusive of Butt-Weld Pipe)

| Nominal Pipe Size | Threads per Inch | Taper on Pitch Line (¾ in/ft) | | Lead in Length of Effective Threads | 60 Degree Angle of Threads, Degrees |
|---|---|---|---|---|---|
| | | Max. | Min. | | |
| 1⁄16, 1⁄8 | 27 | +1⁄8 | −1⁄16 | ±0.003 | ± 2½ |
| 1⁄4, 3⁄8 | 18 | +1⁄8 | −1⁄16 | ±0.003 | ±2 |
| 1⁄2, 3⁄4 | 14 | +1⁄8 | −1⁄16 | ±0.003[a] | ±2 |
| 1, 1¼, 1½, 2 | 11½ | +1⁄8 | −1⁄16 | ±0.003[a] | ±1½ |
| 2½ and larger | 8 | +1⁄8 | −1⁄16 | ±0.003[a] | ±1½ |

[a] The tolerance on lead shall be ±0.003 in. per inch on any size threaded to an effective thread length greater than 1 in.

For tolerances on height of thread, see Table 2.

The limits specified in this table are intended to serve as a guide for establishing limits of the thread elements of taps, dies, and thread chasers. These limits may be required on product threads.

**Table 4. Internal Threads in Pipe Couplings, NPSC for Pressure-Tight Joints with Lubricant or Sealer** *ANSI/ASME B1.20.1-2013 (R2018)*

| Nom. Pipe Size | Thds. per Inch | Minor[a] Dia. | Pitch Diameter[b] | | Nom. Pipe Size | Thds. per Inch | Minor[a] Dia. | Pitch Diameter[b] | |
|---|---|---|---|---|---|---|---|---|---|
| | | Min. | Min. | Max. | | | Min. | Min. | Max. |
| 1/8 | 27 | 0.340 | 0.3701 | 0.3771 | 1 1/2 | 11 1/2 | 1.745 | 1.8142 | 1.8305 |
| 1/4 | 18 | 0.442 | 0.4864 | 0.4968 | 2 | 11 1/2 | 2.219 | 2.2881 | 2.3044 |
| 3/8 | 18 | 0.577 | 0.6218 | 0.6322 | 2 1/2 | 8 | 2.650 | 2.7504 | 2.7739 |
| 1/2 | 14 | 0.715 | 0.7717 | 0.7851 | 3 | 8 | 3.277 | 3.3768 | 3.4002 |
| 3/4 | 14 | 0.925 | 0.9822 | 0.9956 | 3 1/2 | 8 | 3.777 | 3.8771 | 3.9005 |
| 1 | 11 1/2 | 1.161 | 1.2305 | 1.2468 | 4 | 8 | 4.275 | 4.3754 | 4.3988 |
| 1 1/4 | 11 1/2 | 1.506 | 1.5752 | 1.5915 | … | … | … | … | … |

[a] As the American National Standard Pipe Thread form is maintained, the major and minor diameters of the internal thread vary with the pitch diameter. All dimensions are given in inches.

[b] The actual pitch diameter of the straight tapped hole will be slightly smaller than the value given when gaged with a taper plug gage as called for in ANSI/ASME B1.20.1.

**Railing Joint Taper Pipe Threads, NPTR.—**Railing joints require a rigid mechanical thread joint with external and internal taper threads. The external thread is basically the same as the American National Standard Taper Pipe Thread, except that sizes 1/2 through 2 inches are shortened by 3 threads and sizes 2 1/2 through 4 inches are shortened by 4 threads to permit the use of the larger end of the pipe thread. A recess in the fitting covers the last scratch or imperfect threads on the pipe.

**Straight Pipe Threads in Pipe Couplings, NPSC.—**Threads in pipe couplings made in accordance with the ANSI/ASME B1.20.1 specifications are straight (parallel) threads of the same thread form as the American National Standard Taper Pipe Thread. They are used to form pressuretight joints when assembled with an American National Standard external taper pipe thread and made up with lubricant or sealant. These joints are recommended for comparatively low pressures only.

**Straight Pipe Threads for Mechanical Joints, NPSM, NPSL, and NPSH.—**While external and internal taper pipe threads are recommended for pipe joints in practically every service, there are mechanical joints where straight pipe threads are used to advantage. Three types covered by ANSI/ASME B1.20.1 are:

*Loose-Fitting Mechanical Joints with Locknuts (External and Internal), NPSL:* This thread is designed to produce a pipe thread having the largest diameter that it is possible to cut on standard pipe. The dimensions of these threads are given in Table 5. It will be noted that the maximum major diameter of the external thread is slightly greater than the nominal outside diameter of the pipe. The normal manufacturer's variation in pipe diameter provides for this increase.

*Loose-Fitting Mechanical Joints for Hose Couplings (External and Internal), NPSH:* Hose coupling joints are ordinarily made with straight internal and external loose-fitting threads. There are several standards of hose threads having various diameters and pitches. One of these is based on the American National Standard pipe thread and by the use of this thread series, it is possible to join small hose couplings in sizes 1/2 to 4 inches, inclusive, to ends of standard pipe having American National Standard External Pipe Threads, using a gasket to seal the joints. For the hose coupling thread dimensions see *ANSI/ASME Standard Hose Coupling Screw Threads* starting on page 2104.

*Free-Fitting Mechanical Joints for Fixtures (External and Internal), NPSM:* Standard iron, steel, and brass pipe are often used for special applications where there are no internal pressures. Where straight thread joints are required for mechanical assemblies, straight pipe threads are often found more suitable or convenient. Dimensions of these threads are given in Table 5.

**Table 5. American National Standard Straight Pipe Threads for Mechanical Joints, NPSM and NPSL** *ANSI/ASME B1.20.1-2013 (R2018)*

| Nominal Pipe Size | Threads per Inch | Allow-ance | External Thread | | | | Internal Thread | | | |
|---|---|---|---|---|---|---|---|---|---|---|
| | | | Major Diameter | | Pitch Diameter | | Minor Diameter | | Pitch Diameter | |
| | | | Max.[a] | Min. | Max. | Min. | Min.[a] | Max. | Min.[b] | Max. |
| Free-Fitting Mechanical Joints for Fixtures—NPSM | | | | | | | | | | |
| 1/8 | 27 | 0.0011 | 0.397 | 0.390 | 0.3725 | 0.3689 | 0.358 | 0.364 | 0.3736 | 0.3783 |
| 1/4 | 18 | 0.0013 | 0.526 | 0.517 | 0.4903 | 0.4859 | 0.468 | 0.481 | 0.4916 | 0.4974 |
| 3/8 | 18 | 0.0014 | 0.662 | 0.653 | 0.6256 | 0.6211 | 0.603 | 0.612 | 0.6270 | 0.6329 |
| 1/2 | 14 | 0.0015 | 0.823 | 0.813 | 0.7769 | 0.7718 | 0.747 | 0.759 | 0.7784 | 0.7851 |
| 3/4 | 14 | 0.0016 | 1.034 | 1.024 | 0.9873 | 0.9820 | 0.958 | 0.970 | 0.9889 | 0.9958 |
| 1 | 11½ | 0.0017 | 1.293 | 1.281 | 1.2369 | 1.2311 | 1.201 | 1.211 | 1.2386 | 1.2462 |
| 1¼ | 11½ | 0.0018 | 1.638 | 1.626 | 1.5816 | 1.5756 | 1.546 | 1.555 | 1.5834 | 1.5912 |
| 1½ | 11½ | 0.0018 | 1.877 | 1.865 | 1.8205 | 1.8144 | 1.785 | 1.794 | 1.8223 | 1.8302 |
| 2 | 11½ | 0.0019 | 2.351 | 2.339 | 2.2944 | 2.2882 | 2.259 | 2.268 | 2.2963 | 2.3044 |
| 2½ | 8 | 0.0022 | 2.841 | 2.826 | 2.7600 | 2.7526 | 2.708 | 2.727 | 2.7622 | 2.7720 |
| 3 | 8 | 0.0023 | 3.467 | 3.452 | 3.3862 | 3.3786 | 3.334 | 3.353 | 3.3885 | 3.3984 |
| 3½ | 8 | 0.0023 | 3.968 | 3.953 | 3.8865 | 3.8788 | 3.835 | 3.848 | 3.8888 | 3.8988 |
| 4 | 8 | 0.0023 | 4.466 | 4.451 | 4.3848 | 4.3771 | 4.333 | 4.346 | 4.3871 | 4.3971 |
| 5 | 8 | 0.0024 | 5.528 | 5.513 | 5.4469 | 5.4390 | 5.395 | 5.408 | 5.4493 | 5.4598 |
| 6 | 8 | 0.0024 | 6.585 | 6.570 | 6.5036 | 6.4955 | 6.452 | 6.464 | 6.5060 | 6.5165 |
| Loose-Fitting Mechanical Joints for Locknut Connections—NPSL | | | | | | | | | | |
| 1/8 | 27 | … | 0.409 | … | 0.3840 | 0.3805 | 0.362 | … | 0.3863 | 0.3898 |
| 1/4 | 18 | … | 0.541 | … | 0.5038 | 0.4986 | 0.470 | … | 0.5073 | 0.5125 |
| 3/8 | 18 | … | 0.678 | … | 0.6409 | 0.6357 | 0.607 | … | 0.6444 | 0.6496 |
| 1/2 | 14 | … | 0.844 | … | 0.7963 | 0.7896 | 0.753 | … | 0.8008 | 0.8075 |
| 3/4 | 14 | … | 1.054 | … | 1.0067 | 1.0000 | 0.964 | … | 1.0112 | 1.0179 |
| 1 | 11½ | … | 1.318 | … | 1.2604 | 1.2523 | 1.208 | … | 1.2658 | 1.2739 |
| 1¼ | 11½ | … | 1.663 | … | 1.6051 | 1.5970 | 1.553 | … | 1.6106 | 1.6187 |
| 1½ | 11½ | … | 1.902 | … | 1.8441 | 1.8360 | 1.792 | … | 1.8495 | 1.8576 |
| 2 | 11½ | … | 2.376 | … | 2.3180 | 2.3099 | 2.265 | … | 2.3234 | 2.3315 |
| 2½ | 8 | … | 2.877 | … | 2.7934 | 2.7817 | 2.718 | … | 2.8012 | 2.8129 |
| 3 | 8 | … | 3.503 | … | 3.4198 | 3.4081 | 3.344 | … | 3.4276 | 3.4393 |
| 3½ | 8 | … | 4.003 | … | 3.9201 | 3.9084 | 3.845 | … | 3.9279 | 3.9396 |
| 4 | 8 | … | 4.502 | … | 4.4184 | 4.4067 | 4.343 | … | 4.4262 | 4.4379 |
| 5 | 8 | … | 5.564 | … | 5.4805 | 5.4688 | 5.405 | … | 5.4884 | 5.5001 |
| 6 | 8 | … | 6.620 | … | 6.5372 | 6.5255 | 6.462 | … | 6.5450 | 6.5567 |
| 8 | 8 | … | 8.615 | … | 8.5313 | 8.5196 | 8.456 | … | 8.5391 | 8.5508 |
| 10 | 8 | … | 10.735 | … | 10.6522 | 10.6405 | 10.577 | … | 10.6600 | 10.6717 |
| 12 | 8 | … | 12.732 | … | 12.6491 | 12.6374 | 12.574 | … | 12.6569 | 12.6686 |

[a] As the American National Standard Straight Pipe Thread form of thread is maintained, the major and the minor diameters of the internal thread and the minor diameter of the external thread vary with the pitch diameter. The major diameter of the external thread is usually determined by the diameter of the pipe. These theoretical diameters result from adding the depth of the truncated thread ($0.666025 \times p$) to the maximum pitch diameters, and it should be understood that commercial pipe will not always have these maximum major diameters.

[b] This is the same as the pitch diameter at end of internal thread, $E_1$ *Basic*. (See Table 1a.)

All dimensions are given in inches.

*Notes for Free-Fitting Fixture Threads:* The minor diameters of external threads and major diameters of internal threads are those as produced by commercial straight pipe dies and commercial ground straight pipe taps.

The major diameter of the external thread has been calculated on the basis of a truncation of $0.10825p$, and the minor diameter of the internal thread has been calculated on the basis of a truncation of $0.21651p$, to provide no interference at crest and root when product is gaged with gages made in accordance with the Standard.

*Notes for Loose-Fitting Locknut Threads:* The locknut thread is established on the basis of retaining the greatest possible amount of metal thickness between the bottom of the thread and the inside of the pipe. In order that a locknut may fit loosely on the externally threaded part, an allowance equal to the "increase in pitch diameter per turn" is provided, with a tolerance of 1½ turns for both external and internal threads.

**American National Standard Dryseal Pipe Threads for Pressure-Tight Joints.**—Dryseal pipe threads are based on the US (American) pipe thread; however, they differ in that they are designed to seal pressure-tight joints without the necessity of using sealing compounds. To accomplish this, some modification of thread form and greater accuracy in manufacture is required. The roots of both the external and internal threads are truncated slightly more than the crests, i.e., roots have wider flats than crests so that metal-to-metal contact occurs at the crests and roots coincident with, or prior to, flank contact. Thus, as the threads are assembled by wrenching, the roots of the threads crush the sharper crests of the mating threads. This sealing action at both major and minor diameters tends to prevent spiral leakage and makes the joints pressure-tight without the necessity of using sealing compounds, provided that the threads are in accordance with standard specifications and tolerances and are not damaged by galling in assembly. The control of crest and root truncation is simplified by the use of properly designed threading tools. Also, it is desirable that both external and internal threads have full thread height for the length of hand engagement. Where not functionally objectionable, the use of a compatible lubricant or sealant is permissible to minimize the possibility of galling. This is desirable in assembling Dryseal pipe threads in refrigeration and other systems to effect a pressure-tight seal. The crest and root of Dryseal pipe threads may be slightly rounded, but are acceptable if they lie within the truncation limits given in Table 6.

**Table 6. American National Standard Dryseal Pipe Threads—Limits on Crest and Root Truncation** *ANSI/ASME B1.20.3-1976 (R2018)*

| Threads Per Inch | Height of Sharp V Thread (*H*) | Truncation | | | | | | | |
|---|---|---|---|---|---|---|---|---|---|
| | | Minimum | | | | Maximum | | | |
| | | At Crest | | At Root | | At Crest | | At Root | |
| | | Formula | Inch | Formula | Inch | Formula | Inch | Formula | Inch |
| 27 | 0.03208 | 0.047$p$ | 0.0017 | 0.094$p$ | 0.0035 | 0.094$p$ | 0.0035 | 0.140$p$ | 0.0052 |
| 18 | 0.04811 | 0.047$p$ | 0.0026 | 0.078$p$ | 0.0043 | 0.078$p$ | 0.0043 | 0.109$p$ | 0.0061 |
| 14 | 0.06180 | 0.036$p$ | 0.0026 | 0.060$p$ | 0.0043 | 0.060$p$ | 0.0043 | 0.085$p$ | 0.0061 |
| 11½ | 0.07531 | 0.040$p$ | 0.0035 | 0.060$p$ | 0.0052 | 0.060$p$ | 0.0052 | 0.090$p$ | 0.0078 |
| 8 | 0.10825 | 0.042$p$ | 0.0052 | 0.055$p$ | 0.0069 | 0.055$p$ | 0.0069 | 0.076$p$ | 0.0095 |

All dimensions are given in inches. In the formulas, $p$ = pitch.

**Types of Dryseal Pipe Thread.**—American National Standard ANSI/ASME B1.20.3-1976 (R2018) covers four types of standard Dryseal pipe threads:

NPTF, Dryseal US (American) Standard Taper Pipe Thread
PTF-SAE SHORT, Dryseal SAE Short Taper Pipe Thread
NPSF, Dryseal US (American) Standard Fuel Internal Straight Pipe Thread
NPSI, Dryseal US (American) Standard Intermediate Internal Straight Pipe Thread

**Table 7. Recommended Limitation of Assembly among the Various Types of Dryseal Threads**

| External Dryseal Thread | | For Assembly with Internal Dryseal Thread | |
|---|---|---|---|
| Type | Description | Type | Description |
| 1 | NPTF (tapered), ext thd | 1 | NPTF (tapered), int thd |
| | | 2[a,b] | PTF-SAE SHORT (tapered), int thd |
| | | 3[a,c] | NPSF (straight), int thd |
| | | 4[a,c,d] | NPSI (straight), int thd |
| 2[a,e] | PTF-SAE SHORT (tapered) ext thd | 4 | NPSI (straight), int thd |
| | | 1 | NPTF (tapered), int thd |

[a] Pressure-tight joints without the use of a sealant can best be ensured where both components are threaded with NPTF (full length threads), since theoretically interference (sealing) occurs at all threads, but there are two less threads engaged than for NPTF assemblies. When straight internal threads are used, there is interference only at one thread depending on ductility of materials.

[b] PTF-SAE SHORT internal threads are primarily intended for assembly with type 1-NPTF external threads. They are not designed for, and at extreme tolerance limits may not assemble with, type 2- PTF-SAE SHORT external threads.

[c] There is no external straight Dryseal thread.

[d] NPSI internal threads are primarily intended for assembly with type 2-PTF-SAE SHORT external threads but will also assemble with full length type 1 NPTF external threads.

[e] PTF-SAE SHORT external threads are primarily intended for assembly with type 4-NPSI internal threads but can also be used with type 1-NPTF internal threads. They are not designed for, and at extreme tolerance limits may not assemble with, type 2-PTF-SAE SHORT internal threads or type 3-NPSF internal threads.

An assembly with straight internal pipe threads and taper external pipe threads is frequently more advantageous than an all taper thread assembly, particularly in automotive and other allied industries where economy and rapid production are major considerations. Dryseal threads are not used in assemblies in which both components have straight pipe threads.

*NPTF Threads:* This type applies to both external and internal threads and is suitable for pipe joints in practically every type of service. Of all Dryseal pipe threads, NPTF external and internal threads mated are generally conceded to be superior for strength and seal since they have the longest length of thread and, theoretically, interference (sealing) occurs at every engaged thread root and crest. Use of tapered internal threads, such as NPTF or PTF-SAE SHORT in hard or brittle materials having thin sections will minimize the possibility of fracture.

There are two classes of NTPF threads. Class 1 threads are made to interfere (seal) at root and crest when mated, but inspection of crest and root truncation is not required. Consequently, Class 1 threads are intended for applications where close control of tooling is required for conformance of truncation or where sealing is accomplished by means of a sealant applied to the threads.

Class 2 threads are theoretically identical to those made to Class 1. However, inspection of root and crest truncation is required. Consequently, where a sealant is not used, there is more assurance of a pressure-tight seal for Class 2 threads than for Class 1 threads.

*PTF-SAE SHORT Threads:* External threads of this type conform in all respects with NPTF threads except that the thread length has been shortened by eliminating one thread from the small (entering) end. These threads are designed for applications where clearance is not sufficient for the full length of the NPTF threads or for economy of material where the full thread length is not necessary.

Internal threads of this type conform in all respects with NPTF threads, except that the thread length has been shortened by eliminating one thread from the large (entry) end. These threads are designed for thin materials where thickness is not sufficient for the full thread length of the NPTF threads or for economy in tapping where the full thread length is not necessary.

Pressure-tight joints without the use of lubricant or sealer can best be ensured where mating components are both threaded with NPTF threads. This should be considered before specifying PTF-SAE SHORT external or internal threads.

*NPSF Threads:* Threads of this type are straight (cylindrical) instead of tapered and are internal only. They are more economical to produce than tapered internal threads, but when assembled do not offer as strong a guarantee of sealing since root and crest interference will not occur for all threads. NPSF threads are generally used with soft or ductile materials which will tend to adjust at assembly to the taper of external threads, but may be used in hard or brittle materials where the section is thick.

*NPSI Threads:* Threads of this type are straight (cylindrical) instead of tapered, are internal only and are slightly larger in diameter than NPSF threads but have the same tolerance and thread length. They are more economical to produce than tapered threads and may be used in hard or brittle materials where the section is thick or where there is little expansion at assembly with external taper threads. As with NPSF threads, NPSI threads when assembled do not offer as strong a guarantee of sealing as do tapered internal threads.

For more complete specifications for production and acceptance of Dryseal pipe threads, see ANSI/ASME B1.20.3 (Inch) and ANSI/ASME B1.20.4 (Withdrawn) (Metric Translation), and for gaging and inspection, see ANSI/ASME B1.20.5 (Inch) and ANSI/ASME B1.20.6M (Withdrawn) (Metric Translation).

*Designation of Dryseal Pipe Threads:* The standard Dryseal pipe threads are designated by specifying in sequence nominal size, thread series symbol, and class:

*Examples:* 1/8-27 NPTF-1; 1/8-27 PTF-SAE SHORT; and 3/8-18 NPTF-1 AFTER PLATING.

**Table 8. Suggested Tap Drill Sizes for Internal Dryseal Pipe Threads**

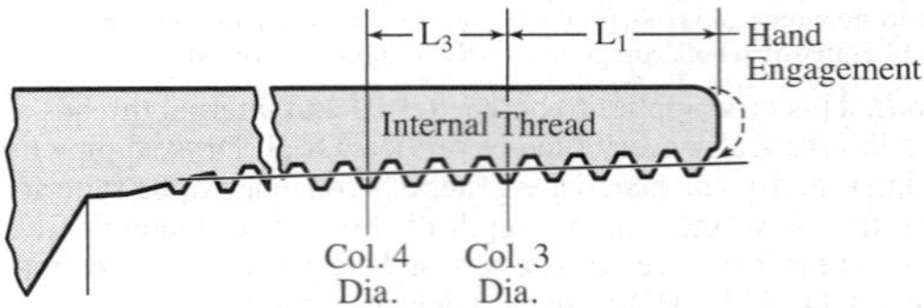


| | | Taper Pipe Thread | | | | Straight Pipe Thread | | |
|---|---|---|---|---|---|---|---|---|
| | | Minor Diameter At Distance | | Drill Size[a] | | Minor Diameter | | |
| Size | Probable Drill Oversize Cut (Mean) | $L_1$ From Large End | $L_1 + L_3$ From Large End | Without Reamer | With Reamer | NPSF | NPSI | Drill Size[a] |
| 1/16-27 | 0.0038 | 0.2443 | 0.2374 | "C" (0.242) | "A" (0.234) | 0.2482 | 0.2505 | "D" (0.246) |
| 1/8-27 | 0.0044 | 0.3367 | 0.3298 | "Q" (0.332) | 21/64 (0.328) | 0.3406 | 0.3429 | "R" (0.339) |
| 1/4-18 | 0.0047 | 0.4362 | 0.4258 | 7/16 (0.438) | 27/64 (0.422) | 0.4422 | 0.4457 | 7/16 (0.438) |
| 3/8-18 | 0.0049 | 0.5708 | 0.5604 | 9/16 (0.562) | 9/16 (0.563) | 0.5776 | 0.5811 | 37/64 (0.578) |
| 1/2-14 | 0.0051 | 0.7034 | 0.6901 | 45/64 (0.703) | 11/16 (0.688) | 0.7133 | 0.7180 | 45/64 (0.703) |
| 3/4-14 | 0.0060 | 0.9127 | 0.8993 | 29/32 (0.906) | 57/64 (0.891) | 0.9238 | 0.9283 | 59/64 (0.922) |
| 1-11½ | 0.0080 | 1.1470 | 1.1307 | 1 9/64 (1.141) | 1 1/8 (1.125) | 1.1600 | 1.1655 | 1 5/32 (1.156) |
| 1¼-11½ | 0.0100 | 1.4905 | 1.4742 | 1 31/64 (1.484) | 1 15/32 (1.469) | … | … | … |
| 1½-11½ | 0.0120 | 1.7295 | 1.7132 | 1 23/32 (1.719) | 1 45/64 (1.703) | … | … | … |
| 2-11½ | 0.0160 | 2.2024 | 2.1861 | 2 3/16 (2.188) | 2 11/64 (2.172) | … | … | … |
| 2½-8 | 0.0180 | 2.6234 | 2.6000 | 2 39/64 (2.609) | 2 37/64 (2.578) | … | … | … |
| 3-8 | 0.0200 | 3.2445 | 3.2211 | 3 15/64 (3.234) | 3 13/64 (3.203) | … | … | … |

[a] Some drill sizes listed may not be standard drills.
All dimensions are given in inches.

**Special Dryseal Threads.—**Where design limitations, economy of material, permanent installation, or other limiting conditions prevail, consideration may be given to using a special Dryseal thread series.

*Dryseal Special Short Taper Pipe Thread, PTF-SPL SHORT:* Threads of this series conform in all respects to PTF-SAE SHORT threads except that the full thread length has been further shortened by eliminating one thread at the small end of internal threads or one thread at the large end of external threads.

*Dryseal Special Extra Short Taper Pipe Thread, PTF-SPL EXTRA SHORT:* Threads of this series conform in all respects to PTF-SAE SHORT threads except that the full thread length has been further shortened by eliminating two threads at the small end of internal threads or two threads at the large end of external threads.

*Limitations of Assembly:* Table 9 applies where Dryseal Special Short or Extra Short Taper Pipe Threads are to be assembled as special combinations.

**Table 9. Assembly Limitations for Special Combinations of Dryseal Threads**

| Thread | May Assemble with[a] | May Assemble with[b] |
|---|---|---|
| PTF SPL SHORT EXTERNAL<br>PTF SPL EXTRA SHORT EXTERNAL | PTF-SAE SHORT INTERNAL<br>NPSF INTERNAL<br>PTF SPL SHORT INTERNAL<br>PTF SPL EXTRA SHORT INTERNAL | NPTF or NPSI INTERNAL |
| PTF SPL SHORT INTERNAL<br>PTF SPL EXTRA SHORT INTERNAL | PTF-SAE SHORT EXTERNAL | NPTF EXTERNAL |

[a] Only when the external thread or the internal thread or both are held closer than the standard tolerance, the external thread toward the minimum and the internal thread toward the maximum pitch diameter to provide a minimum of one turn hand engagement. At extreme tolerance limits the shortened full-thread lengths reduce hand engagement and the threads may not start to assemble.

[b] Only when the internal thread or the external thread or both are held closer than the standard tolerance, the internal thread toward the minimum and the external thread toward the maximum pitch diameter to provide a minimum of two turns for wrench make-up and sealing. At extreme tolerance limits the shortened full-thread lengths reduce wrench make-up and the threads may not seal.

*Dryseal Fine Taper Thread Series, F-PTF:* The need for finer pitches for nominal pipe sizes has brought into use applications of 27 threads per inch to $\frac{1}{4}$- and $\frac{3}{8}$-inch pipe sizes. There may be other needs that require finer pitches for larger pipe sizes. It is recommended that the existing threads per inch be applied to the next larger pipe size for a fine thread series, thus: $\frac{1}{4}$-27, $\frac{3}{8}$-27, $\frac{1}{2}$-18, $\frac{3}{4}$-18, 1-14, 1$\frac{1}{4}$-14, 1$\frac{1}{2}$-14, and 2-14. This series applies to external and internal threads of full length and is suitable for applications where threads finer than NPTF are required.

*Dryseal Special Diameter-Pitch Combination Series, SPL-PTF:* Other applications of diameter-pitch combinations have come into use where taper pipe threads are applied to nominal size thin wall tubing. These combinations are: $\frac{1}{2}$-27, $\frac{5}{8}$-27, $\frac{3}{4}$-27, $\frac{7}{8}$-27, and 1-27. This series applies to external and internal threads of full length and is applicable to thin wall nominal diameter outside tubing.

*Designation of Special Dryseal Pipe Threads:* The designations used for these special dryseal pipe threads are as follows:

$\frac{1}{8}$-27 PTF-SPL SHORT

$\frac{1}{8}$-27 PTF-SPL EXTRA SHORT

$\frac{1}{2}$-27 SPL PTF, OD 0.500

Note that in the last designation the outside diameter of tubing is given.

## British Standard Pipe Threads

**British Standard Pipe Threads for Non-Pressure-Tight Joints.—**The threads in BS EN ISO 228:2003, "Pipe Threads where Pressure-tight Joints are not Made on the Threads" Part 1: Dimensions, Tolerances and Designation and Part 2: Verification by Means of Limit Gauges, are parallel fastening threads that are generally used for fastening purposes such as the mechanical assembly of component parts of fittings, cocks and valves. They are not suitable where pressure-tight joints are made on the threads.

The crests of the basic Whitworth thread form may be truncated to certain limits of size given in the Standard except on internal threads, when they are likely to be assembled with external threads conforming to the requirements of BS 21:1973 (Withdrawn), "Specification for pipe threads for tubes and fittings where pressure-tight joints are made on the threads" (see page 2102).

For external threads two classes of tolerance are provided and for internal, one class. The two classes of tolerance for external threads are Class A and Class B. For economy of manufacture the class B fit should be chosen whenever possible. The class A is reserved for those applications where the closer tolerance is essential. Class A tolerance is an entirely negative value, equivalent to the internal thread tolerance. Class B tolerance is an entirely negative value twice that of class A tolerance. Tables showing limits and dimensions are given in the Standard.

The thread series specified in this Standard shall be designated by the letter "G". A typical reference on a drawing might be "G½", for internal thread; "G½ A", for external thread, class A: and "G ½ B", for external thread, class B. Where no class reference is stated for external threads, that of class B will be assumed. The designation of truncated threads shall have the addition of the letter "T" to the designation, i.e., G ½ T and G ½ BT.

**British Standard Pipe Threads (Non-Pressure-Tight Joints)**
**Metric and Inch Basic Sizes** *BS EN ISO 228-1:2003*

| Nominal Size, Inches | Threads per Inch[a] | Depth of Thread | Major Diameter | Pitch Diameter | Minor Diameter | Nominal Size, Inches | Threads per Inch[a] | Depth of Thread | Major Diameter | Pitch Diameter | Minor Diameter |
|---|---|---|---|---|---|---|---|---|---|---|---|
| 1⁄16 | 28 | 0.581<br>*0.0229* | 7.723<br>*0.3041* | 7.142<br>*0.2812* | 6.561<br>*0.2583* | 1¾ | 11 | 1.479<br>*0.0582* | 53.746<br>*2.1160* | 52.267<br>*2.0578* | 50.788<br>*1.9996* |
| ⅛ | 28 | 0.581<br>*0.0229* | 9.728<br>*0.3830* | 9.147<br>*0.3601* | 8.566<br>*0.3372* | 2 | 11 | 1.479<br>*0.0582* | 59.614<br>*2.3470* | 58.135<br>*2.2888* | 56.656<br>*2.2306* |
| ¼ | 19 | 0.856<br>*0.0337* | 13.157<br>*0.5180* | 12.301<br>*0.4843* | 11.445<br>*0.4506* | 2¼ | 11 | 1.479<br>*0.0582* | 65.710<br>*2.5870* | 64.231<br>*2.5288* | 62.752<br>*2.4706* |
| ⅜ | 19 | 0.856<br>*0.0337* | 16.662<br>*0.6560* | 15.806<br>*0.6223* | 14.950<br>*0.5886* | 2½ | 11 | 1.479<br>*0.0582* | 75.184<br>*2.9600* | 73.705<br>*2.9018* | 72.226<br>*2.8436* |
| ½ | 14 | 1.162<br>*0.0457* | 20.955<br>*0.8250* | 19.793<br>*0.7793* | 18.631<br>*0.7336* | 2¾ | 11 | 1.479<br>*0.0582* | 81.534<br>*3.2100* | 80.055<br>*3.1518* | 78.576<br>*3.0936* |
| ⅝ | 14 | 1.162<br>*0.0457* | 22.911<br>*0.9020* | 21.749<br>*0.8563* | 20.587<br>*0.8106* | 3 | 11 | 1.479<br>*0.0582* | 87.884<br>*3.4600* | 86.405<br>*3.4018* | 84.926<br>*3.3436* |
| ¾ | 14 | 1.162<br>*0.0457* | 26.441<br>*1.0410* | 25.279<br>*0.9953* | 24.117<br>*0.9496* | 3½ | 11 | 1.479<br>*0.0582* | 100.330<br>*3.9500* | 98.851<br>*3.8918* | 97.372<br>*3.8336* |
| ⅞ | 14 | 1.162<br>*0.0457* | 30.201<br>*1.1890* | 29.039<br>*1.1433* | 27.877<br>*1.0976* | 4 | 11 | 1.479<br>*0.0582* | 113.030<br>*4.4500* | 111.551<br>*4.3918* | 110.072<br>*4.3336* |
| 1 | 11 | 1.479<br>*0.0582* | 33.249<br>*1.3090* | 31.770<br>*1.2508* | 30.291<br>*1.1926* | 4½ | 11 | 1.479<br>*0.0582* | 125.730<br>*4.9500* | 124.251<br>*4.8918* | 122.772<br>*4.8336* |
| 1⅛ | 11 | 1.479<br>*0.0582* | 37.897<br>*1.4920* | 36.418<br>*1.4338* | 34.939<br>*1.3756* | 5 | 11 | 1.479<br>*0.0582* | 138.430<br>*5.4500* | 136.951<br>*5.3918* | 135.472<br>*5.3336* |
| 1¼ | 11 | 1.479<br>*0.0582* | 41.910<br>*1.6500* | 40.431<br>*1.5918* | 38.952<br>*1.5336* | 5½ | 11 | 1.479<br>*0.0582* | 151.130<br>*5.9500* | 149.651<br>*5.8918* | 148.172<br>*5.8336* |
| 1½ | 11 | 1.479<br>*0.0582* | 47.803<br>*1.8820* | 46.324<br>*1.8238* | 44.845<br>*1.7656* | 6 | 11 | 1.479<br>*0.0582* | 163.830<br>*6.4500* | 162.351<br>*6.3918* | 160.872<br>*6.3336* |

[a] The thread pitches in millimeters are as follows: 0.907 for 28 threads per inch. 1.337 for 19 threads per inch, 1.814 for 14 threads per inch, and 2.309 for 11 threads per inch.

Each basic metric dimension is given in roman figures (nominal sizes excepted) and each basic inch dimension is shown in italics directly beneath it.

**British Standard Pipe Threads for Pressure-Tight Joints.**—The threads in BS 21:1973 (Withdrawn), "Specification for pipe threads for tubes and fittings where pressure-tight joints are made on the threads," are based on the Whitworth thread form and are specified as:

1) *Jointing threads:* These relate to pipe threads for joints made pressure-tight by the mating of the threads; they include taper external threads for assembly with either taper or parallel internal threads (parallel external pipe threads are not suitable as jointing threads).

2) *Longscrew threads:* These relate to parallel external pipe threads used for longscrews (connectors) specified in BS 1387:1985 (Withdrawn), where a pressure-tight joint is achieved by the compression of a soft material onto the surface of the external thread by tightening a back nut against a socket.

## British Standard External and Internal Pipe Threads (Pressure-Tight Joints) Metric and Inch Dimensions and Limits of Size *BS 21:1973 (Withdrawn)*

| Nominal Size | No. of Threads per Inch[a] | | Basic Diameters at Gage Plane | | | Gage Length | | Number of Useful Threads on Pipe for Basic Gage Length[b] | Tolerance + and − | |
|---|---|---|---|---|---|---|---|---|---|---|
| | | | Major | Pitch | Minor | Basic | Toler-ance (+ and −) | | Gage Plane to Face of Int. Taper Thread | On Diameter of Parallel Int. Threads |
| 1/16 | 28 | { | 7.723 | 7.142 | 6.561 | (4 3/8) | (1) | (7 1/8) | (1 1/4) | 0.071 |
| | | | *0.304* | *0.2812* | *0.2583* | 4.0 | 0.9 | 6.5 | 1.1 | *0.0028* |
| 1/8 | 28 | { | 9.728 | 9.147 | 8.566 | (4 3/8) | (1) | (7 1/8) | (1 1/4) | 0.071 |
| | | | *0.383* | *0.3601* | *0.3372* | 4.0 | 0.9 | 6.5 | 1.1 | *0.0028* |
| 1/4 | 19 | { | 13.157 | 12.301 | 11.445 | (4 1/2) | (1) | (7 1/4) | (1 1/4) | 0.104 |
| | | | *0.518* | *0.4843* | *0.4506* | 6.0 | 1.3 | 9.7 | 1.7 | *0.0041* |
| 3/8 | 19 | { | 16.662 | 15.806 | 14.950 | (4 3/4) | (1) | (7 1/2) | (1 1/4) | 0.104 |
| | | | 0.656 | 0.6223 | 0.5886 | 6.4 | 1.3 | 10.1 | 1.7 | 0.0041 |
| 1/2 | 14 | { | 20.955 | 19.793 | 18.631 | (4 1/2) | (1) | (7 1/4) | (1 1/4) | 0.142 |
| | | | *0.825* | *0.7793* | *0.7336* | 8.2 | 1.8 | 13.2 | 2.3 | *0.0056* |
| 3/4 | 14 | { | 26.441 | 25.279 | 24.117 | (5 1/4) | (1) | (8) | (1 1/4) | 0.142 |
| | | | *1.041* | *0.9953* | *0.9496* | 9.5 | 1.8 | 14.5 | 2.3 | *0.0056* |
| 1 | 11 | { | 33.249 | 31.770 | 30.291 | (4 1/2) | (1) | (7 1/4) | (1 1/4) | 0.180 |
| | | | *1.309* | *1.2508* | *1.1926* | 10.4 | 2.3 | 16.8 | 2.9 | *0.0071* |
| 1 1/4 | 11 | { | 41.910 | 40.431 | 38.952 | (5 1/2) | (1) | (8 1/4) | (1 1/4) | 0.180 |
| | | | *1.650* | *1.5918* | *1.5336* | 12.7 | 2.3 | 19.1 | 2.9 | *0.0071* |
| 1 1/2 | 11 | { | 47.803 | 46.324 | 44.845 | (5 1/2) | (1) | (8 1/4) | (1 1/4) | 0.180 |
| | | | *1.882* | *1.8238* | *1.7656* | 12.7 | 2.3 | 19.1 | 2.9 | *0.0071* |
| 2 | 11 | { | 59.614 | 58.135 | 56.656 | (6 7/8) | (1) | (10 1/8) | (1 1/4) | 0.180 |
| | | | *2.347* | *2.2888* | *2.2306* | 15.9 | 2.3 | 23.4 | 2.9 | *0.0071* |
| 2 1/2 | 11 | { | 75.184 | 73.705 | 72.226 | (7 9/16) | (1 1/2) | (11 9/16) | (1 1/2) | 0.216 |
| | | | *2.960* | *2.9018* | *2.8436* | 17.5 | 3.5 | 26.7 | 3.5 | *0.0085* |
| 3 | 11 | { | 87.884 | 86.405 | 84.926 | (8 15/16) | (1 1/2) | (12 15/16) | (1 1/2) | 0.216 |
| | | | *3.460* | *3.4018* | *3.3436* | 20.6 | 3.5 | 29.8 | 3.5 | *0.0085* |
| 4 | 11 | { | 113.030 | 111.551 | 110.072 | (11) | (1 1/2) | (15 1/2) | (1 1/2) | 0.216 |
| | | | *4.450* | *4.3918* | *4.3336* | 25.4 | 3.5 | 35.8 | 3.5 | *0.0085* |
| 5 | 11 | { | 138.430 | 136.951 | 135.472 | (12 3/8) | (1 1/2) | (17 3/8) | (1 1/2) | 0.216 |
| | | | *5.450* | *5.3918* | *5.3336* | 28.6 | 3.5 | 40.1 | 3.5 | *0.0085* |
| 6 | 11 | { | 163.830 | 162.351 | 160.872 | (12 3/8) | (1 1/2) | (17 3/8) | (1 1/2) | 0.216 |
| | | | *6.450* | *6.3918* | *6.3336* | 28.6 | 3.5 | 40.1 | 3.5 | *0.0085* |

[a] In the British Standard BS 21:1973 (Withdrawn) the thread pitches in millimeters are as follows: 0.907 for 28 threads per inch, 1.337 for 19 threads per inch, 1.814 for 14 threads per inch, and 2.309 for 11 threads per inch.

[b] This is the minimum number of useful threads on the pipe for the basic gage length; for the maximum and minimum gage lengths, the minimum numbers of useful threads are, respectively, greater and less by the amount of tolerance in the column to the left. The design of internally threaded parts shall make allowance for receiving pipe ends of up to the minimum number of useful threads corresponding to the maximum gage length; the minimum number of useful *internal* threads shall be no less than 80 percent of the minimum number of useful external threads for the minimum gage length.

Each basic metric dimension is given in roman figures (nominal sizes excepted) and each basic inch dimension is shown in italics directly beneath it. Figures in ( ) are numbers of turns of thread with metric linear equivalents given beneath. Taper of taper thread is 1 in 16 on diameter.

## Hose Coupling Screw Threads

**ANSI/ASME Standard Hose Coupling Screw Threads.—**Threads for hose couplings, valves, and all other fittings used in direct connection with hose intended for domestic, industrial, and general service in sizes $\frac{1}{2}$, $\frac{5}{8}$, $\frac{3}{4}$, 1, $1\frac{1}{4}$, $1\frac{1}{2}$, 2, $2\frac{1}{2}$, 3, $3\frac{1}{2}$, and 4 inches are covered by ANSI/ASME B1.20.7-1991 (R2018). These threads are designated as follows:

NH— Standard hose coupling threads of full form as produced by cutting or rolling.

NHR— Standard hose coupling threads for garden hose applications where the design utilizes thin walled material which is formed to the desired thread.

NPSH— Standard straight hose coupling thread series in sizes $\frac{1}{2}$ to 4 inches for joining to American National Standard taper pipe threads using a gasket to seal the joint.

Thread dimensions are given in Table 1 and thread lengths in Table 2.

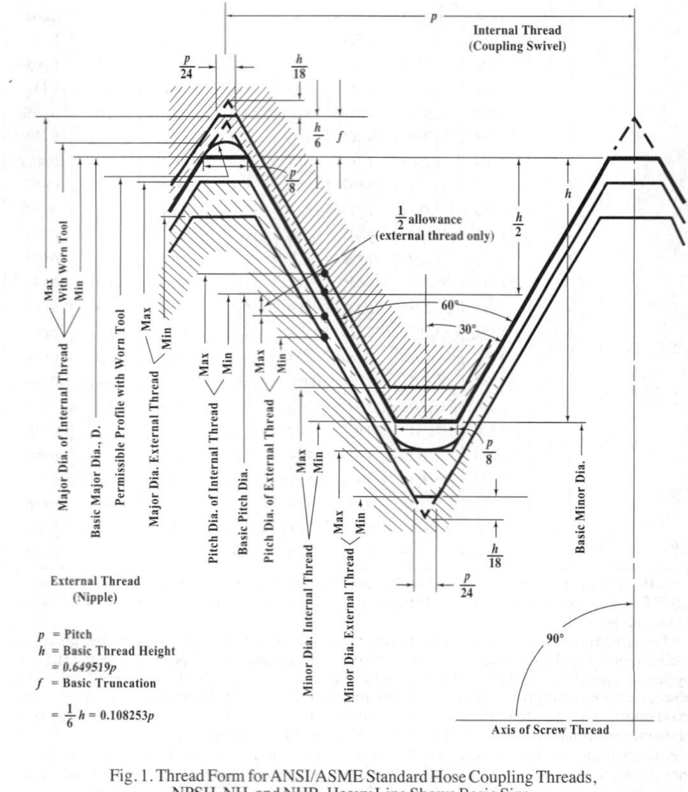


Fig. 1. Thread Form for ANSI/ASME Standard Hose Coupling Threads, NPSH, NH, and NHR. Heavy Line Shows Basic Size.

**Table 1. ANSI/ASME Standard Hose Coupling Threads for NPSH, NH, and NHR Nipples and Coupling Swivels**
*ANSI/ASME B1.20.7-1991 (R2018)*

| Nominal Size of Hose | Threads per Inch | Thread Designation | Pitch | Basic Height of Thread | Nipple (External) Thread | | | | | Coupling (Internal) Thread | | | | |
|---|---|---|---|---|---|---|---|---|---|---|---|---|---|---|
| | | | | | Major Dia. | | Pitch Dia. | | Minor Dia. | Minor Dia. | | Pitch Dia. | | Major Dia. |
| | | | | | Max. | Min. | Max. | Min. | Max. | Min. | Max. | Min. | Max. | Min. |
| $\frac{1}{2}$, $\frac{5}{8}$, $\frac{3}{4}$ | 11.5 | .75-11.5NH | .08696 | .05648 | 1.0625 | 1.0455 | 1.0060 | 0.9975 | 0.9495 | 0.9595 | 0.9765 | 1.0160 | 1.0245 | 1.0725 |
| $\frac{1}{2}$, $\frac{5}{8}$, $\frac{3}{4}$ | 11.5 | .75-11.5NHR | .08696 | .05648 | 1.0520 | 1.0350 | 1.0100 | 0.9930 | 0.9495 | 0.9720 | 0.9930 | 1.0160 | 1.0280 | 1.0680 |
| $\frac{1}{2}$ | 14 | .5-14NPSH | .07143 | .04639 | 0.8248 | 0.8108 | 0.7784 | 0.7714 | 0.7320 | 0.7395 | 0.7535 | 0.7859 | 0.7929 | 0.8323 |
| $\frac{3}{4}$ | 14 | .75-14NPSH | .07143 | .04639 | 1.0353 | 1.0213 | 0.9889 | 0.9819 | 0.9425 | 0.9500 | 0.9640 | 0.9964 | 1.0034 | 1.0428 |
| 1 | 11.5 | 1-11.5NPSH | .08696 | .05648 | 1.2951 | 1.2781 | 1.2396 | 1.2301 | 1.1821 | 1.1921 | 1.2091 | 1.2486 | 1.2571 | 1.3051 |
| $1\frac{1}{4}$ | 11.5 | 1.25-11.5NPSH | .08696 | .05648 | 1.6399 | 1.6229 | 1.5834 | 1.5749 | 1.5269 | 1.5369 | 1.5539 | 1.5934 | 1.6019 | 1.6499 |
| $1\frac{1}{2}$ | 11.5 | 1.5-11.5 NPSH | .08696 | .05648 | 1.8788 | 1.8618 | 1.8223 | 1.8138 | 1.7658 | 1.7758 | 1.7928 | 1.8323 | 1.8408 | 1.8888 |
| 2 | 11.5 | 2-11.5NPSH | .08696 | .05648 | 2.3528 | 2.3358 | 2.2963 | 2.2878 | 2.2398 | 2.2498 | 2.2668 | 2.3063 | 2.3148 | 2.3628 |
| $2\frac{1}{2}$ | 8 | 2.5-8NPSH | .12500 | .08119 | 2.8434 | 2.8212 | 2.7622 | 2.7511 | 2.6810 | 2.6930 | 2.7152 | 2.7742 | 2.7853 | 2.8554 |
| 3 | 8 | 3-8NPSH | .12500 | .08119 | 3.4697 | 3.4475 | 3.3885 | 3.3774 | 3.3073 | 3.3193 | 3.3415 | 3.4005 | 3.4116 | 3.4817 |
| $3\frac{1}{2}$ | 8 | 3.5-8NPSH | .12500 | .08119 | 3.9700 | 3.9478 | 3.8888 | 3.8777 | 3.8076 | 3.8196 | 3.8418 | 3.9008 | 3.9119 | 3.9820 |
| 4 | 8 | 4-8NPSH | .12500 | .08119 | 4.4683 | 4.4461 | 4.3871 | 4.3760 | 4.3059 | 4.3179 | 4.3401 | 4.3991 | 4.4102 | 4.4803 |
| 4 | 6 | 4-6NH (SPL) | .16667 | .10825 | 4.9082 | 4.8722 | 4.7999 | 4.7819 | 4.6916 | 4.7117 | 4.7477 | 4.8200 | 4.8380 | 4.9283 |

All dimensions are given in inches.

Dimensions given for the maximum minor diameter of the nipple are figured to the intersection of the worn tool arc with a centerline through crest and root. The minimum minor diameter of the nipple shall be that corresponding to a flat at the minor diameter of the minimum nipple equal to $\frac{1}{24}p$, and may be determined by subtracting 0.7939$p$ from the minimum pitch diameter of the nipple. (See Fig. 1)

Dimensions given for the minimum major diameter of the coupling correspond to the basic flat, $\frac{1}{8}p$, and the profile at the major diameter produced by a worn tool must not fall below the basic outline. The maximum major diameter of the coupling shall be that corresponding to a flat at the major diameter of the maximum coupling equal to $\frac{1}{24}p$ and may be determined by adding 0.7939$p$ to the maximum pitch diameter of the coupling. (See Fig. 1)

NH and NHR threads are used for garden hose applications. NPSH threads are used for steam, air and all other hose connections to be made up with standard pipe threads. NH (SPL) threads are used for marine applications.

**Table 2. ANSI/ASME Standard Hose Coupling Screw Thread Lengths**
*ANSI/ASME B1.20.7-1991 (R2018)*

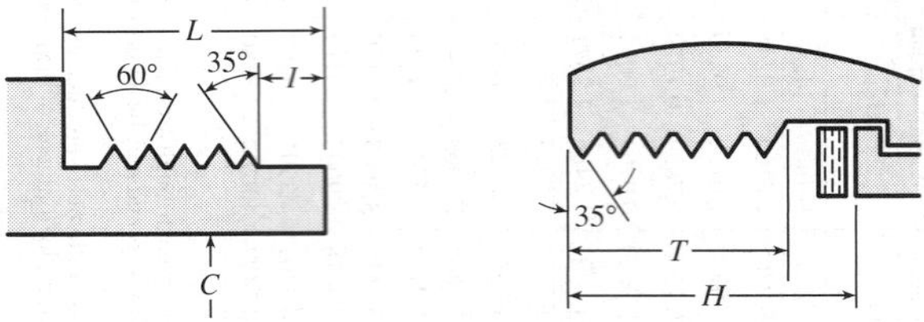


| Nominal Size of Hose | Threads per Inch | I.D. of Nipple, $C$ | Approx. O.D. of Ext. Thd. | Length of Nipple, $L$ | Length of Pilot, $I$ | Depth of Coupl., $H$ | Coupl. Thd. Length, $T$ | Approx. No. Thds. in Length $T$ |
|---|---|---|---|---|---|---|---|---|
| 1/2, 5/8, 3/4 | 11.5 | 25/32 | 1 1/16 | 9/16 | 1/8 | 17/32 | 3/8 | 4 1/4 |
| 1/2, 5/8, 3/4 | 11.5 | 25/32 | 1 1/16 | 9/16 | 1/8 | 17/32 | 3/8 | 4 1/4 |
| 1/2 | 14 | 17/32 | 13/16 | 1/2 | 1/8 | 15/32 | 5/16 | 4 1/4 |
| 3/4 | 14 | 25/32 | 1 1/32 | 9/16 | 1/8 | 17/32 | 3/8 | 5 1/4 |
| 1 | 11.5 | 1 1/32 | 1 9/32 | 9/16 | 5/32 | 17/32 | 3/8 | 4 1/4 |
| 1 1/4 | 11.5 | 1 9/32 | 1 5/8 | 5/8 | 5/32 | 19/32 | 15/32 | 5 1/2 |
| 1 1/2 | 11.5 | 1 17/32 | 1 7/8 | 5/8 | 5/32 | 19/32 | 15/32 | 5 1/2 |
| 2 | 11.5 | 2 1/32 | 2 11/32 | 3/4 | 3/16 | 23/32 | 19/32 | 6 3/4 |
| 2 1/2 | 8 | 2 17/32 | 2 27/32 | 1 | 1/4 | 15/16 | 11/16 | 5 1/2 |
| 3 | 8 | 3 1/32 | 3 15/32 | 1 1/8 | 1/4 | 1 1/16 | 13/16 | 6 1/2 |
| 3 1/2 | 8 | 3 17/32 | 3 31/32 | 1 1/8 | 1/4 | 1 1/16 | 13/16 | 6 1/2 |
| 4 | 8 | 4 1/32 | 4 15/32 | 1 1/8 | 1/4 | 1 1/16 | 13/16 | 6 1/2 |
| 4 | 6 | 4 | 4 29/32 | 1 1/8 | 5/16 | 1 1/16 | 3/4 | 4 1/2 |

All dimensions are given in inches. For thread designation see Table 1.

**American National Fire Hose Connection Screw Thread.—**This thread is specified in the National Fire Protection Association's Standard NFPA No. 194-1974. It covers the dimensions for screw thread connections for fire hose couplings, suction hose couplings, relay supply hose couplings, fire pump suctions, discharge valves, fire hydrants, nozzles, adaptors, reducers, caps, plugs, wyes, siamese connections, standpipe connections, and sprinkler connections.

*Form of Thread:* The basic form of thread is as shown in Fig. 1. It has an included angle of 60 degrees and is truncated top and bottom. The flat at the root and crest of the basic thread form is equal to 1/8 (0.125) times the pitch in inches. The height of the thread is equal to 0.649519 times the pitch. The outer ends of both external and internal threads are terminated by the blunt start or "Higbee Cut" on full thread to avoid crossing and mutilation of thread.

*Thread Designation:* The thread is designated by specifying in sequence the nominal size of the connection, number of threads per inch followed by the thread symbol NH.

Thus, .75-8NH indicates a nominal size connection of 0.75 inch diameter with 8 threads per inch.

*Basic Dimensions:* The basic dimensions of the thread are as given in Table 1.

**Table 1. Basic Dimensions of NH Threads** *NFPA 1963-1993 Edition*

| Nom. Size | Threads per Inch (tpi) | Thread Designation | Pitch, $p$ | Basic Thread Height, $h$ | Minimum Internal Thread Dimensions: Min. Minor Dia. | Basic Pitch Dia. | Basic Major Dia. |
|---|---|---|---|---|---|---|---|
| ¾ | 8 | 0.75-8 NH | 0.12500 | 0.08119 | 1.2246 | 1.3058 | 1.3870 |
| 1 | 8 | 1-8 NH | 0.12500 | 0.08119 | 1.2246 | 1.3058 | 1.3870 |
| 1½ | 9 | 1.5-9 NH | 0.11111 | 0.07217 | 1.8577 | 1.9298 | 2.0020 |
| 2½ | 7.5 | 2.5-7.5 NH | 0.13333 | 0.08660 | 2.9104 | 2.9970 | 3.0836 |
| 3 | 6 | 3-6 NH | 0.16667 | 0.10825 | 3.4223 | 3.5306 | 3.6389 |
| 3½ | 6 | 3.5-6 NH | 0.16667 | 0.10825 | 4.0473 | 4.1556 | 4.2639 |
| 4 | 4 | 4-4 NH | 0.25000 | 0.16238 | 4.7111 | 4.8735 | 5.0359 |
| 4½ | 4 | 4.5-4 NH | 0.25000 | 0.16238 | 5.4611 | 5.6235 | 5.7859 |
| 5 | 4 | 5-4 NH | 0.25000 | 0.16238 | 5.9602 | 6.1226 | 6.2850 |
| 6 | 4 | 6-4 NH | 0.25000 | 0.16238 | 6.7252 | 6.8876 | 7.0500 |
| Nom. Size | Threads per Inch (tpi) | Thread Designation | Pitch, $p$ | External Thread Dimensions (Nipple): Allowance | Max.Major Dia. | Max. Pitch Dia. | Max Minor Dia. |
| ¾ | 8 | 0.75-8 NH | 0.12500 | 0.0120 | 1.3750 | 1.2938 | 1.2126 |
| 1 | 8 | 1-8 NH | 0.12500 | 0.0120 | 1.3750 | 1.2938 | 1.2126 |
| 1½ | 9 | 1.5-9 NH | 0.11111 | 0.0120 | 1.9900 | 1.9178 | 1.8457 |
| 2½ | 7.5 | 2.5-7.5 NH | 0.13333 | 0.0150 | 3.0686 | 2.9820 | 2.8954 |
| 3 | 6 | 3-6 NH | 0.16667 | 0.0150 | 3.6239 | 3.5156 | 3.4073 |
| 3½ | 6 | 3.5-6 NH | 0.16667 | 0.0200 | 4.2439 | 4.1356 | 4.0273 |
| 4 | 4 | 4-4 NH | 0.25000 | 0.0250 | 5.0109 | 4.8485 | 4.6861 |
| 4½ | 4 | 4.5-4 NH | 0.25000 | 0.0250 | 5.7609 | 5.5985 | 5.4361 |
| 5 | 4 | 5-4 NH | 0.25000 | 0.0250 | 6.2600 | 6.0976 | 5.9352 |
| 6 | 4 | 6-4 NH | 0.25000 | 0.0250 | 7.0250 | 6.8626 | 6.7002 |

All dimensions are in inches.

*Thread Limits of Size:* Limits of size for NH external threads are given in Table 2. Limits of size for NH internal threads are given in Table 3.

*Tolerances:* The pitch-diameter tolerances for mating external and internal threads are the same. Pitch-diameter tolerances include lead and half-angle deviations. Lead deviations consuming one-half of the pitch-diameter tolerance are 0.0032 inch for ¾-, 1-, and 1½-inch sizes; 0.0046 inch for 2½-inch size; 0.0052 inch for 3-, and 3½-inch sizes; and 0.0072 inch for 4-, 4½-, 5-, and 6-inch sizes. Half-angle deviations consuming one-half of the pitch-diameter tolerance are 1 degree, 42 minutes for ¾- and 1-inch sizes; 1 degree, 54 minutes for 1½-inch size; 2 degrees, 17 minutes for 2½-inch size; 2 degrees, 4 minutes for 3- and 3½-inch size; and 1 degree, 55 minutes for 4-, 4½-, 5-, and 6-inch sizes.

Tolerances for the external threads are:

Major diameter tolerance = 2 × pitch-diameter tolerance

Minor diameter tolerance = pitch-diameter tolerance + $2h/9$

The minimum minor diameter of the external thread is such as to result in a flat equal to one-third of the $p/8$ basic flat, or $p/24$, at the root when the pitch diameter of the external thread is at its minimum value. The maximum minor diameter is basic, but may be such as results from the use of a worn or rounded threading tool. The maximum minor diameter is shown in Fig. 1 and is the diameter upon which the minor diameter tolerance formula shown above is based.

Tolerances for the internal threads are:

Minor diameter tolerance = 2 × pitch-diameter tolerance

The minimum minor diameter of the internal thread is such as to result in a basic flat, $p/8$, at the crest when the pitch diameter of the thread is at its minimum value.

Major diameter tolerance = pitch-diameter tolerance - $2h/9$

**Table 2. Limits of Size and Tolerances for NH External Threads (Nipples)**
*NFPA 1963, 1993 Edition*

| Nom. Size | Threads per Inch (tpi) | External Thread (Nipple) | | | | | | |
|---|---|---|---|---|---|---|---|---|
| | | Major Diameter | | | Pitch Diameter | | | Minor[a] Dia. |
| | | Max. | Min. | Toler. | Max. | Min. | Toler. | Max. |
| ¾ | 8 | 1.3750 | 1.3528 | 0.0222 | 1.2938 | 1.2827 | 0.0111 | 1.2126 |
| 1 | 8 | 1.3750 | 1.3528 | 0.0222 | 1.2938 | 1.2827 | 0.0111 | 1.2126 |
| 1½ | 9 | 1.9900 | 1.9678 | 0.0222 | 1.9178 | 1.9067 | 0.0111 | 1.8457 |
| 2½ | 7.5 | 3.0686 | 3.0366 | 0.0320 | 2.9820 | 2.9660 | 0.0160 | 2.8954 |
| 3 | 6 | 3.6239 | 3.5879 | 0.0360 | 3.5156 | 3.4976 | 0.0180 | 3.4073 |
| 3½ | 6 | 4.2439 | 4.2079 | 0.0360 | 4.1356 | 4.1176 | 0.0180 | 4.0273 |
| 4 | 4 | 5.0109 | 4.9609 | 0.0500 | 4.8485 | 4.8235 | 0.0250 | 4.6861 |
| 4½ | 4 | 5.7609 | 5.7109 | 0.0500 | 5.5985 | 5.5735 | 0.0250 | 5.4361 |
| 5 | 4 | 6.2600 | 6.2100 | 0.0500 | 6.0976 | 6.0726 | 0.0250 | 5.9352 |
| 6 | 4 | 7.0250 | 6.9750 | 0.0500 | 6.8626 | 6.8376 | 0.0250 | 6.7002 |

[a] Dimensions given for the maximum minor diameter of the nipple are figured to the intersection of the worn tool arc with a center line through crest and root. The minimum minor diameter of the nipple shall be that corresponding to a flat at the minor diameter of the minimum nipple equal to $p/24$ and may be determined by subtracting $11h/9$ (or $0.7939p$) from the minimum pitch diameter of the nipple.

All dimensions are in inches.

**Table 3. Limits of Size and Tolerances for NH Internal Threads (Couplings)**
*NFPA 1963, 1993 Edition*

| Nom. Size | Threads per Inch (tpi) | Internal Thread (Coupling) | | | | | | |
|---|---|---|---|---|---|---|---|---|
| | | Minor Diameter | | | Pitch Diameter | | | Major[a] Dia. |
| | | Min. | Max. | Toler. | Min. | Max. | Toler. | Min. |
| ¾ | 8 | 1.2246 | 1.2468 | 0.0222 | 1.3058 | 1.3169 | 0.0111 | 1.3870 |
| 1 | 8 | 1.2246 | 1.2468 | 0.0222 | 1.3058 | 1.3169 | 0.0111 | 1.3870 |
| 1½ | 9 | 1.8577 | 1.8799 | 0.0222 | 1.9298 | 1.9409 | 0.0111 | 2.0020 |
| 2½ | 7.5 | 2.9104 | 2.9424 | 0.0320 | 2.9970 | 3.0130 | 0.0160 | 3.0836 |
| 3 | 6 | 3.4223 | 3.4583 | 0.0360 | 3.5306 | 3.5486 | 0.0180 | 3.6389 |
| 3½ | 6 | 4.0473 | 4.0833 | 0.0360 | 4.1556 | 4.1736 | 0.0180 | 4.2639 |
| 4 | 4 | 4.7111 | 4.7611 | 0.0500 | 4.8735 | 4.8985 | 0.0250 | 5.0359 |
| 4½ | 4 | 5.4611 | 5.5111 | 0.0500 | 5.6235 | 5.6485 | 0.0250 | 5.7859 |
| 5 | 4 | 5.9602 | 6.0102 | 0.0500 | 6.1226 | 6.1476 | 0.0250 | 6.2850 |
| 6 | 4 | 6.7252 | 6.7752 | 0.0500 | 6.8876 | 6.9126 | 0.0250 | 7.0500 |

[a] Dimensions for the minimum major diameter of the coupling correspond to the basic flat ($p/8$), and the profile at the major diameter produced by a worn tool must not fall below the basic outline. The maximum major diameter of the coupling shall be that corresponding to a flat at the major diameter of the maximum coupling equal to $p/24$ and may be determined by adding $11h/9$ (or $0.7939p$) to the maximum pitch diameter of the coupling.

All dimensions are in inches.

*Gages and Gaging:* Full information on gage dimensions and the use of gages in checking the NH thread are given in NFPA Standard No. 1963, 1993 Edition, published by the National Fire Protection Association, Batterymarch Park, Quincy, MA 02269.

The information and data taken from this Standard are reproduced with the permission of the NFPA.

## OTHER THREADS

### Interference-Fit Threads

**Interference-Fit Threads.**—Interference-fit threads are threads in which the externally threaded member is larger than the internally threaded member when both members are in the free state and that, when assembled, become the same size and develop a holding torque through elastic compression, plastic movement of material, or both. By custom, these threads are designated Class 5.

The data in Table 1, Table 2, and Table 3, which are based on years of research, testing and field study, represent an American standard for interference-fit threads that overcomes the difficulties experienced with previous interference-fit recommendations such as are given in Screw-Thread Standards for Federal Services. These data were adopted as American Standard ASA B1.12-1963. Subsequently, the Standard was revised and issued as American National Standard ANSI B1.12-1972. More recent research conducted by the Portsmouth Naval Shipyard has led to the current revision ANSI/ASME B1.12-1987 (R2018).

The data in Table 1, Table 2, and Table 3 provide dimensions for external and internal interference-fit (Class 5) threads of modified American National form in the Coarse Thread series, sizes $\frac{1}{4}$ inch to $1\frac{1}{2}$ inches. It is intended that interference-fit threads conforming with this Standard will provide adequate torque conditions which fall within the limits shown in Table 3. The minimum torques are intended to be sufficient to ensure that externally threaded members will not loosen in service; the maximum torques establish a ceiling below which seizing, galling, or torsional failure of the externally threaded components is reduced.

Table 1 and Table 2 give external and internal thread dimensions and are based on engagement lengths, external thread lengths, and tapping hole depths specified in Table 3 and in compliance with the design and application data given in the following paragraphs. Table 4 gives the allowances and Table 5 gives the tolerances for pitch, major, and minor diameters for the Coarse Thread Series.

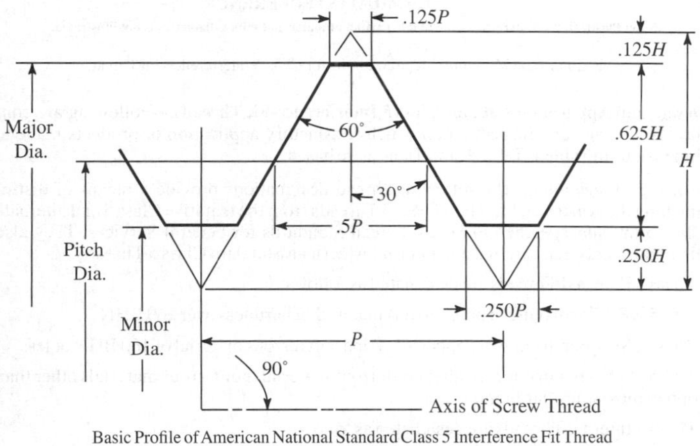


Basic Profile of American National Standard Class 5 Interference Fit Thread

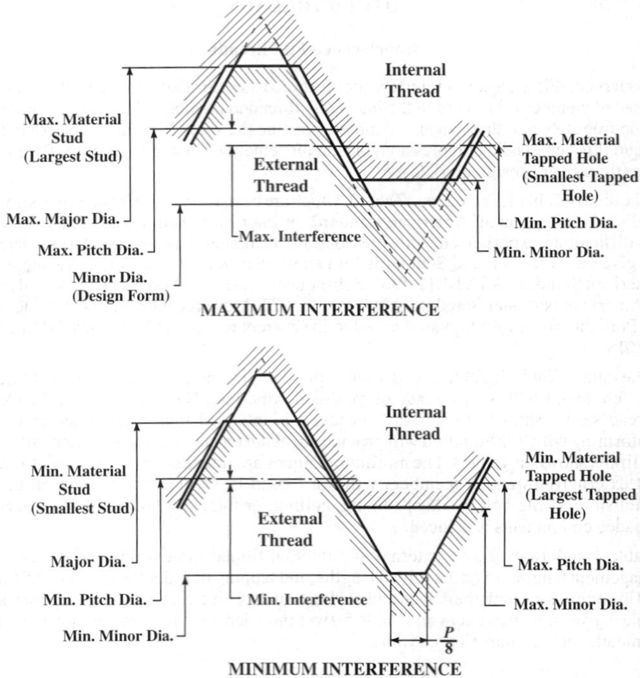


**Note: Plastic flow of interference metal into cavities at major and minor diameters is not illustrated.**

Maximum and Minimum Material Limits for Class 5 Interference-Fit Thread

**Design and Application Data for Class 5 Interference-Fit Threads.—**Following are conditions of usage and inspection on which satisfactory application of products made to dimensions in Table 1, Table 2, and Table 3 are based.

*Thread Designations:* The following thread designations provide a means of distinguishing the American Standard Class 5 Threads from the tentative Class 5 and alternate Class 5 threads, specified in Screw-Thread Standards for Federal Services. They also distinguish between external and internal American Standard Class 5 Threads.

Class 5 External Threads are designated as follows:

NC-5 HF—For driving in hard ferrous material of hardness over 160 BHN.

NC-5 CSF—For driving in copper alloy and soft ferrous material of 160 BHN or less.

NC-5 ONF—For driving in other nonferrous material (nonferrous materials other than copper alloys), any hardness.

Class 5 Internal Threads are designated as follows:

NC-5 IF—Entire ferrous material range.

NC-5 INF—Entire nonferrous material range.

**Table 1. External Thread Dimensions for Class 5 Interference-Fit Threads**
*ANSI/ASME B1.12-1987 (R2018)*

| Nominal Size | Major Diameter, Inches: NC-5 HF for driving in ferrous material with hardness greater than 160 BHN $L_e$ = 1¼ Dia. | | NC-5 CSF for driving in brass and ferrous material with hardness equal to or less than 160 BHN $L_e$ = 1¼ Dia. | | NC-5 ONF for driving in nonferrous except brass (any hardness) $L_e$ = 2½ Dia. | | Pitch Diameter, Inches | | Minor Diameter, Inches |
|---|---|---|---|---|---|---|---|---|---|
| | Max | Min | Max | Min | Max | Min | Max | Min | Max |
| 0.2500-20 | 0.2470 | 0.2418 | 0.2470 | 0.2418 | 0.2470 | 0.2418 | 0.2230 | 0.2204 | 0.1932 |
| 0.3125-18 | 0.3080 | 0.3020 | 0.3090 | 0.3030 | 0.3090 | 0.3030 | 0.2829 | 0.2799 | 0.2508 |
| 0.3750-16 | 0.3690 | 0.3626 | 0.3710 | 0.3646 | 0.3710 | 0.3646 | 0.3414 | 0.3382 | 0.3053 |
| 0.4375-14 | 0.4305 | 0.4233 | 0.4330 | 0.4258 | 0.4330 | 0.4258 | 0.3991 | 0.3955 | 0.3579 |
| 0.5000-13 | 0.4920 | 0.4846 | 0.4950 | 0.4876 | 0.4950 | 0.4876 | 0.4584 | 0.4547 | 0.4140 |
| 0.5625-12 | 0.5540 | 0.5460 | 0.5575 | 0.5495 | 0.5575 | 0.5495 | 0.5176 | 0.5136 | 0.4695 |
| 0.6250-11 | 0.6140 | 0.6056 | 0.6195 | 0.6111 | 0.6195 | 0.6111 | 0.5758 | 0.5716 | 0.5233 |
| 0.7500-10 | 0.7360 | 0.7270 | 0.7440 | 0.7350 | 0.7440 | 0.7350 | 0.6955 | 0.6910 | 0.6378 |
| 0.8750- 9 | 0.8600 | 0.8502 | 0.8685 | 0.8587 | 0.8685 | 0.8587 | 0.8144 | 0.8095 | 0.7503 |
| 1.0000- 8 | 0.9835 | 0.9727 | 0.9935 | 0.9827 | 0.9935 | 0.9827 | 0.9316 | 0.9262 | 0.8594 |
| 1.1250- 7 | 1.1070 | 1.0952 | 1.1180 | 1.1062 | 1.1180 | 1.1062 | 1.0465 | 1.0406 | 0.9640 |
| 1.2500- 7 | 1.2320 | 1.2200 | 1.2430 | 1.2312 | 1.2430 | 1.2312 | 1.1715 | 1.1656 | 1.0890 |
| 1.3750- 6 | 1.3560 | 1.3410 | 1.3680 | 1.3538 | 1.3680 | 1.3538 | 1.2839 | 1.2768 | 1.1877 |
| 1.5000- 6 | 1.4810 | 1.4670 | 1.4930 | 1.4788 | 1.4930 | 1.4788 | 1.4089 | 1.4018 | 1.3127 |

Based on external threaded members being steel ASTM A325 (SAE Grade 5) or better. $L_e$ = length of engagement.

**Table 2. Internal Thread Dimensions for Class 5 Interference-Fit Threads**
*ANSI/ASME B1.12-1987 (R2018)*

| Nominal Size | NC-5 IF Ferrous Material: Minor Dia.[a] | | Tap Drill | NC-5 INF Nonferrous Material: Minor Dia.[a] | | Tap Drill | Pitch Diameter | | Major Diameter |
|---|---|---|---|---|---|---|---|---|---|
| | Min | Max | | Min | Max | | Min | Max | Min |
| 0.2500-20 | 0.196 | 0.206 | 0.2031 | 0.196 | 0.206 | 0.2031 | 0.2175 | 0.2201 | 0.2532 |
| 0.3125-18 | 0.252 | 0.263 | 0.2610 | 0.252 | 0.263 | 0.2610 | 0.2764 | 0.2794 | 0.3161 |
| 0.3750-16 | 0.307 | 0.318 | 0.3160 | 0.307 | 0.318 | 0.3160 | 0.3344 | 0.3376 | 0.3790 |
| 0.4375-14 | 0.374 | 0.381 | 0.3750 | 0.360 | 0.372 | 0.3680 | 0.3911 | 0.3947 | 0.4421 |
| 0.5000-13 | 0.431 | 0.440 | 0.4331 | 0.417 | 0.429 | 0.4219 | 0.4500 | 0.4537 | 0.5050 |
| 0.5625-12 | 0.488 | 0.497 | 0.4921 | 0.472 | 0.485 | 0.4844 | 0.5084 | 0.5124 | 0.5679 |
| 0.6250-11 | 0.544 | 0.554 | 0.5469 | 0.527 | 0.540 | 0.5313 | 0.5660 | 0.5702 | 0.6309 |
| 0.7500-10 | 0.667 | 0.678 | 0.6719 | 0.642 | 0.655 | 0.6496 | 0.6850 | 0.6895 | 0.7565 |
| 0.8750- 9 | 0.777 | 0.789 | 0.7812 | 0.755 | 0.769 | 0.7656 | 0.8028 | 0.8077 | 0.8822 |
| 1.0000- 8 | 0.890 | 0.904 | 0.8906 | 0.865 | 0.880 | 0.8750 | 0.9188 | 0.9242 | 1.0081 |
| 1.1250- 7 | 1.000 | 1.015 | 1.0000 | 0.970 | 0.986 | 0.9844 | 1.0322 | 1.0381 | 1.1343 |
| 1.2500- 7 | 1.125 | 1.140 | 1.1250 | 1.095 | 1.111 | 1.1094 | 1.1572 | 1.1631 | 1.2593 |
| 1.3750- 6 | 1.229 | 1.247 | 1.2344 | 1.195 | 1.213 | 1.2031 | 1.2667 | 1.2738 | 1.3858 |
| 1.5000- 6 | 1.354 | 1.372 | 1.3594 | 1.320 | 1.338 | 1.3281 | 1.3917 | 1.3988 | 1.5108 |

[a] Fourth decimal place is 0 for all sizes.

All dimensions are in inches, unless otherwise specified.

*Externally Threaded Products:* Points of externally threaded components should be chamfered or otherwise reduced to a diameter below the minimum minor diameter of the thread. The limits apply to bare or metallic coated parts. The threads should be free from excessive nicks, burrs, chips, grit or other extraneous material before driving.

**Table 3. Torques, Interferences, and Engagement Lengths for Class 5 Interference-Fit Threads** *ANSI/ASME B1.12-1987 (R2018)*

| | Interference on Pitch Diameter | | Engagement Lengths, External Thread Lengths and Tapped Hole Depths[a] | | | | | | Torque at 1-¼D Engagement in Ferrous Material | |
|---|---|---|---|---|---|---|---|---|---|---|
| | | | In Brass and Ferrous | | | In Nonferrous Except Brass | | | | |
| Nominal Size | Max | Min | $L_e$ | $T_s$ | $T_h$ min | $L_e$ | $T_s$ | $T_h$ min | Max, lb-ft | Min, lb-ft |
| 0.2500-20 | .0055 | .0003 | 0.312 | 0.375 + .125 − 0 | 0.375 | 0.625 | 0.688 + .125 − 0 | 0.688 | 12 | 3 |
| 0.3125-18 | .0065 | .0005 | 0.391 | 0.469 + .139 − 0 | 0.469 | 0.781 | 0.859 + .139 − 0 | 0.859 | 19 | 6 |
| 0.3750-16 | .0070 | .0006 | 0.469 | 0.562 + .156 − 0 | 0.562 | 0.938 | 1.031 + .156 − 0 | 1.031 | 35 | 10 |
| 0.4375-14 | .0080 | .0008 | 0.547 | 0.656 + .179 − 0 | 0.656 | 1.094 | 1.203 + .179 − 0 | 1.203 | 45 | 15 |
| 0.5000-13 | .0084 | .0010 | 0.625 | 0.750 + .192 − 0 | 0.750 | 1.250 | 1.375 + .192 − 0 | 1.375 | 75 | 20 |
| 0.5625-12 | .0092 | .0012 | 0.703 | 0.844 + .208 − 0 | 0.844 | 1.406 | 1.547 + .208 − 0 | 1.547 | 90 | 30 |
| 0.6250-11 | .0098 | .0014 | 0.781 | 0.938 + .227 − 0 | 0.938 | 1.562 | 1.719 + .227 − 0 | 1.719 | 120 | 37 |
| 0.7500-10 | .0105 | .0015 | 0.938 | 1.125 + .250 − 0 | 1.125 | 1.875 | 2.062 + .250 − 0 | 2.062 | 190 | 60 |
| 0.8750-9 | .0016 | .0018 | 1.094 | 1.312 + .278 − 0 | 1.312 | 2.188 | 2.406 + .278 − 0 | 2.406 | 250 | 90 |
| 1.0000-8 | .0128 | .0020 | 1.250 | 1.500 + .312 − 0 | 1.500 | 2.500 | 2.750 + .312 − 0 | 2.750 | 400 | 125 |
| 1.1250-7 | .0143 | .0025 | 1.406 | 1.688 + .357 − 0 | 1.688 | 2.812 | 3.094 + .357 − 0 | 3.095 | 470 | 155 |
| 1.2500-7 | .0143 | .0025 | 1.562 | 1.875 + .357 − 0 | 1.875 | 3.125 | 3.438 + .357 − 0 | 3.438 | 580 | 210 |
| 1.3750-6 | .0172 | .0030 | 1.719 | 2.062 + .419 − 0 | 2.062 | 3.438 | 3.781 + .419 − 0 | 3.781 | 705 | 250 |
| 1.5000-6 | .0172 | .0030 | 1.875 | 2.250 + .419 − 0 | 2.250 | 3.750 | 4.125 + .419 − 0 | 4.125 | 840 | 325 |

[a] $L_e$ = Length of engagement. $T_s$ = External thread length of full form thread. $T_h$ = Minimum depth of full form thread in hole.

All dimensions are inches.

*Materials for Externally Threaded Products:* The length of engagement, depth of thread engagement and pitch diameter in Table 1, Table 2, and Table 3 are designed to produce adequate torque conditions when heat-treated medium-carbon steel products, ASTM A325 (SAE Grade 5) or better, are used. In many applications, case-carburized and non-heat-treated medium-carbon steel products of SAE Grade 4 are satisfactory. SAE Grades 1 and 2 may be usable under certain conditions. This Standard is not intended to cover the use of products made of stainless steel, silicon bronze, brass or similar materials. When such materials are used, the tabulated dimensions will probably require adjustment based on pilot experimental work with the materials involved.

*Lubrication:* For driving in ferrous material, a good lubricant sealer should be used, particularly in the hole. A non-carbonizing type of lubricant (such as a rubber-in-water dispersion) is suggested. The lubricant must be applied to the hole and it may be applied to the male member. In applying it to the hole, care must be taken so that an excess amount of lubricant will not cause the male member to be impeded by hydraulic pressure in a blind hole. Where sealing is involved, the lubricant selected should be insoluble in the medium being sealed.

For driving, in nonferrous material, lubrication may not be needed. The use of medium gear oil for driving in aluminum is recommended. American research has observed that the minor diameter of lubricated tapped holes in nonferrous materials may tend to close in, that is, be reduced in driving; whereas with an unlubricated hole the minor diameter may tend to open up.

*Driving Speed:* This Standard makes no recommendation for driving speed. Some opinion has been advanced that careful selection and control of driving speed is desirable to obtain optimum results with various combinations of surface hardness and roughness. Experience with threads made to this Standard may indicate what limitations should be placed on driving speeds.

**Table 4. Allowances for Coarse Thread Series** *ANSI/ASME B1.12-1987 (R2018)*

| TPI | Difference between Nom. Size and Max Major Dia. of NC-5 HF[a] | Difference between Nom. Size and Max Major Dia. of NC-5 CSF or NC-5 ONF[a] | Difference between Basic Minor Dia. and Min Minor Dia. of NC-5 IF[a] | Difference between Basic Minor Dia. and Min Minor Dia. of NC-5 INF | Max PD Inteference or Neg Allowance, Ext Thread[b] | Difference between Max Minor Dia. and Basic Minor Dia., Ext Thread |
|---|---|---|---|---|---|---|
| 20 | 0.0030 | 0.0030 | 0.000 | 0.000 | 0.0055 | 0.0072 |
| 18 | 0.0045 | 0.0035 | 0.000 | 0.000 | 0.0065 | 0.0080 |
| 16 | 0.0060 | 0.0040 | 0.000 | 0.000 | 0.0070 | 0.0090 |
| 14 | 0.0070 | 0.0045 | 0.014 | 0.000 | 0.0080 | 0.0103 |
| 13 | 0.0080 | 0.0050 | 0.014 | 0.000 | 0.0084 | 0.0111 |
| 12 | 0.0085 | 0.0050 | 0.016 | 0.000 | 0.0092 | 0.0120 |
| 11 | 0.0110 | 0.0055 | 0.017 | 0.000 | 0.0098 | 0.0131 |
| 10 | 0.0140 | 0.0060 | 0.019 | 0.000 | 0.0105 | 0.0144 |
| 9 | 0.0150 | 0.0065 | 0.022 | 0.000 | 0.0116 | 0.0160 |
| 8 | 0.0165 | 0.0065 | 0.025 | 0.000 | 0.0128 | 0.0180 |
| 7 | 0.0180 | 0.0070 | 0.030 | 0.000 | 0.0143 | 0.0206 |
| 6 | 0.0190 | 0.0070 | 0.034 | 0.000 | 0.0172 | 0.0241 |

[a] The allowances in these columns were obtained from industrial research data.

[b] Negative allowance is the difference between the basic pitch diameter and pitch diameter value at maximum material condition.

All dimensions are in inches.

The difference between basic major diameter and internal thread minimum major diameter is $0.075H$ and is tabulated in Table 5.

**Table 5. Tolerances for Pitch Diameter, Major Diameter, and Minor Diameter for Coarse Thread Series** *ANSI/ASME B1.12-1987 (R2018)*

| TPI | PD Tolerance for Ext and Int Threads[a] | Major Dia. Tolerance for Ext Thread[b] | Minor Dia. Tolerance for Int Thread NC-5 IF | Minor Dia. Tolerance for Int Thread NC-5 INF[c] | Tolerance $0.075H$ or $0.065P$ for Tap Major Dia. |
|---|---|---|---|---|---|
| 20 | 0.0026 | 0.0052 | 0.010 | 0.010 | 0.0032 |
| 18 | 0.0030 | 0.0060 | 0.011 | 0.011 | 0.0036 |
| 16 | 0.0032 | 0.0064 | 0.011 | 0.011 | 0.0041 |
| 14 | 0.0036 | 0.0072 | 0.008 | 0.012 | 0.0046 |
| 13 | 0.0037 | 0.0074 | 0.008 | 0.012 | 0.0050 |
| 12 | 0.0040 | 0.0080 | 0.009 | 0.013 | 0.0054 |
| 11 | 0.0042 | 0.0084 | 0.010 | 0.013 | 0.0059 |
| 10 | 0.0045 | 0.0090 | 0.011 | 0.014 | 0.0065 |
| 9 | 0.0049 | 0.0098 | 0.012 | 0.014 | 0.0072 |
| 8 | 0.0054 | 0.0108 | 0.014 | 0.015 | 0.0093 |
| 7 | 0.0059 | 0.0118 | 0.015 | 0.015 | 0.0093 |
| 6 | 0.0071 | 0.0142 | 0.018 | 0.018 | 0.0108 |

[a] National Class 3 pitch diameter tolerance from ASA B1.1-1960.

[b] Twice the NC-3 pitch diameter tolerance.

[c] National Class 3 minor diameter tolerance from ASA B1.1-1960.

All dimensions are in inches.

*Relation of Driving Torque to Length of Engagement:* Torques increase directly as the length of engagement and this increase is proportionately more rapid as size increases. The Standard does not establish recommended breakloose torques.

*Surface Roughness:* Surface roughnesss is not a required measurement. Roughness between 63 and 125 μin Ra is recommended. Surface roughness greater than 125 μin Ra may encourage galling and tearing of threads. Surfaces with roughness less than 63 μin Ra may hold insufficient lubricant and wring or weld together.

*Lead and Angle Variations:* The lead variation values tabulated in Table 6 are the maximum variations from specified lead between any two points not farther apart than the length of the standard GO thread gage. Flank angle variation values tabulated in Table 7 are maximum variations from the basic 30° angle between thread flanks and perpendiculars to the thread axis. The application of these data in accordance with ANSI/ASME B1.3, the screw thread gaging system for dimensional acceptability, is given in the Standard. Lead variation does not change the volume of displaced metal, but it exerts a cumulative unilateral stress on the pressure side of the thread flank. Control of the difference between pitch diameter size and functional diameter size to within one-half the pitch diameter tolerance will hold lead and angle variables to within satisfactory limits. Both the variations may produce unacceptable torque and faulty assemblies.

**Table 6. Maximum Allowable Variations in Lead and Maximum Equivalent Change in Functional Diameter** *ANSI/ASME B1.12-1987 (R2018)*

| Nominal Size | External and Internal Threads | |
|---|---|---|
| | Allowable Variation in Axial Lead (Plus or Minus) | Max Equivalent Change in Functional Dia. (Plus for Ext, Minus for Int) |
| 0.2500-20 | 0.0008 | 0.0013 |
| 0.3125-18 | 0.0009 | 0.0015 |
| 0.3750-16 | 0.0009 | 0.0016 |
| 0.4375-14 | 0.0010 | 0.0018 |
| 0.5000-13 | 0.0011 | 0.0018 |
| 0.5625-12 | 0.0012 | 0.0020 |
| 0.6250-11 | 0.0012 | 0.0021 |
| 0.7500-10 | 0.0013 | 0.0022 |
| 0.8750- 9 | 0.0014 | 0.0024 |
| 1.0000- 8 | 0.0016 | 0.0027 |
| 1.1250- 7 | 0.0017 | 0.0030 |
| 1.2500- 7 | 0.0017 | 0.0030 |
| 1.3750- 6 | 0.0020 | 0.0036 |
| 1.5000- 6 | 0.0020 | 0.0036 |

All dimensions are in inches.

Note: The equivalent change in functional diameter applies to total effect of form errors.

Maximum allowable variation in lead is permitted only when all other form variations are zero.

For sizes not tabulated, maximum allowable variation in lead is equal to 0.57735 times one-half the pitch diameter tolerance.

**Table 7. Maximum Allowable Variation in 30° Basic Half-Angle of External and Internal Screw Threads** *ANSI/ASME B1.12-1987 (R2018)*

| TPI | Allowable Variation in Half-Angle of Thread (Plus or Minus) | TPI | Allowable Variation in Half-Angle of Thread (Plus or Minus) | TPI | Allowable Variation in Half-Angle of Thread (Plus or Minus) |
|---|---|---|---|---|---|
| 32 | 1° 30′ | 14 | 0° 55′ | 8 | 0° 45′ |
| 28 | 1° 20′ | 13 | 0° 55′ | 7 | 0° 45′ |
| 27 | 1° 20′ | 12 | 0° 50′ | 6 | 0° 40′ |
| 24 | 1° 15′ | 11½ | 0° 50′ | 5 | 0° 40′ |
| 20 | 1° 10′ | 11 | 0° 50′ | 4½ | 0° 40′ |
| 18 | 1° 05′ | 10 | 0° 50′ | 4 | 0° 40′ |
| 16 | 1° 00′ | 9 | 0° 50′ | … | … |

## Spark Plug Threads

**British Standard for Spark Plugs BS 45:1952 (Withdrawn).—**This revised British Standard refers solely to spark plugs used in automobiles and industrial spark ignition internal combustion engines. The basic thread form is that of the ISO metric (see page 2049). In assigning tolerances to the threads of the spark plug and the tapped holes, full consideration has been given to the desirability of achieving the closest possible measure of interchangeability between British spark plugs and engines, and those made to the standards of other ISO Member Bodies.

**Basic Thread Dimensions for Spark Plug and Tapped Hole in Cylinder Head**

| Nom. Size | Pitch | Thread | Major Dia. | | Pitch Dia. | | Minor Dia. | |
|---|---|---|---|---|---|---|---|---|
| | | | Max. | Min. | Max. | Min. | Max. | Min. |
| 14 | 1.25 | Plug | 13.937[a] | 13.725 | 13.125 | 12.993 | 12.402 | 12.181 |
| 14 | 1.25 | Hole | | 14.00 | 13.368 | 13.188 | 12.912 | 12.647 |
| 18 | 1.5 | Plug | 17.933[a] | 17.697 | 16.959 | 16.819 | 16.092 | 15.845 |
| 18 | 1.5 | Hole | | 18.00 | 17.216 | 17.026 | 16.676 | 16.376 |

[a] Not specified

All dimensions are given in millimeters.

The tolerance grades for finished spark plugs and corresponding tapped holes in the cylinder head are: for 14 mm size, 6e for spark plugs and 6H for tapped holes which gives a minimum clearance of 0.063 mm; and for 18 mm size, 6e for spark plugs and 6H for tapped holes which gives a minimum clearance of 0.067 mm.

These minimum clearances are intended to prevent the possibility of seizure, as a result of combustion deposits on the bare threads, when removing the spark plugs and applies to both ferrous and nonferrous materials. These clearances are also intended to enable spark plugs with threads in accordance with this Standard to be fitted into existing holes.

**SAE Spark Plug Screw Threads.—**The SAE Standard includes the following sizes: $\frac{7}{8}$-inch nominal diameter with 18 threads per inch: 18-millimeter nominal diameter with a 18-millimeter nominal diameter with 1.5-millimeter pitch; 14-millimeter nominal diameter with a 1.25-millimeter pitch; 10-millimeter nominal diameter with a 1.0 millimeter pitch; $\frac{3}{8}$-inch nominal diameter with 24 threads per inch; and $\frac{1}{4}$-inch nominal diameter with 32 threads per inch. During manufacture, in order to keep the wear on the threading tools within permissible limits, the threads in the spark plug GO (ring) gage should be truncated to the maximum minor diameter of the spark plug; and in the tapped hole GO (plug) gage to the minimum major diameter of the tapped hole.

**SAE Standard Threads for Spark Plugs**

| Size[a] Nom. × Pitch | Major Diameter | | Pitch Diameter | | Minor Diameter | |
|---|---|---|---|---|---|---|
| | Max. | Min. | Max. | Min. | Max. | Min. |
| Spark Plug Threads, mm (inches) | | | | | | |
| M18 × 1.5 | 17.933 | 17.803 | 16.959 | 16.853 | 16.053 | … |
| | (0.7060) | (0.7009) | (0.6677) | (0.6635) | (0.6320) | … |
| M14 × 1.25 | 13.868 | 13.741 | 13.104 | 12.997 | 12.339 | … |
| | (0.5460) | (0.5410) | (0.5159) | (0.5117) | (0.4858) | … |
| M12 × 1.25 | 11.862 | 11.735 | 11.100 | 10.998 | 10.211 | … |
| | (0.4670) | (0.4620) | (0.4370) | (0.4330) | (0.4020) | … |
| M10 × 1.0 | 9.974 | 9.794 | 9.324 | 9.212 | 8.747 | … |
| | (0.3927) | (0.3856) | (0.3671) | (0.3627) | (0.3444) | … |

**SAE Standard Threads for Spark Plugs** *(Continued)*

| Size[a] Nom. × Pitch | Major Diameter Max. | Major Diameter Min. | Pitch Diameter Max. | Pitch Diameter Min. | Minor Diameter Max. | Minor Diameter Min. |
|---|---|---|---|---|---|---|
| Tapped Hole Threads, mm (inches) | | | | | | |
| M18 × 1.5 | … | 18.039 | 17.153 | 17.026 | 16.426 | 16.266 |
| | … | (0.7102) | (0.6753) | (0.6703) | (0.6467) | (0.6404) |
| M14 × 1.25 | … | 14.034 | 13.297 | 13.188 | 12.692 | 12.499 |
| | … | (0.5525) | (0.5235) | (0.5192) | (0.4997) | (0.4921) |
| M12 × 1.25 | … | 12.000 | 11.242 | 11.188 | 10.559 | 10.366 |
| | … | (0.4724) | (0.4426) | (0.4405) | (0.4157) | (0.4081) |
| M10 × 1.0 | … | 10.000 | 9.500 | 9.350 | 9.153 | 8.917 |
| | … | (0.3937) | (0.3740) | (0.3681) | (0,3604) | (0.3511) |

[a] M14 and M18 are preferred for new applications.

In order to keep the wear on the threading tools within permissible limits, the threads in the spark plug GO (ring) gage shall be truncated to the maximum minor diameter of the spark plug, and in the tapped hole GO (plug) gage to the minimum major diameter of the tapped hole. The plain plug gage for checking the minor diameter of the tapped hole shall be the minimum specified. The thread form is that of the ISO metric (see page 2049).



## Lamp Base and Electrical Fixture Threads

**Lamp Base and Socket Shell Threads.—**The "American Standard" threads for lamp base and socket shells are sponsored by the American Society of Mechanical Engineers, the National Electrical Manufacturers' Association and by most of the large manufacturers of products requiring rolled threads on sheet metal shells or parts, such as lamp bases, fuse plugs, attachment plugs, etc. There are five sizes, designated as the "miniature size," the "candelabra size," the "intermediate size," the "medium size" and the "mogul size."

### Rolled Threads for Screw Shells of Electric Sockets and Lamp Bases—American Standard

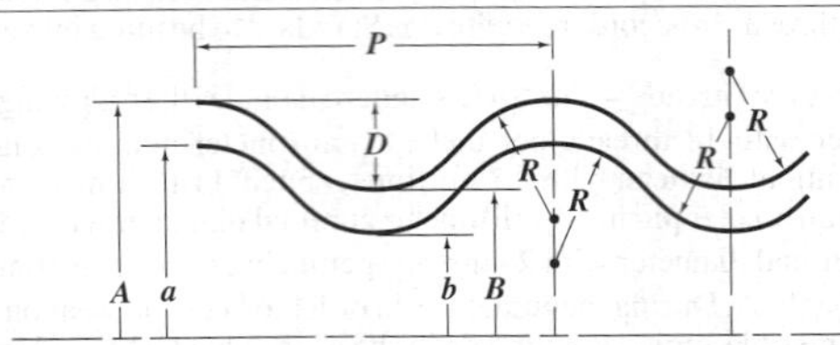


| Size | Threads per Inch | Pitch P | Depth of Thread D | Radius Crest Root R | Major Diameter Max. A | Major Diameter Min. a | Minor Diameter Max. B | Minor Diameter Min. b |
|---|---|---|---|---|---|---|---|---|
| Male or Base Screw Shells Before Assembly | | | | | | | | |
| Miniature | 14 | 0.07143 | 0.020 | 0.0210 | 0.375 | 0.370 | 0.335 | 0.330 |
| Candelabra | 10 | 0.10000 | 0.025 | 0.0312 | 0.465 | 0.460 | 0.415 | 0.410 |
| Intermediate | 9 | 0.11111 | 0.027 | 0.0353 | 0.651 | 0.645 | 0.597 | 0.591 |
| Medium | 7 | 0.14286 | 0.033 | 0.0470 | 1.037 | 1.031 | 0.971 | 0.965 |
| Mogul | 4 | 0.25000 | 0.050 | 0.0906 | 1.555 | 1.545 | 1.455 | 1.445 |
| Socket Screw Shells Before Assembly | | | | | | | | |
| Miniature | 14 | 0.07143 | 0.020 | 0.0210 | 0.3835 | 0.3775 | 0.3435 | 0.3375 |
| Candelabra | 10 | 0.10000 | 0.025 | 0.0312 | 0.476 | 0.470 | 0.426 | 0.420 |
| Intermediate | 9 | 0.11111 | 0.027 | 0.0353 | 0.664 | 0.657 | 0.610 | 0.603 |
| Medium | 7 | 0.14286 | 0.033 | 0.0470 | 1.053 | 1.045 | 0.987 | 0.979 |
| Mogul | 4 | 0.25000 | 0.050 | 0.0906 | 1.577 | 1.565 | 1.477 | 1.465 |

All dimensions are in inches.

*Base Screw Shell Gage Tolerances:* Threaded ring gages—"Go," Max. thread size to minus 0.0003 inch; "Not Go," Min. thread size to plus 0.0003 inch. Plain ring gages—"Go," Max. thread OD to minus 0.0002 inch; "Not Go," Min. thread OD to plus 0.0002 inch.

*Socket Screw Shell Gages:* Threaded plug gages—"Go," Min. thread size to plus 0.0003 inch; "Not Go," Max. thread size to minus 0.0003 inch. Plain plug gages—"Go," Min. minor dia. to plus 0.0002 inch; "Not Go," Max. minor dia. to minus 0.0002 inch.

*Check Gages for Base Screw Shell Gages:* Threaded plugs for checking threaded ring gages—"Go," Max. thread size to minus 0.0003 inch; "Not Go," Min. thread size to plus 0.0003 inch.

**Electric Fixture Thread.—**The special straight electric fixture thread consists of a straight thread of the same pitches as the American standard pipe thread, and having the regular American or US standard form; it is used for caps, etc. The male thread is smaller, and the female thread larger than those of the special straight-fixture pipe threads. The male thread assembles with a standard taper female thread, while the female thread assembles with a standard taper male thread. This thread is used when it is desired to have the joint "make up" on a shoulder. The gages used are straight-threaded limit gages.

## Instrument and Microscope Threads

**British Association Standard Thread (BA).—**This form of thread is similar to the Whitworth thread in that the root and crest are rounded (see illustration). The angle, however, is only 47 degrees 30 minutes and the radius of the root and crest are proportionately larger. This thread is used in Great Britain and, to some extent, in other European countries for very small screws. Its use in the United States is practically confined to the manufacture of tools for export. This thread system was originated in Switzerland as a standard for watch and clock screws, and it is sometimes referred to as the "Swiss small screw thread standard." See also *Swiss Screw Thread.*

This screw thread system is recommended by the British Standards Institution for use in preference to the BSW and BSF systems for all screws smaller than ¼ inch except that the use of the "0" BA thread be discontinued in favor of the ¼-in. BSF. It is further recommended that in the selection of sizes, preference be given to even numbered BA sizes. The thread form is shown by the diagram.

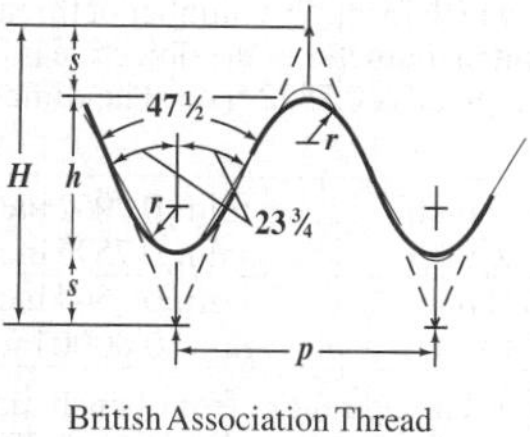


$H = 1.13634 \times p$

$h = 0.60000 \times p$

$r = 0.18083 \times p$

$s = 0.26817 \times p$

British Association Thread

It is a symmetrical V-thread, of 47½ degree included angle, having its crests and roots rounded with equal radii, such that the basic depth of the thread is 0.6000 of the pitch. Where $p$ = pitch of thread, $H$ = depth of V-thread, $h$ = depth of BA thread, $r$ = radius at root and crest of thread, and $s$ = root and crest truncation.

**British Association (BA) Standard Thread, Basic Dimensions**
*BS 93:2008*

| Designation Number | Pitch, mm | Depth of Thread, mm | Bolt and Nut | | | Radius, mm | Threads per Inch (approx.) |
|---|---|---|---|---|---|---|---|
| | | | Major Diameter, mm | Effective Diameter, mm | Minor Diameter, mm | | |
| 0 | 1.0000 | 0.600 | 6.00 | 5.400 | 4.80 | 0.1808 | 25.4 |
| 1 | 0.9000 | 0.540 | 5.30 | 4.760 | 4.22 | 0.1627 | 28.2 |
| 2 | 0.8100 | 0.485 | 4.70 | 4.215 | 3.73 | 0.1465 | 31.4 |
| 3 | 0.7300 | 0.440 | 4.10 | 3.660 | 3.22 | 0.1320 | 34.8 |
| 4 | 0.6600 | 0.395 | 3.60 | 3.205 | 2.81 | 0.1193 | 38.5 |
| 5 | 0.5900 | 0.355 | 3.20 | 2.845 | 2.49 | 0.1067 | 43.0 |
| 6 | 0.5300 | 0.320 | 2.80 | 2.480 | 2.16 | 0.0958 | 47.9 |
| 7 | 0.4800 | 0.290 | 2.50 | 2.210 | 1.92 | 0.0868 | 52.9 |
| 8 | 0.4300 | 0.260 | 2.20 | 1.940 | 1.68 | 0.0778 | 59.1 |
| 9 | 0.3900 | 0.235 | 1.90 | 1.665 | 1.43 | 0.0705 | 65.1 |
| 10 | 0.3500 | 0.210 | 1.70 | 1.490 | 1.28 | 0.0633 | 72.6 |
| 11 | 0.3100 | 0.185 | 1.50 | 1.315 | 1.13 | 0.0561 | 82.0 |
| 12 | 0.2800 | 0.170 | 1.30 | 1.130 | 0.96 | 0.0506 | 90.7 |
| 13 | 0.2500 | 0.150 | 1.20 | 1.050 | 0.90 | 0.0452 | 102 |
| 14 | 0.2300 | 0.140 | 1.00 | 0.860 | 0.72 | 0.0416 | 110 |
| 15 | 0.2100 | 0.125 | 0.90 | 0.775 | 0.65 | 0.0380 | 121 |
| 16 | 0.1900 | 0.115 | 0.79 | 0.675 | 0.56 | 0.0344 | 134 |

*Tolerances and Allowances:* Two classes of bolts and one for nuts are provided: *Close Class bolts* are intended for precision parts subject to stress, no allowance being provided between maximum bolt and minimum nut sizes. *Normal Class bolts* are intended for general commercial production and general engineering use; for sizes 0 to 10 BA, an allowance of 0.025 mm is provided.

**Tolerance Formulas for British Association (BA) Screw Threads**

| | Class or Fit | Tolerance (+ for nuts, – for bolts) | | |
|---|---|---|---|---|
| | | Major Dia. | Effective Dia. | Minor Dia. |
| Bolts | Close Class 0 to 10 BA incl. | $0.15p$ mm | $0.08p + 0.02$ mm | $0.16p + 0.04$ mm |
| | Normal Class 0 to 10 BA incl. | $0.20p$ mm | $0.10p + 0.025$ mm | $0.20p + 0.05$ mm |
| | Normal Class 11 to 16 BA incl. | $0.25p$ mm | $0.10p + 0.025$ mm | $0.20p + 0.05$ mm |
| Nuts | All Classes | | $0.12p + 0.03$ mm | $0.375p$ mm |

In these formulas, $p$ = pitch in millimeters.

**Instrument Makers' Screw Thread System.**—The standard screw system of the Royal Microscopical Society of London, also known as the "Society Thread," is employed for microscope objectives and the nose pieces of the microscope into which these objectives screw. The form of the thread is the standard Whitworth form. The number of threads per inch is 36. There is one size only. The maximum pitch diameter of the objective is 0.7804 inch and the minimum pitch diameter of the nose-piece is 0.7822 inch. The dimensions are as follows:

| | | | |
|---|---|---|---|
| Male thread | outside dia. | max., 0.7982 inch | min., 0.7952 inch |
| | root dia. | max., 0.7626 inch | min., 0.7596 inch |
| Female thread | root of thread | max., 0.7674 inch | min., 0.7644 inch |
| | top of thread | max., 0.8030 inch | min., 0.8000 inch |

The Royal Photographic Society Standard Screw Thread ranges from 1-inch diameter upward. For screws less than 1 inch, the Microscopical Society Standard is used. The British Association thread is another thread system employed on instruments abroad.

**American Microscope Objective Thread (AMO).**—The Standard, ANSI/ASME B1.11-1958 (R2016), describes the American microscope objective thread, AMO, the screw thread form used for mounting a microscope objective assembly to the body or lens turret of a microscope. This screw thread is also recommended for other microscope optical assembles as well as related applications such as photomicrographic equipment. It is based on, and intended to be interchangeable with, the screw thread produced and adopted many years ago by the Royal Microscopical Society of Great Britain, generally known as the RMS thread. While the standard is almost universally accepted as the basic standard for microscope objective mountings, formal recognition has been extremely limited.

The basic thread possesses the overall British Standard Whitworth form. (See *Whitworth Standard Thread Form* starting on page 2089). However, the actual design thread form implementation is based on the WWII era ASA B1.6-1944 "Truncated Whitworth Form" in which the rounded crests and roots are removed. ASA B1.6-1944 was withdrawn in 1951. However, ANSI/ASME B1.11-1958 (R2016) is still active for new design.

*Design Requirements of Microscope Objective Threads:* Due to the inherent longevity of optical equipment and the repeated use to which the objective threads are subjected, the following factors should be considered when designing microscope objective threads:

Adequate clearance to afford protection against binding due to the presence of foreign particles or minor crest damage.

Sufficient depth of thread engagement to assure security in the short lengths of engagement commonly encountered.

Allowances for limited eccentricities so that centralization and squareness of the objective are not influenced by such errors in manufacture.

*Deviation from the Truncated Whitworth Thread Form:* Although ANSI/ASME B1.11-1958 (R2016) is based on the withdrawn ASA B1.6-1944 truncated Whitworth Standard, the previously described design requirements necessitate a deviation from the truncated Whitworth thread form. Some of the more significant modifications are:

A larger allowance on the pitch diameter of the external thread.

Smaller tolerances on the major diameter of the external thread and minor diameter of the internal thread.

The provision of allowances on the major and minor diameters of the external thread.

*Thread Overview:* The thread is a single start type. There is only one class of thread based on a basic major diameter of 0.800 inch and a pitch, $p$, of 0.027778 inch (36 threads per inch). The AMO thread shall be designated on drawings, tools and gages as "0.800-36 AMO." Thread nomenclature, definitions and terminology are based on ANSI/ASME B1.7-2006 (R2021), " Screw Threads: Nomenclature, Definitions, and Letter Symbols."

It should also be noted that ISO 8038-1:1997 "Screw threads for objectives and related nosepieces" is also based on the 0.800 inch, 36 tpi RMS thread form.

*Tolerances and Allowances:* Tolerances are given in Table 2. A positive allowance (minimum clearance) of 0.0018 inch is provided for the pitch diamter $E$, major diameter $D$, and minor diameter, $K$

If interchangeability with full-form Whitworth threads is not required, the allowances for the major and minor diameters are not necessary, because the forms at the root and crest are truncated. In these cases, either both limits or only the maximum limit of the major and minor diameters may be increased by the amount of the allowance, 0.0018 inch.

*Lengths of Engagement:* The tolerances specified in Table 2 are applicable to lengths of engagement ranging from $\frac{1}{8}$ to $\frac{3}{8}$ inch, approximately 15 to 50 percent of the basic diameter. Microscope objective assembles generally have a length of engagement of $\frac{1}{8}$ inch. Lengths exceeding these limits are seldom employed and not covered in this Standard.

*Gage Testing:* Recommended ring and plug testing gage dimensions for the 0.800-36 AMO thread size can be found in ANSI/ASME B1.11-1958 (R2016), Appendix.

*Dimensional Terminology:* Because the active Standard ANSI/ASME B1.11-1958 (R2016) is based on the withdrawn ASA Truncated Whitworth standard, dimensional nomenclature is described below.

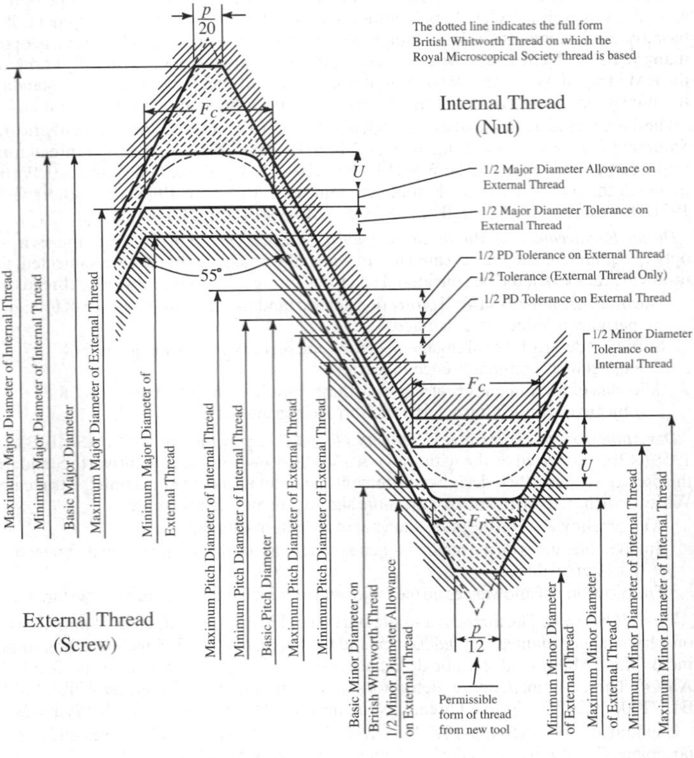


Tolerances, Allowances and Crest Clearances for Microscope Objective Thread (AMO)
ANSI/ASME B1.11-1958 (R2016)

## Table 1. Definitions, Formulas, Basic and Design Dimensions
*ANSI/ASME B1.11-1958 (R2016)*

| Symbol | Property | Formula | Dimension |
|---|---|---|---|
| | Basic Thread Form | | |
| $\alpha$ | Half angle of thread | … | 27°30′ |
| $2\alpha$ | Included angle of thread | … | 55°00′ |
| $n$ | Number of threads per inch | … | 36 |
| $p$ | Pitch | $1/n$ | 0.027778 |
| $H$ | Height of fundamental triangle | $0.960491p$ | 0.026680 |
| $h_b$ | Height of basic thread | $0.640327p$ | 0.0178 |
| $r$ | Radius at crest and root of British Standard Whitworth basic thread (not used) | $0.137329p$ | 0.0038 |

**Table 1.** *(Continued)* **Definitions, Formulas, Basic and Design Dimensions**
*ANSI/ASME B1.11-1958 (R2016)*

| Symbol | Property | Formula | Dimension |
|---|---|---|---|
| Design Thread Form | | | |
| $k$ | Height of truncated Whitworth thread | $h_b - U = 0.566410p$ | 0.0157 |
| $F_c$ | Width of flat at crest | $0.243624p$ | 0.0068 |
| $F_r$ | Width of flat at root | $0.166667p$ | 0.0046 |
| $U$ | Basic truncation of crest from basic Whitworth form | $0.073917p$ | 0.00205 |
| Basic and Design Sizes | | | |
| $D$ | Major diameter, nominal and basic | … | 0.800 |
| $D_n$ | Major diameter of internal thread | $D$ | 0.800 |
| $D_s$ | Major diameter of external thread[a] | D - 2U - G | 0.7941 |
| $E$ | Pitch (effective) diameter, basic | $D - h_b$ | 0.7822 |
| $E_n$ | Pitch (effective) diameter of internal thread | $D - h_b$ | 0.7822 |
| $E_s$ | Pitch (effective) diameter of external thread[b] | $D - h_b$ - G | 0.7804 |
| $K$ | Minor diameter, basic | $D - 2h_b$ | 0.7644 |
| $K_n$ | Minor diameter of internal thread | $D - 2k$ | 0.7685 |
| $K_s$ | Minor diameter of external thread[a] | $D - 2h_b$ - G | 0.7626 |
| $G$ | Allowance at pitch (effective) diameter[a, b] | … | 0.0018 |

[a] An allowance equal to that on the pitch diameter is also provided on the major and minor diameters of the external thread for additional clearance and centralizing.

[b] Allowance (minimum clearance) on pitch (effective) diameter is the same as the British RMS thread.

All dimensions are in inches.

**Table 2. Limits of Size and Tolerances—0.800-36 AMO Thread**
*ANSI/ASME B1.11-1958 (R2016)*

| Element | Major Diameter, *D* | | | Pitch Diameter, *E* | | | Minor Diameter, *K* | | |
|---|---|---|---|---|---|---|---|---|---|
| | Max. | Min. | Tol. | Max. | Min. | Tol. | Max. | Min. | Tol. |
| External thread | 0.7941 | 0.7911 | 0.0030 | 0.7804 | 0.7774 | 0.0030 | 0.7626 | 0.7552[a] | … |
| Internal thread | 0.8092[b] | 0.8000 | … | 0.7852 | 0.7822 | 0.0030 | 0.7715 | 0.7685 | 0.0030 |

[a] Extreme minimum minor diameter produced by a new threading tool having a minimum flat of $p/12 = 0.0023$ inch. This minimum diameter is not controlled by gages but by the form of the threading tool.

[b] Extreme maximum major diameter produced by a new threading tool having a minimum flat of $p/20 = 0.0014$ inch. This maximum diameter is not controlled by gages but by the form of the threading tool.

Tolerances on the internal thread are applied in a plus direction from the basic and design size and tolerances on the external thread are applied in a minus direction from its design (maximum material) size.

All dimensions are in inches.

**Löwenherz or Lowenherz Thread.—**The Löwenherz thread is intended for the fine screws of instruments and is based on the metric system. It is a symmetrical thread, not a buttress thread. The Löwenherz thread has flats at the top and bottom the same as the US standard buttress form, but the angle is 53 degrees 8 minutes. The depth equals 0.75 × the pitch, and the width of the flats at the top and bottom is equal to 0.125 × the pitch. This screw thread is used for the fine threads of measuring instruments, optical apparatus, etc., especially in Germany.

## Löwenherz Thread

| Diameter | | Pitch, Millimeters | Approximate No. of Threads per Inch | Diameter | | Pitch, Millimeters | Approximate No. of Threads per Inch |
|---|---|---|---|---|---|---|---|
| Millimeters | Inches | | | Millimeters | Inches | | |
| 1.0 | 0.0394 | 0.25 | 101.6 | 9.0 | 0.3543 | 1.30 | 19.5 |
| 1.2 | 0.0472 | 0.25 | 101.6 | 10.0 | 0.3937 | 1.40 | 18.1 |
| 1.4 | 0.0551 | 0.30 | 84.7 | 12.0 | 0.4724 | 1.60 | 15.9 |
| 1.7 | 0.0669 | 0.35 | 72.6 | 14.0 | 0.5512 | 1.80 | 14.1 |
| 2.0 | 0.0787 | 0.40 | 63.5 | 16.0 | 0.6299 | 2.00 | 12.7 |
| 2.3 | 0.0905 | 0.40 | 63.5 | 18.0 | 0.7087 | 2.20 | 11.5 |
| 2.6 | 0.1024 | 0.45 | 56.4 | 20.0 | 0.7874 | 2.40 | 10.6 |
| 3.0 | 0.1181 | 0.50 | 50.8 | 22.0 | 0.8661 | 2.80 | 9.1 |
| 3.5 | 0.1378 | 0.60 | 42.3 | 24.0 | 0.9450 | 2.80 | 9.1 |
| 4.0 | 0.1575 | 0.70 | 36.3 | 26.0 | 1.0236 | 3.20 | 7.9 |
| 4.5 | 0.1772 | 0.75 | 33.9 | 28.0 | 1.1024 | 3.20 | 7.9 |
| 5.0 | 0.1968 | 0.80 | 31.7 | 30.0 | 1.1811 | 3.60 | 7.1 |
| 5.5 | 0.2165 | 0.90 | 28.2 | 32.0 | 1.2599 | 3.60 | 7.1 |
| 6.0 | 0.2362 | 1.00 | 25.4 | 36.0 | 1.4173 | 4.00 | 6.4 |
| 7.0 | 0.2756 | 1.10 | 23.1 | 40.0 | 1.5748 | 4.40 | 5.7 |
| 8.0 | 0.3150 | 1.20 | 21.1 | … | … | … | … |

**Swiss Screw Thread.—**This is a thread system originated in Switzerland as a standard for screws used in watch and clock making. The angle between the two sides of the thread is 47 degrees 30 minutes, and the top and bottom of the thread are rounded. This system is used by the British Association as a standard for small screws, and is known as the British Association thread. See *British Association Standard Thread (BA)* on page 2117.

## Historical and Miscellaneous Threads

**Aero-Thread.—**The name "Aero-thread" has been applied to a patented screw thread system that is applicable where the nut or internally threaded part is made from a soft material, such as aluminum or magnesium alloy, for lightness, as in aircraft construction, and where the screw is made from a high-strength steel to provide strength and good wearing qualities. The nut or part containing the internal thread has a 60-degree truncated form of thread. See Fig. 1. The screw, or stud, is provided with a semi-circular thread form, as shown. Between the screw and the nut there is an intermediary part known as a thread lining or insert, made in the form of a helical spring, so that it can be screwed into the nut. The stud is then screwed into the thread formed by the semicircular part of the thread insert. When the screw is provided with a V-form of thread, like the American Standard, frequent loosening and tightening of the screw would cause rapid wear of the softer metal from which the nut is made; furthermore, all the threads might not have an even bearing on the mating threads. By using a thread insert which is screwed into the nut permanently and made from a reasonably hard material, good wearing qualities are obtained. Also, the bearing or load is evenly distributed over the threads of the nut since the insert, being in the form of a spring, can adjust itself to bear on all the thread surfaces.

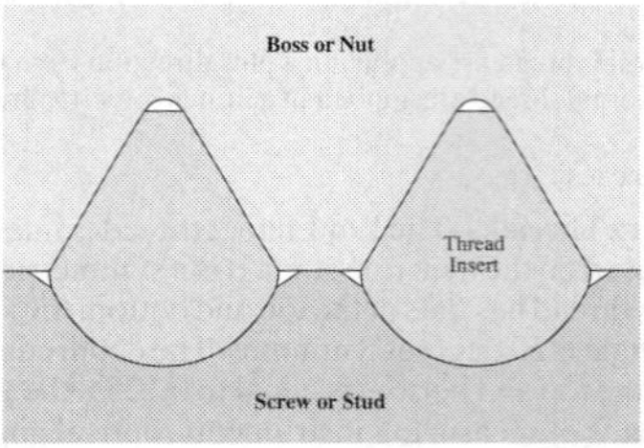


Fig. 1. The Basic Thread Form Used in the Aero-Thread System

**Briggs Pipe Thread.—**The Briggs pipe thread (now known as the American Standard) is used for threaded pipe joints and is the standard for this purpose in the United States. It derives its name from Robert Briggs.

**Casing Thread.—**The standard casing thread of the American Petroleum Institute has an included angle of 60 degrees and a taper of $^3/_4$ inch per foot. The fourteen casing sizes listed in the 1942 revision have outside diameters ranging from $4^1/_2$ to 20 inches. All sizes have 8 threads per inch.

*Rounded Thread Form:* Threads for casing sizes up to $13^3/_8$ inches, inclusive, have rounded crests and roots, and the depth, measured perpendicular to the axis of the pipe, equals $0.626 \times \text{pitch} - 0.007 = 0.07125$ inch.

*Truncated Form:* Threads for the 16-and 20-inch casing sizes have flat crests and roots. The depth equals $0.760 \times \text{pitch} = 0.0950$ inch. This truncated form is designated in the A.P.I. Standard as a "sharp thread."

**Cordeaux Thread.—**The Cordeaux screw thread derives its name from John Henry Cordeaux, an English telegraph inspector who obtained a patent for this thread in 1877. This thread is used for connecting porcelain insulators with their stalks by means of a screw thread on the stalk and a corresponding thread in the insulator. The thread is approximately a Whitworth thread, 6 threads per inch, the diameters most commonly used being $^5/_8$ or $^3/_4$ inch outside diameter of thread; $^5/_8$ inch is almost universally used for telegraph purposes, while a limited number of $^3/_4$-inch sizes are used for large insulators.

**Dardelet Thread.—**The Dardelet patented self-locking thread is designed to resist vibrations and remain tight without auxiliary locking devices. The locking surfaces are the tapered root of the bolt thread and the tapered crest of the nut thread. The nut is free to turn until seated tightly against a resisting surface, thus causing it to shift from the free position (indicated by dotted lines) to the locking position. The locking is due to a wedging action between the tapered crest of the nut thread and the tapered root or binding surface of the bolt thread. This self-locking thread is also applied to set screws and cap screws. The holes must, of course, be threaded with Dardelet taps. The abutment sides of the Dardelet thread carry the major part of the tensile load. The nut is unlocked simply by turning it backward with a wrench. The Dardelet thread can either be cut or rolled, using standard equipment provided with tools, taps, dies, or rolls made to suit the Dardelet thread profile. The included thread angle is 29 degrees; depth $E = 0.3P$; maximum axial movement = $0.28\,P$. The major internal thread diameter (standard series) equals major external thread diameter plus 0.003 inch except for $^1/_4$-inch size which is plus 0.002 inch. The width of both external and internal threads at pitch line equals $0.36\,P$.

**"Drunken" Thread.—**A "drunken" thread, according to prevalent usage of this expression by machinists, etc., is a thread that does not coincide with a true helix or advance uniformly. This irregularity in a taper thread may be due to the fact that in taper turning with the tailstock set over, the work does not turn with a uniform angular velocity, while the cutting tool is advancing along the work longitudinally with a uniform linear velocity. The change in the pitch and the irregularity of the thread is so small as to be imperceptible to the eye, if the taper is slight, but as the tapers increase to, say, $^3/_4$ inch per foot or more, the errors become more pronounced. To avoid this defect, a taper attachment should be used for taper thread cutting.

**Echols Thread.—**Chip room is of great importance in machine taps and tapper taps where the cutting speed is high and always in one direction. The tap as well as the nut to be threaded is liable to be injured, if ample space for the chips to pass away from the cutting edges is not provided. A method of decreasing the number of cutting edges, as well as increasing the amount of chip room, is embodied in the "Echols thread," where every alternate tooth is removed. If a tap has an even number of flutes, the removal of every other tooth in the lands will be equivalent to the removal of the teeth of a continuous thread. It is, therefore, necessary that taps provided with this thread be made with an odd number of lands, so that removing the tooth in alternate lands may result in removing every other tooth in each individual land. Machine taps are often provided with the Echols thread.

**French Thread (SF).—**The French thread has the same form and proportions as the American Standard (formerly US Standard). This French thread is being displaced gradually by the International Metric Thread System.

**Harvey Grip Thread.—**The characteristic feature of this thread is that one side inclines 44 degrees from a line at right angles to the axis, whereas the other side has an inclination of only 1 degree. This form of thread is sometimes used when there is considerable resistance or pressure in an axial direction and when it is desirable to reduce the radial or bursting pressure on the nut as much as possible. See *BUTTRESS THREADS* on page 2081.

**Lloyd & Lloyd Thread.—**The Lloyd & Lloyd screw thread is the same as the regular Whitworth screw thread in which the sides of the thread form an angle of 55 degrees with one another. The top and bottom of the thread are rounded.

**Lock-Nut Pipe Thread.—**The lock-nut pipe thread is a straight thread of the largest diameter which can be cut on a pipe. Its form is the same as the American or Briggs standard taper pipe thread. In general, "Go" gages only are required. These consist of a straight-threaded plug representing the minimum female lock-nut thread, and a straight-threaded ring representing the maximum male lock-nut thread. This thread is used only to hold parts together, or to retain a collar on the pipe. It is never used where a tight threaded joint is required.

**Philadelphia Carriage Bolt Thread.—**This is a screw thread for carriage bolts which is somewhat similar to a square thread, but having rounded corners at the top and bottom. The sides of the thread are inclined to an inclusive angle of 3½ degrees. The width of the thread at the top is 0.53 times the pitch.

**SAE Standard Screw Thread.—**The screw thread standard of the Society of Automotive Engineers (SAE) is intended for use in the automotive industries of the United States. The SAE Standard includes a Coarse series, a Fine series, an 8-thread series, a 12-thread series, a 16-thread series, an Extra-fine series, and a Special-pitch series. The Coarse and Fine series, and also the 8-, 12- and 16-thread series, are identical to the corresponding American Standard series. The Extra-fine and Special-pitch series are SAE Standards only.

The American Standard thread *form* (or the form previously known as the US Standard) is applied to all SAE Standard screw threads. The Extra-fine series has a total of six pitches ranging from 32 down to 16 threads per inch. The 16 threads per inch in the Extra-fine series applies to all diameters from 1¾ up to 6 inches. This Extra-fine series is intended for use on relatively light sections; on parts requiring fine adjustment; where jar and vibration are important factors; when the thickness of a threaded section is relatively small as in tubing, and where assembly is made without the use of wrenches.

The SAE Special pitches include some which are finer than any in the Extra-fine series. The special pitches apply to a range of diameters extending from No. 10 (0.1900 inch) up to 6 inches. Each diameter has a range of pitches varying from five to eight. For example, a ¼-inch diameter has six pitches ranging from 24 to 56 threads per inch, whereas a 6-inch diameter has eight pitches ranging from 4 to 16 threads per inch. These various SAE Standard series are intended to provide adequate screw thread specifications for all uses in the automotive industries.

**Sellers Screw Thread.—**The Sellers screw thread, later known as the 'United States standard thread," and now as the "American Standard," is the most commonly used screw thread in the United States. It was originated by William Sellers, of Philadelphia, and first proposed by him in a paper read before the Franklin Institute, in April, 1864. In 1868, it was adopted by the United States Navy and has since become the generally accepted standard screw thread in the United States.

**Worm Threads.—**The included angle of worm threads range from 29° to 60°; for single-threaded worms 29° is common; multiple-threaded type must have larger helix and thread angles to avoid excessive under-cutting in hobbing the worm-wheel teeth. AGMA recommends 40° included thread angle for triple- and quadruple-thread worms, but many speed reducers and transmissions have 60° thread angles. The 29° angle is the same as the Acme thread, but worm thread depth is greater and widths of the flats at the top and bottom are less. If lead angle is larger than 20°, an increase in included thread angle is desirable. Worm gearing reaches maximum efficiency when lead angle is 45°, thus explaining the 60° thread angle. Thread parts of a 29° worm thread are: $p$ = pitch; $d$ = depth of thread = $0.6866p$; $t$ = width, top of thread = $0.335p$; $b$ = width, bottom of thread = $0.310p$.

# MEASURING SCREW THREADS

## Measuring Screw Threads

**Pitch and Lead of Screw Threads.—**The *pitch* of a screw thread is the distance from the center of one thread to the center of the next thread. This applies no matter whether the screw has a single, double, triple or quadruple thread. The *lead* of a screw thread is the distance the nut will move forward on the screw if it is turned around one full revolution. In a single-threaded screw, the pitch and lead are equal, because the nut would move forward the distance from one thread to the next, if turned around once. In a double-threaded screw, the nut will move forward two threads, or twice the pitch, so that in this case the lead equals twice the pitch. In a triple-threaded screw, the lead equals three times the pitch, and so on.

The word "pitch" is often, although improperly, used to denote the *number of threads per inch*. Screws are spoken of as having a 12-pitch thread, when twelve threads per inch is what is really meant. The number of threads per inch equals 1 divided by the pitch, or expressed as a formula:

$$\text{Number of threads per inch} = \frac{1}{\text{pitch}}$$

The pitch of a screw equals 1 divided by the number of threads per inch, or:

$$\text{Pitch} = \frac{1}{\text{number of threads per inch}}$$

If the number of threads per inch equals 16, the pitch = $\frac{1}{16}$. If the pitch equals 0.05, the number of threads equals 1 ÷ 0.05 = 20. If the pitch is $\frac{2}{5}$ inch, the number of threads per inch equals $1 \div \frac{2}{5} = 2\frac{1}{2}$.

Confusion is often caused by the indefinite designation of multiple-thread screws (double, triple, quadruple, etc.). The expression, "four threads per inch, triple," for example, is not to be recommended. It means that the screw is cut with four triple threads or with twelve threads per inch, if the threads are counted by placing a scale alongside the screw. To cut this screw, the lathe would be geared to cut four threads per inch, but they would be cut only to the depth required for twelve threads per inch. The best expression, when a multiple-thread is to be cut, is to say, in this case, "$\frac{1}{4}$ inch lead, $\frac{1}{12}$ inch pitch, triple thread." For single-threaded screws, only the number of threads per inch and the form of the thread are specified. The word "single" is not required.

**Measuring Screw Thread Pitch Diameters by Thread Micrometers.—**As the pitch or angle diameter of a tap or screw is the most important dimension, it is necessary that the pitch diameter of screw threads be measured, in addition to the outside diameter.

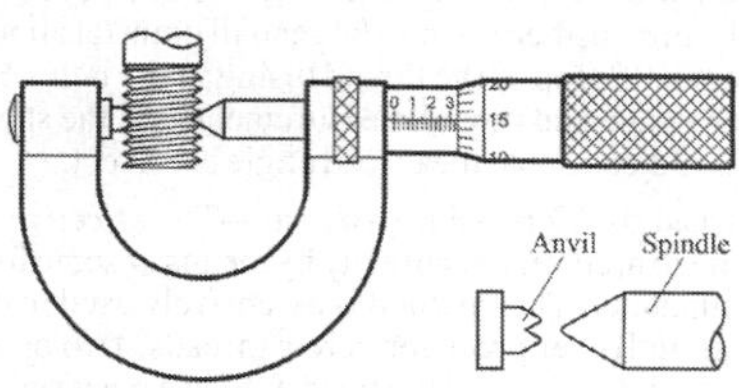


Fig. 1.

One method of measuring in the angle of a thread is by means of a special screw thread micrometer, as shown in the accompanying engraving, Fig. 1. The fixed anvil is W-shaped to engage two thread flanks, and the movable point is cone-shaped so as to enable it to enter the space between two threads, and at the same time be at liberty to revolve. The contact points are on the sides of the thread, as they necessarily must be in order that the pitch diameter may be determined. The cone-shaped point of the measuring screw is slightly rounded so that it will not bear in the bottom of the thread. There is also sufficient clearance at the bottom of the V-shaped anvil to prevent it from bearing on the top of the thread. The movable point is adapted to measuring all pitches, but the fixed anvil is limited in its capacity. To cover the whole range of pitches, from the finest to the coarsest, a number of fixed anvils are therefore required.

To find the theoretical pitch diameter, which is measured by the micrometer, subtract twice the addendum of the thread from the standard outside diameter. The addendum of the thread for the American and other standard threads is given in the section on screw thread systems.

**Ball-Point Micrometers.—**If standard plug gages are available, it is not necessary to actually measure the pitch diameter, but merely to compare it with the standard gage. In this case, a ball-point micrometer, as shown in Fig. 2, may be employed. Two types of ball-point micrometers are ordinarily used. One is simply a regular plain micrometer with ball points made to slip over both measuring points. (See B, Fig. 2.) This makes a kind of combination plain and ball-point micrometer, the ball points being easily removed. These ball points, however, do not fit solidly on their seats, even if they are split, as shown, and are apt to cause errors in measurements. The best, and, in the long run, the cheapest, method is to use a regular micrometer arranged as shown at *A*. Drill and ream out both the end of the measuring screw or spindle and the anvil, and fit ball points into them as shown. Care should be taken to have the ball point in the spindle run true. The holes in the micrometer spindle and anvil and the shanks on the points are tapered to ensure a good fit. The hole *H* in spindle *G* is provided so that the ball point can be easily driven out when a change for a larger or smaller size of ball point is required.

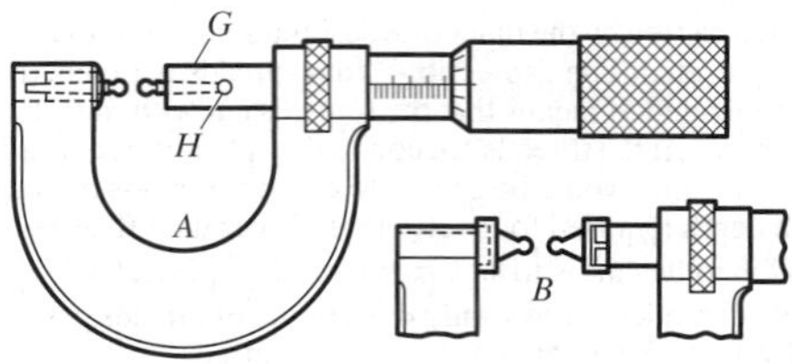


Fig. 2.

A ball-point micrometer may be used for comparing the *angle* of a screw thread, with that of a gage. This can be done by using different sizes of ball points, comparing the size first near the root of the thread, then (using a larger ball point) at about the point of the pitch diameter, and finally near the top of the thread (using in the latter case, of course, a much larger ball point). If the gage and thread measurements are the same at each of the three points referred to, this indicates that the thread angle is correct.

**Measuring Screw Threads by Three-Wire Method.—**The *effective* or *pitch diameter* of a screw thread may be measured very accurately by means of some form of micrometer and three wires of equal diameter. This method is extensively used in checking the accuracy of threaded plug gages and other precision screw threads. Two of the wires are placed in contact with the thread on one side and the third wire in a position diametrically opposite as illustrated by the diagram, (see table *Formulas for Checking Pitch Diameters of Screw Threads* on page 2131) and the dimension over the wires is determined by means of a micrometer. An ordinary micrometer is commonly used but some form of "floating micrometer" is preferable, especially for measuring thread gages and other precision work. The floating micrometer is mounted upon a compound slide so that it can move freely in directions parallel or at right angles to the axis of the screw, which is held in a horizontal position between adjustable centers. With this arrangement the micrometer is held constantly at right angles to the axis of the screw so that only one wire on each side may be used instead of having two on one side and one on the other, as is necessary when using an ordinary micrometer. The pitch diameter may be determined accurately if the correct micrometer reading for wires of a given size is known.

**Classes of Formulas for Three-Wire Measurement.—**Various formulas have been established for checking the pitch diameters of screw threads by measurement over wires of known size. These formulas differ with regard to their simplicity or complexity and resulting accuracy. They also differ in that some show what measurement $M$ over the wires should be to obtain a given pitch diameter $E$, whereas others show the value of the pitch diameter $E$ for a given measurement $M$.

*Formulas for Finding Measurement M:* In using a formula for finding the value of measurement $M$, the required pitch diameter $E$ is inserted in the formula. Then, in cutting or grinding a screw thread, the actual measurement $M$ is made to conform to the calculated value of $M$. Formulas for finding measurement $M$ may be modified so that the basic major or outside diameter is inserted in the formula instead of the pitch diameter; however, the pitch-diameter type of formula is preferable because the pitch diameter is a more important dimension than the major diameter.

*Formulas for Finding Pitch Diameters E:* Some formulas are arranged to show the value of the pitch diameter $E$ when measurement $M$ is known. Thus, the value of $M$ is first determined by measurement and then is inserted in the formula for finding the corresponding pitch diameter $E$. This type of formula is useful for determining the pitch diameter of an existing thread gage or other screw thread in connection with inspection work. The formula for finding measurement $M$ is more convenient to use in the shop or tool room in cutting or grinding new threads, because the pitch diameter is specified on the drawing and the problem is to find the value of measurement $M$ for obtaining that pitch diameter.

**General Classes of Screw Thread Profiles.—**Thread profiles may be divided into three general classes or types as follows:

*Screw Helicoid:* Represented by a screw thread having a straight-line profile in the axial plane. Such a screw thread may be cut in a lathe by using a straight-sided single-point tool, provided the top surface lies in the axial plane.

*Involute Helicoid:* Represented either by a screw thread or a helical gear tooth having an involute profile in a plane perpendicular to the axis. A rolled screw thread, theoretically at least, is an exact involute helicoid.

*Intermediate Profiles:* An intermediate profile that lies somewhere between the screw helicoid and the involute helicoid will be formed on a screw thread either by milling or grinding with a straight-sided wheel set in alignment with the thread groove. The resulting form will approach closely the involute helicoid form. In milling or grinding a thread, the included cutter or wheel angle may either equal the standard thread angle (which is always measured in the axial plane) or the cutter or wheel angle may be reduced to approximate, at least, the thread angle in the normal plane. In practice, all these variations affect the three-wire measurement.

**Accuracy of Formulas for Checking Pitch Diameters by Three-Wire Method.—**The exact measurement $M$ for a given pitch diameter depends upon the lead angle, the thread angle, and the profile or cross-sectional shape of the thread. As pointed out in the preceding paragraph, the profile depends upon the method of cutting or forming the thread. In a milled or ground thread, the profile is affected not only by the cutter or wheel angle, but also by the diameter of the cutter or wheel; hence, because of these variations, an absolutely exact and reasonably simple general formula for measurement $M$ cannot be established; however, if the lead angle is low, as with a standard single-thread screw, and especially if the thread angle is high like a 60-degree thread, simple formulas that are not arranged to compensate for the lead angle are used ordinarily and meet most practical requirements, particularly in measuring 60-degree threads. If lead angles are large enough to greatly affect the result, as with most multiple threads (especially Acme or 29-degree worm threads), a formula should be used that compensates for the lead angle sufficiently to obtain the necessary accuracy.

The formulas that follow include 1) a very simple type in which the effect of the lead angle on measurement $M$ is entirely ignored. This simple formula usually is applicable to the measurement of 60-degree single-thread screws, except possibly when gage-making accuracy is required; 2) formulas that do include the effect of the lead angle but, nevertheless, are approximations and not always suitable for the higher lead angles when extreme accuracy is required; and 3) formulas for the higher lead angles and the most precise classes of work.

Where approximate formulas are applied consistently in the measurement of both thread plug gages and the thread "setting plugs" for ring gages, interchangeability might be secured, assuming that such approximate formulas were universally employed.

**Wire Sizes for Checking Pitch Diameters of Screw Threads.—**In checking screw threads by the 3-wire method, the general practice is to use measuring wires of the so-called "best size." The "best-size" wire is one that contacts at the pitch line or midslope of the thread because then the measurement of the pitch diameter is least affected by an error in the thread angle. In the following formula for determining approximately the "best-size" wire or the diameter for pitch-line contact, $A$ = one-half included angle of thread in the axial plane.

$$\text{Best-size wire} = \frac{0.5 \times \text{pitch}}{\cos A} = 0.5\,\text{pitch} \times \sec A$$

For 60-degree threads, this formula reduces to

$$\text{Best-size wire} = 0.57735 \times \text{pitch}$$

## Diameters of Wires for Measuring American Standard and British Standard Whitworth Screw Threads

| Threads per Inch | Pitch, Inch | Wire Diameters for American Standard Threads | | | Wire Diameters for Whitworth Standard Threads | | |
|---|---|---|---|---|---|---|---|
| | | Max. | Min. | Pitch-Line Contact | Max. | Min. | Pitch-Line Contact |
| 4 | 0.2500 | 0.2250 | 0.1400 | 0.1443 | 0.1900 | 0.1350 | 0.1409 |
| 4½ | 0.2222 | 0.2000 | 0.1244 | 0.1283 | 0.1689 | 0.1200 | 0.1253 |
| 5 | 0.2000 | 0.1800 | 0.1120 | 0.1155 | 0.1520 | 0.1080 | 0.1127 |
| 5½ | 0.1818 | 0.1636 | 0.1018 | 0.1050 | 0.1382 | 0.0982 | 0.1025 |
| 6 | 0.1667 | 0.1500 | 0.0933 | 0.0962 | 0.1267 | 0.0900 | 0.0939 |
| 7 | 0.1428 | 0.1283 | 0.0800 | 0.0825 | 0.1086 | 0.0771 | 0.0805 |
| 8 | 0.1250 | 0.1125 | 0.0700 | 0.0722 | 0.0950 | 0.0675 | 0.0705 |
| 9 | 0.1111 | 0.1000 | 0.0622 | 0.0641 | 0.0844 | 0.0600 | 0.0626 |
| 10 | 0.1000 | 0.0900 | 0.0560 | 0.0577 | 0.0760 | 0.0540 | 0.0564 |
| 11 | 0.0909 | 0.0818 | 0.0509 | 0.0525 | 0.0691 | 0.0491 | 0.0512 |
| 12 | 0.0833 | 0.0750 | 0.0467 | 0.0481 | 0.0633 | 0.0450 | 0.0470 |
| 13 | 0.0769 | 0.0692 | 0.0431 | 0.0444 | 0.0585 | 0.0415 | 0.0434 |
| 14 | 0.0714 | 0.0643 | 0.0400 | 0.0412 | 0.0543 | 0.0386 | 0.0403 |
| 16 | 0.0625 | 0.0562 | 0.0350 | 0.0361 | 0.0475 | 0.0337 | 0.0352 |
| 18 | 0.0555 | 0.0500 | 0.0311 | 0.0321 | 0.0422 | 0.0300 | 0.0313 |
| 20 | 0.0500 | 0.0450 | 0.0280 | 0.0289 | 0.0380 | 0.0270 | 0.0282 |
| 22 | 0.0454 | 0.0409 | 0.0254 | 0.0262 | 0.0345 | 0.0245 | 0.0256 |
| 24 | 0.0417 | 0.0375 | 0.0233 | 0.0240 | 0.0317 | 0.0225 | 0.0235 |
| 28 | 0.0357 | 0.0321 | 0.0200 | 0.0206 | 0.0271 | 0.0193 | 0.0201 |
| 32 | 0.0312 | 0.0281 | 0.0175 | 0.0180 | 0.0237 | 0.0169 | 0.0176 |
| 36 | 0.0278 | 0.0250 | 0.0156 | 0.0160 | 0.0211 | 0.0150 | 0.0156 |
| 40 | 0.0250 | 0.0225 | 0.0140 | 0.0144 | 0.0190 | 0.0135 | 0.0141 |

These formulas are based upon a thread groove of zero lead angle because ordinary variations in the lead angle have little effect on the wire diameter and it is desirable to use one wire size for a given pitch regardless of the lead angle. A theoretically correct solution for finding the *exact* size for pitch-line contact involves the use of cumbersome indeterminate equations with solution by successive trials. The accompanying table gives the wire sizes for both American Standard (formerly, US Standard) and the Whitworth Standard Threads. The following formulas for determining wire diameters do not give the extreme theoretical limits, but the smallest and largest practicable sizes. The diameters in the table are based upon these approximate formulas.

| | |
|---|---|
| American Standard | Smallest wire diameter = 0.56 × pitch<br>Largest wire diameter = 0.90 × pitch<br>Diameter for pitch-line contact = 0.57735 × pitch |
| Whitworth | Smallest wire diameter = 0.54 × pitch<br>Largest wire diameter = 0.76 × pitch<br>Diameter for pitch-line contact = 0.56369 × pitch |

**Measuring Wire Accuracy.—**A set of three measuring wires should have the same diameter within 0.0002 (5.08 μm) inch. To measure the pitch diameter of a screw-thread gage to an accuracy of 0.0001 inch (2.54 μm) by means of wires, it is necessary to know the wire diameters to 0.00002 (0.51 μm) inch. If the diameters of the wires are known only to an accuracy of 0.0001 (2.54 μm) inch, an accuracy better than 0.0003 (7.62 μm) inch in the measurement of pitch diameter cannot be expected. The wires should be accurately finished hardened steel cylinders of the maximum possible hardness without being brittle. The hardness should not be less than that corresponding to a Knoop indentation number of 630. A wire of this hardness can be cut with a file only with difficulty. The surface should not be rougher than the equivalent of a deviation of 3 microinches (0.0762 μm) from a true cylindrical surface.

**Effect of Contact Pressure.—**In measuring screw threads or screw-thread gages by the 3-wire method, variations in contact pressure will result in different readings. The effect of a variation in contact pressure in measuring threads of fine pitches is indicated by the difference in readings obtained with pressures of 2 and 5 pounds (0.91 and 2.27 kg) in checking a thread plug gage having 24 threads per inch. The reading over the wires with 5 pounds (2.27 kg) pressure was 0.00013 inch (3.302 μm) less than with 2 pounds (0.91 kg) pressure. For pitches finer than 20 threads per inch (0.05 inch or 1.27 mm pitch), a pressure of 16 ounces (0.45 kg) is recommended by the National Bureau of Standards, now National Institute of Standards and Technology (NIST). For pitches of 20 threads per inch and coarser, a pressure of 2 ½ pounds (1.13 kg) is recommended.

For Acme threads, the wire presses against the sides of the thread with a pressure of approximately twice that of the measuring instrument. To limit the tendency of the wires to wedge in between the sides of an Acme thread, it is recommended that pitch-diameter measurements be made at 1 pound on 8 threads per inch and finer, and at 2 ½ pounds for pitches coarser than 8 threads per inch (0.125 inch or 3.175 mm pitch).

**Approximate Three-Wire Formulas That Do Not Compensate for Lead Angle.—**A general formula in which the effect of lead angle is ignored is as follows (see accompanying notation used in formulas):

$$M = E - T\cot A + W(1 + \csc A) \qquad (1)$$

This formula can be simplified for any given thread angle and pitch. To illustrate, because $T = 0.5P$, $M = E - 0.5P\cot 30° + W(1 + 2)$, for a 60-degree thread, such as the American Standard,

$$M = E - 0.866025P + 3W$$

The accompanying table contains these simplified formulas for different standard threads. Two formulas are given for each. The upper one is used when the measurement over wires, $M$, is known and the corresponding pitch diameter, $E$, is required; the lower formula gives the measurement $M$ for a specified value of pitch diameter. These formulas are sufficiently accurate for checking practically all standard 60-degree single-thread screws because of the low lead angles, which vary from 1° 11′ to 4° 31′ in the American Standard Coarse-Thread Series.

**Bureau of Standards (now NIST) General Formula.**—Formula (2), which follows, compensates quite largely for the effect of the lead angle. It is from the National Bureau of Standards Handbook H28 (1944), now FED-STD-H28. The formula, however, as here given has been arranged for finding the value of $M$ (instead of $E$).

$$M = E - T \cot A + W(1 + \csc A + 0.5 \tan^2 B \cos A \cot A) \qquad (2)$$

This expression is also found in ANSI/ASME B1.2-1983 (R2017). The Bureau of Standards uses Formula (2) in preference to Formula (1) when the value of $0.5W \tan^2 B \cos A \cot A$ exceeds 0.00015, with the larger lead angles. If this test is applied to American Standard 60-degree threads, it will show that Formula (1) is generally applicable; but for 29-degree Acme or worm threads, Formula (2) (or some other that includes the effect of lead angle) should be employed.

**Notation Used in Formulas for Checking Pitch Diameters by Three-Wire Method**

$A$ = one-half included thread angle in the axial plane

$A_n$ = one-half included thread angle in the normal plane or in plane perpendicular to sides of thread = one-half included angle of cutter when thread is milled ($\tan A_n = \tan A \times \cos B$). (Note: Included angle of milling cutter or grinding wheel may equal the nominal included angle of thread, or may be reduced to whatever normal angle is required to make the thread angle standard in the axial plane. In either case, $A_n$ = one-half cutter angle.)

$B$ = lead angle at pitch diameter = helix angle of thread as measured from a plane perpendicular to the axis, $\tan B = L \div 3.1416E$

$D$ = basic major or outside diameter

$E$ = pitch diameter (basic, maximum, or minimum) for which $M$ is required, or pitch diameter corresponding to measurement $M$

$F$ = angle required in Formulas (4b), (4d), and (4e)

$G$ = angle required in Formula (4)

$H$ = helix angle at pitch diameter and measured from axis = $90° - B$ or $\tan H = \cot B$

$H_b$ = helix angle at $R_b$ measured from axis

$L$ = lead of thread = pitch $P \times$ number of threads $S$

$M$ = dimension over wires

$P$ = pitch = $1 \div$ number of threads per inch

$R_b$ = radius required in Formulas (4) and (4e)

$S$ = number of "starts" or threads on a multiple-threaded worm or screw

$T = 0.5P$ = width of thread in axial plane at diameter $E$

$T_a$ = arc thickness on pitch cylinder in plane perpendicular to axis

$W$ = wire or pin diameter

## Formulas for Checking Pitch Diameters of Screw Threads

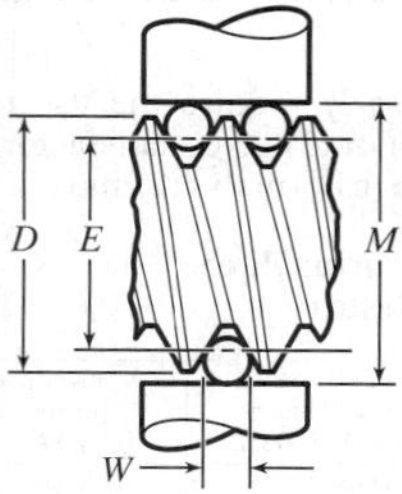


The formulas below do not compensate for the effect of the lead angle upon measurement $M$, but they are sufficiently accurate for checking standard single-thread screws unless exceptional accuracy is required. See accompanying information on effect of lead angle; also matter relating to measuring wire sizes, accuracy required for such wires, and contact or measuring pressure.

The approximate best wire size for pitch-line contact may be obtained by the formula

$$W = 0.5 \times \text{pitch} \times \sec\frac{1}{2}\text{ included thread angle}$$

For 60-degree threads, $W = 0.57735 \times \text{pitch}$.

| Form of Thread | Formulas for determining measurement $M$ corresponding to correct pitch diameter and the pitch diameter $E$ corresponding to a given measurement over wires.[a] |
|---|---|
| American National Standard Unified | When measurement $M$ is known, $E = M + 0.86603P - 3W$<br>When pitch diameter $E$ is used in formula, $M = E - 0.86603P + 3W$<br>The American Standard formerly was known as the US Standard. |
| British Standard Whitworth | When measurement $M$ is known, $E = M + 0.9605P - 3.1657W$<br>When pitch diameter $E$ is used in formula, $M = E - 0.9605P + 3.1657W$ |
| British Association Standard | When measurement $M$ is known, $E = M + 1.1363P - 3.4829W$<br>When pitch diameter $E$ is used in formula, $M = E - 1.1363P + 3.4829W$ |
| Löwenherz Thread | When measurement $M$ is known, $E = M + P - 3.2359W$<br>When pitch diameter $E$ is used in formula, $M = E - P + 3.2359W$ |
| Sharp V-Thread | When measurement $M$ is known, $E = M + 0.86603P - 3W$<br>When pitch diameter $E$ is used in formula, $M = E - 0.86603P + 3W$ |
| International Standard | Use the formula above for the American National Standard Unified Thread. |
| Pipe Thread | See accompanying paragraph on *Buckingham Exact Involute Helicoid Formula Applied to Screw Threads*. |
| Acme and Worm Threads | See Buckingham Formulas page 2135; also *Three-Wire Measurement of Acme and Stub Acme Thread Pitch Diameter*. |
| Buttress Form of Thread | Different forms of buttress threads are used. See paragraph on *Three-Wire Method Applied to Buttress Threads*. |

[a] The wires must be lapped to a uniform diameter and it is very important to insert in the rule or formula the wire diameter as determined by precise means of measurement. Any error will be multiplied. See paragraph on *Wire Sizes for Checking Pitch Diameters of Screw Threads* on page 2128.

**Why Small Thread Angle Affects Accuracy of Three-Wire Measurement.—**In measuring or checking Acme threads, or any others having a comparatively small thread angle $A$, it is particularly important to use a formula that compensates largely, if not entirely, for the effect of the lead angle, especially in all gage and precision work. The effect of the lead angle on the position of the wires and upon the resulting measurement $M$ is much greater in a 29-degree thread than in a higher thread angle such, for example, as a 60-degree thread. This effect results from an increase in the cotangent of the thread angle as this angle becomes smaller. The reduction in the width of the thread groove in the normal plane due to the lead angle causes a wire of given size to rest higher in the groove of a thread having a small thread angle $A$ (like a 29-degree thread) than in the groove of a thread with a larger angle (like a 60-degree American Standard).

*Acme Threads:* Three-wire measurements of high accuracy require the use of Formula (4). For most measurements, however, Formula (2) or (3) gives satisfactory results. The table on page 2138 lists suitable wire sizes for use in Formulas (2) and (4).

**Values of Constants Used in Formulas for Measuring Pitch Diameters of Screws by the Three-wire System**

| No. of Threads per Inch | American Standard Unified and Sharp V-Thread 0.866025*P* | Whitworth Thread 0.9605*P* | No. of Threads per Inch | American Standard Unified and Sharp V-Thread 0.866025*P* | Whitworth Thread 0.9605*P* |
|---|---|---|---|---|---|
| 2¼ | 0.38490 | 0.42689 | 18 | 0.04811 | 0.05336 |
| 2⅜ | 0.36464 | 0.40442 | 20 | 0.04330 | 0.04803 |
| 2½ | 0.34641 | 0.38420 | 22 | 0.03936 | 0.04366 |
| 2⅝ | 0.32992 | 0.36590 | 24 | 0.03608 | 0.04002 |
| 2¾ | 0.31492 | 0.34927 | 26 | 0.03331 | 0.03694 |
| 2⅞ | 0.30123 | 0.33409 | 28 | 0.03093 | 0.03430 |
| 3 | 0.28868 | 0.32017 | 30 | 0.02887 | 0.03202 |
| 3¼ | 0.26647 | 0.29554 | 32 | 0.02706 | 0.03002 |
| 3½ | 0.24744 | 0.27443 | 34 | 0.02547 | 0.02825 |
| 4 | 0.21651 | 0.24013 | 36 | 0.02406 | 0.02668 |
| 4½ | 0.19245 | 0.21344 | 38 | 0.02279 | 0.02528 |
| 5 | 0.17321 | 0.19210 | 40 | 0.02165 | 0.02401 |
| 5½ | 0.15746 | 0.17464 | 42 | 0.02062 | 0.02287 |
| 6 | 0.14434 | 0.16008 | 44 | 0.01968 | 0.02183 |
| 7 | 0.12372 | 0.13721 | 46 | 0.01883 | 0.02088 |
| 8 | 0.10825 | 0.12006 | 48 | 0.01804 | 0.02001 |
| 9 | 0.09623 | 0.10672 | 50 | 0.01732 | 0.01921 |
| 10 | 0.08660 | 0.09605 | 52 | 0.01665 | 0.01847 |
| 11 | 0.07873 | 0.08732 | 56 | 0.01546 | 0.01715 |
| 12 | 0.07217 | 0.08004 | 60 | 0.01443 | 0.01601 |
| 13 | 0.06662 | 0.07388 | 64 | 0.01353 | 0.01501 |
| 14 | 0.06186 | 0.06861 | 68 | 0.01274 | 0.01412 |
| 15 | 0.05774 | 0.06403 | 72 | 0.01203 | 0.01334 |
| 16 | 0.05413 | 0.06003 | 80 | 0.01083 | 0.01201 |

**Constants Used for Measuring Pitch Diameters of Metric Screws by the Three-wire System**

| Pitch in mm | 0.866025*P* in Inches | *W* in Inches | Pitch in mm | 0.866025*P* in Inches | *W* in Inches | Pitch in mm | 0.866025*P* in Inches | *W* in Inches |
|---|---|---|---|---|---|---|---|---|
| 0.2 | 0.00682 | 0.00455 | 0.75 | 0.02557 | 0.01705 | 3.5 | 0.11933 | 0.07956 |
| 0.25 | 0.00852 | 0.00568 | 0.8 | 0.02728 | 0.01818 | 4 | 0.13638 | 0.09092 |
| 0.3 | 0.01023 | 0.00682 | 1 | 0.03410 | 0.02273 | 4.5 | 0.15343 | 0.10229 |
| 0.35 | 0.01193 | 0.00796 | 1.25 | 0.04262 | 0.02841 | 5 | 0.17048 | 0.11365 |
| 0.4 | 0.01364 | 0.00909 | 1.5 | 0.05114 | 0.03410 | 5.5 | 0.18753 | 0.12502 |
| 0.45 | 0.01534 | 0.01023 | 1.75 | 0.05967 | 0.03978 | 6 | 0.20457 | 0.13638 |
| 0.5 | 0.01705 | 0.01137 | 2 | 0.06819 | 0.04546 | 8 | 0.30686 | 0.18184 |
| 0.6 | 0.02046 | 0.01364 | 2.5 | 0.08524 | 0.05683 | … | … | … |
| 0.7 | 0.02387 | 0.01591 | 3 | 0.10229 | 0.06819 | … | … | … |

This table may be used for American National Standard Metric Threads. The formulas for American Standard Unified Threads on page 2131 are used. In the table above, the values of 0.866025$P$ and $W$ are in inches so that the values for $E$ and $M$ calculated from the formulas on page 2131 are also in inches.

**Dimensions Over Wires of Given Diameter for Checking Screw Threads of American National Form (US Standard) and the V-Form**

| Dia. of Thread | No. of Threads per Inch | Wire Dia. Used | Dimension over Wires | | Dia. of Thread | No. of Threads per Inch | Wire Dia. Used | Dimension over Wires | |
|---|---|---|---|---|---|---|---|---|---|
| | | | V-Thread | US Thread | | | | V-Thread | US Thread |
| 1/4 | 18 | 0.035 | 0.2588 | 0.2708 | 7/8 | 8 | 0.090 | 0.9285 | 0.9556 |
| 1/4 | 20 | 0.035 | 0.2684 | 0.2792 | 7/8 | 9 | 0.090 | 0.9525 | 0.9766 |
| 1/4 | 22 | 0.035 | 0.2763 | 0.2861 | 7/8 | 10 | 0.090 | 0.9718 | 0.9935 |
| 1/4 | 24 | 0.035 | 0.2828 | 0.2919 | 15/16 | 8 | 0.090 | 0.9910 | 1.0181 |
| 5/16 | 18 | 0.035 | 0.3213 | 0.3333 | 15/16 | 9 | 0.090 | 1.0150 | 1.0391 |
| 5/16 | 20 | 0.035 | 0.3309 | 0.3417 | 1 | 8 | 0.090 | 1.0535 | 1.0806 |
| 5/16 | 22 | 0.035 | 0.3388 | 0.3486 | 1 | 9 | 0.090 | 1.0775 | 1.1016 |
| 5/16 | 24 | 0.035 | 0.3453 | 0.3544 | 1 1/8 | 7 | 0.090 | 1.1476 | 1.1785 |
| 3/8 | 16 | 0.040 | 0.3867 | 0.4003 | 1 1/4 | 7 | 0.090 | 1.2726 | 1.3035 |
| 3/8 | 18 | 0.040 | 0.3988 | 0.4108 | 1 3/8 | 6 | 0.150 | 1.5363 | 1.5724 |
| 3/8 | 20 | 0.040 | 0.4084 | 0.4192 | 1 1/2 | 6 | 0.150 | 1.6613 | 1.6974 |
| 7/16 | 14 | 0.050 | 0.4638 | 0.4793 | 1 5/8 | 5 1/2 | 0.150 | 1.7601 | 1.7995 |
| 7/16 | 16 | 0.050 | 0.4792 | 0.4928 | 1 3/4 | 5 | 0.150 | 1.8536 | 1.8969 |
| 1/2 | 12 | 0.050 | 0.5057 | 0.5237 | 1 7/8 | 5 | 0.150 | 1.9786 | 2.0219 |
| 1/2 | 13 | 0.050 | 0.5168 | 0.5334 | 2 | 4 1/2 | 0.150 | 2.0651 | 2.1132 |
| 1/2 | 14 | 0.050 | 0.5263 | 0.5418 | 2 1/4 | 4 1/2 | 0.150 | 2.3151 | 2.3632 |
| 9/16 | 12 | 0.050 | 0.5682 | 0.5862 | 2 1/2 | 4 | 0.150 | 2.5170 | 2.5711 |
| 9/16 | 14 | 0.050 | 0.5888 | 0.6043 | 2 3/4 | 4 | 0.150 | 2.7670 | 2.28211 |
| 5/8 | 10 | 0.070 | 0.6618 | 0.6835 | 3 | 3 1/2 | 0.200 | 3.1051 | 3.1670 |
| 5/8 | 11 | 0.070 | 0.6775 | 0.6972 | 3 1/4 | 3 1/2 | 0.200 | 3.3551 | 3.4170 |
| 5/8 | 12 | 0.070 | 0.6907 | 0.7087 | 3 1/2 | 3 1/4 | 0.250 | 3.7171 | 3.7837 |
| 11/16 | 10 | 0.070 | 0.7243 | 0.7460 | 3 3/4 | 3 | 0.250 | 3.9226 | 3.9948 |
| 11/16 | 11 | 0.070 | 0.7400 | 0.7597 | 4 | 3 | 0.250 | 4.1726 | 4.2448 |
| 3/4 | 10 | 0.070 | 0.7868 | 0.8085 | 4 1/4 | 2 7/8 | 0.250 | 4.3975 | 4.4729 |
| 3/4 | 11 | 0.070 | 0.8025 | 0.8222 | 4 1/2 | 2 3/4 | 0.250 | 4.6202 | 4.6989 |
| 3/4 | 12 | 0.070 | 0.8157 | 0.8337 | 4 3/4 | 2 5/8 | 0.250 | 4.8402 | 4.9227 |
| 13/16 | 9 | 0.070 | 0.8300 | 0.8541 | 5 | 2 1/2 | 0.250 | 5.0572 | 5.1438 |
| 13/16 | 10 | 0.070 | 0.8493 | 0.8710 | … | … | … | … | … |

**Buckingham Simplified Formula which Includes Effect of Lead Angle.—**The Formula (3) which follows gives very accurate results for the lower lead angles in determining measurement *M*. However, if extreme accuracy is essential, it may be advisable to use the involute helicoid formulas as explained later.

$$M = E + W(1 + \sin A_n) \quad (3) \quad \text{where} \quad W = \frac{T \times \cos B}{\cos A_n} \quad (3a)$$

Theoretically correct equations for determining measurement *M* are complex and cumbersome to apply. Formula (3) combines simplicity with a degree of accuracy which meets all but the most exacting requirements, particularly for lead angles below 8 or 10 degrees and the higher thread angles. However, the wire diameter used in Formula (3) must conform to that obtained by Formula (3a) to permit a direct solution or one not involving indeterminate equations and successive trials.

*Application of Buckingham Formula:* In the application of Formula (3) to screw or worm threads, two general cases are to be considered.

**Table for Measuring Whitworth Standard Threads by the Three-wire Method**

| Dia. of Thread | No. of Threads per Inch | Dia. of Wire Used | Dia. Measured over Wires | Dia. of Thread | No. of Threads per Inch | Dia. of Wire Used | Dia. Measured over Wires |
|---|---|---|---|---|---|---|---|
| 1/8 | 40 | 0.018 | 0.1420 | 2 1/4 | 4 | 0.150 | 2.3247 |
| 3/16 | 24 | 0.030 | 0.2158 | 2 3/8 | 4 | 0.150 | 2.4497 |
| 1/4 | 20 | 0.035 | 0.2808 | 2 1/2 | 4 | 0.150 | 2.5747 |
| 5/16 | 18 | 0.040 | 0.3502 | 2 5/8 | 4 | 0.150 | 2.6997 |
| 3/8 | 16 | 0.040 | 0.4015 | 2 3/4 | 3 1/2 | 0.200 | 2.9257 |
| 7/16 | 14 | 0.050 | 0.4815 | 2 7/8 | 3 1/2 | 0.200 | 3.0507 |
| 1/2 | 12 | 0.050 | 0.5249 | 3 | 3 1/2 | 0.200 | 3.1757 |
| 9/16 | 12 | 0.050 | 0.5874 | 3 1/8 | 3 1/2 | 0.200 | 3.3007 |
| 5/8 | 11 | 0.070 | 0.7011 | 3 1/4 | 3 1/4 | 0.200 | 3.3905 |
| 11/16 | 11 | 0.070 | 0.7636 | 3 3/8 | 3 1/4 | 0.200 | 3.5155 |
| 3/4 | 10 | 0.070 | 0.8115 | 3 1/2 | 3 1/4 | 0.200 | 3.6405 |
| 13/16 | 10 | 0.070 | 0.8740 | 3 5/8 | 3 1/4 | 0.200 | 3.7655 |
| 7/8 | 9 | 0.070 | 0.9187 | 3 3/4 | 3 | 0.200 | 3.8495 |
| 15/16 | 9 | 0.070 | 0.9812 | 3 7/8 | 3 | 0.200 | 3.9745 |
| 1 | 8 | 0.090 | 1.0848 | 4 | 3 | 0.200 | 4.0995 |
| 1 1/16 | 8 | 0.090 | 1.1473 | 4 1/8 | 3 | 0.200 | 4.2245 |
| 1 1/8 | 7 | 0.090 | 1.1812 | 4 1/4 | 2 7/8 | 0.250 | 4.4846 |
| 1 3/16 | 7 | 0.090 | 1.2437 | 4 3/8 | 2 7/8 | 0.250 | 4.6096 |
| 1 1/4 | 7 | 0.090 | 1.3062 | 4 1/2 | 2 7/8 | 0.250 | 4.7346 |
| 1 5/16 | 7 | 0.090 | 1.3687 | 4 5/8 | 2 7/8 | 0.250 | 4.8596 |
| 1 3/8 | 6 | 0.120 | 1.4881 | 4 3/4 | 2 3/4 | 0.250 | 4.9593 |
| 1 7/16 | 6 | 0.120 | 1.5506 | 4 7/8 | 2 3/4 | 0.250 | 5.0843 |
| 1 1/2 | 6 | 0.120 | 1.6131 | 5 | 2 3/4 | 0.250 | 5.2093 |
| 1 9/16 | 6 | 0.120 | 1.6756 | 5 1/8 | 2 3/4 | 0.250 | 5.3343 |
| 1 5/8 | 5 | 0.120 | 1.6847 | 5 1/4 | 2 5/8 | 0.250 | 5.4316 |
| 1 11/16 | 5 | 0.120 | 1.7472 | 5 3/8 | 2 5/8 | 0.250 | 5.5566 |
| 1 3/4 | 5 | 0.120 | 1.8097 | 5 1/2 | 2 5/8 | 0.250 | 5.6816 |
| 1 13/16 | 5 | 0.120 | 1.8722 | 5 5/8 | 2 5/8 | 0.250 | 5.8066 |
| 1 7/8 | 4 1/2 | 0.150 | 1.9942 | 5 3/4 | 2 1/2 | 0.250 | 5.9011 |
| 1 15/16 | 4 1/2 | 0.150 | 2.0567 | 5 7/8 | 2 1/2 | 0.250 | 6.0261 |
| 2 | 4 1/2 | 0.150 | 2.1192 | 6 | 2 1/2 | 0.250 | 6.1511 |
| 2 1/8 | 4 1/2 | 0.150 | 2.2442 | … | … | … | … |

All dimensions are given in inches.

*Case 1:* The screw thread or worm is to be milled with a cutter having an included angle equal to the nominal or standard thread angle that is assumed to be the angle in the axial plane. For example, a 60-degree cutter is to be used for milling a thread. In this case, the thread angle in the plane of the axis will exceed 60 degrees by an amount increasing with the lead angle. This variation from the standard angle may be of little or no practical importance if the lead angle is small or if the mating nut (or teeth in worm gearing) is formed to suit the thread as milled.

*Case 2:* The screw thread or worm is to be milled with a cutter reduced to whatever normal angle is equivalent to the standard thread angle in the axial plane. For example, a 29-degree Acme thread is to be milled with a cutter having some angle smaller than 29 degrees (the reduction increasing with the lead angle) to make the thread angle standard in the plane of the axis. Theoretically, the milling cutter angle should always be corrected to suit the normal angle; but if the lead angle is small, such correction may be unnecessary.

If the thread is cut in a lathe to the standard angle as measured in the axial plane, Case 2 applies in determining the pin size $W$ and the overall measurement $M$.

In solving all problems under Case 1, angle $A_n$ used in Formulas (3) and (3a) equals one-half the included angle of the milling cutter.

When Case 2 applies, angle $A_n$ for milled threads also equals one-half the included angle of the cutter, but the cutter angle is reduced and is determined as follows:

$$\tan A_n = \tan A \times \cos B$$

The included angle of the cutter or the normal included angle of the thread groove $= 2A_n$. Examples 1 and 2, which follow, illustrate Cases 1 and 2.

*Example 1 (Case 1):* Take, for example, an Acme screw thread that is milled with a cutter having an included angle of 29 degrees; consequently, the angle of the thread exceeds 29 degrees in the axial section.

The outside or major diameter is 3 inches; the pitch, ½ inch; the lead, 1 inch; the number of threads or "starts," 2. Find pin size $W$ and measurement $M$.

Pitch diameter $E = 2.75$; $T = 0.25$; $L = 1.0$; $A_n = 14.50°$ $\tan A_n = 0.258618$; $\sin A_n = 0.25038$; and $\cos A_n = 0.968148$.

$$\tan B = \frac{1.0}{3.1416 \times 2.75} = 0.115749 \qquad B = 6.6025°$$

$$W = \frac{0.25 \times 0.993368}{0.968148} = 0.25651 \text{ inch}$$

$$M = 2.75 + 0.25651 \times (1 + 0.25038) = 3.0707 \text{ inches}$$

*Note:* This value of $M$ is only 0.0001 inch larger than that obtained by using the very accurate involute helicoid Formula (4) discussed on the following page.

*Example 2 (Case 2):* A triple-threaded worm has a pitch diameter of 2.481 inches, pitch of 1.5 inches, lead of 4.5 inches, lead angle of 30 degrees, and nominal thread angle of 60 degrees in the axial plane. Milling cutter angle is to be reduced. $T = 0.75$ inch; $\cos B = 0.866025$; and $\tan A = 0.57735$. Again use Formula (3) to see if it is applicable.

$\tan A_n = \tan A \times \cos B = 0.57735 \times 0.866025 = 0.5000$; hence $A_n = 26.565°$, making the included cutter angle 53.13°, thus $\cos A_n = 0.89443$ and $\sin A_n = 0.44721$.

$$W = \frac{0.75 \times 0.866025}{0.89443} = 0.72618 \text{ inch}$$

$$M = 2.481 + 0.72618 \times (1 + 0.44721) = 3.532 \text{ inches}$$

*Note:* If the value of measurement $M$ is determined by using the following Formula (4) it will be found that $M = 3.515+$ inches; hence the error equals $3.532 - 3.515 = 0.017$ inch approximately, which indicates that Formula (3) is not accurate enough here. The application of this simpler Formula (3) will depend upon the lead angle and thread angle (as previously explained) and upon the class of work.

**Buckingham Exact Involute Helicoid Formula Applied to Screw Threads.—**When extreme accuracy is required in finding measurement $M$ for obtaining a given pitch diameter, the equations that follow, although somewhat cumbersome to apply, have the merit of providing a direct and very accurate solution; consequently, they are preferable to the indeterminate equations and successive trial solutions heretofore employed when extreme precision is required. These equations are exact for involute helical gears and, consequently, give theoretically correct results when applied to a screw thread of the involute helicoidal form; they also give very close approximations for threads having intermediate profiles.

*Helical Gear Equation Applied to Screw Thread Measurement:* In applying the helical gear equations to a screw thread, use either the axial or normal thread angle and the lead angle of the helix. To keep the solution on a practical basis, either thread angle $A$ or $A_n$, as the case may be, is assumed to equal the cutter angle of a milled thread. Actually, the profile of a milled thread will have some curvature in both axial and normal sections; hence angles $A$ and $A_n$ represent the angular approximations of these slightly curved profiles. The equations that follow give the values needed to solve the screw thread problem as a helical gear problem.

$$M = \frac{2R_b}{\cos G} + W \quad (4)$$

$$\tan F = \frac{\tan A}{\tan B} = \frac{\tan A_n}{\sin B} \quad (4a) \qquad R_b = \frac{E}{2}\cos F \quad (4b)$$

$$T_a = \frac{T}{\tan B} \quad (4c) \qquad \tan H_b = \cos F \times \tan H \quad (4d)$$

$$\text{inv } G = \frac{T_a}{E} + \text{inv } F + \frac{W}{2R_b \cos H_b} - \frac{\pi}{S} \quad (4e)$$

The tables of involute functions starting on page 107 provide values for angles from 14 to 51 degrees, used for gear calculations. The formula for involute functions on page 106 may be used to extend this table as required.

*Example:* To illustrate the application of Formula (4) and the supplementary formulas, assume that the number of starts $S = 6$; pitch diameter $E = 0.6250$; normal thread angle $A_n = 20°$; lead of thread $L = 0.864$ inch; $T = 0.072$; $W = 0.07013$ inch.

$$\tan B = \frac{L}{\pi E} = \frac{0.864}{1.9635} = 0.44003 \quad B = 23.751°$$

$$\text{Helix angle } H = 90° - 23.751° = 66.249°$$

$$\tan F = \frac{\tan A_n}{\sin B} = \frac{0.36397}{0.40276} = 0.90369 \quad F = 42.104°$$

$$R_b = \frac{E}{2}\cos F = \frac{0.6250}{2} \times 0.74193 = 0.23185$$

$$T_a = \frac{T}{\tan B} = \frac{0.072}{0.44003} = 0.16362$$

$$\tan H_b = \cos F \tan H = 0.74193 \times 2.27257 = 1.68609 \quad H_b = 59.328°$$

The involute function of $G$ is found next by Formula (4e).

$$\text{inv } G = \frac{0.16362}{0.625} + 0.16884 + \frac{0.07013}{2 \times 0.23185 \times 0.51012} - \frac{3.1416}{6} = 0.20351$$

Since 0.20351 is outside the values for involute functions given in the tables on pages 107 through 110 use the formula for involute functions on page 106 to extend these tables as required. It will be found that 44 deg. 21 min. or 44.350 degrees is the angular equivalent of 0.20351; hence, $G = 44.350$ degrees.

$$M = \frac{2R_b}{\cos G} + W = \frac{2 \times 0.23185}{0.71508} + 0.07013 = 0.71859 \text{ inch}$$

**Accuracy of** Formulas (3) **and** (4) **Compared.—**With the involute helicoid Formula (4) any wire size that makes contact with the flanks of the thread may be used; however, in the preceding example, the wire diameter $W$ was obtained by Formula (3a) in order to compare Formula (4) with (3) . If Example (3) is solved by Formula (3) , $M = 0.71912$; hence the difference between the values of $M$ obtained with Formulas (3) and (4) equals 0.71912 - 0.71859 = 0.00053 inch. The included thread angle in this case is 40 degrees. If

Formulas (3) and (4) are applied to a 29-degree thread, the difference in measurements $M$ or the error resulting from the use of Formulas (3) will be larger. For example, with an Acme thread having a lead angle of about 34 degrees, the difference in values of $M$ obtained by the two formulas equals 0.0008 inch.

**Three-Wire Measurement of Acme and Stub Acme Thread Pitch Diameter.—**For single- and multiple-start Acme and Stub Acme threads having lead angles of less than 5 degrees, the approximate three-wire formula given on page 2129 and the best wire size taken from the table on page 2138 may be used.

Multiple-start Acme and Stub Acme threads commonly have a lead angle of greater than 5 degrees. For these, a direct determination of the actual pitch diameter is obtained by using the formula: $E = M - (C + c)$ in conjunction with the table on page 2139. To enter the table, the lead angle $B$ of the thread to be measured must be known. It is found by the formula: $\tan B = L \div 3.1416E_1$ where $L$ is the lead of the thread and $E_1$ is the nominal pitch diameter. The best wire size is now found by taking the value of $w_1$ as given in the table for lead angle $B$, with interpolation, and dividing it by the number of threads per inch. The value of $(C + c)_1$ given in the table for lead angle $B$ is also divided by the number of threads per inch to get $(C + c)$. Using the best size wires, the actual measurement over wires $M$ is made and the actual pitch diameter $E$ found by using the formula: $E = M - (C + c)$.

*Example:* For a 5 tpi, 4-start Acme thread with a 13.952° lead angle, using three 0.10024-inch wires, $M$ = 1.1498 inches, hence $E$ = 1.1498 − 0.1248 = 1.0250 inches.

Under certain conditions, a wire may contact one thread flank at two points, and it is then advisable to substitute balls of the same diameter as the wires.

**Checking Thickness of Acme Screw Threads.—**In some instances it may be preferable to check the thread thickness instead of the pitch diameter, especially if there is a thread thickness tolerance.

A direct method, applicable to the larger pitches, is to use a vernier gear-tooth caliper for measuring the thickness in the *normal* plane of the thread. This measurement, for an American National Standard General Purpose Acme thread, should be made at a distance below the *basic* outside diameter equal to $p/4$. The thickness at this basic pitch-line depth and in the axial plane should be $p/2 - 0.259 \times$ the pitch diameter allowance from the table on page 2059 with a tolerance of *minus* 0.259 × the pitch diameter tolerance from the table on page 2064. The thickness in the normal plane or plane of measurement is equal to the thickness in the axial plane multiplied by the cosine of the helix angle. The helix angle may be determined from the formula:

$$\text{tangent of helix angle} = \text{lead of thread} \div (3.1416 \times \text{pitch diameter})$$

**Three-Wire Method for Checking Thickness of Acme Threads.—**The application of the 3-wire method of checking the thickness of an Acme screw thread is included in the Report of the National Screw Thread Commission. In applying the 3-wire method for checking thread thickness, the procedure is the same as in checking pitch diameter (see *Three-Wire Measurement of Acme and Stub Acme Thread Pitch Diameter*), although a different formula is required. Assume that $D$ = basic major diameter of screw; $M$ = measurement over wires; $W$ = diameter of wires; $S$ = tangent of helix angle at pitch line; $P$ = pitch; $T$ = thread thickness at depth equal to $0.25P$.

$$T = 1.12931 \times P + 0.25862 \times (M - D) - W \times (1.29152 + 0.48407S^2)$$

This formula transposed to show the correct measurement $M$ equivalent to a given required thread thickness is as follows:

$$M = D + \frac{W \times (1.29152 + 0.48407S^2) + T - 1.12931 \times P}{0.25862}$$

### Wire Sizes for Three-Wire Measurement of Acme Threads with Lead Angles of Less than 5 Degrees

| Threads per Inch | Best Size | Max. | Min. | Threads per Inch | Best Size | Max. | Min. |
|---|---|---|---|---|---|---|---|
| 1 | 0.51645 | 0.65001 | 0.48726 | 5 | 0.10329 | 0.13000 | 0.09745 |
| 1⅓ | 0.38734 | 0.48751 | 0.36545 | 6 | 0.08608 | 0.10834 | 0.08121 |
| 1½ | 0.34430 | 0.43334 | 0.32484 | 8 | 0.06456 | 0.08125 | 0.06091 |
| 2 | 0.25822 | 0.32501 | 0.24363 | 10 | 0.05164 | 0.06500 | 0.04873 |
| 2½ | 0.20658 | 0.26001 | 0.19491 | 12 | 0.04304 | 0.05417 | 0.04061 |
| 3 | 0.17215 | 0.21667 | 0.16242 | 14 | 0.03689 | 0.04643 | 0.03480 |
| 4 | 0.12911 | 0.16250 | 0.12182 | 16 | 0.03228 | 0.04063 | 0.03045 |

Wire sizes are based upon zero helix angle. Best size = 0.51645 × pitch; maximum size = 0.650013 × pitch; minimum size = 0.487263 × pitch.

*Example:* An Acme General Purpose thread, Class 2G, has a 5-inch basic major diameter, 0.5-inch pitch, and 1-inch lead (double thread). Assume the wire size is 0.258 inch. Determine measurement $M$ for a thread thickness $T$ at the basic pitch line of 0.2454 inch. ($T$ is the maximum thickness at the basic pitch line and equals $0.5P$, the basic thickness, −0.259 × allowance from Table 4, page 2064.)

$$M = 5 + \frac{0.258 \times [1.29152 + 0.48407 \times (0.06701)^2] + 0.2454 - 1.12931 \times 0.5}{0.25862}$$

5.056 inches

**Testing Angle of Thread by Three-Wire Method.—**The error in the angle of a thread may be determined by using sets of wires of two diameters, the measurement over the two sets of wires being followed by calculations to determine the amount of error, assuming that the angle cannot be tested by comparison with a standard plug gage, known to be correct. The diameter of the small wires for the American National Standard thread is usually about 0.6 times the pitch and the diameter of the large wires, about 0.9 times the pitch. The total difference between the measurements over the large and small sets of wires is first determined. If the thread is an American National Standard or any other form having an included angle of 60 degrees, the difference between the two measurements should equal three times the difference between the diameters of the wires used. Thus, if the wires are 0.116 and 0.076 inch in diameter, respectively, the difference equals 0.116 − 0.076 = 0.040 inch. Therefore, the difference between the micrometer readings for a standard angle of 60 degrees equals 3 × 0.040 = 0.120 inch for this example. If the angle is incorrect, the amount of error may be determined by the following formula, which applies to any thread regardless of angle:

$$\sin a = \frac{A}{B - A}$$

where $A$ = difference in diameters of the large and small wires used

$B$ = total difference between the measurements over the large and small wires

$a$ = one-half the included thread angle

*Example:* The diameter of the large wires used for testing the angle of a thread is 0.116 inch and of the small wires 0.076 inch. The measurement over the two sets of wires shows a total difference of 0.122 inch instead of the correct difference, 0.120 inch, for a standard angle of 60 degrees when using the sizes of wires mentioned. The amount of error is determined as follows:

$$\sin a = \frac{0.040}{0.122 - 0.040} = \frac{0.040}{0.082} = 0.4878$$

A table of sines shows that this value (0.4878) is the sine of 29 degrees 12 minutes, approximately. Therefore, the angle of the thread is 58 degrees 24 minutes or 1 degree 36 minutes less than the standard angle.

## Best Wire Diameters and Constants for Three-Wire Measurement of Acme and Stub Acme Threads with Large Lead Angles, 1-Inch Axial Pitch

| Lead angle, $B$, deg. | 1-start threads | | 2-start threads | | Lead angle, $B$, deg. | 2-start threads | | 3-start threads | |
|---|---|---|---|---|---|---|---|---|---|
| | $w_1$ | $(C+c)_1$ | $w_1$ | $(C+c)_1$ | | $w_1$ | $(C+c)_1$ | $w_1$ | $(C+c)_1$ |
| 5.0 | 0.51450 | 0.64311 | 0.51443 | 0.64290 | 10.0 | 0.50864 | 0.63518 | 0.50847 | 0.63463 |
| 5.1 | 0.51442 | 0.64301 | 0.51435 | 0.64279 | 10.1 | 0.50849 | 0.63498 | 0.50381 | 0.63442 |
| 5.2 | 0.51435 | 0.64291 | 0.51427 | 0.64268 | 10.2 | 0.50834 | 0.63478 | 0.50815 | 0.63420 |
| 5.3 | 0.51427 | 0.64282 | 0.51418 | 0.64256 | 10.3 | 0.50818 | 0.63457 | 0.50800 | 0.63399 |
| 5.4 | 0.51419 | 0.64272 | 0.51410 | 0.64245 | 10.4 | 0.50802 | 0.63436 | 0.50784 | 0.63378 |
| 5.5 | 0.51411 | 0.64261 | 0.51401 | 0.64233 | 10.5 | 0.40786 | 0.63416 | 0.50768 | 0.63356 |
| 5.6 | 0.51403 | 0.64251 | 0.51393 | 0.64221 | 10.6 | 0.50771 | 0.63395 | 0.50751 | 0.63333 |
| 5.7 | 0.51395 | 0.64240 | 0.51384 | 0.64209 | 10.7 | 0.50755 | 0.63375 | 0.50735 | 0.63311 |
| 5.8 | 0.51386 | 0.64229 | 0.51375 | 0.64196 | 10.8 | 0.50739 | 0.53354 | 0.50718 | 0.63288 |
| 5.9 | 0.51377 | 0.64218 | 0.51366 | 0.64184 | 10.9 | 0.50723 | 0.63333 | 0.50701 | 0.63265 |
| 6.0 | 0.51368 | 0.64207 | 0.51356 | 0.64171 | 11.0 | 0.50707 | 0.63313 | 0.50684 | 0.63242 |
| 6.1 | 0.51359 | 0.64195 | 0.51346 | 0.64157 | 11.1 | 0.50691 | 0.63292 | 0.50667 | 0.63219 |
| 6.2 | 0.51350 | 0.64184 | 0.51336 | 0.64144 | 11.2 | 0.50674 | 0.63271 | 0.50649 | 0.63195 |
| 6.3 | 0.51340 | 0.64172 | 0.41327 | 0.64131 | 11.3 | 0.50658 | 0.63250 | 0.50632 | 0.63172 |
| 6.4 | 0.51330 | 0.64160 | 0.51317 | 0.64117 | 11.4 | 0.50641 | 0.63228 | 0.50615 | 0.63149 |
| 6.5 | 0.51320 | 0.64147 | 0.51306 | 0.64103 | 11.5 | 0.50623 | 0.63206 | 0.50597 | 0.63126 |
| 6.6 | 0.51310 | 0.64134 | 0.51296 | 0.64089 | 11.6 | 0.50606 | 0.63184 | 0.50579 | 0.63102 |
| 6.7 | 0.51300 | 0.64122 | 0.51285 | 0.64075 | 11.7 | 0.50589 | 0.63162 | 0.50561 | 0.63078 |
| 6.8 | 0.51290 | 0.64110 | 0.51275 | 0.64061 | 11.8 | 0.50571 | 0.63140 | 0.50544 | 0.63055 |
| 6.9 | 0.51280 | 0.64097 | 0.51264 | 0.64046 | 11.9 | 0.50553 | 0.63117 | 0.50526 | 0.63031 |
| 7.0 | 0.51270 | 0.64085 | 0.51254 | 0.64032 | 12.0 | 0.50535 | 0.63095 | 0.50507 | 0.63006 |
| 7.1 | 0.51259 | 0.64072 | 0.51243 | 0.64017 | 12.1 | 0.50517 | 0.63072 | 0.50488 | 0.62981 |
| 7.2 | 0.51249 | 0.64060 | 0.51232 | 0.64002 | 12.2 | 0.50500 | 0.63050 | 0.50470 | 0.62956 |
| 7.3 | 0.51238 | 0.64047 | 0.51221 | 0.63987 | 12.3 | 0.50482 | 0.63027 | 0.50451 | 0.62931 |
| 7.4 | 0.51227 | 0.64034 | 0.51209 | 0.63972 | 12.4 | 0.50464 | 0.63004 | 0.50432 | 0.62906 |
| 7.5 | 0.51217 | 0.64021 | 0.51198 | 0.63957 | 12.5 | 0.50445 | 0.62981 | 0.50413 | 0.62881 |
| 7.6 | 0.51206 | 0.64008 | 0.51186 | 0.63941 | 12.6 | 0.50427 | 0.62958 | 0.50394 | 0.62856 |
| 7.7 | 0.51196 | 0.63996 | 0.51174 | 0.63925 | 12.7 | 0.50408 | 0.62934 | 0.50375 | 0.62830 |
| 7.8 | 0.51186 | 0.63983 | 0.51162 | 0.63909 | 12.8 | 0.50389 | 0.62911 | 0.50356 | 0.62805 |
| 7.9 | 0.51175 | 0.63970 | 0.51150 | 0.63892 | 12.9 | 0.50371 | 0.62888 | 0.50336 | 0.62779 |
| 8.0 | 0.51164 | 0.63957 | 0.51138 | 0.63876 | 13.0 | 0.50352 | 0.62865 | | |
| 8.1 | 0.51153 | 0.63944 | 0.51125 | 0.63859 | 13.1 | 0.50333 | 0.62841 | | |
| 8.2 | 0.51142 | 0.63930 | 0.51113 | 0.63843 | 13.2 | 0.50313 | 0.62817 | | |
| 8.3 | 0.51130 | 0.63916 | 0.51101 | 0.63827 | 13.3 | 0.50293 | 0.62792 | | |
| 8.4 | 0.51118 | 0.63902 | 0.51088 | 0.63810 | 13.4 | 0.50274 | 0.62778 | | |
| 8.5 | 0.51105 | 0.63887 | 0.51075 | 0.63793 | 13.5 | 0.50254 | 0.62743 | | |
| 8.6 | 0.51093 | 0.63873 | 0.51062 | 0.63775 | 13.6 | 0.50234 | 0.62718 | | |
| 8.7 | 0.51081 | 0.63859 | 0.51049 | 0.63758 | 13.7 | 0.50215 | 0.62694 | For these 3-start thread values, see table on following page. | |
| 8.8 | 0.51069 | 0.63845 | 0.51035 | 0.63740 | 13.8 | 0.50195 | 0.62670 | | |
| 8.9 | 0.51057 | 0.63831 | 0.51022 | 0.63722 | 13.9 | 0.50175 | 0.62645 | | |
| 9.0 | 0.51044 | 0.63817 | 0.51008 | 0.63704 | 14.0 | 0.50155 | 0.62621 | | |
| 9.1 | 0.51032 | 0.63802 | 0.50993 | 0.63685 | 14.1 | 0.50135 | 0.62596 | | |
| 9.2 | 0.51019 | 0.63788 | 0.50979 | 0.63667 | 14.2 | 0.50115 | 0.62571 | | |
| 9.3 | 0.51006 | 0.63774 | 0.50965 | 0.63649 | 14.3 | 0.50094 | 0.62546 | | |
| 9.4 | 0.50993 | 0.63759 | 0.50951 | 0.63630 | 14.4 | 0.50073 | 0.62520 | | |
| 9.5 | 0.50981 | 0.63744 | 0.50937 | 0.63612 | 14.5 | 0.50051 | 0.62494 | | |
| 9.6 | 0.50968 | 0.63730 | 0.50922 | 0.63593 | 14.6 | 0.50030 | 0.62468 | | |
| 9.7 | 0.50955 | 0.63715 | 0.50908 | 0.63574 | 14.7 | 0.50009 | 0.62442 | | |
| 9.8 | 0.50941 | 0.63700 | 0.50893 | 0.63555 | 14.8 | 0.49988 | 0.62417 | | |
| 9.9 | 0.50927 | 0.63685 | 0.50879 | 0.63537 | 14.9 | 0.49966 | 0.62391 | | |
| 10.0 | 0.50913 | 0.63670 | 0.50864 | 0.63518 | 15.0 | 0.49945 | 0.62365 | | |

All dimensions are in inches.

Values given for $w_1$ and $(C+c)_1$ in table are for 1-inch pitch axial threads. For other pitches, divide table values by number of threads per inch.

*Courtesy of Van Keuren Co.*

**Best Wire Diameters and Constants for Three-Wire Measurement of Acme and Stub Acme Threads with Large Lead Angles—1-Inch Axial Pitch**

| Lead angle, $B$, deg. | 3-start threads | | 4-start threads | | Lead angle, $B$, deg. | 3-start threads | | 4-start threads | |
|---|---|---|---|---|---|---|---|---|---|
| | $w_1$ | $(C+c)_1$ | $w_1$ | $(C+c)_1$ | | $w_1$ | $(C+c)_1$ | $w_1$ | $(C+c)_1$ |
| 13.0 | 0.50316 | 0.62752 | 0.50297 | 0.62694 | 18.0 | 0.49154 | 0.61250 | 0.49109 | 0.61109 |
| 13.1 | 0.50295 | 0.62725 | 0.50277 | 0.62667 | 18.1 | 0.49127 | 0.61216 | 0.49082 | 0.61073 |
| 13.2 | 0.50275 | 0.62699 | 0.50256 | 0.62639 | 18.2 | 0.49101 | 0.61182 | 0.49054 | 0.61037 |
| 13.3 | 0.50255 | 0.62672 | 0.50235 | 0.62611 | 18.3 | 0.49074 | 0.61148 | 0.49027 | 0.61001 |
| 13.4 | 0.50235 | 0.62646 | 0.50215 | 0.62583 | 18.4 | 0.49047 | 0.61114 | 0.48999 | 0.60964 |
| 13.5 | 0.50214 | 0.62619 | 0.50194 | 0.62555 | 18.5 | 0.49020 | 0.61080 | 0.48971 | 0.69928 |
| 13.6 | 0.50194 | 0.62592 | 0.50173 | 0.62526 | 18.6 | 0.48992 | 0.61045 | 0.48943 | 0.60981 |
| 13.7 | 0.50173 | 0.62564 | 0.50152 | 0.62498 | 18.7 | 0.48965 | 0.61011 | 0.48915 | 0.60854 |
| 13.8 | 0.50152 | 0.62537 | 0.50131 | 0.62469 | 18.8 | 0.48938 | 0.60976 | 0.48887 | 0.60817 |
| 13.9 | 0.50131 | 0.62509 | 0.50109 | 0.62440 | 18.9 | 0.48910 | 0.60941 | 0.48859 | 0.60780 |
| 14.0 | 0.50110 | 0.62481 | 0.50087 | 0.62411 | 19.0 | 0.48882 | 0.60906 | 0.48830 | 0.60742 |
| 14.1 | 0.50089 | 0.62453 | 0.50065 | 0.62381 | 19.1 | 0.48854 | 0.60871 | 0.48800 | 0.60704 |
| 14.2 | 0.50068 | 0.62425 | 0.50043 | 0.62351 | 19.2 | 0.48825 | 0.60835 | 0.48771 | 0.60666 |
| 14.3 | 0.50046 | 0.62397 | 0.50021 | 0.62321 | 19.3 | 0.48797 | 0.60799 | 0.48742 | 0.60628 |
| 14.4 | 0.50024 | 0.62368 | 0.49999 | 0.62291 | 19.4 | 0.48769 | 0.60764 | 0.48713 | 0.60590 |
| 14.5 | 0.50003 | 0.62340 | 0.49977 | 0.62262 | 19.5 | 0.48741 | 0.60729 | 0.48684 | 0.60552 |
| 14.6 | 0.49981 | 0.62312 | 0.49955 | 0.62232 | 19.6 | 0.48712 | 0.60693 | 0.48655 | 0.60514 |
| 14.7 | 0.49959 | 0.62883 | 0.49932 | 0.62202 | 19.7 | 0.48638 | 0.60657 | 0.48625 | 0.60475 |
| 14.8 | 0.49936 | 0.62253 | 0.49910 | 0.62172 | 19.8 | 0.48655 | 0.60621 | 0.48596 | 0.60437 |
| 14.9 | 0.49914 | 0.62224 | 0.49887 | 0.62141 | 19.9 | 0.48626 | 0.60585 | 0.48566 | 0.60398 |
| 15.0 | 0.49891 | 0.62195 | 0.49864 | 0.62110 | 20.0 | 0.48597 | 0.60549 | 0.48536 | 0.60359 |
| 15.1 | 0.49869 | 0.62166 | 0.49842 | 0.62080 | 20.1 | … | … | 0.48506 | 0.60320 |
| 15.2 | 0.49846 | 0.62137 | 0.49819 | 0.62049 | 20.2 | … | … | 0.48476 | 0.60281 |
| 15.3 | 0.49824 | 0.62108 | 0.49795 | 0.62017 | 20.3 | … | … | 0.48445 | 0.60241 |
| 15.4 | 0.42801 | 0.62078 | 0.49771 | 0.61985 | 20.4 | … | … | 0.48415 | 0.60202 |
| 15.5 | 0.49778 | 0.62048 | 0.49747 | 0.61953 | 20.5 | … | … | 0.48384 | 0.60162 |
| 15.6 | 0.49754 | 0.62017 | 0.49723 | 0.61921 | 20.6 | … | … | 0.48354 | 0.60123 |
| 15.7 | 0.49731 | 0.61987 | 0.49699 | 0.61889 | 20.7 | … | … | 0.48323 | 0.60083 |
| 15.8 | 0.49707 | 0.61956 | 0.49675 | 0.61857 | 20.8 | … | … | 0.48292 | 0.60042 |
| 15.9 | 0.49683 | 0.61926 | 0.49651 | 0.61825 | 20.9 | … | … | 0.48261 | 0.60002 |
| 16.0 | 0.49659 | 0.61895 | 0.49627 | 0.61793 | 21.0 | … | … | 0.48230 | 0.59961 |
| 16.1 | 0.49635 | 0.61864 | 0.49602 | 0.61760 | 21.1 | … | … | 0.48198 | 0.49920 |
| 16.2 | 0.49611 | 0.61833 | 0.49577 | 0.61727 | 21.2 | … | … | 0.481166 | 0.59879 |
| 16.3 | 0.49586 | 0.61801 | 0.49552 | 0.61694 | 21.3 | … | … | 0.48134 | 0.59838 |
| 16.4 | 0.49562 | 0.61770 | 0.49527 | 0.61661 | 21.4 | … | … | 0.48103 | 0.59797 |
| 16.5 | 0.49537 | 0.61738 | 0.49502 | 0.61628 | 21.5 | … | … | 0.48701 | 0.59756 |
| 16.6 | 0.49512 | 0.61706 | 0.49476 | 0.61594 | 21.6 | … | … | 0.48040 | 0.59715 |
| 16.7 | 0.49488 | 0.61675 | 0.49451 | 0.61560 | 21.7 | … | … | 0.48008 | 0.59674 |
| 16.8 | 0.40463 | 0.61643 | 0.49425 | 0.61526 | 21.8 | … | … | 0.47975 | 0.59632 |
| 16.9 | 0.49438 | 0.61611 | 0.49400 | 0.61492 | 21.9 | … | … | 0.47943 | 0.59590 |
| 17.0 | 0.49414 | 0.61580 | 0.49375 | 0.61458 | 22.0 | … | … | 0.47910 | 0.59548 |
| 17.1 | 0.49389 | 0.61548 | 0.49349 | 0.61424 | 22.1 | … | … | 0.47878 | 0.59507 |
| 17.2 | 0.49363 | 0.61515 | 0.49322 | 0.61389 | 22.2 | … | … | 0.47845 | 0.59465 |
| 17.3 | 0.49337 | 0.61482 | 0.49296 | 0.61354 | 22.3 | … | … | 0.47812 | 0.59422 |
| 17.4 | 0.49311 | 0.61449 | 0.49269 | 0.61319 | 22.4 | … | … | 0.47778 | 0.59379 |
| 17.5 | 0.49285 | 0.61416 | 0.49243 | 0.61284 | 22.5 | … | … | 0.47745 | 0.59336 |
| 17.6 | 0.49259 | 0.61383 | 0.49217 | 0.61250 | 22.6 | … | … | 0.47711 | 0.52993 |
| 17.7 | 0.49233 | 0.61350 | 0.49191 | 0.61215 | 22.7 | … | … | 0.47677 | 0.59250 |
| 17.8 | 0.49206 | 0.61316 | 0.49164 | 0.61180 | 22.8 | … | … | 0.47643 | 0.59207 |
| 17.9 | 0.49180 | 0.61283 | 0.49137 | 0.61144 | 22.9 | … | … | 0.47610 | 0.59164 |
| … | … | … | … | … | 23.0 | … | … | 0.47577 | 0.59121 |

All dimensions are in inches.

Values given for $w_1$ and $(C+c)_1$ in table are for 1-inch pitch axial threads. For other pitches divide table values by number of threads per inch.

*Courtesy of Van Keuren Co.*

**Measuring Taper Screw Threads by Three-Wire Method.**—When the 3-wire method is used in measuring a taper screw thread, the measurement is along a line that is not perpendicular to the axis of the screw thread, the inclination from the perpendicular equaling one-half the included angle of the taper. The formula that follows compensates for this inclination resulting from contact of the measuring instrument surfaces, with two wires on one side and one on the other. The taper thread is measured over the wires in the usual manner except that the single wire must be located in the thread at a point where the effective diameter is to be checked (as described more fully later). The formula shows the dimension equivalent to the correct pitch diameter at this given point. The general formula for taper screw threads follows:

$$M = \frac{E - (\cot a)/2N + W(1 + \csc a)}{\sec b}$$

where $M$ = measurement over the 3 wires
$E$ = pitch diameter
$a$ = one-half the angle of the thread
$N$ = number of threads per inch
$W$ = diameter of wires; and
$b$ = one-half the angle of taper.

This formula is not theoretically correct but it is accurate for screw threads having tapers of $\frac{3}{4}$ inch per foot or less. This general formula can be simplified for a given thread angle and taper. The simplified formula following (in which $P$ = pitch) is for an American National Standard pipe thread:

$$M = \frac{E - (0.866025 \times P) + 3 \times W}{1.00049}$$

Standard pitch diameters for pipe threads will be found in the section "American Pipe Threads," which also shows the location, or distance, of this pitch diameter from the end of the pipe. In using the formula for finding dimension $M$ over the wires, the single wire is placed in whatever part of the thread groove locates it at the point where the pitch diameter is to be checked. The wire must be accurately located at this point. The other wires are then placed on each side of the thread that is diametrically opposite the single wire. If the pipe thread is straight or without taper,

$$M = E - (0.866025 \times P) + 3 \times W$$

*Application of Formula to Taper Pipe Threads:* To illustrate the use of the formula for taper threads, assume that dimension $M$ is required for an American National Standard 3-inch pipe thread gage. Table 1a starting on page 2093 shows that the 3-inch size has 8 threads per inch, or a pitch of 0.125 inch, and a pitch diameter at the gaging notch of 3.3885 inches. Assume that the wire diameter is 0.07217 inch: Then when the pitch diameter is correct

$$M = \frac{3.3885 - (0.866025 \times 0.125) + 3 \times 0.07217}{1.00049} = 3.495 \text{ inches}$$

*Pitch Diameter Equivalent to a Given Measurement Over the Wires:* The formula following may be used to check the pitch diameter at any point along a tapering thread when measurement $M$ over wires of a given diameter is known. In this formula, $E$ = the effective or pitch diameter at the position occupied by the single wire. The formula is not theoretically correct but gives very accurate results when applied to tapers of $\frac{3}{4}$ inch per foot or less.

$$E = 1.00049 \times M + (0.866025 \times P) - 3 \times W$$

*Example:* Measurement $M$ = 3.495 inches at the gaging notch of a 3-inch pipe thread and the wire diameter = 0.07217 inch. Then

$$E = 1.00049 \times 3.495 + (0.866025 \times 0.125) - 3 \times 0.07217 = 3.3885 \text{ inches}$$

*Pitch Diameter at Any Point Along Taper Screw Thread:* When the pitch diameter in any position along a tapering thread is known, the pitch diameter at any other position may be determined as follows:

Multiply the distance (measured along the axis) between the location of the known pitch diameter and the location of the required pitch diameter, by the taper per inch or by 0.0625 for American National Standard pipe threads. Add this product to the known diameter, if the required diameter is at a large part of the taper, or subtract if the required diameter is smaller.

*Example:* The pitch diameter of a 3-inch American National Standard pipe thread is 3.3885 at the gaging notch. Determine the pitch diameter at the small end. The table starting on page 2093 shows that the distance between the gaging notch and the small end of a 3-inch pipe is 0.77 inch. Hence the pitch diameter at the small end = 3.3885 − (0.77 × 0.0625) = 3.3404 inches.

**Three-Wire Method Applied to Buttress Threads.**—The angles of buttress threads vary somewhat, especially on the front or load-resisting side. Formula (1), which follows, may be applied to any angles required. In this formula, $M$ = measurement over wires when *pitch diameter* $E$ is correct; $A$ = included angle of thread and thread groove; $a$ = angle of front face or load-resisting side, measured from a line perpendicular to screw thread axis; $P$ = pitch of thread; and $W$ = wire diameter.

$$M = E - \left[\frac{P}{\tan a + \tan(A - a)}\right] + W\left[1 + \cos\left(\frac{A}{2} - a\right) \times \csc\frac{A}{2}\right] \quad (1)$$

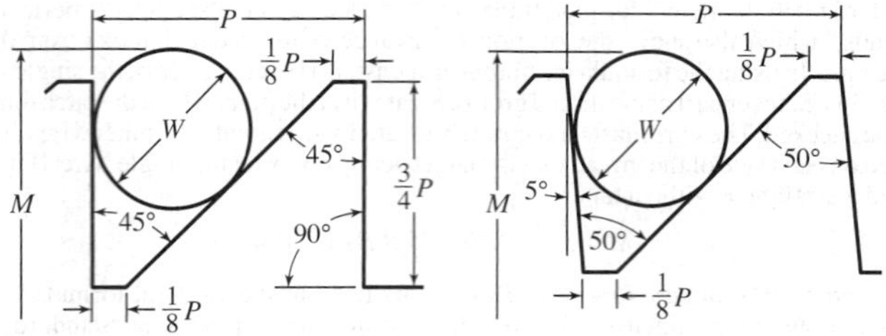


For given angles $A$ and $a$, this general formula may be simplified as shown by Formulas (3) and (4). These simplified formulas contain constants with values depending upon angles $A$ and $a$.

*Wire Diameter:* The wire diameter for obtaining pitch-line contact at the back of a buttress thread may be determined by the following general Formula (2):

$$W = P\left(\frac{\cos a}{1 + \cos A}\right) \quad (2)$$

*45-Degree Buttress Thread:* The buttress thread shown by the diagram at the left, has a front or load-resisting side that is perpendicular to the axis of the screw. Measurement $M$ equivalent to a correct pitch diameter $E$ may be determined by Formula (3):

$$M = E - P + (W \times 3.4142) \quad (3)$$

Wire diameter $W$ for pitch-line contact at back of thread = 0.586 × pitch.

*50-Degree Buttress Thread with Front-face Inclination of 5 Degrees:* This buttress thread form is illustrated by the diagram at the right. Measurement $M$ equivalent to the correct pitch diameter $E$ may be determined by Formula (4):

$$M = E - (P \times 0.91955) + (W \times 3.2235) \qquad (4)$$

Wire diameter $W$ for pitch-line contact at back of thread = 0.606 × pitch. If the width of flat at crest and root = $\frac{1}{8}$ × pitch, depth = 0.69 × pitch.

*American National Standard Buttress Threads ANSI/ASME B1.9-1973 (R2017):* This buttress screw thread has an included thread angle of 52 degrees and a front face inclination of 7 degrees. Measurements $M$ equivalent to a pitch diameter $E$ may be determined by Formula (5):

$$M = E - 0.89064P + 3.15689W + c \qquad (5)$$

The wire angle correction factor $c$ is less than 0.0004 inch for recommended combinations of thread diameters and pitches and may be neglected. Use of wire diameter $W$ = $0.54147P$ is recommended.

**Measurement of Pitch Diameter of Thread Ring Gages.—**The application of direct methods of measurement to determine the pitch diameter of thread ring gages presents serious difficulties, particularly in securing proper contact pressure when a high degree of precision is required. The usual practice is to fit the ring gage to a master setting plug. When the thread ring gage is of correct lead, angle, and thread form, within close limits, this method is quite satisfactory and represents standard American practice. It is the only method available for small sizes of threads. For the larger sizes, various more or less satisfactory methods have been devised, but none of these have found wide application.

**Screw Thread Gage Classification.—**Screw thread gages are classified by their degree of accuracy, that is, by the amount of tolerance afforded the gage manufacturer and the wear allowance, if any.

There are also three classifications according to use: 1) Working gages for controlling production; 2) inspection gages for rejection or acceptance of the finished product; and 3) reference gages for determining the accuracy of the working and inspection gages.

**Gaging Unified Screw Threads.—**ANSI/ASME B1.2-1983 (R2017) covers gaging methods for conformance of Unified screw threads and provides the essential specifications for applicable gages required for Unified inch screw threads.

The standard includes the following gages for *Product Internal Thread:*

*GO Working Thread Plug Gage* for inspecting the maximum-material GO functional limit.

*NOT GO (HI) Thread Plug Gage* for inspecting the NOT GO (HI) functional diameter limit.

*Thread Snap Gage—GO Segments or Rolls* for inspecting the maximum-material GO functional limit.

*Thread Snap Gage—NOT GO (HI) Segments or Rolls* for inspecting the NOT GO (HI) functional diameter limit.

*Thread Snap Gages—Minimum Material: Pitch Diameter Cone Type and Vee and Thread Groove Diameter Type* for inspecting the minimum-material limit pitch diameter.

*Thread-Setting Solid Ring Gage* for setting internal thread indicating and snap gages.

*Plain Plug, Snap, and Indicating Gages* for checking the minor diameter of internal threads.

*Snap and Indicating Gages* for checking the major diameter of internal threads.

*Functional Indicating Thread Gage* for inspecting the maximum-material GO functional limit and size and the NOT GO (HI) functional diameter limit and size.

*Minimum-Material Indicating Thread Gage* for inspecting the minimum-material limit and size.

*Indicating Runout Thread Gage* for inspecting runout of the minor diameter to pitch diameter.

In addition to these gages for product internal threads, the Standard also covers differential gaging and such instruments as pitch micrometers, thread-measuring balls, optical comparator and toolmaker's microscope, profile tracing instrument, surface roughness measuring instrument, and roundness measuring equipment.

The Standard includes the following gages for *Product External Thread:*

*GO Working Thread Ring Gage* for inspecting the maximum-material GO functional limit.

*NOT GO (LO) Thread Ring Gage* for inspecting the NOT GO (LO) functional diameter limit.

*Thread Snap Gage—GO Segments or Rolls* for inspecting the maximum-material GO functional limit.

*Thread Snap Gage—NOT GO (LO) Segments or Rolls* for inspecting the NOT GO (LO) functional diameter limit.

*Thread Snap Gages—Cone and Vee Type and Minimum Material Thread Groove Diameter Type* for inspecting the minimum-material pitch diameter limit.

*Plain Ring and Snap Gages* for checking the major diameter.

*Snap Gage* for checking the minor diameter.

*Functional Indicating Thread Gage* for inspecting the maximum-material GO functional limit and size and the NOT GO (LO) functional diameter limit and size.

*Minimum-Material Indicating Thread Gage* for inspecting the minimum-material limit and size.

*Indicating Runout Gage* for inspecting the runout of the major diameter to the pitch diameter.

*W Tolerance Thread-Setting Plug Gage* for setting adjustable thread ring gages, checking solid thread ring gages, setting thread snap limit gages, and setting indicating thread gages.

*Plain Check Plug Gage for Thread Ring Gage* for verifying the minor diameter limits of thread ring gages after the thread rings have been properly set with the applicable thread-setting plug gages.

*Indicating Plain Diameter Gage* for checking the major diameter.

*Indicating Gage* for checking the minor diameter.

In addition to these gages for product external threads, the Standard also covers differential gaging and such instruments as thread micrometers, thread-measuring wires, optical comparator and toolmaker's microscope, profile tracing instrument, electromechanical lead tester, helical path attachment used with GO type thread indicating gage, helical path analyzer, surface roughness measuring equipment, and roundness measuring equipment.

The Standard lists the following for use of Threaded and Plain Gages for verification of product internal threads:

*Tolerance:* Unless otherwise specified all thread gages which directly check the product thread shall be X tolerance for all classes.

*GO Thread Plug Gages:* GO thread plug gages must enter and pass through the full threaded length of the product freely. The GO thread plug gage is a cumulative check of all thread elements except the minor diameter.

*NOT GO (HI) Thread Plug Gages:* NOT GO (HI) thread plug gages when applied to the product internal thread may engage only the end threads (which may not be representative of the complete thread). Entering threads on product are incomplete and permit gage to start. Starting threads on NOT GO (HI) plugs are subject to greater wear than the remaining threads. Such wear in combination with the incomplete product threads permits further entry of the gage. NOT GO (HI) functional diameter is acceptable when the NOT GO (HI) thread plug gage applied to the product internal thread does not enter more than three complete turns. The gage should not be forced. Special requirements such as exceptionally thin or ductile material, small number of threads, etc., may necessitate modification of this practice.

*GO and NOT GO Plain Plug Gages for Minor Diameter of Product Internal Thread:* (Recommended in Class Z tolerance.) GO plain plug gages must completely enter and pass through the length of the product without force. NOT GO cylindrical plug gage must not enter.

The Standard lists the following for use of Thread Gages for verification of product external threads:

*GO Thread Ring Gages:* Adjustable GO thread ring gages must be set to the applicable W tolerance setting plugs to assure they are within specified limits. The product thread must freely enter the GO thread ring gage for the entire length of the threaded portion. The GO thread ring gage is a cumulative check of all thread elements except the major diameter.

*NOT GO (LO) Thread Ring Gages:* NOT GO (LO) thread ring gages must be set to the applicable W tolerance setting plugs to assure that they are within specified limits. NOT GO (LO) thread ring gages when applied to the product external thread may engage only the end threads (which may not be representative of the complete product thread).

Starting threads on NOT GO (LO) rings are subject to greater wear than the remaining threads. Such wear in combination with the incomplete threads at the end of the product thread permit further entry in the gage. NOT GO (LO) functional diameter is acceptable when the NOT GO (LO) thread ring gage applied to the product external thread does not pass over the thread more than three complete turns. The gage should not be forced. Special requirements such as exceptionally thin or ductile material, small number of threads, etc., may necessitate modification of this practice.

*GO and NOT GO Plain Ring and Snap Gages for Checking Major Diameter of Product External Thread:* The GO gage must completely receive or pass over the major diameter of the product external thread to ensure that the major diameter does not exceed the maximum-material-limit. The NOT GO gage must not pass over the major diameter of the product external thread to ensure that the major diameter is not less than the minimum-material-limit.

Limitations concerning the use of gages are given in the Standard as follows:

*Product threads* accepted by a gage of one type may be verified by other types. It is possible, however, that parts which are near either rejection limit may be accepted by one type and rejected by another. Also, it is possible for two individual limit gages of the same type to be at the opposite extremes of the gage tolerances permitted, and borderline product threads accepted by one gage could be rejected by another. For these reasons, a product screw thread is considered acceptable when it passes a test by any of the permissible gages in ANSI/ASME B1.3 for the gaging system that are within the tolerances.

*Gaging large product external and internal threads* equal to above 6.25-inch nominal size with plain and threaded plug and ring gages presents problems for technical and economic reasons. In these instances, verification may be based on use of modified snap or indicating gages or measurement of thread elements. Various types of gages or measuring devices in addition to those defined in the Standard are available and acceptable when properly correlated to this Standard. Producer and user should agree on the method and equipment used.

**Thread Forms of Gages.—**Thread forms of gages for product internal and external threads are given in Table 1. The Standard ANSI/ASME B1.2-1983 (R2017) also gives illustrations of the thread forms of truncated thread setting plug gages, the thread forms of full-form thread setting plug gages, the thread forms of solid thread setting ring gages, and an illustration that shows the chip groove and removal of partial thread.

**Building Up Worn Plug Gages.—**Plug gages which have been worn under size can be built up by chromium plating and then lapped to size. Any amount of metal up to 0.004 or 0.005 inch can be added to a worn gage. Chromium oxide is used in lapping chromium plated gages, or other parts, to size and for polishing. When the chromium plating of a plug gage has worn under size, it may be removed by subjecting it to the action of muriatic acid. The gage is then built up again by chromium plating and lapped to size. When removing the worn plating the gage should be watched carefully and the action of the acid stopped as soon as the plating has been removed in order to avoid the roughening effect of the acid on the steel.

**Thread Gage Tolerances.—**Gage tolerances of thread plug and ring gages, thread setting plugs, and setting rings for Unified screw threads, designated as W and X tolerances, are given in Table 4. W tolerances represent the highest commercial grade of accuracy and workmanship, and are specified for thread setting gages; X tolerances are larger than W tolerances and are used for product inspection gages. Tolerances for plain gages are given in Table 2.

*Determining Size of Gages:* The three-wire method of determining pitch diameter size of plug gages is recommended for gages covered by American National Standard B1.2, described in Appendix B of the 1983 issue of that Standard.

Size limit adjustments of thread ring and external thread snap gages are determined by their fit on their respective calibrated setting plugs. Indicating gages and thread gages for product external threads are controlled by reference to appropriate calibrated setting plugs.

Size limit adjustments of internal thread snap gages are determined by their fit on their respective calibrated setting rings. Indicating gages and other adjustable thread gages for product internal threads are controlled by reference to appropriate calibrated setting rings or by direct measuring methods.

*Interpretation of Tolerances:* Tolerances on lead, half-angle, and pitch diameter are variations which may be taken independently for each of these elements and may be taken to the extent allowed by respective tabulated dimensional limits. The tabulated tolerance on any one element must not be exceeded, even though variations in the other two elements are smaller than the respective tabulated tolerances.

*Direction of Tolerance on Gages:* At the maximum-material limit (GO), the dimensions of all gages used for final conformance gaging are to be within limits of size of the product thread. At the functional diameter limit, using NOT GO (HI and LO) thread gages, the standard practice is to have the gage tolerance within the limits of size of the product thread.

*Formulas for Limits of Gages:* Formulas for limits of American National Standard Gages for Unified screw threads are given in Table 5. Some constants which are required to determine gage dimensions are tabulated in Table 3.

**Table 1. Thread Forms of Gages for Product Internal and External Threads**

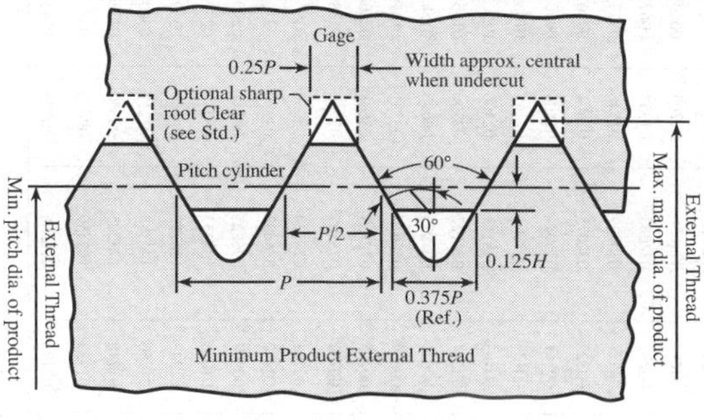


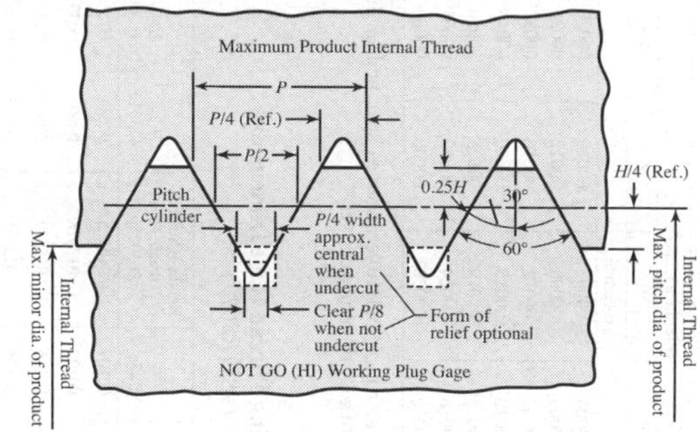


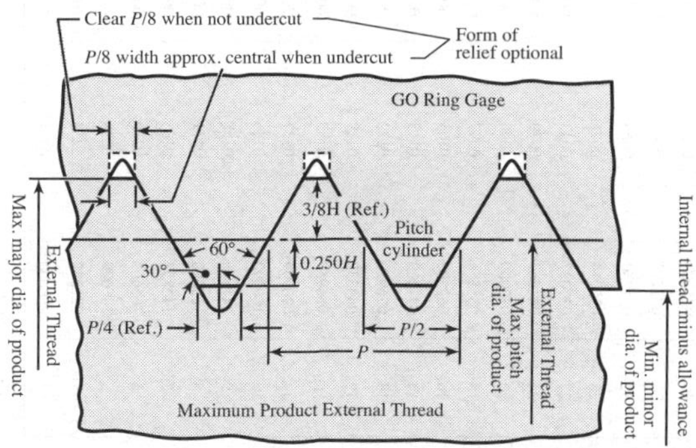


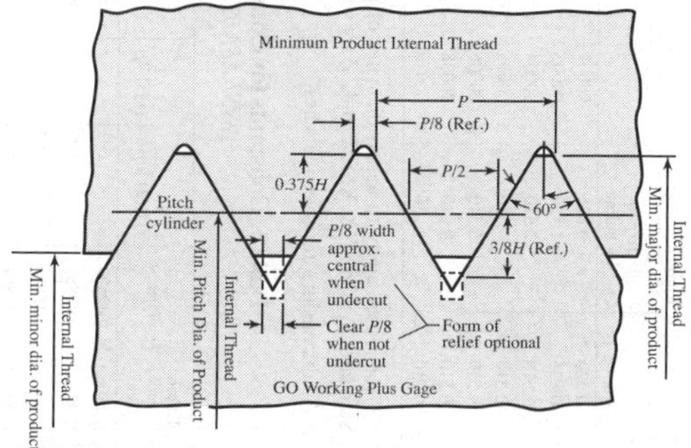

## Table 2. American National Standard Tolerances for Plain Cylindrical Gages
*ANSI/ASME B1.2-1983 (R2017)*

| Size Range | | Tolerance Class[a] | | | | |
|---|---|---|---|---|---|---|
| | | XX | X | Y | Z | ZZ |
| Above | To and Including | Tolerance | | | | |
| 0.020 | 0.825 | .00002 | .00004 | .00007 | .00010 | .00020 |
| 0.825 | 1.510 | .00003 | .00006 | .00009 | .00012 | .00024 |
| 1.510 | 2.510 | .00004 | .00008 | .00012 | .00016 | .00032 |
| 2.510 | 4.510 | .00005 | .00010 | .00015 | .00020 | .00040 |
| 4.510 | 6.510 | .000065 | .00013 | .00019 | .00025 | .00050 |
| 6.510 | 9.010 | .00008 | .00016 | .00024 | .00032 | .00064 |
| 9.010 | 12.010 | .00010 | .00020 | .00030 | .00040 | .00080 |

[a] Tolerances apply to actual diameter of plug or ring. Apply tolerances as specified in the Standard. Symbols XX, X, Y, Z, and ZZ are standard gage tolerance classes.

All dimensions are given in inches.

## Table 3. Constants for Computing Thread Gage Dimensions
*ANSI/ASME B1.2-1983 (R2017)*

| Threads per Inch | Pitch, $p$ | $0.060\sqrt[3]{p^2}+0.017p$ | .05$p$ | .087$p$ | Height of Sharp V-Thread, $H$ = .866025$p$ | $H/2$ = .43301$p$ | $H/4$ = .216506$p$ |
|---|---|---|---|---|---|---|---|
| 80 | .012500 | .0034 | .00063 | .00109 | .010825 | .00541 | .00271 |
| 72 | .013889 | .0037 | .00069 | .00122 | .012028 | .00601 | .00301 |
| 64 | .015625 | .0040 | .00078 | .00136 | .013532 | .00677 | .00338 |
| 56 | .017857 | .0044 | .00089 | .00155 | .015465 | .00773 | .00387 |
| 48 | .020833 | .0049 | .00104 | .00181 | .018042 | .00902 | .00451 |
| 44 | .022727 | .0052 | .00114 | .00198 | .019682 | .00984 | .00492 |
| 40 | .025000 | .0056 | .00125 | .00218 | .021651 | .01083 | .00541 |
| 36 | .027778 | .0060 | .00139 | .00242 | .024056 | .01203 | .00601 |
| 32 | .031250 | .0065 | .00156 | .00272 | .027063 | .01353 | .00677 |
| 28 | .035714 | .0071 | .00179 | .00311 | .030929 | .01546 | .00773 |
| 27 | .037037 | .0073 | .00185 | .00322 | .032075 | .01604 | .00802 |
| 24 | .041667 | .0079 | .00208 | .00361 | .036084 | .01804 | .00902 |
| 20 | .050000 | .0090 | .00250 | .00435 | .043301 | .02165 | .01083 |
| 18 | .055556 | .0097 | .00278 | .00483 | .048113 | .02406 | .01203 |
| 16 | .062500 | .0105 | .00313 | .00544 | .054127 | .02706 | .01353 |
| 14 | .071429 | .0115 | .00357 | .00621 | .061859 | .03093 | .01546 |
| 13 | .076923 | .0122 | .00385 | .00669 | .066617 | .03331 | .01665 |
| 12 | .083333 | .0129 | .00417 | .00725 | .072169 | .03608 | .01804 |
| 11½ | .086957 | .0133 | .00435 | .00757 | .075307 | .03765 | .01883 |
| 11 | .090909 | .0137 | .00451 | .00791 | .078730 | .03936 | .01968 |
| 10 | .100000 | .0146 | .00500 | .00870 | .086603 | .04330 | .02165 |
| 9 | .111111 | .0158 | .00556 | .00967 | .096225 | .04811 | .02406 |
| 8 | .125000 | .0171 | .00625 | .01088 | .108253 | .05413 | .02706 |
| 7 | .142857 | .0188 | .00714 | .01243 | .123718 | .06186 | .03093 |
| 6 | .166667 | .0210 | .00833 | .01450 | .144338 | .07217 | .03608 |
| 5 | .200000 | .0239 | .01000 | .01740 | .173205 | .08660 | .04330 |
| 4½ | .222222 | .0258 | .01111 | .01933 | .192450 | .09623 | .04811 |
| 4 | .250000 | .0281 | .01250 | .02175 | .216506 | .10825 | .05413 |

All dimensions are given in inches unless otherwise specified.

**Table 4. American National Standard Tolerance for GO, HI, and LO Thread Gages for Unified Inch Screw Thread**

| Thds. per Inch | Tolerance on Lead[a]: To & incl. ½ in. Dia. | Tolerance on Lead[a]: Above ½ in. Dia. | Tol. on Thread Half-angle (±), Minutes | Tol. on Major and Minor Diams.[b]: To & incl. ½ in. Dia. | Tol. on Major and Minor Diams.[b]: Above ½ to 4 in. Dia. | Tol. on Major and Minor Diams.[b]: Above 4 in. Dia. | Tolerance on Pitch Diameter[b]: To & incl. ½ in. Dia. | Tolerance on Pitch Diameter[b]: Above ½ to 1½ in. Dia. | Tolerance on Pitch Diameter[b]: Above 1½ to 4 in. Dia. | Tolerance on Pitch Diameter[b]: Above 4 to 8 in. Dia. | Tolerance on Pitch Diameter[b]: Above 8 to 12 in.[c] Dia. |
|---|---|---|---|---|---|---|---|---|---|---|---|
| W GAGES | | | | | | | | | | | |
| 80, 72 | .0001 | .00015 | 20 | .0003 | .0003 | … | .0001 | .00015 | … | … | … |
| 64 | .0001 | .00015 | 20 | .0003 | .0004 | … | .0001 | .00015 | … | … | … |
| 56 | .0001 | .00015 | 20 | .0003 | .0004 | … | .0001 | .00015 | .0002 | … | … |
| 48 | .0001 | .00015 | 18 | .0003 | .0004 | … | .0001 | .00015 | .0002 | … | … |
| 44, 40 | .0001 | .00015 | 15 | .0003 | .0004 | … | .0001 | .00015 | .0002 | … | … |
| 36 | .0001 | .00015 | 12 | .0003 | .0004 | … | .0001 | .00015 | .0002 | … | … |
| 32 | .0001 | .00015 | 12 | .0003 | .0005 | .0007 | .0001 | .00015 | .0002 | .00025 | .0003 |
| 28, 27 | .00015 | .00015 | 8 | .0005 | .0005 | .0007 | .0001 | .00015 | .0002 | .00025 | .0003 |
| 24, 20 | .00015 | .00015 | 8 | .0005 | .0005 | .0007 | .0001 | .00015 | .0002 | .00025 | .0003 |
| 18 | .00015 | .00015 | 8 | .0005 | .0005 | .0007 | .0001 | .00015 | .0002 | .00025 | .0003 |
| 16 | .00015 | .00015 | 8 | .0006 | .0006 | .0009 | .0001 | .0002 | .00025 | .0003 | .0004 |
| 14, 13 | .0002 | .0002 | 6 | .0006 | .0006 | .0009 | .00015 | .0002 | .00025 | .0003 | .0004 |
| 12 | .0002 | .0002 | 6 | .0006 | .0006 | .0009 | .00015 | .0002 | .00025 | .0003 | .0004 |
| 11½ | .0002 | .0002 | 6 | .0006 | .0006 | .0009 | .00015 | .0002 | .00025 | .0003 | .0004 |
| 11 | .0002 | .0002 | 6 | .0006 | .0006 | .0009 | .00015 | .0002 | .00025 | .0003 | .0004 |
| 10 | … | .00025 | 6 | … | .0006 | .0009 | … | .0002 | .0025 | .0003 | .0004 |
| 9 | … | .00025 | 6 | … | .0007 | .0011 | … | .0002 | .00025 | .0003 | .0004 |
| 8 | … | .00025 | 5 | … | .0007 | .0011 | … | .0002 | .00025 | .0003 | .0004 |
| 7 | … | .0003 | 5 | … | .0007 | .0011 | … | .0002 | .00025 | .0003 | .0004 |
| 6 | … | .0003 | 5 | … | .0008 | .0013 | … | .0002 | .00025 | .0003 | .0004 |
| 5 | … | .0003 | 4 | … | .0008 | .0013 | … | … | .00025 | .0003 | .0004 |
| 4½ | … | .0003 | 4 | … | .0008 | .0013 | … | … | .00025 | .0003 | .0004 |
| 4 | … | .0003 | 4 | … | .0009 | .0015 | … | … | .00025 | .0003 | .0004 |
| X GAGES | | | | | | | | | | | |
| 80, 72 | .0002 | .0002 | 30 | .0003 | .0003 | … | .0002 | .0002 | … | … | … |
| 64 | .0002 | .0002 | 30 | .0004 | .0004 | … | .0002 | .0002 | … | … | … |
| 56, 48 | .0002 | .0002 | 30 | .0004 | .0004 | … | .0002 | .0002 | .0003 | … | … |
| 44, 40 | .0002 | .0002 | 20 | .0004 | .0004 | … | .0002 | .0002 | .0003 | … | … |
| 36 | .0002 | .0002 | 20 | .0004 | .0004 | … | .0002 | .0002 | .0003 | … | … |
| 32, 28 | .0003 | .0003 | 15 | .0005 | .0005 | .0007 | .0003 | .0003 | .0004 | .0005 | .0006 |
| 27, 24 | .0003 | .0003 | 15 | .0005 | .0005 | .0007 | .0003 | .0003 | .0004 | .0005 | .0006 |
| 20 | .0003 | .0003 | 15 | .0005 | .0005 | .0007 | .0003 | .0003 | .0004 | .0005 | .0006 |
| 18 | .0003 | .0003 | 10 | .0005 | .0005 | .0007 | .0003 | .0003 | .0004 | .0005 | .0006 |
| 16, 14 | .0003 | .0003 | 10 | .0006 | .0006 | .0009 | .0003 | .0003 | .0004 | .0006 | .0008 |
| 13, 12 | .0003 | .0003 | 10 | .0006 | .0006 | .0009 | .0003 | .0003 | .0004 | .0006 | .0008 |
| 11½ | .0003 | .0003 | 10 | .0006 | .0006 | .0009 | .0003 | .0003 | .0004 | .0006 | .0008 |
| 11, 10 | .0003 | .0003 | 10 | .0006 | .0006 | .0009 | .0003 | .0003 | .0004 | .0006 | .0008 |
| 9 | .0003 | .0003 | 10 | .0007 | .0007 | .0011 | .0003 | .0003 | .0004 | .0006 | .0008 |
| 8, 7 | .0004 | .0004 | 5 | .0007 | .0007 | .0011 | .0004 | .0004 | .0005 | .0006 | .0008 |
| 6 | .0004 | .0004 | 5 | .0008 | .0008 | .0013 | .0004 | .0004 | .0005 | .0006 | .0008 |
| 5, 4½ | .0004 | .0004 | 5 | .0008 | .0008 | .0013 | … | … | .0005 | .0006 | .0008 |
| 4 | .0004 | .0004 | 5 | .0009 | .0009 | .0015 | … | … | .0005 | .0006 | .0008 |

[a] Allowable variation in lead between any two threads not farther apart than the length of the standard gage as shown in ANSI/ASME B47.1. The tolerance on lead establishes the width of a zone, measured parallel to the axis of the thread, within which the actual helical path must lie for the specified length of the thread. Measurements are taken from a fixed reference point, located at the start of the first full thread, to a sufficient number of positions along the entire helix to detect all types of lead variations. The amounts that these positions vary from their basic (theoretical) positions are recorded with due respect to sign. The greatest variation in each direction (±) is selected, and the sum of their values, disregarding sign, must not exceed the tolerance limits specified for W gages.

[b] Tolerances apply to designated size of thread. The application of the tolerances is specified in the Standard.

[c] Above 12 in. the tolerance is directly proportional to the tolerance given in this column below, in the ratio of the diameter to 12 in.

All dimensions are given in inches unless otherwise specified.

**Table 5. Formulas for Limits of American National Standard Gages for Unified Inch Screw Threads** *ANSI/ASME B1.2-1983 (R2017)*

| No. | Thread Gages for External Threads |
|---|---|
| 1 | GO Pitch Diameter = Maximum pitch diameter of external thread. Gage tolerance is *minus*. |
| 2 | GO Minor Diameter = Maximum pitch diameter of external thread minus $H/2$. Gage tolerance is *minus*. |
| 3 | NOT GO (LO) Pitch Diameter (for plus tolerance gage) = Minimum pitch diameter of external thread. Gage tolerance is *plus*. |
| 4 | NOT GO (LO) Minor Diameter = Minimum pitch diameter of external thread minus $H/4$. Gage tolerance is *plus*. |
| | Plain Gages for Major Diameter of External Threads |
| 5 | GO = Maximum major diameter of external thread. Gage tolerance is *minus*. |
| 6 | NOT GO = Minimum major diameter of external thread. Gage tolerance is *plus*. |
| | Thread Gages for Internal Threads |
| 7 | GO Major Diameter = Minimum major diameter of internal thread. Gage tolerance is *plus*. |
| 8 | GO Pitch Diameter = Minimum pitch diameter of internal thread. Gage tolerance is *plus*. |
| 9 | NOT GO (HI) Major Diameter = Maximum pitch diameter of internal thread plus $H/2$. Gage tolerance is *minus*. |
| 10 | NOT GO (HI) Pitch Diameter = Maximum pitch diameter of internal thread. Gage tolerance is *minus*. |
| | Plain Gages for Minor Diameter of Internal Threads |
| 11 | GO = Minimum minor diameter of internal thread. Gage tolerance is *plus*. |
| 12 | NOT GO = Maximum minor diameter of internal thread. Gage tolerance is *minus*. |
| | Full Form nd Truncated Setting Plugs |
| 13 | GO Major Diameter (Truncated Portion) = Maximum major diameter of external thread (= minimum major diameter of full portion of GO setting plug) minus $\left(0.060\sqrt[3]{p^2}+0.017p\right)$. Gage tolerance is *minus*. |
| 14 | GO Major Diameter (Full Portion) = Maximum major diameter of external thread. Gage tolerance is *plus*. |
| 15 | GO Pitch Diameter = Maximum pitch diameter of external thread. Gage tolerance is *minus*. |
| 16 | [a]NOT GO (LO) Major Diameter (Truncated Portion) = Minimum pitch diameter of external thread plus $H/2$. Gage tolerance is *minus*. |
| 17 | NOT GO (LO) Major Diameter (Full Portion) = Maximum major diameter of external thread provided major diameter crest width shall not be less than 0.001 in. (0.0009 in. truncation). Apply W tolerance *plus* for maximum size except that for 0.001 in. crest width apply tolerance *minus*. For the 0.001 in. crest width, major diameter is equal to maximum major diameter of external thread plus 0.216506$p$ minus the sum of external thread pitch diameter tolerance and 0.0017 in. |
| 18 | NOT GO (LO) Pitch Diameter = Minimum pitch diameter of external thread. Gage tolerance is *plus*. |
| | Solid Thread-setting Rings for Snap and Indicating Gages |
| 19 | [b]GO Pitch Diameter = Minimum pitch diameter of internal thread. W gage tolerance is *plus*. |
| 20 | GO Minor Diameter = Minimum minor diameter of internal thread. W gage tolerance is *minus*. |
| 21 | [b]NOT GO (HI) Pitch Diameter = Maximum pitch diameter of internal thread. W gage tolerance is *minus*. |
| 22 | NOT GO (HI) Minor Diameter = Maximum minor diameter of internal thread. W gage tolerance is *minus*. |

[a] Truncated portion is required when optional sharp root profile is used.

[b] Tolerances greater than W tolerance for pitch diameter are acceptable when internal indicating or snap gage can accommodate a greater tolerance and when agreed upon by supplier and user.

See data in Screw Thread Systems section for symbols and dimensions of Unified Screw Threads.

## TAPPING AND THREAD CUTTING

**Selection of Taps.—**For most applications, a standard tap supplied by the manufacturer can be used, but some jobs may require special taps. A variety of standard taps can be obtained. In addition to specifying the size of the tap it is necessary to be able to select the one most suitable for the application at hand.

The elements of standard taps that are varied are: the number of flutes; the type of flute, whether straight, spiral pointed, or spiral fluted; the chamfer length; the relief of the land, if any; the tool steel used to make the tap; and the surface treatment of the tap.

Details regarding the nomenclature of tap elements are given in the section *TAPS* starting on page 957, along with a listing of the standard sizes available.

Factors to consider in selecting a tap include: the method of tapping, by hand or by machine; the material to be tapped and its heat treatment; the length of thread, or depth of the tapped hole; the required tolerance or class of fit; and the production requirement and the type of machine to be used.

The diameter of the hole must also be considered, although this action is usually only a matter of design and the specification of the tap drill size.

*Method of Tapping:* The term *hand tap* is used for both hand and machine taps, and almost all taps can be applied by the hand or machine method. While any tap can be used for hand tapping, those having a concentric land without the relief are preferable. In hand tapping the tool is reversed periodically to break the chip, and the heel of the land of a tap with a concentric land (without relief) will cut the chip off cleanly or any portion of it that is attached to the work, whereas a tap with an eccentric or con-eccentric relief may leave a small burr that becomes wedged between the relieved portion of the land and the work. This wedging creates a pressure towards the cutting face of the tap that may cause it to chip; it tends to roughen the threads in the hole, and it increases the overall torque required to turn the tool. When tapping by machine, however, the tap is usually turned only in one direction until the operation is complete, and an eccentric or con-eccentric relief is often an advantage.

*Chamfer Length:* Three types of hand taps, used both for hand and machine tapping, are available, and they are distinguished from each other by the length of chamfer. *Taper taps* have a chamfer angle that reduces the height about 8–10 teeth; *plug taps* have a chamfer angle with 3–5 threads reduced in height; and *bottoming taps* have a chamfer angle with $1\frac{1}{2}$ threads reduced in height. Since the teeth that are reduced in height do practically all the cutting, the chip load or chip thickness per tooth will be least for a taper tap, greater for a plug tap, and greatest for a bottoming tap.

For most through hole tapping applications it is necessary to use only a plug type tap, which is also most suitable for blind holes where the tap drill hole is deeper than the required thread. If the tap must bottom in a blind hole, the hole is usually threaded first with a plug tap and then finished with a bottoming tap to catch the last threads in the bottom of the hole. Taper taps are used on materials where the chip load per tooth must be kept to a minimum. However, taper taps should not be used on materials that have a strong tendency to work harden, such as the austenitic stainless steels.

*Spiral Point Taps:* Spiral point taps offer a special advantage when machine tapping through holes in ductile materials because they are designed to handle the long continuous chips that form and would otherwise cause a disposal problem. An angular gash is ground at the point or end of the tap along the face of the chamfered threads or lead teeth of the tap. This gash forms a left-hand helix in the flutes adjacent to the lead teeth which causes the chips to flow ahead of the tap and through the hole. The gash is usually formed to produce a rake angle on the cutting face that increases progressively toward the end of the tool. Since the flutes are used primarily to provide a passage for the cutting fluid, they are usually made narrower and shallower thereby strengthening the tool. For tapping thin workpieces short fluted spiral point taps are recommended. They have a spiral point gash along the cutting teeth; the remainder of the threaded portion of the tap has no flute. Most spiral pointed taps are of plug type; however, spiral point bottoming taps are also made.

*Spiral Fluted Taps:* Spiral fluted taps have a helical flute; the helix angle of the flute may be between 15 and 52 degrees and the hand of the helix is the same as that of the threads on the tap. The spiral flute and the rake that it forms on the cutting face of the tap combine to induce the chips to flow backward along the helix and out of the hole. Thus, they are ideally suited for tapping blind holes and they are available as plug and bottoming types. A higher spiral angle should be specified for tapping very ductile materials; when tapping harder materials, chipping at the cutting edge may result and the spiral angle must be reduced.

Holes having a pronounced interruption such as a groove or a keyway can be tapped with spiral fluted taps. The land bridges the interruption and allows the tap to cut relatively smoothly.

*Serial Taps and Close Tolerance Threads:* For tapping holes to close tolerances a set of serial taps is used.

They are usually available in sets of three: the No. 1 tap is undersize and is the first rougher; the No. 2 tap is of intermediate size and is the second rougher; and the No. 3 tap is used for finishing.

The different taps are identified by one, two, and three annular grooves in the shank adjacent to the square. For some applications involving finer pitches only two serial taps are required. Sets are also used to tap hard or tough materials having a high tensile strength, deep blind holes in normal materials, and large coarse threads. A set of more than three taps is sometimes required to produce threads of coarse pitch. Threads to some commercial tolerances, such as American Standard Unified 2B, or ISO Metric 6H, can be produced in one cut using a ground tap; sometimes even closer tolerances can be produced with a single tap. Ground taps are recommended for all close tolerance tapping operations. For much ordinary work, cut taps are satisfactory and more economical than ground taps.

*Tap Steels:* Most taps are made from high-speed steel. The type of tool steel used is determined by the tap manufacturer and is usually satisfactory when correctly applied except in a few exceptional cases. Typical grades of high-speed steel used to make taps are M-1, M-2, M-3, M-42, etc. Carbon tool steel taps are satisfactory where the operating temperature of the tap is low and where a high resistance to abrasion is not required as in some types of hand tapping.

*Surface Treatment:* The life of high-speed steel taps can sometimes be increased significantly by treating the surface of the tap. A very common treatment is oxide coating, which forms a thin metallic oxide coating on the tap that has lubricity and is somewhat porous to absorb and retain oil. This coating reduces the friction between the tap and the work and it makes the surface virtually impervious to rust. It does not increase the hardness of the surface but it significantly reduces or prevents entirely galling, or the tendency of the work material to weld or stick to the cutting edge and to other areas on the tap with which it is in contact. For this reason oxide coated taps are recommended for metals that tend to gall and stick such as non-free cutting low-carbon steels and soft copper. It is also useful for tapping other steels having higher strength properties.

Nitriding provides a very hard and wear resistant case on high-speed steel. Nitrided taps are especially recommended for tapping plastics; they have also been used successfully on a variety of other materials including high strength high alloy steels. However, some caution must be used in specifying nitrided taps because the nitride case is very brittle and may have a tendency to chip.

Chrome plating has been used to increase the wear resistance of taps but its application has been limited because of the high cost and the danger of hydrogen embrittlement which can cause cracks to form in the tool. A flash plate of about .0001 in. or less in thickness is applied to the tap. Chrome-plated taps have been used successfully to tap a variety of ferrous and nonferrous materials including plastics, hard rubber, mild steel, and tool steel. Other surface treatments that have been used successfully to a limited extent are vapor blasting and liquid honing.

*Rake Angle:* For the majority of applications in both ferrous and nonferrous materials the rake angle machined on the tap by the manufacturer is satisfactory. This angle is approximately 5 to 7 degrees. In some instances it may be desirable to alter the rake angle of the tap to obtain beneficial results and Table 1 provides a guide that can be used. In selecting a rake angle from this table, consideration must be given to the size of the tap and the strength of the land. Most standard taps are made with a curved face with the rake angle measured as a chord between the crest and root of the thread. The resulting shape is called a hook angle.

**Table 1. Tap Rake Angles for Tapping Different Materials**

| Material | Rake Angle, Degrees | Material | Rake Angle, Degrees |
|---|---|---|---|
| Cast Iron | 0-3 | Aluminum | 8-20 |
| Malleable Iron | 5-8 | Brass | 2-7 |
| Steel | | Naval Brass | 5-8 |
| AISI 1100 Series | 5-12 | Phosphor Bronze | 5-12 |
| Low-Carbon (up | 5-12 | Tobin Bronze | 5-8 |
| to .25 percent) | | Manganese Bronze | 5-12 |
| Medium-Carbon, Annealed | 5-10 | Magnesium | 10-20 |
| (.30 to .60 percent) | | Monel | 9-12 |
| Heat-Treated, 225-283 Brinell | 0-8 | Copper | 10-18 |
| (.30 to .60 percent) | | Zinc Die Castings | 10-15 |
| High-Carbon and | 0-5 | Plastic | |
| High-Speed | | Thermoplastic | 5-8 |
| Stainless | 8-15 | Thermosetting | 0-3 |
| Titanium | 5-10 | Hard Rubber | 0-3 |

**Cutting Speed.—**The cutting speed for machine tapping is treated in detail on page 1128. It suffices to say here that many variables must be considered in selecting this cutting speed and any tabulation may have to be modified greatly. Where cutting speeds are mentioned in the following section, they are intended only to provide a guideline to show the possible range of speeds that could be used.

**Tapping Specific Materials.—**The work material has a great influence on the ease with which a hole can be tapped. For production work, in many instances, modified taps are recommended; however, for toolroom or short batch work, standard hand taps can be used on most jobs, providing reasonable care is taken when tapping. The following concerns the tapping of metallic materials; information on the tapping of plastics is given on page 605.

*Low-Carbon Steel (Less than 0.15% C):* These steels are very soft and ductile resulting in a tendency for the work material to tear and to weld to the tap. They produce a continuous chip that is difficult to break and spiral pointed taps are recommended for tapping through holes; for blind holes a spiral fluted tap is recommended. To prevent galling and welding, a liberal application of a sulfur base or other suitable cutting fluid is essential and the selection of an oxide coated tap is very helpful.

*Low-Carbon Steels (0.15 to 0.30% C):* The additional carbon in these steels is beneficial as it reduces the tendency to tear and to weld; their machinability is further improved by cold drawing. These steels present no serious problems in tapping provided a suitable cutting fluid is used. An oxide coated tap is recommended, particularly in the lower carbon range.

*Medium-Carbon Steels (0.30 to 0.60% C):* These steels can be tapped without too much difficulty, although a lower cutting speed must be used in machine tapping. The cutting speed is dependent on carbon content and heat treatment. Steels that have a higher carbon content must be tapped more slowly, especially if the heat treatment has produced a pearlitic microstructure. The cutting speed and ease of tapping is significantly improved by heat treating to produce a spheroidized microstructure. A suitable cutting fluid must be used.

*High-Carbon Steels (More than 0.6% C):* Usually these materials are tapped in the annealed or normalized condition although sometimes tapping is done after hardening and tempering to a hardness below 55 RC (Rockwell C scale). Recommendations for tapping after hardening and tempering are given under *High-Tensile Strength Steels*. In the annealed and normalized condition these steels have a higher strength and are more abrasive than steels with a lower carbon content; thus, they are more difficult to tap. The microstructure resulting from the heat treatment has a significant effect on the ease of tapping and the tap life, a spheroidite structure being better in this respect than a pearlitic structure. The rake angle of the tap should not exceed 5 degrees and for the harder materials a concentric tap is recommended. The cutting speed is considerably lower for these steels and an activated sulfur-chlorinated cutting fluid is recommended.

*Alloy Steels:* This classification includes a wide variety of steels, each of which may be heat treated to have a wide range of properties. When annealed and normalized they are similar to medium- to high-carbon steels and usually can be tapped without difficulty, although for some alloy steels a lower tapping speed may be required. Standard taps can be used and for machine tapping a con-eccentric relief may be helpful. A suitable cutting fluid must be used.

*High-Tensile Strength Steels:* Any steel that must be tapped after being heat treated to a hardness range of 40 to 55 RC is included in this classification. Low tap life and excessive tap breakage are characteristics of tapping these materials; those that have a high chromium content are particularly troublesome. Best results are obtained with taps that have concentric lands, a rake angle that is at or near zero degrees, and 6 to 8 chamfered threads on the end to reduce the chip load per tooth. The chamfer relief should be kept to a minimum. The load on the tap should be kept to a minimum by every possible means, including using the largest possible tap drill size; keeping the hole depth to a minimum; avoidance of bottoming holes; and, in the larger sizes, using fine instead of coarse pitches. Oxide coated taps are recommended although a nitrided tap can sometimes be used to reduce tap wear. An active sulfur-chlorinated oil is recommended as a cutting fluid and the tapping speed should not exceed about 10 feet per minute (3.0 m/min).

*Stainless Steels:* Ferritic and martensitic type stainless steels are somewhat like alloy steels that have a high chromium content, and they can be tapped in a similar manner, although a slightly slower cutting speed may have to be used. Standard rake angle oxide coated taps are recommended and a cutting fluid containing molybdenum disulphide is helpful to reduce the friction in tapping. Austenitic stainless steels are very difficult to tap because of their high resistance to cutting and their great tendency to work harden. A work-hardened layer is formed by a cutting edge of the tap and the depth of this layer depends on the severity of the cut and the sharpness of the tool. The next cutting edge must penetrate below the work-hardened layer, if it is to be able to cut. Therefore, the tap must be kept sharp and each succeeding cutting edge on the tool must penetrate below the work-hardened layer formed by the preceding cutting edge. For this reason, a taper tap should not be used, but rather a plug tap having 3 to 5 chamfered threads. To reduce the rubbing of the lands, an eccentric or con-eccentric relieved land should be used and a 10 to 15 degree rake angle is recommended. A tough continuous chip is formed that is difficult to break. To control this chip, spiral pointed taps are recommended for through holes and low-helix angle spiral fluted taps for blind holes. An oxide coating on the tap is very helpful and a sulfur-chlorinated mineral lard oil is recommended, although heavy duty soluble oils have also been used successfully.

*Free Cutting Steels:* There are large numbers of free cutting steels, including free cutting stainless steels, which are also called free machining steels. Sulfur, lead, or phosphorus are added to these steels to improve their machinability. Free machining steels are always easier to tap than their counterparts that do not have the free machining additives. Tool life is usually increased and a somewhat higher cutting speed can be used. The type of tap recommended depends on the particular type of free machining steel and the nature of the tapping operation; usually a standard tap can be used.

*High-Temperature Alloys:* These are cobalt or nickel base nonferrous alloys that cut like austenitic stainless steel, but are often even more difficult to machine. The recommendations given for austenitic stainless steel also apply to tapping these alloys but the rake angle should be 0 to 10 degrees to strengthen the cutting edge. For most applications a nitrided tap or one made from M41, M42, M43, or M44 steel is recommended. The tapping speed is usually in the range of 5–10 ft/min (1.5–3.0 m/min).

*Titanium and Titanium Alloys:* Titanium and its alloys have a low specific heat and a pronounced tendency to weld on to the tool material; therefore, oxide coated taps are recommended to minimize galling and welding. The rake angle of the tap should be from 6 to 10 degrees. To minimize the contact between the work and the tap an eccentric or con-eccentric relief land should be used. Taps having interrupted threads are sometimes helpful. Pure titanium is comparatively easy to tap but the alloys are very difficult. The cutting speed depends on the composition of the alloy and may vary from 10–40 ft/min (3.0–12.2 m/min). Special cutting oils are recommended for tapping titanium.

*Gray Cast Iron:* The microstructure of gray cast iron can vary, even within a single casting, and compositions are used that vary in tensile strength from about 20,000–60,000 psi (138–414 MPa) and 160 to 250 BHN. Thus, cast iron is not a single material, although in general it is not difficult to tap. The cutting speed may vary from 90 ft/min (27.4 m/min) for the softer grades to 30 ft/min (9.1 m/min) for the harder grades. The chip is discontinuous and straight fluted taps should be used for all applications. Oxide coated taps are helpful and gray cast iron can usually be tapped dry, although water soluble oils and chemical emulsions are sometimes used.

*Malleable Cast Iron:* Commercial malleable cast irons are also available having a rather wide range of properties, although within a single casting they tend to be quite uniform. They are relatively easy to tap and standard taps can be used. The cutting speed for ferritic cast irons is 60–90 ft/min (18.3–27.4 m/min), for pearlitic malleable irons 40–50 ft/ min (12.2–15.2 m/min), and for martensitic malleable irons 30–35 ft/min (9.1–10.7 m/min). A soluble oil cutting fluid is recommended except for martensitic malleable iron where a sulfur base oil may work better.

*Ductile or Nodular Cast Iron:* Several classes of nodular iron are used having a tensile strength varying from 60,000–120,000 psi (414–827 MPa). Moreover, the microstructure in a single casting and in castings produced at different times vary rather widely. The chips are easily controlled but have some tendency to weld to the faces and flanks of cutting tools. For this reason oxide coated taps are recommended. The cutting speed may vary from 15 fpm (4.6 m/min) for the harder martensitic ductile irons to 60 fpm (18.3 m/min) for the softer ferritic grades. A suitable cutting fluid should be used.

*Aluminum:* Aluminum and aluminum alloys are relatively soft materials that have little resistance to cutting. The danger in tapping these alloys is that the tap will ream the hole instead of cutting threads, or that it will cut a thread eccentric to the hole. For these reasons, extra care must be taken when aligning the tap and starting the thread. For production tapping a spiral pointed tap is recommended for through holes and a spiral fluted tap for blind holes; preferably these taps should have a 10 to 15 degree rake angle. A lead-screw tapping machine is helpful in cutting accurate threads. A heavy duty soluble oil or a light base mineral oil should be used as a cutting fluid.

*Copper Alloys:* Most copper alloys are not difficult to tap, except beryllium copper and a few other hard alloys. Pure copper is difficult because of its ductility and the ductile continuous chip formed, which can be hard to control. However, with reasonable care and the use of medium heavy duty mineral lard oil it can be tapped successfully. Red brass, yellow brass, and similar alloys containing not more than 35 percent zinc produce a continuous chip. While straight fluted taps can be used for hand tapping these alloys, machine tapping should be done with spiral pointed or spiral fluted taps for through and blind holes respectively. Naval brass, leaded brass, and cast brasses produce a discontinuous chip and a straight fluted tap can be used for machine tapping. These alloys exhibit a tendency to close in on the tap and sometimes an interrupted thread tap is used to reduce the resulting jamming effect. Beryllium copper and the silicon bronzes are the strongest of the copper alloys. Their strength combined with their ability to work harden can cause difficulties in tapping. For these alloys plug type taps should be used and the taps should be kept as sharp as possible. A medium or heavy duty water soluble oil is recommended as a cutting fluid.

**Other Tapping Lubricants.—**The power required in tapping varies considerably with different lubricants. The following lubricants reduce the resistance to the cut when threading forged nuts and hexagon drawn material: stearine oil, lard oil, sperm oil, rape oil, and 10 percent graphite with 90 percent tallow. A mixture of cutting emulsion (soluble oil) with water reduces resistance to threading action well. A few emulsions are almost as good as animal and vegetable oils, but the emulsion used plays an important part; the majority of emulsions do not give good results. A large volume of lubricant gives somewhat better results than a small quantity, especially in the case of the thinner oils. Kerosene, turpentine, and graphite proved unsuitable for tapping steel. Mineral oils not mixed with animal and vegetable oils, and ordinary lubricating and machine oils, are wholly unsuitable.

For aluminum, kerosene is recommended. For tapping cast iron use a strong solution of emulsion; oil has a tendency to make cast-iron chips clog in the flutes, preventing the lubricant from reaching the tap cutting teeth. For tapping copper, milk is a good lubricant.

## Tap Drill and Hole Size

Tapping troubles are sometimes caused by tap drills that are too small in diameter. The tap drill should not be smaller than is necessary to give the required strength to the thread as even a very small decrease in the diameter of the drill will increase the torque required and the possibility of broken taps. Tests have shown that any increase in the percentage of full thread over 60 percent does not significantly increase the strength of the thread. Often, a 55 to 60 percent thread is satisfactory, although 75 percent threads are commonly used to provide an extra measure of safety. The present thread specifications do not always allow the use of the smaller thread depths. However, the specification given on a part drawing must be adhered to and may require smaller minor diameters than might otherwise be recommended.

The depth of the thread in the tapped hole is dependent on the length of thread engagement and on the material. In general, when the engagement length is more than one and one-half times the nominal diameter a 50 or 55 percent thread is satisfactory. Soft ductile materials permit a slightly larger tapping hole than brittle materials such as gray cast iron.

It must be remembered that a twist drill is a roughing tool that may be expected to drill slightly oversize and that some variations in the size of the tapping holes are almost inevitable. When a closer control of the hole size is required it must be reamed. Reaming is recommended for the larger thread diameters and for some fine pitch threads.

For threads of Unified form (see *American National and Unified Screw Thread Forms* on page 1943) the selection of tap drills is covered in Table 2, and the hole size limits are given in Table 3. See the section *Factors Influencing Minor Diameter Tolerances of Tapped Holes* on page 2171. Tap drill sizes for American National Form threads based on 75 percent of full thread depth are given in Table 4 and Table 5. For smaller-size threads the use of slightly larger drills, if permissible, will reduce tap breakage. The selection of tap drills for these threads also may be based on the hole size limits given in Table 3, page 2162, for Unified threads that take lengths of engagement into account. See *Recommended Hole Sizes before Tapping* on page 2172.

The size of the tap drill hole for any desired percentage of full thread depth can be calculated by the formulas below. In these formulas the Percent Full Thread is expressed as a decimal; e.g., 75 percent is expressed as .75. The tap drill size is the size nearest to the calculated hole size.

For American Unified Thread form:

$$\text{Hole Size} = \text{Basic Major Diameter} - \frac{1.08253 \times \text{Percent Full Thread}}{\text{Number of Threads per Inch}}$$

**For ISO Metric threads (all dimensions in millimeters):**

$$\textbf{Hole Size} = \textbf{Basic Major Diameter} - (\textbf{1.08253} \times \textbf{Pitch} \times \textbf{Percent Full Thread})$$

The constant 1.08253 in the above equation represents $5H/8$ where $H$ is the height of a sharp V-thread (see page 1943). The pitch is taken to be 1.

*Hole Sizes for Tapping Unified Miniature Screw Threads* are given on page 2170 and Table 6.

*Theoretical and Tap Drill or Core Hole Sizes for Cold Form Tapping Unified Threads* are given in Table 9, page 2177 and for ISO metric threads are given in Table 10, page 2178.

**Reaming Allowance.—**Reaming is a secondary operation that slightly enlarges previously drilled or bored hole. Reamed holes are very accurate with a smooth surface finish.

The reaming allowance is amount of material left in the hole after drilling. A properly sharpened drill produces a hole that is about 0.002–0.008 inch (0.050–0.203mm) oversize. For accurate diameters, a common practice is to rough out the hole with a smaller drill, then size the hole with the proper drill, leaving the correct reaming allowance.

Use the following table to determine the drill size used prior to reaming the hole.

| Size of Reamer | | Reaming Allowance | | Size of Reamer | | Reaming Allowance | |
|---|---|---|---|---|---|---|---|
| inch | mm | inch | mm | inch | mm | inch | mm |
| $^1/_{32}$–$^1/_8$ | 0.79–3.18 | 0.003–0.005 | 0.076–0.127 | $^3/_8$–$^1/_2$ | 9.53–12.70 | 0.010–0.015 | 0.254–0.381 |
| $^1/_8$–$^1/_4$ | 3.18–6.35 | 0.004–0.008 | 0.101–0.203 | $^1/_2$–$^3/_4$ | 12.70–19.05 | 0.015–0.03 | 0.381–0.762 |
| $^1/_4$–$^3/_8$ | 6.35–9.53 | 0.006–0.010 | 0.152–0.254 | $^3/_4$–1.00 | 19.05–25.40 | 0.03 | 0.762 |

*Example:* A finished bore of 1.000 inches after reaming is required. Select the drill size for predrilling that allows adequate allowance for reaming.

*Solution:* According to the table above, a 1-inch drilled hole needs to be approximately 0.03 inch undersize to allow sufficient allowance for reaming, thus the hole size prior to reaming should be approximately 1.000 − 0.03 = 0.97 inch.

A $^{31}/_{32}$ drill can be expected to produce a slightly oversize hole, say 0.971 to 0.977 inch. The $^{31}/_{32}$ drill provides a reaming allowance of 0.0 23 to 0.029 inch.

**Table 2. Tap Drill Sizes and Percentage of Thread (Unified Threads)**

| Taps Size Threads per Inch | Drill Size | | | | Percentage of Full Thread | Nut Minor Diameter, min[a] max |
|---|---|---|---|---|---|---|
| | No. or Letter | Fractional Inch | mm | Decimal Inch | | |
| 0–80 | … | … | 1.30 | 0.0512 | 54 | 0.0465 |
| | … | … | 1.25 | 0.0492 | 68 | 0.0514 |
| | … | … | 1.20 | 0.0472 | 79 | |
| 1–64 | … | 1⁄16 | … | 0.0625 | 51 | 0.0561 |
| | … | … | 1.55 | 0.0610 | 59 | 0.0623 |
| | 53 | … | … | 0.0595 | 66 | |
| | … | … | 1.50 | 0.0590 | 69 | |
| | … | … | 1.45 | 0.0571 | 77 | |
| 1–72 | 52 | … | … | 0.0635 | 53 | 0.0580 |
| | … | 1⁄16 | … | 0.0625 | 58 | 0.0635 |
| | … | … | 1.55 | 0.0610 | 66 | |
| | 53 | … | … | 0.0595 | 75 | |
| | … | … | 1.50 | 0.0590 | 77 | |
| 2–56 | 49 | … | … | 0.0730 | 56 | 0.0667 |
| | … | … | 1.80 | 0.0709 | 65 | 0.0737 |
| | 50 | … | … | 0.0700 | 69 | |
| | … | … | 1.75 | 0.0689 | 74 | |
| 2–64 | … | … | 1.90 | 0.0748 | 55 | 0.0691 |
| | 49 | … | … | 0.0730 | 64 | 0.0753 |
| | … | … | 1.85 | 0.0728 | 65 | |
| | … | … | 1.80 | 0.0709 | 74 | |
| | 50 | … | … | 0.0700 | 79 | |
| 3–48 | … | … | 2.15 | 0.0846 | 53 | 0.0764 |
| | … | … | 2.10 | 0.0827 | 60 | 0.0845 |
| | 45 | … | … | 0.0820 | 63 | |
| | 46 | … | … | 0.0810 | 66 | |
| | 47 | … | … | 0.0785 | 76 | |
| | … | 5⁄64 | … | 0.0781 | 77 | |
| 3–56 | … | … | 2.20 | 0.0866 | 53 | 0.0797 |
| | 44 | … | … | 0.0860 | 56 | 0.0865 |
| | … | … | 2.15 | 0.0846 | 62 | |
| | … | … | 2.10 | 0.0827 | 70 | |
| | 45 | … | … | 0.0820 | 73 | |
| | 46 | … | … | 0.0801 | 77 | |
| | … | … | 2.05 | 0.0807 | 79 | |
| 4–36 | 43 | … | … | 0.0890 | 64 | 0.0821[b] |
| | 44 | … | … | 0.0860 | 72 | 0.0882[b] |
| 4–40 | … | … | 2.40 | 0.0945 | 54 | 0.0849 |
| | 42 | … | … | 0.0935 | 57 | 0.0939 |
| | … | … | 2.35 | 0.0925 | 60 | |
| | … | … | 2.30 | 0.0905 | 66 | |
| 4–40 | 43 | … | … | 0.0890 | 71 | |
| | … | … | 2.20 | 0.0866 | 78 | |
| 4–48 | … | … | 2.45 | 0.0964 | 57 | 0.0894 |
| | 41 | … | … | 0.0960 | 59 | 0.0968 |
| | … | … | 2.40 | 0.0945 | 64 | |
| | … | 3⁄32 | … | 0.0938 | 68 | |
| | 42 | … | … | 0.0935 | 68 | |
| | … | … | 2.35 | 0.0925 | 72 | |
| | … | … | 2.30 | 0.0905 | 79 | |

**Table 2.** *(Continued)* **Tap Drill Sizes and Percentage of Thread (Unified Threads)**

| Taps Size Threads per Inch | Drill Size | | | | Percentage of Full Thread | Nut Minor Diameter, min[a] max |
|---|---|---|---|---|---|---|
| | No. or Letter | Fractional Inch | mm | Decimal Inch | | |
| 5–40 | … | … | 2.70 | 0.1063 | 57 | 0.0979 |
| | 37 | … | … | 0.1040 | 65 | 0.1062 |
| | … | … | 2.60 | 0.1024 | 70 | |
| | 38 | … | … | 0.1015 | 72 | |
| | 39 | … | … | 0.0995 | 78 | |
| 5–44 | … | … | 2.75 | 0.1083 | 57 | 0.1004 |
| | 36 | … | … | 0.1065 | 62 | 0.1079 |
| | 37 | … | … | 0.1040 | 71 | |
| | … | … | 2.6 | 0.1024 | 76 | |
| | 38 | … | … | 0.1015 | 79 | |
| 6–32 | 33 | … | … | 0.1130 | 62 | 0.1042 |
| | 34 | … | … | 0.1110 | 67 | 0.1130 |
| | … | … | 2.8 | 0.1102 | 68 | |
| | 35 | … | … | 0.1100 | 69 | |
| | … | 7/64 | … | 0.1094 | 70 | |
| | … | … | 2.75 | 0.1083 | 73 | |
| | 36 | … | … | 0.1065 | 77 | |
| 6–40 | 31 | … | … | 0.1200 | 55 | 0.1109 |
| | … | … | 3 | 0.1181 | 61 | 0.1186 |
| | 32 | … | … | 0.1160 | 68 | |
| | … | … | 2.9 | 0.1142 | 73 | |
| | 33 | … | … | 0.1130 | 77 | |
| 8–32 | … | 9/64 | … | 0.1406 | 57 | 0.1302 |
| | 28 | … | … | 0.1405 | 58 | 0.1389 |
| | … | … | 3.5 | 0.1378 | 65 | |
| | 29 | … | … | 0.1360 | 69 | |
| | … | … | 3.4 | 0.1339 | 74 | |
| 8–36 | 27 | … | … | 0.1440 | 55 | 0.1339 |
| | … | … | 3.6 | 0.1417 | 62 | 0.1416 |
| | … | 9/64 | … | 0.1406 | 68 | |
| | 28 | … | … | 0.1405 | 65 | |
| | | … | 3.5 | 0.1378 | 73 | |
| | 29 | … | … | 0.1360 | 78 | |
| 10–24 | 21 | … | … | 0.1590 | 57 | 0.1449 |
| | 22 | … | … | 0.1570 | 61 | 0.1555 |
| | … | 5/32 | … | 0.1563 | 62 | |
| | 23 | … | … | 0.1540 | 66 | |
| | 24 | … | … | 0.1520 | 67 | |
| | 25 | … | … | 0.1496 | 75 | |
| | 26 | … | … | 0.1470 | 79 | |
| 10–32 | 19 | … | … | 0.1660 | 59 | 0.1562 |
| | … | … | 4.1 | 0.1614 | 70 | 0.1641 |
| | 20 | … | … | 0.1610 | 71 | |
| | 21 | … | … | 0.1590 | 76 | |
| 12–24 | 14 | … | … | 0.1820 | 63 | 0.1709 |
| | 15 | … | … | 0.1800 | 67 | 0.1708 |
| | 16 | … | … | 0.1770 | 72 | |
| | 17 | … | … | 0.1730 | 79 | |
| 12–28 | … | 3/16 | … | 0.1875 | 61 | 0.1773 |
| | 13 | … | … | 0.1850 | 67 | 0.1857 |

**Table 2.** *(Continued)* **Tap Drill Sizes and Percentage of Thread (Unified Threads)**

| Taps Size Threads per Inch | Drill Size | | | | Percentage of Full Thread | Nut Minor Diameter, min[a] max |
|---|---|---|---|---|---|---|
| | No. or Letter | Fractional Inch | mm | Decimal Inch | | |
| | 14 | … | … | 0.1820 | 73 | |
| | 15 | … | … | 0.1800 | 78 | |
| 1/4–20 | 4 | … | … | 0.2090 | 63 | 0.1959 |
| | 5 | … | … | 0.2055 | 69 | 0.2067 |
| | 6 | … | … | 0.2040 | 71 | |
| | … | 13/64 | … | 0.2031 | 72 | |
| | 7 | … | … | 0.2010 | 75 | |
| | 8 | … | … | 0.1990 | 79 | |
| 1/4–27 | 2 | … | … | 0.2210 | 61 | 0.2120 |
| | … | … | … | 0.2187 | 65 | 0.2180[b] |
| | 3 | … | … | 0.2130 | 77 | |
| 1/4–28 | 2 | … | … | 0.2210 | 63 | 0.2113 |
| | … | 7/32 | … | 0.2187 | 67 | 0.2190 |
| | … | … | 5.5 | 0.2165 | 72 | |
| | 3 | … | … | 0.2130 | 80 | |
| 1/4–32 | … | … | 5.7 | 0.2244 | 63 | 0.2162 |
| | 2 | … | … | 0.2210 | 71 | 0.2208[b] |
| | … | 7/28 | … | 0.2187 | 77 | |
| 5/16–18 | H | … | … | 0.2660 | 64 | 0.2524 |
| | … | 17/64 | … | 0.2656 | 65 | 0.2630 |
| | G | … | … | 0.2610 | 71 | |
| | F | … | … | 0.2570 | 77 | |
| 5/16–24 | J | … | … | 0.2770 | 66 | 0.2674 |
| | … | … | 7 | 0.2756 | 68 | 0.2754 |
| | I | … | … | 0.2720 | 75 | |
| 5/16–27 | … | 9/32 | … | 0.2812 | 68 | 0.2718 |
| | J | … | … | 0.2770 | 74 | 0.2792[b] |
| 5/16–32 | … | … | 7.3 | 0.2841 | 62 | 0.2787 |
| | … | … | 7.2 | 0.2835 | 75 | 0.2833[b] |
| | … | 9/32 | … | 0.2812 | 77 | |
| 3/8–16 | P | … | … | 0.3230 | 64 | 0.3073 |
| | … | … | 8.1 | 0.3189 | 69 | 0.3182 |
| | O | … | … | 0.3160 | 72 | |
| | … | 5/16 | … | 0.3125 | 77 | |
| 3/8–24 | R | … | … | 0.3390 | 67 | 0.3299 |
| | … | … | 8.5 | 0.3345 | 74 | 0.3372 |
| | Q | … | … | 0.3320 | 79 | |
| 3/8–27 | … | 11/32 | … | 0.3437 | 65 | 0.3347 |
| | R | … | … | 0.3390 | 75 | 0.3416[b] |
| 7/16–14 | … | … | 9.7 | 0.3818 | 60 | 0.3602 |
| | V | … | … | 0.3770 | 65 | 0.3717 |
| | … | 3/8 | … | 0.3750 | 67 | |
| | U | … | … | 0.3680 | 75 | |
| 7/16–20 | X | … | … | 0.3970 | 62 | 0.3834 |
| | … | … | 10 | 0.3937 | 67 | 0.3916 |
| | … | 25/64 | … | 0.3906 | 72 | |
| | W | … | … | 0.3860 | 79 | |
| 7/16–24 | Y | … | … | 0.4040 | 62 | 0.3925 |
| | X | … | … | 0.3970 | 74 | 0.3985[b] |
| 7/16–27 | … | 13/32 | … | 0.4062 | 65 | 0.3982 |

**Table 2.** *(Continued)* **Tap Drill Sizes and Percentage of Thread (Unified Threads)**

| Taps Size Threads per Inch | Drill Size | | | | Percentage of Full Thread | Nut Minor Diameter, min[a] max |
|---|---|---|---|---|---|---|
| | No. or Letter | Fractional Inch | mm | Decimal Inch | | |
| | Y | … | … | 0.4040 | 70 | 0.4043[b] |
| ½–12 | … | 27⁄64 | … | 0.4219 | 72 | 0.4098 |
| | | | | | | 0.4223 |
| ½–13 | … | 7⁄16 | … | 0.4375 | 62 | 0.4167 |
| | … | … | 11 | 0.4331 | 68 | 0.4284 |
| | … | 27⁄64 | … | 0.4219 | 78 | |
| ½–20 | … | 29⁄64 | … | 0.4531 | 72 | 0.4459 |
| | … | [c] | … | 0.4492 | 78 | 0.4537 |
| ½–27 | … | 15⁄32 | … | 0.4687 | 65 | 0.4618 |
| | [c] | … | … | 0.4640 | 75 | 0.4672[b] |
| 9⁄16–12 | … | ½ | … | 0.5000 | 58 | 0.4723 |
| | … | … | 12.5 | 0.4921 | 65 | 0.4843 |
| | … | 31⁄64 | … | 0.4844 | 72 | |
| 9⁄16–18 | … | 33⁄64 | … | 0.5156 | 65 | 0.5024 |
| | … | … | 13 | 0.5118 | 70 | 0.5106 |
| 9⁄16–27 | … | 17⁄32 | … | 0.5312 | 65 | 0.5234 |
| | … | [c] | … | 0.5265 | 75 | 0.5273[b] |
| 5⁄8–11 | … | … | 14 | 0.5512 | 62 | 0.5266 |
| | … | 35⁄64 | … | 0.5469 | 66 | 0.5391 |
| | … | 17⁄32 | … | 0.5312 | 79 | |
| 5⁄8–18 | … | 37⁄64 | … | 0.5781 | 65 | 0.5649 |
| | … | … | 14.5 | 0.5709 | 75 | 0.5730 |
| 5⁄8–27 | … | 19⁄32 | … | 0.5937 | 65 | 0.5860 |
| | … | [c] | … | 0.5890 | 75 | 0.5912 |
| ¾–10 | … | … | 17 | 0.6693 | 62 | 0.6417 |
| | [c] | 21⁄32 | … | 0.6563 | 72 | 0.6545[b] |
| | … | … | 16.5 | 0.6496 | 77 | |
| ¾–16 | … | 45⁄64 | … | 0.7031 | 58 | 0.6823 |
| | … | … | 17.5 | 0.6890 | 75 | 0.6908 |
| | … | 11⁄64 | … | 0.6875 | 77 | |
| ¾–27 | … | 23⁄32 | … | 0.7187 | 65 | 0.7102 |
| | [c] | … | … | 0.7140 | 75 | 0.7164[c] |
| 7⁄8–9 | … | 25⁄32 | … | 0.7813 | 65 | 0.7547 |
| | … | 49⁄64 | … | 0.7656 | 76 | 0.7679 |
| 7⁄8–14 | … | 13⁄16 | … | 0.8125 | 67 | 0.7977 |
| | … | … | 20.5 | 0.8071 | 73 | 0.8068 |
| 7⁄8–18 | … | 53⁄64 | … | 0.8281 | 65 | 0.8149 |
| | [c] | … | … | 0.8210 | 75 | 0.8223[b] |
| 7⁄8–27 | … | 27⁄32 | … | 0.8437 | 65 | 0.8340 |
| | [c] | … | … | 0.8390 | 75 | 0.8406[c] |
| 1–8 | … | … | 23.0 | 0.9055 | 58 | 0.8647 |
| | … | 57⁄64 | … | 0.8906 | 67 | 0.8797 |
| | … | 7⁄8 | … | 0.8750 | 77 | |
| 1–12 | … | 59⁄64 | … | 0.9219 | 72 | 0.9098 |
| | | | | | | 0.9198 |

[a] Unified or American Standard Threads.

[b] Not based on Unified or American Thread Standards.

[c] Special drill required for this size as the next size smaller gives too great a percentage of thread.

## Table 3. Recommended Hole Size Limits Before Tapping Unified Threads

| Thread Size | Classes 1B and 2B | | | | | | | | Class 3B | | | | | | | |
|---|---|---|---|---|---|---|---|---|---|---|---|---|---|---|---|---|
| | Length of Engagement (*D* = Nominal Size of Thread) | | | | | | | | | | | | | | | |
| | To and Including ⅓ *D* | | Above ⅓ *D* to ⅔ *D* | | Above ⅔ *D* to 1½ *D* | | Above 1½ *D* to 3*D* | | To and Including ⅓ *D* | | Above ⅓ *D* to ⅔ *D* | | Above ⅔ *D* to 1½ *D* | | Above 1½ *D* to 3*D* | |
| | Recommended Hole Size Limits | | | | | | | | | | | | | | | |
| | Min[a] | Max | Min | Max | Min | Max[b] | Min | Max | Min[a] | Max | Min | Max | Min | Max[b] | Min | Max |
| 0-80 | 0.0465 | 0.0500 | 0.0479 | 0.0514 | 0.0479 | 0.0514 | 0.0479 | 0.0514 | 0.0465 | 0.0500 | 0.0479 | 0.0514 | 0.0479 | 0.0514 | 0.0479 | 0.0514 |
| 1-64 | 0.0561 | 0.0599 | 0.0585 | 0.0623 | 0.0585 | 0.0623 | 0.0585 | 0.0623 | 0.0561 | 0.0599 | 0.0585 | 0.0623 | 0.0585 | 0.0623 | 0.0585 | 0.0623 |
| 1-72 | 0.0580 | 0.0613 | 0.0596 | 0.0629 | 0.0602 | 0.0635 | 0.0602 | 0.0635 | 0.0580 | 0.0613 | 0.0596 | 0.0629 | 0.0602 | 0.0635 | 0.0602 | 0.0635 |
| 2-56 | 0.0667 | 0.0705 | 0.0686 | 0.0724 | 0.0699 | 0.0737 | 0.0699 | 0.0737 | 0.0667 | 0.0705 | 0.0686 | 0.0724 | 0.0699 | 0.0737 | 0.0699 | 0.0737 |
| 2-64 | 0.0691 | 0.0724 | 0.0707 | 0.0740 | 0.0720 | 0.0753 | 0.0720 | 0.0753 | 0.0691 | 0.0724 | 0.0707 | 0.0740 | 0.0720 | 0.0753 | 0.0720 | 0.0753 |
| 3-48 | 0.0764 | 0.0804 | 0.0785 | 0.0825 | 0.0805 | 0.0845 | 0.0806 | 0.0846 | 0.0764 | 0.0804 | 0.0785 | 0.0825 | 0.0805 | 0.0845 | 0.0806 | 0.0846 |
| 3-56 | 0.0797 | 0.0831 | 0.0814 | 0.0848 | 0.0831 | 0.0865 | 0.0833 | 0.0867 | 0.0797 | 0.0831 | 0.0814 | 0.0848 | 0.0831 | 0.0865 | 0.0833 | 0.0867 |
| 4-40 | 0.0849 | 0.0894 | 0.0871 | 0.0916 | 0.0894 | 0.0939 | 0.0902 | 0.0947 | 0.0849 | 0.0894 | 0.0871 | 0.0916 | 0.0894 | 0.0939 | 0.0902 | 0.0947 |
| 4-48 | 0.0894 | 0.0931 | 0.0912 | 0.0949 | 0.0931 | 0.0968 | 0.0939 | 0.0976 | 0.0894 | 0.0931 | 0.0912 | 0.0949 | 0.0931 | 0.0968 | 0.0939 | 0.0976 |
| 5-40 | 0.0979 | 0.1020 | 0.1000 | 0.1041 | 0.1021 | 0.1062 | 0.1036 | 0.1077 | 0.0979 | 0.1020 | 0.1000 | 0.1041 | 0.1021 | 0.1062 | 0.1036 | 0.1077 |
| 5-44 | 0.1004 | 0.1042 | 0.1023 | 0.1060 | 0.1042 | 0.1079 | 0.1060 | 0.1097 | 0.1004 | 0.1042 | 0.1023 | 0.1060 | 0.1042 | 0.1079 | 0.1060 | 0.1097 |
| 6-32 | 0.104 | 0.109 | 0.106 | 0.112 | 0.109 | 0.114 | 0.112 | 0.117 | 0.1040 | 0.1091 | 0.1066 | 0.1115 | 0.1091 | 0.1140 | 0.1115 | 0.1164 |
| 6-40 | 0.111 | 0.115 | 0.113 | 0.117 | 0.115 | 0.119 | 0.117 | 0.121 | 0.1110 | 0.1148 | 0.1128 | 0.1167 | 0.1147 | 0.1186 | 0.1166 | 0.1205 |
| 8-32 | 0.130 | 0.134 | 0.132 | 0.137 | 0.134 | 0.139 | 0.137 | 0.141 | 0.1300 | 0.1345 | 0.1324 | 0.1367 | 0.1346 | 0.1389 | 0.1367 | 0.1410 |
| 8-36 | 0.134 | 0.138 | 0.136 | 0.140 | 0.138 | 0.142 | 0.140 | 0.144 | 0.1340 | 0.1377 | 0.1359 | 0.1397 | 0.1378 | 0.1416 | 0.1397 | 0.1435 |
| 10-24 | 0.145 | 0.150 | 0.148 | 0.154 | 0.150 | 0.156 | 0.152 | 0.159 | 0.1450 | 0.1502 | 0.1475 | 0.1528 | 0.1502 | 0.1555 | 0.1528 | 0.1581 |
| 10-32 | 0.156 | 0.160 | 0.158 | 0.162 | 0.160 | 0.164 | 0.162 | 0.166 | 0.1560 | 0.1601 | 0.1581 | 0.1621 | 0.1601 | 0.1641 | 0.1621 | 0.1661 |
| 12-24 | 0.171 | 0.176 | 0.174 | 0.179 | 0.176 | 0.181 | 0.178 | 0.184 | 0.1710 | 0.1758 | 0.1733 | 0.1782 | 0.1758 | 0.1807 | 0.1782 | 0.1831 |
| 12-28 | 0.177 | 0.182 | 0.179 | 0.184 | 0.182 | 0.186 | 0.184 | 0.188 | 0.1770 | 0.1815 | 0.1794 | 0.1836 | 0.1815 | 0.1857 | 0.1836 | 0.1878 |
| 12-32 | 0.182 | 0.186 | 0.184 | 0.188 | 0.186 | 0.190 | 0.188 | 0.192 | 0.1820 | 0.1858 | 0.1837 | 0.1877 | 0.1855 | 0.1895 | 0.1873 | 0.1913 |
| ¼-20 | 0.196 | 0.202 | 0.199 | 0.204 | 0.202 | 0.207 | 0.204 | 0.210 | 0.1960 | 0.2013 | 0.1986 | 0.2040 | 0.2013 | 0.2067 | 0.2040 | 0.2094 |
| ¼-28 | 0.211 | 0.216 | 0.213 | 0.218 | 0.216 | 0.220 | 0.218 | 0.222 | 0.2110 | 0.2152 | 0.2131 | 0.2171 | 0.2150 | 0.2190 | 0.2169 | 0.2209 |
| ¼-32 | 0.216 | 0.220 | 0.218 | 0.222 | 0.220 | 0.224 | 0.222 | 0.226 | 0.2160 | 0.2196 | 0.2172 | 0.2212 | 0.2189 | 0.2229 | 0.2206 | 0.2246 |
| ¼-36 | 0.220 | 0.224 | 0.221 | 0.225 | 0.224 | 0.226 | 0.225 | 0.228 | 0.2200 | 0.2243 | 0.2199 | 0.2243 | 0.2214 | 0.2258 | 0.2229 | 0.2273 |
| 5⁄16-18 | 0.252 | 0.259 | 0.255 | 0.262 | 0.259 | 0.265 | 0.262 | 0.268 | 0.2520 | 0.2577 | 0.2551 | 0.2604 | 0.2577 | 0.2630 | 0.2604 | 0.2657 |
| 5⁄16-24 | 0.267 | 0.272 | 0.270 | 0.275 | 0.272 | 0.277 | 0.275 | 0.280 | 0.2670 | 0.2714 | 0.2694 | 0.2734 | 0.2714 | 0.2754 | 0.2734 | 0.2774 |
| 5⁄16-32 | 0.279 | 0.283 | 0.281 | 0.285 | 0.283 | 0.286 | 0.285 | 0.289 | 0.2790 | 0.2817 | 0.2792 | 0.2832 | 0.2807 | 0.2847 | 0.2822 | 0.2862 |
| 5⁄16-36 | 0.282 | 0.286 | 0.284 | 0.288 | 0.285 | 0.289 | 0.287 | 0.291 | 0.2820 | 0.2863 | 0.2824 | 0.2863 | 0.2837 | 0.2877 | 0.2850 | 0.2890 |
| ⅜-16 | 0.307 | 0.314 | 0.311 | 0.318 | 0.314 | 0.321 | 0.318 | 0.325 | 0.3070 | 0.3127 | 0.3101 | 0.3155 | 0.3128 | 0.3182 | 0.3155 | 0.3209 |

**Table 3.** *(Continued)* **Recommended Hole Size Limits Before Tapping Unified Threads**

| Thread Size | Classes 1B and 2B | | | | | | | | Class 3B | | | | | | | |
|---|---|---|---|---|---|---|---|---|---|---|---|---|---|---|---|---|
| | Length of Engagement ($D$ = Nominal Size of Thread) | | | | | | | | | | | | | | | |
| | To and Including 1/3 $D$ | | Above 1/3 $D$ to 2/3 $D$ | | Above 2/3 $D$ to 1 1/2 $D$ | | Above 1 1/2 $D$ to 3$D$ | | To and Including 1/3 $D$ | | Above 1/3 $D$ to 2/3 $D$ | | Above 2/3 $D$ to 1 1/2 $D$ | | Above 1 1/2 $D$ to 3$D$ | |
| | Recommended Hole Size Limits | | | | | | | | | | | | | | | |
| | Min[a] | Max | Min | Max | Min | Max[b] | Min | Max | Min[a] | Max | Min | Max | Min | Max[b] | Min | Max |
| 3/8-24 | 0.330 | 0.335 | 0.333 | 0.338 | 0.335 | 0.340 | 0.338 | 0.343 | 0.3300 | 0.3336 | 0.3314 | 0.3354 | 0.3332 | 0.3372 | 0.3351 | 0.3391 |
| 3/8-32 | 0.341 | 0.345 | 0.343 | 0.347 | 0.345 | 0.349 | 0.347 | 0.351 | 0.3410 | 0.3441 | 0.3415 | 0.3455 | 0.3429 | 0.3469 | 0.3444 | 0.3484 |
| 3/8-36 | 0.345 | 0.349 | 0.346 | 0.350 | 0.347 | 0.352 | 0.349 | 0.353 | 0.3450 | 0.3488 | 0.3449 | 0.3488 | 0.3461 | 0.3501 | 0.3474 | 0.3514 |
| 7/16-14 | 0.360 | 0.368 | 0.364 | 0.372 | 0.368 | 0.376 | 0.372 | 0.380 | 0.3600 | 0.3660 | 0.3630 | 0.3688 | 0.3659 | 0.3717 | 0.3688 | 0.3746 |
| 7/16-20 | 0.383 | 0.389 | 0.386 | 0.391 | 0.389 | 0.395 | 0.391 | 0.397 | 0.3830 | 0.3875 | 0.3855 | 0.3896 | 0.3875 | 0.3916 | 0.3896 | 0.3937 |
| 7/16-28 | 0.399 | 0.403 | 0.401 | 0.406 | 0.403 | 0.407 | 0.406 | 0.410 | 0.3990 | 0.4020 | 0.3995 | 0.4035 | 0.4011 | 0.4051 | 0.4017 | 0.4067 |
| 1/2-13 | 0.417 | 0.426 | 0.421 | 0.430 | 0.426 | 0.434 | 0.430 | 0.438 | 0.4170 | 0.4225 | 0.4196 | 0.4254 | 0.4226 | 0.4284 | 0.4255 | 0.4313 |
| 1/2-12 | 0.410 | 0.414 | 0.414 | 0.424 | 0.414 | 0.428 | 0.424 | 0.433 | 0.4100 | 0.4161 | 0.4129 | 0.4192 | 0.4160 | 0.4223 | 0.4192 | 0.4255 |
| 1/2-20 | 0.446 | 0.452 | 0.449 | 0.454 | 0.452 | 0.457 | 0.454 | 0.460 | 0.4460 | 0.4498 | 0.4477 | 0.4517 | 0.4497 | 0.4537 | 0.4516 | 0.4556 |
| 1/2-28 | 0.461 | 0.467 | 0.463 | 0.468 | 0.466 | 0.470 | 0.468 | 0.472 | 0.4610 | 0.4645 | 0.4620 | 0.4660 | 0.4636 | 0.4676 | 0.4652 | 0.4692 |
| 9/16-12 | 0.472 | 0.476 | 0.476 | 0.486 | 0.476 | 0.490 | 0.486 | 0.495 | 0.4720 | 0.4783 | 0.4753 | 0.4813 | 0.4783 | 0.4843 | 0.4813 | 0.4873 |
| 9/16-18 | 0.502 | 0.509 | 0.505 | 0.512 | 0.509 | 0.515 | 0.512 | 0.518 | 0.5020 | 0.5065 | 0.5045 | 0.5086 | 0.5065 | 0.5106 | 0.5086 | 0.5127 |
| 9/16-24 | 0.517 | 0.522 | 0.520 | 0.525 | 0.522 | 0.527 | 0.525 | 0.530 | 0.5170 | 0.5209 | 0.5186 | 0.5226 | 0.5204 | 0.5244 | 0.5221 | 0.5261 |
| 9/16-28 | 0.524 | 0.528 | 0.526 | 0.531 | 0.528 | 0.532 | 0.531 | 0.535 | 0.5240 | 0.5270 | 0.5245 | 0.5285 | 0.5261 | 0.5301 | 0.5277 | 0.5317 |
| 5/8-11 | 0.527 | 0.536 | 0.532 | 0.541 | 0.536 | 0.546 | 0.541 | 0.551 | 0.5270 | 0.5328 | 0.5298 | 0.5360 | 0.5329 | 0.5391 | 0.5360 | 0.5422 |
| 5/8-12 | 0.535 | 0.544 | 0.540 | 0.549 | 0.544 | 0.553 | 0.549 | 0.558 | 0.5350 | 0.5406 | 0.5377 | 0.5435 | 0.5405 | 0.5463 | 0.5434 | 0.5492 |
| 5/8-18 | 0.565 | 0.572 | 0.568 | 0.575 | 0.572 | 0.578 | 0.575 | 0.581 | 0.5650 | 0.5690 | 0.5670 | 0.5711 | 0.5690 | 0.5730 | 0.5711 | 0.5752 |
| 5/8-24 | 0.580 | 0.585 | 0.583 | 0.588 | 0.585 | 0.590 | 0.588 | 0.593 | 0.5800 | 0.5834 | 0.5811 | 0.5851 | 0.5829 | 0.5869 | 0.5846 | 0.5886 |
| 5/8-28 | 0.586 | 0.591 | 0.588 | 0.593 | 0.591 | 0.595 | 0.593 | 0.597 | 0.5860 | 0.5895 | 0.5870 | 0.5910 | 0.5886 | 0.5926 | 0.5902 | 0.5942 |
| 11/16-12 | 0.597 | 0.606 | 0.602 | 0.611 | 0.606 | 0.615 | 0.611 | 0.620 | 0.5970 | 0.6029 | 0.6001 | 0.6057 | 0.6029 | 0.6085 | 0.6057 | 0.6113 |
| 11/16-24 | 0.642 | 0.647 | 0.645 | 0.650 | 0.647 | 0.652 | 0.650 | 0.655 | 0.6420 | 0.6459 | 0.6436 | 0.6476 | 0.6454 | 0.6494 | 0.6471 | 0.6511 |
| 3/4-10 | 0.642 | 0.653 | 0.647 | 0.658 | 0.653 | 0.663 | 0.658 | 0.668 | 0.6420 | 0.6481 | 0.6449 | 0.6513 | 0.6481 | 0.6545 | 0.6513 | 0.6577 |
| 3/4-12 | 0.660 | 0.669 | 0.665 | 0.674 | 0.669 | 0.678 | 0.674 | 0.683 | 0.6600 | 0.6652 | 0.6626 | 0.6680 | 0.6653 | 0.6707 | 0.6680 | 0.6734 |
| 3/4-16 | 0.682 | 0.689 | 0.686 | 0.693 | 0.689 | 0.696 | 0.693 | 0.700 | 0.6820 | 0.6866 | 0.6844 | 0.6887 | 0.6865 | 0.6908 | 0.6886 | 0.6929 |
| 3/4-20 | 0.696 | 0.702 | 0.699 | 0.704 | 0.702 | 0.707 | 0.704 | 0.710 | 0.6960 | 0.6998 | 0.6977 | 0.7017 | 0.6997 | 0.7037 | 0.7016 | 0.7056 |
| 3/4-28 | 0.711 | 0.716 | 0.713 | 0.718 | 0.716 | 0.720 | 0.718 | 0.722 | 0.7110 | 0.7145 | 0.7120 | 0.7160 | 0.7136 | 0.7176 | 0.7152 | 0.7192 |
| 13/16-12 | 0.722 | 0.731 | 0.727 | 0.736 | 0.731 | 0.740 | 0.736 | 0.745 | 0.7220 | 0.7276 | 0.7250 | 0.7303 | 0.7276 | 0.7329 | 0.7303 | 0.7356 |

**Table 3.** *(Continued)* **Recommended Hole Size Limits Before Tapping Unified Threads**

| Thread Size | Classes 1B and 2B | | | | | | | | Class 3B | | | | | | | |
|---|---|---|---|---|---|---|---|---|---|---|---|---|---|---|---|---|
| | Length of Engagement (*D* = Nominal Size of Thread) | | | | | | | | | | | | | | | |
| | To and Including ⅓ *D* | | Above ⅓ *D* to ⅔ *D* | | Above ⅔ *D* to 1½ *D* | | Above 1½ *D* to 3*D* | | To and Including ⅓ *D* | | Above ⅓ *D* to ⅔ *D* | | Above ⅔*D* to 1½ *D* | | Above 1½ *D* to 3*D* | |
| | Recommended Hole Size Limits | | | | | | | | | | | | | | | |
| | Min[a] | Max | Min | Max | Min | Max[b] | Min | Max | Min[a] | Max | Min | Max | Min | Max[b] | Min | Max |
| 13⁄16-16 | 0.745 | 0.752 | 0.749 | 0.756 | 0.752 | 0.759 | 0.756 | 0.763 | 0.7450 | 0.7491 | 0.7469 | 0.7512 | 0.7490 | 0.7533 | 0.7511 | 0.7554 |
| 13⁄16-20 | 0.758 | 0.764 | 0.761 | 0.766 | 0.764 | 0.770 | 0.766 | 0.772 | 0.7580 | 0.7623 | 0.7602 | 0.7642 | 0.7622 | 0.7662 | 0.7641 | 0.7681 |
| 7⁄8-9 | 0.755 | 0.767 | 0.761 | 0.773 | 0.767 | 0.778 | 0.773 | 0.785 | 0.7550 | 0.7614 | 0.7580 | 0.7647 | 0.7614 | 0.7681 | 0.7647 | 0.7714 |
| 7⁄8-12 | 0.785 | 0.794 | 0.790 | 0.799 | 0.794 | 0.803 | 0.799 | 0.808 | 0.7850 | 0.7900 | 0.7874 | 0.7926 | 0.7900 | 0.7952 | 0.7926 | 0.7978 |
| 7⁄8-14 | 0.798 | 0.806 | 0.802 | 0.810 | 0.806 | 0.814 | 0.810 | 0.818 | 0.7980 | 0.8022 | 0.8000 | 0.8045 | 0.8023 | 0.8068 | 0.8045 | 0.8090 |
| 7⁄8-16 | 0.807 | 0.814 | 0.811 | 0.818 | 0.814 | 0.821 | 0.818 | 0.825 | 0.8070 | 0.8116 | 0.8094 | 0.8137 | 0.8115 | 0.8158 | 0.8136 | 0.8179 |
| 7⁄8-20 | 0.821 | 0.827 | 0.824 | 0.829 | 0.827 | 0.832 | 0.829 | 0.835 | 0.8210 | 0.8248 | 0.8227 | 0.8267 | 0.8247 | 0.8287 | 0.8266 | 0.8306 |
| 7⁄8-28 | 0.836 | 0.840 | 0.838 | 0.843 | 0.840 | 0.845 | 0.843 | 0.847 | 0.8360 | 0.8395 | 0.8370 | 0.8410 | 0.8386 | 0.8426 | 0.8402 | 0.8442 |
| 15⁄16-12 | 0.847 | 0.856 | 0.852 | 0.861 | 0.856 | 0.865 | 0.861 | 0.870 | 0.8470 | 0.8524 | 0.8499 | 0.8550 | 0.8524 | 0.8575 | 0.8550 | 0.8601 |
| 15⁄16-16 | 0.870 | 0.877 | 0.874 | 0.881 | 0.877 | 0.884 | 0.881 | 0.888 | 0.8700 | 0.8741 | 0.8719 | 0.8762 | 0.8740 | 0.8783 | 0.8761 | 0.8804 |
| 15⁄16-20 | 0.883 | 0.889 | 0.886 | 0.891 | 0.889 | 0.895 | 0.891 | 0.897 | 0.8830 | 0.8873 | 0.8852 | 0.8892 | 0.8872 | 0.8912 | 0.8891 | 0.8931 |
| 1-8 | 0.865 | 0.878 | 0.871 | 0.884 | 0.878 | 0.890 | 0.884 | 0.896 | 0.8650 | 0.8722 | 0.8684 | 0.8759 | 0.8722 | 0.8797 | 0.8760 | 0.8835 |
| 1-12 | 0.910 | 0.919 | 0.915 | 0.924 | 0.919 | 0.928 | 0.924 | 0.933 | 0.9100 | 0.9148 | 0.9123 | 0.9173 | 0.9148 | 0.9198 | 0.9173 | 0.9223 |
| 1-14 | 0.923 | 0.931 | 0.927 | 0.934 | 0.931 | 0.938 | 0.934 | 0.942 | 0.9230 | 0.9271 | 0.9249 | 0.9293 | 0.9271 | 0.9315 | 0.9293 | 0.9337 |
| 1-16 | 0.932 | 0.939 | 0.936 | 0.943 | 0.939 | 0.946 | 0.943 | 0.950 | 0.9320 | 0.9366 | 0.9344 | 0.9387 | 0.9365 | 0.9408 | 0.9386 | 0.9429 |
| 1-20 | 0.946 | 0.952 | 0.949 | 0.954 | 0.952 | 0.957 | 0.954 | 0.960 | 0.9460 | 0.9498 | 0.9477 | 0.9517 | 0.9497 | 0.9537 | 0.9516 | 0.9556 |
| 1-28 | 0.961 | 0.966 | 0.963 | 0.968 | 0.966 | 0.970 | 0.968 | 0.972 | 0.9610 | 0.9645 | 0.9620 | 0.9660 | 0.9636 | 0.9676 | 0.9652 | 0.9692 |
| 1 1⁄16-12 | 0.972 | 0.981 | 0.977 | 0.986 | 0.981 | 0.990 | 0.986 | 0.995 | 0.9720 | 0.9773 | 0.9748 | 0.9798 | 0.9773 | 0.9823 | 0.9798 | 0.9848 |
| 1 1⁄16-16 | 0.995 | 1.002 | 0.999 | 1.055 | 1.002 | 1.009 | 1.055 | 1.013 | 0.9950 | 0.9991 | 0.9969 | 1.0012 | 0.9990 | 1.0033 | 1.0011 | 1.0054 |
| 1 1⁄16-18 | 1.002 | 1.009 | 1.005 | 1.012 | 1.009 | 1.015 | 1.012 | 1.018 | 1.0020 | 1.0065 | 1.0044 | 1.0085 | 1.0064 | 1.0105 | 1.0085 | 1.0126 |
| 1 1⁄8-7 | 0.970 | 0.984 | 0.977 | 0.991 | 0.984 | 0.998 | 0.991 | 1.005 | 0.9700 | 0.9790 | 0.9747 | 0.9833 | 0.9789 | 0.9875 | 0.9832 | 0.9918 |
| 1 1⁄8-8 | 0.990 | 1.003 | 0.996 | 1.009 | 1.003 | 1.015 | 1.009 | 1.021 | 0.9900 | 0.9972 | 0.9934 | 1.0009 | 0.9972 | 1.0047 | 1.0010 | 1.0085 |
| 1 1⁄8-12 | 1.035 | 1.044 | 1.040 | 1.049 | 1.044 | 1.053 | 1.049 | 1.058 | 1.0350 | 1.0398 | 1.0373 | 1.0423 | 1.0398 | 1.0448 | 1.0423 | 1.0473 |
| 1 1⁄8-16 | 1.057 | 1.064 | 1.061 | 1.068 | 1.064 | 1.071 | 1.068 | 1.075 | 1.0570 | 1.0616 | 1.0594 | 1.0637 | 1.0615 | 1.0658 | 1.0636 | 1.0679 |
| 1 1⁄8-18 | 1.065 | 1.072 | 1.068 | 1.075 | 1.072 | 1.078 | 1.075 | 1.081 | 1.0650 | 1.0690 | 1.0669 | 1.0710 | 1.0689 | 1.0730 | 1.0710 | 1.0751 |
| 1 1⁄8-20 | 1.071 | 1.077 | 1.074 | 1.079 | 1.077 | 1.082 | 1.079 | 1.085 | 1.0710 | 1.0748 | 1.0727 | 1.0767 | 1.0747 | 1.0787 | 1.0766 | 1.0806 |
| 1 1⁄8-28 | 1.086 | 1.091 | 1.088 | 1.093 | 1.091 | 1.095 | 1.093 | 1.097 | 1.0860 | 1.0895 | 1.0870 | 1.0910 | 1.0886 | 1.0926 | 1.0902 | 1.0942 |
| 1 3⁄16-12 | 1.097 | 1.106 | 1.102 | 1.111 | 1.106 | 1.115 | 1.111 | 1.120 | 1.0970 | 1.1023 | 1.0998 | 1.1048 | 1.1023 | 1.1073 | 1.1048 | 1.1098 |

## Table 3. *(Continued)* Recommended Hole Size Limits Before Tapping Unified Threads

| Thread Size | Classes 1B and 2B | | | | | | | | Class 3B | | | | | | | |
|---|---|---|---|---|---|---|---|---|---|---|---|---|---|---|---|---|
| | Length of Engagement ($D$ = Nominal Size of Thread) | | | | | | | | | | | | | | | |
| | To and Including 1/3 $D$ | | Above 1/3 $D$ to 2/3 $D$ | | Above 2/3 $D$ to 1 1/2 $D$ | | Above 1 1/2 $D$ to 3$D$ | | To and Including 1/3 $D$ | | Above 1/3 $D$ to 2/3 $D$ | | Above 2/3 $D$ to 1 1/2 $D$ | | Above 1 1/2 $D$ to 3$D$ | |
| | Recommended Hole Size Limits | | | | | | | | | | | | | | | |
| | Min[a] | Max | Min | Max | Min | Max[b] | Min | Max | Min[a] | Max | Min | Max | Min | Max[b] | Min | Max |
| 1 3/16-16 | 1.120 | 1.127 | 1.124 | 1.131 | 1.127 | 1.134 | 1.131 | 1.138 | 1.1200 | 1.1241 | 1.1219 | 1.1262 | 1.1240 | 1.1283 | 1.1261 | 1.1304 |
| 1 3/16-18 | 1.127 | 1.134 | 1.130 | 1.137 | 1.134 | 1.140 | 1.137 | 1.143 | 1.1270 | 1.1315 | 1.1294 | 1.1335 | 1.1314 | 1.1355 | 1.1335 | 1.1376 |
| 1 1/4-7 | 1.095 | 1.109 | 1.102 | 1.116 | 1.109 | 1.123 | 1.116 | 1.130 | 1.0950 | 1.1040 | 1.0997 | 1.1083 | 1.1039 | 1.1125 | 1.1082 | 1.1168 |
| 1 1/4-8 | 1.115 | 1.128 | 1.121 | 1.134 | 1.128 | 1.140 | 1.134 | 1.146 | 1.1150 | 1.1222 | 1.1184 | 1.1259 | 1.1222 | 1.1297 | 1.1260 | 1.1335 |
| 1 1/4-12 | 1.160 | 1.169 | 1.165 | 1.174 | 1.169 | 1.178 | 1.174 | 1.183 | 1.1600 | 1.1648 | 1.1623 | 1.1673 | 1.1648 | 1.1698 | 1.1673 | 1.1723 |
| 1 1/4-16 | 1.182 | 1.189 | 1.186 | 1.193 | 1.189 | 1.196 | 1.193 | 1.200 | 1.1820 | 1.1866 | 1.1844 | 1.1887 | 1.1865 | 1.1908 | 1.1886 | 1.1929 |
| 1 1/4-18 | 1.190 | 1.197 | 1.193 | 1.200 | 1.197 | 1.203 | 1.200 | 1.206 | 1.1900 | 1.1940 | 1.1919 | 1.1960 | 1.1939 | 1.1980 | 1.1960 | 1.2001 |
| 1 1/4-20 | 1.196 | 1.202 | 1.199 | 1.204 | 1.202 | 1.207 | 1.204 | 1.210 | 1.1960 | 1.1998 | 1.1977 | 1.2017 | 1.1997 | 1.2037 | 1.2016 | 1.2056 |
| 1 5/16-12 | 1.222 | 1.231 | 1.227 | 1.236 | 1.231 | 1.240 | 1.236 | 1.245 | 1.2220 | 1.2273 | 1.2248 | 1.2298 | 1.2273 | 1.2323 | 1.2298 | 1.2348 |
| 1 5/16-16 | 1.245 | 1.252 | 1.249 | 1.256 | 1.252 | 1.259 | 1.256 | 1.263 | 1.2450 | 1.2491 | 1.2469 | 1.2512 | 1.2490 | 1.2533 | 1.2511 | 1.2554 |
| 1 5/16-18 | 1.252 | 1.259 | 1.256 | 1.262 | 1.259 | 1.265 | 1.262 | 1.268 | 1.2520 | 1.2565 | 1.2544 | 1.2585 | 1.2564 | 1.2605 | 1.2585 | 1.2626 |
| 1 3/8-6 | 1.195 | 1.210 | 1.203 | 1.221 | 1.210 | 1.225 | 1.221 | 1.239 | 1.1950 | 1.2046 | 1.1996 | 1.2096 | 1.2046 | 1.2146 | 1.2096 | 1.2196 |
| 1 3/8-8 | 1.240 | 1.253 | 1.246 | 1.259 | 1.253 | 1.265 | 1.259 | 1.271 | 1.2400 | 1.2472 | 1.2434 | 1.2509 | 1.2472 | 1.2547 | 1.2510 | 1.2585 |
| 1 3/8-12 | 1.285 | 1.294 | 1.290 | 1.299 | 1.294 | 1.303 | 1.299 | 1.308 | 1.2850 | 1.2898 | 1.2873 | 1.2923 | 1.2898 | 1.2948 | 1.2923 | 1.2973 |
| 1 3/8-16 | 1.307 | 1.314 | 1.311 | 1.318 | 1.314 | 1.321 | 1.318 | 1.325 | 1.3070 | 1.3116 | 1.3094 | 1.3137 | 1.3115 | 1.3158 | 1.3136 | 1.3179 |
| 1 3/8-18 | 1.315 | 1.322 | 1.318 | 1.325 | 1.322 | 1.328 | 1.325 | 1.331 | 1.3150 | 1.3190 | 1.3169 | 1.3210 | 1.3189 | 1.3230 | 1.3210 | 1.3251 |
| 1 7/16-12 | 1.347 | 1.354 | 1.350 | 1.361 | 1.354 | 1.365 | 1.361 | 1.370 | 1.3470 | 1.3523 | 1.3498 | 1.3548 | 1.3523 | 1.3573 | 1.3548 | 1.3598 |
| 1 7/16-16 | 1.370 | 1.377 | 1.374 | 1.381 | 1.377 | 1.384 | 1.381 | 1.388 | 1.3700 | 1.3741 | 1.3719 | 1.3762 | 1.3740 | 1.3783 | 1.3761 | 1.3804 |
| 1 7/16-18 | 1.377 | 1.384 | 1.380 | 1.387 | 1.384 | 1.390 | 1.387 | 1.393 | 1.3770 | 1.3815 | 1.3794 | 1.3835 | 1.3814 | 1.3855 | 1.3835 | 1.3876 |
| 1 1/2-6 | 1.320 | 1.335 | 1.328 | 1.346 | 1.335 | 1.350 | 1.346 | 1.364 | 1.3200 | 1.3296 | 1.3246 | 1.3346 | 1.3296 | 1.3396 | 1.3346 | 1.3446 |
| 1 1/2-8 | 1.365 | 1.378 | 1.371 | 1.384 | 1.378 | 1.390 | 1.384 | 1.396 | 1.3650 | 1.3722 | 1.3684 | 1.3759 | 1.3722 | 1.3797 | 1.3760 | 1.3835 |
| 1 1/2-12 | 1.410 | 1.419 | 1.4155 | 1.424 | 1.419 | 1.428 | 1.424 | 1.433 | 1.4100 | 1.4148 | 1.4123 | 1.4173 | 1.4148 | 1.4198 | 1.4173 | 1.4223 |
| 1 1/2-16 | 1.432 | 1.439 | 1.436 | 1.443 | 1.439 | 1.446 | 1.443 | 1.450 | 1.4320 | 1.4366 | 1.4344 | 1.4387 | 1.4365 | 1.4408 | 1.4386 | 1.4429 |
| 1 1/2-18 | 1.440 | 1.446 | 1.443 | 1.450 | 1.446 | 1.452 | 1.450 | 1.456 | 1.4400 | 1.4440 | 1.4419 | 1.4460 | 1.4439 | 1.4480 | 1.4460 | 1.4501 |
| 1 1/2-20 | 1.446 | 1.452 | 1.449 | 1.454 | 1.452 | 1.457 | 1.454 | 1.460 | 1.4460 | 1.4498 | 1.4477 | 1.4517 | 1.4497 | 1.4537 | 1.4516 | 1.4556 |
| 1 9/16-16 | 1.495 | 1.502 | 1.499 | 1.506 | 1.502 | 1.509 | 1.506 | 1.513 | 1.4950 | 1.4991 | 1.4969 | 1.5012 | 1.4990 | 1.5033 | 1.5011 | 1.5054 |
| 1 9/16-18 | 1.502 | 1.509 | 1.505 | 1.512 | 1.509 | 1.515 | 1.512 | 1.518 | 1.5020 | 1.5065 | 1.5044 | 1.5085 | 1.5064 | 1.5105 | 1.5085 | 1.5126 |

**Table 3.** *(Continued)* **Recommended Hole Size Limits Before Tapping Unified Threads**

| Thread Size | Classes 1B and 2B | | | | | | | | Class 3B | | | | | | | |
|---|---|---|---|---|---|---|---|---|---|---|---|---|---|---|---|---|
| | Length of Engagement ($D$ = Nominal Size of Thread) | | | | | | | | | | | | | | | |
| | To and Including ⅓ $D$ | | Above ⅓ $D$ to ⅔ $D$ | | Above ⅔ $D$ to 1½ $D$ | | Above 1½ $D$ to 3$D$ | | To and Including ⅓ $D$ | | Above ⅓ $D$ to ⅔ $D$ | | Above ⅔ $D$ to 1½ $D$ | | Above 1½ $D$ to 3$D$ | |
| | Recommended Hole Size Limits | | | | | | | | | | | | | | | |
| | Min[a] | Max | Min | Max | Min | Max[b] | Min | Max | Min[a] | Max | Min | Max | Min | Max[b] | Min | Max |
| 1⅝-8 | 1.490 | 1.498 | 1.494 | 1.509 | 1.498 | 1.515 | 1.509 | 1.521 | 1.4900 | 1.4972 | 1.4934 | 1.5009 | 1.4972 | 1.5047 | 1.5010 | 1.5085 |
| 1⅝-12 | 1.535 | 1.544 | 1.540 | 1.549 | 1.544 | 1.553 | 1.549 | 1.558 | 1.5350 | 1.5398 | 1.5373 | 1.5423 | 1.5398 | 1.5448 | 1.5423 | 1.5473 |
| 1⅝-16 | 1.557 | 1.564 | 1.561 | 1.568 | 1.564 | 1.571 | 1.568 | 1.575 | 1.5570 | 1.5616 | 1.5594 | 1.5637 | 1.5615 | 1.5658 | 1.5636 | 1.5679 |
| 1⅝-18 | 1.565 | 1.572 | 1.568 | 1.575 | 1.572 | 1.578 | 1.575 | 1.581 | 1.5650 | 1.5690 | 1.5669 | 1.5710 | 1.5689 | 1.5730 | 1.5710 | 1.5751 |
| 1^11^⁄16-16 | 1.620 | 1.627 | 1.624 | 1.631 | 1.627 | 1.634 | 1.631 | 1.638 | 1.6200 | 1.6241 | 1.6219 | 1.6262 | 1.6240 | 1.6283 | 1.6261 | 1.6304 |
| 1^11^⁄16-18 | 1.627 | 1.634 | 1.630 | 1.637 | 1.634 | 1.640 | 1.637 | 1.643 | 1.6270 | 1.6315 | 1.6294 | 1.6335 | 1.6314 | 1.6355 | 1.6335 | 1.6376 |
| 1¾-5 | 1.534 | 1.551 | 1.543 | 1.560 | 1.551 | 1.568 | 1.560 | 1.577 | 1.5340 | 1.5455 | 1.5395 | 1.5515 | 1.5455 | 1.5575 | 1.5515 | 1.5635 |
| 1¾-8 | 1.615 | 1.628 | 1.621 | 1.634 | 1.628 | 1.640 | 1.634 | 1.646 | 1.6150 | 1.6222 | 1.6184 | 1.6259 | 1.6222 | 1.6297 | 1.6260 | 1.6335 |
| 1¾-12 | 1.660 | 1.669 | 1.665 | 1.674 | 1.669 | 1.678 | 1.674 | 1.683 | 1.6600 | 1.6648 | 1.6623 | 1.6673 | 1.6648 | 1.6698 | 1.6673 | 1.6723 |
| 1¾-16 | 1.682 | 1.689 | 1.686 | 1.693 | 1.689 | 1.696 | 1.693 | 1.700 | 1.6820 | 1.6866 | 1.6844 | 1.6887 | 1.6865 | 1.6908 | 1.6886 | 1.6929 |
| 1¾-20 | 1.696 | 1.702 | 1.699 | 1.704 | 1.702 | 1.707 | 1.704 | 1.710 | 1.6960 | 1.6998 | 1.6977 | 1.7017 | 1.6997 | 1.7037 | 1.7016 | 1.7056 |
| 1^13^⁄16-16 | 1.745 | 1.752 | 1.749 | 1.756 | 1.752 | 1.759 | 1.756 | 1.763 | 1.7450 | 1.7491 | 1.7469 | 1.7512 | 1.7490 | 1.7533 | 1.7511 | 1.7554 |
| 1⅞-8 | 1.740 | 1.752 | 1.746 | 1.759 | 1.752 | 1.765 | 1.759 | 1.771 | 1.7400 | 1.7472 | 1.7434 | 1.7509 | 1.7472 | 1.7547 | 1.7510 | 1.7585 |
| 1⅞-12 | 1.785 | 1.794 | 1.790 | 1.799 | 1.794 | 1.803 | 1.799 | 1.808 | 1.7850 | 1.7898 | 1.7873 | 1.7923 | 1.7898 | 1.7948 | 1.7923 | 1.7973 |
| 1⅞-16 | 1.807 | 1.814 | 1.810 | 1.818 | 1.814 | 1.821 | 1.818 | 1.825 | 1.8070 | 1.8116 | 1.8094 | 1.8137 | 1.8115 | 1.8158 | 1.8136 | 1.1879 |
| 1^15^⁄16-16 | 1.870 | 1.877 | 1.874 | 1.881 | 1.877 | 1.884 | 1.881 | 1.888 | 1.8700 | 1.8741 | 1.8719 | 1.8762 | 1.8740 | 1.8783 | 1.8761 | 1.8804 |
| 2-4½ | 1.759 | 1.777 | 1.768 | 1.786 | 1.777 | 1.795 | 1.786 | 1.804 | 1.7590 | 1.7727 | 1.7661 | 1.7794 | 1.7728 | 1.7861 | 1.7794 | 1.7927 |
| 2-8 | 1.865 | 1.878 | 1.871 | 1.884 | 1.878 | 1.890 | 1.884 | 1.896 | 1.8650 | 1.8722 | 1.8684 | 1.8759 | 1.8722 | 1.8797 | 1.8760 | 1.8835 |
| 2-12 | 1.910 | 1.919 | 1.915 | 1.924 | 1.919 | 1.928 | 1.924 | 1.933 | 1.9100 | 1.9148 | 1.9123 | 1.9173 | 1.9148 | 1.9198 | 1.9173 | 1.9223 |
| 2-16 | 1.932 | 1.939 | 1.936 | 1.943 | 1.939 | 1.946 | 1.943 | 1.950 | 1.9320 | 1.9366 | 1.9344 | 1.9387 | 1.9365 | 1.9408 | 1.9386 | 1.9429 |
| 2-20 | 1.946 | 1.952 | 1.949 | 1.954 | 1.952 | 1.957 | 1.954 | 1.960 | 1.9460 | 1.9498 | 1.9477 | 1.9517 | 1.9497 | 1.9537 | 1.9516 | 1.9556 |
| 2^1^⁄16-16 | 1.995 | 2.002 | 2.000 | 2.006 | 2.002 | 2.009 | 2.006 | 2.012 | 1.9950 | 1.9991 | 1.9969 | 2.0012 | 1.9990 | 2.0033 | 2.0011 | 2.0054 |
| 2⅛-8 | 1.990 | 2.003 | 1.996 | 2.009 | 2.003 | 2.015 | 2.009 | 2.021 | 1.9900 | 1.9972 | 1.9934 | 2.0009 | 1.9972 | 2.0047 | 2.0010 | 2.0085 |
| 2⅛-12 | 2.035 | 2.044 | 2.040 | 2.049 | 2.044 | 2.053 | 2.049 | 2.058 | 2.0350 | 2.0398 | 2.0373 | 2.0423 | 2.0398 | 2.0448 | 2.0423 | 2.0473 |
| 2⅛-16 | 2.057 | 2.064 | 2.061 | 2.068 | 2.064 | 2.071 | 2.068 | 2.075 | 2.0570 | 2.0616 | 2.0594 | 2.0637 | 2.0615 | 2.0658 | 2.0636 | 2.0679 |
| 2^3^⁄16-16 | 2.120 | 2.127 | 2.124 | 2.131 | 2.127 | 2.134 | 2.131 | 2.138 | 2.1200 | 2.1241 | 2.1219 | 2.1262 | 2.1240 | 2.1283 | 2.1261 | 2.1304 |
| 2¼-4½ | 2.009 | 2.027 | 2.018 | 2.036 | 2.027 | 2.045 | 2.036 | 2.054 | 2.0090 | 2.0227 | 2.0161 | 2.0294 | 2.0228 | 2.0361 | 2.0294 | 2.0427 |

**Table 3.** *(Continued)* **Recommended Hole Size Limits Before Tapping Unified Threads**

| Thread Size | Classes 1B and 2B | | | | | | | | Class 3B | | | | | | | |
|---|---|---|---|---|---|---|---|---|---|---|---|---|---|---|---|---|
| | Length of Engagement (*D* = Nominal Size of Thread) | | | | | | | | | | | | | | | |
| | To and Including ⅓ *D* | | Above ⅓ *D* to ⅔ *D* | | Above ⅔ *D* to 1½ *D* | | Above 1½ *D* to 3*D* | | To and Including ⅓ *D* | | Above ⅓ *D* to ⅔ *D* | | Above ⅔*D* to 1½ *D* | | Above 1½ *D* to 3*D* | |
| | Recommended Hole Size Limits | | | | | | | | | | | | | | | |
| | Min[a] | Max | Min | Max | Min | Max[b] | Min | Max | Min[a] | Max | Min | Max | Min | Max[b] | Min | Max |
| 2¼-8 | 2.115 | 2.128 | 2.121 | 2.134 | 2.128 | 2.140 | 2.134 | 2.146 | 2.1150 | 2.1222 | 2.1184 | 2.1259 | 2.1222 | 2.1297 | 2.1260 | 2.1335 |
| 2¼-12 | 2.160 | 2.169 | 2.165 | 2.174 | 2.169 | 2.178 | 2.174 | 2.182 | 2.1600 | 2.1648 | 2.1623 | 2.1673 | 2.1648 | 2.1698 | 2.1673 | 2.1723 |
| 2¼-16 | 2.182 | 2.189 | 2.186 | 2.193 | 2.189 | 2.196 | 2.193 | 2.200 | 2.1820 | 2.1866 | 2.1844 | 2.1887 | 2.1865 | 2.1908 | 2.1886 | 2.1929 |
| 2¼-20 | 2.196 | 2.202 | 2.199 | 2.204 | 2.202 | 2.207 | 2.204 | 2.210 | 2.1960 | 2.1998 | 2.1977 | 2.2017 | 2.1997 | 2.2037 | 2.2016 | 2.2056 |
| 2$\frac{5}{16}$-16 | 2.245 | 2.252 | 2.249 | 2.256 | 2.252 | 2.259 | 2.256 | 2.263 | 2.2450 | 2.2491 | 2.2469 | 2.2512 | 2.2490 | 2.2533 | 2.2511 | 2.2554 |
| 2⅜-12 | 2.285 | 2.294 | 2.290 | 2.299 | 2.294 | 2.303 | 2.299 | 2.308 | 2.2850 | 2.2898 | 2.2873 | 2.2923 | 2.2898 | 2.2948 | 2.2923 | 2.2973 |
| 2⅜-16 | 2.307 | 2.314 | 2.311 | 2.318 | 2.314 | 2.321 | 2.318 | 2.325 | 2.3070 | 2.3116 | 2.3094 | 2.3137 | 2.3115 | 2.3158 | 2.3136 | 2.3179 |
| 2$\frac{7}{16}$-16 | 2.370 | 2.377 | 2.374 | 2.381 | 2.377 | 2.384 | 2.381 | 2.388 | 2.3700 | 2.3741 | 2.3719 | 2.3762 | 2.3740 | 2.3783 | 2.3761 | 2.3804 |
| 2½-4 | 2.229 | 2.248 | 2.238 | 2.258 | 2.248 | 2.267 | 2.258 | 2.277 | 2.2290 | 2.2444 | 2.2369 | 2.2519 | 2.2444 | 2.2594 | 2.2519 | 2.2669 |
| 2½-8 | 2.365 | 2.378 | 2.371 | 2.384 | 2.378 | 2.390 | 2.384 | 2.396 | 2.3650 | 2.3722 | 2.3684 | 2.3759 | 2.3722 | 2.3797 | 2.3760 | 2.3835 |
| 2½-12 | 2.410 | 2.419 | 2.415 | 2.424 | 2.419 | 2.428 | 2.424 | 2.433 | 2.4100 | 2.4148 | 2.4123 | 2.4173 | 2.4148 | 2.4198 | 2.4173 | 2.4223 |
| 2½-16 | 2.432 | 2.439 | 2.436 | 2.443 | 2.439 | 2.446 | 2.443 | 2.450 | 2.4320 | 2.4366 | 2.4344 | 2.4387 | 2.4365 | 2.4408 | 2.4386 | 2.4429 |
| 2½-20 | 2.446 | 2.452 | 2.449 | 2.454 | 2.452 | 2.457 | 2.454 | 2.460 | 2.4460 | 2.4498 | 2.4478 | 2.4517 | 2.4497 | 2.4537 | 2.4516 | 2.4556 |
| 2⅝-12 | 2.535 | 2.544 | 2.540 | 2.549 | 2.544 | 2.553 | 2.549 | 2.558 | 2.5350 | 2.5398 | 2.5373 | 2.5423 | 2.5398 | 2.5448 | 2.5423 | 2.5473 |
| 2⅝-16 | 2.557 | 2.564 | 2.561 | 2.568 | 2.564 | 2.571 | 2.568 | 2.575 | 2.5570 | 2.5616 | 2.5594 | 2.5637 | 2.5615 | 2.5658 | 2.5636 | 2.5679 |
| 2¾-4 | 2.479 | 2.498 | 2.489 | 2.508 | 2.498 | 2.517 | 2.508 | 2.527 | 2.4790 | 2.4944 | 2.4869 | 2.5019 | 2.4944 | 2.5094 | 2.5019 | 2.5169 |
| 2¾-8 | 2.615 | 2.628 | 2.621 | 2.634 | 2.628 | 2.640 | 2.634 | 2.644 | 2.6150 | 2.6222 | 2.6184 | 2.6259 | 2.6222 | 2.6297 | 2.6260 | 2.6335 |
| 2¾-12 | 2.660 | 2.669 | 2.665 | 2.674 | 2.669 | 2.678 | 2.674 | 2.683 | 2.6600 | 2.6648 | 2.6623 | 2.6673 | 2.6648 | 2.6698 | 2.6673 | 2.6723 |
| 2¾-16 | 2.682 | 2.689 | 2.686 | 2.693 | 2.689 | 2.696 | 2.693 | 2.700 | 2.6820 | 2.6866 | 2.6844 | 2.6887 | 2.6865 | 2.6908 | 2.6886 | 2.6929 |
| 2⅞-12 | 2.785 | 2.794 | 2.790 | 2.809 | 2.794 | 2.803 | 2.809 | 2.808 | 2.7850 | 2.7898 | 2.7873 | 2.7923 | 2.7898 | 2.7948 | 2.7923 | 2.7973 |
| 2⅞-16 | 2.807 | 2.814 | 2.811 | 2.818 | 2.814 | 2.821 | 2.818 | 2.825 | 2.8070 | 2.8116 | 2.8094 | 2.8137 | 2.8115 | 2.8158 | 2.8136 | 2.8179 |
| 3-4 | 2.729 | 2.748 | 2.739 | 2.758 | 2.748 | 2.767 | 2.758 | 2.777 | 2.7290 | 2.7444 | 2.7369 | 2.7519 | 2.7444 | 2.7594 | 2.7519 | 2.7669 |
| 3-8 | 2.865 | 2.878 | 2.871 | 2.884 | 2.878 | 2.890 | 2.884 | 2.896 | 2.8650 | 2.8722 | 2.8684 | 2.8759 | 2.8722 | 2.8797 | 2.8760 | 2.8835 |
| 3-12 | 2.910 | 2.919 | 2.915 | 2.924 | 2.919 | 2.928 | 2.924 | 2.933 | 2.9100 | 2.9148 | 2.9123 | 2.9173 | 2.9148 | 2.9198 | 2.9173 | 2.9223 |
| 3-16 | 2.932 | 2.939 | 2.936 | 2.943 | 2.939 | 2.946 | 2.943 | 2.950 | 2.9320 | 2.9366 | 2.9344 | 2.9387 | 2.9365 | 2.9408 | 2.9386 | 2.9429 |
| 3⅛-12 | 3.035 | 3.044 | 3.040 | 3.049 | 3.044 | 3.053 | 3.049 | 3.058 | 3.0350 | 3.0398 | 3.0373 | 3.0423 | 3.0398 | 3.0448 | 3.0423 | 3.0473 |
| 3⅛-16 | 3.057 | 3.064 | 3.061 | 3.068 | 3.064 | 3.071 | 3.068 | 3.075 | 3.0570 | 3.0616 | 3.0594 | 3.0637 | 3.0615 | 3.0658 | 3.0636 | 3.0679 |

**Table 3.** *(Continued)* **Recommended Hole Size Limits Before Tapping Unified Threads**

| Thread Size | Classes 1B and 2B | | | | | | | | Class 3B | | | | | | | |
|---|---|---|---|---|---|---|---|---|---|---|---|---|---|---|---|---|
| | Length of Engagement ($D$ = Nominal Size of Thread) | | | | | | | | | | | | | | | |
| | To and Including ⅓ $D$ | | Above ⅓ $D$ to ⅔ $D$ | | Above ⅔ $D$ to 1½ $D$ | | Above 1½ $D$ to 3$D$ | | To and Including ⅓ $D$ | | Above ⅓ $D$ to ⅔ $D$ | | Above ⅔ $D$ to 1½ $D$ | | Above 1½ $D$ to 3$D$ | |
| | Recommended Hole Size Limits | | | | | | | | | | | | | | | |
| | Min[a] | Max | Min | Max | Min | Max[b] | Min | Max | Min[a] | Max | Min | Max | Min | Max[b] | Min | Max |
| 3¼-4 | 2.979 | 2.998 | 2.989 | 3.008 | 2.998 | 3.017 | 3.008 | 3.027 | 2.9790 | 2.9944 | 2.9869 | 3.0019 | 2.9944 | 3.0094 | 3.0019 | 3.0169 |
| 3¼-8 | 3.115 | 3.128 | 3.121 | 3.134 | 3.128 | 3.140 | 3.134 | 3.146 | 3.1150 | 3.1222 | 3.1184 | 3.1259 | 3.1222 | 3.1297 | 3.1260 | 3.1335 |
| 3¼-12 | 3.160 | 3.169 | 3.165 | 3.174 | 3.169 | 3.178 | 3.174 | 3.183 | 3.1600 | 3.1648 | 3.1623 | 3.1673 | 3.1648 | 3.1698 | 3.1673 | 3.1723 |
| 3¼-16 | 3.182 | 3.189 | 3.186 | 3.193 | 3.189 | 3.196 | 3.193 | 3.200 | 3.1820 | 3.1866 | 3.1844 | 3.1887 | 3.1865 | 3.1908 | 3.1886 | 3.1929 |
| 3⅜-12 | 3.285 | 3.294 | 3.290 | 3.299 | 3.294 | 3.303 | 3.299 | 3.299 | 3.2850 | 3.2898 | 3.2873 | 3.2923 | 3.2898 | 3.2948 | 3.2923 | 3.2973 |
| 3⅜-16 | 3.307 | 3.314 | 3.311 | 3.318 | 3.314 | 3.321 | 3.317 | 3.325 | 3.3070 | 3.3116 | 3.3094 | 3.3137 | 3.3115 | 3.3158 | 3.3136 | 3.3179 |
| 3½-4 | 3.229 | 3.248 | 3.239 | 3.258 | 3.248 | 3.267 | 3.258 | 3.277 | 3.2290 | 3.2444 | 3.2369 | 3.2519 | 3.2444 | 3.2594 | 3.2519 | 3.2669 |
| 3½-8 | 3.365 | 3.378 | 3.371 | 2.384 | 3.378 | 3.390 | 3.384 | 3.396 | 3.3650 | 3.3722 | 3.3684 | 3.3759 | 3.3722 | 3.3797 | 3.3760 | 3.3835 |
| 3½-12 | 3.410 | 3.419 | 3.415 | 3.424 | 3.419 | 3.428 | 3.424 | 3.433 | 3.4100 | 3.4148 | 3.4123 | 3.4173 | 3.4148 | 3.4198 | 3.4173 | 3.4223 |
| 3½-16 | 3.432 | 3.439 | 3.436 | 3.443 | 3.439 | 3.446 | 3.443 | 3.450 | 3.4320 | 3.4366 | 3.4344 | 3.4387 | 3.4365 | 3.4408 | 3.4386 | 3.4429 |
| 3⅝-12 | 3.535 | 3.544 | 3.544 | 3.549 | 3.544 | 3.553 | 3.549 | 3.553 | 3.5350 | 3.5398 | 3.5373 | 3.5423 | 3.5398 | 3.5448 | 3.5423 | 3.5473 |
| 3⅝-16 | 3.557 | 3.564 | 3.561 | 3.568 | 3.567 | 3.571 | 3.568 | 3.575 | 3.5570 | 3.5616 | 3.5594 | 3.5637 | 3.5615 | 3.5658 | 3.5636 | 3.5679 |
| 3¾-4 | 3.479 | 3.498 | 3.489 | 3.508 | 3.498 | 3.517 | 3.508 | 3.527 | 3.4790 | 3.4944 | 3.4869 | 3.5019 | 3.4944 | 3.5094 | 3.5019 | 3.5169 |
| 3¾-8 | 3.615 | 3.628 | 3.615 | 3.634 | 3.628 | 3.640 | 3.634 | 3.646 | 3.6150 | 3.6222 | 3.6184 | 3.6259 | 3.6222 | 3.6297 | 3.6260 | 3.6335 |
| 3¾-12 | 3.660 | 3.669 | 3.665 | 3.674 | 3.669 | 3.678 | 3.674 | 3.683 | 3.6600 | 3.6648 | 3.6623 | 3.6673 | 3.6648 | 3.6698 | 3.6673 | 3.6723 |
| ¾-16 | 3.682 | 3.689 | 3.686 | 3.693 | 3.689 | 3.696 | 3.693 | 3.700 | 3.6820 | 3.6866 | 3.6844 | 3.6887 | 3.6865 | 3.6908 | 3.6886 | 3.6929 |
| 3⅞-12 | 3.785 | 3.794 | 3.790 | 3.799 | 3.794 | 3.803 | 3.799 | 3.808 | 3.7850 | 3.7898 | 3.7873 | 3.7923 | 3.7898 | 3.7948 | 3.7923 | 3.7973 |
| 3⅞-16 | 3.807 | 3.814 | 3.811 | 3.818 | 3.814 | 3.821 | 3.818 | 3.825 | 3.8070 | 3.8116 | 3.8094 | 3.8137 | 3.8115 | 3.8158 | 3.8136 | 3.8179 |
| 4-4 | 3.729 | 3.748 | 3.739 | 3.758 | 3.748 | 3.767 | 3.758 | 3.777 | 3.7290 | 3.7444 | 3.7369 | 3.7519 | 3.7444 | 3.7594 | 3.7519 | 3.7669 |
| 4-8 | 3.865 | 3.878 | 3.871 | 3.884 | 3.878 | 3.890 | 3.884 | 3.896 | 3.8650 | 3.8722 | 3.8684 | 3.8759 | 3.8722 | 3.8797 | 3.8760 | 3.8835 |
| 4-12 | 3.910 | 3.919 | 3.915 | 3.924 | 3.919 | 3.928 | 3.924 | 3.933 | 3.9100 | 3.9148 | 3.9123 | 3.9173 | 3.9148 | 3.9198 | 3.9173 | 3.9223 |
| 4-16 | 3.932 | 3.939 | 3.936 | 3.943 | 3.939 | 3.946 | 3.943 | 3.950 | 3.9320 | 3.9366 | 3.9344 | 3.9387 | 3.9365 | 3.9408 | 3.9386 | 3.9429 |
| 4¼-4 | 3.979 | 3.998 | 3.989 | 4.008 | 3.998 | 4.017 | 4.008 | 4.027 | 3.9790 | 3.9944 | 3.9869 | 4.0019 | 3.9944 | 4.0094 | 4.0019 | 4.0169 |
| 4¼-8 | 4.115 | 4.128 | 4.121 | 4.134 | 4.128 | 4.140 | 4.134 | 4.146 | 4.1150 | 4.1222 | 4.1184 | 4.1259 | 4.1222 | 4.1297 | 4.1260 | 4.1335 |
| 4¼-12 | 4.160 | 4.169 | 4.165 | 4.174 | 4.169 | 4.178 | 4.174 | 4.183 | 4.1600 | 4.1648 | 4.1623 | 4.1673 | 4.1648 | 4.1698 | 4.1673 | 4.1723 |
| 4¼-16 | 4.182 | 4.189 | 4.186 | 4.193 | 4.189 | 4.196 | 4.193 | 4.200 | 4.1820 | 4.1866 | 4.1844 | 4.1887 | 4.1865 | 4.1908 | 4.1886 | 4.1929 |
| 4½-4 | 4.229 | 4.248 | 4.239 | 4.258 | 4.248 | 4.267 | 4.258 | 4.277 | 4.2290 | 4.2444 | 4.2369 | 4.2519 | 4.2444 | 4.2594 | 4.2519 | 4.2669 |

**Table 3.** *(Continued)* **Recommended Hole Size Limits Before Tapping Unified Threads**

| Thread Size | Classes 1B and 2B | | | | | | | | Class 3B | | | | | | | |
|---|---|---|---|---|---|---|---|---|---|---|---|---|---|---|---|---|
| | Length of Engagement ($D$ = Nominal Size of Thread) | | | | | | | | | | | | | | | |
| | To and Including ⅓ $D$ | | Above ⅓ $D$ to ⅔ $D$ | | Above ⅔ $D$ to 1½ $D$ | | Above 1½ $D$ to 3$D$ | | To and Including ⅓ $D$ | | Above ⅓ $D$ to ⅔ $D$ | | Above ⅔ $D$ to 1½ $D$ | | Above 1½ $D$ to 3$D$ | |
| | Recommended Hole Size Limits | | | | | | | | | | | | | | | |
| | Min[a] | Max | Min | Max | Min | Max[b] | Min | Max | Min[a] | Max | Min | Max | Min | Max[b] | Min | Max |
| 4½-8 | 4.365 | 4.378 | 4.371 | 4.384 | 4.378 | 4.390 | 4.384 | 4.396 | 4.3650 | 4.3722 | 4.3684 | 4.3759 | 4.3722 | 4.3797 | 4.3760 | 4.3835 |
| 4½-12 | 4.410 | 4.419 | 4.419 | 4.424 | 4.419 | 4.428 | 4.424 | 4.433 | 4.4100 | 4.4148 | 4.4123 | 4.4173 | 4.4148 | 4.4198 | 4.4173 | 4.4223 |
| 4½-16 | 4.432 | 4.439 | 4.437 | 4.444 | 4.439 | 4.446 | 4.444 | 4.455 | 4.4320 | 4.4366 | 4.4344 | 4.4387 | 4.4365 | 4.4408 | 4.4386 | 4.4429 |
| 4¾-8 | 4.615 | 4.628 | 4.621 | 4.646 | 4.628 | 4.640 | 4.646 | 4.646 | 4.6150 | 4.6222 | 4.6184 | 4.6259 | 4.6222 | 4.6297 | 4.6260 | 4.6335 |
| 4¾-12 | 4.660 | 4.669 | 4.665 | 4.674 | 4.669 | 4.678 | 4.674 | 4.683 | 4.6600 | 4.6648 | 4.6623 | 4.6673 | 4.6648 | 4.6698 | 4.6673 | 4.6723 |
| 4¾-16 | 4.682 | 4.689 | 4.686 | 4.693 | 4.689 | 4.696 | 4.693 | 4.700 | 4.6820 | 4.6866 | 4.6844 | 4.6887 | 4.6865 | 4.6908 | 4.6886 | 4.6929 |
| 5-8 | 4.865 | 4.878 | 4.871 | 4.884 | 4.878 | 4.890 | 4.884 | 4.896 | 4.8650 | 4.8722 | 4.8684 | 4.8759 | 4.8722 | 4.8797 | 4.8760 | 4.8835 |
| 5-12 | 4.910 | 4.919 | 4.915 | 4.924 | 4.919 | 4.928 | 4.924 | 4.933 | 4.9100 | 4.9148 | 4.9123 | 4.9173 | 4.9148 | 4.9198 | 4.9173 | 4.9223 |
| 5-16 | 4.932 | 4.939 | 4.936 | 4.943 | 4.939 | 4.946 | 4.943 | 4.950 | 4.9320 | 4.9366 | 4.9344 | 4.9387 | 4.9365 | 4.9408 | 4.9386 | 4.9429 |
| 5¼-8 | 5.115 | 5.128 | 5.121 | 5.134 | 5.128 | 5.140 | 5.134 | 5.146 | 5.1150 | 5.1222 | 5.1184 | 5.1259 | 5.1222 | 5.1297 | 5.1260 | 5.1335 |
| 5¼-12 | 5.160 | 5.169 | 5.165 | 5.174 | 5.169 | 5.178 | 5.174 | 5.183 | 5.1600 | 5.1648 | 5.1623 | 5.1673 | 5.1648 | 5.1698 | 5.1673 | 5.1723 |
| 5¼-16 | 5.182 | 5.189 | 5.186 | 5.193 | 5.189 | 5.196 | 5.193 | 5.200 | 5.1820 | 5.1866 | 5.1844 | 5.1887 | 5.1865 | 5.1908 | 5.1886 | 5.1929 |
| 5½-8 | 5.365 | 5.378 | 5.371 | 5.384 | 5.378 | 5.390 | 5.384 | 5.396 | 5.3650 | 5.3722 | 5.3684 | 5.3759 | 5.3722 | 5.3797 | 5.3760 | 5.3835 |
| 5½-12 | 5.410 | 5.419 | 5.415 | 5.424 | 5.419 | 5.428 | 5.424 | 5.433 | 5.4100 | 5.4148 | 5.4123 | 5.4173 | 5.4148 | 5.4198 | 5.4173 | 5.4223 |
| 5½-16 | 5.432 | 5.439 | 5.436 | 5.442 | 5.439 | 5.446 | 5.442 | 5.450 | 5.4320 | 5.4366 | 5.4344 | 5.4387 | 5.4365 | 5.4408 | 5.4386 | 5.4429 |
| 5¾-8 | 5.615 | 5.628 | 5.621 | 5.634 | 5.628 | 5.640 | 5.634 | 5.646 | 5.6150 | 5.6222 | 5.6184 | 5.6259 | 5.6222 | 5.6297 | 5.6260 | 5.6335 |
| 5¾-12 | 5.660 | 5.669 | 5.665 | 5.674 | 5.669 | 5.678 | 5.674 | 5.683 | 5.6600 | 5.6648 | 5.6623 | 5.6673 | 5.6648 | 5.6698 | 5.6673 | 5.6723 |
| 5¾-16 | 5.682 | 5.689 | 5.686 | 5.693 | 5.689 | 5.696 | 5.693 | 5.700 | 5.6820 | 5.6866 | 5.6844 | 5.6887 | 5.6865 | 5.6908 | 5.6886 | 5.6929 |
| 6-8 | 5.865 | 5.878 | 5.871 | 5.896 | 5.878 | 5.890 | 5.896 | 5.896 | 5.8650 | 5.8722 | 5.8684 | 5.8759 | 5.8722 | 5.8797 | 5.8760 | 5.8835 |
| 6-12 | 5.910 | 5.919 | 5.915 | 5.924 | 5.919 | 5.928 | 5.924 | 5.933 | 5.9100 | 5.9148 | 5.9123 | 5.9173 | 5.9148 | 5.9198 | 5.9173 | 5.9223 |
| 6-16 | 5.932 | 5.939 | 5.935 | 5.943 | 5.939 | 5.946 | 5.943 | 5.950 | 5.9320 | 5.9366 | 5.9344 | 5.9387 | 5.9365 | 5.9408 | 5.9386 | 5.9429 |

[a] This is the minimum minor diameter specified in the thread tables, page 1954.

[b] This is the maximum minor diameter specified in the thread tables, page 1954.

All dimensions are in inches.

For basis of recommended hole size limits see accompanying text.

As an aid in selecting suitable drills, see the listing of American Standard drill sizes starting on page 921 in the twist drill section. For amount of expected drill oversize, see page 950.

## Table 4. Tap Drill Sizes for Threads of American National Form

| Screw Thread | | Commercial Tap Drills[a] | | Screw Thread | | Commercial Tap Drills[a] | |
|---|---|---|---|---|---|---|---|
| Outside Dia. Pitch | Root Dia. | Size or Number | Decimal Equiv. | Outside Dia. Pitch | Root Dia. | Size or Number | Decimal Equiv. |
| 1/16-64 | 0.0422 | 3/64 | 0.0469 | 27 | 0.4519 | 15/32 | 0.4687 |
| 72 | 0.0445 | 3/64 | 0.0469 | 9/16-12 | 0.4542 | 31/64 | 0.4844 |
| 5/64-60 | 0.0563 | 1/16 | 0.0625 | 18 | 0.4903 | 33/64 | 0.5156 |
| 72 | 0.0601 | 52 | 0.0635 | 27 | 0.5144 | 17/32 | 0.5312 |
| 3/32-48 | 0.0667 | 49 | 0.0730 | 5/8-11 | 0.5069 | 17/32 | 0.5312 |
| 50 | 0.0678 | 49 | 0.0730 | 12 | 0.5168 | 35/64 | 0.5469 |
| 7/64-48 | 0.0823 | 43 | 0.0890 | 18 | 0.5528 | 37/64 | 0.5781 |
| 1/8-32 | 0.0844 | 3/32 | 0.0937 | 27 | 0.5769 | 19/32 | 0.5937 |
| 40 | 0.0925 | 38 | 0.1015 | 11/16-11 | 0.5694 | 19/32 | 0.5937 |
| 9/64-40 | 0.1081 | 32 | 0.1160 | 16 | 0.6063 | 5/8 | 0.6250 |
| 5/32-32 | 0.1157 | 1/8 | 0.1250 | 3/4-10 | 0.6201 | 21/32 | 0.6562 |
| 36 | 0.1202 | 30 | 0.1285 | 12 | 0.6418 | 43/64 | 0.6719 |
| 11/64-32 | 0.1313 | 9/64 | 0.1406 | 16 | 0.6688 | 11/16 | 0.6875 |
| 3/16-24 | 0.1334 | 26 | 0.1470 | 27 | 0.7019 | 23/32 | 0.7187 |
| 32 | 0.1469 | 22 | 0.1570 | 13/16-10 | 0.6826 | 23/32 | 0.7187 |
| 13/64-24 | 0.1490 | 20 | 0.1610 | 7/8-9 | 0.7307 | 49/64 | 0.7656 |
| 7/32-24 | 0.1646 | 16 | 0.1770 | 12 | 0.7668 | 51/64 | 0.7969 |
| 32 | 0.1782 | 12 | 0.1890 | 14 | 0.7822 | 13/16 | 0.8125 |
| 15/64-24 | 0.1806 | 10 | 0.1935 | 18 | 0.8028 | 53/64 | 0.8281 |
| 1/4-20 | 0.1850 | 7 | 0.2010 | 27 | 0.8269 | 27/32 | 0.8437 |
| 24 | 0.1959 | 4 | 0.2090 | 15/16-9 | 0.7932 | 53/64 | 0.8281 |
| 27 | 0.2019 | 3 | 0.2130 | 1-8 | 0.8376 | 7/8 | 0.8750 |
| 28 | 0.2036 | 3 | 0.2130 | 12 | 0.8918 | 59/64 | 0.9219 |
| 32 | 0.2094 | 7/32 | 0.2187 | 14 | 0.9072 | 15/16 | 0.9375 |
| 5/16-18 | 0.2403 | F | 0.2570 | 27 | 0.9519 | 31/32 | 0.9687 |
| 20 | 0.2476 | 17/64 | 0.2656 | 1 1/8-7 | 0.9394 | 63/64 | 0.9844 |
| 24 | 0.2584 | I | 0.2720 | 12 | 1.0168 | 1 3/64 | 1.0469 |
| 27 | 0.2644 | J | 0.2770 | 1 1/4-7 | 1.0644 | 1 7/64 | 1.1094 |
| 32 | 0.2719 | 9/32 | 0.2812 | 12 | 1.1418 | 1 11/64 | 1.1719 |
| 3/8-16 | 0.2938 | 5/16 | 0.3125 | 1 3/8-6 | 1.1585 | 1 7/32 | 1.2187 |
| 20 | 0.3100 | 21/64 | 0.3281 | 12 | 1.2668 | 1 19/64 | 1.2969 |
| 24 | 0.3209 | Q | 0.3320 | 1 1/2-6 | 1.2835 | 1 11/32 | 1.3437 |
| 27 | 0.3269 | R | 0.3390 | 12 | 1.3918 | 1 27/64 | 1.4219 |
| 7/16-14 | 0.3447 | U | 0.3680 | 1 5/8-5 1/2 | 1.3888 | 1 29/64 | 1.4531 |
| 20 | 0.3726 | 25/64 | 0.3906 | 1 3/4-5 | 1.4902 | 1 9/16 | 1.5625 |
| 24 | 0.3834 | X | 0.3970 | 1 7/8-5 | 1.6152 | 1 11/16 | 1.6875 |
| 27 | 0.3894 | Y | 0.4040 | 2-4 1/2 | 1.7113 | 1 25/32 | 1.7812 |
| 1/2-12 | 0.3918 | 27/64 | 0.4219 | 2 1/8-4 1/2 | 1.8363 | 1 29/32 | 1.9062 |
| 13 | 0.4001 | 27/64 | 0.4219 | 2 1/4-4 1/2 | 1.9613 | 2 1/32 | 2.0312 |
| 20 | 0.4351 | 29/64 | 0.4531 | 2 3/8-4 | 2.0502 | 2 1/8 | 2.1250 |
| 24 | 0.4459 | 29/64 | 0.4531 | 2 1/2-4 | 2.1752 | 2 1/4 | 2.2500 |

[a] These tap drill diameters allow approximately 75 percent of a full thread to be produced. For small thread sizes in the first column, the use of drills to produce the larger hole sizes shown in Table 3 will reduce defects caused by tap problems and breakage.

**Table 5. Tap Drills for Machine Screws with American National Thread Form**

| Size of Screw | | No. of Threads per Inch | Tap Drills | |
|---|---|---|---|---|
| No. or Dia. | Decimal Equiv. | | Drill Size | Decimal Equiv. |
| 0 | .060 | 80 | 3/64 | .0469 |
| 1 | .073 | 64 | 53 | .0595 |
| | | 72 | 53 | .0595 |
| 2 | .086 | 56 | 50 | .0700 |
| | | 64 | 50 | .0700 |
| 3 | .099 | 48 | 47 | .0785 |
| | | 56 | 45 | .0820 |
| 4 | .112 | 36[a] | 44 | .0860 |
| | | 40 | 43 | .0890 |
| | | 48 | 42 | .0935 |
| 5 | .125 | 40 | 38 | .1015 |
| | | 44 | 37 | 1040 |
| 6 | .138 | 32 | 36 | .1065 |
| | | 40 | 33 | .1130 |
| 8 | .164 | 32 | 29 | .1360 |
| | | 36 | 29 | .1360 |
| 10 | .190 | 24 | 25 | .1495 |
| | | 32 | 21 | .1590 |
| 12 | .216 | 24 | 16 | .1770 |
| | | 28 | 14 | .1820 |
| 14 | .242 | 20[a] | 10 | .1935 |
| | | 24[a] | 7 | .2010 |
| 1/4 | .250 | 20 | 7 | .2010 |
| | | 28 | 3 | .2130 |
| 5/16 | .3125 | 18 | F | .2570 |
| | | 24 | I | .2720 |
| 3/8 | .375 | 16 | 5/16 | .3125 |
| | | 24 | Q | .3320 |
| 7/16 | .4375 | 14 | U | .3680 |
| | | 20 | 25/64 | .3906 |
| 1/2 | .500 | 13 | 27/64 | .4219 |
| | | 20 | 29/64 | .4531 |

[a] These screws are not in the American Standard but are from the former ASME Standard.

**Factors Influencing Minor Diameter Tolerances of Tapped Holes.—**As stated in the Unified screw thread Standard, the principle practical factors that govern minor diameter tolerances of internal threads are tapping difficulties, particularly tap breakage in the small sizes, availability of standard drill sizes in the medium and large sizes, and depth (radial) of engagement. Depth of engagement is related to the stripping strength of the thread assembly, and thus also to the length of engagement. It also has an influence on the tendency toward disengagement of the threads on one side when assembly is eccentric.

The amount of possible eccentricity is one-half of the sum of the pitch diameter allowance and tolerances on both mating threads. For a given pitch, or height of thread, this sum increases with the diameter, and accordingly this factor would require a decrease in minor diameter tolerance with increase in diameter. However, such decrease in tolerance would often require the use of special drill sizes; therefore, to facilitate the use of standard drill sizes, for any given pitch the minor diameter tolerance for Unified thread classes 1B and 2B threads of 1/4 inch diameter and larger is constant, in accordance with a formula given in the American Standard for Unified Screw Threads.

*Effect of Length of Engagement of Minor Diameter Tolerances:* There may be applications where the lengths of engagement of mating threads is relatively short or the combination of materials used for mating threads is such that the maximum minor diameter tolerance given in the Standard (based on a length of engagement equal to the nominal diameter) may not provide the desired strength of the fastening. Experience has shown that for lengths of engagement less than $\frac{2}{3}D$ (the minimum thickness of standard nuts) the minor diameter tolerance may be reduced without causing tapping difficulties. In other applications the length of engagement of mating threads may be long because of design considerations or the combination of materials used for mating threads. As the threads engaged increase in number, a shallower depth of engagement may be permitted and still develop stripping strength greater than the external thread breaking strength. Under these conditions the maximum tolerance given in the Standard should be increased to reduce the possibility of tapping difficulties. The following paragraphs indicate how the aforementioned considerations were taken into account in determining the minor diameter limits for various lengths of engagement given in Table 3.

**Recommended Hole Sizes Before Tapping.**—Recommended hole size limits before threading to provide for optimum strength of fastenings and tapping conditions are shown in Table 3 for classes 1B, 2B, and 3B. The hole size limit before threading, and the tolerances between them, are derived from the minimum and maximum minor diameters of the internal thread given in the dimensional tables for Unified threads in the screw thread section using the following rules:

1) For lengths of engagement in the range to and including $\frac{1}{3}D$, where $D$ equals nominal diameter, the minimum hole size will be equal to the minimum minor diameter of the internal thread and the maximum hole size will be larger by one-half the minor diameter tolerance.

2) For the range from $\frac{1}{3}D$ to $\frac{2}{3}D$, the minimum and maximum hole sizes will each be one quarter of the minor diameter tolerance larger than the corresponding limits for the length of engagement to and including $\frac{1}{3}D$.

3) For the range from $\frac{2}{3}D$ to $1\frac{1}{2}D$ the minimum hole size will be larger than the minimum minor diameter of the internal thread by one-half the minor diameter tolerance and the maximum hole size will be equal to the maximum minor diameter.

4) For the range from $1\frac{1}{2}D$ to $3D$ the minimum and maximum hole sizes will each be one-quarter of the minor diameter tolerance of the internal thread larger than the corresponding limits for the $\frac{2}{3}D$ to $1\frac{1}{2}D$ length of engagement.

From the foregoing it will be seen that the difference between limits in each range is the same and equal to one-half of the minor diameter tolerance given in the Unified screw thread dimensional tables. This is a general rule, except that the minimum differences for sizes below $\frac{1}{4}$ inch are equal to the minor diameter tolerances calculated on the basis of lengths of engagement to and including $\frac{1}{3}D$. Also, for lengths of engagement greater than $\frac{1}{3}D$ and for sizes $\frac{1}{4}$ inch and larger the values are adjusted so that the difference between limits is never less than 0.004 inch.

For diameter-pitch combinations other than those given in Table 3, the foregoing rules should be applied to the tolerances given in the dimensional tables in the screw thread section or the tolerances derived from the formulas given in the Standard to determine the hole size limits.

*Selection of Tap Drills:* In selecting standard drills to produce holes within the limits given in Table 3 it should be recognized that drills have a tendency to cut oversize. The material on pages 949 and 950 may be used as a guide to the expected amount of oversize.

## Table 6. Unified Miniature Screw Threads—Recommended Hole Size Limits Before Tapping

| Thread Size | | Internal Threads | | Lengths of Engagement | | | | | |
|---|---|---|---|---|---|---|---|---|---|
| | | Minor Diameter Limits | | To and including $\frac{2}{3}D$ | | Above $\frac{2}{3}D$ to $1\frac{1}{2}D$ | | Above $1\frac{1}{2}D$ to $3D$ | |
| | | | | Recommended Hole Size Limits | | | | | |
| | Pitch | Min | Max | Min | Max | Min | Max | Min | Max |
| Designation | mm | mm | mm | mm | mm | mm | mm | mm | mm |
| **0.30 UNM** | **0.080** | **0.217** | **0.254** | **0.226** | **0.240** | **0.236** | **0.254** | **0.245** | **0.264** |
| 0.35 UNM | 0.090 | 0.256 | 0.297 | 0.267 | 0.282 | 0.277 | 0.297 | 0.287 | 0.307 |
| **0.40 UNM** | **0.100** | **0.296** | **0.340** | **0.307** | **0.324** | **0.318** | **0.340** | **0.329** | **0.351** |
| 0.45 UNM | 0.100 | 0.346 | 0.390 | 0.357 | 0.374 | 0.368 | 0.390 | 0.379 | 0.401 |
| **0.50 UNM** | **0.125** | **0.370** | **0.422** | **0.383** | **0.402** | **0.396** | **0.422** | **0.409** | **0.435** |
| 0.55 UNM | 0.125 | 0.420 | 0.472 | 0.433 | 0.452 | 0.446 | 0.472 | 0.459 | 0.485 |
| **0.60 UNM** | **0.150** | **0.444** | **0.504** | **0.459** | **0.482** | **0.474** | **0.504** | **0.489** | **0.519** |
| 0.70 UNM | 0.175 | 0.518 | 0.586 | 0.535 | 0.560 | 0.552 | 0.586 | 0.569 | 0.603 |
| **0.80 UNM** | **0.200** | **0.592** | **0.668** | **0.611** | **0.640** | **0.630** | **0.668** | **0.649** | **0.687** |
| 0.90 UNM | 0.225 | 0.666 | 0.750 | 0.687 | 0.718 | 0.708 | 0.750 | 0.729 | 0.771 |
| **1.00 UNM** | **0.250** | **0.740** | **0.832** | **0.763** | **0.798** | **0.786** | **0.832** | **0.809** | **0.855** |
| 1.10 UNM | 0.250 | 0.840 | 0.932 | 0.863 | 0.898 | 0.886 | 0.932 | 0.909 | 0.955 |
| **1.20 UNM** | **0.250** | **0.940** | **1.032** | **0.963** | **0.998** | **0.986** | **1.032** | **1.009** | **1.055** |
| 1.40 UNM | 0.300 | 1.088 | 1.196 | 1.115 | 1.156 | 1.142 | 1.196 | 1.169 | 1.223 |
| Designation | Thds. per inch | inch | inch | inch | inch | inch | inch | inch | inch |
| **0.30 UNM** | **318** | **0.0085** | **0.0100** | **0.0089** | **0.0095** | **0.0093** | **0.0100** | **0.0096** | **0.0104** |
| 0.35 UNM | 282 | 0.0101 | 0.0117 | 0.0105 | 0.0111 | 0.0109 | 0.0117 | 0.0113 | 0.0121 |
| **0.40 UNM** | **254** | **0.0117** | **0.0134** | **0.0121** | **0.0127** | **0.0125** | **0.0134** | **0.0130** | **0.0138** |
| 0.45 UNM | 254 | 0.0136 | 0.0154 | 0.0141 | 0.0147 | 0.0145 | 0.0154 | 0.0149 | 0.0158 |
| **0.50 UNM** | **203** | **0.0146** | **0.0166** | **0.0150** | **0.0158** | **0.0156** | **0.0166** | **0.0161** | **0.0171** |
| 0.55 UNM | 203 | 0.0165 | 0.0186 | 0.0170 | 0.0178 | 0.0176 | 0.0186 | 0.0181 | 0.0191 |
| **0.60 UNM** | **169** | **0.0175** | **0.0198** | **0.0181** | **0.0190** | **0.0187** | **0.0198** | **0.0193** | **0.0204** |
| 0.70 UNM | 145 | 0.0204 | 0.0231 | 0.0211 | 0.0221 | 0.0217 | 0.0231 | 0.0224 | 0.0237 |
| **0.80 UNM** | **127** | **0.0233** | **0.0263** | **0.0241** | **0.0252** | **0.0248** | **0.0263** | **0.0256** | **0.0270** |
| 0.90 UNM | 113 | 0.0262 | 0.0295 | 0.0270 | 0.0283 | 0.0279 | 0.0295 | 0.0287 | 0.0304 |
| **1.00 UNM** | **102** | **0.0291** | **0.0327** | **0.0300** | **0.0314** | **0.0309** | **0.0327** | **0.0319** | **0.0337** |
| 1.10 UNM | 102 | 0.0331 | 0.0367 | 0.0340 | 0.0354 | 0.0349 | 0.0367 | 0.0358 | 0.0376 |
| **1.20 UNM** | **102** | **0.0370** | **0.0406** | **0.0379** | **0.0393** | **0.0388** | **0.0406** | **0.0397** | **0.0415** |
| 1.40 UNM | 85 | 0.0428 | 0.0471 | 0.0439 | 0.0455 | 0.0450 | 0.0471 | 0.0460 | 0.0481 |

As an aid in selecting suitable drills, see the listing of American Standard drill sizes in the twist drill section. Thread sizes in heavy type are preferred sizes.

**Hole Sizes for Tapping Unified Miniature Screw Threads.—**Table 6 indicates the hole size limits recommended for tapping. These limits are derived from the internal thread minor diameter limits given in the American Standard for Unified Miniature Screw Threads ASA B1.10-1958, R1988 and are disposed so as to provide the optimum conditions for tapping. The maximum limits are based on providing a functionally adequate fastening for the most common applications, where the material of the externally threaded member is of a strength essentially equal to or greater than that of its mating part. In applications where, because of considerations other than the fastening, the screw is made of an appreciably weaker material, the use of smaller hole sizes is usually necessary to extend thread engagement to a greater depth on the external thread. Recommended minimum hole sizes are greater than the minimum limits of the minor diameters to allow for the spin-up developed in tapping.

In selecting drills to produce holes within the limits given in Table 6 it should be recognized that drills have a tendency to cut oversize. The material on page 950 may be used as a guide to the expected amount of oversize.

**British Standard Tapping Drill Sizes for Screw and Pipe Threads.—**British Standard BS 1157:1975 provides recommendations for tapping drill sizes for use with fluted taps for various ISO metric, Unified, British Standard fine, British Association, and British Standard Whitworth screw threads as well as British Standard parallel and taper pipe threads.

**Table 7. British Standard Tapping Drill Sizes for ISO Metric Coarse Pitch Series Threads** *BS 1157:1975*

| Nom. Size and Thread Dia. | Standard Drill Sizes[a] | | | | Nom. Size and Thread Dia. | Standard Drill Sizes[a] | | | |
|---|---|---|---|---|---|---|---|---|---|
| | Recommended | | Alternative | | | Recommended | | Alternative | |
| | Size | Theoretical Radial Engagement with Ext. Thread (Percent) | Size | Theoretical Radial Engagement with Ext. Thread (Percent) | | Size | Theoretical Radial Engagement with Ext. Thread (Percent) | Size | Theoretical Radial Engagement with Ext. Thread (Percent) |
| M 1 | 0.75 | 81.5 | 0.78 | 71.7 | M 12 | 10.20 | 83.7 | 10.40 | 74.5[b] |
| M 1.1 | 0.85 | 81.5 | 0.88 | 71.7 | M 14 | 12.00 | 81.5 | 12.20 | 73.4[b] |
| M 1.2 | 0.95 | 81.5 | 0.98 | 71.7 | M 16 | 14.00 | 81.5 | 14.25 | 71.3[c] |
| M 1.4 | 1.10 | 81.5 | 1.15 | 67.9 | M 18 | 15.50 | 81.5 | 15.75 | 73.4[c] |
| M 1.6 | 1.25 | 81.5 | 1.30 | 69.9 | M 20 | 17.50 | 81.5 | 17.75 | 73.4[c] |
| M 1.8 | 1.45 | 81.5 | 1.50 | 69.9 | M 22 | 19.50 | 81.5 | 19.75 | 73.4[c] |
| M 2 | 1.60 | 81.5 | 1.65 | 71.3 | M 24 | 21.00 | 81.5 | 21.25 | 74.7[b] |
| M 2.2 | 1.75 | 81.5 | 1.80 | 72.5 | M 27 | 24.00 | 81.5 | 24.25 | 74.7[b] |
| M 2.5 | 2.05 | 81.5 | 2.10 | 72.5 | M 30 | 26.50 | 81.5 | 26.75 | 75.7[b] |
| M 3 | 2.50 | 81.5 | 2.55 | 73.4 | M 33 | 29.50 | 81.5 | 29.75 | 75.7[b] |
| M 3.5 | 2.90 | 81.5 | 2.95 | 74.7 | M 36 | 32.00 | 81.5 | … | … |
| M 4 | 3.30 | 81.5 | 3.40 | 69.9[b] | M 39 | 35.00 | 81.5 | … | … |
| M 4.5 | 3.70 | 86.8 | 3.80 | 76.1 | M 42 | 37.50 | 81.5 | … | … |
| M 5 | 4.20 | 81.5 | 4.30 | 71.3[b] | M 45 | 40.50 | 81.5 | … | … |
| M 6 | 5.00 | 81.5 | 5.10 | 73.4 | M 48 | 43.00 | 81.5 | … | … |
| M 7 | 6.00 | 81.5 | 6.10 | 73.4 | M 52 | 47.00 | 81.5 | … | … |
| M 8 | 6.80 | 78.5 | 6.90 | 71.7[b] | M 56 | 50.50 | 81.5 | … | … |
| M 9 | 7.80 | 78.5 | 7.90 | 71.7[b] | M 60 | 54.50 | 81.5 | … | … |
| M 10 | 8.50 | 81.5 | 8.60 | 76.1 | M 64 | 58.00 | 81.5 | … | … |
| M 11 | 9.50 | 81.5 | 9.60 | 76.1 | M 68 | 62.00 | 81.5 | … | … |

[a] These tapping drill sizes are for fluted taps only.

[b] For tolerance class 6H and 7H threads only.

[c] For tolerance class 7H threads only.

Drill sizes are given in millimeters.

In the accompanying Table 7, recommended and alternative drill sizes are given for producing holes for ISO metric coarse pitch series threads. These coarse pitch threads are suitable for the large majority of general-purpose applications, and the limits and tolerances for internal coarse threads are given in the table starting on page 2056. It should be noted that Table 7 is for fluted taps only since a fluteless tap will require for the same screw thread a different size of twist drill than will a fluted tap. When tapped, holes produced with drills of the recommended sizes provide for a theoretical radial engagement with the external thread of about 81 percent in most cases. Holes produced with drills of the alternative sizes provide for a theoretical radial engagement with the external thread of about 70 to 75 percent. In some cases, as indicated in Table 7, the alternative drill sizes are suitable only for medium (6H) or for free (7H) thread tolerance classes.

When relatively soft material is being tapped, there is a tendency for the metal to be squeezed down towards the root of the tap thread, and in such instances, the minor diameter of the tapped hole may become smaller than the diameter of the drill employed. Users may wish to choose different tapping drill sizes to overcome this problem or for special purposes, and reference can be made to the pages mentioned above to obtain the minor diameter limits for internal pitch series threads.

Reference should be made to BS 1157:1975 for recommended tapping hole sizes for other types of British Standard screw threads and pipe threads.

**Table 8. British Standard Metric Bolt and Screw Clearance Holes**

*BS 4186:1984 (Withdrawn)*

| Nominal Thread Diameter | Clearance Hole Sizes | | | Nominal Thread Diameter | Clearance Hole Sizes | | |
|---|---|---|---|---|---|---|---|
| | Close Fit Series | Medium Fit Series | Free Fit Series | | Close Fit Series | Medium Fit Series | Free Fit Series |
| 1.6 | 1.7 | 1.8 | 2.0 | 52.0 | 54.0 | 56.0 | 62.0 |
| 2.0 | 2.2 | 2.4 | 2.6 | 56.0 | 58.0 | 62.0 | 66.0 |
| 2.5 | 2.7 | 2.9 | 3.1 | 60.0 | 62.0 | 66.0 | 70.0 |
| 3.0 | 3.2 | 3.4 | 3.6 | 64.0 | 66.0 | 70.0 | 74.0 |
| 4.0 | 4.3 | 4.5 | 4.8 | 68.0 | 70.0 | 74.0 | 78.0 |
| 5.0 | 5.3 | 5.5 | 5.8 | 72.0 | 74.0 | 78.0 | 82.0 |
| 6.0 | 6.4 | 6.6 | 7.0 | 76.0 | 78.0 | 82.0 | 86.0 |
| 7.0 | 7.4 | 7.6 | 8.0 | 80.0 | 82.0 | 86.0 | 91.0 |
| 8.0 | 8.4 | 9.0 | 10.0 | 85.0 | 87.0 | 91.0 | 96.0 |
| 10.0 | 10.5 | 11.0 | 12.0 | 90.0 | 93.0 | 96.0 | 101.0 |
| 12.0 | 13.0 | 14.0 | 15.0 | 95.0 | 98.0 | 101.0 | 107.0 |
| 14.0 | 15.0 | 16.0 | 17.0 | 100.0 | 104.0 | 107.0 | 112.0 |
| 16.0 | 17.0 | 18.0 | 19.0 | 105.0 | 109.0 | 112.0 | 117.0 |
| 18.0 | 19.0 | 20.0 | 21.0 | 110.0 | 114.0 | 117.0 | 122.0 |
| 20.0 | 21.0 | 22.0 | 24.0 | 115.0 | 119.0 | 122.0 | 127.0 |
| 22.0 | 23.0 | 24.0 | 26.0 | 120.0 | 124.0 | 127.0 | 132.0 |
| 24.0 | 25.0 | 26.0 | 28.0 | 125.0 | 129.0 | 132.0 | 137.0 |
| 27.0 | 28.0 | 30.0 | 32.0 | 130.0 | 134.0 | 137.0 | 144.0 |
| 30.0 | 31.0 | 33.0 | 35.0 | 140.0 | 144.0 | 147.0 | 155.0 |
| 33.0 | 34.0 | 36.0 | 38.0 | 150.0 | 155.0 | 158.0 | 165.0 |
| 36.0 | 37.0 | 39.0 | 42.0 | … | … | … | … |
| 39.0 | 40.0 | 42.0 | 45.0 | … | … | … | … |
| 42.0 | 43.0 | 45.0 | 48.0 | … | … | … | … |
| 45.0 | 46.0 | 48.0 | 52.0 | … | … | … | … |
| 48.0 | 50.0 | 52.0 | 56.0 | … | … | … | … |

All dimensions are given in millimeters.

**British Standard Clearance Holes for Metric Bolts and Screws.—**The dimensions of the clearance holes specified in this British Standard BS 4186:1984 (Withdrawn) have been chosen in such a way as to require the use of the minimum number of drills. The recommendations cover three series of clearance holes, namely close fit (H 12), medium fit (H 13), and free fit (H 14) and are suitable for use with bolts and screws specified in the following metric British Standards: BS 3692, ISO metric precision hexagon bolts, screws, and nuts; BS 4168, Hexagon socket screws and wrench keys; BS 4183, Machine screws and machine screw nuts; and BS 4190, ISO metric black hexagon bolts, screws, and nuts. The sizes are in accordance with those given in ISO Recommendation R273, and the range has been extended up to 150 millimeters diameter in accordance with an addendum to that recommendation. The selection of clearance holes sizes to suit particular design requirements can of course be dependent upon many variable factors. It is however felt that the medium fit series should suit the majority of general purpose applications. In the

Standard, limiting dimensions are given in a table which is included for reference purposes only, for use in instances where it may be desirable to specify tolerances.

To avoid any risk of interference with the radius under the head of bolts and screws, it is necessary to countersink slightly all recommended clearance holes in the close and medium fit series. Dimensional details for the radius under the head of fasteners made according to BS 3692 are given on page 1766; those for fasteners to BS 4168 are given on page 1830; those to BS 4183 are given on pages 1804 through 1808.

**Cold Form Tapping.—**Cold form taps do not have cutting edges or conventional flutes; the threads on the tap form the threads in the hole by displacing the metal in an extrusion or swaging process. The threads thus produced are stronger than conventionally cut threads because the grains in the metal are unbroken and the displaced metal is work hardened. The surface of the thread is burnished and has an excellent finish. Although chip problems are eliminated, cold form tapping does displace the metal surrounding the hole and countersinking or chamfering before tapping is recommended. Cold form tapping is not recommended if the wall thickness of the hole is less than two-thirds of the nominal diameter of the thread. If possible, blind holes should be drilled deep enough to permit a cold form tap having a four thread lead to be used as this will require less torque, produce less burr surrounding the hole, and give a greater tool life.

The operation requires 0 to 50 percent more torque than conventional tapping, and the cold form tap will pick up its own lead when entering the hole; thus, conventional tapping machines and tapping heads can be used. Another advantage is the better tool life obtained. The best results are obtained by using a good lubricating oil instead of a conventional cutting oil.

The method can be applied only to relatively ductile metals, such as low-carbon steel, leaded steels, austenitic stainless steels, wrought aluminum, low-silicon aluminum die-casting alloys, zinc die-casting alloys, magnesium, copper, and ductile copper alloys. A higher than normal tapping speed can be used, sometimes by as much as 100 percent.

Conventional tap drill sizes should not be used for cold form tapping because the metal is displaced to form the thread. The cold formed thread is stronger than the conventionally tapped thread, so the thread height can be reduced to 60 percent without much loss of strength; however, the use of a 65 percent thread is strongly recommended. The following formula is used to calculate the theoretical hole size for cold form tapping:

$$\text{US:} \qquad \text{Theoretical hole size} = \text{basic tap OD} - \frac{0.0068 \times \text{percent of full thread}}{\text{threads per inch}}$$

$$\text{Metric:} \quad \textbf{Theoretical hole size} = \textbf{basic tap OD} - \frac{\textbf{Threads pitch} \times \textbf{percent of full thread}}{\textbf{147.06}}$$

The theoretical hole size and the tap drill sizes for American Unified threads are given in Table 9, and Table 10 lists drills for ISO metric threads. Sharp drills should be used to prevent cold-working the walls of the hole, especially on metals that are prone to work hardening. Such damage may cause the torque to increase, possibly stopping the machine or breaking the tap. On materials that can be die cast, cold form tapping can be done in cored holes provided the correct core pin size is used. The core pins are slightly tapered, so the theoretical hole size should be at the position on the pin that corresponds to one-half of the required engagement length of the thread in the hole. The core pins should be designed to form a chamfer on the hole to accept the vertical extrusion.

## Table 9. Theoretical and Tap Drill or Core Hole Sizes for Cold Form Tapping Unified Threads

| Tap Size | Threads Per Inch | Percentage of Full Thread | | | | | | | | |
|---|---|---|---|---|---|---|---|---|---|---|
| | | 75 | | | 65 | | | 55 | | |
| | | Theor. Hole Size | Nearest Drill Size | Dec. Equiv. | Theor. Hole Size | Nearest Drill Size | Dec. Equiv. | Theor. Hole Size | Nearest Drill Size | Dec. Equiv. |
| 0 | 80 | 0.0536 | 1.35 mm | 0.0531 | 0.0545 | … | … | 0.0554 | 54 | 0.055 |
| 1 | 64 | 0.0650 | 1.65 mm | 0.0650 | 0.0661 | … | … | 0.0672 | 51 | 0.0670 |
| | 72 | 0.0659 | 1.65 mm | 0.0650 | 0.0669 | 1.7 mm | 0.0669 | 0.0679 | 51 | 0.0670 |
| 2 | 56 | 0.0769 | 1.95 mm | 0.0768 | 0.0781 | 5⁄64 | 0.0781 | 0.0794 | 2.0 mm | 0.0787 |
| | 64 | 0.0780 | 5⁄64 | 0.0781 | 0.0791 | 2.0 mm | 0.0787 | 0.0802 | … | … |
| 3 | 48 | 0.0884 | 2.25 mm | 0.0886 | 0.0898 | 43 | 0.089 | 0.0913 | 2.3 mm | 0.0906 |
| | 56 | 0.0889 | 43 | 0.089 | 0.0911 | 2.3 mm | 0.0906 | 0.0924 | 2.35 mm | 0.0925 |
| 4 | 40 | 0.0993 | 2.5 mm | 0.0984 | 0.1010 | 39 | 0.0995 | 0.1028 | 2.6 mm | 0.1024 |
| | 48 | 0.0104 | 38 | 0.1015 | 0.1028 | 2.6 mm | 0.1024 | 0.1043 | 37 | 0.1040 |
| 5 | 40 | 0.1123 | 34 | 0.1110 | 0.1140 | 33 | 0.113 | 0.1158 | 32 | 0.1160 |
| | 44 | 0.1134 | 33 | 0.113 | 0.1150 | 2.9 mm | 0.1142 | 0.1166 | 32 | … |
| 6 | 32 | 0.1221 | 3.1 mm | 0.1220 | 0.1243 | … | … | 0.1264 | 3.2 mm | 0.1260 |
| | 40 | 0.1253 | 1⁄8 | 0.1250 | 0.1270 | 3.2 mm | 0.1260 | 0.1288 | 30 | 0.1285 |
| 8 | 32 | 0.1481 | 3.75 mm | 0.1476 | 0.1503 | 25 | 0.1495 | 0.1524 | 24 | 0.1520 |
| | 36 | 0.1498 | 25 | 0.1495 | 0.1518 | 24 | 0.1520 | 0.1537 | 3.9 mm | 0.1535 |
| 10 | 24 | 0.1688 | … | … | 0.1717 | 11⁄64 | 0.1719 | 0.1746 | 17 | 0.1730 |
| | 32 | 0.1741 | 17 | 0.1730 | 0.1763 | … | … | 0.1784 | 4.5 mm | 0.1772 |
| 12 | 24 | 0.1948 | 10 | 0.1935 | 0.1977 | 5.0 mm | 0.1968 | 0.2006 | 5.1 mm | 0.2008 |
| | 28 | 0.1978 | 5.0 mm | 0.1968 | 0.2003 | 8 | 0.1990 | 0.2028 | … | … |
| 1⁄4 | 20 | 0.2245 | 5.7 mm | 0.2244 | 0.2280 | 1 | 0.2280 | 0.2315 | … | … |
| | 28 | 0.2318 | … | … | 0.2343 | A | 0.2340 | 0.2368 | 6.0 mm | 0.2362 |
| 5⁄16 | 18 | 0.2842 | 7.2 mm | 0.2835 | 0.2879 | 7.3 mm | 0.2874 | 0.2917 | 7.4 mm | 0.2913 |
| | 24 | 0.2912 | 7.4 mm | 0.2913 | 0.2941 | M | 0.2950 | 0.2969 | 19⁄64 | 0.2969 |
| 3⁄8 | 16 | 0.3431 | 11⁄32 | 0.3437 | 0.3474 | S | 0.3480 | 0.3516 | … | … |
| | 24 | 0.3537 | 9.0 mm | 0.3543 | 0.3566 | … | … | 0.3594 | 23⁄64 | 0.3594 |
| 7⁄16 | 14 | 0.4011 | … | … | 0.4059 | 13⁄32 | 0.4062 | 0.4108 | … | … |
| | 20 | 0.4120 | Z | 0.413 | 0.4154 | … | … | 0.4188 | … | … |
| 1⁄2 | 13 | 0.4608 | … | … | 0.4660 | … | … | 0.4712 | 12 mm | 0.4724 |
| | 20 | 0.4745 | … | … | 0.4779 | … | … | 0.4813 | … | … |
| 9⁄16 | 12 | 0.5200 | … | … | 0.5257 | … | … | 0.5313 | 17⁄32 | 0.5312 |
| | 18 | 0.5342 | 13.5 mm | 0.5315 | 0.5380 | … | … | 0.5417 | … | … |
| 5⁄8 | 11 | 0.5787 | 37⁄64 | 0.5781 | 0.5848 | … | … | 0.5910 | 15 mm | 0.5906 |
| | 18 | 0.5976 | 19⁄32 | 0.5937 | 0.6004 | … | … | 0.6042 | … | … |
| 3⁄4 | 10 | 0.6990 | … | … | 0.7058 | 45⁄64 | 0.7031 | 0.7126 | … | … |
| | 16 | 0.7181 | 23⁄32 | 0.7187 | 0.7224 | … | … | 0.7266 | … | … |

**Table 10. Tap Drill or Core Hole Sizes for Cold Form Tapping ISO Metric Threads**

| Nominal Size of Tap | Pitch | Recommended Tap Drill Size |
|---|---|---|
| 1.6 mm | 0.35 mm | 1.45 mm |
| 1.8 mm | 0.35 mm | 1.65 mm |
| 2.0 mm | 0.40 mm | 1.8 mm |
| 2.2 mm. | 0.45 mm | 2.0 mm |
| 2.5 mm | 0.45 mm | 2.3 mm |
| 3.0 mm | 0.50 mm | 2.8 mm[a] |
| 3.5 mm | 0.60 mm | 3.2 mm |
| 4.0 mm | 0.70 mm | 3.7 mm |
| 4.5 mm | 0.75 mm | 4.2 mm[a] |
| 5.0 mm | 0.80 mm | 4.6 mm |
| 6.0 mm | 1.00 mm | 5.6 mm[a] |
| 7.0 mm | 1.00 mm | 6.5 mm |
| 8.0 mm | 1.25 mm | 7.4 mm |
| 10.0 mm | 1.50 mm | 9.3 mm |

[a] These diameters are the nearest stocked drill sizes and not the theoretical hole size, and may not produce 60 to 75 percent full thread.

The sizes are calculated to provide 60 to 75 percent of full thread.

**Removing a Broken Tap.—**Broken taps can be removed by electrodischarge machining (EDM), and this method is recommended when available. When an EDM machine is not available, broken taps may be removed by using a tap extractor, which has fingers that enter the flutes of the tap; the tap is backed out of the hole by turning the extractor with a wrench. Sometimes the injection of a small amount of a proprietary solvent into the hole will be helpful. A solvent can be made by diluting about one part nitric acid with five parts water. The action of the proprietary solvent or the diluted nitric acid on the steel loosens the tap so that it can be removed with pliers or with a tap extractor. The hole should be washed out afterwards so that the acid will not continue to work on the part. Another method is to add, by electric arc welding, additional metal to the shank of the broken tap, above the level of the hole. Care must be taken to prevent depositing metal on the threads in the tapped hole. After the shank has been built up, the head of a bolt or a nut is welded to it and then the tap may be backed out.

**Tap Drills for Pipe Taps**

| Size of Tap | Drills for Briggs Pipe Taps | Drills for Whitworth Pipe Taps | Size of Tap | Drills for Briggs Pipe Taps | Drills for Whitworth Pipe Taps | Size of Tap | Drills for Briggs Pipe Taps | Drills for Whitworth Pipe Taps |
|---|---|---|---|---|---|---|---|---|
| 1/8 | 11/32 | 5/16 | 1 1/4 | 1 1/2 | 1 15/32 | 3 1/4 | … | 3 1/2 |
| 1/4 | 7/16 | 27/64 | 1 1/2 | 1 23/32 | 1 25/32 | 3 1/2 | 3 3/4 | 3 3/4 |
| 3/8 | 19/32 | 9/16 | 1 3/4 | … | 1 15/16 | 3 3/4 | … | 4 |
| 1/2 | 23/32 | 11/16 | 2 | 2 3/16 | 2 5/32 | 4 | 4 1/4 | 4 1/4 |
| 5/8 | … | 25/32 | 2 1/4 | … | 2 13/32 | 4 1/2 | 4 3/4 | 4 3/4 |
| 3/4 | 15/16 | 29/32 | 2 1/2 | 2 5/8 | 2 25/32 | 5 | 5 5/16 | 5 1/4 |
| 7/8 | … | 1 1/16 | 2 3/4 | … | 3 1/32 | 5 1/2 | … | 5 3/4 |
| 1 | 1 5/32 | 1 1/8 | 3 | 3 1/4 | 3 9/32 | 6 | 6 3/8 | 6 1/4 |

All dimensions are in inches.

To secure the best results, the hole should be reamed before tapping with a reamer having a taper of 3/4 inch per foot.

**Power for Pipe Taps.—**The power required for driving pipe taps is given in the following table, which includes nominal pipe tap sizes from 2 to 8 inches.

The holes to be tapped were reamed with standard pipe tap reamers before tapping. The horsepower recorded was read off just before the tap was reversed. The table gives the net horsepower, deductions being made for the power required to run the machine without a load. The material tapped was cast iron, except in two instances, where cast steel was tapped. It will be seen that nearly double the power is required for tapping cast steel. The power varies, of course, with the conditions. More power than that indicated in the table will be required if the cast iron is of a harder quality or if the taps are not properly relieved. The taps used in these experiments were of the inserted-blade type, the blades being made of high-speed steel.

**Power Required for Pipe Taps**

| Nominal Tap Size | | Rev. per Min. | Net Power | | Thickness of Metal | | Nominal Tap Size | | Rev. per Min. | Net Power | | Thickness of Metal | |
|---|---|---|---|---|---|---|---|---|---|---|---|---|---|
| inch | mm | | H.P. | kW | inch | mm | inch | mm | | H.P. | kW | inch | mm |
| 2 | 50.80 | 40 | 4.24 | 3.16 | 1⅛ | 28.58 | 3½ | 88.90 | 25.6 | 7.20 | 5.37 | 1¾ | 44.45 |
| 2½ | 63.50 | 40 | 5.15 | 3.84 | 1⅛ | 28.58 | 4 | 101.60 | 18 | 6.60 | 4.92 | 2 | 50.80 |
| [a]2½ | 63.50 | 38.5 | 9.14 | 6.81 | 1⅛ | 28.58 | 5 | 127.00 | 18 | 7.70 | 5.74 | 2 | 50.80 |
| 3 | 76.20 | 40 | 5.75 | 4.29 | 1⅛ | 28.58 | 6 | 152.40 | 17.8 | 8.80 | 6.56 | 2 | 50.80 |
| [a]3 | 76.20 | 38.5 | 9.70 | 7.23 | 1⅛ | 28.58 | 8 | 203.20 | 14 | 7.96 | 5.93 | 2½ | 63.50 |

[a] Tapping cast steel; other tests in cast iron.

**High-Speed CNC Tapping.—**Tapping speed depends on the type of material being cut, the type of cutting tool, the speed and rigidity of the machine, the rigidity of the part-holding fixture, and the proper use of coolants and cutting fluids. When tapping, each revolution of the tool feeds the tap a distance equal to the thread pitch. Both spindle speed and feed per revolution must be accurately controlled so that changes in spindle speed result in a corresponding change in feed rate. If the feed/rev is not right, a stripped thread or broken tap will result. NC/CNC machines equipped with the *synchronous tapping* feature are able to control the tap feed as a function of spindle speed. These machines can use rigid-type tap holders or automatic tapping attachments and are able to control depth very accurately. Older NC machines that are unable to reliably coordinate spindle speed and feed must use a tension-compression type tapping head that permits some variation of the spindle speed while still letting the tap feed at the required rate.

CNC machines capable of synchronous tapping accurately coordinate feed rate and rotational speed so that the tap advances at the correct rate regardless of the spindle speed. A canned tapping cycle (see *Right-Hand Threading Cycle (G84)* and *Left-Hand Threading Cycle (G74)* on page 1357 in the *CNC (Computer Numerical Control) Programming* section) usually controls the operation, and speed and feed are set by the machine operator or part programmer. Synchronized tapping requires reversing the tapping spindle twice for each hole tapped, once after finishing the cut and again at the end of the cycle. Because the rotating mass is fairly large (motor, spindle, chuck or tap holder, and tap), the acceleration and deceleration of the tap are rather slow and a lot of time is lost by this process. The frequent changes in cutting speed during the cut also accelerate tap wear and reduce tap life.

A self-reversing tapping attachment has a forward drive that rotates in the same direction as the machine spindle, a reverse drive that rotates in the opposite direction, and a neutral position in between the two. When a hole is tapped, the spindle feeds at a slightly slower rate than the tap to keep the forward drive engaged until the tap reaches the bottom of the hole. Through holes are tapped by feeding to the desired depth and then retracting the spindle, which engages the tapping-head reverse drive and backs the tap out of the hole—the spindle does not need to be reversed. For tapping blind holes, the spindle is fed to a depth equal to the thread depth minus the self-feed of the tapping attachment. When the spindle is retracted (without reversing), the tap continues to feed forward a short distance (the tapping head self-feed distance) before the reverse drive engages and reverse drives the tap out of the hole. The depth can be controlled to within about ¼ revolution of the tap. The tapping cycle normally used for the self-reversing tap attachment is a standard boring cycle with feed return and no dwell. A typical programming cycle is illustrated with a G85 block on page 1357. The inward feed is set to about 95 percent of the normal tapping feed (i.e., 95 percent of the pitch per revolution). Because the tap is lightweight, tap reversal is almost instantaneous and tapping speed is very fast compared with synchronous tapping.

Tapping speeds are usually given in surface feet per minute (sfm) or the equivalent feet per minute (fpm or ft/min), so a conversion is necessary to get the spindle speed in revolutions per minute. The tapping speed in rpm depends on the diameter of the tap, and is given by the following formula:

$$\text{rpm} = \frac{\text{sfm} \times 12}{d \times 3.14159} = \frac{\text{sfm} \times 3.82}{d}$$

where $d$ is the nominal diameter of the tap in inches. As indicated previously, the feed in in/rev is equal to the thread pitch and is independent of the cutting speed. The feed rate in inches per minute is found by dividing the tapping speed in rpm by the number of threads per inch, or by multiplying the speed in rpm by the pitch or feed per revolution:

$$\text{feed rate (in/min)} = \frac{\text{rpm}}{\text{threads per inch}} = \text{rpm} \times \text{thread pitch} = \text{rpm} \times \text{feed/rev}$$

*Example:* If the recommended tapping speed for 1020 steel is given as 45 to 60 sfm, find the required spindle speed and feed rate for tapping a 1/4-20 UNF thread in 1020 steel.

Assuming that the machine being used is in good condition and rigid, and the tap is sharp, use the higher rate of 60 sfm and calculate the required spindle speed and feed rate as follows:

$$\text{speed} = \frac{60 \times 3.82}{0.25} = 916.8 \approx 920 \text{ rpm} \qquad \text{feed rate} = \frac{920}{20} = 46 \text{ in/min}$$

**Coolant for Tapping.**—Proper use of through-the-tap high-pressure coolant/lubricant can result in increased tap life, increased speed and feed, and more accurate threads. In most chip-cutting processes, cutting fluid is used primarily as a coolant, with lubrication being a secondary but important benefit. Tapping, however, requires a cutting fluid with lubricity as the primary property and coolant as a secondary benefit. Consequently, the typical blend of 5 percent coolant concentrate to 95 percent water is too low for best results. An increased percentage of concentrate in the blend helps the fluid to cling to the tap, providing better lubrication at the cutting interface. A method of increasing the tap lubrication qualities without changing the concentration of the primary fluid blend is to use a cutting fluid dispenser controlled by an M code different from that used to control the high-pressure flood coolant (for example, use an M08 code in addition to M07). The secondary coolant-delivery system applies a small amount of an edge-type cutting fluid (about a drop at a time) directly onto the tap-cutting surfaces providing the lubrication needed for cutting. The edge-type fluid applied in this way clings to the tap, increasing the lubrication effect and ensuring that the cutting fluid becomes directly involved in the cutting action at the shear zone.

High-pressure coolant fed through the tap is important in many high-volume tapping applications. The coolant is fed directly through the spindle or tool holder to the cutting zone, greatly improving the process of chip evacuation and resulting in better thread quality. High-pressure through-the-tap coolant flushes blind holes before the tap enters and can remove chips from the holes after tapping is finished. The flushing action prevents chip recutting by forcing chips through the flutes and back out of the hole, improving the surface of the thread and increasing tap life. By improving lubrication and reducing heat and friction, the use of high-pressure coolant may result in increased tap life up to five times that of conventional tapping and may permit speed and feed increases that reduce overall cycle time.

**Combined Drilling and Tapping.**—A special tool that drills and taps in one operation can save a lot of time by reducing setup and eliminating a secondary operation in some applications. A combination drill and tap can be used for through holes if the length of the fluted drill section is greater than the material thickness, but cannot be used for drilling and tapping blind holes because the tip (drill point) must cut completely through the material before the tapping section begins to cut threads. Drilling and tapping depths up to twice the tool diameter are typical. Determine the appropriate speed by starting the tool at the recommended speed for the tap size and material, and adjust the speed higher or lower to suit the application. Feed during tapping is dependent on the thread pitch. NC/CNC programs can use a fast drilling speed and a slower tapping speed to combine both operations into one and minimize cutting time.

**Relief Angles for Single-Point Thread Cutting Tools.—**The surface finish on threads cut with single-point thread cutting tools is influenced by the relief angles on the tools. The leading and trailing cutting edges that form the sides of the thread, and the cutting edge at the nose of the tool must all be provided with an adequate amount of relief. Moreover, it is recommended that the effective relief angle, $a_e$, for all of these cutting edges be made equal, although the practice in some shops is to use slightly less relief at the trailing cutting edge. While too much relief may weaken the cutting edge, causing it to chip, an inadequate amount of relief will result in rough threads and in a shortened tool life. Other factors that influence the finish produced on threads include the following: the work material; the cutting speed; the cutting fluid used; the method used to cut the thread; and, the condition of the cutting edge.

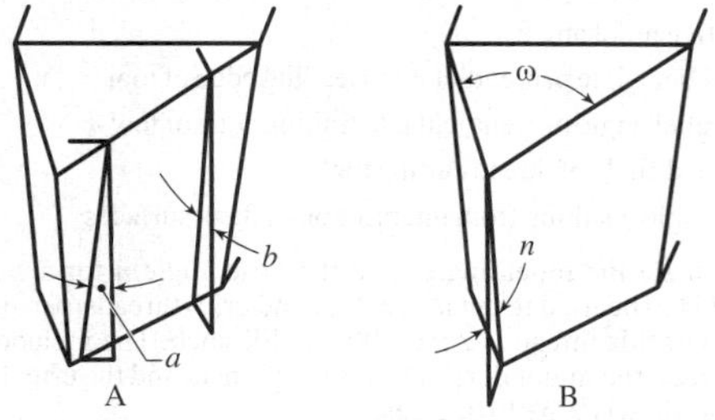


Two similar diagrams showing relationships of various relief angles of thread cutting tools

Relief angles on single-point thread cutting tools are often specified on the basis of experience. While this method may give satisfactory results in many instances, better results can usually be obtained by calculating these angles, using the formulas provided further on. When special high helix angle threads are to be cut, the magnitude of the relief angles should always be calculated. These calculations are based on the effective relief angle, $a_e$; this is the angle between the flank of the tool and the sloping sides of the thread, measured in a direction parallel to the axis of the thread. Recommended values of this angle are 8 to 14 degrees for high-speed steel tools, and 5 to 10 degrees for cemented carbide tools. The larger values are recommended for cutting threads on soft and gummy materials, and the smaller values are for the harder materials, which inherently take a better surface finish. Harder materials also require more support below the cutting edges, which is provided by using a smaller relief angle. These values are recommended for the relief angle below the cutting edge at the nose without any further modification. The angles below the leading and trailing side cutting edges are modified, using the formulas provided. The angles $b$ and $b'$ are the relief angles actually ground on the tool below the leading and trailing side cutting edges respectively; they are measured perpendicular to the side cutting edges. When designing or grinding the thread cutting tool, it is sometimes helpful to know the magnitude of the angle, $n$, for which a formula is provided. This angle would occur only in the event that the tool were ground to a sharp point. It is the angle of the edge formed by the intersection of the flank surfaces.

$$\tan\phi = \frac{\text{lead of thread}}{\pi K} \qquad \tan\phi' = \frac{\text{lead of thread}}{\pi D}$$

$$a = a_e + \phi$$

$$a' = a_e - \phi'$$

$$\tan b = \tan a \cos\frac{1}{2}\omega$$

$$\tan b' = \tan a' \cos\frac{1}{2}\omega$$

$$\tan n = \frac{\tan a - \tan a'}{2\tan\frac{1}{2}\omega}$$

where $\theta$ = helix angle of thread at minor diameter

$\theta'$ = helix angle of thread at major diameter

$K$ = minor diameter of thread

$D$ = major diameter of thread

$a$ = side relief angle parallel to thread axis at leading edge of tool

$a'$ = side relief angle parallel to thread axis at trailing edge of tool

$a_e$ = effective relief angle

$b$ = side relief angle perpendicular to leading edge of tool

$b'$ = side relief angle perpendicular to trailing edge of tool

$\omega$ = included angle of thread cutting tool

$n$ = nose angle resulting from intersection of flank surfaces

*Example:* Calculate the relief angles and the nose angle $n$ for a single-point thread cutting tool that is to be used to cut a 1-inch diameter, 5-threads-per-inch, double Acme thread. The lead of this thread is $2 \times 0.200 = 0.400$ inch. The included angle $\omega$ of this thread is 29 degrees, the minor diameter $K$ is 0.780 inch, and the effective relief angle $a_e$ below all cutting edges is to be 10 degrees.

$$\tan\phi = \frac{\text{lead of thread}}{\pi K} = \frac{0.400}{\pi \times 0.780}$$

$$\phi = 9.27°(9°16')$$

$$\tan\phi' = \frac{\text{lead of thread}}{\pi D} = \frac{0.400}{\pi \times 1.000}$$

$$\phi' = 7.26°(7°15')$$

$$a = a_e + \phi = 10° + 9.27° = 19.27°$$

$$a' = a_e - \phi' = 10° - 7.26° = 2.74°$$

$$\tan b = \tan a \cos\frac{1}{2}\omega = \tan 19.27 \cos 14.5$$

$$b = 18.70°(18°42')$$

$$\tan b' = \tan a' \cos\frac{1}{2}\omega = \tan 2.74 \cos 14.5$$

$$b' = 2.65°(2°39')$$

$$\tan n = \frac{\tan a - \tan a'}{2\tan\frac{1}{2}\omega} = \frac{\tan 19.27 - \tan 2.74}{2\tan 14.5}$$

$$n = 30.26°(30°16')$$

## Lathe Change Gears

**Change Gears for Thread Cutting.—**To determine the change gears to use for cutting a thread of given pitch, first find what number of threads per inch will be cut when gears of the same size are placed on the lead screw and spindle stud, either by trial or by referring to the index plate; then multiply this number, called the "lathe screw constant," by some trial number to obtain the number of teeth in the gear for the spindle stud, and multiply the threads per inch to be cut by the *same* trial number to obtain the number of teeth in the gear for the lead screw. Expressing this rule as a formula:

$$\frac{\text{Trial number} \times \text{lathe screw constant}}{\text{Trial number} \times \text{threads per inch to be cut}} = \frac{\text{teeth in gear on spindle stud}}{\text{teeth in gear on lead screw}}$$

For example, suppose the available change gears supplied with the lathe have 24, 28, 32, 36 teeth, etc., the number increasing by 4 up to 100, and that 10 threads per inch are to be cut in a lathe having a lathe screw constant of 6; then, if the screw constant is written as the numerator, the number of threads per inch to be cut as the denominator of a fraction, and both numerator and denominator are multiplied by some trial number, say, 4, it is found that gears having 24 and 40 teeth can be used. Thus:

$$\frac{6}{10} = \frac{6 \times 4}{10 \times 4} = \frac{24}{40}$$

The 24-tooth gear goes on the spindle stud and the 40-toothgear on the lead screw.

The lathe screw constant is, of course, equal to the number of threads per inch on the lead screw, provided the spindle stud and spindle are geared in the ratio of 1 to 1, which, however, is not always so.

**Compound Gearing.—**To find the change gears used in compound gearing, place the screw constant as the numerator and the number of threads per inch to be cut as the denominator of a fraction; resolve both numerator and denominator into two factors each, and multiply each "pair" of factors by the same number, until values are obtained representing suitable numbers of teeth for the change gears. (One factor in the numerator and one in the denominator make a "pair" of factors.)

*Example:* —$1\frac{3}{4}$ threads per inch are to be cut in a lathe having a screw constant of 8; the available gears have 24, 28, 32, 36, 40 teeth. etc., increasing by 4 up to 100. Following the rule:

$$\frac{8}{1\frac{3}{4}} = \frac{2 \times 4}{1 \times 1\frac{3}{4}} = \frac{(2 \times 36) \times (4 \times 16)}{(1 \times 36) \times (1\frac{3}{4} \times 16)} = \frac{72 \times 64}{36 \times 28}$$

The gears having 72 and 64 teeth are the *driving* gears and those with 36 and 28 teeth are the *driven* gears.

**Fractional Threads.—**Sometimes the lead of a thread is given as a fraction of an inch instead of stating the number of threads per inch. For example, a thread may be required to be cut, having $\frac{3}{8}$ inch lead. The expression "$\frac{3}{8}$ inch lead" should first be transformed to "number of threads per inch." The number of threads per inch (the thread being single) equals:

$$\frac{1}{\frac{3}{8}} = 1 \div \frac{3}{8} = \frac{8}{3} = 2\frac{2}{3}$$

To find the change gears to cut $2\frac{2}{3}$ threads per inch in a lathe having a screw constant 8 and change gears ranging from 24 to 100 teeth, increasing in increments of 4, proceed as below:

$$\frac{8}{2\frac{2}{3}} = \frac{2 \times 4}{1 \times 2\frac{2}{3}} = \frac{(2 \times 36) \times (4 \times 24)}{(1 \times 36) \times (2\frac{2}{3} \times 24)} = \frac{72 \times 96}{36 \times 64}$$

**Change Gears for Metric Pitches.—**When screws are cut in accordance with the metric system, it is the usual practice to give the lead of the thread in millimeters, instead of the number of threads per unit of measurement. To find the change gears for cutting metric threads, when using a lathe having an inch lead screw, first determine the number of threads per inch corresponding to the given lead in millimeters. Suppose a thread of 3 millimeters lead is to be cut in a lathe having an inch lead screw and a screw constant of 6. As there are 25.4 millimeters per inch, the number of threads per inch will equal 25.4 ÷ 3. Place the screw constant as the numerator, and the number of threads per inch to be cut as the denominator:

$$\frac{6}{\frac{25.4}{3}} = 6 \div \frac{25.4}{3} = \frac{6 \times 3}{25.4}$$

The numerator and denominator of this fractional expression of the change gear ratio is next multiplied by some trial number to determine the size of the gears. The first whole number by which 25.4 can be multiplied so as to get a whole number as the result is 5. Thus, 25.4 × 5 = 127. Hence, one gear having 127 teeth is always used when cutting metric threads with an inch lead screw. The other gear required has 90 teeth. Thus:

$$\frac{6 \times 3 \times 5}{25.4 \times 5} = \frac{90}{127}$$

Therefore, the following rule can be used to find the change gears for cutting metric pitches with an inch lead screw:

*Rule:* Place the lathe screw constant multiplied by the lead of the required thread in millimeters multiplied by 5 as the numerator of the fraction and 127 as the denominator. The product of the numbers in the numerator equals the number of teeth for the spindle-stud gear, and 127 is the number of teeth for the lead-screw gear.

If the lathe has a metric pitch lead screw, and a screw having a given number of threads per inch is to be cut, first find the "metric screw constant" of the lathe or the lead of thread in millimeters that would be cut with change gears of equal size on the lead screw and spindle stud; then the method of determining the change gears is simply the reverse of the one already explained for cutting a metric thread with an inch lead screw.

*Rule:* To find the change gears for cutting inch threads with a metric lead screw, place 127 in the numerator and the threads per inch to be cut, multiplied by the metric screw constant multiplied by 5, in the denominator; 127 is the number of teeth on the spindle-stud gear and the product of the numbers in the denominator equals the number of teeth in the lead-screw gear.

**Threads per Inch Obtained with a Given Combination of Gears.—**To determine the number of threads per inch that will be obtained with a given combination of gearing, multiply the lathe screw constant by the number of teeth in the *driven* gear (or by the product of the numbers of teeth in both driven gears of compound gearing), and divide the product thus obtained by the number of teeth in the *driving* gear (or by the product of the two driving gears of a compound train). The quotient equals the number of threads per inch.

**Change Gears for Fractional Ratios.—**When gear ratios cannot be expressed exactly in whole numbers that are within the range of ordinary gearing, the combination of gearing required for the fractional ratio may be determined quite easily, often by the "cancellation method." To illustrate this method, assume that the speeds of two gears are to be in the ratio of 3.423 to 1. The number 3.423 is first changed to $^{3423}/_{1000}$ to clear it of decimals. Then, in order to secure a fraction that can be reduced, 3423 is changed to 3420:

$$\frac{3420}{1000} = \frac{342}{100} = \frac{3 \times 2 \times 57}{2 \times 50} = \frac{3 \times 57}{1 \times 50}$$

Then, multiplying $^3/_1$ by some trial number, say, 24, the following gear combination is obtained:

$$\frac{72}{24} \times \frac{57}{50} = \frac{4104}{1200} = \frac{3.42}{1}$$

As the desired ratio is 3.423 to 1, there is an error of 0.003. When the ratios are comparatively simple, the cancellation method is not difficult and is frequently used; but by the logarithmic method to be described, more accurate results are usually possible.

**Modifying the Quick-Change Gearbox Output.—**On most modern lathes, the gear train connecting the headstock spindle with the lead screw contains a quick-change gearbox. Instead of using different change gears, it is only necessary to position the handles of the gearbox to adjust the speed ratio between the spindle and the lead screw in preparation for cutting a thread. However, a thread sometimes must be cut for which there is no quick-change gearbox setting. It is then necessary to modify the normal, or standard, gear ratio between the spindle and the gearbox by installing modifying change gears to replace the standard gears normally used. Metric and other odd pitch threads can be cut on lathes that have an inch thread lead screw and a quick-change gearbox having only settings for inch threads by using modifying-change gears in the gear train. Likewise, inch threads and other odd pitch threads can be cut on metric lead-screw lathes having a gearbox on which only metric thread settings can be made. Modifying-change gears also can be used for cutting odd pitch threads on lathes having a quick-change gearbox that has both inch and metric thread settings.

The sizes of the modifying-change gears can be calculated by formulas to be given later; they depend on the thread to be cut and on the setting of the quick-change gearbox. Many different sets of gears can be found for each thread to be cut. It is recommended that several calculations be made in order to find the set of gears that is most suitable for installation on the lathe. The modifying-change gear formulas that follow are based on the type of lead screw, i.e., whether the lead screw has inch or metric threads.

*Metric Threads on Inch Lead-Screw Lathes:* A 127-tooth translating gear must be used in the modifying-change gear train in order to be able to cut metric threads on inch lead-screw lathes. The formula for calculating the modifying change gears is:

$$\frac{5 \times \text{gearbox setting in thds/in} \times \text{pitch in mm to be cut}}{127} = \frac{\text{driving gears}}{\text{driven gears}}$$

The numerator and denominator of this formula are multiplied by equal numbers, called trial numbers, to find the gears. If suitable gears cannot be found with one set, then another set of equal trial numbers is used. (Because these numbers are equal, such as 15/15 or 24/24, they are equal to the number one when thought of as a fraction; their inclusion has the effect of multiplying the formula by one, which does not change its value.) It is necessary to select the gearbox setting in threads per inch that must be used to cut the metric thread when using the gears calculated by the formula. One method is to select a quick-change gearbox setting that is close to the actual number of metric threads in a 1-inch length, called the equivalent threads per inch, which can be calculated by the following formula: Equivalent thds/in = 25.4 ÷ pitch in millimeters to be cut.

*Example:* Select the quick-change gearbox setting and calculate the modifying change gears required to set up a lathe having an inch-thread lead screw in order to cut an M12 × 1.75 metric thread.

$$\text{Equivalent thds/in} = \frac{25.4}{\text{pitch in mm to be cut}} = \frac{25.4}{1.75} = 1.45 \text{ (use 14 thds/in)}$$

$$\frac{5 \times \text{gearbox setting in thds/in} \times \text{pitch in mm to be cut}}{127} = \frac{5 \times 14 \times 1.75}{127}$$

$$= \frac{(24) \times 5 \times 14 \times 1.75}{(24) \times 127} = \frac{(5 \times 14) \times (24 \times 1.75 1.75)}{24 \times 127}$$

$$\frac{70 \times 42}{24 \times 127} = \frac{\text{driving gears}}{\text{driven gears}}$$

*Odd Inch Pitch Threads:* The calculation of the modifying change gears used for cutting odd pitch threads that are specified by their pitch in inches involves the sizes of the standard gears, which can be found by counting their teeth. Standard gears are those used to enable the lathe to cut the thread for which the gearbox setting is made; they are the gears that are normally used. The threads on worms used with worm gears are among the odd pitch threads that can be cut by this method. As before, it is usually advisable to calculate the actual number of threads per inch of the odd pitch thread and to select a gearbox setting that is close to this value. The following formula is used to calculate the modifying-change gears to cut odd inch pitch threads:

$$\frac{\text{Standard driving gear} \times \text{pitch to be cut in inches} \times \text{gearbox setting in thds/in}}{\text{Standard driven gear}}$$

$$= \frac{\text{driving gears}}{\text{driven gears}}$$

*Example:* Select the quick-change gearbox setting and calculate the modifying change gears required to cut a thread having a pitch equal to 0.195 inch. The standard driving and driven gears both have 48 teeth. To find equivalent threads per inch:

$$\frac{\text{Thds}}{\text{in.}} = \frac{1}{\text{pitch}} = \frac{1}{0.195} = 5.13 \quad \text{(use 5 thds/in)}$$

$$\frac{\text{Standard driving gear} \times \text{pitch to be cut in inches} \times \text{gearbox setting in thds/in}}{\text{Standard driven gear}}$$

$$= \frac{48 \times 0.195 \times 5}{48} = \frac{(1000) \times 0.195 \times 5}{(1000)} = \frac{195 \times 5}{500 \times 2} = \frac{39 \times 5}{100 \times 2} = \frac{39 \times 5 \times (8)}{50 \times 2 \times 2 \times (8)}$$

$$= \frac{39 \times 40}{50 \times 32} = \frac{\text{driving gears}}{\text{driven gears}}$$

It will be noted that in the second step above, 1000/1000 has been substituted for 48/48. This substitution does not change the ratio. The reason for this substitution is that 1000 × 0.195 = 195, a whole number. Actually, 200/200 might have been substituted because 200 ×0.195 = 39, also a whole number.

The procedure for calculating the modifying gears using the following formulas is the same as illustrated by the two previous examples.

*Odd Threads per Inch on Inch Lead-Screw Lathes:*

$$\frac{\text{Standard driving gear} \times \text{gearbox setting in thds/in}}{\text{Standard driven gear} \times \text{thds/in to be cut}} = \frac{\text{driving gears}}{\text{driven gears}}$$

*Inch Threads on Metric Lead-Screw Lathes:*

$$\frac{127}{5 \times \text{gearbox setting in mm pitch} \times \text{thds/in to be cut}} = \frac{\text{driving gears}}{\text{driven gears}}$$

*Odd Metric Pitch Threads on Metric Lead-Screw Lathes:*

$$\frac{\text{Standard driving gear} \times \text{mm pitch to be cut}}{\text{Standard driven gear} \times \text{gearbox setting in mm pitch}} = \frac{\text{driving gears}}{\text{driven gears}}$$

**Finding Accurate Gear Ratios.—**Tables included in the 23rd and earlier editions of this *Handbook* furnished a series of logarithms of gear ratios as a quick means of finding ratios for all gear combinations having 15 to 120 teeth. The ratios thus determined could be factored into sets of 2, 4, 6, or any other even numbers of gears to provide a desired overall ratio.

Although the method of using logarithms of gear ratios provides results of suitable accuracy for many gear-ratio problems, it does not provide a systematic means of evaluating whether other, more accurate ratios are available. In critical applications, especially in the design of mechanisms using reduction gear trains, it may be desirable to find many or all possible ratios to meet a specified accuracy requirement. The methods best suited to such problems use *Continued Fractions* and *Conjugate Fractions* as explained in the *ADDITIONAL MATERIAL* in the *Machinery's Handbook 32 Digital Edition* and illustrated in the worked-out example provided there for a set of four change gears.

As an example, if an overall reduction of 0.31416 is required, a fraction must be found such that the factors of the numerator and denominator may be used to form a four-gear reduction train in which no gear has more than 120 teeth. By using the method of conjugate fractions, the ratios listed above and their factors are found to be successively closer approximations to the required overall gear ratio.

| Ratio | Numerator Factors | Denominator | Error Factors |
|---|---|---|---|
| 11/35 | 11 | 5×7 | +0.00013 |
| 16/51 | 2×2×2×2 | 3×17 | −0.00043 |
| 27/86 | 3×3×3 | 2×43 | −0.00021 |
| 38/121 | 2×19 | 11×11 | −0.00011 |
| 49/156 | 7×7 | 2×2×3×13 | −0.00006 |
| 82/261 | 2×41 | 3×3×29 | +0.00002 |
| 224/713 | 2×2×2×2×2×7 | 23×31 | +0.000005 |
| 437/1391 | 19×23 | 13×107 | +0.000002 |
| 721/2295 | 7×103 | 3×3×3×5×17 | +0.000001 |
| 1360/4329 | 2×2×2×2×17 | 3×3×13×53 | +0.0000003 |
| 1715/5459 | 5×7×7×7 | 53×103 | +0.0000001 |
| 3927/12500 | 3×7×11×17 | 2×2×5×5×5×5×5 | 0 |

**Lathe Change-Gears.—**To calculate the change gears to cut any pitch on a lathe, the "constant" of the machine must be known. For any lathe, the ratio $C{:}L$ = driver:driven gear, in which $C$ = constant of machine and $L$ = threads per inch.

For example, to find the change gears required to cut 1.7345 threads per inch on a lathe having a constant of 4, the formula:

$$\frac{C}{L} = \frac{4}{1.7345} = 2.306140$$

may be used. The method of conjugate fractions shown on page 5 will find the ratio, 113/49 = 2.306122, which is closer than any other having suitable factors. This ratio is in error by only 2.306140 − 2.306122 = 0.000018. Therefore, the driver should have 113 teeth and the driven gear 49 teeth.

**Relieving Helical-Fluted Hobs.**—Relieving hobs that have been fluted at right angles to the thread is another example of approximating a required change-gear ratio. The usual method is to change the angle of the helical flutes to agree with previously calculated change-gears. The ratio between the hob and the relieving attachment is expressed in the formula:

$$\frac{N}{(C \times \cos^2 \alpha)} = \frac{driver}{driven} \text{gears}$$

and

$$\tan \alpha = \frac{P}{H_c}$$

in which: $N$ = number of flutes in hob; $\alpha$ = helix angle of thread from plane perpendicular to axis; $C$ = constant of relieving attachment; $P$ = axial lead of hob; and $H_c$ = hob pitch circumference, = 3.1416 times pitch diameter.

The constant of the relieving attachment is found on its index plate and is determined by the number of flutes that require equal gears on the change-gear studs. These values will vary with different makes of lathes.

For example, what four change-gears can be used to relieve a helical-fluted worm-gear hob, of 24 diametral pitch, six starts, 13 degrees, 41 minutes helix angle of thread, with eleven helical flutes, assuming a relieving attachment having a constant of 4 is to be used?

$$\frac{N}{(C \times \cos^2 \alpha)} = \frac{11}{(4 \times \cos^2 13°41')} = \frac{11}{(4 \times 0.944045)} = 2.913136$$

Using the conjugate fractions method discussed on page 5, the following ratios are found to provide factors that are successively closer approximations to the required change-gear ratio 2.913136.

| *Numerator/Denominator* | *Ratio* | *Error* |
|---|---|---|
| 67 × 78/(39 × 46) | 2.913043 | −0.000093 |
| 30 × 47/(22 × 22) | 2.913223 | +0.000087 |
| 80 × 26/(21 × 34) | 2.913165 | +0.000029 |
| 27 × 82/(20 × 38) | 2.913158 | +0.000021 |
| 55 × 75/(24 × 59) | 2.913136 | +0.0000004 |
| 74 × 92/(57 × 41) | 2.913136 | +0.00000005 |

## THREAD ROLLING

Screw threads may be formed by rolling either by using some type of thread-rolling machine or by equipping an automatic screw machine or turret lathe with a suitable threading roll. If a thread-rolling machine is used, the unthreaded screw, bolt, or other "blank" is placed (either automatically or by hand) between dies having thread-shaped ridges that sink into the blank, and by displacing the metal, form a thread of the required shape and pitch. The thread-rolling process is applied where bolts, screws, studs, threaded rods, etc., are required in large quantities. Screw threads that are within the range of the rolling process may be produced more rapidly by this method than in any other way. Because of the cold-working action of the dies, the rolled thread is 10 to 20 percent stronger than a cut or ground thread, and the increase may be much higher for fatigue resistance. Other advantages of the rolling process are that no stock is wasted in forming the thread, and the surface of a rolled thread is harder than that of a cut thread, thus increasing wear resistance.

**Thread-Rolling Machine of Flat-Die Type.—**One type of machine that is used extensively for thread rolling is equipped with a pair of flat or straight dies. One die is stationary and the other has a reciprocating movement when the machine is in use. The ridges on these dies, which form the screw thread, incline at an angle equal to the helix angle of the thread. In making dies for precision thread rolling, the threads may be formed either by milling and grinding after heat treatment, or by grinding "from the solid" after heat treating. A vitrified wheel is used.

In a thread-rolling machine, thread is formed in one passage of the work, which is inserted at one end of the dies, either by hand or automatically, and then rolls between the die faces until it is ejected at the opposite end. The relation between the position of the dies and a screw thread being rolled is such that the top of the thread-shaped ridge of one die, at the point of contact with the screw thread, is directly opposite the bottom of the thread groove in the other die at the point of contact. Some form of mechanism ensures starting the blank at the right time and square with the dies.

**Thread-Rolling Machine of Cylindrical-Die Type.—**With machines of this type, the blank is threaded while being rolled between two or three cylindrical dies (depending upon the type of machine) that are pressed into the blank at a rate of penetration adjusted to the hardness of the material, or wall thickness in threading operations on tubing or hollow parts. The dies have ground, or ground and lapped, threads and a pitch diameter that is a multiple of the pitch diameter of the thread to be rolled. As the dies are much larger in diameter than the work, a multiple thread is required to obtain the same lead angle as that of the work. The thread may be formed in one die revolution or even less, or several revolutions may be required (as in rolling hard materials) to obtain a gradual rate of penetration equivalent to that obtained with flat or straight dies if extended to a length of possibly 15 or 20 feet (4.6 or 6 m). Provisions for accurately adjusting or matching the thread rolls to bring them into proper alignment with each other are important features of these machines.

*Two-Roll Type of Machine:* With a two-roll type of machine, the work is rotated between two horizontal power-driven threading rolls and is supported by a hardened rest bar on the lower side. One roll is fed inward by hydraulic pressure to a depth that is governed automatically.

*Three-Roll Type of Machine:* With this machine, the blank to be threaded is held in a "floating position" while being rolled between three cylindrical dies that, through toggle arms, are moved inward at a predetermined rate of penetration until the required pitch diameter is obtained. The die movement is governed by a cam driven through change gears selected to give the required cycle of squeeze, dwell, and release.

**Rate of Production.—**Production rates in thread rolling depend upon the type of machine, the size of both machine and work, and whether the parts to be threaded are inserted by hand or automatically. A reciprocating flat die type of machine, applied to ordinary steels, may thread 30 or 40 parts per minute in diameters ranging from about $\frac{5}{8}$ to

$1\frac{1}{8}$ inch (15.875–28.575 mm), and 150 to 175 per minute in machine screw sizes from No. 10 (.190 inch) to No. 6 (.138 inch). In the case of heat-treated alloy steels in the usual hardness range of 26 to 32 RC (Rockwell C scale), the production may be 30 or 40 per minute or less. With a cylindrical die type of machine, which is designed primarily for precision work and hard metals, 10 to 30 parts per minute are common production rates, the amount depending upon the hardness of material and allowable rate of die penetration per work revolution. These production rates are intended as a general guide only. The diameters of rolled threads usually range from the smallest machine screw sizes up to 1 or $1\frac{1}{2}$ inches (25.4 or 38.1 mm), depending upon the type and size of machine.

**Precision Thread Rolling.—**Both flat and cylindrical dies are used in aeronautical and other plants for precision work. With accurate dies and blank diameters held to close limits, it is practicable to produce rolled threads for American National Standard Class 3 and Class 4 fits. The blank sizing may be by centerless grinding or by means of a die in conjunction with the heading operations. The blank should be round, and, as a general rule, the diameter tolerance should not exceed $\frac{1}{2}$ to $\frac{2}{3}$ the pitch diameter tolerance. The blank diameter should range from the correct size (which is close to the pitch diameter, but should be determined by actual trial), down to the allowable minimum, the tolerance being minus to ensure a correct pitch diameter, even though the major diameter may vary slightly. Precision thread rolling has become an important method of threading alloy steel studs and other threaded parts, especially in aeronautical work where precision and high-fatigue resistance are required. Micrometer screws are also an outstanding example of precision thread rolling. This process has also been applied in tap making, although it is the general practice to finish rolled taps by grinding when the Class 3 and Class 4 fits are required.

**Steels for Thread Rolling.—**Steels vary from soft low-carbon types for ordinary screws and bolts, to nickel, nickel-chromium and molybdenum steels for aircraft studs, bolts, etc., or for any work requiring exceptional strength and fatigue resistance. Typical SAE alloy steels are No. 2330, 3135, 3140, 4027, 4042, 4640 and 6160. The hardness of these steels after heat treatment usually ranges from 26 to 32 RC, with tensile strengths varying from 130,000 to 150,000 psi (896–1034 MPa). While harder materials might be rolled, grinding is more practicable when the hardness exceeds 40 RC. Thread rolling is applicable not only to a wide range of steels but for nonferrous materials, especially if there is difficulty in cutting due to "tearing" the threads.

**Diameter of Blank for Thread Rolling.—**The diameter of the screw blank or cylindrical part upon which a thread is to be rolled should be less than the outside screw diameter by an amount that will just compensate for the metal that is displaced and raised above the original surface by the rolling process. The increase in diameter is approximately equal to the depth of one thread. While there are rules and formulas for determining blank diameters, it may be necessary to make slight changes in the calculated size in order to secure a well-formed thread. Blank diameter should be verified by trial, especially when rolling accurate screw threads. Some stock offers greater resistance to displacement than other stock, owing to greater hardness or tenacity of the metal. The following figures may prove useful in establishing trial sizes. The blank diameters for screws varying from $\frac{1}{4}$ to $\frac{1}{2}$ are from 0.002 to 0.0025 inch (50.8–63.5 μm) larger than the pitch diameter, and for screws varying from $\frac{1}{2}$ to 1 inch (12.7–25.4 mm) or larger, the blank diameters are from 0.0025 to .003 inch (63.5–76.2 μm) larger than the pitch diameter. Blanks which are slightly less than the pitch diameter are intended for bolts, screws, etc., which are to have a comparatively free fit. Blanks for this class of work may vary from 0.002 to 0.003 inch (50.08–76.2 μm) less than the pitch diameter for screw thread sizes varying from $\frac{1}{4}$ to $\frac{1}{2}$ inch (6.35–12.7 μm), and from 0.003 to 0.005 inch (76.2–127 μm) less than the pitch diameter for sizes above $\frac{1}{2}$ inch. If the screw threads are smaller than $\frac{1}{4}$ inch, the blanks are usually from 0.001 to 0.0015 inch (25.4–38.1 μm) less than the pitch diameter for ordinary grades of work.

**Thread Rolling in Automatic Screw Machines.—**Screw threads are sometimes rolled in automatic screw machines and turret lathes when the thread is behind a shoulder so that it cannot be cut with a die. In such cases, the advantage of rolling the thread is that a second operation is avoided. A circular roll is used for rolling threads in screw machines. The roll may be presented to the work either in a tangential direction or radially, either method producing a satisfactory thread. In the former case, the roll gradually comes into contact with the periphery of the work and completes the thread as it passes across the surface to be threaded. When the roll is held in a radial position, it is simply forced against one side until a complete thread is formed. The method of applying the roll may depend upon the relation between the threading operation and other machining operations. Thread rolling in automatic screw machines is generally applied only to brass and other relatively soft metals, owing to the difficulty of rolling threads in steel. Thread rolls made of chrome-nickel steel containing from 0.15 to 0.20 percent of carbon have given fairly good results, however, when applied to steel. A 3 percent nickel steel containing about 0.12 percent carbon has also proved satisfactory for threading brass.

**Factors Governing the Diameter of Thread Rolling.—**The threading roll used in screw machines may be about the same diameter as the screw thread, but for sizes smaller than, say, $^3/_4$ inch (19.05 mm), the roll diameter is some multiple of the thread diameter minus a slight amount to obtain a better rolling action. When the diameters of the thread and roll are practically the same, a single-threaded roll is used to form a single thread on the screw. If the diameter of the roll is made double that of the screw, in order to avoid using a small roll, then the roll must have a double thread. If the thread roll is three times the size of the screw thread, a triple thread is used, and so on. These multiple threads are necessary when the roll diameter is some multiple of the work, in order to obtain corresponding helix angles on the roll and work.

**Diameter of Threading Roll.—**The pitch diameter of a threading roll having a single thread is slightly less than the pitch diameter of the screw thread to be rolled, and in the case of multiple-thread rolls, the pitch diameter is not an exact multiple of the screw thread pitch diameter but is also reduced somewhat. The amount of reduction recommended by one screw machine manufacturer is given by the formula shown at the end of this paragraph. A description of the terms used in the formula is given as follows: $D$ = pitch diameter of threading roll, $d$ = pitch diameter of screw thread, $N$ = number of single threads or "starts" on the roll (this number is selected with reference to diameter of roll desired), $T$ = single depth of thread:

$$D = N\left(d - \frac{T}{2}\right) - T$$

*Example:* Find, by using above formula, the pitch diameter of a double-thread roll for rolling a $^1/_2$-inch American standard screw thread. Pitch diameter $d$ = 0.4500 inch and thread depth $T$ = 0.0499 inch.

$$D = 2\left(0.4500 - \frac{0.0499}{2}\right) - 0.0499 = 0.8001 \text{ inch}$$

**Kind of Thread on Roll and Its Shape.—**The thread (or threads) on the roll should be left-hand for rolling a right-hand thread, and *vice versa*. The roll should be wide enough to overlap the part to be threaded, provided there are clearance spaces at the ends, which should be formed if possible. The thread on the roll should be sharp on top for rolling an American (National) standard form of thread, so that less pressure will be required to displace the metal when rolling the thread. The bottom of the thread groove on the roll may also be left sharp or it may have a flat. If the bottom is sharp, the roll is sunk only far enough into the blank to form a thread having a flat top, assuming that the thread is the American form. The number of threads on the roll (whether double, triple, quadruple, etc.) is selected, as a rule, so that the diameter of the thread roll will be somewhere between $1^1/_4$ and $2^1/_4$ inches (31.75–57.15 mm). In making a thread roll, the ends are beveled at an angle of 45 degrees, to prevent the threads on the ends of the roll from chipping. Precautions should be taken in hardening, because, if the sharp edges are burnt, the roll will be useless. Thread rolls are usually lapped after hardening, by holding them on an arbor in the lathe and using emery and oil on a piece of hard wood. To give good results a thread roll should fit closely in the holder. If the roll is made to fit loosely, it will mar the threads.

**Application of Thread Roll.—**The shape of the work and the character of the operations necessary to produce it, govern, to a large extent, the method employed in applying the thread roll. Some of the points to consider are as follows:

1) Diameter of the part to be threaded.

2) Location of the part to be threaded.

3) Length of the part to be threaded.

4) Relation that the thread rolling operation bears to the other operations.

5) Shape of the part to be threaded, whether straight, tapered or otherwise.

6) Method of applying the support.

When the diameter to be rolled is much smaller than the diameter of the shoulder preceding it, a cross-slide knurl-holder should be used. If the part to be threaded is not behind a shoulder, a holder on the swing principle should be used. When the work is long (greater in length than two-and-one-half times its diameter) a swing roll-holder should be employed, carrying a support. When the work can be cut off after the thread is rolled, a cross-slide roll-holder should be used. The method of applying the support to the work also governs to some extent the method of applying the thread roll. When no other tool is working at the same time as the thread roll, and when there is freedom from chips, the roll can be held more rigidly by passing it under instead of over the work. When passing the roll over the work, there is a tendency to raise the cross-slide. Where the part to be threaded is tapered, the roll can best be presented to the work by holding it in a cross-slide roll-holder.

**Tolerances on Wire for Thread Rolling.—**The wire mills will accept a tolerance specification of plus or minus 0.002 inch (50.8 μm) on the diameter. It is particularly important that this tolerance be maintained on stock used for long screws of small diameter. On screws of short length the material will flow, and if the wire is over size little trouble will be experienced, but in the case of screws having a length greater than ten times their diameter, the material will be confined, and "burning" will take place, if the tolerance is greater than that specified. If the wire is slightly under size, the rolled threads will have a ragged appearance due to the fact that the crest is not fully formed. On screws under the No. 10-24 size, a tolerance of plus or minus 0.001 inch should be adhered to in order to ensure good results.

**Speeds and Feeds for Thread Rolling.—**When the thread roll is made from high-carbon steel and used on brass, a surface speed as high as 200 feet per minute can be used. However, better results are obtained by using a lower speed than this. When the roll is held in a holder attached to the cross-slide, and is presented either tangentially or radially to the work, a considerably higher speed can be used than if it is held in a swing tool. This is due to the lack of rigidity in a holder of the swing type. The feeds to be used when a cross-slide roll-holder is used are given in the upper half of the table "Feeds for Thread Rolling;" the lower half of the table gives the feeds for thread rolling with swing tools. These feeds are applicable for rolling threads without a support, when the root diameter of the blank is not less than five times the double depth of the thread. When the root diameter is less than this, a support should be used. A support should also be used when the width of the roll is more than two-and-one-half times the smallest diameter of the piece to be rolled, irrespective of the pitch of the thread. When the smallest diameter of the piece to be rolled is much less than the root diameter of the thread, the smallest diameter should be taken as the deciding factor for the feed to be used.

## Feeds for Thread Rolling

| Root Dia. of Blank | Number of Threads per Inch | | | | | | | | | | | | | |
|---|---|---|---|---|---|---|---|---|---|---|---|---|---|---|
| | 72 | 64 | 56 | 48 | 44 | 40 | 36 | 32 | 28 | 24 | 22 | 20 | 18 | 14 |
| | Cross-slide Holders—Feed per Revolution in Inches | | | | | | | | | | | | | |
| ⅛ | 0.0045 | 0.0040 | 0.0035 | 0.0030 | 0.0025 | 0.0020 | 0.0015 | 0.0010 | ..... | ..... | ..... | ..... | ..... | ..... |
| 3⁄16 | 0.0050 | 0.0045 | 0.0040 | 0.0035 | 0.0030 | 0.0025 | 0.0020 | 0.0015 | 0.0005 | ..... | ..... | ..... | ..... | ..... |
| ¼ | 0.0055 | 0.0050 | 0.0045 | 0.0040 | 0.0035 | 0.0030 | 0.0025 | 0.0020 | 0.0010 | 0.0005 | 0.0005 | ..... | ..... | ..... |
| 5⁄16 | 0.0060 | 0.0055 | 0.0050 | 0.0045 | 0.0040 | 0.0035 | 0.0030 | 0.0025 | 0.0015 | 0.0010 | 0.0010 | 0.0005 | 0.0005 | ..... |
| ⅜ | 0.0065 | 0.0060 | 0.0055 | 0.0050 | 0.0045 | 0.0040 | 0.0035 | 0.0030 | 0.0020 | 0.0015 | 0.0015 | 0.0010 | 0.0010 | 0.0005 |
| 7⁄16 | 0.0070 | 0.0065 | 0.0060 | 0.0055 | 0.0050 | 0.0045 | 0.0040 | 0.0035 | 0.0025 | 0.0020 | 0.0020 | 0.0015 | 0.0015 | 0.0010 |
| ½ | 0.0075 | 0.0070 | 0.0065 | 0.0060 | 0.0055 | 0.0050 | 0.0045 | 0.0040 | 0.0030 | 0.0025 | 0.0025 | 0.0020 | 0.0020 | 0.0015 |
| ⅝ | 0.0080 | 0.0075 | 0.0070 | 0.0065 | 0.0060 | 0.0055 | 0.0050 | 0.0045 | 0.0035 | 0.0030 | 0.0030 | 0.0025 | 0.0025 | 0.0020 |
| ¾ | 0.0085 | 0.0080 | 0.0075 | 0.0070 | 0.0065 | 0.0060 | 0.0055 | 0.0050 | 0.0040 | 0.0035 | 0.0035 | 0.0030 | 0.0030 | 0.0025 |
| ⅞ | 0.0090 | 0.0085 | 0.0080 | 0.0075 | 0.0070 | 0.0065 | 0.0060 | 0.0055 | 0.0045 | 0.0040 | 0.0040 | 0.0035 | 0.0035 | 0.0030 |
| 1 | 0.0095 | 0.0090 | 0.0085 | 0.0080 | 0.0075 | 0.0070 | 0.0065 | 0.0060 | 0.0050 | 0.0045 | 0.0045 | 0.0040 | 0.0040 | 0.0035 |
| Root Dia. | Swing Holders—Feed per Revolution in Inches | | | | | | | | | | | | | |
| ⅛ | 0.0025 | 0.0020 | 0.0015 | 0.0010 | 0.0005 | ..... | ..... | ..... | ..... | ..... | ..... | ..... | ..... | ..... |
| 3⁄16 | 0.0028 | 0.0025 | 0.0020 | 0.0015 | 0.0008 | 0.0005 | ..... | ..... | ..... | ..... | ..... | ..... | ..... | ..... |
| ¼ | 0.0030 | 0.0030 | 0.0025 | 0.0020 | 0.0010 | 0.0010 | 0.0005 | 0.0005 | 0.0005 | ..... | ..... | ..... | ..... | ..... |
| 5⁄16 | 0.0035 | 0.0035 | 0.0030 | 0.0025 | 0.0015 | 0.0015 | 0.0010 | 0.0010 | 0.0010 | 0.0005 | ..... | ..... | ..... | ..... |
| ⅜ | 0.0040 | 0.0040 | 0.0035 | 0.0030 | 0.0020 | 0.0020 | 0.0015 | 0.0015 | 0.0015 | 0.0010 | 0.0005 | 0.0005 | 0.0005 | ..... |
| 7⁄16 | 0.0045 | 0.0045 | 0.0040 | 0.0035 | 0.0030 | 0.0025 | 0.0020 | 0.0020 | 0.0020 | 0.0015 | 0.0010 | 0.0010 | 0.0010 | ..... |
| ½ | 0.0048 | 0.0048 | 0.0045 | 0.0040 | 0.0035 | 0.0030 | 0.0025 | 0.0025 | 0.0025 | 0.0020 | 0.0015 | 0.0015 | 0.0015 | 0.0005 |
| ⅝ | 0.0050 | 0.0050 | 0.0048 | 0.0043 | 0.0040 | 0.0035 | 0.0030 | 0.0030 | 0.0028 | 0.0025 | 0.0020 | 0.0020 | 0.0018 | 0.0010 |
| ¾ | 0.0055 | 0.0052 | 0.0050 | 0.0045 | 0.0043 | 0.0040 | 0.0035 | 0.0035 | 0.0030 | 0.0028 | 0.0025 | 0.0022 | 0.0020 | 0.0013 |
| ⅞ | 0.0058 | 0.0055 | 0.0052 | 0.0048 | 0.0045 | 0.0043 | 0.0040 | 0.0038 | 0.0032 | 0.0030 | 0.0028 | 0.0025 | 0.0022 | 0.0015 |
| 1 | 0.0060 | 0.0058 | 0.0054 | 0.0050 | 0.0048 | 0.0047 | 0.0043 | 0.0040 | 0.0035 | 0.0032 | 0.0030 | 0.0028 | 0.0025 | 0.0018 |

## THREAD GRINDING

Thread grinding is employed for precision tool and gage work and also in producing certain classes of threaded parts.

Thread grinding may be utilized 1) because of the accuracy and finish obtained; 2) hardness of material to be threaded; and 3) economy in grinding certain classes of screw threads when using modern machines, wheels, and thread-grinding oils.

In some cases pre-cut threads are finished by grinding; but usually, threads are ground "from the solid," being formed entirely by the grinding process. Examples of work include thread gages and taps of steel and tungsten carbide, hobs, worms, lead screws, adjusting or traversing screws, alloy steel studs, etc. Grinding is applied to external, internal, straight, and tapering threads, and to various thread forms.

**Accuracy Obtainable by Thread Grinding.—**With single-edge or single-ribbed wheels it is possible to grind threads on gages to a degree of accuracy that requires but very little lapping to produce a so-called "master" thread gage. As far as lead is concerned, some thread grinding machine manufacturers guarantee to hold the lead within 0.0001 inch per inch (or mm per mm) of thread; and while it is not guaranteed that a higher degree of accuracy for lead is obtainable, it is known that threads have been ground to closer tolerances than this on the lead. Pitch diameter accuracies for either Class 3 or Class 4 fits are obtainable according to the grinding method used; with single-edge wheels, the thread angle can be ground to an accuracy of within two or three minutes in half the angle.

**Wheels for Thread Grinding.—**The wheels used for steel have an aluminous abrasive and, ordinarily, either a resinoid bond or a vitrified bond. The general rule is to use resinoid wheels when extreme tolerances are not required, and it is desirable to form the thread with a minimum number of passes, as in grinding threaded machine parts, such as studs, adjusting screws which are not calibrated, and for some classes of taps. *Resinoid wheels*, as a rule, will hold a fine edge longer than a vitrified wheel but they are more flexible and, consequently, less suitable for accurate work, especially when there is lateral grinding pressure that causes wheel deflection. *Vitrified wheels* are utilized for obtaining extreme accuracy in thread form and lead because they are very rigid and not easily deflected by side pressure in grinding. This rigidity is especially important in grinding pre-cut threads on such work as gages, taps and lead screws. The progressive lead errors in long lead screws, for example, might cause an increasing lateral pressure that would deflect a resinoid wheel. Vitrified wheels are also recommended for internal grinding.

*Diamond Wheels:* Diamond wheels set in a rubber or plastic bond are also used for thread grinding, especially for grinding threads in carbide materials and in other hardened alloys. Thread grinding is now being done successfully on a commercial basis on both taps and gages made from carbides. Gear hobs made from carbides have also been tested with successful results. Diamond wheels are dressed by means of silicon-carbide grinding wheels which travel past the diamond-wheel thread form at the angle required for the flanks of the thread to be ground. The action of the dressing wheels is, perhaps, best described as a "scrubbing" of the bond which holds the diamond grits. Obviously, the silicon-carbide wheels do not dress the diamonds, but they loosen the bond until the diamonds not wanted drop out.

**Thread Grinding with Single-Edge Wheel.—**With this type of wheel, the edge is trued to the cross-sectional shape of the thread groove. The wheel, when new, may have a diameter of 18 or 20 inches (45.7 or 50.8 cm) and, when grinding a thread, the wheel is inclined to align it with the thread groove. On some machines, lead variations are obtained by means of change-gears which transmit motion from the work-driving spindle to the lead screw. Other machines are so designed that a lead screw is selected to suit the lead of thread to be ground and transmits motion directly to the work-driving spindle.

**Wheels with Edges for Roughing and Finishing.—**The "three-ribbed" type of wheel has a roughing edge or rib which removes about two-thirds of the metal. This is followed by an intermediate rib which leaves about 0.005 inch (127 μm) for the third or finishing rib. The accuracy obtained with this triple-edge type compares with that of a single-edge wheel, which means that it may be used for the greatest accuracy obtainable in thread grinding.

When the accuracy required makes it necessary, this wheel can be inclined to the helix angle of the thread, the same as is the single-edge wheel.

The three-ribbed wheel is recommended not only for precision work but for grinding threads which are too long for the multi-ribbed wheel referred to later. It is also well adapted to tap grinding, because it is possible to dress a portion of the wheel adjacent to the finish rib for the purpose of grinding the outside diameter of the thread, as indicated in Fig. 1. Furthermore, the wheel can be dressed for grinding or relieving both crests and flanks at the same time.

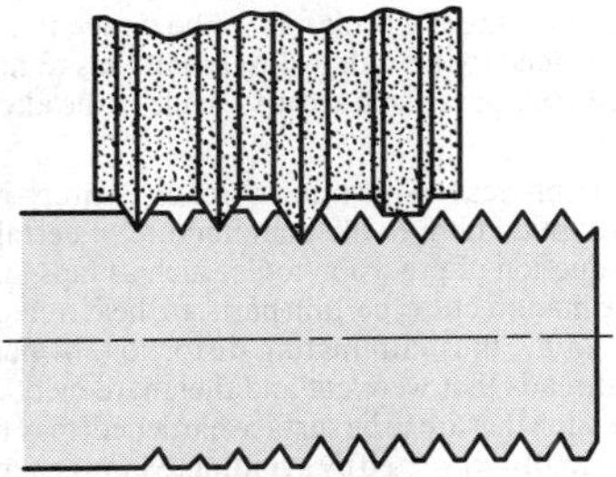

Fig. 1. Wheel with Edges for Roughing and Finishing

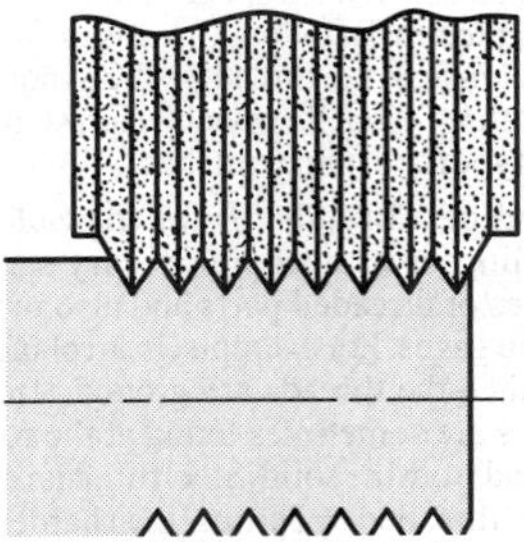

Fig. 2. Multi-Ribbed Type of Thread-Grinding Wheel

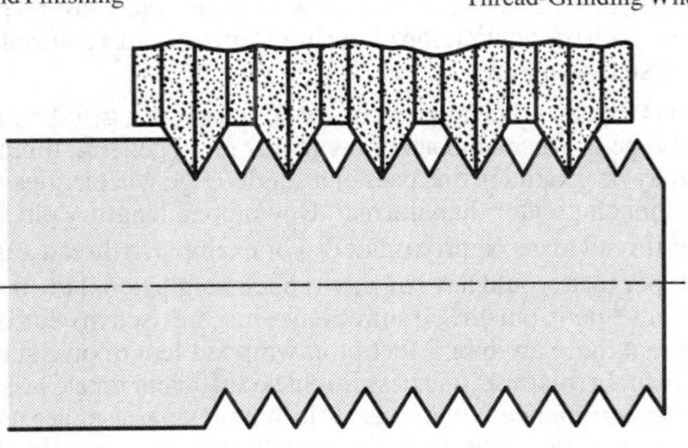

Fig. 3. Alternate-Ribbed Wheel for Grinding the Finer Pitches

**Multi-Ribbed Wheels.—**This type of wheel is employed when rapid production is more important than extreme accuracy, which means that it is intended primarily for the grinding of duplicate parts in manufacturing. A wheel $1\frac{1}{4}$ to 2 inches (3.175–5.08 cm) wide has formed upon its face a series of annular thread-shaped ridges (see Fig. 2); hence, if the length of the thread is not greater than the wheel width, a thread may be ground in one work revolution plus about one-half revolution for feeding in and withdrawing the wheel. The principle of operation is the same as that of thread milling with a multiple type cutter. This type of wheel is not inclined to the lead angle. To obtain a Class 3 fit, the lead angle should not exceed 4 degrees.

It is not practicable to use this form of wheel on thread pitches where the root is less than 0.007 inch (177.8 μm) wide, because of difficulties in wheel dressing. When this method can be applied, it is the fastest means known of producing threads in hardened materials. It is not recommended, however, that thread gages, taps, and work of this character be ground with multi-ribbed wheels. The single-ribbed wheel has a definite field for accurate, small-lot production.

It is necessary, in multi-ribbed grinding, to use more horsepower than is required for single-ribbed wheel grinding. Coarse threads, in particular, may require a wheel motor with two or three times more horsepower than would be necessary for grinding with a single-ribbed wheel.

**Alternate-Ribbed Wheel for Fine Pitches.—**The spacing of ribs on this type of wheel (Fig. 3) equals twice the pitch, so that during the first revolution every other thread groove section is being ground; consequently, about two and one-half work revolutions are required for grinding a complete thread, but the better distribution of cooling oil and resulting increase in work speeds makes this wheel very efficient. This alternate-type of wheel is adapted for grinding threads of fine pitch. Since these wheels cannot be tipped to the helix angle of the thread, they are not recommended for anything closer than Class 3 fits. The "three-ribbed" wheels referred to in a previous paragraph are also made in the alternate type for the finer pitches.

**Grinding Threads "from the Solid".—**The process of forming threads entirely by grinding, or without preliminary cutting, is applied both in the manufacture of certain classes of threaded parts and also in the production of precision tools, such as taps and thread gages. For example, in airplane engine manufacture, certain parts are heat treated and then the threads are ground "from the solid," thus eliminating distortion. Minute cracks are sometimes found at the roots of threads that were cut and then hardened, or ground from the solid. Steel threads of coarse pitch that are to be surface hardened, may be rough threaded by cutting, then hardened and finally corrected by grinding. Many ground thread taps are produced by grinding from the solid after heat treatment. Hardening high-speed steel taps before the thread is formed will ensure there are no narrow or delicate crests to interfere with the application of the high temperature required for uniform hardness and the best steel structure.

**Number of Wheel Passes.—**The number of cuts or passes for grinding from the solid depends upon the type of wheel and accuracy required. In general, threads of 12 or 14 per inch and finer may be ground in one pass of a single-edge wheel unless the "unwrapped" thread length is much greater than normal. Unwrapped length = pitch circumference × total number of thread turns, approximately. For example, a thread gage $1\frac{1}{4}$ inches long with 24 threads per inch would have an unwrapped length equal to 30 × pitch circumference. (If more convenient, outside circumference may be used instead of pitch circumference.) Assume that there are 6 or 7 feet of unwrapped length on a screw thread having 12 threads per inch. In this case, one pass might be sufficient for a Class 3 fit, whereas two passes might be recommended for a Class 4 fit. When two passes are required, too deep a roughing cut may break down the narrow edge of the wheel. To prevent this, try a roughing cut depth equal to about two-thirds the total thread depth, thus leaving one-third for the finishing cut.

**Wheel and Work Rotation.—**When a screw thread, on the side being ground, is moving *upward* or *against* the grinding wheel rotation, less heat is generated and the grinding operation is more efficient than when wheel and work are moving in the same direction on the grinding side; however, to avoid running a machine idle during its return stroke, many screw threads are ground during both the forward and return traversing movements, by reversing the work rotation at the end of the forward stroke. For this reason, thread grinders generally are equipped so that both forward and return work speeds may be changed; they may also be designed to accelerate the return movement when grinding in one direction only.

**Wheel Speeds.**—Wheel speeds should always be limited to the maximum specified on the wheel by the manufacturer. According to the American National Standard Safety Code, resinoid and vitrified wheels are limited to 12,000 surface feet per minute (3657 m/min); however, according to Norton Co., the most efficient speeds are from 9,000 to 10,000 (2743–3048 m/min) for resinoid wheels and 7,500 to 9,500 (2286–2896 m/min) for vitrified wheels. Only tested wheels recommended by the wheel manufacturer should be used. After a suitable surface speed has been established, it should be maintained by increasing the rpm of the wheel, as the latter is reduced in diameter by wear.

Since thread grinding wheels work close to the limit of their stock-removing capacity, some adjustment of the wheel or work speed may be required to get the best results. If the wheel speed is too slow for a given job and excessive heat is generated, try an increase in speed, assuming that such increase is within the safety limits. If the wheel is too soft and the edge wears excessively, again an increase in wheel speed will give the effect of a harder wheel and result in better form-retaining qualities.

**Work Speeds.**—The work speed usually ranges from 3 to 10 fpm (0.9–3.0 m/min). In grinding with a comparatively heavy feed, and a mininum number of passes, the speed may not exceed $2\frac{1}{2}$ or 3 fpm (0.76–0.9 m/min). If very light feeds are employed, as in grinding hardened high-speed steel, the work speed may be much higher than 3 fpm (0.9 m/min) and should be determined by test. If excessive heat is generated by removing stock too rapidly, a work speed reduction is one remedy. If a wheel is working below its normal capacity, an increase in work speed would prevent dulling of the grains and reduce the tendency to heat or "burn" the work. An increase in work speed and reduction in feed may also be employed to prevent burning while grinding hardened steel.

**Truing Grinding Wheels.**—Thread grinding wheels are trued to maintain both the required thread form and an efficient grinding surface. Thread grinders ordinarily are equipped with precision truing devices which function automatically. One type automatically dresses the wheel and also compensates for the slight amount removed in dressing, thus automatically maintaining size control of the work. While truing the wheel, a small amount of grinding oil should be used to reduce diamond wear. Light truing cuts are advisable, especially in truing resinoid wheels which may be deflected by excessive truing pressure. A master former for controlling the path followed by the truing diamond may require a modified profile to prevent distortion of the thread form, especially when the lead angles are comparatively large. Such modification usually is not required for 60-degree threads when the pitches for a given diameter are standard because then the resulting lead angles are less than $4\frac{1}{2}$ degrees. In grinding Acme threads or 29-degree worm threads having lead angles greater than 4 or 5 degrees, modified formers may be required to prevent a bulge in the thread profile. The highest point of this bulge is approximately at the pitch line. A bulge of about 0.001 inch (25.4 μm) may be within allowable limits on some commercial worms but precision worms for gear hobbers, etc., require straight flanks in the axial plane.

*Crushing Method:* Thread grinding wheels are also dressed or formed by the crushing method, which is used in connection with some types of thread grinding machines. When this method is used, the annular ridge or ridges on the wheel are formed by a hardened steel cylindrical dresser or crusher. The crusher has a series of smooth annular ridges which are shaped and spaced like the thread that is to be ground. During the wheel dressing operation, the crusher is positively driven instead of the grinding wheel, and the ridges on the wheel face are formed by the rotating crusher being forced inward.

**Wheel Hardness or Grade.**—Wheel hardness or grade selection is based upon a compromise between efficient cutting and durability of the grinding edge. Grade selection depends on the bond and the character of the work. The following general recommendations are based upon Norton grading.

Vitrified wheels usually range from J to M, and resinoid wheels from R to U. For heat-treated screws or studs and the Unified Standard Thread, try the following. For 8 to

12 threads per inch, grade S resinoid wheel; for 14 to 20 threads per inch, grade T resinoid; for 24 threads per inch and finer, grades T or U resinoid. For high-speed steel taps 4 to 12 threads per inch, grade J vitrified or S resinoid; 14 to 20 threads per inch, grade K vitrified or T resinoid; 24 to 36 threads per inch, grade M vitrified or T resinoid.

**Grain Size.—**A thread grinding wheel usually operates close to its maximum stock-removing capacity, and the narrow edge which forms the root of the thread is the most vulnerable part. In grain selection, the general rule is to use the coarsest grained wheel that will hold its form while grinding a reasonable amount of work. Pitch of thread and quality of finish are two governing factors. Thus, to obtain an exceptionally fine finish, the grain size might be smaller than is needed to retain the edge profile. The usual grain sizes range from 120 to 150. For heat-treated screws and studs with Unified Standard Threads, 100 to 180 is the usual range. For precision screw threads of very fine pitch, the grain size may range from 220 to 320. For high-speed steel taps, the usual range is from 150 to 180 for Unified Standard Threads, and from 80 to 150 for pre-cut Acme threads.

**Thread Grinding by Centerless Method.—**Screw threads may be ground from the solid by the centerless method. A centerless thread grinder is similar in its operating principle to a centerless grinder designed for general work, in that it has a grinding wheel, a regulating or feed wheel (with speed adjustments), and a work-rest. Adjustments are provided to accommodate work of different sizes and for varying the rates of feed. The grinding wheel is a multi-ribbed type, being a series of annular ridges across the face. These ridges conform in pitch and profile with the thread to be ground. The grinding wheel is inclined to suit the helix or lead angle of the thread. In grinding threads on such work as socket type set screws, the blanks are fed automatically and passed between the grinding and regulating wheels in a continuous stream. To illustrate production possibilities, hardened socket set screws of $\frac{1}{4}$-20 size may be ground from the solid at the rate of 60 to 70 per minute and with the wheel operating continuously for 8 hours without redressing. The lead errors of centerless ground screw threads may be limited to 0.0005 inch per inch (or mm per mm) or even less by reducing the production rate. The pitch diameter tolerances are within 0.0002 to 0.0003 inch (5.08–7.62 μm) of the basic size. The grain size for the wheel is selected with reference to the pitch of the thread, the following sizes being recommended: For 11 to 13 threads per inch, 150; for 16 threads per inch, 180; for 18 to 20 threads per inch, 220; for 24 to 28 threads per inch, 320; for 40 threads per inch, 400.

**Principle of Centerless Grinding.—**Centerless grinding is the grinding of cylindrical work without supporting it on centers in the usual way. Two abrasive wheels are mounted so that their peripheries face each other, one of the wheels having its axis so arranged that it can be swung out of parallel with the axis of the other wheel by varying amounts, as required. Between these two abrasive wheels is a work-supporting member equipped with suitable guides. The grinding wheel forces the work downward against the work-rest and also against the regulating wheel. The latter imparts a uniform rotation to the work which has the same peripheral speed as the regulating wheel, the speed of which is adjustable.

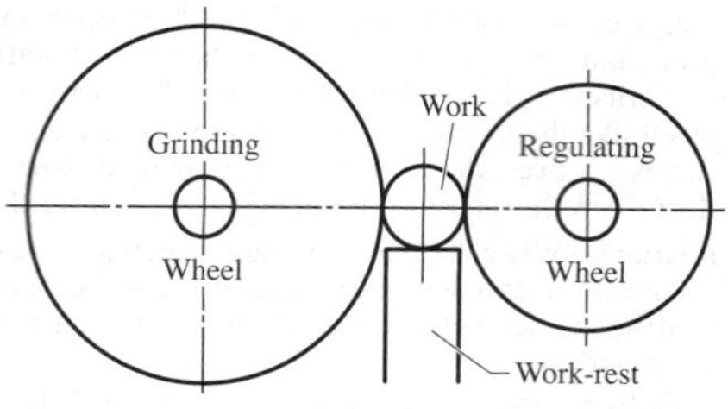


Principle of the Centerless Grinding Process

## THREAD MILLING

**Single-Cutter Method.—**Usually, when a single point cutter is used, the axis of the cutter is inclined an amount equal to the lead angle of the screw thread, in order to locate the cutter in line with the thread groove at the point where the cutting action takes place. Tangent of lead angle = lead of screw thread ÷ pitch circumference of screw.

The helical thread groove is generated by making as many turns around the workpiece diameter as there are pitches in the length of thread to be cut. For example, a 16-pitch thread, 1 inch long, would require 16 turns of the cutter around the work. The single cutter process is especially applicable to the milling of large screw threads of coarse pitch, and either single or multiple threads.

The cutter should revolve as fast as possible without dulling the cutting edges excessively, in order to mill a smooth thread and prevent the unevenness that would result with a slow-moving cutter because of the tooth spaces. As the cutter rotates, the part on which a thread is to be milled is also revolved, but at a very slow rate (a few inches per minute), since this rotation of the work is practically a feeding movement. The cutter is ordinarily set to the full depth of the thread groove and finishes a single thread in one passage, although deep threads of coarse pitch may require two or even three cuts. For fine pitches and short threads, the multiple-cutter method (described in the next paragraph) usually is preferable, because it is more rapid. The milling of taper screw threads may be done on a single-cutter type of machine by traversing the cutter laterally as it feeds along in a lengthwise direction, the same as when using a taper attachment on a lathe.

**Multiple-Cutter Method.—**The multiple cutter for thread milling is practically a series of single cutters, although formed of one solid piece of steel, at least so far as the cutter proper is concerned. The rows of teeth do not lie in a helical path, like the teeth of a hob or tap, but they are annular or without lead. If the cutter had helical teeth the same as a gear hob, it would have to be geared to revolve in a certain fixed ratio with the screw being milled, but a cutter having annular teeth may rotate at any desired cutting speed, while the screw blank is rotated slowly to provide a suitable rate of feed. (The multiple thread milling cutters used are frequently called "hobs," but the term hob should be applied only to cutters having a helical row of teeth like a gear-cutting hob.)

The object in using a multiple cutter instead of a single cutter is to finish a screw thread complete in approximately one revolution of the work, a slight amount of over-travel being allowed to ensure milling the thread to the full depth where the end of cut joins the starting point. The cutter, which is at least one and one half or two threads or pitches wider than the thread to be milled, is fed in to the full thread depth and then either the cutter or screw blank is moved in a lengthwise direction a distance equal to the lead of the thread during one revolution of the work.

The multiple cutter is used for milling comparatively short threads and coarse, medium or fine pitches. The accompanying illustration shows typical examples of external and internal work for which the multiple-cutter type of thread milling has proved very efficient, although its usefulness is not confined to shoulder work and "blind" holes.

In using multiple cutters either for internal or external thread milling, the axis of the cutter is set parallel with the axis of the work, instead of inclining the cutter to suit the lead angle of the thread, as when using a single cutter. Theoretically, this is not the correct position for a cutter, since each cutting edge is revolving in a plane at right angles to the screw's axis while milling a thread groove of helical form. However, as a general rule, interference between the cutter and the thread does not result in a decided change in the standard thread form. Usually the deviation is very slight and may be disregarded except when milling threads which incline considerably relative to the axis like a thread of multiple form and large lead angle. Multiple cutters are suitable for external threads having lead angles under $3\frac{1}{2}$ degrees and for internal threads having lead angles under $2\frac{1}{2}$ degrees. Threads which have steeper sides or smaller included angles than the

American Standard or Whitworth forms have greater limitations on the maximum helix angle and may have to be milled with a single point cutter tilted to the helix angle, assuming that the milling process is preferable to other methods. For instance, in milling an Acme thread which has an included angle between the sides of 29 degrees, there might be considerable interference if a multiple cutter were used, unless the screw thread diameter were large enough in proportion to the pitch to prevent such interference. If an attempt were made to mill a square thread with a multiple cutter, the results would be unsatisfactory owing to the interference.

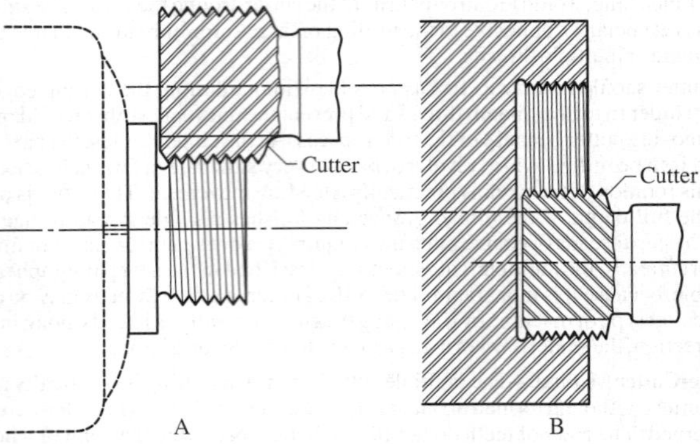


Examples of External and Internal Thread Milling with a Multiple Thread Milling Cutter

Interference between the cutter and work is more pronounced when milling internal threads, because the cutter does not clear itself so well. It is preferable to use as small a cutter as practicable, either for internal or external work, not only to avoid interference, but to reduce the strain on the driving mechanism. Some thread milling cutters, known as "topping cutters," are made for milling the outside diameter of the thread as well as the angular sides and root, but most are made non-tapping.

**Planetary Method.—**The planetary method of thread milling is similar in principle to planetary milling. The part to be threaded is held stationary and the thread milling cutter, while revolving about its own axis, is given a planetary movement around the work in order to mill the thread in one planetary revolution. The machine spindle and the cutter which is held by it is moved longitudinally for thread milling, an amount equal to the thread lead during one planetary revolution. This operation is applicable to both internal and external threads. Other advantages: Thread milling is frequently accompanied by milling operations on other adjoining surfaces, and may be performed with conventional and planetary methods. For example, a machine may be used for milling a screw thread and a concentric cylindrical surface simultaneously. When the milling operation begins, the cutter-spindle feeds the cutter in to the right depth and the planetary movement then begins, thus milling the thread and the cylindrical surface. Thin sharp starting edges are eliminated on threads milled by this method and the thread begins with a smooth gradual approach. One design of machine will mill internal and external threads simultaneously. These threads may be of the same hand or one may be right-hand and the other left-hand. The threads may also be either of the same pitch or of a different pitch, and either straight or tapered.

**Classes of Work for Thread Milling Machines.—**Thread milling machines are used in preference to lathes or taps and dies for certain threading operations.

There are four general reasons why a thread milling machine may be preferred: 1) Because the pitch of the thread is too coarse for cutting with a die; 2) because the milling process is more efficient than using a single-point tool in a lathe; 3) to secure a smoother and more accurate thread than would be obtained with a tap or die; and 4) because the thread is so located relative to a shoulder or other surface that the milling method is superior, if not the only practicable way.

A thread milling machine having a single cutter is especially adapted for coarse pitches, multiple-threaded screws, or any form or size of thread requiring the removal of a relatively large amount of metal, particularly if the pitch of the thread is large in proportion to the screw diameter, since the torsional strain due to the milling process is relatively small. Thread milling often gives a higher rate of production, and a thread is usually finished by means of a single turn of the multiple thread milling cutter around the thread diameter. The multiple-cutter type of thread milling machine frequently comes into competition with dies and taps, and especially self-opening dies and collapsing taps. The use of a multiple cutter is desirable when a thread must be cut close to a shoulder or to the bottom of a shallow recess, although the usefulness of the multiple cutter is not confined to shoulder work and "blind" holes.

**Maximum Pitches of Die-Cut Threads.**—Dies of special design could be constructed for practically any pitch, if the screw blank were strong enough to resist the cutting strains and the size and cost of the die were immaterial; but, as a general rule, when the pitch is coarser than four or five threads per inch, the difficulty of cutting threads with dies increases rapidly, although in a few cases some dies are used successfully on screw threads having two or three threads per inch or less. Much depends upon the design of the die, the finish or smoothness required, and the relation between the pitch of the thread and the diameter of the screw. When the screw diameter is relatively small in proportion to the pitch, there may be considerable distortion due to the twisting strains set up when the thread is being cut. If the number of threads per inch is only one or two less than the standard number for a given diameter, a screw blank ordinarily will be strong enough to permit the use of a die.

**Changing Pitch of Screw Thread Slightly.**—A very slight change in the pitch of a screw thread may be necessary as, for example, when the pitch of a tap is increased a small amount to compensate for shrinkage in hardening. One method of obtaining slight variations in pitch is by means of a taper attachment. This attachment is set at an angle and the work is located at the same angle by adjusting the tailstock center. The result is that the tool follows an angular path relative to the movement of the carriage and, consequently, the pitch of the thread is increased slightly, the amount depending upon the angle to which the work and taper attachment are set. The cosine of this angle, for obtaining a given increase in pitch, equals the standard pitch (which would be obtained with the lathe used in the regular way) divided by the increased pitch necessary to compensate for shrinkage.

If the pitch of a $\frac{3}{4}$-inch American standard screw is to be increased from 0.100 to 0.1005, the cosine of the angle to which the taper attachment and work should be set is found as follows:

$$\text{Cosine of required angle} = \frac{0.100}{0.1005} = 0.9950$$

which is the cosine of 5 degrees 45 minutes, nearly.

**Helix.**—A helix is a curve generated by a point moving about a cylindrical surface (real or imaginary) at a constant rate in the direction of the cylinder's axis. The curvature of a screw thread is one common example of a helical curve.

*Lead of Helix:* The lead of a helix is the distance that it advances in an axial direction, in one complete turn about the cylindrical surface. To illustrate, the lead of a screw thread equals the distance that a thread advances in one turn; it also equals the distance that a nut would advance in one turn.

*Development of Helix:* If one turn of a helical curve were unrolled onto a plane surface (as shown by diagram), the helix would become a straight line forming the hypotenuse of a right-angle triangle. The length of one side of this triangle would equal the circumference of the cylinder with which the helix coincides, and the length of the other side of the triangle would equal the lead of the helix.

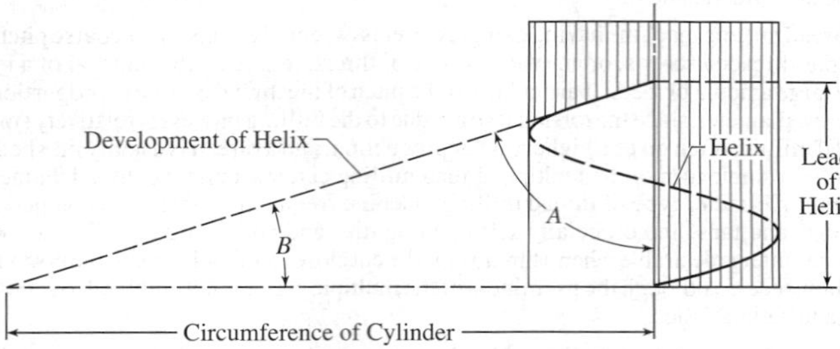


**Helix Angles.**—The triangular development of a helix has one angle *A* subtended by the circumference of the cylinder, and another angle *B* subtended by the lead of the helix. The term "helix angle" applies to angle *A*. For example, the helix angle of a helical gear, according to the general usage of the term, is always angle *A*, because this is the angle used in helical gear-designing formulas. Helix angle *A* would also be applied in milling the helical teeth of cutters, reamers, etc. Angle *A* of a gear or cutter tooth is a measure of its inclination relative to the axis of the gear or cutter.

*Lead Angle:* Angle *B* is applied to screw threads and worm threads and is referred to as the lead angle of the screw thread or worm. This angle *B* is a measure of the inclination of a screw thread from a plane that is perpendicular to the screw thread axis. Angle *B* is called the "lead angle" because it is subtended by the lead of the thread, and to distinguish it from the term "helix angle" as applied to helical gears.

*Finding Helix Angle of Helical Gear:* A helical gear tooth has an infinite number of helix angles, but the angle at the pitch diameter or mid-working depth is the one required in gear designing and gear cutting. This angle *A*, relative to the axis of the gear, is found as follows:

$$\tan \text{helix angle} = \frac{3.1416 \times \text{pitch diameter of gear}}{\text{Lead of gear tooth}}$$

*Finding Lead Angle of Screw Thread:* The lead or helix angle at the pitch diameter of a screw thread usually is required when, for example, a thread milling cutter must be aligned with the thread. This angle measured from a plane perpendicular to the screw thread axis is found as follows:

$$\tan \text{lead angle} = \frac{\text{Lead of screw thread}}{3.1416 \times \text{pitch diameter of screw thread}}$$

## TABLE OF CONTENTS

## GEARS, SPLINES, AND CAMS

### GEARS AND GEARING



### GEARS AND GEARING

*(Continued)*



### HYPOID AND BEVEL GEARING

## TABLE OF CONTENTS

## GEARS, SPLINES, AND CAMS

### HYPOID AND BEVEL GEARING

*(Continued)*



### WORM GEARING



### HELICAL GEARING



### OTHER GEAR TYPES



### OTHER GEAR TYPES

*(Continued)*



### CHECKING GEAR SIZES



### GEAR MATERIALS

TABLE OF CONTENTS

# GEARS, SPLINES, AND CAMS

## SPLINES AND SERRATIONS



## SPLINES AND SERRATIONS

*(Continued)*



## CAMS AND CAM DESIGN

## GEARS AND GEARING

*External spur gears* are cylindrical gears with straight teeth cut parallel to the axes. Gears transmit drive between parallel shafts. Tooth loads produce no axial thrust. Excellent at moderate speeds but tend to be noisy at high speeds. Shafts rotate in opposite directions.

*Internal spur gears* provide compact drive arrangements for transmitting motion between parallel shafts rotating in the same direction.

*Helical gears* are cylindrical gears with teeth cut at an angle to the axes. Provide drive between shafts rotating in opposite directions, with superior load-carrying capacity and quietness than spur gears. Tooth loads produce axial thrust.

*Crossed helical gears* are helical gears that mesh together on non-parallel axes.

*Straight bevel gears* have teeth that are radial toward the apex and are of conical form. Designed to operate on intersecting axes, bevel gears are used to connect two shafts on intersecting axes. The angle between the shafts equals the angle between the two axes of the meshing teeth. End thrust developed under load tends to separate the gears.

*Spiral bevel gears* have curved oblique teeth that contact each other smoothly and gradually from one end of a tooth to the other. Meshing is similar to that of straight bevel gears but is smoother and quieter in use. Left-hand spiral teeth incline away from the axis in an anti-clockwise direction looking on small end of pinion or face of gear; right-hand teeth incline away from axis in clockwise direction. The hand of spiral of the pinion is always opposite to that of the gear and is used to identify the hand of the gear pair. Used to connect two shafts on intersecting axes as with straight bevel gears. The spiral angle does not affect the smoothness and quietness of operation or the efficiency but does affect the direction of the thrust loads created. A left-hand spiral pinion driving clockwise when viewed from the large end of the pinion creates an axial thrust that tends to move the pinion out of mesh.

*Zerol bevel gears* have curved teeth lying in the same general direction as straight bevel teeth but should be considered to be spiral bevel gears with zero spiral angle.

*Hypoid bevel gears* are a cross between spiral bevel gears and worm gears. The axes of hypoid bevel gears are non-intersecting and non-parallel. The distance between the axes is called the offset. The offset permits higher ratios of reduction than is practicable with other bevel gears. Hypoid bevel gears have curved oblique teeth on which contact begins gradually and continues smoothly from one end of the tooth to the other.

*Worm gears* are used to transmit motion between shafts at right angles, that do not lie in a common plane and sometimes to connect shafts at other angles. Worm gears have line tooth contact and are used for power transmission, but the higher the ratio the lower the efficiency.

**Definitions of Gear Terms.—**The following terms are commonly applied to the various classes of gears:

*Active face width* is the dimension of the tooth face width that makes contact with a mating gear.

*Addendum* is the radial or perpendicular distance between the pitch circle and the top of the tooth.

*Arc of action* is the arc of the pitch circle through which a tooth travels from the first point of contact with the mating tooth to the point where contact ceases.

*Arc of approach* is the arc of the pitch circle through which a tooth travels from the first point of contact with the mating tooth to the pitch point.

*Arc of recession* is the arc of the pitch circle through which a tooth travels from its contact with a mating tooth at the pitch point until contact ceases.

*Axial pitch* is the distance parallel to the axis between corresponding sides of adjacent teeth.

*Axial plane* is the plane that contains the two axes in a pair of gears. In a single gear the axial plane is any plane containing the axis and any given point.

*Axial thickness* is the distance parallel to the axis between two pitch line elements of the same tooth.
*Backlash* is the shortest distance between the non-driving surfaces of adjacent teeth when the working flanks are in contact.
*Base circle* is the circle from which the involute tooth curve is generated or developed.
*Base helix angle* is the angle at the base cylinder of an involute gear that the tooth makes with the gear axis.
*Base pitch* is the circular pitch taken on the circumference of the base circles, or the distance along the line of action between two successive and corresponding involute tooth profiles. The *normal base pitch* is the base pitch in the normal plane and the *axial base pitch* is the base pitch in the axial plane.
*Base tooth thickness* is the distance on the base circle in the plane of rotation between involutes of the same pitch.
*Bottom land* is the surface of the gear between the flanks of adjacent teeth.
*Center distance* is the shortest distance between the non-intersecting axes of mating gears, or between the parallel axes of spur gears and parallel helical gears, or the crossed axes of crossed helical gears or worm gears.
*Central plane* is the plane perpendicular to the gear axis in a worm gear, which contains the common perpendicular of the gear and the worm axes. In the usual arrangement with the axes at right angles, it contains the worm axis.
*Chordal addendum* is the radial distance from the circular thickness chord to the top of the tooth, or the height from the top of the tooth to the chord subtending the circular thickness arc.
*Chordal thickness* is the length of the chord subtended by the circular thickness arc. The dimension obtained when a gear tooth caliper is used to measure the tooth thickness at the pitch circle.
*Circular pitch* is the distance on the circumference of the pitch circle in the plane of rotation, between corresponding points of adjacent teeth. The length of the arc of the pitch circle between the centers or other corresponding points of adjacent teeth.
*Circular thickness* is the thickness of the tooth on the pitch circle in the plane of rotation, or the length of arc between the two sides of a gear tooth measured on the pitch circle.
*Clearance* is the radial distance between the top of a tooth and the bottom of a mating tooth space, or the amount by which the dedendum in a given gear exceeds the addendum of its mating gear.
*Contact diameter* is the smallest diameter on a gear tooth with which the mating gear makes contact.
*Contact ratio* is the ratio of the arc of action in the plane of rotation to the circular pitch, and is sometimes thought of as the average number of teeth in contact. This ratio is obtained most directly as the ratio of the length of action to the base pitch.
*Contact ratio - face* is the ratio of the face advance to the circular pitch in helical gears.
*Contact ratio - total* is the ratio of the sum of the arc of action and the face advance to the circular pitch.
*Contact stress* is the maximum compressive stress within the contact area between mating gear tooth profiles. Also called the Hertz stress.
*Cycloid* is the curve formed by the path of a point on a circle as it rolls along a straight line. When such a circle rolls along the outside of another circle the curve is called an *epicycloid*, and when it rolls along the inside of another circle it is called a *hypocycloid*. These curves are used in defining the former American Standard composite Tooth Form.
*Dedendum* is the radial or perpendicular distance between the pitch circle and the bottom of the tooth space.
*Diametral pitch* is the ratio of the number of teeth to the number of inches in the pitch diameter in the plane of rotation, or the number of gear teeth to each inch of pitch diameter. Normal diametral pitch is the diametral pitch as calculated in the normal plane, or the diametral pitch divided by the cosine of the helix angle.

*Efficiency* is the torque ratio of a gear set divided by its gear ratio.

*Equivalent pitch radius* is the radius of curvature of the pitch surface at the pitch point in a plane normal to the pitch line element.

*Face advance* is the distance on the pitch circle that a gear tooth travels from the time pitch point contact is made at one end of the tooth until pitch point contact is made at the other end.

*Fillet radius* is the radius of the concave portion of the tooth profile where it joins the bottom of the tooth space.

*Fillet stress* is the maximum tensile stress in the gear tooth fillet.

*Flank of tooth* is the surface between the pitch circle and the bottom land, including the gear tooth fillet.

*Gear ratio* is the ratio between the numbers of teeth in mating gears.

*Helical overlap* is the effective face width of a helical gear divided by the gear axial pitch.

*Helix angle* is the angle that a helical gear tooth makes with the gear axis at the pitch circle, unless specified otherwise.

*Hertz stress,* see *Contact stress.*

*Highest point of single tooth contact (HPSTC)* is the largest diameter on a spur gear at which a single tooth is in contact with the mating gear.

*Interference* is the contact between mating teeth at some point other than along the line of action.

*Internal diameter* is the diameter of a circle that coincides with the tops of the teeth of an internal gear.

*Internal gear* is a gear with teeth on the inner cylindrical surface.

*Involute* is the curve generally used as the profile of gear teeth. The curve is the path of a point on a straight line as it rolls along a convex base curve, usually a circle.

*Land* The top land is the top surface of a gear tooth and the *bottom land* is the surface of the gear between the fillets of adjacent teeth.

*Lead* is the axial advance of the helix in one complete turn, or the distance along its own axis on one revolution if the gear were free to move axially.

*Length of action* is the distance on an involute line of action through which the point of contact moves during the action of the tooth profile.

*Line of action* is the portion of the common tangent to the base cylinders along which contact between mating involute teeth occurs.

*Lowest point of single tooth contact (LPSTC)* is the smallest diameter on a spur gear at which a single tooth is in contact with its mating gear. Gear set contact stress is determined with a load placed on the pinion at this point.

*Module* is the ratio of the pitch diameter to the number of teeth, normally the ratio of pitch diameter in mm to the number of teeth. Module in the inch system is the ratio of the pitch diameter in inches to the number of teeth.

*Normal plane* is a plane normal to the tooth surfaces at a point of contact and perpendicular to the pitch plane.

*Number of teeth* is the number of teeth contained in a gear.

*Outside diameter* is the diameter of the circle that contains the tops of the teeth of external gears.

*Pitch* is the distance between similar, equally-spaced tooth surfaces in a given direction along a given curve or line.

*Pitch circle* is the circle through the pitch point having its center at the gear axis.

*Pitch diameter* is the diameter of the pitch circle. The operating pitch diameter is the pitch diameter at which the gear operates.

*Pitch plane* is the plane parallel to the axial plane and tangent to the pitch surfaces in any pair of gears. In a single gear, the pitch plane may be any plane tangent to the pitch surfaces.

*Pitch point* is the intersection between the axes of the line of centers and the line of action.

*Plane of rotation* is any plane perpendicular to a gear axis.

*Pressure angle* is the angle between a tooth profile and a radial line at its pitch point. In involute teeth, the pressure angle is often described as the angle between the line of action and the line tangent to the pitch circle. *Standard pressure angles* are established in connection with standard tooth proportions. A given pair of involute profiles will transmit smooth motion at the same velocity ratio when the center distance is changed. Changes in center distance in gear design and gear manufacturing operations may cause changes in pitch diameter, pitch and pressure angle in the same gears under different conditions. Unless otherwise specified, the pressure angle is the *standard pressure angle at the standard pitch diameter.* The *operating pressure angle* is determined by the center distance at which a pair of gears operates. In oblique teeth such as helical and spiral designs, the pressure angle is specified in the transverse, normal or axial planes.

*Principle reference planes* are pitch plane, axial plane and transverse plane, all intersecting at a point and mutually perpendicular.

*Rack:* A rack is a gear with teeth spaced along a straight line, suitable for straight line motion. A basic rack is a rack that is adopted as the basis of a system of interchangeable gears. Standard gear tooth dimensions are often illustrated on an outline of a basic rack.

*Roll angle* is the angle subtended at the center of a base circle from the origin of an involute to the point of tangency of a point on a straight line from any point on the same involute. The radian measure of this angle is the tangent of the pressure angle of the point on the involute.

*Root diameter* is the diameter of the circle that contains the roots or bottoms of the tooth spaces.

*Tangent plane* is a plane tangent to the tooth surfaces at a point or line of contact.

*Tip relief* is an arbitrary modification of a tooth profile where a small amount of material is removed from the involute face of the tooth surface near the tip of the gear tooth.

*Tooth face* is the surface between the pitch line element and the tooth tip.

*Tooth surface* is the total tooth area including the flank of the tooth and the tooth face.

*Total face width* is the dimensional width of a gear blank and may exceed the effective face width as with a double-helical gear where the total face width includes any distance separating the right-hand and left-hand helical gear teeth.

*Transverse plane* is a plane that is perpendicular to the axial plane and to the pitch plane. In gears with parallel axes, the transverse plane and the plane of rotation coincide.

*Trochoid* is the curve formed by the path of a point on the extension of a radius of a circle as it rolls along a curve or line. A trochoid is also the curve formed by the path of a point on a perpendicular to a straight line as the straight line rolls along the convex side of a base curve. By the first definition, a trochoid is derived from the *cycloid*; by the second definition it is derived from the *involute*.

*True involute form diameter* is the smallest diameter on the tooth at which the point of tangency of the involute tooth profile exists. Usually this position is the point of tangency of the involute tooth profile and the fillet curve, and is often referred to as the TIF diameter.

*Undercut* is a condition in generated gear teeth when any part of the fillet curve lies inside a line drawn at a tangent to the working profile at its lowest point. Undercut may be introduced deliberately to facilitate shaving operations, as in preshaving.

*Whole depth* is the total depth of a tooth space, equal to the addendum plus the dedendum and equal to the working depth plus clearance.

*Working depth* is the depth of engagement of two gears, or the sum of their addendums. The standard working distance is the depth to which a tooth extends into the tooth space of a mating gear when the center distance is standard.

Definitions of gear terms are given in AGMA Standards 112.05, 115.01, and 116.01 entitled "Terms, Definitions, Symbols and Abbreviations," "Reference Information—Basic Gear Geometry," and "Glossary—Terms Used in Gearing," respectively; obtainable from American Gear Manufacturers Association, 500 Montgomery St., Alexandria, VA 22314.

## Comparative Sizes and Shape of Gear Teeth

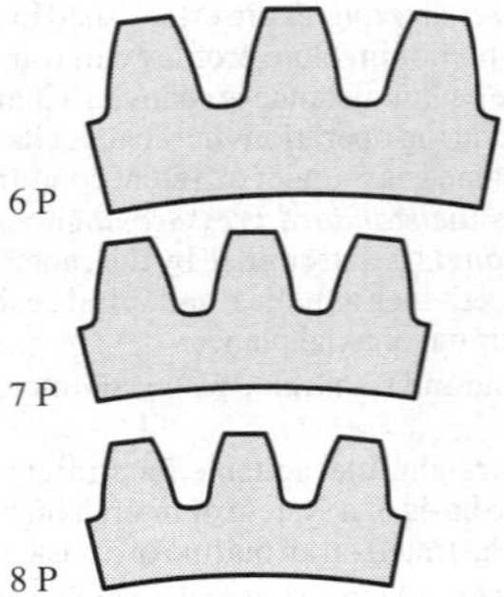


Gear Teeth of Different Diametral Pitch

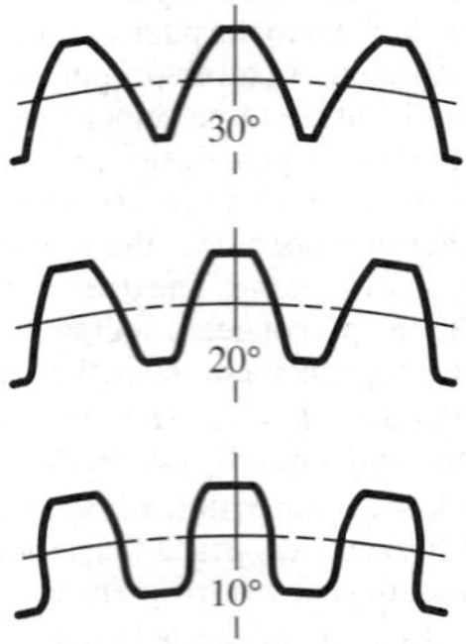


Shapes of Gear Teeth of Different Pressure Angles

## Nomenclature of Gear Teeth

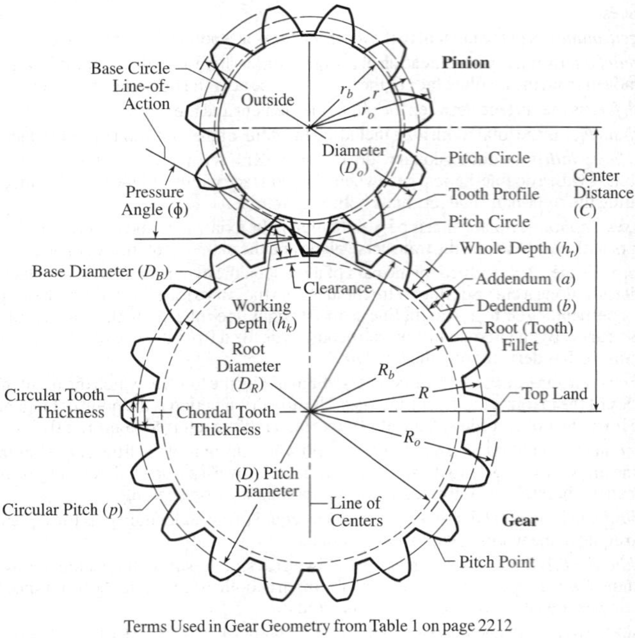


Terms Used in Gear Geometry from Table 1 on page 2212

**Properties of the Involute Curve.—**The involute curve is used almost exclusively for gear-tooth profiles, because of the following important properties.

1) The form or shape of an involute curve depends upon the diameter of the base circle from which it is derived. (If a taut line were unwound from the circumference of a circle—the *base circle* of the involute—the end of that line or any point on the unwound portion would describe an involute curve.)

2) If a gear tooth of involute curvature acts against the involute tooth of a mating gear while rotating at a uniform rate, the angular motion of the driven gear will also be uniform, even though the center-to-center distance is varied.

3) The relative rate of motion between driving and driven gears having involute tooth curves is established by the diameters of their base circles.

4) Contact between intermeshing involute teeth on a driving and driven gear is along a straight line that is tangent to the two base circles of these gears. This is the *line of action*.

5) The point where the line of action intersects the common centerline of the mating involute gears establishes the radii of the pitch circles of these gears; hence true pitch circle diameters are affected by a change in the center distance. (Pitch diameters obtained by dividing the number of teeth by the diametral pitch apply when the center distance equals the total number of teeth on both gears divided by twice the diametral pitch.)

6) The pitch diameters of mating involute gears are directly proportional to the diameters of their respective base circles; thus, if the base circle of one mating gear is three times as large as the other, the pitch circle diameters will be in the same ratio.

7) The angle between the line of action and a line perpendicular to the common centerline of mating gears is the *pressure angle*; hence the pressure angle is affected by any change in the center distance.

8) When an involute curve acts against a straight line (as in the case of an involute pinion acting against straight-sided rack teeth), the straight line is tangent to the involute and perpendicular to its line of action.

9) The pressure angle, in the case of an involute pinion acting against straight-sided rack teeth, is the angle between the line of action and the line of the rack's motion. If the involute pinion rotates at a uniform rate, movement of the rack will also be uniform.

*Nomenclature:*

$\phi$ = Pressure Angle

$a$ = Addendum   $a_G$ = Addendum of Gear   $a_P$ = Addendum of Pinion

$b$ = Dedendum

$c$ = Clearance

$C$ = Center Distance

$D$ = Pitch Diameter   $D_G$ = Pitch Diameter of Gear   $D_P$ = Pitch Diameter of Pinion

$D_B$ = Base Circle Diameter   $D_O$ = Outside Diameter   $D_R$ = Root Diameter

$F$ = Face Width

$h_k$ = Working Depth of Tooth   $h_t$ = Whole Depth of Tooth

$m_G$ = Gear Ratio

$N$ = Number of Teeth   $N_G$ = Number of Teeth in Gear   $N_P$ = Number of Teeth in Pinion

$p$ = Circular Pitch   $P$ = Diametral Pitch

**Diametral and Circular Pitch Systems.—**Gear tooth system standards are established by specifying the tooth proportions of the basic rack. The diametral pitch system is applied to most of the gearing produced in the United States. If gear teeth are larger than about one diametral pitch, it is common practice to use the circular pitch system. The circular pitch system is also applied to cast gearing and it is commonly used in connection with the design and manufacture of worm gearing.

**Pitch Diameters Obtained with Diametral Pitch System.—**The diametral pitch system is arranged to provide a series of standard tooth sizes, the principle being similar to the standardization of screw thread pitches. Inasmuch as there must be a whole number of teeth on each gear, the increase in pitch diameter per tooth varies according to the pitch. For example, the pitch diameter of a gear having, say, 20 teeth of 4 diametral pitch will be 5 inches; 21 teeth, 5¼ inches; and so on, the increase in diameter for each additional tooth being equal to ¼ inch for 4 diametral pitch. Similarly, for 2 diametral pitch the variations for successive numbers of teeth would equal ½ inch, and for 10 diametral pitch the variations would equal $1/10$ inch, etc. Where a given center distance must be maintained and no standard diametral pitch can be used, gears should be designed with reference to the gear set center distance procedure discussed in *Gears for Given Center Distance and Ratio* starting on page 2220.

**Table 1. Formulas for Dimensions of Standard Spur Gears**

| To Find | Formula | To Find | Formula |
|---|---|---|---|
| Base Circle Diameter | $D_B = D\cos\phi$ (1) | Number of Teeth | $N = P \times D$ (2a)<br>$N = \frac{3.1416D}{p}$ (3b) |
| Circular Pitch | $p = \frac{3.1416D}{N}$ (3a)<br>$p = \frac{3.1416}{P}$ (3c) | Outside Diameter (Full-depth Teeth) | $D_O = \frac{N+2}{P}$ (4a)<br>$D_O = \frac{(N+2)p}{3.1416}$ (5b) |
| Center Distance | $C = \frac{N_P(m_G+1)}{2P}$ (5a)<br>$C = \frac{D_P + D_G}{2}$ (5c)<br>$C = \frac{N_G + N_P}{2P}$ (6b)<br>$C = \frac{(N_G + N_P)p}{6.2832}$ (6d) | Outside Diameter (American Standard Stub Teeth) | $D_O = \frac{N+1.6}{P}$ (6a)<br>$D_O = \frac{(N+1.6)p}{3.1416}$ (6c) |
| | | Outside Diameter | $D_O = D + 2a$ (7) |
| Diametral Pitch | $P = \frac{3.1416}{p}$ (8a)<br>$P = \frac{N}{D}$ (9b)<br>$P = \frac{N_P(m_G+1)}{2C}$ (9d) | Pitch Diameter | $D = \frac{N}{P}$ (9a)<br>$D = \frac{Np}{3.1416}$ (9c) |
| | | Root Diameter[a] | $D_R = D - 2b$ (10) |
| Gear Ratio | $m_G = \frac{N_G}{N_P}$ (11) | Whole Depth | $a + b$ (12) |
| | | Working Depth | $a_G + a_P$ (13) |

[a] See also formulas in Table 2 and Table 4 on pages 2212 and 2216.

**Table 2. Formulas for Tooth Parts, 20-and 25-Degree Involute Full-Depth Teeth ANSI Coarse Pitch Spur Gear Tooth Forms** *ANSI B6.1-1968 (R1974)*

| To Find | Diametral Pitch, $P$, Known | Circular Pitch, $p$, Known |
|---|---|---|
| Addendum | $a = 1.000 \div P$ | $a = 0.3183 \times p$ |
| Dedendum (Preferred) | $b = 1.250 \div P$ | $b = 0.3979 \times p$ |
| (Shaved or Ground Teeth)[a] | $b = 1.350 \div P$ | $b = 0.4297 \times p$ |
| Working Depth | $h_k = 2.000 \div P$ | $h_k = 0.6366 \times p$ |
| Whole Depth (Preferred) | $h_t = 2.250 \div P$ | $h_t = 0.7162 \times p$ |
| (Shaved or Ground Teeth) | $h_t = 2.350 \div P$ | $h_t = 0.7480 \times p$ |
| Clearance (Preferred)[b] | $c = 0.250 \div P$ | $c = 0.0796 \times p$ |
| (Shaved or Ground Teeth) | $c = 0.350 \div P$ | $c = 0.1114 \times p$ |

**Table 2.** *(Continued)* **Formulas for Tooth Parts, 20-and 25-Degree Involute Full-Depth Teeth ANSI Coarse Pitch Spur Gear Tooth Forms** *ANSI B6.1-1968 (R1974)*

| To Find | Diametral Pitch, $P$, Known | Circular Pitch, $p$, Known |
|---|---|---|
| Fillet Radius (Rack)[c] | $r_f = 0.300 \div P$ | $r_f = 0.0955 \times p$ |
| Pitch Diameter | $D = N \div P$ | $D = 0.3183 \times N \times p$ |
| Outside Diameter | $D_O = (N+2) \div P$ | $D_O = 0.3183 \times (N+2)p$ |
| Root Diameter (Preferred) | $D_R = (N-2.5) \div P$ | $D_R = 0.3183 \times (N-2.5)p$ |
| (Shaved or Ground Teeth) | $D_R = (N-2.7) \div P$ | $D_R = 0.3183 \times (N-2.7)p$ |
| Circular Thickness—Basic | $t = 1.5708 \div P$ | $t = p \div 2$ |

[a] When gears are preshave cut on a gear shaper the dedendum will usually need to be increased to 1.40/*P* to allow for the higher fillet trochoid produced by the shaper cutter. This is of particular importance on gears of few teeth or if the gear blank configuration requires the use of a small diameter shaper cutter, in which case the dedendum may need to be increased to as much as 1.45/*P*. This should be avoided on highly loaded gears where the consequently reduced *J* factor will increase gear tooth stress excessively.

[b] A minimum clearance of 0.157/*P* may be used for the basic 20-degree and 25-degree pressure angle rack in the case of shallow root sections and use of existing hobs or cutters. However, whenever less than standard clearance is used, the location of the TIF diameter should be determined by the method shown in *True Involute Form Diameter* starting on page 2238. The TIF diameter must be less than the Contact Diameter determined by the method shown on page 2236.

[c] The fillet radius of the basic rack should not exceed 0.235/*P* for a 20-degree pressure angle rack or 0.270/*P* for a 25-degree pressure angle rack for a clearance of 0.157/*P*. The basic rack fillet radius must be *reduced* for teeth with a 25-degree pressure angle having a clearance in excess of 0.250/*P*.

**American National Standard Coarse Pitch Spur Gear Tooth Forms.—**The American National Standard (ANSI B6.1-1968, R1974) provides tooth proportion information on two involute spur gear forms. These two forms are identical except that one has a pressure angle of 20 degrees and a minimum allowable tooth number of 18 while the other has a pressure angle of 25 degrees and a minimum allowable tooth number of 12. (For pinions with fewer teeth, see tooth proportions for long addendum pinions and their mating short addendum gears in Table 7 through Table 9d starting on page 2227.) A gear tooth standard is established by specifying the tooth proportions of the basic rack. Gears made to this standard will thus be conjugate with the specified rack and with each other. The basic rack forms for the 20-degree and 25-degree standard are shown on the following page; basic formulas for these proportions as a function of the gear diametral pitch and also of the circular pitch are given in Table 2. Tooth parts data are given in Table 3.

In recent years, the established standard of almost universal use is the ANSI 20-degree standard spur gear form. It provides a gear with good strength and without fillet undercut in pinions of as few as eighteen teeth. Some more recent applications have required a tooth form of even greater strength and fewer teeth than eighteen. This requirement has stimulated the establishment of the ANSI 25-degree standard. This 25-degree form will give greater tooth strength than the 20-degree standard, will provide pinions of as few as twelve teeth without fillet undercut and will provide a lower contact compressive stress for greater gear set surface durability.

**American National Standard and Former American Standard Gear Tooth Forms**
*ANSI B6.1-1968, (R1974)* **and** *ASA B6.1-1932*

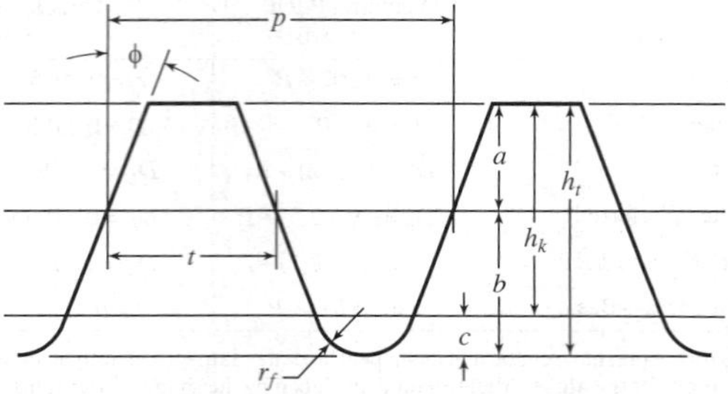


$a$ = addendum $h_k$ = working depth $r_f$ = fillet radius of basic rack
$b$ = dedendum $h_t$ = whole depth $t$ = circular tooth thickness — basic
$c$ = clearance $p$ = circular pitch $\phi$ = pressure angle

Basic Rack of the 20-Degree and 25-Degree Full-Depth Involute Systems

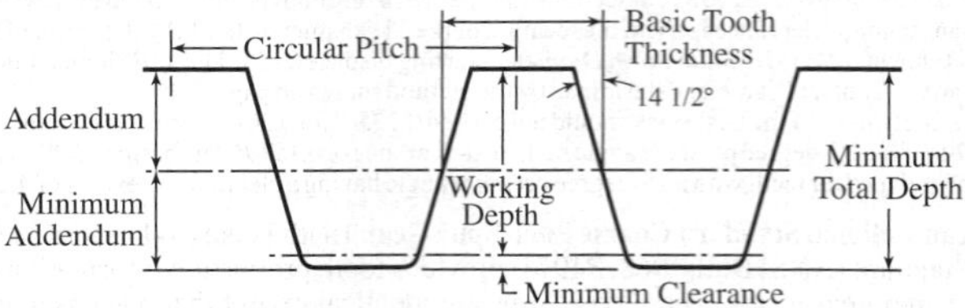


Basic Rack of the 14½-Degree Full-Depth Involute System

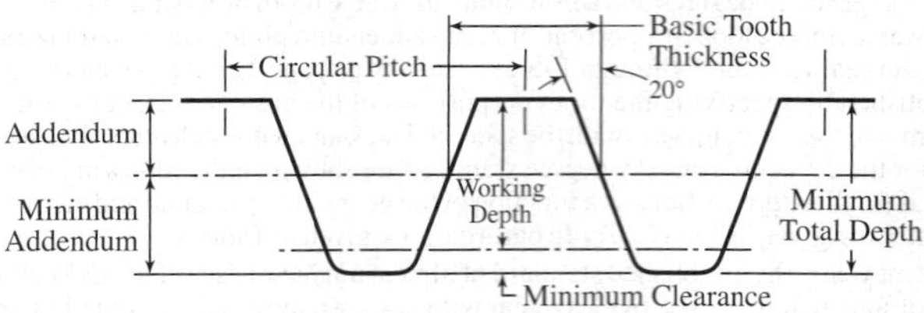


Basic Rack of the 20-Degree Stub Involute System

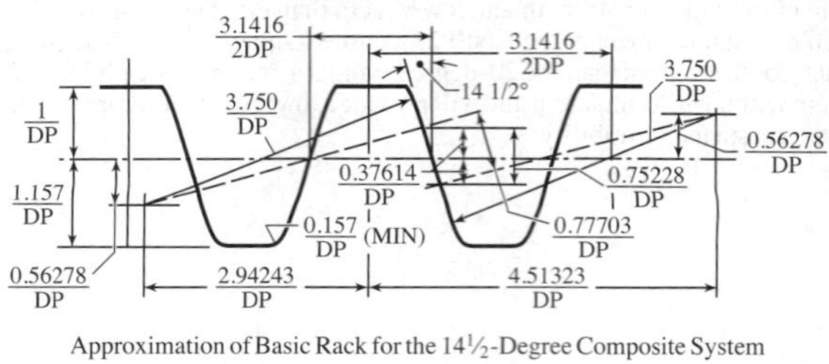


Approximation of Basic Rack for the 14½-Degree Composite System

## Table 3. Gear Tooth Parts for American National Standard Coarse Pitch 20- and 25-Degree Pressure Angle Gears

| Dia. Pitch | Circ. Pitch | Stand. Addend.[a] | Stand. Dedend. | Spec. Dedend.[b] | Min. Dedend. | Stand. F. Rad. | Min. F. Rad. |
|---|---|---|---|---|---|---|---|
| $P$ | $p$ | $a$ | $b$ | $b$ | $b$ | $r_f$ | $r_f$ |
| 0.3142 | 10. | 3.1831 | 3.9789 | 4.2972 | 3.6828 | 0.9549 | 0.4997 |
| 0.3307 | 9.5 | 3.0239 | 3.7799 | 4.0823 | 3.4987 | 0.9072 | 0.4748 |
| 0.3491 | 9. | 2.8648 | 3.5810 | 3.8675 | 3.3146 | 0.8594 | 0.4498 |
| 0.3696 | 8.5 | 2.7056 | 3.3820 | 3.6526 | 3.1304 | 0.8117 | 0.4248 |
| 0.3927 | 8. | 2.5465 | 3.1831 | 3.4377 | 2.9463 | 0.7639 | 0.3998 |
| 0.4189 | 7.5 | 2.3873 | 2.9842 | 3.2229 | 2.7621 | 0.7162 | 0.3748 |
| 0.4488 | 7. | 2.2282 | 2.7852 | 3.0080 | 2.5780 | 0.6685 | 0.3498 |
| 0.4833 | 6.5 | 2.0690 | 2.5863 | 2.7932 | 2.3938 | 0.6207 | 0.3248 |
| 0.5236 | 6. | 1.9099 | 2.3873 | 2.5783 | 2.2097 | 0.5730 | 0.2998 |
| 0.5712 | 5.5 | 1.7507 | 2.1884 | 2.3635 | 2.0256 | 0.5252 | 0.2749 |
| 0.6283 | 5. | 1.5915 | 1.9894 | 2.1486 | 1.8414 | 0.4775 | 0.2499 |
| 0.6981 | 4.5 | 1.4324 | 1.7905 | 1.9337 | 1.6573 | 0.4297 | 0.2249 |
| 0.7854 | 4. | 1.2732 | 1.5915 | 1.7189 | 1.4731 | 0.3820 | 0.1999 |
| 0.8976 | 3.5 | 1.1141 | 1.3926 | 1.5040 | 1.2890 | 0.3342 | 0.1749 |
| 1. | 3.1416 | 1.0000 | 1.2500 | 1.3500 | 1.1570 | 0.3000 | 0.1570 |
| 1.25 | 2.5133 | 0.8000 | 1.0000 | 1.0800 | 0.9256 | 0.2400 | 0.1256 |
| 1.5 | 2.0944 | 0.6667 | 0.8333 | 0.9000 | 0.7713 | 0.2000 | 0.1047 |
| 1.75 | 1.7952 | 0.5714 | 0.7143 | 0.7714 | 0.6611 | 0.1714 | 0.0897 |
| 2. | 1.5708 | 0.5000 | 0.6250 | 0.6750 | 0.5785 | 0.1500 | 0.0785 |
| 2.25 | 1.3963 | 0.4444 | 0.5556 | 0.6000 | 0.5142 | 0.1333 | 0.0698 |
| 2.5 | 1.2566 | 0.4000 | 0.5000 | 0.5400 | 0.4628 | 0.1200 | 0.0628 |
| 2.75 | 1.1424 | 0.3636 | 0.4545 | 0.4909 | 0.4207 | 0.1091 | 0.0571 |
| 3. | 1.0472 | 0.3333 | 0.4167 | 0.4500 | 0.3857 | 0.1000 | 0.0523 |
| 3.25 | 0.9666 | 0.3077 | 0.3846 | 0.4154 | 0.3560 | 0.0923 | 0.0483 |
| 3.5 | 0.8976 | 0.2857 | 0.3571 | 0.3857 | 0.3306 | 0.0857 | 0.0449 |
| 3.75 | 0.8378 | 0.2667 | 0.3333 | 0.3600 | 0.3085 | 0.0800 | 0.0419 |
| 4. | 0.7854 | 0.2500 | 0.3125 | 0.3375 | 0.2893 | 0.0750 | 0.0392 |
| 4.5 | 0.6981 | 0.2222 | 0.2778 | 0.3000 | 0.2571 | 0.0667 | 0.0349 |
| 5. | 0.6283 | 0.2000 | 0.2500 | 0.2700 | 0.2314 | 0.0600 | 0.0314 |
| 5.5 | 0.5712 | 0.1818 | 0.2273 | 0.2455 | 0.2104 | 0.0545 | 0.0285 |
| 6. | 0.5236 | 0.1667 | 0.2083 | 0.2250 | 0.1928 | 0.0500 | 0.0262 |
| 6.5 | 0.4833 | 0.1538 | 0.1923 | 0.2077 | 0.1780 | 0.0462 | 0.0242 |
| 7. | 0.4488 | 0.1429 | 0.1786 | 0.1929 | 0.1653 | 0.0429 | 0.0224 |
| 7.5 | 0.4189 | 0.1333 | 0.1667 | 0.1800 | 0.1543 | 0.0400 | 0.0209 |
| 8. | 0.3927 | 0.1250 | 0.1563 | 0.1687 | 0.1446 | 0.0375 | 0.0196 |
| 8.5 | 0.3696 | 0.1176 | 0.1471 | 0.1588 | 0.1361 | 0.0353 | 0.0185 |
| 9. | 0.3491 | 0.1111 | 0.1389 | 0.1500 | 0.1286 | 0.0333 | 0.0174 |
| 9.5 | 0.3307 | 0.1053 | 0.1316 | 0.1421 | 0.1218 | 0.0316 | 0.0165 |
| 10. | 0.3142 | 0.1000 | 0.1250 | 0.1350 | 0.1157 | 0.0300 | 0.0157 |
| 11. | 0.2856 | 0.0909 | 0.1136 | 0.1227 | 0.1052 | 0.0273 | 0.0143 |
| 12. | 0.2618 | 0.0833 | 0.1042 | 0.1125 | 0.0964 | 0.0250 | 0.0131 |
| 13. | 0.2417 | 0.0769 | 0.0962 | 0.1038 | 0.0890 | 0.0231 | 0.0121 |
| 14. | 0.2244 | 0.0714 | 0.0893 | 0.0964 | 0.0826 | 0.0214 | 0.0112 |
| 15. | 0.2094 | 0.0667 | 0.0833 | 0.0900 | 0.0771 | 0.0200 | 0.0105 |
| 16. | 0.1963 | 0.0625 | 0.0781 | 0.0844 | 0.0723 | 0.0188 | 0.0098 |
| 17. | 0.1848 | 0.0588 | 0.0735 | 0.0794 | 0.0681 | 0.0176 | 0.0092 |
| 18. | 0.1745 | 0.0556 | 0.0694 | 0.0750 | 0.0643 | 0.0167 | 0.0087 |
| 19. | 0.1653 | 0.0526 | 0.0658 | 0.0711 | 0.0609 | 0.0158 | 0.0083 |
| 20. | 0.1571 | 0.0500 | 0.0625 | 0.0675 | 0.0579 | 0.0150 | 0.0079 |

[a] When using equal addendums on pinion and gear the minimum number of teeth on the pinion is 18 and the minimum total number of teeth in the pair is 36 for 20-degree full depth involute tooth form and 12 and 24, respectively, for 25-degree full depth tooth form.

[b] The dedendum in this column is used when the gear tooth is shaved. It allows for the higher fillet cut by a protuberance hob.

The working depth is equal to twice the addendum.

The whole depth is equal to the addendum plus the dedendum.

**Table 4. Tooth Proportions for Fine-Pitch Involute Spur and Helical Gears of 14½-, 20-, and 25-Degree Pressure Angle** *ANSI B6.7-1977*

| Item | Spur | Helical |
|---|---|---|
| Addendum, $a$ | $\frac{1.000}{P}$ | $\frac{1.000}{P_n}$ |
| Dedendum, $b$ | $\frac{1.200}{P} + 0.002\,(\text{min.})$ | $\frac{1.200}{P_n} + 0.002\,(\text{min.})$ |
| Working Depth, $h_k$ | $\frac{2.000}{P}$ | $\frac{2.000}{P_n}$ |
| Whole Depth, $h_t$ | $\frac{2.200}{P} + 0.002\,(\text{min.})$ | $\frac{2.200}{P_n} + 0.002\,(\text{min.})$ |
| Clearance, $c$ (Standard) | $\frac{0.200}{P} + 0.002\,(\text{min.})$ | $\frac{0.200}{P_n} + 0.002\,(\text{min.})$ |
| (Shaved or Ground Teeth) | $\frac{0.350}{P} + 0.002\,(\text{min.})$ | $\frac{0.350}{P_n} + 0.002\,(\text{min.})$ |
| Tooth Thickness, $t$ At Pitch Diameter | $t = \frac{1.5708}{P}$ | $t_n = \frac{1.5708}{P_n}$ |
| Circular Pitch, $p$ | $p = \frac{\pi D}{N}$ or $\frac{\pi d}{n}$ or $\frac{\pi}{P}$ | $p_n = \frac{\pi}{P_n}$ |
| Pitch Diameter Pinion, $d$ | $\frac{n}{P}$ | $\frac{n}{P_n \cos\psi}$ |
| Gear, $D$ | $\frac{N}{P}$ | $\frac{N}{P_n \cos\psi}$ |
| Outside Diameter Pinion, $d_o$ | $\frac{n+2}{P}$ | $\frac{1}{P_n}\left(\frac{n}{\cos\psi} + 2\right)$ |
| Gear, $D_o$ | $\frac{N+2}{P}$ | $\frac{1}{P_n}\left(\frac{N}{\cos\psi} + 2\right)$ |
| Center Distance, $C$ | $\frac{N+n}{2P}$ | $\frac{N+n}{2P_n \cos\psi}$ |

All dimensions are in inches.
$P$ = Transverse Diametral Pitch
$P_n$ = Normal Diametral Pitch
$t_n$ = Normal Tooth Thickness at Pitch Diameter
$p_n$ = Normal Circular Pitch
$\psi$ = Helix Angle
$n$ = Number of pinion teeth
$N$ = Number of gear teeth

**American National Standard Tooth Proportions for Fine-Pitch Involute Spur and Helical Gears.—**The proportions of spur gears in this Standard (ANSI B6.7-1977) follow closely ANSI B6.1-1968, R1974, "Tooth Proportions for Coarse-Pitch Involute Spur Gears." The main difference between fine-pitch and coarse-pitch gears is the greater clearance specified for fine-pitch gears. The increased clearance provides for any foreign material that may tend to accumulate at the bottoms of the teeth and also the relatively larger fillet radius resulting from proportionately greater wear on the tips of fine-pitch cutting tools.

*Pressure Angle:* The standard pressure angle for fine-pitch gears is 20 degrees and is recommended for most applications. For helical gears this pressure angle applies in the *normal* plane. In certain cases, notably sintered or molded gears, or in gearing where greatest strength and wear resistance are desired, a 25-degree pressure angle may be required. However, pressure angles greater than 20 degrees tend to require use of generating tools having very narrow point widths, and higher pressure angles require closer control of center distance when backlash requirements are critical.

In those cases where consideration of angular position or backlash is critical and both pinion and gear contain relatively large numbers of teeth, a $14\frac{1}{2}$-degree pressure angle may be desirable. In general, pressure angles less than 20 degrees require greater amounts of tooth modification to avoid undercutting problems and are limited to larger total numbers of teeth in pinion and gear when operating at a standard center distance. Information Sheet B in the Standard provides tooth proportions for both $14\frac{1}{2}$- and 25-degree pressure angle fine-pitch gears. Table 4 provides tooth proportions for fine-pitch spur and helical gears with $14\frac{1}{2}$-, 20-, and 25-degree pressure angles, and Table 5 provides tooth parts.

*Diametral Pitches:* Diametral pitches preferred are: 20, 24, 32, 40, 48, 64, 72, 80, 96, and 120.

**Table 5. American National Standard Fine Pitch Standard Gear Tooth Parts— $14\frac{1}{2}$-, 20-, and 25-Degree Pressure Angles**

| Diametral Pitch | Circular Pitch | Circular Thickness | Standard Addend. | Standard Dedend. | Special Dedend.[a] |
|---|---|---|---|---|---|
| *P* | *p* | *t* | *a* | *b* | *b* |
| 20 | 0.1571 | 0.0785 | 0.0500 | 0.0620 | 0.0695 |
| 24 | 0.1309 | 0.0654 | 0.0417 | 0.0520 | 0.0582 |
| 32 | 0.0982 | 0.0491 | 0.0313 | 0.0395 | 0.0442 |
| 40 | 0.0785 | 0.0393 | 0.0250 | 0.0320 | 0.0358 |
| 48 | 0.0654 | 0.0327 | 0.0208 | 0.0270 | 0.0301 |
| 64 | 0.0491 | 0.0245 | 0.0156 | 0.0208 | 0.0231 |
| 72 | 0.0436 | 0.0218 | 0.0139 | 0.0187 | 0.0208 |
| 80 | 0.0393 | 0.0196 | 0.0125 | 0.0170 | 0.0189 |
| 96 | 0.0327 | 0.0164 | 0.0104 | 0.0145 | 0.0161 |
| 120 | 0.0262 | 0.0131 | 0.0083 | 0.0120 | 0.0132 |

[a] Based upon clearance for shaved or ground teeth.

The working depth is equal to twice the addendum. The whole depth is equal to the addendum plus the dedendum. For minimum number of teeth see page 2235.

**Other American Spur Gear Standards.—**An appended information sheet in the American National Standard ANSI B6.1-1968, R1974 provides tooth proportion information for three spur gear forms with the notice that they are "not recommended for new designs." These forms are therefore considered to be obsolescent but the information is given on their proportions because they have been used widely in the past. These forms are the $14\frac{1}{2}$-degree full depth form, the 20-degree stub involute form and the $14\frac{1}{2}$-degree composite form which were covered in the former American Standard (ASA B6.1-1932). The basic rack for the $14\frac{1}{2}$-degree full depth form is shown on page 2213; basic formulas for these proportions are given in Table 6.

**Table 6. Formulas for Tooth Parts—Former American Standard Spur Gear Tooth Forms** *ASA B6.1-1932*

| To Find | Diametral Pitch, $P$ Known | Circular Pitch, $p$ Known |
|---|---|---|
| $14\frac{1}{2}$-Degree Involute Full-depth Teeth | | |
| Addendum | $a = 1.000 \div P$ | $a = 0.3183 \times p$ |
| Minimum Dedendum | $b = 1.157 \div P$ | $b = 0.3683 \times p$ |
| Working Depth | $h_k = 2.000 \div P$ | $h_k = 0.6366 \times p$ |
| Minimum Whole Depth | $h_t = 2.157 \div P$ | $h_t = 0.6866 \times p$ |
| Basic Tooth Thickness on Pitch Line | $t = 1.5708 \div P$ | $t = 0.500 \times p$ |
| Minimum Clearance | $c = 0.157 \div P$ | $c = 0.050 \times p$ |
| 20-Degree Involute Stub Teeth | | |
| Addendum | $a = 0.800 \div P$ | $a = 0.2546 \times p$ |
| Minimum Dedendum | $b = 1.000 \div P$ | $b = 0.3183 \times p$ |
| Working Depth | $h_k = 1.600 \div P$ | $h_k = 0.5092 \times p$ |
| Minimum Whole Depth | $h_t = 1.800 \div P$ | $h_t = 0.5729 \times p$ |
| Basic Tooth Thickness on Pitch Line | $t = 1.5708 \div P$ | $t = 0.500 \times p$ |
| Minimum Clearance | $c = 0.200 \div P$ | $c = 0.0637 \times p$ |

*Note:* Radius of fillet equals $1\frac{1}{3} \times$ clearance for $14\frac{1}{2}$-degree full-depth teeth and $1\frac{1}{2} \times$ clearance for 20-degree full-depth teeth.

*Note:* A suitable working tolerance should be considered in connection with all minimum recommendations.

**Fellows Stub Tooth.—**The system of stub gear teeth introduced by the Fellows Gear Shaper Co. is based upon the use of two diametral pitches. One diametral pitch, say, 8, is used as the basis for obtaining the dimensions for the addendum and dedendum, while another diametral pitch, say, 6, is used for obtaining the dimensions of the thickness of the tooth, the number of teeth, and the pitch diameter. Teeth made according to this system are designated as $\frac{6}{8}$ pitch, $\frac{12}{14}$ pitch, etc., the numerator in this fraction indicating the pitch determining the thickness of the tooth and the number of teeth, and the denominator, the pitch determining the depth of the tooth. The clearance is made greater than in the ordinary gear-tooth system and equals 0.25 ÷ denominator of the diametral pitch. The pressure angle is 20 degrees.

This type of stub gear tooth is now used infrequently. Information on the tooth part dimensions may be found in the 18th and earlier editions of *Machinery's Handbook*.

**Basic Gear Dimensions.—**The basic dimensions for all involute spur gears may be obtained using the formulas shown in Table 1. This table is used in conjunction with Table 3 to obtain dimensions for coarse pitch gears and Table 5 to obtain dimensions for fine pitch standard spur gears. To obtain the dimensions of gears that are specified at a standard circular pitch, the equivalent diametral pitch is first calculated by using the formula in Table 1. If the required number of teeth in the pinion ($N_p$) is less than the minimum specified in either Table 3 or Table 5, whichever is applicable, the gears must be proportioned by the long and short addendum method shown on page 2229.

## Formulas for Outside and Root Diameters of Spur Gears that Are Finish-Hobbed, Shaped, or Preshaved

**Notation**

| | |
|---|---|
| $D$ = Pitch Diameter | $a$ = Standard Addendum |
| $D_O$ = Outside Diameter | $b$ = Standard Minimum Dedendum |
| $D_R$ = Root Diameter | $b_s$ = Standard Dedendum |
| $P$ = Diametral Pitch | $b_{ps}$ = Dedendum for Preshaving |

**14½-, 20-, And 25-degree Involute Full-depth Teeth (19$P$ and coarser)[a]**

$$D_O = D + 2a = \frac{N}{P} + \left(2 \times \frac{1}{P}\right)$$

$$D_R = D - 2b = \frac{N}{P} - \left(2 \times \frac{1.157}{P}\right) \qquad \text{(Hobbed)}^b$$

$$D_R = D - 2b_s = \frac{N}{P} - \left(2 \times \frac{1.25}{P}\right) \qquad \text{(Shaped)}^c$$

$$D_R = D - 2b_{ps} = \frac{N}{P} - \left(2 \times \frac{1.35}{P}\right) \qquad \text{(Preshaved)}^d$$

$$D_R = D - 2b_{ps} = \frac{N}{P} - \left(2 \times \frac{1.40}{P}\right) \qquad \text{(Preshaved)}^e$$

**20-degree Involute Fine-pitch Full-depth Teeth (20$P$ and finer)**

$$D_O = D + 2a = \frac{N}{P} + \left(2 \times \frac{1}{P}\right)$$

$$D_R = D - 2b = \frac{N}{P} - 2\left(\frac{1.2}{P} + 0.002\right) \qquad \text{(Hobbed or Shaped)}^f$$

$$D_R = D - 2b_{ps} = \frac{N}{P} - 2\left(\frac{1.35}{P} + 0.002\right) \qquad \text{(Preshaved)}^g$$

**20-degree Involute Stub Teeth[a]**

$$D_O = D + 2a = \frac{N}{P} + \left(2 \times \frac{0.8}{P}\right)$$

$$D_R = D - 2b = \frac{N}{P} - \left(2 \times \frac{1}{P}\right) \qquad \text{(Hobbed)}$$

$$D_R = D - 2b_{ps} = \frac{N}{P} - \left(2 \times \frac{1.35}{P}\right) \qquad \text{(Preshaved)}$$

[a] 14½-degree full-depth and 20-degree stub teeth are not recommended for new designs.

[b] According to ANSI B6.1-1968 a minimum clearance of 0.157/$P$ may be used for the basic 20-degree and 25-degree pressure angle rack in the case of shallow root sections and the use of existing hobs and cutters.

[c] According to ANSI B6.1-1968 the preferred clearance is 0.250/$P$.

[d] According to ANSI B6.1-1968 the clearance for teeth which are shaved or ground is 0.350/$P$.

[e] When gears are preshave cut on a gear shaper the dedendum will usually need to be increased to 1.40/$P$ to allow for the higher fillet trochoid produced by the shaper cutter; this is of particular importance on gears of few teeth or if the gear blank configuration requires the use of a small diameter shaper cutter, in which case the dedendum may need to be increased to as much as 1.45/$P$. This should be avoided on highly loaded gears where the consequently reduced $J$ factor will increase gear tooth stress excessively.

[f] According to ANSI B6.7-1967 the standard clearance is 0.200/$P$ + 0.002 (min.).

[g] According to ANSI B6.7-1967 the clearance for shaved or ground teeth is 0.35[illegible]/$P$ + 0.002 (min.).

**Gears for Given Center Distance and Ratio.—**When it is necessary to use a pair of gears of given ratio at a specified center distance $C_1$, it may be found that no gears of standard diametral pitch will satisfy the center distance requirement. Gears of standard diametral pitch $P$ may need to be redesigned to operate at other than their standard pitch diameter $D$ and standard pressure angle $\phi$. The diametral pitch $P_1$ at which these gears will operate is

$$P_1 = \frac{N_P + N_G}{2C_1} \qquad (1)$$

where $N_p$ = number of teeth in pinion

$N_G$ = number of teeth in gear

and their operating pressure angle $\phi_1$ is

$$\phi_1 = \arccos\left(\frac{P_1}{P}\cos\phi\right) \qquad (2)$$

Thus although the pair of gears are cut to a diametral pitch $P$ and a pressure angle $\phi$, they operate as standard gears of diametral pitch $P_1$ and pressure angle $\phi_1$. The pitch $P$ and pressure angle $\phi$ should be chosen so that $\phi_1$ lies between about 18 and 25 degrees.

The operating pitch diameters of the pinion $D_{p1}$ and of the gear $D_{G1}$ are

$$D_{P1} = \frac{N_P}{P_1} \quad (3a) \qquad \text{and} \qquad D_{G1} = \frac{N_G}{P_1} \quad (3b)$$

The base diameters of the pinion $D_{PB1}$ and of the gear $D_{GB1}$ are

$$D_{PB1} = D_{P1}\cos\phi_1 \quad (4a) \qquad \text{and} \qquad D_{GB1} = D_{G1}\cos\phi_1 \quad (4b)$$

The basic tooth thickness, $t_1$, at the operating pitch diameter for both pinion and gear is

$$t_1 = \frac{1.5708}{P_1} \qquad (5)$$

The root diameters of the pinion $D_{PR1}$ and gear $D_{GR1}$ and the corresponding outside diameters $D_{PO1}$ and $D_{GO1}$ are not standard because each gear is to be cut with a cutter that is not standard for the operating pitch diameters $D_{P1}$ and $D_{G1}$.

The root diameters are

$$D_{PR_1} = \frac{N_P}{P} - 2b_{P_1} \quad (6a) \qquad \text{and} \qquad D_{GR_1} = \frac{N_G}{P} - 2b_{G_1} \quad (6b)$$

where

$$b_{P1} = b_c - \left(\frac{t_{P2} - \frac{1.5708}{P}}{2\tan\phi}\right) \qquad (7a)$$

and

$$b_{G1} = b_c - \frac{t_{G2} - \frac{1.5708}{P}}{2\tan\phi} \qquad (7b)$$

where $b_c$ is the hob or cutter addendum for the pinion and gear.

The tooth thicknesses of the pinion $t_{P2}$ and the gear $t_{G2}$ are

$$t_{P2} = \frac{N_P}{P}\left(\frac{1.5708}{N_P} + \text{inv}\phi_1 - \text{inv}\phi\right) \qquad (8a)$$

$$t_{G2} = \frac{N_G}{P}\left(\frac{1.5708}{N_G} + \text{inv}\phi_1 - \text{inv}\phi\right) \tag{8b}$$

The outside diameter of the pinion $D_{PO}$ and the gear $D_{GO}$ are

$$D_{PO} = 2 \times C_1 - D_{GR1} - 2\left(b_c - \frac{1}{P}\right) \tag{9a}$$

and $$D_{GO} = 2 \times C_1 - D_{PR1} - 2\left(b_c - \frac{1}{P}\right) \tag{9b}$$

*Example:* Design gears of 8 diametral pitch, 20-degree pressure angle, and 28 and 88 teeth to operate at 7.50-inch center distance. The gears are to be cut with a hob of 0.169-inch addendum.

$$P_1 = \frac{28 + 88}{2 \times 7.50} = 7.7333 \tag{1}$$

$$\phi_1 = \arccos\left(\frac{7.7333}{8} \times 0.93969\right) = 24.719° \tag{2}$$

$$D_{P1} = \frac{28}{7.7333} = 3.6207 \text{ in.} \tag{3a}$$

and $$D_{G1} = \frac{88}{7.7333} = 11.3794 \text{ in.} \tag{3b}$$

$$D_{PB1} = 3.6207 \times 0.90837 = 3.2889 \text{ in.} \tag{4a}$$

and $$D_{GB1} = 11.3794 \times 0.90837 = 10.3367 \text{ in.} \tag{4b}$$

$$t_1 = \frac{1.5708}{7.7333} = 0.20312 \text{ in.} \tag{5}$$

$$D_{PR1} = \frac{28}{8} - 2 \times 0.1016 = 3.2968 \text{ in.} \tag{6a}$$

and $$D_{GR1} = \frac{88}{8} - 2 \times (-0.0428) = 11.0856 \text{ in.} \tag{6b}$$

$$b_{P1} = 0.169 - \left(\frac{0.2454 - \frac{1.5708}{8}}{2 \times 0.36397}\right) = 0.1016 \text{ in.} \tag{7a}$$

$$b_{G1} = 0.169 - \left(\frac{0.3505 - \frac{1.5708}{8}}{2 \times 0.36397}\right) = -0.0428 \text{ in.} \tag{7b}$$

$$t_{P2} = \frac{28}{8}\left(\frac{1.5708}{28} + 0.028922 - 0.014904\right) = 0.2454 \text{ in.} \tag{8a}$$

$$t_{G2} = \frac{88}{8}\left(\frac{1.5708}{88} + 0.028922 - 0.014904\right) - 0.3505 \text{ in.} \tag{8b}$$

$$D_{PO1} = 2 \times 7.50 - 11.0856 - 2\left(0.169 - \frac{1}{8}\right) = 3.8264 \text{ in.} \tag{9a}$$

$$D_{GO1} = 2 \times 7.50 - 3.2968 - 2\left(0.169 - \frac{1}{8}\right) = 11.6152 \text{ in.} \tag{9b}$$

**Tooth Thickness Allowance for Shaving.—**Proper stock allowance is important for good results in shaving operations. If too much stock is left for shaving, the life of the shaving tool is reduced and, in addition, shaving time is increased. The following figures represent the amount of stock to be left on the teeth for removal by shaving under average conditions: For diametral pitches of 2 to 4, a thickness of 0.003 to 0.004 inch (0.0762–0.1016 mm)—one-half on each side of the tooth; for 5 to 6 diametral pitch, 0.0025 to 0.0035 inch (0.0635–0.0889 mm); for 7 to 10 diametral pitch, 0.002 to 0.003 inch (0.0508–0.0762 mm); for 11 to 14 diametral pitch, 0.0015 to 0.0020 inch (0.0381–0.0508 mm); for 16 to 18 diametral pitch, 0.001 to 0.002 inch (0.0254–0.0508 mm); for 20 to 48 diametral pitch, 0.0005 to 0.0015 inch (0.0127–0.0381 mm); and for 52 to 72 diametral pitch, 0.0003 to 0.0007 inch (0.00762–0.01778 mm).

The thickness of the gear teeth may be measured in several ways to determine the amount of stock left on the sides of the teeth to be removed by shaving. If it is necessary to measure the tooth thickness during the preshaving operation while the gear is in the gear shaper or hobbing machine, a gear tooth caliper or pins would be employed. Caliper methods of measuring gear teeth are explained in detail on page 2228 for measurements over single teeth, and on page 2317 for measurements over two or more teeth.

When the preshaved gear can be removed from the machine for checking, the center distance method may be employed. In this method, the preshaved gear is meshed without backlash with a gear of standard tooth thickness and the increase in center distance over standard is noted. The amount of total tooth thickness over standard on the preshaved gear can then be determined by the formula: $t_2 = 2 \tan \phi \times d$, where $t_2$ = amount that the total thickness of the tooth exceeds the standard thickness, $\phi$ = pressure angle, and $d$ = amount that the center distance between the two gears exceeds the standard center distance.

**Circular Pitch for Given Center Distance and Ratio.—**When it is necessary to use a pair of gears of given ratio at a specified center distance, it may be found that no gears of standard diametral pitch will satisfy the center distance requirement. Hence, circular pitch gears may be selected. To find the required circular pitch $p$, when the center distance $C$ and total number of teeth $N$ in both gears are known, use the following formula:

$$p = \frac{C \times 6.2832}{N}$$

*Example:* A pair of gears having a ratio of 3 is to be used at a center distance of 10.230 inches. If one gear has 60 teeth and the other 20, what must be their circular pitch?

$$p = \frac{10.230 \times 6.2832}{60 + 20} = 0.8035 \text{ inch}$$

**Circular Thickness of Tooth when Outside Diameter is Standard.—**For a full-depth or stub tooth gear of standard outside diameter, the tooth thickness on the pitch circle (circular thickness or arc thickness) is found by the following formula:

$$t = \frac{1.5708}{P}$$

where $t$ = circular thickness and $P$ = diametral pitch. In Fellows stub tooth gears the diametral pitch used is the numerator of the pitch fraction (for example, 6 if the pitch is 6/8).

*Example 1:* Find the tooth thickness on the pitch circle of a $14\frac{1}{2}$-degree full-depth tooth of 12 diametral pitch.

$$t = \frac{1.5708}{12} = 0.1309 \text{ inch}$$

*Example 2:* Find the tooth thickness on the tooth circle of a 20-degree full-depth involute tooth having a diametral pitch of 5.

$$t = \frac{1.5708}{5} = 0.31416, \text{ say } 0.3142 \text{ inch}$$

The tooth thickness on the pitch circle can be determined very accurately by means of measurement over wires which are located in tooth spaces that are diametrically opposite or as nearly diametrically opposite as possible. Where measurement over wires is not feasible, the circular or arc tooth thickness can be used in determining the chordal thickness which is the dimension measured with a gear tooth caliper.

**Circular Thickness of Tooth when Outside Diameter has been Enlarged.—**When the outside diameter of a small pinion is not standard but is enlarged to avoid undercut and to improve tooth action, the teeth are located farther out radially relative to the standard pitch diameter and consequently the circular tooth thickness at the standard pitch diameter is increased. To find this increased arc thickness the following formula is used, where $t$ = tooth thickness; $e$ = amount outside diameter is increased over standard; $\phi$ = pressure angle; and $p$ = circular pitch at the standard pitch diameter.

$$t = \frac{p}{2} + e \tan \phi$$

*Example:* The outside diameter of a pinion having 10 teeth of 5 diametral pitch and a pressure angle of 14½ degrees is to be increased by 0.2746 inch. The circular pitch equivalent to 5 diametral pitch is 0.6283 inch. Find the arc tooth thickness at the standard pitch diameter.

$$t = \frac{0.6283}{2} + \left(0.2746 \times \tan 14\frac{1}{2}°\right)$$
$$t = 0.3142 + (0.2746 \times 0.25862) = 0.3852 \text{ inch}$$

**Circular Thickness of Tooth when Outside Diameter has been Reduced.—**If the outside diameter of a gear is reduced, as is frequently done to maintain the standard center distance when the outside diameter of the mating pinion is increased, the circular thickness of the gear teeth at the standard pitch diameter will be reduced. This decreased circular thickness can be found by the following formula where $t$ = circular thickness at the standard pitch diameter; $e$ = amount outside diameter is reduced under standard; $\phi$ = pressure angle; and $p$ = circular pitch.

$$t = \frac{p}{2} - e \tan \phi$$

*Example:* The outside diameter of a gear having a pressure angle of 14½ degrees is to be reduced by 0.2746 inch or an amount equal to the increase in diameter of its mating pinion. The circular pitch is 0.6283 inch. Determine the circular tooth thickness at the standard pitch diameter.

$$t = \frac{0.6283}{2} - \left(0.2746 \times \tan 14\frac{1}{2}°\right)$$
$$t = 0.3142 - (0.2746 \times 0.25862) = 0.2432 \text{ inch}$$

**Chordal Thickness of Tooth when Outside Diameter is Standard.—**To find the chordal or straight line thickness of a gear tooth the following formula can be used where $t_c$ = chordal thickness; $D$ = pitch diameter; and $N$ = number of teeth.

$$t_c = D \sin\left(\frac{90°}{N}\right)$$

*Example:* A pinion has 15 teeth of 3 diametral pitch; the pitch diameter is equal to 15 ÷ 3 or 5 inches. Find the chordal thickness at the standard pitch diameter.

$$t_c = 5 \sin\left(\frac{90°}{15}\right) = 5 \sin 6° = 5 \times 0.10453 = 0.5226 \text{ inch}$$

## Chordal Thicknesses and Chordal Addenda of Milled, Full-Depth Gear Teeth and of Gear Milling Cutters

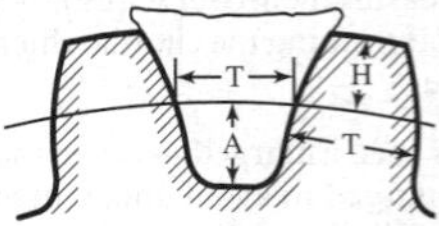


$T$ = chordal thickness of gear tooth and cutter tooth at pitch line;
$H$ = chordal addendum for full-depth gear tooth;
$A$ = chordal addendum of cutter = (2.157 ÷ diametral pitch) − $H$
= (0.6866 × circular pitch) − $H$.

| Diametral Pitch | Dimension | Number of Gear Cutter, and Corresponding Number of Teeth | | | | | | | |
|---|---|---|---|---|---|---|---|---|---|
| | | No. 1 135 Teeth | No. 2 55 Teeth | No. 3 35 Teeth | No. 4 26 Teeth | No. 5 21 Teeth | No. 6 17 Teeth | No. 7 14 Teeth | No. 8 12 Teeth |
| 1 | T | 1.5707 | 1.5706 | 1.5702 | 1.5698 | 1.5694 | 1.5686 | 1.5675 | 1.5663 |
| | H | 1.0047 | 1.0112 | 1.0176 | 1.0237 | 1.0294 | 1.0362 | 1.0440 | 1.0514 |
| 1½ | T | 1.0471 | 1.0470 | 1.0468 | 1.0465 | 1.0462 | 1.0457 | 1.0450 | 1.0442 |
| | H | 0.6698 | 0.6741 | 0.6784 | 0.6824 | 0.6862 | 0.6908 | 0.6960 | 0.7009 |
| 2 | T | 0.7853 | 0.7853 | 0.7851 | 0.7849 | 0.7847 | 0.7843 | 0.7837 | 0.7831 |
| | H | 0.5023 | 0.5056 | 0.5088 | 0.5118 | 0.5147 | 0.5181 | 0.5220 | 0.5257 |
| 2½ | T | 0.6283 | 0.6282 | 0.6281 | 0.6279 | 0.6277 | 0.6274 | 0.6270 | 0.6265 |
| | H | 0.4018 | 0.4044 | 0.4070 | 0.4094 | 0.4117 | 0.4144 | 0.4176 | 0.4205 |
| 3 | T | 0.5235 | 0.5235 | 0.5234 | 0.5232 | 0.5231 | 0.5228 | 0.5225 | 0.5221 |
| | H | 0.3349 | 0.3370 | 0.3392 | 0.3412 | 0.3431 | 0.3454 | 0.3480 | 0.3504 |
| 3½ | T | 0.4487 | 0.4487 | 0.4486 | 0.4485 | 0.4484 | 0.4481 | 0.4478 | 0.4475 |
| | H | 0.2870 | 0.2889 | 0.2907 | 0.2919 | 0.2935 | 0.2954 | 0.2977 | 0.3004 |
| 4 | T | 0.3926 | 0.3926 | 0.3926 | 0.3924 | 0.3923 | 0.3921 | 0.3919 | 0.3915 |
| | H | 0.2511 | 0.2528 | 0.2544 | 0.2559 | 0.2573 | 0.2590 | 0.2610 | 0.2628 |
| 5 | T | 0.3141 | 0.3141 | 0.3140 | 0.3139 | 0.3138 | 0.3137 | 0.3135 | 0.3132 |
| | H | 0.2009 | 0.2022 | 0.2035 | 0.2047 | 0.2058 | 0.2072 | 0.2088 | 0.2102 |
| 6 | T | 0.2618 | 0.2617 | 0.2617 | 0.2616 | 0.2615 | 0.2614 | 0.2612 | 0.2610 |
| | H | 0.1674 | 0.1685 | 0.1696 | 0.1706 | 0.1715 | 0.1727 | 0.1740 | 0.1752 |
| 7 | T | 0.2244 | 0.2243 | 0.2243 | 0.2242 | 0.2242 | 0.2240 | 0.2239 | 0.2237 |
| | H | 0.1435 | 0.1444 | 0.1453 | 0.1462 | 0.1470 | 0.1480 | 0.1491 | 0.1502 |
| 8 | T | 0.1963 | 0.1963 | 0.1962 | 0.1962 | 0.1961 | 0.1960 | 0.1959 | 0.1958 |
| | H | 0.1255 | 0.1264 | 0.1272 | 0.1279 | 0.1286 | 0.1295 | 0.1305 | 0.1314 |
| 9 | T | 0.1745 | 0.1745 | 0.1744 | 0.1744 | 0.1743 | 0.1743 | 0.1741 | 0.1740 |
| | H | 0.1116 | 0.1123 | 0.1130 | 0.1137 | 0.1143 | 0.1151 | 0.1160 | 0.1168 |
| 10 | T | 0.1570 | 0.1570 | 0.1570 | 0.1569 | 0.1569 | 0.1568 | 0.1567 | 0.1566 |
| | H | 0.1004 | 0.1011 | 0.1017 | 0.1023 | 0.1029 | 0.1036 | 0.1044 | 0.1051 |
| 11 | T | 0.1428 | 0.1428 | 0.1427 | 0.1427 | 0.1426 | 0.1426 | 0.1425 | 0.1424 |
| | H | 0.0913 | 0.0919 | 0.0925 | 0.0930 | 0.0935 | 0.0942 | 0.0949 | 0.0955 |
| 12 | T | 0.1309 | 0.1309 | 0.1308 | 0.1308 | 0.1308 | 0.1307 | 0.1306 | 0.1305 |
| | H | 0.0837 | 0.0842 | 0.0848 | 0.0853 | 0.0857 | 0.0863 | 0.0870 | 0.0876 |
| 14 | T | 0.1122 | 0.1122 | 0.1121 | 0.1121 | 0.1121 | 0.1120 | 0.1119 | 0.1118 |
| | H | 0.0717 | 0.0722 | 0.0726 | 0.0731 | 0.0735 | 0.0740 | 0.0745 | 0.0751 |
| 16 | T | 0.0981 | 0.0981 | 0.0981 | 0.0981 | 0.0980 | 0.0980 | 0.0979 | 0.0979 |
| | H | 0.0628 | 0.0632 | 0.0636 | 0.0639 | 0.0643 | 0.0647 | 0.0652 | 0.0657 |
| 18 | T | 0.0872 | 0.0872 | 0.0872 | 0.0872 | 0.0872 | 0.0871 | 0.0870 | 0.0870 |
| | H | 0.0558 | 0.0561 | 0.0565 | 0.0568 | 0.0571 | 0.0575 | 0.0580 | 0.0584 |
| 20 | T | 0.0785 | 0.0785 | 0.0785 | 0.0785 | 0.0784 | 0.0784 | 0.0783 | 0.0783 |
| | H | 0.0502 | 0.0505 | 0.0508 | 0.0511 | 0.0514 | 0.0518 | 0.0522 | 0.0525 |

## Chordal Thicknesses and Chordal Addenda of Milled Full-Depth Gear Teeth and of Gear Milling Cutters *(Continued)*

| Circular Pitch | Dimension | Number of Gear Cutter, and Corresponding Number of Teeth | | | | | | | |
|---|---|---|---|---|---|---|---|---|---|
| | | No. 1 135 Teeth | No. 2 55 Teeth | No. 3 35 Teeth | No. 4 26 Teeth | No. 5 21 Teeth | No. 6 17 Teeth | No. 7 14 Teeth | No. 8 12 Teeth |
| ¼ | T | 0.1250 | 0.1250 | 0.1249 | 0.1249 | 0.1249 | 0.1248 | 0.1247 | 0.1246 |
| | H | 0.0799 | 0.0804 | 0.0809 | 0.0814 | 0.0819 | 0.0824 | 0.0830 | 0.0836 |
| 5⁄16 | T | 0.1562 | 0.1562 | 0.1562 | 0.1561 | 0.1561 | 0.1560 | 0.1559 | 0.1558 |
| | H | 0.0999 | 0.1006 | 0.1012 | 0.1018 | 0.1023 | 0.1030 | 0.1038 | 0.1045 |
| 3⁄8 | T | 0.1875 | 0.1875 | 0.1874 | 0.1873 | 0.1873 | 0.1872 | 0.1871 | 0.1870 |
| | H | 0.1199 | 0.1207 | 0.1214 | 0.1221 | 0.1228 | 0.1236 | 0.1245 | 0.1254 |
| 7⁄16 | T | 0.2187 | 0.2187 | 0.2186 | 0.2186 | 0.2185 | 0.2184 | 0.2183 | 0.2181 |
| | H | 0.1399 | 0.1408 | 0.1416 | 0.1425 | 0.1433 | 0.1443 | 0.1453 | 0.1464 |
| ½ | T | 0.2500 | 0.2500 | 0.2499 | 0.2498 | 0.2498 | 0.2496 | 0.2495 | 0.2493 |
| | H | 0.1599 | 0.1609 | 0.1619 | 0.1629 | 0.1638 | 0.1649 | 0.1661 | 0.1673 |
| 9⁄16 | T | 0.2812 | 0.2812 | 0.2811 | 0.2810 | 0.2810 | 0.2808 | 0.2806 | 0.2804 |
| | H | 0.1799 | 0.1810 | 0.1821 | 0.1832 | 0.1842 | 0.1855 | 0.1868 | 0.1882 |
| 5⁄8 | T | 0.3125 | 0.3125 | 0.3123 | 0.3123 | 0.3122 | 0.3120 | 0.3118 | 0.3116 |
| | H | 0.1998 | 0.2012 | 0.2023 | 0.2036 | 0.2047 | 0.2061 | 0.2076 | 0.2091 |
| 11⁄16 | T | 0.3437 | 0.3437 | 0.3436 | 0.3435 | 0.3434 | 0.3432 | 0.3430 | 0.3427 |
| | H | 0.2198 | 0.2213 | 0.2226 | 0.2239 | 0.2252 | 0.2267 | 0.2283 | 0.2300 |
| ¾ | T | 0.3750 | 0.3750 | 0.3748 | 0.3747 | 0.3747 | 0.3744 | 0.3742 | 0.3740 |
| | H | 0.2398 | 0.2414 | 0.2428 | 0.2443 | 0.2457 | 0.2473 | 0.2491 | 0.2509 |
| 13⁄16 | T | 0.4062 | 0.4062 | 0.4060 | 0.4059 | 0.4059 | 0.4056 | 0.4054 | 0.4050 |
| | H | 0.2598 | 0.2615 | 0.2631 | 0.2647 | 0.2661 | 0.2679 | 0.2699 | 0.2718 |
| 7⁄8 | T | 0.4375 | 0.4375 | 0.4373 | 0.4372 | 0.4371 | 0.4368 | 0.4366 | 0.4362 |
| | H | 0.2798 | 0.2816 | 0.2833 | 0.2850 | 0.2866 | 0.2885 | 0.2906 | 0.2927 |
| 15⁄16 | T | 0.4687 | 0.4687 | 0.4685 | 0.4684 | 0.4683 | 0.4680 | 0.4678 | 0.4674 |
| | H | 0.2998 | 0.3018 | 0.3035 | 0.3054 | 0.3071 | 0.3092 | 0.3114 | 0.3137 |
| 1 | T | 0.5000 | 0.5000 | 0.4998 | 0.4997 | 0.4996 | 0.4993 | 0.4990 | 0.4986 |
| | H | 0.3198 | 0.3219 | 0.3238 | 0.3258 | 0.3276 | 0.3298 | 0.3322 | 0.3346 |
| 1⅛ | T | 0.5625 | 0.5625 | 0.5623 | 0.5621 | 0.5620 | 0.5617 | 0.5613 | 0.5610 |
| | H | 0.3597 | 0.3621 | 0.3642 | 0.3665 | 0.3685 | 0.3710 | 0.3737 | 0.3764 |
| 1¼ | T | 0.6250 | 0.6250 | 0.6247 | 0.6246 | 0.6245 | 0.6241 | 0.6237 | 0.6232 |
| | H | 0.3997 | 0.4023 | 0.4047 | 0.4072 | 0.4095 | 0.4122 | 0.4152 | 0.4182 |
| 1⅜ | T | 0.6875 | 0.6875 | 0.6872 | 0.6870 | 0.6869 | 0.6865 | 0.6861 | 0.6856 |
| | H | 0.4397 | 0.4426 | 0.4452 | 0.4479 | 0.4504 | 0.4534 | 0.4567 | 0.4600 |
| 1½ | T | 0.7500 | 0.7500 | 0.7497 | 0.7495 | 0.7494 | 0.7489 | 0.7485 | 0.7480 |
| | H | 0.4797 | 0.4828 | 0.4857 | 0.4887 | 0.4914 | 0.4947 | 0.4983 | 0.5019 |
| 1¾ | T | 0.8750 | 0.8750 | 0.8746 | 0.8744 | 0.8743 | 0.8737 | 0.8732 | 0.8726 |
| | H | 0.5596 | 0.5633 | 0.5666 | 0.5701 | 0.5733 | 0.5771 | 0.5813 | 0.5855 |
| 2 | T | 1.0000 | 1.0000 | 0.9996 | 0.9994 | 0.9992 | 0.9986 | 0.9980 | 0.9972 |
| | H | 0.6396 | 0.6438 | 0.6476 | 0.6516 | 0.6552 | 0.6596 | 0.6644 | 0.6692 |
| 2¼ | T | 1.1250 | 1.1250 | 1.1246 | 1.1242 | 1.1240 | 1.1234 | 1.1226 | 1.1220 |
| | H | 0.7195 | 0.7242 | 0.7285 | 0.7330 | 0.7371 | 0.7420 | 0.7474 | 0.7528 |
| 2½ | T | 1.2500 | 1.2500 | 1.2494 | 1.2492 | 1.2490 | 1.2482 | 1.2474 | 1.2464 |
| | H | 0.7995 | 0.8047 | 0.8095 | 0.8145 | 0.8190 | 0.8245 | 0.8305 | 0.8365 |
| 3 | T | 1.5000 | 1.5000 | 1.4994 | 1.4990 | 1.4990 | 1.4978 | 1.4970 | 1.4960 |
| | H | 0.9594 | 0.9657 | 0.9714 | 0.9774 | 0.9828 | 0.9894 | 0.9966 | 1.0038 |

**Chordal Thickness of Tooth when Outside Diameter is Special.**—When the outside diameter is larger or smaller than standard the chordal thickness at the standard pitch diameter is found by the following formula where $t_c$ = chordal thickness at the standard pitch diameter $D$; $t$ = circular thickness at the standard pitch diameter of the enlarged pinion or reduced gear being measured.

$$t_c = t - \frac{t^3}{6 \times D^2}$$

*Example 1:* The outside diameter of a pinion having 10 teeth of 5 diametral pitch has been *enlarged* by 0.2746 inch. This enlargement has increased the circular tooth thickness at the standard pitch diameter (as determined by the formula previously given) to 0.3852 inch. Find the equivalent chordal thickness.

$$t_c = 0.3852 - \frac{0.385^3}{6 \times 2^2} = 0.3852 - 0.0024 = 0.3828 \text{ inch}$$

(The error introduced by rounding the circular thickness to three significant figures before cubing it only affects the fifth decimal place in the result.)

*Example 2:* A gear having 30 teeth is to mesh with the pinion in Example 1 and is *reduced* so that the circular tooth thickness at the standard pitch diameter is 0.2432 inch. Find the equivalent chordal thickness.

$$t_c = 0.2432 - \frac{0.243^3}{6 \times 6^2} = 0.2432 - 0.00007 = 0.2431 \text{ inch}$$

**Chordal Addendum.**—In measuring the chordal thickness, the vertical scale of a gear tooth caliper is set to the chordal or "corrected" addendum to locate the caliper jaws at the pitch line (see *Method of setting a gear tooth caliper* on page 2229). The simplified formula which follows may be used in determining the chordal addendum either when the addendum is standard for full-depth or stub teeth or when the addendum is either longer or shorter than standard as in case of an enlarged pinion or a gear which is to mesh with an enlarged pinion and has a reduced addendum to maintain the standard center distance. If $a_c$ = chordal addendum; $a$ = addendum; and $t$ = circular thickness of tooth at pitch diameter $D$; then,

$$a_c = a + \frac{t^2}{4D}$$

*Example 1:* The outside diameter of an 8 diametral pitch 14-tooth pinion with 20-degree full-depth teeth is to be increased by using an enlarged addendum of 1.234 ÷ 8 = 0.1542 inch (see Table 7 on page 2227). The basic tooth thickness of the enlarged pinion is 1.741 ÷ 8 = 0.2176 inch. What is the chordal addendum?

$$\text{Chordal addendum} = 0.1542 + \frac{0.2176^2}{4 \times (14 \div 8)} = 0.1610 \text{ inch}$$

*Example 2:* The outside diameter of a 14½-degree pinion having 12 teeth of 2 diametral pitch is to be enlarged 0.624 inch to avoid undercut (see Table 8 on page 2227), thus increasing the addendum from 0.5000 to 0.8120 inch and the arc thickness at the pitch line from 0.7854 to 0.9467 inch. Then,

$$\text{Chordal addendum of pinion} = 0.8120 + \frac{0.9467^2}{4 \times (12 \div 2)} = 0.8493 \text{ inch}$$

**Table 7. Addendums and Tooth Thicknesses for Coarse-Pitch Long-Addendum Pinions and their Mating Short-Addendum Gears — 20- and 25-Degree Pressure Angles** *ANSI B6.1-1968 (R1974)*

| Number of Teeth in Pinion | Addendum | | Basic Tooth Thickness | | Number of Teeth in Gear |
|---|---|---|---|---|---|
| | Pinion | Gear | Pinion | Gear | |
| $N_P$ | $a_P$ | $a_G$ | $t_P$ | $t_G$ | $N_G$ (min) |
| 20-Degree Involute Full Depth Tooth Form (Less than 20 Diametral Pitch) | | | | | |
| 10 | 1.468 | .532 | 1.912 | 1.230 | 25 |
| 11 | 1.409 | .591 | 1.868 | 1.273 | 24 |
| 12 | 1.351 | .649 | 1.826 | 1.315 | 23 |
| 13 | 1.292 | .708 | 1.783 | 1.358 | 22 |
| 14 | 1.234 | .766 | 1.741 | 1.400 | 21 |
| 15 | 1.175 | .825 | 1.698 | 1.443 | 20 |
| 16 | 1.117 | .883 | 1.656 | 1.486 | 19 |
| 17 | 1.058 | .942 | 1.613 | 1.529 | 18 |
| 25-Degree Involute Full Depth Tooth Form (Less than 20 Diametral Pitch) | | | | | |
| 10 | 1.184 | .816 | 1.742 | 1.399 | 15 |
| 11 | 1.095 | .905 | 1.659 | 1.482 | 14 |

All values are for 1 diametral pitch. For any other sizes of teeth all linear dimensions should be divided by the diametral pitch. Basic tooth thicknesses do not include an allowance for backlash.

**Table 8. Enlarged Pinion and Reduced Gear Dimensions to Avoid Interference Coarse Pitch 14½-Degree Involute Full Depth Teeth**

| Number of Pinion Teeth | Changes in Pinion and Gear Diameters | Circular Tooth Thickness | | Min. No. of Teeth in Mating Gear | |
|---|---|---|---|---|---|
| | | Pinion | Mating Gear | To Avoid Undercut | For Full Involute Action |
| 10 | 1.3731 | 1.9259 | 1.2157 | 54 | 27 |
| 11 | 1.3104 | 1.9097 | 1.2319 | 53 | 27 |
| 12 | 1.2477 | 1.8935 | 1.2481 | 52 | 28 |
| 13 | 1.1850 | 1.8773 | 1.2643 | 51 | 28 |
| 14 | 1.1223 | 1.8611 | 1.2805 | 50 | 28 |
| 15 | 1.0597 | 1.8449 | 1.2967 | 49 | 28 |
| 16 | 0.9970 | 1.8286 | 1.3130 | 48 | 28 |
| 17 | 0.9343 | 1.8124 | 1.3292 | 47 | 28 |
| 18 | 0.8716 | 1.7962 | 1.3454 | 46 | 28 |
| 19 | 0.8089 | 1.7800 | 1.3616 | 45 | 28 |
| 20 | 0.7462 | 1.7638 | 1.3778 | 44 | 28 |
| 21 | 0.6835 | 1.7476 | 1.3940 | 43 | 28 |
| 22 | 0.6208 | 1.7314 | 1.4102 | 42 | 27 |
| 23 | 0.5581 | 1.7151 | 1.4265 | 41 | 27 |
| 24 | 0.4954 | 1.6989 | 1.4427 | 40 | 27 |
| 25 | 0.4328 | 1.6827 | 1.4589 | 39 | 26 |
| 26 | 0.3701 | 1.6665 | 1.4751 | 38 | 26 |
| 27 | 0.3074 | 1.6503 | 1.4913 | 37 | 26 |
| 28 | 0.2447 | 1.6341 | 1.5075 | 36 | 25 |
| 29 | 0.1820 | 1.6179 | 1.5237 | 35 | 25 |
| 30 | 0.1193 | 1.6017 | 1.5399 | 34 | 24 |
| 31 | 0.0566 | 1.5854 | 1.5562 | 33 | 24 |

All dimensions are given in inches and are for 1 diametral pitch. For other pitches divide tabular values by desired diametral pitch.

Add to the standard outside diameter of the pinion the amount given in the second column of the table divided by the desired diametral pitch, and (to maintain standard center distance) subtract the same amount from the outside diameter of the mating gear. Long addendum pinions will mesh with standard gears, but the center distance will be greater than standard.

*Example 3:* The outside diameter of the mating gear for the pinion in Example 3 is to be reduced 0.624 inch. The gear has 60 teeth and the addendum is reduced from 0.5000 to 0.1881 inch (to maintain the standard center distance), thus reducing the arc thickness to 0.6240 inch. Then,

$$\text{Chordal addendum of gear} = 0.1881 + \frac{0.6240^2}{4 \times (60 \div 2)} = 0.1913 \text{ inch}$$

When a gear addendum is reduced as much as the mating pinion addendum is enlarged, the minimum number of gear teeth required to prevent undercutting depends upon the enlargement of the mating pinion. To illustrate, if a $14\frac{1}{2}$-degree pinion with 13 teeth is enlarged 1.185 inches, then the reduced mating gear should have a minimum of 51 teeth to avoid undercut (see Table 8 on page 2227).

**Tables for Chordal Thicknesses and Chordal Addenda of Milled, Full-Depth Teeth.**—Two convenient tables for checking gears with milled, full-depth teeth are given on pages 2224 and 2225. The first shows chordal thicknesses and chordal addenda for the lowest number of teeth cut by gear cutters Nos. 1 through 8, and for the commonly used diametral pitches. The second gives similar data for commonly used circular pitches. In each case the data shown are accurate for the number of gear teeth indicated, but are approximate for other numbers of teeth within the range of the cutter under which they appear in the table. For the higher diametral pitches and lower circular pitches, the error introduced by using the data for any tooth number within the range of the cutter under which it appears is comparatively small. The chordal thicknesses and chordal addenda for gear cutters Nos. 1 through 8 of the more commonly used diametral and circular pitches can be obtained from the table and formulas on pages 2224 and 2225.

**Caliper Measurement of Gear Tooth.**—In cutting gear teeth, the general practice is to adjust the cutter or hob until it grazes the outside diameter of the blank; the cutter is then sunk to the total depth of the tooth space plus whatever slight additional amount may be required to provide the necessary play or backlash between the teeth. (For recommendations concerning backlash and excess depth of cut required, see *Backlash* starting on page 2244.) If the outside diameter of the gear blank is correct, the tooth thickness should also be correct after the cutter has been sunk to the depth required for a given pitch and backlash. However, it is advisable to check the tooth thickness by measuring it, and the vernier gear-tooth caliper (see following illustration) is commonly used in measuring the thickness.

The vertical scale of this caliper is set so that when it rests upon the top of the tooth as shown, the lower ends of the caliper jaws will be at the height of the pitch circle; the horizontal scale then shows the chordal thickness of the tooth at this point. If the gear is being cut on a milling machine or with the type of gear-cutting machine employing a formed milling cutter, the tooth thickness is checked by first taking a trial cut for a short distance at one side of the blank; then the gear blank is indexed for the next space and another cut is taken far enough to mill the full outline of the tooth. The tooth thickness is then measured.

Before the gear-tooth caliper can be used, it is necessary to determine the correct chordal thickness and also the chordal addendum (or "corrected addendum" as it is sometimes called). The vertical scale is set to the chordal addendum, thus locating the ends of the jaws at the height of the pitch circle. The rules or formulas to use in determining the chordal thickness and chordal addendum will depend upon the outside diameter of the gear; for example, if the outside diameter of a small pinion is enlarged to avoid undercut and improve the tooth action, this must be taken into account in figuring the chordal thickness and chordal addendum as shown by the accompanying rules. The detail of a gear tooth included with the gear-tooth caliper illustration represents the chordal thickness *T*, the addendum *S*, and the chordal addendum *H*. For the caliper measurements over two or more teeth see *Checking Spur Gear Size by Chordal Measurement Over Two or More Teeth* starting on page 2317.

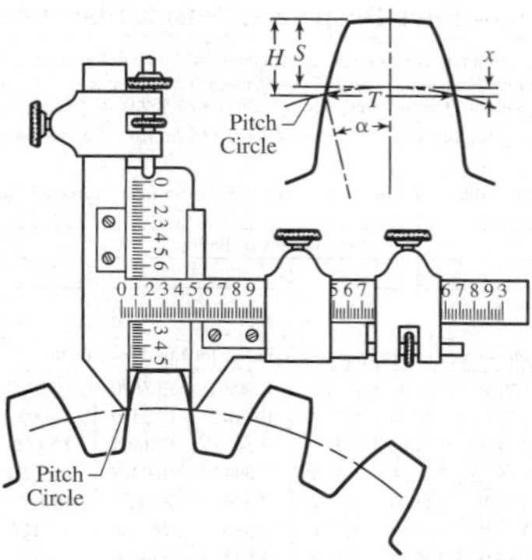


Method of setting a gear tooth caliper

**Selection of Involute Gear Milling Cutter for a Given Diametral Pitch and Number of Teeth.—**When gear teeth are cut by using formed milling cutters, the cutter must be selected to suit both the pitch and the number of teeth, because the shapes of the tooth spaces vary according to the number of teeth. For instance, the tooth spaces of a small pinion are not of the same shape as the spaces of a large gear of equal pitch. Theoretically, there should be a different formed cutter for every tooth number, but such refinement is unnecessary in practice. The involute formed cutters commonly used are made in series of eight cutters for each diametral pitch (see *Series of Involute, Finishing Gear Milling Cutters for Each Pitch*). The shape of each cutter in this series is correct for a certain number of teeth only, but it can be used for other numbers within the limits given. For instance, a No. 6 cutter may be used for gears having from 17 to 20 teeth, but the tooth outline is correct only for 17 teeth or the lowest number in the range, which is also true of the other cutters listed. When this cutter is used for a gear having, say, 19 teeth, too much material is removed from the upper surfaces of the teeth, although the gear meets ordinary requirements. When greater accuracy of tooth shape is desired to ensure smoother or quieter operation, an intermediate series of cutters having half-numbers may be used provided the number of gear teeth is between the number listed for the regular cutters (see *Series of Involute, Finishing Gear Milling Cutters for Each Pitch*).

Involute gear milling cutters are designed to cut a composite tooth form, the center portion being a true involute while the top and bottom portions are cycloidal. This composite form is necessary to prevent tooth interference when milled mating gears are meshed with each other. Because of their composite form, milled gears will not mate satisfactorily enough for high grade work with those of generated, full-involute form Composite form hobs are available, however, which will produce generated gears that mesh with those cut by gear milling cutters.

*Metric Module Gear Cutters:* The accompanying table for selecting the cutter number to be used to cut a given number of teeth may be used also to select metric module gear cutters except that the numbers are designated in reverse order. For example, cutter No. 1, in the metric module system, is used for 12–13 teeth, cutter No. 2 for 14–16 teeth, etc.

**Increasing Pinion Diameter to Avoid Undercut or Interference.—**On coarse-pitch pinions with small numbers of teeth (10 to 17 for 20-degree and 10 and 11 for 25-degree pressure angle involute tooth forms) undercutting of the tooth profile or fillet interference with the tip of the mating gear can be avoided by making certain changes from the standard tooth proportions that are specified in Table 3 on page 2215. These changes consist essentially in increasing the addendum and hence the outside diameter of the pinion and decreasing the addendum and hence the outside diameter of the mating gear.

## Circular Pitch in Gears—Pitch Diameters, Outside Diameters, and Root Diameters

For any particular circular pitch and number of teeth, use the table as shown in the example to find the pitch diameter, outside diameter, and root diameter. *Example:* Pitch diameter for 57 teeth of 6-inch circular pitch = 10 × pitch diameter given under factor for 5 teeth plus pitch diameter given under factor for 7 teeth. (10 × 9.5493) + 13.3690 = 108.862 inches.

Outside diameter of gear equals pitch diameter plus outside diameter factor from next-to-last column in table = 108.862 + 3.8197 = 112.682 inches.

Root diameter of gear equals pitch diameter minus root diameter factor from last column in table = 108.862 − 4.4194 = 104.443 inches.

| Circular Pitch in Inches | Factor for Number of Teeth | | | | | | | | | Outside Dia. Factor | Root Diameter Factor |
|---|---|---|---|---|---|---|---|---|---|---|---|
| | 1 | 2 | 3 | 4 | 5 | 6 | 7 | 8 | 9 | | |
| | Pitch Diameter Corresponding to Factor for Number of Teeth | | | | | | | | | | |
| 6 | 1.9099 | 3.8197 | 5.7296 | 7.6394 | 9.5493 | 11.4591 | 13.3690 | 15.2788 | 17.1887 | 3.8197 | 4.4194 |
| 5 1/2 | 1.7507 | 3.5014 | 5.2521 | 7.0028 | 8.7535 | 10.5042 | 12.2549 | 14.0056 | 15.7563 | 3.5014 | 4.0511 |
| 5 | 1.5915 | 3.1831 | 4.7746 | 6.3662 | 7.9577 | 9.5493 | 11.1408 | 12.7324 | 14.3239 | 3.1831 | 3.6828 |
| 4 1/2 | 1.4324 | 2.8648 | 4.2972 | 5.7296 | 7.1620 | 8.5943 | 10.0267 | 11.4591 | 12.8915 | 2.8648 | 3.3146 |
| 4 | 1.2732 | 2.5465 | 3.8197 | 5.0929 | 6.3662 | 7.6394 | 8.9127 | 10.1859 | 11.4591 | 2.5465 | 2.9463 |
| 3 1/2 | 1.1141 | 2.2282 | 3.3422 | 4.4563 | 5.5704 | 6.6845 | 7.7986 | 8.9127 | 10.0267 | 2.2282 | 2.5780 |
| 3 | 0.9549 | 1.9099 | 2.8648 | 3.8197 | 4.7746 | 5.7296 | 6.6845 | 7.6394 | 8.5943 | 1.9099 | 2.2097 |
| 2 1/2 | 0.7958 | 1.5915 | 2.3873 | 3.1831 | 3.9789 | 4.7746 | 5.5704 | 6.3662 | 7.1620 | 1.5915 | 1.8414 |
| 2 | 0.6366 | 1.2732 | 1.9099 | 2.5465 | 3.1831 | 3.8197 | 4.4563 | 5.0929 | 5.7296 | 1.2732 | 1.4731 |
| 1 7/8 | 0.5968 | 1.1937 | 1.7905 | 2.3873 | 2.9841 | 3.5810 | 4.1778 | 4.7746 | 5.3715 | 1.1937 | 1.3811 |
| 1 3/4 | 0.5570 | 1.1141 | 1.6711 | 2.2282 | 2.7852 | 3.3422 | 3.8993 | 4.4563 | 5.0134 | 1.1141 | 1.2890 |
| 1 5/8 | 0.5173 | 1.0345 | 1.5518 | 2.0690 | 2.5863 | 3.1035 | 3.6208 | 4.1380 | 4.6553 | 1.0345 | 1.1969 |
| 1 1/2 | 0.4775 | 0.9549 | 1.4324 | 1.9099 | 2.3873 | 2.8648 | 3.3422 | 3.8197 | 4.2972 | 0.9549 | 1.1049 |
| 1 7/16 | 0.4576 | 0.9151 | 1.3727 | 1.8303 | 2.2878 | 2.7454 | 3.2030 | 3.6606 | 4.1181 | 0.9151 | 1.0588 |
| 1 3/8 | 0.4377 | 0.8754 | 1.3130 | 1.7507 | 2.1884 | 2.6261 | 3.0637 | 3.5014 | 3.9391 | 0.8754 | 1.0128 |
| 1 5/16 | 0.4178 | 0.8356 | 1.2533 | 1.6711 | 2.0889 | 2.5067 | 2.9245 | 3.3422 | 3.7600 | 0.8356 | 0.9667 |
| 1 1/4 | 0.3979 | 0.7958 | 1.1937 | 1.5915 | 1.9894 | 2.3873 | 2.7852 | 3.1831 | 3.5810 | 0.7958 | 0.9207 |
| 1 3/16 | 0.3780 | 0.7560 | 1.1340 | 1.5120 | 1.8900 | 2.2680 | 2.6459 | 3.0239 | 3.4019 | 0.7560 | 0.8747 |
| 1 1/8 | 0.3581 | 0.7162 | 1.0743 | 1.4324 | 1.7905 | 2.1486 | 2.5067 | 2.8648 | 3.2229 | 0.7162 | 0.8286 |
| 1 1/16 | 0.3382 | 0.6764 | 1.0146 | 1.3528 | 1.6910 | 2.0292 | 2.3674 | 2.7056 | 3.0438 | 0.6764 | 0.7826 |
| 1 | 0.3183 | 0.6366 | 0.9549 | 1.2732 | 1.5915 | 1.9099 | 2.2282 | 2.5465 | 2.8648 | 0.6366 | 0.7366 |
| 15/16 | 0.2984 | 0.5968 | 0.8952 | 1.1937 | 1.4921 | 1.7905 | 2.0889 | 2.3873 | 2.6857 | 0.5968 | 0.6905 |
| 7/8 | 0.2785 | 0.5570 | 0.8356 | 1.1141 | 1.3926 | 1.6711 | 1.9496 | 2.2282 | 2.5067 | 0.5570 | O.6445 |
| 13/16 | 0.2586 | 0.5173 | 0.7759 | 1.0345 | 1.2931 | 1.5518 | 1.8104 | 2.0690 | 2.3276 | 0.5173 | 0.5985 |
| 3/4 | 0.2387 | 0.4475 | 0.7162 | 0.9549 | 1.1937 | 1.4324 | 1.6711 | 1.9099 | 2.1486 | 0.4775 | 0.5524 |
| 11/16 | 0.2188 | 0.4377 | 0.6565 | 0.8754 | 1.0942 | 1.3130 | 1.5319 | 1.7507 | 1.9695 | 0.4377 | 0.5064 |
| 2/3 | 0.2122 | 0.4244 | 0.6366 | 0.8488 | 1.0610 | 1.2732 | 1.4854 | 1.6977 | 1.9099 | 0.4244 | 0.4910 |
| 5/8 | 0.1989 | 0.3979 | 0.5968 | 0.7958 | 0.9947 | 1.1937 | 1.3926 | 1.5915 | 1.7905 | 0.3979 | 0.4604 |
| 9/16 | 0.1790 | 0.3581 | 0.5371 | 0.7162 | 0.8952 | 1.0743 | 1.2533 | 1.4324 | 1.6114 | 0.3581 | 0.4143 |
| 1/2 | 0.1592 | 0.3183 | 0.4775 | 0.6366 | 0.7958 | 0.9549 | 1.1141 | 1.2732 | 1.4324 | 0.3183 | 0.3683 |
| 7/16 | 0.1393 | 0.2785 | 0.4178 | 0.5570 | 0.6963 | 0.8356 | 0.9748 | 1.1141 | 1.2533 | 0.2785 | 0.3222 |
| 3/8 | 0.1194 | 0.2387 | 0.3581 | 0.4775 | 0.5968 | 0.7162 | 0.8356 | 0.9549 | 1.0743 | 0.2387 | 0.2762 |
| 1/3 | 0.1061 | 0.2122 | 0.3183 | 0.4244 | 0.5305 | 0.6366 | 0.7427 | 0.8488 | 0.9549 | 0.2122 | 0.2455 |
| 5/16 | 0.0995 | 0.1989 | 0.2984 | 0.3979 | 0.4974 | 0.5968 | 0.6963 | 0.7958 | 0.8952 | 0.1989 | 0.2302 |
| 1/4 | 0.0796 | 0.1592 | 0.2387 | 0.3183 | 0.3979 | 0.4775 | 0.5570 | 0.6366 | 0.7162 | 0.1592 | 0.1841 |
| 3/16 | 0.0597 | 0.1194 | 0.1790 | 0.2387 | 0.2984 | 0.3581 | 0.4178 | 0.4775 | 0.5371 | 0.1194 | 0.1381 |
| 1/8 | 0.0398 | 0.0796 | 0.1194 | 0.1592 | 0.1989 | 0.2387 | 0.2785 | 0.3183 | 0.3581 | 0.0796 | 0.0921 |
| 1/16 | 0.0199 | 0.0398 | 0.0597 | 0.0796 | 0.0995 | 0.1194 | 0.1393 | 0.1592 | 0.1790 | 0.0398 | 0.0460 |

These changes in outside diameters of pinion and gear do not change the velocity ratio or the procedures in cutting the teeth on a hobbing machine or generating type of shaper or planer.

Data in Table 7 on page 2227 are taken from ANSI Standard B6.1-1968, reaffirmed 1974, and show for 20-degree and 25-degree full-depth standard tooth forms, respectively, the addendums and tooth thicknesses for long addendum pinions and their mating short addendum gears when the number of teeth in the pinion is as given. Similar data for former standard $14\frac{1}{2}$-degree full-depth teeth (20 diametral pitch and coarser) are given in Table 8 on page 2227.

*Example:* A 14-tooth, 20-degree pressure angle pinion of 6 diametral pitch is to be enlarged. What will be the outside diameters of the pinion and a 60-tooth mating gear? If the mating gear is to have the minimum number of teeth to avoid undercut, what will be its outside diameter?

$$D_o\,(\text{pinion}) = \frac{N_P}{P} + 2a = \frac{14}{6} + 2\left(\frac{1.234}{6}\right) = 2.745 \text{ inches}$$

$$D_o\,(\text{gear}) = \frac{N_G}{P} + 2a = \frac{60}{6} + 2\left(\frac{0.766}{6}\right) = 10.255 \text{ inches}$$

For a mating gear with minimum number of teeth to avoid undercut

$$D_o\,(\text{gear}) = \frac{N_G}{P} + 2a = \frac{21}{6} + 2\left(\frac{0.766}{6}\right) = 3.755 \text{ inches}$$

**Series of Involute, Finishing Gear Milling Cutters for Each Pitch**

| Number of Cutter | Will cut Gears from | Number of Cutter | Will cut Gears from |
|---|---|---|---|
| 1 | 135 teeth to a rack | 5 | 21 to 25 teeth |
| 2 | 55 to 134 teeth | 6 | 17 to 20 teeth |
| 3 | 35 to 54 teeth | 7 | 14 to 16 teeth |
| 4 | 26 to 34 teeth | 8 | 12 to 13 teeth |

The regular cutters listed above are used ordinarily.

The cutters listed below (an intermediate series having half numbers) may be used when greater accuracy of tooth shape is essential in cases where the number of teeth is between the numbers for which the regular cutters are intended.

| Number of Cutter | Will cut Gears from | Number of Cutter | Will cut Gears from |
|---|---|---|---|
| $1\frac{1}{2}$ | 80 to 134 teeth | $5\frac{1}{2}$ | 19 to 20 teeth |
| $2\frac{1}{2}$ | 42 to 54 teeth | $6\frac{1}{2}$ | 15 to 16 teeth |
| $3\frac{1}{2}$ | 30 to 34 teeth | $7\frac{1}{2}$ | 13 teeth |
| $4\frac{1}{2}$ | 23 to 25 teeth | … | … |

Roughing cutters are made with No. 1 form only. Dimensions of roughing and finishing cutters are given on page 879. Dimensions of cutters for bevel gears are given on page 880.

*Enlarged Fine-Pitch Pinions:* American Standard ANSI B6.7-1977, Information Sheet A provides a different system for 20-degree pressure angle pinion enlargement than is used for coarse-pitch gears. Pinions with 11 through 23 teeth (9 through 14 teeth for 25-degree pressure angle) are enlarged so that a standard tooth thickness rack with addendum $1.05/P$ will start contact 5° of roll above the base circle radius. The use of $1.05/P$ for the addendum allows for center distance variation and eccentricity of the mating gear outside diameter; the 5° roll angle avoids the fabrication of the involute in the troublesome area near the base circle.

Pinions with less than 11 teeth (9 teeth for 25-degree pressure angle) are enlarged to the extent that the highest point of undercut coincides with the start of contact with the standard rack described previously. The height of undercut considered is that produced by a sharp-cornered 120 pitch hob. Pinions with less than 13 teeth (11 teeth for 25-degree pressure angle) are truncated to provide a top land of $0.275/P$. Data for enlarged pinions may be found in Table 9a, Table 9b, Table 9c, and Table 9d.

**Table 9a. Increase in Dedendum, Δ for 20-, and 25-Degree Pressure Angle Fine-Pitch Enlarged Pinions and Reduced Gears** *ANSI B6.7-1977*

| Diametral Pitch, P | Δ | Diametral Pitch, P | Δ | Diametral Pitch, P | Δ | Diametral Pitch, P | Δ | Diametral Pitch, P | Δ |
|---|---|---|---|---|---|---|---|---|---|
| 20 | 0.0000 | 32 | 0.0007 | 48 | 0.0012 | 72 | 0.0015 | 96 | 0.0016 |
| 24 | 0.0004 | 40 | 0.0010 | 64 | 0.0015 | 80 | 0.0015 | 120 | 0.0017 |

Δ = increase in standard dedendum to provide increased clearance. See footnote to Table 9d.

**Table 9b. Dimensions Required when Using Enlarged, Fine-pitch, 14½-Degree Pressure Angle Pinions** *ANSI B6.7-1977, Information Sheet B*

| Enlarged Pinion | | | Standard Center-distance System (Long and Short Addendum) | | | | Enlarged Center-distance System | | |
|---|---|---|---|---|---|---|---|---|---|
| | | | Reduced Mating Gear | | | | Enlarged Pinion Mating with St'd. Gear | Two Equal Enlarged Mating Pinions[a] | |
| No. of Teeth n | Outside Diameter | Cir. Tooth Thickness at Standard Pitch Dia. | Decrease in Standard Outside Dia.[b] | Cir. Tooth Thickness at Standard Pitch Dia. | Recommended Minimum No. of Teeth N | Contact Ratio, n Mating with N | Increase over St'd. Center Distance | | Contact Ratio of Two Equal Enlarged Mating Pinions |
| 10 | 13.3731 | 1.9259 | 1.3731 | 1.2157 | 54 | 1.831 | 0.6866 | 1.3732 | 1.053 |
| 11 | 14.3104 | 1.9097 | 1.3104 | 1.2319 | 53 | 1.847 | 0.6552 | 1.3104 | 1.088 |
| 12 | 15.2477 | 1.8935 | 1.2477 | 1.2481 | 52 | 1.860 | 0.6239 | 1.2477 | 1.121 |
| 13 | 16.1850 | 1.8773 | 1.1850 | 1.2643 | 51 | 1.873 | 0.5925 | 1.1850 | 1.154 |
| 14 | 17.1223 | 1.8611 | 1.1223 | 1.2805 | 50 | 1.885 | 0.5612 | 1.2223 | 1.186 |
| 15 | 18.0597 | 1.8448 | 1.0597 | 1.2967 | 49 | 1.896 | 0.5299 | 1.0597 | 1.217 |
| 16 | 18.9970 | 1.8286 | 0.9970 | 1.3130 | 48 | 1.906 | 0.4985 | 0.9970 | 1.248 |
| 17 | 19.9343 | 1.8124 | 0.9343 | 1.3292 | 47 | 1.914 | 0.4672 | 0.9343 | 1.278 |
| 18 | 20.8716 | 1.7962 | 0.8716 | 1.3454 | 46 | 1.922 | 0.4358 | 0.8716 | 1.307 |
| 19 | 21.8089 | 1.7800 | 0.8089 | 1.3616 | 45 | 1.929 | 0.4045 | 0.8089 | 1.336 |
| 20 | 22.7462 | 1.7638 | 0.7462 | 1.3778 | 44 | 1.936 | 0.3731 | 0.7462 | 1.364 |
| 21 | 23.6835 | 1.7476 | 0.6835 | 1.3940 | 43 | 1.942 | 0.3418 | 0.6835 | 1.392 |
| 22 | 24.6208 | 1.7314 | 0.6208 | 1.4102 | 42 | 1.948 | 0.3104 | 0.6208 | 1.419 |
| 23 | 25.5581 | 1.7151 | 0.5581 | 1.4265 | 41 | 1.952 | 0.2791 | 0.5581 | 1.446 |
| 24 | 26.4954 | 1.6989 | 0.4954 | 1.4427 | 40 | 1.956 | 0.2477 | 0.4954 | 1.472 |
| 25 | 27.4328 | 1.6827 | 0.4328 | 1.4589 | 39 | 1.960 | 0.2164 | 0.4328 | 1.498 |
| 26 | 28.3701 | 1.6665 | 0.3701 | 1.4751 | 38 | 1.963 | 0.1851 | 0.3701 | 1.524 |
| 27 | 29.3074 | 1.6503 | 0.3074 | 1.4913 | 37 | 1.965 | 0.1537 | 0.3074 | 1.549 |
| 28 | 30.2447 | 1.6341 | 0.2448 | 1.5075 | 36 | 1.967 | 0.1224 | 0.2448 | 1.573 |
| 29 | 31.1820 | 1.6179 | 0.1820 | 1.5237 | 35 | 1.969 | 0.0910 | 0.1820 | 1.598 |
| 30 | 32.1193 | 1.6017 | 0.1193 | 1.5399 | 34 | 1.970 | 0.0597 | 0.1193 | 1.622 |
| 31 | 33.0566 | 1.5854 | 0.0566 | 1.5562 | 33 | 1.971 | 0.0283 | 0.0566 | 1.646 |

[a] If enlarged mating pinions are of unequal size, the center distance is increased by an amount equal to one-half the sum of their increase over standard outside diameters. Data in this column are not given in the Standard.

[b] To maintain standard center distance when using an enlarged pinion, the mating gear diameter must be decreased by the amount of the pinion enlargement.

All dimensions are given in inches and are for 1 diametral pitch. For other pitches divide tabulated dimensions by the diametrical pitch.

**Table 9c. Tooth Proportions Recommended for Enlarging Fine-Pitch Pinions of 20-Degree Pressure Angle— 20 Diametral Pitch and Finer** *ANSI B6.7-1977*

| Enlarged Pinion Dimensions | | | | | Enlarged C.D. System Pinion Mating with Standard Gear | | Standard Center Distance (Long and Short Addendums) Reduced Gear Dimensions | | | | |
|---|---|---|---|---|---|---|---|---|---|---|---|
| Number of Teeth,[a] $n$ | Outside Diameter, $D_{oP}$ | Addendum, $a_P$ | Basic Tooth Thickness, $t_P$ | Dedendum Based on 20 Pitch,[b] $b_P$ | Contact Ratio Two Equal Pinions | Contact Ratio with a 24-Tooth Gear | Addendum, $a_G$ | Basic Tooth Thickness, $t_G$ | Dedendum Based on 20 Pitch,[b] $b_G$ | Recommended Minimum No. of Teeth, $N$ | Contact Ratio $n$ Mating with $N$ |
| 7 | 10.0102 | 1.5051 | 2.14114 | 0.4565 | 0.697 | 1.003 | 0.2165 | 1.00045 | 2.0235 | 42 | 1.079 |
| 8 | 11.0250 | 1.5125 | 2.09854 | 0.5150 | 0.792 | 1.075 | 0.2750 | 1.04305 | 1.9650 | 40 | 1.162 |
| 9 | 12.0305 | 1.5152 | 2.05594 | 0.5735 | 0.893 | 1.152 | 0.3335 | 1.08565 | 1.9065 | 39 | 1.251 |
| 10 | 13.0279 | 1.5140 | 2.01355 | 0.6321 | 0.982 | 1.211 | 0.3921 | 1.12824 | 1.8479 | 38 | 1.312 |
| 11 | 14.0304 | 1.5152 | 1.97937 | 0.6787 | 1.068 | 1.268 | 0.4387 | 1.16222 | 1.8013 | 37 | 1.371 |
| 12 | 15.0296 | 1.5148 | 1.94703 | 0.7232 | 1.151 | 1.322 | 0.4832 | 1.19456 | 1.7568 | 36 | 1.427 |
| 13 | 15.9448 | 1.4724 | 1.91469 | 0.7676 | 1.193 | 1.353 | 0.5276 | 1.22690 | 1.7124 | 35 | 1.457 |
| 14 | 16.8560 | 1.4280 | 1.88235 | 0.8120 | 1.232 | 1.381 | 0.5720 | 1.25924 | 1.6680 | 34 | 1.483 |
| 15 | 17.7671 | 1.3836 | 1.85001 | 0.8564 | 1.270 | 1.408 | 0.6164 | 1.29158 | 1.6236 | 33 | 1.507 |
| 16 | 18.6782 | 1.3391 | 1.81766 | 0.9009 | 1.323 | 1.434 | 0.6609 | 1.32393 | 1.5791 | 32 | 1.528 |
| 17 | 19.5894 | 1.2947 | 1.78532 | 0.9453 | 1.347 | 1.458 | 0.7053 | 1.35627 | 1.5347 | 31 | 1.546 |
| 18 | 20.5006 | 1.2503 | 1.75298 | 0.9897 | 1.385 | 1.482 | 0.7497 | 1.38861 | 1.4903 | 30 | 1.561 |
| 19 | 21.4116 | 1.2058 | 1.72064 | 1.0342 | 1.423 | 1.505 | 0.7942 | 1.42095 | 1.4458 | 29 | 1.574 |
| 20 | 22.3228 | 1.1614 | 1.68839 | 1.0786 | 1.461 | 1.527 | 0.8386 | 1.45320 | 1.4014 | 28 | 1.584 |
| 21 | 23.2340 | 1.1170 | 1.65595 | 1.1230 | 1.498 | 1.548 | 0.8830 | 1.48564 | 1.3570 | 27 | 1.592 |
| 22 | 24.1450 | 1.0725 | 1.62361 | 1.1675 | 1.536 | 1.568 | 0.9275 | 1.51798 | 1.3125 | 26 | 1.598 |
| 23 | 25.0561 | 1.0281 | 1.59127 | 1.2119 | 1.574 | 1.588 | 0.9719 | 1.55032 | 1.2681 | 25 | 1.601 |
| 24 | 26.0000 | 1.0000 | 1.57080 | 1.2400 | 1.602 | 1.602 | 1.0000 | 1.57080 | 1.2400 | 24 | 1.602 |

[a] Caution should be exercised in the use of pinions above the horizontal lines. They should be checked for suitability, particularly in the areas of contact ratio (less than 1.2 is not recommended), center distance, clearance, and tooth strength.

[b] The actual dedendum is calculated by dividing the values in this column by the desired diametral pitch and then adding to the result an amount Δ found in Table 9a. As an example, a 20-degree pressure angle 7-tooth pinion meshing with a 42-tooth gear would have, for 24 diametral pitch, a dedendum of $0.4565 \div 24 + 0.0004 = 0.0194$. The 42-tooth gear would have a dedendum of $2.0235 \div 24 + 0.004 = 0.0847$ inch.

All dimensions are given in inches.

**Table 9d. Tooth Proportions Recommended for Enlarging Fine-Pitch Pinions of 25-Degree Pressure Angle— 20 Diametral Pitch and Finer** *ANSI B6.7-1977, Information Sheet B*

| Enlarged Pinion Dimensions | | | | | Enlarged C.D. System Pinion Mating with Standard Gear | | Standard Center Distance (Long and Short Addendums) Reduced Gear Dimensions | | | | |
|---|---|---|---|---|---|---|---|---|---|---|---|
| Number of Teeth,[a] $n$ | Outside Diameter, $D_{oP}$ | Addendum, $a_P$ | Basic Tooth Thickness, $t_P$ | Dedendum Based on 20 Pitch, Dedendum Based on 20 Pitch,[b] $b_P$ | Contact Ratio Two Equal Pinions | Contact Ratio with a 15-Tooth Gear | Addendum, $a_G$ | Basic Tooth Thickness, $t_G$ | Dedendum Based on 20 Pitch,[b] $b_G$ | Recommended Minimum No. of Teeth, $N$ | Contact Ratio $n$ Mating with $N$ |
| 6 | 8.7645 | 1.3822 | 2.18362 | 0.5829 | 0.696 | 0.954 | 0.3429 | 0.95797 | 1.8971 | 24 | 1.030 |
| 7 | 9.7253 | 1.3626 | 2.10029 | 0.6722 | 0.800 | 1.026 | 0.4322 | 1.04130 | 1.8078 | 23 | 1.108 |
| 8 | 10.6735 | 1.3368 | 2.01701 | 0.7616 | 0.904 | 1.094 | 0.5216 | 1.12459 | 1.7184 | 22 | 1.177 |
| 9 | 11.6203 | 1.3102 | 1.94110 | 0.8427 | 1.003 | 1.156 | 0.6029 | 1.20048 | 1.6371 | 20 | 1.234 |
| 10 | 12.5691 | 1.2846 | 1.87345 | 0.9155 | 1.095 | 1.211 | 0.6755 | 1.26814 | 1.5645 | 19 | 1.282 |
| 11 | 13.5039 | 1.2520 | 1.80579 | 0.9880 | 1.183 | 1.261 | 0.7480 | 1.33581 | 1.4920 | 18 | 1.322 |
| 12 | 14.3588 | 1.1794 | 1.73813 | 1.0606 | 1.231 | 1.290 | 0.8206 | 1.40346 | 1.4194 | 17 | 1.337 |
| 13 | 15.2138 | 1.1069 | 1.67047 | 1.1331 | 1.279 | 1.317 | 0.8931 | 1.47112 | 1.3469 | 16 | 1.347 |
| 14 | 16.0686 | 1.0343 | 1.60281 | 1.2057 | 1.328 | 1.343 | 0.9657 | 1.53878 | 1.2743 | 15 | 1.352 |
| 15 | 17.0000 | 1.0000 | 1.57030 | 1.2400 | 1.358 | 1.358 | 1.0000 | 1.57080 | 1.2400 | 15 | 1.358 |

[a] Caution should be exercised in the use of pinions above the horizontal lines. They should be checked for suitability, particularly in the areas of contact ratio (less than 1.2 is not recommended), center distance, clearance, and tooth strength.

[b] The actual dedendum is calculated by dividing the values in this column by the desired diametral pitch and then adding to the result an amount Δ found in Table 9a. As an example, a 20-degree pressure angle 7-tooth pinion meshing with a 42-tooth gear would have, for 24 diametral pitch, a dedendum of $0.4565 \div 24 + 0.0004 = 0.0194$. The 42-tooth gear would have a dedendum of $2.0235 \div 24 + 0.004 = 0.0847$ inch.

All dimensions are given in inches.

All values are for 1 diametral pitch. For any other sizes of teeth, all linear dimensions should be divided by the diametral pitch.

*Note:* The tables in the ANSI B6.7-1977 Standard also specify Form Diameter, Roll Angle to Form Diameter, and Top Land. These are not shown here. The top land is in no case less than $0.275/P$. The form diameters and the roll angles to form diameter shown in the Standard are the values which should be met with a standard hob when generating the tooth thicknesses shown in the tables. These form diameters provide more than enough length of involute profile for any mating gear smaller than a rack. However, since these form diameters are based on gear tooth generation using standard hobs, they should impose little or no hardship on manufacture except in cases of the most critical quality levels. In such cases, form diameter specifications and master gear design should be based upon actual mating conditions.

**Minimum Number of Teeth to Avoid Undercutting by Hob.—**The data in the above tables give tooth proportions for low numbers of teeth to avoid interference between the gear tooth tip and the pinion tooth flank. Consideration must also be given to possible undercutting of the pinion tooth flank by the hob used to cut the pinion. The minimum number of teeth $N_{min}$ of standard proportion that may be cut without undercut is: $N_{min} = 2P\csc^2\phi\,[a_H - r_t(1 - \sin\phi)]$ where: $a_H$ = cutter addendum; $r_t$ = radius at cutter tip or corners; $\phi$ = cutter pressure angle; and $P$ = diametral pitch.

**Gear to Mesh with Enlarged Pinion.—**Data in the fifth column of Table 8 show minimum number of teeth in a mating gear which can be cut with hob or rack type cutter without undercut, when outside diameter of gear has been reduced an amount equal to the pinion enlargement to retain the standard center distance. To calculate $N$ for the gear, insert addendum $a$ of enlarged mating pinion in the formula $N = 2a \times \csc^2\phi$.

*Example:* A gear is to mesh with a 24-tooth pinion of 1 diametral pitch which has been enlarged 0.4954 inch, as shown by the table. The pressure angle is 14½ degrees. Find minimum number of teeth $N$ for reduced gear.

$$\text{Pinion addendum} = 1 + (0.4954 \div 2) = 1.2477$$

$$\text{Hence,} \quad N = 2 \times 1.2477 \times 15.95 = 39.8 \text{ (use 40)}$$

In the case of fine pitch gears with reduced outside diameters, the recommended minimum numbers of teeth given in Table 9b, Table 9c, and Table 9d are somewhat more than the minimum numbers required to prevent undercutting and are based upon studies made by the *American Gear Manufacturers Association.*

**Standard Center-Distance System for Enlarged Pinions.—**In this system, sometimes referred to as "long and short addendums," the center distance is made standard for the numbers of teeth in pinion and gear. The outside diameter of the gear is decreased by the same amount that the outside of the pinion is enlarged.

The advantages of this system are: 1) No change in center distance or ratio is required; 2) The operating pressure angle remains standard; and 3) A slightly greater contact ratio is obtained than when the center distance is increased.

The disadvantages are 1) The gears as well as the pinion must be changed from standard dimensions; 2) Pinions having fewer than the minimum number of teeth to avoid undercut cannot be satisfactorily meshed together; and 3) In most cases where gear trains include idler gears, the standard center-distance system cannot be used.

**Enlarged Center-Distance System for Enlarged Pinions.—**If an enlarged pinion is meshed with another enlarged pinion or with a gear of standard outside diameter, the center distance must be increased. For fine-pitch gears, it is usually satisfactory to increase the center distance by an amount equal to one-half of the enlargements (see eighth column of Table 9b). This is an approximation as theoretically there is a slight increase in backlash.

The advantages of this system are: 1) Only the pinions need be changed from the standard dimensions; 2) Pinions having fewer than 18 teeth may engage other pinions in this range; 3) The pinion tooth, which is the weaker member, is made stronger by the enlargement; and 4) The tooth contact stress, which controls gear durability, is lowered by being moved away from the pinion base circle.

The disadvantages are: 1) Center distances must be enlarged over the standard; 2) The operating pressure angle increases slightly with different combinations of pinions and gears, which is usually not important; and 3) The contact ratio is slightly smaller than that obtained with the standard center-distance system.

This consideration is of minor importance as in the worst case the loss is approximately only 6 percent.

*Enlarged Pinions Meshing without Backlash:* When two enlarged pinions are to mesh without backlash, their center distance will be greater than the standard and less than that for the enlarged center-distance system. This center distance may be calculated by the formulas given in the following section.

**Center Distance at Which Modified Mating Spur Gears Will Mesh without Backlash.—** When the tooth thickness of one or both of a pair of mating spur gears has been increased or decreased from the standard value ($\pi \div 2P$), the center distance at which they will mesh tightly (without backlash) may be calculated from the following formulas:

$$\text{inv}\,\phi_1 = \text{inv}\,\phi + \frac{P(t+T)-\pi}{n+N} \qquad C = \frac{n+N}{2P} \qquad C_1 = \frac{\cos\phi}{\cos\phi_1} \times C$$

In these formulas, $P$ = diametral pitch; $n$ = number of teeth in pinion; $N$ = number of teeth in gear; $t$ and $T$ are the actual tooth thicknesses of the pinion and gear, respectively, on their standard pitch circles; inv $\phi$ = involute function of standard pressure angle of gears; $C$ = standard center distance for the gears; $C_1$ = center distance at which the gears mesh without backlash; and inv $\phi_1$ = involute function of operating pressure angle when gears are meshed tightly at center distance $C_1$.

*Example:* Calculate the center distance for no backlash when an enlarged 10-tooth pinion of 100 diametral pitch and 20-degree pressure angle is meshed with a standard 30-tooth gear, the circular tooth thickness of the pinion and gear, respectively, being 0.01873 and 0.015708 inch. From the table of involute functions, page 107, inv 20-degrees = 0.014904. Therefore,

$$\text{inv}\,\phi_1 = \text{inv}\,20° + \frac{100(0.01873 + 0.015708) - \pi}{(10+30)} = 0.014904 + \frac{0.34438 - 0.31416}{4}$$

$$\text{inv}\,\phi_1 = 0.022459 \qquad \text{then, from the involute table} \qquad \phi_1 = 22°49'$$

$$C = \frac{n+N}{2P} = \frac{10+30}{2 \times 100} = 0.2000 \text{ inch}$$

$$C_1 = \frac{\cos 20°}{\cos 22°49'} \times 0.2000 = \frac{0.93969}{0.92175} \times 0.2000 = 0.2039 \text{ inch}$$

**Contact Diameter.—**For two meshing gears it is important to know the contact diameter of each. A first gear with number of teeth, $n$, and outside diameter, $d_0$, meshes at a standard center distance with a second gear with number of teeth, $N$, and outside diameter, $D_0$; both gears have a diametral pitch, $P$, and pressure angle, $\phi$, $a$, $A$, $b$, and $B$ are unnamed angles used only in the calculations. The contact diameter, $d_c$, is found by a three-step calculation that can be done by hand using a trigonometric table and a logarithmic table or a desk calculator. Slide rule calculation is not recommended because it is not accurate enough to give good results. The three-step formulas to find the contact diameter, $d_c$, of the first gear are:

$$\cos A = \frac{N\cos\phi}{D_0 \times P} \tag{1}$$

$$\tan b = \tan\phi - \frac{N}{n}(\tan A - \tan\phi) \tag{2}$$

$$d_c = \frac{n\cos\phi}{P\cos b} \tag{3}$$

Similarly the three-step formulas to find the contact diameter, $D_c$, of the second gear are:

$$\cos a = \frac{n\cos\phi}{d_o \times P} \tag{4}$$

$$\tan B = \tan\phi - \frac{n}{N}(\tan a - \tan\phi) \tag{5}$$

$$D_c = \frac{N\cos\phi}{P\cos B} \tag{6}$$

**Contact Ratio.—**The contact ratio of a pair of mating spur gears must be well over 1.0 to assure a smooth transfer of load from one pair of teeth to the next pair as the two gears rotate under load. Because of a reduction in contact ratio due to such factors as tooth deflection, tooth spacing errors, tooth tip breakage, and outside diameter and center distance tolerances, the contact ratio of gears for power transmission as a general rule should not be less than about 1.4. A contact ratio of as low as 1.15 may be used in extreme cases, provided the tolerance effects mentioned above are accounted for in the calculation. The formula for determining the contact ratio, $m_f$, using the nomenclature in the previous section is:

$$m_f = \frac{N}{6.28318}(\tan A - \tan B) \tag{7a}$$

or

$$m_f = \frac{N}{6.28318}(\tan a - \tan b) \tag{7b}$$

or

$$m_f = \frac{\sqrt{R_0^2 - R_B^2} + \sqrt{r_0^2 - r_B^2} - C\sin\theta}{P\cos\theta} \tag{7c}$$

where $R_0$ = outside radius of first gear; $R_B$ = base radius of first gear; $r_0$ = outside radius of second gear; $r_B$ = base radius of second gear; $C$ = center distance; $I$ = pressure angle; and, $p$ = circular pitch.

Both formulas, Equations (7a) and (7b), should give the same answer. It is good practice to use both formulas as a check on the previous calculations.

**Lowest Point of Single Tooth Contact.—**This diameter on the pinion (sometimes referred to as LPSTC is used to find the maximum contact compressive stress (sometimes called the Hertz Stress) of a pair of mating spur gears. The two-step formulas for determining this pinion diameter, $d_L$, using the same nomenclature as in the previous sections with $c$ and $C$ as unnamed angles used only in the calculations are:

$$\tan c = \tan a - \frac{6.28318}{n} \tag{8}$$

$$d_L = \frac{n\cos\phi}{P\cos c} \tag{9}$$

In some cases it is necessary to have a plot of the compressive stress over the whole cycle of contact; in this case the LPSTC for the gear is required also. The similar two-step formulas for this gear diameter are:

$$\tan C = \tan A - \frac{6.28318}{N} \tag{10}$$

$$D_L = \frac{N\cos\phi}{P\cos C} \tag{11}$$

**Maximum Hob Tip Radius.—**The standard gear tooth proportions given by the formulas in Table 2 on page 2212 provide a specified size for the rack fillet radius in the general form of (a constant) × (pitch). For any given standard this constant may vary up to a maximum which it is geometrically impossible to exceed; this maximum constant, $r_c$ (max), is found by the formula:

$$r_c\,(\max) = \frac{0.785398\cos\phi - b\sin\phi}{1 - \sin\phi} \qquad (12)$$

where $b$ is the similar constant in the specified formula for the gear dedendum. The hob tip radius of any standard hob to finish cut any standard gear may vary from zero up to this limiting value.

**Undercut Limit for Hobbed Involute Gears.—**It is well to avoid designing and specifying gears that will have a hobbed trochoidal fillet that undercuts the involute gear tooth profile. This should be avoided because it may cause the involute profile to be cut away up to a point above the required contact diameter with the mating gear so that involute action is lost and the contact ratio reduced to a level that may be too low for proper conjugate action. An undercut fillet will also weaken the beam strength and thus raise the fillet tensile stress of the gear tooth. To assure that the hobbed gear tooth will not have an undercut fillet, the following formula must be satisfied:

$$\frac{b - r_c}{\sin\phi} + r_c \leq 0.5n\sin\phi \qquad (13)$$

where $b$ is the dedendum constant; $r_c$ is the hob or rack tip radius constant; $n$ is the number of teeth in the gear; and $\phi$ is the gear and hob pressure angle. If the gear is not standard or the hob does not roll at the gear pitch diameter, this formula can not be applied and the determination of the expected existence of undercut becomes a considerably more complicated procedure.

**Highest Point of Single Tooth Contact.—**This diameter is used to place the maximum operating load for the determination of the gear tooth fillet stress. The two-step formulas for determining this diameter, $d_H$, of the pinion using the same nomenclature as in the previous sections with $d$ and $D$ as unnamed angles used only in the calculations are:

$$\tan d = \tan b + \frac{6.28318}{n} \qquad (14)$$

$$d_H = \frac{n\cos\phi}{P\cos d} \qquad (15)$$

Similarly for the gear:

$$\tan D = \tan B + \frac{6.28318}{N} \qquad (16)$$

$$D_H = \frac{N\cos\phi}{P\cos D} \qquad (17)$$

**True Involute Form Diameter.—**The point on the gear tooth at which the fillet and the involute profile are tangent to each other should be determined to assure that it lies at a smaller diameter than the required contact diameter with the mating gear. If the TIF diameter is larger than the contact diameter, then fillet interference will occur with severe damage to the gear tooth profile and rough action of the gear set. This two-step calculation is made by using the following two formulas with $e$ and $E$ as unnamed angles used only in the calculations:

$$\tan e = \tan\phi - \frac{4}{n}\left(\frac{b - r_c}{\sin 2\phi} + \frac{r_c}{2\cos\phi}\right) \qquad (18)$$

$$d_{TIF} = \frac{n\cos\phi}{P\cos e} \qquad (19)$$

As in the previous sections, $\phi$ is the pressure angle of the gear; $n$ is the number of teeth in the pinion; $b$ is the dedendum constant; $r_c$ is the rack or hob tip radius constant; $P$ is the gear diametral pitch and $d_{TIF}$ is the true involute form diameter.

Similarly, for the mating gear:

$$\tan E = \tan\phi - \frac{4}{N}\left(\frac{b - r_c}{\sin 2\phi} + \frac{r_c}{2\cos\phi}\right) \tag{20}$$

$$D_{TIF} = \frac{N\cos\phi}{P\cos E} \tag{21}$$

where $N$ is number of teeth in this mating gear and $D_{TIF}$ is the true involute form diameter.

**Profile Checker Settings.—**The actual tooth profile tolerance will need to be determined on high performance gears that operate either at high unit loads or at high pitch-line velocity. This is done on an involute checker, a machine which requires two settings, the gear base radius and the roll angle in degrees to significant points on the involute. From the smallest diameter outward these significant points are: TIF, Contact Diameter, LPSTC, Pitch Diameter, HPSTC, and Outside Diameter.

The base radius is:

$$R_b = \frac{N\cos\phi}{2P} \tag{22}$$

The roll angle, in degrees, at any point is equal to the tangent of the pressure angle at that point multiplied by 57.2958. The following table shows the tangents to be used at each significant diameter.

| Significant Point on Tooth Profile | Pinion | Gear | For Computation |
|---|---|---|---|
| TIF | $\tan e$ | $\tan E$ | (See Formulas (18) & (20)) |
| Contact Dia. | $\tan b$ | $\tan B$ | (See Formulas (2) & (5)) |
| LPSTC | $\tan c$ | $\tan C$ | (See Formulas (8) & (10)) |
| Pitch Dia. | $\tan\phi$ | $\tan\phi$ | ($\phi$ = Pressure angle) |
| HPSTC | $\tan d$ | $\tan D$ | (See Formulas (14) & (16)) |
| Outside Dia. | $\tan a$ | $\tan A$ | (See Formulas (4) & (1)) |

*Example:* Find the significant diameters, contact ratio and hob tip radius for a 10-diametral pitch, 23-tooth, 20-degree pressure angle pinion of 2.5-inch outside diameter if it is to mesh with a 31-tooth gear of 3.3-inch outside diameter.

Thus: $n = 23$

$d_O = 2.5$

$P = 10$

$N = 31$

$D_O = 3.3$

$\phi = 20°$

1) Pinion contact diameter, $d_c$

$$\cos A = \frac{31 \times 0.93969}{3.3 \times 10} \tag{1}$$
$$= 0.88274 \qquad A = 28°1'30''$$

$$\tan b = 0.36397 - \frac{31}{23}(0.53227 - 0.36397) \tag{2}$$
$$= 0.13713 \qquad b = 7°48'26''$$

$$d_c = \frac{23 \times 0.93969}{10 \times 0.99073} \tag{3}$$
$$= 2.1815 \text{ inches}$$

2) Gear contact diameter, $D_c$

$$\cos a = \frac{23 \times 0.93963}{2.5 \times 10} \tag{4}$$
$$= 0.86452 \qquad a = 30°10'20''$$

$$\tan B = 0.36397 - \frac{23}{31}(0.58136 - 0.36937) \tag{5}$$
$$= 0.20267 \qquad B = 11°27'26''$$

$$D_c = \frac{31 \times 0.93969}{10 \times 0.98000} \tag{6}$$
$$= 2.9725 \text{ inches}$$

3) Contact ratio, $m_f$

$$m_f = \frac{31}{6.28318}(0.53227 - 0.20267) \tag{7a}$$
$$= 1.626$$

$$m_f = \frac{23}{6.28318}(0.58136 - 0.13713) \tag{7b}$$
$$= 1.626$$

4) Pinion LPSTC, $d_L$

$$\tan c = 0.58136 - \frac{6.28318}{23} \tag{8}$$
$$= 0.30818 \qquad c = 17°7'41''$$

$$d_L = \frac{23 \times 0.93969}{10 \times 0.95565} \tag{9}$$
$$= 2.2616 \text{ inches}$$

5) Gear LPSTC, $D_L$

$$\tan C = 0.53227 - \frac{6.28318}{31} \tag{10}$$
$$= 0.32959 \qquad C = 18°14'30''$$

$$D_L = \frac{31 \times 0.93969}{10 \times 0.94974} \tag{11}$$
$$= 3.0672 \text{ inches}$$

6) Maximum permissible hob tip radius, $r_c$ (max). The dedendum factor is 1.25.

$$\begin{aligned} r_c\,(\max) &= \frac{0.785398 \times 0.93969 - 1.25 \times 0.34202}{1 - 0.34202} \\ &= 0.4719 \text{ inch} \end{aligned} \tag{12}$$

7) If the hob tip radius $r_c$ is 0.30, determine if the pinion involute is undercut.

$$\begin{gathered} \frac{1.25 - 0.30}{0.34202} + 0.30 \le 0.5 \times 23 \times 0.34202 \\ 3.0776 < 3.9332 \end{gathered} \tag{13}$$

8) Therefore there is no involute undercut.

9) Pinion HPSTC, $d_H$

$$\begin{aligned} \tan d &= 0.13713 + \frac{6.28318}{23} \\ &= 0.41031 \qquad d = 22°18'32'' \end{aligned} \tag{14}$$

$$\begin{aligned} d_H &= \frac{23 \times 0.93969}{10 \times 0.92515} \\ &= 2.3362 \text{ inches} \end{aligned} \tag{15}$$

10) Gear HPSTC, $D_H$

$$\begin{aligned} \tan D &= 0.20267 + \frac{6.28318}{31} \\ &= 0.40535 \qquad D = 22°3'55'' \end{aligned} \tag{16}$$

$$\begin{aligned} D_H &= \frac{31 \times 0.93969}{10 \times 0.92676} \\ &= 3.1433 \text{ inches} \end{aligned} \tag{17}$$

11) Pinion TIF diameter, $d_{TIF}$

$$\begin{aligned} \tan e &= 0.36397 - \frac{4}{23}\left(\frac{1.25 - 0.30}{0.64279} + \frac{0.30}{2 \times 0.93969}\right) \\ &= 0.07917 \qquad e = 4°31'36'' \end{aligned} \tag{18}$$

$$\begin{aligned} d_{TIF} &= \frac{23 \times 0.93969}{10 \times 0.99688} \\ &= 2.1681 \text{ inches} \end{aligned} \tag{19}$$

12) Gear TIF diameter, $D_{TIF}$

$$\begin{aligned} \tan E &= 0.36397 - \frac{4}{31}\left(\frac{1.25 - 0.30}{0.64279} + \frac{0.30}{2 \times 0.93969}\right) \\ &= 0.15267 \qquad E = 8°40'50'' \end{aligned} \tag{20}$$

$$D_{TIF} = \frac{31 \times 0.93969}{10 \times 0.98855} = 2.9468 \text{ inches} \tag{21}$$

**Gear Blanks for Fine-Pitch Gears.**—The accuracy to which gears can be produced is considerably affected by the design of the gear blank and the accuracy to which the various surfaces of the blank are machined. The following recommendations should not be regarded as inflexible rules, but rather as minimum average requirements for gear-blank quality compatible with the expected quality class of the finished gear.

*Design of Gear Blanks:* The accuracy to which gears can be produced is affected by the design of the blank, so the following points of design should be noted: 1) Gears designed with a hole should have the hole large enough that the blank can be adequately supported during machining of the teeth and yet not so large as to cause distortion; 2) Face widths should be wide enough, in proportion to outside diameters, to avoid springing and to permit obtaining flatness in important surfaces; 3) Short bore lengths should be avoided wherever possible. It is feasible, however, to machine relatively thin blanks in stacks, provided the surfaces are flat and parallel to each other; 4) Where gear blanks with hubs are to be designed, attention should be given to the wall sections of the hubs. Too thin a section will not permit proper clamping of the blank during machining operations and may also affect proper mounting of the gear; and 5) Where pinions or gears integral with their shafts are to be designed, deflection of the shaft can be minimized by having the shaft length and shaft diameter well proportioned to the gear or pinion diameter. The foregoing general principles may also be useful when applied to blanks for coarser pitch gears.

**Specifying Spur and Helical Gear Data on Drawings.—**The data that may be shown on drawings of spur and helical gears fall into three groups: The first group consists of data basic to the design of the gear; the second group consists of data used in manufacturing and inspection; and the third group consists of engineering reference data. The accompanying table may be used as a checklist for the various data which may be placed on gear drawings and the sequence in which they should appear.

*Explanation of Terms Used in Gear Specifications:* 1) Number of teeth is the number of teeth in 360 deg of gear circumference. In a sector gear, both the actual number of teeth in the sector and the theoretical number of teeth in 360 deg should be given.

2) Diametral pitch is the ratio of the number of teeth in the gear to the number of inches in the standard pitch diameter. It is used in this standard as a nominal specification of tooth size.

a) Normal diametral pitch is the diametral pitch in the normal plane.

b) Transverse diametral pitch is the diametral pitch in the transverse plane.

c) Module is the ratio of the number of teeth in the gear to the number of mm in the standard pitch diameter.

d) Normal module is the module measured in the normal plane.

e) Transverse module is the module measured in the transverse plane.

3) Pressure angle is the angle between the gear tooth profile and a radial line at the pitch point. It is used in this standard to specify the pressure angle of the basic rack used in defining the gear tooth profile.

a) Normal pressure angle is the pressure angle in the normal plane.

b) Transverse pressure angle is the pressure angle in the transverse plane.

4) Helix angle is the angle between the pitch helix and an element of the pitch cylinder, unless otherwise specified.

a) Hand of helix is the direction in which the teeth twist as they recede from an observer along the axis. A right-hand helix twists clockwise and a left-hand helix twists counterclockwise.

5) Standard pitch diameter is the diameter of the pitch circle. It equals the number of teeth divided by the transverse diametral pitch.

6) Tooth form may be specified as standard addendum, long addendum, short addendum, modified involute or special. If a modified involute or special tooth form is required, a detailed view should be shown on the drawing. If a special tooth form is specified, roll angles must be supplied (see page 2239).

7) Addendum is the radial distance between the standard pitch circle and the outside circle. The actual value depends on the specification of outside diameter.

8) Whole depth is the total radial depth of the tooth space. The actual value is dependent on the specification of outside diameter and root diameter.

9) Maximum calculated circular thickness on the standard pitch circle is the tooth thickness which will provide the desired minimum backlash when the gear is assembled in mesh with its mate on minimum center distance. Control may best be exerted by testing in tight mesh with a master which integrates all errors in the several teeth in mesh through the arc of action as explained on page 2250. This value is independent of the effect of runout.

a) Maximum calculated *normal* circular thickness is the circular tooth thickness in the normal plane which satisfies requirements explained in (9).

10) Gear testing radius is the distance from its axis of rotation to the standard pitch line of a standard master when in intimate contact under recommended pressure on a variable-center-distance running gage. Maximum testing radius should be calculated to provide the maximum circular tooth thickness specified in (9) when checked as explained on page 2250. This value is affected by the runout of the gear. Tolerance on testing radius must be equal to or greater than the total composite error permitted by the quality class specified in (11).

11) Quality class is specified for convenience when talking or writing about the accuracy of the gear.

12) Maximum total composite error, and 13) Maximum tooth-to-tooth composite error. Actual tolerance values (12 and 13) permitted by the quality class (11) are specified in inches to provide machine operator or inspector with tolerances required to inspect the gear.

14) Testing pressure recommendations are given on page 2250. Incorrect testing pressure will result in incorrect measurement of testing radius.

15) Master specifications by tool or code number may be required to call for the use of a special master gear when tooth thickness deviates excessively from standard.

16) Measurement over two 0.xxxx diameter pins may be specified to assist the manufacturing department in determining size at machine for setup only.

17) Outside diameter is usually shown on the drawing of the gear together with other blank dimensions so that it will not be necessary for machine operators to search gear tooth data for this dimension. Since outside diameter is also frequently used in the manufacture and inspection of the teeth, it may be included in the data block with other tooth specifications if preferred. To permit use of topping hobs for cutting gears on which the tooth thickness has been modified from standard, the outside diameter should be related to the specified gear testing radius (10).

18) Maximum root diameter is specified to assure adequate clearance for the outside diameter of the mating gear. This dimension is usually considered acceptable if the gear is checked with a master and meets specifications (10) through (13).

19) Active profile diameter of a gear is the smallest diameter at which the mating gear tooth profile can make contact. Because of difficulties involved in checking, this specification is not recommended for gears finer than 48 pitch.

20) Surface roughness on active profile surfaces may be specified in microinches to be checked by instrument up to about 32 pitch, or by visual comparison in the finer pitch ranges. It is difficult to determine accurately the surface roughness of fine pitch gears. For many commercial applications surface roughness may be considered acceptable on gears which meet the maximum tooth-to-tooth-error specification (13).

21) Mating gear part number may be shown as a convenient reference. If the gear is used in several applications, all mating gears may be listed but usual practice is to record this information in a reference file.

22) Number of teeth in mating gear, and 23) Minimum operating center distance. This information is often specified to eliminate the necessity of getting prints of the mating gear and assemblies for checking the design specifications, interference, backlash, determination of master gear specification, and acceptance or rejection of gears made out of tolerance.

**Data for Spur and Helical Gear Drawings**

| Type of Data | Min. Spur Gear Data | Min. Helical Gear-Data | Add'l Optional Data | Item Number[a] | Data[a] |
|---|---|---|---|---|---|
| Basic Specifications | • | • | | 1 | Number of teeth |
| | • | | | 2 | Diametral pitch or module |
| | | • | | 2a | Normal diametral pitch or module |
| | | | • | 2b | Transverse diametral pitch or module |
| | • | | | 3 | Pressure angle |
| | | • | | 3a | Normal pressure angle |
| | | | • | 3b | Transverse pressure angle |
| | | • | | 4 | Helix angle |
| | | • | | 4a | Hand of helix |
| | • | • | | 5 | Standard pitch diameter |
| | • | • | | 6 | Tooth form |
| | | | • | 7 | Addendum |
| | | | • | 8 | Whole depth |
| | • | | | 9 | Max. calc. circular thickness on std. pitch circle |
| | | • | | 9a | Max. calc. normal circular thickness on std. pitch circle |
| Manufacturing and Inspection | | | • | 10 | Roll angles |
| | • | • | | 11 | A.G.M.A. quality class |
| | • | • | | 12 | Max. total composite error |
| | • | • | | 13 | Max. tooth-to-tooth composite error |
| | | | • | 14 | Testing pressure (Ounces) |
| | • | • | | 15 | Master specification |
| | | | • | 16 | Meas. over two .xxxx dia. pins (For setup only) |
| | • | • | | 17 | Outside diameter (Preferably shown on drawing of gear) |
| | | | • | 18 | Max. root diameter |
| | | | • | 19 | Active profile diameter |
| | | | • | 20 | Surface roughness of active profile |
| Engineering Reference | | | • | 21 | Mating gear part number |
| | | | • | 22 | Number of teeth in mating gear |
| | | | • | 23 | Minimum operating center distance |

[a] An item-by-item explanation of the terms used in this table is given beginning on page 2242.

## Backlash

In general, backlash in gears is play between mating teeth. For purposes of measurement and calculation, backlash is defined as the amount by which a tooth space exceeds the thickness of an engaging tooth. It does not include the effect of center-distance changes of the mountings and variations in bearings. When not otherwise specified, numerical values of backlash are understood to be given on the pitch circles. The general purpose of backlash is to prevent gears from jamming together and making contact on both sides of their teeth simultaneously. Lack of backlash may cause noise, overloading, overheating of the gears and bearings, and even seizing and failure.

Excessive backlash is objectionable, particularly if the drive is frequently reversing, or if there is an overrunning load as in cam drives. On the other hand, specification of an unnecessarily small amount of backlash allowance will increase the cost of gears, because errors in runout, pitch, profile, and mounting must be held correspondingly smaller. Backlash does not affect involute action and usually is not detrimental to proper gear action.

**Determining Proper Amount of Backlash.**—In specifying proper backlash and tolerances for a pair of gears, the most important factor is probably the maximum permissible amount of runout in both gear and pinion (or worm). Next are the allowable errors in profile, pitch, tooth thickness, and helix angle. Backlash between a pair of gears will vary as successive teeth make contact because of the effect of composite tooth errors, particularly runout, and errors in the gear center distances and bearings.

Other important considerations are speed and space for lubricant film. Slow-moving gears, in general, require the least backlash. Fast-moving fine-pitch gears are usually lubricated with relatively light oil, but if there is insufficient clearance for an oil film, and particularly if oil trapped at the root of the teeth cannot escape, heat and excessive tooth loading will occur.

Heat is a factor because gears may operate warmer, and, therefore, expand more, than the housings. The heat may result from oil churning or from frictional losses between the teeth, at bearings or oil seals, or from external causes. Moreover, for the same temperature rise, the material of the gears—for example, bronze and aluminum—may expand more than the material of the housings, usually steel or cast iron.

The higher the helix angle or spiral angle, the more transverse backlash is required for a given normal backlash. The transverse backlash is equal to the normal backlash divided by the cosine of the helix angle.

In designs employing normal pressure angles higher than 20 degrees, special consideration must be given to backlash, because more backlash is required on the pitch circles to obtain a given amount of backlash in a direction normal to the tooth profiles.

Errors in boring the gear housings, both in center distance and alignment, are of extreme importance in determining allowance to obtain the backlash desired. The same is true in the mounting of the gears, which is affected by the type and adjustment of bearings, and similar factors. Other influences in backlash specification are heat treatment subsequent to cutting the teeth, lapping operations, need for recutting, and reduction of tooth thickness through normal wear.

Minimum backlash is necessary for timing, indexing, gun-sighting, and certain instrument gear trains. If the operating speed is very low and the necessary precautions are taken in the manufacture of such gear trains, the backlash may be held to extremely small limits. However, the specification of "zero backlash," so commonly stipulated for gears of this nature, usually involves special and expensive techniques, and is difficult to obtain.

**Table 1. AGMA Recommended Backlash Range for Coarse-Pitch Spur, Helical, and Herringbone Gearing**

| Center Distance (Inches) | Normal Diametral Pitches | | | | |
|---|---|---|---|---|---|
| | 0.5-1.99 | 2-3.49 | 3.5-5.99 | 6-9.99 | 10-19.99 |
| | Backlash, Normal Plane, Inches[a] | | | | |
| Up to 5 | … | … | … | … | .005-.015 |
| Over 5 to 10 | … | … | … | .010-.020 | .010-.020 |
| Over 10 to 20 | … | … | .020-.030 | .015-.025 | .010-.020 |
| Over 20 to 30 | … | .030-.040 | .025-.030 | .020-.030 | … |
| Over 30 to 40 | .040-.060 | .035-.045 | .030-.040 | .025-.035 | … |
| Over 40 to 50 | .050-.070 | .040-.055 | .035-.050 | .030-.040 | … |
| Over 50 to 80 | .060-.080 | .045-.065 | .040-.060 | … | … |
| Over 80 to 100 | .070-.095 | .050-.080 | … | … | … |
| Over 100 to 120 | .080-.110 | … | … | … | … |

[a] Suggested backlash, on nominal centers, measured after rotating to the point of closest engagement. For helical and herringbone gears, divide above values by the cosine of the helix angle to obtain the transverse backlash.

The above backlash tolerances contain allowance for gear expansion due to differential in the operating temperature of the gearing and their supporting structure. The values may be used where the operating temperatures are up to 70 deg F higher than the ambient temperature

For most gearing applications the recommended backlash ranges will provide proper running clearance between engaging teeth of mating gears. Deviation below the minimum or above the maximum values shown, which do not affect operational use of the gearing, should not be cause for rejection.

Definite backlash tolerances on coarse-pitch gearing are to be considered binding on the gear manufacturer only when agreed upon in writing.

Some applications may require less backlash than shown in the above table. In such cases the amount and tolerance should be by agreement between manufacturer and purchaser.

*Recommended Backlash:* In the following tables American Gear Manufacturers Association recommendations for backlash ranges for various kinds of gears are given.* For purposes of measurement and calculation, backlash is defined as the amount by which a tooth space exceeds the thickness of an engaging tooth. When not otherwise specified, numerical values of backlash are understood to be measured at the tightest point of mesh on the pitch circle in a direction normal to the tooth surface when the gears are mounted in their specified position.

*Coarse-Pitch Gears:* Table 1 gives the recommended backlash range for coarse-pitch spur, helical and herringbone gearing. Because backlash for helical and herringbone gears is more conveniently measured in the normal plane, Table 1 has been prepared to show backlash in the normal plane for coarse-pitch helical and herringbone gearing and in the transverse plane for spur gears. To obtain backlash in the transverse plane for helical and herringbone gears, divide the normal plane backlash in Table 1 by the cosine of the helix angle.

**Table 2. AGMA Recommended Backlash Range for Bevel and Hypoid Gears**

| Diametral Pitch | Normal Backlash, Inch | | Diametral Pitch | Normal Backlash, Inch | |
|---|---|---|---|---|---|
| | Quality Numbers 7 through 13 | Quality Numbers 3 through 6 | | Quality Numbers 7 through 13 | Quality Numbers 3 through 6 |
| 1.00 to 1.25 | 0.020-0.030 | 0.045-0.065 | 5.00 to 6.00 | 0.005-0.007 | 0.006-0.013 |
| 1.25 to 1.50 | 0.018-0.026 | 0.035-0.055 | 6.00 to 8.00 | 0.004-0.006 | 0.005-0.010 |
| 1.50 to 1.75 | 0.016-0.022 | 0.025-0.045 | 8.00 to 10.00 | 0.003-0.005 | 0.004-0.008 |
| 1.75 to 2.00 | 0.014-0.018 | 0.020-0.040 | 10.00 to 16.00 | 0.002-0.004 | 0.003-0.005 |
| 2.00 to 2.50 | 0.012-0.016 | 0.020-0.030 | 16.00 to 20.00 | 0.001-0.003 | 0.002-0.004 |
| 2.50 to 3.00 | 0.010-0.013 | 0.015-0.025 | 20 to 50 | 0.000-0.002 | 0.000-0.002 |
| 3.00 to 3.50 | 0.008-0.011 | 0.012-0.022 | 50 to 80 | 0.000-0.001 | 0.000-0.001 |
| 3.50 to 4.00 | 0.007-0.009 | 0.010-0.020 | 80 and finer | 0.000-0.0007 | 0.000-0.0007 |
| 4.00 to 5.00 | 0.006-0.008 | 0.008-0.016 | … | … | … |

Measured at tightest point of mesh

The backlash tolerances given in this table contain allowances for gear expansion due to a differential in the operating temperature of the gearing and their supporting structure. The values may be used where the operating temperature is up to 70° F higher (39° C higher) than the ambient temperature. These backlash values will provide proper running clearances for most gear applications.

The following important factors must be considered in establishing backlash tolerances: a) Center distance tolerance; b) Parallelism of gear axes; c) Side runout or wobble; d) Tooth thickness tolerance; e) Pitch line runout tolerance; f) Profile tolerance; g) Pitch tolerance; h) Lead tolerance; i) Types of bearings and subsequent wear; j) Deflection under load; k) Gear tooth wear; l) Pitch line velocity; m) Lubrication requirements; and n) Thermal expansion of gears and housing.

A tight mesh may result in objectionable gear sound, increased power losses, overheating, rupture of the lubricant film, overloaded bearings and premature gear failure. However, it is recognized that there are some gearing applications where a tight mesh (zero backlash) may be required.

Specifying unnecessarily close backlash tolerances will increase the cost of the gearing. It is obvious from the above summary that the desired amount of backlash is difficult to evaluate. It is, therefore, recommended that when a designer, user or purchaser includes a reference to backlash in a gearing specification and drawing, consultation be arranged with the manufacturer.

* Extracted from Gear Classification Manual, AGMA 390.03 with permission of the publisher, the American Gear Manufacturers Association, 1500 King St., Alexandria, VA 22314.

*Bevel and Hypoid Gears:* Table 2 gives similar backlash range values for bevel and hypoid gears. These are values based upon average conditions for general purpose gearing, but may require modification to meet specific needs.

Backlash on bevel and hypoid gears can be controlled to some extent by axial adjustment of the gears during assembly. However, due to the fact that actual adjustment of a bevel or hypoid gear in its mounting will alter the amount of backlash, it is imperative that the amount of backlash cut into the gears during manufacture is not excessive. Bevel and hypoid gears must always be capable of operation without interference when adjusted for zero backlash. This requirement is imposed by the fact that a failure of the axial thrust bearing might permit the gears to operate under this condition. Therefore, bevel and hypoid gears should never be designed to operate with normal backlash in excess of $0.080/P$ where $P$ is diametral pitch.

*Fine-Pitch Gears:* Table 3 gives similar backlash range values for fine-pitch spur, helical and herringbone gearing.

**Providing Backlash.—**In order to obtain the amount of backlash desired, it is necessary to decrease tooth thicknesses. However, because of manufacturing and assembling inaccuracies not only in the gears but also in other parts, the allowances made on tooth thickness almost always must exceed the desired amount of backlash. Since the amounts of these allowances depend on the closeness of control exercised on all manufacturing operations, no general recommendations for them can be given.

It is customary to make half the allowance for backlash on the tooth thickness of each gear of a pair, although there are exceptions. For example, on pinions having very low numbers of teeth it is desirable to provide all the allowance on the mating gear, so as not to weaken the pinion teeth. In worm gearing, ordinary practice is to provide all of the allowance on the worm which is usually made of a material stronger than that of the worm gear.

In some instances the backlash allowance is provided in the cutter, and the cutter is then operated at the standard tooth depth. In still other cases, backlash is obtained by setting the distance between two tools for cutting the two sides of the teeth, as in straight bevel gears, or by taking side cuts, or by changing the center distance between the gears in their mountings. In spur and helical gearing, backlash allowance is usually obtained by sinking the cutter deeper into the blank than the standard depth. The accompanying table gives the excess depth of cut for various pressure angles.

**Excess Depth of Cut *E* to Provide Backlash Allowance**

| Distribution of Backlash | Pressure Angle ϕ, Degrees | | | | |
|---|---|---|---|---|---|
| | 14½ | 17½ | 20 | 25 | 30 |
| Excess Depth of Cut $E$ to Obtain Circular Backlash $B$[a] | | | | | |
| All on One Gear | 1.93$B$ | 1.59$B$ | 1.37$B$ | 1.07$B$ | 0.87$B$ |
| One-half on Each Gear | 0.97$B$ | 0.79$B$ | 0.69$B$ | 0.54$B$ | 0.43$B$ |
| Excess Depth of Cut $E$ to Obtain Backlash $B_b$, Normal to Tooth Profile | | | | | |
| All on One Gear | 2.00$B_b$ | 1.66$B_b$ | 1.46$B_b$ | 1.[illegible]8$B_b$ | 1.99$B_b$ |
| One-half on Each Gear | 1.00$B_b$ | 0.83$B_b$ | 0.73$B_b$ | 0.59$B_b$ | 0.50$B_b$ |

[a] Circular backlash is the amount by which the width of a tooth space is greater than the thickness of the engaging tooth on the pitch circles. As described in pages 2244 and 2248 this is what is meant by backlash unless otherwise specified.

[b] Backlash measured normal to the tooth profile by inserting a feeler gage between meshing teeth; to convert to circular backlash, $B = B_b \div \cos\phi$.

**Control of Backlash Allowances in Production.—**Measurement of the tooth thickness of gears is perhaps the simplest way of controlling backlash allowances in production.

There are several ways in which this may be done including: 1) chordal thickness measurements as described on page 2226; 2) caliper measurements over two or more teeth as described on page 2317; and 3) measurements over wires.

In this last method, first the theoretical measurement over wires when the backlash allowance is zero is determined by the method described on page 2302; then the amount this measurement must be reduced to obtain a desired backlash allowance is taken from the table on page 2316.

It should be understood, as explained in the section *Measurement of Backlash*, that merely making tooth thickness allowances will not guarantee the amount of backlash in the ready-to-run assembly of two or more gears. Manufacturing limitations will introduce such gear errors as runout, pitch error, profile error, and lead error, and gear-housing errors in both center distance and alignment. All of these make the backlash of the assembled gears different from that indicated by tooth thickness measurements on the individual gears.

**Measurement of Backlash.**—Backlash is commonly measured by holding one gear of a pair stationary and rocking the other back and forth. The movement is registered by a dial indicator having its pointer or finger in a plane of rotation at or near the pitch diameter and in a direction parallel to a tangent to the pitch circle of the moving gear. If the direction of measurement is normal to the teeth, or other than as specified above, it is recommended that readings be converted to the plane of rotation and in a tangent direction at or near the pitch diameter, for purposes of standardization and comparison.

In spur gears, parallel helical gears, and bevel gears, it is immaterial whether the pinion or gear is held stationary for the test. In crossed helical and hypoid gears, readings may vary according to which member is stationary; hence, it is customary to hold the pinion stationary and measure on the gear.

In some instances, backlash is measured by thickness gages or feelers. A similar method utilizes a soft lead wire inserted between the teeth as they pass through mesh. In both methods, it is likewise recommended that readings be converted to the plane of rotation and in a tangent direction at or near the pitch diameter, taking into account the normal pressure angle, and the helix angle or spiral angle of the teeth.

Sometimes backlash in parallel helical or herringbone gears is checked by holding the gear stationary, and moving the pinion axially back and forth, readings being taken on the face or shaft of the pinion, and converted to the plane of rotation by calculation. Another method consists of meshing a pair of gears tightly together on centers and observing the variation from the specified center distance. Such readings should also be converted to the plane of rotation and in a tangent direction at or near the pitch diameter for the reasons previously given.

Measurements of backlash may vary in the same pair of gears, depending on accuracy of manufacturing and assembling. Incorrect tooth profiles will cause a change of backlash at different phases of the tooth action. Eccentricity may cause a substantial difference between maximum and minimum backlash at different positions around the gears. In stating amounts of backlash, it should always be remembered that merely making allowances on tooth thickness does not guarantee the minimum amount of backlash that will exist in assembled gears.

*Fine-Pitch Gears:* The measurement of backlash of fine-pitch gears, when assembled, cannot be made in the same manner and by the same techniques employed for gears of coarser pitches. In the very fine pitches, it is virtually impossible to use indicating devices for measuring backlash. Sometimes a toolmaker's microscope is used for this purpose to good advantage on very small mechanisms.

Another means of measuring backlash in fine-pitch gears is to attach a beam to one of the shafts and measure the angular displacement in inches when one member is held stationary. The ratio of the length of the beam to the nominal pitch radius of the gear or pinion to which the beam is attached gives the approximate ratio of indicator reading to circular backlash. Because of the limited means of measuring backlash between a pair of fine-pitch gears, gear centers and tooth thickness of the gears when cut must be held to very close limits.

Tooth thickness of fine-pitch spur and helical gears can best be checked on a variable-center-distance fixture using a master gear. When checked in this manner, tooth thickness change = 2 × center distance change × tangent of transverse pressure angle, approximately.

**Control of Backlash in Assemblies.**—Provision is often made for adjusting one gear relative to the other, thereby affording complete control over backlash at initial assembly and throughout the life of the gears. Such practice is most common in bevel gearing. It is fairly common in spur and helical gearing when the application permits slight changes between shaft centers. It is practical in worm gearing only for single thread worms with low lead angles. Otherwise faulty contact results.

Another method of controlling backlash quite common in bevel gears and less common in spur and helical gears is to match the high and low spots of the runout gears of one to one ratio and mark the engaging teeth at the point where the runout of one gear cancels the runout of the mating gear.

**Table 3. AGMA Backlash Allowance and Tolerance for Fine-Pitch Spur, Helical, and Herringbone Gearing**

| Backlash Designation | Normal Diametral Pitch Range | Tooth Thinning to Obtain Backlash[a] | | Resulting Approximate Backlash (per Mesh) Normal Plane[b] Inch |
|---|---|---|---|---|
| | | Allowance, per Gear, Inch | Tolerance, per Gear, Inch | |
| A | 20 thru 45 | .002 | 0 to .002 | .004 to .008 |
| | 46 thru 70 | .0015 | 0 to .002 | .003 to .007 |
| | 71 thru 90 | .001 | 0 to .00175 | .002 to .0055 |
| | 91 thru 200 | .00075 | 0 to .00075 | .0015 to .003 |
| B | 20 thru 60 | .001 | 0 to .001 | .002 to .004 |
| | 61 thru 120 | .00075 | 0 to .00075 | .0015 to .003 |
| | 121 thru 200 | .0005 | 0 to .0005 | .001 to .002 |
| C | 20 thru 60 | .0005 | 0 to .0005 | .001 to .002 |
| | 61 thru 120 | .00035 | 0 to .0004 | .0007 to .0015 |
| | 121 thru 200 | .0002 | 0 to .0003 | .0004 to .001 |
| D | 20 thru 60 | .00025 | 0 to .00025 | .0005 to .001 |
| | 61 thru 120 | .0002 | 0 to .0002 | .0004 to .0008 |
| | 121 thru 200 | .0001 | 0 to .0001 | .0002 to .0004 |
| E | 20 thru 60 | Zero[c] | 0 to .00025 | 0 to .0005 |
| | 61 thru 120 | | 0 to .0002 | 0 to .0004 |
| | 121 thru 200 | | 0 to .0001 | 0 to .0002 |

[a] These dimensions are shown primarily for the benefit of the gear manufacturer and represent the amount that the thickness of teeth should be reduced in the pinion and gear below the standard calculated value, to provide for backlash in the mesh. In some cases, particularly with pinions involving small numbers of teeth, it may be desirable to provide for total backlash by thinning the teeth in the gear member only by twice the allowance value shown in column (3). In this case both members will have the tolerance shown in column (4). In some cases, particularly in meshes with a small number of teeth, backlash may be achieved by an increase in basic center distance. In such cases, neither member is reduced by the allowance shown in column (3).

[b] These dimensions indicate the approximate backlash that will occur in a mesh in which each of the mating pairs of gears have the teeth thinned by the amount referred to in Note 1, and are meshed on theoretical centers.

[c] Backlash in gear sets can also be achieved by increasing the center distance above nominal and using the teeth at standard tooth thickness. Class E backlash designation infers gear sets operating under these conditions.

*Backlash* in gears is the play between mating tooth surfaces. For purposes of measurement and calculation, backlash is defined as the amount by which a tooth space exceeds the thickness of an engaging tooth. When not otherwise specified, numerical values of backlash are understood to be measured at the tightest point of mesh on the pitch circle in a direction normal to the tooth surface when the gears are mounted in their specified position.

*Allowance* is the basic amount that a tooth is thinned from basic calculated circular tooth thickness to obtain the required backlash class.

*Tolerance* is the total permissible variation in the circular thickness of the teeth.

**Angular Backlash in Gears.—**When the backlash on the pitch circles of a meshing pair of gears is known, the angular backlash or angular play corresponding to this backlash may be computed from the following formulas.

$$\theta_D = \frac{6875B}{D} \text{ minutes} \qquad \theta_d = \frac{6875B}{d} \text{ minutes}$$

In these formulas, $B$ = backlash between gears, in inches; $D$ = pitch diameter of larger gear, in inches; $d$ = pitch diameter of smaller gear, in inches; $\theta_D$ = angular backlash or angular movement of larger gear in minutes when smaller gear is held fixed and larger gear rocked back and forth; and $\theta_d$ = angular backlash or angular movement of smaller gear, in minutes, when the larger gear is held fixed and the smaller gear rocked back and forth.

**Inspection of Gears.—**Perhaps the most widely used method of determining relative accuracy in a gear is to rotate the gear through at least one complete revolution in intimate contact with a master gear of known accuracy. The gear to be tested and the master gear are mounted on a variable-center-distance fixture and the resulting radial displacements or changes in center distance during rotation of the gear are measured by a suitable device. Except for the effect of backlash, this so-called "composite check" approximates the action of the gear under operating conditions and gives the combined effect of the following errors: runout; pitch error; tooth-thickness variation; profile error; and lateral runout (sometimes called wobble).

*Tooth-to-Tooth Composite Error,* illustrated below, is the error that shows up as flicker on the indicator of a variable-center-distance fixture as the gear being tested is rotated from tooth to tooth in intimate contact with the master gear. Such flicker shows the combined or composite effect of circular pitch error, tooth-thickness variation, and profile error.

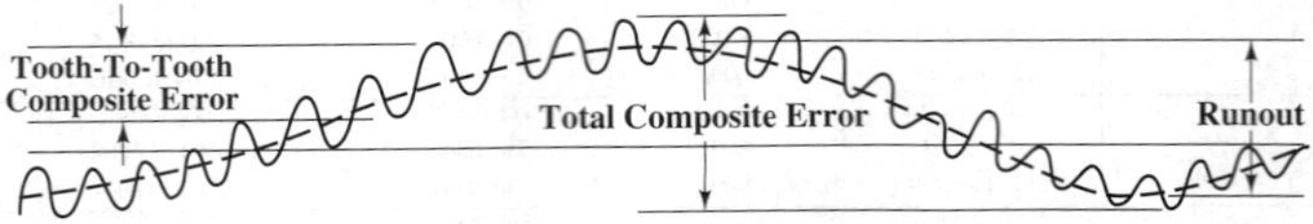


Diagram Showing Nature of Composite Errors

*Total Composite Error,* shown above, is made up of runout, wobble, and the tooth-to-tooth composite error; it is the total center-distance displacement read on the indicating device of the testing fixture, as shown in the accompanying diagram.

**Pressure for Composite Checking of Fine-Pitch Gears.—**In using a variable-center-distance fixture, excessive pressure on fine-pitch gears of narrow face width will result in incorrect readings due to deflection of the teeth. Based on tests, the following checking pressures are recommended for gears of 0.100-inch face width: 20 to 29 diametral pitch, 28 ounces; 30 to 39 pitch, 24 ounces; 40 to 49 pitch, 20 ounces; 50 to 59 pitch, 16 ounces; 60 to 79 pitch, 12 ounces; 80 to 99 pitch, 8 ounces; 100 to 149 pitch, 4 ounces; and 150 and finer pitches, 2 ounces, minimum. These recommended checking pressures are based on the use of antifriction mountings for the movable head of the checking fixture and include the pressure of the indicating device. For face widths less than 0.100 inch, the recommended pressures should be reduced proportionately; for larger widths, no increase is necessary although the force may be increased safely in the proper proportion.

## Internal Gearing

**Internal Spur Gears.—**An internal gear may be proportioned like a standard spur gear turned "outside in" or with addendum and dedendum in reverse positions; however, to avoid interference or improve the tooth form and action, the internal diameter of the gear should be increased and the outside diameter of the mating pinion is also made larger than the size based upon standard or conventional tooth proportions. The extent of these enlargements will be illustrated by means of examples given in the following table, *Rules for Internal Gears – 20-degree Full-Depth Teeth*. The 20-degree involute full-depth tooth form is recommended for internal gears; the 20-degree stub tooth and the $14\frac{1}{2}$-degree full-depth tooth are also used.

**Methods of Cutting Internal Gears.—**Internal spur gears are cut by methods similar in principle to those employed for external spur gears.

They may be cut by one of the following methods: 1) By a generating process, as when using a Fellows gear shaper; 2) by using a formed cutter and milling the teeth; 3) by planing, using a machine of the template or form-copying type (especially applicable to gears of large pitch); and 4) by using a formed tool that reproduces its shape and is given a planing action either on a slotting or a planing type of machine.

Internal gears frequently have a web at one side that limits the amount of clearance space at the ends of the teeth. Such gears may be cut readily on a gear shaper. The most practical method of cutting very large internal gears is on a planer of the form-copying type. A regular spur gear planer is equipped with a special tool holder for locating the tool in the position required for cutting internal teeth.

**Formed Cutters for Internal Gears.—**When formed cutters are used, a special cutter usually is desirable, because the tooth spaces of an internal gear are not the same shape as the tooth spaces of external gearing having the same pitch and number of teeth. This difference is because an internal gear is a spur gear "turned outside in." According to one rule, the standard No. 1 cutter for external gearing may be used for internal gears of 4 diametral pitch and finer, when there are 60 or more teeth. This No. 1 cutter, as applied to external gearing, is intended for all gears having from 135 teeth to a rack. The finer the pitch and the larger the number of teeth, the better the results obtained with a No. 1 cutter. The standard No. 1 cutter is considered satisfactory for jobbing work, and usually when the number of gears to be cut does not warrant obtaining a special cutter, although the use of the No. 1 cutter is not practicable when the number of teeth in the pinion is large in proportion to the number of teeth in the internal gear.

**Arc Thickness of Internal Gear Tooth.—***Rule:* If internal diameter of an internal gear is enlarged as determined by Rules 1 and 2 for Internal Diameters (see *Rules for Internal Gears – 20-degree Full-Depth Teeth*), the arc tooth thickness at the pitch circle equals 1.3888 divided by the diametral pitch, assuming a pressure angle of 20 degrees.

**Arc Thickness of Pinion Tooth.—***Rule:* If the pinion for an internal gear is larger than conventional size (see Outside Diameter of Pinion for Internal Gear, under *Rules for Internal Gears – 20-degree Full-Depth Teeth*), then the arc tooth thickness on the pitch circle equals 1.7528 divided by the diametral pitch, assuming a pressure angle of 20 degrees.

*Note:* For chordal thickness and chordal addendum, see rules and formulas for spur gears.

**Relative Sizes of Internal Gear and Pinion.—**If a pinion is too large or too near the size of its mating internal gear, serious interference or modification of the tooth shape may occur.

*Rule:* For internal gears having a 20-degree pressure angle and full-depth teeth, the difference between the numbers of teeth in gear and pinion should not be less than 12. For teeth of stub form, the smallest difference should be 7 or 8 teeth. For a pressure angle of $14\frac{1}{2}$ degrees, the difference in tooth numbers should not be less than 15.

**Rules for Internal Gears – 20-Degree Full-Depth Teeth**

| To Find | Rule |
|---|---|
| Pitch Diameter | *Rule:* To find the pitch diameter of an internal gear, divide the number of internal gear teeth by the diametral pitch. The pitch diameter of the mating pinion also equals the number of pinion teeth divided by the diametral pitch, the same as for external spur gears. |
| Internal Diameter (Enlarged to Avoid Interference) | *Rule* 1: For internal gears to mesh with pinions having 16 teeth or more, subtract 1.2 from the number of teeth and divide the remainder by the diametral pitch.<br>*Example:* An internal gear has 72 teeth of 6 diametral pitch and the mating pinion has 18 teeth; then<br>$\text{Internal diameter} = \frac{72 - 1.2}{6} = 11.8 \text{ inches}$<br>*Rule* 2: If circular pitch is used, subtract 1.2 from the number of internal gear teeth, multiply the remainder by the circular pitch, and divide the product by 3.1416. |
| Internal Diameter (Based upon Spur Gear Reversed) | *Rule:* If the internal gear is to be designed to conform to a spur gear turned outside in, subtract 2 from the number of teeth and divide the remainder by the diametral pitch to find the internal diameter.<br>*Example:* (Same as Example above.)<br>$\text{Internal diameter} = \frac{72 - 2}{6} = 11.666 \text{ inches}$ |
| Outside Diameter of Pinion for Internal Gear | *Note:* If the internal gearing is to be proportioned like standard spur gearing, use the rule or formula previously given for spur gears in determining the outside diameter. The rule and formula following apply to a pinion that is enlarged and intended to mesh with an internal gear enlarged as determined by the preceding Rules 1 and 2 above.<br>*Rule:* For pinions having 16 teeth or more, add 2.5 to the number of pinion teeth and divide by the diametral pitch.<br>*Example* 1: A pinion for driving an internal gear is to have 18 teeth (full depth) of 6 diametral pitch; then<br>$\text{Outside diameter} = \frac{18 + 2.5}{6} = 3.416 \text{ inches}$<br>By using the rule for external spur gears, the outside diameter = 3.333 inches. |
| Center Distance | *Rule:* Subtract the number of pinion teeth from the number of internal gear teeth and divide the remainder by two times the diametral pitch. |
| Tooth Thickness | See paragraphs, *Arc Thickness of Internal Gear Tooth* and *Effect of Diameter of Cutting on Profile and Pressure Angle of Worms*, on previous page. |

## British Standard for Spur and Helical Gears

BS 436: Part 1:1967 (Withdrawn): Spur and Helical Gears, Basic Rack Form, Pitches and Accuracy for Diametral Pitch Series, now has sections concerned with basic requirements for general tooth form, standard pitches, accuracy and accuracy testing procedures, and the showing of this information on engineering drawings to make sure that the gear manufacturer receives the required data. The latest form of the Standard complies with ISO agreements. The given pitches are in accordance with ISO R54, and the basic rack form and its modifications are in accordance with the ISO R53 "Basic Rack of Cylindrical Gears for General Engineering and for Heavy Engineering Standard."

Five grades of gear accuracy in previous versions are replaced by grades 3 to 12 of the draft ISO Standard. Grades 1 and 2 cover master gears that are not dealt with here. BS 436: Part 1:1967 (Withdrawn) is a companion to the following British Standards:

- BS 235:1987 (Withdrawn), Specification for gears electric traction
- BS 545:1982 (Withdrawn), Specification for bevel gears (machine cut)
- BS 721-1:1963, Specification for Worm Gearing-Imperial Units; BS 721-2:1983 Specification for Worm Gearing- Metric Units
- BS 821:1938 (Withdrawn), Iron Castings for Gears and Gear Blanks (Ordinary, Medium and High Grade)
- BS 978-1:1968, Specification for fine pitch gears-Involute spur and helical gears; BS 978-2:1952, Specification for fine pitch gears-Cycloidal type gears; BS 978-3:1952, Specification for fine pitch gears-Bevel gears
- BS 1807-1:1952 (Withdrawn), Gears for turbines and similar drives-Accuracy; BS 1807-1:1958 (Withdrawn), Gears for turbines and similar drives-Tooth form and pitches
- BS 2519-1:1976 (Withdrawn), Glossary for gears-Geometrical definitions; BS 2519-2:1976, Glossary for gears-Notations
- BS 3027 (Withdrawn), Dimensions for Worm Gear Units
- BS 3696 (Withdrawn), Master Gears

Part 1 of BS 436 applies to external and internal involute spur and helical gears on parallel shafts and having normal diametral pitch of 20 or coarser. The basic rack and tooth form are specified, also first and second preference standard pitches and fundamental tolerances that determine the grades of gear accuracy, and requirements for terminology and notation.

These requirements include: center distance $a$; reference circle diameter $d$, for pinion $d_1$ and wheel $d_2$; tip diameter $d_a$ for pinion $d_{a1}$ and wheel $d_{a2}$; center distance modification coefficient $\gamma$; face width $b$ for pinion $b_1$ and wheel $b_2$; addendum modification coefficient $x$, for pinion $x_1$ and wheel $x_2$; length of arc $l$; diametral pitch $P_t$; normal diametral pitch $p_n$; transverse pitch $p_t$; number of teeth $z$, for pinion $z_1$ and wheel $z_2$; helix angle at reference cylinder $\beta$; pressure angle at reference cylinder $\alpha$; normal pressure angle at reference cylinder $\alpha_n$; transverse pressure angle at reference cylinder $\alpha_t$; and transverse pressure angle, working, $\alpha_{tw}$.

The basic rack tooth profile has a pressure angle of 20. The Standard permits the total tooth depth to be varied within 2.25 to 2.40, so that the root clearance can be increased within the limits of 0.25 to 0.040 to allow for variations in manufacturing processes; and the root radius can be varied within the limits of 0.25 to 0.39. Tip relief can be varied within the limits shown at the right in the illustration.

*Standard normal diametral pitches $P_n$*, BS 436: Part 1:1967 (Withdrawn), are in accordance with ISO R54. The preferred series, rather than the second choice, should be used where possible. Preferred normal diametral pitches for spur and helical gears (second choices in parentheses) are: 20 (18), 16 (14), 12 (11), 10 (9), 8 (7), 6 (5.5), 5 (4.5), 4 (3.5), 3 (2.75), 2.5 (2.25), 2 (1.75), 1.5, 1.25, and 1.

*Information to be Given on Drawings:* British Standard BS 308 (Withdrawn) "Engineering Drawing Practice," specifies data to be included on drawings of spur and helical gears. For all gears the data should include: number of teeth, normal diametral pitch, basic rack tooth form, axial pitch, tooth profile modifications, blank diameter, reference circle diameter, and helix angle at reference cylinder (0 for straight spur gears), tooth thickness at reference cylinder, grade of gear, drawing number of mating gear, working center distance, and backlash.

For single helical gears, the above data should be supplemented with hand and lead of the tooth helix; and for double helical gears, with the hand in relation to a specific part of the face width and the lead of tooth helix.

Inspection instructions should be included, care being taken to avoid conflicting requirements for accuracy of individual elements, and single- and dual-flank testing. Supplementary data covering specific design, manufacturing and inspection requirements or limitations may be needed, together with other dimensions and tolerances, material, heat treatment, hardness, case depth, surface texture, protective finishes, and drawing scale.

**Addendum Modification to Involute Spur and Helical Gears.**—The British Standards Institute guide PD 6457:1970 contains certain design recommendations aimed at making it possible to use standard cutting tools for some sizes of gears. Essentially, the guide covers addendum modification and includes formulas for both English and metric units.

*Addendum Modification* is an enlargement or reduction of gear tooth dimensions that results from displacement of the reference plane of the generating rack from its normal position. The displacement is represented by the coefficient X, X1, or X2, where X is the equivalent dimension for gears of unit module or diametral pitch. The addendum modification establishes a datum tooth thickness at the reference circle of the gear but does not necessarily establish the height of either the reference addendum or the working addendum. In any pair of gears, the datum tooth thicknesses are those that always give zero backlash at the meshing center distance. Normal practice requires allowances for backlash for all unmodified gears.

Taking full advantage of the adaptability of the involute system allows various tooth design features to be obtained. Addendum modification has the following applications: avoiding undercut tooth profiles; achieving optimum tooth proportions and control of the proportion of receding to approaching contact; adapting a gear pair to a predetermined center distance without recourse to nonstandard pitches; and permitting use of a range of working pressure angles using standard geometry tools.

**BS 436: Part 3:1986 (Withdrawn) "Spur and Helical Gears."**—This part provides methods for calculating contact and root bending stresses for metal involute gears, and is somewhat similar to the ANSI/AGMA Standard for calculating stresses in pairs of involute spur or helical gears. Stress factors covered in the British Standard include the following:

*Tangential Force* is the nominal force for contact and bending stresses.

*Zone Factor* accounts for the influence of tooth flank curvature at the pitch point on Hertzian stress.

*Contact Ratio Factor* takes account of the load-sharing influence of the transverse contact ratio and the overlap ratio on the specific loading.

*Elasticity Factor* takes into account the influence of the modulus of elasticity of the material and of Poisson's ratio on the Hertzian stress.

*Basic Endurance Limit* for contact makes allowance for the surface hardness.

*Material Quality* covers the quality of the material used.

*Lubricant Influence, Roughness, and Speed* The lubricant viscosity, surface roughness and pitch line speed affect the lubricant film thickness, which in turn, affects the Hertzian stresses.

*Work Hardening Factor* accounts for the increase in surface durability due to the meshing action.

*Size Factor* covers the possible influences of size on the material quality and its response to manufacturing processes.

*Life Factor* accounts for the increase in permissible stresses when the number of stress cycles is less than the endurance life.

## HYPOID AND BEVEL GEARING

### Hypoid Gears

Hypoid gears are offset and, in effect, are spiral gears whose axes do not intersect but are staggered by an amount decided by the application. Due to the offset, contact between the teeth of the two gears does not occur along a surface line of the cones as it does with spiral bevels having intersecting axes, but along a curve in space inclined to the surface line. The basic solids of the hypoid gear members are not cones, as in spiral bevels, but are hyperboloids of revolution which cannot be projected into the common plane of ordinary flat gears, thus the name hypoid. The visualization of hypoid gears is based on an imaginary flat gear which is a substitute for the theoretically correct helical surface. If certain rules are observed during the calculations to fix the gear dimensions, the errors that result from the use of an imaginary flat gear as an approximation are negligible.

The staggered axes result in meshing conditions that are beneficial to the strength and running properties of the gear teeth. A uniform sliding action takes place between the teeth, not only in the direction of the tooth profile but also longitudinally, producing ideal conditions for movement of lubricants. With spiral gears, great differences in sliding motion arise over various portions of the tooth surface, creating vibration and noise. Hypoid gears are almost free from the problems of differences in these sliding motions and the teeth also have larger curvature radii in the direction of the profile. Surface pressures are thus reduced so that there is less wear and quieter operation.

The teeth of hypoid gears are 1.5 to 2 times stronger than those of spiral bevel gears of the same dimensions, made from the same material. Certain limits must be imposed on the dimensions of hypoid gear teeth so that their proportions can be calculated in the same way as they are for spiral bevel gears. The offset must not be larger than 1/7th of the ring gear outer diameter, and the tooth ratio must not be much less than 4 to 1. Within these limits, the tooth proportions can be calculated in the same way as for spiral bevel gears and the radius of lengthwise curvature can be assumed in such a way that the normal module is a maximum at the center of the tooth face width to produce stabilized tooth bearings.

If the offset is larger or the ratio is smaller than specified above, a tooth form must be selected that is better adapted to the modified meshing conditions. In particular, the curvature of the tooth length curve must be determined with other points in view. The limits are only guidelines since it is impossible to account for all other factors involved, including the pitch line speed of the gears, lubrication, loads, design of shafts and bearings, and the general conditions of operation.

Of the three different designs of hypoid bevel gears now available, the most widely used, especially in the automobile industry, is the Gleason system. Two other hypoid gear systems have been introduced by Oerlikon (Swiss) and Klingelnberg (German). All three methods use the involute gear form, but they have teeth with differing curvatures, produced by the cutting method. Teeth in the Gleason system are arc shaped and their depth tapers. Both the European systems are designed to combine rolling with the sideways motion of the teeth and use a constant tooth depth. Oerlikon uses an epicycloidal tooth form and Klingelnberg uses a true involute form.

With their circular accurate tooth face curves, Gleason hypoid gears are produced with multi-bladed face milling cutters. The gear blank is rolled relative to the rotating cutter to make one inter-tooth groove; then the cutter is withdrawn and returned to its starting position while the blank is indexed into the position for cutting the next tooth. Both roughing and finishing cutters are kept parallel to the tooth root lines, which are at an angle to the gear pitch line. Depending on this angularity, plus the spiral angle, a correction factor must be calculated for both the leading and trailing faces of the gear tooth.

In operation, the convex faces of the teeth on one gear always bear on the concave faces of the teeth on the mating gear. For correct meshing between the pinion and gear wheel, the spiral angles should not vary over the full face width. The tooth form generated is a logarithmic spiral and, as a compromise, the cutter radius is made equal to the mean radius of a corresponding logarithmic spiral.

The involute tooth face curves of the Klingelnberg system gears have constant-pitch teeth cut by (usually) a single-start taper hob. The machine is set up to rotate both the cutter and the gear blank at the correct relative speeds. The surface of the hob is set tangential to a circle radius, which is the gear base circle, from which all the parallel involute curves are struck. To keep the hob size within reasonable dimensions, the cone must lie a minimum distance within the teeth and this requirement governs the size of the module.

Both the module and the tooth depth are constant over the full face width and the spiral angle varies. The cutting speed variations, especially with regard to crown wheels, over the cone surface of the hob, make it difficult to produce a uniform surface finish on the teeth, so a finishing cut is usually made with a truncated hob which is tilted to produce the required amount of crowning automatically, for correct tooth marking and finishing. The dependence of the module, spiral angle and other features on the base circle radius, and the need for suitable hob proportions restrict the gear dimensions and the system cannot be used for gears with a low or zero angle. However, gears can be cut with a large root radius giving teeth of high strength. The favorable geometry of the tooth form gives quieter running and tolerance of inaccuracies in assembly.

Teeth of gears made by the Oerlikon system have elongated epicycloidal form, produced with a face-type rotating cutter. Both the cutter and the gear blank rotate continuously, with no indexing. The cutter head has separate groups of cutters for roughing, outside cutting and inside cutting so that tooth roots and flanks are cut simultaneously, but the feed is divided into two stages. As stresses are released during cutting, there is some distortion of the blank and this distortion will usually be worse for a hollow crown wheel than for a solid pinion.

All the heavy cuts are taken during the first stages of machining with the Oerlikon system and the second stage is used to finish the tooth profile accurately, so distortion effects are minimized. As with the Klingelnberg process, the Oerlikon system produces a variation in spiral angle and module over the width of the face, but unlike the Klingelnberg method, the tooth length curve is cycloidal. It is claimed that, under load, the tilting force in an Oerlikon gear set acts at a point 0.4 times the distance from the small diameter end of the gear and not in the mid-tooth position as in other gear systems, so that the radius is obviously smaller and the tilting moment is reduced, resulting in lower loading of the bearings.

Gears cut by the Oerlikon system have tooth markings of different shape than gears cut by other systems, showing that more of the face width of the Oerlikon tooth is involved in the load-bearing pattern. Thus, the surface loading is spread over a greater area and becomes lighter at the points of contact.

## Bevel Gearing

**Types of Bevel Gears.—**Bevel gears are conical gears, that is, gears in the shape of cones, and are used to connect shafts having intersecting axes. Hypoid gears are similar in general form to bevel gears, but operate on axes that are offset. With few exceptions, most bevel gears may be classified as being either of the straight-tooth type or of the curved-tooth type. The latter type includes spiral bevels, Zerol bevels, and hypoid gears. The following is a brief description of the distinguishing characteristics of the different types of bevel gears.

*Straight Bevel Gears:* The teeth of this most commonly used type of bevel gear are straight but their sides are tapered so that they would intersect the axis at a common point called the pitch cone apex if extended inward. The face cone elements of most straight bevel gears, however, are now made parallel to the root cone elements of the mating gear to obtain uniform clearance along the length of the teeth. The face cone elements of such gears, therefore, would intersect the axis at a point inside the pitch cone. Straight bevel gears are the easiest to calculate and are economical to produce.

Straight bevel gear teeth may be generated for full-length contact or for localized contact. The latter are slightly convex in a lengthwise direction so that some adjustment of the gears during assembly is possible and small displacements due to load deflections can occur without undesirable load concentration on the ends of the teeth. This slight lengthwise rounding of the tooth sides need not be computed in the design but is taken care of automatically in the cutting operation on the newer types of bevel gear generators.

*Zerol Bevel Gears:* The teeth of Zerol bevel gears are curved but lie in the same general direction as the teeth of straight bevel gears. They may be thought of as spiral bevel gears of zero spiral angle and are manufactured on the same machines as spiral bevel gears. The face cone elements of Zerol bevel gears do not pass through the pitch cone apex but instead are approximately parallel to the root cone elements of the mating gear to provide uniform tooth clearance. The root cone elements also do not pass through the pitch cone apex because of the manner in which these gears are cut. Zerol bevel gears are used in place of straight bevel gears when generating equipment of the spiral type but not the straight type is available, and may be used when hardened bevel gears of high accuracy (produced by grinding) are required.

*Spiral Bevel Gears:* Spiral bevel gears have curved oblique teeth on which contact begins gradually and continues smoothly from end to end. They mesh with a rolling contact similar to straight bevel gears. As a result of their overlapping tooth action, however, spiral bevel gears will transmit motion more smoothly than straight bevel or Zerol bevel gears, reducing noise and vibration that become especially noticeable at high speeds.

One of the advantages associated with spiral bevel gears is the complete control of the localized tooth contact. By making a slight change in the radii of curvature of the mating tooth surfaces, the amount of surface over which tooth contact takes place can be changed to suit the specific requirements of each job. Localized tooth contact promotes smooth, quiet running spiral bevel gears, and permits some mounting deflections without concentrating the load dangerously near either end of the tooth. Permissible deflections established by experience are given under the heading *Mountings for Bevel Gears*.

Because their tooth surfaces can be ground, spiral bevel gears have a definite advantage in applications requiring hardened gears of high accuracy. The bottoms of the tooth spaces and the tooth profiles may be ground simultaneously, resulting in a smooth blending of the tooth profile, the tooth fillet, and the bottom of the tooth space. This feature is important from a strength standpoint because it eliminates cutter marks and other surface interruptions that frequently result in stress concentrations.

*Hypoid Gears:* In general appearance, hypoid gears resemble spiral bevel gears, except that the axis of the pinion is offset relative to the gear axis. If there is sufficient offset, the shafts may pass one another thus permitting the use of a compact straddle mounting on the gear and pinion. Whereas a spiral bevel pinion has equal pressure angles and symmetrical profile curvatures on both sides of the teeth, a hypoid pinion properly conjugate to a mating gear having equal pressure angles on both sides of the teeth must have nonsymmetrical profile curvatures for proper tooth action. In addition, to obtain equal arcs of motion for both sides of the teeth, it is necessary to use unequal pressure angles on hypoid pinions. Hypoid gears are usually designed so that the pinion has a larger spiral angle than the gear. The advantage of such a design is that the pinion diameter is increased and is stronger than a corresponding spiral bevel pinion. This diameter increment permits the use of comparatively high ratios without the pinion becoming too small to allow a bore or shank of adequate size. The sliding action along the lengthwise direction of their teeth in hypoid gears is a function of the difference in the spiral angles on the gear and pinion. This sliding effect makes such gears even smoother running than spiral bevel gears. Grinding of hypoid gears can be accomplished on the same machines used for grinding spiral bevel and Zerol bevel gears.

**Applications of Bevel and Hypoid Gears.—**Bevel and hypoid gears may be used to transmit power between shafts at practically any angle and speed. The particular type of gearing best suited for a specific job, however, depends on the mountings and the operating conditions.

*Straight and Zerol Bevel Gears:* For peripheral speeds up to 1000 feet per minute (305 m/min), where maximum smoothness and quietness are not the primary consideration, straight and Zerol bevel gears are recommended. For such applications, plain bearings may be used for radial and axial loads, although the use of antifriction bearings is always preferable. Plain bearings permit a more compact and less expensive design, which is one reason why straight and Zerol bevel gears are much used in differentials. This type of bevel gearing is the simplest to calculate and set up for cutting, and is ideal for small lots where fixed charges must be kept to a minimum.

Zerol bevel gears are recommended in place of straight bevel gears where hardened gears of high accuracy are required, because Zerol gears may be ground; and when only spiral-type equipment is available for cutting bevel gears.

*Spiral Bevel and Hypoid Gears:* Spiral bevel and hypoid gears are recommended for applications where peripheral speeds exceed 1000 feet per minute (305 m/min) or 1000 revolutions per minute. In many instances, they may be used to advantage at lower speeds, particularly where extreme smoothness and quietness are desired. For peripheral speeds above 8000 feet per minute (2438 m/min), ground gears should be used.

For large reduction ratios the use of spiral and hypoid gears will reduce the overall size of the installation because the continuous pitch line contact of these gears makes it practical to obtain smooth performance with a smaller number of teeth in the pinion than is possible with straight or Zerol bevel gears.

Hypoid gears are recommended for industrial applications: when maximum smoothness of operation is desired; for high reduction ratios where compactness of design, smoothness of operation, and maximum pinion strength are important; and for nonintersecting shafts.

Bevel and hypoid gears may be used for both speed-reducing and speed-increasing drives. In speed-increasing drives, however, the ratio should be kept as low as possible and the pinion mounted on antifriction bearings; otherwise bearing friction will cause the drive to lock.

**Notes on the Design of Bevel Gear Blanks.—**The quality of any finished gear is dependent, to a large degree, on the design and accuracy of the gear blank. A number of factors that affect manufacturing economy as well as performance must be considered.

A gear blank should be designed to avoid localized stresses and serious deflections within itself. Sufficient thickness of metal should be provided under the roots of gear teeth to give them proper support. As a general rule, the amount of metal under the root should equal the whole depth of the tooth; this metal depth should be maintained under the small ends of the teeth as well as under the middle. On webless-type ring gears, the minimum stock between the root line and the bottom of tap drill holes should be one-third the tooth depth. For heavily loaded gears, a preliminary analysis of the direction and magnitude of the forces is helpful in the design of both the gear and its mounting. Rigidity is also necessary for proper chucking when cutting the teeth. For this reason, bores, hubs, and other locating surfaces must be in proper proportion to the diameter and pitch of the gear. Small bores, thin webs, or any condition that necessitates excessive overhang in cutting should be avoided.

Other factors to be considered are the ease of machining and, in gears that are to be hardened, proper design to ensure the best hardening conditions. It is desirable to provide a locating surface of generous size on the backs of gears. This surface should be machined or ground square with the bore and is used both for locating the gear axially in assembly and for holding it when the teeth are cut. The front clamping surface must, of course, be flat and parallel to the back surface. In connection with cutting the teeth on Zerol bevel, spiral bevel, and hypoid gears, clearance must be provided for face-mill type cutters; front and rear hubs should not intersect the extended root line of the gear or they will interfere with the path of the cutter. In addition, there must be enough room in the front of the gear for the clamp nut that holds the gear on the arbor, or in the chuck, while cutting the teeth. The same considerations must be given to straight bevel gears that are to be generated using a circular-type cutter instead of reciprocating tools.

**Mountings for Bevel Gears.—**Rigid mountings should be provided for bevel gears to keep the displacements of the gears under operating loads within recommended limits. To align gears properly, care should be taken to ensure accurately machined mountings, properly fitted keys, and couplings that run true and square.

As a result of deflection tests on gears and their mountings, and having observed these same units in service, the *Gleason Works* recommends that the following allowable deflections be used for gears from 6 to 15 inches (15.24–38.10 cm) in diameter: neither the pinion nor the gear should lift or depress more than 0.003 inch (0.076 mm) at the center of the face width; the pinion should not yield axially more than 0.003 inch (0.076 mm) in either direction; and the gear should not yield axially more than 0.003 inch (0.076 mm) in either direction on 1 to 1 ratio gears (miter gears), or near miters, or more than 0.010 inch (0.25 mm) away from the pinion on higher ratios.

When deflections exceed these limits, additional problems are involved in obtaining satisfactory gears. It becomes necessary to narrow and shorten the tooth contacts to suit the more flexible mounting. These changes decrease the bearing area, raise the unit tooth pressure, and reduce the number of teeth in contact, resulting in increased noise and the danger of surface failure as well as tooth breakage.

Spiral bevel and hypoid gears in general should be mounted on antifriction bearings in an oil-tight case. Designs for a given set of conditions may use plain bearings for radial and thrust loads; maintaining gears in satisfactory alignment is usually more easily accomplished with ball or roller bearings.

*Bearing Spacing and Shaft Stiffness:* Bearing spacing and shaft stiffness are extremely important if gear deflections are to be minimized. For both straddle mounted and overhung mounted gears the spread between bearings should never be less than 70 percent of the pitch diameter of the gear. On overhung mounted gears the spread should be at least $2\frac{1}{2}$ times the overhang and, in addition, the shaft diameter should be equal to or preferably greater than the overhang to provide sufficient shaft stiffness. When two spiral bevel or hypoid gears are mounted on the same shaft, the axial thrust should be taken at one place only and near the gear where the greater thrust is developed. Provision should be made for adjusting both the gear and pinion axially in assembly. Details on how this may be accomplished are given in the *Gleason Works* booklet, "Assembling Bevel Gears."

**Cutting Bevel Gear Teeth.—**A correctly formed bevel gear tooth has the same sectional shape throughout its length, but on a uniformly diminishing scale from the large to the small end. The only way to obtain this correct form is by using a generating type of bevel gear cutting machine. This accounts, in part, for the extensive use of generating type gear cutting equipment in the production of bevel gears.

Bevel gears too large to be cut by generating equipment —100 inches (254 cm) or over in diameter—may be produced by a form-copying type of gear planer. A template or former is used to mechanically guide a single cutting tool in the proper path to cut the profile of the teeth. Since the tooth profile produced by this method is dependent on the contour of the template used, it is possible to produce tooth profiles to suit a variety of requirements.

Although generating methods are to be preferred, there are still some cases where straight bevel gears are produced by milling. Milled gears cannot be produced with the accuracy of generated gears and generally are not suitable for use in high-speed applications or where angular motion must be transmitted with a high degree of accuracy. Milled gears are used chiefly as replacement gears in certain applications, and gears which are subsequently to be finished on generating type equipment are sometimes roughed out by milling. Formulas and methods used for the cutting of bevel gears are given in the latter part of this section.

In producing gears by generating methods, the tooth curvature is generated from a straight-sided cutter or tool having an angle equal to the required pressure angle. This tool represents the side of a crown gear tooth. The teeth of a true involute crown gear, however, have sides which are very slightly curved. If the curvature of the cutting tool conforms to that of the involute crown gear, an involute form of bevel gear tooth will be obtained. The use of a straight-sided tool is more practical and results in a very slight change of tooth shape to what is known as the "octoid" form. Both the octoid and involute forms of bevel gear tooth give theoretically correct action.

Bevel gear teeth, like those for spur gears, differ as to pressure angle and tooth proportions. The whole depth and the addendum at the large end of the tooth may be the same as for a spur gear of equal pitch. Most bevel gears, however, both of the straight tooth and spiral-bevel types, have lengthened pinion addendums and shortened gear addendums as in the case of some spur gears, the amount of departure from equal addendums varying with the ratio of gearing. Long addendums on the pinion are used principally to avoid undercut and to increase tooth strength. In addition, where long and short addendums are used, the tooth thickness of the gear is decreased and that of the pinion increased to provide a better balance of strength. See the Gleason Works System for straight and spiral bevel gears and also the British Standard.

**Nomenclature for Bevel Gears.—**The accompanying diagram, Fig. 1a, *Bevel Gear Nomenclature*, illustrates various angles and dimensions referred to in describing bevel gears. In connection with the face angles shown in the diagram, it should be noted that the face cones are made parallel to the root cones of the mating gears to provide uniform clearance along the length of the teeth. See also Fig. 1b, page 2264.

**American Standard for Bevel Gears.—**American Standard ANSI/AGMA 2005-B88, Design Manual for Bevel Gears, replaces AGMA Standards 202.03, 208.03, 209.04, and 330.01, and provides standards for design of straight, zerol, and spiral bevel gears and hypoid gears with information on fabrication, inspection, and mounting. The information covers preliminary design, drawing formats, materials, rating, strength, inspection, lubrication, mountings, and assembly. Blanks for standard taper, uniform depth, duplex taper, and tilted root designs are included so that the material applies to users of Gleason, Klingelnberg, and Oerlikon gear cutting machines.

**Formulas for Dimensions of Milled Bevel Gears.—**As explained earlier, most bevel gears are produced by generating methods. Even so, there are applications for which it may be desired to cut a pair of mating bevel gears by using rotary formed milling cutters. Examples of such applications include replacement gears for certain types of equipment and gears for use in experimental developments.

The tooth proportions of milled bevel gears differ in some respects from those of generated gears, the principal difference being that for milled bevel gears the tooth thicknesses of pinion and gear are made equal, and the addendum and dedendum of the pinion are respectively the same as those of the gear. The rules and formulas in the accompanying table may be used to calculate the dimensions of milled bevel gears with shafts at a right angle, an acute angle, and an obtuse angle.

In the accompanying diagrams, Fig. 1a and Fig. 1b, and list of notations, the various terms and symbols applied to milled bevel gears are as indicated.

$N$ = number of teeth

$P$ = diametral pitch

$p$ = circular pitch

$\alpha$ = pitch cone angle and edge angle

$\Sigma$ = angle between shafts

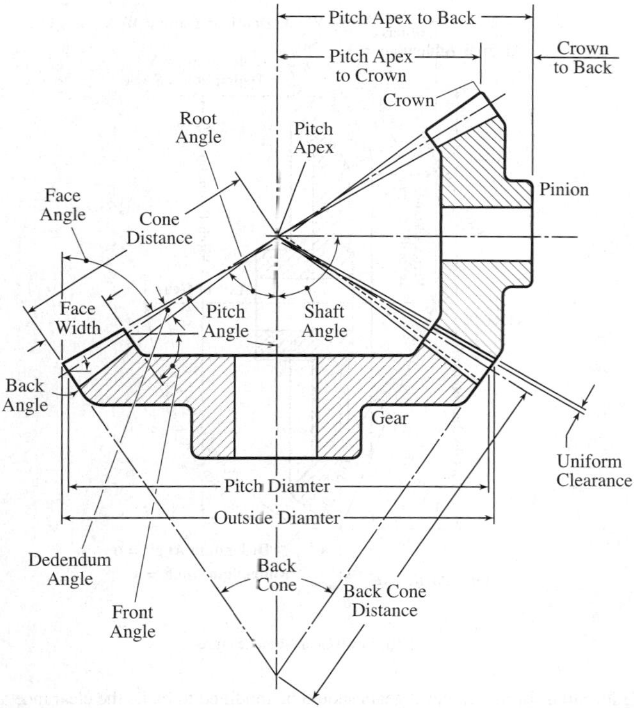


Fig. 1a Bevel Gear Nomenclature

$D$ = pitch diameter
$S$ = addendum
$S + A$ = dedendum ($A$ = clearance)
$W$ = whole depth of tooth
$T$ = thickness of tooth at pitch line
$C$ = pitch cone radius
$F$ = width of face
$s$ = addendum at small end of tooth
$t$ = thickness of tooth at pitch line at small end
$\theta$ = addendum angle
$\phi$ = dedendum angle
$\gamma$ = face angle = pitch cone angle + addendum angle
$\delta$ = angle of compound rest
$\zeta$ = cutting angle
$K$ = angular addendum
$O$ = outside diameter
$J$ = vertex distance
$j$ = vertex distance at small end
$N'$ = number of teeth for which to select cutter

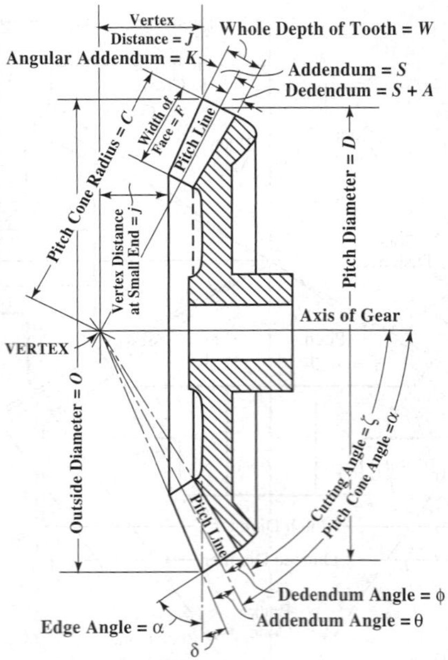


Fig. 1b. Bevel Gear Nomenclature

The formulas for milled bevel gears should be modified to make the clearance at the bottom of the teeth uniform instead of tapering toward the vertex. If this recommendation is followed, then the cutting angle (root angle) should be determined by subtracting the *addendum* angle from the pitch cone angle instead of subtracting the dedendum angle as in the formula given in the table.

**Rules and Formulas for Calculating Dimensions of Milled Bevel Gears**

| To Find | Rule | Formula |
|---|---|---|
| Pitch Cone Angle of Pinion | Divide the sine of the shaft angle by the sum of the cosine of the shaft angle and the quotient obtained by dividing the number of teeth in the gear by the number of teeth in the pinion; this gives the tangent. *Note:* For shaft angles greater than 90° the cosine is negative. | $\tan \alpha_P = \dfrac{\sin \Sigma}{\dfrac{N_G}{N_P} + \cos \Sigma}$ For 90° shaft angle, $\tan \alpha_P = \dfrac{N_P}{N_G}$ |
| Pitch Cone Angle of Gear | Subtract the pitch cone angle of the pinion from the shaft angle. | $\alpha_G = \Sigma - \alpha_P$ |
| Pitch Diameter | Divide the number of teeth by the diametral pitch. | $D = N \div P$ |

**Rules and Formulas for Calculating Dimensions of Milled Bevel Gears** *(Continued)*

| | To Find | Rule | Formula |
|---|---|---|---|
| These dimensions are the same for both gear and pinion. | Addendum | Divide 1 by the diametral pitch. | $S = 1 \div P$ |
| | Dedendum | Divide 1.157 by the diametral pitch. | $S + A = 1.157 \div P$ |
| | Whole Depth of Tooth | Divide 2.157 by the diametral pitch. | $W = 2.157 \div P$ |
| | Thickness of Tooth at Pitch Line | Divide 1.571 by the diametral pitch. | $T = 1.571 \div P$ |
| | Pitch Cone Radius | Divide the pitch diameter by twice the sine of the pitch cone angle. | $C = \frac{D}{2 \times \sin\alpha}$ |
| | Addendum of Small End of Tooth | Subtract the width of face from the pitch cone radius, divide the remainder by the pitch cone radius and multiply by the addendum. | $s = S \times \frac{C - F}{C}$ |
| | Thickness of Tooth at Pitch Line at Small End | Subtract the width of face from the pitch cone radius, divide the remainder by the pitch cone radius and multiply by the thickness of the tooth at pitch line. | $t = T \times \frac{C - F}{C}$ |
| | Addendum Angle | Divide the addendum by the pitch cone radius to get the tangent. | $\tan\theta = \frac{S}{C}$ |
| | Dedendum Angle | Divide the dedendum by the pitch cone radius to get the tangent. | $\tan\phi = \frac{S + A}{C}$ |
| | Face Width (Max.) | Divide the pitch cone radius by 3 or divide 8 by the diametral pitch, whichever gives the smaller value. | $F = \frac{C}{3}$ or $F = \frac{8}{P}$ |
| | Circular Pitch | Divide 3.1416 by the diametral pitch. | $\rho = 3.1416 \div P$ |
| Face Angle | | Add the addendum angle to the pitch cone angle | $\gamma = \alpha + \theta$ |
| Compound Rest Angle for Turning Blank | | Subtract both the pitch cone angle and the addendum angle from 90 degrees. | $\delta = 90° - \alpha - \theta$ |
| Cutting Angle | | Subtract the dedendum angle from the pitch cone angle. | $\zeta = \alpha - \phi$ |
| Angular Addendum | | Multiply the addendum by the cosine of the pitch cone angle. | $K = S \times \cos\alpha$ |
| Outside Diameter | | Add twice the angular addendum to the pitch diameter. | $O = D + 2K$ |
| Vertex or Apex Distance | | Multiply one-half the outside diameter by the cotangent of the face angle. | $J = \frac{O}{2} \times \cot\gamma$ |
| Vertex Distance at Small End of Tooth | | Subtract the width of face from the pitch cone radius; divide the remainder by the pitch cone radius and multiply by the apex distance. | $j = J \times \frac{C - F}{C}$ |
| Number of Teeth for which to Select Cutter | | Divide the number of teeth by the cosine of the pitch cone angle. | $N' = \frac{N}{\cos\alpha}$ |

## Numbers of Formed Cutters Used to Mill Teeth in Mating Bevel Gear and Pinion with Shafts at Right Angles

| Number of Teeth in Gear | Number of Teeth in Pinion → | 12 | 13 | 14 | 15 | 16 | 17 | 18 | 19 | 20 | 21 | 22 | 23 | 24 | 25 | 26 | 27 | 28 |
|---|---|---|---|---|---|---|---|---|---|---|---|---|---|---|---|---|---|---|
| | 12 | 7-7 | … | … | … | … | … | … | … | … | … | … | … | … | … | … | … | … |
| | 13 | 6-7 | 6-6 | … | … | … | … | … | … | … | … | … | … | … | … | … | … | … |
| | 14 | 5-7 | 6-6 | 6-6 | … | … | … | … | … | … | … | … | … | … | … | … | … | … |
| | 15 | 5-7 | 5-6 | 5-6 | 5-5 | … | … | … | … | … | … | … | … | … | … | … | … | … |
| | 16 | 4-7 | 5-7 | 5-6 | 5-6 | 5-5 | … | … | … | … | … | … | … | … | … | … | … | … |
| | 17 | 4-7 | 4-7 | 4-6 | 5-6 | 5-5 | 5-5 | … | … | … | … | … | … | … | … | … | … | … |
| | 18 | 4-7 | 4-7 | 4-6 | 4-6 | 4-5 | 4-5 | 5-5 | … | … | … | … | … | … | … | … | … | … |
| | 19 | 3-7 | 4-7 | 4-6 | 4-6 | 4-6 | 4-5 | 4-5 | 4-4 | … | … | … | … | … | … | … | … | … |
| | 20 | 3-7 | 3-7 | 4-6 | 4-6 | 4-6 | 4-5 | 4-5 | 4-4 | 4-4 | … | … | … | … | … | … | … | … |
| | 21 | 3-8 | 3-7 | 3-7 | 3-6 | 4-6 | 4-5 | 4-5 | 4-5 | 4-4 | 4-4 | … | … | … | … | … | … | … |
| | 22 | 3-8 | 3-7 | 3-7 | 3-6 | 3-6 | 3-5 | 4-5 | 4-5 | 4-4 | 4-4 | 4-4 | … | … | … | … | … | … |
| | 23 | 3-8 | 3-7 | 3-7 | 3-6 | 3-6 | 3-5 | 3-5 | 3-5 | 3-4 | 4-4 | 4-4 | 4-4 | … | … | … | … | … |
| | 24 | 3-8 | 3-7 | 3-7 | 3-6 | 3-6 | 3-6 | 3-5 | 3-5 | 3-4 | 3-4 | 3-4 | 4-4 | 4-4 | … | … | … | … |
| | 25 | 2-8 | 2-7 | 3-7 | 3-6 | 3-6 | 3-6 | 3-5 | 3-5 | 3-5 | 3-4 | 3-4 | 3-4 | 4-4 | 3-3 | … | … | … |
| | 26 | 2-8 | 2-7 | 3-7 | 3-6 | 3-6 | 3-6 | 3-5 | 3-5 | 3-5 | 3-4 | 3-4 | 3-4 | 3-4 | 3-3 | 3-3 | … | … |
| | 27 | 2-8 | 2-7 | 2-7 | 2-6 | 3-6 | 3-6 | 3-5 | 3-5 | 3-5 | 3-4 | 3-4 | 3-4 | 3-4 | 3-4 | 3-3 | 3-3 | … |
| | 28 | 2-8 | 2-7 | 2-7 | 2-6 | 2-6 | 3-6 | 3-5 | 3-5 | 3-5 | 3-4 | 3-4 | 3-4 | 3-4 | 3-4 | 3-3 | 3-3 | 3-3 |
| | 29 | 2-8 | 2-7 | 2-7 | 2-7 | 2-6 | 2-6 | 3-5 | 3-5 | 3-5 | 3-4 | 3-4 | 3-4 | 3-4 | 3-4 | 3-3 | 3-3 | 3-3 |
| | 30 | 2-8 | 2-7 | 2-7 | 2-7 | 2-6 | 2-6 | 2-5 | 2-5 | 3-5 | 3-5 | 3-4 | 3-4 | 3-4 | 3-4 | 3-4 | 3-3 | 3-3 |
| | 31 | 2-8 | 2-7 | 2-7 | 2-7 | 2-6 | 2-6 | 2-6 | 2-5 | 2-5 | 2-5 | 3-4 | 3-4 | 3-4 | 3-4 | 3-4 | 3-3 | 3-3 |
| | 32 | 2-8 | 2-7 | 2-7 | 2-7 | 2-6 | 2-6 | 2-6 | 2-5 | 2-5 | 2-5 | 2-4 | 2-4 | 3-4 | 3-4 | 3-4 | 3-3 | 3-3 |
| | 33 | 2-8 | 2-8 | 2-7 | 2-7 | 2-6 | 2-6 | 2-6 | 2-5 | 2-5 | 2-5 | 2-4 | 2-4 | 2-4 | 3-4 | 3-4 | 3-4 | 3-3 |
| | 34 | 2-8 | 2-8 | 2-7 | 2-7 | 2-6 | 2-6 | 2-6 | 2-5 | 2-5 | 2-5 | 2-4 | 2-4 | 2-4 | 2-4 | 2-4 | 3-4 | 3-3 |
| | 35 | 2-8 | 2-8 | 2-7 | 2-7 | 2-6 | 2-6 | 2-6 | 2-5 | 2-5 | 2-5 | 2-4 | 2-4 | 2-4 | 2-4 | 2-4 | 2-4 | 2-3 |
| | 36 | 2-8 | 2-8 | 2-7 | 2-7 | 2-6 | 2-6 | 2-6 | 2-5 | 2-5 | 2-5 | 2-5 | 2-4 | 2-4 | 2-4 | 2-4 | 2-4 | 2-3 |
| | 37 | 2-8 | 2-8 | 2-7 | 2-7 | 2-6 | 2-6 | 2-6 | 2-5 | 2-5 | 2-5 | 2-5 | 2-4 | 2-4 | 2-4 | 2-4 | 2-4 | 2-3 |
| | 38 | 2-8 | 2-8 | 2-7 | 2-7 | 2-6 | 2-6 | 2-6 | 2-5 | 2-5 | 2-5 | 2-5 | 2-4 | 2-4 | 2-4 | 2-4 | 2-4 | 2-4 |
| | 39 | 2-8 | 2-8 | 2-7 | 2-7 | 2-6 | 2-6 | 2-6 | 2-5 | 2-5 | 2-5 | 2-5 | 2-4 | 2-4 | 2-4 | 2-4 | 2-4 | 2-4 |
| | 40 | 1-8 | 2-8 | 2-7 | 2-7 | 2-6 | 2-6 | 2-6 | 2-5 | 2-5 | 2-5 | 2-5 | 2-4 | 2-4 | 2-4 | 2-4 | 2-4 | 2-4 |
| | 41 | 1-8 | 1-8 | 2-7 | 2-7 | 2-6 | 2-6 | 2-6 | 2-6 | 2-5 | 2-5 | 2-5 | 2-4 | 2-4 | 2-4 | 2-4 | 2-4 | 2-4 |
| | 42 | 1-8 | 1-8 | 2-7 | 2-7 | 2-6 | 2-6 | 2-6 | 2-6 | 2-5 | 2-5 | 2-5 | 2-5 | 2-4 | 2-4 | 2-4 | 2-4 | 2-4 |
| | 43 | 1-8 | 1-8 | 1-7 | 2-7 | 2-6 | 2-6 | 2-6 | 2-6 | 2-5 | 2-5 | 2-5 | 2-5 | 2-4 | 2-4 | 2-4 | 2-4 | 2-4 |
| | 44 | 1-8 | 1-8 | 1-7 | 1-7 | 2-6 | 2-6 | 2-6 | 2-6 | 2-5 | 2-5 | 2-5 | 2-5 | 2-4 | 2-4 | 2-4 | 2-4 | 2-4 |
| | 45 | 1-8 | 1-8 | 1-7 | 1-7 | 1-6 | 2-6 | 2-6 | 2-6 | 2-5 | 2-5 | 2-5 | 2-5 | 2-4 | 2-4 | 2-4 | 2-4 | 2-4 |
| | 46 | 1-8 | 1-8 | 1-7 | 1-7 | 1-7 | 2-6 | 2-6 | 2-6 | 2-5 | 2-5 | 2-5 | 2-5 | 2-4 | 2-4 | 2-4 | 2-4 | 2-4 |
| | 47 | 1-8 | 1-8 | 1-7 | 1-7 | 1-7 | 1-6 | 2-6 | 2-6 | 2-5 | 2-5 | 2-5 | 2-5 | 2-4 | 2-4 | 2-4 | 2-4 | 2-4 |
| | 48 | 1-8 | 1-8 | 1-7 | 1-7 | 1-7 | 1-6 | 1-6 | 2-6 | 2-5 | 2-5 | 2-5 | 2-5 | 2-4 | 2-4 | 2-4 | 2-4 | 2-4 |
| | 49 | 1-8 | 1-8 | 1-7 | 1-7 | 1-7 | 1-6 | 1-6 | 1-6 | 2-5 | 2-5 | 2-5 | 2-5 | 2-4 | 2-4 | 2-4 | 2-4 | 2-4 |
| | 50 | 1-8 | 1-8 | 1-7 | 1-7 | 1-7 | 1-6 | 1-6 | 1-6 | 2-5 | 2-5 | 2-5 | 2-5 | 2-4 | 2-4 | 2-4 | 2-4 | 2-4 |
| | 51 | 1-8 | 1-8 | 1-7 | 1-7 | 1-7 | 1-6 | 1-6 | 1-6 | 1-5 | 2-5 | 2-5 | 2-5 | 2-4 | 2-4 | 2-4 | 2-4 | 2-4 |
| | 52 | 1-8 | 1-8 | 1-7 | 1-7 | 1-7 | 1-6 | 1-6 | 1-6 | 1-5 | 1-5 | 2-5 | 2-5 | 2-4 | 2-4 | 2-4 | 2-4 | 2-4 |
| | 53 | 1-8 | 1-8 | 1-7 | 1-7 | 1-7 | 1-6 | 1-6 | 1-6 | 1-5 | 1-5 | 1-5 | 2-5 | 2-4 | 2-4 | 2-4 | 2-4 | 2-4 |
| | 54 | 1-8 | 1-8 | 1-7 | 1-7 | 1-7 | 1-6 | 1-6 | 1-6 | 1-5 | 1-5 | 1-5 | 1-5 | 2-4 | 2-4 | 2-4 | 2-4 | 2-4 |
| | 55 | 1-8 | 1-8 | 1-7 | 1-7 | 1-7 | 1-6 | 1-6 | 1-6 | 1-5 | 1-5 | 1-5 | 1-5 | 1-4 | 2-4 | 2-4 | 2-4 | 2-4 |

## Numbers of Formed Cutters Used to Mill Teeth in Mating Bevel Gear and Pinion with Shafts at Right Angles *(Continued)*

| Number of Teeth in Gear | Number of Teeth in Pinion | | | | | | | | | | | | | | | | |
|---|---|---|---|---|---|---|---|---|---|---|---|---|---|---|---|---|---|
| | 12 | 13 | 14 | 15 | 16 | 17 | 18 | 19 | 20 | 21 | 22 | 23 | 24 | 25 | 26 | 27 | 28 |
| 56 | 1-8 | 1-8 | 1-7 | 1-7 | 1-6 | 1-6 | 1-6 | 1-6 | 1-5 | 1-5 | 1-5 | 1-5 | 1-4 | 1-4 | 2-4 | 2-4 | 2-4 |
| 57 | 1-8 | 1-8 | 1-7 | 1-7 | 1-6 | 1-6 | 1-6 | 1-6 | 1-5 | 1-5 | 1-5 | 1-5 | 1-4 | 1-4 | 1-4 | 2-4 | 2-4 |
| 58 | 1-8 | 1-8 | 1-7 | 1-7 | 1-6 | 1-6 | 1-6 | 1-6 | 1-5 | 1-5 | 1-5 | 1-5 | 1-4 | 1-4 | 1-4 | 1-4 | 2-4 |
| 59 | 1-8 | 1-8 | 1-7 | 1-7 | 1-6 | 1-6 | 1-6 | 1-6 | 1-5 | 1-5 | 1-5 | 1-5 | 1-5 | 1-4 | 1-4 | 1-4 | 1-4 |
| 60 | 1-8 | 1-8 | 1-7 | 1-7 | 1-6 | 1-6 | 1-6 | 1-6 | 1-5 | 1-5 | 1-5 | 1-5 | 1-5 | 1-4 | 1-4 | 1-4 | 1-4 |
| 61 | 1-8 | 1-8 | 1-7 | 1-7 | 1-6 | 1-6 | 1-6 | 1-6 | 1-5 | 1-5 | 1-5 | 1-5 | 1-5 | 1-4 | 1-4 | 1-4 | 1-4 |
| 62 | 1-8 | 1-8 | 1-7 | 1-7 | 1-6 | 1-6 | 1-6 | 1-6 | 1-5 | 1-5 | 1-5 | 1-5 | 1-5 | 1-4 | 1-4 | 1-4 | 1-4 |
| 63 | 1-8 | 1-8 | 1-7 | 1-7 | 1-6 | 1-6 | 1-6 | 1-6 | 1-5 | 1-5 | 1-5 | 1-5 | 1-5 | 1-4 | 1-4 | 1-4 | 1-4 |
| 64 | 1-8 | 1-8 | 1-7 | 1-7 | 1-6 | 1-6 | 1-6 | 1-6 | 1-5 | 1-5 | 1-5 | 1-5 | 1-5 | 1-4 | 1-4 | 1-4 | 1-4 |
| 65 | 1-8 | 1-8 | 1-7 | 1-7 | 1-7 | 1-6 | 1-6 | 1-6 | 1-6 | 1-5 | 1-5 | 1-5 | 1-5 | 1-4 | 1-4 | 1-4 | 1-4 |
| 66 | 1-8 | 1-8 | 1-7 | 1-7 | 1-7 | 1-6 | 1-6 | 1-6 | 1-6 | 1-5 | 1-5 | 1-5 | 1-5 | 1-4 | 1-4 | 1-4 | 1-4 |
| 67 | 1-8 | 1-8 | 1-7 | 1-7 | 1-7 | 1-6 | 1-6 | 1-6 | 1-6 | 1-5 | 1-5 | 1-5 | 1-5 | 1-4 | 1-4 | 1-4 | 1-4 |
| 68 | 1-8 | 1-8 | 1-7 | 1-7 | 1-7 | 1-6 | 1-6 | 1-6 | 1-6 | 1-5 | 1-5 | 1-5 | 1-5 | 1-4 | 1-4 | 1-4 | 1-4 |
| 69 | 1-8 | 1-8 | 1-7 | 1-7 | 1-7 | 1-6 | 1-6 | 1-6 | 1-6 | 1-5 | 1-5 | 1-5 | 1-5 | 1-4 | 1-4 | 1-4 | 1-4 |
| 70 | 1-8 | 1-8 | 1-7 | 1-7 | 1-7 | 1-6 | 1-6 | 1-6 | 1-6 | 1-5 | 1-5 | 1-5 | 1-5 | 1-4 | 1-4 | 1-4 | 1-4 |
| 71 | 1-8 | 1-8 | 1-7 | 1-7 | 1-7 | 1-6 | 1-6 | 1-6 | 1-6 | 1-5 | 1-5 | 1-5 | 1-5 | 1-4 | 1-4 | 1-4 | 1-4 |
| 72 | 1-8 | 1-8 | 1-7 | 1-7 | 1-7 | 1-6 | 1-6 | 1-6 | 1-6 | 1-5 | 1-5 | 1-5 | 1-5 | 1-4 | 1-4 | 1-4 | 1-4 |
| 73 | 1-8 | 1-8 | 1-7 | 1-7 | 1-7 | 1-6 | 1-6 | 1-6 | 1-6 | 1-5 | 1-5 | 1-5 | 1-5 | 1-4 | 1-4 | 1-4 | 1-4 |
| 74 | 1-8 | 1-8 | 1-7 | 1-7 | 1-7 | 1-6 | 1-6 | 1-6 | 1-6 | 1-5 | 1-5 | 1-5 | 1-5 | 1-4 | 1-4 | 1-4 | 1-4 |
| 75 | 1-8 | 1-8 | 1-7 | 1-7 | 1-7 | 1-6 | 1-6 | 1-6 | 1-6 | 1-5 | 1-5 | 1-5 | 1-5 | 1-4 | 1-4 | 1-4 | 1-4 |
| 76 | 1-8 | 1-8 | 1-7 | 1-7 | 1-7 | 1-6 | 1-6 | 1-6 | 1-6 | 1-5 | 1-5 | 1-5 | 1-5 | 1-4 | 1-4 | 1-4 | 1-4 |
| 77 | 1-8 | 1-8 | 1-7 | 1-7 | 1-7 | 1-6 | 1-6 | 1-6 | 1-6 | 1-5 | 1-5 | 1-5 | 1-5 | 1-4 | 1-4 | 1-4 | 1-4 |
| 78 | 1-8 | 1-8 | 1-7 | 1-7 | 1-7 | 1-6 | 1-6 | 1-6 | 1-6 | 1-5 | 1-5 | 1-5 | 1-5 | 1-4 | 1-4 | 1-4 | 1-4 |
| 79 | 1-8 | 1-8 | 1-7 | 1-7 | 1-7 | 1-6 | 1-6 | 1-6 | 1-6 | 1-5 | 1-5 | 1-5 | 1-5 | 1-4 | 1-4 | 1-4 | 1-4 |
| 80 | 1-8 | 1-8 | 1-7 | 1-7 | 1-7 | 1-6 | 1-6 | 1-6 | 1-6 | 1-5 | 1-5 | 1-5 | 1-5 | 1-4 | 1-4 | 1-4 | 1-4 |
| 81 | 1-8 | 1-8 | 1-7 | 1-7 | 1-7 | 1-6 | 1-6 | 1-6 | 1-6 | 1-5 | 1-5 | 1-5 | 1-5 | 1-4 | 1-4 | 1-4 | 1-4 |
| 82 | 1-8 | 1-8 | 1-7 | 1-7 | 1-7 | 1-6 | 1-6 | 1-6 | 1-6 | 1-5 | 1-5 | 1-5 | 1-5 | 1-4 | 1-4 | 1-4 | 1-4 |
| 83 | 1-8 | 1-8 | 1-7 | 1-7 | 1-7 | 1-6 | 1-6 | 1-6 | 1-6 | 1-5 | 1-5 | 1-5 | 1-5 | 1-4 | 1-4 | 1-4 | 1-4 |
| 84 | 1-8 | 1-8 | 1-7 | 1-7 | 1-7 | 1-6 | 1-6 | 1-6 | 1-6 | 1-5 | 1-5 | 1-5 | 1-5 | 1-4 | 1-4 | 1-4 | 1-4 |
| 85 | 1-8 | 1-8 | 1-7 | 1-7 | 1-7 | 1-6 | 1-6 | 1-6 | 1-6 | 1-5 | 1-5 | 1-5 | 1-5 | 1-4 | 1-4 | 1-4 | 1-4 |
| 86 | 1-8 | 1-8 | 1-7 | 1-7 | 1-7 | 1-6 | 1-6 | 1-6 | 1-6 | 1-5 | 1-5 | 1-5 | 1-5 | 1-4 | 1-4 | 1-4 | 1-4 |
| 87 | 1-8 | 1-8 | 1-7 | 1-7 | 1-7 | 1-6 | 1-6 | 1-6 | 1-6 | 1-5 | 1-5 | 1-5 | 1-5 | 1-4 | 1-4 | 1-4 | 1-4 |
| 88 | 1-8 | 1-8 | 1-7 | 1-7 | 1-7 | 1-6 | 1-6 | 1-6 | 1-6 | 1-5 | 1-5 | 1-5 | 1-5 | 1-4 | 1-4 | 1-4 | 1-4 |
| 89 | 1-8 | 1-8 | 1-7 | 1-7 | 1-7 | 1-6 | 1-6 | 1-6 | 1-6 | 1-5 | 1-5 | 1-5 | 1-5 | 1-4 | 1-4 | 1-4 | 1-4 |
| 90 | 1-8 | 1-8 | 1-7 | 1-7 | 1-7 | 1-6 | 1-6 | 1-6 | 1-6 | 1-5 | 1-5 | 1-5 | 1-5 | 1-4 | 1-4 | 1-4 | 1-4 |
| 91 | 1-8 | 1-8 | 1-7 | 1-7 | 1-7 | 1-6 | 1-6 | 1-6 | 1-6 | 1-5 | 1-5 | 1-5 | 1-5 | 1-4 | 1-4 | 1-4 | 1-4 |
| 92 | 1-8 | 1-8 | 1-7 | 1-7 | 1-7 | 1-6 | 1-6 | 1-6 | 1-6 | 1-5 | 1-5 | 1-5 | 1-5 | 1-4 | 1-4 | 1-4 | 1-4 |
| 93 | 1-8 | 1-8 | 1-7 | 1-7 | 1-7 | 1-6 | 1-6 | 1-6 | 1-6 | 1-5 | 1-5 | 1-5 | 1-5 | 1-4 | 1-4 | 1-4 | 1-4 |
| 94 | 1-8 | 1-8 | 1-7 | 1-7 | 1-7 | 1-6 | 1-6 | 1-6 | 1-6 | 1-5 | 1-5 | 1-5 | 1-5 | 1-4 | 1-4 | 1-4 | 1-4 |
| 95 | 1-8 | 1-8 | 1-7 | 1-7 | 1-7 | 1-6 | 1-6 | 1-6 | 1-6 | 1-5 | 1-5 | 1-5 | 1-5 | 1-4 | 1-4 | 1-4 | 1-4 |
| 96 | 1-8 | 1-8 | 1-7 | 1-7 | 1-7 | 1-6 | 1-6 | 1-6 | 1-6 | 1-5 | 1-5 | 1-5 | 1-5 | 1-4 | 1-4 | 1-4 | 1-4 |
| 97 | 1-8 | 1-8 | 1-7 | 1-7 | 1-7 | 1-6 | 1-6 | 1-6 | 1-6 | 1-5 | 1-5 | 1-5 | 1-5 | 1-4 | 1-4 | 1-4 | 1-4 |
| 98 | 1-8 | 1-8 | 1-7 | 1-7 | 1-7 | 1-6 | 1-6 | 1-6 | 1-6 | 1-5 | 1-5 | 1-5 | 1-5 | 1-4 | 1-4 | 1-4 | 1-4 |
| 99 | 1-8 | 1-8 | 1-7 | 1-7 | 1-7 | 1-6 | 1-6 | 1-6 | 1-6 | 1-5 | 1-5 | 1-5 | 1-5 | 1-4 | 1-4 | 1-4 | 1-4 |
| 100 | 1-8 | 1-8 | 1-7 | 1-7 | 1-7 | 1-6 | 1-6 | 1-6 | 1-6 | 1-5 | 1-5 | 1-5 | 1-5 | 1-4 | 1-4 | 1-4 | 1-4 |

Number of cutter for gear given first, followed by number for pinion. See text, page 2268

**Selecting Formed Cutters for Milling Bevel Gears.**—For milling 14½-degree pressure angle bevel gears, the standard cutter series furnished by manufacturers of formed milling cutters is commonly used. There are 8 cutters in the series for each diametral pitch to cover the full range from a 12-tooth pinion to a crown gear. The difference between formed cutters used for milling spur gears and those used for bevel gears is that bevel gear cutters are thinner because they must pass through the narrow tooth space at the small end of the bevel gear; otherwise the shape of the cutter and, hence, the cutter number are the same.

To select the proper number of cutter to be used when a bevel gear is to be milled, it is necessary, first, to compute what is called the "Number of Teeth, $N'$ for which to Select Cutter." This number of teeth can then be used to select the proper number of bevel gear cutter from the spur gear milling cutter table on page 2231. The value of $N'$ may be computed using the last formula in the table on page 2264.

*Example 1:* What numbers of cutters are required for a pair of bevel gears of 4 diametral pitch and 70 degree shaft angle if the gear has 50 teeth and the pinion 20 teeth?

The pitch cone angle of the pinion is determined by using the first formula in the table on page 2264:

$$\tan \alpha_P = \frac{\sin \Sigma}{\dfrac{N_G}{N_P} + \cos \Sigma} = \frac{\sin 70^\circ}{\dfrac{50}{20} + \cos 70^\circ} = 0.33064;\ \alpha_P = 18^\circ 18'$$

The pitch cone angle of the gear is determined from the second formula in the table on page 2264:

$$\alpha_G = \Sigma - \alpha_P = 70^\circ - 18^\circ 18' = 51^\circ 42'$$

The numbers of teeth N′ for which to select the cutters for the gear and pinion may now be determined from the last formula in the table on page 2264:

$$N' \text{ for the pinion } = \frac{N_P}{\cos \alpha_P} = \frac{20}{\cos 18^\circ 18'} = 21.1 \approx 21 \text{ teeth}$$

$$N' \text{ for the gear } = \frac{N_G}{\cos \alpha_G} = \frac{50}{\cos 51^\circ 42'} = 80.7 \approx 81 \text{ teeth}$$

From the table on page 2231 the numbers of the cutters for pinion and gear are found to be, respectively, 5 and 2.

*Example 2:* The cutters for a pair of bevel gears are required where the gear has 24 teeth and the pinion 12 teeth. The shaft angle is 90 degrees. As in the first example, the formulas given in the table on page 2264 will be used.

$$\tan \alpha_P = N_P \div N_G = 12 \div 24 = 0.5000 \text{ and } \alpha_P = 26^\circ 34'$$

$$\alpha_G = \Sigma - \alpha_P = 90^\circ - 26^\circ 34' = 63^\circ 26'$$

$$N' \text{ for pinion } = 12 \div \cos 26^\circ 34' = 13.4 \approx 13 \text{ teeth}$$

$$N' \text{ for gear } = 24 \div \cos 63^\circ 26' = 53.6 \approx 54 \text{ teeth}$$

And from the table on page 2231 the cutters for pinion and gear are found to be, respectively, 8 and 3.

**Use of Table for Selecting Formed Cutters for Milling Bevel Gears.**—The table beginning on page 2266 gives the numbers of cutters to use for milling various numbers of teeth in the gear and pinion. The table applies only to bevel gears with axes at right angles. Thus, in Example 2 given above, the numbers of the cutters could have been obtained directly by entering the table with the actual numbers of teeth in the gear, 24, and the pinion, 12.

**Offset of Cutter for Milling Bevel Gears.—**When milling bevel gears with a rotary formed cutter, it is necessary to take two cuts through each tooth space with the gear blank slightly off center, first on one side and then on the other, to obtain a tooth of approximately the correct form. The gear blank is also rotated proportionately to obtain the proper tooth thickness at the large and small ends. The amount that the gear blank or cutter should be offset from the central position can be determined quite accurately by the use of the table *Factors for Obtaining Offset for Milling Bevel Gears* in conjunction with the following rule: Find the factor in the table corresponding to the number of cutter used and to the ratio of the pitch cone radius to the face width; then divide this factor by the diametral pitch and subtract the result from half the thickness of the cutter at the pitch line.

**Factors for Obtaining Offset for Milling Bevel Gears**

| No. of Cutter | Ratio of Pitch Cone Radius to Width of Face $\left(\frac{C}{F}\right)$ | | | | | | | | | | | | |
|---|---|---|---|---|---|---|---|---|---|---|---|---|---|
| | $\frac{3}{1}$ | $\frac{3\frac{1}{4}}{1}$ | $\frac{3\frac{1}{2}}{1}$ | $\frac{3\frac{3}{4}}{1}$ | $\frac{4}{1}$ | $\frac{4\frac{1}{4}}{1}$ | $\frac{4\frac{1}{2}}{1}$ | $\frac{4\frac{3}{4}}{1}$ | $\frac{5}{1}$ | $\frac{5\frac{1}{2}}{1}$ | $\frac{6}{1}$ | $\frac{7}{1}$ | $\frac{8}{1}$ |
| 1 | 0.254 | 0.254 | 0.255 | 0.256 | 0.257 | 0.257 | 0.257 | 0.258 | 0.258 | 0.259 | 0.260 | 0.262 | 0.264 |
| 2 | 0.266 | 0.268 | 0.271 | 0.272 | 0.273 | 0.274 | 0.274 | 0.275 | 0.277 | 0.279 | 0.280 | 0.283 | 0.284 |
| 3 | 0.266 | 0.268 | 0.271 | 0.273 | 0.275 | 0.278 | 0.280 | 0.282 | 0.283 | 0.286 | 0.287 | 0.290 | 0.292 |
| 4 | 0.275 | 0.280 | 0.285 | 0.287 | 0.291 | 0.293 | 0.296 | 0.298 | 0.298 | 0.302 | 0.305 | 0.308 | 0.3[illegible]1 |
| 5 | 0.280 | 0.285 | 0.290 | 0.293 | 0.295 | 0.296 | 0.298 | 0.300 | 0.302 | 0.307 | 0.309 | 0.313 | 0.3[illegible]5 |
| 6 | 0.311 | 0.318 | 0.323 | 0.328 | 0.330 | 0.334 | 0.337 | 0.340 | 0.343 | 0.348 | 0.352 | 0.356 | 0.3[illegible]2 |
| 7 | 0.289 | 0.298 | 0.308 | 0.316 | 0.324 | 0.329 | 0.334 | 0.338 | 0.343 | 0.350 | 0.360 | 0.370 | 0.376 |
| 8 | 0.275 | 0.286 | 0.296 | 0.309 | 0.319 | 0.331 | 0.338 | 0.344 | 0.352 | 0.361 | 0.368 | 0.380 | 0.386 |

Note.—For obtaining offset by above table, use formula:

$$\text{Offset} = \frac{T}{2} - \frac{\text{factor from table}}{P}$$

$P$ = diametral pitch of gear to be cut

$T$ = thickness of cutter used, measured at pitch line

To illustrate, what would be the amount of offset for a bevel gear having 24 teeth, 6 diametral pitch, 30-degree pitch cone angle and $1\frac{1}{4}$-inch face or tooth length? In order to obtain a factor from the table, the ratio of the pitch cone radius to the face width must be determined. The pitch cone radius equals the pitch diameter divided by twice the sine of the pitch cone angle = $4 \div (2 \times 0.5) = 4$ inches. As the face width is 1.25, the ratio is $4 \div 1.25$ or about $3\frac{1}{4}$ to 1. The factor in the table for this ratio is 0.280 with a No. 4 cutter, which would be the cutter number for this particular gear. The thickness of the cutter at the pitch line is measured by using a vernier gear tooth caliper. The depth $S + A$ (see Fig. 2; $S$ = addendum; $A$ = clearance) at which to take the measurement equals 1.157 divided by the diametral pitch; thus, $1.157 \div 6 = 0.1928$ inch. The cutter thickness at this depth will vary with different cutters and even with the same cutter as it is ground away, because formed bevel gear cutters are commonly provided with side relief. Assuming that the thickness is 0.1745 inch, and substituting the values in the formula given, we have:

$$\text{Offset} = \frac{0.1745}{2} = \frac{0.280}{6} = 0.0406 \text{ inch}$$

**Adjusting the Gear Blank for Milling.—**After the offset is determined, the blank is adjusted laterally by this amount, and the tooth spaces are milled around the blank. After having milled one side of each tooth to the proper dimensions, the blank is set over in the opposite direction the same amount from a position central with the cutter, and is rotated to line up the cutter with a tooth space at the small end. A trial cut is then taken, which will leave the tooth being milled a little too thick, provided the cutter is thin enough—as it should be—to pass through the small end of the tooth space of the finished gear. This trial tooth is made the proper thickness by rotating the blank toward the cutter. To test the amount of offset, measure the tooth thickness (with a vernier caliper) at the large and small ends. The caliper should be set so that the addendum at the small end is in proper proportion to the addendum at the large end; that is, in the ratio $(C - F)/C$ (see Fig. 2).

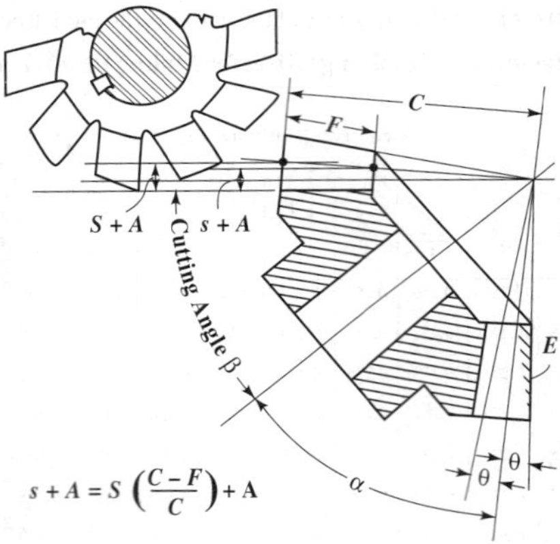


Fig. 2.

In taking these measurements, if the thicknesses at both ends (which should be in this same ratio) are too great, rotate the tooth toward the cutter and take trial cuts until the proper thickness at either the large or small end is obtained. If the large end of the tooth is the right thickness and the small end too thick, the blank was offset too much; inversely, if the small end is correct and the large end too thick, the blank was not set enough off center, and, either way, its position should be changed accordingly. The formula and table previously referred to will enable a properly turned blank to be set accurately enough for general work. The dividing head should be set to the cutting angle β (see Fig. 2), which is found by subtracting the addendum angle θ from the pitch cone angle α. After a bevel gear is cut by the method described, the sides of the teeth at the small end should be filed as indicated by the shade lines at *E*; that is, by filing off a triangular area from the point of the tooth at the large end to the point at the small end, thence down to the pitch line and back diagonally to a point at the large end.

**Circular Thickness, Chordal Thickness, and Chordal Addendum of Milled Bevel Gear Teeth.—** In the formulas that follow, $T$ = circular tooth thickness on pitch circle at large end of tooth; $t$ = circular thickness at small end; $T_c$ and $t_c$ = chordal thickness at large and small ends, respectively; $S_c$ and $s_c$ = chordal addendum at large and small ends, respectively; $D$ = pitch diameter at large end; and $C, F, P, S, s$, and α are as defined on page 2262.

$$T = \frac{1.5708}{P} \qquad T_c = T - \frac{T^3}{6D^2} \qquad S_c = S + \frac{T^2 \cos\alpha}{4D}$$

$$t = \frac{T(C - F)}{C} \qquad t_c = t - \frac{t^3}{6(D - 2F\sin\alpha)^2} \qquad s_c = s + \frac{t^2 \cos\alpha}{4(D - 2F\sin\alpha)}$$

## Typical Steels Used for Bevel Gear Applications

| SAE or AISI No. | Type of Steel | Purchase Specifications: Preliminary Heat Treatment | Purchase Specifications: Brinell Hardness Number | Purchase Specifications: ASTM Grain Size | Remarks |
|---|---|---|---|---|---|
| **Carburizing Steels** | | | | | |
| 1024 | Manganese | Normalize | | | Low alloy—oil quench limited to thin sections |
| 2512 | Nickel Alloy | Normalize—anneal | 163–228 | 5–8 | Aircraft quality |
| 3310<br>3312X | Nickel-Chromium | Normalize, then heat to 1450° F, cool in furnace. Reheat to 1170° F—cool in air | 163–228 | 5–8 | Used for maximum resistance to wear and fatigue |
| 4028 | Molybdenum | Normalize | 163–217 | | Low alloy |
| 4615<br>4620 | Nickel-Molybdenum | Normalize—1700° F–1750° F | 163–217 | 5–8 | Good machining qualities. Well adapted to direct quench—gives tough core with minimum distortion |
| 4815<br>4820 | Nickel-Molybdenum | Normalize | 163–241 | 5–8 | For aircraft and heavily loaded service |
| 5120 | Chromium | Normalize | 163–217 | 5–8 | |
| 8615<br>8620<br>8715<br>8720 | Chromium-Nickel-Molybdenum | Normalize—cool at hammer | 163–217 | 5–8 | Used as an alternate for 4620 |
| **Oil Hardening and Flame Hardening Steels** | | | | | |
| 1141 | Sulfurized free-cutting carbon steel | Normalize—heat-treated | 179–228<br>255–269 | 5 or Coarser | Free-cutting steel used for unhardened gears, oil-treated gears, and for gears to be surface hardened where stresses are low |
| 4140<br>4640 | Chromium-MolybdenumNickel-Molybdenum | For oil hardening, normalize—anneal<br>For surface hardening, Normalize, reheat, quench, and draw | 179–212<br>235–269<br>269–302<br>302–341 | | Used for heat-treated, oil-hardened, and surface-hardened gears. Machine qualities of 4640 are superior to 4140, and it is the preferred steel for flame hardening |
| 6145 | Chromium-Vanadium | Normalize—reheat, quench, and draw | 235–269<br>269–302<br>302–341 | | Fair machining qualities. Used for surface hardened gears when 4640 is not available |
| 8640<br>8739 | Chromium-Nickel-Molybdenum | Same as for 4640 | | | Used as an alternate for 4640 |
| **Nitriding Steels** | | | | | |
| Nitralloy H & G | Special Alloy | Anneal | 163–192 | | Normal hardness range for cutting is 20–28 RC (Rockwell C scale) |

Other steels with qualities equivalent to those listed in the table may also be used.

## WORM GEARING

**Worm Gearing.—**Worm gearing may be divided into two general classes, fine-pitch worm gearing, and coarse-pitch worm gearing. Fine-pitch worm gearing is segregated from coarse-pitch worm gearing for the following reasons:

1) Fine-pitch worms and wormgears are used largely to transmit motion rather than power. Tooth strength except at the coarser end of the fine-pitch range is seldom an important factor; durability and accuracy, as they affect the transmission of uniform angular motion, are of greater importance.

2) Housing constructions and lubricating methods are, in general, quite different for fine-pitch worm gearing.

3) Because fine-pitch worms and wormgears are so small, profile deviations and tooth bearings cannot be measured with the same accuracy as can those of coarse pitches.

4) Equipment generally available for cutting fine-pitch wormgears has restrictions which limit the diameter, the lead range, the degree of accuracy attainable, and the kind of tooth bearing obtainable.

5) Special consideration must be given to top lands in fine-pitch hardened worms and wormgear-cutting tools.

6) Interchangeability and high production are important factors in fine-pitch worm gearing; individual matching of the worm to the gear, as often practiced with coarse-pitch precision worms, is impractical in the case of fine-pitch worm drives.

**American Standard Design for Fine-Pitch Worm Gearing (ANSI B6.9-1977).—** This Standard is intended as a design procedure for fine-pitch worms and wormgears having axes at right angles. It covers cylindrical worms with helical threads, and wormgears hobbed for fully conjugate tooth surfaces. It does not cover helical gears used as wormgears.

*Hobs:* The hob for producing the gear is a duplicate of the mating worm with regard to tooth profile, number of threads, and lead. The hob differs from the worm principally in that the outside diameter of the hob is larger to allow for resharpening and to provide bottom clearance in the wormgear.

*Pitches:* Eight standard axial pitches have been established to provide adequate coverage of the pitch range normally required: 0.030, 0.040, 0.050, 0.065, 0.080, 0.100, 0.130, and 0.160 inch.

Axial pitch is used as a basis for this design standard because: 1) Axial pitch establishes lead which is a basic dimension in the production and inspection of worms; 2) the axial pitch of the worm is equal to the circular pitch of the gear in the central plane; and 3) only one set of change gears or one master lead cam is required for a given lead, regardless of lead angle, on commonly used worm-producing equipment.

*Lead Angles:* Fifteen standard lead angles have been established to provide adequate coverage: 0.5, 1, 1.5, 2, 3, 4, 5, 7, 9, 11, 14, 17, 21, 25, and 30 degrees.

This series of lead angles has been standardized to: 1) Minimize tooling; 2) permit obtaining geometric similarity between worms of different axial pitch by keeping the same lead angle; and 3) take into account the production distribution found in fine-pitch worm gearing applications.

For example, most fine-pitch worms have either one or two threads. This requires smaller increments at the low end of the lead angle series. For the less frequently used thread numbers, proportionately greater increments at the high end of the lead angle series are sufficient.

*Pressure Angle of Worm:* A pressure angle of 20 degrees has been selected as standard for cutters and grinding wheels used to produce worms within the scope of this Standard because it avoids objectionable undercutting regardless of lead angle.

**Table 1. Formulas for Proportions of American Standard Fine-Pitch Worms and Wormgears** *ANSI B6.9-1977*

LETTER SYMBOLS

$P$ = Circular pitch of wormgear

$P$ = Axial pitch of the worm, $P_x$, in the central plane

$P_x$ = Axial pitch of worm

$P_n$ = Normal circular pitch of worm and wormgear = $P_x \cos \lambda = P \cos \psi$

$\lambda$ = Lead angle of worm

$\psi$ = Helix angle of wormgear

$n$ = Number of threads in worm

$N$ = Number of teeth in wormgear

$N = nm_G$

$m_G$ = Ratio of gearing = $N \div n$

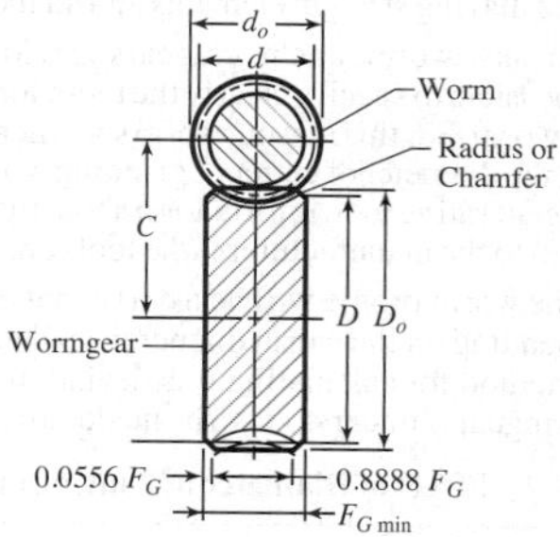


| Item | Formula | Item | Formula |
|---|---|---|---|
| WORM DIMENSIONS | | WORMGEAR DIMENSIONS[a] | |
| Lead | $l = nP_x$ | Pitch Diameter | $D = NP \div \pi = NP_x \div \pi$ |
| Pitch Diameter | $d = l \div (\pi \tan\lambda)$ | Outside Diameter | $D_o = 2C - d + 2a$ |
| Outside Diameter | $d_o = d + 2a$ | Face Width | $F_{G\min} = 1.125 \times \sqrt{(d_o + 2c)^2 - (d_o - 4a)^2}$ |
| Safe Minimum Length of Threaded Portion of Worm[b] | $F_W = \sqrt{D_o^2 - D^2}$ | | |
| DIMENSIONS FOR BOTH WORM AND WORMGEAR | | | |
| Addendum | $a = 0.3183P_n$ | Tooth thickness | $t_n = 0.5P_n$ |
| Whole Depth | $h_t = 0.7003P_n + 0.002$ | Approximate normal pressure angle[c] | $\phi_n = 20$ degrees |
| Working Depth | $h_k = 0.6366P_n$ | | |
| Clearance | $c = h_t - h_k$ | Center distance | $C = 0.5\,(d + D)$ |

All dimensions in inches unless otherwise indicated.

[a] Current practice for fine-pitch worm gearing does not require the use of throated blanks. This results in the much simpler blank shown in the diagram which is quite similar to that for a spur or helical gear. The slight loss in contact resulting from the use of non-throated blanks has little effect on the load-carrying capacity of fine-pitch worm gears. It is sometimes desirable to use topping hobs for producing wormgears in which the size relation between the outside and pitch diameters must be closely controlled. In such cases the blank is made slightly larger than $D_o$ by an amount (usually from 0.010 to 0.020) depending on the pitch. Topped wormgears will appear to have a small throat which is the result of the hobbing operation. For all intents and purposes, the throating is negligible and a blank so made is not to be considered as being a throated blank.

[b] This formula allows a sufficient length for fine-pitch worms.

[c] As stated in the text on page 2272, the actual pressure angle will be slightly greater due to the manufacturing process.

Although the pressure angle of the cutter or grinding wheel used to produce the worm is 20 degrees, the normal pressure angle produced in the worm will actually be slightly greater, and will vary with the worm diameter, lead angle, and diameter of cutter or grinding wheel. A method for calculating the pressure angle change is given under the heading *Effect of Production Method on Worm Profile and Pressure Angle*.

*Pitch Diameter Range of Worms:* The minimum recommended worm pitch diameter is 0.250 inch and the maximum is 2.000 inches.

*Tooth Form of Worm and Wormgear:* The shape of the worm thread in the normal plane is defined as that which is produced by a symmetrical double-conical cutter or grinding wheel having straight elements and an included angle of 40 degrees.

Because worms and wormgears are closely related to their method of manufacture, it is impossible to specify clearly the tooth form of the wormgear without referring to the mating worm. For this reason, worm specifications should include the method of manufacture and the diameter of cutter or grinding wheel used. Similarly, for determining the shape of the generating tool, information about the method of producing the worm threads must be given to the manufacturer if the tools are to be designed correctly.

The worm profile will be a curve that departs from a straight line by varying amounts, depending on the worm diameter, lead angle, and the cutter or grinding wheel diameter. A method for calculating this deviation is given in the Standard. The tooth form of the wormgear is understood to be made fully conjugate to the mating worm thread.

**Effect of Diameter of Cutting on Profile and Pressure Angle of Worms**

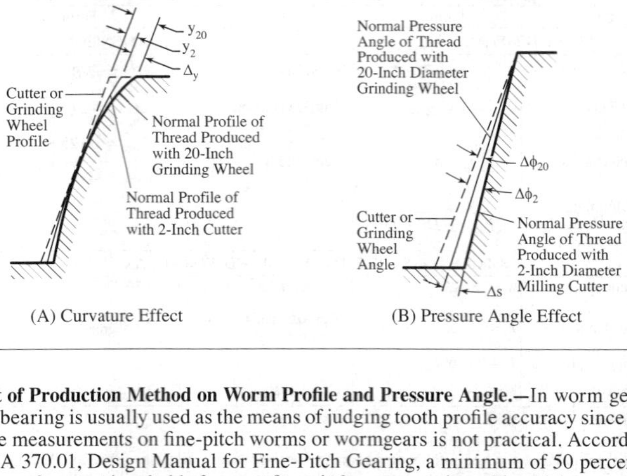


(A) Curvature Effect (B) Pressure Angle Effect

**Effect of Production Method on Worm Profile and Pressure Angle.—**In worm gearing, tooth bearing is usually used as the means of judging tooth profile accuracy since direct profile measurements on fine-pitch worms or wormgears is not practical. According to AGMA 370.01, Design Manual for Fine-Pitch Gearing, a minimum of 50 percent initial area of contact is suitable for most fine-pitch worm gearing, although in some cases, such as when the load fluctuates widely, a more restricted initial area of contact may be desirable.

Except where single-pointed lathe tools, end mills, or cutters of special shape are used in the manufacture of worms, the pressure angle and profile produced by the cutter are different from those of the cutter itself. The amounts of these differences depend on several factors, namely, diameter and lead angle of the worm, thickness and depth of the worm thread, and diameter of the cutter or grinding wheel. The accompanying diagram shows the curvature and pressure angle effects produced in the worm by cutters and grinding wheels, and how the amount of variation in worm profile and pressure angle is influenced by the diameter of the cutting tool used.

**Materials for Worm Gearing.—**Worm gearing, especially for power transmission, should have steel worms and phosphor bronze wormgears. This combination is used extensively. The worms should be hardened and ground to obtain accuracy and a smooth finish.

The phosphor bronze wormgears should contain from 10 to 12 percent of tin. The SAE phosphor gear bronze (No. 65) contains 88–90% copper, 10–12% tin, 0.50% lead, 0.50%

zinc (but with a maximum total lead, zinc and nickel content of 1.0 percent), phosphorous 0.10–0.30%, aluminum 0.005%. The SAE nickel phosphor gear bronze (No. 65 + Ni) contains 87% copper, 11% tin, 2% nickel and 0.2% phosphorous.

**Single-Thread Worm Gears.—**The ratio of the worm speed to the wormgear speed may range from 1.5 or even less up to 100 or more. Worm gearing having high ratios are not very efficient as transmitters of power; nevertheless high as well as low ratios often are required. Since the ratio equals the number of wormgear teeth divided by the number of threads or "starts" on the worm, single-thread worms are used to obtain a high ratio. As a general rule, a ratio of 50 is about the maximum recommended for a single worm and wormgear combination, although ratios up to 100 or higher are possible. When a high ratio is required, it may be preferable to use, in combination, two sets of worm gearing of the multi-thread type in preference to one set of the single-thread type in order to obtain the same total reduction and a higher combined efficiency.

Single-thread worms are comparatively inefficient because of the effect of the low lead angle; consequently, single-thread worms are not used when the primary purpose is to transmit power as efficiently as possible but they may be employed either when a large speed reduction with one set of gearing is necessary, or possibly as a means of adjustment, especially if "mechanical advantage" or self-locking are important factors.

**Multi-Thread Worm Gears.—**When worm gearing is designed primarily for transmitting power efficiently, the lead angle of the worm should be as high as is consistent with other requirements and preferably between, say, 25 or 30 and 45 degrees. This means that the worm must be multi-threaded. To obtain a given ratio, some number of wormgear teeth divided by some number of worm threads must equal the ratio. Thus, if the ratio is 6, combinations such as the following might be used:

$$\frac{24}{4}, \frac{30}{5}, \frac{36}{6}, \frac{42}{7}$$

The numerators represent numbers of wormgear teeth and the denominators, the number of worm threads or "starts." The number of wormgear teeth may not be an exact multiple of the number of threads on a multi-thread worm in order to obtain a "hunting tooth" action.

*Number of Threads or "Starts" on Worm:* The number of threads on the worm ordinarily varies from one to six or eight, depending upon the ratio of the gearing. As the ratio is increased, the number of worm threads is reduced, as a general rule. In some cases, however, the higher of two ratios may also have a larger number of threads. For example, a ratio of $6\frac{1}{5}$ would have 5 threads whereas a ratio of $6\frac{5}{6}$ would have 6 threads. Whenever the ratio is fractional, the number of threads on the worm equals the denominator of the fractional part of the ratio.

**Worm-Gear Cutting.—**The machines used for cutting worm-gears include ordinary milling machines, gear-hobbing machines of the type adapted to cutting either spur, spiral, or worm gearing, and special machines designed expressly for cutting worm-gears The general methods employed are (1) cutting by using a straight hob and a radial feeding movement between hob and gear blank; (2) cutting by feeding a fly cutter tangentially with relation to the worm gear blank; and (3) cutting by feeding a tapering hob tangentially. The fly-cutter method is slow as compared with hobbing but it has two decided advantages: First, a very simple and inexpensive cutter may be used instead of an expensive hob This is of great importance when the number of worm-gears is not large enough to warrant making a hob. Second, with the fly-cutter method, it is possible to produce worm-gears having more accurate teeth than are obtainable by the use of a straight hob. Taper hobs are especially adapted for cutting worm-gears that are to mesh with worms having large helix angles; they are also preferable for worm-gears having large face widths in proportion to the worm diameter. Worm-gear teeth are generated more accurately with a taper hob than with a straight hob that is given a radial feeding movement.

## HELICAL GEARING

**Basic Rules and Formulas for Helical Gear Calculations.**—The rules and formulas in the following table and elsewhere in this article are basic to helical gear calculations. The notation used in the formulas is: $P_n$ = normal diametral pitch of cutter; $D$ = pitch diameter; $N$ = number of teeth; $\alpha$ = helix angle; $\gamma$ = center angle or angle between shafts; $C$ = center distance; $N'$ = number of teeth for which to select a formed cutter for milled teeth; $L$ = lead of tooth helix; $S$ = addendum; $W$ = whole depth; $T_n$ = normal tooth thickness at pitch line; and $O$ = outside diameter.

### Rules and Formulas for Helical Gear Calculations

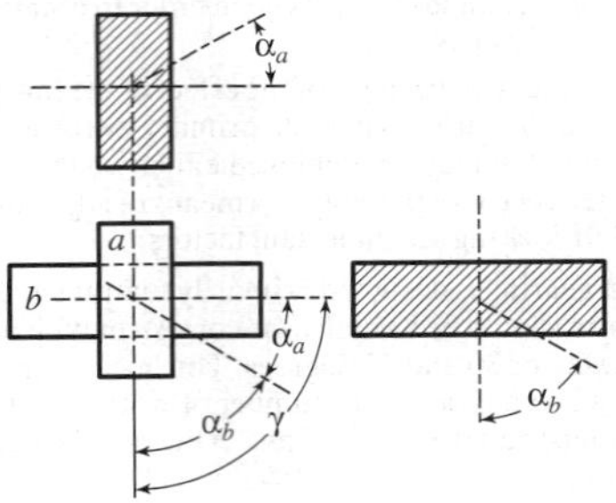


In the formulas, the symbols $D$, $N$, $L$ and a are for either the gear or the pinion. Subscripts $a$ and $b$ refer to the pinion or gear, respectively, in a pair of gears $a$ and $b$.

| No. | To Find | Rule | Formula |
|---|---|---|---|
| 1 | Pitch Diameter | Divide the number of teeth by the product of the normal diameter pitch and the cosine of the helix angle. | $D = \frac{N}{P_n \cos\alpha}$ |
| 2 | Center Distance | Add together the two pitch diameters and divide by 2. | $C = \frac{D_a + D_b}{2}$ |
| 3 | Lead of Tooth Helix | Multiply the pitch diameter by 3.1416 by the cotangent of the helix angle. | $L = \pi D \cot\alpha$ |
| 4 | Addendum | Divide 1 by the normal diametral pitch. | $S = \frac{1}{P_n}$ |
| 5 | Whole Depth of tooth | Divide 2.157 by the normal diametral pitch. | $W = \frac{2.157}{P_n}$ |
| 6 | Normal Tooth Thickness at Pitch Line | Divide 1.5708 by the normal diametral pitch. | $T_n = \frac{1.5708}{P_n}$ |
| 7 | Outside Diameter | Add twice the addendum to the pitch diameter. | $O = D + 2S$ |

**Determining Direction of Thrust.**—The first step in helical gear design is to determine the desired direction of the thrust. When the direction of the thrust has been determined and the relative positions of the driver and driven gears are known, then the direction of helix (right- or left-hand) may be found from the accompanying thrust diagrams, *Directions of Rotation and Resulting Thrust for Parallel Shaft and 90 Degree Shaft Angle Helical Gears*. The diagrams show the directions of rotation and the resulting thrust for parallel-shaft and 90-degree shaft angle helical gears. The thrust bearings are located so as to take the thrust caused by the tooth loads. The direction of the thrust depends on the direction of the helix, the relative positions of driver and driven gears, and the direction of rotation. The thrust may be changed to the opposite direction by changing any one of the three conditions, namely, by changing the hand of the helix, by reversing the direction of rotation, or by exchanging of driver and driven gear positions.

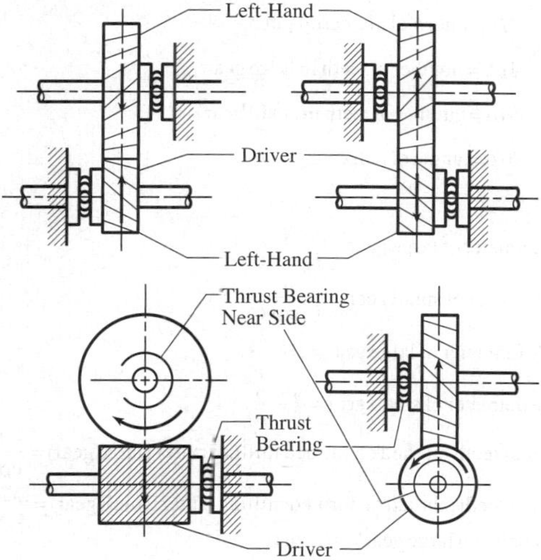


Directions of Rotation and Resulting Thrust for Parallel Shaft and 90 Degree Shaft Angle Helical Gears

**Determining Helix Angles.—**The following rules should be observed for helical gears with shafts at any given angle. If each helix angle is less than the shaft angle, then the sum of the helix angles of the two gears will equal the angle between the shafts, and the helix angle is of the same hand for both gears; if the helix angle of one of the gears is larger than the shaft angle, then the difference between the helix angles of the two gears will be equal to the shaft angle, and the gears will be of opposite hand.

**Pitch of Cutter to be Used.—**The thickness of the cutter at the pitchline for cutting helical gears should equal one-half the *normal* circular pitch. The normal pitch varies with the helix angle, hence, the helix angle must be considered when selecting a cutter. The cutter should be of the same pitch as the *normal* diametral pitch of the gear. This normal pitch is found by dividing the transverse diametral pitch of the gear by the cosine of the helix angle. To illustrate, if the pitch diameter of a helical gear is 6.718 and there are 38 teeth having a helix angle of 45 degrees, the transverse diametral pitch equals 38 divided by 6.718 = 5.656; then the normal diametral pitch equals 5.656 divided by 0.707 = 8. A cutter, then, of 8 diametral pitch is the one to use for this particular gear.

Helical gears should preferably be cut on a generating-type gear cutting machine such as a hobber or shaper. Milling machines are used in some shops when hobbers or shapers are not available or when single, replacement gears are being made. In such instances, the pitch of the formed cutter used in milling a helical gear must not only conform to the normal diametral pitch of the gear, but the cutter number must also be determined. See *Selecting Cutter for Milling Helical Gears* starting on page 2285.

**1. Shafts Parallel, Center Distance Approximate.**—Given or assumed:

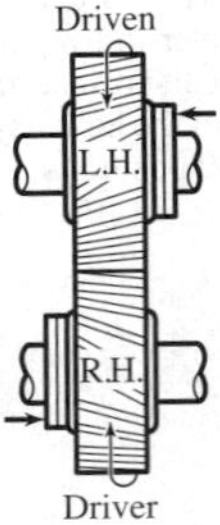


1) Position of gear having right- or left-hand helix, depending upon rotation and direction in which thrust is to be received

2) $C_a$ = approximate center distance

3) $P_n$ = normal diametral pitch

4) $N$ = number of teeth in large gear

5) $n$ = number of teeth in small gear

6) $\alpha$ = angle of helix

To find:

1) $D$ = pitch diameter of large gear $= \dfrac{N}{P_n \cos\alpha}$

2) $d$ = pitch diameter of small gear $= \dfrac{n}{P_n \cos\alpha}$

3) $O$ = outside diameter of large gear $= D + \dfrac{2}{P_n}$

4) $o$ = outside diameter of small gear $= d + \dfrac{2}{P_n}$

5) $T$ = number of teeth marked on formed milling cutter (large gear) $= \dfrac{N}{\cos^3\alpha}$

6) $t$ = number of teeth marked on formed milling cutter (small gear) $= \dfrac{n}{\cos^3\alpha}$

7) $L$ = lead of helix on large gear $= \pi D \cot\alpha$

8) $l$ = lead of helix on small gear $= \pi d \cot\alpha$

9) $C$ = center distance (if not right, vary $\alpha$) $= \frac{1}{2}(D + d)$

*Example:* Given or assumed: 1) See illustration; 2) $C_a$ = 17 inches; 3) $P_n$ = 2; 4) $N$ = 48; 5) $n$ = 20; and 6) $\alpha$ = 20.

To find:

1) $D = \dfrac{N}{P_n \cos\alpha} = \dfrac{48}{2 \times 0.9397} = 25.541$ inches

2) $d = \dfrac{n}{P_n \cos\alpha} = \dfrac{20}{2 \times 0.9397} = 10.642$ inches

3) $O = \dfrac{2}{P_n} = 25.541 + \dfrac{2}{2} = 26.541$ inches

4) $o = d + \dfrac{2}{P_n} = 10.642 + \dfrac{2}{2} = 11.642$ inches

5) $T = \dfrac{N}{\cos^3\alpha} = \dfrac{48}{(0.9397)^3} = 57.8$, say 58 teeth

6) $t = \dfrac{n}{\cos^3\alpha} = \dfrac{20}{(0.9397)^3} = 24.1$, say 24 teeth

7) $L = \pi D \cot\alpha = 3.1416 \times 25.541 \times 2.747 = 220.42$ inches

8) $l = \pi d \cot\alpha = 3.1416 \times 10.642 \times 2.747 = 91.84$ inches

9) $C = \frac{1}{2}(D + d) = \frac{1}{2}(25.541 + 10.642) = 18.091$ inches

**2. Shafts Parallel, Center Distance Exact.**—Given or assumed:

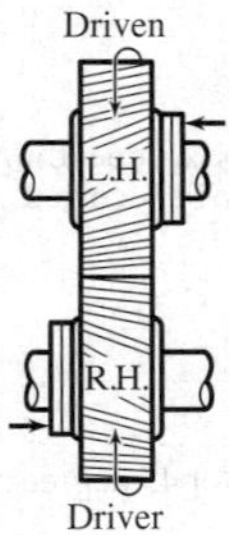


1) Position of gear having right- or left-hand helix, depending upon rotation and direction in which thrust is to be received

2) $C$ = exact center distance

3) $P_n$ = normal diametral pitch (pitch of cutter)

4) $N$ = number of teeth in large gear

5) $n$ = number of teeth in small gear

To find:

1) $\cos\alpha = \dfrac{N+n}{2P_n C}$

2) $D$ = pitch diameter of large gear $= \dfrac{N}{P_n \cos\alpha}$

3) $d$ = pitch diameter of small gear $= \dfrac{n}{P_n \cos\alpha}$

4) $O$ = outside diameter of large gear $= D + \dfrac{2}{P_n}$

5) $o$ = outside diameter of small gear $= d + \dfrac{2}{P_n}$

6) $T$ = number of teeth marked on formed milling cutter (large gear) $= \dfrac{N}{\cos^3\alpha}$

7) $t$ = number of teeth marked on formed milling cutter (small gear) $= \dfrac{n}{\cos^3\alpha}$

8) $L$ = lead of helix (large gear) $= \pi D \cot\alpha$

9) $l$ = lead of helix (small gear) $= \pi d \cot\alpha$

*Example:* Given or assumed: 1) See illustration; 2) $C$ = 18.75 inches; 3) $P_n$ = 4; 4) $N$ = 96; and 5) $n$ = 48.

1) $\cos\alpha = \dfrac{N+n}{2P_n C} = \dfrac{96+48}{2\times 4\times 18.75} = 0.96$, or $\alpha = 16°16'$

2) $D = \dfrac{N}{P_n \cos\alpha} = \dfrac{96}{4\times 0.96} = 25$ inches

3) $d = \dfrac{n}{P_n \cos\alpha} = \dfrac{48}{4\times 0.96} = 12.5$ inches

4) $O = D + \dfrac{2}{P_n} = 25 + \dfrac{2}{4} = 25.5$ inches

5) $o = d + \dfrac{2}{P_n} = 12.5 + \dfrac{2}{4} = 13$ inches

6) $T = \dfrac{N}{\cos^3\alpha} = \dfrac{96}{(0.96)^3} = 108$ teeth

7) $t = \dfrac{n}{\cos^3\alpha} = \dfrac{48}{(0.96)^3} = 54$ teeth

8) $L = \pi D \cot\alpha = 3.1416 \times 25 \times 3.427 = 269.15$ inches

9) $l = \pi d \cot\alpha = 3.1416 \times 12.5 \times 3.427 = 134.57$ inches

**3. Shafts at Right Angles, Center Distance Approximate.—** Sum of helix angles of gear and pinion must equal 90 degrees.

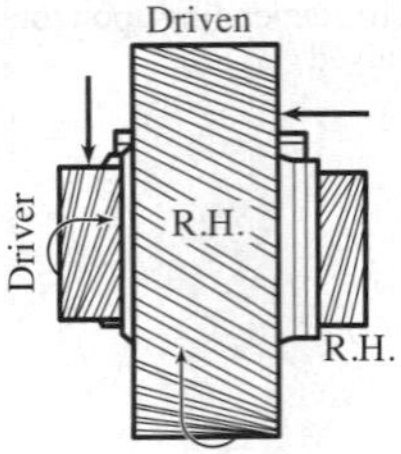


Given or assumed:

1) Position of gear having right- or left-hand helix, depending on rotation and direction in which thrust is to be received

2) $C_a$ = approximate center distance

3) $P_n$ = normal diametral pitch (pitch of cutter)

4) $R$ = ratio of gear to pinion size

5) $n$ = number of teeth in pinion = $\frac{1.41 C_a P_n}{R+1}$ for 45 degrees; and $\frac{2 C_a P_n \cos\alpha \cos\beta}{R \cos\beta + \cos\alpha}$ for any angle

6) $N$ = number of teeth in gear = $Rn$

7) $\alpha$ = angle of helix of gear

8) $\beta$ = angle of helix of pinion

To find:

a) When helix angles are 45 degrees,

1) $D$ = pitch diameter of gear = $\frac{N}{0.70711 P_n}$

2) $d$ = pitch diameter of pinion = $\frac{n}{0.70711 P_n}$

3) $O$ = outside diameter of gear = $D + \frac{2}{P_n}$

4) $o$ = outside diameter of pinion = $d + \frac{2}{P_n}$

5) $T$ = number of formed cutter (gear) = $\frac{N}{0.353}$

6) $t$ = number of formed cutter (pinion) = $\frac{n}{0.353}$

7) $L$ = lead of helix of gear = $\pi D$

8) $l$ = lead of helix of pinion = $\pi d$

9) $C$ = center distance (exact) = $\frac{D + d}{2}$

b) When helix angles are other than 45 degrees

1) $D = \frac{N}{P_n \cos\alpha}$ 2) $d = \frac{n}{P_n \cos\beta}$ 3) $T = \frac{N}{\cos^3\alpha}$

4) $t = \frac{n}{\cos^3\beta}$ 5) $L = \pi D \cot\alpha$ 6) $l = \pi d \cot\beta$

*Example:* Given or assumed: 1) See illustration; 2) $C_a$ = 3.2 inches; 3) $P_n$ = 10; and 4) $R$ = 1.5.

5) $n = \frac{1.41 C_a P_n}{R+1} = \frac{1.41 \times 3.2 \times 10}{1.5 + 1}$ = say 18 teeth.

6) $N = nR = 18 \times 1.5 = 27$ teeth; 7) $\alpha$ = 45 degrees; and 8) $\beta$ = 45 degrees.

To find:

1) $D = \frac{N}{0.70711P_n} = \frac{27}{0.70711 \times 10} = 3.818$ inches

2) $d = \frac{n}{0.70711P_n} = \frac{18}{0.70711 \times 10} = 2\ 545$ inches

3) $O = D + \frac{2}{P_n} = 3.818 + \frac{2}{10} = 4.018$ inches

4) $o = d + \frac{2}{P_n} = 2.545 + \frac{2}{10} = 2.745$ inches

5) $T = \frac{N}{0.353} = \frac{27}{0.353} = 76.5$, say 76 teeth

6) $t = \frac{n}{0.353} = \frac{18}{0.353} = 51$ teeth

7) $L = \pi D = 3.1416 \times 3.818 = 12$ inches

8) $l = \pi d = 3.1416 \times 2.545 = 8$ inches

9) $C = \frac{D + d}{2} = \frac{3.818 + 2.545}{2} = 3.182$ inches

**4A. Shafts at Right Angles, Center Distance Exact.**—Gears have same direction of helix. Sum of the helix angles will equal 90 degrees.

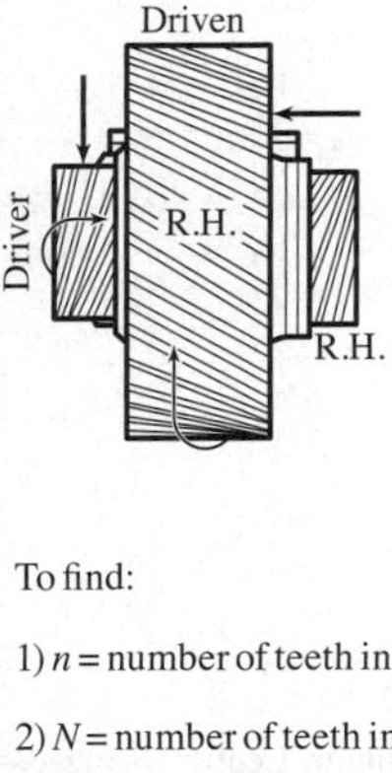


Given or assumed:

1) Position of gear having right- or left-hand helix depending on rotation and direction in which thrust is to be received

2) $P_n$ = normal diametral pitch (pitch of cutter)

3) $R$ = ratio of number of teeth in large gear to number of teeth in small gear

4) $\alpha_a$ = approximate helix angle of large gear

5) $C$ = exact center distance

To find:

1) $n$ = number of teeth in small gear nearest $= 2\,CP_n \sin \alpha_a \div 1 + R \tan \alpha_a$

2) $N$ = number of teeth in large gear $= Rn$

3) $\alpha$ = exact helix angle of large gear, found by trial from $R \sec \alpha + \operatorname{cosec} \alpha = 2\,CP_n \div n$

4) $\beta$ = exact helix angle of small gear $= 90° - \alpha$

5) $D$ = pitch diameter of large gear $= \frac{N}{P_n \cos \alpha}$

6) $d$ = pitch diameter of small gear $= \frac{n}{P_n \cos \beta}$

7) $O$ = outside diameter of large gear $= D + \frac{2}{P_n}$

8) $o$ = outside diameter of small gear $= d + \frac{2}{P_n}$

9) $N'$ and $n'$ = numbers of teeth marked on cutters for large and small gears (see page 2285)

10) $L$ = lead of helix on large gear $= \pi D \cot \alpha$

11) $l$ = lead of helix on small gear $= \pi d \cot \beta$

*Example:* Given or assumed: 1) See illustration; 2) $P_n = 8$; 3) $R = 3$; 4) $\alpha_a = 45$ degrees; and 5) $C = 10$ in.

To find:

1) $n = \frac{2CP_n \sin \alpha_a}{1 + R \tan \alpha_a} = \frac{2 \times 10 \times 8 \times 0.70711}{1 + 3} = 28.25$, say 28 teeth

2) $N = Rn = 3 \times 28 = 84$ teeth

3) $R \sec \alpha + \operatorname{cosec} \alpha = \frac{2CP_n}{n} = \frac{2 \times 10 \times 8}{28} = 5.714$, or $\alpha = 46°6'$

4) $\beta = 90° - \alpha = 90° - 46°6' = 43°54'$

5) $D = \frac{N}{P_n \cos \alpha} = \frac{84}{8 \times 0.6934} = 15.143$ inches

6) $d = \frac{n}{P_n \cos \beta} = \frac{28}{8 \times 0.72055} = 4.857$ inches

7) $O = D + \frac{2}{P_n} = 15.143 + 0.25 = 15.393$ inches

8) $o = d + \frac{2}{P_n} = 4.857 + 0.25 = 5.107$ inches

9) $N' = 275$; $n' = 94$ (see page 2285)

10) $L = \pi D \cot \alpha = 3.1416 \times 15.143 \times 0.96232 = 45.78$ inches

11) $l = \pi d \cot \beta = 3.1416 \times 4.857 \times 1.0392 = 15.857$ inches

**4B. Shafts at Right Angles, Any Ratio, Helix Angle for Minimum Center Distance.—** Diagram similar to 4A. Gears have same direction of helix. The sum of the helix angles will equal 90 degrees.

For any given ratio of gearing $R$ there is a helix angle $\alpha$ for the larger gear and a helix angle $\beta = 90° - \alpha$ for the smaller gear that will make the center distance $C$ a minimum. Helix angle $\alpha$ is found from the formula $\cot \alpha = R^{1/3}$. As an example, using the data found in Case 4A, helix angles $\alpha$ and $\beta$ for minimum center distance would be: $\cot \alpha = R^{1/3} = 1.4422$; $\alpha = 34°44'$ and $\beta = 90° - 34°44' = 55°16'$. Using these helix angles, $D = 12.777$; $d = 6.143$; and $C = 9.460$ from the formulas for $D$ and $d$ given under Case 4A.

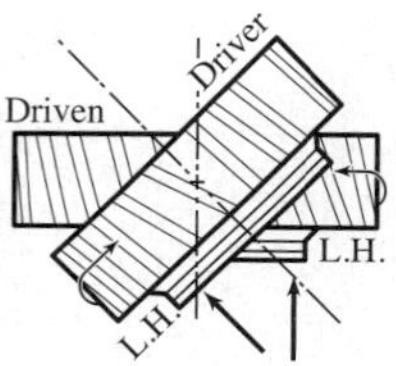


**5. Shafts at Any Angle, Center Distance Approximate.**—The sum of the helix angles of the two gears equals the shaft angle, and the gears are of the same hand, if each angle is less than the shaft angle. The difference between the helix angles equals the shaft angle, and the gears are of opposite hand, if either angle is greater than the shaft angle.

Given or assumed:

1) Hand of helix, depending on rotation and direction in which thrust is to be received

2) $C_a$ = center distance

3) $P_n$ = normal diametral pitch (pitch of cutter)

4) $R$ = ratio of gear to pinion = $\frac{N}{n}$

5) $\alpha$ = angle of helix, gear

6) $\beta$ = angle of helix, pinion

7) $n$ = number of teeth in pinion nearest $\frac{2C_aP_n\cos\alpha\cos\beta}{R\cos\beta + \cos\alpha}$ for any angle

8) and $\frac{2C_aP_n\cos\alpha}{R+1}$ when both angles are equal

9) $N$ = number of teeth in gear = $Rn$

To find:

1) $D$ = pitch diameter of gear = $\frac{N}{P_n\cos\alpha}$

2) $d$ = pitch diameter of pinion = $\frac{n}{P_n\cos\beta}$

3) $O$ = outside diameter of gear = $D + \frac{2}{P_n}$

4) $o$ = outside diameter of pinion = $d + \frac{2}{P_n}$

5) $T$ = number of teeth marked on cutter for gear = $\frac{N}{\cos^3\alpha}$

6) $t$ = number of teeth marked on cutter for pinion = $\frac{n}{\cos^3\beta}$

7) $L$ = lead of helix on gear = $\pi D\cot\alpha$

8) $l$ = lead of helix on pinion = $\pi d\cot\beta$

9) $C$ = actual center distance = $\frac{D-d}{2}$

*Example:* Given or assumed (angle of shafts, 60 degrees):

1) See illustration  2) $C_a$ = 12 inches  3) $P_n = 8$

4) R = 4  5) $\alpha$ = 30 degrees  6) $\beta$ = 30 degrees

5) $n = \frac{2C_aP_n\cos\alpha}{R+1} = \frac{2 \times 12 \times 8 \times 0.86603}{4-1} = 33$ teeth

6) $N = 4 \times 33 = 132$ teeth

To find:

1) $D = \frac{N}{P_n \cos\alpha} = \frac{132}{8 \times 0.86603} = 19.052$ inches

2) $d = \frac{n}{P_n \cos\beta} = \frac{33}{8 \times 0.86603} = 4.763$ inches

3) $O = D + \frac{2}{P_n} = 19.052 + \frac{2}{8} = 19.302$ inches

4) $o = d + \frac{2}{P_n} = 4.763 + \frac{2}{8} = 5.013$ inches

5) $T = \frac{N}{\cos^3\alpha} = \frac{132}{0.65} = 203$ teeth

6) $t = \frac{n}{\cos^3\beta} = \frac{33}{0.65} = 51$ teeth

7) $L = \pi D \cot\alpha = \pi \times 19.052 \times 1.732 = 103.66$ inches

8) $l = \pi d \cot\beta = \pi \times 4.763 \times 1.732 = 25.92$ inches

9) $C = \frac{D+d}{2} = \frac{19.052 + 4.763}{2} = 11.9075$ inches

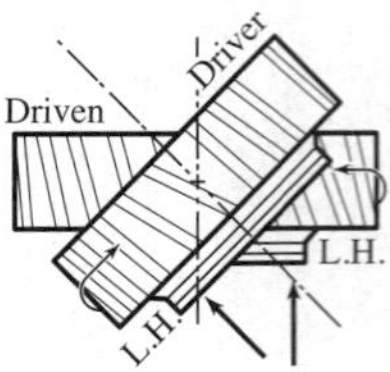


**6. Shafts at Any Angle, Center Distance Exact.—**The sum of the helix angles of the two gears equals the shaft angle, and the gears are of the same hand, if each angle is less than the shaft angle. The difference between the helix angles equals the shaft angle, and the gears are of opposite hand, if either angle is greater than the shaft angle.

Given or assumed:

1) Hand of helix, depending on rotation and direction in which thrust is to be received

2) $C$ = center distance

3) $P_n$ = normal diametral pitch (pitch of cutter)

4) $\alpha_a$ = approximate helix angle of gear

5) $\beta_a$ = approximate helix angle of pinion

6) $R$ = ratio of gear to pinion size $= \frac{N}{n}$

7) $n$ = number of pinion teeth nearest $\frac{2CP_n \cos\alpha_a \cos\beta_a}{R\cos\beta_a + \cos\alpha_a}$

8) $N$ = number of gear teeth $= Rn$

To find:

1) $\alpha$ and $\beta$, exact helix angles, found by trial from $R\sec\alpha + \sec\beta = \frac{2CP_n}{n}$

2) $D$ = pitch diameter of gear = $\dfrac{N}{P_n \cos\alpha}$

3) $d$ = pitch diameter of pinion = $\dfrac{n}{P_n \cos\beta}$

4) $O$ = outside diameter of gear = $D + \dfrac{2}{P_n}$

5) $o$ = outside diameter of pinion = $d + \dfrac{2}{P_n}$

6) $N'$ = number of teeth marked on formed cutter for gear (see below)

7) $n'$ = number of teeth marked on formed cutter for pinion (see below)

8) $L$ = lead of helix on gear = $\pi D \cot\alpha$

9) $l$ = lead of helix on pinion = $\pi d \cot\beta$

**Selecting Cutter for Milling Helical Gears.—**The proper milling cutter to use for *spur* gears depends on the pitch of the teeth and also upon the number of teeth as explained on page 2229 but a cutter for milling helical gears is not selected with reference to the actual number of teeth in the gear, as in spur gearing, but rather with reference to a calculated number $N'$ that takes into account the effect on the tooth profile of lead angle, normal diametral pitch, and cutter diameter.

In the helical gearing examples starting on page 2278 the number of teeth $N'$ on which to base the selection of the cutter has been determined using the approximate formula $N' = N \div \cos^3 \alpha$ or $N' = N \sec^3 \alpha$, where $N$ = the actual number of teeth in the helical gear and $\alpha$ = the helix angle. However, the use of this formula may, where a combination of high helix angle and low tooth number is involved, result in the selection of a higher number of cutter than should actually be used for greatest accuracy. This condition is most likely to occur when the aforementioned formula is used to calculate $N'$ for gears of high helix angle and low number of teeth.

To avoid the possibility of error in choice of cutter number, the following formula, which gives theoretically correct results for all combinations of helix angle and tooth numbers, is to be preferred:

$$N' = N \sec^3\alpha + P_n D_c \tan^2\alpha \qquad (1)$$

where: $N'$ = number of teeth on which to base selection of cutter number from table on page 2231; $N$ = actual number of teeth in helical gear; $\alpha$ = helix angle; $P_n$ = normal diametral pitch of gear and cutter; and $D_c$ = pitch diameter of cutter.

To simplify calculations, Formula (1) may be written as follows:

$$N' = NK + QK' \qquad (2)$$

In this formula, $K$, $K'$ and $Q$ are constants obtained from the tables on page 2286.

*Example:* Helix angle = 30 degrees; number of teeth in helical gear = 15; and normal diametral pitch = 20. From the tables on page 2286 $K$, $K'$, and $Q$ are, respectively, 1.540, 0.333, and 37.80.

$$N' = (15 \times 1.540) + (37.80 \times 0.333) = 23.10 + 12.60$$
$$= 35.70, \text{ say, } 36$$

Hence, from page 2231 select a number 3 cutter. Had the approximate formula been used, then a number 5 cutter would have been selected on the basis of $N' = 23$.

### Factors for Selecting Cutters for Milling Helical Gears

| Helix Angle, α | K | K′ | Helix Angle, α | K | K′ | Helix Angle, α | K | K′ | Helix Angle, α | K | K′ |
|---|---|---|---|---|---|---|---|---|---|---|---|
| 0 | 1.000 | 0 | 16 | 1.127 | 0.082 | 32 | 1.640 | 0.390 | 48 | 3.336 | 1.233 |
| 1 | 1.001 | 0 | 17 | 1.145 | 0.093 | 33 | 1.695 | 0.422 | 49 | 3.540 | 1.323 |
| 2 | 1.002 | 0.001 | 18 | 1.163 | 0.106 | 34 | 1.755 | 0.455 | 50 | 3.767 | 1.420 |
| 3 | 1.004 | 0.003 | 19 | 1.182 | 0.119 | 35 | 1.819 | 0.490 | 51 | 4.012 | 1.525 |
| 4 | 1.007 | 0.005 | 20 | 1.204 | 0.132 | 36 | 1.889 | 0.528 | 52 | 4.284 | 1.638 |
| 5 | 1.011 | 0.008 | 21 | 1.228 | 0.147 | 37 | 1.963 | 0.568 | 53 | 4.586 | 1.761 |
| 6 | 1.016 | 0.011 | 22 | 1.254 | 0.163 | 38 | 2.044 | 0.610 | 54 | 4.925 | 1.894 |
| 7 | 1.022 | 0.015 | 23 | 1.282 | 0.180 | 39 | 2.130 | 0.656 | 55 | 5.295 | 2.039 |
| 8 | 1.030 | 0.020 | 24 | 1.312 | 0.198 | 40 | 2.225 | 0.704 | 56 | 5.710 | 2.198 |
| 9 | 1.038 | 0.025 | 25 | 1.344 | 0.217 | 41 | 2.326 | 0.756 | 57 | 6.190 | 2.371 |
| 10 | 1.047 | 0.031 | 26 | 1.377 | 0.238 | 42 | 2.436 | 0.811 | 58 | 6.720 | 2.561 |
| 11 | 1.057 | 0.038 | 27 | 1.414 | 0.260 | 43 | 2.557 | 0.870 | 59 | 7.321 | 2.770 |
| 12 | 1.068 | 0.045 | 28 | 1.454 | 0.283 | 44 | 2.687 | 0.933 | 60 | 8.000 | 3.000 |
| 13 | 1.080 | 0.053 | 29 | 1.495 | 0.307 | 45 | 2.828 | 1 | 61 | 8.780 | 3.254 |
| 14 | 1.094 | 0.062 | 30 | 1.540 | 0.333 | 46 | 2.983 | 1.072 | 62 | 9.658 | 3.537 |
| 15 | 1.110 | 0.072 | 31 | 1.588 | 0.361 | 47 | 3.152 | 1.150 | 63 | 10.687 | 3.852 |

$K = 1 \div \cos^3 \alpha = \sec^3 \alpha$; $K' = \tan^2 \alpha$

### Outside and Pitch Diameters of Standard Involute-Form Milling Cutters

| Normal Diametral Pitch, $P_n$ | Outside Dia., $D_o$ | Pitch Dia., $D_c$ | $Q = P_n D_c$ | Normal Diametral Pitch, $P_n$ | Outside Dia., $D_o$ | Pitch Dia., $D_c$ | $Q = P_n D_c$ | Normal Diametral Pitch, $P_n$ | Outside Dia., $D_o$ | Pitch Dia., $D_c$ | $Q = P_n D_c$ |
|---|---|---|---|---|---|---|---|---|---|---|---|
| 1 | 8.500 | 6.18 | 6.18 | 6 | 3.125 | 2.76 | 16.56 | 20 | 2.000 | 1.89 | 37.80 |
| 1¼ | 7.750 | 5.70 | 7.12 | 7 | 2.875 | 2.54 | 17.78 | 24 | 1.750 | 1.65 | 39.60 |
| 1½ | 7.000 | 5.46 | 8.19 | 8 | 2.875 | 2.61 | 20.88 | 28 | 1.750 | 1.67 | 46.76 |
| 1¾ | 6.500 | 5.04 | 8.82 | 9 | 2.750 | 2.50 | 22.50 | 32 | 1.750 | 1.68 | 53.76 |
| 2 | 5.750 | 4.60 | 9.20 | 10 | 2.375 | 2.14 | 21.40 | 36 | 1.750 | 1.69 | 60.84 |
| 2½ | 5.750 | 4.83 | 12.08 | 12 | 2.250 | 2.06 | 24.72 | 40 | 1.750 | 1.70 | 68.00 |
| 3 | 4.750 | 3.98 | 11.94 | 14 | 2.125 | 1.96 | 27.44 | 48 | 1.750 | 1.70 | 81.60 |
| 4 | 4.250 | 3.67 | 14.68 | 16 | 2.125 | 1.98 | 31.68 | .. | … | … | … |
| 5 | 3.750 | 3.29 | 16.45 | 18 | 2.000 | 1.87 | 33.66 | .. | … | … | … |

Pitch diameters shown in the table are computed from the formula: $D_c = D_o - 2(1.57 \div P_n)$. This same formula may be used to compute the pitch diameter of a nonstandard outside diameter cutter when the normal diametral pitch $P_n$ and the outside diameter $D_o$ are known.

**Milling the Helical Teeth.—**The teeth of a helical gear are proportioned from the normal pitch and not the circular pitch. The whole depth of the tooth can be found by dividing 2.157 by the normal diametral pitch of the gear, which corresponds to the pitch of the cutter. The thickness of the tooth at the pitch line equals 1.571 divided by the normal diametral pitch. After a tooth space has been milled, the cutter should be prevented from dragging through it when being returned for another cut. This can be done by lowering the blank slightly, or by stopping the machine and turning the cutter to such a position that the teeth will not touch the work. If the gear has teeth coarser than 10 or 12 diametral pitch, it is well to take a roughing and a finishing cut. When pressing a helical gear blank on the arbor, it should be remembered that it is more likely to slip when being milled than a spur gear, because the pressure of the cut, being at an angle, tends to rotate the blank on the arbor.

*Angular Position of Table:* When cutting a helical gear on a milling machine, the table is set to the helix angle of the gear. If the lead of the helical gear is known, but not the helix angle, the helix angle is determined by multiplying the pitch diameter of the gear by 3.1416 and dividing this product by the lead; the result is the tangent of the lead angle which may be obtained from trigonometric tables or a calculator.

**American National Standard Fine-Pitch Teeth for Helical Gears.—**This Standard, ANSI B6.7-1977, provides a 20-degree tooth form for both spur and helical gears of 20 diametral pitch and finer. Formulas for tooth parts are given on page 2216.

*Enlargement of Helical Pinions, 20-Degree Normal Pressure Angle:* Formula (4) and the accompanying graph are based on the use of hobs having sharp corners at their top lands. Pinions cut by shaper cutters may not require as much modification as indicated by (4) or the graph. The number 2.1 appearing in (4) results from the use of a standard tooth thickness rack having an addendum of $1.05/P_n$ which will start contact at a roll angle 5 degrees above the base radius. The roll angle of 5 degrees is also reflected in Formula (4).

To avoid undercutting of the teeth and to provide more favorable contact conditions near the base of the tooth, it is recommended that helical pinions with less than 24 teeth be enlarged in accordance with the following graph and formulas. As with enlarged spur pinions, when an enlarged helical pinion is used it is necessary either to reduce the diameter of the mating gear or to increase the center distance. In the formulas that follow, $\phi_n$ = normal pressure angle; $\phi_t$ = transverse pressure angle; $\psi$ = helix angle of pinion; $P_n$ = normal diametral pitch; $P_t$ = transverse diametral pitch; $d$ = pitch diameter of pinion; $d_o$ = outside diameter of enlarged pinion; $K_h$ = enlargement for full depth pinions of 1 normal diametral pitch; and $n$ = number of teeth in pinion.

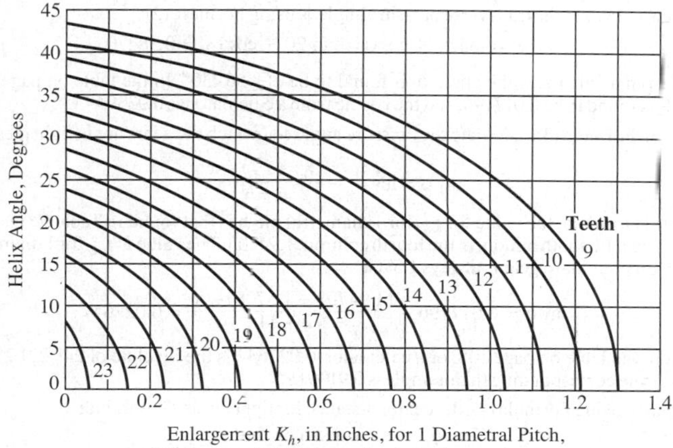


Enlargement $K_h$, in Inches, for 1 Diametral Pitch,
20° Pressure Angle Pinions

To eliminate the need for making the calculations indicated in Formulas (3) and (4), the accompanying graph may be used to obtain the value of $K_h$ directly for full-depth pinions of 20-degree normal pressure angle.

$$P_t = P_n \cos\psi \tag{1}$$

$$d = n \div P_t \tag{2}$$

$$\tan\psi_t = \tan\phi_n \div \cos\psi \tag{3}$$

$$K_h = 2.1 - \frac{n}{\cos\psi}(\sin\phi_t - \cos\phi_t \tan 5°)\sin\phi_t \tag{4}$$

$$d_o = d + \frac{2 + K_h}{P_n} \tag{5}$$

*Example:* Find the outside diameter of a helical pinion having 12 teeth, 32 normal diametral pitch, 20-degree pressure angle, and 18-degree helix angle.

$$P_t = P_n \cos\psi = 32\cos 18° = 32 \times 0.95106 = 30.4339$$

$$d = n \div P_t = 12 \div 30.4339 = 0.3943 \text{ inch}$$

$$K_h = 0.851 \text{ (from graph)}$$

$$d_o = 0.3943 + \frac{2 + 0.851}{32} = 0.4834$$

**Center Distance at Which Modified Mating Helical Gears Mesh without Backlash.—** If the helical pinion in the previous example on page 2288 had been made to standard dimensions, that is, not enlarged, and was in tight mesh with a standard 24-tooth mating gear, the center distance for tight mesh could be calculated from the formula on page 2216:

$$C = \frac{n + N}{2P_n \cos\psi} = \frac{12 + 24}{2 \times 32 \times \cos 18°} = 0.5914 \text{ inch} \tag{1}$$

However, if the pinion is enlarged as in the example and meshed with the same standard 24-tooth gear, then the center distance for tight mesh will be increased. To calculate the new center distance, the following formulas and calculations are required:

First, calculate the transverse pressure angle $\phi_t$ using Formula (2):

$$\tan\phi_t = \tan\phi_n \div \cos\psi = \tan 20° \div \cos 18° = 0.38270 \tag{2}$$

and from a calculator the angle $\phi_t$ is found to be 20° 56′ 30″. In the table on page 107, inv $\phi_t$ is found to be 0.017196, and the cosine from a calculator as 0.93394.

Using Formula (3), calculate the pressure angle $\phi$ at which the gears are in tight mesh:

$$\text{inv}\,\phi = \text{inv}\,\phi_t + \frac{(t_{nP} + t_{nG}) - \pi}{n + N} \tag{3}$$

In this formula, the value for $t_{nP}$ for 1 diametral pitch is that found in Table 9c on page 2233, for a 12-tooth pinion, in the fourth column: 1.94703. The value of $t_{nG}$ for 1 diametral pitch for a standard gear is always 1.5708.

$$\text{inv}\,\phi = 0.017196 + \frac{(1.94703 + 1.5708) - \pi}{12 + 24} = 0.027647$$

From the table on page 108, or a calculator, 0.027647 is the involute of 24° 22′ 7″ and the cosine corresponding to this angle is 0.91091.

Finally, using Formula (4), the center distance for tight mesh, $C'$ is found:

$$C' = \frac{C\cos\phi_t}{\cos\phi} = \frac{0.5914 \times 0.93394}{0.91091} = 0.606 \text{ inch} \tag{4}$$

**Change-Gears for Helical Gear Hobbing.—**If a gear-hobbing machine is not equipped with a differential, there is a fixed relation between the index and feed gears and it is necessary to compensate for even slight errors in the index gear ratio, to avoid excessive lead errors. This may be done readily (as shown by the example to follow) by modifying the ratio of the feed gears slightly, thus offsetting the index gear error and making very accurate leads possible.

*Machine without Differential:* The formulas which follow may be applied in computing the index gear ratio.

$R$ = index-gear ratio

$L$ = lead of gear, inches

$F$ = feed per gear revolution, inch

$K$ = machine constant

$T$ = number of threads on hob

$N$ = number of teeth on gear

$P_n$ = normal diametral pitch

$P_{nc}$ = normal circular pitch

$A$ = helix angle, relative to axis

$M$ = feed gear constant

$$R = \frac{L \div F}{(L \div F) \pm 1} \times \frac{KT}{N} = \frac{L}{L \pm F} \times \frac{KT}{N} = \frac{\text{Driving gear sizes}}{\text{Driven gear sizes}} \qquad (1)$$

Use minus (−) sign in Formulas (1) and (2) when gear and hob are the same "hand" and plus (+) sign when they are of opposite hand; when *climb* hobbing is to be used, reverse this rule.

$$R = \frac{KT}{N \pm \dfrac{P_n \times \sin A \times F}{\pi}} = \frac{KT}{N \pm \dfrac{\sin A \times F}{P_{nc}}} \qquad (2)$$

$$\text{Ratio of feed gears} = \frac{F}{M} \qquad F = \frac{L(NR - KT)}{NR} \qquad (3)$$

$$L = \frac{FNR}{(NR - KT)} = \text{lead obtained with available index and feed gears} \qquad (4)$$

*Note:* If gear and hob are of opposite hand, then in Formulas (3) and (4) change $(NR - KT)$ to $(KT - NR)$. This change is also made if gear and hob are of same hand but *climb* hobbing is used.

*Example:* A right-hand helical gear with 48 teeth of 10 normal diametral pitch, has a lead of 44.0894 inches. The feed is to be 0.035 inch, with whatever slight adjustment may be necessary to compensate for the error in available index gears. $K = 30$ and $M = 0.075$. A single-thread right-hand hob is to be used.

$$R = \frac{44.0894}{44.0894 - 0.035} \times \frac{30 \times 1}{48} = 0.62549654$$

Using the method of *Conjugate Fractions* beginning on page 5, several suitable ratios close to 0.62549654 were found. One of these, $(34 \times 53)/(43 \times 67) = 0.625477264839$, will be used as the index ratio. Other usable ratios and their decimal values were found to be as follows:

$$\frac{32 \times 38}{27 \times 72} = 0.6255144 \qquad \frac{27 \times 42}{42 \times 37} = 0.62548263$$
$$\frac{44 \times 29}{34 \times 60} = 0.6254902 \qquad \frac{26 \times 97}{96 \times 42} = 0.62549603$$
$$\frac{20 \times 41}{23 \times 57} = 0.62547674$$

Index ratio error = 0.62549654 − 0.62547726 = 0.00001928.

Now use Formula (3) to find slight change required in rate of feed. This change compensates sufficiently for the error in available index gears.

*Change in Feed Rate:* Insert in Formula (3) obtainable index ratio.

$$F = \frac{44.0894 \times (48 \times 0.62547726 - 30)}{48 \times 0.62547726} = 0.0336417$$

$$\text{Modified feed gear ratio} = \frac{F}{M} = \frac{0.0336417}{0.075} = 0.448556$$

$$\log 0.448556 = \bar{1}.651817 \qquad \text{log of reciprocal} = 0.348183$$

To find close approximation to modified feed gear ratio, proceed as in finding suitable gears for index ratio, thus obtaining $\frac{106}{71} \times \frac{112}{75}$. Inverting, modified feed gear ratio = $\frac{71}{106} \times \frac{75}{112} = 0.448534$.

Modified feed $F$ = obtainable modified feed ratio × $M$ = 0.448534 × 0.075 = 0.03364 inch. If the feed rate is not modified, even a small error in the index gear ratio may result in an excessive lead error.

*Checking Accuracy of Lead:* The modified feed and obtainable index ratio are inserted in Formula (4). Desired lead = 44.0894 inches. Lead obtained = 44.087196 inches; hence the computed error = 44.0894 − 44.087196 = 0.002204 inch or about 0.00005 inch per inch of lead.

*Machine with Differential:* If a machine is equipped with a differential, the *lead gears* are computed in order to obtain the required helix angle and lead. The instructions of the hobbing machine manufacturer should be followed in computing the lead gears, because the ratio formula is affected by the location of the differential gears. If these gears are *ahead* of the index gears, the lead gear ratio is not affected by a change in the number of teeth to be cut (see Formula (5)); hence, the same lead gears are used when, for example, a gear and pinion are cut on the same machine. In the formulas which follow, the notation is the same as previously given, with these exceptions: $R_d$ = lead gear ratio for machine with differential; $P_a$ = axial or linear pitch of helical gear = distance from center of one tooth to center of next tooth measured parallel to gear axis = total lead $L$ ÷ number of teeth $N$.

$$R_d = \frac{P_a \times T}{K} = \frac{L \times T}{N \times K} = \frac{\pi \times \operatorname{cosec} A \times T}{P_n \times K} = \frac{\text{Driven gear sizes}}{\text{Driving gear sizes}} \qquad (5)$$

The number of hob threads $T$ is included in the formula because double-thread hobs are used sometimes, especially for roughing in order to reduce the hobbing time. Lead gears having a ratio sufficiently close to the required ratio may be determined by using the table of gear ratio logarithms as previously described in connection with the non-differential type of machine. When using a machine equipped with a differential, the effect of a lead-gear ratio error upon the lead of the gear is small in comparison with the effect of an index gear error when using a non-differential type of machine. The lead obtained with a given or obtainable lead gear ratio may be determined by the following formula: $L = (R_d NK) \div T$.

In this formula, $R_d$ represents the ratio obtained with available gears. If the given lead is 44.0894 inches, as in the preceding example, then the desired ratio as obtained with Formula (5) would be 0.9185292 if $K = 1$. Assume that the lead gears selected by using logs of ratios have a ratio of 0.9184704; then this ratio error of 0.0000588 would result in a computed lead error of only 0.000065 inch per inch.

Formula (5), as mentioned, applies to machines having the differential located *ahead* of the index gears. If the differential is located after the index gears, it is necessary to change lead gears whenever the index gears are changed for hobbing a different number of teeth, as indicated by the following formula which gives the lead gear ratio. In this formula, $D$ = pitch diameter.

$$R_d = \frac{L \times T}{K} = \frac{D \times \pi \times T}{K \times \tan A} = \frac{\text{Driven gear sizes}}{\text{Driving gear sizes}} \qquad (6)$$

**General Remarks on Helical Gear Hobbing.—**In cutting teeth having large angles, it is desirable to have the direction of helix of the hob the same as the direction of helix of the gear, or in other words, the gear and the hob of the same "hand." Then the direction of the cut will come against the movement of the blank. At ordinary angles, however, one hob will cut both right- and left-hand gears. In setting up the hobbing machine for helical gears, care should be taken to see that the vertical feed does not trip until the machine has been stopped or the hob has fed down past the finished gear.

## Herringbone Gears

Double helical or herringbone gears are commonly used in parallel-shaft transmissions, especially when a smooth, continuous action (due to the gradual overlapping engagement of the teeth) is essential, as in high-speed drives where the pitch-line velocity may range from about 1000 to 3000 feet per minute (305–914 m/min) in commercial gearing and up to 12,000 feet per minute (3658 m/min) or higher in more specialized installations. These relatively high speeds are encountered in marine reduction gears, in certain speed-reducing and speed-increasing units, and in various other transmissions, particularly in connection with steam turbine and electric motor drives.

**General Classes of Helical Gear Problems.—**There are two general classes of problems. In one, the problem is to design gears capable of transmitting a given amount of power at a given speed, safely and without excessive wear; hence, the required proportions must be determined. In the second, the proportions and speed are known and the power-transmitting capacity is required. The first is the more difficult and common problem.

**Causes of Herringbone Gear Failures.—**Where failure occurs in a herringbone gear transmission, it is rarely due to tooth breakage but usually to excessive wear or sub-surface failures, such as pitting and spalling; hence, it is common practice to base the design of such gears upon durability, or upon tooth pressures which are within the allowable limits for wear. In this connection, it seems to have been well established by tests of both spur gears and herringbone gears, that there is a critical surface pressure value for teeth having given physical properties and coefficient of friction. According to these tests, pressures above the critical value result in rapid wear and a short gear life, whereas when pressures are below the critical, wear is negligible. The yield point or endurance limit of the material marks the critical loading point, and in practical designing a reasonable factor of safety would, of course, be employed.

## OTHER GEAR TYPES

### Elliptic Gears

Gears of this type provide simple means of obtaining a quick-return motion but they present a rather cumbersome manufacturing problem and, as a general rule, it is preferable to obtain quick-return motions by some other type of mechanism. When elliptic gears are used, the two gears that mesh with each other must be equal in size, and each gear must revolve about one of the foci of the ellipse forming the pitch line, as indicated by the diagram, in Fig. 1. By the use of elliptic gears so mounted, it is possible to obtain a variable motion of the driven shaft, because the gear on the driving shaft, while revolving one half of a revolution, will engage with only a small portion of the circumference of the driven gear, while during the other half of its revolution, the driving gear will engage with a great deal more than one-half of the total number of teeth in the driven gear; hence, the cutting stroke of a machine tool, for example, may be made to have a slow motion, while the return stroke is at a rapid rate. The ellipse has two points, each of which is called a *focus,* located as indicated at *A* and *B*. The sum of the distance between the foci and the elliptic curve is constant at all points and is equal to the longer or major axis of the ellipse. On account of this peculiarity of the ellipse, two equal ellipses can be made to mesh with each other during a complete revolution about their axes, if one is mounted on a shaft at its focus *A* and the other at its focus *B*.

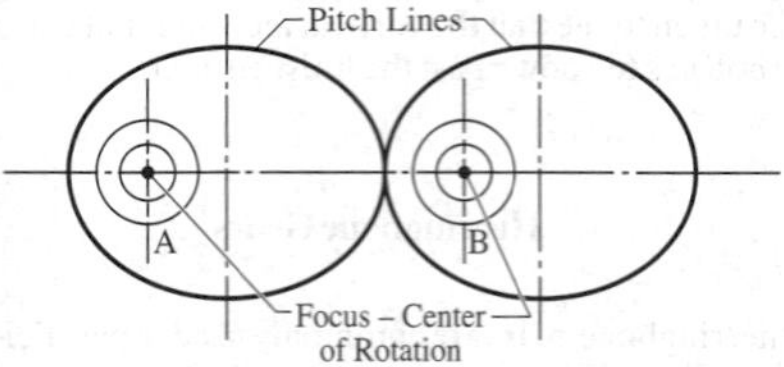


Fig. 1. General Arrangement of Elliptic Gears.

### Planetary Gearing

Planetary or epicyclic gearing provides means of obtaining a compact design of transmission, with driving and driven shafts in line, and a large speed reduction when required. Typical arrangements of planetary gearing are shown by the following diagrams which are accompanied by speed ratio formulas. When planetary gears are arranged as shown by Fig. 5, Fig. 6, Fig. 9 and Fig. 12, the speed of the follower relative to the driver is increased, whereas Fig. 7, Fig. 8, Fig. 10, and Fig. 11 illustrate speed-reducing mechanisms.

**Direction of Rotation.—**In using the following formulas, if the final result is preceded by a minus sign (negative), this indicates that the driver and follower will rotate in opposite directions; otherwise, both will rotate in the same direction.

**Compound Drive.—**The formulas accompanying Fig. 19 through Fig. 22 are for obtaining the speed ratios when there are *two* driving members rotating at different speeds. For example, in Fig. 19, the central shaft with its attached link is one driver. The internal gear $z$, instead of being fixed, is also rotated. In Fig. 22, if $z = 24$, $B = 60$ and $S = 3\frac{1}{2}$, with both drivers rotating in the same direction, then $F = 0$, thus indicating, in this case, the point where a larger value of $S$ will reverse follower rotation.

**Planetary Bevel Gears.—**Two forms of planetary gears of the bevel type are shown in Fig. 23 and Fig. 24. The planet gear in Fig. 23 rotates about a fixed bevel gear at the center of which is the driven shaft. Fig. 24 illustrates the Humpage reduction gear. This is sometimes referred to as cone-pulley back-gearing because of its use within the cone pulleys of certain types of machine tools.

## Ratios of Planetary or Epicyclic Gearing

$D$ = rotation of *driver* per revolution of follower or driven member

$F$ = rotation of *follower* or driven member per revolution of driver. (In Fig. 1 through Fig. 4, $F$ = rotation of planet type follower about its axis.)

$A$ = size of driving gear (use either number of teeth or pitch diameter). Note: When follower derives its motion both from $A$ and from a secondary driving member, $A$ = size of *initial* driving gear, and formula gives speed relationship between $A$ and follower.

$B$ = size of *driven gear or follower* (use either pitch diameter or number of teeth)

$C$ = size of *fixed gear* (use either pitch diameter or number of teeth)

$x$ = size of *planet gear* as shown by diagram (use either pitch diameter or number of teeth)

$y$ = size of *planet gear* as shown by diagram (use either pitch diameter or number of teeth)

$z$ = size of secondary or *auxiliary driving gear*, when follower derives its motion from two driving members

$S$ = rotation of *secondary driver*, per revolution of *initial driver*. $S$ is negative when secondary and initial drivers rotate in opposite directions. (Formulas in which $S$ is used give speed relationship between follower and the initial driver.)

Note: In all cases, if $D$ is known, $F = 1 \div D$, or, if $F$ is known, $D = 1 \div F$.

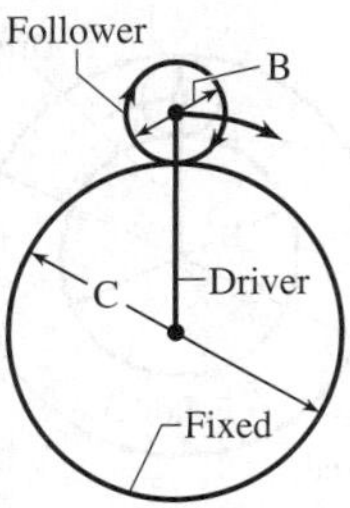


Fig. 1.

$$F = 1 + \frac{C}{B}$$

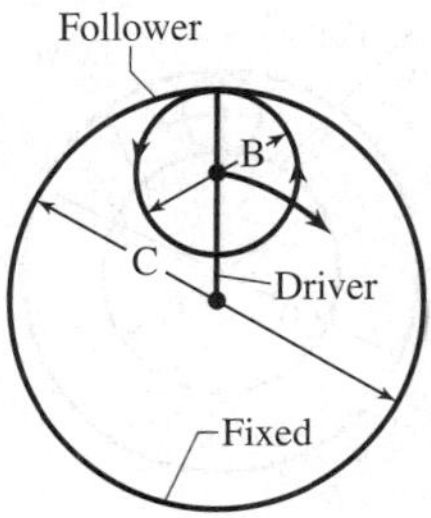


Fig. 2.

$$F = 1 - \frac{C}{B}$$

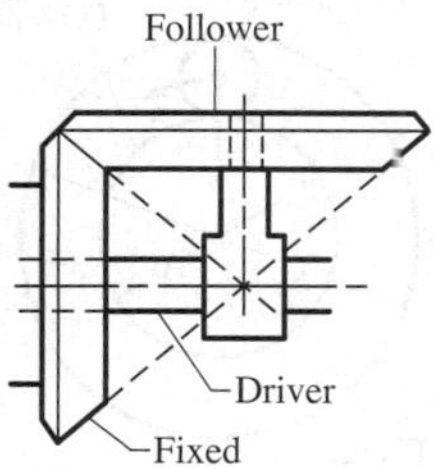


Fig. 3.

$$F = \frac{C}{B}$$

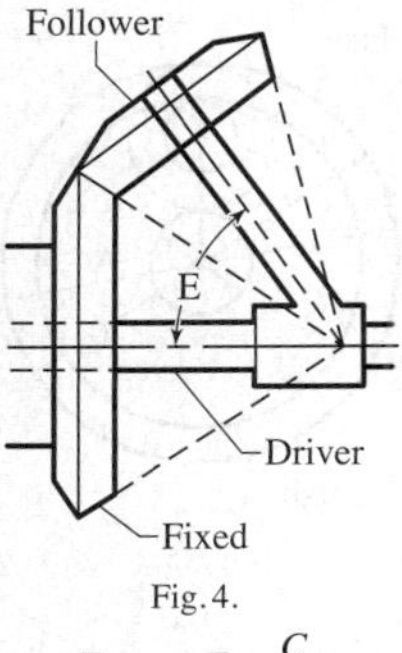


Fig. 4.

$$F = \cos E + \frac{C}{B}$$

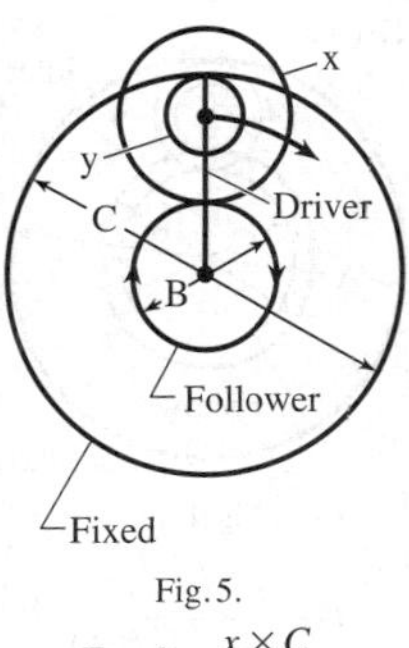


Fig. 5.

$$F = 1 + \frac{x \times C}{y \times B}$$

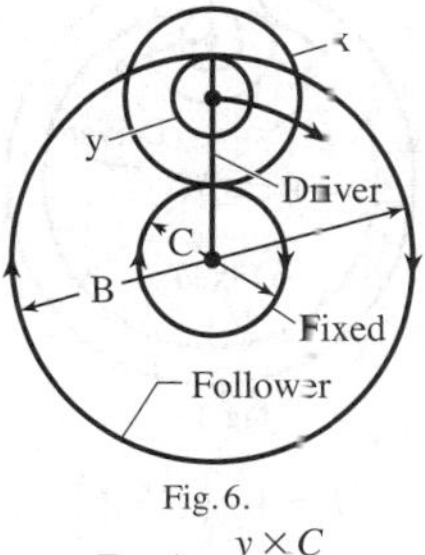


Fig. 6.

$$F = 1 + \frac{y \times C}{x \times B}$$

**Ratios of Planetary or Epicyclic Gearing** *(Continued)*

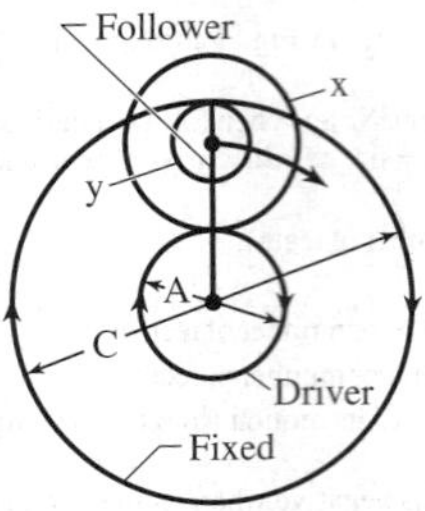


Fig. 7.

$$D = 1 + \frac{x \times C}{y \times A}$$

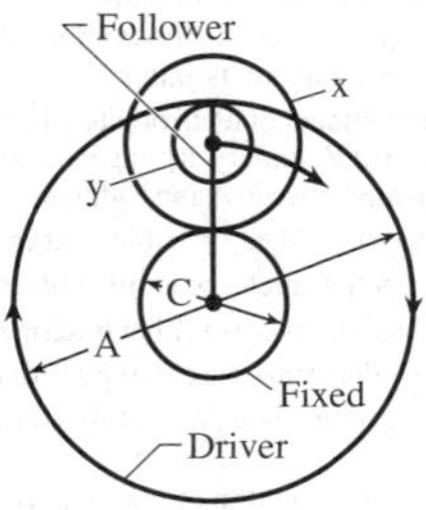


Fig. 8.

$$D = 1 + \frac{y \times C}{x \times A}$$

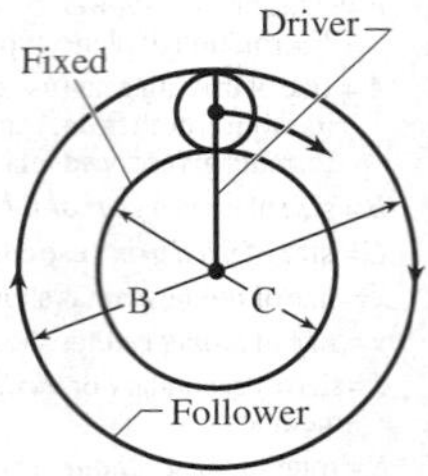


Fig. 9.

$$F = 1 + \frac{C}{B}$$

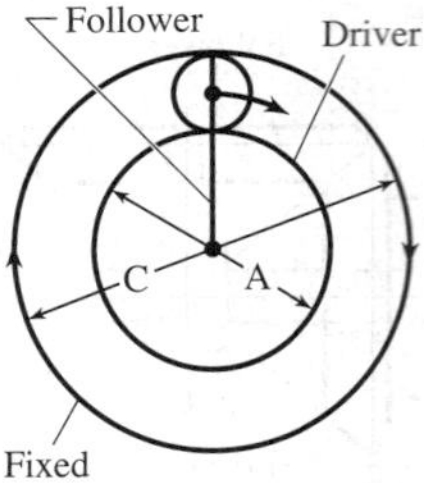


Fig. 10.

$$D = 1 + \frac{C}{A}$$

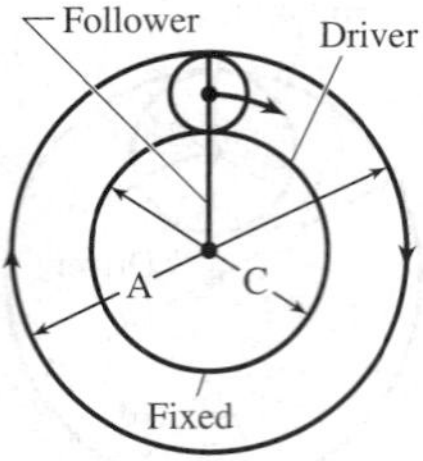


Fig. 11.

$$D = 1 + \frac{C}{A}$$

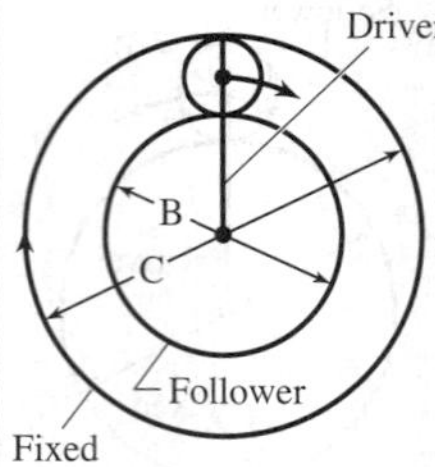


Fig. 12.

$$F = 1 + \frac{C}{B}$$

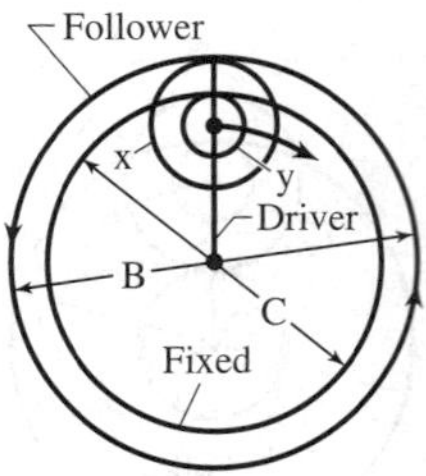


Fig. 13.

$$F = 1 - \left(\frac{C \times x}{y \times B}\right)$$

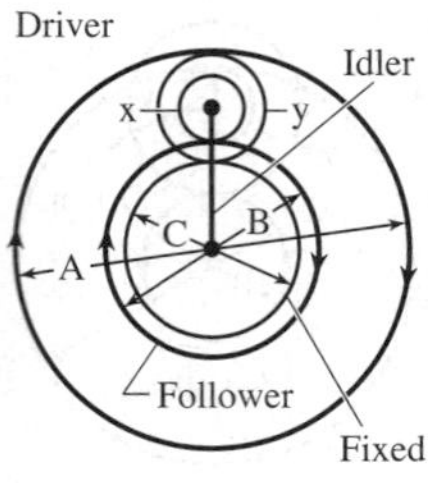


Fig. 14.

$$D = \frac{1 + \frac{C}{A}}{1 - \left(\frac{C \times x}{y \times B}\right)}$$

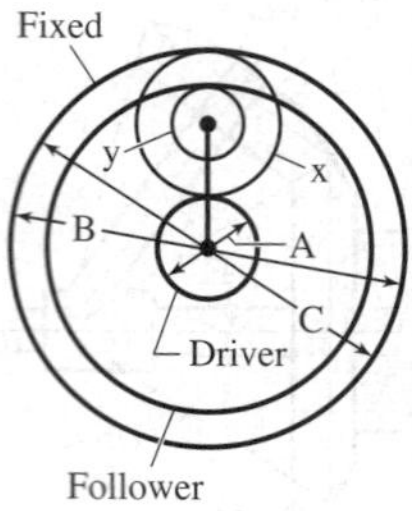


Fig. 15.

$$D = \frac{1 + \frac{C}{A}}{1 - \left(\frac{C \times y}{x \times B}\right)}$$

**Ratios of Planetary or Epicyclic Gearing** *(Continued)*

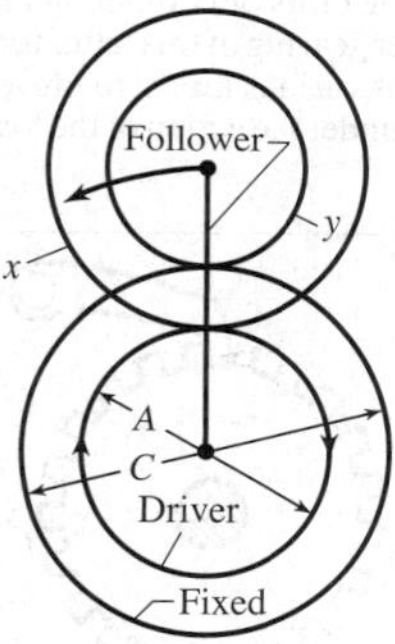


Fig. 16.

$$D = 1 - \left(\frac{C \times x}{y \times A}\right)$$

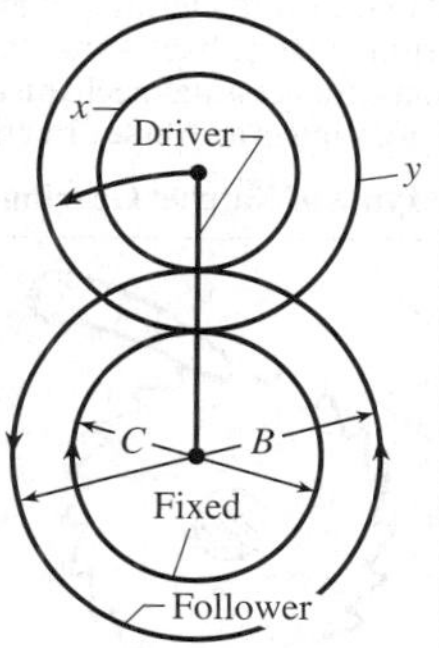


Fig. 17.

$$F = 1 - \left(\frac{C \times x}{y \times B}\right)$$

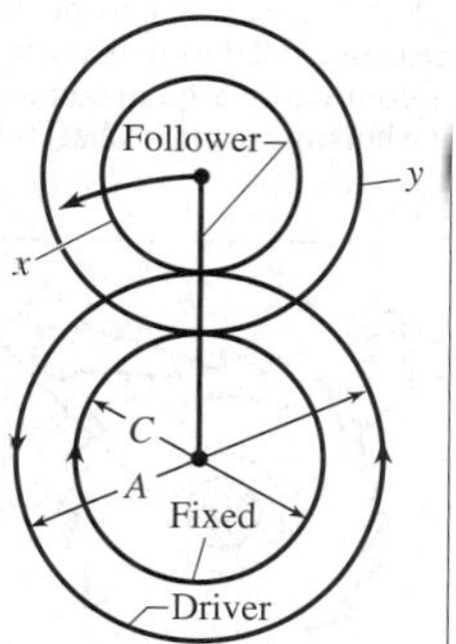


Fig. 18.

$$D = 1 - \left(\frac{C \times x}{y \times A}\right)$$

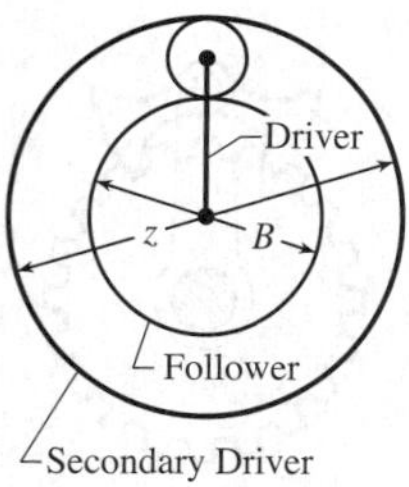


Fig. 19.

$$F = 1 + \frac{z \times (1 - S)}{B}$$

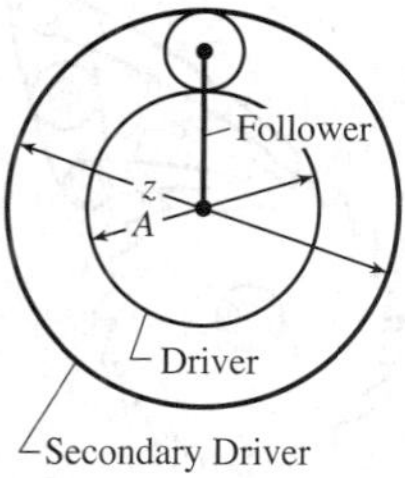


Fig. 20.

$$D = \frac{A + z}{A + (S \times z)}$$

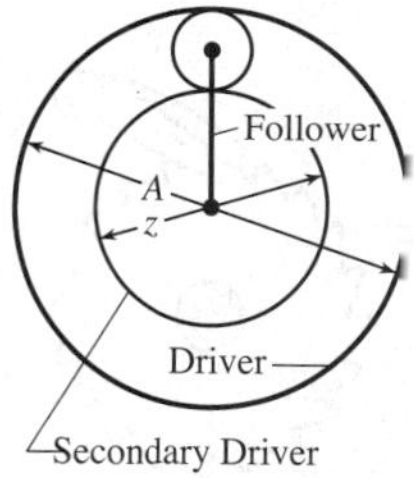


Fig. 21.

$$D = \frac{A + z}{A + (S \times z)}$$

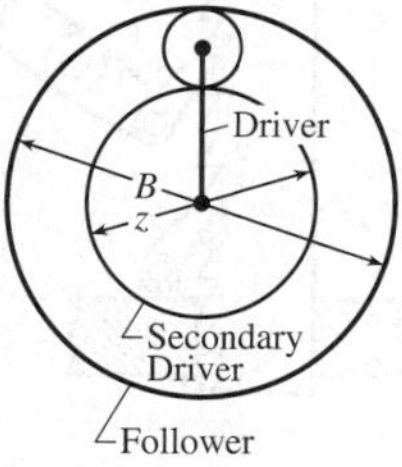


Fig. 22.

$$F = 1 + \frac{z \times (1 - S)}{B}$$

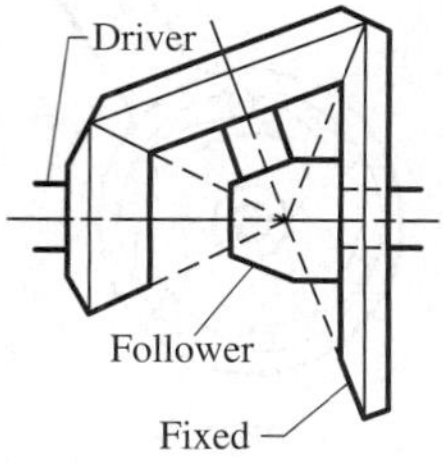


Fig. 23.

$$D = 1 + \frac{C}{A}$$

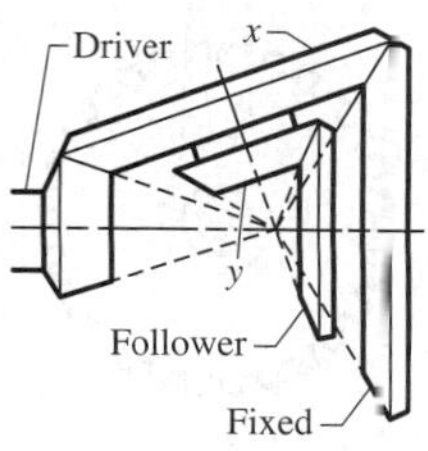


Fig. 24.

$$D = \frac{1 + \frac{C}{A}}{1 - \left(\frac{C \times y}{x \times B}\right)}$$

## Ratchet Gearing

Ratchet gearing may be used to transmit intermittent motion, or its only function may be to prevent the ratchet wheel from rotating backward. Ratchet gearing of this latter form is commonly used in connection with hoisting mechanisms of various kinds, to prevent the hoisting drum or shaft from rotating in a reverse direction under the action of the load.

### Types of Ratchet Gearing

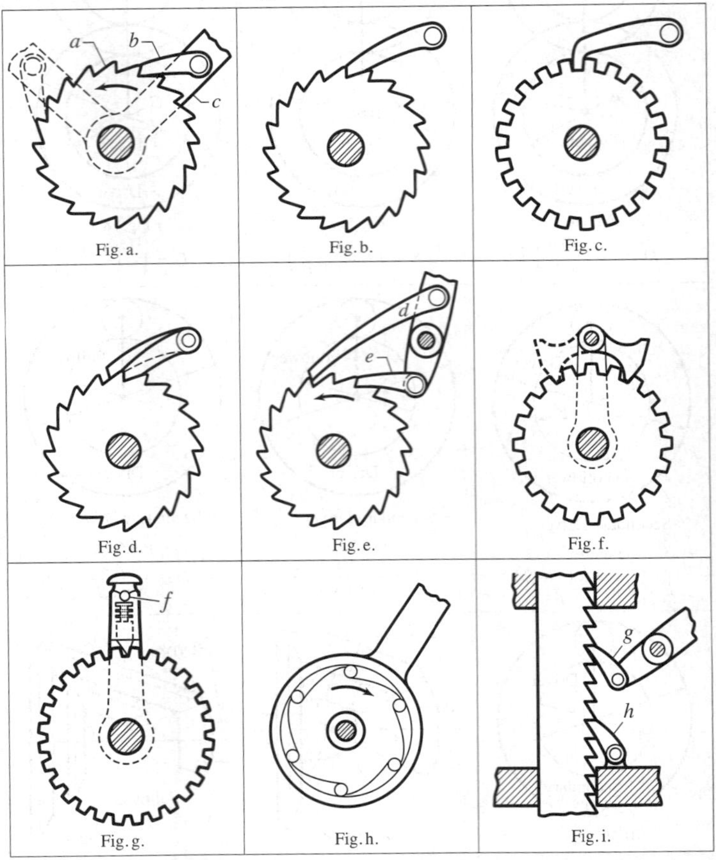


Fig. a. Fig. b. Fig. c.
Fig. d. Fig. e. Fig. f.
Fig. g. Fig. h. Fig. i.

Ratchet gearing in its simplest form consists of a toothed ratchet wheel *a* (see Fig. a), and a pawl or detent *b*, and it may be used to transmit intermittent motion or to prevent relative motion between two parts except in one direction. The pawl *b* is pivoted to lever *c* which, when given an oscillating movement, imparts an intermittent rotary movement to ratchet wheel *a*. Fig. b illustrates another application of the ordinary ratchet and pawl mechanism. In this instance, the pawl is pivoted to a stationary member and its only function is to prevent the ratchet wheel from rotating backward. With the stationary design, illustrated at

Fig. c, the pawl prevents the ratchet wheel from rotating in either direction, so long as it is in engagement with the wheel.

The principle of *multiple-pawl ratchet gearing* is illustrated at Fig. d, which shows the use of two pawls. One of these pawls is longer than the other, by an amount equal to one-half the pitch of the ratchet-wheel teeth, so that the practical effect is that of reducing the pitch one-half. By placing a number of driving pawls side by side and proportioning their lengths according to the pitch of the teeth, a very fine feed can be obtained with a ratchet wheel of comparatively coarse pitch.

This method of obtaining a fine feed from relatively coarse-pitch ratchets may be preferable to the use of single ratchets of fine pitch which, although providing the feed required, may have considerably weaker teeth.

The type of ratchet gearing shown at Fig. e is sometimes employed to impart a rotary movement to the ratchet wheel for both the forward and backward motions of the lever to which the two pawls are attached.

A simple form of *reversing ratchet* is illustrated at Fig. f. The teeth of the wheel are so shaped that either side may be used for driving by simply changing the position of the double-ended pawl, as indicated by the full and dotted lines.

Another form of reversible ratchet gearing for shapers is illustrated at Fig. g. The pawl, in this case, instead of being a pivoted latch, is in the form of a plunger which is free to move in the direction of its axis, but is normally held into engagement with the ratchet wheel by a small spring. When the pawl is lifted and turned one-half revolution, the driving face then engages the opposite sides of the teeth and the ratchet wheel is given an intermittent rotary motion in the opposite direction.

The *frictional type* of ratchet gearing differs from the designs previously referred to, in that there is no positive engagement between the driving and driven members of the ratchet mechanism, the motion being transmitted by frictional resistance. One type of frictional ratchet gearing is illustrated at Fig. h. Rollers or balls are placed between the ratchet wheel and an outer ring which, when turned in one direction, causes the rollers or balls to wedge between the wheel and ring as they move up the inclined edges of the teeth.

Fig. i illustrates one method of utilizing ratchet gearing for moving the driven member in a straight line, as in the case of a lifting jack. The pawl $g$ is pivoted to the operating lever of the jack and does the lifting, whereas the pawl $h$ holds the load while the lifting pawl $g$ is being returned preparatory to another lifting movement.

**Shape of Ratchet Wheel Teeth.**—When designing ratchet gearing, it is important to so shape the teeth that the pawl will remain in engagement when a load is applied. The faces of the teeth which engage the end of the pawl should be in such relation with the center of the pawl pivot that a line perpendicular to the face of the engaging tooth will pass somewhere between the center of the ratchet wheel and the center of the pivot about which the pawl swings. This is true if the pawl *pushes* the ratchet wheel, or if the ratchet wheel *pushes* the pawl. However, if the pawl *pulls* the ratchet wheel or if the ratchet wheel *pulls* the pawl, the perpendicular from the face of the ratchet teeth should fall outside the pawl pivot center. Ratchet teeth may be either cut by a milling cutter having the correct angle, or hobbed in a gear-hobbing machine by the use of a special hob.

**Pitch of Ratchet Wheel Teeth.**—The pitch of ratchet wheels used for holding suspended loads may be calculated by the following formula, in which $P$ = circular pitch, in inches (mm), measured at the outside circumference; $M$ = turning moment acting upon the ratchet wheel shaft, in inch-pounds (N-mm); $L$ = length of tooth face (thickness of ratchet gear), in inches (millimeters); $S$ = safe stress (for steel, 2500 pounds per square inch or 17 MPa when subjected to shock, and 4000 pounds per square inch or 28 MPa when not subjected to shock); $N$ = number of teeth in ratchet wheel; $F$ = a factor the value of which is 50 for ratchet gears with 12 teeth or less, 35 for gears having from 12 to 20 teeth, and 20 for gears having over 20 teeth:

$$P = \sqrt{\frac{FM}{LSN}}$$

This formula has been used in the calculation of ratchet gears for crane design.

**Module System Gear Design.—**The *module* of a gear is equal to the pitch diameter divided by the number of teeth, whereas *diametral pitch* is equal to the number of teeth divided by the pitch diameter. The module system (see accompanying table and diagram) is in general use in countries that have adopted the metric system; hence, the term module is usually understood to mean the pitch diameter *in millimeters* divided by the number of teeth. The module system, however, may also be based on inch measurements and then it is known as the English module to avoid confusion with the metric module. Module is an actual dimension, whereas diametral pitch is only a ratio. Thus, if the pitch diameter of a gear is 50 millimeters and the number of teeth 25, the module is 2, which means that there are 2 millimeters of pitch diameter for each tooth. The table *Tooth Dimensions Based Upon Module System* shows the relation among module, diametral pitch, and circular pitch.

**German Standard Tooth Form for Spur and Bevel Gears** *DIN 867*

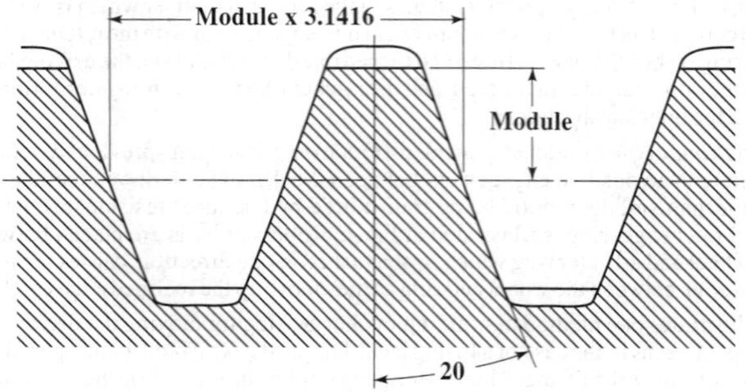


The flanks or sides are straight (involute system) and the pressure angle is 20 degrees. The shape of the root clearance space and the amount of clearance depend upon the method of cutting and special requirements. The amount of clearance may vary from 0.1 × module to 0.3 × module.

| To Find | Module Known | Circular Pitch Known |
|---|---|---|
| Addendum | Equals module | 0.31823 × Circular pitch |
| Dedendum | 1.157 × module*<br>1.167 × module** | 0.3683 × Circular pitch*<br>0.3714 × Circular pitch** |
| Working Depth | 2 × module | 0.6366 × Circular pitch |
| Total Depth | 2.157 × module*<br>2.167 × module** | 0.6866 × Circular pitch*<br>0.6898 × Circulate pitch** |
| Tooth Thickness on Pitch Line | 1.5708 × module | 0.5 × Circular pitch |

Formulas for dedendum and total depth, marked (*) are used when clearance equals 0.157 × module. Formulas marked (**) are used when clearance equals one-sixth module. It is common practice among American cutter manufacturers to make the clearance of metric or module cutters equal to 0.157 × module.

## Tooth Dimensions Based Upon Module System

| Module, *DIN* Standard Series | Equivalent Diametral Pitch | Circular Pitch | | Addendum, Millimeters | Dedendum, Millimeters[a] | Whole Depth,[a] Millimeters | Whole Depth,[b] Millimeters |
|---|---|---|---|---|---|---|---|
| | | Millimeters | Inches | | | | |
| **0.3** | 84.667 | 0.943 | 0.0371 | 0.30 | 0.35 | 0.650 | 0.647 |
| **0.4** | 63.500 | 1.257 | 0.0495 | 0.40 | 0.467 | 0.867 | 0.863 |
| **0.5** | 50.800 | 1.571 | 0.0618 | 0.50 | 0.583 | 1.083 | 1.079 |
| **0.6** | 42.333 | 1.885 | 0.0742 | 0.60 | 0.700 | 1.300 | 1.294 |
| **0.7** | 36.286 | 2.199 | 0.0865 | 0.70 | 0.817 | 1.517 | 1.510 |
| **0.8** | 31.750 | 2.513 | 0.0989 | 0.80 | 0.933 | 1.733 | 1.726 |
| **0.9** | 28.222 | 2.827 | 0.1113 | 0.90 | 1.050 | 1.950 | 1.941 |
| **1** | 25.400 | 3.142 | 0.1237 | 1.00 | 1.167 | 2.167 | 2.157 |
| **1.25** | 20.320 | 3.927 | 0.1546 | 1.25 | 1.458 | 2.708 | 2.697 |
| **1.5** | 16.933 | 4.712 | 0.1855 | 1.50 | 1.750 | 3.250 | 3.236 |
| **1.75** | 14.514 | 5.498 | 0.2164 | 1.75 | 2.042 | 3.792 | 3.774 |
| **2** | 12.700 | 6.283 | 0.2474 | 2.00 | 2.333 | 4.333 | 4.314 |
| **2.25** | 11.289 | 7.069 | 0.2783 | 2.25 | 2.625 | 4.875 | 4.853 |
| **2.5** | 10.160 | 7.854 | 0.3092 | 2.50 | 2.917 | 5.417 | 5.392 |
| **2.75** | 9.236 | 8.639 | 0.3401 | 2.75 | 3.208 | 5.958 | 5.932 |
| **3** | 8.466 | 9.425 | 0.3711 | 3.00 | 3.500 | 6.500 | 6.471 |
| **3.25** | 7.815 | 10.210 | 0.4020 | 3.25 | 3.791 | 7.041 | 7.010 |
| **3.5** | 7.257 | 10.996 | 0.4329 | 3.50 | 4.083 | 7.583 | 7.550 |
| **3.75** | 6.773 | 11.781 | 0.4638 | 3.75 | 4.375 | 8.125 | 8.089 |
| **4** | 6.350 | 12.566 | 0.4947 | 4.00 | 4.666 | 8.666 | 8.628 |
| **4.5** | 5.644 | 14.137 | 0.5566 | 4.50 | 5.25 | 9.750 | 9.707 |
| **5** | 5.080 | 15.708 | 0.6184 | 5.00 | 5.833 | 10.833 | 10.785 |
| **5.5** | 4.618 | 17.279 | 0.6803 | 5.50 | 6.416 | 11.916 | 11.864 |
| **6** | 4.233 | 18.850 | 0.7421 | 6.00 | 7.000 | 13.000 | 12.942 |
| **6.5** | 3.908 | 20.420 | 0.8035 | 6.50 | 7.583 | 14.083 | 14.021 |
| **7** | 3.628 | 21.991 | 0.8658 | 7. | 8.166 | 15.166 | 15.099 |
| **8** | 3.175 | 25.132 | 0.9895 | 8. | 9.333 | 17.333 | 17.256 |
| **9** | 2.822 | 28.274 | 1.1132 | 9. | 10.499 | 19.499 | 19.413 |
| **10** | 2.540 | 31.416 | 1.2368 | 10. | 11.666 | 21.666 | 21.571 |
| **11** | 2.309 | 34.558 | 1.3606 | 11. | 12.833 | 23.833 | 23.728 |
| **12** | 2.117 | 37.699 | 1.4843 | 12. | 14.000 | 26.000 | 25.884 |
| **13** | 1.954 | 40.841 | 1.6079 | 13. | 15.166 | 28.166 | 28.041 |
| **14** | 1.814 | 43.982 | 1.7317 | 14. | 16.332 | 30.332 | 30.198 |
| **15** | 1.693 | 47.124 | 1.8541 | 15. | 17.499 | 32.499 | 32.355 |
| **16** | 1.587 | 50.266 | 1.9790 | 16. | 18.666 | 34.666 | 34.512 |
| **18** | 1.411 | 56.549 | 2.2263 | 18. | 21.000 | 39.000 | 38.825 |
| **20** | 1.270 | 62.832 | 2.4737 | 20. | 23.332 | 43.332 | 43.142 |
| **22** | 1.155 | 69.115 | 2.7210 | 22. | 25.665 | 47.665 | 47.454 |
| **24** | 1.058 | 75.398 | 2.9685 | 24. | 28.000 | 52.000 | 51.768 |
| **27** | 0.941 | 84.823 | 3.339 | 27. | 31.498 | 58.498 | 58.2[illegible]9 |
| **30** | 0.847 | 94.248 | 3.711 | 30. | 35.000 | 65.000 | 64.7[illegible]3 |
| **33** | 0.770 | 103.673 | 4.082 | 33. | 38.498 | 71.498 | 71.131 |
| **36** | 0.706 | 113.097 | 4.453 | 36. | 41.998 | 77.998 | 77.652 |
| **39** | 0.651 | 122.522 | 4.824 | 39. | 45.497 | 84.497 | 84.123 |
| **42** | 0.605 | 131.947 | 5.195 | 42. | 48.997 | 90.997 | 90.594 |
| **45** | 0.564 | 141.372 | 5.566 | 45. | 52.497 | 97.497 | 97.065 |
| **50** | 0.508 | 157.080 | 6.184 | 50. | 58.330 | 108.330 | 107.[illegible]55 |
| **55** | 0.462 | 172.788 | 6.803 | 55. | 64.163 | 119.163 | 118.[illegible]35 |
| **60** | 0.423 | 188.496 | 7.421 | 60. | 69.996 | 129.996 | 129.[illegible]26 |
| **65** | 0.391 | 204.204 | 8.040 | 65. | 75.829 | 140.829 | 140.205 |
| **70** | 0.363 | 219.911 | 8.658 | 70. | 81.662 | 151.662 | 150.997 |
| **75** | 0.339 | 235.619 | 9.276 | 75. | 87.495 | 162.495 | 161.775 |

[a] Dedendum and total depth when clearance = 0.1666 × module, or one-sixth module.
[b] Total depth equivalent to American standard full-depth teeth. (Clearance = 0.157 × module.)

## Rules for Module System of Gearing

| To Find | Rule |
|---|---|
| Metric Module | *Rule 1:* To find the metric module, divide the pitch diameter in millimeters by the number of teeth.<br>*Example 1:* The pitch diameter of a gear is 200 millimeters and the number of teeth, 40; then<br>$\text{Module} = \frac{200}{40} = 5$<br>*Rule 2:* Multiply circular pitch in millimeters by 0.3183.<br>*Example 2:* (Same as Example 1. Circular pitch of this gear equals 15.708 millimeters.)<br>$\text{Module} = 15.708 \times 0.3183 = 5$<br>*Rule 3:* Divide outside diameter in millimeters by the number of teeth plus 2. |
| English Module | *Note:* The module system is usually applied when gear dimensions are expressed in millimeters, but module may also be based on inch measurements.<br>*Rule:* To find the English module, divide pitch diameter in inches by the number of teeth.<br>*Example:* A gear has 48 teeth and a pitch diameter of 12 inches.<br>$\text{Module} = \frac{12}{48} = \frac{1}{4}$ module or 4 diametral pitch |
| Metric Module Equivalent to Diametral Pitch | *Rule:* To find the metric module equivalent to a given diametral pitch, divide 25.4 by the diametral pitch.<br>*Example:* Determine metric module equivalent to 10 diametral pitch.<br>$\text{Equivalent module} = \frac{25.4}{10} = 2.54$<br>*Note:* The nearest standard module is 2.5. |
| Diametral Pitch Equivalent to Metric Module | *Rule:* To find the diametral pitch equivalent to a given module, divide 25.4 by the module. (25.4 = number of millimeters per inch.)<br>*Example:* The module is 12; determine equivalent diametral pitch.<br>$\text{Equivalent diametral pitch} = \frac{25.4}{12} = 2.117$<br>*Note:* A diametral pitch of 2 is the nearest *standard* equivalent. |
| Pitch Diameter | *Rule:* Multiply number of teeth by module.<br>*Example:* The metric module is 8 and the gear has 40 teeth; then<br>$D = 40 \times 8 = 320 \text{ millimeters} = 12.598 \text{ inches}$ |
| Outside Diameter | *Rule:* Add 2 to the number of teeth and multiply sum by the module.<br>*Example:* A gear has 40 teeth and module is 6. Find outside or blank diameter.<br>$\text{Outside diameter} = (40 + 2) \times 6 = 252 \text{ millimeters}$ |

For tooth dimensions, see table *Tooth Dimensions Based Upon Module System*; also formulas in *German Standard Tooth Form for Spur and Bevel Gears DIN 867.*

## Equivalent Diametral Pitches, Circular Pitches, and Metric Modules
## Commonly Used Pitches and Modules in Bold Type

| Diametral Pitch | Circular Pitch, Inches | Module Millimeters | Diametral Pitch | Circular Pitch, Inches | Module Millimeters | Diametral Pitch | Circular Pitch, Inches | Module Millimeters |
|---|---|---|---|---|---|---|---|---|
| **1/2** | 6.2832 | 50.8000 | 2.2848 | **1 3/8** | 11.1170 | 10.0531 | **5/16** | 2.5266 |
| 0.5080 | 6.1842 | **50** | 2.3091 | 1.3605 | **11** | 10.1600 | 0.3092 | **2 1/2** |
| 0.5236 | **6** | 48.5104 | **2 1/2** | 1.2566 | 10.1600 | **11** | 0.2856 | 2.3091 |
| 0.5644 | 5.5658 | **45** | 2.5133 | **1 1/4** | 10.1063 | **12** | 0.2618 | 2.1167 |
| 0.5712 | **5 1/2** | 44.4679 | 2.5400 | 1.2368 | **10** | 12.5664 | **1/4** | 2.0213 |
| 0.6283 | **5** | 40.4253 | **2 3/4** | 1.1424 | 9.2364 | 12.7000 | 0.2474 | **2** |
| 0.6350 | 4.9474 | **40** | 2.7925 | **1 1/8** | 9.0957 | **13** | 0.2417 | 1.9538 |
| 0.6981 | **4 1/2** | 36.3828 | 2.8222 | 1.1132 | **9** | **14** | 0.2244 | 1.8143 |
| 0.7257 | 4.3290 | **35** | **3** | 1.0472 | 8.4667 | **15** | 0.2094 | 1.6933 |
| **3/4** | 4.1888 | 33.8667 | 3.1416 | **1** | 8.0851 | **16** | 0.1963 | 1.5875 |
| 0.7854 | **4** | 32.3403 | 3.1750 | 0.9895 | **8** | 16.7552 | **3/16** | 1.5160 |
| 0.8378 | **3 3/4** | 30.3190 | 3.3510 | **15/16** | 7.5797 | 16.9333 | 0.1855 | **1 1/2** |
| 0.8467 | 3.7105 | **30** | **3 1/2** | 0.8976 | 7.2571 | **17** | 0.1848 | 1.494 |
| 0.8976 | **3 1/2** | 28.2977 | 3.5904 | **7/8** | 7.0744 | **18** | 0.1745 | 1.411 |
| 0.9666 | **3 1/4** | 26.2765 | 3.6286 | 0.8658 | **7** | **19** | 0.1653 | 1.3368 |
| **1** | 3.1416 | 25.4000 | 3.8666 | **13/16** | 6.5691 | **20** | 0.1571 | 1.2700 |
| 1.0160 | 3.0921 | **25** | 3.9078 | 0.8040 | **6 1/2** | **22** | 0.1428 | 1.1545 |
| 1.0472 | **3** | 24.2552 | **4** | 0.7854 | 6.3500 | **24** | 0.1309 | 1.0583 |
| 1.1424 | **2 3/4** | 22.2339 | 4.1888 | **3/4** | 6.0638 | **25** | 0.1257 | 1.0160 |
| **1 1/4** | 2.5133 | 20.3200 | 4.2333 | 0.7421 | **6** | 25.1328 | **1/8** | 1.0106 |
| 1.2566 | **2 1/2** | 20.2127 | 4.5696 | **11/16** | 5.5585 | 25.4000 | 0.1237 | **1** |
| 1.2700 | 2.4737 | **20** | 4.6182 | 0.6803 | **5 1/2** | **26** | 0.1208 | 0.9769 |
| 1.3963 | **2 1/4** | 18.1914 | **5** | 0.6283 | 5.0800 | **28** | 0.1122 | 0.9071 |
| 1.4111 | 2.2263 | **18** | 5.0265 | **5/8** | 5.0532 | **30** | 0.1047 | 0.8467 |
| **1 1/2** | 2.0944 | 16.9333 | 5.0800 | 0.6184 | **5** | **32** | 0.0982 | 0.7937 |
| 1.5708 | **2** | 16.1701 | 5.5851 | **9/16** | 4.5478 | **34** | 0.0924 | 0.7470 |
| 1.5875 | 1.9790 | **16** | 5.6443 | 0.5566 | **4 1/2** | **36** | 0.0873 | 0.7056 |
| 1.6755 | **1 7/8** | 15.1595 | **6** | 0.5236 | 4.2333 | **38** | 0.0827 | 0.6684 |
| 1.6933 | 1.8553 | **15** | 6.2832 | **1/2** | 4.0425 | **40** | 0.0785 | 0.6350 |
| **1 3/4** | 1.7952 | 14.5143 | 6.3500 | 0.4947 | **4** | **42** | 0.0748 | 0.6048 |
| 1.7952 | **1 3/4** | 14.1489 | **7** | 0.4488 | 3.6286 | **44** | 0.0714 | 0.5773 |
| 1.8143 | 1.7316 | **14** | 7.1808 | **7/16** | 3.5372 | **46** | 0.0683 | 0.5522 |
| 1.9333 | **1 5/8** | 13.1382 | 7.2571 | 0.4329 | **3 1/2** | **48** | 0.0654 | 0.5292 |
| 1.9538 | 1.6079 | **13** | **8** | 0.3927 | 3.1750 | **50** | 0.0628 | 0.5080 |
| **2** | 1.5708 | 12.7000 | 8.3776 | **3/8** | 3.0319 | 50.2656 | **1/16** | 0.5053 |
| 2.0944 | **1 1/2** | 12.1276 | 8.4667 | 0.3711 | **3** | 50.8000 | 0.0618 | **1/2** |
| 2.1167 | 1.4842 | **12** | **9** | 0.3491 | 2.8222 | **56** | 0.0561 | 0.4536 |
| **2 1/4** | 1.3963 | 11.2889 | **10** | 0.3142 | 2.5400 | **60** | 0.0524 | 0.4233 |

The module of a gear is the pitch diameter divided by the number of teeth. The module may be expressed in any units; but when no units are stated, it is understood to be in millimeters. The metric module, therefore, equals the pitch diameter in millimeters divided by the number of teeth. To find the metric module equivalent to a given diametral pitch, divide 25.4 by the diametral pitch. To find the diametral pitch equivalent to a given module, divide 25.4 by the module. (25.4 = number of millimeters per inch.)

## CHECKING GEAR SIZES

### Checking Gear Size by Measurement Over Wires or Pins

The wire or pin method of checking gear sizes is accurate, easily applied, and especially useful in shops with limited inspection equipment. Two cylindrical wires or pins of predetermined diameter are placed in diametrically opposite tooth spaces (see diagram). If the gear has an odd number of teeth, the wires are located as nearly opposite as possible, as shown by the diagram at the right. The overall measurement *M* is checked by using any sufficiently accurate method of measurement. The value of measurement *M* when the pitch diameter is correct can be determined easily and quickly by means of the calculated values in the accompanying tables.

**Measurements for Checking External Spur Gears when Wire Diameter Equals 1.728 Divided by Diametral Pitch.—**Table 1 and Table 2 give measurements *M*, in inches, for checking the pitch diameters of external spur gears of 1 diametral pitch. For any other diametral pitch, divide the measurement given in the table by whatever diametral pitch is required. The result shows what measurement *M* should be when the pitch diameter is correct *and there is no allowance for backlash*. The procedure for obtaining a given amount of backlash will be explained later. Table 1 through Table 4 inclusive are based on wire sizes conforming to the Van Keuren standard. For external spur gears, the wire size equals 1.728 divided by the diametral pitch. The wire diameters for various diametral pitches will be found in the left-hand section of Table 5.

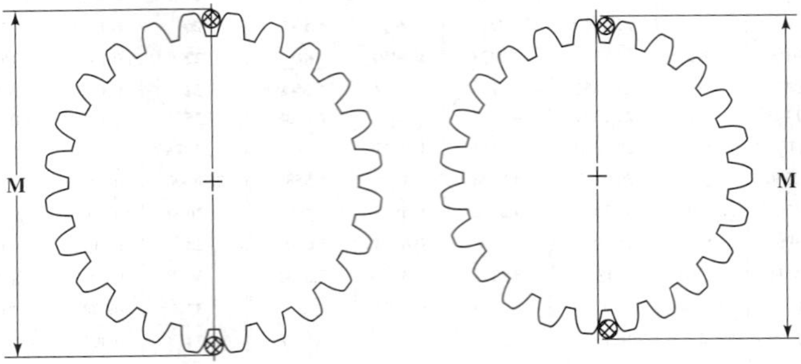


*Even Number of Teeth:* Table 1 is for even numbers of teeth. To illustrate the use of the table, assume that a spur gear has 32 teeth of 4 diametral pitch and a pressure angle of 20 degrees. Table 1 shows that the measurement for 1 diametral pitch is 34.4130; hence, for 4 diametral pitch, the measurement equals 34.4130 ÷ 4 = 8.6032 inches. This dimension is the measurement over the wires when the pitch diameter is correct, provided there is no allowance for backlash. The wire diameter here equals 1.728 ÷ 4 = 0.432 inch (Table 5).

Measurement for even numbers of teeth above 170 and not in Table 1 may be determined as shown by the following example: Assume that number of teeth = 240 and pressure angle = $14\frac{1}{2}$ degrees; then, for 1 diametral pitch, figure at left of decimal point = given No. of teeth + 2 = 240 + 2 = 242. Figure at right of decimal point lies between decimal values given in table for 200 teeth and 300 teeth and is obtained by interpolation. Thus, 240 − 200 = 40 (change to 0.40); 0.5395 − 0.5321 = 0.0074 = difference between decimal values for 300 and 200 teeth; hence, decimal required = 0.5321 + (0.40 × 0.0074) = 0.53506. Total dimension = 242.53506 divided by the diametral pitch required.

*Odd Number of Teeth:* Table 2 is for odd numbers of teeth. Measurement for odd numbers above 171 and not in Table 2 may be determined as shown by the following example: Assume that number of teeth = 335 and pressure angle = 20 degrees; then, for 1 diametral pitch, figure at left of decimal point = given No. of teeth + 2 = 335 + 2 = 337. Figure at right of decimal point lies between decimal values given in table for 301 and 401 teeth. Thus, 335 − 301 = 34 (change to 0.34); 0.4565 − 0.4538 = 0.0027; hence, decimal required = 0.4538 + (0.34 × 0.0027) = 0.4547. Total dimension = 337.4547.

**Table 1. Checking External Spur Gear Sizes by Measurement Over Wires**

**EVEN NUMBERS OF TEETH**

Dimensions in table are for 1 diametral pitch and Van Keuren standard wire sizes. For any other diametral pitch, divide dimension in table by given pitch.

$$\text{Wire or pin diameter} = \frac{1.728}{\text{Diametral Pitch}}$$

| No. of Teeth | Pressure Angle | | | | |
|---|---|---|---|---|---|
| | 14½° | 17½° | 20° | 25° | 30° |
| 6 | 8.2846 | 8.2927 | 8.3032 | 8.3340 | 8.3759 |
| 8 | 10.3160 | 10.3196 | 10.3271 | 10.3533 | 10.3919 |
| 10 | 12.3399 | 12.3396 | 12.3445 | 12.3667 | 12.4028 |
| 12 | 14.3590 | 14.3552 | 14.3578 | 14.3768 | 14.4108 |
| 14 | 16.3746 | 16.3677 | 16.3683 | 16.3846 | 16.4169 |
| 16 | 18.3877 | 18.3780 | 18.3768 | 18.3908 | 18.4217 |
| 18 | 20.3989 | 20.3866 | 20.3840 | 20.3959 | 20.4256 |
| 20 | 22.4087 | 22.3940 | 22.3900 | 22.4002 | 22.4288 |
| 22 | 24.4172 | 24.4004 | 24.3952 | 24.4038 | 24.4315 |
| 24 | 26.4247 | 26.4060 | 26.3997 | 26.4069 | 26.4339 |
| 26 | 28.4314 | 28.4110 | 28.4036 | 28.4096 | 28.4358 |
| 28 | 30.4374 | 30.4154 | 30.4071 | 30.4120 | 30.4376 |
| 30 | 32.4429 | 32.4193 | 32.4102 | 32.4141 | 32.4391 |
| 32 | 34.4478 | 34.4228 | 34.4130 | 34.4159 | 34.4405 |
| 34 | 36.4523 | 36.4260 | 36.4155 | 36.4176 | 36.4417 |
| 36 | 38.4565 | 38.4290 | 38.4178 | 38.4191 | 38.4428 |
| 38 | 40.4603 | 40.4317 | 40.4198 | 40.4205 | 40.4438 |
| 40 | 42.4638 | 42.4341 | 42.4217 | 42.4217 | 42.4447 |
| 42 | 44.4671 | 44.4364 | 44.4234 | 44.4228 | 44.4455 |
| 44 | 46.4701 | 46.4385 | 46.4250 | 46.4239 | 46.4463 |
| 46 | 48.4729 | 48.4404 | 48.4265 | 48.4248 | 48.4470 |
| 48 | 50.4756 | 50.4422 | 50.4279 | 50.4257 | 50.4476 |
| 50 | 52.4781 | 52.4439 | 52.4292 | 52.4265 | 52.4482 |
| 52 | 54.4804 | 54.4454 | 54.4304 | 54.4273 | 54.4487 |
| 54 | 56.4826 | 56.4469 | 56.4315 | 56.4280 | 56.4492 |
| 56 | 58.4847 | 58.4483 | 58.4325 | 58.4287 | 58.4497 |
| 58 | 60.4866 | 60.4496 | 60.4335 | 60.4293 | 60.4501 |
| 60 | 62.4884 | 62.4509 | 62.4344 | 62.4299 | 62.4506 |
| 62 | 64.4902 | 64.4520 | 64.4352 | 64.4304 | 64.4510 |
| 64 | 66.4918 | 66.4531 | 66.4361 | 66.4309 | 66.4513 |
| 66 | 68.4933 | 68.4542 | 68.4369 | 68.4314 | 68.4517 |
| 68 | 70.4948 | 70.4552 | 70.4376 | 70.4319 | 70.4520 |
| 70 | 72.4963 | 72.4561 | 72.4383 | 72.4323 | 72.4523 |
| 72 | 74.4977 | 74.4570 | 74.4390 | 74.4327 | 74.4526 |
| 74 | 76.4990 | 76.4578 | 76.4396 | 76.4331 | 76.4529 |
| 76 | 78.5002 | 78.4586 | 78.4402 | 78.4335 | 78.4532 |
| 78 | 80.5014 | 80.4594 | 80.4408 | 80.4339 | 80.4534 |
| 80 | 82.5026 | 82.4601 | 82.4413 | 82.4342 | 82.4536 |
| 82 | 84.5037 | 84.4608 | 84.4418 | 84.4345 | 84.4538 |
| 84 | 86.5047 | 86.4615 | 86.4423 | 86.4348 | 86.4540 |
| 86 | 88.5057 | 88.4621 | 88.4428 | 88.4351 | 88.4542 |
| 88 | 90.5067 | 90.4627 | 90.4433 | 90.4354 | 90.4544 |

**Table 1.** *(Continued)* **Checking External Spur Gear Sizes by Measurement Over Wires**

**EVEN NUMBERS OF TEETH**

Dimensions in table are for 1 diametral pitch and Van Keuren standard wire sizes. For any other diametral pitch, divide dimension in table by given pitch.

$$\text{Wire or pin diameter} = \frac{1.728}{\text{Diametral Pitch}}$$

| No. of Teeth | Pressure Angle | | | | |
|---|---|---|---|---|---|
| | 14½° | 17½° | 20° | 25° | 30° |
| 90 | 92.5076 | 92.4633 | 92.4437 | 92.4357 | 92.4546 |
| 92 | 94.5085 | 94.4639 | 94.4441 | 94.4359 | 94.4548 |
| 94 | 96.5094 | 96.4644 | 96.4445 | 96.4362 | 96.4550 |
| 96 | 98.5102 | 98.4649 | 98.4449 | 98.4364 | 98.4552 |
| 98 | 100.5110 | 100.4655 | 100.4453 | 100.4367 | 100.4554 |
| 100 | 102.5118 | 102.4660 | 102.4456 | 102.4369 | 102.4555 |
| 102 | 104.5125 | 104.4665 | 104.4460 | 104.4370 | 104.4557 |
| 104 | 106.5132 | 106.4669 | 106.4463 | 106.4372 | 106.4558 |
| 106 | 108.5139 | 108.4673 | 108.4466 | 108.4374 | 108.4560 |
| 108 | 110.5146 | 110.4678 | 110.4469 | 110.4376 | 110.4561 |
| 110 | 112.5152 | 112.4682 | 112.4472 | 112.4378 | 112.4562 |
| 112 | 114.5159 | 114.4686 | 114.4475 | 114.4380 | 114.4563 |
| 114 | 116.5165 | 116.4690 | 116.4478 | 116.4382 | 116.4564 |
| 116 | 118.5171 | 118.4693 | 118.4481 | 118.4384 | 118.4565 |
| 118 | 120.5177 | 120.4697 | 120.4484 | 120.4385 | 120.4566 |
| 120 | 122.5182 | 122.4701 | 122.4486 | 122.4387 | 122.4567 |
| 122 | 124.5188 | 124.4704 | 124.4489 | 124.4388 | 124.4568 |
| 124 | 126.5193 | 126.4708 | 126.4491 | 126.4390 | 126.4569 |
| 126 | 128.5198 | 128.4711 | 128.4493 | 128.4391 | 128.4570 |
| 128 | 130.5203 | 130.4714 | 130.4496 | 130.4393 | 130.4571 |
| 130 | 132.5208 | 132.4717 | 132.4498 | 132.4394 | 132.4572 |
| 132 | 134.5213 | 134.4720 | 134.4500 | 134.4395 | 134.4573 |
| 134 | 136.5217 | 136.4723 | 136.4502 | 136.4397 | 136.4574 |
| 136 | 138.5221 | 138.4725 | 138.4504 | 138.4398 | 138.4575 |
| 138 | 140.5226 | 140.4728 | 140.4506 | 140.4399 | 140.4576 |
| 140 | 142.5230 | 142.4730 | 142.4508 | 142.4400 | 142.4577 |
| 142 | 144.5234 | 144.4733 | 144.4510 | 144.4401 | 144.4578 |
| 144 | 146.5238 | 146.4736 | 146.4512 | 146.4402 | 146.4578 |
| 146 | 148.5242 | 148.4738 | 148.4513 | 148.4403 | 148.4579 |
| 148 | 150.5246 | 150.4740 | 150.4515 | 150.4404 | 150.4580 |
| 150 | 152.5250 | 152.4742 | 152.4516 | 152.4405 | 152.4580 |
| 152 | 154.5254 | 154.4745 | 154.4518 | 154.4406 | 154.4581 |
| 154 | 156.5257 | 156.4747 | 156.4520 | 156.4407 | 156.4581 |
| 156 | 158.5261 | 158.4749 | 158.4521 | 158.4408 | 158.4582 |
| 158 | 160.5264 | 160.4751 | 160.4523 | 160.4409 | 160.4582 |
| 160 | 162.5267 | 162.4753 | 162.4524 | 162.4410 | 162.4583 |
| 162 | 164.5270 | 164.4755 | 164.4526 | 164.4411 | 164.4584 |
| 164 | 166.5273 | 166.4757 | 166.4527 | 166.4411 | 166.4584 |
| 166 | 168.5276 | 168.4759 | 168.4528 | 168.4412 | 168.4585 |
| 168 | 170.5279 | 170.4760 | 170.4529 | 170.4413 | 170.4585 |
| 170 | 172.5282 | 172.4761 | 172.4531 | 172.4414 | 172.4586 |
| 180 | 182.5297 | 182.4771 | 182.4537 | 182.4418 | 182.4589 |
| 190 | 192.5310 | 192.4780 | 192.4542 | 192.4421 | 192.4591 |
| 200 | 202.5321 | 202.4786 | 202.4548 | 202.4424 | 202.4593 |
| 300 | 302.5395 | 302.4831 | 302.4579 | 302.4443 | 302.4606 |
| 400 | 402.5434 | 402.4854 | 402.4596 | 402.4453 | 402.4613 |
| 500 | 502.5458 | 502.4868 | 502.4606 | 502.4458 | 502.4619 |

**Table 2. Checking External Spur Gear Sizes by Measurement Over Wires**

**ODD NUMBERS OF TEETH**

Dimensions in table are for 1 diametral pitch and Van Keuren standard wire sizes. For any other diametral pitch, divide dimension in table by given pitch.

$$\text{Wire or pin diameter} = \frac{1.728}{\text{Diametral Pitch}}$$

| No. of Teeth | Pressure Angle | | | | |
|---|---|---|---|---|---|
| | 14½° | 17½° | 20° | 25° | 30° |
| 7 | 9.1116 | 9.1172 | 9.1260 | 9.1536 | 9.1928 |
| 9 | 11.1829 | 11.1844 | 11.1905 | 11.2142 | 11.2509 |
| 11 | 13.2317 | 13.2296 | 13.2332 | 13.2536 | 13.2882 |
| 13 | 15.2677 | 15.2517 | 15.2639 | 15.2814 | 15.3142 |
| 15 | 17.2957 | 17.2873 | 17.2871 | 17.3021 | 17.3329 |
| 17 | 19.3182 | 19.3072 | 19.3053 | 19.3181 | 19.3482 |
| 19 | 21.3368 | 21.3233 | 21.3200 | 21.3310 | 21.3600 |
| 21 | 23.3524 | 23.3368 | 23.3321 | 23.3415 | 23.3696 |
| 23 | 25.3658 | 25.3481 | 25.3423 | 25.3502 | 25.3775 |
| 25 | 27.3774 | 27.3579 | 27.3511 | 27.3576 | 27.3842 |
| 27 | 29.3876 | 29.3664 | 29.3586 | 29.3640 | 29.3899 |
| 29 | 31.3966 | 31.3738 | 31.3652 | 31.3695 | 31.3948 |
| 31 | 33.4047 | 33.3804 | 33.3710 | 33.3743 | 33.3991 |
| 33 | 35.4119 | 35.3863 | 35.3761 | 35.3786 | 35.4029 |
| 35 | 37.4185 | 37.3916 | 37.3807 | 37.3824 | 37.4063 |
| 37 | 39.4245 | 39.3964 | 39.3849 | 39.3858 | 39.4094 |
| 39 | 41.4299 | 41.4007 | 41.3886 | 41.3889 | 41.4120 |
| 41 | 43.4348 | 43.4047 | 43.3920 | 43.3917 | 43.4145 |
| 43 | 45.4394 | 45.4083 | 45.3951 | 45.3942 | 45.4168 |
| 45 | 47.4437 | 47.4116 | 47.3980 | 47.3965 | 47.4188 |
| 47 | 49.4477 | 49.4147 | 49.4007 | 49.3986 | 49.4206 |
| 49 | 51.4514 | 51.4175 | 51.4031 | 51.4006 | 51.4223 |
| 51 | 53.4547 | 53.4202 | 53.4053 | 53.4024 | 53.4239 |
| 53 | 55.4579 | 55.4227 | 55.4074 | 55.4041 | 55.4254 |
| 55 | 57.4609 | 57.4249 | 57.4093 | 57.4056 | 57.4267 |
| 57 | 59.4637 | 59.4271 | 59.4111 | 59.4071 | 59.428[illegible] |
| 59 | 61.4664 | 61.4291 | 61.4128 | 61.4084 | 61.4292 |
| 61 | 63.4689 | 63.4310 | 63.4144 | 63.4097 | 63.4303 |
| 63 | 65.4712 | 65.4328 | 65.4159 | 65.4109 | 65.4313 |
| 65 | 67.4734 | 67.4344 | 67.4173 | 67.4120 | 67.4323 |
| 67 | 69.4755 | 69.4360 | 69.4186 | 69.4130 | 69.4332 |
| 69 | 71.4775 | 71.4375 | 71.4198 | 71.4140 | 71.4341 |
| 71 | 73.4795 | 73.4389 | 73.4210 | 73.4150 | 73.4349 |
| 73 | 75.4813 | 75.4403 | 75.4221 | 75.4159 | 75.4357 |
| 75 | 77.4830 | 77.4416 | 77.4232 | 77.4167 | 77.4364 |
| 77 | 79.4847 | 79.4428 | 79.4242 | 79.4175 | 79.4371 |
| 79 | 81.4863 | 81.4440 | 81.4252 | 81.4183 | 81.4378 |
| 81 | 83.4877 | 83.4451 | 83.4262 | 83.4190 | 83.4384 |
| 83 | 85.4892 | 85.4462 | 85.4271 | 85.4196 | 85.4390 |
| 85 | 87.4906 | 87.4472 | 87.4279 | 87.4203 | 87.4395 |
| 87 | 89.4919 | 89.4481 | 89.4287 | 89.4209 | 89.4400 |
| 89 | 91.4932 | 91.4490 | 91.4295 | 91.4215 | 91.4405 |
| 91 | 93.4944 | 93.4499 | 93.4303 | 93.4221 | 93.4410 |
| 93 | 95.4956 | 95.4508 | 95.4310 | 95.4227 | 95.4415 |

**Table 2.** *(Continued)* **Checking External Spur Gear Sizes by Measurement Over Wires**

**ODD NUMBERS OF TEETH**

Dimensions in table are for 1 diametral pitch and Van Keuren standard wire sizes. For any other diametral pitch, divide dimension in table by given pitch.

$$\text{Wire or pin diameter} = \frac{1.728}{\text{Diametral Pitch}}$$

| No. of Teeth | Pressure Angle | | | | |
|---|---|---|---|---|---|
| | 14½° | 17½° | 20° | 25° | 30° |
| 95 | 97.4967 | 97.4516 | 97.4317 | 97.4232 | 97.4420 |
| 97 | 99.4978 | 99.4524 | 99.4323 | 99.4237 | 99.4424 |
| 99 | 101.4988 | 101.4532 | 101.4329 | 101.4242 | 101.4428 |
| 101 | 103.4998 | 103.4540 | 103.4335 | 103.4247 | 103.4432 |
| 103 | 105.5008 | 105.4546 | 105.4341 | 105.4252 | 105.4436 |
| 105 | 107.5017 | 107.4553 | 107.4346 | 107.4256 | 107.4440 |
| 107 | 109.5026 | 109.4559 | 109.4352 | 109.4260 | 109.4443 |
| 109 | 111.5035 | 111.4566 | 111.4357 | 111.4264 | 111.4447 |
| 111 | 113.5044 | 113.4572 | 113.4362 | 113.4268 | 113.4450 |
| 113 | 115.5052 | 115.4578 | 115.4367 | 115.4272 | 115.4453 |
| 115 | 117.5060 | 117.4584 | 117.4372 | 117.4275 | 117.4456 |
| 117 | 119.5068 | 119.4589 | 119.4376 | 119.4279 | 119.4459 |
| 119 | 121.5075 | 121.4594 | 121.4380 | 121.4282 | 121.4462 |
| 121 | 123.5082 | 123.4599 | 123.4384 | 123.4285 | 123.4465 |
| 123 | 125.5089 | 125.4604 | 125.4388 | 125.4288 | 125.4468 |
| 125 | 127.5096 | 127.4609 | 127.4392 | 127.4291 | 127.4471 |
| 127 | 129.5103 | 129.4614 | 129.4396 | 129.4294 | 129.4473 |
| 129 | 131.5109 | 131.4619 | 131.4400 | 131.4297 | 131.4476 |
| 131 | 133.5115 | 133.4623 | 133.4404 | 133.4300 | 133.4478 |
| 133 | 135.5121 | 135.4628 | 135.4408 | 135.4302 | 135.4480 |
| 135 | 137.5127 | 137.4632 | 137.4411 | 137.4305 | 137.4483 |
| 137 | 139.5133 | 139.4636 | 139.4414 | 139.4307 | 139.4485 |
| 139 | 141.5139 | 141.4640 | 141.4418 | 141.4310 | 141.4487 |
| 141 | 143.5144 | 143.4644 | 143.4421 | 143.4312 | 143.4489 |
| 143 | 145.5149 | 145.4648 | 145.4424 | 145.4315 | 145.4491 |
| 145 | 147.5154 | 147.4651 | 147.4427 | 147.4317 | 147.4493 |
| 147 | 149.5159 | 149.4655 | 149.4430 | 149.4319 | 149.4495 |
| 149 | 151.5164 | 151.4658 | 151.4433 | 151.4321 | 151.4497 |
| 151 | 153.5169 | 153.4661 | 153.4435 | 153.4323 | 153.4498 |
| 153 | 155.5174 | 155.4665 | 155.4438 | 155.4325 | 155.4500 |
| 155 | 157.5179 | 157.4668 | 157.4440 | 157.4327 | 157.4502 |
| 157 | 159.5183 | 159.4671 | 159.4443 | 159.4329 | 159.4504 |
| 159 | 161.5188 | 161.4674 | 161.4445 | 161.4331 | 161.4505 |
| 161 | 163.5192 | 163.4677 | 163.4448 | 163.4333 | 163.4507 |
| 163 | 165.5196 | 165.4680 | 165.4450 | 165.4335 | 165.4508 |
| 165 | 167.5200 | 167.4683 | 167.4453 | 167.4337 | 167.4510 |
| 167 | 169.5204 | 169.4686 | 169.4455 | 169.4338 | 169.4511 |
| 169 | 171.5208 | 171.4688 | 171.4457 | 171.4340 | 171.4513 |
| 171 | 173.5212 | 173.4691 | 173.4459 | 173.4342 | 173.4514 |
| 181 | 183.5230 | 183.4704 | 183.4469 | 183.4350 | 183.4520 |
| 191 | 193.5246 | 193.4715 | 193.4478 | 193.4357 | 193.4526 |
| 201 | 203.5260 | 203.4725 | 203.4487 | 203.4363 | 203.4532 |
| 301 | 303.5355 | 303.4790 | 303.4538 | 303.4402 | 303.4565 |
| 401 | 403.5404 | 403.4823 | 403.4565 | 403.4422 | 403.4582 |
| 501 | 503.5433 | 503.4843 | 503.4581 | 503.4434 | 503.4592 |

## Table 3. Checking Internal Spur Gear Sizes by Measurement Between Wires

**EVEN NUMBERS OF TEETH**

Dimensions in table are for 1 diametral pitch and Van Keuren standard wire sizes. For any other diametral pitch, divide dimension in table by given pitch.

$$\text{Wire or pin diameter} = \frac{1.44}{\text{Diametral Pitch}}$$

| No. of Teeth | Pressure Angle | | | | |
|---|---|---|---|---|---|
| | 14½° | 17½° | 20° | 25° | 30° |
| 10 | 8.8337 | 8.7383 | 8.6617 | 8.5209 | 8.3966 |
| 12 | 10.8394 | 10.7404 | 10.6623 | 10.5210 | 10.3973 |
| 14 | 12.8438 | 12.7419 | 12.6627 | 12.5210 | 12.3978 |
| 16 | 14.8474 | 14.7431 | 14.6630 | 14.5210 | 14.3982 |
| 18 | 16.8504 | 16.7441 | 16.6633 | 16.5210 | 16.3985 |
| 20 | 18.8529 | 18.7449 | 18.6635 | 18.5211 | 18.3987 |
| 22 | 20.8550 | 20.7456 | 20.6636 | 20.5211 | 20.3989 |
| 24 | 22.8569 | 22.7462 | 22.6638 | 22.5211 | 22.3991 |
| 26 | 24.8585 | 24.7467 | 24.6639 | 24.5211 | 24.3992 |
| 28 | 26.8599 | 26.7471 | 26.6640 | 26.5211 | 26.3993 |
| 30 | 28.8612 | 28.7475 | 28.6641 | 28.5211 | 28.3994 |
| 32 | 30.8623 | 30.7478 | 30.6642 | 30.5211 | 30.3995 |
| 34 | 32.8633 | 32.7481 | 32.6642 | 32.5211 | 32.3995 |
| 36 | 34.8642 | 34.7483 | 34.6643 | 34.5212 | 34.3996 |
| 38 | 36.8650 | 36.7486 | 36.6642 | 36.5212 | 36.3996 |
| 40 | 38.8658 | 38.7488 | 38.6644 | 38.5212 | 38.3997 |
| 42 | 40.8665 | 40.7490 | 40.6644 | 40.5212 | 40.3997 |
| 44 | 42.8672 | 42.7492 | 42.6645 | 42.5212 | 42.3998 |
| 46 | 44.8678 | 44.7493 | 44.6645 | 44.5212 | 44.3998 |
| 48 | 46.8683 | 46.7495 | 46.6646 | 46.5212 | 46.3999 |
| 50 | 48.8688 | 48.7496 | 48.6646 | 48.5212 | 48.3999 |
| 52 | 50.8692 | 50.7497 | 50.6646 | 50.5212 | 50.3999 |
| 54 | 52.8697 | 52.7499 | 52.6647 | 52.5212 | 52.4000 |
| 56 | 54.8701 | 54.7500 | 54.6647 | 54.5212 | 54.4000 |
| 58 | 56.8705 | 56.7501 | 56.6648 | 56.5212 | 56.4001 |
| 60 | 58.8709 | 58.7502 | 58.6648 | 58.5212 | 58.4001 |
| 62 | 60.8712 | 60.7503 | 60.6648 | 60.5212 | 60.4001 |
| 64 | 62.8715 | 62.7504 | 62.6648 | 62.5212 | 62.4001 |
| 66 | 64.8718 | 64.7505 | 64.6649 | 64.5212 | 64.4001 |
| 68 | 66.8721 | 66.7505 | 66.6649 | 66.5212 | 66.4001 |
| 70 | 68.8724 | 68.7506 | 68.6649 | 68.5212 | 68.4001 |
| 72 | 70.8727 | 70.7507 | 70.6649 | 70.5212 | 70.4002 |
| 74 | 72.8729 | 72.7507 | 72.6649 | 72.5212 | 72.4002 |
| 76 | 74.8731 | 74.7508 | 74.6649 | 74.5212 | 74.4002 |
| 78 | 76.8734 | 76.7509 | 76.6649 | 76.5212 | 76.4002 |
| 80 | 78.8736 | 78.7509 | 78.6649 | 78.5212 | 78.4002 |
| 82 | 80.8738 | 80.7510 | 80.6649 | 80.5212 | 80.4002 |
| 84 | 82.8740 | 82.7510 | 82.6649 | 82.5212 | 82.4002 |
| 86 | 84.8742 | 84.7511 | 84.6650 | 84.5212 | 84.4002 |
| 88 | 86.8743 | 86.7511 | 86.6650 | 86.5212 | 86.4003 |
| 90 | 88.8745 | 88.7512 | 88.6650 | 88.5212 | 88.4003 |
| 92 | 90.8747 | 90.7512 | 90.6650 | 90.5212 | 90.4003 |
| 94 | 92.8749 | 92.7513 | 92.6650 | 92.5212 | 92.4003 |

**Table 3.** *(Continued)* **Checking Internal Spur Gear Sizes by Measurement Between Wires**

**EVEN NUMBERS OF TEETH**

Dimensions in table are for 1 diametral pitch and Van Keuren standard wire sizes. For any other diametral pitch, divide dimension in table by given pitch.

$$\text{Wire or pin diameter} = \frac{1.44}{\text{Diametral Pitch}}$$

| No. of Teeth | Pressure Angle | | | | |
|---|---|---|---|---|---|
| | $14\frac{1}{2}°$ | $17\frac{1}{2}°$ | 20° | 25° | 30° |
| 96 | 94.8750 | 94.7513 | 94.6650 | 94.5212 | 94.4003 |
| 98 | 96.8752 | 96.7513 | 96.6650 | 96.5212 | 96.4003 |
| 100 | 98.8753 | 98.7514 | 98.6650 | 98.5212 | 98.4003 |
| 102 | 100.8754 | 100.7514 | 100.6650 | 100.5212 | 100.4003 |
| 104 | 102.8756 | 102.7514 | 102.6650 | 102.5212 | 102.4003 |
| 106 | 104.8757 | 104.7515 | 104.6650 | 104.5212 | 104.4003 |
| 108 | 106.8758 | 106.7515 | 106.6650 | 106.5212 | 106.4003 |
| 110 | 108.8759 | 108.7515 | 108.6651 | 108.5212 | 108.4004 |
| 112 | 110.8760 | 110.7516 | 110.6651 | 110.5212 | 110.4004 |
| 114 | 112.8761 | 112.7516 | 112.6651 | 112.5212 | 112.4004 |
| 116 | 114.8762 | 114.7516 | 114.6651 | 114.5212 | 114.4004 |
| 118 | 116.8763 | 116.7516 | 116.6651 | 116.5212 | 116.4004 |
| 120 | 118.8764 | 118.7517 | 118.6651 | 118.5212 | 118.4004 |
| 122 | 120.8765 | 120.7517 | 120.6651 | 120.5212 | 120.4004 |
| 124 | 122.8766 | 122.7517 | 122.6651 | 122.5212 | 122.4004 |
| 126 | 124.8767 | 124.7517 | 124.6651 | 124.5212 | 124.4004 |
| 128 | 126.8768 | 126.7518 | 126.6651 | 126.5212 | 126.4004 |
| 130 | 128.8769 | 128.7518 | 128.6652 | 128.5212 | 128.4004 |
| 132 | 130.8769 | 130.7518 | 130.6652 | 130.5212 | 130.4004 |
| 134 | 132.8770 | 132.7518 | 132.6652 | 132.5212 | 132.4004 |
| 136 | 134.8771 | 134.7519 | 134.6652 | 134.5212 | 134.4004 |
| 138 | 136.8772 | 136.7519 | 136.6652 | 136.5212 | 136.4004 |
| 140 | 138.8773 | 138.7519 | 138.6652 | 138.5212 | 138.4004 |
| 142 | 140.8773 | 140.7519 | 140.6652 | 140.5212 | 140.4004 |
| 144 | 142.8774 | 142.7519 | 142.6652 | 142.5212 | 142.4004 |
| 146 | 144.8774 | 144.7520 | 144.6652 | 144.5212 | 144.4004 |
| 148 | 146.8775 | 146.7520 | 146.6652 | 146.5212 | 146.4004 |
| 150 | 148.8775 | 148.7520 | 148.6652 | 148.5212 | 148.4005 |
| 152 | 150.8776 | 150.7520 | 150.6652 | 150.5212 | 150.4005 |
| 154 | 152.8776 | 152.7520 | 152.6652 | 152.5212 | 152.4005 |
| 156 | 154.8777 | 154.7520 | 154.6652 | 154.5212 | 154.4005 |
| 158 | 156.8778 | 156.7520 | 156.6652 | 156.5212 | 156.4005 |
| 160 | 158.8778 | 158.7520 | 158.6652 | 158.5212 | 158.4005 |
| 162 | 160.8779 | 160.7520 | 160.6652 | 160.5212 | 160.4005 |
| 164 | 162.8779 | 162.7521 | 162.6652 | 162.5212 | 162.4005 |
| 166 | 164.8780 | 164.7521 | 164.6652 | 164.5212 | 164.4005 |
| 168 | 166.8780 | 166.7521 | 166.6652 | 166.5212 | 166.4005 |
| 170 | 168.8781 | 168.7521 | 168.6652 | 168.5212 | 168.4005 |
| 180 | 178.8783 | 178.7522 | 178.6652 | 178.5212 | 178.4005 |
| 190 | 188.8785 | 188.7522 | 188.6652 | 188.5212 | 188.4005 |
| 200 | 198.8788 | 198.7523 | 198.6652 | 198.5212 | 198.4005 |
| 300 | 298.8795 | 298.7525 | 298.6654 | 298.5212 | 298.4005 |
| 400 | 398.8803 | 398.7527 | 398.6654 | 398.5212 | 398.4006 |
| 500 | 498.8810 | 498.7528 | 498.6654 | 498.5212 | 498.4006 |

## Table 4. Checking Internal Spur Gear Sizes by Measurement Between Wires

| ODD NUMBERS OF TEETH | | | | | |
|---|---|---|---|---|---|
| Dimensions in table are for 1 diametral pitch and Van Keuren standard wire sizes. For any other diametral pitch, divide dimensions in table by given pitch. Wire or pin diameter $= \frac{1.44}{\text{Diametral Pitch}}$ | | | | | |
| No. of Teeth | Pressure Angle | | | | |
| | 14½° | 17½° | 20° | 25° | 30° |
| 7 | 5.6393 | 5.5537 | 5.4823 | 5.3462 | 5.2232 |
| 9 | 7.6894 | 7.5976 | 7.5230 | 7.3847 | 7.2618 |
| 11 | 9.7219 | 9.6256 | 9.5490 | 9.4094 | 9.2867 |
| 13 | 11.7449 | 11.6451 | 11.5669 | 11.4265 | 11.3040 |
| 15 | 13.7620 | 13.6594 | 13.5801 | 13.4391 | 13.3167 |
| 17 | 15.7752 | 15.6703 | 15.5902 | 15.4487 | 15.3265 |
| 19 | 17.7858 | 17.6790 | 17.5981 | 17.4563 | 17.3343 |
| 21 | 19.7945 | 19.6860 | 19.6045 | 19.4625 | 19.3405 |
| 23 | 21.8017 | 21.6918 | 21.6099 | 21.4676 | 21.3457 |
| 25 | 23.8078 | 23.6967 | 23.6143 | 23.4719 | 23.3501 |
| 27 | 25.8130 | 25.7009 | 25.6181 | 25.4755 | 25.3538 |
| 29 | 27.8176 | 27.7045 | 27.6214 | 27.4787 | 27.3571 |
| 31 | 29.8216 | 29.7076 | 29.6242 | 29.4814 | 29.3599 |
| 33 | 31.8251 | 31.7104 | 31.6267 | 31.4838 | 31.3623 |
| 35 | 33.8282 | 33.7128 | 33.6289 | 33.4860 | 33.3645 |
| 37 | 35.8311 | 35.7150 | 35.6310 | 35.4879 | 35.3665 |
| 39 | 37.8336 | 37.7169 | 37.6327 | 37.4896 | 37.3682 |
| 41 | 39.8359 | 39.7187 | 39.6343 | 39.4911 | 39.3698 |
| 43 | 41.8380 | 41.7203 | 41.6357 | 41.4925 | 41.3712 |
| 45 | 43.8399 | 43.7217 | 43.6371 | 43.4938 | 43.3725 |
| 47 | 45.8416 | 45.7231 | 45.6383 | 45.4950 | 45.3737 |
| 49 | 47.8432 | 47.7243 | 47.6394 | 47.4960 | 47.3748 |
| 51 | 49.8447 | 49.7254 | 49.6404 | 49.4970 | 49.3758 |
| 53 | 51.8461 | 51.7265 | 51.6414 | 51.4979 | 51.3768 |
| 55 | 53.8474 | 53.7274 | 53.6422 | 53.4988 | 53.3776 |
| 57 | 55.8486 | 55.7283 | 55.6431 | 55.4996 | 55.3784 |
| 59 | 57.8497 | 57.7292 | 57.6438 | 57.5003 | 57.3792 |
| 61 | 59.8508 | 59.7300 | 59.6445 | 59.5010 | 59.3799 |
| 63 | 61.8517 | 61.7307 | 61.6452 | 61.5016 | 61.3806 |
| 65 | 63.8526 | 63.7314 | 63.6458 | 63.5022 | 63.3812 |
| 67 | 65.8535 | 65.7320 | 65.6464 | 65.5028 | 65.3818 |
| 69 | 67.8543 | 67.7327 | 67.6469 | 67.5033 | 67.3823 |
| 71 | 69.8551 | 69.7332 | 69.6475 | 69.5038 | 69.3828 |
| 73 | 71.8558 | 71.7338 | 71.6480 | 71.5043 | 71.3833 |
| 75 | 73.8565 | 73.7343 | 73.6484 | 73.5048 | 73.3838 |
| 77 | 75.8572 | 75.7348 | 75.6489 | 75.5052 | 75.3842 |
| 79 | 77.8573 | 77.7352 | 77.6493 | 77.5056 | 77.3846 |
| 81 | 79.8584 | 79.7357 | 79.6497 | 79.5060 | 79.3850 |
| 83 | 81.8590 | 81.7361 | 81.6501 | 81.5064 | 81.385[illegible] |
| 85 | 83.8595 | 83.7365 | 83.6505 | 83.5067 | 83.385[illegible] |
| 87 | 85.8600 | 85.7369 | 85.6508 | 85.5071 | 85.386[illegible] |
| 89 | 87.8605 | 87.7373 | 87.6511 | 87.5074 | 87.3864 |
| 91 | 89.8610 | 89.7376 | 89.6514 | 89.5077 | 89.3867 |
| 93 | 91.8614 | 91.7379 | 91.6517 | 91.5080 | 91.3870 |
| 95 | 93.8619 | 93.7383 | 93.6520 | 93.5082 | 93.3873 |
| 97 | 95.8623 | 95.7386 | 95.6523 | 95.5085 | 95.3876 |
| 99 | 97.8627 | 97.7389 | 97.6526 | 97.5088 | 97.3879 |
| 101 | 99.8631 | 99.7391 | 99.6528 | 99.5090 | 99.3881 |
| 103 | 101.8635 | 101.7394 | 101.6531 | 101.5093 | 101.3883 |
| 105 | 103.8638 | 103.7397 | 103.6533 | 103.5095 | 103.3886 |
| 107 | 105.8642 | 105.7399 | 105.6535 | 105.5097 | 105.3888 |
| 109 | 107.8645 | 107.7402 | 107.6537 | 107.5099 | 107.3890 |
| 111 | 109.8648 | 109.7404 | 109.6539 | 109.5101 | 109.3893 |

**Table 4.** *(Continued)* **Checking Internal Spur Gear Sizes by Measurement Between Wires**

**ODD NUMBERS OF TEETH**

Dimensions in table are for 1 diametral pitch and Van Keuren standard wire sizes. For any other diametral pitch, divide dimensions in table by given pitch.

$$\text{Wire or pin diameter} = \frac{1.44}{\text{Diametral Pitch}}$$

| No. of Teeth | Pressure Angle | | | | |
|---|---|---|---|---|---|
| | 14½° | 17½° | 20° | 25° | 30° |
| 113 | 111.8651 | 111.7406 | 111.6541 | 111.5103 | 111.3895 |
| 115 | 113.8654 | 113.7409 | 113.6543 | 113.5105 | 113.3897 |
| 117 | 115.8657 | 115.7411 | 115.6545 | 115.5107 | 115.3899 |
| 119 | 117.8660 | 117.7413 | 117.6547 | 117.5109 | 117.3900 |
| 121 | 119.8662 | 119.7415 | 119.6548 | 119.5110 | 119.3902 |
| 123 | 121.8663 | 121.7417 | 121.6550 | 121.5112 | 121.3904 |
| 125 | 123.8668 | 123.7418 | 123.6552 | 123.5114 | 123.3905 |
| 127 | 125.8670 | 125.7420 | 125.6554 | 125.5115 | 125.3907 |
| 129 | 127.8672 | 127.7422 | 127.6556 | 127.5117 | 127.3908 |
| 131 | 129.8675 | 129.7424 | 129.6557 | 129.5118 | 129.3910 |
| 133 | 131.8677 | 131.7425 | 131.6559 | 131.5120 | 131.3911 |
| 135 | 133.8679 | 133.7427 | 133.6560 | 133.5121 | 133.3913 |
| 137 | 135.8681 | 135.7428 | 135.6561 | 135.5123 | 135.3914 |
| 139 | 137.8683 | 137.7430 | 137.6563 | 137.5124 | 137.3916 |
| 141 | 139.8685 | 139.7431 | 139.6564 | 139.5125 | 139.3917 |
| 143 | 141.8687 | 141.7433 | 141.6565 | 141.5126 | 141.3918 |
| 145 | 143.8689 | 143.7434 | 143.6566 | 143.5127 | 143.3919 |
| 147 | 145.8691 | 145.7436 | 145.6568 | 145.5128 | 145.3920 |
| 149 | 147.8693 | 147.7437 | 147.6569 | 147.5130 | 147.3922 |
| 151 | 149.8694 | 149.7438 | 149.6570 | 149.5131 | 149.3923 |
| 153 | 151.8696 | 151.7439 | 151.6571 | 151.5132 | 151.3924 |
| 155 | 153.8698 | 153.7441 | 153.6572 | 153.5133 | 153.3925 |
| 157 | 155.8699 | 155.7442 | 155.6573 | 155.5134 | 155.3926 |
| 159 | 157.8701 | 157.7443 | 157.6574 | 157.5135 | 157.3927 |
| 161 | 159.8702 | 159.7444 | 159.6575 | 159.5136 | 159.3928 |
| 163 | 161.8704 | 161.7445 | 161.6576 | 161.5137 | 161.3929 |
| 165 | 163.8705 | 163.7446 | 163.6577 | 163.5138 | 163.3930 |
| 167 | 165.8707 | 165.7447 | 165.6578 | 165.5139 | 165.3931 |
| 169 | 167.8708 | 167.7448 | 167.6579 | 167.5139 | 167.3932 |
| 171 | 169.8710 | 169.7449 | 169.6580 | 169.5140 | 169.3933 |
| 181 | 179.8717 | 179.7453 | 179.6584 | 179.5144 | 179.3937 |
| 191 | 189.8721 | 189.7458 | 189.6588 | 189.5148 | 189.3940 |
| 201 | 199.8727 | 199.7461 | 199.6591 | 199.5151 | 199.3944 |
| 301 | 299.8759 | 299.7485 | 299.6612 | 299.5171 | 299.3965 |
| 401 | 399.8776 | 399.7496 | 399.6623 | 399.5182 | 399.3975 |
| 501 | 499.8786 | 499.7504 | 499.6629 | 499.5188 | 499.3981 |

**Table 5. Van Keuren Wire Diameters for Gears**

| External Gears Wire Dia. = 1.728 ÷ D.P. | | | | Internal Gears Wire Dia. = 1.44 ÷ D.P. | | | |
|---|---|---|---|---|---|---|---|
| D.P. | Dia. | D.P. | Dia. | D.P. | Dia. | D.P. | Dia. |
| 2 | 0.86400 | 16 | 0.10800 | 2 | 0.72000 | 16 | 0.09000 |
| 2½ | 0.69120 | 18 | 0.09600 | 2½ | 0.57600 | 18 | 0.08000 |
| 3 | 0.57600 | 20 | 0.08640 | 3 | 0.48000 | 20 | 0.07200 |
| 4 | 0.43200 | 22 | 0.07855 | 4 | 0.36000 | 22 | 0.06545 |
| 5 | 0.34560 | 24 | 0.07200 | 5 | 0.28800 | 24 | 0.06000 |
| 6 | 0.28800 | 28 | 0.06171 | 6 | 0.24000 | 28 | 0.05143 |
| 7 | 0.24686 | 32 | 0.05400 | 7 | 0.20571 | 32 | 0.04500 |
| 8 | 0.21600 | 36 | 0.04800 | 8 | 0.18000 | 36 | 0.04000 |
| 9 | 0.19200 | 40 | 0.04320 | 9 | 0.16000 | 40 | 0.03600 |
| 10 | 0.17280 | 48 | 0.03600 | 10 | 0.14400 | 48 | 0.03000 |
| 11 | 0.15709 | 64 | 0.02700 | 11 | 0.13091 | 64 | 0.02250 |
| 12 | 0.14400 | 72 | 0.02400 | 12 | 0.12000 | 72 | 0.02000 |
| 14 | 0.12343 | 80 | 0.02160 | 14 | 0.10286 | 80 | 0.01800 |

**Measurements for Checking Internal Gears when Wire Diameter Equals 1.44 Divided by Diametral Pitch.**—Table 3 and Table 4 give measurements between wires for checking internal gears of 1 diametral pitch. For any other diametral pitch, divide the measurement given in the table by the diametral pitch required. These measurements are based upon the Van Keuren standard wire size, which, for internal spur gears, equals 1.44 divided by the diametral pitch (see Table 5).

*Even Number of Teeth:* For an even number of teeth above 170 and not in Table 3, proceed as shown by the following example: Assume that the number of teeth = 380 and pressure angle is 14½ degrees; then, for 1 diametral pitch, figure at left of decimal point = given number of teeth − 2 = 380 − 2 = 378. Figure at right of decimal point lies between decimal values given in table for 300 and 400 teeth and is obtained by interpolation. Thus, 380 − 300 = 80 (change to 0.80); 0.8803 − 0.8795 = 0.0008; hence, decimal required = 0.8795 + (0.80 × 0.0008) 0.88014. Total dimension = 378.88014.

*Odd Number of Teeth:* Table 4 is for internal gears having odd numbers of teeth. For tooth numbers above 171 and not in the table, proceed as shown by the following example: Assume that number of teeth = 337 and pressure angle is 14½ degrees; then, for 1 diametral pitch, figure at left of decimal point = given No. of teeth − 2 = 337 − 2 = 335. Figure at right of decimal point lies between decimal values given in table for 301 and 401 teeth and is obtained by interpolation. Thus, 337 − 301 = 36 (change to 0.36); 0.8776 − 0.8759 = 0.0017; hence, decimal required = 0.8759 + (0.36 × 0.0017) = 0.8765. Total dimension = 335.8755.

**Measurements for Checking External Spur Gears when Wire Diameter Equals 1.68 Divided by Diametral Pitch.**—Table 7 and Table 8 give measurements *M*, in inches, for checking the pitch diameters of external spur gears of 1 diametral pitch. For any other diametral pitch, divide the measurement given in the table by whatever diametral pitch is required. The result shows what measurement *M* should be when the pitch diameter is correct and there is no allowance for backlash. The procedure for checking for a given amount of backlash when the diameter of the measuring wires equals 1.68 divided by the diametral pitch is explained under a subsequent heading. Table 7 and Table 8 are based upon wire sizes equal to 1.68 divided by the diametral pitch. The corresponding wire diameters for various diametral pitches are given in Table 6.

**Table 6. Wire Diameters for Spur and Helical Gears Based upon 1.68 Constant**

| Diametral or Normal Diametral Pitch | Wire Diameter | Diametral or Normal Diametral Pitch | Wire Diameter | Diametral or Normal Diametral Pitch | Wire Diameter | Diametral or Normal Diametral Pitch | Wire Diameter |
|---|---|---|---|---|---|---|---|
| 2 | 0.840 | 8 | 0.210 | 18 | 0.09333 | 40 | 0.042 |
| 2½ | 0.672 | 9 | 0.18666 | 20 | 0.084 | 48 | 0.035 |
| 3 | 0.560 | 10 | 0.168 | 22 | 0.07636 | 64 | 0.02625 |
| 4 | 0.420 | 11 | 0.15273 | 24 | 0.070 | 72 | 0.02333 |
| 5 | 0.336 | 12 | 0.140 | 28 | 0.060 | 80 | 0.021 |
| 6 | 0.280 | 14 | 0.120 | 32 | 0.0525 | … | … |
| 7 | 0.240 | 16 | 0.105 | 36 | 0.04667 | … | … |

Pin diameter = 1.68 ÷ diametral pitch for spur gears and 1.68 ÷ normal diametral pitch for helical gears.

To find measurement *M* of an external spur gear using wire sizes equal to 1.68 inches divided by the diametral pitch, the same method is followed in using Table 7 and Table 8 as that outlined for Table 1 and Table 2.

## Table 7. Checking External Spur Gear Sizes by Measurement Over Wires

**EVEN NUMBERS OF TEETH**

Dimensions in table are for 1 diametral pitch and 1.68-inch series wire sizes (a Van Keuren standard). For any other diametral pitch, divide dimension in table by given pitch.

$$\text{Wire or pin diameter} = \frac{1.68}{\text{Diametral Pitch}}$$

| No. of Teeth | Pressure Angle | | | | |
|---|---|---|---|---|---|
| | 14½° | 17½° | 20° | 25° | 30° |
| 6 | 8.1298 | 8.1442 | 8.1600 | 8.2003 | 8.2504 |
| 8 | 10.1535 | 10.1647 | 10.1783 | 10.2155 | 10.2633 |
| 10 | 12.1712 | 12.1796 | 12.1914 | 12.2260 | 12.2722 |
| 12 | 14.1851 | 14.1910 | 14.2013 | 14.2338 | 14.2785 |
| 14 | 16.1964 | 16.2001 | 16.2091 | 16.2397 | 16.2833 |
| 16 | 18.2058 | 18.2076 | 18.2154 | 18.2445 | 18.2871 |
| 18 | 20.2137 | 20.2138 | 20.2205 | 20.2483 | 20.2902 |
| 20 | 22.2205 | 22.2190 | 22.2249 | 22.2515 | 22.2927 |
| 22 | 24.2265 | 24.2235 | 24.2286 | 24.2542 | 24.2949 |
| 24 | 26.2317 | 26.2275 | 26.2318 | 26.2566 | 26.2967 |
| 26 | 28.2363 | 28.2309 | 28.2346 | 28.2586 | 28.2982 |
| 28 | 30.2404 | 30.2339 | 30.2371 | 30.2603 | 30.2996 |
| 30 | 32.2441 | 32.2367 | 32.2392 | 32.2619 | 32.3008 |
| 32 | 34.2475 | 34.2391 | 34.2412 | 34.2632 | 34.3017 |
| 34 | 36.2505 | 36.2413 | 36.2430 | 36.2644 | 36.3026 |
| 36 | 38.2533 | 38.2433 | 38.2445 | 38.2655 | 38.3035 |
| 38 | 40.2558 | 40.2451 | 40.2460 | 40.2666 | 40.3044 |
| 40 | 42.2582 | 42.2468 | 42.2473 | 42.2675 | 42.3051 |
| 42 | 44.2604 | 44.2483 | 44.2485 | 44.2683 | 44.3057 |
| 44 | 46.2624 | 46.2497 | 46.2496 | 46.2690 | 46.3063 |
| 46 | 48.2642 | 48.2510 | 48.2506 | 48.2697 | 48.3068 |
| 48 | 50.2660 | 50.2522 | 50.2516 | 50.2704 | 50.3073 |
| 50 | 52.2676 | 52.2534 | 52.2525 | 52.2710 | 52.3078 |
| 52 | 54.2691 | 54.2545 | 54.2533 | 54.2716 | 54.3082 |
| 54 | 56.2705 | 56.2555 | 56.2541 | 56.2721 | 56.3086 |
| 56 | 58.2719 | 58.2564 | 58.2548 | 58.2726 | 58.3089 |
| 58 | 60.2731 | 60.2572 | 60.2555 | 60.2730 | 60.3093 |
| 60 | 62.2743 | 62.2580 | 62.2561 | 62.2735 | 62.3096 |
| 62 | 64.2755 | 64.2587 | 64.2567 | 64.2739 | 64.3099 |
| 64 | 66.2765 | 66.2594 | 66.2572 | 66.2742 | 66.3102 |
| 66 | 68.2775 | 68.2601 | 68.2577 | 68.2746 | 68.3104 |
| 68 | 70.2785 | 70.2608 | 70.2582 | 70.2749 | 70.3107 |
| 70 | 72.2794 | 72.2615 | 72.2587 | 72.2752 | 72.3109 |
| 72 | 74.2803 | 74.2620 | 74.2591 | 74.2755 | 74.3111 |
| 74 | 76.2811 | 76.2625 | 76.2596 | 76.2758 | 76.3113 |
| 76 | 78.2819 | 78.2631 | 78.2600 | 78.2761 | 78.3115 |
| 78 | 80.2827 | 80.2636 | 80.2604 | 80.2763 | 80.3117 |
| 80 | 82.2834 | 82.2641 | 82.2607 | 82.2766 | 82.3119 |
| 82 | 84.2841 | 84.2646 | 84.2611 | 84.2768 | 84.3121 |
| 84 | 86.2847 | 86.2650 | 86.2614 | 86.2771 | 86.3123 |
| 86 | 88.2854 | 88.2655 | 88.2617 | 88.2773 | 88.3124 |
| 88 | 90.2860 | 90.2659 | 90.2620 | 90.2775 | 90.3126 |
| 90 | 92.2866 | 92.2662 | 92.2624 | 92.2777 | 92.3127 |
| 92 | 94.2872 | 94.2666 | 94.2626 | 94.2779 | 94.3129 |
| 94 | 96.2877 | 96.2670 | 96.2629 | 96.2780 | 96.3130 |
| 96 | 98.2882 | 98.2673 | 98.2632 | 98.2782 | 98.3131 |
| 98 | 100.2887 | 100.2677 | 100.2635 | 100.2784 | 100.3132 |
| 100 | 102.2892 | 102.2680 | 102.2638 | 102.2785 | 102.3134 |
| 102 | 104.2897 | 104.2683 | 104.2640 | 104.2787 | 104.3135 |

**Table 7.** *(Continued)* **Checking External Spur Gear Sizes by Measurement Over Wires**

**EVEN NUMBERS OF TEETH**

Dimensions in table are for 1 diametral pitch and 1.68-inch series wire sizes (a Van Keuren standard). For any other diametral pitch, divide dimension in table by given pitch.

$$\text{Wire or pin diameter} = \frac{1.68}{\text{Diametral Pitch}}$$

| No. of Teeth | Pressure Angle | | | | |
|---|---|---|---|---|---|
| | 14½° | 17½° | 20° | 25° | 30° |
| 104 | 106.2901 | 106.2685 | 106.2642 | 106.2788 | 106.3136 |
| 106 | 108.2905 | 108.2688 | 108.2644 | 108.2789 | 108.3137 |
| 108 | 110.2910 | 110.2691 | 110.2645 | 110.2791 | 110.3138 |
| 110 | 112.2914 | 112.2694 | 112.2647 | 112.2792 | 112.3139 |
| 112 | 114.2918 | 114.2696 | 114.2649 | 114.2793 | 114.3140 |
| 114 | 116.2921 | 116.2699 | 116.2651 | 116.2794 | 116.3141 |
| 116 | 118.2925 | 118.2701 | 118.2653 | 118.2795 | 118.3142 |
| 118 | 120.2929 | 120.2703 | 120.2655 | 120.2797 | 120.3142 |
| 120 | 122.2932 | 122.2706 | 122.2656 | 122.2798 | 122.3143 |
| 122 | 124.2936 | 124.2708 | 124.2658 | 124.2799 | 124.3144 |
| 124 | 126.2939 | 126.2710 | 126.2660 | 126.2800 | 126.3145 |
| 126 | 128.2941 | 128.2712 | 128.2661 | 128.2801 | 128.3146 |
| 128 | 130.2945 | 130.2714 | 130.2663 | 130.2802 | 130.3146 |
| 130 | 132.2948 | 132.2716 | 132.2664 | 132.2803 | 132.3147 |
| 132 | 134.2951 | 134.2718 | 134.2666 | 134.2804 | 134.3147 |
| 134 | 136.2954 | 136.2720 | 136.2667 | 136.2805 | 136.3148 |
| 136 | 138.2957 | 138.2722 | 138.2669 | 138.2806 | 138.3149 |
| 138 | 140.2960 | 140.2724 | 140.2670 | 140.2807 | 140.3149 |
| 140 | 142.2962 | 142.2725 | 142.2671 | 142.2808 | 142.3150 |
| 142 | 144.2965 | 144.2727 | 144.2672 | 144.2808 | 144.315[illegible] |
| 144 | 146.2967 | 146.2729 | 146.2674 | 146.2809 | 146.315[illegible] |
| 146 | 148.2970 | 148.2730 | 148.2675 | 148.2810 | 148.3152 |
| 148 | 150.2972 | 150.2732 | 150.2676 | 150.2811 | 150.3152 |
| 150 | 152.2974 | 152.2733 | 152.2677 | 152.2812 | 152.3153 |
| 152 | 154.2977 | 154.2735 | 154.2678 | 154.2812 | 154.3153 |
| 154 | 156.2979 | 156.2736 | 156.2679 | 156.2813 | 156.3154 |
| 156 | 158.2981 | 158.2737 | 158.2680 | 158.2813 | 158.3155 |
| 158 | 160.2983 | 160.2739 | 160.2681 | 160.2814 | 160.3155 |
| 160 | 162.2985 | 162.2740 | 162.2682 | 162.2815 | 162.3155 |
| 162 | 164.2987 | 164.2741 | 164.2683 | 164.2815 | 164.3156 |
| 164 | 166.2989 | 166.2742 | 166.2684 | 166.2816 | 166.3156 |
| 166 | 168.2990 | 168.2744 | 168.2685 | 168.2816 | 168.3157 |
| 168 | 170.2992 | 170.2745 | 170.2686 | 170.2817 | 170.3157 |
| 170 | 172.2994 | 172.2746 | 172.2687 | 172.2818 | 172.3158 |
| 180 | 182.3003 | 182.2752 | 182.2691 | 182.2820 | 182.3160 |
| 190 | 192.3011 | 192.2757 | 192.2694 | 192.2823 | 192.3161 |
| 200 | 202.3018 | 202.2761 | 202.2698 | 202.2825 | 202.3163 |
| 300 | 302.3063 | 302.2790 | 302.2719 | 302.2839 | 302.3173 |
| 400 | 402.3087 | 402.2804 | 402.2730 | 402.2845 | 402.3178 |
| 500 | 502.3101 | 502.2813 | 502.2736 | 502.2850 | 502.3181 |

*Allowance for Backlash:* Table 1, Table 2, Table 7, and Table 8 give measurements over wires when the pitch diameters are correct and there is no allowance for backlash or play between meshing teeth. Backlash is obtained by cutting the teeth somewhat deeper than standard, thus reducing the thickness. Usually, the teeth of both mating gears are reduced in thickness an amount equal to one-half of the total backlash desired. However, if the pinion is small, it is common practice to reduce the gear teeth the full amount of backlash and the pinion is made to standard size. The changes in measurements *M* over wires, for obtaining backlash in external spur gears, are listed in Table 9.

## Table 8. Checking External Spur Gear Sizes by Measurement Over Wires

**ODD NUMBERS OF TEETH**

Dimensions in table are for 1 diametral pitch and 1.68-inch series wire sizes (a Van Keuren standard). For any other diametral pitch, divide dimension in table by given pitch.

$$\text{Wire or pin diameter} = \frac{1.68}{\text{Diametral Pitch}}$$

| No. of Teeth | Pressure Angle | | | | |
|---|---|---|---|---|---|
| | 14½° | 17½° | 20° | 25° | 30° |
| 5 | 6.8485 | 6.8639 | 6.8800 | 6.9202 | 6.9691 |
| 7 | 8.9555 | 8.9679 | 8.9822 | 9.0199 | 9.0675 |
| 9 | 11.0189 | 11.0285 | 11.0410 | 11.0762 | 11.1224 |
| 11 | 13.0615 | 13.0686 | 13.0795 | 13.1126 | 13.1575 |
| 13 | 15.0925 | 15.0973 | 15.1068 | 15.1381 | 15.1819 |
| 15 | 17.1163 | 17.1190 | 17.1273 | 17.1570 | 17.1998 |
| 17 | 19.1351 | 19.1360 | 19.1432 | 19.1716 | 19.2136 |
| 19 | 21.1505 | 21.1498 | 21.1561 | 21.1832 | 21.2245 |
| 21 | 23.1634 | 23.1611 | 23.1665 | 23.1926 | 23.2334 |
| 23 | 25.1743 | 25.1707 | 25.1754 | 25.2005 | 25.2408 |
| 25 | 27.1836 | 27.1788 | 27.1828 | 27.2071 | 27.2469 |
| 27 | 29.1918 | 29.1859 | 29.1892 | 29.2128 | 29.2522 |
| 29 | 31.1990 | 31.1920 | 31.1948 | 31.2177 | 31.2568 |
| 31 | 33.2053 | 33.1974 | 33.1997 | 33.2220 | 33.2607 |
| 33 | 35.2110 | 35.2021 | 35.2041 | 35.2258 | 35.2642 |
| 35 | 37.2161 | 37.2065 | 37.2079 | 37.2292 | 37.2674 |
| 37 | 39.2208 | 39.2104 | 39.2115 | 39.2323 | 39.2702 |
| 39 | 41.2249 | 41.2138 | 41.2147 | 41.2349 | 41.2726 |
| 41 | 43.2287 | 43.2170 | 43.2174 | 43.2374 | 43.2749 |
| 43 | 45.2323 | 45.2199 | 45.2200 | 45.2396 | 45.2769 |
| 45 | 47.2355 | 47.2226 | 47.2224 | 47.2417 | 47.2788 |
| 47 | 49.2385 | 49.2251 | 49.2246 | 49.2435 | 49.2805 |
| 49 | 51.2413 | 51.2273 | 51.2266 | 51.2452 | 51.2820 |
| 51 | 53.2439 | 53.2294 | 53.2284 | 53.2468 | 53.2835 |
| 53 | 55.2463 | 55.2313 | 55.2302 | 55.2483 | 55.2848 |
| 55 | 57.2485 | 57.2331 | 57.2318 | 57.2497 | 57.2861 |
| 57 | 59.2506 | 59.2348 | 59.2333 | 59.2509 | 59.2872 |
| 59 | 61.2526 | 61.2363 | 61.2347 | 61.2521 | 61.2883 |
| 61 | 63.2545 | 63.2378 | 63.2360 | 63.2532 | 63.2893 |
| 63 | 65.2562 | 65.2392 | 65.2372 | 65.2543 | 65.2902 |
| 65 | 67.2579 | 67.2406 | 67.2383 | 67.2553 | 67.2911 |
| 67 | 69.2594 | 69.2419 | 69.2394 | 69.2562 | 69.2920 |
| 69 | 71.2609 | 71.2431 | 71.2405 | 71.2571 | 71.2928 |
| 71 | 73.2623 | 73.2442 | 73.2414 | 73.2579 | 73.2935 |
| 73 | 75.2636 | 75.2452 | 75.2423 | 75.2586 | 75.2942 |
| 75 | 77.2649 | 77.2462 | 77.2432 | 77.2594 | 77.2949 |
| 77 | 79.2661 | 79.2472 | 79.2440 | 79.2601 | 79.2955 |
| 79 | 81.2673 | 81.2481 | 81.2448 | 81.2607 | 81.2961 |
| 81 | 83.2684 | 83.2490 | 83.2456 | 83.2614 | 83.2967 |
| 83 | 85.2694 | 85.2498 | 85.2463 | 85.2620 | 85.2972 |
| 85 | 87.2704 | 87.2506 | 87.2470 | 87.2625 | 87.2977 |
| 87 | 89.2714 | 89.2514 | 89.2476 | 89.2631 | 89.2982 |
| 89 | 91.2723 | 91.2521 | 91.2482 | 91.2636 | 91.2987 |
| 91 | 93.2732 | 93.2528 | 93.2489 | 93.2641 | 93.2991 |
| 93 | 95.2741 | 95.2534 | 95.2494 | 95.2646 | 95.2996 |

## Table 8. *(Continued)* Checking External Spur Gear Sizes by Measurement Over Wires

**ODD NUMBERS OF TEETH**

Dimensions in table are for 1 diametral pitch and 1.68-inch series wire sizes (a Van Keuren standard). For any other diametral pitch, divide dimension in table by given pitch.

$$\text{Wire or pin diameter} = \frac{1.68}{\text{Diametral Pitch}}$$

| No. of Teeth | Pressure Angle | | | | |
|---|---|---|---|---|---|
| | 14½° | 17½° | 20° | 25° | 30° |
| 95 | 97.2749 | 97.2541 | 97.2500 | 97.2650 | 97.3000 |
| 97 | 99.2757 | 99.2547 | 99.2506 | 99.2655 | 99.3004 |
| 99 | 101.2764 | 101.2553 | 101.2511 | 101.2659 | 101.3008 |
| 101 | 103.2771 | 103.2558 | 103.2516 | 103.2663 | 103.3011 |
| 103 | 105.2778 | 105.2563 | 105.2520 | 105.2667 | 105.3015 |
| 105 | 107.2785 | 107.2568 | 107.2525 | 107.2671 | 107.3018 |
| 107 | 109.2791 | 109.2573 | 109.2529 | 109.2674 | 109.3021 |
| 109 | 111.2798 | 111.2578 | 111.2533 | 111.2678 | 111.3024 |
| 111 | 113.2804 | 113.2583 | 113.2537 | 113.2681 | 113.3027 |
| 113 | 115.2809 | 115.2588 | 115.2541 | 115.2684 | 115.3030 |
| 115 | 117.2815 | 117.2592 | 117.2544 | 117.2687 | 117.3033 |
| 117 | 119.2821 | 119.2596 | 119.2548 | 119.2690 | 119.3036 |
| 119 | 121.2826 | 121.2601 | 121.2552 | 121.2693 | 121.3038 |
| 121 | 123.2831 | 123.2605 | 123.2555 | 123.2696 | 123.304 |
| 123 | 125.2836 | 125.2608 | 125.2558 | 125.2699 | 125.3043 |
| 125 | 127.2841 | 127.2612 | 127.2562 | 127.2702 | 127.3045 |
| 127 | 129.2846 | 129.2615 | 129.2565 | 129.2704 | 129.3048 |
| 129 | 131.2851 | 131.2619 | 131.2568 | 131.2707 | 131.3050 |
| 131 | 133.2855 | 133.2622 | 133.2571 | 133.2709 | 133.3053 |
| 133 | 135.2859 | 135.2626 | 135.2574 | 135.2712 | 135.3055 |
| 135 | 137.2863 | 137.2629 | 137.2577 | 137.2714 | 137.3057 |
| 137 | 139.2867 | 139.3632 | 139.2579 | 139.2716 | 139.3059 |
| 139 | 141.2871 | 141.2635 | 141.2582 | 141.2718 | 141.3060 |
| 141 | 143.2875 | 143.2638 | 143.2584 | 143.2720 | 143.3062 |
| 143 | 145.2879 | 145.2641 | 145.2587 | 145.2722 | 145.3064 |
| 145 | 147.2883 | 147.2644 | 147.2589 | 147.2724 | 147.3066 |
| 147 | 149.2887 | 149.2647 | 149.2591 | 149.2726 | 149.3068 |
| 149 | 151.2890 | 151.2649 | 151.2594 | 151.2728 | 151.3069 |
| 151 | 153.2893 | 153.2652 | 153.2596 | 153.2730 | 153.3071 |
| 153 | 155.2897 | 155.2654 | 155.2598 | 155.2732 | 155.3073 |
| 155 | 157.2900 | 157.2657 | 157.2600 | 157.2733 | 157.3074 |
| 157 | 159.2903 | 159.2659 | 159.2602 | 159.2735 | 159.3076 |
| 159 | 161.2906 | 161.2661 | 161.2604 | 161.2736 | 161.3077 |
| 161 | 163.2909 | 163.2663 | 163.2606 | 163.2738 | 163.3078 |
| 163 | 165.2912 | 165.2665 | 165.2608 | 165.2740 | 165.3080 |
| 165 | 167.2915 | 167.2668 | 167.2610 | 167.2741 | 167.3081 |
| 167 | 169.2917 | 169.2670 | 169.2611 | 169.2743 | 169.3083 |
| 169 | 171.2920 | 171.2672 | 171.2613 | 171.2744 | 171.3084 |
| 171 | 173.2922 | 173.2674 | 173.2615 | 173.2746 | 173.3085 |
| 181 | 183.2936 | 183.2684 | 183.2623 | 183.2752 | 183.3091 |
| 191 | 193.2947 | 193.2692 | 193.2630 | 193.2758 | 193.3097 |
| 201 | 203.2957 | 203.2700 | 203.2636 | 203.2764 | 203.3101 |
| 301 | 303.3022 | 303.2749 | 303.2678 | 303.2798 | 303.3132 |
| 401 | 403.3056 | 403.2774 | 403.2699 | 403.2815 | 403.3147 |
| 501 | 503.3076 | 503.2789 | 503.2711 | 503.2825 | 503.3156 |

**Table 9. Backlash Allowances for External and Internal Spur Gears**

| No. of Teeth | 14½° | | 17½° | | 20° | | 25° | | 30° | |
|---|---|---|---|---|---|---|---|---|---|---|
| | Ext. | Int. | Ext. | Int. | Ext. | Int. | Ext. | Int. | Ext. | Int. |
| 5 | .0019 | .0024 | .0018 | .0024 | .0017 | .0023 | .0015 | .0021 | .0013 | .0019 |
| 10 | .0024 | .0029 | .0022 | .0027 | .0020 | .0026 | .0017 | .0022 | .0015 | .0018 |
| 20 | .0028 | .0032 | .0025 | .0029 | .0023 | .0027 | .0019 | .0022 | .0016 | .0018 |
| 30 | .0030 | .0034 | .0026 | .0030 | .0024 | .0027 | .0020 | .0022 | .0016 | .0018 |
| 40 | .0031 | .0035 | .0027 | .0030 | .0025 | .0027 | .0020 | .0022 | .0017 | .0018 |
| 50 | .0032 | .0036 | .0028 | .0031 | .0025 | .0027 | .0020 | .0022 | .0017 | .0018 |
| 100 | .0035 | .0037 | .0030 | .0031 | .0026 | .0027 | .0021 | .0022 | .0017 | .0017 |
| 200 | .0036 | .0038 | .0031 | .0031 | .0027 | .0027 | .0021 | .0022 | .0017 | .0017 |

*External Gears:* For each 0.001 inch reduction in pitch-line tooth thickness, *reduce* measurement over wires obtained from Table 1, Table 2, Table 7, or Table 8 by the amount shown below.

*Internal Gears:* For each 0.001 inch reduction in pitch-line tooth thickness, *increase* measurement between wires obtained from Table 3 or Table 4 by the amounts shown below.

Backlash on pitch line equals double tooth thickness reduction when teeth of *both* mating gears are reduced. If teeth of *one* gear only are reduced, backlash on pitch line equals amount of reduction.

*Example:* For a 30-tooth, 10-diametral pitch, 20-degree pressure angle external gear, the measurement over wires from Table 1 is 32.4102 ÷ 10. For a backlash of 0.002, this measurement must be reduced by 2 × 0.0024 to 3.2362 or (3.2410 − 0.0048).

**Measurements for Checking Helical Gears Using Wires or Balls.—**Helical gears may be checked for size by using one wire, or ball; two wires, or balls; and three wires, depending on the case at hand. Three wires may be used for measurement of either even or odd tooth numbers provided that the face width and helix angle of the gear permit the arrangement of two wires in adjacent tooth spaces on one side of the gear and a third wire on the opposite side. The wires should be held between flat, parallel plates. The measurement between these plates, and perpendicular to the gear axis, will be the same for both even and odd numbers of teeth because the axial displacement of the wires with the odd numbers of teeth does not affect the perpendicular measurement between the plates. The calculation of measurements over three wires is the same as described for measurements over two wires for even numbers of teeth.

*Measurements over One Wire or One Ball for Even or Odd Numbers of Teeth:* This measurement is calculated by the method for measurement over two wires for even numbers of teeth and the result divided by two to obtain the measurement from over the wire or ball to the center of the gear mounted on an arbor.

*Measurement over Two Wires or Two Balls for Even Numbers of Teeth:* The measurement over two wires (or two balls kept in the same plane by holding them against a surface parallel to the face of the gear) is calculated as follows: First, calculate the pitch diameter of the helical gear from the formula $D$ = Number of teeth divided by the product of the normal diametral pitch and the cosine of the helix angle, $D = N \div (P_n \times \cos\psi)$. Next, calculate the number of teeth, $N_e$, there would be in a spur gear for it to have the same tooth curvature as the helical gear has in the normal plane: $N_e = N/\cos^3\psi$. Next, refer to Table 7 for spur gears with even tooth numbers and find, by interpolation, the *decimal* value of the constant for this number of teeth under the given *normal* pressure angle. Finally, add 2 to this decimal value and divide the sum by the normal diametral pitch $P_n$. The result of this calculation, added to the pitch diameter $D$, is the measurement over two wires or balls.

*Example:* A helical gear has 32 teeth of 6 normal diametral pitch, 20 degree pressure angle, and 23 degree helix angle. Determine the measurement over two wires, $M$, without allowance for backlash.

$D = 32 \div 6 \times \cos 23° = 5.7939$; $N_e = 32 \div \cos^3 23° = 41.027$; and in Table 7, fourth column, the decimal part of the measurement for 40 teeth is 0.2473 and that for 42 teeth is 0.2485.

The decimal part for 41.027 teeth is, by interpolation, $\frac{(41.027-40)}{(42-40)} \times (0.2485 - 0.2473) +$ $0.2473 = 0.2479$; $(0.2479 + 2) \div 6 = 0.3747$; and $M = 0.3747 + 5.7939 = 6.1686$.

This measurement over wires or balls is based upon the use of $1.68/P_n$ wires or balls. If measurements over $1.728/P_n$ diameter wires or balls are preferred, use Table 1 to find the decimal part described above instead of Table 7.

*Measurement Over Two Wires or Two Balls for Odd Numbers of Teeth:* The procedure is similar to that for two wire or two ball measurement for even tooth numbers except that a correction is made in the final $M$ value to account for the wires or balls not being diametrically opposite by one-half tooth interval. In addition, care must be taken to ensure that the balls or wires are kept in a plane of the gear's rotation as described previously.

A helical gear has 13 teeth of 8 normal diametral pitch, $14\frac{1}{2}$ degree pressure angle, and 45 degree helix angle. Determine measurement $M$ without allowance for backlash based upon the use of $1.728/P_n$ balls or wires.

As before, $D = 13/8 \times \cos 45° = 2.2981$; $N_e = 13/\cos^3 45° = 36.770$; and in the second column of Table 1 the *decimal* part of the measurement for 36 teeth is 0.4565 and that for 38 teeth is 0.4603. The decimal part for 36.770 teeth is, by interpolation, $\frac{(36.770-36)}{(38-36)} \times$ $(0.4603 - 0.4565) + 0.4565 = 0.4580$; $(0.4580 + 2)/8 = 0.3073$; and $M = 0.3073 + 2.2981 = 2.6054$. This measurement is correct for three-wire measurements but, for two balls or wires held in the plane of rotation of the gear, $M$ must be corrected as follows:

$$M \text{ corrected} = (M - \text{Ball Dia.}) \times \cos\left(\frac{90°}{N}\right) + \text{Ball Dia.}$$

$$= \left(2.6054 - \frac{1.728}{8}\right) \times \cos\left(\frac{90°}{13}\right) + \frac{1.728}{8} = 2.5880$$

**Checking Spur Gear Size by Chordal Measurement Over Two or More Teeth.—**Another method of checking gear sizes, that is generally available, is illustrated by the diagram accompanying Table 10. A vernier caliper is used to measure the distance $M$ over two or more teeth. The diagram illustrates the measurement over two teeth (or with one intervening tooth space), but three or more teeth might be included, depending upon the pitch. The jaws of the caliper are merely held in contact with the sides or profiles of the teeth and perpendicular to the axis of the gear. Measurement $M$ for involute teeth of the correct size is determined as follows.

*General Formula for Checking External and Internal Spur Gears by Measurement Over Wires:* The following formulas may be used for pressure angles or wire sizes not covered by the tables. In these formulas, $M$ = measurement *over* wires for external gears or measurement *between* wires for internal gears; $D$ = pitch diameter; $T$ = arc tooth thickness on pitch circle; $W$ = wire diameter; $N$ = number of gear teeth; $A$ = pressure angle of gear; $a$ = angle, the cosine of which is required in Formulas (2) and (3).

First determine the involute function of angle $a$ (inv $a$); then the corresponding angle $a$ is found by referring to the tables of involute functions beginning on page 107,

$$\text{inv}\, a = \text{inv}\, A \pm \frac{T}{D} \pm \frac{W}{D\cos A} \mp \frac{\pi}{N} \qquad (1)$$

$$\text{For even numbers of teeth,} \quad M = \frac{D\cos A}{\cos a} \pm W \qquad (2)$$

$$\text{For odd numbers of teeth,} \quad M = \left(\frac{D\cos A}{\cos a}\right)\left(\cos\frac{90°}{N}\right) \pm W \qquad (3)$$

*Note:* In Formulas (1), (2), and (3), use the upper sign for *external* and the lower sign for *internal* gears wherever a ± or ∓ appears in the formulas.

## Table 10. Chordal Measurements over Spur Gear Teeth of 1 Diametral Pitch

Find value of *M* under pressure angle and opposite number of teeth; divide *M* by diametral pitch of gear to be measured and then subtract one-half total backlash to obtain a measurement *M* equivalent to given pitch and backlash. The number of teeth to gage or measure over is shown by Table 11.

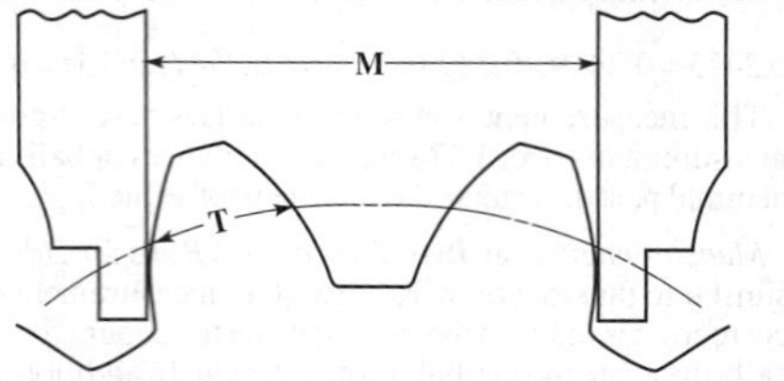


| Number of Gear Teeth | *M* in Inches for 1 D.P. | Number of Gear Teeth | *M* in Inches for 1 D.P. | Number of Gear Teeth | *M* in Inches for 1 D.P. | Number of Gear Teeth | *M* in Inches for 1 D.P. |
|---|---|---|---|---|---|---|---|
| Pressure Angle, 14½ Degrees | | | | | | | |
| 12 | 4.6267 | 37 | 7.8024 | 62 | 14.0197 | 87 | 20.2370 |
| 13 | 4.6321 | 38 | 10.8493 | 63 | 17.0666 | 88 | 23.2838 |
| 14 | 4.6374 | 39 | 10.8547 | 64 | 17.0720 | 89 | 23.2892 |
| 15 | 4.6428 | 40 | 10.8601 | 65 | 17.0773 | 90 | 23.2946 |
| 16 | 4.6482 | 41 | 10.8654 | 66 | 17.0827 | 91 | 23.2999 |
| 17 | 4.6536 | 42 | 10.8708 | 67 | 17.0881 | 92 | 23.3053 |
| 18 | 4.6589 | 43 | 10.8762 | 68 | 17.0934 | 93 | 23.3107 |
| 19 | 7.7058 | 44 | 10.8815 | 69 | 17.0988 | 94 | 23.3160 |
| 20 | 7.7112 | 45 | 10.8869 | 70 | 17.1042 | 95 | 23.3214 |
| 21 | 7.7166 | 46 | 10.8923 | 71 | 17.1095 | 96 | 23.3268 |
| 22 | 7.7219 | 47 | 10.8976 | 72 | 17.1149 | 97 | 23.3322 |
| 23 | 7.7273 | 48 | 10.9030 | 73 | 17.1203 | 98 | 23.3375 |
| 24 | 7.7326 | 49 | 10.9084 | 74 | 17.1256 | 99 | 23.3429 |
| 25 | 7.7380 | 50 | 10.9137 | 75 | 17.1310 | 100 | 23.3483 |
| 26 | 7.7434 | 51 | 13.9606 | 76 | 20.1779 | 101 | 26.3952 |
| 27 | 7.7488 | 52 | 13.9660 | 77 | 20.1833 | 102 | 26.4005 |
| 28 | 7.7541 | 53 | 13.9714 | 78 | 20.1886 | 103 | 26.4059 |
| 29 | 7.7595 | 54 | 13.9767 | 79 | 20.1940 | 104 | 26.4113 |
| 30 | 7.7649 | 55 | 13.9821 | 80 | 20.1994 | 105 | 26.4166 |
| 31 | 7.7702 | 56 | 13.9875 | 81 | 20.2047 | 106 | 26.4220 |
| 32 | 7.7756 | 57 | 13.9929 | 82 | 20.2101 | 107 | 26.4274 |
| 33 | 7.7810 | 58 | 13.9982 | 83 | 20.2155 | 108 | 26.4327 |
| 34 | 7.7683 | 59 | 14.0036 | 84 | 20.2208 | 109 | 26.4381 |
| 35 | 7.7917 | 60 | 14.0090 | 85 | 20.2262 | 110 | 26.4435 |
| 36 | 7.7971 | 61 | 14.0143 | 86 | 20.2316 | … | … |
| Pressure Angle, 20 Degrees | | | | | | | |
| 12 | 4.5963 | 30 | 10.7526 | 48 | 16.9090 | 66 | 23.0653 |
| 13 | 4.6103 | 31 | 10.7666 | 49 | 16.9230 | 67 | 23.0793 |
| 14 | 4.6243 | 32 | 10.7806 | 50 | 16.9370 | 68 | 23.0933 |
| 15 | 4.6383 | 33 | 10.7946 | 51 | 16.9510 | 69 | 23.1073 |
| 16 | 4.6523 | 34 | 10.8086 | 52 | 16.9650 | 70 | 23.1214 |
| 17 | 4.6663 | 35 | 10.8226 | 53 | 16.9790 | 71 | 23.1354 |
| 18 | 4.6803 | 36 | 10.8366 | 54 | 16.9930 | 72 | 23.1494 |
| 19 | 7.6464 | 37 | 13.8028 | 55 | 19.9591 | 73 | 26.1155 |
| 20 | 7.6604 | 38 | 13.8168 | 56 | 19.9731 | 74 | 26.1295 |
| 21 | 7.6744 | 39 | 13.8307 | 57 | 19.9872 | 75 | 26.1435 |
| 22 | 7.6884 | 40 | 13.8447 | 58 | 20.0012 | 76 | 26.1575 |
| 23 | 7.7024 | 41 | 13.8587 | 59 | 20.0152 | 77 | 26.1715 |
| 24 | 7.7165 | 42 | 13.8727 | 60 | 20.0292 | 78 | 26.1855 |
| 25 | 7.7305 | 43 | 13.8867 | 61 | 20.0432 | 79 | 26.1995 |
| 26 | 7.7445 | 44 | 13.9007 | 62 | 20.0572 | 80 | 26.2135 |
| 27 | 7.7585 | 45 | 13.9147 | 63 | 20.0712 | 81 | 26.2275 |
| 28 | 10.7246 | 46 | 16.8810 | 64 | 23.0373 | … | … |
| 29 | 10.7386 | 47 | 16.8950 | 65 | 23.0513 | … | … |

*Table for Determining the Chordal Dimension:* Table 10 gives the chordal dimensions for one diametral pitch when measuring over the number of teeth indicated in Table 11. To obtain any chordal dimension, it is simply necessary to divide chord $M$ in the table (opposite the given number of teeth) by the diametral pitch of the gear to be measured and then subtract from the quotient one-half the total backlash between the mating pair of gears. In cases where a small pinion is used with a large gear and all of the backlash is to be obtained by reducing the gear teeth, the total amount of backlash is subtracted from the chordal dimension of the gear and nothing from the chordal dimension of the pinion. The application of the tables will be illustrated by an example.

**Table 11. Number of Teeth Included in Chordal Measurement**

| Tooth Range for $14\frac{1}{2}°$ Pressure Angle | Tooth Range for 20° Pressure Angle | Number of Teeth to Gage Over | Tooth Range for $14\frac{1}{2}°$ Pressure Angle | Tooth Range for 20° Pressure Angle | Number of Teeth to Gage Over |
|---|---|---|---|---|---|
| 12 to 18 | 12 to 18 | 2 | 63 to 75 | 46 to 54 | 6 |
| 19 to 37 | 19 to 27 | 3 | 76 to 87 | 55 to 63 | 7 |
| 38 to 50 | 28 to 36 | 4 | 88 to 100 | 64 to 72 | 8 |
| 51 to 62 | 37 to 45 | 5 | 101 to 110 | 73 to 81 | 9 |

This table shows the number of teeth to be included between the jaws of the vernier caliper in measuring dimension $M$ as explained in connection with Table 10.

*Example:* Determine the chordal dimension for checking the size of a gear having 30 teeth of 5 diametral pitch and a pressure angle of 20 degrees. A total backlash of 0.008 inch is to be obtained by reducing equally the teeth of both mating gears.

Table 10 shows that chordal distance for 30 teeth of one diametral pitch and a pressure angle of 20 degrees is 10.7526 inches; one-half of the backlash equals 0.004 inch; hence,

$$\text{Chordal dimension} = \frac{10.7526}{5} - 0.004 = 2.1465 \text{ inches}$$

Table 11 shows that this is the chordal dimension when the vernier caliper spans four teeth, this being the number of teeth to gage over whenever gears of 20-degree pressure angle have any number of teeth from 28 to 36, inclusive.

If it is considered necessary to leave enough stock on the gear teeth for a shaving or finishing cut, this allowance is simply added to the chordal dimension of the finished teeth to obtain the required measurement over the teeth for the roughing operation. It may be advisable to place this chordal dimension for rough machining on the detail drawing.

**Formula for Chordal Dimension *M*.—**The required measurement $M$ over spur gear teeth may be obtained by the following formula in which $R$ = pitch radius of gear, $A$ = pressure angle, $T$ = tooth thickness along pitch circle, $N$ = number of gear teeth, $S$ = number of tooth *spaces* between caliper jaws, $F$ = a factor depending on the pressure angle = 0.01109 for $14\frac{1}{2}°$; = 0.01973 for $17\frac{1}{2}°$; = 0.0298 for 20°; = 0.04303 for $22\frac{1}{2}°$; = 0.05995 for 25°. This factor $F$ equals twice the involute function of the pressure angle.

$$M = R \times \cos A \times \left(\frac{T}{R} + \frac{6.2832 \times S}{N} + F\right)$$

*Example:* A spur gear has 30 teeth of 6 diametral pitch and a pressure angle of $14\frac{1}{2}$ degrees. Determine measurement $M$ over three teeth, there being two intervening tooth spaces.

The pitch radius = $2\frac{1}{2}$ inches, the arc tooth thickness equivalent to 6 diametral pitch is 0.2618 inch (if no allowance is made for backlash) and factor $F$ for $14\frac{1}{2}$ degrees = 0.01109 inch.

$$M = 2.5 \times 0.96815 \times \left(\frac{0.2618}{2.5} + \frac{6.2832 \times 2}{30} + 0.01109\right) = 1.2941 \text{ inches}$$

**Checking Enlarged Pinions by Measuring Over Pins or Wires.—**When the teeth of small spur gears or pinions would be undercut if generated by an unmodified straight-sided rack cutter or hob, it is common practice to make the outside diameter larger than standard. The amount of increase in outside diameter varies with the pressure angle and number of teeth, as shown by Table 7 on page 2227. The teeth are always cut to standard depth on a generating type of machine such as a gear hobber or gear shaper; and because the number of teeth and pitch are not changed, the pitch diameter also remains unchanged. The tooth thickness on the pitch circle, however, is increased and wire sizes suitable for standard gears are not large enough to extend above the tops of these enlarged gears or pinions; hence, the Van Keuren wire size recommended for these enlarged pinions equals 1.92 ÷ diametral pitch.

Table 12 gives measurements over wires of this size, for checking full-depth involute gears of 1 diametral pitch. For any other pitch, merely divide the measurement given in the table by the diametral pitch. Table 12 applies to pinions that have been enlarged by the same amounts as given in Table 7 and Table 8, starting on page 2227. These enlarged pinions will mesh with standard gears; but if the standard center distance is to be maintained, reduce the gear diameter below the standard size by as much as the pinion diameter is increased.

**Table 12. Checking Enlarged Spur Pinions by Measurement Over Wires**

Measurements over wires are given in table for 1 diametral pitch. For any other diametral pitch, divide measurement in table by given pitch. Wire size equals 1.92 ÷ diametral pitch.

| Number of Teeth | Outside or Major Diameter (Note 1) | Circular Tooth Thickness (Note 2) | Measurement Over Wires | Number of Teeth | Outside or Major Diameter (Note 1) | Circular Tooth Thickness (Note 2) | Measurement Over Wires |
|---|---|---|---|---|---|---|---|
| $14\frac{1}{2}$-degree full-depth involute teeth: | | | | 20-degree full-depth involute teeth: | | | |
| 10 | 13.3731 | 1.9259 | 13.6186 | 10 | 12.936 | 1.912 | 13.5039 |
| 11 | 14.3104 | 1.9097 | 14.4966 | 11 | 13.818 | 1.868 | 14.3299 |
| 12 | 15.2477 | 1.8935 | 15.6290 | 12 | 14.702 | 1.826 | 15.4086 |
| 13 | 16.1850 | 1.8773 | 16.5211 | 13 | 15.584 | 1.783 | 16.2473 |
| 14 | 17.1223 | 1.8611 | 17.6244 | 14 | 16.468 | 1.741 | 17.2933 |
| 15 | 18.0597 | 1.8449 | 18.5260 | 15 | 17.350 | 1.698 | 18.1383 |
| 16 | 18.9970 | 1.8286 | 19.6075 | 16 | 18.234 | 1.656 | 19.1596 |
| 17 | 19.9343 | 1.8124 | 20.5156 | 17 | 19.116 | 1.613 | 20.0080 |
| 18 | 20.8716 | 1.7962 | 21.5806 | | | | |
| 19 | 21.8089 | 1.7800 | 22.4934 | | | | |
| 20 | 22.7462 | 1.7638 | 23.5451 | | | | |
| 21 | 23.6835 | 1.7476 | 24.4611 | | | | |
| 22 | 24.6208 | 1.7314 | 25.5018 | | | | |
| 23 | 25.5581 | 1.7151 | 26.4201 | | | | |
| 24 | 26.4954 | 1.6989 | 27.4515 | | | | |
| 25 | 27.4328 | 1.6827 | 28.3718 | | | | |
| 26 | 28.3701 | 1.6665 | 29.3952 | | | | |
| 27 | 29.3074 | 1.6503 | 30.3168 | | | | |
| 28 | 30.2447 | 1.6341 | 31.3333 | | | | |
| 29 | 31.1820 | 1.6179 | 32.2558 | | | | |
| 30 | 32.1193 | 1.6017 | 33.2661 | | | | |
| 31 | 33.0566 | 1.5854 | 34.1889 | | | | |

*Note* 1: These enlargements, which are to improve the tooth form and avoid undercut, conform to those given in Table 7 and Table 8, starting on page 2227 where data will be found on the minimum number of teeth in the mating gear.

*Note* 2: The circular or arc thickness is at the standard pitch diameter. The corresponding chordal thickness may be found as follows: Multiply arc thickness by 90 and then divide product by 3.1416 × pitch radius; find sine of angle thus obtained and multiply it by pitch diameter.

## GEAR MATERIALS

**Classification of Gear Steels.—**Gear steels may be divided into two general classes—the plain carbon and the alloy steels. Alloy steels are used to some extent in the industrial field, but heat-treated plain carbon steels are far more common. The use of untreated alloy steels for gears is seldom, if ever, justified, and then, only when heat-treating facilities are lacking. The points to be considered in determining whether to use heat-treated plain carbon steels or heat-treated alloy steels are: Does the service condition or design require the superior characteristics of the alloy steels, or, if alloy steels are not required, will the advantages to be derived offset the additional cost? For most applications, plain carbon steels, heat treated to obtain the best of their qualities for the service intended, are satisfactory and quite economical. The advantages obtained from using heat-treated alloy steels in place of heat-treated plain carbon steels are as follows:

1) Increased surface hardness and depth of hardness penetration for the same carbon content and quench.

2) Ability to obtain the same surface hardness with a less drastic quench and, in the case of some of the alloys, a lower quenching temperature, thus giving less distortion.

3) Increased toughness, as indicated by the higher values of yield point, elongation and reduction of area.

4) Finer grain size, with the resulting higher impact toughness and increased wear resistance.

5) In the case of some of the alloys, better machining qualities or the possibility of machining at higher hardnesses.

**Use of Case-Hardening Steels.—**Each of the two general classes of gear steels may be further subdivided as follows: 1) Case-hardening steels; 2) full-hardening steels; and 3) steels that are heat treated and drawn to a hardness that will permit machining.

The first two—case-hardening and full-hardening steels—are interchangeable for some kinds of service, and the choice is often a matter of personal opinion Case-hardening steels with their extremely hard, fine-grained (when properly treated) case and comparatively soft and ductile core are generally used when resistance to wear is desired. Case-hardening alloy steels have a fairly tough core, but not as tough as that of the full-hardening steels. In order to realize the greatest benefits from the core properties, case-hardened steels should be double-quenched. This is particularly true of the alloy steels, because the benefits derived from their use seldom justify the additional expense, unless the core is refined and toughened by a second quench. The penalty that must be paid for the additional refinement is increased distortion, which may be excessive if the shape or design does not lend itself to the case-hardening process.

**Use of "Thru-Hardening" Steels.—**Thru-hardening (or through-hardening) steels are used when great strength, high endurance limit, toughness, and resistance to shock are required. These qualities are governed by the kind of steel and treatment used. Fairly high surface hardnesses are obtainable in this group, though not so high as those of the case-hardening steels. For that reason, the resistance to wear is not so great as might be obtained, but when wear resistance combined with great strength and toughness is required, this type of steel is superior to the others. Thru-hardening steels become distorted to some extent when hardened, the amount depending upon the steel and quenching medium used. For that reason, thru-hardening steels are not suitable for high-speed gearing where noise is a factor, or for gearing where accuracy is of paramount importance, except, of course, in cases where grinding of the teeth is practicable. The medium and high-carbon percentages require an oil quench, but a water quench may be necessary for the lower carbon contents, in order to obtain the highest physical properties and hardness. The distortion, however, will be greater with the water quench.

**Heat Treatment that Permits Machining.—**When the grinding of gear teeth is not practicable and a high degree of accuracy is required, hardened steels may be drawn or tempered to a hardness that will permit the cutting of the teeth. This treatment gives a highly refined structure, great toughness, and, in spite of the low hardness, excellent wearing qualities. The lower strength is somewhat compensated for by the elimination of the increment loads due to the impacts which are caused by inaccuracies. When steels that have a low degree of hardness penetration from surface to core are treated in this manner, the design cannot be based on the physical properties corresponding to the hardness at the surface. Since the physical properties are determined by the hardness, the drop in hardness from surface to core will give lower physical properties at the root of the tooth, where the stress is greatest. The quenching medium may be either oil, water, or brine, depending on the steel used and hardness penetration desired. The amount of distortion, of course, is immaterial, because the machining is done after heat treating.

**Making Pinion Harder than Gear to Equalize Wear.—**Beneficial results from a wear standpoint are obtained by making the pinion harder than the gear. The pinion, having a lesser number of teeth than the gear, naturally does more work per tooth, and the differential in hardness between the pinion and the gear (the amount being dependent on the ratio) serves to equalize the rate of wear. The harder pinion teeth correct the errors in the gear teeth to some extent by the initial wear and then seem to burnish the teeth of the gear and increase its ability to withstand wear by the greater hardness due to the cold-working of the surface. In applications where the gear ratio is high and there are no severe shock loads, a case-hardened pinion running with an oil-treated gear, treated to a Brinell Hardness at which the teeth may be cut after treating, is an excellent combination. The pinion, being relatively small, is distorted but little, and distortion in the gear is circumvented by cutting the teeth after treatment.

**Forged and Rolled Carbon Steels for Gears.—**These compositions cover steel for gears in three groups, according to heat treatment, as follows:

a) case-hardened gears

b) unhardened gears, not heat treated after machining

c) hardened and tempered gears

Forged and rolled carbon gear steels are purchased on the basis of the requirements as to chemical composition specified in Table 1. Class N steel will normally be ordered in ten point carbon ranges within these limits. Requirements as to physical properties have been omitted, but when they are called for the requirements as to carbon shall be omitted. The steels may be made by either or both the open hearth and electric furnace processes.

**Table 1. Compositions of Forged and Rolled Carbon Steels for Gears**

| Heat Treatment | Class | Carbon | Manganese | Phosphorus | Sulfur |
|---|---|---|---|---|---|
| Case-hardened | C | 0.15–0.25 | 0.40–0.70 | 0.045 max | 0.055 max |
| Untreated | N | 0.25–0.50 | 0.50–0.80 | 0.045 max | 0.055 max |
| Hardened (or untreated) | H | 0.40–0.50 | 0.40–0.70 | 0.045 max | 0.055 max |

**Forged and Rolled Alloy Steels for Gears.—**These compositions cover alloy steel for gears, in two classes according to heat treatment, as follows:

a) case-hardened gears

b) hardened and tempered gears

Forged and rolled alloy gear steels are purchased on the basis of the requirements as to chemical composition specified in Table 2. Requirements as to physical properties have been omitted. The steel shall be made by either or both the open hearth and electric furnace process.

**Table 2. Compositions of Forged and Rolled Alloy Steels for Gears**

| Steel Specification | Chemical Composition[a] | | | | | |
|---|---|---|---|---|---|---|
| | C | Mn | Si | Ni | Cr | Mo |
| AISI 4130 | 0.28–0.30 | 0.40–0.60 | 0.20–0.35 | … | 0.80–1.1 | 0.15–0.25 |
| AISI 4140 | 0.38–0.43 | 0.75–1.0 | 0.20–0.35 | … | 0.80–1.1 | 0.15–0.25 |
| AISI 4340 | 0.38–0.43 | 0.60–0.80 | 0.20–0.35 | 1.65–2.0 | 0.70–90 | 0.20–0.30 |
| AISI 4615 | 0.13–0.18 | 0.45–0.65 | 0.20–0.35 | 1.65–2.0 | … | 0.20–0.30 |
| AISI 4620 | 0.17–0.22 | 0.45–0.65 | 0.20–0.35 | 1.65–2.0 | … | 0.20–0.30 |
| AISI 8615 | 0.13–0.18 | 0.70–0.90 | 0.20–0.35 | 0.40–0.70 | 0.40–0.60 | 0.15–0.25 |
| AISI 8620 | 0.18–0.23 | 0.70–0.90 | 0.20–0.35 | 0.40–0.70 | 0.40–0.60 | 0.15–0.25 |
| AISI 9310 | 0.08–0.13 | 0.45–0.65 | 0.20–0.35 | 3.0–3.5 | 1.0–1.4 | 0.08–0.15 |
| Nitralloy | | | | | | |
| Type N[b] | 0.20–0.27 | 0.40–0.70 | 0.20–0.40 | 3.2–3.8 | 1.0–1.3 | 0.20–0.30 |
| 135 Mod.[b] | 0.38–0.45 | 0.40–0.70 | 0.20–0.40 | … | 1.4–1.8 | 0.30–0.45 |

[a] C = carbon; Mn = manganese; Si = silicon; Ni = nickel; Cr = chromium, and Mo = molybdenum.

[b] Both Nitralloy alloys contain aluminum 0.85–1.2%

**Steel Castings for Gears.—** It is recommended that steel castings for cut gears be purchased on the basis of chemical analysis and that only two types of analysis be used, one for case-hardened gears and the other for both untreated gears and those which are to be hardened and tempered. The steel is to be made by the open hearth, crucible, or electric furnace processes. The converter process is not recognized. Sufficient risers must be provided to secure soundness and freedom from undue segregation. Risers should not be broken off the unannealed castings by force. Where risers are cut off with a torch, the cut should be at least one-half inch above the surface of the castings, and the remaining metal removed by chipping, grinding, or other noninjurious method.

Steel for use in gears should conform to the requirements for chemical composition indicated in Table 3. All steel castings for gears must be thoroughly normalized or annealed, using such temperature and time as will entirely eliminate the characteristic structure of unannealed castings.

**Table 3. Compositions of Cast Steels for Gears**

| Steel Specification | Chemical Composition[a] | | | |
|---|---|---|---|---|
| | C | Mn | Si | |
| SAE-0022 | 0.12–0.22 | 0.50–0.90 | 0.60 Max. | May be carburized |
| SAE-0050 | 0.40–0.50 | 0.50–0.90 | 0.80 Max. | Hardenable 210–250 |

[a] C = carbon; Mn = manganese; Si = silicon.

**Effect of Alloying Metals on Gear Steels.—** The effect of the various alloying elements on steel are here summarized to assist in deciding on the particular kind of alloy steel to use for specific purposes. The characteristics outlined apply only to heat-treated steels. When the effect of the addition of an alloying element is stated, it is understood that reference is made to alloy steels of a given carbon content, compared with a plain carbon steel of the same carbon content.

*Nickel:* The addition of nickel tends to increase the hardness and strength, with but little sacrifice of ductility. The hardness penetration is somewhat greater than that of plain carbon steels. Use of nickel as an alloying element lowers the critical points and produces less distortion, due to the lower quenching temperature. The nickel steels of the case-hardening group carburize more slowly, but the grain growth is less.

*Chromium:* Chromium increases the hardness and strength over that obtained by the use of nickel, though the loss of ductility is greater. Chromium refines the grain and imparts a greater depth of hardness. Chromium steels have a high degree of wear resistance and are easily machined in spite of the fine grain.

*Manganese:* When present in sufficient amounts to warrant the use of the term alloy, the addition of manganese is very effective. It gives greater strength than nickel and a higher degree of toughness than chromium. Owing to its susceptibility to cold-working, it is likely to flow under severe unit pressures. Up to the present time, it has never been used to any great extent for heat-treated gears, but is now receiving an increasing amount of attention.

*Vanadium:* Vanadium has a similar effect to that of manganese—increasing the hardness, strength, and toughness. The loss of ductility is somewhat more than that due to manganese, but the hardness penetration is greater than for any of the other alloying elements. Owing to the extremely fine-grained structure, the impact strength is high; but vanadium tends to make machining difficult.

*Molybdenum:* Molybdenum has the property of increasing the strength without affecting the ductility. For the same hardness, steels containing molybdenum are more ductile than any other alloy steels, and having nearly the same strength, are tougher; in spite of the increased toughness, the presence of molybdenum does not make machining more difficult. In fact, such steels can be machined at a higher hardness than any of the other alloy steels. The impact strength is nearly as great as that of the vanadium steels.

*Chrome-Nickel Steels:* The combination of the two alloying elements chromium and nickel adds the beneficial qualities of both. The high degree of ductility present in nickel steels is complemented by the high strength, finer grain size, deep hardening, and wear-resistant properties imparted by the addition of chromium. The increased toughness makes these steels more difficult to machine than the plain carbon steels, and they are more difficult to heat treat. The distortion increases with the amount of chromium and nickel.

*Chrome-Vanadium Steels:* Chrome-vanadium steels have practically the same tensile properties as the chrome-nickel steels, but the hardening power, impact strength, and wear resistance are increased by the finer grain size. They are difficult to machine and become distorted more easily than the other alloy steels.

*Chrome-Molybdenum Steels:* This group has the same qualities as the straight molybdenum steels, but the hardening depth and wear resistance are increased by the addition of chromium. This steel is very easily heat treated and machined.

*Nickel-Molybdenum Steels:* Nickel-molybdenum steels have qualities similar to chrome-molybdenum steel. The toughness is said to be greater, but the steel is somewhat more difficult to machine.

**Sintered Materials.—**For high production of low and moderately loaded gears, significant production cost savings may be effected by the use of a sintered metal powder. With this material, the gear is formed in a die under high pressure and then sintered in a furnace. The primary cost saving comes from the great reduction in labor cost of machining the gear teeth and other gear blank surfaces. The volume of production must be high enough to amortize the cost of the die and the gear blank must be of such a configuration that it may be formed and readily ejected from the die.

**Bronze and Brass Gear Castings.—**These specifications cover nonferrous metals for spur, bevel, and worm gears, bushings and flanges for composition gears. This material shall be purchased on the basis of chemical composition. The alloys may be made by any approved method.

*Spur and Bevel Gears:* For spur and bevel gears, hard cast bronze is recommended (ASTM B10-18; SAE No. 62; and the well-known 88-10-2 mixture) with the following limits as to composition: Copper, 86 to 89; tin, 9 to 11; zinc, 1 to 3; lead (max), 0.20; iron (max), 0.06 percent. Good castings made from this bronze should have the following minimum physical characteristics: Ultimate strength, 30,000 pounds per square inch; yield point, 15,000 pounds per square inch; elongation in 2 inches, 14 percent.

**Steels for Industrial Gearing**

| Material Specification | Hardness[a] Case RC | Core BHN | Typical Heat Treatment, Characteristics, and Uses |
|---|---|---|---|
| | | | Case-Hardening Steels |
| AISI 1020 AISI 1116 | 55-60 | 160-230 | Carburize, harden, temper at 350° F. For gears that must be wear-resistant. Normalized material is easily machined. Core is ductile but has little strength. |
| AISI 4130 AISI 4140 | 50-55 | 270-370 | Harden, temper at 900° F, Nitride. For parts requiring greater wear resistance than that of through-hardened steels but unable to tolerate the distortion of carburizing. Case is shallow, core is tough. |
| AISI 4615 AISI 4620 | } 55-60 | 170-260 | Carburize, harden, temper at 350° F. For gears requiring high fatigue resistance and strength. |
| AISI 8615 AISI 8620 | } 55-60 | 200-300 | The 86xx series has better machinability. The 20 point steels are used for coarser teeth. |
| AISI 9310 | 58-63 | 250-350 | Carburize, harden, temper at 300° F. Primarily for aerospace gears that are highly loaded and operate at high pitch line velocity and for other gears requiring high reliability under extreme operating conditions. This material is not used at high temperatures. |
| Nitralloy N and Type 135 Mod. (15-N) | 90-94 | 300-370 | Harden, temper at 1200° F, Nitride. For gears requiring high strength and wear resistance that cannot tolerate the distortion of the carburizing process or that operate at high temperatures. Gear teeth are usually finished before nitriding. Care must be exercised in running nitrided gears together to avoid crazing of case-hardened surfaces. |
| | | | Through-Hardening Steels |
| AISI 1045 AISI 1140 | 24-40 | … | Harden and temper to required hardness. Oil quench for lower hardness and water quench for higher hardness For gears of medium and large size requiring moderate strength and wear resistance. Gears that must have consistent, solid sections to withstand quenching. |
| AISI 4140AISI 4340 | 24-40 | … | Harden (oil quench), temper to required hardness. For gears requiring high strength and wear resistance, and high shock loading resistance. Use 41xx series for moderate sections and 43xx series for heavy sections. Gears must have consistent, solid sections to withstand quenching. |

[a] Hardness value scales are: RC for Rockwell C scale, BHN for Brinell Hardness Numbers.

*Worm Gears:* For bronze worm gears, two alternative analyses of phosphor bronze are recommended, SAE No. 65 and No. 63.

SAE No. 65 (called phosphor gear bronze) has the following composition: Copper, 88 to 90; tin, 10 to 12; phosphorus, 0.1 to 0.3; lead, zinc, and impurities (max) 0.5 percent.

Good castings made of this alloy should have the following minimum physical characteristics: Ultimate strength, 35,000 pounds per square inch; yield point, 20,000 pounds per square inch; elongation in 2 inches, 10 percent.

The composition of SAE No. 63 (called leaded gun metal) follows: copper, 86 to 89; tin, 9 to 11; lead, 1 to 2.5; phosphorus (max), 0.25; zinc and impurities (max), 0.50 percent.

Good castings made of this alloy should have the following minimum physical characteristics: Ultimate strength, 30,000 pounds per square inch; yield point, 12,000 pounds per square inch; elongation in 2 inches, 10 percent.

These alloys, especially No. 65, are adapted to chilling for hardness and refinement of grain. No. 65 is to be preferred for use with worms of great hardness and fine accuracy. No. 63 is to be preferred for use with unhardened worms.

*Gear Bushings:* For bronze bushings for gears, SAE No. 64 is recommended of the following analysis: copper, 78.5 to 81.5; tin, 9 to 11; lead, 9 to 11; phosphorus, 0.05 to 0.25; zinc (max), 0.75; other impurities (max), 0.25 percent. Good castings of this alloy should have the following minimum physical characteristics: Ultimate strength, 25,000 pounds per square inch; yield point, 12,000 pounds per square inch; elongation in 2 inches, 8 percent.

*Flanges for Composition Pinions:* For brass flanges for composition pinions ASTM B30-32T, and SAE No. 40 are recommended. This is a good cast red brass of sufficient strength and hardness to take its share of load and wear when the design is such that the flanges mesh with the mating gear. The composition is as follows: copper, 83 to 86; tin, 4.5 to 5.5; lead, 4.5 to 5.5; zinc, 4.5 to 5.5; iron (max) 0.35; antimony (max), 0.25 percent; aluminum, none. Good castings made from this alloy should have the following minimum physical characteristics: ultimate strength, 27,000 pounds per square inch; yield point, 12,000 pounds per square inch; elongation in 2 inches, 16 percent.

**Materials for Worm Gearing.—**The Hamilton Gear & Machine Co. conducted an extensive series of tests on a variety of materials that might be used for worm gears, to ascertain which material is the most suitable. According to these tests chill-cast nickel-phosphor-bronze ranks first in resistance to wear and deformation. This bronze is composed of approximately 87.5 percent copper, 11 percent tin, 1.5 percent nickel, with from 0.1 to 0.2 percent phosphorus. The worms used in these tests were made from SAE-2315, $3\frac{1}{2}$ percent nickel steel, case-hardened, ground, and polished. The Shore scleroscope hardness of the worms was between 80 and 90. This nickel alloy steel was adopted after numerous tests of a variety of steels, because it provided the necessary strength, together with the degree of hardness required.

The material that showed up second best in these tests was a No. 65 SAE bronze. Navy bronze (88-10-2) containing 2 percent zinc, with no phosphorus, and not chilled, performed satisfactorily at speeds of 600 revolutions per minute, but was not sufficiently strong at lower speeds. Red brass (85-5-5) proved slightly better at from 1500 to 1800 revolutions per minute, but would bend at lower speeds, before it would show actual wear.

**Nonmetallic Gearing.—**Nonmetallic or composition gearing is used primarily where quietness of operation at high speed is the first consideration. Nonmetallic materials are also applied very generally to timing gears and numerous other classes of gearing. Rawhide was used originally for nonmetallic gears, but other materials have been introduced that have important advantages. These later materials are sold by different firms under various trade names, such as Micarta, Textolite, Formica, Dilecto, Spauldite, Phenolite, Fibroc, Fabroil, Synthane, Celoron, etc. Most of these gear materials consist of layers of canvas or other material that is impregnated with plastics and forced together under hydraulic pressure, which, in conjunction with the application of heat, forms a dense rigid mass.

Although phenol *resin* gears in general are resilient, they are self-supporting and require no side plates or shrouds unless subjected to a heavy starting torque. The phenol resinoid element protects these gears from vermin and rodents.

The nonmetallic gear materials referred to are generally assumed to have the power-transmitting capacity of cast iron. Although the tensile strength may be considerably less than that of cast iron, the resiliency of these materials enables them to withstand impact and abrasion to a degree that might result in excessive wear of cast-iron teeth. Thus, composition gearing of impregnated canvas has often proved to be more durable than cast iron.

**Application of Nonmetallic Gears.—**The most effective field of use for these nonmetallic materials is for high-speed duty. At low speeds, when the starting torque may be high, or when the load may fluctuate widely, or when high shock loads may be encountered, these nonmetallic materials do not always prove satisfactory. In general, nonmetallic materials should not be used for pitch-line velocities below 600 feet per minute (3.05 m/s).

*Tooth Form:* The best tooth form for nonmetallic materials is the 20-degree stub-tooth system. When only a single pair of gears is involved and the center distance can be varied, the best results will be obtained by making the nonmetallic driving pinion of all-addendum form, and the driven metal gear with standard tooth proportions. Such a drive will carry from 50 to 75 percent greater loads than one of standard tooth proportions.

*Material for Mating Gear:* For durability under load, the use of hardened steel (over 400 BHN or Brinell Hardness Number) for the mating metal gear appears to give the best results. A good second choice for the material of the mating member is cast iron. The use of brass, bronze, or soft steel (under 400 BHN) as a material for the mating member of phenolic laminated gears leads to excessive abrasive wear.

**Power-Transmitting Capacity of Nonmetallic Gears.—**The characteristics of gears made of phenolic laminated materials are so different from those of metal gears that they should be considered in a class by themselves. Because of the low modulus of elasticity, most of the effects of small errors in tooth form and spacing are absorbed at the tooth surfaces by the elastic deformation, and have but little effect on the strength of the gears.

If $S$ = safe working stress for a given velocity lb/in² (MPa)

$S_s$ = allowable static stress lb/in² (MPa)

$V$ = pitch-line velocity in feet per minute (meter/s)

then, the recommended practice of the American Gear Manufacturers' Association,

$$S = S_s \times \left(\frac{150}{200 + V} + 0.25\right) \text{ US Units} \qquad S = S_s \times \left(\frac{0.76}{1.016 + V} + 0.25\right) \text{ SI Units}$$

The value of $S_s$ for phenolic laminated materials is given as 6000 lb/in² (41.36 MPa). The accompanying table gives the safe working stresses $S$ for different pitch-line velocities. When the value of $S$ is known, the horsepower capacity is determined by substituting the value of $S$ for $S_s$ in the appropriate equations in the section on power-transmitting capacity of plastics gears starting on page 607.

**Safe Working Stresses for Nonmetallic Gears**

| Pitch-Line Velocity, *V* | | Safe Working Stress, *S* | | Pitch-Line Velocity, *V* | | Safe Working Stress, *S* | | Pitch-Line Velocity, *V* | | Safe Working Stress, *S* | |
|---|---|---|---|---|---|---|---|---|---|---|---|
| fpm | m/s | lb/in² | MPa | fpm | m/s | lb/in² | MPa | fpm | m/s | lb/in² | MPa |
| 600 | 3.05 | 2625 | 18.10 | 1800 | 9.14 | 1950 | 13.44 | 4000 | 20.32 | 1714 | 11.82 |
| 700 | 3.56 | 2500 | 17.24 | 2000 | 10.16 | 1909 | 13.16 | 4500 | 22.86 | 1691 | 11.66 |
| 800 | 4.06 | 2400 | 16.54 | 2200 | 11.18 | 1875 | 12.93 | 5000 | 25.40 | 1673 | 11.53 |
| 900 | 4.57 | 2318 | 15.98 | 2400 | 12.19 | 1846 | 12.73 | 5500 | 27.94 | 1658 | 11.40 |
| 1000 | 5.08 | 2250 | 15.51 | 2600 | 13.20 | 1821 | 12.56 | 6000 | 30.48 | 1645 | 11.34 |
| 1200 | 6.10 | 2143 | 14.78 | 2800 | 14.22 | 1800 | 12.41 | 6500 | 33.02 | 1634 | 11.27 |
| 1400 | 7.11 | 2063 | 14.22 | 3000 | 15.24 | 1781 | 12.28 | 7000 | 35.56 | 1622 | 11.18 |
| 1600 | 8.13 | 2000 | 13.79 | 3500 | 17.78 | 1743 | 12.02 | 7500 | 38.10 | 1617 | 11.15 |

The tensile strength of the phenolic laminated materials used for gears is slightly less than that of cast iron. These materials are far softer than any metal, and the modulus of elasticity is about one-thirtieth that of steel. In other words, if the tooth load on a steel gear that causes a deformation of 0.001 inch (0.025 mm) were applied to the tooth of a similar gear made of phenolic laminated material, the tooth of the nonmetallic gear would be deformed about $\frac{1}{32}$ inch (0.794 mm). Under these conditions, several things will happen. With all gears, regardless of the theoretical duration of contact, one tooth only will carry the load until the load is sufficient to deform the tooth the amount of the error that may be present. On metal gears, when the tooth has been deformed the amount of the error, the stresses set up in the materials may approach or exceed the elastic limit of the material. Hence, for standard tooth forms and those generated from standard basic racks, it is dangerous to calculate their strength as very much greater than that which can safely be carried on a single tooth. On gears made of phenolic laminated materials, on the other hand, the teeth will be deformed the amount of this normal error without setting up any appreciable stresses in the material, so that the load is actually supported by several teeth.

All materials have their own peculiar and distinct characteristics, so that under certain specific conditions, each material has a field of its own where it is superior to any other. Such fields may overlap to some extent, and only in such overlapping fields are different materials directly competitive. For example, steel is more or less ductile, has a high tensile strength, and a high modulus of elasticity. Cast iron, on the other hand, is not ductile, has a low tensile strength, but a high compressive strength, and a low modulus of elasticity. Hence, when stiffness and high tensile strength are essential, steel is far superior to cast iron. On the other hand, when these two characteristics are unimportant, but high compressive strength and a moderate amount of elasticity are essential, cast iron is superior to steel.

**Preferred Pitch for Nonmetallic Gears.—**The pitch of the gear or pinion should bear a reasonable relation either to the power or speed or to the applied torque, as shown by the accompanying table. The upper half of this table is based upon horsepower (kw) transmitted at a given pitch-line velocity. The lower half gives the torque in pounds-feet (N-m) or the torque at a 1-foot (meter) radius. This torque $T$ for any given horsepower (kw) and speed can be obtained from the following formulas:

$$T = \frac{5252 \times \text{hp}}{\text{rpm}} \text{ pound-feet} \qquad T = \frac{9550 \times \text{kw}}{\text{rpm}} \text{ N-m}$$

**Bore Sizes for Nonmetallic Gears.—**For plain phenolic laminated pinions, that is, pinions without metal end plates, a drive fit of 0.001 inch per inch (or mm/mm) of shaft diameter should be used. For shafts above 2.5 inches (63.5 mm) in diameter, the fit should be constant at 0.0025 to 0.003 inch (0.064–0.076 mm). When metal reinforcing end plates are used, the drive fit should conform to the same standards as used for metal.

The root diameter of a pinion of phenolic laminated type should be such that the minimum distance from the edge of the keyway to the root diameter will be at least equal to the depth of tooth.

**Keyway Stresses for Nonmetallic Gears.—**The keyway stress should not exceed 3000 psi (20.68 MPa) on a plain phenolic laminated gear or pinion. The keyway stress is calculated by the formulas:

$$S = \frac{33000 \times \text{hp}}{V \times A} \text{ psi} \qquad S = \frac{1000 \times \text{kw}}{V \times A} \text{ MPa}$$

where $S$ = unit stress in pounds per square inch (newton per square meter)

$hp$ = horsepower transmitted

$kw$ = kilowatt power transmitted

$V$ = peripheral speed of shaft in feet per minute (meter/sec)

$A$ = square inch (square meter) area of keyway in pinion (length × height)

If the keyway stress formula is expressed in terms of shaft radius $r$ (inch or meter) and revolutions per minute, it will read:

## Preferred Pitches for Nonmetallic Gears

**Applicable both to rawhide and the phenolic laminated types of materials**

| Diametral Pitch for Given Horsepower and Pitch Line Velocities | | | |
|---|---|---|---|
| Horsepower Transmitted | Pitch Line Velocity up to 1000 Feet per Minute | Pitch Line Velocity from 1000 to 2000 Feet per Minute | Pitch Line Velocity over 2000 Feet per Minute |
| ¼-1 | 8-10 | 10-12 | 12-16 |
| 1-2 | 7-8 | 8-10 | 10-12 |
| 2-3 | 6-7 | 7-8 | 8-10 |
| 3-7½ | 5-6 | 6-7 | 7-8 |
| 7½-10 | 4-5 | 5-6 | 6-7 |
| 10-15 | 3-4 | 4-5 | 5-6 |
| 15-25 | 2½-3 | 3-4 | 4-5 |
| 25-60 | 2-2½ | 2½-3 | 3-4 |
| 60-100 | 1¾-2 | 2-2½ | 2½-3 |
| 100-150 | 1½-1¾ | 1¾-2 | 2-2½ |

| Torque in Pounds-Feet for Given Diametral Pitch | | | | | |
|---|---|---|---|---|---|
| Diametral Pitch | Torque in Pounds-feet | | Diametral Pitch | Torque in Pounds-feet | |
| | Minimum | Maximum | | Minimum | Maximum |
| 16 | 1 | 2 | 4 | 50 | 100 |
| 12 | 2 | 4 | 3 | 100 | 200 |
| 10 | 4 | 8 | 2½ | 200 | 450 |
| 8 | 8 | 15 | 2 | 450 | 900 |
| 6 | 15 | 30 | 1½ | 900 | 1800 |
| 5 | 30 | 50 | 1 | 1800 | 3500 |

$$S = \frac{63{,}000 \times \text{hp}}{\text{rpm} \times r \times A}\ \text{psi} \qquad S = \frac{9{,}550 \times \text{kw}}{\text{rpm} \times r \times A}\ \text{N-m}$$

When the design is such that the keyway stresses exceed 3000 psi (20.68 MPa), metal reinforcing end plates may be used. Such end plates should not extend beyond the root diameter of the teeth. The distance from the outer edge of the retaining bolt to the root diameter of the teeth shall not be less than a full tooth depth. The use of drive keys should be avoided, but if required, metal end plates should be used on the pinion to take the wedging action of the key.

For phenolic laminated pinions, the face of the mating gear should be the same or slightly greater than the pinion face.

**Invention of Gear Teeth.—**The invention of gear teeth represents a gradual evolution from gearing of primitive form. The earliest evidence we have of an investigation of the problem of *uniform motion* from toothed gearing and the successful solution of that problem dates from the time of Olaf Roemer, the celebrated Danish astronomer, who in the year 1674, proposed the epicycloidal form to obtain uniform motion. Evidently Robert Willis, professor at the University of Cambridge, was the first to make a practical application of the epicycloidal curve so as to provide for an interchangeable series of gears. Willis gives credit to Charles Étienne Louis Camus for conceiving the idea of interchangeable gears, but claims for himself its first application. The involute tooth was suggested as a theory by early scientists and mathematicians, but it remained for Willis to present it in a practical form. Perhaps the earliest conception of the application of this form of teeth to gears was by Philippe de Lahire, a Frenchman, who considered it, in theory, equally suitable with the epicycloidal for tooth outlines. This was about 1695 and not long after Roemer had first demonstrated the epicycloidal form. The applicability of the involute had been further elucidated by Leonard Euler, a Swiss mathematician, born at Basel, 1707, who is credited by Willis with being the first to suggest it. Willis devised the Willis odontograph for laying out involute teeth.

A pressure angle of 14½ degrees was selected for three different reasons. First, because the sine of 14½ degrees is nearly ¼, making it convenient in calculation; second, because this angle coincided closely with the pressure angle resulting from the usual construction of epicycloidal gear teeth; third, because the angle of the straight-sided involute rack is the same as the 29-degree worm thread.

**Calculating Replacement-Gear Dimensions from Simple Measurements.—**The following Table 1a, Table 1b, and Table 1c provide formulas with which to calculate the dimensions needed to produce replacement spur, bevel, and helical gears when only the number of teeth, the outside diameter, and the tooth depth of the gear to be replaced are known.

For helical gears, exact helix angles can be obtained by the following procedure.

1) Using a common protractor, measure the approximate helix angle *A* at the approximate pitch line.

2) Place sample or its mating gear on the arbor of a gear hobbing machine.

3) Calculate the index and lead gears differentially for the angle obtained by the measurements, and set up the machine as though a gear is to be cut.

4) Attach a dial indicator on an adjustable arm to the vertical swivel head, with the indicator plunger in a plane perpendicular to the gear axis and in contact with the tooth face. Contact may be anywhere between the top and the root of the tooth.

5) With the power shut off, engage the starting lever and traverse the indicator plunger axially by means of the handwheel.

6) If angle *A* is correct, the indicator plunger will not move as it traverses the face width of the gear. If it does move from 0, note the amount. Divide the amount of movement by the width of the gear to obtain the tangent of the angle by which to correct angle *A*, plus or minus, depending on the direction of indicator movement.

**Table 1a. Formulas for Calculating Spur Gear Dimensions**

| Tooth Form and Pressure Angle | Diametral Pitch $P$ | Pitch Diameter $D$ | Circular Pitch $P_c$ | Outside Diameter $O$ | Addendum $J$ | Dedendum $K$ | Whole Tooth Depth $W$ | Clearance $K-J$ | Tooth Thickness on Pitch Circle |
|---|---|---|---|---|---|---|---|---|---|
| American Standard 14½-and 20-degree full depth | $\frac{N+2}{O}$ | $\frac{N}{P}$ | $\frac{3.1416}{P}$ | $\frac{N+2}{P}$ | $\frac{1}{P}$ | $\frac{1.157}{P}$ | $\frac{2.157}{P}$ | $\frac{0.157}{P}$ | $\frac{1.5708}{P}$ |
| American Standard 20-degree stub | $\frac{N+1.6}{O}$ | $\frac{N}{P}$ | $\frac{3.1416}{P}$ | $\frac{N+1.6}{P}$ | $\frac{0.8}{P}$ | $\frac{1}{P}$ | $\frac{1.8}{P}$ | $\frac{0.2}{P}$ | $\frac{1.5708}{P}$ |
| Fellows 20-degree stub | See Note[a] | $\frac{N}{P_N}$ | $\frac{3.1416}{P_N}$ | $\frac{N}{P_N}+\frac{2}{P_D}$ | $\frac{1}{P_D}$ | $\frac{1.25}{P_D}$ | $\frac{2.25}{P_D}$ | $\frac{0.25}{P_D}$ | $\frac{1.5708}{P_N}$ |

[a] In the Fellows stub-tooth system, $P_N$ = diametral pitch in numerator of stub-tooth designation and is used to determine circular pitch and number of teeth, and $P_D$ = diametral pitch in the denominator of stub-tooth designation and is used to determine tooth depth.

$N$ = number of teeth.

**Table 1b. Formulas for Calculating Dimensions of Milled Bevel Gears—90-Degree Shafts**[a]

| Tooth Form and Pressure Angle | Tangent of Pitch Cone Angle of Gear, tan $A$ | Tangent of Pitch Cone Angle of Pinion, tan $a$ | Diametral Pitch[b] of Both Gear and Pinion, $P$ | Outside Diameter of Gear, $O$, or Pinion, $o$ | Pitch-Cone Radius[b] or Cone Distance, $E$ | Tangent of Addendum Angle[b] | Tangent of Dedendum Angle[b] | Cosine of Pitch-Cone Angle[c] of Gear, cos $A$ |
|---|---|---|---|---|---|---|---|---|
| American Standard $14\frac{1}{2}$- and 20-degree full depth | $\frac{N_G}{N_P}$ | $\frac{N_P}{N_G}$ | $\frac{N_G + 2\cos A}{O}$ or $\frac{N_P + 2\cos a}{o}$ | $\frac{N_G + 2\cos A}{P}$ or $\frac{N_P + 2\cos a}{P}$ | $\frac{D}{2\sin A}$ or $\frac{d}{2\sin a}$ | $\frac{2\sin A}{N_a}$ or $\frac{2\sin a}{N_P}$ | $\frac{2.314\sin A}{N_G}$ or $\frac{2.314\sin a}{N_P}$ | $\frac{(P \times O) - N_G}{2}$ |
| American Standard 20-degree stub | $\frac{N_G}{N_P}$ | $\frac{N_P}{N_G}$ | $\frac{N_G + 1.6\cos A}{O}$ or $\frac{N_P + 1.6\cos a}{o}$ | $\frac{N_G + 1.6\cos A}{P}$ or $\frac{N_P + 1.6\cos a}{P}$ | $\frac{D}{2\sin A}$ or $\frac{d}{2\sin a}$ | $\frac{1.6\sin A}{N_G}$ or $\frac{1.6\sin a}{N_P}$ | $\frac{2\sin A}{N_G}$ or $\frac{2\sin a}{N_P}$ | $\frac{(P \times O) - N_G}{1.6}$ |
| Fellows 20-degree stub | $\frac{N_G}{N_P}$ | $\frac{N_P}{N_G}$ | … | $\frac{N_G}{P_N} + \frac{2\cos A}{P_D}$ or $\frac{N_P}{P_N} + \frac{2\cos a}{P_D}$ | $\frac{D}{2\sin A}$ or $\frac{d}{2\sin a}$ | $\frac{2P_N \sin A}{N_G \times P_D}$ or $\frac{2P_N \sin a}{N_P \times P_D}$ | $\frac{2.5P_N \sin A}{N_G \times P_D}$ or $\frac{2.5P_N \sin a}{N_P \times P_D}$ | $\frac{P_D[(O \times P_N) - N_G]}{2P_N}$ |

[a] These formulas do not apply to Gleason System Gearing.

[b] These values are the same for both gear and pinion.

[c] The same formulas apply to the pinion, substituting $N_P$ for $N_G$ and $o$ for $O$.

$N_G$ = number of teeth in gear; $N_P$ = number of teeth in pinion; $O$ = outside diameter of gear; $o$ = outside diameter of pinion; $D$ = pitch diameter of gear = $N_G \div P$; $d$ = pitch diameter of pinion = $N_P \div P$; $P_c$ = circular pitch; $J$ = addendum; $K$ = dedendum; $W$ = whole depth.

See footnote in Table 1a for meaning of $P_N$ and $P_D$. The tooth thickness on the pitch circle is found by means of the formulas in the last column under spur gears.

**Table 1c. Formulas for Calculating Dimensions of Helical Gears**

| Tooth Form and Pressure Angle | Normal Diametral Pitch, $P_N$ | Diametral Pitch, $P$ | Outside Diameter of Blank, $O$ | Pitch Diameter, $D$ | Cosine of Helix Angle, $A$ | Addendum | Dedendum | Whole Depth |
|---|---|---|---|---|---|---|---|---|
| American Standard 14½- and 20-degree full depth | $\frac{N+2\cos A}{O\times\cos A}$ or $\frac{P}{\cos A}$ | $P_N\cos A$ or $\frac{N+2\cos A}{O}$ | $\frac{N+2\cos A}{P_N\cos A}$ or $\frac{N+2\cos A}{P}$ | $\frac{N}{P_N\cos A}$ or $\frac{N}{P}$ | $\frac{P}{P_N}$ or $\frac{N}{O\times P_N-2}$ | $\frac{1}{P_N}$ or $\frac{\cos A}{P}$ | $\frac{1.157}{P_N}$ or $\frac{1.157\cos A}{P}$ | $\frac{2.157}{P_N}$ or $\frac{2.157\cos A}{P}$ |
| American Standard 20-degree stub | $\frac{N+1.6\cos A}{O\times\cos A}$ or $\frac{P}{\cos A}$ | $P_n\cos A$ or $\frac{N+1.6\cos A}{O}$ | $\frac{N+1.6\cos A}{P_N\cos A}$ or $\frac{N+1.6\cos A}{P}$ | $\frac{N}{P_N\cos A}$ or $\frac{N}{P}$ | $\frac{P}{P_N}$ or $\frac{N}{O\times P_N-1.6}$ | $\frac{0.8}{P_N}$ or $\frac{0.8\cos A}{P}$ | $\frac{1}{P_N}$ or $\frac{\cos A}{P}$ | $\frac{1.8}{P_N}$ or $\frac{1.8\cos A}{P}$ |
| Fellows 20-degree stub | … | … | $\frac{N}{(P_N)_N\cos A}+\frac{2}{(P_N)_D}$ | $\frac{N}{(P_N)_N\cos A}$ | $\frac{N}{(P_N)_N\left(O-\frac{2}{(P_N)_D}\right)}$ | $\frac{1}{(P_N)_D}$ | $\frac{1.25}{(P_N)_D}$ | $\frac{2.25}{(P_N)_D}$ |

$P_N$ = normal diametral pitch = normal diametral pitch of cutter or hob used to cut teeth

$P$ = diametral pitch

$O$ = outside diameter of blank

$D$ = pitch diameter

$A$ = helix angle

$N$ = number of teeth

$(P_N)_N$ = normal diametral pitch in numerator of stub-tooth designation, which determines thickness of tooth and number of teeth

$(P_N)_D$ = normal diametral pitch in denominator of stub-tooth designation, which determines the addendum, dedendum, and whole depth

## SPLINES AND SERRATIONS

A splined shaft is one having a series of parallel keys formed integrally with the shaft and mating with corresponding grooves cut in a hub or fitting; this arrangement is in contrast to a shaft having a series of keys or feathers fitted into slots cut into the shaft. The latter construction weakens the shaft to a considerable degree because of the slots cut into it and consequently, reduces its torque-transmitting capacity.

Splined shafts are most generally used in three types of applications: 1) for coupling shafts when relatively heavy torques are to be transmitted without slippage; 2) for transmitting power to slidably-mounted or permanently-fixed gears, pulleys, and other rotating members; and 3) for attaching parts that may require removal for indexing or change in angular position.

Splines having straight-sided teeth have been used in many applications (see SAE Parallel Side Splines for Soft Broached Holes in Fittings); however, the use of splines with teeth of involute profile has steadily increased since 1) involute spline couplings have greater torque-transmitting capacity than any other type; 2) they can be produced by the same techniques and equipment as is used to cut gears; and 3) they have a self-centering action under load even when there is backlash between mating members.

### Involute Splines

**American National Standard Involute Splines*.—**These splines or multiple keys are similar in form to internal and external involute gears. The general practice is to form the external splines either by hobbing, rolling, or on a gear shaper, and internal splines either by broaching or on a gear shaper. The internal spline is held to basic dimensions and the external spline is varied to control the fit. Involute splines have maximum strength at the base, can be accurately spaced and are self-centering, thus equalizing the bearing and stresses, and they can be measured and fitted accurately.

In American National Standard ANSI B92.1-1996, many features of the 1960 and 1970 standards are retained, plus the addition of three tolerance classes, for a total of four. The term "involute serration," formerly applied to involute splines with 45-degree pressure angle, has been deleted and the Standard now includes involute splines with 30-, 37.5-, and 45-degree pressure angles. Tables for these splines have been rearranged accordingly. The term "serration" will no longer apply to splines covered by this Standard.

The Standard has only one fit class for all side fit splines: the former Class 2 fit. Class 1 fit has been deleted because of its infrequent use. The major diameter of the flat root side fit spline has been changed and a tolerance applied to include the range of the 1950 and the 1960 standards. The interchangeability limitations with splines made to previous standards are given later in the section entitled "Interchangeability."

There have been no tolerance nor fit changes to the major diameter fit section.

The Standard recognizes the fact that proper assembly between mating splines is dependent only on the spline being within effective specifications from the tip of the tooth to the form diameter. Therefore, on side fit splines, the internal spline major diameter now is shown as a maximum dimension and the external spline minor diameter is shown as a minimum dimension. The minimum internal major diameter and the maximum external minor diameter must clear the specified form diameter and thus do not need any additional control.

The spline specification tables now include a greater number of tolerance level selections. These tolerance classes were added for greater selection to suit end product needs. The selections differ only in the tolerance as applied to space width and tooth thickness.

* See American National Standard ANSI B92.2M-1980 (R1989), Metric Module Involute Splines; also see page 2353.

The tolerance class used in ASA B5.15-1960 is the basis and is now designated as tolerance Class 5. The newer tolerance classes are based on the following formulas:

$$\text{Tolerance Class 4} = \text{Tolerance Class } 5 \times 0.71$$

$$\text{Tolerance Class 6} = \text{Tolerance Class } 5 \times 1.40$$

$$\text{Tolerance Class 7} = \text{Tolerance Class } 5 \times 2.00$$

All dimensions listed in this Standard are for the finished part. Therefore, any compensation that must be made for operations that take place during processing, such as heat treatment, must be taken into account when selecting the tolerance level for manufacturing.

The Standard has the same internal minimum effective space width and external maximum effective tooth thickness for all tolerance classes and has two types of fit. For tooth side fits, the minimum effective space width and the maximum effective tooth thickness are of equal value. This basic concept makes it possible to have interchangeable assembly between mating splines where they are made to this Standard regardless of the tolerance class of the individual members. A tolerance class "mix" of mating members is thus allowed, which often is an advantage where one member is considerably less difficult to produce than its mate, and the "average" tolerance applied to the two units is such that it satisfies the design need. For instance, assigning a Class 5 tolerance to one member and Class 7 to its mate will provide an assembly tolerance in the Class 6 range. The maximum effective tooth thickness is less than the minimum effective space width for major diameter fits to allow for eccentricity variations.

In the event the fit as provided in this Standard does not satisfy a particular design need and a specific amount of effective clearance or press fit is desired, the change should be made only to the external spline by a reduction or an increase in effective tooth thickness and a like change in actual tooth thickness. The minimum effective space width, in this Standard, is always basic. The basic minimum effective space width should always be retained when special designs are derived from the concept of this Standard.

**Terms Applied to Involute Splines.**—The following definitions of involute spline terms, here listed in alphabetical order, are given in the American National Standard. Some of these terms are illustrated in the diagram in Table 6.

*Active Spline Length* ($L_a$) is the length of spline that contacts the mating spline. On sliding splines, it exceeds the length of engagement.

*Actual Space Width* ($s$) is the circular width on the pitch circle of any single space considering an infinitely thin increment of axial spline length.

*Actual Tooth Thickness* ($t$) is the circular thickness on the pitch circle of any single tooth considering an infinitely thin increment of axial spline length.

*Alignment Variation* is the variation of the effective spline axis with respect to the reference axis (see Fig. 1c).

*Base Circle* is the circle from which involute spline tooth profiles are constructed.

*Base Diameter* ($D_b$) is the diameter of the base circle.

*Basic Space Width* is the basic space width for 30-degree pressure angle splines: half the circular pitch. The basic space width for 37.5- and 45-degree pressure angle splines, however, is greater than half the circular pitch. The teeth are proportioned so that the external tooth, at its base, has about the same thickness as the internal tooth at the form diameter. This proportioning results in greater minor diameters than those of comparable involute splines of 30-degree pressure angle.

*Circular Pitch* ($p$) is the distance along the pitch circle between corresponding points of adjacent spline teeth.

*Depth of Engagement* is the radial distance from the minor circle of the internal spline to the major circle of the external spline, minus corner clearance and/or chamfer depth.

*Diametral Pitch* ($P$) is the number of spline teeth per inch of pitch diameter. The diametral pitch determines the circular pitch and the basic space width or tooth thickness. In conjunction with the number of teeth, it also determines the pitch diameter. (See also Pitch.)

*Effective Clearance* ($c_v$) is the effective space width of the internal spline minus the effective tooth thickness of the mating external spline.

*Effective Space Width* ($S_v$) of an internal spline is equal to the circular tooth thickness on the pitch circle of an imaginary perfect external spline that would fit the internal spline without looseness or interference considering engagement of the entire axial length of the spline. The minimum effective space width of the internal spline is always basic, as shown in Table 3. Fit variations may be obtained by adjusting the tooth thickness of the external spline.

**Three Types of Involute Spline Variations**

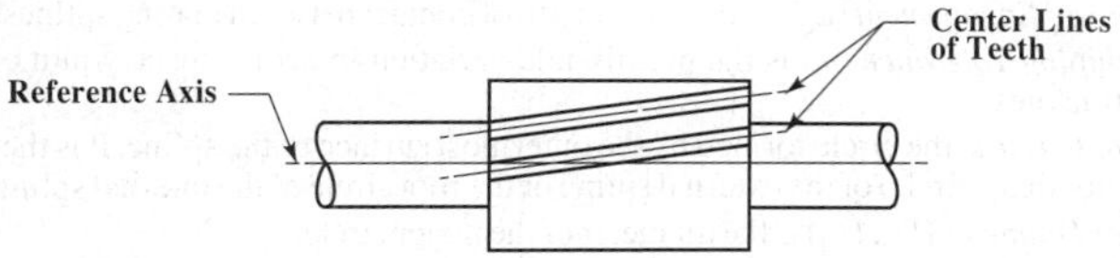


Fig. 1a. Lead Variation

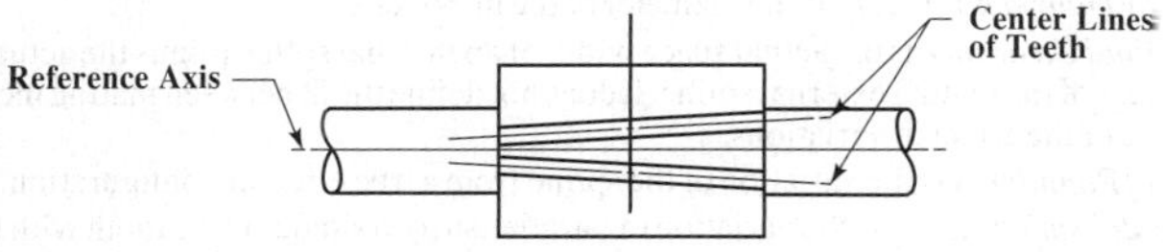


Fig. 1b. Parallelism Variation

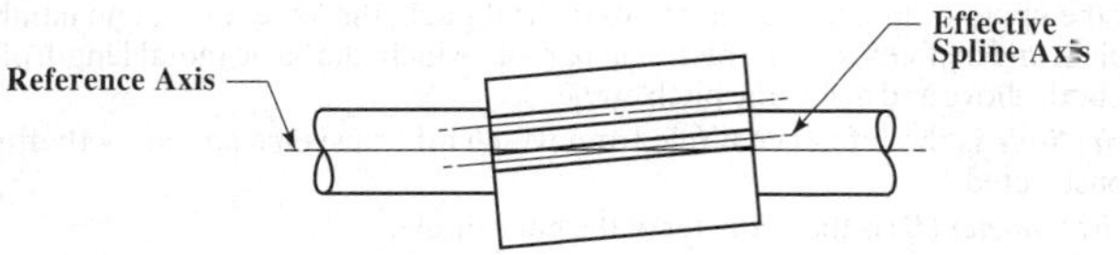


Fig. 1c. Alignment Variation

*Effective Tooth Thickness* ($t_v$) of an external spline is equal to the circular space width on the pitch circle of an imaginary perfect internal spline that would fit the external spline without looseness or interference, considering engagement of the entire axial length of the spline.

*Effective Variation* is the accumulated effect of the spline variations on the fit with the mating part.

*External Spline* is a spline formed on the outer surface of a cylinder.

*Fillet* is the concave portion of the tooth profile that joins the sides to the bottom of the space.

*Fillet Root Splines* are those in which a single fillet in the general form of an arc joins the sides of adjacent teeth.

*Flat Root Splines* are those in which fillets join the arcs of major or minor circles to the tooth sides.

*Form Circle* is the circle which defines the deepest points of involute form control of the tooth profile. This circle along with the tooth tip circle (or start of chamfer circle) determines the limits of tooth profile requiring control. It is located near the major circle on the internal spline and near the minor circle on the external spline.

*Form Clearance* ($c_F$) is the radial depth of involute profile beyond the depth of engagement with the mating part. It allows for looseness between mating splines and for eccentricities between the minor circle (internal), the major circle (external), and their respective pitch circles.

*Form Diameter* ($D_{Fe}, D_{Fi}$) is the diameter of the form circle.

*Internal Spline* is a spline formed on the inner surface of a cylinder.

*Involute Spline* is one having teeth with involute profiles.

*Lead Variation* is the variation of the direction of the spline tooth from its intended direction parallel to the reference axis, also including parallelism and alignment variations (see Fig. 1a). *Note:* Straight (nonhelical) splines have an infinite lead.

*Length of Engagement* ($L_q$) is the axial length of contact between mating splines.

*Machining Tolerance* ($m$) is the permissible variation in actual space width or actual tooth thickness.

*Major Circle* is the circle formed by the outermost surface of the spline. It is the outside circle (tooth tip circle) of the external spline or the root circle of the internal spline.

*Major Diameter* ($D_o, D_{ri}$) is the diameter of the major circle.

*Minor Circle* is the circle formed by the innermost surface of the spline. It is the root circle of the external spline or the inside circle (tooth tip circle) of the internal spline.

*Minor Diameter* ($D_{re}, D_i$) is the diameter of the minor circle.

*Nominal Clearance* is the actual space width of an internal spline minus the actual tooth thickness of the mating external spline. It does not define the fit between mating members, because of the effect of variations.

*Out of Roundness* is the variation of the spline from a true circular configuration.

*Parallelism Variation* is the variation of parallelism of a single spline tooth with respect to any other single spline tooth (see Fig. 1b).

*Pitch* ($P/P_s$) is a combination number of a one-to-two ratio indicating the spline proportions; the upper or first number is the diametral pitch, the lower or second number is the stub pitch and denotes, as that fractional part of an inch, the basic radial length of engagement, both above and below the pitch circle.

*Pitch Circle* is the reference circle from which all transverse spline tooth dimensions are constructed.

*Pitch Diameter* ($D$) is the diameter of the pitch circle.

*Pitch Point* is the intersection of the spline tooth profile with the pitch circle.

*Pressure Angle* ($\phi$) is the angle between a line tangent to an involute and a radial line through the point of tangency. Unless otherwise specified, it is the standard pressure angle.

*Profile Variation* is any variation from the specified tooth profile normal to the flank.

*Spline* is a machine element consisting of integral keys (spline teeth) or keyways (spaces) equally spaced around a circle or portion thereof.

*Standard (Main) Pressure Angle* ($\phi_D$) is the pressure angle at the specified pitch diameter.

*Stub Pitch* ($P_s$) is a number used to denote the radial distance from the pitch circle to the major circle of the external spline and from the pitch circle to the minor circle of the internal spline. The stub pitch for splines is twice the diametral pitch.

*Total Index Variation* is the greatest difference in any two teeth (adjacent or otherwise) between the actual and the perfect spacing of the tooth profiles.

*Total Tolerance* ($m + \lambda$) is the machining tolerance plus the variation allowance.

*Variation Allowance* ($\lambda$) is the permissible effective variation.

**Tooth Proportions.—**There are 17 pitches: 2.5/5, 3/6, 4/8, 5/10, 6/12, 8/16, 10/20, 12/24, 16/32, 20/40, 24/48, 32/64, 40/80, 48/96, 64/128, 80/160, and 128/256. The numerator in this fractional designation is known as the diametral pitch and controls the pitch diameter; the denominator, which is always double the numerator, is known as the stub pitch and controls the tooth depth. For convenience in calculation, only the numerator is used in the formulas given and is designated as *P*. Diametral pitch, as in gears, means the number of teeth per inch of pitch diameter.

Table 1 shows the symbols and Table 2 the formulas for basic tooth dimensions of involute spline teeth of various pitches. Basic dimensions are given in Table 3.

**Table 1. American National Standard Involute Spline Symbols**
*ANSI B92.1-1996*

| | | | |
|---|---|---|---|
| $c_v$ | effective clearance | $M_i$ | measurement between pins, internal spline |
| $c_F$ | form clearance | $N$ | number of teeth |
| $D$ | pitch diameter | $P$ | diametral pitch |
| $D_b$ | base diameter | $P_s$ | stub pitch |
| $D_{ci}$ | pin contact diameter, internal spline | $P/P_s$ | spline pitch |
| $D_{ce}$ | pin contact diameter, external spline | $p$ | circular pitch |
| $D_{Fe}$ | form diameter, external spline | $r_f$ | fillet radius |
| $D_{Fi}$ | form diameter, internal spline | $s$ | actual space width, circular |
| $D_i$ | minor diameter, internal spline | $s_v$ | effective space width, circular |
| $D_o$ | major diameter, external spline | $t$ | actual tooth thickness, circular |
| $D_{re}$ | minor diameter, external spline (root) | $t_v$ | effective tooth thickness, circular |
| $D_{ri}$ | major diameter, internal spline (root) | $\lambda$ | variation allowance |
| $d_e$ | diameter of measuring pin for external spline | $\varepsilon$ | involute roll angle |
| $d_i$ | diameter of measuring pin for internal spline | $\phi$ | pressure angle |
| $K_e$ | change factor for external spline | $\phi_D$ | standard pressure angle |
| $K_i$ | change factor for internal spline | $\phi_{ci}$ | pressure angle at pin contact diameter, internal spline |
| $L$ | spline length | $\phi_{ce}$ | pressure angle at pin contact diameter, external spline |
| $L_a$ | active spline length | $\phi_i$ | pressure angle at pin center, internal spline |
| $L_g$ | length of engagement | $\phi_e$ | pressure angle at pin center, external spline |
| $m$ | machining tolerance | $\phi_F$ | pressure angle at form diameter |
| $M_e$ | measurement over pins, external spline | | |

**Table 2. Formulas for Involute Spline Basic Dimensions** *ANSI B92.1-1996*

| Term | | Symbol | 30 deg $\phi_D$ | | | 37.5 deg $\phi_D$ | 45 deg $\phi_D$ |
|---|---|---|---|---|---|---|---|
| | | | Flat Root Side Fit | Flat Root Major Dia Fit | Fillet Root Side Fit | Fillet Root Side Fit | Fillet Root Side Fit |
| | | | 2.5/5–32/64 Pitch | 3/6–16/32 Pitch | 2.5/5–48/96 Pitch | 2.5/5–48/96 Pitch | 10/20–128/256 Pitch |
| Stub Pitch | | $P_s$ | $2P$ | $2P$ | $2P$ | $2P$ | $2P$ |
| Pitch Diameter | | $D$ | $N/P$ | $N/P$ | $N/P$ | $N/P$ | $N/P$ |
| Base Diameter | | $D_b$ | $D \cos \phi_D$ | $D \cos \phi_D$ | $D \cos \phi_D$ | $D \cos \phi_D$ | $D \cos \phi_D$ |
| Circular Pitch | | $p$ | $\pi/P$ | $\pi/P$ | $\pi/P$ | $\pi/P$ | $\pi/P$ |
| Minimum Effective Space Width | | $s_v$ | $\pi/(2P)$ | $\pi/(2P)$ | $\pi/(2P)$ | $(0.5\pi + 0.1)/P$ | $(0.5\pi + 0.2)/P$ |
| Major Diameter, Internal | | $D_{ri}$ | $(N + 1.35)/P$ | $(N + 1)/P$ | $(N + 1.8)/P$ | $(N + 1.6)/P$ | $(N + 1.4)/P$ |
| Major Diameter, External | | $D_o$ | $(N + 1)/P$ | $(N + 1)/P$ | $(N + 1)/P$ | $(N + 1)/P$ | $(N + 1)/P$ |
| Minor Diameter, Internal | | $D_i$ | $(N - 1)/P$ | $(N - 1)/P$ | $(N - 1)/P$ | $(N - 0.8)/P$ | $(N - 0.6)/P$ |
| Minor Dia. Ext. | 2.5/5 thru 12/24 pitch | $D_{re}$ | $(N - 1.35)/P$ | | $(N - 1.8)/P$ | $(N - 1.3)/P$ | … |
| | 16/32 pitch and finer | | | | $(N - 2)/P$ | | |
| | 10/20 pitch and finer | | | | … | | $(N - 1)/P$ |
| Form Diameter, Internal | | $D_{Fi}$ | $(N + 1)/P - 2c_F$ | $(N + 0.8)/P - 0.004 + 2c_F$ | $(N + 1)/P + 2c_F$ | $(N + 1)/P + 2c_F$ | $(N + 1)/P + 2c_F$ |
| Form Diameter, External | | $D_{Fe}$ | $(N - 1)/P - 2c_F$ | $(N - 1)/P - 2c_F$ | $(N - 1)/P - 2c_F$ | $(N - 0.8)/P - 2c_F$ | $(N - 0.6)/P - 2c_F$ |
| Form Clearance (Radial) | | $c_F$ | 0.001 $D$, with max of 0.010, min of 0.002 | | | | |

$\pi = 3.1415927$

*Note:* All spline specification table dimensions in the Standard are derived from these basic formulas by application of tolerances.

**Table 3. Basic Dimensions for Involute Splines** *ANSI B92.1-1996*

| Pitch, $P/P_s$ | Circular Pitch, $p$ | Min Effective Space Width (BASIC), $S_v$ min | | | Pitch, $P/P_s$ | Circular Pitch, $p$ | Min Effective Space Width (BASIC), $S_v$ min | | |
|---|---|---|---|---|---|---|---|---|---|
| | | 30 deg $\phi$ | 37.5 deg $\phi$ | 45 deg $\phi$ | | | 30 deg $\phi$ | 37.5 deg $\phi$ | 45 deg $\phi$ |
| 2.5/5 | 1.2566 | 0.6283 | 0.6683 | … | 20/40 | 0.1571 | 0.0785 | 0.0835 | 0.0885 |
| 3/6 | 1.0472 | 0.5236 | 0.5569 | … | 24/48 | 0.1309 | 0.0654 | 0.0696 | 0.0738 |
| 4/8 | 0.7854 | 0.3927 | 0.4177 | … | 32/64 | 0.0982 | 0.0491 | 0.0522 | 0.0553 |
| 5/10 | 0.6283 | 0.3142 | 0.3342 | … | 40/80 | 0.0785 | 0.0393 | 0.0418 | 0.0443 |
| 6/12 | 0.5236 | 0.2618 | 0.2785 | … | 48/96 | 0.0654 | 0.0327 | 0.0348 | 0.0369 |
| 8/16 | 0.3927 | 0.1963 | 0.2088 | … | 64/128 | 0.0491 | … | … | 0.0277 |
| 10/20 | 0.3142 | 0.1571 | 0.1671 | 0.1771 | 80/160 | 0.0393 | … | … | 0.0221 |
| 12/24 | 0.2618 | 0.1309 | 0.1392 | 0.1476 | 128/256 | 0.0246 | … | … | 0.0138 |
| 16/32 | 0.1963 | 0.0982 | 0.1044 | 0.1107 | … | … | … | … | .. |

**Tooth Numbers.—**The American National Standard covers involute splines having tooth numbers ranging from 6 to 60 with a 30- or 37.5-degree pressure angle and from 6 to 100 with a 45-degree pressure angle. In selecting the number of teeth for a given spline application, it is well to keep in mind that there are no advantages to be gained by using odd numbers of teeth and that the diameters of splines with odd tooth numbers, particularly internal splines, are troublesome to measure with pins since no two tooth spaces are diametrically opposite each other.

**Types and Classes of Involute Spline Fits.—**Two types of fits are covered by the American National Standard for involute splines: the side fit and the major diameter fit. Dimensional data for flat root side fit, flat root major diameter fit, and fillet root side fit splines are tabulated in this Standard for 30-degree pressure angle splines, but for only the fillet root side fit for 37.5- and 45-degree pressure angle splines.

*Side Fit:* In the side fit, the mating members contact only on the sides of the teeth; major and minor diameters are clearance dimensions. The tooth sides act as drivers and centralize the mating splines.

*Major Diameter Fit:* Mating parts for this fit contact at the major diameter for centralizing. The sides of the teeth act as drivers. The minor diameters are clearance dimensions.

The major diameter fit provides a minimum effective clearance that will allow for contact and location at the major diameter with a minimum amount of location or centralizing effect by the sides of the teeth. The major diameter fit has only one space width and tooth thickness tolerance which is the same as side fit Class 5.

A fillet root may be specified for an external spline, even though it is otherwise designed to the flat root side fit or major diameter fit standard. An internal spline with a fillet root can be used only for the side fit.

**Classes of Tolerances.—**This Standard includes four classes of tolerances on space width and tooth thickness. This has been done to provide a range of tolerances for selection to suit a design need. The classes are variations of the former single tolerance which is now Class 5 and are based on the formulas shown in the footnote of Table 4. All tolerance classes have the same minimum effective space width and maximum effective tooth thickness limits so that a mix of classes between mating parts is possible.

**Table 4. Maximum Tolerances for Space Width and Tooth Thickness of Tolerance Class 5 Splines** *ANSI B92.1-1996*
**(Values shown in ten thousandths; 20 = 0.0020)**

| No. of Teeth | Pitch, $P/P_s$ | | | | | | | |
|---|---|---|---|---|---|---|---|---|
| | 2.5/5 and 3/6 | 4/8 and 5/10 | 6/12 and 8/16 | 10/20 and 12/24 | 16/32 and 20/40 | 24/48 thru 48/96 | 64/128 and 80/160 | 128/256 |
| $N$ | Machining Tolerance, $m$ | | | | | | | |
| 10 | 15.8 | 14.5 | 12.5 | 12.0 | 11.7 | 11.7 | 9.6 | 9.5 |
| 20 | 17.6 | 16.0 | 14.0 | 13.0 | 12.4 | 12.4 | 10.2 | 10.0 |
| 30 | 18.4 | 17.5 | 15.5 | 14.0 | 13.1 | 13.1 | 10.8 | 10.5 |
| 40 | 21.8 | 19.0 | 17.0 | 15.0 | 13.8 | 13.8 | 11.4 | — |
| 50 | 23.0 | 20.5 | 18.5 | 16.0 | 14.5 | 14.5 | — | — |
| 60 | 24.8 | 22.0 | 20.0 | 17.0 | 15.2 | 15.2 | — | — |
| 70 | — | — | — | 18.0 | 15.9 | 15.9 | — | — |
| 80 | — | — | — | 19.0 | 16.6 | 16.6 | — | — |
| 90 | — | — | — | 20.0 | 17.3 | 17.3 | — | — |
| 100 | — | — | — | 21.0 | 18.0 | 18.0 | — | — |
| $N$ | Variation Allowance, λ | | | | | | | |
| 10 | 23.5 | 20.3 | 17.0 | 15.7 | 14.2 | 12.2 | 11.0 | 9.8 |
| 20 | 27.0 | 22.6 | 19.0 | 17.4 | 15.4 | 13.4 | 12.0 | 10.6 |
| 30 | 30.5 | 24.9 | 21.0 | 19.1 | 16.6 | 14.6 | 13.0 | 11.4 |
| 40 | 34.0 | 27.2 | 23.0 | 21.6 | 17.8 | 15.8 | 14.0 | — |
| 50 | 37.5 | 29.5 | 25.0 | 22.5 | 19.0 | 17.0 | — | — |
| 60 | 41.0 | 31.8 | 27.0 | 24.2 | 20.2 | 18.2 | — | — |
| 70 | — | — | — | 25.9 | 21.4 | 19.4 | — | — |
| 80 | — | — | — | 27.6 | 22.6 | 20.6 | — | — |
| 90 | — | — | — | 29.3 | 23.8 | 21.8 | — | — |
| 100 | — | — | — | 31.0 | 25.0 | 23.0 | — | — |
| $N$ | Total Index Variation | | | | | | | |
| 10 | 20 | 17 | 15 | 15 | 14 | 12 | 11 | 10 |
| 20 | 24 | 20 | 18 | 17 | 15 | 13 | 12 | 11 |
| 30 | 28 | 22 | 20 | 19 | 16 | 15 | 14 | 13 |
| 40 | 32 | 25 | 22 | 20 | 18 | 16 | 15 | — |
| 50 | 36 | 27 | 25 | 22 | 19 | 17 | — | — |
| 60 | 40 | 30 | 27 | 24 | 20 | 18 | — | — |
| 70 | — | — | — | 26 | 21 | 20 | — | — |
| 80 | — | — | — | 28 | 22 | 21 | — | — |
| 90 | — | — | — | 29 | 24 | 23 | — | — |
| 100 | — | — | — | 31 | 25 | 24 | — | — |
| $N$ | Profile Variation | | | | | | | |
| All | +7<br>−10 | +6<br>−8 | +5<br>−7 | +4<br>−6 | +3<br>−5 | +2<br>−4 | +2<br>−4 | +2<br>−4 |

| Lead Variation | | | | | | | | | | | | |
|---|---|---|---|---|---|---|---|---|---|---|---|---|
| $L_g$, in. | 0.3 | 0.5 | 1 | 2 | 3 | 4 | 5 | 6 | 7 | 8 | 9 | 10 |
| Variation | 2 | 3 | 4 | 5 | 6 | 7 | 8 | 9 | 10 | 11 | 12 | 13 |

For other tolerance classes: Class 4 = 0.71 × Tabulated value
Class 5 = As tabulated in table
Class 6 = 1.40 × Tabulated value
Class 7 = 2.00 × Tabulated value

**Fillets and Chamfers.—** Spline teeth may have either a flat root or a rounded fillet root.

*Flat Root Splines:* are suitable for most applications. The fillet that joins the sides to the bottom of the tooth space, if generated, has a varying radius of curvature. Specification of this fillet is usually not required. It is controlled by the form diameter, which is the diameter at the deepest point of the desired true involute form (sometimes designated as TIF).

When flat root splines are used for heavily loaded couplings that are not suitable for fillet root spline application, it may be desirable to minimize the stress concentration in the flat root type by specifying an approximate radius for the fillet.

Because internal splines are stronger than external splines due to their broad bases and high pressure angles at the major diameter, broaches for flat root internal splines are normally made with the involute profile extending to the major diameter.

*Fillet Root Splines:* are recommended for heavy loads because the larger fillets provided reduce the stress concentrations. The curvature along any generated fillet varies and cannot be specified by a radius of any given value.

External splines may be produced by generating with a pinion-type shaper cutter or with a hob, or by cutting with no generating motion using a tool formed to the contour of a tooth space. External splines are also made by cold forming and are usually of the fillet root design. Internal splines are usually produced by broaching, by form cutting, or by generating with a shaper cutter. Even when full-tip radius tools are used, each of these cutting methods produces a fillet contour with individual characteristics. Generated spline fillets are curves related to the prolate epicycloid for external splines and the prolate hypocycloid for internal splines. These fillets have a minimum radius of curvature at the point where the fillet is tangent to the external spline minor diameter circle or the internal spline major diameter circle and a rapidly increasing radius of curvature up to the point where the fillet comes tangent to the involute profile.

*Chamfers and Corner Clearance:* In major diameter fits, it is always necessary to provide corner clearance at the major diameter of the spline coupling. This clearance is usually effected by providing a chamfer on the top corners of the external member. This method may not be possible or feasible because of the following:

a) If the external member is roll formed by plastic deformation, a chamfer cannot be provided by the process.

b) A semitopping cutter may not be available.

c) When cutting external splines with small numbers of teeth, a semitopping cutter may reduce the width of the top land to a prohibitive point.

In such conditions, the corner clearance can be provided on the internal spline, as shown in Fig. 2.

When this option is used, the form diameter may fall in the protuberance area.

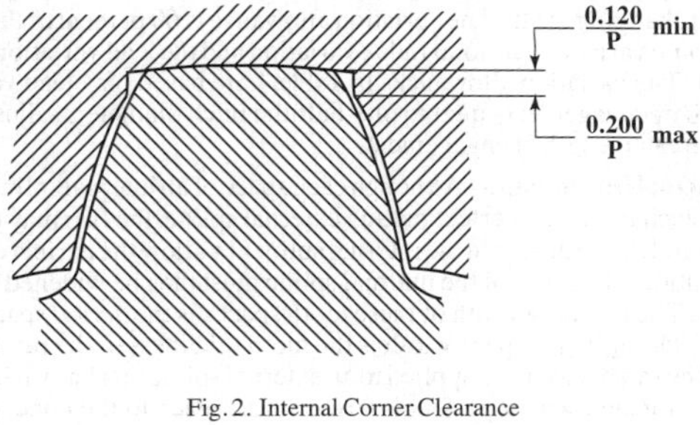


Fig. 2. Internal Corner Clearance

**Spline Variations.—**The maximum allowable variations for involute splines are listed in Table 4.

*Profile Variation:* The reference profile, from which variations occur, passes through the point used to determine the actual space width or tooth thickness. This is either the pitch point or the contact point of the standard measuring pins.

Profile variation is positive in the direction of the space and negative in the direction of the tooth. Profile variations may occur at any point on the profile for establishing effective fits and are shown in Table 4.

*Lead Variations:* The lead tolerance for the total spline length applies also to any portion thereof unless otherwise specified.

*Out of Roundness:* This condition may appear merely as a result of index and profile variations given in Table 4 and requires no further allowance. However, heat treatment and deflection of thin sections may cause out of roundness, which increases index and profile variations. Tolerances for such conditions depend on many variables and are therefore not tabulated. Additional tooth and/or space width tolerance must allow for such conditions.

*Eccentricity:* Eccentricity of major and minor diameters in relation to the effective diameter of side fit splines should not cause contact beyond the form diameters of the mating splines, even under conditions of maximum effective clearance. This Standard does not establish specific tolerances.

Eccentricity of major diameters in relation to the effective diameters of major diameter fit splines should be absorbed within the maximum material limits established by the tolerances on major diameter and effective space width or effective tooth thickness.

If the alignment of mating splines is affected by eccentricity of locating surfaces relative to each other and/or the splines, it may be necessary to decrease the effective and actual tooth thickness of the external splines in order to maintain the desired fit condition. This Standard does not include allowances for eccentric location.

**Effect of Spline Variations.—**Spline variations can be classified as index variations, profile variations, or lead variations.

*Index Variations:* These variations cause the clearance to vary from one set of mating tooth sides to another. Because the fit depends on the areas with minimum clearance, index variations reduce the effective clearance.

*Profile Variations:* Positive profile variations affect the fit by reducing effective clearance. Negative profile variations do not affect the fit but reduce the contact area.

*Lead Variations:* These variations will cause clearance variations and therefore reduce the effective clearance.

*Variation Allowance:* The effect of individual spline variations on the fit (effective variation) is less than their total, because areas of more than minimum clearance can be altered without changing the fit. The variation allowance is 60 percent of the sum of twice the positive profile variation, the total index variation and the lead variation for the length of engagement. The variation allowances in Table 4 are based on a lead variation for an assumed length of engagement equal to one-half the pitch diameter. Adjustment may be required for a greater length of engagement.

**Effective and Actual Dimensions.—**Although each space of an internal spline may have the same width as each tooth of a perfect mating external spline, the two may not fit because of variations of index and profile in the internal spline. To allow the perfect external spline to fit in any position, all spaces of the internal spline must then be widened by the amount of interference. The resulting width of these tooth spaces is the *actual* space width of the internal spline. The *effective* space width is the tooth thickness of the perfect mating external spline. The same reasoning applied to an external spline that has variations of index and profile when mated with a perfect internal spline leads to the concept of effective tooth thickness, which exceeds the actual tooth thickness by the amount of the effective variation.

The effective space width of the internal spline minus the effective tooth thickness of the external spline is the effective clearance and defines the fit of the mating parts. (This statement is strictly true only if high points of mating parts come into contact.) Positive effective clearance represents looseness or backlash. Negative effective clearance represents tightness or interference.

**Space Width and Tooth Thickness Limits.**—The variation of actual space width and actual tooth thickness within the machining tolerance causes corresponding variations of effective dimensions, so that there are four limit dimensions for each component part.

These variations are shown diagrammatically in Table 5.

**Table 5. Specification Guide for Space Width and Tooth Thickness**
*ANSI B92.1-1996*

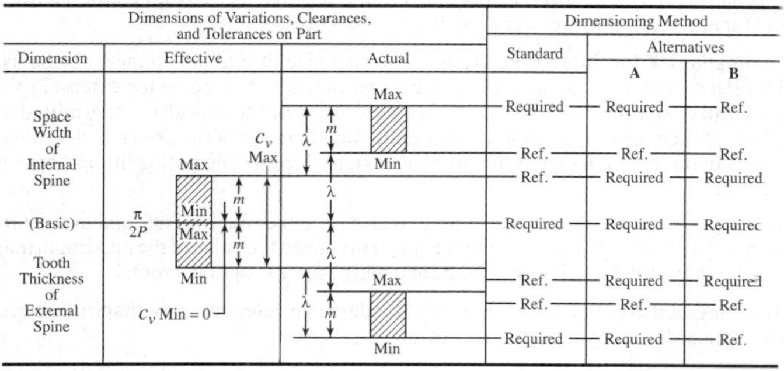


The minimum effective space width is always basic. The maximum effective tooth thickness is the same as the minimum effective space width except for the major diameter fit. The major diameter fit maximum effective tooth thickness is less than the minimum effective space width by an amount that allows for eccentricity between the effective spline and the major diameter. The permissible variation of the effective clearance is divided between the internal and external splines to arrive at the maximum effective space width and the minimum effective tooth thickness. Limits for the actual space width and actual tooth thickness are constructed from suitable variation allowances.

**Use of Effective and Actual Dimensions.**—Each of the four dimensions for space width and tooth thickness shown in Table 5 has a definite function.

*Minimum Effective Space Width and Maximum Effective Tooth Thickness:* These dimensions control the minimum effective clearance, and must always be specified.

*Minimum Actual Space Width and Maximum Actual Tooth Thickness:* These dimensions cannot be used for acceptance or rejection of parts. If the actual space width is less than the minimum without causing the effective space width to be undersized, or if the actual tooth thickness is more than the maximum without causing the effective tooth thickness to be oversized, the effective variation is less than anticipated; such parts are desirable and not defective. The specification of these dimensions as processing reference dimensions is optional. They are also used to analyze undersize effective space width or oversize effective tooth thickness conditions to determine whether or not these conditions are caused by excessive effective variation.

*Maximum Actual Space Width and Minimum Actual Tooth Thickness:* These dimensions control machining tolerance and limit the effective variation. The spread between these dimensions, reduced by the effective variation of the internal and external spline, is the maximum effective clearance. Where the effective variation obtained in machining is appreciably less than the variation allowance, these dimensions must be adjusted in order to maintain the desired fit.

*Maximum Effective Space Width and Minimum Effective Tooth Thickness:* These dimensions define the maximum effective clearance but they do not limit the effective variation. They may be used, in addition to the maximum actual space width and minimum actual tooth thickness, to prevent the increase of maximum effective clearance due to reduction of effective variations. The notation "inspection optional" may be added where maximum effective clearance is an assembly requirement, but does not need absolute control. It will indicate, without necessarily adding inspection time and equipment, that the actual space width of the internal spline must be held below the maximum, or the actual tooth thickness of the external spline above the minimum, if machining methods result in less than the allowable variations. Where effective variation needs no control or is controlled by laboratory inspection, these limits may be substituted for maximum actual space width and minimum actual tooth thickness.

**Combinations of Involute Spline Types.—**Flat root side fit internal splines may be used with fillet root external splines where the larger radius is desired on the external spline for control of stress concentrations. This combination of fits may also be permitted as a design option by specifying, for the minimum root diameter of the external, the value of the minimum root diameter of the fillet root external spline and noting this as "optional root."

A design option may also be permitted to provide either flat root internal or fillet root internal by specifying, for the maximum major diameter, the value of the maximum major diameter of the fillet root internal spline and noting this as "optional root."

**Interchangeability.—**Splines made to this Standard may interchange with splines made to older standards. Exceptions are listed below.

*External Splines:* These external splines will mate with older internal splines as follows:

| Year | Major Dia. Fit | Flat Root Side Fit | Fillet Root Side Fit |
|---|---|---|---|
| 1946 | Yes | No (A)[a] | No (A) |
| 1950[b] | Yes (B) | Yes (B) | Yes (C) |
| 1950[c] | Yes (B) | No (A) | Yes (C) |
| 1957 SAE | Yes | No (A) | Yes (C) |
| 1960 | Yes | No (A) | Yes (C) |

[a] For exceptions A, B, C, see the paragraph on *Exceptions* that follows.

[b] Full dedendum.

[c] Short dedendum.

*Internal Splines:* These will mate with older external splines as follows:

| Year | Major Dia. Fit | Flat Root Side Fit | Fillet Root Side Fit |
|---|---|---|---|
| 1946 | No (D)[a] | No (E) | No (D) |
| 1950 | Yes (F) | Yes | Yes (C) |
| 1957 SAE | Yes (G) | Yes | Yes |
| 1960 | Yes (G) | Yes | Yes |

[a] For exceptions C, D, E, F, G, see the paragraph on *Exceptions* that follows.

**Table 6. Spline Terms, Symbols, and Drawing Data, 30-Degree Pressure Angle, Flat Root Side Fit** *ANSI B92.1-1996*

**30-Deg Pressure Angle**

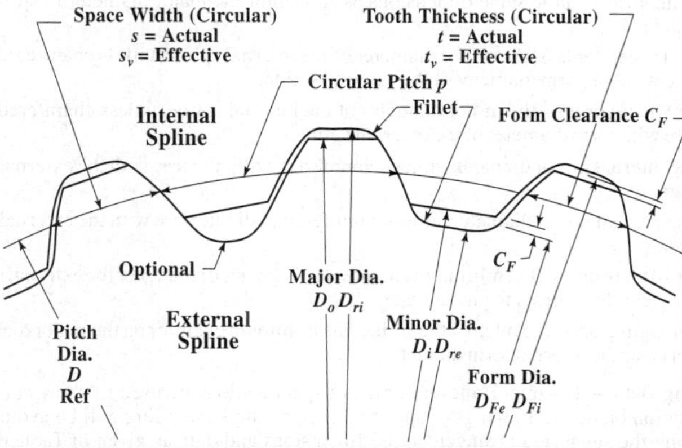


The fit shown is used in restricted areas (as with tubular parts with wall thickness too small to permit use of fillet roots, and to allow hobbing closer to shoulders, etc.) and for economy (when hobbing, shaping, etc., and using shorter broaches for the internal member).

Press fits are not tabulated because their design depends on the degree of tightness desired and must allow for such factors as the shape of the blank, wall thickness, material, hardness, thermal expansion, etc. Close tolerances or selective size grouping may be required to limit fit variations.

**Drawing Data**

| Internal Involute Spline Data | | External Involute Spline Data | |
|---|---|---|---|
| Flat Root Side Fit | | Flat Root Side Fit | |
| Number of Teeth | xx | Number of Teeth | xx |
| Pitch | xx/xx | Pitch | xx/xx |
| Pressure Angle | 30° | Pressure Angle | 30° |
| Base Diameter | x.xxxxxx Ref | Base Diameter | x.xxxxxx Ref |
| Pitch Diameter | x.xxxxxx Ref | Pitch Diameter | x.xxxxxx Ref |
| Major Diameter | x.xxx max | Major Diameter | x.xxx/x.xxx |
| Form Diameter | x.xxx | Form Diameter | x.xxx |
| Minor Diameter | x.xxx/x.xxx | Minor Diameter | x.xxx min |
| Circular Space Width | | Circular Tooth Thickness | |
| Max Actual | x.xxxx | Max Effective | x.xxxx |
| Min Effective | x.xxxx | Min Actual | x.xxxx |
| The following information may be added as required: | | The following information may be added as required: | |
| Max Measurement Between Pins | x.xxx Ref | Min Measurement Over Pins | x.xxxx Ref |
| Pin Diameter | x.xxxx | Pin Diameter | x.xxxx |

The above drawing data are required for the spline specifications. The standard system is shown; for alternate systems, see Table 5. Number of x's indicates number of decimal places normally used.

*Exceptions:*

a) The external major diameter, unless chamfered or reduced, may interfere with the internal form diameter on flat root side fit splines. Internal splines made to the 1957 and 1960 standards had the same dimensions as shown for the major diameter fit splines in this Standard.

b) For 15 teeth or less, the minor diameter of the internal spline, unless chamfered, will interfere with the form diameter of the external spline.

c) For 9 teeth or less, the minor diameter of the internal spline, unless chamfered, will interfere with form diameter of the external spline.

d) The internal minor diameter, unless chamfered, will interfere with the external form diameter.

e) The internal minor diameter, unless chamfered, will interfere with the external form diameter.

f) For 10 teeth or less, the minimum chamfer on the major diameter of the external spline may not clear the internal form diameter.

g) Depending upon the pitch of the spline, the minimum chamfer on the major diameter may not clear the internal form diameter.

**Drawing Data.—**It is important that uniform specifications be used to show complete information on detail drawings of splines. Much misunderstanding will be avoided by following the suggested arrangement of dimensions and data as given in Table 6. The number of x's indicates the number of decimal places normally used. With this tabulated type of spline specifications, it is usually not necessary to show a graphic illustration of the spline teeth.

**Spline Data and Reference Dimensions.—**Spline data are used for engineering and manufacturing purposes. Pitch and pressure angle are not subject to individual inspection.

As used in this standard, *reference* is an added notation or modifier to a dimension, specification, or note when that dimension, specification, or note is:

1) Repeated for drawing clarification.

2) Needed to define a nonfeature datum or basis from which a form or feature is generated.

3) Needed to define a nonfeature dimension from which other specifications or dimensions are developed.

4) Needed to define a nonfeature dimension at which toleranced sizes of a feature are specified.

5) Needed to define a nonfeature dimension from which control tolerances or sizes are developed or added as useful information.

Any dimension, specification, or note that is noted "REF" should not be used as a criterion for part acceptance or rejection.

**Estimating Key and Spline Sizes and Lengths.—**Fig. 3 may be used to estimate the size of American Standard involute splines required to transmit a given torque. It also may be used to find the outside diameter of shafts used with single keys. After the size of the shaft is found, the proportions of the key can be determined from Table 1 on page 2557.

Curve A is for flexible splines with teeth hardened to 55 to 65 RC (Rockwell C scale). For these splines, lengths are generally made equal to or somewhat greater than the pitch diameter for diameters below $1\frac{1}{4}$ inches; on larger diameters, the length is generally one-third to two-thirds the pitch diameter. Curve A also applies for a single key used as a fixed coupling, the length of the key being one to one and one-quarter times the shaft diameter. The stress in the shaft, neglecting stress concentration at the keyway, is about 7500 pounds per square inch. See also *Effect of Keyways on Shaft Strength* starting on page 295.

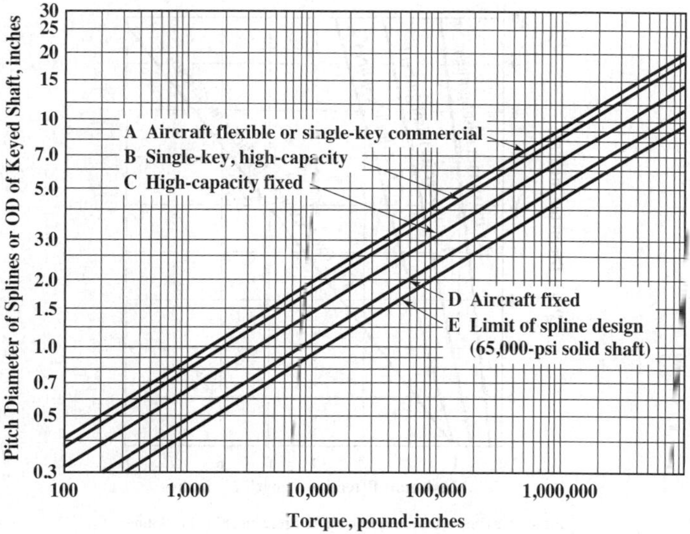


Fig. 3. Chart for Estimating Involute Spline Size Based on Diameter-Torque Relationships

Curve B represents high-capacity single keys used as fixed couplings for stresses of 9500 pounds per square inch, neglecting stress concentration. Key-length is one to one and one-quarter times shaft diameter and both shaft and key are of moderately hard heat-treated steel. This type of connection is commonly used to key commercial flexible couplings to motor or generator shafts.

Curve C is for multiple-key fixed splines with lengths of three-quarters to one and one-quarter times pitch diameter and shaft hardness of 200–300 BHN.

Curve D is for high-capacity splines with lengths one-half to one times the pitch diameter. Hardnesses up to 58 RC (Rockwell C scale) are common and in aircraft applications the shaft is generally hollow to reduce weight.

Curve E represents a solid shaft with 65,000 pounds per square inch shear stress. For hollow shafts with inside diameter equal to three-quarters of the outside diameter the shear stress would be 95,000 pounds per square inch.

*Length of Splines:* Fixed splines with lengths of one-third the pitch diameter will have the same shear strength as the shaft, assuming uniform loading of the teeth; however, errors in spacing of teeth result in only half the teeth being fully loaded. Therefore, for balanced strength of teeth and shaft the length should be two-thirds the pitch diameter. If weight is not important, however, this may be increased to equal the pitch diameter. In the case of flexible splines, long lengths do not contribute to load-carrying capacity when there is misalignment to be accommodated. Maximum effective length for flexible splines may be approximated from Fig. 4.

**Formulas for Torque Capacity of Involute Splines.—**The formulas for torque capacity of 30-degree involute splines given in the following paragraphs are derived largely from an article "When Splines Need Stress Control" by D. W. Dudley, *Product Engineering*, Dec. 23, 1957.

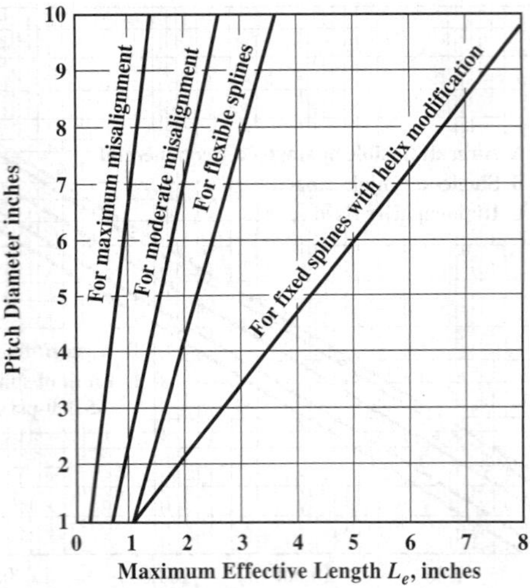


Fig. 4. Maximum Effective Length for Fixed and Flexible Splines

In the formulas that follow the symbols used are as defined on page 2337 with the following additions: $D_h$ = inside diameter of hollow shaft, inches; $K_a$ = application factor from Table 7; $K_m$ = load distribution factor from Table 8; $K_f$ = fatigue life factor from Table 9; $K_w$ = wear life factor from Table 10; $L_e$ = maximum effective length from Fig. 4, to be used in stress formulas even though the actual length may be greater; $T$ = transmitted torque, pound-inches. For fixed splines without helix modification, the effective length $L_e$ should never exceed $5000\ D^{3.5} \div T$.

**Table 7. Spline Application Factors, $K_a$**

| Power Source | Type of Load: Uniform (Generator, Fan) | Light Shock (Oscillating Pumps, etc.) | Intermittent Shock (Actuating Pumps, etc.) | Heavy Shock (Punches, Shears, etc.) |
|---|---|---|---|---|
| | Application Factor, $K_a$ | | | |
| Uniform (Turbine, Motor) | 1.0 | 1.2 | 1.5 | 1.8 |
| Light Shock (Hydraulic Motor) | 1.2 | 1.3 | 1.8 | 2.1 |
| Medium Shock (Internal Combustion Engine) | 2.0 | 2.2 | 2.4 | 2.8 |

**Table 8. Load Distribution Factors, $K_m$, for Misalignment of Flexible Splines**

| Misalignment, inches per inch | Load Distribution Factor, $K_m$[a]: ½-in. Face Width | 1-in. Face Width | 2-in. Face Width | 4-in. Face Width |
|---|---|---|---|---|
| 0.001 | 1 | 1 | 1 | 1½ |
| 0.002 | 1 | 1 | 1½ | 2 |
| 0.004 | 1 | 1½ | 2 | 2½ |
| 0.008 | 1½ | 2 | 2½ | 3 |

[a] For fixed splines, $K_m$ = 1.

**Table 9. Fatigue-Life Factors, $K_f$, for Splines**

| No. of Torque Cycles[a] | Fatigue-Life Factor, $K_f$ | |
|---|---|---|
| | Unidirectional | Fully-reversed |
| 1,000 | 1.8 | 1.8 |
| 10,000 | 1.0 | 1.0 |
| 100,000 | 0.5 | 0.4 |
| 1,000,000 | 0.4 | 0.3 |
| 10,000,000 | 0.3 | 0.2 |

[a] A torque cycle consists of one start and one stop, not the number of revolutions.

**Table 10. Wear Life Factors, $K_w$, for Flexible Splines**

| Number of Revolutions of Spline | Life Factor, $K_w$ | Number of Revolutions of Spline | Life Factor, $K_w$ |
|---|---|---|---|
| 10,000 | 4.0 | 100,000,000 | 1.0 |
| 100,000 | 2.8 | 1,000,000,000 | 0.7 |
| 1,000,000 | 2.0 | 10,000,000,000 | 0.5 |
| 10,000,000 | 1.4 | … | … |

Wear life factors, unlike fatigue life factors given in Table 9, are based on the total number of revolutions of the spline, since each revolution of a flexible spline results in a complete cycle of rocking motion which contributes to spline wear.

*Definitions:* A *fixed* spline is one which is either shrink fitted or loosely fitted but piloted with rings at each end to prevent rocking of the spline which results in small axial movements that cause wear. A *flexible* spline permits some rocking motion such as occurs when the shafts are not perfectly aligned. This flexing or rocking motion causes axial movement and consequently wear of the teeth. Straight-toothed flexible splines can accommodate only small angular misalignments (less than 1 deg.) before wear becomes a serious problem. For greater amounts of misalignment (up to about 5 deg.), crowned splines are preferable to reduce wear and end-loading of the teeth.

*Shear Stress Under Roots of External Teeth:* For a transmitted torque *T*, the torsional shear stress induced in the shaft under the root diameter of an external spline is:

$$S_s = \frac{16TK_a}{\pi D_{re}^3 K_f} \quad \text{for a solid shaft} \qquad (1)$$

$$S_s = \frac{16TD_{re}K_a}{\pi (D_{re}^4 - D_h^4)K_f} \quad \text{for a hollow shaft} \qquad (2)$$

The computed stress should not exceed the values in Table 11.

**Table 11. Allowable Shear, Compressive, and Tensile Stresses for Splines**

| Material | Hardness | | Max. Allowable Stress | | | |
|---|---|---|---|---|---|---|
| | | | Shear Stress, psi | Compressive Stress, psi | | Tensile Stress, psi |
| | Brinell | Rockwell C | | Straight | Crowned | |
| Steel | 160-200 | — | 20,000 | 1,500 | 6,000 | 22,000 |
| | 230-260 | — | 30,000 | 2,000 | 8,000 | 32,000 |
| | 302-351 | 33-38 | 40,000 | 3,000 | 12,000 | 45,000 |
| Surface-hardened Steel | — | 48-53 | 40,000 | 4,000 | 16,000 | 45,000 |
| Case-hardened Steel | — | 58-63 | 50,000 | 5,000 | 20,000 | 55,000 |
| Through-hardened Steel (Aircraft Quality) | — | 42-46 | 45,000 | — | — | 50,000 |

*Shear Stress at the Pitch Diameter of Teeth:* The shear stress at the pitch line of the teeth for a transmitted torque $T$ is:

$$S_s = \frac{4TK_a K_m}{DNL_e t K_f} \quad (3)$$

The factor of 4 in (3) assumes that only half the teeth will carry the load because of spacing errors. For poor manufacturing accuracies, change the factor to 6.

The computed stress should not exceed the values in Table 11.

*Compressive Stresses on Sides of Spline Teeth:* Allowable compressive stresses on splines are very much lower than for gear teeth since non-uniform load distribution and misalignment result in unequal load sharing and end loading of the teeth.

$$\text{For flexible splines, } S_c = \frac{2TK_m K_a}{DNL_e hK_w} \quad (4)$$

$$\text{For fixed splines, } S_c = \frac{2TK_m K_a}{9DNL_e hK_f} \quad (5)$$

In these formulas, $h$ is the depth of engagement of the teeth, which for flat root splines is $0.9/P$ and for fillet root splines is $1/P$, approximately.

The stresses computed from Formulas (4) and (5) should not exceed the values in Table 11.

*Bursting Stresses on Splines:* Internal splines may burst due to three kinds of tensile stress: 1) tensile stress due to the radial component of the transmitted load; 2) centrifugal tensile stress; and 3) tensile stress due to the tangential force at the pitch line causing bending of the teeth.

$$\text{Radial load tensile stress, } S_1 = \frac{T \tan\phi}{\pi D t_w L} \quad (6)$$

where $t_w$ = wall thickness of internal spline = outside diameter of spline sleeve minus spline major diameter, all divided by 2. $L$ = full length of spline.

$$\text{Centrifugal tensile stress, } S_2 = \frac{1.656 \times (\text{rpm})^2 (D_{oi}^2 + 0.212 D_{ri}^2)}{1{,}000{,}000} \quad (7)$$

where $D_{oi}$ = outside diameter of spline sleeve.

$$\text{Beam loading tensile stress, } S_3 = \frac{4T}{D^2 L_e Y} \quad (8)$$

In Equation (8), $Y$ is the Lewis form factor obtained from a tooth layout. For internal splines of 30-deg. pressure angle a value of $Y = 1.5$ is a satisfactory estimate. The factor 4 in (8) assumes that only half the teeth are carrying the load.

The total tensile stress tending to burst the rim of the external member is: $S_t = [K_a K_m (S_1 + S_3) + S_2]/K_f$; and should be less than those in Table 11.

**Crowned Splines for Large Misalignments.**—As mentioned on page 2349, crowned splines can accommodate misalignments of up to about 5 degrees. Crowned splines have considerably less capacity than straight splines of the same size if both are operating with precise alignment. However, when large misalignments exist, the crowned spline has greater capacity.

American Standard tooth forms may be used for crowned external members so that they may be mated with straight internal members of Standard form.

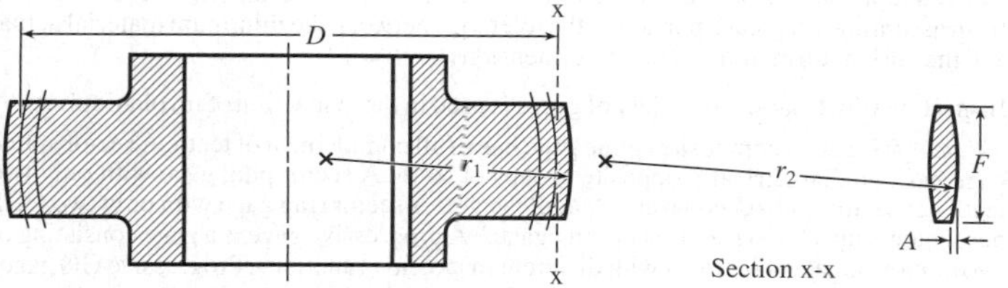


The accompanying diagram of a crowned spline shows the radius of the crown $r_1$; the radius of curvature of the crowned tooth, $r_2$; the pitch diameter of the spline, $D$; the face width, $F$; and the relief or crown height $A$ at the ends of the teeth. The crown height $A$ should always be made somewhat greater than one-half the face width multiplied by the tangent of the misalignment angle. For a crown height $A$, the approximate radius of curvature $r_2$ is $F^2 \div 8A$, and $r_1 = r_2 \tan \phi$, where $\phi$ is the pressure angle of the spline.

For a torque $T$, the compressive stress on the teeth is:

$$S_c = 2290\sqrt{2T \div DNhr_2}$$

and should be less than the value in Table 11.

**Fretting Damage to Splines and Other Machine Elements.—**Fretting is wear that occurs when cyclic loading, such as vibration, causes two surfaces in intimate contact to undergo small oscillatory motions with respect to each other. During fretting, high points or asperities of the mating surfaces adhere to each other and small particles are pulled out, leaving minute, shallow pits and a powdery debris. In steel parts exposed to air, the metallic debris oxidizes rapidly and forms a red, rustlike powder or sludge; hence, the coined designation "fretting corrosion."

Fretting is mechanical in origin and has been observed in most materials, including those that do not oxidize, such as gold, platinum, and nonmetallics; hence, the corrosion accompanying fretting of steel parts is a secondary factor.

Fretting can occur in the operation of machinery subject to motion or vibration or both. It can destroy close fits; the debris may clog moving parts; and fatigue failure may be accelerated because stress levels to initiate fatigue in fretted parts are much lower than for undamaged material. Sites for fretting damage include interference fits; splined, bolted, keyed, pinned, and riveted joints; between wires in wire rope; flexible shafts and tubes; between leaves in leaf springs; friction clamps; small amplitude-of-oscillation bearings; and electrical contacts.

Vibration or cyclic loadings are the main causes of fretting. If these factors cannot be eliminated, greater clamping force may reduce movement but, if not effective, may actually worsen the damage. Lubrication may delay the onset of damage; hard plating or surface hardening methods may be effective, not by reducing fretting, but by increasing the fatigue strength of the material. Plating soft materials having inherent lubricity onto contacting surfaces is effective until the plating wears through.

**Involute Spline Inspection Methods.—**Spline gages are used for routine inspection of production parts.

Analytical inspection, which is the measurement of individual dimensions and variations, may be required:

a) To supplement inspection by gages, for example, where NOT GO composite gages are used in place of NOT GO sector gages and variations must be controlled.

b) To evaluate parts rejected by gages.

c) For prototype parts or short runs where spline gages are not used.

d) To supplement inspection by gages where each individual variation must be restrained from assuming too great a portion of the tolerance between the minimum material actual and the maximum material effective dimensions.

**Inspection with Gages.—**A variety of gages is used in the inspection of involute splines.

*Types of Gages:* A composite spline gage has a full complement of teeth. A sector spline gage has two diametrically opposite groups of teeth. A sector plug gage with only two teeth per sector is also known as a "paddle gage." A sector ring gage with only two teeth per sector is also known as a "snap ring gage." A progressive gage is a gage consisting of two or more adjacent sections with different inspection functions. Progressive GO gages are physical combinations of GO gage members that check consecutively first one feature or one group of features, then their relationship to other features. GO and NOT GO gages may also be combined physically to form a progressive gage.

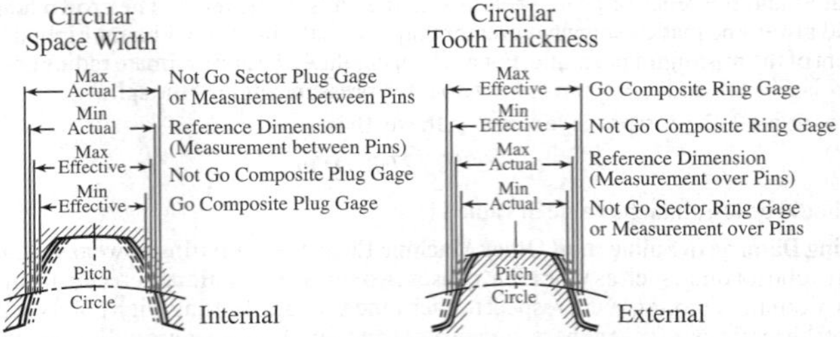


Fig. 5. Space Width and Tooth-Thickness Inspection

*GO and NOT GO Gages:* GO gages are used to inspect maximum material conditions (maximum external, minimum internal dimensions). They may be used to inspect an individual dimension or the relationship between two or more functional dimensions. They control the minimum looseness or maximum interference.

NOT GO gages are used to inspect minimum material conditions (minimum external, maximum internal dimensions), thereby controlling the maximum looseness or minimum interference. Unless otherwise agreed upon, a product is acceptable only if the NOT GO gage does not enter or go on the part. A NOT GO gage can be used to inspect only one dimension. An attempt at simultaneous NOT GO inspection of more than one dimension could result in failure of such a gage to enter or go on (acceptance of part), even though all but one of the dimensions were outside product limits. In the event all dimensions are outside the limits, their relationship could be such as to allow acceptance.

*Effective and Actual Dimensions:* The effective space width and tooth thickness are inspected by means of an accurate mating member in the form of a composite spline gage.

The actual space width and tooth thickness are inspected with sector plug and ring gages, or by measurements with pins.

**Measurements with Pins.—**The actual space width of internal splines, and the actual tooth thickness of external splines, may be measured with pins. These measurements do not determine the fit between mating parts, but may be used as part of the analytic inspection of splines to evaluate the effective space width or effective tooth thickness by approximation.

*Formulas for 2-Pin Measurement Between Pins:* For measurement *between* pins of internal splines using the symbols given on page 2337:

1) Find involute of pressure angle at pin center:

$$\text{inv}\,\phi_i = \frac{s}{D} + \text{inv}\phi_d - \frac{d_i}{D_b}$$

2) Find the value of $\phi_i$ in degrees, in the involute function tables beginning on page 107. Find sec $\phi_i$ = 1/cosine $\phi_i$ in the trig tables, pages 103 through 105, using interpolation to obtain higher accuracy.

3) Compute measurement, $M_i$, between pins:

For even numbers of teeth: $M_i = D_b \sec \phi_i - d_i$

For odd numbers of teeth: $M_i = (D_b \cos 90°/N) \sec \phi_i - d_i$

where: $d_i = 1.7280/P$ for 30° and 37.5° standard pressure angle ($\phi_D$) splines

$d_i = 1.9200/P$ for 45° pressure angle splines

*Example:* Find the measurement between pins for *maximum* actual space width of an internal spline of 30° pressure angle, tolerance class 4, 3/6 diametral pitch, and 20 teeth.

*Solution:* The maximum actual space width to be substituted for $s$ in Step 1 above is obtained as follows: In Table 5, page 2343, the maximum actual space width is the sum of the minimum effective space width (second column) and $\lambda + m$ (third column). The minimum effective space width $s_v$ from Table 2, page 2338, is $\pi/2P = \pi/(2 \times 3)$. The values of $\lambda$ and $m$ from Table 4, page 2340, are, for a class 4 fit, 3/6 diametral pitch, 20-tooth spline: $\lambda = 0.0027 \times 0.71 = 0.00192$; and $m = 0.00176 \times 0.71 = 0.00125$, so that $s = 0.52360 + 0.00192 + 0.00125 = 0.52677$.

Other values required for Step 1 are:

$D = N \div P = 20 \div 3 = 6.66666$

$\text{inv}\,\phi_D = \text{inv}\,30° = 0.053751$ from a calculator

$d_i = 1.7280/3 = 0.57600$

$D_b = D \cos \phi_D = 6.66666 \times 0.86603 = 5.77353$

The computation is made as follows:

1) $\text{inv}\,\phi_i = 0.52677/6.66666 + 0.053751 - 0.57600/5.77353 = 0.03300$

2) From a calculator, $\phi_i = 25°46.18'$ and $\sec \phi_i = 1.11044$

3) $M_i = 5.77353 \times 1.11044 - 0.57600 = 5.8352$ inches

*Formulas for 2-Pin Measurement Over Pins:* For measurement *over* pins of external splines:

1) Find involute of pressure angle at pin center:

$$\text{inv}\,\phi_e = \frac{t}{D} + \text{inv}\,\phi_D + \frac{d_e}{D_b} - \frac{\pi}{N}$$

2) Find the value of $\phi_e$ and sec $\phi_e$ from the involute function tables beginning on page 107.

3) Compute measurement, $M_e$, over pins:

For even numbers of teeth: $M_e = D_b \sec \phi_e + d_e$

For odd numbers of teeth: $M_e = (D_b \cos 90°/N) \sec \phi_e + d_e$

where: $d_e = 1.9200/P$ for all external splines

**American National Standard Metric Module Splines.**—ANSI B92.2M-1980 (R1989) is the American National Standards Institute version of the International Organization for Standardization's involute spline standard. It is not a "soft metric" conversion of any previous, inch-based, standard,* and splines made to this hard metric version are not intended for use with components made to the B92.1 or other, previous standards. The ISO 4156

* A "soft" conversion is one in which dimensions in inches, when multiplied by 25.4 will, after being appropriately rounded off, provide equivalent dimensions in millimeters. In a "hard" system the tools of production, such as hobs, do not bear a usable relation to the tools in another system; i.e., a 10 diametral pitch hob calculates to be equal to a 2.54 module hob in the metric module system, a hob that does not exist in the metric standard.

Standard from which this one is derived is the result of a cooperative effort between the ANSI B92 committee and other members of the ISO/TC 14-2 involute spline committee.

Many of the features of the Standard ANSI B92.1-1996 have been retained such as: 30-, 37.5-, and 45-degree pressure angles; flat root and fillet root side fits; the four tolerance classes 4, 5, 6, and 7; tables for a single class of fit; and the effective fit concept.

Among the major differences are: use of modules of from 0.25 through 10 mm in place of diametral pitch; dimensions in millimeters instead of inches; the "basic rack"; removal of the major diameter fit; and use of ISO symbols in place of those used previously. Also, provision is made for calculating three defined clearance fits.

The Standard recognizes that proper assembly between mating splines is dependent only on the spline being within effective specifications from the tip of the tooth to the form diameter. Therefore, the internal spline major diameter is shown as a maximum dimension and the external spline minor diameter is shown as a minimum dimension. The minimum internal major diameter and the maximum external minor diameter must clear the specified form diameter and thus require no additional control. All dimensions are for the finished part; any compensation that must be made for operations that take place during processing, such as heat treatment, must be considered when selecting the tolerance level for manufacturing.

The Standard provides the same internal minimum effective space width and external maximum effective tooth thickness for all tolerance classes. This basic concept makes possible interchangeable assembly between mating splines regardless of the tolerance class of the individual members, and permits a tolerance class "mix" of mating members. This arrangement is often an advantage when one member is considerably less difficult to produce than its mate, and the "average" tolerance applied to the two units is such that it satisfies the design need. For example, by specifying Class 5 tolerance for one member and Class 7 for its mate, an assembly tolerance in the Class 6 range is provided.

If a fit given in this Standard does not satisfy a particular design need, and a specific clearance or press fit is desired, the change shall be made only to the external spline by a reduction of, or an increase in, the effective tooth thickness and a like change in the actual tooth thickness. The minimum effective space width is always basic and this basic width should always be retained when special designs are derived from the concept of this Standard.

*Spline Terms and Definitions:* The spline terms and definitions given for American National Standard ANSI B92.1-1996 described in the preceding section may be used in regard to ANSI B92.2M-1980 (R1989). The 1980 Standard utilizes ISO symbols in place of those used in the 1970 Standard; these differences are shown in Table 12.

*Dimensions and Tolerances:* Dimensions and tolerances of splines made to the 1980 Standard may be calculated using the formulas given in Table 13. These formulas are for metric module splines in the range of from 0.25 to 10 mm metric module of side-fit design and having pressure angles of 30-, 37.5-, and 45-degrees. The standard modules in the system are: 0.25; 0.5; 0.75; 1; 1.25; 1.5; 1.75; 2; 2.5; 3; 4; 5; 6; 8; and 10. The range of from 0.5 to 10 module applies to all splines except 45-degree fillet root splines; for these, the range of from 0.25 to 2.5 module applies.

*Fit Classes:* Four classes of side fit splines are provided: spline fit class H/h having a minimum effective clearance, $c_v = es = 0$; classes H/f, H/e, and H/d having tooth thickness modifications, *es*, of f, e, and d, respectively, to provide progressively greater effective clearance $c_v$. The tooth thickness modifications h, f, e, and d in Table 14 are fundamental deviations selected from ISO R286, "ISO System of Limits and Fits." They are applied to the external spline by shifting the tooth thickness total tolerance below the basic tooth thickness by the amount of the tooth thickness modification to provide a prescribed minimum effective clearance $c_v$.

**Table 12. Comparison of Symbols Used in *ANSI B92.2M-1980 (R1989)* and Those in *ANSI B92.1-1996***

| Symbol | | | Symbol | | |
|---|---|---|---|---|---|
| B92.2M | B92.1 | Meaning of Symbol | B92.2M | B92.1 | Meaning of Symbol |
| $c$ | … | theoretical clearance | $m$ | … | module |
| $c_v$ | $c_v$ | effective clearance | … | $P$ | diametral pitch |
| $c_F$ | $c_F$ | form clearance | … | $P_s$ | stub pitch = $2P$ |
| $D$ | $D$ | pitch diameter | $P_b$ | … | base pitch |
| $DB$ | $D_b$ | base diameter | $p$ | $p$ | circular pitch |
| $d_{ce}$ | $D_{ce}$ | pin contact diameter, external spline | $\pi$ | $\pi$ | 3.141592654 |
| $d_{ci}$ | $D_{ci}$ | pin contact diameter, internal spline | $rfe$ | $r_f$ | fillet rad., ext. spline |
| $DEE$ | $D_o$ | major dia., ext. spline | $rfi$ | $r_f$ | fillet rad., int. spline |
| $DEI$ | $D_{ri}$ | major dia., int. spline | $E_{bsc}$ | $s_v$ min | basic circular space width |
| $DFE$ | $D_{Fe}$ | form dia., ext. spline | $E_{max}$ | $s$ | max. actual circular space width |
| $DFI$ | $D_{Fi}$ | form dia., int. spline | $E_{min}$ | $s$ | min. actual circular space width |
| $DIE$ | $D_{re}$ | minor dia., ext. spline | $EV$ | $s_v$ | effective circular space width |
| $DII$ | $D_i$ | minor dia., int. spline | $S_{bsc}$ | $t_v$ max | basic circular tooth thickness |
| $DRE$ | $d_e$ | pin dia., ext. spline | $S_{max}$ | $t$ | max. actual circular tooth thickness |
| $DRI$ | $d_i$ | pin dia., int. spline | $S_{min}$ | $t$ | min. actual circular tooth thickness |
| $h_s$ | … | see Fig. 6a, Fig. 6b, Fig. 6c, and Fig. 6d | $SV$ | $t_v$ | effective circular tooth thickness |
| $\lambda$ | $\lambda$ | effective variation | $\alpha$ | $\phi$ | pressure angle |
| INV $\alpha$ | … | involute $\alpha$=tan $\alpha$ − arc $\alpha$ | $\alpha_D$ | $\phi_D$ | standard pressure angle |
| $KE$ | $K_e$ | change factor, ext. spline | $\alpha_{ci}$ | $\phi_{ci}$ | press. angle at pin contact diameter, internal spline |
| $KI$ | $K_i$ | change factor, int. spline | $\alpha_{ce}$ | $\phi_{ce}$ | press. angle at pin contact diameter, external spline |
| $g$ | $L$ | spline length | $\alpha_i$ | $\phi_i$ | press. angle at pin center, internal spline |
| $g_w$ | … | active spline length | $\alpha_e$ | $\phi_e$ | press. angle at pin center, external spline |
| $g\gamma$ | … | length of engagement | $\alpha_{Fe}$ | $\phi_F$ | press. angle at form diameter, external spline |
| $T$ | $m$ | machining tolerance | $\alpha_{Fi}$ | $\phi_F$ | press. angle at form diameter, internal spline |
| $MRE$ | $M_e$ | meas. over 2 pins, ext. spline | $es$ | … | ext. spline cir. tooth thickness modification for required fit class = $c_v$ min (Table 14) |
| $MRI$ | $M_i$ | meas. bet. 2 pins, int. spline | h, f, e, or d | … | tooth thick, size modifiers (called fundamental deviation in ISO R286), Table 14 |
| $Z$ | $N$ | number of teeth | H | … | space width size modifier (called fundamental deviation in ISO R286), Table 14 |

**Table 13. Formulas for Dimensions and Tolerances for All Fit Classes— Metric Module Involute Splines**

| Term | Symbol | Formula: 30-Degree Flat Root 0.5 to 10 module | 30-Degree Fillet Root 0.5 to 10 module | 37.5-Degree Fillet Root 0.5 to 10 module | 45-Degree Fillet Root 0.25 to 2.5 module |
|---|---|---|---|---|---|
| Pitch Diameter | $D$ | $mZ$ | | | |
| Base Diameter | $DB$ | $mZ\cos\alpha_D$ | | | |
| Circular Pitch | $p$ | $\pi m$ | | | |
| Base Pitch | $p_b$ | $\pi m\cos\alpha_D$ | | | |
| Tooth Thick Mod | $es$ | According to selected fit class, H/h, H/f, H/e, or H/d (see Table 14) | | | |
| Min. Maj. Dia. Int. | $DEI$ min | $m(Z+1.5)$ | $m(Z+1.8)$ | $m(Z+1.4)$ | $m(Z+1.2)$ |
| Max. Maj. Dia. Int. | $DEI$ max | $DEI$ min + $(T+\lambda)/\tan\alpha_D$ (see Footnote [a]) | | | |
| Form Dia., Int. | $DFI$ | $m(Z+1)+2c_F$ | $m(Z+1)+2c_F$ | $m(Z+0.9)+2c_F$ | $m(Z+0.8)+2c_F$ |
| Min. Minor Dia., Int. | $DII$ min | $DFE+2c_F$ (see Footnote [b]) | | | |
| Max. Minor Dia., Int. | $DII$ max | $DII$ min + $(0.2m^{0.667}-0.01m^{-0.5})$ (see Footnote [c]) | | | |
| Cir Space Width, | | | | | |
| Basic | $E_{bsc}$ | $0.5\pi m$ | | | |
| Min. Effective | $EV$ min | $0.5\pi m$ | | | |
| Max. Actual | $E$ max | $EV$ min + $(T+\lambda)$ for classes 4, 5, 6, and 7 (see Table 15 for $T+\lambda$) | | | |
| Min. Actual | $E$ min | $EV$ min + $\lambda$ (see text on page 2357 for $\lambda$) | | | |
| Max. Effective | $EV$ max | $E$ max − $\lambda$ (see text on page 2357 for $\lambda$) | | | |
| Max. Major Dia., Ext[d] | $DEE$ max | $m(Z+1)-es/\tan\alpha_D$ | $m(Z+1)-es/\tan\alpha_D$ | $m(Z+0.9)-es/\tan\alpha_D$ | $m(Z+0.8)-es/\tan\alpha_D$ |
| Min. Major Dia., Ext | $DEE$ min | $DEE$ max − $(0.2m^{0.667}-0.01m^{-0.5})$ (see Footnote [c]) | | | |
| Form Dia., External | $DFE$ | $2\times\sqrt{(0.5DB)^2+\left[0.5D\sin\alpha_D-\frac{h_s+(0.5es/\tan\alpha_D)}{\sin\alpha_D}\right]^2}$ | | | |
| Max. Minor Dia., Ext[d] | $DIE$ max | $m(Z-1.5)-es/\tan\alpha_D$ | $m(Z-1.8)-es/\tan\alpha_D$ | $m(Z-1.4)-es/\tan\alpha_D$ | $m(Z-1.2)-es/\tan\alpha_D$ |
| Min. Minor Dia., Ext | $DIE$ min | $DIE$ max − $(T+\lambda)/\tan\alpha_D$ (see Footnote [a]) | | | |
| Cir Tooth Thick, | | | | | |
| Basic | $S_{bsc}$ | $0.5\pi m$ | | | |
| Max. Effective | $SV$ max | $S_{bsc}-es$ | | | |
| Min. Actual | $S$ min | $SV$ max − $(T+\lambda)$ for classes 4, 5, 6, and 7 (see Table 15 for $T+\lambda$) | | | |
| Max. Actual | $S$ max | $SV$ max − $\lambda$ (see text on page 2357 for $\lambda$) | | | |
| Min. Effective | $SV$ min | $S$ min + $\lambda$ (see text on page 2357 for $\lambda$) | | | |
| Total Tolerance on Circular Space Width or Tooth Thickness | $(T+\lambda)$ | See formulas in Table 15 | | | |
| Machining Tolerance on Circular Space Width or Tooth Thickness | $T$ | $T=(T+\lambda)$ from Table 15; $\lambda$ from text on page 2357. | | | |
| Effective Variation Allowed on Circular Space Width or Tooth Thickness | $\lambda$ | See text on page 2357. | | | |
| Form Clearance | $c_F$ | $0.1m$ | | | |
| Rack Dimension | $h_s$ | $0.6m$ (see Fig. 6a) | $0.6m$ (see Fig. 6b) | $0.55m$ (see Fig. 6c) | $0.5m$ (see Fig. 6d) |

[a] Use $(T+\lambda)$ for class 7 from Table 15

[b] For all types of fit, always use the $DFE$ value corresponding to the H/h fit.

[c] Values of $(0.2m^{0.667}-0.01m^{-0.5})$ are as follows: for 10 module, 0.93; for 8 module, 0.80; for 6 module, 0.66; for 5 module, 0.58; for 4 module, 0.50; for 3 module, 0.41; for 2.5 module, 0.36; for 2 module, 0.31; for 1.75 module, 0.28; for 1.5 module, 0.25; for 1.25 module, 0.22; for 1 module, 0.19; for 0.75 module, 0.15; for 0.5 module, 0.11; and for 0.25 module, 0.06.

[d] See Table 17 for values of $es/\tan\alpha_D$.

**Table 14. Tooth Thickness Modification *es* for Selected Spline Fit Classes**

| Pitch Diameter in mm, *D* | External Splines[a] Selected Fit Class | | | | Pitch Diameter in mm, *D* | External Splines[a] Selected Fit Class | | | |
|---|---|---|---|---|---|---|---|---|---|
| | d | e | f | h | | d | e | f | h |
| | Tooth thickness modification (reduction) relative to basic tooth thickness at pitch diameter, *es*, in mm | | | | | Tooth thickness modification (reduction) relative to basic tooth thickness at pitch diameter, *es*, in mm | | | |
| ≤ 3 | 0.020 | 0.014 | 0.006 | 0 | > 120 to 180 | 0.145 | 0.085 | 0.043 | 0 |
| > 3 to 6 | 0.030 | 0.020 | 0.010 | 0 | > 180 to 250 | 0.170 | 0.100 | 0.050 | 0 |
| > 6 to 10 | **0.040** | 0.025 | 0.013 | 0 | > 250 to 315 | 0.190 | 0.110 | 0.056 | 0 |
| > 10 to 18 | 0.050 | 0.032 | 0.016 | 0 | > 315 to 400 | 0.210 | **0.125** | **0.062** | 0 |
| > 18 to 30 | 0.065 | 0.040 | 0.020 | 0 | > 400 to 500 | 0.230 | 0.135 | **0.068** | 0 |
| > 30 to 50 | 0.080 | 0.050 | 0.025 | 0 | > 500 to 630 | 0.260 | 0.145 | **0.076** | 0 |
| > 50 to 80 | 0.100 | 0.060 | 0.030 | 0 | > 630 to 800 | 0.290 | 0.160 | **0.080** | 0 |
| > 80 to 120 | 0.120 | **0.072** | 0.036 | 0 | > 800 to 1000 | 0.320 | **0.170** | **0.086** | 0 |

[a] Internal splines are fit class H and have space width modification from basic space width equal to zero; thus, an H/h fit class has effective clearance $c_v = 0$.

*Note:* The values listed in this table are taken from ISO R286 and have been computed on the basis of the geometrical mean of the size ranges shown. Values in **boldface** type do not comply with any documented rule for rounding but are those used by ISO R286; they are used in this table to comply with established international practice.

*Basic Rack Profiles:* The basic rack profile for the standard pressure angle splines are shown in Fig. 6a, Fig. 6b, Fig. 6c, and Fig. 6d. The dimensions shown are for maximum material condition and for fit class H/h.

**Spline Machining Tolerances and Variations.—**The total tolerance $(T + \lambda)$, Table 15, is the sum of Effective Variation, $\lambda$, and a Machining Tolerance, $T$.

**Table 15. Space Width and Tooth Thickness Total Tolerance, $(T + \lambda)$, in Millimeters**

| Spline Tolerance Class | Formula for Total Tolerance, $(T+\lambda)$ | Spline Tolerance Class | Formula for Total Tolerance, $(T+\lambda)$ |
|---|---|---|---|
| 4 | 10i* + 40i** | 6 | 25i* + 100i** |
| 5 | 16i* + 64i** | 7 | 40i* + 160i** |

In these formulas, i* and i** are tolerance units based upon pitch diameter and tooth thickness, respectively:

$$i^* = 0.001\left(0.45\sqrt[3]{D} + 0.001D\right) \text{ for } D \le 500 \text{ mm}$$

$$= 0.001(0.004D + 2.1) \text{ for } D > 500 \text{ mm}$$

$$i^{**} = 0.001\left(0.45\sqrt[3]{S_{bsc}} + 0.001S_{bsc}\right)$$

*Effective Variation:* The effective variation, $\lambda$, is the combined effect that total index variation, positive profile variation, and tooth alignment variation has on the effective fit of mating involute splines. The effect of the individual variations is less than the sum of the allowable variations because areas of more than minimum clearance can have profile, tooth alignment, or index variations without changing the fit. It is also unlikely that these variations would occur in their maximum amounts simultaneously on the same spline. For this reason, total index variation, total profile variation, and tooth alignment variation are used to calculate the combined effect by the following formula:

$$\lambda = 0.6\sqrt{(F_p)^2 + (f_f)^2 + (F_\beta)^2} \text{ millimeters}$$

The above variation is based upon a length of engagement equal to one-half the pitch diameter of the spline; adjustment of $\lambda$ may be required for a greater length of engagement. Formulas for values of $F_p$, $f_f$, and $F_\beta$ used in the above formula are given in Table 16.

*Machining Tolerance:* A value for machining tolerance may be obtained by subtracting the effective variation, $\lambda$, from the total tolerance $(T + \lambda)$. Design requirements or specific processes used in spline manufacture may require a different amount of machining tolerance in relation to the total tolerance.

**Table 16. Formulas for $F_p$, $f_f$, and $F_\beta$ used to calculate λ**

| Spline Tolerance Class | Total Index Variation, in mm, $F_p$ | Total Profile Variation, in mm, $f_f$ | Total Lead Variation, in mm, $F_\beta$ |
|---|---|---|---|
| 4 | $0.001\left(2.5\sqrt{mZ\frac{\pi}{2}}+6.3\right)$ | $0.001\,[1.6m(1+0.0125Z)+10]$ | $0.001\left(0.8\sqrt{g}+4\right)$ |
| 5 | $0.001\left(3.55\sqrt{mZ\frac{\pi}{2}}+9\right)$ | $0.001\,[2.5m(1+0.0125Z)+16]$ | $0.001\left(1.0\sqrt{g}+5\right)$ |
| 6 | $0.001\left(5\sqrt{mZ\frac{\pi}{2}}+12.5\right)$ | $0.001\,[4m(1+0.0125Z)+25]$ | $0.001\left(1.25\sqrt{g}+6.3\right)$ |
| 7 | $0.001\left(7.1\sqrt{mZ\frac{\pi}{2}}+18\right)$ | $0.001\,[6.3m(1+0.0125Z)+40]$ | $0.001\left(2\sqrt{g}+10\right)$ |

$g$ = length of spline in millimeters.

**Table 17. Reduction, $es/\tan\alpha_D$, of External Spline Major and Minor Diameters Required for Selected Fit Classes**

| Pitch Diameter $D$ in mm | Standard Pressure Angle, in Degrees | | | | | | | | | |
|---|---|---|---|---|---|---|---|---|---|---|
| | 30 | 37.5 | 45 | 30 | 37.5 | 45 | 30 | 37.5 | 45 | All |
| | Classes of Fit | | | | | | | | | |
| | d | | | e | | | f | | | h |
| | $es/\tan\alpha_D$ in millimeters | | | | | | | | | |
| ≤3 | 0.035 | 0.026 | 0.020 | 0.024 | 0.018 | 0.014 | 0.010 | 0.008 | 0.006 | 0 |
| >3 to 6 | 0.052 | 0.039 | 0.030 | 0.035 | 0.026 | 0.020 | 0.017 | 0.013 | 0.010 | 0 |
| > 6 to 10 | 0.069 | 0.052 | 0.040 | 0.043 | 0.033 | 0.025 | 0.023 | 0.017 | 0.013 | 0 |
| > 10 to 18 | 0.087 | 0.065 | 0.050 | 0.055 | 0.042 | 0.032 | 0.028 | 0.021 | 0.016 | 0 |
| > 18 to 30 | 0.113 | 0.085 | 0.065 | 0.069 | 0.052 | 0.040 | 0.035 | 0.026 | 0.020 | 0 |
| > 30 to 50 | 0.139 | 0.104 | 0.080 | 0.087 | 0.065 | 0.050 | 0.043 | 0.033 | 0.025 | 0 |
| > 50 to 80 | 0.173 | 0.130 | 0.100 | 0.104 | 0.078 | 0.060 | 0.052 | 0.039 | 0.030 | 0 |
| > 80 to 120 | 0.208 | 0.156 | 0.120 | 0.125 | 0.094 | 0.072 | 0.062 | 0.047 | 0.036 | 0 |
| > 120 to 180 | 0.251 | 0.189 | 0.145 | 0.147 | 0.111 | 0.085 | 0.074 | 0.056 | 0.043 | 0 |
| > 180 to 250 | 0.294 | 0.222 | 0.170 | 0.173 | 0.130 | 0.100 | 0.087 | 0.065 | 0.050 | 0 |
| > 250 to 315 | 0.329 | 0.248 | 0.190 | 0.191 | 0.143 | 0.110 | 0.097 | 0.073 | 0.056 | 0 |
| > 315 to 400 | 0.364 | 0.274 | 0.210 | 0.217 | 0.163 | 0.125 | 0.107 | 0.081 | 0.062 | 0 |
| > 400 to 500 | 0.398 | 0.300 | 0.230 | 0.234 | 0.176 | 0.135 | 0.118 | 0.089 | 0.068 | 0 |
| > 500 to 630 | 0.450 | 0.339 | 0.260 | 0.251 | 0.189 | 0.145 | 0.132 | 0.099 | 0.076 | 0 |
| > 630 to 800 | 0.502 | 0.378 | 0.290 | 0.277 | 0.209 | 0.160 | 0.139 | 0.104 | 0.080 | 0 |
| > 800 to 1000 | 0.554 | 0.417 | 0.320 | 0.294 | 0.222 | 0.170 | 0.149 | 0.112 | 0.086 | 0 |

These values are used with the applicable formulas in Table 13.

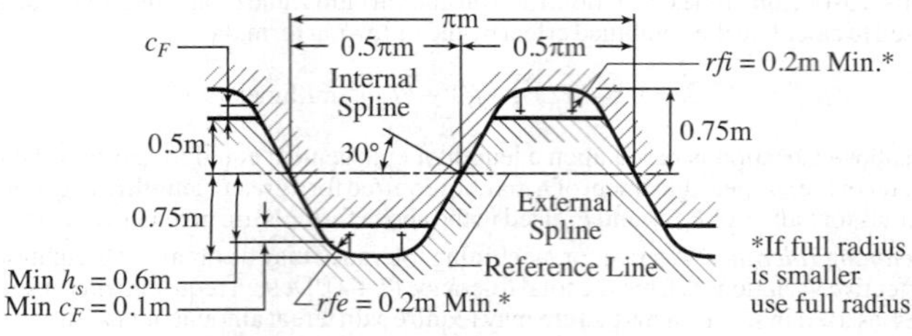


Fig. 6a. Profile of Basic Rack for 30° Flat Root Spline

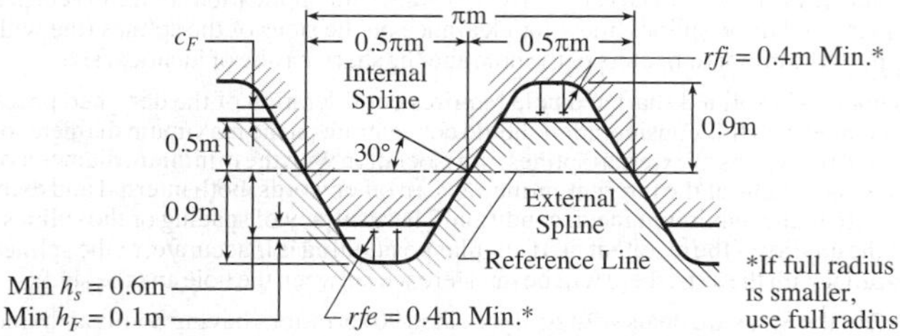


Fig. 6b. Profile of Basic Rack for 30° Fillet Root Spline

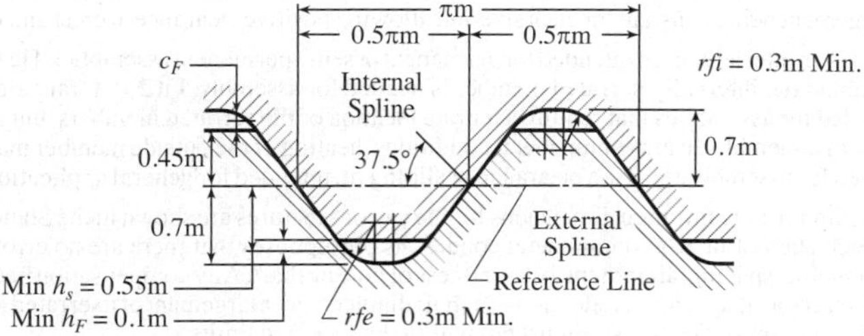


Fig. 6c. Profile of Basic Rack for 37.5° Fillet Root Spline

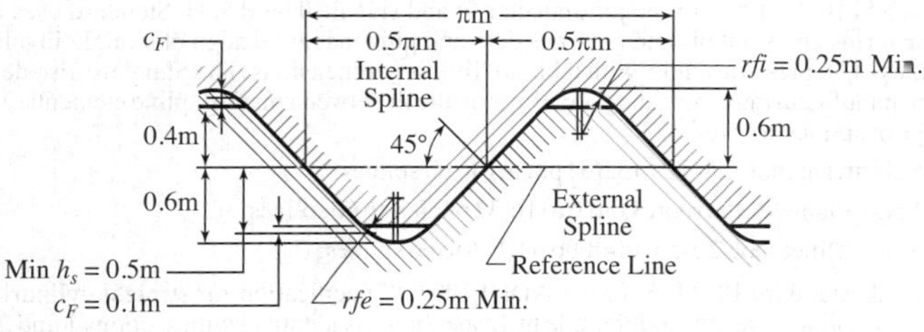


Fig. 6d. Profile of Basic Rack for 45° Fillet Root Spline

**British Standard Straight Splines.**—British Standard BS 2059:1953, "Straight-sided Splines and Serrations," was introduced because of the widespread development and use of splines, and because of the increasing use of involute splines it was necessary to provide a separate standard for straight-sided splines. BS 2059 was prepared on the hole basis, the hole being the constant member, and provided for different fits to be obtained by varying the size of the splined or serrated shaft. Part 1 of the Standard deals with 6 splines only, irrespective of the shaft diameter, with two depths termed shallow and deep. The splines are bottom fitting with top clearance.

The Standard contains three different grades of fit, based on the principle of variations in the diameter of the shaft at the root of the splines, in conjunction with variations in the widths of the splines themselves. Fit 1 represents the condition of closest fit and is designed for minimum backlash. Fit 2 has a positive allowance and is designed for ease of assembly, and Fit 3 has a larger positive allowance for applications that can accept such clearances. All these splines allow for clearance on the sides of the splines (the widths), but in Fit 1, the minor diameters of the hole and the shaft may be of identical size.

Assembly of a splined shaft and hole requires consideration of the designed profile of each member, and this consideration should concentrate on the maximum diameter of the shafts and the widths of external splines, in association with the minimum diameter of the hole and the widths of the internal splineways. In other words, both internal and external splines are in the maximum metal condition. The accuracy of spacing of the splines will affect the quality of the resultant fit. If angular positioning is inaccurate, or the splines are not parallel with the axis, there will be interference between the hole and the shaft.

Part 2 of the Standard deals with straight-sided 90 serrations having nominal diameters from 0.25 to 6.0 inches. Provision is again made for three grades of fits, the basic constant being the serrated hole size. Variations in the fits of these serrations is obtained by varying the sizes of the serrations on the shaft, and the fits are related to flank bearing, the depth of engagement being constant for each size and allowing positive clearance at crest and root.

Fit 1 is an interference fit intended for permanent or semi-permanent assemblies. Heating to expand the internally-serrated member is needed for assembly. Fit 2 is a transition fit intended for assemblies that require accurate location of the serrated members, but must allow disassembly. In maximum metal conditions, heating of the outside member may be needed for assembly. Fit. 3 is a clearance or sliding fit, intended for general applications.

Maximum and minimum dimensions for the various features are shown in the Standard for each class of fit. Maximum metal conditions presupposes that there are no errors of form such as spacing, alignment, or roundness of hole or shaft. Any compensation needed for such errors may require reduction of a shaft diameter or enlargement of a serrated bore, but the measured effective size must fall within the specified limits.

British Standard BS 3550:1963 (Withdrawn), "Involute Splines," is complementary to BS 2059, and the basic dimensions of all the sizes of splines are the same as those in the ANSI B92.1-1996, for major diameter fit and side fit. The British Standard uses the same terms and symbols and provides data and guidance for design of straight involute splines of 30 pressure angle, with tables of limiting dimensions. The Standard also deals with manufacturing errors and their effect on the fit between mating spline elements. The range of splines covered is:

Side fit, flat root, 2.5/5.0 to 32/64 pitch, 6 to 60 splines.

Major diameter, flat root, 3.0/6.0 to 16/32 pitch, 6 to 60 splines.

Side fit, fillet root, 2.5/5.0 to 48/96 pitch, 6 to 60 splines.

British Standard BS 6186: Part 1:ADD1:1993, "Specification for straight cylindrical involute splines: metric module, side fit-Inspection," is identical with sections 1 and 2 of ISO 4156 and with ANSI B92.2M-1980 (R1989) "Straight Cylindrical Involute Splines, Metric Module, Side Fit – Generalities, Dimensions and Inspection."

**SAE Standard Spline Fittings.—**The SAE spline fittings (Table 18 through Table 21 inclusive) have become an established standard for many applications in the agricultural, automotive, machine tool, and other industries. The dimensions given, in inches, apply only to soft broached holes. Dimensions are illustrated in Fig. 7a, Fig. 7b, and Fig. 7c. The tolerances given may be readily maintained by usual broaching methods. The tolerances selected for the large and small diameters may depend upon whether the fit between the mating part, as finally made, is on the large or the small diameter. The other diameter, which is designed for clearance, may have a larger manufactured tolerance. If the final fit between the parts is on the sides of the spline only, larger tolerances are permissible for both the large and small diameters. The spline should not be more than 0.006 inch per foot out of parallel with respect to the shaft axis. No allowance is made for corner radii to obtain clearance. Radii at the corners of the spline should not exceed 0.015 inch.

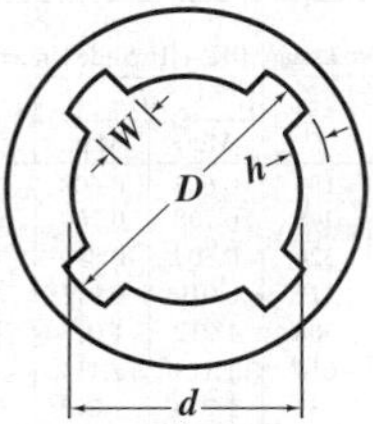

Fig. 7a. 4-Spline Fitting

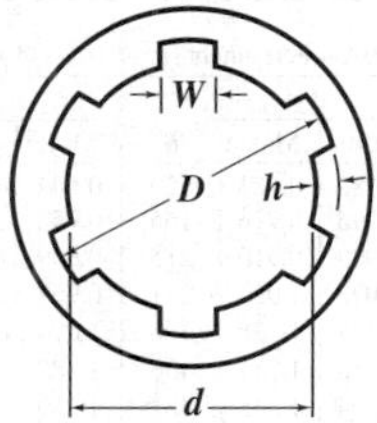

Fig. 7b. 6-Spline Fitting

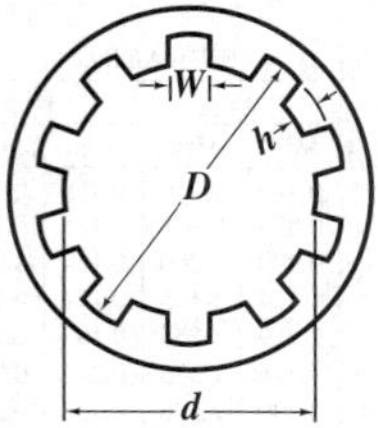

Fig. 7c. 10-Spline Fitting

## Table 18. SAE Standard 4-Spline Fittings

| Nom. Dia. | For All Fits | | | | 4A—Permanent Fit | | | | | 4B—To Slide, No Load | | | | |
|---|---|---|---|---|---|---|---|---|---|---|---|---|---|---|
| | *D* | | *W* | | *d* | | *h* | | | *d* | | *h* | | |
| | Min. | Max. | Min. | Max. | Min. | Max. | Min. | Max. | $T^a$ | Min. | Max. | Min. | Max. | $T^a$ |
| ¾ | 0.749 | 0.750 | 0.179 | 0.181 | 0.636 | 0.637 | 0.055 | 0.056 | 78 | 0.561 | 0.562 | 0.093 | 0.094 | 123 |
| ⅞ | 0.874 | 0.875 | 0.209 | 0.211 | 0.743 | 0.744 | 0.065 | 0.066 | 107 | 0.655 | 0.656 | 0.108 | 0.109 | 167 |
| 1 | 0.999 | 1.000 | 0.239 | 0.241 | 0.849 | 0.850 | 0.074 | 0.075 | 139 | 0.749 | 0.750 | 0.124 | 0.125 | 219 |
| 1⅛ | 1.124 | 1.125 | 0.269 | 0.271 | 0.955 | 0.956 | 0.083 | 0.084 | 175 | 0.843 | 0.844 | 0.140 | 0.141 | 277 |
| 1¼ | 1.249 | 1.250 | 0.299 | 0.301 | 1.061 | 1.062 | 0.093 | 0.094 | 217 | 0.936 | 0.937 | 0.155 | 0.156 | 341 |
| 1⅜ | 1.374 | 1.375 | 0.329 | 0.331 | 1.168 | 1.169 | 0.102 | 0.103 | 262 | 1.030 | 1.031 | 0.171 | 0.172 | 414 |
| 1½ | 1.499 | 1.500 | 0.359 | 0.361 | 1.274 | 1.275 | 0.111 | 0.112 | 311 | 1.124 | 1.125 | 0.186 | 0.187 | 491 |
| 1⅝ | 1.624 | 1.625 | 0.389 | 0.391 | 1.380 | 1.381 | 0.121 | 0.122 | 367 | 1.218 | 1.219 | 0.202 | 0.203 | 577 |
| 1¾ | 1.749 | 1.750 | 0.420 | 0.422 | 1.486 | 1.487 | 0.130 | 0.131 | 424 | 1.311 | 1.312 | 0.218 | 0.219 | 670 |
| 2 | 1.998 | 2.000 | 0.479 | 0.482 | 1.698 | 1.700 | 0.148 | 0.150 | 555 | 1.498 | 1.500 | 0.248 | 0.250 | 875 |
| 2¼ | 2.248 | 2.250 | 0.539 | 0.542 | 1.910 | 1.912 | 0.167 | 0.169 | 703 | 1.685 | 1.687 | 0.279 | 0.281 | 1106 |
| 2½ | 2.498 | 2.500 | 0.599 | 0.602 | 2.123 | 2.125 | 0.185 | 0.187 | 865 | 1.873 | 1.875 | 0.310 | 0.312 | 1365 |
| 3 | 2.998 | 3.000 | 0.720 | 0.723 | 2.548 | 2.550 | 0.223 | 0.225 | 1249 | 2.248 | 2.250 | 0.373 | 0.375 | 1969 |

[a] See note at end of Table 21.

## Table 19. SAE Standard 6-Spline Fittings

| Nom. Dia. | For All Fits | | | | 6A—Permanent Fit | | | 6B—To Slide, No Load | | | 6C—To Slide Under Load | | |
|---|---|---|---|---|---|---|---|---|---|---|---|---|---|
| | *D* | | *W* | | *d* | | | *d* | | | *d* | | |
| | Min. | Max. | Min. | Max. | Min. | Max. | $T^a$ | Min. | Max. | $T^a$ | Min. | Max. | $T^a$ |
| ¾ | 0.749 | 0.750 | 0.186 | 0.188 | 0.674 | 0.675 | 80 | 0.637 | 0.638 | 117 | 0.599 | 0.600 | 152 |
| ⅞ | 0.874 | 0.875 | 0.217 | 0.219 | 0.787 | 0.788 | 109 | 0.743 | 0.744 | 159 | 0.699 | 0.700 | 207 |
| 1 | 0.999 | 1.000 | 0.248 | 0.250 | 0.899 | 0.900 | 143 | 0.849 | 0.850 | 208 | 0.799 | 0.800 | 270 |
| 1⅛ | 1.124 | 1.125 | 0.279 | 0.281 | 1.012 | 1.013 | 180 | 0.955 | 0.956 | 263 | 0.899 | 0.900 | 342 |
| 1¼ | 1.249 | 1.250 | 0.311 | 0.313 | 1.124 | 1.125 | 223 | 1.062 | 1.063 | 325 | 0.999 | 1.000 | 421 |
| 1⅜ | 1.374 | 1.375 | 0.342 | 0.344 | 1.237 | 1.238 | 269 | 1.168 | 1.169 | 393 | 1.099 | 1.100 | 510 |
| 1½ | 1.499 | 1.500 | 0.373 | 0.375 | 1.349 | 1.350 | 321 | 1.274 | 1.275 | 468 | 1.199 | 1.200 | 608 |
| 1⅝ | 1.624 | 1.625 | 0.404 | 0.406 | 1.462 | 1.463 | 376 | 1.380 | 1.381 | 550 | 1.299 | 1.300 | 713 |
| 1¾ | 1.749 | 1.750 | 0.436 | 0.438 | 1.574 | 1.575 | 436 | 1.487 | 1.488 | 637 | 1.399 | 1.400 | 827 |
| 2 | 1.998 | 2.000 | 0.497 | 0.500 | 1.798 | 1.800 | 570 | 1.698 | 1.700 | 833 | 1.598 | 1.600 | 1080 |
| 2¼ | 2.248 | 2.250 | 0.560 | 0.563 | 2.023 | 2.025 | 721 | 1.911 | 1.913 | 1052 | 1.798 | 1.800 | 1367 |
| 2½ | 2.498 | 2.500 | 0.622 | 0.625 | 2.248 | 2.250 | 891 | 2.123 | 2.125 | 1300 | 1.998 | 2.000 | 1688 |
| 3 | 2.998 | 3.000 | 0.747 | 0.750 | 2.698 | 2.700 | 1283 | 2.548 | 2.550 | 1873 | 2.398 | 2.400 | 2430 |

[a] See note at end of Table 21.

## Table 20. SAE Standard 10-Spline Fittings

| Nom. Dia. | For All Fits | | | | 10A—Permanent Fit | | | 10B—To Slide, No Load | | | 10C—To Slide Under Load | | |
|---|---|---|---|---|---|---|---|---|---|---|---|---|---|
| | *D* | | *W* | | *d* | | | *d* | | | *d* | | |
| | Min. | Max. | Min. | Max. | Min. | Max. | *T*[a] | Min. | Max. | *T*[a] | Min. | Max. | *T*[a] |
| ¾ | 0.749 | 0.750 | 0.115 | 0.117 | 0.682 | 0.683 | 120 | 0.644 | 0.645 | 183 | 0.607 | 0.608 | 241 |
| ⅞ | 0.874 | 0.875 | 0.135 | 0.137 | 0.795 | 0.796 | 165 | 0.752 | 0.753 | 248 | 0.708 | 0.709 | 329 |
| 1 | 0.999 | 1.000 | 0.154 | 0.156 | 0.909 | 0.910 | 215 | 0.859 | 0.860 | 326 | 0.809 | 0.810 | 430 |
| 1⅛ | 1.124 | 1.125 | 0.174 | 0.176 | 1.023 | 1.024 | 271 | 0.967 | 0.968 | 412 | 0.910 | 0.911 | 545 |
| 1¼ | 1.249 | 1.250 | 0.193 | 0.195 | 1.137 | 1.138 | 336 | 1.074 | 1.075 | 508 | 1.012 | 1.013 | 672 |
| 1⅜ | 1.374 | 1.375 | 0.213 | 0.215 | 1.250 | 1.251 | 406 | 1.182 | 1.183 | 614 | 1.113 | 1.114 | 813 |
| 1½ | 1.499 | 1.500 | 0.232 | 0.234 | 1.364 | 1.365 | 483 | 1.289 | 1.290 | 732 | 1.214 | 1.215 | 967 |
| 1⅝ | 1.624 | 1.625 | 0.252 | 0.254 | 1.478 | 1.479 | 566 | 1.397 | 1.398 | 860 | 1.315 | 1.316 | 1135 |
| 1¾ | 1.749 | 1.750 | 0.271 | 0.273 | 1.592 | 1.593 | 658 | 1.504 | 1.505 | 997 | 1.417 | 1.418 | 1316 |
| 2 | 1.998 | 2.000 | 0.309 | 0.312 | 1.818 | 1.820 | 860 | 1.718 | 1.720 | 1302 | 1.618 | 1.620 | 1720 |
| 2¼ | 2.248 | 2.250 | 0.348 | 0.351 | 2.046 | 2.048 | 1088 | 1.933 | 1.935 | 1647 | 1.821 | 1.823 | 2176 |
| 2½ | 2.498 | 2.500 | 0.387 | 0.390 | 2.273 | 2.275 | 1343 | 2.148 | 2.150 | 2034 | 2.023 | 2.025 | 2688 |
| 3 | 2.998 | 3.000 | 0.465 | 0.468 | 2.728 | 2.730 | 1934 | 2.578 | 2.580 | 2929 | 2.428 | 2.430 | 3869 |
| 3½ | 3.497 | 3.500 | 0.543 | 0.546 | 3.182 | 3.185 | 2632 | 3.007 | 3.010 | 3987 | 2.832 | 2.835 | 5266 |
| 4 | 3.997 | 4.000 | 0.621 | 0.624 | 3.637 | 3.640 | 3438 | 3.437 | 3.440 | 5208 | 3.237 | 3.240 | 6878 |
| 4½ | 4.497 | 4.500 | 0.699 | 0.702 | 4.092 | 4.095 | 4351 | 3.867 | 3.870 | 6591 | 3.642 | 3.645 | 8705 |
| 5 | 4.997 | 5.000 | 0.777 | 0.780 | 4.547 | 4.550 | 5371 | 4.297 | 4.300 | 8137 | 4.047 | 4.050 | 10746 |
| 5½ | 5.497 | 5.500 | 0.855 | 0.858 | 5.002 | 5.005 | 6500 | 4.727 | 4.730 | 9846 | 4.452 | 4.455 | 13003 |
| 6 | 5.997 | 6.000 | 0.933 | 0.936 | 5.457 | 5.460 | 7735 | 5.157 | 5.160 | 11718 | 4.857 | 4.860 | 15475 |

[a] See note at end of Table 21.

## Table 21. SAE Standard 16-Spline Fittings

| Nom. Dia. | For All Fits | | | | 16A—Permanent Fit | | | 16B—To Slide, No Load | | | 16C—To Slide Under Load | | |
|---|---|---|---|---|---|---|---|---|---|---|---|---|---|
| | *D* | | *W* | | *d* | | | *d* | | | *d* | | |
| | Min. | Max. | Min. | Max. | Min. | Max. | *T*[a] | Min. | Max. | *T*[a] | Min. | Max. | *T*[a] |
| 2 | 1.997 | 2.000 | 0.193 | 0.196 | 1.817 | 1.820 | 1375 | 1.717 | 1.720 | 2083 | 1.617 | 1.620 | 2751 |
| 2½ | 2.497 | 2.500 | 0.242 | 0.245 | 2.273 | 2.275 | 2149 | 2.147 | 2.150 | 3255 | 2.022 | 2.025 | 4299 |
| 3 | 2.997 | 3.000 | 0.291 | 0.294 | 2.727 | 2.730 | 3094 | 2.577 | 2.580 | 4687 | 2.427 | 2.430 | 6190 |
| 3½ | 3.497 | 3.500 | 0.340 | 0.343 | 3.182 | 3.185 | 4212 | 3.007 | 3.010 | 6378 | 2.832 | 2.835 | 8426 |
| 4 | 3.997 | 4.000 | 0.389 | 0.392 | 3.637 | 3.640 | 5501 | 3.437 | 3.440 | 8333 | 3.237 | 3.240 | 11005 |
| 4½ | 4.497 | 4.500 | 0.438 | 0.441 | 4.092 | 4.095 | 6962 | 3.867 | 3.870 | 10546 | 3.642 | 3.645 | 13928 |
| 5 | 4.997 | 5.000 | 0.487 | 0.490 | 4.547 | 4.550 | 8595 | 4.297 | 4.300 | 13020 | 4.047 | 4.050 | 17195 |
| 5½ | 5.497 | 5.500 | 0.536 | 0.539 | 5.002 | 5.005 | 10395 | 4.727 | 4.730 | 15754 | 4.452 | 4.455 | 20806 |
| 6 | 5.997 | 6.000 | 0.585 | 0.588 | 5.457 | 5.460 | 12377 | 5.157 | 5.160 | 18749 | 4.857 | 4.860 | 24760 |

[a] *Torque Capacity of Spline Fittings:* The torque capacities of the different spline fittings are given in the columns headed "*T*." The torque capacity, per inch of bearing length at 1000 pounds pressure per square inch on the sides of the spline, may be determined by the following formula, in which $T$ = torque capacity in inch-pounds per inch of length, $N$ = number of splines, $R$ = mean radius or radial distance from center of hole to center of spline, $h$ = depth of spline: $T = 1000NRh$

## Table 22. Formulas for Determining Dimensions of SAE Standard Splines

| No. of Splines | *W* For All Fits | A Permanent Fit | | B To Slide Without Load | | C To Slide Under Load | |
|---|---|---|---|---|---|---|---|
| | | *h* | *d* | *h* | *d* | *h* | *d* |
| Four | 0.241*D*[a] | 0.075*D* | 0.850*D* | 0.125*D* | 0.750*D* | … | … |
| Six | 0.250*D* | 0.050*D* | 0.900*D* | 0.075*D* | 0.850*D* | 0.100*D* | 0.800*D* |
| Ten | 0.156*D* | 0.045*D* | 0.910*D* | 0.070*D* | 0.860*D* | 0.095*D* | 0.810*D* |
| Sixteen | 0.098*D* | 0.045*D* | 0.910*D* | 0.070*D* | 0.860*D* | 0.095*D* | 0.810*D* |

[a] Four splines for fits *A* and *B* only.

The formulas in the table above give the maximum dimensions for *W*, *h*, and *d*, as listed in Table 18 through Table 21 inclusive.

**Polygon-Type Shaft Connections.—**Involute-form and straight-sided splines are used for both fixed and sliding connections between machine members such as shafts and gears. Polygon-type connections, so called because they resemble regular polygons but with curved sides, may be used similarly. German DIN Standards 32711 and 32712 include data for three- and four-sided metric polygon connections. Data for 11 of the sizes shown in those Standards, but converted to inch dimensions by Stoffel Polygon Systems, are given in the accompanying table.

**Dimensions of Three- and Four-Sided Polygon-Type Shaft Connections**

**Drawing for 3-sided Designs**

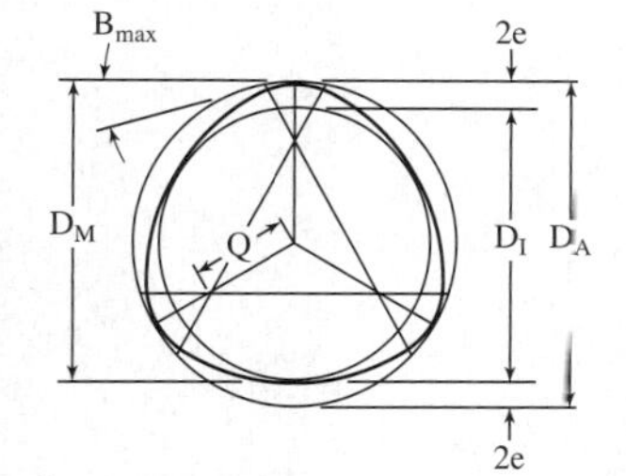


**Drawing for 4-sided Designs**

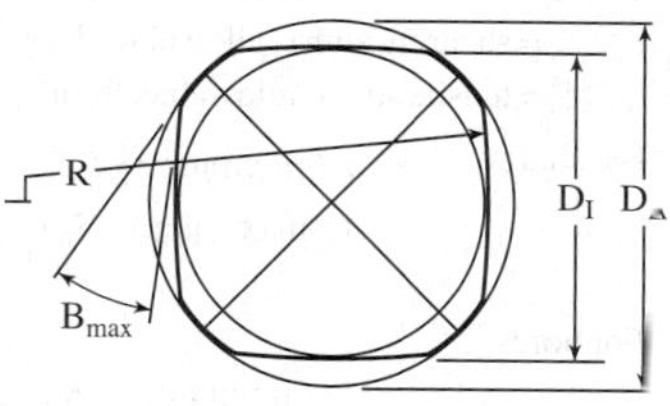


| Three-Sided Designs | | | | | Four-Sided Designs | | | | |
|---|---|---|---|---|---|---|---|---|---|
| Nominal Sizes | | | Design Data | | Nominal Sizes | | | Design Data | |
| $D_A$ (in.) | $D_I$ (in.) | $e$ (in.) | Area ($in^2$) | $Z_P$ ($in^3$) | $D_A$ (in.) | $D_I$ (in.) | $e$ (in.) | Area ($in^2$) | $Z_P$ ($in^3$) |
| 0.530 | 0.470 | 0.015 | 0.194 | 0.020 | 0.500 | 0.415 | 0.075 | 0.155 | 0.014 |
| 0.665 | 0.585 | 0.020 | 0.302 | 0.039 | 0.625 | 0.525 | 0.075 | 0.250 | 0.028 |
| 0.800 | 0.700 | 0.025 | 0.434 | 0.067 | 0.750 | 0.625 | 0.125 | 0.350 | 0.048 |
| 0.930 | 0.820 | 0.027 | 0.594 | 0.108 | 0.875 | 0.725 | 0.150 | 0.470 | 0.075 |
| 1.080 | 0.920 | 0.040 | 0.765 | 0.153 | 1.000 | 0.850 | 0.150 | 0.650 | 0.12 |
| 1.205 | 1.045 | 0.040 | 0.977 | 0.224 | 1.125 | 0.950 | 0.200 | 0.810 | 0.17 |
| 1.330 | 1.170 | 0.040 | 1.208 | 0.314 | 1.250 | 1.040 | 0.200 | 0.980 | 0.22 |
| 1.485 | 1.265 | 0.055 | 1.450 | 0.397 | 1.375 | 1.135 | 0.225 | 1.17 | 0.29 |
| 1.610 | 1.390 | 0.055 | 1.732 | 0.527 | 1.500 | 1.260 | 0.225 | 1.43 | 0.39 |
| 1.870 | 1.630 | 0.060 | 2.378 | 0.850 | 1.750 | 1.480 | 0.250 | 1.94 | 0.64 |
| 2.140 | 1.860 | 0.070 | 3.090 | 1.260 | 2.000 | 1.700 | 0.250 | 2.60 | 0.92 |

Dimensions $Q$ and $R$ shown on the diagrams are approximate and used only for drafting purposes: $Q \approx 7.5e$; $R \approx D_I/2 + 16e$.

Dimension $D_M = D_I + 2e$. Pressure angle $B_{max}$ is approximately $344e/D_M$ degrees for three sides, and $299e/D_M$ degrees for four sides.

Tolerances: ISO H7 tolerances apply to bore dimensions. For shafts, g6 tolerances apply for sliding fits; k7 tolerances for tight fits.

*Choosing Between Three- and Four-Sided Designs:* Three-sided designs are best for applications in which no relative movement between mating components is allowed while torque is transmitted. If a hub is to slide on a shaft while under torque, four-sided designs, which have larger pressure angles $B_{max}$ than those of three-sided designs, are better suited to sliding even though the axial force needed to move the sliding member is approximately 50 percent greater than for comparable involute spline connections.

*Strength of Polygon Connections:* In the formulas that follow,

$H_w$ = hub width, inches $H_t$ = hub wall thickness, inches

$M_b$ = bending moment, pound-inch

$M_t$ = torque, pound-inch

$Z$ = section modulus, bending, $in^3$

$= 0.098 D_M^4 / D_A$ for three sides $= 0.15 D_I^3$ for four sides

$Z_P$ = polar section modulus, torsion, $in^3$

$= 0.196 D_M^4 / D_A$ for three sides $= 0.196 D_I^3$ for four sides

$D_A$ and $D_M$. See table footnotes.

$S_b$ = bending stress, allowable, $lb/in^2$

$S_s$ = shearing stress, allowable, $lb/in^2$

$S_t$ = tensile stress, allowable, $lb/in^2$

For shafts, $M_t \text{ (maximum)} = S_s Z_p$

$M_b \text{ (maximum)} = S_b Z$

For bores,

$$H_t \text{ (minimum)} = K\sqrt{\frac{M_t}{S_t H_w}}$$

in which $K$ = 1.44 for three sides except that if $D_M$ is greater than 1.375 inches, then $K$ = 1.2; $K$ = 0.7 for four sides.

Failure may occur in the hub of a polygon connection if the hoop stresses in the hub exceed the allowable tensile stress for the material used. The radial force tending to expand the rim and cause tensile stresses is calculated from

$$\text{Radial Force, lb} = \frac{2M_t}{D_I n \tan(B_{max} + 11.3)}$$

This radial force acting at $n$ points may be used to calculate the tensile stress in the hub wall using formulas from strength of materials.

*Manufacturing:* Polygon shaft profiles may be produced using conventional machining processes such as hobbing, shaping, contour milling, copy turning, and numerically controlled milling and grinding. Bores are produced using broaches, spark erosion, gear shapers with generating cutters of appropriate form, and, in some instances, internal grinders of special design. Regardless of the production methods used, points on both of the mating profiles may be calculated from the following equations:

$$X = \left(\frac{D_I}{2} + e\right)\cos\alpha - e\cos(n\alpha)\cos\alpha - ne\sin(n\alpha)\sin\alpha$$

$$Y = \left(\frac{D_I}{2} + e\right)\sin\alpha - e\cos(n\alpha)\sin\alpha + ne\sin(n\alpha)\cos\alpha$$

In these equations, $\alpha$ is the angle of rotation of the workpiece from any selected reference position; $n$ is the number of polygon sides, either 3 or 4; $D_I$ is the diameter of the inscribed circle shown on the diagram in the table; and $e$ is the dimension shown on the diagram in the table and may be used as a setting on special polygon grinding machines. The value of $e$ determines the shape of the profile. A value of 0, for example, results in a circular shaft having a diameter of $D_I$. The values of $e$ in the table were selected arbitrarily to provide suitable proportions for the sizes shown.

## CAMS AND CAM DESIGN

**Classes of Cams.—**Cams may, in general, be divided into two classes: uniform motion cams and accelerated motion cams. The uniform motion cam moves the follower at the same rate of speed from the beginning to the end of the stroke; but as the movement is started from zero to the full speed of the uniform motion and stops in the same abrupt way, there is a distinct shock at the beginning and end of the stroke, if the movement is at all rapid. In machinery working at a high rate of speed, therefore, it is important that cams are so constructed that sudden shocks are avoided when starting the motion or when reversing the direction of motion of the follower.

The uniformly accelerated motion cam is suitable for moderate speeds, but it has the disadvantage of sudden changes in acceleration at the beginning, middle and end of the stroke. A cycloidal motion curve cam produces no abrupt changes in acceleration and is often used in high-speed machinery because it results in low noise, vibration and wear. The cycloidal motion displacement curve is so called because it can be generated from a cycloid which is the locus of a point of a circle rolling on a straight line.*

**Cam Follower Systems.—**The three most used cam and follower systems are radial and offset translating roller follower, Fig. 1a and Fig. 1b; and the swinging roller follower, Fig. 1c. When the cam rotates, it imparts a translating motion to the roller followers in Fig. 1a and Fig. 1b and a swinging motion to the roller follower in Fig. 1c. The motion of the follower is, of course, dependent on the shape of the cam; and the following section on displacement diagrams explains how a favorable motion is obtained so that the cam can rotate at high speed without shock.

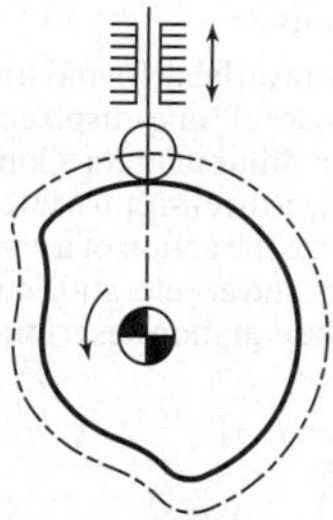

Fig. 1a. Radial Translating Roller Follower

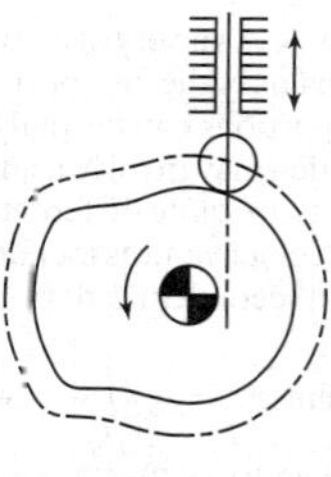

Fig. 1b. Offset Translating Roller Follower

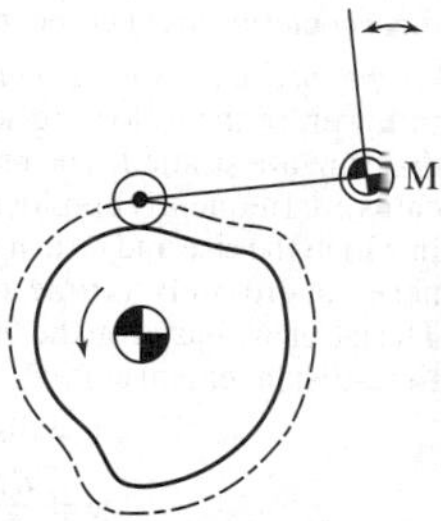


Fig. 1c. Swinging Roller Follower

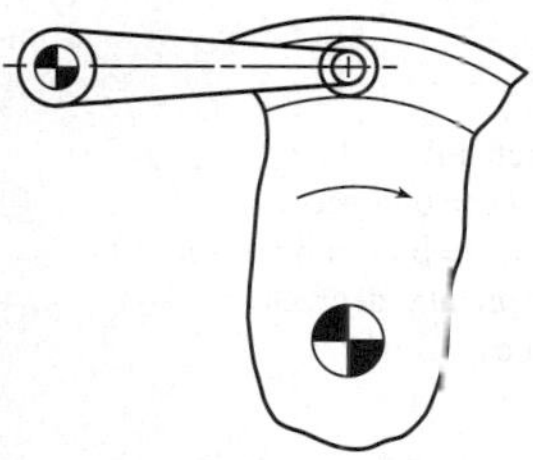

Fig. 2a. Closed-Track Cam

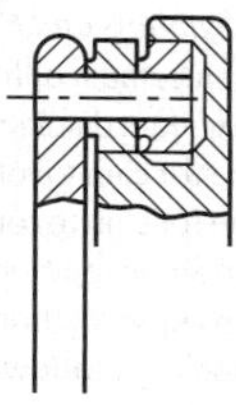

Fig. 2b. Closed-Track Cam with Two Rollers

The arrangements in Fig. 1a, Fig. 1b, and Fig. 1c show open-track cams. In Fig 2a and Fig. 2b the roller is forced to move in a closed track. Open-track cams build smaller than

* Jensen, P. W., *Cam Design and Manufacture*, Industrial Press Inc.

closed-track cams but, in general, springs are necessary to keep the roller in contact with the cam at all times. Closed-track cams do not require a spring and have the advantage of positive drive throughout the rise and return cycle. The positive drive is sometimes required as in the case where a broken spring would cause serious damage to a machine.

**Displacement Diagrams.—**Design of a cam begins with the displacement diagram. A simple displacement diagram is shown in Fig. 3. One cycle means one whole revolution of the cam; i.e., one cycle represents 360°. The horizontal distances $T_1, T_2, T_3, T_4$ are expressed in units of time (seconds); or radians or degrees. The vertical distance, $h$, represents the maximum "rise" or stroke of the follower.

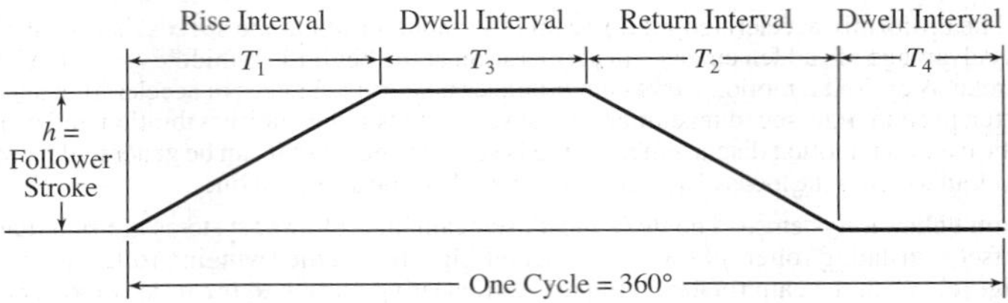


Fig. 3. A Simple Displacement Diagram

The displacement diagram of Fig. 3 is not a very favorable one because the motion from rest (the horizontal lines) to constant velocity takes place instantaneously and this means that accelerations become infinitely large at these transition points.

*Types of Cam Displacement Curves:* A variety of cam curves are available for moving the follower. In the following sections only the rise portions of the total time-displacement diagram are studied. The return portions can be analyzed in a similar manner. Complex cams are frequently employed which may involve a number of rise-dwell-return intervals in which the rise and return aspects are quite different. To analyze the action of a cam it is necessary to study its time-displacement and associated velocity and acceleration curves. The latter are based on the first and second time-derivatives of the equation describing the time-displacement curve:

$$y = \text{displacement} = f(t) \qquad \text{or} \qquad y = f(\phi)$$

$$v = \frac{dy}{dt} = \text{velocity} = \omega\frac{dy}{d\phi}$$

$$a = \frac{d^2y}{dt^2} = \text{acceleration} = \omega^2\frac{d^2y}{d\phi^2}$$

*Meaning of Symbols and Equivalent Relations*

$y$ = displacement of follower, inch (m)

$h$ = maximum displacement of follower, inch (m)

$t$ = time for cam to rotate through angle $\phi$, sec, $= \phi/\omega$, sec

$T$ = time for cam to rotate through angle $\beta$, sec, $= \beta/\omega$, or $\beta/6N$, sec

$\phi$ = cam angle rotation for follower displacement $y$, degrees

$\beta$ = cam angle rotation for total rise $h$, degrees

$v$ = velocity of follower, in/sec (m/s)

$a$ = follower acceleration, in/sec$^2$ (m/s$^2$)

$t/T = \phi/\beta$

$N$ = cam speed, rpm

$\omega$ = angular velocity of cam, degrees/sec $= \beta/T = \phi/t = d\phi/dt = 6N$

$\omega_R$ = angular velocity of cam, radians/sec $= \pi\omega/180$

$W$ = effective weight, lbs (kg)
$g$ = gravitational constant = 386 in/sec$^2$ (9.81 m/s$^2$)
$f(t)$ = means a function of $t$
$f(\phi)$ = means a function of $\phi$
$R_{min}$ = minimum radius to the cam pitch curve, inch (m)
$R_{max}$ = maximum radius to the cam pitch curve, inch (m)
$r_f$ = radius of cam follower roller, inch (m)
$\rho$ = radius of curvature of cam pitch curve (path of center of roller follower), inch (m)
$R_c$ = radius of curvature of actual cam surface, in., (m) = $\rho - r_f$ for convex surface; = $\rho + r_f$ for concave surface.

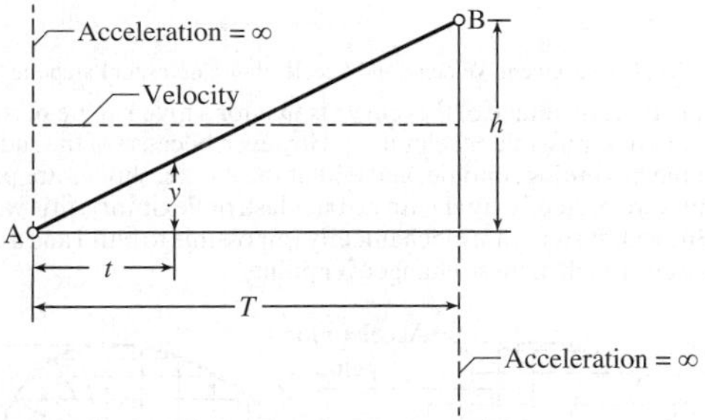


Fig. 4. Cam Displacement, Velocity, and Acceleration Curves for Constant Velocity Motion

Four displacement curves are of the greatest utility in cam design.

*1. Constant-Velocity Motion:* (Fig. 4)

$$y = h\frac{t}{T} \quad \text{or} \quad y = \frac{h\phi}{\beta} \qquad \text{(1a)}$$

$$v = \frac{dy}{dt} = \frac{h}{T} \quad \text{or} \quad v = \frac{h\omega}{\beta} \qquad \text{(1b)} \qquad 0 < t < T$$

$$a = \frac{d^2y}{dt^2} = 0^* \qquad \text{(1c)}$$

* Except at $t = 0$ and $t = T$ where the acceleration is theoretically infinite.

This motion and its disadvantages were mentioned previously. While in the unaltered form shown it is rarely used except in very crude devices, nevertheless, the advantage of uniform velocity is an important one and by modifying the start and finish of the follower stroke this form of cam motion can be utilized. Such modification is explained in the section *Displacement Diagram Synthesis* on page 2369.

*2. Parabolic Motion:* (Fig. 5)

**For $0 \le t \le T/2$ and $0 \le \phi \le \beta/2$**

$y = 2h(t/T)^2 = 2h(\phi/\beta)^2$ (2a)

$v = 4ht/T^2 = 4h\omega\phi/\beta^2$ (2b)

$a = 4h/T^2 = 4h(\omega/\beta)^2$ (2c)

**For $T/2 \le t \le T$ and $\beta/2 \le \phi \le \beta$**

$y = h[1 - 2(1 - t/T)^2] = h[1 - 2(1 - \phi/\beta)^2]$ (2d)

$v = 4h/T(1 - t/T) = (4h\omega/\beta)(1 - \phi/\beta)$ (2e)

$a = -4h/T^2 = -4h(\omega/\beta)^2$ (2f)

Examination of the above formulas shows that the velocity is zero when $t = 0$ and $y = 0$; and when $t = T$ and $y = h$.

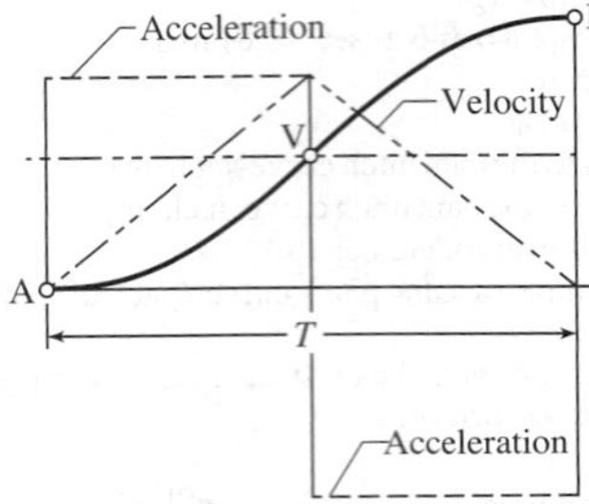


Fig. 5. Cam Displacement, Velocity, and Acceleration Curves for Parabolic Motion

The most important advantage of this curve is that for a given angle of rotation and rise it produces the smallest possible acceleration. However, because of the sudden changes in acceleration at the beginning, middle, and end of the stroke, shocks are produced. If the follower system were perfectly rigid with no backlash or flexibility, this would be of little significance. But such systems are mechanically impossible to build and a certain amount of impact is caused at each of these changeover points.

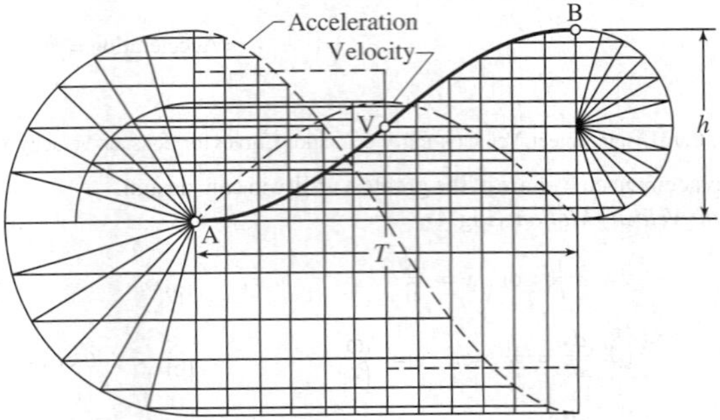


Fig. 6. Cam Displacement, Velocity, and Acceleration Curves for Simple Harmonic Motion

*3. Simple Harmonic Motion:* (Fig. 6)

$$y = \frac{h}{2}\left[1 - \cos\left(\frac{180°t}{T}\right)\right] \quad \text{or} \quad y = \frac{h}{2}\left[1 - \cos\left(\frac{180°\phi}{\beta}\right)\right] \qquad \text{(3a)}$$

$$v = \frac{h}{2} \cdot \frac{\pi}{T}\sin\left(\frac{180°t}{T}\right) \quad \text{or} \quad v = \frac{h}{2} \cdot \frac{\pi\omega}{\beta}\sin\left(\frac{180°\phi}{\beta}\right) \qquad \text{(3b)} \quad \Bigg\} \quad 0 \le t \le T$$

$$a = \frac{h}{2} \cdot \frac{\pi^2}{T^2}\cos\left(\frac{180°t}{T}\right) \quad \text{or} \quad a = \frac{h}{2} \cdot \left(\frac{\pi\omega}{\beta}\right)^2\cos\left(\frac{180°\phi}{\beta}\right) \qquad \text{(3c)}$$

Smoothness in velocity and acceleration during the stroke is the advantage inherent in this curve. However, the instantaneous changes in acceleration at the beginning and end of the stroke tend to cause vibration, noise, and wear. As can be seen from Fig. 6, the maximum acceleration values occur at the ends of the stroke. Thus, if inertia loads are to be overcome by the follower, the resulting forces cause stresses in the members. These forces are in many cases much larger than the externally applied loads.

*4. Cycloidal Motion:* (Fig. 7)

$$y = h\left[\frac{t}{T} - \frac{1}{2\pi}\sin\left(\frac{360°t}{T}\right)\right] \quad \text{or} \quad y = h\left[\frac{\phi}{\beta} - \frac{1}{2\pi}\sin\left(\frac{360°\phi}{\beta}\right)\right] \qquad (4a)$$

$$v = \frac{h}{T}\left[1 - \cos\left(\frac{360°t}{T}\right)\right] \quad \text{or} \quad v = \frac{h\omega}{\beta}\left[1 - \cos\left(\frac{360°\phi}{\beta}\right)\right] \qquad (4b) \quad \} \quad 0 \le i \le T$$

$$a = \frac{2\pi h}{T^2}\sin\left(\frac{360°t}{T}\right) \quad \text{or} \quad a = \frac{2\pi h\omega^2}{\beta^2}\sin\left(\frac{360°\phi}{\beta}\right) \qquad (4c)$$

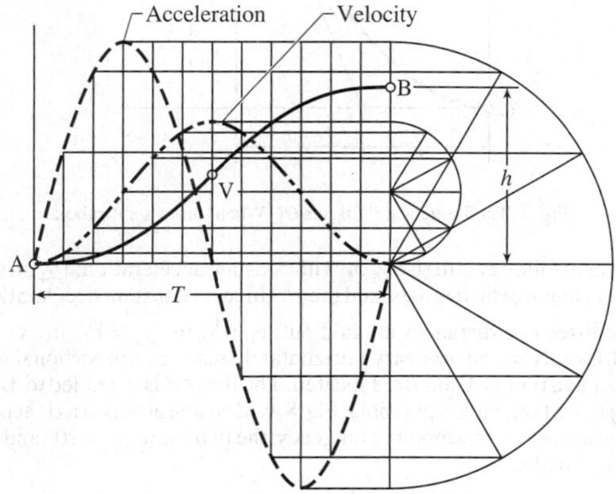


Fig. 7. Cam Displacement, Velocity, and Acceleration Curves for Cycloidal Motion

This time-displacement curve has excellent acceleration characteristics; there are no abrupt changes in its associated acceleration curve. The maximum value of the acceleration of the follower for a given rise and time is somewhat higher than that of the simple harmonic motion curve. In spite of this, the cycloidal curve is used often as a basis for designing cams for high-speed machinery because it results in low levels of noise, vibration, and wear.

**Displacement Diagram Synthesis.—**The straight-line graph shown in Fig. 3 has the important advantage of uniform velocity. This is so desirable that many cams based on this graph are used. To avoid impact at the beginning and end of the stroke, a modification is introduced at these points. There are many different types of modifications possible, ranging from a simple circular arc to much more complicated curves. One of the better curves used for this purpose is the parabolic curve given by Equation (2a). As seen from the derived time graphs, this curve causes the follower to begin a stroke with zero velocity but having a finite and constant acceleration. We must accept the necessity of acceleration, but effort should be made to hold it to a minimum.

*Matching of Constant Velocity and Parabolic Motion Curves:* By matching a parabolic cam curve to the beginning and end of a straight-line cam displacement diagram it is possible to reduce the acceleration from infinity to a finite constant value to avoid impact loads. As illustrated in Fig. 8, it can be shown that for any parabola the vertex of which is at $O$, the tangent to the curve at the point $P$ intersects the line $OQ$ at its midpoint. This means that the tangent at $P$ represents the velocity of the follower at time $X_0$ as shown in Fig. 8. Since the tangent also represents the velocity of the follower over the constant velocity portion of the stroke, the transition from rest to the maximum velocity is accomplished with smoothness.

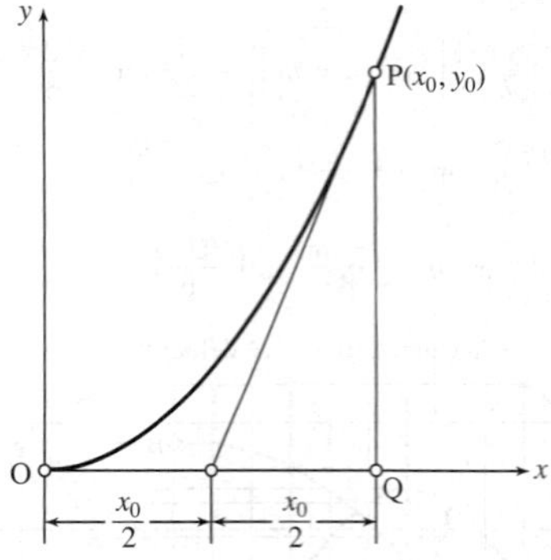


Fig. 8. The Tangent at *P* Bisects *OQ* When Curve is a Parabola

*Example:* A cam follower is to rise ¼ in. with constant acceleration; 1¼ in. with constant velocity, over an angle of 50 degrees; and then ½ in. with constant deceleration.

In Fig. 9 the three rise distances are laid out, $y_1$ = ¼ in., $y_2$ = 1¼ in., $y_3$ = ½ in., and horizontals drawn. Next, an arbitrary horizontal distance $\phi_2$ proportional to 50 degrees is measured off and points *A* and *B* are located. The line *AB* is extended to $M_1$ and $M_2$. By remembering that a tangent to a parabola, Fig. 8, will cut the abscissa axis at point $(X_0/2, 0)$ where $X_0$ is the abscissa of the point of tangency, the two values $\phi_1 = 20°$ and $\phi_3 = 40°$ will be found. Analytically,

$$\frac{M_1E}{\phi_2} = \frac{y_1}{y_2} \quad \frac{\frac{1}{2}\phi_1}{50°} = \frac{0.25}{1.25} \qquad \therefore \phi_1 = 20°$$

$$\frac{FM_2}{\phi_2} = \frac{y_3}{y_2} \quad \frac{\frac{1}{2}\phi_3}{50°} = \frac{0.50}{1.25} \qquad \therefore \phi_3 = 40°$$

In Fig. 9, the portions of the parabola have been drawn in; the details of this operation are as follows:

Assume that accuracy to the nearest thousandth of an inch is desired, and it is decided to plot values for every 5 degrees of cam rotation.

The formula for the acceleration portion of the parabolic curve is:

$$y = \frac{2h}{T^2}t^2 = 2h\left(\frac{\phi}{\beta}\right)^2 \qquad (5)$$

Two different parabolas are involved in this example; one for accelerating the follower during a cam rotation of 20 degrees, the other for decelerating it in 40 degrees, these two being tangent, to opposite ends of the same line *AB*.

In Fig. 9 only the first half of a complete acceleration-deceleration parabolic curve is used to blend with the left end of the straight line *AB*. Therefore, in using the Formula (5) substitute $2y_1$ for *h* and $2\phi_1$ for β so that

$$y = \frac{2h\phi^2}{\beta^2} = \frac{(2)(2y_1)}{(2\phi_1)^2}\phi^2$$

For the right end of the straight line *AB*, the calculations are similar but, in using Formula (5), calculated *y* values are *subtracted* from the *total* rise of the cam $(y_1 + y_2 + y_3)$ to obtain the follower displacement.

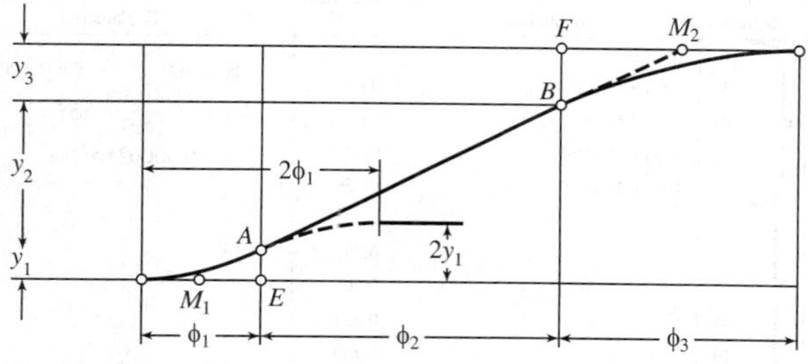


Fig. 9. Matching a Parabola at Each End of Straight Line Displacement Curve *AB* to Provide More Acceptable Acceleration and Deceleration

Table 1 shows the computations and resulting values for the cam displacement diagram described. The calculations are shown in detail so that if equations are programmed for a digital computer, the results can be verified easily. Obviously, the intermediate points are not needed to draw the straight line, but when the cam profile is later to be drawn or cut, these values will be needed since they are to be measured on radial lines.

The matching procedure when using cycloidal motion is exactly the same as for parabolic motion, because parabolic and cycloidal motion have the same maximum velocity for equal rise (or return) and lift angle (or return angle).

**Cam Profile Determination.**—In the cam constructions that follow an artificial device called an *inversion* is used. This represents a mental concept which is very helpful in performing the graphical work. The construction of a cam profile requires the drawing of many positions of the cam with the follower in each case in its related location. However, instead of revolving the cam, it is assumed that the follower rotates around the *fixed* cam. It requires the drawing of many follower positions, but since this is done more or less diagrammatically, it is relatively simple.

As part of the inversion process, the direction of rotation is important. In order to preserve the correct sequence of events, the artificial rotation of the follower must be the reverse of the cam's prescribed rotation. Thus, in Fig. 10 the cam rotation is counterclockwise, whereas the artificial rotation of the follower is clockwise.

*Radial Translating Roller Follower:* The time-displacement diagram for a cam with a radial translating roller follower is shown in Fig. 10(a). This diagram is read from left to right as follows: For 100 degrees of cam shaft rotation the follower rises *h* inches (*AB*), dwells in its upper position for 20 degrees (*BC*), returns over 180 degrees (*CD*), and finally dwells in its lowest position for 60 degrees (*DE*). Then the entire cycle is repeated

Fig. 10(b) shows the cam construction layout with the cam pitch curve as a dot and dash line. To locate a point on this curve, take a point on the displacement curve, as 6′ at the 60-degree position, and project this horizontally to point 6″ on the 0-degree position of the cam construction diagram. Using the center of cam rotation, an arc is struck from point 6″ to intercept the 60-degree position radial line which gives point 6‴ on the cam pitch curve. It will be seen that the smaller circle in the cam construction layout has a radius $R_{min}$ equal to the smallest distance from the center of cam rotation to the pitch curve and, similarly, the larger circle has a radius $R_{max}$ equal to the largest distance to the pitch curve. Thus, the difference in radii of these two circles is equal to the maximum rise *h* of the follower.

**Table 1. Development of Modified Constant Velocity Cam with Parabolic Matching**

| Rise Angle | φ Degrees | Computation | Follower Displacement y | Explanation |
|---|---|---|---|---|
| | 0 | 0 | 0 | |
| | 5 | $0.000625 \times 5^2$ | 0.016 | $\beta = 40°$  $h = 0.500$ |
| $\phi_1 = 20°$ | 10 | $0.00625 \times 10^2$ | 0.063 | $y = \dfrac{(2)(0.500)}{(40)^2}\phi^2$ |
| | 15 | $0.000625 \times 15^2$ | 0.141 | $= 0.000625\phi^2$ |
| | 20 | $0.000625 \times 20^2$ | 0.250 | |
| | 25 | | 0.375 | |
| | 30 | | 0.500 | |
| | 35 | | 0.625 | |
| $\phi_2 = 50°$ | 40 | | 0.750 | |
| | 45 | | 0.875 | 1.250 in. divided into |
| | 50 | | 1.000 | 10 uniform divisions |
| | 55 | | 1.125 | |
| | 60 | | 1.250 | |
| | 65 | | 1.375 | |
| | 70 | | 1.500 | |
| | 75 | $2.000 - (0.0003125 \times 35^2)$ | 1.617 | |
| | 80 | $2.000 - (0.0003125 \times 30^2)$ | 1.719 | $\beta = 80°$  $h = 1.000$ |
| | 85 | $2.000 - (0.0003125 \times 25^2)$ | 1.805 | $y = 2 - \dfrac{(2)(1.000)}{(80°)^2}(110° - \phi)^2$ |
| $\phi_3 = 40°$ | 90 | $2.000 - (0.0003125 \times 20^2)$ | 1.875 | |
| | 95 | $2.000 - (0.0003125 \times 15^2)$ | 1.930 | $= 2 - 0.0003125(110° - \phi)^2$ |
| | 100 | $2.000 - (0.0003125 \times 10^2)$ | 1.969 | See footnote [a] |
| | 105 | $2.000 - (0.0003125 \times 5^2)$ | 1.992 | |
| | 110 | $2.000 - (0.0003125 \times 0^2)$ | 2.000 | |

[a] Since the deceleration portion of a parabolic cam is the same shape as the acceleration portion, but inverted, Formula (5) may be used to calculate the y values by substituting $2y_3$ for $h$ and for $\beta$ and the result subtracted from the total rise ($y_1 + y_2 + y_3$) to obtain the follower displacement.

The cam pitch curve is also the actual profile or working surface when a knife-edged follower is used. To get the profile or working surface for a cam with a roller follower, a series of arcs with centers on the pitch curve and radii equal to the radius of the roller are drawn and the inner envelope drawn tangent to these arcs is the cam working surface or profile shown as a solid line in Fig. 10(b).

***Cam Grinding***: The cams used on gas and gasoline motors, for operating the inlet and exhaust valves, are finished to the correct form by grinding. This grinding may be done in a regular cylindrical grinding machine by using a suitable cam grinding attachment. The general method of grinding cams is by so mounting the cam or camshaft that, while rotating, it will be moved toward and from the grinding wheel by a master cam, the movement causing the cam to be ground to the required form or contour. The master cam is in engagement with a roller which transmits motion to the work-holding fixture. It is evident that cam grinding first involves the generation of master cams, since these must be made to suit each different form of cam that is ground. In modern CNC cam grinding machines, downloaded master cam data replaces the "hard" master cam required on older manual machines.

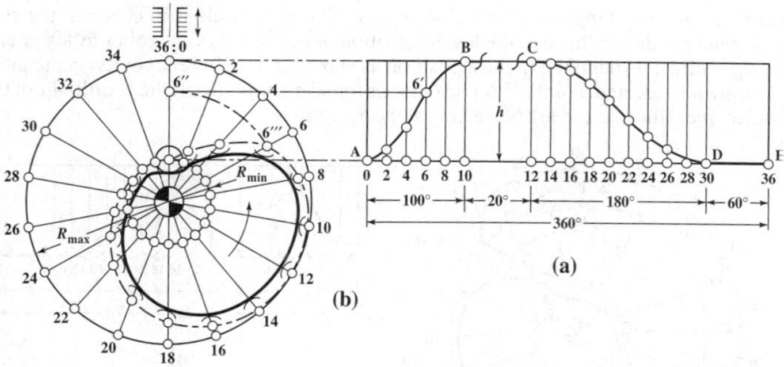


Fig. 10. (a) Time-Displacement Diagram for Cam to be Laid Out;
(b) Construction of Contour of Cam with Radial Translating Roller Follower

*Offset Translating Roller Follower:* Given the time-displacement diagram Fig. 11(a) and an offset follower. The construction of the cam in this case is very similar to the foregoing case and is shown in Fig. 11(b). In this construction it will be noted that the angular position lines are not drawn radially from the cam shaft center but tangent to a circle having a radius equal to the amount of offset of the center line of the cam follower from the center of the cam shaft. For counterclockwise rotation of the cam, points 6′, 6″, and 6‴ are located in succession as indicated.

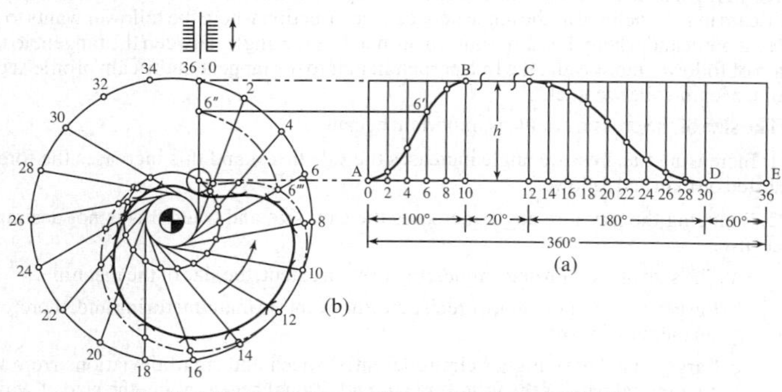


Fig. 11. (a) Time-Displacement Diagram for Cam to be Laid Out;
(b) Construction of Contour of Cam with Offset Translating Roller Follower

*Swinging Roller Follower:* Given the time-displacement diagram Fig. 12(a) and the length of the swinging follower arm $L_f$, it is required that the displacement of the follower center along the circular arc that it describes be equal to the corresponding displacements in the time-displacement diagram. If $\phi_0$ is known, the displacement $h$ of Fig. 12(a) would be found from the formula $h = \pi\phi_0 L_f/180°$; otherwise the maximum rise $h$ of the follower is stepped off on the arc drawn with $M$ as a center and starting at a point on the $R_{min}$ circle. Point $M$ is the actual position of the pivot center of the swinging follower with respect to the cam shaft center. It is again required that the rotation of the cam be counterclockwise and therefore $M$ is considered to have been rotated clockwise around the cam shaft center, whereby the points 2, 4, 6, etc., are obtained as shown in Fig. 12(b). Around each of the pivot points, 2, 4, 6, etc., circular arcs whose radii equal $L_f$ are drawn between the $R_{min}$

and $R_{max}$ circles giving the points 2′, 4′, 6′, etc. The $R_{min}$ circle with center at the cam shaft center is drawn through the lowest position of the center of the roller follower and the $R_{max}$ circle through the highest position as shown. The different points on the pitch curve are now located. Point 6‴, for instance, is found by stepping off the $y_6$ ordinate of the displacement diagram on arc 6′ starting at the $R_{min}$ circle.

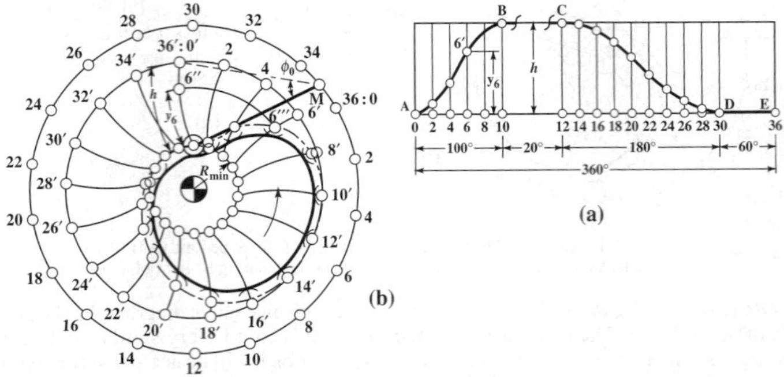

Fig. 12. (a) Time-Displacement Diagram for Cam to be Laid Out; (b) Construction of Contour of Cam with Swinging Roller Follower

**Pressure Angle and Radius of Curvature.—**The pressure angle at any point on the profile of a cam may be defined as the angle between the direction where the follower wants to go at that point and where the cam wants to push it. It is the angle between the tangent to the path of follower motion and the line perpendicular to the tangent of the cam profile at the point of cam-roller contact.

The size of the pressure angle is important because:

1) Increasing the pressure angle increases the side thrust and this increases the forces exerted on cam and follower.

2) Reducing the pressure angle increases the cam size and often this is not desirable because:

a) The size of the cam determines, to a certain extent, the size of the machine.

b) Larger cams require more precise cutting points in manufacturing and, therefore, an increase in cost.

c) Larger cams have higher circumferential speed and small deviations from the theoretical path of the follower cause additional acceleration, the size of which increases with the square of the cam size.

d) Larger cams mean more revolving weight and in high-speed machines this leads to increased vibrations in the machine.

e) The inertia of a large cam may interfere with quick starting and stopping.

The maximum pressure angle $\alpha_m$ should, in general, be kept at or below 30 degrees for translating-type followers and at or below 45 degrees for swinging-type followers. These values are on the conservative side and in many cases may be increased considerably, but beyond these limits trouble could develop and an analysis is necessary.

In the following, graphical methods are described by which a cam mechanism can be designed with translating or swinging roller followers having specified maximum pressure angles for rise and return. These methods are applicable to any kind of time-displacement diagram.

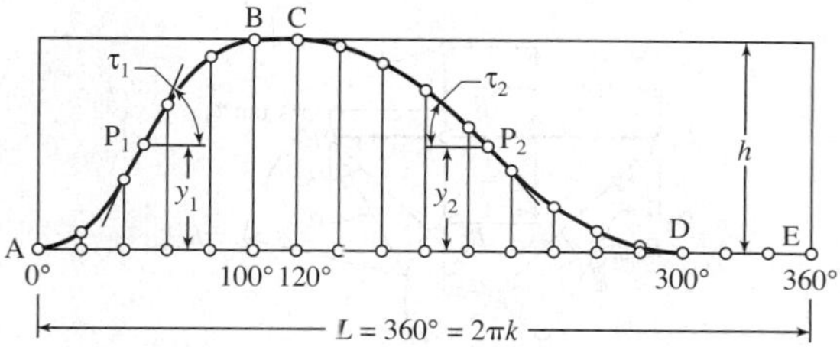


Fig. 13. Displacement Diagram

**Determination of Cam Size for a Radial or an Offset Translating Follower.**—Fig. 13 shows a time-displacement diagram. The maximum displacement is preferably made to scale, but the length of the abscissa, $L$, can be chosen arbitrarily. The distance $L$ from 0 to 360 degrees is measured and is set equal to $2\pi k$ from which

$$k = \frac{L}{2\pi}$$

$k$ is calculated and laid out as length $E$ to $M$ in Fig. 14.

In Fig. 13 the two points $P_1$ and $P_2$ having the maximum angles of slope $\tau_1$ and $\tau_2$ are located by inspection. In this example, $y_1$ and $y_2$ are of equal length.

Angles $\tau_1$ and $\tau_2$ are laid out as shown in Fig. 14, and the points of intersection with a perpendicular to $EM$ at $M$ determine $Q_1$ and $Q_2$. The measured distances

$$MQ_1 = k \tan \tau_1 \quad \text{and} \quad MQ_2 = k \tan \tau_2$$

are laid out in Fig. 15, which is constructed as follows:

Draw a vertical line $R_uR_o$ of length $h$ equal to the stroke of the roller follower, $R_u$ being the lowest position and $R_o$ the highest position of the center of the roller follower. From $R_u$ lay out $R_uR_{y1} = y_1$ and $R_uR_{y2} = y_2$; these are equal lengths in this example. Next, if the rotation of the cam is counterclockwise, lay out $k \tan \tau_1$, to the left, $k \tan \tau_2$ to the right from points $R_{y1}$ and $R_{y2}$, respectively, $R_{y1}$ and $R_{y2}$ being the same point in this case.

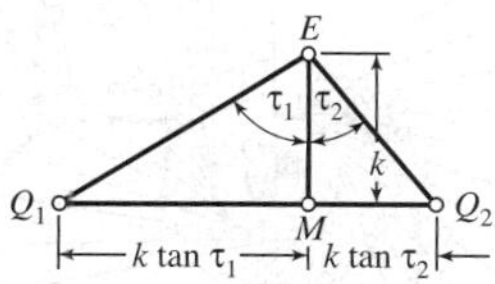


Fig. 14. Construction to Find $k \tan \tau_1$ and $k \tan \tau_2$

The specified maximum pressure angle $\alpha_1$ is laid out at $E_1$ as shown, and a ray (line) $E_1F_1$ is determined. Any point on this ray chosen as the cam shaft center will proportion the cam so that the pressure angle at a point on the cam profile corresponding to point $P_1$, of the displacement diagram will be exactly $\alpha_1$.

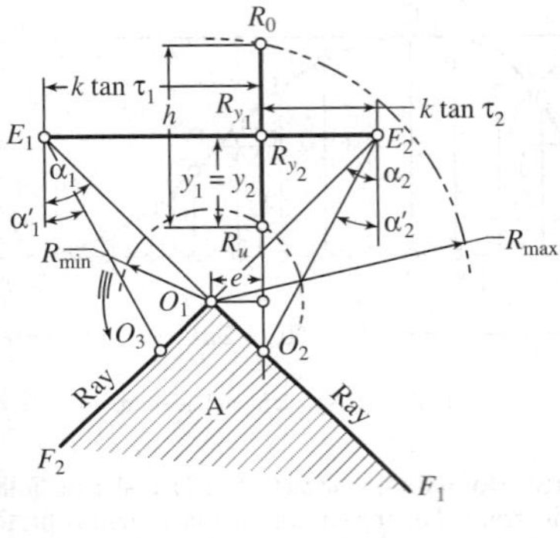


Fig. 15. Finding Proportions of Cam; Offset Translating Follower

The angle $\alpha_2$ is laid out at $E_2$ as shown, and another ray $E_2F_2$ is determined. Similarly, any point on this ray chosen as the cam shaft center will proportion the cam so that the pressure angle at a point on the cam profile corresponding to point $P_2$ of the displacement diagram will be exactly $\alpha_2$.

Any point chosen within the cross-hatched area A as the cam center will yield a cam whose pressure angles at points corresponding to $P_1$ and $P_2$ will not exceed the specified values $\alpha_1$ and $\alpha_2$ respectively. If $O_1$ is chosen as the cam shaft center, the pressure angles on the cam profile corresponding to points $P_1$ and $P_2$ are exactly $\alpha_1$ and $\alpha_2$, respectively. Selection of point $O_1$ also yields the smallest possible cam for the given requirements and requires an offset follower in which $e$ is the offset distance.

If $O_2$ is chosen as the cam shaft center, a radial translating follower is obtained (zero offset). In that case, the pressure angle $\alpha_1$ for the rise is unchanged, whereas the pressure angle for the return is changed from $\alpha_2$ to $\alpha'_2$. That is, the pressure angle on the return stroke is reduced at the point $P_2$. If point $O_3$ had been selected, then $\alpha_2$ would remain unchanged but $\alpha_1$ would be decreased and the offset $e$ increased.

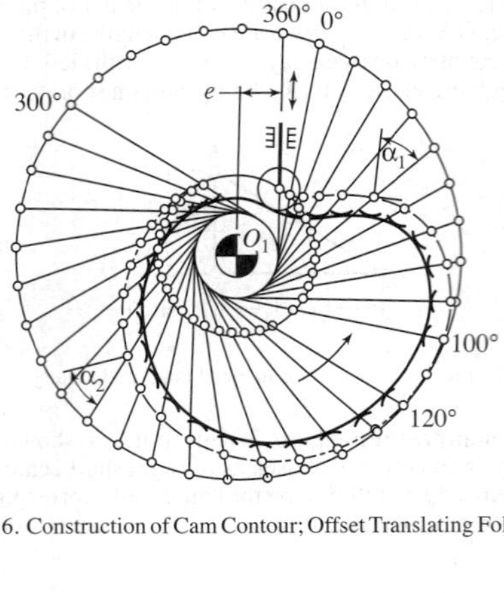


Fig. 16. Construction of Cam Contour; Offset Translating Follower

Fig. 16 shows the shape of the cam when $O_1$ from Fig. 15 is chosen as the cam shaft center, and it is seen that the pressure angle at a point on the cam profile corresponding to point $P_1$ is $\alpha_1$ and at a point corresponding to point $P_2$ is $\alpha_2$.

In the foregoing, a cam mechanism has been so proportioned that the pressure angles $\alpha_1$ and $\alpha_2$ at points on the cam corresponding to points $P_1$ and $P_2$ were obtained. Even though $P_1$ and $P_2$ are the points of greatest slope on the displacement diagram, the pressure angles produced at some other points on the actual cam may be slightly greater.

However, if the pressure angles $\alpha_1$ and $\alpha_2$ are not to be exceeded at any point—i.e., they are to be maximum pressure angles—then $P_1$ and $P_2$ must be selected to be at the locations where these maximum pressure angles occur. If these locations are not known, then the graphical procedure described must be repeated, letting $P_1$ take various positions on the curve for rise ($AB$) and $P_2$ various positions on the return curve ($CD$) and then setting $R_{min}$ equal to the largest of the values determined from the various positions.

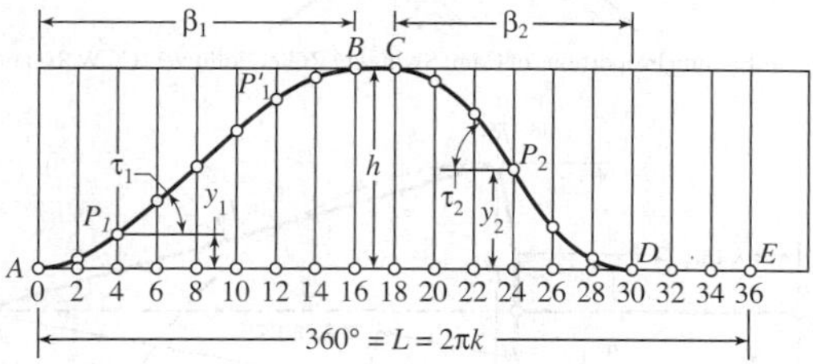


Fig. 17. Displacement Diagram

**Determination of Cam Size for Swinging Roller Follower.**—The proportioning of a cam with swinging roller follower having specific pressure angles at selected points follows the same procedure as that for a translating follower.

*Example:* Given the diagram for the roller displacement along its circular arc, Fig. 17 with $h = 1.95$ in., the periods of rise and fall, respectively, $\beta_1 = 160°$ and $\beta_2 = 120°$, the length of the swinging follower arm $L_f = 3.52$ in., rotation of the cam away from pivot point $M$, and pressure angles $\alpha_1 = \alpha_2 = 45°$ (corresponding to the points $P_1$ and $P_2$ in the displacement diagram). Find the cam proportions.

*Solution:* Distances $k$ tan $\tau_1$ and $k$ tan $\tau_2$ are determined as in the previous example, Fig. 14. In Fig. 18, $R_{y1}$ is determined by making the distance $R_uR_{y1} = y_1$ along the arc $R_uR_o$ and $R_{y2}$ by making $R_uR_{y2} = y_2$. The arc $R_uR_o = h$ and $R_u$ indicates the lowest position of the center of the swinging roller follower and $R_o$ the highest position.

Because the cam (i.e., the surface of the cam as it passes under the follower roller) rotates away from pivot point $M$, $k \tan \tau_1$ is laid out away from $M$, that is, from $R_{y1}$ to $E_1$ and $k \tan \tau_2$ is laid out toward $M$ from $R_{y2}$ to $E_2$. Angle $\alpha_1$ at $E_1$ determines one ray and $\alpha_2$ at $E_2$ another ray, which together subtend an area $A$ having the property that if the cam shaft center is chosen inside this area, the pressure angles at the points of the cam corresponding to $P_1$ and $P_2$ in the displacement diagram will not exceed the given values $\alpha_1$ and $\alpha_2$, respectively. If the cam shaft center is chosen on the ray drawn from $E_1$ at an angle $\alpha_1 = 45°$, the pressure angle $\alpha_1$ on the cam profile corresponding to point $P_1$ will be exactly 45°, and if chosen on the ray from $E_2$, the pressure angle $\alpha_2$ corresponding to $P_2$ will be exactly 45°. If another point, $O_2$ for example, is chosen as the cam shaft center, the pressure angle corresponding to $P_1$ will be $\alpha'_1$ and that corresponding to $P_2$ will be $\alpha_2$.

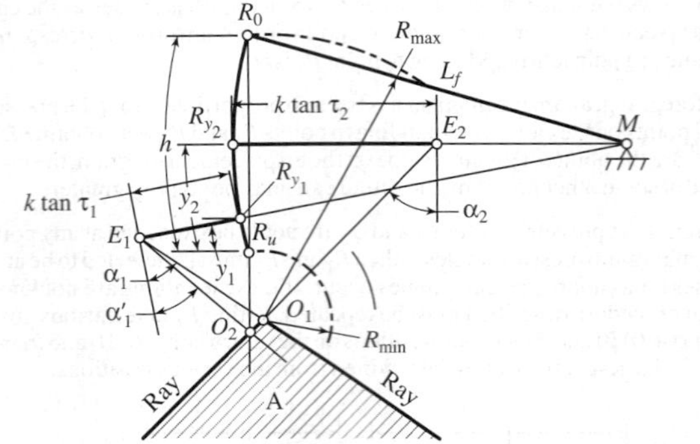


Fig. 18. Finding Proportions of Cam; Swinging Roller Follower (CCW Rotation)

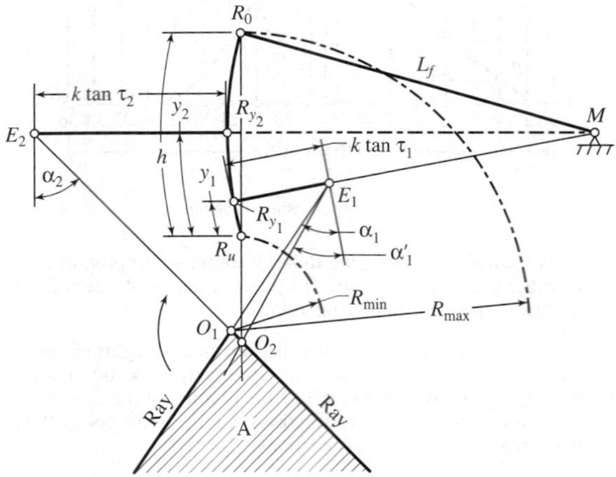


Fig. 19. Finding Proportions of Cam; Swinging Roller Follower (CW Rotation)

Fig. 19 shows the construction for rotation toward pivot point $M$ (clockwise rotation of the cam in this case). The layout of the cam curve is made in a manner similar to that shown previously in Fig. 12.

In this example, the cam mechanism was so proportioned that the pressure angles at certain points (corresponding to $P_1$ and $P_2$) do not exceed certain specified values (namely $\alpha_1$ and $\alpha_2$).

*To make* sure that the pressure angle at *no point* along the cam profile exceeds the specified value, the previous procedure should be repeated for a series of points along the profile.

**Formulas for Calculating Pressure Angles.**—The graphical methods described previously are useful because they permit layout and measurement of pressure angles and radii of curvature of *any* cam profile. For cams of complicated profiles, and especially if the profile cannot be represented by a simple formula, the graphical method may be the only practical solution. However, for some of the standard cam profiles utilizing *radial* translating roller followers, the following formulas may be used to determine key cam dimensions before laying out the cam. These formulas enable the designer to specify the maximum pressure angle (usually 30° or less) and, using the specified value, to calculate the minimum cam size that will satisfy the requirement.

The following symbols are in addition to those starting on page 2366.

$\alpha max$ = specified maximum pressure angle, degrees

$R_{\alpha\, max}$ = radius from cam center to point on pitch curve where $\alpha_{max}$ is located, inches (m)

$\phi_p$ = rise angle, in degrees, corresponding to $\alpha_{max}$ and $R_{\alpha\, max}$

$\alpha$ = pressure angle at any selected point, degrees

$R_\alpha$ = radius from cam center to pitch curve at $\alpha$, inches (m)

$\phi$ = rise angle, in degrees, corresponding to $\alpha$ and $R_\alpha$

*For Uniform Velocity Motion*

$$\alpha = \arctan\left[\frac{180°h}{\pi\beta R_\alpha}\right] \text{ at radius } R_\alpha \text{ to the pitch curve} \qquad (6a)$$

$$\alpha_{max} = \arctan\left[\frac{180°h}{\pi\beta R_{min}}\right] \text{ at radius } R_{min} \text{ of the pitch curve } (\phi = 0\,°). \qquad (6b)$$

If $\alpha_{max}$ is specified, the minimum radius to the lowest point on the pitch curve, $R_{min}$, is:

$$R_{min} = \frac{180°h}{\pi\beta \tan \alpha_{max}} \text{ which corresponds to } \phi = 0\,°. \qquad (6c)$$

*For Parabolic Motion*

$$\alpha = \arctan\left[\frac{720°h\phi}{\pi\beta^2 R_\alpha}\right] \text{ at radius } R_\alpha \text{ to the pitch curve at angle } \phi\text{, where } 0 \le \phi \le \beta / 2 \qquad (7a)$$

$$\text{and } \alpha = \arctan\left[\frac{720°h\,(1 - \phi/\beta)}{\pi\beta R_\alpha}\right] \text{ at radius } R_\alpha \text{ to pitch curve at angle } \phi\text{, where } \beta/2 \le \phi \le \beta.$$

$$\alpha_{max} = \arctan\left[\frac{360°h}{\pi\beta R_\alpha}\right] \text{ which occurs at } \phi\ \beta/2 \text{ and } R_\alpha = R_{min} + h/2 \qquad (7b)$$

If $\alpha_{max}$ is specified, then the minimum radius to the lowest point of the pitch curve is:

$$R_{min} = \left[\frac{360°h}{\pi\beta \tan \alpha_{max}} - \frac{h}{2}\right] \text{ which corresponds to } \phi = 0°. \qquad (7c)$$

*For Simple Harmonic Motion*

$$\alpha = \arctan\left[\frac{90°h}{\beta R_\alpha} \sin\left(\frac{180°\phi}{\beta}\right)\right] \text{ at radius } R_\alpha \text{ to the pitch curve at angle } \phi \qquad (8a)$$

$$\phi_p = \left(\frac{\beta}{180°}\right)\left[\operatorname{arccot}\left(\frac{\beta}{180°} \tan \alpha_{max}\right)\right] = \phi\text{, where pressure angle } \alpha_{max} \text{ occurs} \qquad (8b)$$

$$R_{\alpha max} = \frac{h\left[\sin\left(180°\,\frac{\phi_p}{\beta}\right)\right]^2}{2\cos\left(\frac{180°\phi_p}{\beta}\right)} \text{ at point where } \alpha = \alpha_{max} \text{ and } \phi = \phi_{max} \qquad (8c)$$

$$R_{min} = R_{\alpha max} - \frac{h}{2}\left[1 - \cos\left(\frac{180°\phi_p}{\beta}\right)\right] \tag{8d}$$

*For Cycloidal Motion*

$$\alpha = \arctan\left[\frac{180°}{\pi\beta R_\alpha}\left(1 - \cos\frac{360°\phi}{\beta}\right)\right] \text{ at radius } R_\alpha \text{ to the pitch curve at angle } \phi \tag{9a}$$

$$\phi_p = \frac{\beta}{180°}\left[\operatorname{arccot}\left(\frac{\beta\tan\alpha_{max}}{360°}\right)\right] = \text{value of } \phi \text{ where pressure angle } \alpha_{max} \text{ occurs} \tag{9b}$$

$$R_{\alpha max} = \frac{h}{2\pi}\frac{\left[1 - \cos\left(\frac{360°\phi_p}{\beta}\right)\right]^2}{\sin\left(360°\frac{\phi_p}{\beta}\right)} \text{ at point where } \alpha = \alpha_{max} \text{ and } \phi = \phi_p \tag{9c}$$

$$R_{min} = R_{\alpha max} - h\left[\frac{\phi_p}{\beta} - \frac{1}{2\pi}\sin\left(\frac{360°\phi_p}{\beta}\right)\right] \tag{9d}$$

**Radius of Curvature.—**The minimum radius of curvature of a cam should be kept as large as possible (1) to prevent undercutting of the convex portion of the cam and (2) to prevent too high surface stresses. Fig. 20(a), (b) and (c) illustrate how undercutting occurs.

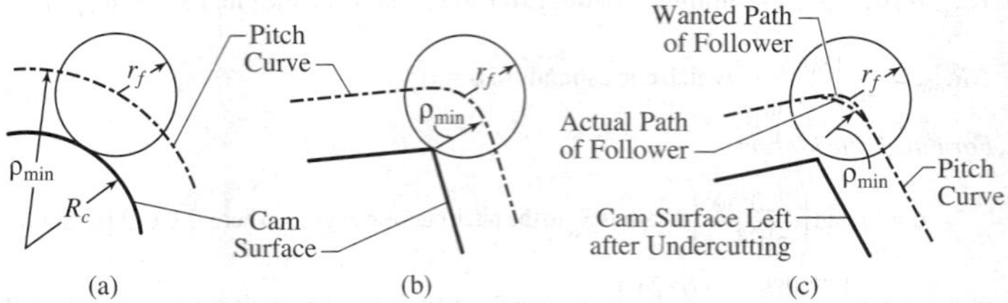


Fig. 20. (a) No Undercutting; (b) Sharp Corner on Cam; (c) Undercutting

In Fig. 20(a) the radius of curvature of the path of the follower is $\rho_{min}$ and the cam will at that point have a radius of curvature $R_c = \rho_{min} - r_f$.

In Fig. 20(b) $\rho_{min} = r_f$ and $R_c = 0$. Therefore, the actual cam will have a sharp corner which in most cases will result in too high surface stresses.

In Fig. 20(c) is shown the case where $\rho_{min} < r_f$. This case is not possible because undercutting will occur and the actual motion of the roller follower will deviate from the desired one as shown.

Undercutting cannot occur at the *concave* portion of the cam profile (working surface), but caution should be exerted in not making the radius of curvature equal to the radius of the roller follower. This condition would occur if there is a cusp on the displacement diagram which, of course, should always be avoided. To enable milling or grinding of *concave* portions of a cam profile, the radius of curvature of concave portions of the cam, $R_c = \rho_{min} + r_f$, must be larger than the radius of the cutter to be used.

The radius of curvature is used in calculating surface stresses (see following section), and may be determined by measurement on the cam layout or, in the case of radial translating followers, may be calculated using the formulas that follow. Although these formulas are exact for radial followers, they may be used for offset and swinging followers to obtain an approximation.

Based upon polar coordinates, the radius of curvature is:

$$\rho = \frac{\left[r^2 + \left(\frac{dr}{d\phi}\right)^2\right]^{\frac{3}{2}}}{r^2 + 2\left(\frac{dr}{d\phi}\right)^2 - r\left(\frac{d^2r}{d\phi^2}\right)} \quad (10)$$

*Positive values (+) indicate convex curve; negative values (−), concave.

In Equation (10), $r = (R_{min} + y)$, where $R_{min}$ is the smallest radius to the pitch curve (see Fig. 12) and $y$ is the displacement of the follower from its lowest position given in terms of $\phi$, the angle of cam rotation. The following formulas for $r$, $dr/d\phi$, and $d^2r/d\phi^2$ may be substituted into Equation (10) to calculate the radius of curvature at any point of the cam pitch curve; however, to determine the possibility of undercutting of the convex portion of the cam, it is the minimum radius of curvature on the convex portion, $\rho_{min}$, that is needed. The minimum radius of curvature occurs, generally, at the point of maximum *negative* acceleration.

*Parabolic motion:*

$$r = R_{min} + h - 2h\left(1 - \frac{\phi}{\beta}\right)^2 \quad (11a)$$

$$\frac{dr}{d\phi} = \frac{720°h}{\pi\beta}\left(1 - \frac{\phi}{\beta}\right) \quad (11b)$$

$$\frac{d^2r}{d\phi^2} = \frac{-4(180°)^2h}{\pi^2\beta^2} \quad (11c)$$

$$\frac{\beta}{2} \le \phi \le \beta$$

These equations are for the deceleration portion of the curve as explained in the footnote to Table 1.

The minimum radius of curvature can occur at either $\phi = \beta/2$ or at $\phi = \beta$, depending on the magnitudes of $h$, $R_{min}$, and $\beta$. Therefore, to determine which is the case, make two calculations using Formula (10), one for $\phi = \beta/2$, and the other for $\phi = \beta$.

*Simple harmonic motion:*

$$r = R_{min} + \frac{h}{2}\left[1 - \cos\left(\frac{180°\phi}{\beta}\right)\right] \quad (12a)$$

$$\frac{dr}{d\phi} = \frac{180°h}{2\beta}\sin\left(\frac{180°\phi}{\beta}\right) \quad (12b)$$

$$\frac{d^2r}{d\phi^2} = \frac{(180°)^2h}{2\beta^2}\cos\left(\frac{180°\phi}{\beta}\right) \quad (12c)$$

$$0 \le \phi \le \beta$$

The minimum radius of curvature can occur at either $\phi = \beta/2$ or at $\phi = \beta$, depending on the magnitudes of $h$, $R_{min}$, and $\beta$. Therefore, to determine which is the case, make two calculations using Formula (10), one for $\phi = \beta/2$, and the other for $\phi = \beta$.

*Cycloidal motion:*

$$r = R_{min} + h\left[\frac{\phi}{\beta} - \frac{1}{2\pi}\sin\left(\frac{360°\phi}{\beta}\right)\right] \quad (13a)$$

$$\frac{dr}{d\phi} = \frac{180°h}{\pi\beta}\left[1 - \cos\left(\frac{360°\phi}{\beta}\right)\right] \quad (13b)$$

$$\frac{d^2r}{d\phi^2} = \frac{2(180°)^2h}{\pi\beta^2}\sin\left(\frac{360°\phi}{\beta}\right) \quad (13c)$$

$$0 \le \phi \le \beta$$

$$\rho_{min} = \frac{\left[(R_{min} + 0.91h)^2 + \left(\frac{180°h}{\pi\beta}\right)^2\right]^{3/2}}{(R_{min} + 0.91h)^2 + 2\left(\frac{180°h}{\pi\beta}\right)^2 + (R_{min} + 0.91h)\left[2(180°)^2 \frac{h}{\pi\beta^2}\right]} \quad (13d)$$

($\rho_{min}$ occurs near $\phi = 0.75\beta$)

*Example:* Given $h = 1$ inch (m), $R_{min} = 2.9$ inch (m), and $\beta = 60°$. Find $\rho_{min}$ for parabolic motion, simple harmonic motion, and cycloidal motion.

*Solution:* $\rho_{min} = 2.02$ inch (m) for parabolic motion, from Equation (10)

$\rho_{min} = 1.8$ inch (m) for simple harmonic motion, from Equation (10)

$\rho_{min} = 1.6$ inch (m) for cycloidal motion, from Equation (13d)

The value of $\rho_{min}$ on *any* cam may also be obtained by measurement on the layout of the cam using a compass.

**Cam Forces, Contact Stresses, and Materials.**—After a cam and follower configuration has been determined, the forces acting on the cam may be calculated or otherwise determined. Next, the stresses at the cam surface are calculated and suitable materials to withstand the stress are selected. If the calculated maximum stress is too great, it will be necessary to change the cam design.

Such changes may include: 1) increasing the cam size to decrease pressure angle and increase the radius of curvature; 2) changing to an offset or swinging follower to reduce the pressure angle; 3) reducing the cam rotation speed to reduce inertia forces; 4) increasing the cam rise angle, $\beta$, during which the rise, $h$, occurs; 5) increasing the thickness of the cam, provided that deflections of the follower are small enough to maintain uniform loading across the width of the cam; and 6) using a more suitable cam curve or modifying the cam curve at critical points.

Although parabolic motion seems to be the best with respect to minimizing the calculated maximum acceleration and, therefore, also the maximum acceleration forces, nevertheless, in the case of high-speed cams, cycloidal motion yields the lower maximum acceleration forces. Thus, it can be shown that owing to the sudden change in acceleration (called *jerk* or *pulse*) in the case of parabolic motion, the actual forces acting on the cam are doubled and sometimes even tripled at high speed, whereas with cycloidal motion, owing to the gradually changing acceleration, the actual dynamic forces are only slightly higher than the theoretical. Therefore, the calculated force due to acceleration should be multiplied by at least a factor of 2 for parabolic and 1.05 for cycloidal motion to provide an allowance for the load-increasing effects of elasticity and backlash.

The main factors influencing cam forces are: 1) displacement and cam speed (forces due to acceleration); 2) dynamic forces due to backlash and flexibility; 3) linkage dimensions which affect weight and weight distribution; 4) pressure angle and friction forces; and 5) spring forces.

The main factors influencing stresses in cams are: 1) radius of curvature for cam and roller; and 2) materials.

*Acceleration Forces:* The formula for the force acting on a translating body given an acceleration $a$ is:

$$R = \frac{Wa}{g} = \frac{Wa}{386} \quad (14)$$

In this formula, $g = 386$ inches/second squared, $a$ = acceleration of $W$ in inches/second squared; $R$ = resultant of all the external forces (except friction) acting on the weight $W$. For cam analysis purposes, $W$, in pounds, consists of the weight of the follower, a portion of the weight of the return spring (1/3), and the weight of the members of the external mechanism against which the follower pushes, for example, the weight of a piston:

$$W = W_f + \frac{1}{3}W_s + W_e \tag{15}$$

where $W$ = equivalent single weight; $W_f$ = follower weight; $W_s$ = spring weight; and $W_e$ = external weight, all in pounds.

*Spring Forces:* The return spring, $K_s$, shown in Fig. 21a must be strong enough to hold the follower against the cam at all times. At high cam speeds the main force attempting to separate the follower from the cam surface is the acceleration force $R$ at the point of maximum *negative* acceleration. Thus, at that point the spring must exert a force $F_s$,

$$F_s = R - W_f - F_e - F_f \tag{16}$$

where $F_e$ = external force resisting motion of follower, and $F_f$ = friction force from follower guide bushings and other sources.

When the follower is at its lowest position ($R_{min}$ in Fig. 21a), it is usual practice to have the spring provide some estimated preload to account for "set" that takes place in a spring after repeated use and to prevent roller sliding at the start of movement.

The required spring constant $K_s$, in pounds per inch of spring deflection, is:

$$K_s = \frac{F_s - \text{preload}}{y_a} \tag{17}$$

where $y_a$ = rise of cam from $R_{min}$ to height at which maximum negative acceleration takes place.

The force, $F_y$, that the spring exerts at any height $y$ above $R_{min}$ is:

$$F_y = yK_s + \text{preload} \tag{18}$$

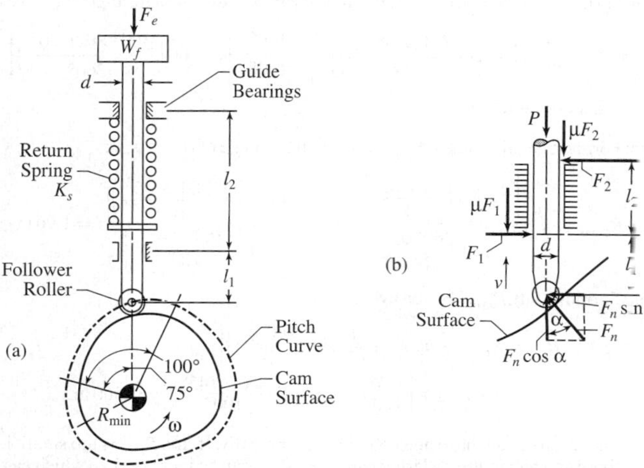


Fig. 21. (a) Radial Translating Follower and Cam System; (b) Force Acting on a Translating Follower

*Pressure Angle and Friction Forces:* As shown in Fig. 21b, the pressure angle of the cam causes a sideways component $F_n \sin \alpha$ which produces friction forces $\mu F_1$ and $\mu F_2$ in the guide bushing. If the follower rod is too flexible, bending of the follower will increase these friction forces. The effect of the friction forces and the pressure angle are taken into account in the formula,

$$F_n = \frac{P}{\cos\alpha - \frac{\mu \sin\alpha}{l_2}(2l_1 + l_2 - \mu d)} \tag{19a}$$

where $\mu$ = coefficient of friction in bushing; $l_1$, $l_2$, and $d$ are as shown in Fig. 21; and $P$ = the sum of all the forces acting down against the upward motion of the follower (acceleration force + spring force + follower weight + external force)

$$P = \frac{W \times a}{386} + (yK_s + \text{preload}) + W_f + F_e \tag{19b}$$

*Cam Torque:* The follower pressing against the cam causes resisting torques during the rise period and assisting torques during the return period. The maximum value of the resisting torque determines the cam drive requirements. Instantaneous torque values may be calculated from

$$T_o = \frac{30vF_n\cos\alpha}{\pi N} = (R_{min} + y)F_n\sin\alpha \tag{20}$$

in which $T_o$ = instantaneous torque in pound-inches.

*Example of Force Analysis:* A radial translating follower system is shown in Fig. 21a. The follower is moved with cycloidal motion over a distance of 1 in. and an angle of lift $\beta$ = 100°. Cam speed $N$ = 900 rpm. The weight of the follower mass, $W_f$, is 2 pounds. Both the spring weight $W_s$ and the external weight $W_e$ are negligible. The follower stem diameter is 0.75 in., $l_1$ = 1.5 in., $l_2$ = 4 in., coefficient of friction $\mu$ = 0.05, external force $F_e$ = 10 lbs, and the pressure angle is not to exceed 30°.

(a) What is the smallest radius $R_{min}$ to the pitch curve?

From Formula (9b) the rise angle $\phi_p$ to where the maximum pressure angle $\alpha_{max}$ exists is:

$$\phi_p = \frac{\beta}{180°}\left[\operatorname{arccot}\left(\frac{\beta\tan\alpha_{max}}{360°}\right)\right] = \frac{100°}{180°}\left[\operatorname{arccot}\left(\frac{100° \times \tan 30°}{360°}\right)\right]$$

$$= 44.94° = 45°$$

From Formula (9c) the radius, $R_{\alpha\,max}$, at which the angle of rise is $\phi_p$ is:

$$R_{\alpha max} = \frac{h}{2\pi}\frac{\left[1 - \cos\left(\frac{360°\phi_p}{\beta}\right)\right]^2}{\sin\left(\frac{360°\phi_p}{\beta}\right)} = \frac{1}{2\pi}\frac{\left[1 - \cos\left(\frac{360° \times 45°}{100°}\right)\right]^2}{\sin\left[\frac{360° \times 45°}{100°}\right]} = 1.96\text{ in.}$$

From Formula (9d), $R_{min}$ is given by

$$R_{min} = R_{\alpha max} - h\left[\frac{\phi_p}{\beta} - \frac{1}{2\pi}\sin\left(\frac{360°\phi_p}{\beta}\right)\right]$$

$$= 1.96 - 1 \times \left[\frac{45°}{100°} - \frac{1}{2\pi}\sin\left(\frac{360° \times 45°}{100°}\right)\right] = 1.560\text{ in.}$$

The same results could have been obtained graphically. If this $R_{min}$ is too small, i.e., if the cam bore and hub require a larger cam, then $R_{min}$ can be increased, in which case the maximum pressure angle will be less than 30°.

(b) If the return spring $K_s$ is specified to provide a preload of 36 lbs when the follower is at $R_{min}$, what is the spring constant required to hold the follower on the cam throughout the cycle?

The follower tends to leave the cam at the point of maximum *negative* acceleration. Fig. 7 shows this to be at $\phi = \frac{3}{4}\beta = 75°$.

From Formula (4c),

$$a = \frac{2\pi h\omega^2}{\beta^2}\sin\left(\frac{360°\phi}{\beta}\right) = \frac{2\pi \times 1 \times (6 \times 900)^2}{(100°)^2}\sin\left(\frac{360° \times 75°}{100°}\right) = -18{,}300 \text{ in/sec}^2$$

From Formulas (14) and (15),

$$R = \frac{Wa}{386} = \frac{\left(W_f + \frac{1}{3}W_s + W_e\right)a}{386} = \frac{(2+0+0)(-18{,}300)}{386} = 95 \text{ lbs (upward)}$$

Using Formula (16) to determine the spring force $F_s$ to hold the follower on the cam,

$$F_s = R - W_f - F_e - F_f$$

as stated on page 2382, the value of $R$ in the above formula should be multiplied by 1.05 for cycloidal motion to provide a factor of safety for dynamic pulses. Thus,

$$F_s = 1.05R - W_f - F_e - F_f = 1.05 \times 95 - 2 - 10 - 0 = 88 \text{ lbs (downward)}$$

The spring constant from Formula (17) is:

$$K_s = \frac{F_s - \text{preload}}{y_a} = \frac{88 - 36}{y_a}$$

and, from Formula (4a), $y_a$ is:

$$y_a = h\left[\frac{\phi}{\beta} - \frac{1}{2\pi}\sin\left(\frac{360°\phi}{\beta}\right)\right] = 1 \times \left[\frac{75°}{100°} - \frac{1}{2\pi}\sin\left(\frac{360° \times 75°}{100°}\right)\right] = 0.909 \text{ in}$$

so that $K_s = (88 - 36)/0.909 = 57$ lb/in.

(c) At the point where the pressure angle $\alpha_{max}$ is 30° ($\phi = 45°$) the rise of the follower is $1.96 - 1.56 = 0.40$ in. What is the normal force, $F_n$, on the cam? From Formulas (19a) and (19b)

$$F_n = \frac{Wa/386 + yK_s + \text{preload} + W_f + F_e}{\cos\alpha - \dfrac{\mu \sin\alpha}{l_2}(2l_1 + l_2 - \mu d)}$$

using $\phi = 45°$, $h = 1$ in., $\beta = 100°$, and $\omega = 6 \times 900$ in Formula (4c) gives $a = 5660$ in/sec$^2$. So that, with $W = 2$ lbs, $y = 0.4$, $K_s = 57$, preload = 36 lbs, $W_f = 2$ lbs, $F_e = 10$ lbs, $\alpha = 30°$, $\mu = 0.05$, $l_1 = 1.5$, $l_2 = 4$, and $d = 0.75$,

$$F_n = \frac{\dfrac{(2 \times 5660)}{386} + 0.4 \times 57 + 36 + 2}{\cos 30° - \dfrac{0.05 \times \sin 30°}{4}(2 \times 1.5 + 4 - 0.05 \times 0.75)} = 110 \text{ lbs}$$

*Note:* If the coefficient of friction had been assumed to be 0, then $F_n = 104$; on the other hand, if the follower is too flexible, so that sidewise bending occurs causing jamming in the bushing, the coefficient of friction may increase to, say, 0.5, in which case the calculated $F_n = 200$ lbs.

(d) Assume that in the manufacture of this cam an error or "bump" resulting from a chattermark or as a result of poor blending occurred, and that this "bump" rose to a height of 0.001 in. in a 1° rise of the cam in the vicinity of $\phi = 45°$. What effect would this bump have on the acceleration force $R$?

One formula that may be used to calculate the change in acceleration caused by such a cam error is:

$$\Delta a = \pm 2e\left(\frac{6N}{\Delta\phi}\right)^2 \tag{21}$$

where $\Delta a$ = change in acceleration,

$e$ = error in inches,

$\Delta\phi$ = width of error in degrees. The plus (+) sign is used for a "bump" and the minus (−) sign for a dent or hollow in the surface

For $e = 0.001$, $\Delta\phi = 1°$, and $N = 900$ rpm,

$$\Delta a = +2 \times 0.001\left(\frac{6 \times 900}{1°}\right)^2 = 58320 \text{ in/sec}^2$$

which is 10 times the acceleration calculated for a perfect cam and would cause sufficient force $F_n$ to damage the cam surface. On high-speed cams, therefore, accuracy is of considerable importance.

(e) What is the cam torque at $\phi = 45°$?

From Formula (20),

$$\begin{aligned} T_o &= (R_{min} + y)F_n \sin\alpha \\ &= (1.56 + 0.4) \times 110 \times \sin 30° = 108 \text{ in.-lbs.} \end{aligned}$$

(f) What is the radius of curvature at $\phi = 45°$?

From Formula (10),

$$\rho = \frac{\left[r^2 + \left(\frac{dr}{d\phi}\right)^2\right]^{\frac{3}{2}}}{r^2 + 2\left(\frac{dr}{d\phi}\right)^2 - r\left(\frac{d^2r}{d\phi^2}\right)}$$

$$r = R_{min} + y = 1.56 + 0.4 = 1.96$$

From Formula (13b),

$$\begin{aligned} \frac{dr}{d\phi} &= \frac{180°h}{\pi\beta}\left[1 - \cos\left(\frac{360°\phi}{\beta}\right)\right] = \frac{180° \times 1}{\pi \times 100°}\left[1 - \cos\left(\frac{360° \times 45°}{100°}\right)\right] \\ &= 1.12 \end{aligned}$$

From Formula (13c),

$$\begin{aligned} \frac{d^2r}{d\phi^2} &= \frac{2(180°)^2h}{\pi\beta^2}\sin\left(\frac{360°\phi}{\beta}\right) = \frac{2 \times (180°)^2 \times 1}{\pi \times (100°)^2}\sin\left(\frac{360° \times 45°}{100°}\right) \\ &= 0.64 \end{aligned}$$

$$\rho = \frac{[(1.96)^2 + (1.12)^2]^{\frac{3}{2}}}{(1.96)^2 + 2(1.12)^2 - 1.96 \times 0.64} = 2.26 \text{ in.}$$

**Calculation of Contact Stresses.—**When a roller follower is loaded against a cam, the compressive stress developed at the surface of contact may be calculated from

$$S_c = 2290\sqrt{\frac{F_n}{b}\left(\frac{1}{r_f} \pm \frac{1}{R_c}\right)} \qquad (22)$$

for a steel roller against a steel cam. For a steel roller on a cast iron cam, use 1850 instead of 2290 in Equation (22).

$S_c$ = maximum calculated compressive stress, psi
$F_n$ = normal load, lb
$b$ = width of cam, inch
$R_c$ = radius of curvature of cam surface, inch
$r_f$ = radius of roller follower, inch

The plus sign in (21) is used in calculating the maximum compressive stress when the roller is in contact with the convex portion of the cam profile and the minus sign is used when the roller is in contact with the concave portion. When the roller is in contact with the straight (flat) portion of the cam profile, $R_c = \infty$ and $1/R_c = 0$. In practice, the greatest compressive stress is most apt to occur when the roller is in contact with that part of the cam profile which is convex and has the smallest radius of curvature.

*Example:* Given the previous cam example, the radius of the roller $r_f = 0.25$ in., the convex radius of the cam $R_c = (2.26–0.25)$ in., the width of contact $b = 0.3$ in., and the normal load $F_n = 110$ lbs. Find the maximum surface compressive stress. From (21),

$$S_c = 2290\sqrt{\frac{110}{0.3}\left(\frac{1}{0.25} + \frac{1}{2.01}\right)} = 93{,}000 \text{ psi}$$

This calculated stress should be less than the allowable stress for the material selected from Table 2.

*Cam Materials:* In considering materials for cams it is difficult to select any single material as being the best for every application. Often the choice is based on custom or the machinability of the material rather than its strength. However, the failure of a cam or roller is commonly due to fatigue, so that an important factor to be considered is the limiting wear load, which depends on the surface endurance limits of the materials used and the relative hardnesses of the mating surfaces.

**Table 2. Cam Materials**

| Cam Materials for Use with Roller of Hardened Steel | Maximum Allowable Compressive Stress, psi |
|---|---|
| Gray-iron casting, ASTM A48-48, Class 20, 160–190 BHN, phosphate-coated | 58,000 |
| Gray-iron casting, ASTM A339-51T, Grade 20, 140–160 BHN | 51,000 |
| Nodular-iron casting, ASTM A339-51T, Grade 80-60-03, 207–241 BHN | 72,000 |
| Gray-iron casting, ASTM A48-48, Class 30, 200–220 BHN | 65,000 |
| Gray-iron casting, ASTM A48-48, Class 35, 225–225 BHN | 78,000 |
| Gray-iron casting, ASTM A48-48, Class 30, heat-treated (Austempered), 225–300 BHN | 90,000 |
| SAE 1020 steel, 130–150 BHN | 82,000 |
| SAE 4150 steel, heat-treated to 270–300 BHN, phosphate-coated | 20,000 |
| SAE 4150 steel, heat-treated to 270–300 BHN | 188,000 |
| SAE 1020 steel, carburized to 0.045 in. depth of case, 50–58 RC | 226,000 |
| SAE 1340 steel, induction hardened to 45–55 RC | 198,000 |
| SAE 4340 steel, induction hardened to 50–55 RC | 226,000 |

Based on United Shoe Machinery Corp. data by Guy J. Talbourdet.

In Table 2 are given maximum permissible compressive stresses (surface endurance limits) for various cam materials when in contact with a roller of hardened steel. The stress values shown are based on 100,000,000 cycles or repetitions of stress for pure rolling. Where the repetitions of stress are considerably greater than 100,000,000, where there is appreciable misalignment, or where there is sliding, more conservative stress figures must be used.

**Layout of Cylinder Cams.—**In Fig. 22 is shown the development of a uniformly accelerated motion cam curve laid out on the surface of a cylindrical cam. This development is necessary for finding the projection on the cylindrical surface, as shown at *KL*. To construct the developed curve, first divide the base circle of the cylinder into, say, twelve equal parts. Set off these parts along line *ag*. Only one-half of the layout has been shown, as the other half is constructed in the same manner, except that the curve is here falling instead of rising. Divide line *aH* into the same number of divisions as the half circle, the divisions being in the proportion 1:3:5:5:3:1. Draw horizontal lines from these division points and vertical lines from *a*, *b*, *c*, etc. The intersections between the two sets of lines are points on the developed cam curve. These points are transferred to the cylindrical surface at the left by projection in the usual manner.

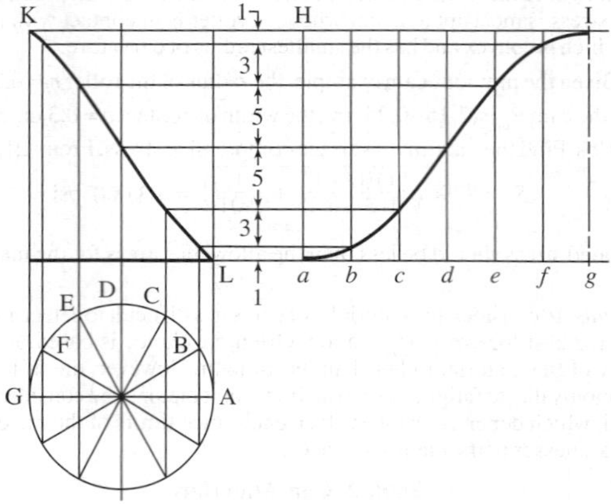


Fig. 22. Development of Cylindrical Cam

**Shape of Rolls for Cylinder Cams.—**The rolls for cylindrical cams working in a groove in the cam should be conical rather than cylindrical in shape, in order that they may rotate freely and without excessive friction. Fig. 23(a) shows a straight roll and groove, the action of which is faulty because of the varying surface speed at the top and bottom of the groove. Fig. 23(b) shows a roll with curved surface. For heavy work, however, the small bearing area is quickly worn down and the roll presses a groove into the side of the cam as well, thus destroying the accuracy of the movement and creating backlash. Fig. 23(c) shows the conical shape which permits a true rolling action in the groove. The amount of taper depends on the angle of spiral of the cam groove. As this angle, as a rule, is not constant for the whole movement, the roll and groove should be designed to meet the requirements on that section of the cam where the heaviest duty is performed. Frequently the cam groove is of a nearly even spiral angle for a considerable length. The method for determining the angle of the roll and groove to work correctly during the important part of the cycle is as follows:

In Fig. 23(d), *b* is the circumferential distance on the surface of the cam that includes the section of the groove for which correct rolling action is required. The throw of the cam for this circumferential movement is *a*. Line *OU* is the development of the movement of the cam roll during the given part of the cycle, and $c$ is the movement corresponding to $b$, but on a circle the diameter of which is equal to that of the cam at the bottom of the groove. With the same throw $a$ as before, the line $OV$ will be the development of the cam at the bottom of the groove. $OU$ then is the length of the helix traveled by the top of the roll, while $OV$ is the travel at the bottom of the groove. If, then, the top width and bottom width of the groove be made proportional to $OU$ and $OV$, the groove will be properly proportioned.

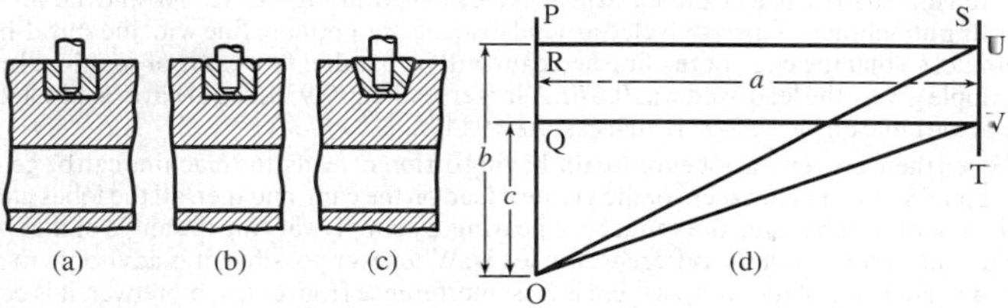


Fig. 23. Shape of Rolls for Cylinder Cams

**Cam Milling.**—Plate cams having a constant rise, such as are used on automatic screw machines, can be cut in a universal milling machine, with the spiral head set at an angle $\alpha$, as shown by the illustration.

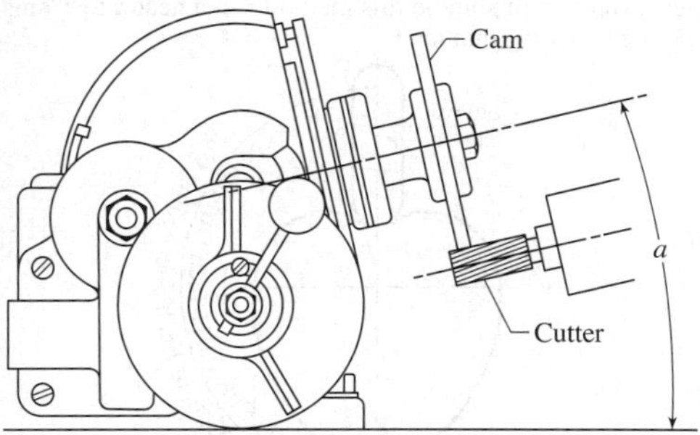


Fig. 24.

When the spiral head is set vertical, the "lead" of the cam (or its rise for one complete revolution) is the same as the lead for which the machine is geared; but when the spiral head and cutter are inclined, any lead or rise of the cam can be obtained, provided it is less than the lead for which the machine is geared, that is, less than the forward feed of the table for one turn of the spiral-head spindle. The cam lead, then, can be varied within certain limits by simply changing the inclination $\alpha$ of the spiral head and cutter. The following formula is for determining this angle of inclination, for a given rise of cam and with the machine geared for a lead, $L$, selected from the tables beginning on page 3554,

$$\sin\alpha = \frac{360° \times r}{\phi \times L}$$

where $\alpha$ = angle to which index head and milling attachment are set from horizontal as shown in the accompanying diagram

$r$ = rise of cam in given part of circumference

$L$ = spiral lead for which milling machine is geared

$\phi$ = angle in which rise is required, expressed in degrees

For example, suppose a cam is to be milled having a rise of 0.125 inch in 300 degrees and that the machine is geared for the smallest possible lead, or 0.670 inch; then:

$$\sin\alpha = \frac{360° \times 0.125}{300° \times 0.670} = 0.2239$$

which is the sine of 12° 56′. Therefore, to secure a rise of 0.125 inch with the machine geared for 0.670 inch lead, the spiral head is elevated to an angle of 12° 56′ and the vertical milling attachment is also swiveled around to locate the cutter in line with the spiral-head spindle, so that the edge of the finished cam will be parallel to its axis of rotation. In the example given, the lead used was 0.670. A larger lead, say 0.930, could have been selected from the table on page 3548. In that case, $\alpha = 9° 17′$.

When there are several lobes on a cam, having different leads, the machine can be geared for a lead somewhat in excess of the greatest lead on the cam, and then all the lobes can be milled without changing the spiral head gearing, by simply varying the angle of the spiral head and cutter to suit the different cam leads. Whenever possible, it is advisable to mill on the under side of the cam, as there is less interference from chips; moreover, it is easier to see any lines that may be laid out on the cam face. To set the cam for a new cut, it is first turned back by operating the handle of the table feed screw, after which the index crank is disengaged from the plate and turned the required amount.

**Simple Method for Cutting Uniform Motion Cams.**—Some cams are laid out with dividers, machined and filed to the line; but for a cam that must advance a certain number of thousandths per revolution of spindle this method is not accurate. Cams are easily and accurately cut in the following manner.

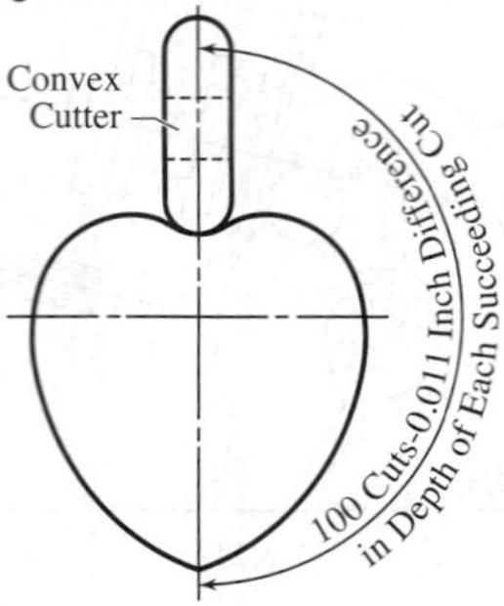


Let it be required to make the heart cam shown in the illustration. The throw of this cam is 1.1 inch. Now, by setting the index on the milling machine to cut 200 teeth and also dividing 1.1 inch by 100, we find that we have 0.011 inch to recede from or advance towards the cam center for each cut across the cam. Placing the cam securely on an arbor, and the latter between the centers of the milling machine, and using a convex cutter set the proper distance from the center of the arbor, make the first cut across the cam. Then, by lowering the milling machine knee 0.011 inch and turning the index pin the proper number of holes on the index plate, take the next cut and so on.

## TABLE OF CONTENTS
## MACHINE ELEMENTS

### PLAIN BEARINGS



### BALL, ROLLER, AND NEEDLE BEARINGS



### BALL, ROLLER, AND NEEDLE BEARINGS

*(Continued)*



### LUBRICATION



MACHINE ELEMENTS

## TABLE OF CONTENTS
## MACHINE ELEMENTS

### LUBRICATION

*(Continued)*



### COUPLINGS, CLUTCHES, BRAKES



### KEYS AND KEYSEATS



### KEYS AND KEYSEATS

*(Continued)*



### FLEXIBLE BELTS AND SHEAVES

TABLE OF CONTENTS

MACHINE ELEMENTS

## FLEXIBLE BELTS AND SHEAVES

*(Continued)*



## TRANSMISSION CHAINS



## BALL AND ACME LEADSCREWS



## ELECTRIC MOTORS



## ADHESIVES AND SEALANTS



## O-RINGS

TABLE OF CONTENTS

MACHINE ELEMENTS

## PLAIN BEARINGS

### Introduction

On the following pages are given data and procedures for designing full-film or hydrodynamically lubricated bearings of the journal and thrust types. However, before proceeding to these design methods, it is useful to first review those bearing aspects concerning the types of bearings available; lubricants and lubrication methods; hardness and surface finish; machining methods; seals; retainers; and typical length-to-diameter ratios for various applications.

The following paragraphs preceding the design sections provide guidance in these matters and suggest modifications in allowable loads when other than full-film operating conditions exist in a bearing.

**Classes of Plain Bearings.—**Bearings that provide sliding contact between mating surfaces fall into three general classes: *radial bearings* that support rotating shafts or journals; *thrust bearings* that support axial loads on rotating members; and *guide* or *slipper bearings* that guide moving parts in a straight line. Radial sliding bearings, more commonly called sleeve bearings, may be of several types, the most usual being the plain full journal bearing, which has 360-degree contact with its mating journal, and the partial journal bearing, which has less than 180-degree contact. This latter type is used when the load direction is constant and has the advantages of simplicity, ease of lubrication, and reduced frictional loss.

The relative motions between the parts of plain bearings may take place: 1) as pure sliding without the benefit of a liquid or gaseous lubricating medium between the moving surfaces such as with the dry operation of nylon or Teflon; 2) with hydrodynamic lubrication in which a wedge or film buildup of lubricating medium is produced, with either whole or partial separation of the bearing surfaces; 3) with hydrostatic lubrication in which a lubricating medium is introduced under pressure between the mating surfaces causing a force opposite to the applied load and a lifting or separation of these surfaces; and 4) with a hybrid form or combination of hydrodynamic and hydrostatic lubrication.

Listed below are some of the advantages and disadvantages of sliding contact (plain) bearings as compared with rolling contact (antifriction) bearings.

*Advantages:* 1) require less space; 2) are quieter in operation; 3) are lower in cost, particularly in high-volume production; 4) have greater rigidity; and 5) their life is generally not limited by fatigue.

*Disadvantages:* 1) have higher frictional properties resulting in higher power consumption; 2) are more susceptible to damage from foreign material in lubrication system; 3) have more stringent lubrication requirements; and 4) are more susceptible to damage from interrupted lubrication supply.

**Types of Journal Bearings.—**Many types of journal bearing configurations have been developed; some of these are shown in Fig. 1.

*Circumferential-groove bearings,* Fig. 1(a), have an oil groove extending circumferentially around the bearing. The oil is maintained under pressure in the groove. The groove divides the bearing into two shorter bearings that tend to run at a slightly greater eccentricity. However, the advantage in terms of stability is slight, and this design is most commonly used in reciprocating-load main and connecting-rod bearings because of the uniformity of oil distribution.

Short cylindrical bearings are a better solution than the circumferential-groove bearing for high-speed, low-load service. Often the bearing can be shortened enough to increase the unit loading to a substantial value, causing the shaft to ride at a position of substantial eccentricity in the bearing. Experience has shown that instability rarely results when the shaft eccentricity is greater that 0.6. Very short bearings are not often used for this type of application, because they do not provide a high temporary rotating-load capacity in the event some unbalance should be created in the rotor during service.

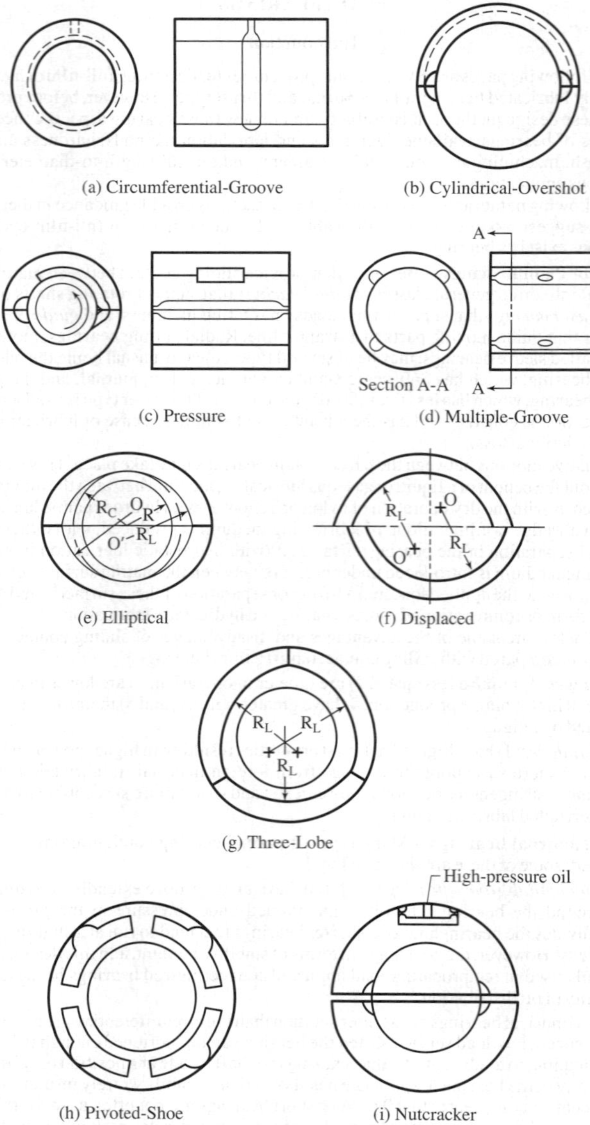


Fig. 1. Typical Shapes of Several Types of Pressure-Fed Bearings.

*Cylindrical-overshot bearings,* Fig. 1(b), are used where surface speeds of 10,000 fpm or more exist, and where additional oil flow is desired to maintain a reasonable bearing temperature. This bearing has a wide circumferential groove extending from one axial oil groove to the other over the upper half of the bearing. Oil is usually admitted to the trailing-edge oil groove. An inlet orifice is used to control the oil flow. Cooler operation results from the elimination of shearing action over a large section of the upper half of the bearing and, to a great extent, from the additional flow of cool oil over the top half of the bearing.

*Pressure bearings,* Fig. 1(c), employ a groove over the top half of the bearing. The groove terminates at a sharp dam about 45 degrees beyond the vertical in the direction of shaft rotation. Oil is pumped into this groove by shear action from the rotation of the shaft and is then stopped by the dam. In high-speed operation, this situation creates a high oil pressure over the upper half of the bearing. The pressure created in the oil groove and surrounding upper half of the bearing increases the load on the lower half of the bearing. This self-generated load increases the shaft eccentricity. If the eccentricity is increased to 0.6 or greater, stable operation under high-speed, low-load conditions can result. The central oil groove can be extended around the lower half of the bearing, further increasing the effective loading. This design has one primary disadvantage: Dirt in the oil will tend to abrade the sharp edge of the dam and impair ability to create high pressures.

*Multiple-groove bearings,* Fig. 1(d), are sometimes used to provide increased oil flow. The interruptions in the oil film also appear to give this bearing some merit as a stable design.

*Elliptical bearings,* Fig. 1(e), are not truly elliptical, but are formed from two sections of a cylinder. This two-piece bearing has a large clearance in the direction of the split and a smaller clearance in the load direction at right angles to the split. At light loads, the shaft runs eccentric to both halves of the bearing, and hence, the elliptical bearing has a higher oil flow than the corresponding cylindrical bearing. Thus, the elliptical bearing will run cooler and will be more stable than a cylindrical bearing.

*Elliptical-overshot bearings* (not shown) are elliptical bearings in which the upper half is relieved by a wide oil groove connecting the axial oil grooves. They are analogous to cylindrical-overshot bearings.

*Displaced elliptical bearings,* Fig. 1(f), shift the centers of the two bearing arcs in both a horizontal and a vertical direction. This design has greater stiffness than a cylindrical bearing, in both horizontal and vertical directions, with substantially higher oil flow. It has not been extensively used, but offers the prospect of high stability and cool operation.

*Three-lobe bearings,* Fig. 1(g), are made up in cross section of three circular arcs. They are most effective as antioil whip bearings when the centers of curvature of each of the three lobes lie well outside the clearance circle that the shaft center can describe within the bearing. Three axial oil-feed grooves are used. It is a more difficult design to manufacture, because it is almost necessary to make it in three parts instead of two. The bore is machined with shims between each of the three parts. The shims are removed after machining is completed.

*Pivoted-shoe bearings,* Fig. 1(h), are one of the most stable bearings. The bearing surface is divided into three or more segments, each of which is pivoted at the center. In operation, each shoe tilts to form a wedge-shaped oil film, thus creating a force tending to push the shaft toward the center of the bearing. For single direction rotation, the shoes are sometimes pivoted near one end and forced toward the shaft by springs.

*Nutcracker bearings,* Fig. 1(i), consist of two cylindrical half-bearings. The upper half-bearing is free to move in a vertical direction and is forced toward the shaft by a hydraulic cylinder. External oil pressure may be used to create load on the upper half of the bearing through the hydraulic cylinder. Or the high-pressure oil may be obtained from the lower half of the bearing by tapping a hole into the high-pressure oil film, thus creating a self-loading bearing. Either type can increase bearing eccentricity to the point where stable operation can be achieved.

**Hydrostatic Bearings.**—Hydrostatic bearings are used when operating conditions require full film lubrication that cannot be developed hydrodynamically. The hydrostatically lubricated bearing, either thrust or radial, is supplied with lubricant under pressure from an external source. Some advantages of the hydrostatic bearing over bearings of other types are: low friction; high load capacity; high reliability; high stiffness; and long life.

Hydrostatic bearings are used successfully in many applications including machine tools, rolling mills, and other heavily loaded slow-moving machinery. However, specialized techniques, including a thorough understanding of hydraulic components external to the bearing package is required. The designer is cautioned against use of this type of bearing without a full knowledge of all aspects of the problem. Determination of the operating performance of hydrostatic bearings is a specialized area of the lubrication field and is described in specialized reference books.

**Guide Bearings.**—This type of bearing is generally used as a positioning device or as a guide to linear motion such as in machine tools. Fig. 2 shows several examples of guideway bearing designs. It is normal for this type of bearing to operate in the boundary lubrication region with either dry, dry film such as molybdenum disulfide ($MoS_2$) or tetrafluorethylene (TFE), grease, oil, or gaseous lubrication. Hydrostatic lubrication is often used to improve performance, reduce wear, and increase stability. This type of design uses pumps to supply air or gas under pressure to pockets designed to produce a bearing film and maintain complete separation of the sliding surfaces.

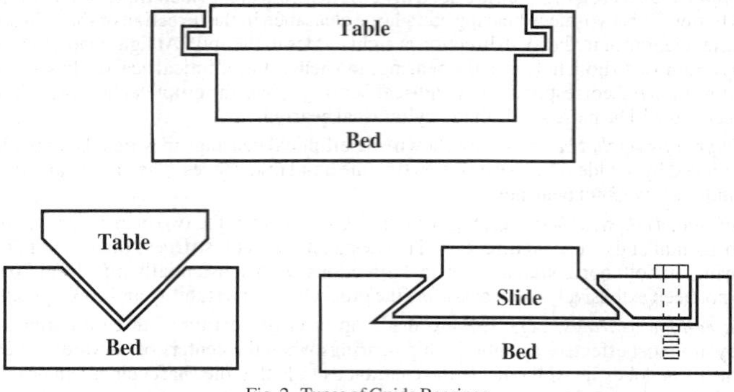


Fig. 2. Types of Guide Bearings

**Design.**—The design of a sliding bearing is generally accomplished in one of two ways: 1) a bearing operating under similar conditions is used as a model or basis from which the new bearing is designed; and 2) in the absence of any previous experience with similar bearings in similar environments, certain assumptions concerning operating conditions and requirements are made and a tentative design prepared based on general design parameters or rules of thumb. Detailed lubrication analysis is then performed to establish design and operating details and requirements.

**Modes of Bearing Operation.**—The load-carrying ability of a sliding bearing depends upon the kind of fluid film that is formed between its moving surfaces. The formation of this film is dependent, in part, on the design of the bearing and, in part, on the speed of rotation. The bearing has three modes or regions of operation designated as *full-film*, *mixed-film*, and *boundary* lubrication with effects on bearing friction, as shown in Fig. 3.

In terms of physical bearing operation these three modes may be further described as follows:

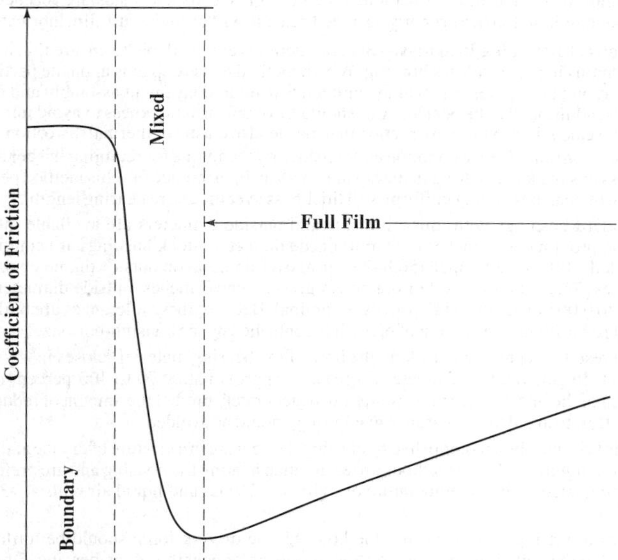


Fig. 3. Three Modes of Bearing Operation.

1) Full-film, or hydrodynamic, lubrication produces a complete physical separation of the sliding surfaces resulting in low friction and long wear-free service life.

To promote full-film lubrication in hydrodynamic operation, the following parameters should be satisfied: a) Lubricant selected has the correct viscosity for the proposed operation; b) proper lubricant flow rates are maintained; c) proper design methods and considerations have been utilized; d) and surface velocity in excess of 25 fpm (7.62 m/min) is maintained.

When full-film lubrication is achieved, a coefficient of friction between 0.001 and 0.005 can be expected.

2) Mixed-film lubrication is a mode of operation between the full-film and boundary modes. With this mode, there is a partial separation of the sliding surfaces by the lubricant film; however, as in boundary lubrication, limitations on surface speed and wear will result. With this type of lubrication, a surface velocity in excess of 10 fpm (3.05 m/min) is required with resulting coefficients of friction of 0.02 to 0.08.

3) Boundary lubrication takes place when the sliding surfaces are rubbing together with only an extremely thin film of lubricant present. This type of operation is acceptable only in applications with oscillating or slow rotary motion. In complete boundary lubrication, the oscillatory or rotary motion is usually less than 10 feet per minute with resulting coefficients of friction of 0.08 to 0.14. These bearings are usually grease lubricated or periodically oil lubricated.

In starting up and accelerating to its operating point, a journal bearing passes through all three modes of operation. At rest, the journal and bearing are in contact, and thus when starting, the operation is in the boundary lubrication region. As the shaft begins to rotate more rapidly and the hydrodynamic film starts to build up, bearing operation enters the region of mixed-film lubrication. When design speeds and loads are reached, the hydrodynamic action in a properly designed bearing will promote full-film lubrication.

**Methods of Retaining Bearings.—**Several methods are available to ensure that a bearing remains in place within a housing. Which method to use depends upon the particular application but requires first that the unit lends itself to convenient assembly and disassembly; additionally, the bearing wall should be of uniform thickness to avoid introduction of weak points in the construction that may lead to elastic or thermal distortion.

*Press or Shrink Fit:* One common and satisfactory technique for retaining the bearing is to press or shrink the bearing in the housing with an interference fit. This method permits the use of bearings having uniform wall thickness over the entire bearing length.

Standard bushings with finished inside and outside diameters are available in sizes up to approximately 5 inches (127 mm) inside diameter. Stock bushings are commonly provided 0.002 to 0.003 inch (50.8–76.2 μm) over nominal on outside diameter sizes of 3 inches (76.2 mm) or less. For diameters greater than 3 inches, outside diameters are 0.003 to 0.005 inch (76.2–127 μm) over nominal. Because these tolerances are built into standard bushings, the amount of press fit is controlled by the housing-bore size.

As a result of a press or shrink fit, the bore of the bearing material "closes in" by some amount. In general, this diameter decrease is approximately 70 to 100 percent of the amount of the interference fit. Any attempt to accurately predict the amount of reduction, in an effort to avoid final clearance machining, should be avoided.

Shrink fits may be accomplished by chilling the bearing in a mixture of dry ice and alcohol, or in liquid air. These methods are easier than heating the housing and are preferred. Dry ice in alcohol has a temperature of −110° F (−79° C) and liquid air boils at −310° F (−190° C).

When a bearing is pressed into the housing, the driving force should be uniformly applied to the end of the bearing to avoid upsetting or peening of the bearing. Of equal importance, the mating surfaces must be clean, smoothly finished, and free of machining imperfections.

*Keying Methods:* A variety of methods can be used to fix the position of the bearing with respect to its housing by "keying" the two together. Possible keying methods are shown in Fig. 4a through Fig. 4f including: a) set screws; b) Woodruff keys; c) bolted bearing flanges; d) threaded bearings; e) dowel pins; and f) housing caps.

Factors to be considered when selecting one of these methods are as follows:

1) Maintaining uniform wall thickness of the bearing material, if possible, especially in the load-carrying region of the bearing.

2) Providing as much contact area as possible between bearing and housing. Mating surfaces should be clean, smooth, and free from imperfections to facilitate heat transfer.

3) Preventing any local deformation of the bearing that might result from the keying method. Machining after keying is recommended.

4) Considering the possibility of bearing distortion resulting from the effect of temperature changes on the particular keying method.

**Methods of Sealing.—**In applications where lubricants or process fluids are utilized in operation, provision must be made normally to prevent leakage to other areas. This provision is made by the use of static and dynamic type sealing devices. In general, three terms are used to describe the devices used for sealing:

*Seal:* A means of preventing migration of fluids, gases, or particles across a joint or opening in a container.

*Packing:* A dynamic seal, used where some form of relative motion occurs between rigid members of an assembly.

*Gaskets:* A static seal, used where there is no relative motion between joined parts.

**Methods of Bearing Retention**

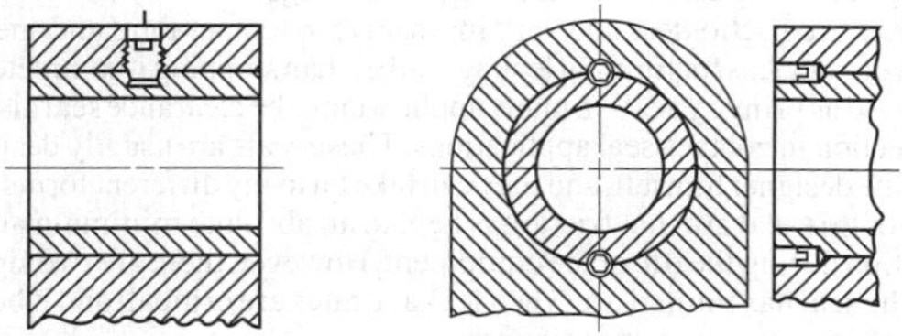

Fig. 4a. Set Screws

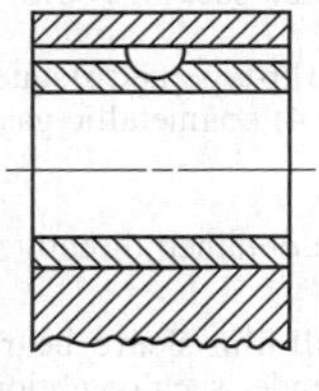

Fig. 4b. Woodruff Key

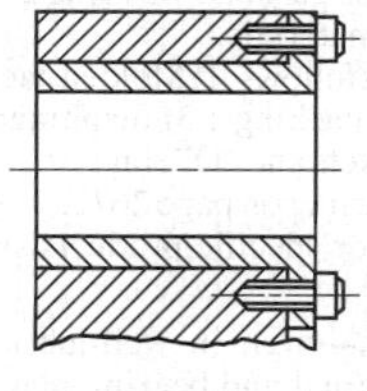

Fig. 4c. Bolts through Flange

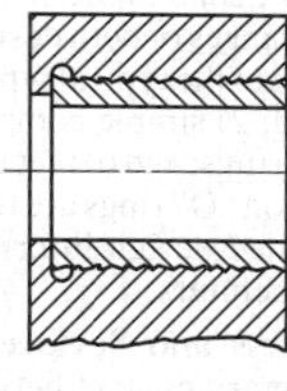

Fig. 4d. Bearing Screwed into Housing

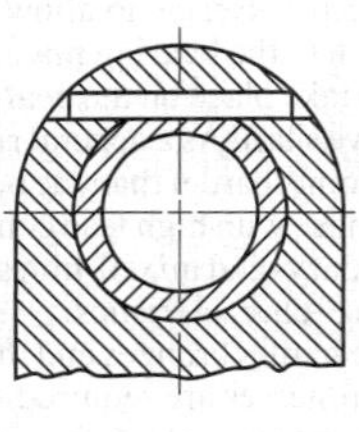

Fig. 4e. Dowel Pin

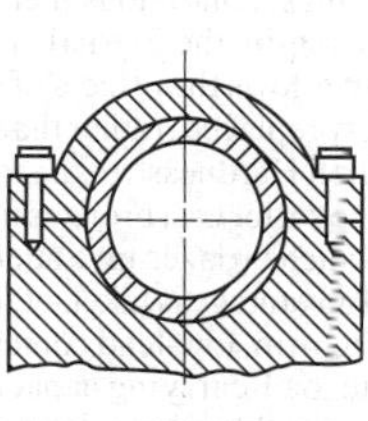

Fig. 4f. Housing Cap

Two major functions must be achieved by all sealing applications: prevent escape of fluid; and prevent migration of foreign matter from the outside.

The first determination in selecting the proper seal is whether the application is static or dynamic. To meet the requirements of a static application there must be no relative motion between the joining parts or between the seal and the mating part. If there is any relative motion, the application must be considered dynamic, and the seal selected accordingly.

Dynamic sealing requires control of fluids leaking between parts with relative motion. Two primary methods are used to this end: positive contact or rubbing seals; and controlled clearance noncontact seals.

*Positive Contact or Rubbing Seals:* These seals are used where positive containment of liquids or gases is required, or where the seal area is continuously flooded. If properly selected and applied, contact seals can provide zero leakage for most fluids. However, because they are sensitive to temperature, pressure, and speed, improper application can result in early failure. These seals are applicable to rotating and reciprocating shafts. In many assemblies, positive-contact seals are available as off-the-shelf items. In other instances, they are custom-designed to the special demands of a particular application. Custom design is offered by many seal manufacturers and, for extreme cases probably offers the best solution to the sealing problem.

*Controlled Clearance Noncontact Seals:* Representative of the controlled-clearance seals, which includes all seals in which there is no rubbing contact between the rotating and stationary members, are throttling bushings and labyrinths. Both types operate by fluid-throttling action in narrow annular or radial passages.

Clearance seals are frictionless and very insensitive to temperature and speed. They are chiefly effective as devices for limiting leakage rather than stopping it completely. Although they are employed as primary seals in many applications, the clearance seal also finds use as auxiliary protection in contact-seal applications. These seals are usually designed into the equipment by the designer himself, and they can take on many different forms.

Advantages of this seal are that friction is kept to an absolute minimum and there is no wear or distortion during the life of the equipment. However, there are two significant disadvantages: The seal has limited use when leakage rates are critical; and it becomes quite costly as the configuration becomes elaborate.

*Static Seals:* Static seals such as gaskets, "O" rings, and molded packings cover very broad ranges of both design and materials.

Some of the typical types are as follows: 1) Molded packings: a) lip type, and b) squeeze-molded; 2) simple compression packings; 3) diaphragm seals; 4) nonmetallic gaskets; 5) "O" rings; and 6) metallic gaskets and "O" rings.

Data on "O" rings are found starting on page 2672.

Detailed design information for specific products should be obtained directly from manufacturers.

**Hardness and Surface Finish.—**Even in well-lubricated full-film sleeve bearings, momentary contact between journal and bearing may occur under such conditions as starting, stopping, or overloading. In mixed-film and boundary-film lubricated sleeve bearings, continuous metal-to-metal contact occurs. Hence, to allow for any necessary wearing-in, the journal is usually made harder than the bearing material. This arrangement allows the effects of scoring or wearing to take place on the bearing, which is more easily replaced, rather than on the more expensive shaft. As a general rule, recommended Brinell Hardness* of the journal is at least 100 points harder than the bearing material.

The softer cast bronzes used for bearings are those with high lead content and very little tin. Such bronzes give adequate service in boundary- and mixed-film applications where full advantage is taken of their excellent "bearing" characteristics.

High-tin, low-lead content cast bronzes are the harder bronzes and these have high ultimate load-carrying capacity: higher journal hardnesses are required with these bearing bronzes. Aluminum bronze, for example, requires a journal hardness in the range of 550 to 600 BHN (Brinell Hardness Number).

In general, harder bearing materials require better alignment and more reliable lubrication to minimize local heat generation if and when the journal touches the shaft. Also, abrasives that find their way into the bearing are a problem for the harder bearing materials and greater care should be taken to exclude them.

*Surface Finish:* Whether bearing operation is complete boundary, mixed film, or fluid film, surface finishes of the journal and bearing must receive careful attention. In applications where operation is hydrodynamic or full-film, peak surface variations should be less than the expected minimum film thickness; otherwise, peaks on the journal surface will contact peaks on the bearing surface, with resulting high friction and temperature rise. Ranges of surface roughness obtained by various finishing methods are: boring, broaching, and reaming, 32 to 64 μin (0.813–1.626 μm), rms; grinding, 16 to 64 μin (0.406–1.626 μm), rms; and fine grinding, 4 to 16 μin (0.102–0.406 μm), rms.

In general, the better surface finishes are required for full-film bearings operating at high eccentricity ratios because full-film lubrication must be maintained with small clearances, and metal-to-metal contact must be avoided. Also, the harder the material, the better the surface finish required. For boundary- and mixed-film applications, surface finish requirements may be somewhat relaxed because bearing wear-in will in time smooth the surfaces.

* Brinell Hardness Number (BHN) also may be identified with the letters HB, HBS for hardness measured with a steel indenter, and HBW for hardness measured with a carbide indenter.

Fig. 5 is a general guide to the ranges required for bearing and journal surface finishes. Selecting a particular surface finish in each range can be simplified by observing the general rule that smoother finishes are required for the harder materials, for high loads, and for high speeds.

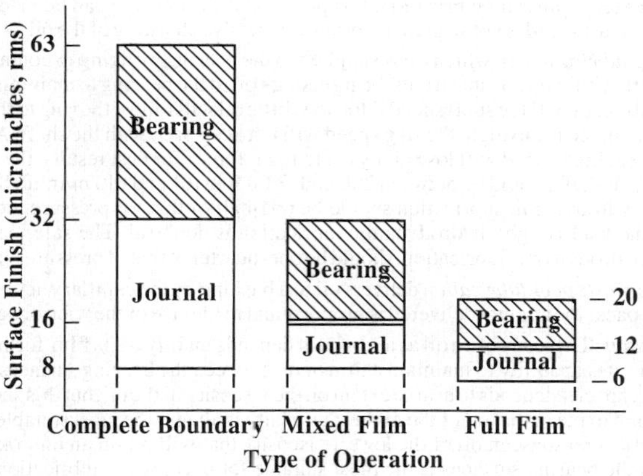


Fig. 5. Recommended Ranges of Surface Finish for the Three Types of Sleeve Bearing Operations.

**Machining Bores.—**The methods most commonly used in finishing journal bearing bores are boring, broaching, reaming, and burnishing.

Broaching is a rapid finishing method providing good size and alignment control when adequate piloting is possible. Soft babbitt materials are particularly compatible with the broaching method. A third finishing method, reaming, facilitates good size and alignment control when piloting is utilized. Reaming can be accomplished both manually or by machine, the machine method being preferred. Burnishing is a fast sizing operation that gives good alignment control, but does not give as good size control as the cutting methods. It is not recommended for soft materials such as babbitt. Burnishing has an ironing effect that gives added seating of the bushing outside diameter in the housing bore; consequently, it is often used for this purpose, especially on a $1/32$-inch (0.794-mm) wall bushing, even if a further sizing operation is to be used subsequently.

Boring of journal bearings provides the best concentricity, alignment, and size control and is the finishing method of choice when close tolerances and clearances are desirable.

**Methods of Lubrication.—**There are numerous ways to supply lubricant to bearings. The more common of these are described in the following.

*Pressure lubrication,* in which an abundance of oil is fed to the bearing from a central groove, single or multiple holes, or axial grooves, is effective and efficient. The moving oil assists in flushing dirt from the bearing and helps keep the bearing cool. In fact, it removes heat faster than other lubricating methods and, therefore, permits thinner oil films and unimpaired load capacities. The oil-supply pressure needed for bushings carrying the basic load is directly proportional to the shaft speed, but for most installations, 50 psi (345 kPa) will be adequate.

*Splash fed* applies to a variety of intermittently lubricated bushings. It includes everything from bearings spattered with oil from the action of other moving parts to bearings regularly dipped in oil. Like oil bath lubrication, splash feeding is practical when the housing can be made oiltight and when the moving parts do not churn the oil. The fluctuating nature of the load and the intermittent oil supply in splash fed applications requires the designer to use experience and judgment when determining the probable load capacity of bearings lubricated in this way.

*Oil bath lubrication,* in which the bushing is submerged in oil, is the most reliable of all methods except pressure lubrication. It is practical if the housing can be made oil tight, and if the shaft speed is not so great as to cause excessive churning of the oil.

*Oil ring lubrication,* in which oil is supplied to the bearing by a ring in contact with the shaft, will, within reasonable limits, bring enough oil to the bearing to maintain hydrodynamic lubrication. If the shaft speed is too low, little oil will follow the ring to the bearing; and, if the speed is too high, the ring speed will not keep pace with the shaft. Also, a ring revolving at high speed will lose oil by centrifugal force. For best results, the peripheral speed of the shaft should be between 200 and 2000 fpm (61 and 610 m/min). Safe load to achieve hydrodynamic lubrication should be one-half of that for pressure fed bearings. Unless the load is light, hydrodynamic lubrication is doubtful. The safe load, then, to achieve hydrodynamic lubrication, should be one-quarter of that of pressure fed bearings.

*Wick or waste pack lubrication* delivers oil to a bushing by the capillary action of a wick or waste pack; the amount delivered is proportional to the size of the wick or pack.

*Lubricants:* The value of an oil as a lubricant depends mainly on its film-forming capacity, that is, its capability to maintain a film of oil between the bearing surfaces. The film-forming capacity depends to a large extent on the viscosity of the oil, but this should not be understood to mean that oil of the highest viscosity is always the most suitable lubricant. For practical reasons, an oil of the lowest viscosity that will retain an unbroken oil film between the bearing surfaces is the most suitable for purposes of lubrication. A higher viscosity than that necessary to maintain the oil film results in a waste of power due to the expenditure of energy necessary to overcome the internal friction of the oil itself.

Fig. 6 provides representative values of viscosity in centipoises for SAE mineral oils. Table 56a on page 2882 is provided as a means of converting viscosities of other units to centipoises.

*Grease* packed in a cavity surrounding the bushing is less adequate than an oil system, but it has the advantage of being more or less permanent. Although hydrodynamic lubrication is possible under certain very favorable circumstances, boundary lubrication is the usual state.

**Lubricant Selection.—**In selecting lubricants for journal bearing operation, several factors must be considered: 1) type of operation (full, mixed, or boundary film) anticipated; 2) surface speed; 3) and bearing loading.

Fig. 7 combines these factors and facilitates general selection of the proper lubricant viscosity range.

As an example of using these curves, consider a lightly loaded bearing operating at 2000 rpm. At the bottom of the figure, locate 2000 rpm and move vertically to intersect the light-load full-film lubrication curve, which indicates an SAE 5 oil.

As a general rule-of-thumb, heavier oils are recommended for high loads and lighter oils for high speeds.

In addition, other than using conventional lubrication oils, journal bearings may be lubricated with greases or solid lubricants. Some of the reasons for use of these lubricants are to:

1) Lengthen the period between relubrication:
2) Avoid contaminating surrounding equipment or material with "leaking" lubricating oil;
3) Provide effective lubrication under extreme temperature ranges;
4) Provide effective lubrication in the presence of contaminating atmospheres; and
5) Prevent intimate metal-to-metal contact under conditions of high unit pressure which might destroy boundary lubricating films.

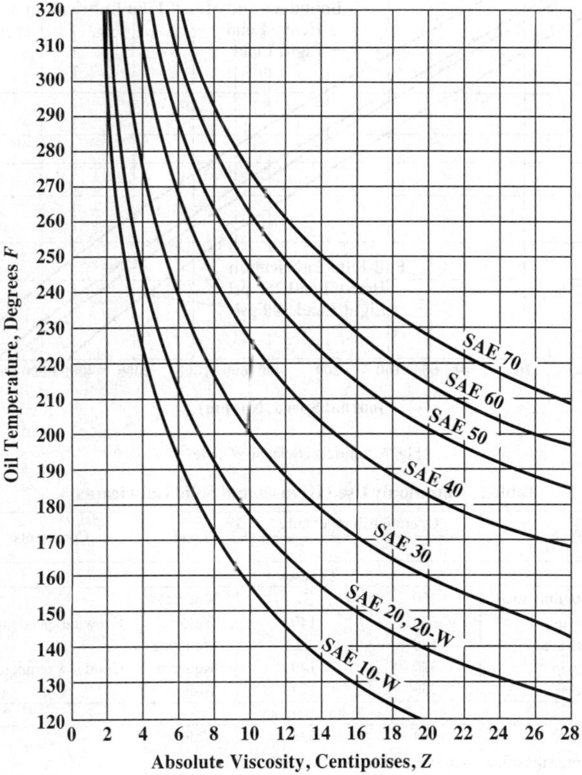


Fig. 6. Viscosity versus Temperature — SAE Oils.

*Greases:* Where full-film lubrication is not possible or is impractical for slow-speed fairly high-load applications, greases are widely used as bearing lubricants. Although full-film lubrication with grease is possible, it is not normally considered since an elaborate pumping system is required to continuously supply a prescribed amount of grease to the bearing. Bearings supplied with grease are usually lubricated periodically. Grease lubrication, therefore, implies that the bearing will operate under conditions of complete boundary lubrication and should be designed accordingly.

Lubricating greases are essentially a combination of a mineral lubricating oil and a thickening agent, which is usually a metallic soap. When suitably mixed, they make excellent bearing lubricants. There are many different types of greases which in general, may be classified according to the soap base used. Information on commonly used greases is shown in Table 1.

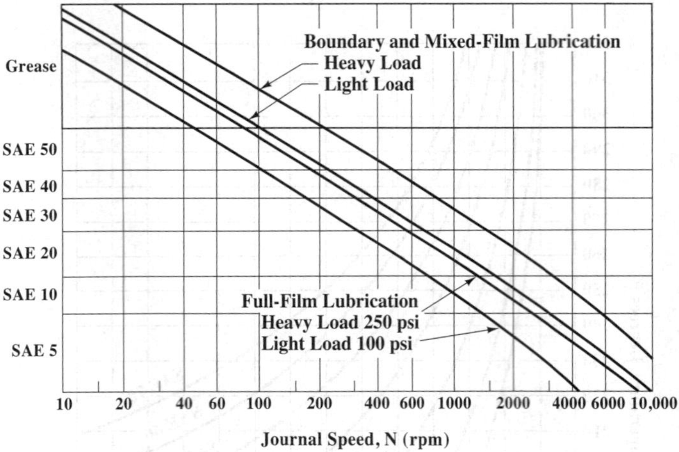


Fig. 7. Lubricant Selection Guide

**Table 1. Commonly Used Greases and Solid Lubricants**

| Type | Operating Temperature | | Load | Comments |
|---|---|---|---|---|
| | °F | °C | | |
| Greases | | | | |
| Calcium or lime soap | 160 | 71 | Moderate | … |
| Sodium soap | 300 | 149 | Wide | For wide speed range |
| Aluminum soap | 180 | 82 | Moderate | … |
| Lithium soap | 300 | 149 | Moderate | Good low temperature |
| Barium soap | 350 | 177 | Wide | … |
| Solid Lubricants | | | | |
| Graphite | 1000 | 538 | Wide | … |
| Molybdenum disulfide | −100 to 750 | −73 to 399 | Wide | … |

Synthetic greases are composed of normal types of soaps but use synthetic hydrocarbons instead of normal mineral oils. They are available in a range of consistencies in both water-soluble and insoluble types. Synthetic greases can accommodate a wide range of variation in operating temperature; however, recommendations on special-purpose greases should be obtained from the lubricant manufacturer.

Application of grease is accomplished by one of several techniques depending upon grease consistency. These classifications are shown in Table 2 along with typical methods of application. Grooves for grease are generally greater in width, up to 1.5 times, than for oil.

Coefficients of friction for grease-lubricated bearings range from 0.08 to 0.16, depending upon consistency of the grease, frequency of lubrication, and type of grease. An average value of 0.12 may be used for design purposes.

*Solid Lubricants:* The need for effective high-temperature lubricants led to the development of several solid lubricants. Essentially, solid lubricants may be described as low-shear-strength solid materials. Their function within a bronze bearing is to act as an intermediary material between sliding surfaces. Since these solids have very low shear

**Table 2. NLGI Consistency Numbers**

| NLGI[a] Consistency No. | Consistency of Grease | Typical Method of Application |
|---|---|---|
| 0 | Semifluid | Brush or gun |
| 1 | Very soft | Pin-type cup or gun |
| 2 | Soft | Pressure gun or centralized pressure system |
| 3 | Light cup grease | Pressure gun or centralized pressure system |
| 4 | Medium cup grease | Pressure gun or centralized pressure system |
| 5 | Heavy cup grease | Pressure gun or hand |
| 6 | Block grease | Hand, cut to fit |

[a] NLGI is National Lubricating Grease Institute

strength, they shear more readily than the bearing material and thereby allow relative motion. So long as solid lubricant remains between the moving surfaces, effective lubrication is provided and friction and wear are reduced to acceptable levels.

Solid lubricants provide the most effective boundary films in terms of reduced friction, wear, and transfer of metal from one sliding component to the other. However, there is a significant deterioration in these desirable properties as the operating temperature of the boundary film approaches the melting point of the solid film. At this temperature the friction may increase by a factor of 5 to 10 and the rate of metal transfer may increase by as much as 1000. What occurs is that the molecules of the lubricant lose their orientation to the surface that exists when the lubricant is solid. As the temperature further increases, additional deterioration sets in with the friction increasing by some additional small amount but the transfer of metal accelerates by an additional factor of 20 or more. The final effect of too high temperature is the same as metal-to-metal contact without benefit of lubricant. These changes, which are due to the physical state of the lubricant, are reversed when cooling takes place.

The effects just described also partially explain why fatty acid lubricants are superior to paraffin base lubricants. The fatty acid lubricants react chemically with the metallic surfaces to form a metallic soap that has a higher melting point than the lubricant itself, the result being that the breakdown temperature of the film, now in the form of a metallic soap is raised so that it acts more like a solid film lubricant than a fluid film lubricant.

## Journal or Sleeve Bearings

Although this type of bearing may take many shapes and forms, there are always three basic components: journal or shaft, bushing or bearing, and lubricant. Fig. 1 shows these components with the nomenclature generally used to describe a journal bearing: $W$ = applied load, $N$ = revolution, $e$ = eccentricity of journal center to bearing center, $\theta$ = attitude angle, which is the angle between the applied load and the point of minimum film thickness, $d$ = diameter of the shaft, $c_d$ = bearing clearance, $d + c_d$ = diameter of the bearing and $h_o$ = minimum film thickness.

**Grooving and Oil Feeding.—**Grooving in a journal bearing has two purposes:

1) to establish and maintain an efficient film of lubricant between the bearing moving surfaces and

2) to provide adequate bearing cooling

The obvious and only practical location for introducing lubricant to the bearing is in a region of low pressure. A typical pressure profile of a bearing is shown by Fig. 2. The arrow $W$ shows the applied load. Typical grooving configurations used for journal bearings are shown in Fig. 3a through Fig. 3e.

**Heat-Radiating Capacity.—**In a self-contained lubrication system for a journal bearing, the heat generated by bearing friction must be removed to prevent continued temperature

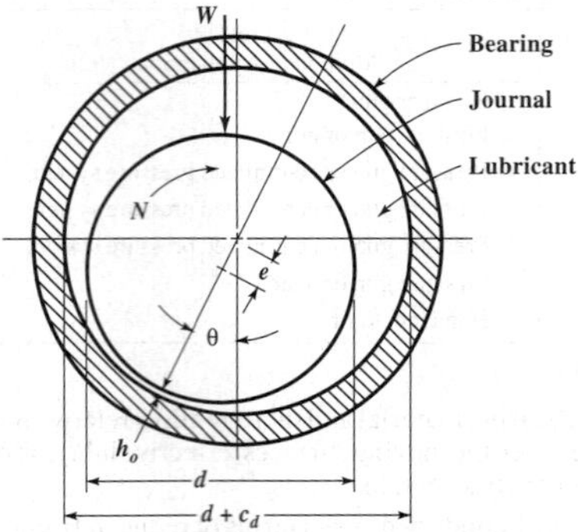


Fig. 1. Basic Components of a Journal Bearing.

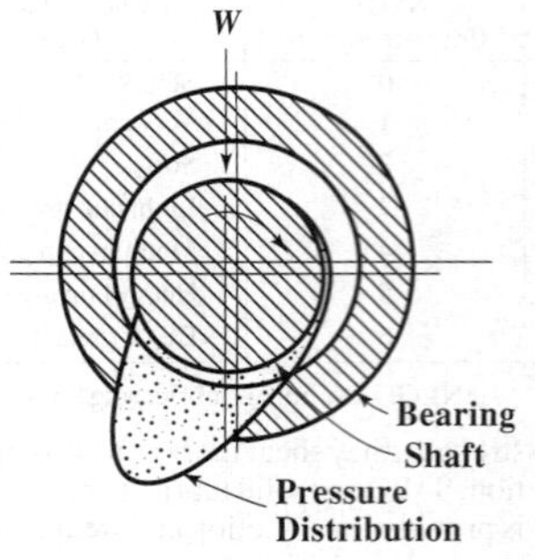


Fig. 2. Typical Pressure Profile of Journal Bearing.

**Types of Journal Bearing Oil Grooving**

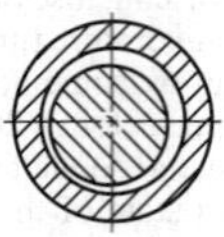

Fig. 3a. Single Inlet Hole

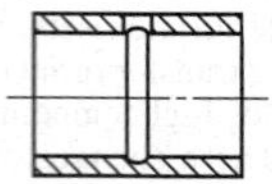

Fig. 3b. Circular Groove

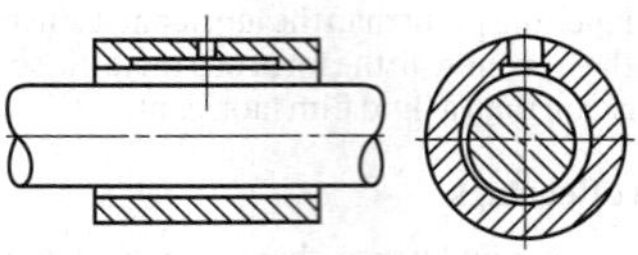

Fig. 3c. Straight Axial Groove

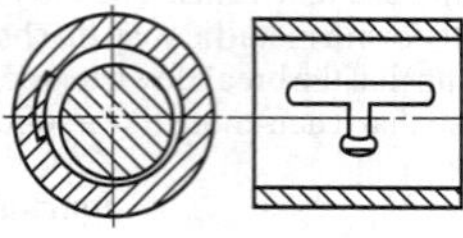

Fig. 3d. Straight Axial Groove with Feeder Groove

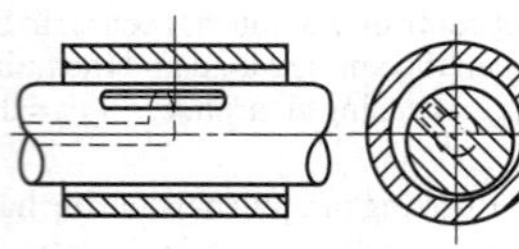

Fig. 3e. Straight Axial Groove in Shaft

rise to an unsatisfactory level. The heat-radiating capacity $H_R$ of the bearing in foot-pounds per minute may be calculated from the formula $H_R = Ld\,Ct_R$ in which $C$ is a constant determined by O. Lasche, and $t_R$ is temperature rise in degrees Fahrenheit.

Values for the product $Ct_R$ may be found from the curves in Fig. 4 for various values of bearing temperature rise $t_R$ and for three operating conditions. In this equation, $L$ = total length of the bearing in inches and $d$ = bearing diameter in inches.

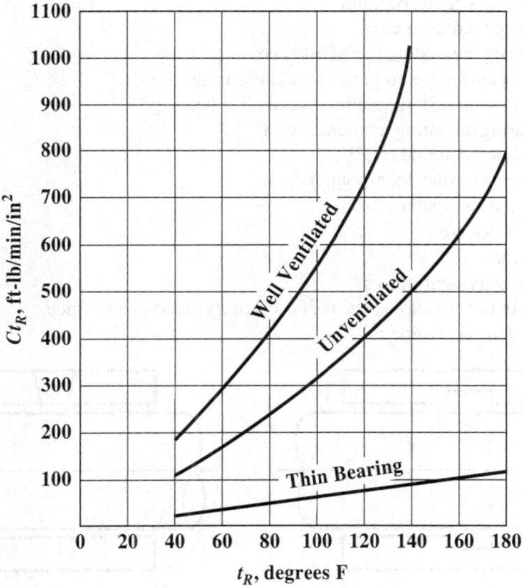


Fig. 4. Heat-Radiating Capacity Factor, $Ct_R$, versus Bearing Temperature Rise, $t_R$ — Journal Bearings.

**Journal Bearing Design Notation.**—The symbols used in the following step-by-step procedure for lubrication analysis and design of a plain sleeve or journal bearing are as follows:

$c$ = specific heat of lubricant, Btu/lb/degree F
$c_d$ = diametral clearance, inches
$C_n$ = capacity number bearing
$d$ = journal diameter, inches
$e$ = eccentricity, inches
$h_o$ = minimum film thickness, inch
$K$ = constants
$l$ = bearing length as defined in Fig. 5, inches
$L$ = actual length of bearing, inches
$m$ = clearance modulus
$N$ = rpm
$p_b$ = unit load, psi
$p_s$ = oil supply pressure, psi
$P_f$ = friction horsepower
$P'$ = bearing pressure parameter
$q$ = flow factor
$Q_1$ = hydrodynamic flow, gpm
$Q_2$ = pressure flow, gpm
$Q$ = total flow, gpm
$Q_{new}$ = new total flow, gpm

$Q_R$ = total flow required, gpm
$r$ = journal radius, inches
$\Delta t$ = actual temperature rise of oil in bearing, °F
$\Delta t_a$ = assumed temperature rise of oil in bearing, °F
$\Delta t_{new}$ = new assumed temperature rise of oil in bearing, °F
$t_b$ = bearing operating temperature, °F
$t_{in}$ = oil inlet temperature, °F
$T_f$ = friction torque, inch-pounds/inch
$T'$ = torque parameter
$W$ = load, pounds
$X$ = factor
$Z$ = viscosity, centipoises
$\in$ = eccentricity ratio — ratio of eccentricity to radial clearance
$\alpha$ = oil density, lbs/inch$^3$

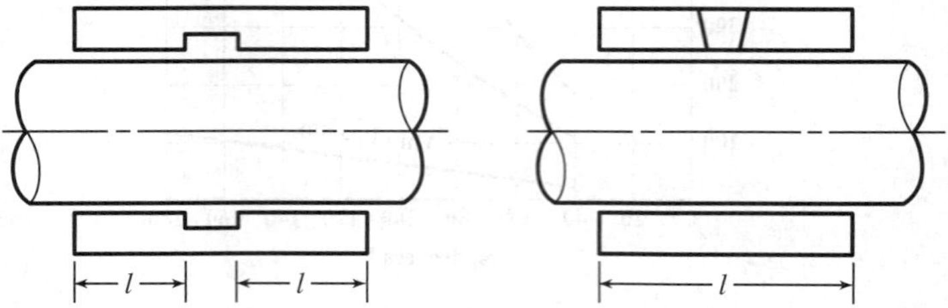


Fig. 5. Length, $l$, of Bearing for Circular Groove Type (left) and Single Inlet Hole Type (right).

**Journal Bearing Lubrication Analysis.—**The following procedure leads to a complete lubrication analysis which forms the basis for the bearing design.

1) *Diameter of bearing d:* This is usually determined by considering strength and/or deflection requirements for the shaft using principles of strength of materials.

2) *Length of bearing L:* This is determined by an assumed $l/d$ ratio in which $l$ may or may not be equal to the overall length, $L$ (see Step 6). Bearing pressure and the possibility of edge loading due to shaft deflection and misalignment are factors to be considered. In general, shaft misalignment resulting from location tolerances and/or shaft deflections should be maintained below 0.0003 inch per inch of length.

3) *Bearing pressure $p_b$:* The unit load in pound per square inch is calculated from the formula:

$$p_b = \frac{W}{Kld}$$

where $K$ = 1 for single oil hole
$K$ = 2 for central groove
$W$ = load, pounds
$l$ = bearing length as defined in Fig. 5, inches
$d$ = journal diameter, inches

Typical unit loads in service are shown in Table 3. These pressures can be used as a safe guide in selection. However, if space limitations impose a higher limit of loading, the complete lubrication analysis and evaluation of material properties will determine acceptability.

**Table 3. Allowable Sleeve Bearing Pressures for Various Classes of Bearings**

| Types of Bearing or Kind of Service | Pressure [a] psi (MPa) | Types of Bearing or Kind of Service | Pressure [a] psi (MPa) |
|---|---|---|---|
| Electric Motor & Generator Bearings (General) | 100–200 (0.69–1.38) | Diesel Engine Rod | 1000–2000 (6.89–13.79) |
| Turbine & Reduction Gears | 100–250 (0.69–1.72) | Wrist Pins | 1800–2000 (12.41–13.79) |
| Heavy Line Shafting | 100–150 (0.69–1.03) | Automotive, Main Bearings | 500–700 (3.45–4.83) |
| Locomotive Axles | 300–350 (2.07–2.41) | Automotive, Rod Bearings | 1500–2500 (10.34–17.24) |
| Light Line Shafting | 15–35 (0.103–0.241) | Centrifugal Pumps | 80–100 (0.55–0.689) |
| Diesel Engine, Main | 800–1500 (5.52–10.34) | Aircraft Rod Bearings | 700–3000 (4.83–20.68) |

[a] These pressures in psi (MPa) of area equal to length times diameter are intended as a general guide only. The allowable unit pressure depends upon operating conditions, especially in regard to lubrication, design of bearings, workmanship, velocity, and nature of load.

4) *Diametral clearance* $c_d$: This is selected on a trial basis from Fig. 6 which shows suggested diametral clearance ranges for various shaft sizes and for two speed ranges. These are *hot* or *operating* clearances so that thermal expansion of journal and bearing to these temperatures must be taken into consideration in establishing machining dimensions. The optimum operating clearance should be determined on the basis of a complete lubrication analysis (see paragraph following Step 23).

5) *Clearance modulus m:* This is calculated from the formula: $m = \frac{c_d}{d}$

6) *Length to diameter ratio l/d:* This is usually between 1 and 2; however, with the modern trend toward higher speeds and more compact units, lower ratios down to 0.3 are used. In shorter bearings there is a consequent reduction in load-carrying capacity due to excessive end or side leakage of lubricant. In longer bearings there may be a tendency towards edge loading. Length *l* for a single oil feed hole is taken as the total length of the bearing as shown in Fig. 5. For a central oil groove length, *l* is taken as one-half the total length.

Typical *l/d* ratio's use for various types of applications are given in Table 4.

7) *Assumed operating temperature* $t_b$: A temperature rise of the lubricant as it passes through the bearing is assumed and the consequent operating temperature in degrees F is calculated from the formula:

$$t_b = t_{in} + \Delta t_a$$

where $t_{in}$ = inlet temperature of oil in °F

$\Delta t_a$ = assumed temperature rise of oil in bearing in °F. An initial assumption of 20° F is usually made.

8) *Viscosity of lubricant Z:* The viscosity in centipoises at the assumed bearing operating temperature is found from the curve in Fig. 6 which shows the viscosity of SAE grade oils versus temperature.

9) *Bearing pressure parameter P′:* This value is required to find the eccentricity ratio and is calculated from the formula:

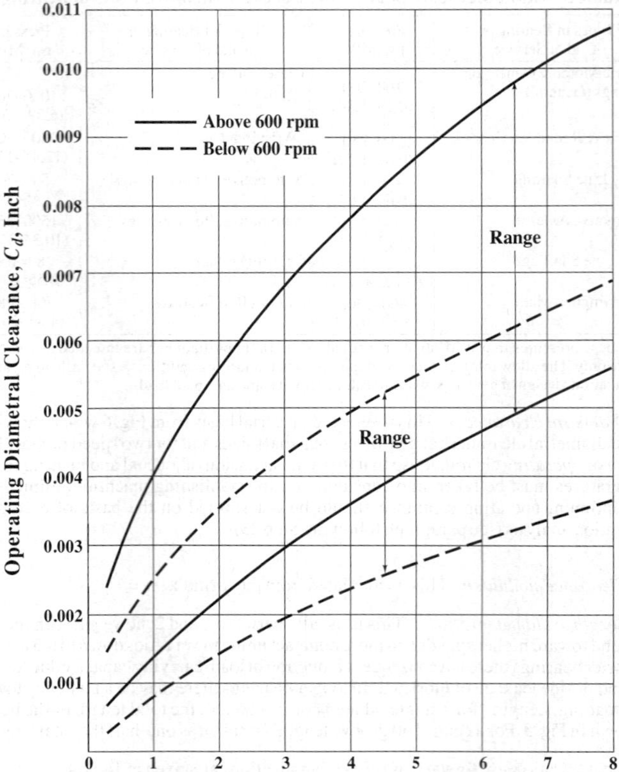


Fig. 6. Operating Diametral Clearance $C_d$ versus Journal Diameter $d$.

**Table 4. Representative $l/d$ Ratios**

| Type of Service | $l/d$ | Type of Service | $l/d$ |
|---|---|---|---|
| Gasoline and diesel engine | | Light shafting | 2.5 to 3.5 |
| main bearings and crankpins | 0.3 to 1.0 | Heavy shafting | 2.0 to 3.0 |
| Generators and motors | 1.2 to 2.5 | Steam engine | |
| Turbogenerators | 0.8 to 1.5 | Main bearings | 1.5 to 2.5 |
| Machine tools | 2.0 to 3.0 | Crank and wrist pins | 1.0 to 1.3 |

$$P' = \frac{6.9(1000m)^2 p_b}{ZN}$$

where $N$ = rpm

10) *Eccentricity ratio* ε: Using $P'$ and $l/d$, the value of $1/(1-\varepsilon)$ is determined from Fig. 7 and from this, ε can be determined.

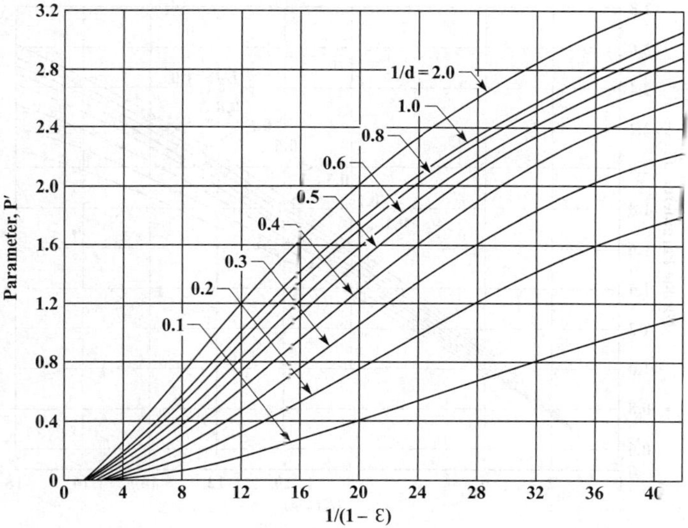


Fig. 7. Bearing Parameter *P′*, versus Eccentricity Ratio 1/(1 – ε) — Journal Bearings.

11) *Torque parameter T′:* This value is obtained from Fig. 8 or Fig. 9 using $1/(1 - \varepsilon)$ and *l*/*d*.

12) *Friction torque T:* This value is calculated from the formula:

$$T = \frac{T'r^2ZN}{6900(1000m)}$$

where $r$ = journal radius, inches

13) *Friction horsepower $P_f$:* This value is calculated from the formula:

$$P_f = \frac{KTNl}{63{,}000}$$

where $K$ = 1 for single oil hole, 2 for central groove.

14) *Factor X:* This factor is used in the calculation of the lubricant flow and can either be obtained from Table 5 or calculated from the formula:

$$X = \frac{0.1837}{\alpha c}$$

where $\alpha$ = oil density in pounds per cubic inch

$c$ = specific heat of lubricant in Btu/lb/°F

15) *Total flow of lubricant required $Q_R$:* This is calculated from the formula:

$$Q_R = \frac{X(P_f)}{\Delta t_a}$$

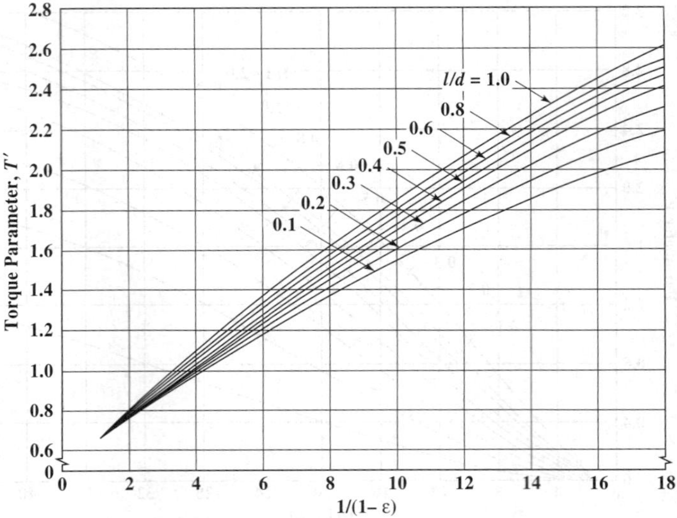


Fig. 8. Torque Parameter, $T'$, versus Eccentricity Ratio, $1/(1-\varepsilon)$ — Journal Bearings.

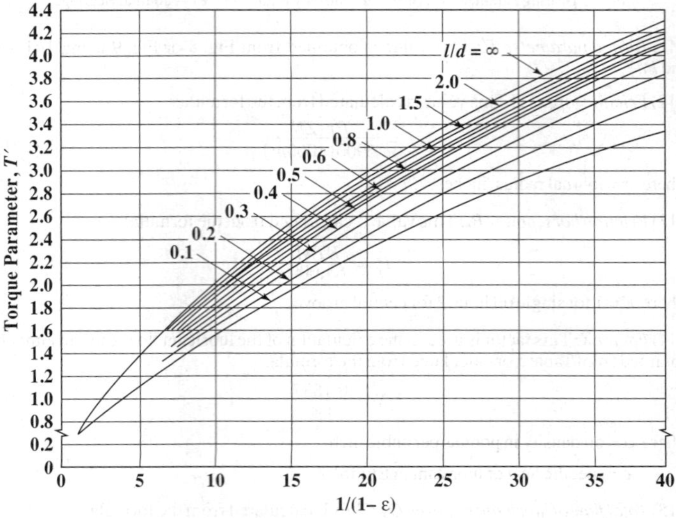


Fig. 9. Torque Parameter, $T'$, versus Eccentricity Ratio, $1/(1-\varepsilon)$ — Journal Bearings.

**Table 5. *X* Factor versus Temperature of Mineral Oils**

| Temperature | *X* Factor |
|---|---|
| 100 | 12.9 |
| 150 | 12.4 |
| 200 | 12.1 |
| 250 | 11.8 |
| 300 | 11.5 |

16) *Bearing capacity number* $C_n$*:* This value is needed to obtain the flow factor and is calculated from the formula:

$$C_n = \left(\frac{l}{d}\right)^2 / 60P'$$

17) *Flow factor* $q$*:* This value is obtained from the curve in Fig. 10.

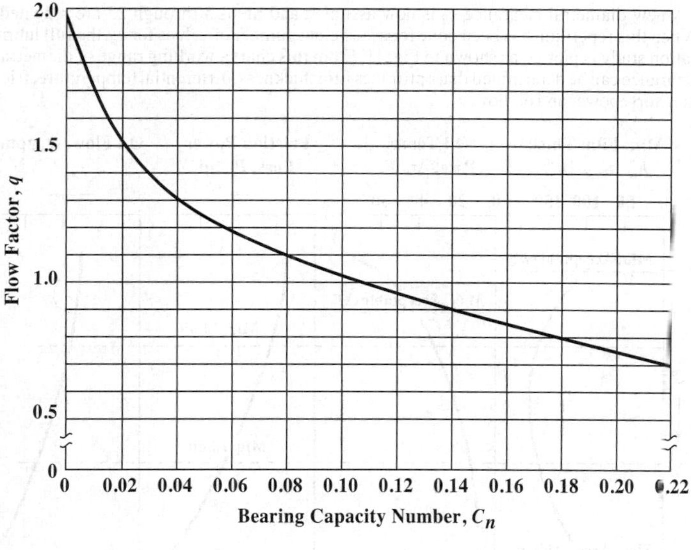


Fig. 10. Flow Factor, $q$, versus Bearing Capacity Number, $C_n$—Journal Bearings.

18) *Hydrodynamic flow of lubricant* $Q_1$*:* This flow in gallons per minute is calculated from the formula:

$$Q_1 = \frac{Nlc_d qd}{294}$$

19) *Pressure flow of lubricant* $Q_2$*:* This flow in gallons per minute is calculated from the formula:

$$Q_2 = \frac{Kp_s c_d^3 d(1 + 1.5\varepsilon^2)}{Zl}$$

where $K = 1.64 \times 10^5$ for single oil hole

$K = 2.35 \times 10^5$ for central groove

$p_s$ = oil supply pressure

20) *Total flow of lubricant Q:* This value is obtained by adding the hydrodynamic flow and the pressure flow.

$$Q = Q_1 + Q_2$$

21) *Bearing temperature rise Δt:* This temperature rise in degrees F is obtained from the formula:

$$\Delta t = \frac{X(P_f)}{Q}$$

22) *Comparison of actual and assumed temperature rises:* At this point if $\Delta t_a$ and $\Delta t$ differ by more than 5° F, Steps 7 through 22 are repeated using a $\Delta t_{new}$ halfway between the former $\Delta t_a$ and $\Delta t$.

23) *Minimum film thickness $h_o$:* When Step 22 has been satisfied, the minimum film thickness in inches is calculated from the formula: $h_o = \frac{1}{2} C_d (1 - \varepsilon)$.

A new diametral clearance $c_d$ is now assumed and Steps 5 through 23 are repeated. When this repetition has been done for a sufficient number of values for $c_d$, the full lubrication study is plotted as shown in Fig. 11. From this chart a working range of diametral clearance can be determined that optimizes film thickness, differential temperature, friction horsepower and oil flow.

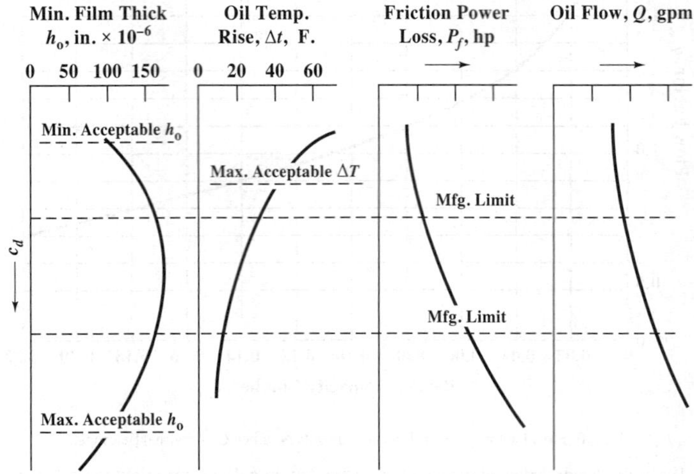


Fig. 11. Example of Lubrication Analysis Curves for Journal Bearing.

**Use of Lubrication Analysis.—**Once the lubrication analysis has been completed and plotted as shown in Fig. 11, the following steps lead to the optimum bearing design, taking into consideration both basic operating requirements and requirements peculiar to the application.

1) Examine the curve (Fig. 11) for minimum film thickness and determine the acceptable range of diametral clearance, $c_d$, based on

a) a minimum of $200 \times 10^{-6}$ inches for small bearings under 1 inch diameter

b) a minimum of $500 \times 10^{-6}$ inches for bearings from 1 to 4 inches diameter

c) a minimum of $750 \times 10^{-6}$ inches for larger bearings.

More conservative designs would increase these requirements

2) Determine the minimum acceptable $c_d$ based on a maximum $\Delta t$ of 40° F from the oil temperature rise curve (Fig. 11).

3) If there are no requirements for maintaining low friction horsepower and oil flow, the possible limits of diametral clearance are now defined.

4) The required manufacturing tolerances can now be placed within this band to optimize $h_o$ as shown by Fig. 11.

5) If oil flow and power loss are a consideration, the manufacturing tolerances may then be shifted within the range permitted by the requirements for $h_o$ and $\Delta t$.

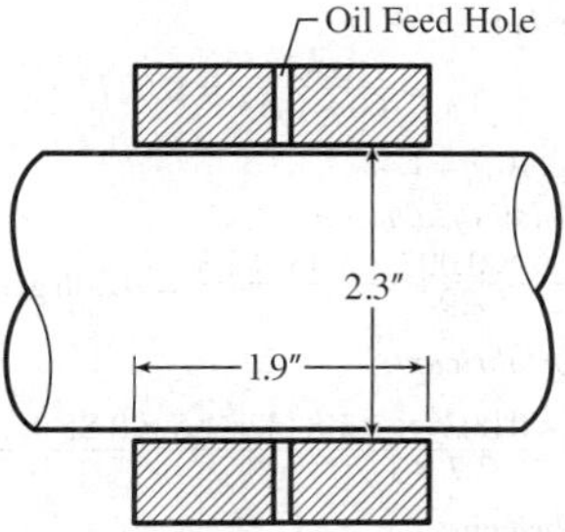


Fig. 12. Full Journal Bearing Example Design.

*Example:* A full journal bearing, Fig. 12, 2.3 inches in diameter and 1.9 inches long, is to carry a load of 6000 pounds at 4800 rpm, using SAE 30 oil supplied at 200° F through a single oil hole at 30 psi. Determine the operating characteristics of this bearing as a function of diametral clearance.

1) *Diameter of bearing,* given as 2.3 inches.

2) *Length of bearing,* given as 1.9 inches.

3) *Bearing pressure:*

$$p_b = \frac{6000}{1 \times 1.9 \times 2.3} = 1372 \text{ lbs. per sq. in.}$$

4) *Diametral clearance:* Assume $c_d$ is equal to 0.003 inch from Fig. 6 on page 2412 for first calculation.

5) *Clearance modulus:* $m = \dfrac{0.003}{2.3} = 0.0013$ inch

6) *Length-to-diameter ratio:*

$$\frac{l}{d} = \frac{1.9}{2.3} = 0.83$$

7) *Assumed operating temperature:* If the temperature rise $\Delta t_a$ is assumed to be 20° F,

$$t_b = 200 + 20 = 220°\text{F}$$

8) *Viscosity of lubricant:* From Fig. 6 on page 2405, Z = 7.7 centipoises

9) *Bearing-pressure parameter:*

$$P' = \frac{6.9 \times 1.3^2 \times 1372}{7.7 \times 4800} = 0.43$$

10) *Eccentricity ratio:* From Fig. 7, $\dfrac{1}{1 - \varepsilon} = 6.8$ and $\varepsilon = 0.85$

11) *Torque parameter:* From Fig. 8, $T' = 1.46$

12) *Friction torque:*

$$T_f = \frac{1.46 \times 1.15^2 \times 7.7 \times 4800}{6900 \times 1.3} = 7.96 \text{ inch-pounds per inch}$$

13) *Friction horsepower:*

$$P_f = \frac{1 \times 7.96 \times 4800 \times 1.9}{63{,}000} = 1.15 \text{ horsepower}$$

14) *Factor X:* From Table 5, $X = 12$, approximately

15) *Total flow of lubricant required:*

$$Q_R = \frac{12 \times 1.15}{20} = 0.69 \text{ gallon per minute}$$

16) *Bearing-capacity number:*

$$C_n = \frac{0.83^2}{60 \times 0.43} = 0.027$$

17) *Flow factor:* From Fig. 10, $q = 1.43$

18) *Actual hydrodynamic flow of lubricant:*

$$Q_1 = \frac{4800 \times 1.9 \times 0.003 \times 1.43 \times 2.3}{294} = 0.306 \text{ gallon per minute}$$

19) *Actual pressure flow of lubricant:*

$$Q_2 = \frac{1.64 \times 10^5 \times 30 \times 0.003^3 \times 2.3 \times (1 + 1.5 \times 0.85^2)}{7.7 \times 1.9} = 0.044 \text{ gallon per min}$$

20) *Actual total flow of lubricant:*

$$Q = 0.306 + 0.044 = 0.350 \text{ gallon per minute}$$

21) *Actual bearing-temperature rise:*

$$\Delta t = \frac{12 \times 1.15}{0.350} = 39.4°\text{F}$$

22) *Comparison of actual and assumed temperature rises:* Because $\Delta t_a$ and $\Delta t$ differ by more than 5° F, a new $\Delta t_a$, midway between these two, of 30° F is assumed and Steps 7 through 22 are repeated.

7a) *Assumed operating temperature:*

$$t_b = 200 + 30 = 230°\text{F}$$

8a) *Viscosity of lubricant:* From Fig. 6, $Z = 6.8$ centipoises

9a) *Bearing-pressure parameter:*

$$P' = \frac{6.9 \times 1.3^2 \times 1372}{6.8 \times 4800} = 0.49$$

10a) *Eccentricity ratio:* From Fig. 7,

$$\frac{1}{1 - \varepsilon} = 7.4$$

and $\varepsilon = 0.86$

11a) *Torque parameter:* From Fig. 8, $T' = 1.53$

12a) *Friction torque:*

$$T_f = \frac{1.53 \times 1.15^2 \times 6.8 \times 4800}{6900 \times 1.3} = 7.36 \text{ inch-pounds per inch}$$

13a) *Friction horsepower:*

$$P_f = \frac{1 \times 7.36 \times 4800 \times 1.9}{63{,}000} = 1.07 \text{ horsepower}$$

14a) *Factor X*: From Table 5, $X = 11.9$ approximately

15a) *Total flow of lubricant required:*

$$Q_R = \frac{11.9 \times 1.07}{30} = 0.42 \text{ gallon per minute}$$

16a) *Bearing-capacity number:*

$$C_n = \frac{0.83^2}{60 \times 0.49} = 0.023$$

17a) *Flow factor:* From Fig. 10, $q = 1.48$

18a) *Actual hydrodynamic flow of lubricant:*

$$Q_1 = \frac{4800 \times 1.9 \times 0.003 \times 1.48 \times 2.3}{294} = 0.317 \text{ gallon per minute}$$

19a) *Pressure flow:*

$$Q_2 = \frac{1.64 \times 10^5 \times 30 \times 0.003^3 \times 2.3 \times (1 + 1.5 \times 0.86^2)}{6.8 \times 1.9} = 0.050 \text{ gallon per minute}$$

20a) *Actual flow of lubricant:*

$$Q_{new} = 0.317 + 0.050 = 0.367 \text{ gallon per minute}$$

21a) *Actual bearing-temperature rise:*

$$\Delta t = \frac{11.9 \times 1.06}{0.367} = 34.4°\text{F}$$

22a) *Comparison of actual and assumed temperature rises:* Now $\Delta t$ and $\Delta t_a$ are within 5° F.

23) *Minimum film thickness:*

$$h_o = \frac{0.003}{2}(1 - 0.86) = 0.00021 \text{ inch}$$

This analysis may now be repeated for other values of $c_d$ determined from Fig. 6 and a complete lubrication analysis performed and plotted as shown in Fig. 11. An operating range for $c_d$ can then be determined to optimize minimum clearance, friction horsepower loss, lubricant flow, and temperature rise.

## Thrust Bearings

As the name implies, thrust bearings are used either to absorb axial shaft loads or to position shafts axially. Brief descriptions of the normal designs for these bearings follow with approximate design methods for each. The generally accepted load ranges for these types of bearings are given in Table 1 and the schematic configurations are shown in Fig. 1.

*The parallel or flat plate thrust bearing* is probably the most frequently used type. It is the simplest and lowest in cost of those considered; however, it is also the least capable of absorbing load, as can be seen from Table 1. It is most generally used as a positioning device where loads are either light or occasional.

*The step bearing,* like the parallel plate, is also a relatively simple design. This type of bearing will accept the normal range of thrust loads and lends itself to low-cost, high-volume production. However, this type of bearing becomes sensitive to alignment as its size increases.

*The tapered land thrust bearing,* as shown in Table 1, is capable of high load capacity. Where the step bearing is generally used for small sizes, the tapered land type can be used in larger sizes. However, it is more costly to manufacture and does require good alignment as size is increased.

*The tilting pad or Kingsbury thrust bearing* (as it is commonly referred to) is also capable of high thrust capacity. Because of its construction it is more costly, but it has the inherent advantage of being able to absorb significant amounts of misalignment.

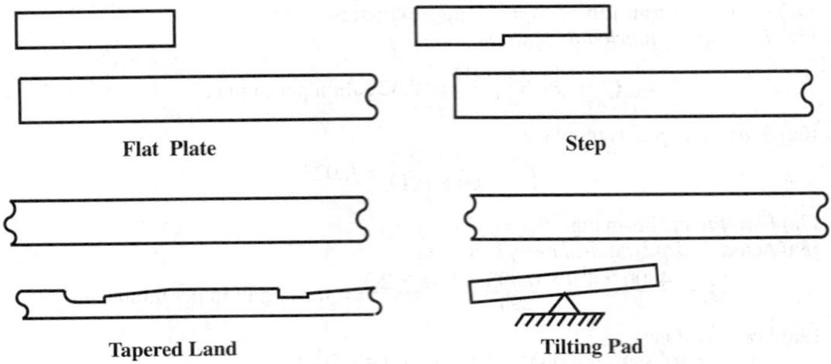


Fig. 1. Types of Thrust Bearings.

**Table 1. Thrust Bearing Loads***

| Type | Normal Unit Loads, lb/in² | Maximum Unit Loads, lb/in² |
|---|---|---|
| Parallel surface | <75 | <150 |
| Step | 200 | 500 |
| Tapered land | 200 | 500 |
| Tilting pad | 200 | 500 |

**Thrust Bearing Design Notation.**—The symbols used in the design procedures that follow for flat plate, step, tapered land, and tilting pad thrust bearings are as follows:

$a$ = radial width of pad, inches
$b$ = circumferential length of pad at pitch line, inches
$b_2$ = pad step length
$B$ = circumference of pitch circle, inches
$c$ = specific heat of oil, Btu/gal/°F
$D$ = diameter, inches
$e$ = depth of step, inch
$f$ = coefficient of friction
$g$ = depth of 45° chamfer, inches
$h$ = film thickness, inch
$i$ = number of pads
$J$ = power loss coefficient
$K$ = film thickness factor
$K_g$ = fraction of circumference occupied by the pads; usually 0.8
$l$ = length of chamfer, inches
$M$ = horsepower per square inch
$N$ = revolutions per minute
$O$ = operating number
$p$ = bearing unit load, psi
$p_s$ = oil-supply pressure, psi
$P_f$ = friction horsepower
$Q$ = total flow, gpm

* Reproduced with permission from Wilcock and Booser, *Bearing Design and Applications*, McGraw-Hill Book Co., © 1957.

$Q_c$ = required flow per chamfer, gpm

$Q^o{}_c$ = uncorrected required flow per chamfer, gpm

$Q_F$ = film flow, gpm

$s$ = oil-groove width

$\Delta t$ = temperature rise, °F

$U$ = velocity, feet per minute

$V$ = effective width-to-length ratio for one pad

$W$ = applied load, pounds

$Y_G$ = oil-flow factor

$Y_L$ = leakage factor

$Y_S$ = shape factor

$Z$ = viscosity, centipoises

$\alpha$ = dimensionless film-thickness factor

$\Delta$ = taper

$\xi$ = kinetic energy correction factor

*Note:* In the following, subscript 1 denotes inside diameter and subscript 2 denotes outside diameter. Subscript *i* denotes inlet and subscript *o* denotes outlet.

**Flat Plate Thrust Bearing Design.**—The following steps define the performance of a flat plate thrust bearing, one section of which is shown in Fig. 2. Although each bearing section is wedge shaped, as shown below right, for the purposes of design calculation, it is considered to be a rectangle with a length *b* equal to the circumferential length along the pitch line of the section being considered, and a width *a* equal to the difference in the external and internal radii.

*General Parameters:* a) From Table 1, the maximum unit load is between 75 and 100 pounds per square inch; and b) the outside diameter is usually between 1.5 and 2.5 times the inside diameter.

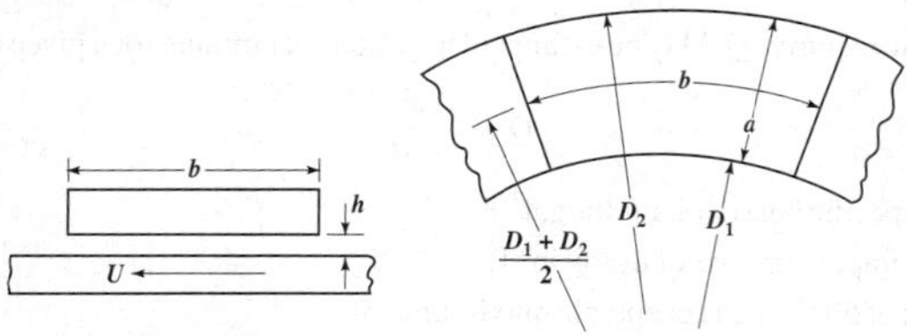


Fig. 2. Basic Elements of Flat Plate Thrust Bearing.*

1) *Inside diameter,* $D_1$. Determined by shaft size and clearance.

2) *Outside diameter,* $D_2$. Calculated by the formula

$$D_2 = \left(\frac{4W}{\pi K_g p} + D_1^2\right)^{\frac{1}{2}}$$

where $W$ = applied load, pounds

$K_g$ = fraction of circumference occupied by pads; usually 0.8

$p$ = bearing unit load, psi

* See footnote on page 2420.

3) *Radial pad width, a.* Equal to one-half the difference between the inside and outside diameters.

$$a = \frac{D_2 - D_1}{2}$$

4) *Pitch line circumference, B.* Found from the pitch diameter.

$$B = \pi(D_2 - a)$$

5) *Number of pads, i.* Assume an oil groove width, *s*. If the length of pad is assumed to be optimum, i.e., equal to its width,

$$i_{app} = \frac{B}{a + s}$$

Take *i* as nearest even number.

6) *Length of pad, b.* If number of pads and oil groove width are known,

$$b = \frac{B - (i \times s)}{i}$$

7) *Actual unit load, p.* Calculated in pounds per square inch based on pad dimensions.

$$p = \frac{W}{iab}$$

8) *Pitch line velocity, U.* Found in feet per minute from

$$U = \frac{BN}{12}$$

where $N$ = rpm

9) *Friction power loss, Pf.* Friction power loss is difficult to calculate for this type of bearing because there is no theoretical method of determining the operating film thickness. However, a good approximation can be made using Fig. 3. From this curve, the value of *M*, horsepower loss per square inch of bearing surface, can be obtained. The total power loss, $P_f$, is then calculated from

$$P_f = iabM$$

10) *Oil flow required, Q.* May be estimated in gallons per minute for a given temperature rise from

$$Q = \frac{42.4P_f}{c\Delta t}$$

where $c$ = specific heat of oil in Btu/gal/°F

$\Delta t$ = temperature rise of the oil in °F

*Note:* A $\Delta$t of 50° F is an acceptable maximum.

Because there is no theoretical method of predicting the minimum film thickness in this type of bearing, only an approximation, based on experience, of the film flow can be made. For this reason and based on practical experience, it is desirable to have a minimum of one-half of the desired oil flow pass through the chamfer.

11) *Film flow, $Q_F$.* Calculated in gallons per minute from

$$Q_F = \frac{(1.5)(10^5)iVh^3p_s}{Z_2}$$

where $V$ = effective width-to-length ratio for one pad, *a*/*b*

$Z_2$ = oil viscosity at outlet temperature

$h$ = film thickness

*Note:* Because *h* cannot be calculated, use $h$ = 0.002 inch.

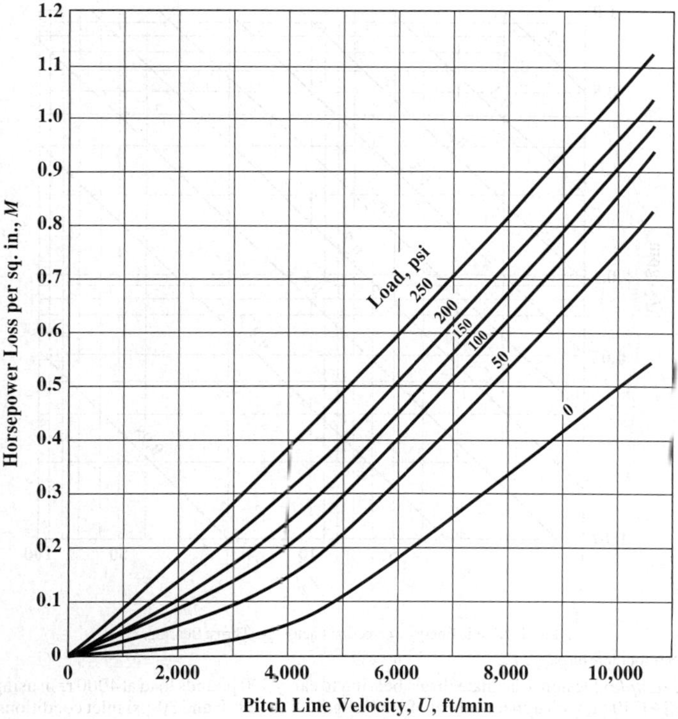


Fig. 3. Friction Power Loss, $M$, versus Peripheral Speed, $U$—Thrust Bearings.[a]

[a]See footnote on page 2420

12) *Required flow per chamfer,* $Q_c$. Readily found from the formula

$$Q_c = \frac{Q}{i}$$

13) *Kinetic energy correction factor,* $\xi$. Found by assuming a chamfer length $l$ and entering Fig. 4 with a value $Z_2 l$ and $Q_c$

14) *Uncorrected required flow per chamfer,* $Q^0{}_c$. Found from the formula

$$Q_c^0 = \frac{Q_c}{\xi}$$

15) *Depth of chamfer, g.* Found from the formula

$$g = \sqrt[4]{\frac{Q_c^0 l Z_2}{4.7410^4 p_s}}$$

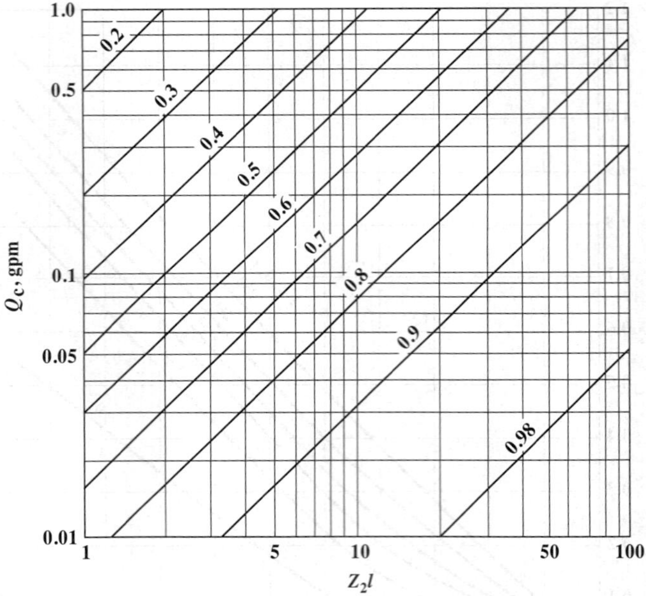


Fig. 4. Kinetic Energy Correction Factor, ξ—Thrust Bearings.[a]

[a]See footnote on page 2420

*Example:* Design a flat plate thrust bearing to carry 900 pounds load at 4000 rpm using an SAE 10 oil with a specific heat of 3.5 Btu/gal/°F at 120° F and 30-psi inlet conditions. The shaft is $2\frac{3}{4}$ inches in diameter and the temperature rise is not to exceed 40° F. Fig. 5 shows the final design of this bearing.

1) *Inside diameter.* Assumed to be 3 inches to clear shaft.

2) *Outside diameter.* Assuming a unit bearing load of 75 pounds per square inch from Table 1,

$$D_2 = \sqrt{\frac{4 \times 900}{\pi \times 0.8 \times 75} + 3^2} = 5.30 \text{ inches}$$

Use $5\frac{1}{2}$ inches.

3) *Radial pad width.*

$$a = \frac{5.5 - 3}{2} = 1.25 \text{ inches}$$

4) *Pitch-line circumference.*

$$B = \pi \times 4.25 = 13.35 \text{ inches}$$

5) *Number of pads.* Assume an oil groove width of $\frac{3}{16}$ inch. If length of pad is assumed to be equal to width of pad, then

$$i_{app} = \frac{13.3}{1.25 + 0.1875} = 9+$$

If the number of pads, $i$, is taken as 10, then

6) *Length of pad.* $b = \dfrac{13.35 - (10 \times 0.875)}{10} = 1.14$ inches

7) *Actual unit load.*

$$p = \frac{900}{10 \times 1.25 \times 1.14} = 63 \text{ psi}$$

8) *Pitch-line velocity.*

$$U = \frac{13.35 \times 4000}{12} = 4{,}430 \text{ ft per min.}$$

9) *Friction power loss.* From Fig. 3, $M = 0.19$

$$P_f = 10 \times 1.25 \times 1.14 \times 0.19 = 2.7 \text{ horsepower}$$

10) *Oil flow required.*

$$Q = \frac{42.4 \times 2.7}{3.5 \times 40} = 0.82 \text{ gallon per minute}$$

(Assuming a temperature rise of 40° F—the maximum allowable according to the given condition—then the assumed operating temperature will be 120° F + 40° F = 160° F and the oil viscosity $Z_2$ is found from Fig. 6 to be 9.6 centipoises.)

11) *Film flow.*

$$Q_F = \frac{1.5 \times 10^5 \times 10 \times 1 \times 0.002^3 \times 30}{9.6} = 0.038 \text{ gpm}$$

Because 0.038 gpm is a very small part of the required flow of 0.82 gpm, the bulk of the flow must be carried through the chamfers.

12) *Required flow per chamfer.* Assume that all the oil flow is to be carried through the chamfers.

$$Q_c = \frac{0.82}{10} = 0.082 \text{ gpm}$$

13) *Kinetic energy correction factor.* If $l$, the length of chamfer is made ⅛ inch, then $Z_2 l = 9.6 \times 1/8 = 1.2$. Entering Fig. 4 with this value and $Q_c = 0.082$,

$$\xi = 0.44$$

14) *Uncorrected required oil flow per chamfer.*

$$Q_c^0 = \frac{0.082}{0.44} = 0.186 \text{ gpm}$$

15) *Depth of chamfer.*

$$g = \sqrt[4]{\frac{0.186 \times 0.125 \times 9.6}{4.74 \times 10^4 \times 30}}$$

$$g = 0.02 \text{ inch}$$

A schematic drawing of this bearing is shown in Fig. 5.

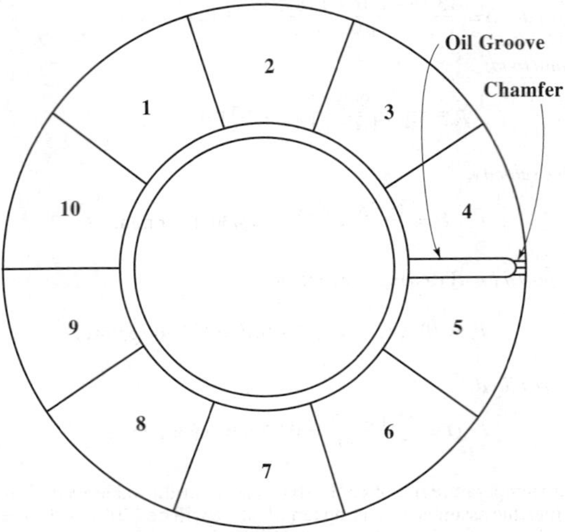


Fig. 5. Flat Plate Thrust Bearing Example Design.*

**Step Thrust Bearing Design.—**The following steps define the performance of a step thrust bearing, one section of which is shown in Fig. 6.

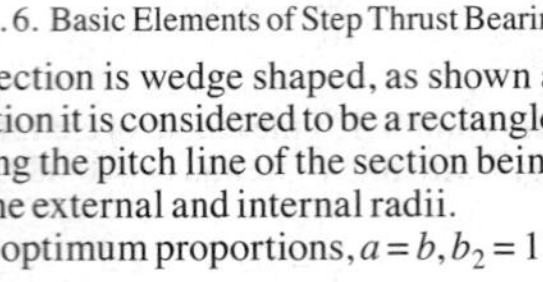


Fig. 6. Basic Elements of Step Thrust Bearing.*

Although each bearing section is wedge shaped, as shown at the right in Fig. 6, for the purposes of design calculation it is considered to be a rectangle with a length $b$ equal to the circumferential length along the pitch line of the section being considered, and a width $a$ equal to the difference in the external and internal radii.

*General Parameters:* For optimum proportions, $a = b, b_2 = 1.2b_1$, and $e = 0.7h$.

1) *Internal diameter,* $D_1$. An internal diameter is assumed that is sufficient to clear the shaft.

2) *External diameter,* $D_2$. A unit bearing pressure is assumed from Table 1 and the external diameter is then found from the formula

$$D_2 = \sqrt{\frac{4W}{\pi K_g p} + D_1^2}$$

3) *Radial pad width, a.* Equal to the difference between the external and internal radii.

$$a = \frac{D_2 - D_1}{2}$$

* See footnote on page 2420

4) *Pitch-line circumference, B.* Found from the formula

$$B = \frac{\pi(D_1 + D_2)}{2}$$

5) *Number of pads, i.* Assume an oil groove width, $s$ (0.062 inch may be taken as a minimum), and find the approximate number of pads, assuming the pad length is equal to $c$. Note that if a chamfer is found necessary to increase the oil flow (see Step 13), the oil groove width should be greater than the chamfer width.

$$i_{app} = \frac{B}{a + s}$$

Then $i$ is taken as the nearest even number.

6) *Length of pad, b.* Readily determined from the number of pads and groove width.

$$b = \frac{B}{i} - s$$

7) *Pitch-line velocity, U.* Found in feet per minute from the formula $U = \frac{BN}{12}$

8) *Film thickness, h.* Found in inches from the formula

$$h = \sqrt{\frac{2.0910^{-9} ia^3 UZ}{W}}$$

9) *Depth of step, e.* According to the general parameter

$$e = 0.7h$$

10) *Friction power loss, $P_f$.* Found from the formula

$$P_f = \frac{7.3510^{-13} ia^2 U^2 Z}{h}$$

11) *Pad step length, $b_2$.* This distance, on the pitch line, from the leading edge of the pad to the step in inches is determined by the general parameters

$$b_2 = \frac{1.2b}{2.2}$$

12) *Hydrodynamic oil flow, Q.* Found in gallons per minute from the formula

$$Q = 6.6510^{-4} iahU$$

13) *Temperature rise, Δt.* Found in degrees F from the formula

$$\Delta t = \frac{42.4P_f}{cQ}$$

If the flow is insufficient, as indicated by too high a temperature rise, chamfers can be added to provide adequate flow as in Steps 12–15 of the flat plate thrust bearing design.

*Example:* Design a step thrust bearing for positioning a $\frac{7}{8}$-inch diameter shaft operating with a 25-pound thrust load and a speed of 5000 rpm. The lubricating oil has a viscosity of 25 centipoises at the operating temperature of 160° F and has a specific heat of 3.4 Btu per gal per °F.

1) *Internal diameter.* Assumed to be 1 inch to clear the shaft.

2) *External diameter.* Because the example is a positioning bearing with low total load, unit load will be negligible and the external diameter is not established by using the formula given in Step 2 of the procedure, but a convenient size is taken to give the desired overall bearing proportions.

$$D_2 = 3 \text{ inches}$$

3) *Radial pad width.*

$$a = \frac{3 - 1}{2} = 1 \text{ inch}$$

4) *Pitch-line circumference.*

$$B = \frac{\pi(3 + 1)}{2} = 6.28 \text{ inches}$$

5) *Number of pads.* Assuming a minimum groove width of 0.062 inch,

$$i_{app} = \frac{6.28}{1 + 0.062} = 5.9$$

Take $i = 6$.

6) *Length of pad.*

$$b = \frac{6.28}{6} - 0.062 = 0.985$$

7) *Pitch-line velocity.*

$$U = \frac{6.28 \times 5{,}000}{12} = 2{,}620 \text{ fpm}$$

8) *Film thickness.*

$$h = \sqrt{\frac{2.0910^{-9} \times 6 \times 1^3 \times 2{,}620 \times 25}{25}} = 0.0057 \text{ inch}$$

9) *Depth of step.*

$$e = 0.7 \times 0.0057 = 0.004 \text{ inch}$$

10) *Power loss.*

$$P_f = \frac{7.3510^{-13} \times 6 \times 1^2 \times 2{,}620^2 \times 25}{0.0057} = 0.133 \text{ hp}$$

11) *Pad step length.*

$$b_2 = \frac{1.2 \times 0.985}{2.2} = 0.537 \text{ inch}$$

12) *Total hydrodynamic oil flow.*

$$Q = 6.6510^{-4} \times 6 \times 1 \times 0.0057 \times 2{,}620 = 0.060 \text{ gpm}$$

13) *Temperature rise.*

$$\Delta t = \frac{42.4 \times 0.133}{3.4 \times 0.060} = 28°\text{F}$$

**Tapered Land Thrust Bearing Design.—**The following steps define the performance of a tapered land thrust bearing, one section of which is shown in Fig. 7. Although each bearing section is wedge shaped, as shown in Fig. 7, right, for the purposes of design calculation, it is considered to be a rectangle with a length $b$ equal to the circumferential length along the pitch line of the section being considered and a width $a$ equal to the difference in the external and internal radii.

*General Parameters:* Usually, the taper extends to only 80 percent of the pad length with the remainder being flat, thus: $b_2 = 0.8b$ and $b_1 = 0.2b$.

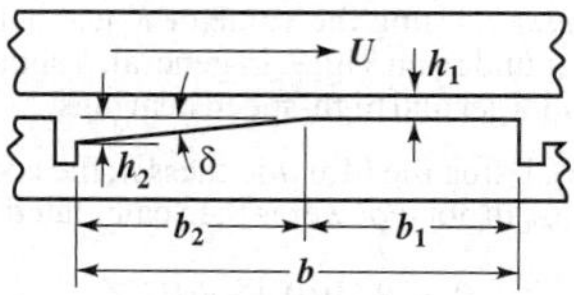


Fig. 7. Basic Elements of Tapered Land Thrust Bearing.*

1) *Inside diameter, $D_1$.* Determined by shaft size and clearance.

2) *Outside diameter, $D_2$.* Calculated by the formula

$$D_2 = \left(\frac{4W}{\pi K_g P_a} + D_1^2\right)^{\frac{1}{2}}$$

where $K_g = 0.8$ or $0.9$ and $W$ = applied load, pounds

$P_a$ = assumed unit load from Table 1, page 2420

3) *Radial pad width, a.* Equal to one-half the difference between the inside and outside diameters.

$$c = \frac{D_2 - D_1}{2}$$

4) *Pitch-line circumference, B.* Found from the mean diameter:

$$B = \frac{\pi(D_1 + D_2)}{2}$$

5) *Number of pads, i.* Assume an oil groove width, *s*, and find the approximate number of pads, assuming the pad length is equal to *a*.

$$i_{app} = \frac{B}{a + s}$$

Then *i* is taken as the nearest even number.

6) *Length of pad, b.* Readily determined because the number of pads and groove width are known.

$$b = \frac{B - is}{i}$$

7) *Taper values, $\delta_1$ and $\delta_2$.* Can be taken from Table 2.

8) *Actual bearing unit load, p.* Calculated in pounds per square inch from the formula

$$p = \frac{W}{iab}$$

9) *Pitch-line velocity, U.* Found in feet per minute at the pitch circle from the formula

$$U = \frac{BN}{12}$$

where $N$ = rpm

10) *Oil leakage factor, $Y_L$.* Found either from Fig. 8 which shows curves for $Y_L$ as functions of the pad width *a* and length of land *b* or from the formula

$$Y_L = \frac{b}{1 + \pi^2 b^2 / 12a^2}$$

11) *Film thickness factor, K.* Calculated using the formula

$$K = \frac{5.75 \times 10^6 p}{U Y_L Z}$$

* See footnote on page 2420

12) *Minimum film thickness, h.* Using the value of $K$ just determined and the selected taper values $\delta_1$ and $\delta_2$, $h$ is found from Fig. 9. In general, $h$ should be 0.001 inch for small bearings and 0.002 inch for larger and high-speed bearings.

13) *Friction power loss, $P_f$.* Using the film thickness $h$, the coefficient $J$ can be obtained from Fig. 10. The friction loss in horsepower is then calculated from the formula

$$P_f = 8.7910^{-13} iabJU^2Z$$

14) *Required oil flow, Q.* May be estimated in gallons per minute for a given temperature rise $\Delta t$ from the formula

$$Q = \frac{42.4P_f}{c\Delta t}$$

where $c$ = specific heat of the oil in Btu/gal/°F

*Note:* A $\Delta t$ of 50° F is an acceptable maximum.

15) *Shape factor, $Y_s$.* Needed to compute the actual oil flow and calculated from

$$Y_S = \frac{8ab}{D_2^2 - D_1^2}$$

16) *Oil flow factor, $Y_G$.* Found from Fig. 11 using $Y_s$ and $D_1/D_2$.

17) *Actual oil film flow, $Q_F$.* The amount of oil in gallons per minute that the bearing film will pass is calculated from the formula

$$Q_F = \frac{8.910^{-4} i\delta_2 D_2^3 NY_G Y_S^2}{D_2 - D_1}$$

18) If the flow is insufficient, the tapers can be increased or chamfers calculated to provide adequate flow, as in Steps 12–15 of the flat plate thrust bearing design procedure.

*Example:* Design a tapered land thrust bearing for 70,000 pounds at 3600 rpm. The shaft diameter is 6.5 inches. The oil inlet temperature is 110° F at 20 psi.

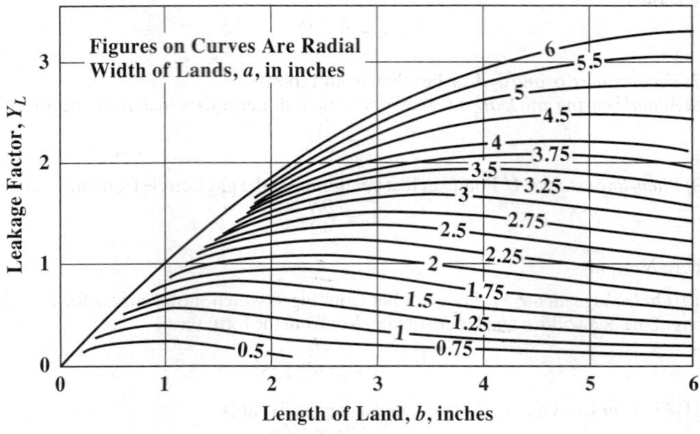


Fig. 8. Leakage Factor, $Y_L$, versus Pad Dimensions $a$ and $b$—Tapered Land Thrust Bearings.*

* See footnote on page 2420.

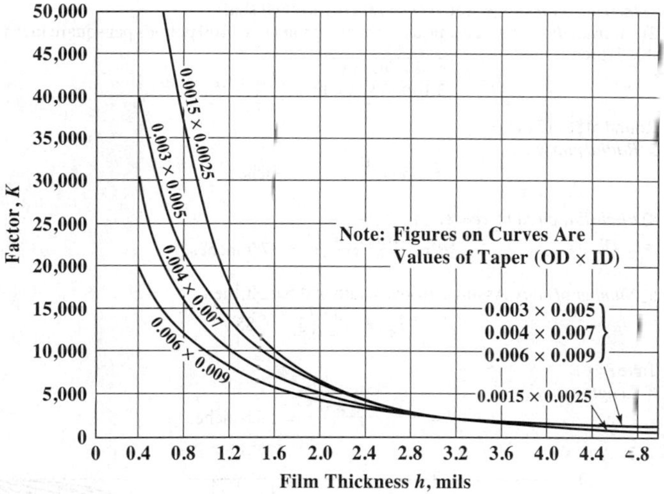


Fig. 9. Thickness, *h*, versus Factor *K*—Tapered Land Thrust Bearings.*

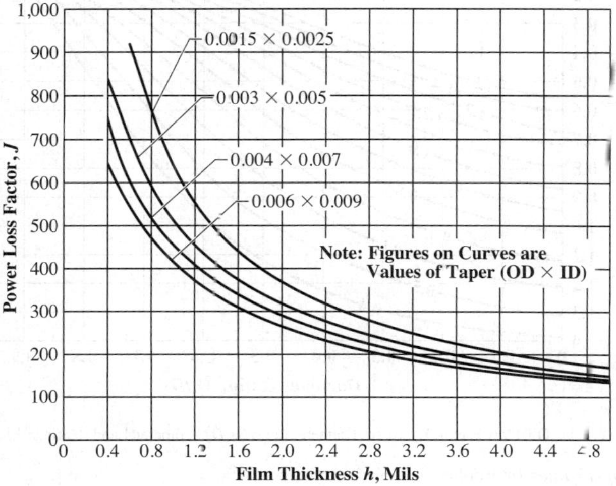


Fig. 10. Power-Loss Coefficient, *J*, versus Film Thickness, *h*—Tapered Land Thrust Bearings.*

A maximum temperature rise of 50° F is acceptable and results in a viscosity of 18 centipoises. Use values of $K_g = 0.9$ and $c = 3.5$ Btu/gal/°F.

* See footnote on page 2420.

1) *Internal diameter.* Assume $D_1 = 7$ inches to clear shaft.

2) *External diameter.* Assume a unit bearing load $p_a$ of 400 pounds per square inch from Table 1, then

$$D_2 = \sqrt{\frac{4 \times 70{,}000}{3.14 \times 0.9 \times 400} + 7^2} = 17.2 \text{ inches}$$

Round off to 17 inches

3) *Radial pad width.*

$$a = \frac{17 - 7}{2} = 5 \text{ inches}$$

4) *Pitch-line circumference.*

$$B = \frac{3.14(17 + 7)}{2} = 37.7 \text{ inches}$$

5) *Number of pads.* Assume groove width of 0.5 inch, then

$$i_{\text{app}} = \frac{37.7}{5 + 0.5} = 6.85$$

Take $i = 6$.

6) *Length of pad.*

$$b = \frac{37.7 - 6 \times 0.5}{6} = 5.78 \text{ inches}$$

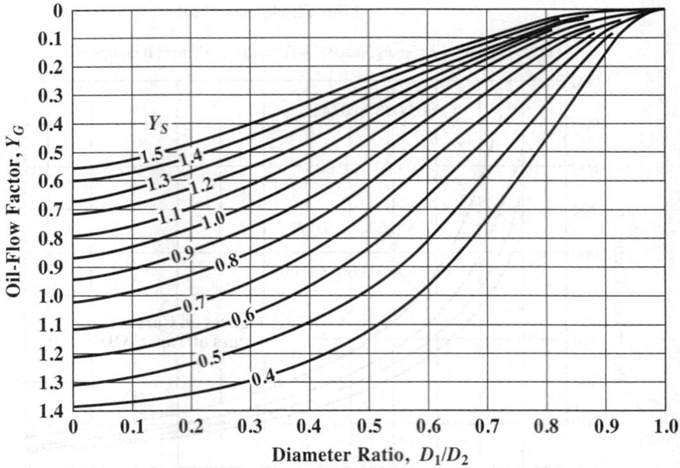


Fig. 11. Oil-Flow Factor, $Y_G$, versus Diameter Ratio $D_1/D_2$—Tapered Land Bearings.*

7) *Taper values.* Interpolate in Table 2 to obtain

$$\delta_1 = 0.008 \text{ inch} \qquad \text{and} \qquad \delta_2 = 0.005 \text{ inch}$$

8) *Actual bearing unit load.*

$$p = \frac{70{,}000}{6 \times 5 \times 5.78} = 404 \text{ psi}$$

* See footnote on page 2420.

9) *Pitch-line velocity.*

$$U = \frac{37.7 \times 3600}{12} = 11{,}300 \text{ ft per min}$$

10) *Oil leakage factor.*

From Fig. 8, $Y_L = 2.75$

11) *Film-thickness factor.*

$$K = \frac{5.75 \times 10^6 \times 404}{11{,}300 \times 2.75 \times 18} = 4150$$

12) *Minimum film thickness.*

From Fig. 9, $h = 2.2$ mils

13) *Friction power loss.* From Fig. 10, $J = 260$, then

$$P_f = 8.79 \times 10^{-13} \times 6 \times 5 \times 5.78 \times 260 \times 11{,}300^2 \times 18 = 91 \text{ hp}$$

14) *Required oil flow.*

$$Q = \frac{42.4 \times 91}{3.5 \times 50} = 22.0 \text{ gpm}$$

See footnote on page 2420.

15) *Shape factor.*

$$Y_S = \frac{8 \times 5 \times 5.78}{17^2 - 7^2} = 0.963$$

16) *Oil-flow factor.*

From Fig. 11, $Y_G = 0.61$

where $D_1/D_2 = 0.41$

17) *Actual oil film flow.*

$$Q_F = \frac{8.9 \times 10^{-4} \times 6 \times 0.005 \times 17^3 \times 3600 \times 0.61 \times 0.963^2}{17 - 7} = 26.7 \text{ gpm}$$

Because calculated film flow exceeds required oil flow, chamfers are not necessary. However, if film flow were less than required, suitable chamfers would be needed.

**Table 2. Taper Values for Tapered Land Thrust Bearings**

| Pad Dimensions, Inches | Taper, Inch | |
|---|---|---|
| $a \times b$ | $\delta_1 = h_2 - h_1$ (at ID) | $\delta_2 = h_2 - h_1$ (at OD) |
| $\frac{1}{2} \times \frac{1}{2}$ | 0.0025 | 0.0015 |
| $1 \times 1$ | 0.005 | 0.003 |
| $3 \times 3$ | 0.007 | 0.004 |
| $7 \times 7$ | 0.009 | 0.006 |

**Tilting Pad Thrust Bearing Design.**—The following steps define the performance of a tilting pad thrust bearing, one section of which is shown in Fig. 12. Although each bearing section is wedge shaped, as shown at the right below, for the purposes of design calculation, it is considered to be a rectangle with a length $b$ equal to the circumferential length along the pitch line of the section being considered and a width $a$ equal to the difference in the external and internal radii, as shown at left in Fig. 12. The optimum location of the pivot also is shown in Fig. 12. If shaft rotation in both directions is required, however, the pivot must be at the midpoint, which results in little or no detrimental effect on the performance.

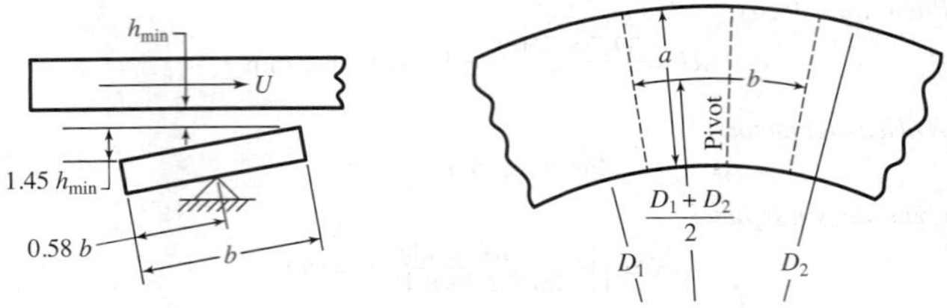


Fig. 12. Basic Elements of Tilting Pad Thrust Bearing.*

1) *Inside diameter, $D_1$.* Determined by shaft size and clearance.

2) *Outside diameter, $D_2$.* Calculated from the formula

$$D_2 = \left(\frac{4W}{\pi K_g p} + D_1^2\right)^{1/2}$$

where $W$ = applied load, pounds

$K_g = 0.8$

$p$ = unit load from Table 1

3) *Radial pad width, a.* Equal to one-half the difference between the inside and outside diameters:

$$a = \frac{D_2 - D_1}{2}$$

4) *Pitch-line circumference, B.* Found from the mean diameter:

$$B = \pi\left(\frac{D_1 + D_2}{2}\right)$$

5) *Number of pads, i.* The number of pads may be estimated from the formula

$$i = \frac{BK_g}{a}$$

Select the nearest even number.

6) *Length of pad, b.* Found from the formula

$$b \cong \frac{BK_g}{i}$$

7) *Pitch-line velocity, U.* Calculated in feet per minute from the formula

$$U = \frac{BN}{12}$$

8) *Bearing unit load, p.* Calculated from the formula

$$p = \frac{W}{iab}$$

9) *Operating number, O.* Calculated from the formula

$$O = \frac{1.4510^{-7}Z_2U}{5pb}$$

where $Z_2$ = viscosity of oil at outlet temperature (inlet temperature plus assumed temperature rise through the bearing).

* See footnote on page 2420.

10) *Minimum film thickness, $h_{min}$.* By using the operating number, the value of $\alpha$ = dimensionless film thickness is found from Fig. 13. Then the actual minimum film thickness is calculated from the formula:

$$h_{min} = \alpha b$$

In general, this value should be 0.001 inch for small bearings and 0.002 inch for larger and high-speed bearings.

11) *Coefficient of friction, f.* Found from Fig. 14.

12) *Friction power loss, $P_f$.* This horsepower loss now is calculated by the formula

$$P_f = \frac{fWU}{33{,}000}$$

13) *Actual oil flow, Q.* This flow over the pad in gallons per minute is calculated from the formula

$$Q = 0.0591\alpha iabU$$

14) *Temperature rise, $\Delta t$.* Found from the formula

$$\Delta t = 0.0217\frac{fp}{\alpha c}$$

where $c$ = specific heat of oil in Btu/gal/°F

If the flow is insufficient, as indicated by too high a temperature rise, chamfers can be added to provide adequate flow, as in Steps 12–15 of the flat plate thrust bearing design.

*Example:* Design a tilting pad thrust bearing for 70,000 pounds thrust at 3600 rpm. The shaft diameter is 6.5 inches and a maximum OD of 15 inches is available. The oil inlet temperature is 110° F and the supply pressure is 20 pounds per square inch. A maximum temperature rise of 50° F is acceptable and results in a viscosity of 18 centipoises. Use a value of 3.5 Btu/gal/°F for *c*.

1) *Inside diameter.* Assume $D_1$ = 7 inches to clear shaft.

2) *Outside diameter.* Given maximum $D_2$ = 15 inches.

3) *Radial pad width.*

$$a = \frac{15 - 7}{2} = 4 \text{ inches}$$

4) *Pitch-line circumference.*

$$B = \pi\left(\frac{7 + 15}{2}\right) = 34.6 \text{ inches}$$

5) *Number of pads.*

$$i = \frac{34.6 \times 0.8}{4} = 6.9$$

Select 6 pads: $i = 6$.

6) *Length of pad.*

$$b = \frac{34.6 \times 0.8}{6} = 4.61 \text{ inches}$$

Make $b$ = 4.75 inches.

7) *Pitch-line velocity.*

$$U = \frac{34.6 \times 3600}{12} = 10{,}400 \text{ ft/min}$$

8) *Bearing unit load.*

$$p = \frac{70{,}000}{6 \times 4 \times 4.75} = 614 \text{ psi}$$

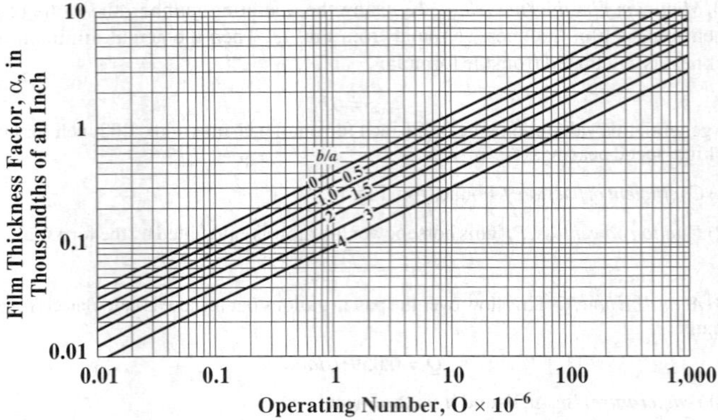


Fig. 13. Dimensionless Minimum Film Thickness, α, versus Operating Number, *O*—Tilting Pad Thrust Bearings.*

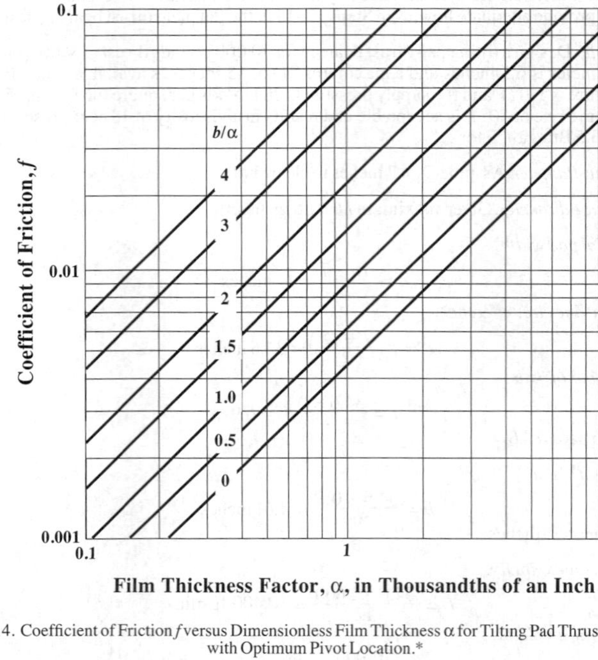


Fig. 14. Coefficient of Friction *f* versus Dimensionless Film Thickness α for Tilting Pad Thrust Bearings with Optimum Pivot Location.*

* See footnote on page 2420.

9) *Operating number.*

$$O = \frac{1.4510^{-7} \times 18 \times 10{,}400}{5 \times 614 \times 4.75} = 1.86 \times 10^{-6}$$

10) *Minimum film thickness.* From Fig. 13, $\alpha = 0.30 \times 10^{-3}$.

$$h_{min} = 0.00030 \times 4.75 = 0.0014 \text{ inch}$$

11) *Coefficient of friction.* From Fig. 14, $f = 0.0036$.

12) *Friction power loss.*

$$P_f = \frac{0.0036 \times 70{,}000 \times 10{,}400}{33{,}000} = 79.4 \text{ hp}$$

13) *Oil flow.*

$$Q = 0.0591 \times 6 \times 0.30 \times 10^{-3} \times 4 \times 4.75 \times 10{,}400 = 21.02 \text{ gpm}$$

14) *Temperature rise.*

$$\Delta t = \frac{0.0217 \times 0.0036 \times 614}{0.30 \times 10^{-3} \times 3.5} = 45.7°\text{F}$$

Because this temperature is less than the 50° F, which is considered as the acceptable maximum, the design is satisfactory.

## Plain Bearing Materials

Materials used for sliding bearings cover a wide range of metals and nonmetals. To make the optimum selection requires a complete analysis of the specific application. The important general categories are: Babbitts, alkali-hardened lead, cadmium alloys, copper lead, aluminum bronze, silver, sintered metals, plastics, wood, rubber, and carbon graphite.

**Properties of Bearing Materials.**—For a material to be used as a plain bearing, it must possess certain physical and chemical properties that permit it to operate properly. If a material does not possess all of these characteristics to some degree, it will not function long as a bearing. It should be noted, however, that few, if any, materials are outstanding in all these characteristics. Therefore, the selection of the optimum bearing material for a given application is at best a compromise to secure the most desirable combination of properties required for that particular usage.

The seven properties generally acknowledged to be the most significant are: 1) Fatigue resistance; 2) embeddability; 3) compatibility; 4) conformability; 5) thermal conductivity; 6) corrosion resistance; and 7) load capacity.

These properties are described as follows:

1) *Fatigue resistance* is the ability of the bearing lining material to withstand repeated applications of stress and strain without cracking, flaking, or being destroyed by some other means.

2) *Embeddability* is the ability of the bearing lining material to absorb or embed within itself any of the larger of the small dirt particles present in a lubrication system. Poor embeddability permits particles circulating around the bearing to score both the bearing surface and the journal or shaft. Good embeddability will permit these particles to be trapped and forced into the bearing surface and out of the way where they can do no harm.

3) *Compatibility or antiscoring tendencies* permit the shaft and bearing to "get along" with each other. It is the ability to resist galling or seizing under conditions of metal-to-metal contact such as at startup. This characteristic is most truly a bearing property, because contact between the bearing and shaft in good designs occurs only at startup.

4) *Conformability* is defined as malleability or as the ability of the bearing material to creep or flow slightly under load, as in the initial stages of running, to permit the shaft and bearing contours to conform with each other or to compensate for nonuniform loading caused by misalignment.

**Table 1. Bearing and Bushing Alloys—Composition, Forms, Characteristics, and Applications** *SAE General Information*

| SAE No. and Alloy Grouping | | Nominal Composition, Percent | Form of Use (1), Characteristics (2), and Applications (3) |
|---|---|---|---|
| Sn-Base Alloys | 11 | Sn, 87.5; Sb, 6.75; Cu, 5.75 | (1) Cast on steel, bronze, or brass backs, or directly in the bearing housing. (2) Soft, corrosion-resistant with moderate fatigue resistance. (3) Main and connecting-rod bearings; motor bushings. Operates with either hard or soft journal. |
| | 12 | Sn, 89; Sb, 7.5; Cu, 3.5 | |
| Pb-Base Alloys | 13 | Pb, 84; Sb, 10; Sn, 6 | (1) SAE 13 and 14 are cast on steel, bronze, or brass, or in the bearing housing; SAE 15 is cast on steel; and SAE 16 is cast into and on a porous sintered matrix, usually copper-nickel bonded to steel. (2) Soft, moderately fatigue-resistant, corrosion-resistant. (3) Main and connecting-rod bearings. Operates with hard or soft journal with good finish. |
| | 14 | Pb, 75; Sb, 15; Sn, 10 | |
| | 15 | Pb, 83; Sb, 15; Sn, 14; As, 1 | |
| | 16 | Pb, 92; Sb, 3.5; Sn, 4.5 | |
| Pb-Sn Overlays | 19 | Pb, 90; Sn, 10 | (1) Electrodeposited as a thin layer on copper-lead or silver bearings faces. (2) Soft, corrosion-resistant. Bearings so coated run satisfactorily against soft shafts throughout the life of the coating. (3) Heavy-duty, high-speed main and connecting-rod bearings. |
| | 190 | Pb, 93; Sn, 7 | |
| Cu-Pb Alloys | 49 | Cu, 76; Pb, 24 | (1) Cast or sintered on steel back with the exception of SAE 481, which is cast on steel back only. (2) Moderately hard. Somewhat subject to oil corrosion. Some oils minimize this; protection with overlay may be desirable. Fatigue resistance good to fairly good. Listed in order of decreasing hardness and fatigue resistance. (3) Main and connecting-rod bearings. The higher lead alloys can be used unplated against a soft shaft, although an overlay is helpful. The lower lead alloys may be used against a hard shaft, or with an overlay against a soft one. |
| | 48 | Cu, 70; Pb, 30 | |
| | 480 | Cu, 65; Pb, 35 | |
| | 481 | Cu, 60; Pb, 40 | |
| Cu-Pb-Sn Alloys | 482 | Cu, 67; Pb, 28; Sn, 5 | (1) Steel-backed and lined with a structure combining sintered copper alloy matrix with corrosion-resistant lead alloy. (2) Moderately hard. Corrosion resistance improved over copper-leads of equal lead content without tin. Fatigue resistance fairly good. Listed in order of decreasing hardness and fatigue resistance. (3) Main and connecting-rod bearings. Generally used without overlay. SAE 484 and 485 may be used with hard or soft shaft, and a hardened or cast shaft is recommended for SAE 482. |
| | 484 | Cu, 55; Pb, 42; Sn, 3 | |
| | 485 | Cu, 46; Pb, 51; Sn, 3 | |
| Al-Base Alloys | 770 | Al, 91.75; Sn, 6.25; Cu, 1; Ni, 1 | (1) SAE 770 cast in permanent molds; work-hardened to improve physical properties. SAE 780 and 782 usually bonded to steel back but is procurable in strip form without steel backing. SAE 781 usually bonded to steel back but can be produced as castings or wrought strip without steel back. (2) Hard, extremely fatigue-resistant, resistant to oil corrosion. (3) Main and connecting-rod bearings. Generally used with suitable overlay. SAE 781 and 782 also used for bushings and thrust bearings with or without overlay. |
| | 780 | Al, 91; Sn, 6; Si, 1.5; Cu, 1; Ni, 0.5 | |
| | 781 | Al, 95; Si, 4; Cd, 1 | |
| | 782 | Al, 95; Cu, 1; Ni, 1; Cd, 3 | |
| Other Cu-Base Alloys | 795 | Cu, 90; Zn, 9.5; Sn, 0.5 | (1) Wrought solid bronze. (2) Hard, strong, good fatigue resistance. (3) Intermediate-load oscillating motion such as tie-rods and brake shafts. |
| | 791 | Cu, 88; Zn, 4; Sn, 4; Pb, 4 | (1) SAE 791, wrought solid bronze; SAE 793, cast on steel back; SAE 798, sintered on steel back. (2) General-purpose bearing material, good shock and load capacity. Resistant to high temperatures. Hard shaft desirable. Less score-resistant than higher lead alloys. (3) Medium to high loads. Transmission bushings and thrust washers. SAE 791 also used for piston pin and 793 and 798 for chassis bushings. |
| | 793 | Cu, 84; Pb, 8; Sn, 4; Zn, 4 | |
| | 798 | Cu, 84; Pb, 8; Sn, 4; Zn, 4 | |
| Other Cu-Base Alloys | 792 | Cu, 80; Sn, 10; Pb, 10 | (1) SAE 792, cast on steel back, SAE 797, sintered on steel back. (2) Has maximum shock and load-carrying capacity of conventional cast bearing alloys; hard, both fatigue- and corrosion-resistant. Hard shaft desirable. (3) Heavy loads with oscillating or rotating motion. Used for piston pins, steering knuckles, differential axles, thrust washers, and wear plates. |
| | 797 | *Cu, 80; Sn, 10; Pb, 10* | |
| | 794 | Cu, 73.5; Pb, 23; Sn, 3.5 | (1) SAE 794, cast on steel back; SAE 799, sintered on steel back. (2) Higher lead content gives improved surface action for higher speeds but results in somewhat less corrosion resistance. (3) Intermediate load application for both oscillating and rotating shafts, that is, rocker-arm bushings, transmissions, and farm implements. |
| | 799 | Cu, 73.5; Pb, 23; Sn, 3.5 | |

5) *High thermal conductivity* is required to absorb and carry away the heat generated in the bearing. This conductivity is most important, not in removing frictional heat generated in the oil film, but in preventing seizures due to hot spots caused by local asperity break-throughs or foreign particles.

6) *Corrosion resistance* is required to resist attack by organic acids that are sometimes formed in oils at operating conditions.

7) *Load capacity or strength* is the ability of the material to withstand the hydrodynamic pressures exerted upon it during operation.

**Babbitt or White Metal Alloys.—**Many different bearing metal compositions are referred to as babbitt metals. The exact composition of the original babbitt metal is not known; however, the ingredients were probably tin, copper, and antimony in approximately the following percentages: 89.3, 3.6, and 7.1. Tin- and lead-base babbitts are probably the best known of all bearing materials. With their excellent embeddability and compatibility characteristics under boundary lubrication, babbitt bearings are used in a wide range of applications including household appliances, automobile and diesel engines, railroad cars, electric motors, generators, steam and gas turbines, and industrial and marine gear units.

**Table 2. White Metal Bearing Alloys—Composition and Properties**
*ASTM B23-00 (R2014)*

| ASTM Alloy[a] Number | Nominal Composition, Percent | | | | Compressive Yield Point,[b] psi | | Ultimate Compressive Strength,[c] psi | | Brinell Hardness Number[d] | | Melting Point °F | Proper Pouring Temperature, °F |
|---|---|---|---|---|---|---|---|---|---|---|---|---|
| | Sn | Sb | Pb | Cu | 68° F | 212° F | 68° F | 212° F | 68° F | 212° F | | |
| 1 | 91.0 | 4.5 | … | 4.5 | 4400 | 2650 | 12,850 | 6950 | 17.0 | 8.0 | 433 | 825 |
| 2 | 89.0 | 7.5 | … | 3.5 | 6100 | 3000 | 14,900 | 8700 | 24.5 | 12.0 | 466 | 795 |
| 3 | 83.33 | 8.33 | … | 8.33 | 6600 | 3150 | 17,600 | 9900 | 27.0 | 14.5 | 464 | 915 |
| 4 | 75.0 | 12.0 | 10.0 | 3.0 | 5550 | 2150 | 16,150 | 6900 | 24.5 | 12.0 | 363 | 710 |
| 5 | 65.0 | 15.0 | 18.0 | 2.0 | 5050 | 2150 | 15,050 | 6750 | 22.5 | 10.0 | 358 | 690 |
| 6 | 20.0 | 15.0 | 63.5 | 1.5 | 3800 | 2050 | 14,550 | 8050 | 21.0 | 10.5 | 358 | 655 |
| 7[e] | 10.0 | 15.0 | bal. | … | 3550 | 1600 | 15,650 | 6150 | 22.5 | 10.5 | 464 | 640 |
| 8[e] | 5.0 | 15.0 | bal. | … | 3400 | 1750 | 15,600 | 6150 | 20.0 | 9.5 | 459 | 645 |
| 10 | 2.0 | 15.0 | 83.0 | … | 3350 | 1850 | 15,450 | 5750 | 17.5 | 9.0 | 468 | 630 |
| 11 | … | 15.0 | 85.0 | … | 3050 | 1400 | 12,800 | 5100 | 15.0 | 7.0 | 471 | 620 |
| 12 | … | 10.0 | 90.0 | … | 2800 | 1250 | 12,900 | 5100 | 14.5 | 6.5 | 473 | 625 |
| 15[f] | 1.0 | 16.0 | bal. | 0.5 | … | … | … | … | 21.0 | 13.0 | 479 | 662 |
| 16 | 10.0 | 12.5 | 77.0 | 0.5 | … | … | … | … | 27.5 | 13.6 | 471 | 620 |
| 19 | 5.0 | 9.0 | 86.0 | … | … | … | 15,600 | 6100 | 17.7 | 8.0 | 462 | 620 |

[a] Data for ASTM alloys 1, 2, 3, 7, 8, and 15 appear in the Appendix of ASTM B23-00 (R2014); the data for alloys 4, 5, 6, 10, 11, 12, 16, and 19 are given in ASTM B23-49. All values are for reference purposes only.

[b] The values for yield point were taken from stress-strain curves at the deformation of 0.125 percent reduction of gage.

[c] The ultimate strength values were taken as the unit load necessary to produce a deformation of 25 percent of the length of the specimen.

[d] These values are the average Brinell Hardness Number of three impressions on each alloy using a 10-mm ball and a 500-kg load applied for 30 seconds.

[e] Also nominal arsenic, 0.45 percent.

[f] Also nominal arsenic, 1 percent.

The compression test specimens were cylinders 1.5 inches in length and 0.5 inch in diameter, machined from chill castings 2 inches in length and 0.75 inch in diameter. The Brinell Hardness tests were made on the bottom face of parallel machined specimens cast in a 2-inch diameter by 0.625-inch deep steel mold at room temperature.

Both the Society of Automotive Engineers and American Society for Testing and Materials have classified white metal bearing alloys. Table 1 and Table 2 give compositions and properties or characteristics for the two classifications.

In small bushings for fractional-horsepower motors and in automotive engine bearings, babbitt is generally used as a thin coating over a flat steel strip. After forming oil distribution grooves and drilling required holes, the strip is cut to size, rolled, and shaped into finished bearing. These bearings are available for shaft diameters from 0.5 to 5 inches (12.7–127 mm). Strip bearings are turned out by the millions yearly in highly automated factories and offer an excellent combination of low cost with good bearing properties.

For larger bearings in heavy-duty equipment, a thicker babbitt is cast on a rigid backing of steel or cast iron. Chemical and electrolytic cleaning of the bearing shell, thorough rinsing, tinning, and then centrifugal casting of the babbitt are desirable for sound bonding of the babbitt to the bearing shell. After machining, the babbitt layer is usually $^1/_2$ to $^1/_4$ inch (12.7–6.35 mm) thick.

Compared to other bearing materials, babbitts generally have lower load-carrying capacity and fatigue strength, are a little higher in cost, and require a more complicated design. Also, their strength decreases rapidly with increasing temperature. These shortcomings can be avoided by using an intermediate layer of high-strength, fatigue-resistant material that is placed between a steel backing and a thin babbitt surface layer. Such composite bearings frequently eliminate any need for using alternate materials having poorer bearing characteristics.

Tin babbitt is composed of 80 to 90 percent tin to which is added about 3 to 8 percent copper and 4 to 14 percent antimony. An increase in copper or antimony produces increased hardness and tensile strength and decreased ductility. However, if the percentages of these alloys are increased above those shown in Table 2, the resulting alloy will have decreased fatigue resistance. These alloys have very little tendency to cause wear to their journals because of their ability to embed dirt. They resist the corrosive effects of acids, are not prone to oil-film failure, and are easily bonded and cast. Two drawbacks are encountered from use of these alloys because they have low fatigue resistance and their hardness and strength drop appreciably at low temperatures.

Lead babbitt compositions generally range from 10 to 15 percent antimony and up to 10 percent tin in combination with the lead. Like tin-base babbitts, these alloys have little tendency to cause wear to their journals, embed dirt well, resist the corrosive effects of acids, are not prone to oil-film failure and are easily bonded and cast. Their chief disadvantages when compared with tin-base alloys are a rather lower strength and a susceptibility to corrosion.

**Cadmium Base.—**Cadmium alloy bearings have a greater resistance to fatigue than babbitt bearings, but their use is very limited due to their poor corrosion resistance. These alloys contain 1 to 15 percent nickel, or 0.4 to 0.75 percent copper, and 0.5 to 2.0 percent silver. Their prime attribute is their high-temperature capability. The load-carrying capacity and relative basic bearing properties are shown in Table 3.

**Copper-Lead.—**Copper-lead bearings are a binary mixture of copper and lead containing from 20 to 40 percent lead. Lead is practically insoluble in copper, so a cast microstructure consists of lead pockets in a copper matrix. A steel backing is commonly used with this material and high volume is achieved either by continuous casting or by powder metallurgy techniques. This material is very often used with an overplate such as lead-tin and lead-tin-copper to increase basic bearing properties. Table 3 provides comparisons of material properties.

The combination of good fatigue strength, high-load capacity, and high-temperature performance has resulted in extensive use of this material for heavy-duty main and connecting-rod bearings as well as moderate-load and speed applications in turbines and electric motors.

**Leaded Bronze and Tin-Bronze.—**Leaded and tin-bronzes contain up to 25 percent lead or approximately 10 percent tin, respectively. Cast leaded bronze bearings offer good compatibility, excellent casting, and easy machining characteristics, low cost, good structural properties and high-load capacity, usefulness as a single material that requires

**Table 3. Properties of Bearing Alloys and Bearing Characteristics Ratings**

| Material | Recommended Shaft Hardness (Brinell Hardness Number) | Load-Carrying Capacity, psi[a] | Maximum Operating Temp., °F[b] | Compatibility[c] | Conformability and Embeddability[c] | Corrosion Resistance[c] | Fatigue Strength[c] |
|---|---|---|---|---|---|---|---|
| Tin-Base Babbitt | 150 or less | 800–1500 | 300 | 1 | 1 | 1 | 5 |
| Lead-Base Babbitt | 150 or less | 800–1200 | 300 | 1 | 1 | 3 | 5 |
| Cadmium Base | 200–250 | 1200–2000 | 500 | 1 | 2 | 5 | 4 |
| Copper-Lead | 300 | 1500–2500 | 350 | 2 | 2 | 5 | 3 |
| Tin-Bronze | 300–400 | 4000+ | 500+ | 3 | 5 | 2 | 1 |
| Lead-Bronze | 300 | 3000–4500 | 450–500 | 3 | 4 | 4 | 2 |
| Aluminum | 300 | 4000+ | 225–300 | 5 | 3 | 1 | 2 |
| Silver-Overplate | 300 | 4000+ | 500 | 2 | 3 | 1 | 1 |
| Trimetal-Overplate | 230 or less | 2000–4000+ | 225–300 | 1 | 2 | 2 | 3 |

[a] 1 psi = 6.8947 kPa

[b] Temp. in °C = (°F − 32)/1.8.

[c] *Note:* 1 is best; 5 is worst.

neither a separate overlay nor a steel backing. Bronzes are available in standard bar stock, sand or permanent molds, investment, centrifugal or continuous casting. Leaded bronzes have better compatibility than tin-bronzes because the spheroids of lead smear over the bearing surface under conditions of inadequate lubrication. These alloys are generally a first choice at intermediate loads and speeds. Table 3 provides comparisons of basic bearing properties of these materials.

**Aluminum.**—Aluminum bearings are either cast solid aluminum, aluminum with a steel backing, or aluminum with a suitable overlay. The aluminum is usually alloyed with small amounts of tin, silicon, cadmium, nickel, or copper, as shown in Table 1. An aluminum bearing alloy with 20 to 30 percent tin alloy and up to 3 percent copper has shown promise as a substitute for bronzes in some industrial applications.

These bearings are best suited for operation with hard journals. Owing to the high thermal expansion of the metal (resulting in diametral contraction when it is confined as a bearing in a rigid housing), large clearances are required, which tend to make the bearing noisy, especially on starting. Overlays of lead-tin, lead, or lead-tin-copper may be applied to aluminum bearings to facilitate their use with soft shafts.

Aluminum alloys are available with properties specifically designed for bearing applications, such as high load-carrying capacity, fatigue strength, and thermal conductivity, in addition to excellent corrosion resistance and low cost.

**Silver.**—Silver bearings were developed for and have an excellent record in heavy-duty applications such as aircraft master rod and diesel engine main bearings. Silver has a higher fatigue rating than any of the other bearing materials; the steel backing used with this material may show evidence of fatigue before the silver. The advent of overlays, or more commonly called overplates, made it possible for silver to be used as a bearing material. Silver by itself does not possess any of the desirable bearing qualities except high fatigue resistance and high thermal conductivity. The overlays such as lead, lead-tin, or lead-indium improve the embeddability and antiscoring properties of silver. The relative basic properties of this material, when used as an overplate, are shown in Table 3.

**Cast Iron.**—Cast iron is an inexpensive bearing material capable of operation at light loads and low speeds, i.e., up to 130 ft/min (40 m/min) and 150 lb/in$^2$ (1.03 MPa). These bearings must be well lubricated and have a rather large clearance so as to avoid scoring from particles torn from the cast iron that ride between bearing and journal. A journal hardness of between 150–250 BHN (Brinell Hardness Number) has been found to be best when using cast-iron bearings.

**Porous Metals.—**Porous metal self-lubricating bearings are usually made by sintering metals such as plain or leaded bronze, iron, and stainless steel. The sintering produces a spongelike structure capable of absorbing fairly large quantities of oil, usually 10–35 percent of the total volume. These bearings are used where lubrication supply is difficult, inadequate, or infrequent. This type of bearing should be flooded from time to time to resaturate the material. Another use of these porous materials is to meter a small quantity of oil to the bearings such as in drip feed systems. The general design operating characteristics of this class of materials are shown in Table 4.

**Table 4. Application Limits — Sintered Metal and Nonmetallic Bearings**

| Bearing Material | Load Capacity | | Maximum Temperature | | Surface Speed, $V_{max}$ (max. fpm) | *PV* Limit *P* = psi load *V* = surface ft/min |
|---|---|---|---|---|---|---|
| | psi | kPa | °F | °C | | |
| Acetal | 1000 | 6895 | 180 | 82 | 1000 | 3000 |
| Graphite (dry) | 600 | 4137 | 750 | 399 | 2500 | 15,000 |
| Graphite (lubricated) | 600 | 4137 | 750 | 399 | 2500 | 150,000 |
| Nylon, Polycarbonate | 1000 | 6895 | 200 | 93 | 1000 | 3000 |
| Nylon composite | … | … | 400 | 204 | … | 16,000 |
| Phenolics | 6000 | 41369 | 200 | 93 | 2500 | 15,000 |
| Porous bronze | 4500 | 31026 | 160 | 71 | 1500 | 50,000 |
| Porous iron | 8000 | 55158 | 160 | 71 | 800 | 50,000 |
| Porous metals | 4000–8000 | 27579–55158 | 150 | 66 | 1500 | 50,000 |
| Virgin Teflon (TFE) | 500 | 3447 | 500 | 260 | 50 | 1000 |
| Reinforced Teflon | 2500 | 17237 | 500 | 260 | 1000 | 10,000–15,000 |
| TFE fabric | 60,000 | 413685 | 500 | 260 | 150 | 25,000 |
| Rubber | 50 | 345 | 150 | 66 | 4000 | 15,000 |
| Maple & Lignum Vitae | 2000 | 13790 | 150 | 66 | 2000 | 15,000 |

1 fpm = 0.3048 m/min; 1 psi = 6.8947 kPa

Table 5a, Table 5b, and Table 5c give the chemical compositions, permissible loads, interference fits, and running clearances of bronze-base and iron-base metal-powder sintered bearings that are specified in the ASTM specifications for oil-impregnated metal-powder sintered bearings (ASTM B438-17 and B439-19).

**Plastics Bearings.—**Plastics are finding increased use as bearing materials because of their resistance to corrosion, quiet operation, ability to be molded into many configurations, and their excellent compatibility, which minimizes or eliminates the need for lubrication. Many plastics are capable of operating as bearings, especially phenolic, tetrafluoroethylene (TFE), and polyamide (nylon) resins. The general application limits for these materials are shown in Table 4.

*Laminated Phenolics:* These composite materials consist of cotton fabric, asbestos, or other fillers bonded with phenolic resin. They have excellent compatibility with various fluids as well as strength and shock resistance. However, precautions must be taken to maintain adequate bearing cooling because the thermal conductivity of these materials is low.

*Nylon:* This material has the widest use for small, lightly loaded applications. It has low frictional properties and requires no lubrication.

*Teflon:* This material, with its exceptional low coefficient of friction, self-lubricating characteristics, resistance to attack by almost any chemicals, and its wide temperature range, is one of the most interesting of the plastics for bearing use. High cost combined with low load capacity cause Teflon to be selected mostly in modified form, where other less expensive materials have proved inadequate for design requirements.

Bearings made of laminated phenolics, nylon, or Teflon are all unaffected by acids and alkalies except if highly concentrated and therefore can be used with lubricants containing dilute acids or alkalies. Water is used to lubricate most phenolic laminate bearings but oil, grease, and emulsions of grease and water are also used. Water and oil are used as lubricants for nylon and Teflon bearings. Almost all types of plastic bearings absorb water and oil to some extent. In some the dimensional change caused by the absorption may be as much as three percent in one direction. This means that bearings have to be treated before use so that proper clearances will be kept. This may be done by boiling in water, for water lubricated bearings. Boiling in water makes bearings swell the maximum amount. Clearances for phenolic bearings are kept at about 0.001 inch per inch (or mm per mm) of diameter on treated bearings. Partially lubricated or dry nylon bearings are given a clearance of 0.004 to 0.006 inch (101.6–152.4 μm) for a 1-inch (25.4-mm) diameter bearing.

*Rubber:* Rubber bearings give excellent performance on propeller shafts and rudders of ships, hydraulic turbines, pumps, sand and gravel washers, dredges and other industrial equipment that handle water or slurries. The resilience of rubber helps to isolate vibration and provide quiet operation, allows running with relatively large clearances and helps to compensate for misalignment. In these bearings a fluted rubber structure is supported by a metal shell. The flutes or scallops in the rubber form a series of grooves through which lubricant or, as generally used, water and foreign material such as sand may pass through the bearing.

**Wood.—**Bearings made from such woods as lignum vitae, rock maple, or oak offer self-lubricating properties, low cost, and clean operation. However, they have frequently been displaced in recent years by various plastics, rubber, and sintered metal bearings. General applications are shown in Table 4.

**Carbon-Graphite.—**Bearings of molded and machined carbon-graphite are used where regular maintenance and lubrication cannot be given. They are dimensionally stable over a wide range of temperatures, may be lubricated if desired, and are not affected by chemicals. These bearings may be used up to temperatures of 700 to 750° F (371–399° C) in air or 1200° F (649° C) in a non-oxidizing atmosphere, and generally are operated at a maximum load of 20 pounds per square inch. In some instances a metal or metal alloy is added to the carbon-graphite composition to improve such properties as compressive strength and density. The temperature limitation depends upon the melting point of the metal or alloy and the maximum load is generally 350 psi (2.4 MPa) when used with no lubrication or 600 psi (4.2 MPa) when used with lubrication.

Normal running clearances for both types of carbon-graphite bearings used with steel shafts and operating at a temperature of less than 200° F (93° C) are as follows: 0.001 inch (0.0254 mm) for bearings of 0.187- to 0.500-inch (4.75–12.7-mm) inside diameter, 0.002 inch (0.0508 mm) for bearings of 0.501- to 1.000-inch (12.73–25.4-mm) inside diameter, 0.003 inch (0.0762 mm) for bearings of 1.001- to 1.250-inch (25.43–31.75-mm) inside diameter 0.004 inch (0.1016 mm) for bearings of 1.251- to 1.500-inch (31.77–38.1-mm) diameter, and 0.005 inch (0.127 mm) for bearings of 1.501- to 2.000-inch (38.13–50.8-mm) inside diameter. Speeds depend upon too many variables to list specifically so it can only be stated here that high loads require a low number of rpm and low loads permit a high number of rpm. Smooth journals are necessary in these bearings as rough ones tend to abrade the bearings quickly. Cast iron and hard chromium-plate steel shafts of 400 BHN (Brinell Hardness Number) and over, and phosphor-bronze shafts over 135 BHN are recommended.

**Table 5a. Copper- and Iron-Base Sintered Bearings (Oil Impregnated)**

*ASTM B438-17, ASTM B439-18, and Appendices*

| Chemical Requirements | | | | | | | | |
|---|---|---|---|---|---|---|---|---|
| Alloying Elements[a] | Percentage Composition | | | | | | | |
| | Copper-Base Bearings | | | | Iron-Base Bearings | | | |
| | Grade 1 | | Grade 2 | | Grades | | | |
| | Class A | Class B | Class A | Class B | 1 | 2 | 3 | 4 |
| Cu | 87.5-99.5 | 87.5-90.5 | 87.5-90.5 | 87.5-90.5 | … | … | 7.0-11.0 | 18.0-22.0 |
| Sn | 9.5-10.5 | 9.5-10.5 | 9.5-10.5 | 9.5-10.5 | … | … | … | … |
| Graphite | 0.1 max. | 1.75 max. | 0.1 max. | 1.75 max. | … | … | … | … |
| Pb | … | … | 2.0-4.0 | 2.0-4.0 | … | … | … | … |
| Fe | 1.0 max. | 1.0 max. | 1.0 max. | 1.0 max. | 96.25 min. | 95.9 min. | Balance[b] | Balance[b] |
| Comb. C[c] | … | … | … | … | 0.25 max. | 0.25-0.60 | … | … |
| Si, max. | … | … | … | … | 0.3 | 0.3 | … | … |
| Al, max. | … | … | … | … | 0.2 | 0.2 | … | … |
| Others | 0.5 max. | 0.5 max. | 1.0 max. | 1.0 max. | 3.0 max. | 3.0 max. | 3.0 max. | 3.0 max. |

[a] Abbreviations used for the alloying elements are as follows: Cu, copper; Fe, iron; Sn, tin; Pb, lead; Zn, zinc; Ni, nickel; Sb, antimony; Si, silicon; Al, aluminum; and C, carbon.

[b] Total of iron plus copper shall be 97 percent, minimum.

[c] Combined carbon (on basis of iron only) may be a metallographic estimate of the carbon in the iron.

| Permissible Loads | | | | | | |
|---|---|---|---|---|---|---|
| Copper-Base Bearings | | | | Iron-Base Bearings | | |
| | Grades 1 & 2 | | | | | |
| | Type 1 | Type 2 | Types 3 & 4 | | Grades 1 & 2 | Grades 3 & 4 |
| Shaft Velocity, fpm | Max. Load, psi | | | Shaft Velocity, fpm | Max. Load, psi | |
| Slow and intermittent | 3200 | 4000 | 4000 | Slow and intermittent | 3600 | 8000 |
| 25 | 2000 | 2000 | 2000 | 25 | 1800 | 3000 |
| 50 to 100 | 500 | 500 | 550 | 50 to 100 | 450 | 700 |
| Over 100 to 150 | 365 | 325 | 365 | Over 100 to 150 | 300 | 400 |
| Over 150 to 200 | 280 | 250 | 280 | Over 150 to 200 | 225 | 300 |
| Over 200 | a | a | a | Over 200 | a | a |

[a] For shaft velocities over 200 fpm, the permissible loads may be calculated as follows: $P = 50{,}000/V$; where $P$ = safe load, psi of projected area; and $V$ = shaft velocity, fpm. With a shaft velocity of less than 50 fpm and a permissible load greater than 1,000 psi, an extreme pressure lubricant should be used; with heat dissipation and removal techniques, higher $PV$ ratings can be obtained.

| Clearances | | | | | | |
|---|---|---|---|---|---|---|
| Press-Fit Clearances | | | Running Clearances[a] | | | |
| Copper- and Iron-Base | | | Copper-Base | | Iron-Base | |
| Bearing OD | Min. | Max. | Shaft Size | Min. Clearance | Shaft Size | Min. Clearance |
| Up to 0.760 | 0.001 | 0.003 | Up to 0.250 | 0.0003 | Up to 0.760 | 0.0008 |
| 0.761-1.510 | 0.0015 | 0.004 | 0.250-0.760 | 0.0005 | 0.761-1.510 | 0.0013 |
| 1.511-2.510 | 0.002 | 0.005 | 0.760-1.510 | 0.0010 | 1.511-2.510 | 0.0018 |
| 2.511-3.010 | 0.002 | 0.006 | 1.510-2.510 | 0.0015 | Over 2.510 | 0.0023 |
| Over 3.010 | 0.002 | 0.007 | Over 2.510 | 0.0020 | | |

[a] Only minimum recommended clearances are listed. It is assumed that ground steel shafting will be used and that all bearings will be oil-impregnated.

**Table 5b. Copper- and Iron-Base Sintered Bearings (Oil Impregnated)**
*ASTM B438-17, ASTM B439-18, and Appendices*

| Commercial Dimensional Tolerances[a,b] | | | | | | | |
|---|---|---|---|---|---|---|---|
| Diameter Tolerance | | Length Tolerance | | Diameter Tolerance | | Length Tolerance | |
| Copper Base | | | | Iron Base | | | |
| Inside or Outside Diameter | Total Diameter Tolerances | Length | Total Length Tolerances | Inside or Outside Diameter | Total Diameter Tolerances | Length | Total Length Tolerances |
| Up to 1 | 0.001 | Up to 1.5 | 0.01 | Up to 0.760 | −0.001 | Up to 1.495 | 0.01 |
| 1 to 1.5 | 0.0015 | 1.5 to 3 | 0.01 | 0.761-1.510 | −0.0015 | 1.496-1.990 | 0.02 |
| 1.5 to 2 | 0.002 | 3 to 4.5 | 0.02 | 1.511-2.510 | −0.002 | 1.991-2.990 | 0.02 |
| 2 to 2.5 | 0.0025 | … | … | 2.511-3.010 | −0.003 | 2.991-4.985 | 0.03 |
| 2.5 to 3 | 0.003 | … | … | 3.011-4.010 | −0.005 | … | … |
| … | … | … | … | 4.011-5.010 | −0.005 | … | … |
| … | … | … | … | 5.011-6.010 | −0.006 | … | … |

**Concentricity Tolerance[a,b,c]**

| Iron Base | | |
|---|---|---|
| Outside Diameter | Max. Wall Thickness | Concentricity Tolerance |
| Up to 1.510 | Up to 0.355 | 0.003 |
| 1.511 to 2.010 | Up to 0.505 | 0.004 |
| 2.011 to 4.010 | Up to 1.010 | 0.005 |
| 4.011 to 5.010 | Up to 1.510 | 0.006 |
| 5.011 to 6.010 | Up to 2.010 | 0.007 |

| Copper Base | | |
|---|---|---|
| Outside Diameter | Length | Concentricity Tolerance |
| Up to 1 | 0 to 1 | 0.000 |
| | 1 to 2 | 0.004 |
| | 2 to 3 | 0.005 |
| 1 to 2 | 0 to 1 | 0.004 |
| | 1 to 2 | 0.005 |
| | 2 to 3 | 0.006 |
| 2 to 3 | 0 to 1 | 0.005 |
| | 1 to 2 | 0.006 |
| | 2 to 3 | 0.007 |

[a] For copper-base bearings with 4-to-1 maximum-length-diameter ratio and a 24-to-1 maximum-length-to-wall-thickness ratio; bearings with greater ratios are not covered here.

[b] For iron-base bearings with a 3-to-1 maximum-length-to-inside diameter ratio and a 20-to-1 maximum-length-to-wall-thickness ratio; bearings with greater ratios are not covered here.

[c] Total indicator reading.

**Table 5c. Copper- and Iron-Base Sintered Bearings (Oil Impregnated)**
*ASTM B438-17, ASTM B439-18, and Appendices*

| Diameter Range | Flange and Thrust Bearings, Diameter, and Thickness Tolerances | | | | Parallelism[a] on Faces, max. | | | |
|---|---|---|---|---|---|---|---|---|
| | Flange Diameter Tolerance | | Flange Thickness Tolerance | | Copper Base | | Iron Base | |
| | Standard | Special | Standard | Special | Standard | Special | Standard | Special |
| 0 to 1½ | ±0.005 | ±0.0025 | ±0.005 | ±0.0025 | 0.003 | 0.002 | 0.005 | 0.003 |
| Over 1½ to 3 | ±0.010 | ±0.005 | ±0.010 | ±0.007 | 0.004 | 0.003 | 0.007 | 0.005 |
| Over 3 to 6 | ±0.025 | ±0.010 | ±0.015 | ±0.010 | 0.005 | 0.004 | 0.010 | 0.007 |

[a] Standard and special tolerances are specified for diameters, thicknesses, and parallelism. Special tolerances should not be specified unless required because they require additional or secondary operations and, therefore, are costlier. Thrust bearings (¼ inch thickness, max.) have a standard thickness tolerance of ±0.005 inch and a special thickness tolerance of ±0.0025 inch for all diameters.

All dimensions in inches except where otherwise noted.

# BALL, ROLLER, AND NEEDLE BEARINGS

## Rolling Contact Bearings

Rolling contact bearings use rolling elements, such as balls or rollers, to carry an impressed load with reduced friction. These bearings are commonly called *antifriction bearings*, since the rolling motion results in less friction than in bearings exhibiting sliding friction. Rolling element bearings are designed to endure the effects of varying speed, temperature, load, lubrication, and maintenance conditions.

Because of their low starting and running capabilities, friction rolling contact bearings are used to reduce the complexity of many systems that otherwise might use journal bearings. These bearings also are desirable due to the precise radial and axial location of the rotating elements, their reduced lubrication requirements, and the ability to function during brief interruptions in normal lubrication.

While many variations are in use throughout industry, rolling contact bearings typically are made of hardened rolling elements and races, usually with a separator to space the rolling elements. The suitability of a rolling element bearing to a particular application is based on several key factors for optimum service, such as available space, loading conditions (magnitude and direction), misalignment, speed, operating temperature, precision required, stiffness, vibration level, contamination level, and lubrication.

Numerous manufacturers produce a complete range of ball and roller bearings in fully interchangeable series with standard dimensions, tolerances, and fits as specified in the bearings standards published by the Anti-Friction Bearing Manufacturers Association (AFBMA). In 1993, this organization changed its name to the American Bearing Manufacturers Association (ABMA). The International Standards Organization (ISO) and other industry organizations also publish applicable standards.

Rolling contact bearings are made to high standards of accuracy and with close metallurgical control. Balls and rollers are normally held to diametral tolerances of .0001 inch (2.54 μm) or less within one bearing. This accuracy is essential to the performance and durability of bearings in limiting runout, providing proper radial and axial clearances, and ensuring smoothness of operation.

The life span of rolling element bearings typically is limited by the fatigue life of the material from which they are made, the lubricant used, and operating and maintenance conditions. In rolling contact fatigue, precise relationships among life, load (radial and thrust, or both), and design characteristics are not predictable. But a statistical function described as the "probability of survival" can be applied to wear estimation, according to equations recommended by the ABMA. Deviations from these formulas should be considered when certain extremes in application conditions—including speed, deflection, temperature, lubrication, and internal geometry—may occur during operation.

**Types of Antifriction Bearings.—**There are two main types of antifriction bearings: *ball bearings* and *roller bearings*. Both types generally have four parts: outer race, inner race, balls or rollers, and cage. In both types, the rolling element bearings support the load via the contact existing between the rolling elements and the raceways.

Ball bearings can operate at higher speeds but have lower load carrying capacity because of their single point contact. Roller bearings provide linear contact, can operate at comparatively lower speeds, and have higher load carrying capacities.

Design variations of both types are available with varying characteristics. While balls as rolling elements are relatively limited in designs variations, different roller designs include cylindrical, tapered, spherical, or needle shapes. There are other modifications of the basic types, but the fundamentals of mounting remain unchanged. Special bearings can be procured with races adapted to specific applications.

**Types of Ball Bearings.—**Most types of ball bearings originate from two basic designs: *radial bearings* and *angular-contact bearings*. Both come in single and double row designs.

## Table 1a. Common Ball Bearing Types

| Type | Mechanism Description | Functionality/Load Capability |
|---|---|---|
| Deep Groove Ball Bearing 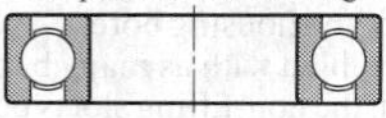 | The most common type of ball bearings, these are self-retaining bearings, with outer and inner rings, and ball and cage assemblies. Available with seals, shields, or open, these bearings produce low noise and need minimal maintenance. | Deep groove bearings can take combined radial and thrust loads where the thrust component is relatively high and can operate at high speeds. Since this type is not self-aligning, the shaft and housing bore must be aligned. |
| Double Row Deep Groove Ball Bearing 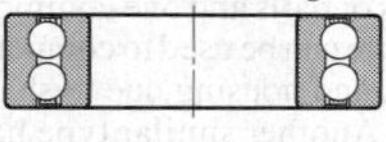 | Double row deep groove ball bearings have two rows of balls in deep, uninterrupted grooves, separated by a spacer to distribute loads evenly, all in a single cage. These bearings do not have seals or shields but may have filling slots. | Double row deep groove ball bearings provide double the load carrying capacity for the same bore and outside diameter. They can carry radial loads, as well as axial loads acting in both directions. |
| Self-Aligning Double Row Deep Groove Ball Bearing  | This self-aligning bearing features two rows of balls and one common, concave-shaped outer raceway. This type produces low noise and vibration. | These bearings provide low friction and high load carrying capacity. They are suitable where considerable shaft deflection or some angular misalignment is expected. |
| Single Row Angular Contact Ball Bearing 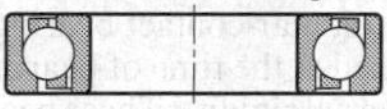 | These bearings have inner and outer raceways that are displaced relative to each other in the direction of the bearing axis. The contact angle, between the contact points of the ball and the raceway in the radial plane, ranges from 15 to 25 degrees. | This type of bearing is designed for high combined radial and thrust loads, where the thrust component may be large and axial deflection must be confined within close limits. |
| Double Row Angular Contact Ball Bearing 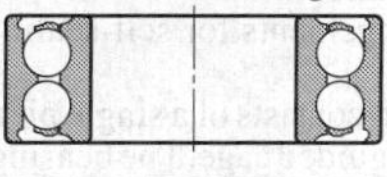 | The double row angular contact bearing is made up of two single row bearings arranged back-to-back to accommodate axial loads in both directions. | Double row angular contact ball bearings work well where high radial and axial loads are acting simultaneously. The axial load carrying capacity for double row angular contact bearing is dependent on the contact angle (i.e., the larger the angle ~ 30°, the higher the axial load to which the bearing can be subjected). |
| Single Direction Thrust Ball Bearing  | Thrust ball bearings consist of two rings (commonly called *shaft* and *housing washers*), balls, and a cage assembly. The shaft washer is placed between the cage and the housing washer, accommodating axial loads on the bearing in one direction. These bearings have low vibration and are easy to maintain. | This type of bearing can accommodate only axial loads (no radial loading) from just one direction and requires a minimum axial load. If the external load is too low, the bearing must be preloaded, usually through a spring mechanism. |
| Double Direction Thrust Ball Bearing 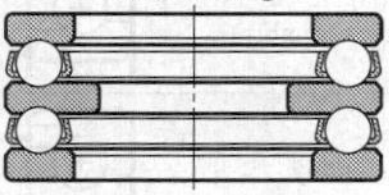 | The components of double direction thrust ball bearings are similar to those of the single direction bearings, but it has three rings (one shaft washer and two housing washers) and double cage assemblies. | The double direction type can accommodate axial loads from both directions. These bearings provide high rigidity and load carrying capacity, and they operate at high speeds. |
| Four-Point Contact Ball Bearing 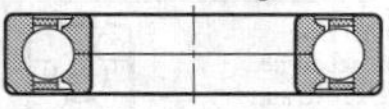 | These separable bearings consist of four contact points between the raceways and balls. They have a single piece outer ring, a radially split inner ring, a ball set, and an outer ring guided cage. The contact angle is around 40 degrees. | Four-point bearings are designed to support axial loads in both directions. This type can handle a combination of radial, moment, and higher thrust loads. |

*Single Row Radial, Non-Filling Slot:* Also known as the *deep-groove* or *Conrad* type, this is probably the most widely used ball bearing, and it is made in many modified forms. This symmetrical bearing can take combined radial and thrust loads in which the thrust component is relatively high. But it is not intended for pure thrust loads. This type is not self-aligning and requires accurate alignment between the shaft and housing bore.

*Single Row Radial, Filling Slot:* Bearings of this type are assembled with as many balls as can be introduced by eccentric displacement of the rings (as in the non-filling slot type), then several more balls are inserted through the loading slot (aided by slight spreading of the rings and heat expansion of the outer ring, if necessary). This type is designed primarily to carry more radial loads, some with a certain level of thrust, but not exceeding 60 percent of the radial load.

*Double Row Radial:* These double row, deep-groove ball bearings have double the load capacity for the same bore and outside diameter. With two rows of balls and one common concave shaped outer raceway, they are self-aligning. This feature can be used to compensate for an appreciable degree of misalignment between shaft and housing due to shaft deflections, mounting inaccuracies, or other common causes. Another similar type has two uninterrupted raceways and two sets of balls.

*Single Row Angular-Contact:* This type is designed for combined radial and thrust loads, where the thrust component may be large and axial deflection must be confined within close limits. A high shoulder on one side of the outer ring takes the thrust, while the shoulder on the other side is just high enough to make the bearing non-separable. Except where used for a pure thrust load in one direction, this type is applied either in pairs (duplex) or with one at each (opposite) end of the shaft.

*Double Row Bearings:* These are, in effect, two single row, angular-contact bearings built as a unit. The internal fit between balls and raceway is fixed at the time of bearing assembly, so it is not dependent upon mounting methods for internal rigidity. These bearings usually have a known amount of internal preload built in for maximum resistance to deflection under combined loads with thrust from either direction. While the balls and races are under compression before an applied external load, this bearing type is very effective for radial loads where bearing deflection must be minimized.

*Other Types:* Modifications of the basic types provide arrangements for self-sealing, location by snap ring, shielding, and so on.

For instance, one special type is the four-point bearing, which consists of a single piece outer ring, a radially split inner ring, a ball set, and an outer ring guided cage. The bearings are separable and designed to support axial loads in both directions.

For more information on ball bearing types, see Table 1a, which illustrates common configurations, with descriptions and functionality/load information. Table 1b and Table 1c provide details of ball bearing elements and corresponding engineering symbols.

**Table 1b. Ball Bearing Elements and Their Symbols**

| Ball Bearings, Single Row, Radial Contact | | | | | |
|---|---|---|---|---|---|
| Symbol | Description | | Symbol | Description | |
| BC | Non-filling slot assembly |  | BH | Non-separable counter-bore assembly |  |
| BL | Filling slot assembly |  | BM | Separable assembly | 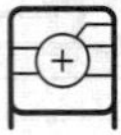 |
| **Ball Bearings, Single Row, Angular Contact[a]** | | | | | |
| Symbol | Description | | Symbol | Description | |
| BN | Non-separable<br>Nominal contact angle: from above 10° to and including 22° | 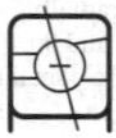 | BAS | Separable inner ring<br>Nominal contact angle: from above 22° to and including 32° | 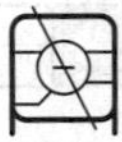 |

## Table 1b. Ball Bearing Elements and Their Symbols *(continued)*

| Ball Bearings, Single Row, Angular Contact[a] | | | | | |
|---|---|---|---|---|---|
| BNS | Separable outer ring<br>Nominal contact angle:<br>from above 10° to and including 22° | 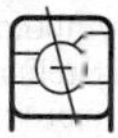 | BT | Non-separable<br>Nominal contact angle:<br>from above 32° to and including 45° |  |
| BNT | Separable inner ring<br>Nominal contact angle:<br>from above 10° to and including 22° | 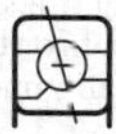 | BY | Two-piece outer ring | 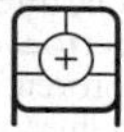 |
| BA | Non-separable<br>Nominal contact angle:<br>from above 22° to and including 32° |  | BZ | Two-piece inner ring |  |
| Ball Bearings, Single Row, Radial Contact, Spherical Outside Surface | | | | | |
| Symbol | Description | | Symbol | Description | |
| BCA | Non-filling slot assembly |  | BLA | Filling slot assembly | 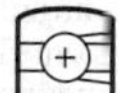 |

[a] A line through the ball contact points forms an acute angle with a perpendicular to the bearing axis.

## Table 1c. Ball Bearing Elements and Their Symbols

| Ball Bearings, Double Row, Radial Contact | | | | | |
|---|---|---|---|---|---|
| Symbol | Description | | Symbol | Description | |
| BF | Filling slot assembly | 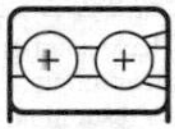 | BHA | Non-separable two-piece outer ring | 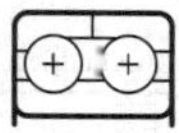 |
| BK | Non-filling slot assembly | 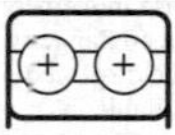 | | | |
| Ball Bearings, Double Row, Angular Contact[a] | | | | | |
| Symbol | Description | | Symbol | Description | |
| BD | Filling slot assembly<br>Vertex of contact angles inside bearing | 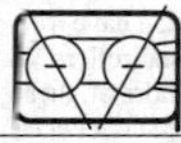 | BG | Non-filling slot assembly<br>Vertex of contact angles outside bearing | 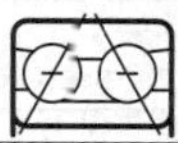 |
| BE | Filling slot assembly<br>Vertex of contact angles outside bearing | 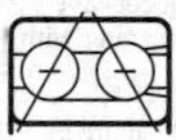 | BAA | Non-separable<br>Vertex of contact angles inside bearing<br>Two-piece outer ring | 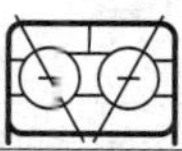 |
| BJ | Non-filling slot assembly<br>Vertex of contact angles inside bearing | 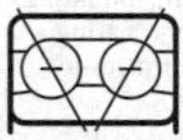 | BVV | Separable<br>Vertex of contact angles outside bearing<br>Two-piece inner ring | 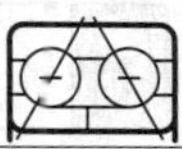 |

| Ball Bearings, Double Row, Self-Aligning[a] | | |
|---|---|---|
| Symbol | Description | |
| BS | Raceway of outer ring spherical | 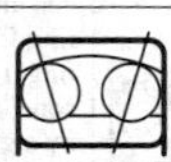 |

[a] A line through the ball contact points forms an acute angle with a perpendicular to the bearing axis.

**Types of Roller Bearings.—**Roller bearings are distinguished by the design of rollers and raceways applied to handle axial, thrust, or combined axial and thrust loads. Table 2a provides illustrations and a summary of common roller bearing types, functionality, and load capabilities. Table 2b and Table 3 include symbology and added details.

*Cylindrical Roller:* These bearings have solid or helically wound hollow cylindrical rollers. The free ring may have a restraining flange to provide restraint to endwise movement in one direction or have no flange so the bearing rings may be displaced axially with respect to each other. Either the rollers or a roller path on the races may be partially crowned to prevent edge loading under slight shaft misalignment. Low friction makes this type suitable for relatively high speeds.

*Barrel Roller:* These bearings have a single row of barrel shaped rollers within a solid outer ring with a concave raceway, and a solid inner ring with two ribs. This type is particularly suitable for high radial shock loads and certain misalignment, with minimal axial load.

*Spherical Roller:* These bearings are usually furnished in a double row, self-aligning mounting. Both rows of barrel shaped rollers have a common spherical outer raceway. This type has a high radial and thrust load carrying capacity and can maintain this capacity under some degree of shaft and housing misalignment.

*Tapered Roller:* In this type, tapered rollers are held in accurate alignment by a guide flange on the inner ring. In these separable bearings, the apexes of the tapered working surfaces of both rollers and races, if extended, are designed to coincide on the bearing axis. These bearings have a high radial and thrust carrying capacity.

**Table 2a. Common Roller Bearing Types**

| Type | Mechanism Description | Functionality/Load Capability |
|---|---|---|
| Cylindrical Roller Bearing 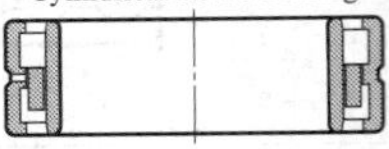 | These floating bearings have solid or helically wound cylindrical rollers. The free ring may have a restraining flange to restrain endwise movement in one direction or have no flange so the bearing rings may be displaced axially. This type is available in single row, double row, and multi-row versions, as well as sealed or unsealed, depending on the application. | This type offers low friction, high radial load carrying capacity, high stiffness, and a long service life operating at high speeds. |
| Barrel Roller Bearing  | These bearings most often have a single row of symmetrical rollers. More bulbous than cylindrical, these barrel shaped rollers are set between a solid outer ring with a concave raceway and a solid inner ring with two ribs. | Barrel roller bearings have poor axial load capacity but are suitable for high radial loads and addressing certain misalignment. Some spherical roller designs are more common and may be used instead. |
| Spherical Roller Bearing 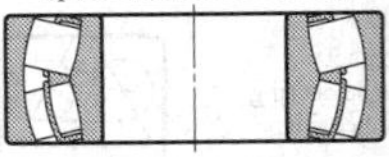 | These bearings usually feature a double row, self-aligning mounting, with both rows of rollers sharing a common spherical outer raceway. | This type of roller bearing has a high radial and thrust load capacity, along with the ability to maintain this capacity under some degree of misalignment of shaft and bearing housing. |
| Tapered Roller Bearing 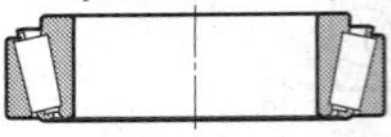 | Tapered roller bearings consist of a cup and cone assembly. The outer ring forms the cup and the inner ring, rollers, and cage make up the cone, all held in accurate alignment by a guide flange on the inner ring. Separable and lubricated, this type is available in single row, two row, and four row designs. | Bearings of this type offer high radial and axial load capacities, particularly at low to intermediate speeds. |

## Table 2b. Cylindrical Roller Bearing Elements and Their Symbols

**Cylindrical Roller Bearing, Single Row, Non-Locating Type**

| Symbol | Description | | Symbol | Description | |
|---|---|---|---|---|---|
| RU | Inner ring without ribs<br>Double-ribbed outer ring<br>Inner ring separable | 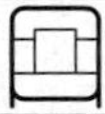 | RNS | Double-ribbed inner ring<br>Outer ring without ribs<br>Outer ring separable<br>Spherical outside surface | 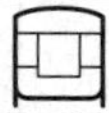 |
| RUP | Inner ring without ribs<br>Double-ribbed outer ring with one loose rib<br>Both rings separable |  | RAB | Inner ring without ribs<br>Single-ribbed outer ring<br>Both rings separable |  |
| RUA | Inner ring without ribs<br>Double-ribbed outer ring<br>Inner ring separable Spherical outside surface | 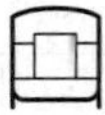 | RM | Inner ring without ribs<br>Rollers located by cage, end-rings or internal snap rings recesses in outer ring<br>Inner ring separable | 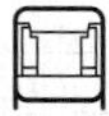 |
| RN | Double-ribbed inner ring<br>Outer ring without ribs<br>Outer ring separable | 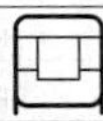 | RNU | Inner ring without ribs<br>Outer ring without ribs<br>Both rings separable | 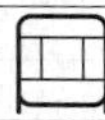 |

**Cylindrical Roller Bearings, Single Row, One-Direction-Locating Type**

| Symbol | Description | | Symbol | Description | |
|---|---|---|---|---|---|
| RR | Single-ribbed inner ring<br>Outer ring with two internal snap rings<br>Inner ring separable |  | RF | Double-ribbed inner ring<br>Single-ribbed outer ring<br>Outer ring separable | 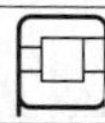 |
| RJ | Single-ribbed inner ring<br>Double-ribbed outer ring<br>Inner ring separable | 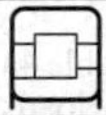 | RS | Single-ribbed inner ring<br>Outer ring with one rib and one internal snap ring<br>Inner ring separable | 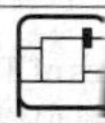 |
| RJP | Single-ribbed inner ring<br>Double-ribbed outer ring with one loose rib<br>Both rings separable | 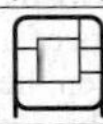 | RAA | Single-ribbed inner ring<br>Single-ribbed outer ring<br>Both rings separable | 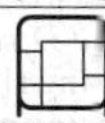 |

**Cylindrical Roller Bearings, Single Row, Two-Direction-Locating Type**

| Symbol | Description | | Symbol | Description | |
|---|---|---|---|---|---|
| RK | Double-ribbed inner ring<br>Outer ring with two internal snap rings<br>Non-separable | 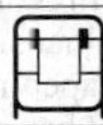 | RY | Double-ribbed inner ring<br>Outer ring with one rib and one internal snap ring<br>Non-separable | 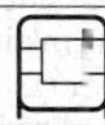 |
| RC | Double-ribbed inner ring<br>Double-ribbed outer ring<br>Non-separable | 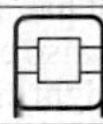 | RCS | Double-ribbed inner ring<br>Double-ribbed outer ring<br>Non-separable<br>Spherical outside surface | 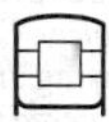 |
| RG | Inner ring, with one rib and one snap ring<br>Double-ribbed outer ring<br>Non-separable | 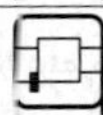 | | | |
| RP | Double-ribbed inner ring<br>Double-ribbed outer ring with one loose rib<br>Outer ring separable | 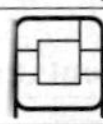 | RT | Double-ribbed inner ring with one loose rib<br>Double-ribbed outer ring<br>Inner ring separable | 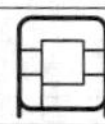 |

**Cylindrical Roller Bearings**

Double Row Non-Locating Type

| Symbol | Description | |
|---|---|---|
| RA | Inner ring without ribs<br>Three integral ribs on outer ring<br>Inner ring separable | 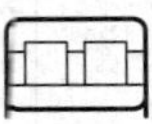 |
| RD | Three integral ribs on inner ring<br>Outer ring without ribs<br>Outer ring separable | 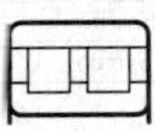 |
| RE | Inner ring without ribs<br>Outer rings without ribs, with two internal snap rings<br>Inner ring separable | 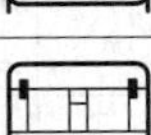 |

Double Row Two-Direction-Locating Type

| Symbol | Description | |
|---|---|---|
| RB | Three integral ribs on inner ring<br>Outer ring without ribs, with two internal snap rings<br>Non-separable | 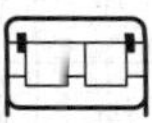 |

Multi-Row Non-Locating Type

| Symbol | Description | |
|---|---|---|
| RV | Inner ring without ribs<br>Double-ribbed outer ring (loose ribs)<br>Both rings separable | 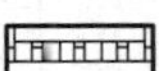 |

**Table 3. Self-Aligning Roller Bearing Elements and Their Symbols**

| Self-Aligning Roller Bearings, Double Row | | |
|---|---|---|
| Symbol | Description | |
| SD | Three integral ribs on inner ring<br>Raceway of outer ring spherical | 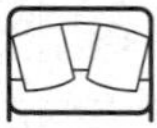 |
| SE | Raceway of outer ring spherical<br>Rollers guided by separate center guide ring in outer ring | 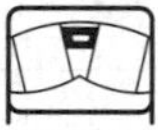 |
| SW | Raceway of inner ring spherical | 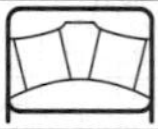 |
| SC | Raceway of outer ring spherical<br>Rollers guided by separate axially floating guide ring on inner ring | 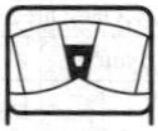 |
| SL | Raceway of outer ring spherical<br>Rollers guided by the cage<br>Two integral ribs on inner ring |  |

| Self-Aligning Roller Bearings Single Row | | |
|---|---|---|
| Symbol | Description | |
| SR | Inner ring with ribs<br>Raceway of outer ring spherical<br>Radial contact |  |
| SA | Raceway of outer ring spherical<br>Angular contact |  |
| SB | Raceway of inner ring spherical<br>Angular contact |  |

**Types of Thrust Bearings.—**These bearings (Table 4a and Table 4b) are designed to take thrust loads alone or combined thrust with radial loads.

*One Direction Ball Thrust:* This type consists of a shaft ring and a flat or spherical housing ring with a single row of balls in between them. These bearings are capable of carrying purely thrust loads in one direction only.

*Two Direction Ball Thrust:* These bearings have a shaft ring with a ball groove in either side, two sets of balls, and two housing rings. This arrangement supports thrust loads in either direction but cannot carry radial loads.

*Spherical Roller Thrust:* Featuring barrel shaped rollers, this type is similar in design to radial spherical roller bearings, except it has a much larger contact angle. With a high thrust load carrying capacity, this design also can carry radial loads.

*Roller Thrust:* In this type, the rollers are straight. Several different arrangements of housing and shaft are used.

*Tapered Roller Thrust:* In this type, the rollers are tapered. Again, there are different versions.

**Table 4a. Thrust Bearing Elements and Their Symbols**

| Symbol | Description | | Symbol | Description | |
|---|---|---|---|---|---|
| Ball Thrust Bearings | | | | | |
| TA<br>TB[a] | Single direction, grooved raceways, flat seats |  | TDA | Double direction, washers with grooved raceways, flat seats | 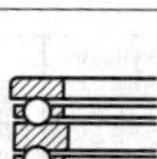 |
| TBF[a] | Single direction, flat washers, flat seats |  | | | |
| Roller Thrust Bearings | | | | | |
| Symbol | Description | | Symbol | Description | |
| TS | Single direction, aligning flat seats, spherical rollers | 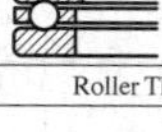 | TPC[a] | Single direction, flat seats, flat races, outside band, cylindrical rollers | 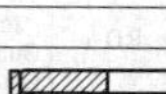 |
| TP | Single direction, flat seats, cylindrical rollers |  | z | Single direction, flat races, aligning seat with aligning washer, cylindrical rollers |  |

[a] Inch-dimensioned only.

**Table 4b. Tapered Thrust Roller Bearings and Their Symbols**

| Tapered Thrust Roller Bearings—Inch | | | |
|---|---|---|---|
| Symbol | Description | Symbol | Description |
| TS | Single row  | TDI | Two row, double cone single cups 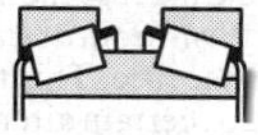 |
| TDO | Two row, double cup single cone adjustable 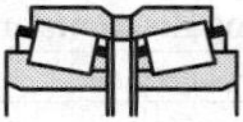 | TNA | Two row, double cup single cone nonadjustable 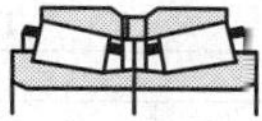 |
| TQD, TQI | Four row, cup adjusted | 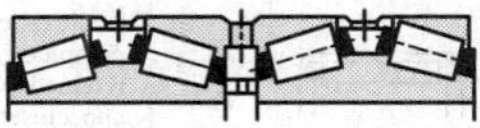 | 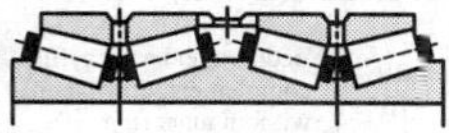 |
| Tapered Roller Thrust Bearings—Metric | | | |
| Symbol | Description | Symbol | Description |
| TS | Single row, straight bore  | TSF | Single row, straight bore, flanged cup  |
| TDO | Double row, straight bore, two single cones, one double cup with lubrication hole and groove 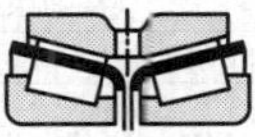 | 2TS | Double row, straight bore, two single cones, two single cups 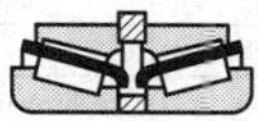 |
| Tapered Thrust Roller Bearings | | | |
| Symbol | Description | | |
| TT | Thrust bearings 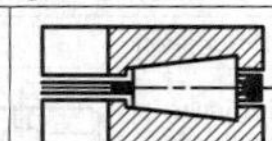 | | |

**Types of Needle Bearings.**—Needle bearings are characterized by their relatively small size rollers, usually not above .25 inch (6.35 mm) in diameter, and a relatively high ratio of length to diameter, usually ranging from about 3 to 1 and 10 to 1. Another characteristic feature of several types of needle bearings is the absence of a cage or separator for retaining the individual rollers. Needle bearings may be divided into three classes: *loose roller*, *outer race and retained roller*, and *non-separable*.

*Loose Roller:* This type of bearing has no integral races or retaining members, the needles being set directly between the shaft and outer bearing bore. Usually, both shaft and outer bearing bore surfaces are hardened; retaining members with smooth unbroken surfaces are provided to prevent endwise movement. This bearing type offers compactness and high radial load capacity.

*Outer Race and Retained Roller:* There are two types of outer race and retained roller bearings. In the *drawn shell* type, the needle rollers are enclosed by a hardened shell, which acts as a retaining member, and an outer race. The needles roll directly on the shaft, the bearing surface of which should be hardened. The capacity for given roller-length and shaft diameter is about two-thirds that of the loose roller type. This type is press-fitted into the housing.

In the *machined race* type, the outer race consists of a heavy machined member. Various modifications of this type provide heavy ends or faces for end location of the needle rollers, or open-end construction with end washers for roller retention, or a cage that maintains alignment of the rollers and is itself held in place by retaining rings. An auxiliary outer member with a spherical seat holding the outer race may be provided for self-alignment. This type is applicable where split housings occur or where a press fit of the bearing into the housing is not possible.

*Non-Separable:* This bearing consists of a non-separable unit of outer race, rollers, and inner race. This type is used where high static or oscillating motion loads are expected (i.e., certain aircraft components) and where both outer and inner races are necessary.

**Table 5. Needle Roller Bearing Elements and Symbols**

| Needle Roller Bearings, Drawn Cup | | | |
|---|---|---|---|
| Symbol[a] | Description | Symbol[a] | Description |
| NIB<br>NB | Needle roller bearing, full complement, drawn cup, without inner ring 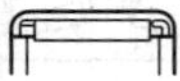 | NIYM<br>NYM | Needle roller bearing, full complement, rollers retained by lubricant, drawn cup, closed end, without inner ring 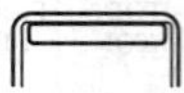 |
| NIBM<br>NBM | Needle roller bearing, full complement, drawn cup, closed end, without inner ring  | NIH<br>NH | Needle roller bearing, with cage, drawn cup, without inner ring 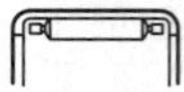 |
| NIY<br>NY | Needle roller bearing, full complement, rollers retained by lubricant, drawn cup, without inner ring  | NIHM<br>NHM | Needle roller bearing, with cage, drawn cup, closed end, without inner ring 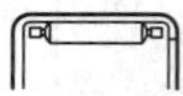 |
| Needle Roller Bearings | | Needle Roller and Cage Assemblies | |
| Symbol[a] | Description | Symbol[a] | Description |
| NIA<br>NA | Needle roller bearing, with cage, machined ring lubrication hole and groove in OD, without inner ring 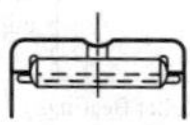 | NIM<br>NM | Needle roller and cage assembly 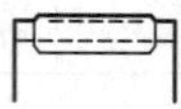 |
| Needle Roller Bearing Inner Rings | | Machined Ring Needle Roller Bearings Type NIA may be used with inch dimensioned inner rings, Type NIR, and Type NA may be used with metric dimensional inner rings, Type NR. | |
| Symbol[a] | Description | | |
| NIR<br>NR | Needle roller bearing inner ring, lubrication hole and groove in bore 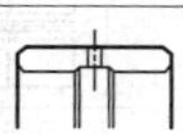 | | |

[a] Symbols with I, as NIB, are inch-dimensioned, and those without the I, as NB, are metric dimensioned.

**Special or Unconventional Types.**—Rolling contact bearings have been developed for many highly specialized applications. They may be constructed of non-corrosive materials, non-magnetic materials, various plastics and ceramics, or even wood. Although materials are chosen to adapt more conventional configurations to difficult applications or environments, even greater ingenuity has been applied in utilizing rolling contact for solving particular problems.

Linear or recirculating bearings are available to provide low friction, accurate location, and simplified lubrication features to such applications as machine ways, axial motion devices, jack screws, steering linkages, collets, and chucks. This type of bearing utilizes the "full-complement" style of loading its rolling elements between "races" or ways without a cage, with each element advancing by the action of the races in the loaded areas, and by contact with the adjacent element in the unloaded areas. The races may be plane surfaces, with suitable interruptions to free the rolling elements, so they can follow a return trough or slot back to the entry point at the start of the race contact area.

Another special application type is the insulated bearing. In motors and alternators, shaft voltage may cause flows of electrical current across the tiny clearances in the bearings. When this happens, resulting arcs can damage the bearing. To interrupt the flow of current, either the drive end or non-drive end bearing is insulated. Now available in most types of rolling element bearings, such bearings with a ceramic insulation layer offer the same outer dimensions and technical characteristics as corresponding uninsulated types.

**Plastics Bearings.—**Acetal resin rollers and balls have gained popularity in applications where abrasive, corrosive, and difficult-to-lubricate conditions exist. Although such bearings do not have the load carrying capacity nor the low friction advantage of their hard steel counterparts, they offer freedom from indentation, wear, and corrosion, while providing significant weight savings.

Of additional value are: 1) their resistance to indentation from shock loads or oscillation; and 2) their self-lubricating properties.

Usually these bearings are not available from stock, but must be designed and produced in accordance with the data made available by the plastics processor.

**Pillow Block and Flanged Housing Bearings.—**Of great interest in the shop and particularly adaptable to "line-shafting" applications are a series of ball and roller bearings supplied with their own housings, adapters, and seals. Often called pre-mounted bearings, they come with a wide variety of flange mountings permitting location on faces parallel to or perpendicular to the shaft axis.

Inner races can be mounted directly on ground shafts, or can be adapter-mounted to "drill-rod" or commercial shafting. For installations sensitive to imbalance and vibration, the use of accurately ground shaft seats is recommended.

Most pillow block designs incorporate self-aligning types of bearings. So they do not require the precision mountings utilized with more normal bearing installations.

**Conventional Bearing Materials.—**Most rolling contact bearings are made with all load carrying members of full hard steel, either through- or case-hardened. For greater reliability, this material is controlled and selected for cleanliness and alloying practices in conformity with rigid specifications in order to reduce anomalies and inclusions that could limit the useful fatigue life. Magnaflux inspection is employed to ensure that elements are free from both material defects and cracks. In addition, a light etch may be employed between rough and finish grinding to allow detection of burns due to heavy stock removal and associated decarburization in finished pieces.

The most common bearing materials are ASTM/SAE/AISI 52100 grade high-carbon chromium steel and ASTM/AISI 440C alloy grade stainless steel used for mild corrosive applications. Carefully controlled processes and alloys are used to manufacture bearing steels for higher temperature applications, as in aerospace. In hybrid ceramic ball bearings, the races and cage may be made of steel, while the balls are ceramic. Ceramic bearings can run faster and cooler with less noise and vibration.

Standard bearings are normally made with cages of free-machining brass or low carbon sulfurized steel. In high-speed applications or where lubrication may be intermittent or marginal, special materials may be employed. Iron-silicon-bronze, laminated phenolics, silver-plating, over-lays, solid-film baked-on coatings, carbon-graphite inserts, and, in extreme cases, sintered or even impregnated materials are used in separators. Commercial bearings usually rely on stamped steel with or without a phosphate treatment; some low cost varieties are found with snap-in plastic or metallic cages.

So long as lubrication is adequate and speeds are both reasonable and steady, the materials and design of the cage are of secondary importance when compared with those of the rolling elements and their contacts with the races. In spite of this tolerance, a good portion of all rolling bearing failures encountered can be traced to cage failures resulting from inadequate lubrication. It can never be overemphasized that *no bearing can be designed to run continuously without lubrication*.

**Standard Method of Bearing Designation.—**The American Bearing Manufacturers Association (ABMA) has adopted a standard identification code that provides a specific designation for each different ball, roller, and needle bearing. Thus, for any given bearing, a uniform designation is provided for manufacturer and user alike, so that the confusion of different company designations can be avoided.

In this identification code there is a "basic number" for each bearing that consists of three elements: a one- to four-digit number to indicate the size of the bore in numbers of millimeters (metric series); a two- or three-letter symbol to indicate the type of bearing; and a two-digit number to identify the dimension series to which the bearing belongs.

In addition to this "basic number" other numbers and letters are added to designate type of tolerance, cage, lubrication, fit up, ring modification, addition of shields, seals, mounting accessories, etc. Thus, a complete designating symbol might be *50BC02JPXE0A10*, for example. The basic number is *50BC02* and the remainder is the supplementary number. For a radial bearing, this latter consists of up to four letters to indicate modification of design, one or two digits to indicate internal fit and tolerances, a letter to indicate lubricants and preservatives, and up to three digits to indicate special requirements.

For a thrust bearing the supplementary number would consist of two letters to indicate modifications of design, one digit to indicate tolerances, one letter to indicate lubricants and preservatives, and up to three digits to indicate special requirements.

For a needle bearing the supplementary number would consist of up to three letters indicating cage material or integral seal information or whether the outer ring has a crowned outside surface and one letter to indicate lubricants or preservatives.

*Dimension Series:* Annular ball, cylindrical roller, and self-aligning roller bearings are made in a series of different outside diameters for every given bore diameter and in a series of different widths for every given outside diameter. Thus, each of these bearings belongs to a dimension series that is designated by a two-digit number such as or, 23, 93, etc. The first digit (8, 0, 1, 2, 3, 4, 5, 6 and 9) indicates the *width series* and the second digit (7, 8, 9, 0, 1, 2, 3, and 4) the *diameter series* to which the bearing belongs. Similar types of identification codes are used for ball and roller thrust bearings and needle roller bearings.

## Bearing Tolerances

**Ball and Roller Bearings.—**In order to provide standards of precision for proper application of ball or roller bearings in all types of equipment, five classes of tolerances have been established by the American Bearing Manufacturers Association for ball bearings, three for cylindrical roller bearings and one for spherical roller bearings. These tolerances are given in Table 6, Table 7, Table 8, Table 9, and Table 10. They are designated as ABEC-1, ABEC-3, ABEC-5, ABEC-7 and ABEC-9 for ball bearings, the ABEC-9 being the most precise, RBEC-1, RBEC-3, and RBEC-5 for roller bearings. In general, bearings to specifications closer than ABEC-1 or RBEC-1 are required because of the need for very precise fits on shaft or housing, to reduce eccentricity or runout of shaft or supported part, or to permit operation at very high speeds. All five classes include tolerances for bore, outside diameter, ring width, and radial runouts of inner and outer rings. ABEC-5, ABEC-7 and ABEC-9 provide added tolerances for parallelism of sides, side runout and groove parallelism with sides.

**Thrust Bearings.—**American Bearing Manufacturers Association and American National Standard tolerance limits for metric single direction thrust ball and roller bearings are given in Table 12. Tolerance limits for single direction thrust ball bearings, inch dimensioned are given in Table 11, and limits for cylindrical thrust roller bearings, inch dimensioned can be found in Table 13.

Only one class of tolerance limits is established for metric thrust bearings.

***Radial* Needle Roller Bearings.—**Tolerance limits for needle roller bearings, drawn cup, without inner ring, inch types NIB, NIBM, NIY, NIYM, NIH, and NIHM are given in Table 14 and limits for metric types NB, NBM, NY, NYM, NH and NHM can be found in Table 15. Standard tolerance limits for needle roller bearings, with cage, machined ring, without inner ring, inch type NIA are given in Table 16 and limits for needle roller bearings inner rings, inch type NIR can be found in Table 17.

## Table 6. ABEC-1 and RBEC-1 Tolerance Limits for Metric Ball and Roller Bearings *ANSI/ABMA 20-2011 (R2020)*

| Basic Inner Ring Bore Diameter, *d* | | $V_{dp}$,[a] max | | | | | |
|---|---|---|---|---|---|---|---|
| mm | | Diameter Series | | | $\Delta_{dmp}$[b] | | $K_{ia}$[c] |
| Over | Incl. | 7, 8, 9 | 0, 1 | 2, 3, 4 | High | Low | max |
| 2.5 | 10 | 10 | 8 | 6 | 0 | −8 | 10 |
| 10 | 18 | 10 | 8 | 6 | 0 | −8 | 10 |
| 18 | 30 | 13 | 10 | 8 | 0 | −10 | 13 |
| 30 | 50 | 15 | 12 | 9 | 0 | −12 | 15 |
| 50 | 80 | 19 | 19 | 11 | 0 | −15 | 20 |
| 80 | 120 | 25 | 25 | 15 | 0 | −20 | 25 |
| 120 | 180 | 31 | 31 | 19 | 0 | −25 | 30 |
| 180 | 250 | 38 | 38 | 23 | 0 | −30 | 40 |
| 250 | 315 | 44 | 44 | 26 | 0 | −35 | 50 |
| 315 | 400 | 50 | 50 | 30 | 0 | −40 | 60 |

[a] Bore diameter variation in a single radial plane.

[b] Single plane mean bore diameter deviation from basic. (For a basically tapered bore, $\Delta_{dmp}$ refers only to the theoretical small end of the bore.)

[c] Radial runout of assembled bearing inner ring.

| Basic Outer Ring Outside Outerside Diameter, *D* | | $V_{Dp}$,[a] max | | | | | | |
|---|---|---|---|---|---|---|---|---|
| | | Open Bearings | | | Capped Bearings[d] | | | |
| mm | | Diameter Series | | | | $\Delta_{Dmp}$[b] | | $K_{ea}$[c] |
| Over | Incl. | 7, 8, 9 | 0, 1 | 2, 3, 4 | 2, 3, 4 | High | Low | max |
| 6 | 18 | 10 | 8 | 6 | 10 | 0 | −8 | 15 |
| 18 | 30 | 12 | 9 | 7 | 12 | 0 | −9 | 15 |
| 30 | 50 | 14 | 11 | 8 | 16 | 0 | −11 | 20 |
| 50 | 80 | 16 | 13 | 10 | 20 | 0 | −13 | 25 |
| 80 | 120 | 19 | 19 | 11 | 26 | 0 | −15 | 35 |
| 120 | 150 | 23 | 23 | 14 | 30 | 0 | −18 | 40 |
| 150 | 180 | 31 | 31 | 19 | 38 | 0 | −25 | 45 |
| 180 | 250 | 38 | 38 | 23 | … | 0 | −30 | 50 |
| 250 | 315 | 44 | 44 | 26 | … | 0 | −35 | 60 |
| 315 | 400 | 50 | 50 | 50 | … | 0 | −40 | 70 |

[a] Outside diameter variation in a single radial plane. Applies before mounting and after removal of internal or external snap ring.

[b] Single plane mean outside diameter deviation from basic.

[c] Radial runout of assembled bearing outer ring.

[d] No values have been established for diameters series 7, 8, 9, 0, and 1.

| Width Tolerances | | | | | | | | | |
|---|---|---|---|---|---|---|---|---|---|
| *d* | | $\Delta_{Bs}$[a] | | | *d* | | $\Delta_{Bs}$[a] | | |
| mm | | All | Normal | Modified[b] | mm | | All | Normal | Modified[b] |
| Over | Incl. | High | Low | | Over | Incl. | High | Low | |
| 2.5 | 10 | 0 | −120 | −250 | 80 | 120 | 0 | −200 | −380 |
| 10 | 18 | 0 | −120 | −250 | 120 | 180 | 0 | −250 | −500 |
| 18 | 30 | 0 | −120 | −250 | 180 | 250 | 0 | −300 | −500 |
| 30 | 50 | 0 | −120 | −250 | 250 | 315 | 0 | −350 | −500 |
| 50 | 80 | 0 | −150 | −380 | 315 | 400 | 0 | −400 | −630 |

[a] Single inner ring width deviation from basic. $\Delta_{Cs}$ (single outer ring width deviation from basic) is identical to $\Delta_{Bs}$ of inner ring of same bearing.

[b] Refers to the rings of single bearings made for paired or stack mounting.

All units are micrometers, unless otherwise indicated. For sizes beyond range of this table, see *ANSI/ABMA 20-2011 (R2020)*. This table does not cover tapered roller bearings.

**Table 7. ABEC-3 AND RBEC-3 Tolerance Limits for Metric Ball and Roller Bearings** *ANSI/ABMA 20-2011 (R2020)*

| Basic Inner Ring Bore Diameter, *d* | | $V_{dp}$,[a] max | | | $\Delta_{dmp}$[b] | | $K_{ia}$[c] |
|---|---|---|---|---|---|---|---|
| mm | | Diameter Series | | | | | |
| Over | Incl. | 7, 8, 9 | 0, 1 | 2, 3, 4 | High | Low | max |
| 2.5 | 10 | 9 | 7 | 5 | 0 | −7 | 6 |
| 10 | 18 | 9 | 7 | 5 | 0 | −7 | 7 |
| 18 | 30 | 10 | 8 | 6 | 0 | −8 | 8 |
| 30 | 50 | 13 | 10 | 8 | 0 | −10 | 10 |
| 50 | 80 | 15 | 15 | 9 | 0 | −12 | 10 |
| 80 | 120 | 19 | 19 | 11 | 0 | −15 | 13 |
| 120 | 180 | 23 | 23 | 14 | 0 | −18 | 18 |
| 180 | 250 | 28 | 28 | 17 | 0 | −22 | 20 |
| 250 | 315 | 31 | 31 | 19 | 0 | −25 | 25 |
| 315 | 400 | 38 | 38 | 23 | 0 | −30 | 30 |

[a] Bore diameter variation in a single radial plane.

[b] Single plane mean bore diameter deviation from basic. (For a basically tapered bore, $\Delta_{dmp}$ refers only to the theoretical small end of the bore.)

[c] Radial runout of assembled bearing inner ring.

| Basic Outer Ring Outside Outerside Diameter, *D* | | $V_{Dp}$,[a] max | | | | $\Delta_{Dmp}$[b] | | $K_{ea}$[c] |
|---|---|---|---|---|---|---|---|---|
| | | Open Bearings | | | Capped Bearings[d] | | | |
| mm | | Diameter Series | | | | | | |
| Over | Incl. | 7, 8, 9 | 0, 1 | 2, 3, 4 | 2, 3, 4 | High | Low | max |
| 6 | 18 | 9 | 7 | 5 | 9 | 0 | −7 | 8 |
| 18 | 30 | 10 | 8 | 6 | 10 | 0 | −8 | 9 |
| 30 | 50 | 11 | 9 | 7 | 13 | 0 | −9 | 10 |
| 50 | 80 | 14 | 11 | 8 | 16 | 0 | −11 | 13 |
| 80 | 120 | 16 | 16 | 10 | 20 | 0 | −13 | 18 |
| 120 | 150 | 19 | 19 | 11 | 25 | 0 | −15 | 20 |
| 150 | 180 | 23 | 23 | 14 | 30 | 0 | −18 | 23 |
| 180 | 250 | 25 | 25 | 15 | … | 0 | −20 | 25 |
| 250 | 315 | 31 | 31 | 19 | … | 0 | −25 | 30 |
| 315 | 400 | 35 | 35 | 21 | … | 0 | −28 | 35 |

[a] Outside diameter variation in a single radial plane. Applies before mounting and after removal of internal or external snap ring.

[b] Single plane mean outside diameter deviation from basic.

[c] Radial runout of assembled bearing outer ring.

[d] No values have been established for diameter series 7, 8, 9, 0, and 1.

| Width Tolerances | | | | | | | | | |
|---|---|---|---|---|---|---|---|---|---|
| *d* | | $\Delta_{Bs}$[a] | | | *d* | | $\Delta_{Bs}$[a] | | |
| mm | | All | Normal | Modified[b] | mm | | All | Normal | Modified[b] |
| Over | Incl. | High | Low | | Over | Incl. | High | Low | |
| 2.5 | 10 | 0 | −120 | −250 | 80 | 120 | 0 | −200 | −380 |
| 10 | 18 | 0 | −120 | −250 | 120 | 180 | 0 | −250 | −500 |
| 18 | 30 | 0 | −120 | −250 | 180 | 250 | 0 | −300 | −500 |
| 30 | 50 | 0 | −120 | −250 | 250 | 315 | 0 | −350 | −500 |
| 50 | 80 | 0 | −150 | −380 | 315 | 400 | 0 | −400 | −630 |

[a] Single inner ring width deviation from basic. $\Delta_{Cs}$ (single outer ring width deviation from basic) is identical to $\Delta_{Bs}$ of *inner ring* of same bearing.

[b] Refers to the rings of single bearings made for paired or stack mounting.

All units are micrometers, unless otherwise indicated. For sizes beyond range of this table, see *ANSI/ABMA 20-2011 (R2020)*. This table does not cover tapered roller bearings.

**Table 8. ABEC-5 and RBEC-5 Tolerance Limits for Metric Ball and Roller Bearings** *ANSI/ABMA 20-2011 (R2020)*

| INNER RING | | | | | | | | | | | | | | |
|---|---|---|---|---|---|---|---|---|---|---|---|---|---|---|
| Inner Ring Bore Basic Dia., $d$ mm | | $V_{dp}$,[a] max Diameter Series | | $\Delta_{dmp}$[b] | | Radial Runout $K_{ia}$ | Ref. Face Runout with Bore $S_d$ | Axial Runout $S_{ia}$[c] | Width | | | |
| | | | | | | | | | $\Delta_{Bs}$[d] | | | $V_{Bs}$[e] |
| Over | Incl. | 7,8,9 | 0,1,2,3,4 | High | Low | max | max | max | All | Normal | Modified[f] | max |
| 2.5 | 10 | 5 | 4 | 0 | −5 | 4 | 7 | 7 | 0 | −40 | −250 | 5 |
| 10 | 18 | 5 | 4 | 0 | −5 | 4 | 7 | 7 | 0 | −80 | −250 | 5 |
| 18 | 30 | 6 | 5 | 0 | −6 | 4 | 8 | 8 | 0 | −120 | −250 | 5 |
| 30 | 50 | 8 | 6 | 0 | −8 | 5 | 8 | 8 | 0 | −120 | −250 | 5 |
| 50 | 80 | 9 | 7 | 0 | −9 | 5 | 8 | 8 | 0 | −150 | −250 | 6 |
| 80 | 120 | 10 | 8 | 0 | −10 | 6 | 9 | 9 | 0 | −200 | −380 | 7 |
| 120 | 180 | 13 | 10 | 0 | −13 | 8 | 10 | 10 | 0 | −250 | −380 | 8 |
| 180 | 250 | 15 | 12 | 0 | −15 | 10 | 11 | 13 | 0 | −300 | −500 | 10 |

[a] Bore ($V_{dp}$) and outside diameter ($V_{Dp}$) variation in a single radial plane.

[b] Single plane mean bore ($\Delta_{dmp}$) and outside diameter ($\Delta_{Dmp}$) deviation from basic. (For a basically tapered bore, $\Delta_{dmp}$ refers only to the theoretical small end of the bore.)

[c] Axial runout of assembled bearing with inner ring $S_{ia}$. Applies to groove-type ball bearings only.

[d] Single bore ($\Delta_{Bs}$) and outer ring ($\Delta_{Cs}$) width variation.

[e] Inner ($V_{Bs}$) and outer ($V_{Cs}$) ring width deviation from basic.

[f] Applies to the rings of single bearings made for paired or stack mounting.

| OUTER RING | | | | | | | | | | | |
|---|---|---|---|---|---|---|---|---|---|---|---|
| Basic Outer Ring Outside Dia., $D$ mm | | $V_{Dp}$,[a,b] max Diameter Series | | $\Delta_{Dmp}$[c] | | Radial Runout $K_{ea}$ | Outside Cylindrical Surface Runout $S_D$[d] | Axial Runout $S_{ea}$[e] | Width | | |
| | | | | | | | | | $\Delta_{Cs}$[f] | | $V_{Cs}$[g] |
| Over | Incl. | 7,8,9 | 0,1,2,3,4 | High | Low | max | max | max | High | Low | max |
| 6 | 18 | 5 | 4 | 0 | −5 | 5 | 8 | 8 | Identical to $\Delta_{Bs}$ of inner ring of same bearing | | 5 |
| 18 | 30 | 6 | 5 | 0 | −6 | 6 | 8 | 8 | | | 5 |
| 30 | 50 | 7 | 5 | 0 | −7 | 7 | 8 | 8 | | | 5 |
| 50 | 80 | 9 | 7 | 0 | −9 | 8 | 8 | 10 | | | 6 |
| 80 | 120 | 10 | 8 | 0 | −10 | 10 | 9 | 11 | | | 8 |
| 120 | 150 | 11 | 8 | 0 | −11 | 11 | 10 | 13 | | | 8 |
| 150 | 180 | 13 | 10 | 0 | −13 | 13 | 10 | 14 | | | 8 |
| 180 | 250 | 15 | 11 | 0 | −15 | 15 | 11 | 15 | | | 10 |

[a] Bore ($V_{dp}$) and outside diameter ($V_{Dp}$) variation in a single radial plane.

[b] No values have been established for capped bearings.

[c] Single plane mean bore ($\Delta_{dmp}$) and outside diameter ($\Delta_{Dmp}$) deviation from basic. (For a basically tapered bore, $\Delta_{dmp}$ refers only to the theoretical small end of the bore.)

[d] Outside cylindrical surface runout with outer ring reference face $S_D$

[e] Axial runout of assembled bearing with outer ring $S_{ea}$.

[f] Single bore ($\Delta_{Bs}$) and outer ring ($\Delta_{Cs}$) width variation.

[g] Inner ($V_{Bs}$) and outer ($V_{Cs}$) ring width deviation from basic.

All units are micrometers, unless otherwise indicated. For sizes beyond range of this table, see *ANSI/ABMA 20-2011 (R2020)*. This table does not cover instrument bearings and tapered roller bearings.

## Table 9. ABEC-7 Tolerance Limits for Metric Ball and Roller Bearings *ANSI/ABMA 20-2011 (R2020)*

| INNER RING | | | | | | | | | | | | | | | | |
|---|---|---|---|---|---|---|---|---|---|---|---|---|---|---|---|---|
| Inner Ring Bore Basic Diameter, $d$ mm | | $V_{dp}$,[a] max Diameter Series | | $\Delta_{dmp}$[b] | | $\Delta_{ds}$[c] | | Radial Runout $K_{ia}$ | Ref. Face Runout with Bore $S_d$ | Axial Runout $S_{ia}$[d] | Width | | | |
| | | | | | | | | | | | $\Delta_{Bs}$[e] | | | $V_{Bs}$[f] |
| Over | Incl. | 7,8,9 | 0,1,2,3,4 | High | Low | High | Low | max | max | max | All | Normal | Modified[g] | max |
| 2.5 | 10 | 4 | 3 | 0 | −4 | 0 | −4 | 2.5 | 3 | 3 | 0 | −40 | −250 | 2.5 |
| 10 | 18 | 4 | 3 | 0 | −4 | 0 | −4 | 2.5 | 3 | 3 | 0 | −80 | −250 | 2.5 |
| 18 | 30 | 5 | 4 | 0 | −5 | 0 | −5 | 3 | 4 | 4 | 0 | −120 | −250 | 2.5 |
| 30 | 50 | 6 | 5 | 0 | −6 | 0 | −6 | 4 | 4 | 4 | 0 | −120 | −250 | 3 |
| 50 | 80 | 7 | 5 | 0 | −7 | 0 | −7 | 4 | 5 | 5 | 0 | −150 | −250 | 4 |
| 80 | 120 | 8 | 6 | 0 | −8 | 0 | −8 | 5 | 5 | 5 | 0 | −200 | −380 | 4 |
| 120 | 180 | 10 | 8 | 0 | −10 | 0 | −10 | 6 | 6 | 7 | 0 | −250 | −380 | 5 |
| 180 | 250 | 12 | 9 | 0 | −12 | 0 | −12 | 8 | 7 | 8 | 0 | −300 | −500 | 6 |

[a] Bore ($V_{dp}$) and outside diameter ($V_{Dp}$) variation in a single radial plane.

[b] Single plane mean bore ($\Delta_{dmp}$) and outside diameter ($\Delta_{Dmp}$) deviation from basic. (For a basically tapered bore, $\Delta_{dmp}$ refers only to the theoretical small end of the bore.)

[c] Single bore ($\Delta_{ds}$) and outside diameter ($\Delta_{Ds}$) deviations from basic. These deviations apply to diameter series 0, 1, 2, 3, and 4 only.

[d] Axial run out of assembled bearing with inner ring $S_{ia}$. Applies to groove-type ball bearings only.

[e] Single bore ($\Delta_{Bs}$) and outer ring ($\Delta_{Cs}$) width deviation from basic.

[f] Inner ($V_{Bs}$) and outer ($V_{Cs}$) ring width variation.

[g] Applies to the rings of single bearings made for paired or stack mounting.

| OUTER RING | | | | | | | | | | | | | | |
|---|---|---|---|---|---|---|---|---|---|---|---|---|---|---|
| Basic Outer Ring Outside Dia., $D$ mm | | $V_{Dp}$,[a,b] max Diameter Series | | $\Delta_{Dmp}$[b] | | $\Delta_{Ds}$[c] | | Radial Runout $K_{ea}$ | Surface Runout $S_D$[d] | Axial Runout $S_{ea}$[f] | Width | | |
| | | | | | | | | | | | $\Delta_{Cs}$[g] | | $V_{Cs}$[h] |
| Over | Incl. | 7,8,9 | 0,1,2,3,4 | High | Low | High | Low | max | max | max | High | Low | max |
| 6 | 18 | 4 | 3 | 0 | −4 | 0 | −4 | 3 | 4 | 5 | | | 2.5 |
| 18 | 30 | 5 | 4 | 0 | −5 | 0 | −5 | 4 | 4 | 5 | | | 2.5 |
| 30 | 50 | 6 | 5 | 0 | −6 | 0 | −6 | 5 | 4 | 5 | Identical to $\Delta_{Bs}$ of inner ring of same bearing | | 2.5 |
| 50 | 80 | 7 | 5 | 0 | −7 | 0 | −7 | 5 | 4 | 5 | | | 3 |
| 80 | 120 | 8 | 6 | 0 | −8 | 0 | −8 | 6 | 5 | 6 | | | 4 |
| 120 | 150 | 9 | 7 | 0 | −9 | 0 | −9 | 7 | 5 | 7 | | | 5 |
| 150 | 180 | 10 | 8 | 0 | −10 | 0 | −10 | 8 | 5 | 8 | | | 5 |
| 180 | 250 | 11 | 8 | 0 | −11 | 0 | −11 | 19 | 7 | 10 | | | 7 |

[a] Bore ($V_{dp}$) and outside diameter ($V_{Dp}$) variation in a single radial plane.

[b] No values have been established for capped bearings.

[c] Single plane mean bore ($\Delta_{dmp}$) and outside diameter ($\Delta_{Dmp}$) deviation from basic. (For a basically tapered bore, $\Delta_{dmp}$ refers only to the theoretical small end of the bore.)

[d] Single bore ($\Delta_{ds}$) and outside diameter ($\Delta_{Ds}$) deviations from basic. These deviations apply to diameter series 0, 1, 2, 3, and 4 only.

[e] Outside cylindrical surface runout outer ring reference face $S_D$

[f] Axial run out of assembled bearing with outer ring $S_{ia}$. Applies to groove-type ball bearings only.

[g] Single bore ($\Delta_{Bs}$) and outer ring ($\Delta_{Cs}$) width deviation from basic.

[h] Inner ($V_{Bs}$) and outer ($V_{Cs}$) ring width variation.

All units are micrometers, unless otherwise indicated. For sizes beyond range of this table, see the *ANSI/ABMA 20-2011 (R2020)*. This table does not cover instrument bearings.

## Table 10. ABEC-9 Tolerance Limits for Metric Ball and Roller Bearing *ANSI/ABMA 20-2011 (R2020)*

| INNER RING | | | | | | | | | | | | | |
|---|---|---|---|---|---|---|---|---|---|---|---|---|---|
| Inner Ring Bore Basic Dia., $d$ mm | | $V_{dp}$,[a] max | $\Delta_{dmp}$[b] | | $\Delta_{ds}$[c] | | Radial Runout $K_{ia}$ | Ref. Face Runout with Bore $S_d$ | Axial Runout of Assembled Bearing with Inner Ring $S_{ia}$[d] | Width | | |
| | | | | | | | | | | $\Delta_{Bs}$[e] | | $V_{Bs}$[f] |
| Over | Incl. | max | High | Low | High | Low | max | max | max | High | Low | max |
| 2.5 | 10 | 2.5 | 0 | −2.5 | 0 | −2.5 | 1.5 | 1.5 | 1.5 | 0 | −40 | 1.5 |
| 10 | 18 | 2.5 | 0 | −2.5 | 0 | −2.5 | 1.5 | 1.5 | 1.5 | 0 | −80 | 1.5 |
| 18 | 30 | 2.5 | 0 | −2.5 | 0 | −2.5 | 2.5 | 1.5 | 2.5 | 0 | −120 | 1.5 |
| 30 | 50 | 2.5 | 0 | −2.5 | 0 | −2.5 | 2.5 | 1.5 | 2.5 | 0 | −120 | 1.5 |
| 50 | 80 | 4 | 0 | −4 | 0 | −4 | 2.5 | 1.5 | 2.5 | 0 | −150 | 1.5 |
| 50 | 80 | 4 | 0 | −4 | 0 | −4 | 2.5 | 1.5 | 2.5 | 0 | −150 | 1.5 |
| 80 | 120 | 5 | 0 | −5 | 0 | −5 | 2.5 | 2.5 | 2.5 | 0 | −200 | 2.5 |
| 120 | 150 | 7 | 0 | −7 | 0 | −7 | 2.5 | 2.5 | 2.5 | 0 | −250 | 2.5 |
| 150 | 180 | 7 | 0 | −7 | 0 | −7 | 5 | 4 | 5 | 0 | −300 | 4 |
| 180 | 250 | 8 | 0 | −8 | 0 | −8 | 5 | 5 | 5 | 0 | −350 | 5 |

[a] Bore ($V_{dp}$) and outside diameter ($V_{Dp}$) variation in a single radial plane.
[b] Single plane mean bore ($\Delta_{dmp}$) and outside diameter ($\Delta_{Dmp}$) deviation from basic. (For a basically tapered bore, $\Delta_{dmp}$ refers to the theoretical small end of the bore.)
[c] Single bore diameter ($\Delta_{ds}$) and outside diameter ($\Delta_{Ds}$) deviation from basic.
[d] Applies to groove-type ball bearings only.
[e] Single bore ($\Delta_{Bs}$) and outer ring ($\Delta_{Cs}$) width variation from basic.
[f] Inner ($V_{Bs}$) and outer ($V_{Cs}$) ring width variation.

| OUTER RING | | | | | | | | | | | | |
|---|---|---|---|---|---|---|---|---|---|---|---|---|
| Basic Outside Diameter, $D$ mm | | $V_{Dp}$[a] | $\Delta_{Dmp}$[b] | | $\Delta_{Ds}$[c] | | Radial Runout $K_{ea}$ | Outside Cylindrical Surface Runout with Outer Ring $S_D$ | Axial Runout of Assembled Bearing with Outer Ring $S_{ea}$ | Width | | |
| | | | | | | | | | | $\Delta_{Cs}$[e] | | $V_{Cs}$[f] |
| Over | Incl. | max | High | Low | High | Low | max | max | max | High | Low | max |
| 6 | 18 | 2.5 | 0 | −2.5 | 0 | −2.5 | 1.5 | 1.5 | 1.5 | Identical to $\Delta_{Bs}$ of inner ring of same bearing | | 1.5 |
| 18 | 30 | 4 | 0 | −4 | 0 | −4 | 2.5 | 1.5 | 2.5 | | | 1.5 |
| 30 | 50 | 4 | 0 | −4 | 0 | −4 | 2.5 | 1.5 | 2.5 | | | 1.5 |
| 50 | 80 | 4 | 0 | −4 | 0 | −4 | 4 | 1.5 | 4 | | | 1.5 |
| 80 | 120 | 5 | 0 | −5 | 0 | −5 | 5 | 2.5 | 5 | | | 1.5 |
| 120 | 150 | 5 | 0 | −5 | 0 | −5 | 5 | 2.5 | 5 | | | 1.5 |
| 150 | 180 | 7 | 0 | −7 | 0 | −7 | 5 | 2.5 | 5 | | | 2.5 |
| 180 | 250 | 8 | 0 | −8 | 0 | −8 | 7 | 4 | 7 | | | 4 |
| 250 | 315 | 8 | 0 | −8 | 0 | −8 | 7 | 5 | 7 | | | 5 |

[a] No values have been established for capped bearings.
All units are micrometers, unless otherwise indicated. For sizes beyond range of this table, see *ANSI/ABMA 20-2011 (R2020)*. This table does not cover instrument bearings.

## Table 11. Tolerance Limits for Single Direction Ball Thrust Bearings— Inch Design *ANSI/ABMA 24.2-1989 (S2020)*[b]

| Bore Diameter[a] $d$, Inches | | Single Plane Mean Bore Dia. Variation, $d$, Inch | | Outside Diameter $D$, Inches | | Single Plane Mean Outside Dia. Variation, $D$, Inch | |
|---|---|---|---|---|---|---|---|
| Over | Incl. | High | Low | Over | Incl. | High | Low |
| 0 | 6.7500 | +0.005 | 0 | 0 | 5.3125 | +0 | −0.002 |
| 6.7500 | 20.0000 | +0.010 | 0 | 5.3125 | 17.3750 | +0 | −0.003 |
| … | … | … | … | 17.3750 | 39.3701 | +0 | −0.004 |

[a] Bore tolerance limits: For bore diameters 0 to 1.8125 inches, inclusive, +0.005, −0.005; over 1.8125 to 12.000 inches, inclusive, +0.010,−0.010; over 12.000 to 20.000, inclusive, +0.0150, −0.0150.
[b] An "S" indicates the standard is maintained under the stabilized maintenance option and satisfies the following criteria: (a) not likely to require revision; (b) not safety or health related; (c) holds status of ANSI/ABMA and has been reaffirmed; (d) at least 10 years have passed since approval or latest revision; and (e) required for use with existing implementations or for reference purposes.

## Table 12. Tolerance Limits for Metric Single Direction Thrust Ball (Type TA) and Roller Type (Type TS) Bearings *ANSI/ABMA 24.1-1989 (S2020)*

| Bore Dia. of Shaft Washer, *d* | | | | | | | | Outside Dia. of Housing Washer, *D* | | | |
|---|---|---|---|---|---|---|---|---|---|---|---|
| mm | | $\Delta dmp$[a] | | $S_i, S_e$[b] | | $\Delta T_{s\ min}$[c] | | mm | | $\Delta Dmp$[a] | |
| Over | Incl. | High | Low | Max | Max | Type TA | Type TS | Over | Incl. | High | Low |
| 18 | 30 | 0 | −10 | 10 | 20 | −250 | … | 10 | 18 | 0 | −11 |
| **30** | **50** | **0** | **−12** | 10 | **20** | −250 | −300 | 18 | 30 | 0 | −13 |
| **50** | **80** | **0** | **−15** | 10 | **20** | −300 | −400 | 30 | 50 | 0 | −16 |
| **80** | **120** | **0** | **−20** | 15 | **25** | −300 | −400 | **50** | **80** | **0** | **−19** |
| **120** | **180** | **0** | **−25** | 15 | **25** | −400 | −500 | **80** | **120** | **0** | **−22** |
| **180** | **250** | **0** | **−30** | 20 | **30** | −400 | −500 | **120** | **180** | **0** | **−25** |
| **250** | **315** | **0** | **−35** | 25 | **40** | −400 | −700 | **180** | **250** | **0** | **−30** |
| **315** | **400** | **0** | **−40** | 30 | **40** | -500 | −700 | **250** | **315** | **0** | **−35** |
| **400** | **500** | **0** | **−45** | 30 | **50** | −500 | −900 | **315** | **400** | **0** | **−40** |
| **500** | **630** | **0** | **−50** | 35 | **60** | −600 | −1200 | **400** | **500** | **0** | **−45** |

[a] Single plane mean bore diameter deviation of central shaft washer ($\Delta_{dmp}$) and outside diameter ($\Delta_{Dmp}$) variation.

[b] Raceway parallelism with the face, housing-mounted ($S_e$) and bore-mounted ($S_i$) race or washer.

[c] Deviation of the actual bearing height.

All dimensions in micrometers, unless otherwise indicated. Tolerances are for normal tolerance class only. For sizes beyond the range of this table and for other tolerance class values, see *ANSI/ABMA 24.1-1989 (S2020)*. All entries apply to type TA bearings; boldface entries also apply to type TS bearings.

## Table 13. Tolerance Limits for Cylindrical Roller Thrust Bearings — Inch Design *ANSI/ABMA 24.2-1989 (S2020)*

| Basic Bore Dia., *d* | | $\Delta dmp$[a] | | $\Delta T_s$[b] | | Basic Outside dia., *D* | | $\Delta Dmp$[c] | |
|---|---|---|---|---|---|---|---|---|---|
| Over | Incl. | Low | High | High | Low | Over | Incl. | High | Low |
| EXTRA LIGHT SERIES—TYPE TP | | | | | | | | | |
| 0 | 0.9375 | +.0040 | +.0060 | +.0050 | −.0050 | 0 | 4.7188 | +0 | −.0030 |
| 0.9375 | 1.9375 | +.0050 | +.0070 | +.0050 | −.0050 | 4.7188 | 5.2188 | +0 | −.0030 |
| 1.9375 | 3.0000 | +.0060 | +.0080 | +.0050 | −.0050 | … | … | … | … |
| 3.0000 | 3.5000 | +.0080 | +.0100 | −.0100 | −.0100 | … | … | … | … |

[a] Single plane mean bore diameter deviation.

[b] Deviation of the actual bearing height, single direction bearing.

[c] Single plane mean outside diameter deviation.

| Basic Bore Diameter, *d* | | $\Delta dmp$[a] | | Basic Outside Diameter, *D* | | Outside Dia., *D* Tolerance Limits | | Basic Bore Diameter, *d* | | $\Delta T_s$ | |
|---|---|---|---|---|---|---|---|---|---|---|---|
| Over | Inc. | High | Low | Over | Incl. | High | Low | Over | Incl. | High | Low |
| LIGHT SERIES—TYPE TP | | | | | | | | | | | |
| 0 | 1.1870 | +0 | −.0005 | 0 | 2.8750 | +.0005 | −0 | 0 | 2.0000 | +0 | −.006 |
| 1.1870 | 1.3750 | +0 | −.0006 | 2.8750 | 3.3750 | +.0007 | −0 | 2.0000 | 3.0000 | +0 | −.008 |
| 1.3750 | 1.5620 | +0 | −.0007 | 3.3750 | 3.7500 | +.0009 | −0 | 3.0000 | 6.0000 | +0 | −.010 |
| 1.5620 | 1.7500 | +0 | −.0008 | 3.7500 | 4.1250 | +.0011 | −0 | 6.0000 | 10.0000 | +0 | −.015 |
| 1.7500 | 1.9370 | +0 | −.0009 | 4.1250 | 4.7180 | +.0013 | −0 | 10.0000 | 18.0000 | +0 | −.020 |
| 1.9370 | 2.1250 | +0 | −.0010 | 4.7180 | 5.2180 | +.0015 | −0 | 18.0000 | 30.0000 | +0 | −.025 |
| 2.1250 | 2.5000 | +0 | −.0011 | … | … | … | … | … | … | … | … |
| 2.2500 | 3.0000 | +0 | −.0012 | … | … | … | … | … | … | … | … |
| 3.0000 | 3.5000 | +0 | −.0013 | … | … | … | … | … | … | … | … |
| HEAVY SERIES—TYPE TP | | | | | | | | | | | |
| 2.0000 | 3.0000 | +0 | −.0010 | 5.0000 | 10.0000 | +.0015 | −0 | 0 | 2.000 | +0 | −.006 |
| 3.0000 | 3.5000 | +0 | −.0012 | 10.0000 | 18.0000 | +.0020 | −0 | 2.000 | 3.000 | +0 | −.008 |
| 3.5000 | 9.0000 | +0 | −.0015 | 18.0000 | 26.0000 | +.0025 | −0 | 3.000 | 6.000 | +0 | −.010 |
| 9.0000 | 12.0000 | +0 | −.0018 | 26.0000 | 34.0000 | +.0030 | −0 | 6.000 | 10.000 | +0 | −.015 |
| 12.0000 | 18.0000 | +0 | −.0020 | 34.0000 | 44.0000 | +.0040 | −0 | 10.000 | 18.000 | +0 | −.020 |
| 18.0000 | 22.0000 | +0 | −.0025 | … | … | … | … | 18.000 | 30.000 | +0 | −.025 |
| 22.0000 | 30.0000 | +0 | −.003 | … | … | … | … | … | … | … | … |
| TYPE TPC | | | | | | | | | | | |
| 0 | 2.0156 | +.010 | −0 | 2.5000 | 4.0000 | +.005 | −.005 | 0 | 2.0156 | +0 | −.008 |
| 2.0156 | 3.0156 | +.010 | −.020 | 4.0000 | 6.0000 | +.006 | −.006 | 2.0156 | 3.0156 | +0 | −.010 |
| 3.0156 | 6.0156 | +.015 | −.020 | 6.0000 | 10.0000 | +.010 | −.010 | 3.0156 | 6.0156 | +0 | −.015 |
| 6.0156 | 10.1560 | +.015 | −.050 | 10.0000 | 18.0000 | +.012 | −.012 | 6.0156 | 10.1560 | +0 | −.020 |

All dimensions are in inches. For Type TR bearings, see *ANSI/ABMA 24.1-1989 (S2020)*.

## Table 14. Tolerance Limits for Needle Roller Bearings, Drawn Cup, Without Inner Ring — Inch Types NIB, NIBM, NIY, NIYM, NIH, and NIHM
*ANSI/ABMA 18.2-1982 (S2013)*

| Ring Gage Bore Diameter[a] | | | Basic Bore Diameter under Needle Rollers, $F_w$ | | Allowable Deviation from $F_w$[a] | | Allowable Deviation from Width, $B$ | |
|---|---|---|---|---|---|---|---|---|
| Basic Outside Diameter, $D$ Inch | | Deviation from $D$ Inch | Inch | | Inch | | Inch | |
| Over | Incl. | | Over | Incl. | Low | High | High | Low |
| 0.1875 | 0.9375 | +0.0005 | | | | | | |
| 0.9375 | 4.0000 | −0.0005 | 0.1875 | 0.6875 | +0.0015 | +0.0024 | +0 | −0.0100 |
| For fitting and mounting practice see Table 23. | | | 0.6875 | 1.2500 | +0.0005 | +0.0014 | +0 | −0.0100 |
| | | | 1.2500 | 1.3750 | +0.0005 | +0.0015 | +0 | −0.0100 |
| | | | 1.3750 | 1.6250 | +0.0005 | +0.0016 | +0 | −0.0100 |
| | | | 1.6250 | 1.8750 | +0.0005 | +0.0017 | +0 | −0.0100 |
| | | | 1.8750 | 2.0000 | +0.0006 | +0.0018 | +0 | −0.0100 |
| | | | 2.0000 | 2.5000 | +0.0006 | +0.0020 | +0 | −0.0100 |
| | | | 2.5000 | 3.5000 | +0.0010 | +0.0024 | +0 | −0.0100 |

[a] The bore diameter under needle rollers can be measured only when bearing is pressed into a ring gage, which rounds and sizes the bearing.

## Table 15. Tolerance Limits for Needle Roller Bearings, Drawn Cup, Without Inner Ring — Metric Types NB, NBM, NY, NYM, NH, and NHM
*ANSI/ABMA 18.1-1982 (S2013)*

| Ring Gage Bore Diameter[a] | | | Basic Bore Diameter under Needle Rollers, $F_w$ | | Allowable Deviation from $F_w$[a] | | Allowable Deviation from Width, $B$ | |
|---|---|---|---|---|---|---|---|---|
| Basic Outside Diameter, $D$ mm | | Deviation from $D$ Micrometers | mm | | Micrometers | | Micrometers | |
| Over | Incl. | | Over | Incl. | Low | High | High | Low |
| 6 | 10 | −16 | 3 | 6 | +10 | +28 | +0 | −250 |
| 10 | 18 | −20 | 6 | 10 | +13 | +31 | +0 | −250 |
| 30 | 50 | −28 | 18 | 30 | +20 | +41 | +0 | −250 |
| 50 | 80 | −33 | 30 | 50 | +25 | +50 | +0 | −250 |
| … | … | … | 50 | 70 | +30 | +60 | +0 | −250 |

[a] The bore diameter under needle rollers can be measured only when bearing is pressed into a ring gage, which rounds and sizes the bearing.

For fitting and mounting practice, see Table 23.

## Table 16. Tolerance Limits for Needle Roller Bearings, With Cage, Machined Ring, Without Inner Ring—Inch Type NIA *ANSI/ABMA 18.2-1982 (S2013)*

| Basic Outside Diameter, $D$ | | Allowable Deviation From $D$ of Single Mean Diameter, $D_{mp}$ | | Basic Bore Diameter under Needle Rollers, $F_w$ | | Allowable Deviation from $F_w$ | | Allowable Deviation from Width, $B$ | |
|---|---|---|---|---|---|---|---|---|---|
| Inch | | Inch | | Inch | | Inch | | Inch | |
| Over | Incl. | High | Low | Over | Incl. | Low | High | High | Low |
| 0.7500 | 2.0000 | +0 | −0.0005 | 0.3150 | 0.7087 | +0.0008 | +0.0017 | +0 | −0.0050 |
| 2.0000 | 3.2500 | +0 | −0.0006 | 0.7087 | 1.1811 | +0.0009 | +0.0018 | +0 | −0.0050 |
| 3.2500 | 4.7500 | +0 | −0.0008 | 1.1811 | 1.6535 | +0.0010 | +0.0019 | +0 | −0.0050 |
| 4.7500 | 7.2500 | +0 | −0.0010 | 1.6535 | 1.9685 | +0.0010 | +0.0020 | +0 | −0.0050 |
| | | | | 1.9685 | 2.7559 | +0.0011 | +0.0021 | +0 | −0.0050 |
| 7.2500 | 10.2500 | +0 | −0.0012 | 2.7559 | 3.1496 | +0.0011 | +0.0023 | +0 | −0.0050 |
| 10.2500 | 11.1250 | +0 | −0.0014 | 3.1496 | 4.0157 | +0.0012 | +0.0024 | +0 | −0.0050 |
| … | … | … | … | 4.0157 | 4.7244 | +0.0012 | +0.0026 | +0 | −0.0050 |
| … | … | … | … | 4.7244 | 6.2992 | +0.0013 | +0.0027 | +0 | −0.0050 |
| … | … | … | … | 6.2992 | 7.0866 | +0.0013 | +0.0029 | +0 | −0.0050 |
| … | … | … | … | 7.0866 | 7.8740 | +0.0014 | +0.0030 | +0 | −0.0050 |
| … | … | … | … | 7.8740 | 9.2520 | +0.0014 | +0.0032 | +0 | −0.0050 |

For fitting and mounting practice, see Table 24.

**Table 17. Tolerance Limits for Needle Roller Bearing Inner Rings—Inch Type NIR**
*ANSI/ABMA 18.2-1982 (S2013)*

| Basic Outside Diameter, *F* | | Allowable Deviation From *F* of Single Mean Diameter, $F_{mp}$ | | Basic Bore Diameter *d* | | Allowable Deviation from *d* of Single Mean Diameter, $d_{mp}$ | | Allowable Deviation from Width, *B* | |
|---|---|---|---|---|---|---|---|---|---|
| Inch | | Inch | | Inch | | Inch | | Inch | |
| Over | Incl. | High | Low | Over | Incl. | High | Low | High | Low |
| 0.3937 | 0.7087 | −0.0005 | −0.0009 | 0.3125 | 0.7500 | +0 | −0.0004 | +0.0100 | +0.0050 |
| 0.7087 | 1.0236 | −0.0007 | −0.0012 | 0.7500 | 2.0000 | +0 | −0.0005 | +0.0100 | +0.0050 |
| 1.0236 | 1.1811 | −0.0009 | −0.0014 | 2.0000 | 3.2500 | +0 | −0.0006 | +0.0100 | +0.0050 |
| 1.1811 | 1.3780 | −0.0009 | −0.0015 | 3.2500 | 4.2500 | +0 | −0.0008 | +0.0100 | +0.0050 |
| 1.3780 | 1.9685 | −0.0010 | −0.0016 | 4.2500 | 4.7500 | +0 | −0.0008 | +0.0150 | +0.0100 |
| 1.9685 | 3.1496 | −0.0011 | −0.0018 | 4.7500 | 7.0000 | +0 | −0.0010 | +0.0150 | +0.0100 |
| 3.1496 | 3.9370 | −0.0013 | −0.0022 | 7.0000 | 8.0000 | +0 | −0.0012 | +0.0150 | +0.0100 |
| 3.9370 | 4.7244 | −0.0015 | −0.0024 | … | … | … | … | … | … |
| 4.7244 | 5.5118 | −0.0015 | −0.0025 | … | … | … | … | … | … |
| 5.5118 | 7.0866 | −0.0017 | −0.0027 | … | … | … | … | … | … |
| 7.0866 | 8.2677 | −0.0019 | −0.0031 | … | … | … | … | … | … |
| 8.2677 | 9.2520 | −0.0020 | −0.0032 | … | … | … | … | … | … |

For fitting and mounting practice, see Table 25.

**Metric Radial Ball and Roller Bearing Shaft and Housing Fits.—**To select the proper fits, it is necessary to consider the type and extent of the load, bearing type, and certain other design and performance requirements.

The required shaft and housing fits are indicated in Table 18 and Table 19. The terms "Light," "Normal," and "Heavy" loads refer to radial loads that are generally within the following limits, with some overlap (*C* being the Basic Load Rating computed in accordance with the ANSI/ABMA Standards):

| Bearing Type | Radial Load | | |
|---|---|---|---|
| | Light | Normal | Heavy |
| Ball | Up to 0.075*C* | From 0.075*C* to 0.15*C* | Over 0.15*C* |
| Cylindrical Roller | Up to 0.075*C* | From 0.075*C* to 0.2*C* | Over 0.15*C* |
| Spherical Roller | Up to 0.075*C* | From 0.070*C* to 0.25*C* | Over 0.15*C* |

*Shaft Fits:* Table 18 indicates the initial approach to shaft fit selection. Note that for most normal applications where the shaft rotates and the radial load direction is constant, an interference fit should be used. Also, the heavier the load, the greater is the required interference. For stationary shaft conditions and constant radial load direction, the inner ring may be moderately loose on the shaft.

For pure thrust (axial) loading, heavy interference fits are not necessary; only a moderately loose to tight fit is needed.

The upper part of Table 20 shows how the shaft diameters for various ANSI shaft limit classifications deviate from the basic bore diameters.

Table 21 gives metric values for the shaft diameter and housing bore tolerance limits given in Table 20.

The lower parts of Table 20 and Table 21 show how housing bores for various ANSI hole limit classifications deviate from the basic shaft outside diameters.

**Table 18. Selection of Shaft Tolerance Classifications for Metric Radial Ball and Roller Bearings of ABEC-1 and RBEC-1 Tolerance Classes** *ANSI/ABMA 7-1995 (S2013)*

| Operating Conditions | | | Ball Bearings mm | Ball Bearings Inch | Cylindrical Roller Bearings mm | Cylindrical Roller Bearings Inch | Spherical Roller Bearings mm | Spherical Roller Bearings Inch | Tolerance Symbol[a] |
|---|---|---|---|---|---|---|---|---|---|
| Inner ring stationary in relation to the direction of the load. | All loads | Inner ring has to be easily displaceable | All diameters | All diameters | All diameters | All diameters | All diameters | All diameters | g6 |
| | | Inner ring does not have to be easily displaceable | All diameters | All diameters | All diameters | All diameters | All diameters | All diameters | h6 |
| | Radial load: | | Nominal Shaft Diameter | | | | | | |
| Direction of load indeterminate or the inner ring rotating in relation to the direction of the load. | LIGHT | | ≤18 | ≤0.71 | | | | | h5 |
| | | | >18 | >0.71 | ≤40 | ≤1.57 | ≤40 | ≤1.57 | j6[b] |
| | | | | | (40)-140 | (1.57)–5.51 | (40)–100 | (1.57)–3.94 | k6[b] |
| | | | | | (140)–320 | (5.51)–12.6 | (100)–320 | (3.94)–12.6 | m6[b] |
| | | | | | (320)–500 | (126)–19.7 | (320)–500 | (126)–19.7 | n6 |
| | | | | | >500 | >19.7 | >500 | >19.7 | p6 |
| | NORMAL | | ≤18 | ≤0.71 | | | | | j5 |
| | | | >18 | >0.71 | ≤40 | ≤1.57 | ≤40 | ≤1.57 | k5 |
| | | | | | (40)–100 | (1.57)–3.94 | (40)–65 | (1.57)–2.56 | m5 |
| | | | | | (100)–140 | (3.94)–5.51 | (65)–100 | (2.56)–3.94 | m6 |
| | | | | | (140)–320 | (5.51)–12.6 | (100)–140 | (3.94)–5.51 | n6 |
| | | | | | (320)–500 | (12.6)–19.7 | (140)–280 | (5.51)–11.0 | p6 |
| | | | | | >500 | >19.7 | (280)–500 | (11.0)–19.7 | r6 |
| | | | | | | | >500 | >19.7 | r7 |
| | HEAVY | | (18)–100 | (0.71)–3.94 | | | | | k5 |
| | | | >100 | >3.94 | ≤40 | ≤1.57 | ≤40 | ≤1.57 | m5 |
| | | | | | (40)-65 | (1.57)–2.56 | (40)–65 | (1.57)–2.56 | m6[b] |
| | | | | | (65)–140 | (2.56)–5.51 | (65)–100 | (2.56)–3.94 | n6[b] |
| | | | | | (140)–200 | (5.51)–7.87 | (100)–140 | (3.94)–5.51 | p6[b] |
| | | | | | (200)–500 | (7.87)–19.7 | (140)–200 | (5.51)–7.87 | r6[b] |
| | | | | | >500 | >19.7 | >200 | >7.87 | r7[b] |
| Pure Thrust Load | | | All diams. | All diams. | Consult Bearing Manufacturer | | | | j6 |

[a] For solid steel shafts. For hollow or nonferrous shafts, tighter fits may be needed.

[b] When greater accuracy is required, use j5, k5, and m5 instead of j6, k6, and m6, respectively.

Numerical values are given in Table 20 and Table 21.

**Table 19. Selection of Housing Tolerance Classifications for Metric Radial Ball and Roller Bearings of ABEC-1 and RBEC-1 Tolerance Classes**

| Design and Operating Conditions | | | | |
|---|---|---|---|---|
| Rotational Conditions | Loading | Outer Ring Axial Displacement Limitations | Other Conditions | Tolerance Classification[a] |
| Outer ring stationary in relation to load direction | Light Normal and Heavy | Outer ring must be easily axially displaceable | Heat input through shaft | G7 |
| | | | Housing split axially | H7[b] |
| | | | Housing not split axially | H6[b] |
| | Shock with temporary complete unloading | Transitional Range[c] | | J6[b] |
| Load direction is indeterminate | Light and normal | | | |
| | Normal and Heavy | | | |
| | Heavy Shock | | Split housing not recommended | K6[b] |
| Outer ring rotating in relation to load direction | Light | | | M6[b] |
| | Normal and Heavy | Outer ring need not be axially displaceable | | N6[b] |
| | Heavy | | Thin wall housing not split | P6[b] |

[a] For cast iron or steel housings. For housings of nonferrous alloys tighter fits may be needed.
[b] Where wider tolerances are permissible, use tolerance classifications P7, N7, M7, K7, J7, and H7, in place of P6, N6, M6, K6, J6, and H6, respectively.
[c] The tolerance zones are such that the outer ring may be either tight or loose in the housing.

**Table 20. Rolling Shaft Diameter and Housing Bore Tolerance Limits** *ANSI/ABMA 7-1995 (S2013)*

| Inches | | mm | | Allowable Deviations of Shaft Diameter from Basic Bore Diameter, Inch | | | | | | | | | | | | |
|---|---|---|---|---|---|---|---|---|---|---|---|---|---|---|---|---|
| Over | Incl. | Over | Incl. | | | | | | | | | | | | | |
| Base Bore Diameter | | | | g6 | h6 | h5 | j5 | j6 | k5 | k6 | m5 | m6 | n6 | p6 | r6 | r7 |
| 0.2362 | 0.3937 | 6 | 10 | −.0002<br>−.0006 | 0<br>−.0004 | 0<br>−.0002 | +.0002<br>−.0001 | +.0003<br>−.0001 | +.0003<br>0 | | +.0005<br>+.0002 | | | | | |
| 0.3937 | 0.7087 | 10 | 18 | −.0002<br>−.0007 | 0<br>−.0004 | 0<br>−.0003 | +.0002<br>−.0001 | +.0003<br>−.0001 | +.0004<br>0 | | +.0006<br>+.0003 | | | | | |
| 0.7087 | 1.1811 | 18 | 30 | −.0003<br>−.0008 | 0<br>−.0005 | | +.0002<br>−.0002 | +.0004<br>−.0002 | +.0004<br>+.0001 | | +.0007<br>+.0003 | | | | | |
| 1.1811 | 1.9685 | 30 | 50 | −.0004<br>−.0010 | 0<br>−.0006 | | +.0002<br>−.0002 | +.0004<br>−.0002 | +.0005<br>+.0001 | +.0007<br>+.0001 | +.0008<br>+.0004 | +.0010<br>+.0004 | | | | |
| 1.9685 | 3.1496 | 50 | 80 | −.0004<br>−.0011 | 0<br>−.0007 | | +.0002<br>−.0003 | +.0005<br>−.0003 | +.0006<br>+.0001 | +.0008<br>+.0001 | +.0009<br>+.0004 | +.0012<br>+.0004 | +.0018<br>+.0009 | | | |
| 3.1496 | 4.7244 | 80 | 120 | −.0005<br>−.0013 | 0<br>−.0009 | | +.0002<br>−.0004 | +.0005<br>−.0004 | +.0007<br>+.0001 | +.0010<br>+.0001 | +.0011<br>+.0005 | +.0014<br>+.0005 | +.0019<br>+.0010 | +.0023<br>+.0015 | | |
| Allowable Deviations of Housing Bore from Basic Outside Diameter of Shaft, Inch | | | | | | | | | | | | | | | | |
| Basic Outside Diameter | | | | G7 | H7 | H6 | J7 | J6 | K6 | K7 | M6 | M7 | N6 | N7 | P6 | P7 |
| 0.7087 | 1.1811 | 18 | 30 | +.0003<br>+.0011 | 0<br>+.0008 | 0<br>+.0005 | −.0004<br>+.0005 | −.0002<br>+.0003 | −.0004<br>+.0001 | −.0006<br>+.0002 | −.0007<br>+.0002 | −.0008<br>0 | −.0009<br>−.0004 | −.0011<br>−.0003 | −.0012<br>−.0007 | −.0014<br>−.0006 |
| 1.1811 | 1.9685 | 30 | 50 | +.0004<br>+.0013 | 0<br>+.0010 | 0<br>+.0006 | −.0004<br>+.0006 | −.0002<br>+.0004 | −.0005<br>+.0001 | −.0007<br>+.0003 | −.0008<br>−.0002 | −.0010<br>0 | −.0011<br>−.0005 | −.0013<br>−.0003 | −.0015<br>−.0008 | −.0017<br>−.0007 |
| 1.9685 | 3.1496 | 50 | 80 | +.0004<br>+.0016 | 0<br>+.0012 | 0<br>+.0007 | −.0005<br>+.0007 | −.0002<br>+.0005 | −.0006<br>+.0002 | −.0008<br>+.0004 | −.0009<br>−.0002 | −.0012<br>0 | −.0013<br>−.0006 | −.0015<br>−.0004 | −.0018<br>−.0010 | −.0020<br>−.0008 |
| 3.1496 | 4.7244 | 80 | 120 | +.0005<br>+.0019 | 0<br>+.0014 | 0<br>+.0009 | −.0005<br>+.0009 | −.0002<br>+.0006 | −.0007<br>+.0002 | −.0010<br>+.0004 | −.0011<br>−.0002 | −.0014<br>0 | −.0015<br>−.0006 | −.0018<br>−.0004 | −.0020<br>−.0012 | −.0023<br>−.0009 |
| 4.7244 | 7.0866 | 120 | 180 | +.0006<br>+.0021 | 0<br>+.0016 | 0<br>+.0010 | −.0006<br>+.0010 | −.0003<br>+.0007 | −.0008<br>+.0002 | −.0011<br>+.0005 | −.0013<br>−.0003 | −.0016<br>0 | −.0018<br>−.0008 | −.0020<br>−.0005 | −.0024<br>−.0014 | −.0027<br>−.0011 |
| 7.0866 | 9.8425 | 180 | 250 | +.0006<br>+.0024 | 0<br>+.0018 | 0<br>+.0011 | −.0006<br>+.0012 | −.0003<br>+.0009 | −.0009<br>+.0002 | −.0013<br>+.0005 | −.0015<br>−.0003 | −.0018<br>0 | −.0020<br>−.0009 | −.0024<br>−.0006 | −.0028<br>−.0016 | −.0031<br>−.0013 |

Based on ANSI/ASME B4.1-1967 (R2020) Preferred Limits and Fits for Cylindrical Parts. Symbols g6, h6, etc., are shaft and G7, H7, etc., hole limits designations. For larger diameters and metric values see *ANSI/ABMA 7-1995 (S2013)*.

**Table 21. Shaft Diameter and Housing Bore Tolerance Limits** *ANSI/ABMA 7-1995 (S2013)*

| Inches | | mm | | Allowable Deviations of Shaft Diameter from Basic Bore Diameter, mm | | | | | | | | | | | | |
|---|---|---|---|---|---|---|---|---|---|---|---|---|---|---|---|---|
| Over | Incl. | Over | Incl. | | | | | | | | | | | | | |
| Base Bore Diameter | | | | g6 | h6 | h5 | j5 | j6 | k5 | k6 | m5 | m6 | n6 | p6 | r6 | r7 |
| 0.2362 | 0.3937 | 6 | 10 | −.005<br>−.014 | 0<br>−.009 | 0<br>−.006 | +.004<br>−.002 | +.007<br>−.002 | +.007<br>−.001 | | +.012<br>+.006 | | | | | |
| 0.3937 | 0.7087 | 10 | 18 | −.006<br>−.017 | 0<br>−.011 | 0<br>−.008 | +.005<br>−.003 | +.008<br>−.003 | +.009<br>+.001 | | +.015<br>+.007 | | | | | |
| 0.7087 | 1.1811 | 18 | 30 | −.007<br>−.020 | 0<br>−.013 | | +.005<br>−.004 | +.009<br>−.004 | +.011<br>+.002 | | +.017<br>+.008 | | | | | |
| 1.1811 | 1.9685 | 30 | 50 | −.009<br>−.025 | 0<br>−.016 | | +.006<br>−.005 | +.011<br>−.005 | +.013<br>+.002 | +.018<br>+.002 | +.020<br>+.009 | +.025<br>+.009 | | | | |
| 1.9685 | 3.1496 | 50 | 80 | −.010<br>−.029 | 0<br>−.019 | | +.006<br>−.007 | +.012<br>−.007 | +.015<br>+.002 | +.021<br>+.002 | +.024<br>+.011 | +.030<br>+.011 | +.039<br>+.020 | | | |
| 3.1496 | 4.7244 | 80 | 120 | −.012<br>−.034 | 0<br>−.022 | | +.006<br>−.009 | +.013<br>−.009 | +.018<br>+.003 | +.025<br>+.003 | +.028<br>+.013 | +.035<br>+.013 | +.045<br>+.023 | +.059<br>+.037 | | |
| Allowable Deviations of Housing Bore from Basic Outside Diameter of Shaft, mm | | | | | | | | | | | | | | | | |
| Basic Outside Diameter | | | | G7 | H7 | H6 | J7 | J6 | K6 | K7 | M6 | M7 | N6 | N7 | P6 | P7 |
| .7086 | 1.1811 | 18 | 30 | +.007<br>+.028 | 0<br>+.021 | 0<br>+.013 | −.009<br>+.012 | −.005<br>+.008 | −.011<br>+.002 | −.015<br>+.006 | −.017<br>−.004 | −.021<br>0 | −.024<br>−.011 | −.028<br>−.007 | −.031<br>−.018 | −.035<br>−.014 |
| 1.1811 | 1.9685 | 30 | 50 | +.009<br>+.034 | 0<br>+.025 | 0<br>+.016 | −.011<br>+.014 | −.006<br>+.010 | −.013<br>+.003 | −.018<br>+.007 | −.020<br>−.004 | −.025<br>0 | −.028<br>−.012 | −.033<br>−.008 | −.037<br>−.021 | −.042<br>−.017 |
| 1.9685 | 3.1496 | 50 | 80 | +.010<br>+.040 | 0<br>+.030 | 0<br>+.019 | −.012<br>+.018 | −.006<br>+.013 | −.015<br>+.004 | −.021<br>+.009 | −.024<br>−.005 | −.030<br>0 | −.033<br>−.014 | −.039<br>−.009 | −.045<br>−.026 | −.051<br>−.021 |
| 3.1496 | 4.7244 | 80 | 120 | +.012<br>+.047 | 0<br>+.035 | 0<br>+.022 | −.013<br>+.022 | −.006<br>+.016 | −.018<br>+.004 | −.025<br>+.010 | −.028<br>−.006 | −.035<br>0 | −.038<br>−.016 | −.045<br>−.010 | −.052<br>−.030 | −.059<br>−.024 |
| 4.7244 | 7.0866 | 120 | 180 | +.014<br>+.054 | 0<br>+.040 | 0<br>+.025 | −.014<br>+.026 | −.007<br>+.018 | −.021<br>+.004 | −.028<br>+.012 | −.033<br>−.008 | −.040<br>0 | −.045<br>−.020 | −.052<br>−.012 | −.061<br>−.036 | −.068<br>−.028 |
| 7.0866 | 9.8425 | 180 | 250 | +.015<br>+.061 | 0<br>+.046 | 0<br>+.029 | −.016<br>+.030 | −.007<br>+.022 | −.024<br>+.005 | −.033<br>+.013 | −.037<br>−.008 | −.046<br>0 | −.051<br>−.022 | −.060<br>−.014 | −.070<br>−.041 | −.079<br>−.033 |

Based on ANSI/ASME B4.1-1967 (R2020) Preferred Limits and Fits for Cylindrical Parts. Symbols g6, h6, etc., are shaft and G7, H7, etc., hole limits designations. For larger diameters and metric values see *ANSI/ABMA 7-1995 (S2013)*.

**Design and Installation Considerations.—**Interference fitting will reduce bearing radial internal clearance, so it is recommended that prospective users consult bearing manufacturers to make certain that the required bearings are correctly specified to satisfy all mounting, environmental and other operating conditions and requirements. This check is particularly necessary where heat sources in associated parts may further diminish bearing clearances in operation.

Standard values of radial internal clearances of radial bearings are listed in ANSI ABMA 20-2011 (R2020).

**Allowance for Axial Displacement.—**Consideration should be given to axial displacement of bearing components owing to thermal expansion or contraction of associated parts. Displacement may be accommodated either by the internal construction of the bearing or by allowing one of the bearing rings to be axially displaceable. For unusual applications consult bearing manufacturers.

**Needle Roller Bearing Fitting and Mounting Practice.—**The tolerance limits required for shaft and housing seat diameters for needle roller bearings with inner and outer rings as well as limits for raceway diameters where inner or outer rings or both are omitted and rollers operate directly upon these surfaces are given in Table 22 through Table 25, inclusive. Unusual design and operating conditions may require a departure from these practices. In such cases, bearing manufacturers should be consulted.

*Needle Roller Bearings, Drawn Cup:* These bearings without inner ring, Types NIB, NB, NIBM, NBM, NIY, NY, NIYM, NYM, NIH, NH, NIHM, NHM, and Inner Rings, Type NIR depend on the housings into which they are pressed for their size and shape. Therefore, the housings must not only have the proper bore dimensions but also must have sufficient strength. Table 22 and Table 23, show the bore tolerance limits for rigid housings such as those made from cast iron or steel of heavy radial section equal to or greater than the ring gage section given in ANSI/ABMA 4-1994 (S2013). The bearing manufacturers should be consulted for recommendations if the housings must be of lower strength materials such as aluminum or even of steel of thin radial section. The shape of the housing bores should be such that when the mean bore diameter of a housing is measured in each of several radial planes, the maximum difference between these mean diameters should not exceed 0.0005 inch (0.013 mm) or one-half the housing bore tolerance limit, if smaller. Also, the radial deviation from circular form should not exceed 0.00025 inch (0.006 mm). The housing bore surface finish should not exceed 125 micro-inches (3.2 micrometers) arithmetical average.

**Table 22. Tolerance Limits for Shaft Raceway and Housing Bore Diameters— Needle Roller Bearings, Drawn Cup, Without Inner Ring, Inch Types NIB, NIBM, NIY, NIYM, NIH, and NIHM** *ANSI/ABMA 18.2-1982 (S2013)*

| Basic Bore Diameter under Needle Rollers, $F_w$ | | Shaft Raceway Diameter[a] Allowable Deviation from $F_w$ | | Basic Outside Diameter, $D$ | | Housing Bore Diameter[a] Allowable Deviation from $D$ | |
|---|---|---|---|---|---|---|---|
| Inch | | Inch | | Inch | | Inch | |
| Over | Incl. | High | Low | Over | Incl. | Low | High |
| OUTER RING STATIONARY RELATIVE TO LOAD | | | | | | | |
| 0.1875 | 1.8750 | +0 | −0.0005 | 0.3750 | 4.0000 | −0.0005 | +0.0005 |
| 1.8750 | 3.5000 | +0 | −0.0006 | … | … | … | … |
| OUTER RING ROTATING RELATIVE TO LOAD | | | | | | | |
| 0.1875 | 1.8750 | −0.0005 | −0.0010 | 0.3750 | 4.0000 | −0.0010 | −0 |
| 1.8750 | 3.5000 | −0.0005 | −0.0011 | … | … | … | … |

[a] See text for additional requirements.

For bearing tolerances, see Table 14.

**Table 23. Tolerance Limits for Shaft Raceway and Housing Bore Diameters— Needle Roller Bearings, Drawn Cup, Without Inner Ring, Metric Types NB, NBM, NY, NYM, NH, and NHM** *ANSI/ABMA 18.1-1982 (S2013)*

| Basic Bore Diameter Under Needle Rollers, $F_w$ | | | | Shaft Raceway Diameter[a] Allowable Deviation from $F_w$ | | Basic Outside Diameter, $D$ | | | | Housing Bore Diameter[a] Allowable Deviation from $D$ | |
|---|---|---|---|---|---|---|---|---|---|---|---|
| OUTER RING STATIONARY RELATIVE TO LOAD | | | | | | | | | | | |
| mm | | Inch | | ANSI h6, Inch | | mm | | Inch | | ANSI N7, Inch | |
| Over | Incl. | Over | Incl. | High | Low | Over | Incl. | Over | Incl. | Low | High |
| 3 | 6 | 0.1181 | 0.2362 | +0 | −0.0003 | 6 | 10 | 0.2362 | 0.3937 | −0.0007 | −0.0002 |
| 6 | 10 | 0.2362 | 0.3937 | +0 | −0.0004 | 10 | 18 | 0.3937 | 0.7087 | −0.0009 | −0.0002 |
| 10 | 18 | 0.3937 | 0.7087 | +0 | −0.0004 | 18 | 30 | 0.7087 | 1.1811 | −0.0011 | −0.0003 |
| 18 | 30 | 0.7087 | 1.1811 | +0 | −0.0005 | 30 | 50 | 1.1811 | 1.9685 | −0.0013 | −0.0003 |
| 30 | 50 | 1.1811 | 1.9685 | +0 | −0.0006 | 50 | 80 | 1.9685 | 3.1496 | −0.0015 | −0.0004 |
| 50 | 80 | 1.9685 | 3.1496 | +0 | −0.0007 | … | … | … | … | … | … |
| OUTER RING ROTATING RELATIVE TO LOAD | | | | | | | | | | | |
| mm | | Inch | | ANSI f6, Inch | | mm | | Inch | | ANSI R7, Inch | |
| Over | Incl. | Over | Incl. | High | Low | Over | Incl. | Over | Incl. | Low | High |
| 3 | 6 | 0.1181 | 0.2362 | −0.0004 | −0.0007 | 6 | 10 | 0.2362 | 0.3937 | −0.0011 | −0.0005 |
| 6 | 10 | 0.2362 | 0.3937 | −0.0005 | −0.0009 | 10 | 18 | 0.3937 | 0.7087 | −0.0013 | −0.0006 |
| 10 | 18 | 0.3937 | 0.7087 | −0.0006 | −0.0011 | 18 | 30 | 0.7087 | 1.1811 | −0.0016 | −0.0008 |
| 18 | 30 | 0.7087 | 1.1811 | −0.0008 | −0.0013 | 30 | 50 | 1.1811 | 1.9685 | −0.0020 | −0.0010 |
| 30 | 50 | 1.1811 | 1.9685 | −0.0010 | −0.0016 | 50 | 65 | 1.9685 | 2.5591 | −0.0024 | −0.0012 |
| 50 | 80 | 1.9685 | 3.1496 | −0.0012 | −0.0019 | 65 | 80 | 2.5591 | 3.1496 | −0.0024 | −0.0013 |

For bearing tolerances, see Table 15.

**Table 24. Tolerance Limits for Shaft Raceway and Housing Bore Diameters— Needle Roller Bearings, With Cage, Machined Ring, Without Inner Ring, Inch Type NIA** *ANSI/ABMA 18.2-1982 (S2013)*

| Basic Bore Diameter under Needle Rollers, $F_w$ | | Shaft Raceway Diameter[a] Allowable Deviation from $F_w$ | | Basic Outside Diameter, $D$ | | Housing Bore Diameter[a] Allowable Deviation from $D$ | |
|---|---|---|---|---|---|---|---|
| OUTER RING STATIONARY RELATIVE TO LOAD | | | | | | | |
| Inch | | ANSI h6, Inch | | Inch | | ANSI H7, Inch | |
| Over | Incl. | High | Low | Over | Incl. | Low | High |
| 0.2362 | 0.3937 | +0 | −0.0004 | 0.3937 | 0.7087 | +0 | +0.0007 |
| 0.3937 | 0.7087 | +0 | −0.0004 | 0.7087 | 1.1811 | +0 | +0.0008 |
| 0.7087 | 1.1811 | +0 | −0.0005 | 1.1811 | 1.9685 | +0 | +0.0010 |
| 1.1811 | 1.9685 | +0 | −0.0006 | 1.9685 | 3.1496 | +0 | +0.0012 |
| 1.9685 | 3.1496 | +0 | −0.0007 | 3.1496 | 4.7244 | +0 | +0.0014 |
| 3.1496 | 4.7244 | +0 | −0.0009 | 4.7244 | 7.0866 | +0 | +0.0016 |
| 4.7244 | 7.0866 | +0 | −0.0010 | 7.0866 | 9.8425 | +0 | +0.0018 |
| 7.0866 | 9.8425 | +0 | −0.0011 | 9.8425 | 12.4016 | +0 | +0.0020 |
| OUTER RING ROTATING RELATIVE TO LOAD | | | | | | | |
| Inch | | ANSI f6, Inch | | Inch | | ANSI N7, Inch | |
| Over | Incl. | High | Low | Over | Incl. | Low | High |
| 0.2362 | 0.3937 | −0.0005 | −0.0009 | 0.3937 | 0.7087 | −0.0009 | −0.0002 |
| 0.3937 | 0.7087 | −0.0006 | −0.0011 | 0.7087 | 1.1811 | −0.0011 | −0.0003 |
| 0.7087 | 1.1811 | −0.0008 | −0.0013 | 1.1811 | 1.9685 | −0.0013 | −0.0003 |
| 1.1811 | 1.9685 | −0.0010 | −0.0016 | 1.9685 | 3.1496 | −0.0015 | −0.0004 |
| *1.9685* | 3.1496 | −0.0012 | −0.0019 | 3.1496 | 4.7244 | −0.0018 | −0.0004 |
| 3.1496 | 4.7244 | −0.0014 | −0.0023 | 4.7244 | 7.0866 | −0.0020 | −0.0005 |
| 4.7244 | 7.0866 | −0.0016 | −0.0027 | 7.0866 | 9.8425 | −0.0024 | −0.0006 |
| 7.0866 | 9.8425 | −0.0020 | −0.0031 | 9.8425 | 11.2205 | −0.0026 | −0.0006 |

[a] See text for additional requirements.

For bearing tolerances, see Table 16.

**Table 25. Tolerance Limits for Shaft Diameters — Needle Roller Bearing Inner Rings, Inch Type NIR (Used with Bearing Type NIA)**
*ANSI/ABMA 18.2-1982 (S2013)*

| Basic Bore, d | | Shaft Diameter[a] | | | |
|---|---|---|---|---|---|
| | | Shaft Rotating Relative to Load, Outer Ring Stationary Relative to Load Allowable Deviation from d | | Shaft Stationary Relative to Load, Outer Ring Rotating Relative to Load Allowable Deviation from d | |
| Inch | | ANSI m5, Inch | | ANSI g6, Inch | |
| Over | Incl. | High | Low | High | Low |
| 0.2362 | 0.3937 | +0.0005 | +0.0002 | −0.0002 | −0.0006 |
| 0.3937 | 0.7087 | +0.0006 | +0.0003 | −0.0002 | −0.0007 |
| 0.7087 | 1.1811 | +0.0007 | +0.0003 | −0.0003 | −0.0008 |
| 1.1811 | 1.9685 | +0.0008 | +0.0004 | −0.0004 | −0.0010 |
| 1.9685 | 3.1496 | +0.0009 | +0.0004 | −0.0004 | −0.0011 |
| 3.1496 | 4.7244 | +0.0011 | +0.0005 | −0.0005 | −0.0013 |
| 4.7244 | 7.0866 | +0.0013 | +0.0006 | −0.0006 | −0.0015 |
| 7.0866 | 9.8425 | +0.0015 | +0.0007 | −0.0006 | −0.0017 |

[a] See text for additional requirements.

For inner ring tolerance limits, see Table 17.

Most needle roller bearings do not use inner rings, but operate directly on the surfaces of shafts. When shafts are used as inner raceways, they should be made of bearing quality steel hardened to 58 RC (Rockwell C scale) minimum. Table 19 and Table 23 show the shaft raceway tolerance limits and Table 25 shows the shaft seat tolerance limits when inner rings are used. However, whether the shaft surfaces are used as inner raceways or as seats for inner rings, the mean outside diameter of the shaft surface in each of several radial planes should be determined. The difference between these mean diameters should not exceed 0.0003 inch (0.008 mm) or one-half the diameter tolerance limit, if smaller. The radial deviation from circular form should not exceed 0.0001 inch (0.0025 mm), for diameters up to and including 1 in. (25.4 mm). Above 1 inch the allowable deviation is 0.0001 times the shaft diameter. The surface finish should not exceed 16 micro-inches (0.4 micrometer) arithmetical average. The housing bore and shaft diameter tolerance limits depend upon whether the load rotates relative to the shaft or the housing.

*Needle Roller Bearing With Cage, Machined Ring, Without Inner Ring:* The following covers needle roller bearings Type NIA and inner rings Type NIR. The shape of the housing bores should be such that when the mean bore diameter of a housing is measured in each of several radial planes, the maximum difference between these mean diameters does not exceed 0.0005 inch (0.013 mm) or one-half the housing bore tolerance limit, if smaller. Also, the radial deviation from circular form should not exceed 0.00025 inch (0.006 mm). The housing bore surface finish should not exceed 125 micro-inches (3.2 micrometers) arithmetical average. Table 25 shows the housing bore tolerance limits.

When shafts are used as inner raceways their requirements are the same as those given above for Needle Roller Bearings, Drawn Cup. Table 24 shows the shaft raceway tolerance limits and Table 25 shows the shaft seat tolerance limits when inner rings are used.

*Needle Roller and Cage Assemblies, Types NIM and NM:* For information concerning boundary dimensions, tolerance limits, and fitting and mounting practice, reference should be made to the standards ANSI/ABMA 18.1-1982 (S2013) and ANSI/ABMA 18.2-1982 (S2013).

## Bearing Mounting Practice

Because of their inherent design and material rigidity, rolling contact bearings must be mounted with careful control of their alignment and runout. Medium-speed or slower (400,000 *DN* values or less where *D* is the bearing bore in millimeters and *N* is the bearing speed in revolutions per minute), and medium to light load (*C*/*P* values of 7 or greater where *C* is the bearing specific dynamic capacity in pounds and *P* is the average bearing load in pounds) applications can endure misalignments equivalent to those acceptable for high-capacity, precision journal bearings utilizing hard bearing materials such as silver, copper-lead, or aluminum. In no case, however, should the maximum shaft deflection exceed .001 inch per inch (or mm per mm) for well-crowned roller bearings, and .003 inch per inch (or mm per mm) for deep-groove ball-bearings. Except for self-aligning ball-bearings and spherical or barrel roller bearings, all other types require shaft alignments with deflections no greater than .0002 inch per inch (or mm per mm). With preloaded ball bearings, this same limit is recommended as a maximum. Close-clearance tapered bearings or thrust bearings of most types require the same shaft alignment also.

Of major importance in achieving reliability in bearings is the location of the races on the shaft and in the housing.

Assembly methods must ensure: 1) that the faces are square before the cavity is closed; 2) that the cover face is square to the shoulder and pulled in evenly; and 3) that it will be located by a face parallel to it when finally seated against the housing.

These requirements are shown in Table 26. In applications not controlled by automatic tooling with closely controlled fixtures and bolt torquing mechanisms, races should be checked for squareness by sweeping with a dial indicator mounted as shown below. For commercial applications with moderate life and reliability requirements, outer race runouts should be held to .0005 inch per inch (or mm per mm) of radius and inner race runout to .0004 inch per inch (or mm per mm) of radius. In preloaded and precision applications, these tolerances must be cut in half. In regard to the question of alignment, it must be recognized that rolling-contact bearings, being made of fully-hardened steel, do not wear in as may certain journal bearings when carefully applied and initially operated. Likewise, rolling contact bearings absorb relatively little deflection when loaded to *C*/*P* values of 6 or less. At such stress levels the rolling element-race deformation is generally not over .0002 inch (5.08 μm). Consequently, proper mounting and control of shaft deflections are imperative for reliable bearing performance. Aside from inadequate lubrication, these factors are the most frequent causes of premature bearing failures.

### Table 26. Commercial Application Alignment Tolerances

| Feature | Location | Tolerance |
|---|---|---|
| **Housing Face Runout** | 1 | Square to shaft center within .0004 inch/inch of radius full indicator reading. |
| **Outer Race Face Runout** | 2 | Square to shaft center within .0004 inch/inch of radius full indicator reading and complementary to the housing runout (not opposed). |
| **Inner Race Face Runout** | 3 | Square to shaft center within .0003 inch/inch of radius full indicator reading. |
| **Cover and Closure Mounting Face Parallelism** | 4 and 5 | Parallel within .001. |
| **Housing Mounting Face Parallelism** | 6 | Parallel within .001 |

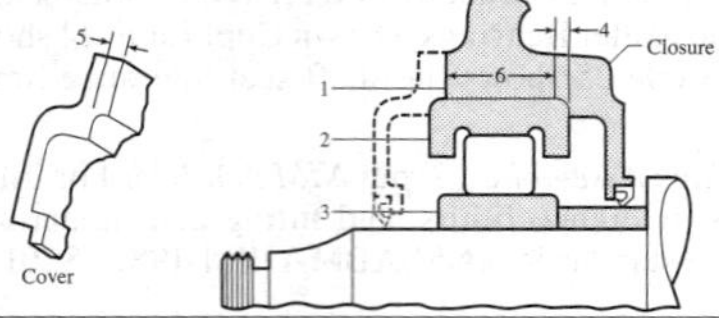

**Mountings for Precision and Quiet-Running Applications.—**In applications of rolling-element bearings where vibration or smoothness of operation is critical, special precautions must be taken to eliminate those conditions which can serve to initiate radial and axial motions. These exciting forces can result in shaft excursions which are in resonance with shaft or housing components over a range of frequencies from well below shaft speed to as much as 100 times above it. The more sensitive the configuration, the greater is the need for precision bearings and mountings to be used.

Precision bearings are normally made to much closer tolerances than standard and therefore benefit from better finishing techniques. Special inspection operations are required, however, to provide races and rolling elements with smoothness and runouts compatible with the needs of the application. Similarly, shafts and housings must be carefully controlled.

Among the important elements to be controlled are shaft, race, and housing roundness; squareness of faces, diameters, shoulders, and rolling paths. Though not readily appreciated, grinding chatter, lobular and compensating out-of-roundness, waviness, and flats of less than 0.0005 inch (0.013 mm) deviation from the average or mean diameter can cause significant roughness. To detect these and ensure the selection of good pieces, three-point electronic indicator inspection must be made.

For ultra-precise or quiet applications, pieces are often checked on a "Talyrond" or a similar continuous recording instrument capable of measuring to within a few millionths of an inch. Though this may seem extreme, it has been found that shaft deformities will be reflected through inner races shrunk onto them. Similarly, tight-fit outer races pick up significant deviations in housings. In many instrument and in missile guidance applications, such deviations and deformities may have to be limited to less than 0.00002 inch (0.508 μm).

In most of these precision applications, bearings are used with rolling elements controlled to less than 5 millionths of an inch deviation from roundness and within the same range for diameter.

Special attention is required both in housing design and in assembly of the bearing to shaft and housing. Housing response to axial excursions forced by bearing wobble (which in itself is a result of out-of-square mounting) has been found to be a major source of small electric and other rotating equipment noise and howl. Stiffer, more massive housings and careful alignment of bearing races can make significant improvements in applications where noise or vibration has been found to be objectionable.

**Squareness and Alignment.—**In addition to the limits for roundness and wall variation of the races and their supports, squareness of end faces and shoulders must be closely controlled. Tolerances of .0001 inch (2.54 μm) full indicator reading per inch of diameter are normally required for end faces and shoulders, with appropriately selected limits for fillet eccentricities. The latter must also fall within specified limits for radii tolerances to prevent interference and the resulting cocking of the race. Reference should be made to the bearing dimension tables which list corner radii for typical bearings.

Shoulders must also be of a sufficient height to ensure proper support for the races, since they are of hardened steel and are less capable of absorbing shock loads and abuse. The general subject of squareness and alignment is of primary importance to the life of rolling element bearings.

The following recommendations for shaft and housing design are given by the New Departure Division of General Motors Corporation:*

"As a rule, there is little trouble experienced with inaccuracies in shafts. Bearings seats and locating shoulders are turned and ground to size with the shaft held on centers and, with ordinary care, there is small chance for serious out-of-roundness or taper. Shaft shoulders should present sufficient surface in contact with the bearing face to assure positive and accurate location."

"Where an undercut must be made for wheel runout in grinding a bearing seat, care should be exercised that no sharp corners are left, for it is at such points that fatigue is most likely to result in shaft breakage. It is best to undercut as little as possible and to have the undercut end in a fillet instead of a sharp corner."

"Where clamping nuts are to be used, it is important to cut the threads as true and square as possible in order to insure even pressure at all points on the bearing inner ring faces when the nuts are set up tight. It is also important not to cut threads so far into the bearing seat as to leave part of the inner ring unsupported or carried on the threads. Excessive deflection is usually the result of improperly designed or undersized machine parts. With a weak shaft, it is possible to seriously affect bearing operation through misalignment due to shaft deflection. Where shafts are comparatively long, the diameter between bearings must be great enough to properly resist bending. In general, the use of more than two bearings on a single shaft should be avoided, owing to the difficulty of securing accurate alignment. With bearings mounted close to each other, this can result in extremely heavy bearing loads."

"Design is as important as careful machining in construction of accurate bearing housings. There should be plenty of metal in the wall sections and large, thin areas should be avoided as much as possible, since they are likely to permit deflection of the boring tool when the housing is being finish-machined."

"Wherever possible, it is best to design a housing so that the radial load placed on the bearing is transmitted as directly as possible to the wall or rib supporting the housing. Diaphragm walls connecting an offset housing to the main wall or side of a machine are apt to deflect unless made thick and well braced."

"When two bearings are to be mounted opposed, but in separate housings, the housings should be so reinforced with fins or webs as to prevent deflection due to the axial load under which the bearings are opposed."

"Where housings are deep and considerable overhang of the boring tool is required, there is a tendency to produce out-of-roundness and taper, unless the tool is very rigid and light finishing cuts are taken. In a too roughly bored housing there is a possibility for the ridges of metal to peen down under load, thus eventually resulting in too loose a fit for the bearing outer ring."

**Soft Metal and Resilient Housings.**—In applications relying on bearing housings made of soft materials (aluminum, magnesium, light sheet metal, etc.) or those which lose their fit because of differential thermal expansion, outer race mounting must be approached in a cautious manner. Of first importance is the determination of the possible consequences of race loosening and turning. In conjunction with this, the type of loading must be considered for it may serve to magnify the effect of race loosening. It must be remembered that generally, balancing processes do not ensure zero unbalance at operating speeds, but rather an "*acceptable*" maximum. This force exerted by the rotating element on the outer race can initiate a precession which will aggravate the race loosening problem by causing further attrition through wear, pounding, and abrasion. Since this force is generally of an order greater than the friction forces in effect between the outer race, housing, and closures (retaining nuts also), no foolproof method can be recommended for securing

* *New Departure Handbook*. Vol. II — 1951.

outer races in housings which deform significantly under load or after appreciable service wear. Though many such "fixes" are offered, the only sure solution is to press the race into a housing of sufficient stiffness with the heaviest fit consistent with the installed and operating clearances. In many cases, inserts, or liners of cast iron or steel, are provided to maintain the desired fit and increase useful life of both bearing and housing.

**Quiet or Vibration-Free Mountings.—**In seeming contradiction is the approach to bearing mountings in which all shaft or rotating element excursions must be isolated from the frame, housing, or supporting structure. Here bearing outer races are often supported on elastomeric or metallic springs. Fundamentally, this is an isolation problem and must be approached with caution to ensure solution of the primary bearing objective — location and restraint of the rotating body, as well as the reduction or elimination of the dynamic problem. Again, the danger of skidding rolling elements must be considered and reference to the resident engineers or sales engineers of the numerous bearing companies is recommended, as this problem generally develops requirements for special, or non-catalog-type bearings.

**General Mounting Precautions.—**Since the last operations involving the bearing application — mounting and closing — have such important effects on bearing performance, durability, and reliability, it must be cautioned that more bearings are abused or "killed" in this early stage of their life than wear out or "die" under conditions for which they were designed. Hammer and chisel "mechanics" invariably handle bearings as though no blow could be too hard, no dirt too abrasive, and no misalignment of any consequence. Proper tools, fixtures, and techniques are a must for rolling bearing application, and it is the responsibility of the design engineer to provide for this in his design, advisory notes, mounting instructions, and service manuals. Nicks, dents, scores, scratches, corrosion staining, and dirt must be avoided if reliability, long life, and smooth running are to be expected of rolling bearings. All manufacturers have pertinent service instructions available for the bearing user. These should be followed for best performance. In a later section, methods for inspecting bearings and descriptions of most common bearing deficiencies will be given.

**Seating Fits for Bearings.—**American Bearing Manufacturers Association (ABMA) standard shaft and housing bearing seat tolerances are given in Table 17 through Table 22, inclusive.

**Clamping and Retaining Methods.—**Various methods of clamping bearings to prevent axial movement on the shaft are employed, one of the most common being a nut screwed on the end of the shaft and held in place by a tongued lock washer (see Table 27). The shaft thread for the clamping nut (see Table 28) should be cut in accurate relation to bearing seats and shoulders if bearing stresses are to be avoided. The threads used are of American National Form, Class 3; special diameters and data for these are given in Table 29 and Table 30.

Where somewhat closer than average accuracy is required, the washers and locknut faces may be obtained ground for closer alignment with the threads. For a high degree of accuracy the shaft threads are ground and a more precise clamping means is employed. Where a bearing inner ring is to be clamped, it is important to provide a sufficiently high shoulder on the shaft to locate the bearing positively and accurately. If the difference between bearing bore and maximum shaft diameter gives a low shoulder which would enter the corner of the radius of the bearing, a shoulder ring that extends above the shoulder and well into the shaft corner is employed. A shoulder ring with snap wire fitting into a groove in the shaft is sometimes used where no locating shaft shoulder is present. A snap ring fitting into a groove is frequently employed to prevent endwise movement of the bearing away from the locating shoulder where tight clamping is not required. Such a retaining ring should not be used where a slot in the shaft surface might lead to fatigue failure. Snap rings are also used to locate the outer bearing ring in the housing. Dimensions of snap rings used for this latter purpose are given in ABMA and ANSI standards.

**Table 27. ABMA Standard Lock Washers (Series W-00) for Ball Bearings and Cylindrical and Spherical Roller Bearings and (Series TW-100) for Tapered Roller Bearings. Inch Design**

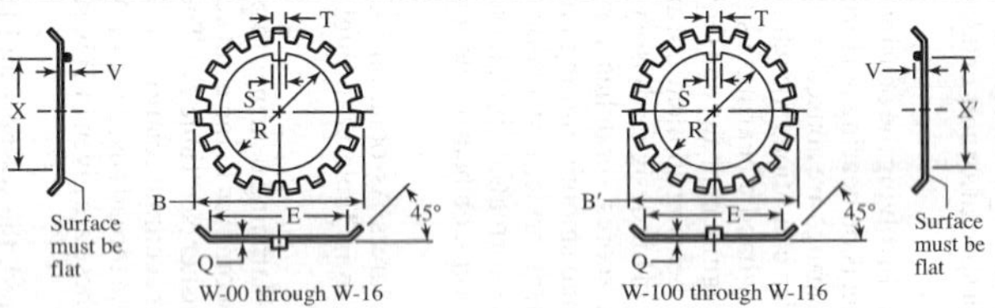


| Type W No. | Q | Type TW No. | Q | Tangs | | Key | | | | | | | | Bore R | | Diameter | | Dia. Over Tangs. Max. | |
|---|---|---|---|---|---|---|---|---|---|---|---|---|---|---|---|---|---|---|---|
| | | | | No. | Width[a] | Project.[a] | Width S | | X | | X′ | | | | | | | | |
| | | | | | T | V | Min. | Max. | Min. | Max. | Min. | Max. | | Min. | Max. | E | Tol. | B | B′ |
| W-00 | .032 | TW-100 | .032 | 9 | .120 | .031 | .110 | .120 | .334 | .359 | .334 | .359 | | .406 | 0.421 | 0.625 | +.015 | 0.875 | 0.891 |
| W-01 | .032 | TW-101 | .032 | 9 | .120 | .031 | .110 | .120 | .412 | .437 | .412 | .437 | | .484 | .499 | 0.719 | +.015 | 1.016 | 1.031 |
| W-02 | .032 | TW-102 | .048 | 11 | .120 | .031 | .110 | .120 | .529 | .554 | .513 | .538 | | .601 | .616 | 0.813 | +.015 | 1.156 | 1.156 |
| W-03 | .032 | TW-103 | .048 | 11 | .120 | .031 | .110 | .120 | .607 | .632 | .591 | .616 | | .679 | .694 | 0.938 | +.015 | 1.328 | 1.344 |
| W-04 | .032 | TW-104 | .048 | 11 | .166 | .031 | .156 | .176 | .729 | .754 | .713 | .738 | | .801 | .816 | 1.125 | +.015 | 1.531 | 1.563 |
| W-05 | .040 | TW-105 | .052 | 13 | .166 | .047 | .156 | .176 | .909 | .939 | .897 | .927 | | .989 | 1.009 | 1.281 | +.015 | 1.719 | 1.703 |
| W-06 | .040 | TW-106 | .052 | 13 | .166 | .047 | .156 | .176 | 1.093 | 1.128 | 1.081 | 1.116 | | 1.193 | 1.213 | 1.500 | +.015 | 1.922 | 1.953 |
| | | TW-065 | .052 | 15 | .166 | … | .156 | .176 | … | … | 1.221 | 1.256 | | 1.333 | 1.353 | 1.813 | +.015 | … | 2.234 |
| W-07 | .040 | TW-107 | .052 | 15 | .166 | .047 | .156 | .176 | 1.296 | 1.331 | 1.284 | 1.319 | | 1.396 | 1.416 | 1.813 | +.015 | 2.250 | 2.250 |
| W-08 | .048 | TW-108 | .062 | 15 | .234 | .047 | .250 | .290 | 1.475 | 1.510 | 1.461 | 1.496 | | 1.583 | 1.603 | 2.000 | +.030 | 2.469 | 2.484 |
| W-09 | .048 | TW-109 | .062 | 17 | .234 | .062 | .250 | .290 | 1.684 | 1.724 | 1.670 | 1.710 | | 1.792 | 1.817 | 2.281 | +.030 | 2.734 | 2.719 |
| W-10 | .048 | TW-110 | .062 | 17 | .234 | .062 | .250 | .290 | 1.884 | 1.924 | 1.870 | 1.910 | | 1.992 | 2.017 | 2.438 | +.030 | 2.922 | 2.922 |
| W-11 | .053 | TW-111 | .062 | 17 | .234 | .062 | .250 | .290 | 2.069 | 2.109 | 2.060 | 2.100 | | 2.182 | 2.207 | 2.656 | ±.030 | 3.109 | 3.094 |
| W-12 | .053 | TW-112 | .072 | 17 | .234 | .062 | .250 | .290 | 2.267 | 2.307 | 2.248 | 2.288 | | 2.400 | 2.425 | 2.844 | +.030 | 3.344 | 3.328 |
| W-13 | .053 | TW-113 | .072 | 19 | .234 | .062 | .250 | .290 | 2.455 | 2.495 | 2.436 | 2.476 | | 2.588 | 2.613 | 3.063 | +.030 | 3.578 | 3.563 |
| W-14 | .053 | TW-114 | .072 | 19 | .234 | .094 | .250 | .290 | 2.658 | 2.698 | 2.639 | 2.679 | | 2.791 | 2.816 | 3.313 | +.030 | 3.828 | 3.813 |
| W-15 | .062 | TW-115 | .085 | 19 | .328 | .094 | .250 | .290 | 2.831 | 2.876 | 2.808 | 2.853 | | 2.973 | 3.003 | 3.563 | +.030 | 4.109 | 4.047 |
| W-16 | .062 | TW-116 | .085 | 19 | .328 | .094 | .313 | .353 | 3.035 | 3.080 | 3.012 | 3.057 | | 3.177 | 3.207 | 3.844 | +.030 | 4.375 | 4.391 |

[a] *Tolerances:* On width, *T*, −.010 inch for Types W-00 to W-03 and TW-100 to TW-103; −.020 inch for W-04 to W-07 and TW-104 to TW-107; −.030 inch for all others shown. On Projection *V*, +.031 inch for all sizes up through W-13 and TW-113; +.062 inch for all others shown.

All dimensions in inches. For dimensions in millimeters, multiply inch values by 25.4 and round result to two decimal places.

For sizes larger than shown here, see ANSI/ABMA 8.2-2023.

**Table 28. ABMA Standard Locknuts (Series N-00) for Ball Bearings and Cylindrical and Spherical Roller Bearings and (Series TN-00) for Tapered Roller Bearings. Inch Design**

Runout and Parallelism of Faces Measured on a Tight Fiting Threaded Arbor.

N-00 to N-06 = .002 Max.
N-07 to AN-15 = .004 Max.

TN-065 to TAN-15 = .002 Max.

Surface Finish Note

TN-065 to TN-11, 100 μin., Max.
TN-12 to TN-15, 120 μin., Max.

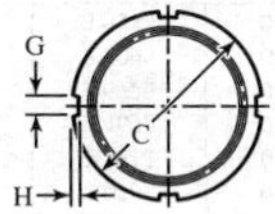


N-00 through AN-15
TN-065 through TAN-15

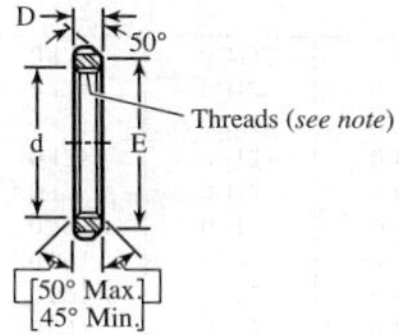


| BB & RB Nut No. | TRB Nut No. | Thds. per Inch | Thread Minor Dia. | | Thread Pitch Dia. | | Thd. Major Dia. *d* | Outside Dia. *C* | Face Dia. *E* | | Slot dimension | | | Thickness *D* | |
|---|---|---|---|---|---|---|---|---|---|---|---|---|---|---|---|
| | | | | | | | | | | | Width *G* | | Height *H* | | |
| | | | Min. | Max. | Min. | Max. | Min. | Max. | Min. | Max. | Min. | Max. | Max. | Min. | Max. |
| N-00 | — | 32 | 0.3572 | 0.3606 | 0.3707 | 0.3733 | 0.391 | 0.755 | .605 | .625 | .120 | .130 | .073 | .209 | .229 |
| N-01 | — | 32 | 0.4352 | 0.4386 | 0.4487 | 0.4513 | 0.469 | 0.880 | .699 | .719 | .120 | .130 | .073 | .303 | .323 |
| N-02 | — | 32 | 0.5522 | 0.5556 | 0.5657 | 0.5687 | 0.586 | 1.005 | .793 | .813 | .120 | .130 | .104 | .303 | .323 |
| N-03 | — | 32 | 0.6302 | 0.6336 | 0.6437 | 0.6467 | 0.664 | 1.130 | .918 | .938 | .120 | .130 | .104 | .334 | .354 |
| N-04 | — | 32 | 0.7472 | 0.7506 | 0.7607 | 0.7641 | 0.781 | 1.380 | 1.105 | 1.125 | .178 | .198 | .104 | .365 | .385 |
| N-05 | — | 32 | 0.9352 | 0.9386 | 0.9487 | 0.9521 | 0.969 | 1.568 | 1.261 | 1.281 | .178 | .198 | .104 | .396 | .416 |
| N-06 | — | 18 | 1.1129 | 1.1189 | 1.1369 | 1.1409 | 1.173 | 1.755 | 1.480 | 1.500 | .178 | .198 | .104 | .396 | .416 |
| | TN-065 | 18 | 1.2524 | 1.2584 | 1.2764 | 1.2804 | 1.312 | 2.068 | 1.793 | 1.813 | .178 | .198 | .104 | — | — |
| N-07 | TN-07 | 18 | 1.3159 | 1.3219 | 1.3399 | 1.3439 | 1.376 | 2.068 | 1.793 | 1.813 | .178 | .198 | .104 | .428 | .448 |
| N-08 | TN-08 | 18 | 1.5029 | 1.5089 | 1.5269 | 1.5314 | 1.563 | 2.255 | 1.980 | 2.000 | .240 | .260 | .104 | .428 | .448 |
| N-09 | TN-09 | 18 | 1.7069 | 1.7129 | 1.7309 | 1.7354 | 1.767 | 2.536 | 2.261 | 2.281 | .240 | .260 | .104 | .428 | .448 |
| N-10 | TN-10 | 18 | 1.9069 | 1.9129 | 1.9309 | 1.9354 | 1.967 | 2.693 | 2.418 | 2.438 | .240 | .260 | .104 | .490 | .510 |
| N-11 | TN-11 | 18 | 2.0969 | 2.1029 | 2.1209 | 2.1260 | 2.157 | 2.974 | 2.636 | 2.656 | .240 | .260 | .135 | .490 | .510 |
| N-12 | TN-12 | 18 | 2.2999 | 2.3059 | 2.3239 | 2.3290 | 2.360 | 3.161 | 2.824 | 2.844 | .240 | .260 | .135 | .521 | .541 |
| N-13 | TN-13 | 18 | 2.4879 | 2.4949 | 2.5119 | 2.5170 | 2.548 | 3.380 | 3.043 | 3.063 | .240 | .260 | .135 | .553 | .573 |
| N-14 | TN-14 | 18 | 2.6909 | 2.6969 | 2.7149 | 2.7200 | 2.751 | 3.630 | 3.283 | 3.313 | .240 | .260 | .135 | .553 | .573 |
| AN-15 | TAN-15 | 12 | 2.8428 | 2.8518 | 2.8789 | 2.8843 | 2.933 | 3.880 | 3.533 | 3.563 | .360 | .385 | .135 | .584 | .604 |

All dimensions in inches. For dimensions in millimeters, multiply inch values, except thread diameters, by 25.4 and round result to two decimal places.

Threads are American National form, Class 3.

Typical steels for locknuts are: AISI, C1015, C1018, C1020, C1025, C1035, C1117, C1118, C1212, C1213, and C1215. Minimum hardness, tensile strength, yield strength and elongation are given in ANSI/ABMA 8.2-2023 which also lists larger sizes of locknuts.

**Table 29. ABMA Standard for Shafts for Locknuts (series N-00) for Ball Bearings and Cylindrical and Spherical Roller Bearings. Inch Design**

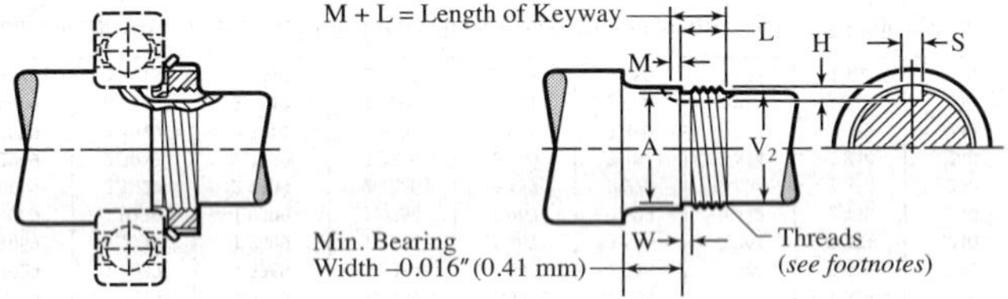


| Locknut Number | Bearing Bore | $V_2$ | Threads[a] | | | | | Relief | | Keyway | | |
|---|---|---|---|---|---|---|---|---|---|---|---|---|
| | | | No. per inch | Major Dia. | Pitch Dia. | Minor Dia. | Length *L* | Dia. *A* | Width *W* | Depth *H* | Width *S* | *M* |
| | | Max. | | Max. | Max. | Max. | Max. | Max. | Max. | Min. | Min. | Min. |
| N-00 | 0.3937 | 0.312 | 32 | 0.391 | 0.3707 | 0.3527 | 0.297 | 0.3421 | 0.078 | 0.062 | 0.125 | 0.094 |
| N-01 | 0.4724 | 0.406 | 32 | 0.469 | 0.4487 | 0.4307 | 0.391 | 0.4201 | 0.078 | 0.062 | 0.125 | 0.094 |
| N-02 | 0.5906 | 0.500 | 32 | 0.586 | 0.5657 | 0.5477 | 0.391 | 0.5371 | 0.078 | 0.078 | 0.125 | 0.094 |
| N-03 | 0.6693 | 0.562 | 32 | 0.664 | 0.6437 | 0.6257 | 0.422 | 0.6151 | 0.078 | 0.078 | 0.125 | 0.094 |
| N-04 | 0.7874 | 0.719 | 32 | 0.781 | 0.7607 | 0.7427 | 0.453 | 0.7321 | 0.078 | 0.078 | 0.188 | 0.094 |
| N-05 | 0.9843 | 0.875 | 32 | 0.969 | 0.9487 | 0.9307 | 0.484 | 0.9201 | 0.078 | 0.094 | 0.188 | 0.125 |
| N-06 | 1.1811 | 1.062 | 18 | 1.173 | 1.1369 | 1.1048 | 0.484 | 1.0942 | 0.109 | 0.094 | 0.188 | 0.125 |
| N-07 | 1.3780 | 1.250 | 18 | 1.376 | 1.3399 | 1.3078 | 0.516 | 1.2972 | 0.109 | 0.094 | 0.188 | 0.125 |
| N-08 | 1.5748 | 1.469 | 18 | 1.563 | 1.5269 | 1.4948 | 0.547 | 1.4842 | 0.109 | 0.094 | 0.312 | 0.125 |
| N-09 | 1.7717 | 1.688 | 18 | 1.767 | 1.7309 | 1.6988 | 0.547 | 1.6882 | 0.141 | 0.094 | 0.312 | 0.156 |
| N-10 | 1.9685 | 1.875 | 18 | 1.967 | 1.9309 | 1.8988 | 0.609 | 1.8882 | 0.141 | 0.094 | 0.312 | 0.156 |
| N-11 | 2.1654 | 2.062 | 18 | 2.157 | 2.1209 | 2.0888 | 0.609 | 2.0782 | 0.141 | 0.125 | 0.312 | 0.156 |
| N-12 | 2.3622 | 2.250 | 18 | 2.360 | 2.3239 | 2.2918 | 0.641 | 2.2812 | 0.141 | 0.125 | 0.312 | 0.156 |
| N-13 | 2.5591 | 2.438 | 18 | 2.548 | 2.5119 | 2.4798 | 0.672 | 2.4692 | 0.141 | 0.125 | 0.312 | 0.156 |
| N-14 | 2.7559 | 2.625 | 18 | 2.751 | 2.7149 | 2.6828 | 0.672 | 2.6722 | 0.141 | 0.125 | 0.312 | 0.250 |
| AN-15 | 2.9528 | 2.781 | 12 | 2.933 | 2.8789 | 2.8308 | 0.703 | 2.8095 | 0.172 | 0.125 | 0.312 | 0.250 |
| AN-16 | 3.1496 | 3.000 | 12 | 3.137 | 3.0829 | 3.0348 | 0.703 | 3.0135 | 0.172 | 0.125 | 0.375 | 0.250 |

[a] Threads are American National form Class 3.

All dimensions in inches. For dimensions in millimeters, multiply inch values, except thread diameters, by 25.4 and round result to two decimal places. See footnote to Table 30 for material other than sttel. For sizes larger than shown, see ANSI/ABMA 8.2-2023.

**Table 30. ABMA Standard for Shafts for Tapered Roller Bearing Locknuts. Inch Design**

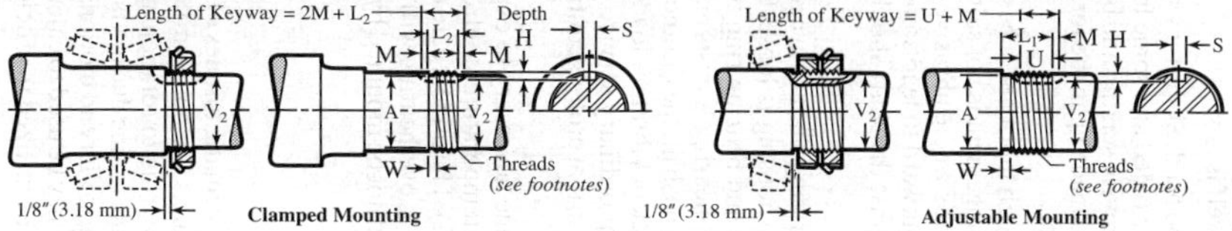


| Locknut Number | Bearing Bore | $V_2$ | Threads[a] | | | | | | Relief | | Keyway | | | |
|---|---|---|---|---|---|---|---|---|---|---|---|---|---|---|
| | | | No. per inch | Major Dia. | Pitch Dia. | Minor Dia. | Length | | Dia. $A$ | Width $W$ | Depth $H$ | Width $S$ | $M$ | $U$ |
| | | | | | | | $L_1$ | $L_2$ | | | | | | |
| | | Max. | | Max. | Max. | Max. | Max. | Max. | Max. | Max. | Max. | Min. | Min. | Min. |
| N-00 | 0.3937 | 0.312 | 32 | 0.391 | 0.3707 | 0.3527 | 0.609 | 0.391 | 0.3421 | 0.078 | 0.094 | 0.125 | 0.094 | 0.469 |
| N-01 | 0.4724 | 0.406 | 32 | 0.469 | 0.4487 | 0.4307 | 0.797 | 0.404 | 0.4201 | 0.078 | 0.094 | 0.125 | 0.094 | 0.562 |
| N-02 | 0.5906 | 0.500 | 32 | 0.586 | 0.5657 | 0.5477 | 0.828 | 0.516 | 0.5371 | 0.078 | 0.094 | 0.125 | 0.094 | 0.594 |
| N-03 | 0.6693 | 0.562 | 32 | 0.664 | 0.6437 | 0.6257 | 0.891 | 0.547 | 0.6151 | 0.078 | 0.078 | 0.125 | 0.094 | 0.625 |
| N-04 | 0.7874 | 0.703 | 32 | 0.781 | 0.7607 | 0.7427 | 0.922 | 0.547 | 0.7321 | 0.078 | 0.094 | 0.188 | 0.094 | 0.625 |
| N-05 | 0.9843 | 0.875 | 32 | 0.969 | 0.9487 | 0.9307 | 1.016 | 0.609 | 0.9201 | 0.078 | 0.125 | 0.188 | 0.125 | 0.719 |
| N-06 | 1.1811 | 1.062 | 18 | 1.173 | 1.1369 | 1.1048 | 1.016 | 0.609 | 1.0942 | 0.109 | 0.125 | 0.188 | 0.125 | 0.719 |
| TN-065 | 1.3750 | 1.188 | 18 | 1.312 | 1.2764 | 1.2443 | 1.078 | 0.641 | 1.2337 | 0.109 | 0.125 | 0.188 | 0.125 | 0.750 |
| TN-07 | 1.3780 | 1.250 | 18 | 1.376 | 1.3399 | 1.3078 | 1.078 | 0.641 | 1.2972 | 0.109 | 0.125 | 0.188 | 0.125 | 0.750 |
| TN-08 | 1.5748 | 1.438 | 18 | 1.563 | 1.5269 | 1.4948 | 1.078 | 0.641 | 1.4842 | 0.109 | 0.125 | 0.312 | 0.125 | 0.750 |
| TN-09 | 1.7717 | 1.656 | 18 | 1.767 | 1.7309 | 1.6988 | 1.078 | 0.641 | 1.6882 | 0.141 | 0.125 | 0.312 | 0.156 | 0.781 |
| TN-10 | 1.9685 | 1.859 | 18 | 1.967 | 1.9309 | 1.8988 | 1.203 | 0.703 | 1.8882 | 0.141 | 0.125 | 0.312 | 0.156 | 0.844 |
| TN-11 | 2.1654 | 2.047 | 18 | 2.157 | 2.1209 | 2.0888 | 1.203 | 0.703 | 2.0782 | 0.141 | 0.125 | 0.312 | 0.156 | 0.844 |
| TN-12 | 2.3622 | 2.250 | 18 | 2.360 | 2.3239 | 2.2918 | 1.297 | 0.766 | 2.2812 | 0.141 | 0.156 | 0.312 | 0.156 | 0.906 |
| TN-13 | 2.5591 | 2.422 | 18 | 2.548 | 2.5119 | 2.4798 | 1.359 | 0.797 | 2.4692 | 0.141 | 0.156 | 0.312 | 0.156 | 0.938 |
| TN-14 | 2.7559 | 2.625 | 18 | 2.751 | 2.7149 | 2.6828 | 1.359 | 0.797 | 2.6722 | 0.141 | 0.156 | 0.312 | 0.250 | 1.000 |
| TAN-15 | 2.9528 | 2.781 | 12 | 2.933 | 2.8789 | 2.8308 | 1.422 | 0.828 | 2.8095 | 0.172 | 0.188 | 0.312 | 0.250 | 1.031 |
| TAN-16 | 3.1496 | 3.000 | 12 | 3.137 | 3.0829 | 3.0348 | 1.422 | 0.828 | 3.0135 | 0.172 | 0.188 | 0.375 | 0.250 | 1.031 |

[a] Threads are American National form Class 3.

All dimensions in inches. For dimensions in millimeters, multiply inch values, except thread diameters, by 25.4 and round results to two decimal places. These data apply to steel. When either the nut or the shaft is made of stainless steel, aluminum, or other material having a tendency to seize, it is recommended that the maximum thread diameter of the shaft, both major and pitch, be reduced by 20 percent of the pitch diameter tolerance listed in the standard. For sizes larger than shown, see ANSI/ABMA 8.2-2023.

**Bearing Closures.—**Shields, seals, labyrinths, and slingers are employed to retain the lubricant in the bearing and to prevent the entry of dirt, moisture, or other harmful substances. The type selected for a given application depends upon the lubricant, shaft, speed, and the atmospheric conditions in which the unit is to operate. The shields or seals may be located in the bearing itself. Shields differ from seals in that they are attached to one bearing race but there is a definite clearance between the shield and the other, usually the inner race. When a shielded bearing is placed in a housing in which the grease space has been filled, the bearing in running will tend to expel excess grease past the shields or to accept grease from the housing when the amount in the bearing itself is low.

Seals of leather, rubber, cork, felt, or plastic composition may be used. Since they must bear against the rotating member, excessive pressure should be avoided and some lubricant must be allowed to flow into the area of contact in order to prevent seizing and burning of the seal and scoring of the rotating member. Some seals are made up in the form of cartridges which can be pressed into the end of the bearing housing.

Leather seals may be used over a wide range of speeds. Although lubricant is best retained with a leather cupped inward toward the bearing, this arrangement is not suitable at high speeds due to danger of burning the leather. At high speeds where abrasive dust is present, the seal should be arranged with the leather cupped outward to lead some lubricant into the contact area. Only light pressure of leather against the shaft should be maintained.

**Bearing Fits.—**The slipping or creeping of a bearing ring on a rotating shaft or in a rotating housing occurs when the fit of the ring on the shaft or in the housing is loose. Such slipping or creeping action may cause rapid wear of both shaft and bearing ring when the surfaces are dry and highly loaded. To prevent this action the bearing is customarily mounted with the rotating ring a press fit and the stationary ring a push fit, the tightness or looseness depending upon the service intended. Thus, where shock or vibratory loads are to be encountered, fits should be made somewhat tighter than for ordinary service. The stationary ring, if correctly fitted, is allowed to creep very slowly so that prolonged stressing of one part of the raceway is avoided.

To facilitate the assembly of a bearing on a shaft it may become necessary to expand the inner ring by heating. This should be done in clean oil or in a temperature-controlled furnace at a temperature of between 200 and 250° F (93 to 121° C). The utmost care must be used to make sure that the temperature does not exceed 250° F as overheating will tend to reduce the hardness of the rings. Prelubricated bearings should not be mounted by this method.

## Design Considerations

**Friction Losses in Rolling Element Bearings.—**The static and kinematic torques of rolling element bearings are generally small and in many applications are not significant. Bearing torque is a measure of the frictional resistance of the bearing to rotation and is the sum of three components: the torque due to the applied load; the torque due to viscous forces in lubricated rolling element bearings; and the torque due to roller end motions, for example, thrust loads against flanges. The friction or torque data may be used to calculate power absorption or heat generation within the bearing and can be utilized in efficiency or system-cooling studies.

Empirical equations have been developed for each of the torque components. These equations are influenced by such factors as bearing load, lubrication environment, and bearing design parameters. These design parameters include sliding friction from contact between the rolling elements and separator surfaces or between adjacent rolling elements; rolling friction from material deformations during the passage of the rolling elements over the race path; skidding or sliding of the Hertzian contact; and windage friction as a function of speed.

Starting or breakaway torques are also of interest in some situations. Breakaway torques tend to be between 1.5 and 1.8 times the running or kinetic torques.

When evaluating the torque requirements of a system under design, it should be noted that other components of the bearing package, such as seals and closures, can increase the overall system torque significantly. Seal torques have been shown to vary from a fraction of the bearing torque to several times that torque. In addition, the torque values given can vary significantly when load, speed of rotation, temperature, or lubrication are outside normal ranges.

For small instrument bearings friction torque has implications more critical than for larger types of bearings. These bearings have three operating friction torques to consider: starting torque, normal running torque, and peak running torque. These torque levels may vary between manufacturers and among lots from a given manufacturer.

Instrument bearings are even more critically dependent on design features — radial play, retainer type, and race conformity — than larger bearings. Typical starting torque values for small bearings are given in Table 31, with values extracted from the New Departure General Catalog.

Finally, if accurate control of friction torque is critical to a particular application, tests of the selected bearings should be conducted to evaluate performance.

**Table 31. Starting Torque — ABEC7**

| Bearing Bore (in.) | Max. Starting Torque (g cm) | Thrust Load (g) | Minimum Radial Play Range (inches) | |
|---|---|---|---|---|
| | | | High Carbon Chrome Steel and All Miniatures | Stainless Steel Except Miniatures |
| 0.125 | 0.10 | 75 | 0.0003–0.0005 | — |
| | 0.14 | 75 | 0.0002–0.0004 | 0.0004–0.0006 |
| | 0.18 | 75 | 0.0001–0.0003 | 0.0003–0.0005 |
| | 0.22 | 75 | 0.0001–0.0003 | 0.0001–0.0003 |
| 0.1875–0.312 | 0.40 | 400 | 0.0005–0.0008 | — |
| | 0.45 | 400 | 0.0004–0.0006 | 0.0005–0.0008 |
| | 0.50 | 400 | 0.0003–0.0005 | 0.0003–0.0005 |
| | 0.63 | 400 | 0.0001–0.0003 | 0.0002–0.0004 |
| 0.375 | 0.50 | 400 | 0.0005–0.0008 | 0.0008–0.0011 |
| | 0.63 | 400 | 0.0004–0.0006 | 0.0005–0.0008 |
| | 0.75 | 400 | 0.0003–0.0005 | 0.0004–0.0006 |
| | 0.95 | 400 | 0.0002–0.0004 | 0.0003–0.0005 |

**Selection of Ball and Roller Bearings.**—As compared with sleeve bearings, ball and roller bearings offer the following advantages: 1) starting friction is low; 2) less axial space is required; 3) relatively accurate shaft alignment can be maintained; 4) both radial and axial loads can be carried by certain types; 5) angle of load application is not restricted; 6) replacement is relatively easy; 7) comparatively heavy overloads can be carried momentarily; 8) lubrication is simple; and 9) design and application can be made with the assistance of bearing supplier engineers.

In selecting a ball or roller bearing for a specific application five choices must be made: 1) the bearing series; 2) the type of bearing; 3) the size of bearing; 4) the method of lubrication; and 5) the type of mounting.

Naturally these considerations are modified or affected by the anticipated operating conditions, expected life, cost, and overhaul philosophy.

It is well to review the possible history of the bearing and its function in the machine it will be applied to, thus: 1) Will it be expected to endure removal and reapplication? 2) Must it be free from maintenance attention during its useful life? 3) Can wear of the housing or shaft be tolerated during the overhaul period? 4) Must it be adjustable to take up wear, or to change shaft location? 5) How accurately can the load spectrum be estimated? and 6) Will it be relatively free from abuse in operation?

Though many cautions could be pointed out, it should always be remembered that inadequate design approaches limit the utilization of rolling element bearings, reduce customer satisfaction, and reduce reliability. Time spent in this stage of design is the most rewarding effort of the bearing engineer, who, here again, can depend on the bearing manufacturers' field organization for assistance.

*Type:* Where loads are low, ball bearings are usually less expensive than roller bearings in terms of unit-carrying capacity. Where loads are high, the reverse is usually true.

For a purely radial load, almost any type of radial bearing can be used, the actual choice being determined by other factors. To support a combination of thrust and radial loads, several types of bearings may be considered. If the thrust load component is large, it may be most economical to provide a separate thrust bearing. When a separate thrust bearing cannot be used due to high speed, lack of space, or other factors, the following types may be considered: angular contact ball bearing, deep groove ball bearing without filling slot, tapered roller bearing with steep contact angle, and self-aligning bearing of the wide type. If movement or deflection in an axial direction must be held to a minimum, then a separate thrust bearing or a preloaded bearing capable of taking considerable thrust load is required. To minimize deflection due to a moment in an axial plane, a rigid bearing such as a double row angular contact type with outwardly converging load lines is required. In such cases, the resulting stresses must be taken into consideration in determining the proper size of the bearing.

For shock loads or heavy loads of short duration, roller bearings are usually preferred.

Special bearing designs may be required where accelerations are usually high as in planetary or crank motions.

Where the problem of excessive shaft deflection or misalignment between shaft and housing is present, a self-aligning type of bearing may be a satisfactory solution.

It should be kept in mind that a great deal of difficulty can be avoided if standard types of bearings are used in preference to special designs, wherever possible.

*Size:* The size of bearing required for a given application is determined by the loads that are to be carried and, in some cases, by the amount of rigidity that is necessary to limit deflection to some specified amount.

The forces to which a bearing will be subjected can be calculated by the laws of engineering mechanics from the known loads, power, operating pressure, etc. Where loads are irregular, varying, or of unknown magnitude, it may be difficult to determine the actual forces. In such cases, empirical determination of such forces, based on extensive experience in bearing design, may be needed to attack the problem successfully. Where such experience is lacking, the bearing manufacturer should be consulted or the services of a bearing expert obtained.

If a ball or roller bearing is to be subjected to a combination of radial and thrust loads, an *equivalent radial load* is computed in the case of radial or angular type bearings and an *equivalent thrust load* is computed in the case of thrust bearings.

**Method of Lubrication.**—If speeds are high, relubrication difficult, the shaft angle other than horizontal, the application environment incompatible with normal lubrication, leakage cannot be tolerated; if other elements of the mechanism establish the lubrication requirements, bearing selection must be made with these criteria as controlling influences. Modern bearing types cover a wide selection of lubrication means. Though the most popular type is the "cartridge" type of sealed grease ball bearing, many applications have requirements which dictate against them. Often, operating environments may subject bearings to temperatures too high for seals utilized in the more popular designs. If minute leakage or the accumulation of traces of dirt at seal lips cannot be tolerated by the application (as in baking industry machinery), then the selections of bearings must be made with other sealing and lubrication systems in mind.

High shaft speeds generally dictate bearing selection based on the need for cooling, the suppression of churning or aeration of conventional lubricants, and most important of all, the inherent speed limitations of certain bearing types. An example of the latter is the effect of cage design and of the roller-end thrust-flange contact on the lubrication requirements in commercial taper roller bearings, which limit the speed they can endure and the thrust load they can carry. Reference to the manufacturers' catalog and application-design manuals is recommended before making bearing selections.

See *Selecting a Suitable Lubricant* on page 2516 for more information on this topic.

**Type of Mounting.—**Many bearing installations are complicated because the best adapted mounting type was not selected. Similarly, performance, reliability, and maintenance operations are restricted because the mounting was not thoroughly considered. There is no universally adaptable bearing for all needs. Careful reviews of the machine requirements should be made before designs are implemented. In many cases complicated machining, redundant shaft and housings, and use of an oversize bearing can be eliminated if the proper bearing in a well-thought-out mounting is chosen.

Advantage should be taken of the many race variations available in "standard" series of bearings. Puller grooves, tapered sleeves, ranged outer races, split races, fully demountable rolling-element and cage assemblies, flexible mountings, hydraulic removal features, relubrication holes and grooves, and many other innovations are available beyond the obvious advantages which are inherent in the basic bearing types.

**Radial and Axial Clearance.—**In designing the bearing mounting, a major consideration is to provide running clearances consistent with the requirements of the application. Race fits must be expected to absorb some of the original bearing clearance so that allowance should be made for approximately 80 percent of the actual interference showing up in the diameter of the race. This will increase for heavy, stiff housings or for extra light series races shrunk onto solid shafts, while light metal housings (aluminum, magnesium, or sheet metal) and tubular shafts with wall sections less than the race wall thickness will cause a lesser change in the race diameter.

Where the application will impose heat losses through housing or shaft, or where a temperature differential may be expected, allowances must be made in the proper direction to ensure proper operating clearance. Some compromises are required in applications where the indicated modification cannot be fully accommodated without endangering the bearing performance at lower speeds, during starting, or under lower temperature conditions than anticipated. Some leeway can be relied on with ball bearings since they can run with moderate preloads (0.0005 inch or 12.7 μm, max.) without affecting bearing life or temperature rise. Roller bearings, however, have a lesser tolerance for preloading, and must be carefully controlled to avoid overheating and resulting self-destruction.

In all critical applications axial and radial clearances should be checked with feeler gages or dial indicators to ensure mounted clearances within tolerances established by the design engineer. Since chips, scores, race misalignment, shaft or housing denting, housing distortion, end cover (closure) off-squareness, and mismatch of rotor and housing and wrong axial and radial clearances in the races, rotor, and housing, as well as variations in temperature can all affect bearing clearances. Careful checks of running clearance is recommended.

For precision applications, taper-sleeve mountings, opposed ball or tapered-roller bearings with adjustable or shimmed closures are employed to provide careful control of radial and/or axial clearances. This practice requires skill and experience as well as the initial assistance of the bearing manufacturer's field engineer.

Tapered bore bearings are often used in applications such as these, again requiring careful and well worked-out assembly procedures. They can be assembled on either tapered shafts or on adapter sleeves. The adapter sleeve is supplied with a lock-nut which is used to advance the race on the tapered sleeve. With the heavier fits normally required to effect the clearance changes compatible with such mountings, hydraulic removal devices are normally recommended.

For the conventional application, with standard fits, clearances provided in the standard bearing are suitable for normal operation. To ensure that the design conditions are "normal," a careful review of the application requirements, environments, operating speed range, anticipated abuses, and design parameters must be made.

**General Bearing Handling Precautions.—**To ensure that rolling element bearings are capable of achieving their design life and that they perform without objectionable noise, temperature rise, or shaft excursions, the following precautions are recommended:

1) Use the best bearing available for the application, consistent with the value of the application. Remember, the cost of the best bearing is generally small compared to the replacement costs of the rotating components that can be destroyed if a bearing fails or malfunctions.

2) If questions arise in designing the bearing application, seek out the assistance of the bearing manufacturer's representative.

3) Handle bearings with care, keeping them in the sealed, original container until ready to use.

4) Follow the manufacturer's instructions in handling and assembling the bearings.

5) Work with clean tools, clean dry hands, and in clean surroundings.

6) Do not wash or wipe bearings prior to installation unless special instructions or requirements have been established to do so.

7) Place unwrapped bearings on clean paper and keep them similarly covered until applied, if they cannot be kept in the original container.

8) Don't use wooden mallets, brittle or chipped tools, or dirty fixtures and tools in mounting bearings.

9) Don't spin uncleaned bearings, nor spin *any* bearing with an air blast.

10) Use care not to scratch or nick bearings.

11) Don't strike or press on race flanges.

12) Use adapters for mounting which provide uniform steady pressure rather than hammering on a drift or sleeve.

13) Make sure that races are started onto shafts and into housings evenly so as to prevent cocking.

14) Inspect shafts and housings before mounting bearing to ensure that proper fits will be maintained.

15) When removing bearings, clean housings, covers, and shafts before exposing the bearings. All dirt can be considered an abrasive, dangerous to the reuse of any rolling bearing.

16) Treat used bearings, which may be reused, as new ones.

17) Protect dismantled bearings from dirt and moisture.

18) Use clean, lint-free rags if bearings are wiped.

19) Wrap bearings in clean, oil-proof paper when not in use.

20) Use clean filtered, water-free Stoddard's solvent or flushing oil to clean bearings.

21) In heating bearings for mounting onto shafts, follow manufacturer's instructions.

22) In assembling bearings onto shafts *never* strike the outer race, or press on it to force the inner race. Apply the pressure on the inner race only. In dismantling follow the same precautions.

23) Do not press, strike, or otherwise force the seal or shield on factory-sealed bearings.

**Bearing Failures, Deficiencies, and Their Origins.—**The general classifications of failures and deficiencies requiring bearing removal are:

1) Overheating due to a) inadequate or insufficient lubrication; b) excessive lubrication; c) grease liquefaction or aeration; d) oil foaming; e) abrasive or corrosive action due to contaminants in bearing; f) distortion of housing due to warping, or out-of-round; g) seal rubbing or failure; h) inadequate or blocked scavenge oil passages; i) inadequate bearing-clearance or bearing-preload; j) race turning; k) cage wear; and l) shaft expansion-loss of bearing or seal clearance.

2) Vibration due to a) shaft misalignment; b) race misalignment; c) dirt or chips in bearing; d) fatigued race or rolling elements; e) race turning; f) rotor unbalance; g) out-of-round shaft; h) housing resonance; i) cage wear; j) flats on races or rolling elements; k) excessive clearance; l) corrosion; m) false-brinelling or indentation of races; n) electrical discharge (similar to corrosion effects); o) mixed rolling element diameters; and p) out-of-square rolling paths in races.

3) Turning on shaft due to a) growth of race due to overheating; b) fretting wear; c) improper initial fit; d) excessive shaft deflection; e) initially coarse shaft finish; and f) seal rub on inner race.

4) Binding of the shaft due to a) lubricant breakdown; b) contamination by abrasive or corrosive matter; c) housing distortion or out-of-round pinching bearing; d) uneven shimming of housing with loss of clearance; e) tight rubbing seals; f) preloaded bearings; g) cocked races; h) loss of clearance due to excessive tightening of adapter; i) thermal expansion of shaft or housing; and j) cage failure.

5) Noisy bearing due to a) lubrication breakdown, inadequate lubrication, stiff grease; b) contamination; c) pinched bearing; d) seal rubbing; e) loss of clearance and preloading; f) bearing slipping on shaft or in housing; g) flattened roller or ball; h) brinelling due to assembly abuse, handling, or shock loads; i) variation in size of rolling elements; j) out-of-round or lobular shaft; k) housing bore waviness; l) chips or scores under bearing race seat; and m) excess misalignment.

6) Displaced shaft due to a) bearing wear; b) improper housing or closure assembly; c) overheated and shifted bearing; d) inadequate shaft or housing shoulder; e) lubrication and cage failure permitting rolling elements to bunch; f) loosened retainer nut or adapter; g) excessive heat application in assembling inner race, causing growth and shifting on shaft; and h) housing pounding out.

7) Lubricant leakage due to a) overfilling of lubricant; b) grease churning due to use of too soft a consistency; c) grease deterioration due to excessive operating temperature; d) operating life longer than grease life (grease breakdown, aeration, and purging); e) seal wear; f) wrong shaft attitude (bearing seals designed for horizontal mounting only); g) seal failure; h) clogged breather; i) oil foaming due to churning or air flow through housing; j) gasket (O-ring) failure or misapplication; k) porous housing or closure; and l) lubricator set at wrong flow rate.

## Load Ratings and Fatigue Life

**Ball and Roller Bearing Life.—**The performance of ball and roller bearings is a function of many variables. These include the bearing design, the characteristics of the material from which the bearings are made, the way in which they are manufactured, as well as many variables associated with their application. The only sure way to establish the satisfactory operation of a bearing selected for a specific application is by actual performance in the application. As this is often impractical, another basis is required to estimate the suitability of a particular bearing for a given application. Two factors are taken into consideration: the bearing fatigue life, and its ability to withstand static loading.

*Life Criterion:* Even if a ball or roller bearing is properly mounted, adequately lubricated, protected from foreign matter and not subjected to extreme operating conditions, it can ultimately fatigue. Under ideal conditions, the repeated stresses developed in the contact areas between the balls or rollers and the raceways eventually can result in the fatigue of the material which manifests itself with the spalling of the load-carrying surfaces. In most applications the fatigue life is the maximum useful life of a bearing.

*Static Load Criterion:* A static load is a load acting on a non-rotating bearing. Permanent deformations appear in balls or rollers and raceways under a static load of moderate magnitude and increase gradually with increasing load. The permissible static load is, therefore, dependent upon the permissible magnitude of permanent deformation. It has been found that for ball and roller bearings suitably manufactured from hardened alloy steel, deformations occurring under maximum contact stress of 4,000 megapascals (580,000 pounds per square inch) acting at the center of contact (in the case of roller bearings, of a uniformly loaded roller) do not greatly impair smoothness or friction. Depending on requirements for smoothness of operation, friction, or sound level, higher or lower static load limits may be tolerated.

**Ball Bearing Types Covered.**—American Bearing Manufacturers Association and American National Standard ANSI/ABMA 9-2015 (R2020) sets forth the method of determining ball bearing Rating Life and Static Load Rating and covers the following types:

1) *Radial, deep groove and angular contact ball bearings* whose inner ring raceways have a cross-sectional radius not larger than 52 percent of the ball diameter and whose outer ring raceways have a cross-sectional radius not larger than 53 percent of the ball diameter.

2) *Radial, self-aligning ball bearings* whose inner ring raceways have cross-sectional radii not larger than 53 percent of the ball diameter.

3) *Thrust ball bearings* whose washer raceways have cross-sectional radii not larger than 54 percent of the ball diameter.

4) *Double row, radial and angular contact ball bearings* and double direction thrust ball bearings are presumed to be symmetrical.

**Limitations for Ball Bearings.**—The following limitations apply:

1) *Truncated contact area.* This Standard* may not be safely applied to ball bearings subjected to loading which causes the contact area of the ball with the raceway to be truncated by the raceway shoulder. This limitation depends strongly on details of bearing design which are not standardized.

2) *Material.* This Standard applies only to ball bearings fabricated from hardened good quality steel.

3) *Types.* The $f_c$ factors specified in the basic load rating formulas are valid only for those ball bearing types specified above.

4) *Lubrication.* The Rating Life calculated according to this Standard is based on the assumption that the bearing is adequately lubricated. The determination of adequate lubrication depends upon the bearing application.

5) *Ring support and alignment.* The Rating Life calculated according to this Standard assumes that the bearing inner and outer rings are rigidly supported and the inner and outer ring axes are properly aligned.

6) *Internal clearance.* The radial ball bearing Rating Life calculated according to this Standard is based on the assumption that only a nominal interior clearance occurs in the mounted bearing at operating speed, load and temperature.

7) *High-speed effects.* The Rating Life calculated according to this Standard does not account for high-speed effects such as ball centrifugal forces and gyroscopic moments. These effects tend to diminish fatigue life. Analytical evaluation of these effects frequently requires the use of high-speed digital computation devices and hence is not covered in the Standard.

* All references to "Standard" are to American Bearing Manufacturers Association and American National Standard "Load Ratings and Fatigue Life for Ball Bearings" ANSI/ABMA 9-2015 (R2020).

8) *Groove radii.* If groove radii are smaller than those specified in the bearing types covered, the ability of a bearing to resist fatigue is not improved: however, it is diminished by the use of larger radii.

**Ball Bearing Rating Life.—**According to the American Bearing Manufacturers Association standards the Rating Life $L_{10}$ of a group of apparently identical ball bearings is the life in millions of revolutions that 90 percent of the group will complete or exceed. For a single bearing, $L_{10}$ also refers to the life associated with 90 percent reliability.

*Radial and Angular Contact Ball Bearings:* The magnitude of the Rating Life $L_{10}$ in millions of revolutions, for a radial or angular contact ball bearing application is given by the formula:

$$L_{10} = \left(\frac{C}{P}\right)^3 \quad (1)$$

where $C$ = basic load rating, newtons (pounds). See Formulas (2), (3a) and (3b)

$P$ = equivalent radial load, newtons (pounds). See Formula (4)

**Table 32. Values of $f_c$ for Radial and Angular Contact Ball Bearings**

| $\frac{D\cos\alpha}{d_m}$ | Single Row Radial Contact; Single and Double Row Angular Contact, Groove Type[a] | | Double Row Radial Contact Groove Type | | Self-Aligning | |
|---|---|---|---|---|---|---|
| | Values of $f_c$ | | | | | |
| | Metric[b] | Inch[c] | Metric[b] | Inch[c] | Metric[b] | Inch[c] |
| 0.05 | 46.7 | 3550 | 44.2 | 3360 | 17.3 | 1310 |
| 0.06 | 49.1 | 3730 | 46.5 | 3530 | 18.6 | 1420 |
| 0.07 | 51.1 | 3880 | 48.4 | 3680 | 19.9 | 1510 |
| 0.08 | 52.8 | 4020 | 50.0 | 3810 | 21.1 | 1600 |
| 0.09 | 54.3 | 4130 | 51.4 | 3900 | 22.3 | 1690 |
| 0.10 | 55.5 | 4220 | 52.6 | 4000 | 23.4 | 1770 |
| 0.12 | 57.5 | 4370 | 54.5 | 4140 | 25.6 | 1940 |
| 0.14 | 58.8 | 4470 | 55.7 | 4230 | 27.7 | 2100 |
| 0.16 | 59.6 | 4530 | 56.5 | 4290 | 29.7 | 2260 |
| 0.18 | 59.9 | 4550 | 56.8 | 4310 | 31.7 | 2410 |
| 0.20 | 59.9 | 4550 | 56.8 | 4310 | 33.5 | 2550 |
| 0.22 | 59.6 | 4530 | 56.5 | 4290 | 35.2 | 2680 |
| 0.24 | 59.0 | 4480 | 55.9 | 4250 | 36.8 | 2790 |
| 0.26 | 58.2 | 4420 | 55.1 | 4190 | 38.2 | 2910 |
| 0.28 | 57.1 | 4340 | 54.1 | 4110 | 39.4 | 3000 |
| 0.30 | 56.0 | 4250 | 53.0 | 4030 | 40.3 | 3060 |
| 0.32 | 54.6 | 4160 | 51.8 | 3950 | 40.9 | 3110 |
| 0.34 | 53.2 | 4050 | 50.4 | 3840 | 41.2 | 3130 |
| 0.36 | 51.7 | 3930 | 48.9 | 3730 | 41.3 | 3140 |
| 0.38 | 50.0 | 3800 | 47.4 | 3610 | 41.0 | 3110 |
| 0.40 | 48.4 | 3670 | 45.8 | 3480 | 40.4 | 3070 |

[a] A. When calculating the basic load rating for a unit consisting of two similar, single row, radial contact ball bearings, in a duplex mounting, the pair is considered as one, double row, radial contact ball bearing.

B. When calculating the basic load rating for a unit consisting of two, similar, single row, angular contact ball bearings in a duplex mounting, "face-to-face" or "back-to-back," the pair is considered as one, double row, angular contact ball bearing.

C. When calculating the basic load rating for a unit consisting of two or more similar, single angular contact ball bearings mounted "in tandem," properly manufactured and mounted for equal load distribution, the rating of the combination is the number of bearings to the 0.7 power times the rating of a single row ball bearing. If the unit may be treated as a number of individually interchangeable single row bearings, this footnote "C" does not apply.

[b] **Use to obtain $C$ in newtons when $D$ is given in mm.**

[c] Use to obtain $C$ in pounds when $D$ is given in inches.

## Table 33. Values of *X* and *Y* for Computing Equivalent Radial Load *P* of Radial and Angular Contact Ball Bearings

| Contact Angle, α | Table Entering Factors[a] | | | | Single Row Bearings[b] $\frac{F_a}{F_r} > e$ | | Double Row Bearings $\frac{F_a}{F_r} \le e$ | | Double Row Bearings $\frac{F_a}{F_r} > e$ | |
|---|---|---|---|---|---|---|---|---|---|---|
| RADIAL CONTACT GROOVE BEARINGS | | | | | | | | | | |
| | $F_a/C_o$ | $F_a/iZD^2$ Metric Units | $F_a/iZD^2$ Inch Units | *e* | *X* | *Y* | *X* | *Y* | *X* | *Y* |
| 0° | 0.014 | 0.172 | 25 | 0.19 | 0.56 | 2.30 | 1 | 0 | 0.56 | 2.30 |
| | 0.028 | 0.345 | 50 | 0.22 | | 1.99 | | | | 1.99 |
| | 0.056 | 0.689 | 100 | 0.26 | | 1.71 | | | | 1.71 |
| | 0.084 | 1.03 | 150 | 0.28 | | 1.56 | | | | 1.55 |
| | 0.11 | 1.38 | 200 | 0.30 | | 1.45 | | | | 1.45 |
| | 0.17 | 2.07 | 300 | 0.34 | | 1.31 | | | | 1.31 |
| | 0.28 | 3.45 | 500 | 0.38 | | 1.15 | | | | 1.15 |
| | 0.42 | 5.17 | 750 | 0.42 | | 1.04 | | | | 1.04 |
| | 0.56 | 6.89 | 1000 | 0.44 | | 1.00 | | | | 1.00 |
| ANGULAR CONTACT GROOVE BEARINGS | | | | | | | | | | |
| | $iF_a/C_o$ | $F_a/ZD^2$ Metric Units | $F_a/ZD^2$ Inch Units | *e* | *X* | *Y* | *X* | *Y* | *X* | *Y* |
| 5° | 0.014 | 0.172 | 25 | 0.23 | For this type use the *X*, *Y*, and *e* values applicable to single row radial contact bearings | | 1 | 2.78 | 0.78 | 3.74 |
| | 0.028 | 0.345 | 50 | 0.26 | | | | 2.40 | | 3.23 |
| | 0.056 | 0.689 | 100 | 0.30 | | | | 2.07 | | 2.78 |
| | 0.085 | 1.03 | 150 | 0.34 | | | | 1.87 | | 2.52 |
| | 0.11 | 1.38 | 200 | 0.36 | | | | 1.75 | | 2.36 |
| | 0.17 | 2.07 | 300 | 0.40 | | | | 1.58 | | 2.13 |
| | 0.28 | 3.45 | 500 | 0.45 | | | | 1.39 | | 1.87 |
| | 0.42 | 5.17 | 750 | 0.50 | | | | 1.26 | | 1.69 |
| | 0.56 | 6.89 | 1000 | 0.52 | | | | 1.21 | | 1.63 |
| 10° | 0.014 | 0.172 | 25 | 0.29 | 0.46 | 1.88 | 1 | 2.18 | 0.75 | 3.06 |
| | 0.029 | 0.345 | 50 | 0.32 | | 1.71 | | 1.98 | | 2.78 |
| | 0.057 | 0.689 | 100 | 0.36 | | 1.52 | | 1.76 | | 2.47 |
| | 0.086 | 1.03 | 150 | 0.38 | | 1.41 | | 1.63 | | 2.20 |
| | 0.11 | 1.38 | 200 | 0.40 | | 1.34 | | 1.55 | | 2.18 |
| | 0.17 | 2.07 | 300 | 0.44 | | 1.23 | | 1.42 | | 2.00 |
| | 0.29 | 3.45 | 500 | 0.49 | | 1.10 | | 1.27 | | 1.79 |
| | 0.43 | 5.17 | 750 | 0.54 | | 1.01 | | 1.17 | | 1.64 |
| | 0.57 | 6.89 | 1000 | 0.54 | | 1.00 | | 1.16 | | 1.63 |
| 15° | 0.015 | 0.172 | 25 | 0.38 | 0.44 | 1.47 | 1 | 1.65 | 0.72 | 2.39 |
| | 0.029 | 0.345 | 50 | 0.40 | | 1.40 | | 1.57 | | 2.28 |
| | 0.058 | 0.689 | 100 | 0.43 | | 1.30 | | 1.46 | | 2.11 |
| | 0.087 | 1.03 | 150 | 0.46 | | 1.23 | | 1.38 | | 2.00 |
| | 0.12 | 1.38 | 200 | 0.47 | | 1.19 | | 1.34 | | 1.93 |
| | 0.17 | 2.07 | 300 | 0.50 | | 1.12 | | 1.26 | | 1.82 |
| | 0.29 | 3.45 | 500 | 0.55 | | 1.02 | | 1.14 | | 1.66 |
| | 0.44 | 5.17 | 750 | 0.56 | | 1.00 | | 1.12 | | 1.63 |
| | 0.58 | 6.89 | 1000 | 0.56 | | 1.00 | | 1.12 | | 1.63 |
| 20° | … | … | … | 0.57 | 0.43 | 1.00 | 1 | 1.09 | 0.70 | 1.63 |
| 25° | … | … | … | 0.68 | 0.41 | 0.87 | 1 | 0.92 | 0.67 | 1.41 |
| 30° | … | … | … | 0.80 | 0.39 | 0.76 | 1 | 0.78 | 0.63 | 1.24 |
| 35° | … | … | … | 0.95 | 0.37 | 0.66 | 1 | 0.66 | 0.60 | 1.07 |
| 40° | … | … | … | 1.14 | 0.35 | 0.57 | 1 | 0.55 | 0.57 | 0.98 |
| Self-aligning Ball Bearings | | | | 1.5 tan α | 0.40 | 0.4 cot α | 1 | 0.42 cot α | 0.65 | 0.65 cot α |

[a] Symbol definitions are given on the following page.

[b] For single row bearings when $F_a/F_r \le e$, use $X = 1$, $Y = 0$. Two similar, single row, angular contact ball bearings mounted face-to-face or back-to-back are considered as one double row, angular contact bearing.

Values of *X*, *Y*, and *e* for a load or contact angle other than shown are obtained by linear interpolation. Values of *X*, *Y*, and *e* do not apply to filling slot bearings for applications in which ball-raceway contact areas project substantially into the filling slot under load. *Symbol Definitions:* $F_a$ is the applied axial load in newtons (pounds); $C_o$ is the static load rating in newtons (pounds) of the bearing under consideration and is found by Formula (20); $i$ is the number of rows of balls in the bearing; $Z$ is the number of balls per row in a radial or angular contact bearing or the number of balls in a single row, single direction thrust bearing; $D$ is the ball diameter in millimeters (inches); and $F_r$ is the applied radial load in newtons (pounds).

For radial and angular contact ball bearings with balls not larger than 25.4 mm (1 inch) in diameter, $C$ is found by the formula:

$$C = f_c (i \cos\alpha)^{0.7} Z^{2/3} D^{1.8} \tag{2}$$

and with balls larger than 25.4 mm (1 inch) in diameter $C$ is found by the formula:

$$\mathbf{C = 3.647 f_c (i \cos\alpha)^{0.7} Z^{2/3} D^{1.4}} \quad \textbf{(metric)} \tag{3a}$$

$$C = f_c (i \cos\alpha)^{0.7} Z^{2/3} D^{1.4} \text{(inch)} \tag{3b}$$

where $f_c$ = a factor which depends on the geometry of the bearing components, the accuracy to which the various bearing parts are made and the material. Values of $f_c$, are given in Table 32

$i$ = number of rows of balls in the bearing

$\alpha$ = nominal contact angle, degrees

$Z$ = number of balls per row in a radial or angular contact bearing

$D$ = ball diameter, mm (inches)

The magnitude of the equivalent radial load, $P$, in newtons (pounds) for radial and angular contact ball bearings, under combined constant radial and constant thrust loads is given by the formula:

$$P = XF_r + YF_a \tag{4}$$

where $F_r$ = the applied radial load in newtons (pounds)

$F_a$ = the applied axial load in newtons (pounds)

$X$ = radial load factor as given in Table 35

$Y$ = axial load factor as given in Table 35

*Thrust Ball Bearings:* The magnitude of the Rating Life $L_{10}$ in millions of revolutions for a thrust ball bearing application is given by the formula:

$$L_{10} = \left(\frac{C_a}{P_a}\right)^3 \tag{5}$$

where $C_a$ = the basic load rating, newtons (pounds). See Formulas (6) to (10)

$P_a$ = equivalent thrust load, newtons (pounds). See Formula (11)

For single row, single and double direction, thrust ball bearing with balls not larger than 25.4 mm (1 inch) in diameter, $C_a$ is found by the formulas:

$$\text{for } \alpha = 90 \text{ degrees,} \quad C_a = f_c Z^{2/3} D^{1.8} \tag{6}$$

$$\text{for } \alpha \neq 90 \text{ degrees,} \quad C_a = f_c (\cos\alpha)^{0.7} Z^{2/3} D^{1.8} \tan\alpha \tag{7}$$

and with balls larger than 25.4 mm (1 inch) in diameter, $C_a$ is found by the formulas:

$$\textbf{for } \boldsymbol{\alpha} \textbf{ = 90 degrees,} \quad \mathbf{C_a = 3.647 f_c Z^{2/3} D^{1.4}} \quad \textbf{(metric)} \tag{8a}$$

$$C_a = f_c Z^{2/3} D^{1.4} \text{(inch)} \tag{8b}$$

$$\textbf{for } \boldsymbol{\alpha \neq} \textbf{ 90 degrees,} \quad \mathbf{C_a = 3.647 f_c (\cos\alpha)^{0.7} Z^{2/3} D^{1.4} \tan\alpha} \quad \textbf{(metric)} \tag{9a}$$

$$C_a = f_c (\cos\alpha)^{0.7} Z^{2/3} D^{1.4} \tan\alpha \text{ (inch)} \tag{9b}$$

where $f_c$ = a factor which depends on the geometry of the bearing components, the accuracy to which the various bearing parts are made, and the material. Values of $f_c$ are given in Table 34

$Z$ = number of balls per row in a single row, single direction thrust ball bearing

$D$ = ball diameter, mm (inches)

$\alpha$ = nominal contact angle, degrees

**Table 34. Values of $f_c$ for Thrust Ball Bearings**

| $\frac{D}{d_m}$ | α = 90° | | | α = 45° | | α = 60° | | α = 75° | |
|---|---|---|---|---|---|---|---|---|---|
| | Metric[a] | Inch[b] | $D \cos \alpha$ | Metric[a] | Inch[b] | Metric[a] | Inch[b] | Metric[a] | Inch[b] |
| 0.01 | 36.7 | 2790 | 0.01 | 42.1 | 3200 | 39.2 | 2970 | 37.3 | 2840 |
| 0.02 | 45.2 | 3430 | 0.02 | 51.7 | 3930 | 48.1 | 3650 | 45.9 | 3490 |
| 0.03 | 51.1 | 3880 | 0.03 | 58.2 | 4430 | 54.2 | 4120 | 51.7 | 3930 |
| 0.04 | 55.7 | 4230 | 0.04 | 63.3 | 4810 | 58.9 | 4470 | 56.1 | 4260 |
| 0.05 | 59.5 | 4520 | 0.05 | 67.3 | 5110 | 62.6 | 4760 | 59.7 | 4540 |
| 0.06 | 62.9 | 4780 | 0.06 | 70.7 | 5360 | 65.8 | 4990 | 62.7 | 4760 |
| 0.07 | 65.8 | 5000 | 0.07 | 73.5 | 5580 | 68.4 | 5190 | 65.2 | 4950 |
| 0.08 | 68.5 | 5210 | 0.08 | 75.9 | 5770 | 70.7 | 5360 | 67.3 | 5120 |
| 0.09 | 71.0 | 5390 | 0.09 | 78.0 | 5920 | 72.6 | 5510 | 69.2 | 5250 |
| 0.10 | 73.3 | 5570 | 0.10 | 79.7 | 6050 | 74.2 | 5630 | 70.7 | 5370 |
| 0.12 | 77.4 | 5880 | 0.12 | 82.3 | 6260 | 76.6 | 5830 | … | … |
| 0.14 | 81.1 | 6160 | 0.14 | 84.1 | 6390 | 78.3 | 5950 | … | … |
| 0.16 | 84.4 | 6410 | 0.16 | 85.1 | 6470 | 79.2 | 6020 | … | … |
| 0.18 | 87.4 | 6640 | 0.18 | 85.5 | 6500 | 79.6 | 6050 | … | … |
| 0.20 | 90.2 | 6854 | 0.20 | 85.4 | 6490 | 79.5 | 6040 | … | … |
| 0.22 | 92.8 | 7060 | 0.22 | 84.9 | 6450 | … | … | … | … |
| 0.24 | 95.3 | 7240 | 0.24 | 84.0 | 6380 | … | … | … | … |
| 0.26 | 97.6 | 7410 | 0.26 | 82.8 | 6290 | … | … | … | … |
| 0.28 | 99.8 | 7600 | 0.28 | 81.3 | 6180 | … | … | … | … |
| 0.30 | 101.9 | 7750 | 0.30 | 79.6 | 6040 | … | … | … | … |
| 0.32 | 103.9 | 7900 | … | … | … | … | … | … | … |
| 0.34 | 105.8 | 8050 | … | … | … | … | … | … | … |

[a] **Use to obtain $C_a$ in newtons when $D$ is given in mm.**

[b] Use to obtain $C_a$ in pounds when $D$ is given in inches.

For thrust ball bearings with two or more rows of similar balls carrying loads in the same direction, the basic load rating, $C_a$, in newtons (pounds) is found by the formula:

$$C_a = (Z_1 + Z_2 + \ldots + Z_n)\left[\left(\frac{Z_1}{C_{a1}}\right)^{10/3} + \left(\frac{Z_2}{C_{a2}}\right)^{10/3} + \ldots + \left(\frac{Z_n}{C_{an}}\right)^{10/3}\right]^{-0.3} \quad (10)$$

where $Z_1, Z_2, \ldots, Z_n$ = number of balls in respective rows of a single direction multi-row thrust ball bearing

$C_{a1}, C_{a2}, \ldots, C_{an}$ = basic load rating per row of a single direction, multi-row thrust ball bearing, each calculated as a single row bearing with $Z_1, Z_2, \ldots, Z_n$ balls, respectively

The magnitude of the equivalent thrust load, $P_a$, in newtons (pounds) for thrust ball bearings with $\alpha \neq 90$ degrees under combined constant thrust and constant radial loads is found by the formula:

$$P_a = XF_r + YF_a \quad (11)$$

where $F_r$ = the applied radial load in newtons (pounds)

$F_a$ = the applied axial load in newtons (pounds)

$X$ = radial load factor as given in Table 35

$Y$ = axial load factor as given in Table 35

**Table 35. Values of $X$ and $Y$ for Computing Equivalent Thrust Load $P_a$ for Thrust Ball Bearings**

| Contact Angle | | Single Direction Bearings | | Double Direction Bearings | | | |
|---|---|---|---|---|---|---|---|
| | | $\frac{F_a}{F_r} > e$ | | $\frac{F_a}{F_r} \le e$ | | $\frac{F_a}{F_r} > e$ | |
| $\alpha$ | $e$ | $X$ | $Y$ | $X$ | $Y$ | $X$ | $Y$ |
| 45° | 1.25 | 0.66 | 1 | 1.18 | 0.59 | 0.66 | 1 |
| 60° | 2.17 | 0.92 | 1 | 1.90 | 0.54 | 0.92 | 1 |
| 75° | 4.67 | 1.66 | 1 | 3.89 | 0.52 | 1.66 | 1 |

For $\alpha = 90°$, $F_r = 0$ and $Y = 1$.

**Roller Bearing Types Covered.—**This Standard* applies to *cylindrical, tapered and self-aligning radial and thrust roller bearings* and to *needle roller bearings*. These bearings are presumed to be within the size ranges shown in the ABMA dimensional standards, of good quality and produced in accordance with good manufacturing practice.

Roller bearings vary considerably in design and execution. Since small differences in relative shape of contacting surfaces may account for distinct differences in load-carrying ability, this Standard does not attempt to cover all design variations, rather it applies to basic roller bearing designs.

The following limitations apply:

1) *Truncated contact area*. This Standard may not be safely applied to roller bearings subjected to application conditions which cause the contact area of the roller with the raceway to be severely truncated by the edge of the raceway or roller.

2) *Stress concentrations*. A cylindrical, tapered or self-aligning roller bearing must be expected to have a basic load rating less than that obtained using a value of $f_c$ taken from Table 36 and Table 37 if, under load, a stress concentration is present in some part of the roller-raceway contact. Such stress concentrations occur in the center of nominal point contacts, at the contact extremities for line contacts and at inadequately blended junctions of a rolling surface profile. Stress concentrations can also occur if the rollers are not accurately guided such as in bearings without cages and bearings not having rigid integral flanges. Values of $f_c$ given in Table 36 and Table 37 are based upon bearings manufactured to achieve optimized contact. For no bearing type or execution will the factor $f_c$ be greater than that obtained in Table 36 and Table 37.

3) *Material*. This Standard applies only to roller bearings fabricated from hardened, good quality steel.

4) *Lubrication*. Rating Life calculated according to this Standard is based on the assumption that the bearing is adequately lubricated. Determination of adequate lubrication depends upon the bearing application.

5) *Ring support and alignment*. Rating Life calculated according to this Standard assumes that the bearing inner and outer rings are rigidly supported, and that the inner and outer ring axes are properly aligned.

6) *Internal clearance*. Radial roller bearing Rating Life calculated according to this Standard is based on the assumption that only a nominal internal clearance occurs in the mounted bearing at operating speed, load, and temperature.

7) *High-speed effects*. The Rating Life calculated according to this Standard does not account for high speed effects such as roller centrifugal forces and gyroscopic moments: These effects tend to diminish fatigue life. Analytical evaluation of these effects frequently requires the use of high-speed digital computation devices and hence, cannot be included.

* All references to "Standard" are to American Bearings Manufacturers Association and American National Standard "Load Ratings and Fatigue Life for Roller Bearings" ANSI/ABMA 11-2014 (R2020).

### Table 36. Values of $f_c$ for Radial Roller Bearings

| $\frac{D\cos\alpha}{d_m}$ | $f_c$ | | $\frac{D\cos\alpha}{d_m}$ | $f_c$ | | $\frac{D\cos\alpha}{d_m}$ | $f_c$ | |
|---|---|---|---|---|---|---|---|---|
| | Metric[a] | Inch[b] | | Metric[a] | Inch[b] | | Metric[a] | Inch[b] |
| 0.01 | 52.1 | 4680 | 0.18 | 88.8 | 7980 | 0.35 | 79.5 | 7140 |
| 0.02 | 60.8 | 5460 | 0.19 | 88.8 | 7980 | 0.36 | 78.6 | 7060 |
| 0.03 | 66.5 | 5970 | 0.20 | 88.7 | 7970 | 0.37 | 77.6 | 6970 |
| 0.04 | 70.7 | 6350 | 0.21 | 88.5 | 7950 | 0.38 | 76.7 | 6890 |
| 0.05 | 74.1 | 6660 | 0.22 | 88.2 | 7920 | 0.39 | 75.7 | 6800 |
| 0.06 | 76.9 | 6910 | 0.23 | 87.9 | 7890 | 0.40 | 74.6 | 6700 |
| 0.07 | 79.2 | 7120 | 0.24 | 87.5 | 7850 | 0.41 | 73.6 | 6610 |
| 0.08 | 81.2 | 7290 | 0.25 | 87.0 | 7810 | 0.42 | 72.5 | 6510 |
| 0.09 | 82.8 | 7440 | 0.26 | 86.4 | 7760 | 0.43 | 71.4 | 6420 |
| 0.10 | 84.2 | 7570 | 0.27 | 85.8 | 7710 | 0.44 | 70.3 | 6320 |
| 0.11 | 85.4 | 7670 | 0.28 | 85.2 | 7650 | 0.45 | 69.2 | 6220 |
| 0.12 | 86.4 | 7760 | 0.29 | 84.5 | 7590 | 0.46 | 68.1 | 6120 |
| 0.13 | 87.1 | 7830 | 0.30 | 83.8 | 7520 | 0.47 | 67.0 | 6010 |
| 0.14 | 87.7 | 7880 | 0.31 | 83.0 | 7450 | 0.48 | 65.8 | 5910 |
| 0.15 | 88.2 | 7920 | 0.32 | 82.2 | 7380 | 0.49 | 64.6 | 5810 |
| 0.16 | 88.5 | 7950 | 0.33 | 81.3 | 7300 | 0.50 | 63.5 | 5700 |
| 0.17 | 88.7 | 7970 | 0.34 | 80.4 | 7230 | … | … | … |

[a] For $\alpha = 0°$, $F_a = 0$ and $X = 1$.

[b] Use to obtain $C$ in pounds when $l_{eff}$ and $D$ are given in inches.

### Table 37. Values of $f_c$ for Thrust Roller Bearings

| $\frac{D\cos\alpha}{d_m}$ | $45° < \alpha < 60°$ | | $60° < \alpha < 75°$ | | $75° \le \alpha < 90°$ | | $\frac{D}{d_m}$ | $\alpha = 90°$ | |
|---|---|---|---|---|---|---|---|---|---|
| | $f_c$ | | | | | | | $f_c$ | |
| | Metric[a] | Inch[b] | Metric[a] | Inch[b] | Metric[a] | Inch[b] | | Metric[a] | Inch[b] |
| 0.01 | 109.7 | 9840 | 107.1 | 9610 | 105.6 | 9470 | 0.01 | 105.4 | 9500 |
| 0.02 | 127.8 | 11460 | 124.7 | 11180 | 123.0 | 11030 | 0.02 | 122.9 | 11000 |
| 0.03 | 139.5 | 12510 | 136.2 | 12220 | 134.3 | 12050 | 0.03 | 134.5 | 12100 |
| 0.04 | 148.3 | 13300 | 144.7 | 12980 | 142.8 | 12810 | 0.04 | 143.4 | 12800 |
| 0.05 | 155.2 | 13920 | 151.5 | 13590 | 149.4 | 13400 | 0.05 | 150.7 | 13200 |
| 0.06 | 160.9 | 14430 | 157.0 | 14080 | 154.9 | 13890 | 0.06 | 156.9 | 14100 |
| 0.07 | 165.6 | 14850 | 161.6 | 14490 | 159.4 | 14300 | 0.07 | 162.4 | 14500 |
| 0.08 | 169.5 | 15200 | 165.5 | 14840 | 163.2 | 14640 | 0.08 | 167.2 | 15100 |
| 0.09 | 172.8 | 15500 | 168.7 | 15130 | 166.4 | 14930 | 0.09 | 171.7 | 15400 |
| 0.10 | 175.5 | 15740 | 171.4 | 15370 | 169.0 | 15160 | 0.10 | 175.7 | 15900 |
| 0.12 | 179.7 | 16120 | 175.4 | 15730 | 173.0 | 15520 | 0.12 | 183.0 | 16300 |
| 0.14 | 182.3 | 16350 | 177.9 | 15960 | 175.5 | 15740 | 0.14 | 189.4 | 17000 |
| 0.16 | 183.7 | 16480 | 179.3 | 16080 | … | … | 0.16 | 195.1 | 17500 |
| 0.18 | 184.1 | 16510 | 179.7 | 16120 | … | … | 0.18 | 200.3 | 18000 |
| 0.20 | 183.7 | 16480 | 179.3 | 16080 | … | … | 0.20 | 205.0 | 18500 |
| 0.22 | 182.6 | 16380 | … | … | … | … | 0.22 | 209.4 | 18800 |
| 0.24 | 180.9 | 16230 | … | … | … | … | 0.24 | 213.5 | 19100 |
| 0.26 | 178.7 | 16030 | … | … | … | … | 0.26 | 217.3 | 19600 |
| 0.28 | … | … | … | … | … | … | 0.28 | 220.9 | 19900 |
| 0.30 | … | … | … | … | … | … | 0.30 | 224.3 | 20100 |

**[a] Use to obtain $C_a$ in newtons when $l_{eff}$ and $D$ are given in mm.**

[b] Use to obtain $C_a$ in pounds when $l_{eff}$ and $D$ are given in inches.

**Roller Bearing Rating Life.**—The Rating Life $L_{10}$ of a group of apparently identical roller bearings is the life in millions of revolutions that 90 percent of the group will complete or exceed. For a single bearing, $L_{10}$ also refers to the life associated with 90 percent reliability.

*Radial Roller Bearings:* The magnitude of the Rating Life, $L_{10}$, in millions of revolutions, for a radial roller bearing application is given by the formula:

$$L_{10} = \left(\frac{C}{P}\right)^{10/3} \qquad (12)$$

where $C$ = the basic load rating in newtons (pounds), see Formula (13); and, $P$ = equivalent radial load in newtons (pounds), see Formula (14).

For radial roller bearings, $C$ is found by the formula:

$$C = f_c\,(il_{eff}\cos\alpha)^{7/9} Z^{3/4} D^{29/27} \qquad (13)$$

where $f_c$ = a factor which depends on the geometry of the bearing components, the accuracy to which the various bearing parts are made, and the material. Maximum values of $f_c$ are given in Table 36

$i$ = number of rows of rollers in the bearing

$l_{eff}$ = effective length, mm (inches) $\alpha$ = nominal contact angle, degrees

$Z$ = number of rollers per row in a radial roller bearing

$D$ = roller diameter, mm (inches) (mean diameter for a tapered roller, major diameter for a spherical roller)

When rollers are longer than 2.5$D$, a reduction in the $f_c$ value must be anticipated. In this case, the bearing manufacturer may be expected to establish load ratings accordingly.

In applications where rollers operate directly on a shaft surface or a housing surface, such a surface must be equivalent in all respects to the raceway it replaces to achieve the basic load rating of the bearing.

When calculating the basic load rating for a unit consisting of two or more similar single row bearings mounted "in tandem," properly manufactured and mounted for equal load distribution, the rating of the combination is the number of bearings to the 7/9 power times the rating of a single row bearing. If, for some technical reason, the unit may be treated as a number of individually interchangeable single row bearings, this consideration does not apply.

The magnitude of the equivalent radial load, $P$, in newtons (pounds), for radial roller bearings, under combined constant radial and constant thrust loads is given by the formula:

$$P = XF_r + YF_a \qquad (14)$$

where $F_r$ = the applied radial load in newtons (pounds)

$F_a$ = the applied axial load in newtons (pounds)

$X$ = radial load factor as given in Table 38

$Y$ = axial load factor as given in Table 38

**Table 38. Values of *X* and *Y* for Computing Equivalent Radial Load *P* for Radial Roller Bearing**

| | $\frac{F_a}{F_r} \le e$ | | $\frac{F_a}{F_r} > e$ | |
|---|---|---|---|---|
| Bearing Type | $X$ | $Y$ | $X$ | $Y$ |
| Self-Aligning and Tapered Roller Bearings[a] $\alpha \neq 0°$ | Single Row Bearings | | | |
| | 1 | 0 | 0.4 | 0.4 cot $\alpha$ |
| | Double Row Bearings[a] | | | |
| | 1 | 0.45 cot $\alpha$ | 0.67 | 0.67 cot $\alpha$ |

[a] For $\alpha = 0°$, $F_a = 0$ and $X = 1$.

$e = 1.5 \tan\alpha$

**Table 39. Typical Bearing Life for Various Design Applications**

| Uses | Design life in hours | Uses | Design life in hours |
|---|---|---|---|
| Agricultural equipment | 3000–6000 | Gearing units | |
| Aircraft equipment | 500–2000 | Automotive | 600–5000 |
| Automotive | | Multipurpose | 8000–15000 |
| Race car | 500–800 | Machine tools | 20000 |
| Light motorcycle | 600–1200 | Rail Vehicles | 15000–25000 |
| Heavy motorcycle | 1000–2000 | Heavy rolling mill | > 50000 |
| Light cars | 1000–2000 | Machines | |
| Heavy cars | 1500–2500 | Beater mills | 20000–30000 |
| Light trucks | 1500–2500 | Briquette presses | 20000–30000 |
| Heavy trucks | 2000–2500 | Grinding spindles | 1000–2000 |
| Buses | 2000–5000 | Machine tools | 10000–30000 |
| Electrical | | Mining machinery | 4000–15000 |
| Household appliances | 1000–2000 | Paper machines | 50000–80000 |
| Motors ≤ ½ hp | 1000–2000 | Rolling mills | |
| Motors ≤ 3 hp | 8000–10000 | Small cold mills | 5000–6000 |
| Motors, medium | 10000–15000 | Large multipurpose mills | 8000–10000 |
| Motors, large | 20000–30000 | Rail vehicle axle | |
| Elevator cables sheaves | 40000–60000 | Mining cars | 5000 |
| Mine ventilation fans | 40000–50000 | Motor rail cars | 16000–20000 |
| Propeller thrust bearings | 15000–25000 | Open-pit mining cars | 20000–25000 |
| Propeller shaft bearings | > 80000 | Streetcars | 20000–25000 |
| Gear drives | | Passenger cars | 26000 |
| Boat gearing units | 3000–5000 | Freight cars | 35000 |
| Gear drives | > 50000 | Locomotive outer bearings | 20000–25000 |
| Ship gear drives | 20000–30000 | Locomotive inner bearings | 30000–40000 |
| Machinery for 8 hour service which are not always fully utilized | 14000–20000 | Machinery for short or intermittent operation where service interruption is of minor importance | 4000–8000 |
| Machinery for 8 hour service which are fully utilized | 20000–30000 | Machinery for intermittent service where reliable operation is of great importance | 8000–14000 |
| Machinery for continuous 24 hour service | 50000–60000 | Instruments and apparatus in frequent use | 0–500 |

Roller bearings are generally designed to achieve optimized contact; however, they usually support loads other than the loading at which optimized contact is maintained. The 10/3 exponent in Rating Life Formulas (12) and (15) was selected to yield satisfactory Rating Life estimates for a broad spectrum from light to heavy loading. When loading exceeds that which develops optimized contact, e.g., loading greater than $C/4$ to $C/2$ or $C_a/4$ to $C_a/2$, the user should consult the bearing manufacturer to establish the adequacy of the Rating Life formulas for the particular application.

*Thrust Roller Bearings:* The magnitude of the Rating Life, $L_{10}$, in millions of revolutions for a thrust roller bearing application is given by the formula:

$$L_{10} = \left(\frac{C_a}{P_a}\right)^{10/3} \tag{15}$$

where $C_a$ = basic load rating, newtons (pounds). See Formulas (16) to (18).

$P_a$ = equivalent thrust load, newtons (pounds). See Formula (19).

For single row, single and double direction, thrust roller bearings, the magnitude of the basic load rating, $C_a$, in newtons (pounds), is found by the formulas:

$$\text{for } \alpha = 90°, \quad C_a = f_c l_{eff}^{7/9} Z^{3/4} D^{29/27} \tag{16}$$

$$\text{for } \alpha \neq 90°, \quad C_a = f_c (l_{eff} \cos\alpha)^{7/9} Z^{3/4} D^{29/27} \tan\alpha \tag{17}$$

where $f_c$ = a factor which depends on the geometry of the bearing components, the accuracy to which the various parts are made, and the material. Values of $f_c$ are given in Table 37.

$l_{eff}$ = effective length, mm (inches)

$Z$ = number of rollers in a single row, single direction, thrust roller bearing

$D$ = roller diameter, mm (inches) (mean diameter for a tapered roller, major diameter for a spherical roller)

$\alpha$ = nominal contact angle, degrees

For thrust roller bearings with two or more rows of rollers carrying loads in the same direction the magnitude of $C_a$ is found by the formula:

$$C_a = (Z_1 l_{eff1} + Z_2 l_{eff2} + \ldots + Z_n l_{effn}) \left\{ \left[\frac{Z_1 l_{eff1}}{C_{a1}}\right]^{9/2} + \left[\frac{Z_2 l_{eff2}}{C_{a2}}\right]^{9/2} + \ldots + \left[\frac{Z_n l_{effn}}{C_{an}}\right]^{9/2} \right\}^{-2/9} \tag{18}$$

where $Z_1, Z_2, \ldots, Z_n$ = the number of rollers in respective rows of a single direction, multi-row bearing

$C_{a1}, C_{a2}, \ldots, C_{an}$ = the basic load rating per row of a single direction, multi-row, thrust roller bearing, each calculated as a single row bearing with $Z_1, Z_2, \ldots, Z_n$ rollers respectively

$l_{eff1}, l_{eff2}, \ldots, l_{effn}$ = effective length, mm (inches), or rollers in the respective rows

In applications where rollers operate directly on a surface supplied by the user, such a surface must be equivalent in all respects to the washer raceway it replaces to achieve the basic load rating of the bearing.

In case the bearing is so designed that several rollers are located on a common axis, these rollers are considered as one roller of a length equal to the total effective length of contact of the several rollers. Rollers as defined above, or portions thereof which contact the same washer-raceway area, belong to one row.

When the ratio of the individual roller effective length to the pitch diameter (at which this roller operates) is too large, a reduction of the $f_c$ value must be anticipated due to excessive slip in the roller-raceway contact.

When calculating the basic load rating for a unit consisting of two or more similar single row bearings mounted "in tandem," properly manufactured and mounted for equal load distribution, the rating of the combination is defined by Formula (18). If, for some technical reason, the unit may be treated as a number of individually interchangeable single row bearings, this consideration does not apply.

The magnitude of the equivalent thrust load, $P_a$, in pounds, for thrust roller bearings with $\alpha$ not equal to 90 degrees under combined constant thrust and constant radial loads is given by the formula:

$$P_a = XF_r + YF_a \tag{19}$$

where $F_r$ = applied radial load, newtons (pounds)

$F_a$ = applied axial load, newtons (pounds)

$X$ = radial load factor as given in Table 40

$Y$ = axial load factor as given in Table 40

**Table 40. Values of $X$ and $Y$ for Computing Equivalent Thrust Load $P_a$ for Thrust Roller Bearings**

| Bearing Type | Single Direction Bearings | | Double Direction Bearings | | | |
|---|---|---|---|---|---|---|
| | $\frac{F_a}{F_r} > e$ | | $\frac{F_a}{F_r} \le e$ | | $\frac{F_a}{F_r} > e$ | |
| | $X$ | $Y$ | $X$ | $Y$ | $X$ | $Y$ |
| Self-Aligning Tapered Thrust Roller Bearings[a] $\alpha \neq 0$ | tan α | 1 | 1.5 tan α | 0.67 | tan α | 1 |

[a] For $\alpha = 90°$, $F_r = 0$ and $Y = 1$.

$e = 1.5 \tan \alpha$

**Life Adjustment Factors.—**In certain applications of ball or roller bearings it is desirable to specify life for a reliability other than 90 percent. In other cases the bearings may be fabricated from special bearing steels such as vacuum-degassed and vacuum-melted steels, and improved processing techniques. Finally, application conditions may indicate other than normal lubrication, load distribution, or temperature. For such conditions a series of life adjustment factors may be applied to the fatigue life formula. This is fully explained in ABMA and American National Standard "Load Ratings and Fatigue Life for Ball Bearings" ANSI/ABMA 9-2015 (R2020) and ABMA and American National Standard "Load Ratings and Fatigue Life for Roller Bearings" ANSI/ABMA 11-2014 (R2020). In addition to consulting these standards it may be advantageous to also obtain information from the bearing manufacturer.

*Life Adjustment Factor for Reliability:* For certain applications, it is desirable to specify life for a reliability greater than 90 percent which is the basis of the Rating Life.

To determine the bearing life of ball or roller bearings for reliability greater than 90 percent, the Rating Life must be adjusted by a factor $a_1$ such that $L_n = a_1 L_{10}$. For a reliability of 95 percent, designated as $L_5$, the life adjustment factor $a_1$ is 0.62; for 96 percent, $L_4$, $a_1$ is 0.53; for 97 percent, $L_3$, $a_1$ is 0.44; for 98 percent, $L_2$, $a_1$ is 0.33; and for 99 percent, $L_1$, $a_1$ is 0.21.

*Life Adjustment Factor for Material:* For certain types of ball or roller bearings which incorporate improved materials and processing, the Rating Life can be adjusted by a factor $a_2$ such that $L_{10}' = a_2 L_{10}$. Factor $a_2$ depends upon steel analysis, metallurgical processes, forming methods, heat treatment, and manufacturing methods in general. Ball and roller bearings fabricated from consumable vacuum remelted steels and certain other special analysis steels, have demonstrated extraordinarily long endurance. These steels are of exceptionally high quality, and bearings fabricated from these are usually considered special manufacture. Generally, $a_2$ values for such steels can be obtained from the bearing manufacturer. However, all of the specified limitations and qualifications for the application of the Rating Life formulas still apply.

*Life Adjustment Factor for Application Condition:* Application conditions which affect ball or roller bearing life include: 1) lubrication; 2) load distribution (including effects of clearance, misalignment, housing and shaft stiffness, type of loading, and thermal gradients); and 3) temperature.

Items 2 and 3 require special analytical and experimental techniques, therefore the user should consult the bearing manufacturer for evaluations and recommendations.

Operating conditions where the factor $a_3$ might be less than 1 include: a) exceptionally low values of $Nd_m$ (rpm times pitch diameter, in mm); e.g., $Nd_m < 10{,}000$; b) lubricant viscosity at less than 70 SSU for ball bearings and 100 SSU for roller bearings at operating temperature; and c) excessively high operating temperatures.

When $a_3$ is less than 1 it may not be assumed that the deficiency in lubrication can be overcome by using an improved steel. When this factor is applied, $L_{10}' = a_3 L_{10}$.

In most ball and roller bearing applications, lubrication is required to separate the rolling surfaces, i.e., rollers and raceways, to reduce the retainer-roller and retainer-land friction and sometimes to act as a coolant to remove heat generated by the bearing.

*Factor Combinations:* A fatigue life formula embodying the foregoing life adjustment factors is $L_{10}' = a_1 a_2 a_3 L_{10}$. Indiscriminate application of the life adjustment factors in this formula may lead to serious overestimation of bearing endurance, since fatigue life is only one criterion for bearing selection. Care must be exercised to select bearings which are of sufficient size for the application.

**Ball Bearing Static Load Rating.—**For ball bearings suitably manufactured from hardened alloy steels, the static radial load rating is that uniformly distributed static radial bearing load which produces a maximum contact stress of 4,000 megapascals (580,000 pounds per square inch). In the case of a single row, angular contact ball bearing, the static radial load rating refers to the radial component of that load which causes a purely radial displacement of the bearing rings in relation to each other. The static axial load rating is that uniformly distributed static centric axial load which produces a maximum contact stress of 4,000 megapascals (580,000 pounds per square inch).

*Radial and Angular Contact Groove Ball Bearings:* The magnitude of the static load rating $C_o$ in newtons (pounds) for radial ball bearings is found by the formula:

$$C_o = f_o iZD^2 \cos\alpha \tag{20}$$

where $f_o$ = a factor for different kinds of ball bearings given in Table 41

$i$ = number of rows of balls in bearing

$Z$ = number of balls per row

$D$ = ball diameter, mm (inches)

$\alpha$ = nominal contact angle, degrees

This formula applies to bearings with a cross sectional raceway groove radius not larger than 0.52$D$ in radial and angular contact groove ball bearing inner rings and 0.53$D$ in radial and angular contact groove ball bearing outer rings and self-aligning ball bearing inner rings.

The load-carrying ability of a ball bearing is not necessarily increased by the use of a smaller groove radius but is reduced by the use of a larger radius than those indicated above.

*Radial or Angular Contact Ball Bearing Combinations:* The basic static load rating for two similar single row radial or angular contact ball bearings mounted side by side on the same shaft such that they operate as a unit (duplex mounting) in "back-to-back" or "face-to-face" arrangement is two times the rating of one single row bearing.

The basic static radial load rating for two or more single row radial or angular contact ball bearings mounted side by side on the same shaft such that they operate as a unit (duplex or stack mounting) in "tandem" arrangement, properly manufactured and mounted for equal load distribution, is the number of bearings times the rating of one single row bearing.

*Thrust Ball Bearings:* The magnitude of the static load rating $C_{oa}$ for thrust ball bearings is found by the formula:

$$C_{oa} = f_o ZD^2 \cos\alpha \tag{21}$$

where $f_o$ = a factor given in Table 41

$Z$ = number of balls carrying the load in one direction

$D$ = ball diameter, mm (inches)

$\alpha$ = nominal contact angle, degrees

This formula applies to thrust ball bearings with a cross sectional raceway radius not larger than 0.54*D*. The load-carrying ability of a bearing is not necessarily increased by use of a smaller radius, but is reduced by use of a larger radius.

*Roller Bearing Static Load Rating:* For roller bearings suitably manufactured from hardened alloy steels, the static radial load rating is that uniformly distributed static radial bearing load which produces a maximum contact stress of 4,000 megapascals (580,000 pounds per square inch) acting at the center of contact of the most heavily loaded rolling element. The static axial load rating is that uniformly distributed static centric axial load which produces a maximum contact stress of 4,000 megapascals (580,000 pounds per square inch) acting at the center of contact of each rolling element.

**Table 41. $f_o$ for Calculating Static Load Rating for Ball Bearings**

| $\frac{D\cos\alpha}{d_m}$ | Radial and Angular Contact Groove Type | | Radial Self-Aligning | | Thrust | |
|---|---|---|---|---|---|---|
| | Metric[a] | Inch[b] | Metric[a] | Inch[b] | Metric[a] | Inch[b] |
| 0.00 | 12.7 | 1850 | 1.3 | 187 | 51.9 | 7730 |
| 0.01 | 13.0 | 1880 | 1.3 | 191 | 52.6 | 7620 |
| 0.02 | 13.2 | 1920 | 1.3 | 195 | 51.7 | 7500 |
| 0.03 | 13.5 | 1960 | 1.4 | 198 | 50.9 | 7380 |
| 0.04 | 13.7 | 1990 | 1.4 | 202 | 50.2 | 7280 |
| 0.05 | 14.0 | 2030 | 1.4 | 206 | 49.6 | 7190 |
| 0.06 | 14.3 | 2070 | 1.5 | 210 | 48.9 | 7090 |
| 0.07 | 14.5 | 2100 | 1.5 | 214 | 48.3 | 7000 |
| 0.08 | 14.7 | 2140 | 1.5 | 218 | 47.6 | 6900 |
| 0.09 | 14.5 | 2110 | 1.5 | 222 | 46.9 | 6800 |
| 0.10 | 14.3 | 2080 | 1.6 | 226 | 46.4 | 6730 |
| 0.11 | 14.1 | 2050 | 1.6 | 231 | 45.9 | 6660 |
| 0.12 | 13.9 | 2020 | 1.6 | 235 | 45.5 | 6590 |
| 0.13 | 13.6 | 1980 | 1.7 | 239 | 44.7 | 6480 |
| 0.14 | 13.4 | 1950 | 1.7 | 243 | 44.0 | 6380 |
| 0.15 | 13.2 | 1920 | 1.7 | 247 | 43.3 | 6280 |
| 0.16 | 13.0 | 1890 | 1.7 | 252 | 42.6 | 6180 |
| 0.17 | 12.7 | 1850 | 1.8 | 256 | 41.9 | 6070 |
| 0.18 | 12.5 | 1820 | 1.8 | 261 | 41.2 | 5970 |
| 0.19 | 12.3 | 1790 | 1.8 | 265 | 40.4 | 5860 |
| 0.20 | 12.1 | 1760 | 1.9 | 269 | 39.7 | 5760 |
| 0.21 | 11.9 | 1730 | 1.9 | 274 | 39.0 | 5650 |
| 0.22 | 11.6 | 1690 | 1.9 | 278 | 38.3 | 5550 |
| 0.23 | 11.4 | 1660 | 2.0 | 283 | 37.5 | 5440 |
| 0.24 | 11.2 | 1630 | 2.0 | 288 | 37.0 | 5360 |
| 0.25 | 11.0 | 1600 | 2.0 | 293 | 36.4 | 5280 |
| 0.26 | 10.8 | 1570 | 2.1 | 297 | 35.8 | 5190 |
| 0.27 | 10.6 | 1540 | 2.1 | 302 | 35.0 | 5080 |
| 0.28 | 10.4 | 1510 | 2.1 | 307 | 34.4 | 4980 |
| 0.29 | 10.3 | 1490 | 2.1 | 311 | 33.7 | 4890 |
| 0.30 | 10.1 | 1460 | 2.2 | 316 | 33.2 | 4810 |
| 0.31 | 9.9 | 1440 | 2.2 | 321 | 32.7 | 4740 |
| 0.32 | 9.7 | 1410 | 2.3 | 326 | 32.0 | 4640 |
| 0.33 | 9.5 | 1380 | 2.3 | 331 | 31.2 | 4530 |
| 0.34 | 9.3 | 1350 | 2.3 | 336 | 30.5 | 4420 |
| 0.35 | 9.1 | 1320 | 2.4 | 341 | 30.0 | 4350 |
| 0.36 | 8.9 | 1290 | 2.4 | 346 | 29.5 | 4270 |
| 0.37 | 8.7 | 1260 | 2.4 | 351 | 28.8 | 4170 |
| 0.38 | 8.5 | 1240 | 2.5 | 356 | 28.0 | 4060 |
| 0.39 | 8.3 | 1210 | 2.5 | 361 | 27.2 | 3950 |
| 0.40 | 8.1 | 1180 | 2.5 | 367 | 26.8 | 3880 |
| 0.41 | 8.0 | 1160 | 2.6 | 372 | 26.2 | 3800 |
| 0.42 | 7.8 | 1130 | 2.6 | 377 | 25.7 | 3720 |
| 0.43 | 7.6 | 1100 | 2.6 | 383 | 25.1 | 3640 |
| 0.44 | 7.4 | 1080 | 2.7 | 388 | 24.6 | 3560 |
| 0.45 | 7.2 | 1050 | 2.7 | 393 | 24.0 | 3480 |

**Table 41.** *(Continued)* $f_o$ **for Calculating Static Load Rating for Ball Bearings**

| $\frac{D\cos\alpha}{d_m}$ | Radial and Angular Contact Groove Type | | Radial Self-Aligning | | Thrust | |
|---|---|---|---|---|---|---|
| | Metric[a] | Inch[b] | Metric[a] | Inch[b] | Metric[a] | Inch[b] |
| 0.46 | 7.1 | 1030 | 2.8 | 399 | 23.5 | 3400 |
| 0.47 | 6.9 | 1000 | 2.8 | 404 | 22.9 | 3320 |
| 0.48 | 6.7 | 977 | 2.8 | 410 | 22.4 | 3240 |
| 0.49 | 6.6 | 952 | 2.9 | 415 | 21.8 | 3160 |
| 0.50 | 6.4 | 927 | 2.9 | 421 | 21.2 | 3080 |

[a] **Use to obtain $C_o$ or $C_{oa}$ in newtons when D is given in mm.**

[b] Use to obtain $C_o$ or $C_{oa}$ in pounds when $D$ is given in inches.

Note: Based on modulus of elasticity = $2.07 \times 10^5$ megapascals ($30 \times 10^6$ pounds per square inch and Poisson's ratio = 0.3.

*Radial Roller Bearings:* The magnitude of the static load rating $C_o$ in newtons (pounds) for radial roller bearings is found by the formulas:

$$C_o = 44\left(1 - \frac{D\cos\alpha}{d_m}\right)iZl_{eff}D\cos\alpha \quad \text{(metric)} \qquad (22a)$$

$$C_o = 6430\left(1 - \frac{D\cos\alpha}{d_m}\right)iZl_{eff}D\cos\alpha \quad \text{(inch)} \qquad (22b)$$

where $D$ = roller diameter, mm (inches); mean diameter for a tapered roller and major diameter for a spherical roller

$d_m$ = mean pitch diameter of the roller complement, mm (inches)

$i$ = number of rows of rollers in bearing

$Z$ = number of rollers per row

$l_{eff}$ = effective length, mm (inches); overall roller length minus roller chamfers or minus grinding undercuts at the ring where contact is shortest

$\alpha$ = nominal contact angle, degrees

*Radial Roller Bearing Combinations:* The static load rating for two similar single row roller bearings mounted side by side on the same shaft such that they operate as a unit is two times the rating of one single row bearing.

The static radial load rating for two or more similar single row roller bearings mounted side by side on the same shaft such that they operate as a unit (duplex or stack mounting) in "tandem" arrangement, properly manufactured and mounted for equal load distribution, is the number of bearings times the rating of one single row bearing.

*Thrust Roller Bearings:* The magnitude of the static load rating $C_{oa}$ in newtons (pounds) for thrust roller bearings is found by the formulas:

$$C_{oa} = 220\left(1 - \frac{D\cos\alpha}{d_m}\right)Zl_{eff}D\sin\alpha \quad \text{(metric)} \qquad (23a)$$

$$C_{oa} = 32150\left(1 - \frac{D\cos\alpha}{d_m}\right)Zl_{eff}D\sin\alpha \quad \text{(inch)} \qquad (23b)$$

where the symbol definitions are the same as for Formulas (22a) and (22b).

*Thrust Roller Bearing Combination:* The static axial load rating for two or more similar single direction thrust roller bearings mounted side by side on the same shaft such that they operate as a unit (duplex or stack mounting) in "tandem" arrangement, properly manufactured and mounted for equal load distribution, is the number of bearings times the rating of one single direction bearing. The accuracy of this formula decreases in the case of single direction bearings when $F_r > 0.44\ F_a \cot\alpha$, where $F_r$ is the applied radial load in newtons (pounds) and $F_a$ is the applied axial load in newtons (pounds).

**Ball Bearing Static Equivalent Load.—**For ball bearings the static equivalent radial load is that calculated static radial load which produces a maximum contact stress equal in magnitude to the maximum contact stress in the actual condition of loading. The static equivalent axial load is that calculated static centric axial load which produces a maximum contact stress equal in magnitude to the maximum contact stress in the actual condition of loading.

*Radial and Angular Contact Ball Bearings:* The magnitude of the static equivalent radial load $P_o$ in newtons (pounds) for radial and angular contact ball bearings under combined thrust and radial loads is the greater of:

$$P_o = X_o F_r + Y_o F_a \qquad (24)$$

$$P_o = F_r \qquad (25)$$

where $X_o$ = radial load factor given in Table 42
$Y_o$ = axial load factor given in Table 42
$F_r$ = applied radial load, newtons (pounds)
$F_a$ = applied axial load, newtons (pounds)

**Table 42. Values of $X_o$ and $Y_o$ for Computing Static Equivalent Radial Load $P_o$ of Ball Bearings**

| Contact Angle | Single Row Bearings[a] $X_o$ | Single Row Bearings[a] $Y_o$[b] | Double Row Bearings $X_o$ | Double Row Bearings $Y_o$[b] |
|---|---|---|---|---|
| RADIAL CONTACT GROOVE BEARINGS[c,a] | | | | |
| $\alpha = 0°$ | 0.6 | 0.5 | 0.6 | 0.5 |
| ANGULAR CONTACT GROOVE BEARINGS | | | | |
| $\alpha = 15°$ | 0.5 | 0.47 | 1 | 0.94 |
| $\alpha = 20°$ | 0.5 | 0.42 | 1 | 0.84 |
| $\alpha = 25°$ | 0.5 | 0.38 | 1 | 0.76 |
| $\alpha = 30°$ | 0.5 | 0.33 | 1 | 0.66 |
| $\alpha = 35°$ | 0.5 | 0.29 | 1 | 0.58 |
| $\alpha = 40°$ | 0.5 | 0.26 | 1 | 0.52 |
| SELF-ALIGNING BEARINGS | | | | |
| … | 0.5 | $0.22 \cot \alpha$ | 1 | $0.44 \cot \alpha$ |

[a] $P_o$ is always $\geq F_r$.
[b] Values of $Y_o$ for intermediate contact angles are obtained by linear interpolation.
[c] Permissible maximum value of $F_a/C_o$ (where $F_a$ is applied axial load and $C_o$ is static radial load rating) depends on the bearing design (groove depth and internal clearance).

*Thrust Ball Bearings:* The magnitude of the static equivalent axial load $P_{oa}$ in newtons (pounds) for thrust ball bearings with contact angle $\alpha \neq 90°$ under combined radial and thrust loads is found by the formula:

$$P_{oa} = F_a + 2.3 F_r \tan \alpha \qquad (26)$$

where the symbol definitions are the same as for Formulas (24) and (25). This formula is valid for all load directions in the case of double direction ball bearings. For single direction ball bearings, it is valid where $F_r/F_a \leq 0.44 \cot \alpha$ and gives a satisfactory but less conservative value of $P_{oa}$ for $F_r/F_a$ up to $0.67 \cot \alpha$.

Thrust ball bearings with $\alpha = 90°$ can support axial loads only. The static equivalent load for this type of bearing is $P_{oa} = F_a$.

**Roller Bearing Static Equivalent Load.—**The static equivalent radial load for roller bearings is that calculated, static radial load which produces a maximum contact stress acting at the center of contact of a uniformly loaded rolling element equal in magnitude to the maximum contact stress in the actual condition of loading. The static equivalent axial load is that calculated, static centric axial load which produces a maximum contact stress acting at the center of contact of a uniformly loaded rolling element equal in magnitude to the maximum contact stress in the actual condition of loading.

*Radial Roller Bearings:* The magnitude of the static equivalent radial load $P_o$ in newtons (pounds) for radial roller bearings under combined radial and thrust loads is the greater of:

$$P_o = X_o F_r + Y_o F_a \tag{27}$$

$$P_o = F_r \tag{28}$$

where $X_o$ = radial factor given in Table 43
$Y_o$ = axial factor given in Table 43
$F_r$ = applied radial load, newtons (pounds)
$F_a$ = applied axial load, newtons (pounds)

**Table 43. Values of $X_o$ and $Y_o$ for Computing Static Equivalent Radial Load $P_o$ for Self-Aligning and Tapered Roller Bearings**

| Bearing Type | Single Row[a] | | Double Row | |
|---|---|---|---|---|
| | $X_o$ | $Y_o$ | $X_o$ | $Y_o$ |
| Self-Aligning and Tapered $\alpha \neq 0$ | 0.5 | 0.22 cot α | 1 | 0.44 cot α |

[a] $P_o$ is always $\geq F_r$.

The static equivalent radial load for radial roller bearings with $\alpha = 0°$ and subjected to radial load only is $P_{or} = F_r$.

Note: The ability of radial roller bearings with $\alpha = 0°$ to support axial loads varies considerably with bearing design and execution. The bearing user should therefore consult the bearing manufacturer for recommendations regarding the evaluation of equivalent load in cases where bearings with $\alpha = 0°$ are subjected to axial load.

*Radial Roller Bearing Combinations:* When calculating the static equivalent radial load for two similar single row angular contact roller bearings mounted side by side on the same shaft such that they operate as a unit (duplex mounting) in "back-to-back" or "face-to-face" arrangement, use the $X_o$ and $Y_o$ values for a double row bearing and the $F_r$ and $F_a$ values for the total loads on the arrangement.

When calculating the static equivalent radial load for two or more similar single row angular contact roller bearings mounted side by side on the same shaft such that they operate as a unit (duplex or stack mounting) in "tandem" arrangement, use the $X_o$ and $Y_o$ values for a single row bearing and the $F_r$ and $F_a$ values for the total loads on the arrangement.

*Thrust Roller Bearings:* The magnitude of the static equivalent axial load $P_{oa}$ in newtons (pounds) for thrust roller bearings with contact angle $\alpha \neq 90°$, under combined radial and thrust loads is found by the formula:

$$P_{oa} = F_a + 2.3 F_r \tan \alpha \tag{29}$$

where $F_a$ = applied axial load, newtons (pounds)
$F_r$ = applied radial load, newtons (pounds)
$\alpha$ = nominal contact angle, degrees

The accuracy of this formula decreases for single direction thrust roller bearings when $F_r > 0.44 F_a \cot \alpha$.

*Thrust Roller Bearing Combinations:* When calculating the static equivalent axial load for two or more thrust roller bearings mounted side by side on the same shaft such that they operate as a unit (duplex or stack mounting) in "tandem" arrangement, use the $F_r$ and $F_a$ values for the total loads acting on the arrangement.

## LUBRICATION

### Friction and Lubrication Theory

Whenever a solid surface moves over another, it must overcome a resistive, opposing force known as *solid friction*. The first stage of solid friction, known as *static friction*, is the frictional resistance that must be overcome to initiate movement of a body at rest. The second stage of frictional resistance, known as *kinetic friction*, is the resistive force of a body in motion as it slides or rolls over another solid body. It is usually smaller in magnitude than static friction.

Although friction varies according to applied load and solid surface roughness, it is unaffected by speed of motion and apparent contact surface area. When viewed under a microscope, a solid surface will appear rough with many *asperities* (peaks and valleys). When two solid surfaces interact without a lubricating medium, full metal-to-metal contact takes place in which the asperity peaks of one solid interferes with asperity peaks of the other solid. When movement is initiated, the asperities collide, causing a rapid increase in heat and the metal peaks to adhere and weld to one another. If the force of motion is great enough, the peaks will plow through each other's surface and the welded areas will shear, causing surface degradation, or wear. In extreme cases, the resistance of the welded solid surfaces can exceed the motive force, causing mechanical seizure.

Some mechanical systems designs, such as brakes, are designed to take advantage of friction. For other systems, such as bearings, this metal-to-metal contact state and level of wear is usually undesirable. To combat such solid friction, heat, wear, and consumed power, a suitable lubricating fluid or fluid film is introduced as an intermediary between the two solid surfaces. Such lubricant applications include a wide range of oils and additives that may be used in bearings, turbines, internal combustion engines, worm and other gears, and between other moving parts, as well as fluids used in hydraulic systems.

Although lubricants themselves are not frictionless, the molecular resistive force of a gas or fluid in motion, known as *fluid friction*, is significantly less than *solid friction*. The level of fluid friction is dependent on the lubricant's *Viscosity* (see page 2507).

**Film Thickness Ratio, λ.**—For all bearings, the working lubricant film thickness has a direct relationship to bearing life. The "working" or specific film thickness ratio lambda λ is defined by dividing the nominal film thickness by the surface roughness, as depicted in Fig. 1.

$$\lambda = \frac{T}{R} \tag{1}$$

where $\lambda$ = specific film thickness
$T$ = nominal film thickness
$R$ = surface roughness

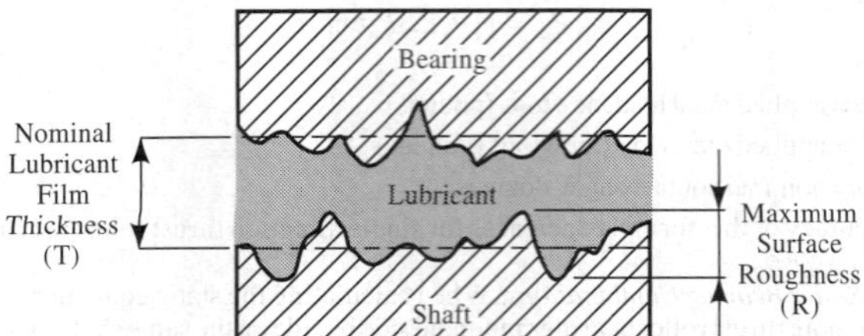


Fig. 1. Determining the Working Film Thickness Ratio Lambda

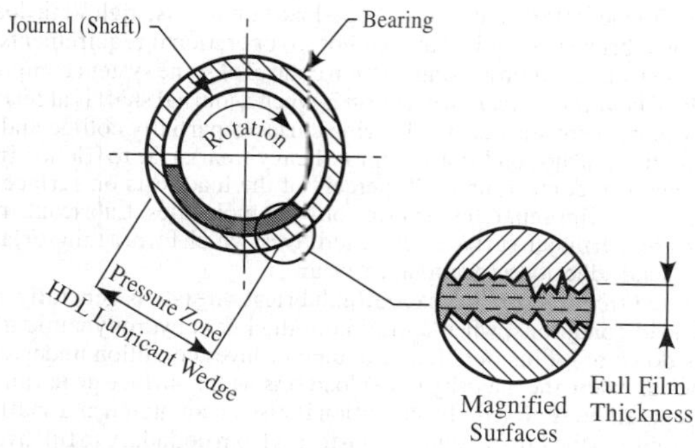


Fig. 2. HDL Hydrodynamic Lubrication of a Journal Bearing

**The Lubrication Film.—**Whenever a plain journal style bearing operates with a fluid film, the coefficient of friction μ or extent of friction reduction will depend on which one of three lubricant film conditions exists between the facing surfaces.

*Hydrodynamic Lubrication (HDL):* Hydrodynamic lubrication is also called *full-film* or *full fluid film lubrication*. In this desired lubrication condition, moving surfaces are fully separated by a hydrodynamically formed wedge, with no contact expected between the asperities of the two surfaces in relative motion, maintaining *sliding friction*. Formation of the oil wedge depends on surface geometry, speed, load, and oil viscosity. Pressure builds up internally due to relative motion between the surfaces and elastic surface deformation.

These conditions are not achieved during startup and shutdown conditions, unless there is pressure on the system to achieve full component separation. This *hydrostatic* or *pre-load* operation may be achieved by using turbine journal bearings or modern engines with automatic shutoff, reducing wear and power requirements.

*Elasto-hydrodynamic lubrication* is formed at rolling contacts, such as rolling element bearings, where surfaces in relative motion converge upon a point (ball bearings) or line (roller bearings) and the bearing material deforms elastically in the small area of contact. A hydrodynamic oil film is formed inside the small volume of deformation; the high pressure inside the deformation results in the oil momentarily changing into a solid in the volume of deformation, temporarily supporting *rolling friction*. Once pressure is released, the lubricant returns to a liquid state.

In the example of plain style bearings, both surfaces are fully separated by a working or specific film thickness lambda λ of more than 2 at the point of pressure distribution loading as shown in Fig. 2. A fluid wedge is created in which the asperities will not collide. Both surfaces are said to be "metal-contact" free at all times.

As the shaft speed accelerates, rotation of the journal acts as a pump, forcing lubricant into the pressure distribution area. Providing the lubricant is of a high enough viscosity, the wedge-shaped lubricant channel will create a load-carrying pressure sufficient to completely separate the two surfaces and support the moving journal. Full film thickness will vary between 5 and 200 microns, depending on speed, load, and viscosity. As the speed increases, so does the lubricating action and ability to carry heavier loads. Inversely, slow speeds do not allow the lubricant wedge to form, causing breakdown of the hydrodynamic action and an undesirable *boundary layer lubrication* state to prevail.

*Boundary Layer Lubrication (BL):* This occurs when any full film lubricant wedge has collapsed, leaving only a residual film of lubricant in its place. Boundary layer lubrication becomes insufficient to prevent metal-to-metal contact from occurring when: (1) the component's surface roughness significantly exceeds the oil film thickness; (2) the equipment is subjected to frequent starts and stops, shock-load conditions, high static loads, or slow speeds, and the lubricant is inadequate; and/or (3) operational requirements dictate the use of low viscosity oil, compromising friction control of some system components.

An example of *boundary layer lubrication* is when a journal shaft is at rest in the bearing and upon subsequent start up, the bearing surfaces partially collide and ride on the thin lubricant film (startup conditions promote heavy wear). Due to friction from viscous drag and mechanical contact, up to 90 percent of the load rests on surface peaks, and boundary anti-wear film separates surfaces only by molecules. Lubricants may include extreme pressure sacrificial, surface-active additives, which form a thin surface laminate that prevents metal adhesion and premature wear.

*Mixed Film Lubrication (MF):* Mixed film lubrication state is generally encountered under shock load conditions, when a minimum thickness hydrodynamic film momentarily breaks down or "thins out" into a boundary layer condition under severe shock load. It can occur where the majority of the load sets on the surface peaks and in cases of incomplete separation. A mixed film condition is also encountered as a shaft accelerates toward full speed and the film thickness transforms from boundary to full hydrodynamic condition.

Choosing too light a viscosity lubricant or where the component's thickness exceeds the applied oil film thickness to some extent can lead to a momentary or full-time mixed film condition. As indicated in Fig. 3, such a mixed film condition occurs when the specific film thickness lambda λ is between 1 and 2.

*Lubricating Film Transition:* Again as depicted in Fig. 3, a boundary layer condition is encountered when the specific film thickness lambda λ ratio is less than 1, a mixed-film state when the lambda λ ratio is between 1 and 2, and hydrodynamic lubrication when the lambda λ ratio is more than 2. Once the lambda λ ratio surpasses 4, relative bearing life is increased four-fold.

To achieve long life while supporting heavy loads, a plain bearing must successfully manage the relationship between load, speed, and lubricant viscosity. If the load and speed change, the lubricant viscosity must be able to compensate for the change.

| Lambda $\lambda < 1$ Boundary Layer Film Lubrication | $1 < \text{Lambda } \lambda < 2$ Mixed Film Lubrication | $2 < \text{Lambda } \lambda < 4$ Hydrodynamic Lubrication |
|---|---|---|

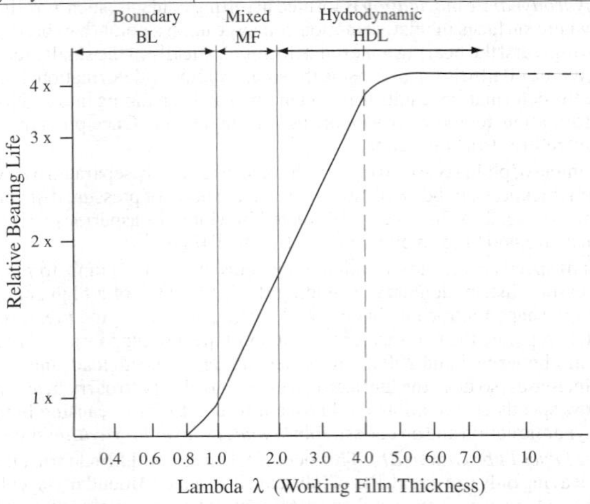


Fig. 3. The Relationship between Film Thickness Ratio Lambda λ and Bearing Life

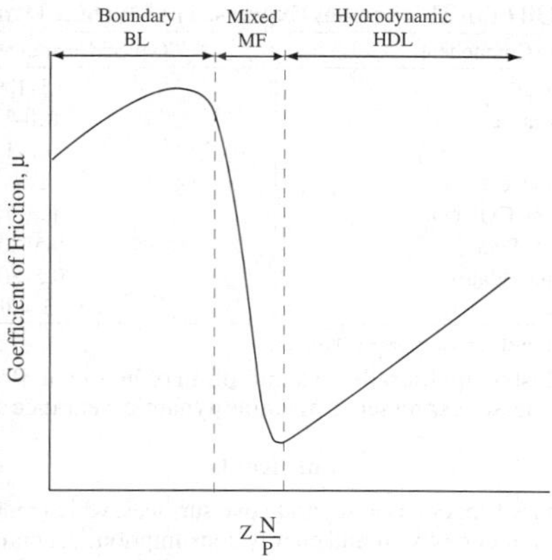


Fig. 4. Stribeck *(ZNP)* Curve

This relationship is shown in the Stribeck or *ZNP* curve illustrated in Fig. 4. Choosing the correct lubricant viscosity allows the bearing to run in the favored hydrodynamic range, producing a low coefficient of friction ranging from 0.002 to 0.005.

Rolling element bearings (point contact) and the rolling section of mating gear teeth (line contact) also favor full hydrodynamic lubrication film. They differ from sliding elements in that rolling elements require considerably less lubrication than their sliding counterparts and that the load is concentrated over a much smaller footprint on a non-conforming surface—small diameter ball or roller "rolling" over or within a much larger diameter raceway. As the ball or roller "rolls" through the load zone, the point of contact experiences a rapid pressure rise causing momentary micro distortion of both the rolling element and race. This area of deformation is named the *Hertzian contact area* (Fig. 5) and is analogous to the contact patch of a properly inflated tire on a moving vehicle. As the loaded section of rolling element moves out of the Hertzian contact area the deformed surface elastically returns to its original shape. The lubricant trapped in the Hertzian contact area benefits greatly from a phenomenon in which a lubricant under pressure will experience a dramatic rise in viscosity and act as a solid lubricant, allowing small amounts of lubricant to provide full film separation under extreme loading conditions. Under these conditions the hydrodynamic film is termed *elastohydrodynamic lubrication* (EHDL), and is unique to point/line contact situations typically found with rolling element bearings and mating gear teeth.

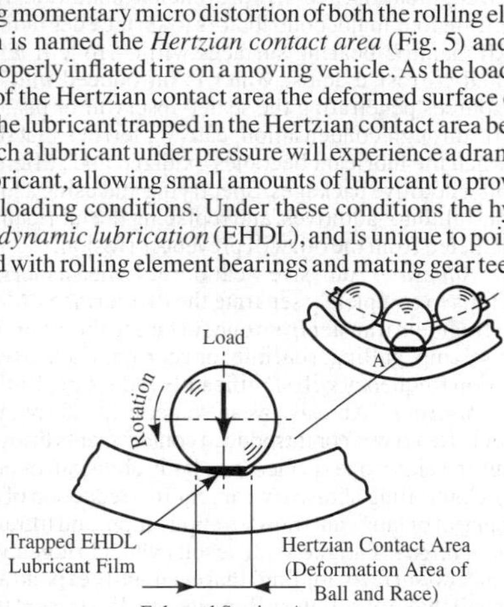


Fig. 5. Hertzian Contact Area Found in Rolling Element Bearing Surfaces

**Table 1. Typical Oil Film Thicknesses Expressed as Machine Dynamic Clearances**

| Machine Component | Typical Clearance in Microns |
|---|---|
| Plain Journal Bearings | 0.5–100 |
| Rolling Element Bearings | 0.1–3 |
| Gears | 0.1–1 |
| Hydraulic Spool to Sleeve | 1–4 |
| Engine Piston Ring to Cylinder | 0.3–7 |
| Engine Rod Bearing (Plain) | 0.5–20 |
| Engine Main Bearing (Plain) | 0.5–80 |
| Pump Piston to Bore | 5–40 |

1 micron = 0.00003937 inch; 25.4 microns = 0.001 inch.

The wheels of industry run literally on a microfilm of lubricant. Practical examples of typical oil film thicknesses expressed in machine dynamic clearance are stated in Table 1.

## Lubricants

A lubricant's primary function is to separate two surfaces, which reduces friction as well as heat; in doing so, it reduces wear and energy consumption. Secondary functions are to reduce impact shock, corrosion, and contamination. A lubricant can be in liquid (oil), solid (grease), or gaseous (oil mist) form and can be formulated from animal, vegetable, hydrocarbon, or synthetic base oil stocks. Adding to each lubricant formulation numerous chemical thickeners, solids, and chemical additives, gives every single manufactured lubricant its own unique signature blend. Lubricant selection is arguably the most influential factor in optimizing component life.

In the 1970's, Dr. Ernest Rabinowicz of MIT performed a landmark study on the effects of lubrication on the Gross National Product (GNP) of the United States. The study concluded that at that time, US manufacturing companies spent over $600 billion US annually to repair damage caused by friction-induced mechanical wear; more importantly, the study determined that over 70 percent of bearing loss of usefulness (failure) is directly attributable to surface degradation, a totally preventable condition. In his study, Rabinowicz determined there are four major contributors to surface degradation:

*Corrosive Wear:* All metallic-bearing surfaces will corrode if left unprotected from contact with water and corrosive acids. Water is introduced into lubricated environments from outside sources penetrating the sealed reservoir or bearing (washout, product contamination), or through condensation, causing ferrous metals to rust. Corrosive acids are produced when the lubricant becomes oxidized and suffers loss or breakdown of its corrosion inhibitor additive packages. Specifying and using a lubricant with rust inhibitors and corrosion-inhibitor additives, and replacing the lubricant in a timely manner when additives are depleted from the oil, will prevent corrosion.

*Mechanical Wear by Adhesion:* Adhesive wear occurs when a lubricant film separating two sliding surfaces fails to completely separate the two surfaces. Metal to metal contact occurs causing metal fragment transfer from one surface to the other. This transfer is commonly referred to as seizing, galling, scuffing, or scoring of surfaces. Correct lubricant viscosity and application frequency will significantly reduce or eliminate adhesive wear.

*Mechanical Wear by Abrasion :* Abrasive wear, sometimes referred to as *cutting wear*, is the result of hard particles from wear or introduced contaminants bridging two moving surfaces, scraping and cutting either one surface (two body abrasion) or both bearing surfaces (three body abrasion). Controlling abrasive wear requires reduction of adhesive wear along with contamination control of lubricant transfer, application, and filtration processes.

*Mechanical Wear by Fatigue:* Fatigue wear results when bridged wear particles cause small surface stress raisers (surface rippling) that eventually expand and break away from the parent metal as a spall (flake or splinter). Repeated cyclic stress at the damaged area accelerates the fatigue wear process. Correct lubricant viscosity choice and contamination control are essential to retard fatigue wear.

In all four types of wear, the primary solution for wear retardation lies in the correct choice of lubricant. Lubricants are categorized into two specific families—oil, and grease. The choice to use either oil or grease will depend upon temperature range, speed of rotation, environment, budget, machine design, bearing and seal design, which operating conditions.

**Lubricating Oil.—**For the majority of industrial applications requiring the separation of moving surfaces, the lubricant of choice continues to be petroleum-based oil, also widely known as mineral oil. Although any liquid will provide a degree of lubrication, hydrocarbon-based petroleum oils provide excellent surface wetting capabilities, water resistance, thermal stability, and sufficient fluid viscosity or "stiffness" to provide full film protection under load—all at an inexpensive price. By adding chemicals, metals, solids, and fillers, mineral-based oil stock can be formulated into an infinite number of tailored lubricating products, including grease. These modified mineral oils can be formulated for virtually any industrial application and widen the lubricant's specification and capabilities even further. Increasingly, however, specially designed synthetic oils and blends are being used for a wide range of applications.

Lubricants are considered in terms of physical properties, including sealing ability, operating temperature, installation/serviceability, service life, and specific application factors, such as component material, as well as environmental agency, manufacturer, and government specifications and regulations. In some cases, using fire-resistant hydraulic oils will not only increase safety but also may reduce insurance premiums.

Chemical properties include viscosity (grade at specific temperature), viscosity index (VI, viscosity/temperature relationship), pour and flash points (low and high temperature volatility), oxidation resistance and thermal stability, specific gravity (density compared to water), and additive response/solvency.

*API Base Oil Grades:* According to the American Petroleum Institute (API), mineral and synthetic based oils can be broken down into five identified oil stock groups, according to the production method, sulfur content, saturation, and viscosity index (See *Viscosity Index (VI)* on page 2510):

Group I: Produced by solvent refining, >0.03% sulfur, <90% saturation, viscosity index 80–120.

Group II: Produced by hydro-processing or hydrocracking, ≤0.03% sulfur, ≥90% saturation, viscosity index 80–120.

Group III: Produced by hydrocracking or catalytic de-waxing, ≤0.03% sulfur, ≥90% saturation, viscosity index ≥120.

Group IV: Produced by the polymerization of ethylene, which is a product derived from crude oil.

Group V: Includes naphthanic or aromatic minerals, silicone, phosphate ester, polyalkylene glycol (PAG), polyolester, and plant and seed oils.

*Viscosity and Standards:* The fundamental defining property for all lubricating oils is viscosity. The starting point for choosing one specific lubricant over another, optimal viscosity selection maximizes lubrication (reducing friction) by creating and maintaining proper fluid film to separate moving parts and/or minimize metal contact, depending on design and application. Viscosity selection is driven by such factors as shape, form, contact surface/smoothness, operating speeds, and material makeup of parts. There also is a relationship between viscosity, oil temperature, and the load-carrying capacity of the oil.

Proper viscosity selection is necessary to ensure the oil performs as required according to the specific application, process, and load requirements. A number of international standards guide viscosity and its classification, as well as other related selection criteria for lubricants:

ANSI/AGMA 9005-F16 (2016), Industrial Gear Lubrication: Provides lubricant classifications guidelines for minimum performance characteristics, as well as generalized application and servicing guidelines for both open and enclosed metallic gearing.

ANSI/ASME F2161-21, Standard Guide for Instrument and Precision Bearing Lubricants: Provides guidelines for choosing lubricants for precision rolling element bearings.

ASTM D2270-10 (R2016), Standard Practice for Calculating Viscosity Index from Kinematic Viscosity at 40° C and 100° C: Covers procedures for calculating viscosity index of petroleum products, such as lubricating oils and related materials, based on kinematic viscosities at specified temperatures.

ISO 11007:2023, Petroleum products and lubricants: Guides determination and consideration of rust-prevention characteristics of lubricating greases: Part 1: Dynamic wet conditions; Part 2: Method with water wash-out.

ISO 3448:1992 (R2018), Industrial liquid lubricants: ISO viscosity classification: Establishes a system of viscosity classification for industrial liquid lubricants and related fluids. This includes mineral oils used as lubricants, hydraulic fluids, and electrical oils, and for other applications.

ISO 6743 and ISO/DIS 8068 (various years), Lubricants, industrial oils, and related products: These multi-part standards cover types and classifications of lubrication applications.

ISO/TS 11366:2011, Petroleum and related products: Guidance for in-servicing of lubricating oils for steam, gas, and combined-cycle turbines. This standard provides insight into how power equipment operators evaluate conditions of oil in various equipment and maintain the oils in proper conditions. The standard also examines how oils deteriorate and includes a meaningful program of sampling and testing of oils in use.

SAE J300-2021, Engine Oil Viscosity Classification: Defines limits for classification of engine lubrication oils "in rhetorical terms only."

SAE J306-2019, Automotive Gear Lubricant Viscosity Classification: Defines limits for classification of automotive engine gear lubrications "in rhetorical terms only."

Additionally, ASME's 2003 publication *Lubrication and Lubricant Selection*, by A.R. Lansdown, provides guidance for the selection of lubricants.

*Flow and Shear:* The viscosity of a fluid is measured as its resistance to flow and shear; resistance caused by fluid friction is set up along the molecular shear planes of the lubricant as depicted in Fig. 6. Thin or light lubricants, such as machine and spindle oils shear at a faster rate than thick lubricants such as gear oils, and are said to be less viscous. Although lower viscosity oil is desirable for reducing energy (less drag), it likely would not be "stiff" or viscous enough to withstand the demands of a heavily loaded gearbox.

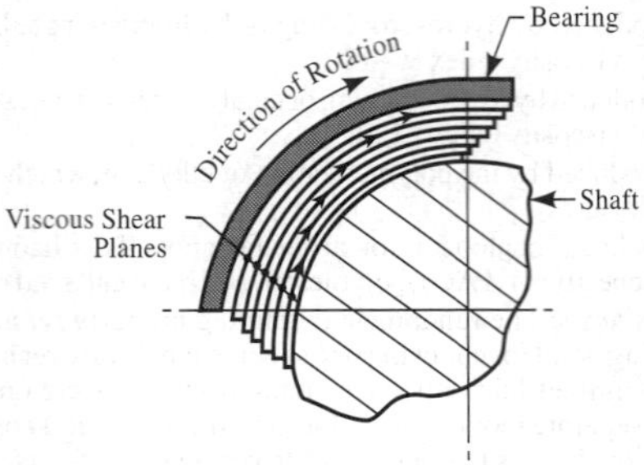


Fig. 6. Viscosity Shear Planes

*Kinematic Viscosity:* Oil viscosity is measured by a variety of classifications. The two generally accepted industrial standards are: Saybolt Universal Seconds or SUS (imperial measure), and ISO VG-centisokes-cSt (metric measure). These two standards rate oil by their *kinematic viscosity* values. The ratings, based on a fluid temperature of 100° F (40° C) and 212° F (100° C), relate the time taken for a fluid to flow through a viscosimeter capillary apparatus and directly measure oil's resistance to flow and shear by the forces of gravity. Other common viscosity classifications and comparison equivalents shown in Table 2 are for comparative purposes only. Grade substitutions are not recommended, due to formulation differences between SAE and ISO grades.

$$cSt = \frac{g/\mathrm{cc}}{\eta(cP)} \text{ at } 60\ ^\circ\mathrm{F} \qquad (2)$$

where $h$ = absolute or dynamic viscosity in centipoise
$g/cc$ = lubricant density (specific gravity)
$cSt$ = kinematic viscosity in centistokes

To convert cSt to SUS at 100° F (40° C), multiply by 4.632
To convert cSt to SUS at 210° F (100° C), multiply by 4.664

*Absolute Viscosity:* The *absolute* or *dynamic* viscosity is measured in poise (metric) or centipoise (cP) and reyn (imperial), where one reyn is equivalent to 68,950 poise. One-poise is equivalent to a one-dyne force required to move a plane surface (shear plane) of unit area a distance of one centimeter with unit speed (one centimeter per second) over a second plane at a unit distance (one centimeter) from it. *Absolute viscosity* is calculated by multiplying the kinematic viscosity value by the density of the lubricant measured at the test temperature, and is the measure of oil's resistance to flow and shear caused by internal friction. Absolute viscosity is the viscosity measured through oil analysis.

$$\eta(cP) = g/cc \text{ at } 60\ ^\circ\mathrm{F} \times cSt \qquad (3)$$

where $h$ = absolute or dynamic viscosity in centipoise
$g/cc$ = lubricant density (specific gravity)
$cSt$ = kinematic viscosity in centistokes

**Table 2. Viscosity Comparison Chart**

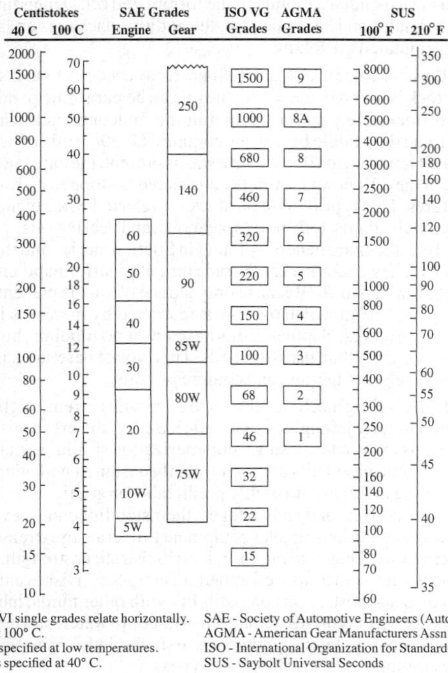


Viscosities based on 95VI single grades relate horizontally.
SAE grades specified at 100° C.
SAE W grades are also specified at low temperatures.
ISO and AGMA Grades specified at 40° C.

SAE - Society of Automotive Engineers (Automotive lubricants)
AGMA - American Gear Manufacturers Assn. (Gear lubricants)
ISO - International Organization for Standardization
SUS - Saybolt Universal Seconds

*Viscosity Index (VI):* Viscosity is dependent on temperature. As oil heats up it becomes thinner or less viscous. Inversely, as oil cools down it becomes thicker or more viscous. This phenomenon dictates that all oils will change their physical properties once they have achieved their working environment temperature. Therefore, before a lubricant viscosity choice can be made, its expected working environment temperature must be known. To engineer for this phenomenon, oil is given a *Viscosity Index*, or *VI* rating, which defines the measure of a lubricant's viscosity change due to temperature change. Higher VI ratings are more desirable, reflecting narrower viscosity change over a standard temperature range. To determine a specific oil's VI rating, its kinematic viscosity is measured at 100° F (40° C) and 212° F (100° C), then its results are compared with two or more series of oils. VI values once ranged between 0 and 100, but recent developments in lubricant technology and additives have allowed this index to raise its upper limit and include a Very High Viscosity Index (VHVI) group. Lubricants are generally classified in four basic VI groups depicted in Table 3.

**Table 3. Viscosity Index Rating**

| VI rating | Viscosity Index Group |
|---|---|
| < 80 | Low (LVI) |
| 80–110 | Medium (MVI) |
| 110–140 | High (HVI) |
| > 140 | Very High (VHVI) |

**Composition of Oil.—**Oil is composed of either a mineral (hydrocarbon based) or synthetic oil base stock to which is added a variety of organic and inorganic compounds that are dissolved or suspended as solids in the formulated oil. Depending on the end use condition the oil formulation is designed for, the additive package can make up from 1% up to 30% of the formulated oil volume.

**Mineral-Based Oils.—**Mineral oils are refined from crude oil stocks. Depending on where the crude stock is found in the world, the oil can be paraffinic or napthenic based.

*Paraffinic* based stocks are generally found in the midcontinental United States, England's North Sea, and the Middle East. They contain a 60/30/10 mix of paraffin/napthene/wax resulting in a VI rating up to 105. Because wax is present, they are known to have a *wax pour point* in which the oil's flow is severely constricted or stopped by wax crystallization at lower temperatures. This type of base oil stock is preferred when blending high quality crankcase oils, hydraulic fluids. turbine oils, gear oils, and bearing oils.

*Napthenic* based oil stocks are generally found in South America and the coastal regions of the United States. They contain a 25/75/trace mix of paraffin/napthene/wax, resulting in a less stable VI rating up to 70. Because only a trace of wax is present they are known as *viscosity pour point oils* in which oil flow is restricted by increases in the lubricant's viscosity at low temperatures. Napthenic oils have lower pour points, higher flash points, and better additive solvency than paraffinic oils. This type of base stock is preferred when blending locomotive oils, refrigerant, and compressor oils.

**Synthetic-Based Oils.—**Originally developed to cope with extreme high temperature situations encountered in early jet engines, synthetic based oil differs from mineral based oil in that its base stock is man-made. Using a polymerization similar to that used in plastics manufacturing, synthetic based oils are scientifically designed with identifiable molecular structures, resulting in fluids with highly predictable properties.

Synthetic lubricants deliver many advantages; their stability under severe high and low temperature operating conditions enables equipment to operate in extreme conditions with a high degree of reliability. They also may provide a higher shear strength, oxidation stability, flash point, and fire resistance, as well as better detergency. Disadvantages can include higher initial and disposal costs; less compatibility with other fluids, lubricants, and seal types; lower solubility with additives; reduced resistance to water; and increased toxicity.

Although there are many different synthetic base stocks, industry is primarily served by the following five common synthetic lubricant types.

*Poly-Alph-Olefins (PAOs):* PAOs, Table 4, are often described as man-made mineral oils (synthesized hydrocarbons) and were amongst the first developed synthetic lubricants for popular use in automotive crankcase oils. They are formulated in a similar molecular structure to that of pure paraffin through the synthesis of ethylene gas molecules into a polymerized uniform structure. They have a wide range of uses that include crankcase oil, gear oil, compressor oil, and turbine oils.

**Table 4. Poly-Alph-Olefins (PAOs)**

| Positive Features | Negative Features |
|---|---|
| Low pour point (down to −90° F or −68° C)<br>High viscosity index, VI > 140<br>High viscosity range<br>Good mineral oil compatibility<br>Excellent corrosion stability | Cost (4–8 × mineral oil cost)<br>Poor additive solubility<br>Poor biodegradability<br>Potential seal compatibility issues |

*Polyalkylene Glycols (PAGs):* PAGs, Table 5, are also known as organic chemical Ucon fluids that possess excellent lubricity and a unique property that causes decomposed or oxidized products to volatilize (clean burn) or become soluble, resulting in no sludge, varnish, or damaging particles to be formed at high temperatures. PAG's are polymers of alkylene oxides and are used for compressor oils, hydraulic oils (water glygols) and severe duty gear oils.

**Table 5. Polyalkylene Glycols (PAGs)**

| Positive Features | Negative Features |
|---|---|
| Low pour point (to −60° F or −51° C)<br>High viscosity index, VI > 150<br>High viscosity range<br>Fair seal compatibility<br>Excellent biodegradability<br>Do not produce sludge or varnish | Cost (4–8 × mineral oil cost)<br>Poor mineral oil compatibility<br>Poor PAO and synthetic ester based oil compatibility |

*Di-Basic Acid Esters (Di-Esters):* Due to their high shear VI stability under extreme temperature, di-esters, Table 6, have become very popular in the aerospace industry. Formulated from the reaction between alcohol and acid-laden oxygen, di-esters originally saw primary use in jet engine oils, but are now used mainly in high temperature compressor oils.

**Table 6. Di-Basic Acid Esters (Di-Esters)**

| Positive Features | Negative Features |
|---|---|
| Low pour point (to −80° F or −62° C)<br>High viscosity index, VI > 150<br>High viscosity range<br>Good mineral oil compatibility<br>Good additive solvency | Cost (4–8 × mineral oil cost)<br>Poor hydrolytic stability<br>Poor seal compatibility issues<br>Fair mineral oil compatibility<br>Poor corrosion stability |

*Polyol-Esters:* Due to their increased thermal stability over di-esters, polyol-esters, Table 7, have now taken over as the preferred oils for gas turbines, jet engines, and 2-cycle oil applications.

**Table 7. Polyol-Esters**

| Positive Features | Negative Features |
|---|---|
| Low pour point (to −95° F or −71° C)<br>High viscosity index, VI >160<br>High viscosity range<br>Good oxidation stability<br>Good mineral oil compatibility<br>Good anti-wear properties<br>Good additive solvency | Cost (10–15 × mineral oil cost)<br>Poor hydrolytic stability<br>Poor seal compatibility issues<br>Fair mineral oil compatibility<br>Poor corrosion stability |

*Silicones:* Silicone lubricants, Table 8, are semi-inorganic compounds formulated to provide the stability of inorganic products, yet retain the versatility of organic products.

Although they have poor lubricity, silicone lubricants find favor in lightly loaded instrument bearings and oils and situations requiring high temperature change and compatibility with plastics. Additives are added to the base stocks to enhance lubricant performance, just as with mineral oils.

**Table 8. Silicone**

| Positive Features | Negative Features |
|---|---|
| Low pour point (to –95° F or –71° C)<br>High viscosity index, VI > 250<br>Very high<br>Viscosity range<br>Very high flash point<br>Good seal compatibility | High Cost (30–100 × mineral oil cost)<br>Poor lubricity<br>Poor mineral oil compatibility<br>Poor biodegradability<br>Poor additive solvency |

**Oil Additives.—**When contact is likely between two bearing surfaces the lubricant should be designed to mitigate the friction through the addition of engineered additives to the base oil. Every manufactured lubricant on the market has its own unique formulation. In effect, it is an engineered liquid, custom built to perform a specific a job in a specific environment. All additives are sacrificial and therefore careful attention to additive package levels through the use of oil analysis will tell the user exactly when to change the oil to prevent damage to the bearing or contact parts. Typically oil additives as shown in Table 9 are used to enhance the existing base oil, add additional properties to the oil, and suppress any undesirable properties the base oil may have.

**Table 9. Oil Additives**

| Enhancement Additives | New Property Additives | Suppressant Additives |
|---|---|---|
| Anti-oxidant<br>Corrosion inhibitor<br>Demulsifier<br>Anti-foam | Extreme Pressure<br>Anti-wear<br>Detergent<br>Dispersant | Pour point depressant<br>Viscosity index improver |

The additive package formulation will depend on the end use. Table 10 references what oil type generally carries what additive package in its formulation.

**Table 10. Additive Package by Oil Type Guide**

| Additive | Bearing Oil | Gear Oil | Turbine Oil | Hydraulic Oil | Compressor Oil | Crankcase Oil | Grease |
|---|---|---|---|---|---|---|---|
| Anti-oxidant | ● | ● | ● | ● | ● | ● | ● |
| Corrosion Inhibitor | ● | ● | ● | ● | ● | ● | ● |
| Demulsifier | ● | ● | ● | ● | ● | | |
| Anti-foam | ● | ● | ● | ● | ● | ● | |
| Extreme Pressure | | ● | | | | | ● |
| Anti-wear | ● | ● | | ● | ● | ● | ● |
| Detergent | | | | | ● | ● | |
| Dispersant | | | | | ● | ● | |
| Pour Point | | ● some | | ● some | ● some | ● some | |
| VI Improver | | | | | | ● | |

*Anti-Oxidants:* Oxygen attacks the base oil, especially at higher temperatures, leading to the formation of sludge, tars, varnish, and corrosive acids; oxidation also can increase viscosity. Anti-oxidant additives can improve the oxidation stability of the oil by more than 10 times; lubricants designed for higher operating temperatures will contain higher levels of antioxidants. Common anti-oxidant additives are typically phenols and aromatic amines. Some other additives have shown multipurpose properties, such as soluble zinc, magnesium, and calcium compounds that reduce wear, as well as act as a detergent.

*Corrosion Inhibitor or Antirust Agents:* Used to form a protective shield against water on ferrous metals, and copper, tin, and lead-based bearing metals. They also act to neutralize corrosive acids or other chemicals that may attack materials.

*Demulsifying Agents:* Stop water from emulsifying with (effectively separating) the oil.

*Antifoaming Agents:* When oil is moved quickly, these agents, usually silicon-based compounds, act to eliminate and retard formation of added air bubbles at the lubricant's surface. (Air bubbles contain oxygen that attack the base oil and cause cavitation in pumps.)

*Extreme Pressure Additives:* Fatty acid, sulfur, phosphorous, and chlorine compounds have been successfully used to reduce extreme pressure damage. Heat activated to bond to metal, these compounds remain bonded until they degrade at specific temperatures. Such additives may produce corrosion in copper or copper alloy materials.

*Anti-Wear Agents:* Solids are employed as added sliding agents where metal-to-metal contact occurs. Most common is zinc dialkyldithiophosphate (ZDDP), which forms a protective layer on metal. Other common materials, such as graphite, molybdenum disulphide (moly), and polytetrafluoroethylene (PTFE), are solid additives used in grease formulations (see Table 11).

*Detergents* are organic metallic soaps of barium, calcium, and magnesium, acting as chemical cleaners to prevent buildup of solid particles, keep surfaces free from deposits, and neutralize harmful combustion acids.

*Dispersants* work in conjunction with detergents to chemically suspend and prevent clustering, settling, and depositing of soot and other dirt particles in the oil and allow them to be extracted by the lubrication system filters.

*Pour Point Depressants* are typically acrylic polymers with short backbones and long-chain molecule groups. These prevent the formation of wax crystals in paraffinic-based mineral oil at low temperatures, allowing it to be more fluid and preventing congelation at colder temperatures.

*Viscosity Improvers:* Sometimes a base oil of inferior quality will require thickeners to assist in achieving the specified viscosity levels over a varied temperature range. Viscosity improvers are also used to prevent the oil from thinning at higher temperatures allowing the manufacturer to build multi-grade lubricants that operate over wider temperature ranges. Viscosity improvers use long-chain organic molecules, such as polymethacrylates and ethylene propylene copolymers, to retard the viscosity shearing and improve an oil's viscosity performance.

**Table 11. Properties of Common Lubricant Solids Additives**

| Solid Additive | Color | Load Capability | Thermal Stability | Average Particle Size | Moisture Sensitivity |
|---|---|---|---|---|---|
| Molybdenum Disulphide (Moly) | Gray-Black | > 100,000 (psi) | < 750° F | < 1–6 micron | Detrimental |
| Graphite | Gray-Black | < 50,000 (psi) | < 1200° F | 2–10 micron | Necessary |
| Polytetrafluoroethylene (PTFE) | White | < 6,000 (psi) | < 500° F | < 1 micron | No Effect |

1 micron = 0.00003937 inch; 1 psi = 6.8947 kPa; Temp. in °C = (°F − 32)/1.8

Solids additives, shown in Table 11, can be added to both mineral and synthetic base oil stocks. In certain high temperature and high-pressure conditions, solids can be mixed with a mineral spirits carrier and applied directly to the bearing surfaces as a dry solid lubricant. The volatile carrier flashes off with the heat and leaves a dry solid film on the bearing surface. These additives are not suitable for liquid lubricants, as they can settle and fall out of solution during storage.

**Temperature Effects on Oil.**—Changes in temperature effect an oil's viscosity and its ability to maintain a load-carrying hydrodynamic film as depicted in Fig. 7. With the exception of silicone-based fluids, which unfortunately have poor lubricating qualities, most oils suffer a dramatic drop in viscosity once the temperature surpasses 100° F (38° C).

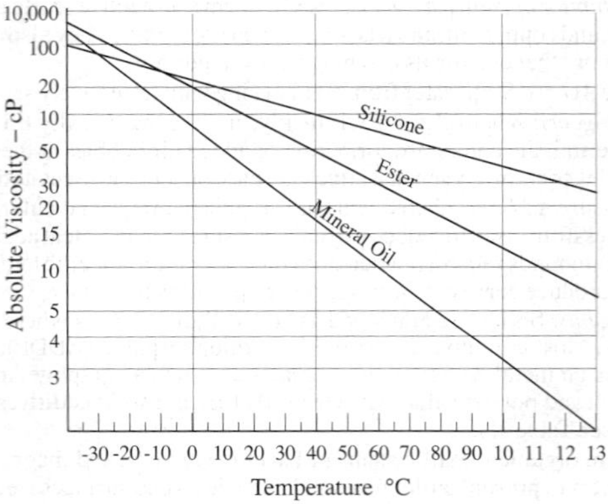


Fig. 7. Temperature Effect on Viscosity for Different Oils

Temperature affects not only the viscosity of the oil, it affects the condition and life expectancy of the oil as shown in Fig. 8. In a common example, for every 17° F (10° C) increase in temperature, oxidation rates double, and effective oil life is halved. However, the reaction kinetics for oxidation may vary dramatically with certain base oils, and the relationship may be polynomial or exponential rather than linear. Operating temperature is the leading indicator in determining oil change out frequencies

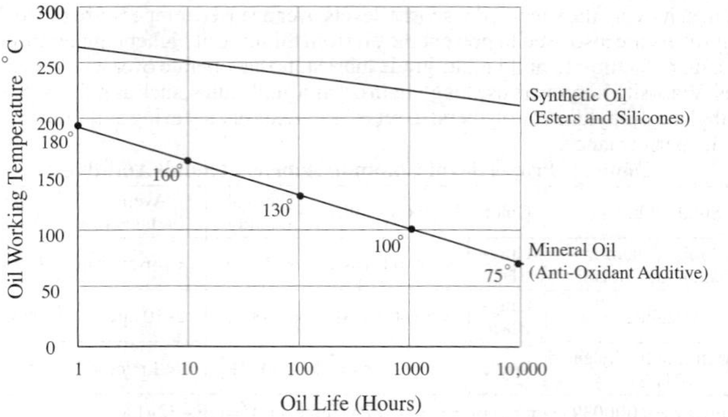


Fig. 8. Expected Oil Life at Varying Operating Temperatures.

*Oxidation* is the leading cause of lubricant failure. It is catalyzed not only by heat, but also by trapped air, water, and copper or iron ions from contamination or wear debris. Fig. 9 shows typical upper and lower working limits for various lubricating oils.

**Lubricating Grease.**—In situations where the containment and continued application of lubricating oil is not practical, lubricating grease is widely used—most specifically in rolling element bearings requiring only periodic lubrication, and slow-speed, high-load boundary lubrication applications. Easier to retain than oil, grease offers lower lubricant losses and good sealing qualities. When utilized in an automatic delivery system, grease can provide full film lubrication.

Grease is a blended mix of the lubricating oil (mineral or synthetic—usually di-ester or silicone based), oil additive package, and fatty acids mixed with metallic alkaline soap

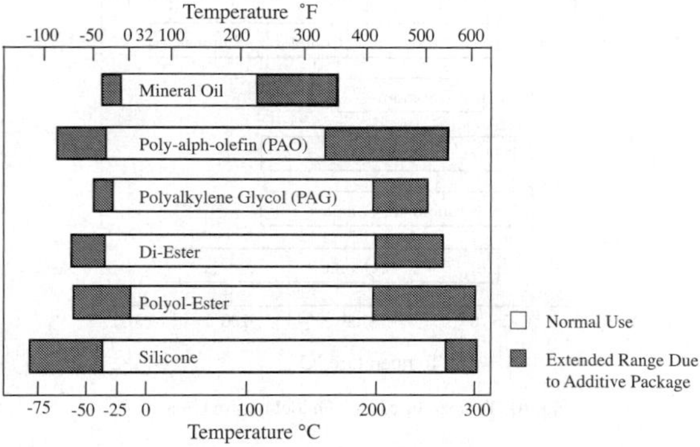


Fig. 9. Temperature Limit Guidelines for Oil

to form the thickening agent. Varying the oil, additive package, and soap blend produces many unique types of grease formulated for a variety of operating conditions. The metallic soap or non-soap greases listed in Table 12 are classified according to their thickener.

Grease works in a similar way to a sponge; as the temperature of the grease rises, the oil bleeds from the soap filler and performs the lubricating function to the balls, raceways, and sliding surfaces. Inversely, once the grease cools down, the oil is soaked back up into the soap filler, which essentially acts as a semi-fluid container for the lubricating oil. An important step in selecting the correct grease is determining if the base oil viscosity is suitable for the application. For example, grease designed for heavily loaded, high temperature applications will probably use a heavy viscosity oil base, whereas general-purpose grease is more likely to use a medium viscosity oil base.

**Table 12. Grease Types and Their Properties**

| Thickener | Appearance | Pump-ability | Heat Resistance | Temperature Range | Water Resistance | Thickener Chemistry |
|---|---|---|---|---|---|---|
| Calcium | Buttery | Fair | Fair | 230° F (110° C) | Excellent | Soap |
| Sodium | Fibrous | Fair | Very Good | 250° F (120° C) | Poor | Soap |
| Calcium Complex | Stringy | Fair | Good | 350° F (175° C) | Very Good | Soap |
| Lithium | Buttery | Excellent | Good | 350° F (175° C) | Excellent | Soap |
| Aluminum Complex | Stringy | Good | Good | 380° F (193° C) | Excellent | Soap |
| Lithium Complex | Buttery | Excellent | Excellent | 375° F (190° C) | Very Good | Soap |
| Barium | Fibrous | Very Good | Excellent | 380° F (193° C) | Excellent | Soap |
| Bentonite | Buttery | Good | Excellent | 500° F (260° C) | Good | Clay |
| Urea | Buttery | Good | Excellent | > 500° F (260° C) | Excellent | Polymer |
| Calcium Sulfonate | Smooth | Very Good | Excellent | > 500° F (260° C) | Excellent | Calcite |

Grease properties may change according to the additive package used

At sustained high temperatures, grease will soften substantially and could leak or drop from the bearing unless rated specifically for high temperature applications. High temperatures rapidly oxidize the lubricant causing the soap to harden; higher temperatures require more frequent application of grease. Lower temperatures can be just as detrimental because the grease "stiffens" considerably as temperatures near –20° F (–30° C). At this temperature the rolling elements no longer rotate and they drag across the raceway. Under heavier loads this effect causes "smearing" of the bearing surfaces leading to premature bearing failure. Grease temperature guidelines by type are shown in Fig. 10.

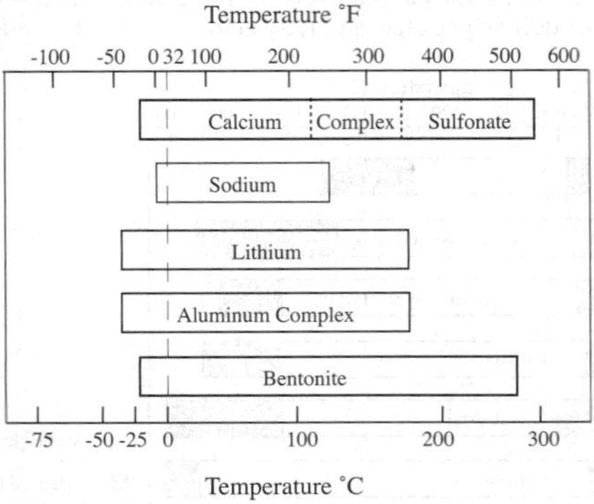


Fig. 10. Temperature Limit Guidelines for Grease

*Grease Classification:* The National Lubricating Grease Institute—NLGI, classifies grease according to a rating standard that measures the consistency of the grease. Using a penetrometer apparatus under laboratory conditions, a conical weight is dropped from a known height into the grease sample, and its depth of penetration is measured after a 5 second time period. The Table 13 rating chart shows that stiffer greases are rated with a higher NLGI code than more fluid grease with higher levels of penetration. Grease consistency largely depends on the type and amount of soap thickener blended in the grease and the oil viscosity—NOT the base oil viscosity alone. Rolling element bearings will use grease in the NLGI 1 to 3 range. Centralized grease lubricating systems favor 0 to 2 NLGI rated grease.

**Table 13. NLGI Grease Consistency Rating Chart**

| NLGI Rating | Description | Penetration Range (0.1mm@77° F) |
|---|---|---|
| 6 | Hard | 85–115 |
| 5 | Very Stiff | 130–160 |
| 4 | Stiff | 175–205 |
| 3 | Medium | 220–250 |
| 2 | Medium Soft | 265–295 |
| 1 | Soft | 310–340 |
| 0 | Very Soft | 355–385 |
| 00 | Semi Fluid | 400–430 |
| 000 | Semi Fluid | 445–475 |

*Grease Additives:* As with oil, grease will also contain solids additives such as graphite, molybdenum disulphide, and PTFE for use in extreme pressure and heavy wear conditions.

## Lubricant Selection and Application

**Selecting a Suitable Lubricant.**—Selecting a suitable lubricant will depend on a number of factors such as type of operation (full film, boundary layer), temperature, speed, load, working environment, and machine design. Machine maintenance requirements and maintenance schedules are not always taken into account in the equipment engineering design process of the lubrication system(s). Careful assessment of the conditions and consultation with a lubricant manufacturer/provider must take place to determine the optimal lubricant choice for each specific application. Table 14 offers general guidelines for lubricant choice when operating conditions are known, and Fig. 11 offers lubricant viscosity guideline choices based on bearing speed in rpm. Once the initial lubricant choice is made, its viscosity must be checked against the specific operating temperature to ensure that the lubricant is suitable for speed, load, and temperature conditions.

**Table 14. General Guidelines for Choosing a Preferred Lubricant Type**

| Condition | Oil | Grease | Solid |
|---|---|---|---|
| Clearances Designed for Oil | ● | | ● |
| Clearances Designed for Grease | | ● | ● |
| High Speed, Low Load | ● | | |
| Low Speed, High Load | | ● | |
| Low Speed, Oscillating Load | | ● | ● |
| High Temperature | ● | ●[a] | |
| Full Film Applications | ● | ● | |
| Boundary Layer Applications | | ● | ● |
| Contaminated Working Environment | | ● | |
| Product Cannot Tolerate Leaks | | ● | ● |
| Closed Gearbox | ● | | |
| Isolated Bearings | | ● | |

[a] Automated delivery system.

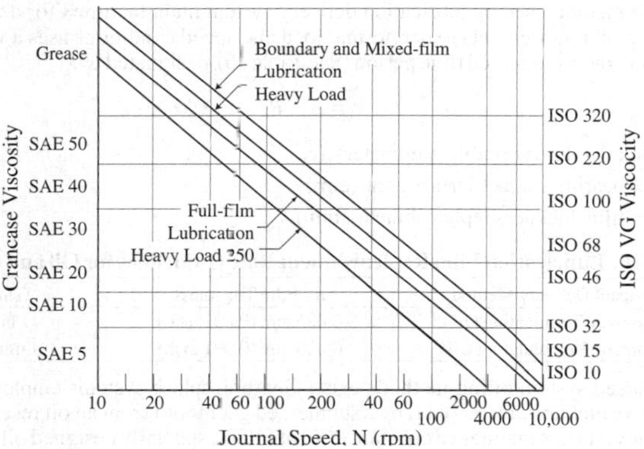


Fig. 11. Lubricant Viscosity Selection Guide based on Bearing Speed in rpm

Lubricant additives deliver different working characteristics to the lubricant. Knowing and documenting a machine or system's lubricant application requirements will facilitate a consolidation of lubricant requirements and assist in determining the optimal lubricant additive package. Table 15 reviews typical lubricated components, and assigns priority guideline ratings against a number of important lubricant functional attributes. This information is a starting point when working with the lubricant manufacturer to enable consolidation of lubricant needs and choose lubricants with suitable additives.

**Table 15. Priority Guideline Ratings of Lubricant Functional Attributes for Different Lubricated Components**

| Lubricant Attribute | Sliding Bearing | Rolling Bearing | Wire Rope, Chain, Open Gears | Closed Gears |
|---|---|---|---|---|
| Friction Reduction | 1 | 2 | 1 | 2 |
| Boundary Lubrication | 1 | 2 | 2 | 3 |
| Cooling Ability | 2 | 2 | 0 | 3 |
| Temperature Range | 1 | 2 | 1 | 2 |
| Corrosion Protection | 1 | 2 | 2 | 1 |
| Seal Out Contaminants | 0 | 2 | 1 | 0 |

0 = Low Priority, 3 = High Priority

**Oil Application.—**Oil lubrication can be broken down into two major categories: terminating (total loss), and recirculating oil systems.

*Terminating Oil Systems:* Terminating oil systems are semi-automated and automated systems that dispense oil at a known rate to the bearing(s) and do not recover the oil. This system can be generally observed in use for lubricating plain bearings, gibs, and slide ways found in small to medium-sized machine tools. Reservoir oil is replenished with new oil on an "as used" basis.

*Recirculating Oil Systems:* Recirculating oil systems pump oil through the bearing(s) on a semi-continuous or continuous cycle, using the oil to cool the bearing as it lubricates. Depending on the system design the oil can be filtered prior to the pump suction inlet, on the pump discharge, and again on the gravity return to the oil storage reservoir.

Recirculating systems are used on every kind of small to very large equipment as long as the oil can be contained; reservoirs retain their original charge of oil, which is changed on a condition or time basis.

A simple method used by lubrication delivery system manufacturers for determining the bearing oil requirements under normal load and speed conditions uses a volumetric requirement over a specified time period (See Table 16), designated by:

$$V = A \times R \tag{4}$$

where $V$ = oil volume in cubic centimeters, (cc)

$A$ = bearing contact surface area, ($cm^2$)

$R$ = film thickness replenishment, (mm)

**Table 16. Lubrication Film Replenishment Rate Guidelines for Oil and Grease**

| Lubricant Delivery Method | $R$ - Film Thickness | Time |
|---|---|---|
| Automatic Terminating Oil | 0.025 mm (0.001 inch) | 1 hour |
| Automatic Recirculating Oil | 0.025 mm (0.001 inch) | 1 minute |

Other closed system oiling methods exist: Gearbox splash systems employ a simple recirculative pickup/transfer of oil by a submerged gear tooth from an oil reservoir bath. Constant level oilers maintain a constant level of oil in a specially designed oil bath bearing housing. Using air over oil technology, oil can be misted and "rifled" into the bearing allowing very high speeds of over 20,000 rpm at light loads.

When replenishing oil reservoirs always use new clean oil of the exact same specification, from the same manufacturer. Mixing different oils of similar specification can cause additive packages to react with one another causing detriment to the bearings. If changing to a new oil specification or manufacturer consult with the new manufacturer for the correct change-out procedure.

**Grease Application.—**Because grease is easy to retain in a bearing housing over a long period of time and because it acts as a seal against contaminants, most rolling element bearings are grease lubricated. For most applications a NLGI 1 or 2 rating grease is used.

The method of grease lubrication will depend on the greased bearing design; bearings can be hand-packed, manually-lubricated with a terminating style grease gun, or automatically greased.

Open rolling bearings are received with a rust inhibiting compound from new and must be pre-packed on assembly—**DO NOT remove bearings from their packaging until ready to use,** and **DO NOT spin dry bearings as this will significantly degrade the life of the bearing.** Shielded or sealed bearings usually arrive pre-packed from the manufacturer—always specify your preferred grease to your bearing supplier when ordering. The initial amount of grease is determined by adjusting the volume according to the known speed and load. For operating temperatures above 180° F (80° C) the bearing is packed to 25 percent of the full pack volume. For temperatures below 180° F (80° C), the guideline for pack volume is shown in Table 17 and is based on the bearing surface speed in operation calculated as:

$$dn \text{ or } Dn = SD \times RPM \tag{5}$$

where $dn$ = bearing surface speed factor, metric, mm

$Dn$ = bearing surface speed factor, US customary, in.

$SD$ = shaft diameter of the bearing bore, mm or in.

$RPM$ = velocity, rotations per minute at full speed

**Table 17. Bearing Packing Guidelines**

| *dn* (mm) | | *Dn* (in.) | | % Full Pack |
|---|---|---|---|---|
| From | To | From | To | |
| 0 | 50,000 | 0 | 2,000 | 100 |
| 50,000 | 100,000 | 2,000 | 4,000 | 75 |
| 100,000 | 150,000 | 4,000 | 6,000 | 50 |
| 150,000 | 200,000 | 6,000 | 8,000 | 33 |
| 200,000+ | | 8,000+ | | 25 |

For vibration applications, do not fill more than 60% of full pack

When hand packing, work the grease with fingers around all the rolling elements; the bearing can be dismantled to make this operation easier. The grease should fill the immediate bearing area. Before renewing grease in an existing bearing, the bearing must be removed and washed in kerosene or any suitable degreasing product. Once clean, the bearing is lightly coated in mineral oil, being careful not to spin the bearing at this point. Once filled with the appropriate amount of grease in the bearing area, the bearing can be hand spun to fling off grease excess, which is wiped away with a lint-free clean cloth. Free spacing in the housing should be filled from 30 to 50 percent. Overfilling bearings with grease is the leading cause of bearing lubrication-related failures. Over greasing causes the lubricant to "churn," which in turn "spikes" the bearing internal temperature, significantly reducing the bearing life using considerably more energy to overcome fluid friction. Bearings designed to be lubricated by a grease handgun or automated delivery system will have a grease port built into the bearing housing and raceway. Grease lubrication intervals will depend on temperature and speed. Fig. 12 provides guidelines for renewing grease based on speed.

Bearings running at high temperature extremes will require more frequent application based on the temperature, load, speed, and type of grease used. When replenishing grease always use new clean grease of the exact same specification, from the same manufacturer. Mixing different greases can lead to compatibility problems causing detriment to the bearings. If changing to a new grease specification or manufacturer, consult with the new manufacturer for the correct change out procedure.

**Lubrication Delivery Methods and Systems.—**Numerous methods and systems are used to deliver oil and grease to bearing points. Automated centralized systems work on the premise of delivering a very small amount of lubrication on an almost continual basis,

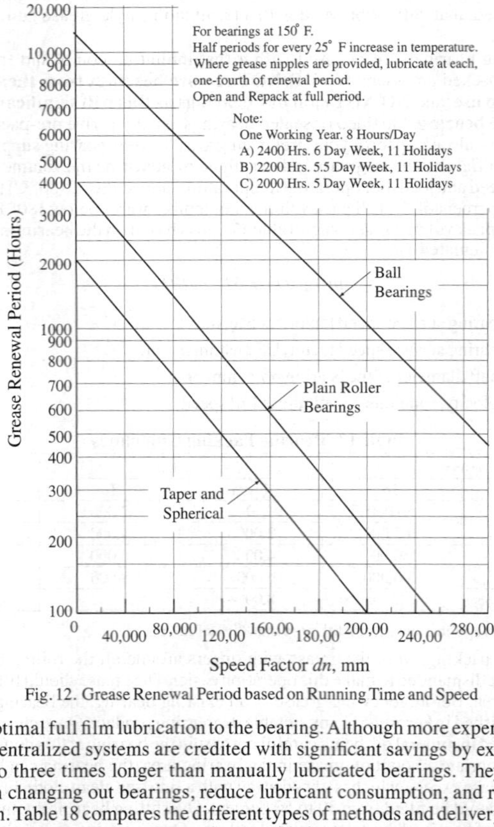


Fig. 12. Grease Renewal Period based on Running Time and Speed

providing optimal full film lubrication to the bearing. Although more expensive initially, automated centralized systems are credited with significant savings by extending bearing life up to three times longer than manually lubricated bearings. They also reduce downtime in changing out bearings, reduce lubricant consumption, and reduce energy consumption. Table 18 compares the different types of methods and delivery systems and some of their features, and can be used as a guide in determining a suitable lubrication delivery approach.

**Table 18. Lubrication Method and System Comparison Guide**

| Feature | Hand Pack | Manual Gun | Single Point | Centralized Total Loss | Centralized Recirculating | Self Contained Splash/Bath | Gravity Fed |
|---|---|---|---|---|---|---|---|
| Oil | | ● | ● | ● | ● | ● | ● |
| Grease | ● | ● | ● | ● | | | |
| Continuous Delivery | | | ● | | ● | ● | ● |
| Cyclic Delivery | | ● | | ● | | | |
| Automatic Control | | | ● | ● | ● | | ● |
| Manual Control | ● | ● | | ● | | | ● |
| Positive Displacement | | ● | ● | ● | ● | | |
| Line Monitor Protection | | | | ● | ● | | |
| # Lube Points | | Unlimited | Unlimited | 20 Min. | 200 Max. | | 20 Max. |

Manual gun delivery systems are commonly known as grease guns and oil guns. These hand-dispensing devices are capable of delivering lubricant at pressures exceeding 15,000 psi (103 MPa), and must be used with extreme caution if the bearing seal is not to be compromised—especially when lubricating from a remote located grease nipple. Bearings manually lubricated with grease and oil guns are lubricated with significantly more lubricant and less frequent applications than automatic centralized lubricated bearings. Manual lubrication results in a high degree of bearing fluid friction and a significant lower life expectancy.

Single point lubricators are self-contained automatic dispensing units that house a lubricant reservoir and can dispense oil or grease to a single bearing or a small number of bearings through a manifold system. Earlier versions of the grease units employed a spring-loaded follower plate that dispensed against a bearing back pressure through a controllable bleed valve; while oil units used gravity (also known as gravity units) to allow oil to drip through a bleed valve at a controlled rate onto a brush or wick device touching the moving shaft or part. Both unit types are refillable and are still available. Modern-day versions are mostly one-time-use units that employ programmable controlled battery operated positive displacement pumps, or electrochemical gas expandable bellows to move the lubricant to the bearing.

Centralized total loss systems employ a pump that can be automatically or manually activated to pump oil (solid or mist) or grease to a series of metering valves mounted at the lubrication point, or in a manifold device piped to the bearing point. These systems are capable of delivering a metered amount of lubricant on a cyclic basis to many hundreds of lubricant points simultaneously. Because the lubricant is not reclaimed at the bearing point, the pump reservoir must be filled with lubricant on a regular basis. This lubrication system is the most common type of found on industrial equipment.

Centralized oil recirculating systems are designed to continually pump a metered amount of oil through each bearing point. The oil is channeled back to the reservoir through a filter system and pumped out again through the distribution system.

Self contained bath and splash installations are "pick-up" type systems that employ oil in a reservoir filled to an engineered level that covers the lowest submersed gear teeth. As the gear moves it picks up the oil and transfers lubricant as each gear engages and disengages. Higher rpm speed causes the lubricant to be splashed high into the gearbox cavity so that is is distributed to all the internal devices.

## Contamination Control

Before an oil lubricant gets to perform its lubrication function at the bearing point, it must often go through a torturous handling process where the oil must be transferred multiple times before it eventually resides in the final application reservoir. The lubricant is shipped from the refinery to the blending station, to the manufacturer's bulk storage tank, to the supplier's storage tank, to the barrel or pail, to the user's storage facility, to the maintenance department, and finally to the machine's reservoir. If the transfer equipment and storage tanks/devices are not dedicated to this exact lubricant type and scrupulously clean, and the oil is not filtered at every transfer point, the virgin oil will be contaminated when placed in the equipment reservoir.

In a study performed by the National Research Council of Canada on bearing failure in primary industries it was found that 82 percent of wear-induced failure was particle-induced failure from dirty lubricants, with the greatest wear caused by particles whose size equaled the oil film thickness Perhaps the greatest contamination enemy for bearings is the ever present silt and its abrasive properties. Fig. 13 shows the Macpherson curve, which depicts the contaminant effect on roller bearing life based on contaminant micron size.

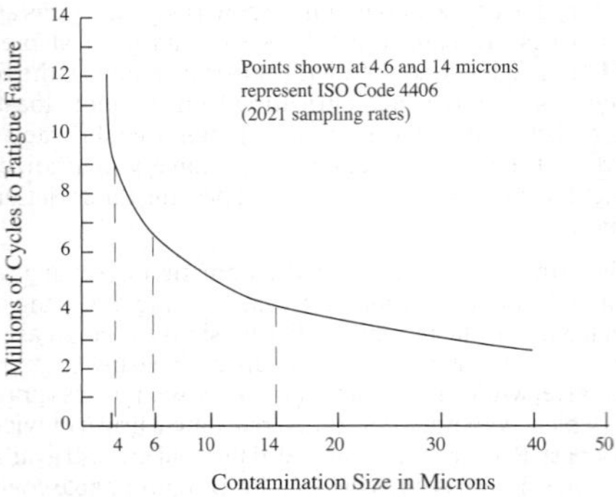


Fig. 13. Macpherson Contamination Effect Curve

The graph in Fig. 13 clearly shows the relationship between bearing life extension and contaminant size. By focusing on controlling contaminates less than 10 microns in size with quality filtration methods, expected bearing life is more than tripled.

**ISO Cleanliness Code.—**When performing a solids lubricant analysis and cleanliness testing, the ISO Code ISO 4406 (1999), "Hydraulic fluid power—Fluids—Method for coding the level of contamination by solid particles," is used as a guide. The number of 4-micron, 6-micron, and 14-micron diameter particles in a 1-ml lubricant sample are counted and compared to a particle concentration range (see Table 19), then assigned a cleanliness code number for each particle count size. The most dangerous particles, between 4 and 6 microns, known as *silting particles*, are determined by the difference in the corresponding ISO 4406 micron values.

**Table 19. ISO Code 4406 (1999)**

| Particles per ml | | | Particles per ml | | |
|---|---|---|---|---|---|
| More Than | Up to and Including | Range Number (R) | More Than | Up to and Including | Range Number (R) |
| 80,000 | 160,000 | 24 | 20 | 40 | 12 |
| 40,000 | 80,000 | 23 | 10 | 20 | 11 |
| 20,000 | 40,000 | 22 | 5 | 10 | 10 |
| 10,000 | 20,000 | 21 | 2.5 | 5 | 9 |
| 5,000 | 10,000 | 20 | 1.3 | 2.6 | 8 |
| 2,500 | 5,000 | 19 | 0.64 | 1.28 | 7 |
| 1,300 | 2,600 | 18 | 0.32 | 0.64 | 6 |
| 640 | 1,280 | 17 | 0.16 | 0.32 | 5 |
| 320 | 640 | 16 | 0.08 | 0.16 | 4 |
| 160 | 320 | 15 | 0.04 | 0.08 | 3 |
| 80 | 160 | 14 | 0.02 | 0.04 | 2 |
| 40 | 80 | 13 | 0.01 | 0.02 | 1 |

*Example:* An ISO code of 21/19/17 would represent findings of between 10,000 to 20,000 4-micron sized particles per ml, between 2,500 and 5,000 6-micron sized particles per ml, and between 640 and 1,280 14-micron sized particles per ml; this sample would be considered very dirty.

Typical cleanliness targets for rolling element bearings would start at 16/14/12 or better, 17/15/12 or better for journal style bearings, 17/14/12 or better for industrial gearboxes, and 15/12 or better for hydraulic fluids.

A study conducted by the British Hydromechanics Research Association (BHRA) looked at the relationship between hydraulic fluid cleanliness and mean time between failure (MTBF) of over 100 hydraulic systems in a variety of industries over a three-year period. The results are seen in Fig. 14 and show that systems that were successful in filtering out and excluding contaminants over 5 microns in size lasted tens of thousands of hours longer between system breakdowns.

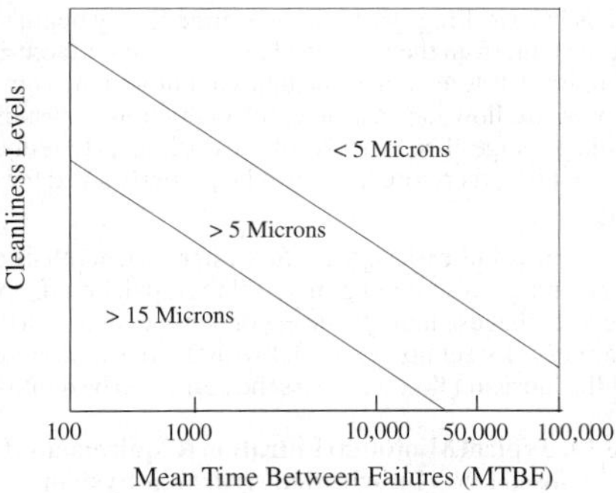


Fig. 14. MTBF versus Cleanliness Levels

Solid particle ingression into a closed lubrication/hydraulic system can come from a variety of sources that include new oil, service and manufacturing debris, improper seals, vents/breathers, filter breakdown, and internal wear generation. For the most part, ingress prevention is all about filtration. Introducing filtered clean new oil into a system will significantly retard the wear process and avoid clogging up breathers and in-line filter systems.

**Water Contamination .—**Water contamination is the other major lubricant contaminant that will significantly degrade the oil's life (Fig. 15). Lubrication fluid typically saturates at 0.04% or 400 ppm, whereas hydraulic fluid (excluding water glycol fluids) saturates at an even lower 0.03% or 300 ppm. Typical water sources are found in the fluid storage areas when lubricants are stored outdoors and subjected to the elements, or stored in continually changing temperatures causing condensation and rust that can be transferred into the equipment's lubrication system.

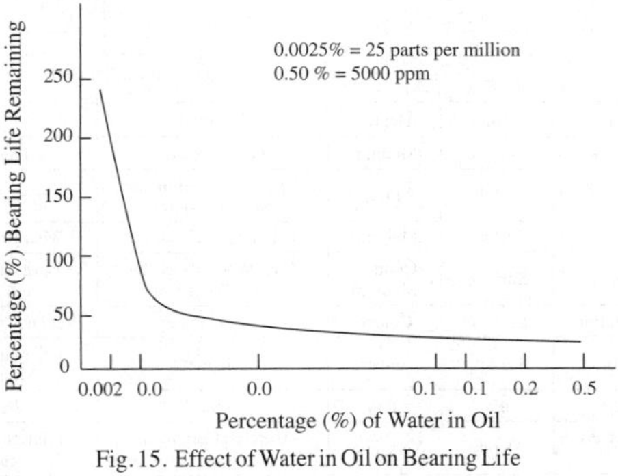


Fig. 15. Effect of Water in Oil on Bearing Life

**Filtration Systems.—**Although contamination cannot be completely eliminated, with diligence and the use of effective filtration techniques and methods the effects of contamination can be seriously mitigated.

Working on the understanding that fluid cleanliness is the basis for contamination control, primary filtration commences on the virgin stock oils prior to the lubricant being placed in the working reservoir or lubrication / hydraulic system.

Once the lubricant is in a working system it will immediately begin to attract contaminants already in the system, from the air, outside sources, manufacturing materials, and wear materials that must be filtered out as the lubricant moves through the system. Table 20 shows a typical pressure flow lubrication delivery, hydraulic system complete with a minimum filter media package. The function of these filters is to reduce operating costs and increase component life; therefore they must be properly sized for the system and be of the highest quality.

There are two basic types of filter design, surface filters and depth filters. Surface filters are the most common and use a screening material to trap debris. Depth filters are deep cleaning filtering devices that use multiple layers of dense materials to "polish" the lubricant. Depth filters are usually set up in parallel with the basic filter system and allow a small percentage of the lubricant flow to bypass the pump and be depth cleaned.

**Table 20. Typical Minimum Filtration Requirements for a Closed Loop Lubrication Hydraulic System**

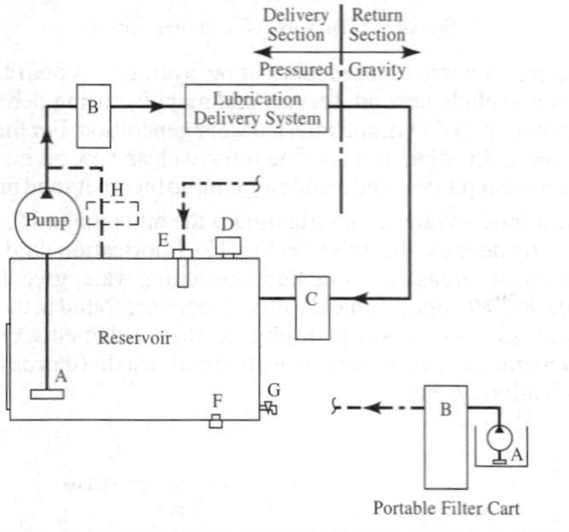


| | Location | Type | Degree | Material | Purpose |
|---|---|---|---|---|---|
| A | Pump Suction | Surface | Medium | Gauze, Paper | Pump Protection |
| B | Pump Delivery Header | Surface | Fine | Felt, Paper, Cellulose, Sintered Metal | Primary System Protection |
| C | Return Line | Surface | Medium | Felt, Paper, Cellulose | Wear Products Protection |
| D | Reservoir Vent | Surface | Course / Medium | Wire, Wool, Paper, Oil Bath Desiccant | Remove Airborne Contaminant and Condensation |
| E | Reservoir Fill Port | Surface | Course | Gauze, Paper | Prevent Course Solids Ingress |
| F | Drain Plug | Surface | Fine | Magnet | Capture Ferrite Wear Metals and Debris |
| G | Drain Valve | n/a | n/a | n/a | Water Removal |
| H | Delivery Bypass Filter | Depth | Very Fine | Carbon, Cellulose, Diamataceous Earth, Felt | Lubricant Deep Cleaning and Polishing |

A diagram and tabulation of the filtration requirements of a closed loop hydraulic system are shown in Table 20. Fluid in the reservoir is sucked up by the pump through the suction filter (A) and pumped into the delivery header line under pressure. If a depth filter option is used, a small percentage, up to 15 percent, of the oil flow is diverted for deep cleaning through a depth filter (H) and sent back to the reservoir for recycling. The lubricant is then forced through the primary pressure filter and allowed to perform its work at the bearing point or hydraulic device before it eventually channels into the system return line under gravity to pass through a low pressure return line filter that takes out any wear materials gathered along the way. Once through the return filter the oil makes its way back into the reservoir. The reservoir is protected against airborne contaminants and condensation by the vent filter, and is protected against ingress of course solids by the fill neck screen filter. Because water is heavier than oil it will settle to the bottom of the tank where most of it can be drained off by opening the drain valve. Metallic debris also settle to the bottom and are captured by the magnetic drain plug at the bottom of the reservoir. As the lubricant oxidizes and breaks down, sludge will form on the bottom of the reservoir, which must be cleaned out periodically manually by removing the reservoir clean-out hatch.

*Filter Efficiency:* Most surface filters are sold in either one-time-use, or cleanable-reusable forms. Depth filters are all one-time-use filters. All filters are performance rated according to the media's particle removal efficiency, known as the filter's filtration ratio, or *beta ratio*. Not all filters are made equal, and they are tested for dirt holding capacity, pressure differential capability, and filter efficiency, using an ISO 4572 Multipass Test Procedure in which fluid is circulated through a mock lube system in a controlled manner. Differential pressure across the test filter element is recorded as contamination is added into the lubricant upstream of the filter. Laser particle sensors determine contamination levels both upstream and downstream of the filter element and the beta ratio is determined using the following formula:

$$B_x = \frac{\#\,\text{Upstream Particulate}}{\#\,\text{Downstream Particulate}} \tag{6}$$

where $B$ = filter filtration ratio
$x$ = specific particle size

*Example:* If 100,000 particles, 10 microns and larger, are counted upstream of the test filter, and 1000 particles are counted after or downstream of the test filter element, the beta ratio would equal:

$$B_{10} = \frac{100{,}000}{1000}$$

Efficiency is determined using the following equation:

$$\text{Efficiency}_x = \left(1 - \frac{1}{B_x}\right) \times 100$$

$$\text{or \% Efficiency} = \left(\frac{B-1}{B}\right) \times 100 \tag{7}$$

$$\text{Efficiency}_{10} = \left(1 - \frac{1}{100}\right) \times 100 = 99\%$$

The higher the beta ratio, the better the capture efficiency of the filter, see Table 21.

**Table 21. Filter Efficiency**

| Beta Ratio at a Specific Particle Size | Filter Efficiency at Same Specific Particle Size | Beta Ratio at a Specific Particle Size | Filter Efficiency at Same Specific Particle Size | Beta Ratio at a Specific Particle Size | Filter Efficiency at Same Specific Particle Size |
|---|---|---|---|---|---|
| 1.01 | 1% | 5 | 80% | 100 | 99% |
| 1.1 | 9% | 10 | 90% | 200 | 99.5% |
| 1.5 | 33% | 20 | 95% | 1000 | 99.9% |
| 2 | 50% | 75 | 98.7% | … | … |

**Food-Grade Lubricants.**—Machine lubricants used in food, beverage, and pharmaceutical industries must perform the same functions as any other technical lubricant, and resist degradation from food, sanitation chemicals, and water/steam, while adhering to international health and safety regulations. In 1944, the US-based National Sanitation Foundation (NSF), was founded; now known as NSF International, it continues with safety and certification programs, specifying protocols, and publishing health and safety standards.

Until 1998, US food-grade lubricants were regulated by the United States Department of Agriculture (USDA) and Food Safety and Inspection Services (FSIS). The design of food-grade grease and oil is now directed by the US Food and Drug Administration's (FDA's) document 21 CFR Part 178.3570 "Lubricants with Incidental Food Contact" (Table 22). In designing or maintaining machines and systems in food processing environments, it is recommended to err on the side of safety, using H1-approved lubricants. Base stocks may be mineral or synthetic oils; common thickening agents include aluminum stearate, aluminum or calcium sulfonate complexes, and organically modified phyllosilicate clays.

Risk and liability for food manufacturers also is guided by the globally recognized Hazard Analysis and Critical Control Point (HACCP) program, which ensures compliance with region-specific regulations and Global Food Safety Initiative (GFSI) protocols. NSF International collaborates with the World Health Organization (WHO) on international food-safe protocols. In addition, these organizations, the National Lubricating Grease Institute (NLGI), and the European Lubricating Grease Institute (ELGI) reference standard DIN V 0010517, 2000-08, "Food-Grade Lubricants–Definitions and Requirements."

During normal operations, the possibility of a lubricant making minor contact with food always exists. Since food-grade lubricants are technically not a food ingredient, this incidental contact cannot lead to contamination of food. To be registered as a lubricant allowed for incidental contact, the product must be formulated in accordance with the US FDA's Code of Federal Regulations Title 21 Section 178.3570 (21 CFR), which gives guidance on food-safe ingredients and includes a list of allowed lubricant chemical compounds and additives, states: (a) Lubricants registered as H1 and/or certified to ISO 21469 standards and with ingredients on the NSF GRAS (Generally Regarded as Safe) list are considered food-safe; (b) Such lubricant needs to be odorless, colorless, and tasteless; and (c) If incidental contact occurs, the amount of lubricant base oils (e.g., mineral oil) that can be present in food is limited to 10 parts per million (ppm).

**Table 22. NSF Food-Grade Lubricant Classes**

| Class | Purpose and Regulation |
|---|---|
| H1 | Food-grade lubricants (must be on the NSF list) considered safe where food or beverage contact is possible in the production line. All contents (base stock, additives) strictly regulated. Maximum level in foodstuffs is 10 ppm. |
| HX-1 | Ingredients for formulation of H1 food-grade lubricants. |
| HT-1 | Heat transfer fluids in cooling and heating systems where food or beverage contact is possible. Strictly regulated. |
| HTX-1 | Ingredients for formulation of HT-1 heat transfer fluids. |
| H2 | Food-safe lubricants for use on processing equipment and machine parts with no direct food contact. |
| HT-2 | Heat transfer fluids. Food contact is not allowed. |
| HTX-2 | Ingredients for formulation of HT-2 heat transfer fluids. |
| H3 | Food-grade soluble oils used for rust protection on equipment peripheral to food processing (hooks, trolleys, etc.) Must be removed before equipment is used around food. |
| 3H | Release agents used to prevent food products from sticking on molds, grills, or other equipment. Maximum level in foodstuffs is 10 ppm. |

## COUPLINGS, CLUTCHES, BRAKES

**Drive Couplings.—**An electric motor, steam turbine, and gas turbine are all examples of primary drivers that power pumps, compressors, or fans by transferring torque. To transmit power, these drivers or prime movers are connected to the driven equipment by drive couplings. Typically, shafts of the systems are collinear during operating condition. Coupling type is chosen based on its ability to function as needed, while absorbing the forces involved in the power transmission. As the weakest link in the system, the coupling should preferably fail before the more expensive drive or driven equipment.

**Connecting Shafts.—**Where misalignment is neither present nor desired, shafts may be connected with a simple rigid coupling, made of two halves, securely bolted together. For certain applications, where couplings transmit up to 150 horsepower, rigid, flange-type couplings of appropriate size (as shown in Table 1) can be used. These are known as safety or protected flange couplings, because the bolt heads and nuts are shrouded by the flange. Typically, such couplings are shielded by sheet metal or other covers.

For fractional horsepower and low-power applications, a setscrew may provide the locked connection between the hub and the shaft. A flat surface on the shaft and locking mechanism using setscrew(s) is recommended. High-power applications require a key and usually two setscrews, with one positioned above the key on each hub (drive and driven).

**Table 1. Safety or Protected Flange Couplings**

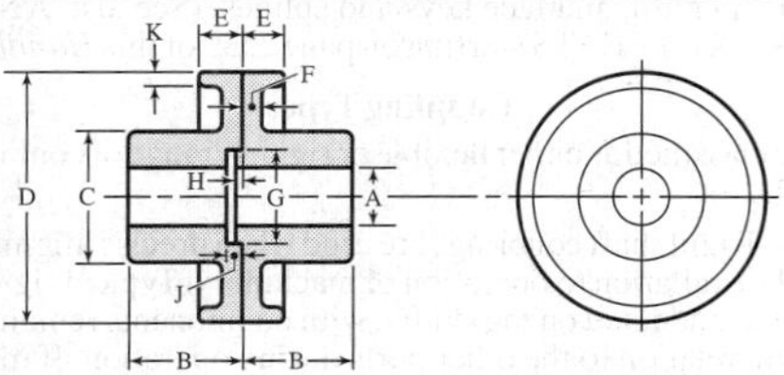


| A | B | C | D | E | F | G | H | J | K | Bolts No. | Bolts Dia. |
|---|---|---|---|---|---|---|---|---|---|---|---|
| 1 | 1 3/4 | 2 1/4 | 4 | 11/16 | 5/16 | 1 1/2 | 1/4 | 9/32 | 1/4 | 5 | 3/8 |
| 1 1/4 | 2 3/16 | 2 3/4 | 5 | 13/16 | 3/8 | 1 7/8 | 1/4 | 9/32 | 1/4 | 5 | 7/16 |
| 1 1/2 | 2 5/8 | 3 3/8 | 6 | 15/16 | 7/16 | 2 1/4 | 1/4 | 9/32 | 1/4 | 5 | 1/2 |
| 1 3/4 | 3 1/16 | 4 | 7 | 1 1/16 | 1/2 | 2 5/8 | 1/4 | 9/32 | 1/4 | 5 | 9/16 |
| 2 | 3 1/2 | 4 1/2 | 8 | 1 3/16 | 9/16 | 3 | 1/4 | 9/32 | 5/16 | 5 | 5/8 |
| 2 1/4 | 3 15/16 | 5 1/8 | 9 | 1 5/16 | 5/8 | 3 3/8 | 1/4 | 9/32 | 5/16 | 5 | 11/16 |
| 2 1/2 | 4 3/8 | 5 5/8 | 10 | 1 7/16 | 11/16 | 3 3/4 | 1/4 | 9/32 | 5/16 | 5 | 3/4 |
| 2 3/4 | 4 13/16 | 6 1/4 | 11 | 1 9/16 | 3/4 | 4 1/8 | 1/4 | 9/32 | 5/16 | 5 | 13/16 |
| 3 | 5 1/4 | 6 3/4 | 12 | 1 11/16 | 13/16 | 4 1/2 | 1/4 | 9/32 | 3/8 | 5 | 7/8 |
| 3 1/4 | 5 11/16 | 7 3/8 | 13 | 1 13/16 | 7/8 | 4 7/8 | 1/4 | 9/32 | 3/8 | 5 | 15/16 |
| 3 1/2 | 6 1/8 | 8 | 14 | 1 15/16 | 15/16 | 5 1/4 | 1/4 | 9/32 | 3/8 | 5 | 1 |
| 3 3/4 | 6 9/16 | 8 1/2 | 15 | 2 1/16 | 1 | 5 5/8 | 1/4 | 9/32 | 3/8 | 5 | 1 1/16 |
| 4 | 7 | 9 | 16 | 2 1/4 | 1 1/8 | 6 | 1/4 | 9/32 | 7/16 | 5 | 1 1/8 |
| 4 1/2 | 7 7/8 | 10 1/4 | 18 | 2 1/2 | 1 1/4 | 6 3/4 | 1/4 | 9/32 | 7/16 | 5 | 1 1/4 |
| 5 | 8 3/4 | 11 1/4 | 20 | 2 3/4 | 1 3/8 | 7 1/2 | 1/4 | 9/32 | 7/16 | 5 | 1 3/8 |
| 5 1/2 | 8 3/4 | 11 1/4 | 20 | 2 3/4 | 1 3/8 | 7 1/2 | 1/4 | 9/32 | 7/16 | 5 | 1 3/8 |
| 6 | 10 1/2 | 12 3/8 | 22 | 2 15/16 | 1 1/2 | 8 1/4 | 5/16 | 11/32 | 1/2 | 5 | 1 7/16 |
| 6 1/2 | 11 3/8 | 13 1/2 | 24 | 3 1/8 | 1 5/8 | 9 | 5/16 | 11/32 | 1/2 | 5 | 1 1/2 |
| 7 | 12 1/4 | 14 5/8 | 26 | 3 1/4 | 1 3/4 | 9 3/4 | 5/16 | 11/32 | 9/16 | 6 | 1 1/2 |
| 7 1/2 | 13 1/8 | 15 3/4 | 28 | 3 7/16 | 1 7/8 | 10 1/2 | 5/16 | 11/32 | 9/16 | 6 | 1 9/16 |
| 8 | 14 | 16 7/8 | 28 | 3 1/2 | 2 | 10 7/8 | 5/16 | 11/32 | 5/8 | 7 | 1 1/2 |
| 8 1/2 | 14 7/8 | 18 | 30 | 3 11/16 | 2 1/8 | 11 1/4 | 5/16 | 11/32 | 5/8 | 7 | 1 9/16 |
| 9 | 15 3/4 | 19 1/8 | 31 | 3 3/4 | 2 1/4 | 11 5/8 | 5/16 | 11/32 | 11/16 | 8 | 1 1/2 |
| 9 1/2 | 16 5/8 | 20 1/4 | 32 | 3 15/16 | 2 3/8 | 12 | 5/16 | 11/32 | 11/16 | 8 | 1 9/16 |
| 10 | 17 1/2 | 21 3/8 | 34 | 4 1/8 | 2 1/2 | 12 3/4 | 5/16 | 11/32 | 3/4 | 8 | 1 5/8 |
| 10 1/2 | 18 3/8 | 22 1/2 | 35 | 4 1/4 | 2 5/8 | 13 1/8 | 5/16 | 11/32 | 3/4 | 10 | 1 5/8 |
| 11 | 19 1/4 | 23 5/8 | 36 | 4 7/16 | 2 3/4 | 13 1/2 | 5/16 | 11/32 | 7/8 | 10 | 1 11/16 |
| 11 1/2 | 20 1/8 | 24 3/4 | 37 | 4 5/8 | 2 7/8 | 13 7/8 | 5/16 | 11/32 | 7/8 | 10 | 1 3/4 |
| 12 | 21 | 25 7/8 | 38 | 4 13/16 | 3 | 14 1/4 | 5/16 | 11/32 | 1 | 10 | 1 13/16 |

The ANSI/AGMA 9002-C14 Standard presents inch dimensions, tolerances, and sizes for straight bores, tapered bores, single keys, and keyways for unmounted industrial flexible couplings. In Class I and II fits, shaft tolerances are −0.0005 inch for shafts from $^1/_2$- to $1^1/_2$-inch diameters and −0.001 inch on larger diameters up to 7 inches. Class I coupling bore tolerances are +0.001 inch on sizes up to $1^1/_2$-inch diameters, then +0.0015 inch up to 7-inch diameters. Class II coupling bore tolerances are +0.002 inch on sizes up to 3-inch diameters, +0.003 inch from $3^1/_4$- to $3^3/_4$-inch diameters, and +0.004 inch on larger diameters up to 7 inches.

ANSI/AGMA 9112-B15 is the metric series equivalent Standard. It includes transitional fits, in addition to the clearance and interference fits. Per best practice, either only inch- or metric-based shaft and key systems should be used.

**Interference Fits and Keys.—** Couplings that use interference fits can provide tight fits to mating parts to help prevent movement and vibrations. Couplings transmitting high torque, or operating at high speeds, or both, may use one or two keys. Good fit of keys in their respective keyways helps to resist movement, reduce vibration, ensure proper transmission of torque, and prevent failure. Standardized sizes for rectangular keys used for couplings range from $^1/_8$ inch wide by $^1/_{16}$ inch high for $^1/_2$-inch diameter shafts to $1^3/_4$ inch wide by $^7/_8$ inch high for 7-inch diameter shafts.

AGMA standards provide recommendations for parallel keys in square, rectangular, and plain tapered forms, for shafts of $^5/_{16}$ through 7 inches diameter, in three classes designated commercial, precision, and fitted. These standards also cover keyway offset, lead, parallelism, finish and radii, and face keys and splines. (See also ANSI and other standards in KEYS AND KEY SEATS starting on page 2539 of this *Handbook*.)

## Coupling Types

Couplings can be classified as either flexible or rigid. A range of common coupling types is shown in Table 2.

**Rigid Couplings.—** Rigid shaft couplings are used when precise alignment of shafts is required from initial installation to operation of machinery. Typical rigid couplings halves are bored in each end and fixed on the shafts, with components remaining in a fixed (unmovable) position in relation to the other parts during operation. If misalignment of the two shafts occurs—due to inaccuracy in mounting, thermal growth during operation, or other abnormal conditions—rigid couplings can exert significant reactive forces on support components of the driving or driven equipment, resulting in unpredictable stresses that can lead to early failure.

**Flexible Couplings.—** Shafts that are out of alignment laterally or angularly can be connected by various designs of flexible couplings. Such couplings also permit some degree of axial movement in one or both shafts. Common types include metallic membrane couplings, elastomeric couplings, and metallic flexible couplings, though there are other types that do not fit in these categories.

*Elastomeric Couplings:* These couplings have resilient materials such as rubber or plastic attached to the flanges to transmit the torque and accommodate certain misalignment between the shafts. While less common these days, coupling hubs can also be connected by links or endless belts made of leather or other strong, pliable material. These non-lubricated couplings provide excellent vibration-damping capability, but are sensitive to chemicals and vulnerable to temperature extremes.

*Metallic Membrane Couplings:* Disks, diaphragms, or the bending of a membrane are used to provide flexibility while still effectively transmitting the torque. This type of coupling generally eliminates the need for lubrication, minimizing maintenance. Additionally, such couplings can transmit higher torque for a given area.

*Metallic Flexible Couplings:* These couplings use toothed hubs, which are engaged by corresponding toothed elements, transmitting power/torque. Such parts may be gears or grids that move against each other to transmit the required torque. These couplings require lubrication, unless one or more of the elements is made of a self-lubricating material. Such couplings do not have good vibration damping capabilities and fatigue/wear are the common cause of failure; however, they maintain high torsional rigidity and allow a considerable degree of angular misalignment.

**Table 2. Common Types of Rigid and Flexible Couplings**

| Coupling Type | Mechanism Description | Applications |
|---|---|---|
| | ***Rigid Couplings*** | |
| Split Muff 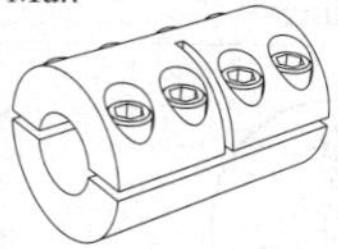 | The sleeve, or muff, of this rigid coupling is made of two halves securely bolted together. Used where shaft misalignment is neither present nor desired, this coupling can simplify assembly. | Vertical pumps, intermediate shafts of turbines, compressors, and conveyors. |
| Flange 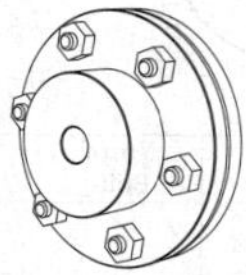 | This rigid coupling uses nuts and bolts to join two flanges that mount on the shaft ends with keys. Flange faces are machined at right angles to the shaft axis; often one face has a projected portion and the other has a matching recess, maintaining alignment. | Agitators, mixers, and some vertical drives. |
| | ***Flexible Couplings*** | |
| Elastomeric 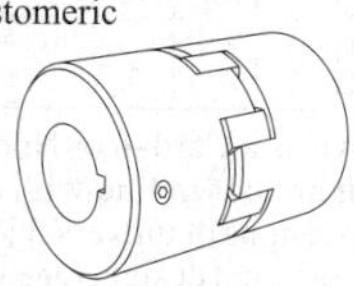 | The elastomeric spider coupling has two hubs with interlocking teeth and a flexible rubber/plastic disc in between. The bored hubs are fitted onto their shafts with keys. This design absorbs shock and can provide electrical isolation. | Pumps, fans, gearboxes, and electrical motor applications (e.g., servo motors and stepping motors). |
| Metallic Membrane 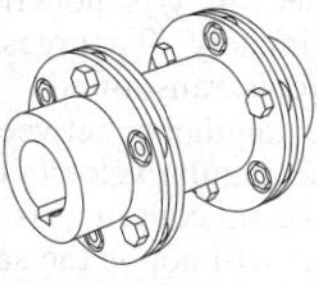 | Metallic membrane couplings use a flexible diaphragm to transmit torque between the shafts. This high-performance coupling is designed to remain torsionally strong under high torque loads, while accommodating shaft misalignment. | Centrifugal compressors, and high-speed and high-power equipment. |
| Gear (Metallic Flexible) 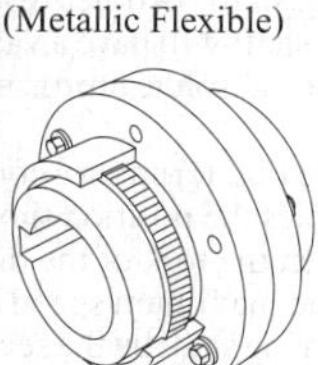 | This coupling utilizes gears between the driving and driven shafts. Two hubs, with crowned external teeth, and two flanges (or sleeves), with internal matching teeth, are bolted together with gaskets or O-rings. This lubricated design can withstand some parallel and angular misalignment and axial displacement. | Gearboxes, compressors, fans, blowers, rolling mills, conveyors, and elevators. |
| Universal Joint 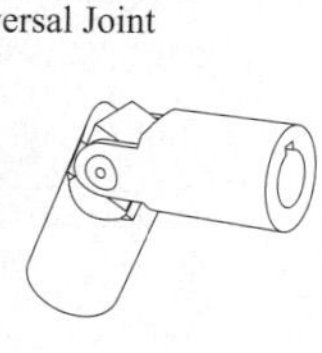 | Universal joint couplings connect and transmit torque between two non-aligned shafts. A pair of hinges, perpendicular to each other, are connected by a cross shaft. These couplings work well where space is restricted and/or high flexibility is needed. | Automotive driveshafts, and agricultural and construction equipment (e.g., lawnmowers, excavators, and cranes). |

**Double-Cone Clamping Couplings.—**As shown in Table 3, double-cone clamping couplings are made in a range of sizes for shafts from $1\frac{7}{16}$ to 6 inches in diameter, and are easily assembled to shafts. These couplings provide an interference fit, but they usually cost more and have larger overall dimensions than regular flanged couplings.

**Table 3. Double-Cone Clamping Couplings**

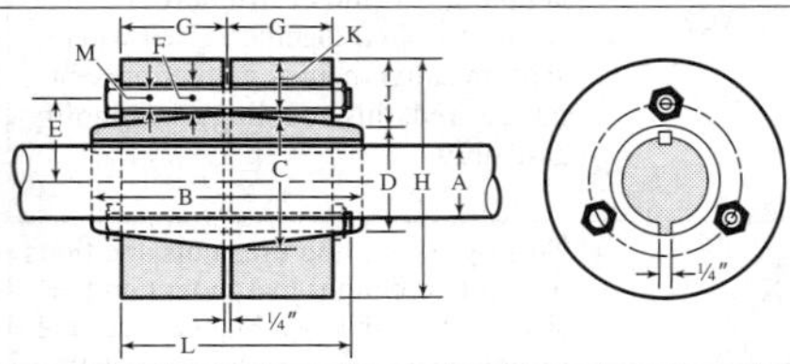


| A | B | C | D | E | F | G | H | J | K | L | M | No. of Bolts | No. of Keys |
|---|---|---|---|---|---|---|---|---|---|---|---|---|---|
| 1 7/16 | 5 1/4 | 2 3/4 | 2 1/8 | 1 5/8 | 5/8 | 2 1/8 | 4 3/4 | 1 1/8 | 1 | 5 | 1/2 | 3 | 1 |
| 1 15/16 | 7 | 3 1/2 | 2 7/8 | 2 1/8 | 5/8 | 2 3/4 | 6 1/4 | 1 1/8 | 1 3/8 | 6 1/4 | 1/2 | 3 | 1 |
| 2 7/16 | 8 3/4 | 4 5/16 | 3 5/8 | 3 | 3/4 | 3 1/2 | 7 13/16 | 1 7/8 | 1 3/4 | 7 7/8 | 5/8 | 3 | 1 |
| 3 | 10 1/2 | 5 1/2 | 4 3/32 | 3 1/2 | 3/4 | 4 3/16 | 9 | 2 1/4 | 2 | 9 1/2 | 5/8 | 3 | 1 |
| 3 1/2 | 12 1/4 | 7 | 5 3/8 | 4 3/8 | 7/8 | 5 1/16 | 11 1/4 | 2 5/8 | 2 1/8 | 11 1/4 | 3/4 | 4 | 1 |
| 4 | 14 | 7 | 5 1/2 | 4 3/4 | 7/8 | 5 1/2 | 12 | 3 3/4 | 2 1/2 | 12 | 3/4 | 4 | 1 |
| 4 1/2 | 15 1/2 | 8 | 6 7/8 | 5 1/4 | 7/8 | 6 3/4 | 13 1/2 | 3 3/4 | 2 3/4 | 14 1/2 | 3/4 | 4 | 1 |
| 5 | 17 | 9 | 7 1/4 | 5 3/4 | 7/8 | 7 | 15 | 3 3/4 | 3 | 15 1/4 | 3/4 | 4 | 1 |
| 5 1/2 | 17 1/2 | 9 1/2 | 7 3/4 | 6 1/4 | 1 | 7 | 15 1/2 | 3 3/4 | 3 | 15 1/4 | 7/8 | 4 | 1 |
| 6 | 18 | 10 | 8 1/4 | 6 3/4 | 1 | 7 | 16 | 3 3/4 | 3 | 15 1/4 | 7/8 | 4 | 2 |

**The Universal Joint.—**This form of coupling, originally known as a Cardan or Hooke's coupling, is used for connecting two shafts the axes of which are not in line with each other, but merely intersect at a point. There are many different designs of universal joints or couplings, which are based on the principle embodied in the original design. One well-known type is shown by the accompanying diagram.

As a rule, a universal joint does not work well if the angle $\alpha$ (see Fig. 1) is more than 45 degrees, and the angle should preferably be limited to approximately 20 degrees or 25 degrees, except when the speed of rotation is slow and little power is transmitted.

*Variation in Angular Velocity of Driven Shaft:* Owing to the angularity between two shafts connected by a universal joint, there is a variation in the angular velocity of one shaft during a single revolution; because of this, the use of universal couplings is sometimes prohibited. Thus, the angular velocity of the driven shaft will not be the same at all points of the revolution as the angular velocity of the driving shaft. In other words, if the driving shaft moves with a uniform motion, then the driven shaft will have a variable motion and, therefore, the universal joint should not be used when absolute uniformity of motion is essential for the driven shaft.

*Determining Maximum and Minimum Velocities:* If shaft A (see Fig. 1) runs at a constant speed, shaft B revolves at maximum speed when shaft A occupies the position shown in the illustration, and the minimum speed of shaft B occurs when the fork of the driving shaft A has turned 90 degrees from the position illustrated. The maximum speed of the driven shaft may be obtained by multiplying the speed of the driving shaft by the secant of angle $\alpha$. The minimum speed of the driven shaft equals the speed of the driver multiplied by cosine $\alpha$. Thus, if the driver rotates at a constant speed of 100 revolutions per minute

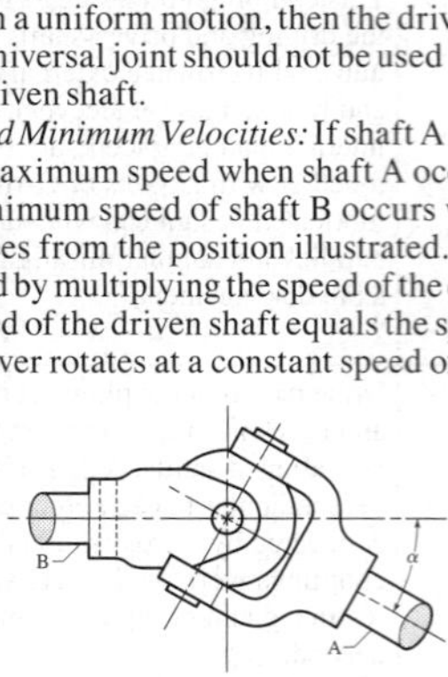


Fig. 1. Standard Universal Joint Design

and the shaft angle is 25 degrees, the maximum speed of the driven shaft is at a rate equal to 1.1034 × 100 = 110.34 rpm. The minimum speed rate equals 0.9063 × 100 = 90.63; hence, the extreme variation equals 110.34 – 90.63 = 19.71 rpm.

**Use of Intermediate Shaft between Two Universal Joints.—**The lack of uniformity in the speed of the driven shaft resulting from the use of a universal coupling, as previously explained, is objectionable for some forms of mechanisms. This variation may be avoided if the two shafts are connected with an intermediate shaft and two universal joints, provided the latter are properly arranged or located. Two conditions are necessary to obtain a constant speed ratio between the driving and driven shafts. First, the shafts must make the same angle with the intermediate shaft; second, the universal joint forks (assuming that the fork design is employed) on the intermediate shaft must be placed relatively so that when the plane of the fork at the left end coincides with the center lines of the intermediate shaft and the shaft attached to the left-hand coupling, the plane of the right-hand fork must also coincide with the center lines of the intermediate shaft and the shaft attached to the right-hand coupling; therefore the driving and the driven shafts may be placed in a variety of positions. One of the most common arrangements is with the driving and driven shafts parallel. The forks on the intermediate shafts should then be placed in the same plane.

This intermediate connecting shaft is frequently made telescoping, and then the driving and driven shafts can be moved independently of each other within certain limits in longitudinal and lateral directions. The telescoping intermediate shaft consists of a rod which enters a sleeve and is provided with a suitable spline, to prevent rotation between the rod and sleeve and permit a sliding movement. This arrangement is applied to various machine tools.

**Knuckle Joints.—**Movement at the joint between two rods may be provided by knuckle joints, for which typical proportions are seen in the table *Proportions of Knuckle Joints* on page 2532.

**Friction Clutches.—**Clutches which transmit motion from the driving to the driven member by the friction between the engaging surfaces are built in many different designs, although practically all of them can be classified under four general types, namely, conical clutches; radially expanding clutches; contracting-band clutches; and friction disk clutches in single and multiple types. There are many modifications of these general classes, some of which combine the features of different types. The proportions of various sizes of cone clutches are given in the table "Cast-iron Friction Clutches."

The multicone friction clutch is a further development of the cone clutch. Instead of having a single cone-shaped surface, there is a series of concentric conical rings which engage annular grooves formed by corresponding rings on the opposite clutch member. The internal-expanding type is provided with shoes which are forced outward against an enclosing drum by the action of levers connecting with a collar free to slide along the shaft. The engaging shoes are commonly lined with wood or other material to increase the coefficient of friction. Disk clutches are based on the principle of multiple-plane friction, and use alternating plates or disks so arranged that one set engages with an outside cylindrical case and the other set with the shaft. When these plates are pressed together by spring pressure, or by other means, motion is transmitted from the driving to the driven members connected to the clutch. Some disk clutches have a few rather heavy or thick plates and others a relatively large number of thinner plates. Clutches of the latter type are common in automobile transmissions. One set of disks may be of soft steel and the other set of phosphor-bronze, or some other combination may be employed. For instance, disks are sometimes provided with cork inserts, or one set or series of disks may be faced with a special friction material such as asbestos-wire fabric, as in "dry plate" clutches, the disks of which are not lubricated like the disks of a clutch having, for example, the steel and phosphor-bronze combination. It is common practice to hold the driving and driven members of friction clutches in engagement by means of spring pressure, although pneumatic or hydraulic pressure may be employed.

## Proportions of Knuckle Joints

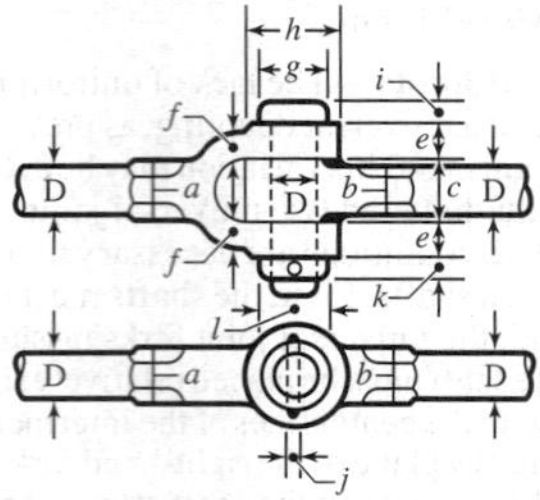


For sizes not given below:

$a = 1.2D$  $h = 2D$
$b = 1.1D$  $i = 0.5D$
$c = 1.2D$  $j = 0.25D$
$e = 0.75D$  $k = 0.5D$
$f = 0.6D$  $l = 1.5D$
$g = 1.5D$

| *D* | *a* | *b* | *c* | *e* | *f* | *g* | *h* | *i* | *j* | *k* | *l* |
|---|---|---|---|---|---|---|---|---|---|---|---|
| ½ | ⅝ | 9⁄16 | ⅝ | ⅜ | 5⁄16 | ¾ | 1 | ¼ | ⅛ | ¼ | ¾ |
| ¾ | ⅞ | ¾ | ⅞ | 9⁄16 | 7⁄16 | 1⅛ | 1½ | ⅜ | 3⁄16 | ⅜ | 1⅛ |
| 1 | 1¼ | 1⅛ | 1¼ | ¾ | ⅝ | 1½ | 2 | ½ | ¼ | ½ | 1½ |
| 1¼ | 1½ | 1⅜ | 1½ | 15⁄16 | ¾ | 1⅞ | 2½ | ⅝ | 5⁄16 | ⅝ | 1⅞ |
| 1½ | 1¾ | 1⅝ | 1¾ | 1⅛ | ⅞ | 2¼ | 3 | ¾ | ⅜ | ¾ | 2¼ |
| 1¾ | 2⅛ | 2 | 2⅛ | 15⁄16 | 11⁄16 | 2⅝ | 3½ | ⅞ | 7⁄16 | ⅞ | 2⅝ |
| 2 | 2⅜ | 2¼ | 2⅜ | 1½ | 13⁄16 | 3 | 4 | 1 | ½ | 1 | 3 |
| 2¼ | 2¾ | 2½ | 2¾ | 111⁄16 | 1⅜ | 3⅜ | 4½ | 1⅛ | 9⁄16 | 1⅛ | 3⅜ |
| 2½ | 3 | 2¾ | 3 | 1⅞ | 1½ | 3¾ | 5 | 1¼ | ⅝ | 1¼ | 3¾ |
| 2¾ | 3¼ | 3 | 3¼ | 21⁄16 | 1⅝ | 4⅛ | 5½ | 1⅜ | 11⁄16 | 1⅜ | 4⅛ |
| 3 | 3⅝ | 3¼ | 3⅝ | 2¼ | 113⁄16 | 4½ | 6 | 1½ | ¾ | 1½ | 4½ |
| 3¼ | 4 | 3⅝ | 4 | 27⁄16 | 2 | 4⅞ | 6½ | 1⅝ | 13⁄16 | 1⅝ | 4⅞ |
| 3½ | 4¼ | 3⅞ | 4¼ | 2⅝ | 2⅛ | 5¼ | 7 | 1¾ | ⅞ | 1¾ | 5¼ |
| 3¾ | 4½ | 4⅛ | 4½ | 213⁄16 | 2¼ | 5⅝ | 7½ | 1⅞ | 15⁄16 | 1⅞ | 5⅝ |
| 4 | 4¾ | 4⅜ | 4¾ | 3 | 2⅜ | 6 | 8 | 2 | 1 | 2 | 6 |
| 4¼ | 5⅛ | 4¾ | 5⅛ | 33⁄16 | 29⁄16 | 6⅜ | 8½ | 2⅛ | 11⁄16 | 2⅛ | 6⅜ |
| 4½ | 5½ | 5 | 5½ | 3⅜ | 2¾ | 6¾ | 9 | 2¼ | 1⅛ | 2¼ | 6¾ |
| 4¾ | 5¾ | 5¼ | 5¾ | 39⁄16 | 2⅞ | 7⅛ | 9½ | 2⅜ | 13⁄16 | 2⅜ | 7⅛ |
| 5 | 6 | 5½ | 6 | 3¾ | 3 | 7½ | 10 | 2½ | 1¼ | 2½ | 7½ |

**Power Transmitting Capacity of Friction Clutches.—**When selecting a clutch for a given class of service, it is advisable to consider any overloads that may be encountered and base the power transmitting capacity of the clutch upon such overloads. When the load varies or is subject to frequent release or engagement, the clutch capacity should be greater than the actual amount of power transmitted. If the power is derived from a gas or gasoline engine, the horsepower rating of the clutch should be 75 or 100 percent greater than that of the engine.

**Power Transmitted by Disk Clutches.—**The approximate amount of power that a disk clutch will transmit may be determined from the following formula, in which $HP$ = horsepower transmitted by the clutch; $\mu$ = coefficient of friction; $r$ = mean radius of engaging surfaces; $F$ = axial force in pounds (spring pressure) holding disks in contact; $N$ = number of frictional surfaces; $S$ = speed of shaft in revolutions per minute:

$$HP = \frac{\mu rFNS}{63{,}000}$$

## Cast-Iron Friction Clutches

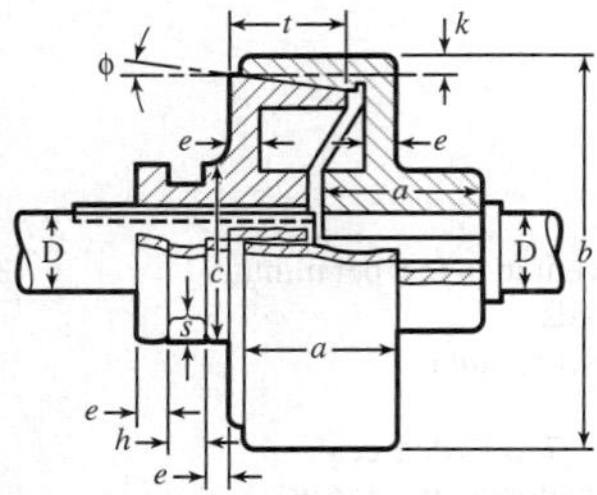


For sizes not given below:

$a = 2D$
$b = 4$ to $8D$
$c = 2\frac{1}{4}D$
$t = 1\frac{1}{2}D$
$e = \frac{3}{8}D$
$h = \frac{1}{2}D$
$s = \frac{5}{16}D$, nearly
$k = \frac{1}{4}D$

*Note:* The angle ϕ of the cone may be from 4 to 10 degrees

| *D* | *a* | *b* | *c* | *t* | *e* | *h* | *s* | *k* |
|---|---|---|---|---|---|---|---|---|
| 1 | 2 | 4-8 | 2 1/4 | 1 1/2 | 3/8 | 1/2 | 5/16 | 1/4 |
| 1 1/4 | 2 1/2 | 5-10 | 2 7/8 | 1 7/8 | 1/2 | 5/8 | 3/8 | 5/16 |
| 1 1/2 | 3 | 6-12 | 3 3/8 | 2 1/4 | 5/8 | 3/4 | 1/2 | 3/8 |
| 1 3/4 | 3 1/2 | 7-14 | 4 | 2 5/8 | 5/8 | 7/8 | 5/8 | 7/16 |
| 2 | 4 | 8-16 | 4 1/2 | 3 | 3/4 | 1 | 5/8 | 1/2 |
| 2 1/4 | 4 1/2 | 9-18 | 5 | 3 3/8 | 7/8 | 1 1/8 | 5/8 | 9/16 |
| 2 1/2 | 5 | 10-20 | 5 5/8 | 3 3/4 | 1 | 1 1/4 | 3/4 | 5/8 |
| 2 3/4 | 5 1/2 | 11-22 | 6 1/4 | 4 1/8 | 1 | 1 3/8 | 7/8 | 11/16 |
| 3 | 6 | 12-24 | 6 3/4 | 4 1/2 | 1 1/8 | 1 1/2 | 7/8 | 3/4 |
| 3 1/4 | 6 1/2 | 13-26 | 7 3/8 | 4 7/8 | 1 1/4 | 1 5/8 | 1 | 13/16 |
| 3 1/2 | 7 | 14-28 | 7 7/8 | 5 1/4 | 1 3/8 | 1 3/4 | 1 | 7/8 |
| 3 3/4 | 7 1/2 | 15-30 | 8 1/2 | 5 5/8 | 1 3/8 | 1 7/8 | 1 1/4 | 15/16 |
| 4 | 8 | 16-32 | 9 | 6 | 1 1/2 | 2 | 1 1/4 | 1 |
| 4 1/4 | 8 1/2 | 17-34 | 9 1/2 | 6 3/8 | 1 5/8 | 2 1/8 | 1 3/8 | 1 1/16 |
| 4 1/2 | 9 | 18-36 | 10 1/4 | 6 3/4 | 1 3/4 | 2 1/4 | 1 3/8 | 1 1/8 |
| 4 3/4 | 9 1/2 | 19-38 | 10 3/4 | 7 1/8 | 1 3/4 | 2 3/8 | 1 1/2 | 1 3/16 |
| 5 | 10 | 20-40 | 11 1/4 | 7 1/2 | 1 7/8 | 2 1/2 | 1 1/2 | 1 1/4 |
| 5 1/4 | 10 1/2 | 21-42 | 11 3/4 | 7 7/8 | 2 | 2 5/8 | 1 5/8 | 1 5/16 |
| 5 1/2 | 11 | 22-44 | 12 3/8 | 8 1/4 | 2 | 2 3/4 | 1 3/4 | 1 3/8 |
| 5 3/4 | 11 1/2 | 23-46 | 13 | 8 5/8 | 2 1/4 | 2 7/8 | 1 3/4 | 1 7/16 |
| 6 | 12 | 24-48 | 13 1/2 | 9 | 2 1/4 | 3 | 1 7/8 | 1 1/2 |

**Frictional Coefficients for Clutch Calculations.—**While the frictional coefficients used by designers of clutches differ somewhat and depend upon variable factors, the following values may be used in clutch calculations: For greasy leather on cast iron about 0.20 or 0.25, leather on metal that is quite oily 0.15; metal and cork on oily metal 0.32; the same on dry metal 0.35; metal on dry metal 0.15; disk clutches having lubricated surfaces 0.10.

**Formulas for Cone Clutches.—**In cone clutch design, different formulas have been developed for determining the horsepower transmitted. These formulas, at first sight, do not seem to agree, there being a variation due to the fact that in some of the formulas the friction clutch surfaces are assumed to engage without slip, whereas, in others, some allowance is made for slip. The following formulas include both of these conditions:

$HP$ = horsepower transmitted
$N$ = revolutions per minute
$r$ = mean radius of friction cone, in inches
$r_1$ = large radius of friction cone, in inches
$r_2$ = small radius of friction cone, in inches
$R_1$ = outside radius of leather band, in inches
$R_2$ = inside radius of leather band, in inches
$V$ = velocity of a point at distance $r$ from the center, in feet per minute
$F$ = tangential force acting at radius $r$, in pounds
$P_n$ = total normal force between cone surfaces, in pounds
$P_s$ = spring force, in pounds
$\alpha$ = angle of clutch surface with axis of shaft = 7 to 13 degrees
$\beta$ = included angle of clutch leather, when developed, in degrees
$f$ = coefficient of friction = 0.20 to 0.25 for greasy leather on iron
$p$ = allowable pressure per square inch of leather band = 7 to 8 pounds
$W$ = width of clutch leather, in inches

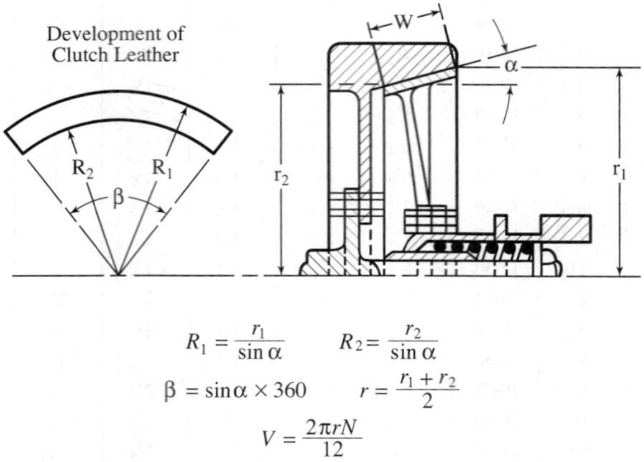


$$R_1 = \frac{r_1}{\sin\alpha} \qquad R_2 = \frac{r_2}{\sin\alpha}$$

$$\beta = \sin\alpha \times 360 \qquad r = \frac{r_1 + r_2}{2}$$

$$V = \frac{2\pi rN}{12}$$

$$F = \frac{HP \times 33{,}000}{V} \qquad W = \frac{P_n}{2\pi rp} \qquad HP = \frac{P_n frN}{63{,}025}$$

For engagement with some slip:

$$P_n = \frac{P_s}{\sin\alpha} \qquad P_s = \frac{HP \times 63{,}025\sin\alpha}{frN}$$

For engagement without slip:

$$P_n = \frac{P_s}{\sin\alpha + f\cos\alpha} \qquad P_s = \frac{HP \times 63{,}025(\sin\alpha + f\cos\alpha)}{frN}$$

**Angle of Cone.—**If the angle of the conical surface of the cone type of clutch is too small, it may be difficult to release the clutch on account of the wedging effect, whereas, if the angle is too large, excessive spring force will be required to prevent slipping. The minimum angle for a leather-faced cone is about 8 or 9 degrees and the maximum angle about 13 degrees. An angle of 12½ degrees appears to be the most common and is generally considered good practice. These angles are given with relation to the clutch axis and are one-half the included angle.

**Magnetic Clutches.**—Many disk and other clutches are operated electromagnetically with the magnetic force used only to move the friction disk(s) and the clutch disk(s) into or out of engagement against spring or other pressure. On the other hand, in a magnetic particle clutch, transmission of power is accomplished by magnetizing a quantity of metal particles enclosed between the driving and the driven components, forming a bond between them. Such clutches can be controlled to provide either a rigid coupling or uniform slip, useful in wire drawing and manufacture of cables.

Another type of magnetic clutch uses eddy currents induced in the input member which interact with the field in the output rotor. Torque transmitted is proportional to the coil current, so precise control of torque is provided. A third type of magnetic clutch relies on the hysteresis loss between magnetic fields generated by a coil in an input drum and a close-fitting cup on the output shaft, to transmit torque. Torque transmitted with this type of clutch also is proportional to coil current, so close control is possible.

Permanent-magnet types of clutches also are available, in which the engagement force is exerted by permanent magnets when the electrical supply to the disengagement coils is cut off. These types of clutches have capacities up to five times the torque-to-weight ratio of spring-operated clutches. In addition, if the controls are so arranged as to permit the coil polarity to be reversed instead of being cut off, the combined permanent magnet and electromagnetic forces can transmit even greater torque.

**Centrifugal and Free-Wheeling Clutches.**—Centrifugal clutches have driving members that expand outward to engage a surrounding drum when speed is sufficient to generate centrifugal force. Free-wheeling clutches are made in many different designs and use balls, cams or sprags, ratchets, and fluids to transmit motion from one member to the other. These types of clutches are designed to transmit torque in only one direction and to take up the drive with various degrees of gradualness up to instantaneously.

**Slipping Clutch/Couplings.**—Where high shock loads are likely to be experienced, a slipping clutch or coupling or both should be used. The most common design uses a clutch plate that is clamped between the driving and driven plates by spring pressure that can be adjusted. When excessive load causes the driven member to slow, the clutch plate surfaces slip, allowing reduction of the torque transmitted. When the overload is removed, the drive is taken up automatically. Switches can be provided to cut off current supply to the driving motor when the driven shaft slows to a preset limit or to signal a warning or both. The slip or overload torque is calculated by taking 150 percent of the normal running torque.

**Wrapped-Spring Clutches.**—For certain applications, a simple steel spring sized so that its internal diameter is a snug fit on both driving and driven shafts will transmit adequate torque in one direction. The tightness of grip of the spring on the shafts increases as the torque transmitted increases. Disengagement can be effected by slight rotation of the spring, through a projecting tang, using electrical or mechanical means, to wind up the spring to a larger internal diameter, allowing one of the shafts to run free within the spring.

Normal running torque $T_r$ in lb-ft = (required horsepower × 5250) ÷ rpm. For heavy shock load applications, multiply by a 200 percent or greater overload factor. (See Motors, factors governing selection.)

The clutch starting torque $T_c$, in lb-ft, required to accelerate a given inertia in a specific time is calculated from the formula:

$$T_c = \frac{WR^2 \times \Delta N}{308t}$$

where $WR^2$ = total inertia encountered by clutch in lb-ft$^2$ ($W$ = weight and $R$ = radius of gyration of rotating part)

$\Delta N$ = final rpm − initial rpm

308 = constant (see *Factors Governing Motor Selection* on page 2658)

$t$ = time to required speed in seconds

*Example 1:* If the inertia is 80 lb-ft², and the speed of the driven shaft is to be increased from 0 to 1500 rpm in 3 seconds, the clutch starting torque in lb-ft is

$$T_c = \frac{80 \times 1500}{308 \times 3} = 130 \text{ lb-ft}$$

The heat $E$, in BTU, generated in one engagement of a clutch can be calculated from the formula:

$$E = \frac{T_c \times WR^2 \times (N_1^2 - N_2^2)}{(T_c - T_1) \times 4.7 \times 10^6}$$

where $WR^2$ = total inertia encountered by clutch in lb-ft²

$N_1$ = final rpm  $N_2$ = initial rpm

$T_c$ = clutch torque in lb-ft  $T_1$ = torque load in lb-ft

*Example 2:* Calculate the heat generated for each engagement under the conditions cited for Example 1.

$$E = \frac{130 \times 80 \times (1500)^2}{(130 - 10) \times 4.7 \times 10^6} = 41.5 \text{ BTU}$$

The preferred location for a clutch is on the high- rather than on the low-speed shaft because a smaller-capacity unit, of lower cost and with more rapid dissipation of heat, can be used. However, the heat generated may also be more because of the greater slippage at higher speeds, and the clutch may have a shorter life. For light-duty applications, such as to a machine tool, where cutting occurs after the spindle has reached operating speed, the calculated torque should be multiplied by a safety factor of 1.5 to arrive at the capacity of the clutch to be used. Heavy-duty applications such as frequent starting of a heavily loaded vibratory-finishing barrel require a safety factor of 3 or more.

**Positive Clutches.**—When the driving and driven members of a clutch are connected by the engagement of interlocking teeth or projecting lugs, the clutch is said to be "positive" to distinguish it from the type in which the power is transmitted by frictional contact. The positive clutch is employed when a sudden starting action is not objectionable and when the inertia of the driven parts is relatively small. The various forms of positive clutches differ merely in the angle or shape of the engaging surfaces. The least positive form is one having planes of engagement which incline backward, with respect to the direction of motion. The tendency of such a clutch is to disengage under load, in which case it must be held in position by axial pressure.

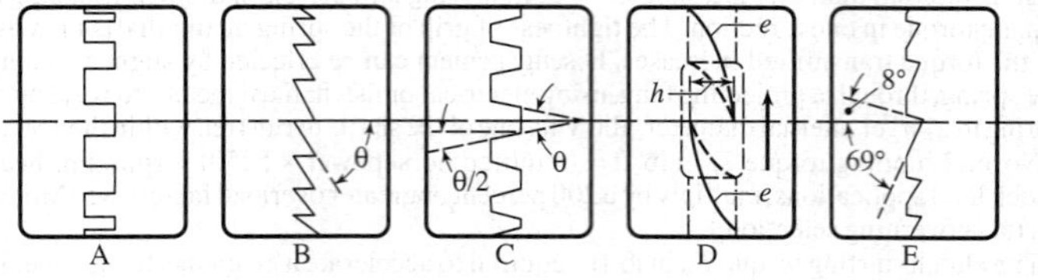


Fig. 1. Types of Clutch Teeth

This pressure may be regulated to perform normal duty, permitting the clutch to slip and disengage when over-loaded. Positive clutches, with the engaging planes parallel to the axis of rotation, are held together to obviate the tendency to jar out of engagement, but they provide no safety feature against over-load. So-called "under-cut" clutches engage more tightly the heavier the load, and are designed to be disengaged only when free from load. The teeth of positive clutches are made in a variety of forms, a few of the more common styles being shown in Fig. 1. Clutch *A* is a straight-toothed type, and *B* has angular or saw-shaped teeth. The driving member of the former can be rotated in either direction the latter is adapted to the transmission of motion in one direction only, but is more readily engaged. The angle θ of the cutter for a saw-tooth clutch *B* is ordinarily 60 degrees. Clutch *C* is similar to *A*, except that the sides of the teeth are inclined to facilitate engagement and disengagement. Teeth of this shape are sometimes used when a clutch is required to run in either direction without backlash. Angle θ is varied to suit requirements and should not exceed 16 or 18 degrees. The straight-tooth clutch *A* is also modified to make the teeth engage more readily, by rounding the corners of the teeth at the top and bottom. Clutch *D* (commonly called a "spiral-jaw" clutch) differs from *B* in that the surfaces *e* are helicoidal. The driving member of this clutch can transmit motion in only one direction

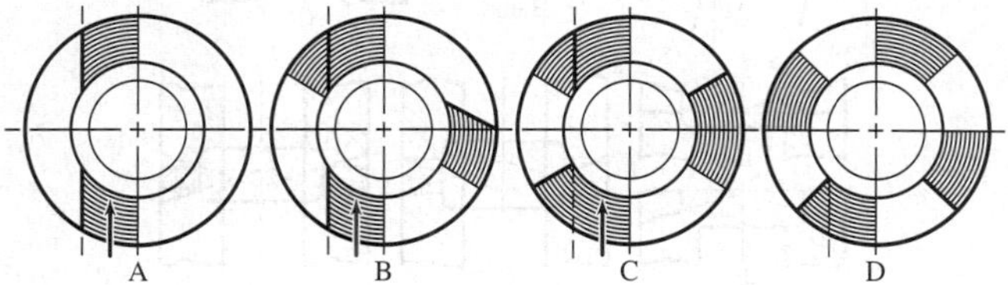


Fig. 2. Diagrammatic View Showing Method of Cutting Clutch Teeth

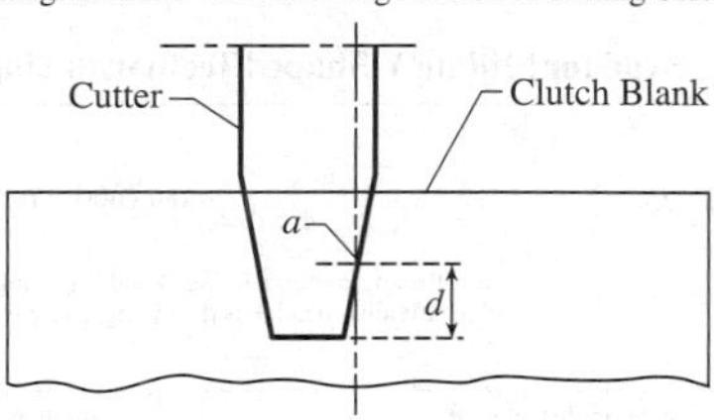


Fig. 3.

Clutches of this type are known as right- and left-hand, the former driving when turning to the right, as indicated by the arrow in the illustration. Clutch *E* is the form used on the back-shaft of the Brown & Sharpe automatic screw machines. The faces of the teeth are radial and incline at an angle of 8 degrees with the axis, so that the clutch can readily be disengaged. This type of clutch is easily operated, with little jar or noise. The 2-inch (50.8-mm) diameter size has 10 teeth. Height of working face, $^1/_8$ inch (3.175 mm).

**Cutting Clutch Teeth.—**A common method of cutting a straight-tooth clutch is indicated by the diagrams *A*, *B* and *C*, Fig. 2, which show the first, second and third cuts required for forming the three teeth. The work is held in the chuck of a dividing head, the latter being set at right angles to the table. A plain milling cutter may be used (unless the corners of the teeth are rounded), the side of the cutter being set to exactly coincide with the centerline. When the number of teeth in the clutch is odd, the cut can be taken clear across the blank as shown, thus finishing the sides of two teeth with one passage of the cutter. When the number of teeth is even, as at *D*, it is necessary to mill all the teeth on one side and then set the cutter for finishing the opposite side. Therefore, clutches of this type commonly have an odd number of teeth. The maximum width of the cutter depends upon the width of the space at the narrow ends of the teeth. If the cutter must be quite narrow in order to pass the narrow ends, some stock may be left in the tooth spaces, which must be removed by a separate cut. If the tooth is of the modified form shown at *C*, Fig. 1, the cutter should be set as indicated in Fig. 3; that is, so that a point a on the cutter at a radial distance *d* equal to one-half the depth of the clutch teeth lies in a radial plane. When it is important to eliminate all backlash, point *a* is sometimes located at a radial distance *d* equal to six-tenths of the depth of the tooth, in order to leave clearance spaces at the bottoms of the teeth; the two clutch members will then fit together tightly. Clutches of this type must be held in mesh.

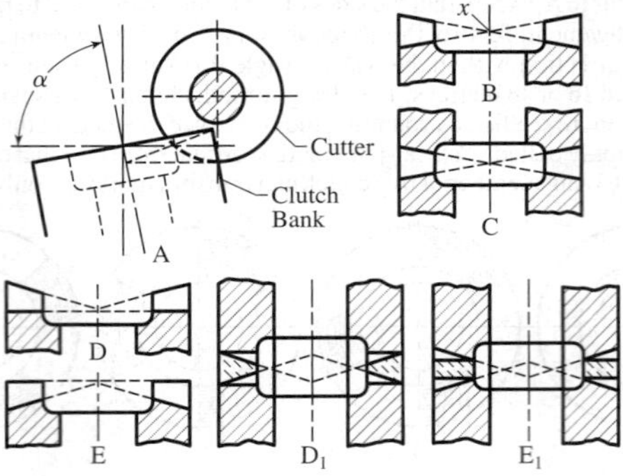


Fig. 4.

**Angle of Dividing Head for Milling V-Shaped Teeth with Single-Angle Cutter**

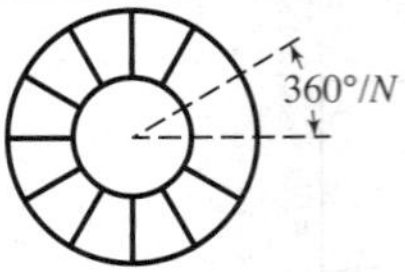


$$\cos\alpha = \frac{\tan(360°/N) \times \cot\theta}{2}$$

α is the angle shown in Fig. 4 and is the angle shown by the graduations on the dividing head. θ is the included angle of a single cutter, see Fig. 1.

| No. of Teeth, *N* | Angle of Single-angle Cutter, θ | | | | | | No. of Teeth, *N* | Angle of Single-angle Cutter, θ | | | | | |
|---|---|---|---|---|---|---|---|---|---|---|---|---|---|
| | 60° | | 70° | | 80° | | | 60° | | 70° | | 80° | |
| | Dividing Head Angle, α | | | | | | | Dividing Head Angle, α | | | | | |
| 5 | 27° | 19.2′ | 55° | 56.3′ | 74° | 15.4′ | 18 | 83° | 58.1′ | 86° | 12.1′ | 88° | 9.67′ |
| 6 | 60 | | 71 | 37.6 | 81 | 13 | 19 | 84 | 18.8 | 86 | 25.1 | 88 | 15.9 |
| 7 | 68 | 46.7 | 76 | 48.5 | 83 | 39.2 | 20 | 84 | 37.1 | 86 | 36.6 | 88 | 21.5 |
| 8 | 73 | 13.3 | 79 | 30.9 | 84 | 56.5 | 21 | 84 | 53.5 | 86 | 46.9 | 88 | 26.5 |
| 9 | 75 | 58.9 | 81 | 13 | 85 | 45.4 | 22 | 85 | 8.26 | 86 | 56.2 | 88 | 31 |
| 10 | 77 | 53.6 | 82 | 24.1 | 86 | 19.6 | 23 | 85 | 21.6 | 87 | 4.63 | 88 | 35.1 |
| 11 | 79 | 18.5 | 83 | 17 | 86 | 45.1 | 24 | 85 | 33.8 | 87 | 12.3 | 88 | 38.8 |
| 12 | 80 | 24.4 | 83 | 58.1 | 87 | 4.94 | 25 | 85 | 45 | 87 | 19.3 | 88 | 42.2 |
| 13 | 81 | 17.1 | 84 | 31.1 | 87 | 20.9 | 26 | 85 | 55.2 | 87 | 25.7 | 88 | 45.3 |
| 14 | 82 | .536 | 84 | 58.3 | 87 | 34 | 27 | 86 | 4.61 | 87 | 31.7 | 88 | 48.2 |
| 15 | 82 | 36.9 | 85 | 21.2 | 87 | 45 | 28 | 86 | 13.3 | 87 | 37.2 | 88 | 50.8 |
| 16 | 83 | 7.95 | 85 | 40.6 | 87 | 54.4 | 29 | 86 | 21.4 | 87 | 42.3 | 88 | 53.3 |
| 17 | 83 | 34.7 | 85 | 57.4 | 88 | 2.56 | 30 | 86 | 28.9 | 87 | 47 | 88 | 55.6 |

*Cutting Saw-Tooth Clutches:* When milling clutches having angular teeth as shown at *B*, Fig. 1, the axis of the clutch blank should be inclined a certain angle α as shown at *A* in Fig. 4. If the teeth were milled with the blank vertical, the tops of the teeth would incline towards the center as at *D*, whereas, if the blank were set to such an angle that the tops of the teeth were square with the axis, the bottoms would incline upwards as at *E*. In either case, the two clutch members would not mesh completely: the engagement of the teeth cut as shown at *D* and *E* would be as indicated at $D_1$ and $E_1$ respectively. As will be seen, when the outer points of the teeth at $D_1$ are at the bottom of the grooves in the opposite member, the inner ends are not together, the contact area being represented by the dotted lines. At $E_1$ the inner ends of the teeth strike first and spaces are left between the teeth around the outside of the clutch. To overcome this objectionable feature, the clutch teeth should be cut as indicated at *B*, or so that the bottoms and tops of the teeth have the same inclination, converging at a central point *x*. The teeth of both members will then engage across the entire width as shown at *C*. The angle α required for cutting a clutch as at *B* can be determined by the following formula in which α equals the required angle, *N* = number of teeth, θ = cutter angle, and 360°/*N* = angle between teeth:

$$\cos\alpha = \frac{\tan(360°/N) \times \cot\theta}{2}$$

The angles α for various numbers of teeth and for 60-, 70- or 80-degree single-angle cutters are given in the table on page 2538. The following table is for double-angle cutters used to cut V-shaped teeth.

**Angle of Dividing Head for Milling V-Shaped Teeth with Double-Angle Cutter**

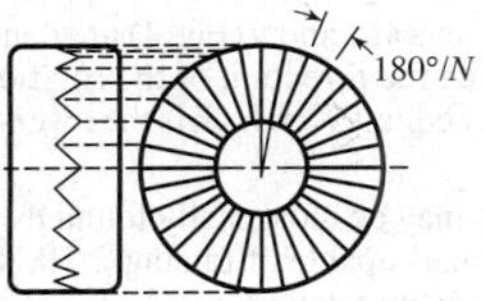


$$\cos\alpha = \frac{\tan(180°/N) \times \cot(\theta/2)}{2}$$

This is the angle (α, Fig. 4) shown by graduations on the dividing head. θ is the included angle of a double-angle cutter, see Fig. 1.

| No. of Teeth, *N* | Included Angle of Cutter, θ | | | | No. of Teeth, *N* | Included Angle of Cutter, θ | | | |
|---|---|---|---|---|---|---|---|---|---|
| | 60° | | 90° | | | 60° | | 90° | |
| | Dividing Head Angle, α | | | | | Dividing Head Angle, α | | | |
| 10 | 73° | 39.4′ | 80° | 39′ | 31 | 84° | 56.9′ | 87° | 5.13′ |
| 11 | 75 | 16.1 | 81 | 33.5 | 32 | 85 | 6.42 | 87 | 1[illegible].6 |
| 12 | 76 | 34.9 | 82 | 18 | 33 | 85 | 15.4 | 87 | 15.8 |
| 13 | 77 | 40.5 | 82 | 55.3 | 34 | 85 | 23.8 | 87 | 20.7 |
| 14 | 78 | 36 | 83 | 26.8 | 35 | 85 | 31.8 | 87 | 25.2 |
| 15 | 79 | 23.6 | 83 | 54 | 36 | 85 | 39.3 | 87 | 29.6 |
| 16 | 80 | 4.83 | 84 | 17.5 | 37 | 85 | 46.4 | 87 | 33.7 |
| 17 | 80 | 41 | 84 | 38.2 | 38 | 85 | 53.1 | 87 | 37.5 |
| 18 | 81 | 13 | 84 | 56.5 | 39 | 85 | 59.5 | 87 | 41.2 |
| 19 | 81 | 41.5 | 85 | 12.8 | 40 | 86 | 5.51 | 87 | 44.7 |
| 20 | 82 | 6.97 | 85 | 27.5 | 41 | 86 | 11.3 | 87 | 48 |
| 21 | 82 | 30 | 85 | 40.7 | 42 | 86 | 16.7 | 87 | 51.2 |
| 22 | 82 | 50.8 | 85 | 52.6 | 43 | 86 | 22 | 87 | 54.2 |
| 23 | 83 | 9.82 | 86 | 3.56 | 44 | 86 | 26.9 | 87 | 57 |
| 24 | 83 | 27.2 | 86 | 13.5 | 45 | 86 | 31.7 | 87 | 59.8 |
| 25 | 83 | 43.1 | 86 | 22.7 | 46 | 86 | 36.2 | 88 | 2.4 |
| 26 | 83 | 57.8 | 86 | 31.2 | 47 | 86 | 40.6 | 88 | 4.91 |
| 27 | 84 | 11.4 | 86 | 39 | 48 | 86 | 44.8 | 88 | 7.32 |
| 28 | 84 | 24 | 86 | 46.2 | 49 | 86 | 48.8 | 88 | 9.63 |
| 29 | 84 | 35.7 | 86 | 53 | 50 | 86 | 52.6 | 88 | 11.8 |
| 30 | 84 | 46.7 | 86 | 59.3 | 51 | 86 | 56.3 | 88 | 14 |

The angles given in the table above are applicable to the milling of V-shaped grooves in brackets, etc., which must have toothed surfaces to prevent the two members from turning relative to each other, except when unclamped for angular adjustment

## Friction Brakes

**Formulas for Band Brakes.**—In any band brake, such as shown in Fig. 1, in the tabulation of formulas, where the brake wheel rotates in a clockwise direction, the tension in that part of the band marked *x* equals $P\frac{1}{e^{\mu\theta}-1}$

The tension in that part marked *y* equals $P\frac{e^{\mu\theta}}{e^{\mu\theta}-1}$.

$P$ = tangential force in pounds at rim of brake wheel

$e$ = base of natural logarithm = 2.71828

$\mu$ = coefficient of friction between the brake band and the brake wheel

$\theta$ = angle of contact of the brake band with the brake wheel expressed in radians

$$1\text{ radian} = \frac{180\text{ deg}}{\pi\text{ radians}} = 57.296\frac{\text{deg}}{\text{radian}}$$

See also *Conversion Tables of Angular Measure* starting on page 98.

For simplicity in the formulas presented, the tensions at *x* and *y* (Fig. 1) are denoted by $T_1$ and $T_2$ respectively, for clockwise rotation. When the direction of the rotation is reversed, the tension in *x* equals $T_2$, and the tension in *y* equals $T_1$, which is the reverse of the tension in the clockwise direction.

The value of the expression $e^{\mu\theta}$ in these formulas may be most easily found by using a handheld calculator of the scientific type; that is, one capable of raising 2.71828 to the power $\mu\theta$. The following example outlines the steps in the calculations.

**Table of Values of $e^{\mu\theta}$**

| Proportion of Contact to Whole Circumference, $\frac{\theta}{2\pi}$ | Steel Band on Cast Iron, $\mu = 0.18$ | Leather Belt on | | | |
|---|---|---|---|---|---|
| | | Wood | Cast Iron | | |
| | | Slightly Greasy; $\mu = 0.47$ | Very Greasy; $\mu = 0.12$ | Slightly Greasy; $\mu = 0.28$ | Damp; $\mu = 0.38$ |
| 0.1 | 1.12 | 1.34 | 1.08 | 1.19 | 1.27 |
| 0.2 | 1.25 | 1.81 | 1.16 | 1.42 | 1.61 |
| 0.3 | 1.40 | 2.43 | 1.25 | 1.69 | 2.05 |
| 0.4 | 1.57 | 3.26 | 1.35 | 2.02 | 2.60 |
| 0.425 | 1.62 | 3.51 | 1.38 | 2.11 | 2.76 |
| 0.45 | 1.66 | 3.78 | 1.40 | 2.21 | 2.93 |
| 0.475 | 1.71 | 4.07 | 1.43 | 2.31 | 3.11 |
| 0.5 | 1.76 | 4.38 | 1.46 | 2.41 | 3.30 |
| 0.525 | 1.81 | 4.71 | 1.49 | 2.52 | 3.50 |
| 0.55 | 1.86 | 5.07 | 1.51 | 2.63 | 3.72 |
| 0.6 | 1.97 | 5.88 | 1.57 | 2.81 | 4.19 |
| 0.7 | 2.21 | 7.90 | 1.66 | 3.43 | 5.32 |
| 0.8 | 2.47 | 10.60 | 1.83 | 4.09 | 6.75 |
| 0.9 | 2.77 | 14.30 | 1.97 | 4.87 | 8.57 |
| 1.0 | 3.10 | 19.20 | 2.12 | 5.81 | 10.90 |

## Formulas for Simple and Differential Band Brakes

$F$ = force in pounds at end of brake handle; $P$ = tangential force in pounds at rim of brake wheel; $e$ = base of natural logarithm = 2.71828; $\mu$ = coefficient of friction between the brake band and the brake wheel; $\theta$ = angle of contact of the brake band with the brake wheel, expressed in radians (1 radian = 57.296 degrees).

$$T_1 = P\frac{1}{e^{\mu\theta}-1} \qquad T_2 = P\frac{e^{\mu\theta}}{e^{\mu\theta}-1}$$

**Simple Band Brake**

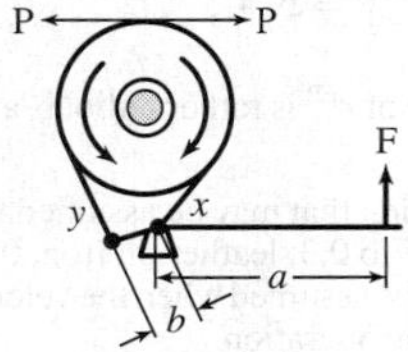


Fig. 1.

For clockwise rotation:

$$F = \frac{bT_2}{a} = \frac{Pb}{a}\left(\frac{e^{\mu\theta}}{e^{\mu\theta}-1}\right)$$

For counter clockwise rotation:

$$F = \frac{bT_1}{a} = \frac{Pb}{a}\left(\frac{1}{e^{\mu\theta}-1}\right)$$

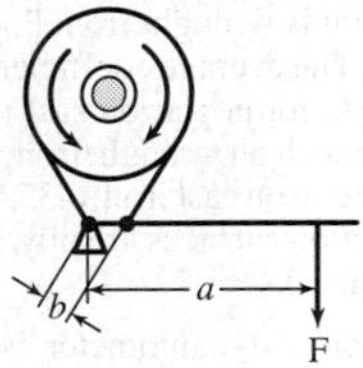


Fig. 2.

For clockwise rotation:

$$F = \frac{bT_1}{a} = \frac{Pb}{a}\left(\frac{1}{e^{\mu\theta}-1}\right)$$

For counter clockwise rotation:

$$F = \frac{bT_2}{a} = \frac{Pb}{a}\left(\frac{e^{\mu\theta}}{e^{\mu\theta}-1}\right)$$

**Differential Band Brake**

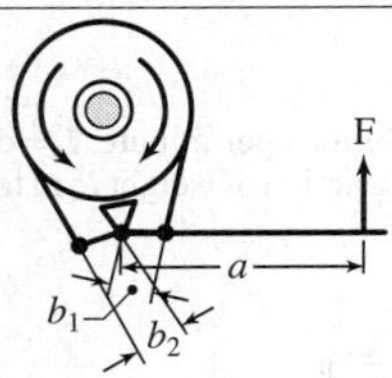


Fig. 3.

For clockwise rotation:

$$F = \frac{b_2T_2 - b_1T_1}{a} = \frac{P}{a}\left(\frac{b_2e^{\mu\theta} - b_1}{e^{\mu\theta}-1}\right)$$

For counter clockwise rotation:

$$F = \frac{b_2T_1 - b_1T_2}{a} = \frac{P}{a}\left(\frac{b_2 - b_1e^{\mu\theta}}{e^{\mu\theta}-1}\right)$$

In this case, if $b_2$ is equal to, or less than, $b_1e^{\mu\theta}$, the force $F$ will be 0 or negative and the band brake works automatically.

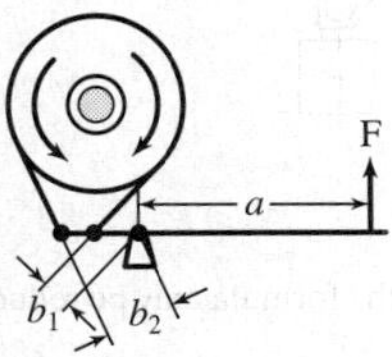


Fig. 4.

For clockwise rotation:

$$F = \frac{b_2T_2 + b_1T_1}{a} = \frac{P}{a}\left(\frac{b_2e^{\mu\theta} + b_1}{e^{\mu\theta}-1}\right)$$

For counter clockwise rotation:

$$F = \frac{b_1T_2 + b_2T_1}{a} = \frac{P}{a}\left(\frac{b_1e^{\mu\theta} + b_2}{e^{\mu\theta}-1}\right)$$

If $b_2 = b_1$, both of the above formulas reduce to $F = \frac{Pb_1}{a}\left(\frac{e^{\mu\theta}-1}{e^{\mu\theta}-1}\right)$.

In this case, the same force $F$ is required for rotation in either direction.

*Example:* In a band brake of the type in Fig. 1, dimension $a$ = 24 inches, and $b$ = 4 inches; force $P$ = 100 pounds; coefficient $\mu$ = 0.2, and angle of contact = 240 degrees, or

$$\theta = \frac{240}{180} \times \pi = 4.18$$

The rotation is clockwise. Find force $F$ required.

$$F = \frac{Pb}{a}\left(\frac{e^{\mu\theta}}{e^{\mu\theta}-1}\right)$$

$$= \frac{100\times 4}{24}\left(\frac{2.71828^{0.2\times 4.18}}{2.71828^{0.2\times 4.18}-1}\right)$$

$$= \frac{400}{24}\times\frac{2.71828^{0.836}}{2.71828^{0.836}-1} = 16.66\times\frac{2.31}{2.31-1} = 29.4$$

If a handheld calculator is not used, determining the value of $e^{\mu\theta}$ is rather tedious, and the table on page 2540 will save calculations.

**Coefficient of Friction in Brakes.—**The coefficients of friction that may be assumed for friction brake calculations are as follows: Iron on iron, 0.25 to 0.3; leather on iron, 0.3; cork on iron, 0.35. Values somewhat lower than these should be assumed when the velocities exceed 400 feet per minute at the beginning of the braking operation.

For brakes where wooden brake blocks are used on iron drums, poplar has proved the best brake block material. The best material for the brake drum is wrought iron. Poplar gives a high coefficient of friction, and is little affected by oil. The average coefficient of friction for poplar brake blocks and wrought-iron drums is 0.6; for poplar on cast iron, 0.35; for oak on wrought iron, 0.5; for oak on cast iron, 0.3; for beech on wrought iron, 0.5; for beech on cast iron, 0.3; for elm on wrought iron, 0.6; and for elm on cast iron, 0.35. The objection to elm is that the friction decreases rapidly if the friction surfaces are oily. The coefficient of friction for elm and wrought iron, if oily, is less than 0.4.

**Calculating Horsepower from Dynamometer Tests.—**When a dynamometer is arranged for measuring the horsepower transmitted by a shaft, as indicated by the diagrammatic view in Fig. 5, the horsepower may be obtained by the formula:

$$HP = \frac{2\pi LPN}{33000}$$

in which $HP$ = horsepower transmitted; $N$ = number of revolutions per minute; $L$ = distance (as shown in illustration) from center of pulley to point of action of weight $P$, in feet; $P$ = weight hung on brake arm or read on scale.

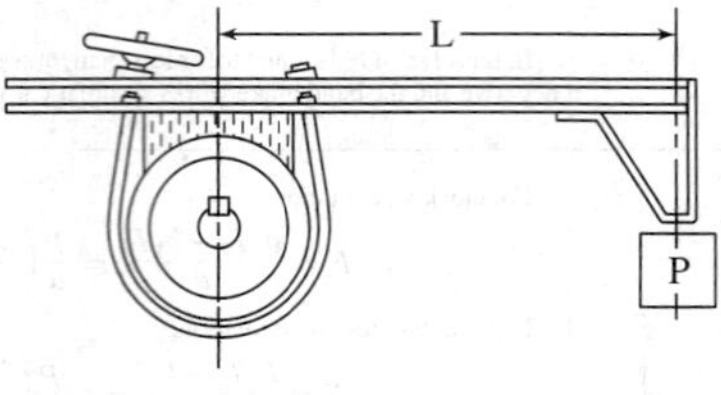


Fig. 5.

By adopting a length of brake arm equal to 5 feet 3 inches, the formula may be reduced to the simple form:

$$HP = \frac{NP}{1000}$$

If a length of brake arm equal to 2 feet $7\frac{1}{2}$ inches is adopted as a standard, the formula takes the form:

$$HP = \frac{NP}{2000}$$

The *transmission* type of dynamometer measures the power by transmitting it through the mechanism of the dynamometer from the apparatus in which it is generated, or to the apparatus in which it is to be utilized. Dynamometers known as *indicators* operate by simultaneously measuring the pressure and volume of a confined fluid. This type may be used for the measurement of the power generated by steam or gas engines or absorbed by refrigerating machinery, air compressors, or pumps. An electrical dynamometer is for measuring the power of an electric current, based on the mutual action of currents flowing in two coils. It consists principally of one fixed and one movable coil, which, in the normal position, are at right angles to each other. Both coils are connected in series, and, when a current traverses the coils, the fields produced are at right angles; hence, the coils tend to take up a parallel position. The movable coil with an attached pointer will be deflected, the deflection measuring directly the electric current.

## Formulas for Block Brakes

$F$ = force in pounds at end of brake handle;
$P$ = tangential force in pounds at rim of brake wheel;
$\mu$ = coefficient of friction between the brake block and brake wheel.

| | |
|---|---|
| 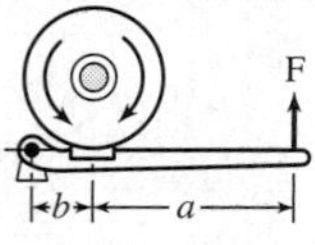 Fig. 1. | Block brake. For rotation in either direction: $F = P\frac{b}{a+b} \times \frac{1}{\mu} = \frac{Pb}{a+b}\left(\frac{1}{\mu}\right)$ |
| 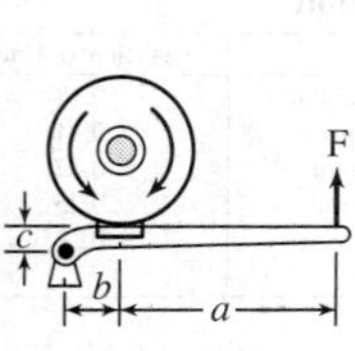 Fig. 2. | Block brake. For clockwise rotation: $F = \frac{\frac{Pb}{\mu} - Pc}{a+b} = \frac{Pb}{a+b}\left(\frac{1}{\mu} - \frac{c}{b}\right)$ For counter clockwise rotation: $F = \frac{\frac{Pb}{\mu} + Pc}{a+b} = \frac{Pb}{a+b}\left(\frac{1}{\mu} + \frac{c}{b}\right)$ |
| 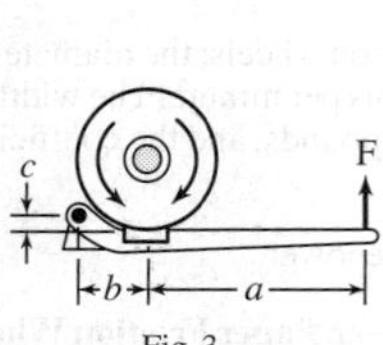 Fig. 3. | Block brake. For clockwise rotation: $F = \frac{\frac{Pb}{\mu} + Pc}{a+b} = \frac{Pb}{a+b}\left(\frac{1}{\mu} + \frac{c}{b}\right)$ For counter clockwise rotation: $F = \frac{\frac{Pb}{\mu} - Pc}{a+b} = \frac{Pb}{a+b}\left(\frac{1}{\mu} - \frac{c}{b}\right)$ |
| 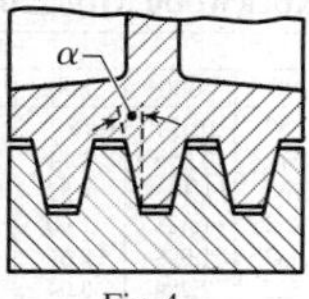 Fig. 4. | The brake wheel and friction block of the block brake are often grooved as shown in Fig. 4. In this case, substitute for $\mu$ in the above equations the value $\frac{\mu}{\sin\alpha + \mu\cos\alpha}$ where $\alpha$ is one-half the angle included by the faces of the grooves. |

## Friction Wheels for Power Transmission

When a rotating member is driven intermittently and the rate of driving does not need to be positive, friction wheels are frequently used, especially when the amount of power to be transmitted is comparatively small. The driven wheels in a pair of friction disks should always be made of a harder material than the driving wheels, so that if the driven wheel should be held stationary by the load, while the driving wheel revolves under its own pressure, a flat spot may not be rapidly worn on the driven wheel. The driven wheels, therefore, are usually made of iron, while the driving wheels are made of or covered with, rubber, paper, leather, wood or fiber. The safe working force per inch of face width of contact for various materials are as follows: Straw fiber, 150; leather fiber, 240; tarred fiber, 240; leather, 150; wood, 100 to 150; paper, 150. Coefficients of friction for different combinations of materials are given in the following table. Smaller values should be used for exceptionally high speeds, or when the transmission must be started while under load.

**Horsepower of Friction Wheels.**—Let $D$ = diameter of friction wheel in inches; $N$ = number of revolutions per minute; $W$ = width of face in inches; $f$ = coefficient of friction; $P$ = force in pounds, per inch width of face. Then:

$$HP = \frac{3.1416 \times D \times N \times P \times W \times f}{33{,}000 \times 12}$$

Assume

$$\frac{3.1416 \times P \times f}{33{,}000 \times 12} = C$$

then,

for $P = 100$ and $f = 0.20$, $C = 0.00016$
for $P = 150$ and $f = 0.20$, $C = 0.00024$
for $P = 200$ and $f = 0.20$, $C = 0.00032$

**Working Values of Coefficient of Friction**

| Materials | Coefficient of Friction | Materials | Coefficient of Friction |
|---|---|---|---|
| Straw fiber and cast iron | 0.26 | Tarred fiber and aluminum | 0.18 |
| Straw fiber and aluminum | 0.27 | Leather and cast iron | 0.14 |
| Leather fiber and cast iron | 0.31 | Leather and aluminum | 0.22 |
| Leather fiber and aluminum | 0.30 | Leather and type metal | 0.25 |
| Tarred fiber and cast iron | 0.15 | Wood and metal | 0.25 |
| Paper and cast iron | 0.20 | | |

The horsepower transmitted is then:

$$HP = D \times N \times W \times C$$

*Example:* Find the horsepower transmitted by a pair of friction wheels; the diameter of the driving wheel is 10 inches, and it revolves at 200 revolutions per minute. The width of the wheel is 2 inches. The force per inch width of face is 150 pounds, and the coefficient of friction 0.20.

$$HP = 10 \times 200 \times 2 \times 0.00024 = 0.96 \text{ horsepower}$$

**Horsepower Which May Be Transmitted by Means of a Clean Paper Friction Wheel of 1-Inch Face When Run Under a Force of 150 Pounds (Rockwood Mfg. Co.)**

| Dia. of Friction Wheel | Revolutions per Minute | | | | | | | | | | |
|---|---|---|---|---|---|---|---|---|---|---|---|
| | 25 | 50 | 75 | 100 | 150 | 200 | 300 | 400 | 600 | 800 | 1000 |
| 4 | 0.023 | 0.047 | 0.071 | 0.095 | 0.142 | 0.190 | 0.285 | 0.380 | 0.571 | 0.76 | 0.95 |
| 6 | 0.035 | 0.071 | 0.107 | 0.142 | 0.214 | 0.285 | 0.428 | 0.571 | 0.856 | 1.14 | 1.42 |
| 8 | 0.047 | 0.095 | 0.142 | 0.190 | 0.285 | 0.380 | 0.571 | 0.761 | 1.142 | 1.52 | 1.90 |
| 10 | 0.059 | 0.119 | 0.178 | 0.238 | 0.357 | 0.476 | 0.714 | 0.952 | 1.428 | 1.90 | 2.38 |
| 14 | 0.083 | 0.166 | 0.249 | 0.333 | 0.499 | 0.666 | 0.999 | 1.332 | 1.999 | 2.66 | 3.33 |
| 16 | 0.095 | 0.190 | 0.285 | 0.380 | 0.571 | 0.761 | 1.142 | 1.523 | 2.284 | 3.04 | 3.80 |
| 18 | 0.107 | 0.214 | 0.321 | 0.428 | 0.642 | 0.856 | 1.285 | 1.713 | 2.570 | 3.42 | 4.28 |
| 24 | 0.142 | 0.285 | 0.428 | 0.571 | 0.856 | 1.142 | 1.713 | 2.284 | 3.427 | 4.56 | 5.71 |
| 30 | 0.178 | 0.357 | 0.535 | 0.714 | 1.071 | 1.428 | 2.142 | 2.856 | 4.284 | 5.71 | 7.14 |
| 36 | 0.214 | 0.428 | 0.642 | 0.856 | 1.285 | 1.713 | 2.570 | 3.427 | 5.140 | 6.85 | 8.56 |
| 42 | 0.249 | 0.499 | 0.749 | 0.999 | 1.499 | 1.999 | 2.998 | 3.998 | 5.997 | 7.99 | 9.99 |
| 48 | 0.285 | 0.571 | 0.856 | 1.142 | 1.713 | 2.284 | 3.427 | 4.569 | 6.854 | 9.13 | 11.42 |
| 50 | 0.297 | 0.595 | 0.892 | 1.190 | 1.785 | 2.380 | 3.570 | 4.760 | 7.140 | 9.52 | 11.90 |

## KEYS AND KEYSEATS

### Metric Square and Rectangular Keys and Keyways

ANSI/ASME B18.25.1M (Withdrawn) covers requirements for square and rectangular parallel keys and keyways intended for both alignment of shafts and hubs, and transmitting torque between shafts and hubs. Keys covered by this Standard have a relatively tight width tolerance. The deviations are less than the basic size. Keys with greater width tolerance and with deviations greater than the basic size are covered by ANSI/ASME B18.25.3M (Withdrawn). All dimensions in these standards are in millimeters (mm).

**Comparison with ISO R773-1969 (Withdrawn) and 2491-1974 (Withdrawn).—**This Standard is based on ISO Standards R773-1969 (Withdrawn), *Rectangular or Square Parallel Keys* and their corresponding keyways, and 2491-1974 (Withdrawn), *Thin Parallel Keys* and their corresponding keyways (dimensions in millimeters). Product manufactured to this Standard will meet the ISO standards. Because of tighter width tolerances in this Standard, products manufactured to the ISO Standard may not meet the requirements of this Standard.

This Standard differs from ISO in that it: a) does not restrict the corners of a key to be chamfered but allows either a chamfer or a radius on the key; and b) specifies a key material hardness rather than a tensile property.

**Tolerances.—**Many of the tolerances shown in Table 1 and Table 2 are from ANSI/ASME B4.2 (ISO 286-1 and ISO 286-2). As a result, in addition to plus-minus tolerances which are common in the United States, some are expressed as plus-plus or minus-minus deviations from the basic size. For further interpretation of these tolerances refer to ANSI/ASME B4.2 or ISO 286.

**Designation.—**Keys conforming to this Standard shall be designated by the following data, preferably in the sequence as follows: a) ASME document number; b) product name; c) nominal size, width ($b$) × height ($h$) × length; d) form; and e) hardness (if other than non-hardened).

*Example:* ANSI/ASME B18.25.1M (Withdrawn) square key 3 × 3 × 15 form B.

ANSI/ASME B18.25.1M (Withdrawn) rectangular key 10 × 6 × 20 form C hardened

**Preferred Lengths and Tolerances.—**Preferred lengths and tolerances of square and rectangular keys are shown below. Tolerances are JS16. To minimize problems due to lack of straightness, key length should be less than 10 times the key width.

| Length | ± Tolerances | Length | ± Tolerances |
|---|---|---|---|
| 6 | 0.38 | 90, 100, 110 | 1.10 |
| 8, 10 | 0.45 | 125, 140, 150, 180 | 1.25 |
| 12, 14, 16, 18 | 0.56 | 200, 220, 250 | 1.45 |
| 20, 22, 25, 28 | 0.65 | 280 | 1.60 |
| 32, 36, 40, 45, 50 | 0.80 | 320, 360, 400 | 1.80 |
| 56, 63, 70, 80 | 0.95 | | |

**Material Requirements.—**Standard steel keys shall have a Vickers hardness of 183 HV minimum. Hardened keys shall be alloy steel through hardened to a Vickers hardness of 390 to 510 HV. When other materials and properties are required, these shall be as agreed upon by the supplier and customer.

**Dimensions and Tolerances.—**Dimensions and tolerances for square and rectangular parallel keys are shown in Table 1. Recommended dimensions and tolerances for keyways are shown in Table 2.

## Figures for Table 1 and Table 3

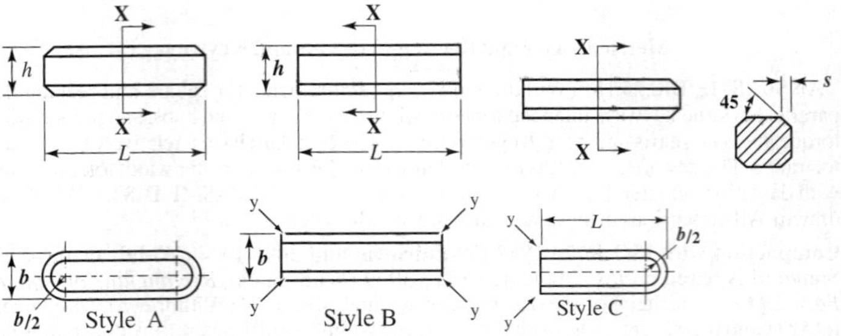


$y$ = sharp edges removed not to exceed $s_{max}$

### Table 1. Dimensions and Tolerances for Metric Square and Rectangular Parallel Keys *ANSI/ASME B18.25.1M-1996 (Withdrawn)*

| Width, $b$ | | Thickness, $h$ | | Chamfer or Radius, $s$ | | Range of Lengths | |
|---|---|---|---|---|---|---|---|
| Basic Size (mm) | Tolerance, h8 | Basic Size | Tolerance, Square, h8 Rectangular, h11 | Minimum | Maximum | From | To[a] |
| Square Keys | | | | | | | |
| 2 | 0 −0.014 | 2 | 0 −0.014 | 0.16 | 0.25 | 6 | 20 |
| 3 | | 3 | | | | 6 | 36 |
| 4 | 0 −0.018 | 4 | 0 −0.018 | | | 8 | 45 |
| 5 | | 5 | | 0.25 | 0.40 | 10 | 56 |
| 6 | | 6 | | | | 14 | 70 |
| Rectangular Keys | | | | | | | |
| 5 | 0 −0.018 | 3 | 0 −0.060 | 0.25 | 0.40 | 10 | 56 |
| 6 | | 4 | 0 −0.075 | | | 14 | 70 |
| 8 | 0 −0.022 | 5 | | | | 18 | 90 |
| | | 7 | 0 −0.090 | | | | |
| 10 | | 6 | 0 −0.075 | 0.40 | 0.60 | 22 | 110 |
| | | 8 | 0 −0.090 | | | | |
| 12 | 0 −0.027 | 6 | 0 −0.075 | | 0.60 | 28 | 110 |
| | | 8 | 0 −0.090 | | | | |
| 14 | | 6 | 0 −0.075 | | | 36 | 160 |
| | | 9 | 0 −0.090 | | | | |
| 16 | | 7 | | | | 45 | 180 |
| | | 10 | | | | | |
| 18 | | 7 | | | | 50 | 200 |
| | | 11 | 0 −0.110 | | | | |

**Table 1.** *(Continued)* **Dimensions and Tolerances for Metric Square and Rectangular Parallel Keys** *ANSI/ASME B18.25.1M-1996 (Withdrawn)*

| Width, *b* | | Thickness, *h* | | Chamfer or Radius, *s* | | Range of Lengths | |
|---|---|---|---|---|---|---|---|
| Basic Size (mm) | Tolerance, h8 | Basic Size | Tolerance, Square, h8 Rectangular, h11 | Minimum | Maximum | From | To[a] |
| Rectangular Keys | | | | | | | |
| 20 | 0 −0.033 | 8 | 0 −0.090 | 0.60 | 0.80 | 56 | 220 |
| | | 12 | 0 −0.110 | | | 63 | 260 |
| 22 | | 6 | 0 −0.075 | | | 70 | 280 |
| | | 14 | 0 −0.110 | | | 80 | 320 |
| 25 | | 9 | 0 −0.090 | | | 90 | 360 |
| | | 14 | 0 −0.110 | | | 100 | 400 |
| 28 | | 10 | 0 −0.090 | | | | |
| | | 16 | 0 −0.110 | | | | |
| 32 | 0 −0.039 | 11 | | | | | |
| | | 18 | | | | | |
| 36 | | 12 | | 1.00 | 1.20 | | |
| | | 20 | 0 −0.110 | | | | |
| 40 | | 22 | | | | | |
| 45 | | 25 | | | | | |
| 50 | | 28 | | | | | |
| 56 | 0 −0.046 | 32 | 0 −0.160 | | | | |
| 63 | | 32 | | 1.60 | 2.00 | | |
| 70 | | 36 | | | | | |
| 80 | | 40 | | | | | |
| 90 | 0 −0.054 | 45 | | 2.50 | 3.00 | | |
| 100 | | 50 | | | | | |

[a] See *Preferred Lengths and Tolerances* starting on page 2545 for preferred maximum length of key.

All dimensions in this Standard are in millimeters (mm).

**Figures for Table 2 and Table 4**

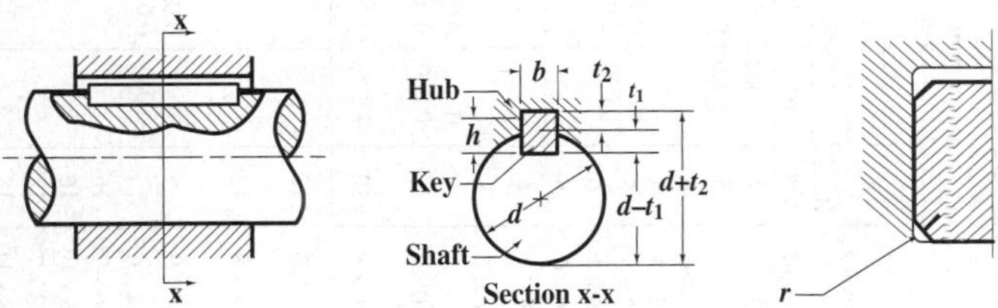

**Table 2. Keyway Dimensions and Tolerances for Metric Square and Rectangular Parallel Keys** *ANSI/ASME B18.25.1M-1996 (Withdrawn)*

| Key size | Keyway | | | | | | | | | | | | | | | | |
|---|---|---|---|---|---|---|---|---|---|---|---|---|---|---|---|---|---|
| | Width. | | | | | | | | | | | Depth | | | | Radius, $r$ | |
| | | Tolerance[a] and Resulting Fits[b] | | | | | | | | | | Shaft, $t_1$ | | Hub, $t_2$ | | | |
| | | Normal Fit | | | | Close Fit | | Free Fit | | | | | | | | | |
| | | Shaft | | Hub | | Shaft and Hub | | Shaft | | Hub | | | | | | | |
| $b \times h$ (mm) | Basic Size[c] | N9 | Fit | JS9 | Fit | P9 | Fit | H9 | Fit | D10 | Fit | Basic Size | Toler-ance | Basic Size | Toler-ance | Min. | Max. |
| 2 × 2 | 2 | −0.004<br>−0.029 | 0.010L<br>0.029T | +0.0125<br>−0.0125 | 0.0265L<br>0.0125T | −0.006<br>−0.031 | 0.008L<br>0.031T | +0.025<br>0 | 0.039L<br>0T | +0.060<br>+0.020 | 0.074L<br>0.020L | 1.2 | +0.1<br>0 | 1 | +0.1<br>0 | 0.08 | 0.16 |
| 3 × 3 | 3 | | | | | | | | | | | 1.8 | | 1.4 | | | |
| 4 × 4 | 4 | 0<br>−0.030 | 0.018L<br>0.030T | +0.0150<br>−0.0150 | 0.033L<br>0.015T | −0.012<br>−0.042 | 0.006L<br>0.042T | +0.030<br>0 | 0.048L<br>0T | +0.078<br>+0.030 | 0.096L<br>0.030L | 2.5 | | 1.8 | | | |
| 5 × 3 | 5 | | | | | | | | | | | 1.8 | | 1.4 | | 0.16 | 0.25 |
| 5 × 5 | 5 | | | | | | | | | | | 3 | | 2.8 | | | |
| 6 × 4 | 6 | | | | | | | | | | | 2.5 | | 1.8 | | | |
| 6 × 6 | 6 | | | | | | | | | | | 3.5 | | 2.8 | | | |
| 8 × 5 | 8 | 0<br>−0.036 | 0.022L<br>0.036T | +0.0180<br>−0.0180 | 0.040L<br>0.018T | −0.015<br>−0.051 | 0.007L<br>0.051T | +0.036<br>0 | 0.058L<br>0T | +0.098<br>+0.040 | 0.120L<br>0.040L | 3 | | 2.8 | | | |
| 8 × 7 | 8 | | | | | | | | | | | 4 | +0.2<br>0 | 3.3 | +0.1<br>0 | | |
| 10 × 6 | 10 | | | | | | | | | | | 3.5 | +0.1<br>0 | 2.8 | +0.1<br>0 | 0.25 | 0.4 |
| 10 × 8 | 10 | | | | | | | | | | | 5 | +0.2<br>0 | 3.3 | +0.2<br>0 | | |
| 12 × 6 | 12 | 0<br>−0.043 | 0.027L<br>0.043T | +0.0215<br>−0.0215 | 0.0485L<br>0.0215T | −0.018<br>−0.061 | 0.009L<br>0.061T | +0.043<br>0 | 0.070L<br>0T | +0.120<br>+0.050 | 0.147L<br>0.050L | 3.5 | +0.1<br>0 | 2.8 | +0.1<br>0 | | |
| 12 × 8 | 12 | | | | | | | | | | | 5 | +0.2<br>0 | 3.3 | +0.2 | | |
| 14 × 6 | 14 | | | | | | | | | | | 3.5 | +0.1<br>0 | 2.8 | +0.1<br>0 | | |
| 14 × 9 | 14 | | | | | | | | | | | 5.5 | +0.2<br>0 | 3.8 | +0.2<br>0 | | |
| 16 × 7 | 16 | | | | | | | | | | | 4 | | 3.3 | | | |
| 16 × 10 | 16 | | | | | | | | | | | 6 | | 4.3 | | | |
| 18 × 7 | 18 | | | | | | | | | | | 4 | | 3.3 | | | |
| 18 × 11 | 18 | | | | | | | | | | | 7 | | 4.4 | | | |

**Table 2.** *(Continued)* **Keyway Dimensions and Tolerances for Metric Square and Rectangular Parallel Keys** *ANSI/ASME B18.25.1M-1996 (Withdrawn)*

| Key size | Keyway | | | | | | | | | | | | | | | | |
|---|---|---|---|---|---|---|---|---|---|---|---|---|---|---|---|---|---|
| | Width. | | | | | | | | | | | Depth | | | | | |
| | | Tolerance[a] and Resulting Fits[b] | | | | | | | | | | Shaft, $t_1$ | | Hub, $t_2$ | | Radius, $r$ | |
| | | Normal Fit | | | | Close Fit | | Free Fit | | | | | | | | | |
| $b \times h$ | Basic | Shaft | | Hub | | Shaft and Hub | | Shaft | | Hub | | Basic | Toler- | Basic | Toler- | | |
| (mm) | Size[c] | N9 | Fit | JS9 | Fit | P9 | Fit | H9 | Fit | D10 | Fit | Size | ance | Size | ance | Min. | Max. |
| 20 × 8 | 20 | | | | | | | | | | | 5 | | 3.3 | | | |
| 20 × 12 | 20 | | | | | | | | | | | 7.5 | | 4.9 | | | |
| 22 × 9 | 22 | | | | | | | | | | | 5.5 | | 3.8 | | | |
| 22 × 14 | 22 | 0 | 0.033L | +0.026 | 0.059L | −0.022 | 0.011L | +0.052 | 0.085L | +0.149 | 0.182L | 9 | | 5.4 | | | |
| 25 × 9 | 25 | −0.052 | 0.052T | −0.026 | 0.026T | −0.074 | 0.074T | 0 | 0T | +0.065 | 0.065L | 5.5 | +0.2 | 3.8 | +0.2 | 0.4 | 0.6[c] |
| 25 × 14 | 25 | | | | | | | | | | | 9 | 0 | 5.4 | 0 | | |
| 28 × 10 | 28 | | | | | | | | | | | 6 | | 4.3 | | | |
| 28 × 16 | 28 | | | | | | | | | | | 10 | | 6.4 | | | |
| 32 × 11 | 32 | | | | | | | | | | | 7[c] | | 4.4 | | | |
| 32 × 18 | 32 | | | | | | | | | | | 11[c] | | 7.4 | | | |
| 36 × 12 | 36 | 0 | 0.039L | +0.031 | 0.070L | −0.026 | 0.013L | +0.062 | 0.101L | +0.180 | 0.219L | 7.5[c] | | 4.9 | | | |
| 36 × 20 | 36 | −0.062 | 0.062T | −0.031 | 0.031T | −0.088 | 0.088T | 0 | 0T | +0.080 | 0.080L | 12 | | 8.4 | | | |
| 40 × 22 | 40 | | | | | | | | | | | 13 | | 9.4 | | 0.7 | 1.0 |
| 45 × 25 | 45 | | | | | | | | | | | 15 | | 10.4 | | | |
| 50 × 28 | 50 | | | | | | | | | | | 17 | | 11.4 | | | |
| 56 × 32 | 56 | | | | | | | | | | | 20 | +0.3 | 12.4 | +0.3 | | |
| 63 × 32 | 63 | 0 | 0.046L | +0.037 | 0.083L | −0.032 | 0.014L | +0.074 | 0.120L | +0.220 | 0.266L | 20 | 0 | 12.4 | 0 | 1.2 | 1.6 |
| 70 × 36 | 70 | −0.074 | 0.074T | −0.037 | 0.037T | −0.106 | 0.106T | 0 | 0T | +0.100 | 0.100L | 22 | | 14.4 | | | |
| 80 × 40 | 80 | | | | | | | | | | | 25 | | 15.4 | | | |
| 90 × 45 | 90 | 0 | 0.054L | +0.0435 | 0.0975L | −0.037 | 0.017L | +0.087 | 0.139L | +0 260 | 0.314L | 28 | | 17.4 | | 2.0 | 2.5 |
| 100 × 50 | 100 | −0.087 | 0.87T | −0.0435 | 0.0435T | −0.1254 | 0.1254T | 0 | 0T | +0.120 | 0.120L | 31 | | 19.5 | | | |

[a] Some of the tolerances are expressed as plus-plus. See *Tolerances* on page 2545 for more information.

[b] Resulting fits: L indicates a clearance between the key and keyway; T indicates an interference between the key and keyway.

[c] Values changed for accuracy from that given in ANSI/ASME 18.25.1M-1996 (Withdrawn).

## Metric Square And Rectangular Keys and Keyways: Width Tolerances and Deviations Greater than Basic Size

ANSI/ASME B18.25.3M-2003 (Withdrawn) covers requirements for square and rectangular parallel keys and keyways intended for both alignment of shafts and hubs, and transmitting torque between shafts and hubs. Keys covered by this Standard have a relatively loose width tolerance. All width tolerances are positive. Keys with minus width tolerances and a smaller tolerance range are covered by ANSI/ASME B18.25.1M-1996 (Withdrawn). Dimensions and tolerances for square and rectangular keys are shown in Table 3. Recommended dimensions and tolerances for keyways are shown in Table 4. All dimensions in this Standard are in millimeters.

**Table 3. Dimensions and Tolerances for Metric Square and Rectangular Parallel Keys** *ANSI/ASME B18.25.3M-2003 (Withdrawn)*

Width Tolerances and Deviations Greater than Basic Size

| Width, *b* | | Thickness, *h* | | Chamfer or Radius, *s* | | Range of Lengths | |
|---|---|---|---|---|---|---|---|
| Basic Size | Tolerance | Basic Size | Tolerance | Min. | Max. | From | To[a] |
| Square Keys | | | | | | | |
| 2 | +0.040 −0.000 | 2 | +0.040 −0.000 | 0.16 | 0.25 | 6 | 20 |
| 3 | | 3 | | | | | 36 |
| 4 | +0.045 −0.000 | 4 | +0.045 −0.000 | | | 8 | 45 |
| 5 | | 5 | | 0.25 | 0.40 | 10 | 56 |
| 6 | | 6 | | | | 14 | 70 |
| Rectangular Keys | | | | | | | |
| 5 | +0.045 −0.000 | 3 | +0.160 −0.000 | 0.25 | 0.40 | 10 | 56 |
| 6 | | 4 | +0.175 −0.000 | | | 14 | 70 |
| 8 | +0.050 −0.000 | 5 | | | | 18 | 90 |
| | | 7 | +0.190 −0.000 | | | | |
| 10 | | 6 | +0.175 −0.000 | 0.40 | 0.60 | 22 | 110 |
| | | 8 | +0.19 −0.000 | | | | |
| 12 | +0.075 −0.000 | 6 | +0.175 −0.000 | 0.40 | 0.60 | 28 | 140 |
| | | 8 | +0.190 −0.000 | | | | |
| 14 | | 6 | +0.175 −0.000 | | | 36 | 160 |
| | | 9 | +0.190 −0.000 | | | | |
| 16 | | 7 | | | | 45 | 180 |
| | | 10 | | | | | |
| 18 | | 7 | | | | 50 | 200 |
| | | 11 | +0.210 −0.000 | | | | |
| 20 | +0.050 −0.033 | 8 | +0.190 −0.000 | 0.60 | 0.80 | 56 | |
| | | 12 | +0.210 −0.000 | | | 63 | |

**Table 3.** *(Continued)* **Dimensions and Tolerances for Metric Square and Rectangular Parallel Keys** *ANSI/ASME B18.25.3M-2003 (Withdrawn)*

Width Tolerances and Deviations Greater than Basic Size

| Width, $b$ | | Thickness, $h$ | | Chamfer or Radius, $s$ | | Range of Lengths | |
|---|---|---|---|---|---|---|---|
| Basic Size | Tolerance | Basic Size | Tolerance | Min. | Max. | From | To[a] |
| 22 | +0.050 −0.033 | 6 | +0.175 −0.000 | 0.60 | 0.80 | 70 | 280 |
| | | 14 | +0.210 −0.000 | | | 80 | 320 |
| 25 | | 9 | +0.210 −0.000 | | | 90 | 360 |
| | | 14 | +0.190 −0.000 | | | 100 | 400 |
| 28 | | 10 | +0.210 −0.000 | | | | |
| | | 16 | +0.280 −0.000 | | | | |
| 32 | +0.090 −0.000 | 11 | | | | | |
| | | 18 | | | | | |
| 36 | | 12 | | 1.00 | 1.20 | | |
| | | 20 | | | | | |
| 40 | | 22 | | | | | |
| 45 | | 25 | | | | | |
| 50 | | 28 | | | | | |
| 56 | +0.125 −0.000 | 32 | +0.310 −0.000 | 1.60 | 2.00 | | |
| 63 | | 32 | | | | | |
| 70 | | 36 | | | | | |
| 80 | | 40 | | 2.50 | 3.00 | | |
| 90 | +0.135 −0.000 | 45 | | | | | |
| 100 | | 50 | | | | | |

[a] See *Preferred Lengths and Tolerances* on page 2551 for preferred maximum length of key except basic width of 2 mm.

**Comparison with ISO R773-1969 and 2491-1974 (Both Withdrawn).—**This Standard has greater tolerances than ISO Standards R773-1969 (Withdrawn) and 2491-1974 (Withdrawn). Product manufactured to this Standard is not interchangeable dimensionally with product manufactured to the ISO Standards nor is product manufactured to the ISO Standards dimensionally interchangeable with product manufactured to this Standard. The ISO Standards do not include hardened keys.

**Preferred Lengths and Tolerances.—**Preferred lengths and tolerances of square and rectangular keys are shown below. Tolerances are JS 16 from ANSI/ASME B4.2. To minimize problems due to lack of straightness, key length should be less than 10 times the key width.

| Length | ±Tolerances | Length | ±Tolerances |
|---|---|---|---|
| 6 | 0.375 | 90, 100, 110 | 1.10 |
| 8, 10 | 0.45 | 125, 140, 160, 180 | 1.25 |
| 12, 14, 16, 18 | 0.55 | 200, 220, 250 | 1.45 |
| 20, 22, 25, 28 | 0.65 | 280 | 1.60 |
| 32, 36, 45, 50 | 0.80 | 320, 360, 400 | 1.80 |
| 56, 63, 70, 80 | 0.95 | | |

**Table 4. Keyway Dimensions and Tolerances for Metric Square and Rectangular Parallel Keys**
*ANSI/ASME B18.25.3M-2003 (Withdrawn)* Width Tolerances and Deviations Greater than Basic Size

| | Keyway Width | | | | | | | | | | | | Keyway Depth | | | | |
|---|---|---|---|---|---|---|---|---|---|---|---|---|---|---|---|---|---|
| | | Tolerance and Resulting Fit[a] | | | | | | | | | | | Shaft, $t_1$ | | Hub, $t_2$ | | Radius, $r$ |
| | | Normal Fit | | | | Close Fit | | Free Fit | | | | | | | | | |
| | | Shaft | | Hub | | Shaft and Hub | | Shaft | | Hub | | | | | | | |
| Key Size, $b \times h$ | Nom-inal | Toler-ance | Fit | Toler-ance | Fit | Toler-ance | Fit | Toler-ance | Fit | Toler-ance | Fit | Nom-inal | Toler-ance | Nomi-nal | Toler-ance | Max. |
| 2×2 | 2 | +0.040 | 0.040L | +0.050 | 0.050L | +0.034 | 0.034L | +0.066 | 0.066L | +0.086 | 0.086L | 1.2 | | 1 | | |
| 3×3 | 3 | +0.010 | 0.030T | +0.025 | 0.015T | −0.008 | 0.032T | +0.040 | 0T | +0.060 | 0.020L | 1.8 | | 1.4 | | 0.16 |
| 4×4 | 4 | | | | | | | | | | | 2.5 | | 1.8 | | |
| 5×3 | 5 | | | | | | | | 0.075L | | | 1.8 | +0.1 | 1.4 | +0.1 | |
| 5×5 | 5 | +0.045 | 0.045L | +0.060 | 0.060L | +0.035 | 0.035L | +0.075 | 0T | +0.105 | 0.105L | 3 | 0 | 2.8 | 0 | |
| 6×4 | 6 | +0.015 | 0.030T | +0.015 | 0.015T | −0.005 | 0.040T | +0.045 | | +0.075 | 0.030L | 2.5 | | 1.8 | | |
| 6×6 | 6 | | | | | | | | | | | 3.5 | | 2.8 | | 0.25 |
| 8×5 | 8 | | | | | | | | | | | 3 | | 2.8 | | |
| 8×7 | 8 | +0.055 | 0.055L | +0.075 | 0.075L | +0.040 | 0.040L | +0.090 | 0.090L | +0.130 | 0.130L | 4 | +0.2 0 | 3.3 | +0.2 0 | |
| 10×6 | 10 | +0.015 | 0.035T | +0.035 | 0.015T | 0.000 | 0.050T | +0.050 | 0T | +0.090 | 0.040L | 3.5 | +0.1 0 | 2.8 | +0.1 0 | |
| 10×8 | 10 | | | | | | | | | | | 5 | +0.2 0 | 3.3 | +0.2 0 | |
| 12×6 | 12 | | | | | | | | | | | 3.5 | +0.1 0 | 2.8 | +0.1 0 | 0.6 |
| 12×8 | 12 | +0.080 | 0.080L | +0.095 | 0.095L | +0.055 | 0.055L | +0.135 | 0.135L | +0.185 | 0.185L | 5 | +0.2 0 | 3.3 | +0.2 0 | |
| 14×6 | 14 | −0.030 | 0.045T | +0.055 | 0.020T | −0.015 | 0.060T | +0.075 | 0T | +0.125 | 0.050L | 3.5 | +0.1 0 | 2.8 | +0.1 0 | |
| 14×9 | 14 | | | | | | | | | | | 5.5 | +0.2 0 | 3.8 | +0.2 0 | |

**Table 4.** *(Continued)* **Keyway Dimensions and Tolerances for Metric Square and Rectangular Parallel Keys**
*ANSI/ASME B18.25.3M-2003 (Withdrawn),* Width Tolerances and Deviations Greater than Basic Size

| Key Size, $b \times h$ | Keyway Width: Nominal | Normal Fit, Shaft: Tolerance | Normal Fit, Shaft: Fit | Normal Fit, Hub: Tolerance | Normal Fit, Hub: Fit | Close Fit, Shaft and Hub: Tolerance | Close Fit, Shaft and Hub: Fit | Free Fit, Shaft: Tolerance | Free Fit, Shaft: Fit | Free Fit, Hub: Tolerance | Free Fit, Hub: Fit | Keyway Depth, Shaft, $t_1$: Nominal | Shaft, $t_1$: Tolerance | Hub, $t_2$: Nominal | Hub, $t_2$: Tolerance | Radius, $r$: Max. |
|---|---|---|---|---|---|---|---|---|---|---|---|---|---|---|---|---|
| 16×7 | 16 | +0.080 −0.030 | 0.080L 0.045T | +0.095 +0.055 | 0.095L 0.020T | +0.055 −0.015 | 0.055L 0.060T | +0.135 +0.075 | 0.135L 0T | +0.185 +0.125 | 0.185L 0.050L | 4 | +0.2 0 | 3.3 | +0.2 0 | 0.6 |
| 16×10 | 16 | | | | | | | | | | | 6 | | 4.3 | | |
| 18×7 | 18 | | | | | | | | | | | 4 | | 3.3 | | |
| 18×11 | 18 | | | | | | | | | | | 7 | | 4.4 | | |
| 20×8 | 20 | +0.085 −0.035 | 0.085L 0.050T | +0.110 +0.060 | 0.110L 0.025T | +0.050 −0.010 | 0.050L 0.075T | +0.135 +0.085 | 0.150L 0T | +0.200 +0.110 | 0.200L 0.065L | 5 | | 3.3 | | |
| 20×12 | 10 | | | | | | | | | | | 7.5 | | 4.9 | | |
| 22×9 | 10 | | | | | | | | | | | 5.5 | | 3.8 | | |
| 22×14 | 22 | | | | | | | | | | | 9 | | 5.4 | | |
| 25×9 | 25 | | | | | | | | | | | 5.5 | | 3.8 | | |
| 25×14 | 25 | | | | | | | | | | | 9 | | 5.4 | | |
| 28×10 | 28 | | | | | | | | | | | 6 | | 4.3 | | |
| 28×16 | 28 | | | | | | | | | | | 10 | | 6.4 | | |
| 32×11 | 32 | +0.110 −0.050 | 0.110L 0.075T | +0.170 +0.090 | 0.170L 0.035T | +0.090 −0.020 | 0.090L 0.105T | +0.200 +0.125 | 0.225L 0T | +0.300 +0.225 | 0.300L 0.100L | 7[b] | | 4.4 | | |
| 32×18 | 32 | | | | | | | | | | | 11[b] | | 7.4 | | |
| 36×12 | 36 | | | | | | | | | | | 7.5[b] | | 4.9 | | 1.6 |
| 56×32 | 56 | | | | | | | | | | | 20 | | 12.4 | | |
| 63×32 | 63 | | | | | | | | | | | | | | | |
| 70×36 | 70 | | | | | | | | | | | 22 | | 14.4 | | |
| 80×40 | 80 | | | | | | | | | | | 25 | | 15.4 | | 2.5 |
| 90×45 | 90 | +0.130 −0.050 | 0.130L 0.085T | +0.180 +0.090 | 0.180L 0.045T | +0.095 −0.015 | 0.095L 0.120T | +0.225 +0.135 | 0.225L 0T | +0.340 +0.255 | 0.340L 0.120L | 28 | | 17.4 | | |
| 100×50 | 100 | | | | | | | | | | | 31 | | 19.5 | | |

[a] In columns labeled "Fit," L indicates the maximum clearance between the key and keyway; the T indicates the maximum interference between the key and keyway.

[b] This value differs from that given in *ANSI/ASME B18.25.3M (Withdrawn)*, which is believed to be inaccurate.

**Tolerances.—**Many of the tolerances shown in Table 3 and Table 4 are from ANSI/ASME B4.2 (ISO 286-1 and ISO 286-2). As a result, in addition to plus-minus tolerances which are common in the United States, some are expressed as plus-plus deviations from the basic size.

**Designation.—**Keys conforming to this Standard shall be designed by the following data, preferably in the sequence shown: a) ASME document number; b) product name; c) nominal size, width ($b$) × height ($h$) × length; d) style; and e) hardness (if other than non-hardened).

Optionally, a part identification number (PIN) per ANSI/ASME B18.24 may be used.

**Material Requirements.—**Same as for ANSI/ASME B18.25.1M-1996 (Withdrawn). See page 2545.

**Metric Keyway Sizes According to Shaft Diameter**

Based on BS 4235: Part 1:1972

| Nominal Shaft Diameter, $d$ | | Key Size, $b \times h$ | Nominal Keyway Width, $b$ | Nominal Shaft Diameter, $d$ | | Key Size, $b \times h$ | Nominal Keyway Width, $b$ |
|---|---|---|---|---|---|---|---|
| Over | Up to and Incl. | | | Over | Up to and Incl. | | |
| Keyways for Square Parallel Keys | | | | *(Cont'd)* Keyways for Rectangular Parallel Keys | | | |
| 6 | 8 | 2×2 | 2 | 85 | 95 | 25×14 | 25 |
| 8 | 10 | 3×3 | 3 | 95 | 110 | 28×16 | 28 |
| 10 | 12 | 4×4 | 4 | 110 | 130 | 32×18 | 32 |
| 12 | 17 | 5×5 | 5 | 130 | 150 | 36×20 | 36 |
| 17 | 22 | 6×6 | 6 | 150 | 170 | 40×22 | 40 |
| Keyways for Rectangular Parallel Keys | | | | 170 | 200 | 45×25 | 45 |
| 22 | 30 | 8×7 | 8 | 200 | 230 | 50×28 | 50 |
| 30 | 38 | 10×8 | 10 | 230 | 260 | 56×32 | 56 |
| 38 | 44 | 12×8 | 12 | 260 | 290 | 63×32 | 63 |
| 44 | 50 | 14×9 | 14 | 290 | 330 | 70×36 | 70 |
| 50 | 58 | 16×10 | 16 | 330 | 380 | 80×40 | 80 |
| 58 | 65 | 18×11 | 18 | 380 | 440 | 90×45 | 90 |
| 65 | 75 | 20×12 | 20 | 440 | 500 | 100×50 | 100 |
| 75 | 85 | 22×14 | 22 | … | … | … | … |

*Note:* This table is NOT part of ANSI/ASME B18.25.1M-1996 (Withdrawn) or ANSI/ASME B18.25.3M-2003 (Withdrawn), and is included for reference only. The selection of the proper size and type of key must rest with the design authority.

## Metric Woodruff Keys and Keyways

This ANSI/ASME B18.25.2M-1996 (Withdrawn) Standard covers requirements for metric Woodruff keys and keyways intended for both alignment of shafts and hubs, and transmitting torque between shafts and hubs. All dimensions in this Standard are in millimeters (mm). Dimensions and tolerances for Woodruff keys are shown in Table 5. Recommended dimensions and tolerances for keyways are shown in Table 6. For inch series Woodruff keys and keyseats, see *ANSI/ASME Standard Woodruff Keys and Keyseats* starting on page 2562.

**Comparison with ISO 3912-1977 (Withdrawn).—**This Standard is based on ISO 3912-1977 (Withdrawn), *Woodruff Keys and Keyways*. However, to improve manufacturability, tolerances are decreased for the keyway width. The resulting fit is approximately the same. Keys manufactured to this Standard are functionally interchangeable with keys manufactured to the ISO Standard. Because of tighter width tolerances in this Standard, products manufactured to the ISO Standard may not meet the requirements of this Standard.

ANSI/ASME B18.25.2M-1996 (Withdrawn) also differs from ISO 3912 in that it: a) does not restrict the corners of a key to be chamfered but allows either a chamfer or a radius on the key; b) specifies a key material hardness rather than a tensile property; and c) specifies h12 rather than h11 for the tolerance of the height of the keys.

**Tolerances.—**Many of the tolerances shown in Table 5 and Table 6 are from ANSI/ASME B4.2, Preferred Metric Limits and Fits (ISO 286-1 and ISO 286-2). As a result in addition to plus-minus tolerances which are common in the United States some are expressed as plus-plus deviations from the basic size.

**Table 5. Dimensions for Metric Woodruff Keys**
*ANSI/ASME B18.25.2M-1996 (Withdrawn)*

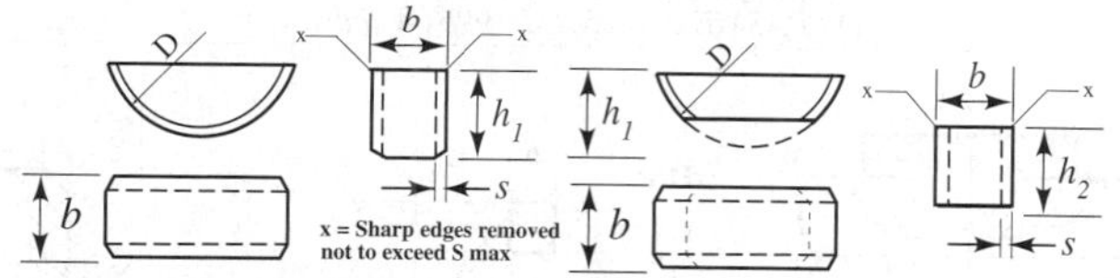

Normal Form (full radius type) Whitney Form (flat bottom type)

| Key Size $b \times h \times D$ | Width $b$ | Width Tolerance | Height $h_1$ | Tolerance h12 | $h_2$ [a] | Tolerance h12 | Diameter, $D$ | Tolerance h12 | Chamfer or Radius, $s$ Min. | Max |
|---|---|---|---|---|---|---|---|---|---|---|
| 1×1.4×4 | 1 | | 1.4 | 0 / −0.10 | 1.1 | | 4 | 0 / −0.120 | | |
| 1.5×2.6×7 | 1.5 | | 2.6 | | 2.1 | 0 / −0.10 | 7 | | | |
| 2×2.6×7 | 2 | | 2.6 | | 2.1 | | 7 | | | |
| 2×3.7×10 | 2 | | 3.7 | 0 / −0.12 | 3.0 | | 10 | 0 / −0.150 | 0.16 | 0.25 |
| 2.5×3.7×10 | 2.5 | | 3.7 | | 3.0 | | 10 | | | |
| 3×5×13 | 3 | | 5.0 | | 4.0 | | 13 | | | |
| 3×6.5×16 | 3 | | 6.5 | | 5.2 | | 16 | 0 / −0.180 | | |
| 4×6.5×16 | 4 | | 6.5 | | 5.2 | | 16 | | | |
| 4×7.5×19 | 4 | 0 / −0.025 | 7.5 | | 6.0 | 0 / −0.12 | 19 | 0 / −0.210 | | |
| 5×6.5×16 | 5 | | 6.5 | 0 / −0.15 | 5.2 | | 16 | 0 / −0.180 | | |
| 5×7.5×19 | 5 | | 7.5 | | 6.0 | | 19 | | 0.25 | 0.40 |
| 5×9×22 | 5 | | 9.0 | | 7.2 | | 22 | | | |
| 6×9×22 | 6 | | 9.0 | | 7.2 | 0 / −0.15 | 22 | 0 / −0.210 | | |
| 6×10×25 | 6 | | 10.0 | | 8.0 | | 25 | | | |
| 8×11×28 | 8 | | 11.0 | | 8.8 | | 28 | | | |
| 10×13×32 | 10 | | 13.0 | 0 / −0.18 | 10.4 | 0 / −0.18 | 32 | 0 / −0.250 | 0.40 | 0.60 |

[a] Height $h_2$ is based on 0.80 times height $h_1$.

**Designation.—**Keys conforming to this Standard shall be designated by the following data, preferably in the sequence as follows: a) ASME document number; b) product name; c) nominal size, width ($b$) × height ($h$) × length; d) form; and e) hardness (if other than non-hardened).

*Example:* ANSI/ASME B18.25.2M (Withdrawn), Woodruff Key 6 × 10 × 25 normal hardened; ANSI/ASME B18.25.2M (Withdrawn), Woodruff Key 3×5×13 Whitney.

**Material Requirements.—**Same as for ANSI/ASME B18.25.2M-1996 (Withdrawn). See page 2545.

**Advantages of Woodruff Keys.—**In the Woodruff key system, half-circular disks of steel are used as keys, the half-circular side of the key being inserted into the keyseat. Part of the key projects and enters into a keyway in the part to be keyed to the shaft in the ordinary way. The advantage of this type of key is that the keyway is easily milled by simply sinking a milling cutter, of the same diameter as the diameter of the stock from which the keys are made, into the shaft. The keys are also very cheaply made, as they are simply cut off from round bar stock and milled apart in the center. Examples of Woodruff keyseat cutters are shown on page 883.

**Table 6. Keyway Dimensions for Metric Woodruff Keys**
*ANSI/ASME B18.25.2M-1996 (Withdrawn)*

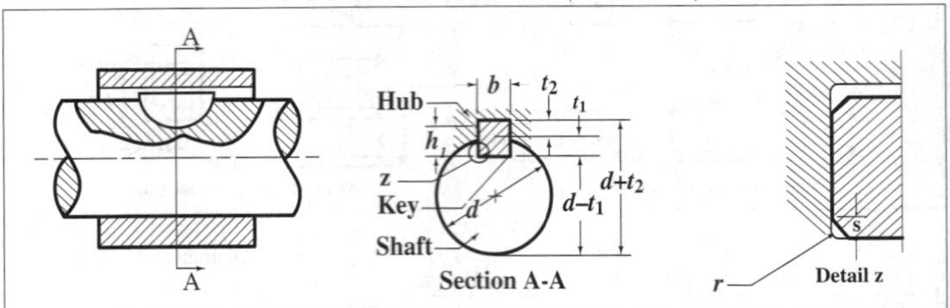


| Key Size[a] $b \times h_1 \times D$ | Width: Basic Size | Width Tolerance[b]: Normal Fit, Shaft N9 | Normal Fit, Hub S9 | Close Fit, Shaft & Hub P9 | Free Fit, Shaft H9 | Free Fit, Hub D10 | Depth: Shaft, $t_1$, Basic Size | Shaft, $t_1$, Tolerance | Hub, $t_2$, Basic Size | Hub, $t_2$, Tolerance | Radius, R, Max. | Radius, R, Min. |
|---|---|---|---|---|---|---|---|---|---|---|---|---|
| 1×1.4×4 | 1 | −0.004 −0.029 | +0.0125 −0.0125 | −0.006 −0.031 | +0.025 0 | +0.60 +0.20 | 1.0 | +0.1 0 | 0.6 | +0.1 0 | 0.16 | 0.08 |
| 1.5×2.6×7 | 1.5 | | | | | | 2.0 | | 0.8 | | | |
| 2×2.6×7 | 2 | | | | | | 1.8 | | 1.0 | | | |
| 2×3.7×10 | 2 | | | | | | 2.9 | | 1.0 | | | |
| 2.5×3.7×10 | 2.5 | | | | | | 2.7 | | 1.2 | | | |
| 3×5×13 | 3 | | | | | | 3.8 | +0.2 0 | 1.4 | | | |
| 3×6.5×16 | 3 | | | | | | 5.3 | | 1.4 | | | |
| 4×6.5×16 | 4 | −0.030 0 | +0.015 −0.015 | −0.012 −0.042 | +0.030 0 | +0.078 +0.030 | 5.0 | | 1.8 | | 0.25 | 0.16 |
| 4×7.5×19 | 4 | | | | | | 6.0 | | 1.8 | | | |
| 5×6.5×16 | 5 | | | | | | 4.5 | | 2.3 | | | |
| 5×7.5×19 | 5 | | | | | | 5.5 | | 2.3 | | | |
| 5×9×22 | 5 | | | | | | 7.0 | +0.3 0 | 2.3 | | | |
| 6×9×22 | 6 | | | | | | 6.5 | | 2.8 | | | |
| 6×10×25 | | | | | | | 7.5 | | 2.8 | +0.2 0 | | |
| 8×11×28 | 8 | 0 −0.036 | +0.018 −0.018 | −0.015 −0.051 | +0.036 0 | +0.098 +0.040 | 8.0 | | 3.3 | | 0.4 | 0.25 |
| 10×13×32 | 10 | | | | | | 10.0 | | 3.3 | | | |

[a] The nominal key diameter is the minimum keyway diameter.

[b] Some of the tolerances are expressed as plus-plus or minus-minus. See *Tolerances* on page 2555 for more informations.

**ANSI Standard Inch Series Keys and Keyseats.—**American National Standard, B17.1 Keys and Keyseats, based on current industry practice, was approved in 1967, and reaffirmed in 2013. This Standard establishes a uniform relationship between shaft sizes and key sizes for parallel and taper keys as shown in Table 1. Other data in this Standard are given in Table 2 and Table 3 through Table 7. The sizes and tolerances shown are for single key applications only.

The following definitions are given in the Standard. *Note:* Inch dimensions converted to metric dimensions (enclosed in parentheses) are not included in the Standard.

*Key:* A demountable machinery part which, when assembled into keyseats, provides a positive means for transmitting torque between the shaft and hub.

*Keyseat:* An axially located rectangular groove in a shaft or hub.

This Standard recognizes that there are two classes of stock for parallel keys used by industry. One is a close, plus toleranced key stock and the other is a broad, negative toleranced bar stock. Based on the use of two types of stock, two classes of fit are shown:

*Class 1:* A clearance or metal-to-metal side fit obtained by using bar stock keys and keyseat tolerances as given in Table 4. This is a relatively free fit and applies only to parallel keys.

*Class 2:* A side fit, with possible interference or clearance, obtained by using key stock and keyseat tolerances as given in Table 4. This is a relatively tight fit.

*Class 3:* This is an interference side fit and is not tabulated in Table 4 since the degree of interference has not been standardized. However, it is suggested that the top and bottom fit range given under Class 2 in Table 4, for parallel keys be used.

**Table 1. Key Size versus Shaft Diameter** *ANSI/ASME B17.1-1967 (R2013)*

| Nominal Shaft Diameter | | Nominal Key Size | | | Normal Keyseat Depth | |
|---|---|---|---|---|---|---|
| | | | Height, *H* | | *H*/2 | |
| Over | To (Incl.) | Width, *W* | Square | Rectangular | Square | Rectangular |
| 5⁄16 | 7⁄16 | 3⁄32 | 3⁄32 | … | 3⁄64 | … |
| 7⁄16 | 9⁄16 | 1⁄8 | 1⁄8 | 3⁄32 | 1⁄16 | 3⁄64 |
| 9⁄16 | 7⁄8 | 3⁄16 | 3⁄16 | 1⁄8 | 3⁄32 | 1⁄16 |
| 7⁄8 | 1 1⁄4 | 1⁄4 | 1⁄4 | 3⁄16 | 1⁄8 | 3⁄32 |
| 1 1⁄4 | 1 3⁄8 | 5⁄16 | 5⁄16 | 1⁄4 | 5⁄32 | 1⁄8 |
| 1 3⁄8 | 1 3⁄4 | 3⁄8 | 3⁄8 | 1⁄4 | 3⁄16 | 1⁄8 |
| 1 3⁄4 | 2 1⁄4 | 1⁄2 | 1⁄2 | 3⁄8 | 1⁄4 | 3⁄16 |
| 2 1⁄4 | 2 3⁄4 | 5⁄8 | 5⁄8 | 7⁄16 | 5⁄16 | 7⁄32 |
| 2 3⁄4 | 3 1⁄4 | 3⁄4 | 3⁄4 | 1⁄2 | 3⁄8 | 1⁄4 |
| 3 1⁄4 | 3 3⁄4 | 7⁄8 | 7⁄8 | 5⁄8 | 7⁄16 | 5⁄16 |
| 3 3⁄4 | 4 1⁄2 | 1 | 1 | 3⁄4 | 1⁄2 | 3⁄8 |
| 4 1⁄2 | 5 1⁄2 | 1 1⁄4 | 1 1⁄4 | 7⁄8 | 5⁄8 | 7⁄16 |
| 5 1⁄2 | 6 1⁄2 | 1 1⁄2 | 1 1⁄2 | 1 | 3⁄4 | 1⁄2 |
| Square Keys preferred for shaft diameters above this line; rectangular keys, below | | | | | | |
| 6 1⁄2 | 7 1⁄2 | 1 3⁄4 | 1 3⁄4 | 1 1⁄2[a] | 7⁄8 | 3⁄4 |
| 7 1⁄2 | 9 | 2 | 2 | 1 1⁄2 | 1 | 3⁄4 |
| 9 | 11 | 2 1⁄2 | 2 1⁄2 | 1 3⁄4 | 1 1⁄4 | 7⁄8 |

[a] Some key standards show 1 1⁄4 inches; preferred height is 1 1⁄2 inches.

All dimensions are given in inches. For larger shaft sizes, see *ANSI/ASME Standard Woodruff Keys and Keyseats.*

*Key Size versus Shaft Diameter:* Shaft diameters are listed in Table 1 for identification of various key sizes and are not intended to establish shaft dimensions, tolerances or selections. For a stepped shaft, the size of a key is determined by the diameter of the shaft at the point of location of the key. Up through $6^1/_2$-inch (165.1-mm) diameter shafts square keys are preferred; rectangular keys are preferred for larger shafts.

If special considerations dictate the use of a keyseat in the hub shallower than the preferred nominal depth shown, it is recommended that the tabulated preferred nominal standard keyseat always be used in the shaft.

*Keyseat Alignment Tolerances:* A tolerance of 0.010 inch (0.254 mm), max is provided for offset (due to parallel displacement of keyseat centerline from centerline of shaft or bore) of keyseats in shaft and bore. The following tolerances for maximum lead (due to angular displacement of keyseat centerline from centerline of shaft or bore and measured at right angles to the shaft or bore centerline) of keyseats in shaft and bore are specified: 0.002 inch (0.0508 mm) for keyseat length up to and including 4 inches (101.6 mm); 0.0005 inch per inch of length (0.0127 mm per mm) for keyseat lengths above 4 inches to and including 10 inches (254 mm); and 0.005 inch (0.127 mm) for keyseat lengths above 10 inches. For the effect of keyways on shaft strength, see *Effect of Keyways on Shaft Strength* on page 295.

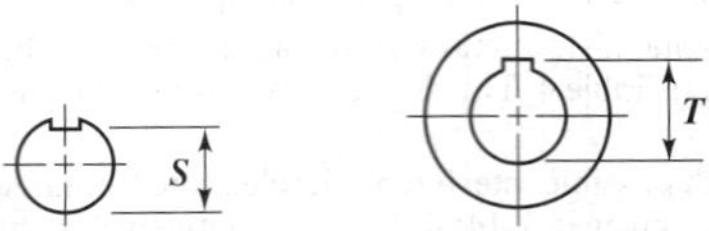


**Table 2. Depth Control Values *S* and *T* for Shaft and Hub**
*ANSI/ASME B17.1-1967 (R2013)*

| Nominal Shaft Diameter | Shafts, Parallel and Taper | | Hubs, Parallel | | Hubs, Taper | |
|---|---|---|---|---|---|---|
| | Square | Rectangular | Square | Rectangular | Square | Rectangular |
| | S | S | T | T | T | T |
| 1/2 | 0.430 | 0.445 | 0.560 | 0.544 | 0.535 | 0.519 |
| 9/16 | 0.493 | 0.509 | 0.623 | 0.607 | 0.598 | 0.582 |
| 5/8 | 0.517 | 0.548 | 0.709 | 0.678 | 0.684 | 0.653 |
| 11/16 | 0.581 | 0.612 | 0.773 | 0.742 | 0.748 | 0.717 |
| 3/4 | 0.644 | 0.676 | 0.837 | 0.806 | 0.812 | 0.781 |
| 13/16 | 0.708 | 0.739 | 0.900 | 0.869 | 0.875 | 0.844 |
| 7/8 | 0.771 | 0.802 | 0.964 | 0.932 | 0.939 | 0.907 |
| 15/16 | 0.796 | 0.827 | 1.051 | 1.019 | 1.026 | 0.994 |
| 1 | 0.859 | 0.890 | 1.114 | 1.083 | 1.089 | 1.058 |
| 1 1/16 | 0.923 | 0.954 | 1.178 | 1.146 | 1.153 | 1.121 |
| 1 1/8 | 0.986 | 1.017 | 1.241 | 1.210 | 1.216 | 1.185 |
| 1 3/16 | 1.049 | 1.080 | 1.304 | 1.273 | 1.279 | 1.248 |
| 1 1/4 | 1.112 | 1.144 | 1.367 | 1.336 | 1.342 | 1.311 |
| 1 5/16 | 1.137 | 1.169 | 1.455 | 1.424 | 1.430 | 1.399 |
| 1 3/8 | 1.201 | 1.232 | 1.518 | 1.487 | 1.493 | 1.462 |
| 1 7/16 | 1.225 | 1.288 | 1.605 | 1.543 | 1.580 | 1.518 |
| 1 1/2 | 1.289 | 1.351 | 1.669 | 1.606 | 1.644 | 1.581 |
| 1 9/16 | 1.352 | 1.415 | 1.732 | 1.670 | 1.707 | 1.645 |
| 1 5/8 | 1.416 | 1.478 | 1.796 | 1.733 | 1.771 | 1.708 |
| 1 11/16 | 1.479 | 1.541 | 1.859 | 1.796 | 1.834 | 1.771 |
| 1 3/4 | 1.542 | 1.605 | 1.922 | 1.860 | 1.897 | 1.835 |
| 1 13/16 | 1.527 | 1.590 | 2.032 | 1.970 | 2.007 | 1.945 |
| 1 7/8 | 1.591 | 1.654 | 2.096 | 2.034 | 2.071 | 2.009 |
| 1 15/16 | 1.655 | 1.717 | 2.160 | 2.097 | 2.135 | 2.072 |
| 2 | 1.718 | 1.781 | 2.223 | 2.161 | 2.198 | 2.136 |
| 2 1/16 | 1.782 | 1.844 | 2.287 | 2.224 | 2.262 | 2.199 |
| 2 1/8 | 1.845 | 1.908 | 2.350 | 2.288 | 2.325 | 2.263 |
| 2 3/16 | 1.909 | 1.971 | 2.414 | 2.351 | 2.389 | 2.326 |
| 2 1/4 | 1.972 | 2.034 | 2.477 | 2.414 | 2.452 | 2.389 |
| 2 5/16 | 1.957 | 2.051 | 2.587 | 2.493 | 2.562 | 2.468 |
| 2 3/8 | 2.021 | 2.114 | 2.651 | 2.557 | 2.626 | 2.532 |
| 2 7/16 | 2.084 | 2.178 | 2.714 | 2.621 | 2.689 | 2.596 |
| 2 1/2 | 2.148 | 2.242 | 2.778 | 2.684 | 2.753 | 2.659 |

**Table 2.** *(Continued)* **Depth Control Values *S* and *T* for Shaft and Hub**
*ANSI/ASME B17.1-1967 (R2013)*

| Nominal Shaft Diameter | Shafts, Parallel and Taper | | Hubs, Parallel | | Hubs, Taper | |
|---|---|---|---|---|---|---|
| | Square | Rectangular | Square | Rectangular | Square | Rectangular |
| | *S* | *S* | *T* | *T* | *T* | *T* |
| 2 9/16 | 2.211 | 2.305 | 2.841 | 2.748 | 2.816 | 2.723 |
| 2 5/8 | 2.275 | 2.369 | 2.905 | 2.811 | 2.880 | 2.786 |
| 2 11/16 | 2.338 | 2.432 | 2.968 | 2.874 | 2.943 | 2.849 |
| 2 3/4 | 2.402 | 2.495 | 3.032 | 2.938 | 3.007 | 2.913 |
| 2 13/16 | 2.387 | 2.512 | 3.142 | 3.017 | 3.117 | 2.992 |
| 2 7/8 | 2.450 | 2.575 | 3.205 | 3.080 | 3.180 | 3.055 |
| 2 15/16 | 2.514 | 2.639 | 3.269 | 3.144 | 3.244 | 3.119 |
| 3 | 2.577 | 2.702 | 3.332 | 3.207 | 3.307 | 3.182 |
| 3 1/16 | 2.641 | 2.766 | 3.396 | 3.271 | 3.371 | 3.246 |
| 3 1/8 | 2.704 | 2.829 | 3.459 | 3.334 | 3.434 | 3.309 |
| 3 3/16 | 2.768 | 2.893 | 3.523 | 3.398 | 3.498 | 3.373 |
| 3 1/4 | 2.831 | 2.956 | 3.586 | 3.461 | 3.561 | 3.436 |
| 3 5/16 | 2.816 | 2.941 | 3.696 | 3.571 | 3.671 | 3.546 |
| 3 3/8 | 2.880 | 3.005 | 3.760 | 3.635 | 3.735 | 3.610 |
| 3 7/16 | 2.943 | 3.068 | 3.823 | 3.698 | 3.798 | 3.673 |
| 3 1/2 | 3.007 | 3.132 | 3.887 | 3.762 | 3.862 | 3.737 |
| 3 9/16 | 3.070 | 3.195 | 3.950 | 3.825 | 3.925 | 3.800 |
| 3 5/8 | 3.134 | 3.259 | 4.014 | 3.889 | 3.989 | 3.864 |
| 3 11/16 | 3.197 | 3.322 | 4.077 | 3.952 | 4.052 | 3.927 |
| 3 3/4 | 3.261 | 3.386 | 4.141 | 4.016 | 4.116 | 3.991 |
| 3 13/16 | 3.246 | 3.371 | 4.251 | 4.126 | 4.226 | 4.101 |
| 3 7/8 | 3.309 | 3.434 | 4.314 | 4.189 | 4.289 | 4.164 |
| 3 15/16 | 3.373 | 3.498 | 4.378 | 4.253 | 4.353 | 4.228 |
| 4 | 3.436 | 3.561 | 4.441 | 4.316 | 4.416 | 4.291 |
| 4 3/16 | 3.627 | 3.752 | 4.632 | 4.507 | 4.607 | 4.482 |
| 4 1/4 | 3.690 | 3.815 | 4.695 | 4.570 | 4.670 | 4.545 |
| 4 3/8 | 3.817 | 3.942 | 4.822 | 4.697 | 4.797 | 4.672 |
| 4 7/16 | 3.880 | 4.005 | 4.885 | 4.760 | 4.860 | 4.735 |
| 4 1/2 | 3.944 | 4.069 | 4.949 | 4.824 | 4.924 | 4.799 |
| 4 3/4 | 4.041 | 4.229 | 5.296 | 5.109 | 5.271 | 5.084 |
| 4 7/8 | 4.169 | 4.356 | 5.424 | 5.236 | 5.399 | 5.211 |
| 4 15/16 | 4.232 | 4.422 | 5.487 | 5.300 | 5.462 | 5.275 |
| 5 | 4.296 | 4.483 | 5.551 | 5.363 | 5.526 | 5.338 |
| 5 3/16 | 4.486 | 4.674 | 5.741 | 5.554 | 5.716 | 5.529 |
| 5 1/4 | 4.550 | 4.737 | 5.805 | 5.617 | 5.780 | 5.592 |
| 5 7/16 | 4.740 | 4.927 | 5.995 | 5.807 | 5.970 | 5.782 |
| 5 1/2 | 4.803 | 4.991 | 6.058 | 5.871 | 6.033 | 5.846 |
| 5 3/4 | 4.900 | 5.150 | 6.405 | 6.155 | 6.380 | 6.130 |
| 5 15/16 | 5.091 | 5.341 | 6.596 | 6.346 | 6.571 | 6.321 |
| 6 | 5.155 | 5.405 | 6.660 | 6.410 | 6.635 | 6.385 |
| 6 1/4 | 5.409 | 5.659 | 6.914 | 6.664 | 6.889 | 6.639 |
| 6 1/2 | 5.662 | 5.912 | 7.167 | 6.917 | 7.142 | 6.892 |
| 6 3/4 | 5.760 | [a]5.885 | 7.515 | [a]7.390 | 7.490 | [a]7.365 |
| 7 | 6.014 | [a]6.139 | 7.769 | [a]7.644 | 7.744 | [a]7.619 |
| 7 1/4 | 6.268 | [a]6.393 | 8.023 | [a]7.898 | 7.998 | [a]7.873 |
| 7 1/2 | 6.521 | [a]6.646 | 8.276 | [a]8.151 | 8.251 | [a]8.126 |
| 7 3/4 | 6.619 | 6.869 | 8.624 | 8.374 | 8.599 | 8.349 |
| 8 | 6.873 | 7.123 | 8.878 | 8.628 | 8.853 | 8.603 |
| 9 | 7.887 | 8.137 | 9.892 | 9.642 | 9.867 | 9.617 |
| 10 | 8.591 | 8.966 | 11.096 | 10.721 | 11.071 | 10.696 |
| 11 | 9.606 | 9.981 | 12.111 | 11.736 | 12.086 | 11.71[illegible] |
| 12 | 10.309 | 10.809 | 13.314 | 12.814 | 13.289 | 12.78[illegible] |
| 13 | 11.325 | 11.825 | 14.330 | 13.830 | 14.305 | 13.80[illegible] |
| 14 | 12.028 | 12.528 | 15.533 | 15.033 | 15.508 | 15.00[illegible] |
| 15 | 13.043 | 13.543 | 16.548 | 16.048 | 16.523 | 16.02[illegible] |

[a] 1¾ × 1½ inch key.

All dimensions are given in inches. See Table 4 for tolerances.

**Table 3. ANSI Standard Plain and Gib Head Keys** *ANSI/ASME B17.1-1967 (R2013)*

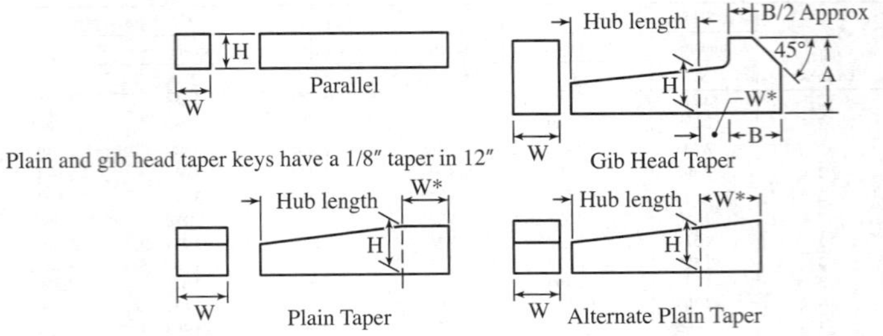


| Key | | | Nominal Key Size Width W | | Tolerance | | | |
|---|---|---|---|---|---|---|---|---|
| | | | Over | To (Incl.) | Width, W | | Height, H | |
| Parallel | Square | Keystock | … | 1¼ | +0.001 | −0.000 | +0.001 | −0.000 |
| | | | 1¼ | 3 | +0.002 | −0.000 | +0.002 | −0.000 |
| | | | 3 | 3½ | +0.003 | −0.000 | +0.003 | −0.000 |
| | | Bar Stock | … | ¾ | +0.000 | −0.002 | +0.000 | −0.002 |
| | | | ¾ | 1½ | +0.000 | −0.003 | +0.000 | −0.003 |
| | | | 1½ | 2½ | +0.000 | −0.004 | +0.000 | −0.004 |
| | | | 2½ | 3½ | +0.000 | −0.006 | +0.000 | −0.006 |
| | Rectangular | Keystock | … | 1¼ | +0.001 | −0.000 | +0.005 | −0.005 |
| | | | 1¼ | 3 | +0.002 | −0.000 | +0.005 | −0.005 |
| | | | 3 | 7 | +0.003 | −0.000 | +0.005 | −0.005 |
| | | Bar Stock | … | ¾ | +0.000 | −0.003 | +0.000 | −0.003 |
| | | | ¾ | 1½ | +0.000 | −0.004 | +0.000 | −0.004 |
| | | | 1½ | 3 | +0.000 | −0.005 | +0.000 | −0.005 |
| | | | 3 | 4 | +0.000 | −0.006 | +0.000 | −0.006 |
| | | | 4 | 6 | +0.000 | −0.008 | +0.000 | −0.008 |
| | | | 6 | 7 | +0.000 | −0.013 | +0.000 | −0.013 |
| Taper | Plain or Gib Head Square or Rectangular | | … | 1¼ | +0.001 | −0.000 | +0.005 | −0.000 |
| | | | 1¼ | 3 | +0.002 | −0.000 | +0.005 | −0.000 |
| | | | 3 | 7 | +0.003 | −0.000 | +0.005 | −0.000 |

| Gib Head Nominal Dimensions | | | | | | | | | | | | | |
|---|---|---|---|---|---|---|---|---|---|---|---|---|---|
| Nominal Key Size Width, W | Square | | | Rectangular | | | Nominal Key Size Width, W | Square | | | Rectangular | | |
| | H | A | B | H | A | B | | H | A | B | H | A | B |
| ⅛ | ⅛ | ¼ | ¼ | 3⁄32 | 3⁄16 | ⅛ | 1 | 1 | 1⅝ | 1⅛ | ¾ | 1¼ | ⅞ |
| 3⁄16 | 3⁄16 | 5⁄16 | 5⁄16 | ⅛ | ¼ | ¼ | 1¼ | 1¼ | 2 | 1 7⁄16 | ⅞ | 1⅜ | 1 |
| ¼ | ¼ | 7⁄16 | ⅜ | 3⁄16 | 5⁄16 | 5⁄16 | 1½ | 1½ | 2⅜ | 1¾ | 1 | 1⅝ | 1⅛ |
| 5⁄16 | 5⁄16 | ½ | 7⁄16 | ¼ | 7⁄16 | ⅜ | 1¾ | 1¾ | 2¾ | 2 | 1½ | 2⅜ | 1¾ |
| ⅜ | ⅜ | ⅝ | ½ | ¼ | 7⁄16 | ⅜ | 2 | 2 | 3½ | 2¼ | 1½ | 2⅜ | 1¾ |
| ½ | ½ | ⅞ | ⅝ | ⅜ | ⅝ | ½ | 2½ | 2½ | 4 | 3 | 1¾ | 2¾ | 2 |
| ⅝ | ⅝ | 1 | ¾ | 7⁄16 | ¾ | 9⁄16 | 3 | 3 | 5 | 3½ | 2 | 3½ | 2¼ |
| ¾ | ¾ | 1¼ | ⅞ | ½ | ⅞ | ⅝ | 3½ | 3½ | 6 | 4 | 2½ | 4 | 3 |
| ⅞ | ⅞ | 1⅜ | 1 | ⅝ | 1 | ¾ | … | … | … | … | … | … | … |

All dimensions are given in inches.

*For locating position of dimension *H*. Tolerance does not apply.

For larger sizes the following relationships are suggested as guides for establishing *A* and *B*: $A = 1.8H$ and $B = 1.2H$.

**Table 4. ASME Standard Fits for Parallel and Taper Keys** *ANSI/ASME B17.1-1967 (R2013)*

| Type of Key | Key Width | | Side Fit | | | Top and Bottom Fit | | | |
|---|---|---|---|---|---|---|---|---|---|
| | | | Width Tolerance | | | Depth Tolerance | | | |
| | Over | To (Incl.) | Key | Key-Seat | Fit Range[a] | Key | Shaft Key-Seat | Hub Key-Seat | Fit Range[a] |
| Class 1 Fit for Parallel Keys | | | | | | | | | |
| Square | … | ½ | +0.000<br>−0.002 | +0.002<br>−0.000 | 0.004 CL<br>0.000 | +0.000<br>−0.002 | +0.000<br>−0.015 | +0.010<br>−0.000 | 0.032 CL<br>0.005 CL |
| | ½ | ¾ | +0.000<br>−0.002 | +0.003<br>−0.000 | 0.005 CL<br>0.000 | +0.000<br>−0.002 | +0.000<br>−0.015 | +0.010<br>−0.000 | 0.032 CL<br>0.005 CL |
| | ¾ | 1 | +0.000<br>−0.003 | +0.003<br>−0.000 | 0.006 CL<br>0.000 | +0.000<br>−0.003 | +0.000<br>−0.015 | +0.010<br>−0.000 | 0.033 CL<br>0.005 CL |
| | 1 | 1½ | +0.000<br>−0.003 | +0.004<br>−0.000 | 0.007 CL<br>0.000 | +0.000<br>−0.003 | +0.000<br>−0.015 | +0.010<br>−0.000 | 0.033 CL<br>0.005 CL |
| | 1½ | 2½ | +0.000<br>−0.004 | +0.004<br>−0.000 | 0.008 CL<br>0.000 | +0.000<br>−0.004 | +0.000<br>−0.015 | +0.010<br>−0.000 | 0.034 CL<br>0.005 CL |
| | 2½ | 3½ | +0.000<br>−0.006 | +0.004<br>−0.000 | 0.010 CL<br>0.000 | +0.000<br>−0.006 | +0.000<br>−0.015 | +0.010<br>−0.000 | 0.036 CL<br>0.005 CL |
| Rectangular | … | ½ | +0.000<br>−0.003 | +0.002<br>−0.000 | 0.005 CL<br>0.000 | +0.000<br>−0.003 | +0.000<br>−0.015 | +0.010<br>−0.000 | 0.033 CL<br>0.005 CL |
| | ½ | ¾ | +0.000<br>−0.003 | +0.003<br>−0.000 | 0.006 CL<br>0.000 | +0.000<br>−0.003 | +0.000<br>−0.015 | +0.010<br>−0.000 | 0.033 CL<br>0.005 CL |
| | ¾ | 1 | +0.000<br>−0.004 | +0.003<br>−0.000 | 0.007 CL<br>0.000 | +0.000<br>−0.004 | +0.000<br>−0.015 | +0.010<br>−0.000 | 0.034 CL<br>0.005 CL |
| | 1 | 1½ | +0.000<br>−0.004 | +0.004<br>−0.000 | 0.008 CL<br>0.000 | +0.000<br>−0.004 | +0.000<br>−0.015 | +0.010<br>−0.000 | 0.034 CL<br>0.005 CL |
| | 1½ | 3 | +0.000<br>−0.005 | +0.004<br>−0.000 | 0.009 CL<br>0.000 | +0.000<br>−0.005 | +0.000<br>−0.015 | +0.010<br>−0.000 | 0.035 CL<br>0.005 CL |
| | 3 | 4 | +0.000<br>−0.006 | +0.004<br>−0.000 | 0.010 CL<br>0.000 | +0.000<br>−0.006 | +0.000<br>−0.015 | +0.010<br>−0.000 | 0.036 CL<br>0.005 CL |
| | 4 | 6 | +0.000<br>−0.008 | +0.004<br>−0.000 | 0.012 CL<br>0.000 | +0.000<br>−0.008 | +0.000<br>−0.015 | +0.010<br>−0.000 | 0.038 CL<br>0.005 CL |
| | 6 | 7 | +0.000<br>−0.013 | +0.004<br>−0.000 | 0.017 CL<br>0.000 | +0.000<br>−0.013 | +0.000<br>−0.015 | +0.010<br>−0.000 | 0.043 CL<br>0.005 CL |
| Class 2 Fit for Parallel and Taper Keys | | | | | | | | | |
| Parallel Square | … | 1¼ | +0.001<br>−0.000 | +0.002<br>−0.000 | 0.002 CL<br>0.001 INT | +0.001<br>−0.000 | +0.000<br>−0.015 | +0.010<br>−0.000 | 0.030 CL<br>0.004 CL |
| | 1¼ | 3 | +0.002<br>−0.000 | +0.002<br>−0.000 | 0.002 CL<br>0.002 INT | +0.002<br>−0.000 | +0.000<br>−0.015 | +0.010<br>−0.000 | 0.030 CL<br>0.003 CL |
| | 3 | 3½ | +0.003<br>−0.000 | +0.002<br>−0.000 | 0.002 CL<br>0.003 INT | +0.003<br>−0.000 | +0.000<br>−0.015 | +0.010<br>−0.000 | 0.030 CL<br>0.002 CL |
| Parallel Rectangular | … | 1¼ | +0.001<br>−0.000 | +0.002<br>−0.000 | 0.002 CL<br>0.001 INT | +0.005<br>−0.005 | +0.000<br>−0.015 | +0.010<br>−0.000 | 0.035 CL<br>0.000 CL |
| | 1¼ | 3 | +0.002<br>−0.000 | +0.002<br>−0.000 | 0.002 CL<br>0.002 INT | +0.005<br>−0.005 | +0.000<br>−0.015 | +0.010<br>−0.000 | 0.035 CL<br>0.000 CL |
| | 3 | 7 | +0.003<br>−0.000 | +0.002<br>−0.000 | 0.002 CL<br>0.003 INT | +0.005<br>−0.005 | +0.000<br>−0.015 | +0.010<br>−0.000 | 0.035 CL<br>0.000 CL |
| Taper | … | 1¼ | +0.001<br>−0.000 | +0.002<br>−0.000 | 0.002 CL<br>0.001 INT | +0.005<br>−0.000 | +0.000<br>−0.015 | +0.010<br>−0.000 | 0.005 CL<br>0.025 INT |
| | 1¼ | 3 | +0.002<br>−0.000 | +0.002<br>−0.000 | 0.002 CL<br>0.002 INT | +0.005<br>−0.000 | +0.000<br>−0.015 | +0.010<br>−0.000 | 0.005 CL<br>0.025 INT |
| | 3 | [b] | +0.003<br>−0.000 | +0.002<br>−0.000 | 0.002 CL<br>0.003 INT | +0.005<br>−0.000 | +0.000<br>−0.015 | +0.010<br>−0.000 | 0.005 CL<br>0.025 INT |

[a] Limits of variation. CL = Clearance; INT = Interference.

[b] To (Incl.) 3½-inch square and 7-inch rectangular key widths.

All dimensions are given in inches. See also text on page 2545.

**Table 5. Suggested Keyseat Fillet Radius and Key Chamfer**
*ANSI/ASME B17.1-1967 (R2013)*

| Keyseat Depth, H/2 | | Fillet Radius | 45 deg. Chamfer | Keyseat Depth, H/2 | | Fillet Radius | 45 deg. Chamfer |
|---|---|---|---|---|---|---|---|
| Over | To (Incl.) | | | Over | To (Incl.) | | |
| 1⁄8 | 1⁄4 | 1⁄32 | 3⁄64 | 7⁄8 | 1 1⁄4 | 3⁄16 | 7⁄32 |
| 1⁄4 | 1⁄2 | 1⁄16 | 5⁄64 | 1 1⁄4 | 1 3⁄4 | 1⁄4 | 9⁄32 |
| 1⁄2 | 7⁄8 | 1⁄8 | 5⁄32 | 1 3⁄4 | 2 1⁄2 | 3⁄8 | 13⁄32 |

All dimensions are given in inches.

**Table 6. Standard Keyseat Tolerances for Electric Motor and Generator Shaft Extensions** *ANSI/ASME B17.1-1967 (R2013)*

| Keyseat Width | | Width Tolerance | Depth Tolerance |
|---|---|---|---|
| Over | To (Incl.) | | |
| … | 1⁄4 | +0.001<br>–0.001 | +0.000<br>–0.015 |
| 1⁄4 | 3⁄4 | +0.000<br>–0.002 | +0.000<br>–0.015 |
| 3⁄4 | 1 1⁄4 | +0.000<br>–0.003 | +0.000<br>–0.015 |

All dimensions are given in inches.

**Table 7. Set Screws for Use Over Keys** *ANSI/ASME B17.1-1967 (R2013)*

| Nom. Shaft Dia. | | Nom. Key Width | Set Screw Dia. | Nom. Shaft Dia. | | Nom. Key Width | Set Screw Dia. |
|---|---|---|---|---|---|---|---|
| Over | To (Incl.) | | | Over | To (Incl.) | | |
| 5⁄16 | 7⁄16 | 3⁄32 | No. 10 | 2 1⁄4 | 2 3⁄4 | 5⁄8 | 1⁄2 |
| 7⁄16 | 9⁄16 | 1⁄8 | No. 10 | 2 3⁄4 | 3 1⁄4 | 3⁄4 | 5⁄8 |
| 9⁄16 | 7⁄8 | 3⁄16 | 1⁄4 | 3 1⁄4 | 3 3⁄4 | 7⁄8 | 3⁄4 |
| 7⁄8 | 1 1⁄4 | 1⁄4 | 5⁄16 | 3 3⁄4 | 4 1⁄2 | 1 | 3⁄4 |
| 1 1⁄4 | 1 3⁄8 | 5⁄16 | 3⁄8 | 4 1⁄2 | 5 1⁄2 | 1 1⁄4 | 7⁄8 |
| 1 3⁄8 | 1 3⁄4 | 3⁄8 | 3⁄8 | 5 1⁄2 | 6 1⁄2 | 1 1⁄2 | 1 |
| 1 3⁄4 | 2 1⁄4 | 1⁄2 | 1⁄2 | … | … | … | … |

All dimensions are given in inches.

These set screw diameter selections are offered as a guide but their use should be dependent upon design considerations.

**ANSI/ASME Standard Woodruff Keys and Keyseats.—**American National Standard B17.2 was approved in 1967, and reaffirmed in 2013. Data from this Standard are shown in Table 8, Table 9, and Table 10.

**Table 8. ANSI Standard Woodruff Keys** *ANSI/ASME B17.2-1967 (R2013)*

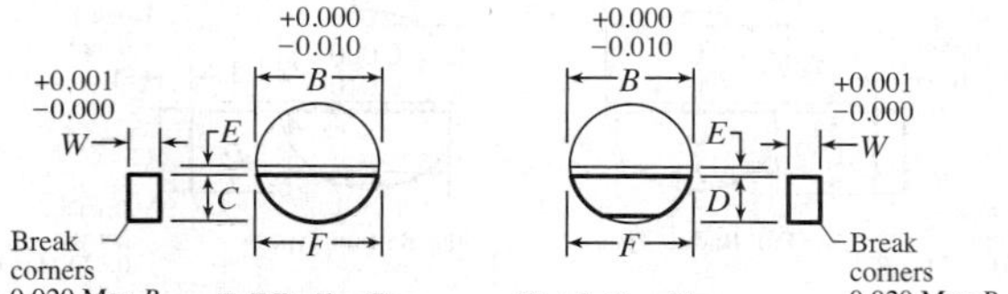


**Full Radius Type** **Flat Bottom Type**

| Key No. | Nominal Key Size W × B | Actual Length F +0.000 −0.010 | Height of Key C Max. | C Min. | D Max. | D Min. | Distance Below Center E |
|---|---|---|---|---|---|---|---|
| 202 | 1/16 × 1/4 | 0.248 | 0.109 | 0.104 | 0.109 | 0.104 | 1/64 |
| 202.5 | 1/16 × 5/16 | 0.311 | 0.140 | 0.135 | 0.140 | 0.135 | 1/64 |
| 302.5 | 3/32 × 5/16 | 0.311 | 0.140 | 0.135 | 0.140 | 0.135 | 1/64 |
| 203 | 1/16 × 3/8 | 0.374 | 0.172 | 0.167 | 0.172 | 0.167 | 1/64 |
| 303 | 3/32 × 3/8 | 0.374 | 0.172 | 0.167 | 0.172 | 0.167 | 1/64 |
| 403 | 1/8 × 3/8 | 0.374 | 0.172 | 0.167 | 0.172 | 0.167 | 1/64 |
| 204 | 1/16 × 1/2 | 0.491 | 0.203 | 0.198 | 0.194 | 0.188 | 3/64 |
| 304 | 3/32 × 1/2 | 0.491 | 0.203 | 0.198 | 0.194 | 0.188 | 3/64 |
| 404 | 1/8 × 1/2 | 0.491 | 0.203 | 0.198 | 0.194 | 0.188 | 3/64 |
| 305 | 3/32 × 5/8 | 0.612 | 0.250 | 0.245 | 0.240 | 0.234 | 1/16 |
| 405 | 1/8 × 5/8 | 0.612 | 0.250 | 0.245 | 0.240 | 0.234 | 1/16 |
| 505 | 5/32 × 5/8 | 0.612 | 0.250 | 0.245 | 0.240 | 0.234 | 1/16 |
| 605 | 3/16 × 5/8 | 0.612 | 0.250 | 0.245 | 0.240 | 0.234 | 1/16 |
| 406 | 1/8 × 3/4 | 0.740 | 0.313 | 0.308 | 0.303 | 0.297 | 1/16 |
| 506 | 5/32 × 3/4 | 0.740 | 0.313 | 0.308 | 0.303 | 0.297 | 1/16 |
| 606 | 3/16 × 3/4 | 0.740 | 0.313 | 0.308 | 0.303 | 0.297 | 1/16 |
| 806 | 1/4 × 3/4 | 0.740 | 0.313 | 0.308 | 0.303 | 0.297 | 1/16 |
| 507 | 5/32 × 7/8 | 0.866 | 0.375 | 0.370 | 0.365 | 0.359 | 1/16 |
| 607 | 3/16 × 7/8 | 0.866 | 0.375 | 0.370 | 0.365 | 0.359 | 1/16 |
| 707 | 7/32 × 7/8 | 0.866 | 0.375 | 0.370 | 0.365 | 0.359 | 1/16 |
| 807 | 1/4 × 7/8 | 0.866 | 0.375 | 0.370 | 0.365 | 0.359 | 1/16 |
| 608 | 3/16 × 1 | 0.992 | 0.438 | 0.433 | 0.428 | 0.422 | 1/16 |
| 708 | 7/32 × 1 | 0.992 | 0.438 | 0.433 | 0.428 | 0.422 | 1/16 |
| 808 | 1/4 × 1 | 0.992 | 0.438 | 0.433 | 0.428 | 0.422 | 1/16 |
| 1008 | 5/16 × 1 | 0.992 | 0.438 | 0.433 | 0.428 | 0.422 | 1/16 |
| 1208 | 3/8 × 1 | 0.992 | 0.438 | 0.433 | 0.428 | 0.422 | 1/16 |
| 609 | 3/16 × 1 1/8 | 1.114 | 0.484 | 0.479 | 0.475 | 0.469 | 5/64 |
| 709 | 7/32 × 1 1/8 | 1.114 | 0.484 | 0.479 | 0.475 | 0.469 | 5/64 |
| 809 | 1/4 × 1 1/8 | 1.114 | 0.484 | 0.479 | 0.475 | 0.469 | 5/64 |
| 1009 | 5/16 × 1 1/8 | 1.114 | 0.484 | 0.479 | 0.475 | 0.469 | 5/64 |
| 610 | 3/16 × 1 1/4 | 1.240 | 0.547 | 0.542 | 0.537 | 0.531 | 5/64 |
| 710 | 7/32 × 1 1/4 | 1.240 | 0.547 | 0.542 | 0.537 | 0.531 | 5/64 |
| 810 | 1/4 × 1 1/4 | 1.240 | 0.547 | 0.542 | 0.537 | 0.531 | 5/64 |
| 1010 | 5/16 × 1 1/4 | 1.240 | 0.547 | 0.542 | 0.537 | 0.531 | 5/64 |
| 1210 | 3/8 × 1 1/4 | 1.240 | 0.547 | 0.542 | 0.537 | 0.531 | 5/64 |
| 811 | 1/4 × 1 3/8 | 1.362 | 0.594 | 0.589 | 0.584 | 0.578 | 3/32 |
| 1011 | 5/16 × 1 3/8 | 1.362 | 0.594 | 0.589 | 0.584 | 0.578 | 3/32 |
| 1211 | 3/8 × 1 3/8 | 1.362 | 0.594 | 0.589 | 0.584 | 0.578 | 3/32 |
| 812 | 1/4 × 1 1/2 | 1.484 | 0.641 | 0.636 | 0.631 | 0.625 | 7/64 |
| 1012 | 5/16 × 1 1/2 | 1.484 | 0.641 | 0.636 | 0.631 | 0.625 | 7/64 |
| 1212 | 3/8 × 1 1/2 | 1.484 | 0.641 | 0.636 | 0.631 | 0.625 | 7/64 |

All dimensions are given in inches.

The key numbers indicate normal key dimensions. The last two digits give the nominal diameter *B* in eighths of an inch and the digits preceding the last two give the nominal width *W* in thirty-seconds of an inch.

**Table 9. ANSI Standard Woodruff Keys** *ANSI/ASME B17.2-1967 (R2013)*

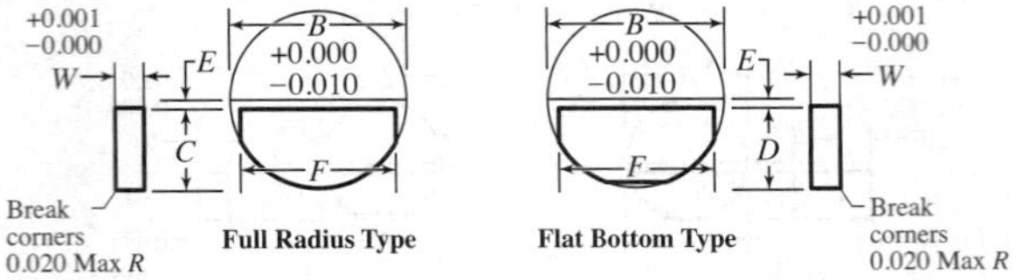


| Key No. | Nominal Key Size $W \times B$ | Actual Length $F$ +0.000 −0.010 | Height of Key $C$ Max. | Height of Key $C$ Min. | Height of Key $D$ Max. | Height of Key $D$ Min. | Distance Below Center $E$ |
|---|---|---|---|---|---|---|---|
| 617-1 | 3/16 × 2 1/8 | 1.380 | 0.406 | 0.401 | 0.396 | 0.390 | 21/32 |
| 817-1 | 1/4 × 2 1/8 | 1.380 | 0.406 | 0.401 | 0.396 | 0.390 | 21/32 |
| 1017-1 | 5/16 × 2 1/8 | 1.380 | 0.406 | 0.401 | 0.396 | 0.390 | 21/32 |
| 1217-1 | 3/8 × 2 1/8 | 1.380 | 0.406 | 0.401 | 0.396 | 0.390 | 21/32 |
| 617 | 3/16 × 2 1/8 | 1.723 | 0.531 | 0.526 | 0.521 | 0.515 | 17/32 |
| 817 | 1/4 × 2 1/8 | 1.723 | 0.531 | 0.526 | 0.521 | 0.515 | 17/32 |
| 1017 | 5/16 × 2 1/8 | 1.723 | 0.531 | 0.526 | 0.521 | 0.515 | 17/32 |
| 1217 | 3/8 × 2 1/8 | 1.723 | 0.531 | 0.526 | 0.521 | 0.515 | 17/32 |
| 822-1 | 1/4 × 2 3/4 | 2.000 | 0.594 | 0.589 | 0.584 | 0.578 | 25/32 |
| 1022-1 | 5/16 × 2 3/4 | 2.000 | 0.594 | 0.589 | 0.584 | 0.578 | 25/32 |
| 1222-1 | 3/8 × 2 3/4 | 2.000 | 0.594 | 0.589 | 0.584 | 0.578 | 25/32 |
| 1422-1 | 7/16 × 2 3/4 | 2.000 | 0.594 | 0.589 | 0.584 | 0.578 | 25/32 |
| 1622-1 | 1/2 × 2 3/4 | 2.000 | 0.594 | 0.589 | 0.584 | 0.578 | 25/32 |
| 822 | 1/4 × 2 3/4 | 2.317 | 0.750 | 0.745 | 0.740 | 0.734 | 5/8 |
| 1022 | 5/16 × 2 3/4 | 2.317 | 0.750 | 0.745 | 0.740 | 0.734 | 5/8 |
| 1222 | 3/8 × 2 3/4 | 2.317 | 0.750 | 0.745 | 0.740 | 0.734 | 5/8 |
| 1422 | 7/16 × 2 3/4 | 2.317 | 0.750 | 0.745 | 0.740 | 0.734 | 5/8 |
| 1622 | 1/2 × 2 3/4 | 2.317 | 0.750 | 0.745 | 0.740 | 0.734 | 5/8 |
| 1228 | 3/8 × 3 1/2 | 2.880 | 0.938 | 0.933 | 0.928 | 0.922 | 13/16 |
| 1428 | 7/16 × 3 1/2 | 2.880 | 0.938 | 0.933 | 0.928 | 0.922 | 13/16 |
| 1628 | 1/2 × 3 1/2 | 2.880 | 0.938 | 0.933 | 0.928 | 0.922 | 13/16 |
| 1828 | 9/16 × 3 1/2 | 2.880 | 0.938 | 0.933 | 0.928 | 0.922 | 13/16 |
| 2028 | 5/8 × 3 1/2 | 2.880 | 0.938 | 0.933 | 0.928 | 0.922 | 13/16 |
| 2228 | 11/16 × 3 1/2 | 2.880 | 0.938 | 0.933 | 0.928 | 0.922 | 13/16 |
| 2428 | 3/4 × 3 1/2 | 2.880 | 0.938 | 0.933 | 0.928 | 0.922 | 13/16 |

All dimensions are given in inches.

The key numbers indicate nominal key dimensions. The last two digits give the nominal diameter *B* in eighths of an inch and the digits preceding the last two give the nominal width *W* in thirty-seconds of an inch.

The key numbers with the −1 designation, while representing the nominal key size have a shorter length *F* and due to a greater distance below center *E* are less in height than the keys of the same number without the −1 designation.

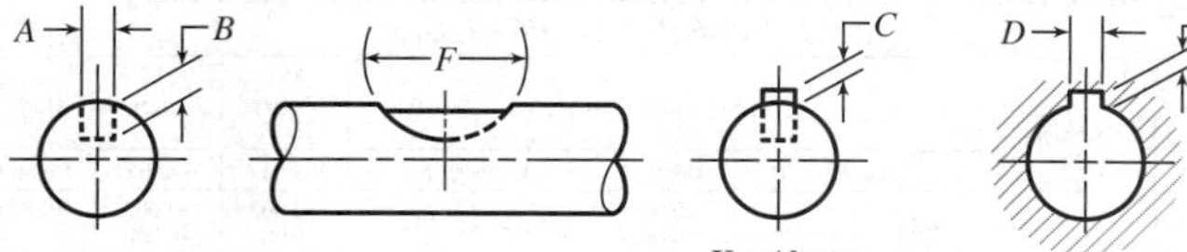


Keyseat—Shaft Key Above Shaft Keyseat—Hub

## Table 10. ANSI Keyseat Dimensions for Woodruff Keys

*ANSI/ASME B17.2-1967 (R2013)*

| Key No. | Nominal Size Key | Keyseat—Shaft | | | | | Key Above Shaft | Keyseat—Hub | |
|---|---|---|---|---|---|---|---|---|---|
| | | Width A[a] | | Depth B | Diameter F | | Height C | Width D | Depth E |
| | | Min. | Max. | +0.005 −0.000 | Min. | Max. | +0.005 −0.005 | +0.002 −0.000 | +0.005 −0.000 |
| 202 | 1/16 × 1/4 | 0.0615 | 0.0630 | 0.0728 | 0.250 | 0.268 | 0.0312 | 0.0635 | 0.0372 |
| 202.5 | 1/16 × 5/16 | 0.0615 | 0.0630 | 0.1038 | 0.312 | 0.330 | 0.0312 | 0.0635 | 0.0372 |
| 302.5 | 3/32 × 5/16 | 0.0928 | 0.0943 | 0.0882 | 0.312 | 0.330 | 0.0469 | 0.0948 | 0.0529 |
| 203 | 1/16 × 3/8 | 0.0615 | 0.0630 | 0.1358 | 0.375 | 0.393 | 0.0312 | 0.0635 | 0.0372 |
| 303 | 3/32 × 3/8 | 0.0928 | 0.0943 | 0.1202 | 0.375 | 0.393 | 0.0469 | 0.0948 | 0.0529 |
| 403 | 1/8 × 3/8 | 0.1240 | 0.1255 | 0.1045 | 0.375 | 0.393 | 0.0625 | 0.1260 | 0.0685 |
| 204 | 1/16 × 1/2 | 0.0615 | 0.0630 | 0.1668 | 0.500 | 0.518 | 0.0312 | 0.0635 | 0.0372 |
| 304 | 3/32 × 1/2 | 0.0928 | 0.0943 | 0.1511 | 0.500 | 0.518 | 0.0469 | 0.0948 | 0.0529 |
| 404 | 1/8 × 1/2 | 0.1240 | 0.1255 | 0.1355 | 0.500 | 0.518 | 0.0625 | 0.1260 | 0.0685 |
| 305 | 3/32 × 5/8 | 0.0928 | 0.0943 | 0.1981 | 0.625 | 0.643 | 0.0469 | 0.0948 | 0.0529 |
| 405 | 1/8 × 5/8 | 0.1240 | 0.1255 | 0.1825 | 0.625 | 0.643 | 0.0625 | 0.1260 | 0.0685 |
| 505 | 5/32 × 5/8 | 0.1553 | 0.1568 | 0.1669 | 0.625 | 0.643 | 0.0781 | 0.1573 | 0.0841 |
| 605 | 3/16 × 5/8 | 0.1863 | 0.1880 | 0.1513 | 0.625 | 0.643 | 0.0937 | 0.1885 | 0.0997 |
| 406 | 1/8 × 3/4 | 0.1240 | 0.1255 | 0.2455 | 0.750 | 0.768 | 0.0625 | 0.1260 | 0.0685 |
| 506 | 5/32 × 3/4 | 0.1553 | 0.1568 | 0.2299 | 0.750 | 0.768 | 0.0781 | 0.1573 | 0.0841 |
| 606 | 3/16 × 3/4 | 0.1863 | 0.1880 | 0.2143 | 0.750 | 0.768 | 0.0937 | 0.1885 | 0.0997 |
| 806 | 1/4 × 3/4 | 0.2487 | 0.2505 | 0.1830 | 0.750 | 0.768 | 0.1250 | 0.2510 | 0.1310 |
| 507 | 5/32 × 7/8 | 0.1553 | 0.1568 | 0.2919 | 0.875 | 0.895 | 0.0781 | 0.1573 | 0.0841 |
| 607 | 3/16 × 7/8 | 0.1863 | 0.1880 | 0.2763 | 0.875 | 0.895 | 0.0937 | 0.1885 | 0.0997 |
| 707 | 7/32 × 7/8 | 0.2175 | 0.2193 | 0.2607 | 0.875 | 0.895 | 0.1093 | 0.2198 | 0.1153 |
| 807 | 1/4 × 7/8 | 0.2487 | 0.2505 | 0.2450 | 0.875 | 0.895 | 0.1250 | 0.2510 | 0.1310 |
| 608 | 3/16 × 1 | 0.1863 | 0.1880 | 0.3393 | 1.000 | 1.020 | 0.0937 | 0.1885 | 0.0997 |
| 708 | 7/32 × 1 | 0.2175 | 0.2193 | 0.3237 | 1.000 | 1.020 | 0.1093 | 0.2198 | 0.1153 |
| 808 | 1/4 × 1 | 0.2487 | 0.2505 | 0.3080 | 1.000 | 1.020 | 0.1250 | 0.2510 | 0.1310 |
| 1008 | 5/16 × 1 | 0.3111 | 0.3130 | 0.2768 | 1.000 | 1.020 | 0.1562 | 0.3135 | 0.1622 |
| 1208 | 3/8 × 1 | 0.3735 | 0.3755 | 0.2455 | 1.000 | 1.020 | 0.1875 | 0.3760 | 0.1935 |
| 609 | 3/16 × 1 1/8 | 0.1863 | 0.1880 | 0.3853 | 1.125 | 1.145 | 0.0937 | 0.1885 | 0.0997 |
| 709 | 7/32 × 1 1/8 | 0.2175 | 0.2193 | 0.3697 | 1.125 | 1.145 | 0.1093 | 0.2198 | 0.1153 |
| 809 | 1/4 × 1 1/8 | 0.2487 | 0.2505 | 0.3540 | 1.125 | 1.145 | 0.1250 | 0.2510 | 0.1310 |
| 1009 | 5/16 × 1 1/8 | 0.3111 | 0.3130 | 0.3228 | 1.125 | 1.145 | 0.1562 | 0.3135 | 0.1622 |
| 610 | 3/16 × 1 1/4 | 0.1863 | 0.1880 | 0.4483 | 1.250 | 1.273 | 0.0937 | 0.1885 | 0.0997 |
| 710 | 7/32 × 1 1/4 | 0.2175 | 0.2193 | 0.4327 | 1.250 | 1.273 | 0.1093 | 0.2198 | 0.1153 |
| 810 | 1/4 × 1 1/4 | 0.2487 | 0.2505 | 0.4170 | 1.250 | 1.273 | 0.1250 | 0.2510 | 0.1310 |
| 1010 | 5/16 × 1 1/4 | 0.3111 | 0.3130 | 0.3858 | 1.250 | 1.273 | 0.1562 | 0.3135 | 0.1622 |
| 1210 | 3/8 × 1 1/4 | 0.3735 | 0.3755 | 0.3545 | 1.250 | 1.273 | 0.1875 | 0.3760 | 0.1935 |
| 811 | 1/4 × 1 3/8 | 0.2487 | 0.2505 | 0.4640 | 1.375 | 1.398 | 0.1250 | 0.2510 | 0.1310 |
| 1011 | 5/16 × 1 3/8 | 0.3111 | 0.3130 | 0.4328 | 1.375 | 1.398 | 0.1562 | 0.3135 | 0.1622 |

**Table 10.** *(Continued)* **ANSI Keyseat Dimensions for Woodruff Keys**
*ANSI/ASME B17.2-1967 (R2013)*

| Key No. | Nominal Size Key | Keyseat—Shaft | | | | | Key Above Shaft | Keyseat—Hub | |
|---|---|---|---|---|---|---|---|---|---|
| | | Width *A*[a] | | Depth *B* | Diameter *F* | | Height *C* | Width *D* | Depth *E* |
| | | Min. | Max. | +0.005 −0.000 | Min. | Max. | +0.005 −0.005 | +0.002 −0.000 | +0.005 −0.000 |
| 1211 | 3⁄8 × 13⁄8 | 0.3735 | 0.3755 | 0.4015 | 1.375 | 1.398 | 0.1875 | 0.3760 | 0.1935 |
| 812 | 1⁄4 × 11⁄2 | 0.2487 | 0.2505 | 0.5110 | 1.500 | 1.523 | 0.1250 | 0.2510 | 0.1310 |
| 1012 | 5⁄16 × 11⁄2 | 0.3111 | 0.3130 | 0.4798 | 1.500 | 1.523 | 0.1562 | 0.3135 | 0.1622 |
| 1212 | 3⁄8 × 11⁄2 | 0.3735 | 0.3755 | 0.4485 | 1.500 | 1.523 | 0.1875 | 0.3760 | 0.1935 |
| 617-1 | 3⁄16 × 21⁄8 | 0.1863 | 0.1880 | 0.3073 | 2.125 | 2.160 | 0.0937 | 0.1885 | 0.0997 |
| 817-1 | 1⁄4 × 21⁄8 | 0.2487 | 0.2505 | 0.2760 | 2.125 | 2.160 | 0.1250 | 0.2510 | 0.1310 |
| 1017-1 | 5⁄16 × 21⁄8 | 0.3111 | 0.3130 | 0.2448 | 2.125 | 2.160 | 0.1562 | 0.3135 | 0.1622 |
| 1217-1 | 3⁄8 × 21⁄8 | 0.3735 | 0.3755 | 0.2135 | 2.125 | 2.160 | 0.1875 | 0.3760 | 0.1935 |
| 617 | 3⁄16 × 21⁄8 | 0.1863 | 0.1880 | 0.4323 | 2.125 | 2.160 | 0.0937 | 0.1885 | 0.0997 |
| 817 | 1⁄4 × 21⁄8 | 0.2487 | 0.2505 | 0.4010 | 2.125 | 2.160 | 0.1250 | 0.2510 | 0.1310 |
| 1017 | 5⁄16 × 21⁄8 | 0.3111 | 0.3130 | 0.3698 | 2.125 | 2.160 | 0.1562 | 0.3135 | 0.1622 |
| 1217 | 3⁄8 × 21⁄8 | 0.3735 | 0.3755 | 0.3385 | 2.125 | 2.160 | 0.1875 | 0.3760 | 0.1935 |
| 822-1 | 1⁄4 × 23⁄4 | 0.2487 | 0.2505 | 0.4640 | 2.750 | 2.785 | 0.1250 | 0.2510 | 0.1310 |
| 1022-1 | 5⁄16 × 23⁄4 | 0.3111 | 0.3130 | 0.4328 | 2.750 | 2.785 | 0.1562 | 0.3135 | 0.1622 |
| 1222-1 | 3⁄8 × 23⁄4 | 0.3735 | 0.3755 | 0.4015 | 2.750 | 2.785 | 0.1875 | 0.3760 | 0.1935 |
| 1422-1 | 7⁄16 × 23⁄4 | 0.4360 | 0.4380 | 0.3703 | 2.750 | 2.785 | 0.2187 | 0.4385 | 0.2247 |
| 1622-1 | 1⁄2 × 23⁄4 | 0.4985 | 0.5005 | 0.3390 | 2.750 | 2.785 | 0.2500 | 0.5010 | 0.2560 |
| 822 | 1⁄4 × 23⁄4 | 0.2487 | 0.2505 | 0.6200 | 2.750 | 2.785 | 0.1250 | 0.2510 | 0.1310 |
| 1022 | 5⁄16 × 23⁄4 | 0.3111 | 0.3130 | 0.5888 | 2.750 | 2.785 | 0.1562 | 0.3135 | 0.1622 |
| 1222 | 3⁄8 × 23⁄4 | 0.3735 | 0.3755 | 0.5575 | 2.750 | 2.785 | 0.1875 | 0.3760 | 0.1935 |
| 1422 | 7⁄16 × 23⁄4 | 0.4360 | 0.4380 | 0.5263 | 2.750 | 2.785 | 0.2187 | 0.4385 | 0.2247 |
| 1622 | 1⁄2 × 23⁄4 | 0.4985 | 0.5005 | 0.4950 | 2.750 | 2.785 | 0.2500 | 0.5010 | 0.2560 |
| 1228 | 3⁄8 × 31⁄2 | 0.3735 | 0.3755 | 0.7455 | 3.500 | 3.535 | 0.1875 | 0.3760 | 0.1935 |
| 1428 | 7⁄16 × 31⁄2 | 0.4360 | 0.4380 | 0.7143 | 3.500 | 3.535 | 0.2187 | 0.4385 | 0.2247 |
| 1628 | 1⁄2 × 31⁄2 | 0.4985 | 0.5005 | 0.6830 | 3.500 | 3.535 | 0.2500 | 0.5010 | 0.2560 |
| 1828 | 9⁄16 × 31⁄2 | 0.5610 | 0.5630 | 0.6518 | 3.500 | 3.535 | 0.2812 | 0.5635 | 0.2872 |
| 2028 | 5⁄8 × 31⁄2 | 0.6235 | 0.6255 | 0.6205 | 3.500 | 3.535 | 0.3125 | 0.6260 | 0.3185 |
| 2228 | 11⁄16 × 31⁄2 | 0.6860 | 0.6880 | 0.5893 | 3.500 | 3.535 | 0.3437 | 0.6885 | 0.3497 |
| 2428 | 3⁄4 × 31⁄2 | 0.7485 | 0.7505 | 0.5580 | 3.500 | 3.535 | 0.3750 | 0.7510 | 0.3810 |

[a] These Width A values were set with the maximum keyseat (shaft) width as that figure which will receive a key with the greatest amount of looseness consistent with assuring the key's sticking in the keyseat (shaft). Minimum keyseat width is that figure permitting the largest shaft distortion acceptable when assembling maximum key in minimum keyseat. Dimensions *A, B, C, D* are taken at side intersection.

All dimensions are given in inches.

The following definitions are given in this Standard:

*Woodruff Key:* A demountable machinery part which, when assembled into key-seats, provides a positive means for transmitting torque between the shaft and hub.

*Woodruff Key Number:* An identification number by which the size of key may be readily determined.

*Woodruff Keyseat—Shaft:* The circular pocket in which the key is retained.

*Woodruff Keyseat—Hub:* An axially located rectangular groove in a hub. (This has been referred to as a keyway.)

*Woodruff Keyseat Milling Cutter:* An arbor type or shank type milling cutter normally used for milling Woodruff keyseats in shafts (see page 883).

**Chamfered Keys and Filleted Keyseats.—**In general practice, chamfered keys and filleted keyseats are not used. However, it is recognized that fillets in keyseats decrease stress concentration at corners. When used, fillet radii should be as large as possible without causing excessive bearing stresses due to reduced contact area between the key and its mating parts. Keys must be chamfered or rounded to clear fillet radii. Values in Table 5 assume general conditions and should be used only as a guide when critical stresses are encountered.

**Depths for Milling Keyseats.—**Table 11 on page 2568 has been compiled to facilitate the accurate milling of keyseats. This table gives the distance $M$ (see illustration accompanying table) between the top of the shaft and a line passing through the upper corners or edges of the keyseat. Dimension $M$ is calculated by the formula: $M = \frac{1}{2}(S - \sqrt{S^2 - E^2})$ where $S$ is diameter of shaft, and $E$ is width of keyseat. A simple approximate formula that gives $M$ to within 0.001 inch is $M = E^2 \div 4S$.

**Keyseating Machines.—**The machines which are designed especially for cutting keyseats or keyways in the hubs of pulleys, gears, etc., are generally known as keyseaters. Machines of this class usually have a base or frame which contains the mechanism for imparting a reciprocating motion to a cutter bar, which moves vertically for cutting a keyseat in the work. There are several types of machines which are used for internal keyseating operations in addition to the machines designed especially for this work. Broaching machines as well as slotters are commonly used, and keyseating is also done to some extent in shapers and planers.

**Other Key Types.—**The *sunk key* is the most common type and is of rectangular section that engages a groove or slot formed both in the shaft and hub of the gear or pulley. The width of an ordinary sunk key ordinarily is equal to about one-fourth of the shaft diameter and the thickness, when a flat key is preferred to the square form, is usually about one-sixth of the shaft diameter; these proportions are varied somewhat by different manufacturers.

The *flat key* is a rectangular shape which bears upon a flat surface formed on one side of the shaft. The *draw* or *gib key* is a sunk key which has a head by means of which it can be removed. The *round tapered key* is simply a taper pin which is driven into a hole that is partly in the shaft and partly in the hub; this form is used for light work. The name *feather* or *spline* is applied to a key which is fixed to either a shaft or hub, as when a gear must be driven by a shaft, but at the same time be free to slide in a lengthwise direction.

The taper of American Standard square and flat keys is 1.8 inch per foot.

The *saddle key* does not enter a slot in the shaft. It has parallel sides and is curved on its under side to fit the shaft. It is slightly tapered on top so that, when it is driven tightly in place, the shaft is held by frictional resistance. This key should be fitted so that it bears lightly on the sides and heavily between the shaft and hub throughout its entire length As the drive with this type of key is not positive, it is only used where there is little power to transmit. It is an inexpensive method of keying, as the shaft does is not machined.

**Effect of Keyways on Shaft Strength.—**See *SHAFTS* starting on page 289 and *Effect of Keyways on Shaft Strength* starting on page 295.

**British Standard Keys and Keyways.—**See *Keys and Keyways* in the *ADDITIONAL MATERIAL* in the *Machinery's Handbook 32 Digital Edition*.

**Cotters.—**A cotter is a form of key that is used to connect rods, etc., that are subjected either to tension or compression or both, the cotter being subjected to shearing stresses at two transverse cross sections. When taper cotters are used for drawing and holding parts together, if the cotter is held in place by the friction between the bearing surfaces, the taper should not be too great. Ordinarily a taper varying from $\frac{1}{4}$ to $\frac{1}{2}$ inch per foot is used for plain cotters. When a set screw or other device is used to prevent the cotter from backing out of its slot, the taper may vary from $1\frac{1}{2}$ to 2 inches per foot.

## Table 11. Finding Depth of Keyseat and Distance from Top of Key to Bottom of Shaft

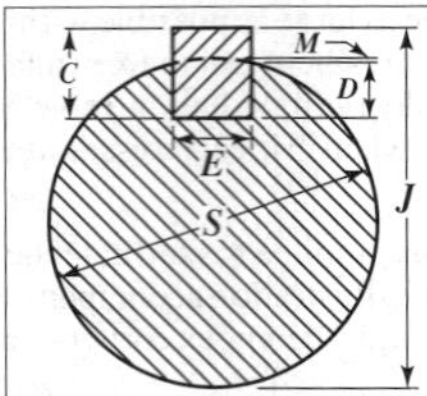


For milling keyseats, the total depth to feed cutter in from outside of shaft to bottom of keyseat is $M + D$, where $D$ is depth of keyseat.

For checking an assembled key and shaft, caliper measurement $J$ between top of key and bottom of shaft is used.

$$J = S - (M + D) + C$$

where $C$ is depth of key. For Woodruff keys, dimensions $C$ and $D$ can be found in Table 8 through Table 10. Assuming shaft diameter $S$ is normal size, the tolerance on dimension $J$ for Woodruff keys in keyslots are + 0.000, −0.010 inch.

| Dia. of Shaft, S Inches | Width of Keyseat, E | | | | | | | | | | | | | | |
|---|---|---|---|---|---|---|---|---|---|---|---|---|---|---|---|
| | 1⁄16 | 3⁄32 | 1⁄8 | 5⁄32 | 3⁄16 | 7⁄32 | 1⁄4 | 5⁄16 | 3⁄8 | 7⁄16 | 1⁄2 | 9⁄16 | 5⁄8 | 11⁄16 | 3⁄4 |
| | Dimension M, Inch | | | | | | | | | | | | | | |
| 0.3125 | .0032 | … | … | … | … | … | … | … | … | … | … | … | … | … | … |
| 0.3437 | .0029 | .0065 | … | … | … | … | … | … | … | … | … | … | … | … | … |
| 0.3750 | .0026 | .0060 | .0107 | … | … | … | … | … | … | … | … | … | … | … | … |
| 0.4060 | .0024 | .0055 | .0099 | … | … | … | … | … | … | … | … | … | … | … | … |
| 0.4375 | .0022 | .0051 | .0091 | … | … | … | … | … | … | … | … | … | … | … | … |
| 0.4687 | .0021 | .0047 | .0085 | .0134 | … | … | … | … | … | … | … | … | … | … | … |
| 0.5000 | .0020 | .0044 | .0079 | .0125 | … | … | … | … | … | … | … | … | … | … | … |
| 0.5625 | … | .0039 | .0070 | .0111 | .0161 | … | … | … | … | … | … | … | … | … | … |
| 0.6250 | … | .0035 | .0063 | .0099 | .0144 | .0198 | … | … | … | … | … | … | … | … | … |
| 0.6875 | … | .0032 | .0057 | .0090 | .0130 | .0179 | .0235 | … | … | … | … | … | … | … | … |
| 0.7500 | … | .0029 | .0052 | .0082 | .0119 | .0163 | .0214 | .0341 | … | … | … | … | … | … | … |
| 0.8125 | … | .0027 | .0048 | .0076 | .0110 | .0150 | .0197 | .0312 | … | … | … | … | … | … | … |
| 0.8750 | … | .0025 | .0045 | .0070 | .0102 | .0139 | .0182 | .0288 | … | … | … | … | … | … | … |
| 0.9375 | … | … | .0042 | .0066 | .0095 | .0129 | .0170 | .0263 | .0391 | … | … | … | … | … | … |
| 1.0000 | … | … | .0039 | .0061 | .0089 | .0121 | .0159 | .0250 | .0365 | … | … | … | … | … | … |
| 1.0625 | … | … | .0037 | .0058 | .0083 | .0114 | .0149 | .0235 | .0342 | … | … | … | … | … | … |
| 1.1250 | … | … | .0035 | .0055 | .0079 | .0107 | .0141 | .0221 | .0322 | .0443 | … | … | … | … | … |
| 1.1875 | … | … | .0033 | .0052 | .0074 | .0102 | .0133 | .0209 | .0304 | .0418 | … | … | … | … | … |
| 1.2500 | … | … | .0031 | .0049 | .0071 | .0097 | .0126 | .0198 | .0288 | .0395 | … | … | … | … | … |
| 1.3750 | … | … | … | .0045 | .0064 | .0088 | .0115 | .0180 | .0261 | .0357 | .0471 | … | … | … | … |
| 1.5000 | … | … | … | .0041 | .0059 | .0080 | .0105 | .0165 | .0238 | .0326 | .0429 | … | … | … | … |
| 1.6250 | … | … | … | .0038 | .0054 | .0074 | .0097 | .0152 | .0219 | .0300 | .0394 | .0502 | … | … | … |
| 1.7500 | … | … | … | … | .0050 | .0069 | .0090 | .0141 | .0203 | .0278 | .0365 | .0464 | … | … | … |
| 1.8750 | … | … | … | … | .0047 | .0064 | .0084 | .0131 | .0189 | .0259 | .0340 | .0432 | .0536 | … | … |
| 2.0000 | … | … | … | … | .0044 | .0060 | .0078 | .0123 | .0177 | .0242 | .0318 | .0404 | .0501 | … | … |
| 2.1250 | … | … | … | … | … | .0056 | .0074 | .0116 | .0167 | .0228 | .0298 | .0379 | .0470 | .0572 | .0684 |
| 2.2500 | … | … | … | … | … | … | .0070 | .0109 | .0157 | .0215 | .0281 | .0357 | .0443 | .0538 | .0643 |
| 2.3750 | … | … | … | … | … | … | … | .0103 | .0149 | .0203 | .0266 | .0338 | .0419 | .0509 | .0608 |
| 2.5000 | … | … | … | … | … | … | … | … | .0141 | .0193 | .0253 | .0321 | .0397 | .0482 | .0576 |
| 2.6250 | … | … | … | … | … | … | … | … | .0135 | .0184 | .0240 | .0305 | .0377 | .0457 | .0547 |
| 2.7500 | … | … | … | … | … | … | … | … | … | .0175 | .0229 | .0291 | .0360 | .0437 | .0521 |
| 2.8750 | … | … | … | … | … | … | … | … | … | .0168 | .0219 | .0278 | .0344 | .0417 | .0498 |
| 3.0000 | … | … | … | … | … | … | … | … | … | … | .0210 | .0266 | .0329 | .0399 | .0476 |

## FLEXIBLE BELTS AND SHEAVES

Flexible belt drives are used in industrial power transmission applications, especially when the speeds of the driver and driven shafts must be different or when shafts must be widely separated. The trend toward higher speed prime movers and the need to achieve a slower, useful driven speed are additional factors favoring the use of belts. Belts have numerous advantages over other means of power transmission; these advantages include overall economy, cleanliness, no need for lubrication, lower maintenance costs, easy installation, dampening of shock loads, and the abilities to be used for clutching and variable-speed power transmission between widely spaced shafts.

### Calculations for Belts and Pulleys

Belt speed plays an important role in the amount of load a friction drive system can transmit. Higher speeds will require higher preloads (increased belt tension) to compensate for the higher centrifugal force. In positive drive (toothed belt) systems, higher speeds generate dynamic forces caused by unavoidable tolerance errors that may result in increased tooth or pin stresses and shorter belt life.

**Pulley Diameters and Drive Ratios.—**Minimum pulley diameters determined by belt manufacturers are based on the minimum radius that a belt can wrap around a pulley without stressing the load-carrying members. For positive drive systems, minimum pulley diameters are also determined by the minimum number of teeth that must be engaged with the sprocket to guarantee the operating load.

Diameters of driving and driven pulleys determine the velocity ratio of the input relative to the output shaft and are derived from the following formulas: for all belt systems, velocity ratio $V = D_{pi}/D_{po}$, and for positive (toothed) drive systems, velocity ratio $V = N_i/N_o$, where $D_{pi}$ is the pitch diameter of the driving pulley, $D_{po}$ is the pitch diameter of the driven pulley, $N_i$ is the number of teeth on the driving pulley, and $N_o$ is the number of teeth on the driven pulley. For most drive systems, a velocity ratio of 8:1 is the largest that should be attempted with a single reduction drive, and 6:1 is a reasonable maximum.

**Wrap Angles and Center-to-Center Distances.—**The radial distance for which the belt is in contact with the pulley surface, or the number of teeth in engagement for positive drive belts, is called the wrap angle. Belt and sprocket combinations should be chosen to ensure a wrap angle of about 120° around the smaller pulley. The wrap angle should not be less than 90°, especially with positive drive belts, because if too few teeth are in engagement, the belt may jump a tooth or pin and timing or synchronization may be lost.

For flat belts, the minimum allowable center-to-center distance (CD) for any belt-and-sprocket combination should be chosen to ensure a minimum wrap angle around the smaller pulley. For high-velocity systems, a good rule of thumb is a minimum CD equal to the sum of the pitch diameter of the larger sprocket and one-half the pitch diameter of the smaller sprocket. This formula ensures a minimum wrap angle of approximately 120°, which is generally sufficient for friction drives and will ensure that positive drive belts do not jump teeth.

**Pulley Center Distances and Belt Lengths.—**Maximum center distances of pulleys should be about 15 to 20 times the pitch diameter of the smaller pulley. Greater spacing requires tight control of the belt tension because a small amount of stretch will cause a large drop in tension. Constant belt tension can be obtained by application of an adjustable tensioning pulley applied to the slack side of the belt. Friction drive systems using flat belts require much more tension than positive drive belt systems.

Belt length can be calculated from: $L = 2C + \pi(D_2 + D_1)/2 + (D_2 - D_1)^2/4C$ for friction drives, and length $L = 2C + \pi(D_2 + D_1)/2 + (D_2 + D_1)^2/4C$ for crossed belt friction belt drives, where $C$ is the center distance, $D_1$ is the pitch diameter of the small pulley and $D_2$

is the pitch diameter of the large pulley. For serrated belt drives, the length determined by use of these equations should be divided by the serration pitch. The belt length must then be adjusted to provide a whole number of serrations.

**Pulley Diameters and Speeds.—**If $D$ = diameter of driving pulley, $d$ = diameter of driven pulley, $S$ = speed of driving pulley, and $s$ = speed of driven pulley:

$$D = \frac{d \times s}{S}, \quad d = \frac{D \times S}{s}, \quad S = \frac{d \times s}{D}, \quad \text{and} \quad s = \frac{D \times S}{d}$$

*Example 1:* If diameter of the driving pulley $D$ is 24 inches, its speed is 100 rpm, and the driven pulley runs at 600 rpm, diameter of the driven pulley, $d = 24 \times 100/600 = 4$ inches.

*Example 2:* If the diameter of the driving pulley $D$ is 60 cm, its speed is 100 rpm, and the driven pulley is to run at 600 rpm, the diameter of the driven pulley, $d = 60 \times 100/600 = 10$ cm.

*Example 3:* If the diameter of the driven pulley $d$ is 36 inches, its required speed is to be 150 rpm, and the speed of the driving pulley is to be 600 rpm, the diameter of the driving pulley $D = 36 \times 150/600 = 9$ inches.

*Example 4:* If diameter of driven pulley $d$ is 4 inches, its required speed is 800 rpm, and diameter of driving pulley $D$ is 26 inches, speed of driving pulley $= 4 \times 800/26 = 123$ rpm.

*Example 5:* If diameter of driven pulley $d$ is 10 cm, its required speed is 800 rpm, and diameter of driving pulley $D$ is 25 cm, speed of the driving pulley $= 10 \times 800/25 = 320$ rpm.

*Example 6:* If the diameter of the driving pulley $D$ is 15 inches and its speed is 180 rpm, and the diameter of the driven pulley $d$ is 9 inches, then the speed of the driven pulley = $15 \times 180/9 = 300$ rpm.

**Pulley Diameters in Compound Drive.—**If speeds of driving and driven pulleys, $A$, $B$, $C$, and $D$ (see illustration) are known, the first step in finding their diameters is to form a fraction with the driving pulley speed as the numerator and the driven pulley speed as the, denominator, and then reduce this fraction to its lowest terms. Resolve the numerator and the denominator into two pairs of factors (a pair being one factor in the numerator and one in the denominator) and, if necessary, multiply each pair by a trial number that will give pulleys of suitable diameters.

*Example 7:* If speed of pulley $A$ is 260 rpm and the required speed of pulley $D$ is 720 rpm, find the diameters of the four pulleys. Reduced to its lowest terms, the fraction 260/720 = 13/36, which represents the required speed ratio. Resolve this ratio 13/36 into two factors:

$$\frac{13}{36} = \frac{1 \times 13}{2 \times 18}$$

Multiply by trial numbers 12 and 1 to get:

$$\frac{(1 \times 12) \times (13 \times 1)}{(2 \times 12) \times (18 \times 1)} = \frac{12 \times 13}{24 \times 18}$$

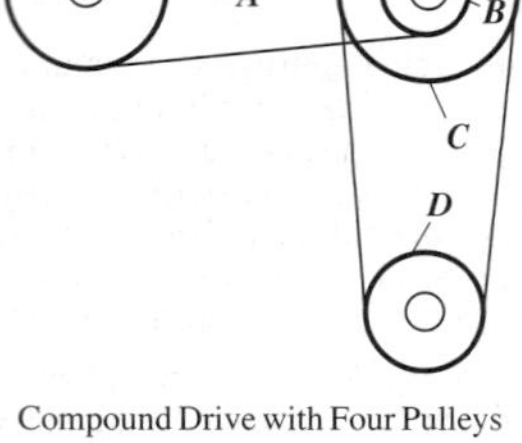


Compound Drive with Four Pulleys

The values 12 and 13 in the numerator represent the diameters of the *driven* pulleys *B* and *D*, and the values 24 and 18 in the denominator represent the diameters of the *driving* pulleys, *A* and *C*, as shown in the illustration.

**Speed of Driven Pulley in Compound Drive.—**If diameters of pulleys *A*, *B*, *C*, and *D* (see illustration above), and speed of pulley *A* are known, the speed of the driven pulley *D* is found from:

$$\frac{\text{driving pulley diameter}}{\text{driven pulley diameter}} \times \frac{\text{driving pulley diameter}}{\text{driven pulley diameter}} \times \text{speed of first driving pulley}$$

*Example 8:* If the diameters of driving pulleys *A* and *C* are 18 and 24 inches, diameters of driven pulleys *B* and *D* are 12 and 13 inches, and the speed of driving pulley *A* is 260 rpm, speed of driven pulley

$$D = \frac{18 \times 24}{12 \times 13} \times 260 = 720 \text{ rpm}$$

**Length of Belt Traversing Three Pulleys.—**The length *L* of a belt traversing three pulleys, as shown in the diagram below, and touching them on one side only, can be found by the following formula.

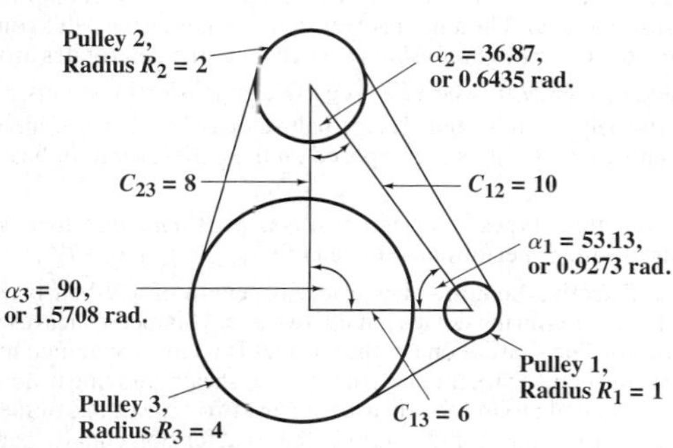


Flat Belt Traversing Three Pulleys

Referring to the diagram, $R_1$, $R_2$, and $R_3$ are the radii of the three pulleys; $C_{12}$, $C_{13}$, and $C_{23}$ are the center distances; and $\alpha_1$, $\alpha_2$, and $\alpha_3$ are the angles, in radians, of the triangle formed by the center distances. Then:

$$L = C_{12} + C_{13} + C_{23} + \frac{1}{2}\left[\frac{(R_2 - R_1)^2}{C_{12}} + \frac{(R_3 - R_1)^2}{C_{13}} + \frac{(R_3 - R_2)^2}{C_{23}}\right] + \pi(R_1 + R_2 + R_3) - (\alpha_1 R_1 + \alpha_2 R_2 + \alpha_3 R_3)$$

*Example 9:* Assume $R_1 = 1, R_2 = 2, R_3 = 4, C_{12} = 10, C_{13} = 6, C_{23} = 8, \alpha_1 = 53.13$ degrees or 0.9273 radian, $\alpha_2 = 36.87$ degrees or 0.6435 radian, and $\alpha_3 = 90$ degrees or 1.5708 radians. Then:

$$L = 10 + 6 + 8 + \frac{1}{2}\left[\frac{(2-1)^2}{10} + \frac{(4-1)^2}{6} + \frac{(4-2)^2}{8}\right] + \pi(1 + 2 + 4) + (0.9273 \times 1 + 0.6435 \times 2 + 1.5708 \times 4)$$
$$= 24 + 1.05 + 21.9911 - 8.4975 = 38.5436$$

**Power Transmitted by Belts.—**With belt drives, the force that produces work acts on the rim of a pulley or sheave and causes it to rotate. Since a belt on a drive must be tight enough to prevent slip, there is a belt pull on both sides of a driven wheel. When a drive is stationary or operating with no power transmitted, the pulls on both sides of the driven wheel are equal. When the drive is transmitting power, however, the pulls are not the same. There is a tight side tension $T_T$ and a slack side tension, $T_S$. The difference between these two pulls $(T_T - T_S)$ is called *effective pull* or *net pull*. This effective pull is applied at the rim of the pulley and is the force that produces work.

Net pull equals horsepower (HP) × 33,000 ÷ belt speed (fpm). Belt speed in fpm can be set by changing the pulley, sprocket, or sheave diameter. The shaft speeds remain the same. Belt speed is directly related to pulley diameter. Double the diameter and the total belt pull is cut in half, reducing the load on the shafts and bearings.

A belt experiences three types of tension as it rotates around a pulley: working tension (tight side – slack side), bending tension, and centrifugal tension.

The *tension ratio* ($R$) equals tight side divided by slack side tension (measured in pounds). The larger $R$ is, the closer a V-belt is to slipping—the belt is too loose. (Synchronous belts do not slip, because they depend on the tooth grip principle.)

In addition to working tension (tight side – slack side), two other tensions are developed in a belt when it is operating on a drive. *Bending tension* $T_B$ occurs when the belt bends around the pulley. One part of the belt is in tension and the other is in compression, so compressive stresses also occur. The amount of tension depends on the belt's construction and the pulley diameter. *Centrifugal tension* ($T_C$) occurs as the belt rotates around the drive and is calculated by $T_C = MV^2$, where $T_C$ is centrifugal tension in pounds, $M$ is a constant dependent on the belt's weight, and $V$ is the belt velocity in feet per minute. Neither the bending nor centrifugal tensions are imposed on the pulley, shaft, or bearing—only on the belt.

Combining these three types of tension results in *peak tension*, which is important in determining the degree of performance or belt life: $T_{peak} = T_T + T_B + T_C$.

**Measuring the Effective Length.—**The effective length of a V-belt is determined by placing the belt on a measuring device having two equal diameter sheaves with standard groove dimensions. The shaft of one of the sheaves is fixed. A specified measuring tension is applied to the housing for the shaft of the other sheave, moving it along a graduated scale. The belt is rotated around the sheaves at least two revolutions of the belt to seat it properly in the sheave grooves and to divide the total tension equally between the two strands of the belt.

The effective length of the belt is obtained by adding the effective (outside) circumference of one of the measuring sheaves to twice the center distance. Synchronous belts are measured in a similar manner.

The following sections cover common belts used in industrial applications for power transmission and specified in Rubber Manufacturers Association (RMA), Mechanical Power Transmission Association (MPTA), and The Rubber Association of Canada (RAC) standards. The information presented does not apply to automotive or agricultural drives, for which other standards exist. The belts covered in this section are narrow, classical, double, and light-duty v-belts, v-ribbed belts, variable-speed belts, 60-degree v-belts, and synchronous (timing) belts.

## Flat Belting

Flat belting was originally made from leather because it was the most durable material available and could easily be cut and joined to make a driving belt suitable for use with cylindrical or domed pulleys. This type of belting was popular because it could be used to transmit high torques over long distances and it was employed in factories to drive many small machines from a large common power source such as a steam engine. As electric motors became smaller, more efficient, and more powerful, and new types of belts and chains were made possible by modern materials and manufacturing processes, flat belts fell out of favor. Flat belts are still used for some drive purposes, but leather has been replaced by other natural and synthetic materials such as urethanes, which can be reinforced by high-strength polyamide or steel fabrics to provide properties such as resistance to stretching. The high modulus of elasticity in these flat belts eliminates the need for periodic retensioning that is usually necessary with V-belts.

Driving belts can be given a coating of an elastomer with a high coefficient of friction, to enable belts to grip pulleys without the degree of tension common with earlier materials. Urethanes are commonly used for driving belts where high resistance to abrasion is required, and also resist attack by chemical solvents of most kinds. Flat belts having good resistance to high temperatures are also available. Typical properties of polyurethane belts include tensile strength up to 40,000 psi (276 MPa), depending on reinforcement type and Shore hardness of 85 to 95. Most polyurethane belts are installed under tension. The amount of tension varies with the belt cross section, being greater for belts of small section. Belt tension can be measured by marking lines 10 inches, or 100 mm, apart on an installed belt, then applying tension until the separation increases by the desired percentage. For 2 percent tension, lines on the tensioned belt would be 10.2 inches, or 102 mm, apart. Mechanical failure may result when belt tensioning is excessive, and 2 to 2.5 percent elongation should be regarded as the limit.

Flat belts offer high load capacities and are capable of transmitting power over long distances, maintaining relative rotational direction, can operate without lubricants, and are generally inexpensive to maintain or replace when worn. Flat belt systems will operate with little maintenance and only periodic adjustment. Because they transmit motion by friction, flat belts have the ability to slip under excessive loads, providing a fail-safe action to guard against malfunctions. This advantage is offset by the problem that friction drives can both slip and creep so that they do not offer exact, consistent velocity ratios nor precision timing between input and output shafts. Flat belts can be made to any desired length, being joined by reliable chemical bonding processes.

Increasing centrifugal force has less effect on the load-carrying capacity of flat belts at high speeds than it has on V-belts, for instance. The low thickness of a flat belt, compared with a V-belt, places its center of gravity near the pulley surface. Flat belts therefore may be run at surface speeds of up to 16,000 or even 20,000 ft/min (81.28 to 101.6 m/s), although ideal speeds are in the range of 3,000 to 10,000 ft/min (15.25 to 50.8 m/s). Elastomeric drive surfaces on flat belts have eliminated the need for belt dressings that were often needed to keep leather belts in place. These surface coatings can also contain antistatic materials. Belt pulley wear and noise are low with flat belts, shock and vibration are damped, and efficiency is generally greater than 98 percent, compared to 96 percent for V-belts.

Driving belt load capacities can be calculated from torque $T = F(d/2)$ and horsepower $HP = T \times rpm/396{,}000$, where $T$ is the torque in in-lb, $F$ is the force transmitted in lb and $d$ is the pulley diameter in inches. Pulley width is usually about 10 percent larger than the belt, and for good tracking, pulleys are often crowned by 0.012–0.10 inch (0.305–2.54 mm) for diameters in the range of 1.5–80 inches (3.8–203 cm).

Before a belt specification is written, the system should be checked for excessive startup and shut-down loads, which sometimes are more than 10 percent above operating conditions. In overcoming such loads, the belt will transmit considerably more force than during normal operation. Large starting and stopping forces will also shorten belt life unless they are taken into account during the design stage.

**Flat Belt Pulleys.—**Flat belt pulleys are usually made of cast iron, fabricated steel, paper, fiber, or various kinds of wood. They may be solid or split and in either case the hub may be split for clamping to the shaft.

Pulley face widths are nominally the same as the widths of the belts they are to carry. The pulley face should be approximately 1 inch (2.54 cm) more than the belt width for belts under 12 inches (30.5 cm) wide, 2 inches (5.1 cm) more for belts from 12 to 24 inches (30.5–61 cm) wide, and 3 inches (7.6 cm) more for belts over 24 inches (61 cm) in width.

Belts may be made to center themselves by the use of crowned pulleys. The usual amount of crowning is $^1/_8$ inch per foot (10.4 mm per meter) of pulley width. Thus, the difference in maximum and minimum radii of a crowned 6-inch (152.4-mm) wide pulley would be $^1/_{16}$ inch (1.59 mm). Crowned pulleys have a rim section either with a convex curve or a flat V form. Flanges on the sides of flat belt pulleys are in general undesirable as the belt tends to crawl against them. Too much crown is undesirable because of the tendency to "break the belt's back." This is particularly true for riding idlers close to driving pulleys where the curvature of the belt changes rapidly from one pulley to the other. Here, the idler should under no circumstances be crowned and the adjacent pulley should have very little crown. Pulleys carrying shifting belts are not crowned.

Open belt drives connecting pulleys on short centers with one pulley considerably larger than the other may be unsatisfactory due to the small angle of wrap on the smaller pulley. This angle may be increased by the use of idler pulleys on one or both sides of the belt.

## V-Belts

**Narrow V-Belts ANSI/RMA IP-22.—**Narrow V-belts serve the same applications as multiple, classical V-belts, but allow for a lighter, more compact drive. Three basic cross sections—3V and 3VX, 5V and 5VX, and 8V—are provided, as shown in Fig. 1. The 3VX and 5VX are molded, notched V-belts that have greater power capacity than conventional belts. Narrow V-belts are specified by cross section and effective length and have top widths ranging from $^3/_8$ to 1 inch (9.525–25.4 mm).

Narrow V-belts usually provide substantial weight and space savings over classical belts. Some narrow belts can transmit up to three times the horsepower of conventional belts in the same drive space, or the same horsepower in one-third to one-half the space. These belts are designed to operate in multiples and are also available in the joined configuration.

*Belt Cross Sections:* Nominal dimensions of the three cross sections are given in Fig. 1.

*Belt Size Designation:* Narrow V-belt sizes are identified by a standard belt number. The first figure of this number followed by the letter V denotes the belt cross section. An X following the V indicates a notched cross section. The remaining figures show the effective belt length in tenths of an inch. For example, the number 5VX1400 designates a notched V-belt with a 5V cross section and an effective length of 140.0 in. Standard effective lengths of narrow V-belts are shown in Table 1.

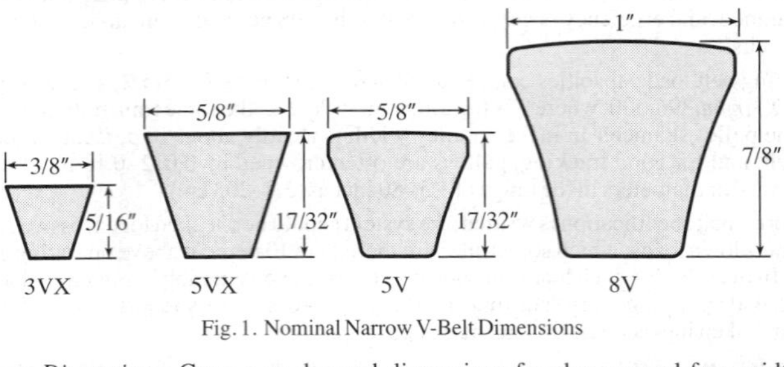


Fig. 1. Nominal Narrow V-Belt Dimensions

*Sheave Dimensions:* Groove angles and dimensions for sheaves and face widths of sheaves for multiple belt drives are given in Table 2a and Table 2b, along with various tolerance values. Standard sheave outside diameters are given in Table 3.

**Table 1. Narrow V-Belt Standard Effective Lengths** *ANSI/RMA IP-22 (1983)*

| Standard Length Designation[a] | Standard Effective Outside Length, Cross Section | | | Permissible Deviation from Standard Length | Matching Limits for One Set | Standard Length Designation[a] | Standard Effective Outside Length, Cross Section | | | Permissible Deviation from Standard Length | Matching Limits for One Set |
|---|---|---|---|---|---|---|---|---|---|---|---|
| | 3V | 5V | 8V | | | | 3V | 5V | 8V | | |
| 250 | 25.0 | … | … | ±0.3 | 0.15 | 1060 | 106.0 | 106.0 | 106.0 | ±0.6 | 0.30 |
| 265 | 26.5 | … | … | ±0.3 | 0.15 | 1120 | 112.0 | 112.0 | 112.0 | ±0.6 | 0.30 |
| 280 | 28.0 | … | … | ±0.3 | 0.15 | 1180 | 118.0 | 118.0 | 118.0 | ±0.6 | 0.30 |
| 300 | 30.0 | … | … | ±0.3 | 0.15 | 1250 | 125.0 | 125.0 | 125.0 | ±0.6 | 0.30 |
| 315 | 31.5 | … | … | ±0.3 | 0.15 | 1320 | 132.0 | 132.0 | 132.0 | ±0.6 | 0.30 |
| 335 | 33.5 | … | … | ±0.3 | 0.15 | 1400 | 140.0 | 140.0 | 140.0 | ±0.6 | 0.30 |
| 355 | 35.5 | … | … | ±0.3 | 0.15 | 1500 | … | 150.0 | 150.0 | ±0.8 | 0.30 |
| 375 | 37.5 | … | … | ±0.3 | 0.15 | 1600 | … | 160.0 | 160.0 | ±0.8 | 0.45 |
| 400 | 40.0 | … | … | ±0.3 | 0.15 | 1700 | … | 170.0 | 170.0 | ±0.8 | 0.45 |
| 425 | 42.5 | … | … | ±0.3 | 0.15 | 1800 | … | 180.0 | 180.0 | ±0.8 | 0.45 |
| 450 | 45.0 | … | … | ±0.3 | 0.15 | 1900 | … | 190.0 | 190.0 | ±0.8 | 0.45 |
| 475 | 47.5 | … | … | ±0.3 | 0.15 | 2000 | … | 200.0 | 200.0 | ±0.8 | 0.45 |
| 500 | 50.0 | 50.0 | … | ±0.3 | 0.15 | 2120 | … | 212.0 | 212.0 | ±0.8 | 0.45 |
| 530 | 53.0 | 53.0 | … | ±0.4 | 0.15 | 2240 | … | 224.0 | 224.0 | ±0.8 | 0.45 |
| 560 | 56.0 | 56.0 | … | ±0.4 | 0.15 | 2360 | … | 236.0 | 236.0 | ±0.8 | 0.45 |
| 600 | 60.0 | 60.0 | … | ±0.4 | 0.15 | 2500 | … | 250.0 | 250.0 | ±0.8 | 0.45 |
| 630 | 63.0 | 63.0 | … | ±0.4 | 0.15 | 2650 | … | 265.0 | 265.0 | ±0.8 | 0.60 |
| 670 | 67.0 | 67.0 | … | ±0.4 | 0.30 | 2800 | … | 280.0 | 280.0 | ±0.8 | 0.60 |
| 710 | 71.0 | 71.0 | … | ±0.4 | 0.30 | 3000 | … | 300.0 | 300.0 | ±0.8 | 0.60 |
| 750 | 75.0 | 75.0 | … | ±0.4 | 0.30 | 3150 | … | 315.0 | 315.0 | ±1.0 | 0.60 |
| 800 | 80.0 | 80.0 | … | ±0.4 | 0.30 | 3350 | … | 335.0 | 335.0 | ±1.0 | 0.60 |
| 850 | 85.0 | 85.0 | … | ±0.5 | 0.30 | 3550 | … | 355.0 | 355.0 | ±1.0 | 0.60 |
| 900 | 90.0 | 90.0 | … | ±0.5 | 0.30 | 3750 | … | … | 375.0 | ±1.0 | 0.60 |
| 950 | 95.0 | 95.0 | … | ±0.5 | 0.30 | 4000 | … | … | 400.0 | ±1.0 | 0.75 |
| 1000 | 100.0 | 100.0 | 100.0 | ±0.5 | 0.30 | 4250 | … | … | 425.0 | ±1.2 | 0.75 |

[a] To specify belt size, use the Standard Length Designation prefixed by the cross section, for example, 5 V850.
All dimensions in inches.

**Table 2a. Narrow V-Belt Standard Sheave and Groove Dimensions** *ANSI/RMA IP-22 (1983)*

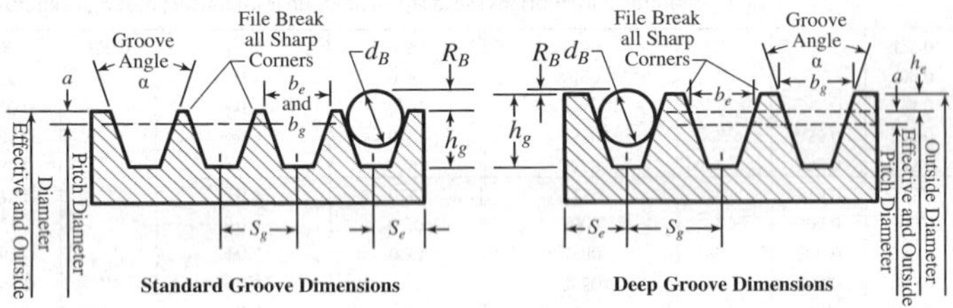


Face Width of Standard and Deep Groove Sheaves = $s_g(N_g - 1) + 2S_e$, where $N_g$ = Number of Grooves

| Cross Section | Standard Groove Outside Diameter | Standard Groove Dimensions: Groove Angle, α, ±0.25 deg | $b_g$ ±0.005 | $b_e$ (Ref) | $h_g$ (Min) | $R_B$ (Min) | $d_B$ ±0.0005 | $S_g$[a] ±0.015 | $S_e$ | Design Factors: Min Recommended Outside Diameter | $2a$ |
|---|---|---|---|---|---|---|---|---|---|---|---|
| 3V | Up through 3.49 | 36 | 0.350 | 0.350 | 0.340 | 0.181 | 0.3438 | 0.406 | 0.344 (+0.099, −0.031) | 2.65 | 0.050 |
| | Over 3.49 up to and including 6.00 | 38 | | | | 0.183 | | | | | |
| | Over 6.00 up to and including 12.00 | 40 | | | | 0.186 | | | | | |
| | Over 12.00 | 42 | | | | 0.188 | | | | | |
| 5V | Up through 9.99 | 38 | 0.600 | 0.600 | 0.590 | 0.329 | 0.5938 | 0.688 | 0.500 +0.125, −0.047) | 7.10 | 0.100 |
| | Over 9.99 up to and including 16.00 | 40 | | | | 0.332 | | | | | |
| | Over 16.00 | 42 | | | | 0.336 | | | | | |
| 8V | Up through 15.99 | 38 | 1.000 | 1.000 | 0.990 | 0.575 | 1.0000 | 1.125 | 0.750 (+0.250, −0.062) | 12.50 | 0.200 |
| | Over 15.99 up to and including 22.40 | 40 | | | | 0.580 | | | | | |
| | Over 22.40 | 42 | | | | 0.585 | | | | | |

[a] See footnote [b] following Table 2b.

**Table 2b. Narrow V-Belt Standard Sheave and Groove Dimensions** *ANSI/RMA IP-22 (1983)*

| | | Deep Groove Dimensions[a] | | | | | | | | Design Factors | | |
|---|---|---|---|---|---|---|---|---|---|---|---|---|
| Cross Section | Deep Groove Outside Diameter | Groove Angle, α, ±0.25 deg | $b_g$ ±0.005 | $b_e$ (Ref) | $h_g$ (Min) | $R_B$ (Min) | $d_B$ ±0.0005 | $S_g$[b] ±0.015 | $S_e$ | Min Recommended Outside Diameter | $2a$ | $2h_e$ |
| 3V | Up through 3.71 | 36 | 0.421 | 0.350 | 0.449 | 0.070 | 0.3438 | 0.500 | 0.375 (+0.094, −0.031) | 2.87 | 0.050 | 0.218 |
| | Over 3.71 up to and including 6.22 | 38 | 0.425 | | | 0.073 | | | | | | |
| | Over 6.22 up to and including 12.22 | 40 | 0.429 | | | 0.076 | | | | | | |
| | Over 12.22 | 42 | 0.434 | | | 0.078 | | | | | | |
| 5V | Up through 10.31 | 38 | 0.710 | 0.600 | 0.750 | 0.168 | 0.5938 | 0.812 | 0.562 (+0.125, −0.047) | 7.42 | 0.100 | 0.320 |
| | Over 10.31 up to and including 16.32 | 40 | 0.716 | | | 0.172 | | | | | | |
| | Over 16.32 | 42 | 0.723 | | | 0.175 | | | | | | |
| 8V | Up through 16.51 | 38 | 1.180 | 1.000 | 1.252 | 0.312 | 1.0000 | 1.312 | 0.844 (+0.250, −0.062) | 13.02 | 0.200 | 0.524 |
| | Over 16.51 up to and including 22.92 | 40 | 1.191 | | | 0.316 | | | | | | |
| | Over 22.92 | 42 | 1.201 | | | 0.321 | | | | | | |

[a] Deep groove sheaves are intended for drives with belt offset such as quarter-turn or vertical shaft drives. They may also be necessary where oscillations in the center distance may occur. Joined belts will not operate in deep groove sheaves.

[b] Summation of the deviations from $S_g$ for all grooves in any one sheave should not exceed ±0.031 in. The variations in pitch diameter between the grooves in any one sheave must be within the following limits: Up through 19.9 in. outside diameter and up through 6 grooves—0.010 in. (add 0.0005 in. for each additional groove). 20.0 in. and over on outside diameter and up through 10 grooves—0.015 in. (add 0.0005 in. for each additional groove). This variation can be obtained by measuring the distance across two measuring balls or rods placed in the grooves diametrically opposite each other. Comparing this "diameter over balls or rods" measurement between grooves will give the variation in pitch diameter.

| Other Sheave Tolerances | | |
|---|---|---|
| Outside Diameter | Radial Runout[a] | Axial Runout[a] |
| Up through 8.0 in. outside diameter ±0.020 in.<br>For each additional inch of outside diameter add ±0.0025 in. | Up through 10.0 in. outside diameter 0.010 in.<br>For each additional inch of outside diameter add 0.0005 in. | Up through 5.0 in. outside diameter 0.005 in.<br>For each additional inch of outside diameter add 0.001 in. |

[a] Total indicator reading.

All dimensions in inches.

**Table 3. Standard Sheave Outside Diameters** *ANSI/RMA IP-22, 1983*

| 3V | | | 5V | | | 8V | | |
|---|---|---|---|---|---|---|---|---|
| Nom | Min | Max | Nom | Min | Max | Nom | Min | Max |
| 2.65 | 2.638 | 2.680 | 7.10 | 7.087 | 7.200 | 12.50 | 12.402 | 12.600 |
| 2.80 | 2.795 | 2.840 | 7.50 | 7.480 | 7.600 | 13.20 | 13.189 | 13.400 |
| 3.00 | 2.953 | 3.000 | 8.00 | 7.874 | 8.000 | 14.00 | 13.976 | 14.200 |
| 3.15 | 3.150 | 3.200 | 8.50 | 8.346 | 8.480 | 15.00 | 14.764 | 15.000 |
| 3.35 | 3.346 | 3.400 | 9.00 | 8.819 | 8.960 | 16.00 | 15.748 | 16.000 |
| 3.55 | 3.543 | 3.600 | 9.25 | 9.291 | 9.440 | 17.00 | 16.732 | 17.000 |
| 3.65 | 3.642 | 3.700 | 9.75 | 9.567 | 9.720 | 18.00 | 17.717 | 18.000 |
| 4.00 | 3.937 | 4.000 | 10.00 | 9.843 | 10.000 | 19.00 | 18.701 | 19.000 |
| 4.12 | 4.055 | 4.120 | 10.30 | 10.157 | 10.320 | 20.00 | 19.685 | 20.000 |
| 4.50 | 4.409 | 4.480 | 10.60 | 10.433 | 10.600 | 21.20 | 20.866 | 21.200 |
| 4.75 | 4.646 | 4.720 | 10.90 | 10.709 | 10.880 | 22.40 | 22.047 | 22.400 |
| 5.00 | 4.921 | 5.000 | 11.20 | 11.024 | 11.200 | 23.60 | 23.622 | 24.000 |
| 5.30 | 5.197 | 5.280 | 11.80 | 11.811 | 12.000 | 24.80 | 24.803 | 25.200 |
| 5.60 | 5.512 | 5.600 | 12.50 | 12.402 | 12.600 | 30.00 | 29.528 | 30.000 |
| 6.00 | 5.906 | 6.000 | 13.20 | 13.189 | 13.400 | 31.50 | 31.496 | 32.000 |
| 6.30 | 6.299 | 6.400 | 14.00 | 13.976 | 14.200 | 35.50 | 35.433 | 36.000 |
| 6.50 | 6.496 | 6.600 | 15.00 | 14.764 | 15.000 | 40.00 | 39.370 | 40.000 |
| 6.90 | 6.890 | 7.000 | 16.00 | 15.748 | 16.000 | 44.50 | 44.094 | 44.800 |
| 8.00 | 7.874 | 8.000 | 18.70 | 18.701 | 19.000 | 50.00 | 49.213 | 50.000 |
| 10.00 | 9.843 | 10.000 | 20.00 | 19.685 | 20.000 | 52.00 | 51.969 | 52.800 |
| 10.60 | 10.433 | 10.600 | 21.20 | 20.866 | 21.200 | 63.00 | 62.992 | 64.000 |
| 12.50 | 12.402 | 12.600 | 23.60 | 23.622 | 24.000 | 71.00 | 70.866 | 72.000 |
| 14.00 | 13.976 | 14.200 | 25.00 | 24.803 | 25.200 | 79.00 | 78.740 | 80.000 |
| 16.00 | 15.748 | 16.000 | 28.00 | 27.953 | 28.400 | 99.00 | 98.425 | 100.000 |
| 19.00 | 18.701 | 19.000 | 31.50 | 31.496 | 32.000 | … | … | … |
| 20.00 | 19.685 | 20.000 | 37.50 | 37.402 | 38.000 | … | … | … |
| 25.00 | 24.803 | 25.200 | 40.00 | 39.370 | 40.000 | … | … | … |
| 31.50 | 31.496 | 32.000 | 44.50 | 44.094 | 44.800 | … | … | … |
| 33.50 | 33.465 | 34.000 | 50.00 | 49.213 | 50.000 | … | … | … |
| … | … | … | 63.00 | 62.992 | 64.000 | … | … | … |
| … | … | … | 71.00 | 70.866 | 72.000 | … | … | … |

All dimensions in inches. The nominal diameters were selected from R40 and R80 preferred numbers (see page 688).

*Minimum Sheave Size:* The recommended minimum sheave size depends on the rpm of the faster shaft. Minimum sheave diameters for each belt cross section are listed in Table 3.

*Cross Section Selection:* The chart (Fig. 2, on page 2579) is a guide to the V-belt cross section to use for any combination of design horsepower and speed of the faster shaft. When the intersection of the design horsepower and speed of the faster shaft falls near a line between two areas on the chart, it is advisable to investigate the possibilities in both areas. Special circumstances (such as space limitations) may lead to a choice of belt cross section different from that indicated in the chart.

*Horsepower Ratings:* The horsepower ratings of narrow V-belts can be calculated using the following formula:

$$HP = d_p r[K_1 - K_2/d_p - K_3(d_p r)^2 - K_4 \log(d_p r)] + K_{SR} r$$

where $d_p$ = the pitch diameter of the small sheave, in.; $r$ = rpm of the faster shaft divided by 1000; $K_{SR}$, speed ratio correction factor (Table 4), and $K_1$, $K_2$, $K_3$, and $K_4$, cross section parameters, are listed in the accompanying Table 5. This formula gives the basic horsepower rating, corrected for the speed ratio. To obtain the horsepower per belt for an arc of contact other than 180° and for belts shorter or longer than average length, multiply the horsepower obtained from this formula by the length correction factor (Table 7) and the arc of contact correction factor (Table 6).

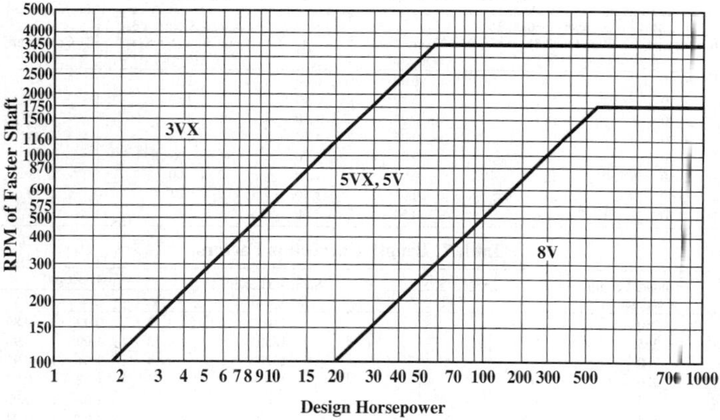


Fig. 2. Selection of Narrow V-Belt Cross Section

## Table 4. Speed Ratio Correction Factors

| Speed Ratio[a] Range | $K_{SR}$ Cross Section 3VX | $K_{SR}$ Cross Section 5VX | Speed Ratio[a] Range | $K_{SR}$ Cross Section 5V | $K_{SR}$ Cross Section 8V |
|---|---|---|---|---|---|
| 1.00–1.01 | 0.0000 | 0.0000 | 1.00–1.01 | 0.0000 | 0.0000 |
| 1.02–1.03 | 0.0157 | 0.0801 | 1.02–1.05 | 0.0963 | 0.4690 |
| 1.04–1.06 | 0.0315 | 0.1600 | 1.06–1.11 | 0.2623 | 1.2780 |
| 1.07–1.09 | 0.0471 | 0.2398 | 1.12–1.18 | 0.4572 | 2.2276 |
| 1.10–1.13 | 0.0629 | 0.3201 | 1.19–1.26 | 0.6223 | 3.0321 |
| 1.14–1.18 | 0.0786 | 0.4001 | 1.27–1.38 | 0.7542 | 3.5747 |
| 1.19–1.25 | 0.0944 | 0.4804 | 1.39–1.57 | 0.8833 | 4.3038 |
| 1.26–1.35 | 0.1101 | 0.5603 | 1.58–1.94 | 0.9941 | 4.8438 |
| 1.36–1.57 | 0.1259 | 0.6405 | 1.95–3.38 | 1.0830 | 5.2767 |
| Over 1.57 | 0.1416 | 0.7202 | Over 3.38 | 1.1471 | 5.5892 |

[a] $D_p/d_p$, where $D_p$ ($d_p$) is the pitch diameter of the large (small) sheave.

## Table 5. Cross Section Correction Factors

| Cross Section | $K_1$ | $K_2$ | $K_3$ | $K_4$ |
|---|---|---|---|---|
| 3VX | 1.1691 | 1.5295 | $1.5229 \times 10^{-4}$ | 0.15960 |
| 5VX | 3.3038 | 7.7810 | $3.6432 \times 10^{-4}$ | 0.43343 |
| 5V | 3.3140 | 10.123 | $5.8758 \times 10^{-4}$ | 0.46527 |
| 8V | 8.6628 | 49.323 | $1.5804 \times 10^{-3}$ | 1.1659 |

*Arc of Contact:* Arc of contact on the small sheave may be determined by the formulas.

*Exact formula:*

$$\text{Arc of Contact (deg)} = 2\cos^{-1}\left(\frac{D_e - d_e}{2C}\right)$$

*Approximate formula:*

$$\text{Arc of Contact (deg)} = 180 - \frac{(D_e - d_e)60}{C}$$

where: $D_e$ = effective diameter of large sheave, inch or mm

$d_e$ = effective diameter of small sheave, inch or mm

$C$ = center distance, inch or mm

### Table 6. Arc of Contact Correction Factors

| $\frac{D_e - d_e}{C}$ | Arc of Contact, θ, on Small Sheave (deg) | Correction Factor | $\frac{D_e - d_e}{C}$ | Arc of Contact, θ, on Small Sheave (deg) | Correction Factor |
|---|---|---|---|---|---|
| 0.00 | 180 | 1.00 | 0.80 | 133 | 0.87 |
| 0.10 | 174 | 0.99 | 0.90 | 127 | 0.85 |
| 0.20 | 169 | 0.97 | 1.00 | 120 | 0.82 |
| 0.30 | 163 | 0.96 | 1.10 | 113 | 0.80 |
| 0.40 | 157 | 0.94 | 1.20 | 106 | 0.77 |
| 0.50 | 151 | 0.93 | 1.30 | 99 | 0.73 |
| 0.60 | 145 | 0.91 | 1.40 | 91 | 0.70 |
| 0.70 | 139 | 0.89 | 1.50 | 83 | 0.65 |

### Table 7. Length Correction Factors

| Standard Length Designation | Cross Section | | | Standard Length Designation | Cross Section | | |
|---|---|---|---|---|---|---|---|
| | 3V | 5V | 8V | | 3V | 5V | 8V |
| 250 | 0.83 | | | 1180 | 1.12 | 0.99 | 0.89 |
| 265 | 0.84 | | | 1250 | 1.13 | 1.00 | 0.90 |
| 280 | 0.85 | | | 1320 | 1.14 | 1.01 | 0.91 |
| 300 | 0.86 | | | 1400 | 1.15 | 1.02 | 0.92 |
| 315 | 0.87 | | | 1500 | | 1.03 | 0.93 |
| 335 | 0.88 | | | 1600 | | 1.04 | 0.94 |
| 355 | 0.89 | | | 1700 | | 1.05 | 0.94 |
| 375 | 0.90 | | | 1800 | | 1.06 | 0.95 |
| 400 | 0.92 | | | 1900 | | 1.07 | 0.96 |
| 425 | 0.93 | | | 2000 | | 1.08 | 0.97 |
| 450 | 0.94 | | | 2120 | | 1.09 | 0.98 |
| 475 | 0.95 | | | 2240 | | 1.09 | 0.98 |
| 500 | 0.96 | 0.85 | | 2360 | | 1.10 | 0.99 |
| 530 | 0.97 | 0.86 | | 2500 | | 1.11 | 1.00 |
| 560 | 0.98 | 0.87 | | 2650 | | 1.12 | 1.01 |
| 600 | 0.99 | 0.88 | | 2800 | | 1.13 | 1.02 |
| 630 | 1.00 | 0.89 | | 3000 | | 1.14 | 1.03 |
| 670 | 1.01 | 0.90 | | 3150 | | 1.15 | 1.03 |
| 710 | 1.02 | 0.91 | | 3350 | | 1.16 | 1.04 |
| 750 | 1.03 | 0.92 | | 3550 | | 1.17 | 1.05 |
| 800 | 1.04 | 0.93 | | 3750 | | | 1.06 |
| 850 | 1.06 | 0.94 | | 4000 | | | 1.07 |
| 900 | 1.07 | 0.95 | | 4250 | | | 1.08 |
| 950 | 1.08 | 0.96 | | 4500 | | | 1.09 |
| 1000 | 1.09 | 0.96 | 0.87 | 4750 | | | 1.09 |
| 1060 | 1.10 | 0.97 | 0.88 | 5000 | | | 1.10 |
| 1120 | 1.11 | 0.98 | 0.88 | … | | | … |

*Number of Belts:* The number of belts required for an application is obtained by dividing the design horsepower by the corrected horsepower rating for one belt.

**Classical V-Belts ANSI/RMA IP-20.—**Classical V-belts are most commonly used in heavy-duty applications and include these standard cross sections: A, AX, B, BX, C, CX, D, and DX (Fig. 3, page 2584). Top widths range from $^1/_2$ to $1^1/_4$ in. and are specified by cross section and nominal length. Classical belts can be teamed in multiples of two or more. These multiple drives can transmit up to several hundred horsepower continuously and absorb reasonable shock loads.

*Belt Cross Sections:* Nominal dimensions of the four cross sections are given in Fig. 3.

*Belt Size Designation:* Classical V-belt sizes are identified by a standard belt number consisting of a letter-numeral combination. The letter identifies the cross section; the numeral identifies the length as shown in Table 8. For example, A60 indicates an A cross section and a standard length designation of 60. An X following the section letter designation indicates a molded notch cross section, for example, AX60.

**Table 8. Classical V-Belt Standard Datum Length** *ANSI/RMA IP-20, 1988*

| Standard Length Designation[a] | Standard Datum Lengths | | | | Permissible Deviations from Std. Datum Length | Matching Limits for One Set |
|---|---|---|---|---|---|---|
| | Cross Section | | | | | |
| | A, AX | B, BX | C, CX | D | | |
| 26 | 27.3 | … | … | … | +0.6, −0.6 | 0.15 |
| 31 | 32.3 | … | … | … | +0.6, −0.6 | 0.15 |
| 35 | 36.3 | 36.8 | … | … | +0.6, −0.6 | 0.15 |
| 38 | 39.3 | 39.8 | … | … | +0.7, −0.7 | 0.15 |
| 42 | 43.3 | 43.8 | … | … | +0.7, −0.7 | 0.15 |
| 46 | 47.3 | 47.8 | … | … | +0.7, −0.7 | 0.15 |
| 51 | 52.3 | 52.8 | 53.9 | … | +0.7, −0.7 | 0.15 |
| 55 | 56.3 | 56.8 | … | … | +0.7, −0.7 | 0.15 |
| 60 | 61.3 | 61.8 | 62.9 | … | +0.7, −0.7 | 0.15 |
| 68 | 69.3 | 69.8 | 70.9 | … | +0.7, −0.7 | 0.30 |
| 75 | 75.3 | 76.8 | 77.9 | … | +0.7, −0.7 | 0.30 |
| 80 | 81.3 | … | … | … | +0.7, −0.7 | 0.30 |
| 81 | … | 82.8 | 83.9 | … | +0.7, −0.7 | 0.30 |
| 85 | 86.3 | 86.8 | 87.9 | … | +0.7, −0.7 | 0.30 |
| 90 | 91.3 | 91.8 | 92.9 | … | +0.8, −0.8 | 0.30 |
| 96 | 97.3 | … | 98.9 | … | +0.8, −0.8 | 0.30 |
| 97 | … | 98.8 | … | … | +0.8, −0.8 | 0.30 |
| 105 | 106.3 | 106.8 | 107.9 | … | +0.8, −0.8 | 0.30 |
| 112 | 113.3 | 113.8 | 114.9 | … | +0.8, −0.8 | 0.30 |
| 120 | 121.3 | 121.8 | 122.9 | 123.3 | +0.8, −0.8 | 0.30 |
| 128 | 129.3 | 129.8 | 130.9 | 131.3 | +0.8, −0.8 | 0.30 |
| 144 | … | 145.8 | 146.9 | 147.3 | +0.8, −0.8 | 0.30 |
| 158 | … | 159.8 | 160.9 | 161.3 | +1.0, −1.0 | 0.45 |
| 173 | … | 174.8 | 175.9 | 176.3 | +1.0, −1.0 | 0.45 |
| 180 | … | 181.8 | 182.9 | 183.3 | +1.0, −1.0 | 0.45 |
| 195 | … | 196.8 | 197.9 | 198.3 | +1.1, −1.1 | 0.45 |
| 210 | … | 211.8 | 212.9 | 213.3 | +1.1, −1.1 | 0.45 |
| 240 | … | 240.3 | 240.9 | 240.8 | +1.3, −1.3 | 0.45 |
| 270 | … | 270.3 | 270.9 | 270.8 | +1.6, −1.6 | 0.60 |
| 300 | … | 300.3 | 300.0 | 300.8 | +1.6, −1.6 | 0.60 |
| 330 | … | … | 330.9 | 330.8 | +2.0, −2.0 | 0.60 |
| 360 | … | … | 380.9 | 360.8 | +2.0, −2.0 | 0.60 |
| 540 | … | … | … | 540.8 | +3.3, −3.3 | 0.90 |
| 390 | … | … | 390.9 | 390.8 | +2.0, −2.0 | 0.75 |
| 420 | … | … | 420.9 | 420.8 | +3.3, −3.3 | 0.75 |
| 480 | … | … | … | 480.8 | +3.3, −3.3 | 0.75 |
| 600 | … | … | … | 600.8 | +3.3, −3.3 | 0.90 |
| 660 | … | … | … | 660.8 | +3.3, −3.3 | 0.90 |

[a] To specify belt size use the Standard Length Designation prefixed by the letter indicating the cross section, e.g., B90.

All dimensions in inches.

*Sheave Dimensions:* Groove angles and dimensions for sheaves and the face widths of sheaves for multiple belt drives are given in Table 9, along with various tolerance values.

## Table 9. Classical V-Belt Sheave and Groove Dimensions *ANSI/RMA IP-20, 1988*

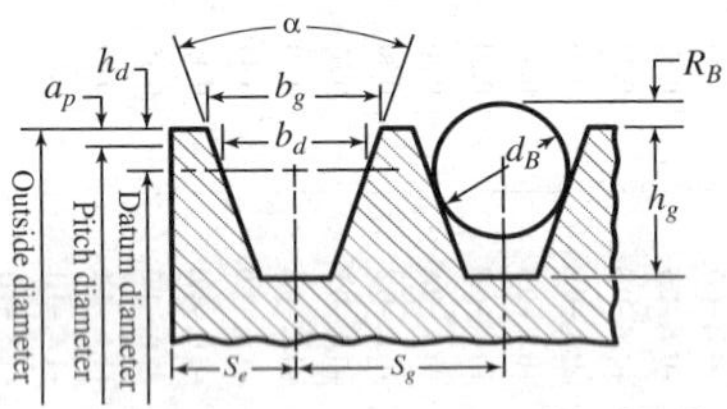


Face width of standard and deep groove sheaves = $S_g (N_g - 1) + 2S_e$, where $N_g$ = number of grooves

| | | Standard Groove Dimensions | | | | | | | | | | | Design Factors | |
|---|---|---|---|---|---|---|---|---|---|---|---|---|---|---|
| Cross Section | Datum[a] Diameter Range | α Groove Angle ±0.33° | $b_d$ Ref | $b_g$ | | $h_g$ Min | $2h_d$ | $R_B$ Min | $d_B$ ±0.0005 | $S_g$[b] ±0.025 | $S_e$ | | Min Recom. Datum Diameter | $2a_p$ |
| A, AX | Through 5.4 | 34 | 0.418 | 0.494 | ±0.005 | 0.460 | 0.250 | 0.148 | 0.4375 (7/16) | 0.625 | 0.375 | +0.090 −0.062 | A 3.0 | 0 |
| | Over 5.4 | 38 | | 0.504 | | | | 0.149 | | | | | AX 2.2 | |
| B, BX | Through 7.0 | 34 | 0.530 | 0.637 | ±0.006 | 0.550 | 0.350 | 0.189 | 0.5625 (9/16) | 0.750 | 0.500 | +0.120 −0.065 | B 5.4 | 0 |
| | Over 7.0 | 38 | | 0.650 | | | | 0.190 | | | | | BX 4.0 | |
| Combination A, AX Belt | Through 7.4[c] | 34 | 0.508[d] | 0.612 | ±0.006 | 0.612 | 0.634[e] | 0.230 | 0.5625 (9/16) | 0.750 | 0.500 | +0.120 −0.065 | A 3.6[c] | 0.37 |
| | Over 7.4 | 38 | | 0.625 | | | 0.602[e] | 0.226 | | | | | AX 2.8 | |
| Combination B, BX Belt | Through 7.4[c] | 34 | | 0.612 | ±0.006 | | 0.333[e] | 0.230 | | | | | B 5.7[c] | −0.01 |
| | Over 7.4 | 38 | | 0.625 | | | 0.334[e] | 0.226 | | | | | BX 4.3 | |
| C, CX | Through 7.99 | 34 | 0.757 | 0.879 | ±0.007 | 0.750 | 0.400 | 0.274 | 0.7812 (25/32) | 1.000 | 0.688 | +0.160 −0.070 | C 9.0 CX 6.8 | 0 |
| | Over 7.99 to and incl. 12.0 | 36 | | 0.887 | | | | 0.276 | | | | | | |
| | Over 12.0 | 38 | | 0.895 | | | | 0.277 | | | | | | |
| D | Through 12.99 | 34 | 1.076 | 1.259 | ±0.008 | 1.020 | 0.600 | 0.410 | 1.1250 (1⅛) | 1.438 | 0.875 | +0.220 −0.080 | 13.0 | 0 |
| | Over 12.99 to and incl. 17.0 | 36 | | 1.271 | | | | 0.410 | | | | | | |
| | Over 17.0 | 38 | | 1.283 | | | | 0.411 | | | | | | |

**Table 9.** *(Continued)* **Classical V-Belt Sheave and Groove Dimensions** *ANSI/RMA IP-20, 1988*

| Deep Groove Dimensions[f] | | | | | | | | | | | | Design Factors | |
|---|---|---|---|---|---|---|---|---|---|---|---|---|---|
| Cross-Section | Datum[a] Dia. Range | α Groove Angle ± 0.33° | $b_g$ Ref | $b_g$ | | $h_g$ Min | $2h_d$ Ref | $R_B$ Min | $d_B$ ± 0.0005 | $S_g$[b] ± 0.025 | $S_e$ | Min Rec. Datum Diameter | $2a_p$ |
| B, BX | Through 7.0 | 34 | 0.530 | 0.747 | ± 0.006 | 0.730 | 0.710 | 0.007 | 0.5625 | 0.875 | 0.562 +0.120 | B 5.4 | 0.36 |
| | Over 7.0 | 38 | | 0.774 | | | | 0.008 | (9⁄16) | | −0.065 | BX 4.0 | |
| C, CX | Through 7.99 | 34 | 0.757 | 1.066 | ± 0.007 | 1.055 | 1.010 | −0.035 | 0.7812 (25⁄32) | 1.250 | 0.812 +0.160 −0.070 | C 9.0 CX 6.8 | 0.61 |
| | Over 7.99 to and incl. 12.0 | 36 | | 1.085 | | | | −0.032 | | | | | |
| | Over 12.0 | 38 | | 1.105 | | | | −0.031 | | | | | |
| D | Through 12.99 | 34 | 1.076 | 1.513 | ±0.008 | 1.435 | 1.430 | −0.010 | 1.1250 (1⅛) | 1.750 | 1.062 +0.220 −0.080 | 13.0 | 0.83 |
| | Over 12.99 to and incl. 17.0 | 36 | | 1.514 | | | | −0.009 | | | | | |
| | Over 17.0 | 38 | | 1.569 | | | | −0.008 | | | | | |

[a] The A/AX, B/BX combination groove should be used when deep grooves are required for A or AX belts.

[b] Summation of the deviations from $S_g$ for all grooves in any one sheave should not exceed ±0.050 in. The variation in datum diameter between the grooves in any one sheave must be within the following limits: Through 19.9 in. outside diameter and through 6 grooves: 0.010 in. (add 0.0005 in. for each additional groove). 20.0 in. and over on outside diameter and through 10 grooves: 0.015 in. (add 0.0005 in. for each additional groove). This variation can be obtained by measuring the distance across two measuring balls or rods placed diametrically opposite each other in a groove. Comparing this "diameter over balls or rods" measurement between grooves will give the variation in datum diameter.

[c] Diameters shown for combination grooves are outside diameters. A specific datum diameter does not exist for either A or B belts in combination grooves.

[d] The $b_d$ value shown for combination grooves is the "constant width" point, but does not represent a datum width for either A or B belts ($2h_d = 0.340$ ref).

[e] $2h_d$ values for combination grooves are calculated based on $b_d$ for A and B grooves.

[f] Deep groove sheaves are intended for drives with belt offset such as quarter-turn or vertical shaft drives. Joined belts will not operate in deep groove sheaves. Also, A and AX joined belts will not operate in A/AX and B/BX combination grooves.

| Other Sheave Tolerances | | |
|---|---|---|
| Outside Diameter | Radial Runout[a] | Axial Runout[a] |
| Through 8.0 in. outside diameter ±0.020 in. For each additional inch of outside diameter add ±0.005 in. | Through 10.0 in. outside diameter 0.010 in. For each additional inch of outside diameter add 0.0005 in. | Through 5.0 in. outside diameter 0.005 in. For each additional inch of outside diameter add 0.001 in. |

[a] Total indicator readings.

A, AX & B, BX Combin. All dimensions in inches.

*Minimum Sheave Size:* The recommended minimum sheave size depends on the rpm of the faster shaft. Minimum sheave diameters for each belt cross section are listed in Table 9.

*Cross Section Selection:* Use the chart (Fig. 4) as a guide to the Classical V-belt cross section for any combination of design horsepower and speed of the faster shaft. When the intersection of the design horsepower and speed of the faster shaft falls near a line between two areas on the chart, the possibilities in both areas should be investigated. Special circumstances (such as space limitations) may lead to a choice of belt cross section different from that indicated in the chart.

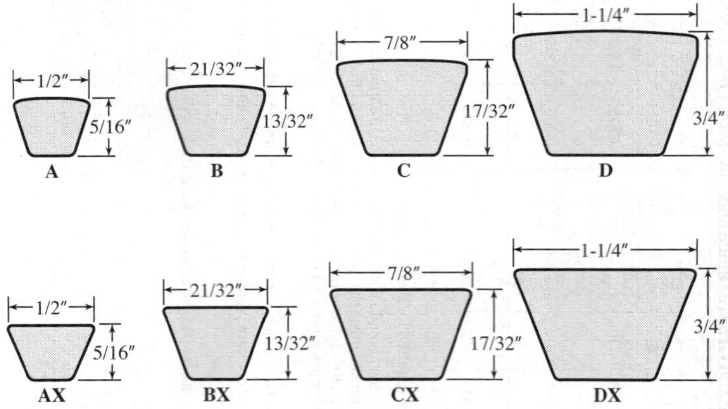


Fig. 3. Classical V-Belt Cross Sections

*Horsepower Ratings:* The horsepower rating formulas for classical V-belts are:

$$\textbf{A}: HP = d_p r\left[1.004 - \frac{1.652}{d_p} - 1.547\times10^{-4}(d_p r)^2 - .0\,2126\log(d_p r)\right] + 1.652r\left(1 - \frac{1}{K_{SR}}\right)$$

$$\textbf{AX}: HP = d_p r\left[1.462 - \frac{2.239}{d_p} - 2.198\times10^{-4}(d_p r)^2 - .0\,4238\log(d_p r)\right] + 2.239r\left(1 - \frac{1}{K_{SR}}\right)$$

$$\textbf{B}: HP = d_p r\left[1.769 - \frac{4.372}{d_p} - 3.081\times10^{-4}(d_p r)^2 - .0\,3658\log(d_p r)\right] + 4.372r\left(1 - \frac{1}{K_{SR}}\right)$$

**BX**: $HP = d_p r\left[2.051 - \frac{3.532}{d_p} - 3.097 \times 10^{-4}(d_p r)^2 - .0\,5735\log(d_p r)\right.$
$$\left. + 3.532r\left(1 - \frac{1}{K_{SR}}\right)\right]$$

**C:** $HP = d_p r\left[3.325 - \frac{12.07}{d_p} - 5.828 \times 10^{-4}(d_p r)^2 - .0\,6886\log(d_p r)\right.$
$$\left. + 12.07r\left(1 - \frac{1}{K_{SR}}\right)\right]$$

**CX**: $HP = d_p r\left[3.272 - \frac{6.655}{d_p} - 5.298 \times 10^{-4}(d_p r)^2 - .0\,8637\log(d_p r)\right.$
$$\left. + 6.655r\left(1 - \frac{1}{K_{SR}}\right)\right]$$

**D**: $HP = d_p r\left[7.160 - \frac{43.21}{d_p} - 1.384 \times 10^{-3}(d_p r)^2 - .1\,454\log(d_p r)\right.$
$$\left. + 43.21r\left(1 - \frac{1}{K_{SR}}\right)\right]$$

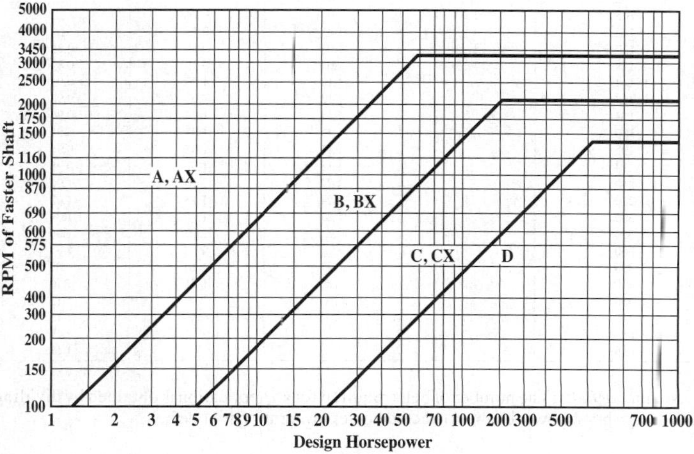


Fig. 4. Selection of Classical V-Belt Cross Sections

In these equations, $d_p$ = pitch diameter of small sheave, in.; $r$ = rpm of the faster shaft divided by 1000; $K_{SR}$ = speed ratio factor given in the accompanying Table 10. These formulas give the basic horsepower rating, corrected for the speed ratio. To obtain the horsepower per belt for an arc of contact other than 180 degrees and for belts shorter or longer than average length, multiply the horsepower obtained from these formulas by the length correction factor (Table 11) and the arc of contact correction factor (Table 12).

**Table 10. Speed Ratio Correction Factors**

| Speed Ratio[a] Range | $K_{SR}$ | Speed Ratio[a] Range | $K_{SR}$ |
|---|---|---|---|
| 1.00–1.01 | 1.0000 | 1.15–1.20 | 1.0586 |
| 1.02–1.04 | 1.0112 | 1.21–1.27 | 1.0711 |
| 1.05–1.07 | 1.0226 | 1.28–1.39 | 1.0840 |
| 1.08–1.10 | 1.0344 | 1.40–1.64 | 1.0972 |
| 1.11–1.14 | 1.0463 | Over 1.64 | 1.1106 |

[a] $D_p/d_p$, where $D_p$ ($d_p$) is the pitch diameter of the large (small) sheave.

**Table 11. Length Correction Factors**

| Std. Length Designation | Cross Section | | | |
|---|---|---|---|---|
| | A, AX | B, BX | C, CX | D |
| 26 | 0.78 | … | … | … |
| 31 | 0.82 | … | … | … |
| 35 | 0.85 | 0.80 | … | … |
| 38 | 0.87 | 0.82 | … | … |
| 42 | 0.89 | 0.84 | … | … |
| 46 | 0.91 | 0.86 | … | … |
| 51 | 0.93 | 0.88 | 0.80 | … |
| 55 | 0.95 | 0.89 | … | … |
| 60 | 0.97 | 0.91 | 0.83 | … |
| 68 | 1.00 | 0.94 | 0.85 | … |
| 75 | 1.02 | 0.96 | 0.87 | … |
| 80 | 1.04 | … | … | … |
| 81 | … | 0.98 | 0.89 | … |
| 85 | 1.05 | 0.99 | 0.90 | … |
| 90 | 1.07 | 1.00 | 0.91 | … |
| 96 | 1.08 | … | 0.92 | … |
| 97 | … | 1.02 | … | … |
| 105 | 1.10 | 1.03 | 0.94 | … |
| 112 | 1.12 | 1.05 | 0.95 | … |
| 120 | 1.13 | 1.06 | 0.96 | 0.88 |
| 128 | 1.15 | 1.08 | 0.98 | 0.89 |
| 144 | … | 1.10 | 1.00 | 0.91 |
| 158 | … | 1.12 | 1.02 | 0.93 |
| 173 | … | 1.14 | 1.04 | 0.94 |
| 180 | … | 1.15 | 1.05 | 0.95 |
| 195 | … | 1.17 | 1.08 | 0.96 |
| 210 | … | 1.18 | 1.07 | 0.98 |
| 240 | … | 1.22 | 1.10 | 1.00 |
| 270 | … | 1.24 | 1.13 | 1.02 |
| 300 | … | 1.27 | 1.15 | 1.04 |
| 330 | … | … | 1.17 | 1.06 |
| 360 | … | … | 1.18 | 1.07 |
| 390 | … | … | 1.20 | 1.09 |
| 420 | … | … | 1.21 | 1.10 |
| 480 | … | … | … | 1.13 |
| 540 | … | … | … | 1.15 |
| 600 | … | … | … | 1.17 |
| 660 | … | … | … | 1.18 |

*Number of Belts:* The number of belts required for an application is obtained by dividing the design horsepower by the corrected horsepower rating for one belt.

*Arc of Contact:* Arc of contact on the small sheave may be determined by the formulas.

*Exact formula:* $\text{Arc of Contact (deg)} = 2\cos^{-1}\left(\frac{D_d - d_d}{2C}\right)$

*Approximate formula:* $\text{Arc of Contact (deg)} = 180 - \left(\frac{(D_d - d_d)60}{C}\right)$

where $D_d$ = datum diameter of large sheave or flat pulley, inch or mm; $d_d$ = datum diameter of small sheave, inch or mm; and, $C$ = distance, inch or mm.

**Table 12. Arc of Contact Correction Factors**

| $\frac{D_d - d_d}{C}$ | Arc of Contact, θ, Small Sheave (deg) | Correction Factor V-V | Correction Factor V-Flat[a] | $\frac{D_d - d_d}{C}$ | Arc of Contact, θ Small Sheave (deg) | Correction Factor V-V | Correction Factor V-Flat[a] |
|---|---|---|---|---|---|---|---|
| 0.00 | 180 | 1.00 | 0.75 | 0.80 | 133 | 0.87 | 0.85 |
| 0.10 | 174 | 0.99 | 0.76 | 0.90 | 127 | 0.85 | 0.85 |
| 0.20 | 169 | 0.97 | 0.78 | 1.00 | 120 | 0.82 | 0.82 |
| 0.30 | 163 | 0.96 | 0.79 | 1.10 | 113 | 0.80 | 0.80 |
| 0.40 | 157 | 0.94 | 0.80 | 1.20 | 106 | 0.77 | 0.77 |
| 0.50 | 151 | 0.93 | 0.81 | 1.30 | 99 | 0.73 | 0.73 |
| 0.60 | 145 | 0.91 | 0.83 | 1.40 | 91 | 0.70 | 0.70 |
| 0.70 | 139 | 0.89 | 0.84 | 1.50 | 83 | 0.65 | 0.65 |

[a] A V-flat drive is one using a small sheave and a large diameter flat pulley.

**Double V-Belts ANSI/RMA IP-21.—**Double V-belts or hexagonal belts are used when power input or takeoff is required on both sides of the belt. Designed for use on "serpentine" drives, which consist of sheaves rotating in opposite directions, the belts are available in AA, BB, CC, and DD cross sections and operate in standard classical sheaves. They are specified by cross section and nominal length.

*Belt Cross Sections:* Nominal dimensions of the four cross sections are given in Fig. 5

*Belt Size Designation:* Double V-belt sizes are identified by a standard belt number, consisting of a letter-numeral combination. The letters identify the cross section; the numbers identify length as shown in Column 1 of Table 13. For example, AA51 indicates an AA cross section and a standard length designation of 51.

**Table 13. Double V-Belt Standard Effective Lengths** *ANSI/RMA IP-21, 1984*

| Standard Length Designation[a] | Standard Effective Length Cross Section AA | BB | CC | DD | Permissible Deviation from Standard Effective Length | Matching Limits for One Set |
|---|---|---|---|---|---|---|
| 51 | 53.1 | 53.9 | … | … | ±0.7 | 0.15 |
| 55 | … | 57.9 | … | … | ±0.7 | 0.15 |
| 60 | 62.1 | 62.9 | … | … | ±0.7 | 0.15 |
| 68 | 70.1 | 70.9 | … | … | ±0.7 | 0.30 |
| 75 | 77.1 | 77.9 | … | … | ±0.7 | 0.30 |
| 80 | 82.1 | … | … | … | ±0.7 | 0.30 |
| 81 | … | 83.9 | 85.2 | … | ±0.7 | 0.30 |
| 85 | 87.1 | 87.9 | 89.2 | … | ±0.7 | 0.30 |
| 90 | 92.1 | 92.9 | 94.2 | … | ±0.8 | 0.30 |
| 96 | 98.1 | … | 100.2 | … | ±0.8 | 0.30 |
| 97 | … | 99.9 | … | … | ±0.8 | 0.30 |
| 105 | 107.1 | 107.9 | 109.2 | … | ±0.8 | 0.30 |
| 112 | 114.1 | 114.9 | 116.2 | … | ±0.8 | 0.30 |
| 120 | 122.1 | 122.9 | 124.2 | 125.2 | ±0.8 | 0.30 |
| 128 | 130.1 | 130.9 | 132.2 | 133.2 | ±0.8 | 0.30 |
| 144 | … | 146.9 | 148.2 | 149.2 | ±0.8 | 0.30 |
| 158 | … | 160.9 | 162.2 | 163.2 | ±1.0 | 0.45 |
| 173 | … | 175.9 | 177.2 | 178.2 | ±1.0 | 0.45 |
| 180 | … | 182.9 | 184.2 | 185.2 | ±1.0 | 0.45 |
| 195 | … | 197.9 | 199.2 | 200.2 | ±1.1 | 0.45 |
| 210 | … | 212.9 | 214.2 | 215.2 | ±1.1 | 0.45 |
| 240 | … | 241.4 | 242.2 | 242.7 | ±1.3 | 0.45 |
| 270 | … | 271.4 | 272.2 | 272.7 | ±1.6 | 0.60 |
| 300 | … | 301.4 | 302.2 | 302.7 | ±1.6 | 0.60 |
| 330 | … | … | 332.2 | 332.7 | ±2.0 | 0.60 |
| 360 | … | … | 362.2 | 362.7 | ±2.0 | 0.60 |

[a] To specify belt size use the Standard Length Designation prefixed by the letters indicating cross section; for example, BB90.

All dimensions in inches.

*Sheave Dimensions:* Groove angles and dimensions for sheaves and face widths of sheaves for multiple belt drives are given in Table 14, along with various tolerance values.

## Table 14. Double V-Belt Sheave and Groove Dimensions *ANSI/RMA IP-21, 1984*

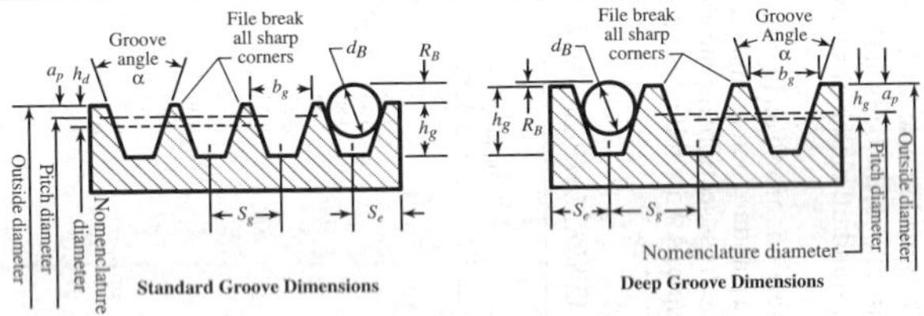


Face width of standard and deep groove sheaves = $s_g(N_g - 1) + 2S_e$, where $N_g$ = number of grooves

| Cross Section | Outside Diameter Range | Groove Angle, α ±0.33° | $b_g$ | | $h_g$ (Min.) | $2h_d$ | $R_B$ (Min.) | $d_B$ ±0.0005 | $S_g$[a] ±0.025 | $S_e$ | | Drive Design Factors: Min. Recomm. Outside Dia. | Drive Design Factors: $2a_p$[b] |
|---|---|---|---|---|---|---|---|---|---|---|---|---|---|
| AA | Up through 5.65 | 34 | 0.494 | ±0.005 | 0.460 | 0.250 | 0.148 | 0.4375 | 0.625 | 0.375 | +0.090 | 3.25 | 0.0 |
| | Over 5.65 | 38 | 0.504 | | | | 0.149 | (7⁄16) | | | −0.062 | | |
| BB | Up through 7.35 | 34 | 0.637 | ±0.006 | 0.550 | 0.350 | 0.189 | 0.5625 | 0.750 | 0.500 | +0.120 | 5.75 | 0.0 |
| | Over 7.35 | 38 | 0.650 | | | | 0.190 | (9⁄16) | | | −0.065 | | |
| AA-BB | Up through 7.35 | 34 | 0.612 | ±0.006 | 0.612 | A = 0.750 | 0.230 | 0.5625 | 0.750 | 0.500 | +0.120 | A = 3.620 | A = 0.370 |
| | Over 7.35 | 38 | 0.625 | | | B = 0.350 | 0.226 | (9⁄16) | | | −0.065 | B = 5.680 | B = −0.070 |
| CC | Up through 8.39 | 34 | 0.879 | } ±0.007 | 0.750 | 0.400 | 0.274 | 0.7812 (25⁄32) | 1.000 | 0.688 | +0.160 −0.070 | 9.4 | 0.0 |
| | Over 8.39 up to and including 12.40 | 36 | 0.887 | | | | 0.276 | | | | | | |
| | Over 12.40 | 38 | 0.895 | | | | 0.277 | | | | | | |
| DD | Up through 13.59 | 34 | 1.259 | } ±0.008 | 1.020 | 0.600 | 0.410 | 1.1250 (1⅛) | 1.438 | 0.875 | +0.220 −0.080 | 13.6 | 0.0 |
| | Over 13.59 up to and including 17.60 | 36 | 1.271 | | | | 0.410 | | | | | | |
| | Over 17.60 | 38 | 1.283 | | | | 0.411 | | | | | | |

**Table 14.** *(Continued)* **Double V-Belt Sheave and Groove Dimensions** *ANSI/RMA IP-21, 1984*

| Deep Groove Dimensions[c] | | | | | | | | | | | | Drive Design Factors | |
|---|---|---|---|---|---|---|---|---|---|---|---|---|---|
| Cross Section | Outside Diameter Range | Groove Angle, α ±0.33° | $b_g$ | | $h_g$ (Min.) | $2h_d$ | $R_B$ (Min.) | $d_B$ ±0.0005 | $S_g$[a] ±0.025 | $S_e$ | | Minimum Recommended Outside Diameter | $2a_p$ |
| AA | Up through 5.96 | 34 | 0.589 | ±0.005 | 0.615 | 0.560 | −0.009 | 0.4375 | 0.750 | 0.438 | +0.090 | 3.56 | 0.310 |
| | Over 5.96 | 38 | 0.611 | | | | −0.008 | (7/16) | | | −0.062 | | |
| BB | Up through 7.71 | 34 | 0.747 | ±0.006 | 0.730 | 0.710 | +0.007 | 0.5625 | 0.875 | 0.562 | +0.120 | 6.11 | 0.360 |
| | Over 7.71 | 38 | 0.774 | | | | +0.008 | (9/16) | | | −0.065 | | |
| CC | Up through 9.00 | 34 | 1.066 | } ±0.007 | 1.055 | 1.010 | −0.035 | 0.7812 (25/32) | 1.250 | 0.812 | +0.160 −0.070 | 10.01 | 0.610 |
| | Over 9.00 up to and including 13.01 | 36 | 1.085 | | | | −0.032 | | | | | | |
| | Over 13.01 | 38 | 1.105 | | | | −0.031 | | | | | | |
| DD | Up through 14.42 | 34 | 1.513 | } ±0.008 | 1.435 | 1.430 | −0.010 | 1.1250 (1 1/8) | 1.750 | 1.062 | +0.220 −0.080 | 14.43 | 0.830 |
| | Over 14.42 up to and including 18.43 | 36 | 1.541 | | | | −0.009 | | | | | | |
| | Over 18.43 | 38 | 1.569 | | | | −0.008 | | | | | | |

[a] Summation of the deviations from $S_g$ for all grooves in any one sheave shall not exceed ±0.050 in. The variation in pitch diameter between the grooves in any one sheave must be within the following limits: Up through 19.9 in. outside diameter and up through 6 grooves: 0.010 in. (add 0.005 in. for each additional groove). 20.0 in. and over on outside diameter and up through 10 grooves: 0.015 in. (add 0.0005 in. for each additional groove). This variation can be obtained easily by measuring the distance across two measuring balls or rods placed diametrically opposite each other in a groove. Comparing this "diameter over balls or rods" measurement between grooves will give the variation in pitch diameter.

[b] The $a_p$ values shown for the A/B combination sheaves are the geometrically derived values. These values may be different from those shown in manufacturer's catalogs.

[c] Deep groove sheaves are intended for drives with belt offset such as quarter-turn or vertical shaft drives.

| Other Sheave Tolerances | | |
|---|---|---|
| Outside Diameter | Radial Runout[a] | Axial Runout[a] |
| Up through 4.0 in. outside diameter ±0.020 in.<br>For each additional inch of outside diameter add ±0.005 in. | Up through 10.0 in. outside diameter ±0.010 in.<br>For each additional inch of outside diameter add 0.0005 in. | Up through 5.0 in. outside diameter 0.005 in.<br>For each additional inch of outside diameter add 0.001 in. |

[a] Total indicator reading.

All dimensions in inches.

*Cross Section Selection:* Use the chart (Fig. 6) as a guide to the double V-belt cross section for any combination of design horsepower and speed of the faster shaft. When the intersection of the design horsepower and speed of the faster shaft falls near a line between two areas on the chart, it is best to investigate the possibilities in both areas. Special circumstances (such as space limitations) may lead to a choice of belt cross section different from that indicated in the chart.

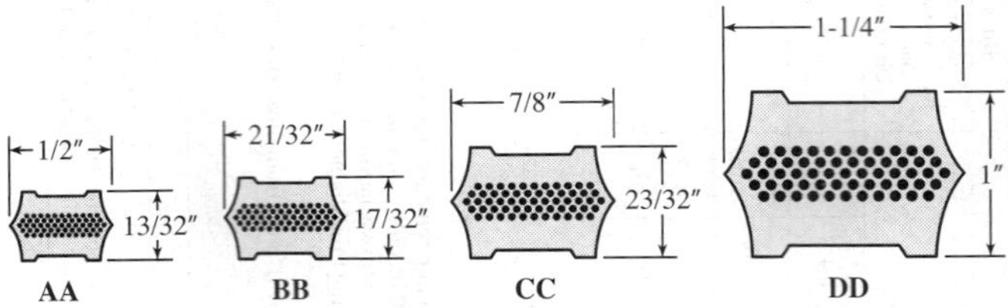


Fig. 5. Double-V Belt Cross Section

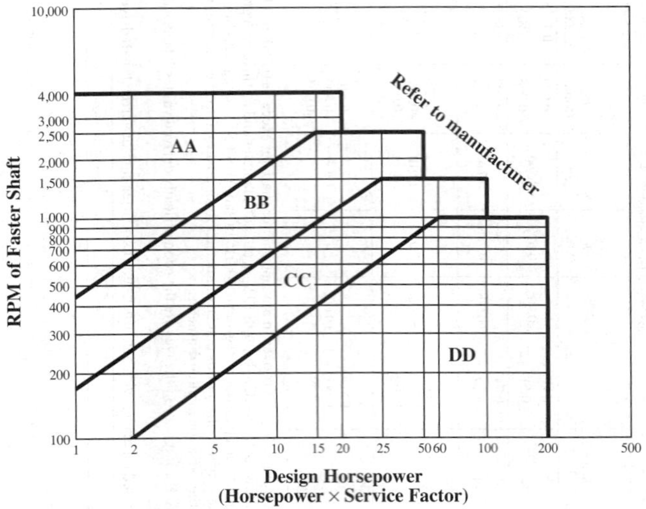


Fig. 6. Selection of Double V-Belt Cross Section

*Effective Diameter Determination:* Fig. 6 shows the relationship of effective diameter, outside diameter, and nomenclature diameter. Nomenclature diameter is used when ordering sheaves for double V-belt drives. The effective diameter is determined as follows:

$$\text{Effective diameter} = \text{Nomenclature diameter} \quad 2h_d - 2a_p$$

The values of $2h_d$ and $2a_p$ are given in Table 14.

*Double V-belt Length Determination:* The effective belt length of a specific drive may be determined by making a scaled layout of the drive. Draw the sheaves in terms of their effective diameters and in the position when a new belt is applied and first brought to driving tension. Next, measure the tangents and calculate the effective arc length ($AL_e$) of each sheave (see Table 15 for a glossary of terms):

$$AL_e = \frac{d_e\theta}{115}$$

The effective length of the belt will then be the sum of the tangents and the connecting arc lengths. Manufacturers may be consulted for mathematical calculation of effective belt length for specific drive applications.

**Table 15. Glossary of Terms for Double V-belt Calculations**

| | | | |
|---|---|---|---|
| $AL_e$ = | Length, arc, effective, in. | $R$ = | Ratio, tight side to slack side tension |
| $2a_p$ = | Diameter, differential, pitch to outside, in. | $R/(R-1)$ = | Factor, tension ratio |
| $d$ = | Diameter, pitch, in. (same as effective diameter) | $r$ = | Angular velocity, faster shaft, rpm/1000 |
| $d_e$ = | Diameter, effective, in. | $S$ = | Speed, belt, fpm/1000 |
| $2h_d$ = | Diameter differential, nomenclature to outside, in. | $T_e$ = | Tension, effective pull, lbf |
| $K_f$ = | Factor, length-flex correction | $T_r$ = | Tension, allowable tight side, lbf |
| $L_e$ = | Length, effective, in. | $T_S$ = | Tension, slack side, lbf |
| $n$ = | Sheaves, number on drive | $T_T$ = | Tension, tight side, lbf |
| $P_d$ = | Power, design, horsepower (transmitted horsepower × service factor) | $\theta$ = | Angle, arc of belt contact, deg |

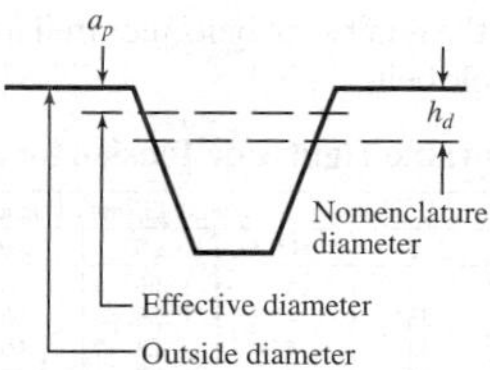


Fig. 7. Effective, Outside, and Nomenclature Sheave Diameters

*Number of Belts Determination:* The number of belts required may be determined on the basis of allowable tight side tension rating ($T_r$) at the most severe sheave. The allowable tight side tensions per belt are given in Table 16 through Table 19, and must be multiplied by the length-flex correction factors ($K_f$) listed in Table 20. To select the allowable tight side tension from the tables for a given sheave, the belt speed and effective diameter of the sheave in question are required.

*Double V-Belt Drive Design Method:* The fourteen drive design steps are as follows:

1) Number the sheaves starting from the driver in the opposite direction to belt rotation; include the idlers.

2) Select the proper service factor for each loaded driven unit.

3) Multiply the horsepower requirement for each loaded driven sheave by the corresponding service factor. This is the design horsepower at each sheave.

4) Calculate driver design horsepower. This hp is equal to the sum of all the driven design horsepower.

5) Calculate belt speed ($S$) in thousands of feet per minute: $S = rd/3.820$.

6) Calculate effective tension ($T_e$) for each loaded sheave: $T_e = 33P_d/S$.

7) Determine minimum $R/(R-1)$ for each loaded sheave from Table 21 using the arc of contact determined from the drive layout.

8) In most drives, slippage will occur first at the driver sheave. Assume this to be true and calculate $T_T$ and $T_S$ for the driver: $T_T = T_e\ [R/(R-1)]$ and $T_s = T_T - T_e$. Use $R/(R-1)$ from Step 7 and $T_e$ from Step 6 for the driver sheave.

9) Starting with the first driven sheave, determine $T_T$ and $T_S$ for each segment of the drive. The $T_T$ for the driver becomes $T_S$ for that sheave and is equal to $T_T - T_e$. Proceed around the drive in like manner.

10) Calculate actual $R/(R-1)$ for each sheave using: $R/(R-1) = T_T/T_e = T_T/(T_T - T_S)$. The $T_T$ and $T_S$ values are for those determined in Step 9. If these values are equal to or greater than those determined in Step 7, the assumption that slippage will first occur at the driver is correct and the next two steps are not necessary. If the value is less, the assumption was not correct, so proceed with Step 11.

11) Take the sheave where the actual value $R/(R-1)$ (Step 10) is less than the minimum, as determined in Step 7, and calculate a new $T_T$ and $T_S$ for this sheave using the minimum $R/(R-1)$ as determined in Step 7: $T_T = T_e\ [R/(R-1)]$ and $T_S = T_T - T_e$.

12) Start with this sheave and recalculate the tension in each segment of the drive as in Step 9.

13) The length-flex factor ($K_f$) is taken from Table 20. Before using this table, calculate the value of $L_e/n$. Be sure to use the appropriate belt cross section column when selecting the correction factor.

14) Beginning with the driver sheave, determine the number of belts ($N_b$) needed to satisfy the conditions at each loaded sheave using: $N_b = T_T/T_rK_f$. Note: $T_T$ is tight side tension as determined in Step 9 or 11 and 12. $T_r$ is allowable tight side tension as shown in Table 18-Table 21. $K_f$ is the length-flex correction factor from Table 20. The sheave that requires the largest number of belts is the number of belts required for the drive. Any fraction of a belt should be treated as a whole belt.

**Table 16. Allowable Tight Side Tension for an AA Section**

| Belt Speed (fpm) | Sheave Effective Diameter (in.) | | | | | | | |
|---|---|---|---|---|---|---|---|---|
| | 3.0 | 3.5 | 4.0 | 4.5 | 5.0 | 5.5 | 6.0 | 6.5 |
| 200 | 30 | 46 | 57 | 66 | 73 | 79 | 83 | 88 |
| 400 | 23 | 38 | 49 | 58 | 65 | 71 | 76 | 80 |
| 600 | 18 | 33 | 44 | 53 | 60 | 66 | 71 | 75 |
| 800 | 14 | 30 | 41 | 50 | 57 | 63 | 67 | 72 |
| 1000 | 12 | 27 | 38 | 47 | 54 | 60 | 65 | 69 |
| 1200 | 9 | 24 | 36 | 45 | 52 | 57 | 62 | 66 |
| 1400 | 7 | 22 | 34 | 42 | 49 | 55 | 60 | 64 |
| 1600 | 5 | 20 | 32 | 40 | 47 | 53 | 58 | 62 |
| 1800 | 3 | 18 | 30 | 38 | 46 | 51 | 56 | 60 |
| 2000 | 1 | 16 | 28 | 37 | 44 | 50 | 54 | 58 |
| 2200 | … | 15 | 26 | 35 | 42 | 48 | 53 | 57 |
| 2400 | … | 13 | 24 | 33 | 40 | 46 | 51 | 55 |
| 2600 | … | 11 | 23 | 31 | 39 | 44 | 49 | 53 |
| 2800 | … | 9 | 21 | 30 | 37 | 43 | 47 | 51 |
| 3000 | … | 8 | 19 | 28 | 35 | 41 | 46 | 50 |
| 3200 | … | 6 | 17 | 26 | 33 | 39 | 44 | 48 |
| 3400 | … | 4 | 16 | 24 | 31 | 37 | 42 | 46 |
| 3600 | … | 2 | 14 | 23 | 30 | 35 | 40 | 44 |
| 3800 | … | 1 | 12 | 21 | 28 | 34 | 38 | 43 |
| 4000 | … | … | 10 | 19 | 26 | 32 | 37 | 41 |
| 4200 | … | … | 8 | 17 | 24 | 30 | 35 | 39 |
| 4400 | … | … | 6 | 15 | 22 | 28 | 33 | 37 |
| 4600 | … | … | 4 | 13 | 20 | 26 | 31 | 35 |
| 4800 | … | … | 2 | 11 | 18 | 24 | 29 | 33 |
| 5000 | … | … | … | 9 | 16 | 22 | 27 | 31 |
| 5200 | … | … | … | 7 | 14 | 20 | 24 | 28 |
| 5400 | … | … | … | 4 | 12 | 17 | 22 | 26 |
| 5600 | … | … | … | 2 | 9 | 15 | 20 | 24 |
| 5800 | … | … | … | … | 7 | 13 | 18 | 22 |

The allowable tight side tension must be evaluated for each sheave in the system (see Step 14). Values must be corrected by $K_f$ from Table 20.

## Table 17. Allowable Tight Side Tension for a BB Section

| Belt Speed (fpm) | Sheave Effective Diameter (in.) | | | | | | | | |
|---|---|---|---|---|---|---|---|---|---|
| | 5.0 | 5.5 | 6.0 | 6.5 | 7.0 | 7.5 | 8.0 | 8.5 | 9.0 |
| 200 | 81 | 93 | 103 | 111 | 119 | 125 | 130 | 135 | 140 |
| 400 | 69 | 81 | 91 | 99 | 107 | 113 | 118 | 123 | 128 |
| 600 | 61 | 74 | 84 | 92 | 99 | 106 | 111 | 116 | 121 |
| 800 | 56 | 68 | 78 | 87 | 94 | 101 | 106 | 111 | 115 |
| 1000 | 52 | 64 | 74 | 83 | 90 | 96 | 102 | 107 | 111 |
| 1200 | 48 | 60 | 71 | 79 | 86 | 93 | 98 | 103 | 107 |
| 1400 | 45 | 57 | 67 | 76 | 83 | 89 | 95 | 100 | 104 |
| 1600 | 42 | 54 | 64 | 73 | 80 | 86 | 92 | 97 | 101 |
| 1800 | 39 | 51 | 61 | 70 | 77 | 84 | 89 | 94 | 98 |
| 2000 | 36 | 49 | 59 | 67 | 74 | 81 | 86 | 91 | 96 |
| 2200 | 34 | 46 | 56 | 64 | 72 | 78 | 84 | 89 | 93 |
| 2400 | 31 | 43 | 53 | 62 | 69 | 75 | 81 | 86 | 90 |
| 2600 | 29 | 41 | 51 | 59 | 67 | 73 | 78 | 83 | 88 |
| 2800 | 26 | 38 | 48 | 57 | 64 | 70 | 76 | 81 | 85 |
| 3000 | 23 | 35 | 45 | 54 | 61 | 68 | 73 | 78 | 82 |
| 3200 | 21 | 33 | 43 | 51 | 59 | 65 | 70 | 75 | 80 |
| 3400 | 18 | 30 | 40 | 49 | 56 | 62 | 68 | 73 | 77 |
| 3600 | 15 | 27 | 37 | 46 | 53 | 59 | 65 | 70 | 74 |
| 3800 | 12 | 24 | 35 | 43 | 50 | 57 | 62 | 67 | 71 |
| 4000 | 9 | 22 | 32 | 40 | 47 | 54 | 59 | 64 | 6[illegible] |
| 4200 | 7 | 19 | 29 | 37 | 45 | 51 | 56 | 61 | 6[illegible] |
| 4400 | 4 | 16 | 26 | 34 | 42 | 48 | 53 | 58 | 6[illegible] |
| 4600 | 1 | 13 | 23 | 31 | 39 | 45 | 50 | 55 | 60 |
| 4800 | … | 10 | 20 | 28 | 35 | 42 | 47 | 52 | 57 |
| 5000 | … | 6 | 16 | 25 | 32 | 39 | 44 | 49 | 53 |
| 5200 | … | 3 | 13 | 22 | 29 | 35 | 41 | 46 | 50 |
| 5400 | … | … | 10 | 18 | 26 | 32 | 38 | 42 | 47 |
| 5600 | … | … | 6 | 15 | 22 | 29 | 34 | 39 | 43 |
| 5800 | … | … | 3 | 11 | 19 | 25 | 31 | 36 | 40 |

The allowable tight side tension must be evaluated for each sheave in the system (see Step 14). Values must be corrected by $K_f$ from Table 20.

## Table 18. Allowable Tight Side Tension for a CC Section

| Belt Speed (fpm) | Sheave Effective Diameter (in.) | | | | | | | | |
|---|---|---|---|---|---|---|---|---|---|
| | 7.0 | 8.0 | 9.0 | 10.0 | 11.0 | 12.0 | 13.0 | 14.0 | 15.0 |
| 200 | 121 | 158 | 186 | 207 | 228 | 244 | 257 | 268 | 278 |
| 400 | 99 | 135 | 164 | 187 | 206 | 221 | 234 | 246 | 256 |
| 600 | 85 | 122 | 151 | 173 | 192 | 208 | 221 | 232 | 242 |
| 800 | 75 | 112 | 141 | 164 | 182 | 198 | 211 | 222 | 232 |
| 1000 | 67 | 104 | 133 | 155 | 174 | 190 | 203 | 214 | 224 |
| 1200 | 60 | 97 | 126 | 149 | 167 | 183 | 196 | 207 | 217 |
| 1400 | 54 | 91 | 120 | 142 | 161 | 177 | 190 | 201 | 211 |
| 1600 | 48 | 85 | 114 | 137 | 155 | 171 | 184 | 196 | 205 |
| 1800 | 43 | 80 | 108 | 131 | 150 | 166 | 179 | 190 | 200 |
| 2000 | 38 | 75 | 103 | 126 | 145 | 160 | 174 | 185 | 195 |
| 2200 | 33 | 70 | 98 | 121 | 140 | 155 | 169 | 180 | 190 |
| 2400 | 28 | 65 | 93 | 116 | 135 | 150 | 164 | 175 | 185 |
| 2600 | 23 | 60 | 88 | 111 | 130 | 145 | 159 | 170 | 180 |
| 2800 | 18 | 55 | 83 | 106 | 125 | 140 | 154 | 165 | 175 |
| 3000 | 13 | 50 | 78 | 101 | 120 | 135 | 149 | 160 | 170 |
| 3200 | 8 | 45 | 73 | 96 | 115 | 130 | 144 | 155 | 165 |
| 3400 | 3 | 39 | 68 | 91 | 110 | 125 | 138 | 150 | 160 |
| 3600 | … | 34 | 63 | 86 | 104 | 120 | 133 | 145 | 154 |
| 3800 | … | 29 | 58 | 80 | 99 | 115 | 128 | 139 | 149 |
| 4000 | … | 24 | 52 | 75 | 94 | 109 | 123 | 134 | 144 |
| 4200 | … | 18 | 47 | 70 | 88 | 104 | 117 | 128 | 138 |
| 4400 | … | 12 | 41 | 64 | 83 | 98 | 112 | 123 | 133 |
| 4600 | … | 7 | 35 | 58 | 77 | 93 | 106 | 117 | 127 |
| 4800 | … | 1 | 29 | 52 | 71 | 87 | 100 | 111 | 121 |
| 5000 | … | … | 23 | 46 | 65 | 81 | 94 | 105 | 115 |
| 5200 | … | … | 17 | 40 | 59 | 75 | 88 | 99 | 109 |
| 5400 | … | … | 11 | 34 | 53 | 68 | 81 | 93 | 103 |
| 5600 | … | … | 5 | 27 | 46 | 62 | 75 | 86 | 96 |
| 5800 | … | … | … | 21 | 40 | 55 | 68 | 80 | 90 |

The allowable tight side tension must be evaluated for each sheave in the system (see Step 14). Values must be corrected by $K_f$ from Table 20.

**Table 19. Allowable Tight Side Tension for a DD Section**

| Belt Speed (fpm) | Sheave Effective Diameter (in.) | | | | | | | | |
|---|---|---|---|---|---|---|---|---|---|
| | 12.0 | 13.0 | 14.0 | 15.0 | 16.0 | 17.0 | 18.0 | 19.0 | 20.0 |
| 200 | 243 | 293 | 336 | 373 | 405 | 434 | 459 | 482 | 503 |
| 400 | 195 | 245 | 288 | 325 | 358 | 386 | 412 | 434 | 455 |
| 600 | 167 | 217 | 259 | 297 | 329 | 358 | 383 | 406 | 426 |
| 800 | 146 | 196 | 239 | 276 | 308 | 337 | 362 | 385 | 405 |
| 1000 | 129 | 179 | 222 | 259 | 291 | 320 | 345 | 368 | 389 |
| 1200 | 114 | 164 | 207 | 244 | 277 | 305 | 331 | 353 | 374 |
| 1400 | 101 | 151 | 194 | 231 | 263 | 292 | 318 | 340 | 361 |
| 1600 | 89 | 139 | 182 | 219 | 251 | 280 | 305 | 328 | 349 |
| 1800 | 78 | 128 | 170 | 207 | 240 | 269 | 294 | 317 | 337 |
| 2000 | 67 | 117 | 159 | 196 | 229 | 258 | 283 | 306 | 326 |
| 2200 | 56 | 106 | 149 | 186 | 218 | 247 | 272 | 295 | 316 |
| 2400 | 45 | 95 | 138 | 175 | 208 | 236 | 262 | 284 | 305 |
| 2600 | 35 | 85 | 128 | 165 | 197 | 226 | 251 | 274 | 294 |
| 2800 | 24 | 74 | 117 | 154 | 187 | 215 | 241 | 263 | 284 |
| 3000 | 14 | 64 | 106 | 144 | 176 | 205 | 230 | 253 | 273 |
| 3200 | 3 | 53 | 96 | 133 | 165 | 194 | 219 | 242 | 263 |
| 3400 | … | 42 | 85 | 122 | 155 | 183 | 209 | 231 | 252 |
| 3600 | … | 31 | 74 | 111 | 144 | 172 | 198 | 220 | 241 |
| 3800 | … | 20 | 63 | 100 | 132 | 161 | 186 | 209 | 230 |
| 4000 | … | 9 | 51 | 89 | 121 | 150 | 175 | 198 | 218 |
| 4200 | … | … | 40 | 77 | 109 | 138 | 163 | 186 | 207 |
| 4400 | … | … | 28 | 65 | 97 | 126 | 152 | 174 | 195 |
| 4600 | … | … | 16 | 53 | 85 | 114 | 139 | 162 | 183 |
| 4800 | … | … | 3 | 40 | 73 | 102 | 127 | 150 | 170 |
| 5000 | … | … | … | 28 | 60 | 89 | 114 | 137 | 158 |
| 5200 | … | … | … | 15 | 47 | 76 | 101 | 124 | 145 |
| 5400 | … | … | … | 1 | 34 | 62 | 88 | 111 | 131 |
| 5600 | … | … | … | … | 20 | 49 | 74 | 97 | 118 |
| 5800 | … | … | … | … | 6 | 35 | 60 | 83 | 104 |

The allowable tight side tension must be evaluated for each sheave in the system (see Step 14). Values must be corrected by $K_f$ from Table 20.

**Table 20. Length-Flex Correction Factors $K_f$**

| $\frac{L_e}{n}$ | Belt Cross Section | | | | $\frac{L_e}{n}$ | Belt Cross Section | | | |
|---|---|---|---|---|---|---|---|---|---|
| | AA | BB | CC | DD | | AA | BB | CC | DD |
| 10 | 0.64 | 0.58 | … | … | 70 | … | 1.03 | 0.95 | 0.91 |
| 15 | 0.74 | 0.68 | … | … | 80 | … | 1.06 | 0.98 | 0.94 |
| 20 | 0.82 | 0.74 | 0.68 | … | 90 | … | 1.09 | 1.00 | 0.96 |
| 25 | 0.87 | 0.79 | 0.73 | 0.70 | 100 | … | 1.11 | 1.03 | 0.99 |
| 30 | 0.92 | 0.84 | 0.77 | 0.74 | 110 | … | … | 1.05 | 1.00 |
| 35 | 0.96 | 0.87 | 0.80 | 0.77 | 120 | … | … | 1.06 | 1.02 |
| 40 | 0.99 | 0.90 | 0.83 | 0.80 | 130 | … | … | 1.08 | 1.04 |
| 45 | 1.02 | 0.93 | 0.86 | 0.82 | 140 | … | … | 1.10 | 1.05 |
| 50 | 1.05 | 0.95 | 0.88 | 0.84 | 150 | … | … | 1.11 | 1.07 |
| 60 | … | 0.99 | 0.92 | 0.88 | … | … | … | … | … |

*Tension Ratings:* The tension rating formulas are:

$$\text{AA} \quad T_r = 118.5 - \frac{318.2}{d} - 0.8380S^2 - 25.76\log S$$

$$\text{BB} \quad T_r = 186.3 - \frac{665.1}{d} - 1.269S^2 - 39.02\log S$$

$$\text{CC} \quad T_r = 363.9 - \frac{2060}{d} - 2.400S^2 - 73.77\log S$$

$$\text{DD} \quad T_r = 783.1 - \frac{7790}{d} - 5.078S^2 - 156.1\log S$$

where $T_r$ = allowable tight side tension for a double-V belt drive, lbf
(not corrected for tension ratio or length-flex correction factor)
$d$ = pitch diameter of small sheave, inch
$S$ = belt speed, fpm/1000

**Table 21. Tension Ratio/Arc of Contact Factors**

| Arc of Contact, θ (deg.) | Design $\frac{R}{R-1}$ | Arc of Contact, θ (deg.) | Design $\frac{R}{R-1}$ |
|---|---|---|---|
| 300 | 1.07 | 170 | 1.28 |
| 290 | 1.08 | 160 | 1.31 |
| 280 | 1.09 | 150 | 1.35 |
| 270 | 1.10 | 140 | 1.40 |
| 260 | 1.11 | 130 | 1.46 |
| 250 | 1.12 | 120 | 1.52 |
| 240 | 1.13 | 110 | 1.60 |
| 230 | 1.15 | 100 | 1.69 |
| 220 | 1.16 | 90 | 1.81 |
| 210 | 1.18 | 80 | 1.96 |
| 200 | 1.20 | 70 | 2.15 |
| 190 | 1.22 | 60 | 2.41 |
| 180 | 1.25 | 50 | 2.77 |

*Minimum Sheave Size:* The recommended minimum sheave size depends on the rpm of the faster shaft. Minimum groove diameters for each belt cross section are listed in Table 14.

**Light Duty V-Belts ANSI/RMA IP-23.—**Light duty V-belts are typically used with fractional horsepower motors or small engines, and are designed primarily for fractional horsepower service. These belts are intended and specifically designed for use with small diameter sheaves and drives of loads and service requirements that are within the capacity of a single belt.

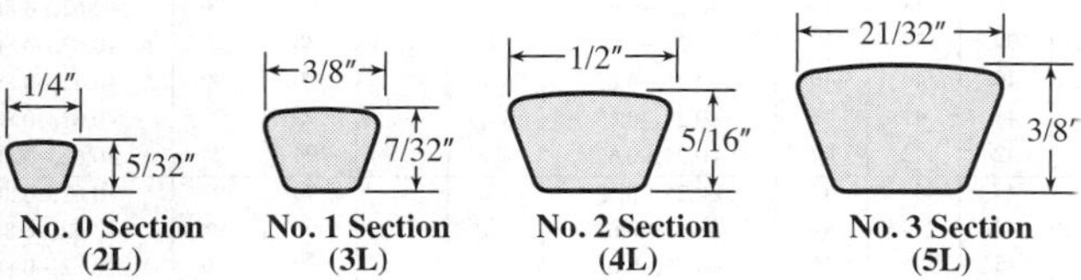


Fig. 8. Light Duty V-Belt Cross Sections

The four belt cross sections and sheave groove sizes are 2L, 3L, 4L, and 5L. The 2L is generally used only by OEMs and is not covered in the RMA standards.

**Belt Cross Sections.—**Nominal dimensions of the four cross sections are given in Fig. 8.

**Belt Size Designation.—**V-belt sizes are identified by a standard belt number, consisting of a letter-numeral combination. The first number and letter identify the cross section; the remaining numbers identify length as shown in Table 22. For example, a 3L520 belt has a 3L cross section and a length of 52.0 in.

**Table 22. Light Duty V-Belt Standard Dimensions** *ANSI/RMA IP-23, 1968*

| Standard Effective Outside Length (in.) | | | | Permissible Deviation From Standard Effective Length (in.) | Standard Effective Outside Length (in.) | | | | Permissible Deviation From Standard Effective Length (in.) |
|---|---|---|---|---|---|---|---|---|---|
| Cross Section | | | | | Cross Section | | | | |
| 2L | 3L | 4L | 5L | | 2L | 3L | 4L | 5L | |
| 8 | … | … | … | +0.12, −0.38 | … | … | 53 | 53 | +0.25, −0.62 |
| 9 | … | … | … | +0.12, −0.38 | … | 54 | 54 | 54 | +0.25, −0.62 |
| 10 | … | … | … | +0.12, −0.38 | … | … | 55 | 55 | +0.25, −0.62 |
| 11 | … | … | … | +0.12, −0.38 | … | 56 | 56 | 56 | +0.25, −0.62 |
| 12 | … | … | … | +0.12, −0.38 | … | … | 57 | 57 | +0.25, −0.62 |
| 13 | … | … | … | +0.12, −0.38 | … | 58 | 58 | 58 | +0.25, −0.62 |
| 14 | 14 | … | … | +0.12, −0.38 | … | … | 59 | 59 | +0.25, −0.62 |
| 15 | 15 | … | … | +0.12, −0.38 | … | 60 | 60 | 60 | +0.25, −0.62 |
| 16 | 16 | … | … | +0.12, −0.38 | … | … | 61 | 61 | +0.31, −0.69 |
| 17 | 17 | … | … | +0.12, −0.38 | … | … | 62 | 62 | +0.31, −0.69 |
| 18 | 18 | 18 | … | +0.12, −0.38 | … | … | 63 | 63 | +0.31, −0.69 |
| 19 | 19 | 19 | … | +0.12, −0.38 | … | … | 64 | 64 | +0.31, −0.69 |
| 20 | 20 | 20 | … | +0.12, −0.38 | … | … | 65 | 65 | +0.31, −0.69 |
| … | 21 | 21 | … | +0.25, −0.62 | … | … | 66 | 66 | +0.31, −0.69 |
| … | 22 | 22 | … | +0.25, −0.62 | … | … | 67 | 67 | +0.31, −0.69 |
| … | 23 | 23 | … | +0.25, −0.62 | … | … | 68 | 68 | +0.31, −0.69 |
| … | 24 | 24 | … | +0.25, −0.62 | … | … | 69 | 69 | +0.31, −0.69 |
| … | 25 | 25 | 25 | +0.25, −0.62 | … | … | 70 | 70 | +0.31, −0.69 |
| … | 26 | 26 | 26 | +0.25, −0.62 | … | … | 71 | 71 | +0.31, −0.69 |
| … | 27 | 27 | 27 | +0.25, −0.62 | … | … | 72 | 72 | +0.31, −0.69 |
| … | 28 | 28 | 28 | +0.25, −0.62 | … | … | 73 | 73 | +0.31, −0.69 |
| … | 29 | 29 | 29 | +0.25, −0.62 | … | … | 74 | 74 | +0.31, −0.69 |
| … | 30 | 30 | 30 | +0.25, −0.62 | … | … | 75 | 75 | +0.31, −0.69 |
| … | 31 | 31 | 31 | +0.25, −0.62 | … | … | 76 | 76 | +0.31, −0.69 |
| … | 32 | 32 | 32 | +0.25, −0.62 | … | … | 77 | 77 | +0.31, −0.69 |
| … | 33 | 33 | 33 | +0.25, −0.62 | … | … | 78 | 78 | +0.31, −0.69 |
| … | 34 | 34 | 34 | +0.25, −0.62 | … | … | 79 | 79 | +0.31, −0.69 |
| … | 35 | 35 | 35 | +0.25, −0.62 | … | … | 80 | 80 | +0.62, −0.88 |
| … | 36 | 36 | 36 | +0.25, −0.62 | … | … | 82 | 82 | +0.62, −0.88 |
| … | 37 | 37 | 37 | +0.25, −0.62 | … | … | 84 | 84 | +0.62, −0.88 |
| … | 38 | 38 | 38 | +0.25, −0.62 | … | … | 86 | 86 | +0.62, −0.88 |
| … | 39 | 39 | 39 | +0.25, −0.62 | … | … | 88 | 88 | +0.62, −0.88 |
| … | 40 | 40 | 40 | +0.25, −0.62 | … | … | 90 | 90 | +0.62, −0.88 |
| … | 41 | 41 | 41 | +0.25, −0.62 | … | … | 92 | 92 | +0.62, −0.88 |
| … | 42 | 42 | 42 | +0.25, −0.62 | … | … | 94 | 94 | +0.62, −0.88 |
| … | 43 | 43 | 43 | +0.25, −0.62 | … | … | 96 | 96 | +0.62, −0.88 |
| … | 44 | 44 | 44 | +0.25, −0.62 | … | … | 98 | 98 | +0.62, −0.88 |
| … | 45 | 45 | 45 | +0.25, −0.62 | … | … | 100 | 100 | +0.62, −0.88 |
| … | 46 | 46 | 46 | +0.25, −0.62 | … | … | … | … | … |
| … | 47 | 47 | 47 | +0.25, −0.62 | … | … | … | … | … |
| … | 48 | 48 | 48 | +0.25, −0.62 | | | | | |
| … | 49 | 49 | 49 | +0.25, −0.62 | … | … | … | … | … |
| … | 50 | 50 | 50 | +0.25, −0.62 | … | … | … | … | … |
| … | … | 51 | 51 | +0.25, −0.62 | … | … | … | … | … |
| … | 52 | 52 | 52 | +0.25, −0.62 | … | … | … | … | … |

All dimensions in inches.

*Sheave Dimensions:* Groove angles and dimensions for sheaves and various sheave tolerances are given in Table 23.

**Table 23. Light Duty V-Belt Sheave and Groove Dimensions**
*ANSI/RMA IP-23, 1968*

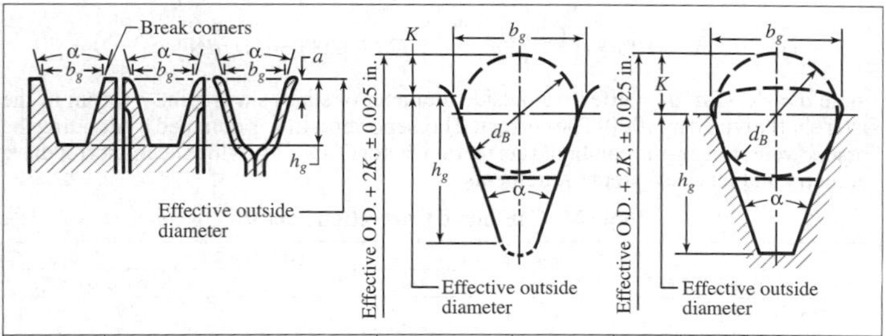


| Belt Section | Effective Outside Diameter: Min. Recomm. | Effective Outside Diameter: Range | α Groove Angle ±0°20' (deg) | $d_B$ Ball Diameter ±0.0005 | 2K | $b_g$ (Ref) | $h_g$ (min) | 2a[a] |
|---|---|---|---|---|---|---|---|---|
| 2L | 0.8 | Less than 1.50 | 32 | 0.2188 | 0.176 | 0.240 | 0.250 | 0.04 |
| | | 1.50 to 1.99 | 34 | | 0.182 | | | |
| | | 2.00 to 2.50 | 36 | | 0.188 | | | |
| | | Over 2.50 | 38 | | 0.194 | | | |
| 3L | 1.5 | Less than 2.20 | 32 | 0.3125 | 0.177 | 0.364 | 0.406 | 0.06 |
| | | 2.20 to 3.19 | 34 | | 0.191 | | | |
| | | 3.20 to 4.20 | 36 | | 0.203 | | | |
| | | Over 4.20 | 38 | | 0.215 | | | |
| 4L | 2.5 | Less than 2.65 | 30 | 0.4375 | 0.299 | 0.490 | 0.490 | 0.10 |
| | | 2.65 to 3.24 | 32 | | 0.316 | | | |
| | | 3.25 to 5.65 | 34 | | 0.331 | | | |
| | | Over 5.65 | 38 | | 0.358 | | | |
| 5L | 3.5 | Less than 3.95 | 30 | 0.5625 | 0.385 | 0.630 | 0.580 | 0.16 |
| | | 3.95 to 4.94 | 32 | | 0.406 | | | |
| | | 4.95 to 7.35 | 34 | | 0.426 | | | |
| | | Over 7.35 | 38 | | 0.461 | | | |

[a] The diameter used in calculating speed ratio and belt speed is obtained by subtracting the 2*a* value from the Effective Outside Diameter of the sheave.

| Other Sheave Tolerances | | | | | |
|---|---|---|---|---|---|
| Outside Diameters | | Outside Diameter Eccentricity[a] | | Groove Side Wobble & Runout[a] | |
| For outside diameters: | | For outside diameters: | | For outside diameters: | |
| Under 6.0 in. | ±0.015 in. | 10.0 in. and under | 0.010 in. | 20.0 in. and under | 0.0015 in. per inch of outside diameter. |
| 6.0 to 12.0 in. | ±0.020 in. | For each additional inch of outside diameter, add 0.0005 in. | | | |
| Over 12.0 in. | ±0.040 in. | | | For each additional inch of outside diameter, add 0.0005 in. | |

[a] Total indicator reading.

All dimensions in inches.

*Horsepower Ratings:* The horsepower ratings for light duty V-belts can be calculated from the following formulas:

$$\text{3L} \quad HP = r\left(\frac{0.2164d^{0.91}}{r^{0.09}} - 0.2324 - 0.0001396r^2d^3\right)$$

$$\text{4L} \quad HP = r\left(\frac{0.4666d^{0.91}}{r^{0.09}} - 0.7231 - 0.0002286r^2d^3\right)$$

$$\text{5L} \quad HP = r\left(\frac{0.7748d^{0.91}}{r^{0.09}} - 1.727 - 0.0003641r^2d^3\right)$$

where $d = d_0 - 2a$; $d_0$ = effective outside diameter of small sheave, in.; $r$ = rpm of the faster shaft divided by 1000. The corrected horsepower rating is obtained by dividing the horsepower rating by the combined correction factor (Table 24), which accounts for drive geometry and service factor requirements.

**Table 24. Combined Correction Factors**

| Type of Driven Unit | Speed Ratio | |
|---|---|---|
| | Less than 1.5 | 1.5 and Over |
| Fans and blowers | 1.0 | 0.9 |
| Domestic laundry machines | 1.1 | 1.0 |
| Centrifugal pumps | 1.1 | 1.0 |
| Generators | 1.2 | 1.1 |
| Rotary compressors | 1.2 | 1.1 |
| Machine tools | 1.3 | 1.2 |
| Reciprocating pumps | 1.4 | 1.3 |
| Reciprocating compressors | 1.4 | 1.3 |
| Woodworking machines | 1.4 | 1.3 |

**V-Ribbed Belts ANSI/RMA IP-26.—**V-ribbed belts are a cross between flat belts and V-belts. The belt is basically flat with V-shaped ribs projecting from the bottom, which guide the belt and provide greater stability than that found in a flat belt. The ribs operate in grooved sheaves.

V-ribbed belts do not have the wedging action of a V-belt and thus operate at higher tensions. This design provides excellent performance in high-speed and serpentine applications, and in drives that utilize small diameter sheaves. The V-ribbed belt comes in five cross sections: H, J, K, L, and M, specified by effective length, cross section and number of ribs.

*Belt Cross Sections:* Nominal dimensions of the five cross sections are given in Table 25.

**Table 25. Nominal Dimensions of V-Ribbed Belt Cross Sections**

*ANSI/RMA IP-26, 1977*

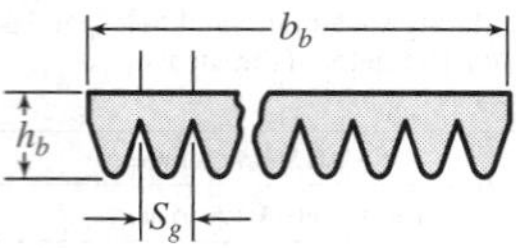


$b_b = N_r \times S_g$, where $N_r$ = number of ribs and $S_g$ is sheave groove spacing

| Cross Section | $h_b$ | $S_g$ | Standard Number of Ribs |
|---|---|---|---|
| H | 0.12 | 0.063 | … |
| J | 0.16 | 0.092 | 4, 6, 10, 16, 20 |
| K | 0.24 | 0.140 | … |
| L | 0.38 | 0.185 | 6, 8, 10, 12, 14, 16, 18, 20 |
| M | 0.66 | 0.370 | 6, 8, 10, 12, 14, 16, 18, 20 |

All dimensions in inches.

**Table 26. V-Ribbed Belt Sheave and Groove Dimensions** *ANSI/RMA IP-26, 1977*

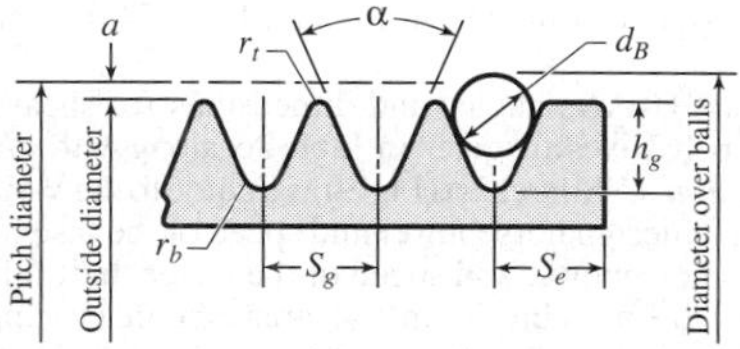


Face width = $S_g (N_g - 1) + 2S_e$, where $N_g$ is number of grooves

| Cross Section | Minimum Recommended Outside Diameter | α Groove Angle ±0.25 (deg) | $S_g$[a] | $r_t$ +0.005, −0.000 | 2a | $r_b$ | $h_g$ (min) | $d_B$ ±0.0005 | $S_e$ |
|---|---|---|---|---|---|---|---|---|---|
| H | 0.50 | 40 | 0.063 ±0.001 | 0.005 | 0.020 | 0.013 +0.000 −0.005 | 0.041 | 0.0469 | 0.080 +0.020 −0.010 |
| J | 0.80 | 40 | 0.092 ±0.001 | 0.008 | 0.030 | 0.015 +0.000 −0.005 | 0.071 | 0.0625 | 0.125 +0.030 −0.015 |
| K | 1.50 | 40 | 0.140 ±0.002 | 0.010 | 0.038 | 0.020 +0.000 −0.005 | 0.122 | 0.1093 | 0.125 +0.050 −0.000 |
| L | 3.00 | 40 | 0.185 ±0.002 | 0.015 | 0.058 | 0.015 +0.000 −0.005 | 0.183 | 0.1406 | 0.375 +0.075 −0.030 |
| M | 7.00 | 40 | 0.370 ±0.003 | 0.030 | 0.116 | 0.030 +0.000 −0.010 | 0.377 | 0.2812 | 0.500 +0.100 −0.040 |

[a] Summation of the deviations from $S_g$ for all grooves in any one sheave shall not exceed ±0.010 in.

| Other Sheave Tolerances[a] | | | | |
|---|---|---|---|---|
| Outside Diameter | | Radial Runout[b] | | Axial Runout[b] |
| Outside Diameter: | | Outside Diameter: | | 0.001 in. per inch of outside diameter |
| Up through 2.9 in. | ±0.010 in. | Up through 2.9 in. | 0.005 in. | |
| Over 2.9 in. to and including 8.0 in. | ±0.020 in. | Over 2.9 in. to and including 10.0 in. | 0.010 in. | |
| For each additional inch of outside diameter over 8.0 in., add ±0.0025 in. | | For each additional inch of outside diameter over 10.0 in., add 0.0005 in. | | |

[a] Variations in pitch diameter between the grooves in any one sheave must be within the following limits: Up through 2.9 in. outside diameter and up through 6 grooves, 0.002 in. (add 0.001 in. for each additional groove); over 2.9 in. to and including 19.9 in. and up through 10 grooves, 0.005 in. (add 0.0002 in. for each additional groove); over 19.9 in. and up through 10 grooves, 0.010 in. (add 0.0005 in. for each additional groove). This variation can be obtained by measuring the distance across two measuring balls or rods placed in the grooves diametrically opposite each other. Comparing this "diameter-over-balls or -rods" measurement between grooves will give the variation in pitch diameter.

[b] Total indicator reading.

All dimensions in inches

*Belt Size Designation:* Belt sizes are identified by a standard belt number, which consists of belt effective length to the nearest tenth of an inch, a letter designating cross section, and the number of ribs. For example, 540L6 signifies a 54.0 in. effective length, L belt, 6 ribs wide.

*Sheave Dimensions:* Groove angles and dimensions for sheaves and face widths of sheaves for multiple belt drives are given in Table 26, along with various tolerance values.

*Cross Section Selection:* Use the chart (Fig. 9) as a guide to the V-ribbed belt cross section for any combination of design horsepower and speed of the faster shaft. When the intersection of the design horsepower and speed of the faster shaft falls near a line between two areas on the chart, the possibilities in both areas should be explored. Special circumstances (such as space limitations) may lead to a choice of belt cross section different from that indicated in the chart. H and K cross sections are not included because of their specialized use. Belt manufacturers should be contacted for specific data.

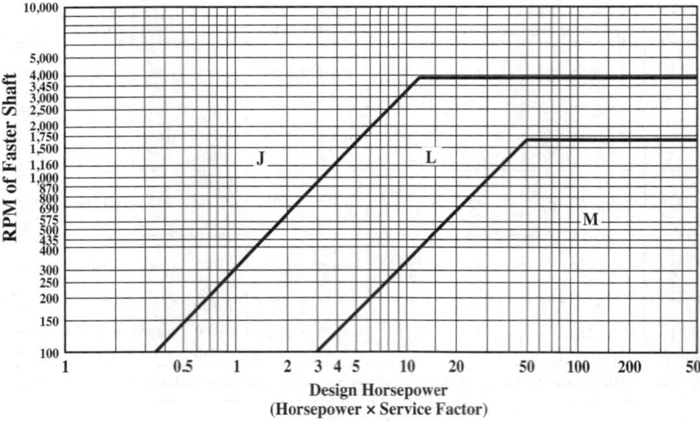


Fig. 9. Selection of V-Ribbed Belt Cross Section

*Horsepower Ratings:* The horsepower rating formulas are:

$$\textbf{J:}\text{HP} = d_p r\left[\frac{0.1240}{(d_p r)^{0.09}} - \frac{0.08663}{d_p} - 0.2318\times10^{-4}(d_p r)^2\right] + .0\ 08663r\left[1 - \frac{1}{K_{SR}}\right]$$

$$\textbf{L:}\text{HP} = d_p r\left[\frac{0.5761}{(d_p r)^{0.09}} - \frac{0.8987}{d_p} - 1.018\times10^{-4}(d_p r)^2\right] + .0\ 8987r\left[1 - \frac{1}{K_{SR}}\right]$$

$$\textbf{M:}\text{HP} = d_p r\left[\frac{1.975}{(d_p r)^{0.09}} - \frac{6.597}{d_p} - 3.922\times10^{-4}(d_p r)^2\right] + .6\ 597r\left[1 - \frac{1}{K_{SR}}\right]$$

In these equations, $d_p$ = pitch diameter of the small sheave, in.; $r$ = rpm of the faster shaft divided by 1000; $K_{SR}$ = speed ratio factor given in the accompanying Table 30. These formulas give the maximum horsepower per rib recommended, corrected for the speed ratio. To obtain the horsepower per rib for an arc of contact other than 180 degrees, and for belts longer or shorter than the average length, multiply the horsepower obtained from these formulas by the length correction factor (Table 28) and the arc of contact correction factor (Table 29).

**Table 27. V-Ribbed Belt Standard Effective Lengths** *ANSI/RMA IP-26, 1977*

| J Cross Section | | | L Cross Section | | | M Cross Section | | |
|---|---|---|---|---|---|---|---|---|
| Standard Length Designation[a] | Standard Effective Length | Permissible Deviation from Standard Length | Standard Length Designation[a] | Standard Effective Length | Permissible Deviation from Standard Length | Standard Length Designation[a] | Standard Effective Length | Permissible Deviation from Standard Length |
| 180 | 18.0 | +0.2, −0.2 | 500 | 50.0 | +0.2, −0.4 | 900 | 90.0 | +0.4, −0.7 |
| 190 | 19.0 | +0.2, −0.2 | 540 | 54.0 | +0.2, −0.4 | 940 | 94.0 | +0.4, −0.8 |
| 200 | 20.0 | +0.2, −0.2 | 560 | 56.0 | +0.2, −0.4 | 990 | 99.0 | +0.4, −0.8 |
| 220 | 22.0 | +0.2, −0.2 | 615 | 61.5 | +0.2, −0.5 | 1060 | 106.0 | +0.4, −0.8 |
| 240 | 24.0 | +0.2, −0.2 | 635 | 63.5 | +0.2, −0.5 | 1115 | 111.5 | +0.4, −0.9 |
| 260 | 26.0 | +0.2, −0.2 | 655 | 65.5 | +0.2, −0.5 | 1150 | 115.0 | +0.4, −0.9 |
| 280 | 28.0 | +0.2, −0.2 | 675 | 67.5 | +0.3, −0.6 | 1185 | 118.5 | +0.4, −0.9 |
| 300 | 30.0 | +0.2, −0.3 | 695 | 69.5 | +0.3, −0.6 | 1230 | 123.0 | +0.4, −1.0 |
| 320 | 32.0 | +0.2, −0.3 | 725 | 72.5 | +0.3, −0.6 | 1310 | 131.0 | +0.5, −1.1 |
| 340 | 34.0 | +0.2, −0.3 | 765 | 76.5 | +0.3, −0.6 | 1390 | 139.0 | +0.5, −1.1 |
| 360 | 36.0 | +0.2, −0.3 | 780 | 78.0 | +0.3, −0.6 | 1470 | 147.0 | +0.6, −1.2 |
| 380 | 38.0 | +0.2, −0.3 | 795 | 79.5 | +0.3, −0.6 | 1610 | 161.0 | +0.6, −1.2 |
| 400 | 40.0 | +0.2, −0.4 | 815 | 81.5 | +0.3, −0.7 | 1650 | 165.0 | +0.6, −1.3 |
| 430 | 43.0 | +0.2, −0.4 | 840 | 84.0 | +0.3, −0.7 | 1760 | 176.0 | +0.7, −1.4 |
| 460 | 46.0 | +0.2, −0.4 | 865 | 86.5 | +0.3, −0.7 | 1830 | 183.0 | +0.7, −1.4 |
| 490 | 49.0 | +0.2, −0.4 | 915 | 91.5 | +0.4, −0.7 | 1980 | 198.0 | +0.8, −1.6 |
| 520 | 52.0 | +0.2, −0.4 | 975 | 97.5 | +0.4, −0.8 | 2130 | 213.0 | +0.8, −1.6 |
| 550 | 55.0 | +0.2, −0.4 | 990 | 99.0 | +0.4, −0.8 | 2410 | 241.0 | +0.9, −1.6 |
| 580 | 58.0 | +0.2, −0.5 | 1065 | 106.5 | +0.4, −0.8 | 2560 | 256.0 | +1.0, −1.8 |
| 610 | 61.0 | +0.2, −0.5 | 1120 | 112.0 | +0.4, −0.9 | 2710 | 271.0 | +1.1, −2.2 |
| 650 | 65.0 | +0.2, −0.5 | 1150 | 115.0 | +0.4, −0.9 | 3010 | 301.0 | +1.2, −2.4 |

[a] To specify belt size, use the standard length designation, followed by the letter indicating belt cross section and the number of ribs desired. For example: 865L10.

All dimensions in inches.

## Table 28. Length Correction Factors

| Std. Length Designation | Cross Section | | | Std. Length Designation | Cross Section | | |
|---|---|---|---|---|---|---|---|
| | J | L | M | | J | L | M |
| 180 | 0.83 | ... | ... | 1230 | ... | 1.08 | 0.94 |
| 200 | 0.85 | ... | ... | 1310 | ... | 1.10 | 0.96 |
| 240 | 0.89 | ... | ... | 1470 | ... | 1.12 | 0.098 |
| 280 | 0.92 | ... | ... | 1610 | ... | 1.14 | 1.00 |
| 320 | 0.95 | ... | ... | 1830 | ... | 1.17 | 1.03 |
| 360 | 0.98 | ... | ... | 1980 | ... | 1.19 | 1.05 |
| 400 | 1.00 | ... | ... | 2130 | ... | 1.21 | 1.06 |
| 440 | 1.02 | ... | ... | 2410 | ... | 1.24 | 1.09 |
| 500 | 1.05 | 0.89 | ... | 2710 | ... | ... | 1.12 |
| 550 | 1.07 | 0.91 | ... | 3010 | ... | ... | 1.14 |
| 610 | 1.09 | 0.93 | ... | 3310 | ... | ... | 1.16 |
| 690 | 1.12 | 0.96 | ... | 3610 | ... | ... | 1.18 |
| 780 | 1.16 | 0.98 | ... | 3910 | ... | ... | 1.20 |
| 910 | 1.18 | 1.02 | 0.88 | 4210 | ... | ... | 1.22 |
| 940 | 1.19 | 1.02 | 0.89 | 4810 | ... | ... | 1.25 |
| 990 | 1.20 | 1.04 | 0.90 | 5410 | ... | ... | 1.28 |
| 1060 | ... | 1.05 | 0.91 | 6000 | ... | ... | 1.30 |
| 1150 | ... | 1.07 | 0.93 | ... | ... | ... | ... |

## Table 29. Arc of Contact Correction Factors

| $\frac{D_o - d_o}{C}$ | Arc of Contact, θ, on Small Sheave, (deg) | Correction Factor |
|---|---|---|
| 0.00 | 180 | 1.00 |
| 0.10 | 174 | 0.98 |
| 0.20 | 169 | 0.97 |
| 0.30 | 163 | 0.95 |
| 0.40 | 157 | 0.94 |
| 0.50 | 151 | 0.92 |
| 0.60 | 145 | 0.90 |
| 0.70 | 139 | 0.88 |
| 0.80 | 133 | 0.85 |
| 0.90 | 127 | 0.83 |
| 1.00 | 120 | 0.80 |
| 1.10 | 113 | 0.77 |
| 1.20 | 106 | 0.74 |
| 1.30 | 99 | 0.71 |
| 1.40 | 91 | 0.67 |
| 1.50 | 83 | 0.63 |

*Number of Ribs:* The number of ribs required for an application is obtained by dividing the design horsepower by the corrected horsepower rating for one rib.

Arc of contact on the small sheave may be determined by the following formulas:

*Exact Formula:* $\text{Arc of Contact (deg)} = 2\cos^{-1}\left(\frac{D_o - d_o}{2C}\right)$

*Approximate Formula:* $\text{Arc of Contact (deg)} = 180 - \frac{(D_o - d_o)60}{C}$

$D_o$ = effective outside diameter of large sheave, inch; $d_o$ = effective outside diameter of small sheave, inch; and, $C$ = center distance, inch.

**Table 30. Speed Ratio Correction Factors**

| Speed Ratio[a] | $K_{SR}$ |
|---|---|
| 1.00 to and incl. 1.10 | 1.0000 |
| Over 1.01 to and incl. 1.04 | 1.0136 |
| Over 1.04 to and incl. 1.08 | 1.0276 |
| Over 1.08 to and incl. 1.12 | 1.0419 |
| Over 1.12 to and incl. 1.18 | 1.0567 |
| Over 1.18 to and incl. 1.24 | 1.0719 |
| Over 1.24 to and incl. 1.34 | 1.0875 |
| Over 1.34 to and incl. 1.51 | 1.1036 |
| Over 1.51 to and incl. 1.99 | 1.1202 |
| Over 1.99 | 1.1373 |

[a] $D_p/d_p$, where $D_p$ ($d_p$) is the pitch diameter of the large (small) sheave.

**Variable-Speed Belts ANSI/RMA IP-25.—**For drives that require more speed variation than can be obtained with conventional industrial V-belts, standard-line variable-speed drives are available. These drives use special wide, thin belts. Package units of standard-line variable-speed belts and sheaves combined with the motor and output gearbox, are available in ranges from approximately ½ through 100 horsepower.

The speed ranges of variable-speed drives can be much greater than those drives using classical V-belts. Speed ranges up to 10:1 can be obtained on lower horsepower units.

This section covers 12 variable-speed belt cross sections and sheave groove sizes designed 1422V, 1922V, 2322V, 1926V, 2926V, 3226V, 2530V, 3230V, 4430V, 4036V, 4436V, and 4836V. The industry supplies many other sizes that are not listed in this section.

*Belt Cross Sections and Lengths:* Nominal dimensions of the 12 cross sections are given in Table 31, and lengths in Table 32.

**Table 31. Normal Variable-Speed Belt Dimensions** *ANSI/RMA IP-25, 1982*

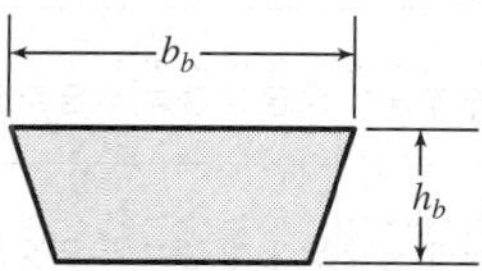


| Cross Section | $b_b$ | $h_b$ | $h_b/b_b$ | Cross Section | $b_b$ | $h_b$ | $h_b/b_b$ |
|---|---|---|---|---|---|---|---|
| 1422V | 0.88 | 0.31 | 0.35 | 2530V | 1.56 | 0.59 | 0.38 |
| 1922V | 1.19 | 0.38 | 0.32 | 3230V | 2.00 | 0.62 | 0.31 |
| 2322V | 1.44 | 0.44 | 0.31 | 4430V | 2.75 | 0.69 | 0.25 |
| 1926V | 1.19 | 0.44 | 0.37 | 4036V | 2.50 | 0.69 | 0.28 |
| 2926V | 1.81 | 0.50 | 0.28 | 4436V | 2.75 | 0.72 | 0.26 |
| 3226V | 2.0 | 0.53 | 0.27 | 4836V | 3.00 | 0.75 | 0.25 |

All dimensions in inches.

**Table 32. Variable-Speed V-Belt Standard Belt Lengths** *ANSI/RMA IP-25, 1982*

| Standard Pitch Length Designation | Standard Effective Lengths | | | | | | | | | | | | Permissible Deviations from Standard Length |
|---|---|---|---|---|---|---|---|---|---|---|---|---|---|
| | Cross Section | | | | | | | | | | | | |
| | 1422V | 1922V | 2322V | 1926V | 2926V | 3226V | 2530V | 3230V | 4430V | 4036V | 4436V | 4836V | |
| 315 | 32.1 | … | … | … | … | … | … | … | … | … | … | … | ±0.7 |
| 335 | 34.1 | … | … | … | … | … | … | … | … | … | … | … | ±0.7 |
| 355 | 36.1 | 36.2 | … | 36.3 | … | … | … | … | … | … | … | … | ±0.7 |
| 375 | 38.1 | 38.2 | … | 38.3 | … | … | … | … | … | … | … | … | ±0.7 |
| 400 | 40.6 | 40.7 | 40.8 | 40.8 | … | … | … | … | … | … | … | … | ±0.7 |
| 425 | 43.1 | 43.2 | 43.3 | 43.3 | … | … | … | … | … | … | … | … | ±0.8 |
| 450 | 45.6 | 45.7 | 45.8 | 45.8 | … | … | … | … | … | … | … | … | ±0.8 |
| 475 | 48.1 | 48.2 | 48.3 | 48.3 | … | … | … | … | … | … | … | … | ±0.8 |
| 500 | 50.6 | 50.7 | 50.8 | 50.8 | … | … | 50.9 | … | … | … | … | … | ±0.8 |
| 530 | 53.6 | 53.7 | 53.8 | 53.8 | 53.9 | … | 53.9 | … | … | … | … | … | ±0.8 |
| 560 | 56.6 | 56.7 | 56.8 | 56.8 | 56.9 | 56.9 | 56.9 | 57.1 | 57.3 | 57.3 | 57.3 | 57.4 | ±0.9 |
| 600 | 60.6 | 60.7 | 60.8 | 60.8 | 60.9 | 60.9 | 60.9 | 61.1 | 61.3 | 61.3 | 61.3 | 61.4 | ±0.9 |
| 630 | 63.6 | 63.7 | 63.8 | 63.8 | 63.9 | 63.9 | 63.9 | 64.1 | 64.3 | 64.3 | 64.3 | 64.4 | ±0.9 |
| 670 | 67.6 | 67.7 | 67.8 | 67.8 | 67.9 | 67.9 | 67.9 | 68.1 | 68.3 | 68.3 | 68.3 | 68.4 | ±0.9 |
| 710 | 71.6 | 71.7 | 71.8 | 71.8 | 71.9 | 71.9 | 71.9 | 72.1 | 72.3 | 72.3 | 72.3 | 72.4 | ±0.9 |
| 750 | 75.6 | 75.7 | 75.8 | 75.8 | 75.9 | 75.9 | 75.9 | 76.1 | 76.3 | 76.3 | 76.3 | 76.4 | ±1.0 |
| 800 | … | 80.7 | 80.8 | 80.8 | 80.9 | 80.9 | 80.9 | 81.1 | 81.3 | 81.3 | 81.3 | 81.4 | ±1.0 |
| 850 | … | 85.7 | 85.8 | 85.8 | 85.9 | 85.9 | 85.9 | 86.1 | 86.3 | 86.3 | 86.3 | 86.4 | ±1.1 |
| 900 | … | 90.7 | 90.8 | 90.8 | 90.9 | 90.9 | 90.9 | 91.1 | 91.3 | 91.3 | 91.3 | 91.4 | ±1.1 |
| 950 | … | 95.7 | 95.8 | 95.8 | 95.9 | 95.9 | 95.9 | 96.1 | 96.3 | 96.3 | 96.3 | 96.4 | ±1.1 |
| 1000 | … | 100.7 | 100.8 | 100.8 | 100.9 | 100.9 | 100.9 | 101.1 | 101.3 | 101.3 | 101.3 | 101.4 | ±1.2 |
| 1060 | … | 106.7 | 106.8 | 106.8 | 106.9 | 106.9 | 106.9 | 107.1 | 107.3 | 107.3 | 107.3 | 107.4 | ±1.2 |
| 1120 | … | 112.7 | 112.8 | 112.8 | 112.9 | 112.9 | 112.9 | 113.1 | 113.3 | 113.3 | 113.3 | 113.4 | ±1.2 |
| 1180 | … | 118.7 | 118.8 | 118.8 | 118.9 | 118.9 | 118.9 | 119.1 | 119.3 | 119.3 | 119.3 | 119.4 | ±1.3 |
| 1250 | … | … | … | … | 125.9 | 125.9 | 125.9 | 126.1 | 126.3 | 126.3 | 126.3 | 126.4 | ±1.3 |
| 1320 | … | … | … | … | … | 132.9 | … | 133.1 | 133.3 | 133.3 | 133.3 | 133.4 | ±1.3 |

All dimensions in inches.

The lengths given in this table are not necessarily available from all manufacturers. Availability should be investigated prior to design commitment.

## Table 33. Variable-Speed Sheave and Groove Dimensions

| | Standard Groove Dimensions | | | | | | | | | Drive Design Factors | | | |
|---|---|---|---|---|---|---|---|---|---|---|---|---|---|
| | Variable | | | | | Companion | | | | | | | |
| Cross Section | α Groove Angle ±0.67 (deg) | $b_g$[a] Closed +0.000 −0.030 | $b_{go}$ Open Max | $h_{gv}$ Min | $S_g$ ±0.03 | α Groove Angle ±0.33 (deg) | $b_g$ ±0.010 | $h_g$ Min | $S_g$ ±0.03 | Min. Recomm. Pitch Diameter | 2a | 2av Max | CL Min |
| 1422V | 22 | 0.875 | 1.63 | 2.33 | 1.82 | 22 | 0.875 | 0.500 | 1.82 | 2.0 | 0.20 | 3.88 | 0.08 |
| 1922V | 22 | 1.188 | 2.23 | 3.14 | 2.42 | 22 | 1.188 | 0.562 | 2.42 | 3.0 | 0.22 | 5.36 | 0.08 |
| 2322V | 22 | 1.438 | 2.71 | 3.78 | 2.89 | 22 | 1.438 | 0.625 | 2.89 | 3.5 | 0.25 | 6.52 | 0.08 |
| 1926V | 26 | 1.188 | 2.17 | 2.65 | 2.36 | 26 | 1.188 | 0.625 | 2.36 | 3.0 | 0.25 | 4.26 | 0.08 |
| 2926V | 26 | 1.812 | 3.39 | 4.00 | 3.58 | 26 | 1.812 | 0.750 | 3.58 | 3.5 | 0.30 | 6.84 | 0.08 |
| 3226V | 26 | 2.000 | 3.75 | 4.41 | 3.96 | 26 | 2.000 | 0.781 | 3.96 | 4.0 | 0.30 | 7.60 | 0.08 |
| 2530V | 30 | 1.562 | 2.81 | 3.01 | 2.98 | 30 | 1.562 | 0.844 | 2.98 | 4.0 | 0.30 | 4.64 | 0.10 |
| 3230V | 30 | 2.000 | 3.67 | 3.83 | 3.85 | 30 | 2.000 | 0.875 | 3.85 | 4.5 | 0.35 | 6.22 | 0.10 |
| 4430V | 30 | 2.750 | 5.13 | 5.23 | 5.38 | 30 | 2.750 | 0.938 | 5.38 | 5.0 | 0.40 | 8.88 | 0.10 |
| 4036V | 36 | 2.500 | 4.55 | 3.95 | 4.80 | 36 | 2.500 | 0.938 | 4.80 | 4.5 | 0.40 | 6.32 | 0.10 |
| 4436V | 36 | 2.750 | 5.03 | 4.33 | 5.30 | 36 | 2.750 | 0.969 | 5.30 | 5.0 | 0.40 | 7.02 | 0.10 |
| 4836V | 36 | 3.000 | 5.51 | 4.72 | 5.76 | 36 | 3.000 | 1.000 | 5.76 | 6.0 | 0.45 | 7.74 | 0.10 |

[a] The effective width ($b_e$), a reference dimension, is the same as the ideal top width of closed variable-speed sheave ($b_g$) and the ideal top width of the companion sheave ($b_g$).

| Other Sheave Tolerances | | | | | |
|---|---|---|---|---|---|
| Outside Diameter | | Radial Runout[a] | | Axial Runout[a] | |
| Up through 4.0 in. outside diameter | ±0.020 in. | Up through 10.0 in. outside diameter | 0.010 in. | Up through 5.0 in. outside diameter | 0.005 in. |
| For each additional inch of outside diameter add ±0.005 in. | | For each additional inch of outside diameter add 0.0005 in. | | For each additional inch of outside diameter add 0.001 in. | |

[a] Total indicator reading.

| Surface Finish | | | |
|---|---|---|---|
| Machined Surface Area | Max Surface Roughness Height, $R_a$ (AA) (μ in.) | Machined Surface Area | Max Surface Roughness Height, $R_a$ (AA) (μ in.) |
| V-Sheave groove sidewalls | 125 | Straight bores with 0.002 in. or less total tolerance | 125 |
| Rim edges and ID, hub ends and OD | 500 | Taper and straight bores with total tolerance over 0.002 in. | 250 |

All dimensions in inches, except where noted.

*Belt Size Designation:* Variable-speed belt sizes are identified by a standard belt number. The first two digits denote the belt top width in sixteenths of an inch; the third and fourth digits indicate the angle of the groove in which the belt is designed to operate. Letter V (for variable) follows the first four digits. The digits after the V indicate pitch length to the nearest 0.1 in. For example, 1422V450 is a belt of 7⁄8 in. (14⁄16 in.) nominal top width designed to operate in a sheave of 22 degree groove angle and having a pitch length of 45.0 in.

*Sheave Groove Data:* A variable-speed sheave is an assembly of movable parts, designed to permit one or both flanges of the sheave to be moved axially causing a radial movement of the variable-speed belt in the sheave groove. This radial movement permits stepless speed variation within the physical limits of the sheave and the belt. A companion sheave may be a solid sheave having a constant diameter and groove profile or another variable sheave. Variable-speed sheave designs should conform to the dimensions in Table 33 and Fig. 10. The included angle of the sheaves, top width, and clearance are boundary dimensions. Groove angles and dimensions of companion sheaves should conform to Table 33 and Fig. 11. Various tolerance values are also given in Table 33.

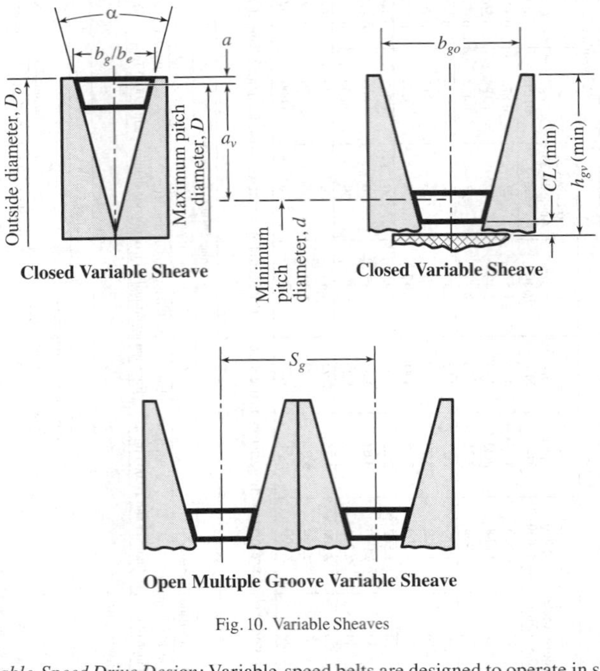


Fig. 10. Variable Sheaves

*Variable-Speed Drive Design:* Variable-speed belts are designed to operate in sheaves that are an assembly of movable parts. The sheave design permits one or both flanges of the sheave to be moved axially, causing a radial movement of the variable-speed belt in the sheave groove. The result is a stepless speed variation within the physical limits of the sheave and the variable-speed belt. Therefore, besides transmitting power, variable-speed belt drives provide speed variation.

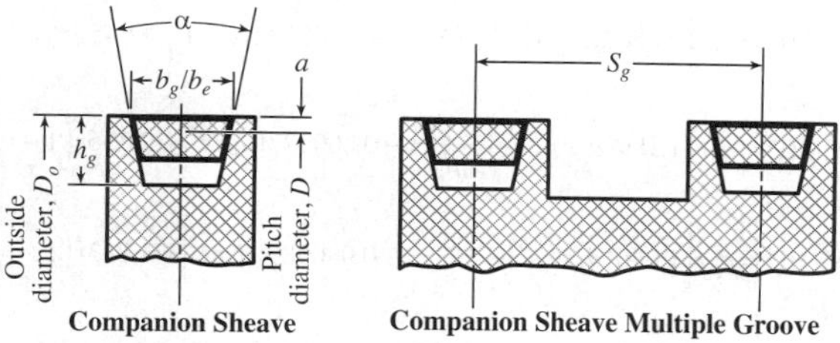


Fig. 11. Companion Sheaves

The factors that determine the amount of pitch diameter change on variable-speed sheaves are belt top width, belt thickness, and sheave angle. This pitch diameter change, combined with the selected operating pitch diameters for a sheave, determines the possible speed variation.

The range of output speeds from a variable-speed sheave drive is established by the companion sheave and is a function of the ratio of the pitch diameter of the companion sheave to the maximum and minimum pitch diameters of the variable sheave. Speed variation is usually obtained by varying the center distance between the two sheaves. This type of drive seldom exceeds a speed variation of 3:1.

For a single variable-speed sheave drive, the speed variation

$$\text{Speed variation} = \frac{\text{PD Max}}{\text{PD Min}}(\text{of variable sheave})$$

For a dual variable-speed sheave drive, which is frequently referred to as a compound drive because both sheaves are variable, the speed variation is

$$\text{Speed variation} = \frac{DR(DN)}{dr(dn)}$$

where $DR$ = max driver PD

$DN$ = max driven PD

$dr$ = min driver PD

$dn$ = min driven PD

With this design, the center distance is generally fixed and speed variation is usually accomplished by mechanically altering the pitch diameter of one sheave. In this type of drive, the other sheave is spring loaded to make an opposite change in the pitch diameter and to provide the correct belt tension. Speed variations of up to 10:1 are common on this type of drive.

*Speed Ratio Adjustment:* All speed ratio changes must be made while the drives are running. Attempting to make adjustments while the unit is stopped creates unnecessary and possibly destructive forces on both the belt and sheaves. In stationary control drives, the belt tension should be released to allow the flanges to adjust without belt force interference.

*Cross Section Selection:* Selection of a variable-speed belt cross section is based on the drive design horsepower and speed variation. Table 33 shows the maximum pitch diameter variation (2*av*) that each cross section can attain.

*Horsepower Ratings:* The general horsepower formulas for variable-speed belts are:

$$1422\text{V} \quad \text{HP} = d_p r\left[0.4907(d_p r)^{-0.09} - \frac{0.8378}{d_p} - 0.000337(d_p r)^2\right] + 0.8378r\left(1 - \frac{1}{K_{CR}}\right)$$

$$1922\text{V} \quad HP = d_p r\left[0.8502(d_p r)^{-0.09} - \frac{1.453}{d_p} - 0.000538(d_p r)^2\right] + 1.453r\left(1 - \frac{1}{K_{SR}}\right)$$

$$2322\text{V} \quad HP = d_p r\left[1.189(d_p r)^{-0.09} - \frac{2.356}{d_p} - 0.000777(d_p r)^2\right] + 2.356r\left(1 - \frac{1}{K_{SR}}\right)$$

$$1926\text{V} \quad HP = d_p r\left[1.046(d_p r)^{-0.09} - \frac{1.833}{d_p} - 0.000589(d_p r)^2\right] + 1.833r\left(1 - \frac{1}{K_{SR}}\right)$$

$$2926\text{V} \quad HP = d_p r\left[1.769(d_p r)^{-0.09} - \frac{4.189}{d_p} - 0.001059(d_p r)^2\right] + 4.189r\left(1 - \frac{1}{K_{SR}}\right)$$

$$3226\text{V} \quad HP = d_p r\left[2.073(d_p r)^{-0.09} - \frac{5.236}{d_p} - 0.001217(d_p r)^2\right] + 5.236r\left(1 - \frac{1}{K_{SR}}\right)$$

$$2530\text{V} \quad HP = d_p r\left[2.395(d_p r)^{-0.09} - \frac{6.912}{d_p} - 0.001148(d_p r)^2\right] + 6.912r\left(1 - \frac{1}{K_{SR}}\right)$$

$$3230\text{V} \quad HP = d_p r\left[2.806(d_p r)^{-0.09} - \frac{7.854}{d_p} - 0.001520(d_p r)^2\right] + 7.854r\left(1 - \frac{1}{K_{SR}}\right)$$

$$4430\text{V} \quad HP = d_p r\left[3.454(d_p r)^{-0.09} - \frac{7.854}{d_p} - 0.002196(d_p r)^2\right] + 9.818r\left(1 - \frac{1}{K_{SR}}\right)$$

$$4036\text{V} \quad HP = d_p r\left[3.566(d_p r)^{-0.09} - \frac{9.687}{d_p} - 0.002060(d_p r)^2\right] + 9.687r\left(1 - \frac{1}{K_{SR}}\right)$$

$$4436\text{V} \quad HP = d_p r\left[4.041(d_p r)^{-0.09} - \frac{11.519}{d_p} - 0.002297(d_p r)^2\right] + 11.519r\left(1 - \frac{1}{K_{SR}}\right)$$

$$4836\text{V} \quad HP = d_p r\left[4.564(d_p r)^{-0.09} - \frac{13.614}{d_p} - 0.002634(d_p r)^2\right] + 13.614r\left(1 - \frac{1}{K_{SR}}\right)$$

In these equations, $d_p$ = pitch diameter of small sheave, in.; $r$ = rpm of faster shaft divided by 1000; $K_{SR}$ = speed ratio factor given in the accompanying Table 34. These formulas give the basic horsepower rating, corrected for the speed ratio. To obtain the horsepower for arcs of contact other than 180 degrees and for belts longer or shorter than average length, multiply the horsepower obtained from these formulas by the arc of contact correction factor (Table 36) and the length correction factor (Table 35).

**Table 34. Speed Ratio Correction Factors**

| Speed Ratio[a] | $K_{SR}$ | Speed Ratio[a] | $K_{SR}$ |
|---|---|---|---|
| 1.00-1.01 | 1.0000 | 1.19-1.24 | 1.0719 |
| 1.02-1.04 | 1.0136 | 1.25-1.34 | 1.0875 |
| 1.05-1.08 | 1.0276 | 1.35-1.51 | 1.1036 |
| 1.09-1.12 | 1.0419 | 1.52-1.99 | 1.1202 |
| 1.13-1.18 | 1.0567 | 2.0 and over | 1.1373 |

[a] $D_p/d_p$, where $D_p$ ($d_p$) is the pitch diameter of the large (small) sheave.

**Table 35. Length Correction Factors**

| Standard Pitch Length Designation | Cross Section | | | | | | | | | | | |
|---|---|---|---|---|---|---|---|---|---|---|---|---|
| | 1422V | 1922V | 2322V | 1926V | 2926V | 3226V | 2530V | 3230V | 4430V | 4036V | 4436V | 4836V |
| 315 | 0.93 | … | … | … | … | … | … | … | … | … | … | … |
| 335 | 0.94 | … | … | … | … | … | … | … | … | … | … | … |
| 355 | 0.95 | 0.90 | … | 0.90 | … | … | … | … | … | … | … | … |
| 375 | 0.96 | 0.91 | … | 0.91 | … | … | … | … | … | … | … | … |
| 400 | 0.97 | 0.92 | 0.90 | 0.92 | … | … | … | … | … | … | … | … |
| 425 | 0.98 | 0.93 | 0.91 | 0.93 | … | … | … | … | … | … | … | … |
| 450 | 0.99 | 0.94 | 0.92 | 0.94 | … | … | … | … | … | … | … | … |
| 475 | 1.00 | 0.95 | 0.93 | 0.95 | … | … | … | … | … | … | … | … |
| 500 | 1.01 | 0.95 | 0.94 | 0.95 | … | … | 0.90 | … | … | … | … | … |
| 530 | 1.02 | 0.96 | 0.95 | 0.96 | 0.92 | … | 0.92 | … | … | … | … | … |
| 560 | 1.03 | 0.97 | 0.96 | 0.97 | 0.93 | 0.92 | 0.93 | 0.91 | 0.90 | 0.91 | 0.91 | 0.92 |
| 600 | 1.04 | 0.98 | 0.97 | 0.98 | 0.94 | 0.93 | 0.94 | 0.93 | 0.92 | 0.93 | 0.92 | 0.93 |
| 630 | 1.05 | 0.99 | 0.98 | 0.99 | 0.95 | 0.94 | 0.95 | 0.94 | 0.93 | 0.94 | 0.93 | 0.94 |
| 670 | 1.06 | 1.00 | 0.99 | 1.00 | 0.97 | 0.95 | 0.96 | 0.95 | 0.94 | 0.95 | 0.95 | 0.95 |
| 710 | 1.07 | 1.01 | 1.00 | 1.01 | 0.98 | 0.96 | 0.98 | 0.96 | 0.96 | 0.96 | 0.96 | 0.96 |
| 750 | 1.08 | 1.02 | 1.01 | 1.02 | 0.99 | 0.98 | 0.99 | 0.97 | 0.97 | 0.97 | 0.97 | 0.98 |
| 800 | … | 1.03 | 1.02 | 1.03 | 1.00 | 0.99 | 1.00 | 0.99 | 0.99 | 0.99 | 0.99 | 0.99 |
| 850 | … | 1.04 | 1.03 | 1.04 | 1.01 | 1.00 | 1.01 | 1.00 | 1.00 | 1.00 | 1.00 | 1.00 |
| 900 | … | 1.05 | 1.04 | 1.05 | 1.02 | 1.01 | 1.02 | 1.01 | 1.01 | 1.01 | 1.01 | 1.01 |
| 950 | … | 1.06 | 1.05 | 1.06 | 1.03 | 1.02 | 1.04 | 1.02 | 1.03 | 1.02 | 1.02 | 1.02 |
| 1000 | … | 1.07 | 1.06 | 1.07 | 1.04 | 1.03 | 1.05 | 1.03 | 1.04 | 1.03 | 1.04 | 1.03 |
| 1060 | … | 1.08 | 1.07 | 1.07 | 1.06 | 1.04 | 1.06 | 1.05 | 1.06 | 1.05 | 1.05 | 1.04 |
| 1120 | … | 1.09 | 1.08 | 1.08 | 1.07 | 1.06 | 1.07 | 1.06 | 1.07 | 1.06 | 1.06 | 1.06 |
| 1180 | … | 1.09 | 1.09 | 1.09 | 1.08 | 1.07 | 1.08 | 1.07 | 1.08 | 1.07 | 1.07 | 1.07 |
| 1250 | … | … | … | … | 1.09 | 1.08 | 1.10 | 1.08 | 1.10 | 1.08 | 1.09 | 1.08 |
| 1320 | … | … | … | … | … | 1.09 | … | 1.09 | 1.11 | 1.09 | 1.10 | 1.09 |

*Rim Speed:* The material and design selected for sheaves must be capable of withstanding the high rim speeds that may occur in variable-speed drives. The rim speed is calculated as follows: Rim speed (fpm) = $(\pi/12)(D_o)$ (rpm).

*Arc of Contact:* Arc of contact on the small sheave may be determined by the formulas:

*Exact Formula:* $\text{Arc of Contact (deg)} = 2\cos^{-1}\left(\frac{D-d}{2C}\right)$

*Approximate Formula:* $\text{Arc of Contact (deg)} = 180 - \frac{(D-d)60}{C}$

where $D$ = Pitch diameter of large sheave or flat pulley, inch or mm

$d$ = Pitch diameter of small sheave, inch or mm

$C$ = Center distance, inch or mm

**Table 36. Arc of Contact Correction Factors**

| $\frac{D-d}{C}$ | Arc of Contact, θ, on Small Sheave, (deg) | Correction Factor | $\frac{D-d}{C}$ | Arc of Contact, θ, on Small Sheave, (deg) | Correction Factor |
|---|---|---|---|---|---|
| 0.00 | 180 | 1.00 | 0.80 | 0.80 | 0.87 |
| 0.10 | 174 | 0.99 | 0.90 | 0.90 | 0.85 |
| 0.20 | 169 | 0.97 | 1.00 | 1.00 | 0.82 |
| 0.30 | 163 | 0.96 | 1.10 | 1.10 | 0.80 |
| 0.40 | 157 | 0.94 | 1.20 | 1.20 | 0.77 |
| 0.50 | 151 | 0.93 | 1.30 | 1.30 | 0.73 |
| 0.60 | 145 | 0.91 | 1.40 | 1.40 | 0.70 |
| 0.70 | 139 | 0.89 | 1.50 | 1.50 | 0.65 |

**60-Degree V-Belts.—**60-degree V-belts are ideal for compact drives. Their 60-degree angle and ribbed top are specifically designed for long life on small diameter sheaves. These belts offer extremely smooth operation at high speeds (in excess of 10,000 rpm) and can be used on drives with high-speed ratios. They are available in 3M, 5M, 7M, and 11M (3, 5, 7, 11 mm) cross sections (top widths) and are commonly found in the joined configuration, which provides extra stability and improved performance. They are specified by cross section and nominal length; for example, a 5M315 designation indicates a belt having a 5 mm cross section and an effective length of 315 mm.

Industry standards have not yet been published for 60-degree V-belts. Belt manufacturers should be contacted for specific applications, specifications, and additional information.

**SAE Standard V-Belts.—**The data for V-belts and pulleys shown in Table 37 cover nine sizes, three of which — 0.250, 0.315, and 0.440 — were added in 1977 to conform to existing practice. This Standard was reaffirmed in 1987.

V-belts are produced in a variety of constructions in a basic trapezoidal shape and are to be dimensioned in such a way that they are functional in pulleys dimensioned as described in the Standard. Standard belt lengths are in increments of $\frac{1}{2}$ inch (12.7 mm) up to and including 80 inches (203.2 cm). Standard lengths above 80 inches up to and including 100 inches (254 cm) are in increments of 1 inch (2.54 cm), without fractions. Standard belt length tolerances are based on the center distance and are as follows: For belt lengths of 50 inches (127 cm) or less, ±0.12 inch; over 50 to 60 inches (127–152.4 cm), inclusive, ±0.16 inch; over 60 to 80 inches (152.4–203.2 cm), inclusive, ±0.19 inch; and over 80 to 100 inches (203.2–254 cm), inclusive, ±0.22 inch.

**Belt Storage and Handling.—**To achieve maximum belt performance, proper belt storage procedures should always be practiced. If belts are not stored properly, their performance can be adversely affected. Four key rules are:

1) Do not store belts on floors unless they are protected by appropriate packaging.

2) Do not store belts near windows where the belts may be exposed to direct sunlight or moisture.

**Table 37. SAE V-Belt and Pulley Dimensions**

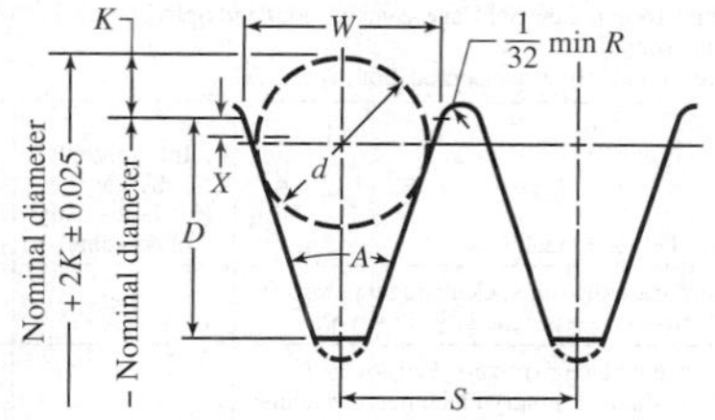


| SAE Size | Recommended Min. Eff Dia[a] | A Groove Angle (deg) ±0.5 | W Eff. Groove Width | D Groove Depth Min | d Ball or Rod Dia (±0.0005) | 2K 2 × Ball Extension | 2X[b] | S Groove[c] Spacing (±0.015) |
|---|---|---|---|---|---|---|---|---|
| 0.250 | 2.25 | 36 | 0.248 | 0.276 | 0.2188 | 0.164 | 0.04 | 0.315 |
| 0.315 | 2.25 | 36 | 0.315 | 0.354 | 0.2812 | 0.222 | 0.05 | 0.413 |
| 0.380 | 2.40 | 36 | 0.380 | 0.433 | 0.3125 | 0.154 | 0.06 | 0.541 |
| 0.440 | 2.75 | 36 | 0.441 | 0.512 | 0.3750 | 0.231 | 0.07 | 0.591 |
| 0.500 | 3.00 | 36 | 0.500 | 0.551 | 0.4375 | 0.314 | 0.08 | 0.661 |
| 11⁄16 | 3.00 | 34 | 0.597 | 0.551 | 0.500 | 0.258 | 0.00 | 0.778 |
| | Over 4.00 | 36 | | | | 0.280 | | |
| | Over 6.00 | 38 | | | | 0.302 | | |
| ¾ | 3.00 | 34 | 0.660 | 0.630 | 0.5625 | 0.328 | 0.02 | 0.841 |
| | Over 4.00 | 36 | | | | 0.352 | | |
| | Over 6.00 | 38 | | | | 0.374 | | |
| ⅞ | 3.50 | 34 | 0.785 | 0.709 | 0.6875 | 0.472 | 0.04 | 0.966 |
| | Over 4.50 | 36 | | | | 0.496 | | |
| | Over 6.00 | 38 | | | | 0.520 | | |
| 1 | 4.00 | 34 | 0.910 | 0.827 | 0.8125 | 0.616 | 0.06 | 1.091 |
| | Over 6.00 | 36 | | | | 0.642 | | |
| | Over 8.00 | 38 | | | | 0.666 | | |

All dimensions in inches.

[a] Pulley effective diameters below those recommended should be used with caution, because power transmission and belt life may be reduced.

[b] The *X* dimension is radial; 2*X* is to be subtracted from the effective diameter to obtain "pitch diameter" for speed ratio calculations.

[c] These values are intended for adjacent grooves of the same effective width (*W*). Choice of pulley manufacture or belt design parameter may justify variance from these values. The *S* dimension should be the same on all multiple groove pulleys in a drive using matched belts. 

3) Do not store belts near electrical devices that may generate ozone (transformers, electric motors, etc.).

4) Do not store belts in areas where solvents or chemicals are present in the atmosphere.

Belts should be stored in a cool, dry environment. When stacked on shelves, the stacks should be short enough to avoid excess weight on the bottom belts, which may cause distortion. When stored in containers, the container size and contents should be sufficiently limited to avoid distortion.

*V-Belts:* A common method is to hang the belts on pegs or pin racks. Very long belts stored this way should use sufficiently large pins or crescent-shaped "saddles" to prevent their weight from causing distortion.

*Joined V-belts, Synchronous Belts, V-Ribbed Belts:* Like V-belts, these belts may be stored on pins or saddles with precautions taken to avoid distortion. However, belts of this

**Table 38. Service Factors for V-Belts**

| Driving Unit | *AC Motors:* Normal Torque, Squirrel Cage, Synchronous and Split Phase.<br>*DC Motors:* Shunt Wound.<br>*Engines:* Multiple Cylinder Internal Combustion. | | |
|---|---|---|---|
| Types of Driven Machines | Intermittent Service (3-5 hours daily or seasonal) | Normal Service (8-10 hours daily) | Continuous Service (16-24 hours daily) |
| Agitators for liquids; Blowers and exhausters; Centrifugal pumps & compressors; Fans up to 10 horsepower; Light duty conveyors | 1.1 | 1.2 | 1.3 |
| Belt conveyors for sand, grain, etc.; Dough mixers; Fans over 10 horsepower; Generators; Line shafts; Laundry machinery; Machine tools; Punches, presses, shears; Printing machinery; Positive displacement rotary pumps; Revolving and vibrating screens | 1.2 | 1.3 | 1.4 |
| Brick machinery; Bucket elevators; Exciters; Piston compressors; Conveyors (drag, pan, screw); Hammer mills; Paper mill beaters; Piston pumps; Positive displacement blowers; Pulverizers; Saw mill and woodworking machinery; Textile machinery | 1.4 | 1.5 | 1.6 |
| Crushers (gyratory, jaw, roll); Mills (ball, rod, tube); Hoists; Rubber calendars, extruders, mills | 1.5 | 1.6 | 1.8 |

| Driving Unit | *AC Motors:* High Torque, High Slip, Repulsion-Induction, Single Phase, Series Wound, Slip Ring.<br>*DC Motors:* Series Wound, Compound Wound.<br>*Engines:* Single Cylinder Internal Combustion.<br>*Line Shafts, Clutches* | | |
|---|---|---|---|
| Types of Driven Machines | Intermittent Service (3-5 hours daily or seasonal) | Normal Service (8-10 hours daily) | Continuous Service (16-24 hours daily) |
| Agitators for liquids; Blowers and exhausters; Centrifugal pumps & compressors; Fans up to 10 horsepower; Light duty conveyors | 1.1 | 1.2 | 1.3 |
| Belt conveyors for sand, grain, etc.; Dough mixers; Fans over 10 horsepower; Generators; Line shafts; Laundry machinery; Machine tools; Punches, presses, shears; Printing machinery; Positive displacement rotary pumps; Revolving and vibrating screens | 1.2 | 1.3 | 1.4 |
| Brick machinery; Bucket elevators; Exciters; Piston compressors; Conveyors (drag, pan, screw); Hammer mills; Paper mill beaters; Piston pumps; Positive displacement blowers; Pulverizers; Saw mill and woodworking machinery; Textile machinery | 1.4 | 1.5 | 1.6 |
| Crushers (gyratory, jaw, roll); Mills (ball, rod, tube); Hoists; Rubber calendars, extruders, mills | 1.5 | 1.6 | 1.8 |

The machines listed above are representative samples only. Select the group listed above whose load characteristics most closely approximate those of the machine being considered.

type up to approximately 120 inch (3.05 m) are normally shipped in a "nested" configuration and should be stored in the same manner. Nests are formed by laying a belt on its side on a flat surface and placing as many belts inside the first belt as possible without undue force. When the nests are tight and are stacked with each rotated 180° from the one below, they may be stacked without damage.

Belts of this type over 120 inch may be "rolled up" and tied for shipment. These rolls may be stacked for easy storage. Care should be taken to avoid small bend radii which could damage the belts.

*Variable-Speed Belts:* Variable-speed belts are more sensitive to distortion than most other belts, and should not be hung from pins or racks but stored on shelves in the sleeves in which they are shipped.

*Service Factors:* Service factors for V-belts are listed in Table 38.

## Synchronous Belts

**Synchronous Belts ANSI/RMA IP-24.—**Synchronous belts are also known as timing or positive-drive belts. These belts have evenly spaced teeth on their surfaces, which mesh with teeth on pulleys or sprockets to produce a positive, no-slip transmission of power. Such designs should not be confused with molded notched V-belts, which transmit power by means of the wedging action of the V-shape. Synchronous belts are used where driven shaft speeds must be synchronized to the rotation of the driver shaft and to eliminate the noise and maintenance problems of chain drives.

*Standard Timing Belts:* Conventional trapezoidal, or rectangular tooth, timing belts come in six cross sections, which relate to the pitch of the belt. Pitch is the distance from center to center of the teeth. The six basic cross sections or pitches are MXL (mini extra light), XL (extra light), L (light), H (heavy), XH (extra heavy), and XXH (double extra heavy) (Fig. 12). Belts are specified by pitch length, cross section (pitch), and width.

Double-sided timing belts have identical teeth on both sides of the belt and are used where synchronization is required from each belt face. They are available in XL, L, and H cross sections.

*Size Designations:* Synchronous belt sizes are identified by a standard number. The first digits specify the belt length to 0.1 inch (2.54 mm) followed by the belt section (pitch) designation. The digits following the belt section designation represent the nominal belt width times 100. For example, an L section belt 30.000 inches pitch length and 0.75 inch in width would be specified as a 300L075 synchronous belt.

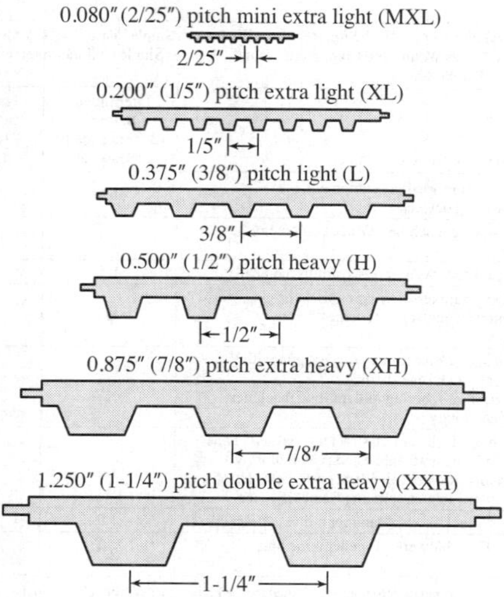


Fig. 12. Standard Synchronous Belt Sections

The RMA nomenclature for double-sided belts is the same as for single-sided belts with the addition of the prefix "D" in front of the belt section. However, some manufacturers use their own designation system for double-sided belts.

*Standard Sections:* Belt sections are specified in terms of pitch. Table 40 gives the Standard Belt Sections and their corresponding pitches.

**Table 39. Service Factors for Synchronous Belt Drives**

| Driving Units | *AC Motors:* Normal Torque, Squirrel Cage, Synchronous and Split Phase. *DC Motors:* Shunt Wound. *Engines:* Multiple Cylinder Internal Combustion. | | |
|---|---|---|---|
| Types of Driven Machines | Intermittent Service (3-5 hours daily or seasonal) | Normal Service (8-10 hours daily) | Continuous Service (16-24 hours daily) |
| Display, Dispensing, Projection, Medical equipment; Instrumentation; Measuring devices | 1.0 | 1.2 | 1.4 |
| Appliances, sweepers, sewing machines; Office equipment; Wood lathes, band saws | 1.2 | 1.4 | 1.6 |
| Conveyors: belt, light package, oven, screens, drums, conical | 1.3 | 1.5 | 1.7 |
| Agitators for liquids; Dough mixers; Drill presses, lathes; Screw machines, jointers; Circular saws, planes; Laundry, paper, printing machinery | 1.4 | 1.6 | 1.8 |
| Agitators for semiliquids; Brick machinery (except pug mills); Conveyor belt: ore, coal, sand; Line shafts; Machine tools: grinder, shaper, boring mill, milling machines; Pumps: centrifugal, gear, rotary | 1.5 | 1.7 | 1.9 |
| Conveyor: apron, pan, bucket, elevator; Extractors, washers; Fans, blowers; centrifugal, induced draft exhausters; Generators & exciters; Hoists, elevators; Rubber calenders, mills, extruders; Saw mill, Textile machinery inc. looms, spinning frames, twisters | 1.6 | 1.8 | 2.0 |
| Centrifuges; Conveyors: flight, screw; Hammer mills; Paper pulpers | 1.7 | 1.9 | 2.1 |
| Brick & clay pug mills; Fans, blowers, propeller mine fans, positive blowers | 1.8 | 2.0 | 2.2 |

| Driving Units | *AC Motors:* High Torque, High Slip, Repulsion-Induction, Single Phase Series Wound and Slip Ring. *DC Motors:* Series Wound and Compound Wound. *Engines:* Single Cylinder Internal Combustion. *Line Shafts. Clutches.* | | |
|---|---|---|---|
| Types of Driven Machines | Intermittent Service (3-5 hours daily or seasonal) | Normal Service (8-10 hours daily) | Continuous Service (16-24 hours daily) |
| Display, dispensing, projection, medical equipment; Instrumentation; Measuring devices | 1.2 | 1.4 | 1.6 |
| Appliances, sweepers, sewing machines; Office equipment; Wood lathes, band saws | 1.4 | 1.6 | 1.8 |
| Conveyors: belt, light package, oven, screens, drums, conical | 1.5 | 1.7 | 1.9 |
| Agitators for liquids; Dough mixers; Drill presses, lathes; Screw machines, jointers; Circular saws, planes; Laundry, paper, printing machinery | 1.6 | 1.8 | 2.0 |
| Agitators for semiliquids; Brick machinery (except pug mills); Conveyor belt: ore, coal, sand; Line shafts; Machine tools: grinder, shaper, boring mill, milling machines; Pumps: centrifugal, gear, rotary | 1.7 | 1.9 | 2.1 |
| Conveyor: apron, pan, bucket, elevator; Extractors, washers; Fans, blowers; centrifugal, induced draft exhausters; Generators & exciters; Hoists, elevators; Rubber calenders, mills, extruders; Saw mill, Textile machinery inc. looms, spinning frames, twisters | 1.8 | 2.0 | 2.2 |
| Centrifuges; Conveyors: flight, screw; Hammer mills; Paper pulpers | 1.9 | 2.1 | 2.3 |
| Brick & clay pug mills; Fans, blowers, propeller mine fans, positive blowers | 2.0 | 2.2 | 2.4 |

Synchronous belts will not slip, and therefore must be belted for the highest loadings anticipated in the system. A minimum service factor of 2.0 is recommended for equipment subject to chocking.

*Pitch Lengths:* Standard belt pitch lengths, belt length designations, and numbers of teeth are shown in Table 42. Belt length tolerances are also given in this table; these tolerances apply to all belt sections and represent the total manufacturing tolerance on belt length.

*Nominal Tooth Dimensions:* Table 40 shows the nominal tooth dimensions for each of the standard belt sections. Tooth dimensions for single- and double-sided belts are identical.

**Table 40. Synchronous Belt Nominal Tooth and Section Dimensions**
*ANSI/RMA IP-24, 1983*

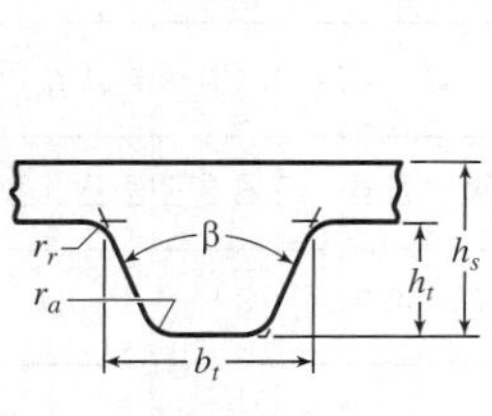


Single-Sided Belts

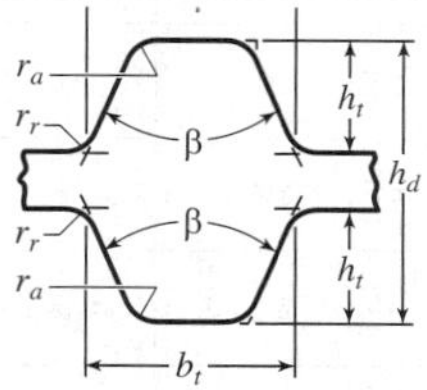


Double-Sided Belts

| Belt Section (Pitch) | Tooth Angle β (deg) | $h_t$ | $b_t$ | $r_a$ | $r_r$ | $h_s$ | $h_d$ | Belt Section (Pitch) | Tooth Angle β (deg) | $h_t$ | $b_t$ | $r_a$ | $r_r$ | $h_s$ | $h_d$ |
|---|---|---|---|---|---|---|---|---|---|---|---|---|---|---|---|
| MXL (0.080) | 40 | 0.020 | 0.045 | 0.005 | 0.005 | 0.045 | … | XXH (1.250) | 40 | 0.375 | 0.750 | 0.060 | 0.090 | 0.62 | … |
| XL (0.200) | 50 | 0.050 | 0.101 | 0.015 | 0.015 | 0.090 | … | DXL (0.200) | 50 | 0.050 | 0.101 | 0.015 | 0.015 | … | 0.120 |
| L (0.375) | 40 | 0.075 | 0.183 | 0.020 | 0.020 | 0.14 | … | DL (0.375) | 40 | 0.075 | 0.183 | 0.020 | 0.020 | … | 0.180 |
| H (0.500) | 40 | 0.090 | 0.241 | 0.040 | 0.040 | 0.16 | … | DH (0.500) | 40 | 0.090 | 0.241 | 0.040 | 0.040 | … | 0.234 |
| XH (0.875) | 40 | 0.250 | 0.495 | 0.047 | 0.062 | 0.44 | … | | | | | | | | |

All dimensions in inches.

**Table 41. Synchronous Belt Standard Pulley and Flange Dimensions**
*ANSI/RMA IP-24, 1983*

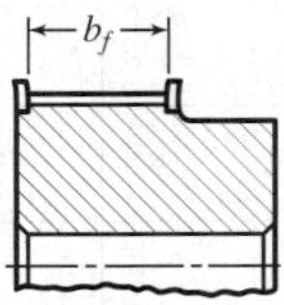


**Flanged Pulley**

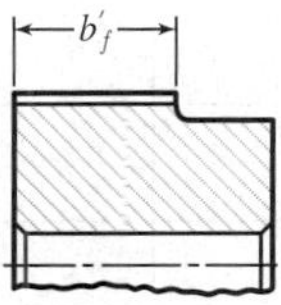


**Unflanged Pulley**

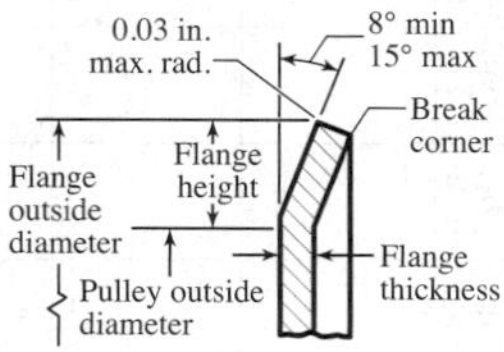


**Flange Dimensions**

| Belt Section | Standard Nominal Pulley Width | Standard Pulley Width Designation | Minimum Pulley Width | | Flange | |
|---|---|---|---|---|---|---|
| | | | Flanged $b_f$ | Unflanged $b'_f$ | Thickness (min) | Height[a] (min) |
| MXL | 0.25 | 025 | 0.28 | 0.35 | 0.023 | 0.020 |
| XL | 0.38 | 037 | 0.41 | 0.48 | 0.029 | 0.040 |
| L | 0.50 | 050 | 0.55 | 0.67 | 0.050 | 0.065 |
| | 0.75 | 075 | 0.80 | 0.92 | | |
| | 1.00 | 100 | 1.05 | 1.17 | | |
| H | 1.00 | 100 | 1.05 | 1.23 | 0.050 | 0.080 |
| | 1.50 | 150 | 1.55 | 1.73 | | |
| | 2.00 | 200 | 2.08 | 2.26 | | |
| | 3.00 | 300 | 3.11 | 3.29 | | |
| XH | 2.00 | 200 | 2.23 | 2.46 | 0.098 | 0.190 |
| | 3.00 | 300 | 3.30 | 3.50 | | |
| | 4.00 | 400 | 4.36 | 4.59 | | |
| XXH | 2.00 | 200 | 2.23 | 2.52 | 0.127 | 0.245 |
| | 3.00 | 300 | 3.30 | 3.59 | | |
| | 4.00 | 400 | 4.36 | 4.65 | | |
| | 5.00 | 500 | 5.42 | 5.72 | | |

[a] Flange outside diameter equals pulley outside diameter plus twice flange height.

**Table 42. Synchronous Belt Standard Pitch Lengths and Tolerances** *ANSI/RMA IP-24, 1983*

| Belt Length Designation | Pitch Length | Permissible Deviation from Standard Length | Number of Teeth for Standard Lengths MXL (0.080) | XL (0.200) | L (0.375) | H (0.500) | XH (0.875) | XXH (1.250) | Belt Length Designation | Pitch Length | Permissible Deviation from Standard Length | Number of Teeth for Standard Lengths MXL (0.080) | XL (0.200) | L (0.375) | H (0.500) | XH (0.875) | XXH (1.250) |
|---|---|---|---|---|---|---|---|---|---|---|---|---|---|---|---|---|---|
| 36 | 3.600 | ±0.016 | 45 | | | | | | 230 | 23.000 | ±0.024 | … | 115 | … | … | | |
| 40 | 4.000 | ±0.016 | 50 | | | | | | 240 | 24.000 | ±0.024 | … | 120 | 64 | 48 | | |
| 44 | 4.400 | ±0.016 | 55 | | | | | | 250 | 25.000 | ±0.024 | … | 125 | … | … | | |
| 48 | 4.800 | ±0.016 | 60 | | | | | | 255 | 25.500 | ±0.024 | … | … | 68 | … | | |
| 56 | 5.600 | ±0.016 | 70 | | | | | | 260 | 26.000 | ±0.024 | … | 130 | … | … | | |
| 60 | 6.000 | ±0.016 | 75 | 30 | | | | | 270 | 27.000 | ±0.024 | … | … | 72 | 54 | | |
| 64 | 6.400 | ±0.016 | 80 | … | | | | | 285 | 28.500 | ±0.024 | … | … | 76 | … | | |
| 70 | 7.000 | ±0.016 | … | 35 | | | | | 300 | 30.000 | ±0.024 | … | … | 80 | 60 | | |
| 72 | 7.200 | ±0.016 | 90 | … | | | | | 322 | 32.250 | ±0.026 | … | … | 86 | … | | |
| 80 | 8.000 | ±0.016 | 100 | 40 | | | | | 330 | 33.000 | ±0.026 | … | … | … | 66 | | |
| 88 | 8.800 | ±0.016 | 110 | … | | | | | 345 | 34.500 | ±0.026 | | | 92 | … | | |
| 90 | 9.000 | ±0.016 | … | 45 | | | | | 360 | 36.000 | ±0.026 | | | … | 72 | | |
| 100 | 10.000 | ±0.016 | 125 | 50 | | | | | 367 | 36.750 | ±0.026 | | | 98 | … | | |
| 110 | 11.000 | ±0.018 | … | 55 | | | | | 390 | 39.000 | ±0.026 | | | 104 | 78 | | |
| 112 | 11.200 | ±0.018 | 140 | … | | | | | 420 | 42.000 | ±0.030 | | | 112 | 84 | | |
| 120 | 12.000 | ±0.018 | … | 60 | … | | | | 450 | 45.000 | ±0.030 | | | 120 | 90 | … | |
| 124 | 12.375 | ±0.018 | … | … | 33 | | | | 480 | 48.000 | ±0.030 | | | 128 | 96 | … | |
| 124 | 12.400 | ±0.018 | 155 | … | … | | | | 507 | 50.750 | ±0.032 | | | … | … | 58 | |
| 130 | 13.000 | ±0.018 | … | 65 | … | | | | 510 | 51.000 | ±0.032 | | | 136 | 102 | … | |
| 140 | 14.000 | ±0.018 | 175 | 70 | … | | | | 540 | 54.000 | ±0.032 | | | 144 | 108 | … | |
| 150 | 15.000 | ±0.018 | … | 75 | 40 | | | | 560 | 56.000 | ±0.032 | | | … | … | 64 | |
| 160 | 16.000 | ±0.020 | 200 | 80 | … | | | | 570 | 57.000 | ±0.032 | | | … | 114 | … | |
| 170 | 17.000 | ±0.020 | … | 85 | … | | | | 600 | 60.000 | ±0.032 | | | 160 | 120 | … | |
| 180 | 18.000 | ±0.020 | 225 | 90 | … | | | | 630 | 63.000 | ±0.034 | | | … | 126 | 72 | |
| 187 | 18.750 | ±0.020 | … | … | 50 | | | | 660 | 66.000 | ±0.034 | | | … | 132 | … | |
| 190 | 19.000 | ±0.020 | … | 95 | … | | | | 700 | 70.000 | ±0.034 | | | | 140 | 80 | 56 |
| 200 | 20.000 | ±0.020 | 250 | 100 | … | | | | 750 | 75.000 | ±0.036 | | | | 150 | … | … |
| 210 | 21.000 | ±0.024 | … | 105 | 56 | | | | 770 | 77.000 | ±0.036 | | | | … | 88 | … |
| 220 | 22.000 | ±0.024 | … | 110 | … | | | | 800 | 80.000 | ±0.036 | | | | 160 | … | 64 |
| 225 | 22.500 | ±0.024 | … | … | 60 | | | | 840 | 84.000 | ±0.038 | | | | … | 96 | … |

All dimensions in inches.

**Table 43. Synchronous Belt Standard Widths and Tolerances**
*ANSI/RMA IP-24, 1983*

| Belt Section | Standard Belt Widths | | Tolerances on Width for Belt Pitch Lengths | | |
|---|---|---|---|---|---|
| | Designation | Dimensions | Up to and including 33 in. | Over 33 in. up to and including 66 in. | Over 66 in |
| MXL (0.080) | 012 | 0.12 | +0.02 −0.03 | … | … |
| | 019 | 0.19 | | | |
| | 025 | 0.25 | | | |
| XL (0.200) | 025 | 0.25 | +0.02 −0.03 | … | … |
| | 037 | 0.38 | | | |
| L (0.375) | 050 | 0.50 | +0.03 −0.03 | +0.03 −0.05 | … |
| | 075 | 0.75 | | | |
| | 100 | 1.00 | | | |
| H (0.500) | 075 | 0.75 | +0.03 −0.03 | +0.03 −0.05 | +0.03 −0.05 |
| | 100 | 1.00 | | | |
| | 150 | 1.50 | | | |
| | 200 | 2.00 | +0.03 −0.05 | +0.05 −0.05 | +0.05 −0.06 |
| | 300 | 3.00 | +0.05 −0.06 | +0.06 −0.06 | +0.06 −0.08 |
| XH (0.875) | 200 | 2.00 | … | +0.19 −0.19 | +0.19 −0.19 |
| | 300 | 3.00 | | | |
| | 400 | 4.00 | | | |
| XXH (1.250) | 200 | 2.00 | … | … | +0.19 −0.19 |
| | 300 | 3.00 | | | |
| | 400 | 4.00 | | | |
| | 500 | 5.00 | | | |

*Widths:* Standard belt widths, width designations, and width tolerances are shown in Table 43.

*Length Determination:* The pitch length of a synchronous belt is determined by placing the belt on a measuring fixture having two pulleys of equal diameter, a method of applying force, and a means of measuring the center distance between the two pulleys. The position of one of the two pulleys is fixed and the other is movable along a graduated scale.

*Synchronous Belt Pulley Diameters:* Table 44 lists the standard pulley diameters by belt section (pitch). Fig. 13 defines the pitch, pitch diameter, outside diameter and pitch line differential.

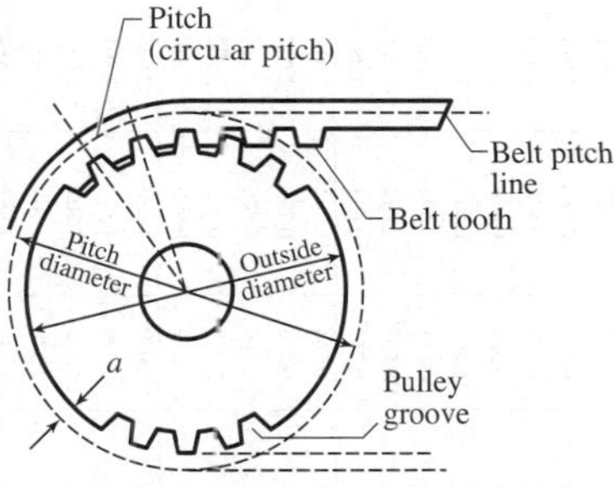


*a* = Pitch line differential

Fig. 13. Synchronous Belt Pulley Dimensions

**Table 44. Synchronous Belt Standard Pulley Diameters** *ANSI/RMA IP-24, 1983*

| Number of Grooves | Belt Section | | | | | | | | | | | |
|---|---|---|---|---|---|---|---|---|---|---|---|---|
| | MXL (0.080) | | XL (0.200) | | L (0.375) | | H (0.500) | | XH (0.875) | | XXH (1.250) | |
| | Diameters | | Diameters | | Diameters | | Diameters | | Diameters | | Diameters | |
| | Pitch | Outside | Pitch | Outside | Pitch | Outside | Pitch | Outside | Pitch | Outside | Pitch | Outside |
| 10 | 0.255 | 0.235 | 0.637 | 0.617 | 1.194[a] | 1.164 | … | … | … | … | … | … |
| 12 | 0.306 | 0.286 | 0.764 | 0.744 | 1.432[a] | 1.402 | … | … | … | … | … | … |
| 14 | 0.357 | 0.337 | 0.891 | 0.871 | 1.671 | 1.641 | 2.228[a] | 2.174 | … | … | … | … |
| 16 | 0.407 | 0.387 | 1.019 | 0.999 | 1.910 | 1.880 | 2.546 | 2.492 | … | … | … | … |
| 18 | 0.458 | 0.438 | 1.146 | 1.126 | 2.149 | 2.119 | 2.865 | 2.811 | 5.013 | 4.903 | 7.162 | 7.042 |
| 20 | 0.509 | 0.489 | 1.273 | 1.253 | 2.387 | 2.357 | 3.183 | 3.129 | 5.570 | 5.460 | 7.958 | 7.838 |
| 22 | 0.560 | 0.540 | 1.401 | 1.381 | 2.626 | 2.596 | 3.501 | 3.447 | 6.127 | 6.017 | 8.754 | 8.634 |
| 24 | 0.611 | 0.591 | 1.528 | 1.508 | 2.865 | 2.835 | 3.820 | 3.766 | 6.685 | 6.575 | 9.549 | 9.429 |
| 26 | 0.662 | 0.642 | … | … | 3.104 | 3.074 | 4.138 | 4.084 | 7.242 | 7.132 | 10.345 | 10.225 |
| 28 | 0.713 | 0.693 | 1.783 | 1.763 | 3.342 | 3.312 | 4.456 | 4.402 | 7.799 | 7.689 | … | … |
| 30 | 0.764 | 0.744 | 1.910 | 1.890 | 3.581 | 3.551 | 4.775 | 4.721 | 8.356 | 8.246 | 11.937 | 11.817 |
| 32 | 0.815 | 0.795 | 2.037 | 2.017 | 3.820 | 3.790 | 5.093 | 5.039 | 8.913 | 8.803 | … | … |
| 34 | 0.866 | 0.846 | … | … | … | … | … | … | … | … | 13.528 | 13.408 |
| 36 | 0.917 | 0.897 | 2.292 | 2.272 | 4.297 | 4.267 | 5.730 | 5.676 | … | … | … | … |
| 40 | 1.019 | 0.999 | 2.546 | 2.526 | 4.775 | 4.745 | 6.366 | 6.312 | 11.141 | 11.031 | 15.915 | 15.795 |
| 42 | 1.070 | 1.050 | 2.674 | 2.654 | … | … | … | … | … | … | … | … |
| 44 | 1.120 | 1.100 | 2.801 | 2.781 | 5.252 | 5.222 | 7.003 | 6.949 | … | … | … | … |
| 48 | 1.222 | 1.202 | 3.056 | 3.036 | 5.730 | 5.700 | 7.639 | 7.585 | 13.369 | 13.259 | 19.099 | 18.979 |
| 60 | 1.528 | 1.508 | 3.820 | 3.800 | 7.162 | 7.132 | 9.549 | 9.495 | 16.711 | 16.601 | 23.873 | 23.753 |
| 72 | 1.833 | 1.813 | 4.584 | 4.564 | 8.594 | 8.564 | 11.459 | 11.405 | 20.054 | 19.944 | 28.648 | 28.528 |
| 84 | … | … | … | … | 10.027 | 9.997 | 13.369 | 13.315 | 23.396 | 23.286 | … | … |
| 90 | … | … | … | … | … | … | … | … | … | … | 35.810 | 35.690 |
| 96 | … | … | … | … | … | … | 15.279 | 15.225 | 26.738 | 26.628 | … | … |
| 120 | … | … | … | … | … | … | 19.099 | 19.045 | 33.423 | 33.313 | … | … |

All dimensions in inches.

* Usually not available in all widths — consult supplier.

*Widths:* Standard pulley widths for each belt section are shown in Table 41. The nominal pulley width is specified in terms of the maximum standard belt width the pulley will accommodate. The minimum pulley width, whether flanged or unflanged, is also shown in Table 41, along with flange dimensions and various pulley tolerances.

*Pulley Size Designation:* Synchronous belt pulleys are designated by the number of grooves, the belt section, and a number representing 100 times the nominal width. For example, a 30 groove L section pulley with a nominal width of 0.75 in. would be designated by 30L075. Pulley tolerances are shown in Table 45.

### Table 45. Pulley Tolerances (All Sections)

| Outside Diameter Range | Outside Diameter Tolerance | Pitch to Pitch Tolerance | |
|---|---|---|---|
| | | Adjacent Grooves | Accumulative Over 90 Degrees |
| Up thru 1.000 | +0.002 −0.000 | ±0.001 | ±0.003 |
| Over 1.000 to and including 2.000 | +0.003 −0.000 | ±0.001 | ±0.004 |
| Over 2.000 to and including 4.000 | +0.004 −0.000 | ±0.001 | ±0.005 |
| Over 4.000 to and including 7.000 | +0.005 −0.000 | ±0.001 | ±0.005 |
| Over 7.000 to and including 12.000 | +0.006 −0.000 | ±0.001 | ±0.006 |
| Over 12.000 to and including 20.000 | +0.007 −0.000 | ±0.001 | ±0.007 |
| Over 20.000 | +0.008 −0.000 | ±0.001 | ±0.008 |

| Radial Runout[a] | Axial Runout[a] |
|---|---|
| For outside diameters 8.0 in. and under 0.005 in. | For outside diameters 1.0 in. and under 0.001 in. |
| For each additional inch of outside diameter add 0.0005 in. | For each additional inch of outside diameter up through 10.0 in., add 0.001 in. |
| | For each additional inch of outside diameter over 10.0 in., add 0.0005 in. |

[a] Total indicator reading.

All dimensions in inches.

*Cross Section Selection:* The chart (Fig. 14) may be used as a guide to the selection of a synchronous belt for any combination of design horsepower and speed of the faster shaft. When the intersection of the design horsepower and speed of the faster shaft falls near a line between two areas on the chart, the possibilities in both areas should be explored. Special circumstances (such as space limitations) may result in selection of a belt cross section different from that indicated in the chart. Belt manufacturers should be contacted for specific data.

*Torque Ratings:* It is customary to use torque load requirements rather than horsepower load when designing drives using the small pitch MXL section belts. These belts operate on small diameters resulting in relatively low belt speeds, so torque is essentially constant for all rpm. The torque rating formulas for MXL sections are:

$$Q_r = d[1.13 - 1.38\times10^{-3}d^2] \text{ for belt width } = 0.12 \text{ in.}$$

$$Q_r = d[1.88 - 2.30\times10^{-3}d^2] \text{ for belt width } = 0.19 \text{ in.}$$

$$Q_r = d[2.63 - 3.21\times10^{-3}d^2] \text{ for belt width } = 0.25 \text{ in.}$$

where $Q_r$ = the maximum torque rating (lbf-in.) for a belt of specified width having six or more teeth in mesh and a pulley surface speed of 6500 fpm or less. Torque ratings for drives with less than six teeth in mesh must be corrected as shown in Table 46. $d$ = pitch diameter of smaller pulley, inch.

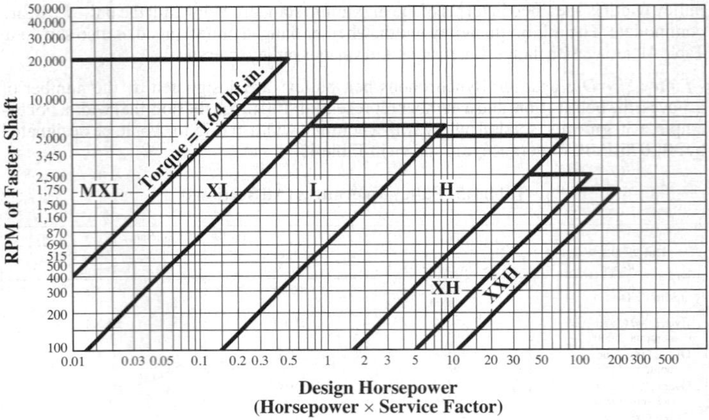


Fig. 14. Selection of Synchronous Belt Cross Section

## Table 46. Teeth in Mesh Factor

| Teeth in Mesh | Factor $K_z$ | Teeth in Mesh | Factor $K_z$ |
|---|---|---|---|
| 6 or more | 1.00 | 3 | 0.40 |
| 5 | 0.80 | 2 | 0.20 |
| 4 | 0.60 | | |

*Horsepower Rating Formulas:* The horsepower rating formulas for synchronous belts, other than the MLX section, are determined from the following formulas, where the number in parentheses is the belt width in inches.

$$\mathrm{XL}(0.38)\ \mathrm{HP} = dr[0.0916 - 7.07\times10^{-5}(dr)^2]$$

$$\mathrm{L}(1.00)\ \mathrm{HP} = dr[0.436 - 3.01\times10^{-4}(dr)^2]$$

$$\mathrm{H}(3.00)\ \mathrm{HP} = dr[3.73 - 1.41\times10^{-3}(dr)^2]$$

$$\mathrm{XH}(4.00)\ \mathrm{HP} = dr[7.21 - 4.68\times10^{-3}(dr)^2]$$

$$\mathrm{XXH}(5.00)\ \mathrm{HP} = dr[11.4 - 7.81\times10^{-3}(dr)^2]$$

where $HP$ = the maximum horsepower rating recommended for the specified standard belt width having six or more teeth in mesh and a pulley surface speed of 6500 fpm or less. Horsepower ratings for drives with less than six teeth in mesh must be corrected as shown in Table 46. $d$ = pitch diameter of smaller pulley, inch; $r$ = rpm of faster shaft divided by 1000. Total horsepower ratings are the same for double-sided as for single-sided belts. Contact manufacturers for percentage of horsepower available for each side of the belt.

*Finding the Required Belt Width:* The belt width should not exceed the small pulley diameter or excessive side thrust will result.

*Torque Rating Method (MXL Section):* Divide the design torque by the teeth in mesh factor to obtain the corrected design torque. Compare the corrected design torque with the torque rating given in Table 47 for the pulley diameter being considered. Select the narrowest belt width that has a torque rating equal to or greater than the corrected design torque.

**Table 47. Torque Rating for MXL Section (0.080 Inch Pitch)**

| Belt Width, (in.) | Rated Torque (lbf-in.) for Small Pulley (Number of Grooves and Pitch Diameter, in.) | | | | | | | | | |
|---|---|---|---|---|---|---|---|---|---|---|
| | 10MXL 0.255 | 12MXL 0.306 | 14MXL 0.357 | 16MXL 0.407 | 18MXL 0.458 | 20MXL 0.509 | 22MXL 0.560 | 24MXL 0.611 | 28MXL 0.713 | 30MXL 0.764 |
| 0.12 | 0.29 | 0.35 | 0.40 | 0.46 | 0.52 | 0.57 | 0.63 | 0.69 | 0.81 | 0.86 |
| 0.19 | 0.48 | 0.58 | 0.67 | 0.77 | 0.86 | 0.96 | 1.05 | 1.15 | 1.34 | 1.44 |
| 0.25 | 0.67 | 0.80 | 0.94 | 1.07 | 1.20 | 1.34 | 1.47 | 1.61 | 1.87 | 2.01 |

*Horsepower Rating Method (XL, L, H, XH, and XXH Sections):* Multiply the horsepower rating for the widest standard belt of the selected section by the teeth in mesh factor to obtain the corrected horsepower rating. Divide the design horsepower by the corrected horsepower rating to obtain the required belt width factor. Compare the required belt width factor with those shown in Table 48. Select the narrowest belt width that has a width factor equal to or greater than the required belt width factor.

**Table 48. Belt Width Factor**

| Belt Section | Belt Width (in.) | | | | | | | | | | | |
|---|---|---|---|---|---|---|---|---|---|---|---|---|
| | 0.12 | 0.19 | 0.25 | 0.38 | 0.50 | 0.75 | 1.00 | 1.50 | 2.00 | 3.00 | 4.00 | 5.00 |
| MXL (0.080) | 0.43 | 0.73 | 1.00 | … | … | … | … | … | … | … | … | … |
| XL (0.200) | … | … | 0.62 | 1.00 | … | … | … | … | … | … | … | … |
| L (0.375) | … | … | … | … | 0.45 | 0.72 | 1.00 | … | … | … | … | … |
| H (0.500) | … | … | … | … | … | 0.21 | 0.29 | 0.45 | 0.63 | 1.00 | … | … |
| XH (0.875) | … | … | … | … | … | … | … | … | 0.45 | 0.72 | 1.00 | … |
| XXH (1.250) | … | … | … | … | … | … | … | … | 0.35 | 0.56 | 0.78 | 1.00 |

*Drive Selection:* Information on design and selection of synchronous belt drives is available in engineering manuals published by belt manufacturers. Manufacturers should be consulted on such matters as preferred stock sizes, desirable speeds, center distances, etc.

*Minimum Pulley Size:* The recommended minimum pulley size depends on the rpm of the faster shaft. Minimum sheave diameters for each belt cross section are listed in Table 44.

*Selection of Flanged Pulleys:* To determine when to use flanged pulleys, consider the following conditions:

1) On all two-pulley drives, the minimum flanging requirements are two flanges on one pulley, or one flange on each pulley on opposite sides.

2) On drives where the center distance is more than eight times the diameter of the small pulley, both pulleys should be flanged on both sides.

3) On vertical shaft drives, one pulley should be flanged on both sides and other pulleys in the system should be flanged on the bottom side only.

4) On drives with more than two pulleys, the minimum flanging requirements are two flanges on every other pulley, or one flange on every pulley, alternating sides around the system.

*Service Factors:* Service factors for synchronous belts in Table 39.

**Gravity Idler.**—The gravity type of idler used in conjunction with many belt drives, consists of a weighted idler pulley, so pivoted from a long arm that the idler runs near the small pulley against the loose side of the belt, in such a way as to increase the arc of contact of the belt on this pulley. These idlers are of special value on difficult drives where there is a large difference in the diameter of the driver and driven pulleys.

# TRANSMISSION CHAINS

## Types of Chains

In addition to the standard roller and inverted tooth types, a wide variety of drive chains of different construction is available. Such chains are manufactured to various degrees of precision ranging from unfinished castings or forgings to chains having certain machined parts. Practically all of these chains as well as standard roller chains can be equipped with attachments to fit them for conveyor use. A few such types are briefly described in the following paragraphs. Detailed information about them can be obtained from the manufacturers.

**Types of Chains.—***Detachable Chains:* The links of this type of chain, which are identical, are easily detachable. Each has a hook-shaped end in which the bar of the adjacent link articulates. These chains are available in malleable iron or pressed steel. The chief advantage is the ease with which any link can be removed.

*Cast Roller Chains:* Cast roller chains are constructed, wholly or partly, of cast metal parts and are available in various styles. In general the rollers and side bars are accurately made castings without machine finish. The links are usually connected by means of forged pins secured by nuts or cotters. Such chains are used for slow speeds and moderate loads, or where the precision of standard roller chains is not required.

*Pintle Chains:* Unlike the roller chain, the pintle chain is composed of hollow-cored cylinders cast or forged integrally with two offset side bars and each link identical. The links are joined by pins inserted in holes in the ends of the side bars and through the cored holes in the adjacent links. Lugs prevent turning of the pins in the side bars ensuring articulation of the chain between the pin and the cored cylinder, allowing for greater force and power transmission.

## Standard Roller Chains

A roller chain is made up of two kinds of links: roller links and pin links alternately spaced throughout the length of the chain as shown in Table 1.

Roller chains are manufactured in several types, each designed for the particular service required. All roller chains are so constructed that the rollers are evenly spaced throughout the chain. The outstanding advantage of this type of chain is the ability of the rollers to rotate when contacting the teeth of the sprocket. Two arrangements of roller chains are in common use: the single-strand type and the multiple-strand type. In the latter type, two or more chains are joined side by side by means of common pins which maintain the alignment of the rollers in the different strands.

**Types of Roller Chains.—***Standard roller chains* are manufactured to the specifications in the American National Standard for precision power transmission roller chains, attachments, and sprockets ANSI/ASME B29.1-2011 (R2022) and, where indicated, the data in the subsequent tables have been taken from this Standard. These roller chains and sprockets are commonly used for the transmission of power in industrial machinery, machine tools, motor trucks, motorcycles, tractors, and similar applications. In tabulating the dimensional information in ANSI/ASME B29.1-2011 (R2022), customary inch-pound units were used. Metric (SI) units are given in separate tabulations in the ANSI/ASME Standard.

*Nonstandard roller chains,* developed individually by various manufacturers prior to the adoption of the ANSI/ASME Standard, are similar in form and construction to standard roller chains but do not conform dimensionally to standard chains. Some sizes are maintained from the originating manufacturers for replacement on existing equipment. They are not recommended for new installations, since their manufacture may be discontinued wihout ongoing support.

**Table 1. Nomenclature for Roller Chain Parts**
*ANSI/ASME B29.1-2011 (R2022)*

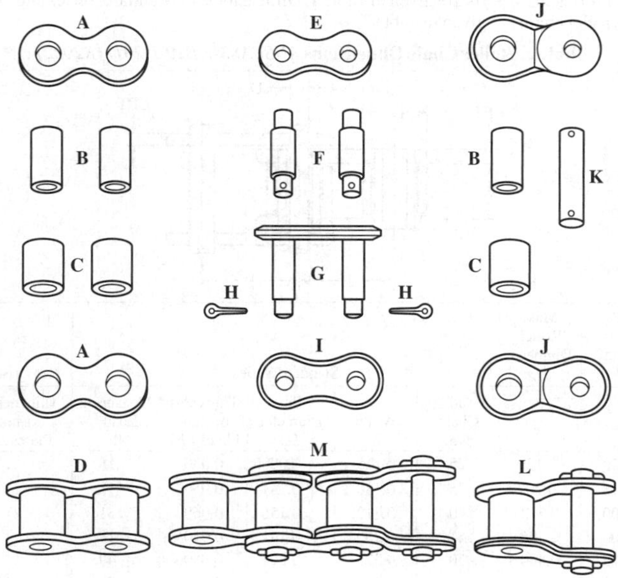


*Roller Link D.* — An inside link consisting of two inside plates, two bushings, and two rollers.

*Pin Link G and E.* — An outside link consisting of two pin-link plates assembled with two pins.

*Inside Plate A.* — One of the plates forming the tension members of a roller link.

*Pin Link Plate E.* — One of the plates forming the tension members of a pin link.

*Pin F.* — A stud articulating within a bushing of an inside link and secured at its ends by the pin-link plates.

*Bushing B.* — A cylindrical bearing in which the pin turns.

*Roller C.* — A ring or thimble which turns over a bushing.

*Assembled Pins G.* — Two pins assembled with one pin-link plate.

*Connecting-Link G and I.* — A pin link having one side plate detachable.

*Connecting-Link Plate I.* — The detachable pin-link plate belonging to a connecting link. It is retained by cotter pins or by a one-piece spring clip (not shown).

*Connecting Link Assembly M.* — A unit designed to connect two roller links.

*Offset Link L.* — A link consisting of two offset plates assembled with a bushing and roller at one end and an offset link pin at the other.

*Offset Plate J.* — One of the plates forming the tension members of the offset link.

*Offset Link Pin K.* — A pin used in offset links.

*Standard double-pitch roller chains* are like standard roller chains, except that their link plates have twice the pitch of the corresponding standard-pitch chain. Their design conforms to specifications in the ANSI/ASME Standard for double-pitch power transmission roller chains and sprockets ANSI/ASME B29.100-2011 (R2021). They are especially useful for low speeds, moderate loads, or long center distances.

## Transmission Roller Chains

**Standard Roller Chain Nomenclature, Dimensions and Loads.—**Standard nomenclature for roller chain parts are given in Table 1. Dimensions for Standard Series and Heavy Series roller chain are given in Table 2.

**Table 2. Roller Chain Dimensions** *ANSI/ASME B29.1-2011 (R2022)*

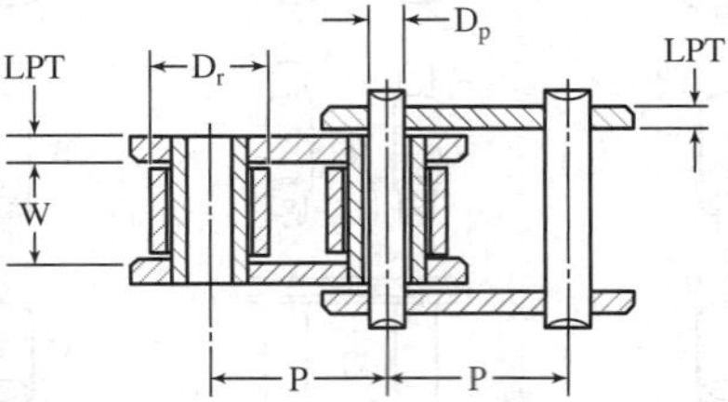


| Pitch *P* | Max. Roller Diameter $D_r$ | Standard Series | | | | | Heavy Series |
|---|---|---|---|---|---|---|---|
| | | Standard Chain No. | Width *W* | Pin Diameter $D_p$ | Thickness of Link Plates *LPT* | Measuring Load,[a] lb | Thickness of Link Plates *LPT* |
| 0.250 | [b]0.130 | 25 | 0.125 | 0.0905 | 0.030 | 18 | … |
| 0.375 | [b]0.200 | 35 | 0.188 | 0.141 | 0.050 | 18 | … |
| 0.500 | 0.312 | 40 | 0.312 | 0.156 | 0.060 | 31 | … |
| 0.500 | 0.306 | 41 | 0.250 | 0.141 | 0.050 | 18 | … |
| 0.625 | 0.400 | 50 | 0.375 | 0.200 | 0.080 | 49 | … |
| 0.750 | 0.469 | 60 | 0.500 | 0.234 | 0.094 | 70 | 0.125 |
| 1.000 | 0.625 | 80 | 0.625 | 0.312 | 0.125 | 125 | 0.156 |
| 1.250 | 0.750 | 100 | 0.750 | 0.375 | 0.158 | 195 | 0.187 |
| 1.500 | 0.875 | 120 | 1.000 | 0.437 | 0.187 | 281 | 0.219 |
| 1.750 | 1.000 | 140 | 1.000 | 0.500 | 0.219 | 383 | 0.250 |
| 2.000 | 1.125 | 160 | 1.250 | 0.562 | 0.250 | 500 | 0.281 |
| 2.250 | 1.406 | 180 | 1.406 | 0.687 | 0.281 | 633 | 0.312 |
| 2.500 | 1.562 | 200 | 1.500 | 0.781 | 0.312 | 781 | 0.375 |
| 3.000 | 1.875 | 240 | 1.875 | 0.937 | 0.375 | 1000 | 0.500 |

[a] For single-strand chain.

[b] Bushing diameter. This size chain has no rollers.

All dimensions are in inches.

*Roller Diameters* $D_r$ are approximately $^5/_8$ *P*.

*The width W* is defined as the distance between the link plates. It is approximately $^5/_8$ of the chain pitch.

*Pin Diameters Dp* are approximately $^5/_{16}$ *P* or $^1/_2$ of the roller diameter.

*Thickness LPT of Inside and Outside Link Plates* for the Standard Series is approximately $^1/_8$ *P*.

*Thickness of Link Plates* for the Heavy Series of any pitch is approximately that of the next larger pitch Standard Series chain.

*Maximum Height of Roller Link Plates* = 0.95 *P*.

*Maximum Height of Pin Link Plates* = 0.82 *P*.

*Maximum Pin Diameter* = nominal pin diameter + 0.0005 inch.

*Minimum Hole in Bushing* = nominal pin diameter + 0.0015 inch.

*Maximum Width of Roller Link* = nominal width of chain + (2.12 × nominal link plate thickness.)

*Minimum Distance between Pin Link Plates* = maximum width of roller link + 0.002 inch.

*Chain Pitch:* Distance in inches between centers of adjacent joint members. Other dimensions are proportional to the pitch.

*Tolerances for Chain Length:* New chains, under standard measuring load, must not be underlength. Overlength tolerance is 0.001/(pitch in inches)$^2$ + 0.015 inch per foot. Length measurements are to be taken over a length of at least 12 inches.

*Measuring Load:* The load in pounds under which a chain should be measured for length. It is equal to 1 percent of the ultimate tensile strength, with a minimum of 18 pounds and a maximum of 1000 pounds for both single and multiple-strand chain.

*Minimum Ultimate Tensile Strength:* For single-strand chain, equal to or greater than 12,500 × (pitch in inches)$^2$ pounds. The minimum tensile strength or breaking strength of a multiple-strand chain is equal to that of a single-strand chain multiplied by the number of strands. Minimum ultimate tensile strength is indicative only of the tensile strength quality of the chain, not the maximum load that can be applied.

**Standard Roller Chain Numbers.—**The right-hand figure in the chain number is zero for roller chains of the usual proportions, 1 for a lightweight chain, and 5 for a rollerless bushing chain. The numbers to the left of the right-hand figure denote the number of $\frac{1}{8}$ inches in the pitch. The letter *H* following the chain number denotes the heavy series; thus the number 80*H* denotes a 1-inch pitch heavy chain. The hyphenated number 2 suffixed to the chain number denotes a double strand, 3 a triple strand, 4 a quadruple strand chain and so on.

*Heavy Series:* These chains, made in $\frac{3}{4}$-inch and larger pitches, have thicker link plates than those of the regular standard. Their value is only in the acceptance of higher loads at lower speeds.

*Light-weight Machinery Chain:* This chain is designated as No. 41. It is $\frac{1}{2}$ inch pitch $\frac{1}{4}$ inch wide; has 0.306-inch diameter rollers and a 0.141-inch pin diameter. The minimum ultimate tensile strength is 1500 pounds.

*Multiple-strand Chain:* This is essentially an assembly of two or more single-strand chains placed side by side with pins that extend through the entire width to maintain alignment of the different strands.

**Types of Sprockets.—**Four different designs or types of roller-chain sprockets are shown by the sectional views, Fig. 1. Type *A* is a plain plate; type *B* has a hub on one side only; type *C*, a hub on both sides; and type *D*, a detachable hub. Also used are shear pin and slip clutch sprockets designed to prevent damage to the drive or to other equipment caused by overloads or stalling.

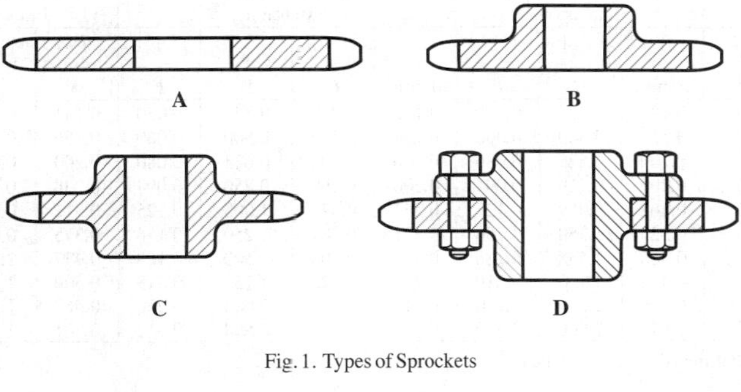


Fig. 1. Types of Sprockets

**Attachments.—**Modifications to standard chain components to adapt the chain for use in conveying, elevating, and timing operations are known as "attachments." The components commonly modified are: 1) the link plates, which are provided with extended lugs which may be straight or bent ; and 2) the chain pins, which are extended in length so as to project substantially beyond the outer surface of the pin link plates.

Hole diameters, thicknesses, hole locations and offset dimensions for straight link and bent link plate extensions and lengths and diameters of extended pins are given in Table 3.

**Table 3. Straight and Bent Link Plate Extensions and Extended Pin Dimensions**
*ANSI/ASME B29.1-2011 (R2022)*

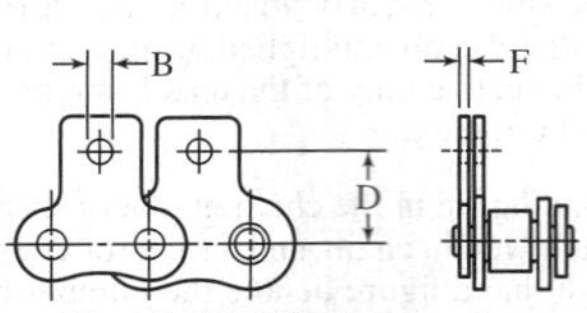


**Straight Link Plate Extension, One Side of Chain**

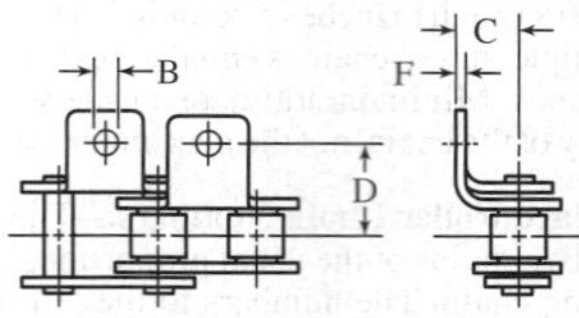


**Bent Link Plate Extension, One Side of Chain**

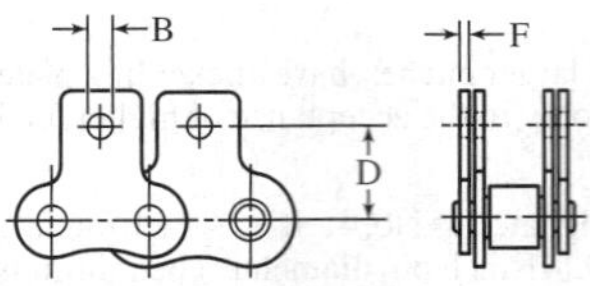


**Straight Link Plate Extension, Both Sides of Chain**

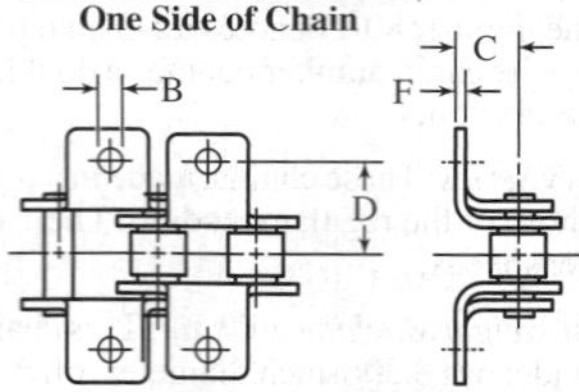


**Bent Link Plate Extension, Both Sides of Chain**

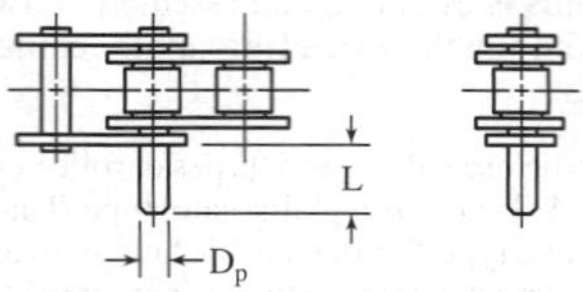


**Extended Pin, One Side of Chain**

| Chain No. | Straight Link Plate Extension | | | Bent Link Plate Extension | | | | Extended Pin | |
|---|---|---|---|---|---|---|---|---|---|
| | *B* min. | *D* | *F* | *B* min. | *C* | *D* | *F* | $D_p$ Nominal | *L* |
| 35 | 0.102 | 0.375 | 0.050 | 0.102 | 0.250 | 0.375 | 0.050 | 0.141 | 0.375 |
| 40 | 0.131 | 0.500 | 0.060 | 0.131 | 0.312 | 0.500 | 0.060 | 0.156 | 0.375 |
| 50 | 0.200 | 0.625 | 0.080 | 0.200 | 0.406 | 0.625 | 0.080 | 0.200 | 0.469 |
| 60 | 0.200 | 0.719 | 0.094 | 0.200 | 0.469 | 0.750 | 0.094 | 0.234 | 0.562 |
| 80 | 0.261 | 0.969 | 0.125 | 0.261 | 0.625 | 1.000 | 0.125 | 0.312 | 0.750 |
| 100 | 0.323 | 1.250 | 0.156 | 0.323 | 0.781 | 1.250 | 0.156 | 0.375 | 0.938 |
| 120 | 0.386 | 1.438 | 0.188 | 0.386 | 0.906 | 1.500 | 0.188 | 0.437 | 1.125 |
| 140 | 0.448 | 1.750 | 0.219 | 0.448 | 1.125 | 1.750 | 0.219 | 0.500 | 1.312 |
| 160 | 0.516 | 2.000 | 0.250 | 0.516 | 1.250 | 2.000 | 0.250 | 0.562 | 1.500 |
| 200 | 0.641 | 2.500 | 0.312 | 0.641 | 1.688 | 2.500 | 0.312 | 0.781 | 1.875 |

All dimensions are in inches.

**Sprocket Classes.—**The American National Standard ANSI/ASME B29.1-2011 (R2022) provides for two classes of sprockets designated as Commercial and Precision. The selection of either is a matter of drive application judgment. The usual moderate- to slow-speed commercial drive is adequately served by Commercial sprockets. Where extreme high speed in combination with high load is involved, or where the drive involves fixed centers, critical timing, or register problems, or close clearance with outside interference, then the use of Precision sprockets may be more appropriate.

As a general guide, drives requiring Type A or Type B lubrication (see page 2644) would be served by Commercial sprockets. Drives requiring Type C lubrication may require Precision sprockets; the manufacturer should be consulted.

**Keys, Keyways, and Set Screws.—**To secure sprockets to the shaft, both keys and set screws should be used. The key is used to prevent rotation of the sprocket on the shaft. Keys should be fitted carefully in the shaft and sprocket keyways to eliminate all backlash, especially on the fluctuating loads. A set screw should be located over a flat key to secure it against longitudinal displacement.

Where a set screw is to be used with a parallel key, the following sizes are recommended by the American Chain Association. For a sprocket bore and shaft diameter in the range of:

$\frac{1}{2}$ through $\frac{7}{8}$ inch, a $\frac{1}{4}$-inch set screw

$\frac{15}{16}$ through $1\frac{3}{4}$ inches, a $\frac{3}{8}$-inch set screw

$1\frac{13}{16}$ through $2\frac{1}{4}$ inches, a $\frac{1}{2}$-inch set screw

$2\frac{5}{16}$ through $3\frac{1}{4}$ inches, a $\frac{5}{8}$-inch set screw

$3\frac{3}{8}$ through $4\frac{1}{2}$ inches, a $\frac{3}{4}$-inch set screw

$4\frac{3}{4}$ through $5\frac{1}{2}$ inches, a $\frac{7}{8}$-inch set screw

$5\frac{3}{4}$ through $7\frac{3}{8}$ inches, a 1-inch set screw

$7\frac{1}{2}$ through $12\frac{1}{2}$ inches, a $1\frac{1}{4}$-inch set screw

**Sprocket Diameters.—**The various diameters of roller chain sprockets are shown in Fig. 2. These are defined as follows.

*Pitch Diameter:* The pitch diameter is the diameter of the pitch circle that passes through the centers of the link pins as the chain is wrapped on the sprocket.

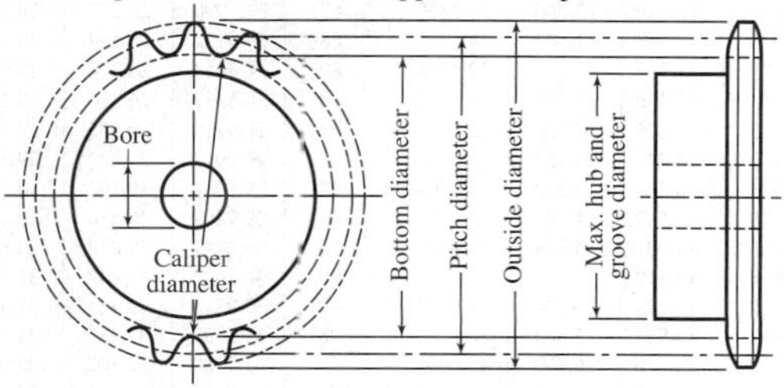


Fig. 2. Sprocket Diameters

Because the chain pitch is measured on a straight line between the centers of adjacent pins, the chain pitch lines form a series of chords of the sprocket pitch circle. Sprocket pitch diameters for 1-inch pitch and for 9 to 108 teeth are given in Table 4. For lower (5 to 8) or higher (109 to 200) numbers of teeth use the following formula in which $P$ = pitch, $N$ = number of teeth: Pitch Diameter = $P \div \sin(180° \div N)$.

*Bottom Diameter:* The bottom diameter is the diameter of a circle tangent to the curve (called the seating curve) at the bottom of the tooth gap. It equals the pitch diameter minus the diameter of the roller.

**Table 4. Roller Chain Sprocket Diameters** *ANSI/ASME B29.1-2011 (R2022)*

These diameters and caliper factors apply only to chain of 1-inch pitch. For any other pitch, multiply the values given below by the pitch.

Caliper Diameter (even teeth) = Pitch Diameter − Roller Diameter

Caliper Diameter (odd teeth) = Caliper factor × Pitch − Roller Diameter

See Table 5 for tolerances on Caliper Diameters.

| No. Teeth[a] | Pitch Diameter | Outside Diameter | | Caliper Factor | No. Teeth[a] | Pitch Diameter | Outside Diameter | | Caliper Factor |
|---|---|---|---|---|---|---|---|---|---|
| | | Turned | Topping Hob Cut | | | | Turned | Topping Hob Cut | |
| 9 | 2.9238 | 3.348 | 3.364 | 2.8794 | 59 | 18.7892 | 19.363 | 19.361 | 18.7825 |
| 10 | 3.2361 | 3.678 | 3.676 | | 60 | 19.1073 | 19.681 | 19.680 | |
| 11 | 3.5495 | 4.006 | 3.990 | 3.5133 | 61 | 19.4255 | 20.000 | 19.998 | 19.4190 |
| 12 | 3.8637 | 4.332 | 4.352 | | 62 | 19.7437 | 20.318 | 20.316 | |
| 13 | 4.1786 | 4.657 | 4.666 | 4.1481 | 63 | 20.0618 | 20.637 | 20.634 | 20.0556 |
| 14 | 4.4940 | 4.981 | 4.982 | | 64 | 20.3800 | 20.956 | 20.952 | |
| 15 | 4.8097 | 5.304 | 5.298 | 4.7834 | 65 | 20.6982 | 21.274 | 21.270 | 20.6921 |
| 16 | 5.1258 | 5.627 | 5.614 | | 66 | 21.0164 | 21.593 | 21.588 | |
| 17 | 5.4422 | 5.949 | 5.930 | 5.4190 | 67 | 21.3346 | 21.911 | 21.907 | 21.3287 |
| 18 | 5.7588 | 6.271 | 6.292 | | 68 | 21.6528 | 22.230 | 22.225 | |
| 19 | 6.0755 | 6.593 | 6.609 | 6.0548 | 69 | 21.9710 | 22.548 | 22.543 | 21.9653 |
| 20 | 6.3924 | 6.914 | 6.926 | | 70 | 22.2892 | 22.867 | 22.861 | |
| 21 | 6.7095 | 7.235 | 7.243 | 6.6907 | 71 | 22.6074 | 23.185 | 23.179 | 22.6018 |
| 22 | 7.0267 | 7.555 | 7.560 | | 72 | 22.9256 | 23.504 | 23.498 | |
| 23 | 7.3439 | 7.876 | 7.877 | 7.3268 | 73 | 23.2438 | 23.822 | 23.816 | 23.2384 |
| 24 | 7.6613 | 8.196 | 8.195 | | 74 | 23.5620 | 24.141 | 24.134 | |
| 25 | 7.9787 | 8.516 | 8.512 | 7.9630 | 75 | 23.8802 | 24.459 | 24.452 | 23.8750 |
| 26 | 8.2962 | 8.836 | 8.829 | | 76 | 24.1984 | 24.778 | 24.770 | |
| 27 | 8.6138 | 9.156 | 9.147 | 8.5992 | 77 | 24.5166 | 25.096 | 25.089 | 24.5116 |
| 28 | 8.9314 | 9.475 | 9.465 | | 78 | 24.8349 | 25.415 | 25.407 | |
| 29 | 9.2491 | 9.795 | 9.782 | 9.2355 | 79 | 25.1531 | 25.733 | 25.725 | 25.1481 |
| 30 | 9.5668 | 10.114 | 10.100 | | 80 | 25.4713 | 26.052 | 26.043 | |
| 31 | 9.8845 | 10.434 | 10.418 | 9.8718 | 81 | 25.7896 | 26.370 | 26.362 | 25.7847 |
| 32 | 10.2023 | 10.753 | 10.736 | | 82 | 26.1078 | 26.689 | 26.680 | |
| 33 | 10.5201 | 11.073 | 11.053 | 10.5082 | 83 | 26.4260 | 27.007 | 26.998 | 26.4213 |
| 34 | 10.8379 | 11.392 | 11.371 | | 84 | 26.7443 | 27.326 | 27.316 | |
| 35 | 11.1558 | 11.711 | 11.728 | 11.1446 | 85 | 27.0625 | 27.644 | 27.635 | 27.0579 |
| 36 | 11.4737 | 12.030 | 12.046 | | 86 | 27.3807 | 27.962 | 27.953 | |
| 37 | 11.7916 | 12.349 | 12.364 | 11.7810 | 87 | 27.6990 | 28.281 | 28.271 | 27.6945 |
| 38 | 12.1095 | 12.668 | 12.682 | | 88 | 28.0172 | 28.599 | 28.589 | |
| 39 | 12.4275 | 12.987 | 13.000 | 12.4174 | 89 | 28.3354 | 28.918 | 28.907 | 28.3310 |
| 40 | 12.7455 | 13.306 | 13.318 | | 90 | 28.6537 | 29.236 | 29.226 | |
| 41 | 13.0635 | 13.625 | 13.636 | 13.0539 | 91 | 28.9719 | 29.555 | 29.544 | 28.9676 |
| 42 | 13.3815 | 13.944 | 13.954 | | 92 | 29.2902 | 29.873 | 29.862 | |
| 43 | 13.6995 | 14.263 | 14.272 | 13.6904 | 93 | 29.6081 | 30.192 | 30.180 | 29.6042 |
| 44 | 14.0175 | 14.582 | 14.590 | | 94 | 29.9267 | 30.510 | 30.499 | |
| 45 | 14.3355 | 14.901 | 14.908 | 14.3269 | 95 | 30.2449 | 30.828 | 30.817 | 30.2408 |
| 46 | 14.6536 | 15.219 | 15.226 | | 96 | 30.5632 | 31.147 | 31.135 | |
| 47 | 14.9717 | 15.538 | 15.544 | 14.9634 | 97 | 30.8815 | 31.465 | 31.454 | 30.8774 |
| 48 | 15.2898 | 15.857 | 15.862 | | 98 | 31.1997 | 31.784 | 31.772 | |
| 49 | 15.6079 | 16.176 | 16.180 | 15.5999 | 99 | 31.5180 | 32.102 | 32.090 | 31.5140 |
| 50 | 15.9260 | 16.495 | 16.498 | | 100 | 31.8362 | 32.421 | 32.408 | |
| 51 | 16.2441 | 16.813 | 16.816 | 16.2364 | 101 | 32.1545 | 32.739 | 32.727 | 32.1506 |
| 52 | 16.5622 | 17.132 | 17.134 | | 102 | 32.4727 | 33.057 | 33.045 | |
| 53 | 16.8803 | 17.451 | 17.452 | 16.8729 | 103 | 32.7910 | 33.376 | 33.363 | 32.7872 |
| 54 | 17.1984 | 17.769 | 17.770 | | 104 | 33.1093 | 33.694 | 33.681 | |
| 55 | 17.5165 | 18.088 | 18.089 | 17.5094 | 105 | 33.4275 | 34.013 | 34.000 | 33.4238 |
| 56 | 17.8347 | 18.407 | 18.407 | | 106 | 33.7458 | 34.331 | 34.318 | |
| 57 | 18.1528 | 18.725 | 18.725 | 18.1459 | 107 | 34.0641 | 34.649 | 34.636 | 34.0604 |
| 58 | 18.4710 | 19.044 | 19.043 | | 108 | 34.3823 | 34.968 | 34.954 | |

[a] For 5-8 and 109-200 teeth see text, pages 2627, 2629.

*Caliper Diameter:* The caliper diameter is the same as the bottom diameter for a sprocket with an even number of teeth. For a sprocket with an odd number of teeth, it is defined as the distance from the bottom of one tooth gap to that of the nearest opposite tooth gap. The caliper diameter for an even tooth sprocket is equal to pitch diameter-roller diameter. The caliper diameter for an odd tooth sprocket is equal to caliper factor-roller diameter. Here, the caliper factor = $PD\cos(90° \div N)$, where $PD$ = pitch diameter and $N$ = number of teeth. Caliper factors for 1-inch pitch and sprockets having 9–108 teeth are given in Table 4. For other tooth numbers use above formula. Caliper diameter tolerances are minus only and are equal to $0.002P\sqrt{N} + 0.006$ inch for the Commercial sprockets and $0.001P\sqrt{N} + 0.003$ inch for Precision sprockets. Tolerances are given in Table 5

**Table 5. Minus Tolerances on the Caliper Diameters of Precision Sprockets**
*ANSI/ASME B29.1-2011 (R2022)*

| | Number of Teeth | | | | |
|---|---|---|---|---|---|
| Pitch | Up to 15 | 16-24 | 25-35 | 36-48 | 49-63 |
| 0.250 | 0.004 | 0.004 | 0.004 | 0.005 | 0.005 |
| 0.375 | 0.004 | 0.004 | 0.004 | 0.005 | 0.005 |
| 0.500 | 0.004 | 0.005 | 0.0055 | 0.006 | 0.0065 |
| 0.625 | 0.005 | 0.0055 | 0.006 | 0.007 | 0.008 |
| 0.750 | 0.005 | 0.006 | 0.007 | 0.008 | 0.009 |
| 1.000 | 0.006 | 0.007 | 0.008 | 0.009 | 0.010 |
| 1.250 | 0.007 | 0.008 | 0.009 | 0.010 | 0.012 |
| 1.500 | 0.007 | 0.009 | 0.0105 | 0.012 | 0.013 |
| 1.750 | 0.008 | 0.010 | 0.012 | 0.013 | 0.015 |
| 2.000 | 0.009 | 0.011 | 0.013 | 0.015 | 0.017 |
| 2.250 | 0.010 | 0.012 | 0.014 | 0.016 | 0.018 |
| 2.500 | 0.010 | 0.013 | 0.015 | 0.018 | 0.020 |
| 3.000 | 0.012 | 0.015 | 0.018 | 0.021 | 0.024 |
| | Number of Teeth | | | | |
| Pitch | 64-80 | 81-99 | 100-120 | 121-143 | 144 up |
| 0.250 | 0.005 | 0.005 | 0.006 | 0.006 | 0.006 |
| 0.375 | 0.006 | 0.006 | 0.006 | 0.007 | 0.007 |
| 0.500 | 0.007 | 0.0075 | 0.008 | 0.0085 | 0.009 |
| 0.625 | 0.009 | 0.009 | 0.009 | 0.010 | 0.011 |
| 0.750 | 0.010 | 0.010 | 0.011 | 0.012 | 0.013 |
| 1.000 | 0.011 | 0.012 | 0.013 | 0.014 | 0.015 |
| 1.250 | 0.013 | 0.014 | 0.016 | 0.017 | 0.018 |
| 1.500 | 0.015 | 0.016 | 0.018 | 0.019 | 0.02 |
| 1.750 | 0.017 | 0.019 | 0.020 | 0.022 | 0.024 |
| 2.000 | 0.019 | 0.021 | 0.023 | 0.025 | 0.027 |
| 2.250 | 0.021 | 0.023 | 0.025 | 0.028 | 0.030 |
| 2.500 | 0.023 | 0.025 | 0.028 | 0.030 | 0.033 |
| 3.000 | 0.027 | 0.030 | 0.033 | 0.036 | 0.039 |

Minus tolerances for Commercial sprockets are twice those shown in this table.

*Outside Diameter:* OD is the diameter over the tips of teeth. Sprocket ODs for 1-in. pitch and 9–108 teeth are given in Table 4. For other tooth numbers the OD may be determined by the following formulas in which $O$ = approximate OD; $P$ = pitch of chain; $N$ = number of sprocket teeth: $O = P[0.6 + \cot(180° \div N)]$, for turned sprocket; $O$ = pitch diameter − roller diameter + 2 × whole depth of topping hob cut, for topping hob cut sprocket.*

* This dimension was added in 1984 as a desirable goal for the future. It should in no way obsolete existing tools or sprockets. The whole depth $WD$ is found from the formula: $WD = \frac{1}{2}D_r + P[0.3 - \frac{1}{2}\tan(90\ \text{deg} \div N_a)]$, where $N_a$ is the intermediate number of teeth for the topping hob. For teeth range 5, $N_a = 5$; 6,6; 7-8, 7.47; 9-11, 9.9; 12-17, 14.07; 18-34, 23.54; 35 and over, 56.

**Table 6. Roller Chain Sprocket Flange Thickness and Tooth Section Profile Dimension** *ANSI/ASME B29.1-2011 (R2022)*

Flange chamfer may be either as in Section "A" or Section "B" or anything in between

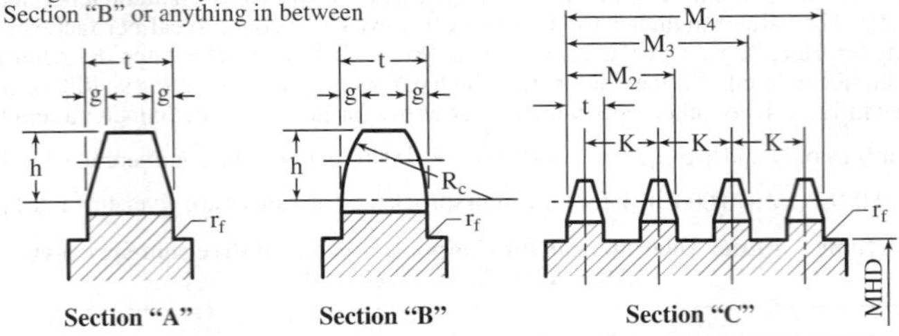


Sprocket Flange Thickness

| Std. Chain No. | Width of Chain, *W* | Maximum Sprocket Flange Thickness, *t* | | | Minus Tolerance on *t* | | Tolerance on *M* | | Max. Variation of *t* on Each Flange | |
|---|---|---|---|---|---|---|---|---|---|---|
| | | Single | Double & Triple | Quad. & Over | Commercial | Precision | Commercial Plus or Minus | Precision Minus Only | Commercial | Precision |
| 25 | 0.125 | 0.110 | 0.106 | 0.096 | 0.021 | 0.007 | 0.007 | 0.007 | 0.021 | 0.004 |
| 35 | 0.188 | 0.169 | 0.163 | 0.150 | 0.027 | 0.008 | 0.008 | 0.008 | 0.027 | 0.004 |
| 41 | 0.250 | 0.226 | … | … | 0.032 | 0.009 | … | … | 0.032 | 0.004 |
| 40 | 0.312 | 0.284 | 0.275 | 0.256 | 0.035 | 0.009 | 0.009 | 0.009 | 0.035 | 0.004 |
| 50 | 0.375 | 0.343 | 0.332 | 0.310 | 0.036 | 0.010 | 0.010 | 0.010 | 0.036 | 0.005 |
| 60 | 0.500 | 0.459 | 0.444 | 0.418 | 0.036 | 0.011 | 0.011 | 0.011 | 0.036 | 0.006 |
| 80 | 0.625 | 0.575 | 0.556 | 0.526 | 0.040 | 0.012 | 0.012 | 0.012 | 0.040 | 0.006 |
| 100 | 0.750 | 0.692 | 0.669 | 0.633 | 0.046 | 0.014 | 0.014 | 0.014 | 0.046 | 0.007 |
| 120 | 1.000 | 0.924 | 0.894 | 0.848 | 0.057 | 0.016 | 0.016 | 0.016 | 0.057 | 0.008 |
| 140 | 1.000 | 0.924 | 0.894 | 0.848 | 0.057 | 0.016 | 0.016 | 0.016 | 0.057 | 0.008 |
| 160 | 1.250 | 1.156 | 1.119 | 1.063 | 0.062 | 0.018 | 0.018 | 0.018 | 0.062 | 0.009 |
| 180 | 1.406 | 1.302 | 1.259 | 1.198 | 0.068 | 0.020 | 0.020 | 0.020 | 0.068 | 0.010 |
| 200 | 1.500 | 1.389 | 1.344 | 1.278 | 0.072 | 0.021 | 0.021 | 0.021 | 0.072 | 0.010 |
| 240 | 1.875 | 1.738 | 1.682 | 1.602 | 0.087 | 0.025 | 0.025 | 0.025 | 0.087 | 0.012 |

Sprocket Tooth Section Profile Dimensions

| Std. Chain No. | Chain Pitch *P* | Depth of Chamfer *h* | Width of Chamfer *g* | Minimum Radius $R_c$ | Transverse Pitch *K* | |
|---|---|---|---|---|---|---|
| | | | | | Standard Series | Heavy Series |
| 25 | 0.250 | 0.125 | 0.031 | 0.265 | 0.252 | … |
| 35 | 0.375 | 0.188 | 0.047 | 0.398 | 0.399 | … |
| 41 | 0.500 | 0.250 | 0.062 | 0.531 | … | … |
| 40 | 0.500 | 0.250 | 0.062 | 0.531 | 0.566 | … |
| 50 | 0.625 | 0.312 | 0.078 | 0.664 | 0.713 | … |
| 60 | 0.750 | 0.375 | 0.094 | 0.796 | 0.897 | 1.028 |
| 80 | 1.000 | 0.500 | 0.125 | 1.062 | 1.153 | 1.283 |
| 100 | 1.250 | 0.625 | 0.156 | 1.327 | 1.408 | 1.539 |
| 120 | 1.500 | 0.750 | 0.188 | 1.593 | 1.789 | 1.924 |
| 140 | 1.750 | 0.875 | 0.219 | 1.858 | 1.924 | 2.055 |
| 160 | 2.000 | 1.000 | 0.250 | 2.124 | 2.305 | 2.437 |
| 180 | 2.250 | 1.125 | 0.281 | 2.392 | 2.592 | 2.723 |
| 200 | 2.500 | 1.250 | 0.312 | 2.654 | 2.817 | 3.083 |
| 240 | 3.000 | 1.500 | 0.375 | 3.187 | 3.458 | 3.985 |

All dimensions are in inches. $r_f$ max = 0.04*P* for max. hub diameter.

**Proportions of Sprockets.—**Typical proportions of single-strand and multiple-strand cast roller chain sprockets, as provided by the American Chain Association, are shown in Table 7. Typical proportions of roller chain bar-steel sprockets, also provided by this association, are shown in Table 8.

**Table 7. Typical Proportions of Single-Strand and Multiple-Strand Cast Roller Chain Sprockets**

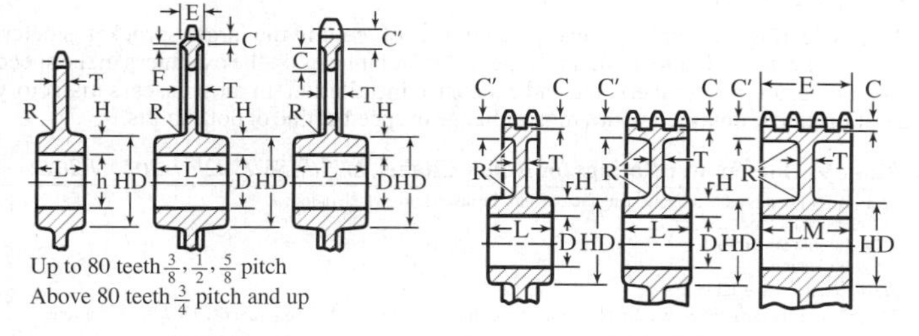


Single-Strand | Multiple-Strand

| Sprocket Web Thickness, *T*, for Various Pitches *P* | | | | | | | | | | | | | | | |
|---|---|---|---|---|---|---|---|---|---|---|---|---|---|---|---|
| Single-Strand | | | | | | | | Multiple-Strand | | | | | | | |
| *P* | *T* | *P* | *T* | *P* | *T* | *P* | *T* | *P* | *T* | *P* | *T* | *P* | *T* | *P* | *T* |
| ⅜ | .312 | ¾ | .437 | 1½ | .625 | 2¼ | 1.000 | ⅜ | .375 | ¾ | .500 | 1½ | .750 | 2¼ | 1.125 |
| ½ | .375 | 1 | .500 | 1¾ | .750 | 2½ | 1.125 | ½ | .406 | 1 | .562 | 1¾ | .875 | 2½ | 1.250 |
| ⅝ | .406 | 1¼ | .562 | 2 | .875 | 3 | 1.250 | ⅝ | .437 | 1¼ | .625 | 2 | 1.000 | 3 | 1.500 |

Formulas for Dimensions of Single and Multiple Sprockets

$H = 0.375 + \frac{D}{6} + 0.01\,PD$

$L = 4H$ for semi-steel castings

$C = 0.5P$

$C' = 0.9P$

$E = 0.625P + 0.93W$

$F = 0.150 + 0.25P$

$G = 2T$

$R = 0.4P$ for single-strand sprockets

$R = 0.5T$ for multiple-strand sprockets

All dimensions in inches. Where $P$ = chain pitch and $W$ = nominal chain width.

**Table 8. Typical Proportions of Roller Chain Bar-Steel Sprockets**

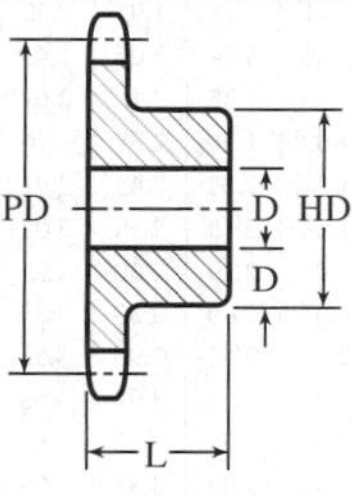


$H = Z + D/6 + 0.01PD$

For *PD* up to 2 inches, $Z = 0.125$ inch; for 2-4 inches, $Z = 0.187$ inch; for 4-6 inches, 0.25 inch; and for over 6 inches 0.375 inch.

Hub length $L = 3.3H$, normally, with a minimum of $2.6H$.

Hub diameter $HD = D + 2H$, but not more than the maximum hub diameter *MHD* given by the formula:

$$MHD = P\left(\cot\frac{180°}{N} - 1\right) - 0.030$$

where

$P$ = Chain pitch, in inches; $N$ = Number of sprocket teeth

When sprocket wheels are designed with spokes, the usual assumptions made in order to determine suitable proportions are as follows: 1) that the maximum torque load acting on a sprocket is the chain tensile strength times the sprocket pitch radius; 2) that the torque load is equally divided between the arms by the rim; and 3) that each arm acts as a cantilever beam.

The arms are generally elliptical in cross section, the major axis twice the minor axis.

**Selection of Chain and Sprockets.—**The smallest applicable pitch of roller chain is desirable for quiet operation and high speed. The horsepower capacity varies with the chain pitch as shown in Table 9 for $^1/_4$, $^3/_8$, and $^1/_2$ in pitch designs. However, short pitch with high working load can often be obtained by the use of multiple-strand chain.

The small sprocket selected must be large enough to accommodate the shaft. Table 10 gives maximum bore and hub diameters consistent with commercial practice for sprockets with up to 25 teeth.

After selecting the small sprocket, the number of teeth in the larger sprocket is determined by the desired ratio of the shaft speed. Overemphasis on the exactness in the speed ratio may result in a cumbersome and expensive installation. In most cases, satisfactory operation can be obtained with a minor change in speed of one or both shafts.

**Table 9. Horsepower Ratings for Roller Chains,** *ANSI/ASME B29.1-2011 (R2022)*

To properly use this table the following factors must be taken into consideration:
1) Service factors
2) Multiple Strand Factors
3) Lubrication

*Service Factors:* See Table 15.

*Multiple Strand Factors:* For two strands, the multiple strand factor is 1.7; for three strands, it is 2.5; and for four strands, it is 3.3.

*Lubrication:*

Required type of lubrication is indicated at the bottom of each roller chain size section of the table. For a description of each type of lubrication, see page 2644.

Type A — Manual or Drip Lubrication
Type B — Bath or Disc Lubrication
Type C — Oil Stream Lubrication

To find the required horsepower table rating, use the following formula:

$$\text{Required hp Table Rating} = \frac{\text{hp to be Transmitted} \times \text{Service Factor}}{\text{Multiple} - \text{Strand Factor}}$$

$^1/_4$-inch Pitch Standard Single-Strand Roller Chain — No. 25

| No. of Teeth Small Spkt. | Revolutions per Minute — Small Sprocket[a] | | | | | | | | | | | | |
|---|---|---|---|---|---|---|---|---|---|---|---|---|---|
| | 50 | 100 | 300 | 500 | 700 | 900 | 1200 | 1500 | 1800 | 2100 | 2500 | 3000 | 3500 |
| | Horsepower Rating | | | | | | | | | | | | |
| 11 | 0.03 | 0.06 | 0.19 | 0.30 | 0.42 | 0.53 | 0.70 | 0.87 | 1.03 | 1.20 | 1.42 | 1.69 | 1.69 |
| 12 | 0.04 | 0.07 | 0.20 | 0.33 | 0.46 | 0.58 | 0.76 | 0.95 | 1.13 | 1.31 | 1.55 | 1.84 | 1.92 |
| 13 | 0.04 | 0.08 | 0.22 | 0.36 | 0.49 | 0.63 | 0.83 | 1.03 | 1.22 | 1.42 | 1.67 | 1.99 | 2.17 |
| 14 | 0.04 | 0.08 | 0.24 | 0.38 | 0.53 | 0.68 | 0.89 | 1.10 | 1.32 | 1.52 | 1.80 | 2.15 | 2.42 |
| 15 | 0.05 | 0.09 | 0.25 | 0.41 | 0.57 | 0.72 | 0.95 | 1.18 | 1.41 | 1.63 | 1.93 | 2.30 | 2.67 |
| 16 | 0.05 | 0.09 | 0.27 | 0.44 | 0.61 | 0.77 | 1.02 | 1.26 | 1.50 | 1.74 | 2.06 | 2.45 | 2.85 |
| 17 | 0.05 | 0.10 | 0.29 | 0.47 | 0.64 | 0.82 | 1.08 | 1.34 | 1.60 | 1.85 | 2.19 | 2.61 | 3.02 |
| 18 | 0.05 | 0.11 | 0.30 | 0.49 | 0.68 | 0.87 | 1.15 | 1.42 | 1.69 | 1.96 | 2.32 | 2.76 | 3.20 |
| 19 | 0.06 | 0.11 | 0.32 | 0.52 | 0.72 | 0.92 | 1.21 | 1.50 | 1.78 | 2.07 | 2.45 | 2.91 | 3.38 |
| 20 | 0.06 | 0.12 | 0.34 | 0.55 | 0.76 | 0.97 | 1.27 | 1.58 | 1.88 | 2.18 | 2.58 | 3.07 | 3.56 |
| 21 | 0.06 | 0.12 | 0.35 | 0.58 | 0.80 | 1.01 | 1.34 | 1.66 | 1.97 | 2.29 | 2.70 | 3.22 | 3.74 |
| 22 | 0.07 | 0.13 | 0.37 | 0.60 | 0.83 | 1.06 | 1.40 | 1.73 | 2.07 | 2.40 | 2.83 | 3.37 | 3.91 |
| 23 | 0.07 | 0.13 | 0.39 | 0.63 | 0.87 | 1.11 | 1.46 | 1.81 | 2.16 | 2.51 | 2.96 | 3.53 | 4.09 |
| 24 | 0.07 | 0.14 | 0.40 | 0.66 | 0.91 | 1.16 | 1.53 | 1.89 | 2.25 | 2.61 | 3.09 | 3.68 | 4.27 |
| 25 | 0.08 | 0.15 | 0.42 | 0.69 | 0.95 | 1.21 | 1.59 | 1.97 | 2.35 | 2.72 | 3.22 | 3.84 | 4.45 |
| 26 | 0.08 | 0.15 | 0.44 | 0.71 | 0.99 | 1.26 | 1.65 | 2.05 | 2.44 | 2.83 | 3.35 | 3.99 | 4.62 |
| 28 | 0.08 | 0.16 | 0.47 | 0.77 | 1.06 | 1.35 | 1.78 | 2.21 | 2.63 | 3.05 | 3.61 | 4.30 | 4.98 |
| 30 | 0.09 | 0.18 | 0.50 | 0.82 | 1.14 | 1.45 | 1.91 | 2.37 | 2.82 | 3.27 | 3.86 | 4.60 | 5.34 |
| 32 | 0.10 | 0.19 | 0.54 | 0.88 | 1.21 | 1.55 | 2.04 | 2.52 | 3.01 | 3.49 | 4.12 | 4.91 | 5.69 |
| 35 | 0.11 | 0.21 | 0.59 | 0.96 | 1.33 | 1.69 | 2.23 | 2.76 | 3.29 | 3.81 | 4.51 | 5.37 | 6.23 |
| 40 | 0.12 | 0.23 | 0.67 | 1.10 | 1.52 | 1.93 | 2.55 | 3.15 | 3.76 | 4.36 | 5.15 | 6.14 | 7.11 |
| 45 | 0.14 | 0.26 | 0.76 | 1.24 | 1.71 | 2.17 | 2.86 | 3.55 | 4.23 | 4.90 | 5.79 | 6.90 | 8.00 |
| | Type A | | | Type B | | | | | | | | | Type C (from 28 teeth; Type B above) |

**Table 9.** *(Continued)* **Horsepower Ratings for Roller Chains,** *ANSI/ASME B29.1-2011 (R2022)*

**3/8-inch Pitch Standard Single-Strand Roller Chain — No. 35**

| No. of Teeth Small Spkt. | Revolutions per Minute — Small Sprocket[a] | | | | | | | | | | | | |
|---|---|---|---|---|---|---|---|---|---|---|---|---|---|
| | 50 | 100 | 240 | 500 | 700 | 900 | 1200 | 1500 | 1800 | 2100 | 2500 | 3000 | 3500 |
| | Horsepower Rating | | | | | | | | | | | | |
| 11 | 0.11 | 0.22 | 0.50 | 1.02 | 1.41 | 1.80 | 2.37 | 2.93 | 3.49 | 4.05 | 3.86 | 2.94 | 2.33 |
| 12 | 0.12 | 0.24 | 0.55 | 1.11 | 1.54 | 1.96 | 2.58 | 3.20 | 3.81 | 4.42 | 4.40 | 3.35 | 2.66 |
| 13 | 0.13 | 0.26 | 0.60 | 1.21 | 1.67 | 2.12 | 2.80 | 3.47 | 4.13 | 4.79 | 4.96 | 3.77 | 3.0[illegible] |
| 14 | 0.14 | 0.28 | 0.64 | 1.30 | 1.80 | 2.29 | 3.01 | 3.73 | 4.45 | 5.15 | 5.55 | 4.22 | 3.35 |
| 15 | 0.15 | 0.30 | 0.69 | 1.39 | 1.92 | 2.45 | 3.23 | 4.00 | 4.76 | 5.52 | 6.15 | 4.68 | 3.71 |
| 16 | 0.16 | 0.32 | 0.73 | 1.49 | 2.05 | 2.61 | 3.44 | 4.26 | 5.08 | 5.89 | 6.77 | 5.15 | 4.09 |
| 17 | 0.17 | 0.34 | 0.78 | 1.58 | 2.18 | 2.77 | 3.66 | 4.53 | 5.40 | 6.26 | 7.40 | 5.64 | 4.48 |
| 18 | 0.18 | 0.36 | 0.83 | 1.67 | 2.31 | 2.94 | 3.87 | 4.80 | 5.72 | 6.63 | 7.83 | 6.15 | 4.88 |
| 19 | 0.19 | 0.38 | 0.87 | 1.76 | 2.44 | 3.10 | 4.09 | 5.06 | 6.03 | 7.00 | 8.27 | 6.67 | 5.29 |
| 20 | 0.20 | 0.40 | 0.92 | 1.86 | 2.56 | 3.26 | 4.30 | 5.33 | 6.35 | 7.36 | 8.71 | 7.20 | 5.72 |
| 21 | 0.21 | 0.42 | 0.96 | 1.95 | 2.69 | 3.43 | 4.52 | 5.60 | 6.67 | 7.73 | 9.14 | 7.75 | 6.[illegible]5 |
| 22 | 0.22 | 0.44 | 1.01 | 2.04 | 2.82 | 3.59 | 4.73 | 5.86 | 6.99 | 8.10 | 9.58 | 8.31 | 6.59 |
| 23 | 0.23 | 0.46 | 1.06 | 2.14 | 2.95 | 3.75 | 4.95 | 6.13 | 7.30 | 8.47 | 10.01 | 8.88 | 7.05 |
| 24 | 0.24 | 0.48 | 1.10 | 2.23 | 3.08 | 3.92 | 5.16 | 6.40 | 7.62 | 8.84 | 10.45 | 9.47 | 7.51 |
| 25 | 0.25 | 0.50 | 1.15 | 2.32 | 3.21 | 4.08 | 5.38 | 6.66 | 7.94 | 9.20 | 10.88 | 10.07 | 7.99 |
| 26 | 0.26 | 0.51 | 1.19 | 2.41 | 3.33 | 4.24 | 5.59 | 6.93 | 8.26 | 9.57 | 11.32 | 10.68 | 8.47 |
| 28 | 0.29 | 0.55 | 1.28 | 2.60 | 3.59 | 4.57 | 6.02 | 7.46 | 8.89 | 10.31 | 12.19 | 11.93 | 9.47 |
| 30 | 0.31 | 0.59 | 1.38 | 2.79 | 3.85 | 4.90 | 6.45 | 8.00 | 9.53 | 11.05 | 13.06 | 13.23 | 10.50 |
| 32 | 0.33 | 0.63 | 1.47 | 2.97 | 4.10 | 5.22 | 6.88 | 8.53 | 10.16 | 11.78 | 13.93 | 14.58 | 1[illegible].57 |
| 35 | 0.36 | 0.69 | 1.61 | 3.25 | 4.49 | 5.71 | 7.53 | 9.33 | 11.11 | 12.89 | 15.23 | 16.67 | 1[illegible].23 |
| 40 | 0.41 | 0.79 | 1.84 | 3.71 | 5.13 | 6.53 | 8.61 | 10.66 | 12.70 | 14.73 | 17.41 | 20.37 | 15.17 |
| 45 | 0.46 | 0.89 | 2.07 | 4.18 | 5.77 | 7.35 | 9.68 | 11.99 | 14.29 | 16.57 | 19.59 | 23.33 | 15.56 |
| | Type A | | Type B | | | | | | | Type C | | | |

**1/2-inch Pitch Standard Single-Strand Roller Chain — No. 40**

| No. of Teeth Small Spkt. | Revolutions per Minute — Small Sprocket[a] | | | | | | | | | | | | |
|---|---|---|---|---|---|---|---|---|---|---|---|---|---|
| | 50 | 100 | 200 | 300 | 500 | 700 | 900 | 1000 | 1200 | 1400 | 1600 | 1800 | 2100 |
| | Horsepower Rating | | | | | | | | | | | | |
| 11 | 0.27 | 0.52 | 1.00 | 1.48 | 2.42 | 3.34 | 4.25 | 4.70 | 5.60 | 6.49 | 5.57 | 4.66 | 3.70 |
| 12 | 0.29 | 0.56 | 1.09 | 1.61 | 2.64 | 3.64 | 4.64 | 5.13 | 6.11 | 7.09 | 6.34 | 5.31 | 4.22 |
| 13 | 0.31 | 0.61 | 1.19 | 1.75 | 2.86 | 3.95 | 5.02 | 5.56 | 6.62 | 7.68 | 7.15 | 5.99 | 4.76 |
| 14 | 0.34 | 0.66 | 1.28 | 1.88 | 3.08 | 4.25 | 5.41 | 5.98 | 7.13 | 8.27 | 7.99 | 6.70 | 5.31 |
| 15 | 0.36 | 0.70 | 1.37 | 2.02 | 3.30 | 4.55 | 5.80 | 6.41 | 7.64 | 8.86 | 8.86 | 7.43 | 5.89 |
| 16 | 0.39 | 0.75 | 1.46 | 2.15 | 3.52 | 4.86 | 6.18 | 6.84 | 8.15 | 9.45 | 9.76 | 8.18 | 6.49 |
| 17 | 0.41 | 0.80 | 1.55 | 2.29 | 3.74 | 5.16 | 6.57 | 7.27 | 8.66 | 10.04 | 10.69 | 8.96 | 7.11 |
| 18 | 0.43 | 0.84 | 1.64 | 2.42 | 3.96 | 5.46 | 6.95 | 7.69 | 9.17 | 10.63 | 11.65 | 9.76 | 7.75 |
| 19 | 0.46 | 0.89 | 1.73 | 2.56 | 4.18 | 5.77 | 7.34 | 8.12 | 9.68 | 11.22 | 12.64 | 10.59 | 8.40 |
| 20 | 0.48 | 0.94 | 1.82 | 2.69 | 4.39 | 6.07 | 7.73 | 8.55 | 10.18 | 11.81 | 13.42 | 11.44 | 9.07 |
| 21 | 0.51 | 0.98 | 1.91 | 2.83 | 4.61 | 6.37 | 8.11 | 8.98 | 10.69 | 12.40 | 14.10 | 12.30 | 9.76 |
| 22 | 0.53 | 1.03 | 2.01 | 2.96 | 4.83 | 6.68 | 8.50 | 9.40 | 11.20 | 12.99 | 14.77 | 13.19 | 10.47 |
| 23 | 0.55 | 1.08 | 2.10 | 3.10 | 5.05 | 6.98 | 8.89 | 9.83 | 11.71 | 13.58 | 15.44 | 14.10 | 11.19 |
| 24 | 0.58 | 1.12 | 2.19 | 3.23 | 5.27 | 7.28 | 9.27 | 10.26 | 12.22 | 14.17 | 16.11 | 15.03 | 11.93 |
| 25 | 0.60 | 1.17 | 2.28 | 3.36 | 5.49 | 7.59 | 9.66 | 10.69 | 12.73 | 14.76 | 16.78 | 15.98 | 12.68 |
| 26 | 0.63 | 1.22 | 2.37 | 3.50 | 5.71 | 7.89 | 10.04 | 11.11 | 13.24 | 15.35 | 17.45 | 16.95 | 13.45 |
| 28 | 0.67 | 1.31 | 2.55 | 3.77 | 6.15 | 8.50 | 10.82 | 11.97 | 14.26 | 16.53 | 18.79 | 18.94 | 15.03 |
| 30 | 0.72 | 1.41 | 2.74 | 4.04 | 6.59 | 9.11 | 11.59 | 12.82 | 15.28 | 17.71 | 20.14 | 21.01 | 16.67 |
| 32 | 0.77 | 1.50 | 2.92 | 4.31 | 7.03 | 9.71 | 12.36 | 13.68 | 16.30 | 18.89 | 21.48 | 23.14 | 18.37 |
| 35 | 0.84 | 1.64 | 3.19 | 4.71 | 7.69 | 10.62 | 13.52 | 14.96 | 17.82 | 20.67 | 23.49 | 26.30 | 21.01 |
| 40 | 0.96 | 1.87 | 3.65 | 5.38 | 8.79 | 12.14 | 15.45 | 17.10 | 20.37 | 23.62 | 26.85 | 30.06 | 25.67 |
| 45 | 1.08 | 2.11 | 4.10 | 6.06 | 9.89 | 13.66 | 17.39 | 19.24 | 22.92 | 26.57 | 30.20 | 33.8[illegible] | 30.63 |
| | Type A | | Type B | | | | | | | | Type C | | |

**Table 9.** *(Continued)* **Horsepower Ratings for Roller Chains,** *ANSI/ASME B29.1-2011 (R2022)*

½-inch Pitch Light Weight Machinery Roller Chain — No. 41

| No. of Teeth Small Spkt. | Revolutions per Minute — Small Sprocket[a] | | | | | | | | | | | | |
|---|---|---|---|---|---|---|---|---|---|---|---|---|---|
| | 10 | 25 | 50 | 100 | 180 | 200 | 300 | 500 | 700 | 900 | 1000 | 1200 | 1400 |
| | Horsepower Rating | | | | | | | | | | | | |
| 11 | 0.03 | 0.07 | 0.15 | 0.28 | 0.50 | 0.55 | 0.81 | 1.33 | 1.84 | 2.34 | 2.25 | 1.71 | 1.36 |
| 12 | 0.03 | 0.08 | 0.16 | 0.31 | 0.54 | 0.60 | 0.89 | 1.45 | 2.00 | 2.55 | 2.57 | 1.95 | 1.55 |
| 13 | 0.04 | 0.09 | 0.17 | 0.34 | 0.59 | 0.70 | 1.04 | 1.69 | 2.34 | 2.97 | 3.23 | 2.20 | 1.75 |
| 14 | 0.04 | 0.10 | 0.19 | 0.36 | 0.63 | 0.57 | 0.83 | 1.31 | 1.77 | 2.22 | 2.44 | 2.46 | 1.95 |
| 15 | 0.04 | 0.10 | 0.20 | 0.39 | 0.68 | 0.75 | 1.11 | 1.81 | 2.50 | 3.19 | 3.53 | 2.73 | 2.17 |
| 16 | 0.05 | 0.11 | 0.21 | 0.41 | 0.73 | 0.80 | 1.18 | 1.93 | 2.67 | 3.40 | 3.76 | 3.01 | 2.39 |
| 17 | 0.05 | 0.12 | 0.23 | 0.44 | 0.77 | 0.85 | 1.26 | 2.05 | 2.84 | 3.61 | 4.00 | 3.29 | 2.61 |
| 18 | 0.05 | 0.12 | 0.24 | 0.46 | 0.82 | 0.90 | 1.33 | 2.18 | 3.00 | 3.82 | 4.23 | 3.59 | 2.85 |
| 19 | 0.05 | 0.13 | 0.25 | 0.49 | 0.86 | 0.95 | 1.41 | 2.30 | 3.17 | 4.04 | 4.47 | 3.89 | 3.09 |
| 20 | 0.06 | 0.14 | 0.27 | 0.52 | 0.91 | 1.00 | 1.48 | 2.42 | 3.34 | 4.25 | 4.70 | 4.20 | 3.33 |
| 21 | 0.06 | 0.14 | 0.28 | 0.54 | 0.95 | 1.05 | 1.55 | 2.54 | 3.51 | 4.46 | 4.94 | 4.52 | 3.59 |
| 22 | 0.06 | 0.15 | 0.29 | 0.57 | 1.00 | 1.10 | 1.63 | 2.66 | 3.67 | 4.67 | 5.17 | 4.85 | 3.85 |
| 23 | 0.07 | 0.16 | 0.30 | 0.59 | 1.04 | 1.15 | 1.70 | 2.78 | 3.84 | 4.89 | 5.41 | 5.18 | 4.11 |
| 24 | 0.07 | 0.16 | 0.32 | 0.62 | 1.09 | 1.20 | 1.78 | 2.90 | 4.01 | 5.10 | 5.64 | 5.52 | 4.38 |
| 25 | 0.07 | 0.17 | 0.33 | 0.64 | 1.13 | 1.25 | 1.85 | 3.02 | 4.17 | 5.31 | 5.88 | 5.87 | 4.66 |
| 26 | 0.07 | 0.18 | 0.34 | 0.67 | 1.18 | 1.30 | 1.92 | 3.14 | 4.34 | 5.52 | 6.11 | 6.23 | 4.94 |
| 28 | 0.08 | 0.19 | 0.37 | 0.72 | 1.27 | 1.40 | 2.07 | 3.38 | 4.67 | 5.95 | 6.58 | 6.96 | 5.52 |
| 30 | 0.08 | 0.20 | 0.40 | 0.77 | 1.36 | 1.50 | 2.22 | 3.63 | 5.01 | 6.37 | 7.05 | 7.72 | 6.13 |
| 32 | 0.09 | 0.22 | 0.42 | 0.82 | 1.45 | 1.60 | 2.37 | 3.87 | 5.34 | 6.80 | 7.52 | 8.50 | 6.75 |
| 35 | 0.10 | 0.24 | 0.46 | 0.90 | 1.59 | 1.76 | 2.59 | 4.23 | 5.84 | 7.44 | 8.23 | 9.80 | 7.72 |
| 40 | 0.11 | 0.27 | 0.53 | 1.03 | 1.81 | 2.01 | 2.96 | 4.83 | 6.68 | 8.50 | 9.40 | 11.20 | 9.43 |
| 45 | 0.13 | 0.31 | 0.60 | 1.16 | 2.04 | 2.26 | 3.33 | 5.44 | 7.51 | 9.56 | 10.58 | 12.60 | 11.25 |
| | Type A | | | | Type B | | | | | | | | |

⅝-inch Pitch Standard Single-Strand Roller Chain — No. 50

| No. of Teeth Small Spkt. | Revolutions per Minute — Small Sprocket[a] | | | | | | | | | | | | |
|---|---|---|---|---|---|---|---|---|---|---|---|---|---|
| | 25 | 50 | 100 | 140 | 200 | 300 | 500 | 700 | 900 | 1200 | 1500 | 1800 | 2100 |
| | Horsepower Rating | | | | | | | | | | | | |
| 11 | 0.27 | 0.52 | 1.00 | 1.39 | 1.95 | 2.88 | 4.70 | 6.50 | 8.27 | 10.24 | 7.33 | 5.58 | 4.42 |
| 12 | 0.29 | 0.56 | 1.09 | 1.51 | 2.13 | 3.14 | 5.13 | 7.09 | 9.02 | 11.67 | 8.35 | 6.35 | 5.04 |
| 13 | 0.31 | 0.61 | 1.19 | 1.64 | 2.31 | 3.40 | 5.56 | 7.68 | 9.77 | 12.88 | 9.42 | 7.16 | 5.69 |
| 14 | 0.34 | 0.66 | 1.28 | 1.76 | 2.48 | 3.67 | 5.99 | 8.27 | 10.53 | 13.87 | 10.52 | 8.01 | 6.35 |
| 15 | 0.36 | 0.70 | 1.37 | 1.89 | 2.66 | 3.93 | 6.41 | 8.86 | 11.28 | 14.86 | 11.67 | 8.88 | 7.05 |
| 16 | 0.39 | 0.75 | 1.46 | 2.02 | 2.84 | 4.19 | 6.84 | 9.45 | 12.03 | 15.85 | 12.86 | 9.78 | 7.76 |
| 17 | 0.41 | 0.80 | 1.55 | 2.14 | 3.02 | 4.45 | 7.27 | 10.04 | 12.78 | 16.85 | 14.08 | 10.71 | 8.50 |
| 18 | 0.43 | 0.84 | 1.64 | 2.27 | 3.19 | 4.71 | 7.70 | 10.63 | 13.53 | 17.84 | 15.34 | 11.67 | 9.26 |
| 19 | 0.46 | 0.89 | 1.73 | 2.39 | 3.37 | 4.98 | 8.12 | 11.22 | 14.28 | 18.83 | 16.64 | 12.66 | 10.05 |
| 20 | 0.48 | 0.94 | 1.82 | 2.52 | 3.55 | 5.24 | 8.55 | 11.81 | 15.04 | 19.82 | 17.97 | 13.67 | 10.85 |
| 21 | 0.51 | 0.98 | 1.92 | 2.65 | 3.73 | 5.50 | 8.98 | 12.40 | 15.79 | 20.81 | 19.34 | 14.71 | 11.67 |
| 22 | 0.53 | 1.03 | 2.01 | 2.77 | 3.90 | 5.76 | 9.41 | 12.99 | 16.54 | 21.80 | 20.73 | 15.77 | 12.52 |
| 23 | 0.55 | 1.08 | 2.10 | 2.90 | 4.08 | 6.02 | 9.83 | 13.58 | 17.29 | 22.79 | 22.16 | 16.89 | 13.38 |
| 24 | 0.58 | 1.13 | 2.19 | 3.02 | 4.26 | 6.28 | 10.26 | 14.18 | 18.04 | 23.78 | 23.62 | 17.97 | 14.26 |
| 25 | 0.60 | 1.17 | 2.28 | 3.15 | 4.44 | 6.55 | 10.69 | 14.77 | 18.79 | 24.77 | 25.11 | 19.11 | 15.16 |
| 26 | 0.63 | 1.22 | 2.37 | 3.28 | 4.61 | 6.81 | 11.12 | 15.36 | 19.55 | 25.76 | 26.64 | 20.26 | 16.08 |
| 28 | 0.67 | 1.31 | 2.55 | 3.53 | 4.97 | 7.33 | 11.97 | 16.54 | 21.05 | 27.75 | 29.77 | 22.65 | 17.97 |
| 30 | 0.72 | 1.41 | 2.74 | 3.78 | 5.32 | 7.86 | 12.83 | 17.72 | 22.55 | 29.73 | 33.01 | 25.11 | 19.93 |
| 32 | 0.77 | 1.50 | 2.92 | 4.03 | 5.68 | 8.38 | 13.68 | 18.90 | 24.06 | 31.71 | 36.37 | 27.67 | 21.96 |
| 35 | 0.84 | 1.64 | 3.19 | 4.41 | 6.21 | 9.16 | 14.97 | 20.67 | 26.31 | 34.68 | 41.60 | 31.65 | 25.11 |
| 40 | 0.96 | 1.88 | 3.65 | 5.04 | 7.10 | 10.47 | 17.10 | 23.63 | 30.07 | 39.64 | 49.11 | 38.67 | 30.68 |
| 45 | 1.08 | 2.11 | 4.10 | 5.67 | 7.98 | 11.78 | 19.24 | 26.58 | 33.83 | 44.59 | 55.24 | 46.14 | 36.61 |
| | Type A | | | Type B | | | | | | Type C | | | |

**Table 9.** *(Continued)* **Horsepower Ratings for Roller Chains,** *ANSI/ASME B29.1-2011 (R2022)*

¾-inch Pitch Standard Single-Strand Roller Chain — No. 60

| No. of Teeth Small Spkt. | Revolutions per Minute — Small Sprocket[a] | | | | | | | | | | | | |
|---|---|---|---|---|---|---|---|---|---|---|---|---|---|
| | 25 | 50 | 100 | 120 | 200 | 300 | 400 | 500 | 600 | 800 | 1000 | 1200 | 1400 |
| | Horsepower Rating | | | | | | | | | | | | |
| 11 | 0.46 | 0.89 | 1.72 | 2.05 | 3.35 | 4.95 | 6.52 | 8.08 | 9.63 | 12.69 | 15.58 | 11.85 | 9.4 |
| 12 | 0.50 | 0.97 | 1.88 | 2.24 | 3.66 | 5.40 | 7.12 | 8.82 | 10.51 | 13.85 | 17.15 | 13.53 | 10.72 |
| 13 | 0.54 | 1.05 | 2.04 | 2.43 | 3.96 | 5.85 | 7.71 | 9.55 | 11.38 | 15.00 | 18.58 | 15.23 | 12.08 |
| 14 | 0.58 | 1.13 | 2.19 | 2.61 | 4.27 | 6.30 | 8.30 | 10.29 | 12.26 | 16.15 | 20.01 | 17.02 | 13.51 |
| 15 | 0.62 | 1.21 | 2.35 | 2.80 | 4.57 | 6.75 | 8.90 | 11.02 | 13.13 | 17.31 | 21.44 | 18.87 | 14.98 |
| 16 | 0.66 | 1.29 | 2.51 | 2.99 | 4.88 | 7.20 | 9.49 | 11.76 | 14.01 | 18.46 | 22.87 | 20.79 | 16.50 |
| 17 | 0.70 | 1.37 | 2.66 | 3.17 | 5.18 | 7.65 | 10.08 | 12.49 | 14.88 | 19.62 | 24.30 | 22.77 | 18.07 |
| 18 | 0.75 | 1.45 | 2.82 | 3.36 | 5.49 | 8.10 | 10.68 | 13.23 | 15.76 | 20.77 | 25.73 | 24.81 | 19.[illegible]9 |
| 19 | 0.79 | 1.53 | 2.98 | 3.55 | 5.79 | 8.55 | 11.27 | 13.96 | 16.63 | 21.92 | 27.16 | 26.91 | 21.35 |
| 20 | 0.83 | 1.61 | 3.13 | 3.73 | 6.10 | 9.00 | 11.86 | 14.70 | 17.51 | 23.08 | 28.59 | 29.06 | 23.[illegible]6 |
| 21 | 0.87 | 1.69 | 3.29 | 3.92 | 6.40 | 9.45 | 12.46 | 15.43 | 18.38 | 24.23 | 30.02 | 31.26 | 24.31 |
| 22 | 0.91 | 1.77 | 3.45 | 4.11 | 6.71 | 9.90 | 13.05 | 16.17 | 19.26 | 25.39 | 31.45 | 33.52 | 26.50 |
| 23 | 0.95 | 1.85 | 3.61 | 4.29 | 7.01 | 10.35 | 13.64 | 16.90 | 20.13 | 26.54 | 32.88 | 35.84 | 28.44 |
| 24 | 0.99 | 1.93 | 3.76 | 4.48 | 7.32 | 10.80 | 14.24 | 17.64 | 21.01 | 27.69 | 34.31 | 38.20 | 30.31 |
| 25 | 1.04 | 2.01 | 3.92 | 4.67 | 7.62 | 11.25 | 14.83 | 18.37 | 21.89 | 28.85 | 35.74 | 40.61 | 32.23 |
| 26 | 1.08 | 2.09 | 4.08 | 4.85 | 7.93 | 11.70 | 15.42 | 19.11 | 22.76 | 30.00 | 37.17 | 43.07 | 34.18 |
| 28 | 1.16 | 2.26 | 4.39 | 5.23 | 8.54 | 12.60 | 16.61 | 20.58 | 24.51 | 32.31 | 40.03 | 47.68 | 38.20 |
| 30 | 1.24 | 2.42 | 4.70 | 5.60 | 9.15 | 13.50 | 17.79 | 22.05 | 26.26 | 34.62 | 42.89 | 51.09 | 42.36 |
| 32 | 1.33 | 2.58 | 5.02 | 5.98 | 9.76 | 14.40 | 18.98 | 23.52 | 28.01 | 36.92 | 45.75 | 54.50 | 46.67 |
| 35 | 1.45 | 2.82 | 5.49 | 6.54 | 10.67 | 15.75 | 20.76 | 25.72 | 30.64 | 40.39 | 50.03 | 59.60 | 53.38 |
| 40 | 1.66 | 3.22 | 6.27 | 7.47 | 12.20 | 18.00 | 23.73 | 29.39 | 35.02 | 46.16 | 57.18 | 68.12 | 65.22 |
| 45 | 1.86 | 3.63 | 7.05 | 8.40 | 13.72 | 20.25 | 26.69 | 33.07 | 39.39 | 51.92 | 64.33 | 76.63 | 77.83 |
| | Type A | | | Type B | | | | | | | Type C | | |

1-inch Pitch Standard Single-Strand Roller Chain — No. 80

| No. of Teeth Small Spkt. | Revolutions per Minute — Small Sprocket[a] | | | | | | | | | | | | |
|---|---|---|---|---|---|---|---|---|---|---|---|---|---|
| | 25 | 50 | 75 | 100 | 200 | 300 | 400 | 500 | 600 | 700 | 800 | 900 | 1000 |
| | Horsepower Ratings | | | | | | | | | | | | |
| 11 | 1.06 | 2.07 | 3.05 | 4.03 | 7.83 | 11.56 | 15.23 | 18.87 | 22.48 | 26.07 | 27.41 | 22.97 | 19.61 |
| 12 | 1.16 | 2.26 | 3.33 | 4.39 | 8.54 | 12.61 | 16.62 | 20.59 | 24.53 | 28.44 | 31.23 | 26.17 | 22.35 |
| 13 | 1.26 | 2.45 | 3.61 | 4.76 | 9.26 | 13.66 | 18.00 | 22.31 | 26.57 | 30.81 | 35.02 | 29.51 | 25.20 |
| 14 | 1.35 | 2.63 | 3.89 | 5.12 | 9.97 | 14.71 | 19.39 | 24.02 | 28.62 | 33.18 | 37.72 | 32.98 | 2[illegible].16 |
| 15 | 1.45 | 2.82 | 4.16 | 5.49 | 10.68 | 15.76 | 20.77 | 25.74 | 30.66 | 35.55 | 40.41 | 36.58 | 3[illegible].23 |
| 16 | 1.55 | 3.01 | 4.44 | 5.86 | 11.39 | 16.81 | 22.16 | 27.45 | 32.70 | 37.92 | 43.11 | 40.30 | 3[illegible].41 |
| 17 | 1.64 | 3.20 | 4.72 | 6.22 | 12.10 | 17.86 | 23.54 | 29.17 | 34.75 | 40.29 | 45.80 | 44.13 | 3[illegible].68 |
| 18 | 1.74 | 3.39 | 5.00 | 6.59 | 12.81 | 18.91 | 24.93 | 30.88 | 36.79 | 42.66 | 48.49 | 48.08 | 4[illegible].05 |
| 19 | 1.84 | 3.57 | 5.28 | 6.95 | 13.53 | 19.96 | 26.31 | 32.60 | 38.84 | 45.03 | 51.19 | 52.15 | 44.52 |
| 20 | 1.93 | 3.76 | 5.55 | 7.32 | 14.24 | 21.01 | 27.70 | 34.32 | 40.88 | 47.40 | 53.88 | 56.32 | 43.08 |
| 21 | 2.03 | 3.95 | 5.83 | 7.69 | 14.95 | 22.07 | 29.08 | 36.03 | 42.92 | 49.77 | 56.58 | 60.59 | 51.73 |
| 22 | 2.13 | 4.14 | 6.11 | 8.05 | 15.66 | 23.12 | 30.47 | 37.75 | 44.97 | 52.14 | 59.27 | 64.97 | 55.47 |
| 23 | 2.22 | 4.33 | 6.39 | 8.42 | 16.37 | 24.17 | 31.85 | 39.46 | 47.01 | 54.51 | 61.97 | 69.38 | 59.30 |
| 24 | 2.32 | 4.52 | 6.66 | 8.78 | 17.09 | 25.22 | 33.24 | 41.18 | 49.06 | 56.88 | 64.66 | 72.40 | 63.21 |
| 25 | 2.42 | 4.70 | 6.94 | 9.15 | 17.80 | 26.27 | 34.62 | 42.89 | 51.10 | 59.25 | 67.35 | 75.42 | 67.20 |
| 26 | 2.51 | 4.89 | 7.22 | 9.52 | 18.51 | 27.32 | 36.01 | 44.61 | 53.14 | 61.62 | 70.05 | 78.43 | 71.27 |
| 28 | 2.71 | 5.27 | 7.77 | 10.25 | 19.93 | 29.42 | 38.78 | 48.04 | 57.23 | 66.36 | 75.44 | 84.47 | 79.65 |
| 30 | 2.90 | 5.64 | 8.33 | 10.98 | 21.36 | 31.52 | 41.55 | 51.47 | 61.32 | 71.10 | 80.82 | 90.50 | 88.33 |
| 32 | 3.09 | 6.02 | 8.89 | 11.71 | 22.78 | 33.62 | 44.32 | 54.91 | 65.41 | 75.84 | 86.21 | 96.53 | 97.31 |
| 35 | 3.38 | 6.58 | 9.72 | 12.81 | 24.92 | 36.78 | 48.47 | 60.05 | 71.54 | 82.95 | 94.29 | 105.58 | 111.31 |
| 40 | 3.87 | 7.53 | 11.11 | 14.64 | 28.48 | 42.03 | 55.40 | 68.63 | 81.76 | 94.80 | 107.77 | 120.67 | 133.51 |
| 45 | 4.35 | 8.47 | 12.49 | 16.47 | 32.04 | 47.28 | 62.32 | 77.21 | 91.98 | 106.65 | 121.24 | 135.75 | 150.20 |
| | Type A | | | Type B | | | | | | | Type C | | |

**Table 9.** *(Continued)* **Horsepower Ratings for Roller Chains,** *ANSI/ASME B29.1-2011 (R2022)*

**1¼-inch Pitch Standard Single-Strand Roller Chain — No. 100**

| No. of Teeth Small Spkt. | Revolutions per Minute — Small Sprocket[a] | | | | | | | | | | | | |
|---|---|---|---|---|---|---|---|---|---|---|---|---|---|
| | 10 | 25 | 50 | 100 | 150 | 200 | 300 | 400 | 500 | 600 | 700 | 800 | 900 |
| | Horsepower Rating | | | | | | | | | | | | |
| 11 | 0.85 | 2.04 | 3.96 | 7.71 | 11.38 | 15.00 | 22.14 | 29.18 | 36.15 | 43.06 | 40.03 | 32.77 | 24.76 |
| 12 | 0.92 | 2.22 | 4.32 | 8.41 | 12.41 | 16.36 | 24.15 | 31.83 | 39.44 | 46.98 | 45.61 | 37.33 | 31.29 |
| 13 | 1.00 | 2.41 | 4.68 | 9.11 | 13.45 | 17.73 | 26.16 | 34.48 | 42.72 | 50.89 | 51.43 | 42.10 | 35.28 |
| 14 | 1.08 | 2.59 | 5.04 | 9.81 | 14.48 | 19.09 | 28.18 | 37.14 | 46.01 | 54.81 | 57.48 | 47.05 | 39.43 |
| 15 | 1.15 | 2.78 | 5.41 | 10.51 | 15.52 | 20.45 | 30.19 | 39.79 | 49.30 | 58.72 | 63.75 | 52.18 | 43.73 |
| 16 | 1.23 | 2.96 | 5.77 | 11.22 | 16.55 | 21.82 | 32.20 | 42.44 | 52.58 | 62.64 | 70.23 | 57.48 | 48.17 |
| 17 | 1.31 | 3.15 | 6.13 | 11.92 | 17.59 | 23.18 | 34.21 | 45.10 | 55.87 | 66.55 | 76.91 | 62.95 | 52.76 |
| 18 | 1.38 | 3.33 | 6.49 | 12.62 | 18.62 | 24.55 | 36.23 | 47.75 | 59.15 | 70.47 | 81.71 | 68.59 | 57.48 |
| 19 | 1.46 | 3.52 | 6.85 | 13.32 | 19.66 | 25.91 | 38.24 | 50.40 | 62.44 | 74.38 | 86.25 | 74.38 | 62.34 |
| 20 | 1.54 | 3.70 | 7.21 | 14.02 | 20.69 | 27.27 | 40.25 | 53.05 | 65.73 | 78.30 | 90.79 | 80.33 | 67.32 |
| 21 | 1.61 | 3.89 | 7.57 | 14.72 | 21.73 | 28.64 | 42.26 | 55.71 | 69.01 | 82.21 | 95.33 | 86.43 | 72.43 |
| 22 | 1.69 | 4.08 | 7.93 | 15.42 | 22.76 | 30.00 | 44.28 | 58.36 | 72.30 | 86.13 | 99.87 | 92.68 | 77.67 |
| 23 | 1.77 | 4.26 | 8.29 | 16.12 | 23.79 | 31.36 | 46.29 | 61.01 | 75.59 | 90.04 | 104.41 | 99.07 | 83.02 |
| 24 | 1.84 | 4.45 | 8.65 | 16.82 | 24.83 | 32.73 | 48.30 | 63.66 | 78.87 | 93.96 | 108.95 | 105.60 | 88.50 |
| 25 | 1.92 | 4.63 | 9.01 | 17.52 | 25.86 | 34.09 | 50.31 | 66.32 | 82.16 | 97.87 | 113.48 | 112.27 | 94.09 |
| 26 | 2.00 | 4.82 | 9.37 | 18.23 | 26.90 | 35.45 | 52.33 | 68.97 | 85.45 | 101.79 | 118.02 | 119.07 | 99.79 |
| 28 | 2.15 | 5.19 | 10.09 | 19.63 | 28.97 | 38.18 | 56.35 | 74.27 | 92.02 | 109.62 | 127.10 | 133.07 | 111.52 |
| 30 | 2.31 | 5.56 | 10.81 | 21.03 | 31.04 | 40.91 | 60.38 | 79.58 | 98.59 | 117.45 | 136.18 | 147.58 | 123.68 |
| 32 | 2.46 | 5.93 | 11.53 | 22.43 | 33.11 | 43.64 | 64.40 | 84.88 | 105.16 | 125.28 | 145.26 | 162.58 | 136.25 |
| 35 | 2.69 | 6.48 | 12.61 | 24.53 | 36.21 | 47.73 | 70.44 | 92.84 | 115.02 | 137.02 | 158.88 | 180.61 | 155.85 |
| 40 | 3.07 | 7.41 | 14.41 | 28.04 | 41.38 | 54.54 | 80.50 | 106.11 | 131.45 | 156.60 | 181.58 | 206.41 | 190.42 |
| 45 | 3.46 | 8.34 | 16.22 | 31.54 | 46.55 | 61.36 | 90.56 | 119.37 | 147.89 | 176.17 | 204.27 | 232.21 | 227.21 |
| | Type A | | | Type B | | | | | | Type C | | | |

**1½-inch Pitch Standard Single-Strand Roller Chain — No. 120**

| No. of Teeth Small Spkt. | Revolutions per Minute — Small Sprocket[a] | | | | | | | | | | | | |
|---|---|---|---|---|---|---|---|---|---|---|---|---|---|
| | 10 | 25 | 50 | 100 | 150 | 200 | 300 | 400 | 500 | 600 | 700 | 800 | 900 |
| | Horsepower Rating | | | | | | | | | | | | |
| 11 | 1.43 | 3.44 | 6.69 | 13.02 | 19.22 | 25.33 | 37.38 | 49.27 | 61.04 | 58.37 | 46.32 | 37.91 | 31.77 |
| 12 | 1.56 | 3.75 | 7.30 | 14.20 | 20.96 | 27.63 | 40.78 | 53.75 | 66.59 | 66.51 | 52.78 | 43.20 | 36.20 |
| 13 | 1.69 | 4.07 | 7.91 | 15.39 | 22.71 | 29.93 | 44.18 | 58.23 | 72.14 | 74.99 | 59.51 | 48.71 | 40.82 |
| 14 | 1.82 | 4.38 | 8.52 | 16.57 | 24.46 | 32.24 | 47.58 | 62.71 | 77.69 | 83.81 | 66.51 | 54.44 | 45.62 |
| 15 | 1.95 | 4.69 | 9.13 | 17.76 | 26.20 | 34.54 | 50.98 | 67.19 | 83.24 | 92.95 | 73.76 | 60.37 | 50.59 |
| 16 | 2.08 | 5.00 | 9.74 | 18.94 | 27.95 | 36.84 | 54.37 | 71.67 | 88.79 | 102.39 | 81.26 | 66.51 | 55.74 |
| 17 | 2.21 | 5.32 | 10.34 | 20.12 | 29.70 | 39.14 | 57.77 | 76.15 | 94.34 | 112.14 | 88.99 | 72.84 | 61.04 |
| 18 | 2.34 | 5.63 | 10.95 | 21.31 | 31.45 | 41.45 | 61.17 | 80.63 | 99.89 | 119.00 | 96.96 | 79.36 | 66.51 |
| 19 | 2.47 | 5.94 | 11.56 | 22.49 | 33.19 | 43.75 | 64.57 | 85.11 | 105.44 | 125.61 | 105.15 | 86.06 | 72.13 |
| 20 | 2.60 | 6.26 | 12.17 | 23.67 | 34.94 | 46.05 | 67.97 | 89.59 | 110.99 | 132.22 | 113.56 | 92.95 | 77.89 |
| 21 | 2.73 | 6.57 | 12.78 | 24.86 | 36.69 | 48.36 | 71.37 | 94.07 | 116.54 | 138.83 | 122.18 | 100.00 | 83.81 |
| 22 | 2.86 | 6.88 | 13.39 | 26.04 | 38.43 | 50.66 | 74.76 | 98.55 | 122.09 | 145.44 | 131.01 | 107.23 | 89.87 |
| 23 | 2.99 | 7.19 | 14.00 | 27.22 | 40.18 | 52.96 | 78.16 | 103.02 | 127.64 | 152.05 | 140.04 | 114.62 | 96.06 |
| 24 | 3.11 | 7.51 | 14.60 | 28.41 | 41.93 | 55.26 | 81.56 | 107.50 | 133.19 | 158.66 | 149.28 | 122.18 | 102.39 |
| 25 | 3.24 | 7.82 | 15.21 | 29.59 | 43.67 | 57.57 | 84.96 | 111.98 | 138.74 | 165.27 | 158.70 | 129.90 | 108.86 |
| 26 | 3.37 | 8.13 | 15.82 | 30.78 | 45.42 | 59.87 | 88.36 | 116.46 | 144.29 | 171.88 | 168.32 | 137.77 | 115.46 |
| 28 | 3.63 | 8.76 | 17.04 | 33.14 | 48.92 | 64.47 | 95.15 | 125.42 | 155.38 | 185.11 | 188.11 | 153.97 | 129.03 |
| 30 | 3.89 | 9.38 | 18.25 | 35.51 | 52.41 | 69.08 | 101.95 | 134.38 | 166.48 | 198.33 | 208.62 | 170.75 | 143.10 |
| 32 | 4.15 | 10.01 | 19.47 | 37.88 | 55.90 | 73.68 | 108.75 | 143.34 | 177.58 | 211.55 | 229.83 | 188.11 | 157.65 |
| 35 | 4.54 | 10.95 | 21.30 | 41.43 | 61.14 | 80.59 | 118.94 | 156.78 | 194.23 | 231.38 | 262.89 | 215.17 | 180.33 |
| 40 | 5.19 | 12.51 | 24.34 | 47.35 | 69.88 | 92.11 | 135.94 | 179.17 | 221.98 | 264.44 | 306.61 | 262.89 | 220.32 |
| 45 | 5.84 | 14.08 | 27.38 | 53.27 | 78.61 | 103.62 | 152.93 | 201.57 | 249.72 | 297.49 | 344.94 | 313.69 | 213.33 |
| | Type A | | | Type B | | | | | | Type C | | | |

[a] For lower or higher rpm, larger chain sizes, and rpm above 3500, see B29.1-2011 (R2022).

For use of table see page 2632.

**Table 10. Recommended Roller Chain Sprocket Maximum Bore and Hub Diameters**

| | Roller Chain Pitch | | | | | | | | | |
|---|---|---|---|---|---|---|---|---|---|---|
| | 3/8 | | 1/2 | | 5/8 | | 3/4 | | 1 | |
| No. of Teeth | Max. Bore | Max. Hub Dia. | Max. Bore | Max. Hub Dia. | Max. Bore | Max. Hub Dia. | Max. Bore | Max. Hub Dia. | Max. Bore | Max. Hub Dia. |
| 11 | 19/32 | 55/64 | 25/32 | 1 11/64 | 31/32 | 1 15/32 | 1 1/4 | 1 49/64 | 1 5/8 | 2 3/8 |
| 12 | 5/8 | 63/64 | 7/8 | 1 21/64 | 1 5/32 | 1 43/64 | 1 9/32 | 2 1/64 | 1 25/32 | 2 45/64 |
| 13 | 3/4 | 1 7/64 | 1 | 1 1/2 | 1 9/32 | 1 7/8 | 1 1/2 | 2 1/4 | 2 | 3 1/64 |
| 14 | 27/32 | 1 15/64 | 1 5/32 | 1 21/32 | 1 5/16 | 2 5/64 | 1 3/4 | 2 1/2 | 2 9/32 | 3 11/32 |
| 15 | 7/8 | 1 23/64 | 1 1/4 | 1 13/16 | 1 17/32 | 2 9/32 | 1 25/32 | 2 3/4 | 2 13/32 | 3 43/64 |
| 16 | 31/32 | 1 15/32 | 1 9/32 | 1 63/64 | 1 11/16 | 2 31/64 | 1 31/32 | 2 63/64 | 2 23/32 | 3 63/64 |
| 17 | 1 3/32 | 1 19/32 | 1 3/8 | 2 9/64 | 1 25/32 | 2 11/16 | 2 7/32 | 3 7/32 | 2 13/16 | 4 5/16 |
| 18 | 1 7/32 | 1 23/32 | 1 17/32 | 2 19/64 | 1 7/8 | 2 57/64 | 2 9/32 | 3 15/32 | 3 1/8 | 4 41/64 |
| 19 | 1 1/4 | 1 27/32 | 1 11/16 | 2 29/64 | 2 1/16 | 3 5/64 | 2 7/16 | 3 45/64 | 3 5/16 | 4 61/64 |
| 20 | 1 9/32 | 1 61/64 | 1 25/32 | 2 5/8 | 2 1/4 | 3 9/32 | 2 11/16 | 3 61/64 | 3 1/2 | 5 9/32 |
| 21 | 1 5/16 | 2 5/64 | 1 25/32 | 2 25/32 | 2 9/32 | 3 31/64 | 2 13/16 | 4 3/16 | 3 3/4 | 5 19/32 |
| 22 | 1 7/16 | 2 13/16 | 1 15/16 | 2 15/16 | 2 7/16 | 3 11/16 | 2 15/16 | 4 7/16 | 3 7/8 | 5 59/64 |
| 23 | 1 9/16 | 2 5/16 | 2 3/32 | 3 3/32 | 2 5/8 | 3 57/64 | 3 1/8 | 4 43/64 | 4 3/16 | 6 15/64 |
| 24 | 1 11/16 | 2 7/16 | 2 1/4 | 3 17/64 | 2 13/16 | 4 5/64 | 3 1/4 | 4 29/32 | 4 9/16 | 6 9/16 |
| 25 | 1 3/4 | 2 9/16 | 2 9/32 | 3 27/64 | 2 27/32 | 4 9/32 | 3 3/8 | 5 5/32 | 4 11/16 | 6 7/8 |
| | Roller Chain Pitch | | | | | | | | | |
| | 1 1/4 | | 1 1/2 | | 1 3/4 | | 2 | | 2 1/2 | |
| No. of Teeth | Max. Bore | Max. Hub Dia. | Max. Bore | Max. Hub Dia. | Max. Bore | Max. Hub Dia. | Max. Bore | Max. Hub Dia. | Max. Bore | Max. Hub Dia. |
| 11 | 1 31/32 | 2 31/32 | 2 5/16 | 3 37/64 | 2 13/16 | 4 11/64 | 3 9/32 | 4 25/32 | 3 15/16 | 5 [illegible]/64 |
| 12 | 2 9/32 | 3 3/8 | 2 3/4 | 4 1/16 | 3 1/4 | 4 3/4 | 3 5/8 | 5 27/64 | 4 23/32 | 6 5[illegible]/64 |
| 13 | 2 17/32 | 3 25/32 | 3 1/16 | 4 35/64 | 3 9/16 | 5 5/16 | 4 1/16 | 6 5/64 | 5 3/32 | 7 3[illegible]/64 |
| 14 | 2 11/16 | 4 3/16 | 3 5/16 | 5 1/32 | 3 7/8 | 5 7/8 | 4 11/16 | 6 23/32 | 5 23/32 | 8 2[illegible]/64 |
| 15 | 3 3/32 | 4 19/32 | 3 3/4 | 5 33/64 | 4 7/16 | 6 29/64 | 4 7/8 | 7 3/8 | 6 1/4 | 9 [illegible]/32 |
| 16 | 3 9/32 | 5 | 4 | 6 | 4 11/16 | 7 1/64 | 5 1/2 | 8 1/64 | 7 | 10 [illegible]/32 |
| 17 | 3 21/32 | 5 13/32 | 4 15/32 | 6 31/64 | 5 1/16 | 7 37/64 | 5 11/16 | 8 21/32 | 7 7/16 | 10 [illegible]/32 |
| 18 | 3 25/32 | 5 51/64 | 4 21/32 | 6 31/32 | 5 5/8 | 8 9/64 | 6 1/4 | 9 5/16 | 8 1/8 | 11 [illegible]/64 |
| 19 | 4 3/16 | 6 13/64 | 4 15/16 | 7 29/64 | 5 11/16 | 8 45/64 | 6 7/8 | 9 61/64 | 9 | 12 [illegible]/16 |
| 20 | 4 19/32 | 6 39/64 | 5 7/16 | 7 15/16 | 6 1/4 | 9 17/64 | 7 | 10 19/32 | 9 3/4 | 13 [illegible]/4 |
| 21 | 4 11/16 | 7 | 5 11/16 | 8 27/64 | 6 13/16 | 9 53/64 | 7 3/4 | 11 15/64 | 10 | 14 [illegible]/64 |
| 22 | 4 7/8 | 7 13/32 | 5 7/8 | 8 57/64 | 7 1/4 | 10 25/64 | 8 3/8 | 11 7/8 | 10 7/8 | 14 [illegible]7/32 |
| 23 | 5 5/16 | 7 13/16 | 6 3/8 | 9 3/8 | 7 7/16 | 10 15/16 | 9 | 12 33/64 | 11 5/8 | 1[illegible] 21/32 |
| 24 | 5 11/16 | 8 13/64 | 6 13/16 | 9 55/64 | 8 | 11 1/2 | 9 5/8 | 13 5/32 | 13 | 1[illegible] 29/64 |
| 25 | 5 23/32 | 8 39/64 | 7 1/4 | 10 11/32 | 8 9/16 | 12 1/16 | 10 1/4 | 13 51/64 | 13 1/2 | 1[illegible] 1/4 |

All dimensions in inches.

For standard key dimensions see pages 2545 through 2558.

*Source:* American Chain Association.

**Center Distance between Sprockets.—**The center-to-center distance between sprockets, as a general rule, should not be less than $1\frac{1}{2}$ times the diameter of the larger sprocket and not less than 30 times the pitch nor more than about 50 times the pitch, although much depends upon the speed and other conditions. A center distance equivalent to 80 pitches may be considered an approved maximum. Very long center distances result in catenary tension in the chain. If roller-chain drives are designed correctly, the center-to-center distance for some transmissions may be so short that the sprocket teeth nearly touch each other, assuming that the load is not too great and the number of teeth is not too small. To avoid interference of the sprocket teeth, the center distance must, of course, be somewhat greater than one-half the sum of the outside diameters of the sprockets. The chain should extend around at least 120 degrees of the pinion circumference, and this minimum amount of contact is obtained for all center distances provided the ratio is less than $3^1/_2$ to 1. Other things being equal, a fairly long chain is recommended in preference to the shortest one allowed by the sprocket diameters, because the rate of chain elongation due to natural wear is inversely proportional to the length, and also because the greater elasticity of the longer strand tends to absorb irregularities of motion and to decrease the effect of shocks.

If possible, the center distance should be adjustable in order to take care of slack due to elongation from wear and this range of adjustment should be at least one and one-half pitches. A little slack is desirable as it allows the chain links to take the best position on the sprocket teeth and reduces the wear on the bearings. Too much sag or an excessive distance between the sprockets may cause the chain to whip up and down — a condition detrimental to smooth running and very destructive to the chain. The sprockets should run in a vertical plane, the sprocket axes being approximately horizontal, unless an idler is used on the slack side to keep the chain in position. The most satisfactory results are obtained when the slack side of the chain is on the bottom.

**Center Distance for a Given Chain Length.**—When the distance between the driving and driven sprockets can be varied to suit the length of the chain, this center distance for a tight chain may be determined by the following formula, in which $c$ = center-to-center distance in inches; $L$ = chain length in pitches; $P$ = pitch of chain; $N$ = number of teeth in large sprocket; $n$ = number of teeth in small sprocket.

$$c = \frac{P}{8}\left(2L - N - n + \sqrt{(2L - N - n)^2 - 0.810(N - n)^2}\right)$$

This formula is approximate, but the error is less than the variation in the length of the best chains. The length $L$ in pitches should be an even number for a roller chain, so that the use of an offset connecting link will not be necessary.

**Idler Sprockets.**—When sprockets have a fixed center distance or are non-adjustable, it may be advisable to use an idler sprocket for taking up the slack. The idler should preferably be placed against the slack side between the two strands of the chain. When a sprocket is applied to the tight side of the chain to reduce vibration, it should be on the lower side and so located that the chain will run in a straight line between the two main sprockets. A sprocket will wear excessively if the number of teeth is too small and the speed too high, because there is impact between the teeth and rollers even though the idler carries practically no load.

**Length of Driving Chains.**—The total length of a block chain should be given in multiples of the pitch, whereas for a roller chain, the length should be in multiples of twice the pitch, because the ends must be connected with an outside and inside link. The length of a chain can be calculated accurately enough for ordinary practice by the use of the following formula, in which $L$ = chain length in pitches; $C$ = center distance in pitches; $N$ = number of teeth in large sprocket; $n$ = number of teeth in small sprocket:

$$L = 2C + \frac{N}{2} + \frac{n}{2} + \left(\frac{N - n}{2\pi}\right)^2 \times \frac{1}{C}$$

## Table 11. ANSI Sprocket Tooth Form for Roller Chains

*ANSI/ASME B29.1-2011 (R2022)*

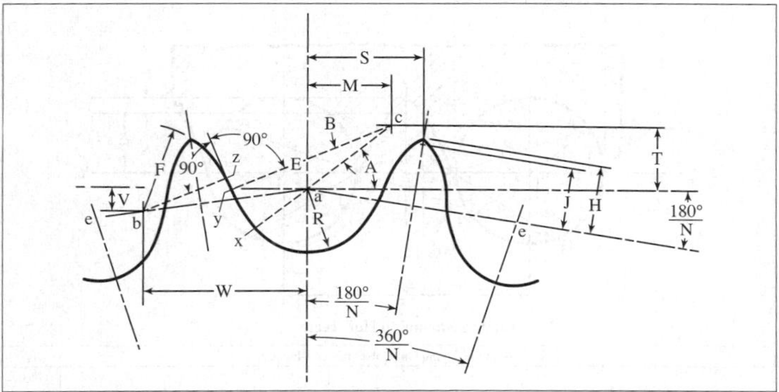


| Seating Curve Data—Inches | | | | | | | | | |
|---|---|---|---|---|---|---|---|---|---|
| $P$ | $D_r$ | Min. $R$ | Min. $D_s$ | $D_s$ Tol.[a] | $P$ | $D_r$ | Min. $R$ | Min. $D_s$ | $D_s$ Tol.[a] |
| 0.250 | 0.130 | 0.0670 | 0.134 | 0.0055 | 1.250 | 0.750 | 0.3785 | 0.757 | 0.0070 |
| 0.375 | 0.200 | 0.1020 | 0.204 | 0.0055 | 1.500 | 0.875 | 0.4410 | 0.882 | 0.0075 |
| 0.500 | 0.306 | 0.1585 | 0.317 | 0.0060 | 1.750 | 1.000 | 0.5040 | 1.008 | 0.0080 |
| 0.500 | 0.312 | 0.1585 | 0.317 | 0.0060 | 2.000 | 1.125 | 0.5670 | 1.134 | 0.0085 |
| 0.625 | 0.400 | 0.2025 | 0.405 | 0.0060 | 2.250 | 1.406 | 0.7080 | 1.416 | 0.0090 |
| 0.750 | 0.469 | 0.2370 | 0.474 | 0.0065 | 2.500 | 1.562 | 0.7870 | 1.573 | 0.0095 |
| 1.000 | 0.625 | 0.3155 | 0.631 | 0.0070 | 3.000 | 1.875 | 0.9435 | 1.887 | 0.0105 |

[a] Plus tolerance only.

$P$ = pitch ($ae$)

$N$ = number of teeth  $D_r$ = nominal roller diameter

$D_s$ = seating curve diameter = $1.005\,D_r + 0.003$ (in inches)

$R = {}^1\!/_2\,D_s$  ($D_s$ has only plus tolerance)

$A = 35° + (60° \div N)$  $B = 18° - (56° \div N)$  $ac = 0.8D_r$

$M = 0.8\,D_r \cos(35° + (60° \div N))$

$T = 0.8\,D_r \sin(35° + (60° \div N))$

$E = 1.3025\,D_r + 0.0015$ (in inches)

*Chord* $xy = (2.605\,D_r + 0.003)\sin(9° - (28° \div N))$ (in inches)

$yz = D_r\,[1.4\sin(17° - (64° \div N)) - 0.8\sin(18° - (56° \div N))]$

Length of a line between $a$ and $b = 1.4\,D_r$

$W = 1.4\,D_r \cos(180° \div N)$  $V = 1.4\,D_r \sin(180° \div N)$

$F = D_r\,[0.8\cos(18° - (56° \div N)) + 1.4\cos(17° - (64° \div N)) - 1.3025] - 0.0015$ inch

$H = \sqrt{F^2 - (1.4D_r - 0.5P)^2}$

$S = 0.5\,P\cos(180° \div N) + H\sin(180° \div N)$

Approximate OD of sprocket when $J$ is $0.3\,P = P\,[0.6 + \cot(180° \div N)]$

OD of sprocket when tooth is pointed $+ P\cot(180° \div N) + \cos(180° \div N)(D_s - D_r) + 2H$

Pressure angle for new chain = $xab = 35° - (120° \div N)$

Minimum pressure angle = $xab - B = 17° - (64° \div N)$;

Average pressure angle = $26° - (92° \div N)$

**Table 12. Standard Hob Design for Roller Chain Sprockets**

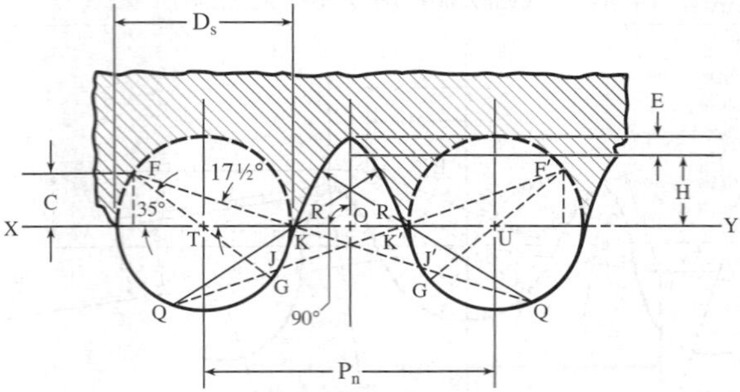


**Section Normal to Hob Teeth**

Data for Laying Out Hob Outlines—Inches

| $P$ | $P_n$ | $H$ | $E$ | O.D. | $W$ | Bore | Keyway | No. Gashes |
|---|---|---|---|---|---|---|---|---|
| ¼ | 0.2527 | 0.0675 | 0.0075 | 2⅝ | 2½ | 1.250 | ¼ × ⅛ | 13 |
| ⅜ | 0.379 | 0.101 | 0.012 | 3⅛ | 2½ | 1.250 | ¼ × ⅛ | 13 |
| ½ | 0.506 | 0.135 | 0.015 | 3⅜ | 2½ | 1.250 | ¼ × ⅛ | 12 |
| ⅝ | 0.632 | 0.170 | 0.018 | 3⅝ | 2½ | 1.250 | 1¼ × ⅛ | 12 |
| ¾ | 0.759 | 0.202 | 0.023 | 3¾ | 2⅞ | 1.250 | ¼ × ⅛ | 11 |
| 1 | 1.011 | 0.270 | 0.030 | 4⅜ | 3¾ | 1.250 | ¼ × ⅛ | 11 |
| 1¼ | 1.264 | 0.337 | 0.038 | 4¾ | 4½ | 1.250 | ¼ × ⅛ | 10 |
| 1½ | 1.517 | 0.405 | 0.045 | 5⅜ | 5¼ | 1.250 | ¼ × ⅛ | 10 |
| 1¾ | 1.770 | 0.472 | 0.053 | 6⅜ | 6 | 1.500 | ⅜ × 3⁄16 | 9 |
| 2 | 2.022 | 0.540 | 0.060 | 6⅞ | 6¾ | 1.500 | ⅜ × 3⁄16 | 9 |
| 2¼ | 2.275 | 0.607 | 0.068 | 8 | 8½ | 1.750 | ⅜ × 3⁄16 | 8 |
| 2½ | 2.528 | 0.675 | 0.075 | 8⅝ | 9⅜ | 1.750 | ⅜ × 3⁄16 | 8 |
| 3 | 3.033 | 0.810 | 0.090 | 9¾ | 11¼ | 2.000 | ½ × 3⁄6 | 8 |

Hobs designed for a given roller diameter ($D_r$) and chain pitch ($P$) will cut any number of teeth.

$P$ = Pitch of Chain

$P_n$ = Normal Pitch of Hob = $1.011P$ inches

$D_s$ = Minimum Diameter of Seating Curve = $1.005D_r + 0.003$ inches

$F$ = Radius Center for Arc $GK$; $TO = OU = P_n \div 2$

$H = 0.27P$ $E = 0.03P$ = Radius of Fillet Circle

$Q$ is located on line passing through $F$ and $J$. Point $J$ is intersection of line $XY$ with circle of diameter $D_s$. $R$ is found by trial and the arc of this radius is tangent to arc $K_G$ at $K$ and to fillet radius.

$OD$ = Outside Diameter = $1.7(\text{Bore} + D_r + 0.7P)$ approx.

$D_h$ = Pitch Diameter = $OD - D_s$; $M$ = Helix Angle; $\sin M = P_n \div \pi D_h$

$L$ = Lead = $P_n \div \cos M$; $W$ = Width = Not less than $2 \times$ Bore, or $6D_r$, or $3.2P$

To the length obtained by this formula, add enough to make a whole number (and for a roller chain, an even number) of pitches. If a roller chain has an odd number of pitches, it will be necessary to use an offset connecting link.

Another formula for obtaining chain length in which $D$ = distance between centers of shafts; $R$ = pitch radius of large sprocket; $r$ = pitch radius of small sprocket: $N$ = number of teeth in large sprocket; $n$ = number of teeth in small sprocket; $P$ = pitch of chain and sprockets; and $l$ = required chain length in inches, is:

$$l = \frac{180° + 2\alpha}{360°}NP + \frac{180° - 2\alpha}{360°}nP + 2D\cos\alpha \quad \text{where} \quad \sin\alpha = \frac{R - r}{D}$$

**Cutting Standard Sprocket Tooth Form.—**The proportions and seating curve data for the standard sprocket tooth form for roller chain are given in Table 11. Either formed or generating types of sprocket cutters may be employed.

*Hobs:* Only one hob will be required to cut any number of teeth for a given pitch and roller diameter. All hobs should be marked with pitch and roller diameter to be cut. Formulas and data for standard hob design are given in Table 12.

*Space Cutters:* Five cutters of this type will be required to cut from 7 teeth up for any given roller diameter. The ranges are, respectively, 7–8, 9–11, 12–17, 18–34, and 35 teeth and over. If less than 7 teeth is necessary, special cutters conforming to the required number of teeth should be used.

The regular cutters are based upon an intermediate number of teeth $N_a$, equal to $2N_1N_2 \div (N_1 + N_2)$ in which $N_1$ = minimum number of teeth and $N_2$ = maximum number of teeth for which cutter is intended; but the topping curve radius $F$ (see diagram in Table 13) is designed to produce adequate tooth height on a sprocket of $N_2$ teeth. The values of $N_a$ for the several cutters are, respectively, 7.47, 9.9, 14.07, 23.54, and 56. Formulas and construction data for space cutter layout are given in Table 13 and recommended cutter sizes are given in Table 14.

**Table 13. Standard Space Cutters for Roller-Chain Sprockets**

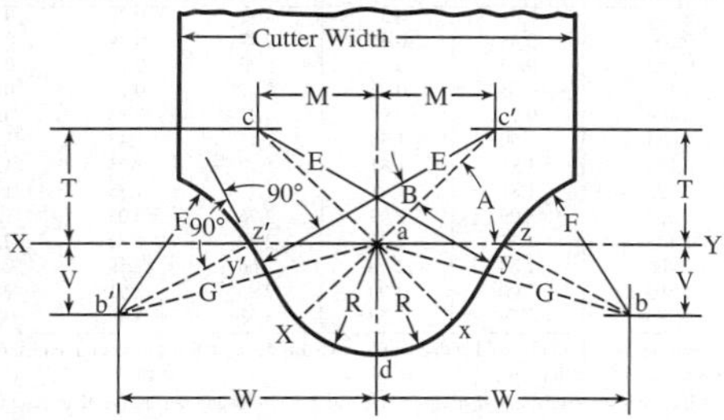


Data for Laying Out Space Cutter

| Range of Teeth | *M* | *T* | *W* | *V* |
|---|---|---|---|---|
| 7–8 | $0.5848D_r$ | $0.5459D_r$ | $1.2790D_r$ | 0.569[illegible]$D_r$ |
| 9–11 | $0.6032D_r$ | $0.5255D_r$ | $1.3302D_r$ | 0.436[illegible]$D_r$ |
| 12–17 | $0.6194D_r$ | $0.5063D_r$ | $1.3694D_r$ | 0.291[illegible]$D_r$ |
| 18–34 | $0.6343D_r$ | $0.4875D_r$ | $1.3947D_r$ | 0.122[illegible]$D_r$ |
| 35 up | $0.6466D_r$ | $0.4710D_r$ | $1.4000D_r$ | 0 |
| **Range of Teeth** | ***F*** | **Chord *xy*** | ***yz*** | **Angle *Yab*** |
| 7–8 | $0.8686D_r - 0.0015$ | $0.2384D_r + 0.0003$ | $0.0618D_r$ | 24° |
| 9–11 | $0.8554D_r - 0.0015$ | $0.2800D_r + 0.0003$ | $0.0853D_r$ | 18°10′ |
| 12–17 | $0.8364D_r - 0.0015$ | $0.3181D_r + 0.0004$ | $0.1269D_r$ | 12° |
| 18–34 | $0.8073D_r - 0.0015$ | $0.3540D_r + 0.0004$ | $0.1922D_r$ | 5° |
| 35 up | $0.7857D_r - 0.0015$ | $0.3850D_r + 0.0004$ | $0.2235D_r$ | 0° |

$E$ (same for all ranges) = $1.3025D_r + 0.0015$; $G$ (same for all ranges) = $1.4D_r$

See Table 14 for recommended cutter sizes.

Angle *Yab* is equal to $180° \div N$ when the cutter is made for a specific number of teeth. For the design of cutters covering a range of teeth, angle *Yab* was determined by layout to ensure chain roller clearance and to avoid pointed teeth on the larger sprockets of each range. It has values as given below for cutters covering the range of teeth shown. The following formulas are for cutters covering the standard ranges of teeth where $N_a$ equals intermediate values given on page 2641.

$$W = 1.4D_r\cos(Yab) \qquad V = 1.4D_r\sin(Yab)$$

$$yz = D_r\left[1.4\sin\left(17° + \frac{116°}{N_a} - Yab\right) - 0.8\sin\left(18° - \frac{56°}{N_a}\right)\right]$$

$$F = D_r\left[0.8\cos\left(18° - \frac{56°}{N_a}\right) + 1.4\cos\left(17° + \frac{116°}{N_a} - Yab\right) - 1.3025\right] - 0.0015\text{ in.}$$

For other points, use the value of $N_a$ for $N$ in the standard formulas in Table 11.

**Table 14. Recommended Space Cutter Sizes for Roller-Chain Sprockets**

| | | Number of Teeth | | | | | |
|---|---|---|---|---|---|---|---|
| | | 6 | 7-8 | 9-11 | 12-17 | 18-34 | 35 up |
| Pitch | Roller Dia. | Cutter Diameter (Minimum) | | | | | |
| 0.250 | 0.130 | 2.75 | 2.75 | 2.75 | 2.75 | 2.75 | 2.75 |
| 0.375 | 0.200 | 2.75 | 2.75 | 2.75 | 2.75 | 2.75 | 2.75 |
| 0.500 | 0.312 | 3.00 | 3.00 | 3.12 | 3.12 | 3.12 | 3.12 |
| 0.625 | 0.400 | 3.12 | 3.12 | 3.25 | 3.25 | 3.25 | 3.25 |
| 0.725 | 0.469 | 3.25 | 3.25 | 3.38 | 3.38 | 3.38 | 3.38 |
| 1.000 | 0.625 | 3.88 | 4.00 | 4.12 | 4.12 | 4.25 | 4.25 |
| 1.250 | 0.750 | 4.25 | 4.38 | 4.50 | 4.50 | 4.62 | 4.62 |
| 1.500 | 0.875 | 4.38 | 4.50 | 4.62 | 4.62 | 4.75 | 4.75 |
| 1.750 | 1.000 | 5.00 | 5.12 | 5.25 | 5.38 | 5.50 | 5.50 |
| 2.000 | 1.125 | 5.38 | 5.50 | 5.62 | 5.75 | 5.88 | 5.88 |
| 2.250 | 1.406 | 5.88 | 6.00 | 6.25 | 6.38 | 6.50 | 6.50 |
| 2.500 | 1.563 | 6.38 | 6.62 | 6.75 | 6.88 | 7.00 | 7.12 |
| 3.000 | 1.875 | 7.50 | 7.75 | 7.88 | 8.00 | 8.00 | 8.25 |
| Pitch | Roller Dia. | Cutter Width (Minimum) | | | | | |
| 0.250 | 0.130 | 0.31 | 0.31 | 0.31 | 0.31 | 0.28 | 0.28 |
| 0.375 | 0.200 | 0.47 | 0.47 | 0.47 | 0.44 | 0.44 | 0.41 |
| 0.500 | 0.312 | 0.75 | 0.75 | 0.75 | 0.75 | 0.72 | 0.69 |
| 0.625 | 0.400 | 0.75 | 0.75 | 0.75 | 0.75 | 0.72 | 0.69 |
| 0.750 | 0.469 | 0.91 | 0.91 | 0.91 | 0.88 | 0.84 | 0.81 |
| 1.000 | 0.625 | 1.50 | 1.50 | 1.47 | 1.47 | 1.41 | 1.34 |
| 1.250 | 0.750 | 1.81 | 1.81 | 1.78 | 1.75 | 1.69 | 1.62 |
| 1.500 | 0.875 | 1.81 | 1.81 | 1.78 | 1.75 | 1.69 | 1.62 |
| 1.750 | 1.000 | 2.09 | 2.09 | 2.06 | 2.03 | 1.97 | 1.88 |
| 2.000 | 1.125 | 2.41 | 2.41 | 2.38 | 2.31 | 2.25 | 2.16 |
| 2.250 | 1.406 | 2.69 | 2.69 | 2.66 | 2.59 | 2.47 | 2.41 |
| 2.500 | 1.563 | 3.00 | 3.00 | 2.94 | 2.91 | 2.75 | 2.69 |
| 3.000 | 1.875 | 3.59 | 3.59 | 3.53 | 3.47 | 3.34 | 3.22 |

Where the same roller diameter is commonly used with chains of two different pitches it is recommended that stock cutters be made wide enough to cut sprockets for both chains.

*Marking of Cutters.*—All cutters are to be marked, giving pitch, roller diameter and range of teeth to be cut.

*Bores for Sprocket Cutters* (recommended practice) are approximately as calculated from the formula:

$$\text{Bore} = 0.7\sqrt{(\text{Width of Cutter} + \text{Roller Diameter} + 0.7\text{ Pitch})}$$

and are equal to 1 inch for ¼- through ¾-inch pitches; 1¼ inches for 1- through 1½-inch for 1¾- through 2¼-inch pitches; 1¾ inches for 2½-inch pitch; and 2 inches for 3-inch pitch.

*Minimum Outside Diameters of Space Cutters* for 35 teeth and over (recommended practice) are approximately as calculated from the formula:

$$\text{Outside Diameter} = 1.2(\text{Bore} + \text{Roller Diameter} + 0.7\text{ Pitch}) + 1\text{ in.}$$

*Shaper Cutters:* Only one will be required to cut any number of teeth for a given pitch and roller diameter. The manufacturer should be referred to for information concerning the cutter form design to be used.

**Sprocket Manufacture.**—Cast sprockets have cut teeth, and the rim, hub face, and bore are machined. The smaller sprockets are generally cut from steel bar stock, and are finished all over. Sprockets are often made from forgings or forged bars. The extent of finishing depends on the particular specifications that are applicable. Many sprockets are made by welding a steel hub to a steel plate. This process produces a one-piece sprocket of desired proportions and one that can be heat treated.

**Sprocket Materials.**—For large sprockets, cast iron is commonly used, especially in drives with large speed ratios, since the teeth of the larger sprocket are subjected to fewer chain engagements in a given time. For severe service, cast steel or steel plate is preferred.

The smaller sprockets of a drive are usually made of steel. With this material the body of the sprocket can be heat treated to produce toughness for shock resistance, and the tooth surfaces can be hardened to resist wear.

Stainless steel or bronze may be used for corrosion resistance, and Formica, nylon or other suitable plastic materials for special applications.

**Roller Chain Drive Ratings.—**In 1961, under auspices of The American Sprocket Chain Manufacturers Association (now called American Chain Association), a joint research program was begun to study pin-bushing interaction at high speeds and to gain further data on the phenomenon of chain joint galling among other research areas. These studies have shown that a separating film of lubricant is formed in chain joints in a manner similar to that found in journal bearings. These developments appear in ANSI/ASME B29.1-2011 (R2022), and are contained in Table 9. The ratings shown in Table 9 are below the galling range.

The horsepower ratings in Table 9 apply to lubricated, single-pitch, single-strand roller chains, both ANSI Standard and Heavy series. To obtain ratings of multiple-strand chains, a multiple-strand factor is applied.

The ratings in Table 9 are based upon: 1) a service factor of 1; 2) a chain length of approximately 100 pitches; 3) use of recommended lubrication methods; and 4) a drive arrangement where two aligned sprockets are mounted on parallel shafts in a horizontal plane.

Under these conditions, approximately 15,000 hours of service life at full load operation may be expected.

**Table 15. Roller Chain Drive Service Factors**

| Type of Driven Load | Type of Input Power | | |
|---|---|---|---|
| | Internal Combustion Engine with Hydraulic Drive | Electric Motor or Turbine | Internal Combustion Engine with Mechanical Drive |
| Smooth | 1.0 | 1.0 | 1.2 |
| Moderate Shock | 1.2 | 1.3 | 1.4 |
| Heavy Shock | 1.4 | 1.5 | 1.7 |

Substantial increases in rated speed loads can be utilized, as when a service life of less than 15,000 hours is satisfactory, or when full load operation is encountered only during a portion of the required service life. Chain manufacturers should be consulted for assistance with any special application requirements.

The horsepower ratings shown in Table 9 relate to the speed of the smaller sprocket and drive selections are made on this basis, whether the drive is speed reducing or speed increasing. Drives with more than two sprockets, idlers, composite duty cycles, or other unusual conditions often require special consideration. Where quietness or extra smooth operation are of special importance, small-pitch chain operating over large diameter sprockets will minimize noise and vibration.

When making drive selection, consideration is given to the loads imposed on the chain by the type of input power and the type of equipment to be driven. Service factors are used to compensate for these loads and the *required* horsepower rating of the chain is determined by the following formula:

$$\text{Required hp Table Rating} = \frac{\text{hp to be Transmitted} \times \text{Service Factor}}{\text{Multiple} - \text{Strand Factor}}$$

*Service Factors:* The service factors in Table 15 are for normal chain loading. For unusual or extremely severe operating conditions not shown in this table, it is desirable to use larger service factors.

*Multiple-Strand Factors:* The horsepower ratings for multiple-strand chains equal single-strand ratings multiplied by these factors:for two strands, a factor of 1.7; for three strands, 2.5; and and for four strands, 3.3.

**Lubrication.**—It has been shown that a separating wedge of fluid lubricant is formed in operating chain joints much like that formed in journal bearings. Therefore, fluid lubricant must be applied to ensure an oil supply to the joints and minimize metal-to-metal contact. If supplied in sufficient volume, lubrication also provides effective cooling and impact damping at higher speeds. For this reason, it is important that lubrication recommendations be followed. *The ratings in* Table 9 *apply only to drives lubricated in the manner specified in this table*.

Chain drives should be protected against dirt and moisture and the oil supply kept free of contamination. Periodic oil change is desirable. A good grade of non-detergent petroleum base oil is recommended. Heavy oils and greases are generally too stiff to enter and fill the chain joints. The following lubricant viscosities are recommended: For temperatures of 20° to 40° F, use SAE 20 lubricant; for 40° to 100°, use SAE 30; for 100° to 120°, use SAE 40; and for 120° to 140°, use SAE 50.

There are three basic types of lubrication for roller chain drives. The recommended type shown in Table 9 as Type A, Type B, or Type C is influenced by the chain speed and the amount of power transmitted. These are *minimum* lubrication requirements and the use of a better type (for example, Type C instead of Type B) is acceptable and may be beneficial. Chain life can vary appreciably depending upon the way the drive is lubricated. The better the chain lubrication, the longer the chain life. For this reason, it is important that the lubrication recommendations be followed when using the ratings given in Table 9. The types of lubrication are as follows:

*Type A — Manual or Drip Lubrication:* In manual lubrication, oil is applied copiously with a brush or spout can at least once every 8 hours of operation. Volume and frequency should be sufficient to prevent overheating of the chain or discoloration of the chain joints. In drip lubrication, oil drops from a drip lubricator are directed between the link plate edges. The volume and frequency should be sufficient to prevent discoloration of the lubricant in the chain joints. Precautions must be taken against misdirection of the drops by windage.

*Type B — Bath or Disc Lubrication:* In bath lubrication, the lower strand of the chain runs through a sump of oil in the drive housing. The oil level should reach the pitch line of the chain at its lowest point while operating. In disc lubrication, the chain operates above the oil level. The disc picks up oil from the sump and deposits it onto the chain, usually by means of a trough. The diameter of the disc should be such as to produce rim speeds of between 600 and 8000 feet per minute.

*Type C — Oil Stream Lubrication:* The lubricant is usually supplied by a circulating pump capable of supplying each chain drive with a continuous stream of oil. The oil should be applied inside the chain loop evenly across the chain width, and directed at the slack strand.

The chain manufacturer should be consulted when it appears desirable to use a type of lubricant other than that recommended.

**Installation and Alignment.**—Sprockets should have the tooth form, thickness, profile, and diameters conforming to ANSI/ASME B29.1. For maximum service life small sprockets operating at moderate to high speeds, or near the rated horsepower, should have hardened teeth. Normally, large sprockets should not exceed 120 teeth.

In general a center distance of 30 to 50 chain pitches is most desirable. The distance between sprocket centers should provide at least a 120 degree chain wrap on the smaller sprocket. Drives may be installed with either adjustable or fixed center distances. Adjustable centers simplify the control of chain slack. Sufficient housing clearance must always be provided for the chain slack to obtain full chain life.

Accurate alignment of shafts and sprocket tooth faces provides uniform distribution of the load across the entire chain width and contributes substantially to optimum drive life.

Shafting, bearings, and foundations should be suitable to maintain the initial alignment. Periodic maintenance should include an inspection of alignment.

**Example of Roller Chain Drive Design Procedure.—**The selection of a roller chain and sprockets for a specific design requirement is best accomplished by a systematic step-by-step procedure such as is used in the following example.

*Example:* Select a roller chain drive to transmit 10 horsepower from a countershaft to the main shaft of a wire drawing machine. The countershaft is $1^{15}/_{16}$-inches diameter and operates at 1000 rpm. The main shaft is also $1^{15}/_{16}$-inches diameter and must operate between 378 and 382 rpm. Shaft centers, once established, are fixed and by initial calculations must be approximately $22\frac{1}{2}$ inches. The load on the main shaft is uneven and presents "peaks," which place it in the heavy shock load category. The input power is supplied by an electric motor. The driving head is fully enclosed and all parts are lubricated from a central system.

*Step 1. Service Factor:* From Table 15 the service factor for heavy shock load and an electric motor drive is 1.5.

*Step 2. Design Horsepower:* The horsepower upon which the chain selection is based (design horsepower) is equal to the specified horsepower multiplied by the service factor, $10 \times 1.5 = 15$ hp.

*Step 3. Chain Pitch and Small Sprocket Size for Single-Strand Drive:* In Table 9 under 1000 rpm, a $^5/_8$-inch pitch chain with a 24-tooth sprocket or a $^3/_4$-inch pitch chain with a 15-tooth sprocket are possible choices.

*Step 4. Check of Chain Pitch and Sprocket Selection:* From Table 10 it is seen that only the 24-tooth sprocket in Step 3 can be bored to fit the $1^{15}/_{16}$-inch diameter main shaft. In Table 9 a $^5/_8$-pitch chain at a small sprocket speed of 1000 rpm is rated at 15.5 hp for a 24-tooth sprocket.

*Step 5. Selection of Large Sprocket:* Since the driver is to operate at 1000 rpm and the driven at a minimum of 378 rpm, the speed ratio 1000/378 = 2.646. Therefore, the large sprocket should have $24 \times 2.646 = 63.5$ (use 63) teeth.

This combination of 24 and 63 teeth will produce a main drive shaft speed of 381 rpm which is within the limitation of 378 to 382 rpm established in the original specification.

*Step 6. Computation of Chain Length:* Since the 24- and 63-tooth sprockets are to be placed on $22\frac{1}{2}$-inch centers, the chain length is determined from the formula:

$$L = 2C + \frac{N}{2} + \frac{n}{2} + \left(\frac{N-n}{2\pi}\right)^2 \times \frac{1}{C}$$

where $L$ = chain length in pitches; $C$ = shaft center distance in pitches; $N$ = number of teeth in large sprocket; and $n$ = number of teeth in small sprocket.

$$L = 2 \times 36 + \frac{63+24}{2} + \left(\frac{63-24}{6.28}\right)^2 \times \frac{1}{36} = 116.57 \text{ pitches}$$

*Step 7. Correction of Center Distance:* Since the chain is to couple at a whole number of pitches, 116 pitches will be used and the center distance recomputed based on this figure using the formula on page 2638 where $c$ is the center distance in inches and $P$ is the pitch.

$$\begin{aligned} c &= \frac{P}{8}\left(2L - N - n + \sqrt{(2L - N - n)^2 - 0.810(N-n)^2}\right) \\ &= \frac{5}{64}\left(2 \times 116 - 63 - 24 + \sqrt{(2 \times 116 - 63 - 24)^2 - 0.810(63-24)^2}\right) \\ &= \frac{5}{64}(145 + 140.69) = 22.32 \text{ inches, say } 22\frac{3}{8} \text{ inches} \end{aligned}$$

# BALL AND ACME LEADSCREWS

## Leadscrew Overview

Leadscrew assemblies convert rotary motion into linear travel and are well suited to precise control using electric motors. Acme leadscrews, much like machine screws, are threaded to allow a mating nut to be pushed or pulled along the axis of the screw as the shaft is rotated. Ball screws are threaded to allow ball bearings to support the attached nut. The nut has raceways that allow the ball bearings to recirculate as the nut travels down the axis of the shaft. Both provide efficient and accurate linear positioning for motion control applications. Each has strengths and weaknesses that makes each type of technology more suitable for different types of applications.

**Ball Screws.**—Ball screws generally require lower-torque motors than similarly sized Acme screws due to their higher efficiency in concerting energy into linear movement. This efficiency is generally above 90 percent compared to 40 percent for Acme screws with plastic nuts and 25 percent for acme screws with bronze-nuts. Ball screws can handle higher loading with high travel rates and lower backlash, though both technologies offer various anti-backlash nut assemblies. Ball screw life is more easily predicted using ball bearing load data. Compared to acme screws, ball screw assemblies exhibit a lower degradation in postitional accuracy over time due to wear on the contact surfaces. Ball screws are more expensive and can be noisier than acme screws. Ball screws may back drive the shaft in vertical orientations.

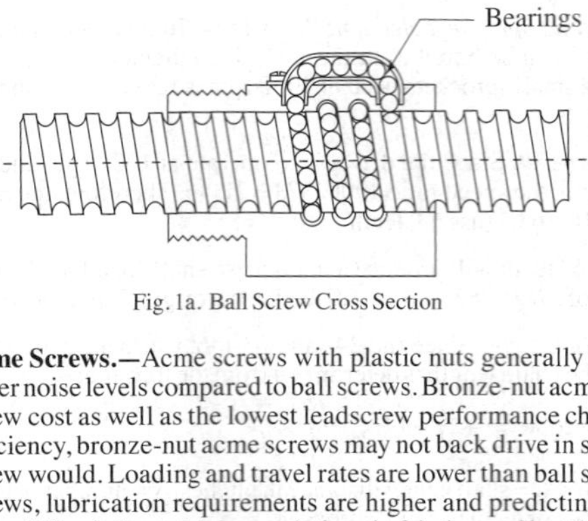


Fig. 1a. Ball Screw Cross Section

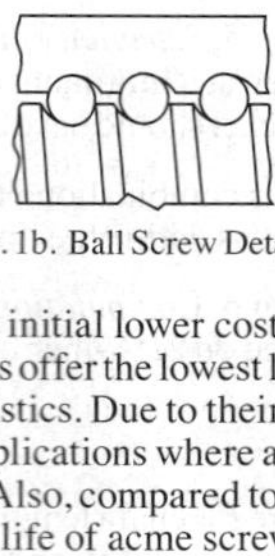

Fig. 1b. Ball Screw Detail

**Acme Screws.**—Acme screws with plastic nuts generally have an initial lower cost and lower noise levels compared to ball screws. Bronze-nut acme screws offer the lowest leadscrew cost as well as the lowest leadscrew performance characteristics. Due to their low efficiency, bronze-nut acme screws may not back drive in some applications where a ball screw would. Loading and travel rates are lower than ball screws. Also, compared to ball screws, lubrication requirements are higher and predicting travel life of acme screws is more difficult due to higher variations in friction and heat build up.

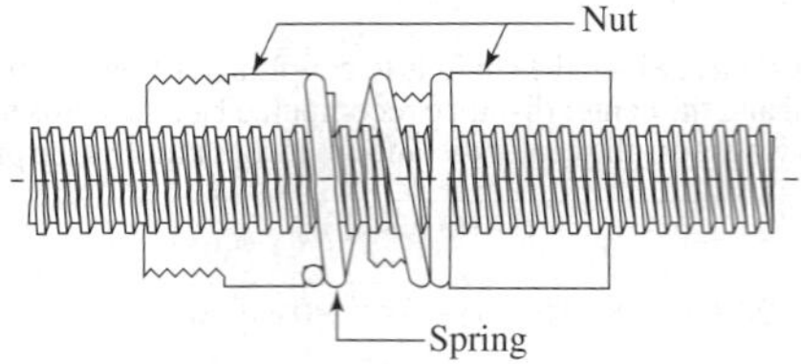


Fig. 2a. Acme Leadscrew with Anti-backlash Nut.

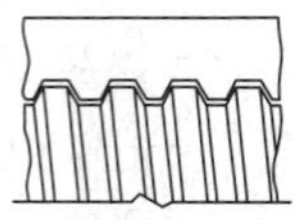

Fig. 2b. Acme Screw Detail

**General Leadscrew Design Considerations.—**The following design criteria should be considered when selecting leadscrews.

*Load:* The load is the weight "as seen by the screw." For vertical applications such as lifting or jacking the load equals the weight of the load being moved. For horizontal applications the "load seen" equals the weight of the load times the coefficient of friction of the supporting media. Another way to determine the load in an actual application is to attach a scale to the load and pull it. Base the load on the moving force required, not on the higher starting (breakaway) force.

*Tension and Compression Load:* For reduced screw thread sizes, design for tension loads whenever possible. Compression load designs must factor in column strength to prevent buckling.

*Back-drive:* When a ball screw or acme nut is forced to move linearly along the shaft by an applied force, or gravity in vertical orientation, it forces or back-drives the shaft making it rotate. With their higher efficiency and lower initial start up (breakway) friction value, ball screw assemblies in vertical orientations may back-drive their shafts.

*Driving Torque:* The driving torque ($T_d$) is the amount of force required to rotate the screw and move a load.

$$T_d = \frac{P \times L}{2\pi e}$$

where $T_d$ = driving torque (lb-in)

$P$ = load (lbs)

$L$ = lead of screw (in/revolution)

$e$ = ball bearing screw efficiency (approximately 0.90 for ball screws, 0.40 for acme screws with plastic nuts, and 0.25 for acme screws with bronze nuts.)

*Accuracy:* Several factors affect accuracy. One factor is the lead deviation in inches per foot. Typical values can range from 0.005 in/ft to 0.001 in/ft. Backlash may also be a consideration in applications that require precise positioning in both directions and in CNC applications. A final consideration is accuracy versus time degradation due to screw wear and ball bearing fatigue. This is especially important in systems that require precise positioning but do not have closed-loop feedback controls. Ball screws have an advantage in long term accuracy since the useful life of ball screws is determined by metal fatigue rather than the ordinary wear characteristics of a conventional screw. Due to rolling contact, very little dimensional change occurs over the life of a ball bearing screw, thus eliminating the need for frequent compensating adjustments.

*Lead:* The distance the nut travels with one rotation of the screw.

*Speed:* When rapid travel is required, design for screws with a high lead to reduce shaft speed to avoid rotational and vibrational side effects.

*Life:* The recommended operating loads for ball screw assemblies are based on a predicted life of one million inches of travel, providing the assembly is lubricated. However, it is possible to operate at greater loads if a reduced life is acceptable. The load should not exceed the maximum static load shown in manufacturer design tables.

*Backlash:* Backlash, end play, positional inaccuracy, or lost motion between the nut and the screw can be eliminated by utilizing a preloaded nut assembly. Backlash can also be reduced, at increased cost, in ball screws by changing the conformity ratio (ball track radius/ball diameter) as well as using ogival (gothic) ball tracks. Fig. 3 illustrates the contact points between a preloaded ball screw assembly and the shaft.

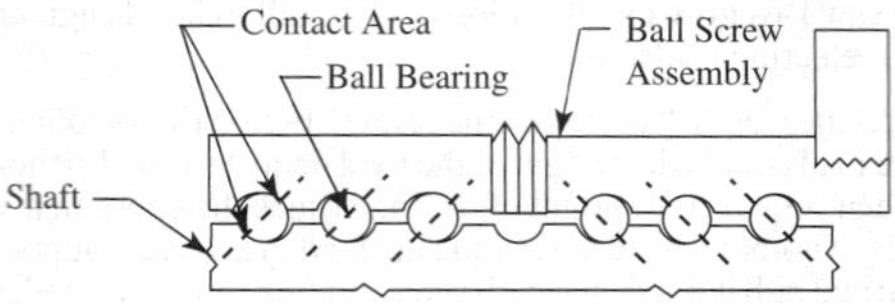


Fig. 3. Contact Points in a Preloaded Anti-Backlash Ball Screw Assembly

*End Supports:* End support type, free or fixed, and the number of end supports for a lead-screw shaft can have a significant impact on the speed and load capacity of the assembly.

*Environment:* Temperature extremes, dirty, dusty conditions, corrosive compounds, and metal chips can cause serious problems with lead-screw assemblies. For applications in such conditions the use of wipers, boots, or both should be considered. Stainless steel or plated assemblies should be considered for corrosive applications. Corrosive environments can also attack lubricants and thus indirectly damage the assemblies.

*Wipers:* Brush type wipers maximize ball screw performance by helping to spread lubricant over the length of the screw and preventing internal ball nut contamination from foreign materials. Bellows and telescopic wipers are recommended for severely "dirty" environments.

*Lubrication:* Ball screws must be lubricated and kept clean. Failure to lubricate can substantially reduce bearing life. Oil and grease can be used but grease is not recommended for low-temperature or high-speed applications.

*PV Value:* Speed and load limits for acme leadscrew assemblies using plastic nuts are generally limited by heat build-up. The factor which affects heat generation in an application is the product of the pressure ($lbs/in^2$) on the nut, and the surface velocity at the thread major diameter. This product is called the *PV* value.

*CNC Considerations:* CNC machine tools require the elimination of backlash and the minimization of elastic deflection to achieve high system stiffness, response, and repeatability of positioning.

**Ball Screw Selection.—**Many types and styles of ball screws and ball screw assemblies are offered by manufacturers. Consult the vendor's specific data sheets for precise design criteria. The selection of the correct ball screw and nut for a particular application involves four factors, all of which are interrelated. These are: the equivalent load measured in pounds, the required life measured in linear inches of travel, travel speed measured in inches per minute, and the length between bearings measured in inches.

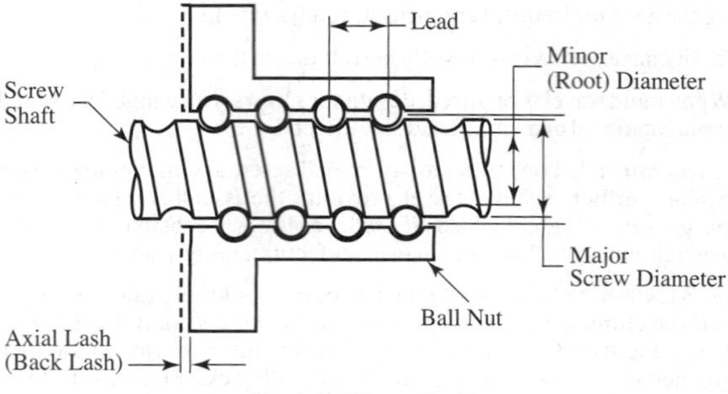


Fig. 4. Ball Screw Detail

*Design Overview:* To design a ball screw assembly many interrelated factors such as load, bearing length, life, and speed must be considered. Changing one parameter will influence others and generally an iterative approach to balancing these requirements is required. An example is the choice of a fine lead versus coarse lead. A fine line lead provides better positioning sensitivity and lower drive torque, but also requires a higher rotary speed to achieve the same linear rate of speed. A coarse lead results in lower rotary speed, but requires a higher drive torque that may require a larger motor and related drive components. Table 1 identifies some of these design interrelationships.

**Table 1. Ball Screw Design Factor Interrelationships**

| Increase In | Affects | How | Increase In | Affects | How |
|---|---|---|---|---|---|
| Screw Length | Critical speed | Decreases | End Mounting Rigidity | Critical speed | Increases |
| | Compression load | Decreases | | Compression load | Increases |
| Screw Diameter | Critical speed | Increases | | System stiffness | Increases |
| | Inertia | Increases | Load | Life | Decreases |
| | Compression load | Increases | Preload | Positioning accuracy | Increases |
| | Stiffness | Increases | | System stiffness | Increases |
| | Spring Rate | Increases | | Drag torque | Increases |
| | Load capacity | Increases | Angular Velocity | Critical speed | Decreases |
| Lead | Drive torque | Increases | Nut length | Load capacity | Increases |
| | Angular velocity | Decreases | | Stiffness | Increases |
| | Load capacity | Increases | Ball diameter | Life | Increases |
| | Positioning accuracy | Decreases | | Stiffness | Increases |
| | Ball diameter | Increases | | Load capacity | Increases |

**Screws, Power Transmission.—**The square form of thread has a somewhat higher efficiency than threads with sloping sides, although when the angle of the thread form is comparatively small, as in the case of an Acme thread, there is little increase in frictional losses. The Acme thread has superseded the square form on many classes of equipment requiring lead screws or other power transmitting screws, because the former has practical advantages in regard to cutting and also in compensating for wear between the screw and nut. Multiple thread screws are much more efficient than single-thread screws, as the efficiency is affected by the helix angle of the thread.

*Force Required to Turn Screw:* In determining the force which must be applied at the end of a given lever-arm in order to turn a screw (or nut surrounding it), there are two conditions to be considered:

1) When rotation is such that the load *resists* the movement of the screw, as in raising a load with a screw jack, $F = L \times \frac{l + 2r\pi\mu}{2r\pi - \mu l} \times \frac{r}{R}$.

2) When rotation is such that the load *assists* the movement of the screw, as in lowering a load, $F = L \times \frac{2r\pi\mu - l}{2r\pi + \mu l} \times \frac{r}{R}$.

In these formulas, $F$ = force applied at end of lever-arm; $L$ = load moved by screw; $R$ = length of lever-arm; $l$ = lead of screw thread; $r$ = mean or pitch radius of screw; $\mu$ = coefficient of friction. If lead $l$ is large in proportion to the diameter so that the helix angle is large, $F$ will have a negative value, indicating that the screw will turn due to the load alone. The efficiency between a screw and nut increases quite rapidly for helix angles up to 10 or 15 degrees (measured from a plane perpendicular to the screw axis). The efficiency remains nearly constant for angles between about 25 and 65 degrees, and the angle of maximum efficiency is between 40 and 50 degrees. A screw will not be self-locking if the efficiency exceeds 50 percent.

## ELECTRIC MOTORS

**Classes of NEMA Standards.—**National Electrical Manufacturers Association (NEMA) Standards, available from NEMA at 1300 North 17th Street, Suite 900, Arlington, VA 22209, www.nema.org, are of two classes: 1) *NEMA Standard,* which relates to a product commercially standardized and subject to repetitive manufacture, which standard has been approved by at least 90 percent of the members of the Subdivision eligible to vote thereon; and 2) *Suggested Standard for Future Design,* which may not have been regularly applied to a commercial product, but which suggests a sound engineering approach to future development and has been approved by at least two-thirds of the members of the Subdivision eligible to vote thereon.

*Authorized Engineering Information* consists of explanatory data and other engineering information of an informative character not falling within the classification of NEMA Standard or Suggested Standard for Future Design.

**Mounting Dimensions and Frame Sizes for Electric Motors.—**Dimensions for foot-mounted electric motors as standardized in the United States by the National Electrical Manufacturers Association (NEMA) include the spacing of bolt holes in the feet of the motor, the distance from the bottom of the feet to the centerline of the motor shaft, the size of the conduit, the length and diameter of shaft, and other dimensions likely to be required by designers or manufacturers of motor-driven equipment. The Standard provides dimensions for face-mounted and flange-mounted motors by means of standard motor frame numbers.

Standard dimensions also are given where the motor is to be mounted upon a belt-tightening base or upon rails.

The NEMA standards also prescribe lettering for dimension drawings, mounting and terminal housing locations and dimensions, symbols and terminal connections, and provision for grounding of field wiring. In addition, the standards give recommended knockout and clearance hole dimensions; tolerances on shaft extension diameters and keyseats; methods of measuring shaft run-out and eccentricity, also face runout of mounting surfaces; and tolerances of face-mounted and flanged-mounted motors.

**Design Letters of Polyphase Integral-Horsepower Motors.—**Designs A, B, C, and D motors are squirrel-cage motors designed to withstand full voltage starting and developing locked-rotor torque and breakdown torque, drawing locked-rotor current, and having a slip as specified below:

*Design A:* Locked-rotor torque as shown in Table 2, breakdown torque as shown in Table 3, locked-rotor current higher than the values shown in Table 1, and a slip at rated load of less than 5 percent. Motors with 10 or more poles may have a slightly greater slip.

**Table 1. NEMA Standard Locked-Rotor Current of 3-Phase 60-Hertz Integral-Horsepower Squirrel-Cage Induction Motors Rated at 230 Volts**

| Horse-power | Locked-rotor Current, Amps. | Design Letters | Horse-power | Locked-rotor Current, Amps. | Design Letters | Horse-power | Locked-rotor Current, Amps. | Design Letters |
|---|---|---|---|---|---|---|---|---|
| ½ | 20 | B, D | 7½ | 127 | B, C, D | 50 | 725 | B, C, D |
| ¾ | 25 | B, D | 10 | 162 | B, C, D | 60 | 870 | B, C, D |
| 1 | 30 | B, D | 15 | 232 | B, C, D | 75 | 1085 | B, C, D |
| 1½ | 40 | B, D | 20 | 290 | B, C, D | 100 | 1450 | B, C, D |
| 2 | 50 | B, D | 25 | 365 | B, C, D | 125 | 1815 | B, C, D |
| 3 | 64 | B, C, D | 30 | 435 | B, C, D | 150 | 2170 | B, C, D |
| 5 | 92 | B, C, D | 40 | 580 | B, C, D | 200 | 2900 | B, C |

*Note:* The locked-rotor current of a motor is the steady-state current taken from the line with the rotor locked and with rated voltage and frequency applied to the motor.

For motors designed for voltages other than 230 volts, the locked-rotor current is inversely proportional to the voltages. For motors larger than 200 hp, see NEMA Standard MG 1-12.34.

## Table 2. NEMA Standard Locked-Rotor Torque of Single-Speed Polyphase 60- and 50-Hertz Squirrel-Cage Integral-Horsepower Motors with Continuous Ratings

| HP | Designs A and B | | | | | | | Design C | | |
|---|---|---|---|---|---|---|---|---|---|---|
| | Synchronous Speed, rpm | | | | | | | | | |
| 60 hertz | 3600 | 1800 | 1200 | 900 | 720 | 600 | 514 | 1800 | 1200 | 900 |
| 50 hertz | 3000 | 1500 | 1000 | 750 | … | … | … | 1500 | 1000 | 750 |
| | Percent of Full-load Torque[a] | | | | | | | | | |
| ½ | … | … | … | 140 | 140 | 115 | 110 | … | … | … |
| ¾ | … | … | 175 | 135 | 135 | 115 | 110 | … | … | … |
| 1 | … | 275 | 170 | 135 | 135 | 115 | 110 | … | … | … |
| 1½ | 175 | 250 | 165 | 130 | 130 | 115 | 110 | … | … | … |
| 2 | 170 | 235 | 160 | 130 | 125 | 115 | 110 | … | … | … |
| 3 | 160 | 215 | 155 | 130 | 125 | 115 | 110 | … | 250 | 225 |
| 5 | 150 | 185 | 150 | 130 | 125 | 115 | 110 | 250 | 250 | 225 |
| 7½ | 140 | 175 | 150 | 125 | 120 | 115 | 110 | 250 | 225 | 200 |
| 10 | 135 | 165 | 150 | 125 | 120 | 115 | 110 | 250 | 225 | 200 |
| 15 | 130 | 160 | 140 | 125 | 120 | 115 | 110 | 225 | 200 | 200 |
| 20 | 130 | 150 | 135 | 125 | 120 | 115 | 110 | 200 for all sizes above 15 hp. | | |
| 25 | 130 | 150 | 135 | 125 | 120 | 115 | 110 | | | |
| 30 | 130 | 150 | 135 | 125 | 120 | 115 | 110 | | | |
| 40 | 125 | 140 | 135 | 125 | 120 | 115 | 110 | | | |
| 50 | 120 | 140 | 135 | 125 | 120 | 115 | 110 | | | |
| 60 | 120 | 140 | 135 | 125 | 120 | 115 | 110 | | | |
| 75 | 105 | 140 | 135 | 125 | 120 | 115 | 110 | For Design D motors, see footnote. | | |
| 100 | 105 | 125 | 125 | 125 | 120 | 115 | 110 | | | |
| 125 | 100 | 110 | 125 | 120 | 115 | 115 | 110 | | | |
| 150 | 100 | 110 | 120 | 120 | 115 | 115 | … | | | |
| 200 | 100 | 100 | 120 | 120 | 115 | … | … | | | |

[a] These values represent the upper limit of application for these motors.

*Note:* The locked-rotor torque of a motor is the minimum torque which it will develop at rest for all angular positions of the rotor, with rated voltage applied at rated frequency.

The locked-rotor torque of Design D, 60- and 50-hertz 4-, 6-, and 8-pole single-speed, polyphase squirrel-cage motors rated 150 hp and smaller, with rated voltage and frequency applied is 275 percent of full-load torque, which represents the upper limit of application for these motors.

For motors larger than 200 hp, see NEMA Standard MG 1-12.37.

## Table 3. NEMA Standard Breakdown Torque of Single-Speed Polyphase Squirrel-Cage, Integral-Horsepower Motors with Continuous Ratings

| Horsepower | Synchronous Speed, rpm | | | | | | |
|---|---|---|---|---|---|---|---|
| 60 hertz | 3600 | 1800 | 1200 | 900 | 720 | 600 | 514 |
| 50 hertz | 3000 | 1500 | 1000 | 750 | … | … | … |
| | Percent of Full Load Torque | | | | | | |
| | Designs A and B[a] | | | | | | |
| ½ | … | … | … | 225 | 200 | 200 | 200 |
| ¾ | … | … | 275 | 220 | 200 | 200 | 200 |
| 1 | … | 300 | 265 | 215 | 200 | 200 | 200 |
| 1½ | 250 | 280 | 250 | 210 | 200 | 200 | 200 |
| 2 | 240 | 270 | 240 | 210 | 200 | 200 | 200 |
| 3 | 230 | 250 | 230 | 205 | 200 | 200 | 200 |
| 5 | 215 | 225 | 215 | 205 | 200 | 200 | 200 |
| 7½ | 200 | 215 | 205 | 200 | 200 | 200 | 200 |
| 10-125, incl. | 200 | 200 | 200 | 200 | 200 | 200 | 200 |
| 150 | 200 | 200 | 200 | 200 | 200 | 200 | … |
| 200 | 200 | 200 | 200 | 200 | 200 | … | … |
| | Design C | | | | | | |
| 3 | … | … | 225 | 200 | … | … | … |
| 5 | … | 200 | 200 | 200 | … | … | … |
| 7½-200, incl. | … | 190 | 190 | 190 | … | … | … |

[a] Design A values are in excess of those shown.

These values represent the upper limit of the range of application for these motors. For above 200 hp, see NEMA Standard MG1-12.38.

*Design B:* Locked-rotor torque as shown in Table 2, breakdown torque as shown in Table 3, locked-rotor current not exceeding that in Table 1, and a slip at rated load of less than 5 percent. Motors with 10 or more poles may have a slightly greater slip.

*Design C:* Locked-rotor torque for special high-torque applications up to values shown in Table 2, breakdown torque up to values shown in Table 3, locked-rotor current not exceeding values shown in Table 1 and a slip at rated load of less than 5 percent.

*Design D:* Locked-rotor torque as indicated in Table 2, locked-rotor current not greater than that shown in Table 1 and a slip at rated load of 5 percent or more.

**Torque and Current Definitions.—**The definitions which follow have been adopted as standard by the National Electrical Manufacturers Association.

*Locked-Rotor or Static Torque:* The locked-rotor torque of a motor is the minimum torque which it will develop at rest for all angular positions of the rotor, with rated voltage applied at rated frequency.

*Breakdown Torque:* The breakdown torque of a motor is the maximum torque which the motor will develop, with rated voltage applied at rated frequency, without an abrupt drop in speed (see Table 4).

*Full-Load Torque:* The full-load torque of a motor is the torque necessary to produce its rated horsepower at full load speed. In pounds at 1-foot radius, it is equal to the horsepower times 5252 divided by the full-load speed.

*Pull-Out Torque:* The pull-out torque of a synchronous motor is the maximum sustained torque which the motor will develop at synchronous speed with rated voltage applied at rated frequency and with normal excitation.

*Pull-In Torque:* The pull-in torque of a synchronous motor is the maximum constant torque under which the motor will pull its connected inertia load into synchronism at rated voltage and frequency, when its field excitation is applied.

*Pull-Up Torque:* The pull-up torque of an alternating current motor is the minimum torque developed by the motor during the period of acceleration from rest to the speed at which breakdown torque occurs. For motors which do not have a definite breakdown torque, the pull-up torque is the minimum torque developed up to rated speed.

*Locked Rotor Current:* The locked rotor current of a motor is the steady-state current taken from the line with the rotor locked and with rated voltage (and rated frequency in the case of alternating-current motors) applied to the motor.

**Table 4. NEMA Standard Breakdown Torque of Polyphase Wound-Rotor Motors with Continuous Ratings — 60- and 50-Hertz**

| Horsepower | Speed, rpm | | | Horsepower | Speed, rpm | | |
|---|---|---|---|---|---|---|---|
| | 1800 | 1200 | 900 | | 1800 | 1200 | 900 |
| | Percent of Full-load Torque | | | | Percent of Full-load Torque | | |
| 1 | … | … | 250 | 7½ | 275 | 250 | 225 |
| 1½ | … | … | 250 | 10 | 275 | 250 | 225 |
| 2 | 275 | 275 | 250 | 15 | 250 | 225 | 225 |
| 3 | 275 | 275 | 250 | 20-200, incl. | 225 | 225 | 225 |
| 5 | 275 | 275 | 250 | … | … | … | … |

These values represent the upper limit of the range of application for these motors.

**Standard Direction of Motor Rotation.—**The standard direction of rotation for all nonreversing direct-current motors, all alternating-current single-phase motors, all synchronous motors, and all universal motors, is counterclockwise when facing that end of the motor opposite the drive.

This rule does not apply to two- and three-phase induction motors, as in most applications the phase sequence of the power lines is rarely known.

**Motor Types According to Variability of Speed.—**Five types of motors classified according to variability of speed are:

*Constant-Speed Motors:* In this type of motor the normal operating speed is constant or practically constant; for example, a synchronous motor, an induction motor with small slip, or a direct-current shunt-wound motor.

*Varying-Speed Motor:* In this type of motor, the speed varies with the load, ordinarily decreasing when the load increases; such as a series-wound or repulsion motor.

*Adjustable-Speed Motor:* In this type of motor, the speed can be varied gradually over a considerable range, but when once adjusted remains practically unaffected by the load; such as a direct-current shunt-wound motor with field resistance control designed for a considerable range of speed adjustment.

The base speed of an adjustable-speed motor is the lowest rated speed obtained at rated load and rated voltage at the temperature rise specified in the rating.

*Adjustable Varying-Speed Motor:* This type of motor is one in which the speed can be adjusted gradually, but when once adjusted for a given load will vary in considerable degree with the change in load; such as a direct-current compound-wound motor adjusted by field control or a wound-rotor induction motor with rheostatic speed control.

*Multispeed Motor:* This type of motor is one which can be operated at any one of two or more definite speeds, each being practically independent of the load; such as a direct-current motor with two armature windings or an induction motor with windings capable of various pole groupings. In the case of multispeed permanent-split capacitor and shaded pole motors, the speeds are dependent upon the load.

**Pull-Up Torque.—**NEMA Standard pull up torques for single-speed, polyphase, squirrel-cage integral-horsepower motors, Designs A and B, with continuous ratings and with rated voltage and frequency applied are as follows: When the locked-rotor torque given in Table 2 is 110 percent or less, the pull-up torque is 90 percent of the locked-rotor torque; when the locked-rotor torque is greater than 110 percent but less than 145 percent, the pull-up torque is 100 percent of full-load torque; and when the locked-rotor torque is 145 percent or more, the pull-up torque is 70 percent of the locked-rotor torque.

For Design C motors, with rated voltage and frequency applied, the pull-up torque is not less than 70 percent of the locked-rotor torque as given in Table 2.

## Types and Characteristics of Electric Motors

**Types of Direct-Current Motors.—**Direct-current motors may be grouped into three general classes: series-wound; shunt-wound; and compound-wound.

In the *series-wound motor* the field windings, which are fixed in the stator frame, and the armature windings, which are placed around the rotor, are connected in series so that all current passing through the armature also passes through the field. In the *shunt-wound motor*, both armature and field are connected across the main power supply so that the armature and field currents are separate. In the *compound-wound motor*, both series and shunt field windings are provided and these may be connected so that the currents in both are flowing in the same direction, called *cumulative compounding*, or so that the currents in each are flowing in opposite directions, called *differential compounding*.

**Characteristics of Series-Wound Direct-Current Motors.—**In the series-wound motor, any increase in load results in more current passing through the armature and the field windings. As the field is strengthened by this increased current, the motor speed decreases. Conversely, as the load is decreased the field is weakened and the speed increases and at very light loads may become excessive. For this reason, series-wound direct-current motors are usually directly connected or geared to the load to prevent "runaway." (A series-wound motor, designated as series-shunt wound, is sometimes provided with a light shunt field winding to prevent dangerously high speeds at light loads.) The increase in armature current with increasing load produces increased torque, so that the series-wound motor is particularly suited to heavy starting duty and where severe overloads may be expected. Its speed may be adjusted by means of a variable resistance placed in series with the motor, but due to variation with load, the speed cannot be held at any constant value. This variation of speed with load becomes greater as the speed is reduced. Series-wound motors are used where the load is practically constant and can easily be controlled by hand. They are usually limited to traction and lifting service.

**Shunt-Wound Direct-Current Motors.**—In the shunt-wound motor, the strength of the field is not affected appreciably by change in the load, so that a fairly constant speed (about 10 to 12 percent drop from no load to full load speed) is obtainable. This type of motor may be used for the operation of machines requiring an approximately constant speed and imposing low starting torque and light overload on the motor.

The shunt-wound motor becomes an adjustable-speed motor by means of field control or by armature control. If a variable resistance is placed in the field circuit, the amount of current in the field windings and hence the speed of rotation can be controlled. As the speed increases, the torque decreases proportionately, resulting in nearly constant horsepower. A speed range of 6 to 1 is possible using field control, but 4 to 1 is more common. Speed regulation is somewhat greater than in the constant-speed shunt-wound motors, ranging from about 15 to 22 percent. If a variable resistance is placed in the armature circuit, the voltage applied to the armature can be reduced and hence the speed of rotation can be reduced over a range of about 2 to 1. With armature control, speed regulation becomes poorer as speed is decreased, and is about 100 percent for a 2 to 1 speed range. Since the current through the field remains unchanged, the torque remains constant.

*Machine Tool Applications:* The adjustable-speed shunt-wound motors are useful on larger machines of the boring mill, lathe, and planer type and are particularly adapted to spindle drives because constant horsepower characteristics permit heavy cuts at low speed and light or finishing cuts at high speed. They have long been used for planer drives because they can provide an adjustable low speed for the cutting stroke and a high speed for the return stroke. Their application has been limited, however, to plants in which direct-current power is available.

**Adjustable-Voltage Shunt-Wound Motor Drive.**—More extensive use of the shunt-wound motor has been made possible by a combination drive that includes a means of converting alternating current to direct current. This conversion may be effected by a self-contained unit consisting of a separately excited direct-current generator driven by a constant-speed alternating-current motor connected to the regular alternating-current line, or by an electronic rectifier with suitable controls connected to the regular alternating-current supply lines. The latter has the advantage of causing no vibration when mounted directly on the machine tool, an important factor in certain types of grinders.

In this type of adjustable-speed, shunt-wound motor drive, speed control is effected by varying the voltage applied to the armature while supplying constant voltage to the field. In addition to providing for the adjustment of the voltage supplied by the conversion unit to the armature of the shunt-wound motor, the amount of current passing through the motor field may also be controlled. In fact, a single control may be provided to vary the motor speed from minimum to base speed (speed of the motor at full load with rated voltage on armature and field) by varying the voltage applied to the armature and from base speed to maximum speed by varying the current flowing through the field. When so controlled, the motor operates at constant torque up to base speed and at constant horsepower above base speed.

*Speed Range:* Speed ranges of at least 20 to 1 below base speed and 4 or 5 to 1 above base speed (a total range of 100 to 1, or more) are obtainable as compared with about 2 to 1 below normal speed and 3 or 4 to 1 above normal speed for the conventional type of control. Speed regulation may be as great as 25 percent at high speeds. Special electronic controls, when used with this type shunt motor drive, make possible maintenance of motor speeds with as little variation as $^1/_2$ to 1 percent of full load speed from full load to no load over a line voltage variation of ± 10 percent and over any normal variation in motor temperature and ambient temperature.

*Applications:* These direct-current, adjustable-voltage drives, as they are sometimes called, have been applied successfully to such machine tools as planers, milling machines, boring mills and lathes, as well as to other industrial machines where wide, stepless speed control, uniform speed under all operating conditions, constant torque acceleration and adaptability to automatic operation are required.

**Compound-Wound Motors.**—In the compound-wound motor, the speed variation due to load changes is much less than in the series-wound motor, but greater than in the shunt-wound motor (ranging up to 25 percent from full load to no load). It has a greater starting torque than the shunt-wound motor, is able to withstand heavier overloads, but has a narrower adjustable speed range. Standard motors of this type have a cumulative-compound winding, the differential-compound winding being limited to special applications. They are used where the starting load is very heavy or where the load changes suddenly and violently as with reciprocating pumps, printing presses and punch presses.

**Types of Polyphase Alternating-Current Motors.**—The most widely used polyphase motors are of the induction type. The *"squirrel cage" induction motor* consists of a wound stator which is connected to an external source of alternating-current power and a laminated steel core rotor with a number of heavy aluminum or copper conductors set into the core around its periphery and parallel to its axis. These conductors are connected together at each end of the rotor by a heavy ring, which provides closed paths for the currents induced in the rotor to circulate. The rotor bars form, in effect, a "squirrel cage" from which the motor takes its name.

*Wound-Rotor Type of Induction Motor:* This type has, in addition to a squirrel cage, a series of coils set into the rotor which are connected through slip-rings to external variable resistors. By varying the resistance of the wound-rotor circuits, the amount of current flowing in these circuits and hence the speed of the motor can be controlled. Since the rotor of an induction motor is not connected to the power supply, the motor is said to operate by transfer action and is analogous to a transformer with a short-circuited secondary that is free to rotate. Induction motors are built with a wide range of speed and torque characteristics which are discussed under "Operating Characteristics of Squirrel-cage Induction Motors."

*Synchronous Motor:* The other type of polyphase alternating-current motor used industrially is the *synchronous motor.* In contrast to the induction motor, the rotor of the synchronous motor is connected to a direct-current supply which provides a field that rotates in step with the alternating-current field in the stator. After having been brought up to synchronous speed, which is governed by the frequency of the power supply and the number of poles in the rotor, the synchronous motor operates at this constant speed throughout its entire load range.

**Operating Characteristics of Squirrel-Cage Induction Motors.**—In general, squirrel-cage induction motors are simple in design and construction and offer rugged service. They are essentially constant-speed motors, their speed changing very little with load and not being subject to adjustment. They are used for a wide range of industrial applications calling for integral horsepower ratings. According to the NEMA (National Electrical Manufacturers Association) Standards, there are four classes of squirrel-cage induction motors designated respectively as *A*, *B*, *C*, and *D*.

*Design A* motors are not commonly used since Design *B* has similar characteristics with the advantage of lower starting current.

*Design B* motors may be designated as a general purpose type suitable for the majority of polyphase alternating-current applications such as blowers, compressors, drill presses, grinders, hammer mills, lathes, planers, polishers, saws, screw machines, shakers, stokers, etc. The starting torque at 1800 rpm is 250 to 275 percent of full load torque for 3 hp (2.24 kW) and below; for 5–75 hp (3.73–56 kW) ratings the starting torque ranges from 185 to 150 percent of full load torque. They have low starting current requirements, usually no more than 5 to 6 times full load current and can be started at full voltage. Their slip (difference between synchronous speed and actual speed at rated load) is relatively low.

*Design C* motors have high starting torque (up to 250 percent of full load torque) but low starting current. They can be started at full voltage. Slip at rated load is relatively low. They are used for compressors requiring a loaded start, heavy conveyors, reciprocating pumps and other applications requiring high starting torque.

*Design D* motors have high slip at rated load, that is, the motor speed drops off appreciably as the load increases, permitting use of the stored energy of a flywheel. They provide heavy starting torque, up to 275 percent of full load torque, are quiet in operation and have relatively low starting current. Applications are for impact, shock and other high peak loads or flywheel drives such as trains, elevators, hoists, punch and drawing presses, shears, etc.

*Design F* motors are no longer standard. They had low starting torque, about 125 percent of full-load torque, and low starting current. They were used to drive machines which required infrequent starting at no load or at very light load.

**Multiple-Speed Induction Motors.**—This type has a number of windings in the stator so arranged and connected that the number of effective poles and hence the speed can be changed. These motors are for the same types of starting conditions as the conventional squirrel-cage induction motors and are available in designs that provide constant horsepower at all rated speeds and in designs that provide constant torque at all rated speeds.

Typical speed combinations obtainable in these motors are 600, 900, 1200 and 1800 rpm; 450, 600, 900 and 1200 rpm; and 600, 720, 900 and 1200 rpm.

Where a gradual change in speed is called for, a wound rotor may be provided in addition to the multiple stator windings.

**Wound-Rotor Induction Motors.**—These motors are designed for applications where extremely low starting current with high starting torque are called for, such as in blowers, conveyors, compressors, fans and pumps. They may be employed for adjustable-varying speed service where the speed range does not extend below 50 percent of synchronous speed, as for steel plate-forming rolls, printing presses, cranes, blowers, stokers, lathes and milling machines of certain types. The speed regulation of a wound rotor induction motor ranges from 5 to 10 percent at maximum speed and from 18 to 30 percent at low speed. They are also employed for reversing service as in cranes, gates, hoists and elevators.

**High-Frequency Induction Motors.**—This type is used in conjunction with frequency changers when very high speeds are desired, as on grinders, drills, routers, portable tools or woodworking machinery. These motors have an advantage over the series-wound or universal type of high-speed motor in that they operate at a relatively constant speed over the entire load range. A motor-generator set, a two-unit frequency converter or a single unit inductor frequency converter may be used to supply three-phase power at the frequency required. The single unit frequency converter may be obtained for delivering any one of a number of frequencies ranging from 360 to 2160 cycles and it is self-driven and self-excited from the general polyphase power supply.

**Synchronous Motors.**—These are widely used in electric timing devices; to drive machines that must operate in synchronism; and also to operate compressors, rolling mills, crushers which are started without load, paper mill screens, shredders, vacuum pumps and motor-generator sets. Synchronous motors have an inherently high power factor and are often employed to make corrections for the low power factor of other types of motors on the same system.

**Types of Single-Phase Alternating-Current Motors.**—Most of the single-phase alternating-current motors are basically induction motors distinguished by different arrangements for starting. (A single-phase induction motor with only a squirrel-cage rotor has no starting torque.) In the *capacitor-start* single-phase motor, an auxiliary winding in the stator is connected in series with a capacitor and a centrifugal switch. During the starting and accelerating period the motor operates as a two-phase induction motor. At about two-thirds full-load speed, the auxiliary circuit is disconnected by the switch and the motor then runs as a single-phase induction motor. In the *capacitor-start, capacitor-run* motor, the auxiliary circuit is arranged to provide high effective capacity for high starting torque and to remain connected to the line but with reduced capacity during the running period. In the *single-value capacitor* or *capacitor split-phase* motor, a relatively small continuously-rated capacitor is permanently connected in one of the two stator windings and the motor both starts and runs like a two-phase motor.

In the *repulsion-start* single-phase motor, a drum-wound rotor circuit is connected to a commutator with a pair of short-circuited brushes set so that the magnetic axis of the rotor winding is inclined to the magnetic axis of the stator winding. The current flowing in this rotor circuit reacts with the field to produce starting and accelerating torques. At about two-thirds full load speed the brushes are lifted, the commutator is short circuited and the motor runs as a single-phase squirrel-cage motor. The *repulsion* motor employs a repulsion winding on the rotor for both starting and running. The *repulsion-induction* motor has an outer winding on the rotor acting as a repulsion winding and an inner squirrel-cage winding. As the motor comes up to speed, the induced rotor current partially shifts from the repulsion winding to the squirrel-cage winding and the motor runs partly as an induction motor.

In the *split-phase* motor, an auxiliary winding in the stator is used for starting with either a resistance connected in series with the auxiliary winding (*resistance-start*) or a reactor in series with the main winding (*reactor-start*).

The *series-wound* single-phase motor has a rotor winding in series with the stator winding as in the series-wound direct-current motor. Since this motor may also be operated on direct current, it is called a *universal* motor.

**Characteristics of Single-Phase Alternating-Current Motors.**—Single-phase motors are used in sizes up to about 7½ horsepower (5.6 kW) for heavy starting duty chiefly in home and commercial appliances for which polyphase power is not available. The *capacitor-start* motor is available in normal starting torque designs for such applications as centrifugal pumps, fans, and blowers and in high-starting torque designs for reciprocating compressors, pumps, loaded conveyors, or belts. The *capacitor-start, capacitor-run* motor is exceptionally quiet in operation when loaded to at least 50 percent of capacity. It is available in low-torque designs for fans and centrifugal pumps and in high-torque designs for applications similar to those of the capacitor-start motor.

The *capacitor split-phase* motor requires the least maintenance of all single-phase motors, but has very low starting torque. Its high maximum torque makes it potentially useful in floor sanders or in grinders where momentary overloads due to excessive cutting pressure are experienced. It is also used for slow-speed direct connected fans.

The *repulsion-start, induction-run* motor has higher starting torque than the capacitor motors, although for the same current, the capacitor motors have equivalent pull-up and maximum torque. Electrical and mechanical noise and the extra maintenance sometimes required are disadvantages. These motors are used for compressors, conveyors and stokers starting under full load. The *repulsion-induction* motor has relatively high starting torque and low starting current. It also has a smooth speed-torque curve with no break and a greater ability to withstand long accelerating periods than capacitor type motors. It is particularly suitable for severe starting and accelerating duty and for high inertia loads such as laundry extractors. Brush noise is, however, continuous.

The *repulsion* motor has no limiting synchronous speed and the speed changes with the load. At certain loads, slight changes in load cause wide changes in speed. A brush shifting arrangement may be provided to adjust the speed which may have a range of 4 to 1 if full rated constant torque is applied but a decreasing range as the torque falls below this value. This type of motor may be reversed by shifting the brushes beyond the neutral point. These motors are suitable for machines requiring constant-torque and adjustable speed.

The *split-phase* and *universal* motors are limited to about $^1/_3$ hp (0.25 kW) ratings and are used chiefly for small appliance and office machine applications.

**Motors with Built-in Speed Reducers.—**Electric motors having built-in speed-changing units are compact and the design of these motorized speed reducers tends to improve the appearance of the machines which they drive. There are several types of these speed reducers; they may be classified according to whether they are equipped with worm gearing, a regular gear train with parallel shafts, or planetary gearing.

The claims made for the worm gearing type of reduction unit are that the drive is quiet in operation and well adapted for use where the slow-speed shaft must be at right angles to the motor shaft and where a high speed ratio is essential.

For very low speeds, the double reduction worm gearing units are suitable. In these units two sets of worm gearing form the gear train, and both the slow-speed shaft and the armature shaft are parallel. The intermediate worm gear shaft can be built to extend from the housing, if required, so as to make two countershaft speeds available on the same unit.

In the parallel-shaft type of speed reducer, the slow-speed shaft is parallel with the armature shaft. The slow-speed shaft is rotated by a pinion on the armature shaft, this pinion meshing with a larger gear on the slow-speed shaft.

Geared motors having built-in speed-changing units are available with constant-mesh change gears for varying the speed ratio.

Planetary gearing permits a large speed reduction with few parts; hence, it is well adapted for geared-head motor units where economy and compactness are essential. The slow-speed shaft is in line with the armature shaft.

## Factors Governing Motor Selection

**Speed, Horsepower, Torque and Inertia Requirements.—**Where more than one speed or a range of speeds are called for, one of the following types of motors may be selected, depending upon other requirements: For direct-current, the standard shunt-wound motor with field control has a 2 to 1 range in some designs; the adjustable speed motor may have a range of from 3 to 1 up to 6 to 1; the shunt motor with adjustable voltage supply has a range up to 20 to 1 or more below base speed and 4 or 5 to 1 above base speed, making a total range of up to 100 to 1 or more. For polyphase alternating current, multi-speed squirrel-cage induction motors have 2, 3 or 4 fixed speeds; the wound-rotor motor has a 2 to 1 range. The two-speed wound-rotor motor has a 4 to 1 range. The brush-shifting shunt motor has a 4 to 1 range. The brush-shifting series motor has a 3 to 1 range; and the squirrel-cage motor with a variable-frequency supply has a very wide range. For single-phase alternating current, the brush-shifting repulsion motor has a $2^1/_2$ to 1 range; the capacitor motor with tapped winding has a 2 to 1 range and the multi-speed capacitor motor has 2 or 3 fixed speeds. Speed regulation (variation in speed from no load to full load) is greatest with motors having series field windings and entirely absent with synchronous motors.

*Horsepower:* Where the load to be carried by the motor is not constant but follows a definite cycle, a horsepower-time curve enables the peak horsepower to be determined as well as the root-mean-square-average horsepower, which indicates the proper motor rating from a heating standpoint. Where the load is maintained at a constant value for a period of from 15 minutes to 2 hours depending on the size, the horsepower rating required will usually not be less than this constant value. When selecting the size of an induction motor, it should be kept in mind that this type of motor operates at maximum efficiency when it is loaded to full capacity. Where operation is to be at several speeds, the horsepower requirement for each speed should be considered.

*Torque:* Starting torque requirements may vary from 10 percent of full load to 250 percent of full load torque depending upon the type of machine being driven. Starting torque may vary for a given machine because of frequency of start, temperature, type and amount of lubricant, etc., and such variables should be taken into account. The motor torque supplied to the machine must be well above that required by the driven machine at all points up to full speed. The greater the excess torque, the more rapid the acceleration. The approximate time required for acceleration from rest to full speed is given by the formula:

$$\text{Time} = \frac{N \times WR^2}{T_a \times 308} \text{ seconds}$$

where $N$ = full load speed in rpm

$T_a$ = torque = average foot-pounds available for acceleration.

$WR^2$ = inertia of rotating part in pounds feet squared ($W$ = weight and $R$ = radius of gyration of rotating part).

308 = combined constant converting minutes into seconds, weight into mass and radius into circumference.

If the time required for acceleration is greater than 20 seconds, special motors or starters may be required to avoid overheating.

The running torque $T_r$ is found by the formula:

$$T_r = \frac{5250 \times \text{HP}}{N} \text{ foot pounds}$$

where $HP$ = horsepower being supplied to the driven machine

$N$ = running speed in rpm

5250 = combined constant converting horsepower to foot-pounds per minute and work per revolution into torque.

The peak horsepower determines the maximum torque required by the driven machine and the motor must have a maximum running torque in excess of this value.

*Inertia:* The inertia or flywheel effect of the rotating parts of a driven machine will, if large, appreciably affect the accelerating time and, hence, the amount of heating in the motor. If synchronous motors are used, the inertia ($WR^2$) of both the motor rotor and the rotating parts of the machine must be known since the pull-in torque (torque required to bring the driven machine up to synchronous speed) varies approximately as the square root of the total inertia of motor and load.

**Space Limitations in Motor Selection.**—If the motor is to become an integral part of the machine which it drives and space is at a premium, a partial motor may be called for. A complete motor is one made up of a stator, a rotor, a shaft, and two end shields with bearings. A *partial motor* is without one or more of these elements. One common type is furnished without drive-end end shield and bearing and is directly connected to the end or side of the machine which it drives, such as the headstock of a lathe. A so-called *shaftless type of motor* is supplied without shaft, end shields or bearings and is intended for built-in application in such units as multiple drilling machines, precision grinders, deep well pumps, compressors and hoists where the rotor is actually made a part of the driven machine. Where a partial motor is used, however, proper ventilation, mounting, alignment and bearings must be arranged for by the designer of the machine to which it is applied.

Sometimes it is possible to use a motor having a smaller frame size and wound with Class $B$ insulation, permitting it to be subjected to a higher temperature rise than the larger-frame Class $A$ insulated motor having the same horsepower rating.

**Temperatures.—**The applicability of a given motor is limited not only by its load-starting and -carrying ability, but also by the temperature which it reaches under load. Motors are given temperature ratings which are based upon the type of insulation (Class A or Class B are the most common) used in their construction and their type of frame (open, semienclosed, or enclosed).

*Insulating Materials:* Class A materials are: cotton, silk, paper, and similar organic materials when either impregnated or immersed in a liquid dielectric; molded and laminated materials with cellulose filler, phenolic resins, and other resins of similar properties; films and sheets of cellulose acetate and other cellulose derivatives of similar properties; and varnishes (enamel) as applied to conductors.

Class B insulating materials are: materials or combinations of materials such as mica, glass fiber, asbestos, etc., with suitable bonding substances. Other materials shown capable of operation at Class B temperatures may be included.

*Ambient Temperature and Allowable Temperature Rise:* Normal ambient temperature is taken to be 40° C (104° F). For open general-purpose motors with Class A insulation, the normal temperature rise on which the performance guarantees are based is 40° C (104° F).

Motors with Class A insulation having protected, semiprotected, drip-proof, or splash-proof, or drip-proof protected enclosures have a 50° C (122° F) rise rating.

Motors with Class A insulation and having totally enclosed, fan-cooled, explosion-proof, waterproof, dust-tight, submersible, or dust-explosion-proof enclosures have a 55° C (131° F) rise rating.

Motors with Class B insulation are permissible for total temperatures up to 110° C (230° F) for open motors and 115° C (239° F) for enclosed motors.

**Motors Exposed to Injurious Conditions.—**Where motors are to be used in locations imposing unusual operating conditions, the manufacturer should be consulted, especially where any of the following conditions apply: exposure to chemical fumes; operation in damp places; operation at speeds in excess of specified overspeed; exposure to combustible or explosive dust; exposure to gritty or conducting dust; exposure to lint; exposure to steam; operation in poorly ventilated rooms; operation in pits, or where entirely enclosed in boxes; exposure to inflammable or explosive gases; exposure to temperatures below 10° C (50° F); exposure to oil vapor; exposure to salt air; exposure to abnormal shock or vibration from external sources; where the departure from rated voltage is excessive; and or where the alternating-current supply voltage is unbalanced.

Improved insulating materials and processes and greater mechanical protection against falling materials and liquids make it possible to use general-purpose motors in many locations where special-purpose motors were previously considered necessary. *Splash-proof motors* having well-protected ventilated openings and specially treated windings are used where they are to be subjected to falling and splashing water or are to be washed down as with a hose. Where climatic conditions are not severe, this type of motor is also successfully used in unprotected outdoor installations.

If the surrounding atmosphere carries abnormal quantities of metallic, abrasive, or non-explosive dust or acid or alkali fumes, a *totally enclosed fan-cooled motor* may be called for. In this type, the motor proper is completely enclosed but air is blown through an outer shell that completely or partially surrounds the inner case. If the dust in the atmosphere tends to pack or solidify and close the air passages of open splash-proof or totally enclosed fan-cooled motors, *totally enclosed (nonventilated) motors* are used. This type, which is limited to low horsepower ratings, is also used for outdoor service in mild or severe climates.

**Table 1. Characteristics and Applications of D.C. Motors, 1–300 Horsepower**

| Type | Starting Duty | Maximum Momentary Running Torque | Speed Regulation | Speed Control[a] | Applications |
|---|---|---|---|---|---|
| Shunt-wound, constant-speed | Medium starting torque. Varies with voltage supplied to armature, and is limited by starting resistor to 125 to 200% full-load torque | 125 to 200%. Limited by commutation | 8 to 12% | Basic speed to 200% basic speed by field control | Drives where starting requirements are not severe. Use constant-speed or adjustable-speed, depending on speed required. Centrifugal pumps, fans, blowers, conveyors, elevators, wood- and metalworking machines |
| Shunt-wound, adjustable speed | | | 10 to 20%, increases with weak fields | Basic speed to 60% basic speed (lower for some ratings) by field control | |
| Shunt-wound, adjustable voltage control | | | Up to 25%. Less than 5% obtainable with special rotating regulator | Basic speed to 2% basic speed and basic speed to 200% basic speed | Drives where wide, stepless speed control, uniform speed, constant-torque acceleration and adaptability to automatic operation are required. Planers, milling machines, boring machines, lathes, etc. |
| Compound wound, constant-speed | Heavy starting torque, Limited by starting resistor to 130 to 260% of full-load torque | 130 to 260%. Limited by commutation | Standard compounding 25%. Depends on amount of series winding | Basic speed to 125% basic speed by field control | Drives requiring high starting torque and fairly constant speed. Pulsating loads. Shears, bending rolls, pumps, conveyors, crushers, etc. |
| Series-wound, varying-speed | Very heavy starting torque. Limited to 300 to 350% full-load torque | 300 to 350%. Limited by commutation | Very high. Infinite no-load speed | From zero to maximum speed, depending on control and load | Drives where very high starting torque is required and speed can be regulated. Cranes, hoists, gates, bridges, car dumpers, etc. |

[a] Minimum speed below basic speed by armature control limited by heating.

**Table 2. Characteristics and Applications of Polyphase AC Motors**

| Polyphase Type | Ratings hp | Speed Regulation | Speed Control | Starting Torque | Breakdown Torque | Applications |
|---|---|---|---|---|---|---|
| General-purpose squirrel cage, normal stg current, normal stg torque. Design B | 0.5 to 200 | Less than 5% | None, except multispeed types, designed for two to four fixed speeds | 100 to 250% of full-load | 200 to 300% of full-load | Constant-speed service where starting torque is not excessive. Fans, blowers, rotary compressors, centrifugal pumps, woodworking machines, machine tools, line shafts |
| Full-voltage starting, high stg torque, normal stg current, squirrel-cage, Design C | 3 to 150 | Less than 5% | None except multispeed types, designed for two to four fixed speeds | 200 to 250% of full-load | 190 to 225% of full-load | Constant-speed service where fairly high starting torque is required at infrequent intervals with starting current of about 500% full-load. Reciprocating pumps and compressors, conveyors, crushers, pulverizers, agitators, etc. |
| Full-voltage starting, high stgtorque, high-slip squirrel cage, Design D | 0.5 to 150 | Drops about 7 to 12% from no load to full load | None, except multispeed types, designed for two to four fixed speeds | 275% of full-load depending on speed and rotor resistance | 275% of full-load Will usually not stall until loaded to its maximum torque, which occurs at standstill | Constant-speed service and high-starting torque if starting not too frequent, and for taking high-peak loads with or without flywheels. Punch presses, die stamping, shears, bulldozers, bailers, hoists, cranes, elevators, etc. |
| Wound-rotor, external-resistance starting | 0.5 to several thousand | With rotor rings short-circuited drops about 3% for large to 5% for small sizes | Speed can be reduced to 50% of normal by rotor resistance. Speed varies inversely as the load | Up to 300% depending on external resistance in rotor circuit and how distributed | 200% when rotor slip rings are short circulated | Where high-starting torque with low-starting current or where limited speed control is required. Fans, centrifugal and plunger pumps, compressors, conveyors, hoists, cranes, ball mills, gate hoists, etc. |
| Synchronous | 25 to several thousand | Constant | None, except special motors designed for two fixed speeds | 40% for slow speed to 160% for medium speed 80% p-f designs. Special high-torque designs | Pull-out torque of unity-p-f motors 170%; 80%-p-f motors 225%. Special designs up to 300% | For constant-speed service, direct connection to slow-speed machines and where power-factor correction is required. |

In addition to these special-purpose motors, there are two types of *explosion-proof motors* designed for hazardous locations. One type is for operation in hazardous dust locations (Class II, Group *G* of the National Electrical Code) and the other is for atmospheres containing explosive vapors and fumes classified as Class I, Group *D* (gasoline, naphtha, alcohols, acetone, lacquer-solvent vapors, natural gas).

## Electric Motor Maintenance

**Electric Motor Inspection Schedule.—**Frequency and thoroughness of inspection depend upon such factors as 1) importance of the motor in the production scheme; 2) percentage of days the motor operates; 3) nature of service; and 4) winding conditions.

The following schedules, recommended by the General Electric Company, and covering both AC and DC motors are based on average conditions in so far as duty and dirt are concerned.

**Weekly Inspection.—**1) *Surroundings*. Check to see if the windings are exposed to any dripping water, acid or alcoholic fumes; also, check for any unusual amount of dust, chips, or lint on or about the motor. See if any boards, covers, canvas, etc., have been misplaced that might interfere with the motor ventilation or jam moving parts.

2) *Lubrication of sleeve-bearing motors*. In sleeve-bearing motors check oil level, if a gage is used, and fill to the specified line. If the journal diameter is less than 2 inches, the motor should be stopped before checking the oil level. For special lubricating systems, such as wool-packed, forced lubrication, flood and disk lubrication, follow instruction book. Oil should be added to bearing housing only when motor is at rest. A check should be made to see if oil is creeping along the shaft toward windings where it may harm the insulation.

3) *Mechanical condition*. Note any unusual noise that may be caused by metal-to-metal contact or any odor as from scorching insulation varnish.

4) *Ball or roller bearings*. Feel ball- or roller-bearing housings for evidence of vibration, and listen for any unusual noise. Inspect for creepage of grease on inside of motor.

5) *Commutators and brushes*. Check brushes and commutator for sparking. If the motor is on cyclic duty it should be observed through several cycles. Note color and surface condition of the commutator. A stable copper oxide-carbon film (as distinguished from a pure copper surface) on the commutator is an essential requirement for good commutation. Such a film may vary in color all the way from copper to straw, chocolate to black. It should be clean and smooth and have a high polish. All brushes should be checked for wear and pigtail connections for looseness. The commutator surface may be cleaned by using a piece of dry canvas or other hard, nonlinting material that is wound around and securely fastened to a wooden stick, and held against the rotating commutator.

6) *Rotors and armatures*. The air gap on sleeve-bearing motors should be checked, especially if they have been recently overhauled. After installing new bearings, make sure that the average reading is within 10 percent, and reading should be less than 0.020 inch (0.51 mm). Check air passages through punchings and make sure they are free of foreign matter.

7) *Windings*. If necessary clean windings by suction or mild blowing. After making sure that the motor is dead, wipe off windings with dry cloth, note evidence of moisture, and see if any water has accumulated in the bottom of frame. Check if any oil or grease has worked its way up to the rotor or armature windings. Clean with carbon tetrachloride in a well-ventilated room.

8) *General*. This is a good time to check the belt, gears, flexible couplings, chain, and sprockets for excessive wear or improper location. The motor starting should be checked to make sure that it comes up to proper speed each time power is applied.

**Monthly or Bimonthly Inspection.—**1) *Windings*. Check shunt, series, and commutating field windings for tightness. Try to move field spools on the poles, as drying out may have caused some play. If this condition exists, a service shop should be consulted. Check motor cable connections for tightness.

2) *Brushes*. Check brushes in holders for fit and free play. Check the brush-spring pressure. Tighten brush studs in holders to take up slack from drying out of washers, making sure that studs are not displaced, particularly on DC motors. Replace brushes that are worn down almost to the brush rivet, examine brush faces for chipped toes or heels, and for heat cracks. Damaged brushes should be replaced immediately.

3) *Commutators*. Examine commutator surface for high bars and high mica, or evidence of scratches or roughness. See that the risers are clean and have not been damaged.

4) *Ball or roller bearings*. On hard-driven, 24-hour service ball- or roller-bearing motors, purge out old grease through drain hole and apply new grease. Check to make sure grease or oil is not leaking out of the bearing housing. If any leakage is present, correct the condition before continuing to operate.

5) *Sleeve bearings*. Check sleeve bearings for wear, including end-play bearing surfaces. Clean out oil wells if there is evidence of dirt or sludge. Flush with lighter oil before refilling.

6) *Enclosed gears*. For motors with enclosed gears, open drain plug and check oil flow for presence of metal scale, sand, or water. If condition of oil is bad, drain, flush, and refill as directed. Rock rotor to see if slack or backlash is increasing.

7) *Loads*. Check loads for changed conditions, bad adjustment, poor handling, or control.

8) *Couplings and other drive details*. Note if belt-tightening adjustment is all used up. Shorten belt if this condition exists. See if belt runs steadily and close to inside (motor edge) of pulley. Chain should be checked for evidence of wear and stretch. Clean inside of chain housing. Check chain-lubricating system. Note inclination of slanting base to make sure it does not cause oil rings to rub on housing.

**Annual or Biannual Inspection.**—1) *Windings*. Check insulation resistance by using either a megohmmeter or a voltmeter having a resistance of about 100 ohms per volt. Check insulation surfaces for dry cracks and other evidence of need for coatings of insulating material. Clean surfaces and ventilating passages thoroughly if inspection shows accumulation of dust. Check for mold or water standing in frame to determine if windings need to be dried out, varnished, and baked.

2) *Air gap and bearings*. Check air gap to make sure that average reading is within 10 percent, provided reading should be less than 0.020 inch (0.51 mm). All bearings, ball, roller, and sleeve should be thoroughly checked and defective ones replaced. Waste-packed and wick-oiled bearings should have waste or wicks renewed, if they have become glazed or filled with metal or dirt, making sure that new waste bears well against shaft.

3) *Rotors (squirrel-cage)*. Check squirrel-cage rotors for broken or loose bars and evidence of local heating. If fan blades are not cast in place, check for loose blades. Look for marks on rotor surface indicating foreign matter in air gap or a worn bearing.

4) *Rotors (wound)*. Clean wound rotors thoroughly around collector rings, washers, and connections. Tighten connections if necessary. If rings are rough, spotted, or eccentric, refer to service shop for refinishing. See that all top sticks or wedges are tight. If any are loose, refer to service shop.

5) *Armatures*. Clean all armature air passages thoroughly if any are obstructed. Look for oil or grease creeping along shaft, checking back to bearing. Check commutator for surface condition, high bars, high mica, or eccentricity. If necessary, remachine the commutator to secure a smooth fresh surface.

6) *Loads*. Read load on motor with instruments at no load, full load, or through an entire cycle, as a check on the mechanical condition of the driven machine.

## ADHESIVES AND SEALANTS

By strict definition, an adhesive is any substance that fastens or bonds materials to be joined (adherends) by means of surface attachment. The bond durability depends on the strength of the adhesive to the substrate (adhesion) and the strength within the adhesive (cohesion). Besides bonding a joint, an adhesive may serve as a seal against foreign matter. When an adhesive performs both bonding and sealing functions, it is usually referred to as an *adhesive sealant*. Joining materials with adhesives offers significant benefits compared with mechanical methods of uniting two materials.

Among these benefits are that an adhesive distributes a load over an area rather than concentrating it at a point, resulting in a more even distribution of stresses. The adhesive bonded joint is therefore more resistant to flexural and vibrational stresses than, for example, a bolted, riveted, or welded joint. Another benefit is that an adhesive forms a seal as well as a bond. This seal prevents the corrosion that may occur with dissimilar metals, such as aluminum and magnesium, or mechanically fastened joints, by providing a dielectric insulation between the substrates. An adhesive also joins irregularly shaped surfaces more easily than does a mechanical fastener. Other benefits include negligible weight addition and virtually no change to part dimensions or geometry.

Most adhesives are available in liquids, gels, pastes, and tape forms. The growing variety of adhesives available can make the selection of the proper adhesive or sealant a challenging experience. In addition to the technical requirements of the adhesive, time and costs are also important considerations. Proper choice of an adhesive is based on knowledge of the suitability of the adhesive or sealant for the particular substrates. Appropriate surface preparation, curing parameters, and matching the strength and durability characteristics of the adhesive to its intended use are essential. The performance of an adhesive-bonded joint depends on a wide range of these factors, many of them quite complex. Adhesive suppliers can usually offer essential expertise in the area of appropriate selection.

Adhesives can be classified as structural or nonstructural. In general, an adhesive can be considered structural when it is capable of supporting heavy loads; nonstructural when it cannot support such loads. Many adhesives and sealants, under various brand names, may be available for a particular bonding application. It is always advisable to check the adhesive manufacturers' information before making an adhesive sealant selection. Also, testing under end-use conditions is always suggested to help ensure bonded or sealed joints meet or exceed expected performance requirements.

Though not meant to be all-inclusive, the following information correlates the features of some successful adhesive compositions available in the marketplace.

### Bonding Adhesives

Reactive-type bonding adhesives are applied as liquids and react (cure) to solids under appropriate conditions. The cured adhesive is either a thermosetting or thermoplastic polymer. These adhesives are supplied as two-component no-mix, two-component mix, and one-component no-mix types, which are discussed in the following paragraphs.

### Two-Component No-Mix Adhesives

**Types of Adhesives.—***Anaerobic (Urethane Methacrylate Ester) Structural Adhesives:* Anaerobic structural adhesives are mixtures of acrylic esters that remain liquid when exposed to air but harden when confined between metal substrates. These adhesives can be used for large numbers of industrial purposes where high reliability of bond joints is required. Benefits include: no mixing is required (no pot-life or waste problems), flexible/durable bonds are made that withstand thermal cycling, have excellent resistance to solvents and severe environments, and rapid cure at room temperatures (eliminating expensive ovens). The adhesives are easily dispensed with automatic equipment. An activator is usually required to be present on one surface to initiate the cure for these adhesives. Applications for these adhesives include bonding of metals, magnets (ferrites), glass, thermosetting plastics, ceramics, and stone.

*Acrylic Adhesives:* Acrylic adhesives are composed of a polyurethane polymer backbone with acrylate end groups. They can be formulated to cure through heat or the use of an activator applied to the substrate surface, but many industrial acrylic adhesives are cured by light. Light-cured adhesives are used in applications where the bond geometry allows light to reach the adhesive and the production rate is high enough to justify the capital expense of a light source. Benefits include: no mixing is required (no pot-life or waste problems); formulations cure (solidify) with activator, heat, or light; the adhesive will bond to a variety of substrates, including metal and most thermoplastics; and tough and durable bonds are produced with a typical resistance to the effects of temperatures up to 180° C. Typical applications include automobile body parts (steel stiffeners), assemblies subjected to paint-baking cycles, speaker magnets to pole plates, and bonding of motor magnets, sheet steel, and many other structural applications. Other applications include bonding glass, sheet metal, magnets (ferrite), thermosetting and thermoplastic plastics, wood, ceramics, and stone.

## Two-Component Mix Adhesives

**Types of Adhesives.—***Epoxy Adhesives:* Two-component epoxy adhesives are well-established adhesives that offer many benefits in manufacturing. The reactive components of these adhesives are separated prior to use, so they usually have a good shelf life without refrigeration. Polymerization begins upon mixing, and a thermoset polymer is formed. Epoxy adhesives cure to form thermosetting polymers made up of a base side with the polymer resin and a second part containing the catalyst. The main benefit of these systems is that the depth of cure is unlimited. As a result, large volume can be filled for work such as potting, without the cure being limited by the need for access to an external influence such as moisture or light to activate the curing process.

For consistent adhesive performance, it is important that the mix ratio remain constant to eliminate variations in adhesive performance. Epoxies can be handled automatically, but the equipment involves initial and maintenance costs. Alternatively, adhesive components can be mixed by hand. However, this approach involves labor costs and the potential for human error. The major disadvantage of epoxies is that they tend to be very rigid and consequently have low peel strength. This lack of peel strength is less of a problem when bonding metal to metal than it is when bonding flexible substrates such as plastics.

Applications of epoxy adhesives include bonding, potting, and coating of metals, bonding of glass, rigid plastics, ceramics, wood, and stone.

*Polyurethane Adhesives:* Like epoxies, polyurethane adhesives are available as two-part systems or as one-component frozen premixes. They are also available as one-part moisture-cured systems. Polyurethane adhesives can provide a wide variety of physical properties. Their flexibility is greater than that of most epoxies. Coupled with the high cohesive strength, this flexibility provides a tough polymer able to achieve better peel strength and lower flexural modulus than most epoxy systems. This superior peel resistance allows use of polyurethanes in applications that require high flexibility. Polyurethanes bond very well to a variety of substrates, though a primer may be needed to prepare the substrate surface. These primers are moisture-reactive and require several hours to react sufficiently for the parts to be used. Such a time requirement may cause a production bottleneck if the bond-strength requirements are such that a primer is needed.

Applications for polyurethane adhesives include bonding of metals, glass, rubber, thermosetting and thermoplastic plastics, and wood.

## One-Component No-Mix Adhesives

**Types of Adhesives.—***Light-Curable Adhesives:* Light-curing systems use a unique curing mechanism. The adhesives contain photoinitiators that absorb light energy and dissociate to form radicals. These radicals then initiate the polymerization of the polymers, oligomers, and monomers in the adhesive. The photoinitiator acts as a chemical solar cell, converting the light energy into chemical energy for the curing process. Typically, these systems are formulated for use with ultraviolet light sources. However, newer products have been formulated for use with visible light sources.

One of the biggest benefits that light-curing adhesives offer to the manufacturer is the elimination of the work time to work-in-progress trade-off, which is embodied in most adhesive systems. With light-curing systems, the user can take as much time as needed to position the part without fear of the adhesive curing. Upon exposure to the appropriate light source, the adhesive then can be fully cured in less than 1 minute, minimizing the costs associated with work in progress. Adhesives that utilize light as the curing mechanism are often one-part systems with good shelf life, which makes them even more attractive for manufacturing use.

Applications for light-curable adhesives include bonding of glass, and glass to metal, tacking of wires, surface coating, thin-film encapsulation, clear substrate bonding, and potting of components,

*Cyanoacrylate Adhesives (Instant Adhesives):* Cyanoacrylates or instant adhesives are often called Superglue™. Cyanoacrylates are one-part adhesives that cure rapidly, as a result of the presence of surface moisture, to form high-strength bonds, when confined between two substrates. Cyanoacrylates have excellent adhesion to many substrates, including most plastics and they achieve fixture strength in seconds and full strength within 24 hours. These qualities make cyanoacrylates suitable for use in automated production environments. They are available in viscosities ranging from water-thin liquids to thixotropic gels.

Because cyanoacrylates are a relatively mature adhesive family, a wide variety of specialty formulations is now available to help the user address difficult assembly problems. One of the best examples is the availability of polyolefin primers, which allow users to obtain high bond strengths on difficult-to-bond plastics such as polyethylene and polypropylene. One common drawback of cyanoacrylates is that they form a very rigid polymer matrix, resulting in very low peel strengths. To address this problem, formulations have been developed that are rubber-toughened. Although the rubber toughening improves the peel strength of the system to some extent, peel strength remains a weak point for this system, and, therefore, cyanoacrylates are poor candidates for joint designs that require high peel resistance. In manufacturing environments with low relative humidity, the cure of the cyanoacrylate can be significantly retarded.

This problem can be addressed in one of two ways. One approach is to use accelerators that deposit active species on the surface to initiate the cure of the product. The other approach is to use specialty cyanoacrylate formulations that have been engineered to be surface-insensitive. These formulations can cure rapidly even on dry or slightly acidic surfaces.

Applications for cyanoacrylate adhesives include bonding of thermoplastic and thermosetting plastics, rubber, metals, wood, and leather, also strain relief of wires.

*Hot-Melt Adhesives:* Hot-melt adhesives are widely used in assembly applications. In general, hot-melt adhesives permit fixturing speeds that are much faster than can be achieved with water- or solvent-based adhesives. Usually supplied in solid form, hot-melt adhesives liquify when exposed to elevated temperatures. After application, they cool quickly, solidifying and forming a bond between two mating substrates. Hot-melt adhesives have been used successfully for a wide variety of adherends and can greatly reduce both the need for clamping and the length of time for curing. Some drawbacks with hot-melt adhesives are their tendency to string during dispensing and relatively low-temperature resistance.

Applications for hot-melt adhesives are bonding of fabrics, wood, paper, plastics, and cardboard.

*Rubber-Based Solvent Cements:* Rubber-based solvent cements are adhesives made by combining one or more rubbers or elastomers in a solvent. These solutions are further modified with additives to improve the tack or stickiness, the degree of peel strength, flexibility, and the viscosity or body. Rubber-based adhesives are used in a wide variety of applications such as contact adhesive for plastics laminates like counter tops, cabinets, desks, and tables. Solvent-based rubber cements have also been the mainstay of the shoe and leather industry for many years.

Applications for rubber-based solvent cements include bonding of plastics laminates, wood, paper, carpeting, fabrics, and leather.

*Moisture-Cured Polyurethane Adhesives:* Like heat-curing systems, moisture-cured polyurethanes have the advantage of a very simple curing process. These adhesives start to cure when moisture from the atmosphere diffuses into the adhesive and initiates the polymerization process. In general, these systems will cure when the relative humidity is above 25 percent, and the rate of cure will increase as the relative humidity increases.

The dependence of these systems on the permeation of moisture through the polymer is the source of their most significant process limitations. As a result of this dependence, depth of cure is limited to between 0.25 and 0.5 in. (6.35 and 12.7 mm). Typical cure times are in the range of 12 to 72 hours. The biggest use for these systems is for windshield bonding in automobile bodies.

Applications for moisture-cured polyurethane adhesives include bonding of metals, glass, rubber, thermosetting and thermoplastic plastics, and wood.

## Retaining Compounds

The term *retaining compounds* is used to describe adhesives used in circumferential assemblies joined by inserting one part into the other. In general, retaining compounds are anaerobic adhesives composed of mixtures of acrylic esters that remain liquid when exposed to air but harden when confined between cylindrical machine components. A typical example is a bearing held in an electric motor housing with a retaining compound. The first retaining compounds were launched in 1963, and the reaction among users of bearings was very strong because these retaining compounds enabled buyers of new bearings to salvage worn housings and minimize their scrap rate.

The use of retaining compounds has many benefits, including elimination of bulk needed for high friction forces, ability to produce more accurate assemblies and to augment or replace press fits, increased strength in heavy press fits, and reduction of machining costs. Use of these compounds also helps in dissipating heat through assembly, and eliminating distortion when installing drill bushings, fretting corrosion and backlash in keys and splines, and bearing seizure during operation.

The major advantages of retaining compounds for structural assemblies are that they require less severe machining tolerances and no securing of parts. Components are assembled quickly and cleanly, and they transmit high forces and torques, including dynamic forces. Retaining compounds also seal, insulate, and prevent micromovements so that neither fretting corrosion nor stress corrosion occurs. The adhesive joint can be taken apart easily after heating above 450° F (230° C) for a specified time.

Applications for retaining compounds include mounting of bearings in housings or on shafts, avoiding distortion of precision tooling and machines, mounting of rotors on shafts, inserting drill jig bushings, retaining cylinder linings, holding oil filter tubes in castings, retaining engine-core plugs, restoring accuracy to worn machine tools, and eliminating keys and set screws.

## Threadlocking

The term *threadlocker* is used to describe adhesives used in threaded assemblies for locking the threaded fasteners by filling the spaces between the nut and bolt threads with a hard, dense material that prevents loosening. In general, thread-lockers are anaerobic adhesives comprising mixtures of acrylic esters that remain liquid when exposed to air but harden when confined between threaded components. A typical example is a mounting bolt on a motor or a pump. Threadlocker strengths range from very low strength (removable) to high strength (permanent).

It is important that the total length of the thread is coated and that there is no restriction to the curing of the threadlocker material. (Certain oils or cleaning systems can impede or even completely prevent the adhesive from curing by anaerobic reaction.) The liquid threadlocker may be applied by hand or with special dispensing devices. Proper coating (wetting) of a thread is dependent on the size of the thread, the viscosity of the adhesive and the geometry of the parts. With blind-hole threads, it is essential that the adhesive be applied all the way to the bottom of the threaded hole. The quantity must be such that after assembly, the displaced adhesive fills the whole length of the thread.

Some threadlocking products cured by anaerobic reaction have a positive influence on the coefficient of friction in the thread. The values are comparable with those of oiled bolts. Prestress and installation torque therefore can be defined exactly. This property allows threadlocking products cured by anaerobic reaction to be integrated into automated production lines using existing assembly equipment. The use of thread-lockers has many benefits including ability to lock and seal all popular bolt and nut sizes with all industrial finishes, and to replace mechanical locking devices. The adhesive can seal against most industrial fluids and will lubricate threads so that the proper clamp load is obtained. The materials also provide vibration-resistant joints that require handtool dismantling for servicing, prevent rusting of threads, and cure (solidify) without cracking or shrinking.

The range of applications includes such uses as locking and sealing nuts on hydraulic pistons, screws on vacuum cleaner bell housings, track bolts on bulldozers, hydraulic-line fittings, screws on typewriters, oil-pressure switch assembly, screws on carburetors, rocker nuts, machinery driving keys, and on construction equipment.

## Sealants

The primary role of a sealant composition is the prevention of leakage from or access by dust, fluids, and other materials to assembly structures. Acceptable leak rates can range from a slight drip to bubbletight to molecular diffusion through the base materials. Equipment users in the industrial market want trouble-free operation, but it is not always practical to specify zero leak rates. Factors influencing acceptable leak rates are toxicity, product or environmental contamination, combustibility, economics, and personnel considerations. All types of fluid seals perform the same basic function: they seal the process fluid (gas, liquid, or vapor) and keep it where it belongs. A general term for these assembly approaches is gasketing. Many products are being manufactured that are capable of sealing a variety of substrates.

**Types of Sealants.—***Anaerobic Formed-in-Place Gasketing Materials:* Mechanical assemblies that require the joining of metal-to-metal flange surfaces have long been designed with prefabricated, precut materials required to seal the imperfect surfaces of the assembly. Numerous gasket materials that have been used to seal these assemblies include paper, cork, asbestos, wood, metals, dressings, and even plastics. Fluid seals are divided into static and dynamic systems, depending on whether or not the parts move in relationship to each other. Flanges are classed as static systems, although they may be moved relative to each other by vibration, temperature, and/or pressure changes, shocks, and impacts.

The term *anaerobic formed-in-place gasketing* is used to describe sealants that are used in flanged assemblies to compensate for surface imperfections of metal-to-metal components by filling the space between the substrates with a flexible, nonrunning material. In general, anaerobic formed-in-place gaskets are sealants made up of mixtures of acrylic esters that remain liquid when exposed to air but harden when confined between components. A typical example is sealing two halves of a split crankcase.

The use of anaerobic formed-in-place gaskets has many benefits, including the ability to seal all surface imperfections, allow true metal-to-metal contact, eliminate compression set and fastener loosening, and add structural strength to assemblies. These gaskets also help improve torque transmission between bolted flange joints, eliminate bolt retorquing needed with conventional gaskets, permit use of smaller fasteners and lighter flanges, and provide for easy disassembly and cleaning.

Applications in which formed-in-place gasketing can be used to produce leakproof joints include pipe flanges, split crankcases, pumps, compressors, power takeoff covers, and axle covers. These types of gaskets may also be used for repairing damaged conventional gaskets and for coating soft gaskets.

*Silicone Rubber Formed-in-Place Gasketing:* Another type of formed-in-place gasket uses room-temperature vulcanizing (RTV) silicone rubbers. These materials are one-component sealants that cure on exposure to atmospheric moisture. They have excellent properties for vehicle use such as flexibility, low volatility, good adhesion, and high resistance to most automotive fluids. The materials will also withstand temperatures up to 600° F (320° C) for intermittent operation.

RTV silicones are best suited for fairly thick section (gap) gasketing applications where flange flexing is greatest. In the form of a very thin film, for a rigid metal-to-metal seal, the cured elastomer may abrade and eventually fail under continual flange movement. The RTV silicone rubber does not unitize the assembly, and it requires relatively clean, oil-free surfaces for sufficient adhesion and leakproof seals.

Because of the silicone's basic polymeric structure, RTV silicone elastomers have several inherent characteristics that make them useful in a wide variety of applications. These properties include outstanding thermal stability at temperatures from 400 to 600° F (204 to 320° C), and good low-temperature flexibility at −85 to −165° F (−65 to −115° C). The material forms an instant seal, as is required of all liquid gaskets, and will fill large gaps up to 0.250 in. (6.35 mm) for stamped metal parts and flanges. The rubber also has good stability in ultraviolet light and excellent weathering resistance.

Applications for formed-in-place RTV silicones in the automotive field are valve, camshaft and rocker covers, manual transmission (gearbox) flanges, oil pans, sealing panels, rear axle housings, timing chain covers, and window plates. The materials are also used on oven doors and flues.

## Tapered Pipe-Thread Sealing

Thread sealants are used to prevent leakage of gases and liquids from pipe joints. All joints of this type are considered to be dynamic because of vibration, changing pressures, or changing temperatures.

Several types of sealants are used on pipe threads including noncuring pipe dopes, which are one of the oldest methods of sealing the spiral leak paths of threaded joints. In general, pipe dopes are pastes made from oils and various fillers. They lubricate joints and jam threads but provide no locking advantage. They also squeeze out under pressure, and have poor solvent resistance. Noncuring pipe dopes are not suitable for use on straight threads.

Another alternative is solvent-drying pipe dopes, which are an older method of sealing tapered threaded joints. These types of sealant offer the advantages of providing lubrication and orifice jamming and they also extrude less easily than noncuring pipe dopes. One disadvantage is that they shrink during cure as the solvents evaporate and fittings must be retorqued to minimize voids. These materials generally lock the threaded joint together by friction. A third type of sealer is the trapped elastomer supplied in the form of a thin tape incorporating polytetrafluorethylene (PTFE). This tape gives a good initial seal and resists chemical attack, and is one of the only materials used for sealing systems that will seal against oxygen gas.

Some other advantages of PTFE are that it acts as a lubricant, allows for high torquing, and has a good resistance to various solvents. Some disadvantages are that it may not provide a true seal between the two threaded surfaces, and it lubricates in the off direction so it may allow fittings to loosen. In dynamic joints, tape may allow creep, resulting in leakage over time. The lubrication effect may allow overtightening, which can add stress or lead to breakage. Tape also may be banned in some hydraulic systems due to shredding, which may cause clogging of key orifices.

**Anaerobic Pipe Sealants.—**The term *anaerobic pipe sealants* is used to describe anaerobic sealants used in tapered threaded assemblies for sealing and locking threaded joints. Sealing and locking are accomplished by filling the space between the threads with the sealant. In general, these pipe sealants are anaerobic adhesives consisting of mixtures of acrylic esters that remain liquid when exposed to air but harden when confined between threaded components to form an insoluble tough plastics. The strength of anaerobic pipe sealants is between that of elastomers and yielding metal.

Clamp loads need be only tight enough to prevent separation in use. Because they develop strength by curing after they are in place, these sealants are generally forgiving of tolerances, tool marks, and slight misalignment. These sealants are formulated for use on metal substrates. If the materials are used on plastics, an activator or primer should be used to prepare the surfaces.

Among the advantages of these anaerobic sealers are that they lubricate during assembly, they seal regardless of assembly torque, and they make seals that correspond with the burst rating of the pipe. They also provide controlled disassembly torque, do not cure outside the joint, and are easily dispensed on the production line. These sealants also have the lowest cost per sealed fitting. Among the disadvantages are that the materials are not suitable for oxygen service, for use with strong oxidizing agents, or for use at temperatures above 200° C (392° F). The sealants also are typically not suitable for diameters over M80 or 80 mm (approximately 3 inches).

The many influences faced by pipe joints during service should be known and understood at the design stage, when sealants are selected. Sealants must be chosen for reliability and long-term quality. Tapered pipe threads must remain leak-free under the severest vibration and chemical attack, also under heat and pressure surges.

Applications of aerobic sealants are found in industrial plant fluid power systems, the textile industry, chemical processing, utilities and power generation facilities, petroleum refining, and in marine, automotive, and industrial equipment. The materials are also used in the pulp and paper industries, in gas compression and distribution, and in waste-treatment facilities.

# O-RINGS

## Overview

An O-ring is a one-piece molded seal that distorts when squeezed between two parts to fill a gap. Most O-rings are toroidal (a round donut shape with a hole in the middle) and made from highly resilient materials, such as thermoplastic elastomers. Some O-rings are made from materials that permanently deform in use, such as PTFE (polytetrafluoroethylene).

To create the sealing geometry, an O-ring is contained in a cylindrical or planar groove cut into one part. An O-ring gland is composed of the groove in one part and a sealing surface in the other part. If the two parts fit together with no gap, the groove and gland depths are the same. From a sealing standpoint, it usually does not matter which part in the cylindrical or planar joint carries the gland or groove. Ease of manufacture and/or assembly normally drives that decision. When properly installed in a gland, an O-ring is slightly deformed, so that the naturally round cross section is squeezed diametrically out of round prior to application of pressure. This compression ensures that under static conditions, the O-ring is in forceful contact (or "interferes with") the sealing surfaces of both parts enclosing it. The resiliency of the material ensures adequate sealing force at zero applied pressure and allows some dimensional variation within the gland.

When pressure is applied, it tends to force the O-ring across the gland, causing the O-ring to further deform and increase the sealing force, as in Fig. 1(a). As additional pressure is applied, the O-ring deforms into a D shape and can flow into the clearance gap between the two parts, as in Fig. 1(b). When the applied pressure changes direction (sometimes referred to as *cycling*), the O-ring can be forced back and forth within the gland. In such cases, a stiffer or abrasion-resistant O-ring, backup rings, or a second O-ring seal (one for each direction) may be needed. (Other seal types also are available that are better suited to pressure-reversal applications.) In gaseous applications, the permeability of the O-ring material must be considered in estimating leakage rates and avoiding explosive decompression failure of the O-ring, particularly at high pressures.

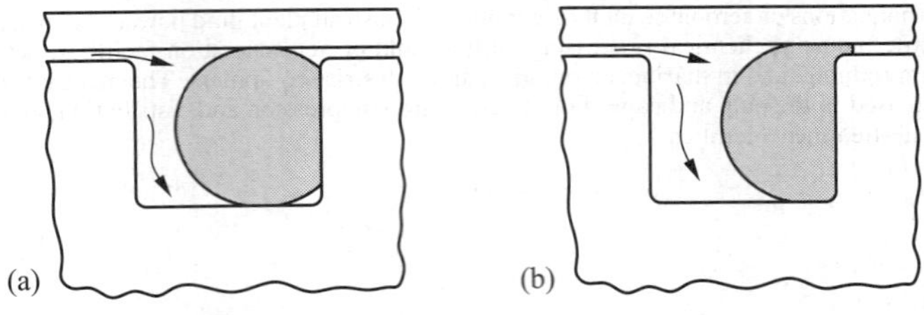


Fig. 1. Effect of Fluid Pressure on the Cross Section of an O-Ring in a Gland

**O-Ring Standards and Designations.—**O-ring size is based on two parameters: inner *diameter* and cross-sectional diameter. However, to completely characterize an O-ring, both the size and the material must be specified. Manufacturers often use different designations to identify O-ring compounds.

Table 1 provides some prevailing standards for general-purpose inch and metric O-rings and gland designs. Note that previous O-ring metric standards BS 1806, DIN 3771, and JIS B 2401 have been superseded by the ISO standards shown in this table.

**Table 1. Commonly Referenced O-Ring Standards**

| Application | Units | Standard | Title |
|---|---|---|---|
| O-ring dimensions and tolerances | inch | SAE AS568 | Aerospace Size Standard for O-rings |
| | mm | ISO 3601-1 | Fluid power systems – O-rings – Part 1: Inside diameters, cross-sections, tolerances and designation codes |
| General use piston or rod gland design, static and dynamic applications | inch | SAE AS4716 | Gland Design, O-ring and Other Seals |
| | mm | ISO 3601-2 | Fluid power systems – O-rings – Part 2: Housing dimensions for general applications |
| Static face seal (axial) gland design | inch | SAE AS6235 | Face Seal Gland Design, Static, O-ring and Other Seals for Aerospace Hydraulic and Pneumatic Applications |
| | mm | ISO 3601-2 | Fluid power systems – O-rings – Part 2: Housing dimensions for general applications |
| Backup rings | inch | SAE *(various)* | *(Backup rings are contained in seperate SAE standards based on material and style.)* |
| | mm | ISO 3601-4 | Fluid power systems – O-rings – Part 4: Anti-extrusion rings (backup rings) |

*Inch System O-Rings:* The inch-basis SAE standard O-ring sizes are assigned three-digit size codes. A dash precedes the size code. The first digit refers to cross-sectional diameter, and the next two digits refer to inner diameter; however, the numbers that make up the size code are not derivative of the actual inch measurements. For example, for an SAE standard O-ring with designation -230-8307, the O-ring size code is -230, which specifies a cross-sectional diameter of 0.139 in. and inner diameter of 2.484 inch. The complete designation number indicates that the O-ring is made from the vendor-specific material designated by material code 8307.

*Metric General-Purpose O-Rings:* According to ISO specifications, O-rings for general use are functionally equivalent to those found in SAE AS568F when used in the same glands. Size codes for these use the same three-digit numbers set by the SAE Standard. But the ISO size code is completed by an added tolerance class designation (A or B). Tolerance class A O-rings, which are equivalent to those in the SAE (inch) Standard, are suitable for aerospace or general applications requiring tight tolerances. Even in applications where tolerance needs are not as tight, class A O-rings are most often used. Tolerance class B O-rings, which are less tight and considered appropriate for general applications, are not as common in industry.

To give an ISO example of an SAE equivalent O-ring: for a part with nominal inside diameter of 63.09 mm and nominal cross-sectional diameter of 3.53 mm, the ISO size code is -230A, and the full designation is -230A-63.09x3.53-N. The final letter N is the grade. Grade indicates the inspection standards required per ISO 3601-3: Grade N = general purpose, S = special, and CS = critical service.

It is important to note that, as with SAE designations, ISO standard designations do not identify the material. A manufacturer-specific compound identifier must be appended or called out separately.

*Metric Aerospace O-Rings:* Another group of O-rings in the ISO Standard is intended specifically for aerospace applications. These are designed for use in metric glands. This is advantageous when close fits are required or when working with metric bores and pistons. Codes for aerospace sizes differ from those for general applications; they are preceded by the letter A followed by a dash. The size code then begins with a cross section size identifier, designated from A to E; the next four digits give the inside diameter in tenths of a millimeter. There is only one tolerance class in this group, so a tolerance class identifier is not included, but the inspection standard per ISO 3601-3 is added to the end of the code. For example, a critical service aerospace O-ring with a 3.55 mm cross section and a 63 mm inside diameter has a size code of C0630 and the full designation is A-C0630-63.00x3.55-CS. In this case, the grade code is CS. This series of O-ring sizes are not covered here, but size codes and dimensional information can be found in ISO standards and supplier catalogs.

**O-Ring Size Selection.—**O-ring sizing involves determining the best inside and cross-sectional diameter for an application. Choice of inner diameter usually is determined by the basic needs of the assembly being sealed: the groove for the seal must fit in one of the assembled parts, leave enough material on the part to provide sufficient stiffness, and allow for installation. Deciding which part will bear the groove and choosing the O-ring is heavily influenced by cost and assembly considerations.

There are more variables involved in determining cross-sectional diameter that can substantially impact sealing effectiveness. Smaller cross section O-rings have the advantages of lighter weight, compact footprint, better resistance to explosive decompression, lower friction in dynamic sealing applications, and better physical performance for many materials. Disadvantages of smaller cross sections include higher likeliness of leaks due to surface defects in the gland, less stability in reciprocating applications, higher vulnerability to compression set, higher percentage swell (expansion due to fluid exposure or heat), and the need for tighter gland tolerances.

When working with an assembly containing thin or fragile parts, compression force of the O-ring also must be taken into account. Compression force depends on O-ring material properties, cross section, length (mean circumference), and amount of squeeze induced in the O-ring cross section by the gland.

SAE and ISO standards indicate appropriate sizes for various applications. A greater range of sizes are suitable for static applications than for dynamic applications. There also are differences in size ranges for internal pressure (pushing out from the O-ring's center) and external pressure (pushing inward from outside the O-ring) axial face seal applications. Fig. 2, Fig. 3, and the tables on the following pages show size ranges and relevant gland dimensions for SAE and ISO standard size codes.

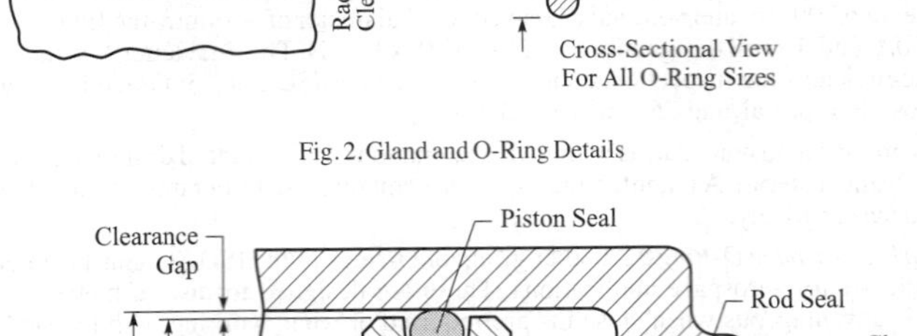


Fig. 2. Gland and O-Ring Details

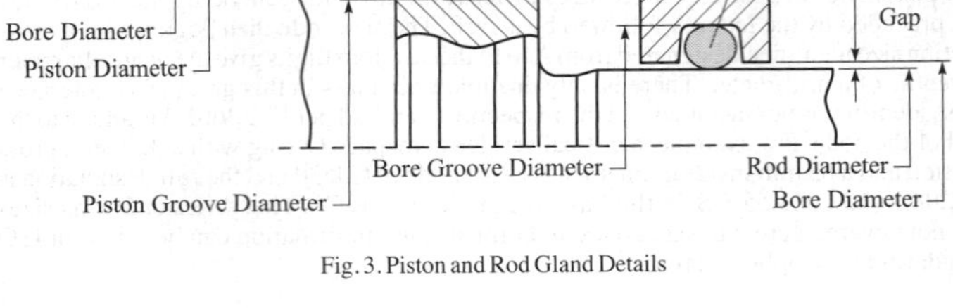


Fig. 3. Piston and Rod Gland Details

## Table 2. Inch O-Ring Sizes and Gland Data by Size Range

*(Based on SAE AS4716C-2017, AS6235A-2015, and 70 Shore A O-Rings)*

| O-Ring | | | Groove Inside Corner Radii | Piston Seal & Rod Seal | | | | | | | | | Axial Face Seal | | |
|---|---|---|---|---|---|---|---|---|---|---|---|---|---|---|---|
| Size Code | Cross-Section Diameter | ± Tol. | | Recommended Range of O-Ring Sizes | | Groove Eccentricity | Size Sub-Range for Groove Width | Groove Width *G* | | | Recommended Range of O-Ring Sizes | | Groove Width *G* | Groove Depth *F* | Installed Axial Squeeze (No Swell, No Radii, Straight Sides) |
| | Nominal | | Min/Max | Static | Dynamic | Max | | No Backup Min/Max | 1 Backup Ring Min/Max | 2 Backup Rings Min/Max | Internal Pressure | External Pressure | Min/Max | Min/Max | Min/Max |
| -0XX | 0.070 | 0.003 | 0.005 | 004 to 028 | 004 to 012 | 0.002 | 004 to 009 | 0.098 | 0.154 | 0.210 | 008 to 028 | 006 to 028 | 0.098 | 0.056 | 10% |
| | | | 0.015 | | | | | 0.103 | 0.164 | 0.220 | | | | | |
| | | | 0.005 | | | | 010 to 028 | 0.094 | 0.150 | 0.207 | | | 0.103 | 0.060 | 23% |
| | | | 0.015 | | | | | 0.099 | 0.160 | 0.217 | | | | | |
| -1XX | 0.103 | 0.003 | 0.005 | 104 to 149 | 104 to 116 | 0.002 | | 0.141 | 0.183 | 0.245 | 108 to 149 | 106 to 149 (exc. 108) | 0.141 | 0.086 | 10% |
| | | | 0.015 | | | | | | | | | | 0.151 | 0.090 | |
| | | | 0.005 | | | | | 0.151 | 0.193 | 0.255 | | 0.108 | 0.141 | 0.086 | 19% |
| | | | 0.015 | | | | | | | | | | 0.146 | 0.090 | |
| -2XX | 0.139 | 0.004 | 0.010 | 210 to 247 | 210 to 222 | 0.003 | | 0.188 | 0.235 | 0.304 | 210 to 247 | | 0.188 | 0.117 | 10% |
| | | | 0.025 | | | | | 0.198 | 0.245 | 0.314 | | | 0.198 | 0.121 | 18% |
| -3XX | 0.210 | 0.005 | 0.020 | 325 to 349 | 325 to 349 | 0.004 | | 0.281 | 0.334 | 0.424 | 325 to 349 | | 0.281 | 0.178 | 10% |
| | | | 0.035 | | | | | 0.291 | 0.344 | 0.434 | | | 0.291 | 0.184 | 17% |
| -4XX | 0.275 | 0.006 | 0.020 | 425 to 460 | 425 to 460 | 0.005 | | 0.375 | 0.475 | 0.579 | 425 to 460 | | 0.375 | 0.234 | 10% |
| | | | 0.035 | | | | | 0.385 | 0.485 | 0.589 | | | 0.385 | 0.242 | 17% |

Note: Blank table cells indicate O-ring sizes that are not recommended for the given application.

**Table 3. Metric (mm) O-Ring Sizes and Gland Data by Size Range**
*(Based on ISO 3601-2-2016 and 70 Shore A O-Rings)*

| O-Ring | | | Piston Seal & Rod Seal | | | | | | | Axial Face Seal | | | | | |
|---|---|---|---|---|---|---|---|---|---|---|---|---|---|---|---|
| Size Code | Cross-Section Diameter | ± Tol. | Groove Inside Corner Radii | Recommended Range of O-Ring Sizes | | Groove Eccentricity | Groove Width *G* | | | Recommended Range of O-Ring Sizes | | Groove Width *G* | | Groove Depth *F* | Installed Axial Squeeze (No Swell, No Radii, Straight Sides) |
| | | | | | | | No Backup | 1 Backup Ring | 2 Backup Rings | Internal Pressure | External Pressure | Liquid | Gas & Vac. | | |
| | Nominal | | Min/Max | Static | Dynamic | Max | Min/Max | Min/Max | Min/Max | | | Min/Max | Min/Max | Min/Max | Min/Max |
| -0XXA | 1.780 | 0.080 | 0.2 | 004 to 050 | 006 to 012 | If Groove or Rod Dia. > 50: E = 0.050 Else: 0.025 | 2.8 | 4.2 | 5.6 | 011 to 050 | 004 to 050 | 3.2 | 2.9 | 1.3 | 18% |
| | | | 0.4 | 004 to 027 | | | 3.05 | 4.45 | 5.85 | | | 3.4 | 3.1 | 1.4 | 30% |
| -1XXA | 2.620 | 0.090 | 0.2 | 104 to 178 | 104 to 116 | | 3.8 | 5.2 | 6.6 | 112 to 178 | 102 to 178 | 4 | 3.6 | 2 | 17% |
| | | | 0.4 | 102 to 134 | | | 4.05 | 5.45 | 6.85 | | | 4.2 | 3.8 | 2.1 | 26% |
| -2XXA | 3.530 | 0.100 | 0.4 | 201 to 284 | 210 to 222 | | 5 | 6.4 | 7.8 | 212 to 284 | 201 to 284 | 5.3 | 4.8 | 2.7 | 18% |
| | | | 0.8 | 201 to 231 | 210 to 221 (Gas) 210 to 222 (Liq.) | | 5.25 | 6.65 | 8.05 | | | 5.5 | 5 | 2.8 | 26% |
| -3XXA | 5.330 | 0.130 | 0.4 | 309 to 395 | 309 to 349 | | 7.2 | 9 | 10.9 | 318 to 395 | 309 to 395 | 7.6 | 7 | 4.2 | 17% |
| | | | 0.8 | 309 to 349 | 325 to 331 (Gas) 325 to 336 (Liq.) | | 7.45 | 9.25 | 11.15 | | | 7.8 | 7.2 | 4.3 | 23% |
| -4XXA | 6.990 | 0.150 | 0.4 | 425 to 475 | 425 to 460 | | 9.5 | 12.3 | 15.1 | 425 to 475 | | 9 | 8.5 | 5.7 | 15% |
| | | | 0.8 | 425 to 429 | | | 9.75 | 12.55 | 15.35 | | | 9.2 | 8.7 | 5.8 | 20% |

**Table 4. Inch Axial Face Seal Data, Recommended Sizes in Each Range**
*(Based on SAE AS6235A-2015 and 70 Shore A O-Rings)*

| O-Ring | | | | Axial Face Seal Groove | | | | | |
|---|---|---|---|---|---|---|---|---|---|
| | | | | Internal Pressure | | | External Pressure | | |
| Size Code | Cross-Section Diameter | Inner Diameter | | Groove Outer Diameter | Cramping [a] | Gland Fill (No Swell, Straight Sides, No Radii) | Groove Inner Diameter | Stretch | Gland Fill (No Swell, Straight Sides, No Radii) |
| | Nominal | Nominal | ± Tol. | Max/Min | | Min/Max | Min/Max | | Min/Max |
| -006 | 0.070 | 0.114 | 0.005 | | | | 0.107<br>0.112 | 4.1% | 46%<br>72% |
| -008 | | 0.176 | | 0.328<br>0.323 | 0.2% | 60%<br>86% | 0.173<br>0.178 | 5.0% | 48%<br>71% |
| -028 | | 1.364 | 0.013 | 1.504<br>1.499 | −0.7% | 56%<br>79% | 1.364<br>1.369 | 0.8% | 54%<br>76% |
| -106 | 0.103 | 0.174 | 0.005 | | | | 0.169<br>0.174 | 4.9% | 47%<br>67% |
| -108 | | 0.237 | | 0.452<br>0.447 | −0.9% | 64%<br>83% | 0.235<br>0.240 | 4.9% | 49%<br>67% |
| -149 | | 2.800 | 0.022 | 3.006<br>3.001 | −0.4% | 57%<br>75% | 2.800<br>2.805 | 0.5% | 55%<br>72% |
| -210 | 0.139 | 0.734 | 0.010 | 1.019<br>1.014 | −0.9% | 62%<br>78% | 0.734<br>0.739 | 2.3% | 54%<br>70% |
| -247 | | 4.609 | 0.030 | 4.887<br>4.877 | −0.4% | 59%<br>74% | 4.609<br>4.619 | 0.4% | 57%<br>73% |
| -325 | 0.210 | 1.475 | 0.015 | 1.899<br>1.894 | −1.0% | 63%<br>77% | 1.475<br>1.480 | 1.6% | 58%<br>70% |
| -349 | | 4.475 | 0.030 | 4.895<br>4.885 | −0.5% | 61%<br>74% | 4.475<br>4.485 | 0.6% | 59%<br>72% |
| -425 | 0.275 | 4.475 | 0.033 | 5.025<br>5.015 | −0.6% | 61%<br>73% | 4.475<br>4.485 | 0.7% | 58%<br>70% |
| -460 | | 15.475 | 0.070 | 16.025<br>16.015 | −0.2% | 60%<br>72% | 15.475<br>15.485 | 0.2% | 59%<br>71% |

Note: Blank table cells indicate O-ring sizes that are not recommended for the given application.
[a] Positive values indicate no cramping.

**Table 5. Metric (mm) Axial Face Seal Data, Recommended Sizes in Each Range**
*(Based on ISO 3601-2-2016 and 70 Shore A O-Rings)*

| O-Ring | | | | Axial Face Seal Groove | | | | | | | |
|---|---|---|---|---|---|---|---|---|---|---|---|
| | | | | Internal Pressure | | | | External Pressure | | | |
| Size Code | Cross-Section Diameter | Inner Diameter | | Groove Outer Diameter | Cramping | Gland Fill Min/Max (No Swell, Straight Sides, No Radii) | | Groove Inner Diameter | Stretch | Gland Fill Min/Max (No Swell, Straight Sides, No Radii) | |
| | Nominal | Nominal | ± Tol. | Min/Max | | Liquid Min/Max | Gas & Vac Min/Max | Min/Max | | Liquid Min/Max | Gas & Vac Min/Max |
| -004A | 1.78 | 1.78 | 0.13 | | | | | 1.78 / 1.80 | 1.1% | 31% / 49% | 36% / 58% |
| -011A | | 7.65 | | 11.20 / 11.24 | −3.1% | 56% / 79% | 59% / 84% | 7.70 / 7.72 | 0.9% | 40% / 58% | 45% / 66% |
| -050A | | 133.07 | 0.94 | 136.62 / 136.66 | −0.3% | 48% / 67% | 52% / 73% | 134.37 / 134.39 | 1.0% | 46% / 65% | 51% / 71% |
| -102A | 2.62 | 1.24 | 0.13 | | | | | 1.25 / 1.27 | 2.4% | 38% / 56% | 45% / 67% |
| -112A | | 12.37 | | 17.60 / 17.64 | −2.5% | 63% / 81% | 67% / 87% | 12.48 / 12.50 | 1.1% | 50% / 67% | 57% / 76% |
| -178A | | 247.32 | 1.40 | 252.55 / 252.59 | −0.2% | 57% / 73% | 63% / 81% | 248.82 / 248.84 | 0.6% | 56% / 72% | 62% / 80% |
| -201A | 3.53 | 4.34 | 0.13 | | | | | 4.37 / 4.39 | 1.2% | 46% / 61% | 54% / 71% |
| -212A | | 21.82 | 0.25 | 28.88 / 28.92 | −2.1% | 64% / 79% | 69% / 85% | 22.03 / 22.05 | 1.1% | 54% / 68% | 61% / 77% |
| -284A | | 456.06 | 2.16 | 463.12 / 463.16 | −0.1% | 60% / 73% | 66% / 80% | 457.56 / 457.58 | 0.3% | 59% / 72% | 65% / 80% |
| -318A | 5.33 | 24.77 | 0.25 | 35.43 / 35.47 | −2.3% | 68% / 80% | 72% / 85% | 25.00 / 25.02 | 1.0% | 57% / 69% | 63% / 76% |
| -395A | | 658.88 | 3.05 | 669.54 / 669.58 | −0.1% | 63% / 74% | 68% / 80% | 660.38 / 660.40 | 0.2% | 63% / 73% | 68% / 80% |
| -425A | 6.99 | 113.67 | 0.84 | 127.63 / 127.67 | −0.7% | 70% / 80% | 73% / 84% | 114.78 / 114.80 | 1.0% | 66% / 77% | 71% / 82% |
| -475A | | 658.88 | 3.05 | 672.84 / 672.88 | −0.1% | 69% / 79% | 73% / 83% | 660.38 / 660.40 | 0.2% | 68% / 78% | 72% / 83% |

Note: Blank table cells indicate O-ring sizes that are not recommended for the given application.

## Table 6. Inch Piston Seal Data, Recommended Sizes in Each Range

*(Based on SAE AS4716C-2017 and 70 Shore A O-Rings)*

| O-Ring | | | | Piston Gland | | | | | | Installed O-Ring | |
|---|---|---|---|---|---|---|---|---|---|---|---|
| Size Code | Cross-Section Diameter | Inner Diameter | | Pneumatic & Hydraulic Application | Bore Diameter | Piston Groove Diameter | Piston Diameter | Diametrical Clearance | Gland Depth Including Eccentricity & Side Load | Stretch | Squeeze Including Eccentricity & Side Load |
| | Nominal | Nominal | ± Tol. | | Min/Max | Max/Min | Max/Min | Min/Max | Max/Min | Min/Max | Min/Max |
| -004 | 0.070 | 0.070 | 0.005 | Static & Dynamic | 0.190 | 0.076 | 0.188 | 0.002 | 0.061 | 0% | 4% * |
| | | | | | 0.191 | 0.075 | 0.187 | 0.004 | 0.055 | 17% | 25% |
| -012 | | 0.364 | 0.005 | | 0.485 | 0.375 | 0.483 | 0.002 | 0.059 | 1% | 9% |
| | | | | | 0.486 | 0.374 | 0.482 | 0.004 | 0.053 | 4% | 27% |
| -028 | | 1.364 | 0.013 | Static | 1.491 | 1.381 | 1.489 | 0.002 | 0.061 | 0% | 8% |
| | | | | | 1.493 | 1.379 | 1.488 | 0.005 | 0.053 | 2% | 28% |
| -104 | 0.103 | 0.112 | 0.005 | Static & Dynamic | 0.297 | 0.128 | 0.295 | 0.002 | 0.089 | 9% | 5% |
| | | | | | 0.298 | 0.127 | 0.294 | 0.004 | 0.082 | 20% | 20% |
| -116 | | 0.737 | 0.009 | | 0.925 | 0.751 | 0.923 | 0.002 | 0.093 | 0% | 5% |
| | | | | | 0.927 | 0.749 | 0.922 | 0.004 | 0.085 | 3% | 20% |
| -149 | | 2.800 | 0.022 | Static | 2.993 | 2.819 | 2.990 | 0.003 | 0.094 | 0% | 5% |
| | | | | | 2.995 | 2.817 | 2.988 | 0.007 | 0.084 | 1% | 21% |
| -210 | 0.139 | 0.734 | 0.010 | Static & Dynamic | 0.991 | 0.750 | 0.989 | 0.002 | 0.127 | 1% | 4% |
| | | | | | 0.993 | 0.748 | 0.988 | 0.005 | 0.118 | 4% | 17% |
| -222 | | 1.484 | 0.015 | | 1.741 | 1.499 | 1.739 | 0.002 | 0.127 | 0% | 4% |
| | | | | | 1.743 | 1.497 | 1.738 | 0.005 | 0.118 | 2% | 17% |
| -247 | | 4.609 | 0.030 | Static | 4.868 | 4.626 | 4.864 | 0.004 | 0.129 | 0% | 4% |
| | | | | | 4.870 | 4.624 | 4.862 | 0.008 | 0.117 | 1% | 19% |
| -325 | 0.210 | 1.475 | 0.015 | Static & Dynamic | 1.867 | 1.495 | 1.864 | 0.003 | 0.193 | 0% | 4% |
| | | | | | 1.869 | 1.493 | 1.863 | 0.006 | 0.182 | 2% | 15% |
| -349 | | 4.475 | 0.030 | | 4.868 | 4.496 | 4.864 | 0.004 | 0.194 | 0% | 4% |
| | | | | | 4.870 | 4.494 | 4.862 | 0.008 | 0.181 | 1% | 16% |
| -425 | 0.275 | 4.475 | 0.033 | Static & Dynamic | 4.974 | 4.497 | 4.970 | 0.004 | 0.249 | 0% | 7% |
| | | | | | 4.977 | 4.494 | 4.968 | 0.009 | 0.233 | 1% | 17% |
| -460 | | 15.475 | 0.070 | | 15.974 | 15.497 | 15.970 | 0.004 | 0.250 | 0% | 7% |
| | | | | | 15.978 | 15.494 | 15.967 | 0.011 | 0.233 | 1% | 17% |

* This minimum squeeze is less than the recommended minimum of 0.005 inches.

## Table 7. Metric (mm) Dynamic Piston Seal Data, Recommended Sizes in Each Range

*(Based on ISO 3601-2-2016 and 70 Shore A O-Rings)*

| O-Ring | | | | Piston Gland | | | | | | Installed O-Ring | |
|---|---|---|---|---|---|---|---|---|---|---|---|
| Size Code | Cross-Section Diameter | Inner Diameter | | Dynamic Application | Bore Diameter | Piston Groove Diameter | Piston Diameter | Diametrical Clearance | Gland Depth Including Eccentricity & Side Load | Installed Stretch | Squeeze Including Eccentricity & Side Load |
| | Nominal | Nominal | ± Tol. | | Min/Max | Max/Min | Max/Min | Min/Max | Max/Min | Min/Max | Min/Max |
| -006A | 1.78 | 2.90 | 0.13 | Pneumatic | 5.650 | 3.050 | 5.640 | 0.010 | 1.346 | 0% | 16% |
| | | | | | 5.668 | 3.042 | 5.628 | 0.040 | 1.277 | 10% | 31% |
| | | | | Hydraulic | 5.740 | 3.050 | 5.730 | 0.010 | 1.391 | 0% | 13% |
| | | | | | 5.758 | 3.042 | 5.718 | 0.040 | 1.322 | 10% | 29% |
| -012A | 1.78 | 9.25 | 0.13 | Pneumatic | 12.270 | 9.500 | 12.254 | 0.016 | 1.446 | 1% | 12% |
| | | | | | 12.297 | 9.491 | 12.236 | 0.061 | 1.356 | 4% | 26% |
| | | | | Hydraulic | 12.220 | 9.540 | 12.204 | 0.016 | 1.401 | 2% | 15% |
| | | | | | 12.247 | 9.531 | 12.186 | 0.061 | 1.311 | 5% | 29% |
| -104A | 2.62 | 2.84 | 0.13 | Pneumatic | 7.280 | 2.990 | 7.267 | 0.013 | 2.197 | 0% | 7% |
| | | | | | 7.302 | 2.984 | 7.252 | 0.050 | 2.119 | 10% | 21% |
| | | | | Hydraulic | 7.080 | 2.990 | 7.067 | 0.013 | 2.097 | 0% | 12% |
| | | | | | 7.102 | 2.984 | 7.052 | 0.050 | 2.019 | 10% | 25% |
| -116A | 2.62 | 18.72 | 0.23 | Pneumatic | 23.440 | 19.170 | 23.420 | 0.020 | 2.218 | 1% | 10% |
| | | | | | 23.473 | 19.137 | 23.399 | 0.074 | 2.102 | 4% | 22% |
| | | | | Hydraulic | 23.330 | 19.270 | 23.310 | 0.020 | 2.113 | 2% | 14% |
| | | | | | 23.363 | 19.237 | 23.289 | 0.074 | 1.997 | 4% | 25% |
| -210A | 3.53 | 18.64 | 0.25 | Pneumatic | 25.170 | 19.320 | 25.150 | 0.020 | 3.017 | 2% | 9% |
| | | | | | 25.203 | 19.268 | 25.129 | 0.074 | 2.892 | 5% | 19% |
| | | | | Hydraulic | 24.970 | 19.320 | 24.950 | 0.020 | 2.917 | 2% | 12% |
| | | | | | 25.003 | 19.268 | 24.929 | 0.074 | 2.792 | 5% | 22% |
| -222A | 3.53 | 37.69 | 0.38 | Pneumatic | 44.730 | 38.990 | 44.695 | 0.035 | 2.978 | 2% | 10% |
| | | | | | 44.769 | 38.928 | 44.680 | 0.089 | 2.833 | 5% | 20% |
| | | | | Hydraulic | 44.530 | 38.990 | 44.505 | 0.025 | 2.878 | 2% | 13% |
| | | | | | 44.569 | 38.928 | 44.480 | 0.089 | 2.733 | 5% | 23% |
| -309A | 5.33 | 10.46 | 0.13 | Pneumatic & Hydraulic | 19.500 | 10.840 | 19.480 | 0.020 | 4.418 | 2% | 12% |
| | | | | | 19.533 | 10.797 | 19.459 | 0.074 | 4.297 | 5% | 20% |
| -349A | 5.33 | 113.67 | 0.76 | | 125.420 | 116.940 | 125.377 | 0.043 | 4.413 | 2% | 13% |
| | | | | | 125.483 | 116.853 | 125.337 | 0.146 | 4.174 | 4% | 22% |
| -425A | 6.99 | 113.67 | 0.84 | Pneumatic & Hydraulic | 128.500 | 117.020 | 128.457 | 0.043 | 5.913 | 2% | 11% |
| | | | | | 128.563 | 116.933 | 128.417 | 0.146 | 5.674 | 4% | 19% |
| -460A | 6.99 | 393.07 | 1.78 | | 414.450 | 403.150 | 414.382 | 0.068 | 5.915 | 2% | 11% |
| | | | | | 414.547 | 402.995 | 414.319 | 0.228 | 5.560 | 3% | 21% |

## Table 8. Metric (mm) Static Piston Seal Data, Recommended Sizes in Each Range

*(Based on ISO 3601-2-2016 and 70 Shore A O-Rings)*

| O-Ring | | | | Piston Gland | | | | | | Installed O-Ring | |
|---|---|---|---|---|---|---|---|---|---|---|---|
| Size Code | Cross-Section Diameter | Inner Diameter | | Static Application | Bore Diameter | Piston Groove Diameter | Piston Diameter | Diametrical Clearance | Gland Depth Including Eccentricity & Side Load | Installed Stretch | Squeeze Including Eccentricity & Side Load |
| | Nominal | Nominal | ± Tol. | | Min/Max | Max/Min | Max/Min | Min/Max | Max/Min | Min/Max | Min/Max |
| -004A | 1.78 | 1.78 | 0.13 | Pneumatic & Hydraulic | 4.520 | 1.930 | 4.510 | 0.010 | 1.327 | 1% | 14% |
| | | | | | 4.538 | 1.924 | 4.498 | 0.040 | 1.284 | 17% | 30% |
| -050A | 1.78 | 133.07 | 0.94 | | 139.290 | 136.940 | 139.247 | 0.043 | 1.329 | 2% | 20% |
| | | | | | 139.353 | 136.840 | 139.207 | 0.146 | 1.134 | 4% | 38% |
| -102A | 2.62 | 1.24 | 0.13 | Pneumatic & Hydraulic | 5.380 | 1.380 | 5.370 | 0.010 | 2.032 | 0% | 9% |
| | | | | | 5.398 | 1.374 | 5.358 | 0.040 | 1.989 | 24% | 26% |
| -178A | 2.62 | 247.32 | 1.40 | | 257.690 | 254.010 | 257.634 | 0.056 | 2.040 | 2% | 18% |
| | | | | | 257.771 | 253.880 | 257.582 | 0.189 | 1.786 | 3% | 33% |
| -201A | 3.53 | 4.34 | 0.13 | Pneumatic & Hydraulic | 9.910 | 4.530 | 9.897 | 0.013 | 2.741 | 1% | 16% |
| | | | | | 9.932 | 4.500 | 9.882 | 0.050 | 2.676 | 8% | 26% |
| -284A | 3.53 | 456.06 | 2.16 | | 472.780 | 467.780 | 472.712 | 0.068 | 2.740 | 2% | 18% |
| | | | | | 472.877 | 467.625 | 472.649 | 0.228 | 2.435 | 3% | 32% |
| -309A | 5.33 | 10.46 | 0.13 | Pneumatic & Hydraulic | 19.000 | 10.830 | 18.980 | 0.020 | 4.160 | 2% | 17% |
| | | | | | 19.033 | 10.787 | 18.959 | 0.074 | 4.065 | 5% | 24% |
| -395A | 5.33 | 658.88 | 3.05 | | 683.370 | 675.650 | 683.290 | 0.080 | 4.165 | 2% | 18% |
| | | | | | 683.495 | 675.450 | 683.210 | 0.285 | 3.780 | 3% | 30% |
| -425A | 6.99 | 113.67 | 0.84 | Pneumatic & Hydraulic | 127.800 | 117.020 | 127.757 | 0.043 | 5.538 | 2% | 17% |
| | | | | | 127.863 | 116.933 | 127.717 | 0.146 | 5.349 | 4% | 24% |
| -475A | 6.99 | 658.88 | 3.05 | | 686.510 | 675.650 | 686.430 | 0.080 | 5.735 | 2% | 14% |
| | | | | | 686.635 | 675.450 | 686.350 | 0.285 | 5.350 | 3% | 24% |

## Table 9. Inch Rod Seal Data, Recommended Sizes in Each Range

*(Based on SAE AS4716C-2017 and 70 Shore A O-Rings)*

| O-Ring | | | | Rod Lead-In Chamfer Depth 15°–30° | Rod Gland | | | | | | Installed O-Ring | | |
|---|---|---|---|---|---|---|---|---|---|---|---|---|---|
| Size Code | Cross-Section Diameter | Inner Diameter | | | Pneumatic & Hydraulic Application | Bore Diameter | Bore Groove Diameter | Rod Diameter | Diametrical Clearance | Gland Depth Including Eccentricity & Side Load | Installed Stretch | Cramping [a] | Squeeze Including Eccentricity & Side Load |
| | Nominal | Nominal | ± Tol. | Minimum | | Min/Max | Min/Max | Max/Min | Min/Max | Max/Min | Min/Max | | Min/Max |
| -004 | 0.070 | 0.070 | 0.005 | 0.018 | Static & Dynamic | 0.078 | 0.190 | 0.076 | 0.002 | 0.061 | 0% | -4.0% | 4% * |
| | | | | | | 0.079 | 0.191 | 0.075 | 0.004 | 0.055 | 17% | | 25% |
| -012 | | 0.070 | 0.005 | | | 0.375 | 0.484 | 0.373 | 0.002 | 0.060 | 1% | -1.6% | 9% |
| | | | | | | 0.376 | 0.485 | 0.372 | 0.004 | 0.053 | 4% | | 27% |
| -028 | | 1.364 | 0.013 | 0.018 | Static | 1.375 | 1.483 | 1.373 | 0.002 | 0.061 | 0% | 0.0% | 8% |
| | | | | | | 1.376 | 1.485 | 1.371 | 0.005 | 0.053 | 2% | | 28% |
| -104 | 0.103 | 0.112 | 0.005 | 0.021 | Static & Dynamic | 0.125 | 0.295 | 0.123 | 0.002 | 0.090 | 4% | -3.6% | 10% |
| | | | | | | 0.126 | 0.296 | 0.122 | 0.004 | 0.084 | 15% | | 21% |
| -116 | | 0.112 | 0.005 | | | 0.750 | 0.923 | 0.748 | 0.002 | 0.093 | 0% | -0.3% | 5% |
| | | | | | | 0.751 | 0.925 | 0.746 | 0.005 | 0.085 | 3% | | 20% |
| -149 | | 2.800 | 0.022 | 0.021 | Static | 2.814 | 2.985 | 2.811 | 0.003 | 0.093 | 0% | 0.3% | 5% |
| | | | | | | 2.816 | 2.987 | 2.809 | 0.007 | 0.084 | 1% | | 21% |
| -210 | 0.139 | 0.734 | 0.010 | 0.023 | Static & Dynamic | 0.750 | 0.989 | 0.748 | 0.002 | 0.127 | 0% | -0.3% | 4% |
| | | | | | | 0.751 | 0.991 | 0.746 | 0.005 | 0.118 | 3% | | 17% |
| -222 | | 0.734 | 0.010 | | | 1.500 | 1.740 | 1.498 | 0.002 | 0.127 | 0% | 0.2% | 4% |
| | | | | | | 1.501 | 1.742 | 1.496 | 0.005 | 0.118 | 2% | | 17% |
| -247 | | 4.609 | 0.030 | 0.023 | Static | 4.626 | 4.864 | 4.622 | 0.004 | 0.129 | 0% | 0.4% | 4% |
| | | | | | | 4.628 | 4.866 | 4.620 | 0.008 | 0.117 | 1% | | 19% |
| -325 | 0.210 | 1.475 | 0.015 | 0.029 | Static & Dynamic | 1.500 | 1.870 | 1.498 | 0.002 | 0.193 | 0% | 0.1% | 4% |
| | | | | | | 1.502 | 1.872 | 1.496 | 0.006 | 0.182 | 3% | | 15% |
| -349 | | 4.475 | 0.030 | | | 4.500 | 4.869 | 4.497 | 0.003 | 0.194 | 0% | 0.3% | 5% |
| | | | | | | 4.502 | 4.871 | 4.495 | 0.007 | 0.182 | 1% | | 16% |
| -425 | 0.275 | 4.475 | 0.033 | 0.043 | Static & Dynamic | 4.501 | 4.974 | 4.497 | 0.004 | 0.249 | 0% | -0.1% | 7% |
| | | | | | | 4.503 | 4.977 | 4.494 | 0.009 | 0.233 | 1% | | 17% |
| -460 | | 15.475 | 0.070 | | | 15.501 | 15.974 | 15.497 | 0.004 | 0.250 | 0% | 0.2% | 7% |
| | | | | | | 15.504 | 15.978 | 15.494 | 0.010 | 0.233 | 1% | | 17% |

* This minimum squeeze is less than the recommended minimum of 0.005 inches.

[a] Positive values indicate no cramping.

**Table 10. Metric (mm) Dynamic Rod Seal Data, Recommended Sizes in Each Range**
*(Based on ISO 3601-2-2016 and 70 Shore A O-Rings)*

| O-Ring | | | | Rod Lead-In Chamfer Depth 15°–20° | Rod Gland | | | | | | Installed O-Ring | | |
|---|---|---|---|---|---|---|---|---|---|---|---|---|---|
| Size Code | Cross-Section Diameter | Inner Diameter | | | Dynamic Application | Bore Diameter | Bore Groove Diameter | Rod Diameter | Diametrical Clearance | Gland Depth Including Eccentricity & Side Load | Installed Stretch | Cramping[a] | Squeeze Including Eccentricity & Side Load |
| | Nominal | Nominal | ± Tol. | Minimum | | Min/Max | Max/Min | Max/Min | Min/Max | Max/Min | Min/Max | | Min/Max |
| -006A | 1.78 | 2.90 | 0.13 | 0.3 | Pneumatic | 3.160 | 5.930 | 3.150 | 0.010 | 1.429 | 4% | −3.9% | 9% |
| | | | | | | 3.178 | 5.900 | 3.138 | 0.040 | 1.349 | 14% | | 26% |
| | | | | | Hydraulic | 3.160 | 5.850 | 3.150 | 0.010 | 1.404 | 4% | −4.7% | 10% |
| | | | | | | 3.178 | 5.880 | 3.138 | 0.040 | 1.324 | 14% | | 27% |
| -012A | | 9.25 | 0.13 | 0.3 | Pneumatic & Hydraulic | 9.500 | 12.353 | 9.487 | 0.013 | 1.478 | 1% | −1.3% | 10% |
| | | | | | | 9.522 | 12.310 | 9.472 | 0.050 | 1.382 | 4% | | 25% |
| -104A | 2.62 | 2.84 | 0.13 | 0.4 | Pneumatic & Hydraulic | 3.030 | 7.196 | 3.020 | 0.010 | 2.127 | 1% | −7.4% | 10% |
| | | | | | | 3.048 | 7.160 | 3.008 | 0.040 | 2.044 | 11% | | 24% |
| -116A | | 18.72 | 0.23 | 0.4 | | 19.300 | 23.552 | 19.280 | 0.020 | 2.196 | 2% | 0.0% | 11% |
| | | | | | | 19.333 | 23.500 | 19.259 | 0.074 | 2.071 | 4% | | 22% |
| -210A | 3.53 | 18.64 | 0.25 | 0.5 | Pneumatic & Hydraulic | 19.300 | 25.052 | 19.280 | 0.020 | 2.946 | 2% | −0.8% | 11% |
| | | | | | | 19.333 | 25.000 | 19.259 | 0.074 | 2.821 | 5% | | 21% |
| -221A | | 36.09 | 0.30 | 0.5 | | 36.800 | 42.562 | 36.775 | 0.025 | 2.963 | 1% | −0.2% | 12% |
| | | | | | | 36.839 | 42.500 | 36.750 | 0.089 | 2.818 | 3% | | 22% |
| -222A | | 37.69 | 0.38 | 0.5 | Hydraulic | 38.500 | 44.162 | 38.475 | 0.025 | 2.913 | 1% | 0.0% | 13% |
| | | | | | | 38.539 | 44.100 | 38.450 | 0.089 | 2.768 | 3% | | 23% |
| -325A | 5.33 | 37.47 | 0.38 | 0.8 | Pneumatic & Hydraulic | 38.500 | 47.462 | 38.475 | 0.025 | 4.563 | 2% | −0.1% | 10% |
| | | | | | | 38.539 | 47.400 | 38.450 | 0.089 | 4.418 | 4% | | 18% |
| -331A | | 56.52 | 0.46 | 0.8 | | 57.550 | 66.454 | 57.520 | 0.030 | 4.560 | 1% | 0.0% | 10% |
| | | | | | | 57.596 | 66.380 | 57.490 | 0.106 | 4.367 | 3% | | 19% |
| -336A | | 72.39 | 0.51 | 0.8 | Hydraulic | 73.610 | 82.277 | 73.580 | 0.030 | 4.442 | 1% | 0.0% | 13% |
| | | | | | | 73.656 | 82.190 | 73.550 | 0.106 | 4.242 | 2% | | 22% |

[a] Positive values indicate no cramping.

## Table 11. Metric (mm) Static Rod Seal Data, Recommended Sizes in Each Range

*(Based on ISO 3601-2-2016 and 70 Shore A O-Rings)*

| O-Ring | | | | Rod Lead-In Chamfer Depth 15°–20° | Rod Gland | | | | | | Installed O-Ring | | |
|---|---|---|---|---|---|---|---|---|---|---|---|---|---|
| Size Code | Cross-Section Diameter | Inner Diameter | | | Static Application | Bore Diameter | Bore Groove Diameter | Rod Diameter | Diametrical Clearance | Gland Depth Including Eccentricity & Side Load | Stretch | Cramping [a] | Squeeze Including Eccentricity & Side Load |
| | Nominal | Nominal | ± Tol. | Minimum | | Min/Max | Max/Min | Max/Min | Min/Max | Max/Min | Min/Max | | Min/Max |
| -004A | 1.78 | 1.78 | 0.13 | 0.3 | Pneumatic & Hydraulic | 1.950 | 4.590 | 1.944 | 0.006 | 1.343 | 1% | −9.1% | 13% |
| | | | | | | 1.964 | 4.560 | 1.934 | 0.030 | 1.298 | 18% | | 27% |
| -027A | | 133.07 | 0.94 | 0.3 | | 33.700 | 36.162 | 33.675 | 0.025 | 1.301 | 1% | 0.0% | 22% |
| | | | | | | 33.739 | 36.100 | 33.650 | 0.089 | 1.181 | 3% | | 36% |
| -102A | 2.62 | 1.24 | 0.13 | 0.4 | Pneumatic & Hydraulic | 1.400 | 5.260 | 1.394 | 0.006 | 1.953 | 1% | −14.7% | 12% |
| | | | | | | 1.414 | 5.230 | 1.384 | 0.030 | 1.908 | 26% | | 29% |
| -134A | | 247.32 | 1.40 | 0.4 | | 48.200 | 51.974 | 48.175 | 0.025 | 1.956 | 1% | 0.0% | 21% |
| | | | | | | 48.239 | 51.900 | 48.150 | 0.089 | 1.831 | 3% | | 32% |
| -201A | 3.53 | 4.34 | 0.13 | 0.5 | Pneumatic & Hydraulic | 4.560 | 10.233 | 4.550 | 0.010 | 2.868 | 2% | −7.6% | 12% |
| | | | | | | 4.578 | 10.190 | 4.538 | 0.040 | 2.806 | 8% | | 22% |
| -231A | | 456.06 | 2.16 | 0.5 | | 67.500 | 72.614 | 67.470 | 0.030 | 2.640 | 1% | 0.0% | 21% |
| | | | | | | 67.546 | 72.540 | 67.440 | 0.106 | 2.497 | 3% | | 31% |
| -309A | 5.33 | 10.46 | 0.13 | 0.8 | Pneumatic & Hydraulic | 10.800 | 19.552 | 10.784 | 0.016 | 4.424 | 2% | −5.7% | 12% |
| | | | | | | 10.827 | 19.500 | 10.766 | 0.061 | 4.337 | 4% | | 19% |
| -349A | | 658.88 | 3.05 | 0.8 | | 115.500 | 123.310 | 115.464 | 0.036 | 4.003 | 1% | 0.0% | 22% |
| | | | | | | 115.554 | 123.210 | 115.429 | 0.125 | 3.828 | 2% | | 29% |
| -425A | 6.99 | 113.674 | 0.84 | 1.0 | Pneumatic & Hydraulic | 115.700 | 126.500 | 115.664 | 0.036 | 5.498 | 1% | 0.0% | 18% |
| | | | | | | 115.754 | 126.400 | 115.629 | 0.125 | 5.323 | 3% | | 25% |
| -429A | | 658.88 | 3.05 | 1.0 | | 128.500 | 139.110 | 128.457 | 0.043 | 5.419 | 1% | 0.0% | 19% |
| | | | | | | 128.563 | 139.010 | 128.417 | 0.146 | 5.224 | 2% | | 26% |

[a] Positive values indicate no cramping.

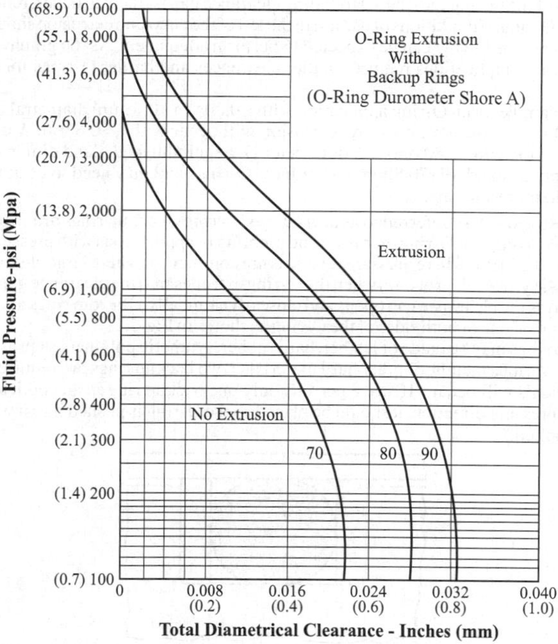


Fig. 4. Extrusion Potential of O-Rings as a Function of Hardness and Clearance Gap

**Clearance Gap.**—When there is relative motion between two parts enclosing an O-ring, there must be a clearance gap between them (see Fig. 2 and Fig. 3). This broadly applies to piston and rod seals. Axial face seal applications also may have a clearance gap, but generally they do not.

For a piston gland (see Fig. 3), the radial clearance gap is the distance between the piston surface and the bore wall. For rod glands (again, see Fig. 3), the gap is the distance between the rod surface and the bore wall. The radial clearance gap on one side is at its maximum value when the piston or rod is side-loaded and touches the bore (zero clearance gap) on the other side. This gap often is expressed in terms of diametral clearance, which is twice the nominal radial clearance; diametral clearance is the same, regardless of piston or rod position.

If the radial clearance gap between the sealing surface and the gland edge is too large, or if the fluid pressure exceeds the deformation limits of the O-ring, it will extrude into the gap, reducing the effective life of the seal. This is especially problematic for soft silicone O-rings, where a 50 percent reduction in the standard clearance gap may be recommended. To prevent O-ring extrusion, reduce the clearance gap by modifying the gland dimensions, lower the system operating pressure, or install backup rings. A harder O-ring material may be used but could result in higher drag, increase the tendency of the seal to leak at low pressures, and necessitate additional dimensional changes.

Fig. 4 indicates effective diametral clearance values that may be used, depending on fluid pressure and O-ring hardness. If conditions of use fall to the right of the curve, extrusion of the O-ring into the surrounding clearance gap is likely to occur, reducing the life of the O-ring. If conditions of use fall to the left of the curve, no extrusion should occur, in which case the O-ring can be expected to perform adequately. As the graph indicates, high-pressure applications require smaller clearances and harder O-rings for effective sealing.

As an example, in an O-ring application with a 0.004 in. (0.1 mm) diametral clearance and 2500 psi (17.24 MPa) pressure, extrusion will occur with a 70 Shore A durometer O-ring but not with an 80 Shore A durometer O-ring. Published gland dimensions normally apply to standard (70 Shore A) hardness O-rings and may need to be adjusted for use with harder materials.

**Backup Rings.—**Also referred to as *anti-extrusion rings,* backup rings are strongly recommended to prevent O-ring extrusion and nibbling in applications with pressures above 1500 psi (10.3 MPa), where pressure reversals are common, or where large clearance gaps are necessary. (*Nibbling* occurs when the O-ring extrudes into the clearance gap and then is pinched as the clearance gap opens and closes dynamically.) Backup rings are installed in the gland on one or both sides of the O-ring as shown in Fig. 5.

Backup rings may be made of leather, hard rubber, synthetic polymers such as phenolic or PTFE, various metals, or other hard materials. Two backup rings are required if pressure reversals will occur. (If space permits only one backup ring in an application with constant pressure direction, it should be placed so the O-ring is pushed into it when pressure is applied.)

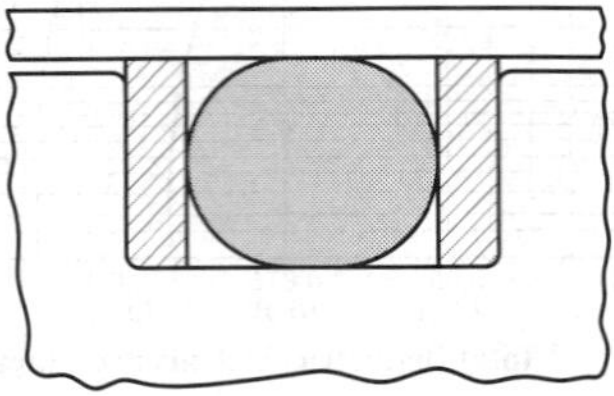

Fig. 5. Preferred High-Pressure Configuration Including Backup Rings

**Groove Width and Slope.—**Groove widths are standardized to create a gland that approaches 85 percent filled when the O-ring is installed, assuming no swell, straight (perpendicular) walls, and no radii or edge breaks. If one or more backup rings are used, groove width must be increased as needed. If nonstandard gland dimensions are used in a customized application, gland fill must be evaluated to ensure groove width is sufficient. Groove widths for many O-ring sizes are shown in Table 2 and Table 3.

Perpendicular groove walls are best for preventing extrusion or nibbling, but sloped walls often facilitate machining and may be specified as a cost-saving measure. Groove walls sloped outward up to 5 degrees from perpendicular generally are acceptable in low-pressure applications (less than 1500 psi or 10.34 MPa).

**O-Ring Stretch, Cramping, and Cross Section Reduction.—***Stretch* occurs when the inner diameter of an installed and squeezed O-ring is smaller than that of the housing groove or rod. When an elastomeric O-ring is stretched, its cross-sectional area decreases. *Cramping,* also called *cramming,* occurs when the outer diameter of an installed and squeezed O-ring is larger than that of the housing groove or bore.

Interference caused by stretch or cramping provides some zero-pressure sealing force, stabilizes the O-ring, and works to hold it in place. Stretch and cramping allowance often is traded against the economic value of using standard stock and tooling sizes. Excess stretch or cramping can be successful; it can be found in the gland dimensional standards for some O-ring sizes. (Relevant stretch and cramping equations can be found for each gland type beginning on page 2688.)

**O-Ring Squeeze.—**The installed squeeze of an O-ring has a great effect on its sealing ability and effective life. Squeeze is the difference, often expressed as a percent, between the stretched O-ring cross-sectional diameter and the gland depth *F* (see Fig. 2). Any stretch applied to the O-ring by a radial gland will reduce the O-ring's cross-sectional diameter and affect squeeze. High squeeze values are used with very soft O-rings or with high pressures, while low squeeze values should be favored in dynamic applications where friction (heat and drag) is a concern. Sealing against vacuum also usually requires higher squeeze, with values between 10 and 30 percent not unusual.

Target squeeze will depend on O-ring size, material, and application parameters. For most elastomers and applications, the O-ring gland should ensure a minimum squeeze of 0.005 in. (0.13 mm) regardless of O-ring size to overcome the elastomers' tendency to permanently deform a very small amount. Standard general squeeze target ranges can be found in the sections specific to each gland type. O-rings with smaller cross sections tolerate higher squeeze better than larger cross section O-rings, but compression set or rupture can become a problem for most materials if squeeze is above 30 percent. (Relevant squeeze equations can be found for each gland type on page 2688.)

For inch system static seals, many vendors recommend gland dimensions that adhere to the SAE AS5857 Standard. This Standard targets greater squeeze values than the general-use standards and applies to low-temperature, low-swell, and vacuum applications. It falls to the designer to decide if the greater squeeze values are necessary for a given application. Many O-ring suppliers have developed nonstandard glands; designers also may customize gland dimensions to optimize a particular application.

**Gland Fill.—**When installed, an O-ring compresses slightly and distorts into free space within the gland. Additional expansion or swelling also may occur due to contact of the O-ring with fluid or heat. The gland must be large enough to accommodate maximum expansion of the O-ring, to prevent it from extruding into the clearance gap or rupturing the assembly. In a dynamic application, any extruded O-ring material will quickly wear and fray, severely limiting seal life.

When designing a gland for a squeeze value that is higher than standard, gland fill must be carefully evaluated. Groove width can be increased to accommodate the squeezed volume of the O-ring, plus 15 percent to accommodate swell. This corresponds to an 85 percent gland fill target, assuming straight groove sides, no O-ring swell, and no internal or external radii. When significant swell is expected, further examination of gland fill must be undertaken. (Relevant gland fill equations can be found for each gland type below.)

**Axial Face Seal Glands.—**When an O-ring is compressed between two flat parts, this forms an axial face seal. This type of gland is always static, and it normally has no clearance gap between parts. A groove is cut into the face of one of the parts to house the O-ring, with gland fill considered when customizing gland dimensions. Groove diameter will depend on which direction the pressure or vacuum is coming from. If a face seal is expected to experience reversals in the direction of pressure, an O-ring is not the best solution; a flat gasket or other type of seal should be investigated.

*Internal Pressure Axial Face Seals:* If the pressure differential tends to expand the O-ring away from the center, this is considered internal pressure. When the gland is closed, the O-ring deforms as discussed below, with the outer diameter interfering (called *cramping* or *cramming*) in this case. Once squeezed, the O-ring will be in good contact with the outer wall of the gland, and the internal pressure should not cause O-ring movement. For SAE and ISO standard glands, the groove outer diameter maximum is often set equal to the O-ring nominal outer diameter; then the groove is adjusted to fulfill squeeze and gland fill targets. Using nominal dimensions, the SAE system allows up to 1 percent cramping, while the ISO system allows up to 5 percent cramping (reduces to 3 percent for inner diameters > 250 mm). Larger cramping values can work in some situations.

*External Pressure Axial Face Seals:* If the pressure differential tends to collapse the O-ring inward toward the center, this is considered external pressure. The inner diameter of the face groove should be specified to stretch the squeezed O-ring. SAE and ISO standard glands set the inner diameter minimum as equal to the O-ring nominal inner diameter. The SAE system allows up to 5 percent stretch, while the ISO system allows up to 8 percent stretch of the squeezed O-ring inner diameter. Larger stretch values can work in some applications.

*Axial Face Gland Calculations:* When working with standard gland dimensions, the designer usually begins with a preferred cross section and inner diameter. Standards and catalogs can then be used to determine appropriate groove dimensions. These dimensions can be customized if desired, but the resulting O-ring stretch/cramping, squeeze, and gland fill must be evaluated to check for potential problems. Standard gland dimensions include some violations of standard recommended squeeze ranges. In the standards, target squeeze ranges are as follows:

| | | |
|---|---|---|
| SAE Sizes -0XX: 10–25% | SAE Sizes -1XX and -2XX: 10–20% | |
| SAE Sizes -3XX: 10–18% | SAE Sizes -4XX: 10–17% | |
| ISO Sizes -0XXA: 21–36% | ISO Sizes -1XXA: 19–30% | ISO Sizes -2XXA: 17–26% |
| ISO Sizes -3XXA: 15–23% | ISO Sizes -4XXA: 13–20% | |

In axial face glands, most O-ring stretch/cramping occurs after it is squeezed and widens to interfere with (deform against) the groove wall. When an axial face gland is closed, the O-ring cross section is squeezed into an oval shape with length *L*. Assuming that the nominal mean diameter of the O-ring does not move, the squeezed O-ring inner diameter reduces and the outer diameter expands. This causes the O-ring to stretch and/or cramp in the gland, putting it in good contact with the groove walls and preventing movement of the O-ring. The following equations, which apply to Fig. 2, can be used to calculate stretch/cramping, squeeze, and gland fill for many elastomeric O-rings:

*Definition of variables:*

$ID$ = O-ring inner diameter

$W$ = O-ring cross section diameter

$AF$ = gland depth

$WR$ = reduced O-ring cross section in axial gland

$IDG$ = face groove inner diameter

$ODG$ = face groove outer diameter

$AG$ = groove width

$L$ = length of axially squeezed O-ring cross section

$ASID$ = axially squeezed O-ring inner diameter

$ASOD$ = axially squeezed O-ring outer diameter

$GVol$ = groove volume

$OVol$ = O-ring volume

$Fill\%$ = gland fill percent

$AStretch\%$ = inner diameter stretch (external pressure)

$Sq\%$ = axial O-ring squeeze percentage

$ACramp\%$ = outer diameter cramping (internal pressure)

For Axial Glands:

$$WR_{min} = W_{min}$$

$$WR_{max} = W_{max}$$

$$L_{max} = \frac{\pi}{4AF_{min}}\left(W_{nom}^2 - AF_{min}^2\right) + AF_{min}$$

$$L_{min} = \frac{\pi}{4AF_{max}}\left(W_{nom}^2 - AF_{max}^2\right) + AF_{max}$$

$$ASID_{max} = ID_{nom} + W_{nom} - L_{min}$$

$$ASID_{min} = ID_{nom} + W_{nom} - L_{max}$$

$$ASOD_{max} = ID_{nom} + W_{nom} + L_{max}$$

$$ASOD_{min} = ID_{nom} + W_{nom} + L_{min}$$

$$AStretch\%_{max} = ((IDG_{max} - ASID_{min})/ASID_{min}) \times 100\%$$

$$AStretch\%_{min} = ((IDG_{min} - ASID_{max})/ASID_{max}) \times 100\%$$

$$ACramp\%_{max} = ((ODG_{min} - ASOD_{max})/ASOD_{max}) \times 100\%$$

$$ACramp\%_{min} = ((ODG_{max} - ASOD_{min})/ASOD_{min}) \times 100\%$$

$$AF_{max} = AF_{max} \text{ (given)}$$

$$AF_{min} = AF_{min} \text{ (given)}$$

$$Sq\%_{max} = ((WR_{max} - AF_{min})/W_{max}) \times 100\%$$

$$Sq\%_{min} = ((WR_{min} - AF_{max})/W_{min}) \times 100\%$$

$$OVol_{max} = \left(\frac{\pi^2}{4}\right) \times \left(ID_{max} + W_{max}\right) \times \left(W_{max}^2\right)$$

$$OVol_{min} = \left(\frac{\pi^2}{4}\right) \times \left(ID_{min} + W_{min}\right) \times \left(W_{min}^2\right)$$

$$Vol_{max} = \left(\frac{\pi}{4}\right) \times \left(ODG_{max}^2 - IDG_{min}^2\right) \times AF_{max}$$

$$GVol_{min} = \left(\frac{\pi}{4}\right) \times \left(ODG_{min}^2 - IDG_{max}^2\right) \times AF_{min}$$

$$Fill\%_{max} = (OVol_{max} / GVol_{min}) \times 100\%$$

$$Fill\%_{min} = (OVol_{min} / GVol_{max}) \times 100\%$$

*Example:* Calculate stretch/cramping, squeeze, and gland fill for a -116 size O-ring (-116A in metric, shown in parentheses) in a standard axial face gland with external liquid pressure.

Given: Use gland Standard SAE AS6235A-2015 (ISO 3601-2-2016):

| | | | |
|---|---|---|---|
| $ID_{max} = 0.746$ | $ID_{min} = 0.728$ | $(ID_{max} = 18.95)$ | $(ID_{min} = 18.49)$ |
| $W_{max} = 0.106$ | $W_{min} = 0.1$ | $(W_{max} = 2.71)$ | $(W_{min} = 2.53)$ |
| $WR_{min} = W_{min} = 0.1$ | $WR_{max} = W_{max} = 0.106$ | $(WR_{min} = W_{min} = 2.53)$ | $(WR_{max} = W_{max} = 2.71)$ |
| $IDG_{max} = 0.742$ | $IDG_{min} = 0.737$ | $(IDG_{max} = 18.9)$ | $(IDG_{min} = 18.84)$ |
| $AF_{max} = 0.09$ | $AF_{min} = 0.086$ | $(AF_{max} = 2.1)$ | $(AF_{min} = 2)$ |
| $AG_{min} = 0.141$ | $AG_{max} = 0.151$ | $(AG_{min} = 4)$ | $(AG_{max} = 4.2)$ |

Calculate:

| | | | |
|---|---|---|---|
| $L_{max} = 0.115$ | $L_{min} = 0.112$ | $(L_{max} = 3.125)$ | $(L_{min} = 3.018)$ |
| $SID_{max} = 0.728$ | $SID_{min} = 0.725$ | $(SID_{max} = 18.322)$ | $(SID_{min} = 18.215)$ |
| $Stretch\%_{max} = 2.4\%$ | $Stretch\%_{min} = 1.2\%$ | $(Stretch\%_{max} = 3.8\%)$ | $(Stretch\%_{min} = 2.9\%)$ |
| $ODG_{max} = 1.044$ | $ODG_{min} = 1.019$ | $(ODG_{max} = 27.3)$ | $(ODG_{min} = 26.85)$ |
| $Sq\%_{max} = 19\%$ | $Sq\%_{min} = 10\%$ | $(Sq\%_{max} = 26\%)$ | $(Sq\%_{min} = 17\%)$ |
| $OVol_{max} = 0.024$ | $OVol_{min} = 0.02$ | $(OVol_{max} = 392.5)$ | $(OVol_{min} = 331.98)$ |
| $GVol_{max} = 0.039$ | $GVol_{min} = 0.033$ | $(GVol_{max} = 643.31)$ | $(GVol_{min} = 571.15)$ |
| $Fill\%_{max} = 72\%$ | $Fill\%_{min} = 53\%$ | $(Fill\%_{max} = 69\%)$ | $(Fill\%_{min} = 52\%)$ |

Validate for the -116 SAE O-Ring:

| | | |
|---|---|---|
| $Stretch\%_{max} = 2.4\%$ | $Stretch\%_{min} = 1.2\%$ | Well within SAE limit of 5%. |
| $Sq\%_{min} = 10\%$ | $Sq\%_{max} = 19\%$ | Meets SAE recommendation of 10–20% for this cross section. |
| $Fill\%_{max} = 72\%$ | $Fill\%_{min} = 53\%$ | Within SAE target maximum of 85%. |

Validate for the -116A ISO O-Ring:

| | | |
|---|---|---|
| $(Stretch\%_{max} = 3.8\%)$ | $(Stretch\%_{min} = 2.9\%)$ | Well within ISO limit of 8%. |
| $(Sq\%_{min} = 17\%)$ | $(Sq\%_{max} = 26\%)$ | ISO recommends 19–30% for this cross section. This slight lower limit violation invites caution in critical applications but is not uncommon. |
| $(Fill\%_{max} = 69\%)$ | $(Fill\%_{min} = 52\%)$ | Within ISO target maximum of 85%. |

*PTFE O-Rings in Axial Face Seals:* Axial face seals can work with very stiff O-ring materials. Solid PTFE seals are often chosen because they are highly chemically inert and less costly than perfluorinated elastomer seals with similar chemical resistance. Glands for solid PTFE O-rings often target a squeeze of 10 to 20 percent for O-rings with cross-sectional diameters up to 0.210 in. (5.33 mm), and 10 to 15 percent for cross sections 0.210 to 0.275 in. (5.33 to 6.99 mm).

It is common practice to dimension the groove so that the groove's mean diameter (inner diameter plus one cross-sectional diameter) is equal to the O-ring's free-state mean diameter. Deliberate contact with the inner or outer groove walls also can be targeted as with elastomeric O-rings, but stretch and cramping should be zero or nearly zero. A groove with a full-radius floor is best to support the O-ring and control deformation. The groove width is normally 10 percent greater than the O-ring cross-sectional diameter. If a clearance gap is present, it must not exceed 0.005 in. (0.13 mm).

PTFE O-rings and their glands are extremely vulnerable to nicks, scratches, and particle contamination. These seals must be handled with care and installed in a clean environment.

If stretch is required, a PTFE O-ring can be heated up to 212° F (100° C) with boiling water or a dry heat source. If the gland is reopened after the O-ring is squeezed, the O-ring should be replaced, as PTFE will not rebound from compression and may not seal a second time.

*Axial Face Gland Dimensions:* Axial face seal grooves should have edge breaks or radii on the external edges to protect the O-ring from sharp edges during installation. Recommended edge breaks are 0.005–0.01 in. or 0.13–0.25 mm. Flatness must be tightly controlled for both mating parts, with 0.0004 in. per 4 in. (0.01 mm per 100 mm) a good target. Parallelism of the groove floor to the mating face also should be tightly controlled. A target parallelism of 0.0008 in. (0.02 mm) is normally sufficient. Ideally, groove walls should be vertical but may slope from vertical by up to 5 degrees to aid manufacturing. Table 2, Table 3, Table 4, and Table 5 contain a sampling of gland data for standard size ranges for axial face sealing. Many O-ring suppliers recommend axial gland dimensions according to the SAE AS5857 Standard, which recommends greater squeeze and is suitable for low-temperature, low-swell, and vacuum applications.

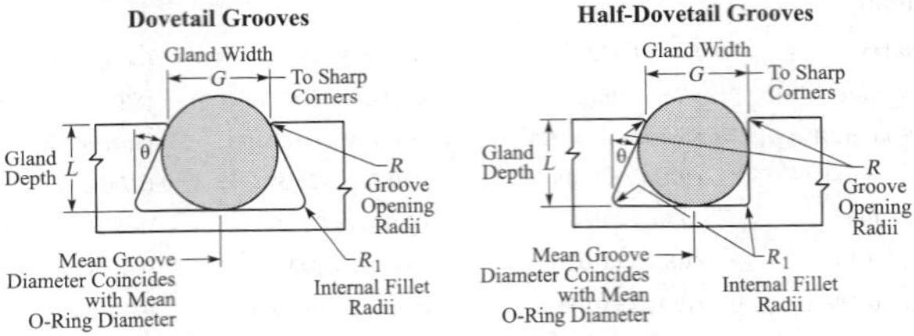


Fig. 6. Dovetail and Half-Dovetail Grooves

**Dovetail Grooves.—**It is sometimes necessary to retain an O-ring in a static face groove more securely than by using the O-ring stretch alone. In these cases, if sticky O-ring grease is not a practical solution, a dovetail or half-dovetail groove (see Fig. 6) can be used. This is an expensive option, and it requires tight tolerances. Entry radii are critical for preventing installation damage and extrusion.

A half-dovetail groove is appropriate in cases where the applied pressure does not change direction. Be sure to orient a half-dovetail so that the pressure differential forces the O-ring against the vertical side of the groove. Full dovetail grooves can be used in applications with pressure reversals. The nominal mean diameter of these grooves should be the same as the nominal mean diameter of the O-ring (see Fig. 6 and Table 12).

Dovetail and half-dovetail grooves are not recommended for O-ring cross sections smaller than 0.139 in. (3.53 mm). These groove shapes also should not be used with rigid O-rings, because installation would cause permanent deformation. These glands can easily fail due to O-ring swell, and can cause problems in vacuum applications where the groove can release the trapped volume of air contained in its corner(s).

**Radial (Piston or Rod) Glands.—**Piston and rod seals can be static or dynamic, pneumatic or hydraulic. Different gland dimensions and O-ring size ranges may be required for different applications. Radial O-ring grooves for piston or rod seals should have edge breaks on the external edges to protect the O-ring during installation. Recommended edge breaks are 0.005–0.01 in. or 0.1–0.3 mm. Rod seals also will require a lead-in chamfer on the rod to facilitate assembly; piston seals will need an equivalent lead-in to the housing bore. Pneumatic radial seals are typically limited to pressures less than 150 psi (1 MPa), while hydraulic seals can perform at much higher pressures.

*Rod Glands:* Rod gland designs tend to begin with a known rod diameter. Rod glands contain the O-ring in a groove cut into the housing bore, and the groove diameter should be sized to cramp the O-ring outer diameter. The O-ring is often simultaneously stretched by the rod. Here the groove maximum outer diameter often is set equal to the nominal O-ring diameter as a starting point. Alternatively, using the minimum O-ring outside diameter and the

**Table 12. Dovetail and Half-Dovetail Groove Dimensions for Static Face Seals**
*(Zero Clearance Gap in Gland, 70 Shore A O-Rings)*

| O-Ring | | | Groove Details | | | Dovetail Groove | | | Half-Dovetail Groove | | |
|---|---|---|---|---|---|---|---|---|---|---|---|
| Size Code | Units | Cross-Section Nominal | Wall Angle θ ± 1° | Internal Radii | Opening Radii | Squeeze | Depth *L* | Width *G* | Squeeze | Depth *L* | Width *G* |
| -0XX | inch | 0.070 | 24% | 0.016 | 0.005 | 27% | 0.050–0.052 | 0.055–0.059 | 23% | 0.053–0.055 | 0.064–0.066 |
| -0XXA | mm | 1.78 | | 0.40 | 0.13 | | 1.27–1.32 | 1.40–1.50 | | 1.35–1.40 | 1.63–1.68 |
| -1XX | inch | 0.103 | 24% | 0.016 | 0.010 | 21% | 0.081–0.083 | 0.083–0.087 | 19% | 0.083–0.085 | 0.095–0.097 |
| -1XXA | mm | 2.62 | | 0.40 | 0.25 | | 2.06–2.11 | 2.11–2.21 | | 2.11–2.16 | 2.41–2.46 |
| -2XX | inch | 0.139 | 24% | 0.031 | 0.010 | 20% | 0.111–0.113 | 0.113–0.117 | 18% | 0.113–0.115 | 0.124–0.128 |
| -2XXA | mm | 3.53 | | 0.80 | 0.25 | | 2.87–2.92 | 2.87–2.97 | | 2.87–2.92 | 3.15–3.25 |
| -3XX | inch | 0.210 | 24% | 0.031 | 0.015 | 18% | 0.171–0.173 | 0.171–0.175 | 17% | 0.173–0.176 | 0.190–0.193 |
| -3XXA | mm | 5.33 | | 0.80 | 0.40 | | 4.52–4.57 | 4.01–4.11 | | 4.40–4.47 | 4.83–4.90 |
| -4XX | inch | 0.275 | 24% | 0.063 | 0.015 | 16% | 0.231–0.234 | 0.231–0.235 | 15% | 0.234–0.238 | 0.255–0.257 |
| -4XXA | mm | 6.99 | | 1.60 | 0.40 | | 5.89–5.94 | 5.54–5.64 | | 5.95–6.05 | 6.48–6.53 |

Note: Sizes -0XX to -1XXA are not recommended for these grooves.

Note: Sizes -2XX to -2XXA should be used with caution in these grooves.

maximum housing diameter, target a maximum cramping of 1 percent of nominal O-ring dimensions at nominal squeeze in the SAE system or 5 percent in the ISO system for O-rings with an inner diameter less than 250 mm. Larger diameter metric O-rings are allowed only 3 percent cramping. Stretch on the rod is normally limited to 5 percent. It is sometimes acceptable to violate these limits, and standards commonly include such cases for smaller sizes.

*Piston Glands:* To design a piston gland, it is best to start with a known housing bore diameter. Piston glands have a groove cut into the outer diameter of a piston, where the O-ring is stretched slightly around the groove diameter and squeezed radially between the groove floor and housing bore. One exception to this is a pneumatic "floating" piston gland, discussed later. In most piston seal applications, limit maximum installed stretch of most elastomeric O-rings to 5 percent for long seal life. This can be exceeded for some small sizes, and standard gland dimensions include such exceptions.

*Radial Gland Calculations:* Designers usually begin a radial seal design with a preferred cross section and inner diameter selected to work with a given rod or bore. Standards and catalogs can then be used to determine appropriate gland dimensions. In the standards, target squeeze ranges are as follows:

| | | |
|---|---|---|
| SAE Sizes -0XX: 5–30% | SAE Sizes -1XX: 5–21% | SAE Sizes -2XX: 4–20% |
| SAE Sizes -3XX: 4–17% | SAE Sizes -4XX: 7–17% | |
| ISO Sizes -0XXA: 14–35% (static) | 10–24% (pneumatic reciprocating) | 13–27% (hydraulic reciprocating) |
| ISO Sizes -1XXA: 13–30% (static) | 8–22% (pneumatic reciprocating) | 12–24% (hydraulic reciprocating) |
| ISO Sizes -2XXA: 13–30% (static) | 7–20% (pneumatic reciprocating) | 11–22% (hydraulic reciprocating) |
| ISO Sizes -3XXA: 12–28% (static) | 7–18% (pneumatic reciprocating) | 11–20% (hydraulic reciprocating) |
| ISO Sizes -4XXA: 10–25% (static) | 6–17% (pneumatic reciprocating) | 9–19% (hydraulic reciprocating) |

Radial gland depth used to calculate squeeze should take into account any clearance gap and groove eccentricity *E*. The radial clearance gap maximum and minimum are seen when the piston or rod is side-loaded and touching the wall of the bore. While SAE standards include these factors in squeeze range calculations and recommendations, ISO standards do not. Calculation by either method is valid, as long as it is compared with recommended ranges calculated the same way. (Calculated squeeze values given in tables here take clearance gap and eccentricity into account; as a result, some squeeze values shown may be outside ISO recommended ranges.)

Gland data for appropriate size ranges are given in Table 2, Table 3, and Table 6 to Table 11. These dimensions can be customized if desired, but the resulting O-ring stretch/cramping, squeeze, and gland fill must be evaluated to check for potential problems. Precise calculation methods for a radially stretched O-ring cross section are included in the standards. As an approximation, Poisson's ratio (μ) with a value of 0.5 (for rubber) can be used as shown below. The following equations, which apply to Fig. 2 and Fig. 3, can be used to calculate stretch/cramping, squeeze, and gland fill for many elastomeric O-rings:

*Definition of variables:*

*ID* = O-ring inner diameter

*G* = groove width

*PISTON* = outer diameter of the piston

*ROD* = outer diameter of a rod

*BG* = diameter of a bore groove for a rod gland

*Sq%* = O-ring squeeze percentage

*GVol* = groove volume

*Fill%* = gland fill percent

μ = Poisson's ratio = 0.5 (rubber)

*E* = groove eccentricity

*W* = O-ring cross section diameter

*F* = gland depth including clearance gap and groove eccentricity

*PG* = diameter of a piston groove

*BORE* = inner diameter of the housing bore

*Stretch%* = percent O-ring stretch once installed in the groove

*ACramp%* = outer diameter cramping

*OVol* = O-ring volume

*WR* = reduced cross section of a stretched O-ring

*AR%* = approx. cross section reduction of stretched O-ring

Note: ISO standards omit the second two terms (clearance gap and eccentricity) in calculating *F*, affecting *Squeeze%*.

Piston Seal:

$Stretch\%_{min} = ((PG_{min} - ID_{max})/ID_{max}) \times 100\%$

$Stretch\%_{max} = ((PG_{max} - ID_{min})/ID_{min}) \times 100\%$

$F_{max} = ((BORE_{max} - PG_{min})/2) + ((BORE_{max} - PISTON_{min})/2) + (E_{max}/2)$

$F_{min} = ((BORE_{min} - PG_{max})/2) - ((BORE_{min} - PISTON_{min})/2) - (E_{max}/2)$

$$GVol_{max} = \left(\frac{\pi}{4}\right) \times \left(BORE^2_{max} - PG^2_{min}\right) \times G_{max}$$

$$GVol_{min} = \left(\frac{\pi}{4}\right) \times \left(BORE^2_{min} - PG^2_{max}\right) \times G_{min}$$

Rod Seal:

$Stretch\%_{min} = ((ROD_{min} - ID_{max})/ID_{max}) \times 100\%$

$Stretch\%_{max} = ((ROD_{max} - ID_{min})/ID_{min}) \times 100\%$

$F_{max} = ((BG_{max} - ROD_{min})/2) + ((BORE_{max} - ROD_{min})/2) + (E_{max}/2)$

$F_{min} = ((BG_{min} - ROD_{max})/2) - ((BORE_{max} - ROD_{max})/2) - (E_{max}/2)$

$$GVol_{max} = \left(\frac{\pi}{4}\right) \times \left(BG^2_{max} - ROD^2_{min}\right) \times G_{max}$$

$$GVol_{min} = \left(\frac{\pi}{4}\right) \times \left(BG^2_{min} - ROD^2_{max}\right) \times G_{min}$$

All Radial Seals:

$AR\%_{min} = (Stretch\%_{min} \times \mu)$

$AR\%_{max} = (Stretch\%_{max} \times \mu)$

$WR_{max} = W_{max} - ((AR\%_{min}/100) \times W_{max})$

$WR_{min} = W_{min} - ((AR\%_{max}/100) \times W_{min})$

$Sq\%_{min} = ((WR_{min} - F_{max})/W_{min}) \times 100\%$

$Sq\%_{max} = ((WR_{max} - F_{min})/W_{max}) \times 100\%$

$$OVol_{max} = \left(\frac{\pi^2}{4}\right) \times \left(ID_{max} + W_{max}\right) \times \left(W^2_{max}\right)$$

$$OVol_{min} = \left(\frac{\pi^2}{4}\right) \times \left(ID_{min} + W_{min}\right) \times \left(W^2_{min}\right)$$

$Fill\%_{max} = (OVol_{max} / GVol_{min}) \times 100\%$

$Fill\%_{min} = (OVol_{min} / GVol_{max}) \times 100\%$

As an example, calculate stretch/cramping, squeeze, and gland fill for an SAE size -116 O-ring (-116A in ISO, shown in parentheses below) in a standard pneumatic reciprocating piston gland:

Given: Use gland Standard SAE AS4716C-2017 (ISO 3601-2-2016):

| | | | |
|---|---|---|---|
| $ID_{max} = 0.746$ | $ID_{min} = 0.728$ | $(ID_{max} = 18.95)$ | $(ID_{min} = 18.49)$ |
| $W_{max} = 0.106$ | $W_{min} = 0.100$ | $(W_{max} = 2.71)$ | $(W_{min} = 2.53)$ |
| $G_{min} = 0.141$ | $G_{max} = 0.151$ | $(G_{min} = 4)$ | $(G_{max} = 4.2)$ |
| $PG_{max} = 0.751$ | $PG_{min} = 0.749$ | $(PG_{max} = 19.17)$ | $(PG_{min} = 19.137)$ |
| $BORE_{max} = 0.927$ | $BORE_{min} = 0.925$ | $(BORE_{max} = 23.473)$ | $(BORE_{min} = 23.44)$ |
| $PISTON_{max} = 0.923$ | $PISTON_{min} = 0.922$ | $(PISTON_{max} = 23.42)$ | $(PISTON_{min} = 23.399)$ |
| $E_{max} = 0.002$ | | $(E_{max} = 0.025)$ | |

Calculate:

| | | | |
|---|---|---|---|
| $F_{max} = 0.093$ | $F_{min} = 0.085$ | $(F_{max} = 2.218)$ | $(F_{min} = 2.102)$ |
| $Stretch\%_{max} = 3.2\%$ | $Stretch\%_{min} = 0.4\%$ | $(Stretch\%_{max} = 3.7\%)$ | $(Stretch\%_{min} = 1.0\%)$ |
| $AR\%_{max} = 1.6\%$ | $AR\%_{min} = 0.2\%$ | $(AR\%_{max} = 1.8\%)$ | $(AR\%_{min} = 0.5\%)$ |
| $WR_{max} = 0.106$ | $WR_{min} = 0.100$ | $(WR_{max} = 2.697)$ | $(WR_{min} = 2.518)$ |
| $Sq\%_{min} = 7.3\%$ | $Sq\%_{max} = 20.1\%$ | $(Sq\%_{min} = 11.9\%)$ | $(Sq\%_{max} = 21.9\%)$ |
| $GVol_{max} = 0.035$ | $GVol_{min} = 0.032$ | $(GVol_{max} = 609.45)$ | $(GVol_{min} = 571.60)$ |
| $Fill\%_{max} = 73.1\%$ | $Fill\%_{min} = 57.7\%$ | $(Fill\%_{max} = 68.7\%)$ | $(Fill\%_{min} = 54.5\%)$ |

Validate for the -116 SAE O-Ring:

| | | |
|---|---|---|
| $Stretch\%_{max} = 3.2\%$ | $Stretch\%_{min} = 0.4\%$ | Well within SAE limit of 5%. |
| $Sq\%_{min} = 7.3\%$ | $Sq\%_{max} = 20.1\%$ | Meets SAE recommendation of 5–21% for this cross section. |
| $Fill\%_{max} = 73.1\%$ | $Fill\%_{min} = 57.7\%$ | Within SAE target maximum of 85%. |

Validate for the -116A ISO O-Ring:

| | | |
|---|---|---|
| $(Stretch\%_{max} = 3.7\%)$ | $(Stretch\%_{min} = 1\%)$ | Well within ISO limit of 5%. |
| $(Sq\%_{min} = 11.9\%)$ | $(Sq\%_{max} = 21.9\%)$ | Meets ISO recommendation of 8–22% for this cross section. |
| $(Fill\%_{min} = 68.7\%)$ | $(Fill\%_{min} = 54.5\%)$ | Within ISO target maximum of 85%. |

*Floating Piston Glands:* In pneumatic reciprocating applications, where friction (manifesting as drag and/or heat) is a critical consideration, a "floating" piston gland design can be beneficial because it applies zero stretch to the O-ring. The gland is located on the piston and is overly deep to avoid contact with the O-ring and allow some radial flotation. Groove width is reduced by about 20 percent to help prevent axial movement in the absence of stretch. All compression of the O-ring happens at the outer diameter, where it interferes slightly (called "cramping") with the housing bore. The amount of recommended cramping increases with O-ring cross section. Some recommended cramping values for floating piston glands are as follows:

Here, floating piston gland cramping = O-ring nominal outside diameter − housing bore diameter minimum

Sizes -006 to -012: Cramping = 0.033–0.040 in. (0.8–1.0 mm)

Sizes -104 to -116: Cramping = 0.036–0.060 in. (0.9–1.5 mm)

Sizes -201 to -222: Cramping = 0.059–0.080 in. (1.5–2.0 mm)

Sizes -309 to -349: Cramping = 0.082–0.122 in. (2.1–3.1 mm)

Sizes -425 to -460: Cramping = 0.138–0.173 in. (3.5–4.4 mm)

Floating piston glands are suitable for low-pressure hydraulics below 200 psi (1.38 MPa) that allow for some leakage. Pneumatic piston glands also can be designed to float, but floating pneumatic rod glands have drag problems due to rod-induced stretch and therefore are not recommended. Each time the direction of pressure changes and the O-ring moves from one side of the groove to the other, fluid or gas can travel under and around the seal until the O-ring is again in contact with a groove wall. Lubrication is needed for pneumatic applications.

*Rotary Motion Radial Seals:* Lip seals and other engineered seals generally work best in rotary applications, but O-rings made of wear-resistant, hard material may be used with caution. Rotary O-ring applications are configured with the groove in the housing bore to prevent centrifugal expansion of the O-ring. The shaft should rotate inside the O-ring, and squeeze must be kept very low. Groove diameter normally is sized 5 percent smaller than the O-ring outer diameter to prevent rotation of the O-ring and cramp it against the shaft. The shaft must be supported with bearings, with the O-ring simply acting as a seal. This application requires reliable lubrication and heat management.

**O-Ring Lubrication.—**Dynamic seals typically rely on system fluid to lubricate the seal as it rubs against the facing part. Installation instructions for O-rings in dynamic applications often suggest coating the O-ring in either the process fluid if applicable, or an acceptable seal lubricant. Lubrication issues can arise when sealing a dry, volatile, or crystallizing fluid. In pneumatic applications, lubrication is provided either by using lubricated air or with periodic lubrication of the seals. Reciprocating seals maintain lubrication film thickness best at high piston velocities.

**Gland Surface Quality.—**All gland surfaces should be free of burrs, nicks, and scratches; the chosen machining method should not leave tool marks that cut across the gland. To seal properly, the gland and O-ring surfaces must be clean of debris. Surface quality and cleanliness matter most with small cross section O-rings and/or stiff materials. In a static seal, an O-ring made from softer material can be used to compensate for a rough gland. It should be noted that the two systems of measurement recommend nonequivalent maximum surface roughness average (Ra) for various cases, as follows:

| | | |
|---|---|---|
| Face seals, sealing surfaces: | SAE: 16 μin. | ISO: 1.6 μm |
| Face seals, contact wall of groove: | SAE: 32 μin. | ISO: 1.6 μm |
| Face seals, non-contact wall of groove: | SAE: 63 μin. | ISO: 1.6 μm |
| Piston and rod seals, sealing surfaces: | SAE: Liquids: 32 μin. (Gases: 16 μin.) | ISO: Dynamic: 0.4 μm (Static: 1.6 μm) |
| Piston and rod seals, groove walls: | SAE: Dynamic: 32 μin. (Static: 63 μin.) | ISO: 1.6 μm |

Surface finishes of less than 5 μin. (0.13 μm) Ra are too smooth to ensure long seal life because they wipe too cleanly, causing the O-ring to wear against the housing in the absence of a lubricating film. The best quality surfaces are honed, burnished, or hard chromium-plated. Soft and stringy metals, such as aluminum, brass, bronze, Monel, or free machining stainless steel, should not be used in contact with moving seals.

**O-Ring Installation.—**During installation, O-rings must not traverse across sharp edges or external threads. All edges encountered by the O-ring must have chamfers or radii. In radial glands, a generous lead-in chamfer of 15 to 20 degrees with a rounded edge is necessary to prevent damage, although an installation mandrel still may be needed in some cases. Never twist or roll an O-ring, and pay special attention during installation of O-rings with a large ratio of inner diameter to cross-sectional diameter. Avoid stretching the inner diameter of an elastomeric O-ring more than 50 percent, and certainly not more than 100 percent during installation. Rigid O-rings normally should not be stretched, though some materials may allow gentle expansion through applied heat for installation purposes.

The preferred bore materials are steel and cast iron, and pistons should be softer than the bore to avoid scratching them. The bore sections should be thick enough to resist expansion and contraction under pressure so that the radial clearance gap remains constant, reducing the chance of damage to the O-ring by extrusion and nibbling. Some compatibility problems may occur when O-rings are used with plastics parts because certain compounding ingredients may attack the plastics, causing crazing of the plastics surface.

**O-Rings Used as Drive Belts.—**O-rings are frequently used as driving belts in round bottom or V-grooves with light tension for low-power drive elements. Special compounds are available with high resistance to stress relaxation and fatigue for these applications. Best service is obtained in drive belt applications when the initial belt tension is between 80 and 200 psi and the initial installed stretch is between 8 and 25 percent of the circumferential length. Most of the compounds used for drive belts operate best between 10 and 15 percent stretch, although polyurethane has good service life when stretched as much as 20 to 25 percent.

**Table 13. Typical O-Ring Compounds**

| | |
|---|---|
| Nitrile | General-purpose compound for use with most petroleum oils, greases, gasoline, alcohols and glycols, LP gases, propane and butane fuels. Also for food service to resist vegetable and animal fats. Effective temperature range is about –40° to 250° F (–40 to 121° C). Excellent compression set, tear and abrasion resistance, but poor resistance to ozone, sunlight and weather. Higher-temperature nitrile compounds with similar properties are also available. |
| Hydrogenated Nitrile | Similar to general-purpose nitrile compounds with improved high-temperature performance, resistance to aging, and petroleum product compatibility. |
| Polychloroprene (Neoprene) | General-purpose compound with low compression set and good resistance to elevated temperatures. Good resistance to sunlight, ozone, and weathering, and fair oil resistance. Frequently used for refrigerator gases such as Freon. Effective temperature range is about –40° to 250° F (–40 to 121° C). |
| Ethylene Propylene | General-purpose compound with excellent resistance to polar fluids such as water, steam, ketones, and phosphate esters, and brake fluids, but not resistant to petroleum oils and solvents. Excellent resistance to ozone and flexing. Recommended for belt-drive applications. Continuous duty service in temperatures up to 250° F (121° C). |
| Silicone | Widest temperature range (–150° to 500° F or –101 to 260° C) and best low-temperature flexibility of all elastomeric compounds. Not recommended for dynamic applications, due to low strength, or for use with most petroleum oils. Shrinkage characteristics similar to organic rubber, allowing existing molds to be used. |
| Polyurethane | Toughest of the elastomers used for O-rings, characterized by high tensile strength, excellent abrasion resistance, and tear strength. Compression set and heat resistance are inferior to nitrile. Suitable for hydraulic applications that anticipate abrasive contaminants and shock loads. Temperature service range of –65° to 212° F (–54 to 100° C). |
| Fluorosilicone | Wide temperature range (–80° to 450° F or –62 to 232° C) for continuous duty and excellent resistance to petroleum oils and fuels. Recommended for static applications only, due to limited strength and low abrasion resistance. |
| Polyacrylate | Heat resistance better than nitrile compounds, but inferior low temperature, compression set, and water resistance. Often used in power steering and transmission applications due to excellent resistance to oil, automatic transmission fluids, oxidation, and flex cracking. Temperature service range of –20° to 300° F (–29 to 149° C). |
| Fluorocarbon (Viton) | General-purpose compound suitable for applications requiring resistance to aromatic or halogenated solvents or to high temperatures (–20° to 500° F or –29 to 260° C with limited service to 600° F or 316° C). Outstanding resistance to blended aromatic fuels, straight aromatics, and halogenated hydrocarbons and other petroleum products. Good resistance to strong acids (temperature range in acids (–20° to 250° F or –29 to 121° C), but not effective for use with very hot water, steam, and brake fluids. |

**O-Ring Materials.**—Thousands of O-ring compounds have been formulated for specific applications. Some of the most common types of compounds and their typical applications are given in Table 13. The Shore A durometer is the standard instrument used for measuring the hardness of elastomeric compounds. The softest O-rings are 50 and 60 Shore A and stretch more easily, exhibit lower breakout friction, seal better on rough surfaces, and need less clamping pressure than harder rings. For a given squeeze, the higher the durometer hardness of a ring, the greater the associated friction because a greater compressive force is exerted by hard rings than soft rings.

The most widely used rings are medium-hard O-rings with 70 Shore A hardness, which have the best wear resistance and frictional properties for running seals. Applications that involve oscillating or rotary motion frequently use 80 Shore A materials. Rings with a hardness above 85 Shore A often leak more because of less effective wiping action. These harder rings have a greater resistance to extrusion, but for small sizes may break easily during installation. O-ring hardness varies inversely with temperature, but when used for continuous service at high temperatures, the hardness may eventually increase after an initial softening of the compound.

O-ring compounds have thermal coefficients of expansion in the range of 7 to 20 times that of metal components, so shrinkage or expansion with temperature change can pose problems of leakage past the seal at low temperatures and excessive pressures at high temperatures when a ring is installed in a tight-fitting groove. Likewise, when an O-ring is immersed in a fluid, the compound usually absorbs some of the fluid and consequently increases in volume. Manufacturer's data give volumetric increase data for compounds completely immersed in various fluids. For confined rings (those with only a portion of the ring exposed to fluid), the size increase may be considerably lower than for rings completely immersed in fluid. Certain fluids can also cause ring shrinkage during "idle" periods, i.e., when the seal has a chance to dry out. If this shrinkage is more than 3 to 4 percent, the seal may leak.

Excessive swelling due to fluid contact and high temperatures softens all compounds approximately 20 to 30 Shore A points from room temperature values and designs should anticipate the expected operating conditions. At low temperatures, swelling may be beneficial because fluid absorption may make the seal more flexible. However, the combination of low temperature and low pressure makes a seal particularly difficult to maintain. A soft compound should be used to provide a resilient seal at low temperatures. Below −65° F, only compounds formulated with silicone are useful; other compounds are simply too stiff, especially for use with air and other gases.

Compression set is another material property and a very important sealing factor. It is a measure of the shape memory of the material, that is, the ability to regain shape after being deformed. Compression set is a ratio, expressed as a percentage, of the unrecovered to original thickness of an O-ring compressed for a specified period of time between two heated plates and then released. O-rings with excessive compressive set will fail to maintain a good seal because, over time, the ring will be unable to exert the necessary compressive force (squeeze) on the enclosing walls. Swelling of the ring due to fluid contact tends to increase the squeeze and may partially compensate for the loss due to compression set. Generally, compression set varies by compound and ring cross-sectional diameter, and increases with the operating temperature.

## ROLLED STEEL, WIRE, SHEET METAL, WIRE ROPE

### Rolled Steel Sections

**Lengths of Angles Bent to Circular Shape.—**To calculate the length of an angle-iron used either inside or outside of a tank or smokestack, the following table of constants may be used: Assume, for example, that a stand-pipe, 20 feet inside diameter, is provided with a 3 by 3 by 3⁄8 inch angle-iron on the inside at the top. The circumference of a circle 20 feet in diameter is 754 inches. From the table of constants, find the constant for a 3 by 3 by 3⁄8 inch angle-iron, which is 4.319. The length of the angle then is 754 − 4.319 = 749.681 inches. Should the angle be on the outside, add the constant instead of subtracting it; thus, 754 + 4.319 = 758.319 inches.

| Size of Angle | Const. | Size of Angle | Const. | Size of Angle | Const. |
|---|---|---|---|---|---|
| 1⁄4 × 2 × 2 | 2.879 | 5⁄16 × 3 × 3 | 4.123 | 1⁄2 × 5 × 5 | 6.804 |
| 5⁄16 × 2 × 2 | 3.076 | 3⁄8 × 3 × 3 | 4.319 | 3⁄8 × 6 × 6 | 7.461 |
| 3⁄8 × 2 × 2 | 3.272 | 1⁄2 × 3 × 3 | 4.711 | 1⁄2 × 6 × 6 | 7.854 |
| 1⁄4 × 2 1⁄2 × 2 1⁄2 | 3.403 | 3⁄8 × 3 1⁄2 × 3 1⁄2 | 4.843 | 3⁄4 × 6 × 6 | 8.639 |
| 5⁄16 × 2 1⁄2 × 2 1⁄2 | 3.600 | 1⁄2 × 3 1⁄2 × 3 1⁄2 | 5.235 | 1⁄2 × 8 × 8 | 9.949 |
| 3⁄8 × 2 1⁄2 × 2 1⁄2 | 3.796 | 3⁄8 × 4 × 4 | 5.366 | 3⁄4 × 8 × 8 | 10.734 |
| 1⁄2 × 2 1⁄2 × 2 1⁄2 | 4.188 | 1⁄2 × 4 × 4 | 5.758 | 1 × 8 × 8 | 11.520 |
| 1⁄4 × 3 × 3 | 3.926 | 3⁄8 × 5 × 5 | 6.414 | … | … |

**Standard Designations of Rolled Steel Shapes.—**Through a joint effort, the American Iron and Steel Institute (AISI) and the American Institute of Steel Construction (AISC) have changed most of the designations for their hot-rolled structural steel shapes. The present designations, standard for steel producing and fabricating industries, should be used when designing, detailing, and ordering steel. The accompanying Table 1 compares the present designations with the previous descriptions.

**Table 1. Hot-Rolled Structural Steel Shape Designations (AISI and AISC)**

| Present Designation | Type of Shape | Previous Designation |
|---|---|---|
| W 24 × 76 | W shape | 24 WF 76 |
| W 14 × 26 | W shape | 14 B 26 |
| S 24 × 100 | S shape | 24 I 100 |
| M 8 × 18.5 | M shape | 8 M 18.5 |
| M 10 × 9 | M shape | 10 JR 9.0 |
| M 8 × 34.3 | M shape | 8 × 8 M 34.3 |
| C 12 × 20.7 | American Standard Channel | 12 [20.7 |
| MC 12 × 45 | Miscellaneous Channel | 12 × 4 [45.0 |
| MC 12 × 10.6 | Miscellaneous Channel | 12 JR [10.6 |
| HP 14 × 73 | HP shape | 14 BP 73 |
| L 6 × 6 × 3⁄4 | Equal Leg Angle | ∠ 6 × 6 × 3⁄4 |
| L 6 × 4 × 5⁄8 | Unequal Leg Angle | ∠ 6 × 4 × 5⁄8 |
| WT 12 × 38 | Structural Tee cut from W shape | ST 12 WF 38 |
| WT 7 × 13 | Structural Tee cut from W shape | ST 7 B 13 |
| St 12 × 50 | Structural Tee cut from S shape | ST 12 I 50 |
| MT 4 × 9.25 | Structural Tee cut from M shape | ST 4 M 9.25 |
| MT 5 × 4.5 | Structural Tee cut from M shape | ST 5 JR 4.5 |
| MT 4 × 17.15 | Structural Tee cut from M shape | ST 4 M 17.15 |
| PL 1⁄2 × 18 | Plate | PL 18 × 1⁄2 |
| Bar 1 | Square Bar | Bar 1 |
| Bar 1 1⁄4 ∅ | Round Bar | Bar 1 1⁄4 ∅ |
| Bar 2 1⁄2 × 1⁄2 | Flat Bar | Bar 2 1⁄2 × 1⁄2 |
| Pipe 4 Std. | Pipe | Pipe 4 Std. |
| Pipe 4 X-Strong | Pipe | Pipe 4 X-Strong |
| Pipe 4 XX-Strong | Pipe | Pipe 4 XX-Strong |
| TS 4 × 4 × .375 | Structural Tubing: Square | Tube 4 × 4 × .375 |
| TS 5 × 3 × .375 | Structural Tubing: Rectangular | Tube 5 × 3 × .375 |
| TS 3 OD × .250 | Structural Tubing: Circular | Tube 3 OD × .250 |

Data taken from the "Manual of Steel Construction," 8th Edition, 1980, with permission of the American Institute of Steel Construction.

## Table 2a. Steel Wide-Flange Sections

Wide-flange sections are designated, in order, by a section letter, nominal depth of the member in inches, and the nominal weight in pounds per foot; thus:

**W 18 × 64**

indicates a wide-flange section having a nominal depth of 18 inches, and a nominal weight per foot of 64 pounds. Actual geometry for each section can be obtained from the values below.

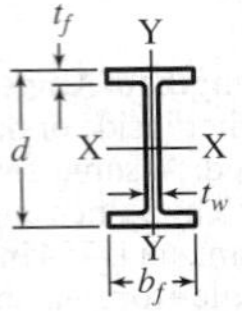


| Designation | Area, $A$ inch² | Depth, $d$ inch | Flange Width, $b_f$ inch | Flange Thickness, $t_f$ inch | Web Thickness, $t_w$ inch | Axis XX $I$ inch⁴ | Axis XX $S$ inch³ | Axis XX $r$ inch | Axis YY $I$ inch⁴ | Axis YY $S$ inch³ | Axis YY $r$ inch |
|---|---|---|---|---|---|---|---|---|---|---|---|
| [a]W 27 × 178 | 52.3 | 27.81 | 14.085 | 1.190 | 0.725 | 6990 | 502 | 11.6 | 555 | 78.8 | 3.26 |
| × 161 | 47.4 | 27.59 | 14.020 | 1.080 | 0.660 | 6280 | 455 | 11.5 | 497 | 70.9 | 3.24 |
| × 146 | 42.9 | 27.38 | 13.965 | 0.975 | 0.605 | 5630 | 411 | 11.4 | 443 | 63.5 | 3.21 |
| × 114 | 33.5 | 27.29 | 10.070 | 0.930 | 0.570 | 4090 | 299 | 11.0 | 159 | 31.5 | 2.18 |
| × 102 | 30.0 | 27.09 | 10.015 | 0.830 | 0.515 | 3620 | 267 | 11.0 | 139 | 27.8 | 2.15 |
| × 94 | 27.7 | 26.92 | 9.990 | 0.745 | 0.490 | 3270 | 243 | 10.9 | 124 | 24.8 | 2.12 |
| × 84 | 24.8 | 26.71 | 9.960 | 0.640 | 0.460 | 2850 | 213 | 10.7 | 106 | 21.2 | 2.07 |
| W 24 × 162 | 47.7 | 25.00 | 12.955 | 1.220 | 0.705 | 5170 | 414 | 10.4 | 443 | 68.4 | 3.05 |
| × 146 | 43.0 | 24.74 | 12.900 | 1.090 | 0.650 | 4580 | 371 | 10.3 | 391 | 60.5 | 3.01 |
| × 131 | 38.5 | 24.48 | 12.855 | 0.960 | 0.605 | 4020 | 329 | 10.2 | 340 | 53.0 | 2.97 |
| × 117 | 34.4 | 24.26 | 12.800 | 0.850 | 0.550 | 3540 | 291 | 10.1 | 297 | 46.5 | 2.94 |
| × 104 | 30.6 | 24.06 | 12.750 | 0.750 | 0.500 | 3100 | 258 | 10.1 | 259 | 40.7 | 2.91 |
| × 94 | 27.7 | 24.31 | 9.065 | 0.875 | 0.515 | 2700 | 222 | 9.87 | 109 | 24.0 | 1.98 |
| × 84 | 24.7 | 24.10 | 9.020 | 0.770 | 0.470 | 2370 | 196 | 9.79 | 94.4 | 20.9 | 1.95 |
| × 76 | 22.4 | 23.92 | 8.990 | 0.680 | 0.440 | 2100 | 176 | 9.69 | 82.5 | 18.4 | 1.92 |
| × 68 | 20.1 | 23.73 | 8.965 | 0.585 | 0.415 | 1830 | 154 | 9.55 | 70.4 | 15.7 | 1.87 |
| × 62 | 18.2 | 23.74 | 7.040 | 0.590 | 0.430 | 1550 | 131 | 9.23 | 34.5 | 9.80 | 1.38 |
| × 55 | 16.2 | 23.57 | 7.005 | 0.505 | 0.395 | 1350 | 114 | 9.11 | 29.1 | 8.30 | 1.34 |
| W 21 × 147 | 43.2 | 22.06 | 12.510 | 1.150 | 0.720 | 3630 | 329 | 9.17 | 376 | 60.1 | 2.95 |
| × 132 | 38.8 | 21.83 | 12.440 | 1.035 | 0.650 | 3220 | 295 | 9.12 | 333 | 53.5 | 2.93 |
| × 122 | 35.9 | 21.68 | 12.390 | 0.960 | 0.600 | 2960 | 273 | 9.09 | 305 | 49.2 | 2.92 |
| × 111 | 32.7 | 21.51 | 12.340 | 0.875 | 0.550 | 2670 | 249 | 9.05 | 274 | 44.5 | 2.90 |
| × 101 | 29.8 | 21.36 | 12.290 | 0.800 | 0.500 | 2420 | 227 | 9.02 | 248 | 40.3 | 2.89 |
| × 93 | 27.3 | 21.62 | 8.420 | 0.930 | 0.580 | 2070 | 192 | 8.70 | 92.9 | 22.1 | 1.84 |
| × 83 | 24.3 | 21.43 | 8.355 | 0.835 | 0.515 | 1830 | 171 | 8.67 | 81.4 | 19.5 | 1.83 |
| × 73 | 21.5 | 21.24 | 8.295 | 0.740 | 0.455 | 1600 | 151 | 8.64 | 70.6 | 17.0 | 1.81 |
| × 68 | 20.0 | 21.13 | 8.270 | 0.685 | 0.430 | 1480 | 140 | 8.60 | 64.7 | 15.7 | 1.80 |
| × 62 | 18.3 | 20.99 | 8.240 | 0.615 | 0.400 | 1330 | 127 | 8.54 | 57.5 | 13.9 | 1.77 |
| × 57 | 16.7 | 21.06 | 6.555 | 0.650 | 0.405 | 1170 | 111 | 8.36 | 30.6 | 9.35 | 1.35 |
| × 50 | 14.7 | 20.83 | 6.530 | 0.535 | 0.380 | 984 | 94.5 | 8.18 | 24.9 | 7.64 | 1.30 |
| × 44 | 13.0 | 20.66 | 6.500 | 0.450 | 0.350 | 843 | 81.6 | 8.06 | 20.7 | 6.36 | 1.26 |
| W 18 × 119 | 35.1 | 18.97 | 11.265 | 1.060 | 0.655 | 2190 | 231 | 7.90 | 253 | 44.9 | 2.69 |
| × 106 | 31.1 | 18.73 | 11.200 | 0.940 | 0.590 | 1910 | 204 | 7.84 | 220 | 39.4 | 2.66 |
| × 97 | 28.5 | 18.59 | 11.145 | 0.870 | 0.535 | 1750 | 188 | 7.82 | 201 | 36.1 | 2.65 |
| × 86 | 25.3 | 18.39 | 11.090 | 0.770 | 0.480 | 1530 | 166 | 7.77 | 175 | 31.6 | 2.63 |
| × 76 | 22.3 | 18.21 | 11.035 | 0.680 | 0.425 | 1330 | 146 | 7.73 | 152 | 27.6 | 2.61 |
| × 71 | 20.8 | 18.47 | 7.635 | 0.810 | 0.495 | 1170 | 127 | 7.50 | 60.3 | 15.8 | 1.70 |
| × 65 | 19.1 | 18.35 | 7.590 | 0.750 | 0.450 | 1070 | 117 | 7.49 | 54.8 | 14.4 | 1.69 |
| × 60 | 17.6 | 18.24 | 7.555 | 0.695 | 0.415 | 984 | 108 | 7.47 | 50.1 | 13.3 | 1.69 |
| × 55 | 16.2 | 18.11 | 7.530 | 0.630 | 0.390 | 890 | 98.3 | 7.41 | 44.9 | 11.9 | 1.67 |
| × 50 | 14.7 | 17.99 | 7.495 | 0.570 | 0.355 | 800 | 88.9 | 7.38 | 40.1 | 10.7 | 1.65 |
| × 46 | 13.5 | 18.06 | 6.060 | 0.605 | 0.360 | 712 | 78.8 | 7.25 | 22.5 | 7.43 | 1.29 |
| × 40 | 11.8 | 17.90 | 6.015 | 0.525 | 0.315 | 612 | 68.4 | 7.21 | 19.1 | 6.35 | 1.27 |
| × 35 | 10.3 | 17.70 | 6.000 | 0.425 | 0.300 | 510 | 57.6 | 7.04 | 15.3 | 5.12 | 1.22 |

[a] Consult the AISC Manual, noted above, for W steel shapes having nominal depths greater than 27 inches.

*Symbols:* $I$ = moment of inertia; $S$ = section modulus; $r$ = radius of gyration.

Data taken from the "Manual of Steel Construction," 8th Edition, 1980, with permission of the American Institute of Steel Construction.

## Table 2b. Steel Wide-Flange Sections

Wide-flange sections are designated, in order, by a section letter, nominal depth of the member in inches, and the nominal weight in pounds per foot. Thus:

**W 16 × 78**

indicates a wide-flange section having a nominal depth of 16 inches, and a nominal weight per foot of 78 pounds. Actual geometry for each section can be obtained from the values below.

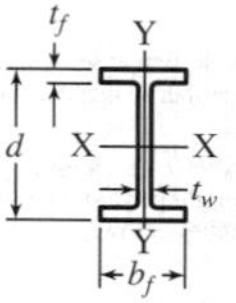


| Designation | Area, $A$ inch² | Depth, $d$ inch | Flange: Width, $b_f$ inch | Flange: Thickness, $t_f$ inch | Web Thickness, $t_w$ inch | Axis XX: $I$ inch⁴ | Axis XX: $S$ inch³ | Axis XX: $r$ inch | Axis YY: $I$ inch⁴ | Axis YY: $S$ inch³ | Axis YY: $r$ inch |
|---|---|---|---|---|---|---|---|---|---|---|---|
| W 16 × 100 | 29.4 | 16.97 | 10.425 | 0.985 | 0.585 | 1490 | 175 | 7.10 | 186 | 35.7 | 2.51 |
| × 89 | 26.2 | 16.75 | 10.365 | 0.875 | 0.525 | 1300 | 155 | 7.05 | 163 | 31.4 | 2.49 |
| × 77 | 22.6 | 16.52 | 10.295 | 0.760 | 0.455 | 1110 | 134 | 7.00 | 138 | 26.9 | 2.47 |
| × 67 | 19.7 | 16.33 | 10.235 | 0.665 | 0.395 | 954 | 117 | 6.96 | 119 | 23.2 | 2.46 |
| × 57 | 16.8 | 16.43 | 7.120 | 0.715 | 0.430 | 758 | 92.2 | 6.72 | 43.1 | 12.1 | 1.60 |
| × 50 | 14.7 | 16.26 | 7.070 | 0.630 | 0.380 | 659 | 81.0 | 6.68 | 37.2 | 10.5 | 1.59 |
| × 45 | 13.3 | 16.13 | 7.035 | 0.565 | 0.345 | 586 | 72.7 | 6.65 | 32.8 | 9.34 | 1.57 |
| × 40 | 11.8 | 16.01 | 6.995 | 0.505 | 0.305 | 518 | 64.7 | 6.63 | 28.9 | 8.25 | 1.57 |
| × 36 | 10.6 | 15.86 | 6.985 | 0.430 | 0.295 | 448 | 56.5 | 6.51 | 24.5 | 7.00 | 1.52 |
| × 31 | 9.12 | 15.88 | 5.525 | 0.440 | 0.275 | 375 | 47.2 | 6.41 | 12.4 | 4.49 | 1.17 |
| × 26 | 7.68 | 15.69 | 5.500 | 0.345 | 0.250 | 301 | 38.4 | 6.26 | 9.59 | 3.49 | 1.12 |
| W 14 × 730 | 215.0 | 22.42 | 17.890 | 4.910 | 3.070 | 14300 | 1280 | 8.17 | 4720 | 527 | 4.69 |
| × 665 | 196.0 | 21.64 | 17.650 | 4.520 | 2.830 | 12400 | 1150 | 7.98 | 4170 | 472 | 4.62 |
| × 605 | 178.0 | 20.92 | 17.415 | 4.160 | 2.595 | 10800 | 1040 | 7.80 | 3680 | 423 | 4.55 |
| × 550 | 162.0 | 20.24 | 17.200 | 3.820 | 2.380 | 9430 | 931 | 7.63 | 3250 | 378 | 4.49 |
| × 500 | 147.0 | 19.60 | 17.010 | 3.500 | 2.190 | 8210 | 838 | 7.48 | 2880 | 339 | 4.43 |
| × 455 | 134.0 | 19.02 | 16.835 | 3.210 | 2.015 | 7190 | 756 | 7.33 | 2560 | 304 | 4.38 |
| × 426 | 125.0 | 18.67 | 16.695 | 3.035 | 1.875 | 6600 | 707 | 7.26 | 2360 | 283 | 4.34 |
| × 398 | 117.0 | 18.29 | 16.590 | 2.845 | 1.770 | 6000 | 656 | 7.16 | 2170 | 262 | 4.3 |
| × 370 | 109.0 | 17.92 | 16.475 | 2.660 | 1.655 | 5440 | 607 | 7.07 | 1990 | 241 | 4.27 |
| × 342 | 101.0 | 17.54 | 16.360 | 2.470 | 1.540 | 4900 | 559 | 6.98 | 1810 | 221 | 4.24 |
| × 311 | 91.4 | 17.12 | 16.230 | 2.260 | 1.410 | 4330 | 506 | 6.88 | 1610 | 199 | 4.20 |
| × 283 | 83.3 | 16.74 | 16.110 | 2.070 | 1.290 | 3840 | 459 | 6.79 | 1440 | 179 | 4.17 |
| × 257 | 75.6 | 16.38 | 15.995 | 1.890 | 1.175 | 3400 | 415 | 6.71 | 1290 | 161 | 4.13 |
| × 233 | 68.5 | 16.04 | 15.890 | 1.720 | 1.070 | 3010 | 375 | 6.63 | 1150 | 145 | 4.10 |
| × 211 | 62.0 | 15.72 | 15.800 | 1.560 | 0.980 | 2660 | 338 | 6.55 | 1030 | 130 | 4.07 |
| × 193 | 56.8 | 15.48 | 15.710 | 1.440 | 0.890 | 2400 | 310 | 6.50 | 931 | 119 | 4.05 |
| × 176 | 51.8 | 15.22 | 15.650 | 1.310 | 0.830 | 2140 | 281 | 6.43 | 838 | 107 | 4.02 |
| × 159 | 46.7 | 14.98 | 15.565 | 1.190 | 0.745 | 1900 | 254 | 6.38 | 748 | 96.2 | 4.00 |
| × 145 | 42.7 | 14.78 | 15.500 | 1.090 | 0.680 | 1710 | 232 | 6.33 | 677 | 87.3 | 3.98 |

*Symbols:* $I$ = moment of inertia; $S$ = section modulus; $r$ = radius of gyration.

Data taken from the "Manual of Steel Construction," 8th Edition, 1980, with permission of the American Institute of Steel Construction.

## Table 2c. Steel Wide-Flange Sections

Wide-flange sections are designated, in order, by a section letter, nominal depth of the member in inches, and the nominal weight in pounds per foot. Thus:

**W 14 × 38**

indicates a wide-flange section having a nominal depth of 14 inches, and a nominal weight per foot of 38 pounds. Actual geometry for each section can be obtained from the values below.

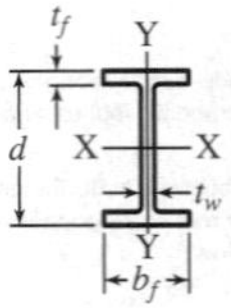


| Designation | Area, $A$ inch² | Depth, $d$ inch | Flange Width, $b_f$ inch | Flange Thickness, $t_f$ inch | Web Thickness, $t_w$ inch | Axis XX $I$ inch⁴ | Axis XX $S$ inch³ | Axis XX $r$ inch | Axis YY $I$ inch⁴ | Axis YY $S$ inch³ | Axis YY $r$ inch |
|---|---|---|---|---|---|---|---|---|---|---|---|
| W 14 × 132 | 38.8 | 14.66 | 14.725 | 1.030 | 0.645 | 1530 | 209 | 6.28 | 548 | 74.5 | 3.76 |
| × 120 | 35.3 | 14.48 | 14.670 | 0.940 | 0.590 | 1380 | 190 | 6.24 | 495 | 67.5 | 3.74 |
| × 109 | 32.0 | 14.32 | 14.605 | 0.860 | 0.525 | 1240 | 173 | 6.22 | 447 | 61.2 | 3.73 |
| × 99 | 29.1 | 14.16 | 14.565 | 0.780 | 0.485 | 1110 | 157 | 6.17 | 402 | 55.2 | 3.71 |
| × 90 | 26.5 | 14.02 | 14.520 | 0.710 | 0.440 | 999 | 143 | 6.14 | 362 | 49.9 | 3.70 |
| × 82 | 24.1 | 14.31 | 10.130 | 0.855 | 0.510 | 882 | 123 | 6.05 | 148 | 29.3 | 2.48 |
| × 74 | 21.8 | 14.17 | 10.070 | 0.785 | 0.450 | 796 | 112 | 6.04 | 134 | 26.6 | 2.48 |
| × 68 | 20.0 | 14.04 | 10.035 | 0.720 | 0.415 | 723 | 103 | 6.01 | 121 | 24.2 | 2.46 |
| × 61 | 17.9 | 13.89 | 9.995 | 0.645 | 0.375 | 640 | 92.2 | 5.98 | 107 | 21.5 | 2.45 |
| × 53 | 15.6 | 13.92 | 8.060 | 0.660 | 0.370 | 541 | 77.8 | 5.89 | 57.7 | 14.3 | 1.92 |
| × 48 | 14.1 | 13.79 | 8.030 | 0.595 | 0.340 | 485 | 70.3 | 5.85 | 51.4 | 12.8 | 1.91 |
| × 43 | 12.6 | 13.66 | 7.995 | 0.530 | 0.305 | 428 | 62.7 | 5.82 | 45.2 | 11.3 | 1.89 |
| × 38 | 11.2 | 14.10 | 6.770 | 0.515 | 0.310 | 385 | 54.6 | 5.87 | 26.7 | 7.88 | 1.55 |
| × 34 | 10.0 | 13.98 | 6.745 | 0.455 | 0.285 | 340 | 48.6 | 5.83 | 23.3 | 6.91 | 1.53 |
| × 30 | 8.85 | 13.84 | 6.730 | 0.385 | 0.270 | 291 | 42.0 | 5.73 | 19.6 | 5.82 | 1.49 |
| × 26 | 7.69 | 13.91 | 5.025 | 0.420 | 0.255 | 245 | 35.3 | 5.65 | 8.91 | 3.54 | 1.08 |
| × 22 | 6.49 | 13.74 | 5.000 | 0.335 | 0.230 | 199 | 29.0 | 5.54 | 7.00 | 2.80 | 1.04 |
| W 12 × 336 | 98.8 | 16.82 | 13.385 | 2.955 | 1.775 | 4060 | 483 | 6.41 | 1190 | 177 | 3.47 |
| × 305 | 89.6 | 16.32 | 13.235 | 2.705 | 1.625 | 3550 | 435 | 6.29 | 1050 | 159 | 3.42 |
| × 279 | 81.9 | 15.85 | 13.140 | 2.470 | 1.530 | 3110 | 393 | 6.16 | 937 | 143 | 3.38 |
| × 252 | 74.1 | 15.41 | 13.005 | 2.250 | 1.395 | 2720 | 353 | 6.06 | 828 | 127 | 3.34 |
| × 230 | 67.7 | 15.05 | 12.895 | 2.070 | 1.285 | 2420 | 321 | 5.97 | 742 | 115 | 3.31 |
| × 210 | 61.8 | 14.71 | 12.790 | 1.900 | 1.180 | 2140 | 292 | 5.89 | 664 | 104 | 3.28 |
| × 190 | 55.8 | 14.38 | 12.670 | 1.735 | 1.060 | 1890 | 263 | 5.82 | 589 | 93.0 | 3.25 |
| × 170 | 50.0 | 14.03 | 12.570 | 1.560 | 0.960 | 1650 | 235 | 5.74 | 517 | 82.3 | 3.22 |
| × 152 | 44.7 | 13.71 | 12.480 | 1.400 | 0.870 | 1430 | 209 | 5.66 | 454 | 72.8 | 3.19 |
| × 136 | 39.9 | 13.41 | 12.400 | 1.250 | 0.790 | 1240 | 186 | 5.58 | 398 | 64.2 | 3.16 |
| × 120 | 35.3 | 13.12 | 12.320 | 1.105 | 0.710 | 1070 | 163 | 5.51 | 345 | 56.0 | 3.13 |
| × 106 | 31.2 | 12.89 | 12.220 | 0.990 | 0.610 | 933 | 145 | 5.47 | 301 | 49.3 | 3.11 |
| × 96 | 28.2 | 12.71 | 12.160 | 0.900 | 0.550 | 833 | 131 | 5.44 | 270 | 44.4 | 3.09 |
| × 87 | 25.6 | 12.53 | 12.125 | 0.810 | 0.515 | 740 | 118 | 5.38 | 241 | 39.7 | 3.07 |
| × 79 | 23.2 | 12.38 | 12.080 | 0.735 | 0.470 | 662 | 107 | 5.34 | 216 | 35.8 | 3.05 |
| × 72 | 21.1 | 12.25 | 12.040 | 0.670 | 0.430 | 597 | 97.4 | 5.31 | 195 | 32.4 | 3.04 |
| × 65 | 19.1 | 12.12 | 12.000 | 0.605 | 0.390 | 533 | 87.9 | 5.28 | 174 | 29.1 | 3.02 |
| × 58 | 17.0 | 12.19 | 10.010 | 0.640 | 0.360 | 475 | 78.0 | 5.28 | 107 | 21.4 | 2.51 |
| × 53 | 15.6 | 12.06 | 9.995 | 0.575 | 0.345 | 425 | 70.6 | 5.23 | 95.8 | 19.2 | 2.48 |
| × 50 | 14.7 | 12.19 | 8.080 | 0.640 | 0.370 | 394 | 64.7 | 5.18 | 56.3 | 13.9 | 1.96 |
| × 45 | 13.2 | 12.06 | 8.045 | 0.575 | 0.335 | 350 | 58.1 | 5.15 | 50.0 | 12.4 | 1.94 |
| × 40 | 11.8 | 11.94 | 8.005 | 0.515 | 0.295 | 310 | 51.9 | 5.13 | 44.1 | 11.0 | 1.93 |
| × 35 | 10.3 | 12.50 | 6.560 | 0.520 | 0.300 | 285 | 45.6 | 5.25 | 24.5 | 7.47 | 1.54 |
| × 30 | 8.79 | 12.34 | 6.520 | 0.440 | 0.260 | 238 | 38.6 | 5.21 | 20.3 | 6.24 | 1.52 |
| × 26 | 7.65 | 12.22 | 6.490 | 0.380 | 0.230 | 204 | 33.4 | 5.17 | 17.3 | 5.34 | 1.51 |
| × 22 | 6.48 | 12.31 | 4.030 | 0.425 | 0.260 | 156 | 25.4 | 4.91 | 4.66 | 2.31 | 0.847 |
| × 19 | 5.57 | 12.16 | 4.005 | 0.350 | 0.235 | 130 | 21.3 | 4.82 | 3.76 | 1.88 | 0.822 |
| × 16 | 4.71 | 11.99 | 3.990 | 0.265 | 0.220 | 103 | 17.1 | 4.67 | 2.82 | 1.41 | 0.773 |
| × 14 | 4.16 | 11.91 | 3.970 | 0.225 | 0.200 | 88.6 | 14.9 | 4.62 | 2.36 | 1.19 | 0.753 |

Data taken from the "Manual of Steel Construction," 8th Edition, 1980, with permission of the American Institute of Steel Construction.

## Table 2d. Steel Wide-Flange Sections

Wide-flange sections are designated, in order, by a section letter, nominal depth of the member in inches, and the nominal weight in pounds per foot; thus:

**W 8 × 67**

indicates a wide-flange section having a nominal depth of 8 inches, and a nominal weight per foot of 67 pounds. Actual geometry for each section can be obtained from the values below.

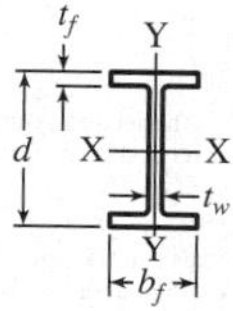


| Designation | Area, $A$ inch² | Depth, $d$ inch | Flange: Width, $b_f$ inch | Flange: Thickness, $t_f$ inch | Web Thickness, $t_w$ inch | Axis XX: $I$ inch⁴ | Axis XX: $S$ inch³ | Axis XX: $r$ inch | Axis YY: $I$ inch⁴ | Axis YY: $S$ inch³ | Axis YY: $r$ inch |
|---|---|---|---|---|---|---|---|---|---|---|---|
| W 10 × 112 | 32.9 | 11.36 | 10.415 | 1.250 | 0.755 | 716 | 126 | 4.66 | 236 | 45.3 | 2.68 |
| × 100 | 29.4 | 11.10 | 10.340 | 1.120 | 0.680 | 623 | 112 | 4.60 | 207 | 40.0 | 2.65 |
| × 88 | 25.9 | 10.84 | 10.265 | 0.990 | 0.605 | 534 | 98.5 | 4.54 | 179 | 34.8 | 2.63 |
| × 77 | 22.6 | 10.60 | 10.190 | 0.870 | 0.530 | 455 | 85.9 | 4.49 | 154 | 30.1 | 2.60 |
| × 68 | 20.0 | 10.40 | 10.130 | 0.770 | 0.470 | 394 | 75.7 | 4.44 | 134 | 26.4 | 2.59 |
| × 60 | 17.6 | 10.22 | 10.080 | 0.680 | 0.420 | 341 | 66.7 | 4.39 | 116 | 23.0 | 2.57 |
| × 54 | 15.8 | 10.09 | 10.030 | 0.615 | 0.370 | 303 | 60.0 | 4.37 | 103 | 20.6 | 2.5[illegible] |
| × 49 | 14.4 | 9.98 | 10.000 | 0.560 | 0.340 | 272 | 54.6 | 4.35 | 93.4 | 18.7 | 2.54 |
| × 45 | 13.3 | 10.10 | 8.020 | 0.620 | 0.350 | 248 | 49.1 | 4.32 | 53.4 | 13.3 | 2.0[illegible] |
| × 39 | 11.5 | 9.92 | 7.985 | 0.530 | 0.315 | 209 | 42.1 | 4.27 | 45.0 | 11.3 | 1.98 |
| × 33 | 9.71 | 9.73 | 7.960 | 0.435 | 0.290 | 170 | 35.0 | 4.19 | 36.6 | 9.20 | 1.94 |
| × 30 | 8.84 | 10.47 | 5.810 | 0.510 | 0.300 | 170 | 32.4 | 4.38 | 16.7 | 5.75 | 1.37 |
| × 26 | 7.61 | 10.33 | 5.770 | 0.440 | 0.260 | 144 | 27.9 | 4.35 | 14.1 | 4.89 | 1.35 |
| × 22 | 6.49 | 10.17 | 5.750 | 0.360 | 0.240 | 118 | 23.2 | 4.27 | 11.4 | 3.97 | 1.33 |
| × 19 | 5.62 | 10.24 | 4.020 | 0.395 | 0.250 | 96.3 | 18.8 | 4.14 | 4.29 | 2.14 | 0.874 |
| × 17 | 4.99 | 10.11 | 4.010 | 0.330 | 0.240 | 81.9 | 16.2 | 4.05 | 3.56 | 1.78 | 0.844 |
| × 15 | 4.41 | 9.99 | 4.000 | 0.270 | 0.230 | 68.9 | 13.8 | 3.95 | 2.89 | 1.45 | 0.810 |
| × 12 | 3.54 | 9.87 | 3.960 | 0.210 | 0.190 | 53.8 | 10.9 | 3.90 | 2.18 | 1.10 | 0.785 |
| W 8 × 67 | 19.7 | 9.00 | 8.280 | 0.935 | 0.570 | 272 | 60.4 | 3.72 | 88.6 | 21.4 | 2.[illegible]2 |
| × 58 | 17.1 | 8.75 | 8.220 | 0.810 | 0.510 | 228 | 52.0 | 3.65 | 75.1 | 18.3 | 2.[illegible]0 |
| × 48 | 14.1 | 8.50 | 8.110 | 0.685 | 0.400 | 184 | 43.3 | 3.61 | 60.9 | 15.0 | 2.08 |
| × 40 | 11.7 | 8.25 | 8.070 | 0.560 | 0.360 | 146 | 35.5 | 3.53 | 49.1 | 12.2 | 2.04 |
| × 35 | 10.3 | 8.12 | 8.020 | 0.495 | 0.310 | 127 | 31.2 | 3.51 | 42.6 | 10.6 | 2.03 |
| × 31 | 9.13 | 8.00 | 7.995 | 0.435 | 0.285 | 110 | 27.5 | 3.47 | 37.1 | 9.27 | 2.02 |
| × 28 | 8.25 | 8.06 | 6.535 | 0.465 | 0.285 | 98.0 | 24.3 | 3.45 | 21.7 | 6.63 | 1.62 |
| × 24 | 7.08 | 7.93 | 6.495 | 0.400 | 0.245 | 82.8 | 20.9 | 3.42 | 18.3 | 5.63 | 1.61 |
| × 21 | 6.16 | 8.28 | 5.270 | 0.400 | 0.250 | 75.3 | 18.2 | 3.49 | 9.77 | 3.71 | 1.26 |
| × 18 | 5.26 | 8.14 | 5.250 | 0.330 | 0.230 | 61.9 | 15.2 | 3.43 | 7.97 | 3.04 | 1.23 |
| × 15 | 4.44 | 8.11 | 4.015 | 0.315 | 0.245 | 48.0 | 11.8 | 3.29 | 3.41 | 1.70 | 0.876 |
| × 13 | 3.84 | 7.99 | 4.000 | 0.255 | 0.230 | 39.6 | 9.91 | 3.21 | 2.73 | 1.37 | 0.843 |
| × 10 | 2.96 | 7.89 | 3.940 | 0.205 | 0.170 | 30.8 | 7.81 | 3.22 | 2.09 | 1.06 | 0.841 |
| W 6 × 25 | 7.34 | 6.38 | 6.080 | 0.455 | 0.320 | 53.4 | 16.7 | 2.70 | 17.1 | 5.61 | 1.52 |
| × 20 | 5.87 | 6.20 | 6.020 | 0.365 | 0.260 | 41.4 | 13.4 | 2.66 | 13.3 | 4.41 | 1.50 |
| × 16 | 4.74 | 6.28 | 4.030 | 0.405 | 0.260 | 32.1 | 10.2 | 2.60 | 4.43 | 2.20 | 0.966 |
| × 15 | 4.43 | 5.99 | 5.990 | 0.260 | 0.230 | 29.1 | 9.72 | 2.56 | 9.32 | 3.11 | [illegible].46 |
| × 12 | 3.55 | 6.03 | 4.000 | 0.280 | 0.230 | 22.1 | 7.31 | 2.49 | 2.99 | 1.50 | 0.918 |
| × 9 | 2.68 | 5.90 | 3.940 | 0.215 | 0.170 | 16.4 | 5.56 | 2.47 | 2.19 | 1.11 | 0.905 |
| W 5 × 19 | 5.54 | 5.15 | 5.030 | 0.430 | 0.270 | 26.2 | 10.2 | 2.17 | 9.13 | 3.63 | 1.28 |
| × 16 | 4.68 | 5.01 | 5.000 | 0.360 | 0.240 | 21.3 | 8.51 | 2.13 | 7.51 | 3.00 | 1.27 |
| W 4 × 13 | 3.83 | 4.16 | 4.060 | 0.345 | 0.280 | 11.3 | 5.46 | 1.72 | 3.86 | 1.90 | 1.00 |

*Symbols:* I = moment of inertia; S = section modulus; r = radius of gyration.

Data taken from the "Manual of Steel Construction," 8th Edition, 1980, with permission of the American Institute of Steel Construction.

## Table 3. Steel S Sections

"S" is the section symbol for "I" beams. S shapes are designated, in order, by their section letter, actual depth in inches, and nominal weight in pounds per foot. Thus:

**S 5 × 14.75**

indicates an S shape (or I beam) having a depth of 5 inches and a nominal weight of 14.75 pounds per foot.

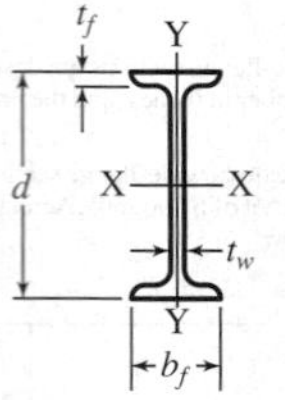


| Designation | Area $A$ inch² | Depth, $d$ inch | Flange Width, $b_f$ inch | Flange Thickness, $t_f$ inch | Web Thickness, $t_w$ inch | Axis XX $I$ inch⁴ | Axis XX $S$ inch³ | Axis XX $r$ inch | Axis YY $I$ inch⁴ | Axis YY $S$ inch³ | Axis YY $r$ inch |
|---|---|---|---|---|---|---|---|---|---|---|---|
| S 24 × 121 | 35.6 | 24.50 | 8.050 | 1.090 | 0.800 | 3160 | 258 | 9.43 | 83.3 | 20.7 | 1.53 |
| × 106 | 31.2 | 24.50 | 7.870 | 1.090 | 0.620 | 2940 | 240 | 9.71 | 77.1 | 19.6 | 1.57 |
| ×100 | 29.3 | 24.00 | 7.245 | 0.870 | 0.745 | 2390 | 199 | 9.02 | 47.7 | 13.2 | 1.27 |
| × 90 | 26.5 | 24.00 | 7.125 | 0.870 | 0.625 | 2250 | 187 | 9.21 | 44.9 | 12.6 | 1.30 |
| × 80 | 23.5 | 24.00 | 7.000 | 0.870 | 0.500 | 2100 | 175 | 9.47 | 42.2 | 12.1 | 1.34 |
| S 20 × 96 | 28.2 | 20.30 | 7.200 | 0.920 | 0.800 | 1670 | 165 | 7.71 | 50.2 | 13.9 | 1.33 |
| × 86 | 25.3 | 20.30 | 7.060 | 0.920 | 0.660 | 1580 | 155 | 7.89 | 46.8 | 13.3 | 1.36 |
| × 75 | 22.0 | 20.00 | 6.385 | 0.795 | 0.635 | 1280 | 128 | 7.62 | 29.8 | 9.32 | 1.16 |
| × 66 | 19.4 | 20.00 | 6.255 | 0.795 | 0.505 | 1190 | 119 | 7.83 | 27.7 | 8.85 | 1.19 |
| S 18 × 70 | 20.6 | 18.00 | 6.251 | 0.691 | 0.711 | 926 | 103 | 6.71 | 24.1 | 7.72 | 1.08 |
| × 54.7 | 16.1 | 18.00 | 6.001 | 0.691 | 0.461 | 804 | 89.4 | 7.07 | 20.8 | 6.94 | 1.14 |
| S 15 × 50 | 14.7 | 15.00 | 5.640 | 0.622 | 0.550 | 486 | 64.8 | 5.75 | 15.7 | 5.57 | 1.03 |
| × 42.9 | 12.6 | 15.00 | 5.501 | 0.622 | 0.411 | 447 | 59.6 | 5.95 | 14.4 | 5.23 | 1.07 |
| S 12 × 50 | 14.7 | 12.00 | 5.477 | 0.659 | 0.687 | 305 | 50.8 | 4.55 | 15.7 | 5.74 | 1.03 |
| × 40.8 | 12.0 | 12.00 | 5.252 | 0.659 | 0.462 | 272 | 45.4 | 4.77 | 13.6 | 5.16 | 1.06 |
| × 35 | 10.3 | 12.00 | 5.078 | 0.544 | 0.428 | 229 | 38.2 | 4.72 | 9.87 | 3.89 | 0.980 |
| × 31.8 | 9.35 | 12.00 | 5.000 | 0.544 | 0.350 | 218 | 36.4 | 4.83 | 9.36 | 3.74 | 1.00 |
| S 10 × 35 | 10.3 | 10.00 | 4.944 | 0.491 | 0.594 | 147 | 29.4 | 3.78 | 8.36 | 3.38 | 0.901 |
| × 25.4 | 7.46 | 10.00 | 4.661 | 0.491 | 0.311 | 124 | 24.7 | 4.07 | 6.79 | 2.91 | 0.954 |
| S 8 × 23 | 6.77 | 8.00 | 4.171 | 0.426 | 0.441 | 64.9 | 16.2 | 3.10 | 4.31 | 2.07 | 0.798 |
| × 18.4 | 5.41 | 8.00 | 4.001 | 0.426 | 0.271 | 57.6 | 14.4 | 3.26 | 3.73 | 1.86 | 0.831 |
| S 7 × 20 | 5.88 | 7.00 | 3.860 | 0.392 | 0.450 | 42.4 | 12.1 | 2.69 | 3.17 | 1.64 | 0.734 |
| × 15.3 | 4.50 | 7.00 | 3.662 | 0.392 | 0.252 | 36.7 | 10.5 | 2.86 | 2.64 | 1.44 | 0.766 |
| S 6 × 17.25 | 5.07 | 6.00 | 3.565 | 0.359 | 0.465 | 26.3 | 8.77 | 2.28 | 2.31 | 1.30 | 0.675 |
| × 12.5 | 3.67 | 6.00 | 3.332 | 0.359 | 0.232 | 22.1 | 7.37 | 2.45 | 1.82 | 1.09 | 0.705 |
| S 5 × 14.75 | 4.34 | 5.00 | 3.284 | 0.326 | 0.494 | 15.2 | 6.09 | 1.87 | 1.67 | 1.01 | 0.620 |
| × 10 | 2.94 | 5.00 | 3.004 | 0.326 | 0.214 | 12.3 | 4.92 | 2.05 | 1.22 | 0.809 | 0.643 |
| S 4 × 9.5 | 2.79 | 4.00 | 2.796 | 0.293 | 0.326 | 6.79 | 3.39 | 1.56 | 0.903 | 0.646 | 0.569 |
| × 7.7 | 2.26 | 4.00 | 2.663 | 0.293 | 0.193 | 6.08 | 3.04 | 1.64 | 0.764 | 0.574 | 0.581 |
| S 3 × 7.5 | 2.21 | 3.00 | 2.509 | 0.260 | 0.349 | 2.93 | 1.95 | 1.15 | 0.586 | 0.468 | 0.516 |
| × 5.7 | 1.67 | 3.00 | 2.330 | 0.260 | 0.170 | 2.52 | 1.68 | 1.23 | 0.455 | 0.390 | 0.522 |

Data taken from the "Manual of Steel Construction," 8th Edition, 1980, with permission of the American Institute of Steel Construction.

## Table 4. American Standard Steel Channels

American Standard Channels are designated, in order, by a section letter, actual depth in inches, and nominal weight per foot in pounds. Thus:

**C 7 × 14.75**

indicates an American Standard Channel with a depth of 7 inches and a nominal weight of 14.75 pounds per foot.

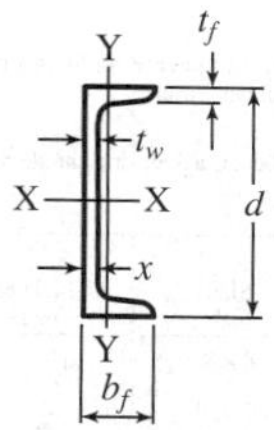


| | | | Flange | | | Axis XX | | | Axis YY | | | |
|---|---|---|---|---|---|---|---|---|---|---|---|---|
| Designation | Area, $A$ inch² | Depth, $d$ inch | Width, $b_f$ inch | Thickness, $t_f$ inch | Web Thickness, $t_w$ inch | $I$ inch⁴ | $S$ inch³ | $r$ inch | $I$ inch⁴ | $S$ inch³ | $r$ inch | $x$ inch |
| C 15 × 50 | 14.7 | 15.00 | 3.716 | 0.650 | 0.716 | 404 | 53.8 | 5.24 | 11.0 | 3.78 | 0.867 | 0.798 |
| × 40 | 11.8 | 15.00 | 3.520 | 0.650 | 0.520 | 349 | 46.5 | 5.44 | 9.23 | 3.37 | 0.886 | 0.777 |
| × 33.9 | 9.96 | 15.00 | 3.400 | 0.650 | 0.400 | 315 | 42.0 | 5.62 | 8.13 | 3.11 | 0.904 | 0.787 |
| C 12 × 30 | 8.82 | 12.00 | 3.170 | 0.501 | 0.510 | 162 | 27.0 | 4.29 | 5.14 | 2.06 | 0.763 | 0.674 |
| × 25 | 7.35 | 12.00 | 3.047 | 0.501 | 0.387 | 144 | 24.1 | 4.43 | 4.47 | 1.88 | 0.780 | 0.674 |
| × 20.7 | 6.09 | 12.00 | 2.942 | 0.501 | 0.282 | 129 | 21.5 | 4.61 | 3.88 | 1.73 | 0.799 | 0.698 |
| C 10 × 30 | 8.82 | 10.00 | 3.033 | 0.436 | 0.673 | 103 | 20.7 | 3.42 | 3.94 | 1.65 | 0.669 | 0.649 |
| × 25 | 7.35 | 10.00 | 2.886 | 0.436 | 0.526 | 91.2 | 18.2 | 3.52 | 3.36 | 1.48 | 0.676 | 0.617 |
| × 20 | 5.88 | 10.00 | 2.739 | 0.436 | 0.379 | 78.9 | 15.8 | 3.66 | 2.81 | 1.32 | 0.692 | 0.606 |
| × 15.3 | 4.49 | 10.00 | 2.600 | 0.436 | 0.240 | 67.4 | 13.5 | 3.87 | 2.28 | 1.16 | 0.713 | 0.634 |
| C 9 × 20 | 5.88 | 9.00 | 2.648 | 0.413 | 0.448 | 60.9 | 13.5 | 3.22 | 2.42 | 1.17 | 0.642 | 0.583 |
| × 15 | 4.41 | 9.00 | 2.485 | 0.413 | 0.285 | 51.0 | 11.3 | 3.40 | 1.93 | 1.01 | 0.661 | 0.586 |
| × 13.4 | 3.94 | 9.00 | 2.433 | 0.413 | 0.233 | 47.9 | 10.6 | 3.48 | 1.76 | 0.962 | 0.669 | 0.601 |
| C 8 × 18.75 | 5.51 | 8.00 | 2.527 | 0.390 | 0.487 | 44.0 | 11.0 | 2.82 | 1.98 | 1.01 | 0.599 | 0.565 |
| × 13.75 | 4.04 | 8.00 | 2.343 | 0.390 | 0.303 | 36.1 | 9.03 | 2.99 | 1.53 | 0.854 | 0.615 | 0.553 |
| × 11.5 | 3.38 | 8.00 | 2.260 | 0.390 | 0.220 | 32.6 | 8.14 | 3.11 | 1.32 | 0.781 | 0.625 | 0.571 |
| C 7 × 14.75 | 4.33 | 7.00 | 2.299 | 0.366 | 0.419 | 27.2 | 7.78 | 2.51 | 1.38 | 0.779 | 0.564 | 0.532 |
| × 12.25 | 3.60 | 7.00 | 2.194 | 0.366 | 0.314 | 24.2 | 6.93 | 2.60 | 1.17 | 0.703 | 0.571 | 0.525 |
| × 9.8 | 2.87 | 7.00 | 2.090 | 0.366 | 0.210 | 21.3 | 6.08 | 2.72 | 0.968 | 0.625 | 0.581 | 0.540 |
| C 6 × 13 | 3.83 | 6.00 | 2.157 | 0.343 | 0.437 | 17.4 | 5.80 | 2.13 | 1.05 | 0.642 | 0.525 | 0.514 |
| × 10.5 | 3.09 | 6.00 | 2.034 | 0.343 | 0.314 | 15.2 | 5.06 | 2.22 | 0.866 | 0.564 | 0.529 | 0.499 |
| × 8.2 | 2.40 | 6.00 | 1.920 | 0.343 | 0.200 | 13.1 | 4.38 | 2.34 | 0.693 | 0.492 | 0.537 | 0.511 |
| C 5 × 9 | 2.64 | 5.00 | 1.885 | 0.320 | 0.325 | 8.90 | 3.56 | 1.83 | 0.632 | 0.450 | 0.489 | 0.478 |
| × 6.7 | 1.97 | 5.00 | 1.750 | 0.320 | 0.190 | 7.49 | 3.00 | 1.95 | 0.479 | 0.378 | 0.493 | 0.484 |
| C 4 × 7.25 | 2.13 | 4.00 | 1.721 | 0.296 | 0.321 | 4.59 | 2.29 | 1.47 | 0.433 | 0.343 | 0.450 | 0.459 |
| × 5.4 | 1.59 | 4.00 | 1.584 | 0.296 | 0.184 | 3.85 | 1.93 | 1.56 | 0.319 | 0.283 | 0.449 | 0.457 |
| C 3 × 6 | 1.76 | 3.00 | 1.596 | 0.273 | 0.356 | 2.07 | 1.38 | 1.08 | 0.305 | 0.268 | 0.416 | 0.455 |
| × 5 | 1.47 | 3.00 | 1.498 | 0.273 | 0.258 | 1.85 | 1.24 | 1.12 | 0.247 | 0.233 | 0.410 | 0.438 |
| × 4.1 | 1.21 | 3.00 | 1.410 | 0.273 | 0.170 | 1.66 | 1.10 | 1.17 | 0.197 | 0.202 | 0.404 | 0.436 |

*Symbols:* $I$ = moment of inertia; $S$ = section modulus; $r$ = radius of gyration; $x$ = distance from center of gravity of section to outer face of structural shape.

Data taken from the "Manual of Steel Construction," 8th Edition, 1980, with permission of the American Institute of Steel Construction.

## Table 5. Steel Angles with Equal Legs

These angles are commonly designated by section symbol, width of each leg, and thickness, thus:

**L 3 × 3 × 1/4**

indicates a 3 × 3-inch angle of 1/4-inch thickness.

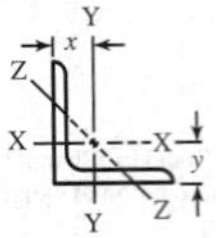


| Size inch | Thickness inch | Weight per Foot lb. | Area inch² | XX and YY Axes: $I$ inch⁴ | XX and YY Axes: $r$ inch | XX and YY Axes: $x$ or $y$ inch | Axis ZZ: $r$ inch |
|---|---|---|---|---|---|---|---|
| 8 × 8 | 1 1/8 | 56.9 | 16.7 | 98.0 | 2.42 | 2.41 | 1.56 |
| | 1 | 51.0 | 15.0 | 89.0 | 2.44 | 2.37 | 1.56 |
| | 7/8 | 45.0 | 13.2 | 79.6 | 2.45 | 2.32 | 1.57 |
| | 3/4 | 38.9 | 11.4 | 69.7 | 2.47 | 2.28 | 1.58 |
| | 5/8 | 32.7 | 9.61 | 59.4 | 2.49 | 2.23 | 1.58 |
| | 9/16 | 29.6 | 8.68 | 54.1 | 2.50 | 2.21 | 1.59 |
| | 1/2 | 26.4 | 7.75 | 48.6 | 2.50 | 2.19 | 1.59 |
| 6 × 6 | 1 | 37.4 | 11.00 | 35.5 | 1.80 | 1.86 | 1.17 |
| | 7/8 | 33.1 | 9.73 | 31.9 | 1.81 | 1.82 | 1.17 |
| | 3/4 | 28.7 | 8.44 | 28.2 | 1.83 | 1.78 | 1.17 |
| | 5/8 | 24.2 | 7.11 | 24.2 | 1.84 | 1.73 | 1.18 |
| | 9/16 | 21.9 | 6.43 | 22.1 | 1.85 | 1.71 | 1.18 |
| | 1/2 | 19.6 | 5.75 | 19.9 | 1.86 | 1.68 | 1.18 |
| | 7/16 | 17.2 | 5.06 | 17.7 | 1.87 | 1.66 | 1.19 |
| | 3/8 | 14.9 | 4.36 | 15.4 | 1.88 | 1.64 | 1.19 |
| | 5/16 | 12.4 | 3.65 | 13.0 | 1.89 | 1.62 | 1.20 |
| 5 × 5 | 7/8 | 27.2 | 7.98 | 17.8 | 1.49 | 1.57 | .973 |
| | 3/4 | 23.6 | 6.94 | 15.7 | 1.51 | 1.52 | .975 |
| | 5/8 | 20.0 | 5.86 | 13.6 | 1.52 | 1.48 | .978 |
| | 1/2 | 16.2 | 4.75 | 11.3 | 1.54 | 1.43 | .983 |
| | 7/16 | 14.3 | 4.18 | 10.0 | 1.55 | 1.41 | .986 |
| | 3/8 | 12.3 | 3.61 | 8.74 | 1.56 | 1.39 | .990 |
| | 5/16 | 10.3 | 3.03 | 7.42 | 1.57 | 1.37 | .994 |
| 4 × 4 | 3/4 | 18.5 | 5.44 | 7.67 | 1.19 | 1.27 | .778 |
| | 5/8 | 15.7 | 4.61 | 6.66 | 1.20 | 1.23 | .779 |
| | 1/2 | 12.8 | 3.75 | 5.56 | 1.22 | 1.18 | .782 |
| | 7/16 | 11.3 | 3.31 | 4.97 | 1.23 | 1.16 | .785 |
| | 3/8 | 9.8 | 2.86 | 4.36 | 1.23 | 1.14 | .788 |
| | 5/16 | 8.2 | 2.40 | 3.71 | 1.24 | 1.12 | .791 |
| | 1/4 | 6.6 | 1.94 | 3.04 | 1.25 | 1.09 | .795 |
| 3 1/2 × 3 1/2 | 1/2 | 11.1 | 3.25 | 3.64 | 1.06 | 1.06 | .683 |
| | 7/16 | 9.8 | 2.87 | 3.26 | 1.07 | 1.04 | .684 |
| | 3/8 | 8.5 | 2.48 | 2.87 | 1.07 | 1.01 | .687 |
| | 5/16 | 7.2 | 2.09 | 2.45 | 1.08 | .990 | .690 |
| | 1/4 | 5.8 | 1.69 | 2.01 | 1.09 | .968 | .694 |
| 3 × 3 | 1/2 | 9.4 | 2.75 | 2.22 | .898 | .932 | .584 |
| | 7/16 | 8.3 | 2.43 | 1.99 | .905 | .910 | .585 |
| | 3/8 | 7.2 | 2.11 | 1.76 | .913 | .888 | .587 |
| | 5/16 | 6.1 | 1.78 | 1.51 | .922 | .865 | .589 |
| | 1/4 | 4.9 | 1.44 | 1.24 | .930 | .842 | .592 |
| | 3/16 | 3.71 | 1.09 | .962 | .939 | .820 | .596 |
| 2 1/2 × 2 1/2 | 1/2 | 7.7 | 2.25 | 1.23 | .739 | .806 | .487 |
| | 3/8 | 5.9 | 1.73 | .984 | .753 | .762 | .487 |
| | 5/16 | 5.0 | 1.46 | .849 | .761 | .740 | .489 |
| | 1/4 | 4.1 | 1.19 | .703 | .769 | .717 | .491 |
| | 3/16 | 3.07 | .902 | .547 | .778 | .694 | .495 |
| 2 × 2 | 3/8 | 4.7 | 1.36 | .479 | .594 | .636 | .389 |
| | 5/16 | 3.92 | 1.15 | .416 | .601 | .614 | .390 |
| | 1/4 | 3.19 | .938 | .348 | .609 | .592 | .391 |
| | 3/16 | 2.44 | .715 | .272 | .617 | .569 | .394 |
| | 1/8 | 1.65 | .484 | .190 | .626 | .546 | .398 |

Data taken from the "Manual of Steel Construction," 8th Edition, 1980, with permission of the American Institute of Steel Construction.

## Table 6. Steel Angles with Unequal Legs

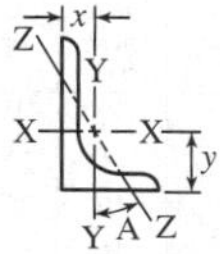


These angles are commonly designated by section symbol, width of each leg, and thickness, thus:

$$L\,7 \times 4 \times \tfrac{1}{2}$$

indicates a 7 × 4-inch angle of ½-inch thickness.

| Size inch | Thick-ness inch | Weight per Ft. lb. | Area inch² | Axis XX | | | | Axis YY | | | Axis ZZ | | |
|---|---|---|---|---|---|---|---|---|---|---|---|---|---|
| | | | | *I* inch⁴ | *S* inch³ | *r* inch | *y* inch | *I* inch⁴ | *S* inch³ | *r* inch | *x* inch | *r* inch | Tan *A* |
| 9 × 4 | ⅝ | 26.3 | 7.73 | 64.9 | 11.5 | 2.90 | 3.36 | 8.32 | 2.65 | 1.04 | .858 | .847 | .215 |
| | 9⁄16 | 23.8 | 7.00 | 59.1 | 10.4 | 2.91 | 3.33 | 7.63 | 2.41 | 1.04 | .834 | .850 | .218 |
| | ½ | 21.3 | 6.25 | 53.2 | 9.34 | 2.92 | 3.31 | 6.92 | 2.17 | 1.05 | .810 | .854 | .220 |
| 8 × 6 | 1 | 44.2 | 13.0 | 80.8 | 15.1 | 2.49 | 2.65 | 38.8 | 8.92 | 1.73 | 1.65 | 1.28 | .543 |
| | ⅞ | 39.1 | 11.5 | 72.3 | 13.4 | 2.51 | 2.61 | 34.9 | 7.94 | 1.74 | 1.61 | 1.28 | .547 |
| | ¾ | 33.8 | 9.94 | 63.4 | 11.7 | 2.53 | 2.56 | 30.7 | 6.92 | 1.76 | 1.56 | 1.29 | .551 |
| | ⅝ | 28.5 | 8.36 | 54.1 | 9.37 | 2.54 | 2.52 | 26.3 | 5.88 | 1.77 | 1.52 | 1.29 | .554 |
| | 9⁄16 | 25.7 | 7.56 | 49.3 | 8.95 | 2.55 | 2.50 | 24.0 | 5.34 | 1.78 | 1.50 | 1.30 | .556 |
| | ½ | 23.0 | 6.75 | 44.3 | 8.02 | 2.56 | 2.47 | 21.7 | 4.79 | 1.79 | 1.47 | 1.30 | .558 |
| | 7⁄16 | 20.2 | 5.93 | 39.2 | 7.07 | 2.57 | 2.45 | 19.3 | 4.23 | 1.80 | 1.45 | 1.31 | .560 |
| 8 × 4 | 1 | 37.4 | 11.0 | 69.6 | 14.1 | 2.52 | 3.05 | 11.6 | 3.94 | 1.03 | 1.05 | .846 | .247 |
| | ¾ | 28.7 | 8.44 | 54.9 | 10.9 | 2.55 | 2.95 | 9.36 | 3.07 | 1.05 | .953 | .852 | .258 |
| | 9⁄16 | 21.9 | 6.43 | 42.8 | 8.35 | 2.58 | 2.88 | 7.43 | 2.38 | 1.07 | .882 | .861 | .265 |
| | ½ | 19.6 | 5.75 | 38.5 | 7.49 | 2.59 | 2.86 | 6.74 | 2.15 | 1.08 | .859 | .865 | .267 |
| 7 × 4 | ¾ | 26.2 | 7.69 | 37.8 | 8.42 | 2.22 | 2.51 | 9.05 | 3.03 | 1.09 | 1.01 | .860 | .224 |
| | ⅝ | 22.1 | 6.48 | 32.4 | 7.14 | 2.24 | 2.46 | 7.84 | 2.58 | 1.10 | .963 | .865 | .229 |
| | ½ | 17.9 | 5.25 | 26.7 | 5.81 | 2.25 | 2.42 | 6.53 | 2.12 | 1.11 | .917 | .872 | .235 |
| | ⅜ | 13.6 | 3.98 | 20.6 | 4.44 | 2.27 | 2.37 | 5.10 | 1.63 | 1.13 | .870 | .880 | .240 |
| 6 × 4 | ⅞ | 27.2 | 7.98 | 27.7 | 7.15 | 1.86 | 2.12 | 9.75 | 3.39 | 1.11 | 1.12 | .857 | .421 |
| | ¾ | 23.6 | 6.94 | 24.5 | 6.25 | 1.88 | 2.08 | 8.68 | 2.97 | 1.12 | 1.08 | .860 | .428 |
| | ⅝ | 20.0 | 5.86 | 21.1 | 5.31 | 1.90 | 2.03 | 7.52 | 2.54 | 1.13 | 1.03 | .864 | .435 |
| | 9⁄16 | 18.1 | 5.31 | 19.3 | 4.83 | 1.90 | 2.01 | 6.91 | 2.31 | 1.14 | 1.01 | .866 | .438 |
| | ½ | 16.2 | 4.75 | 17.4 | 4.33 | 1.91 | 1.99 | 6.27 | 2.08 | 1.15 | .987 | .870 | .440 |
| | 7⁄16 | 14.3 | 4.18 | 15.5 | 3.83 | 1.92 | 1.96 | 5.60 | 1.85 | 1.16 | .964 | .873 | .443 |
| | ⅜ | 12.3 | 3.61 | 13.5 | 3.32 | 1.93 | 1.94 | 4.90 | 1.60 | 1.17 | .941 | .877 | .446 |
| | 5⁄16 | 10.3 | 3.03 | 11.4 | 2.79 | 1.94 | 1.92 | 4.18 | 1.35 | 1.17 | .918 | .882 | .448 |
| 6 × 3½ | ½ | 15.3 | 4.50 | 16.6 | 4.24 | 1.92 | 2.08 | 4.25 | 1.59 | .972 | .833 | .759 | .344 |
| | ⅜ | 11.7 | 3.42 | 12.9 | 3.24 | 1.94 | 2.04 | 3.34 | 1.23 | .988 | .787 | .676 | .350 |
| | 5⁄16 | 9.8 | 2.87 | 10.9 | 2.73 | 1.95 | 2.01 | 2.85 | 1.04 | .996 | .763 | .772 | .352 |
| 5 × 3½ | ¾ | 19.8 | 5.81 | 13.9 | 4.28 | 1.55 | 1.75 | 5.55 | 2.22 | .977 | .996 | .748 | .464 |
| | ⅝ | 16.8 | 4.92 | 12.0 | 3.65 | 1.56 | 1.70 | 4.83 | 1.90 | .991 | .951 | .751 | .472 |
| | ½ | 13.6 | 4.00 | 9.99 | 2.99 | 1.58 | 1.66 | 4.05 | 1.56 | 1.01 | .906 | .755 | .479 |
| | 7⁄16 | 12.0 | 3.53 | 8.90 | 2.64 | 1.59 | 1.63 | 3.63 | 1.39 | 1.01 | .883 | .758 | .482 |
| | ⅜ | 10.4 | 3.05 | 7.78 | 2.29 | 1.60 | 1.61 | 3.18 | 1.21 | 1.02 | .861 | .762 | .486 |
| | 5⁄16 | 8.7 | 2.56 | 6.60 | 1.94 | 1.61 | 1.59 | 2.72 | 1.02 | 1.03 | .838 | .766 | .489 |
| | ¼ | 7.0 | 2.06 | 5.39 | 1.57 | 1.62 | 1.56 | 2.23 | .830 | 1.04 | .814 | .770 | .492 |
| 5 × 3 | ⅝ | 15.7 | 4.61 | 11.4 | 3.55 | 1.57 | 1.80 | 3.06 | 1.39 | .815 | .796 | .644 | .349 |
| | ½ | 12.8 | 3.75 | 9.45 | 2.91 | 1.59 | 1.75 | 2.58 | 1.15 | .829 | .750 | .648 | .357 |

**Table 6.** *(Continued)* **Steel Angles with Unequal Legs**

| Size inch | Thickness inch | Weight per Ft. lb. | Area inch² | Axis XX and Axis YY | | | | Axis YY | | | Axis ZZ | | |
|---|---|---|---|---|---|---|---|---|---|---|---|---|---|
| | | | | $I$ inch⁴ | $S$ inch³ | $r$ inch | $y$ inch | $I$ inch⁴ | $S$ inch³ | $r$ inch | $x$ inch | $r$ inch | Tan $A$ |
| 5×3 | 7/16 | 11.3 | 3.31 | 8.43 | 2.58 | 1.60 | 1.73 | 2.32 | 1.02 | .837 | .727 | .651 | .361 |
| | 3/8 | 9.8 | 2.86 | 7.37 | 2.24 | 1.61 | 1.70 | 2.04 | .888 | .845 | .704 | .654 | .364 |
| | 5/16 | 8.2 | 2.40 | 6.26 | 1.89 | 1.61 | 1.68 | 1.75 | .753 | .853 | .681 | .658 | .368 |
| | 1/4 | 6.6 | 1.94 | 5.11 | 1.53 | 1.62 | 1.66 | 1.44 | .614 | .861 | .657 | .663 | .371 |
| 4×3½ | 5/8 | 14.7 | 4.30 | 6.37 | 2.35 | 1.22 | 1.29 | 4.52 | 1.84 | 1.03 | 1.04 | .719 | .745 |
| | 1/2 | 11.9 | 3.50 | 5.32 | 1.94 | 1.23 | 1.25 | 3.79 | 1.52 | 1.04 | 1.00 | .722 | .750 |
| | 7/16 | 10.6 | 3.09 | 4.76 | 1.72 | 1.24 | 1.23 | 3.40 | 1.35 | 1.05 | .978 | .724 | .753 |
| | 3/8 | 9.1 | 2.67 | 4.18 | 1.49 | 1.25 | 1.21 | 2.95 | 1.17 | 1.06 | .955 | .727 | .755 |
| | 5/16 | 7.7 | 2.25 | 3.56 | 1.26 | 1.26 | 1.18 | 2.55 | .994 | 1.07 | .932 | .730 | .757 |
| | 1/4 | 6.2 | 1.81 | 2.91 | 1.03 | 1.27 | 1.16 | 2.09 | .808 | 1.07 | .909 | .734 | .759 |
| 4×3 | 5/8 | 13.6 | 3.98 | 6.03 | 2.30 | 1.23 | 1.37 | 2.87 | 1.35 | .849 | .871 | .637 | .534 |
| | 1/2 | 11.1 | 3.25 | 5.05 | 1.89 | 1.25 | 1.33 | 2.42 | 1.12 | .864 | .827 | .639 | .543 |
| | 7/16 | 9.8 | 2.87 | 4.52 | 1.68 | 1.25 | 1.30 | 2.18 | .992 | .871 | .804 | .641 | .547 |
| | 3/8 | 8.5 | 2.48 | 3.96 | 1.46 | 1.26 | 1.28 | 1.92 | .866 | .879 | .782 | .644 | .551 |
| | 5/16 | 7.2 | 2.09 | 3.38 | 1.23 | 1.27 | 1.26 | 1.65 | .734 | .887 | .759 | .647 | .554 |
| | 1/4 | 5.8 | 1.69 | 2.77 | 1.00 | 1.28 | 1.24 | 1.36 | .599 | .896 | .736 | .651 | .558 |
| 3½×3 | 1/2 | 10.2 | 3.00 | 3.45 | 1.45 | 1.07 | 1.13 | 2.33 | 1.10 | .881 | .875 | .621 | .714 |
| | 7/16 | 9.1 | 2.65 | 3.10 | 1.29 | 1.08 | 1.10 | 2.09 | .975 | .889 | .853 | .622 | .718 |
| | 3/8 | 7.9 | 2.30 | 2.72 | 1.13 | 1.09 | 1.08 | 1.85 | .851 | .897 | .830 | .625 | .721 |
| | 5/16 | 6.6 | 1.93 | 2.33 | .954 | 1.10 | 1.06 | 1.58 | .722 | .905 | .808 | .627 | .724 |
| | 1/4 | 5.4 | 1.56 | 1.91 | .776 | 1.11 | 1.04 | 1.30 | .589 | .914 | .785 | .631 | .727 |
| 3½×2½ | 1/2 | 9.4 | 2.75 | 3.24 | 1.41 | 1.09 | 1.20 | 1.36 | .760 | .704 | .705 | .534 | .486 |
| | 7/16 | 8.3 | 2.43 | 2.91 | 1.26 | 1.09 | 1.18 | 1.23 | .677 | .711 | .682 | .535 | .491 |
| | 3/8 | 7.2 | 2.11 | 2.56 | 1.09 | 1.10 | 1.16 | 1.09 | .592 | .719 | .660 | .537 | .496 |
| | 5/16 | 6.1 | 1.78 | 2.19 | .927 | 1.11 | 1.14 | .939 | .504 | .727 | .637 | .540 | .501 |
| | 1/4 | 4.9 | 1.44 | 1.80 | .755 | 1.12 | 1.11 | .777 | .412 | .735 | .614 | .544 | .506 |
| 3×2½ | 1/2 | 8.5 | 2.50 | 2.08 | 1.04 | .913 | 1.00 | 1.30 | .744 | .722 | .750 | .520 | .667 |
| | 7/16 | 7.6 | 2.21 | 1.88 | .928 | .920 | .978 | 1.18 | .664 | .729 | .728 | .521 | .672 |
| | 3/8 | 6.6 | 1.92 | 1.66 | .810 | .928 | .956 | 1.04 | .581 | .736 | .706 | .522 | .676 |
| | 5/16 | 5.6 | 1.62 | 1.42 | .688 | .937 | .933 | .898 | .494 | .744 | .683 | .525 | .680 |
| | 1/4 | 4.5 | 1.31 | 1.17 | .561 | .945 | .911 | .743 | .404 | .753 | .661 | .528 | .684 |
| | 3/16 | 3.39 | .996 | .907 | .430 | .954 | .888 | .577 | .310 | .761 | .638 | .533 | .688 |
| 3×2 | 1/2 | 7.7 | 2.25 | 1.92 | 1.00 | .924 | 1.08 | .672 | .474 | .546 | .583 | .428 | .414 |
| | 7/16 | 6.8 | 2.00 | 1.73 | .894 | .932 | 1.06 | .609 | .424 | .553 | .561 | .429 | .421 |
| | 3/8 | 5.9 | 1.73 | 1.53 | .781 | .940 | 1.04 | .543 | .371 | .559 | .539 | .430 | .428 |
| | 5/16 | 5.0 | 1.46 | 1.32 | .664 | .948 | 1.02 | .740 | .317 | .567 | .516 | .432 | .435 |
| | 1/4 | 4.1 | 1.19 | 1.09 | .542 | .957 | .993 | .392 | .260 | .574 | .493 | .435 | .440 |
| | 3/16 | 3.07 | .902 | .842 | .415 | .966 | .970 | .307 | .200 | .583 | .470 | .439 | .446 |
| 2½×2 | 3/8 | 5.3 | 1.55 | .912 | .547 | .768 | .831 | .514 | .363 | .577 | .581 | .420 | .614 |
| | 5/16 | 4.5 | 1.31 | .788 | .466 | .776 | .809 | .446 | .310 | .584 | .559 | .422 | .620 |
| | 1/4 | 3.62 | 1.06 | .654 | .381 | .784 | .787 | .372 | .254 | .592 | .537 | .424 | .626 |
| | 3/16 | 2.75 | .809 | .509 | .293 | .793 | .764 | .291 | .196 | .600 | .514 | .427 | .631 |

*Symbols:* $I$ = moment of inertia; $S$ = section modulus; $r$ = radius of gyration; $x$ = distance from center of gravity of section to outer face of structural shape.

Data taken from the "Manual of Steel Construction," 8th Edition, 1980, with permission of the American Institute of Steel Construction.

## Table 7. Aluminum Association Standard Structural Shapes

**I-Beams**

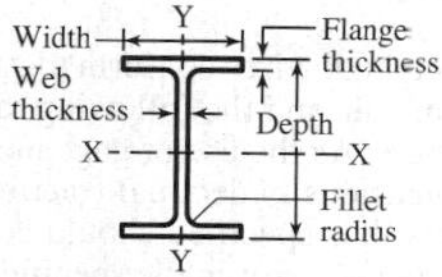


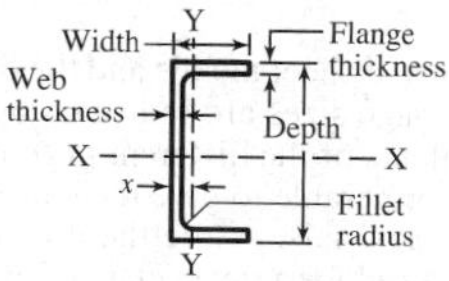


**Channels**

| Depth | Width | Weight per Foot | Area | Flange Thickness | Web Thickness | Fillet Radius | Axis XX *I* | Axis XX *S* | Axis XX *r* | Axis YY *I* | Axis YY *S* | Axis YY *r* | *x* |
|---|---|---|---|---|---|---|---|---|---|---|---|---|---|
| inch | inch | lb. | $inch^2$ | inch | inch | inch | $inch^4$ | $inch^3$ | inch | $inch^4$ | $inch^3$ | inch | inch |
| | | | | | | | I-BEAMS | | | | | | |
| 3.00 | 2.50 | 1.637 | 1.392 | 0.20 | 0.13 | 0.25 | 2.24 | 1.49 | 1.27 | 0.52 | 0.42 | 0.61 | ... |
| 3.00 | 2.50 | 2.030 | 1.726 | 0.26 | 0.15 | 0.25 | 2.71 | 1.81 | 1.25 | 0.68 | 0.54 | 0.63 | ... |
| 4.00 | 3.00 | 2.311 | 1.965 | 0.23 | 0.15 | 0.25 | 5.62 | 2.81 | 1.69 | 1.04 | 0.69 | 0.73 | ... |
| 4.00 | 3.00 | 2.793 | 2.375 | 0.29 | 0.17 | 0.25 | 6.71 | 3.36 | 1.68 | 1.31 | 0.87 | 0.74 | ... |
| 5.00 | 3.50 | 3.700 | 3.146 | 0.32 | 0.19 | 0.30 | 13.94 | 5.58 | 2.11 | 2.29 | 1.31 | 0.85 | ... |
| 6.00 | 4.00 | 4.030 | 3.427 | 0.29 | 0.19 | 0.30 | 21.99 | 7.33 | 2.53 | 3.10 | 1.55 | 0.95 | ... |
| 6.00 | 4.00 | 4.692 | 3.990 | 0.35 | 0.21 | 0.30 | 25.50 | 8.50 | 2.53 | 3.74 | 1.87 | 0.97 | ... |
| 7.00 | 4.50 | 5.800 | 4.932 | 0.38 | 0.23 | 0.30 | 42.89 | 12.25 | 2.95 | 5.78 | 2.57 | 1.08 | ... |
| 8.00 | 5.00 | 6.181 | 5.256 | 0.35 | 0.23 | 0.30 | 59.69 | 14.92 | 3.37 | 7.30 | 2.92 | 1.18 | ... |
| 8.00 | 5.00 | 7.023 | 5.972 | 0.41 | 0.25 | 0.30 | 67.78 | 16.94 | 3.37 | 8.55 | 3.42 | 1.20 | ... |
| 9.00 | 5.50 | 8.361 | 7.110 | 0.44 | 0.27 | 0.30 | 102.02 | 22.67 | 3.79 | 12.22 | 4.44 | 1.31 | ... |
| 10.00 | 6.00 | 8.646 | 7.352 | 0.41 | 0.25 | 0.40 | 132.09 | 26.42 | 4.24 | 14.78 | 4.93 | 1.42 | ... |
| 10.00 | 6.00 | 10.286 | 8.747 | 0.50 | 0.29 | 0.40 | 155.79 | 31.16 | 4.22 | 18.03 | 6.01 | 1.44 | ... |
| 12.00 | 7.00 | 11.672 | 9.925 | 0.47 | 0.29 | 0.40 | 255.57 | 42.60 | 5.07 | 26.90 | 7.69 | 1.65 | ... |
| 12.00 | 7.00 | 14.292 | 12.153 | 0.62 | 0.31 | 0.40 | 317.33 | 52.89 | 5.11 | 35.48 | 10.14 | 1.71 | ... |
| | | | | | | | CHANNELS | | | | | | |
| 2.00 | 1.00 | 0.577 | 0.491 | 0.13 | 0.13 | 0.10 | 0.288 | 0.288 | 0.766 | 0.045 | 0.064 | 0.303 | 0.298 |
| 2.00 | 1.25 | 1.071 | 0.911 | 0.26 | 0.17 | 0.15 | 0.546 | 0.546 | 0.774 | 0.139 | 0.178 | 0.391 | 0.471 |
| 3.00 | 1.50 | 1.135 | 0.965 | 0.20 | 0.13 | 0.25 | 1.41 | 0.94 | 1.21 | 0.22 | 0.22 | 0.47 | 0.49 |
| 3.00 | 1.75 | 1.597 | 1.358 | 0.26 | 0.17 | 0.25 | 1.97 | 1.31 | 1.20 | 0.42 | 0.37 | 0.55 | 0.62 |
| 4.00 | 2.00 | 1.738 | 1.478 | 0.23 | 0.15 | 0.25 | 3.91 | 1.95 | 1.63 | 0.60 | 0.45 | 0.64 | 0.65 |
| 4.00 | 2.25 | 2.331 | 1.982 | 0.29 | 0.19 | 0.25 | 5.21 | 2.60 | 1.62 | 1.02 | 0.69 | 0.72 | 0.78 |
| 5.00 | 2.25 | 2.212 | 1.881 | 0.26 | 0.15 | 0.30 | 7.88 | 3.15 | 2.05 | 0.98 | 0.64 | 0.72 | 0.73 |
| 5.00 | 2.75 | 3.089 | 2.627 | 0.32 | 0.19 | 0.30 | 11.14 | 4.45 | 2.06 | 2.05 | 1.14 | 0.88 | 0.95 |
| 6.00 | 2.50 | 2.834 | 2.410 | 0.29 | 0.17 | 0.30 | 14.35 | 4.78 | 2.44 | 1.53 | 0.90 | 0.80 | 0.79 |
| 6.00 | 3.25 | 4.030 | 3.427 | 0.35 | 0.21 | 0.30 | 21.04 | 7.01 | 2.48 | 3.76 | 1.76 | 1.05 | 1.12 |
| 7.00 | 2.75 | 3.205 | 2.725 | 0.29 | 0.17 | 0.30 | 22.09 | 6.31 | 2.85 | 2.10 | 1.10 | 0.88 | 0.84 |
| 7.00 | 3.50 | 4.715 | 4.009 | 0.38 | 0.21 | 0.30 | 33.79 | 9.65 | 2.90 | 5.13 | 2.23 | 1.13 | 1.20 |
| 8.00 | 3.00 | 4.147 | 3.526 | 0.35 | 0.19 | 0.30 | 37.40 | 9.35 | 3.26 | 3.25 | 1.57 | 0.96 | 0.93 |
| 8.00 | 3.75 | 5.789 | 4.923 | 0.41 | 0.25 | 0.35 | 52.69 | 13.17 | 3.27 | 7.13 | 2.82 | 1.20 | 1.22 |
| 9.00 | 3.25 | 4.983 | 4.237 | 0.35 | 0.23 | 0.35 | 54.41 | 12.09 | 3.58 | 4.40 | 1.89 | 1.02 | 0.93 |
| 9.00 | 4.00 | 6.970 | 5.927 | 0.44 | 0.29 | 0.35 | 78.31 | 17.40 | 3.63 | 9.61 | 3.49 | 1.27 | 1.25 |
| 10.00 | 3.50 | 6.136 | 5.218 | 0.41 | 0.25 | 0.35 | 83.22 | 16.64 | 3.99 | 6.33 | 2.56 | 1.10 | 1.02 |
| 10.00 | 4.25 | 8.360 | 7.109 | 0.50 | 0.31 | 0.40 | 116.15 | 23.23 | 4.04 | 13.02 | 4.47 | 1.35 | 1.34 |
| 12.00 | 4.00 | 8.274 | 7.036 | 0.47 | 0.29 | 0.40 | 159.76 | 26.63 | 4.77 | 11.03 | 3.86 | 1.25 | 1.14 |
| 12.00 | 5.00 | 11.822 | 10.053 | 0.62 | 0.35 | 0.45 | 239.69 | 39.95 | 4.88 | 25.74 | 7.60 | 1.60 | 1.61 |

Structural sections are available in 6061-T6 aluminum alloy. Data supplied by The Aluminum Association.

## Wire and Sheet Metal Gages

The thicknesses of sheet metals and the diameters of wires conform to various gaging systems. These gage sizes are indicated by numbers, and the following tables give the decimal equivalents of the different gage numbers. As the use of gage numbers can be confusing, it is preferable to give the exact dimensions in decimal fractions of an inch when ordering materials. While the dimensions thus specified should conform to the gage ordinarily used for a given class of material, any error in the specification due, for example, to the use of a table having "rounded off" or approximate equivalents, will be apparent to the manufacturer at the time the order is placed. Furthermore, the decimal method of indicating wire diameters and sheet metal thicknesses has the advantage of being self-explanatory, whereas arbitrary gage numbers are not. The decimal system of indicating gage sizes is now mostly used, and gage numbers are gradually becoming obsolete. Unfortunately, there is considerable variation in the use of different gages. For example, a gage ordinarily used for copper, brass, and other nonferrous materials, may at times be used for steel, and vice versa. The gages specified in the following are the ones ordinarily employed for the materials mentioned, but there are some minor exceptions and variations throughout different industries.

**Wire Gages.—**The wire gage system used by most steel producers in the United States is known by the name Steel Wire Gage or, to distinguish it from the Standard Wire Gage (S.W.G.) used in Great Britain, it is called the United States Steel Wire Gage. It is the same as the Washburn and Moen, American Steel and Wire Company, and Roebling wire gages. The name has the official sanction of the Bureau of Standards at Washington but is not legally effective. The only wire gage which has been recognized in Acts of Congress is the Birmingham Gage (also known as Stub's Iron Wire). The Birmingham Gage is, however, nearly obsolete in both the United States and Great Britain, where it originated. Copper and aluminum wires are specified in decimal fractions. They were formerly universally specified in the United States by the American or Brown & Sharpe Wire Gage. Music spring steel wire, one of the highest quality wires of several types used for mechanical springs, is specified by the piano or music wire gage.

In Great Britain one wire gage has been legalized. This is called the Standard Wire Gage (S.W.G.), formerly called Imperial Wire Gage.

**Gages for Rods.—**Steel wire rod sizes are designated by fractional or decimal parts of an inch and by the gage numbers of the United States Steel Wire Gage. Copper and aluminum rods are specified by decimal fractions and fractions. Drill rod may be specified in decimal fractions but in the carbon and alloy tool steel grades may also be specified in the Stub's Steel Wire Gage and in the high-speed steel drill rod grade may be specified by the Morse Twist Drill Gage (Manufacturers' Standard Gage for Twist Drills). For gage numbers with corresponding decimal equivalents see the tables of American Standard Straight Shank Twist Drills, for example, page 921, and Table 5a on page 2716.

**Gages for Wall Thicknesses of Tubing.—**At one time the Birmingham or Stub's Iron Wire Gage was used to specify the wall thickness of the following classes of tubing: seamless brass, seamless copper, seamless steel, and aluminum. The Brown & Sharpe Wire Gage was used for brazed brass and brazed copper tubing. Wall thicknesses are now specified by decimal parts of an inch but the wall thickness of steel pressure tubes and steel mechanical tubing may be specified by the Birmingham or Stub's Iron Wire Gage. In Great Britain the Standard Wire Gage (S.W.G.) is used to specify the wall thickness of some kinds of steel tubes.

## Table 1. Wire Gages in Approximate Decimals of an Inch

| No. of Wire Gage | American Wire or Brown & Sharpe Gage | Steel Wire Gage (US)[a] | British Standard Wire Gage (Imperial Wire Gage) | Music or Piano Wire Gage | Birmingham or Stub's Iron Wire Gage | Stub's Steel Wire Gage | No. of Wire Gage | Stub's Steel Wire Gage |
|---|---|---|---|---|---|---|---|---|
| 7/0 | … | 0.4900 | 0.5000 | … | … | … | 51 | 0.066 |
| 6/0 | 0.5800 | 0.4615 | 0.4640 | 0.004 | … | … | 52 | 0.063 |
| 5/0 | 0.5165 | 0.4305 | 0.4320 | 0.005 | 0.5000 | … | 53 | 0.058 |
| 4/0 | 0.4600 | 0.3938 | 0.4000 | 0.006 | 0.4540 | … | 54 | 0.055 |
| 3/0 | 0.4096 | 0.3625 | 0.3720 | 0.007 | 0.4250 | … | 55 | 0.050 |
| 2/0 | 0.3648 | 0.3310 | 0.3480 | 0.008 | 0.3800 | … | 56 | 0.045 |
| 1/0 | 0.3249 | 0.3065 | 0.3240 | 0.009 | 0.3400 | … | 57 | 0.042 |
| 1 | 0.2893 | 0.2830 | 0.3000 | 0.010 | 0.3000 | 0.227 | 58 | 0.041 |
| 2 | 0.2576 | 0.2625 | 0.2760 | 0.011 | 0.2840 | 0.219 | 59 | 0.040 |
| 3 | 0.2294 | 0.2437 | 0.2520 | 0.012 | 0.2590 | 0.212 | 60 | 0.039 |
| 4 | 0.2043 | 0.2253 | 0.2320 | 0.013 | 0.2380 | 0.207 | 61 | 0.038 |
| 5 | 0.1819 | 0.2070 | 0.2120 | 0.014 | 0.2200 | 0.204 | 62 | 0.037 |
| 6 | 0.1620 | 0.1920 | 0.1920 | 0.016 | 0.2030 | 0.201 | 63 | 0.036 |
| 7 | 0.1443 | 0.1770 | 0.1760 | 0.018 | 0.1800 | 0.199 | 64 | 0.035 |
| 8 | 0.1285 | 0.1620 | 0.1600 | 0.020 | 0.1650 | 0.197 | 65 | 0.033 |
| 9 | 0.1144 | 0.1483 | 0.1440 | 0.022 | 0.1480 | 0.194 | 66 | 0.032 |
| 10 | 0.1019 | 0.1350 | 0.1280 | 0.024 | 0.1340 | 0.191 | 67 | 0.031 |
| 11 | 0.0907 | 0.1205 | 0.1160 | 0.026 | 0.1200 | 0.188 | 68 | 0.030 |
| 12 | 0.0808 | 0.1055 | 0.1040 | 0.029 | 0.1090 | 0.185 | 69 | 0.029 |
| 13 | 0.0720 | 0.0915 | 0.0920 | 0.031 | 0.0950 | 0.182 | 70 | 0.027 |
| 14 | 0.0641 | 0.0800 | 0.0800 | 0.033 | 0.0830 | 0.180 | 71 | 0.026 |
| 15 | 0.0571 | 0.0720 | 0.0720 | 0.035 | 0.0720 | 0.178 | 72 | 0.024 |
| 16 | 0.0508 | 0.0625 | 0.0640 | 0.037 | 0.0650 | 0.175 | 73 | 0.023 |
| 17 | 0.0453 | 0.0540 | 0.0560 | 0.039 | 0.0580 | 0.172 | 74 | 0.022 |
| 18 | 0.0403 | 0.0475 | 0.0480 | 0.041 | 0.0490 | 0.168 | 75 | 0.020 |
| 19 | 0.0359 | 0.0410 | 0.0400 | 0.043 | 0.0420 | 0.164 | 76 | 0.018 |
| 20 | 0.0320 | 0.0348 | 0.0360 | 0.045 | 0.0350 | 0.161 | 77 | 0.016 |
| 21 | 0.0285 | 0.0318 | 0.0320 | 0.047 | 0.0320 | 0.157 | 78 | 0.015 |
| 22 | 0.0253 | 0.0286 | 0.0280 | 0.049 | 0.0280 | 0.155 | 79 | 0.014 |
| 23 | 0.0226 | 0.0258 | 0.0240 | 0.051 | 0.0250 | 0.153 | 80 | 0.013 |
| 24 | 0.0201 | 0.0230 | 0.0220 | 0.055 | 0.0220 | 0.151 | … | … |
| 25 | 0.0179 | 0.0204 | 0.0200 | 0.059 | 0.0200 | 0.148 | … | … |
| 26 | 0.0159 | 0.0181 | 0.0180 | 0.063 | 0.0180 | 0.146 | … | … |
| 27 | 0.0142 | 0.0173 | 0.0164 | 0.067 | 0.0160 | 0.143 | … | … |
| 28 | 0.0126 | 0.0162 | 0.0149 | 0.071 | 0.0140 | 0.139 | … | … |
| 29 | 0.0113 | 0.0150 | 0.0136 | 0.075 | 0.0130 | 0.134 | … | … |
| 30 | 0.0100 | 0.0140 | 0.0124 | 0.080 | 0.0120 | 0.127 | … | … |
| 31 | 0.00893 | 0.0132 | 0.0116 | 0.085 | 0.0100 | 0.120 | … | … |
| 32 | 0.00795 | 0.0128 | 0.0108 | 0.090 | 0.0090 | 0.115 | … | … |
| 33 | 0.00708 | 0.0118 | 0.0100 | 0.095 | 0.0080 | 0.112 | … | … |
| 34 | 0.00630 | 0.0104 | 0.0092 | 0.100 | 0.0070 | 0.110 | … | … |
| 35 | 0.00561 | 0.0095 | 0.0084 | 0.106 | 0.0050 | 0.108 | … | … |
| 36 | 0.00500 | 0.0090 | 0.0076 | 0.112 | 0.0040 | 0.106 | … | … |
| 37 | 0.00445 | 0.0085 | 0.0068 | 0.118 | … | 0.103 | … | … |
| 38 | 0.00396 | 0.0080 | 0.0060 | 0.124 | … | 0.101 | … | … |
| 39 | 0.00353 | 0.0075 | 0.0052 | 0.130 | … | 0.099 | … | … |
| 40 | 0.00314 | 0.0070 | 0.0048 | 0.138 | … | 0.097 | … | … |
| 41 | 0.00280 | 0.0066 | 0.0044 | 0.146 | … | 0.095 | … | … |
| 42 | 0.00249 | 0.0062 | 0.0040 | 0.154 | … | 0.092 | … | … |
| 43 | 0.00222 | 0.0060 | 0.0036 | 0.162 | … | 0.088 | … | … |
| 44 | 0.00198 | 0.0058 | 0.0032 | 0.170 | … | 0.085 | … | … |
| 45 | 0.00176 | 0.0055 | 0.0028 | 0.180 | … | 0.081 | … | … |
| 46 | 0.00157 | 0.0052 | 0.0024 | … | … | 0.079 | … | … |
| 47 | 0.00140 | 0.0050 | 0.0020 | … | … | 0.077 | … | … |
| 48 | 0.00124 | 0.0048 | 0.0016 | … | … | 0.075 | … | … |
| 49 | 0.00111 | 0.0046 | 0.0012 | … | … | 0.072 | … | … |
| 50 | 0.00099 | 0.0044 | 0.0010 | … | … | 0.069 | … | … |

[a] Also known as Washburn and Moen, American Steel and Wire Co. and Roebling wire gages. A greater selection of sizes is available and is specified by what are known as split gage numbers. They can be recognized by ½ fractions which follow the gage number; i.e., 4½. The decimal equivalents of split gage numbers are in the Steel Products Manual entitled: *Wire and Rods, Carbon Steel* published by the American Iron and Steel Institute, Washington, DC.

## Wire Gauge, Diameter, and Area of Solid Copper Wire[a]

| Gauge | Diameter (in.) | Area (in²) | Gauge | Diameter (in.) | Area (in²) |
|---|---|---|---|---|---|
| 0000 | 0.460000 | 0.1661901110 | 22 | 0.025350 | 0.0005047141 |
| 000 | 0.409600 | 0.1317678350 | 23 | 0.022570 | 0.0004000853 |
| 00 | 0.364800 | 0.1045199453 | 24 | 0.020100 | 0.0003173084 |
| 0 | 0.324900 | 0.0829065680 | 25 | 0.017900 | 0.0002516492 |
| 1 | 0.289300 | 0.0657334432 | 26 | 0.015940 | 0.0001995566 |
| 2 | 0.257600 | 0.0521172188 | 27 | 0.014200 | 0.0001583676 |
| 3 | 0.229400 | 0.0413310408 | 28 | 0.012640 | 0.0001254826 |
| 4 | 0.204300 | 0.0327813057 | 29 | 0.011260 | 0.0000995787 |
| 5 | 0.181900 | 0.0259869262 | 30 | 0.010030 | 0.0000790117 |
| 6 | 0.162000 | 0.0206119720 | 31 | 0.008928 | 0.0000626034 |
| 7 | 0.144300 | 0.0163539316 | 32 | 0.007950 | 0.0000496391 |
| 8 | 0.128500 | 0.0129686799 | 33 | 0.007080 | 0.0000393691 |
| 9 | 0.114400 | 0.0102787798 | 34 | 0.006305 | 0.0000312219 |
| 10 | 0.101900 | 0.0081552613 | 35 | 0.005615 | 0.0000247622 |
| 11 | 0.090740 | 0.0064667648 | 36 | 0.005000 | 0.0000196349 |
| 12 | 0.080810 | 0.0051288468 | 37 | 0.004453 | 0.0000155738 |
| 13 | 0.071960 | 0.0040669780 | 38 | 0.003965 | 0.0000123474 |
| 14 | 0.064080 | 0.0032250357 | 39 | 0.003531 | 0.0000097923 |
| 15 | 0.057070 | 0.0025580278 | 40 | 0.003145 | 0.0000077684 |
| 16 | 0.050820 | 0.0020284244 | 41 | 0.002800 | 0.0000061575 |
| 17 | 0.045260 | 0.0016088613 | 42 | 0.002490 | 0.0000048695 |
| 18 | 0.040300 | 0.0012755562 | 43 | 0.002220 | 0.0000038708 |
| 19 | 0.035890 | 0.0010116643 | 44 | 0.001970 | 0.0000030480 |
| 20 | 0.031960 | 0.0008022377 | 45 | 0.001760 | 0.0000024328 |
| 21 | 0.028460 | 0.0006361497 | 46 | 0.001570 | 0.0000019359 |

[a] Gauge is American Wire Gauge (AWG). For a given gauge, $d = 0.005 \times 92^{\frac{36 - \text{Gauge}}{39}}$

## Hypodermic Needle Dimensions

| | Needle Nominal Outside Diameter | | | Needle Nominal Inside Diameter | | |
|---|---|---|---|---|---|---|
| Gauge[a] | mm | in. | Tolerance (in.) | mm | in. | Tolerance (in.) |
| 6 | 5.156 | 0.2030 | | 4.394 | 0.1730 | |
| 7 | 4.572 | 0.1800 | | 3.810 | 0.1500 | ±0.0030 |
| 8 | 4.191 | 0.1650 | | 3.429 | 0.1350 | |
| 9 | 3.759 | 0.1480 | | 2.997 | 0.1180 | |
| 10 | 3.404 | 0.1340 | ±0.0010 | 2.692 | 0.1060 | |
| 11 | 3.048 | 0.1200 | | 2.388 | 0.0940 | |
| 12 | 2.769 | 0.1090 | | 2.159 | 0.0850 | ±0.0020 |
| 13 | 2.413 | 0.0950 | | 1.803 | 0.0710 | |
| 14 | 2.108 | 0.0830 | | 1.600 | 0.0630 | |
| 15 | 1.829 | 0.0720 | | 1.372 | 0.0540 | |
| 16 | 1.651 | 0.0650 | | 1.194 | 0.0470 | |
| 17 | 1.473 | 0.0580 | ±0.0005 | 1.067 | 0.0420 | ±0.0015 |
| 18 | 1.270 | 0.0500 | | 0.838 | 0.0330 | |
| 19 | 1.067 | 0.0420 | | 0.686 | 0.0270 | |
| 20 | 0.902 | 0.0355 | | 0.584 | 0.0230 | |
| 21 | 0.813 | 0.0320 | | 0.495 | 0.0195 | |
| 22 | 0.711 | 0.0280 | | 0.394 | 0.0155 | |
| 22s | 0.711 | 0.0280 | | 0.140 | 0.0055 | |
| 23 | 0.635 | 0.0250 | | 0.318 | 0.0125 | |
| 24 | 0.559 | 0.0220 | | 0.292 | 0.0115 | |
| 25 | 0.508 | 0.0200 | | 0.241 | 0.0095 | |
| 25s | 0.508 | 0.0200 | | 0.140 | 0.0055 | |
| 26 | 0.457 | 0.0180 | +0.0005 −0.0000 | 0.241 | 0.0095 | +0.0015 −0.0000 |
| 26s | 0.467 | 0.0184 | | 0.114 | 0.0045 | |
| 27 | 0.406 | 0.0160 | | 0.191 | 0.0075 | |
| 28 | 0.356 | 0.0140 | | 0.165 | 0.0065 | |
| 29 | 0.330 | 0.0130 | | 0.165 | 0.0065 | |
| 30 | 0.305 | 0.0120 | | 0.140 | 0.0055 | |
| 31 | 0.254 | 0.0100 | | 0.114 | 0.0045 | |
| 32 | 0.229 | 0.0090 | | 0.089 | 0.0035 | |
| 33 | 0.203 | 0.0080 | | 0.089 | 0.0035 | |

[a] Gauge is Birmingham Wire Gauge (same as Stub's Iron Wire Gauge)

**Strength and Stiffness of Perforated Metals.—**It is common practice to use perforated metals in equipment enclosures to provide cooling by the flow of air or fluids. If the perforated material is to serve also as a structural member, then calculations of stiffness and strength must be made that take into account the effect of the perforations on the strength of the panels.

The accompanying table provides equivalent or effective values of the yield strength $S^*$; modulus of elasticity $E^*$; and Poisson's ratio $\nu^*$ of perforated metals in terms of the values for solid material. The $S^*/S$ and $E^*/E$ ratios, given in the accompanying table for the standard round hole staggered pattern, can be used to determine the safety margins or deflections for perforated metal use as compared to the unperforated metal for any geometry or loading condition.

Perforated material has different strengths depending on the direction of loading; therefore, values of $S^*/S$ in the table are given for the width (strongest) and length (weakest) directions. Also, the effective elastic constants are for plane stress conditions and apply to the in-plane loading of thin perforated sheets; the bending stiffness is greater. However, since most loading conditions involve a combination of bending and stretching, it is more convenient to use the same effective elastic constants for these combined loading conditions. The plane stress effective elastic constants given in the table can be conservatively used for all loading conditions.

**Mechanical Properties of Materials Perforated with Round Holes in IPA Standard Staggered Hole Pattern**

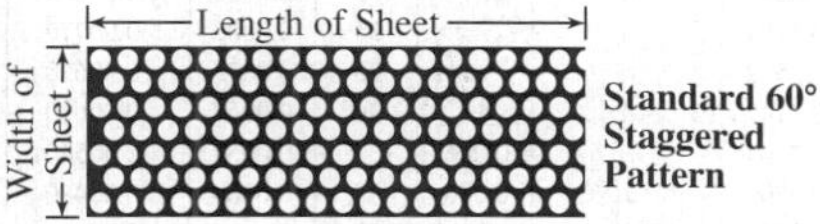


| IPA No. | Perforation Dia. (in.) | Center Distance (in.) | Open Area (%) | $S^*/S$ Width (in.) | $S^*/S$ Length (in.) | $E^*/E$ | $\nu^*$ |
|---|---|---|---|---|---|---|---|
| 100 | 0.020 | (625) | 20 | 0.530 | 0.465 | 0.565 | 0.32 |
| 106 | 1/16 | 1/8 | 23 | 0.500 | 0.435 | 0.529 | 0.33 |
| 107 | 5/64 | 7/64 | 46 | 0.286 | 0.225 | 0.246 | 0.38 |
| 108 | 5/64 | 1/8 | 36 | 0.375 | 0.310 | 0.362 | 0.35 |
| 109 | 3/32 | 5/32 | 32 | 0.400 | 0.334 | 0.395 | 0.34 |
| 110 | 3/32 | 3/16 | 23 | 0.500 | 0.435 | 0.529 | 0.33 |
| 112 | 1/10 | 5/32 | 36 | 0.360 | 0.296 | 0.342 | 0.35 |
| 113 | 1/8 | 3/16 | 40 | 0.333 | 0.270 | 0.310 | 0.36 |
| 114 | 1/8 | 7/32 | 29 | 0.428 | 0.363 | 0.436 | 0.33 |
| 115 | 1/8 | 1/4 | 23 | 0.500 | 0.435 | 0.529 | 0.33 |
| 116 | 5/32 | 7/32 | 46 | 0.288 | 0.225 | 0.249 | 0.38 |
| 117 | 5/32 | 1/4 | 36 | 0.375 | 0.310 | 0.362 | 0.35 |
| 118 | 3/16 | 1/4 | 51 | 0.250 | 0.192 | 0.205 | 0.42 |
| 119 | 3/16 | 5/16 | 33 | 0.400 | 0.334 | 0.395 | 0.34 |
| 120 | 1/4 | 5/16 | 58 | 0.200 | 0.147 | 0.146 | 0.47 |
| 121 | 1/4 | 3/8 | 40 | 0.333 | 0.270 | 0.310 | 0.36 |
| 122 | 1/4 | 7/16 | 30 | 0.428 | 0.363 | 0.436 | 0.33 |
| 123 | 1/4 | 1/2 | 23 | 0.500 | 0.435 | 0.529 | 0.33 |
| 124 | 3/8 | 1/2 | 51 | 0.250 | 0.192 | 0.205 | 0.42 |
| 125 | 3/8 | 9/16 | 40 | 0.333 | 0.270 | 0.310 | 0.36 |
| 126 | 3/8 | 5/8 | 33 | 0.400 | 0.334 | 0.395 | 0.34 |
| 127 | 7/16 | 5/8 | 45 | 0.300 | 0.239 | 0.265 | 0.38 |
| 128 | 1/2 | 11/16 | 47 | 0.273 | 0.214 | 0.230 | 0.39 |
| 129 | 9/16 | 3/4 | 51 | 0.250 | 0.192 | 0.205 | 0.42 |

Value in parentheses specifies holes per square inch instead of center distance. $S^*/S$ = ratio of yield strength of perforated to unperforated material; $E^*/E$ = ratio of modulus of elasticity of perforated to unperforated material; $\nu^*$ = Poisson's ratio for given percentage of open area.

IPA is Industrial Perforators Association.

**Sheet Metal Gages.—**Thicknesses of steel sheets given in Table 2 are based upon a weight of 41.82 pounds per square foot per inch of thickness, which is known as the Manufacturers' Standard Gage for Sheet Steel. This gage differs from the older United States Standard Gage for iron and steel sheets and plates, established by Congress in 1893, based upon a weight of 40 pounds per square foot per inch of thickness which is the weight of wrought-iron plate.

**Table 2. Sheet Metal Gages in Approximate Decimals of an Inch**

| Gage No. | Steel Gage[a] | B.G[b]. | Galvanized Sheet | Zinc Gage | Gage No. | Steel Gage[a] | B.G.[b] | Galvanized Sheet | Zinc Gage |
|---|---|---|---|---|---|---|---|---|---|
| 15/0 | ... | 1.000 | ... | ... | 20 | 0.0359 | 0.0392 | 0.0396 | 0.070 |
| 14/0 | ... | 0.9583 | ... | ... | 21 | 0.0329 | 0.0349 | 0.0366 | 0.080 |
| 13/0 | ... | 0.9167 | ... | ... | 22 | 0.0299 | 0.03125 | 0.0336 | 0.090 |
| 12/0 | ... | 0.8750 | ... | ... | 23 | 0.0269 | 0.02782 | 0.0306 | 0.100 |
| 11/0 | ... | 0.8333 | ... | ... | 24 | 0.0239 | 0.02476 | 0.0276 | 0.125 |
| 10/0 | ... | 0.7917 | ... | ... | 25 | 0.0209 | 0.02204 | 0.0247 | ... |
| 9/0 | ... | 0.7500 | ... | ... | 26 | 0.0179 | 0.01961 | 0.0217 | ... |
| 8/0 | ... | 0.7083 | ... | ... | 27 | 0.0164 | 0.01745 | 0.0202 | ... |
| 7/0 | ... | 0.6666 | ... | ... | 28 | 0.0149 | 0.01562 | 0.0187 | ... |
| 6/0 | ... | 0.6250 | ... | ... | 29 | 0.0135 | 0.01390 | 0.0172 | ... |
| 5/0 | ... | 0.5883 | ... | ... | 30 | 0.0120 | 0.01230 | 0.0157 | ... |
| 4/0 | ... | 0.5416 | ... | ... | 31 | 0.0105 | 0.01100 | 0.0142 | ... |
| 3/0 | ... | 0.5000 | ... | ... | 32 | 0.0097 | 0.00980 | 0.0134 | ... |
| 2/0 | ... | 0.4452 | ... | ... | 33 | 0.0090 | 0.00870 | ... | ... |
| 1/0 | ... | 0.3964 | ... | ... | 34 | 0.0082 | 0.00770 | ... | ... |
| 1 | ... | 0.3532 | ... | ... | 35 | 0.0075 | 0.00690 | ... | ... |
| 2 | ... | 0.3147 | ... | ... | 36 | 0.0067 | 0.00610 | ... | ... |
| 3 | 0.2391 | 0.2804 | ... | 0.006 | 37 | 0.0064 | 0.00540 | ... | ... |
| 4 | 0.2242 | 0.2500 | ... | 0.008 | 38 | 0.0060 | 0.00480 | ... | ... |
| 5 | 0.2092 | 0.2225 | ... | 0.010 | 39 | ... | 0.00430 | ... | ... |
| 6 | 0.1943 | 0.1981 | ... | 0.012 | 40 | ... | 0.00386 | ... | ... |
| 7 | 0.1793 | 0.1764 | ... | 0.014 | 41 | ... | 0.00343 | ... | ... |
| 8 | 0.1644 | 0.1570 | 0.1681 | 0.016 | 42 | ... | 0.00306 | ... | ... |
| 9 | 0.1495 | 0.1398 | 0.1532 | 0.018 | 43 | ... | 0.00272 | ... | ... |
| 10 | 0.1345 | 0.1250 | 0.1382 | 0.020 | 44 | ... | 0.00242 | ... | ... |
| 11 | 0.1196 | 0.1113 | 0.1233 | 0.024 | 45 | ... | 0.00215 | ... | ... |
| 12 | 0.1046 | 0.0991 | 0.1084 | 0.028 | 46 | ... | 0.00192 | ... | ... |
| 13 | .0897 | 0.0882 | 0.0934 | 0.032 | 47 | ... | 0.00170 | ... | ... |
| 14 | 0.0747 | 0.0785 | 0.0785 | 0.036 | 48 | ... | 0.00152 | ... | ... |
| 15 | 0.0673 | 0.0699 | 0.0710 | 0.040 | 49 | ... | 0.00135 | ... | ... |
| 16 | 0.0598 | 0.0625 | 0.0635 | 0.045 | 50 | ... | 0.00120 | ... | ... |
| 17 | 0.0538 | 0.0556 | 0.0575 | 0.050 | 51 | ... | 0.00107 | ... | ... |
| 18 | .0478 | 0.0495 | 0.0516 | 0.055 | 52 | ... | 0.00095 | ... | ... |
| 19 | 0.0418 | 0.0440 | 0.0456 | 0.060 | ... | ... | ... | ... | ... |

[a] Manufacturers' Standard Gage for Sheet Steel

[b] B.G. is the Birmingham Gage for sheets and hoops.

*The United States Standard Gage* (not shown above) for iron and steel sheets and plates was established by Congress in 1893 and was primarily a *weight* gage rather than a thickness gage. The equivalent thicknesses were derived from the weight of wrought iron. The weight per cubic foot was taken at 480 pounds, thus making the weight of a plate 12 inches square and 1 inch thick, 40 pounds. In converting weight to equivalent thickness, gage tables formerly published contained thicknesses equivalent to the basic weights just mentioned. For example, a No. 3 US gage represents a wrought-iron plate having a weight of 10 pounds per square foot; hence, if the weight per square foot per inch thick is 40 pounds, the plate thickness for a No. 3 gage = 10 ÷ 40 = 0.25 inch, which was the original thickness equivalent for this gage number. Because this and the other thickness equivalents were derived from the weight of wrought iron, they are not correct for steel.

Zinc sheets are usually ordered by specifying decimal thickness although a zinc gage exists and is shown in Table 2.

Most sheet metal products in Great Britain are specified by the British Standard Wire Gage (Imperial Wire Gage). Black iron and steel sheet and hooping, and galvanized flat and corrugated steel sheet, however, are specified by the Birmingham Gage (B.G.), which was legalized in 1914, and are also shown in Table 2. This Birmingham Gage should not be confused with the Birmingham or Stub's Iron Wire Gage mentioned previously.

Thicknesses of aluminum, copper, and copper-base alloys were formerly designated by the American or Brown & Sharpe Wire Gage but now are specified in decimals or fractions of an inch. American National Standard B32.1-1952 (Withdrawn) entitled Preferred Thicknesses for Uncoated Thin Flat Metals (see accompanying Table 3) gives thicknesses that are based on the 20- and 40-series of preferred numbers in American National Standard Preferred Numbers — ANSI Z17.1 (Withdrawn) (see *Handbook* page 688) and are applicable to uncoated, thin, flat metals and alloys. Each number of the 20-series is approximately 12 percent greater than the next smaller one, and each number of the 40-series is approximately 6 percent greater than the next smaller one.

**Table 3. Preferred Thicknesses for Uncoated Metals and Alloys— Under 0.250 Inch in Thickness** *ANSI/ASME B32.1-1952 (Withdrawn)*

| Preferred Thickness, Inches | | | | | | | |
|---|---|---|---|---|---|---|---|
| Based on 20-Series | Based on 40-Series | Based on 20-Series | Based on 40-Series | Based on 20-Series | Based on 40-Series | Based on 20-Series | Based on 40-Series |
| … | 0.236 | 0.100 | 0.100 | … | 0.042 | 0.018 | 0.018 |
| 0.224 | 0.224 | … | 0.095 | 0.040 | 0.040 | … | 0.017 |
| … | 0.212 | 0.090 | 0.090 | … | 0.038 | 0.016 | 0.016 |
| 0.200 | 0.200 | … | 0.085 | 0.036 | 0.036 | … | 0.015 |
| … | 0.190 | 0.080 | 0.080 | … | 0.034 | 0.014 | 0.014 |
| 0.180 | 0.180 | … | 0.075 | 0.032 | 0.032 | … | 0.013 |
| … | 0.170 | 0.071 | 0.071 | … | 0.030 | 0.012 | 0.012 |
| 0.160 | 0.160 | … | 0.067 | 0.028 | 0.028 | 0.011 | 0.011 |
| … | 0.150 | 0.063 | 0.063 | … | 0.026 | 0.010 | 0.010 |
| 0.140 | 0.140 | … | 0.060 | 0.025 | 0.025 | 0.009 | 0.009 |
| … | 0.132 | 0.056 | 0.056 | … | 0.024 | 0.008 | 0.008 |
| 0.125 | 0.125 | … | 0.053 | 0.022 | 0.022 | 0.007 | 0.007 |
| … | 0.118 | 0.050 | 0.050 | … | 0.021 | 0.006 | 0.006 |
| 0.112 | 0.112 | … | 0.048 | 0.020 | 0.020 | 0.005 | 0.005 |
| … | 0.106 | 0.045 | 0.045 | … | 0.019 | 0.004 | 0.004 |

The American National Standard ANSI/ASME B32.1-1952 (Withdrawn) lists preferred thicknesses that are based on the 20- and 40-series of preferred numbers and states that those based on the 40-series should provide adequate coverage. However, where intermediate thicknesses are required, the Standard recommends that thicknesses be based on the 80-series of preferred numbers (see *Handbook* page 688).

Thicknesses for copper and copper-base alloy flat products below $1/4$ inch thick are specified by the 20-series of American National Standard Preferred Numbers given in ANSI/ASME B32.1. Although the table in ANSI/ASME B32.1 gives only the 20- and 40-series of numbers, it states that when intermediate thicknesses are required they should be selected from thicknesses based on the 80-series of numbers (see *Handbook* page 688).

**Metric Sizes for Flat Metal Products.**—American National Standard B32.100-2016 establishes a preferred series of metric thicknesses, widths, and lengths for flat metal products of rectangular cross section; the thickness and width values are also applicable to base metals that may be coated in later operations. The Standard also establishes a preferred series of metric sizes for round and hexagonal metal products.

The First Choice of sizes of this Standard are derived from Renard's R5 series of preferred numbers (see *American National Standard for Preferred Numbers* on page 688, also ANSI/ASME B4.2 or ISO 497) in which each preferred number is approximately 60 percent larger than the preceding number in the series. The Second Choice series are derived from the R10 series in which each number is approximately 25 percent larger than the preceding, and Third Choice sizes are generally from the R20 series in which each number is approximately 12.5 percent larger than the preceding number.

Table 4a gives preferred metric diameters from 1 to 500 millimeters for round metal products; Table 4b lists the preferred thicknesses; Table 4c lists the preferred widths. Wherever possible, sizes should be selected from the First Choice series shown in the tables. Second and Third Choice series are also shown. For preferred metric diameters less than 1 millimeter, preferred across flat metric sizes of square and hexagon metal products, preferred across flat metric sizes of rectangular metal products, and preferred metric lengths of metal products, reference should be made to the B32.100-2016 Standard.

**Table 4a. American National Standard Preferred Metric Diameters of Round Metal Product** *ANSI/ASME B32.100-2016*

| Basic Size, mm | | | Basic Size, mm | | |
|---|---|---|---|---|---|
| First Choice | Second Choice | Third Choice | First Choice | Second Choice | Third Choice |
| 1 | … | … | 25 | … | … |
| … | … | 1.1 | … | … | 26, 28 |
| … | 1.2 | … | … | 30 | … |
| … | … | 1.4 | … | … | 32, 35, 36, 38 |
| 1.6 | … | … | 40 | … | … |
| … | … | 1.8 | … | … | 42, 45, 48 |
| … | 2 | … | … | 50 | … |
| … | … | 2.2, 2.4 | … | … | 55, 56 |
| 2.5 | … | … | 60 | … | … |
| … | … | 2.6, 2.8 | … | … | 63, 65, 70, 75 |
| … | 3 | … | … | 80 | … |
| … | … | 3.2, 3.5, 3.8 | … | … | 85, 90, 95 |
| 4 | … | … | 100 | … | … |
| … | … | 4.5, 4.8 | … | … | 105, 110 |
| … | 5 | … | … | 120 | … |
| … | … | 5.5 | … | … | 125, 130, 140, 150 |
| 6 | … | … | 160 | … | … |
| … | … | 6.5, 7, 7.5 | … | … | 170, 180, 190 |
| … | 8 | … | … | 200 | … |
| … | … | 8.5, 9, 9.5 | … | … | 220, 240 |
| 10 | … | … | 250 | … | … |
| … | … | 11 | … | … | 260, 280 |
| … | 12 | … | … | 300 | … |
| … | … | 13, 14, 15 | … | … | 320, 340, 350, 360, 380 |
| 16 | … | … | 400 | … | … |
| … | … | 17, 18, 19 | … | … | 420, 440, 450, 480 |
| … | 20 | … | … | 500 | … |
| … | … | 21, 22, 23, 24 | … | … | … |

## Table 4b. Preferred Metric Thicknesses (mm) for All Flat Metal Products *ANSI/ASME B32.100-2016*

| First Choice | Second Choice | Third Choice | First Choice | Second Choice | Third Choice | First Choice | Second Choice | Third Choice |
|---|---|---|---|---|---|---|---|---|
| … | 0.050 | … | 1.6 | … | … | 25 | … | … |
| 0.060 | … | … | … | … | 1.7 | … | … | 28 |
| … | 0.080 | … | … | … | 1.8 | … | 30 | … |
| 0.10 | … | … | … | 2 | … | … | … | 32 |
| … | … | 0.11 | … | … | 2.2 | … | … | 35 |
| … | 0.12 | … | … | … | 2.3 | … | … | 38 |
| … | … | 0.14 | 2.5 | … | … | 40 | … | … |
| 0.16 | … | … | … | … | 2.6 | … | … | 45 |
| … | … | 0.18 | … | … | 2.8 | … | 50 | … |
| … | 0.20 | … | … | 3 | … | … | … | 55 |
| … | … | 0.22 | … | … | 3.2 | 60 | … | … |
| 0.25 | … | … | … | … | 3.5 | … | … | 70 |
| … | … | 0.28 | … | … | 3.8 | … | 80 | … |
| … | 0.30 | … | 4 | … | … | … | … | 90 |
| … | … | 0.35 | … | … | 4.5 | … | … | … |
| 0.40 | … | … | … | 5 | … | 100 | … | … |
| … | … | 0.45 | … | … | 5.5 | … | … | 110 |
| … | 0.50 | … | 6 | … | … | … | 120 | … |
| … | … | 0.55 | … | … | 7 | … | … | 130 |
| 0.60 | … | … | … | 8 | … | … | … | 140 |
| … | … | 0.65 | … | … | 9 | … | … | 150 |
| … | … | 0.70 | 10 | … | … | 160 | … | … |
| … | … | 0.75 | … | … | 11 | … | … | 180 |
| … | 0.80 | … | … | 12 | … | … | 200 | … |
| … | … | 0.90 | … | … | 14 | 250 | … | … |
| 1.0 | … | … | … | … | 15 | … | 300 | … |
| … | … | 1.1 | 16 | … | … | … | … | … |
| … | 1.2 | … | … | … | 18 | … | … | … |
| … | … | 1.4 | … | 20 | … | … | … | … |
| … | … | 1.5 | … | … | 22 | … | … | … |

## Table 4c. Preferred Metric Widths (mm) for All Flat Metal Products *ANSI/ASME B32.100-2016*

| First Choice | Second Choice | Third Choice | First Choice | Second Choice | Third Choice | First Choice | Second Choice | Third Choice |
|---|---|---|---|---|---|---|---|---|
| … | 2 | … | 60 | … | … | 400 | … | … |
| 2.5 | … | … | … | … | 70 | … | 500 | … |
| … | 3 | … | … | 80 | … | 600 | … | … |
| 4 | … | … | … | … | 90 | … | … | 700 |
| … | 5 | … | 100 | … | … | … | 800 | … |
| 6 | … | … | … | … | 110 | … | … | 900 |
| … | 8 | … | … | 120 | … | 1000 | … | … |
| 10 | … | … | … | … | 130 | … | 1200 | … |
| … | 12 | … | … | 140 | … | … | … | 1500 |
| 16 | … | … | … | … | 150 | 1600 | … | … |
| … | 20 | … | 160 | … | … | … | 2000 | … |
| 25 | 30 | … | … | … | 180 | 2500 | … | … |
| … | … | 35 | … | 200 | … | … | 3000 | … |
| 40 | … | … | … | … | 225 | … | … | 3500 |
| … | … | 45 | 250 | … | … | 4000 | … | … |
| … | 50 | … | … | … | 280 | … | 5000 | … |
| … | … | 55 | … | 300 | … | … | … | … |

All dimensions are in millimeters.

**Table 5a. Decimal Equivalent of Numbered Drill Sizes**

| Drill Number | Decimal | | Drill Number | Decimal | | Drill Number | Decimal | | Drill Number | Decimal | |
|---|---|---|---|---|---|---|---|---|---|---|---|
| | Inch | mm | | Inch | mm | | Inch | mm | | Inch | mm |
| 1 | 0.2280 | 5.791 | 26 | 0.1470 | 3.734 | 51 | 0.0670 | 1.702 | 76 | 0.0200 | 0.508 |
| 2 | 0.2210 | 5.613 | 27 | 0.1440 | 3.658 | 52 | 0.0635 | 1.613 | 77 | 0.0180 | 0.457 |
| 3 | 0.2130 | 5.410 | 28 | 0.1405 | 3.569 | 53 | 0.0595 | 1.511 | 78 | 0.0160 | 0.406 |
| 4 | 0.2090 | 5.309 | 29 | 0.1360 | 3.454 | 54 | 0.0550 | 1.397 | 79 | 0.0145 | 0.368 |
| 5 | 0.2055 | 5.220 | 30 | 0.1285 | 3.264 | 55 | 0.0520 | 1.321 | 80 | 0.0135 | 0.343 |
| 6 | 0.2040 | 5.182 | 31 | 0.1200 | 3.048 | 56 | 0.0465 | 1.181 | 81 | 0.0130 | 0.330 |
| 7 | 0.2010 | 5.105 | 32 | 0.1160 | 2.946 | 57 | 0.0430 | 1.092 | 82 | 0.0125 | 0.318 |
| 8 | 0.1990 | 5.054 | 33 | 0.1130 | 2.870 | 58 | 0.0420 | 1.067 | 83 | 0.0120 | 0.305 |
| 9 | 0.1960 | 4.978 | 34 | 0.1110 | 2.819 | 59 | 0.0410 | 1.041 | 84 | 0.0115 | 0.292 |
| 10 | 0.1935 | 4.915 | 35 | 0.1100 | 2.794 | 60 | 0.0400 | 1.016 | 85 | 0.0110 | 0.280 |
| 11 | 0.1910 | 4.851 | 36 | 0.1065 | 2.705 | 61 | 0.0390 | 0.991 | 86 | 0.0105 | 0.267 |
| 12 | 0.1890 | 4.800 | 37 | 0.1040 | 2.642 | 62 | 0.0380 | 0.965 | 87 | 0.0100 | 0.254 |
| 13 | 0.1850 | 4.700 | 38 | 0.1015 | 2.578 | 63 | 0.0370 | 0.940 | 88 | 0.0095 | 0.241 |
| 14 | 0.1820 | 4.623 | 39 | 0.0995 | 2.527 | 64 | 0.0360 | 0.914 | 89 | 0.0091 | 0.231 |
| 15 | 0.1800 | 4.572 | 40 | 0.0980 | 2.489 | 65 | 0.0350 | 0.889 | 90 | 0.0087 | 0.221 |
| 16 | 0.1770 | 4.496 | 41 | 0.0960 | 2.438 | 66 | 0.0330 | 0.838 | 91 | 0.0083 | 0.211 |
| 17 | 0.1730 | 4.394 | 42 | 0.0935 | 2.375 | 67 | 0.0320 | 0.813 | 92 | 0.0079 | 0.200 |
| 18 | 0.1695 | 4.305 | 43 | 0.0890 | 2.261 | 68 | 0.0310 | 0.787 | 93 | 0.0075 | 0.190 |
| 19 | 0.1660 | 4.216 | 44 | 0.0860 | 2.184 | 69 | 0.0292 | 0.742 | 94 | 0.0071 | 0.180 |
| 20 | 0.1610 | 4.089 | 45 | 0.0820 | 2.083 | 70 | 0.0280 | 0.711 | 95 | 0.0067 | 0.170 |
| 21 | 0.1590 | 4.039 | 46 | 0.0810 | 2.057 | 71 | 0.0260 | 0.660 | 96 | 0.0063 | 0.160 |
| 22 | 0.1570 | 3.988 | 47 | 0.0785 | 1.994 | 72 | 0.0250 | 0.635 | 97 | 0.0059 | 0.150 |
| 23 | 0.1540 | 3.912 | 48 | 0.0760 | 1.930 | 73 | 0.0240 | 0.610 | … | … | … |
| 24 | 0.1520 | 3.861 | 49 | 0.0730 | 1.854 | 74 | 0.0225 | 0.572 | … | … | … |
| 25 | 0.1495 | 3.797 | 50 | 0.0700 | 1.778 | 75 | 0.0210 | 0.533 | … | … | … |

**Table 5b. Decimal Equivalent of Letter Drill Sizes**

| Drill Size | Decimal | | Drill Size | Decimal | | Drill Size | Decimal | | Drill Size | Decimal | | Drill Size | Decimal | |
|---|---|---|---|---|---|---|---|---|---|---|---|---|---|---|
| | Inch | mm | | Inch | mm | | Inch | mm | | Inch | mm | | Inch | mm |
| A | 0.234 | 5.944 | G | 0.261 | 6.629 | M | 0.295 | 7.493 | S | 0.348 | 8.839 | Y | 0.404 | 10.262 |
| B | 0.238 | 6.045 | H | 0.266 | 6.756 | N | 0.302 | 7.671 | T | 0.358 | 9.093 | Z | 0.413 | 10.490 |
| C | 0.242 | 6.147 | I | 0.272 | 6.909 | O | 0.316 | 8.026 | U | 0.368 | 9.347 | … | … | … |
| D | 0.246 | 6.248 | J | 0.277 | 7.036 | P | 0.323 | 8.204 | V | 0.377 | 9.576 | … | … | … |
| E | 0.250 | 6.350 | K | 0.281 | 7.137 | Q | 0.332 | 8.433 | W | 0.386 | 9.804 | … | … | … |
| F | 0.257 | 6.528 | L | 0.290 | 7.366 | R | 0.339 | 8.611 | X | 0.397 | 10.084 | … | … | … |

**Wire Drawing History.—**Wire drawing is the process used for producing wire by drawing a wire or rod of larger diameter through a plate or die provided with a hole which reduces the size to the desired dimension. Previous to wire drawing, wire was made by hammering or beating metal into thin sheets or plates, which were cut into continuous strips. These strips were then afterwards rounded by hammering. There is evidence that wire was produced by a simple hammering process as early as 2000 B.C., and perhaps earlier. The history of the art does not clearly indicate when simple hammering was superseded by the method of drawing through a die, but records are available to show that wire drawing on a commercial scale was practiced in France in 1270, in Germany in 1350, and in England in 1466. The first wire-drawing mill in America was built in 1775, in Norwich, Connecticut, by Nathaniel Niles, who was granted a loan of $1500 by the court for this purpose.

**Shaped Wire.—**Shaped wire is any wire not round in section; it may be square, rectangular, hexagonal, triangular, half-round, oval, etc. Some sections of shaped wire may be either rolled or drawn. Shaped wire is a specialty, subject to agreement between purchaser and manufacturer as to size, finish, temper, and weight of coils and packages.

## Strength and Properties of Wire Rope

**Wire Rope Construction.—**Essentially, a wire rope is made up of a number of strands laid helically about a metallic or nonmetallic core. Each strand consists of a number of wires also laid helically about a metallic or nonmetallic center. Various types of wire rope have been developed to meet a wide range of uses and operating conditions. These types are distinguished by the kind of core; the number of strands; the number, sizes, and arrangement of the wires in each strand; and the way in which the wires and strands are wound or laid about each other. The following descriptive material is based largely on information supplied by the Bethlehem Steel Co.

*Rope Wire Materials:* Materials used in the manufacture of rope wire are, in order of increasing strength: iron, phosphor bronze, traction steel, plow steel, improved plow steel, and bridge rope steel. Iron wire rope is largely used for low-strength applications such as elevator ropes not used for hoisting, and for stationary guy ropes.

Phosphor bronze wire rope is used occasionally for elevator governor-cable rope and for certain marine applications as life lines, clearing lines, wheel ropes and rigging.

Traction steel wire rope is used primarily as hoist rope for passenger and freight elevators of the traction drive type, an application for which it was specifically designed.

Ropes made of galvanized wire or wire coated with zinc by the electrodeposition process are used in certain applications where additional protection against rusting is required. As will be noted from the tables of wire-rope sizes and strengths, the breaking strength of galvanized wire rope is 10 percent less than that of ungalvanized (bright) wire rope. Bethanized (zinc-coated) wire rope can be furnished to bright wire rope strength when so specified.

Galvanized carbon steel, tinned carbon steel, and stainless steel are used for small cords and strands ranging in diameter from $^1/_{64}$ to $^3/_8$ inch and larger.

Marline clad wire rope has each strand wrapped with a layer of tarred marline. The cladding provides hand protection for workers and wear protection for the rope.

*Rope Cores:* Wire-rope cores are made of fiber, cotton, asbestos, polyvinyl plastic, a small wire rope (independent wire-rope core), a multiple-wire strand (wire-strand core) or a cold-drawn wire-wound spring.

*Fiber* (manila or sisal) is the type of core most widely used when loads are not too great. It supports the strands in their relative positions and acts as a cushion to prevent nicking of the wires lying next to the core.

*Cotton* is used for small ropes such as sash cord and aircraft cord.

*Asbestos cores* can be furnished for certain special operations where the rope is used in oven operations.

*Polyvinyl plastics cores* are offered for use where exposure to moisture, acids, or caustics is excessive.

*Wire-strand core* often referred to as WSC, consists of a multiple-wire strand that may be the same as one of the strands of the rope. It is smoother and more solid than the independent wire rope core and provides a better support for the rope strands.

The *independent wire rope core*, often referred to as IWRC, is a small 6 × 7 wire rope with a wire-strand core and is used to provide greater resistance to crushing and distortion of the wire rope. For certain applications it has the advantage over a wire-strand core in that it stretches at a rate closer to that of the rope itself.

Wire ropes with wire-strand cores are, in general, less flexible than wire ropes with independent wire-rope or nonmetallic cores.

Ropes with metallic cores are rated $7^1/_2$ percent stronger than those with nonmetallic cores.

*Wire-Rope Lay:* The lay of a wire rope is the direction of the helical path in which the strands are laid and, similarly, the lay of a strand is the direction of the helical path in which the wires are laid. If the wires in the strand or the strands in the rope form a helix similar to the threads of a right-hand screw, i.e., they wind around to the right, the lay is called right-hand and, conversely, if they wind around to the left, the lay is called left-hand. In the *regular lay*, the wires in the strands are laid in the opposite direction to the lay of the strands in the rope. In right-regular lay, the strands are laid to the right and the wires to the left. In left-regular lay, the strands are laid to the left, the wires to the right. In *Lang lay*, the wires and strands are laid in the same direction, i.e., in right Lang lay, both the wires and strands are laid to the right and in left Lang they are laid to the left.

Alternate lay ropes having alternate right and left laid strands are used to resist distortion and prevent clamp slippage, but because other advantages are missing, have limited use.

The regular lay wire rope is most widely used and right regular lay rope is customarily furnished. Regular lay rope has less tendency to spin or untwist when placed under load and is generally selected where long ropes are employed and the loads handled are frequently removed. Lang lay ropes have greater flexibility than regular lay ropes and are more resistant to abrasion and fatigue.

In preformed wire ropes the wires and strands are preshaped into a helical form so that when laid to form the rope they tend to remain in place. In a non-preformed rope, broken wires tend to "wicker out" or protrude from the rope and strands that are not seized tend to spring apart. Preforming also tends to remove locked-in stresses, lengthen service life, and make the rope easier to handle and to spool.

*Strand Construction:* Various arrangements of wire are used in the construction of wire rope strands. In the simplest arrangement six wires are grouped around a central wire thus making seven wires, all of the same size. Other types of construction known as "filler-wire," Warrington, Seale, etc. make use of wires of different sizes. Their respective patterns of arrangement are shown diagrammatically in the table of wire weights and strengths.

**Specifying Wire Rope.**—In specifying wire rope the following information will be required: length, diameter, number of strands, number of wires in each strand, type of rope construction, grade of steel used in rope, whether preformed or not preformed, type of center, and type of lay. The manufacturer should be consulted in selecting the best type of wire rope for a new application.

**Properties of Wire Rope.**—Important properties of wire rope are strength, wear resistance, flexibility, and resistance to crushing and distortion.

*Strength:* The strength of wire rope depends upon its size, kind of material of which the wires are made and their number, the type of core, and whether the wire is galvanized or not. Strengths of various types and sizes of wire ropes are given in the accompanying tables together with appropriate factors to apply for ropes with steel cores and for galvanized wire ropes.

*Wear Resistance:* When wire rope must pass back and forth over surfaces that subject it to unusual wear or abrasion, it must be specially constructed to give satisfactory service.

Such construction may make use of 1) relatively large outer wires; 2) Lang lay in which wires in each strand are laid in the same direction as the strand; and 3) flattened strands.

The object in each type is to provide a greater outside surface area to take the wear or abrasion. From the standpoint of material, improved plow steel has not only the highest tensile strength but also the greatest resistance to abrasion in regularly stocked wire rope.

*Flexibility:* Wire rope that undergoes repeated and severe bending, such as in passing around small sheaves and drums, must have a high degree of flexibility to prevent premature breakage and failure due to fatigue. Greater flexibility in wire rope is obtained by 1) using small wires in larger numbers; 2) using Lang lay; and 3) preforming, that is, the wires and strands of the rope are shaped during manufacture to fit the position they will assume in the finished rope.

*Resistance to Crushing and Distortion:* Where wire rope is to be subjected to transverse loads that may crush or distort it, care should be taken to select a type of construction that will stand up under such treatment.

Wire rope designed for such conditions may have 1) large outer wires to spread the load per wire over a greater area; and 2) an independent wire core or a high-carbon cold-drawn wound spring core.

**Standard Classes of Wire Rope.—**Wire rope is commonly designated by two figures, the first indicating the number of strands and the second, the number of wires per strand, as: 6 × 7, a six-strand rope having seven wires per strand, or 8 × 19, an eight-strand rope having 19 wires per strand. When such numbers are used as designations of standard wire rope classes, the second figure in the designation may be purely nominal in that the number of wires per strand for various ropes in the class may be slightly less or slightly more than the nominal as will be seen from the following brief descriptions. (For ropes with a wire strand core, a second group of two numbers may be used to indicate the construction of the wire core, as 1 × 21, 1 × 43, and so on.)

*6 × 7 Class (Standard Coarse Laid Rope):* Wire ropes in this class are for use where resistance to wear, as in dragging over the ground or across rollers, is an important requirement. Heavy hauling, rope transmissions, and well drilling are common applications. These wire ropes are furnished in right regular lay and occasionally in Lang lay. The cores may be of fiber, independent wire rope, or wire strand. Since this class is a relatively stiff type of construction, these ropes should be used with large sheaves and drums. Because of the small number of wires, a larger factor of safety may be called for.

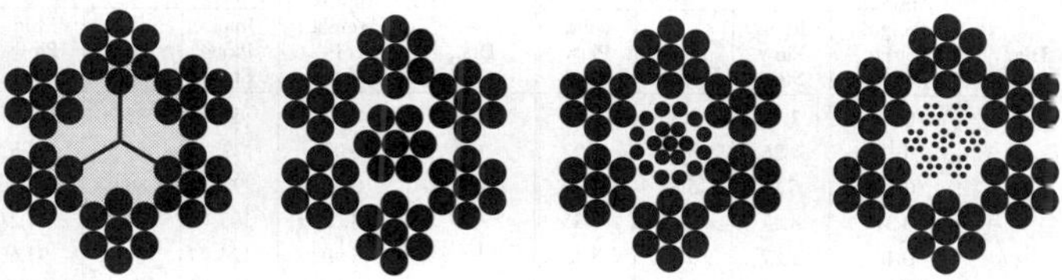

Fig. 1a. 6×7 with Fiber Core | Fig. 1b. 6×7 with 1×7 WSC | Fig. 1c. 6×7 with 1×19 WSC | Fig. 1d. 6×7 with IWRC

As shown in Fig. 1a through Fig. 1d, this class includes a 6 × 7 construction with fiber core: a 6 × 7 construction with 1 × 7 wire strand core (sometimes called 7 × 7); a 6 × 7 construction with 1 × 19 wire strand core; and a 6 × 7 construction with independent wire rope core. Table 1 provides strength and weight data for this class.

Two special types of wire rope in this class are: aircraft cord, a 6 × 6 or 7 × 7 Bethanized wire rope of high tensile strength and sash cord, a 6 × 7 iron rope used for a variety of purposes where strength is not an important factor.

**Table 1. Weights and Strengths of 6 × 7 (Standard Coarse Laid) Wire Ropes, Preformed and Not Preformed**

| Dia., Inches | Approx. Weight per Ft., Pounds | Breaking Strength, Tons of 2000 Lbs. | | | Dia., Inches | Approx. Weight per Ft., Pounds | Breaking Strength, Tons of 2000 Lbs. | | |
|---|---|---|---|---|---|---|---|---|---|
| | | Impr. Plow Steel | Plow Steel | Mild Plow Steel | | | Impr. Plow Steel | Plow Steel | Mild Plow Steel |
| ¼ | 0.094 | 2.64 | 2.30 | 2.00 | ¾ | 0.84 | 22.7 | 19.8 | 17.2 |
| 5⁄16 | 0.15 | 4.10 | 3.56 | 3.10 | ⅞ | 1.15 | 30.7 | 26.7 | 23.2 |
| ⅜ | 0.21 | 5.86 | 5.10 | 4.43 | 1 | 1.50 | 39.7 | 34.5 | 30.0 |
| 7⁄16 | 0.29 | 7.93 | 6.90 | 6.00 | 1⅛ | 1.90 | 49.8 | 43.3 | 37.7 |
| ½ | 0.38 | 10.3 | 8.96 | 7.79 | 1¼ | 2.34 | 61.0 | 53.0 | 46.1 |
| 9⁄16 | 0.48 | 13.0 | 11.3 | 9.82 | 1⅜ | 2.84 | 73.1 | 63.6 | 55.3 |
| ⅝ | 0.59 | 15.9 | 13.9 | 12.0 | 1½ | 3.38 | 86.2 | 75.0 | 65.2 |

For ropes with steel cores, add 7½ percent to above strengths.

For galvanized ropes, deduct 10 percent from above strengths.

*Source:* Rope diagrams, Bethlehem Steel Co. All data, US Simplified Practice Recommendation 198-50.

*6 × 19 Class (Standard Hoisting Rope)*: This rope is the most popular and widely used class. Ropes in this class are furnished in regular or Lang lay and may be obtained preformed or not preformed. Cores may be of fiber, independent wire rope, or wire strand. As can be seen from Table 2 and Fig. 2a through Fig. 2h, there are four common types: 6 × 25 filler wire construction with fiber core (not illustrated), independent wire core, or wire strand core (1 × 25 or 1 × 43); 6 × 19 Warrington construction with fiber core; 6 × 21 filler wire construction with fiber core; and 6 × 19, 6 × 21, and 6 × 17 Seale construction with fiber core.

**Table 2. Weights and Strengths of 6 × 19 (Standard Hoisting) Wire Ropes, Preformed and Not Preformed**

| Dia., Inches | Approx. Weight per Ft., Pounds | Breaking Strength, Tons of 2000 Lbs. | | | Dia., Inches | Approx. Weight per Ft., Pounds | Breaking Strength, Tons of 2000 Lbs. | | |
|---|---|---|---|---|---|---|---|---|---|
| | | Impr. Plow Steel | Plow Steel | Mild Plow Steel | | | Impr. Plow Steel | Plow Steel | Mild Plow Steel |
| ¼ | 0.10 | 2.74 | 2.39 | 2.07 | 1¼ | 2.50 | 64.6 | 56.2 | 48.8 |
| 5⁄16 | 0.16 | 4.26 | 3.71 | 3.22 | 1⅜ | 3.03 | 77.7 | 67.5 | 58.8 |
| ⅜ | 0.23 | 6.10 | 5.31 | 4.62 | 1½ | 3.60 | 92.0 | 80.0 | 69.6 |
| 7⁄16 | 0.31 | 8.27 | 7.19 | 6.25 | 1⅝ | 4.23 | 107 | 93.4 | 81.2 |
| ½ | 0.40 | 10.7 | 9.35 | 8.13 | 1¾ | 4.90 | 124 | 108 | 93.6 |
| 9⁄16 | 0.51 | 13.5 | 11.8 | 10.2 | 1⅞ | 5.63 | 141 | 123 | 107 |
| ⅝ | 0.63 | 16.7 | 14.5 | 12.6 | 2 | 6.40 | 160 | 139 | 121 |
| ¾ | 0.90 | 23.8 | 20.7 | 18.0 | 2⅛ | 7.23 | 179 | 156 | … |
| ⅞ | 1.23 | 32.2 | 28.0 | 24.3 | 2¼ | 8.10 | 200 | 174 | … |
| 1 | 1.60 | 41.8 | 36.4 | 31.6 | 2½ | 10.00 | 244 | 212 | … |
| 1⅛ | 2.03 | 52.6 | 45.7 | 39.8 | 2¾ | 12.10 | 292 | 254 | … |

The 6 × 25 filler wire with fiber core not illustrated.

For ropes with steel cores, add 7½ percent to above strengths.

For galvanized ropes, deduct 10 percent from above strengths.

*Source:* Rope diagrams, Bethlehem Steel Co. All data, US Simplified Practice Recommendation 198-50.

*6 × 37 Class (Extra Flexible Hoisting Rope):* For a given size of rope, the component wires are of smaller diameter than those in the two classes previously described and hence have less resistance to abrasion. Ropes in this class are furnished in regular and Lang lay with fiber core or independent wire rope core, preformed or not preformed.

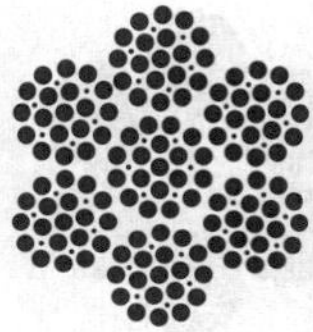

Fig. 2a.
6 × 25 Filler Wire
with WSC (1 × 25)

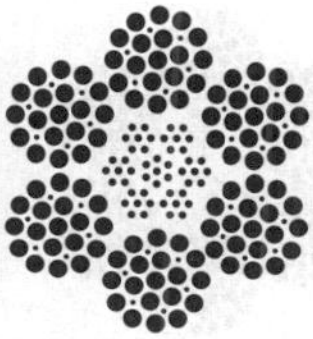

Fig. 2b.
6 × 25 Filler Wire
with IWRC

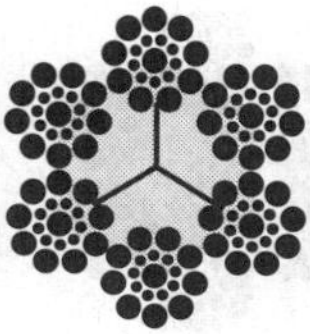

Fig. 2c.
6 × 19 Seale
with Fiber Core

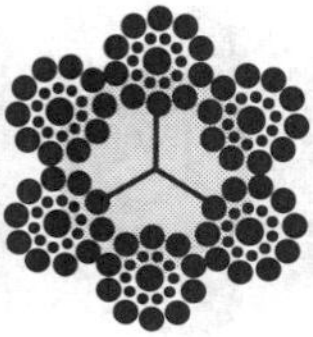

Fig. 2d.
6 × 21 Seale
with Fiber Core

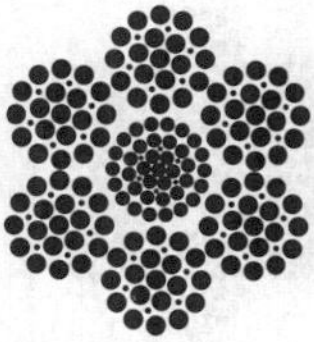

Fig. 2e.
6 × 25 Filler Wire
with WSC (1 × 43)

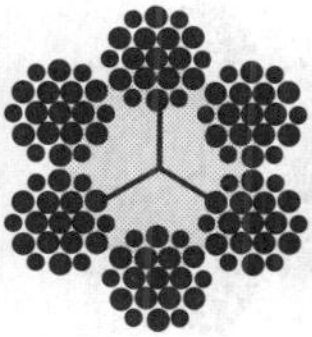

Fig. 2f.
6 × 19 Warrington
with Fiber Core

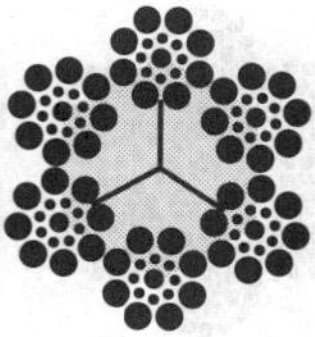

Fig. 2g.
6 × 17 Seale
with Fiber Core

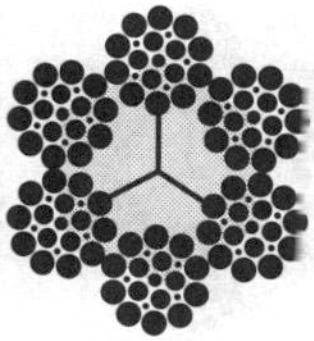

Fig. 2h.
6 × 21 Filler Wire
with Fiber Core

**Table 3. Weights and Strengths of 6 × 37 (Extra Flexible Hoisting) Wire Ropes, Preformed and Not Preformed**

| Dia., Inches | Approx. Weight per Ft., Pounds | Breaking Strength, Tons of 2000 Lbs. | | Dia., Inches | Approx. Weight per Ft., Pounds | Breaking Strength, Tons of 2000 Lbs. | |
|---|---|---|---|---|---|---|---|
| | | Impr. Plow Steel | Plow Steel | | | Impr. Plow Steel | Plow Steel |
| ¼ | 0.10 | 2.59 | 2.25 | 1½ | 3.49 | 87.9 | 76.4 |
| 5⁄16 | 0.16 | 4.03 | 3.50 | 1⅝ | 4.09 | 103 | 89.3 |
| ⅜ | 0.22 | 5.77 | 5.02 | 1¾ | 4.75 | 119 | 103 |
| 7⁄16 | 0.30 | 7.82 | 6.80 | 1⅞ | 5.45 | 136 | 118 |
| ½ | 0.39 | 10.2 | 8.85 | 2 | 6.20 | 154 | 134 |
| 9⁄16 | 0.49 | 12.9 | 11.2 | 2⅛ | 7.00 | 173 | 150 |
| ⅝ | 0.61 | 15.8 | 13.7 | 2¼ | 7.85 | 193 | 168 |
| ¾ | 0.87 | 22.6 | 19.6 | 2½ | 9.69 | 236 | 205 |
| ⅞ | 1.19 | 30.6 | 26.6 | 2¾ | 11.72 | 284 | 247 |
| 1 | 1.55 | 39.8 | 34.6 | 3 | 14.0 | 335 | 291 |
| 1⅛ | 1.96 | 50.1 | 43.5 | 3¼ | 16.4 | 390 | 339 |
| 1¼ | 2.42 | 61.5 | 53.5 | 3½ | 19.0 | 449 | 390 |
| 1⅜ | 2.93 | 74.1 | 64.5 | … | … | … | … |

For ropes with steel cores, add $7^1/_2$ percent to above strengths.

For galvanized ropes, deduct 10 percent from above strengths.

*Source:* Rope diagrams, Bethlehem Steel Co. All data, US Simplified Practice Recommendation 198-50.

As shown in Table 3 and Fig. 3a through Fig. 3h, there are four common types: 6 × 29 filler wire construction with fiber core and 6 × 36 filler wire construction with independent wire rope core, a special rope for construction equipment; 6 × 35 (two operations) construction with fiber core and 6 × 41 Warrington Seale construction with fiber core, a standard crane rope in this class of rope construction; 6 × 41 filler wire construction with fiber core or

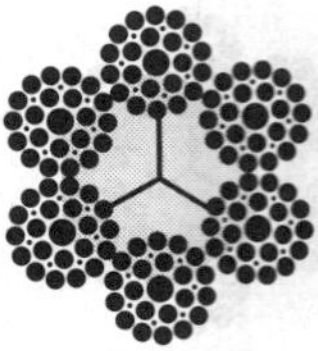

Fig. 3a.
6×29 Filler Wire
with Fiber Core

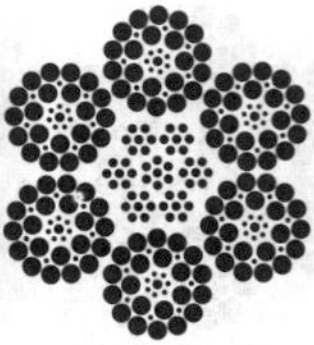

Fig. 3b.
6×36 Filler Wire
with IWRC

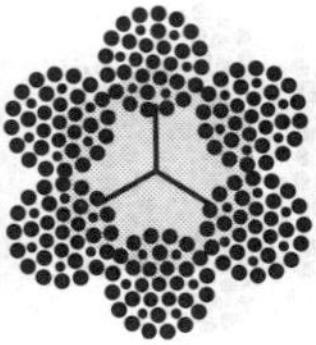

Fig. 3c.
6×35 with
Fiber Core

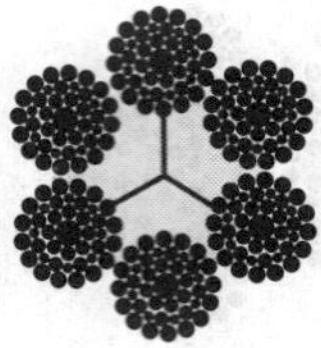

Fig. 3d.
6×41 Warrington-Seale
with Fiber Core

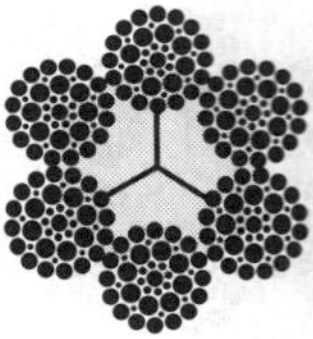

Fig. 3e.
6×41 Filler Wire
with Fiber Core

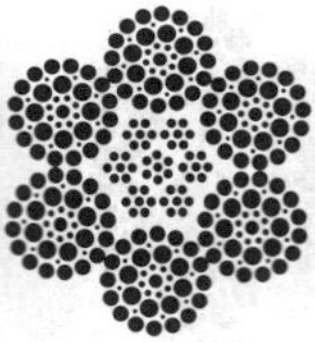

Fig. 3f.
6×41 Filler Wire
with IWRC

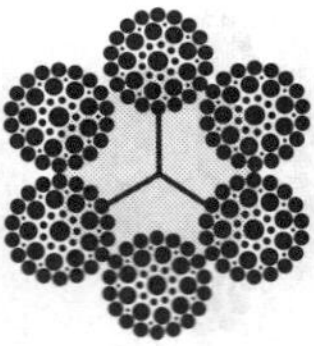

Fig. 3g.
6×46 Filler Wire
with Fiber Core

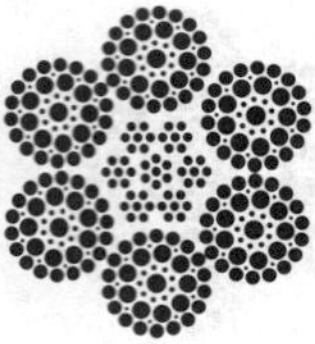

Fig. 3h.
6×46 Filler Wire
with IWRC

independent wire core, a special large shovel rope usually furnished in Lang lay; and 6 × 46 filler wire construction with fiber core or independent wire rope core, a special large shovel and dredge rope.

*8× 19 Class (Special Flexible Hoisting Rope):* This rope is stable and smooth-running, and is especially suitable, because of its flexibility, for high-speed operation with reverse bends. Ropes in this class are available in regular lay with fiber core.

As shown in Table 4 and Fig. 4a through Fig. 4d, there are four common types: 8 × 25 filler wire construction, the most flexible but the least wear resistant rope of the four types; Warrington type in 8 × 19 construction, less flexible than the 8 × 25; 8 × 21 filler wire construction, less flexible than the Warrington; and Seale type in 8 × 19 construction, which has the greatest wear resistance of the four types but is also the least flexible.

**Table 4. Weights and Strengths of 8 × 19 (Special Flexible Hoisting) Wire Ropes, Preformed and Not Preformed**

| Dia., Inches | Approx. Weight per Ft., Pounds | Breaking Strength, Tons of 2000 Lbs. | | Dia., Inches | Approx. Weight per Ft., Pounds | Breaking Strength, Tons of 2000 Lbs. | |
|---|---|---|---|---|---|---|---|
| | | Impr. Plow Steel | Plow Steel | | | Impr. Plow Steel | Plow Steel |
| ¼ | 0.09 | 2.35 | 2.04 | ¾ | 0.82 | 20.5 | 17.8 |
| 5⁄16 | 0.14 | 3.65 | 3.18 | ⅞ | 1.11 | 27.7 | 24.1 |
| ⅜ | 0.20 | 5.24 | 4.55 | 1 | 1.45 | 36.0 | 31.3 |
| 7⁄16 | 0.28 | 7.09 | 6.17 | 1⅛ | 1.84 | 45.3 | 39.4 |
| ½ | 0.36 | 9.23 | 8.02 | 1¼ | 2.27 | 55.7 | 48.4 |
| 9⁄16 | 0.46 | 11.6 | 10.1 | 1⅜ | 2.74 | 67.1 | 58.3 |
| ⅝ | 0.57 | 14.3 | 12.4 | 1½ | 3.26 | 79.4 | 69.1 |

For ropes with steel cores, add 7½ percent to above strengths.
For galvanized ropes, deduct 10 percent from above strengths.
*Source:* Rope diagrams, Bethlehem Steel Co. All data, US Simplified Practice Recommendation 198-50.

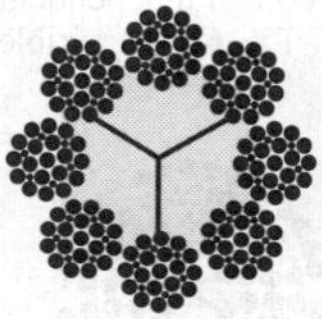

Fig. 4a.
8 × 25 Filler Wire
with Fiber Core

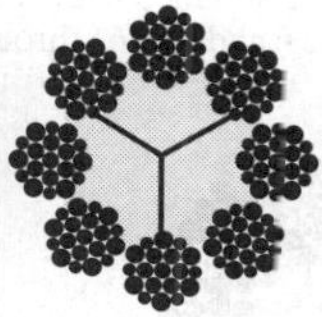

Fig. 4b.
8 × 19 Warrington
with Fiber Core

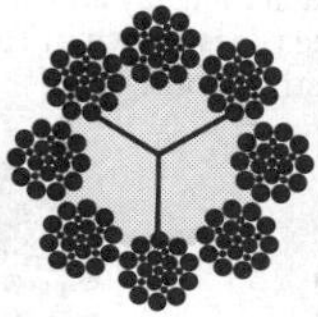

Fig. 4c.
8 × 21 Filler Wire
with Fiber Core

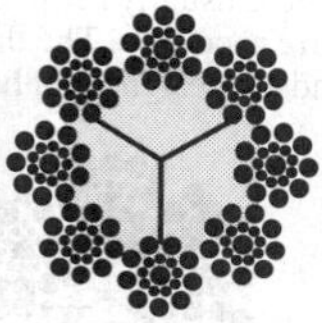

Fig. 4d.
8 × 19 Seale
with Fiber Core

Also in this class, but not shown in Table 4, are elevator ropes made of traction steel and iron.

*18 × 7 Non-Rotating Wire Rope:* This rope is specially designed for use where a minimum of rotating or spinning is called for, especially in the lifting or lowering of free loads with a single-part line. It has an inner layer composed of 6 strands of 7 wires each laid in left Lang lay over a fiber core and an outer layer of 12 strands of 7 wires each laid in right regular lay. The combination of opposing lays tends to prevent rotation when the rope is stretched. However, to avoid any tendency to rotate or spin, loads should be kept to at least one-eighth and preferably one-tenth of the breaking strength of the rope. Weights and strengths are shown in Table 5.

**Table 5. Weights and Strengths of Standard 18 × 7 Nonrotating Wire Rope, Preformed and Not Preformed**

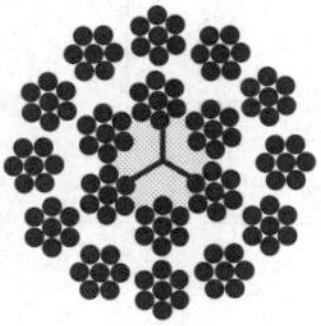

Fig. 5.
18 × 7 Non-Rotating Rope

**Recommended Sheave and Drum Diameters**

| | |
|---|---|
| Single layer on drum | 36 rope diameters |
| Multiple layers on drum | 48 rope diameters |
| Mine service | 60 rope diameters |

| Dia., Inches | Approx. Weight per Ft., Pounds | Breaking Strength, Tons of 2000 Lbs. | | Dia., Inches | Approx. Weight per Ft., Pounds | Breaking Strength, Tons of 2000 Lbs. | |
|---|---|---|---|---|---|---|---|
| | | Impr. Plow Steel | Plow Steel | | | Impr. Plow Steel | Plow Steel |
| 3⁄16 | 0.061 | 1.42 | 1.24 | 7⁄8 | 1.32 | 29.5 | 25.7 |
| 1⁄4 | 0.108 | 2.51 | 2.18 | 1 | 1.73 | 38.3 | 33.3 |
| 5⁄16 | 0.169 | 3.90 | 3.39 | 1 1⁄8 | 2.19 | 48.2 | 41.9 |
| 3⁄8 | 0.24 | 5.59 | 4.86 | 1 1⁄4 | 2.70 | 59.2 | 51.5 |
| 7⁄16 | 0.33 | 7.58 | 6.59 | 1 3⁄8 | 3.27 | 71.3 | 62.0 |
| 1⁄2 | 0.43 | 9.85 | 8.57 | 1 1⁄2 | 3.89 | 84.4 | 73.[illegible] |
| 9⁄16 | 0.55 | 12.4 | 10.8 | 1 5⁄8 | 4.57 | 98.4 | 85.5 |
| 5⁄8 | 0.68 | 15.3 | 13.3 | 1 3⁄4 | 5.30 | 114 | 98.3 |
| 3⁄4 | 0.97 | 21.8 | 19.0 | … | … | … | … |

For galvanized ropes, deduct 10 percent from above strengths.

*Source:* Rope diagrams, sheave and drum diameters, and data for 3⁄16, 1⁄4 and 5⁄16-inch sizes, Bethlehem Steel Co. All other data, US Simplified Practice Recommendation 198-50.

*Flattened Strand Wire Rope:* The wires forming the strands of this type of rope are wound around triangular centers so that a flattened outer surface is provided with a greater area than in the regular round rope to withstand severe conditions of abrasion. The triangular shape of the strands also provides superior resistance to crushing. Flattened strand wire rope is usually furnished in Lang lay and may be obtained with fiber core or independent wire rope core. The three types shown in Table 6 and Fig. 6a through Fig. 6c are flexible and are designed for hoisting work.

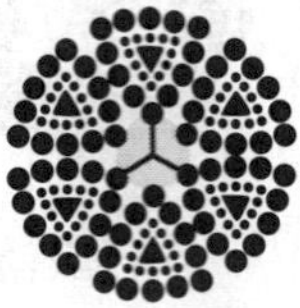

Fig. 6a.
6 × 25 with Fiber Core

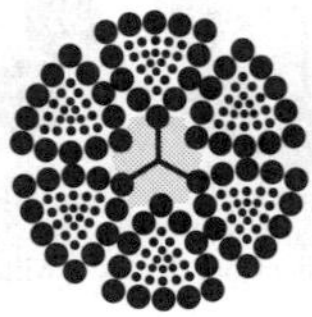

Fig. 6b.
6 × 30 with Fiber Core

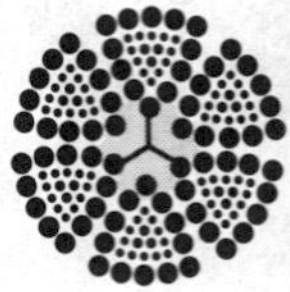

Fig. 6c.
6 × 27 with Fiber Core

**Table 6. Weights and Strengths of Flattened Strand Wire Rope, Preformed and Not Preformed**

| Dia., Inches | Approx. Weight per Ft., Pounds | Breaking Strength, Tons of 2000 Lbs. | | Dia., Inches | Approx. Weight per Ft., Pounds | Breaking Strength, Tons of 2000 Lbs. | |
|---|---|---|---|---|---|---|---|
| | | Impr. Plow Steel | Mild Plow Steel | | | Impr. Plow Steel | Mild Plow Steel |
| ⅜[a] | 0.25 | 6.71 | … | 1⅜ | 3.40 | 85.5 | … |
| ½[a] | 0.45 | 11.8 | 8.94 | 1½ | 4.05 | 101 | … |
| 9⁄16[a] | 0.57 | 14.9 | 11.2 | 1⅝ | 4.75 | 118 | … |
| ⅝ | 0.70 | 18.3 | 13.9 | 1¾ | 5.51 | 136 | … |
| ¾ | 1.01 | 26.2 | 19.8 | 2 | 7.20 | 176 | … |
| ⅞ | 1.39 | 35.4 | 26.8 | 2¼ | 9.10 | 220 | … |
| 1 | 1.80 | 46.0 | 34.8 | 2½ | 11.2 | 269 | … |
| 1⅛ | 2.28 | 57.9 | 43.8 | 2¾ | 13.6 | 321 | … |
| 1¼ | 2.81 | 71.0 | 53.7 | … | … | … | … |

[a] These sizes in Type B only.

Type H is not in US Simplified Practice Recommendation.

*Source:* Rope diagrams, Bethlehem Steel Co. All other data, US Simplified Practice Recommendation 198-50.

*Flat Wire Rope:* This type of wire rope is made up of a number of four-strand rope units placed side by side and stitched together with soft steel sewing wire. These four-strand units are alternately right and left lay to resist warping, curling, or rotating in service. Weights and strengths are shown in Table 7.

**Simplified Practice Recommendations.—**Because the total number of wire rope types is large, manufacturers and users have agreed upon and adopted a US Simplified Practice Recommendation to provide a simplified listing of those kinds and sizes of wire rope which are most commonly used and stocked. These, then, are the types and sizes which are most generally available. Other types and sizes for special or limited uses also may be found in individual manufacturer's catalogs.

**Sizes and Strengths of Wire Rope.—**The data shown in Table 1 through Table 7 have been taken from US Simplified Practice Recommendation 198-50 but do not include those wire ropes shown in that Simplified Practice Recommendation which are intended primarily for marine use.

*Wire Rope Diameter:* The diameter of a wire rope is the diameter of the circle that will just enclose it, hence when measuring the diameter with calipers, care must be taken to obtain the largest outside dimension, taken across the opposite strands, rather than the smallest dimension across opposite "valleys" or "flats." It is standard practice for the nominal diameter to be the minimum with all tolerances taken on the plus side. Limits for diameter as well as for minimum breaking strength and maximum pitch are given in Federal Specification for Wire Rope, RR-R—571a.

*Wire Rope Strengths:* The strength figures shown in the accompanying tables have been obtained by a mathematical derivation based on actual breakage tests of wire rope and represent from 80 to 95 percent of the total strengths of the individual wires, depending upon the type of rope construction.

**Table 7. Weights and Strengths of Standard Flat Wire Rope, Not Preformed**

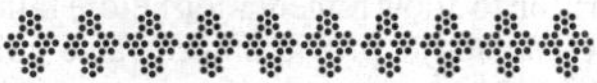

Flat Wire Rope

This rope consists of a number of 4-strand rope units placed side by side and stitched together with soft steel sewing wire.

| Width and Thickness, Inches | No. of Ropes | Approx. Weight per Ft., Pounds | Breaking Strength, Tons of 2000 Lbs. | | Width and Thickness, Inches | No. of Ropes | Approx. Weight per Ft., Pounds | Breaking Strength Tons of 2000 Lbs. | |
|---|---|---|---|---|---|---|---|---|---|
| | | | Plow Steel | Mild Plow-Steel | | | | Plow Steel | Mild Plow Steel |
| ¼ × 1½ | 7 | 0.69 | 16.8 | 14.6 | ½ × 4 | 9 | 3.16 | 81.8 | 71.2 |
| ¼ × 2 | 9 | 0.88 | 21.7 | 18.8 | ½ × 4½ | 10 | 3.82 | 90.9 | 79.1 |
| ¼ × 2½ | 11 | 1.15 | 26.5 | 23.0 | ½ × 5 | 12 | 4.16 | 109 | 94.9 |
| ¼ × 3 | 13 | 1.34 | 31.3 | 27.2 | ½ × 5½ | 13 | 4.50 | 118 | 103 |
| | | | | | ½ × 6 | 14 | 4.85 | 127 | 111 |
| 5/16 × 1½ | 5 | 0.77 | 18.5 | 16.0 | ½ × 7 | 16 | 5.85 | 145 | 126 |
| 5/16 × 2 | 7 | 1.05 | 25.8 | 22.4 | | | | | |
| 5/16 × 2½ | 9 | 1.33 | 33.2 | 28.8 | ⅝ × 3½ | 6 | 3.40 | 85.8 | 74.5 |
| 5/16 × 3 | 11 | 1.61 | 40.5 | 35.3 | ⅝ × 4 | 7 | 3.95 | 100 | 87.[illegible] |
| 5/16 × 3½ | 13 | 1.89 | 47.9 | 41.7 | ⅝ × 4½ | 8 | 4.50 | 114 | 99.5 |
| 5/16 × 4 | 15 | 2.17 | 55.3 | 48.1 | ⅝ × 5 | 9 | 5.04 | 129 | 112 |
| | | | | | ⅝ × 5½ | 10 | 5.59 | 143 | 124 |
| ⅜ × 2 | 6 | 1.25 | 31.4 | 27.3 | ⅝ × 6 | 11 | 6.14 | 157 | 137 |
| ⅜ × 2½ | 8 | 1.64 | 41.8 | 36.4 | ⅝ × 7 | 13 | 7.23 | 186 | 162 |
| ⅜ × 3 | 9 | 1.84 | 47.1 | 40.9 | ⅝ × 8 | 15 | 8.32 | 214 | 186 |
| ⅜ × 3½ | 11 | 2.23 | 57.5 | 50.0 | | | | | |
| ⅜ × 4 | 12 | 2.44 | 62.7 | 54.6 | ¾ × 5 | 8 | 6.50 | 165 | 143 |
| ⅜ × 4½ | 14 | 2.83 | 73.2 | 63.7 | ¾ × 6 | 9 | 7.31 | 185 | 16[illegible] |
| ⅜ × 5 | 15 | 3.03 | 78.4 | 68.2 | ¾ × 7 | 10 | 8.13 | 206 | 17[illegible] |
| ⅜ × 5½ | 17 | 3.42 | 88.9 | 77.3 | ¾ × 8 | 11 | 9.70 | 227 | 19[illegible] |
| ⅜ × 6 | 18 | 3.63 | 94.1 | 81.9 | | | | | |
| | | | | | ⅞ × 5 | 7 | 7.50 | 190 | 165 |
| ½ × 2½ | 6 | 2.13 | 54.5 | 47.4 | ⅞ × 6 | 8 | 8.56 | 217 | 188 |
| ½ × 3 | 7 | 2.47 | 63.6 | 55.4 | ⅞ × 7 | 9 | 9.63 | 244 | 212 |
| ½ × 3½ | 8 | 2.82 | 72.7 | 63.3 | ⅞ × 8 | 10 | 10.7 | 271 | 2[illegible]6 |

*Source:* Rope diagrams, Bethlehem Steel Co. All data, US Simplified Practice Recommendation 198-50.

**Safe Working Loads and Factors of Safety.—**The maximum load for which a wire rope is to be used should take into account such associated factors as friction, load caused by bending around each sheave, acceleration and deceleration, and, if a long length of rope is to be used for hoisting, the weight of the rope at its maximum extension. The condition of the rope — whether new or old, worn or corroded — and type of attachments should also be considered.

Factors of safety for standing rope usually range from 3 to 4; for operating rope, from 5 to 12. Where there is the element of hazard to life or property, higher values are used.

## SHAFT ALIGNMENT

### Introduction

Shaft alignment is the positioning of the rotational centers of two or more shafts, so that the shafts are co-linear when the machines are operating. The purpose of shaft alignment is to optimize equipment efficiency and increase the operating life span of rotating machinery. Shaft misalignment will result in high axial and radial vibrations, high temperatures in casings, bearings, or lubricants; lubricant leakage; loose, broken, or missing coupling bolts or foundation bolts; and cracks in shafts or other parts. Modern vibration spectrum analyzers can detect misalignment early on to allow remediation before failure.

ANSI/ASA S2.75-2017, "Shaft Alignment Methodology," provides acceptable guidelines for shaft alignment of rotating machinery. The Standard has three parts: Part 1-2017 (R2020), "General Principles, Methods, Practices, and Tolerances" establishes the methodology for measurement, analysis, and correction of the misalignment of shafts coupled by means of a flexible coupling supported by bearing systems; Part 2-2017 (R2020), "Vocabulary," defines terms; Part 3-2021 discusses vertically oriented rotating shafts. This standard addresses critical alignment principles and modeling, machine movement detection and calculations, pipe strain, tolerances for acceptable relative shaft position, and repeatability.

**Misalignment.**—While ANSI/ASA S2.75 fills a need for setting shaft alignment tolerances and best practices, there already are defined limits for shaft-to-shaft alignment of coupled machines. These limits are based on two measures of misalignment: *angularity* and *offset.*

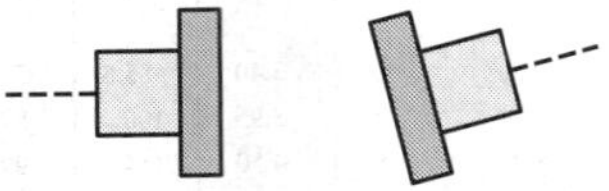

Fig. 1. Shafts in Angular Misaligned Position

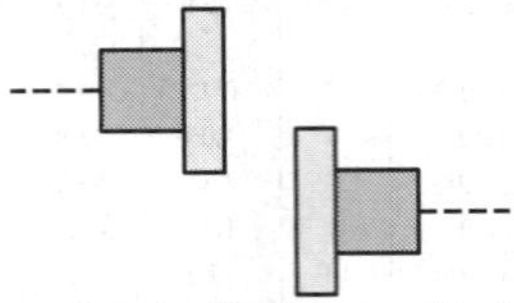

Fig. 2. Shafts in Offset Misaligned Position

*Angular Misalignment:* This is the difference in the slope of one shaft, as compared to the slope of the other shaft, in units expressed as rise/run. The run (distance along the shaft) is measured in inches and the rise (perpendicular distance to the shafts) is measured in mils. The process of correcting this type of alignment problem may be referred to as gap or face alignment.

*Offset Misalignment:* This is the distance between the shaft centers of rotation as measured at the plane of power transmission or coupling center. The units of measurement are in mils.

**Correcting Misalignment.**—There are four alignment considerations to be carefully measured and corrected: vertical angularity, vertical offset, horizontal angularity, and horizontal offset. Proper shaft alignment is extremely critical when shafts are running at higher speeds. Thus, as shaft speeds increase, the allowable limits of misalignment decrease. Values in Table 1 may be used as a general guide for acceptable limits of misalignment.

**Table 1. Misalignment Tolerance Guide**

| | Offset Misalignment (Mils) | | Angular Misalignment (mils/inch) | |
|---|---|---|---|---|
| RPM | Excellent | Acceptable | Excellent | Acceptable |
| 600 | ±2.00 | ±4.00 | 0.80 | 1.25 |
| 900 | ±1.50 | ±3.00 | 0.70 | 1.00 |
| 1200 | ±1.25 | ±2.00 | 0.50 | 0.80 |
| 1800 | ±1.0 | ±1.50 | 0.30 | 0.50 |
| 3600 | ±0.50 | ±0.75 | 0.20 | 0.30 |
| ≥4000 | ±0.50 | ±0.75 | 0.10 | 0.25 |

Source: *Engineering Design for Manufacturability, Volume 1*, by Kelly L. Bramble, Monroe, Georgia: Engineers Edge, ©2006–2015, www.engineersedge.com.

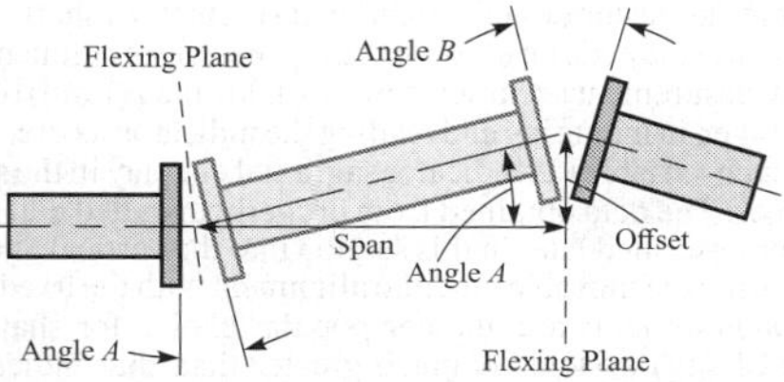


Fig. 3. Using Angles Between Hubs to Measure Misalignment

While the terms offset and angularity are commonly used to describe alignment conditions, a more accurate method is to use the angles between the adjacent shaft hubs and the coupling mechanical link (CML) that connects those hubs. These angles are generally measured at the two flex planes, such that:

$$\text{angle } A \text{ (or } B) = \frac{offset}{span}\left(\frac{mils}{inch}\right)$$

where the CML connects to each of the two hubs as Angle A and B (as shown in Fig. 3).

ANSI/ASA S2.75-2017, Part 1, "General Principles, Methods, Practices, and Tolerances," specifies "alignment quality grades" (AL#.#), which permit classification of shaft alignment tolerances for common machinery and are expressed in units of mils/in (μm/mm). There are three appropriate levels: AL4.5 = minimal; AL2.2 = standard; and AL1.2 = precise. Alignment grades are chosen based on machine construction and operating conditions. The tolerance angle *T* is the maximum angle at each flex plane and is related to the ratio of offset at the flex plane to flex plane separation (offset/span). This allowable residual misalignment for a given alignment quality grade may be calculated as:

$$T = \frac{AL}{\sqrt{\frac{RPM}{1000}}} + 1$$

where AL is the acceptable alignment quality grade,
RPM is the maximum operating rotating speed of the machinery,

Units of the tolerance angle are typically in $\frac{\mu m\ offset}{mm\ span}$ or $\frac{mils\ offset}{inch\ span}$

When the shafts of two machines require alignment, the process generally calls for one of the machines to be permanently mounted and the other one to be movable. The fixed unit usually is the driven machinery, such as the pump in a pump-motor pair. The second machine (usually the motor) is moved into approximate alignment (by eye and straight edge, for example) in preparation for measurements that will determine the magnitude and direction of repositioning required to place it in final alignment with the fixed machine. (Effectively, the movable machine's shaft is aligned with the shaft of the fixed machinery.) The position of the movable machine is adjusted vertically by adding and/or removing shims from under its feet, and horizontally by making small lateral moves as needed, until the desired alignment is obtained.

### Shaft Alignment with Dial Indicators

The material that follows describes the process of shaft alignment when dial indicators are used to measure the alignment data.

**Instruments and Methods.—**Numbers of instruments are available for making shaft measurements and calculating moves, but the most important requirement for any shaft alignment system is repeatability of the readings. Dial indicators and lasers are two choice measuring systems.

*Dial indicators* provide accurate and reliable measurement of shaft alignment. They are the most useful because they can be used to measure bearing alignment, shaft runout, and soft foot directly. Measurement accuracy down to 0.001 inch (1 mil) or 0.025 mm may be achieved if care is taken in mounting and reading the indicators correctly, and controlling or accounting for such variables as indicator sag, axial endplay in the shaft, and vibration from outside sources. The data obtained from properly installed dial indicators are converted by equations described later in this section into the vertical and horizontal movements required to bring the movable unit into alignment with the fixed unit.

*Laser measurement systems* are another popular choice for shaft alignment work, although the cost of such systems is much greater than dial indicators. Accuracy of 0.0001 inch (0.0025 mm) or greater is possible, and setup and operation is generally faster and simpler than with dial indicators. Many laser systems can perform some or all of the calculations required to obtain the horizontal and vertical moves. Lasers may not be safe for use in explosive environments.

**Runout Check.—**A runout test of either or both couplings is important only if there is a runout problem. Seldom will the runout of a pump or motor coupling be enough to detect with the eye alone. Standard practice is to position a rim and face indicator setup on the uncoupled couplings (one at a time) and roll said couplings through several 360° turns while monitoring the indicators. The indicator bracket should be firmly attached to a static object close to the pump or motor coupling being checked. Usually a magnetic base designed to hold indicators is used for this purpose if the two couplings involved are more than five or six inches apart. For an ordinary pump and motor, however, it is frequently convenient to mount the indicator jig to the motor to check the runout on the pump and then vice-versa for the motor.

Fig. 4 shows a profile view on the left and a motor- or pump-eye view (right) of a typical runout check setup involving the use of indicators attached to a magnetic base. Assume the rim indicator to have been adjusted to zero, the coupling being checked is rotated several full turns. Repeated turns deliver repeated readings of a maximum minus of −0.XXX inch at one extreme, and a maximum plus of 0.XXX inch at the other extreme. Assuming there is no paint or rust, etc. involved, at least three possibilities exist.

1) the coupling was bored off-center
2) the shaft is bent
3) a combination of the above two conditions

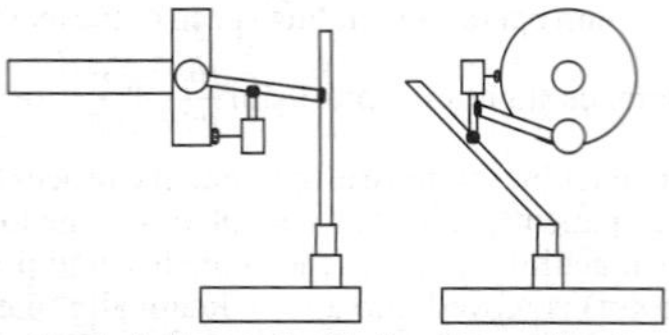

Fig. 4. Indicators Set Up for a Runout Check
In field applications, mount the face dial as near as possible to the rim indicator.
The face indicator is drawn at the 6 o'clock position in both views for clarity.

Concerning Item 3) in the previous list, it is possible (however unlikely) for a coupling that was bored off center and would have shown up as coupling runout with a rim indicator test except that a slight bend in the shaft has canceled out what would normally have been evident due to the off-center bored coupling. This odd situation can cause vibration even when a standard test proves there is no misalignment. Coupling runout can also be caused by too large a coupling bore, with the slack being taken up by set screws.

*Interpreting the Indicators:* The actual amount of runout registered by the rim indicator is *half of the difference between the plus and minus reading extremes*.

*Example:* A rim indicator is mounted as shown in Fig. 4 and set on zero. Then, several full turns of the shaft reveals a maximum minus of −0.006 inch and a maximum plus of +0.012 inch on the indicator. The total variation is 0.018 inch and half of 0.018 is 0.009. Thus, the result of Items 1), 2), or 3) is 0.009 inch of eccentricity, or 0.018 inch TIR (total indicator reading) of coupling runout. *Note:* Much less coupling runout than is described above is usually expected.

If the face indicator shows zero at all points around a coupling while the rim indicator moves between +0.XXX inch and −0.XXX inch, it is probable that the coupling was bored off-center. If the rim indicator shows practically zero movement while the face indicator moves between −0.XXX inch and +0.XXX inch, the cause could be Item 2) or 3) above, or the coupling was bored essentially centered, but at an angle other than perpendicular to the face of the coupling, as is shown in Fig. 5.

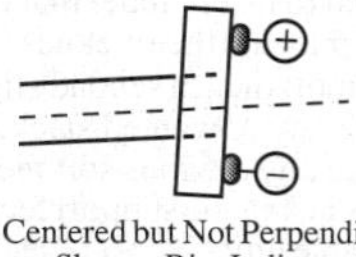


Fig. 5. Coupling Bore Centered but Not Perpendicular to Coupling Face
Face Indicator Shown, Rim Indicator Not Shown

In Fig. 5, both the indicators shown will alternate from plus to minus and back again as the coupling being checked is rotated. Even if the shaft centerline goes through the center of the coupling, it will do so at an angle other than perpendicular to the face of the coupling. The result is serious coupling runout on the face indicator and less serious runout on the rim indicator.

Face runout can easily show up when there is none. If a machine has sleeve bearings and if the shaft is not axially restrained by a thrust bearing during the runout test, severe face runout will likely be indicated when little, if any, exists.

*Note:* Unacceptable runout of either or both couplings should be of little concern in the alignment process. The extended centerlines of the two shafts can be aligned even when there is serious coupling runout. A runout check is mainly to have recorded information if excessive vibration occurs.

**Soft Legs.**—One of the most often overlooked precautions relevant to motor alignment is the firming up of the soft legs. For the motor in Fig. 6, one leg (D) is $^1/_2$ inch short. In alignment jargon, the short leg is said to be *soft*. If a $^1/_2$ inch shim were placed under the short leg, all four motor feet would firmly support the motor. On the other hand, if the motor were tightened down without the use of proper shims, its support would be very spongy. While indicator readings may vary, bolting down electric motors with soft legs can result in high vibration readings two times the line frequency of the incoming power.

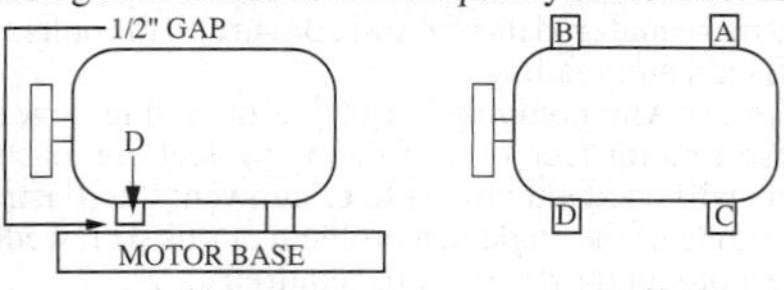


Fig. 6. Profile (a) and Overhead (b) Views of Motor with Soft Foot

On an ordinary electric motor any of the four motor feet may be either raised or lowered via the placement or removal of shims, as applicable, to render the motor equally supported by all four feet. Geometry may allow a certain freedom of choice, so it is practical to be very selective about which motor foot or feet will receive additional shims during the firming up of the motor.

When a soft leg involves only a few thousandths of an inch, it is much less obvious than in Fig. 6. By using what is known to some mechanics as the rock-a-bye method, it is possible to detect a slight amount of instability in the motor supports. With the bolts inserted, but only partly threaded in, alternately apply heel-of-the-hands pressure on or above two motor feet at a diagonal. If no rock-a-bye is evident when heel-of-the-hands pressure is alternately applied above feet A and D, switch to B and C. If rock-a-bye is found, remove applicable shims if possible or add shims under one or both of the unstable feet.

For the $^1/_2$ inch (12.7 mm) gap shown in Fig. 6, downward heel-of-the-hands pressure at points A and D would yield more movement than the indicators could handle. On more subtle soft feet, it takes an experienced feel to detect only a few mils of rock-a-bye at a motor foot. When feel becomes too little to be certain, it is time to apply indicators.

On units where 0.010 inch (0.254 mm) or more of movement is evident, a tapered wedge gauge can be inserted under the motor foot as the motor is gently rocked. When the motor will no longer rock, the approximate thickness of the wedge gauge under the motor foot will be the amount of shim material to insert under that foot. If a single foot requires more shim thickness than is desirable, put half the thickness under the soft foot and the other half under the opposite diagonal foot; then, a second slight amount of shim placed under one of the two feet should remove any unwanted slack. During the rock-a-bye test, trial and error shimming will eliminate any obvious soft feet in a few tries. The careful step-by-step elimination of unwanted slack in the support feet of a motor or other movable unit should result in a very stable movable unit.

On larger motors, leverage proportional to resistance should be used, involving anything from a small pry-bar to hydraulics. When the rock-a-bye method gets so involved, it is advisable to have dial indicators mounted at the applicable motor feet and/or at the motor coupling, with the indicator button on top of the pump coupling or comparable static object, to register any movement.

A heavy duty version of the rock-a-bye method is to have all the hold-down bolts tight and then loosen them one at a time while monitoring the indicators.

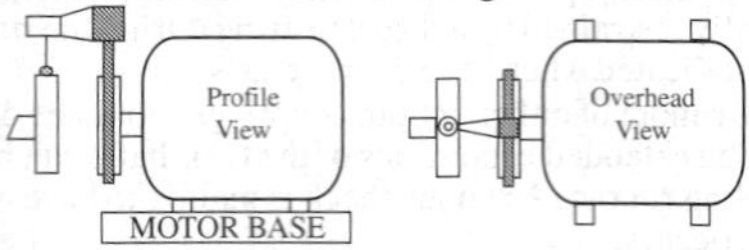


Fig. 7. Rim Indicator Mounted to Test for Soft Feet

Fig. 7 shows a profile view (left) and overhead view (right) of the motor previously shown. In this figure, the indicator illustrated is mounted exactly as the rim indicator will later be mounted for checking alignment, and any soft leg activity will register the same as in the line-up process. Once a motor gets a clean bill of health from this soft foot check, it is safe to proceed with the normal procedures of motor alignment. Mount the indicators in the most convenient manner that will apply to the problem at hand and select the proper formula as discussed in the material that follows. Be sure all the bolts are tight before making the first (vertical) indicator reading

*Other Mounting Issues:* Any combination of the mounting conditions illustrated in Fig. 8 will produce soft motor feet, even if the motor feet are machined properly. The motor, base, or shims will need alterations to circumvent the illustrated problems. Depending on the magnitude of the angle in a problem of this sort, wedge-shaped shims or machine work on the motor or the base may be required.

Sometimes on relatively unimportant equipment, it is permissible to simply fold shims in various thicknesses to take up the slack. This method should not be used as common practice; but as an emergency measure, it is very effective.

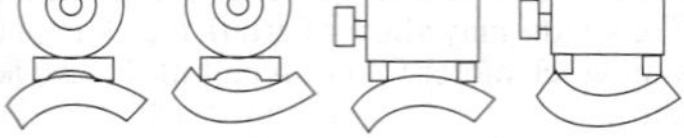

Fig. 8. Mounting Conditions That Produce Soft Motor Feet.

**Mounting Dial Indicators, Perpendicular versus Otherwise.—**When indicators are mounted to get a rim reading on a coupling, the indicator stem should be set to aim straight through the center of the shaft, and at a perpendicular to the axis of the shaft. This center shot at a perpendicular does not have to be absolutely perfect, but it should be close enough so that a closer than casual visual inspection can find no flaw. This is close enough for practical purposes.

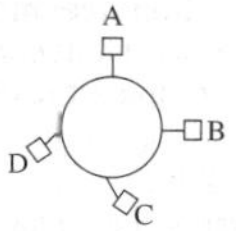


Fig. 9. Indicator Stem Orientation, Correct at A and B, Incorrect at C and D

In Fig. 9 observe the rim indicator settings at positions A, B, C, and D. Indicators at positions A and B are set correctly, indicators at C and D are not. A slight tilt in the mounting of the indicators will make too little difference to matter, but any tilt over a few degrees should be avoided.

Just as important, the face indicator should be set perpendicular to the coupling face with the centerline of the indicator stem parallel to the axis of the fixed unit coupling shaft.

Fig. 10a shows a setup with rim and face indicators set at odd angles that will cause incorrect readings. Note the difference between the arrangement of Fig. 10a, and the correct one shown in Fig. 10b.

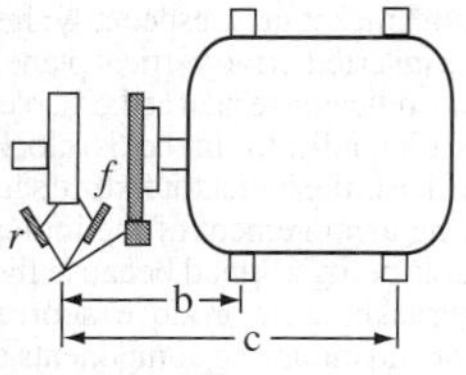


Fig. 10a. Incorrect Positioning of Dial Indicators

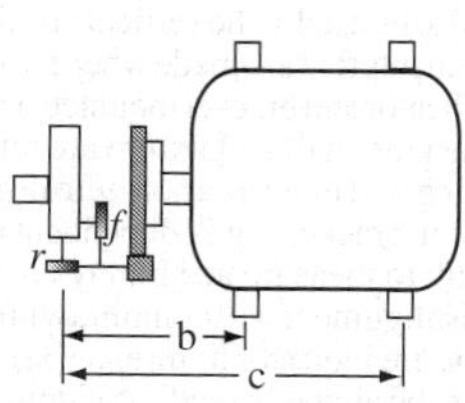


Fig. 10b. Correct Positioning of Dial Indicators

**Preparations.—**Before starting an alignment process several important factors need to be considered, among them the following. Is temperature rise in the coupled machinery a consideration? What is the indicator sag of the dial indicators in use? Is coupling runout a consideration? Are there improperly supported or "soft feet" in the movable machine? Are the motor feet, jack bolts, and shims clean? Can the movable unit be moved laterally far enough to permit alignment, or is movement restricted by bolts ("bolt bound") or nearby obstructions? Is the driving unit powered off and locked out?

*Thermal Effects:* Metals increase in length as temperature increases, and decrease in length as temperature decreases. Rotating equipment at rest slowly attains the temperature of the surrounding environment, the *ambient* or room temperature. In service, machinery such as electric motors and fuel-burning engines generate heat and increase in temperature during operation; other machinery such as pumps may warm or cool during operation.

If initial alignment adjustments are made when machines are at *ambient* or room temperature, thermal movement will occur in various parts of the machines as they reach operating temperature. The various dimensions will not increase or decrease uniformly. In a motor-pump combination, for example, the motor shaft height above the base may increase while the pump shaft height may decrease.

The amount of thermal expansion can be estimated by the formula $\Delta e = \alpha H \Delta T$:

where $\Delta e$ = change in height of shaft

$H$ = vertical height or distance between shaft centerline and shims

$\alpha$ = expansion coefficient of material in μin/in-°F (for stainless steel 7.4, mild steel 6.3, cast iron 5.9, bronze 10, and aluminum 12.6). *Note:* 1 μin/in-°F = 1.8 μm/m-°C

$\Delta T$ = the difference between operating temperature and ambient (initial measurement) temperature

*Example 1(a):* A pump is to be installed and aligned with a motor. The cast iron pump delivers fresh water at 34° F and is at present at 75° F. The pump shaft centerline is 12 inches above the baseline of the pump. The carbon steel motor is also at 75° F and is expected to rise to 140° F when in operation. The motor shaft centerline is 12 inches above the base of the motor. At operating temperature, compare the expected thermal growth of the pump and motor shaft centerlines.

For the pump, the change in shaft height:

$$\Delta e = \alpha H \Delta T = 0.0000059 \times 12 \times (34 - 75) = -0.002903 \text{ inch}$$

For the motor, the change in shaft height is:

$$\Delta e = \alpha H \Delta T = 0.0000063 \times 12 \times (140 - 75) = 0.004914 \text{ inch}$$

These calculations indicate that the pump and motor shafts, if they were in alignment at ambient temperature, would be misaligned at operating temperature by approximately 0.0078 inch difference in elevation. Thermal movement also can be the result of improperly isolated and/or supported heated process piping. This should be checked if thermal growth is still an issue after making alignment changes.

*Indicator Bracket Sag:* Indicator sag, indicator bracket sag, or jig sag are all terms commonly used to describe the effect on indicator readings of the weight of the dial indicator(s) in combination with the length, weight, and orientation of the indicator bracket. Dial indicators are designed to be used in the vertical position and indicator bracket sag is especially significant in measurements that are made when the dial indicator is inverted in the vertical plane. Consider a rim indicator and bracket mounted on a shaft with the indicator zeroed at the 12 o'clock position; when the shaft and fixture are rotated to put the dial indicator in the 6 o'clock position, the reading will indicate a non-zero negative value, which is the indicator fixture sag factor.

Indicator bracket sag is dependent on the mounting arrangement of the indicators, and is difficult to measure accurately on equipment that is being aligned because the effect of shaft misalignment will combine with the indicator bracket sag to produce incorrect results.

To measure indicator fixture sag, set up the indicator and mounting components on the machinery to be aligned exactly as if doing alignment, and tighten everything down. Remove the bracket and indicator from the machine and remount in exactly the same manner on a rigid fixture, such as a piece of pipe supported by V-blocks. Zero the indicator at the 12 o'clock (reference) position, then without changing the setup carefully rotate the fixture and read the indicator at the 6 o'clock (reading) position. At the reading position, the indicator displays the total indicator reading (TIR); this value is the *total sag factor.* The actual sag in the indicator bracket is equal to one half of the total indicator reading.

When the indicator bracket sag has been determined, it can be accounted for in measurements taken during the alignment procedure. One procedure is to add the indicator bracket sag factor to the indicator value when the indicator is "zeroed" at the reference position, so that the sag factor is accounted for throughout subsequent calculations. For example, if an indicator is zeroed at the 12 o'clock position and reads 0.006 inch at the 6 o'clock position, set the indicator to positive 0.006 at the 12 o'clock reference position to account for the sag factor in further measurements. If the reference position used for checking indicator sag is the 6 o'clock position (indicator zeroed at 6 o'clock position), the positive reading found at the 12 o'clock position is the sag factor that must be subtracted from indicator value when the indicator is zeroed at the 6 o'clock (reference) position.

**Alignment Procedures for Machinery.**—The *face and rim method* and the *reverse dial indicator method* are the techniques most commonly used for aligning machinery shafts. Common shaft alignment procedure is to establish one machine, usually the driven machine, as fixed, and the other machine as movable. The shaft of the movable machine is positioned, as required, to align with the shaft of the fixed one. Most aligning specialists recommend vertical alignment first, and then horizontal alignment. It is important to verify in advance that the movable unit has sufficient horizontal clearance and freedom of movement to obtain the required horizontal position before starting on the vertical alignment. Also axial positioning of the driven shaft must ensure there is no preload on its bearings, whether antifriction or sleeve, or their life will be drastically reduced.

*Vertical Moves:* Vertical adjustments are accomplished by adding or removing shims from under the machine feet. Use as few shims as possible. Too many shims under a foot may cause it to act like a soft foot by creating a spring effect; usually a maximum of four shims are allowed. Shims should be measured by micrometer before placing. If possible, apply a single shim that is the total thickness required under the foot. Insert the shims all the way until the shim slot bottoms out on the bolt, then pull the shim back about a quarter inch before tightening the bolts. If alignment is not possible with four shims, then use fewer, amounting to the required thickness. Tighten bolts to their prescribed torque each time after placing shims to avoid any movement when working on the other side of the machine.

Soft feet correction allows all foot bolts to be loose at the same time for adjusting the vertical machine height. Once the machine is lifted, remove all shims and add or subtract the number of shims required for the needed vertical adjustment. (Be cautious not to get confused among a lot of loose shims.) Re-torque all foot bolts and check vertical alignment. Once it is within specification, loosen the foot bolts to allow for side-to-side movement for horizontal alignment.

*Horizontal Moves:* It is preferable to utilize a base plate with jackbolts for horizontal moves. If jack bolts are not available, a small hydraulic jack, pipe clamps, or pony clamps can be used. The most accurate method for measuring the horizontal move is to place dial indicators around the machine at convenient locations, such as at machine feet. The indicators can be located anywhere convenient as long as the dimensions used in the alignment calculations reflect the actual position at which measurements are made. When the indicators are mounted, adjust the dials to zero, and move the unit the distance and direction indicated by the calculations. Once the horizontal move has been completed, a final set of readings should be taken to verify both vertical and horizontal alignment.

*Safety:* Shaft alignment should not be performed when machines are running. Before shaft alignment, all sources of power to the machine should be off and verified. Energy sources must be locked out by electrical controls; steam valves, gas main valve, and fuel oil valve must be shut down before shaft alignment.

**Rim and Face Alignment Procedure.—**Fig. 11 and Fig. 12 show dial indicator arrangements for rim and face alignment that are typical of those used in the remainder of this section. Fig. 11 illustrates a chain indicator jig, but it could be of any type. The arrangement in Fig. 12 represents an equivalent setup. In each figure, two dial indicators are shown, one for the rim reading and the other for the face reading. Darkened squares, triangles, or circles are simply to indicate attachment points.

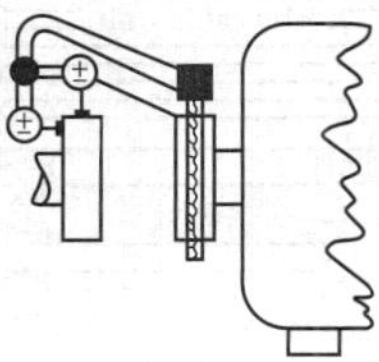

Fig. 11.

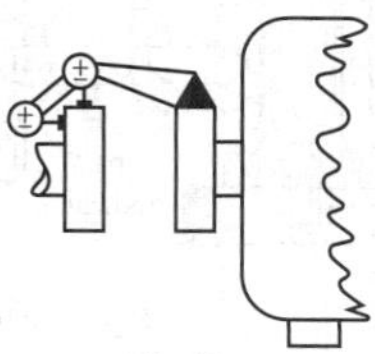

Fig. 12.

Starting on page 2737, a number of different indicator arrangements are illustrated in Fig. 1a through Fig. 16a. Each arrangement is associated with a specific equation and a chart that facilitates understanding the dial and face readings and the corresponding tasks of determining where shims need to be added or removed and what horizontal moves are required. To use this system, the mechanic needs to set up the rim and face indicators in the most convenient manner, then turn to the figure of the matching indicator arrangement and use the formulas provided.

*Example Aligning a Motor with a Pump by the Rim and Face Method:* A 10-horsepower (7.5 kW) electric motor needs to be precision aligned to a specific pump. Both the pump and motor are to be mounted on a base that when finished will look similar to Fig. 13.

Not shown are the intake and discharge piping, electrical fixtures, and other incidentals including shims under the motor feet.

The shims are omitted to visually explain why the motor will usually have a little less distance from the motor shaft centerline to the base than the pump has. This set up is desirable because with the motor shaft a little lower than the pump shaft, the motor can be shimmed up until both shafts share a common elevation. If the motor shaft is higher than the pump shaft, it can be very difficult to alter the position of the pump because of the rigidly connected piping.

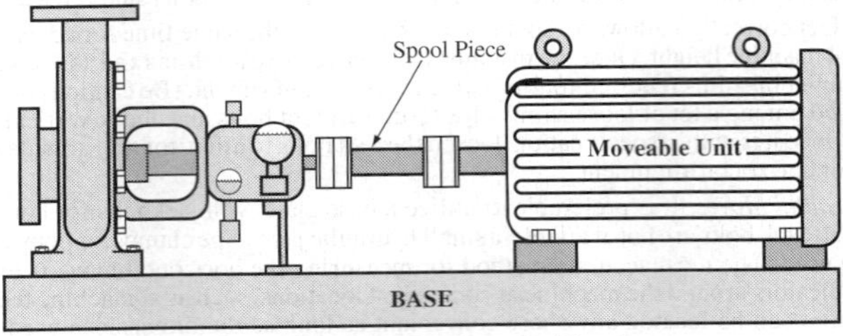

Fig. 13. Pump and Motor are to be Aligned

The motor in Fig. 13 is in its approximate finished position about 8 inches (203.2 mm) from the pump. The spool piece shown in Fig. 13 is to be bolted between the motor and pump couplings after alignment is complete. The spool piece (not shown in Fig. 14) is 8 inches long and there is an $8^1/_8$ inch (206.4 mm) space between pump and motor couplings.

Fig. 14 shows a viable mounting of dial indicators in the process of procuring readings from the zero setting at 12 o'clock (broken lines) to the 6 o'clock position (solid lines). Some calibrated eyeballing, a scale measure and some straight edging are used to align the motor initially to the pump.

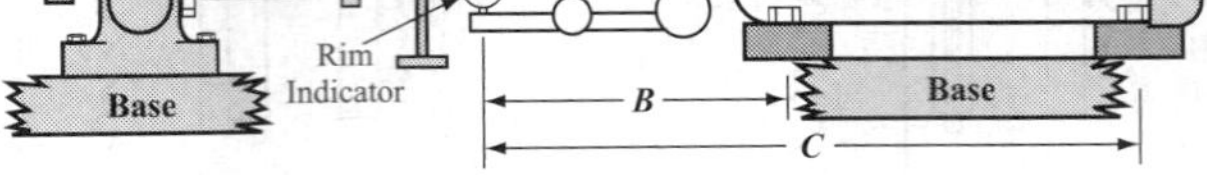

Fig. 14. Alignment Setup for Pump and Motor

There are many ways to mount and adjust dial indicators to measure the remaining misalignment between the two units. Several are correct and only a few are optimum. The mounting of the pair of dial indicators between the motor and pump shown in Fig. 14 is as good as any and commonly used. Other arrangements are shown starting on page 2737.

Although there seem to be four dial indicators in Fig. 14, there are only two. The two indicators with broken lines are drawn at the initial 12 o'clock position where they are zeroed with about half the indicator stems exposed, (both pointers are shown at 12 o'clock). Then the motor and pump shafts are both turned 180° to the 6 o'clock position (those indicators are drawn with solid lines, and the pointers are shown set to the 3 o'clock and 9 o'clock positions in Fig. 14). In real life, the pointers may be aimed anywhere on the dial face, but the dial will rotate 360°, so the zero can be adjusted to align with the pointer.

*Solution for Vertical Alignment:* Once a motor has been aligned with straight edges and such, it is probably close enough to use dial indicators. The dials most often used have a stem range of just over 0.200 inch (roughly $^7/_{32}$ inch). These dial indicators (usually two) are positioned with the buttons of the indicators against the rim and face of the pump coupling respectively, as shown in Fig. 14. The seasoned millwright/mechanic will adjust the indicator jig/bracket in such manner that the stem on each of the indicators has freedom to travel inward or outward about 0.100 inch. This adjustment is made at the 12 o'clock position, shown by broken lines in Fig. 14.

A good arrangement is for the rim indicator in Fig. 14 to register a minus *r* reading at the 6 o'clock position and the face indicator to register a plus *f* reading (see bottom right of Fig. 1b on page 2738). Readings on both indicators show that the motor needs to be raised involving the simple task of determining how much shim material to add under each foot For the indicator arrangement in Fig. 14. the required shim thickness can be determined by using the equations below.

$$F = \frac{\pm r}{-2} + \frac{\pm f \times B}{A} \qquad \text{and} \qquad R = \frac{\pm r}{-2} + \frac{\pm f \times C}{A}$$

where $F$ = ±change in shim thickness needed under the front feet
$R$ = ±change in shim thickness needed under the rear feet
$r$ = plus or minus rim indicator reading
$f$ = plus or minus face indicator reading
$A$ = diameter of face indicator path on coupling
$B$ = distance from rim indicator path to front feet
$C$ = distance from rim indicator path to rear feet

The equations use the gathered data to calculate the correct shim placement. With either equation, a positive solution indicates that shims equal to the calculated value are added, and a negative solution indicates that shims equal to the calculated value are removed.

The rim and face indicator setup depicted in Fig. 14 is one of the sixteen arrangements described in the material that follows. Just plug the variables into the formulas that match the current indicator format and calculate the shims required.

*Solution for Horizontal Alignment:* Fig. 15 shows a top view of the motor in Fig. 14, with indicators in the reading positions for horizontal alignment. The rim and face indicators are zeroed at the position 180° opposed to that shown in Fig. 15. The directions corresponding to horizontal movement of the motor in Fig. 15 are indicated by east and west arrows. Magnetic-mount dial indicators (not shown) may be set up to monitor the horizontal movements. The dimensions *B* and *C* represent the distances from the rim indicator path to the locations of the monitoring indicators or jackbolts.

If jackbolts (or monitor indicators) are positioned on the west side as indicated in Fig 15, the *B* and *C* dimensions are used in the equations. If jackbolts (or monitor indicators) are positioned at a different location on the east side (for example, jackbolts shown by dashed lines in Fig. 15), the east-side set of dimensions, $B'$ and $C'$, are used in the equations.

The equations are the same as previously used in the vertical alignment procedure Per the indicator arrangement in Fig. 15, a positive solution demands a move eastward, and a negative solution requires a westward move.

$F$ = ±east-west movement needed at the front monitoring location
$R$ = ±east-west movement needed at the rear monitoring location
$r$ = plus or minus rim indicator reading
$f$ = plus or minus face indicator reading
$A$ = diameter of face indicator path
$B$ = distance from rim indicator path to front monitoring location
$C$ = distance from rim indicator path to rear monitoring location

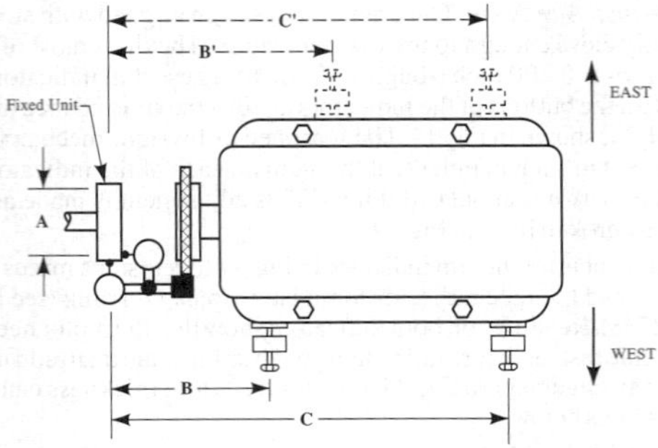


Fig. 15. Top View of Motor from Fig. 14

*Side to Side Adjustment Using Jackbolts:* Avoid using jackbolts that are longer than necessary. Make sure internal and external threads are clean and lubricated enough to be functional. Then, if heavy resistance is encountered when adjusting jackbolts, it indicates a mechanically-bound condition, not a rusty jackbolt thread. Before advancing jackbolts on one side of the movable unit, back off the jackbolts on the opposite side about $1\frac{1}{2}$ turns to provide enough space to make the movement without creating a mechanical bind.

Jackbolts should not be left in a "bind" against the movable unit to assist in keeping the movable unit in alignment. After alignment is complete, the jackbolts should be removed and stored for future use. This action will prevent jackbolt corrosion, and leave the jackbolt nuts vacant to receive a tap and lubrication, prior to being used again.

On movable units outfitted with jackbolts, one complete (360°) turn of a jackbolt will advance the thread an amount equal to the pitch of the screw thread. For inch-threaded screws, pitch equals 1 divided by the number of threads per inch (1 ÷ TPI). The axial travel per head flat is the pitch divided by the number of flats on the bolt head.

*Example:* A jackbolt has a $\frac{1}{2}$-13 thread. Find the axial travel per revolution and per flat for square-head and hex-head bolts. The thread has 13 threads per inch.

$$\text{advance per revolution (full turn) of bolt} = \frac{1}{\text{TPI}} = \frac{1}{13} = 0.0769 \text{ inch}$$

$$\text{advance per flat (1/4 turn) of square-head bolt} = \frac{1}{\text{TPI} \times 4} = \frac{1}{(13 \times 4)} = 0.0192 \text{ inch}$$

$$\text{advance per flat (1/6 turn) of hex-head bolt} = \frac{1}{\text{TPI} \times 6} = \frac{1}{(13 \times 6)} = 0.0128 \text{ inch}$$

*Example:* A motor needs to be moved 0.064 inch east at the front jackbolt and 0.091 east at the rear. The square-head jackbolts are $\frac{5}{8}$-11. How many turns of the front and rear jackbolts are need to bring the unit into approximate alignment?

$$\text{one full turn of jackbolt} = \frac{1}{11} = 0.0909 \text{ inch}$$

$$\text{full turns at front jackbolt} = \frac{0.064}{0.0909} = 0.7040 \approx \frac{3}{4} \text{ turn or 3 flats}$$

$$\text{full turns at rear jackbolt} = \frac{0.091}{0.0909} = 1.0011 \text{ turn} \approx 1 \text{ turn or 4 flats}$$

## Practical Rim and Face Dial Indicator Arrangements for Shaft Alignment

This phase of millwright/machinist work is usually referred to as motor-to-pump alignment. The true objective is to correctly align the axis of one shaft to the axis of another by any of a rather wide range of dial indicator mountings, depending on a myriad of varying situations. These motor to pump scenarios can be duplicated on almost any shaft that is axially in close proximity to another shaft that has to share a common axis. Addressed here is standard rim and face pump alignment.

There are sixteen indicator arrangements to select from in this rim and face section. Units are aligned as closely as possible with a straight edge before using the dial indicators. At the first indicator reading, a neat sketch of the scenario is made, along with a carefully measured dimensioning schedule such as that shown in Fig. 1a. Based on indicator readings, select the applicable block from the second illustration, Fig. 1b, that matches the current plus or minus (±) indicator reading. The dashed lines in Fig. 1b represent the centerline of the movable unit shaft. The two heavy round dots on the dashed lines (to the right of the couplings) represent points directly above or below the movable unit shim locations, and indicate that the movable unit must be raised or lowered at that location. The diagrams will help clarify the results of the shim calculations made using the appropriate formulas. Be certain to use minus signs any time they are called for in the formulas.

*Note:* In all the rim and face formats that follow, the axial location of the rim indicator always dictates the beginning point of the dimensions that are used in the various formulas. Also, the *dial indicator arrangements in* Fig. 1a *through* Fig. 16a *are shown in the reading position*, after having been zeroed at the position 180° opposite that shown and then rotated into the position shown.

**1st Arrangement.—**Fig. 1a illustrates the most universally accepted method of mounting dial indicators for shaft alignment, and it is also the same arrangement that was used in the preceding example, illustrated by Fig. 14 and Fig. 15.

When a reading of the rim and face indicators has been made, find the diagram in Fig. 1b that corresponds to the signs (positive or negative) of the readings obtained (positive rim value and negative face value, for example). The appropriate diagram will provide a visual clue to the meaning of the indicator readings, and indicate the direction that the motor feet must be moved to obtain alignment of the movable shaft with the fixed shaft.

Finally, use the equations adjacent to Fig. 1a to calculate the exact thickness of shims that need to be added or removed at each foot.

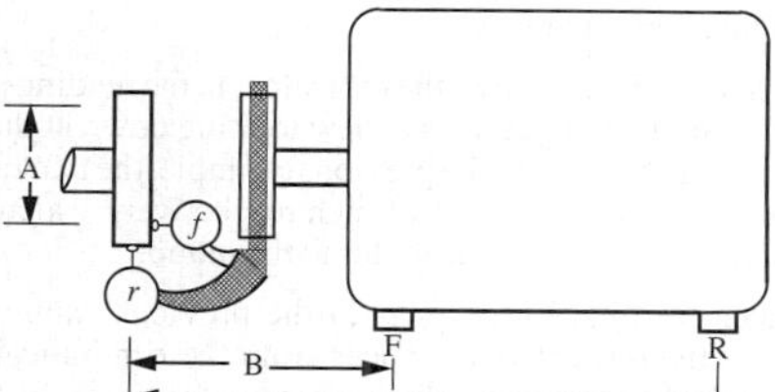


$$F = \frac{\pm r}{-2} + \frac{\pm f \times B}{A}$$

$$R = \frac{\pm r}{-2} + \frac{\pm f \times C}{A}$$

Fig. 1a. Dial Indicator Arrangement 1

In the rim and face format in Fig. 1a, *r* and *f* represent the values of the rim and face indicators in their reading positions. The F and R below the motor are the front and rear feet, corresponding to the *F* and *R* values in the formulas that represent the needed plus or minus (±) shim packs. The formulas adjacent to Fig. 1a, are used to calculate the necessary addition or subtraction of shims required on the front or rear feet.

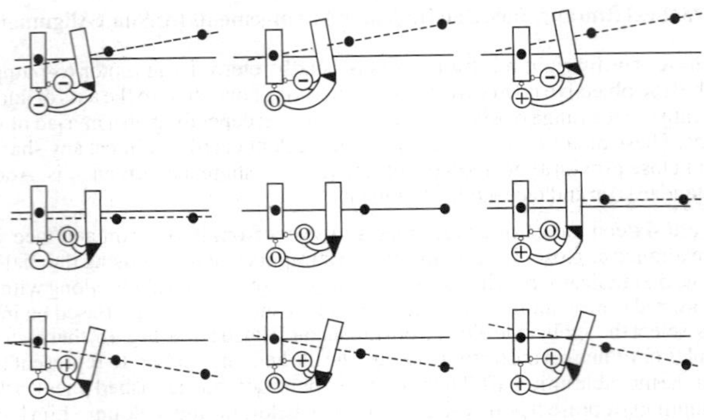

Fig. 1b. Interpreting Dial Indicator Readings for Indicator Arrangement 1

Indicators are shown in the reading position. Solid lines represent the desired location for the dashed lines (movable unit). The plus, minus, and zero readings on the indicators are represented by one of these nine scenarios. Depending on the magnitude of the variables, the generic graph gives a mental picture of what the formulas should yield.

*Example 1(a), Vertical Alignment:* Calculate the necessary shims for a pump and motor alignment where dimensions are $A = 4$ inches, $B = 6.5$ inches, and $C = 18$ inches. The two indicator dials read $r$ = minus (−) 0.014 inch and $f$ = plus (+) 0.021 inch respectively.

*Solution:* The lower left block in Fig. 1b corresponds to a negative rim reading and a positive face reading. The figure indicates that both the front and rear feet will require shims, and that the rear feet will require more shims than the front feet. Shims on front and rear feet will be as follows:

$$F = \frac{\pm r}{-2} + \frac{\pm f \times B}{A} = \frac{-0.014}{-2} + \frac{0.021 \times 6.5}{4} = 0.007 + 0.034125 = 0.041 \text{ inch}$$

$$R = \frac{\pm r}{-2} + \frac{\pm f \times C}{A} = \frac{-0.014}{-2} + \frac{0.021 \times 18}{4} = 0.007 + 0.0945 = 0.102 \text{ inch}$$

The calculations indicate that 0.041 inch of shims is required under both front feet, and 0.102 inch of shims is required under both rear feet.

*Example 1(b), Horizontal Alignment:* Fig. 1a. shows the indicators in the reading position for horizontal alignment, if the figure is thought of as a view looking down at the motor from the top. Assuming the same data obtained in the previous example, the two indicator dials read $r$ = minus (−) 0.014 inch and $f$ = plus (+) 0.021 inch respectively. Calculate the necessary horizontal movement required to align the motor to the pump.

*Solution:* The equations adjacent to Fig. 1a apply, as in the previous example. The dimensions $B$ and $C$ in Fig. 1a are the respective distances from the rim indicator path on the coupling to the locations at the front and rear of the motor, not necessarily the feet, where the horizontal adjustments will be made. These locations could be the front and rear jackbolts, or they could be magnetic base monitoring indicators placed conveniently near the front and rear feet, as discussed previously with respect to Fig. 15.

The diagram in the lower left corner of Fig. 1b indicates the direction of the move when the figure is viewed as if looking at the motor from the top. If $B$ and $C$ are 6.5 and 18 inches, respectively, then the equations indicate that the front and rear of the motor must be moved in the same direction, +0.041 inch and +0.102 inch respectively, to attain alignment.

**2nd Arrangement.—**Rim and face indicators are placed as shown in Fig. 2a. This arrangement is commonly used when the space between couplings is limited.

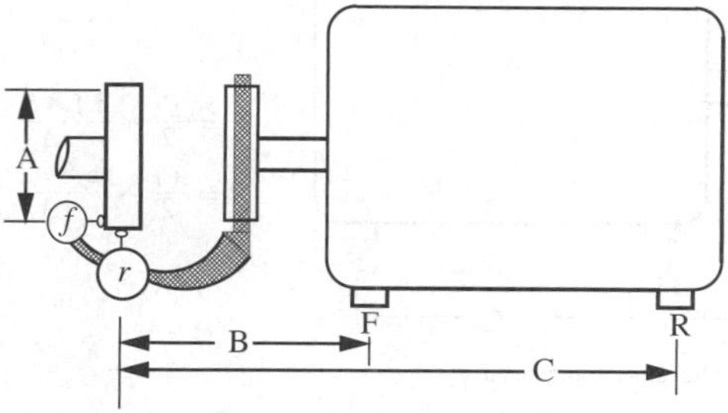


$$F = \frac{\pm r}{-2} + \frac{\pm f \times B}{-A}$$

$$R = \frac{\pm r}{-2} + \frac{\pm f \times C}{-A}$$

Fig. 2a. Dial Indicator Arrangement 2

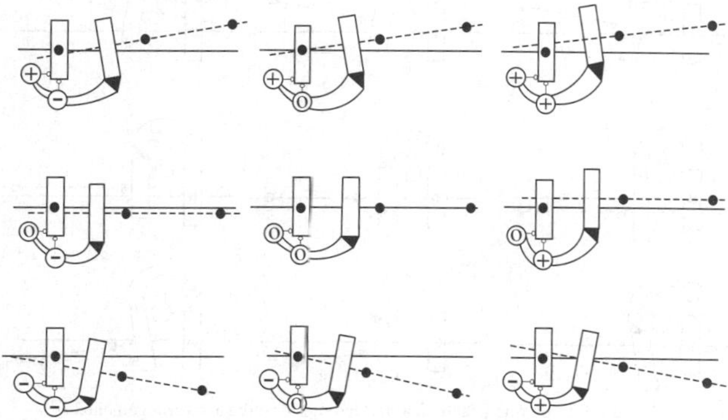

Fig. 2b. Interpreting Dial Indicator Readings for Indicator Arrangement 2

Indicators are shown in the reading position. Solid lines represent the desired location for the dashed lines (movable unit). The plus, minus, and zero readings on the indicators are represented by one of these nine scenarios. The generic graph gives a picture of what the formulas should yield.

*Example:* Calculate the shims necessary for a pump and motor alignment where dimensions are $A = 4.75$ inches, $B = 7.75$ inches, and $C = 17.25$ inches. The two indicator dials read $r =$ plus (+) 0.004 inch and $f =$ minus (−) 0.022 inch respectively.

*Solution:* The lower right block in Fig. 2b corresponds to a positive rim reading and a negative face. The diagram indicates that shims will be required under both the front and rear feet. Calculations confirm this, as follows:

$$\begin{aligned} F &= \frac{\pm r}{-2} + \frac{\pm f \times B}{-A} \\ &= \frac{0.004}{-2} + \frac{-0.022 \times 7.75}{-4.75} \\ &= -0.002 + 0.036 \\ &= 0.034 \text{ inch} \end{aligned}$$

$$\begin{aligned} R &= \frac{\pm r}{-2} + \frac{\pm f \times C}{-A} \\ &= \frac{0.004}{-2} + \frac{-0.022 \times 17.25}{-4.75} \\ &= -0.002 + 0.07989 \\ &= 0.078 \text{ inch} \end{aligned}$$

**3rd Arrangement.**—Place rim and face indicators as shown in Fig. 3a.

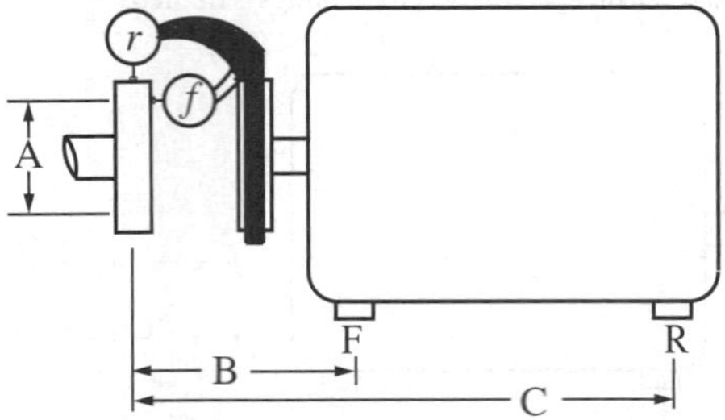


$$F = \frac{\pm r}{2} + \frac{\pm f \times B}{-A}$$

$$R = \frac{\pm r}{2} + \frac{\pm f \times C}{-A}$$

Fig. 3a. Dial Indicator Arrangement 3

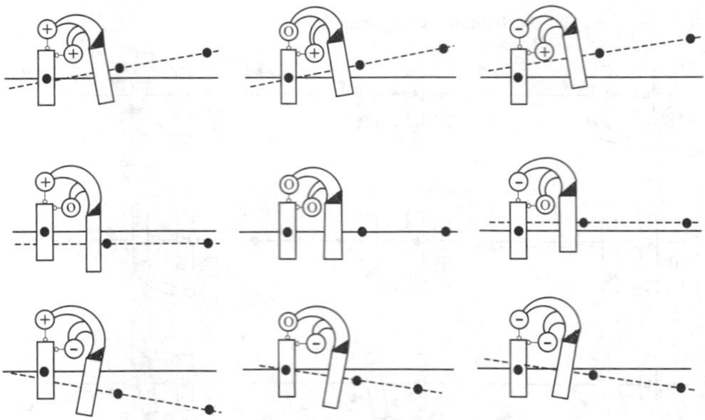

Fig. 3b. Interpreting Dial Indicator Readings for Indicator Arrangement 3

Solid lines represent the desired location for the dashed lines (movable unit). The plus, minus, and zero readings on the indicators are represented by one of these nine scenarios. Depending on the magnitude of the variables, the generic graph gives a mental picture of what the formulas should yield.

*Example:* Calculate the shims necessary for a pump and motor alignment where dimensions are $A = 4.25$ inches, $B = 8.875$ inches, and $C = 28.75$ inches. The two indicators read $r =$ plus (+)0.042 inch and $f =$ minus (−) 0.036 inch respectively.

*Solution:* Only the lower left block in Fig. 3b corresponds to a negative face reading and a positive rim reading. The diagram indicates that shims will be required under both front and rear feet. The required shim thickness for front and rear feet is calculated as follows:

$$\begin{aligned} F &= \frac{\pm r}{2} + \frac{\pm f \times B}{-A} \\ &= \frac{0.042}{2} + \frac{-0.036 \times 8.875}{-4.25} \\ &= 0.021 + 0.075 \\ &= 0.096 \text{ inch} \end{aligned}$$

$$\begin{aligned} R &= \frac{\pm r}{2} + \frac{\pm f \times C}{-A} \\ &= \frac{0.042}{2} + \frac{-0.036 \times 28.75}{-4.25} \\ &= 0.021 + 0.244 \\ &= 0.265 \text{ inch} \end{aligned}$$

**4th Arrangement.—**Place rim and face indicators as shown in Fig. 4a.

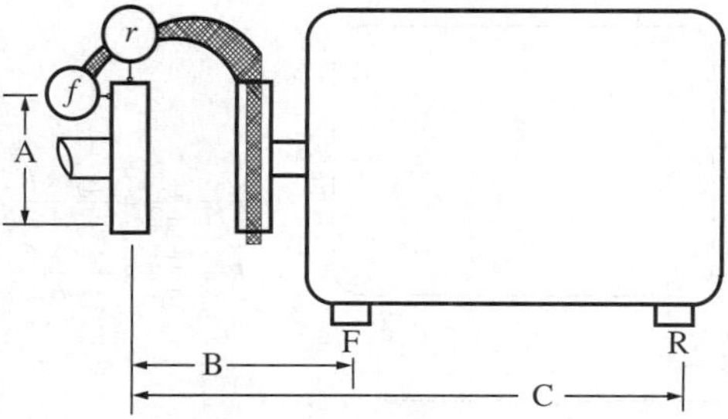


$$F = \frac{\pm r}{2} + \frac{\pm f \times B}{A}$$

$$R = \frac{\pm r}{2} + \frac{\pm f \times C}{A}$$

Fig. 4a. Dial Indicator Arrangement 4

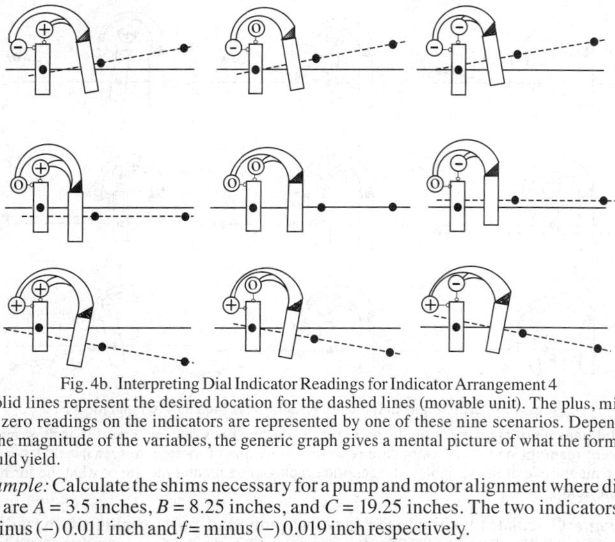

Fig. 4b. Interpreting Dial Indicator Readings for Indicator Arrangement 4

Solid lines represent the desired location for the dashed lines (movable unit). The plus, minus, and zero readings on the indicators are represented by one of these nine scenarios. Depending on the magnitude of the variables, the generic graph gives a mental picture of what the formulas should yield.

*Example:* Calculate the shims necessary for a pump and motor alignment where dimensions are $A = 3.5$ inches, $B = 8.25$ inches, and $C = 19.25$ inches. The two indicators read $r =$ minus (–) 0.011 inch and $f =$ minus (–) 0.019 inch respectively.

*Solution:* According to the top left block in Fig. 4b, the negative rim and negative face values indicate that the motor needs to have shims removed from beneath both the front and rear feet. The calculations that follow confirm that 0.05 inch needs to be removed from beneath the front feet, and 0.11 inch from below the rear feet. This adjusting may be difficult if the motor is sitting on its base without shims.

$$\begin{aligned} F &= \frac{\pm r}{2} + \frac{\pm f \times B}{A} \\ &= \frac{-0.011}{2} + \frac{-0.019 \times 8.25}{3.5} \\ &= -0.0055 + (-0.04479) \\ &= -0.050 \text{ inch} \end{aligned}$$

$$\begin{aligned} R &= \frac{\pm r}{2} + \frac{\pm f \times C}{A} \\ &= \frac{-0.011}{2} + \frac{-0.019 \times 19.25}{3.5} \\ &= -0.0055 + (-0.1045) \\ &= -0.110 \text{ inch} \end{aligned}$$

**5th Arrangement.**—Place rim and face indicators as shown in Fig. 5a. In the figure, notice that the origin of the dimensions *B* and *C* is the line of the rim indicator.

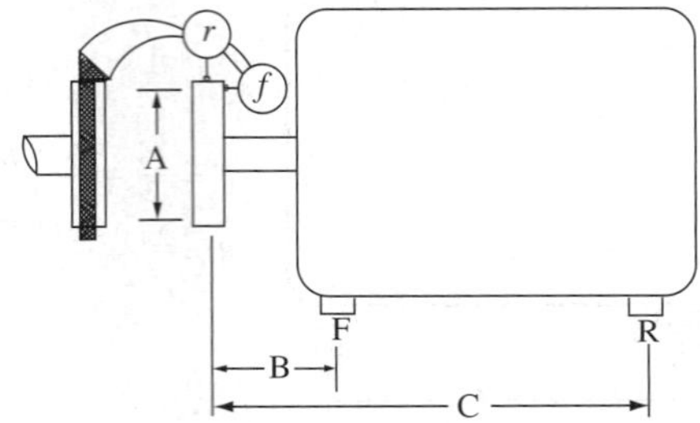


$$F = \frac{\pm r}{-2} + \frac{\pm f \times B}{A}$$

$$R = \frac{\pm r}{-2} + \frac{\pm f \times C}{A}$$

Fig. 5a. Dial Indicator Arrangement 5

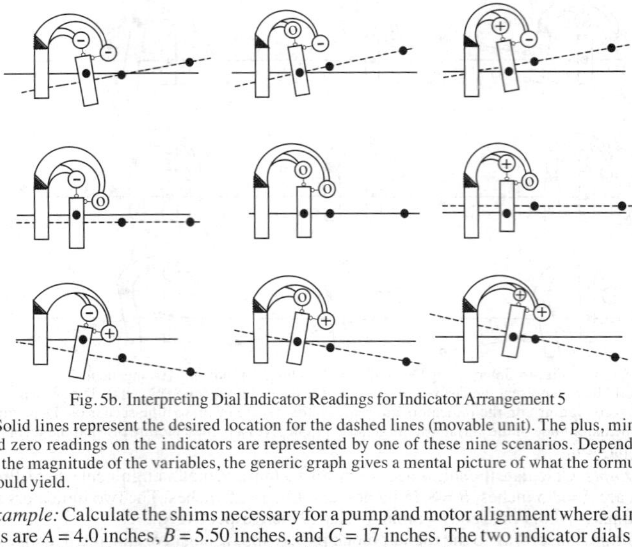

Fig. 5b. Interpreting Dial Indicator Readings for Indicator Arrangement 5

Solid lines represent the desired location for the dashed lines (movable unit). The plus, minus, and zero readings on the indicators are represented by one of these nine scenarios. Depending on the magnitude of the variables, the generic graph gives a mental picture of what the formulas should yield.

*Example:* Calculate the shims necessary for a pump and motor alignment where dimensions are $A = 4.0$ inches, $B = 5.50$ inches, and $C = 17$ inches. The two indicator dials read $r$ = plus (+) 0.042 inch and $f$ = minus (−) 0.036 inch respectively.

*Solution:* The top right block in Fig. 5b corresponds to the positive rim value and a negative face value of the indicators, and indicates that both ends of the motor need to be moved down. This requirement is confirmed by the negative shim numbers that provide the exact distance as follows:

$$\begin{aligned} F &= \frac{\pm r}{-2} + \frac{\pm f \times B}{A} \\ &= \frac{0.042}{-2} + \frac{-0.036 \times 5.50}{4.0} \\ &= -0.021 + (-0.0495) \\ &= -0.0705 \text{ inch} \end{aligned}$$

$$\begin{aligned} R &= \frac{\pm r}{-2} + \frac{\pm f \times C}{A} \\ &= \frac{0.042}{-2} + \frac{-0.036 \times 17.0}{4.0} \\ &= -0.021 + (-0.153) \\ &= -0.174 \text{ inch} \end{aligned}$$

**6th Arrangement.—**Place rim and face indicators as shown in Fig. 6a.

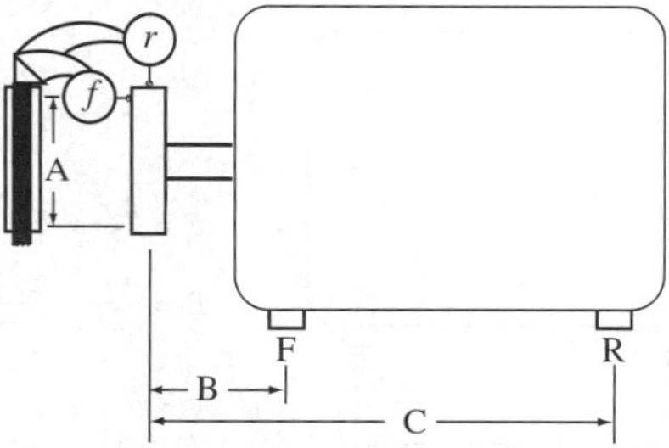


$$F = \frac{\pm r}{-2} + \frac{\pm f \times B}{-A}$$

$$R = \frac{\pm r}{-2} + \frac{\pm f \times C}{-A}$$

Fig. 6a. Dial Indicator Arrangement 6

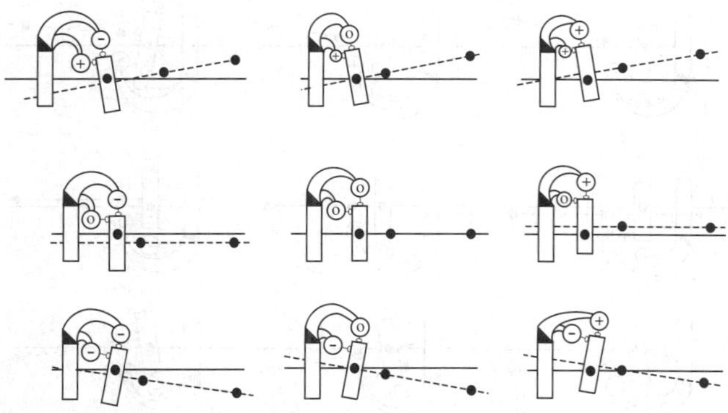

Fig. 6b. Interpreting Dial Indicator Readings for Indicator Arrangement 6

Solid lines represent the desired location for the dashed lines (movable unit). The plus, minus, and zero readings on the indicators are represented by one of these nine scenarios. Depending on the magnitude of the variables, the generic graph gives a mental picture of what the formulas should yield.

*Example:* Calculate the shims necessary for a pump and motor alignment where dimensions are $A = 4.75$ inches, $B = 6.25$ inches, and $C = 31.75$ inches. The two indicator dials: $r$ = minus (–) 0.024 inch and $f$ = plus (+) 0.005 inch respectively.

*Solution:* The top left block in Fig. 6b describes the situation of negative rim and positive face values. The generic graph suggests that both the front and rear feet need some shims to be removed, but the calculations below require that some positive movement of the front feet is required. If in doubt, follow the numbers; the graphs are not drawn to scale. The formulas call for 0.005 inch added under the front feet and the rear feet to be lowered by 0.021 inch:

$$\begin{aligned} F &= \frac{\pm r}{-2} + \frac{\pm f \times B}{-A} \\ &= \frac{-0.024}{-2} + \frac{0.005 \times 6.25}{-4.75} \\ &= 0.012 - 0.006578 \\ &= 0.005 \text{ inch} \end{aligned}$$

$$\begin{aligned} R &= \frac{\pm r}{-2} + \frac{\pm f \times C}{-A} \\ &= \frac{-0.024}{-2} + \frac{0.005 \times 31.75}{-4.75} \\ &= 0.012 + (-0.03342) \\ &= -0.021 \text{ inch} \end{aligned}$$

**7th Arrangement.**—Place rim and face indicators as shown in Fig. 7a.

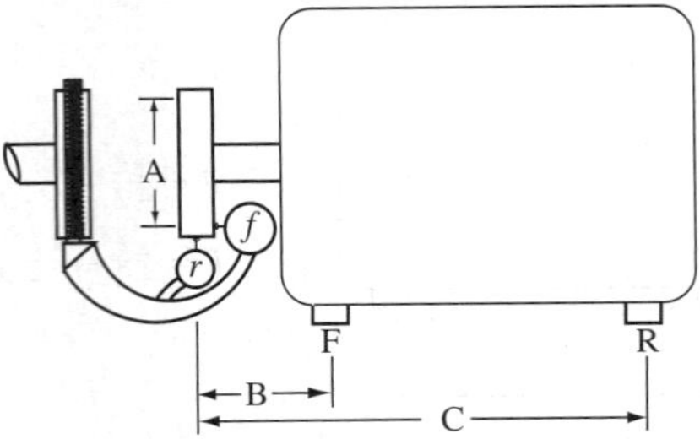


$$F = \frac{\pm r}{2} + \frac{\pm f \times B}{-A}$$

$$R = \frac{\pm r}{2} + \frac{\pm f \times C}{-A}$$

Fig. 7a. Dial Indicator Arrangement 7

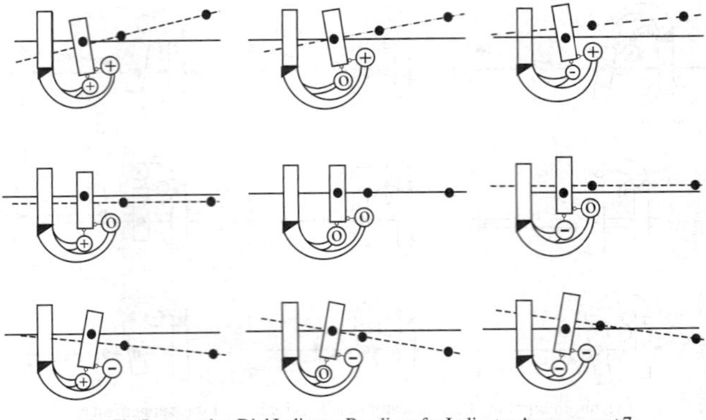

Fig. 7b. Interpreting Dial Indicator Readings for Indicator Arrangement 7

Solid lines represent the desired location for the dashed lines (movable unit). The plus, minus, and zero readings on the indicators are represented by one of these nine scenarios. Depending on the magnitude of the variables, the generic graph gives a mental picture of what the formulas should yield.

*Example:* Calculate the shims necessary for a pump and motor alignment where dimensions are $A$ =12.375 inches, $B$ =14.875 inches, and $C$ = 47.75 inches. The two indicator dials: $r$ = minus (−) 0.002 inch and $f$ = plus (+) 0.001 inch respectively.

*Solution:* This motor is almost in alignment. The upper right block of Fig. 7b corresponds to the negative rim and positive face condition. The graph suggests both the front and rear motor feet be lowered, depending on the magnitude of the variables. The formula solution dictates exactly how much the movable unit needs to be lowered at both front and rear shim points. Shims on front and rear feet will be as follows:

$$\begin{aligned} F &= \frac{\pm r}{2} + \frac{\pm f \times B}{-A} \\ &= \frac{-0.002}{2} + \frac{0.001 \times 14.875}{-12.375} \\ &= -0.001 + (-0.001) \\ &= -0.002 \text{ inch} \end{aligned}$$

$$\begin{aligned} R &= \frac{\pm r}{2} + \frac{\pm f \times C}{-A} \\ &= \frac{-0.002}{2} + \frac{0.001 \times 47.75}{-12.375} \\ &= -0.001 + (-0.003858) \\ &= -0.005 \text{ inch} \end{aligned}$$

**8th Arrangement.**—Place rim and face indicators as shown in Fig. 8a.

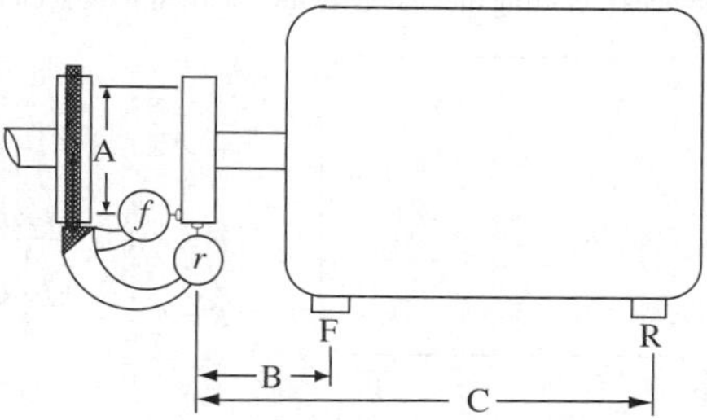


$$F = \frac{\pm r}{2} + \frac{\pm f \times B}{A}$$

$$R = \frac{\pm r}{2} + \frac{\pm f \times C}{A}$$

Fig. 8a. Dial Indicator Arrangement 8

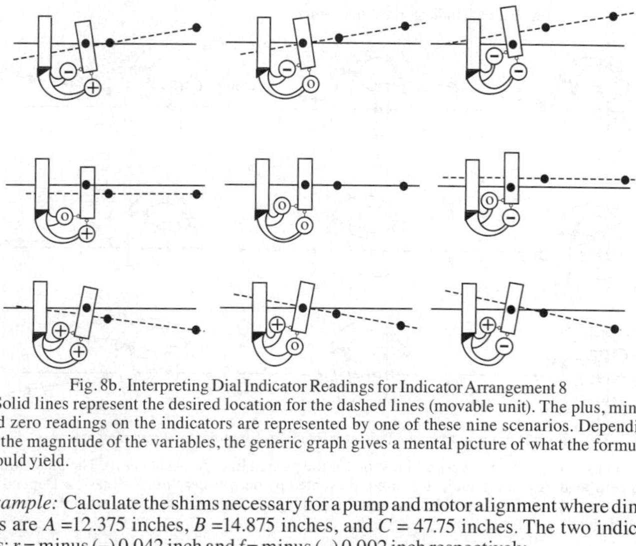

Fig. 8b. Interpreting Dial Indicator Readings for Indicator Arrangement 8
Solid lines represent the desired location for the dashed lines (movable unit). The plus, minus, and zero readings on the indicators are represented by one of these nine scenarios. Depending on the magnitude of the variables, the generic graph gives a mental picture of what the formulas should yield.

*Example:* Calculate the shims necessary for a pump and motor alignment where dimensions are $A$ =12.375 inches, $B$ =14.875 inches, and $C$ = 47.75 inches. The two indicator dials: $r$ = minus (−) 0.042 inch and $f$ = minus (−) 0.002 inch respectively.

*Solution:* In Fig. 8b, the top right block conforms to negative rim and negative face values. The graph suggests both the front and rear motor feet be lowered. The formula dictates exactly how much the movable unit is to be lowered at both front and rear shim points. Shims removed on front and rear feet will be as follows:

$$F = \frac{\pm r}{2} + \frac{\pm f \times B}{A} = \frac{-0.042}{2} + \frac{-0.002 \times 14.875}{12.375} = -0.021 + (-0.002) = -0.023 \text{ inch}$$

$$R = \frac{\pm r}{2} + \frac{\pm f \times C}{A} = \frac{-0.042}{2} + \frac{-0.002 \times 47.75}{12.375} = -0.021 + (-0.0077) = -0.029 \text{ inch}$$

**9th Arrangement.—**Place rim and face indicators as shown in Fig. 9a. The driven unit here could be any of several devices including the "catch" drum for a centrifugal clutch.

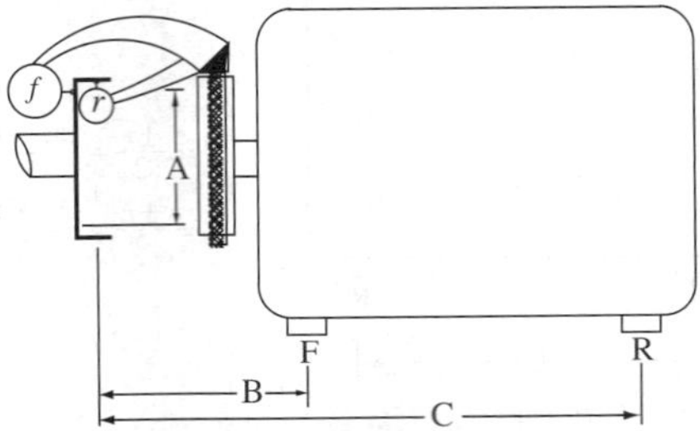


$$F = \frac{\pm r}{-2} + \frac{\pm f \times B}{A}$$

$$R = \frac{\pm r}{-2} + \frac{\pm f \times C}{A}$$

Fig. 9a. Dial Indicator Arrangement 9

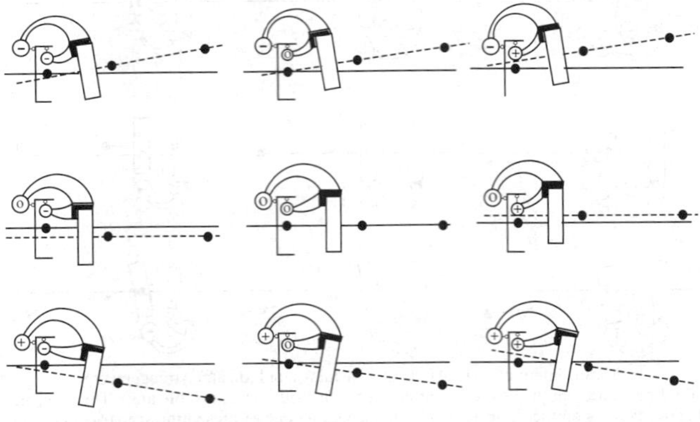

Fig. 9b. Interpreting Dial Indicator Readings for Indicator Arrangement 9

Solid lines represent the desired location for the dashed lines (movable unit). The plus, minus, and zero readings on the indicators are represented by one of these nine scenarios. Depending on the magnitude of the variables, the generic graph gives a mental picture of what the formulas should yield.

*Example:* Calculate the shims necessary for a pump and motor alignment where dimensions are $A = 10.375$ inches, $B = 22.0$ inches, and $C = 48.0$ inches. The two indicator dials read $r = 0.00$ inch and $f =$ plus (+) 0.017 inch respectively.

*Solution:* In Fig. 9b, the center block of the bottom row represents zero rim and positive face values. The graph suggests both the front and rear motor feet be moved up by adding shims, depending on the magnitude of the variables. The formula solution below dictates that the front feet be moved up 0.036 inch and the rear feet be moved up 0.079 inch:

$$\begin{aligned} F &= \frac{\pm r}{-2} + \frac{\pm f \times B}{A} \\ &= \frac{0.00}{-2} + \frac{0.017 \times 22}{10.375} \\ &= -0.00 + 0.036 \\ &= 0.036 \text{ inch} \end{aligned}$$

$$\begin{aligned} R &= \frac{\pm r}{-2} + \frac{\pm f \times C}{A} \\ &= \frac{0.00}{-2} + \frac{0.017 \times 48}{10.375} \\ &= -0.00 + 0.079 \\ &= 0.079 \text{ inch} \end{aligned}$$

**10th Arrangement.**—Fig. 10a illustrates an arrangement used for taking two separate readings with one dial indicator.

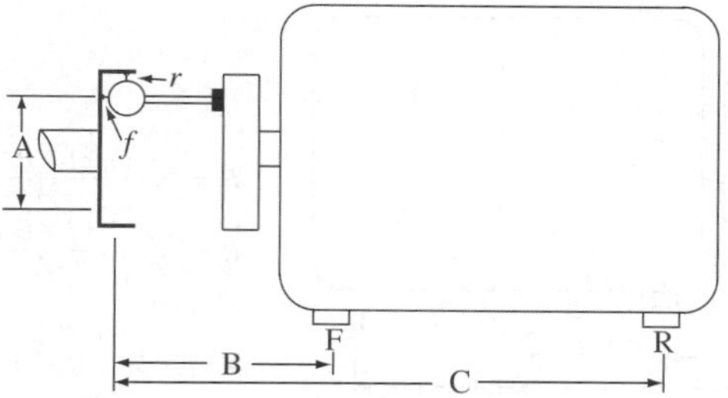


$$F = \frac{\pm r}{-2} + \frac{\pm f \times B}{-A}$$

$$R = \frac{\pm r}{-2} + \frac{\pm f \times C}{-A}$$

Fig. 10a. Dial Indicator Arrangement 10

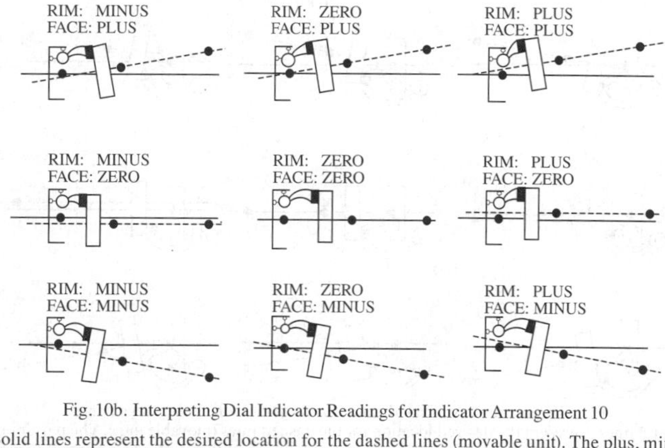


Fig. 10b. Interpreting Dial Indicator Readings for Indicator Arrangement 10

Solid lines represent the desired location for the dashed lines (movable unit). The plus, minus, and zero readings on the indicators are represented by one of these nine scenarios. Depending on the magnitude of the variables, the generic graph gives a mental picture of what the formulas should yield.

*Example:* Calculate the shims necessary for a pump and motor alignment where dimensions are $A = 5.50$ inches, $B = 14.0$ inches, and $C = 35.0$ inches. The indicator readings are $r$ = plus (+) 0.030 inch and $f$ = plus (+) 0.022 inch, respectively.

*Solution:* In Fig. 10b locate the block illustrating the positive rim and positive face condition. The graph suggests both the front and rear motor feet be lowered. The formula solution dictates exactly how much the movable unit is to be lowered at both front and rear shim points:

$$F = \frac{\pm r}{-2} + \frac{\pm f \times B}{-A} = \frac{0.030}{-2} + \frac{0.022 \times 14}{-5.5} = -0.015 + (-0.056) = -0.071 \text{ inch}$$

$$R = \frac{\pm r}{-2} + \frac{\pm f \times C}{-A} = \frac{0.030}{-2} + \frac{0.022 \times 35}{-5.5} = -0.015 + (-0.140) = -0.155 \text{ inch}$$

**11th Arrangement.**—Place rim and face indicators as shown in Fig. 11a.

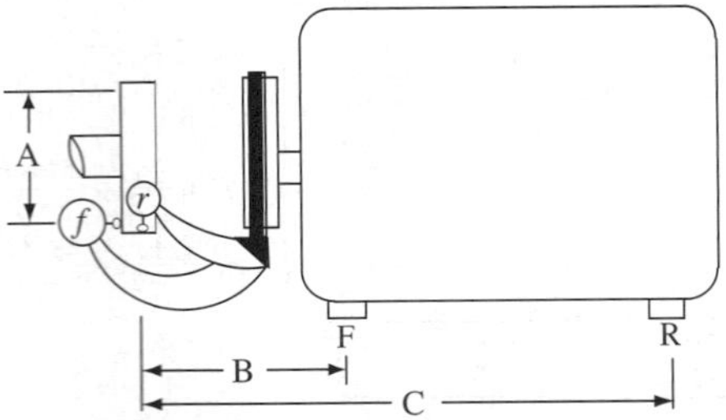


$$F = \frac{\pm r}{2} + \frac{\pm f \times B}{-A}$$

$$R = \frac{\pm r}{2} + \frac{\pm f \times C}{-A}$$

Fig. 11a. Dial Indicator Arrangement 11

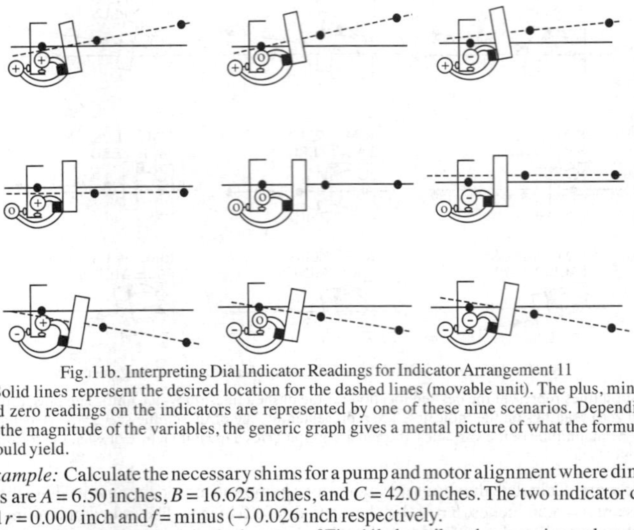

Fig. 11b. Interpreting Dial Indicator Readings for Indicator Arrangement 11

Solid lines represent the desired location for the dashed lines (movable unit). The plus, minus, and zero readings on the indicators are represented by one of these nine scenarios. Depending on the magnitude of the variables, the generic graph gives a mental picture of what the formulas should yield.

*Example:* Calculate the necessary shims for a pump and motor alignment where dimensions are $A = 6.50$ inches, $B = 16.625$ inches, and $C = 42.0$ inches. The two indicator dials read $r = 0.000$ inch and $f =$ minus (−) 0.026 inch respectively.

*Solution:* The center block in the last row of Fig. 11b describes the zero rim and negative face value condition. The motor is too low because of angularity (face reading), but the elevation is correct at the axial position of the rim indicator reading. The graph suggests both the front and rear motor feet be moved up. The formula solution dictates exactly how much the movable unit is raised at both front and rear shim points:

$$F = \frac{\pm r}{2} + \frac{\pm f \times B}{-A} = \frac{0.00}{2} + \frac{-0.026 \times 16.625}{-6.50} = 0.00 + 0.0665 = 0.0665 \text{ inch}$$

$$R = \frac{\pm r}{2} + \frac{\pm f \times C}{-A} = \frac{0.00}{2} + \frac{-0.026 \times 42}{-6.50} = 0.00 + 0.168 = 0.168 \text{ inch}$$

**12th Arrangement.—**The set up in Fig. 12a could be one indicator with two stems, or one indicator taking separate face and rim readings at the same location.

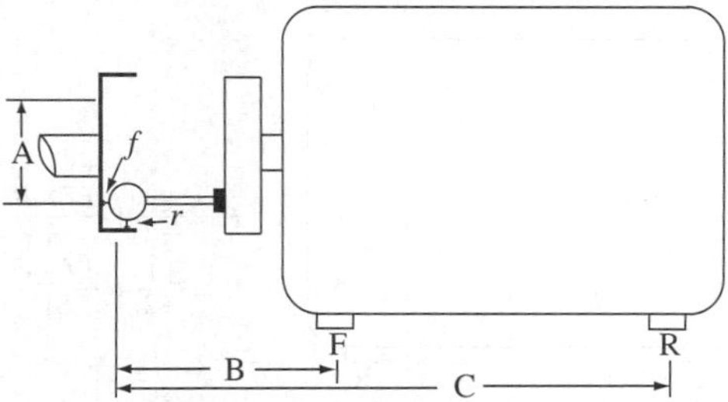


$$F = \frac{\pm r}{2} + \frac{\pm f \times B}{A}$$
$$R = \frac{\pm r}{2} + \frac{\pm f \times C}{A}$$

Fig. 12a. Dial Indicator Arrangement 12

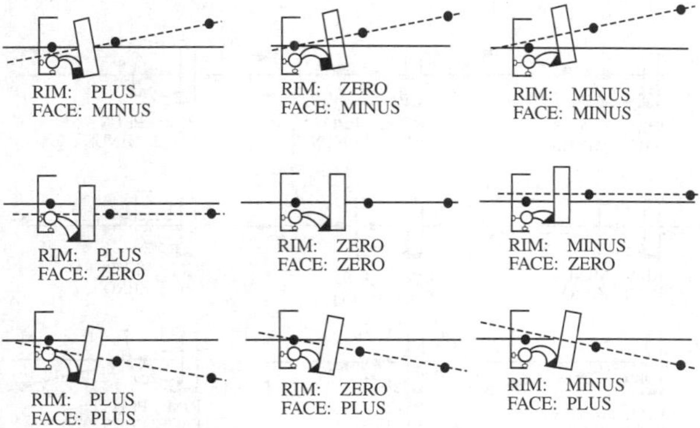


Fig. 12b. Interpreting Dial Indicator Readings for Indicator Arrangement 12

Solid lines represent the desired location for the dashed lines (movable unit). The plus, minus, and zero readings on the indicators are represented by one of these nine scenarios. Depending on the magnitude of the variables, the graph gives a mental picture of what the formulas should yield.

*Example:* Calculate the shims necessary for a pump and motor alignment where dimensions are $A = 23.0$ inches, $B = 41.0$ inches, and $C = 96.0$ inches. The two indicator readings are $r$ = plus (+)0.004 inch and $f$ = minus (−)0.003 inch, respectively.

*Solution:* For practical purposes, this unit is already within tolerance by most standards. However, the block in the upper left corner of Fig. 12b illustrates the positive rim and negative face condition. The graph suggests both the front and rear motor feet be lowered. Removal of 0.003 inch of shim at the front and 0.011 inch at the rear feet will reduce misalignment to less than a mil on the next reading.

$$\begin{aligned} F &= \frac{\pm r}{2} + \frac{\pm f \times B}{A} \\ &= \frac{0.004}{2} + \frac{-0.003 \times 41}{23.0} \\ &= 0.002 + (-0.005) \\ &= -0.003 \text{ inch} \end{aligned}$$

$$\begin{aligned} R &= \frac{\pm r}{2} + \frac{\pm f \times C}{A} \\ &= \frac{0.004}{2} + \frac{-0.003 \times 96}{23.0} \\ &= 0.002 + (-0.013) \\ &= -0.011 \text{ inch} \end{aligned}$$

**13th Arrangement.—**In Fig. 13a, the indicator jig is mounted on the static unit and two separate readings are taken from surfaces of the inside of a centrifugal clutch drum on the movable unit.

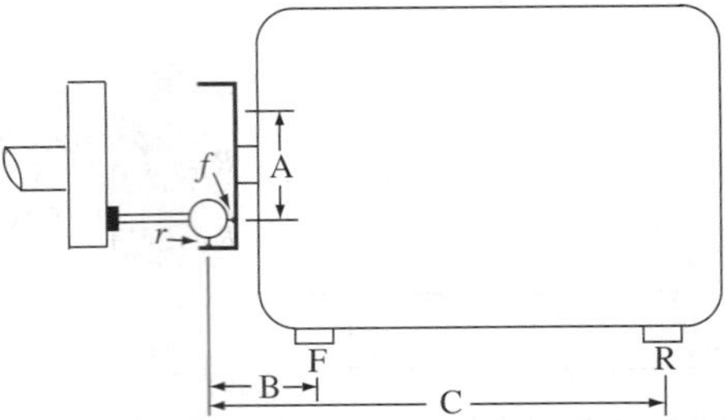


$$F = \frac{\pm r}{-2} + \frac{\pm f \times B}{A}$$

$$R = \frac{\pm r}{-2} + \frac{\pm f \times C}{A}$$

Fig. 13a. Dial Indicator Arrangement 13

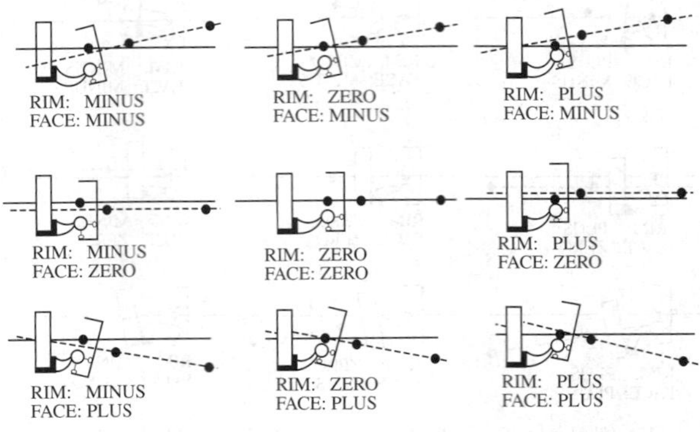


Fig. 13b. Interpreting Dial Indicator Readings for Indicator Arrangement 13

Solid lines represent the desired location for the dashed lines (movable unit). The plus, minus, and zero readings on the indicators are represented by one of these nine scenarios. Depending on the magnitude of the variables, the generic graph gives a mental picture of what the formulas should yield.

*Example:* Calculate the shims necessary for a pump and motor alignment where dimensions are $A = 23.0$ inches, $B = 20.0$ inches, and $C = 75.0$ inches. The indicator readings are $r =$ plus (+) 0.004 inch and $f =$ minus (−) 0.003 inch, respectively.

*Solution:* The positive rim and negative face condition is illustrated in the top right block of Fig. 13b. The graph suggests that both the front and rear motor feet needs to be lowered. The formula solution dictates exactly how much the movable unit is to be lowered at both front and rear shim points:

$$\begin{aligned} F &= \frac{\pm r}{-2} + \frac{\pm f \times B}{A} \\ &= \frac{0.004}{-2} + \frac{-0.003 \times 20}{23.0} \\ &= -0.002 + (-0.0026) \\ &= -0.005 \text{ inch} \end{aligned} \qquad \begin{aligned} R &= \frac{\pm r}{-2} + \frac{\pm f \times C}{A} \\ &= \frac{0.004}{-2} + \frac{-0.003 \times 75}{23.0} \\ &= -0.002 + (-0.00978) \\ &= -0.012 \text{ inch} \end{aligned}$$

**14th Arrangement.—**Place rim and face indicators as shown in Fig. 14a.

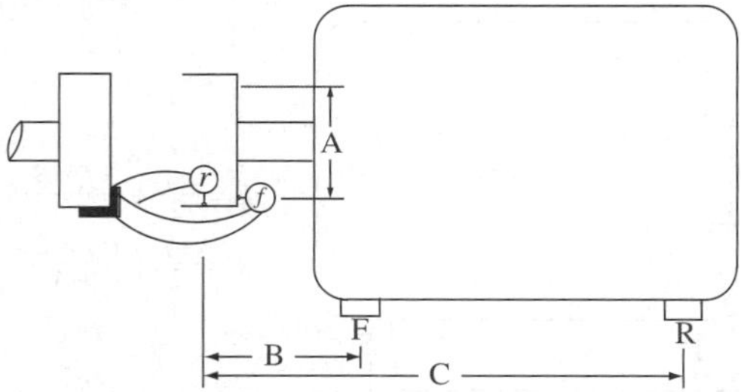


$$F = \frac{\pm r}{-2} + \frac{\pm f \times B}{-A}$$

$$R = \frac{\pm r}{-2} + \frac{\pm f \times C}{-A}$$

Fig. 14a. Dial Indicator Arrangement 14

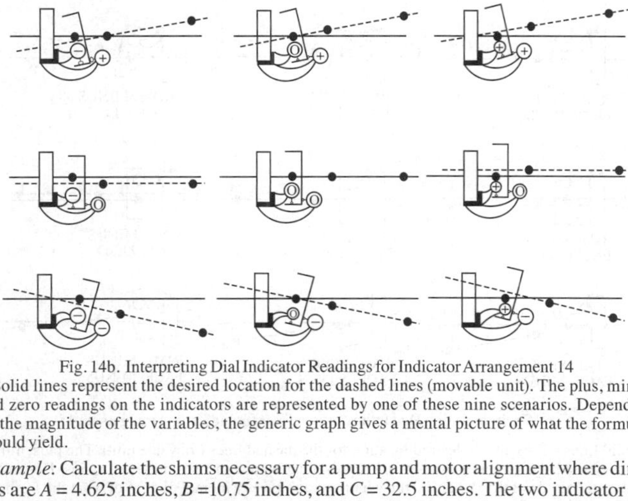

Fig. 14b. Interpreting Dial Indicator Readings for Indicator Arrangement 14

Solid lines represent the desired location for the dashed lines (movable unit). The plus, minus, and zero readings on the indicators are represented by one of these nine scenarios. Depending on the magnitude of the variables, the generic graph gives a mental picture of what the formulas should yield.

*Example:* Calculate the shims necessary for a pump and motor alignment where dimensions are $A = 4.625$ inches, $B = 10.75$ inches, and $C = 32.5$ inches. The two indicator dials read $r =$ minus (−) 0.014 inch and $f =$ plus (+) 0.036 inch respectively.

*Solution:* In Fig. 14b, the negative rim and positive face condition is shown in the top left block. The graph suggests that both the front and rear motor feet should be lowered, depending on the magnitude of the variables. The formula solution dictates exactly how much the movable unit is to be lowered at both front and rear shim points:

$$\begin{aligned} F &= \frac{\pm r}{-2} + \frac{\pm f \times B}{-A} \\ &= \frac{-0.014}{-2} + \frac{0.036 \times 10.75}{-4.625} \\ &= 0.007 + (-0.0837) \\ &= -0.077 \text{ inch} \end{aligned} \qquad \begin{aligned} R &= \frac{\pm r}{-2} + \frac{\pm f \times C}{-A} \\ &= \frac{-0.014}{-2} + \frac{0.036 \times 32.5}{-4.625} \\ &= 0.007 + (-0.25297) \\ &= -0.246 \text{ inch} \end{aligned}$$

**15th Arrangement.—**Place rim and face indicators as shown in Fig. 15a.

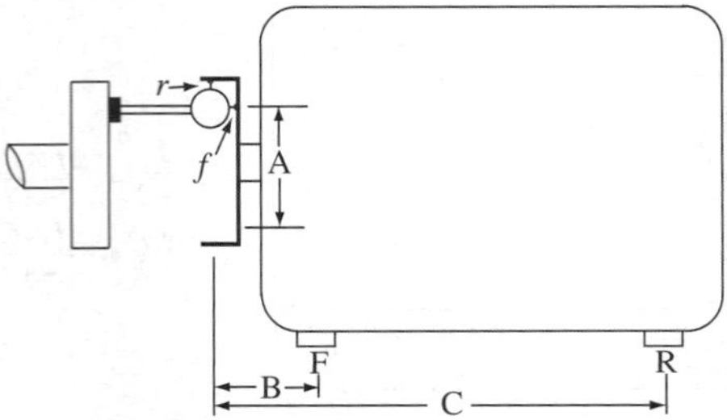


$$F = \frac{\pm r}{2} + \frac{\pm f \times B}{-A}$$

$$R = \frac{\pm r}{2} + \frac{\pm f \times C}{-A}$$

Fig. 15a. Dial Indicator Arrangement 15

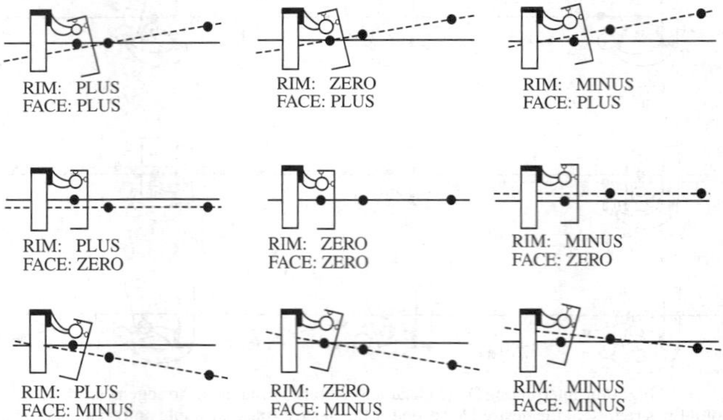


Fig. 15b. Interpreting Dial Indicator Readings for Indicator Arrangement 15

Solid lines represent the desired location for the dashed lines (movable unit). The plus, minus, and zero readings on the indicators are represented by one of these nine scenarios. Depending on the magnitude of the variables, the generic graph gives a mental picture of what the formulas should yield.

*Example:* Calculate the shims necessary for a pump and motor alignment where dimensions are $A = 6.0$ inches, $B = 38.0$ inches, and $C = 84.0$ inches. The two indicator dials reads $r$ = minus (–) 0.002 inch and $f$ = plus (+) 0.003 inch respectively.

*Solution:* The negative rim and positive face condition is found in the top right corner of Fig. 15b. The graph suggests that both the front and rear motor feet need to be lowered. The formula solution dictates exactly how much the movable unit is to be lowered at both front and rear shim points:

$$\begin{aligned} F &= \frac{\pm r}{2} + \frac{\pm f \times B}{-A} \\ &= \frac{-0.002}{2} + \frac{0.003 \times 38.0}{-6.0} \\ &= -0.001 + (-0.019) \\ &= -0.020 \text{ inch} \end{aligned}$$

$$\begin{aligned} R &= \frac{\pm r}{2} + \frac{\pm f \times C}{-A} \\ &= \frac{-0.002}{2} + \frac{0.003 \times 84.0}{-6.0} \\ &= -0.001 + (-0.042) \\ &= -0.043 \text{ inch} \end{aligned}$$

**16th Arrangement.**—Place rim and face indicators as shown in Fig. 16a.

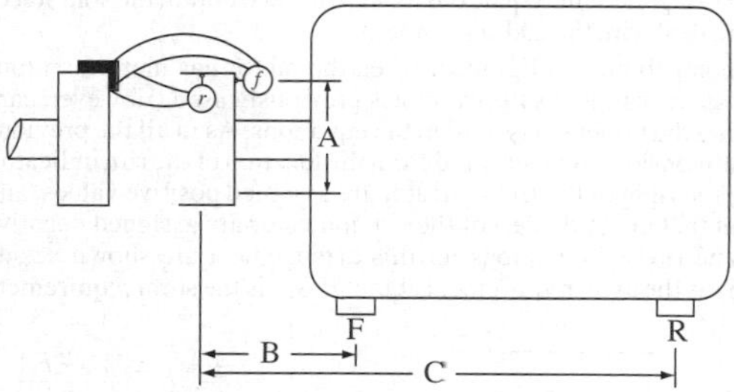


$$F = \frac{\pm r}{2} + \frac{\pm f \times B}{A}$$

$$R = \frac{\pm r}{2} + \frac{\pm f \times C}{A}$$

Fig. 16a. Dial Indicator Arrangement 16

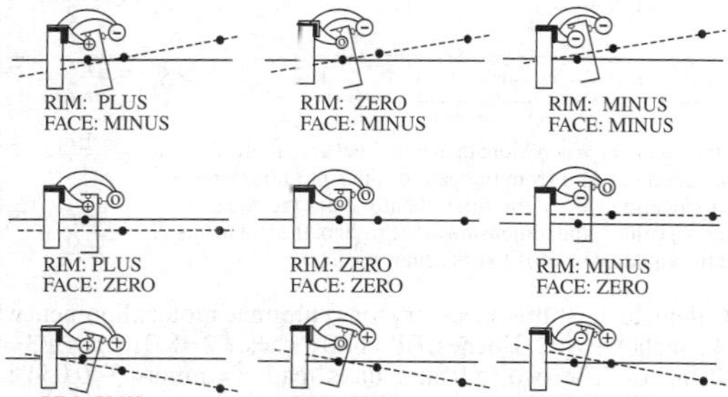


Fig. 16b. Interpreting Dial Indicator Readings for Indicator Arrangement 16
Solid lines represent the desired location for the dashed lines (movable unit). The plus, minus, and zero readings on the indicators are represented by one of these nine scenarios. Depending on the magnitude of the variables, the graph gives a mental picture of what the formulas should yield.

*Example:* Calculate the shims necessary for a pump and motor alignment where dimensions are $A = 16.0$ inches, $B = 26.0$ inches, and $C = 54.0$ inches. The two indicator dials read $r =$ minus (−) 0.002 inch and $f =$ minus (−) 0.006 inch respectively.

*Solution:* The rim and face numbers indicate that the whole movable machine is one mil high (from the rim indicator, $^r/_2 = {}^{0.002}/_2 = 0.001$), and the angularity from the face indicator say that for every 16 inches of axial length from the rim indicator there is an upward incline toward the rear of 0.006 inch. In Fig. 16b, locate the block with negative rim and negative face values. The graph suggests both the front and rear motor feet be lowered. The formula solution dictates that the movable unit is to be lowered at both front and rear shim points:

$$\begin{aligned} F &= \frac{\pm r}{2} + \frac{\pm f \times B}{A} \\ &= \frac{-0.002}{2} + \frac{-0.006 \times 26}{16} \\ &= -0.001 + (-0.00975) \\ &= -0.011 \text{ inch} \end{aligned}$$

$$\begin{aligned} R &= \frac{\pm r}{2} + \frac{\pm f \times C}{A} \\ &= \frac{-0.002}{2} + \frac{-0.006 \times 54.0}{16} \\ &= -0.001 + (-0.02025) \\ &= -0.021 \text{ inch} \end{aligned}$$

**Rim and Face Alignment Method Extended to Motors with Extra Feet.**—Fig. 17 illustrates an alignment problem that is complicated by the fact that the motor has ten feet. The indicator arrangement used matches the arrangement illustrated in Fig. 14a. In this problem the solution is obtained by using the same equations as were used to obtain the solution of Fig. 14a, adapted slightly to deal with the additional feet.

Shim calculations for motor to pump alignment when the motor has more than four feet are performed in the same manner as the examples previously used. However, care must be used in determining the dimensions used in the equations. As in all the previous examples, the horizontal dimensions are measured from the location of the rim indicator. Distances to motor feet to the right of the rim indicator are assigned positive values, and distances measured to feet that are to the left of the rim indicator are assigned negative values in the formulas. The revised equations for this arrangement are shown beside Fig. 17. In the equations, $S_F$ is the shim requirement at foot F, $S_{F1}$ is the shim requirement at foot F1, etc.

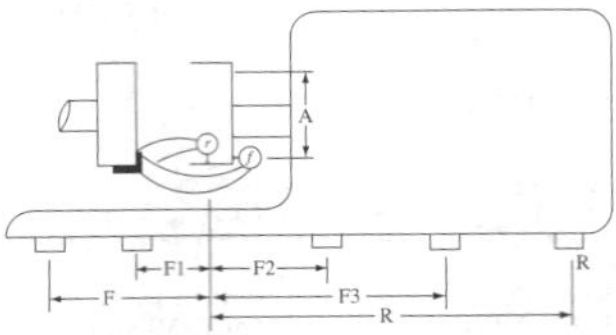


Fig. 17. Shaft Alignment When More than Four Feet are Involved Dimensions are measured from the path of the rim indicator. Dimensions to locations left of the rim indicator enter the equations as minus (–) values, and dimension to the right of the rim indicator enter the equations as positive (+) quantities.

$$S_F = \frac{\pm r}{-2} + \frac{\pm f \times (-F)}{-A}$$

$$S_{F1} = \frac{\pm r}{-2} + \frac{\pm f \times (-F1)}{-A}$$

$$S_{F2} = \frac{\pm r}{-2} + \frac{\pm f \times F2}{-A}$$

$$S_{F3} = \frac{\pm r}{-2} + \frac{\pm f \times F3}{-A}$$

$$S_R = \frac{\pm r}{-2} + \frac{\pm f \times R}{-A}$$

*Example:* Calculate the shims necessary for a pump and motor alignment where dimensions are $A = 4.5$ inches, $F = 22.0$ inches, $F1 = 10.0$ inches, $F2 = 8.0$ inches, $F3 = 24.0$ inches, and $R = 46.0$ inches. The two indicator dials read $r$ = minus (–)0.054 inch and $f$ = plus (+)0.032 inch, respectively.

*Solution, Vertical Alignment:*

$$S_F = \frac{\pm r}{-2} + \frac{\pm f \times (-F)}{-A} = \frac{-0.054}{-2} + \frac{0.032 \times (-22)}{-4.5} = 0.027 + 0.156 = 0.183 \text{ inch}$$

$$S_{F1} = \frac{\pm r}{-2} + \frac{\pm f \times (-F1)}{-A} = \frac{-0.054}{-2} + \frac{0.032 \times (-10)}{-4.5} = 0.027 + 0.071 = 0.098 \text{ inch}$$

$$S_{F2} = \frac{\pm r}{-2} + \frac{\pm f \times F2}{-A} = \frac{-0.054}{-2} + \frac{0.032 \times 8}{-4.5} = 0.027 + (-0.156) = -0.030 \text{ inch}$$

$$S_{F3} = \frac{\pm r}{-2} + \frac{\pm f \times F3}{-A} = \frac{-0.054}{-2} + \frac{0.032 \times 24}{-4.5} = 0.027 + (-0.171) = -0.144 \text{ inch}$$

$$S_R = \frac{\pm r}{-2} + \frac{\pm f \times R}{-A} = \frac{-0.054}{-2} + \frac{0.032 \times 46}{-4.5} = 0.027 + (-0.327) = -0.300 \text{ inch}$$

The calculations show that more shims are required under the two pairs of front feet, $F$ and $F1$, and need to be reduced for the other three pairs of rear feet. The top left figure in Fig. 14b (negative rim and positive face indicator readings) confirms that the numerical solution appears reasonable.

*Solution, Horizontal Alignment:* The horizontal alignment solution follows the same procedure as the vertical solution above, and will use the same equations if the indicator set up shown in Fig. 17 is used.

The dimensions $F$, $F1$, $F2$, $F3$, and $R$ are the distances from the rim indictor path to the respective locations where the move occurs, whether it be the location of the hold-down bolts, or another location where a monitoring dial indicator is mounted. See *Solution for Horizontal Alignment* on page 2735 for additional information on this subject.

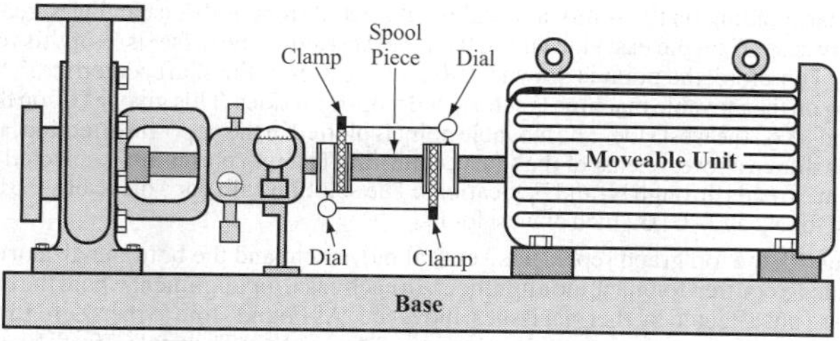


Fig. 18. Interpreting Dial Indicator Readings for Indicator Arrangement 1

**Reverse Dial Alignment Method.—**While the rim and face method involves one dial reading the pump coupling, the reverse dial method includes an additional dial reading the rim of the motor coupling. This method does not require removing coupling components, a time saver that reduces wear and tear on coupling components. But the reverse dial method only works if both shafts can be rotated. And the axial position of the shafts must be correct before proceeding to prevent bearing preloading.

The preliminary work—correcting soft foot, rough aligning the two machines, calibrating bar sag, etc.—is as previously discussed. The dials are mounted as shown in Fig. 18, and the shafts are rotated together, with readings gathered for top, bottom, east, and west positions of the two dials. These readings can be plotted on graph paper or in a spreadsheet to find the vertical and horizontal movements needed for the motor feet. Mathematical formulas can also be used to calculate the moves.

*Example:* Calculate the shims necessary for a pump and motor alignment, where the dimensions are: $D_1 = 8$ inches between the dials; $D_2 = 12.5$ inches from first dial to near foot ($N_f$), and $D_3 = 27.5$ inches from first dial to far foot ($F_f$). The readings below have been corrected for bar sag. East readings are to the left, facing the back of the motor looking towards the pump.

| Stationary | Movable |
|---|---|
| Top 0 | Top 0 |
| east 5 (stationary machine); west −15 | east 14 (movable machine); west 6 |
| TIR bottom −10 (offset = −5) | TIR bottom 20 (offset = 10) |
| horizontal offset = −10 west | horizontal offset = −8, which is east for movable machine |

The bottom reading represents the total indicator reading (TIR). Actual offset is ½ TIR. The stationary machine bottom reading is plotted up for + (positive) and down for – (negative). The movable side bottom reading is plotted opposite: down for + and up for –.

In this example and Fig. 19, the dial indicator readings and machine movements (the vertical scale) are in thousandths of inches, and the length dimensions (horizontal) are in inches. (The units are not important, as long as they are consistent; millimeters and centimeters could be used for the same results.)

The stationary dial reading calls for a point −5, or 0.005 inch down. The moveable dial reading calls for a point 20 down, or 0.010 inch down. Those points are connected and extended from the dials to the far foot. The vertical distance to the stationary shaft line for a zero thermal growth alignment is the amount of shim required under both feet. Add 0.013 inch under the near feet ($N_f$); add 0.020 inch under the far feet ($F_f$).

The horizontal move requires readings to be referenced from the rear of the stationary machine, looking down the shaft to the movable machine. The dial readings on the stationary machine are read as positive towards the east and negative toward the west.

The east reading on the stationary dial is subtracted from both east and west readings. This gives a "0" on the east side and −20 on the west side. The offset is ½ of this reading or −10. Therefore, the point is drawn 0.010 inch west from the shaft centerline. The east reading on the movable machine is subtracted from both sides. This gives a "0" on the east side and −8 on the west side. The movable point is plotted in the opposite direction, and the point is drawn above, or east of the shaft centerline. The two points are connected with a line that extends through $N_f$ and $F_f$ locations. The solution calls for a move of −0.011 inch or west for $N_f$, and −0.032 inch or west for $F_f$.

In Fig. 19, the top graph represents vertical movement and the bottom half horizontal movement required for machine alignment. In each case, for alignment where the components are not affected by thermal issues, move the "As Found" line to the "Zero Line." To correct for heat expansion, move the "As Found" line to the "Thermal Offset" line; when hot, the machine will "grow" to the "Zero Line."

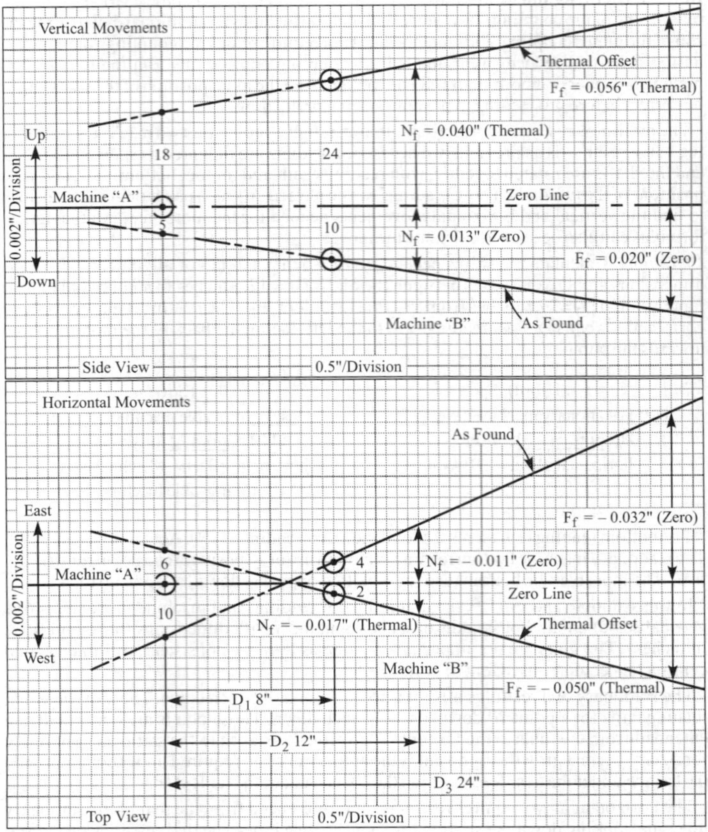


Fig. 19. Graphical Representation of Reverse Dial Indicator Alignment

*Math Solution for Zero Thermal Growth:* $_S$ = stationary, $_M$ = Movable, E = east, W = west, V = vertical

*Vertical:*

Stationary side factor – $V_1 = b_s/2$

Movable side factor – $V_2 = b_m/2$

$N_f - ((V_2 \times D_2)/D_1) - V_1 = +13$ (add)

$F_f = ((V_2 \times D_3)/D_1) - V_1 = +20$ (add)

*Horizontal:*

Stationary side factor – $V_1 = (W_s - E_m)/2$

Movable side factor – $V_2 = (W_m - E_m)/2$

$N_f = ((V_2 \times D_2)/D_1) - V_1 = -11$ (west)

$F_f = ((V_2 \times D_3)/D_1) - V_1 = -32$ (west)

A reverse dial indicator alignment can also be used to calculate a preset thermal offset if known. These are the readings for the ambient thermal growth offset values.

Top 0
East 12 (stationary machine); west 24
TIR bottom 36 (offset = 18)

Top 0
East –26 (movable machine); west –22
TIR bottom –48 (offset = –24)

*Math Solution with Thermal Offset:* $_S$ = stationary, $_M$ = movable, E = east, W = west, V=vertical, $_P$ = present state, $_O$ = desired thermal offset

*Vertical:*

$V_1 = (bsp\text{-}bso)/2$

$V_2 = (bmp\text{-}bmo)/2$

$N_f - ((V_2 \times D_2)/D_1) - V_1 = 40$

$F_f = ((V_2 \times D_3)/D_1) - V_1 = 56$

*Horizontal:*

$V_1 = ((wps - eps) - (Wos - Eos))/2$

$V_2 = ((wpm - epm) - (wos - eos))/2$

$N_f = ((V_2 \times D_2)/D_1) - V_1 = -17$

$F_f = ((V_2 \times D_3)/D_1) - V_1 = -50$

**Laser Alignment.**—The most precise method of machine shaft alignment is to use the components in a standard laser alignment system. The screen of the handheld readout automatically graphs the shaft centerlines with required shimming information. Once a mechanic or millwright has learned the graphic method for performing reverse dial analysis, they should be able to quickly learn to use a laser tool for this calculation.

Thermal growth is accurately calculated by mounting the laser head to a bracket on one machine and the corresponding prism to a bracket on the other. Zero both when the machines are cold. As the machines come up to normal operating temperature, the laser system will identify the precise vertical and horizontal change in the movable machine shaft centerline. The laser alignment tool will allow for these numbers to be input as the thermal offset numbers needed to have a "zero line" alignment after the machinery is at normal operating temperature.

**Choosing Alignment Methods.**—Methods for calculating thermal growth can be as elementary as taking thermal imaging temperatures of the machine casing and entering these into spreadsheets provided by the original equipment manufacturer.

In lieu of provided templates, the dial indictor methods for machine alignment are both easy to learn. The rim and face method is the only method that works when only one shaft can be rotated. The reverse dial indicator method enables couplings to stay assembled, with the axial shaft positions remaining correct, plus this method is easy to graph.

Potential issues of these methods include coupling face errors, axial movement of shafts, the dial rotation can be difficult to view, sag must be compensated for in calculations, dial reading errors, and angled dials giving false readings. Recording readings as dials rotate also requires good lighting and mirrors.

Pros of laser alignment include digital readout, requiring only 45 degrees of rotation, the internal GPS in the laser automatically tracks position, vertical and side-to-side measurements can be performed simultaneously, and these measurements can be programmed to include thermal offset. Cons include the initial investment cost, though this has been decreasing, and the fact that laser systems may not be safe in certain applications.

# FLUID POWER

## Introduction

A fluid is a substance which deforms continuously when subjected to a shear stress. Fluid power systems convert mechanical or potential energy into fluid energy, and usually perform work with the fluid energy. The fluid used can be either a liquid or a gas. When liquids are used, it is a hydraulic system, and when a gas is used it is a pneumatic system.

**Properties of Liquids and Gases.**—One of the main differences between liquids and gases has to do with compressibility. Compressibility is a measure of the relative volume change of a substance with a change in pressure. Liquids are only slightly compressible, and are usually considered to be incompressible for the purposes of fluid power calculations. Gases, on the other hand, are very compressible under normal conditions.

Specific weight is the weight of a fluid per unit of volume. Mass density of a fluid is the mass contained within a unit of volume. When a fluid is pressurized, its density and specific weight increase, and when a fluid is heated, its density and specific weight decrease. Liquids are affected very little by these factors, while gases are very much affected. Specific gravity is the ratio between specific weight or density at actual versus standard conditions. Specific gravity changes with specific weight or density. Standard conditions are taken to be 4° C for water, and 0° C for air. Other fluids may have different standard conditions. Standard symbols and equations for these properties are shown below. Table 1, Table 2 and Table 3 contain the properties of some common fluids.

### Fluid Property Symbols and Equations

Specific weight: $lbf/ft^3$, $N/m^3$ $\qquad \gamma = \rho g = \frac{w}{V}$

Density $\qquad \rho = \frac{m}{V}$

slugs/ft$^3$, kg/m$^3$ $\qquad \rho$

lbm/ft$^3$ $\qquad \rho_{lbm} = \rho g_c$

Gravitational Constant: lbm-ft/lbf-s$^2$ $\qquad g_c = 32.2$

Specific Gravity: dimensionless $\qquad SG = \frac{\gamma}{\gamma_c} = \frac{\rho}{\rho_c}$

The control specific weight $\gamma_c$ for liquids is water, and for gases is air.

The density of dry air at 32 degrees F and atmospheric pressure (29.92 inches of mercury or 14.70 pounds per square inch) is 0.08073 pound per cubic foot.

The density of air at any other temperature or pressure is

$$\rho = \frac{1.325 \times B}{T}$$

in which $\rho$ = density in pounds per cubic foot; $B$ = height of barometric pressure in inches of mercury; $T$ = absolute temperature in degrees Rankine. (When using pounds as a unit care must be exercised to differentiate between pounds mass and pounds force. See *Acceleration of Gravity g Used in Mechanics Formulas* on page 152.

The absolute zero from which all temperatures must be counted when dealing with the weight and volume of gases is assumed to be −459.7 degrees F. Hence, to obtain the absolute temperature $T$ used in the preceding formula, add the value 459.7 to the temperature observed on a regular Fahrenheit thermometer.

In obtaining the value of $B$, 1 inch of mercury at 32 degrees F may be taken as equal to a pressure of 0.491 pound per square inch.

## Table 1. Density of Air at Different Pressures and Temperatures

| Temp. of Air, °F | Gage Pressure, Pounds | | | | | | | | | | | | | | |
|---|---|---|---|---|---|---|---|---|---|---|---|---|---|---|---|
| | 0 | 5 | 10 | 20 | 30 | 40 | 50 | 60 | 80 | 100 | 120 | 150 | 200 | 250 | 300 |
| | Density in Pounds per Cubic Foot | | | | | | | | | | | | | | |
| −20 | 0.0900 | 0.1205 | 0.1515 | 0.2125 | 0.274 | 0.336 | 0.397 | 0.458 | 0.580 | 0.702 | 0.825 | 1.010 | 1.318 | 1.625 | 1.930 |
| −10 | 0.0882 | 0.1184 | 0.1485 | 0.2090 | 0.268 | 0.328 | 0.388 | 0.448 | 0.567 | 0.687 | 0.807 | 0.989 | 1.288 | 1.588 | 1.890 |
| 0 | 0.0864 | 0.1160 | 0.1455 | 0.2040 | 0.263 | 0.321 | 0.380 | 0.438 | 0.555 | 0.672 | 0.790 | 0.968 | 1.260 | 1.553 | 1.850 |
| 10 | 0.0846 | 0.1136 | 0.1425 | 0.1995 | 0.257 | 0.314 | 0.372 | 0.429 | 0.543 | 0.658 | 0.774 | 0.947 | 1.233 | 1.520 | 1.810 |
| 20 | 0.0828 | 0.1112 | 0.1395 | 0.1955 | 0.252 | 0.307 | 0.364 | 0.420 | 0.533 | 0.645 | 0.757 | 0.927 | 1.208 | 1.489 | 1.770 |
| 30 | 0.0811 | 0.1088 | 0.1366 | 0.1916 | 0.246 | 0.301 | 0.357 | 0.412 | 0.522 | 0.632 | 0.742 | 0.908 | 1.184 | 1.460 | 1.735 |
| 40 | 0.0795 | 0.1067 | 0.1338 | 0.1876 | 0.241 | 0.295 | 0.350 | 0.404 | 0.511 | 0.619 | 0.727 | 0.890 | 1.161 | 1.431 | 1.701 |
| 50 | 0.0780 | 0.1045 | 0.1310 | 0.1839 | 0.237 | 0.290 | 0.343 | 0.396 | 0.501 | 0.607 | 0.713 | 0.873 | 1.139 | 1.403 | 1.668 |
| 60 | 0.0764 | 0.1025 | 0.1283 | 0.1803 | 0.232 | 0.284 | 0.336 | 0.388 | 0.493 | 0.596 | 0.700 | 0.856 | 1.116 | 1.376 | 1.636 |
| 80 | 0.0736 | 0.0988 | 0.1239 | 0.1738 | 0.224 | 0.274 | 0.324 | 0.374 | 0.473 | 0.572 | 0.673 | 0.824 | 1.074 | 1.325 | 1.573 |
| 100 | 0.0710 | 0.0954 | 0.1197 | 0.1676 | 0.215 | 0.264 | 0.312 | 0.360 | 0.455 | 0.551 | 0.648 | 0.794 | 1.035 | 1.276 | 1.517 |
| 120 | 0.0680 | 0.0921 | 0.1155 | 0.1618 | 0.208 | 0.255 | 0.302 | 0.348 | 0.440 | 0.533 | 0.626 | 0.767 | 1.001 | 1.234 | 1.465 |
| 140 | 0.0663 | 0.0889 | 0.1115 | 0.1565 | 0.201 | 0.246 | 0.291 | 0.336 | 0.426 | 0.516 | 0.606 | 0.742 | 0.968 | 1.194 | 1.416 |
| 150 | 0.0652 | 0.0874 | 0.1096 | 0.1541 | 0.198 | 0.242 | 0.286 | 0.331 | 0.419 | 0.508 | 0.596 | 0.730 | 0.953 | 1.175 | 1.392 |
| 175 | 0.0626 | 0.0840 | 0.1054 | 0.1482 | 0.191 | 0.233 | 0.275 | 0.318 | 0.403 | 0.488 | 0.573 | 0.701 | 0.914 | 1.128 | 1.337 |
| 200 | 0.0603 | 0.0809 | 0.1014 | 0.1427 | 0.184 | 0.225 | 0.265 | 0.305 | 0.388 | 0.470 | 0.552 | 0.674 | 0.879 | 1.084 | 1.287 |
| 225 | 0.0581 | 0.0779 | 0.0976 | 0.1373 | 0.177 | 0.216 | 0.255 | 0.295 | 0.374 | 0.452 | 0.531 | 0.649 | 0.846 | 1.043 | 1.240 |
| 250 | 0.0560 | 0.0751 | 0.0941 | 0.1323 | 0.170 | 0.208 | 0.247 | 0.284 | 0.360 | 0.436 | 0.513 | 0.627 | 0.817 | 1.007 | 1.197 |
| 275 | 0.0541 | 0.0726 | 0.0910 | 0.1278 | 0.164 | 0.201 | 0.238 | 0.274 | 0.348 | 0.421 | 0.494 | 0.605 | 0.789 | 0.972 | 1.155 |
| 300 | 0.0523 | 0.0707 | 0.0881 | 0.1237 | 0.159 | 0.194 | 0.230 | 0.265 | 0.336 | 0.407 | 0.478 | 0.585 | 0.762 | 0.940 | 1.118 |
| 350 | 0.0491 | 0.0658 | 0.0825 | 0.1160 | 0.149 | 0.183 | 0.216 | 0.249 | 0.316 | 0.382 | 0.449 | 0.549 | 0.715 | 0.883 | 1.048 |
| 400 | 0.0463 | 0.0621 | 0.0779 | 0.1090 | 0.140 | 0.172 | 0.203 | 0.235 | 0.297 | 0.360 | 0.423 | 0.517 | 0.674 | 0.831 | 0.987 |
| 450 | 0.0437 | 0.0586 | 0.0735 | 0.1033 | 0.133 | 0.163 | 0.192 | 0.222 | 0.281 | 0.340 | 0.399 | 0.488 | 0.637 | 0.786 | 0.934 |
| 500 | 0.0414 | 0.0555 | 0.0696 | 0.978 | 0.126 | 0.154 | 0.182 | 0.210 | 0.266 | 0.322 | 0.379 | 0.463 | 0.604 | 0.746 | 0.885 |
| 550 | 0.0394 | 0.0528 | 0.0661 | 0.930 | 0.120 | 0.146 | 0.173 | 0.200 | 0.253 | 0.306 | 0.359 | 0.440 | 0.573 | 0.749 | 0.841 |
| 600 | 0.0376 | 0.0504 | 0.0631 | 0.883 | 0.114 | 0.139 | 0.165 | 0.190 | 0.241 | 0.292 | 0.343 | 0.419 | 0.547 | 0.675 | 0.801 |

**Table 2. Mass Density and Specific Weight of Water at Selected Temperatures**

| Temperature | | Mass Density | | Specific Weight | |
|---|---|---|---|---|---|
| °F | °C | slug/ft³ | kg/m³ | lbf/ft³ | N/m³ |
| 32 | 0 | 1.94 | 999.835 | 62.42 | 9810 |
| 50 | 10 | 1.94 | 999.835 | 62.41 | 9800 |
| 60 | 15.6 | 1.938 | 998.804 | 62.37 | 9800 |
| 70 | 21.1 | 1.936 | 997.773 | 62.31 | 9790 |
| 80 | 26.7 | 1.934 | 996.743 | 62.23 | 9780 |
| 100 | 37.8 | 1.927 | 993.135 | 62.02 | 9740 |
| 120 | 48.9 | 1.918 | 988.497 | 61.74 | 9700 |
| 140 | 60 | 1.908 | 983.343 | 61.37 | 9640 |
| 160 | 71.1 | 1.896 | 977.158 | 60.98 | 9580 |
| 180 | 82.2 | 1.883 | 970.458 | 60.55 | 9510 |
| 200 | 93.3 | 1.869 | 963.243 | 60.12 | 9440 |
| 212 | 100 | 1.86 | 958.605 | 59.83 | 9400 |

**Table 3. Properties of Common Fluids at Atmospheric Pressure**

| Fluid | Temperature | | Density | | Specific Weight | | Specific Gravity |
|---|---|---|---|---|---|---|---|
| | °F | °C | slug/ft³ | kg/m³ | lbf/ft³ | N/m³ | |
| Air | 68 | 20 | $2.33 \times 10^{-3}$ | 1.205 | 0.075 | 11.8 | 1 |
| Nitrogen | 68 | 20 | $2.26 \times 10^{-3}$ | 1.165 | … | … | 0.9669 |
| Water | 68 | 20 | 1.936 | 998.2 | 62.4 | $9.81 \times 10^3$ | 1 |
| Seawater | 68 | 20 | 1.987 | 1024 | 63.9 | $10.03 \times 10^3$ | 1.025 |

**Pressure.—**Pressure, $p$, is defined as the average force per unit area, $A$, exerted by a fluid.

$$p = \frac{F}{A}$$

The pressure at a point in a static fluid is the same in all directions, and the same at all points in that fluid. The pressure exerted by a fluid on a solid surface is always perpendicular to that surface. Pressure measured relative to atmospheric pressure is called gauge pressure.

Pressure measured relative to a perfect vacuum is called absolute pressure. The relationship between absolute pressure and gauge pressure is given in the following equation:

$$P_a = P_g + P_{atm}$$

where $P_a$ = 14.696 psi (US); and, $\mathbf{P_a}$ **= 101.325 kPa (SI)**.

The change in pressure in a non-moving liquid due to a change in elevation is dependent on the specific weight of the fluid and can be calculated using the following equations. Elevation is always taken to be positive in the upward direction relative to the reference plane. The pressure caused by elevation is independent of the shape and size of the container and its free surface.

$p = \gamma h$ where $\gamma$ is specific weight (lbf/ft³ or N/m³), $h$ is elevation (ft or m), and $p$ is pressure (lbf/ft² or N/m²)

$p = \dfrac{\gamma h}{144}$ where $\gamma$ is specific weight (lbf/ft³), $h$ is elevation (ft), and $p$ is pressure (psi)

For ideal gases, the relationship between pressure, temperature, and specific volume can be found in the following equations. Specific volume is the volume occupied by a unit mass of a fluid.

$$PV = 1545.3nT \quad \text{For fps units}$$
$$PV = 8314nT \quad \text{For SI units}$$

in which $P$ = absolute pressure in pounds per square foot or Pa ($N/m^2$)

$V$ = volume in cubic feet or cubic meter

$T$ = absolute temperature in degrees R or degrees K

$n$ = number of pound moles or kg moles. A mole is the mass of substance, in appropriate units, divided by its molecular weight. The equations above are general forms that apply to any gas that behaves the ideal gas law.

**Viscosity.**—Viscosity is a liquid property that indicates its resistance to flow. It is a measure of a fluid's resistance to a shearing force. High viscosity fluids require a greater force to shear at a given rate than low viscosity fluids. This is the most important property of fluids when designing and analyzing fluid power systems. An ideal fluid has zero viscosity, and gases have nearly zero viscosity. For most fluids, viscosity varies significantly with temperature. It is relatively unaffected by changes in pressure, unless they are extreme. Newtonian fluids are those whose viscosity is independent of velocity gradient.

Viscosity index is the change in viscosity relative to change in temperature. Absolute, or dynamic, viscosity relates shearing stress in a fluid to the velocity gradient. Kinematic viscosity is the ratio of a fluid's absolute viscosity to its mass density. Absolute viscosity is often measured in centipoise (cP), and kinematic viscosity in centistokes (cSt).

Equations and conversion factors for absolute and kinematic viscosity are given below. The absolute and kinematic viscosity of various fluids can be found in Table 4.

*Absolute viscosity:* $\mu = \dfrac{F\,dy}{A\,dv}$ lbf-s/ft² or N-s/m² (Pa-s)

Units may be expressed in centipoise, cP

$cP = 2.09 \times 10^{-5}$ lbf-s/ft²

$cP = 1 \times 10^{-3}$ Pa-s

*Kinematic viscosity:* $\upsilon = \dfrac{\mu}{\rho}$ ft²/s or m²/s

Units may be expressed in centistokes, cSt

$cSt = 1.07 \times 10^{-5}$ ft²/s

$cSt = 1 \times 10^{-6}$ m²/s

**Table 4. Viscosity of Common Fluids at Atmospheric Pressure**

| Fluid | Absolute Viscosity | | | Kinematic Viscosity | | |
|---|---|---|---|---|---|---|
| | lbf-s/ft² | Pa-s | cP | ft²/s | m²/s | cSt |
| Air | $3.75 \times 10^{-7}$ | $1.80 \times 10^{-5}$ | 0.018 | $1.64 \times 10^{-4}$ | $1.52 \times 10^{-5}$ | 15.24 |
| Nitrogen | $3.75 \times 10^{-7}$ | $1.80 \times 10^{-5}$ | 0.018 | $1.64 \times 10^{-4}$ | $1.52 \times 10^{-5}$ | 15.24 |
| Water | $2.09 \times 10^{-5}$ | $1.00 \times 10^{-3}$ | 1 | $1.07 \times 10^{-5}$ | $1.00 \times 10^{-6}$ | 1 |
| Seawater | $2.26 \times 10^{-5}$ | $1.08 \times 10^{-3}$ | 1.08 | $1.13 \times 10^{-5}$ | $1.05 \times 10^{-6}$ | 1.15 |
| Mercury | $3.34 \times 10^{-5}$ | $1.60 \times 10^{-3}$ | 1.6 | $1.27 \times 10^{-6}$ | $1.18 \times 10^{-7}$ | 0.118 |

Properties based on temperature 68° F or 20° C.

## Fluid Circuits

Fluid circuits contain compressed and flowing fluid, either liquid or gas, and convey it to perform work. They usually employ a valve to stop and/or control the direction of flow.

**Standard Fluid Circuit Components.—**Fluid circuits typically include fluid preparation components, valves, actuators, and connecting lines. Pneumatic circuits also normally include provisions for gas to be exhausted to the atmosphere, through a silencer, to a reclassifying filter, or to an exhaust manifold where it can be ducted elsewhere. Quick exhausting of gas is essential for fast pneumatic circuit response. Hydraulic circuits are generally closed, that is, recirculating fluid rather than discharging it.

**Fluid Circuit Types.—**There are two main fluid circuit types, which are classified in terms of how the valve functions.

*Open and Closed Center Circuits:* Open center circuits permit nearly unrestricted flow through the valve and back to the source when the valve is de-energized. Closed center circuits block flow through the valve when the valve is de-energized. This type of circuit is recommended when you have multiple circuits on a single supply.

*Flow Control Circuits:* To control the speed of an actuator in a pneumatic circuit, use flow controls. Flow controls always generate some heat in hydraulic circuits, so with hydraulics they should be avoided where possible and other methods used. There are three types of flow control circuits: meter-in, meter-out, and bleed-off (or bypass). Pneumatic and some hydraulic systems use meter-in and meter-out circuits, while many hydraulic circuits use bleed-off types.

Meter-in circuits have a flow control valve before an actuator. Meter-in control can be used for actuator speed control. For hydraulic systems, a regulated bypass circuit is recommended rather than a restrictive flow control to reduce generated heat.

Meter-out circuits are best for pneumatic actuator speed control, and are required for applications with large negative loads on actuators. For example when lowering a heavy weight, a meter-out circuit is best. Flow control is placed after the actuator to restrict flow coming out. Because the positive side of the actuator sees full system pressure and flow, the mechanical system is very stable under loading. Care must be taken to avoid overloading actuators when using meter-out circuits.

**Fluid Component Symbols and Circuit Diagrams.—**Fluid power circuits are documented through the use of circuit diagrams or schematics. The components in a circuit are represented symbolically and connected with lines representing pipe or tubing. The symbols are used to identify and represent the function of the components. Symbols for fluid power components have been standardized by ANSI Y32.10 and ISO 1219. Some common symbols can be found in Table 5.

Directional control valve symbols show the valve's positions, methods of actuation, flow paths, and ports. The valve positions are represented by boxes. The methods of left and right valve actuation are represented by symbols on each end of the valve symbol. The flow path for each position is shown within its respective position box. Ports are marked on the position box that corresponds to the de-energized state of the valve.

**Table 5. Selected ANSI Fluid Component Symbols**

| | | | |
|---|---|---|---|
| ———— Solid Line<br>Main Line Conductor, Physical Outline, or Shaft | — — — — Dashed Line<br>Pilot Line for Control | - - - - - - Dotted Line<br>Exhaust or Drain Line | —— - —— - —— Center Line<br>Enclosure Outline |
| 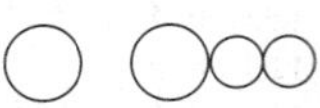 Circle<br>Device, Connector, or Component | 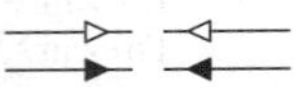 Triangle<br>Indicates Flow Direction[a]: (See footnote [a]) |  Square<br>Reservoir (Closed, Open) | 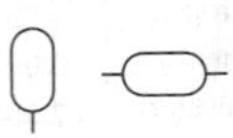 Oval<br>Accumulator or Receiver |

**Table 5.** *(Continued)* **Selected ANSI Fluid Component Symbols**

| | | | |
|---|---|---|---|
| 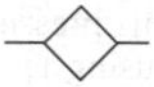<br>Diamond<br>Fluid Conditioner<br>(Filter, Strainer) | <br>Arrow Through Symbol at 45°<br>Indicates Component Can<br>Be Adjusted | 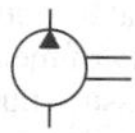<br>Pump, Fixed Capacity<br>(One direction of flow) | <br>Pump, Variable Capacity<br>(Two directions of flow) |
| 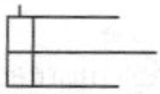<br>Single Acting Air Cylinder | 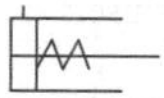<br>Single Acting Spring<br>Return Air Cylinder | 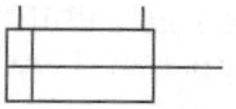<br>Double Acting<br>Air Cylinder | <br>Double Acting Rotary<br>Actuator |
| 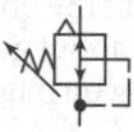<br>Pressure Regulator,<br>Adjustable and Relieving | <br>Pressure Gauge | <br>Temperature Gauge | <br>Cooler |
| <br>Check Valve | <br>Shut-off Valve | <br>Flow Control Valve,<br>Adjustable | 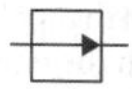<br>Pressure Control Valve |
| 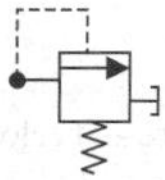<br>Pressure Relief Valve | 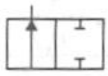<br>Directional Valve<br>Two Port, Two Position<br>(2/2) Two Way,<br>Normally Open | 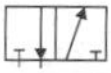<br>Directional Valve<br>Three Port, Two Position<br>(3/2) Two Way,<br>Normally Closed | <br>Directional Valve<br>Five Port, Two Position<br>(5/2) Four Way,<br>Normally Open |
| 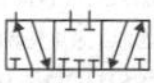<br>Directional Valve<br>Five Port, Three Position<br>(5/3) Four Way,<br>Normally Closed | 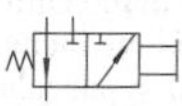<br>Directional Valve<br>Two Position, Three Way,<br>Normally Open, With<br>Manual Actuation and<br>Spring Return | 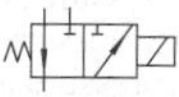<br>Directional Valve<br>Two Position, Three Way,<br>Normally Open, With<br>Single Solenoid and<br>Spring Return | 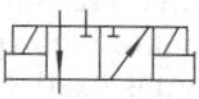<br>Directional Valve<br>Two Position, Three Way,<br>Normally Open, With<br>Double Solenoids and<br>Manual Overrides |

[a] A pneumatic direction symbol is shown as an outline; a hydraulic symbol is filled.

**Fluid Circuit Analysis Considerations.**—When designing or analyzing a fluid circuit, one must consider fluid type, flow rate, velocity, pressure, and temperature. Fluid type and temperature will determine how it behaves as it is manipulated. Excessive velocities can lead to noise, vibration, and stresses. Pressure and flow rate will determine what size and type of pipe, tube, and components can be used. Also important to consider is the structural integrity of the system, including the weight of components, connections, and fluid. Every component in a fluid circuit, with the exceptions of pumps and compressors, will cause energy to be lost from the fluid. Friction, gravity, directional change, and restriction are all causes of lost energy. A careful analysis of energy added and lost from a fluid circuit must be completed as part of the design process.

**Safety Notes.**—When designing fluid circuits, safety should be of paramount importance. The logic of all circuits should be checked for proper performance during startup, shutdown, pressure loss, emergency stops, and other foreseeable events. Before selecting any control element, it is important to first know the safety category for the function being controlled by the fluid circuit. The manufacturer can assist in selecting components with the proper safety functions and ratings. It is important to note that any compressed gas other than air is an asphyxiation hazard. Air quality monitoring devices should be used when the volume of available gas has the potential to create an unsafe environment in the event of a leak. Some fluids can be incompatible with materials used in the circuit. Check material compatibility of seals, tubing, and other components. Check the temperature ratings of any components against the expected temperature of the fluid application.

## Pressure and Flow

Fluids are often delivered at a much higher pressure than needed. This pressure can be measured through a variety of means and brought down to a set level using a pressure regulator. When delivery pressure fluctuates, the installation of a regulator set at a lower pressure can stabilize pressure downstream for the fluid power components in the circuit.

**Pressure Generation.—**Pressure is raised in fluids in a variety of ways, the most common being pumps, compressors, and diffusers. Diffusers are sometimes contained within pumps, and are commonly part of HVAC systems.

*Compressors, Fans, Blowers, and Pumps:* Liquids have their pressures increased by pumps, while gases have their pressures increased by compressors, fans, and blowers. Compressors and some types of fans and blowers are similar to pumps, but the compression of gases is subject to the laws of thermodynamics. More information about gas compressors, fans, and blowers can be found in the section *Pneumatics* starting on page 2819.

*Diffusers:* Diffusers use a gradual enlargement of a pipe to convert velocity energy into pressure. Diffusers are used both to raise pressure and to decrease velocity of a fluid. They are used in pipelines to reduce the energy loss when the pipe goes through an enlargement. More on the energy loss of sudden versus gradual enlargement can be found later in this text. The pressure increase generated by a diffuser can be calculated using the following equation:

$$\Delta p = p_o - p_i = \gamma\left[\frac{(v_i^2 - v_o^2)}{2g} - h_l\right]$$

where $p_o$ and $v_o$ are pressure and velocity at the outlet, $p_i$ and $v_i$ are pressure and velocity at the inlet, and $h_l$ is the energy loss through the diffuser.

**Pressure Measurement.—**Pressure is measured in the United States in terms of pounds per square inch (psi). Where pressures are quite low, inches of water gauge (in$H_2O$, or in. w.g.) is used. In SI units, pressure is measured in pascals (Pa), kilopascals (kPa), and Megapascals (MPa). Other units of pressure are bars and mm$H_2O$. Pressure is usually expressed relative to atmospheric pressure (gauge pressure), as opposed to absolute pressure. Negative gauge pressure is a vacuum. Some common pressure unit conversions are given in the following table.

**Pressure Conversions**

| US | SI |
|---|---|
| 1 in$H_2O$ = 0.036 psi | 1 mm$H_2O$ = 9.81 Pa |
| 1 psi = 6895 Pa | 1 Pa = 0.000145 Psi |
| 1 psi = 144 psf | 1 Pa = 1 N/m2 |
| 1 atm = 14.7 psi | 1 bar = 100,000 Pa |
| 1 atm = 1.01 bar | 1 bar = 0.99 atm |
| 1 inHg = 0.49 psi | 1 mmHg = 133 Pa |

*Manometers:* Manometers are devices which expose a static column of relatively dense fluid (mercury and oils are commonly used) to a differential pressure. These are used for liquid pressure measurement. In many cases, one end is exposed to a pressurized fluid while the other is left open to atmospheric pressure. The change in elevation of the column of fluid can be used to calculate the pressure of the fluid being measured by using the relationship between pressure and elevation.

Commercial manometers are calibrated and can be purchased in a variety of ranges. Piezometer tubes are types of manometers used for relatively low pressures. They consist of a vertical glass tube inserted into the wall of a pipe or pressure vessel. Its end is open to atmosphere or some known pressure reservoir. The height of the liquid in the tube is read against a calibrated scale and is related to pressure in the pipe or vessel.

*Barometers:* Barometers are used to measure atmospheric pressure. They operate on the same principle as manometers, where the change in height of a column of gauge fluid (commonly mercury) is used to calculate a change in pressure. Commercial barometers are calibrated and generally measure in the range between 25.5 and 32.5 inches (650–825 mm) of mercury. Corresponding pressures in pound per square inch and pound per square foot are given in Table 6.

**Table 6. Relation Between Barometric Pressure, and Pressures in Pounds per Square Inch and Square Foot**

| Barometer, Inches | Pressure in Psi[a] | Pressure in Psf[a] | Barometer, Inches | Pressure in Psi[a] | Pressure in Psf[a] | Barometer, Inches | Pressure in Psi[a] | Pressure in Psf[a] |
|---|---|---|---|---|---|---|---|---|
| 28.00 | 13.75 | 1980 | 29.25 | 14.36 | 2068 | 30.50 | 14.98 | 2156 |
| 28.25 | 13.87 | 1997 | 29.50 | 14.48 | 2086 | 30.75 | 15.10 | 2174 |
| 28.50 | 13.99 | 2015 | 29.75 | 14.61 | 2103 | 31.00 | 15.22 | 2192 |
| 28.75 | 14.12 | 2033 | 30.00 | 14.73 | 2121 | 31.25 | 15.34 | 2210 |
| 29.00 | 14.24 | 2050 | 30.25 | 14.85 | 2139 | … | … | … |

[a] Psi is pound per square inch; Psf is pound per square foot

*Pressure Gauges:* The most common method of measuring pressure in a fluid power circuit is by using a pressure gauge. There are gauges for liquids and gauges for gases. There are a variety of methods of construction of pressure gauges, including elastic diaphragm, dead weight, magnetic, and Bourdon tube gauges. Commercial gauges are normally calibrated to read zero at standard atmospheric pressure and temperature. Pressure gauges are available in a variety of ranges and units.

Some are capable of reading vacuum, or negative gauge pressure. Liquid filled gauges provide more stable readouts when vibration or pulsation is present in the system. The use of a snubber is recommended when fluid pressure pulsates.

*Pressure Transducers:* These are used when the display needs to be decoupled from the measurement device. Transducers convert pressures to electrical signals which can be transmitted to remote displays. Pressure transducers can be linear differential transformers, piezoelectric, quartz resonators, or solid state devices.

**Pressure Regulation.**—Fluids are often delivered at pressures higher than needed. Pressure regulators are used where pressure needs to be lowered and controlled. There are regulators for liquids and others for gases. A pressure regulator is a device generally comprised of three elements; a pressure reducing or restrictive element, a sensing element, and a force generator (usually a spring). In operation, the force generator opens the valve. Pressure introduced into the inlet port then flows through the valve and applies pressure to the sensing device (usually a diaphragm or piston). The pressure acts on the sensing element to produce a force which opposes the opening force and closes the valve. Regulators may be manually or electronically adjusted, and may include a pressure gauge or sensor.

In some cases, pressure may fluctuate due to variable supply and/or demand. This may be undesirable where work is being performed. A pressure reservoir with a regulator after it is a desirable source for most fluid circuits because the combination can buffer pressure variations and supply multiple circuits.

When choosing a pressure regulator, there are many factors to consider. The allowable ranges of inlet and outlet pressures must be known when sizing a regulator. The maximum flow rate and any porting requirements are also factors. It is also important to consider the chemical properties of the fluid before determining the best materials for your application. There are a wide range of materials available to suit all applications. A non-relieving regulator is preferred for use with hazardous, explosive, or expensive gases since a relieving regulator will vent excess pressure to atmosphere.

The temperature range that the regulator will see is also a factor. Extreme operating temperatures may affect flow capacity and/or the spring rate.

**Flow Measurement.**—There are many ways to measure the flow of liquid or gas. Measurement can be in terms of velocity, volume flow rate, or mass flow rate. When selecting a flow meter, one must consider the viscosity and chemical compatibility of the fluid being measured, the desired output units, the range of flow values being measured, the desired accuracy, and the maximum pressure encountered by the unit. The ratio of the maximum to the minimum flow rate a meter can measure within the expected accuracy is called "turndown."

When designing a pipe system, it is often beneficial to run a bypass system of pipes with valves to allow service of a flow meter without interrupting system flow. Flow meters should be placed in straight sections of pipe for best measurement results. The length of straight pipe before and after each meter will depend on its type, the pipe diameter, and the type of nearest fitting or bend.

*Pitot Tubes:* Some pitot tubes are available to measure liquids and others are made for gases. These devices measure fluid flow velocity directly as a function of the height of a stationary column of fluid pushed out of the flow stream into a narrow tube. Most of the fluid flows past the device, but a small amount is captured in the static tube. Pitot tubes can be permanently installed in a pipe or duct, or used as a portable device for insertion as needed. The location of the capture tip of the pitot tube is critical, since flow velocity in a pipe varies across its cross section. Pitot-static tubes have ports perpendicular to the fluid flow direction to measure the static pressure of the fluid. Velocity is then calculated using the following equation:

$$v = \sqrt{\frac{2(p_t - p_s)}{\rho}}$$

where $v$ is fluid velocity at the collection point; $p_s$ is static pressure; and $p_t$ is the stagnation pressure measured as a function of displacement of the fluid in the tube.

*Cup Anemometer:* Air velocity is often measured with one of these devices. They consist of several cups arranged about a shaft that is free to spin. As air flows perpendicularly to the cups, the device rotates at a speed dependent on the air flow velocity.

*Hot Wire Anemometer:* These devices employ a thin wire inserted into a gas stream. The wire is heated by current passing through it, and the gas flowing by tends to dissipate this heat. Some hot wire anemometers apply a fixed current, and thus the change in resistance of the wire (related to temperature change) is related to flow velocity. Others vary the current to maintain a fixed temperature, and thus the current required is related to flow velocity. Hot wire anemometers tend to break when used with liquids, so hot film probes are used in liquid applications.

*Venturi Meters:* These flow-through meters measure the pressure drop across a restriction and relate that to volume flow rate. The restriction is called a "throat" and has a gradual contraction followed by a gradual expansion. Venturi meters are more expensive than orifice meters, but cause less pressure loss in the system. Overall pressure loss is generally 5 to 20 percent of the measured differential pressure. Many venturi meters are available for measuring liquids, and others are meant for gases. They are suitable for contaminated fluids and slurries. The lengths of straight pipe required upstream and downstream of a Venturi meter for accurate flow measurement are given in ISO 5167-1. Flow rate $Q$ through a venturi is calculated using the following equation:

$$Q = C\left(\sqrt{\frac{2\Delta p}{\rho}}\right)\frac{A_1}{\sqrt{\left(\frac{A_1}{A_2}\right)^2 - 1}}$$

where $\Delta p$ is the measured differential pressure, $A_1$ is the area at the upstream pressure tap, and $A_2$ is the area at the throat. Coefficient $C$ is generally 1 for ideal fluids, and 0.90 to 0.98 for viscous fluids.

*Orifice Plates and Meters:* These are the most widely used method of flow measurement in industry. Similar to venturi meters, these flow- through devices measure pressure drop across a restriction. The restriction is generally a hole in a plate placed across the flow path. This causes sudden contraction and expansion. The placement of the pressure measurement point is critical and dependent on the shape of the orifice. These meters generally create a larger energy loss than venturi meters due to the turbulence created by sudden changes in diameter, but are very cost-effective. The relevant Standard covering orifice plates is ISO 5167. Flow through an orifice meter can be calculated using the following equation.

$$Q = C_f A_o \sqrt{\frac{2\Delta p}{\rho}}$$

where $A_o$ is the area of the orifice. The coefficient $C_f$ is generally between 0.6 and 0.9, and depends on the Reynolds number as well as pipe and orifice diameter. Tabulations of coefficient $C_f$ can be found in various reference sources.

*Flow Nozzle Meters:* Similar to venturi and orifice meters, these flow-through devices measure pressure drop caused by flow restriction. Flow nozzles have a smooth elliptical inlet leading to a throat section with a sharp outlet. The geometry of these nozzles has been highly standardized by ASME and ISO. The nozzle tends to have a well-rounded intake, and therefore creates a much lower energy loss than an orifice meter. This higher efficiency means greater flow capacity when compared to most other differential meters of the same size. Energy loss does tend to be higher than with venturi tubes. Nozzle meters are available in standard and long radius configurations, with different discharge coefficients. Flow is calculated similarly to orifice and venturi meters, using Bernoulli's equation and the appropriate coefficient.

*Variable Area Flow Meters:* These devices are extremely common in industry, and are available for measuring liquids or gases. There are tube and vane types. Tube meters have a weighted float that is suspended in the flowing fluid. The clearance space between the float and the tube forms an annular passage or orifice. Since the tube is tapered, the area of this orifice is larger when the float is near the top than it is when the float is near the bottom. Volumetric flow is measured based on the height the float is carried to. As a result, these meters must be installed vertically. Vane type meters use a vane to obstruct a bend in the pipe. It is normally closed, and fluid flow swings it open. Both tube and vane meters are for use with clean fluids at moderate pressures. Some types allow the float or vane to be viewed against a calibrated scale. In those cases, the fluid must be clear enough to see the float. Other types of variable area flow meters are suitable for high pressures and opaque fluids, and use sensors to detect the position of the float or vane and display the result remotely. Due to the restriction caused by the float or vane in the fluid stream, some energy is lost by fluid flowing through these meters.

*Elbow Meters:* These meters are simply a bend in a pipe through which the fluid flows. These meters are sometimes the only choice where space constraints do not allow the other types of meters. Pressure differential is measured between a point on the outside radius and a point on the inside radius. Because the fluid is moving faster along the outside radius, it will be generate higher pressure.

*Turbine Flow Meters:* These meters allow fluid flowing through to turn a turbine rotor. The speed at which the rotor turns is indicative of the fluid velocity. The speed of the rotor is mechanically or electronically monitored. Because the turbine obstructs the fluid stream to some extent, some energy is lost due to restriction. These meters can be used for liquids or gases. Care must be taken when using these meters in a fluid that is non-lubricating, since the bearings that support the rotor may prematurely wear. Turbine flowmeters are less accurate at low flow rates due to drag, and should be operated above 5 percent of rated flow. Use in dirty fluids should generally be avoided to reduce the possibility of wear and damage. Two-phase flow and abrupt transitions from gas to liquid should be avoided because this can damage the mechanism.

*Vortex Shedding Flow Meters:* These are flow-through meters used for both liquids and gases. Placement of a blunt obstruction, called a *bluff body*, in the fluid stream causes vortices to be shed at a frequency dependent on the flow velocity. The shedding frequency is independent of fluid properties such as density, viscosity, conductivity, etc., however the flow must be turbulent for vortex shedding to occur. Sensors electronically monitor the shedding vortices. The obstruction need not be very large, and varies between manufacturers. Therefore, the energy lost by the fluid varies greatly from device to device. The advantage of this type of meter is that it is suitable for a wide range of fluids and gases. However, if a fluid includes substances likely to stick to or coat the bluff body, the meter's characteristics will change and accuracy will be compromised. The minimum measured flow rate should be at least twice the minimum rate detectable by the meter. The maximum flow capacity of the meter should be at least five times the anticipated maximum flow rate. The energy loss through a vortex meter is about half that of an orifice plate. The installed cost of vortex meters is competitive with that of orifice meters in sizes under six inches.

*Magnetic Flow Meters:* For use with conducting fluids only, magnetic flow meters use electrodes to detect the voltage generated in a fluid as it moves between two electromagnetic coils. The voltage generated is dependent on fluid velocity. Since there is nothing obstructing flow, these flow-through meters cause less energy loss than obstructing types. Another advantage of these meters is their suitability for a wide variety of fluids because their function is not affected by temperature, viscosity, or specific gravity.

*Ultrasonic Flow Meters:* These flow-through devices are attached to the outside of a pipe or tube, and therefore do not restrict flow or cause energy loss in the fluid. There are meters for gases and for fluids. Ultrasonic signals are transmitted through the pipe and fluid. The time it takes the signal to reach the receiver depends on the flow rate of the fluid. Some ultrasonic flow meters measure across the pipe, while others measure axially. A special type of ultrasonic flow meter, called a "Doppler" meter, is used for fluids with particulate content which may obstruct traversing ultrasonic signals. These meters measure the time it takes for signals to be reflected back from the fluid.

*Positive Displacement Flow Meters:* These are not flow-through devices, but instead allow fluid to fill a volumetrically calibrated chamber before the fluid is released. These meters can utilize a wide variety of mechanisms to capture and release volumes of fluid, one of the most common being diaphragm types typical to home gas meters. Other designs incorporate gears, pistons, and nutating disks (wobble plates). These meters are most often used for liquids, but can be used for gases as well.

*Coriolis Mass Flow Meters:* These flow-through devices measure mass flow rate directly, as opposed to measuring volume flow rate or velocity. By routing fluid through looping geometry, Coriolis force is generated and measured with sensors. Measurement results depend on the vibration of the tubes, so changes in tube rigidity due to temperature and pressure changes will affect calibration of the meter. These factors can be compensated for, but pitting, cracking, coating, erosion, or corrosion of the tubes cannot. Some energy is lost as the looping geometry causes changes in direction and velocity. These devices can be outfitted to monitor density of the fluid.

*Thermal Mass Flow Meters:* These devices are used almost exclusively for gas flow applications. Resistance temperature detectors (RTD) exhibit a linear change in resistance with respect to temperature. To measure mass flow rate, two RTD probes are used. One probe measures the fluid temperature, while the second probe is maintained at a higher temperature. As fluid flows by the heated probe, it dissipates the heat. The energy required to maintain the heated probe temperature is proportional to mass flow rate of the fluid. Some fluid energy is lost due to the obstruction of the probes in the fluid stream.

**Flow Control.—**Flow control is accomplished with manual or automatic valves, for both liquids and gases. Monitoring and control are often combined into a single device, in which flow is measured and the results are used to control a proportional valve to regulate flow. There are many types of flow controllers available, including manual needle valves and fully automated flow controllers. When density and viscosity is known and constant, and temperature is known, mass flow control can be performed. Mass flow controllers are usually used with gases, but some are available for liquids. Commercial flow control devices range in accuracy and flow capacity, and often contain proprietary control software and flow measurement technology.

**Coefficient of Velocity, $C_v$.—**$C_v$ is defined as the flow rate in US gallons per minute (gpm) of water at 60° F with a pressure drop across the device of 1 psi. This is directly applicable to hydraulic components, but $C_v$ is also applied to pneumatic devices as an indication of their flow capacity. This value is normally used when matching valves to actuators, or sizing other fluid power devices.

Flow capacity of valves in a circuit must be equal to or greater than flow required by the devices they supply. For actuators, higher $C_v$ values can also indicate faster movement capability.

$C_v$ is typically used in the United States and calculated using English units. The SI system has its own flow coefficient, $K_v$, which is calculated using metric units. These two values are not equivalent. $K_v$ is defined as the flow rate in cubic meters per hour ($m^3/h$) of water at a temperature of 16° C, with a pressure drop across the valve of 1 bar.

$C_v$ is commonly used by manufacturers for both English and metric products, and is normally provided in catalogs. The equations used to calculate $C_v$ and $K_v$ are listed below. The equations for liquids are simpler than those for gases, because gases are compressible.

*Conversion:* $C_v = 1.157K_v$ $\qquad K_v = 0.864C_v$

### $C_v$ and $K_v$ Equations for Liquids

for <u>*Non-choked Turbulent Flow*</u> when $\Delta p < F_L^2(p_i - F_F P_v)$

$$C_v = Q\sqrt{\frac{SG}{\Delta p}} \qquad\qquad K_v = Q\sqrt{\frac{SG}{\Delta p}}$$

for <u>*Choked Turbulent Flow*</u> when $\Delta p \geq F_L^2(p_i - F_F P_v)$

$$C_v = \frac{Q}{F_L}\sqrt{\frac{SG}{p_i - F_F P_v}} \qquad\qquad K_v = \frac{Q}{F_L}\sqrt{\frac{SG}{p_i - F_F P_v}}$$

for <u>*Laminar or Transitional Flow*</u>

$$C_v = \frac{Q}{F_R}\sqrt{\frac{SG}{\Delta p}} \qquad\qquad K_v = \frac{Q}{F_R}\sqrt{\frac{SG}{\Delta p}}$$

### $C_v$ and $K_v$ Equations for Gases

for Non-choked Turbulent Flow when $x < F_\gamma x_T$

$$C_v = \frac{Q}{1360 p_i Y}\sqrt{\frac{SG p_i T_i Z}{\Delta p}} \qquad K_v = \frac{Q}{482 p_i Y}\sqrt{\frac{SG p_i T_i Z}{\Delta p}}$$

for Choked Turbulent Flow when $x \geq F_\gamma x_T$

$$C_v = \frac{Q}{907 p_i}\sqrt{\frac{SG p_i T_i Z}{F_\gamma \Delta p}} \qquad K_v = \frac{Q}{321.5 p_i}\sqrt{\frac{SG p_i T_i Z}{F_\gamma \Delta p}}$$

for Laminar or Transitional Flow

$$C_v = \frac{Q}{N_1 F_R}\sqrt{\frac{M T_i}{\Delta p (p_i + p_o)}} \qquad K_v = \frac{Q}{N_2 F_R}\sqrt{\frac{M T_i}{\Delta p (p_i + p_o)}}$$

$N_1 = 1730$ for $T_s = 0°$ C $\qquad N_2 = 4920$ for $T_s = 0°$ C

$N_1 = 1840$ for $T_s = 15°$ C $\qquad N_2 = 5200$ for $T_s = 15°$ C

The following definitions apply to equations for $C_v$ and $K_v$ given above:

$Q$ = flow in gpm for liquids, and flow in SCFM for gasses

SG = specific gravity

$\Delta p = p_i - p_o$ is the pressure differential between upstream and downstream, $p_i$ is inlet absolute pressure (psia or bar), and $p_o$ is outlet absolute pressure (psia or bar)

$F_F$ = liquid critical pressure ratio factor (ranges between 0.68 to 0.96)

$F_L$ = liquid pressure recovery factor. $F_L$ depends on valve geometry and ranges between 0.6 for ball valves and 0.9 for globe valves

$P_v$ = absolute vapor pressure of the liquid at inlet temperature

$x = \frac{\Delta p}{p_i}$ is the ratio of the pressure differential to the inlet absolute pressure

$F_\gamma$ = specific heat ratio factor. $F_\gamma = 1$ for air, otherwise $F_\gamma = \frac{\gamma}{1.4}$

$x_T$ = choked flow pressure differential ratio factor, experimentally determined, that varies from 0.3 to 0.9 depending on valve geometry

$Y$ = expansion factor for gases. $Y = 1 - \frac{x}{3F_\gamma x_T}$ for normal flow, and ranges from $Y = 1$ to $Y = 0.667$ at choked flow

$Z$ = compressibility factor (assume $Z = 1$ for ideal gasses)

$F_R$ = Reynolds number factor that depends on valve characteristics, and is limited to $F_R \leq 1$

$M$ = molecular mass of fluid in kg/kg-mol or lb/lb-mol

$T_i$ = absolute inlet temperature for the gas, °C

$T_s$ = absolute reference temperature for the gas, °C

For additional information on flow equations see International Electrotechnical Commission Standard IEC 60534.

## Pipe, Tubing, and Fittings

Design of pipe systems must be undertaken with great care. Safety, performance, reliability, and cost are all significantly affected by selection of appropriate pipe sizes, materials, and connection methods. For many applications, the design must comply with codes and standards. Organizations such as ISO, ASTM, NSF, NFPA, AWWA, AFSA, and IAPMO should be consulted when working with pipe. Tubing systems are also covered by standards in cases of sanitary, food, medical, and high pressure applications.

**Metal Pipe.—**Metal pipe is ideal for applications where structural strength is needed. Metal pipe is also resistant to high temperatures and is non-combustible. Extremely low temperatures should be avoided. Metal pipe systems may need proper grounding to prevent shocks and accelerated corrosion. A wide range of materials is available.

*Materials:* Steel is the most commonly used metal pipe material for general use. Steel pipe is highly resistant to shock and vibration from water hammer or external sources. To carry a given pressure, steel pipe can be made with a thinner wall than pipes made from other materials, and as a result the pipe will have a larger flow capacity. Carbon steel pipes are susceptible to corrosion and oxidation, so additional allowance should be made when calculating flow capacity. Galvanized steel can inhibit atmospheric corrosion to some extent, but the coating can break down when exposed to constant moisture. Stainless steel pipes cost more, but are resistant to corrosion and oxidation. Stainless steel is also compatible with a wide variety of chemicals.

Ductile iron pipe is commonly used for water and sewer applications. It combines the physical strength of mild steel with the long life of gray cast iron. It has superseded cast iron pipe in the water handling industry. In addition to the benefits of long life, corrosion resistance, high structural strength, and tight joints, ductile iron is also readily machinable. Although somewhat corrosion resistant when in contact with water, these pipes are often internally lined or coated to inhibit corrosion further.

Copper pipe is commonly used in household and industrial supply lines for both liquids and gases. It costs more than steel pipe, but is much easier to join. It is extremely thermally conductive, which makes it a top choice for heat exchangers and other thermal applications. Copper resists corrosion, is unaffected by UV radiation, and inhibits the growth of bacteria. It will corrode in the presence of high or low pH liquids.

Connecting pipes and fittings of different metals is not recommended because galvanic corrosion to occur. Protective coatings should be employed where necessary to prevent contamination or failure of the joint.

*Schedules and Classes:* Schedule number designations for steel pipe, previously called "Iron Pipe Sizes (IPS)" are an indicator of wall thickness. Schedules range from 10 to 160, with higher numbers designating thicker walls. Schedules 40 and 80 are the most commonly used. From the IPS system, designations like STD, XS or XHY (extra heavy or strong), and XXS or XXHY (double extra strong or heavy) are still used. For a given pipe size, all schedules have the same outside diameter.

Higher schedule pipes of the same diameter have smaller flow areas. The nominal pipe size (NPS) is given, along with the schedule or IPS designation to indicate the full dimensional characteristics of the pipe.

Ductile iron pipe ranges in size from 3 to 64 inches and in pressure classes from 150 psi to 350 psi, although custom sizes and pressure ratings can be obtained.

Copper pipe is classified as copper tube. Information on copper tube can be found in the section *Metal Tube* starting on page 2780. Copper pipe is suitable for water and steam service at pressures up to 250 psi (1750 kPa) and temperatures up to 406° F (208° C).

*Sizes:* Nominal sizes for commercial steel pipe are generally given in inches. Actual diameter of the pipe can be different from the nominal designation, so refer to Table 7 for actual dimensions of common pipe sizes up to 6 inches in diameter. This is the Nominal

Pipe Size (NPS) system. ISO has designated metric equivalent pipes for customary US pipe sizes. As with US pipe, metric pipe dimensions often differ from the nominal size designations.

Sizes of ductile iron pipe follow the Ductile Iron Pipe Sizes (DIPS) Standard. These sizes and dimensions can be found in Table 8. While the maximum length of flanged pipes is generally around 20 feet, the minimum length varies according to pipe size and manufacturer. The minimum class thickness for ductile iron flanged pipe to be threaded is Special Thickness Class 53 for sizes up through 54 and pressure class 350 for 60 and 64 sizes. Greater pipe wall thicknesses are available.

*Pressure Ratings:* When working with ductile iron pipe applications, the designer should choose the appropriate class of pipe to carry the anticipated working pressure.

When working with other types of metal pipe, wall thickness must be selected or calculated based on working pressure. ANSI/ASME Standard B31.1-1998 "Power Piping" details a method of calculating the minimum wall thickness for pipes. The maximum working pressure for a pipe is often provided by the manufacturer, and can be calculated using US standard units in the following equation:

$$p_{max} = \frac{2S(0.875t - A)}{D - 2Y(0.875t - A)}$$

where $S$ is the allowable tensile stress in psi, $t$ is the wall thickness in inches, $D$ is the outer diameter in inches, $A$ is an allowance for corrosion and threading, and $Y$ is a temperature factor given in Table 9. A factor of safety should be used when selecting allowable stress.

Allowance $A$ is added to compensate for corrosion, erosion, threads, and bends, which thin the walls of the pipe. Allowance $A = 0.05$ inch for threaded steel up to $3/8$ inch OD and for unthreaded steel up to $1/2$ inch OD. $A = 0.065$ inch for unthreaded steel larger than $3/8$ inch OD. For threaded steel larger than $3/8$ inch OD, $A$ is equal to the thread depth.

Wall thickness based on pressure requirement can be calculated using the following equation and Table 9. This is for straight pipe, and wall thickness tolerances should be taken into account.

$$t = \frac{pD_o}{2(S + pY)} + A$$

where $D_o$ is the outside diameter of the pipe, $p$ is internal pressure, $Y$ is the temperature coefficient given in Table 9, $A$ is the allowance discussed above, and $S$ is the maximum allowable stress in the material. $S$ should be modified to account for welds and joints according to the standards, and a factor of safety should be used.

The temperature at which material properties are given must be considered. Many hydraulic systems run at a temperature of approximately 100° F (38° C). A factor of safety of 3 to 4 should be used when calculating pressure ratings. If the tube is carrying liquid, the pressure rating must be derated to account for the pressure spikes that occur during valve closures. Derating for pressure spikes (water hammer) is often 20 percent. Derating to account for fittings is usually also done, with 25 percent additional reduction a typical value.

*Connections:* There are many types of connections for metal pipe, the most common of which are threaded, flanged, welded, and soldered.

Threaded connections on pipes normally use tapered threads per the standard for either American National Pipe Thread (NPT) or British Standard Pipe Thread (BSP). Metric pipe uses ISO standard pipe threads. These threads are detailed in *PIPE AND HOSE THREADS* starting on page 2092. Sealing is accomplished through thread interlocking and the use of a sealing compound or PTFE tape. The compound or tape lubricates the threads and ensures that spiral leak paths through the threads are blocked. Tape must be used in just the right amount to be effective. Too much tape can cause the joint to fail. Tape should not be used on bull plugs on heat exchangers. Tape is available in several

## Table 7. Dimensions of Steel Pipe, Common Sizes, US Customary Units

| Nominal Pipe Size inch | Pipe O.D. inch | Wall Thickness inch | Pipe I.D. inch | Weight per Foot lbf | Iron Pipe Size (IPS) | Schedule | Gallon Capacity Per Linear Foot |
|---|---|---|---|---|---|---|---|
| ½ | 0.8400 | 0.1090 | 0.6220 | 0.8500 | STD | 40 | 0.0158 |
| | | 0.1470 | 0.5460 | 1.0900 | XHY | 80 | 0.0122 |
| | | 0.1880 | 0.4640 | 1.3100 | … | 160 | 0.0088 |
| | | 0.2940 | 0.2520 | 1.7200 | XXHY | … | 0.0026 |
| ¾ | 1.0500 | 0.1130 | 0.8240 | 1.1300 | STD | 40 | 0.0277 |
| | | 0.1540 | 0.7420 | 1.4800 | XHY | 80 | 0.0225 |
| | | 0.2190 | 0.6120 | 1.9500 | … | 160 | 0.0153 |
| | | 0.3080 | 0.4340 | 2.4400 | XXHY | … | 0.0077 |
| 1 | 1.3150 | 0.1330 | 1.0490 | 1.6800 | STD | 40 | 0.0449 |
| | | 0.1790 | 0.9570 | 2.1700 | XHY | 80 | 0.0374 |
| | | 0.2500 | 0.8150 | 2.8500 | … | 160 | 0.0271 |
| | | 0.3580 | 0.5990 | 3.6600 | XXHY | … | 0.0146 |
| 1 ¼ | 1.6600 | 0.1400 | 1.3800 | 2.2700 | STD | 40 | 0.0777 |
| | | 0.1910 | 1.2780 | 3.0000 | XHY | 80 | 0.0666 |
| | | 0.2500 | 1.1600 | 3.7700 | … | 160 | 0.0549 |
| | | 0.3820 | 0.8960 | 5.2200 | … | … | 0.0328 |
| 1 ½ | 1.9000 | 0.1450 | 1.6100 | 2.7200 | STD | 40 | 0.1058 |
| | | 0.2000 | 1.5000 | 3.6300 | XHY | 80 | 0.0918 |
| | | 0.2810 | 1.3380 | 4.8600 | … | 160 | 0.0730 |
| | | 0.4000 | 1.1000 | 6.4100 | … | … | 0.0494 |
| 2 | 2.3750 | 0.1090 | 2.1570 | 2.6400 | … | 10 | 0.1898 |
| | | 0.1540 | 2.0670 | 3.6600 | STD | 40 | 0.1743 |
| | | 0.2180 | 1.9390 | 5.0300 | XHY | 80 | 0.1534 |
| | | 0.2500 | 1.8750 | 5.6800 | … | … | 0.1434 |
| | | 0.2810 | 1.8130 | 6.2900 | … | … | 0.1344 |
| | | 0.3440 | 1.6870 | 7.4700 | … | 160 | 0.1161 |
| | | 0.4360 | 1.5030 | 9.0400 | XXHY | | 0.0922 |
| 2 ½ | 2.8750 | 0.1200 | 2.6350 | 3.5300 | … | 10 | 0.2833 |
| | | 0.2030 | 2.4690 | 5.8000 | STD. | 40 | 0.2487 |
| | | 0.2160 | 2.4430 | 6.1400 | … | … | 0.2435 |
| | | 0.2500 | 2.3750 | 7.0200 | … | … | 0.2301 |
| | | 0.2760 | 2.3230 | 7.6700 | XHY | 80 | 0.2202 |
| | | 0.3750 | 2.1250 | 10.0200 | … | 160 | 0.1842 |
| | | 0.5520 | 1.7710 | 13.7100 | XXXHY | … | 0.1280 |
| 3 | 3.5000 | 0.1200 | 3.2600 | 4.3400 | … | … | 0.4336 |
| | | 0.1560 | 3.1880 | 5.5800 | … | … | 0.4147 |
| | | 0.1720 | 3.1560 | 6.1200 | … | … | 0.4064 |
| | | 0.1880 | 3.1240 | 6.6600 | … | … | 0.3984 |
| | | 0.2160 | 3.0680 | 7.5800 | STD. | 40 | 0.3840 |
| | | 0.2500 | 3.0000 | 8.6900 | … | … | 0.3672 |
| | | 0.2810 | 2.9380 | 9.6700 | … | … | 0.3521 |
| | | 0.3000 | 2.9000 | 10.2600 | XHY | 80 | 0.3431 |
| | | 0.4380 | 2.6240 | 14.3400 | … | 160 | 0.2811 |
| | | 0.6000 | 2.3000 | 18.6000 | XXHY | … | 0.2158 |
| 3 ½ | 4.0000 | 0.1200 | 3.7600 | 4.9800 | … | 10 | 0.5768 |
| | | 0.2260 | 3.5480 | 9.1200 | STD. | 140 | 0.5136 |
| | | 0.2500 | 3.5000 | 10.0200 | … | … | 0.4998 |
| | | 0.2810 | 3.4380 | 11.1700 | … | … | 0.4821 |
| | | 0.3180 | 3.3640 | 12.5200 | XHY | 80 | 0.4617 |
| | | 0.6360 | 2.728 | 22.8700 | XXHY | … | 0.3085 |

**Table 7.** *(Continued)* **Dimensions of Steel Pipe, Common Sizes, US Customary Units**

| Nominal Pipe Size inch | Pipe O.D. inch | Wall Thickness inch | Pipe I.D. inch | Weight per Foot lbf | Iron Pipe Size (IPS) | Schedule | Gallon Capacity Per Linear Foot |
|---|---|---|---|---|---|---|---|
| 4 | 4.5000 | 0.1200 | 4.2600 | 5.6200 | … | 10 | 0.7404 |
| | | 0.1560 | 4.1880 | 7.2400 | … | … | 0.7156 |
| | | 0.1880 | 4.1240 | 8.6700 | … | … | 0.6942 |
| | | 0.2030 | 4.0940 | 9.3200 | … | … | 0.6838 |
| | | 0.2190 | 4.0620 | 10.0200 | … | … | 0.6725 |
| | | 0.2370 | 4.0260 | 10.8000 | STD. | 40 | 0.6613 |
| | | 0.2500 | 4.0000 | 11.3600 | … | … | 0.6528 |
| | | 0.2810 | 3.9380 | 12.6700 | … | 60 | 0.6326 |
| | | 0.3120 | 3.8760 | 13.9700 | … | … | 0.6126 |
| | | 0.3370 | 3.8260 | 15.0000 | XHY | 80 | 0.5972 |
| | | 0.4380 | 3.6240 | 19.0200 | … | 120 | 0.5361 |
| | | 0.5310 | 3.4380 | 22.5300 | … | 160 | 0.4822 |
| | | 0.6740 | 3.1520 | 27.5700 | XXHY | … | 0.4054 |
| 5 | 5.5630 | 0.1880 | 5.1870 | 10.8000 | … | … | 1.0979 |
| | | 0.2190 | 5.1250 | 12.5100 | … | … | 1.0716 |
| | | 0.2580 | 5.0470 | 14.6300 | STD. | 40 | 1.0391 |
| | | 0.2810 | 5.0010 | 15.8700 | … | … | 1.0204 |
| | | 0.3120 | 4.9390 | 17.5100 | … | … | 0.9965 |
| | | 0.3440 | 4.8750 | 19.1900 | … | … | 0.9696 |
| | | 0.3750 | 4.8130 | 20.8000 | XHY | 80 | 0.9449 |
| | | 0.5000 | 4.5630 | 27.0600 | … | 120 | 0.8495 |
| | | 0.6250 | 4.3130 | 32.9900 | … | 160 | 0.7590 |
| | | 0.7500 | 4.0630 | 38.5900 | XXHY | … | 0.6734 |
| 6 | 6.6250 | 0.1090 | 6.4070 | 7.5900 | … | 5 | 1.6748 |
| | | 0.1340 | 6.3570 | 9.3000 | … | 10 | 1.6488 |
| | | 0.1560 | 6.3130 | 10.7900 | … | … | 1.6260 |
| | | 0.1880 | 6.2490 | 12.9400 | … | … | 1.5937 |
| | | 0.2030 | 6.2190 | 13.9400 | … | … | 1.5780 |
| | | 0.2190 | 6.1870 | 15.0000 | … | … | 1.5620 |
| | | 0.2500 | 6.1250 | 17.0400 | … | … | 1.5306 |
| | | 0.2800 | 6.0650 | 18.9900 | STD. | 40 | 1.5008 |
| | | 0.3120 | 6.0010 | 21.0600 | … | … | 1.4688 |
| | | 0.3440 | 5.9370 | 23.1000 | … | … | 1.4391 |
| | | 0.3750 | 5.8750 | 25.0500 | … | … | 1.4082 |
| | | 0.4320 | 5.7610 | 28.6000 | XHY | 80 | 1.3541 |
| | | 0.5000 | 5.6250 | 32.7400 | … | … | 1.2909 |
| | | 0.5620 | 5.5010 | 36.4300 | … | 120 | 1.2346 |
| | | 0.6250 | 5.3750 | 40.0900 | … | … | 1.1787 |
| | | 0.7190 | 5.1870 | 45.3900 | … | 160 | 1.0977 |
| | | 0.8640 | 4.8970 | 53.2100 | XXHY | … | 0.9784 |
| | | 1.0000 | 4.6250 | 60.1300 | … | … | 0.8727 |
| | | 1.1250 | 4.3750 | 66.1400 | … | … | 0.7809 |

thicknesses. Thicker tape should be used on higher pressure or larger size installations. Always follow the manufacturer's recommendations when using liquid thread compounds, and always verify chemical compatibility with the process fluid.

Threaded connections are specified by the nominal pipe size and the type of thread (NPT, BSP or other).

Threaded pipe connections require the rotation of one of the pipes in the joint. When this is not practical, a threaded union may be used. Unions are comprised of two halves that form a flange connection.

Flanged connections are commonly used on large piping systems. For process and utilities, the two most common flange standards are ANSI B16.5 and BS 1560. Wellhead flanges are specified by API 6A. Flanges of different standards are not normally joined. There are three commonly used types of flange facings: Ring Type Joint (RTJ), Raised

## Table 8. Ductile Iron Pipe Specifications

| Nom. Size | Outside Diameter inches | Pressure Class, psi | | | | | Special Thickness Class | | | | | | |
|---|---|---|---|---|---|---|---|---|---|---|---|---|---|
| | | 150 | 200 | 250 | 300 | 350 | 50 | 51 | 52 | 53 | 54 | 55 | 56 |
| | | Nominal Wall Thickness, inch | | | | | Nominal Wall Thickness, inch | | | | | | |
| 3 | 3.96 | … | … | … | … | 0.25 | … | 0.25 | 0.28 | 0.31 | 0.34 | 0.37 | 0.40 |
| 4 | 4.80 | … | … | … | … | 0.25 | … | 0.26 | 0.29 | 0.32 | 0.35 | 0.38 | 0.41 |
| 6 | 6.90 | … | … | … | … | 0.25 | 0.25 | 0.28 | 0.31 | 0.34 | 0.37 | 0.40 | 0.43 |
| 8 | 9.05 | … | … | … | … | 0.25 | 0.27 | 0.30 | 0.33 | 0.36 | 0.39 | 0.42 | 0.45 |
| 10 | 11.10 | … | … | … | … | 0.26 | 0.29 | 0.32 | 0.35 | 0.38 | 0.41 | 0.44 | 0.47 |
| 12 | 13.20 | … | … | … | … | 0.28 | 0.31 | 0.34 | 0.37 | 0.40 | 0.43 | 0.46 | 0.49 |
| 14 | 15.30 | … | … | 0.28 | 0.30 | 0.31 | 0.33 | 0.36 | 0.39 | 0.42 | 0.45 | 0.48 | 0.51 |
| 16 | 17.40 | … | … | 0.30 | 0.32 | 0.34 | 0.34 | 0.37 | 0.40 | 0.43 | 0.46 | 0.49 | 0.52 |
| 18 | 19.50 | … | … | 0.31 | 0.34 | 0.36 | 0.35 | 0.38 | 0.41 | 0.44 | 0.47 | 0.50 | 0.53 |
| 20 | 21.60 | … | … | 0.33 | 0.36 | 0.38 | 0.36 | 0.39 | 0.42 | 0.45 | 0.48 | 0.51 | 0.54 |
| 24 | 25.80 | … | 0.33 | 0.37 | 0.40 | 0.43 | 0.38 | 0.41 | 0.44 | 0.47 | 0.50 | 0.53 | 0.56 |
| 30 | 32.00 | 0.34 | 0.38 | 0.42 | 0.45 | 0.49 | 0.39 | 0.43 | 0.47 | 0.51 | 0.55 | 0.59 | 0.63 |
| 36 | 38.30 | 0.38 | 0.42 | 0.47 | 0.51 | 0.56 | 0.43 | 0.48 | 0.53 | 0.58 | 0.63 | 0.68 | 0.73 |
| 42 | 44.50 | 0.41 | 0.47 | 0.52 | 0.57 | 0.63 | 0.47 | 0.53 | 0.59 | 0.65 | 0.71 | 0.77 | 0.83 |
| 48 | 50.80 | 0.46 | 0.52 | 0.58 | 0.64 | 0.70 | 0.51 | 0.58 | 0.65 | 0.72 | 0.79 | 0.86 | 0.93 |
| 54 | 57.56 | 0.51 | 0.58 | 0.65 | 0.72 | 0.79 | 0.57 | 0.65 | 0.73 | 0.81 | 0.89 | 0.97 | 1.05 |
| 60 | 61.61 | 0.54 | 0.61 | 0.68 | 0.76 | 0.83 | … | … | … | … | … | … | … |
| 64 | 65.67 | 0.56 | 0.64 | 0.72 | 0.80 | 0.87 | … | … | … | … | … | … | … |

*Casting Tolerances*: Sizes 3 to 8, 0.05 inch; sizes 10 to 12, 0.06 inch; sizes 14 to 42, 0.07 inch; size 48, 0.08 inch; sizes 54 to 64, 0.09 inch.

## Table 9. Temperature Coefficient, *Y*

| Material | Temperature | | | | | | | |
|---|---|---|---|---|---|---|---|---|
| | ≤900° F ≤482° C | 950° F 510° C | 1000° F 538° C | 1050° F 566° C | 1100° F 593° C | 1150° F 621° C | 1200° F 649° C | 1250+°F 677+°C |
| Cast Iron | 0 | 0 | 0 | 0 | 0 | 0 | 0 | 0 |
| Nonferrous | 0 | 0 | 0 | 0 | 0 | 0 | 0 | 0 |
| Ferritic Steel | 0.4 [a] | 0.5 | 0.7 | 0.7 | 0.7 | 0.7 | 0.7 | 0.7 |
| Austenitic Steel | 0.4 [a] | 0.4 | 0.4 | 0.4 | 0.5 | 0.7 | 0.7 | 0.7 |
| Nickel Alloys: N06617, N08800, N08810, N08825 | 0.4 | 0.4 | 0.4 | 0.4 | 0.4 | 0.4 | 0.5 | 0.7 |

[a] If $D_o \div t < 6$, calculate $Y = \dfrac{D_i}{D_i + D_o}$.

Face (RF), and Flat Face (FF). RTJ facings use ring type metal gaskets and are for the most severe duties, such as with high pressure gas piping. Raised face flanges have a nonmetallic gasket that fits within the bolts of the flange.

The face has a series of grooves into which the gasket flows upon compression. A flat face flange also uses a series of grooves, but the nonmetallic gasket fits over the entire face of the flange. Flat face flanges are used for the lowest pressure duties. The way in which the flange is attached to the pipe determines the flange type. Weld-Neck (WN) flanges are butt-welded to the pipe, and are used on high pressure and hazardous applications. Socket Weld (SW) flanges are fillet welded onto the pipe and are also suitable for high pressures, but are limited to pipe sizes of 1½ inch or less. Slip-On Weld (SO) flanges are used for low pressure, low hazard applications. The pipe is welded both to the hub and the bore of the flange. This weld configuration makes it impractical to radiographically inspect the welds. Composite lap joint flanges are used to save material expense on alloy pipe systems, where a backing flange of an inexpensive material is added to a hub welded to the pipe. Swivel ring flanges, like composite lap joint flanges, use a hub welded to the pipe. A swivel ring fits over the hub for the purpose of bolting the joint together. The swivel ring flange facilitates flange alignment. There are a number of compact flange designs that include tapered sections or proprietary designs. Always comply with the manufacturer's specifications when using this type of flange connection.

Flanges are specified using the nominal pipe size, type and facing, pressure class, material, schedule, and applicable standard. Since surface condition is critical to flange sealing, all surfaces must be carefully inspected before use, and all scratches, rust, and burrs removed. Gasket selection must be done with consideration of temperature, internal pressure, cycling conditions, vibration, material compatibility, erosive/corrosive potential of the media, and potential to leach substances into the fluid. Gaskets may be ring-type, spiral wound, or sheet. They must be inspected carefully before use for damage, and should never be trimmed or re-used. Bolts must be of the proper material and grade for the joint, and tightened gradually according to standard procedure. Lubricant of the proper type should be used if recommended, and a torque wrench should be employed upon final tightening to ensure proper compression.

Welded and soldered pipe connections should be made according to applicable standards and safety precautions. Steel pipe is often welded, while copper pipe is typically joined by soldering. In soldered pipe joints, capillary fittings are used.

*Expansion with Temperature:* Metal pipe expands and contracts with temperature by an amount dependent on both temperature change and material. This must be considered when designing pipe systems where temperature will change between installation and operation, or when the fluids pumped will be above or below installation temperature. Use the factors in the following table to calculate length increase or decrease for each pipe run.

**Linear Expansion and Contraction Factors per 100 Feet of Pipe**

| Temperature Change, °F | Pipe Material | | | | |
|---|---|---|---|---|---|
| | Steel | Copper | PVC | FRP | PP and PVDF |
| 0 | 0 | 0 | 0 | 0 | 0 |
| 20 | 0.15 | 0.25 | 0.62 | 0.26 | 2.00 |
| 40 | 0.30 | 0.45 | 1.30 | 0.52 | 4.00 |
| 60 | 0.46 | 0.65 | 2.20 | 0.78 | 6.00 |
| 80 | 0.61 | 0.87 | 2.80 | 1.05 | 8.00 |
| 100 | 0.77 | 1.10 | 3.50 | 1.31 | 10.00 |
| 120 | 0.92 | 1.35 | 4.25 | 1.57 | 12.00 |
| 140 | 1.08 | 1.57 | 4.80 | 1.83 | 14.00 |
| 160 | 1.24 | 1.77 | 5.50 | 2.09 | 16.00 |
| 180 | 1.40 | 2.00 | 6.30 | 2.35 | 18.00 |
| 200 | 1.57 | 2.25 | 7.12 | 2.62 | 20.00 |

Multiply length of pipe by the table factor and divide by 100 for the increase or decrease in length.

**Plastic Pipe.—**Plastic pipe is lighter and less expensive than metal pipe, and can be more chemically inert. It is generally non-conductive and thermally insulative. Due to its smooth inner surfaces, it can offer better flow characteristics than metal pipe. However, plastic pipe can be affected by ultraviolet radiation and extreme temperatures. The most common plastic pipe is suitable for nonflammable liquids up to 150 psi (1000 kPa) and temperatures up to 140° F (60° C). Plastic pipe expands with temperature much more than metal pipe.

*Materials:* Many types of plastics are available as plastic pipe. Polyvinyl chloride (PVC), acrylonitrile butadiene styrene (ABS), and polyethylene (PE) are the most common in the United States. Crosslinked polyethylene (PEX) is common in Europe and gaining popularity in the United States.

PVC and chlorinated PVC (CPVC) are related, but CPVC can withstand a wider range of temperatures. PVC is limited to 140° F while CPVC can be used up to 190° F. PVC has inherently superior fire retarding properties due to its chlorine content. The ignition temperature of PVC is as high as 455° C. PVC and CPVC are highly resistant to oxidation and are very durable. PVC is resistant to acids, bases, and most inorganic chemicals. PVC swells or dissolves in aromatic hydrocarbons, ketones, and cyclic ethers. These plastics crack when frozen, and CPVC requires more mechanical support than PVC.

ABS pipes are used mainly in applications where rigidity and impact strength are needed. ABS can be tailored to a wide range of needs by altering its chemical composition. It is suitable for water transport and has good resistance to a wide range of diluted inorganic acids, organic acids, salts, animal fats and oils. ABS is not resistant to organic solvents, alcohol, gasoline, acetic acid or vegetable oils. ABS is resistant to temperatures up to 70° C, but its pressure capacity must be derated with elevated temperature. ABS is non-toxic and is used for food and beverage applications.

PE pipes can be joined through heat fusion, and are common in municipal water applications. They are extremely slippery when wet. PE Water Service lines are typically ½ to 3 nominal diameter pipe or tubing in accordance with AWWA C901, ASTM D2239, ASTM D3035, or ASTM D2737. PE water distribution lines are typically 4 to 12 nominal diameter pipe in accordance with AWWA C906, ASTM D3035, or ASTM F714. Install PE pressure pipe in accordance with ASTM D2774. PE pipe may be cold bent. The allowable bend ratio is determined by the pipe diameter (*D*) and the Standard Dimension Ratio (*SDR*) in accordance with Table 10.

**Table 10. Cold Bend Radius, Polyethylene Pipe**

| Standard Dimension Ratio, *SDR* | Minimum Bend Radius to Centerline of Pipe |
|---|---|
| 7 through 9 | 20 × Pipe Outside Diameter |
| 11 through 13.5 | 25 × Pipe Outside Diameter |
| 17 through 21 | 27 × Pipe Outside Diameter |
| Fitting in Bend | 100 × Pipe Outside Diameter |

PEX is gaining popularity for residential water service because it is flexible, does not corrode, is easy to work and connect, and is resistant to scale and chlorine. It is made from crosslinked HDPE and subject to ASTM F876. It can be bent according to table, as well as used with compression tube fittings. Standard PEX does admit some oxygen, so an added oxygen barrier may be needed in some applications. PEX can be used at temperatures from below freezing to almost boiling.

**Table 11. Bend Radius, PEX Pipe or Tube**

| PEX Size, inch | Outside Diameter, inch | Minimum Bend Radius, inch |
|---|---|---|
| ⅜ | 0.500 | 4 |
| ½ | 0.625 | 5 |
| ⅝ | 0.750 | 6 |
| ¾ | 0.875 | 7 |
| 1 | 1.125 | 9 |

*Schedules:* Schedule number designations for plastic pipe are an indicator of wall thickness. Schedules 40 and 80 are the most commonly used, and 120 is also widely available. For a given nominal pipe size, all schedules have the same outside diameter. Therefore, higher schedule pipes of the same diameter have smaller flow areas. The nominal pipe size

(NPS) is given, along with the schedule or IPS designation to indicate the full dimensional characteristics of the pipe.

*Sizes:* Plastic pipe size designations are often tied to the outside diameter, as they are for steel pipe according to the NPS system. PVC has its own ASTM pipe sizing standard, while CPVC is available in either ASTM standard sizing or CPVC-CTS (Copper Tube Standard). Dimensions for schedule 40, 80, and 120 PVC pipe are given in Table 12. Alternative sizing methods are the Standard Inside Dimension Ratio (SIDR) method and the Standard Dimension Ratio (SDR) method. The SDR method uses OD and wall thickness, and SIDR uses ID and wall thickness. The SIDR method is useful when designing flow-critical applications. The SDR method is useful when pressure rating is of primary concern. PEX is most commonly sized as tube, which is covered in the section *Plastic Tube* starting on page 2782.

**Table 12. PVC Pipe Dimensions, Schedule 40, 80, and 120, Sizes Up To 12″**

| PVC Pipe Size, inch | Outside Diameter, inch | Inside Diameter, inch | | |
|---|---|---|---|---|
| | | SCH 40 | SCH 80 | SCH 120 |
| ⅛ | 0.405 | 0.249 | 0.195 | … |
| ¼ | 0.54 | 0.344 | 0.282 | … |
| ⅜ | 0.675 | 0.473 | 0.403 | … |
| ½ | 0.84 | 0.602 | 0.526 | 0.48 |
| ¾ | 1.05 | 0.804 | 0.722 | 0.69 |
| 1 | 1.315 | 1.029 | 0.936 | 0.891 |
| 1¼ | 1.66 | 1.36 | 1.255 | 1.204 |
| 1½ | 1.9 | 1.59 | 1.476 | 1.423 |
| 2 | 2.375 | 2.047 | 1.913 | 1.845 |
| 2½ | 2.875 | 2.445 | 2.29 | 2.239 |
| 3 | 3.5 | 3.042 | 2.864 | 2.758 |
| 3½ | 4 | 3.521 | 3.326 | … |
| 4 | 4.5 | 3.998 | 3.786 | 3.574 |
| 5 | 5.563 | 5.016 | 4.768 | … |
| 6 | 6.625 | 6.031 | 5.709 | 5.434 |
| 8 | 8.625 | 7.942 | 7.565 | 7.189 |
| 10 | 10.75 | 9.976 | 9.493 | … |
| 12 | 12.75 | 11.889 | 11.294 | … |

*Pressure Ratings:* Pressure rating of plastic pipe is highly dependent on material, as well as wall thickness. Most commercially available plastic pipe will be rated by the manufacturer. PEX pressure ratings are 160 psi at 74° F and 100 psi at 180° F.

Pressure ratings for polyethylene pipes depend on the material formulation. Pressure ratings of standard sizes of PVC pipe are given in Table 13. Pressure ratings must be derated with elevated temperature. Derating factors for PVC pipe are given in Table 14. Further derating must be done to account for water hammer, and this is often 20 percent.

*Connections:* Some plastic pipe is threaded for use with threaded fittings. Threading normally follows the NPT standard. If the pipe is not threaded, a variety of other connection methods may be used. PVC pipe is usually bonded into socket fittings using solvent cement. PE pipe can be joined using heat fusion and electrofusion. In some cases, tube fittings may be used to connect plastic pipe.

*Expansion with Temperature:* Plastic pipe expands and contracts with temperature by an amount dependent on both temperature change and material. This must be considered

**Table 13. Pressure Ratings of Schedule 40, 80, 120 PVC Pipe Up To 12″ at 73.4° F**

| Pipe Size | Maximum Water Pressure (psig) | | |
|---|---|---|---|
| | SCH 40 | SCH 80 | SCH 120 |
| ⅛ | 810 | 1230 | … |
| ¼ | 780 | 1130 | … |
| ⅜ | 620 | 920 | … |
| ½ | 600 | 850 | 1010 |
| ¾ | 480 | 690 | 770 |
| 1 | 450 | 630 | 720 |
| 1¼ | 370 | 520 | 600 |
| 1½ | 330 | 470 | 540 |
| 2 | 280 | 400 | 470 |
| 2½ | 300 | 420 | 470 |
| 3 | 260 | 370 | 440 |
| 3½ | 240 | 350 | … |
| 4 | 220 | 320 | 430 |
| 5 | 190 | 290 | … |
| 6 | 180 | 280 | 370 |
| 8 | 160 | 250 | 380 |
| 10 | 140 | 230 | … |
| 12 | 130 | 230 | … |

**Table 14. Pressure Derating Factors for PVC Pipe**

| Temperature | | | | | | | | | |
|---|---|---|---|---|---|---|---|---|---|
| | °F | 73.4 | 80 | 90 | 100 | 110 | 120 | 130 | 140 |
| | °C | 23 | 26.7 | 32.2 | 37.8 | 43.3 | 48.9 | 54.4 | 60 |
| Derating Factor | | 1 | 0.88 | 0.75 | 0.62 | 0.5 | 0.4 | 0.3 | 0.22 |

when designing pipe systems where temperature will change between installation and operation, or when the fluids pumped will be above or below installation temperature. Plastics expand much more than metals, so this may become a factor in selecting one over the other. Use the factors in Table 15 to calculate length increase or decrease for each pipe run. To use the tabulated factors, use the equation $\Delta L = \alpha L_0 \Delta T$ where $L_0$ is the length at the installation temperature, and $\Delta T$ is the difference between the installation temperature and the temperature in question.

**Table 15. Coefficient of Linear Expansion (α) for Common Pipe Materials**

| Material | Coefficient of Linear Expansion | |
|---|---|---|
| | $10^{-6}$ in/in °F | $10^{-6}$ m/m °C |
| Carbon Steel | 6.5 | 11.7 |
| Copper | 9.3 | 16.8 |
| Stainless Steel | 9.9 | 17.8 |
| ABS | 35 | 63 |
| HDPE | 67 | 120 |
| PE | 83 | 150 |
| CPVC | 44 | 79 |

**Pipe Fittings.**—Pipe fittings are used to connect pipes together either permanently or non-permanently. Pipe systems should be designed to minimize the number of fittings because they cause pressure loss, require pressure derating, and are a potential source of leaks.

*Materials:* Pipe fittings are available in a variety of materials, and should be selected for compatibility with the pipe material and the substance being transported. Dissimilar metals should be avoided to prevent galvanic corrosion. Plastic fittings are also available for use with plastic pipe.

*Types:* Pipe fittings may be threaded, slip, or interference type connections. Interference connections, like compression or stab fittings, are used with plastic pipe and with metal and plastic tubing. PEX pipe is typically connected using either a copper crimp ring or expansion fittings with outer expansion rings. In the crimping method, a copper crimp ring is put over the pipe near the end, a copper fitting is inserted into the pipe end, and a crimping tool is used to crimp the ring over the pipe and fitting. In the expansion fitting method, an expansion tool is used to open the diameter of the pipe, into which an expansion fitting is inserted. The pipe shrinks over the fitting as it cools, and then a plastic expansion ring is pressed over the pipe and fitting. Compression fittings can be used as an alternative; that method is discussed in the section *Tube Fittings* on page 2785.

The most common types of pipe fittings are illustrated in Table 16. These examples are by no means exhaustive of the types and shapes available.

**Table 16. Common Pipe Fittings**

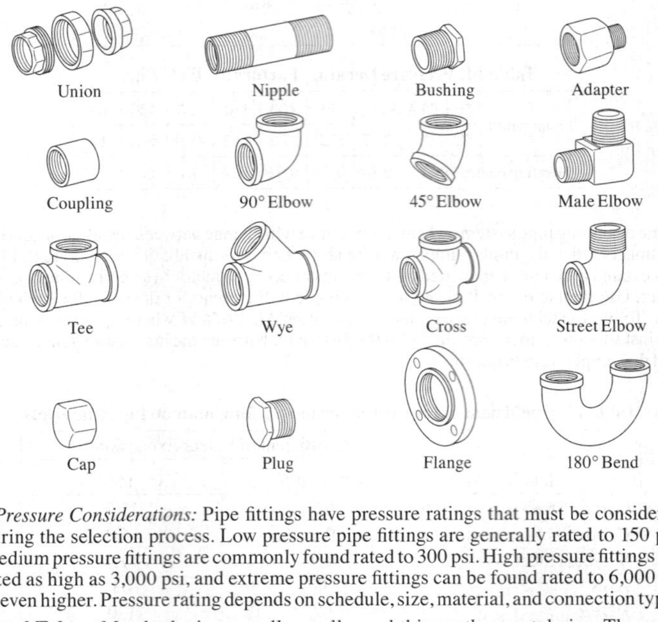


*Pressure Considerations:* Pipe fittings have pressure ratings that must be considered during the selection process. Low pressure pipe fittings are generally rated to 150 psi. Medium pressure fittings are commonly found rated to 300 psi. High pressure fittings are rated as high as 3,000 psi, and extreme pressure fittings can be found rated to 6,000 psi or even higher. Pressure rating depends on schedule, size, material, and connection type.

**Metal Tube.**—Metal tube is generally smaller and thinner than metal pipe. The same considerations in terms of material properties, mechanical stresses, and environmental factors apply to both tube and pipe.

*Material:* Steel tubing is usually made from carbon steel (plain or galvanized) or stainless steels (304 or 316). It is used in a variety of applications to carry liquids or gases. Stainless steel is known for its corrosion resistance when exposed to a variety of materials.

Copper tubing is typically offered as either soft annealed or hard drawn. Annealed tubing is easy to bend, while drawn tubing is stiff. Drawn tubing has a higher pressure rating than annealed. Flare fittings may be installed on annealed temper tube, but not on drawn temper; both types can be soldered or brazed. Copper tube types and applications are given in Table 17.

**Table 17. Copper Tubing Types and Applications**

| Tube Type | Standard | Applications | Commercially Available Sizes |
|---|---|---|---|
| K | ASTM B88 | Domestic Water<br>Fire Protection<br>Solar<br>Fuel/Fuel Oil<br>HVAC<br>Compressed Air<br>Natural Gas, LP Gas<br>Vacuum | • Straight Lengths: ¼ inch to 12 inch<br>• Coils: ¼ inch to 2 inch |
| L | ASTM B88 | Domestic Water<br>Fire Protection<br>Solar<br>Fuel/Fuel Oil<br>Natural Gas, LP Gas<br>HVAC<br>Compressed Air<br>Vacuum | • Straight Lengths: ¼ inch to 12 inch<br>• Coils: ¼ inch to 2 inch |
| M | ASTM B88 | Domestic Water<br>Fire Protection<br>Solar<br>Fuel/Fuel Oil<br>HVAC<br>Vacuum | • Straight Lengths: ¼ inch to 12 inch |
| DWV | ASTM B306 | Drain, Waste, Vent<br>HVAC<br>Solar | • Straight Lengths: 1¼ inch to 8 inch |
| ACR | ASTM B280 | Air Conditioning<br>Refrigeration<br>Natural Gas, LP Gas<br>Compressed Air | • Straight Lengths: ⅜ inch to 4⅛ inch<br>• Coils: ⅛ inch to 1⅝ inch |
| OXY/MED | ASTM B819 | Medical Gas<br>Compressed Medical Air<br>Vacuum | • Straight Lengths: ¼ inch to 8 inch |

*Sizes:* Steel tubing is highly standardized, with designations of outside diameter and wall thickness. Imperial and metric sizes are available. A selection of Imperial sizes and pressure ratings of stainless steel tubing is given in Table 18.

The system of Copper Tube Sizes (CTS) depends on type. Types K, L, M, DVW (Drain Waste Vent), and OXY/MED all have an outside diameter ⅛ inch larger than the nominal size designation. Type ACR (Air Conditioning & Refrigeration) tubing outside diameter is equal to the nominal size designation. Dimensions of copper tube are found in Table 19.

*Pressure Ratings:* Pressure ratings for stainless steel tubes were given in Table 18. Allowable maximum pressure for metal tubes can be calculated as it was for metal pipes.

**Table 18. Stainless Steel Tube Specifications, Selected Sizes**

| Tube OD (in.) | Wall Thickness (in.) | Working Pressure (psig) |
|---|---|---|
| 1⁄16 | 0.014 | 8100 |
| | 0.020 | 12000 |
| 1⁄8 | 0.028 | 8500 |
| | 0.035 | 10900 |
| 1⁄4 | 0.035 | 5100 |
| | 0.049 | 7500 |
| | 0.065 | 10200 |
| 3⁄8 | 0.035 | 3300 |
| | 0.049 | 4800 |
| | 0.065 | 6500 |
| 1⁄2 | 0.035 | 2600 |
| | 0.049 | 3700 |
| | 0.065 | 5100 |
| | 0.083 | 6700 |
| 5⁄8 | 0.049 | 2900 |
| | 0.065 | 4000 |
| 3⁄4 | 0.049 | 2400 |
| | 0.065 | 3300 |
| 1 | 0.083 | 3100 |
| 1 1⁄4 | 0.095 | 2800 |
| | 0.12 | 3600 |
| 1 1⁄2 | 0.12 | 3000 |
| | 0.134 | 3400 |
| 2 | 0.134 | 2500 |
| | 0.188 | 3600 |

Alternatively, the maximum working pressure of metal tubing can be calculated using the Barlow formula for thin-walled cylinders as given in the following equation. This formula is typically used for steel tubing connected with flare fittings.

$$p_{max} = \frac{2St}{D}$$

where $S$ is the ultimate tensile strength, $t$ is wall thickness, and $D$ is outer diameter. A factor of safety between 1.5 and 10 should be applied.

If the tube is carrying liquid, the pressure rating must be derated to account for the pressure spikes that occur during valve closures. The amount of reduction of working pressure is called the "water hammer factor" and is typically 20 percent. The types of fittings being used must be carefully considered when working with elevated pressures, and further derating of the system may be required. For pressure ratings of copper tube, consult the "Copper Tube Handbook," available from the Copper Development Association, Inc.

**Plastic Tube.—**Plastic tube is often selected for its flexibility, clarity, or non-conductivity. When selecting from the wide variety of available tubing materials, one must consider the operating temperature range in the application, the media being transported, the environment in which the tubing will be located, the pressures involved, whether there will be vacuum, whether there are any standards or certifications needed, whether standard tubing tolerances will be acceptable, and whether the tubing must be transparent. Temperature must be considered not only for extremes, but also for slight variations from 73° F (22° C) because most plastic tubing expands and softens even with slight changes in temperature.

**Table 19. Copper Tube and Pipe Dimensions**

| Nominal Size | Actual Outer Diameter, inch | Type K | Type L | Type M | Type DWV | ACR Actual Outer Diameter, inch | ACR Actual I.D., inch (Temper) |
|---|---|---|---|---|---|---|---|
| | | Actual Inner Diameter, inch | | | | | |
| ⅛ | … | … | … | … | … | 0.125 | 0.065 (A) |
| 3⁄16 | … | … | … | … | … | 0.187 | 0.128 (A) |
| ¼ | 0.375 | 0.305 | 0.315 | … | … | 0.250 | 0.190 (A) |
| 5⁄16 | … | … | … | … | … | 0.312 | 0.248 (A) |
| ⅜ | 0.500 | 0.402 | 0.430 | 0.450 | … | 0.375 | 0.311 (A), 0.315 (D) |
| ½ | 0.625 | 0.527 | 0.545 | 0.569 | … | 0.500 | 0.436 (A), 0.430 (D) |
| ⅝ | 0.750 | 0.652 | 0.666 | | … | 0.625 | 0.555 (A), 0.545 (D) |
| ¾ | 0.875 | 0.745 | 0.785 | 0.811 | … | 0.750 | 0.680 (A), 0.666 (A, D |
| ⅞ | … | … | … | … | … | 0.875 | 0.785 (A, D) |
| 1 | 1.125 | 0.995 | 1.025 | 1.055 | … | … | … |
| 1⅛ | … | … | … | … | … | 1.125 | 1.025 (A, D) |
| 1¼ | 1.375 | 1.245 | 1.265 | 1.291 | 1.295 | … | … |
| 1⅜ | … | … | … | … | … | 1.375 | 1.265 (A, D) |
| 1½ | 1.625 | 1.481 | 1.505 | 1.527 | 1.541 | … | … |
| 1⅝ | … | … | … | … | … | 1.625 | 1.505 (A,D) |
| 2 | 2.125 | 1.959 | 1.985 | 2.009 | 2.041 | … | … |
| 2⅛ | … | … | … | … | … | 2.125 | 1.985 (D) |
| 2½ | 2.625 | 2.435 | 2.465 | 2.495 | … | … | … |
| 2⅝ | … | … | … | … | … | 2.625 | 2.465 (D) |
| 3 | 3.125 | 2.907 | 2.945 | 2.981 | 3.030 | … | … |
| 3⅛ | … | … | … | … | … | 3.125 | 2.945 (D) |
| 3½ | 3.625 | 3.385 | 3.425 | 3.459 | … | … | … |
| 3⅝ | … | … | … | … | … | 3.625 | 3.425 (D) |
| 4 | 4.125 | 3.857 | 3.905 | 3.935 | 4.009 | … | … |
| 4⅛ | … | … | … | … | … | 4.125 | 3.905 (D) |
| 5 | 5.125 | 4.805 | 4.875 | 4.907 | 4.981 | … | … |
| 6 | 6.125 | 5.741 | 5.845 | 5.881 | 5.959 | … | … |
| 8 | 8.125 | 7.583 | 7.725 | 7.785 | 7.907 | … | … |
| 10 | 10.125 | 9.449 | 9.625 | 9.701 | … | … | … |
| 12 | 12.125 | 11.315 | 11.565 | 11.617 | … | … | … |

*Materials:* Some of the more popular tubing materials are polyvinyl chloride (PVC), low density polyethylene (LDPE), high density polyethylene (HDPE), polypropylene (PP), nylon (PA 6, 6/6, 11, 12), silicone, polyurethane, fluoropolymers (PTFE, FEP, PFA, PVDF, ETFE), and polyetheretherketone (PEEK). PEX tube is also common in Europe for water applications.

PVC can be rigid or flexible, and can be transparent. Exposure to water can cause clouding of clear PVC over time. Flexibility will depend on the amount of plasticizer present, so a variety of durometers are available. Flexible PVC is rated for fairly low pressures, unless reinforced with braiding.

LDPE is low cost, light weight, relatively transparent, and flexible. It is suitable for air, water, and some chemicals. Caution should be used when barbed fittings are employed because LDPE has poor environmental crack resistance, which can be accentuated when solvents are present.

HDPE is similar to LDPE in terms of material compatibility. It is fairly opaque, and is more dense and rigid than LDPE.

Polypropylene has high impact resistance combined with good stiffness and excellent chemical resistance. PP is often mixed with polyethylene or rubber to improve its impact and low temperature properties. PP is more heat resistant than PE, with an operating range up to 60° C.

Nylons can be used at elevated temperatures. They cost more than PE and PP, but exhibit superior abrasion resistance and stiffness. Types 11 and 12 are more flexible than type 6, and have better chemical resistance to acids and bases.

Silicone tubing is suitable for elevated temperatures and can be sterilized using heat. It is extremely soft and flexible, and is non-toxic. There are different ways to cure silicone tubing, so material properties may vary.

Polyurethane tubing is flexible and can be relatively transparent. This type of tubing has good abrasion resistance and tensile strength. It is flexible at low temperatures and holds its properties over time. There are two types of polyurethane: esters and ethers. Esters have good chemical and oil resistance. Ethers have good water compatibility and are suitable for outdoor applications. Both can be made FDA compliant.

Fluoropolymers are chemically inert and often chosen for harsh applications or high purity applications. They are also suitable for high temperatures in the 300–500° F (148–260° C) range. Rigidity varies with composition, but they are generally stiff but bendable. FEP is flexible and transparent.

PEEK is fairly rigid but bendable, has excellent mechanical stability and chemical compatibility, and can be used to 100° C. It is often used where stainless steel tubing has material compatibility issues.

PEX tubing is of the same material and properties as PEX piping, discussed in the section *Plastic Pipe* starting on page 2776. It comes in a variety of colors, or a natural hazy clear state.

There are many more materials than those listed. In addition, there are many brand-name and proprietary formulations.

*Sizes:* Most plastic tubing is sized according to OD, ID, and wall thickness. A wide variety of sizes are available for each material. PEX tubing is sized according to the Copper Tube Size standard for outer diameter. Standard PEX dimensions are given in Table 20.

**Table 20. Standard PEX Dimensions**

| Nominal PEX Size inch | Outside Diameter inch | Inside Diameter inch | Wall Thickness, inch |
|---|---|---|---|
| 3⁄8 | 0.500 | 0.350 | 0.070 |
| 1⁄2 | 0.625 | 0.475 | 0.070 |
| 5⁄8 | 0.750 | 0.574 | 0.083 |
| 3⁄4 | 0.875 | 0.677 | 0.097 |
| 1 | 1.125 | 0.863 | 0.125 |

All dimensions are in inches.

*Pressure Ratings:* Care must be taken when using plastic tubing because pressure ratings are generally much lower than with metal tubing. In addition, if vacuum is to be used in the system, the tubing must be vacuum rated to ensure that it does not collapse.

Pressure rating will depend on material, diameter, and wall thickness. Consult the manufacturer's pressure rating when specifying plastic tubing.

**Tube Fittings.—**Tube fittings are used to connect tubing to tubing, or tubing to pipes or other objects. Tube fittings are generally non-permanent connections, although they may permanently deform the end of the tube.

*Materials:* Fitting materials should be chosen for compatibility with the tubing and the fluid being transported. Combining dissimilar metals should be avoided to prevent galvanic corrosion. Metal fittings and plastic fittings are widely available. Some plastic fittings have metal ferrules, and some metal fittings can have elastomeric seals. It is critical to check material compatibility with all components of a fitting. Plastic fittings are used with plastic tubing that is softer than the fitting material. Steel fittings are used with steel or plastic tubing. Brass fittings are used with copper, aluminum, and plastic tubing.

*Types:* There are many types of tube fittings, including proprietary and name-brand fittings. The most common general types are barbed, compression, flare, and push-in.

*Barbed fittings* are used with soft plastic tubing which can expand enough to fit over the barb end but have elastic memory enough to seal over the barb or barbs. A clamp should be employed to hold the tube on if pressures are moderate. Pressure in the system should not exceed 150 psi when using barbed connections. To specify a barbed fitting connection, one must know the tube inner diameter, process pressure, material compatibility, and whether the tubing is suitable for use with barbed fittings.

*Compression fittings* are used on metal tubing and stiff plastic tubing. Metal and plastic compression fittings are available. They are suitable for high pressures when constructed out of high strength material and combined with a high pressure rated tube. There are numerous designs available, but all include at least one ferrule, a nut, and a fitting body. When the nut is tightened, the ferrule swages the tube to form a pressure tight seal. The tube does not need to rotate during tightening. Compression fittings can be re-used if care is taken not to damage the ferrule or tube end. To specify a compression fitting, one must know the tube outer diameter, wall thickness, process pressure, and material compatibility requirements. To ensure proper sealing, the tube should be clean and cut square without burrs. The tube material must always be softer than the fitting material. Plastic compression fittings are used with softer plastic tubing, and stainless compression fittings are generally used with stainless tubing. Brass fittings can be used with firmer plastic, aluminum, and copper tubing. Carbon steel compression fittings are used with carbon steel tubing.

*Flare fittings* are typically used with metal tubing which is forced over a flared seat and secured with a nut to create a leak-tight seal. They are similar to compression fittings but have a sleeve instead of a ferrule. A flaring tool is used on the end of the tubing to create the proper flare for the fitting being used. These fittings are suitable for high pressures, even greater than with compression fittings. The most common flare fitting standards in use today are the 45° SAE fitting, and the 37° JIC fitting. The 37° style generally has a higher pressure rating for a given size tubing. The 45° style is generally used for low pressure refrigeration applications, while the 37° style is used for high pressure hydraulic applications. SAE and JIC fittings are incompatible due to the different flare angles. Flare fittings can be reused if care is taken not to damage the flared end of the tubing. Stainless steel fittings are used with flareable seamless stainless tubing. Brass fittings are used with flareable seamless copper. Carbon steel fittings are used with flareable welded and seamless carbon steel tubing. Aluminum fittings are used with seamless aluminum tubing.

*Push-in fittings* are popular for pneumatic connections, but are also made for liquid applications. These fittings have an internal gripping ring and usually an o-ring. The tube must be cut square and be burr-free. A leak-free seal is created when the end of the tube is pushed into the fitting. To release the tubing, the end of the fitting must be compressed axially to release the gripping mechanism. These fittings normally have elastomeric, plastic, and stainless steel components, regardless of the body material, so material compatibility can become an issue. These fittings can be selected to accept plastic or metal tubing, and pressure ratings vary. Pressure capability of push-in fittings is generally less than compression fittings but greater than or equivalent to barbed fittings.

Many other fittings and couplings are available. Couplings attach to the ends of two tubes and have a leak-free means of connecting the halves of the coupling together. Couplings are available with and without shut-off valves to prevent spillage when disconnected.

## Flow and Energy Loss in Pipe, Tube, and Fittings

Fluids experience energy loss due to friction when flowing through pipes and fittings. The energy lost is dependent on the length and diameter of the pipe/tube, the surface conditions, and the number and shape of any bends or restrictions.

**Flow Rates.—**The principal of continuity for constant flow dictates that the volumetric flow rate (or mass flow rate) remains the same along a pipe, even if the pipe expands, contracts, or is restricted. This means that velocity changes with flow area. One can calculate the velocity before or after a change in flow area if the velocity on the other side of the change is known.

$$Q_i = A_i v_i = Q_o = A_o v_o$$

where $A$ is the cross sectional flow area, $Q$ is volumetric flow rate, and $v$ is velocity.

The behavior of a fluid, and its energy losses through pipe and fittings, is dependent upon whether it follows a laminar, transitional, or turbulent flow regime. To determine what flow regime is present, the dimensionless Reynolds number is calculated with the following equations. If the Reynolds number is less than 2000, the flow is laminar. If the Reynolds number is greater than 4000, flow is turbulent. Between those values lies the transitional flow regime. Note that the Reynolds number is dependent on fluid viscosity.

$$R_e = \frac{\rho v D}{\mu}$$

where $D$ is the flow diameter, $\rho$ is the density, $v$ is the mean velocity, $\mu$ is the absolute viscosity of the fluid.

$$R_e = \frac{vD}{\upsilon} = \frac{QD}{A\upsilon}$$

where $Q$ is the volumetric flow rate, $A$ is the cross sectional area of flow, and $\upsilon$ is the kinematic viscosity.

For general liquid piping applications, the US Army Corps of Engineers manual "Liquid Process Piping" recommends a flow velocity of 7 ± 3 ft/s (2.1 ±0.9 m/s) with a maximum velocity of 7 ft/s at discharge points including pump suction lines and drains.

All formulas for calculating the mean velocity of flow through a pipe or tube are approximate. The following is a formula for low viscosity liquids that will yield results within 5 to 10 percent of actual if applied to carefully laid pipe in fair condition:

$$v = C\sqrt{\frac{hd}{L + 54d}}$$

where $C$ is taken from Table 21, $h$ is total head in feet, $d$ is the inside diameter of a round pipe in feet, and $L$ is the total length of the pipe in feet. Length, where pressure drop due to bends and fittings is a factor, is equivalent length. See *Energy Loss in Fittings and Valves* on page 2792, and *Energy Loss in Fittings and Valves* on page 2792 for more on this topic.

For air, the following velocity equation can be used to estimate flow:

$$v = \sqrt{\frac{25,000\ dp}{L}} \qquad p = \frac{Lv^2}{25,000\ d}$$

where $v$ = velocity of air in feet per second

$p$ = loss of pressure due to flow through the pipes in ounces per square inch

$d$ = inside diameter of pipe in inches

$L$ = length of pipe in feet

**Table 21. Values of Coefficient $C$**

| Dia. of Pipe | | | Dia. of Pipe | | | Dia. of Pipe | | |
|---|---|---|---|---|---|---|---|---|
| Feet | Inches | $C$ | Feet | Inches | $C$ | Feet | Inches | $C$ |
| 0.1 | 1.2 | 23 | 0.8 | 9.6 | 46 | 3.5 | 42 | 64 |
| 0.2 | 2.4 | 30 | 0.9 | 10.8 | 47 | 4.0 | 48 | 66 |
| 0.3 | 3.6 | 34 | 1.0 | 12.0 | 48 | 5.0 | 60 | 68 |
| 0.4 | 4.8 | 37 | 1.5 | 18.0 | 53 | 6.0 | 72 | 70 |
| 0.5 | 6.0 | 39 | 2.0 | 24.0 | 57 | 7.0 | 84 | 72 |
| 0.6 | 7.2 | 42 | 2.5 | 30.0 | 60 | 8.0 | 96 | 74 |
| 0.7 | 8.4 | 44 | 3.0 | 36.0 | 62 | 10.0 | 120 | 77 |

The quantity of air discharged in cubic feet per second is the product of the velocity as obtained from the preceding formula and the area of the pipe in square feet. The horsepower required to drive air through a pipe equals the volume of air in cubic feet per second multiplied by the pressure in pounds per square foot, and this product divided by 550.

**Volume of Air Transmitted Through Pipes, in CFM**

| Velocity of Air in Feet per Second | Actual Inside Diameter of Pipe, Inches | | | | | | | | | |
|---|---|---|---|---|---|---|---|---|---|---|
| | 1 | 2 | 3 | 4 | 6 | 8 | 10 | 12 | 16 | 24 |
| 1 | 0.33 | 1.31 | 2.95 | 5.2 | 11.8 | 20.9 | 32.7 | 47.1 | 83.8 | 188 |
| 2 | 0.65 | 2.62 | 5.89 | 10.5 | 23.6 | 41.9 | 65.4 | 94.2 | 167.5 | 377 |
| 3 | 0.98 | 3.93 | 8.84 | 15.7 | 35.3 | 62.8 | 98.2 | 141.4 | 251.3 | 565 |
| 4 | 1.31 | 5.24 | 11.78 | 20.9 | 47.1 | 83.8 | 131.0 | 188.0 | 335.0 | 754 |
| 5 | 1.64 | 6.55 | 14.7 | 26.2 | 59.0 | 104.0 | 163.0 | 235.0 | 419.0 | 942 |
| 6 | 1.96 | 7.85 | 17.7 | 31.4 | 70.7 | 125.0 | 196.0 | 283.0 | 502.0 | 1131 |
| 7 | 2.29 | 9.16 | 20.6 | 36.6 | 82.4 | 146.0 | 229.0 | 330.0 | 586.0 | 1319 |
| 8 | 2.62 | 10.50 | 23.5 | 41.9 | 94.0 | 167.0 | 262.0 | 377.0 | 670.0 | 1508 |
| 9 | 2.95 | 11.78 | 26.5 | 47.0 | 106.0 | 188.0 | 294.0 | 424.0 | 754.0 | 1696 |
| 10 | 3.27 | 13.1 | 29.4 | 52.0 | 118.0 | 209.0 | 327.0 | 471.0 | 838.0 | 1885 |
| 12 | 3.93 | 15.7 | 35.3 | 63.0 | 141.0 | 251.0 | 393.0 | 565.0 | 1005.0 | 2262 |
| 15 | 4.91 | 19.6 | 44.2 | 78.0 | 177.0 | 314.0 | 491.0 | 707.0 | 1256.0 | 2827 |
| 18 | 5.89 | 23.5 | 53.0 | 94.0 | 212.0 | 377.0 | 589.0 | 848.0 | 1508.0 | 3393 |
| 20 | 6.55 | 26.2 | 59.0 | 105.0 | 235.0 | 419.0 | 654.0 | 942.0 | 1675.0 | 3770 |
| 24 | 7.86 | 31.4 | 71.0 | 125.0 | 283.0 | 502.0 | 785.0 | 1131.0 | 2010.0 | 4524 |
| 25 | 8.18 | 32.7 | 73.0 | 131.0 | 294.0 | 523.0 | 818.0 | 1178.0 | 2094.0 | 4712 |
| 28 | 9.16 | 36.6 | 82.0 | 146.0 | 330.0 | 586.0 | 916.0 | 1319.0 | 2346.0 | 5278 |
| 30 | 9.80 | 39.3 | 88.0 | 157.0 | 353.0 | 628.0 | 982.0 | 1414.0 | 2513.0 | 5655 |

**Flow of Compressed Air in Pipes.—**When there is a comparatively small difference of pressure at the two ends of a pipe, the volume of flow in cubic feet per minute is found by the formula:

$$V = 58\sqrt{\frac{pd^5}{WL}}$$

where $V$ = volume of air in cubic feet per minute

$p$ = difference in pressure at the two ends of the pipe in pounds per square inch

$d$ = inside diameter of pipe in inches

$W$ = weight in pounds of one cubic foot of entering air

$L$ = length of pipe in feet

**Flow in Branched (Parallel) Pipes.**—Pipe systems commonly have parallel sections to allow some fluid to bypass a component. The principle of continuity dictates that in a steady flow situation, total flow into a branched system is equal to total flow exiting the branched system. The following are the governing equations for branching pipe systems.

$$Q_i = A_1 v_1 + A_2 v_2$$

where $Q_i$ is the total flow rate entering the branching system. $A_1$ and $v_1$ are the cross sectional area and flow velocity of the first branch, and subscript 2 refers to the second branch.

Head loss is equal in all branches and equal to the total head loss across the network:

$$h_L = h_{L1} = h_{L2}$$

**Energy Loss in Pipes.**—As fluid moves through straight pipe and tube, some energy is lost to friction. The relationship between pressures and velocities at two ends of a pipe is governed by Bernoulli's equation:

$$\frac{p_1}{\gamma_1} + \frac{v_1^2}{2g} + z_1 = \frac{p_2}{\gamma_2} + \frac{v_2^2}{2g} + z_2$$

where $p$ is pressure, $v$ is velocity, $z$ is elevation, and $g$ is acceleration due to gravity.

For inviscid, incompressible liquid flow moving through a pipe, Bernoulli's equation simplifies because the specific weight of the fluid is assumed to be the same at both points of interest.

*Darcy-Weisbach Equation and the Moody Diagram:* This equation can be used to calculate friction loss in Newtonian liquids for all flow regimes. It requires the use of a friction factor, which is dimensionless and can be determined using the equation for laminar flow, the Colebrook equation for turbulent flow, or graphically through the use of the Moody diagram (Fig. 1). The Darcy-Weisbach equation is:

$$h_L = \frac{fLv^2}{2gD}$$

where $h_L$ is head loss in units of length, $L$ is pipe length, $v$ is average fluid velocity, $D$ is inside diameter of the pipe, and $f$ is the friction factor.

The friction factor for laminar flow is $f = \frac{64}{R_e}$. The Colebrook equation for turbulent flow is $\frac{1}{\sqrt{f}} = -2\log_{10}\left(\frac{\varepsilon}{3.7D} + \frac{2.51}{R_e\sqrt{f}}\right)$ where ε is the pipe roughness, and $D$ is the inside diameter. This applies when the pipe is completely full of fluid. Calculators are available online for this factor, or it can be solved numerically.

Moody's diagram was generated experimentally. A chart of surface roughness of common pipe and tube materials can be found in Table 22 for use with the turbulent flow equation and Moody diagram.

**Table 22. Surface Roughness of Common Pipe and Tube**

| Material | Surface Roughness | | Material | Surface Roughness | |
|---|---|---|---|---|---|
| | ft × 10⁻⁶ | m × 10⁻⁶ | | ft × 10⁻⁶ | m × 10⁻⁶ |
| Glass | 0.0003 | 0.0001 | Ductile Iron (Coated) | 400 | 120 |
| Plastic | 0.0003–0.0023 | 0.0001–0.0007 | Concrete (Smooth) | 400 | 120 |
| Drawn Metal Tubing | 5 | 2 | Galvanized Iron | 500 | 150 |
| Carbon Steel | 150 | 46 | Ductile Iron (Uncoated) | 800 | 240 |
| Wrought Iron | 150 | 46 | Cast Iron | 850 | 260 |
| Rubber Tubing | 230 | 70 | Riveted Steel | 6000 | 1800 |

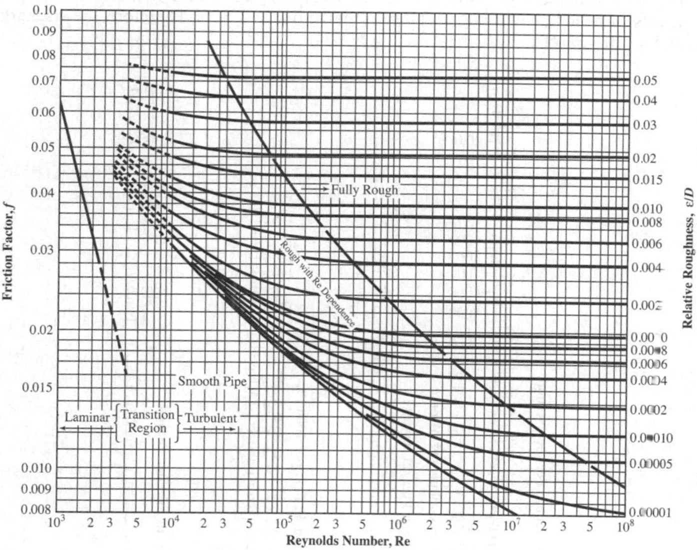


Fig. 1. Moody Diagram

*Hazen-Williams Equation for Water Flow:* For the special case of 60° F water in pipes between 2 inch and 6 feet in diameter, at velocities less than 10 ft/s, the Hazen-Williams formula or nomograph (Fig. 2) can be used. The coefficient values of common pipes can be found in Table 23. The nomograph is read by aligning a straightedge with the known values and reading the unknowns at the intersection points. The Hazen-Williams formula for US units is:

$$v = 1.32 C_h R^{0.63} \left(\frac{h_L}{L}\right)^{0.54}$$

where $v$ is average velocity (ft/s), $R$ is hydraulic radius (ft), $L$ is length of pipe (ft), $h_L$ is head loss (ft), and $C_h$ is the Hazen-Williams coefficient (see Table 23). The Hazen-Williams formula for SI units, with $R$, $h$, and $L$ in meters, and $v$ in m/s is:

$$v = 0.85 C_h R^{0.63} \left(\frac{h_L}{L}\right)^{0.54}$$

A solution of the Hazen-Williams equations may be found by use of the nomograph in Fig. 2, based on the Hazen-Williams formula and using $C_h = 100$. For larger or smaller values of $C_h$, the discharge or velocity obtained from the nomograph is multiplied by the ratio of the given value of $C_h$ to 100. If the discharge or velocity is given, it should be multiplied by the ratio of 100 to the known value of $C_h$ before the nomograph is used.

*Example:* By using Fig. 2, determine the discharge, in cubic feet per second (cfs), from a 12-inch pipe for which $C_h = 120$ when the loss of head is 5 feet per 1000 feet.

*Solution:* The discharge corresponding to the given diameter and loss of head for a value of $C_h = 100$ is found first. A straightedge passing through 12 on the diameter line and 5 on the loss-of-head line will intersect the discharge line at 2.5 cfs. Therefore, the discharge is $2.5 \times 120 = 3.0$ cfs.

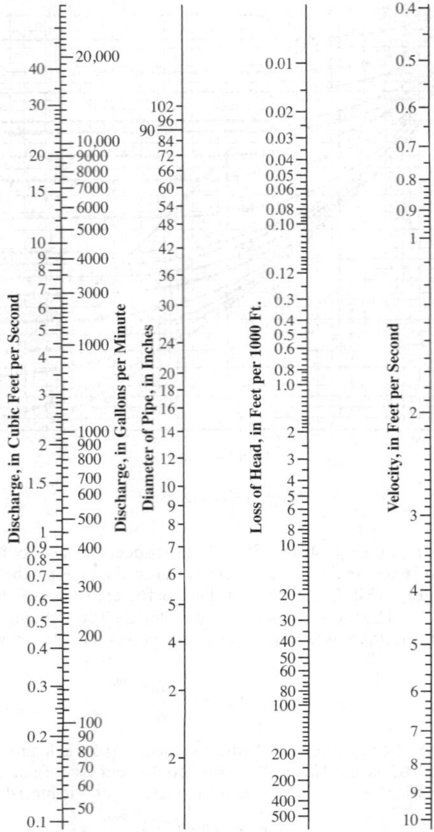


Fig. 2. Nomograph for Hazen-Williams equations in which $C_h = 100$.

**Table 23. Hazen-Williams Coefficient, Selected Values**

| Pipe Type | $C_h$ | Pipe Type | $C_h$ |
|---|---|---|---|
| Corrugated Metal | 60 | Copper | 130–140 |
| ABS Plastic | 130 | Drawn metal | 130–140 |
| Aluminum | 130–150 | Ductile iron pipe | 140 |
| Brass | 130–140 | Ductile iron, cement lined | 120 |
| Cast iron | 100 | Fiberglass | 150 |
| new unlined | 130 | Galvanized iron | 120 |
| 10 years old | 107–113 | Glass | 130 |
| 20 years old | 89–100 | Lead | 130–140 |
| 40 years old | 64–83 | Plastic | 130–150 |
| asphalt coated | 100 | Polyethylene | 140 |
| bituminous lined | 140 | PVC, CPVC | 150 |
| cement lined | 140 | Steel new unlined | 140–150 |
| sea-coated | 120 | | |
| Concrete | 100–140 | | |
| lined, steel forms | 140 | | |
| lined, wooden forms | 120 | | |

*Energy Loss for Gas Flow:* Bernoulli's equation can be applied to examine pressure loss for gas flow through pipes. Gas flowing through pipes can be considered incompressible and treated like a liquid if the pressure ratio is small enough. Pressure ratio is calculated using the following equation. Engineering judgment should be used when applying this equation. In different industries, different values are used to make the decision how to treat gas flow.

$$PR\% = 100\left(\frac{p_1 - p_2}{p_1}\right)$$

where $p$ is pressure at the point of interest and $PR\%$ is the pressure ratio.

Pressure drop for gas flow in pipes can be calculated most easily if certain assumptions are made. One can assume flow is isothermal, the gas is perfect, the pipe is straight and horizontal, the friction factor is constant, and flow is steady. This simplifies the pressure drop equation to the following form:

$$p_i^2 - p_o^2 = Z_m RT\left(\frac{\dot{m}}{A}\right)^2\left(f\frac{L}{D}\right)$$

where $p_i$ and $p_o$ are the inlet and outlet pressures, $Z_m$ is the mean compressibility factor, $R$ is the gas constant, $T$ is temperature, $A$ is pipe cross sectional area, $f$ is the friction factor, $L$ is pipe length, $D$ is inner diameter, and $\dot{m}$ is mass flow rate.

Alternative methods of calculation include the Weymouth and Panhandle formula, and the Renouard equation for natural gas flow.

**Energy Loss in Pipe and Tube Bends.—**The energy lost by fluid in a pipe bend is caused by both friction and momentum changes as the direction of flow changes. This loss depends on the bend angle, the curvature ratio (bend mean radius divided by pipe diameter, $R/D$), the roughness of the pipe, and the angle of bend. To compute the energy lost by liquids in bends, use the following equations and Table 24.

$$h_l = K_b\frac{V^2}{2g} \qquad K_b = f\frac{L_e}{D}$$

where the head loss, $h_l$, is expressed in units of length; $V$ is the average velocity in the pipe; and $K_b$ is the bend loss coefficient; $f$ is the friction factor given in the Moody diagram, (Fig. 1); and $L_e/D$ is given in Table 24. $K_b$ increases with relative roughness (or screwed connections) and bend angle, but decreases with the curvature ratio.

**Table 24. Bend Equivalent Length Ratio, $L_e/D$**

| Bend Type | Bend Angle | $R/D$ | $L_e/D$ |
|---|---|---|---|
| Radius Bend | 90° | 1 | 20 |
| | | 2 | 12 |
| | | 3 | 12 |
| | | 4 | 14 |
| | | 6 | 17 |
| | | 8 | 24 |
| | | 10 | 30 |
| | | 12 | 34 |
| | | 16 | 42 |
| Miter Bend | 30° | … | 8 |
| | 45° | … | 15 |
| | 60° | … | 25 |
| | 90° | … | 60 |
| Close Return Bend | 180° | … | 50 |

$R/D$ is the bend mean radius divided by pipe diameter

The loss of head due to a bend in the pipe is most frequently given as the equivalent length of straight pipe, which would cause the same loss in head as the bend. Experiments show that a right-angle bend should have a radius of about three times the diameter of the pipe. Assuming this curvature, then, if $d$ is the diameter of the pipe in inches, and $L$ is the length of straight pipe in feet that causes the same loss of head as the bend in the pipe, the following formula gives the equivalent length of straight pipe that should be added to simulate a right-angle bend:

$$L = 4d \div 3$$

**Energy Loss in Fittings and Valves.**—There are two standard ways to calculate the energy lost in fittings and valves: the resistance coefficient method, and the equivalent length method. The method chosen will depend on what information is provided with the fitting to be analyzed.

*Resistance Coefficient Method:* Resistance coefficients are commonly reported by the manufacturers of liquid fittings and valves. The following equations and Table 25 (or manufacturer's data) can be used to calculate friction loss in terms of resistance coefficient. There are methods that use up to 3 different resistance coefficients to improve the accuracy of the calculation.

$$h_l = K_F\left(\frac{v^2}{2g}\right) \qquad K_F = f\frac{L_e}{D}$$

where $v$ is average velocity, $h_l$ is head loss in units of length, $K_F$ is the resistance coefficient for the fitting or valve, $f$ is the friction factor given in the Moody diagram, Fig. 1, and $L_e/D$ is the equivalent length ratio given in Table 25.

**Table 25. Equivalent Length Ratio ($L_e/D$) for Liquid Fittings and Valves**

| Fitting or Valve Type | $L_e/D$ | Fitting or Valve Type | $L_e/D$ |
|---|---|---|---|
| Elbow | 30 | Swing Check Valve | 100 |
| Street Elbow | 50 | Ball Check Valve | 150 |
| Long Radius Elbow | 20 | Foot Valve (hinged) | 75 |
| 45° Elbow | 16 | Ball Valve | 3 |
| Close Return Bend | 50 | Gate Valve | 8 |
| Tee (Flow thru Run) | 20 | Gate Valve, 1/2 Closed | 160 |
| Tee (Flow thru Branch) | 60 | Butterfly Valve (< 8″) | 45 |
| | | Globe Valve | 340 |

Valves fully open unless otherwise specified.

*Equivalent Length Method:* Pressure loss can be calculated for systems based on the equivalent length of straight pipe representing all pipe, fittings, and valves in the circuit. Equivalent lengths of some common fittings and valves for liquids can be found in Table 26.

**Table 26. Equivalent Length in Feet of Selected Pipe Fittings and Valves for Liquid**

| Nominal Pipe Size (in.) | Elbows | | | | | Standard Tee | | |
|---|---|---|---|---|---|---|---|---|
| | 90° Std. | 45° Std. | 90° Long Radius | 90° Street | 45° Street | Square Corner | Flow thru Run | Flow thru Branch |
| ¼ | 0.9 | 0.5 | 0.6 | 1.5 | 0.8 | 1.7 | 0.6 | 1.8 |
| ½ | 1.6 | 0.8 | 1.0 | 2.6 | 1.3 | 3.0 | 1.0 | 4.0 |
| ¾ | 2.1 | 1.1 | 1.4 | 3.4 | 1.8 | 3.9 | 1.4 | 5.1 |
| 1 | 2.6 | 1.4 | 1.7 | 4.4 | 2.3 | 5.0 | 1.7 | 6.0 |
| 1¼ | 3.5 | 1.8 | 2.3 | 5.8 | 3.0 | 6.5 | 2.3 | 6.9 |
| 1½ | 4.0 | 2.1 | 2.7 | 6.7 | 3.5 | 7.6 | 2.7 | 8.1 |
| 2 | 5.5 | 2.8 | 4.3 | 8.6 | 4.5 | 9.8 | 4.3 | 12.0 |
| 2½ | 6.2 | 3.3 | 5.1 | 10.3 | 5.4 | 11.7 | 5.1 | 14.3 |
| 3 | 7.7 | 4.1 | 6.3 | 12.8 | 6.6 | 14.6 | 6.3 | 16.3 |
| 4 | 10.1 | 5.4 | 8.3 | 16.8 | 8.7 | 19.1 | 8.3 | 22.1 |
| 6 | 15.2 | 8.1 | 12.5 | 25.3 | 13.1 | 28.8 | 12.5 | 32.2 |
| 8 | 20.0 | 10.6 | 16.5 | 33.3 | 17.3 | 37.9 | 16.5 | 39.9 |
| 10 | 25.1 | 13.4 | 20.7 | 41.8 | 21.7 | 47.6 | 20.7 | 50.1 |
| 12 | 29.8 | 15.9 | 24.7 | 49.7 | 25.9 | 56.7 | 24.7 | 59.7 |

Energy loss, in terms of pressure, for gas flow through fittings is often found using tabulated values of pressure drops for fittings of different sizes. The equivalent length method can also be used, with pressure loss calculated in terms of the total equivalent length of all pipe and fittings. Table 27 includes some equivalent lengths of selected pneumatic fittings and valves.

**Table 27. Selected Pneumatic Fittings and Valves, Equivalent Lengths**

| Pipe Size, inch | Elbow | Long Radius Elbow | Close Return Bend | Gate Valve | Globe Valve |
|---|---|---|---|---|---|
| ½ | 0.84 | 0.41 | 1.30 | 0.31 | 2.50 |
| ¾ | 1.20 | 0.57 | 1.80 | 0.44 | 3.50 |
| 1 | 1.60 | 0.77 | 2.30 | 0.57 | 4.70 |
| 1½ | 2.60 | 1.30 | 3.90 | 0.98 | 7.80 |
| 2 | 3.60 | 1.70 | 5.30 | 1.30 | 10.60 |
| 3 | 5.70 | 3.00 | 8.50 | 2.10 | 17.10 |
| 4 | 7.90 | 3.90 | 11.80 | 3.00 | 23.70 |

**Entrance, Exit, Enlargement, and Contraction Losses for Subsonic Flow.—**Any change in diameter of a pipe will cause energy losses in the fluid.

*Sudden Enlargement:* Sudden enlargement of a pipe or tube causes minor losses due to turbulence. The friction loss, in terms of head, due to sudden enlargement can be approximated using the following equations:

$$h_l = \frac{(v_i - v_o)^2}{2g} = K\left(\frac{v_i^2}{2g}\right)$$

$$K = \left[1 - \left(\frac{A_i}{A_o}\right)\right]^2$$

where head loss $h_l$ is measured in units of length; $v_i$ is velocity entering the enlargement; $v_o$ is velocity leaving the enlargement; $K$ is the loss coefficient for sudden enlargement; $A_i$ and $A_o$ are the areas before and after the enlargement. The equation for $K$ is only valid if velocity at the inlet is less than 4 ft/sec (1.2 m/s). For faster flow rates, a graph of experimentally determined $K$ values must be used.

*Gradual Enlargement:* Gradual enlargement causes less energy loss than sudden enlargement because it causes less turbulence. Diffusers are gradual enlargement devices. Velocity decreases while pressure increases. Gradual enlargements generally have a conical section that joins two different pipe diameters. The divergence angle is the included angle in the cone. The following two equations give approximations of head loss in a gradual enlargement. $K_L$ is the Gibson loss coefficient (see Fig. 3). $K_D$ is the resistance coefficient (see Fig. 4). Both depend on the ratio of areas and the angle of divergence. The formulas to calculate this loss ignore the effects of friction inside the conical section. If the enlargement is very gradual, the effects of friction can become significant. Therefore, an included angle of 6° or more is recommended in the conical section to mitigate the effects of friction. For area ratios up to 6:1 with angles of divergence of 40° or more, head loss may be less with a sudden expansion than with a gradual one. These two effects are clearest on the graph of $K_L$ (Fig. 3).

$$h_l = K_D\frac{v_i^2}{2g} \qquad h_l = K_L\frac{(v_i^2 - v_o^2)}{2g}$$

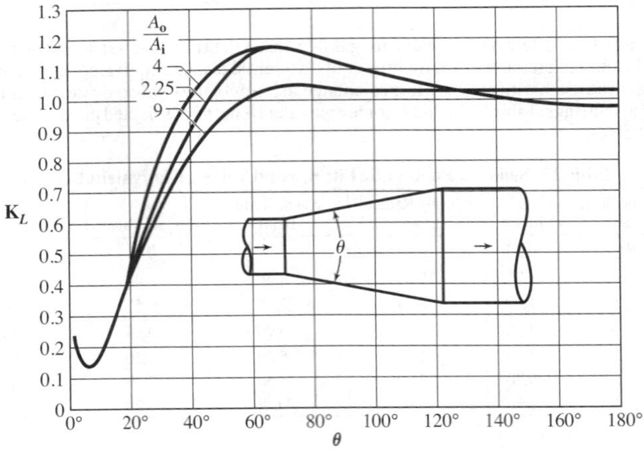


Fig. 3. Head Loss Coefficient $K_L$, Conical Diffuser

*Sudden Contraction:* Sudden contraction causes energy loss due to turbulence. This head loss can be calculated by using the following equation and Fig. 5. Because loss varies with the square of velocity, small changes in flow can greatly affect energy loss.

$$h_l = K_C\frac{v_o^2}{2g}$$

where $v_o$ is the velocity at the outlet and $K_C$ is the loss coefficient. $D_i/D_o$ is the ratio of the inlet diameter to the outlet diameter.

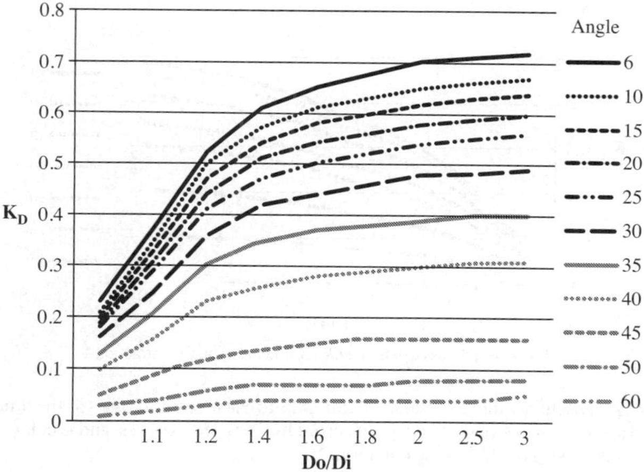


Fig. 4. Head Loss Coefficient $K_D$, Conical Diffuser

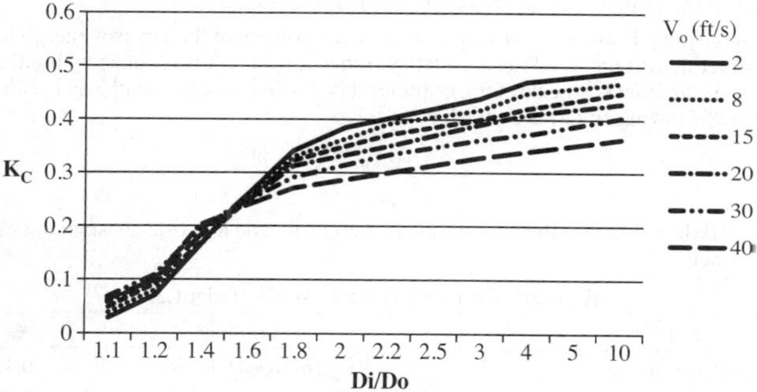


Fig. 5. Head Loss Coefficient $K_C$, Sudden Contraction

*Gradual Contraction:* Gradual contraction, like gradual enlargement, reduces losses by reducing turbulence. Like with gradual enlargement, if the angle is shallow enough, friction losses become significant. Included angles greater than 15° avoid this effect. Good performance can be achieved with included angles as high as 60°. To approximate head loss due to gradual contraction, use the following equation and Fig. 6.

$$h_l = K_G \frac{v_o^2}{2g}$$

where $v_o$ is the velocity at the outlet and $K_G$ is the loss coefficient. $D_i/D_o$ is the ratio of the inlet diameter to the outlet diameter. By rounding the exposed edge of the contraction, one can reduce the head loss even further.

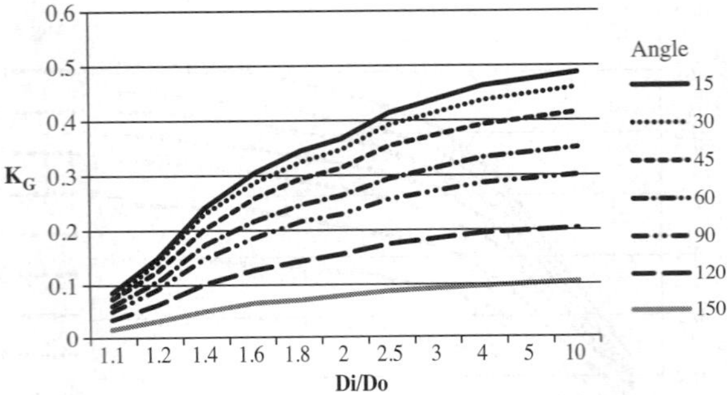


Fig. 6. Head Loss Coefficient $K_G$, Conical Gradual Contraction

*Exit Loss:* Discharge into to a static volume of fluid causes minor losses as the fluid loses its kinetic energy as its velocity drops to zero. This is called exit loss, and is calculated in terms of head using the following equation:

$$h_l = \frac{v_i^2}{2g}$$

where $v_i$ is the velocity in the pipe before entering the reservoir.

*Entrance Loss:* Entrance into a pipe from a static volume of fluid causes energy loss due to contraction turbulence. This is called entrance loss, and it is reduced by breaking the corners of the inlet with chamfers, or preferably, radii. Entrance head loss is calculated using the following equation and Table 28.

$$h_l = K_E \frac{v_o^2}{2g}$$

where $v_o$ is the velocity in the pipe exiting the reservoir, and $K_E$ is the resistance coefficient for entrance.

**Table 28. Entrance Head Loss Coefficient, $K_E$**

| Condition | $K_E$ | Condition | $K_E$ |
|---|---|---|---|
| Rounded Inlet, $R/D_o$: | | Square-edged Inlet | 0.5 |
| 0.02 | 0.28 | Chamfered Inlet | 0.25 |
| 0.04 | 0.24 | Inward Projecting Pipe | 1 |
| 0.06 | 0.15 | | |
| 0.1 | 0.09 | | |
| >0.15 | 0.04 | | |

## Fluid Power Valves

There are many types of pneumatic and hydraulic valves. Valves can be normally open or normally closed. Normally open valves allow flow when the valve is not energized. Normally closed valves block flow when not energized. Valves may be actuated manually or automatically. Automatic valves typically change state using solenoids, often assisted by energy from either internal or external pilot pressure. An external pilot supply should be used to control valve when the supply pressure to the valve is insufficient to shift the valve or when the medium being controlled by the valve would damage the shifting mechanism. Automatic valves can be equipped with manual overrides which permit the operator to actuate the valve manually.

**Valve Types.**—There are many valves types commercially available. The most common types are check valves, foot valves, ball valves, gate valves, globe valves, butterfly valves relief valves, and directional control valves.

*Check Valves and Foot Valves:* Check valves are constructed to allow flow in one direction and stop flow in the other direction. This type of valve is essential in pipe and tube systems where backflow could cause damage or flooding. The two most common types of check valves are ball type and swing type. Ball check valves create a higher energy loss due to restriction around the ball. Check valves have a "minimum cracking pressure" that is the pressure required to open the valve and allow flow. This may be expressed as a minimum pressure or a minimum flow velocity, depending on the application.

Foot valves are commonly used on the intake lines of pumps that draw fluid from an open reservoir. They are comprised of a check valve outfitted with a low-loss strainer to keep out debris. The strainer should be cleaned regularly to prevent suction line restriction.

*Restriction Valves:* There are several types of valves that are used to reduce flow or to stop flow entirely. Ball valves contain a sphere with a hole through it inside the valve body. When the sphere is turned, the through-hole can be aligned to allow full flow or to obstruct flow partially or completely. These valves are fully actuated with one quarter turn, and are typically used for on/off control rather than throttling. Gate valves have a blade or gate that is cranked into or out of the flow path to allow full flow, partial flow, or no flow. Globe valves are true throttling valves, and can be used even in high flow applications. Due to their design, they cause a significantly higher pressure drop than other types of valves, even when fully open. Butterfly valves consist of a disk that can be turned to fully or partially block flow. When fully open, the disk is still in the flow path, but is turned to present its edge to the flow.

*Relief Valves:* Pressure relief valves are used to protect the fluid circuit and its components from dangerously high pressures. A pressure relief valve opens to relieve pressure at a predetermined set point. These valves can be selected such that they close again when the system pressure has returned to a safe level. Since pressure relief valves are safety devices, their design and use is covered by various codes. The information and calculations required to properly size pressure relief valves is quite extensive and beyond the scope of this text. Most major pressure relief valve manufacturers also offer sizing software. In liquid pumping systems, the relief valve should discharge to the suction reservoir in a way that prevents gases from entering the suction line to the pump.

*Directional Control:* Directional control valves are used to change the direction of flow. They can also be used to stop and start flow. Directional control valves are functionally described in terms of the number of ports, the number of positions (usually 2 or 3), the number of paths (passing and non-passing) the fluid can take through the valve, and the valve's de-energized state. The number of paths the fluid can take through the valve is said to be the number of "ways." Three way valves have one output and four way valves have two outputs. Generally, three way valves operate single acting linear actuators and four way valves operate double acting linear actuators. Some of the most commonly used directional control valves are:

- Two port, two position (2/2) two way valves have one inlet and one outlet port. They perform a basic on/off function.
- Three port, two position (3/2) two way valves have one inlet, one outlet, and one exhaust port. These valves switch flow between inlet and exhaust, and are commonly used to pressurize and exhaust a single actuator supply line.
- Five port, two position (5/2) four way valves are typically used to control pressure and exhaust to two output lines. This is ideal for controlling double acting cylinders. 5/2 valves are also available with four ports. Five ported valves have separate exhaust

ports for each supply port. Flow controls can be installed on each exhaust port and thus the speed of an actuator can be controlled in each direction.

- Five port, three position (5/3) four way valves also can control pressure and exhaust to two output lines. These valves often have the center position normally closed when the valve is de-energized. This valve is also excellent for control of double acting cylinders. The closed position makes it possible to 'jog' actuators.

Directional control valves can be either manually controlled or automatically controlled. Manual control valves are outfitted with buttons, levers, or knobs. Automatic control valves are electrically operated solenoids. Solenoid valves are most commonly actuated using spools or poppets. These valves can be single or double acting. Single acting valves normally have spring returns. Double acting valves use fluid pressure on either side of the valve to actuate back and forth.

Automatic control valves can be either direct acting or pilot operated. Direct acting solenoid valves open and close the valve. They are used in systems with low flow capacities or low pressure differentials across the valve. The sealing surface that opens and closes the main valve orifice is connected to the solenoid plunger. Direct acting valves can operate with a zero pressure differential, or up to the maximum rated pressure differential (MOPD) regardless of fluid pressure. Pressure drop across the valve is not required to hold the valve open.

Pilot operated valves use system fluid pressure to open and close the valve. In a normally closed valve, the sealing surface is opened when the pilot is energized. When a normally open valve is energized, fluid pressure builds up behind the sealing surface, forcing it to close the valve. Pneumatic solenoid valves are generally pilot operated. When vacuum or air with insufficient pressure is the medium, separate pilot lines must be provided with sufficient positive pressure to actuate the valve.

**Energy Loss in Valves.**—Energy is lost in all valves due to restriction, friction, and changes in direction of flow. The type of valve and its specific geometry will determine the amount of energy lost. Manufacturers will generally provide equivalent length of straight pipe values or resistance coefficients for their valves. Equivalent length is added to the pipe circuit when calculating pressure drop through the system. Resistance coefficient $K$ is used in previous equations for friction loss. Some values for equivalent lengths are given in Table 26 and Table 27.

**Valve Selection and Sizing.**—Valves should be selected by first considering their function and selecting a type. Compatibility with the fluid must be carefully evaluated, since most valves include seals and mechanical components of different materials. The temperature and pressure range of the system should be taken into account. The cleanliness of the fluid matters when selecting valves with small or sensitive internal components. A fluid filter (with its own pressure drop) may be required. Pressure drop across the valve should be examined at the operating flow rate to determine if it meets the valve's minimum pressure differential requirement in the case of pilot operated valves.

Valves are usually sized based on Coefficient of Velocity ($C_v$). These values are normally listed in manufacturer's catalogs. $C_v$ and its calculation is covered in *Coefficient of Velocity, Cv* on page 2769. Using $C_v$ to size valves is industry standard practice, but be aware that it is only an approximation. Energy loss in the circuit due to pipe and tubing runs, fittings, and other components will cause deviation of system performance form ideal. Valve switching response time can also be a factor to consider when selecting a valve. To select the proper size valve for an actuator to perform a specific task, first calculate the $C_v$ of the actuator, and then select a valve that has a corresponding or greater $C_v$.

## Fluid Power Actuators

**Linear Actuators.—**When a gas, typically air, is used in a linear actuator, it is called a gas cylinder or air cylinder. When a liquid is used, it is typically called a hydraulic ram or cylinder. These are essentially the same in form and function. Linear actuators can be either single or double acting.

*Single Acting Cylinders:* Single acting actuators with spring return use a spring to push the actuator to its free state when fluid pressure is removed. The free state of a single acting cylinder is usually fully retracted. A fluid pressure differential is used to extend the cylinder. Single acting cylinders use half the fluid of double acting cylinders. Single acting cylinders can be actuated with three way valves.

*Double Acting Cylinders:* These cylinders require pressure differentials on both the extend and retract strokes. When pressure equalizes across the cylinder, the state of the actuator becomes indeterminate. Double acting cylinders are most commonly actuated using four way valves. Double acting cylinders use twice the fluid of single acting cylinders, which matters most in pneumatic applications where that air is exhausted and lost.

*Selection of Linear Actuators:* Linear actuators are available with a variety of options. In addition to single and double acting versions, some cylinders are available with a magnet embedded in the piston to enable position sensing. For applications with side loading a double rod cylinder or external bearing may be needed. When a cylinder is guiding a load that should not rotate, anti- rotation options such as a non-round rod or integral guide rods are commonly used. For heavy loads or high speeds, internal or external end cushions are usually offered to soften the impacts and allow the cylinder to withstand higher kinetic energy movements.

*Force Exerted by Linear Actuators:* The force exerted by a linear actuator is calculated using the following equation:

$$F = \eta p A$$

where $\eta_{\text{hydraulic}} = 0.9$ and $\eta_{\text{pneumatic}} = 1$, and $A$ is the effective area of the cylinder. Effective area on the retract stroke is the piston area minus the rod area.

It is good practice to oversize the load-carrying capacity of an actuator by 25 percent to ensure smooth operation and account for losses in the system. For high-speed operation, oversizing by 50 percent is not uncommon. For a spring return cylinder, the force on the retraction stroke will be equal to the spring force rather than a function of air pressure. For applications where rapid retraction speed is critical, double acting cylinders may be the best choice.

*Speed of Linear Actuators:* The speed achieved by an unloaded linear actuator depends on the effective area of the cylinder and the flow rate of the fluid. Actuation velocity can be calculated for an unloaded cylinder using the following equation:

$$v = \frac{231Q}{A}$$

where $v$ is velocity in inches per minute, $Q$ is flow rate in gallons per minute, and $A$ is piston active area in square inches. Active area is reduced by the area of the rod when a rod is present on the acting side of the cylinder.

Most manufacturers provide graphs of actuator speed versus loading for their products. Fluid supply can be characterized by Coefficient of Velocity ($C_v$) factor. Valves and circuits with a higher $C_v$ will be capable of moving an actuator faster because they can provide more air flow.

If the $C_v$ of a valve and other system components is known, it can be used to determine the time required to extend or retract a cylinder. It is common to oversize the speed of an actuator by 25 percent in critical applications to account for losses.

*Kinetic Energy and Linear Actuators:* Kinetic energy is generated when a load is moved. When a moving load is stopped, the kinetic energy is absorbed by the mechanism that stops it. Actuators can have either internal or external stops. External stops can generally withstand greater impacts than internal stops. Manufacturers normally publish the kinetic energy limits of their actuators.

*Safe Loading of Linear Actuators:* Rod failure and piston wear is of concern when linear actuators are used in compression or are subjected to side loading. Side loading can be a result of external forces, eccentrically mounted loads, or mounting and operating a cylinder at some angle rather than vertical. It is critical to align connections to a cylinder such that side loading due to misalignment or eccentric loading is minimized. Compliant connections are often used to eliminate overcontraint and resultant loading. When rod bending or buckling is a risk, the use of external bearings or double rod cylinders can help reduce the risk. For horizontal cylinder installations, center trunnion mounting can help reduce the side loading on the cylinder. Another way to reduce premature failure due to side loading is to limit the extension stroke of a cylinder so that it cannot fully extend.

The potential for buckling must be evaluated for any cylinder in compression. To increase the buckling strength of a cylinder, increase the piston rod diameter. Critical buckling load is calculated as shown in the following equation.

Equivalent length is used to calculate critical buckling load rather than actual rod length. The equivalent length is dependent on the end conditions of the cylinder. The length $L$ used to calculate equivalent length is assumed to be the distance at maximum extension between the exposed end of the piston rod and the nearest point of fixation.

$$F_{cr} = \frac{\pi^2 EI}{L_e^2} \qquad I = \frac{\pi r^2}{4}$$

where $F_{cr}$ is the critical buckling load, $E$ is the modulus of elasticity for the rod material, and $L_e$ is given in Table 29.

**Table 29. Equivalent Length, Linear Actuator**

| Actuator and Load Fixations | Equivalent Length $L_e$ |
|---|---|
| Actuator Rigidly Mounted, Load Free | $2L$ |
| Actuator and Load Pin Connected | $L$ |
| Actuator Rigidly Mounted, Load Pinned | $0.7L$ |
| Actuator and Load Rigidly Mounted | $0.5L$ |

**Intensifiers and Hydraulic Presses.**—Intensifiers use the energy in a relatively large volume (work cylinder) of moderate pressure fluid to displace a smaller piston (ram). The two pistons are usually mechanically linked, so their stroke is the same. Intensifiers will hold a fixed pressure without additional input energy. The intensifier ratio, output pressure, and stroke are found using the following equations.

$$R_I = \frac{p_o}{p_i} \qquad p_o = \frac{p_i A_i}{A_o} \qquad S_1 = \frac{V_w + V_C}{A_r} + l$$

$$V_C = 0.1 V_{oil} \frac{p_r}{1000} \qquad S_2 = \frac{V_o + V_C}{A_r} + X + l$$

where $R_I$ = the intensifier ratio

$p_o$ and $p_i$ = output and input pressures

$A_o$ and $A_i$ = the two different piston areas

$S_1$ = the stroke of a single pressure intensifier

$V_w$ = the volume of the work cylinder

$V_C = 0.1 V_{oil}\frac{p_r}{1000}$ is the compressibility allowance, where $p_r$ is the pressure in the high pressure circuit, and $V_{oil}$ is the total volume of oil in the high pressure circuit

$l$ = any pretravel of the work cylinder required

$A_r$ = the area of the ram

$S_2$ = the stroke of a dual pressure intensifier

$V_o$ = the oil volume required to move the work cylinder through the high pressure portion of the stroke

$X$ = the stroke required to close the high pressure seal on the intensifier (often assumed to be 2 inches)

Intensifiers can be either pneumatically driven or fully hydraulic. Fully hydraulic intensifiers are more efficient than air driven ones. They operate on the same principle, with liquid used on both sides of the actuator. Commercial hydraulic intensifiers are generally powered by electric motors or diesel engines. Hydraulic intensifiers can be either single or dual pressure. A single pressure intensifier is used when the work cylinder requires a high pressure for the entire stroke. A dual pressure intensifier is recommended if the high pressure is required only during the last portion of the work cylinder stroke. For best economy, select the smallest driving cylinder bore size for a given intensifier ratio. Minimize pressure losses by keeping lines short and by sizing lines and fittings as large as practical. Size the lines so that fluid velocity does not exceed 7 ft/sec. Self-bleeding boosters be used wherever possible to ensure that air does not build up in the system and cause "spongy" operation.

Hydraulic presses, like intensifiers, use two pistons of different diameters. They operate on a shared fluid volume. When a force is applied to the small cylinder to create pressure, that pressure is reflected in the large cylinder. When the large cylinder is exposed to high pressure, it can exert a lot more force than was applied to the small cylinder. The mechanical advantage of the press is the ratio of output to input force. Hydraulic presses are subject to the following equations:

$$p_1 = p_2 = \frac{F_1}{A_1} = \frac{F_2}{A_2} \qquad F_1 s_1 = F_2 s_2 \qquad R = \frac{F_1}{F_2} = \frac{A_1}{A_2} = \frac{s_1}{s_2}$$

where $F$ is force, $p$ is pressure, and $A$ is area, $s$ is stroke, and $R$ is the gear ratio of the press. Note that the stroke of the smaller diameter cylinder will be greater than that of the larger diameter cylinder.

**Rotary Actuators.—**These actuators use fluid pressure to induce rotary motion, and tend to be double acting. Rotary actuators can have either internal or external stops. External stops can generally withstand greater impacts than internal stops. Stops on rotary actuators are often adjustable. Rotary actuators can be hydraulic, pneumatic, or a combination of both.

*Types:* Rotary actuators come in two main types: vane and rack and pinion. Vane type actuators offer the smallest package and zero backlash, but are usually capable of only up to 270° of rotation. They employ a rotary vane mechanism to turn the actuator under fluid pressure. Rack and pinion actuators use one or more linear actuators to move racks back and forth, thereby turning a pinion shaft. They can be outfitted to have multiple stopping positions, and are capable of 360° rotation. Many packaging options are available for both vane and rack and pinion types, with many mounting and output shaft options.

*Sizing Rotary Actuators:* When sizing rotary actuators, one must know the accuracy and actuation time required, the required angle of rotation, any resisting torque, and the moment of inertia of any mass loading about the center of rotation. Torque values for both directions of rotation as a function of input pressure are normally provided by manufacturers. Manufacturers sometimes publish the $C_v$ of their actuators, or it can be calculated using the rotary swept volume. Kinetic energy is of particular concern with rotary actuators. Manufacturers normally publish the kinetic energy limits of their actuators. Although cushions are normally provided as part of rotary actuators, it is recommended to use proportional valves to decelerate rotating loads where loading is significant. Kinetic energy is calculated using the following equation:

$$KE = \frac{1}{2}I\omega^2$$

where ω is the angular velocity (peak) in radians/sec, and $I$ is the mass moment of inertia of the load carried on the rotary actuator shaft.

**Shock Absorbers and Hydraulic Dampers.—**These devices are special types of linear actuators that are used to control applied loads. Pneumatic and hydraulic types are commercially available. All have frequency limitations that should be examined when sizing them for an application.

*Pneumatic Shock Absorbers and Gas Springs:* Pneumatic shock absorbers and gas springs use the compressibility of gas (usually nitrogen) to resist an applied force. Shock absorbers are used to decelerate an applied load over a distance, and are generally quite small. Gas springs are used mostly to counterbalance loads and often have very long strokes. Some gas springs have intermediate holding positions.

Shock absorbers are normally used when no return assistance is required and no restriction of velocity is required. There are some available with an orifice in the piston that provides velocity control. Shock absorbers often have a spring to provide a return function and assist in decelerating loads. They also usually have a mechanical stop, such as an adjustable nut, to prevent overcompression.

Shock absorbers are sized according to impact speed, stroke, and energy absorbing capacity. The energy absorbed will be the sum of the kinetic energy of the moving load and any external thrust provided by a driving mechanism. Vertical applications will also need to resist the weight of the applied load. Kinetic and thrust energies are calculated using the following equations:

$$E_{total} = KE + E_t$$

$KE = \frac{1}{2}mv^2$ = Kinetic energy of a linear impact, where $m$ is mass and $v$ is velocity.

$E_t = Fs$ = Thrust energy of a linear impact, where $F$ is applied force and $s$ is the shock absorber stroke.

$KE = \frac{1}{2}I\omega^2$ = Kinetic energy of a rotary impact, where $I$ is moment of inertia and is angular velocity.

$E_t = \frac{Ts}{R}$ = Thrust energy of a rotary impact, where $T$ is applied torque and $R$ is radius from the moment center and impact point.

$KE = mgh$ = Kinetic energy of a falling weight impact, where $m$ is mass and $h$ is height.

$E_t = mgs$ = Thrust energy of a falling weight impact.

Gas springs are used when return assistance or load support (counterbalance) is required throughout the range of motion. If the gas spring is the quick return type, the force increases as the piston moves. If an orifice is present in the piston through which the gas can escape, they can limit velocity and provide a more constant force over the full stroke.

Gas springs can be either push or pull (tension or traction) type, depending on where the gas chamber is relative to the piston rod. Gas springs usually have a reservoir of oil near the rod outlet of the cylinder that is used to lubricate the rod seal. They are designed to be stored and installed with the rod pointing down, at an angle of 60° or less from vertical to ensure that the seal is lubricated. The oil also serves to dampen the end of stroke. When they cannot be installed rod down, special gas springs with provisions for lubrication are required. Due to their long, narrow form factor, side loading should be avoided.

Sizing gas springs is done with consideration of the required stroke, compressed length, diameter, compressed force, and extended force. Extended force is normally given at a point short of full extension because they should never be fully extended in operation to minimize the chances of buckling. Hysteresis is the difference between the actual and rated output force of a spring caused by friction. *K* factor, or progression, is the spring rate of a gas spring, and is calculated based on the following equation:

$$K = \frac{F_2}{F_1}$$ where $F_2$ is force at the compressed position, and $F_1$ is the uncompressed force.

Progression occurs because as the rod enters the chamber during compression, it reduces the available volume for the gas to occupy. This causes a pressure increase.

*Hydraulic Dampers (Hydraulic Shock Absorbers):* These devices are used to safely control the speed of applied loads. Dampers consist of a piston in a pressure tube containing a hydraulic fluid. When force is applied to the piston rod to move the piston, the damping fluid must pass through orifices in the piston. As a result, the damping force is dependent on the piston speed. As the fluid is forced through the orifices, kinetic energy is converted into thermal energy. Using valving on the orifices, it is possible to regulate the damping forces in each direction independently. Hydraulic dampers are usually self-contained, sealed units. They can be single or double acting. In some designs, the damping speed is adjustable over a range by turning the rod while fully extended. Hydraulic dampers are sized according to diameter, stroke length, and required damping characteristics. When extreme speeds and forces are required, cavitation/foaming of the oil can introduce gas bubbles into the chamber and cause a drop in damping capacity.

## Hydraulics

**Hydraulic Fluids.—**Hydraulic systems use liquids as their medium. Water is a common hydraulic medium in general cases. When actuators are involved, petroleum based oils are the most common type of hydraulic fluid, but silicones, synthetics, and water based fluids are also common. Hydraulic fluids can be broken down in these major categories: hydraulic fluids based on mineral oils and related hydrocarbons (ISO 11158), environmentally acceptable hydraulic fluids (ISO 15380), fire- resistant, water-free hydraulic fluids (ISO 12922), and fire-resistant, water-containing hydraulic fluids (ISO 12922).

When selecting a hydraulic fluid, one must consider viscosity, viscosity index, lubricity, compressibility, flammability, foaming resistance, resistance to rust/oxidation, resistance to bacterial growth, pour point, flash point, chemical compatibility with system seals/components, cleanliness, cost, and environmental impact. The manufacturer of the hydraulic components should be consulted as well.

Viscosity at the required temperatures is the most important selection factor. Many petroleum products are graded according to the ISO Viscosity Classification System, approved by the Organization for International Standards (ISO). Each ISO viscosity grade number corresponds to the mid-point of a viscosity range expressed in centistokes (cSt) at 40° C. A chart of viscosity related to the various systems of grading fluids is given in Fig. 7.

**Fluid Conditioning and Stabilization.—**To provide useful work, fluid must have energy added in the form of pressure. Pumps are usually used to pressurize fluid, unless gravity can perform the needed energy. Strainers and filters are needed to keep contaminants out

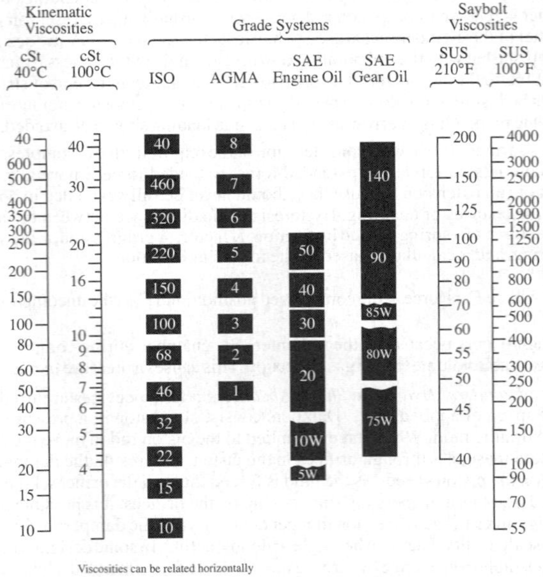


Fig. 7. Hydraulic Fluid Viscosity Chart

The viscosity/temperature relationships are based on 95 VI oils and are usable only for mono grade engine oils, gear oils and other 95 VI oils. Crankcase oils and gear oils are based on 100° C viscosity. The "W" grades are classified mono grade engine oils, gear oils and other 95 VI oils. ISO oils and AGMA grades are based on 40° C viscosity.

of sensitive components, and accumulators are used to stabilize the fluid pressure and act as reservoirs. In some cases, cooling must be provided, either passively or actively.

*Strainers and Filters:* Despite being closed systems, hydraulic circuits can become contaminated with particles due to component wear and incoming fluid contaminants. Filters are generally installed after a pump. Installation in the suction line, while more protective, would cause pressure loss in the suction line and jeopardize pump efficiency. An additional coarse strainer that creates very little pressure drop is often placed in the suction line to protect the pump. The pressure drop induced by a commercial filter is normally provided by the manufacturer.

Hydraulic filters must be capable of handling the pressures and temperatures encountered in the system.

Chemical compatibility of the filter membrane, body, and seals must also be considered. The filter must be rated for the maximum system flow rate and must be able to handle the viscosity of fluid used in the system. The allowable particle size that can pass through the filter will determine the membrane type. Hydraulic filters are available as in-line, in-tank, and other configurations. Regular maintenance will be required, so filters must be chosen and installed to facilitate membrane replacement.

*Liquid Pulsation (Water Hammer):* Valves and reciprocating pumps are often the culprits when water hammer occurs. Water hammer is a pressure spike in a fluid circuit caused by a sudden stoppage of flow. A valve closing can cause a shock wave in the liquid, resulting in noise. Reciprocating pumps cause acceleration and deceleration of the fluid that result in pressure variations. Hydraulic water hammer can be severe enough to burst pipes. Pulsation dampeners are recommended when water hammer is likely. An equation for the pressure rise that can occur due to water hammer follows:

$$\Delta p = \frac{Q}{20d^2}\sqrt{\mathrm{SG}b}$$

where $p$ is pressure in psi, $Q$ is flow rate in GPM, SG is specific gravity, $b$ is bulk modulus in psi, and $d$ is pipe inside diameter, in inches, upstream of the point of interest.

Accumulators and pulsation dampeners are used to store pressurized hydraulic fluid, which contains potential energy. This fluid and its energy can then be released to smooth out intermittent system cycles, dampen pulsations, absorb shocks, and compensate for leaks in the system.

*Accumulators:* Pressurization against an accumulation of fluid can be accomplished through weight loading, spring loading, or compressed gas loading. Hydro-pneumatic, or gas loaded, accumulators are the most common. The three most common of this class are piston, bladder, and diaphragm accumulators. In these, a piston, bladder, or diaphragm separates the fluid from the compressed gas.

When a hydro-pneumatic accumulator has a closed volume of gas, the change in volume of hydraulic fluid is equal to the change in volume of the gas. The equation for polytropic processes can then be used to calculate accumulator volume or pressure, depending on which values are known. The polytropic exponent, n, depends on the type of process. For slow gas expansion and compression processes which occur almost isothermally, the polytropic exponent can be assumed to be $n = 1$. For rapid processes, the adiabatic exponent can be taken as $n = 1.4$ for Nitrogen and air. For pressures above 3000 psi the real gas behavior deviates considerably from the ideal one, which reduces the effective fluid volume. In such cases the adiabatic exponent used is greater than 1.4.

The equation for polytropic processes is:

$$p_0 V_0^n = p_1 V_1^n = p_2 V_2^n = \text{Constant}$$

where $p_0$ is the gas precharge pressure, $p_1$ is the minimum working pressure, and $p_2$ is the maximum working pressure. $V_0$, $V_1$, and $V_2$ are the gas volumes in each of the states.

For low pressure applications of less than 150 psi, absolute gas pressures must always be used in the equation for required gas volume. These equations are assuming ideal gas behavior. Correction factors must be used to take into account real gas behavior. Those correction factors can range from 1 to 2 depending on the pressures involved.

$$V_0 = \frac{\Delta V}{\left(\frac{p_0}{p_1}\right)^{\frac{1}{n}} - \left(\frac{p_0}{p_2}\right)^{\frac{1}{n}}}$$

where $V_0$ is the precharge volume of gas, and $\Delta V = V_1 - V_2$

Precharge pressure for energy storage is generally about 90 percent of system minimum working pressure. For shock absorption, precharge is usually 60 to 90 percent of median working pressure; for pulsation dampening, precharge is usually 60 to 80 percent of median working pressure. A small amount of fluid should remain inside the accumulator at minimum working pressure to prevent the mechanical components from rubbing or impacting. Therefore the precharge pressure should be slightly lower than the minimum working system pressure.

*Pulsation Dampeners and Snubbers:* Pulsating flow can wreak havoc with hydraulic devices in a circuit. A pulsation dampener is a type of accumulator that absorbs these pressure variations and stabilizes the downstream flow velocity. They absorb only the portion of fluid accelerated above the mean flow rate, store the excess fluid momentarily, and then discharge it during the portion of the cycle where flow rate drops below mean. These devices are less effective at higher frequencies. For best results, dampeners are placed as close as possible to a pump or source of pulsation. Snubbers are generally placed before gauges. Suction stabilizing dampeners are placed soon after the supply reservoir to protect a pump from acceleration head variations.

The diaphragm and bladder types of pulsation dampener are extremely common. The diaphragm or bladder is pressurized on one side with a gas (nitrogen is common) to about 70 to 80 percent of system peak pressure. A pulsation dampener can be as simple as a section of elastic tubing in a circuit that can expand and contract as pressure builds and subsides. Sometimes pulsations can be dampened by placing a restriction in the line to limit fluid flow. This method is less preferred because dangerous levels of pressure may occur behind the restriction.

Gauge snubbers vary in their method of dampening. Less expensive ones often restrict flow using a porous disc, while more expensive snubbers use a piston mechanism to dampen pulsations.

The most common application for a pulsation dampener involves a reciprocating piston pump system. Since the dampener must absorb and store the volume of fluid above the mean discharge volume of the pump, this excess fluid volume is the sizing criterion. A simplex pump (one piston) displaces a volume of fluid above mean equal to about 60 percent of total displacement. A duplex pump (two pistons) displaces a fluid volume above mean equal to about 30 percent of total displacement. A triplex pump displaces a fluid volume above mean equal to about 4 percent of total displacement. Piston displacement is calculated by multiplying face area by stroke. The following equation is used to calculate the volume of gas required for the pulsation dampener.

$$V_g = \frac{V_p}{p_g X C_p}$$

where $V_g$ is the volume of gas required, $V_p$ is the piston displacement volume, $p_g$ is the gas precharge percent (as a decimal), $X$ is the level of dampening desired (as a decimal), and $C_p$ is the pump constant. The level of dampening is the residual peak to peak pulsations, such that lower values correspond to more damping.

The precharge percentage of system pressure is typically 70 to 80 percent, as mentioned earlier, so as a decimal this would be expressed as 0.7 to 0.8. The level of dampening desired is expressed as the percentage of residual peak to peak pulsations. So, for 5 percent residual pulsations, this would be expressed in decimal form as 0.05. Conservative values for the pump constant are 1.5 (Simplex), 2 (Duplex), and 7 (Triplex).

Snubbers, on the other hand, are sized primarily based on the pressure range they are going to experience in operation. The pressure spikes must be quantified when selecting a snubber.

**Head.**—The energy added to a liquid by a pump or by gravity is often quantified in terms of head. Head has units of length, usually feet or meters. Pressure can be calculated from head, and vice-versa, through use of the following equations:

$p = 0.434 \times h \times \text{SG}$ where $h$ is head in feet, $p$ is pressure in psi

$p = 0.0981 \times h \times \text{SG}$ Where $h$ is head in meters, $p$ is pressure in bar

$h = 2.31 p \times \text{SG}$ where $h$ is head in feet, $p$ is pressure in psi

$h = 10.197 p \div \text{SG}$ where $h$ is head in meters, $p$ is pressure in bar

For water at 70° F (20° C), head and pressure equivalents can be found in Table 30 and Table 31.

**Table 30. Pressure in Pounds per Square Inch for Different Heads of Water**

| Head, ft | 0 | 1 | 2 | 3 | 4 | 5 | 6 | 7 | 8 | 9 |
|---|---|---|---|---|---|---|---|---|---|---|
| 0 | … | 0.43 | 0.87 | 1.30 | 1.73 | 2.16 | 2.60 | 3.03 | 3.46 | 3.90 |
| 10 | 4.33 | 4.76 | 5.20 | 5.63 | 6.06 | 6.49 | 6.93 | 7.36 | 7.79 | 8.23 |
| 20 | 8.66 | 9.09 | 9.53 | 9.96 | 10.39 | 10.82 | 11.26 | 11.69 | 12.12 | 12.56 |
| 30 | 12.99 | 13.42 | 13.86 | 14.29 | 14.72 | 15.15 | 15.59 | 16.02 | 16.45 | 16.89 |
| 40 | 17.32 | 17.75 | 18.19 | 18.62 | 19.05 | 19.48 | 19.92 | 20.35 | 20.78 | 21.22 |
| 50 | 21.65 | 22.08 | 22.52 | 22.95 | 23.38 | 23.81 | 24.25 | 24.68 | 25.11 | 25.55 |
| 60 | 25.98 | 26.41 | 26.85 | 27.28 | 27.71 | 28.14 | 28.58 | 29.01 | 29.44 | 29.88 |
| 70 | 30.31 | 30.74 | 31.18 | 31.61 | 32.04 | 32.47 | 32.91 | 33.34 | 33.77 | 34.21 |
| 80 | 34.64 | 35.07 | 35.51 | 35.94 | 36.37 | 36.80 | 37.24 | 37.67 | 38.10 | 38.54 |
| 90 | 38.97 | 39.40 | 39.84 | 40.27 | 40.70 | 41.13 | 41.57 | 42.00 | 42.43 | 42.87 |

**Table 31. Heads of Water in Feet Corresponding to Certain Pressures in Pounds per Square Inch**

| Pressure, lb/in² | 0 | 1 | 2 | 3 | 4 | 5 | 6 | 7 | 8 | 9 |
|---|---|---|---|---|---|---|---|---|---|---|
| 0 | … | 2.3 | 4.6 | 6.9 | 9.2 | 11.5 | 13.9 | 16.2 | 18.5 | 20.8 |
| 10 | 23.1 | 25.4 | 27.7 | 30.0 | 32.3 | 34.6 | 36.9 | 39.3 | 41.6 | 43.9 |
| 20 | 46.2 | 48.5 | 50.8 | 53.1 | 55.4 | 57.7 | 60.0 | 62.4 | 64.7 | 67.0 |
| 30 | 69.3 | 71.6 | 73.9 | 76.2 | 78.5 | 80.8 | 83.1 | 85.4 | 87.8 | 90.[illegible] |
| 40 | 92.4 | 94.7 | 97.0 | 99.3 | 101.6 | 103.9 | 106.2 | 108.5 | 110.8 | 113.2 |
| 50 | 115.5 | 117.8 | 120.1 | 122.4 | 124.7 | 127.0 | 129.3 | 131.6 | 133.9 | 136.3 |
| 60 | 138.6 | 140.9 | 143.2 | 145.5 | 147.8 | 150.1 | 152.4 | 154.7 | 157.0 | 159.3 |
| 70 | 161.7 | 164.0 | 166.3 | 168.6 | 170.9 | 173.2 | 175.5 | 177.8 | 180.1 | 182.4 |
| 80 | 184.8 | 187.1 | 189.4 | 191.7 | 194.0 | 196.3 | 198.6 | 200.9 | 203.2 | 205.5 |
| 90 | 207.9 | 210.2 | 212.5 | 214.8 | 217.1 | 219.4 | 221.7 | 224.0 | 226.3 | 228.6 |

Bernoulli's equation is commonly used in the design of hydraulic systems. In terms of head, it takes the following form. This equation assumes steady flow, so additional terms are used in cases of pulsating flow as with reciprocating pumps.

$$h_1 = h_2 + h_L \quad \text{or} \quad z_1 + \frac{v_1^2}{2g} + \frac{p_1}{\gamma} = z_2 + \frac{v_2^2}{2g} + \frac{p_2}{\gamma} + h_L$$

where $h_L$ is minor loss of head due to viscous and turbulent shear stresses and other factors. The terms of this equation are explored further in this section.

*Elevation Head:* This is the vertical distance between the point of concern and the free surface of the fluid. This head is considered positive if the free surface is above the point of concern, and negative if the point of concern is above the free surface. No equation is required. When applied to a suction line, when the pump is above the fluid free surface (negative head) this is often referred to as "suction lift." Elevation head is often represented by the letter $z$ or as $h_z$.

*Pressure Head:* The equivalent head imparted by the pressure applied to the fluid at its free surface. This is calculated using the following equation:

$h_p = \frac{p}{\gamma}$ where $p$ is applied absolute pressure and $\gamma$ is specific weight.

*Static Head:* This term is used to include static effects like elevation and pressure head, while excluding dynamic effects like friction and velocity head.

*Dynamic or Velocity Head:* The head imparted by the specific kinetic energy of the fluid. This value is proportional to the square of the fluid's velocity. Velocity head is calculated using the following equation:

$$h_v = \frac{v^2}{2g}$$

*Acceleration Head:* This head is critical when evaluating the suction side of reciprocating positive displacement pumps and other pumps with pulsating flow. It is also important to consider during startup of a kinetic pump. Acceleration head is subtracted from the *NPSHA*, which is covered in the section *NPSHR and NPSHA* starting on page 2813.

Acceleration head is the head required to accelerate the fluid in the line. It must be considered when evaluating cavitation risk at the inlet and peak loads on the pump components at discharge. It is calculated using the following equation and Table 32.

$$h_a = \frac{LVNC}{gk}$$

where $h_a$ is acceleration head in feet, *L* is the length of the suction line in feet, *V* is the average liquid velocity in the suction line (ft/sec), *N* is the speed of the pump in RPM, *g* is the gravitational constant, and *C* and *k* are given in Table 32.

**Table 32. Acceleration Head Constants**

| Pump Type | Constant *C* | Fluid Type | Constant *k* |
|---|---|---|---|
| Single Acting Simplex | 0.4 | Non-Compressible (degassed) | 1.4 |
| Single Acting Duplex | 0.2 | Most Liquids | 1.5 |
| Double Acting Duplex | 0.115 | Compressible | 2.5 |
| Triplex | 0.066 | | |
| Quintuplex | 0.04 | | |
| Septuplex | 0.028 | | |

*Friction Head:* The head required to overcome friction effects in the line due to surface roughness and restrictions. Restrictions can include fittings, valves, and other devices. Velocity of the fluid is directly proportional to friction head. Friction head is calculated as the sum of all energy losses incurred in the pipe system and fittings. These equations were given in section *Energy Loss in Pipes* on page 2788, and following pages.

*Vapor Pressure Head:* This head term ($h_{vp}$) is used on the suction line to evaluate the potential for cavitation. It is the head equal to the vapor pressure of the liquid being used, and is calculated using the equation for pressure head. The vapor pressure is equal to the saturation pressure at a given temperature. The vapor pressure head of water at 68° F (20° C) is 0.84 ft (2.338 kPa). When the pressure of a liquid drops below its vapor pressure, it boils.

**Discharge from Tanks.—**To calculate the velocity of discharge of liquid from a tank the following equations can be used:

For a non-pressurized tank,

$$v_o = C_V\sqrt{2gh}$$

where $v_o$ is velocity through the discharge orifice, *h* is the height of the free surface of liquid in the tank above the orifice, and $C_V$ is coefficient of velocity for the discharge orifice (see Table 33).

For a tank at gauge pressure $p$,

$$v_o = C_V\sqrt{2g\left[\left(\frac{p}{\gamma}\right)+h\right]}$$

The time required to drain a tank can be calculated using the following equation if the tank has a constant cross-sectional area and is not replenished. Strategies to reduce turbulence as a fluid exiting a tank into a mouthpiece or discharge tube include employing a re-entrant mouthpiece that protrudes into the tank, or using a convergent-divergent mouthpiece.

$$t = \frac{2A_t(\sqrt{z_1}-\sqrt{z_2})}{C_d A_o \sqrt{2g}}$$

where $t$ is time in seconds, $A_t$ is the cross-sectional area of the tank, $A_o$ is the cross-sectional area of the discharge orifice, $z_1$ and $z_2$ are the initial and final liquid heights in the tank, and $C_d$ is the coefficient of discharge for the type of orifice used (see Table 33).

**Table 33. Discharge and Velocity Coefficients**

| Orifice Type | $C_d$ | $C_V$ |
|---|---|---|
| Sharp Edged | 0.62 | 0.98 |
| Round Edged | 0.98 | 0.98 |
| Short Tube (Fluids separates from walls) | 0.61 | 0.61 |
| Sharp Tube (No fluid separation) | 0.82 | 0.82 |
| Sharp Tube with Rounded Entrance | 0.97 | 0.98 |
| Re-entrant Tube, $L < 0.5d$ | 0.54 | 0.99 |
| Re-entrant Tube, $2 \leq L \leq 3$ | 0.72 | 0.72 |
| Smooth, Well-Tapered Nozzle | 0.98 | 0.99 |

**Pumps.**—Pumps are used to move fluid. More generally, pumps add pressure energy to fluid through a variety of means. Pumps fall into two general classifications: positive displacement pumps and kinetic (dynamic) pumps. Some pumps are meant to be submerged completely in the fluid being pumped, while others are not.

*Positive Displacement Pumps:* Positive displacement pumps move a fixed amount of fluid during each discharge stroke or revolution. Flow rate is directly related to the speed of the moving parts of the pump, and for many of these pumps flow pulsates as the cavity or cavities discharge and refill. These pumps are suitable for low to moderate flow rates, for moving highly viscous fluids, and generating very high pressures. Because of the high pressure capability of these pumps, they tend to be sturdily built and the use of a pressure relief valve is recommended in fluid circuits employing these pumps. Positive displacement pumps can be further broken down into reciprocating and rotary types. The most common types of reciprocating pumps are: bladder, diaphragm, peristaltic, and piston or plunger pumps. Commonly encountered rotary types are: gear, lobe, vane, screw, and progressive cavity pumps. Metering pumps are positive displacement pumps that have a precisely controlled discharge volume. This volume is normally adjustable through stroke adjustment.

Bladder pumps deliver pulsing flow, usually by inflating and deflating a bladder arranged axially along a pumping chamber. Diaphragm pumps deliver pulsing flow from a single cavity, but can be arranged in multiples to provide smoother flow. Peristaltic pumps deliver pulsing flow, and have the advantage of completely containing the fluid within a section of tubing that runs through the pump mechanism. The pump uses a series of rollers to progressively pinch the tube and move the fluid ahead of the pinch point.

Piston pumps and plunger pumps deliver pulsing flow. Axial piston pumps are designed with multiple pistons arranged axially to the motor. A tilted swashplate turns and actuates the pistons in sequence. More pistons in the pump produce smoother flow. Radial piston pumps have pistons arranged radially around a central cam that actuates them in sequence. Reciprocating piston and plunger pumps used for fluid transfer are typically either single acting or double acting, which means the piston delivers fluid on one or both directions of movement. These are driven with a crank arrangement or other reciprocating mechanism. A simplex pump has one cylinder, and a duplex pump has two cylinders. Arrangements with more cylinders are possible.

Gear pumps deliver pulsing flow at high pressures and fairly high flow capacities. They rely on close clearances to capture the fluid. This means that the fluid being moved must be free of particulate matter. These pumps perform best with viscous fluids. Lobe pumps deliver pulsing flow. They are similar to gear pumps, but the lobes in mesh do not contact one another. These pumps are often used to move fluid with large particulate inclusions. The lobes are often nonmetallic, and are suitable for handling delicate items in a fluid stream. Vane pumps deliver pulsing flow at moderately high pressures. They are comprised of a rotating set of vanes that capture and release fluid inside a round housing.

Screw pumps deliver smooth flow at moderately high pressures. They are commonly found in single, double, and triple screw configurations. Single screw pumps are sometimes called progressing-cavity pumps. Direct acting pumps are driven by a fluid with a differential pressure, and generally use a piston. They can use air, steam, or a liquid as the driver. These pumps are suitable for high pressure, low flow applications. They have the disadvantage of consuming high volumes of drive fluid.

*Kinetic Pumps:* Kinetic pumps generate momentum in a fluid, which is then transformed into pressure. These pumps do not have a fixed discharge volume, and therefore do not impart a pulsation to the fluid being moved. The most common type of kinetic pump is the centrifugal pump. Other types of kinetic pumps include air lifts, hydraulic rams, turbine pumps, and jet pumps

Centrifugal pumps use a one or more spinning impellers to accelerate a fluid radially from the inlet. The fluid is then slowed by a spiral volute, converting the velocity energy into pressure energy. Centrifugal pumps can be radial, axial, and mixed flow types. The pump specific speed determines which would be the best type for a given application. Some centrifugal pumps are configured to handle fluids with solids. In those cases, a grinder can be integrated into the pump. Centrifugal pumps are used mainly in high flow applications with moderate heads. Multistage versions are suitable for high head applications. The size of a centrifugal pump is designated by the discharge port size, inlet port size, impeller diameter, and number of stages. Therefore, a 1 x 2 x 5 pump would be a single stage pump with a 1 inch discharge, a 2 inch inlet, and a 5 inch impeller. If the pump case is supported at the centerline of the shaft, it is said to be "centerline mounted." This mounting configuration is ideal for high temperature applications because shaft alignment with the driver is maintained as the case expands due to heat. Foot mounted pumps are supported under the case, and frame mounted pumps are supported under the bearing housing.

Air lift pumps use a stream of air bubbles injected into the submerged base of a pipe to entrain the fluid. The fluid rises with the air bubbles. These pumps are used when only light suction is needed.

Hydraulic ram pumps operate cyclically to deliver pulsing flow. A hydraulic ram is typically comprised of a spring or weight loaded normally open valve, and a check valve. Flowing water with kinetic energy creates pressure against the normally open valve until it closes. This elevated pressure sends fluid through the check valve to a higher elevation. The check valve then prevents the fluid from reversing. Once pressure is relieved, the normally open valve opens once more and the cycle begins again.

Turbine pumps are comprised of a series of progressive impellers. Each stage, or impeller, increases the energy of the fluid. Jet pumps are commonly used in home water well systems. These pumps are comprised of a centrifugal pump located above, with a jet assembly down in the well. The centrifugal pump pressurizes the water, which is then passed through a nozzle. The jet created by the nozzle entrains additional water and lifts it up toward the pump inlet. In the suction line there is also a diffuser, which slows the water and converts velocity to pressure energy.

*The Effect of Viscosity on Pumps:* Viscosity affects positive displacement pumps, resulting in higher friction head values for a given flow rate. When centrifugal pumps handle fluids more viscous than water, the pumping efficiency decreases and the flow for a given head also decreases. Centrifugal pump performance is typically rated at values achieved when pumping cool water. For computing centrifugal pump performance, some correction factors for viscous fluids can be found in Fig. 8. To use the chart, enter at capacity, go straight up to the head curve, over horizontally to the viscosity curve, and then straight up to the correction factor curve.The correction factors are applied according to the following equations:

$Q_v = C_Q Q_{water}$ where $Q_v$ is the corrected viscous capacity, $C_Q$ is the capacity correction factor, and $Q_{water}$ is the flow rate achieved while pumping cool water.

$h_v = C_h h_{water}$ where $h_v$ is the corrected viscous head, $C_h$ is the head correction factor, and $h_{water}$ is the head achieved while pumping cool water.

$\eta_v = C_h \eta_{water}$ where $\eta_v$ is the corrected viscous efficiency, $C_h$ is the efficiency correction factor, and $\eta_{water}$ is the efficiency achieved while pumping cool water.

$P_v = \dfrac{P_{water}}{C_\eta}$ where $P_v$ is the corrected viscous pumping power, and $P_{water}$ is the power achieved while pumping cool water.

*Pump Capacity:* Flow rate through a pump is known as capacity. For positive displacement pumps, flow rate is simply related to cavity volume and cycle rate. For centrifugal pumps, a change in impeller diameter or speed results in a capacity change according to the affinity laws discussed later in *Affinity and Similarity* on page 2817.

Centrifugal pumps are intended to be used at or near rated capacity. When operated below this capacity, the pump experiences higher axial thrust loads, elevated temperatures, and there is a potential for gases to become trapped in the pump case. Another effect, called *recirculation*, can also occur in which flow reversals occur at the inlet or discharge tips of the impeller. This can result in pressure pulsations which can damage the pump. In cases where demand is likely to fall below pump capacity, a parallel pump arrangement can be considered, or a bypass installed in the discharge line of the pump to allow the pump to operate at higher flow than demanded.

*Total Suction Head:* This is the sum of all the heads imparted on the fluid in the suction line supplying a pump. It includes static and dynamic heads, unless otherwise specified. The general equation for total suction head is: $h_{t(s)} = h_z + h_p + h_v - h_f$ where $h_z$ is static head, $h_p$ is pressure head, $h_v$ is velocity head, and $h_f$ is friction head. All are taken with respect to the suction line. Acceleration head can also be subtracted when dealing with reciprocating pumps.

*Total Discharge Head:* This is the sum of all the heads imparted on the fluid in the discharge line exiting a pump. The general equation for total discharge head is: $h_{t(d)} = h_z + h_p + h_v + h_f$. Here the terms are taken with respect to the discharge line. Acceleration head can also be added for reciprocating pumps.

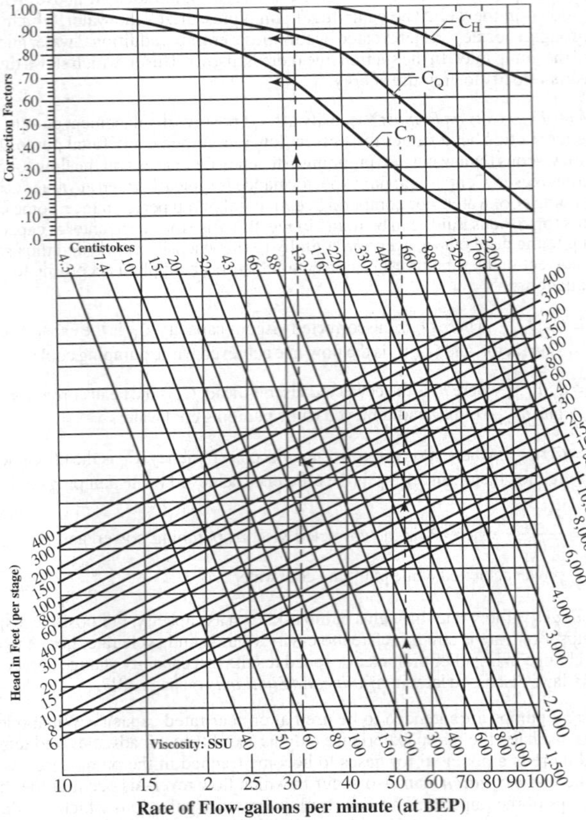


Fig. 8. Centrifugal Pump Viscosity Correction Factors

SSU are Saybolt Universal Seconds. ASTM D2161 covers conversion of kinematic viscosity to SSU at a given temperature. BEP is *Best Efficiency Point*.

*System Head or Total Dynamic Head (TDH):* This is equal to the difference between total discharge head and total suction head: $h_t = h_{t(d)} - h_{t(s)}$

For centrifugal pumps, a change in impeller diameter or speed results in a change in head according to the affinity laws discussed in *Affinity and Similarity* on page 2817.

Using the following equation, head can be calculated from impeller and fluid speeds for radial and mixed-flow pumps where the incoming fluid velocity is nearly zero.

$$h_t = \frac{\eta_i v_i v_f}{g}$$

where $\eta_i$ is impeller efficiency (typically 0.85 – 0.95), $v$ is tangential impeller velocity at a given radius, and $v_f$ is the tangential velocity of the fluid.

*NPSHR and NPSHA: Net Positive Suction Head Required* is the minimum fluid energy required at the pump inlet for normal operation. *NPSHR* is measured when the pump total head is reduced by 3 percent due to cavitation. *NPSHR* depends on flow rate as well as the characteristics of the pump and suction line. This value is normally provided by pump manufacturers for one or more flow rates. To some extent, *NPSHR* depends on the characteristics of the valves used on the pump inlet and discharge ports. If *NPSHR* for one flow rate is known, it can be calculated for another flow rate using the equation:

$$\frac{NPSHR_2}{NPSHR_1} = \left(\frac{Q_2}{Q_1}\right)^2$$

*Net Positive Suction Head Available* is the actual fluid energy at the pump inlet, which can be calculated for the suction line as the difference between total suction head and the vapor pressure head using the following equation. For reciprocating pumps, acceleration head ($h_a$) must be accounted for as well.

$$NPSHA = h_z + h_p + h_v - h_f - h_{vp} - h_a$$

where all terms are taken with respect to the suction line.

*Cavitation:* Cavitation occurs when the pressure of the fluid in the inlet of a pump drops below its vapor pressure for a given temperature. A fluid whose pressure drops below its vapor pressure boils, or vaporizes, degrading pumping performance. The symptoms of cavitation include noise, vibration, and pitting of pump surfaces. In extreme cases, structural damage can occur. If *NPSHA* < *NPSHR*, the fluid will cavitate excessively and the pump will not run as expected. An additional margin is recommended, with recommendations ranging from 10 to 200 percent depending on application. If the liquid has been exposed to gas other than its own vapor, additional margin of *NPSHR* over *NPSHA* is needed to prevent cavitation. If *NPSHR* is not yet known, the designer must aim for a *NPSHA* as high as reasonably possible, and certainly above zero.

Cavitation can be caused by a variety of conditions, including: high pump speeds, high fluid temperatures, low suction head, and fluids with high vapor pressures. When working at high altitudes, the effect of altitude on vapor pressure must be taken into account when calculating suction head. If *NPSHA* is zero or negative, the suction head is by definition below the fluid's vapor pressure. To increase *NPSHA* and prevent cavitation, the inlet pressure can be increased through a variety of methods: pressurizing the fluid source or using a booster pump, raising the fluid source, reducing the flow rate, cooling the fluid, shortening and/or widening the suction line, and reducing suction line losses due to friction or turns. Additional strategies include adding a suction line stabilizer (pulsation dampener) to the suction line, or adding a section of highly flexible hose to the suction line to allow some dampening.

*Hydraulic (Pump) Power:* Hydraulic power, or pump power, is a function of mass flow rate and total dynamic head (added head). This is the energy imparted to the fluid per unit time by the pump. Typical units for power are horsepower or kilowatts. The equations for hydraulic power are given below. These equations assume the density of water is 62.4 lbm/ft$^3$ (1000 kg/m$^3$).

**Hydraulic Kilowatt**

In the equations that follow, $Q$ is in l/s; $\dot{m}$ is in kg/s; and $\dot{V}$ is in m$^3$/s.

$$KW = 9.81\frac{h_t Q(\mathrm{SG})}{1000} = 9.81\frac{h_t \dot{m}}{1000} = 9.81 h_t \dot{V}(\mathrm{SG}) \text{ where } h_t \text{ is in meters}$$

$$KW = \frac{\Delta p Q}{1000} = \frac{\Delta p \dot{m}}{1000(\mathrm{SG})} = \Delta p \dot{V} \text{ where } \Delta p \text{ is in kPa}$$

$$KW = \frac{WQ(\mathrm{SG})}{1000} = \frac{W\dot{m}}{1000} = W\dot{V}(\mathrm{SG}) \text{ where } W \text{ is in J/kg}$$

**Hydraulic Horsepower**

In the equations that follow, $Q$ is in gal/min; $\dot{m}$ is in lbm/s; and $\dot{V}$ is in ft$^3$/s.

$$HP = \frac{h_t Q(\mathrm{SG})}{3956} = \frac{h_t \dot{m} g}{550 g_c} = \frac{h_t \dot{V}(\mathrm{SG})}{8.814} \text{ where } h_t \text{ is in feet}$$

$$HP = \frac{\Delta p Q}{1714} = \frac{\Delta p \dot{m} g}{238.3(\mathrm{SG}) g_c} = \frac{\Delta p \dot{V}}{3.819} \text{ where } \Delta p \text{ is in psi}$$

$$HP = \frac{\Delta p Q}{246800} = \frac{\Delta p \dot{m} g}{34320(\mathrm{SG}) g_c} = \frac{\Delta p \dot{V}}{550} \text{ where } \Delta p \text{ is in psf}$$

$$HP = \frac{WQ(\mathrm{SG})}{3956} = \frac{W\dot{m}}{550} = \frac{W\dot{V}(\mathrm{SG})}{8.814} \text{ where } W \text{ is in ft-lbf/lbm}$$

For centrifugal pumps, a change in impeller diameter or speed results in a change in power according to the affinity laws discussed in *Affinity and Similarity* on page 2817. Required power for centrifugal pumps is directly proportional to the specific gravity of the liquid being pumped. It is influenced also by viscosity.

*Pump Efficiency:* Pump efficiency is the ratio of the actual power delivered by the pump driver (usually a motor) and the power the pump ideally delivers to the fluid. Losses occur in the system due to friction between the fluid and the pump, as well as friction and/or slip within the pump mechanism. Since hydraulic power is a function of mass flow rate, efficiencies of pumps are often given in terms of flow rate, speed, or specific speed. Drive power to a pump is often called "brake horsepower" (bhp). Pump efficiency is defined by the following equation:

$$\eta_p = \frac{P_{hydraulic}}{P_{brake}}$$

where $P_{hydraulic}$ is the power delivered to the fluid, and $P_{brake}$ is the drive power to the pump.

For positive displacement pumps, volumetric efficiency is often used to measure performance. This is the ratio of volumetric flow rate input to displacement per unit time delivered by the pump. Torsional efficiency is also commonly used for positive displacement pumps. This is the ratio of actual to ideal torque required to drive the pump. Overall efficiency of a positive displacement pump is the ratio of actual energy delivered, to the input power to the pump.

Volumetric efficiency is calculated using the following equation:

$$\eta_V = 1 - \left(\Delta p \beta \frac{C}{D}\right) + L_V$$

where $\beta$ is the compressibility factor of the fluid, $C/D$ is the pumping chamber clearance to displacement ratio, and $L_V$ is the fluid loss to slippage back through the pump valves before they can seal. $\Delta p = p_d - p_s$ where $p_d$ is pressure at discharge and $p_s$ is pressure at suction.

Centrifugal pumps are intended for use near their best efficiency. When operated below peak efficiency, a phenomenon called "recirculation" can occur, which causes flow reversal at the inlet or discharge tips of the impeller. This can be very detrimental to performance and damaging to the pump.

*Specific Speed:* Centrifugal pumps are characterized in terms of specific speed. Specific speed is the speed of an ideal pump needed to raise a unit of volume through a unit of head in a unit of time. It assumes the ideal pump is geometrically similar to the pump being evaluated. Specific speed is used to determine the geometry (type of impeller) needed for a given set of input and output requirements. It is calculated using the following equations where $n$ is rotation in RPM:

$$\eta_s = \frac{n\sqrt{Q}}{h_t^{0.75}}$$ where $Q$ is in gal/min. and $h_t$ is total dynamic head in feet.

$$\eta_s = \frac{n\sqrt{\dot{V}}}{h_t^{0.75}}$$ where $\dot{V}$ is in m$^3$/sec. and $h_t$ is in meters.

Low specific speeds produce the largest heads. In US units, specific speeds range from approximately 500 to 1500 RPM for radial vane pumps, 1500–4000 for Francis vane, 4000–9000 for mixed flow impeller designs, and above 9000 for axial flow impellers. When specific speed is low for a given style, efficiency suffers, and a multi-stage pump should be used. Once specific speed has been calculated from a given head, drive speed, and flow rate, the appropriate type of impeller can be selected. In addition, pump efficiency, *NPSHR*, maximum speed, and suction lift can be determined using specific speed. It is good practice to select a centrifugal pump that has the highest specific speed, since that will yield the smallest pump.

*Critical Speed:* Critical speed is related to static deflection of the shaft in the horizontal position. When operated within about 20 percent of critical speed, the pump will vibrate excessively and damage can occur. The wet critical speed must be well above the operating speed of the pump in order to ensure smooth running. Centrifugal pumps have a "dry" critical speed and a "wet" critical speed. Wet critical speed is normally higher (smaller deflections experienced) than dry critical speed. As the pump shaft support components (bearings or bushings) wear, the wet critical speed drops. When a pump is operated away from its best efficiency, higher deflections are likely, causing critical speed to drop. Dry critical speed can be determined using the following equations:

$$n_c = \frac{187.7}{\sqrt{y}}$$

where $y$ is shaft deflection in inches, and $n_c$ is critical speed in RPM. Shaft deflection is typically limited to 0.006 inches in commercial pumps. This leads of a critical speed value of 2423 RPM.

$$n_c = \frac{946}{\sqrt{y}}$$

where $y$ is shaft deflection in mm. A maximum shaft deflection of 0.15mm leads to a critical speed of 2442 RPM.

*Pump Shaft Loading:* Piston pumps carry the entire piston compression force on their drive rod. The axial load on the rod is easily calculated as force = pressure/area. Buckling of the load bearing members and/or piston rod may also need consideration. Many pumps are rated based on the maximum discharge plunger load that can be carried when the suction pressure is zero. If the piston is massive, or pump speeds high, the force to accelerate the piston must be taken into account.

For radially driven pumps, the torque on the motor shaft driving a pump depends upon the type of pump being driven. Pumps which produce pulsating flow must accelerate the fluid during each cycle, and therefore experience variable shaft torque. Centrifugal pumps experience a steady torque on the shaft, which can be calculated either from brake power and speed or from the change in momentum of the fluid being pumped. Torque can be calculated using the following equations:

$$T = \frac{63025P}{n}$$

where $T$ is torque in in-lbf, $P$ is brake horsepower, and $n$ is rotation in RPM.

$$T = \frac{9549P}{n}$$

where $T$ is torque in N-m, $P$ is brake power in kW, and $n$ is rotation in RPM.

$$T = \frac{\dot{m}}{g_c}(v_{t(d)}r - v_{t(i)}r_i)$$

where $T$ is torque, $\dot{m}$ is mass flow rate, $v_{t(d)}$ and $v_{t(i)}$ are tangential velocity at the discharge and inlet, $r$ is the radius of the impeller, and $r_i$ is the radius at the inlet. When the inlet is central to the impeller, the tangential velocity at the inlet is zero.

If a pump is driven using a belt or chain, overhung loads on both the motor and the pump shafts should be calculated. Additional bearing support or larger shafts may be required.

In centrifugal pumps and some rotary positive displacement pumps, axial thrust is generated that must be accommodated by the pump drive shaft. In some multi-stage centrifugal pumps, mounting the impellers in opposing directions can balance the thrust load and reduce axial load transferred to the drive shaft. In other cases, a hydraulic balancing device is used. This type of device is piped to the suction line or reservoir and exerts an opposing force to the thrust generated by the impellers.

For axially driven reciprocating pumps, the axial loading on the drive shaft must be considered, as well as radial loading, moment loading, and torque. Axial and radial/moment loading on the motor shaft can be calculated using simple trigonometric relationships. Torque required to axially drive the motor will depend on how the pump is being driven. In some cases, a lead screw or equivalent mechanism is used and the lead screw equation for torque is applicable.

*Pump Curves:* Pump performance can be shown graphically by plotting various performance parameters as a function of flow rate. These graphs are usually supplied by pump manufacturers and usually include efficiency, power requirement, dynamic head, and *NPSHR*. These values are given as a function of the pump alone, and normally are measured in a way that minimizes the influence of suction/discharge line effects. Pump curves are also normally generated using water as the fluid. Since hydraulic power is a function of the fluid's specific gravity, care must be taken when planning to pump a fluid with a different specific gravity. In those cases, the pump's power rating should be multiplied by the specific gravity of the planned fluid.

The performance of a centrifugal pump is generally based on capacity (gal/min or $m^3$/h), added fluid head (ft or m), input power (bhp of kW), efficiency (%), and speed (rpm). It is common to plot centrifugal pump curves for a range of impeller diameters together for a fixed speed. This allows selection of the diameter when looking at operating points.

*System Head Curves:* System performance curves display the influence of suction and discharge line configurations. On a system head curve, head requirements of the system are plotted versus capacity. The curve represents a total static head plus friction losses. When system head is variable, as with changing reservoir levels or when there is switching between lines through valving, multiple system head curves will be represented on the graph.

*Operating Point:* The operating point is the intersection of the pump curve and the system curve on a graph that sets head against flow rate. From this graph, the flow rate and head of the pump/system combination can be easily determined. An example of the operating point on a graph can be seen in Fig. 9. A pump/system combination will have multiple viable operating points if there are multiple pump curves given for different speeds and/or geometries. The designer should choose the operating point with the highest pump efficiency. Another case where multiple operating points are seen is in systems where the system head varies and multiple system head curves are present. The designer should ensure that the pump can meet the system requirements at all operating points. When dealing with centrifugal pumps, the pump curve chosen for the operating point will in turn determine the speed and/or impeller. It is good practice to select a centrifugal pump with the largest impeller diameter that will meet the application needs while operating at or near its maximum rated speed.

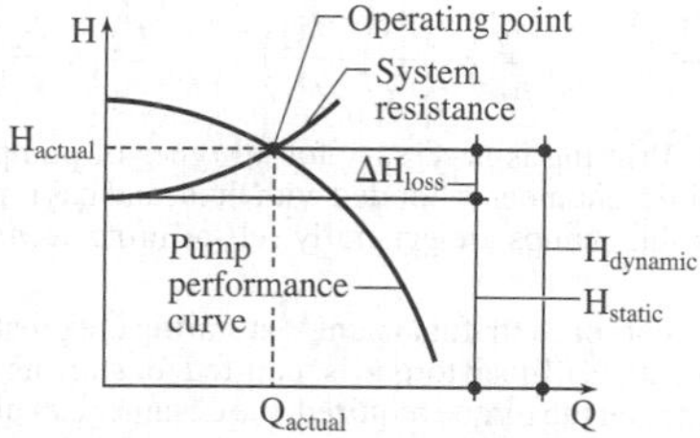


Fig. 9. Operating Point of a Pump

The operating point of a pump in a system can be changed by throttling the discharge line. The suction line should never be throttled because it could lead to cavitation. When a pump is throttled, the operating point moves along the pump curve as the flow rate is reduced. Operating point can also be changed through the use of a variable-speed drive. Changing the viscosity of the fluid being pumped also moves the operating point in the direction of less flow for a given head.

*Combining Pumps in Series or Parallel:* When one pump discharges into the suction line of another pump, this is a series system. Pumps in series are capable of producing higher discharge heads than single pumps. A slight increase in flow rate will also be seen when pumps are combined in series. To generate a pump curve for pumps in series, the head of each pump is added at each flow rate to create a new curve. This curve is then graphed together with the system curve to find a new operating point. Multistage pumps contain multiple impellers in one pump body to achieve a similar effect as a series of pumps.

A system of pumps in parallel experiences an increase primarily in flow capacity, with only a slight increase in head. Parallel pump systems have the advantage of increased reliability, and the ability to efficiently adjust to variations in demand. When demand varies, flow rate moves away from the operating point, resulting in decreased efficiency.

When two or more parallel pumps are used, they can be engaged and disengaged to keep the system flow rate near demand.

*Affinity and Similarity:* Comparing centrifugal pumps can be done if one assumes the efficiencies of the two pumps are the same. This becomes less true as the difference in size between pumps grows. Note that as the specific speed increases, the accuracy of the affinity equations decreases. To estimate the efficiency of a very different size pump, the following equation can be used.

$$\left(\frac{D_L}{D_S}\right)^n = \frac{1-\eta_S}{1-\eta_L}$$

where subscript $L$ denotes the larger pump, and $S$ denotes the smaller pump. $D$ is impeller diameter and $n$ is speed.

The affinity equations (assuming equal efficiencies) which allow comparison between pumps are as follows:

- When the impeller diameter is constant while the speed changes:

$$\frac{n_2}{n_1} = \frac{Q_2}{Q_1} \qquad \left(\frac{n_2}{n_1}\right)^2 = \frac{h_2}{h_1} = \left(\frac{Q_2}{Q_1}\right)^2 \qquad \left(\frac{n_2}{n_1}\right)^3 = \frac{P_2}{P_1} = \left(\frac{Q_2}{Q_1}\right)^3$$

- When the speed is constant and the impeller size changes:

$$\frac{D_2}{D_1} = \frac{Q_2}{Q_1} \qquad \left(\frac{D_2}{D_1}\right)^2 = \frac{h_2}{h_1} \qquad \left(\frac{D_2}{D_1}\right)^3 = \frac{P_2}{P_1}$$

- When flow is assumed to be turbulent and the percentage of wide open flow is assumed to be the same, similarity can be used as follows:

$$\frac{Q_2}{Q_1} = \frac{n_2}{n_1}\left(\frac{D_2}{D_1}\right)^3 \qquad \frac{h_2}{h_1} = \left(\frac{n_2}{n_1}\right)^2\left(\frac{D_2}{D_1}\right)^2 \qquad \frac{P_2}{P_1} = \frac{\rho_2}{\rho_1}\left(\frac{n_2}{n_1}\right)^3\left(\frac{D_2}{D_1}\right)^5$$

*Startup Conditions:* Priming is necessary for all types of pumps to begin operation. During priming, the inlet chamber is flooded with fluid, and gas is purged from the pump mechanism. Submersible pumps are generally self-priming, as are some types of non-submersible pumps.

The torque requirements of centrifugal pumps at startup are greatly different from running torque requirements. Full-load torque is required for starting against full discharge pressure. To reduce the startup torque required, the designer can pipe the discharge back to the suction line, and close it off from any head inherent in the running discharge line. This equalizes discharge and suction pressure and eliminates the load on the pump during startup. Using valves, this loop can be closed off and the pump discharge connected to the discharge pipe system once the pump gets up to running speed.

*Piping Considerations:* To ensure peak performance of hydraulic systems, some points to consider are:

- Make piping runs as short and direct as possible.
- Design for minimum fluid velocities and accelerations.
- Use long radius bends rather than elbows to redirect piping where possible.
- Design pipe systems to preclude the collection of vapor bubbles, except where venting is provided.
- Include pulsation dampeners and suction stabilizers where acceleration head is high or the fluid needs degassing. A suction stabilizer is a device placed on the suction line of a pump that both absorbs pulsations and separates free gases from the liquid being pumped.
- Always ensure that *NPSHA* >> *NPSHR*.
- Include strainers only when absolutely necessary, and ensure regular maintenance is conducted.
- Provide a relief valve that is adequately sized.
- Pipe a bypass line around the pump with valves so that the pump may be started with low load. This must be done slowly to avoid shock to the system.
- In systems with high static discharge heads, install a check valve at the pump discharge to prevent static pressure from loading the pump during startup.

## Pneumatics

Pneumatic systems use compressed and flowing gas, usually air, to perform work. Many pneumatic components, such as valves, actuators, and compressors share similarities with components used with liquids. In many cases, pneumatic components and calculations can be found in earlier sections.

**Gas Consumption Calculation and Measurement.—**It is common practice to exhaust air used in pneumatic circuits to atmosphere through silencers, or to an exhaust manifold and/or filter. Since that air is lost, and compression of air costs money, air consumption should be evaluated and limited as much as possible when designing a pneumatic circuit. Silencers are essential to limit noise produced by exhausting air.

In US units, the volumetric flow rate of gases is normally expressed in cubic feet per minute (cfm). Velocity in US units is normally expressed in feet per minute. SI units for gas volumetric flow are cubic meters per second and meters per second for velocity. Liters per second is used for volumetric flows in low ranges.

*Gas Consumption:* Compressors and devices using compressed air are often rated in terms of free air delivery (f.a.d.), which is the quantity of air used per unit time assuming the air is at standard atmospheric pressure and temperature. These values are 14.7 psia (0.101 MPa) and 60° F (15° C). When conditions differ from standard, the following equation can be used:

$$Q_a = \frac{Q_s p_s T_a}{(p_{atm} + p_a)T_s}$$

where $Q_a$ is actual volumetric flow rate, $Q_s$ is flow rate at standard conditions, $p_s$ is standard absolute atmospheric pressure, $p_{atm}$ is actual absolute atmospheric pressure, $p_a$ is actual gauge pressure, $T_a$ and $T_s$ are actual and standard absolute temperature (°R or K).

This equation is often used in another form to calculate the flow of free air required to supply a given compressed air flow. When the free air is at standard temperature, this takes the form of the following equation:

$$Q_s = \frac{Q_a(p_{atm} + p_a)}{p_s}$$

where $Q_s$ is the free air flow at standard temperature, $Q_a$ is the compressed air flow, $p_a$ is actual gage pressure, and $p_s$ is the absolute standard pressure. The ratio of absolute pressure to absolute standard pressure is known as the *compression ratio.* In US units, the free flow is normally given in units of SCFM and compressed air flow is in units of CFM.

**Fans and Blowers.—**Fans and blowers add energy to air. When static pressure is very low, fans are typically used. Blowers are appropriate when static pressure is higher.

*Types of Fans and Blowers:* Axial propeller fans have a number of blades arranged around a spinning shaft. They move air axially through themselves with very low pressure and velocity increase. Axial fans are normally used to move large quantities of air against static pressures below 0.4 psig (3kPa). A diffuser can be added after a fan to increase static pressure output. Axial propeller fans are the least expensive of the fan types, but are not as efficient as centrifugal fans. Vane-axial blowers are similar to propeller fans, but have highly optimized casing geometry that creates much higher pressures than propeller fans.

Centrifugal fans and blowers use a rotating impeller to accelerate air radially outward from a central intake. Significant velocity energy is added, which is then converted to pressure energy when the air is slowed using a volute. The inlet and volute geometry is usually organized so that air enters and exits the centrifugal device axially. Scroll casings are arranged such that the air exiting the device is radial. Centrifugal fans are appropriate up to a static pressure of approximately 2 psi (15 kPa). Forward-curved centrifugal fans (squirrel cage fans) are the most common type, but with an efficiency of approximately 75 percent they are the least efficient of all the centrifugal designs. Backward-curved centrifugal fans are less likely to overload than forward-curved fans, and their increased efficiency over forward-curved centrifugals makes up for their increased cost. Radial blade fans are commonly found in industrial cooling applications where a large volume of air is required at moderate pressure. Centrifugal blowers look more like centrifugal pumps than fans. The impeller is typically gear-driven. The two most common types of centrifugal blowers are pressure blowers and volume blowers. Pressure blowers are designed to draw or push air at high pressures, rated in static pressure water gauge (SPWG). Volume blowers are designed to draw or push larger volumes of air, rated in cubic feet of air per minute (CFM), at lower static pressures than pressure blowers.

Cross-flow fans and blowers have air enter and exit the impeller tangentially. The pressure produced by these impellers is low. Mixed-flow fans and blowers use impellers that impart both axial and radial momentum to a gas. These impellers can be housed in either axial or scroll-type casings to output either axial or radial flow. Positive displacement blowers are essentially positive displacement pumps used for moving air. Positive displacement pumps are described earlier in this text.

*Fan and Blower Calculations:* When selecting a fan or blower, one must consider the cleanliness and type of gas being moved, the pressure differential or velocity needed, the volumetric flow rate required, the energy loss in the ductwork or system, the noise produced, and the cost.

Fan capacity is a measure of the maximum volumetric flow rate of the fan or blower. Total pressure of a fan or blower is the difference between the inlet and outlet pressures, and it is also the sum of static pressure and velocity pressure. Static and velocity pressures are often measured in inches of water (in. w.g.). Fan static, velocity, and total pressures at standard conditions can be calculated using the following equations:

$$p_{s(\text{in. w. g.})} = \frac{p_{psig}}{0.0361}$$

$$p_{v(\text{in. w.g.})} = \left(\frac{v_{fpm}}{4005}\right)^2$$

$$p_t = p_s + p_v = \frac{p}{\gamma} + \frac{\rho v^2}{2}$$

where $p_s$ is static pressure, and the specific weight of water is assumed to be 0.0361 lbf/in$^3$; $p_v$ is velocity pressure; $v_{fpm}$ is velocity in fpm; and $p_t$ is total pressure.

When a fan is connected to ductwork, the pressure drop through the duct and fittings should be taken into account using the following equations, by using the Moody diagram given in Fig. 1, or by using friction loss graphs like those published by ASHRAE. The Darcy-Weisbach equation (see *Darcy-Weisbach Equation and the Moody Diagram* on page 2788) should be used for "nonstandard" duct type such as flex duct.

$$p_{f(\text{in. w.g.})} = \frac{(3.9\times10^{-9})v_{fpm}^{2.43}L_{ft}}{Q_{CFM}^{0.61}}$$

where $p_f$ is friction pressure of air moving through ducts. This assumes clean, galvanized, circular ducts with a specific roughness of 0.0005 ft.

$$p_{f(\text{in. w.g.})} = K\left(\frac{v_{fpm}}{4005}\right)^2$$

where $p_f$ is friction pressure of air moving through fittings. The loss coefficient $K$ can be found in Table 34.

**Table 34. Loss Coefficients, Air Duct Fittings**

| Fitting Type | Geometry | | $K$ (upstream) | $K$ (downstream) |
|---|---|---|---|---|
| Sudden Expansion | $\frac{A_{upstream}}{A_{downstream}} =$ | 0.2 | 0.64 | |
| | | 0.4 | 0.36 | |
| | | 0.6 | 0.16 | |
| | | 0.8 | 0.04 | |
| Sudden Contraction | $\frac{A_{downstream}}{A_{upstream}} =$ | 0.2 | | 0.32 |
| | | 0.4 | | 0.25 |
| | | 0.6 | | 0.16 |
| | | 0.8 | | 0.06 |
| Gradual Contraction | Included Angle | 40° | 0.012 | |
| | | 80° | 0.032 | |
| | | 120° | 0.07 | |
| 90° Elbow | $r/d =$ | 1.25 | 0.55 | |
| | | 1.5 | 0.39 | |
| | | 1.75 | 0.32 | |
| | | 2 | 0.27 | |
| | | 2.25 | 0.26 | |
| | | 2.5 | 0.22 | |
| | | 2.75 | 0.26 | |
| 45° Elbow | $r/d =$ | 1.25 | 0.275 | |
| | | 1.5 | 0.195 | |
| | | 1.75 | 0.16 | |
| | | 2 | 0.135 | |
| | | 2.25 | 0.13 | |
| | | 2.5 | 0.11 | |
| | | 2.75 | 0.13 | |
| Straight Miter | Two Piece | | 1.2 | |
| | Three Piece | | 0.34 | |

Compressibility factor is the ratio of total pressure that would be developed with an incompressible fluid to the total pressure developed with a compressible gas. When fan total pressure is less than 12 in. w.g., the compressibility factor can be assumed to be equal to 1 (incompressible flow). Power output is the product of capacity, total pressure, and compressibility factor. Total efficiency is the ratio of power output to power input. Static efficiency is the product of total efficiency and the ratio of static pressure to total pressure. These values can be calculated using the following equations. Fan efficiency grades have been established to classify fan performance.

$$P_o = K_p Q p_t$$

where $P_o$ is output power imparted to the gas, $K_p$ is compressibility factor, and $p_t$ is fan total pressure.

$$\eta_t = \frac{P_o}{P_B}$$

where $P_B$ is power supplied to the driver.

$$\eta_t = \frac{Q_{CFM} p_{t(\text{in. w.g.})}}{6356 P_B} \qquad \eta_s = \frac{Q_{CFM} p_{s(\text{in. w.g.})}}{6356 P_B}$$

The fan laws of similarity and affinity rely on the fact that two geometrically similar impellers of different sizes are homologous. These laws can be used to compare similar fans, and are expressed in the following equations. It is essential to make sure that the fans under consideration have the same efficiency. When one fan is compared at different speeds:

$$\frac{Q_2}{Q_1} = \frac{n_2}{n_1} \qquad \frac{p_2}{p_1} = \left(\frac{n_2}{n_1}\right)^2 \qquad \frac{P_{o2}}{P_{o1}} = \frac{P_{B2}}{P_{B1}} = \left(\frac{n_2}{n_1}\right)^3$$

When two similar fans are compared:

$$\frac{Q_2}{Q_1} = \left(\frac{n_2}{n_1}\right)\left(\frac{D_2}{D_1}\right)^3 \qquad \frac{P_2}{P_1} = \left(\frac{p_2 Q_2}{p_1 Q_1}\right)$$

Sizing of fans and blowers is similar in process to that of pumps, where graphs are commonly used. The performance graph is usually a constant-speed plot with capacity on the X-axis and a variety of performance factors on the Y-axis. As part of the selection process, a graph of the system is also generated, with capacity as the X-axis and head (energy loss in the system) on the Y-axis. As with pumps, the operating point of a fan is found at the intersection of the performance graph and the system graph. When multiple system curves are applicable, as with variable duct geometry or variable static pressure, there will be multiple operating points. When multiple performance curves are available, as with adjustable impeller vane angle, damper control, or variable speed, there will be multiple operating points. A slightly undersized fan is preferable to an oversized fan because of lower cost and increased stability.

Operating a fan or blower at much lower than rated capacity can result in pulsations and instability due to a variety of factors. This should be avoided. To ensure stable fan and blower operation, the engineer should select an operating point where the slope of the pressure/capacity performance curve has opposite sign as the system curve. This is usually the case, except at or above design capacity. Pulsation can be avoided by selecting an operating point to the right of the surge point. If this is not possible, speed control, impeller geometry control, inlet dampers, exhaust dampers, and bleed lines may be employed. Changing a fan's speed cannot be used to stabilize a fan operating in an unstable condition. According to affinity, if an operating point for a fan is in an unstable region to the left of the peak, it will remain in an unstable region at any speed.

Multiple fans may be called for in some applications to avoid pressure buildup in a space or to fit a space constraint. When demand or system conditions are highly variable, multiple fans may serve the need better than a single fan. Because operating point is based on capacity, if capacity varies the operating point will move along the performance curve and may depart from reasonable efficiency for the fan or blower. In these cases, one fan is sized for one operating point, and additional fans are called into service as needed to provide additional capacity. Fans or blowers in parallel and series follow the same additive rules as pumps. Instability can occur in parallel fan systems when the combined performance curve intersects the system curve at a point where the individual operating points are widely separated. In that condition, the fans may suddenly exchange loading and shock the mechanism and/or ductwork. To avoid this, fans in parallel must be carefully matched to the system.

**Compressors.—**When significant air pressure must be generated, in excess of 10 psig *(68.9 kPa)*, a compressor is employed. The most common pressures range from 60 to 110 psig (413.7–758.4 kPa). There are many standards covering compressors, including those by ANSI, ASME, API, SAE, and CAGI.

*Compression of Air:* The formula for the relationship between pressure, temperature, and volume of air just given indicates that when the pressure remains constant the volume is directly proportional to the absolute temperature. If the temperature remains constant, the volume is inversely proportional to the absolute pressure. Theoretically, air (as well as other gases) can be expanded or compressed according to different laws.

*Adiabatic Expansion or Compression* takes place when the air is expanded or compressed without transmission of heat to or from it, as, for example, if the air could be expanded or compressed in a cylinder of an absolutely nonconducting material.

Let: $P_1$ = initial absolute pressure in pounds per square foot
$V_1$ = initial volume in cubic feet
$T_1$ = initial absolute temperature in degrees R
$P_2$ = absolute pressure in pounds per square foot, after compression
$V_2$ = volume in cubic feet, after compression
$T_2$ = absolute temperature in degrees R, after compression

Then:

$$\frac{V_2}{V_1} = \left(\frac{P_1}{P_2}\right)^{0.71} \qquad \frac{P_2}{P_1} = \left(\frac{V_1}{V_2}\right)^{1.41} \qquad \frac{T_2}{T_1} = \left(\frac{V_1}{V_2}\right)^{0.41}$$

$$\frac{V_2}{V_1} = \left(\frac{T_1}{T_2}\right)^{2.46} \qquad \frac{P_2}{P_1} = \left(\frac{T_2}{T_1}\right)^{3.46} \qquad \frac{T_2}{T_1} = \left(\frac{P_2}{P_1}\right)^{0.29}$$

These formulas are also applicable if all pressures are in pounds per square inch; if all volumes are in cubic inches; or if any other consistent set of units is used for pressure or volume.

*Example 1:* A volume of 165 cubic feet of air, at a pressure of 15 pounds per square inch, is compressed adiabatically to a pressure of 80 pounds per square inch. What will be the volume at this pressure?

$$V_2 = V_1\left(\frac{P_1}{P_2}\right)^{0.71} = 165\left(\frac{15}{80}\right)^{0.71} = 50 \text{ cubic feet, approx.}$$

*Isothermal Expansion or Compression* takes place when a gas is expanded or compressed with an addition or transmission of sufficient heat to maintain a constant temperature.

Let: $P_1$ = initial absolute pressure in pounds per square foot
$V_1$ = initial volume in cubic feet
$P_2$ = absolute pressure in pounds per square foot, after compression
$V_2$ = volume in cubic feet, after compression
$R$ = 53.3
$T$ = temperature in degrees Rankine maintained during isothermal expansion or contraction

Then:

$$P_1 \times V_1 = P_2 \times V_2 = RT$$

*Example 2:* The same volume of air as in Example 1 is compressed isothermally from 15 to 80 pounds per square inch. What will be the volume after compression?

$$V_2 = \frac{P_1 \times V_1}{P_2} = \frac{15 \times 165}{80} = 31 \text{ cubic feet}$$

*Work Required in Compression of Air:* The total work required for compression and expulsion of air, adiabatically compressed, is:

$$\text{Total work in foot-pounds} = 2.44 P_1 V_1\left[\left(\frac{P_2}{P_1}\right)^{0.29} - 1\right]$$

$P_1$ = initial absolute pressure in pounds per square foot; $P_2$ = absolute pressure in pounds per square foot, after compression; and, $V_1$ = initial volume in cubic feet.

The total work required for isothermal compression is:

$$\text{Total work in foot-pounds} = P_1 V_1 \log_e \frac{V_1}{V_2}$$

in which $P_1$, $P_2$, and $V_1$ denote the same quantities as in the previous equation, and $V_2$ = volume of air in cubic feet, after compression.

The work required to compress air isothermally, that is, when the heat of compression is removed as rapidly as produced, is considerably less than the work required for compressing air adiabatically, or when all the heat is retained. In practice, neither of these two theoretical extremes is obtainable, but the power required for air compression is about the median between the powers that would be required for each. The accompanying table gives the average number of foot-pounds of work required to compress air.

*Horsepower Required to Compress Air:* In the accompanying tables is given the horsepower required to compress one cubic foot of free air per minute (isothermally and adiabatically) from atmospheric pressure (14.7 pounds per square inch) to various gage pressures, for one-, two-, and three-stage compression. The formula for calculating the horsepower required to compress, adiabatically, a given volume of free air to a given pressure is:

$$HP = \frac{144NPVn}{33{,}000(n-1)}\left[\left(\frac{P_2}{P}\right)^{\frac{n-1}{Nn}} - 1\right]$$

where $N$ = number of stages in which compression is accomplished

$P$ = atmospheric pressure in pounds per square inch

$P_2$ = absolute terminal pressure in pounds per square inch

$V$ = volume of air, in cubic feet, compressed per minute, at atmospheric pressure

$n$ = exponent of the compression curve = 1.41 for adiabatic compression

For different methods of compression and for one cubic foot of air per minute, this formula may be simplified as follows:

For one-stage compression: $HP = 0.015P(R^{0.29} - 1)$

For two-stage compression: $HP = 0.030\,P(R^{0.145} - 1)$

For three-stage compression: $HP = 0.045\,P(R^{0.0975} - 1)$

For four-stage compression: $HP = 0.060\,P(R^{0.0725} - 1)$

In these latter formulas $R = \frac{P_2}{P}$ = number of atmospheres to be compressed

The formula for calculating the horsepower required to compress isothermally a given volume of free air to a given pressure is:

$$HP = \frac{144PV}{33000}\left(\log_e \frac{P_2}{P}\right)$$

Natural logarithms are obtained by multiplying common logarithms by 2.30259 or by using a handheld calculator.

The work and power required for air compression can be found in Table 35 to Table 38.

Stress and heat are what limits the maximum compression of any single stage of a compressor. The volumetric efficiency drops as pressure rise increases, further limiting the maximum compression of a single stage. Multi-stage compressors are used when a single stage cannot develop enough pressure. Multi-stage axial compressors use a series of axial fans to deliver very large flow rates. Perfect intercooling is impractical, so it is important to note that for every 10° F temperature increase, the horsepower requirement increases nearly 1%.

**Table 35. Foot-pounds of Work Required in Compression of Air
Initial Pressure = 1 atmosphere = 14.7 Pounds per Square Inch**

| Gage Pressure, lb/in² | Isothermal Compression | Adiabatic Compression | Actual Compression | Gage Pressure, lb/in² | Isothermal Compression | Adiabatic Compression | Actual Compression |
|---|---|---|---|---|---|---|---|
| | Foot-pounds Required per Cubic Foot of Air at Initial Pressure | | | | Foot-pounds Required per Cubic Foot of Air at Initial Pressure | | |
| 5 | 619.6 | 649.5 | 637.5 | 55 | 3393.7 | 4188.9 | 3870.8 |
| 10 | 1098.2 | 1192.0 | 1154.6 | 60 | 3440.4 | 4422.8 | 4029.8 |
| 15 | 1488.3 | 1661.2 | 1592.0 | 65 | 3577.6 | 4645.4 | 4218.2 |
| 20 | 1817.7 | 2074.0 | 1971.4 | 70 | 3706.3 | 4859.6 | 4398.1 |
| 25 | 2102.6 | 2451.6 | 2312.0 | 75 | 3828.0 | 5063.9 | 4569.5 |
| 30 | 2353.6 | 2794.0 | 2617.8 | 80 | 3942.9 | 5259.7 | 4732.9 |
| 35 | 2578.0 | 3111.0 | 2897.8 | 85 | 4051.5 | 5450.0 | 4890.1 |
| 40 | 2780.8 | 3405.5 | 3155.6 | 90 | 4155.7 | 5633.1 | 5042.1 |
| 45 | 2966.0 | 3681.7 | 3395.4 | 95 | 4254.3 | 5819.3 | 5187.3 |
| 50 | 3136.2 | 3942.3 | 3619.8 | 100 | 4348.1 | 5981.2 | 5327.9 |

**Table 36. Horsepower Required to Compress Air, Single-Stage Compression**

Horsepower required to compress one cubic foot of free air per minute (isothermally and adiabatically) from atmospheric pressure (14.7 pounds per square inch) to various gage pressures.
Single-Stage Compression, initial temperature of air, 60° F, jacket cooling not considered.

| Gage Pressure, Pounds | Absolute Pressure, Pounds | Number of Atmospheres | Isothermal Compression: Mean Effective Pressure[a] | Isothermal Compression: Horsepower | Adiabatic Compression: Mean Effective Pressure,[a] Theoretical | Adiabatic Compression: Mean Eff. Pressure plus 15% Friction | Adiabatic Compression: Horsepower, Theoretical | Adiabatic Compression: Horsepower plus 15% Friction |
|---|---|---|---|---|---|---|---|---|
| 5 | 19.7 | 1.34 | 4.13 | 0.018 | 4.46 | 5.12 | 0.019 | 0.022 |
| 10 | 24.7 | 1.68 | 7.57 | 0.033 | 8.21 | 9.44 | 0.036 | 0.04 |
| 15 | 29.7 | 2.02 | 11.02 | 0.048 | 11.46 | 13.17 | 0.050 | 0.057 |
| 20 | 34.7 | 2.36 | 12.62 | 0.055 | 14.30 | 16.44 | 0.062 | 0.071 |
| 25 | 39.7 | 2.70 | 14.68 | 0.064 | 16.94 | 19.47 | 0.074 | 0.085 |
| 30 | 44.7 | 3.04 | 16.30 | 0.071 | 19.32 | 22.21 | 0.084 | 0.095 |
| 35 | 49.7 | 3.38 | 17.90 | 0.078 | 21.50 | 24.72 | 0.094 | 0.108 |
| 40 | 54.7 | 3.72 | 19.28 | 0.084 | 25.53 | 27.05 | 0.103 | 0.118 |
| 45 | 59.7 | 4.06 | 20.65 | 0.090 | 25.40 | 29.21 | 0.111 | 0.127 |
| 50 | 64.7 | 4.40 | 21.80 | 0.095 | 27.23 | 31.31 | 0.119 | 0.136 |
| 55 | 69.7 | 4.74 | 22.95 | 0.100 | 28.90 | 33.23 | 0.126 | 0.145 |
| 60 | 74.7 | 5.08 | 23.90 | 0.104 | 30.53 | 35.10 | 0.133 | 0.153 |
| 65 | 79.7 | 5.42 | 24.80 | 0.108 | 32.10 | 36.91 | 0.140 | 0.151 |
| 70 | 84.7 | 5.76 | 25.70 | 0.112 | 33.57 | 38.59 | 0.146 | 0.158 |
| 75 | 89.7 | 6.10 | 26.62 | 0.116 | 35.00 | 40.25 | 0.153 | 0.175 |
| 80 | 94.7 | 6.44 | 27.52 | 0.120 | 36.36 | 41.80 | 0.159 | 0.182 |
| 85 | 99.7 | 6.78 | 28.21 | 0.123 | 37.63 | 43.27 | 0.164 | 0.189 |
| 90 | 104.7 | 7.12 | 28.93 | 0.126 | 38.89 | 44.71 | 0.169 | 0.195 |
| 95 | 109.7 | 7.46 | 29.60 | 0.129 | 40.11 | 46.12 | 0.175 | 0.201 |
| 100 | 114.7 | 7.80 | 30.30 | 0.132 | 41.28 | 47.46 | 0.180 | 0.207 |
| 110 | 124.7 | 8.48 | 31.42 | 0.137 | 43.56 | 50.09 | 0.190 | 0.218 |
| 120 | 134.7 | 9.16 | 32.60 | 0.142 | 45.69 | 52.53 | 0.199 | 0.229 |
| 130 | 144.7 | 9.84 | 33.75 | 0.147 | 47.72 | 54.87 | 0.208 | 0.239 |
| 140 | 154.7 | 10.52 | 34.67 | 0.151 | 49.64 | 57.08 | 0.216 | 0.249 |
| 150 | 164.7 | 11.20 | 35.59 | 0.155 | 51.47 | 59.18 | 0.224 | 0.258 |
| 160 | 174.7 | 11.88 | 36.30 | 0.158 | 53.70 | 61.80 | 0.234 | 0.269 |
| 170 | 184.7 | 12.56 | 37.20 | 0.162 | 55.60 | 64.00 | 0.242 | 0.278 |
| 180 | 194.7 | 13.24 | 38.10 | 0.166 | 57.20 | 65.80 | 0.249 | 0.286 |
| 190 | 204.7 | 13.92 | 38.80 | 0.169 | 58.80 | 67.70 | 0.256 | 0.294 |
| 200 | 214.7 | 14.60 | 39.50 | 0.172 | 60.40 | 69.50 | 0.263 | 0.303 |

[a] Mean Effective Pressure (MEP) is defined as that single pressure rise, above atmospheric, which would require the same horsepower as the actual varying pressures during compression.

## Table 37. Horsepower Required to Compress Air, Two-Stage Compression

Horsepower required to compress one cubic foot of free air per minute (isothermally and adiabatically) from atmospheric pressure (14.7 pounds per square inch) to various gage pressures.

Two-Stage Compression, initial temperature of air, 60° F, jacket cooling not considered.

| Gage Pressure, Pounds | Absolute Pressure, Pounds | Number of Atmospheres | Correct Ratio of Cylinder Volumes | Intercooler Gage Pressure | Isothermal Compression: Mean Effective Pressure[a] | Isothermal Compression: Horsepower | Adiabatic Compression: Mean Eff. Pressure,[a] Theoretical | Adiabatic Compression: Mean Eff. Pressure plus 15 per cent Friction | Adiabatic Compression: Horsepower, Theoretical | Adiabatic Compression: HP plus 15 per cent Friction | Percentage of Saving over One-stage Compression |
|---|---|---|---|---|---|---|---|---|---|---|---|
| 50 | 64.7 | 4.40 | 2.10 | 16.2 | 21.80 | 0.095 | 24.30 | 27.90 | 0.106 | 0.123 | 10.9 |
| 60 | 74.7 | 5.08 | 2.25 | 18.4 | 23.90 | 0.104 | 27.20 | 31.30 | 0.118 | 0.136 | 11.3 |
| 70 | 84.7 | 5.76 | 2.40 | 20.6 | 25.70 | 0.112 | 29.31 | 33.71 | 0.128 | 0.147 | 12.3 |
| 80 | 94.7 | 6.44 | 2.54 | 22.7 | 27.52 | 0.120 | 31.44 | 36.15 | 0.137 | 0.158 | 13.8 |
| 90 | 104.7 | 7.12 | 2.67 | 24.5 | 28.93 | 0.126 | 33.37 | 38.36 | 0.145 | 0.167 | 14.2 |
| 100 | 114.7 | 7.80 | 2.79 | 26.3 | 30.30 | 0.132 | 35.20 | 40.48 | 0.153 | 0.176 | 15.0 |
| 110 | 124.7 | 8.48 | 2.91 | 28.1 | 31.42 | 0.137 | 36.82 | 42.34 | 0.161 | 0.185 | 15.2 |
| 120 | 134.7 | 9.16 | 3.03 | 29.8 | 32.60 | 0.142 | 38.44 | 44.20 | 0.168 | 0.193 | 15.6 |
| 130 | 144.7 | 9.84 | 3.14 | 31.5 | 33.75 | 0.147 | 39.86 | 45.83 | 0.174 | 0.200 | 16.3 |
| 140 | 154.7 | 10.52 | 3.24 | 32.9 | 34.67 | 0.151 | 41.28 | 47.47 | 0.180 | 0.207 | 16.7 |
| 150 | 164.7 | 11.20 | 3.35 | 34.5 | 35.59 | 0.155 | 42.60 | 48.99 | 0.186 | 0.214 | 16.9 |
| 160 | 174.7 | 11.88 | 3.45 | 36.1 | 36.30 | 0.158 | 43.82 | 50.39 | 0.191 | 0.219 | 18.4 |
| 170 | 184.7 | 12.56 | 3.54 | 37.3 | 37.20 | 0.162 | 44.93 | 51.66 | 0.196 | 0.225 | 19.0 |
| 180 | 194.7 | 13.24 | 3.64 | 38.8 | 38.10 | 0.166 | 46.05 | 52.95 | 0.201 | 0.231 | 19.3 |
| 190 | 204.7 | 13.92 | 3.73 | 40.1 | 38.80 | 0.169 | 47.16 | 54.22 | 0.206 | 0.236 | 19.5 |
| 200 | 214.7 | 14.60 | 3.82 | 41.4 | 39.50 | 0.172 | 48.18 | 55.39 | 0.210 | 0.241 | 20.1 |
| 210 | 224.7 | 15.28 | 3.91 | 42.8 | 40.10 | 0.174 | 49.35 | 56.70 | 0.216 | 0.247 | … |
| 220 | 234.7 | 15.96 | 3.99 | 44.0 | 40.70 | 0.177 | 50.30 | 57.70 | 0.220 | 0.252 | … |
| 230 | 244.7 | 16.64 | 4.08 | 45.3 | 41.30 | 0.180 | 51.30 | 59.10 | 0.224 | 0.257 | … |
| 240 | 254.7 | 17.32 | 4.17 | 46.6 | 41.90 | 0.183 | 52.25 | 60.10 | 0.228 | 0.262 | … |
| 250 | 264.7 | 18.00 | 4.24 | 47.6 | 42.70 | 0.186 | 52.84 | 60.76 | 0.230 | 0.264 | … |
| 260 | 274.7 | 18.68 | 4.32 | 48.8 | 43.00 | 0.188 | 53.85 | 62.05 | 0.235 | 0.270 | … |
| 270 | 284.7 | 19.36 | 4.40 | 50.0 | 43.50 | 0.190 | 54.60 | 62.90 | 0.238 | 0.274 | … |
| 280 | 294.7 | 20.04 | 4.48 | 51.1 | 44.00 | 0.192 | 55.50 | 63.85 | 0.242 | 0.278 | … |
| 290 | 304.7 | 20.72 | 4.55 | 52.2 | 44.50 | 0.194 | 56.20 | 64.75 | 0.246 | 0.282 | … |
| 300 | 314.7 | 21.40 | 4.63 | 53.4 | 45.80 | 0.197 | 56.70 | 65.20 | 0.247 | 0.283 | … |
| 350 | 364.7 | 24.80 | 4.98 | 58.5 | 47.30 | 0.206 | 60.15 | 69.16 | 0.262 | 0.301 | … |
| 400 | 414.7 | 28.20 | 5.31 | 63.3 | 49.20 | 0.214 | 63.19 | 72.65 | 0.276 | 0.317 | … |
| 450 | 464.7 | 31.60 | 5.61 | 67.8 | 51.20 | 0.223 | 65.93 | 75.81 | 0.287 | 0.329 | … |
| 500 | 514.7 | 35.01 | 5.91 | 72.1 | 52.70 | 0.229 | 68.46 | 78.72 | 0.298 | 0.342 | … |

[a] Mean Effective Pressure (MEP) is defined as that single pressure rise, above atmospheric, which would require the same horsepower as the actual varying pressures during compression.

## Table 38. Horsepower Required to Compress Air, Three-Stage Compression

Horsepower required for compressing one cubic foot of free air per minute (isothermally and adiabatically) from atmospheric pressure (14.7 pounds per square inch) to various gage pressures.
Three-stage Compression, initial temperature of air, 60° F, jacket-cooling not considered.

| Gage Pressure, Pounds | Absolute Pressure, Pounds | Number of Atmospheres | Correct Ratio of Cylinder Volumes | Intercooler Gage Pressure, First and Second Stages | Isothermal Compression | | Adiabatic Compression | | | | Percentage of Saving over Two-stage Compression |
|---|---|---|---|---|---|---|---|---|---|---|---|
| | | | | | Mean Effective Pressure[a] | Horsepower | Mean Eff. Pressure,[a] Theoretical | Mean Eff. Pressure plus 15 per cent Friction | Horsepower Theoretical | HP plus 15 per cent Friction | |
| 100 | 114.7 | 7.8 | 1.98 | 14.4-42.9 | 30.30 | 0.132 | 33.30 | 38.30 | 0.145 | 0.167 | 5.23 |
| 150 | 164.7 | 11.2 | 2.24 | 18.2-59.0 | 35.59 | 0.155 | 40.30 | 46.50 | 0.175 | 0.202 | 5.92 |
| 200 | 214.7 | 14.6 | 2.44 | 21.2-73.0 | 39.50 | 0.172 | 45.20 | 52.00 | 0.196 | 0.226 | 6.67 |
| 250 | 264.7 | 18.0 | 2.62 | 23.8-86.1 | 42.70 | 0.186 | 49.20 | 56.60 | 0.214 | 0.246 | 6.96 |
| 300 | 314.7 | 21.4 | 2.78 | 26.1-98.7 | 45.30 | 0.197 | 52.70 | 60.70 | 0.229 | 0.264 | 7.28 |
| 350 | 364.7 | 24.8 | 2.92 | 28.2-110.5 | 47.30 | 0.206 | 55.45 | 63.80 | 0.242 | 0.277 | 7.6[illegible] |
| 400 | 414.7 | 28.2 | 3.04 | 30.0-121.0 | 49.20 | 0.214 | 58.25 | 66.90 | 0.253 | 0.292 | 8.3[illegible] |
| 450 | 464.7 | 31.6 | 3.16 | 31.8-132.3 | 51.20 | 0.223 | 60.40 | 69.40 | 0.263 | 0.302 | 8.3[illegible] |
| 500 | 514.7 | 35.0 | 3.27 | 33.4-142.4 | 52.70 | 0.229 | 62.30 | 71.70 | 0.273 | 0.314 | 8.33 |
| 550 | 564.7 | 38.4 | 3.38 | 35.0-153.1 | 53.75 | 0.234 | 65.00 | 74.75 | 0.283 | 0.326 | 8.8[illegible] |
| 600 | 614.7 | 41.8 | 3.47 | 36.3-162.3 | 54.85 | 0.239 | 66.85 | 76.90 | 0.291 | 0.334 | 8.85 |
| 650 | 664.7 | 45.2 | 3.56 | 37.6-171.5 | 56.00 | 0.244 | 67.90 | 78.15 | 0.296 | 0.340 | 9.[illegible]2 |
| 700 | 714.7 | 48.6 | 3.65 | 38.9-180.8 | 57.15 | 0.249 | 69.40 | 79.85 | 0.303 | 0.348 | 9.[illegible]8 |
| 750 | 764.7 | 52.0 | 3.73 | 40.1-189.8 | 58.10 | 0.253 | 70.75 | 81.40 | 0.309 | 0.355 | ... |
| 800 | 814.7 | 55.4 | 3.82 | 41.4-199.5 | 59.00 | 0.257 | 72.45 | 83.25 | 0.315 | 0.362 | ... |
| 850 | 864.7 | 58.8 | 3.89 | 42.5-207.8 | 60.20 | 0.262 | 73.75 | 84.90 | 0.321 | 0.369 | ... |
| 900 | 914.7 | 62.2 | 3.95 | 43.4-214.6 | 60.80 | 0.265 | 74.80 | 86.00 | 0.326 | 0.375 | ... |
| 950 | 964.7 | 65.6 | 4.03 | 44.6-224.5 | 61.72 | 0.269 | 76.10 | 87.50 | 0.331 | 0.381 | ... |
| 1000 | 1014.7 | 69.0 | 4.11 | 45.7-233.3 | 62.40 | 0.272 | 77.20 | 88.80 | 0.336 | 0.383 | ... |
| 1050 | 1064.7 | 72.4 | 4.15 | 46.3-238.3 | 63.10 | 0.275 | 78.10 | 90.10 | 0.340 | 0.391 | ... |
| 1100 | 1114.7 | 75.8 | 4.23 | 47.5-248.3 | 63.80 | 0.278 | 79.10 | 91.10 | 0.344 | 0.396 | ... |
| 1150 | 1164.7 | 79.2 | 4.30 | 48.5-256.8 | 64.40 | 0.281 | 80.15 | 92.20 | 0.349 | 0.401 | ... |
| 1200 | 1214.7 | 82.6 | 4.33 | 49.0-261.3 | 65.00 | 0.283 | 81.00 | 93.15 | 0.353 | 0.405 | ... |
| 1250 | 1264.7 | 86.0 | 4.42 | 50.3-272.3 | 65.60 | 0.286 | 82.00 | 94.30 | 0.357 | 0.411 | ... |
| 1300 | 1314.7 | 89.4 | 4.48 | 51.3-280.8 | 66.30 | 0.289 | 82.90 | 95.30 | 0.362 | 0.416 | ... |
| 1350 | 1364.7 | 92.8 | 4.53 | 52.0-287.3 | 66.70 | 0.291 | 84.00 | 96.60 | 0.366 | 0.421 | ... |
| 1400 | 1414.7 | 96.2 | 4.58 | 52.6-293.5 | 67.00 | 0.292 | 84.60 | 97.30 | 0.368 | 0.423 | ... |
| 1450 | 1464.7 | 99.6 | 4.64 | 53.5-301.5 | 67.70 | 0.295 | 85.30 | 98.20 | 0.371 | 0.426 | ... |
| 1500 | 1514.7 | 103.0 | 4.69 | 54.3-309.3 | 68.30 | 0.298 | 85.80 | 98.80 | 0.374 | 0.430 | ... |
| 1550 | 1564.7 | 106.4 | 4.74 | 55.0-317.3 | 68.80 | 0.300 | 86.80 | 99.85 | 0.378 | 0.434 | ... |
| 1600 | 1614.7 | 109.8 | 4.79 | 55.8-323.3 | 69.10 | 0.302 | 87.60 | 100.80 | 0.382 | 0.438 | ... |

[a] Mean Effective Pressure (MEP) is defined as that single pressure rise, above atmospheric, which would require the same horsepower as the actual varying pressures during compression.

*Compressor Types:* Compressors can be centrifugal, multi-stage centrifugal, multi-stage axial, or positive displacement. Centrifugal compressors are similar to centrifugal blowers. Positive displacement compressors are essentially positive displacement pumps, used to deliver high pressures at moderate flow rates. These compressors can employ a variety of mechanisms, like pistons, lobes, screws, and vanes. Small positive displacement compressors are often sold to homes and businesses.

Positive displacement compressors compress a fixed amount of gas within a closed chamber by decreasing its volume. Volumetric efficiency is affected by leakage and mechanical losses. Positive displacement compressors tend to be selected for low to moderate flow applications with high compression requirements. Compressor types in this category include: reciprocating piston, rotary vane, rolling piston, rotary twin screw, and orbiting scroll compressors. Screw compressors are commonly used in the range between 100 and 150 psig. Reciprocating compressors can be either single acting or double acting, and are commonly used for pressures in excess of 200 psig. These types of compressors deliver a pulsating flow of gas, unlike other types of compressors. Multi-stage positive displacement compressors with intercooling are common at pressures exceeding 80 psig.

Many state laws and general good practice requires the use of an appropriately sized relief valve ahead of the first stop valve in positive displacement compressors. This relief valve should be set to release at 125 percent of the maximum discharge pressure of the compressor, or at the maximum working pressure of the compressor, whichever is lower.

Dynamic compressors impart momentum to the gas by means of a spinning rotor. This momentum energy is converted to pressure energy when the gas is slowed again. This is a constant flow process. Dynamic compressors are subject to the fan laws, and the specific speed is used to select the geometry. Dynamic compressors tend to be selected for moderate to high flow applications with low to moderate compression requirements. Multi-stage dynamic compressors can provide high flow and high compression.

*Compressor Sizing:* When sizing a compressor for a given application, one must analyze the piping system to the point of use to account for any pressure drops between it and the compressor. Any pressure loss must be added to the compressor range to make sure the desired pressure is delivered at the point of use. If the pressure drop is significant (exceeding 10 percent) the change in specific weight of the gas should be taken into account.

The maximum capacity of a compressor should exceed the maximum mean air consumption of the system.

Because the maximum capacity of an air compressor will exceed the minimum air consumption in the system, the compressor must modulate its capacity during normal work. This will cause variation of system pressure, or pressure pulsation. As a result, a receiver should be installed after the compressor to stabilize system pressure.

Like with fans and pumps, compressors are usually sized by graphing the characteristic curve. Discharge pressure, head, or pressure ratio is graphed against inlet flow rate. For dynamic compressors, the graph includes values at various speeds and efficiencies. The system curve(s) is then plotted over the characteristic curve(s) and the intersection is the operating point. Multiple operating points may be encountered. Each stage of compression will have its own characteristic curve, and in an ideal case each stage will operate at a high efficiency.

For dynamic compressors, inlet pressure variation has a direct effect on horsepower. To correctly size the compressor drive, the full range of inlet pressures must be considered as well as discharge pressures. Dynamic compressors are subject to a minimum flow rate called the surge limit. Surge is seen to the left of the characteristic curve(s). The demand limits of the application must be considered and operation below the surge limit prevented.

**Gas Conditioning and Stabilization.**—In order to provide consistent, clean, dry gas supply; dryers, filters, oilers, receivers, silencers, and exhaust cleaners are used to condition and stabilize pressurized gas.

*Filters:* Compressed gas used to perform work normally needs to be very clean to avoid fouling the devices in the system. Filters of various types are employed to clean compressed air. Most manufacturers offer a range of high efficiency coalescing compressed air filters for the removal of water and oil aerosols, solid particulates and micro-organisms, and dust filters for the removal of dry particulate and micro-organisms from industrial compressed air systems. Filters are selected and sized based on what contaminant is to be removed, output cleanliness level, pressure capacity, flow capacity, pressure drop, and mounting method. Filters require regular preventative maintenance.

*Dryers:* Compressed air supplies for pneumatic systems normally have moisture removed to prevent fouling and corrosion of the system components. For this purpose, compressed air dryers are employed. Dryer type and size is selected according to required air quality, pressure, dew point, and delivery volume. There are several types of commercial dryers available: refrigerated, desiccant, membrane, and deliquescent.

A refrigerated dryer uses a compressor, heat exchanger, and refrigerant to pull liquid out of the gas. Normally a coalescing prefilter is installed immediately upstream of a refrigerated dryer to remove lubricating oil and other contaminants that have the potential to foul the dryer's heat exchangers.

A desiccant dryer passes compressed air through a pressure vessel filled with a media such as activated alumina, silica gel, molecular sieve or other desiccant material which pulls water out of the gas. Normally the desiccant can be regenerated.

A membrane dryer requires no electricity to operate. Water vapor permeates a membrane of hollow fibers and is swept out with redirected dry air to exhaust, while the bulk of the dry air is passed back into the pressurized system.

Normally a coalescing prefilter is installed immediately upstream of a membrane dryer to remove lubricating oil and other contaminants that have potential to foul the dryer's membrane.

A deliquescent dryer typically consists of a pressure vessel filled with a hygroscopic media that absorbs water vapor. The media gradually dissolves, or deliquesces. The vessel must be regularly drained and new media must be added.

*Lubricators:* In some cases a small amount of oil or other lubricant is added to the compressed air supply to lubricate components like actuators. Manufacturers will specify whether their components are "oil-free" or not.

Lubricators are often combined with filters and regulators to form filter-regulator-lubricator (FRL) units. Lubricators usually introduce a fine mist of lubricant into the air stream with particles in the micron range.

*Receivers:* Most compressed gas systems have gas introduced in a pulsating supply. Receivers are essentially pressure vessels that act as reservoirs for the compressed gas. These reservoirs act to stabilize the system pressure by absorbing pressure pulsations. They also stabilize the system at times of fluctuating demand by supplying excess gas when demand peaks, and storing it when demand drops. Receivers are most effective when placed close to the pressure source in the system. A pressure regulator is placed after the receiver to provide constant pressure to the downstream components. Secondary receivers are often placed near high demand to further stabilize the system pressure. Receivers may have valves at the bottom where collected condensate may be drained.

*Silencers and Exhaust Cleaners:* When air is the gas employed, it is usually exhausted to atmosphere after it is used to perform work. When compressed air is released, significant noise can be produced. Silencers or mufflers are used to minimize this effect. The most common silencers are sintered metal, and they are very effective. Silencers are chosen based on port size, pressure, flow requirement, construction, material, and noise reduction level. Exhaust cleaners may be required in some cases where contaminants are present. The proper cleaning system will depend on the contaminant type and level, flow rate, pressure, and construction.

**Nozzles.**—These devices include a restriction through which a gas passes from a region of higher pressure to a region of lower pressure, typically atmospheric. Nozzles are used to accelerate and direct a gas to perform some function. Analysis of nozzles is greatly simplified if one assumes that there is zero heat transfer during discharge, or the discharge is adiabatic. Real nozzles are not adiabatic, but this assumption is generally safe to make. Flow rate of gas through a nozzle is calculated using the following equation:

$$\dot{m} = \rho v_n A_n$$

where $A_n$ is the cross sectional area of the nozzle, $v_n$ is velocity through the nozzle, and $\dot{m}$ is the mass flow rate. In the following equations, $v_n$ and $p_n$ are nozzle outlet velocity and exit pressure ($p_n$ = atmospheric normally), $v_t$ and $p_t$ are source (tank) velocity and pressure, $\gamma$ is specific weight, and $k$ is the adiabatic exponent given in Table 39. Pressures are absolute.

**Table 39. Adiabatic Gas Exponent and Critical Pressure Ratio**

| Gas | $k$ | Critical Pressure Ratio | Gas | $k$ | Critical Pressure Ratio |
|---|---|---|---|---|---|
| Air | 1.4 | 0.528 | Nitrogen | 1.41 | 0.527 |
| Ammonia | 1.32 | 0.542 | Oxygen | 1.4 | 0.528 |
| Carbon Dioxide | 1.3 | 0.546 | Propane | 1.15 | 0.574 |

Flow rate through a converging nozzle reaches a maximum at a critical pressure ratio. At this ratio, the nozzle velocity is equal to the speed of sound for that gas at those conditions. For subsonic flow, an increase in nozzle cross sectional area causes flow velocity to decrease. For supersonic flow, an increase in nozzle area causes flow velocity to increase. To calculate the critical pressure ratio use the following equation:

$$\left(\frac{p'_n}{p_t}\right)_c = \left(\frac{2}{k+1}\right)^{\frac{k}{k-1}}$$

If a converging nozzle's pressure ratio ($p_n/p_t$) is greater than the critical pressure ratio, use the following equation to calculate flow.

$$W = A_n\sqrt{\frac{p_t\gamma_t 2gk}{k+1}\left(\frac{2}{k+1}\right)^{\frac{2}{k-1}}}$$

where $W$ is the weight flow rate.

If a converging nozzle is not operating above the critical pressure ratio, then velocity is equal to sonic velocity and the following equations are used:

Pressure at the nozzle throat is equal to the critical pressure:

$$p_n = p'_n = p_t\left(\frac{2}{k+1}\right)^{\frac{k}{k+1}}$$

$$\dot{m} = A_n\sqrt{kp_i\rho_i}\left(\frac{2}{k+1}\right)^{\frac{k+1}{2(k-1)}}$$

where subscript $i$ indicates inlet conditions, $\dot{m}$ is the mass flow in kg/s, $A$ is nozzle area in $m^2$, and $p$ is pressure in Pa.

Flow rate faster than the speed of sound is possible with a converging-diverging nozzle. Analysis of this type of nozzle is beyond the scope of this text.

*Velocity of Escaping Compressed Air:* For a large container with a small orifice or hole from which the air escapes, the velocity of escape (theoretical) may be calculated from the formula:

$$v_2 = \sqrt{2g \cdot \frac{k}{k-1} \cdot 53.3(459.7 + F)\left[1 - \left(\frac{p_2}{p_1}\right)^{\frac{k-1}{k}}\right]}$$

In this formula, $v_2$ = velocity of escaping air in feet per second; $g$ = acceleration due to gravity, 32.16 feet per second squared; $k$ = 1.41 for adiabatic expansion or compression of air; $F$ = temperature, degrees F; $p_2$ = atmospheric pressure = 14.7 pounds per square inch; and $p_1$ = pressure of air in container, pounds per square inch. In applying the preceding formula, when the ratio $p_2/p_1$ approximately equals 0.53, under normal temperature conditions at sea level, the escape velocity $v_2$ will be equal to the velocity of sound. Increasing the pressure $p_1$ will not increase the velocity of escaping air beyond this limiting velocity unless a special converging diverging nozzle design is used rather than an orifice.

The accompanying Table 40 provides velocity of escaping air for various values of $p_1$. These values were calculated from the preceding formula simplified by substituting the appropriate constants:

$$v_2 = 108.58\sqrt{(459.7 + F)\left[1 - \left(\frac{14.7}{p_1}\right)^{0.29}\right]}$$

**Table 40. Velocity of Escaping Air at 70-Degrees F**

| Pressure Above Atmospheric Pressure | | | Theoretical Velocity, Feet per Second | Pressure Above Atmospheric Pressure | | | Theoretical Velocity, Feet per Second |
|---|---|---|---|---|---|---|---|
| Atmo-spheres | Inches Mercury | lbs per sq. in. | | Atmo-spheres | Inches Mercury | lbs per sq. in. | |
| 0.010 | 0.30 | 0.147 | 134 | 0.408 | 12.24 | 6.00 | 769 |
| 0.068 | 2.04 | 1.00 | 344 | 0.500 | 15.00 | 7.35 | 833 |
| 0.100 | 3.00 | 1.47 | 413 | 0.544 | 16.33 | 8.00 | 861 |
| 0.136 | 4.08 | 2.00 | 477 | 0.612 | 18.37 | 9.00 | 900 |
| 0.204 | 6.12 | 3.00 | 573 | 0.680 | 20.41 | 10.0 | 935 |
| 0.272 | 8.16 | 4.00 | 650 | 0.816 | 24.49 | 12.0 | 997 |
| 0.340 | 10.20 | 5.00 | 714 | 0.884 | 26.53 | 13.0 | 1025 |

The theoretical velocities in Table 40 must be reduced by multiplying by a "coefficient of discharge," which varies with the orifice and the pressure. The coefficients in Table 41 are used for orifices in thin plates and short tubes.

**Table 41. Coefficients of Discharge**

| Type of Orifice | Pressures in Atmospheres Above Atmospheric Pressure | | | |
|---|---|---|---|---|
| | 0.01 | 0.1 | 0.5 | 1 |
| Orifice in thin plate | 0.65 | 0.64 | 0.57 | 0.54 |
| Orifice in short tube | 0.83 | 0.82 | 0.71 | 0.67 |

## Vacuum

Vacuum is any pressure lower than atmospheric. It can be measured in psi or Pa as a negative gauge pressure, or as a positive absolute pressure.Other common units for vacuum are inches of mercury (inHg), millimeters of mercury (mmHg), millibar, and torr. A perfect vacuum has zero absolute pressure, or −14.7 psi (−29.92 inHg). Flow in low to medium vacuum (down to $10^{-4}$ torr) is characterized as viscous flow, while in high (below $10^{-4}$ torr) and ultra-high vacuum (below $10^{-8}$ torr) it is characterized as molecular flow.

**Tubing and Fittings for Vacuum Applications.—**Tubing and fittings used for vacuum applications must be rated for vacuum use to prevent collapse. When working with vacuum, the resistance to flow created by tubing and fittings is discussed in terms of conductance. Conductance can be calculated as follows:

$$C = \frac{Q}{\Delta p}$$

where $C$ is conductance in l/sec, $p$ is pressure in torr, and $Q$ is flow rate in torr–l/sec. For dry air at 20° C in the molecular flow region of vacuum, estimate conductance of long tubes as follows:

$$C = 75\frac{D^3}{L}$$

where $L$ is tube length in inches, $D$ is tube inner diameter in inches, and $C$ is in l/sec.

Effective pumping speed at the end of a tube with conductance $C$ can be calculated using the following equation:

$$\frac{1}{S_{eff}} = \frac{1}{C} + \frac{1}{S}$$

where $S$ is the pumping speed of the pump by itself, and $C$ is conductance of the tube.

The conductance of tubes in series and parallel can be combined according to the following equations.

Conductance changes with pressure, so effective pumping speed is best calculated piecemeal at incremental pressures.

$C_p = C_1 + C_2$ where $C_p$ is the conductance of tubes 1 and 2 in parallel.

$\frac{1}{C_s} = \frac{1}{C_1} + \frac{1}{C_2}$ where $C_s$ is the conductance of tubes 1 and 2 in series.

**Vacuum Pumps and Generators.—**Vacuum pumps are essentially compressors in reverse. Where deep or continuous vacuum is required on a large evacuation volume, it is common to use a fast evacuation pump to remove the gas, and then a deep vacuum pump is used to reach the desired vacuum level. Vacuum deeper than −14.7psi (−29.92 inHg) is not possible.

Vacuum generators usually use a venturi to generate vacuum when compressed air is supplied. Single and multi- stage units are widely available to generate vacuum as deep as −25 inHg (−85 kPa). Suction flow for vacuum generators is moderate, with values of 7 SCFM (200 l/min) commercially available. Vacuum generators consume compressed air, so these tend to be more costly to run than vacuum pumps. Vacuum generators are sometimes combined into devices called "vacuum ejectors" which provide a puff of positive air pressure when the vacuum is halted. This provides a positive release function when picking and placing components using vacuum.

*Types of Vacuum Pumps:* Vacuum pumps can employ pistons, vanes, and lobes like compressors or pumps. There are also turbomolecular pumps, diffuser pumps, chemical reaction pumps, gettering pumps, ion pumps, and cryogenic pumps. Vacuum pumps tend to fall into two main categories: transfer pumps and capture pumps.

Coarse vacuum pumps used for rapid evacuation are commonly rotary vane-type positive displacement pumps. These pumps are usually flooded with oil, which lubricates, cools, and seals the pump. The oil is discharged with the gas. To achieve deep vacuum, vane pumps in series are sometimes used, as are screw pumps. Where the presence of oil is objectionable, multi-stage oil-free pumps are employed, but are not as efficient as oil sealed pumps.

Turbomolecular pumps in general use rotors to accelerate gas molecules, similar to centrifugal pumps. Drag pumps are an example of this type of vacuum pump.

Diffuser, or jet pumps are the most commonly used type of vacuum pump for high vacuum applications. In these pumps, a high velocity jet of fluid is used to transfer momentum to gas in a channel. This type of pump can sustain a vacuum level up to 1 torr, but must be used in combination with an evacuation pump. Multi-stage diffuser pumps are available.

Cryogenic pumps cool gases until they condense into a liquid or solid. In practice, a cooled pump is used, rather than cooling an entire reservoir or chamber. Helium is often used as the refrigerant, and cooling to temperatures of 10 Kelvin are possible.

Gettering pumps use specially treated surfaces, called "gettering" surfaces. In vacuum, these surfaces are capable of holding greater amounts of gas than would occupy a section of free space of the same cross sectional area. Ion pumps use electron bombardment to ionize inert gases that would not normally be affected by gettering surfaces.

*Selection Factors:* The type of pump or generator required by an application depends on whether the system must provide rapid evacuation, long term maintenance of vacuum, high suction flow rate, and/or deep vacuum. Diffuser pumps cannot be exposed to gas at atmospheric pressure once they are at operating temperature. To prevent the risk of damage, a system of valves must be engineered to prevent this condition. Filters should always be used with vacuum generators to protect them from contamination.

*Sizing Vacuum Pumps and Generators:* Vacuum pump performance is rated in terms of throughput, pumping speed, and vacuum pressure capability. They are rated for either intermittent or continuous duty. Throughput is the product of volumetric flow rate and pressure. Pumping speed has units of volume/time and is the mean volume flow through the cross section of the inlet port of a vacuum pump.

Determination of the pumping speed is described in ISO 21360-1. Vacuum pump performance curves are plots of pumping speed as a function of inlet pressure. The lowest pressure that can be achieved by a vacuum pump can be calculated using the following equation:

$$p_{\min} = p_0 e^{\left(\frac{-S_{eff}t}{V}\right)} + \frac{Q_o}{S_{eff}} + \frac{Q_D}{S_{eff}} + \frac{Q_K}{S_{eff}}$$

where $p_{\min}$ is the deepest vacuum achievable, $p_o$ is the initial pressure, $t$ is time, $S_{eff}$ is the effective pumping speed, and $V$ is the evacuation volume. $Q_o$, $Q_D$, and $Q_K$ are outgassing, diffusion, and permeation flow rates.

The time required to reach a certain pressure depends on the pumping speed, the volume being evacuated, and the initial pressure. It can be calculated using the following equation. Effective pumping speed changes as the pressure decreases, so the following equations are best applied piecemeal using pumping speed values at increments as the pressure varies.

$$t = \frac{V}{S_{eff}} \ln\left(\frac{p_0}{p_1}\right)$$

where $t$ is pumpdown time, $V$ is volume, $S_{eff}$ is effective pumping speed, $p_0$ is initial pressure (absolute), and $p_1$ is final pressure.

Minimum achievable pressure is affected by ambient air pressure according to the following relationship:

$$p'_{min} = p_{atm}\frac{p_r}{29.92}$$

where $p'_{min}$ is the adjusted achievable vacuum (psi), $p_{atm}$ is atmospheric pressure (psi), and $p_r$ is the rated achievable vacuum (inHg). Most vacuum pumps are rated at sea level atmospheric pressure.

Vacuum generators are sized according to the level of vacuum required and the required suction flow rate.

Vacuum level generated is often expressed as a percentage of input pressure. Input tube diameter and air pressure must be known. A variety of options are available, including ejection, vacuum monitoring, and valving. Multi-stage generators are used to create deeper vacuum than single stage units. Some vacuum generators are optimized for high flow, while others for deep vacuum. Deep vacuum is desirable when picking up a heavy object with a leak-free seal. High suction flow is desirable when picking up objects where a leak-free seal is not possible.

*Vacuum Accessories:* Many accessories are used with vacuum pumps and generators to filter, regulate, control, exploit, measure, and stabilize vacuum. It is important to note that vacuum rated greases must be used when grease is exposed to vacuum.

*Filters:* Filters are critical when using vacuum, because contaminants can be sucked into the vacuum lines and damage the pump, generator, valve, or other components. Selection is based on body type, tube diameter, and maximum particle size that can pass through the filter. Fine filters are typically used with vacuum systems. They must be regularly replaced or cleaned to prevent clogging.

*Regulators:* Like pressure regulators, these devices are essential for limiting and stabilizing vacuum pressure in a circuit. They may be manually or automatically controlled. These devices are outfitted with a gauge or digital readout. Vacuum regulators are sized with respect to range and accuracy of measurement, maximum flow rate, and port size.

*Valves:* Many pressure valves can be used to control vacuum. However, if the valve uses air pressure to cause switching, an external pilot line with positive pressure must be provided.

*Grippers:* There are a wide variety of grippers available for the purpose of picking up objects with vacuum. Their primary purpose is to provide a leak-free seal to maximize the effect of the vacuum system. Compliant grippers include suction cups, bellows, and vacuum pads of all shapes and sizes. Miniature valves are often used in applications with multiple grippers to restrict flow if no part is present.

*Measurement:* Vacuum gauges and switches are used to measure the level of vacuum. They may be analog or digital, and measure in absolute or gauge pressure.

*Level Compensators:* These compliant devices are used when multiple grippers are employed in picking up a part. They provide some degree of shock and vibration dampening. Level compensators are available in a wide variety of sizes and stroke lengths.

# TABLE OF CONTENTS

## MEASURING UNITS

### SYMBOLS AND ABBREVIATIONS



### MEASURING UNITS



### US SYSTEM AND METRIC SYSTEM CONVERSIONS



### US SYSTEM AND METRIC SYSTEM CONVERSIONS

*(Continued)*



UNITS

## SYMBOLS AND ABBREVIATIONS

### Greek Letters and Standard Abbreviations

The Greek letters are frequently used in mathematical expressions and formulas. The Greek alphabet is given below.

| | | | | | | | | | | | |
|---|---|---|---|---|---|---|---|---|---|---|---|
| Α | α | Alpha | Η | η | Eta | Ν | ν | Nu | Τ | τ | Tau |
| Β | β | Beta | Θ | θ ϑ | Theta | Ξ | ξ | Xi | ϒ | υ | Upsilon |
| Γ | γ | Gamma | Ι | ι | Iota | Ο | ο | Omicron | Φ | ϕ | Phi |
| Δ | δ | Delta | Κ | κ | Kappa | Π | π | Pi | Χ | χ | Chi |
| Ε | ε | Epsilon | Λ | λ | Lambda | R | ρ | Rho | Ψ | ψ | Psi |
| Ζ | ζ | Zeta | Μ | μ | Mu | Σ | σ ς | Sigma | Ω | ω | Omega |

### Abbreviations for Scientific and Engineering Terms

*ANSI/ASME Y14.38-2019*

| | |
|---|---|
| Absolute | abs |
| Alternating current | ac |
| Ampere | amp |
| Ampere-hour | amp hr |
| **Angstrom unit** | **Å** |
| Antilogarithm | antilog |
| Arithmetic average | aa |
| Atmosphere | atm |
| Atomic weight | at wt |
| Avoirdupois | avdp |
| Barometer | baro |
| Board-foot (foot board measure) | fbm |
| Boiler pressure | bopress |
| Boiling point | bp |
| Brinell Hardness Number | HBW or BH or BHN |
| British thermal unit | Btu |
| Bushel | bu |
| **Calorie** | **cal** |
| Candle | cd |
| Center to center | c to c |
| Centimeter | cm |
| **Centimeter-gram-second (system)** | **cgs** |
| **Chemical** | **chem** |
| **Chemically pure** | **cp** |
| Circular | circ |
| **Circular mil** | **cmil** |
| **Coefficient** | **coef** |
| **Cologarithm** | **colog** |
| **Concentrate** | **conc** |
| **Conductivity** | **cndct** |
| **Constant** | **const** |
| **Cord** | **cd** |
| Cosecant | csc |
| Cosine | cos |
| **Cost, insurance, and freight** | **cif** |
| Cotangent | cot |
| **Counter electromotive force** | **cemf** |
| Cubic | cu |
| Cubic centimeter | $cm^3$ or cc |
| Cubic foot | $ft^3$ or cu ft |
| Cubic feet per second | $ft^3$ or cfs |
| Cubic inch | $in^3$ or cu in |
| Cubic meter | $m^3$ or cu m |
| Cubic millimeter | $mm^3$ or cumm |
| Cubic yard | $yd^3$ or cu yd |
| **Current density** | **cd** |
| Cylinder | cyl |

| | |
|---|---|
| Decibel | dB |
| Degree | deg or ° |
| Degree Centigrade | °C |
| Degree Fahrenheit | °F |
| Degree Kelvin | K |
| Diameter | dia |
| Direct current | dc |
| Dozen | doz |
| Dram | dr |
| Efficiency | eff |
| Electric | elec |
| Electromotive force | emf |
| Elevation | el |
| Engine | eng |
| Engineer | engr |
| Engineering | engrg |
| Equation | eq |
| External | ext |
| Fluid | fl |
| Foot | ft |
| Foot-candle | fc |
| Foot-lambert | fL or fl |
| Foot per minute | fpm |
| Foot per second | fps |
| Foot-pound | ft-lb |
| Foot-pound-second (system) | fps |
| Free on board | fob |
| Freezing point | fp |
| Frequency | freq |
| Fusion point | fnpt |
| Gallon | gal |
| Gallon per minute | gpm |
| Gallon per second | gps |
| Grain | gr |
| Gram | g |
| Greatest common divisor | gcd |
| High pressure | hp |
| Horsepower | hp |
| Horsepower-hour | hp-hr |
| Hour | h or hr |
| Hyperbolic cosine | cosh |
| Hyperbolic sine | sinh |
| Hyperbolic tangent | tanh |
| Inch | in |
| Inch per second | in/s or ips |
| Inch-pound | in-lb |

## Abbreviations for Scientific and Engineering Terms

*ANSI/ASME Y14.38-2019 (Continued)*

| | | | |
|---|---|---|---|
| Indicated horsepower-hour | ihp | Pound-force foot | $lb_f \cdot ft$ or lb-ft |
| Intermediate pressure | ip | Pound-force inch | $lb_f \cdot in$ or lb-in |
| Internal | intl | Pound-force per square foot | $lb_f/ft^2$ or psf |
| Kilovolt-ampere hour | KVA-h or kVah | Pound-force per square inch | $lb_f/in^2$ or psi |
| Kilowatt-hour meter | kwhm | Pound per horsepower | lb/hp or php |
| Latitude | lat | Power factor | pf |
| Least common multiple | lcm | Quart | qt |
| Liquid | liq | Reactive volt-ampere meter | rva |
| Logarithm (common) | log | Revolution per minute | r/min or rpm |
| Logarithm (natural) | ln | Revolution per second | r/s or rps |
| Low pressure | lp | Root mean square | rms |
| Lumen per watt | lm/W or lpw | Round | rnd |
| Magnetomotive force | mmf | Secant | sec |
| Mathematics (ical) | math | Second | s or sec |
| Maximum | max | Sine | sin |
| Mean effective pressure | mep | Specific gravity | sp gr |
| Melting point | mp | Specific heat | sp ht |
| Meter | m | Square | sq |
| Meter-kilogram-second (system) | mks | Square centimeter | $cm^2$ or sq cm |
| **Microfarad** | **μF** | Square foot | $ft^2$ or sq ft |
| Mile | mi | Square inch | $in^2$ or sq in |
| Mile per hour | mi/h or mph | Square kilometer | $km^2$ or sq km |
| **Milliampere** | **mA** | Standard | std |
| Minimum | min | Tangent | tan |
| Molecular weight | mol wt | Temperature | temp |
| Molecule | mol | Tensile strength | ts |
| National Electrical Code | NEC | Versed sine | vers |
| Ounce | oz | Volt | V |
| Ounce-inch | oz-in | Watt | W |
| Pennyweight | dwt | Watt-hour | Wh |
| Pint | pt | Week | wk |
| Potential | pot | Weight | wt |
| Potential difference | pd | Yard | yd |
| Pound | lb | | |

Alternative abbreviations conforming to the practice of the International Electrotechnical Commission.

| | | | | | | | |
|---|---|---|---|---|---|---|---|
| Ampere | A | Kilovolt-ampere | kVA | Microfarad | μF | Ohm | Ω |
| Ampere-hour | Ah | Kilowatt | kW | Microwatt | μW | Volt | V |
| Coulomb | C | Kilowatt-hour | kWh | Milliampere | mA | Volt-ampere | VA |
| Farad | F | Megawatt | MW | Millifarad | mF | Volt-coulomb | VC |
| Henry | H | Megaohm | MΩ | Millihenry | mH | Watt | W |
| Joule | J | Microampere | μA | Millivolt | mV | Watthour | Wh |
| Kilovolt | kV | | | | | | |

Note: The bold terms and abbreviations are from previous versions of ANSI/ASME Y14.38-2019 or are otherwise widely used industry acronyms not included in the current standard.

In the two tables above, only the most commonly used terms have been included. These forms are recommended for those whose familiarity with the terms used makes possible a maximum of abbreviations. For others, less contracted combinations made up from this list may be used. For example, the list gives the abbreviation of the term "feet per second" as "fps." To some, however, ft per sec will be more easily understood.

Abbreviations should be used sparingly and only where their meaning will be clear. If there is any doubt, then spell out the term or unit of measurement.

The following points are good practice when preparing engineering documentation. Terms denoting units of measurement should be abbreviated in text only when preceded by the amounts indicated in numerals: "several inches," "one inch," "12 in." A sentence should not begin with a numeral followed by an abbreviation. The use of conventional signs for abbreviations in text should be avoided: use "lb," not "#" or "in," not ″.

Symbols for the chemical elements are listed in the table on page 362.

## Mathematical Signs and Commonly Used Abbreviations

| | | | | |
|---|---|---|---|---|
| + | Plus (sign of addition) | | $\pi$ | Pi (3.1416) |
| + | Positive | | $\Sigma$ | Sigma (sign of summation) |
| − | Minus (sign of subtraction) | | $\omega$ | Omega |
| − | Negative | | $g$ | Acceleration due to gravity (32.16 ft/s² or 9.81 m/s²) |
| $\pm (\mp)$ | Plus or minus (minus or plus) | | $i$ (or $j$) | Imaginary quantity $(\sqrt{-1})$ |
| $\times$ | Multiplied by (multiplication sign) | | sin | Sine |
| $\cdot$ | Multiplied by (multiplication sign) | | cos | Cosine |
| $\div$ | Divided by (division sign) | | tan | Tangent |
| / | Divided by (division sign) | | cot | Cotangent |
| : | Is to (in proportion) | | sec | Secant |
| = | Equals | | csc | Cosecant |
| $\neq$ | Is not equal to | | vers | Versed sine |
| $\equiv$ | Is identical to | | covers | Coversed sine |
| $\cong$ or $\approx$ | Approximately equals | | $\sin^{-1} a$ arcsin $a$ or asin $a$ | Angle the sine of which is $a$ |
| > | Greater than | | $(\sin a)^{-1}$ | Reciprocal of sin $a$ ($1 \div \sin a$) |
| < | Less than | | $\sin^n x$ | $n$th power of sin $x$ |
| $\geq$ | Greater than or equal to | | $\sinh x$ | Hyperbolic sine of $x$ |
| $\leq$ | Less than or equal to | | $\cosh x$ | Hyperbolic cosine of $x$ |
| $\rightarrow$ | Approaches as a limit; implies | | $\Delta$ | Delta (increment of) |
| $\propto$ | Varies directly as | | $\delta$ | Delta (variation of) |
| $\therefore$ | Therefore | | $d$ | Differential (in calculus) |
| :: | Equals (in proportion) | | $\partial$ | Partial differentiation (in calculus) |
| $\sqrt{a}$ | Square root of $a$ | | $\int$ | Integral (in calculus) |
| $\sqrt[3]{a}$ | Cube root of $a$ | | $\int_b^a$ | Integral between the limits $a$ and $b$ |
| $\sqrt[4]{a}$ | 4th root of $a$ | | $n!$ | Factorial, $n(n-1)...(2)(1)$; $0! = 1$; $1! = 1$ *Ex:* $5! = 5 \times 4 \times 3 \times 2 \times 1 = 120$ |
| $\sqrt[n]{a}$ | $n$th root of $a$ (also $a^{1/n}$) | | $\angle$ | Angle |
| $a^2$ | $a$ squared (2nd power of $a$) | | ∟ | Right angle |
| $a^3$ | $a$ cubed (3rd power of $a$) | | $\perp$ | Perpendicular to |
| $a^4$ | 4th power of $a$ | | $\Delta$ | Triangle |
| $a^n$ | $n$th power of $a$ | | ○ | Circle |
| $a^{-n}$ | $1 \div a^n$ | | ▱ | Parallelogram |
| $\frac{1}{n}$ | Reciprocal value of $n$ | | ° | Degree (circular arc or temperature) |
| log | Logarithm | | ′ | Minutes or feet |
| $\log_e$ | Natural or Napierian logarithm | | ″ | Seconds or inches |
| ln | Natural or Napierian logarithm | | $a'$ | $a$ prime |
| $e$ | Base of natural logarithms (2.71828) | | $a''$ | $a$ double prime |
| lim | Limit value (of an expression) | | $a_1$ | $a$ sub one |
| $\infty$ | Infinity | | $a_2$ | $a$ sub two |
| $\alpha$ | Alpha | | $a_n$ | $a$ sub $n$ |
| $\beta$ | Beta | commonly used to denote angles | ( ) | Parentheses |
| $\gamma$ | Gamma | | [ ] | Brackets |
| $\theta$ | Theta | | { } | Braces |
| $\phi$ | Phi | | $\lvert K \rvert$ | Absolute value of $K$, size of $K$ irrespective of sign |
| $\mu$ | Mu (coefficient of friction) | | | |

## Letter Symbols for Mechanics and Time-Related Phenomena

*ANSI/ASME Y10.3M-1984 (Withdrawn)*

| | | | |
|---|---|---|---|
| Acceleration, angular | α (alpha) | Height | $h$ |
| Acceleration, due to gravity | $g$ | Inertia, moment of | $I$ or $J$ |
| Acceleration, linear | $a$ | Inertia, polar (area) moment of[a] | $J$ |
| Amplitude[a] | $A$ | Inertia, product (area) moment of[a] | $I_{xy}$ |
| | | Length | $L$ or $l$ |
| Angle | α (alpha), β (beta), γ (gamma), θ (theta), ϕ (phi), ψ (psi) | Load per unit distance[a] | $q$ or $w$ |
| | | Load, total[a] | $P$ or W |
| | | Mass | $m$ |
| | | Moment of force, including bending moment | $M$ |
| | | Neutral axis, distance to extreme fiber from[a] | $c$ |
| Angle, solid | Ω (omega) | Period | $T$ |
| Angular frequency | ω (omega) | Poisson's ratio | μ (mu) or ν (nu) |
| Angular momentum | $L$ | Power | $P$ |
| Angular velocity | ω (omega) | Pressure, normal force per unit area | $p$ |
| Arc length | $s$ | Radius | $r$ |
| Area | $A$ | Revolutions per unit of time | $n$ |
| Axes, through any point[a] | $X$-$X$, $Y$-$Y$, or $Z$-$Z$ | Second moment of area (second axial moment of area) | $I_a$ |
| Bulk modulus | $K$ | Second polar moment of area | $I_P$ or $J$ |
| Breadth (width) | $b$ | Section modulus | $Z$ |
| Coefficient of expansion, linear[a] | α (alpha) | Shear force in beam section[a] | $V$ |
| Coefficient of friction | μ (mu) | Spring constant (load per unit deflection)[a] | $k$ |
| Concentrated load (same as force) | $F$ | Statical moment of any area about a given axis[a] | $Q$ |
| Deflection of beam, max[a] | δ (delta) | Strain, normal | ε (epsilon) |
| Density | ρ (rho) | Strain, shear | γ (gamma) |
| Depth | $d$, δ (delta), or $t$ | Stress, concentration factor[a] | $K$ |
| Diameter | $D$ or $d$ | Stress, normal | σ (sigma) |
| Displacement[a] | $u$, $v$, $w$ | Stress, shear | τ (tau) |
| Distance, linear[a] | $s$ | Temperature, absolute[b] | $T$, or θ (theta) |
| Eccentricity of application of load[a] | $e$ | Temperature[b] | $t$, or θ (theta) |
| Efficiency[a] | η (eta) | Thickness | $d$, δ (delta), or $t$ |
| Elasticity, modulus of | $E$ | Time | $t$ |
| Elasticity, modulus of, in shear | $G$ | Torque | $T$ |
| Elongation, total[a] | δ (delta) | Velocity, linear | $v$ |
| Energy, kinetic | $E_k$, $K$, $T$ | Volume | $V$ |
| Energy, potential | $E_P$, $V$, or Φ (phi) | Wavelength | λ (lambda) |
| Factor of safety[a] | $N$ or $n$ | Weight | $W$ |
| Force or load, concentrated | $F$ | Weight per unit volume | γ (gamma) |
| Frequency | $f$ | Work | $W$ |
| Gyration, radius of[a] | $k$ | | |

[a] Not specified in Standard

[b] Specified in ANSI Y10.4-1982 (R1988)

## MEASURING UNITS

### Metric Systems of Measurement

A metric system of measurement was first established in France in the years following the French Revolution, and various systems of metric units have been developed since that time. All metric unit systems are based, at least in part, on the International Metric Standards, which are the meter and kilogram, or decimal multiples or submultiples of these standards.

In 1795, a metric system called the centimeter-gram-second (cgs) system was proposed, and it was adopted in France in 1799. In 1873, the British Association for the Advancement of Science recommended the use of the cgs system, and since then it has been widely used in all branches of science throughout the world. From the base units in the cgs system are derived the following:

*Unit of velocity* = 1 centimeter per second

*Acceleration due to gravity (at Paris)* = 981 centimeters per second per second

*Unit of force* = 1 dyne = $\frac{1}{981}$ gram

*Unit of work* = 1 erg = 1 dyne-centimeter

*Unit of power* = 1 watt = 10,000,000 ergs per second

Another metric system, called the MKS (meter-kilogram-second) system of units, was proposed by Professor G. Giorgi in 1902. In 1935, the International Electrotechnical Commission (IEC) accepted his recommendation that this system of units of mechanics be linked with the electromagnetic units by the adoption of a fourth base unit. In 1950, the IEC adopted the ampere, the unit of electric current, as the fourth unit, and the MKSA system thus came into being.

A gravitational system of metric units, known as the technical system, is based on the meter, the kilogram as a force, and the second. It has been widely used in engineering. Because the standard of force is defined as the weight of the mass of the standard kilogram, the fundamental unit of force varies due to the difference in gravitational pull at different locations around the earth. By international agreement, a standard value for acceleration due to gravity was chosen (9.81 meters per second squared) that for all practical measurements is approximately the same as the local value at the point of measurement.

**International System of Units (SI).**—The Conférence Générale des Poids et Mesures (CGPM) is the body responsible for all international matters concerning the metric system, which was originally based on the four MKSA units for: length (*meter*), time (*second*), mass (*kilogram*), and electric current (*ampere*). In 1954, a unit for temperature (*kelvin*) and another for luminous intensity (*candela*) were added.

Over the years, these standards have been refined and redefined. In 1960, the CGPM formally renamed the metric system the *Système International d'Unites* (abbreviated SI in all languages). In 1971, the 14th CGPM adopted a seventh base unit, the *mole*, for the amount of substance in a physical sample. In 2011, CGPM stated its intention of defining all seven base units with "the explicit-constant formulation," where all units would be defined indirectly through their relation to a fundamental physical constant, as had been done for the speed of light. For several more years, only three of the seven *base units*—the meter, candela, and second—were as yet defined in terms of fundamental physical constants. But in May 2019, the other four base units—the mole, ampere, kelvin, and kilogram—were redefined, no longer based on physical objects, but on universal physical constants.

**Table 1. SI Base Units**

| Quantity | Name | Symbol |
|---|---|---|
| Length | meter | m |
| Mass | kilogram | kg |
| Time | second | s |
| Electric Current | ampere | A |
| Luminous Intensity | candela | cd |
| Thermodynamic Temperature | kelvin | K |
| Amount of Substance | mole | mol |

## Definitions of Base Units

*Meter (m):* The base unit of length, equal to the path length traveled by light in a vacuum during a time interval of $3.3356410 \times 10^{-9}$ s (derived from the precise measurement of light's speed, $c = 299{,}792{,}458$ m/s). This value was adopted in 1983, based on the newly defined value for the speed of light, made possible by the technology developed at the National Institute of Standards and Technology (NIST).

*Kilogram (kg):* The base unit of mass, related to the most precise measurement to date of the Planck constant $h$, which is the energy carried by a photon to its frequency, and measured as $6.6260702 \times 10^{-34}$ kg m$^2$/s$^2$. It is with respect to this invariant of nature that the precise standard for the kilogram is given.

*Second (s):* The duration of $9.1926318 \times 10^9$ periods of the radiation, corresponding to the transition between the two hyperfine levels of the ground state of the cesium-133 atom, is the element used for the atomic clock due to its stability.

*Ampere (A):* The base unit of electric current, measured as current flow per unit of time. One ampere (*amp*) is equal to one coulomb per second. One elementary electric charge (e) is equal to $1.60218 \times 10^{-19}$ coulomb (that is, $1\ C = 6.241 \times 10^{18}$ e).

*Candela (cd):* The base unit for photometry, which is the science of measuring light perceived by the human eye. The candela (cd) is the luminous intensity, in a given direction, of a source that emits monochromatic radiation of frequency $540 \times 10^{12}$ hertz. The human eye is most sensitive to this frequency (greenish-yellow light). One cd has a radiant intensity of 1/683 watt per steradian.

*Kelvin (K):* The base unit of thermodynamic temperature, defined by taking the fixed numerical value of the Boltzmann constant $k$ to be $1.380649 \times 10^{-23}$ J·K$^{-1}$.

*Mole (mol):* The base unit for the amount of a chemical substance consisting of exactly $6.02214076 \times 10^{23}$ constitutive particles. This number is the Avogadro constant $N_A = 6.02214076 \times 10^{23}$ mol$^{-1}$. (Previously, 1 mole was linked to the number of atoms in 12.1 g of the carbon-12 isotope.)

From the base units, come the SI derived units, as shown in Table 2.

**Table 2. SI Derived Units**

| Quantity | Name | Symbol |
|---|---|---|
| Area | square meter | m$^2$ |
| Volume | cubic meter | m$^3$ |
| Speed, Velocity | meter per second | m/s |
| Acceleration | meter per second squared | m/s$^2$ |
| Wave Number | reciprocal meter | m$^{-1}$ |
| Mass Density | kilogram per cubic meter | kg/m$^3$ |
| Specific Volume | cubic meter per kilogram | m$^3$/kg |
| Current Density | ampere per square meter | A/m$^3$ |
| Magnetic Field Strength | ampere per meter | A/m |
| Amount-of-Substance Concentration | mole per cubic meter | mol/m$^3$ |
| Luminance | candle per square meter | cd/m$^2$ |
| Mass Fraction | kilogram per kilogram | kg/kg |

**Table 3. SI Derived Units with Special Names and Symbols**

| Quantity | Name and Derived Units | Symbol |
|---|---|---|
| Plane Angle | radian, m/m | rad |
| Solid Angle | steradian, $m^2/m^2$ | sr |
| Frequency | hertz, $s^{-1}$ | Hz |
| Force | newton, $kg \cdot m \cdot s^{-2}$ | N |
| Pressure, Stress | pascal, $kg \cdot m^{-1} \cdot s^{-2}$ | Pa |
| Energy, Work, Heat | joule, $kg \cdot m^2 \cdot s^{-2}$ | J |
| Power, Radiant Flux | watt, $kg \cdot m^2 \cdot s^{-3}$ | W |
| Quantity of Electric Charge | coulomb, A·s | C |
| Electric Potential Difference | volt, $kg \cdot m^2 \cdot s^{-3} \cdot A^{-1}$ | V |
| Capacitance | farad, $s^4 \cdot A^2 \cdot m^{-2} \cdot kg^{-1}$ | F |
| Electric Resistance | ohm, $kg \cdot m^2 \cdot s^{-3} \cdot A^{-2}$ | Ω |
| Electric Conductance | siemens, $kg^{-1} \cdot m^{-2} \cdot s^3 \cdot A^2$ | S |
| Magnetic Flux | weber, $kg \cdot m^2 \cdot s^{-2} \cdot A^{-1}$ | Wb |
| Magnetic Flux Density | tesla, $kg \cdot s^{-2} \cdot A^{-1}$ | T |
| Inductance | henry, $kg \cdot m^2 \cdot s^{-2} \cdot A^{-2}$ | H |
| Celsius Temperature | degree Celsius, K – 273.15 | °C |
| Luminous Flux | lumen, cd·sr | lm |
| Illuminance | lux, $cd \cdot sr \cdot m^{-2}$ | lx |
| Activity of Radionuclide | becquerel, $s^{-1}$ | Bq |
| Absorbed Dose | gray, $J \cdot kg^{-1}$ | Gy |
| Dose Equivalent | sievert, $m^2 \cdot s^{-2}$ | Sv |
| Catalytic Activity | katal, $mol \cdot s^{-1}$ | kat |

**Table 4. SI Derived Units that Include SI Derived Units with Special Names**

| Quantity | Name | Symbol |
|---|---|---|
| Dynamic Viscosity | pascal-second | Pa·s |
| Moment of Force | newton-meter | N·m |
| Angular Velocity | radian per second | rad/s |
| Angular Acceleration | radian per second squared | $rad/s^2$ |
| Heat Flux Density, Irradiance | watt per square meter | $W/m^2$ |
| Heat Capacity, Entropy | joule per kelvin | J/K |
| Specific Heat Capacity, Specific Entropy | joule per kilogram-kelvin | J/kg·K |
| Specific Energy | joule per kilogram | J/kg |
| Energy Density | joule per cubic meter | $J/m^3$ |
| Thermal Conductivity | watt per meter-kelvin | W/m·K |
| Electric Field Strength | volt per meter | V/m |
| Electric Charge Density | coulomb per cubic meter | $C/m^3$ |
| Electric Flux Density | coulomb per square meter | $C/m^2$ |
| Permittivity | farad per meter | F/m |
| Permeability | henry per meter | H/m |
| Molar Energy | joule per mole | J/mol |
| Molar Entropy, Molar Heat Capacity | joule per mole-kelvin | J/mol·K |
| Exposure (X-rays and γ rays) | coulomb per kilogram | C/kg |
| Absorbed Dose Rate | gray per second | Gy/s |
| Radian Intensity | watt per steradian | W/sr |
| Radiance | watt per square meter-steradian | $W/m^2 \cdot sr$ |

**Table 5. Units Outside SI, Accepted for Use with SI**

| Name | Symbol | Value in SI Units |
|---|---|---|
| minute | min | 1 min = 60 s |
| hour | h | 1 h = 60 min = 3600 s |
| day | d | 1 d = 24 h = 1440 min = 86400 s |
| liter | L | 1 L = 1 $dm^3$ = $10^{-3}$ m |
| metric ton | t | 1 t = $10^3$ kg =2205 lb |
| bel | B | 1 B = 10 dB |
| degree (angle) | ° | 1° = π/180 rad |
| minute (angle) | ′ | 1′ = (1/60)° = (π/10800) rad |
| second (angle) | ″ | 1″ = (1/60)′ = (π/6480000) rad |
| electron volt | eV | 1 eV = 1.60218 × $10^{-19}$ J |
| unified atomic mass unit | Da or u | 1 u = 1.66054 × $10^{-27}$ kg |
| astronomical unit | au | 1 au = 1.49598 × $10^{11}$ m |
| nautical mile | nmi | 1 nmi = 1852 m |
| knot | kn | 1 kn = 1 nmi·$h^{-1}$ = 0.514444 m·$s^{-}$ |
| are | a | 1 a = 100 $m^2$ |
| hectare | ha | 1 ha = 100 a = $10^4$ $m^2$ |
| bar | bar | 1 bar = $10^2$ kPa = $10^5$ Pa |
| ångström | Å | 1 Å = 0.1 nm = $10^{-10}$ m |
| curie | Ci | 1 Ci = 3.7 × $10^{10}$ Bq |
| roentgen | R | 1 R = 2.58 × $10^{-4}$ C·$kg^{-1}$ |
| rad | rad | 1 rad = $10^{-2}$ Gy |
| rem | rem | 1 rem = $10^2$ Sv |

**Table 6. SI Prefixes**

| Factor | Name | Symbol | Factor | Name | Symbol |
|---|---|---|---|---|---|
| $10^1$ | deca | da | $10^{-1}$ | deci | d |
| $10^2$ | hecto | h | $10^{-2}$ | centi | c |
| $10^3$ | kilo | k | $10^{-3}$ | milli | m |
| $10^6$ | mega | M | $10^{-6}$ | micro | μ |
| $10^9$ | giga | G | $10^{-9}$ | nano | n |
| $10^{12}$ | tera | T | $10^{-12}$ | pico | p |
| $10^{15}$ | peta | P | $10^{-15}$ | femto | f |
| $10^{18}$ | exa | E | $10^{-18}$ | atto | a |

## Standard of Length and the US Customary Unit System

Among all units of measure, the history of standard of length traces a clear path from the less scientific approach of physical object standards used in past centuries to the today's precise standards, based on physical constants on an atomic level.

The primary Imperial yard was set by the British Weights and Measures Act of 1824. But it was partially destroyed in a fire in 1834, and replaced by a new standard, made of an alloy of copper, tin, and zinc. Between 1845 and 1855, forty copies of the Imperial yard were cast. Bronze yard No. 11 went to the United States, an exact copy of the British Imperial yard, in both form and material.

By an Act of Congress, in 1866, the US legally recognized the meter as a standard of length equal to 39/39.37 = 0.9144 yard; for commercial purposes, 1 meter = 39.37 inches.

The Treaty of the Meter was signed in 1875, and the International Bureau of Weights and Measures (BIPM) in Severs, France, created 30 prototype line standards of platinum iridium. In 1889, United States received Nos. 27 and 21 prototype meters from BIPM. No. 27 was used as a standard for the US metric system from 1889 to 1960, but then a new standard was adopted, based not on an object whose dimensions could change over time but on a physical constant related the number of wavelengths krypton-86 emits during a specific energy level transition. In 1983 the meter was again redefined, described now in terms of the speed of light. In 2019, the standard was once more updated, based on an even more precise measurement of the speed of light. Prototype meter No. 27 is now kept in the NIST museum.

According to US 1988 Law on Metrification, the country would phase out the imperial and US customary unit, replacing them with their metric equivalents. As of 2018, the use of the metric system in trade and commerce was mandatory in all countries except the United States, Myanmar, and Liberia.

The meter has gone through several restandardizations. In 1983, the modern base unit was set as that length equal to the path length traveled by light in a vacuum during a time interval of $3.3356410 \times 10^{-9}$ second. The new meter represents the most advanced ideas of physical measurement, much like the second is based on the Cs 133 atomic clock.

Each new meter represented the most advanced ideas of standards, so it seemed that greater stability, as well as higher accuracy, would be secured by accepting the international meter as a fundamental standard of length. The director of national standard laboratory of the United States entered into agreement effective July 1, 1959, whereby the equivalent of 1 yard = 0.9144 meter and 1 avoirdupois pound = 0.45359237 kilogram, where adapted. In the United States, the SI metric system of units is predominantly used in the fields of physics, chemistry, and other sciences; however, US customary units are still widely used in industry and by the general public.

**Fundamental Constants**

| Name | Symbol | USCS units | SI units |
|---|---|---|---|
| Avogadro's number | $N_A$ | | $6.022 \times 10^{23}$ mol$^{-1}$ |
| Boltzman constant | $k$ | $5.65 \times 10^{-24}$ ft·lb$_f$/°R | $1.38065 \times 10^{-23}$ J/K |
| Faraday Constant | $F$ | | 96,485 C/mol |
| Gravitational constant | $g$ | 32.174 lb$_m$-ft/lb$_f$-sec$^2$ | 9.80667 m/sec$^2$ |
| Gravitational constant | $G$ | $5.65 \times 10^{-24}$ ft·lb$_f$/°R | $6.672 \times 10^{-11}$ N·m$^2$/kg$^2$ |
| Specific gas constant | $R$ | 53.3 ft·lb$_f$/lb$_m$·°R | 287 J/kg·K |
| Universal gas constant | $R$ | 1545 ft·lb$_f$/lbmol·°R | 8314 J/kmol·K |
| Volume (molal ideal gas) | $V$ | 359 ft$^3$/lbmol | 22.41 m$^3$/kmol |
| Pressure, atmospheric | $P$ | 14.696 lb$_f$/in$^2$ | 101,330 Pa (N/m$^2$) |
| Temperature, standard | $T$ | 32° F | 0° C |
| Density | | | |
| Air at 32° F (0° C) | | 0.0805 lb$_m$/ft$^3$ | 1.29 kg/m$^3$ |
| Air at 70° F (20° C), 1 atm | | 0.0749 lb$_m$/ft$^3$ | 1.20 kg/m$^3$ |
| Sea water | | 64 lb$_m$/ft$^3$ | 1025 kg/m$^3$ |
| Fresh water | | 62.4 lb$_m$/ft$^3$ | 1000 kg/m$^3$ |
| Mercury | | 849 lb$_m$/ft$^3$ | 13600 kg/m$^3$ |
| Earth | | 345 lb$_m$/ft$^3$ | 5520 kg/m$^3$ |

# US SYSTEM AND METRIC SYSTEM CONVERSIONS

## Units of Length

### Table 1. Linear Measure Conversion Factors

**Metric**

1 *kilometer (km)* =
- **1000** meters
- **100,000** centimeters
- **1,000,000** millimeters
- 0.539956 nautical mile
- 0.621371 mile
- 1093.61 yards
- 3280.83 feet
- 39,370.08 inches

1 *meter (m)* =
- **10** decimeters
- **100** centimeters
- **1000** millimeters
- 1.09361 yards
- 3.28084 feet
- 39.37008 inches

1 *decimeter (dm)* = **10** centimeters

1 *centimeter (cm)* =
- **0.01** meter
- **10** millimeters
- 0.0328 foot
- 0.3937 inch

1 *millimeter (mm)* =
- **0.001** meter
- **0.1** centimeter
- **1000** microns
- 0.03937 inch

1 *micrometer* or *micron (μm)* =
- **0.000001** meter = one millionth meter
- **0.0001** centimeter
- **0.001** millimeter
- 0.00003937 inch
- 39.37 micro-inches

**US Customary**

1 *mile (mi)* =
- 0.868976 nautical mile
- **1760** yards
- **5280** feet
- **63,360** inches
- **1.609344** kilometers
- **1609.344** meters
- **160,934.4** centimeters
- **1,609,344** millimeters

1 *yard (yd)* =
- **3** feet
- **36** inches
- **0.9144** meter
- **91.44** centimeter
- **914.4** millimeter

1 *foot (international) (ft)* =
- **12** inches = $\frac{1}{3}$ yard
- **0.3048** meter
- **30.48** centimeter
- **304.8** millimeters

1 *survey foot* =
- 1.000002 international feet
- $\frac{12}{39.37}$ = 0.3048006096012 meter

1 *inch (in)* =
- **1000** mils
- **1,000,000** micro-inches
- **2.54** centimeters
- **25.4** millimeters
- **25,400** microns

1 *mil* =
- **0.001** inch
- **1000** micro-inches
- **0.0254** millimeters

1 *micro-inch (μin)* =
- **0.000001** inch = one millionth inch
- **0.0254** micrometer (micron)

*Note*: Figures in **Bold** indicate exact conversion values

**Surveyors Measure**

1 *mile* = **8** furlongs = **80** chains

1 *furlong* = **10** chains = **220** yards

1 *chain* =
- **4** rods = **22** yards = **66** feet = **100** links

1 *rod* =
- **5.5** yards = **16.5** feet = **25** links
- 5.0292 meters

1 *link* = 7.92 inches

1 *span* = 9 inches

1 *hand* = 4 inches

**Nautical Measure**

1 *league* = 3 nautical miles

1 *nautical mile* =
- 1.1508 statute miles
- **6076.11549** feet
- 1.8516 kilometers

1 *fathom* = **2** yards = **6** feet

1 *knot* = nautical unit of speed =
- 1 nautical mile per hour
- 1.1508 statute miles per hour
- 1.8516 kilometers per hour

**Table 1.** *(Continued)* **Linear Measure Conversion Factors**

*One degree at the equator* =
60 nautical miles
69.047 statute miles
111.098 kilometers

*One minute at the equator* =
1 nautical mile
1.1508 statute miles
1.8516 kilometers

*360 degrees at the equator* =
circumference at equator
21,600 nautical miles
24,856.8 statute miles
39,995.4 kilometers

**Table 2. Circular and Angular Measure Conversion Factors**

*Circumference of circle* =
360 degrees = $2\pi$ radian = 6.283185 radian

1 *quadrant* = 90 degrees = $\pi/2$ radians = 1.570796 radians

1 *radian* = 57.2957795 degrees

1 *degree* (°) = 60 minutes = 3600 seconds = $\pi/180$ radian = 0.017453 radian

1 *minute* (′) = 60 seconds = 0.016667 degree = 0.000291 radian

$\pi$ = 3.141592654

**Table 3. Feet and Inches to Inches Conversion**

| Inches → / Feet ↓ | 0 | 1 | 2 | 3 | 4 | 5 | 6 | 7 | 8 | 9 | 10 | 11 |
|---|---|---|---|---|---|---|---|---|---|---|---|---|
| | Inches | | | | | | | | | | | |
| 0 | 0 | 1 | 2 | 3 | 4 | 5 | 6 | 7 | 8 | 9 | 10 | 11 |
| 1 | 12 | 13 | 14 | 15 | 16 | 17 | 18 | 19 | 20 | 21 | 22 | 23 |
| 2 | 24 | 25 | 26 | 27 | 28 | 29 | 30 | 31 | 32 | 33 | 34 | 35 |
| 3 | 36 | 37 | 38 | 39 | 40 | 41 | 42 | 43 | 44 | 45 | 46 | 47 |
| 4 | 48 | 49 | 50 | 51 | 52 | 53 | 54 | 55 | 56 | 57 | 58 | 59 |
| 5 | 60 | 61 | 62 | 63 | 64 | 65 | 66 | 67 | 68 | 69 | 70 | 71 |
| 6 | 72 | 73 | 74 | 75 | 76 | 77 | 78 | 79 | 80 | 81 | 82 | 83 |
| 7 | 84 | 85 | 86 | 87 | 88 | 89 | 90 | 91 | 92 | 93 | 94 | 95 |
| 8 | 96 | 97 | 98 | 99 | 100 | 101 | 102 | 103 | 104 | 105 | 106 | 107 |
| 9 | 108 | 109 | 110 | 111 | 112 | 113 | 114 | 115 | 116 | 117 | 118 | 119 |
| 10 | 120 | 121 | 122 | 123 | 124 | 125 | 126 | 127 | 128 | 129 | 130 | 131 |
| 20 | 240 | 241 | 242 | 243 | 244 | 245 | 246 | 247 | 248 | 249 | 250 | 251 |
| 30 | 360 | 361 | 362 | 363 | 364 | 365 | 366 | 367 | 368 | 369 | 370 | 371 |
| 40 | 480 | 481 | 482 | 483 | 484 | 485 | 486 | 487 | 488 | 489 | 490 | 491 |
| 50 | 600 | 601 | 602 | 603 | 604 | 605 | 606 | 607 | 608 | 609 | 610 | 611 |
| 60 | 720 | 721 | 722 | 723 | 724 | 725 | 726 | 727 | 728 | 729 | 730 | 731 |
| 70 | 840 | 841 | 842 | 843 | 844 | 845 | 846 | 847 | 848 | 849 | 850 | 851 |
| 80 | 960 | 961 | 962 | 963 | 964 | 965 | 966 | 967 | 968 | 969 | 970 | 971 |
| 90 | 1080 | 1081 | 1082 | 1083 | 1084 | 1085 | 1086 | 1087 | 1088 | 1089 | 1090 | 1091 |
| 100 | 1200 | 1201 | 1202 | 1203 | 1204 | 1205 | 1206 | 1207 | 1208 | 1209 | 1210 | 1211 |

*Example:* A tape measure reads 17 feet 8 inches. How many inches is this? *Solution:* Read down the first column of Table 3 to find 10 ft 0 inch = 120 inches. Next, find the intersection of the 7 ft row and the 8 inch column to get 92 inches. Add both results to get 120 inches + 92 inches = 212 inches.

**Table 4. Inches to Feet and Yards Conversion**

| inch | feet | yard | inch | feet | yard | inch | feet | yard | inch | feet | yard | inch | feet | yard |
|---|---|---|---|---|---|---|---|---|---|---|---|---|---|---|
| 100 | 8.3333 | 2.7778 | 10 | 0.8333 | 0.2778 | 1 | 0.0833 | 0.0278 | 0.1 | 0.0083 | 0.0028 | 0.01 | 0.0008 | 0.0003 |
| 200 | 16.6667 | 5.5556 | 20 | 1.6667 | 0.5556 | 2 | 0.1667 | 0.0556 | 0.2 | 0.0167 | 0.0056 | 0.02 | 0.0017 | 0.0006 |
| 300 | 25 | 8.3333 | 30 | 2.5 | 0.8333 | 3 | 0.25 | 0.0833 | 0.3 | 0.025 | 0.0083 | 0.03 | 0.0025 | 0.0008 |
| 400 | 33.3333 | 11.1111 | 40 | 3.3333 | 1.1111 | 4 | 0.3333 | 0.1111 | 0.4 | 0.0333 | 0.0111 | 0.04 | 0.0033 | 0.0011 |
| 500 | 41.6667 | 13.8889 | 50 | 4.1667 | 1.3889 | 5 | 0.4167 | 0.1389 | 0.5 | 0.0417 | 0.0139 | 0.05 | 0.0042 | 0.0014 |
| 600 | 50 | 16.6667 | 60 | 5 | 1.6667 | 6 | 0.5 | 0.1667 | 0.6 | 0.05 | 0.0167 | 0.06 | 0.005 | 0.0017 |
| 700 | 58.3333 | 19.4444 | 70 | 5.8333 | 1.9444 | 7 | 0.5833 | 0.1944 | 0.7 | 0.0583 | 0.0194 | 0.07 | 0.0058 | 0.0019 |
| 800 | 66.6667 | 22.2222 | 80 | 6.6667 | 2.2222 | 8 | 0.6667 | 0.2222 | 0.8 | 0.0667 | 0.0222 | 0.08 | 0.0067 | 0.0022 |
| 900 | 75 | 25.0000 | 90 | 7.5 | 2.5000 | 9 | 0.75 | 0.2500 | 0.9 | 0.075 | 0.0250 | 0.09 | 0.0075 | 0.0025 |
| 1000 | 83.3333 | 27.7778 | 100 | 8.3333 | 2.7778 | 10 | 0.8333 | 0.2778 | 1 | 0.0833 | 0.0278 | 0.1 | 0.0083 | 0.0028 |

**Table 5. Fractional Inches to Decimal Feet for 0 to 1 Foot**

| → Inches ↓ | 0 | 1 | 2 | 3 | 4 | 5 | 6 | 7 | 8 | 9 | 10 | 11 |
|---|---|---|---|---|---|---|---|---|---|---|---|---|
| | Feet | | | | | | | | | | | |
| 0 | 0.0000 | 0.0833 | 0.1667 | 0.2500 | 0.3333 | 0.4167 | 0.5000 | 0.5833 | 0.6667 | 0.7500 | 0.8333 | 0.9167 |
| 1/64 | 0.0013 | 0.0846 | 0.1680 | 0.2513 | 0.3346 | 0.4180 | 0.5013 | 0.5846 | 0.6680 | 0.7513 | 0.8346 | 0.9180 |
| 1/32 | 0.0026 | 0.0859 | 0.1693 | 0.2526 | 0.3359 | 0.4193 | 0.5026 | 0.5859 | 0.6693 | 0.7526 | 0.8359 | 0.9193 |
| 3/64 | 0.0039 | 0.0872 | 0.1706 | 0.2539 | 0.3372 | 0.4206 | 0.5039 | 0.5872 | 0.6706 | 0.7539 | 0.8372 | 0.9206 |
| 1/16 | 0.0052 | 0.0885 | 0.1719 | 0.2552 | 0.3385 | 0.4219 | 0.5052 | 0.5885 | 0.6719 | 0.7552 | 0.8385 | 0.9219 |
| 5/64 | 0.0065 | 0.0898 | 0.1732 | 0.2565 | 0.3398 | 0.4232 | 0.5065 | 0.5898 | 0.6732 | 0.7565 | 0.8398 | 0.9232 |
| 3/32 | 0.0078 | 0.0911 | 0.1745 | 0.2578 | 0.3411 | 0.4245 | 0.5078 | 0.5911 | 0.6745 | 0.7578 | 0.8411 | 0.9245 |
| 7/64 | 0.0091 | 0.0924 | 0.1758 | 0.2591 | 0.3424 | 0.4258 | 0.5091 | 0.5924 | 0.6758 | 0.7591 | 0.8424 | 0.9258 |
| 1/8 | 0.0104 | 0.0938 | 0.1771 | 0.2604 | 0.3438 | 0.4271 | 0.5104 | 0.5938 | 0.6771 | 0.7604 | 0.8438 | 0.9271 |
| 9/64 | 0.0117 | 0.0951 | 0.1784 | 0.2617 | 0.3451 | 0.4284 | 0.5117 | 0.5951 | 0.6784 | 0.7617 | 0.8451 | 0.9284 |
| 5/32 | 0.0130 | 0.0964 | 0.1797 | 0.2630 | 0.3464 | 0.4297 | 0.5130 | 0.5964 | 0.6797 | 0.7630 | 0.8464 | 0.9297 |
| 11/64 | 0.0143 | 0.0977 | 0.1810 | 0.2643 | 0.3477 | 0.4310 | 0.5143 | 0.5977 | 0.6810 | 0.7643 | 0.8477 | 0.9310 |
| 3/16 | 0.0156 | 0.0990 | 0.1823 | 0.2656 | 0.3490 | 0.4323 | 0.5156 | 0.5990 | 0.6823 | 0.7656 | 0.8490 | 0.9323 |
| 13/64 | 0.0169 | 0.1003 | 0.1836 | 0.2669 | 0.3503 | 0.4336 | 0.5169 | 0.6003 | 0.6836 | 0.7669 | 0.8503 | 0.9336 |
| 7/32 | 0.0182 | 0.1016 | 0.1849 | 0.2682 | 0.3516 | 0.4349 | 0.5182 | 0.6016 | 0.6849 | 0.7682 | 0.8516 | 0.9349 |
| 15/64 | 0.0195 | 0.1029 | 0.1862 | 0.2695 | 0.3529 | 0.4362 | 0.5195 | 0.6029 | 0.6862 | 0.7695 | 0.8529 | 0.9362 |
| 1/4 | 0.0208 | 0.1042 | 0.1875 | 0.2708 | 0.3542 | 0.4375 | 0.5208 | 0.6042 | 0.6875 | 0.7708 | 0.8542 | 0.9375 |
| 17/64 | 0.0221 | 0.1055 | 0.1888 | 0.2721 | 0.3555 | 0.4388 | 0.5221 | 0.6055 | 0.6888 | 0.7721 | 0.8555 | 0.9388 |
| 9/32 | 0.0234 | 0.1068 | 0.1901 | 0.2734 | 0.3568 | 0.4401 | 0.5234 | 0.6068 | 0.6901 | 0.7734 | 0.8568 | 0.9401 |
| 19/64 | 0.0247 | 0.1081 | 0.1914 | 0.2747 | 0.3581 | 0.4414 | 0.5247 | 0.6081 | 0.6914 | 0.7747 | 0.8581 | 0.9414 |
| 5/16 | 0.0260 | 0.1094 | 0.1927 | 0.2760 | 0.3594 | 0.4427 | 0.5260 | 0.6094 | 0.6927 | 0.7760 | 0.8594 | 0.9427 |
| 21/64 | 0.0273 | 0.1107 | 0.1940 | 0.2773 | 0.3607 | 0.4440 | 0.5273 | 0.6107 | 0.6940 | 0.7773 | 0.8607 | 0.9440 |
| 11/32 | 0.0286 | 0.1120 | 0.1953 | 0.2786 | 0.3620 | 0.4453 | 0.5286 | 0.6120 | 0.6953 | 0.7786 | 0.8620 | 0.9453 |
| 23/64 | 0.0299 | 0.1133 | 0.1966 | 0.2799 | 0.3633 | 0.4466 | 0.5299 | 0.6133 | 0.6966 | 0.7799 | 0.8633 | 0.9466 |
| 3/8 | 0.0313 | 0.1146 | 0.1979 | 0.2813 | 0.3646 | 0.4479 | 0.5313 | 0.6146 | 0.6979 | 0.7813 | 0.8646 | 0.9479 |
| 25/64 | 0.0326 | 0.1159 | 0.1992 | 0.2826 | 0.3659 | 0.4492 | 0.5326 | 0.6159 | 0.6992 | 0.7826 | 0.8659 | 0.9492 |
| 13/32 | 0.0339 | 0.1172 | 0.2005 | 0.2839 | 0.3672 | 0.4505 | 0.5339 | 0.6172 | 0.7005 | 0.7839 | 0.8672 | 0.9505 |
| 27/64 | 0.0352 | 0.1185 | 0.2018 | 0.2852 | 0.3685 | 0.4518 | 0.5352 | 0.6185 | 0.7018 | 0.7852 | 0.8685 | 0.9518 |
| 7/16 | 0.0365 | 0.1198 | 0.2031 | 0.2865 | 0.3698 | 0.4531 | 0.5365 | 0.6198 | 0.7031 | 0.7865 | 0.8698 | 0.9531 |
| 29/64 | 0.0378 | 0.1211 | 0.2044 | 0.2878 | 0.3711 | 0.4544 | 0.5378 | 0.6211 | 0.7044 | 0.7878 | 0.8711 | 0.9544 |
| 15/32 | 0.0391 | 0.1224 | 0.2057 | 0.2891 | 0.3724 | 0.4557 | 0.5391 | 0.6224 | 0.7057 | 0.7891 | 0.8724 | 0.9557 |
| 31/64 | 0.0404 | 0.1237 | 0.2070 | 0.2904 | 0.3737 | 0.4570 | 0.5404 | 0.6237 | 0.7070 | 0.7904 | 0.8737 | 0.9570 |
| 1/2 | 0.0417 | 0.1250 | 0.2083 | 0.2917 | 0.3750 | 0.4583 | 0.5417 | 0.6250 | 0.7083 | 0.7917 | 0.8750 | 0.9583 |
| 33/64 | 0.0430 | 0.1263 | 0.2096 | 0.2930 | 0.3763 | 0.4596 | 0.5430 | 0.6263 | 0.7096 | 0.7930 | 0.8763 | 0.9596 |
| 17/32 | 0.0443 | 0.1276 | 0.2109 | 0.2943 | 0.3776 | 0.4609 | 0.5443 | 0.6276 | 0.7109 | 0.7943 | 0.8776 | 0.9609 |
| 35/64 | 0.0456 | 0.1289 | 0.2122 | 0.2956 | 0.3789 | 0.4622 | 0.5456 | 0.6289 | 0.7122 | 0.7956 | 0.8789 | 0.9622 |
| 9/16 | 0.0469 | 0.1302 | 0.2135 | 0.2969 | 0.3802 | 0.4635 | 0.5469 | 0.6302 | 0.7135 | 0.7969 | 0.8802 | 0.9635 |
| 37/64 | 0.0482 | 0.1315 | 0.2148 | 0.2982 | 0.3815 | 0.4648 | 0.5482 | 0.6315 | 0.7148 | 0.7982 | 0.8815 | 0.9648 |
| 19/32 | 0.0495 | 0.1328 | 0.2161 | 0.2995 | 0.3828 | 0.4661 | 0.5495 | 0.6328 | 0.7161 | 0.7995 | 0.8828 | 0.9661 |
| 39/64 | 0.0508 | 0.1341 | 0.2174 | 0.3008 | 0.3841 | 0.4674 | 0.5508 | 0.6341 | 0.7174 | 0.8008 | 0.8841 | 0.9674 |
| 5/8 | 0.0521 | 0.1354 | 0.2188 | 0.3021 | 0.3854 | 0.4688 | 0.5521 | 0.6354 | 0.7188 | 0.8021 | 0.8854 | 0.9688 |
| 41/64 | 0.0534 | 0.1367 | 0.2201 | 0.3034 | 0.3867 | 0.4701 | 0.5534 | 0.6367 | 0.7201 | 0.8034 | 0.8867 | 0.9701 |
| 21/32 | 0.0547 | 0.1380 | 0.2214 | 0.3047 | 0.3880 | 0.4714 | 0.5547 | 0.6380 | 0.7214 | 0.8047 | 0.8880 | 0.9714 |
| 43/64 | 0.0560 | 0.1393 | 0.2227 | 0.3060 | 0.3893 | 0.4727 | 0.5560 | 0.6393 | 0.7227 | 0.8060 | 0.8893 | 0.9727 |
| 11/16 | 0.0573 | 0.1406 | 0.2240 | 0.3073 | 0.3906 | 0.4740 | 0.5573 | 0.6406 | 0.7240 | 0.8073 | 0.8906 | 0.9740 |
| 45/64 | 0.0586 | 0.1419 | 0.2253 | 0.3086 | 0.3919 | 0.4753 | 0.5586 | 0.6419 | 0.7253 | 0.8086 | 0.8919 | 0.9753 |
| 23/32 | 0.0599 | 0.1432 | 0.2266 | 0.3099 | 0.3932 | 0.4766 | 0.5599 | 0.6432 | 0.7266 | 0.8099 | 0.8932 | 0.9766 |
| 47/64 | 0.0612 | 0.1445 | 0.2279 | 0.3112 | 0.3945 | 0.4779 | 0.5612 | 0.6445 | 0.7279 | 0.8112 | 0.8945 | 0.9779 |
| 3/4 | 0.0625 | 0.1458 | 0.2292 | 0.3125 | 0.3958 | 0.4792 | 0.5625 | 0.6458 | 0.7292 | 0.8125 | 0.8958 | 0.9792 |
| 49/64 | 0.0638 | 0.1471 | 0.2305 | 0.3138 | 0.3971 | 0.4805 | 0.5638 | 0.6471 | 0.7305 | 0.8138 | 0.8971 | 0.9805 |
| 25/32 | 0.0651 | 0.1484 | 0.2318 | 0.3151 | 0.3984 | 0.4818 | 0.5651 | 0.6484 | 0.7318 | 0.8151 | 0.8984 | 0.9818 |
| 51/64 | 0.0664 | 0.1497 | 0.2331 | 0.3164 | 0.3997 | 0.4831 | 0.5664 | 0.6497 | 0.7331 | 0.8164 | 0.8997 | 0.9831 |
| 13/16 | 0.0677 | 0.1510 | 0.2344 | 0.3177 | 0.4010 | 0.4844 | 0.5677 | 0.6510 | 0.7344 | 0.8177 | 0.9010 | 0.9844 |
| 53/64 | 0.0690 | 0.1523 | 0.2357 | 0.3190 | 0.4023 | 0.4857 | 0.5690 | 0.6523 | 0.7357 | 0.8190 | 0.9023 | 0.9857 |
| 27/32 | 0.0703 | 0.1536 | 0.2370 | 0.3203 | 0.4036 | 0.4870 | 0.5703 | 0.6536 | 0.7370 | 0.8203 | 0.9036 | 0.9870 |
| 55/64 | 0.0716 | 0.1549 | 0.2383 | 0.3216 | 0.4049 | 0.4883 | 0.5716 | 0.6549 | 0.7383 | 0.8216 | 0.9049 | 0.9883 |
| 7/8 | 0.0729 | 0.1563 | 0.2396 | 0.3229 | 0.4063 | 0.4896 | 0.5729 | 0.6563 | 0.7396 | 0.8229 | 0.9063 | 0.9896 |
| 57/64 | 0.0742 | 0.1576 | 0.2409 | 0.3242 | 0.4076 | 0.4909 | 0.5742 | 0.6576 | 0.7409 | 0.8242 | 0.9076 | 0.9909 |
| 29/32 | 0.0755 | 0.1589 | 0.2422 | 0.3255 | 0.4089 | 0.4922 | 0.5755 | 0.6589 | 0.7422 | 0.8255 | 0.9089 | 0.9922 |
| 59/64 | 0.0768 | 0.1602 | 0.2435 | 0.3268 | 0.4102 | 0.4935 | 0.5768 | 0.6602 | 0.7435 | 0.8268 | 0.9102 | 0.9935 |
| 15/16 | 0.0781 | 0.1615 | 0.2448 | 0.3281 | 0.4115 | 0.4948 | 0.5781 | 0.6615 | 0.7448 | 0.8281 | 0.9115 | 0.9948 |
| 61/64 | 0.0794 | 0.1628 | 0.2461 | 0.3294 | 0.4128 | 0.4961 | 0.5794 | 0.6628 | 0.7461 | 0.8294 | 0.9128 | 0.9961 |
| 31/32 | 0.0807 | 0.1641 | 0.2474 | 0.3307 | 0.4141 | 0.4974 | 0.5807 | 0.6641 | 0.7474 | 0.8307 | 0.9141 | 0.9974 |
| 63/64 | 0.0820 | 0.1654 | 0.2487 | 0.3320 | 0.4154 | 0.4987 | 0.5820 | 0.6654 | 0.7487 | 0.8320 | 0.9154 | 0.9987 |
| 1 | 0.0833 | 0.1667 | 0.2500 | 0.3333 | 0.4167 | 0.5000 | 0.5833 | 0.6667 | 0.7500 | 0.8333 | 0.9167 | 1.0000 |

*Example:* Convert 78¾ inches to feet. *Solution:* From Table 4, find 70 inches = 5.8333 feet and add to that 8¾ inches = 0.7292 feet found in Table 5 at the intersection of the ¾ inch row and the 8 inch column. Thus, 78¾ inches = 5.8333 + 0.7292 = 6.5625 feet.

## Table 6. Feet to Inches Conversion

| feet | inch | feet | inch | feet | inch | feet | inch | feet | inch | feet | inch | feet | inch |
|---|---|---|---|---|---|---|---|---|---|---|---|---|---|
| 100 | 1200 | 10 | 120 | 1 | 12 | 0.1 | 1.2 | 0.01 | 0.12 | 0.001 | 0.012 | 0.0001 | 0.0012 |
| 200 | 2400 | 20 | 240 | 2 | 24 | 0.2 | 2.4 | 0.02 | 0.24 | 0.002 | 0.024 | 0.0002 | 0.0024 |
| 300 | 3600 | 30 | 360 | 3 | 36 | 0.3 | 3.6 | 0.03 | 0.36 | 0.003 | 0.036 | 0.0003 | 0.0036 |
| 400 | 4800 | 40 | 480 | 4 | 48 | 0.4 | 4.8 | 0.04 | 0.48 | 0.004 | 0.048 | 0.0004 | 0.0048 |
| 500 | 6000 | 50 | 600 | 5 | 60 | 0.5 | 6 | 0.05 | 0.6 | 0.005 | 0.06 | 0.0005 | 0.006 |
| 600 | 7200 | 60 | 720 | 6 | 72 | 0.6 | 7.2 | 0.06 | 0.72 | 0.006 | 0.072 | 0.0006 | 0.0072 |
| 700 | 8400 | 70 | 840 | 7 | 84 | 0.7 | 8.4 | 0.07 | 0.84 | 0.007 | 0.084 | 0.0007 | 0.0084 |
| 800 | 9600 | 80 | 960 | 8 | 96 | 0.8 | 9.6 | 0.08 | 0.96 | 0.008 | 0.096 | 0.0008 | 0.0096 |
| 900 | 10800 | 90 | 1080 | 9 | 108 | 0.9 | 10.8 | 0.09 | 1.08 | 0.009 | 0.108 | 0.0009 | 0.0108 |
| 1000 | 12000 | 100 | 1200 | 10 | 120 | 1 | 12 | 0.1 | 1.2 | 0.01 | 0.12 | 0.001 | 0.012 |

## Table 7. Fractional and Decimal Inches to Millimeters, Exact[a] Values

| Fractional Inch | Decimal Inches | Millimeters | Fractional Inches | Decimal Inches | Millimeters |
|---|---|---|---|---|---|
| 1/64 | 0.015625 | 0.396875 | | 0.511811024 | 13 |
| 1/32 | 0.03125 | 0.79375 | 33/64 | 0.515625 | 13.096875 |
| | 0.039370079 | 1 | 17/32 | 0.53125 | 13.49375 |
| 3/64 | 0.046875 | 1.190625 | 35/64 | 0.546875 | 13.890625 |
| 1/16 | 0.0625 | 1.5875 | | 0.551181102 | 14 |
| 5/64 | 0.078125 | 1.984375 | 9/16 | 0.5625 | 14.2875 |
| | 0.078740157 | 2 | 37/64 | 0.578125 | 14.684375 |
| 1/12 | 0.08$\overline{33}$[b] | 2.11$\overline{66}$ | 7/12 | 0.58$\overline{33}$ | 14.81$\overline{66}$ |
| 3/32 | 0.09375 | 2.38125 | | 0.590551181 | 15 |
| 7/64 | 0.109375 | 2.778125 | 19/32 | 0.59375 | 15.08125 |
| | 0.118110236 | 3 | 39/64 | 0.609375 | 15.478125 |
| 1/8 | 0.125 | 3.175 | 5/8 | 0.625 | 15.875 |
| 9/64 | 0.140625 | 3.571875 | | 0.62992126 | 16 |
| 5/32 | 0.15625 | 3.96875 | 41/64 | 0.640625 | 16.271875 |
| | 0.157480315 | 4 | 21/32 | 0.65625 | 16.66875 |
| 1/6 | 0.1$\overline{66}$ | 4.2$\overline{33}$ | 2/3 | 0.$\overline{66}$ | 16.9$\overline{33}$ |
| 11/64 | 0.171875 | 4.365625 | | 0.669291339 | 17 |
| 3/16 | 0.1875 | 4.7625 | 43/64 | 0.671875 | 17.065625 |
| | 0.196850394 | 5 | 11/16 | 0.6875 | 17.4625 |
| 13/64 | 0.203125 | 5.159375 | 45/64 | 0.703125 | 17.859375 |
| 7/32 | 0.21875 | 5.55625 | | 0.708661417 | 18 |
| 15/64 | 0.234375 | 5.953125 | 23/32 | 0.71875 | 18.25625 |
| | 0.236220472 | 6 | 47/64 | 0.734375 | 18.653125 |
| 1/4 | 0.25 | 6.35 | | 0.748031496 | 19 |
| 17/64 | 0.265625 | 6.746875 | 3/4 | 0.75 | 19.05 |
| | 0.275590551 | 7 | 49/64 | 0.765625 | 19.446875 |
| 9/32 | 0.28125 | 7.14375 | 25/32 | 0.78125 | 19.84375 |
| 19/64 | 0.296875 | 7.540625 | | 0.787401575 | 20 |
| 5/16 | 0.3125 | 7.9375 | 51/64 | 0.796875 | 20.240625 |
| | 0.31496063 | 8 | 13/16 | 0.8125 | 20.6375 |
| 21/64 | 0.328125 | 8.334375 | | 0.826771654 | 21 |
| 1/3 | 0.$\overline{33}$ | 8.4$\overline{66}$ | 53/64 | 0.828125 | 21.034375 |
| 11/32 | 0.34375 | 8.73125 | 27/32 | 0.84375 | 21.43125 |
| | 0.354330709 | 9 | 55/64 | 0.859375 | 21.828125 |
| 23/64 | 0.359375 | 9.128125 | | 0.866141732 | 22 |
| 3/8 | 0.375 | 9.525 | 7/8 | 0.875 | 22.225 |
| 25/64 | 0.390625 | 9.921875 | 57/64 | 0.890625 | 22.621875 |
| | 0.393700787 | 10 | | 0.905511811 | 23 |
| 13/32 | 0.40625 | 10.31875 | 29/32 | 0.90625 | 23.01875 |
| 5/12 | 0.41$\overline{66}$ | 10.58$\overline{33}$ | 11/12 | 0.91$\overline{66}$ | 23.28$\overline{33}$ |
| 27/64 | 0.421875 | 10.715625 | 59/64 | 0.921875 | 23.415625 |
| | 0.433070866 | 11 | 15/16 | 0.9375 | 23.8125 |
| 7/16 | 0.4375 | 11.1125 | | 0.94488189 | 24 |
| 29/64 | 0.453125 | 11.509375 | 61/64 | 0.953125 | 24.209375 |
| 15/32 | 0.46875 | 11.90625 | 31/32 | 0.96875 | 24.60625 |
| | 0.472440945 | 12 | | 0.984251969 | 25 |
| 31/64 | 0.484375 | 12.303125 | 63/64 | 0.984375 | 25.003125 |
| 1/2 | 0.5 | 12.7 | | | |

[a] Reading left to right from all "Fractional" and "Decimal Inches" columns to "Millimeters" columns: conversion values are exact. Reading right to left from the "Millimeters" columns: whole number millimeters conversions to "Decimal Inches" are rounded to 9 decimal places.

[b] A bar over a group of digits indicates they repeat indefinitely; for example, 0.08$\overline{33}$ = 0.08333..

## Table 8a. Inches to Millimeters Conversion

| inch | mm | inch | mm | inch | mm | inch | mm | inch | mm | inch | mm |
|---|---|---|---|---|---|---|---|---|---|---|---|
| 10 | 254.00000 | 1 | 25.40000 | 0.1 | 2.54000 | .01 | 0.25400 | 0.001 | 0.02540 | 0.0001 | 0.00254 |
| 20 | 508.00000 | 2 | 50.80000 | 0.2 | 5.08000 | .02 | 0.50800 | 0.002 | 0.05080 | 0.0002 | 0.00508 |
| 30 | 762.00000 | 3 | 76.20000 | 0.3 | 7.62000 | .03 | 0.76200 | 0.003 | 0.07620 | 0.0003 | 0.00762 |
| 40 | 1,016.00000 | 4 | 101.60000 | 0.4 | 10.16000 | .04 | 1.01600 | 0.004 | 0.10160 | 0.0004 | 0.01016 |
| 50 | 1,270.00000 | 5 | 127.00000 | 0.5 | 12.70000 | .05 | 1.27000 | 0.005 | 0.12700 | 0.0005 | 0.01270 |
| 60 | 1,524.00000 | 6 | 152.40000 | 0.6 | 15.24000 | .06 | 1.52400 | 0.006 | 0.15240 | 0.0006 | 0.01524 |
| 70 | 1,778.00000 | 7 | 177.80000 | 0.7 | 17.78000 | .07 | 1.77800 | 0.007 | 0.17780 | 0.0007 | 0.01778 |
| 80 | 2,032.00000 | 8 | 203.20000 | 0.8 | 20.32000 | .08 | 2.03200 | 0.008 | 0.20320 | 0.0008 | 0.02032 |
| 90 | 2,286.00000 | 9 | 228.60000 | 0.9 | 22.86000 | .09 | 2.2860 | 0.009 | 0.22860 | 0.0009 | 0.02286 |
| 100 | 2,540.00000 | 10 | 254.00000 | 1.0 | 25.40000 | .10 | 2.54000 | 0.010 | 0.25400 | 0.0010 | 0.02540 |

All values in this table are exact. For inches to centimeters, shift decimal point in mm column one place to left and read centimeters; thus, for example, 40 in. = 1016 mm = 101.6 cm.

## Table 8b. Millimeters to Inches Conversion

| mm | inch | mm | inch | mm | inch | mm | inch | mm | inch | mm | inch |
|---|---|---|---|---|---|---|---|---|---|---|---|
| 100 | 3.93701 | 10 | 0.39370 | 1 | 0.03937 | 0.1 | 0.00394 | 0.01 | .000039 | 0.001 | 0.00004 |
| 200 | 7.87402 | 20 | 0.78740 | 2 | 0.07874 | 0.2 | 0.00787 | 0.02 | .00079 | 0.002 | 0.00008 |
| 300 | 11.81102 | 30 | 1.18110 | 3 | 0.11811 | 0.3 | 0.01181 | 0.03 | .00118 | 0.003 | 0.00012 |
| 400 | 15.74803 | 40 | 1.57480 | 4 | 0.15748 | 0.4 | 0.01575 | 0.04 | .00157 | 0.004 | 0.00016 |
| 500 | 19.68504 | 50 | 1.96850 | 5 | 0.19685 | 0.5 | 0.01969 | 0.05 | .00197 | 0.005 | 0.00020 |
| 600 | 23.62205 | 60 | 2.36220 | 6 | 0.23622 | 0.6 | 0.02362 | 0.06 | .00236 | 0.006 | 0.00024 |
| 700 | 27.55906 | 70 | 2.75591 | 7 | 0.27559 | 0.7 | 0.02756 | 0.07 | .00276 | 0.007 | 0.00028 |
| 800 | 31.49606 | 80 | 3.14961 | 8 | 0.31496 | 0.8 | 0.03150 | 0.08 | .00315 | 0.008 | 0.00031 |
| 900 | 35.43307 | 90 | 3.54331 | 9 | 0.35433 | 0.9 | 0.03543 | 0.09 | .00354 | 0.009 | 0.00035 |
| 1,000 | 39.37008 | 100 | 3.93701 | 10 | 0.39370 | 1.0 | 0.03937 | 0.10 | .00394 | 0.010 | 0.00039 |

Based on 1 inch = 25.4 millimeters, exactly. For centimeters to inches, shift decimal point of centimeter value one place to right and enter mm column; thus, for example, 70 cm = 700 mm = 27.55906 inches.

## Table 9. Feet to Millimeters Conversion

| feet | mm | feet | mm | feet | mm | feet | mm | feet | mm |
|---|---|---|---|---|---|---|---|---|---|
| 100 | 30,480 | 10 | 3,048 | 1 | 304.8 | 0.1 | 30.48 | 0.01 | 3.048 |
| 200 | 60,960 | 20 | 6,096 | 2 | 609.6 | 0.2 | 60.96 | 0.02 | 6.096 |
| 300 | 91,440 | 30 | 9,144 | 3 | 914.4 | 0.3 | 91.44 | 0.03 | 9.144 |
| 400 | 121,920 | 40 | 12,192 | 4 | 1,219.2 | 0.4 | 121.92 | 0.04 | 12.192 |
| 500 | 152,400 | 50 | 15,240 | 5 | 1,524.0 | 0.5 | 152.40 | 0.05 | 15.240 |
| 600 | 182,880 | 60 | 18,288 | 6 | 1,828.8 | 0.6 | 182.88 | 0.06 | 18.288 |
| 700 | 213,360 | 70 | 21,336 | 7 | 2,133.6 | 0.7 | 213.36 | 0.07 | 21.336 |
| 800 | 243,840 | 80 | 24,384 | 8 | 2,438.4 | 0.8 | 243.84 | 0.08 | 24.384 |
| 900 | 274,320 | 90 | 27,432 | 9 | 2,743.2 | 0.9 | 274.32 | 0.09 | 27.432 |
| 1,000 | 304,800 | 100 | 30,480 | 10 | 3,048.0 | 1.0 | 304.80 | 0.10 | 30.480 |

Based on 1 inch = 25.4 millimeters, exactly. All values in this table are exact.

*Example 1:* Convert 293 feet, $5\frac{47}{64}$ inches to mm.

| | | | | |
|---|---|---|---|---|
| 200 ft | | = | 60,960.0 | mm |
| 90 ft | | = | 27,432.0 | mm |
| 3 ft | | = | 914.4 | mm |
| 5 | inch | = | 127.0 | mm |
| $\frac{47}{64}$ | inch | = | 18.653 | mm |
| 293 ft $5\frac{47}{64}$ | inch | = | 89,452.053 | mm |

*Example 2:* Convert 71.86 feet to mm.

| | | | | |
|---|---|---|---|---|
| 70.0 | feet | = | 21,336.0 | mm |
| 1.0 | feet | = | 304.8 | mm |
| 0.80 | feet | = | 243.84 | mm |
| 0.06 | feet | = | 18.288 | mm |
| 71.86 | feet | = | 21,902.928 | mm |

**Table 10. Mixed Fractional Inches to Millimeters Conversion for 0 to 41 Inches in 1/64-Inch Increments**

| → Inches↓ | 0 | 1 | 2 | 3 | 4 | 5 | 6 | 7 | 8 | 9 | 10 | 20 | 30 | 40 |
|---|---|---|---|---|---|---|---|---|---|---|---|---|---|---|
| | Millimeters | | | | | | | | | | | | | |
| 0 | 0 | 25.4 | 50.8 | 76.2 | 101.6 | 127.0 | 152.4 | 177.8 | 203.2 | 228.6 | 254.0 | 508.0 | 762.0 | 1016.0 |
| 1/64 | 0.396875 | 25.796875 | 51.196875 | 76.596875 | 101.996875 | 127.396875 | 152.796875 | 178.196875 | 203.596875 | 228.996875 | 254.396875 | 508.396875 | 762.396875 | 1016.396875 |
| 1/32 | 0.79375 | 26.19375 | 51.59375 | 76.99375 | 102.39375 | 127.79375 | 153.19375 | 178.59375 | 203.99375 | 229.39375 | 254.79375 | 508.79375 | 762.79375 | 1016.79375 |
| 3/64 | 1.190625 | 26.590625 | 51.990625 | 77.390625 | 102.790625 | 128.190625 | 153.590625 | 178.990625 | 204.390625 | 229.790625 | 255.190625 | 509.190625 | 763.190625 | 1017.190625 |
| 1/16 | 1.5875 | 26.9875 | 52.3875 | 77.7875 | 103.1875 | 128.5875 | 153.9875 | 179.3875 | 204.7875 | 230.1875 | 255.5875 | 509.5875 | 763.5875 | 1017.5875 |
| 5/64 | 1.984375 | 27.384375 | 52.784375 | 78.184375 | 103.584375 | 128.984375 | 154.384375 | 179.784375 | 205.184375 | 230.584375 | 255.984375 | 509.984375 | 763.984375 | 1017.984375 |
| 3/32 | 2.38125 | 27.78125 | 53.18125 | 78.58125 | 103.98125 | 129.38125 | 154.78125 | 180.18125 | 205.58125 | 230.98125 | 256.38125 | 510.38125 | 764.38125 | 1018.38125 |
| 7/64 | 2.778125 | 28.178125 | 53.578125 | 78.978125 | 104.378125 | 129.778125 | 155.178125 | 180.578125 | 205.978125 | 231.378125 | 256.778125 | 510.778125 | 764.778125 | 1018.778125 |
| 1/8 | 3.175 | 28.575 | 53.975 | 79.375 | 104.775 | 130.175 | 155.575 | 180.975 | 206.375 | 231.775 | 257.175 | 511.175 | 765.175 | 1019.175 |
| 9/64 | 3.571875 | 28.971875 | 54.371875 | 79.771875 | 105.171875 | 130.571875 | 155.971875 | 181.371875 | 206.771875 | 232.171875 | 257.571875 | 511.571875 | 765.571875 | 1019.571875 |
| 5/32 | 3.96875 | 29.36875 | 54.76875 | 80.16875 | 105.56875 | 130.96875 | 156.36875 | 181.76875 | 207.16875 | 232.56875 | 257.96875 | 511.96875 | 765.96875 | 1019.96875 |
| 11/64 | 4.365625 | 29.765625 | 55.165625 | 80.565625 | 105.965625 | 131.365625 | 156.765625 | 182.165625 | 207.565625 | 232.965625 | 258.365625 | 512.365625 | 766.365625 | 1020.365625 |
| 3/16 | 4.7625 | 30.1625 | 55.5625 | 80.9625 | 106.3625 | 131.7625 | 157.1625 | 182.5625 | 207.9625 | 233.3625 | 258.7625 | 512.7625 | 766.7625 | 1020.7625 |
| 13/64 | 5.159375 | 30.559375 | 55.959375 | 81.359375 | 106.759375 | 132.159375 | 157.559375 | 182.959375 | 208.359375 | 233.759375 | 259.159375 | 513.159375 | 767.159375 | 1021.159375 |
| 7/32 | 5.55625 | 30.95625 | 56.35625 | 81.75625 | 107.15625 | 132.55625 | 157.95625 | 183.35625 | 208.75625 | 234.15625 | 259.55625 | 513.55625 | 767.55625 | 1021.55625 |
| 15/64 | 5.953125 | 31.353125 | 56.753125 | 82.153125 | 107.553125 | 132.953125 | 158.353125 | 183.753125 | 209.153125 | 234.553125 | 259.953125 | 513.953125 | 767.953125 | 1021.953125 |
| 1/4 | 6.35 | 31.75 | 57.15 | 82.55 | 107.95 | 133.35 | 158.75 | 184.15 | 209.55 | 234.95 | 260.35 | 514.35 | 768.35 | 1022.35 |
| 17/64 | 6.746875 | 32.146875 | 57.546875 | 82.946875 | 108.346875 | 133.746875 | 159.146875 | 184.546875 | 209.946875 | 235.346875 | 260.746875 | 514.746875 | 768.746875 | 1022.746875 |
| 9/32 | 7.14375 | 32.54375 | 57.94375 | 83.34375 | 108.74375 | 134.14375 | 159.54375 | 184.94375 | 210.34375 | 235.74375 | 261.14375 | 515.14375 | 769.14375 | 1023.14375 |
| 19/64 | 7.540625 | 32.940625 | 58.340625 | 83.740625 | 109.140625 | 134.540625 | 159.940625 | 185.340625 | 210.740625 | 236.140625 | 261.540625 | 515.540625 | 769.540625 | 1023.540625 |
| 5/16 | 7.9375 | 33.3375 | 58.7375 | 84.1375 | 109.5375 | 134.9375 | 160.3375 | 185.7375 | 211.1375 | 236.5375 | 261.9375 | 515.9375 | 769.9375 | 1023.9375 |
| 21/64 | 8.334375 | 33.734375 | 59.134375 | 84.534375 | 109.934375 | 135.334375 | 160.734375 | 186.134375 | 211.534375 | 236.934375 | 262.334375 | 516.334375 | 770.334375 | 1024.334375 |
| 11/32 | 8.73125 | 34.13125 | 59.53125 | 84.93125 | 110.33125 | 135.73125 | 161.13125 | 186.53125 | 211.93125 | 237.33125 | 262.73125 | 516.73125 | 770.73125 | 1024.73125 |
| 23/64 | 9.128125 | 34.528125 | 59.928125 | 85.328125 | 110.728125 | 136.128125 | 161.528125 | 186.928125 | 212.328125 | 237.728125 | 263.128125 | 517.128125 | 771.128125 | 1025.128125 |
| 3/8 | 9.525 | 34.925 | 60.325 | 85.725 | 111.125 | 136.525 | 161.925 | 187.325 | 212.725 | 238.125 | 263.525 | 517.525 | 771.525 | 1025.525 |
| 25/64 | 9.921875 | 35.321875 | 60.721875 | 86.121875 | 111.521875 | 136.921875 | 162.321875 | 187.721875 | 213.121875 | 238.521875 | 263.921875 | 517.921875 | 771.921875 | 1025.921875 |
| 13/32 | 10.31875 | 35.71875 | 61.11875 | 86.51875 | 111.91875 | 137.31875 | 162.71875 | 188.11875 | 213.51875 | 238.91875 | 264.31875 | 518.31875 | 772.31875 | 1026.31875 |
| 27/64 | 10.715625 | 36.115625 | 61.515625 | 86.915625 | 112.315625 | 137.715625 | 163.115625 | 188.515625 | 213.915625 | 239.315625 | 264.715625 | 518.715625 | 772.715625 | 1026.715625 |
| 7/16 | 11.1125 | 36.5125 | 61.9125 | 87.3125 | 112.7125 | 138.1125 | 163.5125 | 188.9125 | 214.3125 | 239.7125 | 265.1125 | 519.1125 | 773.1125 | 1027.1125 |
| 29/64 | 11.509375 | 36.909375 | 62.309375 | 87.709375 | 113.109375 | 138.509375 | 163.909375 | 189.309375 | 214.709375 | 240.109375 | 265.509375 | 519.509375 | 773.509375 | 1027.509375 |
| 15/32 | 11.90625 | 37.30625 | 62.70625 | 88.10625 | 113.50625 | 138.90625 | 164.30625 | 189.70625 | 215.10625 | 240.50625 | 265.90625 | 519.90625 | 773.90625 | 1027.90625 |
| 31/64 | 12.303125 | 37.703125 | 63.103125 | 88.503125 | 113.903125 | 139.303125 | 164.703125 | 190.103125 | 215.503125 | 240.903125 | 266.303125 | 520.303125 | 774.303125 | 1028.303125 |
| 1/2 | 12.7 | 38.1 | 63.5 | 88.9 | 114.3 | 139.7 | 165.1 | 190.5 | 215.9 | 241.3 | 266.7 | 520.7 | 774.7 | 1028.7 |

**Table 10.** *(Continued)* **Mixed Fractional Inches to Millimeters Conversion for 0 to 41 Inches in $^{1}/_{64}$-Inch Increments**

| → Inches↓ | 0 | 1 | 2 | 3 | 4 | 5 | 6 | 7 | 8 | 9 | 10 | 20 | 30 | 40 |
|---|---|---|---|---|---|---|---|---|---|---|---|---|---|---|
| | Millimeters | | | | | | | | | | | | | |
| 33/64 | 13.096875 | 38.496875 | 63.896875 | 89.296875 | 114.696875 | 140.096875 | 165.496875 | 190.896875 | 216.296875 | 241.696875 | 267.096875 | 521.096875 | 775.096875 | 1029.096875 |
| 17/32 | 13.49375 | 38.89375 | 64.29375 | 89.69375 | 115.09375 | 140.49375 | 165.89375 | 191.29375 | 216.69375 | 242.09375 | 267.49375 | 521.49375 | 775.49375 | 1029.49375 |
| 35/64 | 13.890625 | 39.290625 | 64.690625 | 90.090625 | 115.490625 | 140.890625 | 166.290625 | 191.690625 | 217.090625 | 242.490625 | 267.890625 | 521.890625 | 775.890625 | 1029.890625 |
| 9/16 | 14.2875 | 39.6875 | 65.0875 | 90.4875 | 115.8875 | 141.2875 | 166.6875 | 192.0875 | 217.4875 | 242.8875 | 268.2875 | 522.2875 | 776.2875 | 1030.2875 |
| 37/64 | 14.684375 | 40.084375 | 65.484375 | 90.884375 | 116.284375 | 141.684375 | 167.084375 | 192.484375 | 217.884375 | 243.284375 | 268.684375 | 522.684375 | 776.684375 | 1030.684375 |
| 19/32 | 15.08125 | 40.48125 | 65.88125 | 91.28125 | 116.68125 | 142.08125 | 167.48125 | 192.88125 | 218.28125 | 243.68125 | 269.08125 | 523.08125 | 777.08125 | 1031.08125 |
| 39/64 | 15.478125 | 40.878125 | 66.278125 | 91.678125 | 117.078125 | 142.478125 | 167.878125 | 193.278125 | 218.678125 | 244.078125 | 269.478125 | 523.478125 | 777.478125 | 1031.478125 |
| 5/8 | 15.875 | 41.275 | 66.675 | 92.075 | 117.475 | 142.875 | 168.275 | 193.675 | 219.075 | 244.475 | 269.875 | 523.875 | 777.875 | 1031.875 |
| 41/64 | 16.271875 | 41.671875 | 67.071875 | 92.471875 | 117.871875 | 143.271875 | 168.671875 | 194.071875 | 219.471875 | 244.871875 | 270.271875 | 524.271875 | 778.271875 | 1032.271875 |
| 21/32 | 16.66875 | 42.06875 | 67.46875 | 92.86875 | 118.26875 | 143.66875 | 169.06875 | 194.46875 | 219.86875 | 245.26875 | 270.66875 | 524.66875 | 778.66875 | 1032.66875 |
| 43/64 | 17.065625 | 42.465625 | 67.865625 | 93.265625 | 118.665625 | 144.065625 | 169.465625 | 194.865625 | 220.265625 | 245.665625 | 271.065625 | 525.065625 | 779.065625 | 1033.065625 |
| 11/16 | 17.4625 | 42.8625 | 68.2625 | 93.6625 | 119.0625 | 144.4625 | 169.8625 | 195.2625 | 220.6625 | 246.0625 | 271.4625 | 525.4625 | 779.4625 | 1033.4625 |
| 45/64 | 17.859375 | 43.259375 | 68.659375 | 94.059375 | 119.459375 | 144.859375 | 170.259375 | 195.659375 | 221.059375 | 246.459375 | 271.859375 | 525.859375 | 779.859375 | 1033.859375 |
| 23/32 | 18.25625 | 43.65625 | 69.05625 | 94.45625 | 119.85625 | 145.25625 | 170.65625 | 196.05625 | 221.45625 | 246.85625 | 272.25625 | 526.25625 | 780.25625 | 1034.25625 |
| 47/64 | 18.653125 | 44.053125 | 69.453125 | 94.853125 | 120.253125 | 145.653125 | 171.053125 | 196.453125 | 221.853125 | 247.253125 | 272.653125 | 526.653125 | 780.653125 | 1034.653125 |
| 3/4 | 19.05 | 44.45 | 69.85 | 95.25 | 120.65 | 146.05 | 171.45 | 196.85 | 222.25 | 247.65 | 273.05 | 527.05 | 781.05 | 1035.05 |
| 49/64 | 19.446875 | 44.846875 | 70.246875 | 95.646875 | 121.046875 | 146.446875 | 171.846875 | 197.246875 | 222.646875 | 248.046875 | 273.446875 | 527.446875 | 781.446875 | 1035.446875 |
| 25/32 | 19.84375 | 45.24375 | 70.64375 | 96.04375 | 121.44375 | 146.84375 | 172.24375 | 197.64375 | 223.04375 | 248.44375 | 273.84375 | 527.84375 | 781.84375 | 1035.84375 |
| 51/64 | 20.240625 | 45.640625 | 71.040625 | 96.440625 | 121.840625 | 147.240625 | 172.640625 | 198.040625 | 223.440625 | 248.840625 | 274.240625 | 528.240625 | 782.240625 | 1036.240625 |
| 13/16 | 20.6375 | 46.0375 | 71.4375 | 96.8375 | 122.2375 | 147.6375 | 173.0375 | 198.4375 | 223.8375 | 249.2375 | 274.6375 | 528.6375 | 782.6375 | 1036.6375 |
| 53/64 | 21.034375 | 46.434375 | 71.834375 | 97.234375 | 122.634375 | 148.034375 | 173.434375 | 198.834375 | 224.234375 | 249.634375 | 275.034375 | 529.034375 | 783.034375 | 1037.034375 |
| 27/32 | 21.43125 | 46.83125 | 72.23125 | 97.63125 | 123.03125 | 148.43125 | 173.83125 | 199.23125 | 224.63125 | 250.03125 | 275.43125 | 529.43125 | 783.43125 | 1037.43125 |
| 55/64 | 21.828125 | 47.228125 | 72.628125 | 98.028125 | 123.428125 | 148.828125 | 174.228125 | 199.628125 | 225.028125 | 250.428125 | 275.828125 | 529.828125 | 783.828125 | 1037.828125 |
| 7/8 | 22.225 | 47.625 | 73.025 | 98.425 | 123.825 | 149.225 | 174.625 | 200.025 | 225.425 | 250.825 | 276.225 | 530.225 | 784.225 | 1038.225 |
| 57/64 | 22.621875 | 48.021875 | 73.421875 | 98.821875 | 124.221875 | 149.621875 | 175.021875 | 200.421875 | 225.821875 | 251.221875 | 276.621875 | 530.621875 | 784.621875 | 1038.621875 |
| 29/32 | 23.01875 | 48.41875 | 73.81875 | 99.21875 | 124.61875 | 150.01875 | 175.41875 | 200.81875 | 226.21875 | 251.61875 | 277.01875 | 531.01875 | 785.01875 | 1039.01875 |
| 59/64 | 23.415625 | 48.815625 | 74.215625 | 99.615625 | 125.015625 | 150.415625 | 175.815625 | 201.215625 | 226.615625 | 252.015625 | 277.415625 | 531.415625 | 785.415625 | 1039.415625 |
| 15/16 | 23.8125 | 49.2125 | 74.6125 | 100.0125 | 125.4125 | 150.8125 | 176.2125 | 201.6125 | 227.0125 | 252.4125 | 277.8125 | 531.8125 | 785.8125 | 1039.8125 |
| 61/64 | 24.209375 | 49.609375 | 75.009375 | 100.409375 | 125.809375 | 151.209375 | 176.609375 | 202.009375 | 227.409375 | 252.809375 | 278.209375 | 532.209375 | 786.209375 | 1040.209375 |
| 31/32 | 24.60625 | 50.00625 | 75.40625 | 100.80625 | 126.20625 | 151.60625 | 177.00625 | 202.40625 | 227.80625 | 253.20625 | 278.60625 | 532.60625 | 786.60625 | 1040.60625 |
| 63/64 | 25.003125 | 50.403125 | 75.803125 | 101.203125 | 126.603125 | 152.003125 | 177.403125 | 202.803125 | 228.203125 | 253.603125 | 279.003125 | 533.003125 | 787.003125 | 1041.003125 |
| 1 | 25.4 | 50.8 | 76.2 | 101.6 | 127 | 152.4 | 177.8 | 203.2 | 228.6 | 254 | 279.4 | 533.4 | 787.4 | 1041.4 |

Based on 1 inch = 25.4 millimeters, exactly. All values in this table are exact. *Example:* Convert $21^{23}/_{64}$ inches to millimeters. *Solution:* From the first page of this table, find 20 inches = 508.0 millimeters and add to that $1^{23}/_{64}$ inches = 34.528125 millimeters found at the intersection of the 1- inch column and the row containing $^{23}/_{64}$ inch. Thus, $21^{23}/_{64}$ inches = 508.0 + 34.528125 = 542.528125 mm, exactly.

## Table 11. Decimals of an Inch to Millimeters Conversion

| → Inches ↓ | 0.000 | 0.001 | 0.002 | 0.003 | 0.004 | 0.005 | 0.006 | 0.007 | 0.008 | 0.009 |
|---|---|---|---|---|---|---|---|---|---|---|
| | Millimeters | | | | | | | | | |
| 0.000 | … | 0.0254 | 0.0508 | 0.0762 | 0.1016 | 0.1270 | 0.1524 | 0.1778 | 0.2032 | 0.2286 |
| 0.010 | 0.2540 | 0.2794 | 0.3048 | 0.3302 | 0.3556 | 0.3810 | 0.4064 | 0.4318 | 0.4572 | 0.4826 |
| 0.020 | 0.5080 | 0.5334 | 0.5588 | 0.5842 | 0.6096 | 0.6350 | 0.6604 | 0.6858 | 0.7112 | 0.7366 |
| 0.030 | 0.7620 | 0.7874 | 0.8128 | 0.8382 | 0.8636 | 0.8890 | 0.9144 | 0.9398 | 0.9652 | 0.9906 |
| 0.040 | 1.0160 | 1.0414 | 1.0668 | 1.0922 | 1.1176 | 1.1430 | 1.1684 | 1.1938 | 1.2192 | 1.2446 |
| 0.050 | 1.2700 | 1.2954 | 1.3208 | 1.3462 | 1.3716 | 1.3970 | 1.4224 | 1.4478 | 1.4732 | 1.4986 |
| 0.060 | 1.5240 | 1.5494 | 1.5748 | 1.6002 | 1.6256 | 1.6510 | 1.6764 | 1.7018 | 1.7272 | 1.7526 |
| 0.070 | 1.7780 | 1.8034 | 1.8288 | 1.8542 | 1.8796 | 1.9050 | 1.9304 | 1.9558 | 1.9812 | 2.0066 |
| 0.080 | 2.0320 | 2.0574 | 2.0828 | 2.1082 | 2.1336 | 2.1590 | 2.1844 | 2.2098 | 2.2352 | 2.2606 |
| 0.090 | 2.2860 | 2.3114 | 2.3368 | 2.3622 | 2.3876 | 2.4130 | 2.4384 | 2.4638 | 2.4892 | 2.5146 |
| 0.100 | 2.5400 | 2.5654 | 2.5908 | 2.6162 | 2.6416 | 2.6670 | 2.6924 | 2.7178 | 2.7432 | 2.7686 |
| 0.110 | 2.7940 | 2.8194 | 2.8448 | 2.8702 | 2.8956 | 2.9210 | 2.9464 | 2.9718 | 2.9972 | 3.0226 |
| 0.120 | 3.0480 | 3.0734 | 3.0988 | 3.1242 | 3.1496 | 3.1750 | 3.2004 | 3.2258 | 3.2512 | 3.2766 |
| 0.130 | 3.3020 | 3.3274 | 3.3528 | 3.3782 | 3.4036 | 3.4290 | 3.4544 | 3.4798 | 3.5052 | 3.5306 |
| 0.140 | 3.5560 | 3.5814 | 3.6068 | 3.6322 | 3.6576 | 3.6830 | 3.7084 | 3.7338 | 3.7592 | 3.7846 |
| 0.150 | 3.8100 | 3.8354 | 3.8608 | 3.8862 | 3.9116 | 3.9370 | 3.9624 | 3.9878 | 4.0132 | 4.0386 |
| 0.160 | 4.0640 | 4.0894 | 4.1148 | 4.1402 | 4.1656 | 4.1910 | 4.2164 | 4.2418 | 4.2672 | 4.2926 |
| 0.170 | 4.3180 | 4.3434 | 4.3688 | 4.3942 | 4.4196 | 4.4450 | 4.4704 | 4.4958 | 4.5212 | 4.5466 |
| 0.180 | 4.5720 | 4.5974 | 4.6228 | 4.6482 | 4.6736 | 4.6990 | 4.7244 | 4.7498 | 4.7752 | 4.8006 |
| 0.190 | 4.8260 | 4.8514 | 4.8768 | 4.9022 | 4.9276 | 4.9530 | 4.9784 | 5.0038 | 5.0292 | 5.0546 |
| 0.200 | 5.0800 | 5.1054 | 5.1308 | 5.1562 | 5.1816 | 5.2070 | 5.2324 | 5.2578 | 5.2832 | 5.3086 |
| 0.210 | 5.3340 | 5.3594 | 5.3848 | 5.4102 | 5.4356 | 5.4610 | 5.4864 | 5.5118 | 5.5372 | 5.5626 |
| 0.220 | 5.5880 | 5.6134 | 5.6388 | 5.6642 | 5.6896 | 5.7150 | 5.7404 | 5.7658 | 5.7912 | 5.8166 |
| 0.230 | 5.8420 | 5.8674 | 5.8928 | 5.9182 | 5.9436 | 5.9690 | 5.9944 | 6.0198 | 6.0452 | 6.0706 |
| 0.240 | 6.0960 | 6.1214 | 6.1468 | 6.1722 | 6.1976 | 6.2230 | 6.2484 | 6.2738 | 6.2992 | 6.3246 |
| 0.250 | 6.3500 | 6.3754 | 6.4008 | 6.4262 | 6.4516 | 6.4770 | 6.5024 | 6.5278 | 6.5532 | 6.5786 |
| 0.260 | 6.6040 | 6.6294 | 6.6548 | 6.6802 | 6.7056 | 6.7310 | 6.7564 | 6.7818 | 6.8072 | 6.8326 |
| 0.270 | 6.8580 | 6.8834 | 6.9088 | 6.9342 | 6.9596 | 6.9850 | 7.0104 | 7.0358 | 7.0612 | 7.0866 |
| 0.280 | 7.1120 | 7.1374 | 7.1628 | 7.1882 | 7.2136 | 7.2390 | 7.2644 | 7.2898 | 7.3152 | 7.3406 |
| 0.290 | 7.3660 | 7.3914 | 7.4168 | 7.4422 | 7.4676 | 7.4930 | 7.5184 | 7.5438 | 7.5692 | 7.5946 |
| 0.300 | 7.6200 | 7.6454 | 7.6708 | 7.6962 | 7.7216 | 7.7470 | 7.7724 | 7.7978 | 7.8232 | 7.8486 |
| 0.310 | 7.8740 | 7.8994 | 7.9248 | 7.9502 | 7.9756 | 8.0010 | 8.0264 | 8.0518 | 8.0772 | 8.1026 |
| 0.320 | 8.1280 | 8.1534 | 8.1788 | 8.2042 | 8.2296 | 8.2550 | 8.2804 | 8.3058 | 8.3312 | 8.3566 |
| 0.330 | 8.3820 | 8.4074 | 8.4328 | 8.4582 | 8.4836 | 8.5090 | 8.5344 | 8.5598 | 8.5852 | 8.6106 |
| 0.340 | 8.6360 | 8.6614 | 8.6868 | 8.7122 | 8.7376 | 8.7630 | 8.7884 | 8.8138 | 8.8392 | 8.8646 |
| 0.350 | 8.8900 | 8.9154 | 8.9408 | 8.9662 | 8.9916 | 9.0170 | 9.0424 | 9.0678 | 9.0932 | 9.1186 |
| 0.360 | 9.1440 | 9.1694 | 9.1948 | 9.2202 | 9.2456 | 9.2710 | 9.2964 | 9.3218 | 9.3472 | 9.3726 |
| 0.370 | 9.3980 | 9.4234 | 9.4488 | 9.4742 | 9.4996 | 9.5250 | 9.5504 | 9.5758 | 9.6012 | 9.6266 |
| 0.380 | 9.6520 | 9.6774 | 9.7028 | 9.7282 | 9.7536 | 9.7790 | 9.8044 | 9.8298 | 9.8552 | 9.8806 |
| 0.390 | 9.9060 | 9.9314 | 9.9568 | 9.9822 | 10.0076 | 10.0330 | 10.0584 | 10.0838 | 10.1092 | 10.1346 |
| 0.400 | 10.1600 | 10.1854 | 10.2108 | 10.2362 | 10.2616 | 10.2870 | 10.3124 | 10.3378 | 10.3632 | 10.3886 |
| 0.410 | 10.4140 | 10.4394 | 10.4648 | 10.4902 | 10.5156 | 10.5410 | 10.5664 | 10.5918 | 10.6172 | 10.6426 |
| 0.420 | 10.6680 | 10.6934 | 10.7188 | 10.7442 | 10.7696 | 10.7950 | 10.8204 | 10.8458 | 10.8712 | 10.8966 |
| 0.430 | 10.9220 | 10.9474 | 10.9728 | 10.9982 | 11.0236 | 11.0490 | 11.0744 | 11.0998 | 11.1252 | 11.1506 |
| 0.440 | 11.1760 | 11.2014 | 11.2268 | 11.2522 | 11.2776 | 11.3030 | 11.3284 | 11.3538 | 11.3792 | 11.4046 |
| 0.450 | 11.4300 | 11.4554 | 11.4808 | 11.5062 | 11.5316 | 11.5570 | 11.5824 | 11.6078 | 11.6332 | 11.6586 |
| 0.460 | 11.6840 | 11.7094 | 11.7348 | 11.7602 | 11.7856 | 11.8110 | 11.8364 | 11.8618 | 11.8872 | 11.9126 |
| 0.470 | 11.9380 | 11.9634 | 11.9888 | 12.0142 | 12.0396 | 12.0650 | 12.0904 | 12.1158 | 12.1412 | 12.1666 |
| 0.480 | 12.1920 | 12.2174 | 12.2428 | 12.2682 | 12.2936 | 12.3190 | 12.3444 | 12.3698 | 12.3952 | 12.4206 |
| 0.490 | 12.4460 | 12.4714 | 12.4968 | 12.5222 | 12.5476 | 12.5730 | 12.5984 | 12.6238 | 12.6492 | 12.6746 |
| 0.500 | 12.7000 | 12.7254 | 12.7508 | 12.7762 | 12.8016 | 12.8270 | 12.8524 | 12.8778 | 12.9032 | 12.9286 |

**Table 11.** *(Continued)* **Decimals of an Inch to Millimeters Conversion**

| → Inches ↓ | 0.000 | 0.001 | 0.002 | 0.003 | 0.004 | 0.005 | 0.006 | 0.007 | 0.008 | 0.009 |
|---|---|---|---|---|---|---|---|---|---|---|
| | Millimeters | | | | | | | | | |
| 0.510 | 12.9540 | 12.9794 | 13.0048 | 13.0302 | 13.0556 | 13.0810 | 13.1064 | 13.1318 | 13.1572 | 13.1826 |
| 0.520 | 13.2080 | 13.2334 | 13.2588 | 13.2842 | 13.3096 | 13.3350 | 13.3604 | 13.3858 | 13.4112 | 13.4366 |
| 0.530 | 13.4620 | 13.4874 | 13.5128 | 13.5382 | 13.5636 | 13.5890 | 13.6144 | 13.6398 | 13.6652 | 13.6906 |
| 0.540 | 13.7160 | 13.7414 | 13.7668 | 13.7922 | 13.8176 | 13.8430 | 13.8684 | 13.8938 | 13.9192 | 13.9446 |
| 0.550 | 13.9700 | 13.9954 | 14.0208 | 14.0462 | 14.0716 | 14.0970 | 14.1224 | 14.1478 | 14.1732 | 14.1986 |
| 0.560 | 14.2240 | 14.2494 | 14.2748 | 14.3002 | 14.3256 | 14.3510 | 14.3764 | 14.4018 | 14.4272 | 14.4526 |
| 0.570 | 14.4780 | 14.5034 | 14.5288 | 14.5542 | 14.5796 | 14.6050 | 14.6304 | 14.6558 | 14.6812 | 14.7066 |
| 0.580 | 14.7320 | 14.7574 | 14.7828 | 14.8082 | 14.8336 | 14.8590 | 14.8844 | 14.9098 | 14.9352 | 14.9606 |
| 0.590 | 14.9860 | 15.0114 | 15.0368 | 15.0622 | 15.0876 | 15.1130 | 15.1384 | 15.1638 | 15.1892 | 15.2146 |
| 0.600 | 15.2400 | 15.2654 | 15.2908 | 15.3162 | 15.3416 | 15.3670 | 15.3924 | 15.4178 | 15.4432 | 15.4686 |
| 0.610 | 15.4940 | 15.5194 | 15.5448 | 15.5702 | 15.5956 | 15.6210 | 15.6464 | 15.6718 | 15.6972 | 15.7226 |
| 0.620 | 15.7480 | 15.7734 | 15.7988 | 15.8242 | 15.8496 | 15.8750 | 15.9004 | 15.9258 | 15.9512 | 15.9766 |
| 0.630 | 16.0020 | 16.0274 | 16.0528 | 16.0782 | 16.1036 | 16.1290 | 16.1544 | 16.1798 | 16.2052 | 16.2306 |
| 0.640 | 16.2560 | 16.2814 | 16.3068 | 16.3322 | 16.3576 | 16.3830 | 16.4084 | 16.4338 | 16.4592 | 16.4846 |
| 0.650 | 16.5100 | 16.5354 | 16.5608 | 16.5862 | 16.6116 | 16.6370 | 16.6624 | 16.6878 | 16.7132 | 16.7386 |
| 0.660 | 16.7640 | 16.7894 | 16.8148 | 16.8402 | 16.8656 | 16.8910 | 16.9164 | 16.9418 | 16.9672 | 16.9926 |
| 0.670 | 17.0180 | 17.0434 | 17.0688 | 17.0942 | 17.1196 | 17.1450 | 17.1704 | 17.1958 | 17.2212 | 17.2466 |
| 0.680 | 17.2720 | 17.2974 | 17.3228 | 17.3482 | 17.3736 | 17.3990 | 17.4244 | 17.4498 | 17.4752 | 17.5006 |
| 0.690 | 17.5260 | 17.5514 | 17.5768 | 17.6022 | 17.6276 | 17.6530 | 17.6784 | 17.7038 | 17.7292 | 17.7546 |
| 0.700 | 17.7800 | 17.8054 | 17.8308 | 17.8562 | 17.8816 | 17.9070 | 17.9324 | 17.9578 | 17.9832 | 18.0086 |
| 0.710 | 18.0340 | 18.0594 | 18.0848 | 18.1102 | 18.1356 | 18.1610 | 18.1864 | 18.2118 | 18.2372 | 18.2626 |
| 0.720 | 18.2880 | 18.3134 | 18.3388 | 18.3642 | 18.3896 | 18.4150 | 18.4404 | 18.4658 | 18.4912 | 18.5166 |
| 0.730 | 18.5420 | 18.5674 | 18.5928 | 18.6182 | 18.6436 | 18.6690 | 18.6944 | 18.7198 | 18.7452 | 18.7706 |
| 0.740 | 18.7960 | 18.8214 | 18.8468 | 18.8722 | 18.8976 | 18.9230 | 18.9484 | 18.9738 | 18.9992 | 19.0246 |
| 0.750 | 19.0500 | 19.0754 | 19.1008 | 19.1262 | 19.1516 | 19.1770 | 19.2024 | 19.2278 | 19.2532 | 19.2786 |
| 0.760 | 19.3040 | 19.3294 | 19.3548 | 19.3802 | 19.4056 | 19.4310 | 19.4564 | 19.4818 | 19.5072 | 19.5326 |
| 0.770 | 19.5580 | 19.5834 | 19.6088 | 19.6342 | 19.6596 | 19.6850 | 19.7104 | 19.7358 | 19.7612 | 19.7866 |
| 0.780 | 19.8120 | 19.8374 | 19.8628 | 19.8882 | 19.9136 | 19.9390 | 19.9644 | 19.9898 | 20.0152 | 20.0406 |
| 0.790 | 20.0660 | 20.0914 | 20.1168 | 20.1422 | 20.1676 | 20.1930 | 20.2184 | 20.2438 | 20.2692 | 20.2946 |
| 0.800 | 20.3200 | 20.3454 | 20.3708 | 20.3962 | 20.4216 | 20.4470 | 20.4724 | 20.4978 | 20.5232 | 20.5486 |
| 0.810 | 20.5740 | 20.5994 | 20.6248 | 20.6502 | 20.6756 | 20.7010 | 20.7264 | 20.7518 | 20.7772 | 20.8026 |
| 0.820 | 20.8280 | 20.8534 | 20.8788 | 20.9042 | 20.9296 | 20.9550 | 20.9804 | 21.0058 | 21.0312 | 21.0566 |
| 0.830 | 21.0820 | 21.1074 | 21.1328 | 21.1582 | 21.1836 | 21.2090 | 21.2344 | 21.2598 | 21.2852 | 21.3106 |
| 0.840 | 21.3360 | 21.3614 | 21.3868 | 21.4122 | 21.4376 | 21.4630 | 21.4884 | 21.5138 | 21.5392 | 21.5646 |
| 0.850 | 21.5900 | 21.6154 | 21.6408 | 21.6662 | 21.6916 | 21.7170 | 21.7424 | 21.7678 | 21.7932 | 21.8186 |
| 0.860 | 21.8440 | 21.8694 | 21.8948 | 21.9202 | 21.9456 | 21.9710 | 21.9964 | 22.0218 | 22.0472 | 22.0726 |
| 0.870 | 22.0980 | 22.1234 | 22.1488 | 22.1742 | 22.1996 | 22.2250 | 22.2504 | 22.2758 | 22.3012 | 22.3266 |
| 0.880 | 22.3520 | 22.3774 | 22.4028 | 22.4282 | 22.4536 | 22.4790 | 22.5044 | 22.5298 | 22.5552 | 22.5806 |
| 0.890 | 22.6060 | 22.6314 | 22.6568 | 22.6822 | 22.7076 | 22.7330 | 22.7584 | 22.7838 | 22.8092 | 22.8346 |
| 0.900 | 22.8600 | 22.8854 | 22.9108 | 22.9362 | 22.9616 | 22.9870 | 23.0124 | 23.0378 | 23.0632 | 23.0886 |
| 0.910 | 23.1140 | 23.1394 | 23.1648 | 23.1902 | 23.2156 | 23.2410 | 23.2664 | 23.2918 | 23.3172 | 23.3426 |
| 0.920 | 23.3680 | 23.3934 | 23.4188 | 23.4442 | 23.4696 | 23.4950 | 23.5204 | 23.5458 | 23.5712 | 23.5966 |
| 0.930 | 23.6220 | 23.6474 | 23.6728 | 23.6982 | 23.7236 | 23.7490 | 23.7744 | 23.7998 | 23.8252 | 23.8506 |
| 0.940 | 23.8760 | 23.9014 | 23.9268 | 23.9522 | 23.9776 | 24.0030 | 24.0284 | 24.0538 | 24.0792 | 24.1046 |
| 0.950 | 24.1300 | 24.1554 | 24.1808 | 24.2062 | 24.2316 | 24.2570 | 24.2824 | 24.3078 | 24.3332 | 24.3586 |
| 0.960 | 24.3840 | 24.4094 | 24.4348 | 24.4602 | 24.4856 | 24.5110 | 24.5364 | 24.5618 | 24.5872 | 24.6126 |
| 0.970 | 24.6380 | 24.6634 | 24.6888 | 24.7142 | 24.7396 | 24.7650 | 24.7904 | 24.8158 | 24.8412 | 24.8666 |
| 0.980 | 24.8920 | 24.9174 | 24.9428 | 24.9682 | 24.9936 | 25.0190 | 25.0444 | 25.0698 | 25.0952 | 25.1206 |
| 0.990 | 25.1460 | 25.1714 | 25.1968 | 25.2222 | 25.2476 | 25.2730 | 25.2984 | 25.3238 | 25.3492 | 25.3746 |
| 1.000 | 25.4000 | … | … | … | … | … | … | … | … | … |

Based on 1 inch = 25.4 millimeters, exactly. All values in this table are exact. Use Table 8a to obtain whole inch and other decimal equivalents to add to decimal equivalents above. *Example:* Convert 10.9983 in. to mm. *Solution:* 10.9983 in. = 254.0 + 25.3492 + 0.00762 = 279.35682 mm.

**Table 12. Millimeters to Inches Conversion**

| → Millimeters ↓ | 0 | 1 | 2 | 3 | 4 | 5 | 6 | 7 | 8 | 9 |
|---|---|---|---|---|---|---|---|---|---|---|
| | Inches | | | | | | | | | |
| 0 | … | 0.03937 | 0.07874 | 0.11811 | 0.15748 | 0.19685 | 0.23622 | 0.27559 | 0.31496 | 0.35433 |
| 10 | 0.39370 | 0.43307 | 0.47244 | 0.51181 | 0.55118 | 0.59055 | 0.62992 | 0.66929 | 0.70866 | 0.74803 |
| 20 | 0.78740 | 0.82677 | 0.86614 | 0.90551 | 0.94488 | 0.98425 | 1.02362 | 1.06299 | 1.10236 | 1.14173 |
| 30 | 1.18110 | 1.22047 | 1.25984 | 1.29921 | 1.33858 | 1.37795 | 1.41732 | 1.45669 | 1.49606 | 1.53543 |
| 40 | 1.57480 | 1.61417 | 1.65354 | 1.69291 | 1.73228 | 1.77165 | 1.81102 | 1.85039 | 1.88976 | 1.92913 |
| 50 | 1.96850 | 2.00787 | 2.04724 | 2.08661 | 2.12598 | 2.16535 | 2.20472 | 2.24409 | 2.28346 | 2.32283 |
| 60 | 2.36220 | 2.40157 | 2.44094 | 2.48031 | 2.51969 | 2.55906 | 2.59843 | 2.63780 | 2.67717 | 2.71654 |
| 70 | 2.75591 | 2.79528 | 2.83465 | 2.87402 | 2.91339 | 2.95276 | 2.99213 | 3.03150 | 3.07087 | 3.11024 |
| 80 | 3.14961 | 3.18898 | 3.22835 | 3.26772 | 3.30709 | 3.34646 | 3.38583 | 3.42520 | 3.46457 | 3.50394 |
| 90 | 3.54331 | 3.58268 | 3.62205 | 3.66142 | 3.70079 | 3.74016 | 3.77953 | 3.81890 | 3.85827 | 3.89764 |
| 100 | 3.93701 | 3.97638 | 4.01575 | 4.05512 | 4.09449 | 4.13386 | 4.17323 | 4.21260 | 4.25197 | 4.29134 |
| 110 | 4.33071 | 4.37008 | 4.40945 | 4.44882 | 4.48819 | 4.52756 | 4.56693 | 4.60630 | 4.64567 | 4.68504 |
| 120 | 4.72441 | 4.76378 | 4.80315 | 4.84252 | 4.88189 | 4.92126 | 4.96063 | 5.00000 | 5.03937 | 5.07874 |
| 130 | 5.11811 | 5.15748 | 5.19685 | 5.23622 | 5.27559 | 5.31496 | 5.35433 | 5.39370 | 5.43307 | 5.47244 |
| 140 | 5.51181 | 5.55118 | 5.59055 | 5.62992 | 5.66929 | 5.70866 | 5.74803 | 5.78740 | 5.82677 | 5.86614 |
| 150 | 5.90551 | 5.94488 | 5.98425 | 6.02362 | 6.06299 | 6.10236 | 6.14173 | 6.18110 | 6.22047 | 6.25984 |
| 160 | 6.29921 | 6.33858 | 6.37795 | 6.41732 | 6.45669 | 6.49606 | 6.53543 | 6.57480 | 6.61417 | 6.65354 |
| 170 | 6.69291 | 6.73228 | 6.77165 | 6.81102 | 6.85039 | 6.88976 | 6.92913 | 6.96850 | 7.00787 | 7.04724 |
| 180 | 7.08661 | 7.12598 | 7.16535 | 7.20472 | 7.24409 | 7.28346 | 7.32283 | 7.36220 | 7.40157 | 7.44094 |
| 190 | 7.48031 | 7.51969 | 7.55906 | 7.59843 | 7.63780 | 7.67717 | 7.71654 | 7.75591 | 7.79528 | 7.83465 |
| 200 | 7.87402 | 7.91339 | 7.95276 | 7.99213 | 8.03150 | 8.07087 | 8.11024 | 8.14961 | 8.18898 | 8.22835 |
| 210 | 8.26772 | 8.30709 | 8.34646 | 8.38583 | 8.42520 | 8.46457 | 8.50394 | 8.54331 | 8.58268 | 8.62205 |
| 220 | 8.66142 | 8.70079 | 8.74016 | 8.77953 | 8.81890 | 8.85827 | 8.89764 | 8.93701 | 8.97638 | 9.01575 |
| 230 | 9.05512 | 9.09449 | 9.13386 | 9.17323 | 9.21260 | 9.25197 | 9.29134 | 9.33071 | 9.37008 | 9.40945 |
| 240 | 9.44882 | 9.48819 | 9.52756 | 9.56693 | 9.60630 | 9.64567 | 9.68504 | 9.72441 | 9.76378 | 9.80315 |
| 250 | 9.84252 | 9.88189 | 9.92126 | 9.96063 | 10.0000 | 10.0394 | 10.0787 | 10.1181 | 10.1575 | 10.1969 |
| 260 | 10.2362 | 10.2756 | 10.3150 | 10.3543 | 10.3937 | 10.4331 | 10.4724 | 10.5118 | 10.5512 | 10.5906 |
| 270 | 10.6299 | 10.6693 | 10.7087 | 10.7480 | 10.7874 | 10.8268 | 10.8661 | 10.9055 | 10.9449 | 10.9843 |
| 280 | 11.0236 | 11.0630 | 11.1024 | 11.1417 | 11.1811 | 11.2205 | 11.2598 | 11.2992 | 11.3386 | 11.3780 |
| 290 | 11.4173 | 11.4567 | 11.4961 | 11.5354 | 11.5748 | 11.6142 | 11.6535 | 11.6929 | 11.7323 | 11.7717 |
| 300 | 11.8110 | 11.8504 | 11.8898 | 11.9291 | 11.9685 | 12.0079 | 12.0472 | 12.0866 | 12.1260 | 12.1654 |
| 310 | 12.2047 | 12.2441 | 12.2835 | 12.3228 | 12.3622 | 12.4016 | 12.4409 | 12.4803 | 12.5197 | 12.5591 |
| 320 | 12.5984 | 12.6378 | 12.6772 | 12.7165 | 12.7559 | 12.7953 | 12.8346 | 12.8740 | 12.9134 | 12.9528 |
| 330 | 12.9921 | 13.0315 | 13.0709 | 13.1102 | 13.1496 | 13.1890 | 13.2283 | 13.2677 | 13.3071 | 13.3465 |
| 340 | 13.3858 | 13.4252 | 13.4646 | 13.5039 | 13.5433 | 13.5827 | 13.6220 | 13.6614 | 13.7008 | 13.7402 |
| 350 | 13.7795 | 13.8189 | 13.8583 | 13.8976 | 13.9370 | 13.9764 | 14.0157 | 14.0551 | 14.0945 | 14.1339 |
| 360 | 14.1732 | 14.2126 | 14.2520 | 14.2913 | 14.3307 | 14.3701 | 14.4094 | 14.4488 | 14.4882 | 14.5276 |
| 370 | 14.5669 | 14.6063 | 14.6457 | 14.6850 | 14.7244 | 14.7638 | 14.8031 | 14.8425 | 14.8819 | 14.9213 |
| 380 | 14.9606 | 15.0000 | 15.0394 | 15.0787 | 15.1181 | 15.1575 | 15.1969 | 15.2362 | 15.2756 | 15.3150 |
| 390 | 15.3543 | 15.3937 | 15.4331 | 15.4724 | 15.5118 | 15.5512 | 15.5906 | 15.6299 | 15.6693 | 15.7087 |
| 400 | 15.7480 | 15.7874 | 15.8268 | 15.8661 | 15.9055 | 15.9449 | 15.9843 | 16.0236 | 16.0630 | 16.1024 |
| 410 | 16.1417 | 16.1811 | 16.2205 | 16.2598 | 16.2992 | 16.3386 | 16.3780 | 16.4173 | 16.4567 | 16.4961 |
| 420 | 16.5354 | 16.5748 | 16.6142 | 16.6535 | 16.6929 | 16.7323 | 16.7717 | 16.8110 | 16.8504 | 16.8898 |
| 430 | 16.9291 | 16.9685 | 17.0079 | 17.0472 | 17.0866 | 17.1260 | 17.1654 | 17.2047 | 17.2441 | 17.2835 |
| 440 | 17.3228 | 17.3622 | 17.4016 | 17.4409 | 17.4803 | 17.5197 | 17.5591 | 17.5984 | 17.6378 | 17.6772 |
| 450 | 17.7165 | 17.7559 | 17.7953 | 17.8346 | 17.8740 | 17.9134 | 17.9528 | 17.9921 | 18.0315 | 18.0709 |
| 460 | 18.1102 | 18.1496 | 18.1890 | 18.2283 | 18.2677 | 18.3071 | 18.3465 | 18.3858 | 18.4252 | 18.4646 |
| 470 | 18.5039 | 18.5433 | 18.5827 | 18.6220 | 18.6614 | 18.7008 | 18.7402 | 18.7795 | 18.8189 | 18.8583 |
| 480 | 18.8976 | 18.9370 | 18.9764 | 19.0157 | 19.0551 | 19.0945 | 19.1339 | 19.1732 | 19.2126 | 19.2520 |
| 490 | 19.2913 | 19.3307 | 19.3701 | 19.4094 | 19.4488 | 19.4882 | 19.5276 | 19.5669 | 19.6063 | 19.6457 |

**Table 12.** *(Continued)* **Millimeters to Inches Conversion**

| → Millimeters ↓ | 0 | 1 | 2 | 3 | 4 | 5 | 6 | 7 | 8 | 9 |
|---|---|---|---|---|---|---|---|---|---|---|
| | Inches | | | | | | | | | |
| 500 | 19.6850 | 19.7244 | 19.7638 | 19.8031 | 19.8425 | 19.8819 | 19.9213 | 19.9606 | 20.0000 | 20.0394 |
| 510 | 20.0787 | 20.1181 | 20.1575 | 20.1969 | 20.2362 | 20.2756 | 20.3150 | 20.3543 | 20.3937 | 20.4331 |
| 520 | 20.4724 | 20.5118 | 20.5512 | 20.5906 | 20.6299 | 20.6693 | 20.7087 | 20.7480 | 20.7874 | 20.8268 |
| 530 | 20.8661 | 20.9055 | 20.9449 | 20.9843 | 21.0236 | 21.0630 | 21.1024 | 21.1417 | 21.1811 | 21.2205 |
| 540 | 21.2598 | 21.2992 | 21.3386 | 21.3780 | 21.4173 | 21.4567 | 21.4961 | 21.5354 | 21.5748 | 21.6142 |
| 550 | 21.6535 | 21.6929 | 21.7323 | 21.7717 | 21.8110 | 21.8504 | 21.8898 | 21.9291 | 21.9685 | 22.0079 |
| 560 | 22.0472 | 22.0866 | 22.1260 | 22.1654 | 22.2047 | 22.2441 | 22.2835 | 22.3228 | 22.3622 | 22.4016 |
| 570 | 22.4409 | 22.4803 | 22.5197 | 22.5591 | 22.5984 | 22.6378 | 22.6772 | 22.7165 | 22.7559 | 22.7953 |
| 580 | 22.8346 | 22.8740 | 22.9134 | 22.9528 | 22.9921 | 23.0315 | 23.0709 | 23.1102 | 23.1496 | 23.1890 |
| 590 | 23.2283 | 23.2677 | 23.3071 | 23.3465 | 23.3858 | 23.4252 | 23.4646 | 23.5039 | 23.5433 | 23.5827 |
| 600 | 23.6220 | 23.6614 | 23.7008 | 23.7402 | 23.7795 | 23.8189 | 23.8583 | 23.8976 | 23.9370 | 23.9764 |
| 610 | 24.0157 | 24.0551 | 24.0945 | 24.1339 | 24.1732 | 24.2126 | 24.2520 | 24.2913 | 24.3307 | 24.3701 |
| 620 | 24.4094 | 24.4488 | 24.4882 | 24.5276 | 24.5669 | 24.6063 | 24.6457 | 24.6850 | 24.7244 | 24.7638 |
| 630 | 24.8031 | 24.8425 | 24.8819 | 24.9213 | 24.9606 | 25.0000 | 25.0394 | 25.0787 | 25.1181 | 25.1575 |
| 640 | 25.1969 | 25.2362 | 25.2756 | 25.3150 | 25.3543 | 25.3937 | 25.4331 | 25.4724 | 25.5118 | 25.5512 |
| 650 | 25.5906 | 25.6299 | 25.6693 | 25.7087 | 25.7480 | 25.7874 | 25.8268 | 25.8661 | 25.9055 | 25.9449 |
| 660 | 25.9843 | 26.0236 | 26.0630 | 26.1024 | 26.1417 | 26.1811 | 26.2205 | 26.2598 | 26.2992 | 26.3386 |
| 670 | 26.3780 | 26.4173 | 26.4567 | 26.4961 | 26.5354 | 26.5748 | 26.6142 | 26.6535 | 26.6929 | 26.7323 |
| 680 | 26.7717 | 26.8110 | 26.8504 | 26.8898 | 26.9291 | 26.9685 | 27.0079 | 27.0472 | 27.0866 | 27.1260 |
| 690 | 27.1654 | 27.2047 | 27.2441 | 27.2835 | 27.3228 | 27.3622 | 27.4016 | 27.4409 | 27.4803 | 27.5197 |
| 700 | 27.5591 | 27.5984 | 27.6378 | 27.6772 | 27.7165 | 27.7559 | 27.7953 | 27.8346 | 27.8740 | 27.9134 |
| 710 | 27.9528 | 27.9921 | 28.0315 | 28.0709 | 28.1102 | 28.1496 | 28.1890 | 28.2283 | 28.2677 | 28.3071 |
| 720 | 28.3465 | 28.3858 | 28.4252 | 28.4645 | 28.5039 | 28.5433 | 28.5827 | 28.6220 | 28.6614 | 28.7008 |
| 730 | 28.7402 | 28.7795 | 28.8189 | 28.8583 | 28.8976 | 28.9370 | 28.9764 | 29.0157 | 29.0551 | 29.0945 |
| 740 | 29.1339 | 29.1732 | 29.2126 | 29.2520 | 29.2913 | 29.3307 | 29.3701 | 29.4094 | 29.4488 | 29.4882 |
| 750 | 29.5276 | 29.5669 | 29.6063 | 29.6457 | 29.6850 | 29.7244 | 29.7638 | 29.8031 | 29.8425 | 29.8819 |
| 760 | 29.9213 | 29.9606 | 30.0000 | 30.0394 | 30.0787 | 30.1181 | 30.1575 | 30.1969 | 30.2362 | 30.2756 |
| 770 | 30.3150 | 30.3543 | 30.3937 | 30.4331 | 30.4724 | 30.5118 | 30.5512 | 30.5906 | 30.6299 | 30.6693 |
| 780 | 30.7087 | 30.7480 | 30.7874 | 30.8268 | 30.8661 | 30.9055 | 30.949 | 30.9843 | 31.0236 | 31.0630 |
| 790 | 31.1024 | 31.1417 | 31.1811 | 31.2205 | 31.2598 | 31.2992 | 31.3386 | 31.3780 | 31.4173 | 31.4567 |
| 800 | 31.4961 | 31.5354 | 31.5748 | 31.6142 | 31.6535 | 31.6929 | 31.7323 | 31.7717 | 31.8110 | 31.8504 |
| 810 | 31.8898 | 31.9291 | 31.9685 | 32.0079 | 32.0472 | 32.0866 | 32.1260 | 32.1654 | 32.2047 | 32.2441 |
| 820 | 32.2835 | 32.3228 | 32.3622 | 32.4016 | 32.4409 | 32.4803 | 32.5197 | 32.5591 | 32.5984 | 32.6378 |
| 830 | 32.6772 | 32.7165 | 32.7559 | 32.7953 | 32.8346 | 32.8740 | 32.9134 | 32.9528 | 32.9921 | 33.03[illegible]5 |
| 840 | 33.0709 | 33.1102 | 33.1496 | 33.1890 | 33.2283 | 33.2677 | 33.3071 | 33.3465 | 33.3858 | 33.4252 |
| 850 | 33.4646 | 33.5039 | 33.5433 | 33.5827 | 33.6220 | 33.6614 | 33.7008 | 33.7402 | 33.7795 | 33.8139 |
| 860 | 33.8583 | 33.8976 | 33.9370 | 33.9764 | 34.0157 | 34.0551 | 34.0945 | 34.1339 | 34.1732 | 34.2126 |
| 870 | 34.2520 | 34.2913 | 34.3307 | 34.3701 | 34.4094 | 34.4488 | 34.4882 | 34.5276 | 34.5669 | 34.6063 |
| 880 | 34.6457 | 34.6850 | 34.7244 | 34.7638 | 34.8031 | 34.8425 | 34.8819 | 34.9213 | 34.9606 | 35.0000 |
| 890 | 35.0394 | 35.0787 | 35.1181 | 35.1575 | 35.1969 | 35.2362 | 35.2756 | 35.3150 | 35.3543 | 35.3937 |
| 900 | 35.4331 | 35.4724 | 35.5118 | 35.5512 | 35.5906 | 35.6299 | 35.6693 | 35.7087 | 35.7480 | 35.7874 |
| 910 | 35.8268 | 35.8661 | 35.9055 | 35.9449 | 35.9843 | 36.0236 | 36.0630 | 36.1024 | 36.1417 | 36.1811 |
| 920 | 36.2205 | 36.2598 | 36.2992 | 36.3386 | 36.3780 | 36.4173 | 36.4567 | 36.4961 | 36.5354 | 36.5748 |
| 930 | 36.6142 | 36.6535 | 36.6929 | 36.7323 | 36.7717 | 36.8110 | 36.8504 | 36.8898 | 36.9291 | 36.9685 |
| 940 | 37.0079 | 37.0472 | 37.0866 | 37.1260 | 37.1654 | 37.2047 | 37.2441 | 37.2835 | 37.3228 | 37.3622 |
| 950 | 37.4016 | 37.409 | 37.4803 | 37.5197 | 37.5591 | 37.5984 | 37.6378 | 37.6772 | 37.7165 | 37.7559 |
| 960 | 37.7953 | 37.8346 | 37.8740 | 37.9134 | 37.9528 | 37.9921 | 38.0315 | 38.0709 | 38.1102 | 38.[illegible]496 |
| 970 | 38.1800 | 38.2283 | 38.2677 | 38.3071 | 38.3465 | 38.3858 | 38.4252 | 38.4646 | 38.5039 | 38.5433 |
| 980 | 38.5827 | 38.6220 | 38.6614 | 38.7008 | 38.7402 | 38.7795 | 38.8189 | 38.8583 | 38.8976 | 38.9370 |
| 990 | 38.9764 | 39.0157 | 39.0551 | 39.0945 | 39.1339 | 39.1732 | 39.2126 | 39.2520 | 39.2913 | 39.3307 |
| 1000 | 39.3701 | … | … | … | … | … | … | … | … | .. |

Based on 1 inch = 25.4 millimeters, exactly.

**Table 13a. Microinches to Micrometers (Microns) Conversion**

| → Microinches ↓ | 0 | 1 | 2 | 3 | 4 | 5 | 6 | 7 | 8 | 9 |
|---|---|---|---|---|---|---|---|---|---|---|
| | Micrometers (microns) | | | | | | | | | |
| 0 | 0 | 0.0254 | 0.0508 | 0.0762 | 0.1016 | 0.127 | 0.1524 | 0.1778 | 0.2032 | 0.2286 |
| 10 | 0.254 | 0.2794 | 0.3048 | 0.3302 | 0.3556 | 0.381 | 0.4064 | 0.4318 | 0.4572 | 0.4826 |
| 20 | 0.508 | 0.5334 | 0.5588 | 0.5842 | 0.6096 | 0.635 | 0.6604 | 0.6858 | 0.7112 | 0.7366 |
| 30 | 0.762 | 0.7874 | 0.8128 | 0.8382 | 0.8636 | 0.889 | 0.9144 | 0.9398 | 0.9652 | 0.9906 |
| 40 | 1.016 | 1.0414 | 1.0668 | 1.0922 | 1.1176 | 1.143 | 1.1684 | 1.1938 | 1.2192 | 1.2446 |
| 50 | 1.27 | 1.2954 | 1.3208 | 1.3462 | 1.3716 | 1.397 | 1.4224 | 1.4478 | 1.4732 | 1.4986 |
| 60 | 1.524 | 1.5494 | 1.5748 | 1.6002 | 1.6256 | 1.651 | 1.6764 | 1.7018 | 1.7272 | 1.7526 |
| 70 | 1.778 | 1.8034 | 1.8288 | 1.8542 | 1.8796 | 1.905 | 1.9304 | 1.9558 | 1.9812 | 2.0066 |
| 80 | 2.032 | 2.0574 | 2.0828 | 2.1082 | 2.1336 | 2.159 | 2.1844 | 2.2098 | 2.2352 | 2.2606 |
| 90 | 2.286 | 2.3114 | 2.3368 | 2.3622 | 2.3876 | 2.413 | 2.4384 | 2.4638 | 2.4892 | 2.5146 |
| 100 | 2.54 | 2.5654 | 2.5908 | 2.6162 | 2.6416 | 2.667 | 2.6924 | 2.7178 | 2.7432 | 2.7686 |
| 110 | 2.794 | 2.8194 | 2.8448 | 2.8702 | 2.8956 | 2.921 | 2.9464 | 2.9718 | 2.9972 | 3.0226 |
| 120 | 3.048 | 3.0734 | 3.0988 | 3.1242 | 3.1496 | 3.175 | 3.2004 | 3.2258 | 3.2512 | 3.2766 |
| 130 | 3.302 | 3.3274 | 3.3528 | 3.3782 | 3.4036 | 3.429 | 3.4544 | 3.4798 | 3.5052 | 3.5306 |
| 140 | 3.556 | 3.5814 | 3.6068 | 3.6322 | 3.6576 | 3.683 | 3.7084 | 3.7338 | 3.7592 | 3.7846 |
| 150 | 3.81 | 3.8354 | 3.8608 | 3.8862 | 3.9116 | 3.937 | 3.9624 | 3.9878 | 4.0132 | 4.0386 |
| 160 | 4.064 | 4.0894 | 4.1148 | 4.1402 | 4.1656 | 4.191 | 4.2164 | 4.2418 | 4.2672 | 4.2926 |
| 170 | 4.318 | 4.3434 | 4.3688 | 4.3942 | 4.4196 | 4.445 | 4.4704 | 4.4958 | 4.5212 | 4.5466 |
| 180 | 4.572 | 4.5974 | 4.6228 | 4.6482 | 4.6736 | 4.699 | 4.7244 | 4.7498 | 4.7752 | 4.8006 |
| 190 | 4.826 | 4.8514 | 4.8768 | 4.9022 | 4.9276 | 4.953 | 4.9784 | 5.0038 | 5.0292 | 5.0546 |
| 200 | 5.08 | 5.1054 | 5.1308 | 5.1562 | 5.1816 | 5.207 | 5.2324 | 5.2578 | 5.2832 | 5.3086 |
| 210 | 5.334 | 5.3594 | 5.3848 | 5.4102 | 5.4356 | 5.461 | 5.4864 | 5.5118 | 5.5372 | 5.5626 |
| 220 | 5.588 | 5.6134 | 5.6388 | 5.6642 | 5.6896 | 5.715 | 5.7404 | 5.7658 | 5.7912 | 5.8166 |
| 230 | 5.842 | 5.8674 | 5.8928 | 5.9182 | 5.9436 | 5.969 | 5.9944 | 6.0198 | 6.0452 | 6.0706 |
| 240 | 6.096 | 6.1214 | 6.1468 | 6.1722 | 6.1976 | 6.223 | 6.2484 | 6.2738 | 6.2992 | 6.3246 |
| 250 | 6.35 | 6.3754 | 6.4008 | 6.4262 | 6.4516 | 6.477 | 6.5024 | 6.5278 | 6.5532 | 6.5786 |
| 260 | 6.604 | 6.6294 | 6.6548 | 6.6802 | 6.7056 | 6.731 | 6.7564 | 6.7818 | 6.8072 | 6.8326 |
| 270 | 6.858 | 6.8834 | 6.9088 | 6.9342 | 6.9596 | 6.985 | 7.0104 | 7.0358 | 7.0612 | 7.0866 |
| 280 | 7.112 | 7.1374 | 7.1628 | 7.1882 | 7.2136 | 7.239 | 7.2644 | 7.2898 | 7.3152 | 7.3406 |
| 290 | 7.366 | 7.3914 | 7.4168 | 7.4422 | 7.4676 | 7.493 | 7.5184 | 7.5438 | 7.5692 | 7.5946 |
| 300 | 7.62 | 7.6454 | 7.6708 | 7.6962 | 7.7216 | 7.747 | 7.7724 | 7.7978 | 7.8232 | 7.8486 |
| 310 | 7.874 | 7.8994 | 7.9248 | 7.9502 | 7.9756 | 8.001 | 8.0264 | 8.0518 | 8.0772 | 8.1026 |
| 320 | 8.128 | 8.1534 | 8.1788 | 8.2042 | 8.2296 | 8.255 | 8.2804 | 8.3058 | 8.3312 | 8.3566 |
| 330 | 8.382 | 8.4074 | 8.4328 | 8.4582 | 8.4836 | 8.509 | 8.5344 | 8.5598 | 8.5852 | 8.6106 |
| 340 | 8.636 | 8.6614 | 8.6868 | 8.7122 | 8.7376 | 8.763 | 8.7884 | 8.8138 | 8.8392 | 8.8646 |
| 350 | 8.89 | 8.9154 | 8.9408 | 8.9662 | 8.9916 | 9.017 | 9.0424 | 9.0678 | 9.0932 | 9.1186 |
| 360 | 9.144 | 9.1694 | 9.1948 | 9.2202 | 9.2456 | 9.271 | 9.2964 | 9.3218 | 9.3472 | 9.3726 |
| 370 | 9.398 | 9.4234 | 9.4488 | 9.4742 | 9.4996 | 9.525 | 9.5504 | 9.5758 | 9.6012 | 9.6266 |
| 380 | 9.652 | 9.6774 | 9.7028 | 9.7282 | 9.7536 | 9.779 | 9.8044 | 9.8298 | 9.8552 | 9.8806 |
| 390 | 9.906 | 9.9314 | 9.9568 | 9.9822 | 10.0076 | 10.033 | 10.0584 | 10.0838 | 10.1092 | 10.1346 |
| 400 | 10.16 | 10.1854 | 10.2108 | 10.2362 | 10.2616 | 10.287 | 10.3124 | 10.3378 | 10.3632 | 10.3886 |
| 410 | 10.414 | 10.4394 | 10.4648 | 10.4902 | 10.5156 | 10.541 | 10.5664 | 10.5918 | 10.6172 | 10.6426 |
| 420 | 10.668 | 10.6934 | 10.7188 | 10.7442 | 10.7696 | 10.795 | 10.8204 | 10.8458 | 10.8712 | 10.8966 |
| 430 | 10.922 | 10.9474 | 10.9728 | 10.9982 | 11.0236 | 11.049 | 11.0744 | 11.0998 | 11.1252 | 11.1506 |
| 440 | 11.176 | 11.2014 | 11.2268 | 11.2522 | 11.2776 | 11.303 | 11.3284 | 11.3538 | 11.3792 | 11.4046 |
| 450 | 11.43 | 11.4554 | 11.4808 | 11.5062 | 11.5316 | 11.557 | 11.5824 | 11.6078 | 11.6332 | 11.6586 |
| 460 | 11.684 | 11.7094 | 11.7348 | 11.7602 | 11.7856 | 11.811 | 11.8364 | 11.8618 | 11.8872 | 11.9126 |
| 470 | 11.938 | 11.9634 | 11.9888 | 12.0142 | 12.0396 | 12.065 | 12.0904 | 12.1158 | 12.1412 | 12.1666 |
| 480 | 12.192 | 12.2174 | 12.2428 | 12.2682 | 12.2936 | 12.319 | 12.3444 | 12.3698 | 12.3952 | 12.4206 |
| 490 | 12.446 | 12.4714 | 12.4968 | 12.5222 | 12.5476 | 12.573 | 12.5984 | 12.6238 | 12.6492 | 12.6746 |
| 500 | 12.7 | 12.7254 | 12.7508 | 12.7762 | 12.8016 | 12.827 | 12.8524 | 12.8778 | 12.9032 | 12.9286 |

Use the small table below to convert microinches to micrometers for ranges higher than given in the main table above. Appropriate quantities chosen from both tables are simply added to obtain the higher converted value:

| μin. | μm | μin. | μm | μin. | μm | μin. | μm | μin. | μm | μin. | μm |
|---|---|---|---|---|---|---|---|---|---|---|---|
| 600 | 15.24 | 800 | 20.32 | 1000 | 25.4 | 1500 | 38.1 | 2100 | 53.34 | 2700 | 68.58 |
| 700 | 17.78 | 900 | 22.86 | 1200 | 30.48 | 1800 | 45.72 | 2400 | 60.96 | 3000 | 76.2 |

Both tables based on 1 microinch = 0.0254 micrometer, exactly. All values in both parts of this table are exact; figures to the right of the last place figures are all zeros.

*Example:* Convert 1375 μin. to μm:

| | | | |
|---|---|---|---|
| From lower portion of Table 13a: | 1200 μin. | = | 30.48 μm |
| From upper portion of Table 13a: | 175 μin. | = | 4.445 μm |
| | 1375 μin. | = | 34.925 μm |

## Table 13b. Micrometers (Microns) to Microinches Conversion

| Microns ↓ \ → | 0 | 0.01 | 0.02 | 0.03 | 0.04 | 0.05 | 0.06 | 0.07 | 0.08 | 0.09 |
|---|---|---|---|---|---|---|---|---|---|---|
| | Microinches | | | | | | | | | |
| 0.00 | 0.0000 | 0.3937 | 0.7874 | 1.1811 | 1.5748 | 1.9685 | 2.3622 | 2.7559 | 3.1496 | 3.5433 |
| 0.10 | 3.9370 | 4.3307 | 4.7244 | 5.1181 | 5.5118 | 5.9055 | 6.2992 | 6.6929 | 7.0866 | 7.4803 |
| 0.20 | 7.8740 | 8.2677 | 8.6614 | 9.0551 | 9.4488 | 9.8425 | 10.2362 | 10.6299 | 11.0236 | 11.4173 |
| 0.30 | 11.8110 | 12.2047 | 12.5984 | 12.9921 | 13.3858 | 13.7795 | 14.1732 | 14.5669 | 14.9606 | 15.3543 |
| 0.40 | 15.7480 | 16.1417 | 16.5354 | 16.9291 | 17.3228 | 17.7165 | 18.1102 | 18.5039 | 18.8976 | 19.2913 |
| 0.50 | 19.6850 | 20.0787 | 20.4724 | 20.8661 | 21.2598 | 21.6535 | 22.0472 | 22.4409 | 22.8346 | 23.2283 |
| 0.60 | 23.6220 | 24.0157 | 24.4094 | 24.8031 | 25.1969 | 25.5906 | 25.9843 | 26.3780 | 26.7717 | 27.1654 |
| 0.70 | 27.5591 | 27.9528 | 28.3465 | 28.7402 | 29.1339 | 29.5276 | 29.9213 | 30.3150 | 30.7087 | 31.1024 |
| 0.80 | 31.4961 | 31.8898 | 32.2835 | 32.6772 | 33.0709 | 33.4646 | 33.8583 | 34.2520 | 34.6457 | 35.0394 |
| 0.90 | 35.4331 | 35.8268 | 36.2205 | 36.6142 | 37.0079 | 37.4016 | 37.7953 | 38.1890 | 38.5827 | 38.9764 |
| 1.00 | 39.3701 | 39.7638 | 40.1575 | 40.5512 | 40.9449 | 41.3386 | 41.7323 | 42.1260 | 42.5197 | 42.9134 |
| 1.10 | 43.3071 | 43.7008 | 44.0945 | 44.4882 | 44.8819 | 45.2756 | 45.6693 | 46.0630 | 46.4567 | 46.8504 |
| 1.20 | 47.2441 | 47.6378 | 48.0315 | 48.4252 | 48.8189 | 49.2126 | 49.6063 | 50.0000 | 50.3937 | 50.7874 |
| 1.30 | 51.1811 | 51.5748 | 51.9685 | 52.3622 | 52.7559 | 53.1496 | 53.5433 | 53.9370 | 54.3307 | 54.7244 |
| 1.40 | 55.1181 | 55.5118 | 55.9055 | 56.2992 | 56.6929 | 57.0866 | 57.4803 | 57.8740 | 58.2677 | 58.6614 |
| 1.50 | 59.0551 | 59.4488 | 59.8425 | 60.2362 | 60.6299 | 61.0236 | 61.4173 | 61.8110 | 62.2047 | 62.5984 |
| 1.60 | 62.9921 | 63.3858 | 63.7795 | 64.1732 | 64.5669 | 64.9606 | 65.3543 | 65.7480 | 66.1417 | 66.5354 |
| 1.70 | 66.9291 | 67.3228 | 67.7165 | 68.1102 | 68.5039 | 68.8976 | 69.2913 | 69.6850 | 70.0787 | 70.4724 |
| 1.80 | 70.8661 | 71.2598 | 71.6535 | 72.0472 | 72.4409 | 72.8346 | 73.2283 | 73.6220 | 74.0157 | 74.4094 |
| 1.90 | 74.8031 | 75.1969 | 75.5906 | 75.9843 | 76.3780 | 76.7717 | 77.1654 | 77.5591 | 77.9528 | 78.3465 |
| 2.00 | 78.7402 | 79.1339 | 79.5276 | 79.9213 | 80.3150 | 80.7087 | 81.1024 | 81.4961 | 81.8898 | 82.2835 |
| 2.10 | 82.6772 | 83.0709 | 83.4646 | 83.8583 | 84.2520 | 84.6457 | 85.0394 | 85.4331 | 85.8268 | 86.2205 |
| 2.20 | 86.6142 | 87.0079 | 87.4016 | 87.7953 | 88.1890 | 88.5827 | 88.9764 | 89.3701 | 89.7638 | 90.1575 |
| 2.30 | 90.5512 | 90.9449 | 91.3386 | 91.7323 | 92.1260 | 92.5197 | 92.9134 | 93.3071 | 93.7008 | 94.0945 |
| 2.40 | 94.4882 | 94.8819 | 95.2756 | 95.6693 | 96.0630 | 96.4567 | 96.8504 | 97.2441 | 97.6378 | 98.0315 |
| 2.50 | 98.4252 | 98.8189 | 99.2126 | 99.6063 | 100.0000 | 100.3937 | 100.7874 | 101.1811 | 101.5748 | 101.9685 |
| 2.60 | 102.3622 | 102.7559 | 103.1496 | 103.5433 | 103.9370 | 104.3307 | 104.7244 | 105.1181 | 105.5118 | 105.9055 |
| 2.70 | 106.2992 | 106.6929 | 107.0866 | 107.4803 | 107.8740 | 108.2677 | 108.6614 | 109.0551 | 109.4488 | 109.8425 |
| 2.80 | 110.2362 | 110.6299 | 111.0236 | 111.4173 | 111.8110 | 112.2047 | 112.5984 | 112.9921 | 113.3858 | 113.7795 |
| 2.90 | 114.1732 | 114.5669 | 114.9606 | 115.3543 | 115.7480 | 116.1417 | 116.5354 | 116.9291 | 117.3228 | 117.7165 |
| 3.00 | 118.1102 | 118.5039 | 118.8976 | 119.2913 | 119.6850 | 120.0787 | 120.4724 | 120.8661 | 121.2598 | 121.6535 |
| 3.10 | 122.0472 | 122.4409 | 122.8346 | 123.2283 | 123.6220 | 124.0157 | 124.4094 | 124.8031 | 125.1969 | 125.5906 |
| 3.20 | 125.9843 | 126.3780 | 126.7717 | 127.1654 | 127.5591 | 127.9528 | 128.3465 | 128.7402 | 129.1339 | 129.5276 |
| 3.30 | 129.9213 | 130.3150 | 130.7087 | 131.1024 | 131.4961 | 131.8898 | 132.2835 | 132.6772 | 133.0709 | 133.4646 |
| 3.40 | 133.8583 | 134.2520 | 134.6457 | 135.0394 | 135.4331 | 135.8268 | 136.2205 | 136.6142 | 137.0079 | 137.4016 |
| 3.50 | 137.7953 | 138.1890 | 138.5827 | 138.9764 | 139.3701 | 139.7638 | 140.1575 | 140.5512 | 140.9449 | 141.3386 |
| 3.60 | 141.7323 | 142.1260 | 142.5197 | 142.9134 | 143.3071 | 143.7008 | 144.0945 | 144.4882 | 144.8819 | 145.2756 |
| 3.70 | 145.6693 | 146.0630 | 146.4567 | 146.8504 | 147.2441 | 147.6378 | 148.0315 | 148.4252 | 148.8189 | 149.2126 |
| 3.80 | 149.6063 | 150.0000 | 150.3937 | 150.7874 | 151.1811 | 151.5748 | 151.9685 | 152.3622 | 152.7559 | 153.1496 |
| 3.90 | 153.5433 | 153.9370 | 154.3307 | 154.7244 | 155.1181 | 155.5118 | 155.9055 | 156.2992 | 156.6929 | 157.0866 |
| 4.00 | 157.4803 | 157.8740 | 158.2677 | 158.6614 | 159.0551 | 159.4488 | 159.8425 | 160.2362 | 160.6299 | 161.0236 |
| 4.10 | 161.4173 | 161.8110 | 162.2047 | 162.5984 | 162.9921 | 163.3858 | 163.7795 | 164.1732 | 164.5669 | 164.9606 |
| 4.20 | 165.3543 | 165.7480 | 166.1417 | 166.5354 | 166.9291 | 167.3228 | 167.7165 | 168.1102 | 168.5039 | 168.8976 |
| 4.30 | 169.2913 | 169.6850 | 170.0787 | 170.4724 | 170.8661 | 171.2598 | 171.6535 | 172.0472 | 172.4409 | 172.8346 |
| 4.40 | 173.2283 | 173.6220 | 174.0157 | 174.4094 | 174.8031 | 175.1969 | 175.5906 | 175.9843 | 176.3780 | 176.7717 |
| 4.50 | 177.1654 | 177.5591 | 177.9528 | 178.3465 | 178.7402 | 179.1339 | 179.5276 | 179.9213 | 180.3150 | 180.7087 |
| 4.60 | 181.1024 | 181.4961 | 181.8898 | 182.2835 | 182.6772 | 183.0709 | 183.4646 | 183.8583 | 184.2520 | 184.6457 |
| 4.70 | 185.0394 | 185.4331 | 185.8268 | 186.2205 | 186.6142 | 187.0079 | 187.4016 | 187.7953 | 188.1890 | 188.5827 |
| 4.80 | 188.9764 | 189.3701 | 189.7638 | 190.1575 | 190.5512 | 190.9449 | 191.3386 | 191.7323 | 192.1260 | 192.5197 |
| 4.90 | 192.9134 | 193.3071 | 193.7008 | 194.0945 | 194.4882 | 194.8819 | 195.2756 | 195.6693 | 196.0630 | 196.4567 |
| 5.00 | 196.8504 | 197.2441 | 197.6378 | 198.0315 | 198.4252 | 198.8189 | 199.2126 | 199.6063 | 200.0000 | 200.3937 |

The table given below can be used with the preceding main table to obtain higher converted values, simply by adding appropriate quantities chosen from each table:

| μm | μin. | μm | μin. | μm | μin. | μm | μin. | μm | μin. |
|---|---|---|---|---|---|---|---|---|---|
| 10 | 393.7008 | 20 | 787.4016 | 30 | 1181.1024 | 40 | 1574.8032 | 50 | 1968.5039 |
| 15 | 590.5512 | 25 | 984.2520 | 35 | 1378.9528 | 45 | 1771.6535 | 55 | 2165.3543 |

Both portions of Table 13b are based on 1 microinch = 0.0254 micrometer, exactly.

*Example:* Convert 23.55 μm to μin.:

| | | | |
|---|---|---|---|
| From above table: | 20.00 μm | = | 787.4016 μin |
| From main table: | 3.55 μm | = | 139.7638 μin |
| | 23.55 μm | = | 927.1654 μin |

**Table 14a. Feet to Meters Conversion**

| feet | meters | feet | meters | feet | meters | feet | meters | feet | meters |
|---|---|---|---|---|---|---|---|---|---|
| 100 | 30.48 | 10 | 3.048 | 1 | 0.3048 | 0.1 | 0.03048 | 0.01 | 0.003048 |
| 200 | 60.96 | 20 | 6.096 | 2 | 0.6096 | 0.2 | 0.06096 | 0.02 | 0.006096 |
| 300 | 91.44 | 30 | 9.144 | 3 | 0.9144 | 0.3 | 0.09144 | 0.03 | 0.009144 |
| 400 | 121.92 | 40 | 12.192 | 4 | 1.2192 | 0.4 | 0.12192 | 0.04 | 0.012192 |
| 500 | 152.4 | 50 | 15.24 | 5 | 1.524 | 0.5 | 0.1524 | 0.05 | 0.01524 |
| 600 | 182.88 | 60 | 18.288 | 6 | 1.8288 | 0.6 | 0.18288 | 0.06 | 0.018288 |
| 700 | 213.36 | 70 | 21.336 | 7 | 2.1336 | 0.7 | 0.21336 | 0.07 | 0.021336 |
| 800 | 243.84 | 80 | 24.384 | 8 | 2.4384 | 0.8 | 0.24384 | 0.08 | 0.024384 |
| 900 | 274.32 | 90 | 27.432 | 9 | 2.7432 | 0.9 | 0.27432 | 0.09 | 0.027432 |
| 1,000 | 304.8 | 100 | 30.48 | 10 | 3.048 | 1.0 | 0.3048 | 0.10 | 0.03048 |

1 ft = 0.3048 m, exactly

**Table 14b. Meters to Feet Conversion**

| meters | feet | meters | feet | meters | feet | meters | feet | meters | feet |
|---|---|---|---|---|---|---|---|---|---|
| 100 | 328.084 | 10 | 32.808 | 1 | 3.281 | 0.1 | 0.328 | 0.01 | 0.033 |
| 200 | 656.168 | 20 | 65.617 | 2 | 6.562 | 0.2 | 0.656 | 0.02 | 0.066 |
| 300 | 984.252 | 30 | 98.425 | 3 | 9.843 | 0.3 | 0.984 | 0.03 | 0.098 |
| 400 | 1,312.336 | 40 | 131.234 | 4 | 13.123 | 0.4 | 1.312 | 0.04 | 0.131 |
| 500 | 1,640.420 | 50 | 164.042 | 5 | 16.404 | 0.5 | 1.640 | 0.05 | 0.164 |
| 600 | 1,968.504 | 60 | 196.850 | 6 | 19.685 | 0.6 | 1.969 | 0.06 | 0.197 |
| 700 | 2,296.588 | 70 | 229.659 | 7 | 22.966 | 0.7 | 2.297 | 0.07 | 0.230 |
| 800 | 2,624.672 | 80 | 262.467 | 8 | 26.247 | 0.8 | 2.625 | 0.08 | 0.262 |
| 900 | 2,952.756 | 90 | 295.276 | 9 | 29.528 | 0.9 | 2.953 | 0.09 | 0.295 |
| 1,000 | 3,280.840 | 100 | 328.084 | 10 | 32.808 | 1.0 | 3.281 | 0.10 | 0.328 |

1 m = 3.280840 ft

**Table 15a. Miles to Kilometers Conversion**

| miles | km | miles | km | miles | km | miles | km | miles | km |
|---|---|---|---|---|---|---|---|---|---|
| 1,000 | 1,609.34 | 100 | 160.93 | 10 | 16.09 | 1 | 1.61 | 0.1 | 0.16 |
| 2,000 | 3,218.69 | 200 | 321.87 | 20 | 32.19 | 2 | 3.22 | 0.2 | 0.32 |
| 3,000 | 4,828.03 | 300 | 482.80 | 30 | 48.28 | 3 | 4.83 | 0.3 | 0.48 |
| 4,000 | 6,437.38 | 400 | 643.74 | 40 | 64.37 | 4 | 6.44 | 0.4 | 0.64 |
| 5,000 | 8,046.72 | 500 | 804.67 | 50 | 80.47 | 5 | 8.05 | 0.5 | 0.80 |
| 6,000 | 9,656.06 | 600 | 965.61 | 60 | 96.56 | 6 | 9.66 | 0.6 | 0.97 |
| 7,000 | 11,265.41 | 700 | 1,126.54 | 70 | 112.65 | 7 | 11.27 | 0.7 | 1.13 |
| 8,000 | 12,874.75 | 800 | 1,287.48 | 80 | 128.75 | 8 | 12.87 | 0.8 | 1.29 |
| 9,000 | 14,484.10 | 900 | 1,448.41 | 90 | 144.84 | 9 | 14.48 | 0.9 | 1.45 |
| 10,000 | 16,093.44 | 1,000 | 1,609.34 | 100 | 160.93 | 10 | 16.09 | 1.0 | 1.61 |

1 mile = 1.609344 km, exactly

**Table 15b. Kilometers to Miles Conversion**

| km | miles | km | miles | km | miles | km | miles | km | miles |
|---|---|---|---|---|---|---|---|---|---|
| 1,000 | 621.37 | 100 | 62.14 | 10 | 6.21 | 1 | 0.62 | 0.1 | 0.06 |
| 2,000 | 1,242.74 | 200 | 124.27 | 20 | 12.43 | 2 | 1.24 | 0.2 | 0.12 |
| 3,000 | 1,864.11 | 300 | 186.41 | 30 | 18.64 | 3 | 1.86 | 0.3 | 0.19 |
| 4,000 | 2,485.48 | 400 | 248.55 | 40 | 24.85 | 4 | 2.49 | 0.4 | 0.25 |
| 5,000 | 3,106.86 | 500 | 310.69 | 50 | 31.07 | 5 | 3.11 | 0.5 | 0.31 |
| 6,000 | 3,728.23 | 600 | 372.82 | 60 | 37.28 | 6 | 3.73 | 0.6 | 0.37 |
| 7,000 | 4,349.60 | 700 | 434.96 | 70 | 43.50 | 7 | 4.35 | 0.7 | 0.43 |
| 8,000 | 4,970.97 | 800 | 497.10 | 80 | 49.71 | 8 | 4.97 | 0.8 | 0.50 |
| 9,000 | 5,592.34 | 900 | 559.23 | 90 | 55.92 | 9 | 5.59 | 0.9 | 0.56 |
| 10,000 | 6,213.71 | 1,000 | 621.37 | 100 | 62.14 | 10 | 6.21 | 1.0 | 0.62 |

1 km = 0.6213712 mile

## Units of Area

### Table 16. Square Measure and Conversion Factors

**Metric System**

1 *square kilometer* ($km^2$) =
- **100** hectares
- **1,000,000** square meters
- 0.3861 square mile
- 247.1 acres

1 *hectare (ha)* =
- **0.01** square kilometer
- **100** ares
- **10,000** square meters
- 2.471 acres
- 107,639 square feet

1 *are (a)* =
- **0.0001** square kilometer
- **100** square meters
- 0.0247 acre
- 1076.4 square feet

1 *square meter* ($m^2$) =
- **0.000001** square kilometer
- **100** square decimeters
- **10000** square centimeters
- **1,000,000** square millimeters
- 10.764 square feet
- 1.196 square yards

1 *square decimeter* ($dm^2$) =
- **100** square centimeters

1 *square centimeter* ($cm^2$) =
- **0.0001** square meter
- **100** square millimeters
- 0.001076 square foot
- 0.155 square inch

1 *square millimeter* ($mm^2$) =
- **0.01** square centimeter
- **1,000,000** square microns
- 0.00155 square inch

1 *square micrometer (micron)* ($\mu m^2$) =
- **$1 \times 10^{-12}$** square meter
- **0.000001** square millimeter
- $1 \times 10^{-9}$ square inch
- 1549.997 square micro-inch

**US System**

1 *square mile* ($mi^2$) =
- **640** acres
- **6400** square chains
- 2.5899 square kilometers

1 *acre* =
- **10** square chains
- **4840** square yards
- **43,560** square feet
- a square, 208.71 feet on a side
- 0.4046856 hectare
- 40.47 ares
- 4046.856 square meters

1 *square chain* =
- **16** square rods
- **484** square yards
- **4356** square feet

1 *square rod* =
- **30.25** square yards
- **272.25** square feet
- **625** square links

1 *square yard* ($yd^2$) =
- **9** square feet
- **1296** square inches
- **0.83612736** square meter
- **8361.2736** square centimeter
- **836,127.36** square millimeter

1 *square foot* ($ft^2$) =
- 0.111111 square yard
- **144** square inches
- **0.09290304** square meter
- **929.0304** square centimeters
- **92,903.04** square millimeters

1 *square inch* ($in^2$) =
- 0.0007716 square yard
- 0.006944 square foot
- 0.00064516 square meter
- **6.4516** square centimeters
- **645.16** square millimeters

1 *square mil* ($mil^2$) =
- **0.000001** square inch
- **0.00064516** square millimeter

1 *square micro-inch* ($\mu in^2$) =
- **$1 \times 10^{-12}$** square inch
- **0.00064516** square micrometer (micron)

*Note*: Figures in **Bold** indicate exact conversion values

### Measure Used for Diameters and Areas of Electric Wires

1 *circular inch* =
- area of 1-inch diameter circle
- $\pi/4$ square inch
- 0.7854 square inch
- 5.067 square centimeters
- 1,000,000 circular mils

1 *circular mil* =
- area of 0.001-inch diameter circle
- $\pi/4$ square mil

1 *square inch* =
- 1.2732 circular inches
- 1,273,239 circular mils

## Table 17a. Square Inches to Square Centimeters Conversion

| inch² | cm² | inch² | cm² | inch² | cm² | inch² | cm² | inch² | cm² |
|---|---|---|---|---|---|---|---|---|---|
| 100 | 645.16 | 10 | 64.516 | 1 | 6.4516 | 0.1 | 0.64516 | 0.01 | 0.064516 |
| 200 | 1,290.32 | 20 | 129.032 | 2 | 12.9032 | 0.2 | 1.29032 | 0.02 | 0.129032 |
| 300 | 1,935.48 | 30 | 193.548 | 3 | 19.3548 | 0.3 | 1.93548 | 0.03 | 0.135489 |
| 400 | 2,580.64 | 40 | 258.064 | 4 | 25.8064 | 0.4 | 2.58064 | 0.04 | 0.258064 |
| 500 | 3,225.80 | 50 | 322.58 | 5 | 32.258 | 0.5 | 3.2258 | 0.05 | 0.32258 |
| 600 | 30,870.96 | 60 | 387.096 | 6 | 38.7096 | 0.6 | 3.87096 | 0.06 | 0.387096 |
| 700 | 4,516.12 | 70 | 451.612 | 7 | 45.1612 | 0.7 | 4.51612 | 0.07 | 0.451612 |
| 800 | 5,161.28 | 80 | 516.128 | 8 | 51.6128 | 0.8 | 5.16128 | 0.08 | 0.516128 |
| 900 | 5,806.44 | 90 | 580.644 | 9 | 58.0644 | 0.9 | 5.80644 | 0.09 | 0.580644 |
| 1,000 | 6,451.60 | 100 | 645.16 | 10 | 64.516 | 1.0 | 6.4516 | 0.10 | 0.64516 |

Based on 1 inch = 2.54 centimeters, exactly; 1 $inch^2 = 6.4516\ cm^2$, exactly.

## Table 17b. Square Centimeters to Square Inches Conversion

| cm² | inch² | cm² | inch² | cm² | inch² | cm² | inch² | cm² | inch² |
|---|---|---|---|---|---|---|---|---|---|
| 100 | 15.500 | 10 | 1.550 | 1 | 0.155 | 0.1 | 0.016 | 0.01 | 0.002 |
| 200 | 31.000 | 20 | 3.100 | 2 | 0.310 | 0.2 | 0.031 | 0.02 | 0.003 |
| 300 | 46.500 | 30 | 4.650 | 3 | 0.465 | 0.3 | 0.047 | 0.03 | 0.005 |
| 400 | 62.000 | 40 | 6.200 | 4 | 0.620 | 0.4 | 0.062 | 0.04 | 0.006 |
| 500 | 77.500 | 50 | 7.750 | 5 | 0.75 | 0.5 | 0.078 | 0.05 | 0.008 |
| 600 | 93.000 | 60 | 9.300 | 6 | 0.930 | 0.6 | 0.093 | 0.06 | 0.009 |
| 700 | 108.500 | 70 | 10.850 | 7 | 1.085 | 0.7 | 0.109 | 0.07 | 0.011 |
| 800 | 124.000 | 80 | 12.400 | 8 | 1.240 | 0.8 | 0.124 | 0.08 | 0.012 |
| 900 | 139.500 | 90 | 13.950 | 9 | 1.395 | 0.9 | 0.140 | 0.09 | 0.014 |
| 1,000 | 155.000 | 100 | 15.500 | 10 | 1.550 | 1.0 | 0.155 | 0.10 | 0.016 |

Based on 1 inch = 2.54 centimeters, exactly; 1 $cm^2 = 0.1550003\ inch^2$.

## Table 18a. Square Feet to Square Meters Conversion

| ft² | m² | ft² | m² | ft² | m² | ft² | m² | ft² | m² |
|---|---|---|---|---|---|---|---|---|---|
| 1,000 | 92.903 | 100 | 9.290 | 10 | 0.929 | 1 | 0.093 | 0.1 | 0.009 |
| 2,000 | 185.806 | 200 | 18.581 | 20 | 1.858 | 2 | 0.186 | 0.2 | 0.019 |
| 3,000 | 278.709 | 300 | 27.871 | 30 | 2.787 | 3 | 0.279 | 0.3 | 0.028 |
| 4,000 | 371.612 | 400 | 37.161 | 40 | 3.716 | 4 | 0.372 | 0.4 | 0.037 |
| 5,000 | 464.515 | 500 | 46.452 | 50 | 4.645 | 5 | 0.465 | 0.5 | 0.046 |
| 6,000 | 557.418 | 600 | 55.742 | 60 | 5.574 | 6 | 0.557 | 0.6 | 0.056 |
| 7,000 | 650.321 | 700 | 65.032 | 70 | 6.503 | 7 | 0.650 | 0.7 | 0.065 |
| 8,000 | 743.224 | 800 | 74.322 | 80 | 7.432 | 8 | 0.743 | 0.8 | 0.074 |
| 9,000 | 836.127 | 900 | 83.613 | 90 | 8.361 | 9 | 0.836 | 0.9 | 0.084 |
| 10,000 | 929.030 | 1,000 | 92.903 | 100 | 9.290 | 10 | 0.929 | 1.0 | 0.093 |

Based on 1 inch = 2.54 centimeters, exactly; 1 $ft^2 = 0.09290304\ m^2$, exactly.

## Table 18b. Square Meters to Square Feet Conversion

| m² | ft² | m² | ft² | m² | ft² | m² | ft² | m² | ft² |
|---|---|---|---|---|---|---|---|---|---|
| 100 | 1,076.39 | 10 | 107.64 | 1 | 10.76 | 0.1 | 1.08 | 0.01 | 0.11 |
| 200 | 2,152.78 | 20 | 215.28 | 2 | 21.53 | 0.2 | 2.15 | 0.02 | 0.22 |
| 300 | 3,229.17 | 30 | 322.92 | 3 | 32.29 | 0.3 | 3.23 | 0.03 | 0.32 |
| 400 | 4,305.56 | 40 | 430.56 | 4 | 43.06 | 0.4 | 4.31 | 0.04 | 0.43 |
| 500 | 5,381.96 | 50 | 538.20 | 5 | 53.82 | 0.5 | 5.38 | 0.05 | 0.54 |
| 600 | 6,458.35 | 60 | 645.83 | 6 | 64.58 | 0.6 | 6.46 | 0.06 | 0.65 |
| 700 | 7,534.74 | 70 | 753.47 | 7 | 75.35 | 0.7 | 7.53 | 0.07 | 0.75 |
| 800 | 8,611.13 | 80 | 861.11 | 8 | 86.11 | 0.8 | 8.61 | 0.08 | 0.86 |
| 900 | 9,687.52 | 90 | 968.75 | 9 | 96.88 | 0.9 | 9.69 | 0.09 | 0.97 |
| 1,000 | 10,763.91 | 100 | 1,076.39 | 10 | 107.64 | 1.0 | 10.76 | 0.10 | 1.08 |

Based on 1 inch = 2.54 centimeters, exactly; 1 $m^2 = 10.76391\ ft^2$.

## Table 19a. Square Yards to Square Meters Conversion

| $yd^2$ | $m^2$ | $yd^2$ | $m^2$ | $yd^2$ | $m^2$ | $yd^2$ | $m^2$ | $yd^2$ | $m^2$ |
|---|---|---|---|---|---|---|---|---|---|
| 1000 | 836.12736 | 100 | 83.612736 | 10 | 8.3612736 | 1 | 0.83612736 | 0.1 | 0.083612736 |
| 2000 | 1672.25472 | 200 | 167.225472 | 20 | 16.7225472 | 2 | 1.67225472 | 0.2 | 0.167225472 |
| 3000 | 2508.38208 | 300 | 250.838208 | 30 | 25.0838208 | 3 | 2.50838208 | 0.3 | 0.250838208 |
| 4000 | 3344.50944 | 400 | 334.450944 | 40 | 33.4450944 | 4 | 3.34450944 | 0.4 | 0.334450944 |
| 5000 | 4180.6368 | 500 | 418.06368 | 50 | 41.806368 | 5 | 4.1806368 | 0.5 | 0.41806368 |
| 6000 | 5016.76416 | 600 | 501.676416 | 60 | 50.1676416 | 6 | 5.01676416 | 0.6 | 0.501676416 |
| 7000 | 5852.89152 | 700 | 585.289152 | 70 | 58.5289152 | 7 | 5.85289152 | 0.7 | 0.585289152 |
| 8000 | 6689.01888 | 800 | 668.901888 | 80 | 66.8901888 | 8 | 6.68901888 | 0.8 | 0.668901888 |
| 9000 | 7525.14624 | 900 | 752.514624 | 90 | 75.2514624 | 9 | 7.52514624 | 0.9 | 0.752514624 |
| 10000 | 8361.2736 | 1000 | 836.12736 | 100 | 83.612736 | 10 | 8.3612736 | 1 | 0.83612736 |

Based on 1 inch = 2.54 centimeters, exactly; 1 $yd^2$ = 0.83612736 $m^2$, exactly

## Table 19b. Square Meters to Square Yards Conversion

| $m^2$ | $yd^2$ | $m^2$ | $yd^2$ | $m^2$ | $yd^2$ | $m^2$ | $yd^2$ | $m^2$ | $yd^2$ |
|---|---|---|---|---|---|---|---|---|---|
| 1000 | 1195.990046 | 100 | 119.5990046 | 10 | 11.95990046 | 1 | 1.195990046 | 0.1 | 0.119599005 |
| 2000 | 2391.980093 | 200 | 239.1980093 | 20 | 23.91980093 | 2 | 2.391980093 | 0.2 | 0.239198009 |
| 3000 | 3587.970139 | 300 | 358.7970139 | 30 | 35.87970139 | 3 | 3.587970139 | 0.3 | 0.358797014 |
| 4000 | 4783.960185 | 400 | 478.3960185 | 40 | 47.83960185 | 4 | 4.783960185 | 0.4 | 0.478396019 |
| 5000 | 5979.950232 | 500 | 597.9950232 | 50 | 59.79950232 | 5 | 5.979950232 | 0.5 | 0.597995023 |
| 6000 | 7175.940278 | 600 | 717.5940278 | 60 | 71.75940278 | 6 | 7.175940278 | 0.6 | 0.717594028 |
| 7000 | 8371.930324 | 700 | 837.1930324 | 70 | 83.71930324 | 7 | 8.371930324 | 0.7 | 0.837193032 |
| 8000 | 9567.92037 | 800 | 956.792037 | 80 | 95.6792037 | 8 | 9.56792037 | 0.8 | 0.956792037 |
| 9000 | 10763.91042 | 900 | 1076.391042 | 90 | 107.6391042 | 9 | 10.76391042 | 0.9 | 1.076391042 |
| 10000 | 11959.90046 | 1000 | 1195.990046 | 100 | 119.5990046 | 10 | 11.95990046 | 1 | 1.195990046 |

Based on 1 inch = 2.54 centimeters, exactly; 1 $m^2$ = 1.195990046 $yd^2$.

## Table 20a. Acres to Hectares Conversion

| → | 0 | 10 | 20 | 30 | 40 | 50 | 60 | 70 | 80 | 90 |
|---|---|---|---|---|---|---|---|---|---|---|
| acres ↓ | hectares | | | | | | | | | |
| 0 | … | 4.047 | 8.094 | 12.141 | 16.187 | 20.234 | 24.281 | 28.328 | 32.375 | 36.422 |
| 100 | 40.469 | 44.515 | 48.562 | 52.609 | 56.656 | 60.703 | 64.750 | 68.797 | 72.843 | 76.890 |
| 200 | 80.937 | 84.984 | 89.031 | 93.078 | 97.125 | 101.171 | 105.218 | 109.265 | 113.312 | 117.359 |
| 300 | 121.406 | 125.453 | 129.499 | 133.546 | 137.593 | 141.640 | 145.687 | 149.734 | 153.781 | 157.827 |
| 400 | 161.874 | 165.921 | 169.968 | 174.015 | 178.062 | 182.109 | 186.155 | 190.202 | 194.249 | 198.296 |
| 500 | 202.343 | 206.390 | 240.437 | 214.483 | 218.530 | 222.577 | 226.624 | 230.671 | 234.718 | 238.765 |
| 600 | 242.811 | 246.858 | 250.905 | 254.952 | 258.999 | 263.046 | 267.092 | 271.139 | 275.186 | 279.233 |
| 700 | 283.280 | 287.327 | 291.374 | 295.420 | 299.467 | 303.514 | 307.561 | 311.608 | 315.655 | 319.702 |
| 800 | 323.748 | 327.795 | 331.842 | 335.889 | 339.936 | 343.983 | 348.030 | 352.076 | 356.123 | 360.170 |
| 900 | 364.217 | 368.264 | 372.311 | 376.358 | 380.404 | 384.451 | 388.498 | 392.545 | 396.592 | 400.639 |
| 1000 | 404.686 | … | … | … | … | … | … | … | … | … |

1 acre = 0.4046856 hectare

## Table 20b. Hectares to Acres Conversion

| → | 0 | 10 | 20 | 30 | 40 | 50 | 60 | 70 | 80 | 90 |
|---|---|---|---|---|---|---|---|---|---|---|
| hectares ↓ | acres | | | | | | | | | |
| 0 | … | 24.71 | 49.42 | 74.13 | 98.84 | 123.55 | 148.26 | 172.97 | 197.68 | 222.39 |
| 100 | 247.11 | 271.82 | 296.53 | 321.24 | 345.95 | 370.66 | 395.37 | 420.08 | 444.79 | 469.50 |
| 200 | 494.21 | 518.92 | 543.63 | 568.34 | 593.05 | 617.76 | 642.47 | 667.18 | 691.90 | 716.61 |
| 300 | 741.32 | 766.03 | 790.74 | 815.45 | 840.16 | 864.87 | 889.58 | 914.29 | 939.00 | 963.71 |
| 400 | 988.42 | 1013.13 | 1037.84 | 1062.55 | 1087.26 | 1111.97 | 1136.68 | 1161.40 | 1186.11 | 1210.82 |
| 500 | 1235.53 | 1260.24 | 1284.95 | 1309.66 | 1334.37 | 1359.08 | 1383.79 | 1408.50 | 1433.21 | 1457.92 |
| 600 | 1482.63 | 1507.34 | 1532.05 | 1556.76 | 1581.47 | 1606.19 | 1630.90 | 1655.61 | 1680.32 | 1705.03 |
| 700 | 1729.74 | 1754.45 | 1779.16 | 1803.87 | 1828.58 | 1853.29 | 1878.00 | 1902.71 | 1927.42 | 1952.13 |
| 800 | 1976.84 | 2001.55 | 2026.26 | 2050.97 | 2075.69 | 2100.40 | 2125.11 | 2149.82 | 2174.53 | 2199.24 |
| 900 | 2223.95 | 2248.66 | 2273.37 | 2298.08 | 2322.79 | 2347.50 | 2372.21 | 2396.92 | 2421.63 | 2446.34 |
| 1000 | 2471.05 | … | … | … | … | … | … | … | … | … |

1 hectare = 2.471054 acres

## Units of Volume

### Table 21. Cubic Measure and Conversion Factors

**Metric System**

1 *cubic meter* ($m^3$) =
- **1000** cubic decimeters (liters)
- **1,000,000** cubic centimeters
- 1.30795 cubic yards
- 35.314667 cubic feet
- 61,023.74 cubic inches
- 264.17205 US gallons
- 219.96925 British Imperial gallons

1 *liter* (*l*) *or* 1 *cubic decimeter* ($dm^3$) =
- **1** liter = volume of 1 kg water at 39.2° F
- **0.001** cubic meter
- **1000** cubic centimeters
- **10** deciliters
- 0.03531466 cubic foot
- 61.023744 cubic inches
- 0.2642 US gallon
- 0.21997 British Imperial gallon
- 1.0566882 US quarts
- 33.814 US fluid ounces

1 *cubic centimeter* ($cm^3$) =
- **0.001** liter
- **1000** cubic millimeters
- 0.061024 cubic inch

1 *cubic millimeter* = **0.001** cubic centimeters

1 *hectoliter* (*hl*) = **100** liters

1 *deciliter* (*dl*) = **10** centiliters

1 *centiliter* (*cl*) = **10** milliliters

**US System**

1 *cubic yard* ($yd^3$) =
- **27** cubic feet
- 201.97403 US gallons
- **46,656** cubic inch
- 0.7646 cubic meter

1 *cubic foot* ($ft^3$) =
- **1728** cubic inches
- 7.4805 US gallons
- 6.23 British Imperial gallons
- 0.02831685 cubic meter
- 28.31685 liters

1 *cubic inch* ($in^3$) =
- 0.55411256 US fluid ounce
- **16.387064** cubic centimeters

**Shipping Measure**

For measuring internal capacity of a vessel:
- 1 *register ton* = 100 cubic feet

For measurement of cargo:
- 1 *shipping ton* =
  Approximately 40 cubic feet of merchandise is considered a shipping ton, unless that bulk would weigh more than 2000 pounds, in which case the freight charge may be based upon weight
- 40 *cubic feet* =
  - 32.143 US bushels
  - 31.16 Imperial bushels

**British (Imperial) Liquid and Dry Measure**

1 *British Imperial gallon* =
- 0.1605 cubic foot
- 277.42 cubic inches
- 1.2009 US gallon
- **160** Imperial fluid ounces
- **4** Imperial quarts
- **8** Imperial pints
- **4.54609** liters

1 *quart* =
- **2** Imperial pints
- **8** Imperial gills
- **40** Imperial fluid ounces
- 69.354 cubic inches
- **1.1365225** liters

1 *pint* =
- **4** Imperial gills
- **20** Imperial fluid ounces
- 34.678 cubic inches
- **568.26125** milliliters

1 *gill* =
- **5** Imperial fluid ounces
- 8.669 cubic inches
- 142.07 milliliters

**US Liquid Measure**

1 *US gallon* =
- 0.13368 cubic foot
- **231** cubic inches
- **128** US fluid ounces
- **4** US quarts
- **8** US pints
- 0.8327 British Imperial gallon
- **3.785411784** liters

1 *quart* =
- **2** US pints
- **8** US gills
- **32** US fluid ounces
- **57.75** cubic inches
- 0.9463529 liters

1 *pint* =
- **4** US gills
- **16** US fluid ounces
- **28.875** cubic inches
- 473.176 milliliters

1 *gill* =
- **1/2** cup = **4** US fluid ounces
- **7.21875** cubic inches
- 118.29 milliliters

*Note*: Figures in **Bold** indicate exact conversion values

## Table 21. *(Continued)* Cubic Measure and Conversion Factors

**British (Imperial) Liquid and Dry Measure**

1 *British Imperial fluid ounce* =
 1.733871 cubic inch
 $^1/_{160}$ British Imperial gallon
 28.41306 milliliters

1 *British Imperial bushel* =
 **8** Imperial gallons = 1.284 cubic feet
 2219.36 cubic inches

**US Dry Measure**

1 *bushel (US or Winchester struck bushel)* =
 1.2445 cubic feet
 **2150.42** cubic inches
 a cylinder 18.5 inches dia., 8 inches deep
 a cylinder 47.0 cm dia., 20.3 cm deep

1 *bushel* = **4** pecks = **32** quarts = **64** pints

1 *peck* = **8** quarts = **16** pints

1 *dry quart* = **2** pints =
 **67.200625** cubic inches
 1.101221 liters

1 *heaped bushel* = 1¼ struck bushel

1 *cubic foot* = 0.8036 struck bushel

**Apothecaries' Fluid Measure**

1 *US fluid ounce* =
 **1.8046875** cubic inches
 $^1/_{128}$ US gallon
 **8** drachms
 0.02957353 liter
 29.57353 milliliters

1 *fluid drachm* = **60** minims

**Old Liquid Measure**

1 *barrel (bbl)* = 31½ gallons

1 *hogshead* = 2 barrels = 63 gallons

1 *pipe* or *butt* = 2 hogsheads = 4 barrels = 126 gallons

1 *tierce* = 42 gallons

1 *puncheon* = 2 tierces = 84 gallons

1 *tun* = 2 pipes = 3 puncheons

**Other Cubic Measure**

The following are used for wood and masonry:

1 *cord of wood* = 4 × 4 × 8 feet = 128 cubic feet

1 *perch of masonry* =
 16½ × 1½ × 1 foot = 24¾ cubic feet

**Barrel Measure**

1 *drum* =
 **55** US gallon
 7.3524 cubic feet
 208.19765 liters

1 *petroleum barrel (bo)* =
 **42** US gallons
 5.614583 cubic feet
 158.98729 liters

*Note*: Figures in **Bold** indicate exact conversion values

## Table 22a. Cubic Inches to Cubic Centimeters Conversion

| inch³ | cm³ | inch³ | cm³ | inch³ | cm³ | inch³ | cm³ | inch³ | cm³ |
|---|---|---|---|---|---|---|---|---|---|
| 100 | 1,638.71 | 10 | 163.87 | 1 | 16.39 | 0.1 | 1.64 | 0.01 | 0.16 |
| 200 | 3,277.41 | 20 | 327.74 | 2 | 32.77 | 0.2 | 3.28 | 0.02 | 0.33 |
| 300 | 4,916.12 | 30 | 491.61 | 3 | 49.16 | 0.3 | 4.92 | 0.03 | 0.49 |
| 400 | 6,554.82 | 40 | 655.48 | 4 | 65.55 | 0.4 | 6.55 | 0.04 | 0.66 |
| 500 | 8,193.53 | 50 | 819.35 | 5 | 81.94 | 0.5 | 8.19 | 0.05 | 0.82 |
| 600 | 9,832.24 | 60 | 983.22 | 6 | 98.32 | 0.6 | 9.83 | 0.06 | 0.98 |
| 700 | 11,470.94 | 70 | 1,147.09 | 7 | 114.71 | 0.7 | 11.47 | 0.07 | 1.15 |
| 800 | 13,109.65 | 80 | 1,310.96 | 8 | 131.10 | 0.8 | 13.11 | 0.08 | 1.31 |
| 900 | 14,748.35 | 90 | 1,474.84 | 9 | 147.48 | 0.9 | 14.75 | 0.09 | 1.47 |
| 1,000 | 16,387.06 | 100 | 1,638.71 | 10 | 163.87 | 1.0 | 16.39 | 0.10 | 1.64 |

Based on 1 inch = 2.54 centimeters, exactly. 1 inch³ = 16.387064 cm³, exactly

## Table 22b. Cubic Centimeters to Cubic Inches Conversion

| cm³ | inch³ | cm³ | in³ | cm³ | inch³ | cm³ | in³ | cm³ | in³ |
|---|---|---|---|---|---|---|---|---|---|
| 1,000 | 61.024 | 100 | 6.102 | 10 | 0.610 | 1 | 0.061 | 0.1 | 0.006 |
| 2,000 | 122.048 | 200 | 12.205 | 20 | 1.220 | 2 | 0.122 | 0.2 | 0.012 |
| 3,000 | 183.071 | 300 | 18.307 | 30 | 1,831 | 3 | 0.183 | 0.3 | 0.018 |
| 4,000 | 244.095 | 400 | 24.410 | 40 | 2.441 | 4 | 0.244 | 0.4 | 0.024 |
| 5,000 | 305.119 | 500 | 30.512 | 50 | 3.051 | 5 | 0.305 | 0.5 | 0.031 |
| 6,000 | 366.143 | 600 | 36.614 | 60 | 3.661 | 6 | 0.366 | 0.6 | 0.037 |
| 7,000 | 427.166 | 700 | 42.717 | 70 | 4.272 | 7 | 0.427 | 0.7 | 0.043 |
| 8,000 | 488.190 | 800 | 48.819 | 80 | 4.882 | 8 | 0.488 | 0.8 | 0.049 |
| 9,000 | 549.214 | 900 | 54.921 | 90 | 5.492 | 9 | 0.549 | 0.9 | 0.055 |
| 10,000 | 610.238 | 1,000 | 61.024 | 100 | 6.102 | 10 | 0.610 | 1.0 | 0.06 |

Based on 1 inch = 2.54 centimeters, exactly. 1 cm³ = 0.06102376 inch³

**Table 23a. Cubic Feet to Cubic Meters Conversion**

| $ft^3$ | $m^3$ | $ft^3$ | $m^3$ | $ft^3$ | $m^3$ | $ft^3$ | $m^3$ | $ft^3$ | $m^3$ |
|---|---|---|---|---|---|---|---|---|---|
| 1,000 | 28.317 | 100 | 2.832 | 10 | 0.283 | 1 | 0.028 | 0.1 | 0.003 |
| 2,000 | 56.634 | 200 | 5.663 | 20 | 0.566 | 2 | 0.057 | 0.2 | 0.006 |
| 3,000 | 84.951 | 300 | 8.495 | 30 | 0.850 | 3 | 0.085 | 0.3 | 0.008 |
| 4,000 | 113.267 | 400 | 11.327 | 40 | 1.133 | 4 | 0.113 | 0.4 | 0.011 |
| 5,000 | 141.584 | 500 | 14.158 | 50 | 1.416 | 5 | 0.142 | 0.5 | 0.014 |
| 6,000 | 169.901 | 600 | 16.990 | 60 | 1.699 | 6 | 0.170 | 0.6 | 0.017 |
| 7,000 | 198.218 | 700 | 19.822 | 70 | 1.982 | 7 | 0.198 | 0.7 | 0.020 |
| 8,000 | 226.535 | 800 | 22.653 | 80 | 2.265 | 8 | 0.227 | 0.8 | 0.023 |
| 9,000 | 254.852 | 900 | 25.485 | 90 | 2.549 | 9 | 0.255 | 0.9 | 0.025 |
| 10,000 | 283.168 | 1,000 | 28.317 | 100 | 2.832 | 10 | 0.283 | 1.0 | 0.028 |

Based on 1 inch = 2.54 centimeters, exactly. 1 $ft^3$ = 0.02831685 $m^3$

**Table 23b. Cubic Meters to Cubic Feet Conversion**

| $m^3$ | $ft^3$ | $m^3$ | $ft^3$ | $m^3$ | $ft^3$ | $m^3$ | $ft^3$ | $m^3$ | $ft^3$ |
|---|---|---|---|---|---|---|---|---|---|
| 100 | 3,531.47 | 10 | 353.15 | 1 | 35.31 | 0.1 | 3.53 | 0.01 | 0.35 |
| 200 | 7,062.93 | 20 | 706.29 | 2 | 70.63 | 0.2 | 7.06 | 0.02 | 0.71 |
| 300 | 10,594.40 | 30 | 1,059.44 | 3 | 105.94 | 0.3 | 10.59 | 0.03 | 1.06 |
| 400 | 14,125.86 | 40 | 4,412.59 | 4 | 141.26 | 0.4 | 14.13 | 0.04 | 1.41 |
| 500 | 17,657.33 | 50 | 1,756.73 | 5 | 176.57 | 0.5 | 17.66 | 0.05 | 1.77 |
| 600 | 21,188.80 | 60 | 2,118.88 | 6 | 211.89 | 0.6 | 21.19 | 0.06 | 2.12 |
| 700 | 24,720.26 | 70 | 2,472.03 | 7 | 247.20 | 0.7 | 24.72 | 0.07 | 2.47 |
| 800 | 28,251.73 | 80 | 2,825.17 | 8 | 282.52 | 0.8 | 28.25 | 0.08 | 2.83 |
| 900 | 31,783.19 | 90 | 3,178.32 | 9 | 317.83 | 0.9 | 31.78 | 0.09 | 3.18 |
| 1,000 | 35,314.66 | 100 | 3,531.47 | 10 | 353.15 | 1.0 | 35.311 | 0.10 | 3.53 |

Based on 1 inch = 2.54 centimeters, exactly. 1 $m^3$ = 35.31466 $ft^3$

**Table 24a. Cubic Feet to Liters Conversion**

| $ft^3$ | liters | $ft^3$ | liters | $ft^3$ | liters | $ft^3$ | liters | $ft^3$ | liters |
|---|---|---|---|---|---|---|---|---|---|
| 100 | 2,831.68 | 10 | 283.17 | 1 | 28.32 | 0.1 | 2.83 | 0.01 | 0.28 |
| 200 | 5,663.37 | 20 | 566.34 | 2 | 56.63 | 0.2 | 5.66 | 0.02 | 0.57 |
| 300 | 8,495.06 | 30 | 849.51 | 3 | 84.95 | 0.3 | 8.50 | 0.03 | 0.85 |
| 400 | 11,326.74 | 40 | 1,132.67 | 4 | 113.27 | 0.4 | 11.33 | 0.04 | 1.13 |
| 500 | 14,158.42 | 50 | 1,415.84 | 5 | 141.58 | 0.5 | 14.16 | 0.05 | 1.42 |
| 600 | 16,990.11 | 60 | 1,699.01 | 6 | 169.90 | 0.6 | 16.99 | 0.06 | 1.70 |
| 700 | 19,821.80 | 70 | 1,982.18 | 7 | 198.22 | 0.7 | 19.82 | 0.07 | 1.98 |
| 800 | 22,653.48 | 80 | 2,263.35 | 8 | 226.53 | 0.8 | 22.65 | 0.08 | 2.27 |
| 900 | 25,485.16 | 90 | 2,548.52 | 9 | 254.85 | 0.9 | 25.49 | 0.09 | 2.55 |
| 1,000 | 28,316.85 | 100 | 2,831.68 | 10 | 283.17 | 1.0 | 28.32 | 0.10 | 2.83 |

1 $ft^3$ = 28.31685 liters

**Table 24b. Liters to Cubic Feet Conversion**

| liters | $ft^3$ | liters | $ft^3$ | liters | $ft^3$ | liters | $ft^3$ | liters | $ft^3$ |
|---|---|---|---|---|---|---|---|---|---|
| 1,000 | 35.315 | 100 | 3.531 | 10 | 0.353 | 1 | 0.035 | 0.1 | 0.004 |
| 2,000 | 70.629 | 200 | 7.063 | 20 | 0.706 | 2 | 0.071 | 0.2 | 0.007 |
| 3,000 | 105.944 | 300 | 10.594 | 30 | 1.059 | 3 | 0.106 | 0.3 | 0.011 |
| 4,000 | 141.259 | 400 | 14.126 | 40 | 1.413 | 4 | 0.141 | 0.4 | 0.014 |
| 5,000 | 176.573 | 500 | 17.657 | 50 | 1.766 | 5 | 0.177 | 0.5 | 0.018 |
| 6,000 | 211.888 | 600 | 21.189 | 60 | 2.119 | 6 | 0.212 | 0.6 | 0.021 |
| 7,000 | 247.203 | 700 | 24.720 | 70 | 2.472 | 7 | 0.247 | 0.7 | 0.025 |
| 8,000 | 282.517 | 800 | 28.252 | 80 | 2.825 | 8 | 0.283 | 0.8 | 0.028 |
| 9,000 | 317.832 | 900 | 31.783 | 90 | 3.178 | 9 | 0.318 | 0.9 | 0.032 |
| 10,000 | 353.147 | 1,000 | 35.315 | 100 | 3.531 | 10 | 0.353 | 1.0 | 0.035 |

1 liter = 0.03531466 $ft^3$

## Table 25a. UK (Imperial) Gallons to Liters Conversion

| Imp. gals | 0 | 1 | 2 | 3 | 4 | 5 | 6 | 7 | 8 | 9 |
|---|---|---|---|---|---|---|---|---|---|---|
| | liters | | | | | | | | | |
| 0 | … | 4.546 | 9.092 | 13.638 | 18.184 | 22.730 | 27.277 | 31.823 | 36.369 | 40.915 |
| 10 | 45.461 | 50.007 | 54.553 | 59.099 | 63.645 | 68.191 | 72.737 | 77.284 | 81.830 | 86.376 |
| 20 | 90.922 | 95.468 | 100.014 | 104.560 | 109.106 | 113.652 | 118.198 | 122.744 | 127.291 | 131.837 |
| 30 | 136.383 | 140.929 | 145.475 | 150.021 | 154.567 | 159.113 | 163.659 | 168.205 | 172.751 | 177.298 |
| 40 | 181.844 | 186.390 | 190.936 | 195.482 | 200.028 | 204.574 | 209.120 | 213.666 | 218.212 | 222.759 |
| 50 | 227.305 | 231.851 | 236.397 | 240.943 | 245.489 | 250.035 | 254.581 | 259.127 | 263.673 | 268.219 |
| 60 | 272.766 | 277.312 | 281.858 | 286.404 | 290.950 | 295.496 | 300.042 | 304.588 | 309.134 | 313.680 |
| 70 | 318.226 | 322.773 | 327.319 | 331.865 | 336.411 | 340.957 | 345.503 | 350.049 | 354.595 | 359.141 |
| 80 | 363.687 | 368.233 | 372.780 | 377.326 | 381.872 | 386.418 | 390.964 | 395.510 | 400.056 | 404.602 |
| 90 | 409.148 | 413.694 | 418.240 | 422.787 | 427.333 | 431.879 | 436.425 | 440.971 | 445.517 | 450.063 |
| 100 | 454.609 | 459.155 | 463.701 | 468.247 | 472.794 | 477.340 | 481.886 | 486.432 | 490.978 | 495.524 |

1 UK gallon = 4.546092 liters

## Table 25b. Liters to UK (Imperial) Gallons Conversion

| liters | 0 | 1 | 2 | 3 | 4 | 5 | 6 | 7 | 8 | 9 |
|---|---|---|---|---|---|---|---|---|---|---|
| | Imperial gallons | | | | | | | | | |
| 0 | … | 0.220 | 0.440 | 0.660 | 0.880 | 1.100 | 1.320 | 1.540 | 1.760 | 1.980 |
| 10 | 2.200 | 2.420 | 2.640 | 2.860 | 3.080 | 3.300 | 3.520 | 3.739 | 3.959 | 4.179 |
| 20 | 4.399 | 4.619 | 4.839 | 5.059 | 5.279 | 5.499 | 5.719 | 5.939 | 6.159 | 6.379 |
| 30 | 6.599 | 6.819 | 7.039 | 7.259 | 7.479 | 7.699 | 7.919 | 8.139 | 8.359 | 8.579 |
| 40 | 8.799 | 9.019 | 9.239 | 9.459 | 9.679 | 9.899 | 10.119 | 10.339 | 10.559 | 10.778 |
| 50 | 10.998 | 11.218 | 11.438 | 11.658 | 11.878 | 12.098 | 12.318 | 12.538 | 12.758 | 12.978 |
| 60 | 13.198 | 13.418 | 13.638 | 13.858 | 14.078 | 14.298 | 14.518 | 14.738 | 14.958 | 15.178 |
| 70 | 15.398 | 15.618 | 15.838 | 16.058 | 16.278 | 16.498 | 16.718 | 16.938 | 17.158 | 17.378 |
| 80 | 17.598 | 17.818 | 18.037 | 18.257 | 18.477 | 18.697 | 18.917 | 19.137 | 19.357 | 19.577 |
| 90 | 19.797 | 20.017 | 20.237 | 20.457 | 20.677 | 20.897 | 21.117 | 21.337 | 21.557 | 21.777 |
| 100 | 21.997 | 22.217 | 22.437 | 22.657 | 22.877 | 23.097 | 23.317 | 23.537 | 23.757 | 23.977 |

1 liter = 0.2199692 UK gallons

## Table 26a. US Gallons to Liters Conversion

| gals | liters | gals | liters | gals | liters | gals | liters | gals | liters |
|---|---|---|---|---|---|---|---|---|---|
| 1,000 | 3,785.41 | 100 | 378.54 | 10 | 37.85 | 1 | 3.79 | 0.1 | 0.38 |
| 2,000 | 7,570.82 | 200 | 757.08 | 20 | 75.71 | 2 | 7.57 | 0.2 | 0.76 |
| 3,000 | 11,356.24 | 300 | 1,135.62 | 30 | 113.56 | 3 | 11.36 | 0.3 | 1.14 |
| 4,000 | 15,141.65 | 400 | 1,514.16 | 40 | 151.42 | 4 | 15.14 | 0.4 | 1.51 |
| 5,000 | 18,927.06 | 500 | 1,892.71 | 50 | 189.27 | 5 | 18.93 | 0.5 | 1.89 |
| 6,000 | 22,712.47 | 600 | 2,271.25 | 60 | 227.12 | 6 | 22.71 | 0.6 | 2.27 |
| 7,000 | 26,497.88 | 700 | 2,649.79 | 70 | 264.98 | 7 | 26.50 | 0.7 | 2.65 |
| 8,000 | 30,283.30 | 800 | 3,028.33 | 80 | 302.83 | 8 | 30.28 | 0.8 | 3.03 |
| 9,000 | 34,068.71 | 900 | 3,406.87 | 90 | 340.69 | 9 | 34.07 | 0.9 | 3.41 |
| 10,000 | 37,854.12 | 1,000 | 3,785.41 | 100 | 378.54 | 10 | 37.85 | 1.0 | 3.79 |

1 US gallon = 3.785412 liters

## Table 26b. Liters to US Gallons Conversion

| liters | gals | liters | gals | liters | gals | liters | gals | liters | gals |
|---|---|---|---|---|---|---|---|---|---|
| 1,000 | 264.17 | 100 | 26.42 | 10 | 2.64 | 1 | 0.26 | 0.1 | 0.03 |
| 2,000 | 528.34 | 200 | 52.83 | 20 | 5.28 | 2 | 0.53 | 0.2 | 0.05 |
| 3,000 | 792.52 | 300 | 79.25 | 30 | 7.93 | 3 | 0.79 | 0.3 | 0.08 |
| 4,000 | 1,056.69 | 400 | 105.67 | 40 | 10.57 | 4 | 1.06 | 0.4 | 0.11 |
| 5,000 | 1,320.86 | 500 | 132.09 | 50 | 13.21 | 5 | 1.32 | 0.5 | 0.13 |
| 6,000 | 1,585.03 | 600 | 158.50 | 60 | 15.85 | 6 | 1.59 | 0.6 | 0.16 |
| 7,000 | 1,849.20 | 700 | 184.92 | 70 | 18.49 | 7 | 1.85 | 0.7 | 0.18 |
| 8,000 | 2,113.38 | 800 | 211.34 | 80 | 21.13 | 8 | 2.11 | 0.8 | 0.21 |
| 9,000 | 2,377.55 | 900 | 237.75 | 90 | 23.78 | 9 | 2.38 | 0.9 | 0.24 |
| 10,000 | 2,641.72 | 1,000 | 264.17 | 100 | 26.42 | 10 | 2.64 | 1.0 | 0.26 |

1 liter = 0.2641720 US gallon

### Table 27a. US Fluid Ounces to Milliliters Conversion

| oz | mL | oz | mL | oz | mL | oz | mL | oz | mL |
|---|---|---|---|---|---|---|---|---|---|
| 100 | 2957.353 | 10 | 295.7353 | 1 | 29.57353 | 0.1 | 2.957353 | 0.01 | 0.2957353 |
| 200 | 5914.706 | 20 | 591.4706 | 2 | 59.14706 | 0.2 | 5.914706 | 0.02 | 0.5914706 |
| 300 | 8872.059 | 30 | 887.2059 | 3 | 88.72059 | 0.3 | 8.872059 | 0.03 | 0.8872059 |
| 400 | 11829.412 | 40 | 1182.9412 | 4 | 118.29412 | 0.4 | 11.829412 | 0.04 | 1.1829412 |
| 500 | 14786.765 | 50 | 1478.6765 | 5 | 147.86765 | 0.5 | 14.786765 | 0.05 | 1.4786765 |
| 600 | 17744.118 | 60 | 1774.4118 | 6 | 177.44118 | 0.6 | 17.744118 | 0.06 | 1.7744118 |
| 700 | 20701.471 | 70 | 2070.1471 | 7 | 207.01471 | 0.7 | 20.701471 | 0.07 | 2.0701471 |
| 800 | 23658.824 | 80 | 2365.8824 | 8 | 236.58824 | 0.8 | 23.658824 | 0.08 | 2.3658824 |
| 900 | 26616.177 | 90 | 2661.6177 | 9 | 266.16177 | 0.9 | 26.616177 | 0.09 | 2.6616177 |
| 1000 | 29573.53 | 100 | 2957.353 | 10 | 295.7353 | 1 | 29.57353 | 0.1 | 2.957353 |

1 US fluid ounce = 29.57353 milliliters

### Table 27b. Milliliters to US Fluid Ounces Conversion

| mL | oz | mL | oz | mL | oz | mL | oz | mL | oz |
|---|---|---|---|---|---|---|---|---|---|
| 100 | 3.3814 | 10 | 0.33814 | 1 | 0.033814 | 0.1 | 0.0033814 | 0.01 | 0.00033814 |
| 200 | 6.7628 | 20 | 0.67628 | 2 | 0.067628 | 0.2 | 0.0067628 | 0.02 | 0.00067628 |
| 300 | 10.1442 | 30 | 1.01442 | 3 | 0.101442 | 0.3 | 0.0101442 | 0.03 | 0.00101442 |
| 400 | 13.5256 | 40 | 1.35256 | 4 | 0.135256 | 0.4 | 0.0135256 | 0.04 | 0.00135256 |
| 500 | 16.907 | 50 | 1.6907 | 5 | 0.16907 | 0.5 | 0.016907 | 0.05 | 0.0016907 |
| 600 | 20.2884 | 60 | 2.02884 | 6 | 0.202884 | 0.6 | 0.0202884 | 0.06 | 0.00202884 |
| 700 | 23.6698 | 70 | 2.36698 | 7 | 0.236698 | 0.7 | 0.0236698 | 0.07 | 0.00236698 |
| 800 | 27.0512 | 80 | 2.70512 | 8 | 0.270512 | 0.8 | 0.0270512 | 0.08 | 0.00270512 |
| 900 | 30.4326 | 90 | 3.04326 | 9 | 0.304326 | 0.9 | 0.0304326 | 0.09 | 0.00304326 |
| 1000 | 33.814 | 100 | 3.3814 | 10 | 0.33814 | 1 | 0.033814 | 0.1 | 0.0033814 |

1 milliliter = 0.003814 US fluid ounce

## Units of Volumetric Flow Rate

### Table 28a. Volume Flow per Second Conversion

| To Convert ↓ | Multiply By Factor To Obtain → | cm$^3$/sec | meter$^3$/sec | foot$^3$/sec | liter/sec | gallon/sec (US) | gallon/sec (UK) |
|---|---|---|---|---|---|---|---|
| cm$^3$/sec | | 1 | **1 × 10$^{-6}$** | 3.531 × 10$^{-5}$ | **0.001** | 2.642 × 10$^{-4}$ | 2.19969 × 10$^{-4}$ |
| meter$^3$/sec | | **1 × 10$^6$** | 1 | 35.31466 | **1,000** | 264.172 | 219.9692 |
| foot$^3$/sec | | 28,316.846 | 0.028316 | 1 | 28.3168 | 7.480519 | 6.22883 |
| liter/sec | | **1000** | **0.001** | 0.0353146 | 1 | 0.264172 | 0.21996 |
| gallon/sec (US) | | 3,785.412 | 3.7854 × 10$^{-3}$ | 0.133368 | 3.785412 | 1 | 0.8326739 |
| gallon/sec (UK) | | 4,546.092 | 4.546 × 10$^{-3}$ | 0.1605432 | 4.546092 | 1.2009504 | 1 |

### Table 28b. Volume Flow per Minute Conversion

| To Convert ↓ | Multiply By Factor To Obtain → | foot$^3$/min | liter/min | gallon/min (US) | gallon/min (UK) |
|---|---|---|---|---|---|
| foot$^3$/min | | 1 | 28.316846 | 7.480519 | 6.2288327 |
| liter/min | | 0.035314 | 1 | 0.264172 | 0.2199692 |
| gallon/min (US) | | 0.133680 | 3.785412 | 1 | 0.832673 |
| gallon/min (UK) | | 0.1605437 | 4.546092 | 1.20095 | 1 |

**Pitot Tube.—**A pitot tube is a small, transparent, open tube bent at a right angle. It is a hollow tube that is placed longitudinally in the direction of fluid flow, allowing the flow to enter one end at the fluid's velocity of approach. When the fluids enter the pitot tube, it comes to a stop, and all of the velocity head is converted to pressure head. The difference between the total and static energies is the kinetic energy of the fluid. The velocity of the fluid can be calculated by using the Bernoulli equation.

$$\frac{p_1}{\rho} + \frac{v_1^2}{2} = \frac{p_2}{\rho} \qquad v_1 = \sqrt{\frac{2(p_2 - p_1)}{\rho}} \text{ (SI)} \qquad v_1 = \sqrt{\frac{2(p_2 - p_1)g_c}{\rho}} \text{ (US)}$$

## Units of Mass and Weight

### Table 29. Mass and Weight Conversion Factors

**Metric System**

1 *metric ton (t)* =
**1000** kilograms
2204.6223 pounds
0.9842 gross or long ton (of 2240 pounds)
0.9072 net or short ton (of 2000 pounds)

1 *kilogram (kg)* =
**1000** grams = **10** hectograms
2.2046 pounds
35.274 ounces avoirdupois

1 *hectogram (hg)* = **10** dekagrams

1 *dekagram (dag)* = **10** grams

1 *gram (g)* =
**10** decigrams
0.0022046 pound
0.03215 ounce Troy
0.03527 ounce avoirdupois
15.432 grains

1 *decigram (dg)* = **10** centigrams

1 *centigram (cg)* = **10** milligrams

**Avoirdupois or Commercial Weight**

1 *gross or long ton* =
**2240** pounds
1.016 metric ton
1016 kilograms

1 *net or short ton* = **2000** pounds

1 *pound* = **16** ounces
**7000** grains
**0.45359237** kilogram
453.6 grams

1 *ounce* =
$^1/_{16}$ pound
16 drachms
**437.5** grains
28.3495 grams
0.2780139 newton

1 *grain Avoirdupois* =
**1** grain apothecaries' weight =
**1** grain Troy weight
0.064799 gram

**Troy Weight**

Used for Weighing Gold and Silver

1 *pound Troy* =
12 ounces Troy = 5760 grains
$^{144}/_{175}$ avoirdupois pound

1 *ounce Troy* =
20 pennyweights = 480 grains
31.103 grams

1 *pennyweight* = 24 grains

1 *grain Troy* =
**1** grain avoirdupois
**1** grain apothecaries' weight
0.0648 gram

1 *carat (used in weighing diamonds)* =
3.086 grains
**200** milligrams = $^1/_5$ gram

1 *gold karat* = $^1/_{24}$ proportion pure gold

**Apothecaries' Weight**

1 *pound* = 12 ounces = 5760 grains

1 *ounce* =
8 drachms = 480 grains
31.103 grams

1 *drachm* = 3 scruples = 60 grains

1 *scruple* = 20 grains

**Old Weight Measures**

Measures for weight seldom used in the United States:

1 *gross or long ton* = 20 hundred-weights

1 *hundred-weight* = 4 quarters = 112 pounds

1 *quarter* = 28 pounds

1 *stone* = 14 pounds

1 *quintal* = 100 pounds

*Note*: Figures in **Bold** indicate exact conversion values

### Table 30a. Pounds to Kilograms Conversion

| lb | kg | lb | kg | lb | kg | lb | kg | lb | kg |
|---|---|---|---|---|---|---|---|---|---|
| 1,000 | 453.59 | 100 | 45.36 | 10 | 4.54 | 1 | 0.45 | 0.1 | 0.05 |
| 2,000 | 907.18 | 200 | 90.72 | 20 | 9.07 | 2 | 0.91 | 0.2 | 0.09 |
| 3,000 | 1,360.78 | 300 | 136.08 | 30 | 13.61 | 3 | 1.36 | 0.3 | 0.14 |
| 4,000 | 1,814.37 | 400 | 181.44 | 40 | 18.14 | 4 | 1.81 | 0.4 | 0.18 |
| 5,000 | 2,267.96 | 500 | 226.80 | 50 | 22.68 | 5 | 2.27 | 0.5 | 0.23 |
| 6,000 | 2,721.55 | 600 | 272.16 | 60 | 27.22 | 6 | 2.72 | 0.6 | 0.27 |
| 7,000 | 3,175.15 | 700 | 317.51 | 70 | 31.75 | 7 | 3.18 | 0.7 | 0.32 |
| 8,000 | 3,628.74 | 800 | 362.87 | 80 | 36.29 | 8 | 3.63 | 0.8 | 0.36 |
| 9,000 | 4,082.33 | 900 | 408.23 | 90 | 40.82 | 9 | 4.08 | 0.9 | 0.41 |
| 10,000 | 4,535.92 | 1,000 | 453.59 | 100 | 45.36 | 10 | 4.54 | 1.0 | 0.45 |

1 pound = 0.4535924 kilogram

## Table 30b. Kilograms to Pounds Conversion

| kg | lb | kg | lb | kg | lb | kg | lb | kg | lb |
|---|---|---|---|---|---|---|---|---|---|
| 1,000 | 2,204.62 | 100 | 220.46 | 10 | 22.05 | 1 | 2.20 | 0.1 | 0.22 |
| 2,000 | 4,409.24 | 200 | 440.92 | 20 | 44.09 | 2 | 4.41 | 0.2 | 0.44 |
| 3,000 | 6,613.87 | 300 | 661.39 | 30 | 66.14 | 3 | 6.61 | 0.3 | 0.66 |
| 4,000 | 8,818.49 | 400 | 881.85 | 40 | 88.18 | 4 | 8.82 | 0.4 | 0.88 |
| 5,000 | 11,023.11 | 500 | 1,102.31 | 50 | 110.23 | 5 | 11.02 | 0.5 | 1.10 |
| 6,000 | 13,227.73 | 600 | 1,322.77 | 60 | 132.28 | 6 | 13.23 | 0.6 | 1.32 |
| 7,000 | 15,432.35 | 700 | 1,543.24 | 70 | 154.32 | 7 | 15.43 | 0.7 | 1.54 |
| 8,000 | 17,636.98 | 800 | 1,763.70 | 80 | 176.37 | 8 | 17.64 | 0.8 | 1.76 |
| 9,000 | 19,841.60 | 900 | 1,984.16 | 90 | 198.42 | 9 | 19.84 | 0.9 | 1.98 |
| 10,000 | 22,046.22 | 1,000 | 2,204.62 | 100 | 220.46 | 10 | 22.05 | 1.0 | 2.20 |

1 kilogram = 2.204622 pounds

## Table 31a. Ounces to Grams Conversion

| oz | g | oz | g | oz | g | oz | g | oz | g |
|---|---|---|---|---|---|---|---|---|---|
| 10 | 283.50 | 1 | 28.35 | 0.1 | 2.83 | 0.01 | 0.28 | 0.001 | 0.03 |
| 20 | 566.99 | 2 | 56.70 | 0.2 | 5.67 | 0.02 | 0.57 | 0.002 | 0.06 |
| 30 | 850.49 | 3 | 85.05 | 0.3 | 8.50 | 0.03 | 0.85 | 0.003 | 0.09 |
| 40 | 1,133.98 | 4 | 113.40 | 0.4 | 11.34 | 0.04 | 1.13 | 0.004 | 0.11 |
| 50 | 1,417.48 | 5 | 141.75 | 0.5 | 14.17 | 0.05 | 1.42 | 0.005 | 0.14 |
| 60 | 1,700.97 | 6 | 170.10 | 0.6 | 17.01 | 0.06 | 1.70 | 0.006 | 0.17 |
| 70 | 1,984.47 | 7 | 198.45 | 0.7 | 19.84 | 0.07 | 1.98 | 0.007 | 0.20 |
| 80 | 2,267.96 | 8 | 226.80 | 0.8 | 22.68 | 0.08 | 2.27 | 0.008 | 0.23 |
| 90 | 2,551.46 | 9 | 255.15 | 0.9 | 25.51 | 0.09 | 2.55 | 0.009 | 0.26 |
| 100 | 2,834.95 | 10 | 283.50 | 1.0 | 28.35 | 0.10 | 2.83 | 0.010 | 0.28 |

1 ounce = 28.34952 grams

## Table 31b. Grams to Ounces Conversion

| g | oz | g | oz | g | oz | g | oz | g | oz |
|---|---|---|---|---|---|---|---|---|---|
| 100 | 3.527 | 10 | 0.353 | 1 | 0.035 | 0.1 | 0.004 | 0.01 | 0.000 |
| 200 | 7.055 | 20 | 0.705 | 2 | 0.071 | 0.2 | 0.007 | 0.02 | 0.001 |
| 300 | 10.582 | 30 | 1.058 | 3 | 0.106 | 0.3 | 0.011 | 0.03 | 0.001 |
| 400 | 14.110 | 40 | 1.411 | 4 | 0.141 | 0.4 | 0.014 | 0.04 | 0.001 |
| 500 | 17.637 | 50 | 1.764 | 5 | 0.176 | 0.5 | 0.018 | 0.05 | 0.002 |
| 600 | 21.164 | 60 | 2.116 | 6 | 0.212 | 0.6 | 0.021 | 0.06 | 0.002 |
| 700 | 24.692 | 70 | 2.469 | 7 | 0.247 | 0.7 | 0.025 | 0.07 | 0.002 |
| 800 | 28.219 | 80 | 2.822 | 8 | 0.282 | 0.8 | 0.028 | 0.08 | 0.003 |
| 900 | 31.747 | 90 | 3.175 | 9 | 0.317 | 0.9 | 0.032 | 0.09 | 0.003 |
| 1,000 | 35.274 | 100 | 3.527 | 10 | 0.353 | 1.0 | 0.035 | 0.10 | 0.004 |

1 gram = 0.03527397 ounce

## Table 32. Density Conversion Factors

| To Convert ↓ | Multiply By This Factor To Obtain → | gram/mL | gram/cm³ | kg/m³ | lb/inch³ | lb/feet³ | lb/gallon (US) | ton/yard³ |
|---|---|---|---|---|---|---|---|---|
| grams/mL | | 1 | 1 | **1000** | 0.036128 | 62.43 | 8.345 | 0.8428 |
| grams/cm³ | | 1 | 1 | **1000** | 0.036128 | 62.43 | 8.345 | 0.8428 |
| kilogram/m³ | | **0.001** | **0.001** | 1 | $3.6128 \times 10^{-5}$ | 0.06243 | 0.008345 | $8.428 \times 10^{-4}$ |
| lb/inch³ | | 27.67788 | 27.67788 | 27677.83 | 1 | 1728.0 | 230.9718 | 23.32687 |
| lb/feet³ | | 0.01602 | 0.01602 | 16.02 | $5.787 \times 10^{-4}$ | 1 | 0.1337 | 0.01349 |
| lb/gallon (US) | | 0.11983 | 0.11983 | 119.83 | 0.004329 | 7.481126 | 1 | 0.10099 |
| ton/yard³ | | 1.18652 | 1.18652 | 1186.52 | 0.042869 | 74.07451 | 9.9015 | 1 |

**Table 33a. Pounds per Cubic Inch to Grams per Cubic Centimeter Conversion**

| lb/in³ | g/cm³ | lb/in³ | g/cm³ | lb/in³ | g/cm³ | lb/in³ | g/cm³ | lb/in³ | g/cm³ |
|---|---|---|---|---|---|---|---|---|---|
| 100 | 2,767.99 | 10 | 276.80 | 1 | 27.68 | 0.1 | 2.77 | 0.01 | 0.28 |
| 200 | 5,535.98 | 20 | 553.60 | 2 | 55.36 | 0.2 | 5.54 | 0.02 | 0.55 |
| 300 | 8,303.97 | 30 | 830.40 | 3 | 83.04 | 0.3 | 8.30 | 0.03 | 0.83 |
| 400 | 11,071.96 | 40 | 1,107.20 | 4 | 110.72 | 0.4 | 11.07 | 0.04 | 1.11 |
| 500 | 13,839.95 | 50 | 1,384.00 | 5 | 138.40 | 0.5 | 13.84 | 0.05 | 1.38 |
| 600 | 16,607.94 | 60 | 1,660.79 | 6 | 166.08 | 0.6 | 16.61 | 0.06 | 1.66 |
| 700 | 19,375.93 | 70 | 1,937.59 | 7 | 193.76 | 0.7 | 19.38 | 0.07 | 1.94 |
| 800 | 22,143.92 | 80 | 2,214.39 | 8 | 221.44 | 0.8 | 22.14 | 0.08 | 2.21 |
| 900 | 24,911.91 | 90 | 2,491.19 | 9 | 249.12 | 0.9 | 24.91 | 0.09 | 2.49 |
| 1,000 | 27,679.90 | 100 | 2,767.99 | 10 | 276.80 | 1.0 | 27.68 | 0.10 | 2.77 |

1 lb/in³ = 27.67990 g/cm³

**Table 33b. Grams per Cubic Centimeter to Pounds per Cubic Inch Conversion**

| g/cm³ | lb/in³ | g/cm³ | lb/in³ | g/cm³ | lb/in³ | g/cm³ | lb/in³ | g/cm³ | lb/in³ |
|---|---|---|---|---|---|---|---|---|---|
| 1,000 | 36.127 | 100 | 3.613 | 10 | 0.361 | 1 | 0.036 | 0.1 | 0.004 |
| 2,000 | 72.255 | 200 | 7.225 | 20 | 0.723 | 2 | 0.072 | 0.2 | 0.007 |
| 3,000 | 108.382 | 300 | 10.838 | 30 | 1.084 | 3 | 0.108 | 0.3 | 0.011 |
| 4,000 | 144.509 | 400 | 14.451 | 40 | 1.445 | 4 | 0.145 | 0.4 | 0.014 |
| 5,000 | 180.636 | 500 | 18.064 | 50 | 1.806 | 5 | 0.181 | 0.5 | 0.018 |
| 6,000 | 216.764 | 600 | 21.676 | 60 | 2.168 | 6 | 0.217 | 0.6 | 0.022 |
| 7,000 | 252.891 | 700 | 25.289 | 70 | 2.529 | 7 | 0.253 | 0.7 | 0.025 |
| 8,000 | 289.018 | 800 | 28.902 | 80 | 2.890 | 8 | 0.289 | 0.8 | 0.029 |
| 9,000 | 325.146 | 900 | 32.515 | 90 | 3.251 | 9 | 0.325 | 0.9 | 0.033 |
| 10,000 | 361.273 | 1,000 | 36.127 | 100 | 3.613 | 10 | 0.361 | 1.0 | 0.036 |

1 g/cm³ = 0.03612730 lb/in³

**Table 34a. Pounds per Cubic Foot to Kilograms per Cubic Meter Conversion**

| lb/ft³ | kg/m³ | lb/ft³ | kg/m³ | lb/ft³ | kg/m³ | lb/ft³ | kg/m³ | lb/ft³ | kg/m³ |
|---|---|---|---|---|---|---|---|---|---|
| 100 | 1,601.85 | 10 | 160.18 | 1 | 16.02 | 0.1 | 1.60 | 0.01 | 0.16 |
| 200 | 3,203.69 | 20 | 320.37 | 2 | 32.04 | 0.2 | 3.20 | 0.02 | 0.32 |
| 300 | 4,805.54 | 30 | 480.55 | 3 | 48.06 | 0.3 | 4.81 | 0.03 | 0.48 |
| 400 | 6,407.38 | 40 | 640.74 | 4 | 64.07 | 0.4 | 6.41 | 0.04 | 0.64 |
| 500 | 8,009.23 | 50 | 800.92 | 5 | 80.09 | 0.5 | 8.01 | 0.05 | 0.80 |
| 600 | 9,611.08 | 60 | 961.11 | 6 | 96.11 | 0.6 | 9.61 | 0.06 | 0.96 |
| 700 | 11,212.92 | 70 | 1,121.29 | 7 | 112.13 | 0.7 | 11.21 | 0.07 | 1.12 |
| 800 | 12,814.77 | 80 | 1,281.48 | 8 | 128.15 | 0.8 | 12.81 | 0.08 | 1.28 |
| 900 | 14,416.61 | 90 | 1,441.66 | 9 | 144.17 | 0.9 | 14.42 | 0.09 | 1.44 |
| 1,000 | 16,018.46 | 100 | 1,601.85 | 10 | 160.18 | 1.0 | 16.02 | 0.10 | 1.60 |

1 lb/ft³ = 16.01846 kg/m³

**Table 34b. Kilograms per Cubic Meter to Pounds per Cubic Foot Conversion**

| kg/m³ | lb/ft³ | kg/m³ | lb/ft³ | kg/m³ | lb/ft³ | kg/m³ | lb/ft³ | kg/m³ | lb/ft³ |
|---|---|---|---|---|---|---|---|---|---|
| 1,000 | 62.428 | 100 | 6.243 | 10 | 0.624 | 1 | 0.062 | 0.1 | 0.0[illegible]6 |
| 2,000 | 124.856 | 200 | 12.486 | 20 | 1.249 | 2 | 0.125 | 0.2 | 0.0[illegible]2 |
| 3,000 | 187.284 | 300 | 18.728 | 30 | 1.873 | 3 | 0.187 | 0.3 | 0.0[illegible]9 |
| 4,000 | 249.712 | 400 | 24.971 | 40 | 2.497 | 4 | 0.250 | 0.4 | 0.0[illegible]5 |
| 5,000 | 312.140 | 500 | 31.214 | 50 | 3.121 | 5 | 0.312 | 0.5 | 0.0[illegible]1 |
| 6,000 | 374.568 | 600 | 37.457 | 60 | 3.746 | 6 | 0.375 | 0.6 | 0.0[illegible]7 |
| 7,000 | 436.996 | 700 | 43.700 | 70 | 4.370 | 7 | 0.437 | 0.7 | 0.0[illegible]4 |
| 8,000 | 499.424 | 800 | 49.942 | 80 | 4.994 | 8 | 0.499 | 0.8 | 0.0[illegible]0 |
| 9,000 | 561.852 | 900 | 56.185 | 90 | 5.619 | 9 | 0.562 | 0.9 | 0.0[illegible]6 |
| 10,000 | 624.280 | 1,000 | 62.428 | 100 | 6.243 | 10 | 0.624 | 1.0 | 0.0[illegible]2 |

1 kg/m³ = 0.06242797 lb/ft³

## Units of Pressure and Stress

### Table 35. Pressure and Stress Conversion Factors

1 *kilogram per sq. millimeter* ($kg_f/mm^2$) =
1422.32 pounds per square inch

1 *kilogram per sq. centimeter* ($kg_f/cm^2$) =
14.223 pounds per square inch

1 *bar* =
**1,000,000** dynes per square centimeter
**1000** millibars
**100** kilopascals
750.06168 torr
1.0197162 kg force per $cm^2$
14.50377 pounds per square inch
29.529983 inches of mercury at 0° C
10,197.162 mm water at 4° C
33.455256 feet of water at 4° C

1 *millibar* =
**100,000** dynes per square centimeter
**100** pascal

1 *torr* =
**760** millimeters mercury
1⁄760 atmosphere
133.224 pascal
1.333224 millibar

1 *pound per square inch* =
144 pounds per square foot
0.068 atmosphere
2.042 inches of mercury at 62° F
27.7 inches of water at 62° F
2.31 feet of water at 62° F
0.0703 kilogram per square centimeter
6.894757 kilopascals
6894.757 pascal

1 *atmosphere* =
30 inches of mercury at 62° F
14.7 pounds per square inch
2116.3 pounds per square foot
33.95 feet of water at 62° F

1 *foot of water at 62° F* =
62.355 pounds per square foot
0.433 pound per square inch

1 *inch of mercury at 62° F* =
1.132 foot of water
13.58 inches of water
0.491 pound per square inch

1 *inch of water* =
0.0735559 inch mercury at 0° C
1.8683205 torr
0.5780367 ounce force per square inch
0.0024583 atmosphere

### Table 36a. Pounds per Square Inch to Kilograms per Square Centimeter Conversion

| lb/in² | kg/cm² | lb/in² | kg/cm² | lb/in² | kg/cm² | lb/in² | kg/cm² | lb/in² | kg/cm² |
|---|---|---|---|---|---|---|---|---|---|
| 1,000 | 70.307 | 100 | 7.031 | 10 | 0.703 | 1 | 0.070 | 0.1 | 0.007 |
| 2,000 | 140.614 | 200 | 14.061 | 20 | 1.406 | 2 | 0.141 | 0.2 | 0.014 |
| 3,000 | 210.921 | 300 | 21.092 | 30 | 2.109 | 3 | 0.211 | 0.3 | 0.021 |
| 4,000 | 281.228 | 400 | 28.123 | 40 | 2.812 | 4 | 0.281 | 0.4 | 0.028 |
| 5,000 | 351.535 | 500 | 35.153 | 50 | 3.515 | 5 | 0.352 | 0.5 | 0.035 |
| 6,000 | 421.842 | 600 | 42.184 | 60 | 4.218 | 6 | 0.422 | 0.6 | 0.042 |
| 7,000 | 492.149 | 700 | 49.215 | 70 | 4.921 | 7 | 0.492 | 0.7 | 0.049 |
| 8,000 | 562.456 | 800 | 56.246 | 80 | 5.625 | 8 | 0.562 | 0.8 | 0.056 |
| 9,000 | 632.763 | 900 | 63.276 | 90 | 6.328 | 9 | 0.633 | 0.9 | 0.063 |
| 10,000 | 703.070 | 1,000 | 70.307 | 100 | 7.031 | 10 | 0.703 | 1.0 | 0.070 |

1 lb/in² = 0.07030697 kg/cm²

### Table 36b. Kilograms per Square Centimeter to Pounds per Square Inch Conversion

| kg/cm² | lb/in² | kg/cm² | lb/in² | kg/cm² | lb/in² | kg/cm² | lb/in² | kg/cm² | lb/in² |
|---|---|---|---|---|---|---|---|---|---|
| 100 | 1,422.33 | 10 | 142.23 | 1 | 14.22 | 0.1 | 1.42 | 0.01 | 0.14 |
| 200 | 2,844.67 | 20 | 284.47 | 2 | 28.45 | 0.2 | 2.84 | 0.02 | 0.28 |
| 300 | 4,267.00 | 30 | 426.70 | 3 | 42.67 | 0.3 | 4.27 | 0.03 | 0.43 |
| 400 | 5,689.34 | 40 | 568.93 | 4 | 56.89 | 0.4 | 5.69 | 0.04 | 0.57 |
| 500 | 7,111.67 | 50 | 711.17 | 5 | 71.12 | 0.5 | 7.11 | 0.05 | 0.71 |
| 600 | 8,534.00 | 60 | 853.40 | 6 | 85.34 | 0.6 | 8.53 | 0.06 | 0.85 |
| 700 | 9,956.34 | 70 | 995.63 | 7 | 99.56 | 0.7 | 9.96 | 0.07 | 1.00 |
| 800 | 11,378.67 | 80 | 1,137.87 | 8 | 113.79 | 0.8 | 11.38 | 0.08 | 1.14 |
| 900 | 12,801.01 | 90 | 1,280.10 | 9 | 128.01 | 0.9 | 12.80 | 0.09 | 1.28 |
| 1,000 | 14,223.34 | 100 | 1,422.33 | 10 | 142.23 | 1.0 | 14.22 | 0.10 | 1.42 |

1 kg/cm² = 14.22334 lb/in²

**Table 37a. Pounds per Square Foot to Kilograms per Square Meter Conversion**

| lb/ft² | kg/m² | lb/ft² | kg/m² | lb/ft² | kg/m² | lb/ft² | kg/m² | lb/ft² | kg/m² |
|---|---|---|---|---|---|---|---|---|---|
| 1,000 | 4,882.43 | 100 | 488.24 | 10 | 48.82 | 1 | 4.88 | 0.1 | 0.49 |
| 2,000 | 9,764.86 | 200 | 976.49 | 20 | 97.65 | 2 | 9.76 | 0.2 | 0.98 |
| 3,000 | 14,647.29 | 300 | 1,464.73 | 30 | 146.47 | 3 | 14.65 | 0.3 | 1.46 |
| 4,000 | 19,529.72 | 400 | 1,952.97 | 40 | 195.30 | 4 | 19.53 | 0.4 | 1.95 |
| 5,000 | 24,412.14 | 500 | 2,441.21 | 50 | 244.12 | 5 | 24.41 | 0.5 | 2.44 |
| 6,000 | 29,294.57 | 600 | 2,929.46 | 60 | 292.95 | 6 | 29.29 | 0.6 | 2.93 |
| 7,000 | 34,177.00 | 700 | 3,417.70 | 70 | 341.77 | 7 | 34.18 | 0.7 | 3.42 |
| 8,000 | 39,059.43 | 800 | 3,905.94 | 80 | 390.59 | 8 | 39.06 | 0.8 | 3.91 |
| 9,000 | 43,941.86 | 900 | 4,394.19 | 90 | 439.42 | 9 | 43.94 | 0.9 | 4.39 |
| 10,000 | 48,824.28 | 1,000 | 4,882.43 | 100 | 488.24 | 10 | 48.82 | 1.0 | 4.88 |

1 lb/ft² = 4.882429 kg/m²

**Table 37b. Kilograms per Square Meter to Pounds per Square Foot Conversion**

| kg/m² | lb/ft² | kg/m² | lb/ft² | kg/m² | lb/ft² | kg/m² | lb/ft² | kg/m² | lb/ft² |
|---|---|---|---|---|---|---|---|---|---|
| 1,000 | 204.82 | 100 | 20.48 | 10 | 2.05 | 1 | 0.20 | 0.1 | 0.02 |
| 2,000 | 409.63 | 200 | 40.96 | 20 | 4.10 | 2 | 0.41 | 0.2 | 0.04 |
| 3,000 | 614.45 | 300 | 61.44 | 30 | 6.14 | 3 | 0.61 | 0.3 | 0.06 |
| 4,000 | 819.26 | 400 | 81.93 | 40 | 8.19 | 4 | 0.82 | 0.4 | 0.08 |
| 5,000 | 1,024.08 | 500 | 102.41 | 50 | 10.24 | 5 | 1.02 | 0.5 | 0.10 |
| 6,000 | 1,228.90 | 600 | 122.89 | 60 | 12.29 | 6 | 1.23 | 0.6 | 0.12 |
| 7,000 | 1,433.71 | 700 | 143.37 | 70 | 14.34 | 7 | 1.43 | 0.7 | 0.14 |
| 8,000 | 1,638.53 | 800 | 163.85 | 80 | 16.39 | 8 | 1.64 | 0.8 | 0.16 |
| 9,000 | 1,843.34 | 900 | 184.33 | 90 | 18.43 | 9 | 1.84 | 0.9 | 0.18 |
| 10,000 | 2,048.16 | 1,000 | 204.82 | 100 | 20.48 | 10 | 2.05 | 1.0 | 0.20 |

1 kg/m² = 0.2048161 lb/ft²

**Table 38a. Pounds Per Square Inch to Kilopascals Conversion**

| → lb/in² ↓ | 0 | 1 | 2 | 3 | 4 | 5 | 6 | 7 | 8 | 9 |
|---|---|---|---|---|---|---|---|---|---|---|
| | kilopascals | | | | | | | | | |
| 0 | … | 6.895 | 13.790 | 20.684 | 27.579 | 34.474 | 41.369 | 48.263 | 55.158 | 62.053 |
| 10 | 68.948 | 75.842 | 82.737 | 89.632 | 96.527 | 103.421 | 110.316 | 117.211 | 124.106 | 131.000 |
| 20 | 137.895 | 144.790 | 151.685 | 158.579 | 165.474 | 172.369 | 179.264 | 186.158 | 193.053 | 199.948 |
| 30 | 206.843 | 213.737 | 220.632 | 227.527 | 234.422 | 241.316 | 248.211 | 255.106 | 262.001 | 268.896 |
| 40 | 275.790 | 282.685 | 289.580 | 296.475 | 303.369 | 310.264 | 317.159 | 324.054 | 330.948 | 337.843 |
| 50 | 344.738 | 351.633 | 358.527 | 365.422 | 372.317 | 379.212 | 386.106 | 393.001 | 399.896 | 406.791 |
| 60 | 413.685 | 420.580 | 427.475 | 434.370 | 441.264 | 448.159 | 455.054 | 461.949 | 468.843 | 475.738 |
| 70 | 482.633 | 489.528 | 496.423 | 503.317 | 510.212 | 517.107 | 524.002 | 530.896 | 537.791 | 544.686 |
| 80 | 551.581 | 558.475 | 565.370 | 572.265 | 579.160 | 586.054 | 592.949 | 599.844 | 606.739 | 613.633 |
| 90 | 620.528 | 627.423 | 634.318 | 641.212 | 648.107 | 655.002 | 661.897 | 668.791 | 675.686 | 682.581 |
| 100 | 689.476 | 696.370 | 703.265 | 710.160 | 717.055 | 723.949 | 730.844 | 737.739 | 744.634 | 751.529 |

1 lb/in² = 6.894757 kPa. *Note:* 1 kilopascal = 1 kilonewton/meter².

**Table 38b. Kilopascals to Pounds Per Square Inch Conversion**

| → kPa ↓ | 0 | 1 | 2 | 3 | 4 | 5 | 6 | 7 | 8 | 9 |
|---|---|---|---|---|---|---|---|---|---|---|
| | lb/in² | | | | | | | | | |
| 0 | … | 0.145 | 0.290 | 0.435 | 0.580 | 0.725 | 0.870 | 1.015 | 1.160 | 1.305 |
| 10 | 1.450 | 1.595 | 1.740 | 1.885 | 2.031 | 2.176 | 2.321 | 2.466 | 2.611 | 2.756 |
| 20 | 2.901 | 3.046 | 3.191 | 3.336 | 3.481 | 3.626 | 3.771 | 3.916 | 4.061 | 4.206 |
| 30 | 4.351 | 4.496 | 4.641 | 4.786 | 4.931 | 5.076 | 5.221 | 5.366 | 5.511 | 5.656 |
| 40 | 5.802 | 5.947 | 6.092 | 6.237 | 6.382 | 6.527 | 6.672 | 6.817 | 6.962 | 7.107 |
| 50 | 7.252 | 7.397 | 7.542 | 7.687 | 7.832 | 7.977 | 8.122 | 8.267 | 8.412 | 8.557 |
| 60 | 8.702 | 8.847 | 8.992 | 9.137 | 9.282 | 9.427 | 9.572 | 9.718 | 9.863 | 10.008 |
| 70 | 10.153 | 10.298 | 10.443 | 10.588 | 10.733 | 10.878 | 11.023 | 11.168 | 11.313 | 11.458 |
| 80 | 11.603 | 11.748 | 11.893 | 12.038 | 12.183 | 12.328 | 12.473 | 12.618 | 12.763 | 12.908 |
| 90 | 13.053 | 13.198 | 13.343 | 13.489 | 13.634 | 13.779 | 13.924 | 14.069 | 14.214 | 14.359 |
| 100 | 14.504 | 14.649 | 14.794 | 14.939 | 15.084 | 15.229 | 15.374 | 15.519 | 15.664 | 15.809 |

1 kPa = 0.1450377 lb/in². *Note:* 1 kilopascal = 1 kilonewton/meter².

## Table 39. Pressure and Stress Conversion Factors

| To Convert ↓ / Multiply By This Factor To Obtain → | atmosphere | pascal ($N/m^2$) | dyne/cm² | bar | kg/cm² | kg/m² | psi (lb/inch²) | pound/ft² | inch of water | inch of mercury | millimeter of mercury | ton/ft² (short) |
|---|---|---|---|---|---|---|---|---|---|---|---|---|
| atmosphere | 1 | 101325 | $1.0133 \times 10^6$ | 1.01325 | 1.03319076 | 10,331.9076 | 14.6959488 | 2,116.216 | 407.1893 | 29.9212 | 760 | 0.9597354 |
| pascal ($N/m^2$) | $9.8692 \times 10^{-6}$ | 1 | **10** | $\mathbf{1 \times 10^{-5}}$ | $1.01968 \times 10^{-5}$ | 0.101968 | 0.00014504 | 0.02088 | 0.004019 | 0.0002953 | 0.0075 | $9.472 \times 10^{-6}$ |
| dyne/cm² | $9.8692 \times 10^{-7}$ | **0.1** | 1 | $\mathbf{1 \times 10^{-6}}$ | $1.01968 \times 10^{-6}$ | 0.0101968 | $1.4504 \times 10^{-5}$ | 0.002088 | 0.000402 | $2.95 \times 10^{-5}$ | 0.00075 | $9.472 \times 10^{-7}$ |
| bar | 0.98692327 | $\mathbf{1 \times 10^5}$ | $\mathbf{1 \times 10^6}$ | 1 | 1.01968 | 10194.8 | 14.5037256 | 2088.5434 | 401.8646 | 29.5299 | 750.06168 | 0.9471852 |
| kilogram/cm² | 0.96784111 | 98,069.982 | 980,699.83 | 0.9807 | 1 | **10000** | 14.2232691 | 2048.6123 | 394.0945 | 28.9653 | 735.58536 | 0.9289043 |
| kilogram/meter² | $9.6787 \times 10^{-5}$ | 9.80699 | 98.06998 | $9.807 \times 10^{-5}$ | **0.0001** | 1 | 0.001422 | 0.204823 | 0.039409 | 0.002896 | 0.0735585 | $9.289 \times 10^{-5}$ |
| psi (lb/inch²) | 0.06804596 | 6,894.7572 | 68,947.573 | 0.068947 | 0.07029148 | 703.0446 | 1 | **144** | 27.70768 | 2.03602 | 51.71493 | 0.0653061 |
| pound/ft² | $4.7254 \times 10^{-4}$ | 47.88025 | 478.80258 | 0.000478 | 0.00048813 | 4.88225 | 0.006944 | 1 | 0.19241 | 0.014139 | 0.3591314 | 0.0004535 |
| inch of water | 0.00245586 | 248.8400 | 2488.4003 | 0.002488 | 0.00253690 | 25.3737 | 0.036091 | 5.19713 | 1 | 0.073482 | 1.866453 | 0.002356 |
| inch of mercury | 0.03342112 | 3386.3949 | 33,863.949 | 0.033863 | 0.03452401 | 345.3039 | 0.491153 | 70.72632 | 13.6087 | 1 | **25.4** | 0.0320754 |
| mm of mercury | 0.00131579 | 133.32236 | 1333.22368 | 0.001333 | 0.00135921 | 13.594615 | 0.019336 | 2.784495 | 0.53577 | 0.03937 | 1 | 0.0012628 |
| ton/ft² (short) | 0.94508279 | 95760.514 | 957605.14 | 0.957605 | 0.9764854 | 9764.854 | 13.88888 | 2000 | 384.8277 | 28.27801 | 718.2616 | 1 |

## Units of Force

### Table 40. Force Conversion Factors

| To Convert ↓ / Multiply By This Factor To Obtain → | dyne | gram force | joule/cm | newton | $kg_f$ | $lb_f$ | kip | poundal | ounce force |
|---|---|---|---|---|---|---|---|---|---|
| dyne | 1 | 0.00101968 | **0.001** | **0.00001** | $1.01968 \times 10^{-6}$ | $2.24809 \times 10^{-6}$ | $2.2481 \times 10^{-9}$ | $7.233013 \times 10^{-5}$ | $3.59694 \times 10^{-5}$ |
| gram force | 980.7 | 1 | 0.9807 | 0.009807 | **0.001** | 0.0022047 | $2.2047 \times 10^{-6}$ | 0.0709341 | 0.03527521 |
| joule/cm | **1000** | 1.0196798 | 1 | **0.01** | 0.00101968 | 0.002248 | $2.2481 \times 10^{-6}$ | 0.0723301 | 0.03596942 |
| newton | $\mathbf{1 \times 10^5}$ | 101.96798 | **100** | 1 | 0.101967982 | 0.2248089 | $2.2481 \times 10^{-4}$ | 7.23301 | 3.596942 |
| kg force | $9.807 \times 10^5$ | **1000** | 980.7 | 9.807 | 1 | 2.2047 | 0.0022047 | 70.934129 | 35.2752102 |
| $lb_f$ | $4.4482 \times 10^5$ | 453.57627 | 444.822 | 4.44822 | 0.45357626 | 1 | **0.001** | 32.174038 | 16 |
| kip | $4.4482 \times 10^8$ | $4.5357 \times 10^5$ | $4.4482 \times 10^5$ | 4448.2224 | 453.5762688 | **1000** | 1 | 32174.038 | 16000 |
| poundal | 13825.50 | 14.097586 | 13.8255 | 0.1382555 | 0.014097586 | 0.0310809 | $3.1081 \times 10^{-5}$ | 1 | 0.497296 |
| ounce force | 27801.39 | 28.348519 | 27.8013 | 0.278013 | 0.02834852 | 0.06250 | $6.25 \times 10^{-5}$ | 2.010877 | 1 |

Figures in **bold face** indicate the conversion is exact

## Table 41a. Pounds Force to Newtons Conversion

| lbf → ↓ | 0 | 1 | 2 | 3 | 4 | 5 | 6 | 7 | 8 | 9 |
|---|---|---|---|---|---|---|---|---|---|---|
| | newtons | | | | | | | | | |
| 0 | … | 4.448 | 8.896 | 13.345 | 17.793 | 22.241 | 26.689 | 31.138 | 35.586 | 40.034 |
| 10 | 44.482 | 48.930 | 53.379 | 57.827 | 62.275 | 66.723 | 71.172 | 75.620 | 80.068 | 84.516 |
| 20 | 88.964 | 93.413 | 97.861 | 102.309 | 106.757 | 111.206 | 115.654 | 120.102 | 124.550 | 128.998 |
| 30 | 133.447 | 137.895 | 142.343 | 146.791 | 151.240 | 155.688 | 160.136 | 164.584 | 169.032 | 173.481 |
| 40 | 177.929 | 182.377 | 186.825 | 191.274 | 195.722 | 200.170 | 204.618 | 209.066 | 213.515 | 217.963 |
| 50 | 222.411 | 226.859 | 231.308 | 235.756 | 240.204 | 244.652 | 249.100 | 253.549 | 257.997 | 262.445 |
| 60 | 266.893 | 271.342 | 275.790 | 280.238 | 284.686 | 289.134 | 293.583 | 298.031 | 302.479 | 306.927 |
| 70 | 311.376 | 315.824 | 320.272 | 324.720 | 329.168 | 333.617 | 338.065 | 342.513 | 346.961 | 351.410 |
| 80 | 355.858 | 360.306 | 364.754 | 369.202 | 373.651 | 378.099 | 382.547 | 386.995 | 391.444 | 395.892 |
| 90 | 400.340 | 404.788 | 409.236 | 413.685 | 418.133 | 422.581 | 427.029 | 431.478 | 435.926 | 440.374 |
| 100 | 444.822 | 449.270 | 453.719 | 458.167 | 462.615 | 467.063 | 471.512 | 475.960 | 480.408 | 484.856 |

1 pound force = 4.448222 newtons

## Table 41b. Newtons to Pounds Force Conversion

| N → ↓ | 0 | 1 | 2 | 3 | 4 | 5 | 6 | 7 | 8 | 9 |
|---|---|---|---|---|---|---|---|---|---|---|
| | pounds force | | | | | | | | | |
| 0 | … | 0.22481 | 0.44962 | 0.67443 | 0.89924 | 1.12404 | 1.34885 | 1.57366 | 1.79847 | 2.02328 |
| 10 | 2.24809 | 2.47290 | 2.69771 | 2.92252 | 3.14732 | 3.37213 | 3.59694 | 3.82175 | 4.04656 | 4.27137 |
| 20 | 4.49618 | 4.72099 | 4.94580 | 5.17060 | 5.39541 | 5.62022 | 5.84503 | 6.06984 | 6.29465 | 6.51946 |
| 30 | 6.74427 | 6.96908 | 7.19388 | 7.41869 | 7.64350 | 7.86831 | 8.09312 | 8.31793 | 8.54274 | 8.76755 |
| 40 | 8.99236 | 9.21716 | 9.44197 | 9.66678 | 9.89159 | 10.1164 | 10.3412 | 10.5660 | 10.7908 | 11.0156 |
| 50 | 11.2404 | 11.4653 | 11.6901 | 11.9149 | 12.1397 | 12.3645 | 12.5893 | 12.8141 | 13.0389 | 13.2637 |
| 60 | 13.4885 | 13.7133 | 13.9382 | 14.1630 | 14.3878 | 14.6126 | 14.8374 | 15.0622 | 15.2870 | 15.5118 |
| 70 | 15.7366 | 15.9614 | 16.1862 | 16.4110 | 16.6359 | 16.8607 | 17.0855 | 17.3103 | 17.5351 | 17.7599 |
| 80 | 17.9847 | 18.2095 | 18.4343 | 18.6591 | 18.8839 | 19.1088 | 19.3336 | 19.5584 | 19.7832 | 20.0080 |
| 90 | 20.2328 | 20.4576 | 20.6824 | 20.9072 | 21.1320 | 21.3568 | 21.5817 | 21.8065 | 22.0313 | 22.2561 |
| 100 | 22.4809 | 22.7057 | 22.9305 | 23.1553 | 23.3801 | 23.6049 | 23.8297 | 24.0546 | 24.2794 | 24.5042 |

1 newton = 0.2248089 pound force

# Units of Moment and Torque

## Table 42. Bending Moment or Torque Conversion Factors

| To Convert ↓ | → Multiply By Factor | dyne-centimeter | kilogram-meter | newton-millimeter | newton-meter | ounce-inch | pound-foot |
|---|---|---|---|---|---|---|---|
| dyne-centimeter | | 1 | **$1 \times 10^{-7}$** | **0.0001** | **$1 \times 10^{-7}$** | $1.416 \times 10^{-5}$ | $7.375 \times 10^{-8}$ |
| kilogram-meter | | $9.80665 \times 10^{7}$ | 1 | 9806.65 | 9.80665 | 1388.78818707 | 7.233271722 |
| newton-millimeter | | **10,000** | 0.000101968 | 1 | **0.001** | 0.14161193 | 0.000737562 |
| newton-meter | | **$1 \times 10^{7}$** | 0.101967982 | **1000** | 1 | 141.61192894 | 0.737562121 |
| ounce-inch | | 70615.52 | 0.000720052 | 7.061552 | 0.007061552 | 1 | 0.005208333 |
| pound-feet | | 13,558,180 | 0.138250025 | 1355.818 | 1.355818 | 192 | 1 |

Figures in **bold face** indicate the conversion is exact

## Table 43a. Pound-Inches to Newton-Meters Conversion

| lbf-in | N•m | lbf-in | N•m | lbf-in | N•m | lbf-in | N•m | lbf-in | N•m |
|---|---|---|---|---|---|---|---|---|---|
| 100 | 11.298 | 10 | 1.130 | 1 | 0.113 | 0.1 | 0.011 | 0.01 | 0.001 |
| 200 | 22.597 | 20 | 2.260 | 2 | 0.226 | 0.2 | 0.023 | 0.02 | 0.002 |
| 300 | 33.895 | 30 | 3.390 | 3 | 0.339 | 0.3 | 0.034 | 0.03 | 0.003 |
| 400 | 45.194 | 40 | 4.519 | 4 | 0.452 | 0.4 | 0.045 | 0.04 | 0.005 |
| 500 | 56.492 | 50 | 5.649 | 5 | 0.565 | 0.5 | 0.056 | 0.05 | 0.006 |
| 600 | 67.791 | 60 | 6.779 | 6 | 0.678 | 0.6 | 0.068 | 0.06 | 0.007 |
| 700 | 79.089 | 70 | 7.909 | 7 | 0.791 | 0.7 | 0.079 | 0.07 | 0.008 |
| 800 | 90.388 | 80 | 9.039 | 8 | 0.904 | 0.8 | 0.090 | 0.08 | 0.009 |
| 900 | 101.686 | 90 | 10.169 | 9 | 1.017 | 0.9 | 0.102 | 0.09 | 0.010 |
| 1000 | 112.985 | 100 | 11.298 | 10 | 1.130 | 1.0 | 0.113 | 0.10 | 0.011 |

1 pound-inch = 0.1129848 newton-meter

**Table 43b. Newton-Meters to Pound-Inches Conversion**

| N•m | $lb_f$-in | N•m | $lb_f$-in | N•m | $lb_f$-in | N•m | $lb_f$-in | N•m | $lb_f$-in |
|---|---|---|---|---|---|---|---|---|---|
| 100 | 885.07 | 10 | 88.51 | 1 | 8.85 | 0.1 | 0.89 | 0.01 | 0.09 |
| 200 | 1770.15 | 20 | 177.01 | 2 | 17.70 | 0.2 | 1.77 | 0.02 | 0.18 |
| 300 | 2655.22 | 30 | 265.52 | 3 | 26.55 | 0.3 | 2.66 | 0.03 | 0.27 |
| 400 | 3540.30 | 40 | 354.03 | 4 | 35.40 | 0.4 | 3.54 | 0.04 | 0.35 |
| 500 | 4425.37 | 50 | 442.54 | 5 | 44.25 | 0.5 | 4.43 | 0.05 | 0.44 |
| 600 | 5310.45 | 60 | 531.04 | 6 | 53.10 | 0.6 | 5.31 | 0.06 | 0.53 |
| 700 | 6195.52 | 40 | 619.55 | 7 | 61.96 | 0.7 | 6.20 | 0.07 | 0.62 |
| 800 | 7080.60 | 80 | 708.06 | 8 | 70.81 | 0.8 | 7.08 | 0.08 | 0.71 |
| 900 | 7965.67 | 90 | 796.57 | 9 | 79.66 | 0.9 | 7.97 | 0.09 | 0.80 |
| 1000 | 8850.75 | 100 | 885.07 | 10 | 88.51 | 1.0 | 8.85 | 0.10 | 0.89 |

1 newton-meter = 8.850748 pound-inches

**Poundal.—**The expression "poundal" is sometimes used in connection with calculations in mechanics. Many mechanical handbooks, however, do not define it, because of its limited use. A poundal is a unit of force and is defined as that force which, acting on a mass of one pound for one second, produces a velocity of one foot per second. A foot-poundal is a unit of energy equal to the energy resulting when a force of one poundal acts through a distance of one foot. In order to reduce foot-poundals to foot-pounds, multiply the number of foot-poundals by 0.03108. Dividing the number of foot-poundals by 32.16 (acceleration due to gravity) will also give foot-pounds.

## Units of Energy, Power, and Heat

**Table 44a. Energy Conversion Factors**

1 *horsepower-hour* =
- 0.746 kilowatt-hour
- 1,980,000 foot-pounds
- 2545 Btu (British thermal units)
- 2.64 pounds of water evaporated at 212° F
- 17 pounds of water raised from 62° to 212° F

1 *kilowatt-hour* =
- 1000 watt-hours
- 1.34 horsepower-hour
- 2,655,200 foot-pounds
- 3,600,000 joules
- 3415 Btu
- 3.54 pounds of water evaporated at 212° F
- 22.8 pounds of water raised from 62° to 212° F

**Table 44b. Power Conversion Factors**

1 *horsepower* =
- 746 watts
- 0.746 kilowatt
- 33,000 foot-pounds/minute
- 550 foot-pounds/second
- 2545 Btu/hour
- 42.4 Btu/minute
- 0.71 Btu/second
- 2.64 pounds of water evaporated per hour at 212° F

1 *kilowatt* =
- 1000 watts
- 1.34 horsepower
- 2,654,200 foot-pounds/hour
- 44,200 foot-pounds/minute
- 737 foot-pounds/second
- 3415 Btu/hour
- 57 Btu/minute
- 0.95 Btu/second
- 3.54 pounds of water evaporated per hour at 212° F

1 *watt* =
- 1 joule/second
- 0.00134 horsepower
- 0.001 kilowatt
- 3.42 Btu/hour
- 44.22 foot-pounds/minute
- 0.74 foot-pounds/second
- 0.0035 pound of water evaporated per hour at 212° F

**Table 44c. Heat Conversion Factors**

1 *Btu (British thermal unit)* =
- 1052 watt-seconds
- 778 foot-pounds
- 0.252 kilogram-calorie
- 0.000292 kilowatt-hour
- 0.000393, horsepower-hour
- 0.00104 pound of water evaporated at 212° F

1 *kilogram calorie* = 3.968 Btu

1 *foot-pound* =
- 1.36 joules
- 0.000000377 kilowatt-hour
- 0.00129 Btu
- 0.0000005 horsepower-hour

1 *kilogram-meter* =
- 7.233 foot-pounds

1 *joule* =
- 1 watt-second
- 0.00000078 kilowatt-hour
- 0.00095 Btu
- 0.74 foot-pound

1 *therm* =
- 100,000 Btu (US)
- 29.3 kilowatt-hour
- 105.5 megajoule

## Table 45a. British Thermal Units to Foot-Pounds

| Btu | ft-lb | Btu | ft-lb | Btu | ft-lb | Btu | ft-lb | Btu | ft-lb |
|---|---|---|---|---|---|---|---|---|---|
| 100 | 77,826 | 10 | 7,783 | 1 | 778 | 0.1 | 78 | 0.01 | 8 |
| 200 | 155,652 | 20 | 15,565 | 2 | 1,557 | 0.2 | 156 | 0.02 | 16 |
| 300 | 233,478 | 30 | 23,348 | 3 | 2,335 | 0.3 | 233 | 0.03 | 23 |
| 400 | 311,304 | 40 | 31,130 | 4 | 3,113 | 0.4 | 311 | 0.04 | 31 |
| 500 | 389,130 | 50 | 38,913 | 5 | 3,891 | 0.5 | 389 | 0.05 | 39 |
| 600 | 466,956 | 60 | 46,696 | 6 | 4,670 | 0.6 | 467 | 0.06 | 47 |
| 700 | 544,782 | 70 | 54,478 | 7 | 5,448 | 0.7 | 545 | 0.07 | 54 |
| 800 | 622,608 | 80 | 62,261 | 8 | 6,226 | 0.8 | 623 | 0.08 | 62 |
| 900 | 700,434 | 90 | 70,043 | 9 | 7,004 | 0.9 | 700 | 0.09 | 70 |
| 1,000 | 778,260 | 100 | 77,826 | 10 | 7,783 | 1.0 | 778 | 0.10 | 78 |

1 Btu = 778.26 ft-lb, conversion factor defined by International Steam Table Conference, 1929.

## Table 45b. Foot-Pounds to British Thermal Units

| ft-lb | Btu | ft-lb | Btu | ft-lb | Btu | ft-lb | Btu | ft-lb | Btu |
|---|---|---|---|---|---|---|---|---|---|
| 10,000 | 12.849 | 1,000 | 1.285 | 100 | 0.128 | 10 | 0.013 | 1 | 0.001 |
| 20,000 | 25.698 | 2,000 | 2.570 | 200 | 0.257 | 20 | 0.026 | 2 | 0.003 |
| 30,000 | 38.548 | 3,000 | 3.855 | 300 | 0.385 | 30 | 0.039 | 3 | 0.004 |
| 40,000 | 51.397 | 4,000 | 5.140 | 400 | 0.514 | 40 | 0.051 | 4 | 0.005 |
| 50,000 | 64.246 | 5,000 | 6.425 | 500 | 0.642 | 50 | 0.064 | 5 | 0.006 |
| 60,000 | 77.095 | 6,000 | 7.710 | 600 | 0.771 | 60 | 0.077 | 6 | 0.008 |
| 70,000 | 89.944 | 7,000 | 8.994 | 700 | 0.899 | 70 | 0.090 | 7 | 0.009 |
| 80,000 | 102.794 | 8,000 | 10.279 | 800 | 1.028 | 80 | 0.103 | 8 | 0.010 |
| 90,000 | 115.643 | 9,000 | 11.564 | 900 | 1.156 | 90 | 0.116 | 9 | 0.012 |
| 100,000 | 128.492 | 10,000 | 12.849 | 1,000 | 1.285 | 100 | 0.128 | 10 | 0.013 |

1 ft-lb = 0.00128492 Btu, conversion factor defined by International Steam Table Conference 1929.

## Table 46a. British Thermal Units to Kilojoules

| Btu→ ↓ | 0 | 100 | 200 | 300 | 400 | 500 | 600 | 700 | 800 | 900 |
|---|---|---|---|---|---|---|---|---|---|---|
| | kilojoules | | | | | | | | | |
| 0 | … | 105.51 | 211.01 | 316.52 | 422.02 | 527.53 | 633.03 | 738.54 | 844.04 | 949.55 |
| 1000 | 1055.06 | 1160.56 | 1266.07 | 1371.57 | 1477.08 | 1582.58 | 1688.09 | 1793.60 | 1899.10 | 2004.61 |
| 2000 | 2110.11 | 2215.62 | 2321.12 | 2426.63 | 2532.13 | 2637.64 | 2743.15 | 2848.65 | 2954.16 | 3059.66 |
| 3000 | 3165.17 | 3270.67 | 3376.18 | 3481.68 | 3587.19 | 3692.70 | 3798.20 | 3903.71 | 4009.21 | 4114.72 |
| 4000 | 4220.22 | 4325.73 | 4431.24 | 4536.74 | 4642.25 | 4747.75 | 4853.26 | 4958.76 | 5064.27 | 5169.77 |
| 5000 | 5275.28 | 5380.79 | 5486.29 | 5591.80 | 5697.30 | 5802.81 | 5908.31 | 6013.82 | 6119.32 | 6224.83 |
| 6000 | 6330.34 | 6435.84 | 6541.35 | 6646.85 | 6752.36 | 6857.86 | 6963.37 | 7068.88 | 7174.38 | 7279.89 |
| 7000 | 7385.39 | 7490.90 | 7596.40 | 7701.91 | 7807.41 | 7912.92 | 8018.43 | 8123.93 | 8229.44 | 8334.94 |
| 8000 | 8440.45 | 8545.95 | 8651.46 | 8756.96 | 8862.47 | 8967.98 | 9073.48 | 9178.99 | 9284.49 | 9390.00 |
| 9000 | 9495.50 | 9601.01 | 9706.52 | 9812.02 | 9917.53 | 10023.0 | 10128.5 | 10234.0 | 10339.5 | 10445.1 |
| 10,000 | 10550.6 | … | … | … | … | … | … | … | … | … |

1 Btu = 1055.056 joules

## Table 46b. Kilojoules to British Thermal Units

| kJ→ ↓ | 0 | 100 | 200 | 300 | 400 | 500 | 600 | 700 | 800 | 900 |
|---|---|---|---|---|---|---|---|---|---|---|
| | British Thermal Units | | | | | | | | | |
| 0 | … | 94.78 | 189.56 | 284.35 | 379.13 | 473.91 | 568.69 | 663.47 | 758.25 | 853.04 |
| 1000 | 947.82 | 1042.60 | 1137.38 | 1232.16 | 1326.94 | 1421.73 | 1516.51 | 1611.29 | 1706.07 | 1800.85 |
| 2000 | 1895.63 | 1990.42 | 2085.20 | 2179.98 | 2274.76 | 2369.54 | 2464.32 | 2559.11 | 2653.89 | 2748.67 |
| 3000 | 2843.45 | 2938.23 | 3033.01 | 3127.80 | 3222.58 | 3317.36 | 3412.14 | 3506.92 | 3601.70 | 3696.49 |
| 4000 | 3791.27 | 3886.05 | 3980.83 | 4075.61 | 4170.39 | 4265.18 | 4359.96 | 4454.74 | 4549.52 | 4644.30 |
| 5000 | 4739.08 | 4833.87 | 4928.65 | 5023.43 | 5118.21 | 5212.99 | 5307.78 | 5402.56 | 5497.34 | 5592.12 |
| 6000 | 5686.90 | 5781.68 | 5876.47 | 5971.25 | 6066.03 | 6160.81 | 6255.59 | 6350.37 | 6445.16 | 6539.94 |
| 7000 | 6634.72 | 6729.50 | 6824.28 | 6919.06 | 7013.85 | 7108.63 | 7203.41 | 7298.19 | 7392.97 | 7487.75 |
| 8000 | 7582.54 | 7677.32 | 7772.10 | 7866.88 | 7961.66 | 8056.44 | 8151.23 | 8246.01 | 8340.79 | 8435.57 |
| 9000 | 8530.35 | 8625.13 | 8719.92 | 8814.70 | 8909.48 | 9004.26 | 9099.04 | 9193.82 | 9288.61 | 9383.39 |
| 10,000 | 9478.17 | … | … | … | … | … | … | … | … | … |

1 joule = 0.0009478170 Btu

**Table 47a. Horsepower to Kilowatts Conversion**

| hp | kW | hp | kW | hp | kW | hp | kW | hp | kW |
|---|---|---|---|---|---|---|---|---|---|
| 1,000 | 745.7 | 100 | 74.6 | 10 | 7.5 | 1 | 0.7 | 0.1 | 0.07 |
| 2,000 | 1,491.4 | 200 | 149.1 | 20 | 14.9 | 2 | 1.5 | 0.2 | 0.15 |
| 3,000 | 2,237.1 | 300 | 223.7 | 30 | 22.4 | 3 | 2.2 | 0.3 | 0.22 |
| 4,000 | 2,982.8 | 400 | 298.3 | 40 | 29.8 | 4 | 3.0 | 0.4 | 0.30 |
| 5,000 | 3,728.5 | 500 | 372.8 | 50 | 37.3 | 5 | 3.7 | 0.5 | 0.37 |
| 6,000 | 4,474.2 | 600 | 447.4 | 60 | 44.7 | 6 | 4.5 | 0.6 | 0.45 |
| 7,000 | 5,219.9 | 700 | 522.0 | 70 | 52.2 | 7 | 5.2 | 0.7 | 0.52 |
| 8,000 | 5,965.6 | 800 | 596.6 | 80 | 59.7 | 8 | 6.0 | 0.8 | 0.60 |
| 9,000 | 6,711.3 | 900 | 671.1 | 90 | 67.1 | 9 | 6.7 | 0.9 | 0.67 |
| 10,000 | 7,457.0 | 1,000 | 745.7 | 100 | 74.6 | 10 | 7.5 | 1.0 | 0.75 |

1 hp = 0.7456999 kW, based on 1 horsepower = 550 foot-pounds per second.

**Table 47b. Kilowatts to Horsepower Conversion**

| kW | hp | kW | hp | kW | hp | kW | hp | kW | hp |
|---|---|---|---|---|---|---|---|---|---|
| 1,000 | 1,341.0 | 100 | 134.1 | 10 | 13.4 | 1 | 1.3 | 0.1 | 0.13 |
| 2,000 | 2,682.0 | 200 | 268.2 | 20 | 26.8 | 2 | 2.7 | 0.2 | 0.27 |
| 3,000 | 4,023.1 | 300 | 402.3 | 30 | 40.2 | 3 | 4.0 | 0.3 | 0.40 |
| 4,000 | 5,364.1 | 400 | 536.4 | 40 | 53.6 | 4 | 5.4 | 0.4 | 0.54 |
| 5,000 | 6,705.1 | 500 | 670.5 | 50 | 67.1 | 5 | 6.7 | 0.5 | 0.67 |
| 7,000 | 9,387.2 | 700 | 938.7 | 70 | 93.9 | 7 | 9.4 | 0.7 | 0.94 |
| 8,000 | 10,728.2 | 800 | 1,072.8 | 80 | 107.3 | 8 | 10.7 | 0.8 | 1.07 |
| 9,000 | 12,069.2 | 900 | 1,206.9 | 90 | 120.7 | 9 | 12.1 | 0.9 | 1.21 |
| 10,000 | 13,410.2 | 1,000 | 1,341.0 | 100 | 134.1 | 10 | 13.4 | 1.0 | 1.34 |

1 kW = 1.341022 hp, based on 1 horsepower = 550 foot-pounds per second.

**Table 48a. Foot-Pounds to Joules Conversion**

| ft-lb→ ↓ | 0 | 1 | 2 | 3 | 4 | 5 | 6 | 7 | 8 | 9 |
|---|---|---|---|---|---|---|---|---|---|---|
| | joules | | | | | | | | | |
| 0 | … | 1.356 | 2.712 | 4.067 | 5.423 | 6.779 | 8.135 | 9.491 | 10.847 | 12.202 |
| 10 | 13.558 | 14.914 | 16.270 | 17.626 | 18.981 | 20.337 | 21.693 | 23.049 | 24.405 | 25.761 |
| 20 | 27.116 | 28.472 | 29.828 | 31.184 | 32.540 | 33.895 | 35.251 | 36.607 | 37.963 | 39.319 |
| 30 | 40.675 | 42.030 | 43.386 | 44.742 | 46.098 | 47.454 | 48.809 | 50.165 | 51.521 | 52.877 |
| 40 | 54.233 | 55.589 | 56.944 | 58.300 | 59.656 | 61.012 | 62.368 | 63.723 | 65.079 | 66.435 |
| 50 | 67.791 | 69.147 | 70.503 | 71.858 | 73.214 | 74.570 | 75.926 | 77.282 | 78.637 | 79.993 |
| 60 | 81.349 | 82.705 | 84.061 | 85.417 | 86.772 | 88.128 | 89.484 | 90.840 | 92.196 | 93.551 |
| 70 | 94.907 | 96.263 | 97.619 | 98.975 | 100.331 | 101.686 | 103.042 | 104.398 | 105.754 | 107.110 |
| 80 | 108.465 | 109.821 | 111.177 | 112.533 | 113.889 | 115.245 | 116.600 | 117.956 | 119.312 | 120.668 |
| 90 | 122.024 | 123.379 | 124.735 | 126.091 | 127.447 | 128.803 | 130.159 | 131.514 | 132.870 | 134.226 |
| 100 | 135.582 | 136.938 | 138.293 | 139.649 | 141.005 | 142.361 | 143.717 | 145.073 | 146.428 | 147.784 |

1 foot-pound = 1.355818 joules

**Table 48b. Joules to Foot-Pounds Conversion**

| J→ ↓ | 0 | 1 | 2 | 3 | 4 | 5 | 6 | 7 | 8 | 9 |
|---|---|---|---|---|---|---|---|---|---|---|
| | foot-pounds | | | | | | | | | |
| 0 | … | 0.7376 | 1.4751 | 2.2127 | 2.9502 | 3.6878 | 4.4254 | 5.1629 | 5.9005 | 6.6381 |
| 10 | 7.3756 | 8.1132 | 8.8507 | 9.5883 | 10.3259 | 11.0634 | 11.8010 | 12.5386 | 13.2761 | 14.0137 |
| 20 | 14.7512 | 15.4888 | 16.2264 | 16.9639 | 17.7015 | 18.4391 | 19.1766 | 19.4142 | 20.6517 | 21.3893 |
| 30 | 22.1269 | 22.8644 | 23.6020 | 24.3395 | 25.0771 | 25.8147 | 26.5522 | 27.2898 | 28.0274 | 28.7649 |
| 40 | 29.5025 | 30.2400 | 30.9776 | 31.7152 | 32.4527 | 33.1903 | 33.9279 | 34.6654 | 35.4030 | 36.1405 |
| 50 | 36.8781 | 37.6157 | 38.3532 | 39.0908 | 39.8284 | 40.5659 | 41.3035 | 42.0410 | 42.7786 | 43.5162 |
| 60 | 44.2537 | 44.9913 | 45.7289 | 46.4664 | 47.2040 | 47.9415 | 48.6791 | 49.4167 | 50.1542 | 50.8918 |
| 70 | 51.6293 | 52.3669 | 53.1045 | 53.8420 | 54.5796 | 55.3172 | 56.0547 | 56.7923 | 57.5298 | 58.2674 |
| 80 | 59.0050 | 59.7425 | 60.4801 | 61.2177 | 61.9552 | 62.6928 | 63.4303 | 64.1679 | 64.9055 | 65.6430 |
| 90 | 66.3806 | 67.1182 | 67.8557 | 68.5933 | 69.3308 | 70.0684 | 70.8060 | 71.5435 | 72.2811 | 73.0186 |
| 100 | 73.7562 | 74.4938 | 75.2313 | 75.9689 | 76.7065 | 77.4440 | 78.1816 | 78.9191 | 79.6567 | 80.3943 |

1 joule = 0.7375621 foot-pound

## Table 49. Power Conversion Factors

| To Convert ↓ / Multiply By This Factor To Obtain → | horsepower | watts | kilowatts | hp (metric) | kg$_f$-m/s | ft-lb$_f$/s | ft-lb$_f$/min | calories/sec | Btu/sec | Btu/hr |
|---|---|---|---|---|---|---|---|---|---|---|
| horsepower | 1 | 745.699 | 0.745699 | 1.0138681 | 76.04 | **550** | **33000** | 178.1 | 0.7068 | 2544.48 |
| watts | 0.00134024 | 1 | **0.001** | 0.0013596 | 0.1019714 | 0.7375630 | 44.253727 | 0.2388363 | 0.0009478 | 3.4122 |
| kilowatts | 1.34102365 | **1000** | 1 | 1.3596196 | 101.9713158 | 737.563011 | 44253.727270 | 238.836025 | 0.9478344 | 3412.20 |
| hp (metric) | 0.9863215 | 735.499 | 0.735499 | 1 | **75** | 542.476857 | 32548.61114 | 175.663869 | 0.6971321 | 2509.6754 |
| kg$_f$-m/s | 0.01315097 | 9.8066 | 0.0098067 | 0.0133334 | 1 | 7.2330352 | 433.982114 | 2.3421883 | 0.0092951 | 33.4623 |
| ft-lb$_f$/s | 0.00181818 | 1.35581 | 0.0013558 | 0.0018434 | 0.1382545 | 1 | **60** | 0.3238181 | 0.0012851 | 4.6263 |
| ft-lb$_f$/min | $3.0303 \times 10^{-5}$ | 0.02259 | $2.2596 \times 10^{-5}$ | $3.07233 \times 10^{-5}$ | 0.0023042 | 0.0166667 | 1 | 0.0053969 | $2.1418 \times 10^{-5}$ | 0.077105 |
| calories/sec | 0.00561482 | 4.18696 | 0.0041869 | 0.0056927 | 0.4269512 | 3.0881527 | 185.288916 | 1 | 0.0039686 | 14.2868 |
| Btu/sec | 1.41482739 | 1055.035 | 1.0550353 | 1.4344484 | 107.5834748 | 778.155065 | 46689.3039 | **252** | 1 | **3600** |
| Btu/hr | 0.0003930 | 0.29306 | 0.0002931 | 0.0003985 | 0.0298843 | 0.2161542 | 12.969251 | 0.069994 | 0.0002778 | 1 |

Figures in **bold face** indicate the conversion is exact

## Table 50. Energy and Work Conversion Factors

| To Convert ↓ / Multiply By This Factor To Obtain → | joules | ft-lb$_f$ | ft-poundal | Btu | kg-m | calories | watt-hour | erg | therm | hp-hours | hp-hours (m) |
|---|---|---|---|---|---|---|---|---|---|---|---|
| joules | 1 | 0.73756 | 23.7303 | 0.0009478 | 0.101972 | 0.2388458 | 0.00027778 | **$1 \times 10^{7}$** | $9.478 \times 10^{-9}$ | $3.725 \times 10^{-7}$ | $3.7764 \times 10^{-7}$ |
| ft-lb$_f$ | 1.355818 | 1 | 32.1740 | 0.00128506 | 0.138255 | 0.3238316 | 0.00037661 | $1.356 \times 10^{7}$ | $1.285 \times 10^{-8}$ | $5.0505 \times 10^{-7}$ | $5.1201 \times 10^{-7}$ |
| ft-poundal | 0.04214 | 0.03108 | 1 | $3.994 \times 10^{-5}$ | 0.0042971 | 0.010065 | $1.1705 \times 10^{-5}$ | $4.214 \times 10^{5}$ | $3.994 \times 10^{-10}$ | $1.5697 \times 10^{-8}$ | $1.5914 \times 10^{-8}$ |
| Btu | 1055.055 | 778.1692 | 25036.8174 | 1 | 107.5875 | 252 | 0.29307071 | $1.055 \times 10^{10}$ | $1 \times 10^{-5}$ | 0.0003930 | 0.0003984 |
| kg-m | 9.80665 | 7.233013 | 232.714987 | 0.00929524 | 1 | 2.342278 | 0.00272416 | $9.807 \times 10^{7}$ | $9.294 \times 10^{-8}$ | $3.653 \times 10^{-6}$ | $3.703 \times 10^{-6}$ |
| calories | 4.1868 | 3.088025 | 99.35427 | 0.00396832 | 0.42691934 | 1 | 0.001163 | $4.187 \times 10^{7}$ | $3.968 \times 10^{-8}$ | $1.5596 \times 10^{-6}$ | $1.5811 \times 10^{-6}$ |
| watt-hour | **3600** | 2655.2237 | 85429.168 | 3.4121416 | 367.09783 | 859.845227 | 1 | $3.6 \times 10^{10}$ | $3.412 \times 10^{-5}$ | 0.001341 | 0.0013595 |
| erg | **$1 \times 10^{-7}$** | $7.375 \times 10^{-8}$ | $2.373 \times 10^{-6}$ | $9.478 \times 10^{-11}$ | $1.0197 \times 10^{-8}$ | $2.3884 \times 10^{-8}$ | $2.778 \times 10^{-11}$ | 1 | $9.478 \times 10^{-16}$ | $3.725 \times 10^{-14}$ | $3.776 \times 10^{-14}$ |
| therm | $1.055 \times 10^{8}$ | $7.781 \times 10^{7}$ | $2.503 \times 10^{7}$ | **$1 \times 10^{5}$** | $1.0758 \times 10^{7}$ | $2.5196 \times 10^{7}$ | 29307.222 | $1.055 \times 10^{15}$ | 1 | 39.3020 | 39.843655 |
| hp-hours | $2.6845 \times 10^{6}$ | $1.9799 \times 10^{6}$ | $6.3704 \times 10^{7}$ | 2544.4150 | $2.7374 \times 10^{5}$ | $6.4118 \times 10^{5}$ | 745.6944 | $2.685 \times 10^{13}$ | 0.025444 | 1 | 1.0137839 |
| hp-hours (m) | $2.648 \times 10^{6}$ | $1.953 \times 10^{6}$ | $6.2837 \times 10^{7}$ | 2509.8197 | $2.70 \times 10^{5}$ | $6.3246 \times 10^{5}$ | 735.555 | $2.648 \times 10^{13}$ | 0.025098 | 0.9864034 | 1 |

Figures in **bold face** indicate the conversion is exact

## Table 51. Thermal Conductance Conversion Factors

| To Convert ↓ / Multiply By This Factor To Obtain → | Btu-ft/(h-ft²-°F) | Btu-in/(h-ft²-°F) | Btu-in/(sec-ft²-°F) | cal/(cm-s-°C) | kcal/(cm-s-°C) | kcal/(m-h-°C) | erg/(cm-s-°C) | joules/(m-h-°C) | watt/(ft-°C) | watt/(m-°K) |
|---|---|---|---|---|---|---|---|---|---|---|
| Btu-ft/(h-ft²-°F) | 1 | **12** | 0.00333333 | 0.00413385 | $4.13386 \times 10^{-6}$ | 1.488188976 | 173076.378 | 6230.0055 | 0.5274738 | 1.73056 |
| Btu-in/(h-ft²-°F) | 0.083333 | 1 | 0.000277778 | 0.00034448 | $3.44448 \times 10^{-7}$ | 0.124015748 | 14423.0315 | 519.25573 | 0.04395615 | 0.14421 |
| Btu-in/(sec-ft²-°F) | **300** | **3600** | 1 | 1.24001574 | 0.001240157 | 446.4566929 | $5.1925 \times 10^{7}$ | $1.8693 \times 10^{6}$ | 158.24214 | 519.167 |
| cal/(cm-s-°C) | 241.9047 | 2902.8571 | 0.806349 | 1 | **0.001** | **360** | $4.1868 \times 10^{7}$ | $1.507 \times 10^{6}$ | 127.598424 | 418.63 |
| kcal/(cm-s-°C) | $2.419 \times 10^{5}$ | $2.902 \times 10^{6}$ | 806.3492 | **1000** | 1 | **360000** | $4.1868 \times 10^{10}$ | $1.507 \times 10^{9}$ | $1.276 \times 10^{5}$ | $4.1863 \times 10^{5}$ |
| kcal/(m-h-°C) | 0.671957 | 8.063349 | 0.00223985 | 0.00277778 | $2.77778 \times 10^{-6}$ | 1 | 116300 | 4186.8 | 0.35444 | 1.16286 |
| erg/(cm- s-°C) | $5.7778 \times 10^{-6}$ | $6.933 \times 10^{-5}$ | $1.92593 \times 10^{-8}$ | $2.3884 \times 10^{-8}$ | $2.3884 \times 10^{-11}$ | $8.5984 \times 10^{-6}$ | 1 | 0.036 | $3.0476 \times 10^{-6}$ | $1 \times 10^{-5}$ |
| joules/(m-h-°C) | $1.6051 \times 10^{-4}$ | 0.00192616 | $5.35045 \times 10^{-7}$ | $6.6354 \times 10^{-7}$ | $6.6354 \times 10^{-10}$ | 0.000238874 | 27.781095 | 1 | $8.4666 \times 10^{-5}$ | $2.7777 \times 10^{-4}$ |
| watt/(ft-°C) | 1.895828 | 22.75 | 0.006319429 | 0.00783708 | $7.83709 \times 10^{-6}$ | 2.821351461 | 328123.1749 | 11811.024 | 1 | 3.28 |
| watt/(m-°K) | 0.5778486 | 6.934183 | 0.001926162 | 0.002388744 | $2.38874 \times 10^{-6}$ | 0.859947925 | **$1 \times 10^{5}$** | **3600** | 0.304878 | 1 |

Figures in **bold face** indicate the conversion is exact

**Conduction.**—Whenever the molecules of a working substance, whether liquid, solid, or vapor, are restrained so that no appreciable relative translatory motion occurs among them, the kinetic energies of the various molecules will be largely due to vibration. If a temperature difference exists in the working substance, some adjacent molecules will necessarily be at different temperatures, hence will possess different degrees of vibratory motion. In this case the molecule which is vibrating most rapidly will transfer some of its motion to the slower-moving molecule next to it, the one then undergoing a decrease in temperature and the other an increase. In this way, thermal energy will be transferred by the mechanism of conduction from the region of higher to the region of lower temperature. The process will continue spontaneously until the entire system has reached a uniform equilibrium temperature.

In contrast to radiation, conduction only occurs when a working substance is present and when the molecules of that working substance retain practically fixed positions with respect to one another. Thus, conductive heat flow would always occur through solids, but would take place in liquids and vapors only if special conditions prevented or greatly reduced the normal translatory motion of the molecules within these materials.

**Fuel Oil, Coal, and Gas.**—One gallon of fuel oil equals 13.1 pounds of coal, equals 160 cubic feet of natural gas. One barrel of fuel oil equals 0.278 ton of coal, equals 6806 cubic feet of natural gas. One pound of fuel oil equals 1.75 pounds of coal, equals 21.3 cubic feet of natural gas. One pound of coal equals 0.763 gallon of oil, equals 12.2 cubic feet of natural gas. One ton of coal equals 3.6 barrels of oil, equals 24,500 cubic feet of natural gas. The heating value of the average mid-continent fuel oil having a Baume gravity of 26.9 is 19,376 British thermal units per pound of oil, and 143,950 British thermal units per gallon of oil. The specific gravity and the heat value may be expressed approximately by means of a simple formula, as follows: BTU per pound = 18,650 + 40 × (Degrees Baume – 10).

## Units of Temperature

There are two thermometer scales in general use: the Fahrenheit (°F), which is used in the United States and in other countries still using the English system of units, and the Celsius (°C) or Centigrade used throughout the rest of the world.

In the Fahrenheit thermometer, the freezing point of water is marked at 32 degrees on the scale and the boiling point, at atmospheric pressure, at 212 degrees. The distance between these two points is divided into 180 degrees. On the Celsius scale, the freezing point of water is at 0 degree and the boiling point at 100 degrees. The following formulas may be used for converting temperatures given on any one of the scales to the other scale:

$$°F = \frac{9}{5}°C + 32 \qquad C = \frac{5}{9}(°F - 32)$$

Tables on the pages that follow can be used to convert degrees Celsius into degrees Fahrenheit or vice versa, and to convert the absolute temperatures of Kelvin and Rankine scales. In the event that the conversions are not covered in the tables, use those applicable portions of the formulas given above and in Table 52 for converting.

**Table 52. Temperature Conversion Fomulas**

| To Convert | To | Use Formula | To Convert | To | Use Formula |
|---|---|---|---|---|---|
| | K, $t_K$ | $t_K = t_C + 273.15$ | | °C, $t_C$ | $t_C = t_K - 273.15$ |
| Celsius, $t_C$ | °F, $t_F$ | $t_F = 1.8t_C + 32$ | Kelvin, $t_K$ | °F, $t_F$ | $t_F = 1.8t_K - 459.67$ |
| | °R, $t_R$ | $t_R = 9(t_C + 273.15)/5$ | | °R, $t_R$ | $t_R = 9/5 \times t_K$ |
| | K, $t_K$ | $t_K = (t_F + 459.67)/1.8$ | | K, $t_K$ | $t_K = 5/9 \times t_R$ |
| Fahrenheit, $t_F$ | °C, $t_C$ | $t_C = (t_F - 32)/1.8$ | Rankine, $t_R$ | °C, $t_C$ | $t_C = 5/9 \times t_R - 273.15$ |
| | °R, $t_R$ | $t_R = t_F + 459.67$ | | °F, $t_F$ | $t_F = t_R - 459.67$ |

Kelvin temperatures are expressed by K without use of the degree symbol (°).

**Absolute Temperature and Absolute Zero.**—A point has been determined on the thermometer scale, by theoretical considerations, that is called the *absolute zero* and beyond which a further decrease in temperature is inconceivable. This point is located at −273.15 degrees Celsius or −459.67 degrees F. A temperature reckoned from this point, instead of from the zero on the ordinary thermometers, is called absolute temperature. The absolute temperature scale corresponding to Celsius is the Kelvin scale; the absolute scale corresponding to Fahrenheit is the Rankine. 0 K and 0° R are both absolute zero. Conversions from Celsius and Fahrenheit are obtained as follows:

$$K = °C + 273.15 \qquad °R = °F + 459.67$$

**Measures of the Quantity of Thermal Energy.**—The unit of quantity of thermal energy used in the United States is the British thermal unit, which is the quantity of heat or thermal energy required to raise the temperature of one pound of pure water one degree F. (American National Standard abbreviation, Btu; conventional British symbol, B.Th.U.) The French thermal unit, or *kilogram calorie,* is the quantity of heat or thermal energy required to raise the temperature of one kilogram of pure water one degree C. One kilogram calorie = 3.968 British thermal units = 1000 gram calories. The number of foot-pounds of mechanical energy equivalent to one British thermal unit is called the *mechanical equivalent of heat* and equals 778 foot-pounds.

In the modern metric or SI system of units, the unit for thermal energy is the *joule* (J); a commonly used multiple being the kilojoule (kJ), or 1000 joules. See page 2840 for an explanation of the SI System. One kilojoule = 0.9478 Btu. Also in the SI System, the *watt* (W), equal to joule per second (J/s), is used for power, where 1 watt = 3.412 Btu per hour.

**Table 53. °C → °F and °R Temperature Conversion °F → °C and K**

| K | °C | | °F | °R | K | °C | | °F | °R | K | °C | | °F | °R |
|---|---|---|---|---|---|---|---|---|---|---|---|---|---|---|
| 0.0 | −273.2 | **−459.7** | ... | ... | 261.5 | −11.7 | **11** | 51.8 | 511.5 | 293.7 | 20.6 | **69** | 156.2 | 615.9 |
| 5.4 | −267.8 | **−450** | ... | ... | 262.0 | −11.1 | **12** | 53.6 | 513.3 | 294.3 | 21.1 | **70** | 158.0 | 617.7 |
| 10.9 | −262.2 | **−440** | ... | ... | 262.6 | −10.6 | **13** | 55.4 | 515.1 | 294.8 | 21.7 | **71** | 159.8 | 619.5 |
| 16.5 | −256.7 | **−430** | ... | ... | 263.2 | −10.0 | **14** | 57.2 | 516.9 | 295.4 | 22.2 | **72** | 161.6 | 621.3 |
| 22.0 | −251.1 | **−420** | ... | ... | 263.7 | −9.4 | **15** | 59.0 | 518.7 | 295.9 | 22.8 | **73** | 163.4 | 623.1 |
| 27.6 | −245.6 | **−410** | ... | ... | 264.3 | −8.9 | **16** | 60.8 | 520.5 | 296.5 | 23.3 | **74** | 165.2 | 624.9 |
| 33.2 | −240.0 | **−400** | ... | ... | 264.8 | −8.3 | **17** | 62.6 | 522.3 | 297.0 | 23.9 | **75** | 167.0 | 626.7 |
| 38.7 | −234.4 | **−390** | ... | ... | 265.4 | −7.8 | **18** | 64.4 | 524.1 | 297.6 | 24.4 | **76** | 168.8 | 628.5 |
| 44.3 | −228.9 | **−380** | ... | ... | 265.9 | −7.2 | **19** | 66.2 | 525.9 | 298.2 | 25.0 | **77** | 170.6 | 630.3 |
| 49.8 | −223.3 | **−370** | ... | ... | 266.5 | −6.7 | **20** | 68.0 | 527.7 | 298.7 | 25.6 | **78** | 172.4 | 632.1 |
| 55.4 | −217.8 | **−360** | ... | ... | 267.0 | −6.1 | **21** | 69.8 | 529.5 | 299.3 | 26.1 | **79** | 174.2 | 633.9 |
| 60.9 | −212.2 | **−350** | ... | ... | 267.6 | −5.6 | **22** | 71.6 | 531.3 | 299.8 | 26.7 | **80** | 176.0 | 635.7 |
| 66.5 | −206.7 | **−340** | ... | ... | 268.2 | −5.0 | **23** | 73.4 | 533.1 | 300.4 | 27.2 | **81** | 177.8 | 637.5 |
| 72.0 | −201.1 | **−330** | ... | ... | 268.7 | −4.4 | **24** | 75.2 | 534.9 | 300.9 | 27.8 | **82** | 179.6 | 639.3 |
| 77.6 | −195.6 | **−320** | ... | ... | 269.3 | −3.9 | **25** | 77.0 | 536.7 | 301.5 | 28.3 | **83** | 181.4 | 641.1 |
| 83.2 | −190.0 | **−310** | ... | ... | 269.8 | −3.3 | **26** | 78.8 | 538.5 | 302.0 | 28.9 | **84** | 183.2 | 642.9 |
| 88.7 | −184.4 | **−300** | ... | ... | 270.4 | −2.8 | **27** | 80.6 | 540.3 | 302.6 | 29.4 | **85** | 185.0 | 644.7 |
| 94.3 | −178.9 | **−290** | ... | ... | 270.9 | −2.2 | **28** | 82.4 | 542.1 | 303.2 | 30.0 | **86** | 186.8 | 646.5 |
| 99.8 | −173.3 | **−280** | ... | ... | 271.5 | −1.7 | **29** | 84.2 | 543.9 | 303.7 | 30.6 | **87** | 188.6 | 648.3 |
| 103.6 | −169.5 | **−273.2** | −459.7 | 0.0 | 272.0 | −1.1 | **30** | 86.0 | 545.7 | 304.3 | 31.1 | **88** | 190.4 | 650.1 |
| 105.4 | −167.8 | **−270** | −454.0 | 5.7 | 272.6 | −0.6 | **31** | 87.8 | 547.5 | 304.8 | 31.7 | **89** | 192.2 | 651.9 |
| 110.9 | −162.2 | **−260** | −436.0 | 23.7 | 273.2 | 0.0 | **32** | 89.6 | 549.3 | 305.4 | 32.2 | **90** | 194.0 | 653.7 |
| 116.5 | −156.7 | **−250** | −418.0 | 41.7 | 273.7 | 0.6 | **33** | 91.4 | 551.1 | 305.9 | 32.8 | **91** | 195.8 | 655.5 |
| 122.0 | −151.1 | **−240** | −400.0 | 59.7 | 274.3 | 1.1 | **34** | 93.2 | 552.9 | 306.5 | 33.3 | **92** | 197.6 | 657.3 |
| 127.6 | −145.6 | **−230** | −382.0 | 77.7 | 274.8 | 1.7 | **35** | 95.0 | 554.7 | 307.0 | 33.9 | **93** | 199.4 | 659.1 |
| 133.2 | −140.0 | **−220** | −364.0 | 95.7 | 275.4 | 2.2 | **36** | 96.8 | 556.5 | 307.6 | 34.4 | **94** | 201.2 | 660.9 |
| 138.7 | −134.4 | **−210** | −346.0 | 113.7 | 275.9 | 2.8 | **37** | 98.6 | 558.3 | 308.2 | 35.0 | **95** | 203.0 | 662.7 |
| 144.3 | −128.9 | **−200** | −328.0 | 131.7 | 276.5 | 3.3 | **38** | 100.4 | 560.1 | 308.7 | 35.6 | **96** | 204.8 | 664.5 |
| 149.8 | −123.3 | **−190** | −310.0 | 149.7 | 277.0 | 3.9 | **39** | 102.2 | 561.9 | 309.3 | 36.1 | **97** | 206.6 | 666.3 |
| 155.4 | −117.8 | **−180** | −292.0 | 167.7 | 277.6 | 4.4 | **40** | 104.0 | 563.7 | 309.8 | 36.7 | **98** | 208.4 | 668.1 |
| 160.9 | −112.2 | **−170** | −274.0 | 185.7 | 278.2 | 5.0 | **41** | 105.8 | 565.5 | 310.4 | 37.2 | **99** | 210.2 | 669.9 |
| 166.5 | −106.7 | **−160** | −256.0 | 203.7 | 278.7 | 5.6 | **42** | 107.6 | 567.3 | 310.9 | 37.8 | **100** | 212.0 | 671.7 |
| 172.0 | −101.1 | **−150** | −238.0 | 221.7 | 279.3 | 6.1 | **43** | 109.4 | 569.1 | 311.5 | 38.3 | **101** | 213.8 | 673.5 |
| 177.6 | −95.6 | **−140** | −220.0 | 239.7 | 279.8 | 6.7 | **44** | 111.2 | 570.9 | 312.0 | 38.9 | **102** | 215.6 | 675.3 |
| 183.2 | −90.0 | **−130** | −202.0 | 257.7 | 280.4 | 7.2 | **45** | 113.0 | 572.7 | 312.6 | 39.4 | **103** | 217.4 | 677.1 |
| 188.7 | −84.4 | **−120** | −184.0 | 275.7 | 280.9 | 7.8 | **46** | 114.8 | 574.5 | 313.2 | 40.0 | **104** | 219.2 | 678.9 |
| 194.3 | −78.9 | **−110** | −166.0 | 293.7 | 281.5 | 8.3 | **47** | 116.6 | 576.3 | 313.7 | 40.6 | **105** | 221.0 | 680.7 |
| 199.8 | −73.3 | **−100** | −148.0 | 311.7 | 282.0 | 8.9 | **48** | 118.4 | 578.1 | 314.3 | 41.1 | **106** | 222.8 | 682.5 |
| 205.4 | −67.8 | **−90** | −130.0 | 329.7 | 282.6 | 9.4 | **49** | 120.2 | 579.9 | 314.8 | 41.7 | **107** | 224.6 | 684.3 |
| 210.9 | −62.2 | **−80** | −112.0 | 347.7 | 283.2 | 10.0 | **50** | 122.0 | 581.7 | 315.4 | 42.2 | **108** | 226.4 | 686.1 |
| 216.5 | −56.7 | **−70** | −94.0 | 365.7 | 283.7 | 10.6 | **51** | 123.8 | 583.5 | 315.9 | 42.8 | **109** | 228.2 | 687.9 |
| 222.0 | −51.1 | **−60** | −76.0 | 383.7 | 284.3 | 11.1 | **52** | 125.6 | 585.3 | 316.5 | 43.3 | **110** | 230.0 | 689.7 |
| 227.6 | −45.6 | **−50** | −58.0 | 401.7 | 284.8 | 11.7 | **53** | 127.4 | 587.1 | 317.0 | 43.9 | **111** | 231.8 | 691.5 |
| 233.2 | −40.0 | **−40** | −40.0 | 419.7 | 285.4 | 12.2 | **54** | 129.2 | 588.9 | 317.6 | 44.4 | **112** | 233.6 | 693.3 |
| 238.7 | −34.4 | **−30** | −22.0 | 437.7 | 285.9 | 12.8 | **55** | 131.0 | 590.7 | 318.2 | 45.0 | **113** | 235.4 | 695.1 |
| 244.3 | −28.9 | **−20** | −4.0 | 455.7 | 286.5 | 13.3 | **56** | 132.8 | 592.5 | 318.7 | 45.6 | **114** | 237.2 | 696.9 |
| 249.8 | −23.3 | **−10** | 14.0 | 473.7 | 287.0 | 13.9 | **57** | 134.6 | 594.3 | 319.3 | 46.1 | **115** | 239.0 | 698.7 |
| 255.4 | −17.8 | **0** | 32.0 | 491.7 | 287.6 | 14.4 | **58** | 136.4 | 596.1 | 319.8 | 46.7 | **116** | 240.8 | 700.5 |
| 255.9 | −17.2 | **1** | 33.8 | 493.5 | 288.2 | 15.0 | **59** | 138.2 | 597.9 | 320.4 | 47.2 | **117** | 242.6 | 702.3 |
| 256.5 | −16.7 | **2** | 35.6 | 495.3 | 288.7 | 15.6 | **60** | 140.0 | 599.7 | 320.9 | 47.8 | **118** | 244.4 | 704.1 |
| 257.0 | −16.1 | **3** | 37.4 | 497.1 | 289.3 | 16.1 | **61** | 141.8 | 601.5 | 321.5 | 48.3 | **119** | 246.2 | 705.9 |
| 257.6 | −15.6 | **4** | 39.2 | 498.9 | 289.8 | 16.7 | **62** | 143.6 | 603.3 | 322.0 | 48.9 | **120** | 248.0 | 707.7 |
| 258.2 | −15.0 | **5** | 41.0 | 500.7 | 290.4 | 17.2 | **63** | 145.4 | 605.1 | 322.6 | 49.4 | **121** | 249.8 | 709.5 |
| 258.7 | −14.4 | **6** | 42.8 | 502.5 | 290.9 | 17.8 | **64** | 147.2 | 606.9 | 323.2 | 50.0 | **122** | 251.6 | 711.3 |
| 259.3 | −13.9 | **7** | 44.6 | 504.3 | 291.5 | 18.3 | **65** | 149.0 | 608.7 | 323.7 | 50.6 | **123** | 253.4 | 713.1 |
| 259.8 | −13.3 | **8** | 46.4 | 506.1 | 292.0 | 18.9 | **66** | 150.8 | 610.5 | 324.3 | 51.1 | **124** | 255.2 | 714.9 |
| 260.4 | −12.8 | **9** | 48.2 | 507.9 | 292.6 | 19.4 | **67** | 152.6 | 612.3 | 324.8 | 51.7 | **125** | 257.0 | 716.7 |
| 260.9 | −12.2 | **10** | 50.0 | 509.7 | 293.2 | 20.0 | **68** | 154.4 | 614.1 | 325.4 | 52.2 | **126** | 258.8 | 718.5 |

**Table 53.** *(Continued)* **°C → °F and °R Temperature Conversion °F → °C and K**

| K | °C | | °F | °R | K | °C | | °F | °R | K | °C | | °F | °R |
|---|---|---|---|---|---|---|---|---|---|---|---|---|---|---|
| 325.9 | 52.8 | **127** | 260.6 | 720.3 | 357.6 | 84.4 | **184** | 363.2 | 822.9 | 741.5 | 468.3 | **875** | 1607.0 | 2066.7 |
| 326.5 | 53.3 | **128** | 262.4 | 722.1 | 358.2 | 85.0 | **185** | 365.0 | 824.7 | 755.4 | 482.2 | **900** | 1652.0 | 2111.7 |
| 327.0 | 53.9 | **129** | 264.2 | 723.9 | 358.7 | 85.6 | **186** | 366.8 | 826.5 | 769.3 | 496.1 | **925** | 1697.0 | 2156.7 |
| 327.6 | 54.4 | **130** | 266.0 | 725.7 | 359.3 | 86.1 | **187** | 368.6 | 828.3 | 783.2 | 510.0 | **950** | 1742.0 | 2201.7 |
| 328.2 | 55.0 | **131** | 267.8 | 727.5 | 359.8 | 86.7 | **188** | 370.4 | 830.1 | 797.0 | 523.9 | **975** | 1787.0 | 2246.7 |
| 328.7 | 55.6 | **132** | 269.6 | 729.3 | 360.4 | 87.2 | **189** | 372.2 | 831.9 | 810.9 | 537.8 | **1000** | 1832.0 | 2291.7 |
| 329.3 | 56.1 | **133** | 271.4 | 731.1 | 360.9 | 87.8 | **190** | 374.0 | 833.7 | 838.7 | 565.6 | **1050** | 1922.0 | 2381.7 |
| 329.8 | 56.7 | **134** | 273.2 | 732.9 | 361.5 | 88.3 | **191** | 375.8 | 835.5 | 866.5 | 593.3 | **1100** | 2012.0 | 2471.7 |
| 330.4 | 57.2 | **135** | 275.0 | 734.7 | 362.0 | 88.9 | **192** | 377.6 | 837.3 | 894.3 | 621.1 | **1150** | 2102.0 | 2561.7 |
| 330.9 | 57.8 | **136** | 276.8 | 736.5 | 362.6 | 89.4 | **193** | 379.4 | 839.1 | 922.0 | 648.9 | **1200** | 2192.0 | 2651.7 |
| 331.5 | 58.3 | **137** | 278.6 | 738.3 | 363.2 | 90.0 | **194** | 381.2 | 840.9 | 949.8 | 676.7 | **1250** | 2282.0 | 2741.7 |
| 332.0 | 58.9 | **138** | 280.4 | 740.1 | 363.7 | 90.6 | **195** | 383.0 | 842.7 | 977.6 | 704.4 | **1300** | 2372.0 | 2831.7 |
| 332.6 | 59.4 | **139** | 282.2 | 741.9 | 364.3 | 91.1 | **196** | 384.8 | 844.5 | 1005.4 | 732.2 | **1350** | 2462.0 | 2921.7 |
| 333.2 | 60.0 | **140** | 284.0 | 743.7 | 364.8 | 91.7 | **197** | 386.6 | 846.3 | 1033.2 | 760.0 | **1400** | 2552.0 | 3011.7 |
| 333.7 | 60.6 | **141** | 285.8 | 745.5 | 365.4 | 92.2 | **198** | 388.4 | 848.1 | 1060.9 | 787.8 | **1450** | 2642.0 | 3101.7 |
| 334.3 | 61.1 | **142** | 287.6 | 747.3 | 365.9 | 92.8 | **199** | 390.2 | 849.9 | 1088.7 | 815.6 | **1500** | 2732.0 | 3191.7 |
| 334.8 | 61.7 | **143** | 289.4 | 749.1 | 366.5 | 93.3 | **200** | 392.0 | 851.7 | 1116.5 | 843.3 | **1550** | 2822.0 | 3281.7 |
| 335.4 | 62.2 | **144** | 291.2 | 750.9 | 367.0 | 93.9 | **201** | 393.8 | 853.5 | 1144.3 | 871.1 | **1600** | 2912.0 | 3371.7 |
| 335.9 | 62.8 | **145** | 293.0 | 752.7 | 367.6 | 94.4 | **202** | 395.6 | 855.3 | 1172.0 | 898.9 | **1650** | 3002.0 | 3461.7 |
| 336.5 | 63.3 | **146** | 294.8 | 754.5 | 368.2 | 95.0 | **203** | 397.4 | 857.1 | 1199.8 | 926.7 | **1700** | 3092.0 | 3551.7 |
| 337.0 | 63.9 | **147** | 296.6 | 756.3 | 368.7 | 95.6 | **204** | 399.2 | 858.9 | 1227.6 | 954.4 | **1750** | 3182.0 | 3641.7 |
| 337.6 | 64.4 | **148** | 298.4 | 758.1 | 369.3 | 96.1 | **205** | 401.0 | 860.7 | 1255.4 | 982.2 | **1800** | 3272.0 | 3731.7 |
| 338.2 | 65.0 | **149** | 300.2 | 759.9 | 369.8 | 96.7 | **206** | 402.8 | 862.5 | 1283.2 | 1010.0 | **1850** | 3362.0 | 3821.7 |
| 338.7 | 65.6 | **150** | 302.0 | 761.7 | 370.4 | 97.2 | **207** | 404.6 | 864.3 | 1310.9 | 1037.8 | **1900** | 3452.0 | 3911.7 |
| 339.3 | 66.1 | **151** | 303.8 | 763.5 | 370.9 | 97.8 | **208** | 406.4 | 866.1 | 1338.7 | 1065.6 | **1950** | 3542.0 | 4001.7 |
| 339.8 | 66.7 | **152** | 305.6 | 765.3 | 371.5 | 98.3 | **209** | 408.2 | 867.9 | 1366.5 | 1093.3 | **2000** | 3632.0 | 4091.7 |
| 340.4 | 67.2 | **153** | 307.4 | 767.1 | 372.0 | 98.9 | **210** | 410.0 | 869.7 | 1394.3 | 1121.1 | **2050** | 3722.0 | 4181.7 |
| 340.9 | 67.8 | **154** | 309.2 | 768.9 | 372.6 | 99.4 | **211** | 411.8 | 871.5 | 1422.0 | 1148.9 | **2100** | 3812.0 | 4271.7 |
| 341.5 | 68.3 | **155** | 311.0 | 770.7 | 373.2 | 100.0 | **212** | 413.6 | 873.3 | 1449.8 | 1176.7 | **2150** | 3902.0 | 4361.7 |
| 342.0 | 68.9 | **156** | 312.8 | 772.5 | 377.6 | 104.4 | **220** | 428.0 | 887.7 | 1477.6 | 1204.4 | **2200** | 3992.0 | 4451.7 |
| 342.6 | 69.4 | **157** | 314.6 | 774.3 | 383.2 | 110.0 | **230** | 446.0 | 905.7 | 1505.4 | 1232.2 | **2250** | 4082.0 | 4541.7 |
| 343.2 | 70.0 | **158** | 316.4 | 776.1 | 388.7 | 115.6 | **240** | 464.0 | 923.7 | 1533.2 | 1260.0 | **2300** | 4172.0 | 4631.7 |
| 343.7 | 70.6 | **159** | 318.2 | 777.9 | 394.3 | 121.1 | **250** | 482.0 | 941.7 | 1560.9 | 1287.8 | **2350** | 4262.0 | 4721.7 |
| 344.3 | 71.1 | **160** | 320.0 | 779.7 | 408.2 | 135.0 | **275** | 527.0 | 986.7 | 1588.7 | 1315.6 | **2400** | 4352.0 | 4811.7 |
| 344.8 | 71.7 | **161** | 321.8 | 781.5 | 422.0 | 148.9 | **300** | 572.0 | 1031.7 | 1616.5 | 1343.3 | **2450** | 4442.0 | 4901.7 |
| 345.4 | 72.2 | **162** | 323.6 | 783.3 | 435.9 | 162.8 | **325** | 617.0 | 1076.7 | 1644.3 | 1371.1 | **2500** | 4532.0 | 4991.7 |
| 345.9 | 72.8 | **163** | 325.4 | 785.1 | 449.8 | 176.7 | **350** | 662.0 | 1121.7 | 1672.0 | 1398.9 | **2550** | 4622.0 | 5081.7 |
| 346.5 | 73.3 | **164** | 327.2 | 786.9 | 463.7 | 190.6 | **375** | 707.0 | 1166.7 | 1699.8 | 1426.7 | **2600** | 4712.0 | 5171.7 |
| 347.0 | 73.9 | **165** | 329.0 | 788.7 | 477.6 | 204.4 | **400** | 752.0 | 1211.7 | 1727.6 | 1454.4 | **2650** | 4802.0 | 5261.7 |
| 347.6 | 74.4 | **166** | 330.8 | 790.5 | 491.5 | 218.3 | **425** | 797.0 | 1256.7 | 1755.4 | 1482.2 | **2700** | 4892.0 | 5351.7 |
| 348.2 | 75.0 | **167** | 332.6 | 792.3 | 505.4 | 232.2 | **450** | 842.0 | 1301.7 | 1783.2 | 1510.0 | **2750** | 4982.0 | 544[illegible].7 |
| 348.7 | 75.6 | **168** | 334.4 | 794.1 | 519.3 | 246.1 | **475** | 887.0 | 1346.7 | 1810.9 | 1537.8 | **2800** | 5072.0 | 553[illegible].7 |
| 349.3 | 76.1 | **169** | 336.2 | 795.9 | 533.2 | 260.0 | **500** | 932.0 | 1391.7 | 1838.7 | 1565.6 | **2850** | 5162.0 | 562[illegible].7 |
| 349.8 | 76.7 | **170** | 338.0 | 797.7 | 547.0 | 273.9 | **525** | 977.0 | 1436.7 | 1866.5 | 1593.3 | **2900** | 5252.0 | 571[illegible].7 |
| 350.4 | 77.2 | **171** | 339.8 | 799.5 | 560.9 | 287.8 | **550** | 1022.0 | 1481.7 | 1894.3 | 1621.1 | **2950** | 5342.0 | 5801.7 |
| 350.9 | 77.8 | **172** | 341.6 | 801.3 | 574.8 | 301.7 | **575** | 1067.0 | 1526.7 | 1922.0 | 1648.9 | **3000** | 5432.0 | 5891.7 |
| 351.5 | 78.3 | **173** | 343.4 | 803.1 | 588.7 | 315.6 | **600** | 1112.0 | 1571.7 | 2033.2 | 1760.0 | **3200** | 5792.0 | 6251.7 |
| 352.0 | 78.9 | **174** | 345.2 | 804.9 | 602.6 | 329.4 | **625** | 1157.0 | 1616.7 | 2144.3 | 1871.1 | **3400** | 6152.0 | 6611.7 |
| 352.6 | 79.4 | **175** | 347.0 | 806.7 | 616.5 | 343.3 | **650** | 1202.0 | 1661.7 | 2255.4 | 1982.2 | **3600** | 6512.0 | 6971.7 |
| 353.2 | 80.0 | **176** | 348.8 | 808.5 | 630.4 | 357.2 | **675** | 1247.0 | 1706.7 | 2366.5 | 2093.3 | **3800** | 6872.0 | 7331.7 |
| 353.7 | 80.6 | **177** | 350.6 | 810.3 | 644.3 | 371.1 | **700** | 1292.0 | 1751.7 | 2477.6 | 2204.4 | **4000** | 7232.0 | 7691.7 |
| 354.3 | 81.1 | **178** | 352.4 | 812.1 | 658.2 | 385.0 | **725** | 1337.0 | 1796.7 | 2588.7 | 2315.6 | **4200** | 7592.0 | 8051.7 |
| 354.8 | 81.7 | **179** | 354.2 | 813.9 | 672.0 | 398.9 | **750** | 1382.0 | 1841.7 | 2699.8 | 2426.7 | **4400** | 7952.0 | 8411.7 |
| 355.4 | 82.2 | **180** | 356.0 | 815.7 | 685.9 | 412.8 | **775** | 1427.0 | 1886.7 | 2810.9 | 2537.8 | **4600** | 8312.0 | 8771.7 |
| 355.9 | 82.8 | **181** | 357.8 | 817.5 | 699.8 | 426.7 | **800** | 1472.0 | 1931.7 | 2922.0 | 2648.9 | **4800** | 8672.0 | 9131.7 |
| 356.5 | 83.3 | **182** | 359.6 | 819.3 | 713.7 | 440.6 | **825** | 1517.0 | 1976.7 | 3033.2 | 2760.0 | **5000** | 9032.0 | 9491.7 |
| 357.0 | 83.9 | **183** | 361.4 | 821.1 | 727.6 | 454.4 | **850** | 1562.0 | 2021.7 | … | … | … | … | … |

Table converts °C → °F and °R, or °F → °C and K. Find "convert from" temperature in **bold** column and read result from °F and °R or °C and K columns. *Example 1:* 183° C = 361.4° F and 821.1° R. *Example 2:* 183° F = 83.9° C and 357.0 K.

## Units of Velocity and Acceleration

### Table 54. Velocity Conversion Factors

| To Convert ↓ / Multiply By Factor To Obtain → | cm/sec | m/sec | km/hr | ft/sec | ft/min | ft/hr | knot[a] | mile/hr |
|---|---|---|---|---|---|---|---|---|
| cm/sec | 1 | **0.01** | **0.036** | 0.032808 | 1.9685 | 118.110236 | 0.01944 | 0.02237 |
| m/sec | **100** | 1 | **3.6** | 3.2808 | 196.8504 | 11811.0236 | 1.94384 | 2.236936 |
| km/hr | 27.77778 | 0.27778 | 1 | 0.911344 | 54.6806 | 3280.8399 | 0.53995 | 0.621371 |
| ft/sec | 30.48 | 0.3048 | 1.09728 | 1 | **60** | **3600** | 0.59248 | 0.681818 |
| ft/min | 0.5080 | 0.00508 | 0.018288 | 0.016667 | 1 | **60** | $9.8 \times 10^{-3}$ | 0.011364 |
| ft/hr | 0.008467 | $8.47 \times 10^{-5}$ | $3.05 \times 10^{-4}$ | $2.78 \times 10^{-4}$ | 0.01666 | 1 | $1.6 \times 10^{-4}$ | $1.89 \times 10^{-4}$ |
| knot | 51.444 | 0.51444 | 1.852 | 1.687808 | 101.2686 | 6076.11549 | 1 | 1.15167 |
| mile/hr | 44.704 | 0.447040 | 1.609344 | 1.466667 | **88** | **5280** | 0.8689 | 1 |

[a] Knot means nautical miles per hour

Figures in **bold face** indicate the conversion is exact

### Table 55. Acceleration Conversion Factors

| To Convert ↓ / Multiply By Factor To Obtain → | cm/sec² | m/sec² | km/hr² | feet/sec² | ft/hr² | knot/sec | miles/hr² |
|---|---|---|---|---|---|---|---|
| cm/sec² | 1 | **0.01** | 129.6 | 0.0328 | $4.252 \times 10^{5}$ | 0.0194384 | 80.529 |
| m/sec² | **100** | 1 | 12960 | 3.280 | $4.252 \times 10^{7}$ | 1.943844 | 8052.970 |
| km/hr² | 0.007716 | $7.72 \times 10^{-5}$ | 1 | $2.532 \times 10^{-4}$ | 3280.84 | 0.0001499 | 0.6213 |
| ft/sec² | 30.48 | 0.3048 | 3950.20 | 1 | $1.296 \times 10^{7}$ | 0.592483 | 2454.545 |
| ft/hr² | $2.35 \times 10^{-6}$ | $2.35 \times 10^{-5}$ | $3.048 \times 10^{-4}$ | $7.716 \times 10^{-8}$ | 1 | $4.571 \times 10^{-8}$ | $1.893 \times 10^{-4}$ |
| knot/sec | 51.44444 | 0.514444 | 6667.2 | 1.687809 | $2.187 \times 10^{7}$ | 1 | 4142.8060 |
| mile/hr² | 0.0124 | 0.000124 | 1.609 | $4.074 \times 10^{-4}$ | **5280** | 0.00024138 | 1 |

Figures in **bold face** indicate the conversion is exact

## Units of Viscosity

### Table 56a. Oil Viscosity Conversion Factors

| To Convert ↓ | | Multiply By This Factor To Obtain → Poise (*P*) | Centipoise (*Z*) | Reyn (μ) | Stoke (*S*) | Centistoke (*v*) |
|---|---|---|---|---|---|---|
| Poise (*P*) | $\frac{\text{dyne-s}}{\text{cm}^2} = \frac{\text{gram mass}}{\text{cm-s}}$ | 1 | 100 | $1.45 \times 10^{-5}$ | $\frac{1}{\rho}$ | $\frac{100}{\rho}$ |
| Centipoise (*Z*) | $\frac{\text{dyne-s}}{100\ \text{cm}^2} = \frac{\text{gram mass}}{100\ \text{cm-s}}$ | 0.01 | 1 | $1.45 \times 10^{-7}$ | $\frac{0.01}{\rho}$ | $\frac{1}{\rho}$ |
| Reyn (μ) | $\frac{\text{lb force-s}}{\text{in}^2}$ | $6.9 \times 10^{4}$ | $6.9 \times 10^{6}$ | 1 | $\frac{6.9 \times 10^{4}}{\rho}$ | $\frac{6.9 \times 10^{6}}{\rho}$ |
| Stoke (*S*) | $\frac{\text{cm}^2}{\text{s}}$ | ρ | 100ρ | $1.45 \times 10^{-5}\ \rho$ | 1 | 100 |
| Centistoke (*v*) | $\frac{\text{cm}^2}{100\text{s}}$ | 0.01ρ | ρ | $1.45 \times 10^{-7}\ \rho$ | 0.01 | 1 |

### Table 56b. Additional Viscosity Conversion Factors

| Multiply | By | To Obtain | Multiply | By | To Obtain |
|---|---|---|---|---|---|
| centipoise | **0.001** | pascal-second (Pa-s) | pascal-second | **1000** | centipoise |
| centistoke | **0.000001** | meter²/second (m²/s) | pascal-second | **10** | poise |
| stoke | **0.0001** | meter²/second (m²/s) | poise | **0.1** | pascal-second (Pa-s) |

ρ = Specific gravity of the oil.

Figures in **bold face** indicate the conversion is exact

## Units of Moment of Inertia and Momentum

### Table 57. Moment of Inertia Conversion Factors

| Multiply | By | To Obtain |
|---|---|---|
| Moment of Inertia and Section Modulus | | |
| moment of inertia [kg•m²] | 23.73036 | pound-foot² |
| moment of inertia [kg•m²] | 3417.171 | pound-inch² |
| moment of inertia [lb•ft²] | 0.04214011 | kilogram-meter² (kg•m²) |
| moment of inertia [lb•inch²] | 0.0002926397 | kilogram-meter² (kg•m²) |
| moment of section [foot⁴] | 0.008630975 | meter⁴ (m⁴) |
| moment of section [inch⁴] | 41.62314 | centimeter⁴ |
| moment of section [meter⁴] | 115.8618 | foot⁴ |
| moment of section [centimeter⁴] | 0.02402510 | inch⁴ |
| section modulus [foot³] | 0.02831685 | meter³ (m³) |
| section modulus [inch³] | 0.00001638706 | meter³ (m³) |
| section modulus [meter³] | 35.31466 | foot³ |
| section modulus [meter³] | 61,023.76 | inch³ |

### Table 58. Momentum Conversion Factors

| Multiply | By | To Obtain |
|---|---|---|
| Momentum | | |
| kilogram-meter/second | 7.233011 | pound-foot/second |
| kilogram-meter/second | 86.79614 | pound-inch/second |
| pound-foot/second | 0.1382550 | kilogram-meter/second (kg•m/s) |
| pound-inch/second | 0.01152125 | kilogram-meter/second (kg•m/s) |

## Miscellaneous Measuring Units

1 *great gross* = 12 gross = 144 dozen
1 *gross* = 12 dozen = 144 units
1 *dozen* = 12 units

1 *quire* = 24 sheets
1 *ream* = 20 quires = 480 sheets
1 *ream printing paper* = 500 sheets
1 *score* = 20 units

**Ohm's Law.**—The following figure represents basic electrical relationships. This chart has been formatted in such a way that each variable has been related to the other three variables. This figure is simply for reference.

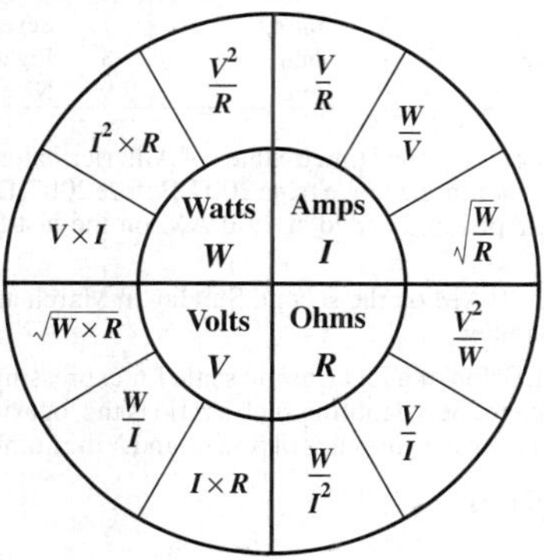


*Key to variables:*
$V$ = Voltage (Volts)
$R$ = Resistance (Ohms)
$I$ = Current (Amps)
$W$ = Power (Watts)

Circular Model of Electrical Relations

**Windchill Temperature.—**Windchill temperature is a measure of the combined cooling effect of wind and temperature. The formula below is used for calculating windchill by the National Weather Service (NWS) in the United States:

$$\text{Windchill Temperature } °F = 35.74 + 0.6215T - 35.75(V^{0.16}) + 0.4275T(V^{0.16})$$

where $T$ = air temperature in °F

$V$ = wind speed in miles per hour (mph) measured at NWS standard height, 33 feet

The formula calculates the chilling effect of wind on the human body at 5 feet above the ground and assumes no influence caused by sunlight (that is, as if the temperature and wind speed measurements were made at night in clear sky conditions). Windchill temperature is only defined for temperatures at or below 50 degrees F and wind speeds above 3 mph. Bright sunshine may increase the windchill temperature by 10 to 18 degrees F.

**Wind Velocity to Pressure**

| *V* | *P* | *V* | *P* | *V* | *P* |
|---|---|---|---|---|---|
| 5 | 0.1 | 35 | 4.9 | 65 | 16.9 |
| 10 | 0.4 | 40 | 6.4 | 70 | 19.6 |
| 15 | 0.9 | 45 | 8.1 | 75 | 22.5 |
| 20 | 1.6 | 50 | 10.0 | 80 | 25.6 |
| 25 | 2.5 | 55 | 12.1 | 100 | 40.0 |
| 30 | 3.6 | 60 | 14.4 | | |

The formula is $P = 0.004V^2$ where $V$ = wind velocity, mph; $P$ = pressure, lbs/sq ft.

**Phonetic Alphabet**

| | | | | | | | |
|---|---|---|---|---|---|---|---|
| A | Alfa | J | Juliett | S | Sierra | 1 | One |
| B | Bravo | K | Kilo | T | Tango | 2 | Two |
| C | Charlie | L | Lima | U | Uniform | 3 | Three |
| D | Delta | M | Mike | V | Victor | 4 | Four |
| E | Echo | N | November | W | Whiskey | 5 | Five |
| F | Foxtrot | O | Oscar | X | X-Ray | 6 | Six |
| G | Golf | P | Papa | Y | Yankee | 7 | Seven |
| H | Hotel | Q | Quebec | Z | Zulu | 8 | Eight |
| I | India | R | Romeo | 0 | Zero | 9 | Nine |

**Daylight Savings Time.—**The Congress of the United States of America extended Daylight Savings Time (DST) by about a month, effective from 2007. Before 2007, DST began at 2:00 AM on the first Sunday in April and ended at 2:00 AM on the last Sunday in October.

Beginning 2007, DST begins at 2:00 AM on the second Sunday in March and ends at 2:00 AM on the first Sunday in November.

**Bel.—**The *bel* is the fundamental division of a logarithmic scale for expressing the ratio of two amounts of power. The number of bels denoting such a ratio is the logarithm to the base 10 of this ratio. Thus, if $P_1$ and $P_2$ are two amounts of power, and $N$ the number of bels denoting their ratio, then $N = \log \frac{P_1}{P_2}$ bels.

The *decibel* is one-tenth of a bel and is commonly abbreviated as db. This unit is used extensively in the measurement of sound volume in telephone and radio transmission and reception, and in noise measurements of various kinds.

# INDEX

## A

# B

# C

# D

# E

# F

# G

# H

# I

# J



# K

# L

# M

# N

## Q



## R

# S

# T

# U

# V

## W

# Y



# Z

## Force

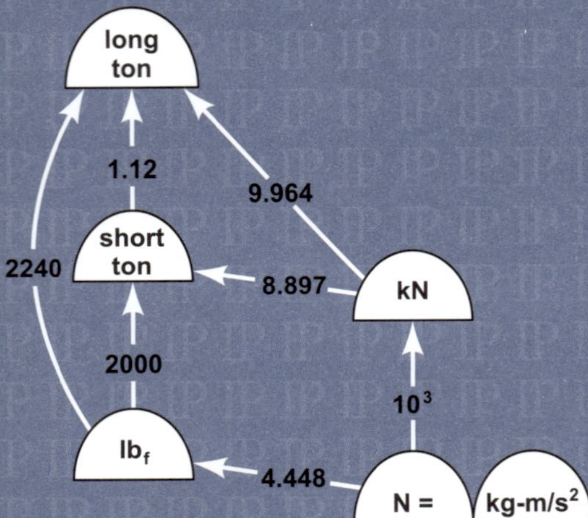


## Pressure

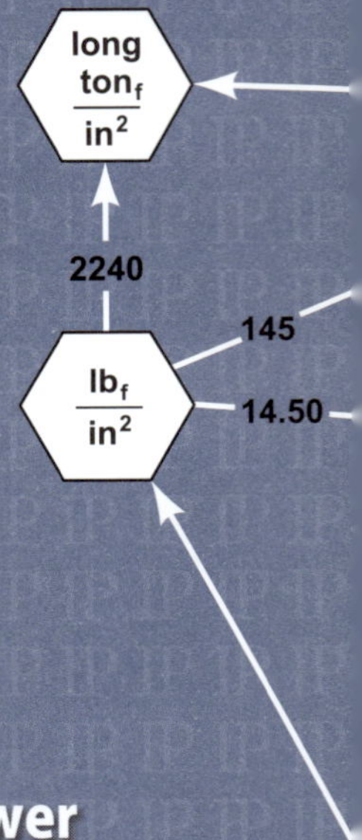


## Energy, Work — Power

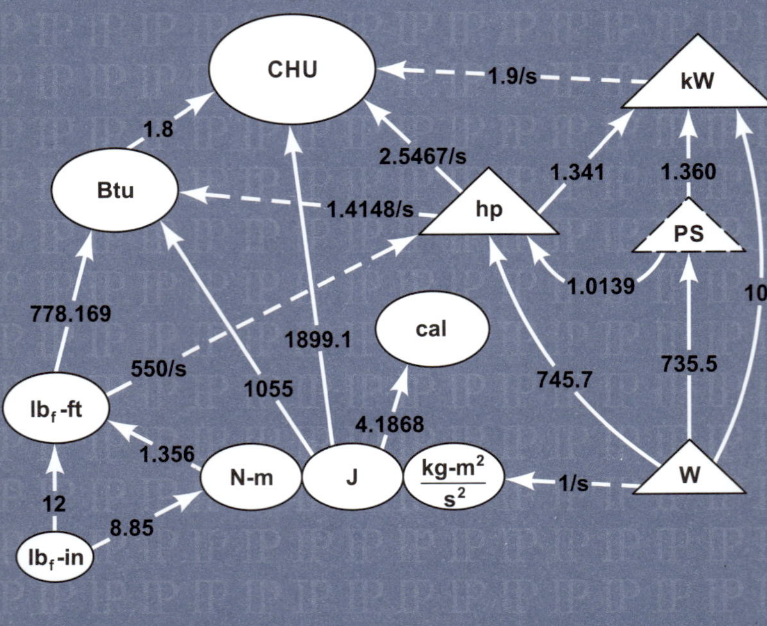

# USAGE

This edition of the *Machinery's Handbook Guide* should be used in conjunction with the *Machinery's Handbook, 32nd Edition*, which is available in two print sizes, as well as digitally.

This *Guide* is designed to aid in the most efficient use of the *Machinery's Handbook* and to reinforce the extensive information it provides. Hundreds of examples and test questions with answer keys are provided to aid in comprehension of selected topics and guide the use of tables, formulas, and general data in the *Handbook*. This companion volume is designed for use, along with the *32nd Edition*, in general, engineering, and trade schools; home study courses; apprenticeships; and other practical and professional enrichment.

To facilitate usage of this excellent learning and teaching text, *Machinery's Handbook Guide* includes updated page references throughout, both to text and other elements within this edition of the *Guide*, and to the corresponding material in the *Machinery's Handbook, 32nd Edition,* and the *Machinery's Handbook 32 Digital Edition*.

# Machinery's Handbook Guide

*A Guide to Using*
*Tables, Formulas, & More*
*in the 32nd Edition*


By John M. Amiss, Franklin D. Jones,
Henry H. Ryffel, and Christopher J. McCauley

Laura Brengelman, Editor



2024
Industrial Press, Inc.

**INDUSTRIAL PRESS, INC.**
1 Chestnut Street
South Norwalk, Connecticut 06854 U.S.A.
Phone: 203-956-5593
Toll-Free: 888-528-7852
Email: info@industrialpress.com

Title: *Machinery's Handbook Guide: A Guide to Using Tables, Formulas, & More in the 32nd Edition*
Authors: John H. Amiss, Franklin D. Jones, Henry H. Ryffel, and Christopher J. McCauley
Library of Congress Control Number: 2023949275



ISBN PRINT: 978-0-8311-5032-7
ISBN ePDF: 978-0-8311-9722-3
ISBN ePub: 978-0-8311-9723-0
ISBN: eMobi: 978-0-8311-9823-7

For ISBNs of other digital products and packages including this title, see books.industrialpress.com.



Printed and bound at Thomson Press

**MACHINERY'S HANDBOOK GUIDE**
**TO USING THE 32ND EDITION**
**Second Printing**

books.industrialpress.com
ebooks.industrialpress.com

# TABLE OF CONTENTS

# TABLE OF CONTENTS

# TABLE OF CONTENTS

# TABLE OF CONTENTS

# TABLE OF CONTENTS

# THE PURPOSE OF THIS BOOK

An engineering handbook is essential equipment for practically all engineers, machine designers, drafters, tool engineers, and skilled mechanics in machine shops and toolrooms. Such a book, with its tables and general data, saves time and labor. To make the best use of any handbook, however, the user must know how to apply the text, tables, formulas, and other data as needed.

One purpose of this *Guide*, which is based on the *Machinery's Handbook*, is to show—through extracted explanations, examples, solutions, and practice questions and answers—typical applications of *Handbook* information, as well as to familiarize engineering students and other users with the *Handbook's* contents. To this end, cross references to material in the *Machinery's Handbook, 32nd Edition,* and in the *Machinery's Handbook 32 Digital Edition* are interspersed throughout this *Guide*. Another objective is to provide test questions and drill work that will enable *Handbook* users, through practice, to both increase their technical knowledge and learn how to obtain needed information quickly and easily.

The *Guide* is available as a paperback, as a standalone eBook, and as part of the *Machinery's Handbook 32 Digital Edition*. In the *Digital Edition* package, the *Guide* includes "live" clickable links to pages, and specific tables, diagrams, and figures in the *Machinery's Handbook* and the *Digital Edition*–only material.

The *Machinery's Handbook*, as with most handbooks, serves as a primary reference, presenting information in condensed form so that a large variety of subjects can be covered in a single volume. Because of this condensed treatment, the practical application of some parts may not always be readily apparent, especially to those who have had little experience in engineering. Therefore, the topical discussions, examples, and questions in this companion volume are not only intended to supplement some of the *Handbook's* content. It is our hope that use of the *Guide* will inform, instruct, and also stimulate interest both in those parts that are referred to regularly and in other interesting and potentially valuable information, even if it is not used that often.

**Laura Brengelman**
**Editor**

## SECTION 1

## DIMENSIONS AND AREAS OF CIRCLES AND SPHERES

*Machinery's Handbook* pages **78** and **86**

Circumferences of circles are used in calculating speeds of rotating machine parts, including drills, reamers, milling cutters grinding wheels, gears, and pulleys. These speeds are variously referred to as surface speed, circumferential speed, and peripheral speed; meaning, for each, the distance that a point on the surface or circumference would travel in one minute. This distance usually is expressed as feet per minute. Circumferences are also required in calculating the circular pitch of gears, laying out involute curves, finding the lengths of arcs, and solving many geometrical problems. Letters from the Greek alphabet frequently are used to designate angles, and the Greek letter π (pi) is always used to indicate the ratio between the circumference and the diameter of a circle:

$$\pi = 3.14159265... = \frac{\text{circumference of circle}}{\text{diameter of circle}}$$

For most practical purposes the value of $\pi = 3.1416$ may be used.

*Example 1:* Find the circumference and area of a circle whose diameter is 8 inches.

On *Handbook* page **78**, the circumference $C$ of a circle is given as $3.1416d$. Therefore, $3.1416 \times 8 = 25.1328$ inches.

On the same page, the area is given as $0.7854d^2$. Therefore, $A$ (area) $= 0.7854 \times 8^2 = 0.7854 \times 64 = 50.2656$ square inches.

*Example 2:* From page **84** of the *Handbook*, the area of a cylindrical surface equals $S = 3.1416dh$. For a diameter of 8 inches and a height of 10 inches, the area is $3.1416 \times 8 \times 10 = 251.328$ square inches.

*Example 3:* For the cylinder in **Example 2** but with the area of both ends included, the total area is the sum of the area found in **Example 2** plus two times the area found in **Example 1**.

Thus, 251.328 + 2 × 50.2656 = 351.8592 square inches. The same result could have been obtained by using the formula for total area: $A = 3.1416 \times d \times (\frac{1}{2}\, d + h) = 3.1416 \times 8 \times (\frac{1}{2} \times 8 + 10) = 351.8592$ square inches. Also see the formulas given page **84** of the *Handbook*.

*Example 4:* If the circumference of a tree is 96 inches, what is its diameter? Since the circumference of a circle $C = 3.1416d$, $96 = 3.1416d$, so $d = 96 \div 3.1416 = 30.558$ inches.

*Example 5:* The tables starting on page **1074** of the *Handbook* provide values of revolutions per minute required to produce various cutting speeds for workpieces of selected diameters. How are these speeds calculated? Cutting speed in feet per minute is calculated by multiplying the circumference in feet of a workpiece by the rpm of the spindle: cutting speed in fpm = circumference in feet × rpm. By transposing this formula as explained in *Formulas and Their Rearrangement* starting on page **13**,

$$\text{rpm} = \frac{\text{cutting speed, fpm}}{\text{circumference in feet}}$$

For a 3-inch diameter workpiece ($\frac{1}{4}$-foot diameter) and for a cutting speed of 40 fpm, rpm = $40 \div (3.1416 \times \frac{1}{4}) = 50.92 = 51$ rpm, approximately, which is the same as the value given on page **1074** of the *Handbook*.

**Area of Square Inscribed in Circle.**—The area of a square inscribed in a circle can be found by drawing a circle and dividing it by two diameters drawn at right angles through the center.

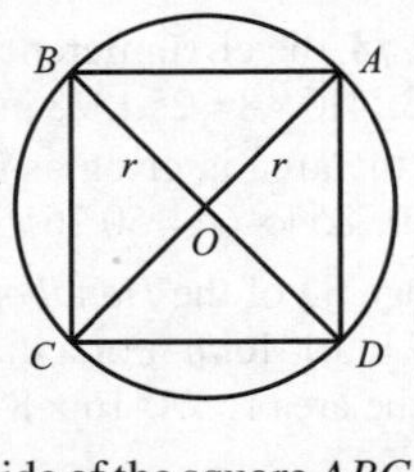


Line $\overline{AB}$ forms one side of the square $ABCD$, and the length of $\overline{AB}$ can be found using the right-triangle formula $a^2 + b^2 = c^2$, *Handbook* page **90**, where lengths $a$ and $b$ are equal to the length of radius $r$, and $c$ is the length of line $\overline{AB}$.

Therefore,

$$c^2 = a^2 + b^2$$

$$(\overline{AB})^2 = r^2 + r^2 = 2r^2$$

Because the sides of a square are of equal length, the area of the square inscribed in the circle is $\overline{AB} \times \overline{AB} = (\overline{AB})^2 = 2r^2$.

**Spheres.—***Handbook* page **86** gives formulas for calculating spherical areas and volumes. Additional formulas are given starting on page **51**.

*Example 6:* If the diameter of a sphere is $24\frac{5}{8}$ inches, what is the volume, given the formula:

$$\text{Volume} = 0.5236d^3$$

The cube of $24\frac{5}{8}$ = 14,932.369; hence, the volume of this sphere = $0.5236 \times 14{,}932.369 = 7818.5$ cubic inches.

*Example 7:* If the sphere in **Example 6** is hollow and $\frac{1}{4}$ inch thick, what is its weight?

The volume can be obtained using the formula for $G_v$ in the table starting on *Handbook* page **51**, and the density of steel from the table of specific gravity on *Handbook* page **375**.

$$G_v = \frac{4\pi}{3}(R_1^3 - R_2^3) = \frac{\pi}{6}(d_1^3 - d_2^3)$$

$$\text{volume of sphere wall} = \frac{\pi}{6}(24.625^3 - 24.125^3) = 466.65 \text{ in}^3$$

$$466.65 \text{ in}^3 \times \frac{1 \text{ ft}^3}{1728 \text{ in}^3} = 0.27 \text{ ft}^3$$

$$\text{weight of sphere} = 491 \times 0.27 = 132.6 \text{ lb}$$

## PRACTICE EXERCISES FOR SECTION 1

(See *Answers to Practice Exercises for Section 1* on page **236**)

1) Find the area and circumference of a circle 10 mm in diameter

2) On *Handbook* page **1076**, for a 5-mm diameter tool or workpiece rotating at 318 rpm, the corresponding cutting speed is given as 5 meters per minute. Check this value.

3) For a cylinder 100 mm in diameter and 10 mm high, what is the surface area not including the top or bottom?

4) A steel column carrying a load of 10,000 pounds has a diameter of 10 inches. What is the pressure on the floor in pounds per square inch (psi)?

5) What is the ratio of the area of a square of any size to the area of a circle having the same diameter as one side of the square?

6) What is the ratio of the area of a circle to the area of a square inscribed in that circle?

7) The drilling speed for cast iron is assumed to be 70 feet per minute. Find the time required to drill two holes in each of 500 castings if each hole has a diameter of $^3/_4$ inch and is 1 inch deep. Use 0.010-inch feed and allow $^1/_4$ minute per hole for setup.

8) Find the weight of a cast-iron column 10 inches in diameter and 10 feet high. Cast iron weighs 0.26 pound per cubic inch.

9) If machine steel has a tensile strength of 55,000 psi, what should be the diameter of a rod to support 36,000 pounds if the safe working stress is assumed to be one-fifth of the tensile strength?

10) Moving the circumference of a 16-inch automobile flywheel 2 inches moves the camshaft through how many degrees? (The camshaft rotates at one-half the flywheel speed.)

11) What is the area within a circle that surrounds a square inscribed in the circle (shaded area)?

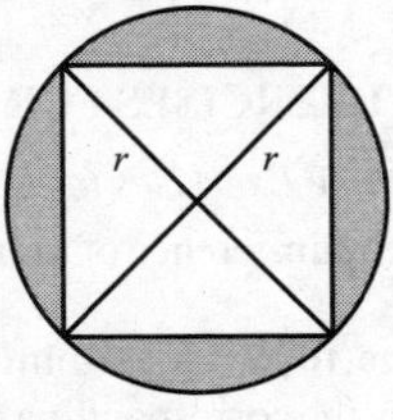


12) A hydraulic cylinder, rated for 2000 psi, is said to have a 5-inch bore and a 2.5-inch piston diameter. If operated at the rated pressure, what forces could be developed on extension and return strokes?

13) Find the number of gallons of oil in a tank 6 feet in diameter and 12 feet long if the tank is in a horizontal position, and the oil measures 2 feet deep.

14) Find the surface areas of the following spheres, the diameters of which are: $1\frac{1}{2}$; $3\frac{3}{8}$; 65; $20\frac{3}{4}$.

15) Find the volume of each sphere in the above exercise.

16) The volume of a sphere is 1,802,725 cubic inches. What are its surface area and diameter?

## SECTION 2

## CHORDS, SEGMENTS, AND HOLE CIRCLES

*Machinery's Handbook* pages **50**, **79**, **85**, **702**, and **708–720**

A chord of a circle is the distance along a straight line from one point to any other point on the circumference. A segment of a circle is that part or area between a chord and the arc it intercepts. The lengths of chords and the dimensions and areas of segments are often required in mechanical work.

**Lengths of Chords.—**The table of chords, *Handbook* **720**, can be applied to a circle of any diameter as explained and illustrated by examples on pages **711** and **720**. The table is given to six decimal places so that it can be used in connection with precision-tool work. Additional related formulas are given on page **702**.

*Example 1:* A circle has 56 equal divisions, and the chordal distance from one division to the next is 2.156 inches. What is the diameter of the circle?

The chordal length in the table for 56 divisions and a diameter of 1 equals 0.05607; therefore, in this example,

$$2.156 = 0.05607 \times \text{Diameter}$$

$$\text{Diameter} = \frac{2.156}{0.05607} = 38.452 \text{ inches}$$

*Example 2:* A drill jig is to have eight holes equally spaced around a circle 6 inches in diameter. How can the chordal distance between adjacent holes be determined when the table on *Handbook* page **720** is not available?

One-half the angle between the radial center lines of adjacent holes = 180 ÷ number of holes. If the sine of this angle is multiplied by the diameter of the circle, the product equals the chordal distance. In this example, we have 180 ÷ 8 = 22.5 degrees. The sine of 22.5 degrees from a calculator is 0.38268; hence, the chordal distance = 0.38268 × 6 = 2.296 inches. The result is the same as would be obtained with the table on *Handbook* page **720** because the figures in the column "Length of the Chord" represent the sines of angles equivalent to 180 divided by the different numbers of spaces.

**Use of the Table of Segments of Circles—*Handbook* page 79.—** This table is of the unit type in that the values all apply to a radius of 1. As explained above the table, the value for any other radius can be obtained by multiplying the figures in the table by the given radius. For areas, the *square* of the given radius is used. Thus, the unit type of table is universal in its application.

*Example 3:* Find the area of a segment of a circle, the center angle of which is 57 degrees, and the radius $2\frac{1}{2}$ inches.

First locate 57 degrees in the center angle column; opposite this figure in the area column will be found 0.0781. Since the area is required, this number is multiplied by the square of $2\frac{1}{2}$. Thus, $0.0781 \times (2\frac{1}{2})^2 = 0.488$ square inch.

*Example 4:* A cylindrical oil tank is $4\frac{1}{2}$ feet in diameter and 10 feet long, and is in a horizontal position. When the depth of the oil is 3 feet, 8 inches, what is the number of gallons of oil?

The total capacity of the tank equals $0.7854 \times (4\frac{1}{2})^2 \times 10 = 159$ cubic feet. One US gallon equals 0.1337 cubic foot (see *Handbook* page **2862**); hence, the total capacity of the tank equals $159 \div 0.1337 = 1190$ gallons.

The unfilled area at the top of the tank is a segment having a height of 10 inches or $\frac{10}{27}$ (0.37037) of the tank radius. The nearest decimal equivalent to $\frac{10}{27}$ in Column *h* of the table starting on page **79** is 0.3707; hence, the number of cubic feet in the segment-shaped space $= (27^2 \times 0.401 \times 120) \div 1728 = 20.3$ cubic feet and $20.3 \div 0.1337 = 152$ gallons. Therefore, when the depth of oil is 3 feet, 8 inches, there are $1190 - 152 = 1038$ gallons. (See also *Handbook* page **68** for additional information on the capacity of cylindrical tanks.)

*Example 5:* Use the tank from **Example 4** and the table on *Handbook* page **79** to estimate the height of fuel in the tank when the tank contains 150 gallons.

When the tank contains 150 gallons, it is $^{150}/_{1190} \times 100 = 12.6\%$ full. In the table starting on page **79**, locate the value in the $^{A}/_{\pi}$ column closest to 12.6 and find the corresponding value of $h$. For $^{A}/_{\pi} = 12.4$, $h = .36392$, and the approximate height of fuel in the tank is $h \times r = 0.36392 \times 2.25 \times 12 = 9.83$ inches.

**Hole Circle Coordinates.—** It is often necessary to manually locate the positions of equally spaced holes around a given hole circle diameter. Pages **712–719** of the *Handbook* provide tables with constants derived using equations on *Handbook* pages **708–710**. These constants can be used to calculate the precise $x$ and $y$ position of each hole.

Type "A" hole circles, as shown in **Fig. 1a** on *Handbook* page **708** and **Fig. 1** below, can be identified by hole number 1 being positioned at the top of the hole circle diameter, with the other holes numbered in a counterclockwise direction. The $x$ and $y$ coordinates of each hole are calculated by multiplying the hole circle diameter by the constants in **Table 1a** on *Handbook* pages **712–713**. The reference point for the $x$ and $y$ coordinates is located at the center of the hole circle diameter.

*Example 6:* **Fig. 1** shows 7 equally spaced holes on a hole circle diameter of 1.125″ with each hole's $x$ and $y$ coordinates given in the included table at right. The position of hole number 7 is calculated using inch coordinates to 3 decimal places:

$x = 0.39092$ (constant) × 1.125 (hole circle diameter) = 0.440

$y = -0.31174$ (constant) × 1.125 (hole circle diameter) = −0.351

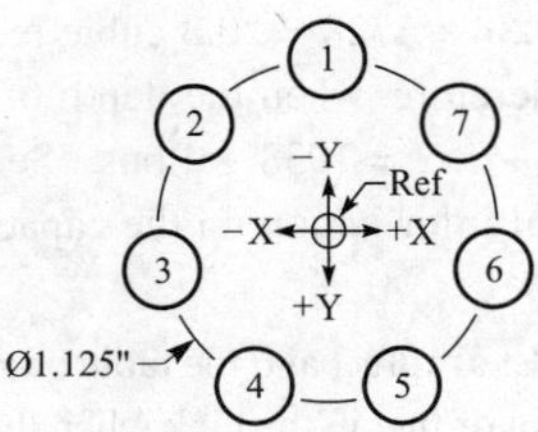


| # | x | y |
|---|---|---|
| 1 | 0.000 | –0.563 |
| 2 | –0.440 | –0.351 |
| 3 | –0.548 | 0.125 |
| 4 | –0.244 | 0.507 |
| 5 | 0.244 | 0.507 |
| 6 | 0.548 | 0.125 |
| 7 | 0.440 | –0.351 |

**Fig. 1.**

Type "A" hole circles, as shown in **Fig. 1b** on *Handbook* page **708** and in **Fig. 3** below, again can be identified by hole number 1 at the top of the hole circle diameter, with the other holes numbered in a counterclockwise direction. The $x$ and $y$ coordinates of each hole are calculated by multiplying the hole circle diameter by the constants in **Table 1b** on *Handbook* pages **714–715**.

As shown in **Fig. 2**, X0, Y0 is located at the intersection of the vertical $x$ coordinate line and the horizontal $y$ coordinate line, with both lines tangent to the hole circle diameter.

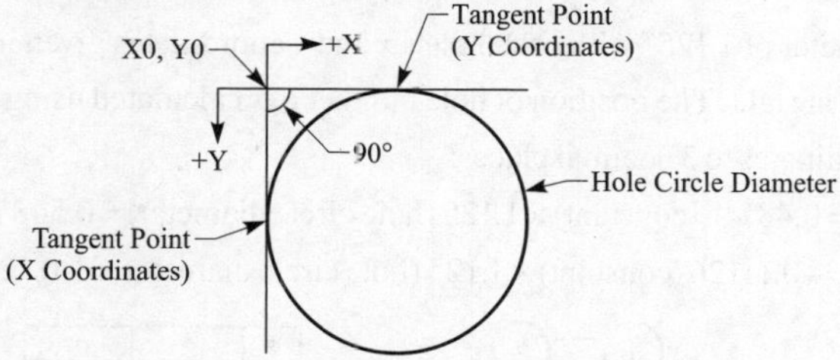


**Fig. 2.**

*Example 7:* **Fig. 3** shows 7 equally spaced holes on a hole circle diameter of 1.125″ with each hole's $x$ and $y$ coordinates given in the table at right. The position of hole number 4 is calculated using inch coordinates to 3 decimal places:

$x = 0.28306$ (constant) × 1.125 (hole circle diameter) = 0.318

$y = 0.95048$ (constant) × 1.125 (hole circle diameter) = 1.069

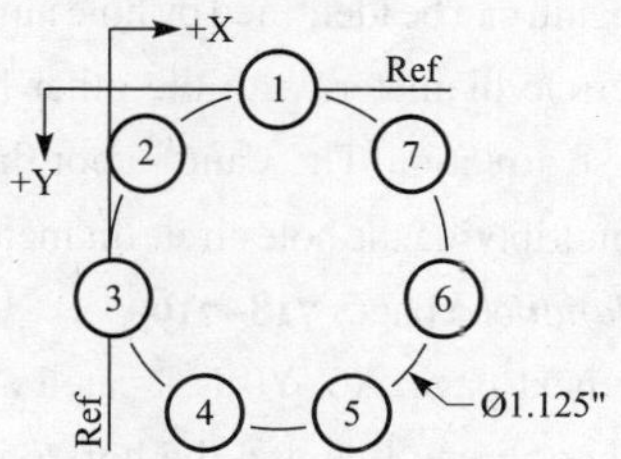


| # | $x$ | $y$ |
|---|---|---|
| 1 | 0.563 | 0.000 |
| 2 | 0.123 | 0.212 |
| 3 | 0.014 | 0.688 |
| 4 | 0.318 | 1.069 |
| 5 | 0.807 | 1.069 |
| 6 | 1.111 | 0.688 |
| 7 | 1.002 | 0.212 |

**Fig. 3.**

Type "B" hole circles, as shown in **Fig. 2a** on *Handbook* page **709** and in **Fig. 4** below, can be identified by hole number 1 being positioned at the top left of the hole circle diameter, with the other holes numbered in a counterclockwise direction. The *x* and *y* coordinates of each hole are calculated by multiplying the hole circle diameter by the constants in **Table 2a** on *Handbook* pages **716–717**. The reference point for the x and y coordinates is located at the center of the hole circle diameter.

*Example 8:* **Fig. 4** shows 7 equally spaced holes on a hole circle diameter of 1.125″ with each hole's *x* and *y* coordinates given in the table at right. The position of hole number 6 is calculated using inch coordinates to 3 decimal places:

$x = 0.48746$ (constant) × 1.125 (hole circle diameter) = 0.548

$y = -0.11126$ (constant) × 1.125 (hole circle diameter) = −0.125

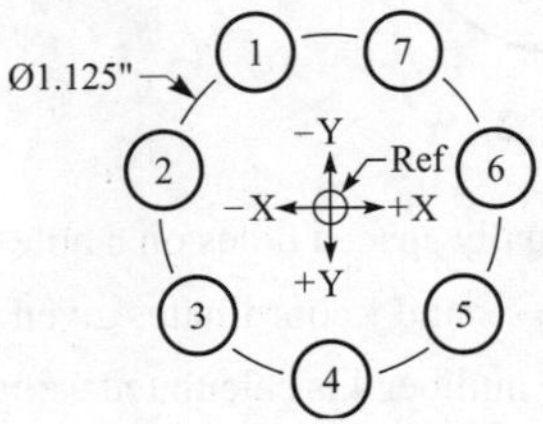


| # | *x* | *y* |
|---|---|---|
| 1 | −0.244 | −0.507 |
| 2 | −0.548 | −0.125 |
| 3 | −0.440 | 0.351 |
| 4 | 0.000 | 0.563 |
| 5 | 0.440 | 0.351 |
| 6 | 0.548 | −0.125 |
| 7 | 0.244 | −0.507 |

**Fig. 4.**

Type "B" hole circles, as shown in **Fig. 2b** on *Handbook* page **709** and in **Fig. 6** on the next page, again can be identified by hole number 1 at the top left of the hole circle diameter, with the other holes numbered in a counterclockwise direction. The *x* and *y* coordinates of each hole are calculated by multiplying the hole circle diameter by the constants in **Table 2b** on *Handbook* pages **718–719**.

As shown in **Fig. 5** on the next page, X0, Y0 is located at the intersection of the vertical *x* coordinate line and the horizontal *y* coordinate line, with both lines tangent to the hole circle diameter.

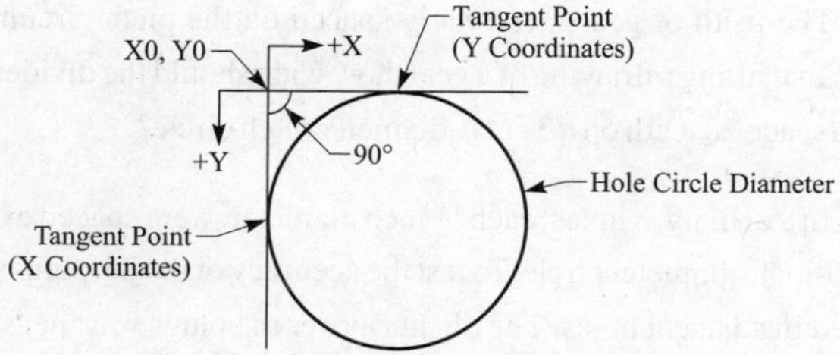


**Fig. 5.**

*Example 9:* **Fig. 6** shows 7 equally spaced holes on a hole circle diameter of 1.125″ with each hole's $x$ and $y$ coordinates given in the table at right. The position of hole number 2 is calculated using inch coordinates to 3 decimal places:

$x = 0.01254$ (constant) × 1.125 (hole circle diameter) = 0.014

$y = 0.38874$ (constant) × 1.125 (hole circle diameter) = 0.437

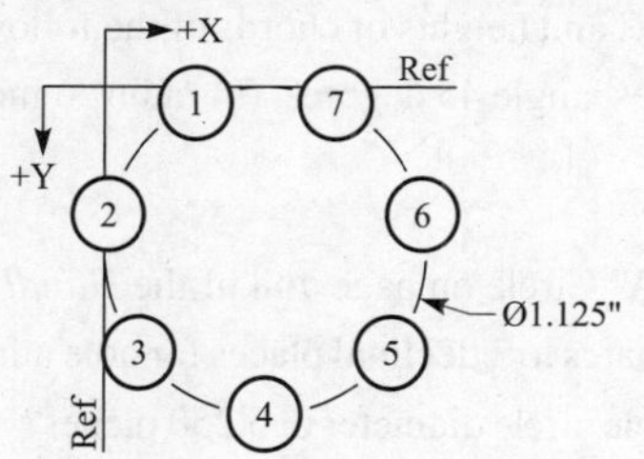


| # | $x$ | $y$ |
|---|---|---|
| 1 | 0.318 | 0.056 |
| 2 | 0.014 | 0.437 |
| 3 | 0.123 | 0.913 |
| 4 | 0.563 | 1.125 |
| 5 | 1.002 | 0.913 |
| 6 | 1.111 | 0.437 |
| 7 | 0.807 | 0.056 |

**Fig. 6.**

## PRACTICE EXERCISES FOR SECTION 2

(See *Answers to Practice Exercises for Section 2* on page **236**)

1) Find the lengths of chords when the number of divisions of a circumference and the radii are as follows: 30 and 4; 14 and $2\frac{1}{2}$; 18 and $3\frac{1}{2}$.

2) Find the chordal distance between the graduations for thousandths on the following dial indicators: (a) Starrett has 100 divisions and $1\frac{3}{8}$-inch dial. (b) Brown & Sharpe has 100 divisions and $1\frac{3}{4}$-inch dial. (c) Ames has 50 divisions and $1\frac{5}{8}$-inch dial.

3) The teeth of gears are evenly spaced on the pitch circumference. In making a drawing of a gear, how wide should the dividers be set to space 28 teeth on a 3-inch-diameter pitch circle?

4) In a drill jig, 8 holes, each $\frac{1}{2}$ inch diameter, were spaced evenly on a 6-inch-diameter circle. To test the accuracy of the jig, plugs were placed in adjacent holes. The distance over the plugs was measured by micrometer. What should be the micrometer reading?

5) In the preceding problem, what should be the distance over plugs placed in alternate holes?

6) What is the length of the arc of contact of a belt over a pulley 2 feet, 3 inches in diameter if the arc of contact is 215 degrees?

7) Find the areas, lengths, and heights of chords of the following segments: (a) radius 2 inches, angle 45 degrees; (b) radius 6 inches, angle 27 degrees.

8) Using **Fig. 1a** Type "A" Circle on page **708** of the *Handbook*, what are the *x* and *y* coordinates to 3 decimal places for hole number 8 in a 9 hole circle, with a hole circle diameter of 5.250 inches?

9) Using **Fig. 1b** Type "A" Circle on page **708** of the *Handbook*, what are the *x* and *y* coordinates to 3 decimal places for hole number 4 in a 5 hole circle, with a hole circle diameter of 3.625 inches?

10) Using **Fig. 2a** Type "B" Circle on page **709** of the *Handbook*, what are the *x* and *y* coordinates to 3 decimal places for hole number 11 in a 17 hole circle, with a hole circle diameter of 4.835 inches?

11) Using **Fig. 2b** Type "B" Circle on page **709** of the *Handbook*, what are the *x* and *y* coordinates to 3 decimal places for hole number 3 in a 5 hole circle, with a hole circle diameter of 2.000 inches?

## SECTION 3

## FORMULAS AND THEIR REARRANGEMENT

*Machinery's Handbook* page **21**

A formula may be defined as a mathematical rule expressed by signs and symbols instead of in actual words. In formulas, letters are used to represent numbers or quantities, the term "quantity" being used to designate any number involved in a mathematical process. The use of letters in formulas, in place of the actual numbers, simplifies the solution of problems and makes it possible to condense into small space the information that otherwise would be imparted by long and cumbersome rules. The figures or values for a given problem are inserted in the formula according to the requirements in each specific case. When the values are thus inserted, in place of the letters, the result or answer is obtained by ordinary arithmetical methods. There are two reasons why a formula is preferable to a rule expressed in words. 1) The formula is more concise, it occupies less space, and it is possible to see at a glance the whole meaning of the rule laid down. 2) It is easier to remember a brief formula than a long rule, and it is, therefore, of greater value and convenience.

*Example 1:* In spur gears, the outside diameter of the gear can be found by adding 2 to the number of teeth and dividing the sum obtained by the diametral pitch of the gear. This rule can be expressed very simply by a formula. Assume that we write $D$ for the outside diameter of the gear, $N$ for the number of teeth, and $P$ for the diametral pitch. Then the formula would be:

$$D = \frac{N+2}{P}$$

This formula reads exactly as the rule given above. It says that the outside diameter ($D$) of the gear equals 2 added to the number of teeth ($N$), and this sum is divided by the pitch ($P$).

If the number of teeth in a gear is 16 and the diametral pitch 6, then simply put these figures in the place of *N* and *P* in the formula, and solve for the outside diameter as in ordinary arithmetic.

$$D = \frac{16+2}{6} = \frac{18}{6} = 3 \text{ inches}$$

*Example 2:* The formula for the horsepower generated by a steam engine is as follows:

$$H = \frac{P \times L \times A \times N}{33{,}000}$$

in which $H$ = indicated horsepower of engine;
$P$ = mean effective pressure on piston in pounds per square inch;
$L$ = length of piston stroke in feet;
$A$ = area of piston in square inches;
$N$ = number of strokes of piston per minute.

Assume that $P = 90$, $L = 2$, $A = 320$, and $N = 110$; what would be the horsepower?

If we insert the given values in the formula, we have:

$$H = \frac{90 \times 2 \times 320 \times 110}{33{,}000} = 192$$

From the examples given, we may formulate the following general rule: *In formulas, each letter stands for a certain dimension or quantity; when using a formula for solving a problem, replace the letters in the formula by the values given for a certain problem, and find the required answer as in ordinary arithmetic.*

**Omitting Multiplication Signs in Formulas.**—In formulas, the sign for multiplication (×) is often left out between letters the values of which are to be multiplied. Thus $AB$ means $A \times B$, and the formula $H = \frac{P \times L \times A \times N}{33{,}000}$ can also be written $H = \frac{PLAN}{33{,}000}$.

If $A = 3$, and $B = 5$, then: $AB = A \times B = 3 \times 5 = 15$.

It is only the multiplication sign (×) that can be thus left out between the symbols or letters in a formula. All other signs must be indicated the same as in arithmetic. However, the product of two numbers *does* require the multiplication sign, such as $23 \times 4$; this also may be written with parentheses, as (23)(4).

As a general rule, the figure in an expression such as "3*A*" is written first and is known as the *coefficient* of *A*.

**Rearrangement of Formulas.**—A formula can be rearranged or "transposed" to determine the values represented by different letters of the formula. To illustrate by a simple example, the formula for determining the speed ($s$) of a driven pulley, when its diameter ($d$) and the diameter ($D$) and speed ($S$) of the driving pulley are known, is as follows: $s = (S \times D)/d$. If the speed of the driven pulley is known, and the problem is to find its diameter or the value of $d$ instead of $s$, this formula can be rearranged or changed. Thus: $d = (S \times D)/s$

Rearranging a formula in this way is governed by four general rules.

*Rule 1.* An independent term preceded by a plus sign (+) may be transposed to the other side of the equals sign (=) if the plus sign is changed to a minus sign (−).

*Rule 2.* An independent term preceded by a minus sign may be transposed to the other side of the equals sign if the minus sign is changed to a plus sign.

As an illustration of these rules, if $A = B - C$, then $C = B - A$, and if $A = C + D - B$, then $B = C + D - A$. That the foregoing are correct may be proved by substituting numerical values for the different letters and then transposing them as shown.

*Rule 3.* A term that multiplies all the other terms on one side of the equals sign may be moved to the other side if it is made to divide all the terms on that side.

As an illustration of this rule, if $A = BCD$, then $A/(BC) = D$ or according to the common arrangement $D = A/(BC)$. Suppose, in the preceding formula, that $B = 10$, $C = 5$, and $D = 3$; then $A = 10 \times 5 \times 3 = 150$ and $150/(10 \times 5) = 3$.

*Rule 4.* A term that divides all the other terms on one side of the equals sign may be moved to the other side if it is made to multiply all the terms on that side.

To illustrate, if $s = SD/d$, then $sd = SD$, and, according to *Rule 3*, $d = SD/s$. This formula may also be rearranged for determining the values of $S$ and $D$; thus $ds/D = S$, and $ds/S = D$.

If, in the rearrangement of formulas, minus signs precede quantities, the signs may be changed to obtain positive rather than minus quantities. All the signs on both sides of the equals sign or on both sides of the equation may be changed. For example, if $-2A = -B + C$, then $2A = B - C$. The same result would be obtained by placing all the terms on the opposite side of the equals sign, which involves changing signs. For instance, if $-2A = -B + C$, then $B - C = 2A$.

**Fundamental Laws Governing Rearrangement.**—After a few fundamental laws that govern any formula or equation are understood, its solution usually is very simple. An equation states that one quantity equals another quantity. So long as both parts of the equation are treated exactly alike, the values remain equal. Thus, in the equation $A = \frac{1}{2}ab$, which states that the area $A$ of a triangle equals one-half the product of the base $a$ times the altitude $b$, each side of the equation would remain equal if we added the same amount: $A + 6 = \frac{1}{2}ab + 6$; or we could subtract an equal amount from both sides: $A - 8 = \frac{1}{2}ab - 8$; or multiply both parts by the same number: $7A = 7(\frac{1}{2}ab)$; or we could divide both parts by the same number, and we would still have a true equation.

One formula for the total area $T$ of a cylinder is:

$$T = 2\pi r^2 + 2\pi rh$$

where: $r$ = radius and $h$ = height of the cylinder. Suppose we want to solve this equation for $h$. Transposing the part that does not contain $h$ to the other side by changing its sign, we get: $2\pi rh = T - 2\pi r^2$. To obtain $h$, we can divide both sides of the equation by any quantity that will leave $h$ on the left-hand side; thus:

$$\frac{2\pi rh}{2\pi r} = \frac{T - 2\pi r^2}{2\pi r}$$

It is clear that, in the left side of the equation, the $2\pi r$ will cancel out, leaving: $h = (T - 2\pi r^2)/(2\pi r)$. The expression $2\pi r$ on the right side cannot be cancelled because it is not an independent factor, since the numerator equals the difference between $T$ and $2\pi r^2$.

*Example 3:* Rearrange the formula for a trapezoid (*Handbook* page **72**) to obtain $h$.

$$A = \frac{(a+b)h}{2}$$

$2A = (a+b)h$ (multiply both sides of the equation by 2)

$(a+b)h = 2A$ (transpose both sides so as to get the multiple of $h$ on the left side)

$\frac{(a+b)h}{a+b} = \frac{2A}{a+b}$ (divide both sides by $a + b$)

$h = \frac{2A}{a+b}$ (cancel $a + b$ from the left side)

*Example 4:* The formula for determining the radius of a sphere (*Handbook* page **86**) is as follows:

$$r = \sqrt[3]{\frac{3V}{4\pi}}$$

Rearrange to obtain a formula for finding the volume $V$.

$r^3 = \frac{3V}{4\pi}$ (cube each side)

$4\pi r^3 = 3V$ (multiply each side by $4\pi$)

$3V = 4\pi r^3$ (transpose both sides)

$\frac{3V}{3} = \frac{4\pi r^3}{3}$ (divide each side by 3)

$V = \frac{4\pi r^3}{3}$ (cancel 3 from the left side)

The procedure has been shown in detail to indicate the underlying principles involved. Rearrangement could be simplified somewhat by direct application of the rules previously given.

To illustrate:

$$r^3 = \frac{3V}{4\pi} \quad \text{(cube each side)}$$

$$4\pi r^3 = 3V \quad \text{(applying } \textit{Rule 4} \text{ move } 4\pi \text{ to the left side)}$$

$$\frac{4\pi r^3}{3} = V \quad \text{(move 3 to the left side—} \textit{Rule 3}\text{)}$$

This final equation would, of course, be reversed to locate $V$ on the left side of the equals sign, as this is the usual position for whatever letter represents the quantity or value to be determined.

*Example 5:* Determine the diameter of cylinder and length of stroke of a steam engine to deliver 150 horsepower. The mean effective steam pressure is 75 pounds, and the number of strokes per minute is 120. The length of the stroke is to be 1.4 times the diameter of the cylinder.

First, insert the known values into the horsepower formula (**Example 2**, page **14**):

$$150 = \frac{75 \times L \times A \times 120}{33{,}000} = \frac{3 \times L \times A}{11}$$

The last expression is found by cancellation.

Assume now that the diameter of the cylinder in inches equals $D$. Then, $L = 1.4D/12 = 0.117D$ according to the requirements in the problem; the divisor 12 is introduced to change the inches to feet ($L$ being in feet in the horsepower formula). The area $A = D^2 \times 0.7854$. If we insert these values in the last expression in our formula, we have:

$$150 = \frac{3 \times 0.117D \times 0.7854D^2}{11} = \frac{0.2757D^3}{11}$$

$$0.2757D^3 = 150 \times 11 = 1650$$

$$D^3 = \frac{1650}{0.2757} \qquad D = \sqrt[3]{\frac{1650}{0.2757}} = \sqrt[3]{5984.8} = 18.15$$

The diameter of the cylinder should be about 18¼ inches, and the length of the stroke $18.15 \times 1.4 = 25.41$, or about 25½ inches.

**Solving Equations or Formulas by Trial.**—One of the equations used for spiral gear calculations, when the shafts are at right angles, the ratios are unequal, and the center distance must be exact, is as follows:

$$R\sec\alpha + \csc\alpha = \frac{2CP_n}{n}$$

In this equation

$R$ = ratio of number of teeth in large gear to number in small gear

$C$ = exact center distance

$P_n$ = normal diametral pitch

$n$ = number of teeth in small gear

The exact spiral angle $\alpha$ of the large gear is found by trial using the equation just given.

Equations of this form are solved by trial by selecting an angle assumed to be approximately correct and inserting the secant and cosecant of this angle in the equation, adding the values thus obtained, and comparing the sum with the known value to the right of the equals sign in the equation. An example will show this more clearly. By using the problem given in the *Example* near the top of *Handbook* page **2282** to illustrate, $R = 3$; $C = 10$; $P_n = 8$; $n = 28$.

Hence, the whole expression $\frac{2CP_n}{n} = \frac{2 \times 10 \times 8}{28} = 5.714$ from which it follows that:

$$R\sec\alpha + \csc\alpha = 5.714$$

In the problem given, the spiral angle required is 45 degrees. The spiral gears, however, would not meet all the conditions given in the problem if the angle could not be slightly modified. To determine whether the angle should be greater or smaller than 45 degrees, insert the values of the secant and cosecant of 45 degrees in the formula. The secant of 45 degrees is 1.4142, and the cosecant is 1.4142. Then,

$$3 \times 1.4142 + 1.4142 = 5.6568$$

The value 5.6568 is too small, as it is less than 5.714, which is the required value. Hence, try 46 degrees. The secant of 46 degrees is 1.4395, and the cosecant, 1.3902.

Then,

$$3 \times 1.4395 + 1.3902 = 5.7087$$

Obviously, an angle of 46 degrees is also too small. Proceed, therefore, to try an angle of 46 degrees, 30 minutes. This angle will be found too great. Similarly 46 degrees, 15 minutes, if tried, will be found too great, and by repeated trials it will finally be found that an angle of 46 degrees, 6 minutes, the secant of which is 1.4422, and the cosecant, 1.3878, meets the requirements. Then,

$$3 \times 1.4422 + 1.3878 = 5.7144$$

which is as close to the required value as necessary.

In general, when an equation must be solved by the trial-and-error method, all the known quantities may be written on the right side of the equal sign and all the unknown quantities on the left side. A value is assumed for the unknown quantity. This value is substituted in the equation, and all the values thus obtained on the left side are added. In general, if the result is greater than the values on the right side, the assumed value of the unknown quantity is too great. If the result obtained is smaller than the sum of the known values, the assumed value for the unknown quantity is too small. By thus adjusting the value of the unknown quantity until the left side of the equation with the assumed value of the unknown quantity will just equal the known quantities on the right side of the equal sign, the correct value of the unknown quantity may be determined.

**Derivation of Formulas.—**Most formulas in engineering handbooks are given without showing how they have been derived or originated, because engineers and designers usually want only the final results; moreover, such derivations would require considerable additional space, and they belong in textbooks rather than in handbooks, which are primarily works of reference. Although *Machinery's Handbook* contains thousands of formulas derived from published Standards and other special formulas, it is apparent that no handbook can include every kind of formula, because a great many formulas apply only to local designing or manufacturing problems. Such special formulas are derived by engineers and designers for their own use. The exact methods of deriving formulas are based upon mathematical principles as they are related to the particular factors that apply. A few examples will be given to show how several different types of special formulas have been derived.

*Example 6:* The problem is to deduce the general formula for finding the point of intersection of two tapers with reference to measured diameters on those tapers. In **Fig. 1**,

$L$ = distance between the two measured diameters, $D$ and $d$

$X$ = the required distance from one measured diameter to the intersection of tapers

$a$ = angle of long taper as measured from center line

$a_1$ = angle of short taper as measured from center line

Then,

$$E = \frac{D-d}{2} = Z + Y$$

$$Z = (L - X)\tan a_1$$

$$Y = X \tan a$$

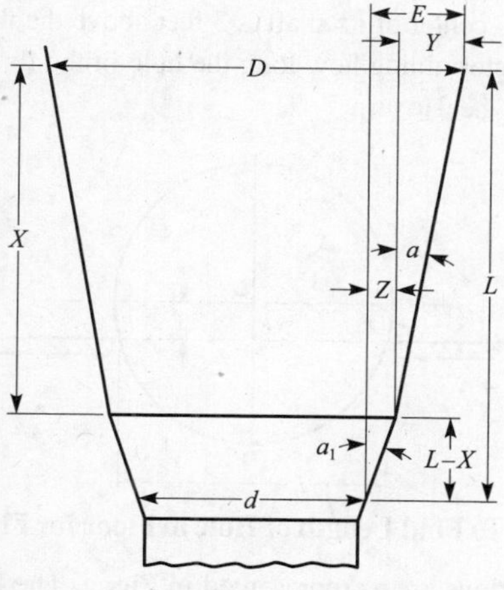


**Fig. 1. To Find Dimension *X* from a Given Diameter *D* to the Intersection of Two Conical Surfaces**

Therefore:

$$\frac{D-d}{2} = (L - X)\tan a_1 + X \tan a$$

and

$$D - d = 2\tan a_1 (L - X) + 2X \tan a \qquad (1)$$

But

$$2\tan a_1 = T_1 \qquad \text{and} \qquad 2\tan a = T$$

in which $T$ and $T_1$ represent the long and short tapers per inch, respectively.

Therefore, from **Equation (1)**,

$$D - d = T_1 (L - X) + TX$$

$$D - d = T_1 L - T_1 X + TX$$

$$X(T_1 - T) = T_1 L - (D - d)$$

$$X = \frac{T_1 L - (D - d)}{T_1 - T}$$

*Example 7:* A flywheel is 16 feet in diameter (outside measurement), and the center of its shaft is 3 feet above the floor. Derive a formula for determining how long the hole in the floor must be to permit the flywheel to turn.

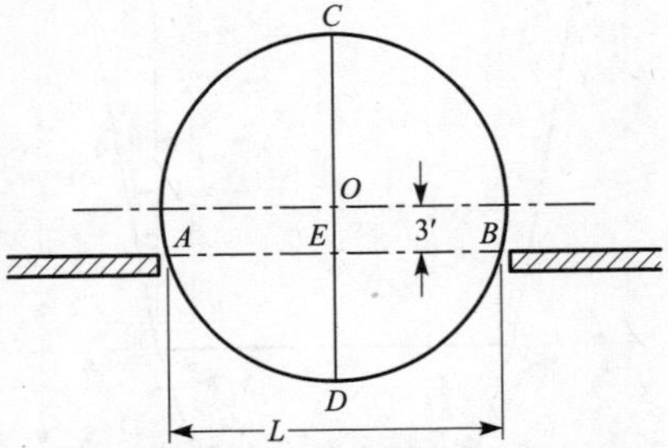


**Fig. 2. To Find Length of Hole in Floor for Flywheel**

The conditions are as represented in **Fig. 2**. The line $AB$ is the floor level and is a chord of the arc $ABD$; it is parallel to the horizontal diameter through the center $O$. $CD$ is the vertical diameter and is perpendicular to $AB$. It is shown in geometry that the diameter $CD$ bisects the chord $AB$ at the point of intersection $E$.

One of the most useful theorems of geometry is that when a diameter bisects a chord, the product of the two parts of the diameter is equal to the square of one half the chord; in other words,

$(AE)^2 = ED \times EC$. If $AB$ is represented by $L$ and $OE$ by $a$, $ED = r - a$ and $EC = r + a$, in which $r$ = the radius $OC$; hence,

$$\left(\frac{L}{2}\right)^2 = (r-a)(r+a) = r^2 - a^2$$

$$\frac{L}{2} = \sqrt{r^2 - a^2} \text{ and } L = 2\sqrt{r^2 - a^2}$$

By substituting the values given,

$$L = 2\sqrt{8^2 - 3^2} = 14.8324 \text{ feet} = 14 \text{ feet, } 10 \text{ inches}$$

The length of the hole, therefore, should be at least 15 feet, to allow sufficient clearance.

**Empirical Formulas.**—Many formulas used in engineering calculations cannot be established fully by mathematical derivation but must be based upon actual tests instead of relying upon mere theories or assumptions that might introduce excessive errors. These formulas are known as *empirical formulas*. Usually such a formula contains a constant (or constants) that represents the result of the tests; consequently, the value obtained by the formula is consistent with these tests or with actual practice.

A simple example of an empirical formula is found in the following formula for the breaking load of chain in pounds, given on *Machinery's Handbook 32 Digital Edition* page **3793**:

$$W = 54{,}000D^2$$

This particular formula contains the constant 54,000, which was established by tests, and the formula is used to obtain the breaking load of wrought-iron crane chains to which a factor of safety of 3, 4, or 5 is then applied to obtain the working load.

*Handbook* page **289** contains an example of an empirical formula based upon experiments made with power-transmitting shafts. This formula gives the diameter of shaft required to prevent excessive twisting during transmission of power.

**Parentheses.**—Two important rules related to the use of parentheses are based upon the principles of positive and negative numbers:

1) If a parenthesis is preceded by a + sign, it may be removed, and the terms within the parentheses will retain their signs.

$$a + (b - c) = a + b - c$$

2) If a parenthesis is preceded by a – sign, it may be removed, but only if the signs preceding each of the terms inside of the parentheses are changed (+ changed to –, and – to +). Multiplication and division signs are not affected.

$$a - (b - c) = a - b + c$$

$$a - (-b + c) = a + b - c$$

Knowledge of algebra is not necessary to make successful use of formulas of the general type such as are found in engineering handbooks; it is only necessary to understand thoroughly the use of letters or symbols in place of numbers, and to be well versed in the methods, rules, and processes of ordinary arithmetic. Knowledge of algebra becomes necessary only where a general rule or formula that gives the answer to a problem directly is not available. In other words, algebra is useful in *developing* or originating a general rule or formula, but the formula can be *used* without recourse to algebraic processes.

**Constants.**—A constant is a value that does not change or is not variable. Constants at one stage of a mathematical investigation may be variables at another stage, but an *absolute constant* has the same value under all circumstances. The ratio of the circumference to the diameter of a circle, or 3.1416, is a simple example of an absolute constant. In a common formula used for determining the indicated horsepower of a reciprocating steam engine, the product of the mean effective pressure in pounds per square inch, the length of the stroke in feet, the area of the piston in square inches, and the number of piston strokes per minute is divided by the constant 33,000, which represents the number of foot-pounds of work per minute equivalent to 1 horsepower. Constants occur in many mathematical formulas.

**Mathematical Signs and Abbreviations.**—Every division of mathematics has its traditions, customs, and signs that are frequently of ancient origin. Hence, we encounter Greek letters in many problems (see *Handbook* page **2836**) where it would seem that English letters would do as well or better. Many of the signs and abbreviations on *Handbook* pages **2836** to **2838** will be used frequently. They should, therefore, be understood.

**Conversion Tables.**—It may sometimes be necessary to convert English units of measurement into metric units and vice versa. The tables provided at the back of the *Handbook* will be found useful in this connection. A table of metric conversion factors is also provided in this book starting on page **268**.

## PRACTICE EXERCISES FOR SECTION 3

(See *Answers to Practice Exercises for Section 3* on page **237**)

1) An approximate formula for determining the horsepower $H$ of automobile engines is: $H = D^2SN/3$, where $D$ = diameter of bore, inches; $S$ = length of stroke, inches; and $N$ = number of cylinders. Find the horsepower of the following automobile engine: a) bore, $3\frac{1}{2}$ inches; stroke, $4\frac{1}{4}$ inches; cylinders, 6. b) By using the reciprocal of 3, how could this formula be stated?

2) Using the right-angle triangle formula: $C = \sqrt{a^2 + b^2}$, where $a$ = one leg of the triangle, $b$ = the other leg, and $C$ = the hypotenuse, find the hypotenuse of a right triangle whose legs are 16 inches and 63 inches.

3) The formula for finding the blank diameter of a cylindrical shell is: $D = \sqrt{d \times (d + 4h)}$, where $D$ = blank diameter, $d$ = diameter of the shell, and $h$ = height of the shell. Find the diameter of the blank to form a cylindrical shell of 3 inches in diameter and 2 inches high.

4) If $D$ = diagonal of a cube, $d$ = diagonal of face of a cube, $s$ = side of a cube, and $V$ = volume of a cube, then $d = \sqrt{2D^2/3}$; $s = \sqrt{D^2/3}$; and $V = s^3$. Find the side, volume of a cube, and diagonal of the face of a cube if the diagonal of the cube is 10.

5) The area of an equilateral triangle equals one fourth of the square of the side times the square root of 3, or

$$A = (S^2/4)\sqrt{3} = 0.43301S^2$$

Find the area of an equilateral triangle whose side is 14.5 inches.

6) The formula for the volume of a sphere is: $4\pi r^3/3$ or $\pi d^3/6$. What constants may be used in place of $4\pi/3$ and $\pi/6$?

7) The formula for the volume of a solid ring is $2\pi^2 Rr^2$, where $r$ = radius of cross section and $R$ = radius from the center of the ring to the center of the cross section. Find the volume of a solid ring made from 2-inch round stock if the mean diameter of the ring is 6 inches.

8) Explain these signs: $\pm$, $>$, $<$, $\sin^{-1}a$, tan, $\angle$, $\sqrt[4]{\ }$, log, $\theta$, $\beta$, ::

9) The area $A$ of a trapezoid (see *Handbook* page 72) is found by the formula:

$$A = \frac{(a+b)h}{2}$$

Transpose the formula for determining width $a$.

10) $R = \sqrt{r^2 + s^2/4}$; solve for $r$.

11) $P = 3.1416\sqrt{2(a^2 + b^2)}$; solve for $a$.

12) $\cos A = \sqrt{1 - \sin^2 A}$; solve for $\sin A$.

13) $a/\sin A = b/\sin B$; solve for $a, b, \sin A, \sin B$.

## SECTION 4

## SPREADSHEET CALCULATIONS

Spreadsheet computer programs or spreadsheets are versatile, powerful tools for doing repetitive or complicated algebraic calculations. They are used in diverse technological fields including manufacturing, design, and finance. Spreadsheets blend the power of high-level computer languages with the simplicity of hand calculators. They are ideal for doing "what-if" calculations, such as changing a problem's parameters and comparing the new result to the initial answer. The visual nature of spreadsheets allows the user to grasp quickly and simultaneously the interaction of many variables in a given problem.

Generally only 5 to 10 percent of a spreadsheet program functionality needs to be understood to begin doing productive spreadsheet calculations. Since the underlying concepts of all spreadsheets are the same, it is easy to transfer this basic understanding from one spreadsheet program to another with a minimal learning curve. Only a small percentage of the actual spreadsheet commands will be covered in this section, but understanding these core concepts will allow the reader to do productive work immediately.

There are many varieties of spreadsheet programs. It is impossible to cover all these spreadsheet programs individually in this brief overview. The formulas listed below are for conceptual understanding and may not work when plugged directly into a particular program. The user should consult the spreadsheet's manual or built-in help system for examples. Generally for any given topic a spreadsheet's help system will list a properly constructed example of what the user is trying to do. The reader can use this as a guide and template to get started.

**Basic Spreadsheet Concepts.**—To begin using spreadsheets, several key spreadsheet concepts must be understood.

*Cell Content:* The basic calculating unit of all spreadsheets are cells. Cells may either contain formulas, which are discussed further on; or numbers, words, dates, percentages, and currency. A cell normally has to be formatted using the spreadsheet's cell format commands to display its contents correctly. The formatting usually does not affect the internal representation of the cell, e.g. the actual value of the number. For example, a cell formatted as a *percentage* such as "12%" would actually contain a value of "0.12" in the cell. If the cell were left unformatted "0.12" would be displayed. A cell formatted for currency would display "3.4" as "$3.40,"

| *Number* | *Currency* | *Text* | *Percentage* |
|---|---|---|---|
| 12.7854 | $12.05 | Feed Rate | 12% *or* 0.12 |

Cells containing numbers may be formatted to display an arbitrary level of precision. Again the displayed precision has no affect on actual calculations. For example, the contents of a particular cell containing "3.1415" could be formatted to display "3.141" or "3.14" or "3". Regardless of what is displayed "3.1415" will be used internally by the program for all calculations that refer to that cell.

Formatting cells, while not absolutely necessary, is usually a good idea for several reasons. Formatted cells help others understand your spreadsheet. "12%" is easily identifiable as an interest rate, "0.12" is not. Formatting can also help to avoid input mistakes in large spreadsheets, such as accidently placing an interest rate percentage in a payment currency-formatted cell. The interest rate will be displayed as "$0.12," immediately telling the user something is wrong. For quick "back-of-the-envelope" calculations, formatting can be dispensed with to save time.

*Cell Address:* In addition to content, cells also have addresses. A cell address is created by combining the column and row names of that cell. In the spreadsheet in **Table 1a**, *Parts* would have an address of *A1*, *Machine 2* would be *C1*, and "13.76" would be *B3*. Spreadsheets use these cell addresses to combine and manipulate the cell contents using formulas.

**Table 1a. Machine Cost Spreadsheet (Display)**

| | A | B | C | D |
|---|---|---|---|---|
| 1 | Parts | Machine 1 | Machine 2 | Total |
| 2 | Motor | 12.89 | $18.76 | $31.65 |
| 3 | Controls | 13.76 | $19.56 | $33.32 |
| 4 | Chassis | 15 | $21.87 | $36.87 |
| 5 | Rebate | −7.5 | −$10.00 | −$17.50 |
| 6 | Total | 34.15 | $50.19 | $84.34 |

*Formulas:* Instead of containing values, a cell may have a formula assigned to it. Spreadsheets use these formulas to manipulate, combine, and chain cells mathematically. The specific format or syntax for properly constructing a formula varies from spreadsheet to spreadsheet. The two most common formula construction techniques are illustrated using the spreadsheet in **Table 1b**.

**Table 1b. Machine Cost Spreadsheet (Formulas)**

| | A | B | C | D |
|---|---|---|---|---|
| 1 | Parts | Machine 1 | Machine 2 | Total |
| 2 | Motor | 12.89[a] | $18.76 | = +B2+C2[b]<br>= $31.65 |
| 3 | Controls | 13.76[a] | $19.56 | = Sum(B3:C3)[b]<br>= $33.32 |
| 4 | Chassis | 15[a] | $21.87 | = Sum(B4:C4)[b]<br>= $36.87 |
| 5 | Rebate | −7.5[a] | −$10.00 | = Sum (B5:C5)[b]<br>= −$17.50 |
| 6 | Total | = +B2+B3+B4 +B5[b]<br>= Sum(B2:B5)<br>= 34.15[a] | = Sum(C2:C5)[b]<br>= $50.19 | = Sum(D2:D5)[b,c]<br>= Sum(B6:C6)[d]<br>= $84.34 |

[a] This cell is unformatted. This does not change the value of the intermediate calculations or final results.

[b] Cells cannot contain more than one value or formula. The double values and formulas listed in this cell are for illustration only and would not be allowed in a working spreadsheet.

[c] Sum of the machine *Parts*.

[d] Sum of *Machine 1* and *Machine 2*.

*Cell by Cell:* Each cell is added, subtracted, multiplied, or divided individually. For example in **Table 1b**, the total cost of *Machine 1* would be the values of each individual part cost in column *B* added vertically in cell *B6*.

*B6*=+ B2+B3+B4+B5=$34.15

*Sum Function:* For long columns or rows of cells, individual cell addition becomes cumbersome. Built-in functions simplify manipulating multiple cells by applying a specific function, like addition, over a range of cells. All spreadsheets have a summation or *Sum* function that adds all the cells that are called out in the function's address range. The *Sum* function adds cells horizontally or vertically. Again in **Table 1b**, the total cost of *Machine 1* using the *Sum* function would be:

*B6*=Sum(B2:B5)=$34.15

Either method yields the same result and may be used interchangeably. The cell-by-cell method must be used for cells that are not aligned horizontally or vertically. The compact *Sum* method is useful for long chains or ranges of cells. Spreadsheets contain many, many built-in functions that work with math, text strings, dates, etc.

*Adding Formulas:* Cells containing formulas can themselves be combined, i.e. formulas containing formulas. In **Table 1b**, the total of the *motor* parts (row 2) for *Machine 1* and *Machine 2*, is calculated by the formula in cell *D2*, the total of the *control* parts *D3*, the total of all *chassis* parts *D4*, and the total of the *rebates* in *D5*. These formulas are summed together vertically in the first formula in cell *D6* to get the total cost of all the parts, in this case $84.34. Note that a spreadsheet cell may only contain one formula or value. The multiple formulas in *D6* are for illustration only.

Alternatively, the cost of *Machine 1*, *B6* and *Machine 2*, *C6* could be added together horizontally to get the cost of all the machines, which, in this case, equals the cost of all parts $84.34. This illustrates that it is possible to set up a spreadsheet to find a solution in more than one way. In this case the total cost of all machines was calculated by adding the parts' subtotals or the individual machines' subtotals.

*Positive and Negative:* Spreadsheets usually display negative numbers with a minus sign (–) in front of them. Sometimes a negative cell number may be formatted to display parentheses around a number instead of a minus sign. For example, –12.874 would be equivalent to (12.874). As with general formatting, this has no effect on the actual cell value.

It is extremely important to treat positive and negative cell values consistently. For example, cell values representing a loan amount of $22,000 and a payment of $500 might be entered as +$22,000 and –$500 if you are receiving a loan or –$22,000 and +$500 if you are loaning the money to someone. Switching one of the signs will create an error in the spreadsheet.

Generally it doesn't matter how positive and negative numbers are assigned, so long as the user is consistent throughout the spreadsheet and the people using the spreadsheet understand the positive-negative frame of reference. Failure to be consistent will lead to errors in your results.

*Basic Mathematical Operators:* Spreadsheets generally use the following conventions for basic mathematical operators. These operators may be applied to cell values or cell formulas.

**Basic Spreadsheet Mathematical Operators**

| Function | Operator | Function | Operator |
|---|---|---|---|
| Add | + | Divide | / |
| Subtract | – | Square | ^2 |
| Multiply | * | Square Root | ^.5 |
| Grouping | ((5+B2)/A2) – (6*((9+16)^0.5)) | | |

Consult the spreadsheet's help system to properly construct other mathematical operations, such as sine, cosine, tangent, logarithms, etc.

*Built-In Functions:* As previously mentioned, spreadsheets contain many built-in functions to aid the user in setting up equations. For example, most spreadsheets have built-in interest functions sometimes referred to as Time Value of Money or *TVM* equations. Generally the names of the variables in the built-in equations do not always exactly match the generally accepted mathematical names used in particular field, such as economics.

To illustrate this point, let's compare the *TVM* terms found in *Variables* on *Handbook* page **138** to the variable names found in a spreadsheet's future value (*FV*) built-in function. Then we'll redo the *Compound Interest* problem found on *Handbook* page **139**.

*Example 1, Compound Interest:* At 10 percent interest compounded annually for 3 years, a principal amount *P* of $1000 becomes a sum $F = 1000(1 + 10/100)^3 = \$1,331.93$.

To solve this problem using a spreadsheet, use the future value, *FV* built-in equation *FV*(*Rate*, *Nper*, *Pmt*, *Pv*) where

*FV* = *F* or the future value of the amount owed or received.

*Rate* = *I* or nominal annual interest rate per period. In this yearly case, divide by 1; for monthly payments, divide by 12.

*Nper* = *n* or number of interest periods—in this case, 3. If the interest were compounded monthly, then *Nper* = 3 years × 12 periods /yr. = 36 periods

*Pmt* = *R* or the payments made or received. For a compound interest loan, *Pmt* = $0.00

*PV* = *P* or principal amount lent or borrowed.

To solve, plug in the appropriate values. Again note that leaving column *B* unformatted or formatting column *C* makes no difference for the final answer, but does make it easier to understand the spreadsheet values.

**Table 2. Compound Interest Calculations Spreadsheet**

| | A | B | C | D |
|---|---|---|---|---|
| 1 | | Value | Value | |
| 2 | *Rate* | .1[a] | 10%[b] | |
| 3 | *Nper* | 3[a] | 3[b] | |
| 4 | *Pmt* | 0[a] | $0.00[b] | |
| 5 | *PV* | −1000[a,c] | −$1,000.00[b,c] | |
| 6 | *FV* | = *FV*(B2,B3,B4,B5) = 1,331.93[a] | = $1,331.93[b] | |

[a] Unformatted cell.
[b] Formatted cell.
[c] This number is negative because you are loaning the money out to collect interest.

**Advanced Spreadsheet Concepts.**—One of the great strengths of spreadsheets is their ability to quickly and easily do what-if calculations. The two key concepts required to do this are cell content and formula "copying and pasting" and "relative and absolute" cell addressing.

*Copying and Pasting:* Spreadsheets allow cells to be moved, or copied and pasted into new locations. Since a chain of cells can represent a complete problem and solution, copying these chains and pasting them repeatedly into adjacent areas allows several experimental what-if scenarios to be set up. It is then easy to vary the initial conditions of the problem and compare the results side by side. This is illustrated in the following example.

*Example 2, What-If Compound Interest Comparison:* Referring back to the compound interest problem in **Example 1**, compare the effects of different interest rates from three banks using the same loan amount and loan period. The banks offer a 10%, 11%, and 12% rate. In the spreadsheet, enter 10%, 11%, and 12% into *B2*, *C2*, and *D2*, respectively. Instead of typing in the initial amounts and formulas for the other values for other banks, type them in once in *B3*, *B4*, *B5*, and *B6*. Copy these cells one column over, into column *C* and column *D*. The spreadsheet will immediately solve all three interest rate solutions.

**Table 3. Interest Calculations Spreadsheet Using Relative Addressing**

| | A | B | C | D | E |
|---|---|---|---|---|---|
| 1 | Term | Bank A | Bank B | Bank C | |
| 2 | *Rate* | 10% | 11% | 12% | 4 cells above "relative" to **E5** |
| 3 | *Nper* | 3 | 3 | 3 | 3 |
| 4 | *Pmt* | $0.00 | $0.00 | $0.00 | 2 |
| 5 | *PV* | –$1,000 | –$1,000 | –$1,000 | 1 |
| 6 | *FV* | =*FV*(**B2**,**B3**, **B4**,**B5**) =$1,331.93 | =*FV*(**C2**,**C3**, **C4**,**C5**) =$1,367.63 | =*FV*(**D2**,**D3**, **D4**,**D5**) =$1,404.93 | Cell **E5** |

*Relative versus Absolute Address:* In row 6 of **Table 3**, notice how the *FV* function cell addresses changed as they were copied from column *B* and pasted into the columns *C* and *D*. The formula cell addresses were changed from **B** to **C** in column *C* and **B** to **D** in column *D*. This is known as *relative addressing*. Instead of the formulas pointing to the original or *absolute* locations in the *B* column they were changed by the spreadsheet program as they were pasted to match a cell location with the same relative distance and direction as the original cell. To clarify, in column *E*, the cell *E2* is 4 cells up relative to *E5*. Relative addressing while pasting allows spreadsheets users to easily copy and paste multiple copies of a series of calculations. This easy what-if functionality is a cornerstone of spreadsheet usefulness.

*Absolute Addressing:* For large complicated spreadsheets the user may want to examine several what-if conditions while varying one basic parameter. For this type of problem it is useful to use *absolute addressing*. There are several formats for creating absolute addresses. Some spreadsheets require a "$" be placed in front of each address. The relative address "*B2*" would become an absolute address when entered as "*$B$2*". When a formula with an absolute address is copied and pasted the copied formula refers to the same cell as the original. The power of this is best illustrated by an example. Formulas can use a combination of relative and absolute addressing, as in $B2 or B$2, where the absolute portion of the cell address is preceded by the $ symbol.

*Example 3, Absolute and Relative Addressing:* Suppose that in **Example 1** we wanted to find the future value of $1,000, $1,500 and $2,000 for 10% and 11% interest rates. Using the previous example as a starting point we enter values for *Rate*, *Nper*, *Pmt*, and *Pv*. We also enter the function *FV* into cell *B6*. This time we enter the absolute address *$B$2* for the *Rate* variable. Now when we copy cell *B6* into *C6* and *D6*, the Rate variable continues to point to cell B2 (absolute addresses) while the other variables *Nper*, *Pmt*, and *Pv* point to locations in columns *C* and *D* (relative addresses).

From the **Table 4a** we find the future value for different starting amounts for a 10% rate. We change cell *B2* from 10% to 11%, and the spreadsheet updates all the loan calculations based on the new interest rate.

**Table 4a. 10% Interest Rate Calculations Spreadsheet Using Absolute Addressing**

| | A | B | C | D |
|---|---|---|---|---|
| 1 | Term | Loan Amount A | Loan Amount B | Loan Amount C |
| 2 | *Rate* | 10% | | |
| 3 | *Nper* | 5 | 4 | 3 |
| 4 | *Pmt* | $0.00 | $0.00 | $0.00 |
| 5 | *PV* | –$1,000 | –$1,500 | –$2,000 |
| 6 | *FV* | =*FV*(**$B$2**,B3, B4,B5) =$1,610.51 | =*FV*(**$B$2**,C3, C4,C5) =$2,196.15 | =*FV*(**$B$2**,D3, D4,D5) =$2,662.00 |

These new values are displayed in **Table 4b**. All we had to do was change one cell to try a new what-if. By combining relative and absolute addresses, we were able to compare the effects of three different loan amounts using two interest rates by changing one cell value.

**Table 4b. 11% Interest Rate Calculations Spreadsheet Using Absolute Addressing**

| | A | B | C | D |
|---|---|---|---|---|
| 1 | Term | Loan Amount A | Loan Amount B | Loan Amount C |
| 2 | *Rate* | 11% | | |
| 3 | *Nper* | 5 | 4 | 3 |
| 4 | *Pmt* | $0.00 | $0.00 | $0.00 |
| 5 | *PV* | –$1,000 | –$1,500 | –$2,000 |
| 6 | *FV* | =*FV*(**$B$2**,B3, B4,B5) =$1,685.06 | =*FV*(**$B$2**,C3, C4,C5) =$2,277.11 | =*FV*(**$B$2**,D3, D4,D5) =$2,735.26 |

*Other Capabilities:* In addition to mathematical manipulations, most spreadsheets can create graphs, work with dates and text strings, link results to other spreadsheets, and create conditional programming algorithms, to name a few advanced capabilities. While these features may be useful in some cases, many real-world problems can be solved with spreadsheets using simple operators and concepts.

## PRACTICE EXERCISES FOR SECTION 4

(See *Answers to Practice Exercises for Section 4* on page **237**)

1) Use a spreadsheet to format a cell in different ways. Enter the number 0.34 in the first cell. Using the spreadsheet menu bar and online help, change the formatting of the cell to display this number as a percentage, a dollar amount, and then back to a general number.

2) Create a multiplication table. Enter the numbers 1 through 10 in the first column (A), and in the first row (1). In cell B2 enter the formula B1 × A2. In cell B3, enter the formula B1 × A3. Repeat this operation down the column. The last cell in column B should have the formula B1 × A10. For column C, formulas are C1 × A2, C1 × A3, C1 × A4, ..., C1 × A10. Complete entering formulas for columns C through J in the same manner. Use your spreadsheet to look up the value of 2 × 2, 5 × 7, and 8 × 9.

| | A | B | C | D | E | F | G | H | I | J |
|---|---|---|---|---|---|---|---|---|---|---|
| 1 | 1 | 2 | 3 | 4 | 5 | 6 | 7 | 8 | 9 | 10 |
| 2 | 2 | | | | | | | | | |
| 3 | 3 | | | | | | | | | |
| 4 | 4 | | | | | | | | | |
| 5 | 5 | | | | | | | | | |
| 6 | 6 | | | | | | | | | |
| 7 | 7 | | | | | | | | | |
| 8 | 8 | | | | | | | | | |
| 9 | 9 | | | | | | | | | |
| 10 | 10 | | | | | | | | | |

3) Write a single formula that can be entered into cell B2 of the previous problem and then copied into each blank cell (B2...J10) to complete the multiplication table (*Hint:* Use relative and absolute addressing). Explain how the formula changes as it is copied into different locations.

4) Use a spreadsheet to recreate **Table 1b** on page **29**. Make sure to format currency cells where required.

5) Using your spreadsheet's online help for guidance, recreate the compound interest calculation in **Table 2** on page **32** using the spreadsheet's future value interest rate function. Make sure to format currency and percentage cells correctly.

6) Using the spreadsheet you created in the previous question, calculate the future value of $2,500 compounded annually for 12 years at 7.5% interest. What would the future value be if the interest was compounded monthly?

7) An equation for windchill temperature is given on *Handbook* page **2884**. Build a windchill temperature table. What is the windchill temperature for: a) 15° F and 20 mph wind, b) 55° F and 20 mph wind, c) 5° F and 5 mph wind?

8) The windchill equation on *Handbook* page **2884** contains two sets of parentheses. Are these parentheses required when entering the formula into a spreadsheet? Why?

## SECTION 5

## CALCULATIONS INVOLVING LOGARITHMS

*Machinery's Handbook* page **31**

The purpose of logarithms is to facilitate and shorten calculations involving multiplication and division, obtaining the powers of numbers, and extracting the roots of numbers. By means of logarithms, long multiplication problems become simple addition of logarithms; cumbersome division problems are easily solved by simple subtraction of logarithms; the fourth root or, say, the 10.4th root of a number can be extracted easily; and any number can be raised to the twelfth power as readily as it can be squared.

The availability of inexpensive computers and handheld calculators has eliminated much of the need to use logarithms for such purposes; there are, however, many applications in which the logarithm of a number is used to obtain the solution of a problem. For example, in the section *Compound Interest* on page **139** of the *Handbook*, there is a formula to find the number of years $n$ required for a sum of money to grow a specified amount. The example accompanying the formula shows the calculations that include the logarithms 3, 2.69897, and 0.025306, which correspond to the numbers 1000, 500, and 1.06, respectively. These logarithms were obtained directly from a handheld calculator and are the common or *Briggs* system of logarithms, which have a base 10. Any other system of logarithms, such as that of base $e$ ($e$ = 2.71828…), could have been used with the same result. Base $e$ logarithms are sometimes called *natural logarithms*.

There are other types of problems in which logarithms of a specific base, usually 10 or $e$, must be used to obtain the correct result. On the logarithm keys of most calculators, base 10 logs are identified by the word "log" and those of base $e$ are referred to as "ln".

In the common or Briggs system of logarithms, which is used ordinarily, the base of the logarithms is 10; that is, the logarithm is the *exponent* that would be affixed to 10 to produce the number corresponding to the logarithm. To illustrate:

$$\text{Logarithm of } 10 = 1 \text{ because } 10^1 = 10$$

$$\text{Logarithm of } 100 = 2 \text{ because } 10^2 = 100$$

$$\text{Logarithm of } 1000 = 3 \text{ because } 10^3 = 1000$$

In each case, it will be seen that the exponent of 10 equals the logarithm of the number. The logarithms of all numbers between 10 and 100 equal 1 plus some fraction. For example: The logarithm of 20 = 1.301030.

The logarithms of all numbers between 100 and 1000 = 2 plus some fraction; between 1000 and 10,000 = 3 plus some fraction; and so on. The tables of logarithms in engineering handbooks give only this fractional part of a logarithm, which is called the *mantissa*. The whole number part of a logarithm, which is called the *characteristic*, is not given in the tables because it can easily be determined by simple rules. The logarithm of 350 is 2.544068. The whole number 2 is the characteristic (see *Handbook* page **32**) and the decimal part 0.544068, the mantissa, is found in the table (*Machinery's Handbook 32 Digital Edition* page **3175**). Logarithms can most easily be found by using a scientific calculator.

**Principles Governing the Application of Logarithms.**—When logarithms are used, the product of two numbers can be obtained as follows: Add the logarithms of the two numbers; the sum equals the logarithm of the product. For example: The logarithm of 10 (commonly abbreviated log 10) equals 1; log 100 = 2; 2 + 1 = 3, which is the logarithm of 1000 and the product of 100 × 10.

Logarithms would not be used for such a simple example of multiplication; these particular numbers are employed merely to illustrate the principle involved.

For division by logarithms, subtract the logarithm of the divisor from the logarithm of the dividend to obtain the logarithm of the quotient. To use another simple example, divide 1000 by 100 using logarithms. The respective logarithms of these numbers are 3 and 2, and the difference equals the logarithm of the quotient 10.

In using logarithms to raise a number to any power, simply multiply the logarithm of the number by the exponent of the number; the product equals the logarithm of the power. To illustrate, find the value of $10^3$ using logarithms. The logarithm of 10 = 1 and the exponent is 3; hence, $3 \times 1 = 3$ = log of 1000; hence, $10^3 = 1000$.

To extract any root of a number, merely divide the logarithm of this number by the index of the root; the quotient is the logarithm of the root. Thus, to obtain the cube root of 1000, divide 3 (log 1000) by 3 (index of root); the quotient equals 1, which is the logarithm of 10. Therefore,

$$\sqrt[3]{1000} = 10$$

Logarithms are of great value in many engineering and shop calculations because they make it possible to readily solve cumbersome or difficult problems that otherwise would require complicated formulas or higher mathematics. Keep constantly in mind that logarithms are merely exponents. Any number might be the base of a system of logarithms. Thus, if 2 were selected as a base, then the logarithm of 256 would equal 8 because $2^8 = 256$. However, unless otherwise mentioned, the term "logarithm" is used to apply to the common or Briggs system, which has 10 for a base.

The tables of common logarithms are found on *Machinery's Handbook 32 Digital Edition* pages **3175** and **3176**. The natural logarithms, *Machinery's Handbook 32 Digital Edition* pages **3177** and **3178**, are based upon the number 2.71828. These logarithms are used in higher mathematics and also in connection with the formula to determine the mean effective pressure of steam in engine cylinders.

**Finding the Logarithms of Numbers.**—There is nothing complicated about the use of logarithms, but a little practice is required to readily locate the logarithm of a given number or to reverse this process and find the number corresponding to a given logarithm. These corresponding numbers are sometimes called *antilogarithms*.

Carefully study the rules for finding logarithms given on *Handbook* page **31** and in the section *Logarithms* starting on page **3169**, *Machinery's Handbook 32 Digital Edition*. Although the characteristic or whole-number part of a logarithm is easily determined, the following table will assist the beginner in memorizing the rules.

**Sample Numbers and Their Characteristics**

| Number | Characteristic | Number | Characteristic |
|---|---|---|---|
| 0.008 | −3 | 88 | 1 |
| 0.08 | −2 | 888 | 2 |
| 0.8 | −1 | 8888 | 3 |
| 8.0 | 0 | 88888 | 4 |

For common logarithms: For numbers greater than or equal to 1, the characteristic is 1 less than the number of places to the left of the decimal point. For numbers smaller than 1 and greater than 0, the characteristic is negative and its numerical value is 1 more than the number of zeros immediately to the right of the decimal point.

*Example 1:* Use of the table of numbers and their characteristics.

What number corresponds to the log $\overline{2}.55145$? (Note that the bar over a digit is used when the characteristic is negative and the mantissa is positive.) Find 0.551450 in the log tables to correspond to 356. From the table of characteristics, note that a −2 characteristic calls for one zero in front of the first integer; hence, 0.0356 is the number corresponding to the log $\overline{2}.55145$. Evaluating logarithms with negative characteristics is explained more thoroughly later.

*Example 2:* Find the logarithm of 46.8.

The mantissa of this number is 0.670246. When there are two whole-number places, the characteristic is 1; hence, the log of 46.8 is 1.670246.

After a little practice with the above table, you will become familiar enough with the rules governing the characteristic so that referring to the table will no longer be necessary.

**Obtaining More Accurate Values than Given Directly by Tables.**—The method of using the tables of logarithms to obtain more accurate values than are given directly, by means of interpolation, is explained on *Machinery's Handbook 32 Digital Edition* page **3170**. These instructions should be read carefully in order to understand the procedure in connection with the following example:

*Example 3:*

$$\frac{76824 \times 52.076}{435.21} =$$

$$\log 76824 = 4.88549 \qquad \log \text{numerator} = 6.60213$$

$$\log 52.076 = \underline{1.71664} \qquad -\log 435.21 = \underline{2.63870}$$

$$\log \text{numerator} = 6.60213 \qquad \log \text{quotient} = 3.96343$$

The number corresponding to the logarithm 3.96343 is 9192.4. The logarithms just given for the dividend and divisor are obtained by interpolation from the log table in the following manner:

In the log tables on *Machinery's Handbook 32 Digital Edition* page **3175**, find the mantissa corresponding to the first three digits of the number 76824 and the mantissa of the next higher 3-digit number in the table, 769. The mantissa of 76824 is the mantissa of 768 plus $^{24}/_{100}$ times the difference between the mantissas of 769 and 768.

$$\text{Mantissa } 769 = .885926$$

$$\text{Mantissa } 768 = \underline{.885361}$$

$$\text{Difference} = .000565$$

Thus, log 76824 = 0.24 × 0.000565 + log 76800 = 4.885497. The characteristic 4 is obtained as previously illustrated in the table on page **41**. By again using interpolation, as explained in the *Handbook*, the corrected mantissas are found for the logarithms of 52.076 and 435.21.

After obtaining the logarithm of the quotient, which is 3.96343, interpolation is again used to determine the corresponding number more accurately than would be possible otherwise. The mantissa .96343 (see *Machinery's Handbook 32 Digital Edition* page **3176**) is found in the table between 0.963316 and 0.963788, the mantissas corresponding to 919 and 920, respectively.

$$0.963788 - 0.963316 = 0.000472$$

$$0.96343 - 0.963316 = 0.000114$$

Note that the first line gives the difference between the two mantissas nearest .96343, and the second line gives the difference between the mantissa of the quotient and the nearest smaller mantissa in the *Handbook* table. The characteristic 3 in the quotient 3.96343 indicates 4 digits before the decimal point in the answer, and thus the number sought is $9190 + {}^{114}/_{472}\,(9200 - 9190) = 9192.4$.

**Changing Form of Logarithm Having Negative Characteristic.—** The characteristic is frequently rearranged for easier manipulation. Note that 8 − 8 is the same as 0; hence, the log of 4.56 could be stated as 0.658965 or 8.658965 − 8. Similarly, the log of 0.075 = $\bar{2}$.875061 or 8.875061 − 10 or 7.875061 − 9. Any similar arrangement could be made, as determined by case in multiplication or division.

*Example 4:*

$$\sqrt[3]{0.47} = ?$$

$$\log 0.47 = \bar{1}.672098 \text{ or } 8.672098 - 9$$

$$\log \sqrt[3]{0.47} = (8.672098 - 9) \div 3 = 2.890699 \div 3 = \bar{1}.89070$$

In the first line above, 9 − 9 was added to log 0.47 because 3 (the index of the root) will divide evenly into 9; 11 − 12 or 5 − 6 could have been used as well. (Refer also to **Example 7** on *Machinery's Handbook 32 Digital Edition* page **3172**. The procedure differs from that just described, but the same result is obtained.)

To find the number corresponding to $\bar{1}$.89070, locate the nearest mantissa. Mantissa .890421 is found in the table and corresponds to 777. The $\bar{1}$ characteristic indicates that the decimal point immediately precedes the first integer; therefore, the number equivalent to the log 1.89070 is 0.777. If desired, additional accuracy can be obtained by interpolation, as explained previously. Thus, $\sqrt[3]{0.47} = 0.777$ .

**Cologarithms.—**The cologarithm of a number is the logarithm of the reciprocal of that number. The resulting *cologs* have no properties different from those of ordinary logarithms, but they enable division to be done by addition because the addition of a colog is the same as the subtraction of a logarithm.

*Example 5:* $\dfrac{742 \times 6.31}{55 \times 0.92} = ?$

Note that this problem could be stated as 742 × 6.31 × 1/55 × 1/0.92. Then the logs of each number could be added because the process is one of multiplication only.

Log 1/55 can be obtained readily in two ways:

$$\begin{array}{rrll} \log 1/55 = & \log 1 - \log 55 & & \\ \log 1 = & 10.000000 & -10 & \\ -\log 55 = & -1.740363 & & \\ \hline & 8.259637 & -10 & = \bar{2}.259637 \end{array}$$

or

$$\log 1/55 = \log 0.0181818 \text{ (see reciprocals)}$$
$$\log 0.0181818 = \bar{2}.25964$$

This number $\bar{2}.259637$ is called the colog of 55; hence, to find the colog of any number, subtract the logarithm of that number from 10.000000 – 10; this is the same as dividing 1 by the number whose colog is sought.

To find the colog of 0.92. subtract log 0.92 (or $\bar{1}.963788$) from 10.00000 – 10; thus:

$$\begin{array}{rrll} & 10.000000 & -10 & \\ \log 0.92 = & \bar{1}.963788 & & \\ \hline \text{colog } 0.92 = & 9.963788 & -10 & = 0.036212 \end{array}$$

(In subtracting negative characteristics, change the sign of the lower one and add.)

Another method is to use log 0.92 = $\bar{1}.96379$ or 9.96379 – 10, and proceed as above:

$$\begin{array}{rrrl} & & 10.000000 & -10 \\ \log 0.92 = & \bar{1}.963788 = & 9.963788 & -10 \\ \hline \text{colog } 0.92 & = & 0.036212 & \end{array}$$

**Example 5** may then be solved by adding logs; thus:

$$\begin{array}{lcr} \log 742 & = & 2.870404 \\ \log 6.31 & = & 0.800029 \\ \text{colog } 55 & = & \bar{2}.259637 \\ \text{colog } 0.92 & = & 0.036212 \\ \hline \log \text{quotient} & = & 1.966282 \end{array}$$

The number corresponding to the logarithm of the quotient = 92.53.

*Example 6:* The initial absolute pressure of the steam in a steam engine cylinder is 120 pounds per square inch (psi); the length of the stroke is 26 inches; the clearance, $1\frac{1}{2}$ inches; and the period of admission, measured from the beginning of the stroke, 8 inches. Find the mean effective pressure.

The mean effective pressure is found by the formula:

$$p = \frac{P(1 + \log_e R)}{R}$$

in which $p$ = mean effective pressure in pounds per square inch

$P$ = initial absolute pressure in pounds per square inch

$R$ = ratio of expansion, which in turn is found from the formula:

$$R = \frac{L + C}{l + C}$$

in which $L$ = length of stroke in inches

$l$ = period of admission in inches

$C$ = clearance in inches

The given values are $P = 120$, $L = 26$, $l = 8$, and $C = 1$. By inserting the last three values in the formula for $R$, we have:

$$R = \frac{26 + 1\frac{1}{2}}{8 + 1\frac{1}{2}} = \frac{27.5}{9.5} = 2.89$$

We now insert the value of $P$ and the found value of $R$ in the formula for $p$:

$$p = \frac{120(1 + \log_e 2.89)}{2.89}$$

The logarithm may be found from tables or a calculator. The natural logarithm for 2.89 is 1.061257 (see *Machinery's Handbook 32 Digital Edition* page **3177**). Inserting this value in the formula, we have:

$$p = \frac{120(1 + 1.061257)}{2.89} = \frac{120 \times 2.061257}{2.89} = 85.6 \text{ lb/in}^2$$

## PRACTICE EXERCISES FOR SECTION 5

(See *Answers to Practice Exercises for Section 5* on page **238**)

1) What are the rules governing the characteristics?

2) Find the mantissas of: 762, 478, 26, 0.0098, 6743, and 24.82.

3) What are the characteristics of the numbers just given?

4) What numbers could correspond to the following mantissas: 0.085016, 0.88508, and 0.22763?

5) (a) If the characteristic of each of the mantissas just given is 1, what would the corresponding numbers be? (b) Using the following characteristics (2, 0, 3) for each mantissa, find the antilogarithms or corresponding numbers.

6) log 765.4 = ?; log 87.2 = ?; log 0.00874 = ?

7) What are the antilogarithms of: 2.89894, 1.24279, 0.18013, and 2.68708?

8) Find by interpolation the logarithm of: 75186 and 42.037.

9) Find the numbers corresponding to the following logarithms: 1.82997 and 0.67712.

10) $(2.71)^5 = ?$ $(4.23)^{2.5} = ?$

11) $\sqrt{97.65} = ?$ $\sqrt[5]{4687} = ?$ $\sqrt[2.3]{44.5} = ?$

12) $\dfrac{62876 \times 54.2 \times 0.0326}{1728 \times 231} = ?$

13) $(2/19)^7 = ?$

14) $(9.16)^{2.47} = ?$

15) $\sqrt[3]{\dfrac{(75)^2 \times (5.23)^{2/3}}{0.00036 \times \sqrt{51.7}}} =$

16) The area of a circular sector $= 0.008727ar^2$ where $a$ = angle in degrees and $r$ = radius of the circle. Find the area of a circular sector the radius of which is 6.25 inches and the central angle is 42° 15′.

17) The diameter of a lineshaft carrying pulleys may be found from the formula: $d = \sqrt[3]{53.5\text{hp/rpm}}$. Find the diameter of shafting necessary to transmit 50 hp at 250 rpm.

18) The horsepower of a steam engine is found from the formula: $hp = PLAN/33000$, where

$P$ = mean effective pressure in pounds per square inch
$L$ = length of stroke in feet
$A$ = area of piston in square inches
$N$ = number of strokes per minute = revolutions per minute $\times 2$

Find the horsepower of a steam engine if the pressure is 120 pounds, stroke is 18 inches, piston is 10 inches in diameter, and the number of revolutions per minute is 125.

19) Can the tables of logarithms be used for addition and subtraction?

20) Can logarithms be used to solve gear-ratio problems?

## SECTION 6

## DIMENSIONS, AREAS, AND VOLUMES OF FIGURES

*Machinery's Handbook* pages **37** to **88**

The formulas given for the solution of different problems relating to the areas of surfaces and volumes of various geometrical figures are derived from plane and solid geometry. For purposes of shop mathematics, all that is necessary is to select the appropriate figure and use the formula given. Keep in mind the applicable tables that are provided and use them in the solution of the formulas whenever such usage can be done to advantage.

Many rules may be developed directly from the table for polygons on *Handbook* page **77**. These rules will permit easy solution of nearly every problem involving a regular polygon. For instance, in the first "*A*" columns at the left, $A/S^2 = 7.6942$ for a decagon (10-sided polygon); by transposition, $S = \sqrt{A \div 7.6942}$. In the first "*R*" column, $R = 1.3066S$ for an octagon (8-sided polygon); hence, $S = R \div 1.3066$.

Given the frequent occurrence of such geometrical figures as squares, hexagons, spheres, and spherical segments in shop calculations, such tables dealing with these figures are very useful.

*Example 1:* A rectangle 12 inches long has an area of 120 square inches; what is the length of its diagonal?

The area of a rectangle equals the product of the two sides; hence, the unknown side of this rectangle equals $^{120}/_{12} = 10$ inches.

$$\text{Length of diagonal} = \sqrt{12^2 + 10^2} = \sqrt{244} = 15.6205$$

*Example 2:* If the diameter of a sphere, the diameter of the base, and the height of a cone are all equal, find the volume of the sphere if the volume of the cone is 250 cubic inches.

The formula on *Handbook* page **85** for the volume of a cone shows that the value for $250 = 0.2618d^2h$, in which $d$ = diameter of cone base and $h$ = vertical height of cone; hence,

$$d^2 = \frac{250}{0.2618h}$$

Since in this example $d$ and $h$ are equal,

$$d^3 = \frac{250}{0.2618}$$

and

$$d = \sqrt[3]{\frac{250}{0.2618}} = 9.8474 \text{ inches}$$

By referring to the formula on *Handbook* page **86**, the volume of a sphere $= 0.5236d^3 = 0.5236 \times (9.8474)^3 = 500$ cubic inches.

In solving the following exercises, first, construct the figure carefully, and then apply the formula. Use the examples in the *Handbook* as models.

## PRACTICE EXERCISES FOR SECTION 6

(See *Answers to Practice Exercises for Section 6* on page **239**)

1) Find the volume of a cylinder having a base radius of 12.5 and a height of 16.3 inches.

2) Find the area of a triangle with sides that are 12, 14, and 18 inches in length.

3) Find the volume of a torus or circular ring made from $1\frac{1}{2}$ inch round stock with an outside diameter of 14 inches.

4) A bar of hexagonal screw stock measures 0.750 inch per side. What is the largest diameter that can be turned from this bar?

5) Using the prismoidal formula (*Handbook* page **66**), find the volume of the frustum of a regular triangular pyramid if its lower base is 6 inches per side, upper base 2 inches per side, and height 3 inches. (Use the table on *Handbook* page **77** for areas. The side of the midsection equals one-half the sum of one side of the lower base and one side of the upper base.)

6) What is the diameter of a circle the area of which is equivalent to that of a spherical zone whose radius is 4 inches and height 2 inches?

7) Find the volume of a steel ball $^3/_8$ inch in diameter.

8) What is the length of the side of a cube if the volume equals the volume of a frustum of a pyramid with square bases, 4 inches and 6 inches per side, and 3 inches high?

9) Find the volume of a bronze bushing if its inside diameter is 1 inch, outside diameter is $1^1/_2$ inches, and length is 2 inches.

10) Find the volume of a hollow sphere with an outside diameter of 10 inches and an inside diameter of 6 inches.

11) Find the area of a 10-equal-sided polygon inscribed in a 6-inch diameter circle.

12) What is the radius of a fillet if its chord is 2 inches? What is its area?

13) Find the area of the conical surface and volume of a frustum of a cone if the diameter of its lower base is 3 feet, diameter of upper base 1 foot, and height 3 feet.

14) Find the total area of the sides and the volume of a triangular prism 10 feet high, with a base width of 8 feet.

15) The diagonal of a square is 16 inches. What is the length of its side?

16) How many gallons can be contained in a barrel having the following dimensions: height $2^1/_2$ feet, bottom diameter 18 inches, and bilge diameter 21 inches? (The sides are formed to the arc of a circle.)

17) Find the area of a sector of a circle if the radius is 8 inches and the central angle is 32 degrees.

18) Find the height of a cone if its volume is 17.29 cubic inches and the radius of its base is 4 inches.

19) Find the volume of a rectangular pyramid having a base 4 × 5 inches and height 6 inches.

20) Find the distance across the corners of both hexagons and squares when the distance across in each case is: $\frac{1}{2}$, $1\frac{5}{8}$, $3\frac{3}{10}$, 5, and 8.

21) The diagonal of one square is 2.0329 and of the other square is 4.6846. Find the lengths of the sides of both squares.

22) In measuring the distance over plugs in a die that has six $\frac{3}{4}$-inch holes equally spaced on a circle, what should be the micrometer reading over opposite plugs if the distance over alternate plugs is $4\frac{1}{2}$ inches?

23) To what diameter should a shaft be turned in order to mill on one end a hexagon 2 inches on a side? An octagon 2 inches on a side?

## SECTION 7

## GEOMETRICAL PROPOSITIONS AND CONSTRUCTIONS

*Machinery's Handbook* pages **56** to **65**

Geometry is the branch of mathematics that deals with the relations of lines, angles, surfaces, and solids. Plane geometry treats the relations of lines, angles, and surfaces in one plane only, and since this branch of geometry is of special importance in mechanical work, the various propositions or fundamental principles are given in the *Handbook*, as well as various problems or constructions. This information is particularly useful in mechanical drafting and in solving problems about measurement.

*Example 1:* A segment-shaped casting (see **Fig. 1**) has a chordal length of 12 inches, and the height of the chord is 2 inches; determine by the application of a geometrical principle the radius *R* of the segment.

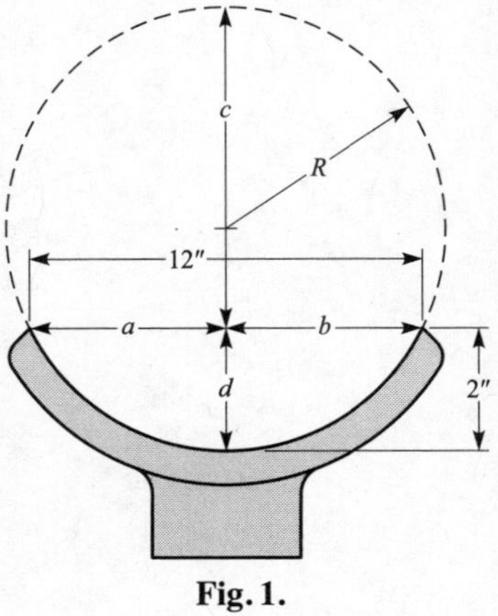


**Fig. 1.**

This problem may be solved by the application of the third geometrical proposition given on *Handbook* page **60**. In this example, one chord consists of two sections $a$ and $b$, each 6 inches long; the other intersecting chord consists of one section $d$, 2 inches long; and the length of section $c$ is to be determined in order to find radius $R$. Since $a \times b = c \times d$, it follows that:

$$c = \frac{a \times b}{d} = \frac{6 \times 6}{2} = 18 \quad \text{inches}$$

therefore,

$$R = \frac{c + d}{2} = \frac{18 + 2}{2} = 10 \quad \text{inches}$$

In this example, one chordal dimension, $c + d$ = the diameter; but the geometrical principle given in the *Handbook* applies regardless of the relative lengths of the intersecting chords.

*Example 2:* The center lines of three holes in a jig plate form a triangle. The angle between two of these intersecting center lines is 52 degrees. Another angle between adjacent center lines is 63 degrees. What is the third angle?

This problem is solved by application of the first geometrical principle on *Handbook* page **56**. The unknown angle = 180 – (63 + 52) = 65 degrees.

*Example 3:* The center lines of four holes in a jig plate form a four-sided figure. Three of the angles between the different intersecting center lines are 63 degrees, 105 degrees, and 58 degrees, respectively. What is the fourth angle?

According to the geometrical principle at the middle of *Handbook* page **58**, the unknown angle = 360 – (63 + 105 + 58) = 134 degrees.

*Example 4:* The centers of three holes are located on a circle. The angle between the radial center lines of the first and second holes is 22 degrees, and the center-to-center distance measured along the circle is $2\frac{1}{2}$ inches. The angle between the second and third holes is 44 degrees. What is the center-to-center distance along the circle?

This problem is solved by application of the fourth principle on *Handbook* page **60**. Since the lengths of the arcs are proportional to the angles, the center distance between the second and third holes = $(44 \times 2\frac{1}{2})/22 = 5$ inches. (See also rules governing proportion starting on *Handbook* page **9**.)

The following practice exercises relate to the propositions and constructions given and should be answered without the aid of the *Handbook*.

## PRACTICE EXERCISES FOR SECTION 7

(See *Answers to Practice Exercises for Section 7* on page **239**)

1) If any two angles of a triangle are known, how can the third angle be determined?

2) State three instances where one triangle is equal to another.

3) When are triangles similar?

4) What is the purpose of proving triangles similar?

5) If a triangle is equilateral, what follows?

6) What are the properties of the bisector of any angle of an equilateral triangle?

7) What is an isosceles triangle?

8) How do the size of an angle and the length of a side of a triangle relate to each other?

9) Can you draw a triangle whose sides are 5, 6, and 11 inches?

10) What is the length of the hypotenuse of a right-angled triangle with sides measuring 12 and 16 inches?

11) What is the value of the exterior angle of any triangle?

12) What are the relations of angles formed by two intersecting lines?

13) Draw two intersecting straight lines and a circle tangent to these lines.

14) Construct a right-angled triangle given the hypotenuse and one side.

15) When are the areas of two parallelograms equal?

16) When are the areas of two triangles equal?

17) If a radius of a circle is perpendicular to a chord, what follows?

18) What is the relation between the radius and tangent of a circle?

19) What lines pass through the point of tangency of two tangent circles?

20) What are the attributes to two tangents drawn to a circle from an external point?

21) What is the value of an angle between a tangent and a chord drawn from the point of tangency?

22) Are all angles equal if their vertices are on the circumference of a circle and they are subtended by the same chord?

23) If two chords intersect within a circle, what is the value of the product of their respective segments?

24) How can a right angle be drawn using a semicircle?

25) What does the length of circular arcs in the same circle depend on?

26) What are the circumferences and areas of two circles proportional to?

## SECTION 8

## TRIGONOMETRY: FUNCTIONS OF ANGLES

*Machinery's Handbook* page **89** to **106**

The basis of trigonometry is proportion. If the sides of any angle are indefinitely extended and perpendiculars from various points on one side are drawn to intersect the other side, right triangles will be formed, and the ratios of the respective sides and hypotenuses will be identical. If the base of the smallest triangle thus formed is 1 inch, and the height is $\frac{1}{2}$ inch (see **Fig. 1**), the ratio between these sides is $1 \div \frac{1}{2} = 2$ or $\frac{1}{2} \div 1 = \frac{1}{2}$ depending upon how the ratio is stated. If the next triangle is measured, the ratio between the base and altitude will likewise be either 2 or $\frac{1}{2}$, and this will always be true for any number of triangles, if the angle remains unchanged. For example, $3 \div 1\frac{1}{2} = 2$ and $4\frac{1}{2} \div 2\frac{1}{4} = 2$ or $1\frac{1}{2} \div 3 = \frac{1}{2}$ and $2\frac{1}{4} \div 4\frac{1}{2} = \frac{1}{2}$.

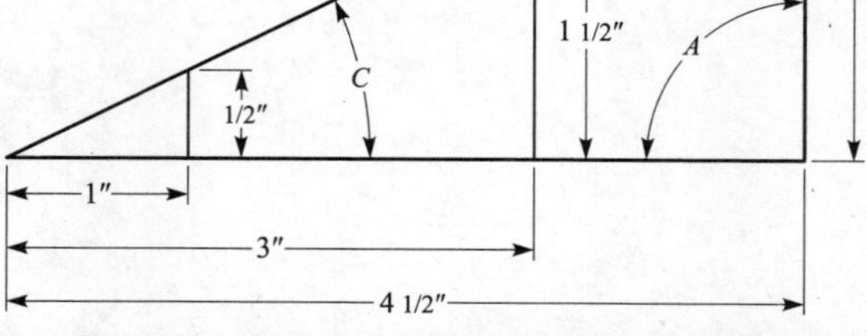


**Fig. 1. For a Given Angle, the Ratio of the Base to the Altitude Is the Same for All Triangle Sizes**

This relationship explains why rules can be developed to find the length of any side of a right triangle when the angle and one side are known or to find the angle when any two sides are known. Since there are two relations between any two sides of a right triangle, there can be, therefore, a total of six ratios with three sides. These ratios are defined and explained in the *Handbook*. Refer to pages **90** and **91** and note explanations of the terms *side adjacent*, *side opposite*, and *hypotenuse*.

The abbreviations of the trigonometric functions begin with a small letter and are not followed by periods.

**Functions of Angles and Use of Trigonometric Tables.**—On page **91** of the *Handbook* are the rules for determining the functions of angles. These rules, which should be memorized, may also be expressed as simple formulas:

$$\text{sine} = \frac{\text{side opposite}}{\text{hypotenuse}} \qquad \text{cosecant} = \frac{\text{hypotenuse}}{\text{side opposide}}$$

$$\text{cosine} = \frac{\text{side adjacent}}{\text{hypotenuse}} \qquad \text{secant} = \frac{\text{hypotenuse}}{\text{side adjacent}}$$

$$\text{tangent} = \frac{\text{side opposite}}{\text{side adjacent}} \qquad \text{cotangent} = \frac{\text{side adjacent}}{\text{side opposite}}$$

Note that these functions are arranged in pairs to include sine and cosecant, cosine and secant, tangent and cotangent, and that each pair consists of a function and its reciprocal. Also, note that these functions are merely ratios: the sine is the ratio of the *side opposite* to the *hypotenuse*, cosine is the ratio of the *side adjacent* to the *hypotenuse*, etc. Tables of trigonometric functions are, therefore, tables of ratios, and these functions can be obtained easily and quickly from most pocket calculators. For example, tan 20° 30′ = 0.37388; this means that in any right triangle having an acute angle of 20° 30′, the side opposite that angle is equal in length to 0.37388 times the length of the side adjacent. Cos 50° 22′ = 0.63787; this means that in any right triangle having an angle of 50° 22′, if the hypotenuse equals a certain length, say, 8, the side adjacent to the angle will equal 0.63787 × 8 or 5.10296.

Referring to **Fig. 1**, tan angle $C = 2\frac{1}{4} \div 4\frac{1}{2} = 1\frac{1}{2} \div 3 = \frac{1}{2} \div 1 = 0.5$; therefore, for this particular angle $C$, the *side opposite* is always equal to 0.5 times *side adjacent*, thus: $1 \times 0.5 = \frac{1}{2}$; $3 \times 0.5 = 1\frac{1}{2}$; and $4\frac{1}{2} \times 0.5 = 2\frac{1}{4}$. The side opposite angle $B$ equals $4\frac{1}{2}$; hence, tan $B = 4\frac{1}{2} \div 2\frac{1}{4} = 2$.

**Finding Angle Equivalent to Given Function.—**After determining the tangent of angle $C$ or of angle $B$, the values of these angles can be determined readily. As tan $C$ = 0.5, find the numbers nearest to this in the tangent column on *Handbook* page **104**: 0.498582, corresponding to 26 degrees, 30 minutes, and 0.502219 corresponding to the angle 26 degrees, 40 minutes. Because 0.5 is approximately midway between 0.498582 and 0.502219, angle $C$ can be accurately estimated as 26 degrees, 35 minutes. This degree of accuracy is usually sufficient; however, improved accuracy may be obtained by interpolation, as explained in the examples to follow.

Since angle $A$ = 90 degrees, and, as the sum of three angles of a triangle always equals 180 degrees, it is evident that angle $C + B$ = 90 degrees; therefore, $B$ = 90 degrees minus 26 degrees, 35 minutes = 63 degrees, 25 minutes. The table on *Handbook* page **104** also shows that tan 63 degrees, 25 minutes is midway between 1.991164 and 2.005690, or approximately 2 within 0.0002.

Note that for angles 45° to 90°, *Handbook* pages **103–105**, the table is used by reading from the bottom up, using the function labels across the bottom of the table, as explained on *Handbook* page **101**.

In the foregoing example, the tangent is used to determine the unknown angles because the known sides are the side adjacent and the side opposite the unknown angles, which are the sides required for determining the tangent. If the side adjacent and the length of hypotenuse had been given instead, the unknown angles might have been determined by first finding the cosine because the cosine equals the side adjacent divided by the hypotenuse.

The acute angles (like $B$ and $C$, **Fig. 1**) of any right triangle must be complementary, so the function of any angle equals the cofunction of its complement; thus, the sine of angle $B$ = the cosine of angle $C$; the tangent of angle $B$ = the cotangent of angle $C$; etc. Thus, tan $B = 4\frac{1}{2} \div 2\frac{1}{4}$ and cotangent $C$ also equals $4\frac{1}{2} \div 2\frac{1}{4}$. The tangent of 20° 30′ = 0.37388, which also equals the cotangent of 20° 30′. For this reason, it is only necessary to calculate the trigonometric ratios to 45° when making a table of trigonometric functions for angles between 45° and 90°, and this is why the functions of angles between 45° and 90° are located in the table by reading it backwards or in reverse order, as previously mentioned.

*Example 1:* Find the tangent of 44 degrees, 59 minutes.

Following instructions given on page **101** of the *Handbook*, find 44 degrees, 50 minutes, and 45 degrees, 0 minutes at the bottom of page **105**; and find their respective tangents, 0.994199 and 1.0000000, in the column labeled "tan" across the top of the table. The tangent of 44° 59′ is 0.994199 + 0.9 × (1 − 0.994199) = 0.99942.

*Example 2:* Find the tangent of 45 degrees, 5 minutes.

At the bottom of *Handbook* page **105**, and above "tan" at the bottom right of the table, are the tangents of 45° 0′ and 45° 10′, 1.000000 and 1.005835, respectively. The required tangent is midway between these two values and can be found from 1.000000 + 0.5 × (1.005835 − 1) = 1.00292.

**How to Find More Accurate Functions and Angles than Those Given in the Table.**—In the *Handbook*, the values of trigonometric functions are given for degrees and 10-minute increments; hence, if the given angle is in degrees, minutes, and seconds, the value of the function is determined from the nearest given values by interpolation.

*Example 3:* Determine the sine of 14° 22′ 26″. It is evident that this value lies between the sine of 14° 20′ and the sine of 14° 30′.

Sine 14° 20′ = 0.247563 and sine 14° 30′ = 0.250380; the difference = 0.250389 − 0.247563 = 0.002817. Consider this difference as a whole number (2817) and multiply it by a fraction with its numerator being the number of additional minutes and fractions of minutes (number of seconds divided by 60) in the given angle (2 + $^{26}/_{60}$), and its denominator being the number of minutes in the interval between 14° 20′ and the sine of 14° 30′.

Thus, $(2 + {}^{26}\!/_{60})/10 \times 2817 = [(2 \times 60) + 26]/(10 \times 60) \times 2817 =$ 685.47; hence, by adding 0.000685 to sine of 14° 20′, we find that sine 14° 22′ 26″ = 0.247563 + 0.000685 = 0.24825.

The correction value (represented in this example by 0.000685) is *added* to the function of the *smaller* angle nearest the given angle in dealing with sines or tangents, but this correction value is *subtracted* in dealing with cosines or cotangents.

*Example 4:* Find the angle whose cosine is 0.27052.

The table of trigonometric functions shows that the desired angle is between 74° 10′ and 74° 20′ because the cosines of these angles are, respectively, 0.272840 and 0.270040. The difference = 0.272840 – 0.270040 = 0.00280′. From the cosine of the smaller angle (i.e., the larger cosine) or 0.272840, subtract the given cosine; thus, 0.272840 – 0.27052 = 0.00232; hence 232/280 × 10 = 8.28571′ or the number of minutes to add to the smaller angle to obtain the required angle. Thus, the angle for a cosine of 0.27052 is 74° 18.28571′, or 74° 18′ 17″. Angles corresponding to given sines, tangents, or cotangents may be determined by the same method.

**Trigonometric Functions of Angles Greater than 90 Degrees.**—In obtuse triangles, one angle is greater than 90 degrees, and the *Handbook* tables can be used for finding the functions of angles larger than 90 degrees, but the angle must be first expressed in terms of an angle less than 90 degrees.

The sine of an angle greater than 90 degrees but less than 180 degrees equals the sine of an angle that is the difference between 180 degrees and the given angle.

*Example 5:* Find the sine of 118 degrees.

sin 118° = sin (180° – 118°) = sin 62°. Refer to page **104** to see that the sine given for 62 degrees is 0.882948.

The cosine, tangent, and cotangent of an angle greater than 90 but less than 180 degrees equals, respectively, the cosine, tangent, and cotangent of the difference between 180 degrees and the given angle; but the angular function has a negative value and must be preceded by a minus sign.

*Example 6:* Find tan 123 degrees, 20 minutes.

tan 123° 20′ = –tan (180° – 123° 20′) = –tan 56° 40′ = –1.520426

*Example 7:* Find csc 150 degrees.

Cosecant, abbreviated "csc" or "cosec," equals 1/sin, and is positive for angles 90 to 180 degrees (see *Handbook* page **102**):

$\csc 15° = 1/\sin(180° - 150°) = 1/\sin 30° = 1/0.5 = 2.0$

In the calculation of triangles, it is very important to include the minus sign in connection with the sines, cosines, tangents, and cotangents of angles greater than 90 degrees. *Signs of Trigonometric Functions, Fractions of* $\pi$*, and Degree-Radian Conversion* (**Fig. 4** on page **102** of the *Handbook*) shows clearly the negative and positive values of different functions and angles between 0 and 360 degrees. *Useful Relationships Among Angles* (**Table 1** on page **102**) is also helpful in determining the function, sign, and angle less than 90 degrees that is equivalent to the function of an angle greater than 90 degrees.

**Use of Functions for Laying Out Angles.**—Trigonometric functions may be used for laying out angles accurately either on drawings or in connection with template work, etc. The following example illustrates the general method:

*Example 8:* Construct or lay out an angle of 27 degrees, 29 minutes by using its sine instead of a protractor.

First, draw two lines at right angles, as in **Fig. 2**, and to any convenient length. Using a calculator, find the sine of 27 degrees, 29 minutes: 0.46149. If there is enough space, lay out your figure to an enlarged scale to obtain greater accuracy. Assume that the scale is 10 to 1: therefore, multiply the sine of the angle by 10, obtaining 4.6149 or about $4^{39}/_{64}$. Set the dividers or the compass to this dimension and with *a* (**Fig. 2**) as a center, draw an arc, thus obtaining one side of the triangle *ab*. Now set the compass to 10 inches (since the scale is 10 to 1) and, with *b* as the center, describe an arc so as to obtain intersection *c*. The hypotenuse of the triangle is now drawn through the intersections *c* and *b*, thus obtaining an angle *C* of 27 degrees, 29 minutes within fairly close limits. The angle *C*, laid out in this way, equals 27 degrees, 29 minutes because:

$$\frac{\text{side opposite}}{\text{hypotenuse}} = \frac{4.6149}{10} = 0.46149 = \sin 27°29'$$

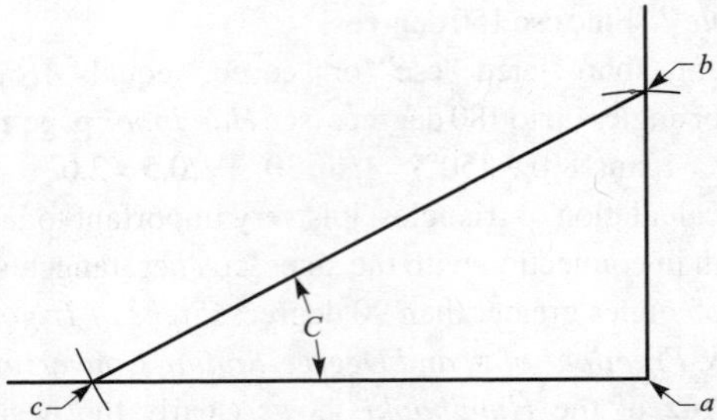


**Fig. 2. Method of Laying Out Angle by Using Its Sine**

**Tables of Functions Used in Conjunction with Formulas.–** When milling keyways, it is often desirable to know the total depth from the outside of the shaft to the bottom of the keyway. With this depth known, the cutter can be fed down to the required depth without taking any measurements other than that indicated by the graduations on the machine. To determine the total depth, it is necessary to calculate the height of the arc, which is designated as dimension *A* in **Fig. 3**. The formula usually employed to determine *A* for a given diameter of shaft *D* and width of key *W* is as follows:

$$A = \frac{D}{2} - \sqrt{\left(\frac{D}{2}\right)^2 - \left(\frac{W}{2}\right)^2}$$

Another formula, which is simpler than the one above, is used in conjunction with a calculator, as follows:

$$A = \frac{D}{2} \times \text{ versed sine of an angle whose sine is } \frac{W}{D}$$

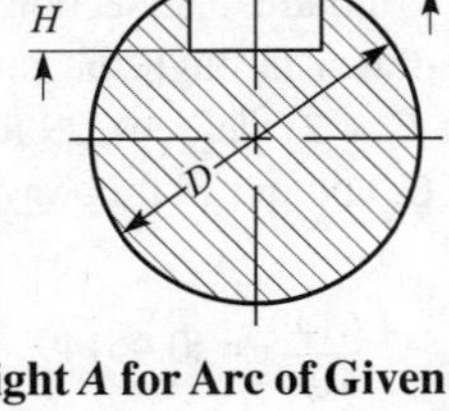


**Fig. 3. To Find Height *A* for Arc of Given Radius and Width *W***

*Example 9:* To illustrate the application of this formula, let it be required to find the height $A$ when the shaft diameter is $\frac{7}{8}$ inch and the width $W$ of the key is $\frac{7}{32}$ inch. Then, $W/D = (\frac{7}{32})/(\frac{7}{8}) = \frac{7}{32} \times \frac{8}{7} = 0.25$. Using the formula at the bottom of *Handbook* page **106** for versed sin $\theta = 1 - \cos\theta$, and a calculator, the angle corresponding to sin 0.25 = 14.4775 degrees, or 14 degrees, 28 minutes, 39 seconds. The cosine of this angle is 0.9682, and subtracting this value from 1 gives 0.03175 for the versed sine. Then, the height of the circular segment $A = D/2 \times 0.03175 = (7 \times 0.03175)/(8 \times 2) = 0.01389$, so the total depth of the keyway equals dimension $H$ plus 0.01389 inch.

## PRACTICE EXERCISES FOR SECTION 8

(See *Answers to Practice Exercises for Section 8* on page **240**)

1) How should a scientific pocket calculator be used to solve triangles?

2) Explain the meaning of sin 30° = 0.50000.

3) Find sin 18° 26′ 30″, tan 27° 16′ 15″, and cos 32° 55′ 17″.

4) Find the angles that correspond to the following tangents: 0.52035 and 0.13025; to the following cosines: 0.06826 and 0.66330.

5) Give two rules for finding the *side opposite* a given angle.

6) Give two rules for finding the *side adjacent* to a given angle.

7) Explain the following terms: equilateral, isosceles, acute angle, obtuse angle, and oblique angle.

8) What is meant by complement, side adjacent, and side opposite?

9) Can the elements referred to in Exercise 8 be used in solving an isosceles triangle?

10) Construct by use of tangents an angle of 42° 20′.

11) Construct by use of sines an angle of 68° 15′.

12) Construct by use of cosines an angle of 55° 5′.

## SECTION 9

## SOLUTION OF RIGHT-ANGLE TRIANGLES

*Machinery's Handbook* pages **93–94**

A thorough knowledge of the solution of triangles or trigonometry is essential in drafting, layout work, bench work, and for convenient and rapid operation of some machine tools. Calculations concerning gears, screw threads, dovetails, angles, tapers, solution of polygons, gage design, cams, dies, and general inspection work are dependent upon trigonometry. Many geometrical problems may be solved more rapidly by trigonometry than by geometry.

In shop trigonometry, it is not necessary to develop and memorize the various rules and formulas, but it is essential that the six trigonometric functions be mastered thoroughly. Remember that a thorough, working knowledge of trigonometry depends upon drill work; hence a large number of problems should be solved.

The various formulas for the solution of right-angle triangles are given on *Handbook* page **93**; and examples showing their application are given on page **94**. These formulas may, of course, be applied to a large variety of practical problems in drafting rooms, tool rooms, and machine shops, as indicated by the following examples.

Whenever two sides of a right-angle triangle are given, the third side can always be found by a simple arithmetical calculation, as shown by the second and third examples on *Handbook* page **94**. To find the angles, however, it is necessary to use tables of sines, cosines, tangents, and cotangents, or a calculator, and, if only one side and one of the acute angles are given, the natural trigonometric functions (trig tables, pages **103–105**) must be used for finding the lengths of the other sides.

*Example 1:* The Jarno taper is 0.600 inch per foot for all numbers. What is the included angle?

As the angle measured from the axis or center line is 0.600 ÷ 2 = 0.300 inch per foot, the tangent of one-half the included angle = 0.300 ÷ 12 = 0.25 = tan 1° 26′; hence, the included angle = 2° 52′. A more direct method is to find the angle whose tangent equals the taper per foot divided by 24, as explained on *Handbook* page **699**.

*Example 2:* Determine the width *W* (see **Fig. 1**) of a cutter for milling a splined shaft having 6 splines 0.312 inch wide and a diameter *B* of 1.060 inches.

Dimension *W* may be computed by using the following formula:

$$W = \sin\left(\frac{\frac{360°}{N} - 2a}{2}\right) \times B$$

in which *N* = number of splines; *B* = diameter of body or of the shaft at the root of the spline groove.

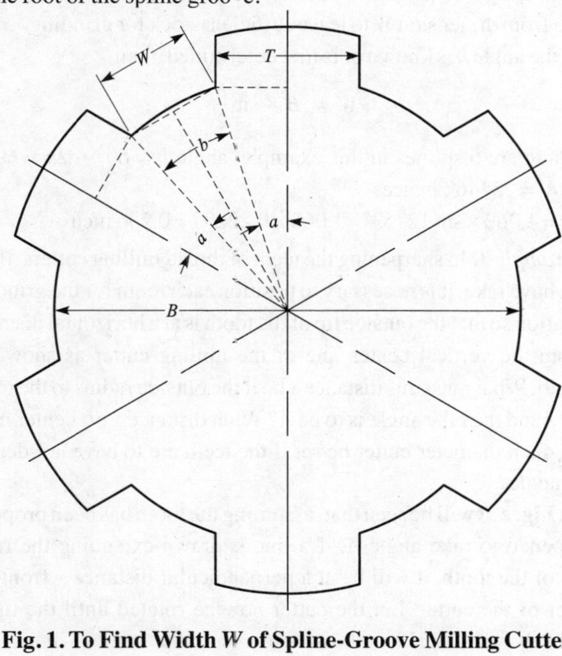


**Fig. 1. To Find Width *W* of Spline-Groove Milling Cutter**

Angle $a$ must first be computed, as follows:

$$\sin a = \frac{T}{2} \div \frac{B}{2} \quad \text{or} \quad \sin a = \frac{T}{B}$$

where $T$ = width of spline; $B$ = diameter at the root of the spline groove. In this example,

$$\sin a = \frac{0.312}{1.060} = 0.29434$$

$$a = 17°7'; \text{ hence}$$

$$W = \left( \frac{\sin \frac{360°}{6} - 2 \times 17°7'}{2} \right) \times 1.060 = 0.236 \text{ inch}$$

This formula has also been used frequently in connection with broach design, but it is capable of a more general application. If the splines are to be ground on the sides, suitable deduction must be made from dimension $W$ to leave sufficient stock for grinding.

If the angle $b$ is known or is first determined, then

$$W = B \times \sin\frac{b}{2}$$

As there are 6 splines in this example, angle $b = 60° - 2a = 60° - 34°\,14' = 25°\,46'$; hence,

$W = 1.060 \times \sin 12°\,53' = 1.060 \times 0.22297 = 0.236$ inch

*Example 3:* In sharpening the teeth of thread milling cutters, if the teeth have rake, it is necessary to position each tooth for the grinding operation so that the outside tip of the tooth is at a horizontal distance $x$ from the vertical center line of the milling cutter as shown in **Fig. 2b**. What must this distance $x$ be if the outside radius to the tooth tip is $r$ and the rake angle is to be $A$? What distance $x$ off center must a $4\frac{1}{2}$-inch diameter cutter be set if the teeth are to have a 3-degree rake angle?

In **Fig. 2**, it will be seen that, assuming the tooth has been properly sharpened to rake angle $A$, if a line is drawn extending the front edge of the tooth, it will be at a perpendicular distance $x$ from the center of the cutter. Let the cutter now be rotated until the tip of the tooth is at a horizontal distance $x$ from the vertical center line of the cutter as shown in **Fig. 2b**. It will be noted that an extension of the front edge of the cutter is still at perpendicular distance $x$ from the center of the cutter, indicating that the cutter face is parallel to the vertical center line or is itself vertical, which is the desired position for sharpening using a vertical wheel. Thus, $x$ is the proper offset distance for grinding the tooth to rake angle $A$ if the radius to the tooth tip is $r$. Since $r$ is the hypotenuse, and $x$ is one side of a right-angled triangle,

$$x = r\sin A$$

For a cutter diameter of $4\frac{1}{2}$ inches and a rake angle of 3 degrees,

$$x = (4.5 \div 2)\sin 3° = 2.25 \times 0.05234$$
$$= 0.118 \text{ inch}$$

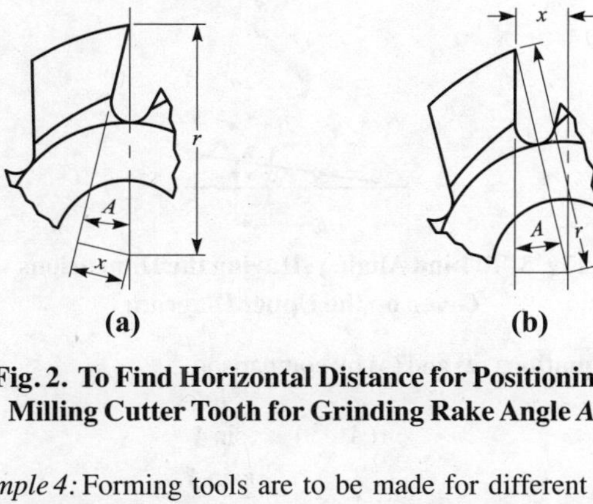


**Fig. 2. To Find Horizontal Distance for Positioning Milling Cutter Tooth for Grinding Rake Angle *A***

*Example 4:* Forming tools are to be made for different sizes of poppet valve heads, and a general formula is required for finding angle $x$ from dimensions given in **Fig. 3**.

The values for $b$, $h$, and $r$ can be determined easily from the given dimensions. Angle $x$ can then be found in the following manner:

Referring to the lower diagram in **Fig. 3**,

$$\tan A = \frac{h}{b} \quad (1)$$

$$c = \frac{h}{\sin A} \quad (2)$$

Also,

$$c = \frac{r}{\sin B} = \frac{r}{\sin(A-x)} \tag{3}$$

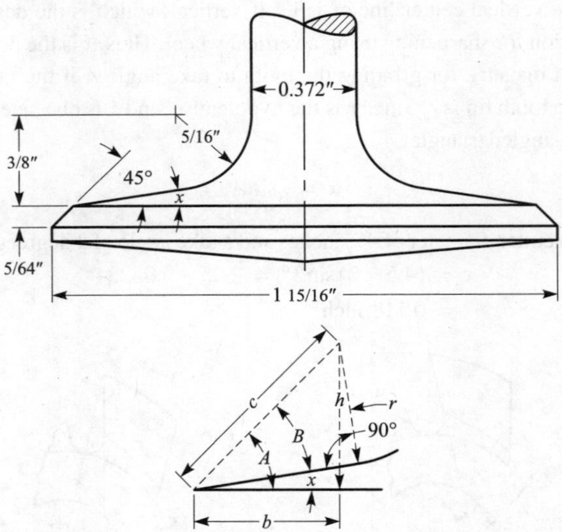


**Fig. 3. To Find Angle *x*, Having the Dimensions Given on the Upper Diagram**

From **Equations (2)** and **(3)** by comparison,

$$\frac{r}{\sin(A-x)} = \frac{h}{\sin A} \tag{4a}$$

$$\sin(A-x) = \frac{r \sin A}{h} \tag{4b}$$

From the dimensions given, it is obvious that $b = 0.392125$ inch, $h = 0.375$ inch, and $r = 0.3125$ inch. Substituting these values in **Equations (1)** and **(4b)** and solving, angle $A$ will be found to be 43 degrees, 43 minutes, and angle $(A - x)$ to be 35 degrees, 10 minutes. By subtracting these two values, angle $x$ will be found to equal 8 degrees, 33 minutes.

*Example 5:* In tool designing, it is frequently necessary to determine the length of a tangent to two circles. In **Fig. 4**, $R$ = radius of large circle = $^{13}/_{16}$ inch; $r$ = radius of small circle = $^{3}/_{8}$ inch; $W$ = center distance between circles = $1^{11}/_{16}$ inches.

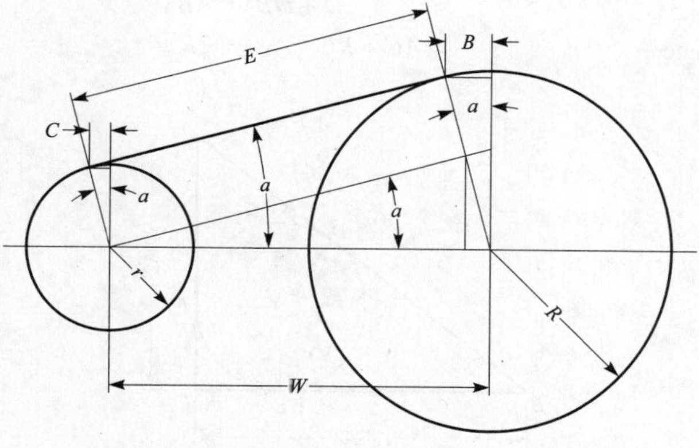


**Fig. 4. To Find Dimension *E* or Distance Between Points of Tangency**

With the values given, find the following: $E$ = length of tangent, $B$ = length of horizontal line from point of tangency on large circle to the vertical line, and $C$ = length of horizontal line from point of tangency on small circle to the vertical center line.

$$\sin a = \frac{R - r}{W} = \frac{13/16 - 3/8}{1\frac{11}{16}} = 0.25925$$

$$\text{Angle } a = 15°1' \text{ nearly}$$

$$E = W \cos a = 1\frac{11}{16} \times 0.9658 = 1.63 \text{ inches}$$

$$B = R \sin a \qquad \text{and} \qquad C = r \sin a$$

*Example 6:* A circle is inscribed in a right triangle having the dimensions shown in **Fig. 5**. Find the radius of the circle.

In **Fig. 5**, $BD = BE$ and $AD = AF$, because tangents drawn to a circle from the same point are equal. $EC = CF$, and $EC$ = radius $OF$. Then, let $R$ = radius of inscribed circle. $AC - R = AD$ and $BC - R = DB$.

Adding,

$$AC + BC - 2R = AD + DB$$
$$AD + DB = AB$$

hence,

$$AC + BC - AB = 2R$$

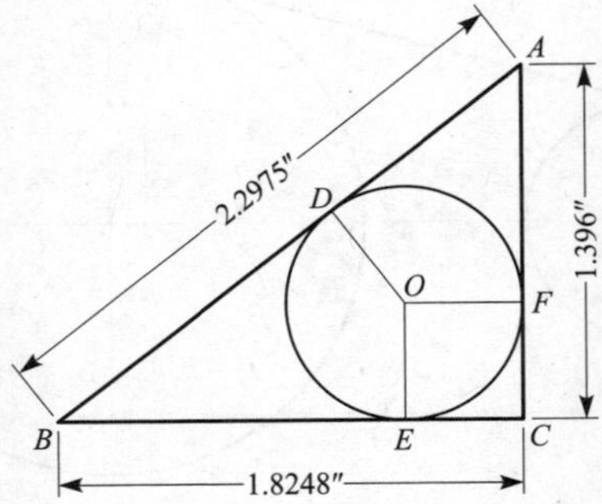


**Fig. 5. To Find Radius of Circle Inscribed in Triangle**

Stated as a rule: *The diameter of a circle inscribed in a right triangle is equal to the difference between the lengths of the hypotenuse and the sum of the lengths of the other sides.* Substituting the given dimensions, we have 1.396 + 1.8248 − 2.2975 = 0.9233 = 2$R$ and $R$ = 0.4616.

*Example 7:* A part is to be machined to an angle $b$ of 30 degrees (**Fig. 6**) by using a vertical forming tool with a clearance angle $a$ of 10 degrees. Calculate the angle of the forming tool as measured in a plane $Z$-$Z$, which is perpendicular to the front or clearance surface of the tool.

Assume that $B$ represents the angle in plane $Z$-$Z$.

$$\tan B = \frac{Y}{X} \text{ and } Y = y \times \cos a \qquad (1)$$

Also,

$$y = X \times \tan b \text{ and } X = \frac{y}{\tan b} \qquad (2)$$

Now, substituting the values of $Y$ and $X$ in **Equation (1)**:

$$\tan B = \frac{y \times \cos a}{\dfrac{y}{\tan b}}$$

Clearing this equation of fractions,

$$\tan B = \cos a \times \tan b$$

In this example, $\tan B = 0.98481 \times 0.57735 = 0.56858$; hence, $B = 29°37'$ nearly.

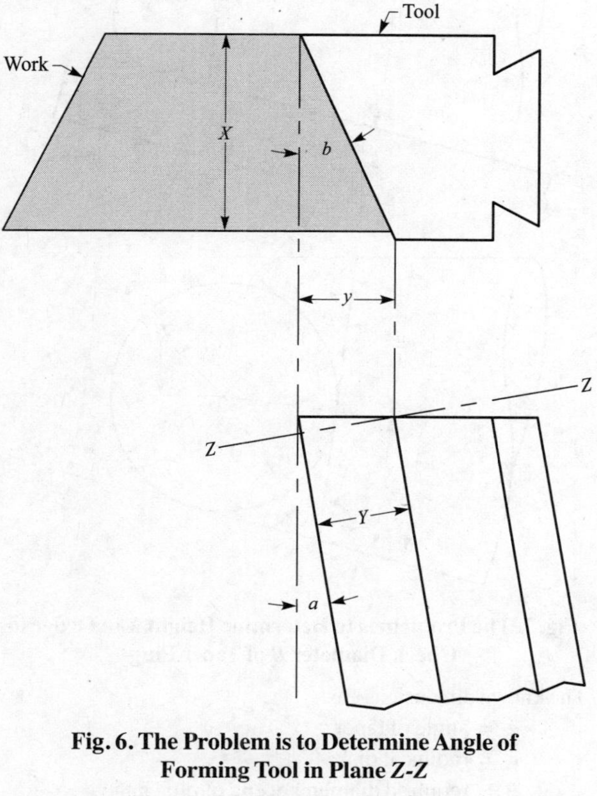


**Fig. 6. The Problem is to Determine Angle of Forming Tool in Plane *Z-Z***

*Example 8:* A method of checking the diameter at the small end of a taper plug gage is shown by **Fig. 7**. The gage is first mounted on a sine-bar so that the top of the gage is parallel with the surface plate. A disk of known radius $r$ is then placed in the corner formed by the end of the plug gage and the top side of the sine-bar. Now by determining the difference $X$ in height between the top of the gage and the top edge of the disk, the accuracy of the diameter $B$ can be checked readily. Derive formulas for determining dimension $X$.

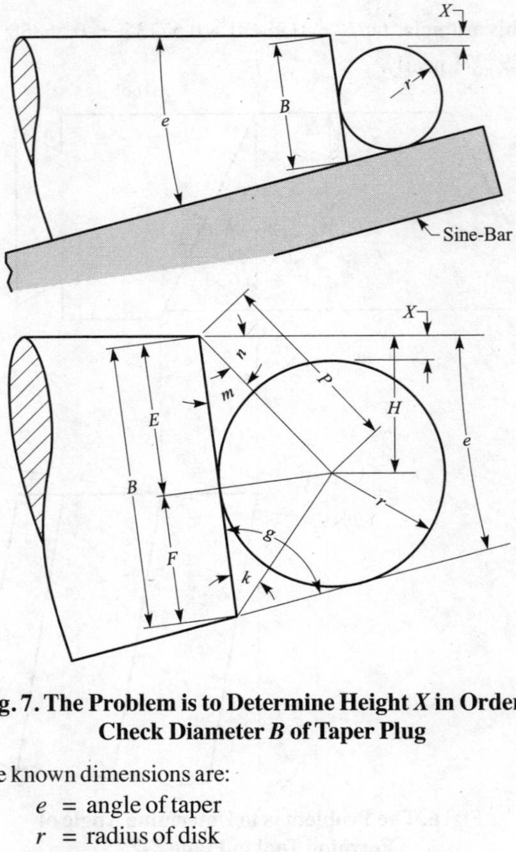


**Fig. 7. The Problem is to Determine Height *X* in Order to Check Diameter *B* of Taper Plug**

The known dimensions are:

$e$ = angle of taper
$r$ = radius of disk
$B$ = required diameter at end of plug gage
$g$ = 90 degrees $- \frac{1}{2} e$
$k = \frac{1}{2} g$

By trigonometry,

$$F = \frac{r}{\tan k}; E = B - F\text{; and } \tan m = \frac{r}{E}$$

Also,

$$P = \frac{r}{\sin m}; n = g - m; \text{and } H = P\sin n$$

Therefore, $X = H - r$ or $r - H$, depending on whether or not the top edge of the disk is above or below the top of the plug gage. In **Fig. 7**, the top of the disk is below the top surface of the plug gage, so it is evident that $X = H - r$.

To illustrate the application of these formulas, assume that $e$ = 6 degrees, $r$ = 1 inch, and $B$ = 2.400 inches. The dimension $X$ is then found as follows:

$$g = 90 - 6/2 = 87°; \text{and } k = 43° 30'$$

By trigonometry,

$$F = \frac{1}{0.9896} = 1.0538''; E = 2.400 - 1.0538 = 1.3462 \text{ inches}$$

$$\tan m = \frac{1}{1.3462} = 0.74283 \text{ and } m = 36° 36' 22''$$

$$P = \frac{1}{0.59631} = 1.6769''; n = 87° - 36°36'22'' = 50°23'38''$$

and $$H = 1.6769 \times 0.77044 = 1.2920 \text{ inches}$$

Therefore, $$X = H - r = 1.2920 - 1 = 0.2920 \text{ inch}$$

The disk here is below the top surface of the plug gage; hence, the formula $X = H - r$ was applied.

*Example 9:* In **Fig. 8**, $a = 1\frac{1}{4}$ inches, $h$ = 4 inches, and angle $A$ = 12 degrees. Find dimension $x$ and angle $B$.

Draw an arc through points $E$, $F$, and $G$, as shown, with $r$ as a radius. According to a well-known theorem of geometry, which is given on *Handbook* page **59**, if an angle at the circumference of a circle, between two chords, is subtended by the same arc as the angle at the center, between two radii, then the angle at the circumference is equal to one-half the angle at the center. This being true, angle $C$ is twice the magnitude of angle $A$, and angle $D$ = angle $A$ = 12 degrees.

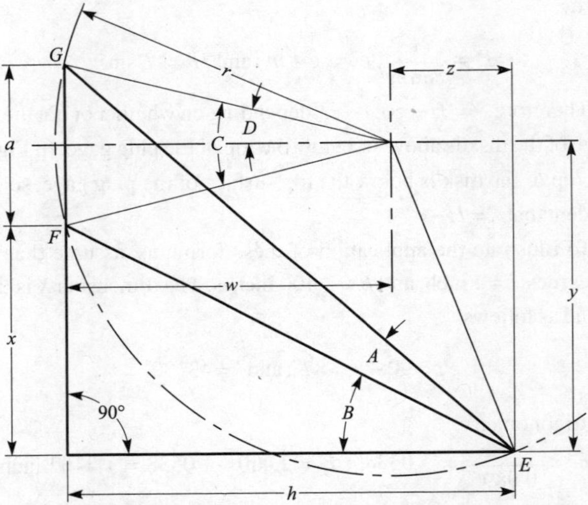


**Fig. 8. Find Dimension *x* and Angle *B*, Given *a*, *h*, and Angle *A***

Thus,

$$r = \frac{a}{2\sin D} = \frac{1.25}{2 \times 0.20791} = 3.0061$$

$$w = \frac{a}{2}\cot D = 0.625 \times 4.7046 = 2.9404$$

and

$$z = h - w = 4 - 2.9404 = 1.0596$$

Now

$$y = \sqrt{r^2 - z^2} = \sqrt{7.9138505} = 2.8131$$

and

$$x = y - \frac{a}{2} = 2.8131 - 0.625 = 2.1881 \text{ inches}$$

Finally,

$$\tan B = \frac{x}{h} = \frac{2.1881}{4} = 0.54703$$

and

$$B = 28 \text{ degrees, } 40 \text{ minutes, } 47 \text{ seconds}$$

*Example 10:* A steel ball is placed inside a taper gage as shown in **Fig. 9**. If the angle of the taper, length of taper, radius of ball, and its position in the gage are known, how can the end diameters *X* and *Y* of the gage be determined by measuring dimension *C*?

The ball should be of such size as to project above the face of the gage. Although not necessary, this projection is preferable, as it permits the required measurements to be obtained more readily. After measuring the distance *C*, the calculation of dimension *X* is as follows: First obtain dimension *A*, which equals *R* multiplied by csc *a*. Then adding *R* to *A* and subtracting *C*, we obtain dimension *B*. Dimension *X* may then be obtained by multiplying 2*B* by the tangent of angle *a*. The formulas for *X* and *Y* can therefore be written as follows:

$$X = 2(R\csc a + R - C)\tan a$$
$$= 2(R\sec a + 2\tan a(R - C))$$
$$Y = X - 2T\tan a$$

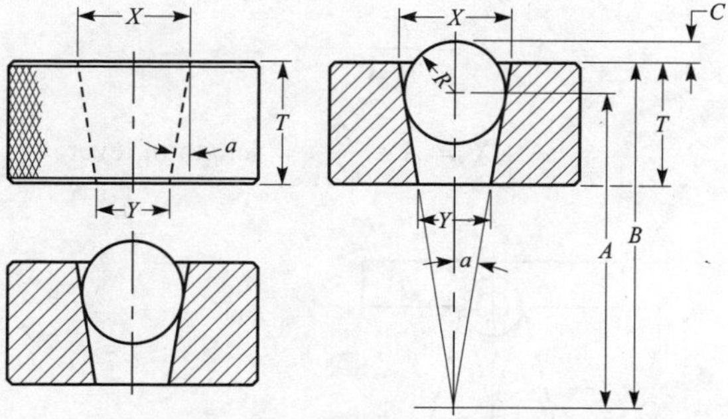


**Fig. 9. Checking Dimensions *X* and *Y* by Using One Ball of Given Size**

If, in **Fig. 9**, angle $a = 9$ degrees, $T = 1.250$ inches, $C = 0.250$ inch and $R = 0.500$ inch, what are the dimensions *X* and *Y*? Applying the formula,

$$X = 2 \times 0.500 \times 1.0125 + 2 \times 0.15838\,(0.500 - 0.250)$$

By solving this equation, $X = 1.0917$ inches. Then

$$Y = 1.0917 - (2.500 \times 0.15838) = 0.6957$$

*Example 11:* In designing a motion of the type shown in **Fig. 10**, it is essential, usually, to have link *E* swing equally above and below the center line *M-M*. A mathematical solution of this problem follows. In the illustration, *G* represents the machine frame; *F*, a lever shown in extreme positions; *E*, a link; and *D*, a slide. The distances *A* and *B* are fixed, and the problem is to obtain $A + X$, or the required length of the lever. In the right triangle:

$$A + X = \sqrt{(A - X)^2 + \left(\frac{B}{2}\right)^2}$$

Squaring, we have:

$$A^2 + 2AX + X^2 = A^2 - 2AX + X^2 + \frac{B^2}{4}$$

$$4AX = \frac{B^2}{4}$$

$$X = \frac{B^2}{16A}$$

$$A + X = A + \frac{B^2}{16A} = \text{length of lever}$$

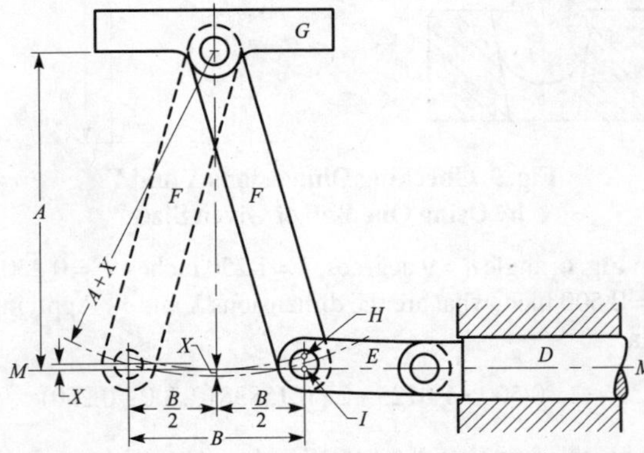


**Fig. 10. Determining Length *F* so that Link *E* Will Swing Equally Above and Below the Center Line**

To illustrate the application of this formula, assume that the length of a lever is required when the distance $A = 10$ inches, and the stroke $B$ of the slide is 4 inches.

$$\text{Length of lever} = A + \frac{B^2}{16A} = 10 + \frac{16}{16 \times 10}$$

$$= 10.100 \text{ inches}$$

Thus, it is evident that the pin in the lower end of the lever will be 0.100 inch below the center line *M-M* when half the stroke has been made, and at each end of the stroke, the pin will be 0.100 inch above this center line.

*Example 12:* The spherical hubs of bevel gears are checked by measuring the distance *x* (**Fig. 11**) over a ball or plug placed against a plug gage that fits into the bore. Determine this distance *x*.

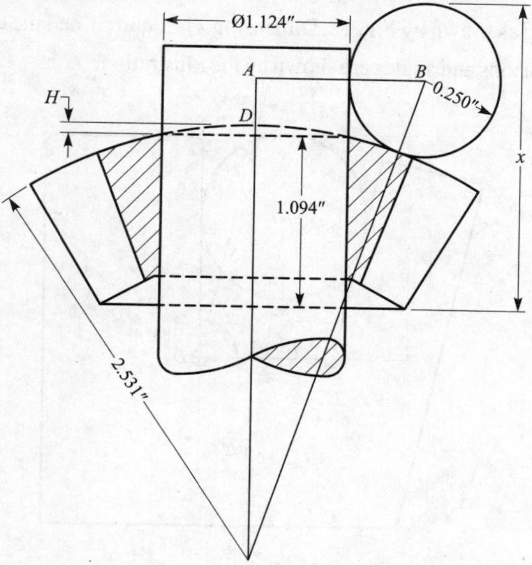


**Fig. 11. Method of Checking the Spherical Hub of a Bevel Gear with Plug Gages**

First find $H$ by means of the formula for circular segments on *Handbook* page **74**.

$$H = 2.531 - 1/2\sqrt{4 \times (2.531)^2 - (1.124)^2} = 0.0632 \text{ inch}$$

$$AB = \frac{1.124}{2} + 0.25 = 0.812 \text{ inch}$$

$$BC = 2.531 + 0.25 = 2.781 \text{ inches}$$

Applying the formula on *Handbook* page **93** for a right triangle with side $a$ and hypotenuse $c$ known,

$$AC = \sqrt{(2.781)^2 - (0.812)^2} = 2.6599 \text{ inches}$$

$$AD = AC - DC = 2.6599 - 2.531 = 0.1289 \text{ inch}$$

$$x = 1.094 + 0.0632 + 0.1289 + 0.25 = 1.536 \text{ inches}$$

*Example 13:* The accuracy of a gage is to be checked by placing a ball or plug between the gage jaws and measuring to the top of the ball or plug as shown by **Fig. 12**. Dimension $x$ is required, and the known dimensions and angles are shown by the illustration.

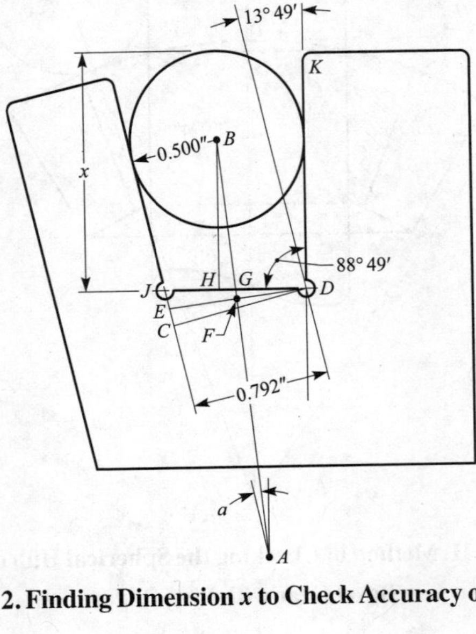


**Fig. 12. Finding Dimension $x$ to Check Accuracy of Gage**

One-half of the included angle between the gage jaws equals one-half of $13° \times 49'$ or $6° \times 54\frac{1}{2}'$, and the latter equals angle $a$.

$$AB = \frac{0.500}{\sin 6°54\frac{1}{2}'} = 4.1569 \quad \text{inches}$$

$DE$ is perpendicular to $AB$ and angle $CDE$ = angle $a$; hence,

$$DE = \frac{CD}{\cos 6°54\frac{1}{2}'} = \frac{0.792}{\cot 6°54\frac{1}{2}'} = 0.79779 \text{ inch}$$

$$AF = \frac{DE}{2} \times \cot 6°54\frac{1}{2}' = 3.2923 \text{ inches}$$

$$\text{Angle } CDK = 90° + 13°49' = 103°49'$$

$$\text{Angle } CDJ = 103°49' - 88°49' = 15°$$

$$\text{Angle } EDJ = 15° - 6°54\frac{1}{2}' = 8°5\frac{1}{2}'$$

$$GF = \frac{DE}{2} \times \tan 8°5\frac{1}{2}' = 0.056711 \text{ inch}$$

$$\text{Angle } HBG = \text{angle } EDJ = 8°5\frac{1}{2}'$$

$$BG = AB - (GF + AF) = 0.807889 \text{ inch}$$

$$BH = BG \times \cos 8°5\frac{1}{2}' = 0.79984 \text{ inch}$$

$$x = BH + 0.500 = 1.2998 \text{ inches}$$

If surface $JD$ is parallel to the bottom surface of the gage, the distance between these surfaces might be added to $x$ to make it possible to use a height gage from a surface plate.

**Helix Angles of Screw Threads, Hobs, and Helical Gears.—**The terms "helical" and "spiral" often are used interchangeably in drafting rooms and shops, although the two curves are entirely different. As the illustration on *Handbook* page **65** shows, every point on a helix is equidistant from the axis, and the curve advances at a uniform rate around a cylindrical area. The helix is illustrated by the springs shown on *Handbook* page **311**. A spiral is flat like a clock spring. A spiral may be defined mathematically as a curve having a constantly increasing radius of curvature.

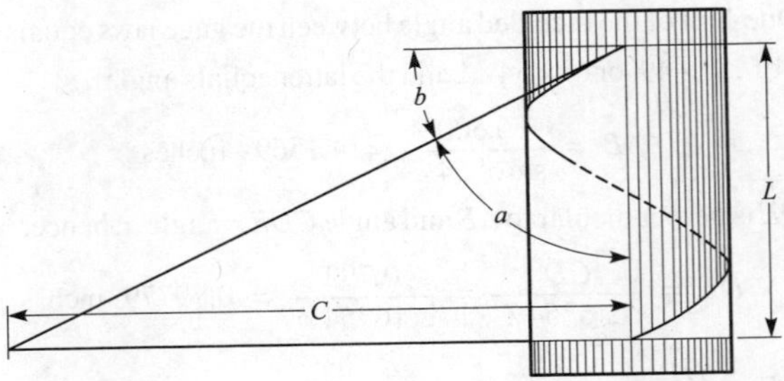


**Fig. 13. Helix Represented by a Triangular Piece of Paper Wound Upon a Cylinder**

If a piece of paper is cut in the form of a right triangle and wrapped around a cylinder (as indicated by **Fig. 13**), the hypotenuse will form a helix. The curvature of a screw thread represents a helix. From the properties of a right triangle, simple formulas can be derived for determining helix angles. Thus, if the circumference of a part is divided by the lead or distance that the helix advances axially in one turn, the quotient equals the tangent of the helix angle as measured from the axis. The angles of helical curves usually (but not always) are measured from the axis. The helix angle of a helical or spiral gear is measured from the axis, but the helix angle of a screw thread is measured from a plane perpendicular to the axis. In a helical gear, the angle is $a$ (**Fig. 13**), whereas for a screw thread, the angle is $b$; hence, for helical gears, tan $a$ of helix angle = $C/L$; for screw threads, tan $b$ of helix angle = $L/C$. The helix angle of a hob, such as is used for gear cutting, also is measured as indicated at $b$ and often is known as the *end angle*, because it is measured from the plane of the end surface of the hob. In calculating helix angles of helical gears, screw threads, and bobs, the pitch circumference is used.

*Example 14:* If the pitch diameter of a helical gear = 3.818 inches and the lead = 12 inches, what is the helix angle?

Tan helix angle = $(3.818 \times 3.1416) / 12 = 1$ very nearly; hence the angle = 45 degrees.

## PRACTICE EXERCISES FOR SECTION 9

(See *Answers to Practice Exercises for Section 9* on page **241**)

1) The No. 4 Morse taper is 0.6233 inch per foot; calculate the included angle.

2) ANSI Standard pipe threads have a taper of $^3/_4$ inch per foot. What is the angle on each side of the center line?

3) To what dimension should the dividers be set to space 8 holes evenly on a circle of 6 inches diameter?

4) Explain the derivation of the formula

$$W = \sin\left(\frac{\frac{360°}{N} - 2a}{2}\right) \times B$$

For notation, see **Example 2** and **Fig. 1** on page **65**.

5) The top of a male dovetail is 4 inches wide. If the angle is 55 degrees, and the depth is $^5/_8$ inch, what is the width at the bottom of the dovetail?

6) Angles may be laid out accurately by describing an arc with a radius of given length and then determining the length of a chord of this arc. In laying out an angle of 25 degrees, 20 minutes, using a radius of 8 inches, what should the length of the chord opposite the named angle be?

7) What is the largest square that may be milled on the end of a $2^1/_2$-inch bar of round stock?

8) A guy wire from a smoke stack is 120 feet long. How high is the stack if the wire is attached 10 feet from the top and makes an angle of 57 degrees with the stack?

9) In laying out a master jig plate, it is required that holes *F* and *H*, **Fig. 14**, shall be on a straight line that is $1^3/_4$ inch distant from hole *E*. The holes must also be on lines making, respectively, 40- and 50-degree angles with line *EG*, drawn at right angles to the sides of the jig plate through *E*, as shown in the figure. Find the dimensions *a*, *b*, *c*, and *d*.

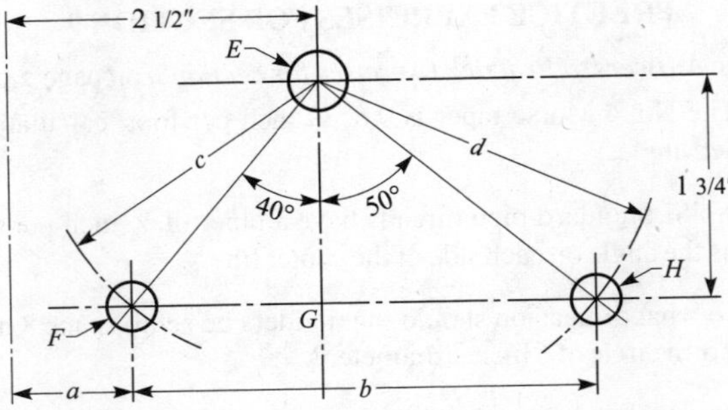


**Fig. 14. Find Dimensions *a*, *b*, *c*, and *d***

10) **Fig. 15** shows a template for locating a pump body on a milling fixture, with the inside contour of the template corresponding with the contour of the pump flange. Find the angle *a* from the values given.

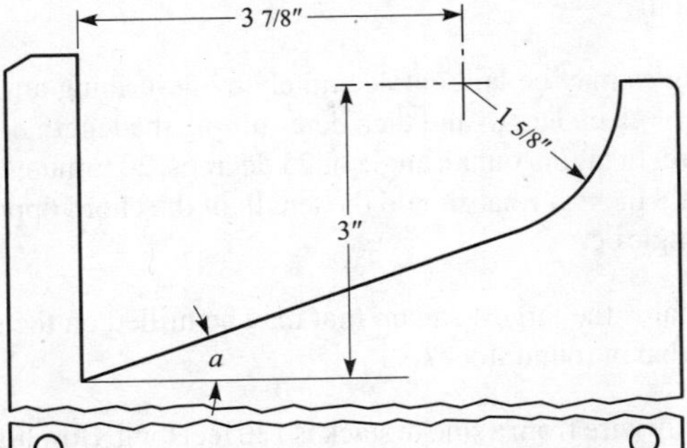


**Fig. 15. Find an Angle *a* Having the Dimensions Given**

11) Find the chordal distances as measured over plugs placed in holes located at different radii in the taximeter drive ring shown in **Fig. 16**. All holes are $^7/_{32}$ inch diameter; the angle between the center line of each pair of holes is 60 degrees.

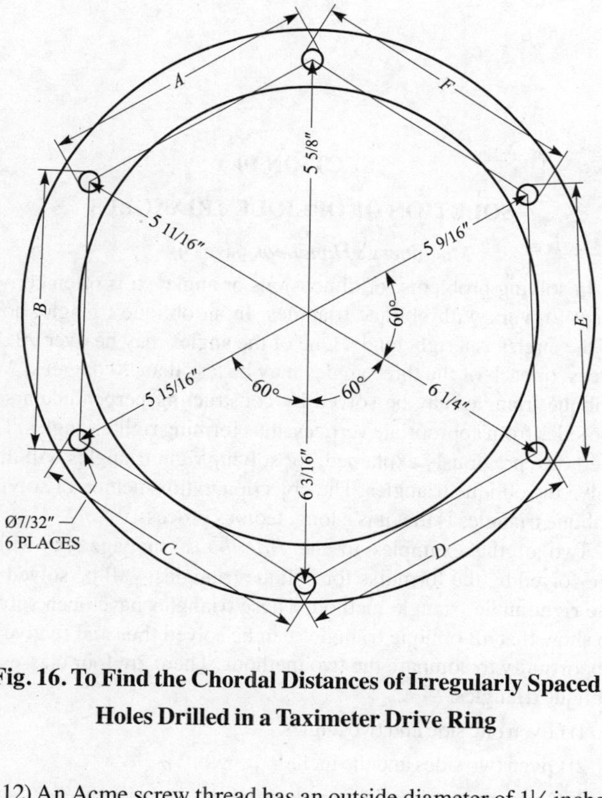


**Fig. 16. To Find the Chordal Distances of Irregularly Spaced Holes Drilled in a Taximeter Drive Ring**

12) An Acme screw thread has an outside diameter of $1\frac{1}{4}$ inches and has 6 threads per inch. Find the helix angle using the pitch diameter as a base. Also find the helix angle if a double thread is cut on the screw.

13) What is the lead of the flutes in a $\frac{7}{8}$-inch drill if the helix angle, measured from the center line of the drill, is 27° 30′?

14) A 4-inch diameter milling cutter has a lead of 68.57 inches. What is the helix angle measured from the axis?

## SECTION 10

## SOLUTION OF OBLIQUE TRIANGLES

*Machinery's Handbook* pages **95–97**

In solving problems for dimensions or angles, it is often convenient to work with oblique triangles. In an oblique triangle, none of the angles is a right angle. One of the angles may be over 90 degrees, or each of the three angles may be less than 90 degrees. Any oblique triangle may be solved by constructing perpendiculars to the sides from appropriate vertices, thus forming right triangles. The methods, previously explained, for solving right triangles will then solve the oblique triangles. The objection to this method of solving oblique triangles is that it is a long, tedious process.

Two of the examples in the *Handbook* on page **95**, which are solved by the formulas for oblique triangles, will be solved by the right-angle triangle method. These triangles have been solved to show that all oblique triangles can be solved thus and to give an opportunity to compare the two methods. There are four classes of oblique triangles:

1) Given one side and two angles

2) Given two sides and the included angle

3) Given two sides and the angle opposite one of them

4) Given the three sides

*Example 1:* Solve the first example on *Handbook* page **95** by the right-angle triangle method. By referring to **Fig. 1** on the next page:

$$\text{Angle } C = 180° - (62° + 80°) = 38°$$

Draw a line $DC$ perpendicular to $AB$.
In the right triangle $BDC$, $DC/BC = \sin 62°$.

$$\frac{DC}{5} = 0.88295; \; DC = 5 \times 0.88295 = 4.41475$$

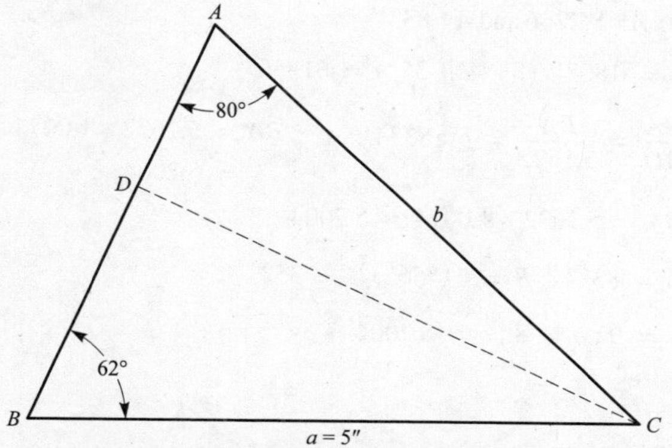


**Fig. 1. Oblique Triangle Solved by Right-Angle Triangle Method**

Angle $BCD = 90° - 62° = 28°; DCA = 38° - 28° = 10°$

$$\frac{BD}{5} = \cos 62°;\ BD = 5 \times 0.46947 = 2.34735$$

In triangle $ADC, AC/DC = \sec 10°$.

$$AC = 4.41475 \times 1.0154 = 4.4827$$

$$\frac{AD}{4.41475} = \tan 10°;\ AD = 4.41475 \times 0.17633 = 0.7785$$

$$\text{and } AB = AD + BD = 0.7785 + 2.34735 = 3.1258$$

$$C = 38°;\ b = 4.4827;\ c = 3.1258$$

*Example 2:* Apply the right-angle triangle method to the solution of the second example on *Handbook* page **95**.

Referring to **Fig. 2**, draw a line $BD$ perpendicular to $CA$.

In the right triangle $BDC, BD/9 = \sin 35°$.

$$BD = 9 \times 0.57358 = 5.16222$$

$$\frac{CD}{9} = \cos 35°;\ CD = 9 \times 0.81915 = 7.37235$$

$$DA = 8 - 7.37235 = 0.62765$$

In the right triangle $BDA, \dfrac{BD}{DA} = \dfrac{5.16222}{0.62765} = \tan A$.

$\tan A = 8.2246$ and $A = 83°4'$

$$B = 180° - (83°4' + 35°) = 61°56'$$

$$\frac{BA}{BD} = \frac{BA}{5.1622} = \csc 83°4'; \qquad BA = 5.1622 \times 1.0074 = 5.2004$$

$$BA = 5.1622 \times 1.0074 = 5.2004$$

$$A = 83°4' ; B = 61°56' ; C = 35°$$

$$a = 9 ; b = 8 ; c = 5.2004$$

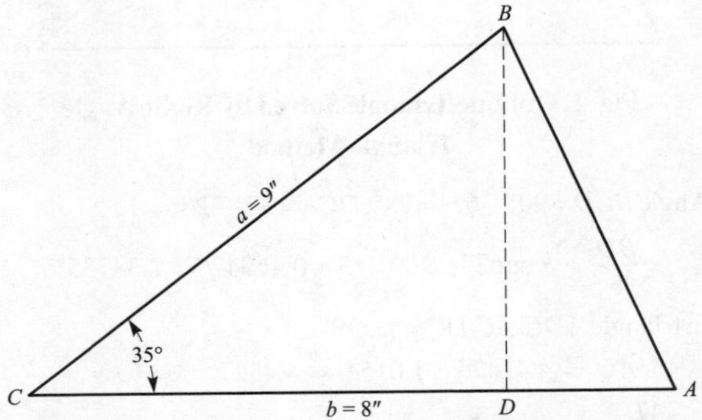


**Fig. 2. Another Example of the Right-Angle Triangle Solution of an Oblique Triangle Equation**

**Use of Formulas for Oblique Triangles.—**Oblique triangles are not encountered as frequently as right triangles, and, therefore, the methods of solving the latter may be fresh in the memory whereas methods for solving the former may be forgotten. All the formulas involved in the solution of the four classes of oblique triangles are derived from (1) the law of sines, (2) the law of cosines, and (3) the sum of angles of a triangle equal 180°.

The law of sines is that, in any triangle, the lengths of the sides are proportional to the sines of the opposite angles. (See diagrams on *Handbook* page **95**.)

$$\frac{a}{\sin A} = \frac{b}{\sin B} = \frac{c}{\sin C} \qquad (1)$$

Solving this equation, we get:

$\dfrac{a}{\sin A} = \dfrac{b}{\sin B}$; then $a \sin B = b \sin A$ and

$a = \dfrac{b \sin A}{\sin B}$; $\sin B = \dfrac{b \sin A}{a}$

$b = \dfrac{a \sin B}{\sin A}$ ; $\sin A = \dfrac{a \sin B}{b}$

In like manner, $\dfrac{a}{\sin A} = \dfrac{c}{\sin C}$ and

$a \sin C = c \sin A$; hence $\sin A = \dfrac{a \sin C}{c}$

and $\dfrac{b}{\sin B} = \dfrac{c}{\sin C}$ or $b \sin C = c \sin B$

Thus, twelve formulas may be derived. As a general rule, only **Formula (1)** is remembered, and special formulas are derived from it as required.

The law of cosines states that, in any triangle, the square of any side equals the sum of the squares of the other two sides minus twice their product multiplied by the cosine of the angle between them. These relations are stated as the following formulas:

$$a^2 = b^2 + c^2 - 2bc \cos A \quad \text{or} \quad a = \sqrt{b^2 + c^2 - 2bc \cos A} \qquad (1)$$

$$b^2 = a^2 + c^2 - 2ac \cos B \quad \text{or} \quad b = \sqrt{a^2 + c^2 - 2ac \cos B} \qquad (2)$$

$$c^2 = a^2 + b^2 - 2ab \cos C \quad \text{or} \quad c = \sqrt{a^2 + b^2 - 2ab \cos C} \qquad (3)$$

By solving **(1)**, $a^2 = b^2 + c^2 - 2bc \cos A$ for $\cos A$,

$$2bc \cos A = b^2 + c^2 - a^2 \quad \text{(transposing)}$$

$$\cos A = \frac{b^2 + c^2 - a^2}{2bc}$$

In like manner, formulas for $\cos B$ and $\cos C$ may be found.

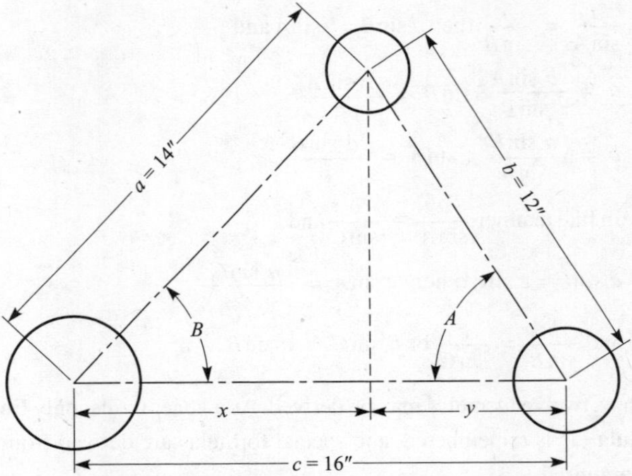


**Fig. 3. Finding Dimensions *x* and *y* from Known Values**

*Example 3:* A problem quite often encountered in layout work is illustrated in **Fig. 3**. Find the dimensions *x* and *y* between the holes, with these dimensions being measured from the intersection of the perpendicular line with the center line of the two lower holes. The three center-to-center distances are the only known values.

The first method might be to find the angle *A* (or *B*) by using a formula such as:

$$\cos A = \frac{b^2 + c^2 - a^2}{2bc}$$

and then solve the right triangle for *y* by the formula

$$y = b \cos A$$

**Formulas (1)** and **(2)** can be combined as follows:

$$y = \frac{b^2 + c^2 - a^2}{2c}$$

The value of *x* can be determined in a similar manner.

The second solution of this problem involves the following geometrical proposition: In any oblique triangle where the three sides are known, the ratio of the length of the base to the sum of the other two sides equals the ratio of the difference between the length of the two sides to the difference between the lengths $x$ and $y$. Therefore, if $a = 14$, $b = 12$, and $c = 16$ inches, then

$$c:(a+b) = (a-b):(x-y)$$

$$16:26 = 2:(x-y)$$

$$(x-y) = \frac{26 \times 2}{16} = 3\tfrac{1}{4} \text{ inches}$$

$$x = \frac{(x+y)+(x-y)}{2} = \frac{16 + 3\tfrac{1}{4}}{2} = 9.625 \text{ inches}$$

$$y = \frac{(x+y)-(x-y)}{2} = \frac{16 - 3\tfrac{1}{4}}{2} = 6.375 \text{ inches}$$

**When Angles Have Negative Values.**—In the solution of oblique triangles with one angle larger than 90 degrees, it is sometimes necessary to use angles whose functions are negative. (Review *Handbook* pages **3** and **129**.) Notice that for angles between 90 and 180 degrees, the cosine, tangent, cotangent, and secant are negative.

*Example 4:* In **Fig. 4**, two sides and the angle between them are shown. Find angles $A$ and $B$. (See *Handbook* page **95**.)

$$\tan A = \frac{4 \times \sin 20°}{3 - 4 \times \cos 20°} = \frac{4 \times 0.34202}{3 - 4 \times 0.93969} = \frac{1.36808}{3 - 3.75876}$$

In the denominator of the fraction above, the number to be subtracted from 3 is greater than 3; the difference will thus be negative. Hence:

$$\tan A = \frac{1.36808}{3 - 3.75876} = \frac{1.36808}{-0.75876} = -1.80305$$

The final result is negative because a positive number (1.36808) is divided by a negative number (–0.75876). The tangents of angles greater than 90 degrees and smaller than 180 degrees are negative. To illustrate an angle whose tangent is negative, enter the value −1.80305 in the calculator and find the corresponding angle, which is −60.986558 degrees or −60 degrees, 59 minutes, 59 seconds. Because the tangent is negative, angle *A* must be subtracted from 180 degrees, giving 119.01344 degrees or 119 degrees, 0 minutes, 49 seconds as the angle. Now angle *B* is found from the formula,

$$B = 180° - (A + C) = 180° - (119°0'11'' + 20°)$$
$$= 180° - 139°0'11'' = 40°59'49''$$

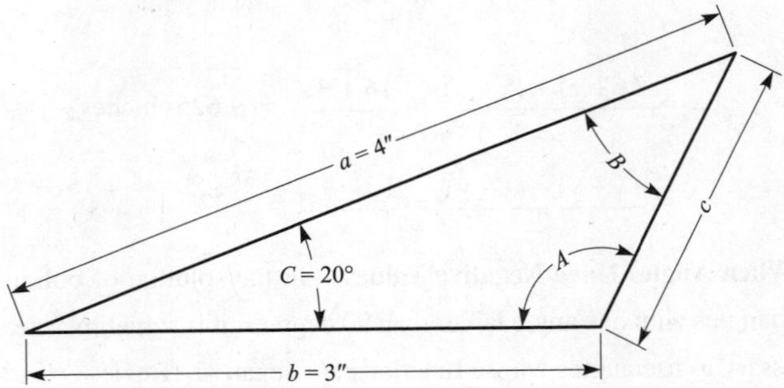


**Fig. 4. Finding Angles *A* and *B* from the Dimensions Given**

**When Either of Two Triangles Conforms to the Given Dimensions.—** When two sides and the angle opposite one of the given sides are known, *if the side opposite the given angle is shorter than the other given side*, two triangles can be drawn, having sides of the required length (as shown by **Fig. 5**) and the required angle opposite one of the sides. The lengths of the two known sides of each triangle are 8 and 9 inches, and the angle opposite the 8-inch side is 49 degrees, 27 minutes in each triangle; but it will be seen that the angle *B* of the lower triangle is very much larger than the corresponding angle of the upper triangle, and there is a great difference in the area. When two sides and one of the opposite angles are given, the problem can have two possible solutions when (and only when) the side opposite the given angle is shorter than the other given side. When the triangle to be calculated is drawn to scale, it is possible to determine from the shape of the triangle which of the two solutions applies.

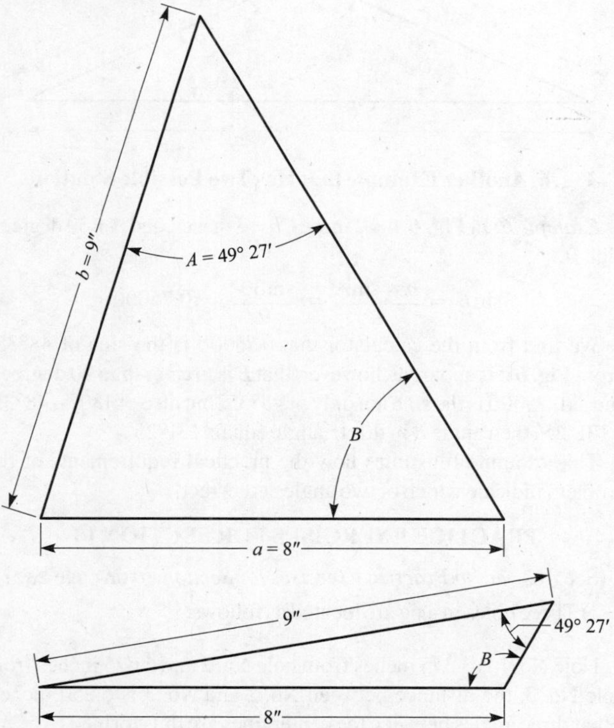


**Fig. 5. Example of Two Possible Solutions of the Same Problem, Which Is to Find Angle *B***

*Example 5:* Find angle $B$, from the formula $\sin B = (b \sin A)/a$, where $b = 9$ inches; $A = 49$ degrees, 27 minutes; and $a$ is the side opposite angle $A = 8$ inches.

$$\sin B = 9 \times 0.75984/8 = 0.85482 = \sin 58°44'34'' \quad \text{or}$$
$$\sin B = 121°15'36''$$

The practical requirements of the problem doubtless will indicate which of the two triangles shown in **Fig. 5** is the correct one.

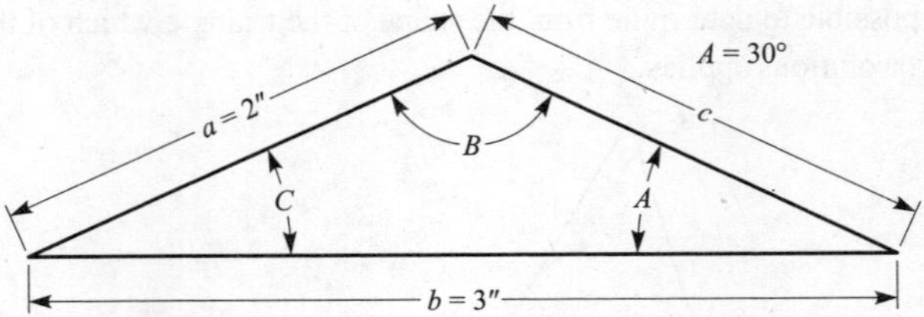


**Fig. 6. Another Example that Has Two Possible Solutions**

*Example 6:* In **Fig. 6**, $a = 2$ inches, $b = 3$ inches, and $A = 30$ degrees. Find $B$.

$$\sin B = \frac{b \times \sin A}{a} = \frac{\sin 30°}{2} = 0.75000$$

We find from the calculator that 0.75000 is the sine of 48°35′. From **Fig. 6** it is apparent, however, that $B$ is greater than 90 degrees, and as 0.75000 is the sine not only of 48°35′, but also of 180° − 48°35′ = 131°25′, then angle $B$ in this triangle equals 131°25′.

This example illustrates how the practical requirements of the problem indicate which of two angles is correct.

## PRACTICE EXERCISES FOR SECTION 10

(See *Answers to Practice Exercises for Section 10* on page **242**)

1) Three holes in a jig are located as follows:

Hole No. 1 is 3.375 inches from hole No. 2 and 5.625 inches from hole No. 3; the distance between No. 2 and No. 3 is 6.250 inches. What three angles between the center lines are thus formed?

2) In **Fig. 7**, a triangle has one side measuring 6.5 feet, and two angles $A$ and $C$ are 78 and 73 degrees, respectively. Find angle $B$, sides $b$ and $c$, and the area.

3) In **Fig. 8**, side $a$ equals 3.2 inches; angle $A$, 118 degrees; and angle $B$, 40 degrees. Find angle $C$, sides $b$ and $c$, and the area.

4) In **Fig. 9**, side $b$ = 0.3 foot, angle $B$ = 35°40′, and angle $C$ = 24°10′. Find angle $A$, sides $a$ and $c$, and the area.

5) Give two general rules for finding the areas of triangles.

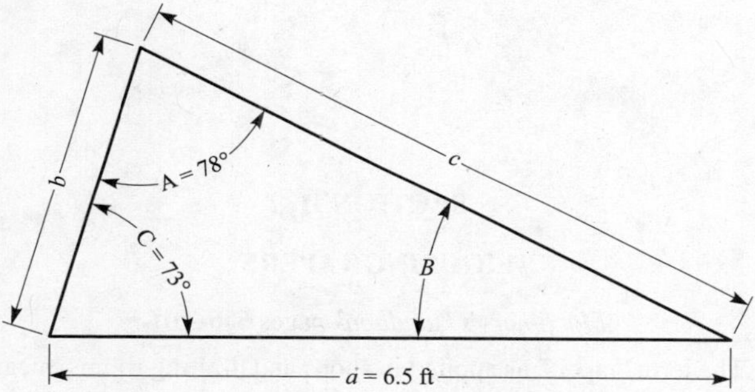


**Fig. 7. Example for Practice Exercise No. 2**

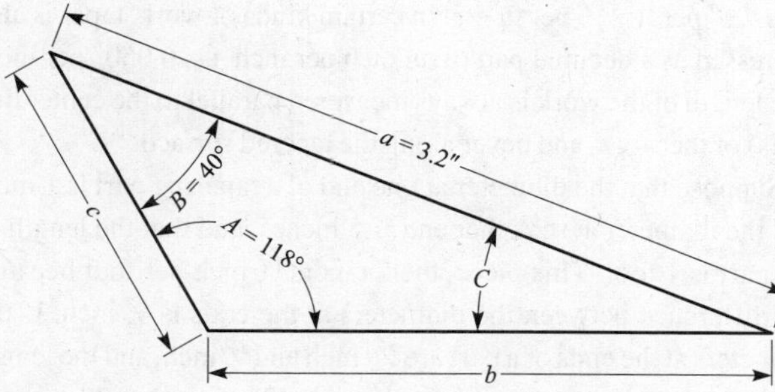


**Fig. 8. Example for Practice Exercise No. 3**

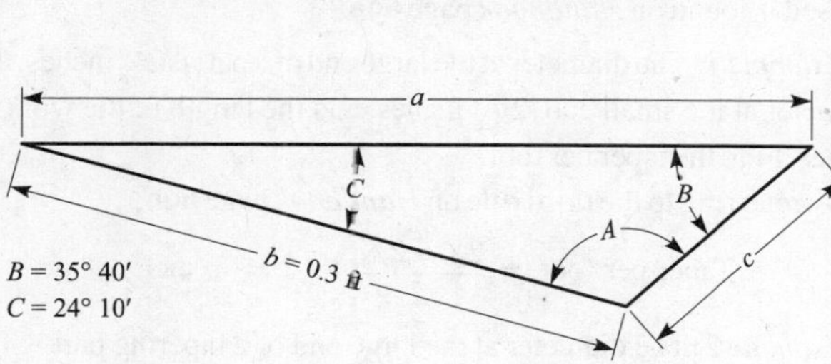


**Fig. 9. Example for Practice Exercise No. 4**

## SECTION 11

## FIGURING TAPERS

*Machinery's Handbook* pages **696–701**

The term "taper," as applied in shops and drafting rooms, means the difference between the large and small dimensions where the increase in size is uniform. Since tapering parts generally are conical, taper means the difference between the large and small diameters. Taper is ordinarily expressed as a certain number of inches per foot; thus, $\frac{1}{2}''$ per ft; $\frac{3}{4}''$ per ft; etc. In certain kinds of work, taper is also expressed as a decimal part of an inch per inch, as: 0.050″ per inch. The length of the work is always measured parallel to the center line (axis) of the work, and never along the tapered surface.

Suppose that the diameter at one end of a tapering part is 1 inch, and the diameter at the other end, 1.5 inches, and that the length of the part is 1 foot. This piece, then, tapers $\frac{1}{2}$ inch per foot because the difference between the diameters at the ends is $\frac{1}{2}$ inch. If the diameters at the ends of a part are $\frac{7}{16}$ inch and $\frac{1}{2}$ inch, and the length is 1 inch, this piece tapers $\frac{1}{16}$ inch per inch. The usual problems met when figuring tapers may be divided into seven classes. The rule to be used is found on *Handbook* page **699**.

*Example 1:* The diameter at the large end of a part is $2\frac{5}{8}$ inches, the diameter at the small end, $2\frac{3}{16}$ inches, and the length of the work, 7 inches. Find the taper per foot.

By referring to the third rule on *Handbook* page **699**,

$$\text{Taper per foot} = \frac{2\frac{5}{8} - 2\frac{3}{16}}{7} \times 12 = \tfrac{3}{4}\text{ inch}$$

*Example 2:* The diameter at the large end of a tapering part is $1\frac{5}{8}$ inches, the length is $3\frac{1}{2}$ inches, and the taper is $\frac{3}{4}$ inch per foot. The problem is to find the diameter at the small end.

By applying the fourth rule on *Handbook* page **699**,

$$\text{Diameter at small end} = 1\tfrac{5}{8} - \left(\frac{3/4}{12} \times 3\tfrac{1}{2}\right) = 1\tfrac{13}{32}$$

*Example 3:* What is the length of the taper if the two end diameters are 2.875 inches and 2.542 inches, the taper being 1 inch per foot?

By applying the sixth rule on *Handbook* page **699**,

$$\text{Distance between the two diameters} = \frac{2.875 - 2.542}{1} \times 12$$
$$= 4 \text{ inches nearly}$$

*Example 4:* If the length of the taper is 10 inches, and the taper is ¾ inch per foot, what is the taper in the given length?

By applying the last rule on *Handbook* page **699**,

$$\text{Taper in given length} = \frac{3/4}{12} \times 10 = 0.625 \text{ inch}$$

*Example 5:* The small diameter is 1.636 inches, the length of the work is 5 inches, and the taper is ¼ inch per foot; what is the large diameter?

By referring to the fifth rule on *Handbook* page **699**,

$$\text{Diameter at large end} = \left(\frac{1/4}{12} \times 5\right) + 1.636 = 1.740 \text{ inches}$$

*Example 6:* Sketch *A*, **Fig. 1**, shows a part used as a clamp bolt. The diameter, 3¼ inches, is given 3 inches from the large end of the taper. The total length of the taper is 10 inches. The taper is ⅜ inch per foot. Find the diameter at the large and small ends of the taper.

First find the diameter of the large end using the fifth rule on *Handbook* page **699**.

$$\text{Diameter at large end} = \left(\frac{3/8}{12} \times 3\right) + 3\tfrac{1}{4} = 3\tfrac{11}{32} \text{ inches}$$

To find the diameter at the small end, use the fourth rule on *Handbook* page **699**.

$$\text{Diameter at small end} = 3\tfrac{11}{32} - \left(\frac{3/8}{12} \times 10\right) = 3\tfrac{1}{32} \text{ inches}$$

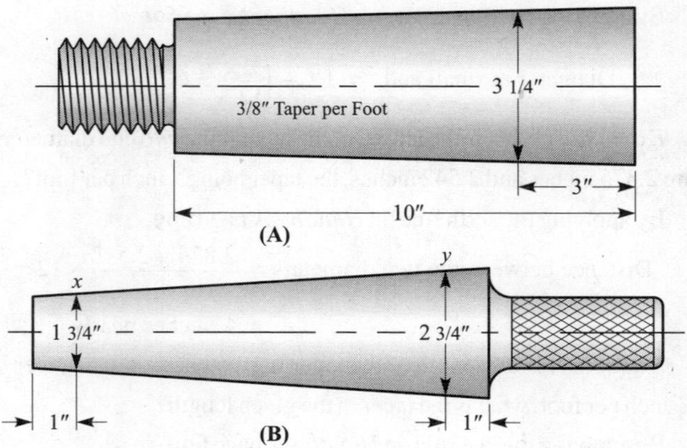


**Fig. 1. Illustrations for Example 6 and Example 7**

*Example 7:* At *B*, **Fig. 1**, is shown a taper master gage intended for inspecting taper-ring gages of various dimensions. The smallest diameter of the smallest ring gage is $1\frac{3}{4}$ inches, and the largest diameter of the largest ring gage is $2\frac{3}{4}$ inches. The taper is $1\frac{1}{2}$ inches per foot. It is required that the master gage extend 1 inch outside of the ring gages at both the small and the large ends, when these ring gages are tested. How long should the taper be on the master gage?

The sixth rule on *Handbook* page **699** may be applied here.

$$\text{Distance between the two diameters} = \frac{2\frac{3}{4} - 1\frac{3}{4}}{1\frac{1}{2}} \times 12$$

$$= 8 \text{ inches}$$

$$\text{Total length of taper} = 8 + 2 = 10 \text{ inches}$$

**Table for Converting Taper per Foot to Degrees.**—Some types of machines, such as milling machines, are graduated in degrees, making it necessary to convert the taper per foot to the corresponding angle in degrees. This conversion is quickly done by means of the table, *Handbook* page **698**.

*Example 8:* If a taper of $1\frac{1}{2}$ inches per foot is to be milled on a piece of work, at what angle must the machine table be set if the taper is measured from the axis of the work?

By referring to the table on *Handbook* page **698**, the angle corresponding to a taper of $1\frac{1}{2}$ inches to the foot is 3° 34′ 35″ as measured from the center line.

Note that the taper per foot varies directly as *the tangent of one-half the included angle.* Two mistakes frequently made in figuring tapers are assuming that the taper per foot varies directly as the included angle or that it varies directly as the tangent of the included angle. In order to verify this point, refer to the table on *Handbook* page **698**, where it will be seen that the included angle for a taper of 4 inches per foot (18° 55′ 29″) is not twice the included angle for a taper of 2 inches per foot (9° 31′ 38″). Neither is the tangent of 18° 55′ 29″ (0.3428587) twice the tangent of 9° 31′ 38″ (0.1678311).

**Tapers for Machine-Tool Spindles.**—The holes in machine-tool spindles, for receiving tool shanks, arbors, and centers, are tapered to ensure a tight grip, accuracy of location, and to facilitate removal of arbors, cutters, etc. The most common tapers are the Morse, the Brown & Sharpe, and the Jarno. The Morse has been very generally adopted for drilling-machine spindles. Most engine-lathe spindles also have the Morse taper, but some lathes have the Jarno or a modification of it, and others, a modified Morse taper, which is longer than the standard. A standard milling-machine spindle was adopted in 1927 by the milling-machine manufacturers of the National Machine Tool Builders' Association. A comparatively steep taper of $3\frac{1}{2}$ inches per foot was adopted in connection with this standard spindle to ensure instant release of arbors. Prior to the adoption of the standard spindle, the Brown & Sharpe taper was used for practically all milling machines and is also the taper for dividing-head spindles. There is considerable variation in grinding machine spindles. The Brown & Sharpe taper is the most common, but the Morse and the Jarno have also been used. Tapers of $\frac{5}{8}$ inch per foot and $\frac{3}{4}$ inch per foot also have been used to some extent on miscellaneous classes of machines requiring a taper hole in the spindle.

## PRACTICE EXERCISES FOR SECTION 11

(See *Answers to Practice Exercises for Section 11* on page **242**)

1) What tapers, per foot, are used with the following tapers: a) Morse taper; b) Jarno taper; c) milling-machine spindle; d) and taper pin?

2) What is the taper per foot on a part if the included angle is 10° 30′; 55° 45′?

3) In setting up a taper gage like that shown on *Handbook* page **697**, what should be the center distance between 1.75-inch and 2-inch disks to check either the taper per foot or angle of a No. 4 Morse taper?

4) If it is required to check an angle of 14½°, using two disks in contact, and the smaller disk is 1-inch diameter, what should the diameter of the larger disk be?

5) What should be the center distance, using disks of 2-inch and 3-inch diameter, to check an angle of 18° 30′ if the taper is measured from one side?

6) In grinding a reamer shank to fit a standard No. 2 Morse taper gage, it was found that the reamer stopped ⅜ inch short of going into the gage to the gage mark. How much should be ground off the diameter?

7) A milling machine arbor has a shank 6½ inches long with a No. 10 Brown & Sharpe taper. What is the total taper in this length?

8) A taper bushing for a grinding machine has a small inside diameter of ⅞ inch. It is 3 inches long with ½-inch taper per toot. Find the large inside diameter.

9) If a 5-inch sine-bar is used for finding the angle of the tapering bloc *A* (**Fig. 2**), and the heights of the sine-bar plug are as shown, find the corresponding angle *a* by means of the instructions beginning on *Handbook* page **694**.

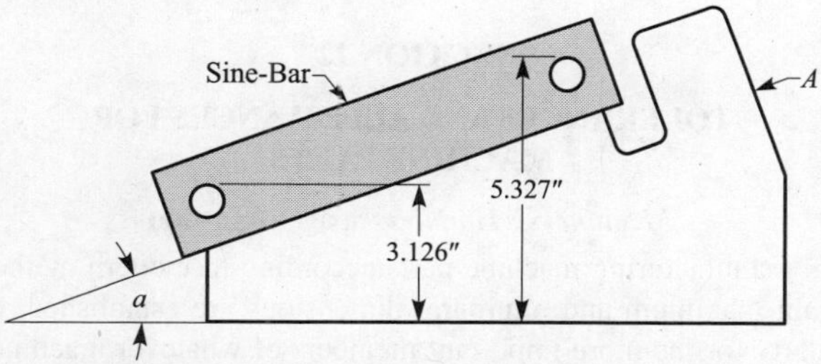


**Fig. 2. Finding Angle *a* by Means of a Sine-Bar and Handbook Instructions**

## SECTION 12

## TOLERANCES AND ALLOWANCES FOR MACHINE PARTS

*Machinery's Handbook* pages **641–690**

In manufacturing machine parts according to modern methods, certain maximum and minimum dimensions are established, particularly for the more important members of whatever machine or mechanism is to be constructed. These limiting dimensions serve two purposes: they prevent both unnecessary accuracy and excessive inaccuracies. A certain degree of accuracy is essential to the proper functioning of the assembled parts of a mechanism, but it is useless and wasteful to make parts more precise than needed to meet practical requirements. Hence, the use of proper limiting dimensions promotes efficiency in manufacturing and ensures standards of accuracy and quality that are consistent with the functions of the different parts of a mechanical device.

Parts made to specified limits usually are considered interchangeable or capable of use without selection, but there are several degrees of interchangeability in machinery manufacture. Strictly speaking, interchangeability consists of making the different parts of a mechanism so uniform in size and contour that each part of a certain model will fit any mating part of the same model, regardless of the lot to which it belongs or when it was made. However, as often defined, interchangeability consists in making each part fit any mating part in a certain series; that is, the interchangeability exists only in the same series. Selective assembly is sometimes termed interchangeability, but it involves a selection or sorting of parts as explained later. It will be noted that the strict definition of interchangeability does not imply that the parts must always be assembled without handwork, although that is usually considered desirable. It does mean, however, that when whatever process finishes the mating parts, they must assemble and function properly without fitting individual parts one to the other.

When a machine having interchangeable parts has been installed, possibly at some distant point, a broken part can readily be replaced by a new one sent by the manufacturer, but this feature is secondary as compared with the increased efficiency in manufacturing on an interchangeable basis. To make parts interchangeable, it is necessary to use gages and measuring tools, to provide some system of inspection, and to adopt suitable tolerances. Whether absolute interchangeability is practicable or not may depend upon the tolerances adopted, the relation between the different parts, and their form.

**Meanings of "Limit," "Tolerance," and "Allowance"**—The terms "limit" and "tolerance" and "allowance" are often used interchangeably, but each of these three terms has a distinct meaning and refers to different dimensions. As shown by **Fig. 1**, the *limits* of a hole or shaft are its diameters. *Tolerance* is the difference between two *limits* and limiting dimensions of a given part, and the term means that a certain amount of error is tolerated for practical reasons. *Allowance* is the difference between limiting dimensions on mating parts that are to be assembled either loosely or tightly, depending upon the amount allowed for the fit.

*Example 1:* Limits and fits for cylindrical parts are given starting on page **647** in the *Handbook*. These data provide a series of standard types and classes of fits. From **Table 8a** on page **654**, establish limits of size and clearance for a 2-inch diameter hole and shaft for a class RC-1 fit (hole H5, shaft g4).

$$\text{Max. hole} = 2 + 0.0005 = 2.0005$$
$$\text{Min. hole} = 2 - 0 = 2$$
$$\text{Max. shaft} = 2 - 0.0004 = 1.9996$$
$$\text{Min. shaft} = 2 - 0.0007 = 1.9993$$
$$\text{Min. allow.} = \text{min. hole} - \text{max. shaft} = 2 - 1.996 = 0.0004$$
$$\text{Max. allow.} = \text{max. hole} - \text{min. shaft}$$
$$= 2.0005 - 1.9993 = 0.0012$$

*Example 2:* Beginning on *Handbook* page **1952**, there are tables of dimensions for the Standard Unified Screw Thread Series—Class 1A, 2A, and 3A and B Fits. Determine the pitch-diameter tolerance of both screw and nut and the minimum and maximum allowance between screw and nut at the pitch diameter, assuming that the nominal diameter is 1 inch, the pitch is 8 threads per inch, and the fits are Class 2A and 2B for screw and nut, respectively.

**Differences Among "Limit," "Tolerance," and "Allowance"**

<table>
<tr><td rowspan="2">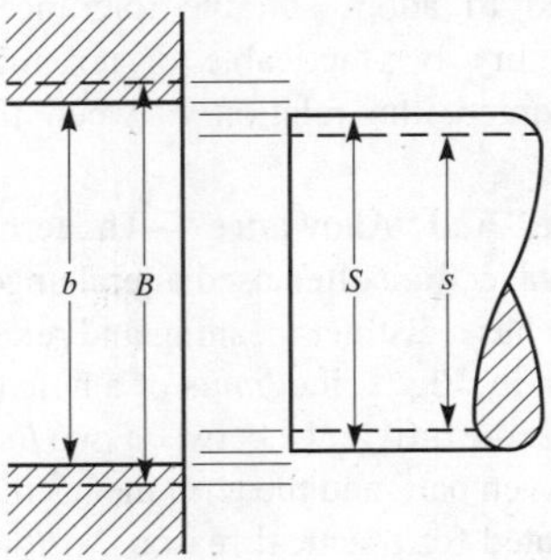

**Fig. 1.**</td><td>Limits</td><td>$B$ = maximum limit of bore<br>$b$ = minimum limit of bore<br>$S$ = maximum limit of shaft<br>$s$ = minimum limit of shaft</td></tr>
<tr><td>Tolerances</td><td>$B - b$ = maximum tolerance of bore<br>$S - s$ = maximum tolerance of shaft</td></tr>
<tr><td colspan="3">Allowances</td></tr>
<tr><td colspan="3">$B - s$ = maximum allowance, or if $s$ is greater than $B$ (as for tight or forced fits) then $s - B$ = minimum allowance for fit.<br>$b - S$ = minimum allowance, or if S is greater than b (as for tight or forced fits) then S − b = maximum allowance for fit.</td></tr>
</table>

The maximum pitch diameter or limit of the screw = 0.9168, and the minimum pitch diameter = 0.9100; hence, the tolerance = 0.9168 − 0.9100 = 0.0068 inch. The nut tolerance = 0.9276 − 0.9100 = 0.0176 inch. The maximum allowance for medium fit = maximum pitch diameter of nut − minimum pitch diameter of screw = 0.9276 − 0.9168 = 0.0108 inch. The minimum allowance = minimum pitch diameter of nut − maximum pitch diameter of screw = 0.9188 − 0.9168 = 0.0020.

**Relation of Tolerances to Limiting Dimensions and How Basic Size Is Determined.**—The absolute limits of the various dimensions and surfaces indicate danger points, in as much as parts made beyond these limits are unserviceable. A careful analysis of a mechanism shows that one of these danger points is more sharply defined than the other. For example, a certain stud must always assemble into a certain hole. If the stud is made beyond its maximum limit, it may be too large to assemble. If it is made beyond its minimum limit, it may be too loose or too weak to function. The absolute maximum limit in this case may cover a range of 0.001 inch, whereas the absolute minimum limit may have a range of at least 0.004 inch. In this case the maximum limit is the more sharply defined.

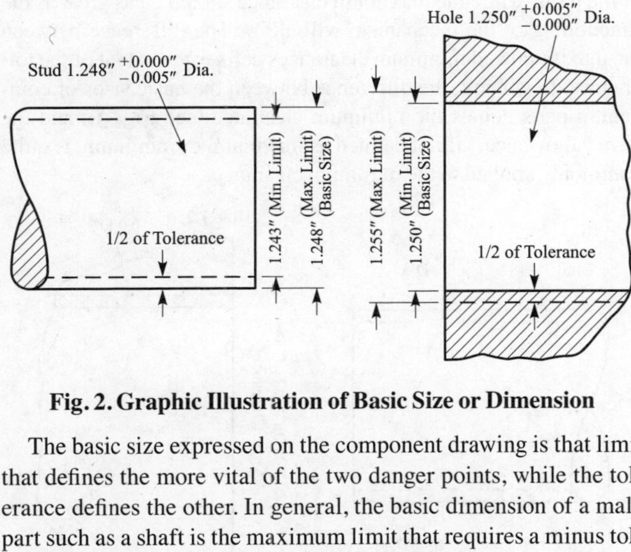


**Fig. 2. Graphic Illustration of Basic Size or Dimension**

The basic size expressed on the component drawing is that limit that defines the more vital of the two danger points, while the tolerance defines the other. In general, the basic dimension of a male part such as a shaft is the maximum limit that requires a minus tolerance. Similarly, the basic dimension of a female part is the minimum limit requiring a plus tolerance, as shown in **Fig. 2**. There are, however, dimensions that define neither a male nor a female surface, such as, for example, dimensions for the location of holes. In a few such instances, a variation in one direction is less dangerous than a variation in the other. Under these conditions, the basic dimension represents the danger point, and the unilateral tolerance permits a variation only in the less dangerous direction. At other times, the conditions are such that any variation from a fixed point in either direction is equally dangerous. The basic size then represents this fixed point, and tolerances on the drawing are bilateral and extend equally in both directions. (See *Handbook* page **641** for explanation of unilateral and bilateral tolerances.)

**When Allowance Provides Clearance Between Mating Parts.—** When one part must fit freely into another part like a shaft in its bearing, the allowance between the shaft and bearing represents a clearance space. It is evident that the amount of clearance varies widely for different classes of work. The minimum clearance should be as small as will permit the ready assembly and operation of the parts, while the maximum clearance should be as great as the functioning of the mechanism will allow. The difference between the maximum and minimum clearances defines the extent of the tolerances. In general, the difference between the basic sizes of companion parts equals the minimum clearance (see **Fig. 3**), and the term "allowance," if not defined as maximum or minimum, is quite commonly applied to the minimum clearance.

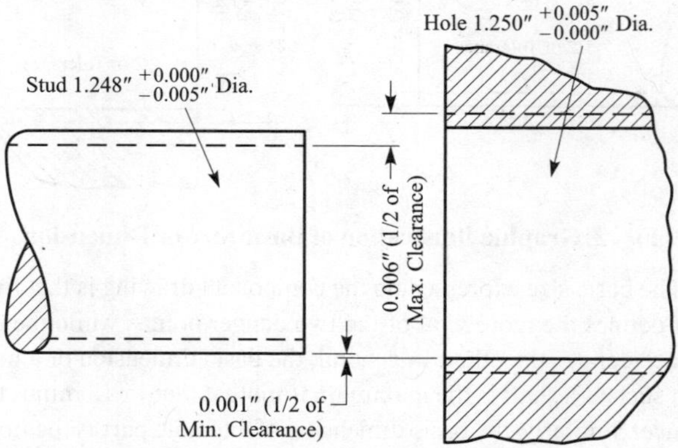


**Fig. 3. Graphic Illustration of Maximum and Minimum Clearance**

**When Interference of Metal Is the Result of Allowance.—** If a shaft or pin is larger in diameter than the hole into which it is forced, there is, of course, interference between the two parts. The metal surrounding the hole is expanded and compressed as the shaft or other part is forced into place.

Engine crankpins, car axles, and various other parts are assembled in this way (see paragraph *Allowance for Forced Fits*,

*Handbook* page **643**). The force and shrink fits in **Table 11** (starting on *Handbook* page **659**) all represent interference of metal.

If interchangeable parts are to be forced together, the minimum interference establishes the danger point. Thus, for force fits, the basic dimension of the shaft or pin is the minimum limit requiring a plus tolerance, and the basic dimension of the hole is the maximum limit requiring a minus tolerance, see **Fig. 4**.

**Obtaining Allowance by Selection of Mating Parts.**—The term "selective assembly" is applied to a method of manufacturing that is similar in many of its details to interchangeable manufacturing. In selective assembly, the mating parts are sorted according to size and assembled or interchanged with little or no further machining or hand work.

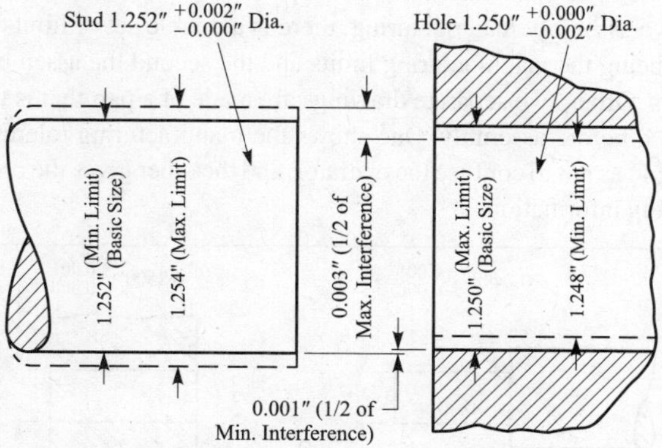


**Fig. 4. Illustration of the Meaning of Maximum and Minimum Interference**

The chief purpose of manufacturing by selective assembly is the production of large quantities of duplicate parts as economically as possible. As a general rule, the smaller the tolerances, the more exacting and expensive will be the manufacturing processes. However, it is possible to use comparatively large tolerances and then reduce them, in effect, by selective assembly, provided the quantity of parts is large enough to make such selective fitting possible. To illustrate, **Fig. 5** shows a plug or stud that has a plus tolerance of 0.001 inch and a hole that also has a plus tolerance of 0.001 inch. Assume that this tolerance of 0.001 inch represents the normal size variation on each part when manufactured efficiently. With this tolerance, a minimum plug in a maximum hole would have a clearance 0.2510 – 0.2498 = 0.0012 inch, and a maximum plug in a minimum hole would have a *metal interference* of 0.2508 – 0.2500 = 0.0008 inch. Suppose, however, that the clearance required for these parts must range from zero to 0.0004 inch. This reduction can be obtained by dividing both plugs and holes into five groups. (See below.) Any studs in Group *A*, for example, will assemble in any hole in Group *A*, but the studs in one group will not assemble properly in the holes in another group. When the largest stud in Group *A* is assembled in the smallest hole in Group *A*, the clearance equals zero. When the smallest stud in Group *A* is assembled in the largest hole in Group *A*, the clearance equals 0.0004 inch. Thus, in selective assembly manufacturing, there is a double set of limits, the first being the manufacturing limits and the second the assembling limits. Often, two separate drawings are made of a part that is to be graded before assembly. One shows the manufacturing tolerances only, so as not to confuse the operator, and the other gives the proper grading information.

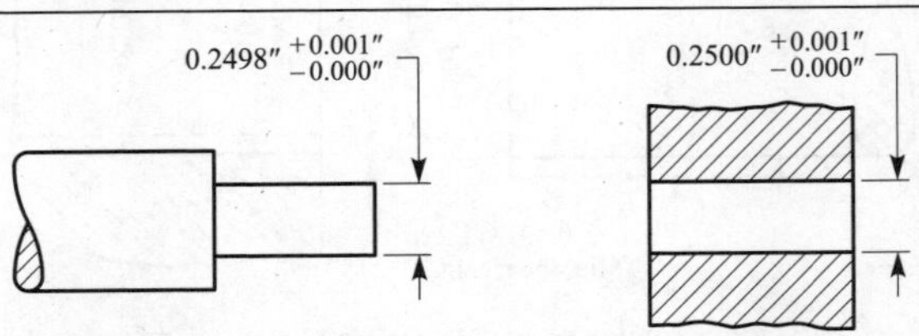


**Fig. 5. Information Placed on Drawings in Selective Assembly Manufacturing to Facilitate Grading of Parts**

| | | | | | |
|---|---|---|---|---|---|
| GROUP A | 0.2498″ | +0.002″ −0.0000″ | GROUP A | 0.2500″ | +0.002″ −0.0000″ |
| GROUP B | 0.2500″ | +0.002″ −0.0000″ | GROUP B | 0.2502″ | +0.002″ −0.0000″ |
| GROUP C | 0.2502″ | +0.002″ −0.002″ | GROUP C | 0.2504″ | +0.002″ −0.002″ |
| GROUP D | 0.2504″ | +0.0000″ −0.002″ | GROUP D | 0.2506″ | +0.0000″ −0.002″ |
| GROUP E | 02506″ | +0.0000″ −0.002″ | GROUP E | 0.2508″ | +0.0000″ −0.002″ |

*Example 3:* Force and shrink fit data are given in **Table 11**, page **659** of the *Handbook*. Establish the limits of size and interference of the hole and shaft for a Class FN-1 fit of 2-inch diameter.

$$\text{Max. hole} = 2 + 0.0007 = 2.0007; \text{ min. hole} = 2 - 0 = 2$$

$$\text{Max. shaft} = 2 + 0.0018 = 2.0018; \text{ min. shaft} = 2 + 0.0013 = 2.0013$$

In the second column of the table, the minimum and maximum interferences are given as 0.0006 and 0.0018 inch, respectively, for a FN-1 fit of 2-inch diameter. For a *selected* fit, shafts are selected that are 0.0012 inch larger than the mating holes; that is, for any mating pair, the shaft is larger than the hole by an amount midway between the minimum (0.0006-inch) and maximum (0.0018 inch) interference.

**Dimensioning Drawings to Ensure Obtaining Required Tolerances.**—In dimensioning the drawings of parts requiring tolerances, there are certain fundamental rules that should be applied.

*Rule 1:* In interchangeable manufacturing there is only one dimension (or group of dimensions) in the same straight line that can be controlled within fixed tolerances. This dimension is the distance between the cutting surface of the tool and the locating or registering surface of the part being machined. Therefore, it is incorrect to locate any point or surface with tolerances from more than one point in the same straight line.

*Rule 2:* Dimensions should be given between those points that it is essential to hold in a specific relation to each other. Most dimensions, however, are relatively unimportant in this respect. It is good practice to establish common location points in each plane and give, as far as possible, all such dimensions from these points.

*Rule 3:* Basic dimensions given on component drawings for interchangeable parts should be, except for force fits and other unusual conditions, the *maximum metal* size (maximum shaft or plug and minimum hole). The direct comparison of the basic sizes should check the danger zone, which is the minimum clearance condition in most instances. It is evident that these sizes are the most important ones, as they control interchangeability, and should be the first determined. Once established, they should remain fixed if the mechanism functions properly, and the design is unchanged.

The direction of the tolerances, then, would be that which recedes from the danger zone. In most instances, this directionality means that the direction of the tolerances is that which will increase the clearance. For force fits, basic dimensions determine the minimum interference, and tolerances limit the maximum interference.

*Rule 4:* Dimensions must not be duplicated between the same points. The duplication of dimensions causes much needless trouble, due to changes being made in one place and not in the others. It is easier to search a drawing to find a dimension than it is to have them duplicated and more readily found but inconsistent.

*Rule 5:* As far as possible, the dimensions on comparison parts should be given from the same relative locations. Such a procedure assists in detecting interferences and other improper conditions.

In attempting to work in accordance with general laws or principles, one other elementary rule should always be kept in mind. Special requirements need special consideration. The following detailed examples are given to illustrate the application of the five rules and to indicate results of their violation.

**Violations of Rules for Dimensioning.—Fig. 6** shows a very common method of dimensioning a part such as the stud shown, but one that is bad practice as it violates the first and second rules. The dimensions given for the diameters are correct, so they are eliminated from the discussion. The dimensions given for the various lengths are wrong: First, because they give no indication as to the essential lengths; second, because of several possible sequences of operations, some of which would not maintain the specified conditions.

**Fig. 7** shows one possible sequence of operations indicated alphabetically. If we first finish the dimension *a* and then finish *b*, the dimension *c* will be within the specified limits. However, the dimension *c* is then superfluous. **Fig. 8** gives another possible sequence of operations. If we first establish *a*, and then *b*, the dimension *c* may vary 0.030 instead of 0.010 inch as specified in **Fig. 6**. **Fig. 9** gives a third possible sequence of operations. If we first finish the overall length *a*, and then the length of the body *b*, the stem *c* may vary 0.030 inch instead of 0.010 inch as specified in **Fig. 6**.

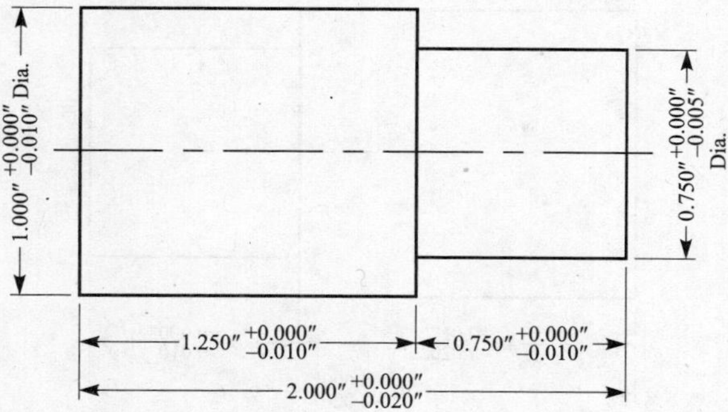


**Fig. 6. Common but Incorrect Method of Dimensioning**

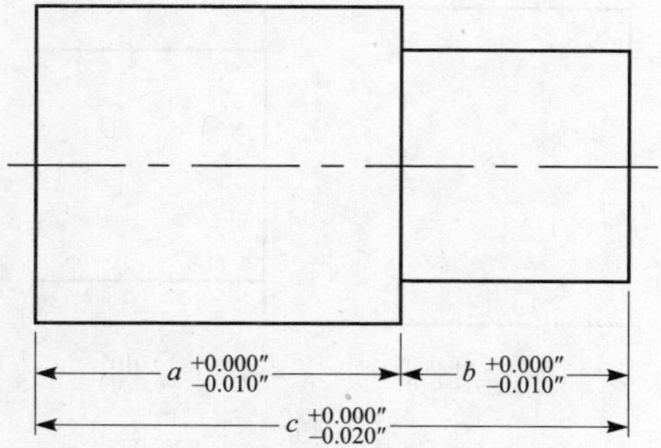


**Fig. 7. One Interpretation of Dimensioning in Fig. 6**

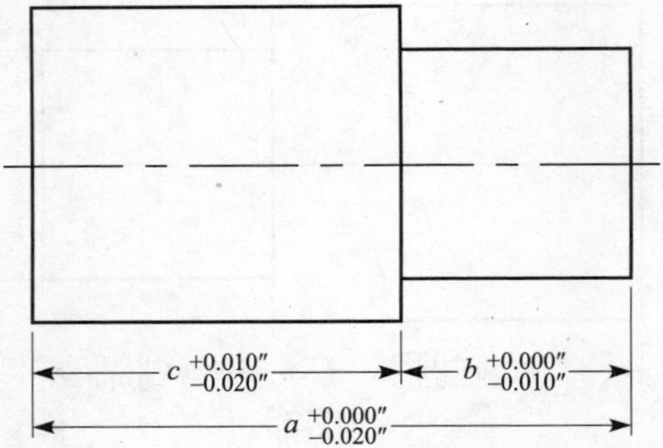


**Fig. 8. Another Interpretation of Dimensioning in Fig. 6**

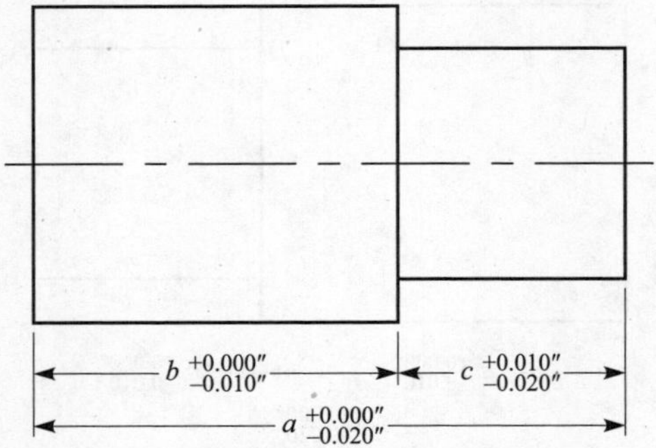


**Fig. 9. A Third Interpretation of Dimensioning in Fig. 6**

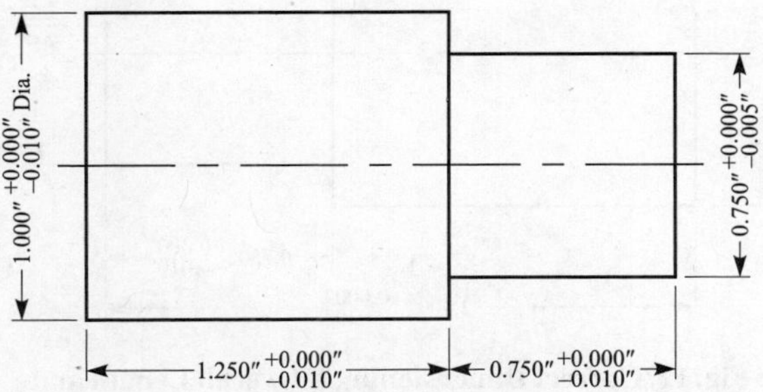


**Fig. 10. Correct Dimensioning if Length of Body and Length of Stem Are Most Important**

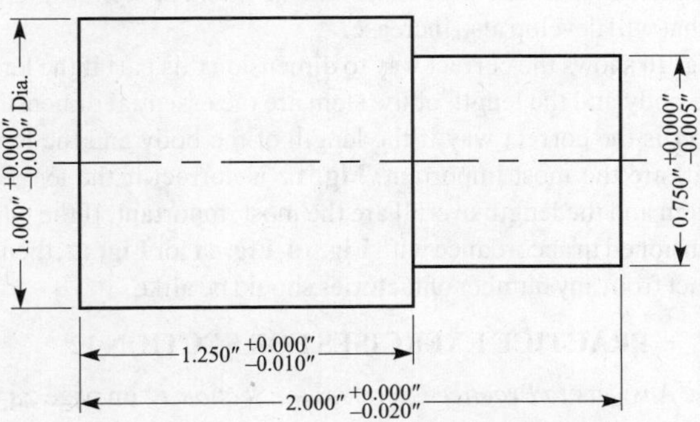


**Fig. 11. Correct Dimensioning if Length of Body and Overall Length Are Most Important**

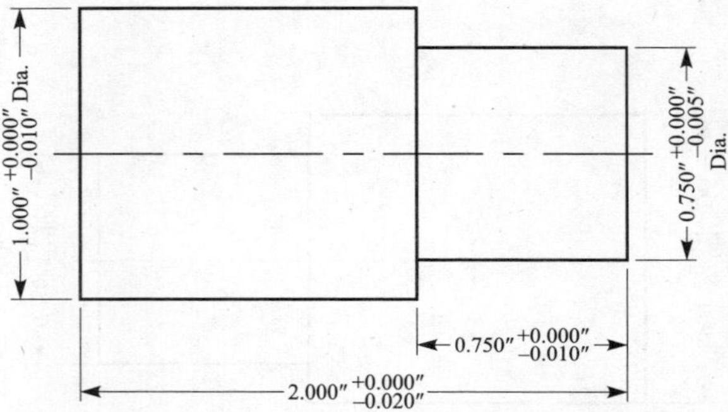


**Fig. 12. Correct Dimensioning if Overall Length and Length of Stem Are Most Important**

If three different plants were manufacturing this part, each one using a different sequence of operations, it is evident from the foregoing that a different product would be received from each plant. The example given is the simplest one possible. As the parts become more complex, and the number of dimensions increases, the number of different combinations possible and the extent of the variations in size that will develop also increase.

**Fig. 10** shows the correct way to dimension this part if the length of the body and the length of the stem are the essential dimensions. **Fig. 11** is the correct way if the length of the body and the length overall are the most important. **Fig. 12** is correct if the length of the stem and the length overall are the most important. If the part is dimensioned in accordance with **Fig. 10**, **Fig. 11**, or **Fig. 12**, then the product from any number of factories should be alike.

## PRACTICE EXERCISES FOR SECTION 12

(See *Answers to Practice Exercises for Section 12* on page **243**)

1) What factors influence the allowance for a forced fit?

2) What is the general practice in applying tolerances to center distances between holes?

3) A 2-inch shaft is to have a tolerance of 0.003 inch on the diameter. Show, by examples, three ways of expressing the shaft dimensions.

4) In what respect does a bilateral tolerance differ from a unilateral tolerance? Give an example that demonstrates this difference.

5) What is the relationship between gagemaker's tolerance and workplace tolerance?

6) Name the different classes of fits for screw thread included in the American Standards.

7) How does the Unified screw for screw threads differ from the former American Standard with regard to clearance between mating parts? With regard toward working tolerance?

8) Under what conditions is one limiting dimension or "limit" also a basic dimension?

9) According to table at the bottom of *Handbook* page **650**, broaching will produce work within tolerance grades 5 through 8. What does this mean in terms of thousandths of an inch, considering a 1-inch diameter broached hole?

10) Does surface roughness affect the ability to work within the tolerance grades specified in Exercise 10?

## SECTION 13
## USING STANDARDS DATA AND INFORMATION

(References to Standards appear throughout the *Machinery's Handbook, 32nd Edition*)

Standards are needed in metalworking manufacturing to establish dimensional and physical property limits for parts that are to be interchangeable. Standards make it possible for essential parts, such as nuts, screws, bolts, splines, gears, etc., to be manufactured at different times and places with the assurance that they will meet assembly requirements. Standards are also needed for commonly used tools, such as twist drills, reamers, milling cutters, etc., so that only a given number of sizes need be made available to cover a given range and to ensure adequate performance. Also, performance standards often are established to make sure that machines and equipment will satisfy their application requirements.

A specific standard may be established by a company on a limited basis for its own use. An industry may find that a standardized set of specifications is needed, and its member companies working through their trade association come to an agreement as to what requirements should be included. Sometimes, industry standards sponsored by a trade association or an engineering society become acceptable by a wide range of consumers, manufacturers, and government agencies as national standards and are made available through a national authority such as the American National Standards Institute (ANSI) and the American Society of Mechanical Engineers (ASME). More countries have been deciding that standards should be universal and are working to this end through the International Organization for Standardization (ISO) and other international bodies.

In the United States and some other English-speaking countries, there are two systems of measurement in use: the inch system and the metric system. As a result, standards for, say, bolts, nuts, and screws have been developed for both inch and metric dimensions as will be found in *Machinery's Handbook*. However, an increasing number of multinational corporations and their local suppliers are finding it prohibitively expensive to operate with two systems of measurements and standards. Thus, in order to use available expertise in one plant location, a machine may be designed in an "inch" nation only to be produced later in a "metric" country or vice versa. This situation generates additional costs in the conversion of drawings, substitution of equivalent standard steel sizes and fasteners, and conversion of testing and material specifications, etc. Because of these problems, more and more official Standards are being developed and published in the United States and throughout the world that are based, wherever practicable, upon established ISO and other authoritative organizational Standards.

In the *Handbook*, the user will find that a large number of both inch and metric Standards data and information are provided. It should be noted that at the head of each table of data derived from a specific Standard, the source is given in parentheses, such as (ANSI/ASME B18.3-2012). ANSI indicates the American National Standards Institute and ASME stands for the American Society of Mechanical Engineers; B18.3 is the identifying number of the Standard; and 2012 is the date the Standard was published, or revised, and became effective. If there is an additional year after the publishing year with an "R" in front of it, the Standard was "reaffirmed" during that year. And even when Standards are officially withdrawn, they may still be referenced for important content.

Generally, new products are produced to the metric Standards; older products and replacement parts for them may require reference to older inch Standards, and some products, such as inch-unit pipe threads, are considered as standard for the near future because of widespread use throughout the world.

**Important Objectives of Standardization.**—The purpose of standardization is to manufacture goods for less direct and indirect costs and to provide finished products that meet the demands of the marketplace. A more detailed description of the objectives could be as follows:

*Lower the production costs when the aim is to:*

1) Facilitate and systematize the work of skilled designers;

2) Ensure optimum selection of materials, components, and semi-finished products;

3) Reduce stocks of materials, semifinished products, and finished products;

4) Minimize the number of different products sold; and

5) Facilitate and reduce the cost of procurement of purchased goods.

*Meet the demands of the market place, when the objective is to:*

1) Conform to regulations imposed by government and trade organizations;

2) Stay within safety regulations set forth by governments; and

3) Facilitate interchangeability requirements with existing products.

**Standardization Technique.**—The two commonly used basic principles for the preparation of a Standard are:

1) Analytical standardization–Standard developed from scratch.

2) Conservative standardization–Standard based, so far as is possible, on existing practice.

In practice, it appears that a Standard cannot be prepared completely by one or the other of the two methods but emerges from a compromise between the two. The goal of the standardization technique, then, should be to utilize the basic material and the rules and the aids available in such a way that a valid and practical compromise solution is reached.

The basic material could consist of such items as former company Standards, vendor catalog data, national and international Standards, requirements of the company's customers, and competitor's material. Increasingly important are the national and international Standards in existence on the subject; they should always play an important part in any conservative standardization work. For example, it would be foolish to create a new metric Standard without first considering some existing European metric Standards.

**Standards Information in the *Handbook*.**—Among the many kinds of material and data to be found in the *Handbook*, the user will note that extensive coverage is given to Standards of several types: American National Standards, British Standards, ISO Standards, engineering society Standards, trade association Standards, and, in certain instances, company product Standards. Both inch and metric system Standards are given wherever appropriate. Inch dimension Standards sometimes are provided only for use during transition to metric Standards or to provide information for the manufacture of replacement parts.

In selecting Standards to be presented in the *Handbook*, the editors have chosen those Standards most appropriate to the needs of *Handbook* users. Text, illustrations, formulas, tables of data, and examples have been arranged in the order best suitable for direct and quick use. As an example of this type of presentation, the section on bevel gearing, starting on *Handbook* page **2258**, begins with text material that provides the basis for understanding information presented in the AGMA Standards; the illustrations on *Handbook* pages **2263–2264** provide visual definition of essential parts and dimensions of a bevel gear; the formulas on *Handbook* page **2252** show how to calculate dimensions of milled bevel gears; the tables starting on *Handbook* page **2266** give numbers of formed cutters used to mill teeth in mating bevel gear and pinion sets with shafts at right angles; and finally, the worked-out examples beginning on *Handbook* page **2268** give a step-by-step procedure for selecting formed cutters for milling bevel gears. Also, where combinations of tables and formulas are given, the formulas have been arranged in the best sequence for computation with the aid of a pocket calculator.

**"Soft" Conversion of Inch to Metric Dimensions.**—The dimensions of certain products, when specified in inches, may be converted to metric dimensions, or vice versa, by multiplying by the appropriate conversion factor so that the parts can be fabricated either to inch or to the equivalent metric dimensions and still be fully interchangeable. Such a conversion is called a *soft* conversion. An example of a soft conversion is available on *Handbook* page **2476**, which gives the inch dimensions of standard lockwashers for ball bearings. The footnote to the table indicates that multiplication of the tabulated inch dimensions by 25.4 and rounding the results to two decimal places will provide the equivalent metric dimensions.

**"Hard" Metric or Inch Standard Systems.**—In a *hard* system, those dimensions in the system that have been standardized cannot be converted to another dimensional system that has been standardized independently of the first system. As stated in the footnote on page **2353** of the *Handbook*, "In a 'hard' system the tools of production, such as hobs, do not bear a usable relation to the tools in another system; i.e., a 10 diametral pitch hob calculates to be equal to a 2.54 module hob in the metric module system, a hob that does not exist in the metric standard."

**Interchangeability of Parts Made to Revised Standards.—** Where a Standard has been revised, there may still remain some degree of interchangeability between older parts and those made to the new Standard. As an example, starting on page **2344** of the *Handbook*, there are two tables showing which of the internal and external involute splines made to older Standards will mate with those made to newer Standards.

## PRACTICE EXERCISES FOR SECTION 13

(See *Answers to Practice Exercises for Section 13* on page **244**)

1) What is the breaking strength of a 6 × 7 fiber-core wire rope $\frac{1}{4}$ inch in diameter if the rope material is mild plow steel?

2) What factor of safety should be applied to the rope in Exercise 1?

3) How many carbon steel balls of $\frac{1}{4}$-inch diameter would weigh 1 lb? How would this information be obtained without the table?

4) For a 1-inch diameter of shaft, what size square key is appropriate?

5) Find the hole size needed for a $\frac{5}{32}$-inch standard cotter pin.

6) Find the limits of size for a 0.1250-inch diameter hardened and ground dowel pin.

7) For a 3AM1-17 retaining ring (snap ring), what is the maximum allowable speed of rotation?

8) Find the hole size required for a type AB steel thread-forming screw of number 6 size in 0.105-inch-thick stainless steel.

## SECTION 14

## STANDARD SCREW AND PIPE THREADS

*Machinery's Handbook* pages **1943–2150**

Different screw thread forms and standards have been originated and adopted at various times, either because they were considered superior to other forms or because of the special requirements of screws used for a certain class of work.

A standard thread conforms to an adopted Standard with regard to the form or contour of the thread itself and as to the pitches or numbers of threads per inch for different screw diameters.

The United States Standard formerly used in the United States was replaced by an American Standard having the same thread form as the former Standard and a more extensive series of pitches, as well as tolerances and allowances for different classes of fits. This American Standard was revised in 1949 to include a Unified Thread Series, which was established to obtain screw thread interchangeability among the United Kingdom, Canada, and the United States.

The Standard was revised again in 1959. The Unified threads are now generally used in the United States and the former American Standard threads are now used only in certain applications where the changeover in tools, gages, and manufacturing has not been completed. The differences between Unified and the former National Standard threads are explained on pages **1943** and **1950** in the *Handbook*.

As may be seen in the table on *Handbook* page **1953**, the Unified series of screw threads consists of three Standard series having graded pitches (UNC, UNF, and UNEF) and eight Standard series of uniform (constant) pitch. In addition to these Standard series. There are places in the table beginning on *Handbook* page **1954** where special threads (UNS) are listed. These UNS threads are for use only if Standard series threads do not meet requirements.

*Example 1:* The table on *Handbook* page **1981** shows that the pitch diameter of a 2-inch screw thread is 1.8557 inches. What is meant by the term "pitch diameter" as applied to a screw thread and how is it determined?

According to a definition given in connection with American Standard screw threads, the *pitch diameter* of a straight (nontapering) screw thread is the diameter of an imaginary cylinder, the surface of which would pass through the threads at such points as to make equal the width of the threads and the width of the spaces cut by the surface of the cylinder.

The basic pitch diameter equals the basic major (outside) diameter minus two times the addendum of the external thread (*Handbook* page **1952**), so the basic pitch diameter for the 2-inch example, with $4\frac{1}{2}$ threads per inch, is $2.00 - 2 \times 0.07217 = 1.8557$ inches.

*Example 2:* The tensile strength of a bolt $3\frac{1}{2}$ inches in diameter at a stress of 6000 pounds per square inch may be calculated by means of the formulas on *Handbook* page **1668**. This formula uses the largest diameter of the bolt, avoiding the need to take account of the reduced diameter at the thread root, and gives a tensile strength of 35,175 pounds for the conditions noted.

If the second formula on page **1668**, based on the area of the smallest diameter, is used for the same bolt and stress, and the diameter of the thread root is taken as 3.1 inches, then the tensile strength is calculated as 40,636 pounds. The difference in these formulas is that the first uses a slightly greater factor of safety than the second, taking account of possible variations in thread depth.

*Example 3: Handbook* page **2131** gives formulas for checking the pitch diameter of screw threads by the three-wire method (when effect of lead angle is ignored). Show how these formulas have been derived using the one for the American National Standard Unified thread as an example.

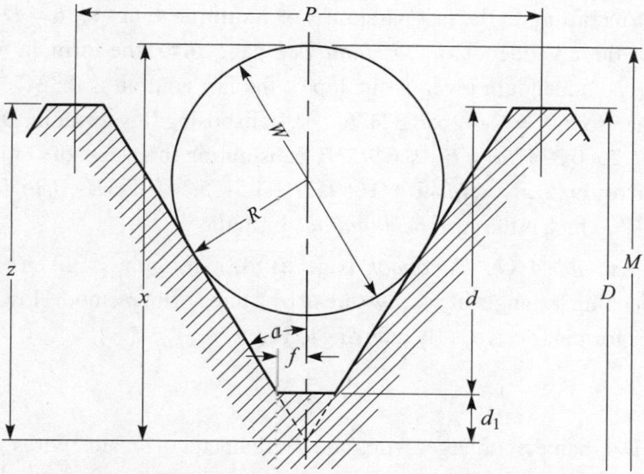


**Fig. 1. Diagram Illustrating the Derivation of Formulas for Three-Wire Measurements of Screw Thread Pitch Diameters**

It is evident from **Fig. 1** that:

$$M = D - 2z + 2x \tag{1}$$

$$x = R + \frac{R}{\sin a} \text{ and } 2x = 2R + \frac{2R}{0.5}\text{; hence,}$$

$$2x = \frac{(2 \times 0.5 + 2)R}{0.5} = \frac{3R}{0.5} = 6R = 3W$$

$$z = d + d_1 = 0.6495P + f \times \cot\alpha$$

$$f = 0.0625P\text{; therefore,}$$

$$z = 0.6495P + 0.10825P = 0.75775P$$

If, in **Formula (1)**, we substitute the value of $2z$ or $2 \times 0.75775P$ and the value of $2x$, we have:

$$M = D - 1.5155 \times P + 3W \tag{2}$$

This **Formula (2)** is the one found in previous editions of the *Handbook*. In the 22nd and subsequent editions of the *Handbook* use of the outside diameter $D$ in **Formula (2)** above was eliminated to provide a formula in terms of the pitch diameter $E$. Such a formula is useful for finding the wire measurement corresponding to the actual pitch diameter, whether it be correct, undersize, or oversize.

According to the last paragraph of **Example 1**, above, $E = D - 2 \times$ thread addendum. On *Handbook* page **1952**, the formula for thread addendum given at the top of the last column is $0.32476P$. Therefore, $E = D - 2 \times 0.32476P$, or, transposing this formula, $D = E + 2 \times 0.32476P = E + 0.64952P$. Substituting this value of $D$ into **Formula (2)** gives: $M = E + 0.64952P - 1.5155P + 3W = E - 0.8660P + 3W$, which is the current *Handbook* formula.

*Example 4:* On *Handbook* page **2138**, a formula is given for checking the angle of a screw thread by a three-wire method. How is this formula derived? By referring to **Fig. 2**,

$$\sin a = \frac{W}{S} \qquad (1)$$

If $D$ = diameter of larger wires and $d$ = diameter of smaller wires,

$$W = \frac{D - d}{2}$$

If $B$ = difference in measurement over wires, then the difference $S$ between the centers of the wires is:

$$S = \frac{B - (D - d)}{2}$$

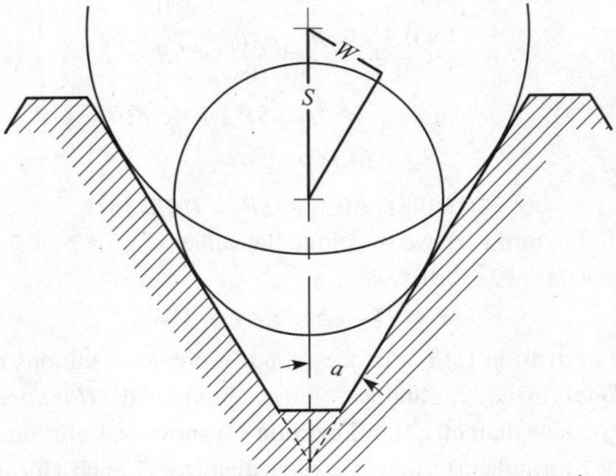


**Fig. 2. Diagram Illustrating the Derivation of Formula for Checking the Thread Angle by the Three-Wire System**

By inserting these expressions for $W$ and $S$ in **Formula (1)** and canceling, the formula given in the *Handbook* is obtained if $A$ is substituted for $D-d$.

$$\sin a = \frac{A}{B-A}$$

*Example 5:* A vernier gear-tooth caliper (like the one shown on *Handbook* page **2229**) is to be used for checking the width of an Acme screw by measuring squarely across or perpendicular to the thread. Since standard screw thread dimensions are in the plane of the axis, how is the width square or normal to the sides of the thread determined? Assume that the width is to be measured at the pitch line and that the number of threads per inch is two.

The table on *Handbook* page **2060** shows that for two threads per inch, the depth is 0.260 inch; hence, if the measurement is to be at the pitch line, the vertical scale of the caliper is set to $(0.260 - 0.010) \div 2 = 0.125$ inch. The pitch equals

$$\frac{1}{\text{No. of threads per inch}} = \tfrac{1}{2}\text{ inch}$$

The width $A$, **Fig. 3**, in the plane of the axis equals $\frac{1}{2}$ the pitch, or $\frac{1}{4}$ inch. The width $B$ perpendicular to the sides of the thread = width in axial plane × cosine helix angle.

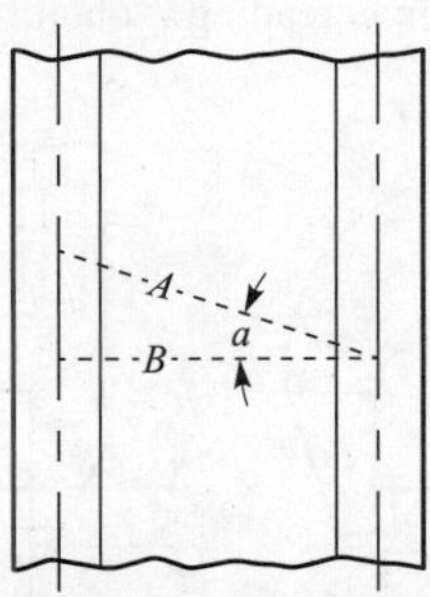


**Fig. 3. Determining the Width Perpendicular to the Sides of a Thread at the Pitch Line**

(The helix angle, which equals angle $a$, is based upon the pitch diameter and is measured from a plane perpendicular to the axis of the screw thread.) The width $A$ in the plane of the axis represents the hypotenuse of a right triangle, and the required width *B* equals the side adjacent; hence width $B = A \times$ cosine of helix angle. The angle of the thread itself (29° for an Acme Thread) does not affect the solution.

**Width of Flat End of Unified Screw Thread and American Standard Acme Screw Thread Tools.—**The widths of the flat or end of the threading tool for either of these threads may be measured by using a micrometer as illustrated at **(A)** in **Fig. 4**. In measuring the thread tool, a scale is held against the spindle and anvil of the micrometer, and the end of the tool is placed against this scale. The micrometer is then adjusted to the position shown and 0.2887 inch subtracted from the reading for an American Standard screw-thread tool. For American Standard Acme threads, 0.1293 inch is subtracted from the micrometer reading to obtain the width of the tool point. The constants (0.2887 and 0.1293), which are subtracted from the micrometer reading, are only correct when the micrometer spindle has the usual diameter of 0.25 inch.

An ordinary gear-tooth vernier caliper also may be used for testing the width of a thread tool point, as illustrated at **(B)** in **Fig. 4**. If the measurement is made at a vertical distance *x* of $\frac{1}{4}$ inch from the points of the caliper jaws, the constants previously given for American Standard caliper reading to obtain the actual width of the cutting end of the tool.

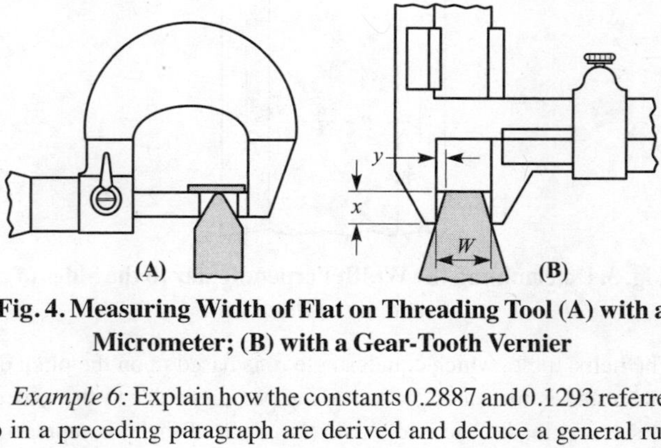


**Fig. 4. Measuring Width of Flat on Threading Tool (A) with a Micrometer; (B) with a Gear-Tooth Vernier**

*Example 6:* Explain how the constants 0.2887 and 0.1293 referred to in a preceding paragraph are derived and deduce a general rule applicable regardless of the micrometer spindle diameter or vertical dimension $x$, **Fig. 4**.

The dimension $x$ (which also is equivalent to the micrometer spindle diameter) represents one side of a right triangle (the side adjacent), having an angle of $29 \div 2 = 14$ degrees and 30 minutes, in the case of an Acme thread. The side opposite or $y$ = side adjacent × tangent = dimension $x \times \tan 14° 30'$.

If $x$ equals 0.25 inch, then side opposite or $y = 0.25 \times 0.25862 = 0.06465$; hence, the caliper reading minus $2 \times 0.06465$ = width of the flat end ($2 \times 0.06465 = 0.1293$ = constant).

The same result would be obtained by multiplying 0.25862 by $2x$; hence, the following rule: To determine the width of the end of the threading tool, by the general method illustrated in **Fig. 4**, multiply twice the dimension $x$ (or spindle diameter in the case of the micrometer) by the tangent of one-half the thread-tool angle, and subtract this product from the width $w$ to obtain the width at the end of the tool.

*Example 7:* A gear-tooth vernier caliper is to be used for measuring the width of the flat of an American Standard external screw-thread tool. The vertical scale is set to $^1/_8$ inch (corresponding to the dimension $x$, **Fig. 4**). How much is subtracted from the reading on the horizontal scale to obtain the width of the flat end of the tool?

$$^1/_8 \times 2 \times \tan 30° = {}^1/_4 \times 0.57735 = 0.1443 \text{ inch}$$

Hence, the width of the flat equals $w$, **Fig. 4**, minus 0.1443. This width should be equal to one-eighth of the pitch of the thread to be cut, since this is the width of flat at the minimum minor diameter of American Standard external screw threads.

## PRACTICE EXERCISES FOR SECTION 14

(See *Answers to Practice Exercises for Section 14* on page **244**)

1) What form of screw thread is most commonly used (*a*) in the United States? (*b*) in Britain?

2) What is the meaning of abbreviations 3″- 4NC-2?

3) What are the advantages of an Acme thread compared to a square thread?

4) For what reason would a Stub Acme thread be preferred in some applications?

5) Find the pitch diameters of the following screw threads of American Standard Unified form: $^1/_4$ – 28 (meaning $^1/_4$-inch diameter and 28 threads per inch); $^3/_4$ – 10?

6) How much taper is used on a standard pipe thread?

7) Under what conditions are straight, or nontapering, pipe threads used?

8) In cutting a taper thread, what is the proper position for the lathe tool?

9) If a lathe is used for cutting a British Standard pipe thread, in what position is the tool set?

10) A thread tool is to be ground for cutting an Acme thread having 4 threads per inch; what is the correct width of the tool at the end?

11) What are the common shop and toolroom methods of checking the pitch diameters of American Standard screw threads requiring accuracy?

12) In using the formula on *Handbook* page **2131** for measuring an American Standard screw thread by the three-wire method, why should the constant 0.86603 be multiplied by the pitch before adding it to measurement *M*, even if not enclosed by parentheses?

13) What is the difference between the pitch and the lead (*a*) of a double thread? (*b*) of a triple thread?

14) In using a lathe to cut American Standard Unified threads, what should be the truncations of the tool points and the thread depths for the following pitches: 0.1, 0.125, 0.2, and 0.25 inch?

15) In using the three-wire method of measuring a screw thread, what is the micrometer reading for a $^3/_4$ - 12 special thread of American Standard form if the wires have a diameter of 0.070 inch?

16) Are most nuts made to the United States Standard dimensions?

17) Is there, at the present time, a manufacturing Standard for bolts and nuts?

18) The American Standard for machine screws includes a coarse-thread series and a fine-thread series as shown by the tables starting on *Handbook* page **1981**. Which series is commonly used?

19) How is the length (*a*) of a flat head or countersunk type of machine screw measured? (*b*) of a fillister head machine screw?

20) What size tap drill should be used for an American Standard machine screw of No. 10 size, 24 threads per inch?

21) What is the diameter of a No. 10 drill?

22) Is a No. 6 drill larger than a No. 16?

23) What is the relation between the letter size drills and the numbered sizes?

24) Why is it common practice to use tap drills that leave about $^3/_4$ of the full thread depth after tapping, as shown by the tables starting on page **2170** in the *Handbook*?

25) What form of a screw thread is used on (*a*) machine screws? (*b*) cap screws?

26) What Standard governs the pitches of cap screw threads?

27) What form of thread is used on the National Standard fire hose couplings? How many standards diameters are there?

28) In what way do hand taps differ from machine screw taps?

29) What are tapper taps?

30) The diameter of a $^3/_4$ – 10 American Standard Thread is to be checked by the three-wire method. What is the largest size wire that can be used?

31) Why is the advance of some threading dies positively controlled by a lead screw instead of relying upon the die to lead itself?

32) What is the included angle of the heads of American Standard (*a*) flat head machine screws? (*b*) flat head cap screws? (*c*) flat head wood screws?

# SECTION 15

## PROBLEMS IN MECHANICS

*Machinery's Handbook* pages **151–172**

In the design of machines or other mechanical devices, it is often necessary to deal with the actions of forces and their effects. For example, the problem may be to determine what force is equivalent to two or more forces acting in the same plane but in different directions. Another type of problem is to determine the change in the magnitude of a force resulting from the application of mechanical appliances, such as levers, pulleys, and screws used either separately or in combination. It also may be necessary to determine the magnitude of a force needed to proportion machine parts to resist the force safely; or, possibly, to ascertain if the force is enough to perform a given amount of work. Determining the amount of energy stored in a moving body or its capacity to perform work, and the power developed by mechanical apparatus, or the rate at which work is performed, are additional examples of problems frequently encountered in developing mechanical appliances. The section in *Machinery's Handbook* on mechanics, beginning on page **151**, deals with fundamental principles and formulas applicable to various mechanical problems.

**The Moment of a Force.**—The tendency of a force acting upon a body is, in general, to produce either a motion of translation (that is, to cause every part of the body to move in a straight line) or to produce a motion of rotation. A moment, in mechanics, is the measure of the turning effect of a force that tends to produce rotation. For example, suppose a force acts upon a body that is supported by a pivot. Unless the line of action of the force happens to pass through the pivot, the body will tend to rotate. Its tendency to rotate, moreover, will depend upon two things: (1) the magnitude of the force acting, and (2) the distance of the force from the pivot, *measuring along a line at right angles to the line of action of the force*. (See **Fig. 10** and the accompanying text on *Handbook* page **156**.)

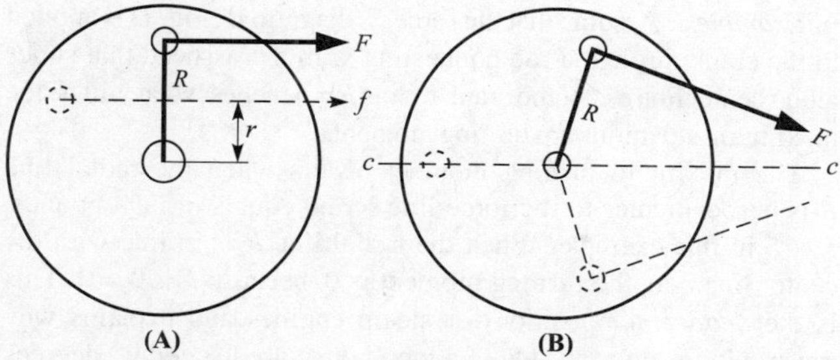


**Fig. 1. Example of the Turning Moment of a Crank Disk from Zero to Maximum**

*Example 1:* A force $F$ of 300 pounds is applied to a crank disk (diagram A, **Fig. 1**) in the direction of the arrow. If the radius $R$ = 5 inches, what is the turning moment? Also, determine how much the turning moment is reduced when the crankpin is in the position shown by the dashed lines, assuming that the force is along line $f$ and that $r = 2\frac{1}{2}$ inches.

When the crankpin is in the position illustrated by the solid lines, the maximum turning moment is obtained, and it equals $F \times R = 300 \times 5 = 1500$ inch-pounds or pound-inches. When the crankpin is in the position shown by the dashed lines, the turning moment is reduced one-half and equals $f \times r = 300 \times 2\frac{1}{2} = 750$ inch-pounds.

*Note:* Foot-pound is the unit for measurement of work and is in common use in horsepower calculations. However, torque, or turning moment, is also a unit of measurement of work. To differentiate between these two similar terms, which have the same essential meaning, it is convenient to express torque in terms of *pound-feet* (or *pound-inches*). This reversal of word sequence will readily indicate the different meanings of the two terms for units of measurement—the unit of horsepower and the unit of turning moment. A strong reason for expressing the unit of turning moment as *pound-inches* (rather than as *foot-pounds*) is because the dimensions of shafts and other machine parts ordinarily are stated in inches.

*Example 2:* Assume that the force $F$ (diagram B, **Fig. 1**) is applied to the crank through a rod connecting with a crosshead that slides along center line *c-c*. If the crank radius $R = 5$ inches, what will be the maximum and minimum turning moments?

The maximum turning moment occurs when the radial line $R$ is perpendicular to the force line $F$ and equals in inch-pounds, $F \times 5$ in this example. When the radial line $R$ is in line with the center line *c-c*, the turning moment is 0, because $F \times 0 = 0$. This is the *deadcenter* position for steam engines and explains why the crankpins on each side of a locomotive are located 90 degrees apart, or, in such a position that the maximum turning moment, approximately, occurs when the turning moment is zero on the opposite side. With this arrangement, it is always possible to start the locomotive since only one side at a time can be in the dead-center position.

**The Principle of Moments in Mechanics.**—When two or more forces act upon a rigid body and tend to turn it about an axis, then, for equilibrium to exist, the sum of the moments of the forces that tend to turn the body in one direction must be equal to the sum of the moments of those that tend to turn it in the opposite direction about the same axis.

*Example 3:* In **Fig. 2**, a lever 30 inches long is pivoted at the fulcrum $F$. At the right, and 10 inches from $F$, is a weight, $B$, of 12 pounds tending to turn the bar in a right-hand direction about its fulcrum $F$. At the left end, 12 inches from $F$, the weight $A$, of 4 pounds tends to turn the bar in a left-hand direction, while weight $C$, at the other end, 18 inches from $F$, has a like effect, through the use of the string and pulley $P$.

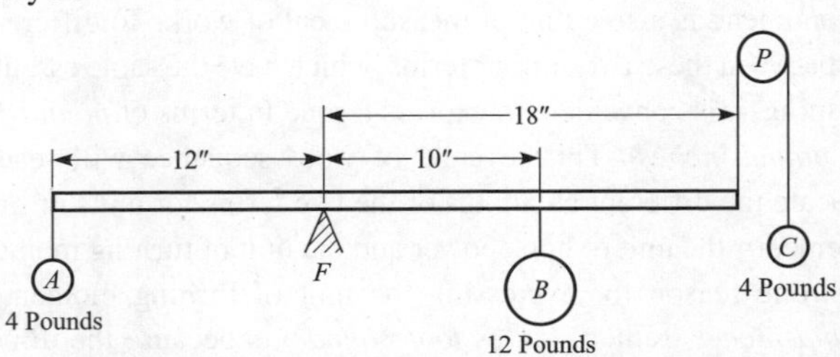


**Fig. 2. Lever in Equilibrium Because the Turning Moment of a Crank Disk Varies from Zero to Maximum**

Taking moments about *F*, which is the center of rotation, we have:

Moment of *B* = 10 × 12 = 120 inch-pounds

Opposed to this are the moments of *A* and *C*:

| | | | | |
|---|---|---|---|---|
| Moment of *A* | = | 4 × 12 | = | 48 inch-pounds |
| Moment of *C* | = | 4 × 18 | = | 72 inch-pounds |
| Sum of negative numbers | | | = | 120 inch-pounds |

Hence, the moments are equal, and, if we suppose, for simplicity, that the lever is weightless, it will balance or be in equilibrium. Should weight *A* be increased, the negative moments would be greater, and the lever would turn to the left, while if *B* should be increased or its distance from *F* be made greater, the lever would turn to the right. (See *Handbook* **Fig. 10** and the accompanying text on page **156**.)

*Example 4:* Another application of the principle of moments is given in **Fig. 3**. A beam of uniform cross section, weighing 200 pounds, rests upon two supports, *R* and *S*, that are 12 feet apart. The weight of the beam is considered to be concentrated at its center of gravity *G*, at a distance of 6 feet from each support. Each of the supports reacts or pushes upward, with a combined force equal to the downward pressure of the beam.

To make this clear, suppose two people take hold of the beam, one at each end, and that the supports are withdrawn. Then, in order to hold the beam in position, the two people must together lift or pull upward an amount equal to the weight of the beam and its load, or 250 pounds. Placing the supports in position again, and resting the beam upon them, does not change the conditions. The weight of the beam acts downward, and the supports react by an equal amount.

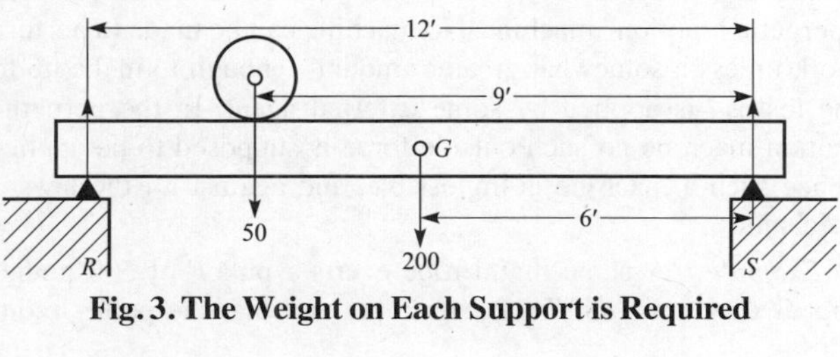


**Fig. 3. The Weight on Each Support is Required**

Now, to solve the problem, assume the beam to be pivoted at one support, say, at *S*. The forces or weights of 50 pounds and 200 pounds tend to rotate the beam in a left-hand direction about this point, while the reaction of *R* in an upward direction tends to give it a right-hand rotation. As the beam is balanced and has no tendency to rotate, it is in equilibrium, and the opposing moments of these forces must balance; hence, taking moments,

$$\begin{aligned} 9 \times 50 &= \phantom{0}450 \text{ pound-feet} \\ 6 \times 200 &= \underline{1200 \text{ pound-feet}} \\ \text{Sum of negative numbers} &= 1650 \text{ pound-feet} \end{aligned}$$

By letting *R* represent the reaction of support,

$$\text{Moment of } R = R \times 12 = \text{pound-feet}$$

By the principle of moments, $R \times 12 = 1650$. That is, if *R*, the quantity that we wish to obtain, is multiplied by 12, the result will be 1650; hence, to obtain *R*, divide 1650 by 12. Therefore, $R = 137.5$ pounds, which is also the weight of that end of the beam. As the total load is 250 pounds, the weight at the other end must be $250 - 137.5 = 112.5$ pounds.

**The Principle of Work in Mechanics.—** Another principle of more importance than the principle of moments, even in the study of machine elements, is the principle of work. According to this principle (neglecting frictional or other losses), the applied force, multiplied by the distance through which it moves, equals the resistance overcome, multiplied by the distance through which it is overcome. The principle of work may also be stated as follows:

$$\text{Work put in} = \text{lost work} + \text{work done by machine}$$

This principle holds absolutely in every case. It applies equally to a simple lever, the most complex mechanism, or to a so-called "perpetual motion" machine. No machine can be made to perform work unless a somewhat greater amount—enough to make up for the losses—is applied by some external agent. In the perpetual motion machine no such outside force is supposed to be applied, hence such a machine is impossible, and against all the laws of mechanics.

*Example 5:* Assume that a rope exerts a pull *F* of 500 pounds (upper diagram, *Handbook* page **171**) and that the pulley radius

$R = 10$ inches and the drum radius $r = 5$ inches. How much weight $W$ can be lifted (ignoring frictional losses) and upon what mechanical principle is the solution based?

According to one of the formulas accompanying the diagram at the top of *Handbook* page **171**,

$$W = \frac{F \times R}{r} = \frac{500 \times 10}{5} = 1000 \text{ pounds}$$

This formula (and the others for finding the values of $F$, $R$, etc.) agrees with the principle of moments, and with the principle of work. The principle of moments will be applied first.

The moment of the force $F$ about the center of the pulley, which corresponds to the fulcrum of a lever, is $F$ multiplied by the perpendicular distance $R$, it being a principle of geometry that a radius is perpendicular to a line drawn tangent to a circle, at the point of tangency. Also, the opposing moment of $W$ is $W \times r$. Hence, by the principle of moments,

$$F \times R = W \times r$$

Now, for comparison, we will apply the principle of work. Assuming this principle to be true, force $F$ multiplied by the distance traversed by this force or by a given point on the rim of the large pulley should equal the resistance $W$ multiplied by the distance that the load is raised. In one revolution, force $F$ passes through a distance equal to the circumference of the pulley, which is equal to $2 \times 3.1416 \times R = 6.2832 \times R$, and the hoisting rope passes through a distance equal to $2 \times 3.1416 \times r$. Hence, by the principle of work,

$$6.2832 \times F \times R = 6.2832 \times W \times r$$

The statement simply shows that $F \times R$ multiplied by 6.2832 equals $W \times r$ multiplied by the same number, and it is evident therefore, that the equality will not be altered by canceling the 6.2832 and writing:

$$F \times R = W \times r$$

However, this statement is the same as that obtained by applying the principle of moments; hence, we see that the principle of moments and the principle of work are in harmony.

The basis of operation of a train of wheels is a continuation of the principle of work. For example, in the gear train represented by the diagram at the bottom of *Handbook* page **171**, the continued product of the applied force $F$ and the radii of the driven wheels equals the continued product of the resistance $W$ and the radii of the drivers. In calculations, the pitch diameters or the numbers of teeth in gear wheels may be used instead of the radii.

**Efficiency of a Machine or Mechanism.**—The efficiency of a machine is the ratio of the power delivered by the machine to the power received by it. For example, the efficiency of an electric motor is the ratio between the power delivered by the motor to the machinery it drives and the power it receives from the generator. Assume, for example, that a motor receives 50 kilowatts from the generator, but that the output of the motor is only 47 kilowatts. Then, the efficiency of the motor is 47 ÷ 50 = 94 percent. The efficiency of a machine tool is the ratio of the power consumed at the cutting tool to the power delivered by the driving belt. The efficiency of gearing is the ratio between the power obtained from the driven shaft to the power used by the driving shaft. Generally speaking, the efficiency of any machine or mechanism is the ratio of the output of power to the input. The percentage of power representing the difference between the input and output has been dissipated through frictional and other mechanical losses.

*Mechanical Efficiency:* If $E$ represents the energy that a machine transforms into useful work or delivers at the driven end, and $L$ equals the energy loss through friction or dissipated in other ways, then,

$$\text{Mechanical efficiency} = \frac{E}{E+L}$$

In this equation, the total energy $F + L$ is assumed to be the amount of energy that is transformed into useful and useless work. The actual total amount of energy, however, may be considerably larger than the amount represented by $E + L$. For example, in a steam engine, there are heat losses due to radiation and steam condensation, and considerable heat energy supplied to an internal combustion engine is dissipated either through the cooling water or direct to the atmosphere. In other classes of mechanical and electrical machinery, the total energy is much larger than that represented by the amount transformed into useful and useless work.

*Absolute Efficiency:* If $E_1$ equals the full amount of energy or the true total, then,

$$\text{Absolute efficiency} = \frac{E}{E_1}$$

It is evident that absolute efficiency of a prime mover, such as a steam or gas engine, will be much lower than the mechanical efficiency. Ordinarily, the term *efficiency* as applied to engines and other classes of machinery means the mechanical efficiency. The *mechanical efficiency* of reciprocating steam engines may vary from 85 to 95 percent, but the *thermal efficiency* may range from 5 to 25 percent, the smaller figure representing noncondensing engines of the cheaper class and the higher figure the best types.

*Example 6:* Assume that a motor driving through a compound train of gearing (see **Fig. 4**) is to lift a weight $W$ of 1000 pounds. The pitch radius $R$ = 6 inches; $R_1$ = 8 inches; pitch radius of pinion $r$ = 2 inches; and radius of winding drum $r_1 = 2\frac{1}{2}$ inches. What motor horsepower will be required if the frictional loss in the gear train and bearings is assumed to be 10 percent? The pitch-line velocity of the motor pinion $M$ is 1200 feet per minute.

The problem is to determine first the tangential force $F$ required at the pitch line of the motor pinion; then, the equivalent horsepower is easily found. According to the formula at the bottom of *Handbook* page **171**, which does not take into account frictional losses,

$$F = \frac{1000 \times 2 \times 2\frac{1}{2}}{6 \times 8} = 104 \text{ pounds}$$

The pitch-line velocity of the motor pinion is 1200 feet per minute and, as the friction loss is assumed to be 10 percent, the mechanical efficiency equals 90 ÷ (90 + 10) = 0.90 or 90 percent as commonly written; thus,

$$\text{Horsepower} = \frac{104 \times 1200}{33{,}000 \times 0.90} = 4\frac{1}{4} \text{ approximately}$$

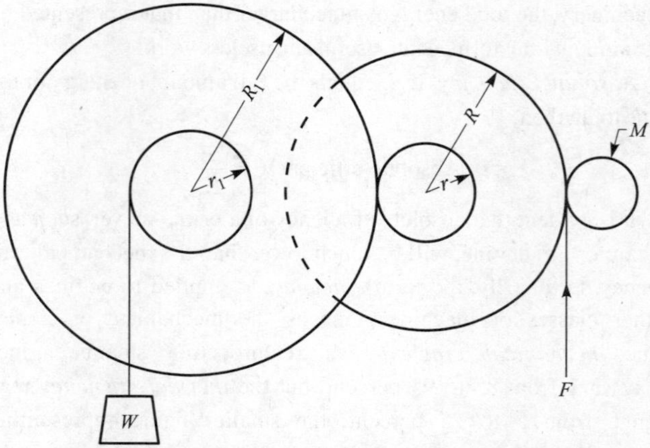


**Fig. 4. Determining the Power Required for Lifting a Weight by Means of a Motor and a Compound Train of Gearing**

*Example 7:* In designing practice, a motor of horsepower, or larger, might be selected for the drive referred to in **Example 6** (depending upon conditions) to provide extra power should it be needed. However, to illustrate the procedure, assume that the gear train is to be modified so that the calculated horsepower will be 4 instead of $4\frac{1}{4}$; conditions otherwise are the same as in **Example 6**.

$$F = \frac{33{,}000 \times 4}{1200} = 110 \text{ pounds}$$

Hence, since $W = 1000$ pounds,

$$1000 = \frac{110 \times 0.90 \times R \times R_1}{r \times r_1}$$

Insert any values for the pitch radii $R$, $R_1$, etc., that will balance the equation, so that the right-hand side equals 1000, at least approximately. Several trial solutions may be necessary to obtain a total of about 1000, at the same time, secure properly proportional gears that meet other requirements of the design. Suppose the same radii are used here, except $R_1$, which is increased from 8 to $8\frac{1}{2}$ inches.

Then,

$$\frac{110 \times 0.90 \times 6 \times 8\frac{1}{2}}{2 \times 2\frac{1}{2}} = 1000 \text{ approximately}$$

This example shows that the increase in the radius of the last driven gear from 8 to $8\frac{1}{2}$ inches makes it possible to use the 4-horsepower motor. The hoisting speed has been decreased somewhat, and the center distance between the gears has been increased. These changes might or might not be objectionable in actual designing practice, depending upon the particular requirements.

**Force Required to Turn a Screw Used for Elevating or Lowering Loads.**—In determining the force that must be applied at the end of a given lever arm in order to turn a screw (or nut surrounding it), there are two conditions to be considered: (1) when rotation is such that the load *resists* the movement of the screw, as in raising a load with a screw jack; (2) when rotation is such that the load *assists* the movement of the screw, as in lowering a load. The formulas at the bottom of the table on *Handbook* page **172** apply to both these conditions. When the load resists the screw movement, use the formula "for motion in a direction opposite to $Q$." When the load assists the screw movement, use the formula "for motion in the same direction as $Q$."

If the lead of the thread is large in proportion to the diameter so that the helix angle is large, the force $F$ may have a negative value, which indicates that the screw will turn due to the load alone, unless resisted by a force that is great enough to prevent rotation of a nonlocking screw.

*Example 8:* A screw is to be used for elevating a load $Q$ of 6000 pounds. The pitch diameter is 4 inches, the lead is 0.75 inch, and the coefficient of friction $\mu$ between screw and nut is assumed to be 0.150. What force $F$ will be required at the end of a lever arm $R$ of 10 inches? In this example, the load is in the direction opposite to the arrow $Q$ (see the applicable diagram on *Handbook* page **172**).

$$F = 6000 \times \frac{0.75 + 6.2832 \times 0.150 \times 2}{6.2832 \times 2 - 0.150 \times 0.75} \times \frac{2}{10}$$

$$= 254 \text{ pounds}$$

*Example 9:* What force $F$ will be required to lower a load of 6000 pounds using the screw referred to in **Example 8**? In this case, the load assists in turning the screw; hence,

$$F = 6000 \times \frac{6.2832 \times 0.150 \times 2 - 0.75}{6.2832 \times 2 + 0.150 \times 0.75} \times \frac{2}{10} = 107 \text{ pounds}$$

**Coefficients of Friction for Screws and Their Efficiency.—** According to experiments Professor Kingsbury made with square-threaded screws, a friction coefficient μ of 0.10 is about right for pressures less than 3000 pounds per square inch (psi) and velocities above 50 feet per minute, assuming that fair lubrication is maintained. If the pressures vary from 3000 to 10,000 psi, a coefficient of 0.15 is recommended for low velocities. The coefficient of friction varies with lubrication and the materials used for the screw and nut. For pressures of 3000 psi and by using heavy machinery oil as a lubricant, the coefficients were as follows: Mild steel screw and cast-iron nut, 0.132; mild-steel nut, 0.147; cast-brass nut, 0.127. For pressures of 10,000 psi using a mild-steel screw, the coefficients were, for a cast-iron nut, 0.136; for a mild-steel nut, 0.141; for a cast-brass nut, 0.136. For dry screws, the coefficient may be 0.3 to 0.4 or higher.

Frictional resistance is proportional to the normal pressure, and for a thread of angular form, the increase in the coefficient of friction is equivalent practically to $\mu \sec\beta$, in which β equals one-half the included thread angle; hence, for a 60-degree thread, a coefficient of 1.155μ may be used. The square form of thread has a somewhat higher efficiency than threads with sloping sides, although when the angle of the thread form is comparatively small, as in an Acme thread, there is little increase in frictional losses. Multiple-thread screws are much more efficient than single-thread screws, as the efficiency is affected by the helix angle of the thread.

The efficiency between a screw and nut increases quite rapidly for helix angles up to 10 to 15 degrees (measured from a plane perpendicular to the screw axis). The efficiency remains nearly constant for angles between about 25 and 65 degrees, and the angle of maximum efficiency is between 40 and 50 degrees. A screw will not be self-locking if the efficiency exceeds 50 percent. For example, the screw of a jack or other lifting or hoisting appliance would turn under the action of the load if the efficiency were over 50 percent. It is evident that maximum efficiency for power transmission screws often is impractical, as for example, when the smaller helix angles are required to permit moving a given load by the application of a smaller force or turning moment than would be needed for a multiple-screw thread.

In determining the efficiency of a screw and a nut, the helix angle of the thread and the coefficient of friction are the important factors. If $E$ equals the efficiency, $A$ equals the helix angle, measured from a plane perpendicular to the screw axis, and $\mu$ equals the coefficient of friction between the screw thread and nut, then the efficiency may be determined by the following formula, which does not take into account any additional friction losses, such as may occur between a thrust collar and its bearing surfaces:

$$E = \frac{\tan A(1 - \mu \tan A)}{\tan A + \mu}$$

This formula would be suitable for a screw having ball-bearing thrust collars. Where collar friction should be taken into account, a fair approximation may be obtained by changing the denominator of the foregoing formula to $\tan A + 2\mu$. Otherwise, the formula remains the same.

**Angles and Angular Velocity Expressed in Radians.**—There are three systems generally used to indicate the sizes of angles, which are ordinarily measured by the number of degrees in the arc subtended by the sides of the angle. Thus, if the arc subtended by the sides of the angle equals one-sixth of the circumference, the angle is said to be 60 degrees. Angles are also designated as multiples of a right angle. As an example, the sum of the interior angles of any polygon equals the number of sides less two, times two right angles. Thus the sum of the interior angles of an octagon equals $(8-2)\times 2\times 90 = 6\times 180 = 1080$ degrees. Hence each interior angle equals $1080 \div 8 = 135$ degrees.

A third method of designating the size of an angle is very helpful in certain problems. This method makes use of radians. A radian is defined as a central angle, the subtended arc of which equals the radius of the arc.

By using the symbols on *Handbook* page **176**, $v$ may represent the length of an arc as well as the velocity of a point on the periphery of a body. Then, according to the definition of a radian: $\omega = v/r$, or the angle in radians equals the length of the arc divided by the radius. Both the length of the arc and the radius must, of course, have the same unit of measurement—both must be in feet or inches or centimeters, etc. By rearranging the preceding equation:

$$v = \omega r \quad \text{and} \quad r = \frac{v}{\omega}$$

These three formulas will solve practically every problem involving radians.

The circumference of a circle equals $\pi d$ or $2\pi r$, which equals $6.2832r$, which indicates that a radius is contained in a circumference 6.2832 times; hence there are 6.2832 radians in a circumference. Since a circumference represents 360 degrees, 1 radian equals $360 \div 6.2832 = 57.2958$ degrees. Since 57.2958 degrees = 1 radian, 1 degree = 1 radian $\div$ 57.2958 = 0.01745 radian.

*Example 10:* 2.5 radians equal how many degrees? One radian = 57.2958 degrees; hence, 2.5 radians = $57.2958 \times 2.5 = 143.239$ degrees.

*Example 11:* $22° 31' 12'' =$ how many radians? 12 seconds $= {}^{12}/_{60} = {}^{1}/_{5} = 0.2$ minute; $31.2' \div 60 = 0.52$ degree. One radian = 57.3 degrees approximately. $22.52° = 22.52 + 57.3 = 0.393$ radian.

*Example 12:* In the figure on *Handbook* page **79**, let $l = v =$ 30 inches; and radius $r = 50$ inches; find the central angle $\omega = v/r = {}^{30}/_{50} = {}^{3}/_{5} = 0.6$ radian.

$$57.2958 \times 0.6 = 34°22.6'$$

*Example 13:* ${}^{3\pi}/_{4}$ radians equal how many degrees? $2\pi$ radians $=360°$; $\pi$ radians $= 180°$. ${}^{3\pi}/_{4} = {}^{3}/_{4} \times 180 = 135$ degrees.

*Example 14:* A 20-inch grinding wheel has a surface speed of 6000 feet per minute. What is the angular velocity?

The radius ($r$) = $^{10}/_{12}$ foot; the velocity ($n$) in feet per second = $^{6000}/_{60}$; hence,

$$\omega = \frac{6000}{60 \times {}^{10}/_{12}} = 120 \text{ radians per second}$$

*Example 15:* Use the table on *Handbook* page **98** to solve **Example 11**.

$$\begin{aligned} 20° &= 0.349066 \text{ radian} \\ 2° &= 0.034907 \text{ radian} \\ 31' &= 0.009018 \text{ radian} \\ 12'' &= 0.000058 \text{ radian} \\ \hline 22°31'12'' &= 0.393049 \text{ radian} \end{aligned}$$

*Example 16:* 7.23 radians equals how many degrees? On *Handbook* page **99**, find:

$$\begin{aligned} 7.0 \text{ radians} &= 401° \, 4' \, 14'' \\ 0.2 \text{ radian} &= 11° \, 27' \, 33'' \\ 0.03 \text{ radian} &= 1° \, 43' \, 8'' \\ \hline 7.23 \text{ radians} &= 414° \, 14' \, 55'' \end{aligned}$$

## PRACTICE EXERCISES FOR SECTION 15

(See *Answers to Practice Exercises for Section 15* on page **246**)

1) In what respect does a foot-pound differ from a pound?

2) If a 100-pound weight is dropped, how much energy will it be capable of exerting after falling 10 feet?

3) Can the force of a hammer blow be expressed in pounds?

4) If a 2-pound hammer is moving 30 feet per second, what is its kinetic energy?

5) If this 2-pound hammer drives a nail into a $^1/_4$-inch board, what is the average force of the blow?

6) What relationship is there between the muzzle velocity of a projectile fired upward and the velocity with which the projectile strikes the ground?

7) What is the difference between the composition of forces and the resolution of forces?

8) If four equal forces act along lines 90 degrees apart through a given point, what is the shape of the corresponding polygon of forces?

9) Skids are to be employed for transferring boxed machinery from one floor to the floor above. If these skids are inclined at an angle of 35 degrees, what force in pounds, applied parallel to the skids, will be required to slide a boxed machine weighing 2500 pounds up the incline, assuming that the coefficient of friction is 0.20?

10) Refer to Exercise 9. If the force or pull were applied in a horizontal direction instead of in line with the skids, what increase, if any, would be required?

11) Will the boxed machine referred to in #9 above slide down the skids by gravity?

12) At what angle will the skids need to be for the boxed machine to start sliding due to gravity?

13) What name is applied to the angle that marks the dividing line between sliding and nonsliding when a body is placed on an inclined plane?

14) How is the "angle of repose" determined?

15) What figure or value is commonly used in engineering calculations for acceleration due to gravity?

16) Is the value commonly used for acceleration due to gravity strictly accurate for any locality?

17) A flywheel 3 feet in diameter has a rim speed of 1200 feet per minute, and another flywheel 6 feet in diameter has the same rim speed. Will the rim stress or the force tending to burst the larger flywheel be greater than the force in the rim of the smaller flywheel?

18) What factors of safety are commonly used in designing flywheels?

19) Does the stress in the rim of a flywheel increase in proportion to the rim velocity?

20) What is generally considered the maximum safe speed for the rim of a solid or one-piece cast-iron flywheel?

21) Why is a well-constructed wood flywheel better adapted to higher speeds than one made of cast iron?

22) What is the meaning of the term "critical speed" as applied to a rotating body?

23) How is angular velocity generally expressed?

24) What is a radian, and how is its angle indicated?

25) How many degrees are there in 2.82 radians?

26) How many degrees are in the following radians: $\pi/3$ and $2\pi/5$.

27) Reduce to radians: 63°; 45°32′; 6°37′46″; 22°22′22″.

28) Find the angular velocity in radians per second of the following: 157 rpm; 275 rpm; 324 rpm.

29) Why do the values in the *l* column starting on *Handbook* page **79** equal those in the radian column on page **98**?

30) If the length of the arc of a sector is 4⅞ inches, and the radius is 6⅞ inches, find the central angle.

31) A 12-inch grinding wheel has a surface speed of a mile a minute. Find its angular velocity and its revolutions per minute.

32) The radius of a circle is 1½ inches, and the central angle is 60 degrees. Find the length of the arc.

33) If an angle of 34°12′ subtends an arc of 16.25 inches, find the radius of the arc.

## SECTION 16

## STRENGTH OF MATERIALS

*Machinery's Handbook* pages **200–221**

The *Strength of Materials* section of *Machinery's Handbook* contains fundamental formulas and data for use in proportioning parts that are common to almost every type of machine or mechanical structure. In designing machine parts, factors other than strength often are of vital importance. For example, some parts are made much larger than required for strength alone to resist extreme vibrations, deflection, or wear; consequently, many machine parts cannot be designed merely by mathematical or strength calculations, and their proportions should, if possible, be based upon experience or upon similar designs that have proved successful. It is evident that no engineering handbook can take into account the endless variety of requirements relating to all types of mechanical apparatus, and it is necessary for the designer to determine these local requirements for each, but, even when the strength factor is secondary due to some other requirement, the strength, especially of the more important parts, should be calculated, in many instances, merely to prove that it will be sufficient.

In designing for strength, the part is so proportioned that the maximum working stress likely to be encountered will not exceed the strength of the material by a suitable margin. The design is accomplished by the use of a factor of safety. The relationship between the working stress $s_w$, the strength of the material, $S_m$, and the factor of safety, $f_s$ is given by **Equation (1)** on page **205** of the *Handbook*:

$$s_w = \frac{S_m}{f_s} \qquad \text{(a)}$$

The value selected for the strength of the material, $S_m$ depends on the type of material, whether failure is expected to occur because of tensile, compressive, or shear stress, and on whether the stresses are constant, fluctuating, or are abruptly applied as with shock loading. In general, the value of $S_m$ is based on yield strength for ductile materials, ultimate strength for brittle materials, and fatigue strength for parts subject to cyclic stresses. Moreover, the value for $S_m$ must be for the temperature at which the part operates. Values of $S_m$ for common materials at 68° F can be obtained from the tables in *Machinery's Handbook* from pages **427** and **508**. Factors from the table given on *Handbook* page **390**, *Influence of Temperature on the Strength of Metals*, can be used to convert strength values at 68° F to values applicable at elevated temperatures. For heat-treated carbon and alloy-steel parts, see data starting on *Handbook* page **421**.

The factor of safety depends on the relative importance of reliability, weight, and cost. General recommendations are given in the *Handbook* on page **205**.

Working stress is dependent on the shape of the part, hence on a stress concentration factor, and on a nominal stress associated with the way in which the part is loaded. Equations and data for calculating nominal stresses, stress concentration factors, and working stresses are given starting on *Handbook* page **206**.

*Example 1:* Determine the allowable working stress for a part that is to be made from SAE 1112 free-cutting steel; the part is loaded in such a way that failure is expected to occur in tension when the yield strength has been exceeded. A factor of safety of 3 is to be used.

From the table, *Strength Data for Iron and Steel*, on page **427** of the *Handbook*, a value of 30,000 pounds per square inch (psi) is selected for the strength of the material, $S_m$. Working stress $s_w$ is calculated from **Equation (a)** as follows:

$$s_w = \frac{30{,}000}{3} = 10{,}000 \text{ psi}$$

**Finding Diameter of a Bar to Resist Safely Under a Given Load.—**Assume that a direct tension load, $F$, is applied to a bar such that the force acts along the longitudinal axis of the bar. From *Handbook* page **210**, the following equation is given for calculating the nominal stress:

$$\sigma = \frac{F}{A} \tag{b}$$

where $A$ is the cross-sectional area of the bar. **Equation (2)** on *Handbook* page **206** related the nominal stress to the stress concentration factor, $K$, and working stress, $s_w$:

$$s_w = K\sigma \tag{c}$$

Combining **Equation (a)**, **(b)**, and **(c)** results in the following:

$$\frac{S_m}{Kf_s} = \frac{F}{A} \tag{d}$$

*Example 2:* A structural steel bar supports in tension a load of 40,000 pounds. The load is gradually applied and then, after having reached its maximum value, is gradually removed. Find the diameter of round bar required.

According to the table on *Handbook* page **427**, the yield strength of structural steel is 33,000 psi. Suppose that a factor of safety of 3 and a stress concentration factor of 1.1 are used. Then, inserting known values in **Equation (d)**:

$$\frac{33{,}000}{1.1 \times 3} = \frac{40{,}000}{A};\ A = \frac{40{,}000 \times 3.3}{33{,}000};\ A = 4 \text{ square inches}$$

Hence, the cross-section of the bar must be about 4 square inches. As the bar is circular in section, the diameter must then be about $2\frac{1}{4}$ inches.

**Diameter of a Bar to Resist Compression.**—If a short bar is subjected to compression in such a way that the line of application of the load coincides with the longitudinal axis of the bar, the formula for nominal stress is the same as for direct tension loading. **Equation (b)** and hence **Equation (d)** also may be applied to direct compression loading.

*Example 3:* A short structural-steel bar supports in compression a load of 40,000 pounds. (See **Fig. 1**.) The load is steady. Find the diameter of the bar required.

From page **427** in the *Handbook*, the yield strength of structural steel is 33,000 psi. If a stress concentration factor of 1.1 and a factor of safety of 2.5 are used, then, substituting values into **Equation (d)**:

$$\frac{33{,}000}{1.1 \times 2.5} = \frac{40{,}000}{A}; \quad A = 3.33 \text{ square inches}$$

The diameter of a bar, the cross-section of which is 3.33 square inches, is about $2\frac{1}{16}$ inches.

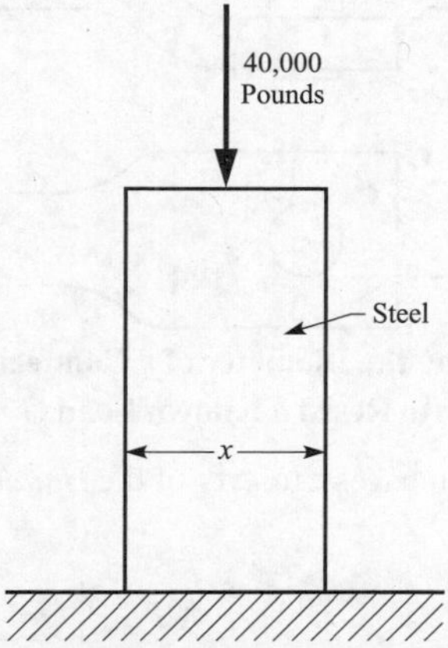


**Fig. 1. Calculating Diameter *x* to Support a Given Load Safely**

According to a general rule, the simple formulas that apply to compression should be used only if the length of the member being compressed is not greater than 6 times the least cross-sectional dimension. For example, these formulas should be applied to round bars only when the length of the bar is less than 6 times the diameter. If the bar is rectangular, the formulas should be applied only to bars having a length less than 6 times the shortest side of the rectangle. When bars are longer than this, a compressive stress causes a sidewise bending action, and an even distribution of the compression stresses over the total area of the cross-section should no longer be depended upon. Special formulas for long bars or columns will be found on *Handbook* page 277; see also text beginning on page 275, *Strength of Columns or Struts*.

**Diameter of a Pin to Resist Shearing Stress.—**The pin *E* shown in the illustration, **Fig. 2**, is subjected to shear. Parts *G* and *B* are held together by the pin and tend to shear it off at C and D. The areas resisting the shearing action are equal to the pin at these points.

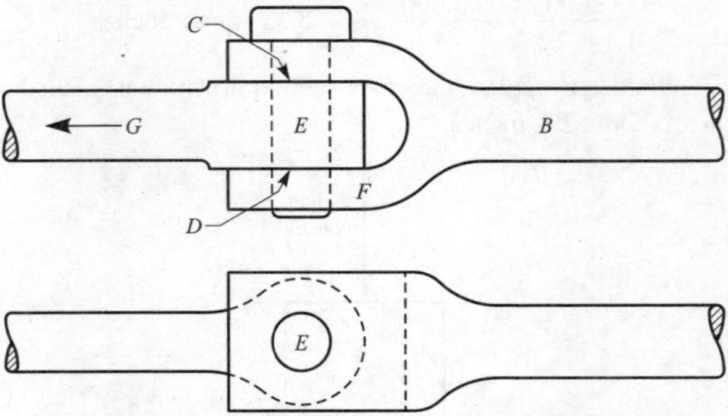


**Fig. 2. Finding the Diameter of a Connecting-Rod Pin to Resist a Known Load *G***

From **Table 2** on pages **210–211** of the *Handbook*, the equation for direct shear is:

$$\tau = \frac{F}{A} \qquad \text{(e)}$$

$\tau$ is a simple stress related to the working stress, $s_w$, by **Equation (3)** on *Handbook* page **206**:

$$s_w = K\tau \qquad \text{(f)}$$

where $K$ is a stress concentration factor. Combining **Equation (a)**, **(e)**, and **(f)** gives **Equation (d)** on page **146**, where $S_m$ is, of course, the shearing strength of the material.

If a pin is subjected to shear as in **Fig. 2**, so that two surfaces, as at C and D, must fail by shearing before breakage occurs, the areas of both surfaces must be taken into consideration when calculating the strength. The pin is then said to be in *double shear*. If the lower part F of connecting rod B were removed, so that member G were connected with B by a pin subjected to shear at C only, the pin would be said to be in *single shear*.

*Example 4:* Assume that in **Fig. 2** the load at G pulling on the connecting rod is 20,000 pounds. The material of the pin is SAE

1025 steel. The load is applied in such a manner that shocks are liable to occur. Find the required dimensions for the pin.

Since the pins are subjected to shock loading, the nominal stress resulting from the application of the 20,000-pound load must be assumed to be twice as great (see *Handbook* starting on page **272**) as it would be if the load were gradually applied or steady. From *Handbook* page **427**, the ultimate strength in shear for SAE 1025 steel is 75 percent of 60,000 or 45,000 psi. A factor of safety of 3 and a stress concentration factor of 1.8 are to be used. By substituting values into **Equation (d)**:

$$\frac{45,000}{1.8 \times 3} = \frac{2 \times 20,000}{A}; \quad A = \frac{10.8 \times 20,000}{45,000}$$

$$= 4.8 \text{ square inch}$$

As the pin is in double shear, that is, as there are two surfaces *C* and *D* over which the shearing stress is distributed, each surface must have an area of one-half the total shearing area *A*. Then, the cross-sectional area of the pin will be 2.4 square inches, and the diameter of the pin, to give a cross-sectional area of 2.4 square inches, must be $1\frac{3}{4}$ inches.

**Beams and Stresses to Which They Are Subjected.**—Parts of machines and structures subjected to bending are known mechanically as *beams*. Hence, in this sense, a lever fixed at one end and subjected to a force at its other end, a rod supported at both ends and subjected to a load at its center, or the overhanging arm of a jib crane would all be known as beams.

The stresses in a beam are principally tension and compression stresses. If a beam is supported at the ends, and a load rests upon the upper side, the lower fibers will be stretched by the bending action and will be subjected to a tensile stress, while the upper fibers will be compressed and be subjected to a compressive stress. There will be a slight lengthening of the fibers in the lower part of the beam, while those on the upper side will be somewhat shorter, depending upon the amount of deflection. If we assume that the beam is either round or square in cross-section, there will be a layer or surface through its center line, which will be neither in compression nor in tension.

This surface is known as the neutral surface. The stresses of the individual layers or fibers of the beam will be proportional to their distances from the neutral surface, the stresses being greater the farther away from the neutral surface the fiber is located. Hence, there is no stress on the fibers in the neutral surface, but there is a maximum tension on the fibers at the extreme lower side and a maximum compression on the fibers at the extreme upper side of the beam. In calculating the strength of beams, it is, therefore, only necessary to determine that the fibers of the beam that are at the greatest distance from the neutral surface are not stressed beyond the safe working stress of the material. If this condition exists, all the other parts of the section of the beam are not stressed beyond the safe working stress of the material.

In addition to the tension and compression stresses, a loaded beam is also subjected to a stress that tends to shear it. This shearing stress depends upon the magnitude and kind of load. In most instances, the shearing action can be ignored for metal beams, especially if the beams are long and the loads far from the supports. If the beams are very short and the load quite close to a support, then the shearing stress may become equal to or greater than the tension or compression stresses in the beam, and the beam should then be calculated for shear.

**Beam Formulas.**—The bending action of a load upon a beam is called the *bending moment*. For example, in **Fig. 3** the load $P$ acting downward on the free end of the cantilever beam has a moment or bending action about the support at $A$ equal to the load multiplied by its distance from the support. The bending moment is commonly expressed in inch-pounds, the load being expressed in pounds and the lever arm or distance from the support in inches. The length of the lever arm should always be measured in a direction at right angles to the direction of the load. Thus, in the second example in **Fig. 3**, the bending moment is not $P \times a$, but is $P \times l$, because $l$ is measured in a direction at right angles to the direction of the load $P$.

The property of a beam to resist the bending action or the bending moment is called the *moment of resistance* of the beam. It is evident that the bending moment must be equal to the moment of resistance. The moment of resistance, in turn, is equal to the stress in the fiber farthest away from the neutral plane multiplied by the section modulus. The *section modulus* is a factor that depends upon the shape and size of the cross-section of a beam and is given for different cross-sections in all engineering handbooks. (See table, *Moments of Inertia, Section Moduli, and Radii of Gyration* starting on *Handbook* page **235**.) The section modulus, in turn, equals the moment of inertia of the cross-section, divided by the distance from the neutral surface to the most extreme fiber. The moment of inertia formulas for various cross-sections also will be found in the table just mentioned.

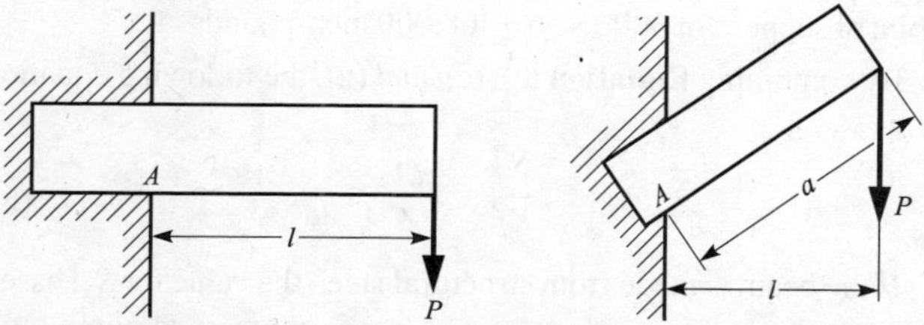


**Fig. 3. Diagrams Illustrating Principle of Bending Moments**

The following formula on *Handbook* page **211** may be given as the fundamental formula for bending of beams:

$$\sigma = \pm\frac{M}{Z} = \pm\frac{My}{I} \qquad \text{(g)}$$

The moment of inertia $I$ is a property of the cross-section that determines its relative strength. In calculations of strength of materials, a handbook is necessary because of the tabulated formulas and data relating to section moduli and moments of inertia, areas of cross-sections, etc., to be found therein.

There are many different ways in which a beam can be supported and loaded, and the bending moment caused by a given load varies greatly according to whether the beam is supported at one end only or at both ends, also whether it is freely supported at the ends or is held firmly. The load may be equally distributed over the full length of the beam or may be applied at one point either in the center or near to one or the other of the supports. The point where stress is maximum is generally called the critical point. The stress at the critical point equals bending moment divided by section modulus.

Formulas for determining the stresses at the critical points will be found in the table of beam formulas, starting on *Handbook* page **251**.

*Example 5:* A rectangular steel bar 2 inches thick and firmly built into a wall, as shown in **Fig. 4**, is to support 3000 pounds at its outer end 36 inches from the wall. What would be the necessary depth *h* of the beam to support this weight safely?

The bending moment equals the load times the distance from the point of support, or 3000 × 36 = 108,000 inch-pounds.

By combining **Equation (a)**, **(c)**, and **(g)**, the following equation is obtained:

$$\frac{S_m}{Kf_s} = \frac{M}{Z} \qquad \text{(h)}$$

If the beam is made from structural steel, the value for $S_m$, based on yield strength, from page **427** in the *Handbook*, is 33,000 psi. By using a stress concentration factor of 1.1 and a factor of safety of 2.5, values may be inserted into the above equation:

$$\frac{33{,}000}{1.1 \times 2.5} = \frac{108{,}000}{Z}; \quad Z = \frac{2.75 \times 108{,}000}{33{,}000}; \quad Z = 9 \text{ inches}^3$$

The section modulus for a rectangle equals $bd^2/6$, in which *b* is the length of the shorter side and *d* of the longer side of the rectangle (see *Handbook* page **236**), hence, $Z = bd^2/6$.

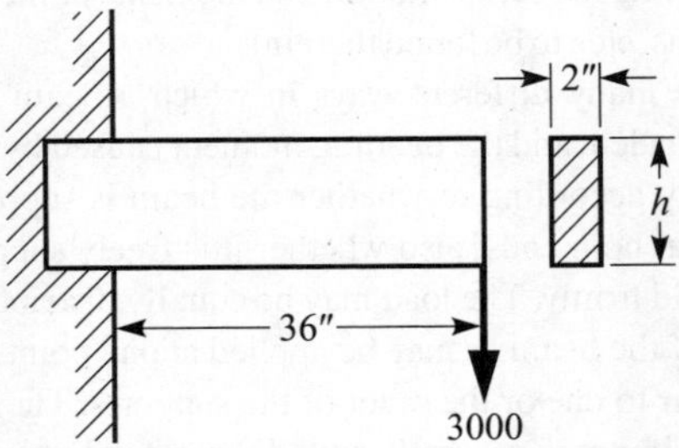


**Fig. 4. Determining the Depth *h* of a Beam to Support a Known Weight**

But $Z = 9$ and $b = 2$. Inserting these values into the formula, we have:

$$9 = \frac{2d^2}{6}$$

from which $d^2 = 27$, and $d = 5.2$ inches. This value $d$ corresponds to dimension $h$ in **Fig. 4**. Hence, the required depth of the beam to support a load of 3000 pounds at the outer end with a factor of safety of 3 would be 5.2 inches.

In calculating beams having either rectangular or circular cross-sections, the formulas on *Handbook* pages **262–263** are convenient to use. A beam loaded as shown by **Fig. 4** is similar to the first diagram on *Handbook* page **262**. If the formula on this page in the *Handbook* for determining height $h$ is applied to **Example 5**, **Fig. 4**, then,

$$h = \sqrt{\frac{6lW}{bf}} \quad \sqrt{\frac{6 \times 36 \times 3000}{2 \times 12{,}000}} = 5.2 \text{ inches}$$

In the above calculation the stress value $f$ is equivalent to $S_m/Kf_s$.

*Example 6:* A steel I-beam is to be used as a crane trolley track. This I-beam is to be supported at the ends, and the unsupported span is 20 feet long. The maximum load is 6000 pounds, and the nominal stress is not to exceed 10,000 pounds per square inch. Determine the size of the standard I-beam; also determine the maximum deflection when the load is at the center of the beam.

The foregoing conditions are represented by Case 2, *Handbook* page **251**. A formula for the stress at the critical point is $Wl/4Z$. As explained on *Handbook* page **250**, all dimensions are in inches, and the minus sign preceding a formula merely denotes compression of the upper fibers and tension in the lower fibers.

By inserting the known values in the formula:

$$10{,}000 = \frac{6000 \times 240}{4Z}; \text{ hence}$$

$$Z = \frac{6000 \times 240}{10{,}000 \times 4} = 36$$

The table of standard I-beams on *Handbook* page **2702** shows that a 12-inch I-beam, which weighs 31.8 pounds per foot, has a section modulus of 36.4.

The formula for maximum deflection (see *Handbook* starting on page **251**, Case 2) is $Wl^3/48EI$. According to the table on *Handbook* page **427**, the modulus of elasticity ($E$) of structural steel is 29,000,000.

As $Z$ = moment of inertia $I \div$ distance from neutral axis to extreme fiber (see *Handbook* page **250**), then for a 12-inch I-beam $I = 6Z = 216$; hence,

$$\text{Maximum deflection} = \frac{6000 \times (240)^3}{48 \times 29{,}000{,}000 \times 216} = 0.27 \text{ inch}$$

*Example 7:* All conditions are the same as in **Example 6**, except that the maximum deflection at the "critical point," or center of the I-beam, must not exceed $\frac{1}{8}$ inch. What size I-beam is required?

To meet the requirement regarding deflection,

$$\frac{1}{8} = \frac{Wl^3}{48EI}; \quad \text{therefore,}$$

$$I = \frac{8Wl^3}{48E} = \frac{8 \times 6000 \times (240)^3}{48 \times 29{,}000{,}000} = 476$$

If $x$ = distance from neutral axis to most remote fiber ($\frac{1}{2}$ beam depth in this case), then $Z = I/x$, and the table on *Handbook* page **2702** shows that a 15-inch, 50-pound I-beam should be used because it has a section modulus of 64.8 and $476/7.5 = 63.5$ nearly.

If 476 were divided by 6 ($\frac{1}{2}$ depth of a 12-inch I-beam), the result would be much higher than the section modulus of any standard 12-inch I-beam ($476 \div 6 = 79.3$); moreover, $576 \div 9 = 53$, which shows that an 18-inch I-beam is larger than is necessary because the lightest beam of this size has a section modulus of 81.9.

*Example 8:* If the speed of a motor is 1200 revolutions per minute and if its driving pinion has a pitch diameter of 3 inches, determine the torsional moment to which the pinion shaft is subjected, assuming that 10 horsepower is being transmitted.

If $W$ = tangential load in pounds, $H$ = the number of horsepower, and $V$ = pitch-line velocity in feet per minute,

$$W = \frac{33,000 \times H}{V}$$
$$= \frac{33,000 \times 10}{943} = 350 \text{ pounds}$$

The torsional moment = $W \times$ pitch radius of pinion = $350 \times 1.5$ = 525 pound-inches (or inch-pounds).

*Example 9:* If the pinion referred to in **Example 8** drives a gear having a pitch diameter of 12 inches, to what torsional or turning moment is the gear shaft subjected?

The torque or torsional moment in any case = pitch radius of gear × tangential load. The latter is the same for both gear and pinion; hence, torsional moment of gear = $350 \times 6 = 2100$ inch-pounds.

The torsional moment or the turning effect of a force that tends to produce rotation depends upon (1) the magnitude of the force acting, and (2) the distance of the force from the axis of rotation, measuring along a line at right angles to the line of action of the force.

**Strength in Plastic and Polymer Composite Parts.**—Various plastic materials are broadly used and provide desirable properties. Commonly used plastics, such as polyurethane and silicone, tend to be *viscoelastic*, displaying properties between those of elastic solids and viscous fluids, exhibiting a nonlinear stress-strain relationship. When these materials are stressed by sustained loads, they often deform or *creep*; composition, exposure time, and temperature affect the rate at which this can occur. After such deformation, the object will not return to its original shape and dimensions, causing a reduction in material strength. For more on mechanical, thermal, and electrical properties of plastics, see *Handbook* pages **561–580**; manufacturing for plastics is discussed on pages **580–591**; plastics parts design is discussed on pages **591–618**.

*Polymer Composites:* For applications requiring higher strength-to-weight ratios, good corrosion resistance, and reduced vibrational effects, products may be engineered with a matrix containing oriented composite plies or other polymer fibers for added reinforcement. The majority of the strength and stiffness of most composites is provided by the fiber reinforcements, while the matrix is a continuous material that holds the fiber reinforcements together. Typical loadings of fiber reinforcements are on the order of 50 percent by component volume and 75 percent by component mass, with the balance going to the matrix material.

Formulas for calculating density, maximum modulus, and maximum strength of polymer composites can be found on *Handbook* page **586**.

A composite's material properties can be estimated by considering the volume fraction ($f$) of the fiber reinforcements, as well as the orientation of the fiber reinforcements relative to the load direction. The density ($\rho_C$), maximum modulus ($E_C$), and maximum strength ($\sigma_C$) of the composite in the direction of the reinforcements are estimated using a rule of mixtures, where the properties for the reinforcements and matrix are indicated by the subscripts $R$ and $M$, respectively:

$$\rho_C = f\rho_R + (1-f)\rho_M$$
$$E_C = fE_R + (1-f)E_M$$
$$\sigma_C = f\sigma_R + (1-f)\sigma_M$$

*Example 10:* Use these formulas to determine the density, maximum modulus, and maximum strength of a composite with 55 percent volume fraction of carbon fibers embedded within an epoxy thermoset matrix:

Carbon properties: density = 1.81g/cm$^3$, tensile modulus = 242 GPa, tensile strength = 4.1 GPa

Epoxy properties: density = 1.35g/cm$^3$, tensile modulus = 35 GPa, tensile strength = 0.56 GPa

$$\rho_C = 0.55 \times \frac{1.81g}{cm^3} + (1 - 0.55) \times \frac{1.35g}{cm^3} = \frac{1.60g}{cm^3}$$

$$E_C = 0.55 \times 242\ GPa + (1 - 0.55) \times 35\ GPa = 148.85\ GPa$$

$$\sigma_C = 0.55 \times 4.1\ GPa + (1 - 0.55) \times 0.56\ GPa = 2.51\ GPa$$

**Sustainability Considerations.**—Added to the challenges of selecting materials with the appropriate mechanical and thermal strength and durability, fine-tuning design parameters, and optimizing manufacturing and machining methods, is the ever more important emphasis on sustainable materials and processes. For an overview of some common materials and calculations related to energy consumption, production cost, carbon footprint, and more, see pages **361–366** of the *Handbook*.

## PRACTICE EXERCISES FOR SECTION 16

(See *Answers to Practice Exercises for Section 16* on page **248**)

1) What is a "factor of safety," and why are different factors used in machine design?

2) If the ultimate strength of a steel rod is 60,000 pounds per square inch, and the factor of safety is 5, what is the equivalent working stress?

3) If a steel bar must withstand a maximum pull of 9000 pounds and if the maximum nominal stress must not exceed 12,000 pounds per square inch, what diameter bar is required?

4) Is a steel rod stronger when at ordinary room temperature or when heated to 500° F?

5) What is the meaning of the term "elastic limit"?

6) Approximately what percentages of copper and zinc in brass result in the greatest tensile strength?

7) If four 10-foot-long pipes are to be used to support a water tank installation weighing 100,000 pounds, what diameter standard-weight pipe is required?

8) Under the elastic limit of common metals, how is stress correlated to strain?

9) When plastics are exhibiting high loads, what common phenomena occurs? What happens when the temperature also increases with this high loading?

10) What are the two major components of polymer composites? What component provides the tensile strength?

## SECTION 17

## DESIGN OF SHAFTS AND KEYS FOR POWER TRANSMISSION

*Machinery's Handbook* pages **289–297** and pages **2545–2568**

This section is a review of the general procedure in designing shafts to resist both torsional and combined torsional and bending stresses. The diameter of a shaft through which power is transmitted depends, for a given shaft material, upon the amount and kind of stress or stresses to which the shaft is subjected. To illustrate the general procedure, we shall assume first that the shaft is subjected only to a uniform torsional or twisting stress and that there is no additional bending stress that needs to be considered in determining the diameter.

*Example 1:* A lineshaft carrying pulleys located close to the bearings is to transmit 50 horsepower at 1200 revolutions per minute. If the load is applied gradually and is steady, what diameter steel shaft is required, assuming that the pulleys are fastened to the shaft by means of keys and that the bending stresses caused by the pull of the belts are negligible?

According to the former American Standard Association's Code for the Design of Transmission Shafting, the diameter of shaft required to meet the stated conditions can be determined by using **Formula (16b)**, *Handbook* page **294**.

$$D = B \times \sqrt[3]{\frac{321,000 K_t P}{S_s N}}$$

In this formula, $D$ = required shaft diameter in inches; $B$ = a factor, which for solid shafts is taken as 1; $K_t$ = combined shock and fatigue factor; $P$ = maximum horsepower transmitted by shaft; $S_s$ = maximum allowable torsional shearing stress in pounds per square inch; and $N$ = shaft speed in revolutions per minute.

From **Table 1** on *Handbook* page **295**, $K_t$ = 1.0 for gradually applied and steady loads, and from **Table 2** (on the same page) the recommended maximum allowable working stress for "Commercial Steel" shafting with keyways subjected to pure torsion loads is 6000 pounds per square inch. By substituting in the formula,

$$D = 1 \times \sqrt[3]{\frac{321{,}000 \times 1.0 \times 50}{6000 \times 1200}} = 1.306 \text{ inches}$$

The nearest standard-size transmission shafting from the table on *Handbook* page **293** is $1\frac{7}{16}$ inches.

*Example 2:* If, in **Example 1**, the shaft diameter had been determined by using **Formula (5b)**, *Handbook* page **289**, what would the result have been and why?

$$D = \sqrt[3]{\frac{53.5P}{N}} = \sqrt[3]{\frac{53.5 \times 50}{1200}} = 1.306 \text{ inches}$$

This formula gives the same shaft diameter as was previously determined because it is a simplified form of the first formula used and contains the same values of $K_t$ and $S_s$, but combined as the single constant 53.5 . For lineshafts carrying pulleys under conditions ordinarily encountered, this simplified formula is usually quite satisfactory; but, where conditions of shock loading are known to exist, it is safer to use **Formula (16b)**, *Handbook* page **294**, which takes such conditions into account.

**Shafts Subjected to Combined Stresses.**—The preceding formulas are based on the assumption that the shaft is subjected to torsional stresses only. However, many shafts must withstand stresses that result from combinations of torsion, bending, and shock loading. In such conditions it is necessary to use formulas that take such stresses into account.

*Example 3:* Suppose that, after the lineshaft in **Example 1** was installed, it became necessary to relocate a machine that was being driven by one of the pulleys on the shaft. Because of the new machine location, it was necessary to move the pulley on the lineshaft farther away from the nearest bearing, and, as a result, a bending moment of 2000 inch-pounds was introduced. Is the

$1\frac{7}{16}$-inch diameter shaft sufficient to take this additional stress, or will it be necessary to relocate the bearing to provide better support?

Since there are now both bending and torsional loads acting on the shaft, **Formula (18b)**, *Handbook* page **294** should be used to compute the required shaft diameter. This diameter is then compared with the $1\frac{7}{16}$-inch diameter previously determined.

$$D = B \times \sqrt[3]{\frac{5.1}{p_t}\sqrt{(K_m M)^2 + \left(\frac{63{,}000 K_t P}{N}\right)^2}}$$

In this formula $B$, $K_t$, $P$, and $N$ are quantities previously defined and $p_t$ = maximum allowable shearing stress under combined loading conditions in pounds per square inch; $K_m$ = combined shock and fatigue factor; and $M$ = maximum bending moment in inch-pounds.

From **Table 1** on *Handbook* page **295**, $K_m$ = 1.5 for gradually applied and steady loads and from **Table 2** (on the same page), $p_t$ = 6000 pounds per square inch. By substituting in the formula,

$$D = 1 \times \sqrt[3]{\frac{5.1}{6000}\sqrt{(1.5 \times 2000)^2 + \left(\frac{63{,}000 \times 1 \times 50}{1200}\right)^2}}$$

$$= \sqrt[3]{\frac{5.1}{6000}\sqrt{9000000 + 6{,}890{,}625}} = \sqrt[3]{\frac{5.1}{6000} \times 3986}$$

$$= \sqrt[3]{3.388} = 1.502 \text{ inches or about } 1\tfrac{1}{2} \text{ inches}$$

This diameter is larger than the $1\frac{7}{16}$-inch diameter used for the shaft in **Example 1**, so it will be necessary to relocate the bearing closer to the pulley, thus reducing the bending moment. The $1\frac{7}{16}$-inch diameter shaft will then be able to operate within the allowable working stress for which it was originally designed.

**Design of Shafts to Resist Torsional Deflection.—** Shafts must often be proportioned not only to provide the strength required to transmit a given torque, but also to prevent torsional deflection (twisting) through a greater angle than has been found satisfactory for a given type of service. This requirement is particularly true for machine shafts and machine tool spindles.

For ordinary service, it is customary that the angle of twist of machine shafts be limited to $\frac{1}{10}$ degree per foot of shaft length, and, for machine shafts subject to load reversals, $\frac{1}{20}$ degree per foot of shaft length. As explained in the *Handbook*, the usual design procedure for shafting that is to have a specified maximum angular deflection is to compute the diameter of shaft required based on both deflection and strength considerations and then to choose the larger of the two diameters thus determined.

*Example 4:* A 6-foot-long feed shaft is to transmit a torque of 200 inch-pounds. If there are no bending stresses, and the shaft is to be limited to a torsional deflection of $\frac{1}{20}$ degree per foot of length, what diameter shaft should be used? The shaft is to be made of cold drawn steel and is to be designed for a maximum working stress of 6000 pounds per square inch in torsion.

The diameter of shaft required for a maximum angular deflection α is given by **Formula (13)**, *Handbook* page **291**.

$$D = 4.9\sqrt[4]{\frac{Tl}{G\alpha}}$$

In this formula $T$ = applied torque in inch-pounds; $l$ = length of shaft in inches; $G$ = torsional modulus of elasticity, which, for steel, is 11,500,000 pounds per square inch; and $\alpha$ = angular deflection of shaft in degrees.

In the problem at hand, $T = 200$ inch-pounds; $l = 6 \times 12 = 72$ inches; and $\alpha = 6 \times 1/20 = 0.3$ degree.

$$D = 4.9\sqrt[4]{\frac{200 \times 72}{11{,}500{,}000 \times 0.3}} = 4.9\sqrt[4]{0.0041739}$$

$$= 4.9 \times 0.254 = 1.24 \text{ inches}$$

The diameter of the shaft based on strength considerations is obtained by using **Formula (3a)**, *Handbook* page **289**.

$$D = \sqrt[3]{\frac{5.1T}{S_s}} = \sqrt[3]{\frac{5.1 \times 200}{6000}} = \sqrt[3]{0.17} = 0.55 \text{ inch}$$

From the above calculations, the diameter based on torsional deflection considerations is the larger of the two values obtained, so the nearest standard diameter, $1\frac{1}{4}$ inches, should be used.

**Selection of Key Size Based on Shaft Size.**—Keys are generally proportioned in relation to shaft diameter instead of in relation to torsional load to be transmitted because of practical reasons, such as standardization of keys and shafts. Standard sizes are listed in the table *Key Size versus Shaft Diameter ANSI/ASME B17.1-1967 (R2013)* on *Handbook* page **2557**. Dimensions of both square and rectangular keys are given, but for shaft diameters up to and including $6\frac{1}{2}$ inches, square keys are preferred. For larger shafts, rectangular keys are commonly used.

Two rules that base key length on shaft size are: (1) $L = 1.5D$ and (2) $L = 0.3D^2 \div T$, where $L$ = length of key, $D$ = diameter of shaft, and $T$ = key thickness.

If the keyset is to have fillets, and the key is to be chamfered, suggested dimensions for these modifications are given on *Handbook* page **2562**. If a set screw is to be used over the key, suggested sizes are given in the table on *Handbook* page **2562**.

*Example 5:* If the maximum torque output of a 2-inch diameter shaft is to be transmitted to a keyed pulley, what should be the proportions of the key?

According to the table on *Handbook* page **2557**, a $\frac{1}{2}$-inch square key would be preferred. If a rectangular key were selected, its dimensions would be $\frac{1}{2}$ inch by $\frac{3}{8}$ inch. According to rule 1 above, its length would be 3 inches.

The key and keyseat may be proportioned so as to provide a clearance or an interference fit. The table on *Handbook* page **2561** gives tolerances for widths and depths of keys and keyseats to provide Class 1 (clearance) and Class 2 (interference) fits. An additional Class 3 (interference) fit, which has not been standardized, is mentioned on *Handbook* page **2557** together with suggested tolerances.

**Keys Proportioned According to Transmitted Torque.**—As previously stated, if key sizes are based on shaft diameter, the dimensions of the key sometimes will be excessive, usually when a gear or pulley transmits only a portion of the total torque capacity of the shaft to which it is keyed. If excessively large keys are to be avoided, it may be advantageous to base the determination on the torque to be transmitted rather than on the shaft diameter and to use the dimensions thus determined as a guide in selecting a standard-size key.

A key proportioned to transmit a specified torque may fail in service either by shearing or by crushing, depending on the proportions of the key and the manner in which it is fitted to the shaft and hub. The best proportions for a key are those that make it equally resistant to failure by shearing and by crushing. The safe torque in inch-pounds that a key will transmit, based on the allowable shearing stress of the key material, may be found from the formula:

$$T_s = L \times W \times \frac{D}{2} \times S_s \qquad (1)$$

The safe torque based on the allowable compressive stress of the key material is found from the formula:

$$T_c = L \times \frac{H}{2} \times \frac{D}{2} \times S_c \qquad (2)$$

(For Woodruff keys the amount that the key projects above the shaft is substituted for $H/2$.)

In these formulas, $T_s$ = safe torque in shear; $T_c$ = safe torque in compression; $S_s$ = allowable shearing stress; $S_c$ = allowable compressive stress; $L$ = key length in inches; $W$ = key width in inches; $H$ = key thickness in inches; and $D$ = shaft diameter in inches.

To satisfy the condition that the key be equally resistant to shearing and crushing, $T_s$ should equal $T_c$. Thus, by equating **Formulas (1)** and **(2)**, it is found that the width of the keyway in terms of the height of the keyway is:

$$W = \frac{HS_c}{2S_s} \qquad (3)$$

For the type of steel commonly used in making keys, the allowable compressive stress *Sc* may be taken as twice the allowable shearing stress *Ss* of the material if the key is properly fitted on all four sides. By substituting $Sc = 2Ss$ in **Formula (3)** it will be found that $W = H$, so that for equal strength in compression and shear a square key should be used.

If a rectangular key is used, and the thickness *H* is less than the width *W*, then the key will be weaker in compression than in shear so that it is sufficient to check the torque capacity of the key using **Formula (2)**.

*Example 6:* A 3-inch shaft is to deliver 100 horsepower at 200 revolutions per minute through a gear keyed to the shaft. If the hub of the gear is 4 inches long, what size key, equally strong in shear and compression, should be used? The allowable compressive stress in the shaft is not to exceed 16,000 pounds per square inch and the key material has an allowable compressive stress of 20,000 pounds per square inch and an allowable shearing stress of 15,000 pounds per square inch.

The first step is to decide on the length of the key. Since the hub of the gear is 4 inches long, a key of the same length may be used. The next step is to determine the torque that the key will have to transmit. By using **Formula (2)**, *Handbook* page **289**,

$$T = \frac{63{,}000P}{N} = \frac{63{,}000 \times 100}{200} = 31{,}500 \text{ inch-pounds}$$

To determine the width of the key, based on the allowable shearing stress of the key material, **Equation (1)** above is used.

$$T_s = L \times W \times \frac{D}{2} \times S_s$$

$$31{,}500 = 4 \times W \times \frac{D}{2} \times 15{,}000$$

or

$$W = \frac{31{,}500 \times 2}{15{,}000 \times 4 \times 3} = 0.350, \text{ say, } \tfrac{3}{8} \text{ inch}$$

In using **Equation (2)** to determine the thickness of the key, however, it should be noted that, if the shaft material has a different allowable compressive stress than the key material, then the lower of the two values should be used. The shaft material then has the lower allowable compressive stress, and the keyway in the shaft would fail by crushing before the key would fail. Therefore,

$$T_c = L \times \frac{H}{2} \times \frac{D}{2} \times S_c$$

$$31{,}500 = 4 \times \frac{H}{2} \times \frac{3}{2} \times 16{,}000$$

$$H = \frac{31{,}500 \times 2 \times 2}{16{,}000 \times 4 \times 3} = 0.656 = \frac{21}{32} \text{ inch}$$

Therefore, the dimensions of the key for equal resistance to failure by shearing and crushing are $^3/_8$ inch wide, $^{21}/_{32}$ inch thick, and 4 inches long. If, for some reason, it is desirable to use a key shorter than 4 inches, say, 2 inches, then it will be necessary to increase both the width and thickness by a factor of 4 ÷ 2 if equal resistance to shearing and crushing is to be maintained. Thus the width would be $^3/_8 \times {}^4/_2 = {}^3/_4$ inch, and the thickness would be $^{21}/_{32} \times {}^4/_2 = 1^5/_{16}$ inch for a 2-inch-long key.

**Set-Screws Used to Transmit Torque.—**For certain applications it is common practice to use set-screws to transmit torque because they are relatively inexpensive to install and permit axial adjustment of the member mounted on the shaft. However, set-screws depend primarily on friction and the shearing force at the point of the screw, so they are not especially well-suited for high torques or where sudden load changes take place.

One rule for determining the proper size of a set-screw states that the diameter of the screw should equal $^5/_{16}$ inch plus one-eighth the shaft diameter. The holding power of set-screws selected by this rule can be checked using the formula on page **1834** of the *Handbook*.

## PRACTICE EXERCISES FOR SECTION 17

(See *Answers to Practice Exercises for Section 17* on page **248**)

1) What is the polar section modulus of a 2-inch diameter shaft?

2) Using the information in the note at the bottom of page **3242** in the *Machinery's Handbook 32 Digital Edition*, redo #1 above using the tables *Section Moduli and Moments of Inertia for Round Shafts* starting on page **3242**.

3) If a 3-inch shaft is subjected to a torsional or twisting moment of 32,800 pound-inches, what is the equivalent torsional or shearing stress?

4) Is the shaft referred to in #2 above subjected to an excessive torsional stress?

5) If a 10-horsepower motor operating at its rated capacity connects by a belt with a 16-inch pulley on the driving shaft of a machine, what is the load tangential to the pulley rim and the

resulting twisting moment on the shaft, assuming that the rim speed of the driven pulley is 600 feet per minute?

6) How is the maximum distance between bearings for steel line-shafting determined?

7) What are "gib-head" keys, and why are they used on some classes of work?

8) What is the distinctive feature of Woodruff keys?

9) What are the advantages of Woodruff keys?

10) If a $^3/_8$-inch wide keyseat is to be milled into a $1^1/_2$-inch diameter shaft and if the keyseat depth is $^3/_{16}$ inch (as measured at one side), what is the depth from the top surface of the shaft or the amount to sink the cutter after it grazes the top of the shaft?

## SECTION 18

## SPLINES

*Machinery's Handbook* pages **2333–2362**

This section of the *Handbook* shows how to calculate the dimensions of involute splines and how to provide specifications for manufacturing drawings. Many types of mechanical connections between shafts and hubs are available for both fixed and sliding applications. Among these connections are the ordinary key and keyway (*Handbook* pages **2545–2568**), multiple keys and keyways, three- and four-lobed polygon shaft and hub connections, and involute splines of both inch dimension and metric module sizes.

The major advantages of involute splines are that they may be manufactured on the same equipment used to manufacture gears, they may be used for fixed and interference fit connections as well as for sliding connections, and they are stronger than most other connections with the exception of polygon-shaped members.

The section in the *Handbook* on involute splines, pages **2333–2359**, provides tables, data, formulas, and diagrams for American Standard splines made to both inch and metric module systems. Both systems share common definitions of terms, although the symbols used to identify dimensions and angles may differ, as shown on *Handbook* page **2354**. The two systems do not provide for interchangeability of parts; the new metric module Standard is the American National Standards Institute version of the International Standards Organization involute spline Standard, which is based upon metric, not inch, dimensions.

*Example 1:* A metric module involute spline pair is required to meet the following specification: pressure angle $\alpha_D = 30°$; module $m = 5$; number of teeth $Z = 32$; fit class = H/h; tolerance class 5 for both the internal and external splines; flat-root design for both members; length of engagement of the splines is 100 mm.

**Table 13** beginning on *Handbook* page **2356** provides all the formulas necessary to calculate the dimensions of these splines.

Pitch diameter:

$$D = mZ = 5 \times 32 = 160 \text{ mm} \quad (1)$$

Base diameter:

$$\begin{aligned} DB &= mZ\cos\alpha_D = 160 \times \cos\alpha_D = 160 \times \cos 30° \\ &= 160 \times 0.86603 = 138.5641 \text{ mm} \end{aligned} \quad (2)$$

Circular pitch:

$$p = \pi m = 3.1416 \times 5 = 15.708 \quad (3)$$

Base pitch:

$$p_b = \pi m \cos\alpha_D = \pi \times 5 \times 0.86603 = 13.60350 \quad (4)$$

Tooth thickness modification:

$$es = 0 \quad (5)$$

in accordance with footnote to **Table 14**, *Handbook* page **2357**, and the Fit Classes paragraph on page **2354** that refers to H/h fits.

Minimum major diameter, internal spline,

$$DEI \text{ min} = m(Z + 1.8) = 5 \times (32 + 1.8) = 169.000 \quad (6)$$

Maximum major diameter, internal spline,

$$\begin{aligned} DEI \text{ max} &= DEI \text{ min} + (T + \lambda)/(\tan\alpha_D) \\ &= 169.000 + 0.248/\tan 30° \\ &= 169.4295 \text{ mm} \end{aligned} \quad (7)$$

In this last calculation, the value of $(T + \lambda) = 0.248$ for class 7 was calculated using the formula in **Table 15**, *Handbook* page **2357**, as follows:

$$\begin{aligned} i^* &= 0.001(0.45\sqrt[3]{D} + 0.001D) \\ &= 0.001(0.45\sqrt[3]{160} + 0.001 \times 160) \\ &= 0.00260 \end{aligned} \quad (8a)$$

$$\begin{aligned} i^{**} &= 0.001(0.45\sqrt[3]{7.85398} + 0.001 \times 7.85398) \\ &= 0.00090 \end{aligned} \quad (8b)$$

In **Equation (8b)**, 7.85398 is the value of $S_{bsc}$ calculated from the formula $S_{bsc} = 0.5\pi m$ given in the table starting on *Handbook* page **2356**.

$$\begin{aligned}(T+\lambda) &= 40i^* + 160i^* \\ &= 40 \times 0.00260 + 160 \times 0.00090 \qquad (8c) \\ &= 0.248 \text{ mm}\end{aligned}$$

Form diameter, internal spline,

$$\begin{aligned}DFI &= m(Z+1) + 2c_F \\ &= 5(32+1) + 2 \times 0.1m \\ &= 5(32+1) + 2 \times 0.1 \times 5 \qquad (9) \\ &= 166 \text{ mm}\end{aligned}$$

In the above calculation the value of $c_F = 0.1m$ is taken from the diagram on *Handbook* page **2358**, and the corresponding formula for form clearance on *Handbook* page **2356**. Minimum minor diameter, internal spline,

$$\begin{aligned}DII \min &= DFE + 2c_F \\ &= 154.3502 + 2 \times 0.1 \times 5 \qquad (10) \\ &= 155.3502 \text{ mm}\end{aligned}$$

The *DFE* value of 154.3502 used in this calculation was calculated from the formula on *Handbook* page **2356** as follows: $DB = 138.564$ from step (**2**); $D = 160$ from step (**1**); $h_s = 0.6m = 3.0$ from the last formula in the table starting on *Handbook* page **2356**; $es = 0$ from step (**5**); sin 30° = 0.50000; tan 30° = 0.57735. Therefore,

$$DFE = 2 \times \sqrt{(0.5 \times 138.564)^2 + \left[0.5 \times 160 \times 0.50000 - \frac{0.6 \times 5 + \left(\frac{0.5 \times 0}{0.57735}\right)}{0.50000}\right]^2} \qquad (11)$$

$$= 154.3502$$

Maximum minor diameter, internal spline,

$$
\begin{aligned}
DII \max &= DII \min + (0.2m^{0.667} - 0.1m^{-0.5}) \\
&= 155.3502 + 0.58 \qquad (12) \\
&= 155.9302 \text{ mm}
\end{aligned}
$$

The value 0.58 used in this calculation comes from the footnote *c* to the table on *Handbook* page **2356**. Circular space width, basic,

$$E_{bsc} = 0.5\pi m = 0.5 \times 3.1416 \times 5 = 7.854 \text{ mm} \qquad (13)$$

Circular space width, minimum effective,

$$EV \min = E_{bsc} = 7.854 \text{ mm} \qquad (14)$$

Circular space width, maximum actual,

$$
\begin{aligned}
E \max &= EV \min + (T + \lambda) \\
&= 7.854 + 0.0992 \quad \text{from step } \mathbf{(16c)} \qquad (15) \\
&= 7.9532 \text{ mm}
\end{aligned}
$$

The value of $(T + \lambda)$ calculated in step **(16c)** is based upon class 5 fit stated at the beginning of the example. The value calculated in step **(8c)**, on the other hand, is based upon class 7 fit as required by the formula in step **(7)**. For class 5 fit, using the formula given in **Table 15**, *Handbook* page **2357**:

$$i^* = 0.00260 \quad \text{from step } \mathbf{(8a)} \qquad (16a)$$

$$i^{**} = 0.00090 \quad \text{from step } \mathbf{(8b)} \qquad (16b)$$

$$
\begin{aligned}
(T + \lambda) &= 16i^* + 64i^{**} = 16 \times 0.00260 + 64 \times 0.00090 \\
&= 0.0992 \text{ mm}
\end{aligned} \qquad (16c)
$$

Circular space width, minimum actual,

$$E \min = EV \min + \lambda = 7.854 + 0.045 = 7.899 \text{ mm} \qquad (17)$$

The value of $\lambda$ used in this formula was calculated from the formulas for class 5 fit in **Table 16**, on *Handbook* page **2358**, and the formula in the text on *Handbook* page **2357** as follows:

$$F_p = 0.001(3.55\sqrt{5 \times 32 \times 3.1416/2} + 9) = 0.065 \text{ mm} \quad (18a)$$

$$f_f = 0.001[2.5 \times 5(1 + 0.0125 \times 32) + 16] = 0.034 \text{ mm} \quad (18b)$$

$$F_\beta = 0.001(1 \times \sqrt{100} \times 5) = 0.015 \text{ mm} \quad (18c)$$

$$\lambda = 0.6\sqrt{(0.065)^2 + (0.034)^2 + (0.015)^2} = 0.045 \text{ mm} \quad (18d)$$

Circular space width, maximum effective,

$$\begin{aligned} EV \max &= E \max - \lambda \qquad \textbf{(15)} \qquad\qquad \textbf{(18d)} \\ &= 7.9532 \text{ from step (15)} - 0.045 \text{ from step (18d)} \qquad (19) \\ &= 7.9082 \text{ mm} \end{aligned}$$

Maximum major diameter, external spline,

$$\begin{aligned} DEE \max &= m(Z+1) - es/\tan\alpha_D = 5(32+1) - 0 \\ &= 165 \text{ mm} \end{aligned} \quad (20)$$

The value 0 in this last calculation is from **Table 17**, *Handbook* page **2358**, for h class fit.

Minimum major diameter, external spline, is calculated using the results of step **(20)** and footnote **c** on *Handbook* page **2356**,

$$\begin{aligned} &DEE \min = DEE \max - (0.2m^{0.667} - 0.01m^{-0.5}) \\ &= 165 - 0.58 = 164.42 \text{ mm} \end{aligned} \quad (21)$$

Maximum minor diameter, external spline,

$$\begin{aligned} DIE \max &= m(Z - 1.8) - es/\tan\alpha_D \\ &= 5(32 - 1.8) - 0 \qquad (22) \\ &= 151 \text{ mm} \end{aligned}$$

The value 0 in this calculation is from **Table 17**, *Handbook* page **2358**, for h class fit.

Minimum minor diameter, external spline, is calculated using the results of steps **(22)** and **(7)**,

$$\begin{aligned} DIE \text{ min} &= DIE \text{ max} - (T + \lambda)/\tan\alpha_D \\ &= 151 - 0.248/\tan 30° \\ &= 151 - 0.4295 \\ &= 150.570 \text{ mm} \end{aligned} \tag{23}$$

Circular tooth thickness, basic, has been taken from step **(13)**,

$$S_{bsc} = 7.854 \text{ mm} \tag{24}$$

Circular tooth thickness, maximum effective, is calculated using the results of steps **(13)** and **(5)**,

$$\begin{aligned} SV \text{ max} &= S_{bsc} - es \\ &= 7.854 - 0 \\ &= 7.854 \text{ mm} \end{aligned} \tag{25}$$

Circular tooth thickness, minimum actual, is calculated using the results of steps **(25)** and **(16c)**,

$$S \text{ min} = SV \text{ max} - (T + \lambda) = 7.854 - 0.0992 = 7.7548 \text{ mm} \tag{26}$$

Circular tooth thickness, maximum actual, is calculated using the results of steps **(25)** and **(18d)**,

$$\begin{aligned} S \text{ max} &= SV \text{ max} - \lambda \\ &= 7.854 - 0.045 \\ &= 7.809 \text{ mm} \end{aligned} \tag{27}$$

Circular tooth thickness, minimum effective, is calculated using the results of steps **(26)** and **(18d)**,

$$\begin{aligned} SV \text{ min} &= S \text{ min} + \lambda \\ &= 7.754 + 0.045 \\ &= 7.799 \text{ mm} \end{aligned} \tag{28}$$

*Example 2:* As explained on *Handbook* page **2351**, spline gages are used for routine inspection of production parts. However, as part of an analytical procedure to evaluate effective space width or effective tooth thickness, measurements with pins are often used. Measurements with pins are also used for checking the actual space width and tooth thickness of splines during the machining process. Such measurements help in making the necessary size adjustments both during the setup process and as manufacturing proceeds. For the splines calculated in **Example 1**, what are the pin measurements for the tooth thickness and space width?

The maximum space width for the internal spline is 7.953 mm from step **(15)** in **Example 1**. The minimum tooth thickness for the external spline is 7.755 mm from step **(26)**.

*Handbook* page **2352** gives a method for calculating pin measurements for splines. This procedure was developed for inch-dimension splines. However, it may be used for metric module splines simply by replacing *P* wherever it appears in a formula by 1/*m*; and by using millimeters instead of inches as dimensional units throughout.

For two-pin measurement *between* pins for the *internal* spline, steps **1)**, **2)**, and **3)** starting on *Handbook* page **2352** are used as follows:

$$\begin{aligned} \text{inv}\ \phi_i &= 7.953/160 + \text{inv}\ 30° - 8.64/138.564 \\ &= 0.049706 + 0.053751 - 0.062354 = 0.041103 \end{aligned} \tag{1}$$

The numbers used in this calculation are taken from the results in **Example 1** except for the involute of 30°, which is from the table on page **108** of the *Handbook*, and 8.64 is the diameter of the wire as calculated from the formula on *Handbook* page **2353**, 1.7280/*P* in which 1/*m* has been substituted for *P* to give 1.7280*m* = 1.7280 × 5 = 8.64. Note that the symbols on page **2352** are not the same as those used in **Example 1**. This is because the metric Standard for involute splines uses different symbols for the same dimensions. The table on page **2355** of the *Handbook* shows how these different symbols compare.

The value of inv $\phi_i$ = 0.041103 is used to enter the table on *Handbook* page **108** to find, by interpolation,

$$\phi_i = 27°36'20'' \tag{2}$$

From a calculator find

$$\sec 27°36'20'' = 1.1285 \tag{3}$$

Calculate the measurement between wires:

$$\begin{aligned} M_i &= D_b \sec\phi_i - d_i = 138.564 \times 1.1285 - 8.64 \\ &= 147.729 \text{ mm} \end{aligned} \tag{4}$$

For two-pin measurement *over* the teeth of *external* splines, steps **1)**, **2)**, and **3)** on *Handbook* pages **2352–2353** are used as follows:

$$\text{inv}\ \phi_e = 7.755/160 + 0.053751 + 9.6/138.564 - 3.1416/32 \qquad (5)$$
$$= 0.073327$$

Therefore, from *Handbook* page **109**, $\phi_e = 32° 59'$ and, from a calculator, sec $32° 59' = 1.1921$. From the formula in step **3)** on *Handbook* page **2353**:

$$M_e = 138.564 \times 1.1921 + 9.6 = 174.782 \text{ mm} \qquad (6)$$

The pin diameter 9.6 in this calculation was calculated from the formula in step **3)** on *Handbook* page **2353** by substituting $1/m$ for $P$ in the formula $d_e = 1.9200/P = 1.9200m$.

**Specifying Spline Data on Drawings.—**As stated on *Handbook* page **2346**, if the data specified on a spline drawing are suitably arranged and presented in a consistent manner, it is usually not necessary to provide a graphic illustration of the spline teeth. **Table 6** on *Handbook* page **2345** illustrates a flat-root spline similar to the one in **Example 1** except that it is an inch-dimension spline. The method of presenting drawing data for metric module splines differs somewhat from that shown on page **2345** in that the number of decimal places used for metric spline data is sometimes less than that for the corresponding inch-dimension system.

*Example 3:* How much of the data calculated or given in **Example 1** and **Example 2** should be presented on the spline drawing?

For the internal spline, the data required to manufacture the spline should be presented as follows, including the number of decimal places shown:

**Example of Internal Involute Spline Drawing Data**

| Flat-Root Side Fit | Tolerance Class 5H |
|---|---|
| Number of Teeth | 32 |
| Module | 5 |
| Pressure Angle | 30 deg |
| Base Diameter | 138.5641 REF |

**Example of Internal Involute Spline Drawing Data** *(Continued)*

| Flat-Root Side Fit | Tolerance Class 5H |
|---|---|
| Pitch Diameter | 160.0000 REF |
| Major Diameter | 169.42 Max |
| Form Diameter | 166.00 |
| Minor Diameter | 155.35/155.93 |
| *Circular Space Width:* | |
| Max. Actual | 7.953 |
| Min. Effective | 7.854 |
| *Max. Measurement Between Pins:* | 147.729 REF |
| Pin Diameter | 8.640 |

For the external spline, the data required to manufacture the spline should be presented as follows:

**Example of External Involute Spline Drawing Data**

| Flat-Root Side Fit | Tolerance Class 5H |
|---|---|
| Number of Teeth | 32 |
| Module | 5 |
| Pressure Angle | 30 deg |
| Base Diameter | 138.5641 REF |
| Pitch Diameter | 160.0000 REF |
| Major Diameter | 164.42/165.00 |
| Form Diameter | 154.35 |
| Minor Diameter | 150.57 MIN |
| *Circular Space Width:* | |
| Max. Actual | 7.854 |
| Min. Effective | 7.809 |
| *Min. Measurement Over Pins:* | 74.782 REF |
| Pin Diameter | 9.6 |

## PRACTICE EXERCISES FOR SECTION 18

(See *Answers to Practice Exercises for Section 18* on page **249**)

1) What is the difference between a "soft" conversion of a standard and a "hard" system?

2) The Standard for metric module splines does not include a major diameter fit. What Standard provides for a major diameter fit?

3) What is an involute serration and is it still called this in American Standards?

4) What are some of the advantages of involute splines?

5) What is the meaning of the term "effective tooth thickness"?

6) What advantage is there in using an odd number of spline teeth?

7) If a spline connection is misaligned, fretting can occur at certain combinations of torque, speed, and misalignment angle. Is there any method for diminishing such damage?

8) For a given design of spline is there a method for estimating the torque capacity based upon wear? Based on shearing stress?

9) What does REF following a dimension of a spline mean?

10) Why are fillet-root splines sometimes preferred over flat-root splines?

## SECTION 19

## DESIGNING AND CUTTING GEARS

*Machinery's Handbook* pages **2206–2332**

In the design of gearing, there may be three distinct types of problems. These are: (1) determining the relative sizes of two or more gears to obtain a given speed or series of speeds; (2) determining the pitch of the gear teeth so that they will be strong enough to transmit a given amount of power; and (3) calculating the dimensions of a gear of a given pitch, such as the outside diameter, the depth of the teeth, and other dimensions needed in cutting the gear.

When the term "diameter" is applied to a spur gear, the pitch diameter is generally referred to and not the outside diameter. In calculating the speeds of gearing, the pitch diameters are used and not the outside diameters, because when gears are in mesh, the imaginary pitch circles roll in contact with each other.

**Calculating Gear Speeds.**—The simple rules for calculating the speeds of pulleys beginning on *Handbook* page **2569** may be applied to gearing, provided either the pitch diameters of the gears or the numbers of teeth are substituted for the pulley diameters. Information on gear speeds, especially as applied to compound trains of gearing, also will be found in the section dealing with lathe change gears beginning on *Handbook* page **2183**. When gear trains must be designed to secure unusual or fractional gear ratios, the directions beginning on *Handbook* page **2184** will be found very useful. A practical application of these methods is shown by examples beginning on *Handbook* page **2188**.

Planetary or epicyclic gearing is an increasingly important class of power transmission in various industries because of compactness, efficiency, and versatility. The rules for calculating rotational speeds and ratios are different from those for other types of gearing. Formulas for the most commonly used types of planetary gears are provided on *Handbook* pages **2293–2295**.

*Example 1:* The following example illustrates the method of calculating the speed of a driven shaft in a combination belt and gear drive when the diameters of the pulleys and the pitch diameters of the gears are known, and the number of revolutions per minute (rpm) of the driving shaft is given. If driving pulley *A*, **Fig. 1**, is 16 inches in diameter, and driven pulley *B*, 6 inches in diameter, and the pitch diameter of driving gear *C* is 12 inches, driving gear *D* is 14 inches, driven gear *E*, 7 inches, driven gear *F*, 6 inches, and driving pulley *A* makes 60 rpm, determine the number of rpm of *F*.

$$\frac{16 \times 12 \times 14}{6 \times 7 \times 6} \times 60 = 640 \text{ rpm}$$

The calculations required in solving problems of this kind can be simplified if the gears are considered as pulleys having diameters equal to their pitch diameters. When this is done, the rules that apply to compound belt drives can be used in determining the speed or size of the gears or pulleys.

Substituting the numbers of teeth in each gear for the pitch diameter gives the same result as when the pitch diameters are used.

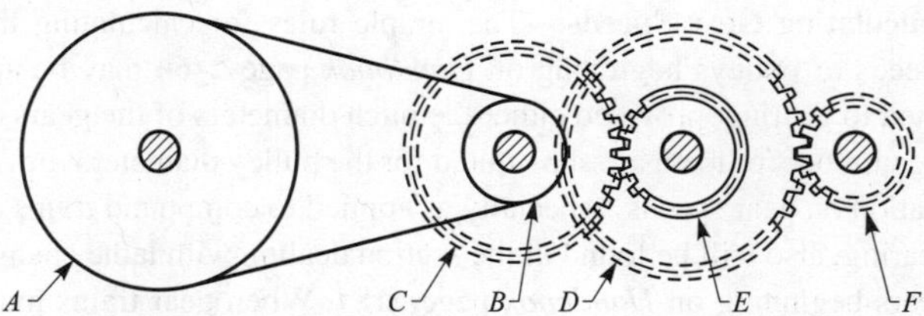


**Fig. 1. Combination Pulley and Compound Gear Drive**

*Example 2:* If driving spur gear *A* (**Fig. 2**) makes 336 rpm and has 42 teeth, driven spur gear *B*, 21 teeth, driving bevel gear *C*, 33 teeth, driven bevel gear *D*, 24 teeth, driving worm *E*, one thread, and driven worm-wheel *F*, 42 teeth, determine the number of rpm of *F*.

When a combination of spur, bevel, and wormgearing is employed to transmit motion and power from one shaft to another, the speed of the driven shaft can be found by the following method: Consider the worm as a gear having one tooth if it is single-threaded and as a gear having two teeth if double-threaded, etc. The speed of the driving shaft can then be found by applying the rules for ordinary compound spur gearing. In this example,

$$\frac{42 \times 33 \times 1}{21 \times 24 \times 42} \times 336 = 22 \text{ rpm}$$

If the pitch diameters of the gears are used instead of the number of teeth in making calculations, the worm should be considered as a gear having a pitch diameter of 1 inch if single-threaded, and 2 inches if a double-threaded worm, etc.

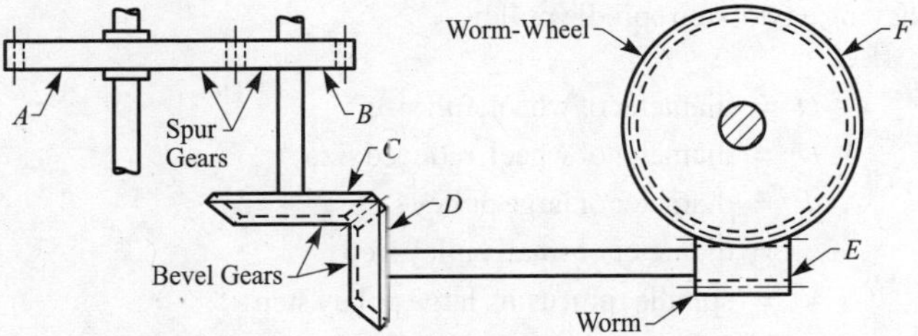


**Fig. 2. Combination of Spur, Bevel, and Wormgearing**

*Example 3:* If a worm is triple-threaded and makes 180 rpm, and the worm-wheel is required to make 5 rpm, determine the number of teeth in the worm-wheel.

*Rule:* Multiply the number of threads in the worm by its number of rpm, and divide the product by the number of rpm of the worm-wheel. By applying this rule,

$$\frac{3 \times 180}{5} = 108 \text{ teeth}$$

*Example 4:* A 6-inch grinding machine with a spindle speed of 1773 rpm, for a recommended peripheral speed of 6500 feet per minute (as figured for a full-size 14-inch wheel for this size of machine), has two steps on the spindle pulley; the large step is 5.5 inches in diameter and the small step, 4 inches. What should be the minimum diameter of the wheel before the belt is shifted to the smaller step in order to select a peripheral wheel speed of 6500 feet per minute?

As the spindle makes 1773 rpm when the belt is on the large pulley, its speed with the belt on the smaller pulley may be determined as follows: 5.5:4 = $x$:1773, or (5.5 × 1773)/4 = 2438 rpm, approximately. To obtain the same peripheral speed as when the belt is on the large pulley, the diameters of the grinding wheel should be 14:$x$ = 2438:1773, or (14 × 1773)/2438 = 10.18 inches. Therefore, when the grinding wheel has been worn down to a diameter of 10.18 inches, or approximately $10\frac{3}{16}$ inches, the spindle belt should be shifted to the smaller step of the spindle pulley to obtain a peripheral speed of 6500 feet per minute. The method used in this example may be reduced to a formula for use with any make of grinding machine having a two-step spindle pulley.

Let

$D$ = diameter of wheel, full size
$D_1$ = diameter of wheel, reduced size
$d$ = diameter of large pulley step
$d_1$ = diameter of small pulley step
$V$ = spindle rpm, using large pulley step
$v$ = spindle rpm, using small pulley step

Then,

$$v = \frac{dV}{d_1};\quad D_1 = \frac{DV}{v}$$

*Example 5:* Planetary gear sets are widely used in power transmission because of their compactness and relatively high efficiency when properly designed. The simple planetary configuration shown in **Fig. 10** on *Handbook* page **2294** is typical of high-efficiency designs. If $A = 20$ and $C = 40$, what is the rotation of the driver $D$ per revolution of the follower?

Using the formula given on *Handbook* page **2294**,

$$D = 1 + \frac{C}{A} = 1 + \frac{40}{20} = 3$$

*Example 6:* If, in **Example 5**, the diameter of the fixed gear is doubled to $C = 80$, what effect does that produce in the rotation of the drive $D$?

$$D = 1 + \frac{80}{20} = 5$$

Note that doubling the size of the fixed gear *C* does not double the ratio or the driver speed of the gear set because the overall ratio is always plus the ratio of *C* to *A*.

*Example 7:* The compound type of planetary gear shown in **Fig. 13** on *Handbook* page **2294** can provide high revolution ratios, although the efficiency decreases as the ratio increases. What is the rotation of the follower *F* when $B = 61$, $C = 60$, $x = 19$, and $y = 20$?

$$F = 1 - \left(\frac{C \times x}{y \times B}\right) = 1 - \left(\frac{60 \times 19}{20 \times 61}\right) = 1 - \frac{57}{61} = 0.06557$$

*Example 8:* In **Example 7**, what is the rotation of the driver per revolution of the follower?

$$\text{Driver} = \frac{1}{\text{follower}} = \frac{1}{0.06557} = 15.25$$

Note that in compound planetary gear drives the sum of meshing tooth pairs must be equal for proper meshing. Thus, $C + y = x + B$.

**Diametral Pitch of a Gear.**—The diametral pitch represents the number of gear teeth for each inch of pitch diameter and, therefore, equals the number of teeth divided by the pitch diameter. The term "diametral pitch" as applied to bevel gears has the same meaning as with spur gears. This method of basing the pitch on the relation between the number of teeth and the pitch diameter is used almost exclusively in connection with cut gearing and to some extent for cast gearing. The term "circular pitch," which refers to the distance between the centers of adjacent teeth measured along the pitch circle, is used for cast gearing but very little for cut gearing except very large sizes. If 3.1416 is divided by the diametral pitch, the quotient equals the circular pitch, or, if the circular pitch is known, the diametral pitch may be found by dividing 3.1416 by the circular pitch. The pitch of the gear teeth may depend primarily upon the strength required to transmit a given amount of power.

**Power-Transmitting Capacity of Bevel Gears.**—The design of bevel gears to meet a set of operating conditions is best accomplished in four steps: (1) determine the design load upon which the bevel gear sizes will be based; (2) using design literature and charts available from gear manufacturers and distributors, select approximate gear and pinion sizes to satisfy the load requirements; (3) determine the maximum safe tooth load, based on gear geometry and material, using manufacturer's and/or AGMA formulas; and (4) determine the safe horsepower capacity of the gears, based on safe tooth load and tooth surface durability. The horsepower capacity of the gears should meet or exceed the design load requirements. To check the capacity of an existing bevel-gear drive, only steps (3) and (4) are necessary.

**Dimensions and Angles Required in Producing Gears.—**Many of the rules and formulas given in the gear section of the *Handbook* beginning on page **2206** are used in determining tooth dimensions, gear-blank sizes, and angles in bevel, helical, and wormgearing. These dimensions or angles are required on the working drawings used in connection with machining operations, such as turning gear blanks and cutting the teeth.

*Example 9:* If a spur gear is to have 40 teeth of 8 diametral pitch, to what diameter should the blank be turned? By applying **Example (4a)**, *Handbook* page **2212**, (40 + 2)/8 = 5.25 inches. Therefore, the outside diameter of this gear or the diameter to which the blank would be turned is $5\frac{1}{4}$ inches.

For internal spur gears, the inside diameter to which the gear blank would be bored may be obtained by subtracting 2 from the number of teeth and dividing the remainder by the diametral pitch.

*Example 10:* A sample spur gear has 22 teeth, and the outside diameter, or diameter measured across the tops of the teeth, is 6 inches. Determine the diametral pitch. According to **Example (4a)**, *Handbook* page **2212**,

$$D_o = \frac{N+2}{P}$$

Hence,

$$P = \frac{N+2}{D_o} = \frac{22+2}{6} = 4 \text{ diametral pitch}$$

The table, *Handbook* page **2212**, also shows that when the sample gear has American Standard Stub teeth, **Example (6a)** should be used to determine the outside diameter, or diametral pitch.

*Example 11:* A 25-degree involute full-depth spur gear is to be produced by hobbing. How is the hob tip radius found?

Per *Handbook* pages **2237–2238**, the maximum hob tip radius, $r_c$ (max), is found by the formula:

$$r_c\,(\max) = \frac{0.785398\cos\phi - b\sin\phi}{1 - \sin\phi}$$

where $\phi$ is the pressure angle, here, 25°, and $b$ is the dedendum constant, which is 1.250 according to **Table 2** on *Handbook* page **2212**. Thus,

$$r_c\,(\max) = \frac{0.785398 \times 0.90631 - 1.25 \times 0.42262}{1 - 0.42262}$$

$$= 0.3179 \text{ inch for a 1-diametral-pitch gear}$$

*Example 12:* If a 20-degree involute full-depth pinion having 24 teeth of 6 diametral pitch is to mesh with a rack, determine the whole depth of the rack teeth and the linear pitch of the teeth.

The teeth of a rack are of the same proportions as the teeth of a spur gear or pinion that is intended to mesh with the rack; hence the pitch of the rack teeth is equal to the circular pitch of the pinion and is found by dividing 3.1416 by the diametral pitch.

The pitch = 3.1416 ÷ 6 = 0.5236 inch = linear pitch of a rack to mesh with a pinion of 6 diametral pitch. This dimension (0.5236) represents the distance that the cutter would be indexed when milling rack teeth or the distance that the planer tool would be moved for cutting successive teeth if a planer were used. The whole depth of a full-depth rack tooth of 20-degree pressure angle equals 2.157 divided by the diametral pitch of the meshing gear, or the whole depth equals the circular pitch multiplied by 0.6866. Here, the circular pitch is 0.5236, and the whole depth equals 0.5236 × 0.6866 = 0.3595 inch.

*Example 13:* If the teeth of a spur gear are to be cut to a certain diametral pitch, is it possible to obtain any diameter that may be desired? Thus, if the diametral pitch is 4, is it possible to make the pitch diameter $5\frac{1}{8}$ inches?

The diametral pitch system is so arranged as to provide a series of tooth sizes, just as the pitches of screw threads are standardized. In as much as there must be a whole number of teeth in each gear, it is apparent that gears of a given pitch vary in diameter according to the number of teeth. Suppose, for example, that a series of gears is of 4 diametral pitch. Then the pitch diameter of a gear having, say, 20 teeth will be 5 inches; 21 teeth, $5\frac{1}{4}$ inches; 22 teeth, $5\frac{1}{2}$ inches, and so on. It will be seen that the increase in diameter for each additional tooth is equal to $\frac{1}{4}$ inch for 4 diametral pitch. Similarly, for 2 diametral pitch, the variations for successive numbers of teeth would equal $\frac{1}{2}$ inch, and for 10 diametral pitch the variations would equal $\frac{1}{10}$ inch, etc.

The center-to-center distance between two gears is equal to one-half the total number of teeth in the gears divided by the diametral pitch. It may be desirable at times to have a center distance that cannot be obtained exactly by any combination of gearing of given diametral pitch, but this condition is unusual, and, ordinarily, the designer of a machine can alter the center distance whatever slight amount may be required for gearing of the desired ratio and pitch. By using a standard system of pitches, all calculations are simplified, and it is also possible to obtain the benefits of standardization in the manufacturing of gears and gear-cutters.

**Proportioning Spur Gears When Center Distance Is Fixed.—**If the center-to-center distance between two shafts is fixed, and it is desired to use gears of a certain pitch, the number of teeth in each gear for a given speed may be determined as follows: Since the gears must be of a certain pitch, the total number of teeth available should be determined and then the number of teeth in the driving and the driven gears. The total number of teeth equals twice the product of the center distance multiplied by the diametral pitch. If the center distance is 6 inches, and the diametral pitch 2, the total number of teeth equals $6 \times 2 \times 10 = 120$ teeth. The next step is to find the number of teeth in the driving and the driven gears for a given rate of speed.

*Rule:* Divide the speed of the driving gear in revolutions per minute by the speed of the driven gear and add one to the quotient. Next divide the total number of teeth in both gears by the sum previously obtained, and the quotient will equal the number of teeth required in the driving gear. This number subtracted from the total number of teeth will equal the number of teeth required in the driven gear.

*Example 14:* If the center-to-center distance is 6 inches, and the diametral pitch is 10, the total number of teeth available will be 120. If the speeds of the driving and the driven gears are to be 100 and 60 revolutions per minute, respectively, find the number of teeth for each gear.

$^{100}/_{60} = 1^2/_3$ and $1^2/_3 + 1 = 2^2/_3$

$120 \div 2^2/_3 = {}^{120}/_1 \times {}^3/_8 = 45 =$ number of teeth in driving gear

The number of teeth in the driven gear equals 120 − 45 = 75 teeth.

When the center distance and the velocity ratios are fixed by some essential construction of a machine, it is often impossible to use standard-diametral-pitch gear teeth. If cast gears are to be used, it does not matter so much, as a pattern maker can lay out the teeth according to the pitch desired, but if cut gears are required, an effort should be made to alter the center distance so that standard-diametral-pitch cutters can be used since these are usually carried in stock.

**Dimensions in Generated Bevel Gears.**—*Example 15:* Find all the dimensions and angles necessary to manufacture a pair of straight bevel gears if the number of teeth in the pinion is 16, the number of teeth in the mating gear is 49, the diametral pitch is 5, and the face width is 1.5 inches. The gears are to have a 20-degree pressure angle, a 90-degree shaft angle, and must be in accordance with the Gleason System.

On page **186** of this *Guide*, **Table 1** gives formulas for Gleason System 20-degree pressure angle straight bevel gears with 90-degree shaft angle. These formulas are given in the same order as is normally used in computation. Computations of the gear dimensions should be arranged as shown in the table on the following pages to establish a consistent procedure when calculations for bevel gears are required frequently.

Given:

| | | | |
|---|---|---|---|
| Number of pinion teeth, $n$ | = | 16 | (1) |
| Number of gear teeth, $N$ | = | 49 | (2) |
| Diametral pitch, $P$ | = | 5 | (3) |
| Face width, $F$ | = | 1.5 | (4) |
| Pressure angle, $\phi = 20°$ | = | 20° | (5) |
| Shaft angle, $\Sigma = 90°$ | = | 90° | (6) |

Sample calculations for **(7)** to **(24)** are given in **Table 2** on page **188**.

**Table 1. Formulas for Gleason System 20-Degree Straight Bevel Gears—90-Degree Shaft Angle**

| To Find | | Formula | |
|---|---|---|---|
| No. | Item | Pinion | Gear |
| 7 | Working Depth | $h_k = \dfrac{2.000}{P}$ | Same as pinion |
| 8 | Whole Depth | $h_t = \dfrac{2.188}{P} + 0.002$ | Same as pinion |
| 9 | Pitch Diameter | $d = \dfrac{n}{P}$ | $D = \dfrac{N}{P}$ |
| 10 | Pitch Angle | $\gamma = \tan^{-1}\dfrac{n}{N}$ | $\Gamma = 90° - \gamma$ |
| 11 | Cone Distance | $A_O = \dfrac{D}{2\sin\Gamma}$ | Same as pinion |
| 12 | Circular Pitch | $p = \dfrac{3.1416}{P}$ | Same as pinion |
| 13 | Addendum | $a_p = h_t - a_G$ | $a_G = \dfrac{0.540}{P} + \dfrac{0.460}{P\left(\dfrac{N}{n}\right)^2}$ |
| 14 | Dedendum[a] | $b_p = \dfrac{2.188}{P} - a_p$ | $b_G = \dfrac{2.188}{P} - a_G$ |
| 15 | Clearance | $c = h_t - h_k$ | Same as pinion |
| 16 | Dedendum Angle | $\delta_P = \tan^{-1}\dfrac{b_p}{A_O}$ | $\delta_G = \tan^{-1}\dfrac{b_G}{A_O}$ |
| 17 | Face Angle of Blank | $\gamma_O = \gamma + \delta_G$ | $\Gamma_O = \Gamma + \delta_p$ |
| 18 | Root Angle | $\gamma_r = \gamma - \delta_p$ | $\Gamma_R = \Gamma - \delta_G$ |
| 19 | Outside Diameter | $d_O = d + 2a_p\cos\gamma$ | $D_O = D + 2a_G\cos\Gamma$ |

**Table 1.** *(Continued)* **Formulas for Gleason System 20-Degree Straight Bevel Gears—90-Degree Shaft Angle**

| To Find | | Formula | |
|---|---|---|---|
| No. | Item | Pinion | Gear |
| 20 | Pitch Apex to Crown | $x_O = \frac{D}{2} - a_p \sin\gamma$ | $X_O = \frac{d}{2} - a_G \sin\Gamma$ |
| 21 | Circular Thickness | $t = p - T$ | $T = \frac{p}{2} - (a_p - a_G)\tan\phi - \frac{K}{P}$<br>$K = \text{(Chart 1)}$ |
| 22 | Backlash | $B =$ (See table on *Handbook* page **2245**) | |
| 23 | Chordal Thickness | $t_c = t - \frac{t^3}{6d^2} - \frac{B}{2}$ | $T_c = T - \frac{T^3}{6D^2} - \frac{B}{2}$ |
| 24 | Chordal Addendum | $a_{cp} = a_p + \frac{t^2 \cos\gamma}{4d}$ | $a_{CG} = a_G + \frac{T^2 \cos\gamma}{4D}$ |
| 25 | Tooth Angle | $\frac{3438}{A_O}\left(\frac{t}{2} + b_p \tan\phi\right)$ minutes | $\frac{3438}{A_O}\left(\frac{T}{2} + b_G \tan\phi\right)$ minutes |
| 26 | Limit Point Width | $\frac{A_O - F}{A_O}(T - 2b_p \tan\phi) - 0.0015$ | $\frac{A_O - F}{A_O}(t - 2b_G \tan\phi) - 0.0015$ |

[a] The actual dedendum will be 0.002 inches greater than calculated due to tool advance

All linear dimensions are in inches.

The tooth angle (Item 25, **Table 1**) is a machine setting and is only computed if a Gleason two-tool type straight bevel gear generator is to be used. Calculations continue on page **188**.

**Dimensions of Milled Bevel Gears.**—As explained on *Handbook* page **2262**, the tooth proportions of milled bevel gears differ in some respects from those of generated bevel gears. To take these differences into account, a separate table of formulas is given on *Handbook* page **2264** for use in calculating dimensions of milled bevel gears.

*Example 16:* Compute the dimensions and angles of a pair of mating bevel gears that are to be cut on a milling machine using rotary formed milling cutters if the data given are as follows:

| | | |
|---|---|---|
| Number of pinion teeth | = | 15 |
| Number of gear teeth | = | 60 |
| Diametral pitch | = | 3 |
| Face width | = | 1.5 |
| Pressure angle | = | $14\frac{1}{2}°$ |
| Shaft angle | = | 90° |

## Table 2. Calculations of Dimensions for Example 15

| | Dimension | Pinion | Gear |
|---|---|---|---|
| (7) | Working depth | $2.000/5 = 0.400$ | Same as Pinion |
| (8) | Whole depth | $2.188/5 + 0.002 = 0.440$ | Same as Pinion |
| (9) | Pitch diameter | $^{16}/_{5} = 3.2000$ | $^{49}/_{5} = 9.8000$ |
| (10) | Pitch angle | $\tan^{-1}(^{16}/_{49}) = 18° 5'$ | $90° - 18°5' = 71°55'$ |
| (11) | Cone distance | $9.8000/(2 \times \sin 71°55') = 5.1546$ | Same as pinion |
| (12) | Circular pitch | $3.1416/5 = 0.6283$ | Same as pinion |
| (13) | Addendum | $0.400 - 0.118 = 0.282$ | $0.540/5 + 0.460/(5(49/16)^2) = 0.118$ |
| (14) | Dedendum | $2.188/5 - 0.282 = 0.1554$ | $2.188/5 - 0.118 = 0.3196$ |
| (15) | Clearance | $0.440 - 0.400 = 0.040$ | Same as pinion |
| (16) | Dedendum angle | $\tan^{-1}(0.1536/5.1546) = 1°42'$ | $\tan^{-1}(0.3214/5.1546) = 3°34'$ |
| (17) | Face angle of blank | $18°5' + 3°34' = 21°39'$ | $71°55' + 1°42' = 73°37'$ |
| (18) | Root angle | $18°5' \angle 1°42' = 16°23'$ | $71°55' \angle 3°34' = 68°21'$ |
| (19) | Outside diameter | $3.2000 + 2 \times 0.282 \cos 18°5' = 3.735$ | $9.8000 + 2 \times 0.118 \cos 71°55' = 9.875$ |
| (20) | Pitch apex to crown | $9.8000/2 - 0.284 \sin 18°5' = 4.812$ | $3.2000/2 - 0.118 \sin 71°55' = 1.488$ |
| (21) | Circular thickness | $0.6283 - 0.2467 = 0.3816$ | $0.6283/2 - (0.284 - 0.118)\tan 20° - (0.038(\text{chart } 1))/5 = 0.2467$ |
| (22) | Backlash | 0.006 | 0.006 |
| (23) | Chordal thickness | $0.3816 - \frac{(0.3816)^3}{6 \times (3.2000)^2} - \frac{0.006}{2} = 0.378$ | $0.2467 - \frac{(0.2467)^3}{6 \times (9.8000)^2} - \frac{0.006}{2} = 0.244$ |
| (24) | Chordal addendum | $0.282 + \frac{0.3816^2 \cos 18°5'}{4 \times 3.2000} = 0.293$ | $0.118 + \frac{0.2467^2 \cos 71°55'}{4 \times 9.8000} = 0.118$ |

By using the formulas on *Handbook* page **2264**,

$$\tan \alpha_p = 15 \div 60 = 0.25 = \tan 14° 2.2', \text{say}, 14° 2'$$

$$\alpha_G = 90° - 14° 2' = 75° 57.8', \text{say}, 75° 58'$$

$$D_p = 15 \div 3 = 5.0000 \text{ inches}$$

$$D_G = 60 \div 3 = 20.0000 \text{ inches}$$

$$S = 1 \div 3 = 0.3333 \text{ inch}$$

$$S + A = 1.157 \div 3 = 0.3857 \text{ inch}$$

$$W = 2.157 \div 3 = 0.7190 \text{ inch}$$

$$T = 1.571 \div 3 = 0.5236 \text{ inch}$$

$$C = \frac{5.000}{2 \times 0.24249} = 10.308 \text{ inches}$$

(In determining $C$, the sine of unrounded value of $\alpha_p$, 14° 2.2′, is used.)

$$F = 8 \div 3 = 2\tfrac{2}{3}, \text{say}, 2\tfrac{5}{8} \text{ inches}$$

$$s = 0.3333 \times \frac{10.308 - 2\tfrac{5}{8}}{10.308} = 0.2484 \text{ inch}$$

$$t = 0.5236 \times \frac{10.308 - 2\tfrac{5}{8}}{10.308} = 0.3903 \text{ inch}$$

$$\tan\theta = 0.3333 \div 10.308 = \tan 1° 51'$$

$$\tan\phi = 0.3857 \div 10.308 = \tan 2° 9'$$

$$\gamma_P = 14° 2' + 1° 51' = 15° 53'$$

$$\gamma_G = 75° 58' + 1° 51' = 77° 49'$$

$$\delta_P = 90° - 15° 53' = 74° 7'$$

$$\delta_G = 90° - 77° 49' = 12° 11'$$

$$\zeta_P = 14° 2' + 2° 9' = 11° 53'$$

$$\zeta_G = 75° 582' + 2° 9' = 73° 49'$$

$$K_P = 0.3333 \times 0.97015 = 0.3234 \text{ inch}$$

$$K_G = 0.3333 \times 0.24249 = 0.0808 \text{ inch}$$

$$O_P = 5.000 + 2 \times 0.3234 = 5.6468 \text{ inches}$$

$$O_G = 20.000 + 2 \times 0.0808 = 20.1616 \text{ inches}$$

$$J_P = \frac{5.6468}{2} \times 3.5144 = 9.9226 \text{ inches}$$

$$J_G = \frac{20.1616}{2} \times 0.21590 = 2.1764 \text{ inches}$$

$$j_p = 9.9226 \times \frac{10.3097 - 2\tfrac{5}{8}}{10.3097} = 7.3961 \text{ inches}$$

$$j_g = 2.1764 \times \frac{10.3097 - 2\tfrac{5}{8}}{10.3097} = 1.6222 \text{ inches}$$

$$N'_P = \frac{15}{0.97015} = 15.4\text{, say, 15 teeth}$$

$$N'_G = \frac{60}{0.24249} = 247 \text{ teeth}$$

If these gears are to have uniform clearance at the bottom of the teeth, in accordance with the recommendation given in the last paragraph on *Handbook* page **2262**, then the cutting angles $\zeta_P$ and $\zeta_G$ should be determined by subtracting the addendum angle from the pitch cone angles. Thus,

$$\zeta_P = 14°2' - 1°51' = 12°11'$$
$$\zeta_G = 75°58' - 1°51' = 74°7'$$

**Selection of Formed Cutters for Bevel Gears.**—*Example 17:* In **Example 16**, the numbers of teeth for which to select the cutters were calculated as 15 and 247 for the pinion and gear, respectively. Therefore, as explained on page **2268** of the *Handbook*, the cutters selected from the table on page **2231** are the No. 7 and the No. 1 cutters. As further noted on page **2268**, bevel gear milling cutters may be selected directly from the table beginning on page **2266**, when the shaft angle is 90 degrees, instead of using the computed value of $N'$ to enter the table on page **2231**. Thus, for a 15-tooth pinion and a 60-tooth gear, the table on page **2266** shows that the numbers of the cutters to use are 1 and 7 for gear and pinion, respectively.

**Pitch of Hob for Helical Gears.—***Example 18:* A helical gear that is to be used for connecting shafts has 83 teeth, a helix angle of 7 degrees, and a pitch diameter of 47.78 inches. Determine the pitch of hob to use in cutting this gear.

As explained on *Handbook* page **2277**, the normal diametral pitch and the pitch of the hob are determined as follows: the transverse diametral pitch equals 83 ÷ 47.78 = 1.737. The cosine of the helix angle of the gear (7 degrees) is 0.99255; hence the normal diametral pitch equals 1.737 ÷ 0.99255 = 1.75; therefore, a hob of $1\frac{3}{4}$ diametral pitch should be used. This hob is the same as would be used for spur gears of $1\frac{3}{4}$ diametral pitch, and it will cut any spur or helical gear of that pitch regardless of the number of teeth, provided $1\frac{3}{4}$ is the diametral pitch of the spur gear and the normal diametral pitch of the helical gear.

**Determining Contact Ratio.—**As pointed out on *Handbook* page **2237**, if a smooth transfer of load is to be obtained from one pair of teeth to the next pair of teeth as two mating gears rotate under load, the contact ratio must be well over 1.0. Usually, this ratio should be 1.4 or more, although in extreme cases it may be as low as 1.15.

*Example 19:* Find the contact ratio for a pair of 18-diametral pitch, 20-degree pressure gears, one having 36 teeth and the other 90 teeth. From **Formula (1)** given on *Handbook* page **2236**:

$$\cos A = \frac{90 \times \cos 20°}{5.111 \times 18} = \frac{90 \times 0.93969}{91.9998} = 0.91926 \quad \text{and}$$

$$A = 23°11'$$

From **Formula (4)** given on *Handbook* page **2236**:

$$\cos a = \frac{36 \times \cos 20°}{2.111 \times 18} = \frac{36 \times 0.93969}{37.9998} = 0.89024 \quad \text{and}$$

$$a = 27°6'$$

From **Formula (5)** given on *Handbook* page **2237**:

$$\tan B = \tan 20° - \frac{36}{90}(\tan 27°6' - \tan 20°)$$

$$= 0.36397 - \frac{36}{90}(0.51172 - 0.36397) = 0.30487$$

From **Formula (7a)** given on *Handbook* page **2237**, the contact ratio $m_f$ is found:

$$m_f = \frac{90}{6.28318}(0.42826 - 0.30487)$$

$$= 1.77$$

which is satisfactory.

**Dimensions Required When Using Enlarged Fine-Pitch Pinions.—** On *Handbook* pages **2232–2234**, there are tables of dimensions for enlarged fine-pitch pinions. These tables show how much the dimensions of enlarged pinions must differ from standard when the number of teeth is small, and undercutting of the teeth is to be avoided.

*Example 20:* If a 10- and a 31-tooth mating pinion and gear of 20 diametral pitch and $14\frac{1}{2}°$ pressure angle have both been enlarged to avoid undercutting of the teeth, what increase over the standard center distance is required?

$$\text{Standard center distance} = \frac{n + N}{2P} = \frac{10 + 31}{2 \times 20} = 1.0250 \text{ inches}$$

The amount by which the center distance must be increased over standard can be obtained by taking the sum of the amounts shown in the eighth column of **Table 9b** on *Handbook* page **2232** and dividing this sum by the diametral pitch. Thus, the increase over the standard center distance is (0.6866 + 0.0283)/20 = 0.0357 inch.

*Example 21:* At what center distance would the gears in **Example 20** have to be meshed if there were to be no backlash?

Obtaining the two thicknesses of both gears at the standard pitch diameters from **Table 9b** on *Handbook* page **2232**, dividing them by 20, and using the formulas on *Handbook* page **2236**:

$$\text{inv}\phi_1 = \text{inv}14\tfrac{1}{2}° + \frac{20(0.09630 + 0.07927) - 3.1416}{10 + 31}$$

The involute of $14\frac{1}{2}°$ is found on *Handbook* page **107** to be 0.0055448. Therefore,

$$\text{inv}\phi_1 = 0.0055448 + 0.0090195 = 0.0145643$$

By referring again to the table on *Handbook* page **107**:

$$\phi_1 = 19°51'6''$$

$$C = \frac{n+N}{2P} = \frac{10+31}{2 \times 20} = 1.025 \text{ inch}$$

$$C_1 = \frac{\cos 14\tfrac{1}{2}°}{\cos 19°51'6''} \times 1.025 = \frac{0.96815}{0.94057} \times 1.025 = 1.0551 \text{ inch}$$

**End Thrust of Helical Gears Applied to Parallel Shafts.—** *Example 22:* The diagrams on *Handbook* pages **2278–2279** show the application of helical or spiral gears to parallel shaft drives. If a force of 7 horsepower is to be transmitted at a pitch-line velocity of 200 feet per minute, determine the end thrust in pounds, assuming that the helix angle of the gear is 15 degrees.

To determine the end thrust of helical gearing as applied to parallel shafts, first calculate the tangential load on the gear teeth.

$$\text{Tangential load} = \frac{33{,}000 \times 7}{200} = 1155 \text{ pounds}$$

(This formula is derived from the formulas for power given on *Handbook* page **187**.)

The axial or end thrust may now be determined approximately by multiplying the tangential load by the tangent of the tooth angle. Thus, in this instance, the thrust = 1155 × tan 15 degrees = about 310 pounds. (Note that this formula agrees with the one on *Handbook* page **170** for determining force *P* parallel to base of inclined plane.) The end thrust obtained by this calculation will be somewhat greater than the actual end thrust, because frictional losses in the shaft bearings, etc., have not been taken into account, although a test on a helical gear set, with a motor drive, showed that the actual thrust of the $7\tfrac{1}{2}$-degree helical gears tested was not much below the values calculated as just explained.

According to most textbooks, the maximum angle for single helical gears should be about 20 degrees, although one prominent manufacturer mentions that the maximum angle for industrial drives ordinarily does not exceed 10 degrees, and this will give quiet running without excessive end thrust. On some of the heavier single helical gearing used for street railway transmissions, etc., an angle of 7 degrees is employed.

**Dimensions of Wormgear Blank and the Gashing Angle.—** *Example 23:* A wormgear having 45 teeth is to be driven by a double-threaded worm having an outside diameter of 2½ inches and a lead of 1 inch, the linear pitch being ½ inch. The throat diameter and throat radius of the wormgear are required as well as the angle for gashing the blank.

The throat diameter $D_t$ equals the pitch diameter $D$ plus twice the addendum $A$; thus, $D_t = D + 2A$. The addendum of the worm thread equals the linear pitch multiplied by 0.3183, and here, $0.5 \times 0.3183 = 0.1591$ inch. The pitch diameter of the wormgear $= 45 \times 0.5 \div 3.1416 = 7.162$ inches; hence, the throat diameter equals $7.162 + 2 \times 0.1591 = 7.48$ inches.

The radius of the wormgear throat is found by subtracting twice the addendum of the worm thread from ½ the outside diameter of the worm. The addendum of the worm thread equals 0.1591 inch, and the radius of the throat, therefore, equals $(2.5 \div 2) - 2 \times 0.1591 = 0.931$ inch.

When a wormgear is hobbed in a milling machine, gashes are milled before the hobbing operation. The table must be swiveled around while gashing, the amount depending upon the relation between the lead of the worm thread and the pitch circumference. The first step is to find the circumference of the pitch circle of the worm. The pitch diameter equals the outside diameter minus twice the addendum of the worm thread; hence, the pitch diameter equals $2.5 - 2 \times 0.1591 = 2.18$ inches, and the pitch circumference equals $2.18 \times 3.1416 = 6.848$ inches.

Next, divide the lead of the worm thread by the pitch circumference to obtain the tangent of the desired angle, and then refer to a table of tangents or a calculator to determine what this angle is. For this example, it is $1 \div 6.848 = 0.1460$, which is the tangent of 8⅓ degrees from its normal position.

**Change Gear Ratio for Diametral-Pitch Worms.—** *Example 24:* In cutting worms to a given diametral pitch, the ratio of the change gears is 22 × threads per inch / 7 × diametral pitch.

The reason why the constants 22 and 7 are used in determining the ratio of change-gears for cutting worm threads is because 22⁄7 equals, very nearly, 3.1416, which is the circular pitch equivalent to diametral pitch.

Assume that the diametral pitch of the wormgear is 5, and the lathe screw constant is 4. (See *Handbook* page **2183** for the meaning of "lathe screw constant.") Then, $(4 \times 22)/(5 \times 7) = 88/35$. If this simple combination of gearing were used, the gear on the stud would have 88 teeth and the gear on the lead screw, 35 teeth. Of course, any other combination of gearing having this same ratio could be used, as, for example, the following compound train of gearing: $(24 \times 66)/(30 \times 21)$.

If the lathe screw constant is 4, as previously assumed, then the number of threads per inch obtained with gearing having a ratio of $88/35 = (4 \times 35)/88 = 1.5909$; hence, the pitch of the worm thread equals $1 \div 1.5909 = 0.6284$ inch, which is the circular pitch equivalent to 5 diametral pitch, correct to within 0.0001 inch.

**Bearing Loads Produced by Bevel Gears.**—In applications where bevel gears are used, not only must the gears be proportioned with regard to the power to be transmitted, but also the bearings supporting the gear shafts must be of adequate size and design to sustain the radial and thrust loads that will be imposed on them. Assuming that suitable gear and pinion proportions have been selected, the next step is to compute the loads needed to determine whether or not adequate bearings can be provided. To find the loads on the bearings, first, use the formulas on the following pages to compute the tangential, axial, and separating components of the load on the tooth surfaces. Second, use the principle of moments, together with the components determined in the first step, to find the radial loads on the bearings. To illustrate the procedure, the following example will be used.

*Example 25:* A 16-tooth left-hand spiral pinion rotating clockwise at 1800 rpm transmits 71 horsepower to a 49-tooth mating gear. If the pressure angle is 20 degrees, the spiral angle is 35 degrees, the face width is 1.5 inches, and the diametral pitch is 5, what are the radial and thrust loads that govern the selection of bearings?

In **Fig. 3**, the locations of the bearings for the gear shafts are shown. It should be noted that distances *J*, *K*, *L*, and *M* are measured from the center line of the bearings and from the midfaces of the gears at their mean pitch diameters. In this example, it will be assumed that these distances are given and are as follows: $J = 3.5$ inches; $K = 2.5$ inches; $L = 1.5$ inches; and $M = 5.0$ inches.

Also given:

| | | | |
|---|---|---|---|
| Number of pinion teeth, $n$ | = | 16 | (1) |
| Number of gear teeth, $N$ | = | 49 | (2) |
| Diametral pitch, $P$ | = | 5 | (3) |
| Face width, $F$ | = | 1.5 | (4) |
| Pressure angle, $\phi = 20°$ | = | 20° | (5) |
| Shaft angle, $\Sigma = 90°$ | = | 90° | (6) |

**Table 3. Formulas for Gleason System 20-Degree Pressure Angle, Spiral Bevel Gears—90-Degree Shaft Angle**

| No | Item | Formula | |
|---|---|---|---|
| | | Pinion | Gear |
| 7 | Working Depth | $h_k = \frac{1.700}{P}$ | Same as pinion |
| 8 | Whole Depth | $h_t = \frac{2.188}{P}$ | Same as pinion |
| 9 | Pitch Diameter | $d = \frac{n}{P}$ | $D = \frac{N}{P}$ |
| 10 | Pitch Angle | $\gamma = \tan^{-1}\frac{n}{N}$ | $\Gamma = 90° - \gamma$ |
| 11 | Cone Distance | $A_O = \frac{D}{2\sin\Gamma}$ | Same as pinion |
| 12 | Circular Pitch | $p = \frac{3.1416}{P}$ | Same as pinion |
| 13 | Addendum | $a_p = h_k - a_G$ | $a_G = \frac{0.540}{P} + \frac{0.390}{P\left(\frac{N}{n}\right)^2}$ |
| 14 | Dedendum | $b_p = h_t - a_p$ | $b_G = h_t - a_G$ |
| 15 | Clearance | $c = h_t - h_k$ | Same as pinion |
| 16 | Dedendum Angle | $\delta_P = \tan^{-1}\frac{b_p}{A_O}$ | $\delta_G = \tan^{-1}\frac{b_G}{A_O}$ |
| 17 | Face Angle of Blank | $\gamma_O = \gamma + \delta_G$ | $\Gamma_O = \Gamma + \delta_p$ |

**Table 3.** *(Continued)* **Formulas for Gleason System 20-Degree Pressure Angle, Spiral Bevel Gears—90-Degree Shaft Angle**

| No | Item | Formula | |
|---|---|---|---|
| | | Pinion | Gear |
| 18 | Root Angle | $\gamma_R = \gamma - \delta_p$ | $\Gamma_R = \Gamma - \delta_G$ |
| 19 | Outside Diameter | $d_O = d + 2a_p \cos\gamma$ | $D_O = D + 2a_G \cos\Gamma$ |
| 20 | Pitch Apex to Crown | $x_O = \frac{D}{2} - a_p \sin\gamma$ | $X_O = \frac{d}{2} - a_G \sin\Gamma$ |
| 21 | Circular Thickness | $t = p - T$ | $T = \frac{(1.5708 - K)}{P} - \frac{\tan\phi}{\cos\psi}(a_p - a_G)$ |
| 22 | Backlash[a] | E = (See table on *Handbook* page **2245**) | |

[a] When the gear is cut spread-blade, all the backlash is taken from the pinion thickness. When both members are cut single-side, each thickness is reduced by half of the backlash.

All linear dimensions are in inches.

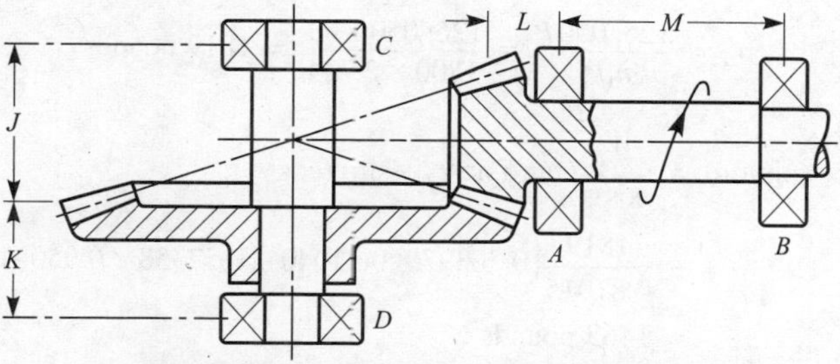


**Fig. 3. Diagram Showing Location of Bearings for Bevel Gear Drive in Example 25**

Other quantities that will be required in the solution of this example are the pitch diameter, pitch angle, and mean pitch diameter of both the gear and pinion. These are computed using formulas given in **Table 3** on the previous page as follows:

By using Formula 9 in **Table 3**,

Pitch dia. of pinion $d = 3.2$ inches

Pitch dia. of gear $D = 9.8$ inches

By using Formula 10 in **Table 3**,

Pitch angle of pinion $\gamma = 18° 5'$

Pitch angle of gear $\Gamma = 71° 55'$

By using the formula given below,

Mean pitch diameter of pinion

$$d_m = d - F\sin\gamma$$
$$= 3.2 - 1.5 \times 0.31040$$
$$= 2.734 \text{ inches}$$

Mean pitch diameter of gear

$$D_m = D - F\sin\Gamma$$
$$= 9.8 - 1.5 \times 0.95061$$
$$= 8.374 \text{ inches}$$

The first step in determining the bearing loads is to compute the tangential $W_t$, axial $W_x$, and separating $W_s$, components of the tooth load, using the formulas that follow.

$$W_t = \frac{126,050P}{nd_m} = \frac{126,050 \times 71}{1800 \times 2.734} = 1819 \text{ pounds}$$

$$W_x(\text{pinion}) = \frac{W_t}{\cos\psi}(\tan\phi\sin\gamma_d + \sin\psi\cos\gamma_d)$$
$$= \frac{1819}{0.81915}(0.36397 \times 0.31040 + 0.57358 \times 0.95061)$$
$$= 1462 \text{ pounds}$$

$$W_x\ (\text{gear}) = \frac{W_t}{\cos\psi}(\tan\phi\sin\gamma_D - \sin\psi\cos\gamma_D)$$
$$= \frac{1819}{0.81915}(0.36397 \times 0.95061 - 0.57358 \times 0.31040)$$
$$= 373 \text{ pounds}$$

$$W_s(\text{pinion}) = \frac{W_t}{\cos\psi}(\tan\phi\cos\gamma_d - \sin\psi\cos\gamma_d)$$
$$= \frac{1819}{0.81915}(0.36397 \times 0.95061 - 0.57358 \times 0.31040)$$
$$= 373 \text{ pounds}$$

$$W_s \text{ (gear)} = \frac{W_t}{\cos\psi}(\tan\phi\cos\gamma_D + \sin\psi\cos\gamma_D)$$
$$= \frac{1819}{0.81915}(0.36397 \times 0.31040 + 0.57358 \times 0.95061)$$
$$= 1462 \text{ pounds}$$

The axial thrust load on the bearings is equal to the axial component of the tooth load $W_x$. Since thrust loads are always taken up at only one mounting point, either bearing *A* or bearing *B* must be a bearing capable of taking a thrust of 1462 pounds, and either bearing *C* or bearing *D* must be capable of taking a thrust of 373 pounds.

The next step is to determine the magnitudes of the radial loads on the bearings *A*, *B*, *C*, and *D*. For an overhung mounted gear, or pinion, it can be shown, using the principle of moments, that the radial load on bearing *A* is:

$$R_A = \frac{1}{M}\sqrt{[W_t(L+M)]^2 + [W_s(L+M) - W_x r]^2} \qquad (1)$$

And the radial load on bearing *B* is:

$$R_B = \frac{1}{M}\sqrt{(W_t L)^2 + (W_s L - W_x r)^2} \qquad (2)$$

For a *straddle mounted gear* or pinion the radial load on bearing *C* is:

$$R_C = \frac{1}{J+K}\sqrt{(W_t K)^2 + (W_s K - W_x r)^2} \qquad (3)$$

And the radial load on bearing *D* is:

$$R_D = \frac{1}{J+K}\sqrt{(W_t J)^2 + (W_s J + W_x r)^2} \qquad (4)$$

In these formulas, *r* is the mean pitch radius of the gear or pinion.

These formulas will now be applied to the gear and pinion bearings in the example. An overhung mounting is used for the pinion, so **Formulas (1)** and **(2)** are used to determine the radial loads on the pinion bearings:

$$R_A = \frac{1}{5}\sqrt{[1819(1.5+5)]^2 + [373(1.5+5) - 1462 \times 1.367]^2}$$

$$= 2365 \text{ pounds}$$

$$R_B = \frac{1}{5}\sqrt{(1819 \times 1.5)^2 + [373 \times 1.5 - 1462 \times 1.367]^2}$$

$$= 618 \text{ pounds}$$

Because of the straddle mounting used for the gear, **Formulas (3)** and **(4)** are used to determine the radial loads on the gear bearings:

$$R_C = \frac{1}{3.5+2.5}\sqrt{(1819 \times 2.5)^2 + (1462 \times 2.5 - 373 \times 4.187)^2}$$

$$= 833 \text{ pounds}$$

$$R_D = \frac{1}{3.5+2.5}\sqrt{(1819 \times 3.5)^2 + (1462 \times 3.5 + 373 \times 4.187)^2}$$

$$= 1533 \text{ pounds}$$

These radial loads, and the thrust loads previously computed, are then used to select suitable bearings from manufacturers' catalogs.

It should be noted, in applying **Formulas (1)** to **(4)**, that if both gear and pinion had overhung mountings, then **Formulas (1)** and **(2)** would have been used for both; if both gear and pinion had straddle mountings, then **Formulas (3)** and **(4)** would have been used for both. In any arrangement, the dimensions and loads for the corresponding member must be used. Also, in applying the formulas, the computed values of $W_x$ and $W_s$, if they are negative, must be used in accordance with the rules applicable to negative numbers.

**Gear Strength Calculations.—**Methods of calculating the strength and power capacity for gears used in all types of applications are provided in American Gear Manufacturers Association (AGMA) Standards. These Standards are revised as needed by improvements in gear materials, calculation methods, and increased field experience with typical designs and application factors.

AGMA Standard 2001-B88, "Fundamental Rating Factors and Calculation Methods for Involute Spur and Helical Gear Teeth," is a revision of, and supersedes, AGMA 218.01.

The AGMA Standard presents general formulas for rating the pitting resistance and the bending strength of spur and helical involute gear teeth. It is intended to establish a common base for rating various types of gears for differing applications and to encourage the maximum practical degree of uniformity and consistency between rating practices in the gear industry. The Standard provides the basis from which more detailed AGMA Application Standards are developed and is a means for calculation of approximate ratings in the absence of such Standards. Where applicable AGMA Standards exist, they should be used in preference to this Standard. Where no application Standard exists, numerical values may be estimated for the factors used in the general equations presented in the Standard. The values of these factors may vary significantly, depending on the application, system effects, gear accuracy, manufacturing practice, and definition of what constitutes gear failure.

Information on geometry factors used in pitting resistance independent strength calculations for AGMA 908-B89, "Geometry Factors for Determining the Pitting Resistance and Bending Strength of Spur, Helical, and Herringbone Gear Teeth," is used in conjunction with AGMA 2001-B88 formulas.

## PRACTICE EXERCISES FOR SECTION 19

(See *Answers to Practice Exercises for Section 19* on page **250**)

1) A spur gear of 6 diametral pitch has an outside diameter of 3.3333 inches. How many teeth has it? What is the pitch diameter? What is the tooth thickness measured along the pitch circle?

2) A gear of 6 diametral pitch has 14 teeth. Find the outside diameter, the pitch diameter, and the addendum.

3) When is the 25-degree tooth form standard preferred?

4) What dimension does a gear-tooth vernier caliper measure?

5) What are the principal 20-degree pressure angle tooth dimensions for the following diametral pitches: 4; 6; 8; 18?

6) Give the important $14\frac{1}{2}$ degree pressure angle tooth dimensions for the following circular pitches: $\frac{1}{2}$ inch; $\frac{3}{4}$ inch; $\frac{9}{16}$ inch.

7) What two principal factors are taken into consideration in determining the power-transmitting capacity of spur gears?

8) The table on *Handbook* page **2231** shows that a No. 8 formed cutter (involute system) would be used for milling either a 12- or 13-tooth pinion, whereas a No. 7 would be used for tooth numbers from 14 to 16, inclusive. If the pitch is not changed, why is it necessary to use different cutter numbers?

9) Are hobs made in series or numbers for each pitch similar to formed cutters?

10) If the teeth of a gear have a $\frac{6}{8}$ pitch, what name is applied to the tooth form?

11) A stub-tooth gear has $\frac{8}{10}$ pitch. What do the figures 8 and 10 indicate?

12) What is the module of a gear?

13) Explain the use of the table of chordal thicknesses on *Handbook* page **2224**.

14) Give the dimensions of a 20-degree stub tooth of 12 pitch.

15) What are the recommended diametral pitches for fine-pitch standard gears?

16) What tooth numbers could be used in pairs of gears having the following ratios: 0.2642; 0.9615?

17) What amount of backlash is provided for general-purpose gearing, and how is the excess depth of cut to obtain it calculated?

18) What diametral pitches correspond to the following modules: 2.75; 4; 8?

19) Can bevel gears be cut by formed milling cutters?

20) Can the formed cutters used for cutting spur gears also be used for bevel gears?

21) What is the pitch angle of a bevel gear?

22) What does the term "miter" mean as applied to bevel gears?

23) What is the difference between the terms "whole depth" and "working depth" as applied to gear teeth?

24) Why do preshaved gears have a greater dedendum than gears that are finish-hobbled?

25) Are gear teeth of 8 diametral pitch larger or smaller than teeth of 4 diametral pitch, and how do these two pitches compare in regard to tooth depth and thickness?

26) Where is the pitch diameter of a bevel gear measured?

27) What is the relation between the circular pitch of a wormgear and the linear pitch of the mating worm?

28) In what respect does the helix angle of a worm differ from the helix angle of a helical or spiral gear?

29) How do the terms "pitch" and "lead," as applied to a worm, compare with the same terms as applied to screw threads?

30) Why is the outside diameter of a hob for cutting a wormgear somewhat larger than the outside diameter of the worm?

31) Why are triple, quadruple, or other multiple-threaded worms used when an efficient transmission is required?

32) In designing worm drives having multi-threaded worms, it is common practice to select a number of wormgear teeth that is not an exact multiple of the number of worm threads. Why is this done? When should this practice be avoided?

33) Explain the following terms used in connection with helical or spiral gears: "transverse diametral pitch" and "normal diametral pitch." What is the relation between these?

34) Are helical gear calculations based upon diametral pitch or circular pitch?

35) Can helical gears be cut with the formed cutters used for spur gears?

36) In spiral gearing, the tangent of the tooth or helix angle = the circumference ÷ lead. Is this circumference calculated from the outside diameter, the pitch diameter, or the root diameter?

37) What advantages are claimed for gearing of the herringbone type?

# SECTION 20

## SPEEDS, FEEDS, AND MACHINING POWER

*Machinery's Handbook* pages **1065–1141** and **1144–1183**

Metal-cutting operations such as turning and drilling may not be as productive as they could be unless the material removal rate is at or near the maximum permitted by the available power of the machine. It is not always possible to use the machine's full power owing to limitations imposed by a combination of part configuration, part material, tool material, surface finish and tolerance requirements, coolant employed, and tool life. However, even with such restrictions, it is practical to find a combination of depth of cut, feed rate, and cutting speed to achieve the best production rate for the job at hand.

The information on *Handbook* pages **1065–1141** is useful in determining how to get the most out of machining operations. The tabular data are based on actual shop experience and extensive testing in machining laboratories. A list of machining data tables is given on *Handbook* page **1078**, and these tables are referred to in the following.

Machining operations such as milling, drilling, and turning using very small tooling requires special consideration as such tools easily break, even at conservative values of cutting speed, feed, and depth of cut of conventional machining. The information on *Handbook* pages **1144–1183** explores the requirements and techniques of sucessful micromachining operations.

Most materials can be machined over a wide range of speeds; however, there is usually a narrower spread of speeds within which the most economical results are obtained. This narrower spread is determined by the economical tool life for the job at hand as, for example, when a shorter tool life is tolerable the speed can be increased. On the other hand, if tool life is too short, causing excessive down time, then speed can be reduced to lengthen tool life.

To select the best cutting conditions for machining a part the following procedure may be followed:

1) Select the maximum depth of cut consistent with the job.

2) Select the maximum feed rate that can be used consistent with such job requirements as surface finish and the rigidity of the cutting tool, workpiece, and the machine tool. Use **Table 15a** on *Handbook* page **1110** to assist in feed selection for milling. When possible, use the combined feed/speed portions of the tables to select two pairs of feed and speed data and determine the spindle speed as illustrated by **Example 1**.

3) If the combined feed/speed data are not used, select the cutting speed and determine the spindle speed (for turning also use **Table 5a** on *Handbook* page **1091**). This order of selection is based on the laws governing tool life; i.e., the life of a cutting tool is affected most by the cutting speed, then by the feed, and least by the depth of cut.

By using the same order of selection, when very heavy cuts are to be taken, the cutting speed that will utilize the maximum power available on the machine tool can be estimated by using a rearrangement of the machining power formulas on *Handbook* pages **1133–1137**. These formulas are used together with those on *Handbook* pages **1072** and **1097** which are used when taking ordinary cuts, as well as heavy cuts. Often, the available power on the machine will limit the size of the cut that can be taken. The maximum depth of cut and feed should then be used and the cutting speed adjusted to utilize the maximum available power. When the cutting speed determined in this manner is equal to or less than recommended, the maximum production and the best possible tool life will be achieved. When the estimated cutting speed is greater than recommended, the depth of cut or feed may be increased, but the cutting speed should not be increased beyond the value that will provide a reasonable tool life.

*Example 1:* An ASTM Class 25 (160–180 Brinell Hardness Number, BHN, which also may be expressed as HB) gray-iron casting is to be turned on a geared head lathe using a cemented carbide cutting tool. The heaviest cut will be 0.250 inch (6.35 mm) deep, taken on an 8-inch (203.2-mm) diameter of the casting; a feed rate of 0.020 in/rev (0.51 mm/rev) is selected for this cut. Calculate the spindle speed of the lathe, and estimate the power required to take this cut.

Locate the selected work material in **Table 4a** (*Handbook* page **1089**), and select the feed/speed pairs that correspond to the chosen cutter material. For an uncoated carbide tool, the given feed/speed pairs are: optimum 28/240, and average 13/365.

Factors to correct for feed and depth of cut are found in **Table 5a** (page **1091**). First, determine the ratios of $^{\text{chosen feed}}/_{\text{optimum feed}} = {}^{20}/_{28} = 0.71$ and $V_{avg}/V_{opt} = {}^{365}/_{240} = 1.52$, then, by estimation or interpolation, determine $F_f$ and $F_d$, and calculate $V$ and $N$ as follows:

$F_f = 1.22; F_d = 0.86$

$V = V_{opt} \times F_f \times F_d = 240 \times 1.22 \times 0.86 = 252$ ft/min

$$N = \frac{12V}{\pi D} = \frac{12 \times 252}{\pi \times 8} = 120 \text{ rpm}$$

See tables from *Handbook* pages **1133** on to estimate the power requirements using: $K_p = 0.52$ (**Table 1a**), $C = 0.90$ (**Table 2**), $Q = 12Vfd$ (**Table 5**), $W = 1.30$ (**Table 3**), and $E = 0.80$ (**Table 4**).

$Q = 12Vfd = 12 \times 252 \times 0.020 \times 0.250 = 15.12$ in³/min

$$P_m = \frac{K_p CQW}{E} = \frac{0.52 \times 0.90 \times 15.12 \times 1.30}{0.80} = 11.5 \text{ hp}$$

The equivalent results, expressed in the metric system, can be obtained by converting the cutting speed $V$, the metal removal rate $Q$, and the power at the motor $P_m$ into metric units using factors found starting on page **2845** of the *Handbook*, as illustrated in the following.

$V = 252$ ft/min $= 252 \times 0.3 = 76$ m/min

$Q = 15.12$ in³/min $= 15.12 \times 16.4 \div 60 = 4.13$ cm³/s

$P_m = 11.5$ hp $= 11.5 \times 0.745 = 8.6$ kw

Alternatively, if metric units are used throughout the problem, $F_f$ and $F_d$ are determined in the same manner as above. However, if $V$ is in meters per minute, and $D$ and $d$ are in millimeters, then $N = 1000V/\pi D$, and $Q = Vfd/60$.

*Example 2:* If the lathe in **Example 1** has only a 10-hp motor, estimate the cutting speed and spindle speed that will utilize the maximum available power. Use inch units only.

$$Q_{max} = \frac{P_m E}{K_p C W} = \frac{10 \times 0.80}{0.52 \times 0.90 \times 1.30} \quad \left(P_m = \frac{K_p CQW}{E}\right)$$

$$= 13.15 (\text{in}^3/\text{min})$$

$$V = \frac{Q_{max}}{12fd} = \frac{13.15}{12 \times 0.020 \times 0.250} \qquad (Q = 12Vfd)$$

$$= 219 \text{ fpm}$$

$$N = \frac{12V}{\pi D} = \frac{12 \times 219}{\pi \times 8} = 105 \text{ rpm}$$

*Example 3:* A slab milling operation is to be performed on 120–140 BHN AISI 1020 steel using a 3-inch diameter high-speed steel plain milling cutter having 8 teeth. The width of this cut is 2 inches; the depth is 0.250 inch, and the feed rate is 0.004 in/tooth. Estimate the power at the motor required to take this cut (except for **Table 11**, refer to tables starting on *Handbook* page **1135**).

$V = 110$ fpm (**Table 11**, page **1101**) $Q = f_m wd$ (**Table 5**)
$K_p = 0.69$ (**Table 1b**) $W = 1.10$ (**Table 3**)
$C = 1.25$ (**Table 2**) $E = 0.80$ (**Table 4**)

$$N = \frac{12V}{\pi D} = \frac{12 \times 110}{\pi \times 3} = 140 \text{ rpm}$$

$$f_m = f_t n_t N = 0.004 \times 8 \times 140 = 4.5 \text{ in/min}$$

$$P_m = \frac{K_p CQW}{E} = \frac{0.69 \times 1.25 \times 2.25 \times 1.10}{0.80} = 2.67 \text{ hp}$$

*Example 4:* A 16-inch-diameter cemented carbide face milling cutter having 18 teeth is to be used to take a 14-inch-wide and 0.125-inch-deep cut on an H12 tool steel die block having a hardness of 250–275 BHN. The feed used will be 0.008 in/tooth, and the milling machine has a 20-hp motor. Using the tables starting on *Handbook* page **1133**, estimate the cutting speed and the spindle speed to be used that will utilize the maximum horsepower available on the machine.

$K_p = 0.98$ fpm (**Table 1a**) $W = 1.25$ (**Table 3**)
$C = 1.08$ (**Table 2**) $E = 0.80$ (**Table 4**)
$Q = f_m wd$ (**Table 5**)

$$Q_{max} = \frac{P_m E}{K_p C W} = \frac{20 \times 0.80}{0.98 \times 1.08 \times 1.25} \qquad \left(P_m = \frac{K_p C Q W}{E}\right)$$

$$= 12.1 \ \ (\text{in}^3/\text{min})$$

$$f_m = \frac{Q_{max}}{wd} = \frac{12}{14 \times 0.125} \qquad (Q = f_m w d)$$

$$= 6.9 \ \ \text{in/min; use } 7 \text{ in/min}$$

$$N = \frac{f_m}{f_t n_t} = \frac{7}{0.008 \times 18} \qquad (f_m = f_t n_t N)$$

$$= 48.6 \ \ \text{rpm; use 50 rpm}$$

$$V = \frac{\pi D N}{12} = \frac{\pi \times 16 \times 50}{12} = 209 \text{ fpm}$$

Formulas for estimating the thrust, torque, and power for drilling are given on *Handbook* page **1139**. Thrust is the force required to push or feed the drill when drilling. This force can be very large. It is sometimes helpful to know the magnitude of this force and the torque exerted by the drill when designing drill jigs or work-holding fixtures; it is essential to have this information as well as the power required to drill when designing machine tools on which drilling operations are to be performed. In the ordinary shop, it is often helpful to be able to estimate the power required to drill larger holes in order to determine if the operation is within the capacity of the machine to be used.

*Example 5:* Estimate the thrust, torque, and power at the motor required to drill a ¾-inch-diameter hole in a part made from AISI 1117 steel, using a conventional twist drill and a feed rate of 0.008 in/rev. (Except for **Table 17**, see *Handbook* pages **1136** on for tables.)

| | |
|---|---|
| $K_d$ = 12,000 (**Table 6**) | $B$ = 1.355 (**Table 7**) |
| $F_f$ = 0.021 (**Table 8**) | $J$ = 0.030 (**Table 7**) |
| $F_T$ = 0.794 (**Table 9**) | $E$ = 0.80 (**Table 4**) |
| $F_M$ = 0.596 (**Table 9**) | $W$ = 1.30 (**Table 3**) |
| $A$ = 1.085 (**Table 7**) | $V$ = 101 fpm (**Table 17**, page **1117**) |

$$\begin{aligned}
T &= 2K_dF_fF_TBW + K_dd^2JW \\
&= 2\times 12{,}000\times 0.021\times 0.794\times 1.355\times 1.30 + 12{,}000\times \\
&\quad 0.75^2\times 0.030\times 1.30 \\
&= 968 \text{ lb} \\
M &= K_dF_fF_MAW \\
&= 12{,}000\times 0.021\times 0.596\times 1.085\times 1.30 \\
&= 212 \text{ in-lb} \\
N &= \frac{12V}{\pi D} = \frac{12\times 101}{\pi\times 0.750} = 514 \text{ rpm} \\
P_c &= \frac{MN}{63{,}025} = \frac{212\times 514}{63{,}025} = 1.73 \text{ hp} \\
P_m &= \frac{P_c}{E} = \frac{1.73}{0.80} = 2.16 \text{ hp}
\end{aligned}$$

## PRACTICE EXERCISES FOR SECTION 20

(See *Answers to Practice Exercises for Section 20* on page **252**)

1) Calculate the spindle speeds for turning $\frac{1}{2}$-inch and 4-inch bars made from the following steels, using a high-speed steel cutting tool and the cutting conditions given as follows:

| Steel Designation | Feed, in/rev | Depth of Cut, inch |
|---|---|---|
| AISI 1108, Cold Drawn | 0.012 | 0.062 |
| 12L13, 150–200 BHN | 0.008 | 0.250 |
| 1040, Hot Rolled | 0.015 | 0.100 |
| 1040, 375–425 BHN | 0.015 | 0.100 |
| 41L40, 200–250 BHN | 0.015 | 0.100 |
| 4140, Hot Rolled | 0.015 | 0.100 |
| O2, Tool Steel | 0.012 | 0.125 |
| M2, Tool Steel | 0.010 | 0.200 |

2) Calculate the spindle speeds for turning 6-inch-diameter sections of the following materials, using a cemented carbide cutting tool and the cutting conditions given in the table on the next page.

| Material | Feed, in/rev | Depth of Cut, inch |
|---|---|---|
| AISI 1330, 200 BHN | 0.030 | 0.150 |
| 201 Stainless Steel, Cold Drawn | 0.012 | 0.100 |
| ASTM Class 50 Gray Cast Iron | 0.016 | 0.125 |
| 6A1-4V Titanium Alloy | 0.018 | 0.188 |
| Waspaloy | 0.020 | 0.062 |

3) A 200 BHN AISI 1030 forged steel shaft is being turned at a constant spindle speed of 400 rpm, using a cemented carbide cutting tool. The as-forged diameters of the shaft are 1½, 3, and 4 inches. Calculate the cutting speeds (fpm) at these diameters, and check to see if they are within the recommended cutting speed.

4) A 75-mm diameter bar of cold-drawn wrought aluminum is to be turned with a high-speed steel cutting tool, using a cutting speed of 180 m/mm. Calculate the spindle speed that should be used.

5) Calculate the spindle speed required to mill a 745 nickel-silver part using a ½-inch end milling cutter.

6) An AISI 4118 part having a hardness of 200 BHN is to be machined on a milling machine. Calculate the spindle speeds for each of the operations below and the milling-machine table feed rates for Operations a) and b).

a) Face mill top surface, using an 8-inch diameter cemented-carbide face milling cutter having 2 teeth. (Use $f_t = 0.008$ in/tooth.)

b) Mill ¼-inch-deep slot, using a ¾-inch-diameter two-fluted high-speed steel end milling cutter.

c) Drill a 23⁄64-inch hole.

d) Ream the hole ⅜ inch, using HSS reamer.

7) A 3-inch-diameter high-speed steel end milling cutter having 12 teeth is used to mill a piece of D2 high carbon, high chromium cold-work tool steel having a hardness of 220 BHN. The spindle speed used is 75 rpm, and the milling-machine table feed rate is 10 in/mm. Check the cutting conditions with respect to the recommended values, and make recommendations for improvements, if possible.

8) A 100–150 BHN low-carbon steel casting is to be machined with a 12-inch-diameter cemented carbide face milling cutter having 14 teeth, using a spindle speed of 60 rpm and a table-feed rate of 5 in/mm. Check these cutting conditions and recommend improvements, if possible.

9) Estimate the cutting speed and the power at the cutter and at the motor required to turn 210 BHN AISI 1040 steel in a geared head lathe, using an uncoated carbide tool, a depth of cut of 0.125 inches, a feed of 0.015 in/rev, and efficiency $E$ of 0.80.

10) A 165 BHN A286 high-temperature alloy, or superalloy, is to be turned on a 3-hp geared head lathe using a cemented carbide cutting tool. The depth of cut selected is 0.100 inch, and the feed is 0.020 in/rev. Estimate the cutting speed that will utilize the maximum power available on the lathe.

11) An AISI 8642 steel having a hardness of 210 BHN is to be milled with a 6-inch-diameter cemented carbide face milling cutter having 8 teeth on a 10-hp milling machine. The depth of cut is to be 0.200 inch, the width is 4 inches, and the feed is to be 0.010 in/tooth. Estimate the cutting speed that will utilize the maximum power available on the machine.

12) Estimate the thrust, torque, and power at the motor required to drill 200 BHN steel using the following drill sizes, feeds, and spindle speeds.

| Drill Size | Feed | Spindle Speed |
|---|---|---|
| $\frac{1}{4}$ in. | 0.0005 in/rev | 1500 rpm |
| $\frac{1}{2}$ in. | 0.002 in/rev | 750 rpm |
| 1 in. | 0.008 in/rev | 375 rpm |
| 19 mm | 0.15 mm/rev | 500 rpm |

13) Estimate the thrust, torque, and power at the motor for the 1-inch drill in Exercise 12 if the drill is ground to have a split point.

14) Describe the general characteristics of high-speed steels that make them suitable for use as cutting-tool materials.

15) What guidelines should be followed in selecting a grade of cemented carbide?

16) How does the cutting speed, feed, and depth of cut influence tool life?

17) List the steps for selecting the cutting conditions in their correct order and explain why.

18) What are the advantages of coated carbides, and how should they be used?

19) Name the factors that must be considered when selecting a cutting speed for tapping.

20) Why is it important to calculate the table-feed rate for milling?

21) Name the factors that affect the basic feed rate for milling.

22) When should the power required to take a cut be estimated? Why?

23) Name the factors that affect the power constant, $K_p$. This constant is unaffected by what?

24) Why is it necessary to have a separate method for estimating the drilling thrust, torque, and power?

25) Why are traditional speeds and feeds generally inappropriate for micromachining?

26) What is pecking or peck drilling and why is it used?

## SECTION 21

## CNC (COMPUTER NUMERICAL CONTROL) PROGRAMMING

*Machinery's Handbook* pages **1338–1387**

Numerical control (NC) was defined by the Electronic Industries Association as "a system in which actions are controlled by the direct insertion of numerical data at some point. The system must automatically interpret at least some portion of these data." From this concept evolved modern manufacturing methods using computer numerical control programming, better known as *CNC programming*.

CNC manufacturing can be defined as using coded instructions (expressed with letters in the English alphabet, numerals, and symbols) submitted to and processed by a programmable machine control system, which then operates a machine's tools to modify or create a workpiece. All instructions must be written in a logical order, in a specified format called the *program structure*, taking into account the CNC machine's setup. Individual program instructions include all required machine activities: tool selection, motion control, spindle speed, cutting feed rates, coolant flow, various adjustments, and so on. The part program can be stored and used repeatedly to produce the same part with minimal human interaction.

The CNC programmer is responsible for gathering and integrating all pertinent information involved in developing safe, efficient, error-free, production-ready part programs for CNC machines installed in the shop. The programmer needs to know the machines and tools, visualize all tool motions, and, understanding the materials and part parameters, recognize any restricting factors. Even with the help of today's automated software, knowledge of programming is essential, as is effective communication among collaborators on any parts project.

Information in the *Handbook*, starting on page **1338**, gives an overview of the most important CNC concepts and operations. Additionally, material on numerical control programming from previous editions of the *Handbook*, starting on *Machinery's Handbook 32 Digital Edition* page **3430**, is arranged by subject matter for ease of reference. (Note that as with all *Digital Edition*–only material, this section has not been updated.) A related older discussion—of the Automatic Programmed Tool (APT) language, its use in part programming, and examples of computational and geometric programs—also can be found starting on *Machinery's Handbook 32 Digital Edition* page **3455**, though APT programming of this type is no longer required with today's CNC machines.

**CNC Coordinate Geometry.**—Key to working with CNC principles is understanding the *geometry* of machine tools. This encompasses the relationship between the machine data, part data, and tool data, which includes setup (see *Handbook* page **1339**).

As with all CAD/CAM systems (see *CAD/CAM* on *Handbook* page **1377**), CNC is based on the same principles of a system of coordinates that defines the location of a point in two-dimensional (2D) or three-dimensional space (3D). A system of coordinates is founded on the concept of two perpendicular lines (named axes "X" and "Y") intersecting at a point called the *origin*, where both coordinates have a value of zero (X0, Y0). Both lines are divided into equal units of measurement as shown in **Fig. 1**.

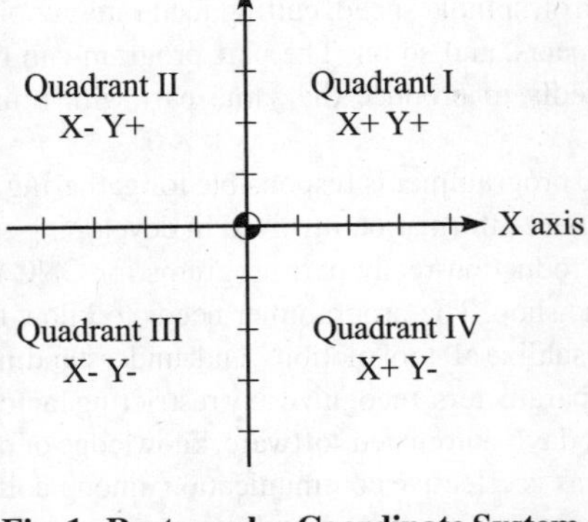


**Fig. 1. Rectangular Coordinate System**

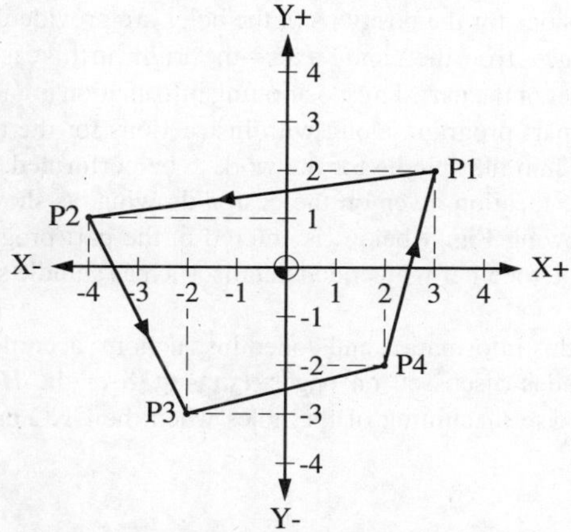


**Fig. 2. Absolute Coordinates and Incremental Linear Motion**

A point can be defined in a plane (any two axes = 2D) or in space (three axes = 3D). For CNC work, the *rectangular coordinate system*—also known as the *Cartesian coordinate system*—is the most commonly used. It is based on three standard axes: X,Y,Z. *Points* are defined as locations with a distance from the origin defined by projecting a line at 90 degrees to each axis, forming a visual rectangle. For example, in **Fig. 2**, P1 is defined as X3.0, Y2.0, P2 is defined as X-4.0, Y1.0, P3 as X-2.0,Y-3.0, and P4 as X2.0, Y-2.0.

Once the point locations are established, the programmer can use these points to specify the center of individual holes to be machined or as endpoints for a continuous contour. These points also can be connected by providing specific motion instructions, as a toolpath between given points.

**Point-to-Point Programming.**—Point-to-point programming is covered beginning on *Handbook* page **1354**. As an example of the use of CNC for point-to-point part programs, consider the rectangular plate shown in **Fig. 3**, in which it is required to machine eight holes as shown.

Dimensions for the positions of the holes are provided based on their distances from the *X* and *Y* axes—the origin, in this case, located at the center of the part. This positioning information is entered into the CNC part program, along with instructions for the tooling to be loaded into the spindle for the work to be performed. The hole coordinate location given on the design drawing, as shown in the table following **Fig. 3** below, is entered in the part programming manuscript, together with coded details, such as spindle speed and feed rates.

All of this information and coded instructions, according to the special codes discussed on pages **1343–1348** of the *Handbook*, direct precise machining of the holes when the CNC program is executed.

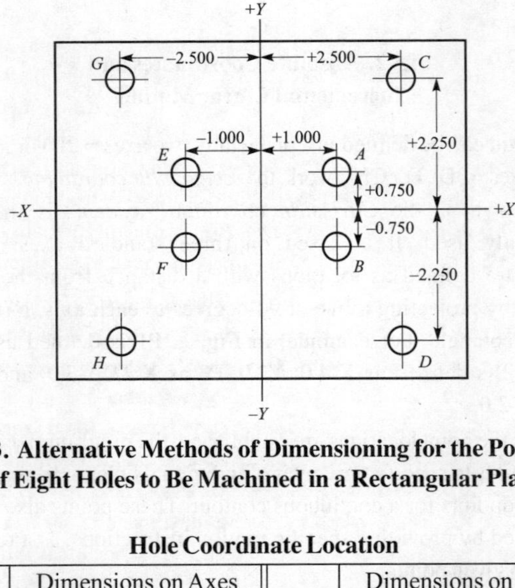


**Fig. 3. Alternative Methods of Dimensioning for the Positions of Eight Holes to Be Machined in a Rectangular Plate**

**Hole Coordinate Location**

| Point | Dimensions on Axes *X* | Dimensions on Axes *Y* | Point | Dimensions on Axes *X* | Dimensions on Axes *Y* |
|---|---|---|---|---|---|
| *A* | + 1.000 | + 0.750 | *E* | – 1.000 | + 0.750 |
| *B* | + 1.000 | – 0.750 | *F* | – 1.000 | – 0.750 |
| *C* | + 2.500 | + 2.250 | *G* | – 2.500 | + 2.250 |
| *D* | + 2.500 | – 2.250 | *H* | – 2.500 | – 2.250 |

**Absolute and Incremental Programming.**—A motion dimension along the axes *X*, *Y*, and *Z* for milling and axes *X* and *Z* for turning (as well as any parallel or rotary axis) can be programmed either as an *absolute location* measured from part zero (origin) or as an *incremental distance* (*U* and *W* for turning) with direction measured from the current tool position. The absolute method of programming is selected by the G90 command, and the incremental method is selected by the G91 command as described on *Handbook* page **1347**. (Each command cancels the other.)

The benefit of absolute programming is that one change of a point location requires only one change in the program; the same change in incremental programming requires two changes. The main benefit of incremental programming is that it can be used for toolpath repetition, typically in subprograms or macros.

**Continuous-Path Programming.**—Surfaces along or at angles to the axes and curved surfaces are produced by *continuous-path* or *contouring* programs. (See the section *Calculations for Contouring* on page **1359** of the *Handbook*.) These programs coordinate two or more machine motions simultaneously, so that the movement of the workpiece relative to the cutting tool generates the required shape.

Angular shapes are generated by *straight line* or *linear interpolation* programs that coordinate movements of two slides to produce the required angular path. Circular arcs can be generated by means of a *circular interpolation* program that controls the slide movements automatically to produce a curved path.

**Linear Interpolation.**—In *linear interpolation* mode (programmed in G01 mode with a cutting feed rate in effect), the tool moves along a linear path, directly from one point to the next, traversing the shortest distance, as described on *Handbook* page **1350**. (This is a departure from *rapid motion* mode, programmed in G00 mode, which instead selects the path for the fastest motion.) The word "interpolation" defines a mathematical process that translates into a synchronized feed rate for all programmed axes; this is automatically calculated and applied by the control system.

**Circular Interpolation.**—*Circular interpolation* applies to a cutting toolpath moving along an arc or a circle. The cutting motion along an arc is two-dimensional and takes place in a single plane. The plane in which the arc is positioned must be determined initially.

(No arc or circle exists as a 3D entity; it becomes a helix.) Depending on the required outcome, the desired plane can be defined using one of three G-codes (G17, G18, or G19), as described on *Handbook* page **1350**. CNC mills usually default to the *X-Y* axes (G17) upon startup.

Programming an arc requires a direction of motion in the plane (G02 = clockwise; G03 = counterclockwise), a start point (typically, the current block), an end point (typically, the target point at the end of the motion), and the arc radius. The arc radius can be specified directly (for example, R1.5), or it can be defined by special vectors, using I-J-K addresses (see **Fig. 4** and the definitions given below).

The definition of vectors I, J, and K is the same for all three axes, where X and I, Y and J, and Z and K are related (paired): the vector I-J-K is the incremental distance and direction from the start point of the arc to the arc center, measured along axis X-Y-Z respectively.

A typical milling application in the plane XY is shown in **Fig. 4**. Current plane selection determines the arc vectors. Some controls use I-J-K as an absolute location. The start point of the arc is defined as the endpoint of the previous motion.

When specifying a direct radius R, there are some limitations. Radius R (positive) can only be programmed for arcs up to and including a 180-degree arc sweep angle. If the sweep angle exceeds 180 degrees but is less than 360 degrees, R– (negative) must be used. For a full circle (360 degrees arc sweep angle), the address R cannot be used at all. Instead, a two-vector combination must be used; there is no need to break arc motion at the quadrants.

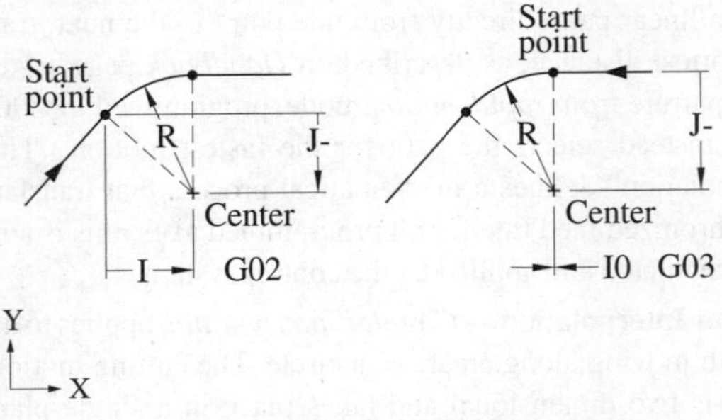


**Fig. 4. Arc Vectors**

## PRACTICE EXERCISES FOR SECTION 21

(See *Answers to Practice Exercises for Section 21* on page **254**)

1) List the different methods of program development.

2) What type of actions are controlled in CNC programs?

3) What is the address (code beginning) used to preset or to prepare the control system for a certain desired condition, or to a certain mode or a state of operation?

4) Which of the following applications of CNC is most common? (a) grinding, (b) turning, (c) broaching, or (d) milling and turning?

5) What is the default (at machine startup) feed rate on CNC lathes measured in? (a) cutting speed, (b) inches per revolution in/rev or mm/rev, (c) inches per minute in/min or mm/min, or (d) constant surface speed?

6) What other program word is necessary when programming G01 linear interpolation? (a) S, (b) F, (c) T, or (d) H?

7) When programming an arc, which letters identify the arc center location? (a) X, Y, and Z, (b) A, B, and C, (c) I, J, and K, or (d) Q and P?

8) The letter address H is used to indicate a tool length offset register number. Which preparatory function is it used in conjunction with? (a) G54, (b) G43, (c) G42, or (d) G41?

9) When programming an arc, an additional method exists that does not use I, J, and K. Which program word is used? (a) A, (b) B, (c) C, or (d) R?

10) What is the program word "N" that identifies a program block called?

11) The advantage of using G96 Constant Cutting Speed in turning is that as the diameter changes (position of the tool changes in relation to the centerline), the r/min increases or decreases to accomplish the programmed cutting speed. True or false?

12) What is the difference between cutter radius offset and cutter compensation?

13) Which G-Code is used to cancel Cutter Diameter Compensation? (a) G40, (b) G41, (c) G42, or (d) G43?

14) Fixed cycles, sometimes called canned cycles, are a series of preset sequences of toolpath motions, allowing consecutive motions to be performed with a single block of program. What do the codes G98 and G99 represent?

15) In the first column below various fixed or canned cycles are shown. In the second column are some preparatory codes. Match the functions with the codes.

| | |
|---|---|
| a. Spot-drilling cycle | 1. G80 |
| b. Peck-drilling cycle (deep hole) | 2. G81 |
| c. Boring cycle, spindle rotating on withdrawal at feed rate | 3. G85 |
| d. Drilling cycle | 4. G84 |
| e. Right-hand threading cycle | 5. G82 |
| f. Fixed cycle cancellation | 6. G83 |

16) A parametric subroutine is used exclusively for describing the path around the outside of a part. True or false?

17) What is a G word?

18) Explain the rule that describes the orientation and directions of the motions of slides and spindles on a machine tool.

19) With computer-aided manufacturing (CAM), the design model from CAD is used to identify specific cutting tool data and create cutting toolpaths for CNC machining. True or false?

20) The simulation and verification of the final output of CNC code requires input of accurate setup data to perform effectively. Name some of the common criteria required.

## SECTION 22

## THE METRIC SYSTEM

*Machinery's Handbook* contains a considerable amount of metric material in terms of texts, tables, and formulas. This material is included because much of the world now uses the metric system, also known as the Système International (SI), and the movement in that direction continues in all countries that intend to compete in the international marketplace, including the United States.

An explanation of the SI metric system is found on *Handbook* pages **151–154** and **2840–2843**. A brief history is given of the development of this system, and a description is provided for each of its seven basic units. Factors and prefixes for forming decimal multiples and submultiples of the SI units also are shown. Another table lists SI units with complex names and symbols for them.

Tables of SI units and conversion factors appear on pages **2845–2883**. Factors are provided for converting English units to metric units, or vice versa, and cover units of length, area, volume (including capacity), velocity, acceleration, flow, mass, density, force, force per unit length, bending moment or torque, moment of inertia, section modulus, momentum, pressure, stress, energy, work, power, and viscosity. By using the factors in these tables, it is a simple matter of multiplication to convert from one system of units to the other. Where the conversion factors are exact, they are given to only 3 or 4 significant figures, but where they are not exact they are given to 7 significant figures to permit the maximum degree of accuracy to be obtained that is ordinarily required in the metalworking field.

To avoid the need to use some of the conversion factors, various conversion tables are given starting on page **2845**. The tables for length conversion on pages **2845–2858** will probably be the most frequently used. Two different types of tables are shown. The two tables on page **2849** facilitate converting lengths up to 100 inches into millimeters, in steps of one ten-thousandth of an inch; and up to 1000 millimeters to inches, in steps of a thousandth of a millimeter.

The table starting on page **2850** enables converting fractions and mixed number lengths up to 41 inches into millimeters, in steps of one sixty-fourth of an inch.

To make possible such a wide range in a compact table, the reader often must take two or more numbers from the table and add them together, as explained in the accompanying text. The tables starting on pages **2852** and **2854** have a much more limited range of conversion for inches to millimeters and millimeters to inches. However, these table have the advantage of being direct-reading; that is, only a single value is taken from the table, and no addition is required.

For those who are engaged in design work where it is necessary to do computations in the fields of mechanics and strength of materials, a considerable amount of guidance will be found for the use of metric units. Thus, beginning on *Handbook* page **151**, the use of the metric SI system in mechanics calculations is explained in detail. In succeeding pages, **boldface type** is used to highlight references to metric units in the combined Mechanics and Strength of Materials section. Metric formulas are provided also, to parallel the formulas for English units.

As another example, on page **212**, it is explained in boldface type that SI metric units can be applied in the calculations in place of the English units of measurement without changes to the formulas for simple stresses.

The reader also should be aware that certain tables in the *Handbook*, such as that on page **80**, which gives values for segments of circles for a radius = 1, can be used for either English or metric units, as indicated directly under the table heading. There are other instances, however, where separate tables are needed, such as are shown for the conversion of cutting speed in feet per minute into revolutions per minute on pages **1074–1075**, and cutting speed in meters per minute into revolutions per minute on pages **1076–1077**.

The metric material in the *Handbook* will provide considerable useful data and assistance to engineers and technicians who are required to use metric units of measurements. It is strongly suggested that all readers, whether or not they are using metric units at the present time, become familiar with the SI System by reading the explanatory material in the *Handbook* and by studying the SI units and the ways of converting English units to them.

**SI Base Units and Definitions.**—*Meter (m):* The base unit of length, equal to the path length traveled by light in a vacuum during a time interval of $3.3356410 \times 10^{-9}$ s (derived from the precise measurement of light's speed, $c = 299{,}792{,}458$ m/s). This value was adopted in 1983, based on the newly defined value for the speed of light, made possible by the technology developed at the National Institute of Standards and Technology (NIST).

*Kilogram (kg):* The base unit of mass, related to the most precise measurement to date of the Planck constant $h$, which is the energy carried by a photon to its frequency, and measured as $6.6260702 \times 10^{-34}$ kg $m^2/s^2$. It is with respect to this invariant of nature that the precise standard for the kilogram is given.

*Second (s):* The duration of $9.1926318 \times 10^9$ periods of the radiation, corresponding to the transition between the two hyperfine levels of the ground state of the cesium-133 atom, is the element used for the atomic clock due to its stability.

*Ampere (A):* The base unit of electric current, measured as current flow per unit of time. One ampere (*amp*) is equal to one coulomb per second. One elementary electric charge (e) is equal to $1.60218 \times 10^{-19}$ coulomb (that is, 1 C = $6.241 \times 10^{18}$ e).

*Candela (cd):* The base unit for photometry, which is the science of measuring light perceived by the human eye. The candela (cd) is the luminous intensity, in a given direction, of a source that emits monochromatic radiation of frequency $540 \times 10^{12}$ hertz. The human eye is most sensitive to this frequency (greenish-yellow light). One cd has a radiant intensity of 1/683 watt per steradian.

*Kelvin (K):* The base unit of thermodynamic temperature, defined by taking the fixed numerical value of the Boltzmann constant $k$ to be $1.380649 \times 10^{-23}$ $J{\cdot}K^{-1}$.

*Mole (mol):* The base unit for the amount of a chemical substance consisting of exactly $6.02214076 \times 10^{23}$ constitutive particles. This

**Table 1. SI Base Units**

| Quantity | Name | Symbol |
|---|---|---|
| Length | meter | m |
| Mass | kilogram | kg |
| Time | second | s |
| Electric Current | ampere | A |
| Luminous Intensity | candela | cd |
| Thermodynamic Temperature | kelvin | K |
| Amount of Substance | mole | mol |

number is the Avogadro constant $N_A = 6.02214076 \times 10^{23}$ mol$^{-1}$. (Previously, 1 mole was linked to the number of atoms in 12.1 g of the carbon-12 isotope.)

**SI Derived Units.**—From the base units, come the SI derived units, as shown in **Table 2**. See also the following tables, *Units of Measure and Conversion Factors* on page **268** of this *Guide*, and the *MEASURING UNITS* section of the *Handbook*, starting on page **2835**.

## Table 2. SI Derived Units

| Quantity | Name | Symbol |
|---|---|---|
| Area | square meter | $m^2$ |
| Volume | cubic meter | $m^3$ |
| Speed, Velocity | meter per second | m/s |
| Acceleration | meter per second squared | $m/s^2$ |
| Wave Number | reciprocal meter | $m^{-1}$ |
| Mass Density | kilogram per cubic meter | $kg/m^3$ |
| Specific Volume | cubic meter per kilogram | $m^3/kg$ |
| Current Density | ampere per square meter | $A/m^3$ |
| Magnetic Field Strength | ampere per meter | A/m |
| Amount-of-Substance Concentration | mole per cubic meter | $mol/m^3$ |
| Luminance | candle per square meter | $cd/m^2$ |
| Mass Fraction | kilogram per kilogram | kg/kg |

## Table 3. SI Derived Units with Special Names and Symbols

| Quantity | Name and Derived Units | Symbol |
|---|---|---|
| Plane Angle | radian, m/m | rad |
| Solid Angle | steradian, $m^2/m^2$ | sr |
| Frequency | hertz, $s^{-1}$ | Hz |
| Force | newton, $kg{\cdot}m{\cdot}s^{-2}$ | N |
| Pressure, Stress | pascal, $kg{\cdot}m^{-1}{\cdot}s^{-2}$ | Pa |
| Energy, Work, Heat | joule, $kg{\cdot}m^2{\cdot}s^{-2}$ | J |
| Power, Radiant Flux | watt, $kg{\cdot}m^2{\cdot}s^{-3}$ | W |
| Quantity of Electric Charge | coulomb, A·s | C |
| Electric Potential Difference | volt, $kg{\cdot}m^2{\cdot}s^{-3}{\cdot}A^{-1}$ | V |
| Capacitance | farad, $s^4{\cdot}A^2{\cdot}m^{-2}{\cdot}kg^{-1}$ | F |
| Electric Resistance | ohm, $kg{\cdot}m^2{\cdot}s^{-3}{\cdot}A^{-2}$ | Ω |
| Electric Conductance | siemens, $kg^{-1}{\cdot}m^{-2}{\cdot}s^3{\cdot}A^2$ | S |
| Magnetic Flux | weber, $kg{\cdot}m^2{\cdot}s^{-2}{\cdot}A^{-1}$ | Wb |
| Magnetic Flux Density | tesla, $kg{\cdot}s^{-2}{\cdot}A^{-1}$ | T |
| Inductance | henry, $kg{\cdot}m^2{\cdot}s^{-2}{\cdot}A^{-2}$ | H |
| Celsius Temperature | degree Celsius, K – 273.15 | °C |
| Luminous Flux | lumen, cd·sr | lm |
| Illuminance | lux, $cd{\cdot}sr{\cdot}m^{-2}$ | lx |
| Activity of Radionuclide | becquerel, $s^{-1}$ | Bq |
| Absorbed Dose | gray, $J{\cdot}kg^{-1}$ | Gy |
| Dose Equivalent | sievert, $m^2{\cdot}s^{-2}$ | Sv |
| Catalytic Activity | katal, $mol{\cdot}s^{-1}$ | kat |

## Table 4. SI Derived Units that Include SI Derived Units with Special Names

| Quantity | Name | Symbol |
|---|---|---|
| Dynamic Viscosity | pascal-second | Pa·s |
| Moment of Force | newton-meter | N·m |
| Angular Velocity | radian per second | rad/s |
| Angular Acceleration | radian per second squared | $rad/s^2$ |
| Heat Flux Density, Irradiance | watt per square meter | $W/m^2$ |
| Heat Capacity, Entropy | joule per kelvin | J/K |
| Specific Heat Capacity, Specific Entropy | joule per kilogram-kelvin | J/kg·K |
| Specific Energy | joule per kilogram | J/kg |
| Energy Density | joule per cubic meter | $J/m^3$ |
| Thermal Conductivity | watt per meter-kelvin | W/m·K |
| Electric Field Strength | volt per meter | V/m |
| Electric Charge Density | coulomb per cubic meter | $C/m^3$ |
| Electric Flux Density | coulomb per square meter | $C/m^2$ |
| Permittivity | farad per meter | F/m |
| Permeability | henry per meter | H/m |
| Molar Energy | joule per mole | J/mol |
| Molar Entropy, Molar Heat Capacity | joule per mole-kelvin | J/mol·K |
| Exposure (X-rays and γ rays) | coulomb per kilogram | C/kg |
| Absorbed Dose Rate | gray per second | Gy/s |
| Radian Intensity | watt per steradian | W/sr |
| Radiance | watt per square meter-steradian | $W/m^2 \cdot sr$ |

## Table 5. Units Outside SI, Accepted for Use with SI

| Name | Symbol | Value in SI Units |
|---|---|---|
| minute | min | 1 min = 60 s |
| hour | h | 1 h = 60 min = 3600 s |
| day | d | 1 d = 24 h = 1440 min = 86400 s |
| liter | L | $1 L = 1 dm^3 = 10^{-3} m$ |
| metric ton | t | $1 t = 10^3$ kg =2205 lb |
| bel | B | 1 B = 10 dB |
| degree (angle) | ° | $1° = \pi/180$ rad |
| minute (angle) | ′ | $1' = (1/60)° = (\pi/10800)$ rad |
| second (angle) | ″ | $1'' = (1/60)' = (\pi/6480000)$ rad |
| electron volt | eV | $1 eV = 1.60218 \times 10^{-19}$ J |
| unified atomic mass unit | Da or u | $1 u = 1.66054 \times 10^{-27}$ kg |
| astronomical unit | au | $1 au = 1.49598 \times 10^{11}$ m |
| nautical mile | nmi | 1 nmi = 1852 m |
| knot | kn | $1 kn = 1 nmi \cdot h^{-1} = 0.514444 m \cdot s^{-1}$ |
| are | a | $1 a = 100 m^2$ |
| hectare | ha | $1 ha = 100 a = 10^4 m^2$ |
| bar | bar | $1 bar = 10^2 kPa = 10^5$ Pa |
| ångström | Å | $1 Å = 0.1 nm = 10^{-10}$ m |
| curie | Ci | $1 Ci = 3.7 \times 10^{10}$ Bq |
| roentgen | R | $1 R = 2.58 \times 10^{-4} C \cdot kg^{-1}$ |
| rad | rad | $1 rad = 10^{-2}$ Gy |
| rem | rem | $1 rem = 10^2$ Sv |

## Table 6. SI Prefixes

| Factor | Name | Symbol | Factor | Name | Symbol |
|---|---|---|---|---|---|
| $10^1$ | deca | da | $10^{-1}$ | deci | d |
| $10^2$ | hecto | h | $10^{-2}$ | centi | c |
| $10^3$ | kilo | k | $10^{-3}$ | milli | m |
| $10^6$ | mega | M | $10^{-6}$ | micro | μ |
| $10^9$ | giga | G | $10^{-9}$ | nano | n |
| $10^{12}$ | tera | T | $10^{-12}$ | pico | p |
| $10^{15}$ | peta | P | $10^{-15}$ | femto | f |
| $10^{18}$ | exa | E | $10^{-18}$ | atto | a |

## SECTION 23

## GENERAL REVIEW QUESTIONS

(See *Answers to General Review Questions* on page **256**)

1) If a regular polygon of 20 sides is to have an area of 100 square inches what formula may be used to calculate the length of one side of the polygon?

2) What does the number of a Jarno taper indicate?

3) What is the general rule for determining the direction in which to apply tolerances?

4) Why is 1 horsepower equivalent to 33,000 foot-pounds of work per minute? Why not 30,000 or some other number?

5) What is the chief element in the composition of babbitt metals?

6) If the pitch of a stub-tooth gear is $^8/_{10}$, what is the tooth depth?

7) What does the figure 8 mean if the pitch of a stub-tooth gear is $^8/_{10}$?

8) Explain how to determine the diametral pitch of a spur gear from a sample gear.

9) If a sample gear is cut to circular pitch, how can this pitch be determined?

10) What gage is used for seamless tubing, and does it apply to all metals?

11) How does the strength of iron wire rope compare with steel rope?

12) Is the friction between two bearing surfaces proportional to the pressure?

13) If the surfaces are well lubricated, upon what does frictional resistance depend?

14) What is the general rule for subtracting a negative number from a positive number? For example, 8 – (–4) =?

15) Is 1 meter longer than 1 yard?

16) On *Handbook* page **2874**, two of the equivalents of horsepower-hour are: 1,980,000 foot-pounds and 2.64 pounds of water evaporated at 212° F. How is this relationship between work and heat established?

17) Are "extra strong" and "double extra strong" wrought or steel pipe larger in diameter than standard-weight pipe?

18) In the design of plain bearings, what is the general relationship between surface finish and hardness of journal?

19) Are the nominal sizes of wrought or steel pipe ever designated by giving the outside diameter?

20) What are the advantages of plastics pipe?

21) Will charcoal ignite at a lower temperature than dry pine?

22) What general classes of steel are referred to as "stainless"?

23) What are free-cutting steels?

24) Does the nominal length of a file include the tang? For example, is a 12-inch file 12 inches long overall?

25) Is steel heavier (denser) than cast iron?

26) What is meant by specific heat?

27) What is the specific gravity (a) of solid bodies, (b) of liquids, (c) of gases?

28) A system of four-digit designations for wrought aluminum and aluminum alloys was adopted by The Aluminum Association in 1954. What do the various digits signify?

29) What alloys are known as "red brass," and how do they compare with "yellow brass"?

30) What is the difference between adiabatic expansion or compression and isothermal expansion or compression?

31) Are the sizes of all small twist drills designated by numbers?

32) Why are steel tools frequently heated in molten baths to harden them?

33) In hardening tool steel, what is the best temperature for refining the grain of the steel?

34) In cutting a screw thread on a tap, assume that the pitch is to be increased from 0.125 inch to 0.1255 inch to compensate for shrinkage in hardening. How can this be done?

35) What is the general rule for reading a vernier scale (a) for linear measurements; (b) for angular measurements?

36) The end of a shaft is to be turned to a taper of $^3/_8$ inch per foot for a length of inches without leaving a shoulder at the end of the cut. How is the diameter of the small end determined?

37) Is there a simple way of converting the function of 90° plus an angle to the function of the angle itself?

38) What decimal part of a degree is 53 minutes?

39) If $10x - 5 = 3x + 16$, what is the value of $x$?

40) Approximately what angle is required for a cone clutch to prevent either slipping or excessive wedging action?

41) What is the coefficient of friction?

42) Is Stub's steel wire gage used for the same purpose as Stub's iron wire gage?

43) Why are some ratchet mechanisms equipped with two pawls of different lengths?

44) How does the modulus of elasticity affect the application of flat belts?

45) What is the effect of centrifugal force on flat and V-belts?

46) Is the ultimate strength of a crane or hoisting chain equal to twice the ultimate strength of the bar or rod used for making the links?

47) How would you determine the size of chain required for lifting a given weight?

48) If a shaft $3^1/_2$ inches in diameter is to be turned at a cutting speed of 90 feet per minute, what number of revolutions per minute will be required?

49) In lapping by the "wet method," what kind of lubricant is preferable (a) with a steel lap, (b) with a cast-iron lap?

50) What is the meaning of the terms "right-hand" and "left-hand" as applied to helical or spiral gears, and how is the "hand" of the gear determined?

51) Are mating helical or spiral gears always made to the same hand?

52) How would you determine the total weight of 100 feet of $1^1/_2$-inch standard-weight pipe?

53) What is the difference between casehardening and pack-hardening?

54) What is the nitriding process of heat-treating steel?

55) What is the difference between single-cut and double-cut files?

56) For general purposes, what is the usual height of work benches?

57) What do the terms "major diameter" and "minor diameter" mean as applied to screw threads in connection with the American Standard?

58) Is the present SAE Standard for screw threads the same as the Unified and American Standard?

59) Does the machinability of steel depend only upon its hardness?

60) Is there any direct relationship between the hardness of steel and its strength?

61) What is the millimeter equivalent of $^{33}/_{64}$ of an inch?

62) How is the sevolute function of an angle calculated?

63) What is the recommended cutting speed in feet per minute for turning normalized AISI 4320 alloy steel with a hardness of 250 BHN, when using a coated, tough carbide tool?

64) The diametral pitch of a spur gear equals the number of teeth divided by pitch diameter. Is the diametral pitch of the cutter or hob for a helical or spiral gear determined in the same way?

65) Why are casehardening steels preferred for some gears and what special heat treatment is recommended?

66) Are the symbols for dimensions and angles used in spline calculations the same for both inch-dimension and metric module involute splines?

67) What kind of bearing surface and tool insert rake are provided by an indexable insert tool holder?

68) Is it necessary in making ordinary working drawings of gears to lay out the tooth curves? Why?

69) In milling plate cams on a milling machine, how is the cam rise varied other than by changing the gears between the dividing head and feed screw?

70) How is the angle of the dividing head spindle determined for milling plate cams?

71) How is the center-to-center distance between two gears determined if the number of teeth and diametral pitch are known?

72) How is the center-to-center distance determined for internal gears?

73) In the failure of riveted joints, rivets may fail through one or two cross-sections or by crushing. How may plates fail?

74) What gage is used in Britain to designate wire sizes?

75) What is a transmission dynamometer?

76) What is the advantage of a dynamometer for measuring power?

77) If a beam supported at each end is uniformly loaded throughout its length, will its load capacity exceed that of a similar beam loaded at the center only?

78) Is there any relationship between Brinell hardness and tensile strength of steel?

79) Is the outside diameter of a 2-inch pipe about 2 inches?

80) The hub of a lever 10 inches long is secured to a 1-inch shaft by a taper pin. If the maximum pull at the end of the lever equals 60 pounds, what pin diameter is required? (Give mean diameter or diameter at center.)

81) What are the two laws that form the basis of all formulas relating to the solution of triangles?

82) What are the sine and the cosine of the angle 45 degrees?

83) How is the pressure of water in pounds per square inch determined for any depth?

84) When calculating the basic load rating for a unit consisting of two bearings mounted in tandem, is the rated load of the combination equal to 2 times the capacity of a single bearing?

85) If a machine producing 50 parts per day is replaced by a machine that produces 100 parts per day, what is the percentage of increase?

86) If production is decreased from 100 to 50, what is the percentage of reduction?

87) What kind of steel is used ordinarily for springs in the automotive industry?

88) What is the heat-treating process known as "normalizing"?

89) What are the important Standards applicable to electric motors?

90) Is there an American Standard for section linings to represent different materials on drawings?

91) Is the taper per foot of the Morse Standard uniform for all numbers or sizes?

92) Is there more than one way to remove a tap that has broken in the hole during tapping?

93) The center-to-center distance between two bearings for gears is to be 10 inches, with a tolerance of 0.005 inch. Should this tolerance be (a) unilateral and plus, (b) unilateral and minus, (c) bilateral?

94) How are the available pitch-diameter tolerances for Acme screw threads obtained?

95) On *Handbook* page **1403**, there is a rule for determining the pressure required for punching circular holes into steel sheets or plates. Why is the product of the hole diameter and stock thickness multiplied by 80 to obtain the approximate pressure in tons?

96) What gage is used in the United States for cold-rolled sheet steel?

97) What gage is used for brass wire and is the same gage used for brass sheets?

98) Is the term "babbitt metal" applied to a single composition?

99) What are the chief elements in high-grade babbitt metal?

100) How many bars of stock 20 feet long will be needed to make 20,000 dowel-pins 2 inches long if the tool for cutting them off is 0.100 inch wide?

101) What is the melting point and density of cast iron; steel; lead; copper; nickel?

102) What lubricant is recommended for machining aluminum?

103) What relief angles are recommended for turning copper, brass, bronze, and aluminum?

104) Why is stock annealed between drawing operations in producing parts in drawing dies?

105) When is it advisable to mill screw threads?

106) How does a fluted chucking reamer differ from a rose chucking reamer?

107) What kind of material is commonly used for gage blocks?

108) What grade of gage blocks is used for shop standards?

109) What is the "lead" of a milling machine?

110) The table in *Machinery's Handbook 32 Digital Edition* page **3573** shows that a lead of 9.625 inches will be obtained if the numbers of teeth in the *driven* gears are 44 and 28 and the numbers of teeth on the *driving* gears 32 and 40. Prove that this lead of 9.625 inches is correct.

111) Use the prime number and factor table beginning on *Handbook* page **15** to reduce the following fractions to their lowest terms: $^{210}/_{462}$; $^{2765}/_{6405}$; $^{741}/_{1131}$.

112) If a bevel gear and a spur gear each have 30 teeth of 4 diametral pitch, how do the tooth sizes compare?

113) For what types of work are the following machinists' files used: (a) flat files? (b) half round files? (c) hand files? (d) knife files? (e) general-purpose files? (f) pillar files?

114) Referring to the illustration on *Handbook* page **697**, what is the dimension *x* over the rods used for measuring the dovetail slide if *a* is 4 inches, angle α is 60 degrees, and the diameter of the rods used is $^5/_8$ inch?

115) Determine the diameter of the bar or rod for making the links of a single chain required to lift safely a load of 6 tons.

116) Why will a helical gear have a greater tendency to slip on an arbor while the teeth are being milled than when milling a straight tooth gear?

117) What is meant by "trepanning"?

118) When is a removable or "slip" bushing used in a jig?

119) What are the relative ratings and properties of an H43 molybdenum high-speed tool steel?

120) What systematic procedure may be used in designing a roller chain drive to meet certain requirements as to horsepower, center distance, etc.?

121) In the solution of oblique triangles having two sides and the angle opposite one of the sides known, it is possible to have no solution or more than one solution. Under what condition will there be no solution?

122) What gear steels would you use (1) for casehardened gears? (2) for fully hardened gears? (3) for gears that are to be machined after heat treatment?

123) Is it practicable to tap holes and obtain (1) Class 2B fits? (2) Class 3B fits?

124) What is the maximum safe operating speed of an organic bonded type grinding wheel when used in a bench grinder?

125) What is the recommended type of diamond wheel and abrasive specification for internal grinding?

126) Is there a standard direction of rotation for all types of nonreversing electric motors?

127) What conditions are best suited for grease-lubricated antifriction bearings? What materials make up the grease used in this application?

128) In the example on *Handbook* page **2182**, the side relief angle at the leading edge of the single-point Acme thread cutting tool was calculated to be 19.27°, or 19°16′, which provides an effective relief angle ($a_e$) between the flank of the tool and the side of the thread of 10° at the minor diameter. What is the effective relief angle of this tool at the pitch diameter ($E$) and at the major diameter ($D$)? The pitch diameter of the thread is 0.900 inch, the major diameter is 1.000 inch, and the lead of the thread is 0.400 inch.

129) Helical flute milling cutters having eccentric relief are known to provide better support of the cutting edge than cutters ground with straight or concave relief. For a 1-inch-diameter milling cutter having a 35-degree helix angle, what is the measured indicator drop according to the methods described beginning on *Handbook* page **889** if the radial relief angle is to be 7°?

130) On *Handbook* page **2442**, **Table 4** shows that TFE fabric bearings have a load capacity of 60,000 pounds per square inch. Also shown in the table is a PV limit of 25,000 for this material. At what maximum surface speed in feet per minute can this material operate when the load is 60,000 pounds per square inch (psi)?

131) Is there a Standard for shaft diameter and housing bore tolerance limits that applies to rolling element bearings?

132) In designing an aluminum bronze plain bearing, what hardness should the steel journal have?

133) Steel balls are usually sold by the pound. How many pounds will provide 100 balls of $^{13}/_{32}$-inch-diameter carbon steel?

134) If a 3AM1-18 steel retaining ring were used on a rotating shaft, what is the maximum allowable speed of rotation?

135) What procedure applies to 3-wire measurements of Acme threads when the lead angle is greater than 5 degrees?

136) Twelve $1\frac{1}{2}$-inch-diameter rods are to be packed in a tube. What is the minimum inside diameter of the tube?

137) A four-wheel dolly supports a rack that holds six 5-gallon bottles filled with water that must be moved up a ramp onto the bed of a truck. What force is required to pull the loaded dolly up the ramp into the truck if the ramp is 14 feet long and the truck bed is 44 inches high? Neglect the weight of the dolly and rack, and friction in the dolly wheels. What if the ramp was 8 feet long?

## SECTION 24

## ANSWERS TO EXERCISES AND REVIEW QUESTIONS

All references are to *Machinery's Handbook, 32nd Edition* and *Machinery's Handbook 32 Digital Edition* page numbers

**Answers to** *Practice Exercises for Section 1 (see* **page 3***)*

| Number of Exercise | Answers (Or where information is given in the *Handbook*) |
|---|---|
| 1 | 78.54 $mm^2$; 31.416 mm |
| 2 | 4.995 or 5, approx. |
| 3 | 3141.6 $mm^2$ |
| 4 | 127.3 psi |
| 5 | 1.27 |
| 6 | 1.5708 |
| 7 | 8 hours, 50 minutes |
| 8 | 2450.448 pounds |
| 9 | $2\frac{1}{16}$ inches |
| 10 | 7 degrees, 10 minutes |
| 11 | $(\pi-2)r^2 = 1.1416r^2$ |
| 12 | Formula on *Handbook 32 Digital Edition* page **3770** |
| 13 | 740 gallons, approximately |
| 14 | Formula on *Handbook* page **86** |
| 15 | Formula on *Handbook* page **86** |
| 16 | Formula on *Handbook* page **86** |

**Answers to** *Practice Exercises for Section 2 (see* **page 11***)*

| Number of Exercise | Answers (Or where information is given in the *Handbook*) |
|---|---|
| 1 | *Handbook* pages **711** and **720** |
| 2 | (a) 0.043 inch, (b) 0.055 inch, (c) 0.102 inch |
| 3 | 0.336 inch |
| 4 | 2.796 inches |
| 5 | 4.743 inches |
| 6 | 4.221 feet |
| 7 | *Handbook* page **70** and **79** |
| 8 | $x = 2.585$, $y = -0.456$ |
| 9 | $x = 2.878$, $y = 3.279$ |
| 10 | $x = 1.629$, $y = 1.787$ |
| 11 | $x = 1.000$, $y = 2.000$ |

**Answers to** *Practice Exercises for Section 3* (*see* **page 25**)

| Number of Exercise | Answers (Or where information is given in the *Handbook*) |
|---|---|
| 1 | (a) 104 horsepower; (b) if reciprocal is used, $H = 0.33\ D^2 SN$ |
| 2 | 65 inches |
| 3 | 5.74 inches |
| 4 | Side $s$ = 5.77 inches; diagonal $d$ = 8.165 inches, and volume = 192.1 cubic inches |
| 5 | 91.0408 square inches |
| 6 | 4.1888 and 0.5236 |
| 7 | 59.217 cubic inches |
| 8 | *Handbook* page **2838** |
| 9 | $a = \frac{2A}{h} - b$ |
| 10 | $r = \sqrt{R^2 - \frac{s^2}{4}}$ |
| 11 | $a = \sqrt{\frac{(P/\pi)^2}{2} - b^2}$ |
| 12 | $\sin A = \sqrt{1 - \cos^2 A}$ |
| 13 | $a = \frac{b \times \sin A}{\sin B}$; $b = \frac{a \times \sin B}{\sin A}$<br>$\sin A = \frac{a \times \sin B}{b}$ $\sin B = \frac{b \times \sin A}{a}$ |

**Answers to** *Practice Exercises for Section 4* (*see* **page 36**)

| Number of Exercise | Answers (Or where information is given in the *Handbook*) |
|---|---|
| 2 | 4; 35; 72 |
| 3 | \$A2 × B\$1. The dollar sign (\$) indicates an absolute reference to the row or column it precedes. For \$A2, the referenced cell is in column A, and row changes are relative to starting row 2. For B\$1, the referenced cell is in row 1, and the column changes relative to column B. |
| 6 | \$5,954.45; \$6,131.81 |

**Answers to** *Practice Exercises for Section 4 (Continued)*

| Number of Exercise | Answers (Or where information is given in the *Handbook*) |
|---|---|
| 7 | −2; undefined; −16 |
| 8 | No, operations with exponents are performed before multiplication, division, addition, or subtraction. Refer to spreadsheet documentation for operator precedence rules. |

**Answers to** *Practice Exercises for Section 5 (see* **page 46***)*

| Number of Exercise | Answers (Or where information is given in the *Handbook*) |
|---|---|
| 1 | *Guide* page **41** |
| 2 | Calculator, or table beginning on *Handbook 32 Digital Edition* page **3175** |
| 3 | 2; 2; 1; $\overline{3}$; 3; 1 |
| 4 | As location of decimal point is indicated by characteristic, which is not given, the number might be 7082, 708.20, 70.82, 7.082, 0.7082, 0.07082, etc.; 7675, 767.5, etc.; 1689, 168.9, etc. |
| 5 | (a) 70.82; 76.75; 16.89; (b) 708.2; 767.5, 168.9; 7.082, 7.675, 1.689; 7082, 7675, 1689 |
| 6 | 2.88389; 1.94052; $\overline{3}$.94151 |
| 7 | 792.4; 17.49; 1.514; 486.5 |
| 8 | 4.87614; 1.62363 |
| 9 | 67.603; 4.7547 |
| 10 | 146.17; 36.8 |
| 11 | 9.88; 5.422; 5.208 |
| 12 | 0.2783 |
| 13 | 0.0000001432 |
| 14 | 237.6 |
| 15 | 187.08 |
| 16 | 14.403 square inches |
| 17 | 2.203 or, say, 2¼ inches |
| 18 | 107 horsepower |
| 19 | No |
| 20 | Yes, see *Handbook* page **2187** |

**Answers to** *Practice Exercises for Section 6 (see* **page 49***)*

| Number of Exercise | Answers (Or where information is given in the *Handbook*) |
|---|---|
| 1 | 8001.3 cubic inches |
| 2 | 83.905 square inches |
| 3 | 69.395 cubic inches |
| 4 | 1.299 inches |
| 5 | 22.516 cubic inches |
| 6 | 8 inches |
| 7 | 0.0276 cubic inch |
| 8 | 4.2358 inches |
| 9 | 1.9635 cubic inches |
| 10 | 410.5024 cubic inches |
| 11 | 26.4501 square inches |
| 12 | Radius, 1.4142 inches; area, 0.43 square inch |
| 13 | Area, 19.869 square feet; volume, 10.2102 cubic feet |
| 14 | Area, 240 square feet; volume, 277.12 cubic feet |
| 15 | 11.3137 inches |
| 16 | 41.03 gallons |
| 17 | 17.872 square gallons |
| 18 | 1.032 inches |
| 19 | 40 cubic inches |
| 20 | Table on *Handbook* page **82** |
| 21 | Table on *Handbook* page **82** |
| 22 | 5.0801 inches |
| 23 | 4 inches; 5226 inches |

**Answers to** *Practice Exercises for Section 7 (see* **page 54***)*

| Number of Exercise | Answers (Or where information is given in the *Handbook*) |
|---|---|
| 1 | *Handbook* page **56** |
| 2 | *Handbook* page **56** |
| 3 | *Handbook* page **56** |
| 4 | *Handbook* page **56** |
| 5 | *Handbook* page **57** |
| 6 | *Handbook* page **57** |
| 7 | *Handbook* page **57** |
| 8 | *Handbook* page **57** |
| 9 | *Handbook* page **57** |
| 10 | *Handbook* page **57** |
| 11 | *Handbook* page **58** |
| 12 | *Handbook* page **58** |

**Answers to** *Practice Exercises for Section 7 (Continued)*

| Number of Exercise | Answers (Or where information is given in the *Handbook*) |
|---|---|
| 13 | *Handbook* page **58** |
| 14 | *Handbook* page **57** |
| 15 | *Handbook* page **58** |
| 16 | *Handbook* page **58** |
| 17 | *Handbook* page **58** |
| 18 | *Handbook* page **59** |
| 19 | *Handbook* page **59** |
| 20 | *Handbook* page **59** |
| 21 | *Handbook* page **59** |
| 22 | *Handbook* page **59** |
| 23 | *Handbook* page **60** |
| 24 | *Handbook* page **60** |
| 25 | *Handbook* page **60** |
| 26 | *Handbook* page **60** |

**Answers to** *Practice Exercises for Section 8 (see* **page 63***)*

| Number of Exercise | Answers (Or where information is given in the *Handbook*) |
|---|---|
| 1 | See *Handbook* pages **93–97** |
| 2 | In any right-angle triangle having an acute angle of 30 degrees, the side opposite that angle equals 0.5 × hypotenuse. |
| 3 | Sine = 0.31634; tangent = 0.51549; cosine = 0.83942 |
| 4 | Angles equivalent to tangents are 27°29′24″ and 7°25′16″; angles equivalent to cosines are 86°5′8″ and 48°26′52″ |
| 5 | Rule 1: Side opposite = hypotenuse × sine; Rule 2: Side opposite = side adjacent × tangent |
| 6 | Rule 1: Side adjacent = hypotenuse × cosine; Rule 2: Side adjacent = side opposite × cotangent |
| 7 | *Handbook* page **89** |
| 8 | *Handbook* page **91** |

**Answers to** *Practice Exercises for Section 8 (Continued)*

| Number of Exercise | Answers (Or where information is given in the *Handbook*) |
|---|---|
| 9 | Yes. This can be done after dividing the isosceles triangle into two right-angle triangles. |

**Answers to** *Practice Exercises for Section 9 (see* **page 81***)*

| Number of Exercise | Answers (Or where information is given in the *Handbook*) |
|---|---|
| 1 | 2 degrees, 58 minutes |
| 2 | 1 degree, 47 minutes |
| 3 | 2.296 inches, as shown by the table on *Handbook* page **720** |
| 4 | $360°/N - 2a$ = angle intercepted by width $W$. The sine of $\frac{1}{2}$ this angle; $\frac{1}{2}B = \frac{1}{2}W$ hence, this sine $\times B = W$ |
| 5 | 3.1247 inches |
| 6 | 3.5085 inches |
| 7 | 1.7677 inches |
| 8 | 75 feet approximately |
| 9 | $a = 1.0316$ inches; $b = 3.5540$ inches; $c = 2.2845$ inches; $d = 2.7225$ inches |
| 10 | $a = 18°22'$. For solution of similar problem, see *Guide*, **Example 4** of Section 8, page **60** |
| 11 | $A = 5.8758''$; $B = 6.0352''$; $C = 6.2851''$; $D = 6.4378''$; $E = 6.1549''$; $F = 5.8127''$. apply formula on *Handbook* page **95** |
| 12 | 2°37′33″; 5°15′6″ |
| 13 | 5.2805 inches |
| 14 | 10 degrees, 23 minutes |

**Answers to** *Practice Exercises for Section 10 (see* **page 92***)*

| Number of Exercise | Answers (Or where information is given in the *Handbook*) |
|---|---|
| 1 | 84°; 63°31′; 32°29′ |
| 2 | $B = 29°$; $b = 3.222$ feet; $c = 6.355$ feet; area = 10.013 square feet |
| 3 | $C = 22°$; $b = 2.33$ inches; $c = 1.358$ inches; area = 1.396 square inches |
| 4 | $A = 120°10′$; $a = 0.445$ foot; $c = 0.211$ foot; area = 0.027 square feet |
| 5 | The area of a triangle equals one-half the product of two of its sides multiplied by the sine of the angle between them. The area of a triangle may also be found by taking one-half of the product of the base and the altitude. |

**Answers to** *Practice Exercises for Section 11 (see* **page 98***)*

| Number of Exercise | Answers (Or where information is given in the *Handbook*) |
|---|---|
| 1 | *Handbook* page **1001** for Morse<br>*Handbook* page **1012** for Jarno<br>*Handbook* page **1012** for milling machine<br>*Handbook* page **1893** for taper pins |
| 2 | 2.205 inches; 12.694 inches |
| 3 | 4.815 inches. *Handbook* page **701**<br>$C = \frac{D-d}{2} \times \frac{\sqrt{1+\frac{T^2}{24}}}{\frac{T}{24}}$<br>$= \frac{2-1.75}{2} \times \frac{\sqrt{1+\left(\frac{0.62326}{24}\right)^2}}{\frac{0.62326}{24}}$ |
| 4 | 1.289 inches. *Handbook* page **702** |

**Answers to** *Practice Exercises for Section 11 (Continued)*

| Number of Exercise | Answers (Or where information is given in the *Handbook*) |
|---|---|
| 5 | 3.110 inches. *Handbook* page **695** |
| 6 | 0.0187 inch |
| 7 | 0.2796 inch |
| 8 | 1.000 inch |
| 9 | 26 degrees, 7 minutes |

**Answers to** *Practice Exercises for Section 12 (see* **page 112***)*

| Number of Exercise | Answers (Or where information is given in the *Handbook*) |
|---|---|
| 1 | *Handbook* pages **643**, **645** |
| 2 | *Handbook* page **642** |
| 3 | *Handbook* page **642** |
| 4 | *Handbook* page **641** |
| 5 | *Handbook* page **674** |
| 6 | *Handbook* page **1950** |
| 7 | *Handbook* pages **1943**, **1954** |
| 8 | When the tolerance is unilateral |
| 9 | It means that a tolerance of 0.0004 to 0.0012 inch could normally be worked to. See table on *Handbook* page **649** |
| 10 | Yes. See *Handbook* page **802** |

**Answers to** *Practice Exercises for Section 13* (*see* **page 118**)

| Number of Exercise | Answers (Or where information is given in the *Handbook*) |
|---|---|
| 1 | 4000 pounds. *Handbook* page **2720** |
| 2 | *Handbook* page **2725** |
| 3 | 430 balls. *Handbook 32 Digital Edition* page **3714**. To calculate, use density ρ from *Handbook* page **371** $$\frac{\text{balls}}{\text{lb}} = \frac{1}{\rho V} = \frac{6}{\rho \pi d^3}$$ |
| 4 | ¼ inch. *Handbook* page **2557** |
| 5 | 0.172 inch. *Handbook* page **1936** |
| 6 | 0.1251 to 0.1252. *Handbook* page **1885** |
| 7 | 24,000 rpm. *Handbook* page **1904** |
| 8 | 0.128 inch. *Handbook* page **1846** |

**Answers to** *Practice Exercises for Section 14* (*see* **page 125**)

| Number of Exercise | Answers (Or where information is given in the *Handbook*) |
|---|---|
| 1 | Both countries have used the Unified Standard, but Britain is changing to the ISO Metric. See *Handbook* page **1943** and page **2047** |
| 2 | The symbol is used to specify an American Standard screw thread 3 inches in diameter, 4 threads per inch or the coarse series, and Class 2 fit. |
| 3 | An Acme thread is stronger, easier to cut with a die, and more readily engaged by a split nut used with a lead screw. |
| 4 | The Stub Acme form of thread is preferred for those applications where a coarse thread of shallow depth is required. |
| 5 | See tables, *Handbook* pages **1981–1982** |
| 6 | ¾ inch per foot measured on the diameter-American and British Standards |

**Answers to** *Practice Exercises for Section 14 (Continued)*

| Number of Exercise | Answers (Or where information is given in the *Handbook*) |
|---|---|
| 7 | *Handbook* page **2096** |
| 8 | Center line of tool is set square to axis of screw thread. |
| 9 | Present practice is to set center line of tool square to axis of pipe. |
| 10 | See formulas for $F_{rn}$ and $F_{rs}$, *Handbook* page **2068** |
| 11 | By three-wire method or by use of special micrometers. See *Handbook* pages **2125–2146** |
| 12 | Two quantities to be multiplied are the same as if enclosed by parentheses. See instructions about order of operations, *Handbook* page **4** |
| 13 | (a) Lead of double thread equals twice the pitch; (b) lead of triple thread equals three times the pitch. See *Handbook* page **2125** |
| 14 | See *Handbook* page **1952** |
| 15 | 0.8337 inch. See *Handbook* page **2133** |
| 16 | No. Bulk of production is made to American Standard dimensions given in *Handbook*. |
| 17 | This Standard has been superseded by the American Standard. |
| 18 | Most machine screws (about 80% of the production) have the coarse series of pitches. |
| 19 | (a) Length includes head; (b) Length does not include head |
| 20 | No. 25. See table, *Handbook* page **2171** |
| 21 | 0.1935 inch. See table, *Handbook* page **921** |
| 22 | Yes. The diameters decrease as the numbers increase. |
| 23 | The numbered sizes range in diameter from 0.0059 to 0.228 inch, and the letter sizes from 0.234 to 0.413 inch. See *Handbook* pages **919–929** |

**Answers to** *Practice Exercises for Section 14 (Continued)*

| Number of Exercise | Answers (Or where information is given in the *Handbook*) |
|---|---|
| 24 | A thread of $^3/_4$ standard depth has sufficient strength, and tap breakage is reduced. |
| 25 | (a) and (b) the American Standard Unified form |
| 26 | Cap-screws are made in the same pitches as the Coarse-, Fine-, and 8- thread series of the American Standard, class 2A. |
| 27 | For thread form, see *Handbook* page **2104**. There are seven standard diameters as shown on page **2105**. |
| 28 | *Handbook* page **957** |
| 29 | *Handbook* page **957** |
| 30 | 0.90 × pitch. See *Handbook* page **2128** |
| 31 | To reduce errors in the finished thread |
| 32 | Included angle is 82° for each |

**Answers to** *Practice Exercises for Section 15 (see* **page 141***)*

| Number of Exercise | Answers (Or where information is given in the *Handbook*) |
|---|---|
| 1 | A foot-pound in mechanics is a unit of work and is the work equivalent to raising 1 pound 1 foot high. |
| 2 | 1000 foot-pounds |
| 3 | Only as an average value. See *Handbook* page **184** |
| 4 | 28 foot-pounds. See *Handbook* pages **182** and **184** |
| 5 | 1346 pounds |
| 6 | Neglecting air resistance, the muzzle velocity is the same as the velocity with which the projectile strikes the ground. See *Handbook* page **176** |
| 7 | See *Handbook* page **157** |

**Answers to** *Practice Exercises for Section 15 (Continued)*

| Number of Exercise | Answers (Or where information is given in the *Handbook*) |
|---|---|
| 8 | Square |
| 9 | 1843 pounds approximately |
| 10 | The pull will have been increased from 1843 pounds to about 2617 pounds. See *Handbook* page **170** |
| 11 | Yes |
| 12 | About 11 degrees |
| 13 | The angle of repose |
| 14 | The coefficient of friction equals the tangent of the angle of repose. |
| 15 | 32.16 feet per second$^2$ |
| 16 | No. 32.16 feet per second$^2$ is the value at sea level at a latitude of about 40 degrees, but this figure is commonly used. See *Handbook* page **152** |
| 17 | No. The rim stress is independent of the diameter and depends upon the velocity. |
| 18 | 10 to 13. See *Handbook 32 Digital Edition* page **3236** |
| 19 | No. The increase in stress is proportional to the square of the rim velocity. |
| 20 | 110 feet per second or approximately 1.25 miles per minute |
| 21 | Because the strength of wood is greater in proportion to its weight than cast iron. |
| 22 | See *Handbook* page **198** |
| 23 | In radians per second |
| 24 | A radian equals the angle subtended by the arc of circle; this angle is 57.3 degrees nearly. |
| 25 | *Handbook* page **99** |
| 26 | 60 degrees; 72 degrees |
| 27 | *Handbook* page **99** |

**Answers to** *Practice Exercises for Section 15 (Continued)*

| Number of Exercise | Answers (Or where information is given in the *Handbook*) |
|---|---|
| 28 | *Handbook* page **177** (see *Guide* page **141** for example illustrating method of using tables) |
| 29 | Length of arc = radians × radius. As radius = 1 in the table segments, $l$ = radians |
| 30 | 40 degrees, 37 minutes, 42 seconds |
| 31 | 176 radians per second; 1680.7 revolutions per minute |
| 32 | 1.5705 inches |
| 33 | 27.225 inches |

**Answers to** *Practice Exercises for Section 16 (see* **page 157***)*

| Number of Exercise | Answers (Or where information is given in the *Handbook*) |
|---|---|
| 1 | *Handbook* page **205** |
| 2 | 12,000 pounds |
| 3 | 1 inch |
| 4 | *Handbook* page **508** |
| 5 | *Handbook* page **201** |
| 6 | *Handbook* page **508** |
| 7 | 3-inch diameter. See *Handbook* page **280** |
| 8 | Linearly |
| 9 | Plastics under high loads can creep, or slowly deform, due to their viscous nature. At higher temperatures, plastics are likely to become more viscous, and creep is more rapid. See *Handbook* page **568** |
| 10 | Matrix and fiber reinforcements. The fiber reinforcements provide most of the tensile strength. See *Handbook* page **586** |

**Answers to** *Practice Exercises for Section 17 (see* **page 165***)*

| Number of Exercise | Answers (Or where information is given in the *Handbook*) |
|---|---|
| 1 | 1.568 (See formula on *Handbook* page **246**) |
| 2 | 1.571 |
| 3 | 6200 pounds per square inch approximately |

**Answers to** *Practice Exercises for Section 17 (Continued)*

| Number of Exercise | Answers (Or where information is given in the *Handbook*) |
|---|---|
| 4 | It depends upon the class of service. See *Handbook* page **293** |
| 5 | Tangential load = 550 pounds; twisting moment = 4400 inch-pounds |
| 6 | See formulas on *Handbook* page **292** |
| 7 | The head is useful for withdrawing the key, especially when it is not possible to drive against the inner end. See *Handbook* page **2560** |
| 8 | Key is segment-shaped and fits into circular keyseat. See *Handbook* pages **2563–2564** |
| 9 | These keys are inexpensive to make from round bar stock, and keyseats are easily formed by milling. |
| 10 | 0.211 inch. See table, *Handbook* page **2568** |

**Answers to** *Practice Exercises for Section 18 (see* **page 176***)*

| Number of Exercise | Answers (Or where information is given in the *Handbook*) |
|---|---|
| 1 | See text and footnote on *Handbook* page **2353**; see also *Guide* page **117** |
| 2 | American Standard B92.1, *Handbook* pages **2333** and **2339** |
| 3 | See text, *Handbook* page **2333** |
| 4 | See text, *Handbook* page **2333** |
| 5 | See definitions, *Handbook* page **2335** |
| 6 | None. See text, *Handbook* page **2339** |
| 7 | Yes, a crowned spline permits small amount of misalignment. See *Handbook* page **2350** |
| 8 | The torque capacity of splines may be calculated using the formulas and charts on *Handbook* page **2347–2351** |
| 9 | *Handbook* page **2346** |
| 10 | The fillet radius permits heavier loading and effects greater fatigue resistance than flat roots through absence of stress raisers. |

**Answers to** *Practice Exercises for Section 19 (see* **page 201***)*

| Number of Exercise | Answers (Or where information is given in the *Handbook*) |
|---|---|
| 1 | 18 teeth; 3 inches; 0.2618 inch |
| 2 | 2.666 inches; 2.333 inches; 0.166 inches |
| 3 | *Handbook* pages **2216–2217** |
| 4 | Chordal thickness at intersections of pitch circle with sides of tooth |
| 5 | **Table 3**, *Handbook* page **2215** |
| 6 | Calculate using **Table 5**, *Handbook* page **2217** |
| 7 | Surface durability stress and tooth fillet tensile stress are the two principal factors to be found in determining the power-transmitting capacity of spur gears. |
| 8 | Because the tooth shape varies as the number of teeth is changed |
| 9 | No. One hob may be used for all tooth numbers, and the same applies to any generating process. |
| 10 | Stub |
| 11 | *Handbook* (see *Fellows Stub Tooth* on page **2218**) |
| 12 | *Handbook* page **2298** |
| 13 | *Handbook* page **2228** |
| 14 | *Handbook* page **2218** |
| 15 | See table on *Handbook* page **2217** |
| 16 | *Handbook* page **2187** |
| 17 | *Handbook* pages **2244–2249** |
| 18 | *Handbook* page **2299** |
| 19 | Yes, but accurate tooth form is obtained only by a generating process. |
| 20 | See paragraph on *Handbook* page **2268** |
| 21 | *Handbook* page **2262** |
| 22 | When the numbers of teeth in both the pinion and the gear are the same, the pitch angle being 45 degrees for each. |

**Answers to** *Practice Exercises for Section 19 (Continued)*

| Number of Exercise | Answers (Or where information is given in the *Handbook*) |
|---|---|
| 23 | The whole depth minus the clearance between the bottom of a tooth space and the end of a mating tooth = the working depth. |
| 24 | See *Handbook* page **2222** |
| 25 | See *Handbook* pages **2210** and **2212** |
| 26 | See *Handbook* page **2263** |
| 27 | Circular pitch of gear equals linear pitch of worm. |
| 28 | Helix angle or lead angle of worm is measured from a plane perpendicular to the axis; helix angle of a helical gear is measured from the axis. |
| 29 | These terms each have the same meaning. |
| 30 | To provide a grinding allowance and to increase hob life over repeated sharpening |
| 31 | See explanation beginning on *Handbook 32 Digital Edition* page **3677** |
| 32 | *Handbook 32 Digital Edition* page **3677** |
| 33 | *Handbook* page **2277** |
| 34 | Normal diameter pitch is commonly used. |
| 35 | Yes (see *Handbook* page **2277**), but the hobbing process is generally applied. |
| 36 | Pitch diameter |
| 37 | *Handbook* page **2291** |

**Answers to** *Practice Exercises for Section 20* (*see* **page 209**)

| Number of Exercise | Answers (Or where information is given in the *Handbook*) |
|---|---|
| 1 | AISI 1108 CD $\frac{1}{2}$ in. dia. = 1008 rpm<br>12L13, 150–200 BHN : = 1192 rpm<br>1040, Hot Rolled : = 611 rpm<br>1040, 375–425 BHN : = 214 rpm<br>41L40, 200–250 BHN : = 718 rpm<br>4140, Hot Rolled : = 611 rpm<br>O2, Tool Steel : = 535 rpm<br>M2, Tool Steel : = 497 rpm<br>AISI 1108 CD 4 in. dia. = 126 rpm<br>12L13, 150–200 BHN : = 149 rpm<br>1040, Hot Rolled : = 576 rpm<br>1040, 375–425 BHN : = 27 rpm<br>41L40, 200–250 BHN : = 90 rpm<br>4140, Hot Rolled : = 76 rpm<br>O2, Tool Steel : = 67 rpm<br>M2, Tool Steel : = 62 rpm |
| 2 | AISI 1330, 200 BHN : 153 rpm<br>201 Stainless Steel, CD : 345 rpm<br>ASTM Class 50 Gray Cast Iron : 145 rpm<br>6A1-4V Titanium Alloy : 52 rpm<br>Waspaloy : 20 rpm<br>(V = 60 fpm) |
| 3 | $1\frac{1}{2}$ in. dia.: 157 fpm—OK<br>3 in. dia. : 314 fpm—OK<br>4 in. dia. : 419 fpm—Too Fast |
| 4 | 764 rpm |
| 5 | 840 rpm ($V$ = 110 fpm) |
| 6 | Operation:<br>1: $N$ = 167 rpm; $f_m$ = 13 in./min.<br>2: $N$ = 127 rpm; $f_m$ = 2.0 in./min.<br>3: $N$ = 744 rpm<br>4: $N$ = 458 rpm |

**Answers to** *Practice Exercises for Section 20 (Continued)*

| Number of Exercise | Answers (Or where information is given in the *Handbook*) |
|---|---|
| 7 | Existing operation:<br>$V = 59$ fpm (too fast)<br>$f_t = 0.011$ in/tooth (too severe)<br>Change to:<br>$V = 40$ fpm $N = 50$ rpm<br>$f_t = 0.006$ in/tooth; $f_m = 3.6$ in/min |
| 8 | Existing operation:<br>V = 188 fpm (too slow)<br>$f_t = 0.006$ in/tooth (too slow)<br>Change to:<br>$V = 375$ fpm $N = 120$ rpm<br>$f_t = 0.012$ in/tooth; $f_m = 520$ in/min |
| 9 | $V = 414$ fpm, $P_c = 9.0$ hp, $P_m = 11.24$ hp |
| 10 | $V = 104$ fpm |
| 11 | $V = 205$ fpm<br>($Q_{max} = 8.55$ in$^3$/min.;<br>$f_m = 10.5$ in/min;<br>$N = 131$ rpm) |
| 12 | ¼ in: $T = 123$ lb; $M = 6.38$ in-lb;<br>$P_m = 0.19$ hp<br>½ in: $T = 574$ lb; $M = 68$ in-lb;<br>$P_m = 1.0$ hp<br>1 in: $T = 2712$ lb; $M = 711$ in-lb;<br>$P_m = 5.3$ hp<br>19 mm: $T = 7244$ N; $M = 37.12$ N-m;<br>$P_m = 2.43$ kw |
| 13 | $T = 1473$ lb; $M = 655$ in-lb; $P_m = 4.9$ hp |
| 14 | *Handbook* page **1065** |
| 15 | *Handbook* page **1066** |
| 16 | *Handbook* page **1070** |
| 17 | *Handbook* page **1070** |
| 18 | *Handbook* pages **851** and **1067** |
| 19 | *Handbook* pages **1128** and **1130** |
| 20 | *Handbook* pages **1096** and **1099** |
| 21 | *Handbook* page **1096** |

**Answers to** *Practice Exercises for Section 20 (Continued)*

| Number of Exercise | Answers (Or where information is given in the *Handbook*) |
|---|---|
| 22 | *Handbook* page **1133** |
| 23 | *Handbook* pages **1133–1134** |
| 24 | *Handbook* page **1139** |
| 25 | *Handbook* page **1145** |
| 26 | *Handbook* page **1177** |

**Answers to** *Practice Exercises for Section 21 (see* **page 219***)*

| Number of Exercise | Answers (Or where information is given in the *Handbook*) |
|---|---|
| 1 | Manual programming (also known as G-code programming), computer-aided manufacturing (also known as CAM programming), shop-floor (also known as conversational programming). See *Handbook* pages **1341–1342** |
| 2 | All machine activities: tool selection, motion control of axes, spindle speed, cutting feed rates, coolant flow, and more. See *Handbook* page **1338** |
| 3 | The address "G" identifies a preparatory command. See *Handbook* page **1344** |
| 4 | (d) milling and turning. See *Handbook* pages **1338–1387** |
| 5 | (b) Inches per revolution in/rev or mm/rev. See *Handbook* page **1348** |
| 6 | (b) F. See *Handbook* pages **1348** and **1350** |
| 7 | (c) I, J, and K. See *Handbook* pages **1350–1351** |
| 8 | (b) G43. See *Handbook* pages **1352–1353** |
| 9 | (d) R. See *Handbook* pages **1350–1351** |
| 10 | Sequence number. See *Handbook* page **1344** |
| 11 | True. See *Handbook* pages **1347–1348** |
| 12 | Cutter radius offset is an adjustment parallel to one of the axes. Cutter compensation is an adjustment that is normal to the part, whether or not the adjustment is parallel to an axis. See *Handbook* pages **1353–1354** |

**Answers to** *Practice Exercises for Section 21 (Continued)*

| Number of Exercise | Answers (Or where information is given in the *Handbook*) |
|---|---|
| 13 | (a) G40. See *Handbook* page **1353** |
| 14 | G98 = Initial level. G99 = R-level (return level). See *Handbook* pages **1354–1359** |
| 15 | a. Spot-drilling cycle — 5. G82<br>b. Peck-drilling cycle (deep hole) — 6. G83<br>c. Boring cycle, spindle rotating on withdrawal at feed rate — 3. G85<br>d. Drilling cycle — 2. G81<br>e. Right-hand threading cycle — 4. G84<br>f. Fixed cycle cancellation — 1. G80<br>See *Handbook* pages **1354–1359** |
| 16 | False. See *Handbook* page **1365** |
| 17 | A G word is a preparatory code consisting of the address G and two digits used to tell the control system to accept the remainder of the block in the required way. See *Handbook* page **1344** |
| 18 | The "right-hand rule" says that if a right-hand is laid palm up on the table of a vertical milling machine, the thumb will point in the positive *X* direction, the forefinger in the positive *Y* direction, and the erect middle finger in the positive *Z* direction. See *Handbook* pages **1374–1375** |
| 19 | True. See *Handbook* page **1383** |
| 20 | The machine type (mill, lathe, etc.), stock/raw material dimensions, part material, specific data related to the cutting tool, and work holding considerations. See *Handbook* page **1384** |

**Answers to** *General Review Questions (see* **page 227***)*

| Number of Exercise | Answers (Or where information is given in the *Handbook*) |
|---|---|
| 1 | *Handbook* page **77** gives the formula for length of side *S* in terms of the given area *A*. |
| 2 | The diameter of each end and the length of the taper; see explanation on *Handbook* page **1002**, also the table on page **1012** |
| 3 | Tolerance is applied in whatever direction is likely to be the least harmful; see *Handbook* page **643** |
| 4 | It is said that James Watt found, by experiment, that an average carthorse can develop 22,000 foot-pounds per minute, and added 50 percent to ensure good measure to purchasers of his engines ($22{,}000 \times 1.50 = 33{,}000$). |
| 5 | Tin in the high grades, and lead in the lower grades |
| 6 | Same depth as ordinary gear of 10 diametral pitch |
| 7 | The tooth thickness and the number of teeth are the same as an ordinary gear of 8 diametral pitch. |
| 8 | Add 2 to the number of teeth and divide by the outside diameter. |
| 9 | Multiply the outside diameter by 3.1416 and divide the product by the number of teeth plus 2. |
| 10 | Birmingham or Stub's iron wire gage is used for seamless steel, brass, copper, and aluminium tubing. |
| 11 | Iron wire rope has the least strength of all wire rope materials |
| 12 | If surfaces are well-lubricated, the friction is almost independent of the pressure, but if the surfaces are unlubricated, the friction is directly proportional to the normal pressure except for the higher pressures. |
| 13 | It depends very largely upon temperature. See *Handbook* section, *Lubricated Surfaces* on page **166** |

**Answers to** *General Review Questions (Continued)*

| Number of Exercise | Answers (Or where information is given in the *Handbook*) |
|---|---|
| 14 | 8 – (–4) = 12. See rules for positive and negative numbers, *Handbook* page **3** |
| 15 | Yes. One meter equals 3.2808 feet; see other equivalents on *Handbook* page **2845** |
| 16 | Experiments have shown that there is a definite relationship between heat and work and that 1 British thermal unit equals 778 foot-pounds. To change 1 pound of water at 212° F into steam at that temperature requires about 966 British thermal units, or 966 × 788 = about 751,600 foot-pounds; hence, the number of pounds of water evaporated 212° F, equivalent to 1 horsepower-hour = 1,980,000 ÷ 751,600 = 2.64 pounds of water as given in *Handbook*, page **2874** |
| 17 | No. The thickness of the pipe is increased by reducing the inside diameter; compare thickness in the table on *Handbook 32 Digital Edition* page **3746** |
| 18 | As a general rule, smoother finishes are required for harder materials, for high loads, and for high speeds. See *Handbook* page **2402** |
| 19 | Yes. The so-called "O.D. pipe" begins, usually, with the $\frac{1}{4}$-inch size. See *Handbook 32 Digital Edition* page **3751** |
| 20 | It is light in weight and resists deterioration from corrosive or caustic fluids. See *Handbook 32 Digital Edition* page **3751** |
| 21 | Yes. About 140 degrees lower. See *Handbook* page **371** |
| 22 | Low-carbon alloy steels with high chromium content. See *Handbook* page **392** |
| 23 | Low-carbon steels containing 0.20% sulfur or less and usually from 0.90 to 1.20% manganese. See *Handbook* page **407** |

**Answers to** *General Review Questions (Continued)*

| Number of Exercise | Answers (Or where information is given in the *Handbook*) |
|---|---|
| 24 | No. The nominal length of a file indicates the distance from the point to the "heel" and does not include the tang. |
| 25 | Yes. See table, *Handbook* page **375** |
| 26 | Specific heat is a ratio of the amount of heat required to raise the temperature of a certain weight of substance 1° F to the amount of heat required to raise the temperature of an equivalent of water 1° F. See *Handbook* page **367** |
| 27 | (a) and (b) A number indicating how a given volume of the material or liquid compares in weight with an equal volume of water. (c) A number indicating a comparison in weight with an equal volume of air. See *Handbook* pages **375–376** |
| 28 | The first digit identifies the alloy type; the second, the impurity control; etc. See *Handbook* page **529** |
| 29 | Red brass contains 84 to 86% copper, about 5% tin, 5% lead, and 5% zinc, whereas yellow brass contains 62 to 67% copper, about 30% zinc, 1.5 to 3.5% lead, and not even 1% tin. See UNS Designations on *Handbook* pages **510**, **525** |
| 30 | See *Handbook 32 Digital Edition* pages **3261** and **3264** |
| 31 | No. Twenty-six sizes ranging from 0.234 to 0.413 inch are indicated by capital letters of the alphabet (see table, *Handbook* page **921–928**). Fractional sizes are also listed in manufacturers' catalogues beginning either at $\frac{1}{32}$ inch, $\frac{1}{16}$ inch, or $\frac{1}{8}$ inch, the smallest size varying with different firms. |
| 32 | To ensure uniform heating at a given temperature and protect the steel against oxidation. See *Handbook* page **469** |

**Answers to** *General Review Questions (Continued)*

| Number of Exercise | Answers (Or where information is given in the *Handbook*) |
|---|---|
| 33 | Hardening temperatures vary for different steels; see critical temperatures and how they are determined, *Handbook* pages **468–469** |
| 34 | Set the taper attachment to an angle the cosine of which equals 0.125 ÷ 0.1255. See *Handbook* page **2201** |
| 35 | See *Handbook* page **691** |
| 36 | Divide $\frac{3}{4}$ by 12; multiply the taper per inch found by 5 and subtract the result from the large diameter. See rules for figuring tapers, *Handbook* page **697** |
| 37 | Yes. See "Useful Relationships Among Angles," *Handbook* page **102** |
| 38 | 0.8833. See *Handbook* page **99** |
| 39 | $x = 3$ |
| 40 | About $12\frac{1}{2}$ degrees. See *Handbook* page **2533** |
| 41 | Ratio between resistance to the motion of a body due to friction, and the perpendicular pressure between the sliding and fixed surfaces. See formula, *Handbook* page **166** |
| 42 | No. Stub's steel wire gage applies to tool steel rod and wire, and the most important applications of Stub's iron wire gage (also known as Birmingham) are to seamless tubing, steel strips, and telephone and telegraph wire. |
| 43 | If the difference between the length of the pawls equals one-half of the pitch of the ratchet wheel teeth, the practical effect is that of reducing the pitch of one-half. See *Ratchet Gearing* starting on *Handbook* page **2296** |
| 44 | The high modulus of elasticity eliminates the need for periodic retensioning that is normally required with V-belts. See *Handbook* page **299** |

**Answers to** *General Review Questions (Continued)*

| Number of Exercise | Answers (Or where information is given in the *Handbook*) |
|---|---|
| 45 | Increasing centrifugal force has less effect on flat belts because of the low center of gravity. See *Handbook Flat Belting* starting on page **2572** |
| 46 | The ultimate strength is less due to bending action. See *Handbook 32 Digital Edition* page **3793**, and also see table, *Close-Link Hoisting, Sling and Crane Chain* on page **3797** |
| 47 | Refer to *Handbook 32 Digital Edition* page **3797** |
| 48 | Multiply 90 by 12 and divide the circumference of the shaft to obtain rpm. See cutting speed calculations, *Handbook* pages **1072–1074** |
| 49 | (a) Lard oil; (b) gasoline |
| 50 | If the teeth advance around the gear to the right, as viewed from one end, the gear is right-handed; and, if they advance to the left, it is a left-hand gear. See illustrations, *Handbook* page **2276** |
| 51 | No. They may be opposite hand depending upon the helix angle. See *Handbook* pages **2276–2277** |
| 52 | Multiply the total length by the weight per foot for plain end and coupled pipe, given in the table on *Handbook* page **2773** |
| 53 | The processes are similar but the term *pack-hardening* usually is applied to the casehardening of tool steel. See *Handbook* pages **479–480** |
| 54 | A gas process of surface hardening. See *Handbook* page **480** |
| 55 | See definitions for these terms given on *Handbook* page **1039** |

**Answers to** *General Review Questions (Continued)*

| Number of Exercise | Answers (Or where information is given in the *Handbook*) |
|---|---|
| 56 | About 34 inches, but the height may vary from 32 to 36 inches for heavy and light assembling, respectively. |
| 57 | Major diameter is the same as outside diameter, and the minor diameter is the same as root diameter. See definitions, on *Handbook* page **1947** |
| 58 | The SAE Standards conform, in general, with the Unified and American Standard Screw Thread Series as revised in 1959 and may, therefore, be considered to be virtually the same. |
| 59 | See information on work materials, *Handbook* page **1065** |
| 60 | Yes. See *Handbook* page **467** and page **505** |
| 61 | 13.097 millimeters. See the table on *Handbook* page **2848**, which give millimeter equivalents of inch fractions and decimals. |
| 62 | The sevolute of an angle is obtained by subtracting the involute of the angle from the secant of that angle. See *Handbook* page **106**. The involute functions of angles are found in the tables beginning on *Handbook* page **107** |
| 63 | In **Table 1** starting on *Handbook* page **1083**, two feed-speed pairs are given, *Opt*. 28 in/tooth, 685 ft/min, and *Avg*., 13 in/tooth, 960 ft/min. These feed-speed pairs represent values for optimum and average conditions respectively and are intended as a starting point and generally require adjustment for actual cutting conditions. It is important to understand the factors that affect the cutting speed and feed, as covered in the footnote to **Table 1**, and *How to Use the Tables* which begins on *Handbook* page **1078** |

**Answers to** *General Review Questions (Continued)*

| Number of Exercise | Answers (Or where information is given in the *Handbook*) |
|---|---|
| 64 | No. First determine the diametral pitch the same as for a spur gear; then divide this "real diametral pitch" by the cosine of "real diametral pitch" by the cosine of the helix angle to obtain the "normal diametral pitch," which is the pitch of the cutter. See *Handbook* page **2277** |
| 65 | Casehardening steels can have hard, fine-grained surfaces and a soft, ductile core giving good strength combined with wear resistance. See *Handbook* page **2321** |
| 66 | Not in every instance. See *Handbook* page **2353** |
| 67 | A cemented carbide seat provides a flat bearing surface and a positive-, negative-, or neutral-rake orientation to the tool insert. See *Handbook* page **829** |
| 68 | No. The size of the gear blank, the pitch of the teeth, and depth of cut are sufficient for the operator in the shop. The tooth curvature is the result of the gear-cutting process. Tooth curves on the working drawing are of no practical value |
| 69 | By changing the inclination of the dividing head spindle. See *Handbook* page **2389** |
| 70 | See formula and example on *Handbook* page **2389** |
| 71 | Divide the total number of teeth in both gears by twice the diametral pitch to obtain the theoretical center-to-center distance. See formula in the table of *Formulas for Dimensions of Standard Spur Gears*, *Handbook* page **2212** |
| 72 | Subtract number of teeth on pinion from number of teeth on gear and divide the remainder by two times the diametral pitch. See Rule at bottom of *Handbook* page **2252** |
| 73 | See *Handbook* page **1865** |

**Answers to** *General Review Questions (Continued)*

| Number of Exercise | Answers (Or where information is given in the *Handbook*) |
|---|---|
| 74 | The Standard Wire Gage (S.W.G.), also known as the Imperial Wire Gage and as the English Legal Standard, is used in Britain for all wires. |
| 75 | A simple type of apparatus for measuring power |
| 76 | With a dynamometer, the actual amount of power delivered may be determined; that is, the power input minus losses. See *Handbook* page **2542** |
| 77 | The uniformly loaded beam has double the load capacity of a beam loaded at the center only. See formulas, *Handbook* page **251** |
| 78 | Refer to *Handbook* page **467** for graph of SAE-determined relationships. |
| 79 | No. The nominal size of steel pipe, except for sizes above 12 inches, is approximately equal to the inside diameter. See tables, *Handbook 32 Digital Edition* pages **3746** and **3748** |
| 80 | 0.357 inch. See formula, *Handbook* page **221** |
| 81 | The laws of sines and cosines are stated on *Handbook* page **91** |
| 82 | Both the sine and cosine of 45 degrees are 0.70711. |
| 83 | Multiply depth in feet by 0.4335 |
| 84 | No. See *Handbook* page **2492** |
| 85 | 100% |
| 86 | 50% |
| 87 | Various steels are used, depending on kind of spring. See *Handbook* page **403** |
| 88 | Normalizing is a special annealing process.The steel is heated above the critical range and allowed to cool in still air at ordinary temperature, *Handbook* page **479**. Normalizing temperatures for steels are given on *Handbook* pages **485–486** |

**Answers to** *General Review Questions (Continued)*

| Number of Exercise | Answers (Or where information is given in the *Handbook*) |
|---|---|
| 89 | The National Electrical Manufacturers Association (NEMA) Standards for mounting dimensions, frame sizes, horsepower, and speed ratings. See section beginning on *Handbook* page **2650** |
| 90 | Yes. The American Standard drafting room practice includes section lining, etc. See *Handbook* page **623** |
| 91 | No. There are different tapers per foot, ranging from 0.5986 to 0.6315 inch. See table on *Handbook* page **1001** |
| 92 | Yes. See *Handbook* page **2178** |
| 93 | Unilateral and plus. See *Handbook* page **642** |
| 94 | See table, *Handbook* page **2064** |
| 95 | If $D$ = diameter of hole in inches; $T$ = stock thickness in inches; shearing strength of steel = 51,000 pounds per square inch, then tonnage for punching = $51{,}000D\pi T/2000 = 80DT$ |
| 96 | See *Handbook* pages **2712–2713** |
| 97 | The Brown & Sharpe or American wire gage is used for each. See *Handbook* pages **2708–2713** |
| 98 | No, this name is applied to several compositions that vary widely. |
| 99 | Antimony and copper |
| 100 | 177 nearly. See table on *Handbook* page **1228** |
| 101 | See *Handbook* pages **362**, **371**, **375** |
| 102 | See *Handbook* page **1238** |
| 103 | See *Handbook* page **824** |
| 104 | See *Handbook 32 Digital Edition* page **3503** |
| 105 | See *Handbook* page **2201** |

**Answers to** *General Review Questions (Continued)*

| Number of Exercise | Answers (Or where information is given in the *Handbook*) |
|---|---|
| 106 | See *Handbook* page **897** |
| 107 | Steel, chromium-plated steel, chromium carbide, tungsten carbide, and other materials. See *Handbook* page **724** |
| 108 | See text on *Handbook* page **724** |
| 109 | The lead of a milling machine equals lead of helix or spiral milled when gears of equal size are placed on feed screw and wormgear stud; see rule for finding lead on *Handbook 32 Digital Edition* page **3567** |
| 110 | Multiply product of driven gears by lead of machine and divide by product of driving gears. If lead of machine is 10, divide 10 times product of driven gears by product of drivers. |
| 111 | $^5/_{11}$; $^{79}/_{183}$; $^{19}/_{29}$ |
| 112 | The whole depth and tooth thickness at the large ends of the bevel-gear teeth are the same as the whole depth and thickness of spur-gear teeth of the same pitch. |
| 113 | See Text on *Handbook* page **1040** |
| 114 | 5.7075 inches |
| 115 | Use the formula (found on *Handbook 32 Digital Edition* page **3793**) for finding the breaking load, which in this case is taken as three times the actual load. Transposing, $D = \sqrt{\dfrac{6 \times 2000 \times 3}{54,000}} = 0.816$, say, $^7/_8$ inch diameter |
| 116 | Because the direction of the cutter thrust tends to cause the gear to rotate upon the arbor. See *Handbook Milling the Helical Teeth* on page **2286** |

**Answers to** *General Review Questions (Continued)*

| Number of Exercise | Answers (Or where information is given in the *Handbook*) |
|---|---|
| 117 | Trepanning describes use of a fly-cutter or circular toothed cutter to cut a groove to the full depth of a plate, producing a hole of the required size. See *Handbook 32 Digital Edition* page **3488** |
| 118 | Chiefly when a hole is to be tapped or reamed after drilling. See *Handbook 32 Digital Edition* page **3389** |
| 119 | See table on *Handbook* page **445** |
| 120 | See *Handbook* page **2645** |
| 121 | See *Handbook* page **96** |
| 122 | See *Handbook* pages **2321–2322** |
| 123 | See *Handbook* page **985** |
| 124 | See table *Handbook* page **1301** |
| 125 | See table *Handbook* page **1285** |
| 126 | Motor rotation has been standardized by the National Electrical Manufacturers Association. See *Handbook* page **2652** |
| 127 | Grease-lubricated bearings are intended for slow-speed, moderate-to-high load applications. The grease is comprised of a mineral oil and a thickening agent, such as a metallic soap. See *Lubrication Selection* starting on *Handbook* page **2404**. |

**Answers to** *General Review Questions (Continued)*

| Number of Exercise | Answers (Or where information is given in the *Handbook*) |
|---|---|
| 128 | To solve this problem, the helix angle $\phi$ of the thread at the pitch and major diameters must be found, which is accomplished by substituting these diameters (*E* and *D*) for the minor diameters (*K*) in the formula for $\phi$. Thus, at the pitch diameter: $\tan\phi = \frac{\text{lead of thread}}{\pi E} = \frac{0.400}{\pi \times 0.900}$ $\phi = 8.052° = 8°3'$ $a = a_e + \phi$ $a_e = a - \phi = 19°16' - 8°3' = 11°13'$ At the major diameter: $\tan\phi = \frac{\text{lead of thread}}{\pi D} = \frac{0.400}{\pi \times 1.000}$ $\phi = 7.256° = 7°15'$ $a_e = a - \phi = 19°16' - 7°15' = 12°1'$ |
| 129 | 0.0037 inch |
| 130 | $^5/_{12}$ foot (5 inches) per minute obtained by dividing 25,000 by 60,000. Note that this speed is considerably less than maximum surface speed at any load to prevent excess heat and wear |
| 131 | Yes. See **Table 14**, *Handbook* page **2463**, and following tables |
| 132 | 550–600 BHN (Brinell Hardness Number) (See *Handbook* page **2402**) |
| 133 | 1 pound. See *Handbook 32 Digital Edition* page **3714** |
| 134 | 23,000 rpm. See *Handbook* page **1904** |
| 135 | See *Handbook* page **2137** |
| 136 | See footnote, **Table 2**, *Handbook 32 Digital Edition* page **3210** |
| 137 | See *Handbook* pages **2862** and **375**, and table on *Handbook* page **169**. Rope tension on 14-foot ramp = 64.3 lb; rope tension on 8-foot ramp= 111.1 lb |

## SECTION 25

## UNITS OF MEASURE AND CONVERSION FACTORS

In the table of conversion factors that follows, the symbols for SI units, multiples, and submultiples are given in parentheses in the right-hand column. The symbol "*a*" following a number indicates that the conversion factor is exact.

**Table 1. Metric Conversion Factors**

| Multiply | By | To Obtain |
|---|---|---|
| | Length | |
| centimeter | 0.03280840 | foot |
| centimeter | 0.3937008 | inch |
| fathom | 1.8288[a] | meter (m) |
| foot | 0.3048[a] | meter (m) |
| foot | 30.48[a] | centimeter (cm) |
| foot | 304.8[a] | millimeter (mm) |
| inch | 0.0254[a] | meter (m) |
| inch | 2.54[a] | centimeter (cm) |
| inch | 25.4[a] | millimeter (mm) |
| kilometer | 0.6213712 | mile [US statute] |
| meter | 39.37008 | inch |
| meter | 0.5468066 | fathom |
| meter | 3.280840 | foot |
| meter | 0.1988388 | rod |
| meter | 1.093613 | yard |
| meter | 0.0006213712 | mile [US statute] |
| microinch | 0.0254[a] | micrometer [micron] (μm) |
| micrometer [micron] | 39.37008 | microinch |
| mile [US statute] | 1609.344[a] | meter (m) |
| mile [US statute] | 1.609344[a] | kilometer (km) |
| millimeter | 0.003280840 | foot |
| millimeter | 0.03937008 | inch |
| rod | 5.0292[a] | meter (m) |
| yard | 0.9144[a] | meter (m) |
| | Area | |
| acre | 4046.856 | meter$^2$ (m$^2$) |
| acre | 0.4046856 | hectare |
| centimeter$^2$ | 0.1550003 | inch$^2$ |
| centimeter$^2$ | 0.001076391 | foot$^2$ |
| foot$^2$ | 0.09290304[a] | meter$^2$ (m$^2$) |
| foot$^2$ | 929.0304[a] | centimeter$^2$ (cm$^2$) |
| foot$^2$ | 92,903.04[a] | millimeter$^2$ (mm$^2$) |

## Table 1. Metric Conversion Factors *(Continued)*

| Multiply | By | To Obtain |
|---|---|---|
| hectare | 2.471054 | acre |
| inch$^2$ | 645.16[a] | millimeter$^2$ (mm$^2$) |
| inch$^2$ | 6.4516[a] | centimeter$^2$ (cm$^2$) |
| inch$^2$ | 0.00064516[a] | meter$^2$ (m$^2$) |
| meter$^2$ | 1550.003 | inch$^2$ |
| meter$^2$ | 10.763910 | foot$^2$ |
| meter$^2$ | 1.195990 | yard$^2$ |
| meter$^2$ | 0.0002471054 | acre |
| mile$^2$ | 2.5900 | kilometer$^2$ |
| millimeter$^2$ | 0.00001076391 | foot$^2$ |
| millimeter$^2$ | 0.001550003 | inch$^2$ |
| yard$^2$ | 0.8361274 | meter$^2$ (m$^2$) |
| | Volume (including Capacity) | |
| centimeter$^3$ | 0.06102376 | inch$^3$ |
| foot$^3$ | 28.31685 | liter |
| foot$^3$ | 28.31685 | liter |
| gallon [UK liquid] | 0.004546092 | meter$^3$ (m$^3$) |
| gallon [UK liquid] | 4.546092 | liter |
| gallon [US liquid] | 0.003785412 | meter$^3$ (m$^3$) |
| gallon [US liquid] | 3.785412 | liter |
| inch$^3$ | 16,387.06 | millimeter$^3$ (mm$^3$) |
| inch$^3$ | 16.38706 | centimeter$^3$ (cm$^3$) |
| inch$^3$ | 0.00001638706 | meter$^3$ (m$^3$) |
| liter | 0.001[a] | meter$^3$ (m$^3$) |
| liter | 0.2199692 | gallon [UK liquid] |
| liter | 0.2641720 | gallon [US liquid] |
| liter | 0.03531466 | foot$^3$ |
| meter$^3$ | 219.9692 | gallon [UK liquid] |
| meter$^3$ | 264.1720 | gallon [US liquid] |
| meter$^3$ | 35.31466 | foot$^3$ |
| meter$^3$ | 1.307951 | yard$^3$ |
| meter$^3$ | 1000.[a] | liter |
| meter$^3$ | 61,023.76 | inch$^3$ |
| millimeter$^3$ | 0.00006102376 | inch$^3$ |
| quart [US liquid] | 0.946 | liter |
| quart [UK liquid] | 1.136 | liter |
| yard$^3$ | 0.7645549 | meter$^3$ (m$^3$) |
| | Velocity, Acceleration, and Flow | |
| centimeter/second | 1.968504 | foot/minute |
| centimeter/second | 0.03280840 | foot/second |
| centimeter/minute | 0.3937008 | inch/minute |
| foot/hour | 0.00008466667 | meter/second (m/s) |
| foot/hour | 0.00508[a] | meter/minute |
| foot/hour | 0.3048[a] | meter/hour |
| foot/minute | 0.508[a] | centimeter/second |
| foot/minute | 18.288[a] | meter/hour |
| foot/minute | 0.3048[a] | meter/minute |
| foot/minute | 0.00508[a] | meter/second (m/s) |

## Table 1. Metric Conversion Factors *(Continued)*

| Multiply | By | To Obtain |
|---|---|---|
| foot/second | 30.48[a] | centimeter/second |
| foot/second | 18.288[a] | meter/minute |
| foot/second | 0.3048[a] | meter/second (m/s) |
| foot/second$^2$ | 0.3048[a] | meter/second$^2$ (m/s$^2$) |
| foot$^3$/minute | 28.31685 | liter/minute |
| foot$^3$/minute | 0.0004719474 | meter$^3$/second (m$^3$/s) |
| gallon [US liquid]/min. | 0.003785412 | meter$^3$/minute |
| gallon [US liquid]/min. | 0.00006309020 | meter$^3$/second (m$^3$/s) |
| gallon [US liquid]/min. | 0.06309020 | liter/second |
| gallon [US liquid]/min. | 3.785412 | liter/minute |
| gallon [UK liquid]/min. | 0.004546092 | meter$^3$/minute |
| gallon [UK liquid]/min. | 0.00007576820 | meter$^3$/second (m$^3$/s) |
| inch/minute | 25.4[a] | millimeter/minute |
| inch/minute | 2.54[a] | centimeter/minute |
| inch/minute | 0.0254[a] | meter/minute |
| inch/second$^2$ | 0.0254[a] | meter/second$^2$ (m/s$^2$) |
| kilometer/hour | 0.6213712 | mile/hour [US statute] |
| liter/minute | 0.03531466 | foot$^3$/minute |
| liter/minute | 0.2641720 | gallon [US liquid]/minute |
| liter/second | 15.85032 | gallon [US liquid]/minute |
| mile/hour | 1.609344[a] | kilometer/hour |
| millimeter/minute | 0.03937008 | inch/minute |
| meter/second | 11,811.02 | foot/hour |
| meter/second | 196.8504 | foot/minute |
| meter/second | 3.280840 | foot/second |
| meter/second$^2$ | 3.280840 | foot/second$^2$ |
| meter/second$^2$ | 39.37008 | inch/second$^2$ |
| meter/minute | 3.280840 | foot/minute |
| meter/minute | 0.05468067 | foot/second |
| meter/minute | 39.37008 | inch/minute |
| meter/hour | 3.280840 | foot/hour |
| meter/hour | 0.05468067 | foot/minute |
| meter$^3$/second | 2118.880 | foot$^3$/minute |
| meter$^3$/second | 13,198.15 | gallon [UK liquid]/minute |
| meter$^3$/second | 15,850.32 | gallon [US liquid]/minute |
| meter$^3$/minute | 219.9692 | gallon [UK liquid]/minute |
| meter$^3$/minute | 264.1720 | gallon [US liquid]/minute |
| | Mass and Density | |
| grain [$^1/_{7000}$ 1b avoirdupois] | 0.06479891 | gram (g) |
| gram | 15.43236 | grain |
| gram | 0.001[a] | kilogram (kg) |
| gram | 0.03527397 | ounce [avoirdupois] |
| gram | 0.03215074 | ounce [Troy] |
| gram/centimeter$^3$ | 0.03612730 | pound/inch$^3$ |
| hundredweight [long] | 50.80235 | kilogram (kg) |
| hundredweight [short] | 45.35924 | kilogram (kg) |

## Table 1. Metric Conversion Factors *(Continued)*

| Multiply | By | To Obtain |
|---|---|---|
| kilogram | 1000.[a] | gram (g) |
| kilogram | 35.27397 | ounce [avoirdupois] |
| kilogram | 32.15074 | ounce [Troy] |
| kilogram | 2.204622 | pound [avoirdupois] |
| kilogram | 0.06852178 | slug |
| kilogram | 0.0009842064 | ton [long] |
| kilogram | 0.001102311 | ton [short] |
| kilogram | 0.001[a] | ton [metric] |
| kilogram | 0.001[a] | tonne |
| kilogram | 0.01968413 | hundredweight [long] |
| kilogram | 0.02204622 | hundredweight [short] |
| kilogram/meter$^3$ | 0.06242797 | pound/foot$^3$ |
| kilogram/meter$^3$ | 0.01002242 | pound/gallon [UK liquid] |
| kilogram/meter$^3$ | 0.008345406 | pound/gallon [US liquid] |
| ounce [avoirdupois] | 28.34952 | gram (g) |
| ounce [avoirdupois] | 0.02834952 | kilogram (kg) |
| ounce [Troy] | 31.10348 | gram (g) |
| ounce [Troy] | 0.03110348 | kilogram (kg) |
| pound [avoirdupois] | 0.4535924 | kilogram (kg) |
| pound/foot$^3$ | 16.01846 | kilogram/meter$^3$ (kg/m$^3$) |
| pound/inch$^3$ | 27.67990 | gram/centimeter$^3$ (g/cm$^3$) |
| pound/gal [US liquid] | 119.8264 | kilogram/meter$^3$ (kg/m$^3$) |
| pound/gal [UK liquid] | 99.77633 | kilogram/meter$^3$ (kg/m$^3$) |
| slug | 14.59390 | kilogram (kg) |
| ton [long 2240 lb] | 1016.047 | kilogram (kg) |
| ton [short 2000 lb] | 907.1847 | kilogram (kg) |
| ton [metric] | 1000.[a] | kilogram (kg) |
| ton [metric] | 0.9842 | ton [long 2240 lb] |
| ton [metric] | 1.1023 | ton [short 2000 lb] |
| tonne | 1000.[a] | kilogram (kg) |
| | Force and Force/Length | |
| dyne | 0.00001[a] | newton (N) |
| kilogram-force | 9.806650[a] | newton (N) |
| kilopound | 9.806650[a] | newton (N) |
| newton | 0.1019716 | kilogram-force |
| newton | 0.1019716 | kilopound |
| newton | 0.2248089 | pound-force |
| newton | 100,000.[a] | dyne |
| newton | 7.23301 | poundal |
| newton | 3.596942 | ounce-force |
| newton/meter | 0.005710148 | pound/inch |
| newton/meter | 0.06852178 | pound/foot |
| ounce-force | 0.2780139 | newton (N) |
| pound-force | 4.448222 | newton (N) |
| poundal | 0.1382550 | newton (N) |
| pound/inch | 175.1268 | newton/meter (N/m) |
| pound/foot | 14.59390 | newton/meter (N/m) |

## Table 1. Metric Conversion Factors *(Continued)*

| Multiply | By | To Obtain |
|---|---|---|
| **Bending Moment or Torque** | | |
| dyne-centimeter | 0.0000001[a] | newton-meter (N · m) |
| kilogram-meter | 9.806650[a] | newton-meter (N · m) |
| ounce-inch | 7.061552 | newton-millimeter |
| ounce-inch | 0.007061552 | newton-meter (N · m) |
| newton-meter | 0.7375621 | pound-foot |
| newton-meter | 10,000,000.[a] | dyne-centimeter |
| newton-meter | 0.1019716 | kilogram-meter |
| newton-meter | 141.6119 | ounce-inch |
| newton-millimeter | 0.1416119 | ounce-inch |
| pound-foot | 1.355818 | newton-meter (N · m) |
| **Moment of Inertia and Section Modulus** | | |
| moment of inertia [kg · m$^2$] | 23.73036 | pound-foot$^2$ |
| moment of inertia [kg · m$^2$] | 3417.171 | pound-inch$^2$ |
| moment of inertia [lb · ft$^2$] | 0.04214011 | kilogram-meter$^2$ (kg · m$^2$) |
| moment of inertia [lb · inch$^2$] | 0.0002926397 | kilogram-meter$^2$ (kg · m$^2$) |
| moment of section [foot$^4$] | 0.008630975 | meter$^4$ (m$^4$) |
| moment of section [inch$^4$] | 41.62314 | centimeter$^4$ |
| moment of section [meter$^4$] | 115.8618 | foot$^4$ |
| moment of section [cm$^4$] | 0.02402510 | inch$^4$ |
| section modulus [foot$^3$] | 0.02831685 | meter$^3$ (m$^3$) |
| section modulus [inch$^3$] | 0.00001638706 | meter$^3$ (m$^3$) |
| section modulus [meter$^3$] | 35.31466 | foot$^3$ |
| section modulus [meter$^3$] | 61,023.76 | inch$^3$ |
| **Momentum** | | |
| kilogram-meter/second | 7.233011 | pound-foot/second |
| kilogram-meter/second | 86.79614 | pound-inch/second |
| pound-foot/second | 0.1382550 | kilogram-meter/second (kg · m/s) |
| pound-inch/second | 0.01152125 | kilogram-meter/second (kg · m/s) |
| **Pressure and Stress** | | |
| atmosphere [14.6959 lb/inch$^2$] | 101,325. | pascal (Pa) |
| bar | 100,000.[a] | pascal (Pa) |
| bar | 14.50377 | pound/inch$^2$ |
| bar | 100,000.[a] | newton/meter$^2$ (N/m$^2$) |
| hectobar | 0.6474898 | ton [long]/inch$^2$ |
| kilogram/centimeter$^2$ | 14.22334 | pound/inch$^2$ |
| kilogram/meter$^2$ | 9.806650[a] | newton/meter$^2$ (N/m$^2$) |
| kilogram/meter$^2$ | 9.806650[a] | pascal (Pa) |
| kilogram/meter$^2$ | 0.2048161 | pound/foot$^2$ |
| kilonewton/meter$^2$ | 0.1450377 | pound/inch$^2$ |
| newton/centimeter$^2$ | 1.450377 | pound/inch$^2$ |
| newton/meter$^2$ | 0.00001[a] | bar |
| newton/meter$^2$ | 1.0[a] | pascal (Pa) |
| newton/meter$^2$ | 0.0001450377 | pound/inch$^2$ |
| newton/meter$^2$ | 0.1019716 | kilogram/meter$^2$ |
| newton/millimeter$^2$ | 145.0377 | pound/inch$^2$ |

**Table 1. Metric Conversion Factors** *(Continued)*

| Multiply | By | To Obtain |
|---|---|---|
| pascal | 0.00000986923 | atmosphere |
| pascal | 0.00001[a] | bar |
| pascal | 0.1019716 | kilogram/meter$^2$ |
| pascal | 1.0[a] | newton/meter$^2$ (N/m$^2$) |
| pascal | 0.02088543 | pound/foot$^2$ |
| pascal | 0.0001450377 | pound/inch$^2$ |
| pound/foot$^2$ | 4.882429 | kilogram/meter$^2$ |
| pound/foot$^2$ | 47.88026 | pascal (Pa) |
| pound/inch$^2$ | 0.06894757 | bar |
| pound/inch$^2$ | 0.07030697 | kilogram/centimeter$^2$ |
| pound/inch$^2$ | 0.6894757 | newton/centimeter$^2$ |
| pound/inch$^2$ | 6.894757 | kilonewton/meter$^2$ |
| pound/inch$^2$ | 6894.757 | newton/meter$^2$ (N/m$^2$) |
| pound/inch$^2$ | 0.006894757 | newton/millimeter$^2$ (N/mm$^2$) |
| pound/inch$^2$ | 6894.757 | pascal (Pa) |
| ton [long]/inch$^2$ | 1.544426 | hectobar |
| | Energy and Work | |
| Btu [International Table] | 1055.056 | joule (J) |
| Btu [mean] | 1055.87 | joule (J) |
| calorie [mean] | 4.19002 | joule (J) |
| foot-pound | 1.355818 | joule (J) |
| foot-poundal | 0.04214011 | joule (J) |
| joule | 0.0009478170 | Btu [International Table] |
| joule | 0.0009470863 | Btu [mean] |
| joule | 0.2386623 | calorie [mean] |
| joule | 0.7375621 | foot-pound |
| joule | 23.73036 | foot-poundal |
| joule | 0.9998180 | joule [International US] |
| joule | 0.9999830 | joule [US legal, 1948] |
| joule [International US] | 1.000182 | joule (J) |
| joule [US legal, 1948] | 1.000017 | joule (J) |
| joule | .0002777778 | watt-hour |
| watt-hour | 3600.[a] | joule (J) |
| | Power | |
| Btu [International Table]/hour | 0.2930711 | watt (W) |
| foot-pound/hour | 0.0003766161 | watt (W) |
| foot-pound/minute | 0.02259697 | watt (W) |
| horsepower [550 ft-lb/s] | 0.7456999 | kilowatt (kW) |
| horsepower [550 ft-lb/s] | 745.6999 | watt (W) |
| horsepower [electric] | 746.[a] | watt (W) |
| horsepower [metric] | 735.499 | watt (W) |
| horsepower [UK] | 745.70 | watt (W) |
| kilowatt | 1.341022 | horsepower [550 ft-lb/s] |
| watt | 2655.224 | foot-pound/hour |
| watt | 44.25372 | foot-pound/minute |
| watt | 0.001341022 | horsepower [550 ft-lb/s] |
| watt | 0.001340483 | horsepower [electric] |
| watt | 0.001359621 | horsepower [metric] |
| watt | 0.001341022 | horsepower [UK] |
| watt | 3.412141 | Btu [International Table]/hour |

## Table 1. Metric Conversion Factors *(Continued)*

| Multiply | By | To Obtain |
|---|---|---|
| | Viscosity | |
| poise | 0.1[a] | pascal-second (Pa · s) |
| centipoise | 0.001[a] | pascal-second (Pa · s) |
| stoke | 0.0001[a] | meter²/second ($m^2/s$) |
| centistoke | 0.000001[a] | meter²/second ($m^2/s$) |
| meter²/second | 1,000,000.[a] | centistoke |
| meter²/second | 10,000.[a] | stoke |
| pascal-second | 1000.[a] | centipoise |
| pascal-second | 10.[a] | poise |
| | Temperature | |
| temperature Celsius, $t_C$ | temperature Kelvin, $t_K$ | $t_K = t_C + 273.15$ |
| temperature Fahrenheit, $t_F$ | temperature Kelvin, $t_K$ | $t_K = (t_F + 459.67)/1.8$ |
| temperature Celsius, $t_C$ | temperature Fahrenheit, $t_F$ | $t_F = 1.8\,t_C + 32$ |
| temperature Fahrenheit, $t_F$ | temperature Celsius, $t_C$ | $t_C = (t_F - 32)/1.8$ |
| temperature Kelvin, $t_K$ | temperature Celsius, $t_C$ | $t_C = t_K - 273.15$ |
| temperature Kelvin, $t_K$ | temperature Fahrenheit, $t_F$ | $t_F = 1.8\,t_K - 459.67$ |
| temperature Kelvin, $t_K$ | temperature Rankine, $t_R$ | $t_R = 9/5\,t_K$ |
| temperature Rankine, $t_R$ | temperature Kelvin, $t_K$ | $t_K = 5/9\,t_R$ |

[a] The figure is exact.

## Table 2. Circular and Angular Measure Conversion Factors

*Circumference of circle* = 360 degrees = $2\pi$ radian = 6.283185 radian

1 *degree* (°) = 60 minutes = 3600 seconds = $\pi/180$ radian = 0.017453 radian

1 *quadrant* = 90 degrees = $\pi/2$ radians = 1.570796 radians

1 *minute* (′) = 60 seconds = 0.016667 degree = 0.000291 radian

1 *radian* = 57.2957795 degrees

$\pi = 3.141592654$

# Table 3. Linear Measure Conversion Factors

**Metric**

1 *kilometer (km)* =
- **1000** meters
- **100,000** centimeters
- **1,000,000** millimeters
- 0.539956 nautical mile
- 0.621371 mile
- 1093.61 yards
- 3280.83 feet
- 39,370.08 inches

1 *meter (m)* =
- **10** decimeters
- **100** centimeters
- **1000** millimeters
- 1.09361 yards
- 3.28084 feet
- 39.37008 inches

1 *decimeter (dm)* = **10** *centimeters*

1 *centimeter (cm)* =
- **0.01** meter
- **10 millimeters**
- 0.0328 foot
- 0.3937 inch

1 *millimeter (mm)* =
- **0.001** meter
- **0.1** centimeter
- **1000** microns
- 0.03937 inch

1 *micrometer or micron (μm)* =
- **0.000001** meter = one millionth meter
- **0.0001** centimeter
- **0.001** millimeter
- 0.00003937 inch
- 39.37 micro-inches

**US Customary**

1 *mile (mi)* =
- 0.868976 nautical mile
- **1760** yards
- **5280** feet
- **63,360** inches
- **1.609344** kilometers
- **1609.344** meters
- **160,934.4** centimeters
- **1,609,344** millimeters

1 *yard (yd)* =
- **3** feet
- **36** inches
- **0.9144** meter
- **91.44** centimeter
- **914.4** millimeter

1 *foot (international) (ft)* =
- **12** inches = $\frac{1}{3}$ yard
- **0.3048** meter
- **30.48** centimeter
- **304.8** millimeters

1 *survey foot* =
- 1.000002 international feet
- $\frac{12}{39.37}$ = 0.3048006096012 meter

1 *inch (in)* =
- **1000** mils
- **1,000,000** micro-inches
- **2.54** centimeters
- **25.4** millimeters
- **25,400** microns

1 *mil* =
- **0.001** inch
- **1000** micro-inches
- **0.0254** millimeters

1 *micro-inch (μin)* =
- **0.000001** inch = one millionth inch
- **0.0254** micrometer (micron)

**Surveyors Measure**

1 *mile* = **8** furlongs = **80** chains

1 *furlong* = **10** chains = **220** yards

1 *chain* =
- **4** rods = **22** yards = **66** feet = **100** links

1 *rod* =
- **5.5** yards = **16.5** feet = **25** links
- 5.0292 meters

1 *link* = 7.92 inches

1 *span* = 9 inches

1 *hand* = 4 inches

**Nautical Measure**

1 *league* = 3 nautical miles

1 *nautical mile* =
- 1.1508 statute miles
- **6076.11549** feet
- 1.8516 kilometers

1 *fathom* = **2** yards = **6** feet

1 *knot* = nautical unit of speed =
- 1 nautical mile per hour
- 1.1508 statute miles per hour
- 1.8516 kilometers per hour

*One degree at the equator* =
- 60 nautical miles
- 69.047 statute miles
- 111.098 kilometers

*One minute at the equator* =
- 1 nautical mile
- 1.1508 statute miles
- 1.8516 kilometers

*360 degrees at the equator* =
- circumference at equator
- 21,600 nautical miles
- 24,856.8 statute miles
- 39,995.4 kilometers

*Note*: Figures in **Bold** indicate exact conversion values

# Table 4. Square Measure and Conversion Factors

### Metric System

1 *square kilometer* (*km*$^2$) =
- **100** hectares
- **1,000,000** square meters
- 0.3861 square mile
- 247.1 acres

1 *hectare* (*ha*) =
- **0.01** square kilometer
- **100 ares**
- **10,000** square meters
- 2.471 acres
- 107,639 square feet

1 *are* (*a*) =
- **0.0001** square kilometer
- **100** square meters
- 0.0247 acre
- 1076.4 square feet

1 *square meter* (*m*$^2$) =
- **0.000001** square kilometer
- **100** square decimeters
- **10000** square centimeters
- **1,000,000** square millimeters
- 10.764 square feet
- 1.196 square yards

1 *square decimeter* (*dm*$^2$) =
- **100** square centimeters

1 *square centimeter* (*cm*$^2$) =
- **0.0001** square meters
- **100** square millimeters
- **0.001076** square foot
- **0.155** square inch

1 *square millimeter* (*mm*$^2$) =
- **0.01** square centimeters
- **1,000,000** square microns
- 0.00155 square inch

1 *square micrometer* (*micron*) ($\mu m^2$) =
- **1 × 10$^{-12}$** square meter
- **0.000001** square millimeters
- 1 × 10$^{-9}$ square inch
- 1549.997 square micro-inch

### US Customary System

1 *square mile* (*mi*$^2$) =
- **640** acres
- **6400** square chains
- 2.5899 square kilometers

1 *acre* =
- **10** square chains
- **4840** square yards
- **43,560** square feet
- a square, 208.71 feet on a side
- 0.4046856 hectare
- 40.47 ares
- 4046.856 square meters

1 *square chain* =
- **16** square rods
- **484** square yards
- **4356** square feet

1 *square rod* =
- **30.25** square yards
- **272.25** square feet
- **625** square links

1 *square yard* (*yd*$^2$) =
- **9** square feet
- **1296** square inches
- **0.83612736** square meter
- **8361.2736** square centimeter
- **836,127.36** square millimeter

1 *square foot* (*ft*$^2$) =
- 0.111111 square yard
- **144** square inches
- **0.09290304** square meter
- **929.0304** square centimeters
- **92,903.04** square millimeters

1 *square inch* (*in*$^2$) =
- 0.0007716 square yard
- 0.006944 square foot
- 0.00064516 square meter
- **6.4516** square centimeters
- **645.16** square millimeters

1 *square mil* (*mil*$^2$) =
- **0.000001** square inch
- **0.00064516** square millimeter

### Measure Used for Diameters and Areas of Electric Wires

1 *circular inch* =
- area of 1-inch diameter circle
- $\pi/4$ square inch
- 0.7854 square inch
- 5.067 square centimeters
- 1,000,000 circular mils

1 *circular mil* =
- area of 0.001-inch diameter circle
- $\pi/4$ square mil

1 *square inch* =
- 1.2732 circular inches
- 1,273,239 circular mils

*Note*: Figures in **Bold** indicate exact conversion values

# Table 5. Cubic Measure and Conversion Factors

### Metric System

1 *cubic meter* ($m^3$) =
- **1000** cubic decimeters (liters)
- **1,000,000** cubic centimeters
- 1.30795 cubic yards
- 35.314667 cubic feet
- 61,023.74 cubic inches
- 264.17205 US gallons
- 219.96925 British Imperial gallons

1 *liter* (*l*) *or 1 cubic decimeter* ($dm^3$) =
- **1** liter = volume of 1 kg water at 39.2° F
- **0.001** cubic meter
- **1000** cubic centimeters
- **10** deciliters
- 0.03531466 cubic foot
- 61.023744 cubic inches
- 0.2642 US gallon
- 0.21997 British Imperial gallon
- 1.0566882 US quarts
- 33.814 US fluid ounces

1 *cubic centimeter* ($cm^3$) =
- **0.001** liter
- **1000** cubic millimeters
- 0.061024 cubic inch

1 *cubic millimeter* = **0.001** cubic centimeters
1 *hectoliter* (*hl*) = **100** liters
1 *deciliter* (*dl*) = **10** centiliters
1 *centiliter* (*cl*) = **10** milliliters

### British (Imperial) Liquid and Dry Measure

1 *British Imperial gallon* =
- 0.1605 cubic foot
- 277.42 cubic inches
- 1.2009 US gallon
- **160** Imperial fluid ounces
- **4** Imperial quarts
- **8** Imperial pints
- **4.54609** liters

1 *quart* =
- **2** Imperial pints
- **8** Imperial gills
- **40** Imperial fluid ounces
- 69.354 cubic inches
- **1.1365225** liters

1 *pint* =
- **4** Imperial gills
- **20** Imperial fluid ounces
- 34.678 cubic inches
- **568.26125** milliliters

1 *gill* =
- **5** Imperial fluid ounces
- 8.669 cubic inches
- 142.07 milliliters

### US Customary System

1 *cubic yard* ($yd^3$) =
- **27** cubic feet
- 201.97403 US gallons
- **46,656** cubic inch
- 0.7646 cubic meter

1 *cubic foot* ($ft^3$) =
- **1728** cubic inches
- 7.4805 US gallons
- 6.23 British Imperial gallons
- 0.02831685 cubic meter
- 28.31685 liters

1 *cubic inch* ($in^3$) =
- 0.55411256 US fluid ounce
- **16.387064** cubic centimeters

### Shipping Measure

For measuring internal capacity of a vessel:

1 *register ton* = 100 cubic feet

For measurement of cargo:

1 *shipping ton* =

Approximately 40 cubic feet of merchandise is considered a shipping ton, unless that bulk would weigh more than 2000 pounds, in which case the freight charge may be based upon weight

40 *cubic feet* = 32.143 US bushels = 31.16 Imperial bushels

### US Liquid Measure

1 *US gallon* =
- 0.13368 cubic foot
- **231** cubic inches
- **128** US fluid ounces
- **4** US quarts
- **8** US pints
- 0.8327 British Imperial gallon
- **3.785411784** liters

1 *quart* =
- **2** US pints
- **8** US gills
- **32** US fluid ounces
- **57.75** cubic inches
- 0.9463529 liters

1 *pint* =
- **4** US gills
- **16** US fluid ounces
- **28.875** cubic inches
- 473.176 milliliters

1 *gill* =
- **1/2** cup = **4** US fluid ounces
- **7.21875** cubic inches
- 118.29 milliliters

*Note*: Figures in **Bold** indicate exact conversion values

## Table 5. Cubic Measure and Conversion Factors *(Continued)*

**British (Imperial) Liquid and Dry Measure**

1 *British Imperial fluid ounce* =
1.733871 cubic inch
$\frac{1}{160}$ British Imperial gallon
28.41306 milliliters

1 *British Imperial bushel* =
**8** Imperial gallons = 1.284 cubic feet
2219.36 cubic inches

**US Dry Measure**

1 *bushel (US or Winchester struck bushel)* =
1.2445 cubic feet
**2150.42** cubic inches
a cylinder 18.5 inches dia., 8 inches deep
a cylinder 47.0 cm dia., 20.3 cm deep

1 *bushel* = **4** pecks = **32** quarts = **64** pints

1 *peck* = **8** quarts = **16** pints

1 *dry quart* = **2** pints =
**67.200625** cubic inches
1.101221 liters

1 *heaped bushel* = $1\frac{1}{4}$ struck bushel

1 *cubic foot* = 0.8036 struck bushel

**Apothecaries' Fluid Measure**

1 *US fluid ounce* =
**1.8046875** cubic inches
$\frac{1}{128}$ US gallon
**8** drachms
0.02957353 liter
29.57353 milliliters

1 *fluid drachm* = **60** minims

**Old Liquid Measure**

1 *barrel (bbl)* = $31\frac{1}{2}$ gallons

1 *hogshead* = 2 barrels = 63 gallons

1 *pipe* or *butt* = 2 hogsheads = 4 barrels = 126 gallons

1 *tierce* = 42 gallons

1 *puncheon* = 2 tierces = 84 gallons

1 *tun* = 2 pipes = 3 puncheons

**Other Cubic Measure**

The following are used for wood and masonry:

1 *cord of wood* = 4 × 4 × 8 feet = 128 cubic feet

1 *perch of masonry* =
$16\frac{1}{2} \times 1\frac{1}{2} \times 1$ foot = $24\frac{3}{4}$ cubic feet

**Barrel Measure**

1 *drum* =
**55** US gallon
7.3524 cubic feet
208.19765 liters

1 *petroleum barrel (bo)* =
**42** US gallons
5.614583 cubic feet
158.98729 liters

## Table 6. Mass and Weight Conversion Factors

**Metric System**

1 *metric ton (t)* =
**1000** kilograms
2204.6223 pounds
0.9842 gross or long ton (of 2240 pounds)
0.9072 net or short ton (of 2000 pounds)

1 *kilogram (kg)* =
**1000** grams = **10** hectograms
2.2046 pounds
35.274 ounces avoirdupois

1 *hectogram (hg)* = **10** dekagrams

1 *dekagram (dag)* = **10** grams

1 *gram (g)* =
**10** decigrams
0.0022046 pound
0.03215 ounce Troy
0.03527 ounce avoirdupois
15.432 grains

1 *decigram (dg)* = **10** centigrams

1 *centigram (cg)* = **10** milligrams

**Avoirdupois or Commercial Weight**

1 *gross or long ton* =
**2240** pounds
1.016 metric ton
1016 kilograms

1 *net or short ton* = **2000** pounds

1 *pound* =
**16** ounces
**7000** grains
**0.45359237** kilogram
453.6 grams

1 *ounce* =
$\frac{1}{16}$ pound
16 drachms
**437.5** grains
28.3495 grams
0.2780139 newton

1 *grain Avoirdupois* =
**1** grain apothecaries' weight =
**1** grain Troy weight
0.064799 gram

*Note*: Figures in **Bold** indicate exact conversion values

## Table 6. Mass and Weight Conversion Factors *(Continued)*

**Troy Weight**

Used for Weighing Gold and Silver

1 *pound Troy* =
- 12 ounces Troy = 5760 grains
- $^{144}/_{175}$ avoirdupois pound

1 *ounce Troy* =
- 20 pennyweights = 480 grains
- 31.103 grams

1 *pennyweight* = 24 grains

1 *grain Troy* =
- **1** grain avoirdupois
- **1** grain apothecaries' weight
- 0.0648 gram

1 *carat (used in weighing diamonds)* =
- 3.086 grains
- **200** milligrams = $^1/_5$ gram

1 *gold karat* = $^1/_{24}$ proportion pure gold

**Apothecaries' Weight**

1 *pound* = 12 ounces = 5760 grains

1 *ounce* =
- 8 drachms = 480 grains
- 31.103 grams

1 *drachm* = 3 scruples = 60 grains

1 *scruple* = 20 grains

**Old Weight Measures**

Measures for weight seldom used in the United States:

1 *gross or long ton* = 20 hundred-weights

1 *hundred-weight* = 4 quarters = 112 pounds

1 *quarter* = 28 pounds

1 *stone* = 14 pounds

1 *quintal* = 100 pounds

## Table 7. Pressure and Stress Conversion Factors

1 *kilogram per sq. millimeter* ($kg_f/mm^2$) =
- 1422.32 pounds per square inch

1 *kilogram per sq. centimeter* ($kg_f/cm^2$) =
- 14.223 pounds per square inch

1 *bar* =
- **1,000,000** dynes per square centimeter
- **1000** millibars
- **100** kilopascals
- 750.06168 torr
- 1.0197162 kg force per $cm^2$
- 14.50377 pounds per square inch
- 29.529983 inches of mercury at 0° C
- 10,197.162 mm water at 4° C
- 33.455256 feet of water at 4° C

1 *millibar* =
- **100,000** dynes per square centimeter
- **100** pascal

1 *torr* =
- **760** millimeters mercury
- $^1/_{760}$ atmosphere
- 133.224 pascal
- 1.333224 millibar

1 *pound per square inch* =
- 144 pounds per square foot
- 0.068 atmosphere
- 2.042 inches of mercury at 62° F
- 27.7 inches of water at 62° F
- 2.31 feet of water at 62° F
- 0.0703 kilogram per square centimeter
- 6.894757 kilopascals
- 6894.757 pascal

1 *atmosphere* =
- 30 inches of mercury at 62° F
- 14.7 pounds per square inch
- 2116.3 pounds per square foot
- 33.95 feet of water at 62° F

1 *foot of water at 62° F* =
- 62.355 pounds per square foot
- 0.433 pound per square inch

1 *inch of mercury at 62° F* =
- 1.132 foot of water
- 13.58 inches of water
- 0.491 pound per square inch

1 *inch of water* =
- 0.0735559 inch mercury at 0° C
- 1.8683205 torr
- 0.5780367 ounce force per square inch
- 0.0024583 atmosphere

*Note*: Figures in **Bold** indicate exact conversion values

## Table 8. Temperature Conversion Fomulas

| To Convert | To | Use Formula |
|---|---|---|
| Celsius, $t_C$ | K, $t_K$ | $t_K = t_C + 273.15$ |
| | °F, $t_F$ | $t_F = 1.8\,t_C + 32$ |
| | °R, $t_R$ | $t_R = 9(t_C + 273.15)/5$ |
| Fahrenheit, $t_F$ | K, $t_K$ | $t_K = (t_F + 459.67)/1.8$ |
| | °C, $t_C$ | $t_C = (t_F - 32)/1.8$ |
| | °R, $t_R$ | $t_R = t_F + 459.67$ |
| Kelvin, $t_K$ | °C, $t_C$ | $t_C = t_K - 273.15$ |
| | °F, $t_F$ | $t_F = 1.8\,t_K - 459.67$ |
| | °R, $t_R$ | $t_R = 9/5 \times t_K$ |
| Rankine, $t_R$ | K, $t_K$ | $t_K = 5/9 \times t_R$ |
| | °C, $t_C$ | $t_C = 5/9 \times t_R - 273.15$ |
| | °F, $t_F$ | $t_F = t_R - 459.67$ |

Kelvin temperatures are expressed by K without use of the degree symbol (°).

## Table 9. Energy Conversion Factors

1 *horsepower-hour =*
0.746 kilowatt-hour
1,980,000 foot-pounds
2545 Btu (British thermal units)
2.64 pounds of water evaporated at 212° F
17 pounds of water raised from 62° to 212° F

1 *kilowatt-hour =*
1000 watt-hours
1.34 horsepower-hour
2,655,200 foot-pounds
3,600,000 joules
3415 Btu
3.54 pounds of water evaporated at 212° F
22.8 pounds of water raised from 62° to 212° F

## Table 10. Power Conversion Factors

1 *horsepower =*
746 watts
0.746 kilowatt
33,000 foot-pounds/minute
550 foot-pounds/second
2545 Btu/hour
42.4 Btu/minute
0.71 Btu/second
2.64 pounds of water evaporated per hour at 212° F

1 *kilowatt =*
1000 watts
1.34 horsepower
2,654,200 foot-pounds/hour
44,200 foot-pounds/minute
737 foot-pounds/second
3415 Btu/hour
57 Btu/minute
0.95 Btu/second
3.54 pounds of water evaporated per hour at 212° F

1 *watt =*
1 joule/second
0.00134 horsepower
0.001 kilowatt
3.42 Btu/hour
44.22 foot-pounds/minute
0.74 foot-pounds/second
0.0035 pound of water evaporated per hour at 212° F

## Table 11. Heat Conversion Factors

1 *Btu (British thermal unit) =*
1052 watt-seconds
778 foot-pounds
0.252 kilogram-calorie
0.000292 kilowatt-hour
0.000393, horsepower-hour
0.00104 pound of water evaporated at 212° F

1 *kilogram calorie =*
3.968 Btu

1 *foot-pound =*
1.36 joules
0.000000377 kilowatt-hour
0.00129 Btu
0.0000005 horsepower-hour

1 *kilogram-meter =*
7.233 foot-pounds

1 *joule =*
1 watt-second
0.00000078 kilowatt-hour
0.00095 Btu
0.74 foot-pound

1 *therm =*
100,000 Btu (US)
29.3 kilowatt-hour
105.5 megajoule

# INDEX

## A

## B



## C

# E



# F

# G

# H

## N



## O

# P



# Q

## R



## S

## U



## V



## W

# NOTES

# NOTES

# Volume

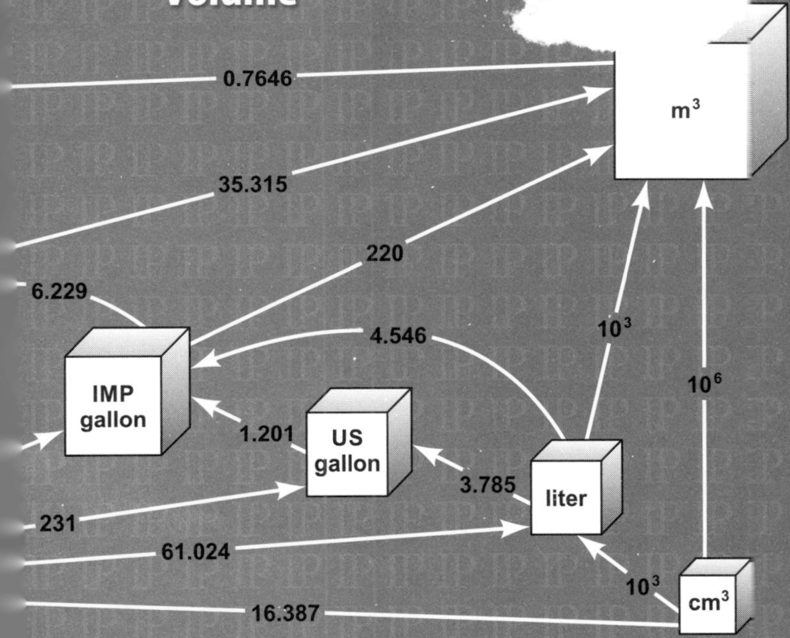


kW = kilowatt
lb = pound
M = mega
m = meter
mm = millimeter
MPa = megapascal
N = newton
oz = ounce
Pa = pascal
PS = pferdestarke (metric hp)
s = second
US = United States (gallon)
W = watt
yd = yard

**Example:** Convert 10 $m^2$ to $yd^2$

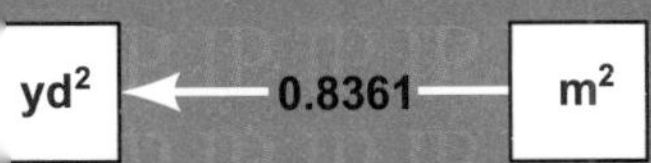


10 divided by 0.8361 = 11.96 $yd^2$

**Example:** Convert 10 $yd^2$ to $m^2$

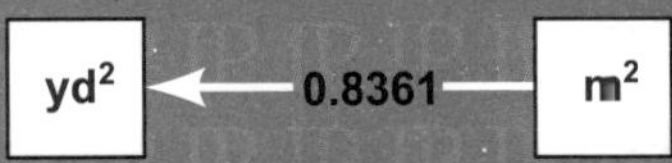


10 multiplied by 0.8361 = 8.36 $m^2$

***Continued on the inside back cover...***

# Machinery's Handbook

## $32^{nd}$ Edition

# REAL NUMBERS AND THEIR OPERATIONS

## Real Numbers

Most mathematical computation is performed in the *real number system*. The universal set of the "reals" includes the subsets: *naturals, whole numbers, integers, rationals,* and *irrationals*. The naturals (also called *counting numbers*): {1, 2, 3, . . .} are included in the whole numbers: {0, 1, 2, 3, . . .}, which are included in the integers (or signed whole numbers): {. . . , –2, –1, 0, 1, 2, . . .}. And all of these subsets are included in the rationals.

Rational numbers, including integers, can be written in fraction form. Since all fractions can be divided numerator by denominator, their decimal form either terminates or repeats. Examples of rational numbers: –4/1, 3/5 = 0.6, 1/3 = 0.333. . . .

The only set in the real numbers larger than the naturals that does not contain any of the other sets is the irrationals. These are *not* expressible as ratios. An irrational number's decimal representation does not terminate and it has no pattern of repetition. Examples of irrational numbers are roots that cannot be simplified, such as $\sqrt{6}$ and $\sqrt[3]{70}$, as well as quantities like $\pi$ and the natural log base $e$. The entire real number set is the union of the rationals and the irrationals.

**Properties of Real Numbers.**—Though often obvious and followed almost automatically, the properties of real numbers are critical to mathematical reasoning. These properties justify various steps in solving algebraic problems, such as those in this *Handbook*. Equivalence properties (symmetry, reflexivity, transitivity) and operational properties of numbers are summarized here.

*Equivalence Properties:* The properties of equivalence relations are the basis of equation solving.

Reflexive: $a = a$.
Symmetric: If $a = b$, then $b = a$.
Transitive: If $a = b$ and $b = c$, then $a = c$.
Substitution: If $a = b$, then $a$ may be replaced by $b$ in any equation or expression.

*Operational Properties:* These concern addition, subtraction, multiplication, and division, as summarized in the table below.

| *Property* | *Addition* | *Multiplication* |
|---|---|---|
| Commutative: | $a + b = b + a$ | $a \times b = b \times a$ |
| Associative: | $(a + b) + c = a + (b + c)$ | $(a \times b) \times c = a \times (b \times c)$ |
| Identity: | $a + 0 = 0 + a = a$ | $1 \times a = a \times 1 = a$ |
| Inverse: | $a + (-a) = 0$ | $a \times 1/a = 1$ |

*Other Properties:*

| | | |
|---|---|---|
| Distributive of multiplication over addition: | $a \times (b + c) = (a \times b) + (a \times c)$ | $(a + b) \times c = (a \times c) + (b \times c)$ |
| Zero property of multiplication: | If $a \times b = 0$, then either $a = 0$ or $b = 0$ | |
| Zero property of division: | If $a/b = 0$, then $a = 0$ $(b \neq 0)$ | |

**Integers (Signed Numbers).**—Positive whole numbers extend to the right of zero on the number line. Negative whole numbers extend to the left of zero. Together with zero, these make up the *integers* (sometimes called *signed numbers*): {. . . , –2, –1, 0, 1, 2, . . .}.

The sciences (as well as economics and other fields) deal with negative as well as non-negative quantities. Temperature is an obvious example; so is land altitude, which can be above, at, or below sea level. Angles can be negative, too, as explained in

previous example, production increased from 280 to 300, then the percent change would be:
$\frac{\text{final} - \text{original amount}}{\text{original}} \times 100 = \frac{300-280}{280} \times 100 = \frac{20}{280} \times 100 = 7.14\%$.

The denominator is always the original amount. Percent change also can be negative. If production decreased from 280 to 245, the percent change would be: $\frac{245-280}{280} \times 100 = \frac{-35}{280} \times 100 = -12.5\%$.

**Powers and Roots**

*Powers:* The *square* or *second power* of a number (or quantity) is the product of that number multiplied by itself. Thus, the square of 9 is $9 \times 9$. The square of a number is indicated by the *exponent* $^2$, thus: $9^2 = 9 \times 9 = 81$.

The *cube* or *third power* of a number $n$ is the product $n \times n \times n$, or $n^3$. Thus, the cube of 4 is $4 \times 4 \times 4 = 64$, and is written $4^3$.

In general, the $n$th power of $a$ is written $a^n$, where $a$ is the *base* and $n$ is the *exponent*.

*Roots:* The *square root* of a given number is the positive number which, when multiplied by itself, will produce the given number. The square root of 16 (written $\sqrt{16}$) is 4 because $4 \times 4 = 16$. The other root of 16 is −4, but the use of the square root symbol indicates the positive (principal) square root only.

Similarly, the *cube root* of a given number is the number which, when used as a factor three times, will produce the given number. Thus, the cube root of 64 (written $\sqrt[3]{64}$) is 4 because $4 \times 4 \times 4 = 64$.

In general, the $n$th root of $a$ is written as $\sqrt[n]{a}$ or $a^{1/n}$.

**Properties of Exponents**

$a^n a^m = a^{n+m}$ $\qquad \frac{a^n}{a^m} = a^{(n-m)}$ $\qquad (a^m)^n = a^{mn}$ $\qquad (ab)^m = a^m b^m$

$\left(\frac{a}{b}\right)^n = \frac{a^n}{b^n}$ $\qquad a^{m/n} = (a^{1/n})^m$ or $(a^m)^{1/n}$ $\qquad a^{-n} = \frac{1}{a^n}$ $\qquad \frac{1}{a^{-n}} = a^n$

$a^0 = 1 \ (a \neq 0)$ $\qquad a^{1/n} = \sqrt[n]{a}$ $\qquad a^{m/n} = (\sqrt[n]{a})^m = \left(\sqrt[n]{a^m}\right)$ unless $a < 0$, and $m$ and $n$ are both even

*Examples:*

$3^1 3^2 = 3^{1+2} = 3^3 = 27$ $\qquad (x)(x^3) = x^{(1+3)} = x^4$

$\frac{5^4}{5^2} = 5^{4-2} = 5^2 = 25$ $\qquad \frac{x^9}{x^6} = x^{(9-6)} = x^3$

$(2^4)^2 = 2^{(4)(2)} = 2^8 = 256$ $\qquad (x^3)^3 = x^{(3)(3)} = x^9$

$(9x)^2 = 9^2 x^2 = 81x^2$ $\qquad (ab^4)^2 = a^2 b^8$

$32^{3/5} = (32^{1/5})^3 = (\sqrt[5]{32})^3 = 2^3 = 8$ $\qquad 4^{-3} = \frac{1}{4^3} = \frac{1}{64}$

$\frac{1}{2^{-5}} = 2^5 = 32$ $\qquad 9x^0 = 9(1) = 9$

$$\frac{(2.5 \times 10^2) \times (4.698 \times 10^3) \times (3.9 \times 10^{-4})}{(4.3678 \times 10^4) \times (2 \times 10^{-3}) \times (1.47 \times 10^{-2})}$$

$$= \frac{(2.5 \times 4.698 \times 3.9)(10^{2+3-4})}{(4.3678 \times 2 \times 1.47)(10^{4-3-2})} = \frac{45.8055 \times 10}{12.8413 \times 10^{-1}}$$

$$= 3.5670 \times 10^{1-(-1)} = 3.5670 \times 10^2 = 356.70 \text{ (rounded)}$$

**Factorial Notation.—**A factorial is a mathematical shortcut denoted by the symbol **!** following a number (for example, 3! is "three factorial"). $n!$ is found by multiplying together all the positive integers less than or equal to the factorial number $n$. Zero factorial (0!) is defined as 1. For example: $3! = 1 \times 2 \times 3 = 6$; $4! = 1 \times 2 \times 3 \times 4 = 24$; $7! = 1 \times 2 \times 3 \times 4 \times 5 \times 6 \times 7 = 5040$; etc. Factorial notation is used in certain areas, including probability and analysis. The following two topics (permutations and combinations) relate to probability and statistics.

**Permutation.—**A permutation is an arrangement of objects of a set into a sequence or order. In mathematics, the number of arrangements of $n$ objects is given by $n!$. For example, 4 objects can be arranged 4! ways, that is, $4 \times 3 \times 2 \times 1 = 24$ ways. The number of ways $r$ objects can be arranged (that is, ordered) from a set of $n$ is given by the *permutation* formula $_nP_r = \frac{n!}{(n-r)!}$

*Example:* How many ways can the letters X, Y, and Z be arranged?

*Solution:* Three objects ($r = 3$) out of a set of 3 ($n = 3$) are being arranged. The numbers of possible arrangements for the three letters are $3!/(3-3)! = (3 \times 2 \times 1)/1 = 6$. Listing them is not difficult, since there are so few: XYZ, XZY, YXZ, YZX, ZXY, ZYX.

*Example:* There are 10 people participating in a foot race. How many arrangements of first, second, and third place winners are there?

*Solution:* Here $r$ is 3 and $n$ is 10. The number of possible arrangements of winners are:

$$_{10}P_3 = \frac{10!}{(10-3)!} = \frac{10!}{7!} = 10 \times 9 \times 8 = 720$$

**Combination.—**This is the number of ways $r$ objects can be chosen from $n$ in a way that order does not matter. It is expressed as "$n$ choose $r$." There are fewer combinations than permutations of $r$ objects out of $n$, since it does not matter in what order the three objects are chosen. So in a combination, choosing ABC is the same as choosing ACB or BAC and so on. The formula is $_nC_r = \frac{n!}{(n-r)!r!}$

*Example:* How many possible sets of 6 numbers can be picked with no regard for order from the numbers 1 to 52?

*Solution:* Here $r$ is 6 and $n$ is 52. So the possible number of combinations is:

$$_{52}C_6 = \frac{52!}{(52-6)!6!} = \frac{52!}{46!6!} = \frac{52 \times 51 \times 50 \times 49 \times 48 \times 47}{1 \times 2 \times 3 \times 4 \times 5 \times 6} = 20{,}358{,}520$$

**Prime Factorization of Numbers.—**Tables of prime numbers and factors of numbers are particularly useful for calculations involving change-gear ratios for compound gearing, dividing heads, gear-generating machines, and mechanical designs having gear trains.

*Definition: p* is a *factor* of a number $n$ if the division $n/p$ leaves no remainder. Thus, any number $n$ has factors of itself and 1, because $n/n = 1$ and $n/1 = n$. Other factors of a number are found as follows:

2 is a factor of any even number. Thus, $28 = 2 \times 14$, and $210 = 2 \times 105$.

3 is a factor of any number where the sum of its digits is divisible by 3. Thus, 3 is a factor of 2397, because $2 + 3 + 9 + 7 = 21$, and $21 \div 3 = 7$.

**Table 1a. Prime Number and Factor Table for 1 to 1199**

| From<br>To | 0<br>100 | 100<br>200 | 200<br>300 | 300<br>400 | 400<br>500 | 500<br>600 | 600<br>700 | 700<br>800 | 800<br>900 | 900<br>1000 | 1000<br>1100 | 1100<br>1200 |
|---|---|---|---|---|---|---|---|---|---|---|---|---|
| 1 | **P** | **P** | 3 | 7 | **P** | 3 | **P** | **P** | 3 | 17 | 7 | 3 |
| 2 | **P** | 2 | 2 | 2 | 2 | 2 | 2 | 2 | 2 | 2 | 2 | 2 |
| 3 | **P** | **P** | 7 | 3 | 13 | **P** | 3 | 19 | 11 | 3 | 17 | **P** |
| 5 | **P** | 3 | 5 | 5 | 3 | 5 | 5 | 3 | 5 | 5 | 3 | 5 |
| 7 | **P** | **P** | 3 | **P** | 11 | 3 | **P** | 7 | 3 | **P** | 19 | 3 |
| 9 | 3 | **P** | 11 | 3 | **P** | **P** | 3 | **P** | **P** | 3 | **P** | **P** |
| 11 | **P** | 3 | **P** | **P** | 3 | 7 | 13 | 3 | **P** | **P** | 3 | 11 |
| 13 | **P** | **P** | 3 | **P** | 7 | 3 | **P** | 23 | 3 | 11 | **P** | 3 |
| 15 | 3 | 5 | 5 | 3 | 5 | 5 | 3 | 5 | 5 | 3 | 5 | 5 |
| 17 | **P** | 3 | 7 | **P** | 3 | 11 | **P** | 3 | 19 | 7 | 3 | **P** |
| 19 | **P** | 7 | 3 | 11 | **P** | 3 | **P** | **P** | 3 | **P** | **P** | 3 |
| 21 | 3 | 11 | 13 | 3 | **P** | **P** | 3 | 7 | **P** | 3 | **P** | 19 |
| 23 | **P** | 3 | **P** | 17 | 3 | **P** | 7 | 3 | **P** | 13 | 3 | **P** |
| 25 | 5 | 5 | 3 | 5 | 5 | 3 | 5 | 5 | 3 | 5 | 5 | 5 |
| 27 | 3 | **P** | **P** | 3 | 7 | 17 | 3 | **P** | **P** | 3 | 13 | 7 |
| 29 | **P** | 3 | **P** | 7 | 3 | 23 | 17 | 3 | **P** | **P** | 3 | **P** |
| 31 | **P** | **P** | 3 | **P** | **P** | 3 | **P** | 17 | 3 | 7 | **P** | 3 |
| 33 | 3 | 7 | **P** | 3 | **P** | 13 | 3 | **P** | 7 | 3 | **P** | 11 |
| 35 | 5 | 3 | 5 | 5 | 3 | 5 | 5 | 3 | 5 | 5 | 3 | 5 |
| 37 | **P** | **P** | 3 | **P** | 19 | 3 | 7 | 11 | 3 | **P** | 17 | 3 |
| 39 | 3 | **P** | **P** | 3 | **P** | 7 | 3 | **P** | **P** | 3 | **P** | 17 |
| 41 | **P** | 3 | **P** | 11 | 3 | **P** | **P** | 3 | 29 | **P** | 3 | 7 |
| 43 | **P** | 11 | 3 | 7 | **P** | 3 | **P** | **P** | 3 | 23 | 7 | 3 |
| 45 | 3 | 5 | 5 | 3 | 5 | 5 | 3 | 5 | 5 | 3 | 5 | 5 |
| 47 | **P** | 3 | 13 | **P** | 3 | **P** | **P** | 3 | 7 | **P** | 3 | 31 |
| 49 | 7 | **P** | 3 | **P** | **P** | 3 | 11 | 7 | 3 | 13 | **P** | 3 |
| 51 | 3 | **P** | **P** | 3 | 11 | 19 | 3 | **P** | 23 | 3 | **P** | **P** |
| 53 | **P** | 3 | 11 | **P** | 3 | 7 | **P** | 3 | **P** | **P** | 3 | **P** |
| 55 | 5 | 5 | 3 | 5 | 5 | 3 | 5 | 5 | 3 | 5 | 5 | 3 |
| 57 | 3 | **P** | **P** | 3 | **P** | **P** | 3 | **P** | **P** | 3 | 7 | 13 |
| 59 | **P** | 3 | 7 | **P** | 3 | 13 | **P** | 3 | **P** | 7 | 3 | 19 |
| 61 | **P** | 7 | 3 | 19 | **P** | 3 | **P** | **P** | 3 | 31 | **P** | 3 |
| 63 | 3 | **P** | **P** | 3 | **P** | **P** | 3 | 7 | **P** | 3 | **P** | **P** |
| 65 | 5 | 3 | 5 | 5 | 3 | 5 | 5 | 3 | 5 | 5 | 3 | 5 |
| 67 | **P** | **P** | 3 | **P** | **P** | 3 | 23 | 13 | 3 | **P** | 11 | 3 |
| 69 | 3 | 13 | **P** | 3 | 7 | **P** | 3 | **P** | 11 | 3 | **P** | 7 |
| 71 | **P** | 3 | **P** | 7 | 3 | **P** | 11 | 3 | 13 | **P** | 3 | **P** |
| 73 | **P** | **P** | 3 | **P** | 11 | 3 | **P** | **P** | 3 | 7 | 29 | 3 |
| 75 | 3 | 5 | 5 | 3 | 5 | 5 | 3 | 5 | 5 | 3 | 5 | 5 |
| 77 | 7 | 3 | **P** | 13 | 3 | **P** | **P** | 3 | **P** | **P** | 3 | 11 |
| 79 | **P** | **P** | 3 | **P** | **P** | 3 | 7 | 19 | 3 | 11 | 13 | 3 |
| 81 | 3 | **P** | **P** | 3 | 13 | 7 | 3 | 11 | **P** | 3 | 23 | **P** |
| 83 | **P** | 3 | **P** | **P** | 3 | 11 | **P** | 3 | **P** | **P** | 3 | 7 |
| 85 | 5 | 5 | 3 | 5 | 5 | 3 | 5 | 5 | 3 | 5 | 5 | 3 |
| 87 | 3 | 11 | 7 | 3 | **P** | **P** | 3 | **P** | **P** | 3 | **P** | **P** |
| 89 | **P** | 3 | 17 | **P** | 3 | 19 | 13 | 3 | 7 | 23 | 3 | 29 |
| 91 | 7 | **P** | 3 | 17 | **P** | 3 | **P** | 7 | 3 | **P** | **P** | 3 |
| 93 | 3 | **P** | **P** | 3 | 17 | **P** | 3 | 13 | 19 | 3 | **P** | **P** |
| 95 | 5 | 3 | 5 | 5 | 3 | 5 | 5 | 3 | 5 | 5 | 3 | 5 |
| 97 | **P** | **P** | 3 | **P** | 7 | 3 | 17 | **P** | 3 | **P** | **P** | 3 |
| 99 | 3 | **P** | 13 | 3 | **P** | **P** | 3 | 17 | 29 | 3 | 7 | 11 |

### Table 2a. Prime Numbers from 2 to 6869

| | | | | | | | | | | | | |
|---|---|---|---|---|---|---|---|---|---|---|---|---|
| 2 | 347 | 773 | 1259 | 1753 | 2293 | 2803 | 3389 | 3929 | 4517 | 5099 | 5689 | 6277 |
| 3 | 349 | 787 | 1277 | 1759 | 2297 | 2819 | 3391 | 3931 | 4519 | 5101 | 5693 | 6287 |
| 5 | 353 | 797 | 1279 | 1777 | 2309 | 2833 | 3407 | 3943 | 4523 | 5107 | 5701 | 6299 |
| 7 | 359 | 809 | 1283 | 1783 | 2311 | 2837 | 3413 | 3947 | 4547 | 5113 | 5711 | 6301 |
| 11 | 367 | 811 | 1289 | 1787 | 2333 | 2843 | 3433 | 3967 | 4549 | 5119 | 5717 | 6311 |
| 13 | 373 | 821 | 1291 | 1789 | 2339 | 2851 | 3449 | 3989 | 4561 | 5147 | 5737 | 6317 |
| 17 | 379 | 823 | 1297 | 1801 | 2341 | 2857 | 3457 | 4001 | 4567 | 5153 | 5741 | 6323 |
| 19 | 383 | 827 | 1301 | 1811 | 2347 | 2861 | 3461 | 4003 | 4583 | 5167 | 5743 | 6329 |
| 23 | 389 | 829 | 1303 | 1823 | 2351 | 2879 | 3463 | 4007 | 4591 | 5171 | 5749 | 6337 |
| 29 | 397 | 839 | 1307 | 1831 | 2357 | 2887 | 3467 | 4013 | 4597 | 5179 | 5779 | 6343 |
| 31 | 401 | 853 | 1319 | 1847 | 2371 | 2897 | 3469 | 4019 | 4603 | 5189 | 5783 | 6353 |
| 37 | 409 | 857 | 1321 | 1861 | 2377 | 2903 | 3491 | 4021 | 4621 | 5197 | 5791 | 6359 |
| 41 | 419 | 859 | 1327 | 1867 | 2381 | 2909 | 3499 | 4027 | 4637 | 5209 | 5801 | 6361 |
| 43 | 421 | 863 | 1361 | 1871 | 2383 | 2917 | 3511 | 4049 | 4639 | 5227 | 5807 | 6367 |
| 47 | 431 | 877 | 1367 | 1873 | 2389 | 2927 | 3517 | 4051 | 4643 | 5231 | 5813 | 6373 |
| 53 | 433 | 881 | 1373 | 1877 | 2393 | 2939 | 3527 | 4057 | 4649 | 5233 | 5821 | 6379 |
| 59 | 439 | 883 | 1381 | 1879 | 2399 | 2953 | 3529 | 4073 | 4651 | 5237 | 5827 | 6389 |
| 61 | 443 | 887 | 1399 | 1889 | 2411 | 2957 | 3533 | 4079 | 4657 | 5261 | 5839 | 6397 |
| 67 | 449 | 907 | 1409 | 1901 | 2417 | 2963 | 3539 | 4091 | 4663 | 5273 | 5843 | 6421 |
| 71 | 457 | 911 | 1423 | 1907 | 2423 | 2969 | 3541 | 4093 | 4673 | 5279 | 5849 | 6427 |
| 73 | 461 | 919 | 1427 | 1913 | 2437 | 2971 | 3547 | 4099 | 4679 | 5281 | 5851 | 6449 |
| 79 | 463 | 929 | 1429 | 1931 | 2441 | 2999 | 3557 | 4111 | 4691 | 5297 | 5857 | 6451 |
| 83 | 467 | 937 | 1433 | 1933 | 2447 | 3001 | 3559 | 4127 | 4703 | 5303 | 5861 | 6469 |
| 89 | 479 | 941 | 1439 | 1949 | 2459 | 3011 | 3571 | 4129 | 4721 | 5309 | 5867 | 6473 |
| 97 | 487 | 947 | 1447 | 1951 | 2467 | 3019 | 3581 | 4133 | 4723 | 5323 | 5869 | 6481 |
| 101 | 491 | 953 | 1451 | 1973 | 2473 | 3023 | 3583 | 4139 | 4729 | 5333 | 5879 | 6491 |
| 103 | 499 | 967 | 1453 | 1979 | 2477 | 3037 | 3593 | 4153 | 4733 | 5347 | 5881 | 6521 |
| 107 | 503 | 971 | 1459 | 1987 | 2503 | 3041 | 3607 | 4157 | 4751 | 5351 | 5897 | 6529 |
| 109 | 509 | 977 | 1471 | 1993 | 2521 | 3049 | 3613 | 4159 | 4759 | 5381 | 5903 | 6547 |
| 113 | 521 | 983 | 1481 | 1997 | 2531 | 3061 | 3617 | 4177 | 4783 | 5387 | 5923 | 6551 |
| 127 | 523 | 991 | 1483 | 1999 | 2539 | 3067 | 3623 | 4201 | 4787 | 5393 | 5927 | 6553 |
| 131 | 541 | 997 | 1487 | 2003 | 2543 | 3079 | 3631 | 4211 | 4789 | 5399 | 5939 | 6563 |
| 137 | 547 | 1009 | 1489 | 2011 | 2549 | 3083 | 3637 | 4217 | 4793 | 5407 | 5953 | 6569 |
| 139 | 557 | 1013 | 1493 | 2017 | 2551 | 3089 | 3643 | 4219 | 4799 | 5413 | 5981 | 6571 |
| 149 | 563 | 1019 | 1499 | 2027 | 2557 | 3109 | 3659 | 4229 | 4801 | 5417 | 5987 | 6577 |
| 151 | 569 | 1021 | 1511 | 2029 | 2579 | 3119 | 3671 | 4231 | 4813 | 5419 | 6007 | 6581 |
| 157 | 571 | 1031 | 1523 | 2039 | 2591 | 3121 | 3673 | 4241 | 4817 | 5431 | 6011 | 6599 |
| 163 | 577 | 1033 | 1531 | 2053 | 2593 | 3137 | 3677 | 4243 | 4831 | 5437 | 6029 | 6607 |
| 167 | 587 | 1039 | 1543 | 2063 | 2609 | 3163 | 3691 | 4253 | 4861 | 5441 | 6037 | 6619 |
| 173 | 593 | 1049 | 1549 | 2069 | 2617 | 3167 | 3697 | 4259 | 4871 | 5443 | 6043 | 6637 |
| 179 | 599 | 1051 | 1553 | 2081 | 2621 | 3169 | 3701 | 4261 | 4877 | 5449 | 6047 | 6653 |
| 181 | 601 | 1061 | 1559 | 2083 | 2633 | 3181 | 3709 | 4271 | 4889 | 5471 | 6053 | 6659 |
| 191 | 607 | 1063 | 1567 | 2087 | 2647 | 3187 | 3719 | 4273 | 4903 | 5477 | 6067 | 6661 |
| 193 | 613 | 1069 | 1571 | 2089 | 2657 | 3191 | 3727 | 4283 | 4909 | 5479 | 6073 | 6673 |
| 197 | 617 | 1087 | 1579 | 2099 | 2659 | 3203 | 3733 | 4289 | 4919 | 5483 | 6079 | 6679 |
| 199 | 619 | 1091 | 1583 | 2111 | 2663 | 3209 | 3739 | 4297 | 4931 | 5501 | 6089 | 6689 |
| 211 | 631 | 1093 | 1597 | 2113 | 2671 | 3217 | 3761 | 4327 | 4933 | 5503 | 6091 | 6691 |
| 223 | 641 | 1097 | 1601 | 2129 | 2677 | 3221 | 3767 | 4337 | 4937 | 5507 | 6101 | 6701 |
| 227 | 643 | 1103 | 1607 | 2131 | 2683 | 3229 | 3769 | 4339 | 4943 | 5519 | 6113 | 6703 |
| 229 | 647 | 1109 | 1609 | 2137 | 2687 | 3251 | 3779 | 4349 | 4951 | 5521 | 6121 | 6709 |
| 233 | 653 | 1117 | 1613 | 2141 | 2689 | 3253 | 3793 | 4357 | 4957 | 5527 | 6131 | 6719 |
| 239 | 659 | 1123 | 1619 | 2143 | 2693 | 3257 | 3797 | 4363 | 4967 | 5531 | 6133 | 6733 |
| 241 | 661 | 1129 | 1621 | 2153 | 2699 | 3259 | 3803 | 4373 | 4969 | 5557 | 6143 | 6737 |
| 251 | 673 | 1151 | 1627 | 2161 | 2707 | 3271 | 3821 | 4391 | 4973 | 5563 | 6151 | 6761 |
| 257 | 677 | 1153 | 1637 | 2179 | 2711 | 3299 | 3823 | 4397 | 4987 | 5569 | 6163 | 6763 |
| 263 | 683 | 1163 | 1657 | 2203 | 2713 | 3301 | 3833 | 4409 | 4993 | 5573 | 6173 | 6779 |
| 269 | 691 | 1171 | 1663 | 2207 | 2719 | 3307 | 3847 | 4421 | 4999 | 5581 | 6197 | 6781 |
| 271 | 701 | 1181 | 1667 | 2213 | 2729 | 3313 | 3851 | 4423 | 5003 | 5591 | 6199 | 6791 |
| 277 | 709 | 1187 | 1669 | 2221 | 2731 | 3319 | 3853 | 4441 | 5009 | 5623 | 6203 | 6793 |
| 281 | 719 | 1193 | 1693 | 2237 | 2741 | 3323 | 3863 | 4447 | 5011 | 5639 | 6211 | 6803 |
| 283 | 727 | 1201 | 1697 | 2239 | 2749 | 3329 | 3877 | 4451 | 5021 | 5641 | 6217 | 6823 |
| 293 | 733 | 1213 | 1699 | 2243 | 2753 | 3331 | 3881 | 4457 | 5023 | 5647 | 6221 | 6827 |
| 307 | 739 | 1217 | 1709 | 2251 | 2767 | 3343 | 3889 | 4463 | 5039 | 5651 | 6229 | 6829 |
| 311 | 743 | 1223 | 1721 | 2267 | 2777 | 3347 | 3907 | 4481 | 5051 | 5653 | 6247 | 6833 |
| 313 | 751 | 1229 | 1723 | 2269 | 2789 | 3359 | 3911 | 4483 | 5059 | 5657 | 6257 | 6841 |
| 317 | 757 | 1231 | 1733 | 2273 | 2791 | 3361 | 3917 | 4493 | 5077 | 5659 | 6263 | 6857 |
| 331 | 761 | 1237 | 1741 | 2281 | 2797 | 3371 | 3919 | 4507 | 5081 | 5669 | 6269 | 6863 |
| 337 | 769 | 1249 | 1747 | 2287 | 2801 | 3373 | 3923 | 4513 | 5087 | 5683 | 6271 | 6869 |

*Example*: Three similar trinomials, only two are factorable:

$x^2 + 2x - 3 = (x+3)(x-1)$, since $(3)(-1) = -3$ and $3 + -1 = 2$

$x^2 - 2x - 3 = (x-3)(x+1)$, since $(-3)(1) = -3$ and $-3 + 1 = -2$

$x^2 - 2x + 3$ Does not factor over the real numbers, since no two integers have a product of $+3$ and a sum of $-2$

*Examples*: Factoring by reverse FOIL, along with the explanation:

$x^2 - 8x - 20 = (x-10)(x+2)$ The two integers whose product is –20 and whose sum is –3 are –10 and 2.

$x^2 - 21x + 20 = (x-20)(x-1)$ The two integers whose product is 20 and whose sum is –21 are –20 and –1.

Reverse *FOIL for polynomials of Form* $ax^2 + bx + c$: If the leading coefficient is other than 1, reverse FOIL works, but a different tactic is needed. Sometimes called "magic factoring," the procedure is shown in this example:

$$3x^2 + 13x + 12, \quad a = 3, b = 13, c = 12$$

*Step 1*: $ac = (3)(12) = 36$

*Step 2*: Write all factor pairs of $ac = 36$: $1, 36; 2, 18; 3, 12; 4, 9; 6, 6$

*Step 3*: Choose the factor pair whose sum is $b = 13$: $4, 9$

*Step 4*: Rewrite polynomial with middle term $13x$ as $4x + 9x$: $3x^2 + 4x + 9x + 12$

*Step 5*: Group first two terms and second two terms in parentheses: $(3x^2 + 4x) + (9x + 12)$

*Step 6*: Take the common factor out of each group: $x(3x + 4) + 3(3x + 4)$

*Step 7*: Take the common *binomial* factor out of each large term: $(3x + 4)(x + 3)$ Correct factorization

*Step 8*: Check the factorization found $3x^2 + 9x + 4x + 12 = 3x^2 + 13x + 12$

*Special Products*: Certain binomials are factored according to formulas, which can be checked by multiplying. The *difference of squares* is perhaps the most important because it comes up so often in applications, as do many quadratic (second-degree) polynomials. It comes from multiplying the *conjugate pair* of binomials, $(a + b)$ with $(a - b)$ to get $a^2 - b^2$.

Difference of squares: $a^2 - b^2 = (a + b)(a - b)$

Difference of cubes: $a^3 - b^3 = (a - b)(a^2 + ab + b^2)$

Sum of cubes: $a^3 + b^3 = (a + b)(a^2 - ab + b^2)$

Square of a sum: $(a + b)^2 = (a + b)(a + b) = a^2 + 2ab + b^2$

Difference of a sum: $(a - b)^2 = (a - b)(a - b) = a^2 - 2ab + b^2$

*Note*: *The sum of squares* $a^2 + b^2$ *is not a factorable binomial over the set of real numbers*. However, it is factorable over the set of complex numbers (see *Complex Numbers* on page 54).

*Examples*: Several special product factorizations are shown:

| Difference of squares: $a^2 - b^2$ | Differerence of cubes: $a^3 - b^3$ | Sum of cubes: $a^3 + b^3$ |
|---|---|---|
| $x^2 - 81 = (x+9)(x-9)$ | $x^3 - 1 = (x-1)(x^2 + x + 1)$ | $x^3 + 1 = (x+1)(x^2 - x + 1)$ |
| $100 - 49x^2 = (10 + 7x)(10 - 7x)$ | $8 - x^3 = (2 - x)(4 + 2x + x^2)$ | $27 + x^3 = (3 + x)(9 - 3x + x^2)$ |

Factorization of all other cases is done using the *quadratic formula*, as shown in the next section.

is $x^2 - r^2 = 0$, and then factor as the difference of squares (see *Polynomials* on page 23): $(x+r)(x-r)=0$, hence, $x=-r$ and $r$.

Verifying this to be the factorization of the difference of squares is simply a matter of applying the distributive property of multiplication over addition to each term:

$$(x+r)(x-r) = xx - xr + xr - rr = x^2 - r^2$$

As previously explained, this process is called expanding by FOIL, for the **F**irst, **O**uter, **I**nner, and **L**ast terms of each binomial, which are multiplied in that order to get the product. From the zero property of multiplication,

$$(x+r)(x-r)=0 \quad \text{implies} \quad x+r=0 \quad \text{or} \quad x-r=0, \text{thus } x=-r \quad \text{or} \quad x=r.$$

If a quadratic cannot be factored as a difference of squares, then factorization is approached as shown in *Factoring Polynomials* on page 24. Given a quadratic equation in the form $ax^2 + bx + c = 0$, first obtain the product $ac$ from the coefficients $a$ and $c$; then determine two numbers whose product is $ac$ and whose sum is $b$.

*Example:* Find the solution to $x^2 - 5x + 6 = 0$ by factoring.

In this example, $a=1, b=-5, c=6$ and $ac=(1)(6)=6$. The factors of 6 whose sum is $-5$ are $-2$ and $-3$. The equation is factored as $x^2 - 5x + 6 = (x-2)(x-3) = 0$. Then by the zero property of multiplication, the roots of the equation are $x=2$ and $x=3$. The parabola intersects the $x$-axis at these two values of $x$.

A more difficult example has a leading coefficient of $x$ other than 1.

*Example:* Factor $8x^2 + 22x + 5 = 0$ and find the values of $x$ that satisfy the equation.

*Solution:* Here, $a=8, b=22$, and $c=5$. Therefore, $ac = 8\times 5 = 40$, and $ac$ is positive, so we are looking for two factors of $ac$, namely $f_1$ and $f_2$, such that $f_1 + f_2 = 22$.

The possible combinations of numbers with product of 40 are 20 and 2, 8 and 5, 4 and 10, and 40 and 1. The requirements that satisfy a sum of 22 are 20 and 2, since $20\times 2 = 40$ and $20 + 2 = 22$. Hence:

$$\begin{aligned} 8x^2 + 22x + 5 &= 0 \\ 8x^2 + 20x + 2x + 5 &= 0 \\ 4x(2x+5) + 1(2x+5) &= 0 \\ (2x+5)(4x+1) &= 0 \end{aligned}$$

On the second line, the common factor of $4x$ in the first two terms is factored out, so the common binomial factor of $2x + 5$ is then apparent, to be factored out of the larger terms. Checking the answer is a matter of simply remultiplying the factors to produce the original expression.

Because the product of the two factors equals zero, each of the factors also equals zero. Thus, $2x + 5 = 0$ and $4x + 1 = 0$. Rearranging and solving, $x = -5/2$ or $x = -1/4$.

*Example:* Factor $8x^2 + 3x - 5 = 0$ and find the solutions of the equation.

*Solution:* Here $a = 8, b = 3, c = -5$, and $ac = 8\times(-5) = -40$. The required numbers must have a product of $-40$ and a sum of 3.

As in the previous example, the possible combinations are 20 and $-2$, $-20$ and 2, $-8$ and 5, 8 and $-5$, 40 and $-1$, and $-40$ and 1. Only 8 and $-5$ satisfy the requirements because $8\times(-5) = -40$, and $8 + (-5) = 3$. Notice that $3x$ in the first line is thus written as $8x - 5x$ in the second line, making it possible to rearrange and simplify the expression.

$$\begin{aligned} 8x^2 + 3x - 5 &= 0 \\ 8x^2 + 8x - 5x - 5 &= 0 \\ 8x(x+1) - 5(x+1) &= 0 \\ (x+1)(8x-5) &= 0 \end{aligned}$$

Solving, for $x + 1 = 0$, $x = -1$; and, for $8x - 5 = 0$, $x = 5/8$.

**Geometric Sequence Formulas**

| To Find | Given | Use Formula |
|---|---|---|
| $a$ | $l \quad n \quad r$ | $a = \dfrac{l}{r^{n-1}}$ |
| | $n \quad r \quad S$ | $a = \dfrac{(r-1)S}{r^n - 1}$ |
| | $l \quad r \quad S$ | $a = lr - (r-1)S$ |
| | $l \quad n \quad S$ | $a(S-a)^{n-1} = l(S-l)^{n-1}$ |
| $l$ | $a \quad n \quad r$ | $l = ar^{n-1}$ |
| | $a \quad r \quad S$ | $l = \dfrac{1}{r}[a + (r-1)S]$ |
| | $a \quad n \quad S$ | $l(S-l)^{n-1} = a(S-a)^{n-1}$ |
| | $n \quad r \quad S$ | $l = \dfrac{S(r-1)r^{n-1}}{r^n - 1}$ |
| $n$ | $a \quad l \quad r$ | $n = \dfrac{\log l - \log a}{\log r} + 1$ |
| | $a \quad r \quad S$ | $n = \dfrac{\log[a + (r-1)S] - \log a}{\log r}$ |
| | $a \quad l \quad S$ | $n = \dfrac{\log l - \log a}{\log(S-a) - \log(S-l)} + 1$ |
| | $l \quad r \quad S$ | $n = \dfrac{\log l - \log[lr - (r-1)S]}{\log r} + 1$ |
| $r$ | $a \quad l \quad n$ | $r = \sqrt[n-1]{\dfrac{l}{n}}$ |
| | $a \quad n \quad S$ | $r^n = \dfrac{Sr}{a} + \dfrac{a-S}{a}$ |
| | $a \quad l \quad S$ | $r = \dfrac{S-a}{S-l}$ |
| | $l \quad n \quad S$ | $r^n = \dfrac{Sr^{n-1}}{S-l} - \dfrac{l}{S-l}$ |
| $S$ | $a \quad n \quad r$ | $S = \dfrac{a(r^n - 1)}{r-1}$ |
| | $a \quad l \quad r$ | $S = \dfrac{lr - a}{r-1}$ |
| | $a \quad l \quad n$ | $S = \dfrac{\sqrt[n-1]{l^n} - \sqrt[n-1]{a^n}}{\sqrt[n-1]{l} - \sqrt[n-1]{a}}$ |
| | $l \quad n \quad r$ | $S = \dfrac{l(r^n - 1)}{(r-1)r^{n-1}}$ |

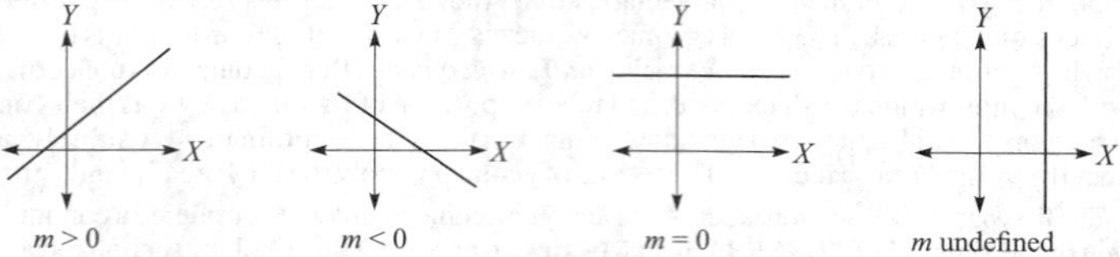


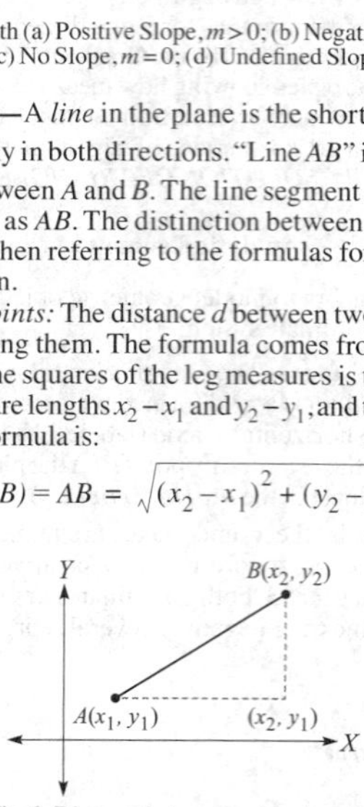

*Point-Point Form:* Substituting $m = \dfrac{y_2 - y_1}{x_2 - x_1}$ into the point-slope form and rearranging terms gives $y - y_1 = \frac{y_1 - y_2}{x_1 - x_2}(x - x_1)$, the *point-point form*.

*Slope-Intercept Form:* Another often-used form is $y = mx + b$, where $m$ is the slope and $b$ is the $y$-intercept. (That is, when $x = 0$, then $y = m(0) + b = b$.) If the previous example is rearranged and solved for $y$, then the *slope-intercept form* is arrived at: $y = -\frac{2}{3}(x-4)+1 = -\frac{2}{3}x + \frac{8}{3} + 1 = -\frac{2}{3}x + \frac{11}{3}$.

*Standard Form of a Line:* In vector representations (see *Complex Numbers* on page 54), lines usually take the standard form, $Ax + By = C$, where $A$, $B$, $C$. For example, $7x - 3y = 4$ is a line in standard form, with $A = 7$, $B = -3$, and $C = 4$.

*Example 4:* What is the standard form equation of a line $AB$ between points $A(4, 5)$ and $B(7, 8)$?

*Solution:* Using the point-point form of the line, where (4,5) is $(x_1, y_1)$ and (7,8) is $(x_2, y_2)$,

$$y - y_1 = \frac{y_1 - y_2}{x_1 - x_2}(x - x_1)$$

$$y - 5 = \frac{5-8}{4-7}(x-4)$$

$$y - 5 = x - 4$$

$$x - y = -1$$

*Example 5:* Find the slope-intercept equation of a line passing through the points (3, 2) and (5, 6). The $y$-intercept is the intersection point of the line with the $y$-axis.

*Solution:* First, find the slope:

$$m = \frac{\Delta y}{\Delta x} = \frac{6-2}{5-3} = \frac{4}{2} = 2$$

The slope-intercept form of the line is $y = 2x + b$, and the value of the constant $b$ can be determined by substituting the coordinates of a point on the line into the general form. Using the coordinates of the point (3,2) gives $2 = (2)(3) + b$ and rearranging, $b = 2 - 6 = -4$. As a check, using another point on the line, (5, 6), yields the same result, $y = 6 = (2)(5) + b$ and $b = 6 - 10 = -4$.

The equation of the line, therefore, is $y = 2x - 4$, indicating that line $y = 2x - 4$ intersects the $y$-axis at point (0,–4), the $y$-intercept.

*Example 6:* Use the point-slope form to find the equation of the line passing through the point (3, 2) and having a slope of 2.

$$y - 2 = 2(x - 3)$$

$$y = 2x - 6 + 2$$

$$y = 2x - 4$$

The slope of this line is positive and crosses the $y$-axis at the $y$-intercept, point (0,–4).

*Parallel Lines:* The two lines, $l_1$ and $l_2$, are parallel if they have the same slope, that is, if $m_1 = m_2$.

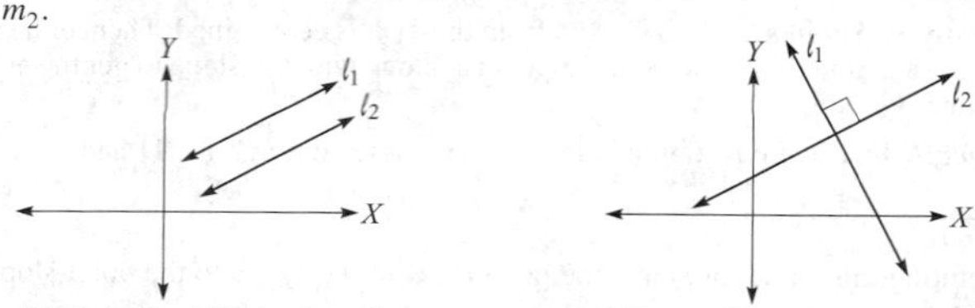


Fig. 5. (a) Parallel lines, $l_1$ and $l_2$. (b) Perpendicular lines, $l_1$ and $l_2$.

*Solution:* Segment area $T+S$:

$$\phi = \cos^{-1}\left(\frac{X_2}{A}\right) = 1.51072\text{ rad}, \quad Y_2 = B\sin\phi = 7\sin(1.51072\text{ rad}) = 10.57504$$

$$T+S = AB\cos^{-1}\left(\frac{X_2}{A}\right) - X_2Y_2 = 10\times7\times0.9268-(6.0041\times5.5978)$$

$$= 64.876-33.6097 = 31.266\ \text{cm}^2$$

*Example 4:* Find the area of elliptical segment $V$ if major radius of ellipse is 4 inches, minor radius is 3 inches, dimension $X = 2.3688$ inches, dimension $Y = 2.4231$ inches.

*Solution:* Segment area $V$:

$$R_2 = \sqrt{AB}, \quad (R_2)^2 = AB = 3\times4 = 12$$

$$V = (R_2)^2\sin^{-1}\left(\frac{X}{A}\right) - XY = 12\sin^{-1}\left(\frac{2.3688}{4}\right) - (2.3688\times2.4231)$$

$$= 7.6054-5.7398 = 1.3656\ \text{in}^2$$

**Four-Arc Oval Approximating an Ellipse*.**—The method of constructing an approximate ellipse by circular arcs, described on page 64, fails when the ratio of the major to minor diameter is 4 or greater. Additionally, it is reported that the method always produces a somewhat larger minor axis than intended. The method described below presents an alternative.

An oval that approximates an ellipse, illustrated in Fig. 7, can be constructed from the following equations:

$$r = \frac{B^2}{2A}\left(\frac{A}{B}\right)^{0.38} \qquad (1)$$

where $A$ and $B$ are dimensions of the major and minor axis, respectively, and $r$ is the radius of the curve at the long ends.

The radius $R$ and its location are found from Equations (2) and (3):

$$X = \frac{\frac{A^2}{4} - Ar + Br - \frac{B^2}{4}}{B-2r} \qquad (2) \qquad\qquad R = \frac{B}{2} + X \qquad (3)$$

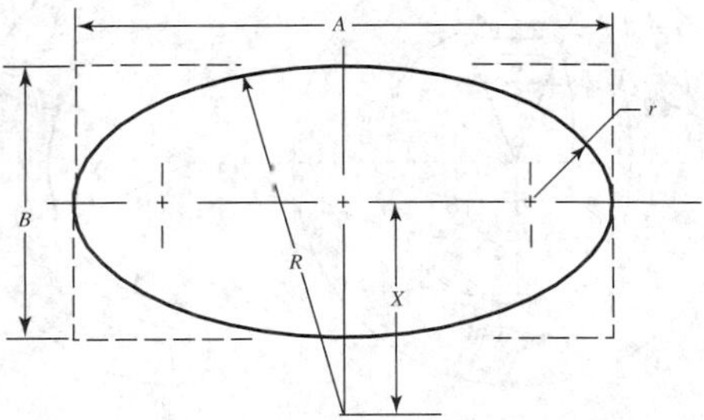


Fig. 7. Four-Arc Oval Ellipse

To make an oval thinner or fatter than what is given, select a smaller or larger radius $r$ is chosen than that calculated by Equation (1), and then $X$ and $R$ are found using Equations (2) and (3).

* *Four-Arc Oval* material contributed by Manfred K. Brueckner

 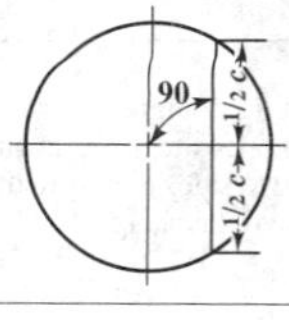  

 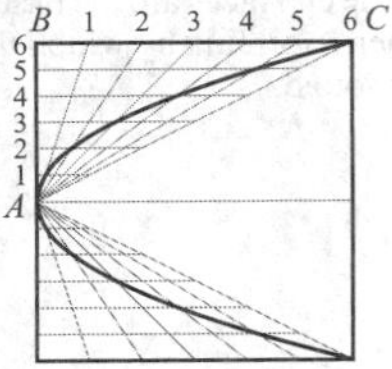  

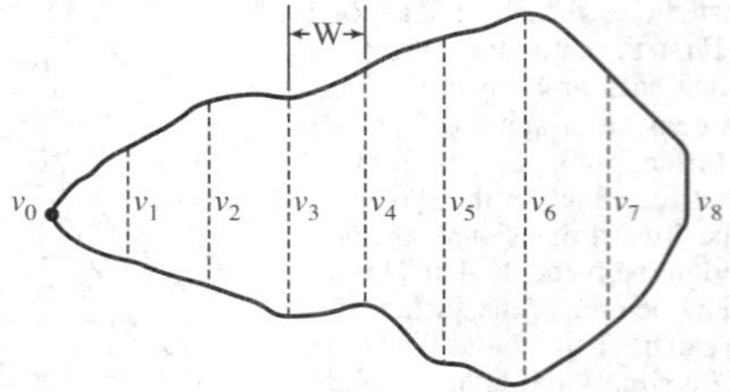


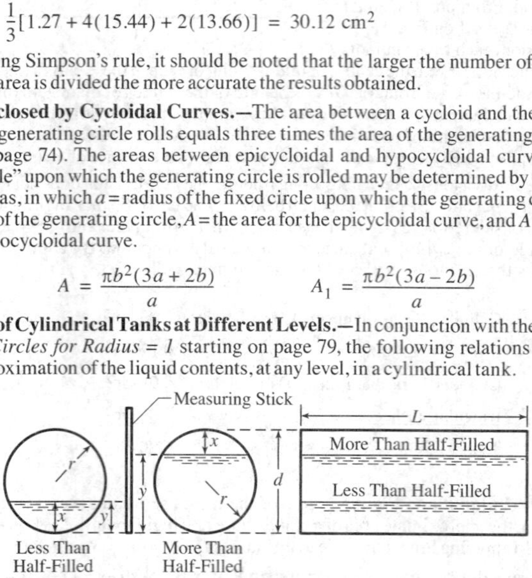

*Segment of a Circle:*

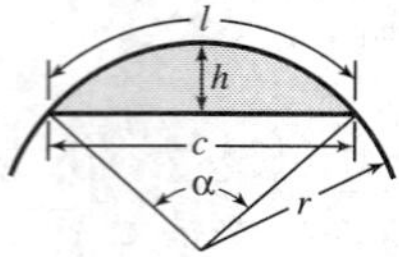


$A$ = area  $l$ = length of arc  $\alpha$ = angle, in degrees

$$c = 2\sqrt{h(2r-h)} \qquad A = \tfrac{1}{2}[rl - c(r-h)]$$

$$r = \frac{c^2 + 4h^2}{8h} \qquad l = 0.01745 r\alpha$$

$$h = r - \tfrac{1}{2}\sqrt{4r^2 - c^2} = r[1 - \cos(\alpha/2)] \qquad \alpha = \frac{57.296\, l}{r}$$

See also *Segments of a Circle* starting on page 78.

*Example:* The radius $r$ is 60 inches and the height $h$ is 8 inches. Find the length of the chord $c$.

$$c = 2\sqrt{h(2r-h)} = 2\sqrt{8 \times (2 \times 60 - 8)} = 2\sqrt{896} = 2 \times 29.93 = 59.86 \text{ in.}$$

*Example:* If $c = 16$, and $h = 6$ inches, what is the radius of the circle of which the segment is a part?

$$r = \frac{c^2 + 4h^2}{8h} = \frac{16^2 + 4 \times 6^2}{8 \times 6} = \frac{256 + 144}{48} = \frac{400}{48} = 8\tfrac{1}{3} \text{ in.}$$

*Cycloid:*

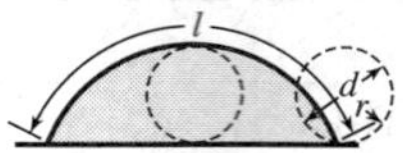


$$\text{Area} = A = 3\pi r^2 = 9.4248 r^2 = 2.3562 d^2$$
$$= 3 \times \text{area of generating circle}$$

$$\text{Length of cycloid} = l = 8r = 4d$$

See also *Areas Enclosed by Cycloidal Curves* on page 68.

*Example:* The diameter of the generating circle of a cycloid is 6 inches. Find the length $l$ of the cycloidal curve and the area enclosed between the curve and the base line.

$$l = 4d = 4 \times 6 = 24 \text{ in} \qquad A = 2.3562 d^2 = 2.3562 \times 6^2 = 84.82 \text{ in}^2$$

*Circular Ring (Annulus):*

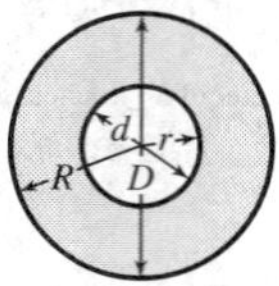


$$\text{Area} = A = \pi(R^2 - r^2) = 3.1416(R^2 - r^2)$$
$$= 3.1416(R + r)(R - r)$$
$$= 0.7854(D^2 - d^2) = 0.7854(D + d)(D - d)$$

*Example:* Let the outside diameter $D = 12$ centimeters and the inside diameter $d = 8$ centimeters. Find the area of the ring.

$$A = 0.7854(D^2 - d^2) = 0.7854(12^2 - 8^2) = 0.7854(144 - 64) = 0.7854 \times 80$$
$$= 62.83 \text{ cm}^2$$

By the alternative formula:

$$A = 0.7854(D + d)(D - d) = 0.7854(12 + 8)(12 - 8) = 0.7854 \times 20 \times 4$$
$$= 62.83 \text{ cm}^2$$

*Sector of Circular Ring:*

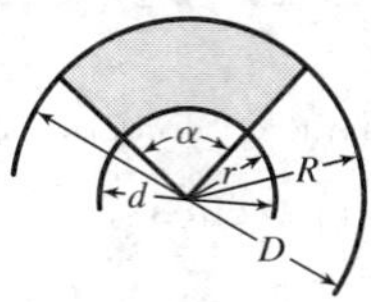


$A$ = area,  $\alpha$ = central angle, in degrees

$$A = \frac{\alpha\pi}{360}(R^2 - r^2) = 0.00873\alpha(R^2 - r^2)$$

$$= \frac{\alpha\pi}{4 \times 360}(D^2 - d^2) = 0.00218\alpha(D^2 - d^2)$$

*Example:* Find the area, if the outside radius $R = 5$ inches, the inside radius $r = 2$ inches, and $\alpha = 72$ degrees.

$$A = 0.00873\alpha(R^2 - r^2) = 0.00873 \times 72(5^2 - 2^2)$$
$$= 0.6286(25 - 4) = 0.6286 \times 21 = 13.2 \text{ in}^2$$

**Segments of Circles for Radius = 1.**—Formulas for segments of circles are given on pages 74 and 78. When central angle α and radius $r$ are known, the following table can be used to find the length of arc $l$, height of segment $h$ ($x$ in the discussion on page 74), chord length $c$, and segment area $A$. Column $A/\pi$ is the ratio of segment area $A$ to the area of a circle with radius $r = 1$, in percent.

When angle α and radius $r$ are not known, but segment height $h$ and chord length $c$ are known, ratio $h/c$ can be used to find α, $l$, and $A$ by linear interpolation. Radius $r$ is found by the formula on page 74 or 78. The value of $l$ is then multiplied by the radius $r$ and the area $A$ by $r^2$.

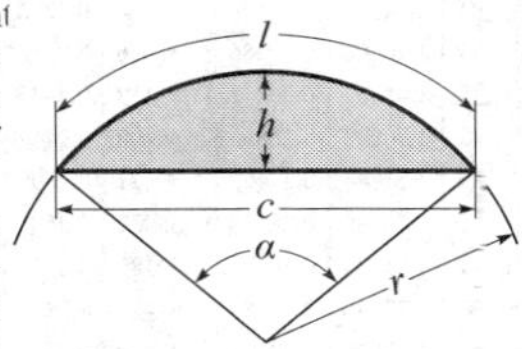


Angle α can be found thus with an accuracy of about 0.001 degree; arc length $l$ with an error of about 0.02 percent; and area $A$ with an error ranging from about 0.02 percent for the highest entry value of $h/c$ to about 1 percent for values of $h/c$ of about 0.050. For lower values of $h/c$, and where greater accuracy is required, area $A$ should be found by the formula on page 74.

*Example:* A 3-foot diameter cylindrical tank, mounted horizontally, contains fuel. What is the fuel depth, in inches, when the tank is 20% full? *Solution:* Locate 20% in table column $A/\pi$%. The depth equals $h$ multiplied by the radius: $hr = 0.50758 \times 1.5 \times 12 = 9.14$ inches

**Segments of Circles for Radius = 1** (US Customary or Metric Units)

| α, Deg. | l | h | c | Area A | A/π % | h/c | α, Deg. | l | h | c | Area A | A/π % | h/c |
|---|---|---|---|---|---|---|---|---|---|---|---|---|---|
| 1 | 0.01745 | 0.00004 | 0.01745 | 0.0000 | 0.0 | 0.00218 | 41 | 0.71558 | 0.06333 | 0.70041 | 0.0298 | 0.9 | 0.09[illegible]41 |
| 2 | 0.03491 | 0.00015 | 0.03490 | 0.0000 | 0.0 | 0.00436 | 42 | 0.73304 | 0.06642 | 0.71674 | 0.0320 | 1.0 | 0.09[illegible]67 |
| 3 | 0.05236 | 0.00034 | 0.05235 | 0.0000 | 0.0 | 0.00655 | 43 | 0.75049 | 0.06958 | 0.73300 | 0.0342 | 1.1 | 0.09[illegible]93 |
| 4 | 0.06981 | 0.00061 | 0.06980 | 0.0000 | 0.0 | 0.00873 | 44 | 0.76794 | 0.07282 | 0.74921 | 0.0366 | 1.2 | 0.09[illegible]19 |
| 5 | 0.08727 | 0.00095 | 0.08724 | 0.0001 | 0.0 | 0.01091 | 45 | 0.78540 | 0.07612 | 0.76537 | 0.0391 | 1.2 | 0.09[illegible]46 |
| 6 | 0.10472 | 0.00137 | 0.10467 | 0.0001 | 0.0 | 0.01309 | 46 | 0.80285 | 0.07950 | 0.78146 | 0.0418 | 1.3 | 0.10[illegible]73 |
| 7 | 0.12217 | 0.00187 | 0.12210 | 0.0002 | 0.0 | 0.01528 | 47 | 0.82030 | 0.08294 | 0.79750 | 0.0445 | 1.4 | 0.10[illegible]00 |
| 8 | 0.13963 | 0.00244 | 0.13951 | 0.0002 | 0.0 | 0.01746 | 48 | 0.83776 | 0.08645 | 0.81347 | 0.0473 | 1.5 | 0.10[illegible]28 |
| 9 | 0.15708 | 0.00308 | 0.15692 | 0.0003 | 0.0 | 0.01965 | 49 | 0.85521 | 0.09004 | 0.82939 | 0.0503 | 1.6 | 0.10[illegible]56 |
| 10 | 0.17453 | 0.00381 | 0.17431 | 0.0004 | 0.0 | 0.02183 | 50 | 0.87266 | 0.09369 | 0.84524 | 0.0533 | 1.7 | 0.11[illegible]85 |
| 11 | 0.19199 | 0.00460 | 0.19169 | 0.0006 | 0.0 | 0.02402 | 51 | 0.89012 | 0.09741 | 0.86102 | 0.0565 | 1.8 | 0.1[illegible]314 |
| 12 | 0.20944 | 0.00548 | 0.20906 | 0.0008 | 0.0 | 0.02620 | 52 | 0.90757 | 0.10121 | 0.87674 | 0.0598 | 1.9 | 0.1[illegible]543 |
| 13 | 0.22689 | 0.00643 | 0.22641 | 0.0010 | 0.0 | 0.02839 | 53 | 0.92502 | 0.10507 | 0.89240 | 0.0632 | 2.0 | 0.1[illegible]773 |
| 14 | 0.24435 | 0.00745 | 0.24374 | 0.0012 | 0.0 | 0.03058 | 54 | 0.94248 | 0.10899 | 0.90798 | 0.0667 | 2.1 | 0.1[illegible]004 |
| 15 | 0.26180 | 0.00856 | 0.26105 | 0.0015 | 0.0 | 0.03277 | 55 | 0.95993 | 0.11299 | 0.92350 | 0.0704 | 2.2 | 0.1[illegible]235 |
| 16 | 0.27925 | 0.00973 | 0.27835 | 0.0018 | 0.1 | 0.03496 | 56 | 0.97738 | 0.11705 | 0.93894 | 0.0742 | 2.4 | 0.1[illegible]466 |
| 17 | 0.29671 | 0.01098 | 0.29562 | 0.0022 | 0.1 | 0.03716 | 57 | 0.99484 | 0.12118 | 0.95432 | 0.0781 | 2.5 | 0.1[illegible]698 |
| 18 | 0.31416 | 0.01231 | 0.31287 | 0.0026 | 0.1 | 0.03935 | 58 | 1.01229 | 0.12538 | 0.96962 | 0.0821 | 2.6 | 0.1[illegible]931 |
| 19 | 0.33161 | 0.01371 | 0.33010 | 0.0030 | 0.1 | 0.04155 | 59 | 1.02974 | 0.12964 | 0.98485 | 0.0863 | 2.7 | 0.1[illegible]164 |
| 20 | 0.34907 | 0.01519 | 0.34730 | 0.0035 | 0.1 | 0.04374 | 60 | 1.04720 | 0.13397 | 1.00000 | 0.0906 | 2.9 | 0.1[illegible]397 |
| 21 | 0.36652 | 0.01675 | 0.36447 | 0.0041 | 0.1 | 0.04594 | 61 | 1.06465 | 0.13837 | 1.01508 | 0.0950 | 3.0 | 0.1[illegible]632 |
| 22 | 0.38397 | 0.01837 | 0.38162 | 0.0047 | 0.1 | 0.04814 | 62 | 1.08210 | 0.14283 | 1.03008 | 0.0996 | 3.2 | 0.1[illegible]866 |
| 23 | 0.40143 | 0.02008 | 0.39874 | 0.0053 | 0.2 | 0.05035 | 63 | 1.09956 | 0.14736 | 1.04500 | 0.1043 | 3.3 | 0.1[illegible]101 |
| 24 | 0.41888 | 0.02185 | 0.41582 | 0.0061 | 0.2 | 0.05255 | 64 | 1.11701 | 0.15195 | 1.05984 | 0.1091 | 3.5 | 0.1[illegible]337 |
| 25 | 0.43633 | 0.02370 | 0.43288 | 0.0069 | 0.2 | 0.05476 | 65 | 1.13446 | 0.15661 | 1.07460 | 0.1141 | 3.6 | 0.1[illegible]574 |
| 26 | 0.45379 | 0.02563 | 0.44990 | 0.0077 | 0.2 | 0.05697 | 66 | 1.15192 | 0.16133 | 1.08928 | 0.1192 | 3.8 | 0.1[illegible]811 |
| 27 | 0.47124 | 0.02763 | 0.46689 | 0.0086 | 0.3 | 0.05918 | 67 | 1.16937 | 0.16611 | 1.10387 | 0.1244 | 4.0 | 0.1[illegible]048 |
| 28 | 0.48869 | 0.02970 | 0.48384 | 0.0096 | 0.3 | 0.06139 | 68 | 1.18682 | 0.17096 | 1.11839 | 0.1298 | 4.1 | 0.1[illegible]287 |
| 29 | 0.50615 | 0.03185 | 0.50076 | 0.0107 | 0.3 | 0.06361 | 69 | 1.20428 | 0.17587 | 1.13281 | 0.1353 | 4.3 | 0.1[illegible]525 |
| 30 | 0.52360 | 0.03407 | 0.51764 | 0.0118 | 0.4 | 0.06583 | 70 | 1.22173 | 0.18085 | 1.14715 | 0.1410 | 4.5 | 0.[illegible]765 |
| 31 | 0.54105 | 0.03637 | 0.53448 | 0.0130 | 0.4 | 0.06805 | 71 | 1.23918 | 0.18588 | 1.16141 | 0.1468 | 4.7 | 0.[illegible]6005 |
| 32 | 0.55851 | 0.03874 | 0.55127 | 0.0143 | 0.5 | 0.07027 | 72 | 1.25664 | 0.19098 | 1.17557 | 0.1528 | 4.9 | 0.[illegible]6246 |
| 33 | 0.57596 | 0.04118 | 0.56803 | 0.0157 | 0.5 | 0.07250 | 73 | 1.27409 | 0.19614 | 1.18965 | 0.1589 | 5.1 | 0.[illegible]6488 |
| 34 | 0.59341 | 0.04370 | 0.58474 | 0.0171 | 0.5 | 0.07473 | 74 | 1.29154 | 0.20136 | 1.20363 | 0.1651 | 5.3 | 0.[illegible]6730 |
| 35 | 0.61087 | 0.04628 | 0.60141 | 0.0186 | 0.6 | 0.07696 | 75 | 1.30900 | 0.20665 | 1.21752 | 0.1715 | 5.5 | 0.[illegible]6973 |
| 36 | 0.62832 | 0.04894 | 0.61803 | 0.0203 | 0.6 | 0.07919 | 76 | 1.32645 | 0.21199 | 1.23132 | 0.1781 | 5.7 | 0.[illegible]7216 |
| 37 | 0.64577 | 0.05168 | 0.63461 | 0.0220 | 0.7 | 0.08143 | 77 | 1.34390 | 0.21739 | 1.24503 | 0.1848 | 5.9 | 0.[illegible]7461 |
| 38 | 0.66323 | 0.05448 | 0.65114 | 0.0238 | 0.8 | 0.08367 | 78 | 1.36136 | 0.22285 | 1.25864 | 0.1916 | 6.1 | 0.[illegible]7706 |
| 39 | 0.68068 | 0.05736 | 0.66761 | 0.0257 | 0.8 | 0.08592 | 79 | 1.37881 | 0.22838 | 1.27216 | 0.1986 | 6.3 | 0.[illegible]7952 |
| 40 | 0.69813 | 0.06031 | 0.68404 | 0.0277 | 0.9 | 0.08816 | 80 | 1.39626 | 0.23396 | 1.28558 | 0.2057 | 6.5 | 0.[illegible]8199 |

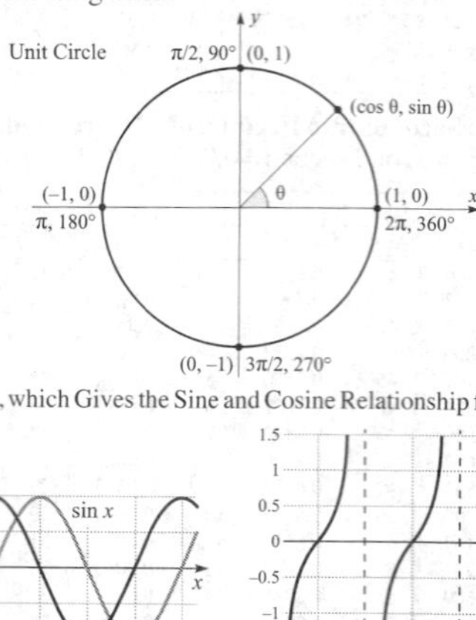


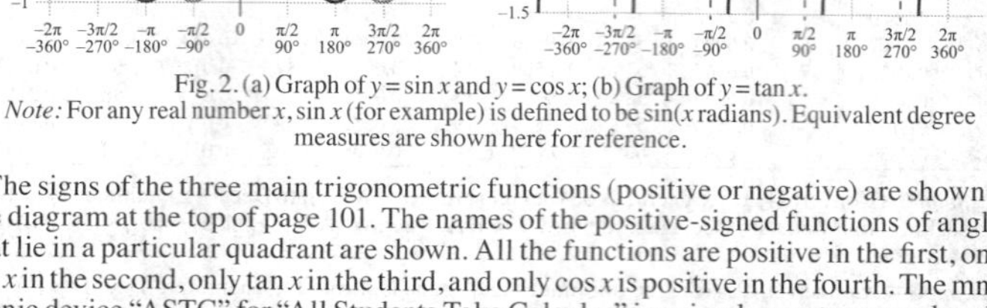

## Involute Function Values for Angles from 32 to 41 Degrees

| | Degrees | | | | | | | | |
|---|---|---|---|---|---|---|---|---|---|
| | 32 | 33 | 34 | 35 | 36 | 37 | 38 | 39 | 40 |
| Minutes | Involute Function Values | | | | | | | | |
| 0 | 0.066364 | 0.073449 | 0.081097 | 0.089342 | 0.098224 | 0.107782 | 0.118061 | 0.129106 | 0.140968 |
| 1 | 0.066478 | 0.073572 | 0.081229 | 0.089485 | 0.098378 | 0.107948 | 0.118238 | 0.129297 | 0.141173 |
| 2 | 0.066591 | 0.073695 | 0.081362 | 0.089628 | 0.098532 | 0.108113 | 0.118416 | 0.129488 | 0.141378 |
| 3 | 0.066705 | 0.073818 | 0.081494 | 0.089771 | 0.098686 | 0.108279 | 0.118594 | 0.129679 | 0.141584 |
| 4 | 0.066820 | 0.073941 | 0.081627 | 0.089914 | 0.098840 | 0.108445 | 0.118773 | 0.129870 | 0.141789 |
| 5 | 0.066934 | 0.074064 | 0.081760 | 0.090058 | 0.098994 | 0.108611 | 0.118951 | 0.130062 | 0.141995 |
| 6 | 0.067048 | 0.074188 | 0.081894 | 0.090201 | 0.099149 | 0.108777 | 0.119130 | 0.130254 | 0.142201 |
| 7 | 0.067163 | 0.074312 | 0.082027 | 0.090345 | 0.099303 | 0.108943 | 0.119309 | 0.130446 | 0.142408 |
| 8 | 0.067277 | 0.074435 | 0.082161 | 0.090489 | 0.099458 | 0.109110 | 0.119488 | 0.130639 | 0.142614 |
| 9 | 0.067392 | 0.074559 | 0.082294 | 0.090633 | 0.099614 | 0.109277 | 0.119667 | 0.130832 | 0.14282[illegible] |
| 10 | 0.067507 | 0.074684 | 0.082428 | 0.090777 | 0.099769 | 0.109444 | 0.119847 | 0.131025 | 0.143028 |
| 11 | 0.067622 | 0.074808 | 0.082562 | 0.090922 | 0.099924 | 0.109611 | 0.120027 | 0.131218 | 0.143236 |
| 12 | 0.067738 | 0.074932 | 0.082697 | 0.091067 | 0.100080 | 0.109779 | 0.120207 | 0.131411 | 0.14344[illegible] |
| 13 | 0.067853 | 0.075057 | 0.082831 | 0.091211 | 0.100236 | 0.109947 | 0.120387 | 0.131605 | 0.14365[illegible] |
| 14 | 0.067969 | 0.075182 | 0.082966 | 0.091356 | 0.100392 | 0.110114 | 0.120567 | 0.131799 | 0.14385[illegible] |
| 15 | 0.068084 | 0.075307 | 0.083101 | 0.091502 | 0.100549 | 0.110283 | 0.120748 | 0.131993 | 0.144063 |
| 16 | 0.068200 | 0.075432 | 0.083235 | 0.091647 | 0.100705 | 0.110451 | 0.120929 | 0.132187 | 0.144275 |
| 17 | 0.068316 | 0.075557 | 0.083371 | 0.091793 | 0.100862 | 0.110619 | 0.121110 | 0.132381 | 0.144485 |
| 18 | 0.068432 | 0.075683 | 0.083506 | 0.091938 | 0.101019 | 0.110788 | 0.121291 | 0.132576 | 0.144694 |
| 19 | 0.068549 | 0.075808 | 0.083641 | 0.092084 | 0.101176 | 0.110957 | 0.121473 | 0.132771 | 0.144903 |
| 20 | 0.068665 | 0.075934 | 0.083777 | 0.092230 | 0.101333 | 0.111126 | 0.121655 | 0.132966 | 0.1451[illegible]3 |
| 21 | 0.068782 | 0.076060 | 0.083913 | 0.092377 | 0.101490 | 0.111295 | 0.121837 | 0.133162 | 0.145323 |
| 22 | 0.068899 | 0.076186 | 0.084049 | 0.092523 | 0.101648 | 0.111465 | 0.122019 | 0.133358 | 0.145533 |
| 23 | 0.069016 | 0.076312 | 0.084185 | 0.092670 | 0.101806 | 0.111635 | 0.122201 | 0.133553 | 0.145743 |
| 24 | 0.069133 | 0.076439 | 0.084321 | 0.092816 | 0.101964 | 0.111805 | 0.122384 | 0.133750 | 0.1459[illegible]4 |
| 25 | 0.069250 | 0.076565 | 0.084458 | 0.092963 | 0.102122 | 0.111975 | 0.122567 | 0.133946 | 0.1461[illegible]5 |
| 26 | 0.069367 | 0.076692 | 0.084594 | 0.093111 | 0.102280 | 0.112145 | 0.122750 | 0.134143 | 0.146376 |
| 27 | 0.069485 | 0.076819 | 0.084731 | 0.093258 | 0.102439 | 0.112316 | 0.122933 | 0.134339 | 0.1465[illegible]7 |
| 28 | 0.069602 | 0.076946 | 0.084868 | 0.093406 | 0.102598 | 0.112486 | 0.123117 | 0.134537 | 0.1467[illegible]9 |
| 29 | 0.069720 | 0.077073 | 0.085005 | 0.093553 | 0.102757 | 0.112657 | 0.123300 | 0.134734 | 0.1470[illegible]0 |
| 30 | 0.069838 | 0.077200 | 0.085142 | 0.093701 | 0.102916 | 0.112829 | 0.123484 | 0.134931 | 0.147222 |
| 31 | 0.069956 | 0.077328 | 0.085280 | 0.093849 | 0.103075 | 0.113000 | 0.123668 | 0.135129 | 0.147435 |
| 32 | 0.070075 | 0.077455 | 0.085418 | 0.093998 | 0.103235 | 0.113172 | 0.123853 | 0.135327 | 0.147647 |
| 33 | 0.070193 | 0.077583 | 0.085555 | 0.094146 | 0.103395 | 0.113343 | 0.124037 | 0.135525 | 0.147860 |
| 34 | 0.070312 | 0.077711 | 0.085693 | 0.094295 | 0.103555 | 0.113515 | 0.124222 | 0.135724 | 0.148073 |
| 35 | 0.070430 | 0.077839 | 0.085832 | 0.094443 | 0.103715 | 0.113688 | 0.124407 | 0.135923 | 0.148[illegible]86 |
| 36 | 0.070549 | 0.077968 | 0.085970 | 0.094593 | 0.103875 | 0.113860 | 0.124592 | 0.136122 | 0.148[illegible]00 |
| 37 | 0.070668 | 0.078096 | 0.086108 | 0.094742 | 0.104036 | 0.114033 | 0.124778 | 0.136321 | 0.148714 |
| 38 | 0.070788 | 0.078225 | 0.086247 | 0.094891 | 0.104196 | 0.114205 | 0.124964 | 0.136520 | 0.148[illegible]28 |
| 39 | 0.070907 | 0.078354 | 0.086386 | 0.095041 | 0.104357 | 0.114378 | 0.125150 | 0.136720 | 0.149[illegible]42 |
| 40 | 0.071026 | 0.078483 | 0.086525 | 0.095190 | 0.104518 | 0.114552 | 0.125336 | 0.136920 | 0.149[illegible]57 |
| 41 | 0.071146 | 0.078612 | 0.086664 | 0.095340 | 0.104680 | 0.114725 | 0.125522 | 0.137120 | 0.149[illegible]72 |
| 42 | 0.071266 | 0.078741 | 0.086804 | 0.095490 | 0.104841 | 0.114899 | 0.125709 | 0.137320 | 0.149[illegible]87 |
| 43 | 0.071386 | 0.078871 | 0.086943 | 0.095641 | 0.105003 | 0.115073 | 0.125896 | 0.137521 | 0.150002 |
| 44 | 0.071506 | 0.079000 | 0.087083 | 0.095791 | 0.105165 | 0.115247 | 0.126083 | 0.137722 | 0.150218 |
| 45 | 0.071626 | 0.079130 | 0.087223 | 0.095942 | 0.105327 | 0.115421 | 0.126270 | 0.137923 | 0.150434 |
| 46 | 0.071747 | 0.079260 | 0.087363 | 0.096093 | 0.105489 | 0.115595 | 0.126457 | 0.138124 | 0.150650 |
| 47 | 0.071867 | 0.079390 | 0.087503 | 0.096244 | 0.105652 | 0.115770 | 0.126645 | 0.138326 | 0.150866 |
| 48 | 0.071988 | 0.079520 | 0.087644 | 0.096395 | 0.105814 | 0.115945 | 0.126833 | 0.138528 | 0.15[illegible]083 |
| 49 | 0.072109 | 0.079651 | 0.087784 | 0.096546 | 0.105977 | 0.116120 | 0.127021 | 0.138730 | 0.15[illegible]299 |
| 50 | 0.072230 | 0.079781 | 0.087925 | 0.096698 | 0.106140 | 0.116296 | 0.127209 | 0.138932 | 0.15[illegible]517 |
| 51 | 0.072351 | 0.079912 | 0.088066 | 0.096850 | 0.106304 | 0.116471 | 0.127398 | 0.139134 | 0.15[illegible]734 |
| 52 | 0.072473 | 0.080043 | 0.088207 | 0.097002 | 0.106467 | 0.116647 | 0.127587 | 0.139337 | 0.15[illegible]952 |
| 53 | 0.072594 | 0.080174 | 0.088348 | 0.097154 | 0.106631 | 0.116823 | 0.127776 | 0.139540 | 0.152169 |
| 54 | 0.072716 | 0.080306 | 0.088490 | 0.097306 | 0.106795 | 0.116999 | 0.127965 | 0.139743 | 0.152388 |
| 55 | 0.072838 | 0.080437 | 0.088631 | 0.097459 | 0.106959 | 0.117175 | 0.128155 | 0.139947 | 0.152606 |
| 56 | 0.072960 | 0.080569 | 0.088773 | 0.097611 | 0.107123 | 0.117352 | 0.128344 | 0.140151 | 0.152825 |
| 57 | 0.073082 | 0.080700 | 0.088915 | 0.097764 | 0.107288 | 0.117529 | 0.128534 | 0.140355 | 0.153044 |
| 58 | 0.073204 | 0.080832 | 0.089057 | 0.097917 | 0.107452 | 0.117706 | 0.128725 | 0.140559 | 0.1[illegible]3263 |
| 59 | 0.073326 | 0.080964 | 0.089200 | 0.098071 | 0.107617 | 0.117883 | 0.128915 | 0.140763 | 0.1[illegible]3482 |
| 60 | 0.073449 | 0.081097 | 0.089342 | 0.098224 | 0.107782 | 0.118061 | 0.129106 | 0.140968 | 0.1[illegible]3702 |

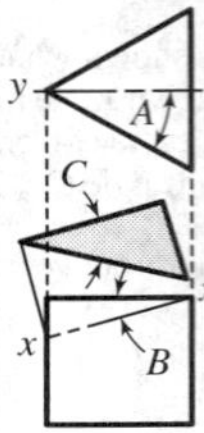


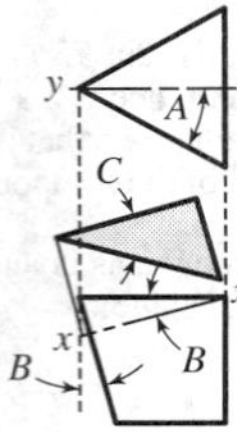


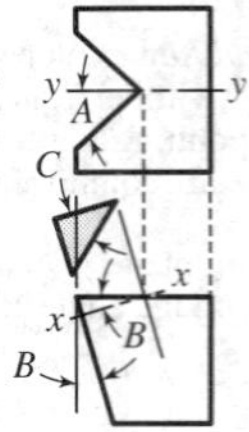


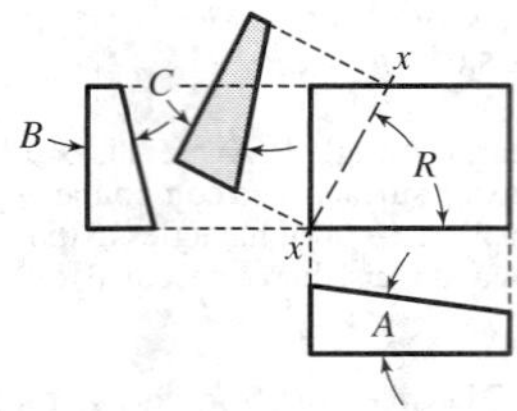


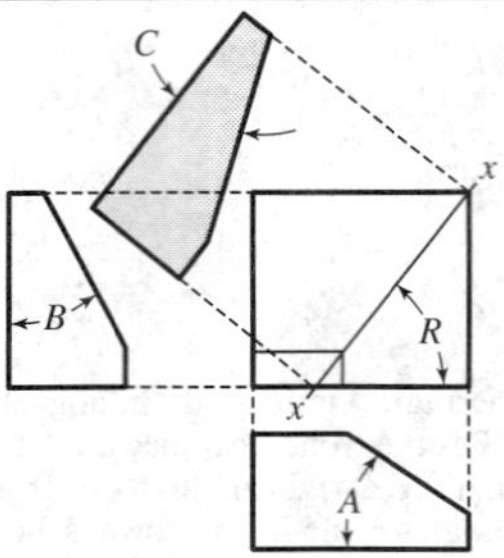


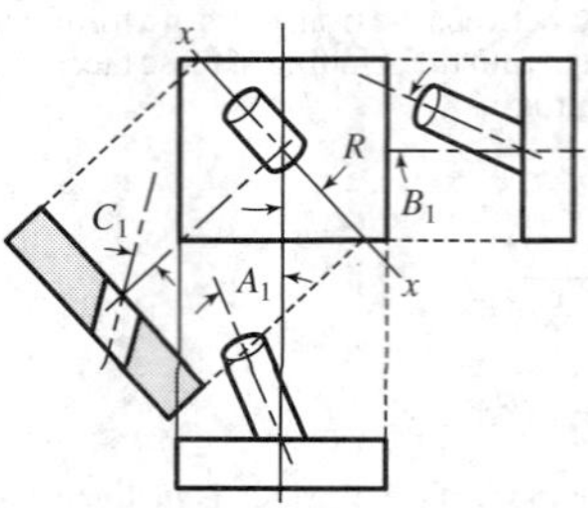

## Matrix Operations

**Matrix Addition and Subtraction.—**The sum or difference of matrices is determined simply by adding or subtracting the corresponding elements of each matrix.

So, matrix $C$ is the result of adding or subtracting matrices $A$ and $B$; the entries are combined as follows:

$$a_{ij}+b_{ij}=c_{ij} \quad \text{or} \quad a_{ij}-b_{ij}=c_{ij}$$

Matrices must be the same size, $m \times n$, to be combined this way. That is, $A_{mn}+B_{mn}=C_{mn}$; $A_{mn}-B_{mn}=C_{mn}$.

An efficient way to indicate both + and – is with the symbol ±. Thus, in the matrix display below, $a_{ij} \pm b_{ij}$ covers both matrix operations.

*Example 1:*

$$\begin{bmatrix} 4 & 6 & -5 \\ 5 & -7 & 8 \\ -8 & 6 & -7 \end{bmatrix} + \begin{bmatrix} 8 & -2 & 6 \\ -6 & 9 & 5 \\ 9 & -2 & 2 \end{bmatrix} = \begin{bmatrix} (4+8) & (6-2) & (-5+6) \\ (5-6) & (-7+9) & (8+5) \\ (-8+9) & (6-2) & (-7+2) \end{bmatrix} = \begin{bmatrix} 12 & 4 & 1 \\ -1 & 2 & 13 \\ 1 & 4 & -5 \end{bmatrix}$$

**Matrix Multiplication.—**Two matrices can be multiplied *only* if the number of columns in the first matrix equals the number of rows of the second matrix. For example, a $1 \times 3$ matrix can multiplied by a $3 \times 2$ matrix, but not the other way around. Or a $2 \times 4$ by a $4 \times 3$, in that order. In general, an $m \times n$ and an $n \times p$ can be multiplied; the product matrix is an $m \times p$ matrix. Matrix multiplication is not commutative, that is, $A \times B$ is not necessarily equal to $B \times A$.

The steps in matrix multiplication are shown in the instructive example below for a general $3 \times 2$ matrix multiplied by a $2 \times 1$ matrix. The result is a $3 \times 1$ matrix.

$$\begin{pmatrix} 1 & 4 \\ -3 & 1 \\ 2 & -1 \end{pmatrix} \times \begin{pmatrix} 0 \\ -2 \end{pmatrix} = \begin{pmatrix} 1 \cdot 0 + 4 \cdot -2 \\ -3 \cdot 0 + 1 \cdot -2 \\ 2 \cdot 0 + -1 \cdot -2 \end{pmatrix} = \begin{pmatrix} -8 \\ -2 \\ 2 \end{pmatrix}$$

In general, *each entry* $(AB)_{ij}$ in the product matrix (where $i$ is the row number and $j$ is the column number) is equal to $a_{i1}b_{1j}+a_{i2}b_{2j}+\ldots+a_{in}b_{nj}$. For example:

$$\underset{\substack{3 \times 2 \\ \text{matrix}}}{\begin{pmatrix} a_{11} & a_{12} \\ a_{21} & a_{22} \\ a_{31} & a_{32} \end{pmatrix}} \times \underset{\substack{2 \times 2 \\ \text{matrix}}}{\begin{pmatrix} b_{11} & b_{12} \\ b_{21} & b_{22} \end{pmatrix}} = \underset{\substack{3 \times 2 \\ \text{matrix}}}{\begin{pmatrix} a_{11}b_{11}+a_{12}b_{21} & a_{11}b_{12}+a_{12}b_{22} \\ a_{21}b_{11}+a_{22}b_{21} & a_{21}b_{12}+a_{22}b_{22} \\ a_{31}b_{11}+a_{32}b_{21} & a_{31}b_{12}+a_{32}b_{22} \end{pmatrix}}$$

*Example 2:*

$$\begin{bmatrix} 1 & 2 & 3 \\ 4 & 5 & 6 \\ 3 & 2 & 1 \end{bmatrix} \times \begin{bmatrix} 7 & 8 & 9 \\ 1 & 2 & 3 \\ 4 & 5 & 7 \end{bmatrix} = \begin{bmatrix} (1 \cdot 7+2 \cdot 1+3 \cdot 4) & (1 \cdot 8+2 \cdot 2+3 \cdot 5) & (1 \cdot 9+2 \cdot 3+3 \cdot 7) \\ (4 \cdot 7+5 \cdot 1+6 \cdot 4) & (4 \cdot 8+5 \cdot 2+6 \cdot 5) & (4 \cdot 9+5 \cdot 3+6 \cdot 7) \\ (3 \cdot 7+2 \cdot 1+1 \cdot 4) & (3 \cdot 8+2 \cdot 2+1 \cdot 5) & (3 \cdot 9+2 \cdot 3+1 \cdot 7) \end{bmatrix}$$

$$= \begin{bmatrix} (7+2+12) & (8+4+15) & (9+6+21) \\ (28+5+24) & (32+10+30) & (36+15+42) \\ (21+2+4) & (24+4+5) & (27+6+7) \end{bmatrix} = \begin{bmatrix} 21 & 27 & 36 \\ 57 & 72 & 93 \\ 27 & 33 & 40 \end{bmatrix}$$

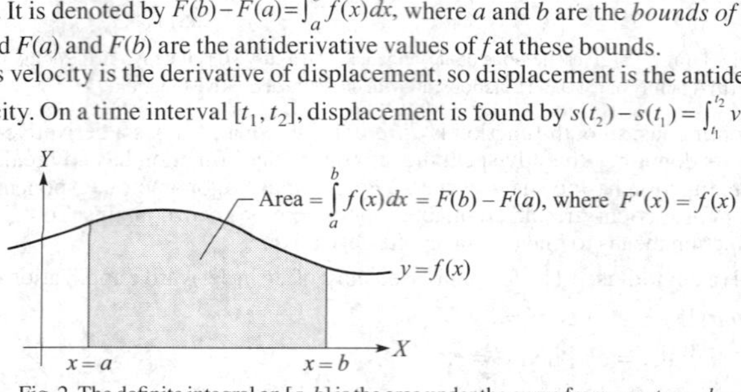

**Table of Derivatives and Integrals** *(Continued)*

| Derivatives | Integrals |
|---|---|
| $\frac{d}{dx}(\cos x) = -\sin x$ | $\int \sin x\,dx = -\cos x + C$ |
| $\frac{d}{dx}(\tan x) = \sec^2 x$ | $\int \tan x\,dx = -\ln\lvert\cos x\rvert + C$ |
| $\frac{d}{dx}(\cot x) = -\csc^2 x$ | $\int \cot x\,dx = \ln\lvert\sin x\rvert + C$ |
| $\frac{d}{dx}(\sec x) = \sec x \tan x$ | $\int \sin^2 x\,dx = -\frac{1}{4}\sin(2x) + \frac{1}{2}x + C$ |
| $\frac{d}{dx}(\csc x) = -\csc x \cot x$ | $\int \cos^2 x\,dx = \frac{1}{4}\sin(2x) + \frac{1}{2}x + C$ |
| $\frac{d}{dx}(\sin^{-1} x) = \frac{1}{\sqrt{1-x^2}} \quad x \neq 1, -1$ | $\int \frac{dx}{\sqrt{b^2 - x^2}} = \sin^{-1}\left(\frac{x}{b}\right) + C$ |
| $\frac{d}{dx}(\cos^{-1} x) = \frac{-1}{\sqrt{1-x^2}} \quad x \neq 1, -1$ | $\int \frac{dx}{\sqrt{x^2 - b^2}} = \ln\left(\frac{x}{b} + \frac{\sqrt{x^2 - b^2}}{b}\right) + C$ |
| $\frac{d}{dx}(\tan^{-1} x) = \frac{1}{1 + x^2}$ | $\int \frac{dx}{b^2 + x^2} = \frac{1}{b}\tan^{-1}\left(\frac{x}{b}\right) + C$ |
| $\frac{d}{dx}(\cot^{-1} x) = \frac{-1}{1 + x^2}$ | $\int \frac{dx}{b^2 - x^2} = \frac{1}{b}\tan^{-1}\left(\frac{x}{b}\right) = \frac{-1}{2b}\log\left[\frac{\lvert x-b\rvert}{\lvert x+b\rvert}\right] + C$ |
| $\frac{d}{dx}(\sec^{-1} x) = \frac{1}{x\sqrt{x^2 - 1}} \quad x \neq 1, -1$ | $\int \frac{dx}{x^2 - b^2} = -\frac{1}{b}\cos^{-1}\left(\frac{x}{b}\right) = \frac{1}{2b}\log\left[\frac{\lvert x-b\rvert}{\lvert x+b\rvert}\right] + C$ |
| $\frac{d}{dx}(\csc^{-1} x) = \frac{-1}{x\sqrt{x^2 - 1}} \quad x \neq 1, -1$ | $\int \frac{dx}{ax^2 + bx + c} = \frac{2}{\sqrt{4ac - b^2}}\tan^{-1}\left[\frac{2ax + b}{\sqrt{4ac - b^2}}\right] + C$ |
| $\frac{d}{dx}(\ln(\sin x)) = \cot x$ | $\int e^{ax}\sin(bx)\,dx = \frac{a\sin(bx) - b\cos(bx)}{a^2 + b^2}e^{ax} + C$ |
| $\frac{d}{dx}(\ln(\cos x)) = -\tan x$ | $\int e^{ax}\cos(bx)\,dx = \frac{a\cos(bx) + b\sin(bx)}{a^2 + b^2}e^{ax} + C$ |
| $\frac{d}{dx}(\ln(\tan x)) = \frac{2}{\sin(2x)}$ | $\int \frac{1}{\sin x}dx = \ln\lvert\csc x - \cot x\rvert + C$ |
| $\frac{d}{dx}(\ln(\cot x)) = \frac{-2}{\sin(2x)}$ | $\int \frac{1}{\cos x}dx = \ln\lvert\sec x + \tan x\rvert + C$ |
| $\frac{d}{dx}(\ln(ax + b)) = \frac{a}{ax + b}$ | $\int \frac{1}{1 + \cos x}dx = \csc x - \cot x + C$ |

## Depreciation

Depreciation is an accounting method that allocates the cost of an asset over its useful life. A machine may decline in value because it is wearing out and no longer performing its function as well as when it was new. As an economics technique, depreciation spreads the purchase price of an asset or other property over a number of years. Tax regulations do not allow the cost of an asset to be treated as a deductible expense in the year of purchase; rather, a portion of the expense is allocated to each year of the asset's depreciation period. This yearly allocation is referred to as *depreciation*.

**Straight Line Depreciation.**—Straight line depreciation is a constant depreciation charge over the period of life. If $P$ is principal value, $L$ is salvage value and $n$ is period of life, then depreciation is calculated:

Depreciated value of an asset at year $n$: $A = P(1 - x) + L$

Depreciation at $x$th year: $$D_x = \frac{P - L}{n}$$

Book value after $x$ years: $$BV_x = \frac{(P - L)(n - x)}{n} + L$$

After tax depreciation recovery: $$ATDR = TR\left(\frac{P - L}{n}\right)\frac{(1 + i)^n - 1}{i(1 + i)^n}$$

**Sum of the Years Digits.**—Another method for allocating the cost of an asset minus its salvage value over useful life is *sum of the years digits depreciation*. This method results in charges that are greater than straight line depreciation charges during the early years of an asset and smaller charges near the end period.

Depreciation at $x$th year: $$D_x = \frac{2(P - L)(n - x + 1)}{n(n + 1)}$$

Book value after $x$ years: $$BV_x = P - (P - L)(2n - x + 1)\frac{x}{n(n + 1)}$$

**Double Declining Balance Method.**—A constant depreciation is applied to the book value of a property.

Depreciation at $x$th year: $$D_x = 2\left(\frac{P}{n}\right)\left(\frac{n - 2}{n}\right)^{(x - 1)}$$

Book value after $x$ years: $$BV_x = P\left(\frac{n - 2}{n}\right)^x$$

**Statutory Depreciation System.**—A depreciation method used for income tax purposes is *accelerated cost recovery system* (ACRS) depreciation. The first step in ACRS is to determine the property class of the asset being depreciated. All personal property falls into one of six classes. Reading across Table 3 after identifying the correct property class, principal $P$ for year $x$ is substituted into the following formula:

Depreciation at $x$th year: $$D_x = P \times Depreciation\ Factor$$

*Example, Annual Cost Considering Salvage Value:* If in the previous example the salvage value of the equipment installed was $5,000 at the end of 10 years, what is the effect on the annual cost of the proposed investment of $15,000?

*Solution:* The only item in the annual cost of the previous example that will be affected is the capital recovery amount of $2,442. The following formula gives the amount of annual capital recovery when salvage value is considered:

$$A = (P - L)\frac{i(1+i)^n}{(1+i)^n - 1} + Li + AE$$

$$= (15000 - 5000)\frac{(0.10)(1+0.10)^{10}}{(1+0.10)^{10} - 1} + 5000(0.10) + 5100 = 7227.45$$

This amount is $315 less than the previous annual cost of $7,542 for the proposed investment.

**Rate of Return.**—This is the estimated interest rate produced by an investment. *Rate of return* (*ROR*) is the interest rate at which benefits are equivalent to costs. It is defined as the interest rate paid on the unpaid balance of a loan in such a way that the payment schedule makes the unpaid loan balance equal zero when the final payment is made. It may be computed by finding the interest rate in such a way that the estimated expenditures are equal to the capital gain. *Net Present Worth* = 0, or equivalently, *PW of benefits* – *PW of costs* = 0

$$\frac{(1+ROR)^n - 1}{ROR(1+ROR)^n}(AR - AE) + \frac{L}{(1+ROR)^n} = P$$

The rate of return can only be calculated through trial and error. To find out the present worth, a reasonable interest rate is selected and the present worth is calculated. Then another rate is chosen and the present worth is calculated. The value of the *ROR* is interpolated or extrapolated to find the zero value of present worth.

**Benefit-Cost Ratio.**—This is the ratio of present worth of benefit to present worth of cost. This method is applied to municipal project evaluations where benefits ($B$) and costs ($C$) accrue to different segments of the community. A project is considered acceptable if the ratio equals or exceeds 1. For fixed input, $B/C \geq 1$ is maximized, and for fixed output, $B/C \geq 1$ is maximized, and if neither input nor output is fixed, to compute the incremental benefit-cost ratio ($\Delta B/\Delta C$), choose $\Delta B/\Delta C \geq 1$.

*Example:* To build a bridge over a river costs $1,200,000, with benefits of $2,000,000, and disbenefits of $500,000. (a) What is the benefit-cost ratio? (b) What is the excess of benefits over costs?

*Solution:* (a) The benefit-cost ratio $B/C$ is regarded as $(B - D)/C$. Here $B$ = benefits, $D$ = disbenefits, and $C$ = cost.

Benefit-cost ratio: (2,000,000 – 500,000)/1,200,000 = 1.25

As the ratio is greater than 1, the project is worth undertaking.

(b) The excess of benefits over cost = 2,000,000 – 1,200,000 – 500,000 = 300,000.

**Payback Period.**—This is the period required to obtain benefits or cost savings equal to the initial investment made. For example, if a piece of equipment costs $10,000 and the company can obtain cost savings of $2,500 per year by procuring that equipment, the payback period can be said to be $10,000 divided by $2,500, which is 4 years. Payback period is a quick way of evaluating whether an investment is worth making. For example, a plant manager who has a payback period of 5 years in her mind might think the $10,000 investment is worth making.

Payback period ignores the rate of interest or assumes that rate to be 0%. Also, payback period ignores benefits coming in a non-linear manner over the life of the equipment.

synonymous and expressions such as $WV^2 / 2g$ are used without pondering the distinction. Nevertheless, on reflection it seems odd that $g$ should appear in a formula which has nothing to do with gravity at all. In fact the $g$ used here is not the true, local value of the acceleration due to gravity, but an arbitrary standard value chosen as part of the definition of the pound force and is more properly designated $g_0$ (page 152). Its function is not to indicate the strength of the local gravitational field but to convert from one unit to another.

In SI the unit of mass is the *kilogram*, and the unit of force (and weight) is the *newton*.

The following are typical statements in dynamics expressed in SI units:

A force of $R$ newtons acting on a mass of $M$ kilograms produces an acceleration of $R/M$ m/s$^2$. The kinetic energy of a mass of $M$ kg moving with velocity $V$ m/s is $\frac{1}{2} MV^2$ kg (m/s)$^2$ or $\frac{1}{2} MV^2$ joules. The work done by a force of $R$ newtons moving a distance $L$ meters is $RL$ N·m, or $RL$ joules. If this work were converted entirely into kinetic energy we could write $RL = \frac{1}{2} MV^2$ and it is instructive to consider the units. Remembering that the N is the same as the kg·m/s$^2$, we have (kg·m/s$^2$) m = kg (m/s)$^2$, which is clearly correct. It will be noted that $g$ does not appear anywhere in these statements.

In contrast, in many branches of engineering where the weight of a body is important, rather than its mass, using SI units, $g$ does appear where formerly it was absent. Thus, if a rope hangs vertically supporting a mass of $M$ kilograms the tension in the rope is $Mg$ N. Here $g$ is acceleration due to gravity, and its units are m/s$^2$. The ordinary numerical value of 9.81 will be sufficiently accurate for most purposes on earth. The expression is still valid elsewhere, for example, on the moon, provided the proper value of $g$ is used. The maximum tension the rope can safely withstand (and other similar properties) will also be specified in terms of the newton, so that direct comparison may be made with the tension predicted.

Words like load and weight have to be used with greater care. In everyday language we might say "a lift carries a load of five people of average weight 70 kg," but in precise technical language we say that if the average mass is 70 kg, then the average weight is $70g$ N, and the total load (that is force) on the lift is $350g$ N.

If the lift starts to rise with acceleration $a$ m/s$^2$, the load becomes $350\ (g + a)$ N; both $g$ and $a$ have units of m/s$^2$, the mass is in kg, so the load is in terms of kg·m/s$^2$, which is the same as the newton.

*Pressure and stress:* These quantities are expressed in terms of force per unit area. In SI the unit is the *pascal* (Pa), which expressed in terms of SI derived and base units is the newton per meter squared (N/m$^2$). The pascal is very small—equivalent to $0.15 \times 10^{-3}$ lb/in$^2$ — hence the kilopascal (kPa = 1000 pascals) and the megapascal (MPa = $10^6$ pascals) may be more convenient multiples in practice. Thus, note: 1 N/mm$^2$ = 1 MN/m$^2$ = 1 MPa.

In addition to the pascal, the *bar*, a non-SI unit, is in use in the field of pressure measurement in some countries, including England. Thus, in view of existing practice, the International Committee of Weights and Measures (CIPM) decided in 1969 to retain this unit for a limited time for use with those of SI. 1 bar = $10^5$ pascals and 1 hectobar = $10^7$ pascals.

## Force Systems

**Scalar and Vector Quantities.**—The quantities dealt with in mechanics are of two kinds according to whether magnitude alone or direction as well as magnitude must be known in order to completely specify them. Quantities such as time, volume and density are completely specified when their magnitude is known. Such quantities are called *scalar* quantities. Quantities such as force, velocity, acceleration, moment, and displacement must, in order to be specified completely, include direction as well as magnitude, are called *vector* quantities.

**Graphical Representation of Forces.**—A force has three characteristics which, when known, determine it. They are *direction, point of application*, and *magnitude*. The direction of a force is the direction it tends to move the body upon which it acts. The point of application is the place on the line of action where the force is applied. Forces may conveniently be represented by straight lines and arrow heads. The arrow head indicates the direction of the force, and the length of the line, its magnitude to any suitable scale. The point of application may be at any point on the line, but it is generally convenient to assume it to be at one end. In Fig. 1, a force is supposed to act along line $AB$ in a direction from left to right. The length of line $AB$ shows the magnitude of the force. If point $A$ is the point of application, the force is exerted as

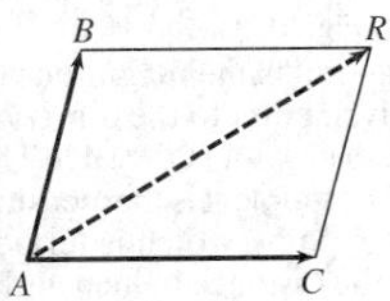




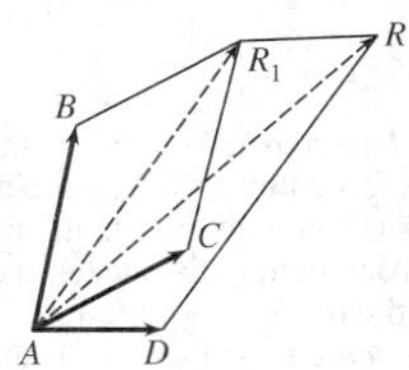

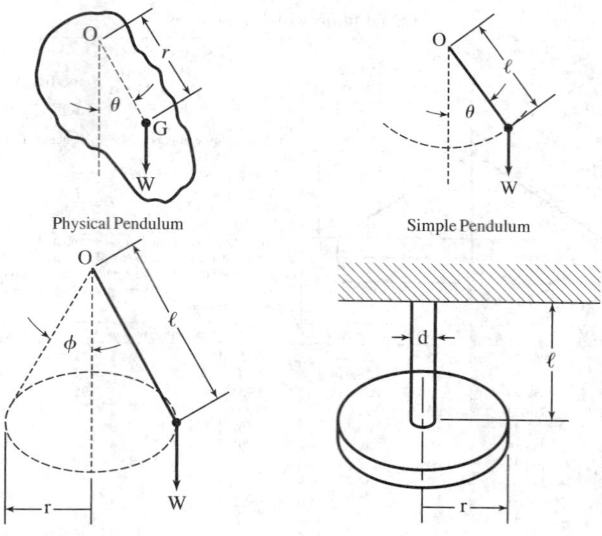

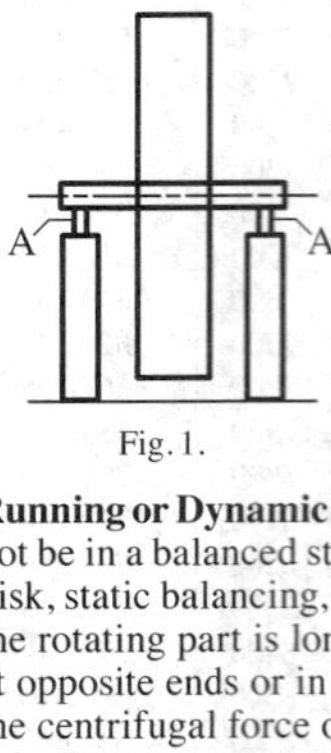


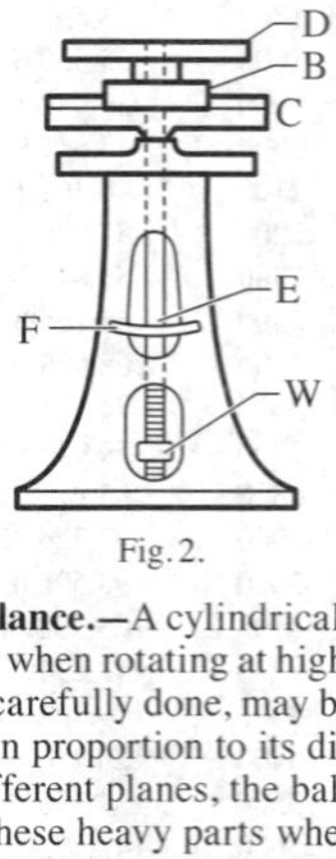


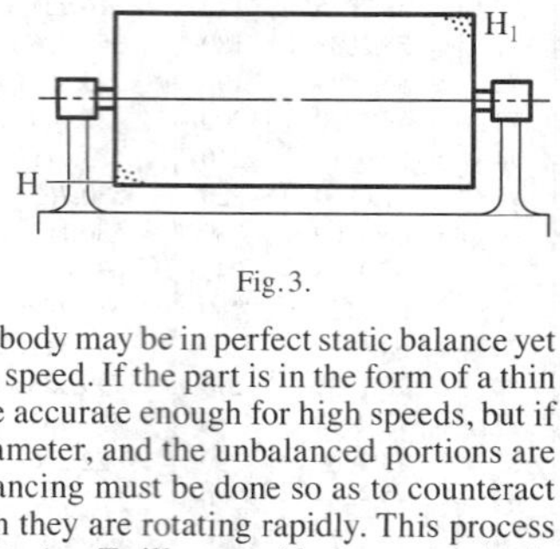

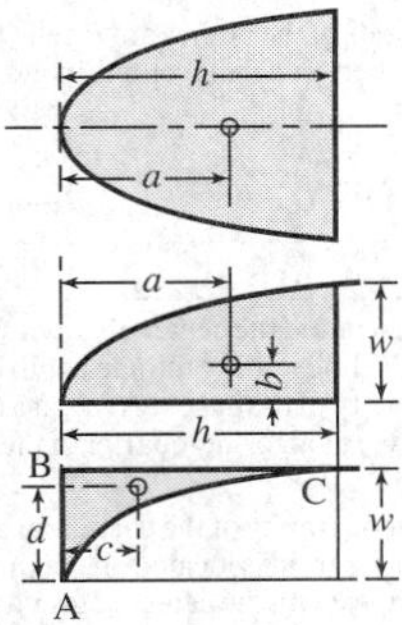


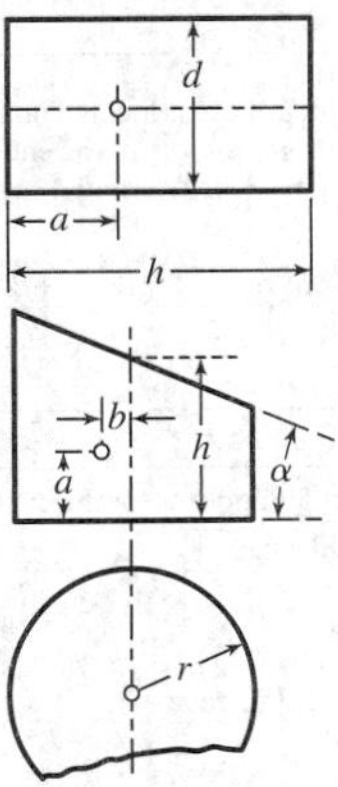


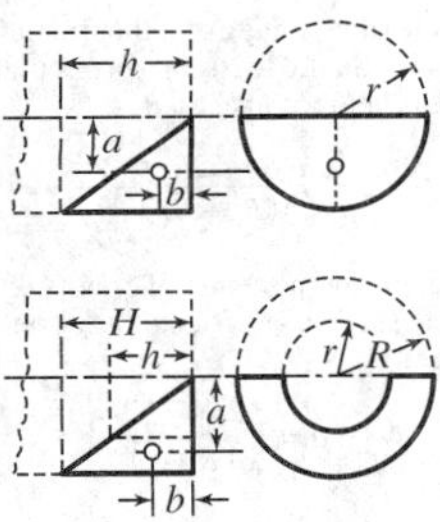

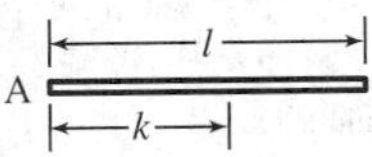


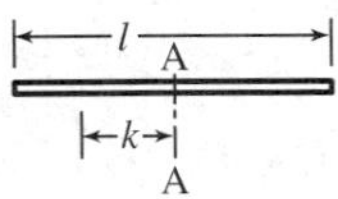


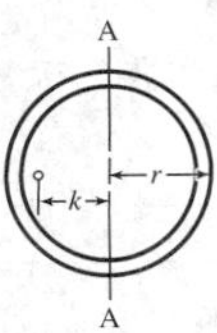


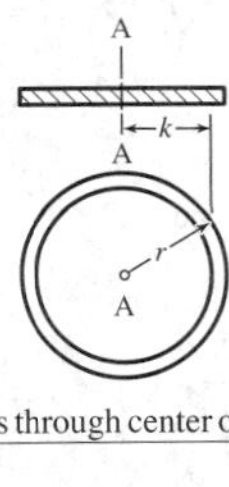


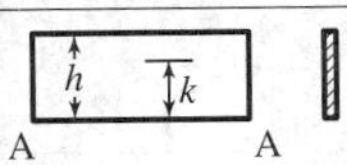


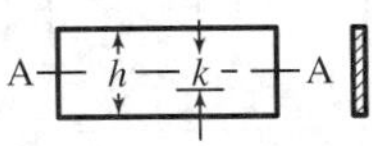

*Sphere:*

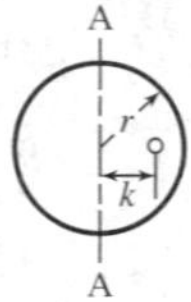


$k = 0.6325r$

$k^2 = \frac{2}{5}r^2$

Axis its diameter

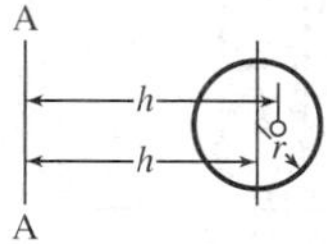


$k = \sqrt{a^2 + \frac{2}{5}r^2}$

$k^2 = a^2 + \frac{2}{5}r^2$

Axis at a distance

*Hollow Sphere and Thin Spherical Shell:*

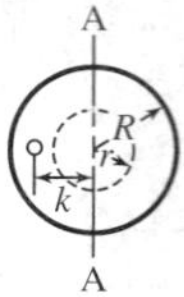


$k = 0.6325\sqrt{\frac{R^5 - r^5}{R^3 - r^3}}$

$k^2 = \frac{2(R^5 - r^5)}{5(R^3 - r^3)}$

Hollow Sphere
Axis its diameter

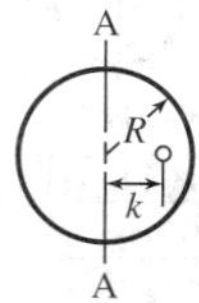


$k = 0.8165r$

$k^2 = \frac{2}{3}r^2$

Thin Spherical Shell

*Ellipsoid and Paraboloid:*

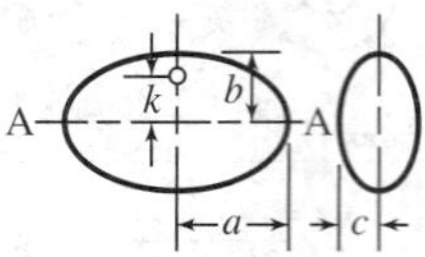


$k = \frac{\sqrt{b^2 + c^2}}{\sqrt{5}}$

$k^2 = \frac{1}{5}(b^2 + c^2)$

Ellipsoid
Axis through center

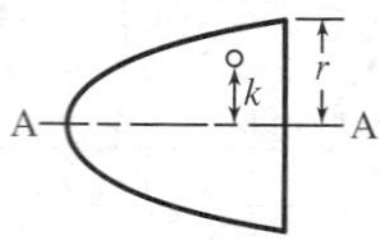


$k = 0.5773r$

$k^2 = \frac{1}{3}r^2$

Paraboloid
Axis through center

**Center and Radius of Oscillation.—**If a body oscillates about a horizontal axis which does not pass through its center of gravity, there will be a point on the line drawn from the center of gravity, perpendicular to the axis, the motion of which will be the same as if the whole mass were concentrated at that point. This point is called the *center of oscillation*. The *radius of oscillation* is the distance between the center of oscillation and the point of suspension. In a straight line, or in a bar of small diameter suspended at one end and oscillating about it, the center of oscillation is at $^2/_3$ the length of the rod from the end by which it is suspended.

When the vibrations are perpendicular to the plane of the figure, and the figure is suspended by the vertex of an angle or its uppermost point, the radius of oscillation of an isosceles triangle is equal to $^3/_4$ of the height of the triangle; of a circle, $^5/_8$ of the diameter; of a parabola, $^5/_7$ of the height.

If the vibrations are in the plane of the figure, then the radius of oscillation of a circle equals $^3/_4$ of the diameter; of a rectangle suspended at the vertex of one angle, $^2/_3$ of the diagonal.

**Center of Percussion.—**For a body that moves without rotation, the resultant of all the forces acting on the body passes through the center of gravity. On the other hand, for a body that rotates about some *fixed axis*, the resultant of all the forces acting on it does not pass through the center of gravity of the body but through a point called the *center of*

 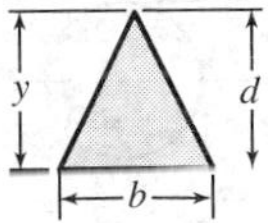 

**Moments of Inertia, Section Moduli, and Radii of Gyration** *(Continued)*

| Section | Area of Section, $A$ | Distance from Neutral Axis to Extreme Fiber, $y$ | Moment of Inertia, $I$ | Section Modulus, $Z = \frac{I}{y}$ | Radius of Gyration, $k = \sqrt{\frac{I}{A}}$ |
|---|---|---|---|---|---|
| 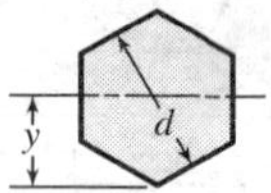 | $\frac{3d^2 \tan 30°}{2} = 0.866d^2$ | $\frac{d}{2\cos 30°} = 0.577d$ | $\frac{A}{12}\left[\frac{d^2(1+2\cos^2 30°)}{4\cos^2 30°}\right] = 0.06d^4$ | $\frac{A}{6.9}\left[\frac{d(1+2\cos^2 30°)}{4\cos^2 30°}\right] = 0.104d^3$ | $\sqrt{\frac{d^2(1+2\cos^2 30°)}{48\cos^2 30°}} = 0.264d$ |
| 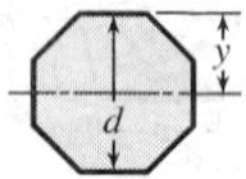 | $2d^2 \tan 22\frac{1}{2}° = 0.828d^2$ | $\frac{d}{2}$ | $\frac{A}{12}\left[\frac{d^2(1+2\cos^2 22\frac{1}{2}°)}{4\cos^2 22\frac{1}{2}°}\right] = 0.055d^4$ | $\frac{A}{6}\left[\frac{d(1+2\cos^2 22\frac{1}{2}°)}{4\cos^2 22\frac{1}{2}°}\right] = 0.109d^3$ | $\sqrt{\frac{d^2(1+2\cos^2 22\frac{1}{2}°)}{48\cos^2 22\frac{1}{2}°}} = 0.257d$ |
| Circular, Elliptical, and Circular Arc Sections | | | | | |
| 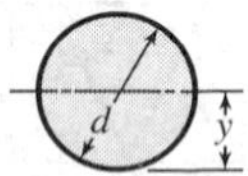 | $\frac{\pi d^2}{4} = 0.7854d^2$ | $\frac{d}{2}$ | $\frac{\pi d^4}{64} = 0.049d^4$ | $\frac{\pi d^3}{32} = 0.098d^3$ | $\frac{d}{4}$ |
| 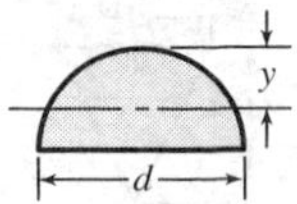 | $\frac{\pi d^2}{8} = 0.393d^2$ | $\frac{(3\pi - 4)d}{6\pi} = 0.288d$ | $\frac{(9\pi^2 - 64)d^4}{1152\pi} = 0.007d^4$ | $\frac{(9\pi^2 - 64)d^3}{192(3\pi - 4)} = 0.024d^3$ | $\frac{\sqrt{(9\pi^2 - 64)d^2}}{12\pi} = 0.132d$ |
| 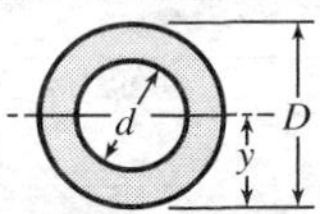 | $\frac{\pi(D^2 - d^2)}{4} = 0.7854(D^2 - d^2)$ | $\frac{D}{2}$ | $\frac{\pi(D^4 - d^4)}{64} = 0.049(D^4 - d^4)$ | $\frac{\pi(D^4 - d^4)}{32D} = 0.098\frac{D^4 - d^4}{D}$ | $\frac{\sqrt{D^2 + d^2}}{4}$ |

**Moments of Inertia, Section Moduli, and Radii of Gyration** *(Continued)*

| Section | Area of Section, $A$ | Distance from Neutral Axis to Extreme Fiber, $y$ | Moment of Inertia, $I$ | Section Modulus, $Z = \frac{I}{y}$ | Radius of Gyration, $k = \sqrt{\frac{I}{A}}$ |
|---|---|---|---|---|---|
| 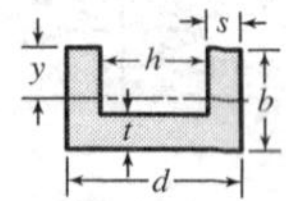  | $bd - h(b-t)$ | $b - \frac{2b^2s + ht^2}{2bd - 2h(b-t)}$ | $\frac{2sb^3 + ht^3}{3} - A(b-y)^2$ | $\frac{I}{y}$ | $\sqrt{\frac{I}{A}}$ |
| | | T-Sections | | | |
| 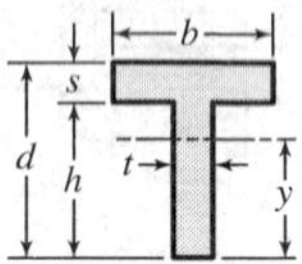  | $bs + ht$ | $d - \frac{d^2t + s^2(b-t)}{2(bs+ht)}$ | $\frac{1}{3}[ty^3 + b(d-y)^3 - (b-t)(d-y-s)^3]$ | $\frac{I}{y}$ | $\sqrt{\frac{1}{3(bs+ht)}[ty^3 + b(d-y)^3 - (b-t)(d-y-s)^3]}$ |
| 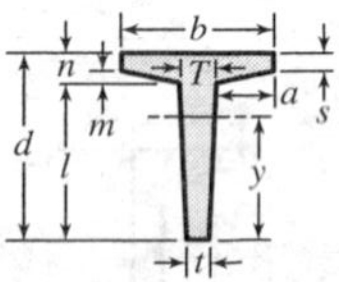  | $\frac{l(T+t)}{2} + Tn + a(s+n)$ | $d - [3s^2(b-T) + 2am(m+3s) + 3Td^2 - l(T-t)(3d-l)] \div 6A$ | $\frac{1}{12}[l^3(T+3t) + 4bn^3 - 2am^3] - A(d-y-n)^2$ | $\frac{I}{y}$ | $\sqrt{\frac{I}{A}}$ |
| 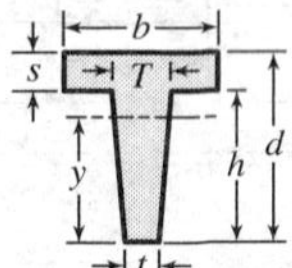  | $bs + \frac{h(T+t)}{2}$ | $d - [3bs^2 + 3ht(d+s) + h(T-t)(h+3s)] \div 6A$ | $\frac{1}{12}[4bs^3 + h^3(3t+T)] - A(d-y-s)^2$ | $\frac{I}{y}$ | $\sqrt{\frac{I}{A}}$ |

**Moments of Inertia, Section Moduli, and Radii of Gyration** *(Continued)*

| Section | Area of Section, $A$ | Distance from Neutral Axis to Extreme Fiber, $y$ | Moment of Inertia, $I$ | Section Modulus, $Z = \frac{I}{y}$ | Radius of Gyration, $k = \sqrt{\frac{I}{A}}$ |
|---|---|---|---|---|---|
| 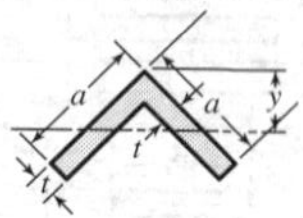 | $t(2a-t)$ | $\frac{a^2+at-t^2}{2(2a-t)\cos 45°}$ | $\frac{A}{12}[7(a^2+b^2)-12y^2]$ $-2ab^2(a-b)$ in which $b=(a-t)$ | $\frac{I}{y}$ | $\sqrt{\frac{I}{A}}$ |
| 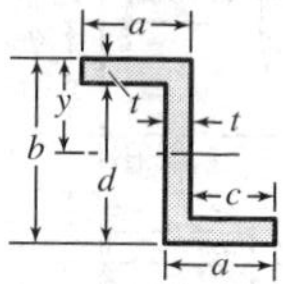 | $t[b+2(a-t)]$ | $\frac{b}{2}$ | $\frac{ab^3-c(b-2t)^3}{12}$ | $\frac{ab^3-c(b-2t)^3}{6b}$ | $\sqrt{\frac{ab^3-c(b-2t)^3}{12t[b+2(a-t)]}}$ |
| 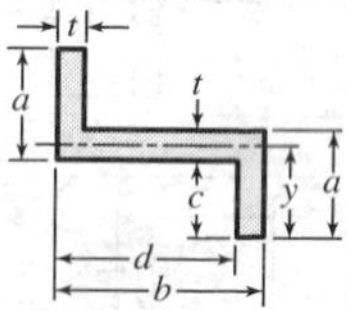 | $t[b+2(a-t)]$ | $\frac{2a-t}{2}$ | $\frac{b(a+c)^3-2c^3d-6a^2cd}{12}$ | $\frac{b(a+c)^3-2c^3d-6a^2cd}{6(2a-t)}$ | $\sqrt{\frac{b(a+c)^3-2c^3d-6a^2cd}{12t[b+2(a-t)]}}$ |
| 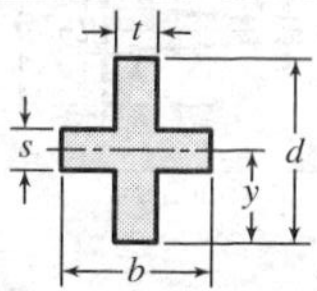 | $dt+s(b-t)$ | $\frac{d}{2}$ | $\frac{td^3+s^3(b-t)}{12}$ | $\frac{td^3-s^3(b-t)}{6d}$ | $\sqrt{\frac{td^3+s^3(b-t)}{12[td+s(b-t)]}}$ |

## Table 1. *(Continued)* Stresses and Deflections in Beams

| Type of Beam | Stresses: General Formula for Stress at any Point | Stresses: Stresses at Critical Points | Deflections: General Formula for Deflection at any Point[a] | Deflections: Deflections at Critical Points[a] |
|---|---|---|---|---|
| **Case 4. — Supported at Both Ends, Two Symmetrical Loads** | | | | |
| 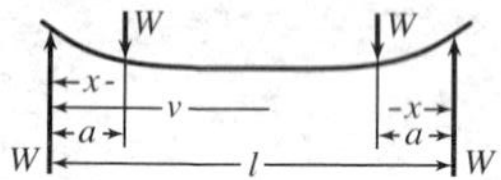  | Between each support and adjacent load, $s=-\frac{Wx}{Z}$<br>Between loads, $s=-\frac{Wa}{Z}$ | Stress at each load, and at all points between, $-\frac{Wa}{Z}$ | Between each support and adjacent load, $y=\frac{Wx}{6EI}[3a(l-a)-x^2]$<br>Between loads, $y=\frac{Wa}{6EI}[3v(l-v)-a^2]$ | Maximum deflection at center, $\frac{Wa}{24EI}(3l^2-4a^2)$<br>Deflection at loads $\frac{Wa^2}{6EI}(3l-4a)$ |
| **Case 5. — Both Ends Overhanging Supports Symmetrically, Uniform Load** | | | | |
| 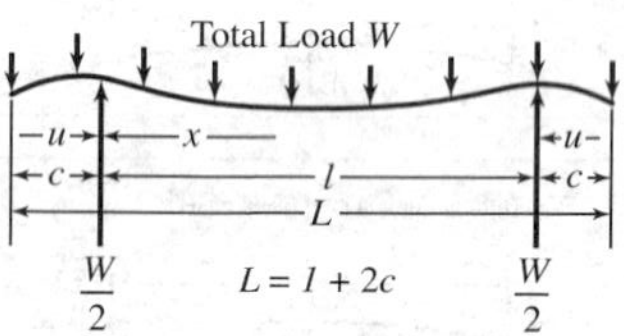  | Between each support and adjacent end, $s=\frac{W}{2Zl}(c-u)^2$<br>Between supports, $s=\frac{W}{2ZL}[c^2-x(l-x)]$ | Stress at each support, $\frac{Wc^2}{2ZL}$<br>Stress at center, $\frac{W}{2ZL}\left(c^2-\frac{1}{4}l^2\right)$<br>If cross section is constant, the greater of these is the maximum stress.<br>If $l$ is greater than $2c$, the stress is zero at points $\sqrt{\frac{1}{4}l^2-c^2}$ on both sides of the center.<br>If cross section is constant and if $l=2.828c$, the stresses at supports and center are equal and opposite, and are $\pm\frac{WL}{46.62Z}$ | Between each support and adjacent end, $y=\frac{Wu}{24EIL}[6c^2(l+u)-u^2(4c-u)-l^3]$<br>Between supports, $y=\frac{Wx(l-x)}{24EIL}[x(l-x)+l^2-6c^2]$ | Deflection at ends, $\frac{Wc}{24EIL}[3c^2(c+2l)-l^3]$<br>Deflection at center, $\frac{Wl^2}{384EIL}(5l^2-24c^2)$<br>If $l$ is between $2c$ and $2.449c$, there are maximum upward deflections at points $\sqrt{3\left(\frac{1}{4}l^2-c^2\right)}$ on both sides of the center, which are, $-\frac{W}{96EIL}(6c^2-l^2)^2$ |

**Table 1.** *(Continued)* **Stresses and Deflections in Beams**

| Type of Beam | Stresses: General Formula for Stress at any Point | Stresses: Stresses at Critical Points | Deflections: General Formula for Deflection at any Point[a] | Deflections: Deflections at Critical Points[a] |
|---|---|---|---|---|
| Case 6. — Both Ends Overhanging Supports Unsymmetrically, Uniform Load | | | | |
| 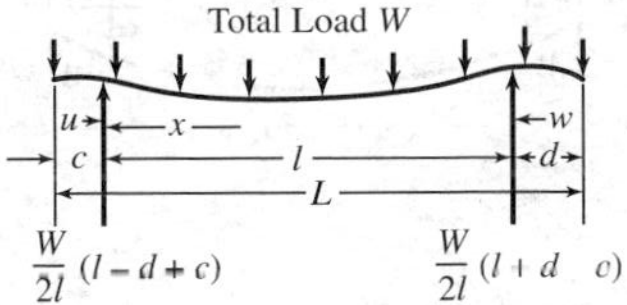  | For overhanging end of length $c$, $s=\frac{W}{2ZL}(c-u)^2$ <br> Between supports, $s=\frac{W}{2ZL}\left\{c^2\left(\frac{l-x}{l}\right)+d^2\frac{x}{l}-x(l-x)\right\}$ <br> For overhanging end of length $d$, $s=\frac{W}{2ZL}(d-w)^2$ | Stress at support next to end of length $c$, $\frac{Wc^2}{2ZL}$ <br> Critical stress between supports is at $x=\frac{l^2+c^2-d^2}{2l}=x_1$ and is $\frac{W}{2ZL}(c^2-x_1^2)$ <br> Stress at support next to end of length $d$, $\frac{Wd^2}{2ZL}$ <br> If cross section is constant, the greatest of these three is the maximum stress. <br> If $x_1>c$, the stress is zero at points $\sqrt{x_1^2-c^2}$ on both sides of $x=x_1$. | For overhanging end of length $c$, $y=\frac{Wu}{24EIL}[2l(d^2+2c^2)+6c^2u-u^2(4c-u)-l^3]$ <br> Between supports, $y=\frac{Wx(l-x)}{24EIL}\{x(l-x)+l^2-2(d^2+c^2)-\frac{2}{l}[d^2x+c^2(l-x)]\}$ <br> For overhanging end of length $d$, $y=\frac{Ww}{24EIL}[2l(c^2+2d^2)+6d^2w-w^2(4d-w)-l^3]$ | Deflection at end $c$, $\frac{Wc}{24EIL}[2l(d^2+2c^2)+3c^3-l^3]$ <br> Deflection at end $d$, $\frac{Wd}{24EIL}[2l(c^2+2d^2)+3d^3-l^3]$ <br> This case is so complicated that convenient general expressions for the critical deflections between supports cannot be obtained. |
| Case 7. — Both Ends Overhanging Supports, Load at any Point Between | | | | |
| 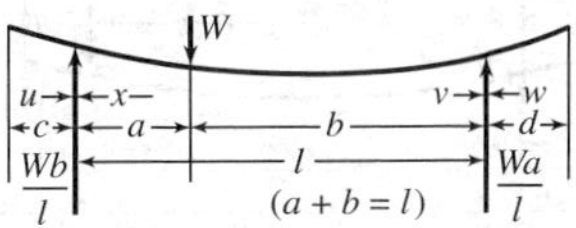  | Between supports: <br> For segment of length $a$, $s=-\frac{Wbx}{Zl}$ <br> For segment of length $b$, $s=-\frac{Wav}{Zl}$ <br> Beyond supports $s=0$. | Stress at load, $-\frac{Wab}{Zl}$ <br> If cross section is constant, this is the maximum stress. | Between supports, same as Case 3. <br> For overhanging end of length $c$, $y=-\frac{Wabu}{6EIl}(l+b)$ <br> For overhanging end of length $d$, $y=-\frac{Wabw}{6EIl}(l+a)$ | Between supports, same as Case 3. <br> Deflection at end $c$, $-\frac{Wabc}{6EIl}(l+b)$ <br> Deflection at end $d$, $-\frac{Wabd}{6EIl}(l+a)$ |

**Table 1.** *(Continued)* **Stresses and Deflections in Beams**

| Type of Beam | Stresses: General Formula for Stress at any Point | Stresses: Stresses at Critical Points | Deflections: General Formula for Deflection at any Point[a] | Deflections: Deflections at Critical Points[a] |
|---|---|---|---|---|
| Case 22. — Continuous Beam, with Two Equal Spans, Uniform Load | | | | |
| Total Load on Each Span, $W$ 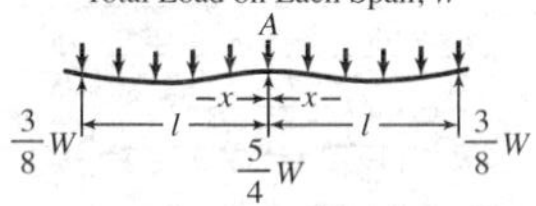 | $s = \frac{W(l-x)}{2Zl}\left(\frac{1}{4}l - x\right)$ | Maximum stress at point $A$, $\frac{Wl}{8Z}$ <br> Stress is zero at $x = \frac{1}{4}l$ <br> Greatest negative stress is at $x = \frac{5}{8}l$ and is, $-\frac{9}{128}\frac{Wl}{Z}$ | $y = \frac{Wx^2(l-x)}{48EIl}(3l - 2x)$ | Maximum deflection is at $x = 0.5785l$, and is $\frac{Wl^3}{185EI}$ <br> Deflection at center of span, $\frac{Wl^3}{192EI}$ <br> Deflection at point of greatest negative stress, at $x = \frac{5}{8}l$ is $\frac{Wl^3}{187EI}$ |
| Case 23. — Continuous Beam, with Two Equal Spans, Equal Loads at Center of Each | | | | |
| 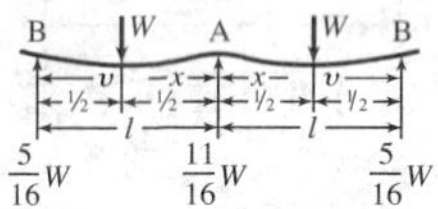 | Between point $A$ and load, $s = \frac{W}{16Z}(3l - 11x)$ <br> Between point $B$ and load, $s = -\frac{5}{16}\frac{Wv}{Z}$ | Maximum stress at point $A$, $\frac{3}{16}\frac{Wl}{Z}$ <br> Stress is zero at $x = \frac{3}{11}l$ <br> Greatest negative stress at center of span, $-\frac{5}{32}\frac{Wl}{Z}$ | Between point $A$ and load, $y = \frac{Wx^2}{96EI}(9l - 11x)$ <br> Between point $B$ and load, $y = \frac{Wv}{96EI}(3l^2 - 5v^2)$ | Maximum deflection is at $v = 0.4472l$, and is $\frac{Wl^3}{107.33EI}$ <br> Deflection at load, $\frac{7}{768}\frac{Wl^3}{EI}$ |

*Example:* Find the thickness of metal required in the hemi-spherical end of a cylindrical vessel, 2 feet in diameter, subjected to an internal pressure of 500 pounds per square inch.

*Solution:* The material is mild steel and a tensile stress of 10,000 psi is allowable.

$$t = \frac{500 \times 2 \times 12}{4 \times 10{,}000} = 0.3 \text{ inch}$$

***Example:* A similar example using metric SI units is as follows: find the thickness of metal required in the hemi-spherical end of a cylindrical vessel, 750 mm in diameter, subjected to an internal pressure of 3 newtons/mm$^2$. The material is mild steel and a tensile stress of 70 newtons/mm$^2$ is allowable.**

$$t = \frac{3 \times 750}{4 \times 70} = 8.04 \text{ mm}$$

If the radius of curvature of the domed head of a boiler or container subjected to internal pressure is made equal to the diameter of the boiler, the thickness of the cylindrical shell and of the spherical head should be made the same. For example, if a boiler is 3 feet in diameter, the radius of curvature of its head should also be 3 feet, if material of the same thickness is to be used and the stresses are to be equal in both the head and cylindrical portion.

**Collapsing Pressure of Cylinders and Tubes Subjected to External Pressures.—**The following formulas may be used for finding the collapsing pressures of lap-welded Bessemer steel tubes:

$$P = 86{,}670\frac{t}{D} - 1386 \qquad (29)$$

$$P = 50{,}210{,}000\left(\frac{t}{D}\right)^3 \qquad (30)$$

in which $P$ = collapsing pressure in psi; $D$ = outside diameter of tube or cylinder in inches; $t$ = thickness of wall in inches.

Formula (29) is for values of $P$ greater than 580 pounds per square inch, and Formula (30) is for values of $P$ less than 580 pounds per square inch. These formulas are substantially correct for all lengths of pipe greater than six diameters between transverse joints that tend to hold the pipe to a circular form. The pressure $P$ found is the actual collapsing pressure, and a suitable factor of safety must be used. Ordinarily, a factor of safety of 5 is sufficient. In cases where there are repeated fluctuations of the pressure, vibration, shocks and other stresses, a factor of safety of from 6 to 12 should be used.

**If metric SI units are used the formulas are:**

$$P = 597.6\frac{t}{D} - 9.556 \qquad (31)$$

$$P = 346{,}200\left(\frac{t}{D}\right)^3 \qquad (32)$$

**where $P$ = collapsing pressure in newtons per square millimeter; $D$ = outside diameter of tube or cylinder in millimeters; and $t$ = thickness of wall in millimeters. Formula (31) is for values of $P$ greater than 4 N/mm$^2$, and Formula (32) is for values of $P$ less than 4 N/mm$^2$.**

The table *Tubes Subjected to External Pressure* is based upon the requirements of the Steam Boat Inspection Service of the Department of Commerce and Labor and gives the permissible working pressures and corresponding minimum wall thickness for long, plain, lap-welded and seamless steel flues subjected to external pressure only. The table thicknesses have been calculated from the formula:

## Diameters of Finished Shafting (former American Standard *ASA B17.1*)

| Diameters, Inches | | Minus Tolerances, Inches[a] | Diameters, Inches | | Minus Tolerances, Inches[a] | Diameters, Inches | | Minus Tolerances, Inches[a] |
|---|---|---|---|---|---|---|---|---|
| Transmission Shafting | Machinery Shafting | | Transmission Shafting | Machinery Shafting | | Transmission Shafting | Machinery Shafting | |
| | ½ | 0.002 | | 1 13/16 | 0.003 | | 3¾ | 0.004 |
| | 9/16 | 0.002 | | 1⅞ | 0.003 | | 3⅞ | 0.004 |
| | ⅝ | 0.002 | 1 15/16 | 1 15/16 | 0.003 | 3 15/16 | 4 | 0.004 |
| | 11/16 | 0.002 | | 2 | 0.003 | | 4¼ | 0.005 |
| | ¾ | 0.002 | | 2 1/16 | 0.004 | 4 7/16 | 4½ | 0.005 |
| | 13/16 | 0.002 | | 2⅛ | 0.004 | | 4¾ | 0.005 |
| | ⅞ | 0.002 | 2 3/16 | 2 3/16 | 0.004 | 4 15/16 | 5 | 0.005 |
| 15/16 | 15/16 | 0.002 | | 2¼ | 0.004 | | 5¼ | 0.005 |
| | 1 | 0.002 | | 2 5/16 | 0.004 | 5 7/16 | 5½ | 0.005 |
| | 1 1/16 | 0.003 | | 2⅜ | 0.004 | | 5¾ | 0.005 |
| | 1⅛ | 0.003 | 2 7/16 | 2 7/16 | 0.004 | 5 15/16 | 6 | 0.005 |
| 1 3/16 | 1 3/16 | 0.003 | | 2½ | 0.004 | | 6¼ | 0.006 |
| | 1¼ | 0.003 | | 2⅝ | 0.004 | 6½ | 6½ | 0.006 |
| | 1 5/16 | 0.003 | | 2¾ | 0.004 | | 6¾ | 0.006 |
| | 1⅜ | 0.003 | 2 15/16 | 2⅞ | 0.004 | 7 | 7 | 0.006 |
| 1 7/16 | 1 7/16 | 0.003 | | 3 | 0.004 | | 7¼ | 0.006 |
| | 1½ | 0.003 | | 3⅛ | 0.004 | 7½ | 7½ | 0.006 |
| | 1 9/16 | 0.003 | | 3¼ | 0.004 | | 7¾ | 0.006 |
| | 1⅝ | 0.003 | | 3⅜ | 0.004 | 8 | 8 | 0.006 |
| 1 11/16 | 1 11/16 | 0.003 | 3 7/16 | 3½ | 0.004 | … | … | … |
| | 1¾ | 0.003 | | 3⅝ | 0.004 | … | … | … |

[a] *Note:*—These tolerances are *negative* or minus and represent the maximum allowable variation *below* the exact nominal size. For instance the maximum diameter of the $1^{15}/_{16}$ inch shaft is 1.938 inch and its minimum allowable diameter is 1.935 inch. Stock lengths of finished transmission shafting shall be: 16, 20 and 24 feet.

**Design of Transmission Shafting.**—The following guidelines for the design of shafting for transmitting a given amount of power under various conditions of loading are based upon formulas given in the former American Standard ASA B17c Code for the Design of Transmission Shafting. These formulas are based on the *maximum-shear theory* of failure which assumes that the elastic limit of a *ductile* ferrous material in shear is practically one-half its elastic limit in tension. This theory agrees, very nearly, with the results of tests on ductile materials and has gained wide acceptance in practice.

The formulas given apply in all shaft designs including shafts for special machinery. The limitation of these formulas is that they provide only for the strength of shafting and are not concerned with the torsional or lineal deformations which may, in shafts used in machine design, be the controlling factor (see *Torsional Deflection of Circular Shafts* on page 291 and *Linear Deflection of Shafting* on page 292 for deflection considerations). In the formulas that follow,

$B = \sqrt[3]{1 \div (1 - K^4)}$ (see Table 3)

$D$ = outside diameter of shaft in inches

$D_1$ = inside diameter of a hollow shaft in inches

$K_m$ = shock and fatigue factor to be applied in every case to the computed bending moment (see Table 1)

$K_t$ = combined shock and fatigue factor to be applied in every case to the computed torsional moment (see Table 1)

$M$ = maximum bending moment in inch-pounds

$N$ = revolutions per minute

$P$ = maximum power to be transmitted by the shaft in horsepower

(see Table 6.) The bending stress $S_b$ at section B should be compared with allowable stresses for torsion springs and with the elastic limit of the material in tension (See Fig. 7 through Fig. 10.)

*Stresses in Cross-Over Hooks:* Results of tests on springs having a normal average index show that the cross-over hooks last longer than regular hooks. These results may not occur on springs of small index or if the cross-over bend is made too sharply.

In as much as both types of hooks have the same bending stress, it would appear that the fatigue life would be the same. However, the large bend radius of the regular hooks causes some torsional stresses to coincide with the bending stresses, thus explaining the earlier breakages. If sharper bends were made on the regular hooks, the life should then be the same as for cross-over hooks.

**Table 6. Formula for Bending Stress at Section B**

<table>
<tr><th colspan="2">Type of Hook</th><th>Stress in Bending</th></tr>
<tr><td>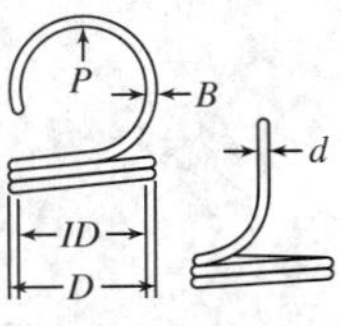

Regular Hook</td><td>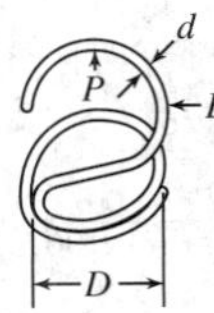

Cross-over Hook</td><td>$S_b = \frac{5PD^2}{IDd^3}$</td></tr>
</table>

*Stresses in Half Hooks:* The formulas for regular hooks can also be used for half hooks, because the smaller bend radius allows for the increase in stress. It will therefore be observed that half hooks have the same stress in bending as regular hooks.

Frequently overlooked facts by many designers are that one full hook deflects an amount equal to one half a coil and each half hook deflects an amount equal to one tenth of a coil. Allowances for these deflections should be made when designing springs. Thus, an extension spring, with regular full hooks and having 10 coils, will have a deflection equal to 11 coils, or 10 percent more than the calculated deflection.

**Extension Spring Design.**—The available space in a product or assembly usually determines the limiting dimensions of a spring, but the wire size, number of coils, and initial tension are often unknown.

*Example:* An extension spring is to be made from spring steel ASTM A229, with regular hooks as shown in Fig. 17. Calculate the wire size, number of coils and initial tension.

*Note:* Allow about 20 to 25 percent of the 9 pound load for initial tension, say 2 pounds, and then design for a 7 pound load (not 9 pounds) at $\frac{5}{8}$ inch deflection. Also use lower stresses than for a compression spring to allow for overstretching during assembly and to obtain a safe stress on the hooks. Proceed as for compression springs, but locate a load in the tables somewhat higher than the 9 pound load.

*Method 1, Using Table:* From Table 5 locate $\frac{3}{4}$ inch outside diameter in the left column and move to the right to locate a load *P* of 13.94 pounds. A deflection *f* of 0.212 inch appears above this figure. Moving vertically from this position to the top of the column a suitable wire diameter of 0.0625 inch is found.

The remaining design calculations are completed as follows:

*Step 1:* The stress with a load of 7 pounds is obtained as follows:

The 7 pound load is 50.2 percent of the 13.94 pound load. Therefore, the stress *S* at 7 pounds = 0.502 percent × 100,000 = 50,200 pounds per square inch.

*Step 2:* The 50.2 percent figure is also used to determine the deflection per coil *f*: 0.502 percent × 0.212 = 0.1062 inch.

From Fig. 8,

$$\frac{D_a}{2} = \frac{D}{2} - (a + b) \qquad (5)$$

where $a = t\sin\alpha$ and $b = r\cos\alpha$. Substitution of $a$ and $b$ values into Equation (5) gives:

$$\frac{D_a}{2} = \frac{D}{2} - (t\sin\alpha + r\cos\alpha) \qquad (6)$$

The cone angle $\alpha$ is found from:

$$\tan\alpha = \frac{h}{\frac{D}{2} - \frac{d}{2}} = \frac{2h}{D-d} \qquad \alpha = \operatorname{atan}\left(\frac{2h}{D-d}\right) \qquad (7)$$

Substituting $\alpha$ from Equation (7) and $r = \frac{t}{6}$ into Equation (6) gives:

$$\frac{D_a}{2} = \frac{D}{2} - t\left\{\sin\left[\operatorname{atan}\left(\frac{2h}{D-d}\right)\right] + \frac{1}{6}\cos\left[\operatorname{atan}\left(\frac{2h}{D-d}\right)\right]\right\} \qquad (8)$$

Finally,

$$D_a = D - 2t\left\{\sin\left[\operatorname{atan}\left(\frac{2h}{D-d}\right)\right] + \frac{1}{6}\cos\left[\operatorname{atan}\left(\frac{2h}{D-d}\right)\right]\right\} \qquad (9)$$

Substituting $D_a$ from Equation (9) for $D$ in Equation (4) yields Equation (10), that provides better accuracy for calculating disc spring forces.

$$F = \frac{4 \cdot E \cdot s}{(1-\mu^2)K_1 \cdot D_a^2}\left[\left(h - \frac{s}{2}\right) \cdot (h - s) \cdot t + t^3\right] \qquad (10)$$

The constant $K_1$ depends on disc spring outside diameter $D$, inside diameter $d$, and their ratio $\delta = D/d$:

$$K_1 = \frac{\left(\frac{\delta - 1}{\delta}\right)^2}{\pi\left(\frac{\delta + 1}{\delta - 1} - \frac{2}{\ln\delta}\right)} \qquad (11)$$

Table 2 compares the spring force of a series of disc springs deflected by 75 percent of their cone height, i.e., $s = 0.75h$, as determined from manufacturers catalogs calculated in accordance with Equation (4), calculated forces by use of Equation (10), and measured forces.

**Table 2. Comparison Between Calculated and Measured Disc Spring Forces**

| Disc Spring Catalog Item | *Schnorr* Handbook for Disc Springs | *Christian Bauer* Disc Spring Handbook | *Key Bellevilles* Disc Spring Catalog | Spring Force Calculated by Equation (10) | Measured Disc Spring Force |
|---|---|---|---|---|---|
| 50 – 22.4 – 2.5<br>$S$ = 1.05 mm | 8510 N<br>1913 lbf | 8510 N<br>1913 lbf | 8616 N<br>1937 lbf | 9020 N<br>2028 lbf | 9563 N<br>2150 lbf |
| 60 – 30.5 – 2.5<br>$S$ = 1.35 mm | 8340 N<br>1875 lbf | 8342 N<br>1875 lbf | 8465 N<br>1903 lbf | 8794 N<br>1977 lbf | 8896 N<br>2000 lbf |
| 60 – 30.5 – 3<br>$S$ = 1.275 mm | 13200 N<br>2967 lbf | 13270 N<br>2983 lbf | 13416 N<br>3016 lbf | 14052 N<br>3159 lbf | 13985 N<br>3144 lbf |
| 70 – 35.5 – 3<br>$S$ = 1.575 mm | 12300 N<br>2765 lbf | 12320 N<br>2770 lbf | 12397 N<br>2787 lbf | 12971 N<br>2916 lbf | 13287 N<br>2987 lbf |
| 70 – 35.5 – 3.5<br>$S$ = 1.35 mm | | 16180 N<br>3637 lbf | | 17170 N<br>3860 lbf | 17304 N<br>3890 lbf |

Comparison made at 75 percent deflection, in newtons (N) and pounds (lbf)

The difference between disc spring forces calculated by Equation (10) and the measured forces varies from −5.7 percent (maximum) to +0.5 percent (minimum). Disc spring forces calculated by Equation (4) and shown in manufacturers catalogs are less than measured forces by −11 percent (maximum) to −6 percent (minimum).

**Force Generated by Disc Spring with Contact Surfaces.—**Some of disc springs in *Group 2* and all disc springs in *Group 3* are manufactured with small contact (load-bearing) surfaces or flats in addition to the corner radii. These flats provide better contact between disc springs, but, at the same time, they reduce the springs outside diameter and generate higher spring force because in Equation (4) force $F$ is inversely proportional to the square of outside diameter $D^2$. To compensate for the undesired force increase, the disc spring thickness is reduced from $t$ to $t'$. Thickness reduction factors $t'/t$ are approximately 0.94 for disc spring series A and B, and approximately 0.96 for series C springs. With such reduction factors, the disc spring force at 75 percent deflection is the same as for equivalent disc spring without contact surfaces. Equation (12), which is similar to Equation (10), has an additional constant $K_4$ that correlates the increase in spring force due to contact surfaces. If disc springs do not have contact surfaces, then $K_4^2 = K_4 = 1$.

$$F = \frac{4 \cdot E \cdot K_4^2 \cdot s}{(1-\mu^2) \cdot K_1 \cdot D_a^2}\left[K_4^2 \cdot \left(h' - \frac{s}{2}\right) \cdot (h' - s) \cdot (t' + t')^3\right] \tag{12}$$

where $t'$ = reduced thickness of a disc spring

$h'$ = cone height adjusted to reduced thickness: $h' = H - t'$ $(h' > h)$

$K_4$ = constant applied to disc springs with contact surfaces.

$K_4^2$ can be calculated as follows:

$$K_4^2 = \frac{-b + \sqrt{b^2 - 4ac}}{2a} \tag{13}$$

where $a = t'(H - 4t' + 3t)(5H - 8t' + 3t)$; $b = 32(t')^3$; and, $c = -t[5(H-t)^2 + 32t^2]$.

**Disc Spring Functional Stresses.—**Disc springs are designed for both static and dynamic load applications. In static load applications, disc springs may be under constant or fluctuating load conditions that change up to 5,000 or 10,000 cycles over long time intervals. Dynamic loads occur when disc springs are under continuously changing deflection between pre-load (approximately 15 to 20 percent of the cone height) and the maximum deflection values over short time intervals. Both static and dynamic loads cause compressive and tensile stresses. The position of critical stress points on a disc spring cross section are shown in Fig. 9.

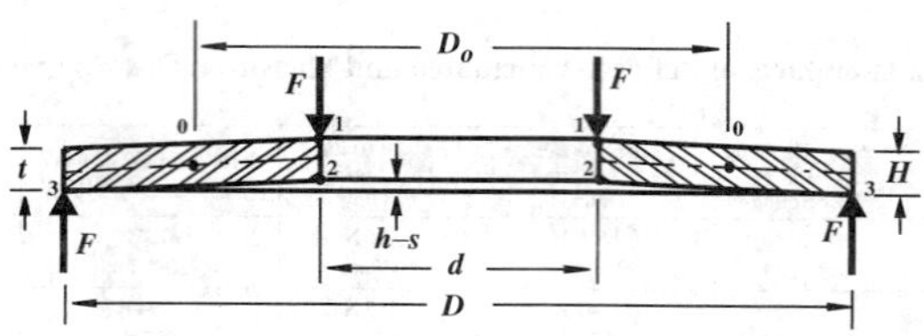


Fig. 9. Critical Stress Points
*s* is Deflection of Spring by Force *F*; *h* − *s* is a Cone Height of Loaded Disc Spring.

Compressive stresses are acting at points 0 and 1, that are located on the top surface of the disc spring. Point 0 is located on the cross-sectional midpoint diameter, and point 1 is located on the top inside diameter. Tensile stresses are acting at points 2 and 3, which are located on the bottom surface of the disc spring. Point 2 is on the bottom inside diameter, and point 3 is on the bottom outside diameter. The following equations are used to

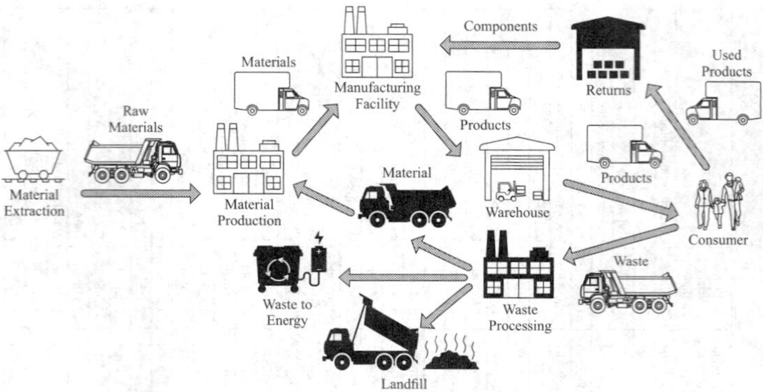


Fig. 1. Materials Lifecycle

The lifecycle of materials is depicted in Fig. 1. Raw materials are extracted or otherwise recovered from waste processing streams. Primary material production involves refinement of these raw materials, blending of alloys, and other treatments to provide feedstocks with known properties. Materials recovery from waste streams provides significant energy savings, often as much as 70 percent if recycled material sources are available.

Feedstocks are provided to manufacturing facilities where components are made and assembled into products. Significant additional energy is typically added during manufacture and assembly. Energy and cost may be reduced by minimizing material handling and shipping.

**Production Cost and Carbon Footprint.—**The amount of energy is related to the type of manufacturing process, properties of the processed material, and details of the machine design and operation. The minimum total embodied energy, E, for a component is:

$$E_{Component} = (E_{Material} = V_{Material} \times e_{Material}) + \sum (E_{Process} = P_{Process} \times t_{Process})$$

Where:

$E_{Material}$ is the embodied energy of the material

$V_{Material}$ is the volume of the material used per component

$e_{Material}$ is the embodied energy of the material as measured in MJ/m$^3$ per Table 2

$P_{process}$ is the average power of each manufacturing process

$t_{process}$ is the processing time for each manufacturing process.

While extraction and refining processes vary for different materials, material ecology research indicates a strong correlation between embodied energy and carbon footprint with a representative value of 0.5 kg $CO_2$ per MJ of embodied energy. The associated energy cost can also be estimated from the cost of energy given the economic value per kilowatt hour (kWh).

*Example 1:* Estimate the embodied energy for production of a 3/8-16 inch hex head bolt made from a 1.5 inch long, $^3/_4$ inch diameter rod of A36 steel. The bolt is produced using (1) a 75-ton cold forging press drawing 20 HP, and (2) a 50-ton threading machine drawing 12 HP, and then (3) annealed for 3 hours in an oven, drawing 10,000 BTUs per hour. The production rate is 1,000 parts per hour.

the listed quality requirement for the processes. For example, 62 percent of the black walnut pieces planed came out perfect, but only 34 percent of the pieces run on the shaper achieved good to excellent results.

**Table 3. Machinability and Related Properties of Various Domestic Hardwoods**

| Type of Wood | Planing | Shaping | Turning | Boring | Mortising | Sanding |
|---|---|---|---|---|---|---|
| | Quality Required | | | | | |
| | Perfect | Good to Excellent | Fair to Excellent | Good to Excellent | Fair to Excellent | Good to Excellent |
| Alder, red | 61 | 20 | 88 | 64 | 52 | … |
| Ash | 75 | 55 | 79 | 94 | 58 | 75 |
| Aspen | 26 | 7 | 65 | 78 | 60 | … |
| Basswood | 64 | 10 | 68 | 76 | 51 | 17 |
| Beech | 83 | 24 | 90 | 99 | 92 | 49 |
| Birch | 63 | 57 | 80 | 97 | 97 | 34 |
| Birch, paper | 47 | 22 | … | … | … | … |
| Cherry, black | 80 | 80 | 88 | 100 | 100 | … |
| Chestnut | 74 | 28 | 87 | 91 | 70 | 64 |
| Cottonwood | 21 | 3 | 70 | 70 | 52 | 19 |
| Elm, soft | 33 | 13 | 65 | 94 | 75 | 66 |
| Hackberry | 74 | 10 | 77 | 99 | 72 | … |
| Hickory | 76 | 20 | 84 | 100 | 98 | 80 |
| Magnolia | 65 | 27 | 79 | 71 | 32 | 37 |
| Maple, bigleaf | 52 | 56 | 8 | 100 | 80 | … |
| Maple, hard | 54 | 72 | 82 | 99 | 95 | 38 |
| Maple, soft | 41 | 25 | 76 | 80 | 34 | 37 |
| Oak, red | 91 | 28 | 84 | 99 | 95 | 81 |
| Oak, white | 87 | 35 | 85 | 95 | 99 | 83 |
| Pecan | 88 | 40 | 89 | 100 | 98 | … |
| Sweetgum | 51 | 28 | 86 | 92 | 53 | 23 |
| Sycamore | 22 | 12 | 85 | 98 | 96 | 21 |
| Tanoak | 80 | 39 | 81 | 100 | 100 | … |
| Tupelo, black | 48 | 32 | 75 | 82 | 24 | 21 |
| Tupelo, water | 55 | 52 | 79 | 62 | 33 | 34 |
| Walnut, black | 62 | 34 | 91 | 100 | 98 | … |
| Willow | 52 | 5 | 58 | 71 | 24 | 24 |
| Yellow-poplar | 70 | 13 | 81 | 87 | 63 | 19 |

The data above represent the percentage of pieces attempted that meet the quality requirement listed.

**Nominal and Minimum Sizes of Sawn Lumber**

| Type of Lumber | Thickness (inches) | | | Face Widths (inches) | | |
|---|---|---|---|---|---|---|
| | Nominal, $T_n$ | Dry | Green | Nominal, $W_n$ | Dry | Green |
| Boards | 1 | ¾ | 25/32 | 2 to 4 | $W_n - 1/2$ | $W_n - 7/16$ |
| | 1¼ | 1 | 1 1/32 | 5 to 7 | $W_n - 1/2$ | $W_n - 3/8$ |
| | 1½ | 1¼ | 1 9/32 | 8 to 16 | $W_n - 3/4$ | $W_n - 1/2$ |
| Dimension Lumber | 2 | 1½ | 1 9/16 | 2 to 4 | $W_n - 1/2$ | $W_n - 7/16$ |
| | 2½ | 2 | 2 1/16 | 5 to 6 | $W_n - 1/2$ | $W_n - 3/8$ |
| | 3 | 2½ | 2 9/16 | 8 to 16 | $W_n - 3/4$ | $W_n - 1/2$ |
| | 3½ | 3 | 3 1/16 | … | … | … |
| | 4 | 3½ | 3 9/16 | … | … | … |
| | 4½ | 4 | 4 1/16 | … | … | … |
| Timbers | 5 and up | … | $T_n - 1/2$ | 5 and up | … | $W_n - 1/2$ |

*Source:* National Forest Products Association: *Design Values for Wood Construction*. Moisture content: dry lumber ≤ 19%; green lumber > 19%. Dimension lumber refers to lumber 2 to 4 inches thick (nominal) and 2 inches or greater in width. Timbers refers to lumber of approximately square cross section, 5 × 5 inches or larger, and a width no more than 2 inches greater than the thickness.

Propeller shafts, 2340, 2345, 4140
Races, ball-bearing, 52100
Ring gears, 3115, 3120, 4119
Rings, snap, 1060, 1070, 1090
Rivets, rimmed*
Rod and wire, killed*
Rod, cold-heading, 1035
Roller bearings, 4815
Rollers for bearings, 52100
Screws and bolts, 1035
Screw stock
- Bessemer, 1111, 1112, 1113
- open-hearth, 1115

Screws, heat-treated, 2330
Seat springs, 10956
Shafts
- axle, 1045
- cyanide-hardening, 5140
- heavy-duty, 4340, 6150, 4615, 4620
- oil-hardening, 5150
- propeller, 2340, 2345, 4140
- transmission, 4140

Sheets and strips, rimmed*
Snap rings, 1060, 1070, 1090
Spline shafts, 1045, 1320, 2340, 2345, 3115, 3120, 3135, 3140, 4023
Spring clips, 1060
Springs
- coil, 1095, 4063, 6150
- clutch, 1060
- cushion, 1060
- hard-drawn coiled, 1066
- leaf, 1085, 1095, 4063, 4068, 9260, 6150
- oil-hardening, 5150
- oil-tempered wire, 1066
- seat, 1095
- valve, 1060

Spring wire, 1045
- hard-drawn, 1055
- oil-tempered, 1055

Stainless irons, 51210, 51710
Steel
- cold-rolled, 1070
- cold-heading, 30905
- free-cutting carbon, 1111, 1113
- free-cutting chrome-nickel, 30615
- free-cutting manganese, 1132
- minimum distortion, 4615, 4620, 4640
- soft ductile, 30905

Steering arms, 4042
Steering-arm bolts, 3130
Steering knuckles, 3141
Steering-knuckle pins, 4815, 4820
Tacks, rimmed*
Thrust washers, 1060
- oil-hardened, 5150

Transmission shafts, 4140
Tubing, 1040
- front axle, 4140
- seamless, 1030
- welded, 1020

Universal joints, 1145
Valve springs, 1060
Washers, lock, 1060
Welded structures, 30705
Wire and rod, killed*
Wire
- cold-heading, rimmed*
- hard-drawn spring, 1045, 1055
- music, 1085
- oil-tempered spring, 1055

Wrist-pins, automobile, 1020
Yokes, 1145

**Carbon Steels.—***SAE Steels 1006, 1008, 1010, 1015:* These steels are the lowest carbon steels of the plain carbon type and are selected where cold formability is the primary requisite of the user. They are produced both as rimmed and killed steels. Rimmed steel is used for sheet, strip, rod, and wire where excellent surface finish or good drawing qualities are required, such as body and fender stock, hoods, lamps, oil pans, and other deep-drawn and -formed products. This steel is also used for cold-heading wire for tacks, and rivets and low carbon wire products. Killed steel (usually aluminum killed or special killed) is used for difficult stampings, or where nonaging properties are needed. Killed steels (usually silicon killed) should be used in preference to rimmed steel for forging or heat-treating applications.

**Table 10. Expected Minimum Mechanical Properties of Cold-Drawn Carbon-Steel Rounds, Squares, and Hexagons**

| Size, in. | As Cold-Drawn: Tensile Strength Ultimate (1000 lb/in²) | As Cold-Drawn: Tensile Strength Yield (1000 lb/in²) | As Cold-Drawn: Elongation in 2 inches, Percent | As Cold-Drawn: Reduction in Area, Percent | As Cold-Drawn: Hardness, BHN | Cold-Drawn Followed by Low-Temperature Stress Relief: Tensile Strength Ultimate (1000 lb/in²) | Low-Temp.: Tensile Strength Yield (1000 lb/in²) | Low-Temp.: Elongation in 2 inches, Percent | Low-Temp.: Reduction in Area, Percent | Low-Temp.: Hardness, BHN | Cold-Drawn Followed by High-Temperature Stress Relief: Tensile Strength Ultimate (1000 lb/in²) | High-Temp.: Tensile Strength Yield (1000 lb/in²) | High-Temp.: Elongation in 2 inches, Percent | High-Temp.: Reduction in Area, Percent | High-Temp.: Hardness, BHN |
|---|---|---|---|---|---|---|---|---|---|---|---|---|---|---|---|
| AISI 1018 and 1025 Steels | | | | | | | | | | | | | | | |
| ⅝ - ⅞ | 70 | 60 | 18 | 40 | 143 | … | … | … | … | … | 65 | 45 | 20 | 45 | 131 |
| Over ⅞ - 1¼ | 65 | 55 | 16 | 40 | 131 | … | … | … | … | … | 60 | 45 | 20 | 45 | 121 |
| Over 1¼ - 2 | 60 | 50 | 15 | 35 | 121 | … | … | … | … | … | 55 | 45 | 16 | 40 | 111 |
| Over 2 - 3 | 55 | 45 | 15 | 35 | 111 | … | … | … | … | … | 50 | 40 | 15 | 40 | 101 |
| AISI 1117 and 1118 Steels | | | | | | | | | | | | | | | |
| ⅝ - ⅞ | 75 | 65 | 15 | 40 | 149 | 80 | 70 | 15 | 40 | 163 | 70 | 50 | 18 | 45 | 143 |
| Over ⅞ - 1¼ | 70 | 60 | 15 | 40 | 143 | 75 | 65 | 15 | 40 | 149 | 65 | 50 | 16 | 45 | 131 |
| Over 1¼ - 2 | 65 | 55 | 13 | 35 | 131 | 70 | 60 | 13 | 35 | 143 | 60 | 50 | 15 | 40 | 121 |
| Over 2 - 3 | 60 | 50 | 12 | 30 | 121 | 65 | 55 | 12 | 35 | 131 | 55 | 45 | 15 | 40 | 111 |
| AISI 1035 Steel | | | | | | | | | | | | | | | |
| ⅝ - ⅞ | 85 | 75 | 13 | 35 | 170 | 90 | 80 | 13 | 35 | 179 | 80 | 60 | 16 | 45 | 163 |
| Over ⅞ - 1¼ | 80 | 70 | 12 | 35 | 163 | 85 | 75 | 12 | 35 | 170 | 75 | 60 | 15 | 45 | 149 |
| Over 1¼ - 2 | 75 | 65 | 12 | 35 | 149 | 80 | 70 | 12 | 35 | 163 | 70 | 60 | 15 | 40 | 143 |
| Over 2 - 3 | 70 | 60 | 10 | 30 | 143 | 75 | 65 | 10 | 30 | 149 | 65 | 55 | 12 | 35 | 131 |
| AISI 1040 and 1140 Steels | | | | | | | | | | | | | | | |
| ⅝ - ⅞ | 90 | 80 | 12 | 35 | 179 | 95 | 85 | 12 | 35 | 187 | 85 | 65 | 15 | 45 | 170 |
| Over ⅞ - 1¼ | 85 | 75 | 12 | 35 | 170 | 90 | 80 | 12 | 35 | 179 | 80 | 65 | 15 | 45 | 163 |
| Over 1¼ - 2 | 80 | 70 | 10 | 30 | 163 | 85 | 75 | 10 | 30 | 170 | 75 | 60 | 15 | 40 | 149 |
| Over 2 - 3 | 75 | 65 | 10 | 30 | 149 | 80 | 70 | 10 | 30 | 163 | 70 | 55 | 12 | 35 | 143 |

## Table 9. Cold-Work Tool Steels

| AISI | Group | | D2 | D3 | D4 | D5 | D7 | A2 | A3 | A4 | A6 | A7 | A8 | A9 | A10 | O1 | O2 | O6 | O7 |
|---|---|---|---|---|---|---|---|---|---|---|---|---|---|---|---|---|---|---|---|
| Identifying Chemical Composition and Typical Heat-Treatment Data | | | | | | | | | | | | | | | | | | | |
| | Group | | High-Carbon, High-Chromium Types | | | | | Medium-Alloy, Air-Hardening Types | | | | | | | | Oil-Hardening Types | | | |
| | Types | | D2 | D3 | D4 | D5 | D7 | A2 | A3 | A4 | A6 | A7 | A8 | A9 | A10 | O1 | O2 | O6 | O7 |
| Identifying Chemical Elements in Percent | C | | 1.50 | 2.25 | 2.25 | 1.50 | 2.35 | 1.00 | 1.25 | 1.00 | 0.70 | 2.25 | 0.55 | 0.50 | 1.35 | 0.90 | 0.90 | 1.45 | 1.20 |
| | Mn | | … | … | … | … | … | … | … | 2.00 | 2.00 | … | … | … | 1.80 | 1.00 | 1.60 | … | … |
| | Si | | … | … | … | … | … | … | … | … | … | … | … | … | 1.25 | … | … | 1.00 | … |
| | W | | … | … | … | … | … | … | … | … | … | 1.00 | 1.25 | … | … | 0.50 | … | … | 1.75 |
| | Mo | | 1.00 | … | 1.00 | 1.00 | 1.00 | 1.00 | 1.00 | 1.00 | 1.25 | 1.00 | 1.25 | 1.40 | 1.50 | … | … | 0.25 | … |
| | Cr | | 12.00 | 12.00 | 12.00 | 12.00 | 12.00 | 5.00 | 5.00 | 1.00 | 1.00 | 5.25 | 5.00 | 5.00 | … | 0.50 | … | … | 0.75 |
| | V | | 1.00 | … | … | … | 4.00 | … | 1.00 | … | … | 4.75 | … | 1.00 | … | … | … | … | … |
| | Co | | … | … | … | 3.00 | … | … | … | … | … | … | … | … | … | … | … | … | … |
| | Ni | | … | … | … | … | … | … | … | … | … | … | … | 1.50 | 1.80 | … | … | … | … |
| Heat-Treatment Data | Hardening Temperature Range | °F | 1800–1875 | 1700–1800 | 1775–1850 | 1800–1875 | 1850–1950 | 1700–1800 | 1750–1850 | 1500–1600 | 1525–1600 | 1750–1800 | 1800–1850 | 1800–1875 | 1450–1500 | 1450–1500 | 1400–1475 | 1450–1500 | 1550–1525 |
| | | °C | 982–1024 | 927–982 | 968–1010 | 982–1024 | 1010–1066 | 927–982 | 954–1010 | 816–871 | 829–871 | 954–982 | 982–1010 | 982–1024 | 788–816 | 788–816 | 760–802 | 788–816 | 843–829 |
| | Quenching Medium | | Air | Oil | Air | Air | Air | Air | Air | Air | Air | Air | Air | Air | Air | Oil | Oil | Oil | Oil |
| | Tempering Temperature Range | °F | 400–1000 | 400–1000 | 400–1000 | 400–1000 | 300–1000 | 350–1000 | 350–1000 | 350–800 | 300–800 | 300–1000 | 350–1100 | 950–1150 | 350–800 | 350–500 | 350–500 | 350–600 | 350–550 |
| | | °C | 204–538 | 204–538 | 204–538 | 204–538 | 149–538 | 177–538 | 177–538 | 177–427 | 149–427 | 149–538 | 177–593 | 510–621 | 177–427 | 177–260 | 177–260 | 177–316 | 177–288 |
| | Approx. Tempered Hardness, RC (Rockwell C scale) | | 61–54 | 61–54 | 61–54 | 61–54 | 65–58 | 62–57 | 65–57 | 62–54 | 60–54 | 67–57 | 60–50 | 56–35 | 62–55 | 62–57 | 62–57 | 63–58 | 64–58 |
| Relative Ratings of Properties (A = greatest to E = least) | | | | | | | | | | | | | | | | | | | |
| Characteristics in Heat Treatment | Safety in Hardening | | A | C | A | A | A | A | A | A | A | A | A | A | A | B | B | B | B |
| | Depth of Hardening | | A | A | A | A | A | A | A | A | A | A | A | A | A | B | B | B | B |
| | Resistance to Decarburization | | B | B | B | B | B | B | B | A/B | A/B | B | B | B | A/B | A | A | A | A |
| | Stability of Shape in Heat Treatment | | A | B | A | A | A | A | A | A | A | A | A | A | A | B | B | B | B |
| Service Properties | Machinability | | E | E | E | E | E | D | D | D/E | D/E | E | D | D | C/D | C | C | B | C |
| | Hot Hardness | | C | C | C | C | C | C | C | D | D | C | C | C | D | E | E | E | E |
| | Wear Resistance | | B/C | B | B | B/C | A | C | B | C/D | C/D | A | C/D | C/D | C | D | D | D | D |
| | Toughness | | E | E | E | E | E | D | D | D | D | E | C | C | D | D | D | D | C |

to use the middle of the temperature range. All other cutting tools, such as drills, countersinks, taps, milling cutters, reamers, broaches, and form tools, should be hardened in the middle of the range shown. For certain tools, such as slender taps, cold punches, and blanking and trimming dies, where greater toughness to resist shocks is required, the lower end of the hardening temperature range should be used.

**Table 4. Heat Treatment of Molybdenum High-Speed Steels**

| Heat-Treating Operation | Molybdenum–Tungsten Types Ia and Ib[a] Temp. Deg. F | Molybdenum–Tungsten Types Ia and Ib[a] Temp. Deg. C | Molybdenum–Vanadium Type II Temp. Deg. F | Molybdenum–Vanadium Type II Temp. Deg. C | Tungsten–Molybdenum Type III Temp. Deg. F | Tungsten–Molybdenum Type III Temp. Deg. C |
|---|---|---|---|---|---|---|
| Forging | 1850–2000 | 1010–1093 | 1850–2000 | 1010–1093 | 1900–2050 | 1038–1121 |
| Not below | 1600 | 871 | 1600 | 871 | 1600 | 871 |
| Annealing | 1450–1550 | 788–843 | 1450–1550 | 788–843 | 1450–1550 | 788–843 |
| Strain relief | 1150–1350 | 621–732 | 1150–1350 | 621–732 | 1150–1350 | 621–732 |
| Preheating | 1250–1500 | 677–816 | 1250–1500 | 677–816 | 1250–1550 | 677–843 |
| Hardening[b] | 2150–2250[a] | 1177–1232[a] | 2150–2250 | 1177–1232 | 2175–2275 | 1191–1246 |
| Salt | 2150–2225 | 1177–1218 | 2150–2225 | 1177–1218 | 2150–2250 | 1177–1232 |
| Tempering | 950–1100 | 510–593 | 950–1100 | 510–593 | 950–1100 | 510–593 |

[a] For similar working conditions, Type Ib steel requires slightly higher hardening heat than Type Ia.
[b] The higher side of the hardening range should be used for large sections, and the lower side for small sections.

Molybdenum high-speed steels can be pack-hardened following the same practice as is used for tungsten high-speed steels but keeping on the lower side of the hardening range (approximately 1850° F or 1010° C). Special surface treatments such as nitriding by immersion in molten cyanide that are used for tungsten high-speed steels are also applicable to molybdenum high-speed tools.

When heated in an open fire or in furnaces without atmosphere control, these steels do not sweat like 18-4-1 steels; consequently, determining the proper time in the high-heat chamber is a matter of experience. This time approximates that used with 18-4-1 steels, although it may be slightly longer when the lower part of the hardening range is used. Much can be learned by preliminary hardening of test pieces and checking on the hardness fracture and structure. It is difficult to give the exact heating time because it is affected by temperature, type of furnace, size and shape, and furnace atmosphere. Rate of heat transfer is most rapid in salt baths and slowest in controlled-atmosphere furnaces with high carbon monoxide content.

**Quenching and Tempering of Molybdenum High-Speed Tools.—**Quenching may be done in oil, air, or molten bath. To reduce the possibility of breakage and undue distortion of intricately shaped tools, it is advisable to quench in a molten bath at approximately 1100° F (593° C). The tool also may be quenched in oil and removed while still red, or at approximately 1100° F (593° C). The tool is then cooled in air to room temperature and tempered immediately to avoid cracking.

When straightening is necessary, it should be done after quenching and before cooling to room temperature prior to tempering.

To temper, the tools should be reheated slowly and uniformly to 950 to 1100° F (510 to 593° C). For general work, 1050° F (566° C) is most common. The tools should be held at this temperature at least 1 hour. Two hours is a safer minimum, and 4 hours is maximum. The time and temperature depend on the hardness and toughness required. Where tools are subjected to more or less shock, multiple temperings are suggested.

**Protective Coatings for Molybdenum Steels.—**To protect the surface from oxidation during heat treatment, borax may be sprinkled lightly over the steel when the latter is heated in a furnace to a low temperature, 1200 to 1400° F (649 to 760° C). Small tools may be rolled in a box of borax before heating. Another method more suitable for finished tools is to apply the borax or boric acid in the form of a supersaturated water solution. The tools

*-T2, annealed (cast products only):* Designates a type of annealing treatment used to improve ductility and increase dimensional stability of castings.

*-T3, solution heat treated and then cold-worked:* Applies to products that are cold-worked to improve strength, or in which the effect of cold-work in flattening or straightening is recognized in applicable specifications.

*-T4, solution heat treated and naturally aged to a substantially stable condition:* Applies to products that are not cold-worked after solution heat treatment, or in which the effect of cold-work in flattening or straightening may not be recognized in applicable specifications.

*-T5, artificially aged only:* Applies to products that are artificially aged after an elevated-temperature rapid-cool fabrication process, such as casting or extrusion, to improve mechanical properties or dimensional stability, or both.

*-T6, solution heat treated and then artificially aged:* Applies to products that are not cold-worked after solution heat treatment, or in which the effect of cold-work in flattening or straightening may not be recognized in applicable specifications.

*-T7, solution heat treated and then stabilized:* Applies to products that are stabilized to carry them beyond the point of maximum hardness, providing control of growth or residual stress or both.

*-T8, solution heat treated, cold-worked, and then artificially aged:* Applies to products that are cold-worked to improve strength, or in which the effect of cold-work in flattening or straightening is recognized in applicable specifications.

*-T9, solution heat treated, artificially aged, and then cold-worked:* Applies to products that are cold-worked to improve strength.

*-T10, artificially aged and then cold-worked:* Applies to products that are artificially aged after an elevated-temperature rapid-cool fabrication process, such as casting or extrusion, and then cold-worked to improve strength.

Additional digits may be added to designations -T1 through -T10 to indicate a variation in treatment that significantly alters the characteristics of the product. These may be arbitrarily assigned and registered with The Aluminum Association for an alloy and product to indicate a specific treatment or specific mechanical property limits.

These additional digits have been assigned for wrought products in all alloys:

*-T__51, stress-relieved by stretching:* Applies to products that are stress-relieved by stretching the following amounts after solution heat treatment:

| | |
|---|---|
| Plate | $1\frac{1}{2}$ to 3 percent permanent set |
| Rod, Bar and Shapes | 1 to 3 percent permanent set |
| Drawn tube | 0.5 to 3 percent permanent set |

Applies directly to plate and rolled or cold-finished rod and bar.

These products receive no further straightening after stretching.

Applies to extruded rod and bar shapes and tube when designated as follows:

*-T__510 :* Products that receive no further straightening after stretching.

*-T__511:* Products that receive minor straightening after stretching to comply with standard tolerances.

*-T__52, stress-relieved by compressing:* Applies to products that are stress-relieved by compressing after solution heat treatment to produce a nominal permanent set of $2\frac{1}{2}$ percent.

*-T__54, stress-relieved by combined stretching and compressing:* Applies to die forgings that are stress relieved by restriking cold in the finish die.

The following two-digit -T temper designations have been assigned for wrought products in all alloys:

**Table 7b.** *(Continued)* **Typical Mechanical Properties of Wrought Aluminum Alloys**

| Alloy and Temper | Tension: Strength (ksi) Ultimate | Tension: Strength (ksi) Yield | Tension: Elongation in 2 inches (%) 1/16-inch Thick Specimen | Tension: Elongation in 2 inches (%) 1/2-inch Diameter Specimen | Brinell Hardness Number (500 kg load, 10-mm ball) | Ultimate Shearing Strength (ksi) | Endurance Limit[a] (ksi) |
|---|---|---|---|---|---|---|---|
| 6066-O | 22 | 12 | | 18 | 43 | 14 | |
| 6066-T4, T451 | 52 | 30 | | 18 | 90 | 29 | |
| 6066-T6, T651[b] | 57 | 52 | | 12 | 120 | 34 | 16 |
| 6070-T6 | 55 | 51 | 10 | | | 34 | 14 |
| 6101-H111 | 14 | 11 | | | | | |
| 6101-T6 | 32 | 28 | 15 | | 71 | 20 | |
| 6262-T9 | 58 | 55 | | 10 | 120 | 35 | 13 |
| 6351-T4 | 36 | 22 | 20 | | | | |
| 6351-T6 | 45 | 41 | 14 | | 95 | 29 | 13 |
| 6463-T1 | 22 | 13 | 20 | | 42 | 14 | 10 |
| 6463-T5 | 27 | 21 | 12 | | 60 | 17 | 10 |
| 6463-T6 | 35 | 31 | 12 | | 74 | 22 | 10 |
| 7049-T73 | 75 | 65 | | 12 | 135 | 44 | |
| 7049-T7352 | 75 | 63 | | 11 | 135 | 43 | |
| 7050-T73510, T73511 | 72 | 63 | | 12 | | | |
| 7050-T7451[f] | 76 | 68 | | 11 | | 44 | |
| 7050-T7651 | 80 | 71 | | 11 | | 47 | |
| 7075-O | 33 | 15 | 17 | 16 | 60 | 22 | |
| 7075-T6, T651 | 83 | 73 | 11 | 11 | 150 | 48 | 23 |
| Alclad 7075-O | 32 | 14 | 17 | | | 22 | |
| Alclad 7075-T6, T651 | 76 | 67 | 11 | | | 46 | |
| 7178-O | 33 | 15 | 15 | 16 | | | |
| 7178-T6, T651 | 88 | 78 | 10 | 11 | | | |
| 7178-T76, T7651 | 83 | 73 | | 11 | | | |
| Alclad 7178-O | 32 | 14 | 16 | | | | |
| Alclad 7178-T6, T651 | 81 | 71 | 10 | | | | |
| 8176-H24 | 17 | 14 | 15 | | | 10 | |

[a] Based on 500,000,000 cycles of completely reversed stress using the R. R. Moore type of machine and specimen.

[b] 1350-O wire should have an elongation of approximately 23 percent in 10 inches.

[c] 1350-H19 wire should have an elongation of approximately 1.5 percent in 10 inches.

[d] Tempers T361 and T861 were formerly designated T36 and T86, respectively.

[e] Based on $10^7$ cycles using flexural type testing of sheet specimens.

[f] T7451, although not previously registered, has appeared in the literature and in some specifications as T73651.

The data given in this table are intended only as a basis for comparing alloys and tempers, and should not be specified as engineering requirements or used for design purposes. The indicated typical mechanical properties for all except O temper material are higher than the specified minimum properties. For O temper products, typical ultimate and yield values are slightly lower than specified (maximum) values.

*Source: Aluminum Standards and Data.* Courtesy of the Aluminum Association.

**Heat-Treatability of Wrought Aluminum Alloys.**—In high-purity form, aluminum is soft and ductile. Most commercial uses, however, require greater strength than pure aluminum affords. This extra strength is achieved in aluminum first by the addition of other elements to produce various alloys, which singly or in combination impart strength to the metal. Further strengthening is possible by means that classify the alloys roughly into two categories, non-heat-treatable and heat-treatable.

*Non-Heat-Treatable Alloys:* The initial strength of alloys in this group depends upon the hardening effect of elements such as manganese, silicon, iron and magnesium, singly or

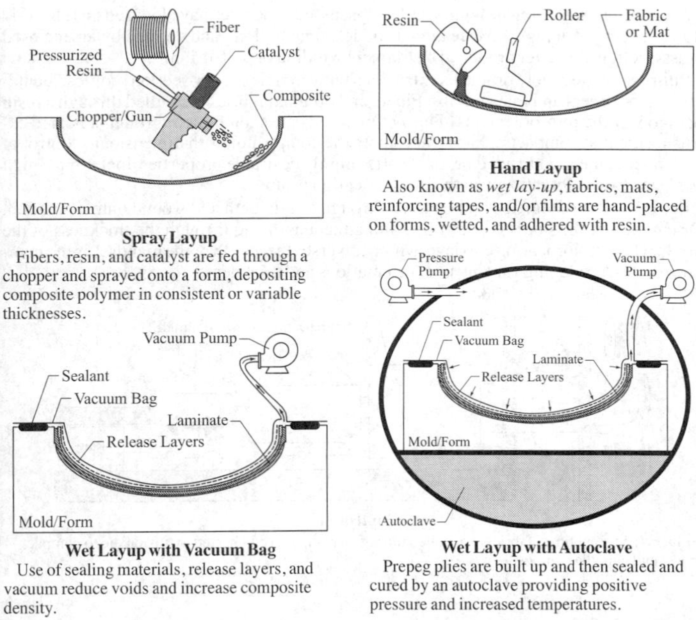


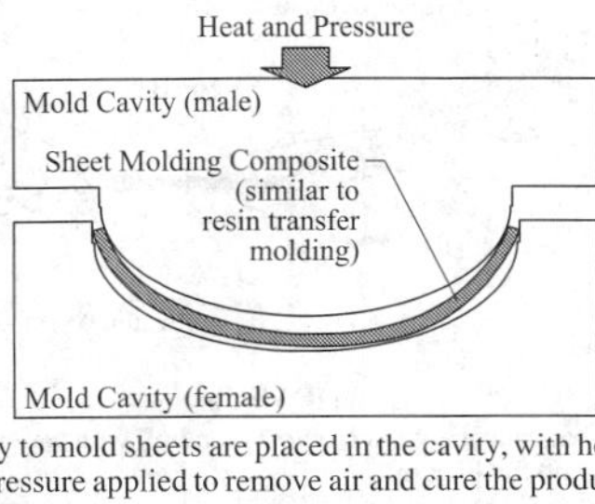

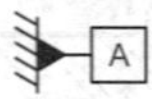 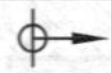 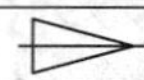 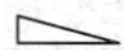 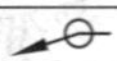 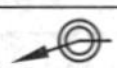 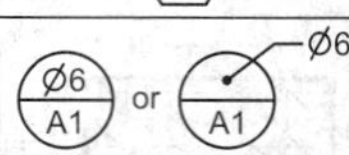 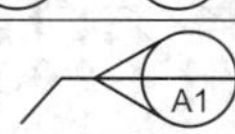 

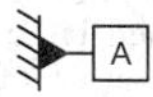 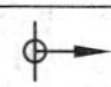  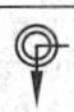 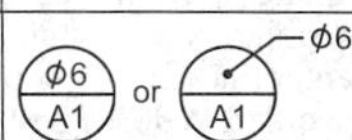 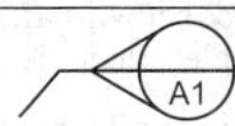 

When unilateral tolerances are used, one of the following three methods should be used to express them:

1) Specify, limiting dimensions only as

Diameter of hole: 2.250, 2.252

Diameter of shaft: 2.249, 2.247

2) One limiting size may be specified with its tolerances as

Diameter of hole: 2.250 + 0.002, −0.000

Diameter of shaft: 2.249 + 0.000, −0.002

3) The nominal size may be specified for both parts, with a notation showing both allowance and tolerance, as

Diameter of hole: $2\frac{1}{4}$ + 0.002, −0.000

Diameter of shaft: $2\frac{1}{4}$ − 0.001, −0.003

Bilateral tolerances should be specified as such, usually with plus and minus tolerances of equal amount. An example of the expression of bilateral tolerances is

$$2 \pm 0.001 \quad \text{or} \quad 2\,^{+0.001}_{-0.001}$$

**Application of Tolerances.—**According to common practice, tolerances are applied in such a way as to show the permissible amount of dimensional variation in the direction that is less dangerous. When a variation in either direction is equally dangerous, a bilateral tolerance should be given. When a variation in one direction is more dangerous than a variation in another, a unilateral tolerance should be given in the less dangerous direction.

For nonmating surfaces, or atmospheric fits, the tolerances may be bilateral or unilateral, depending entirely upon the nature of the variations that develop in manufacture. On mating surfaces, with few exceptions, the tolerances should be unilateral.

Where tolerances are required on the distances between holes, usually they should be bilateral, as variation in either direction is normally equally dangerous. The variation in the distance between shafts carrying gears, however, should always be unilateral and plus; otherwise, the gears might run too tight. A slight increase in the backlash between gears is seldom of much importance.

One exception to the use of unilateral tolerances on mating surfaces occurs when tapers are involved; either bilateral or unilateral tolerances may then prove advisable, depending upon conditions. These tolerances should be determined in the same manner as the tolerances on the distances between holes. When a variation either in or out of the position of the mating taper surfaces is equally dangerous, the tolerances should be bilateral. When a variation in one direction is of less danger than a variation in the opposite direction, the tolerance should be unilateral and in the less dangerous direction.

**Locating Tolerance Dimensions.—**Only one dimension in the same straight line can be controlled within fixed limits. That dimension is the distance between the cutting surface of the tool and the locating or registering surface of the part being machined. Therefore, it is incorrect to locate any point or surface with tolerances from more than one point in the same straight line.

Every part of a mechanism must be located in each plane. Every operating part must be located with proper operating allowances. After such requirements of location are met, all other surfaces should have liberal clearances. Dimensions should be given between those points or surfaces that it is essential to hold in a specific relation to each other. This restriction applies particularly to those surfaces in each plane that control the location of other component parts. Many dimensions are relatively unimportant in this respect. It is good practice to establish a common locating point in each plane and give, as far as possible, all such dimensions from these common locating points. The locating points on the drawing, the locating or registering points used for machining the surfaces and the locating points for measuring should all be identical.

**Table 2. Factors for Calculating Shrinkage Fit Allowances for Hollow Steel Shafts and Steel Hubs**

| $\frac{D_2}{D_1}$ | $\frac{D_1}{D_0}$ | $C$[a] | $\frac{D_2}{D_1}$ | $\frac{D_1}{D_0}$ | $C$[a] | $\frac{D_2}{D_1}$ | $\frac{D_1}{D_0}$ | $C$[a] |
|---|---|---|---|---|---|---|---|---|
| 1.5 | 2.0 | 0.455 | 2.4 | 2.0 | 0.760 | 3.4 | 2.0 | 0.876 |
| | 2.5 | 0.357 | | 2.5 | 0.597 | | 2.5 | 0.689 |
| | 3.0 | 0.313 | | 3.0 | 0.523 | | 3.0 | 0.602 |
| | 3.5 | 0.288 | | 3.5 | 0.481 | | 3.5 | 0.555 |
| 1.6 | 2.0 | 0.509 | 2.6 | 2.0 | 0.793 | 3.6 | 2.0 | 0.888 |
| | 2.5 | 0.400 | | 2.5 | 0.624 | | 2.5 | 0.698 |
| | 3.0 | 0.350 | | 3.0 | 0.546 | | 3.0 | 0.611 |
| | 3.5 | 0.322 | | 3.5 | 0.502 | | 3.5 | 0.562 |
| 1.8 | 2.0 | 0.599 | 2.8 | 2.0 | 0.820 | 3.8 | 2.0 | 0.900 |
| | 2.5 | 0.471 | | 2.5 | 0.645 | | 2.5 | 0.707 |
| | 3.0 | 0.412 | | 3.0 | 0.564 | | 3.0 | 0.619 |
| | 3.5 | 0.379 | | 3.5 | 0.519 | | 3.5 | 0.570 |
| 2.0 | 2.0 | 0.667 | 3.0 | 2.0 | 0.842 | 4.0 | 2.0 | 0.909 |
| | 2.5 | 0.524 | | 2.5 | 0.662 | | 2.5 | 0.715 |
| | 3.0 | 0.459 | | 3.0 | 0.580 | | 3.0 | 0.625 |
| | 3.5 | 0.422 | | 3.5 | 0.533 | | 3.5 | 0.576 |
| 2.2 | 2.0 | 0.718 | 3.2 | 2.0 | 0.860 | … | … | … |
| | 2.5 | 0.565 | | 2.5 | 0.676 | | … | … |
| | 3.0 | 0.494 | | 3.0 | 0.591 | | … | … |
| | 3.5 | 0.455 | | 3.5 | 0.544 | | … | … |

[a] Values of factor $C$ for hollow steel shafts of external and internal diameters $D_1$ and $D_0$, respectively, and steel hubs of nominal external diameter $D_2$.

**Table 3. Factors for Calculating Shrinkage Fit Allowances for Hollow Steel Shafts and Cast-Iron Hubs**

| $\frac{D_2}{D_1}$ | $\frac{D_1}{D_0}$ | $C$[a] | $\frac{D_2}{D_1}$ | $\frac{D_1}{D_0}$ | $C$[a] | $\frac{D_2}{D_1}$ | $\frac{D_1}{D_0}$ | $C$[a] |
|---|---|---|---|---|---|---|---|---|
| 1.5 | 2.0 | 0.468 | 2.4 | 2.0 | 0.798 | 3.4 | 2.0 | 0.926 |
| | 2.5 | 0.368 | | 2.5 | 0.628 | | 2.5 | 0.728 |
| | 3.0 | 0.322 | | 3.0 | 0.549 | | 3.0 | 0.637 |
| | 3.5 | 0.296 | | 3.5 | 0.506 | | 3.5 | 0.587 |
| 1.6 | 2.0 | 0.527 | 2.6 | 2.0 | 0.834 | 3.6 | 2.0 | 0.941 |
| | 2.5 | 0.414 | | 2.5 | 0.656 | | 2.5 | 0.740 |
| | 3.0 | 0.362 | | 3.0 | 0.574 | | 3.0 | 0.647 |
| | 3.5 | 0.333 | | 3.5 | 0.528 | | 3.5 | 0.596 |
| 1.8 | 2.0 | 0.621 | 2.8 | 2.0 | 0.864 | 3.8 | 2.0 | 0.953 |
| | 2.5 | 0.488 | | 2.5 | 0.679 | | 2.5 | 0.749 |
| | 3.0 | 0.427 | | 3.0 | 0.594 | | 3.0 | 0.656 |
| | 3.5 | 0.393 | | 3.5 | 0.547 | | 3.5 | 0.603 |
| 2.0 | 2.0 | 0.696 | 3.0 | 2.0 | 0.888 | 4.0 | 2.0 | 0.964 |
| | 2.5 | 0.547 | | 2.5 | 0.698 | | 2.5 | 0.758 |
| | 3.0 | 0.479 | | 3.0 | 0.611 | | 3.0 | 0.663 |
| | 3.5 | 0.441 | | 3.5 | 0.562 | | 3.5 | 0.610 |
| 2.2 | 2.0 | 0.753 | 3.2 | 2.0 | 0.909 | … | … | … |
| | 2.5 | 0.592 | | 2.5 | 0.715 | | … | … |
| | 3.0 | 0.518 | | 3.0 | 0.625 | | … | … |
| | 3.5 | 0.477 | | 3.5 | 0.576 | | … | … |

[a] Values of factor $C$ for hollow steel shafts and cast-iron hubs. Notation as in Table 2.

**Table 6. Standard Tolerances** *ANSI/ASME B4.1-1967 (R2020)*

| Nominal Size, Inches | | Grade | | | | | | | | | |
|---|---|---|---|---|---|---|---|---|---|---|---|
| | | 4 | 5 | 6 | 7 | 8 | 9 | 10 | 11 | 12 | 13 |
| Over | To | Tolerances in thousandths of an inch[a] | | | | | | | | | |
| 0 | 0.12 | 0.12 | 0.15 | 0.25 | 0.4 | 0.6 | 1.0 | 1.6 | 2.5 | 4 | 6 |
| 0.12 | 0.24 | 0.15 | 0.20 | 0.3 | 0.5 | 0.7 | 1.2 | 1.8 | 3.0 | 5 | 7 |
| 0.24 | 0.40 | 0.15 | 0.25 | 0.4 | 0.6 | 0.9 | 1.4 | 2.2 | 3.5 | 6 | 9 |
| 0.40 | 0.71 | 0.2 | 0.3 | 0.4 | 0.7 | 1.0 | 1.6 | 2.8 | 4.0 | 7 | 10 |
| 0.71 | 1.19 | 0.25 | 0.4 | 0.5 | 0.8 | 1.2 | 2.0 | 3.5 | 5.0 | 8 | 12 |
| 1.19 | 1.97 | 0.3 | 0.4 | 0.6 | 1.0 | 1.6 | 2.5 | 4.0 | 6 | 10 | 16 |
| 1.97 | 3.15 | 0.3 | 0.5 | 0.7 | 1.2 | 1.8 | 3.0 | 4.5 | 7 | 12 | 18 |
| 3.15 | 4.73 | 0.4 | 0.6 | 0.9 | 1.4 | 2.2 | 3.5 | 5 | 9 | 14 | 22 |
| 4.73 | 7.09 | 0.5 | 0.7 | 1.0 | 1.6 | 2.5 | 4.0 | 6 | 10 | 16 | 25 |
| 7.09 | 9.85 | 0.6 | 0.8 | 1.2 | 1.8 | 2.8 | 4.5 | 7 | 12 | 18 | 28 |
| 9.85 | 12.41 | 0.6 | 0.9 | 1.2 | 2.0 | 3.0 | 5.0 | 8 | 12 | 20 | 30 |
| 12.41 | 15.75 | 0.7 | 1.0 | 1.4 | 2.2 | 3.5 | 6 | 9 | 14 | 22 | 35 |
| 15.75 | 19.69 | 0.8 | 1.0 | 1.6 | 2.5 | 4 | 6 | 10 | 16 | 25 | 40 |
| 19.69 | 30.09 | 0.9 | 1.2 | 2.0 | 3 | 5 | 8 | 12 | 20 | 30 | 50 |
| 30.09 | 41.49 | 1.0 | 1.6 | 2.5 | 4 | 6 | 10 | 16 | 25 | 40 | 60 |
| 41.49 | 56.19 | 1.2 | 2.0 | 3 | 5 | 8 | 12 | 20 | 30 | 50 | 80 |
| 56.19 | 76.39 | 1.6 | 2.5 | 4 | 6 | 10 | 16 | 25 | 40 | 60 | 100 |
| 76.39 | 100.9 | 2.0 | 3 | 5 | 8 | 12 | 20 | 30 | 50 | 80 | 125 |
| 100.9 | 131.9 | 2.5 | 4 | 6 | 10 | 16 | 25 | 40 | 60 | 100 | 160 |
| 131.9 | 171.9 | 3 | 5 | 8 | 12 | 20 | 30 | 50 | 80 | 125 | 200 |
| 171.9 | 200 | 4 | 6 | 10 | 16 | 25 | 40 | 60 | 100 | 160 | 250 |

[a] All tolerances above the heavy line are in accordance with American-British-Canadian (ABC) agreements.

**Table 7. Relation of Machining Processes to Tolerance Grades**
*ANSI/ASME B4.1-1967 (R2020)*

This chart may be used as a general guide to determine the machining processes that will, under normal conditions, produce work within the tolerance grades indicated.

(See also *Relation of Surface Roughness to Tolerances* starting on page 802.)

| MACHINING OPERATION | TOLERANCE GRADES 4 | 5 | 6 | 7 | 8 | 9 | 10 | 11 | 12 | 13 |
|---|---|---|---|---|---|---|---|---|---|---|
| Lapping & Honing | X | X | | | | | | | | |
| Cylindrical Grinding | | X | X | X | | | | | | |
| Surface Grinding | | X | X | X | X | | | | | |
| Diamond Turning | | X | X | X | | | | | | |
| Diamond Boring | | X | X | X | | | | | | |
| Broaching | | X | X | X | X | | | | | |
| Reaming | | | X | X | X | X | X | | | |
| Turning | | | | X | X | X | X | X | X | X |
| Boring | | | | | X | X | X | X | X | X |
| Milling | | | | | | | X | X | X | X |
| Planing & Shaping | | | | | | | X | X | X | X |
| Drilling | | | | | | | X | X | X | X |

The letter symbols used to designate standard fits are as follows:

*RC* = Running or Sliding Clearance Fit *LC* = Locational Clearance Fit

*LT* = Transition Clearance or Interference Fit

*LN* = Locational Interference Fit

*FN* = Force or Shrink Fit

## Table 8a. American National Standard Running and Sliding Fits *ANSI/ASME B4.1-1967 (R2020)*

| Nominal Size Range, Inches | Class RC 1 | | | Class RC 2 | | | Class RC 3 | | | Class RC 4 | | |
|---|---|---|---|---|---|---|---|---|---|---|---|---|
| | | Standard Tolerance Limits | | | Standard Tolerance Limits | | | Standard Tolerance Limits | | | Standard Tolerance Limits | |
| | Clearance[a] | Hole H5 | Shaft g4 | Clearance[a] | Hole H6 | Shaft g5 | Clearance[a] | Hole H7 | Shaft f6 | Clearance[a] | Hole H8 | Shaft f7 |
| Over - To | Values shown below are in thousandths of an inch | | | | | | | | | | | |
| 0 - 0.12 | 0.1 | +0.2 | −0.1 | 0.1 | +0.25 | −0.1 | 0.3 | +0.4 | −0.3 | 0.3 | +0.6 | −0.3 |
| | 0.45 | 0 | −0.25 | 0.55 | 0 | −0.3 | 0.95 | 0 | −0.55 | 1.3 | 0 | −0.7 |
| 0.12 - 0.24 | 0.15 | +0.2 | −0.15 | 0.15 | +0.3 | −0.15 | 0.4 | +0.5 | −0.4 | 0.4 | +0.7 | −0.4 |
| | 0.5 | 0 | −0.3 | 0.65 | 0 | −0.35 | 1.2 | 0 | −0.7 | 1.6 | 0 | −0.9 |
| 0.24 - 0.40 | 0.2 | +0.25 | −0.2 | 0.2 | +0.4 | −0.2 | 0.5 | +0.6 | −0.5 | 0.5 | +0.9 | −0.5 |
| | 0.6 | 0 | −0.35 | 0.85 | 0 | −0.45 | 1.5 | 0 | −0.9 | 2.0 | 0 | −1.1 |
| 0.40 - 0.71 | 0.25 | +0.3 | −0.25 | 0.25 | +0.4 | −0.25 | 0.6 | +0.7 | −0.6 | 0.6 | +1.0 | −0.6 |
| | 0.75 | 0 | −0.45 | 0.95 | 0 | −0.55 | 1.7 | 0 | −1.0 | 2.3 | 0 | −1.3 |
| 0.71 - 1.19 | 0.3 | +0.4 | −0.3 | 0.3 | +0.5 | −0.3 | 0.8 | +0.8 | −0.8 | 0.8 | +1.2 | −0.8 |
| | 0.95 | 0 | −0.55 | 1.2 | 0 | −0.7 | 2.1 | 0 | −1.3 | 2.8 | 0 | −1.6 |
| 1.19 - 1.97 | 0.4 | +0.4 | −0.4 | 0.4 | +0.6 | −0.4 | 1.0 | +1.0 | −1.0 | 1.0 | +1.6 | −1.0 |
| | 1.1 | 0 | −0.7 | 1.4 | 0 | −0.8 | 2.6 | 0 | −1.6 | 3.6 | 0 | −2.0 |
| 1.97 - 3.15 | 0.4 | +0.5 | −0.4 | 0.4 | +0.7 | −0.4 | 1.2 | +1.2 | −1.2 | 1.2 | +1.8 | −1.2 |
| | 1.2 | 0 | −0.7 | 1.6 | 0 | −0.9 | 3.1 | 0 | −1.9 | 4.2 | 0 | −2.4 |
| 3.15 - 4.73 | 0.5 | +0.6 | −0.5 | 0.5 | +0.9 | −0.5 | 1.4 | +1.4 | −1.4 | 1.4 | +2.2 | −1.4 |
| | 1.5 | 0 | −0.9 | 2.0 | 0 | −1.1 | 3.7 | 0 | −2.3 | 5.0 | 0 | −2.8 |
| 4.73 - 7.09 | 0.6 | +0.7 | −0.6 | 0.6 | +1.0 | −0.6 | 1.6 | +1.6 | −1.6 | 1.6 | +2.5 | −1.6 |
| | 1.8 | 0 | −1.1 | 2.3 | 0 | −1.3 | 4.2 | 0 | −2.6 | 5.7 | 0 | −3.2 |
| 7.09 - 9.85 | 0.6 | +0.8 | −0.6 | 0.6 | +1.2 | −0.6 | 2.0 | +1.8 | −2.0 | 2.0 | +2.8 | −2.0 |
| | 2.0 | 0 | −1.2 | 2.6 | 0 | −1.4 | 5.0 | 0 | −3.2 | 6.6 | 0 | −3.8 |
| 9.85 - 12.41 | 0.8 | +0.9 | −0.8 | 0.8 | +1.2 | −0.8 | 2.5 | +2.0 | −2.5 | 2.5 | +3.0 | −2.5 |
| | 2.3 | 0 | −1.4 | 2.9 | 0 | −1.7 | 5.7 | 0 | −3.7 | 7.5 | 0 | −4.5 |
| 12.41 - 15.75 | 1.0 | +1.0 | −1.0 | 1.0 | +1.4 | −1.0 | 3.0 | +2.2 | −3.0 | 3.0 | +3.5 | −3.0 |
| | 2.7 | 0 | −1.7 | 3.4 | 0 | −2.0 | 6.6 | 0 | −4.4 | 8.7 | 0 | −5.2 |
| 15.75 - 19.69 | 1.2 | +1.0 | −1.2 | 1.2 | +1.6 | −1.2 | 4.0 | +2.5 | −4.0 | 4.0 | +4.0 | −4.0 |
| | 3.0 | 0 | −2.0 | 3.8 | 0 | −2.2 | 8.1 | 0 | −5.6 | 10.5 | 0 | −6.5 |

## Table 9a. American National Standard Clearance Locational Fits *ANSI/ASME B4.1-1967 (R2020)*

| Nominal Size Range, Inches | Class LC 1 | | | Class LC 2 | | | Class LC 3 | | | Class LC 4 | | | Class LC 5 | | |
|---|---|---|---|---|---|---|---|---|---|---|---|---|---|---|---|
| | | Standard Tolerance Limits | | | Standard Tolerance Limits | | | Standard Tolerance Limits | | | Standard Tolerance Limits | | | Standard Tolerance Limits | |
| Over To | Clearance[a] | Hole H6 | Shaft h5 | Clearance[a] | Hole H7 | Shaft h6 | Clearance[a] | Hole H8 | Shaft h7 | Clearance[a] | Hole H10 | Shaft h9 | Clearance[a] | Hole H7 | Shaft g6 |
| | Values shown below are in thousandths of an inch | | | | | | | | | | | | | | |
| 0- 0.12 | 0 | +0.25 | 0 | 0 | +0.4 | 0 | 0 | +0.6 | 0 | 0 | +1.6 | 0 | 0.1 | +0.4 | −0.1 |
| | 0.45 | 0 | −0.2 | 0.65 | 0 | −0.25 | 1 | 0 | −0.4 | 2.6 | 0 | −1.0 | 0.75 | 0 | −0.35 |
| 0.12- 0.24 | 0 | +0.3 | 0 | 0 | +0.5 | 0 | 0 | +0.7 | 0 | 0 | +1.8 | 0 | 0.15 | +0.5 | −0.15 |
| | 0.5 | 0 | −0.2 | 0.8 | 0 | −0.3 | 1.2 | 0 | −0.5 | 3.0 | 0 | −1.2 | 0.95 | 0 | −0.45 |
| 0.24- 0.40 | 0 | +0.4 | 0 | 0 | +0.6 | 0 | 0 | +0.9 | 0 | 0 | +2.2 | 0 | 0.2 | +0.6 | −0.2 |
| | 0.65 | 0 | −0.25 | 1.0 | 0 | −0.4 | 1.5 | 0 | −0.6 | 3.6 | 0 | −1.4 | 1.2 | 0 | −0.6 |
| 0.40- 0.71 | 0 | +0.4 | 0 | 0 | +0.7 | 0 | 0 | +1.0 | 0 | 0 | +2.8 | 0 | 0.25 | +0.7 | −0.25 |
| | 0.7 | 0 | −0.3 | 1.1 | 0 | −0.4 | 1.7 | 0 | −0.7 | 4.4 | 0 | −1.6 | 1.35 | 0 | −0.65 |
| 0.71- 1.19 | 0 | +0.5 | 0 | 0 | +0.8 | 0 | 0 | +1.2 | 0 | 0 | +3.5 | 0 | 0.3 | +0.8 | −0.3 |
| | 0.9 | 0 | −0.4 | 1.3 | 0 | −0.5 | 2 | 0 | −0.8 | 5.5 | 0 | −2.0 | 1.6 | 0 | −0.8 |
| 1.19- 1.97 | 0 | +0.6 | 0 | 0 | +1.0 | 0 | 0 | +1.6 | 0 | 0 | +4.0 | 0 | 0.4 | +1.0 | −0.4 |
| | 1.0 | 0 | −0.4 | 1.6 | 0 | −0.6 | 2.6 | 0 | −1 | 6.5 | 0 | −2.5 | 2.0 | 0 | −1.0 |
| 1.97- 3.15 | 0 | +0.7 | 0 | 0 | +1.2 | 0 | 0 | +1.8 | 0 | 0 | +4.5 | 0 | 0.4 | +1.2 | −0.4 |
| | 1.2 | 0 | −0.5 | 1.9 | 0 | −0.7 | 3 | 0 | −1.2 | 7.5 | 0 | −3 | 2.3 | 0 | −1.1 |
| 3.15- 4.73 | 0 | +0.9 | 0 | 0 | +1.4 | 0 | 0 | +2.2 | 0 | 0 | +5.0 | 0 | 0.5 | +1.4 | −0.5 |
| | 1.5 | 0 | −0.6 | 2.3 | 0 | −0.9 | 3.6 | 0 | −1.4 | 8.5 | 0 | −3.5 | 2.8 | 0 | −1.4 |
| 4.73- 7.09 | 0 | +1.0 | 0 | 0 | +1.6 | 0 | 0 | +2.5 | 0 | 0 | +6.0 | 0 | 0.6 | +1.6 | −0.6 |
| | 1.7 | 0 | −0.7 | 2.6 | 0 | −1.0 | 4.1 | 0 | −1.6 | 10.0 | 0 | −4 | 3.2 | 0 | −1.6 |
| 7.09- 9.85 | 0 | +1.2 | 0 | 0 | +1.8 | 0 | 0 | +2.8 | 0 | 0 | +7.0 | 0 | 0.6 | +1.8 | −0.6 |
| | 2.0 | 0 | −0.8 | 3.0 | 0 | −1.2 | 4.6 | 0 | −1.8 | 11.5 | 0 | −4.5 | 3.6 | 0 | −1.8 |
| 9.85- 12.41 | 0 | +1.2 | 0 | 0 | +2.0 | 0 | 0 | +3.0 | 0 | 0 | +8.0 | 0 | 0.7 | +2.0 | −0.7 |
| | 2.1 | 0 | −0.9 | 3.2 | 0 | −1.2 | 5 | 0 | −2.0 | 13.0 | 0 | −5 | 3.9 | 0 | −1.9 |
| 12.41- 15.75 | 0 | +1.4 | 0 | 0 | +2.2 | 0 | 0 | +3.5 | 0 | 0 | +9.0 | 0 | 0.7 | +2.2 | −0.7 |
| | 2.4 | 0 | −1.0 | 3.6 | 0 | −1.4 | 5.7 | 0 | −2.2 | 15.0 | 0 | −6 | 4.3 | 0 | −2.1 |
| 15.75- 19.69 | 0 | +1.6 | 0 | 0 | +2.5 | 0 | 0 | +4 | 0 | 0 | +10.0 | 0 | 0.8 | +2.5 | −0.8 |
| | 2.6 | 0 | −1.0 | 4.1 | 0 | −1.6 | 6.5 | 0 | −2.5 | 16.0 | 0 | −6 | 4.9 | 0 | −2.4 |

Tolerance limits given in body of Table 9a and Table 9b are added or subtracted to basic size (as indicated by + or − sign) to obtain maximum and minimum sizes of mating parts. All data above heavy lines are in accordance with American-British-Canadian (ABC) agreements. Symbols H6, H7, s6, etc., are hole and shaft designations in ABC system.

## Table 11. Standard Force and Shrink Fits *ANSI/ASME B4.1-1967 (R2020)*

| Nominal Size Range, Inches Over–To | Class FN 1 Interference[a] | Class FN 1 Standard Tolerance Limits Hole H6 | Class FN 1 Standard Tolerance Limits Shaft | Class FN 2 Interference[a] | Class FN 2 Standard Tolerance Limits Hole H7 | Class FN 2 Standard Tolerance Limits Shaft s6 | Class FN 3 Interference[a] | Class FN 3 Standard Tolerance Limits Hole H7 | Class FN 3 Standard Tolerance Limits Shaft t6 | Class FN 4 Interference[a] | Class FN 4 Standard Tolerance Limits Hole H7 | Class FN 4 Standard Tolerance Limits Shaft u6 | Class FN 5 Interference[a] | Class FN 5 Standard Tolerance Limits Hole H8 | Class FN 5 Standard Tolerance Limits Shaft x7 |
|---|---|---|---|---|---|---|---|---|---|---|---|---|---|---|---|
| | Values shown below are in thousandths of an inch | | | | | | | | | | | | | | |
| 0- 0.12 | 0.05 | +0.25 | +0.5 | 0.2 | +0.4 | +0.85 | | | | 0.3 | +0.4 | +0.95 | 0.3 | +0.6 | +1.3 |
| | 0.5 | 0 | +0.3 | 0.85 | 0 | +0.6 | | | | 0.95 | 0 | +0.7 | 1.3 | 0 | +0.9 |
| 0.12- 0.24 | 0.1 | +0.3 | +0.6 | 0.2 | +0.5 | +1.0 | | | | 0.4 | +0.5 | +1.2 | 0.5 | +0.7 | +1.7 |
| | 0.6 | 0 | +0.4 | 1.0 | 0 | +0.7 | | | | 1.2 | 0 | +0.9 | 1.7 | 0 | +1.2 |
| 0.24- 0.40 | 0.1 | +0.4 | +0.75 | 0.4 | +0.6 | +1.4 | | | | 0.6 | +0.6 | +1.6 | 0.5 | +0.9 | +2.0 |
| | 0.75 | 0 | +0.5 | 1.4 | 0 | +1.0 | | | | 1.6 | 0 | +1.2 | 2.0 | 0 | +1.4 |
| 0.40- 0.56 | 0.1 | +0.4 | +0.8 | 0.5 | +0.7 | +1.6 | | | | 0.7 | +0.7 | +1.8 | 0.6 | +1.0 | +2.3 |
| | 0.8 | 0 | +0.5 | 1.6 | 0 | +1.2 | | | | 1.8 | 0 | +1.4 | 2.3 | 0 | +1.6 |
| 0.56- 0.71 | 0.2 | +0.4 | +0.9 | 0.5 | +0.7 | +1.6 | | | | 0.7 | +0.7 | +1.8 | 0.8 | +1.0 | +2.5 |
| | 0.9 | 0 | +0.6 | 1.6 | 0 | +1.2 | | | | 1.8 | 0 | +1.4 | 2.5 | 0 | +1.8 |
| 0.71- 0.95 | 0.2 | +0.5 | +1.1 | 0.6 | +0.8 | +1.9 | | | | 0.8 | +0.8 | +2.1 | 1.0 | +1.2 | +3.0 |
| | 1.1 | 0 | +0.7 | 1.9 | 0 | +1.4 | | | | 2.1 | 0 | +1.6 | 3.0 | 0 | +2.2 |
| 0.95- 1.19 | 0.3 | +0.5 | +1.2 | 0.6 | +0.8 | +1.9 | 0.8 | +0.8 | +2.1 | +1.0 | +0.8 | +2.3 | 1.3 | +1.2 | +3.3 |
| | 1.2 | 0 | +0.8 | 1.9 | 0 | +1.4 | 2.1 | 0 | +1.6 | 2.3 | 0 | +1.8 | 3.3 | 0 | +2.5 |
| 1.19- 1.58 | 0.3 | +0.6 | +1.3 | 0.8 | +1.0 | +2.4 | 1.0 | +1.0 | +2.6 | 1.5 | +1.0 | +3.1 | 1.4 | +1.6 | +4.0 |
| | 1.3 | 0 | +0.9 | 2.4 | 0 | +1.8 | 2.6 | 0 | +2.0 | 3.1 | 0 | +2.5 | 4.0 | 0 | +3.0 |
| 1.58- 1.97 | 0.4 | +0.6 | +1.4 | 0.8 | +1.0 | +2.4 | 1.2 | +1.0 | +2.8 | 1.8 | +1.0 | +3.4 | 2.4 | +1.6 | +5.0 |
| | 1.4 | 0 | +1.0 | 2.4 | 0 | +1.8 | 2.8 | 0 | +2.2 | 3.4 | 0 | +2.8 | 5.0 | 0 | +4.0 |
| 1.97- 2.56 | 0.6 | +0.7 | +1.8 | 0.8 | +1.2 | +2.7 | 1.3 | +1.2 | +3.2 | 2.3 | +1.2 | +4.2 | 3.2 | +1.8 | +6.2 |
| | 1.8 | 0 | +1.3 | 2.7 | 0 | +2.0 | 3.2 | 0 | +2.5 | 4.2 | 0 | +3.5 | 6.2 | 0 | +5.0 |
| 2.56- 3.15 | 0.7 | +0.7 | +1.9 | 1.0 | +1.2 | +2.9 | 1.8 | +1.2 | +3.7 | 2.8 | +1.2 | +4.7 | 4.2 | +1.8 | +7.2 |
| | 1.9 | 0 | +1.4 | 2.9 | 0 | +2.2 | 3.7 | 0 | +3.0 | 4.7 | 0 | +4.0 | 7.2 | 0 | +6.0 |
| 3.15- 3.94 | 0.9 | +0.9 | +2.4 | 1.4 | +1.4 | +3.7 | 2.1 | +1.4 | +4.4 | 3.6 | +1.4 | +5.9 | 4.8 | +2.2 | +8.4 |
| | 2.4 | 0 | +1.8 | 3.7 | 0 | +2.8 | 4.4 | 0 | +3.5 | 5.9 | 0 | +5.0 | 8.4 | 0 | +7.0 |
| 3.94- 4.73 | 1.1 | +0.9 | +2.6 | 1.6 | +1.4 | +3.9 | 2.6 | +1.4 | +4.9 | 4.6 | +1.4 | +6.9 | 5.8 | +2.2 | +9.4 |
| | 2.6 | 0 | +2.0 | 3.9 | 0 | +3.0 | 4.9 | 0 | +4.0 | 6.9 | 0 | +6.0 | 9.4 | 0 | +8.0 |

**Table 12. Standard Interference Locational Fits**
*ANSI/ASME B4.1-1967 (R2020)*

| Nominal Size Range, Inches | Class LN 1 | | | Class LN 2 | | | Class LN 3 | | |
|---|---|---|---|---|---|---|---|---|---|
| | Limits of Interference | Standard Limits | | Limits of Interference | Standard Limits | | Limits of Interference | Standard Limits | |
| | | Hole H6 | Shaft n5 | | Hole H7 | Shaft p6 | | Hole H7 | Shaft r6 |
| Over To | Values shown below are given in thousandths of an inch | | | | | | | | |
| 0- 0.12 | 0<br>0.45 | +0.25<br>0 | +0.45<br>+0.25 | 0<br>0.65 | +0.4<br>0 | +0.65<br>+0.4 | 0.1<br>0.75 | +0.4<br>0 | +0.75<br>+0.5 |
| 0.12- 0.24 | 0<br>0.5 | +0.3<br>0 | +0.5<br>+0.3 | 0<br>0.8 | +0.5<br>0 | +0.8<br>+0.5 | 0.1<br>0.9 | +0.5<br>0 | +0.9<br>+0.6 |
| 0.24- 0.40 | 0<br>0.65 | +0.4<br>0 | +0.65<br>+0.4 | 0<br>1.0 | +0.6<br>0 | +1.0<br>+0.6 | 0.2<br>1.2 | +0.6<br>0 | +1.2<br>+0.8 |
| 0.40- 0.71 | 0<br>0.8 | +0.4<br>0 | +0.8<br>+0.4 | 0<br>1.1 | +0.7<br>0 | +1.1<br>+0.7 | 0.3<br>1.4 | +0.7<br>0 | +1.4<br>+1.0 |
| 0.71- 1.19 | 0<br>1.0 | +0.5<br>0 | +1.0<br>+0.5 | 0<br>1.3 | +0.8<br>0 | +1.3<br>+0.8 | 0.4<br>1.7 | +0.8<br>0 | +1.7<br>+1.2 |
| 1.19- 1.97 | 0<br>1.1 | +0.6<br>0 | +1.1<br>+0.6 | 0<br>1.6 | +1.0<br>0 | +1.6<br>+1.0 | 0.4<br>2.0 | +1.0<br>0 | +2.0<br>+1.4 |
| 1.97- 3.15 | 0.1<br>1.3 | +0.7<br>0 | +1.3<br>+0.8 | 0.2<br>2.1 | +1.2<br>0 | +2.1<br>+1.4 | 0.4<br>2.3 | +1.2<br>0 | +2.3<br>+1.6 |
| 3.15- 4.73 | 0.1<br>1.6 | +0.9<br>0 | +1.6<br>+1.0 | 0.2<br>2.5 | +1.4<br>0 | +2.5<br>+1.6 | 0.6<br>2.9 | +1.4<br>0 | +2.9<br>+2.0 |
| 4.73- 7.09 | 0.2<br>1.9 | +1.0<br>0 | +1.9<br>+1.2 | 0.2<br>2.8 | +1.6<br>0 | +2.8<br>+1.8 | 0.9<br>3.5 | +1.6<br>0 | +3.5<br>+2.5 |
| 7.09- 9.85 | 0.2<br>2.2 | +1.2<br>0 | +2.2<br>+1.4 | 0.2<br>3.2 | +1.8<br>0 | +3.2<br>+2.0 | 1.2<br>4.2 | +1.8<br>0 | +4.2<br>+3.0 |
| 9.85- 12.41 | 0.2<br>2.3 | +1.2<br>0 | +2.3<br>+1.4 | 0.2<br>3.4 | +2.0<br>0 | +3.4<br>+2.2 | 1.5<br>4.7 | +2.0<br>0 | +4.7<br>+3.5 |
| 12.41- 15.75 | 0.2<br>2.6 | +1.4<br>0 | +2.6<br>+1.6 | 0.3<br>3.9 | +2.2<br>0 | +3.9<br>+2.5 | 2.3<br>5.9 | +2.2<br>0 | +5.9<br>+4.5 |
| 15.75- 19.69 | 0.2<br>2.8 | +1.6<br>0 | +2.8<br>+1.8 | 0.3<br>4.4 | +2.5<br>0 | +4.4<br>+2.8 | 2.5<br>6.6 | +2.5<br>0 | +6.6<br>+5.0 |

All data in this table are in accordance with American-British-Canadian (ABC) agreements.

Limits for sizes above 19.69 inches are not covered by ABC agreements but are given in the ANSI/ASME Standard.

Symbols H7, p6, etc., are hole and shaft designations in the ABC system.

Tolerance limits given in body of table are added or subtracted to basic size (as indicated by + or – sign) to obtain maximum and minimum sizes of mating parts.

## ANSI/ASME Preferred Metric Limits and Fits

ANSI/ASME B4.2-1978 (R2020) describes the ISO system of metric limits and fits for mating parts as approved for general engineering usage in the United States.

It establishes 1) the designation symbols used to define dimensional limits on drawings, material stock, related tools, gages, etc.; 2) the preferred basic sizes (first and second choices); 3) the preferred tolerance zones (first, second, and third choices); 4) the preferred limits and fits for sizes (first choice only) up to and including 500 millimeters; and 5) the definitions of related terms.

The general terms "hole" and "shaft" can also be taken to refer to the space containing or contained by two parallel faces of any part, such as the width of a slot or the thickness of a key.

**Definitions.—**The most important terms relating to limits and fits are shown in Fig. 1 and are defined as follows:

*Basic Size:* The size to which limits of deviation are assigned. The basic size is the same for both members of a fit. For example, it is designated by the numbers 40 in 40H7.

*Deviation:* The algebraic difference between a size and the corresponding basic size

**Table 2.** *(Continued)* **American National Standard Preferred Hole Basis Metric Clearance Fits** *ANSI/ASME B4.2-1978 (R2020)*

| Basic Size[a] | | Loose Running Hole H11 | Loose Running Shaft c11 | Loose Running Fit[b] | Free Running Hole H9 | Free Running Shaft d9 | Free Running Fit[b] | Close Running Hole H8 | Close Running Shaft f7 | Close Running Fit[b] | Sliding Hole H7 | Sliding Shaft g6 | Sliding Fit[b] | Locational Clearance Hole H7 | Locational Clearance Shaft h6 | Locational Clearance Fit[b] |
|---|---|---|---|---|---|---|---|---|---|---|---|---|---|---|---|---|
| 30 | Max | 30.130 | 29.890 | 0.370 | 30.052 | 29.935 | 0.169 | 30.033 | 29.980 | 0.074 | 30.021 | 29.993 | 0.041 | 30.021 | 30.000 | 0.034 |
| | Min | 30.000 | 29.760 | 0.110 | 30.000 | 29.883 | 0.065 | 30.000 | 29.959 | 0.020 | 30.000 | 29.980 | 0.007 | 30.000 | 29.987 | 0.000 |
| 40 | Max | 40.160 | 39.880 | 0.440 | 40.062 | 39.920 | 0.204 | 40.039 | 39.975 | 0.089 | 40.025 | 39.991 | 0.050 | 40.025 | 40.000 | 0.041 |
| | Min | 40.000 | 39.720 | 0.120 | 40.000 | 39.858 | 0.080 | 40.000 | 39.950 | 0.025 | 40.000 | 39.975 | 0.009 | 40.000 | 39.984 | 0.000 |
| 50 | Max | 50.160 | 49.870 | 0.450 | 50.062 | 49.920 | 0.204 | 50.039 | 49.975 | 0.089 | 50.025 | 49.991 | 0.050 | 50.025 | 50.000 | 0.041 |
| | Min | 50.000 | 49.710 | 0.130 | 50.000 | 49.858 | 0.080 | 50.000 | 49.950 | 0.025 | 50.000 | 49.975 | 0.009 | 50.000 | 49.984 | 0.000 |
| 60 | Max | 60.190 | 59.860 | 0.520 | 60.074 | 59.900 | 0.248 | 60.046 | 59.970 | 0.106 | 60.030 | 59.990 | 0.059 | 60.030 | 60.000 | 0.049 |
| | Min | 60.000 | 59.670 | 0.140 | 60.000 | 59.826 | 0.100 | 60.000 | 59.940 | 0.030 | 60.000 | 59.971 | 0.010 | 60.000 | 59.981 | 0.000 |
| 80 | Max | 80.190 | 79.850 | 0.530 | 80.074 | 79.900 | 0.248 | 80.046 | 79.970 | 0.106 | 80.030 | 79.990 | 0.059 | 80.030 | 80.000 | 0.049 |
| | Min | 80.000 | 79.660 | 0.150 | 80.000 | 79.826 | 0.100 | 80.000 | 79.940 | 0.030 | 80.000 | 79.971 | 0.010 | 80.000 | 79.981 | 0.000 |
| 100 | Max | 100.220 | 99.830 | 0.610 | 100.087 | 99.880 | 0.294 | 100.054 | 99.964 | 0.125 | 100.035 | 99.988 | 0.069 | 100.035 | 100.000 | 0.057 |
| | Min | 100.000 | 99.610 | 0.170 | 100.000 | 99.793 | 0.120 | 100.000 | 99.929 | 0.036 | 100.000 | 99.966 | 0.012 | 100.000 | 99.978 | 0.000 |
| 120 | Max | 120.220 | 119.820 | 0.620 | 120.087 | 119.880 | 0.294 | 120.054 | 119.964 | 0.125 | 120.035 | 119.988 | 0.069 | 120.035 | 120.000 | 0.057 |
| | Min | 120.000 | 119.600 | 0.180 | 120.000 | 119.793 | 0.120 | 120.000 | 119.929 | 0.036 | 120.000 | 119.966 | 0.012 | 120.000 | 119.978 | 0.000 |
| 160 | Max | 160.250 | 159.790 | 0.710 | 160.100 | 159.855 | 0.345 | 160.063 | 159.957 | 0.146 | 160.040 | 159.986 | 0.079 | 160.040 | 160.000 | 0.065 |
| | Min | 160.000 | 159.540 | 0.210 | 160.000 | 159.755 | 0.145 | 160.000 | 159.917 | 0.043 | 160.000 | 159.961 | 0.014 | 160.000 | 159.975 | 0.000 |
| 200 | Max | 200.290 | 199.760 | 0.820 | 200.115 | 199.830 | 0.400 | 200.072 | 199.950 | 0.168 | 200.046 | 199.985 | 0.090 | 200.046 | 200.000 | 0.075 |
| | Min | 200.000 | 199.470 | 0.240 | 200.000 | 199.715 | 0.170 | 200.000 | 199.904 | 0.050 | 200.000 | 199.956 | 0.015 | 200.000 | 199.971 | 0.000 |
| 250 | Max | 250.290 | 249.720 | 0.860 | 250.115 | 249.830 | 0.400 | 250.072 | 249.950 | 0.168 | 250.046 | 249.985 | 0.090 | 250.046 | 250.000 | 0.075 |
| | Min | 250.000 | 249.430 | 0.280 | 250.000 | 249.715 | 0.170 | 250.000 | 249.904 | 0.050 | 250.000 | 249.956 | 0.015 | 250.000 | 249.971 | 0.000 |
| 300 | Max | 300.320 | 299.670 | 0.970 | 300.130 | 299.810 | 0.450 | 300.081 | 299.944 | 0.189 | 300.052 | 299.983 | 0.101 | 300.052 | 300.000 | 0.084 |
| | Min | 300.000 | 299.350 | 0.330 | 300.000 | 299.680 | 0.190 | 300.000 | 299.892 | 0.056 | 300.000 | 299.951 | 0.017 | 300.000 | 299.968 | 0.000 |
| 400 | Max | 400.360 | 399.600 | 1.120 | 400.140 | 399.790 | 0.490 | 400.089 | 399.938 | 0.208 | 400.057 | 399.982 | 0.111 | 400.057 | 400.000 | 0.093 |
| | Min | 400.000 | 399.240 | 0.400 | 400.000 | 399.650 | 0.210 | 400.000 | 399.881 | 0.062 | 400.000 | 399.946 | 0.018 | 400.000 | 399.964 | 0.000 |
| 500 | Max | 500.400 | 499.520 | 1.280 | 500.155 | 499.770 | 0.540 | 500.097 | 499.932 | 0.228 | 500.063 | 499.980 | 0.123 | 500.063 | 500.000 | 0.103 |
| | Min | 500.000 | 499.120 | 0.480 | 500.000 | 499.615 | 0.230 | 500.000 | 499.869 | 0.068 | 500.000 | 499.940 | 0.020 | 500.000 | 499.960 | 0.000 |

[a] The sizes shown are first-choice basic sizes (see Table 1, page 689). Preferred fits for other sizes can be calculated from data given in ANSI/ASME B4.2-1978 (R2020).

[b] All fits shown in this table have clearance.

All dimensions are in millimeters.

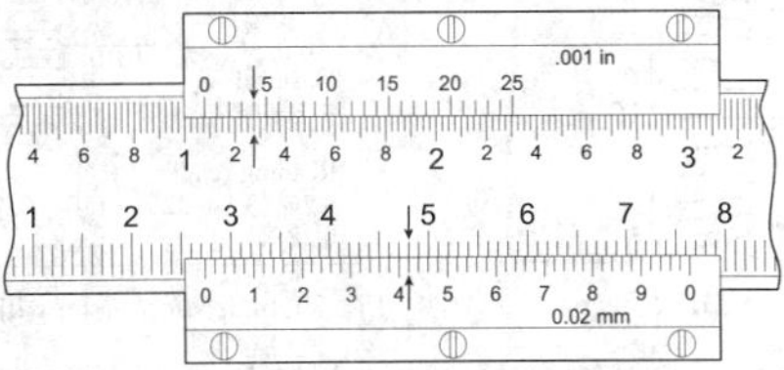


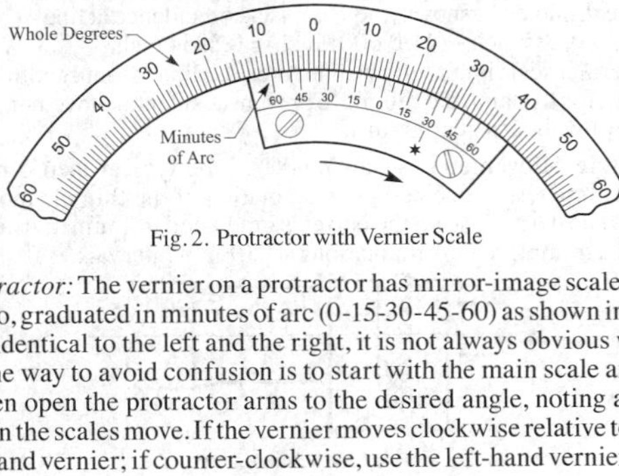

To check for taper, the shaft is supported in the V-block and the dial indicator is used to measure the maximum height over the shaft at various positions along its length, as shown in Fig. 4b, without turning the workpiece. Again, the shaft should be marked with the reading positions and values, and with the direction of the taper, and a written record should be made of the amount and direction of any taper discovered.

Checking for a bent shaft requires that the shaft be clocked at the shoulder and at the farther end, as shown in Fig. 4c. For a second check, the shaft is rotated only 90° or a quarter turn. When the recorded readings are compared with those from the ovality and taper checks, the three conditions can be distinguished.

To detect a curved or bowed condition, the shaft should be suspended in two V-blocks with only about $\frac{1}{8}$ inch of each end in each V. Alternatively, the shaft can be placed between centers. The shaft is then clocked at several points, as shown in Fig. 4d, but preferably not at those locations used for the ovality, taper, or crookedness checks. If the single element due to curvature is to be distinguished from the effects of ovality, taper, and crookedness, and its value assessed, great care must be taken to differentiate between the conditions detected by the measurements.

Finally, the amount of eccentricity between one shaft diameter and another may be tested by the setup shown in Fig. 4e. With the indicator plunger in contact with the smaller diameter, close to the shoulder, the shaft is rotated in the V-block and the indicator needle position is monitored to find the maximum and minimum readings.

Curvature, ovality, or crookedness conditions may tend to cancel each other, as shown in Fig. 5, and one or more of these degrees of defectiveness may add themselves to the true eccentricity readings, depending on their angular positions. Fig. 5a shows, for instance, how crookedness and ovality tend to cancel each other, and also shows their effect in falsifying the reading for eccentricity. As the same shaft is turned in the V-block to the position shown in Fig. 5b, the maximum curvature reading could tend to cancel or reduce the maximum eccentricity reading. Where maximum readings for ovality, curvature, or crookedness occur at the same angular position, their values should be subtracted from the eccentricity reading to arrive at a true picture of the shaft condition. Confirmation of eccentricity readings may be obtained by reversing the shaft in the V-block, as shown in Fig. 5c, and clocking the larger diameter of the shaft.

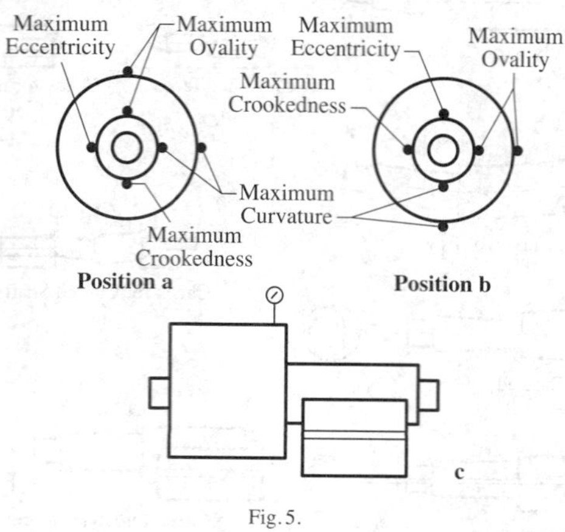


Fig. 5.

**Out-of-Roundness—Lobing.**—With the imposition of finer tolerances and the development of improved measurement methods, it has become apparent that no hole, cylinder, or

where $n$ = number of holes in circle; $D$ = diameter of hole circle; $\theta$ = angle between adjacent holes; $x_H$ = $x$ coordinate at position of hole number $H$; and, $y_H$ = $y$ coordinate at position of hole number $H$.

*Example 2(a):* Calculate the hole coordinates for the 5-hole circle shown in Fig. 2a when circle diameter = 1. Compare the results to the data in Table 2a.

| Hole | $\theta = 360/5 = 72°$ $\theta/2 = 36°$ $D = 1$ | |
|---|---|---|
| 1 | $x_1 = -1/2 \times \sin(36) = -0.29389$ | $y_1 = -1/2 \times \cos(36) = -0.40451$ |
| 2 | $x_2 = -1/2 \times \sin(108) = -0.47553$ | $y_2 = -1/2 \times \cos(108) = 0.15451$ |
| 3 | $x_3 = -1/2 \times \sin(180) = 0.00000$ | $y_3 = -1/2 \times \cos(180) = 0.50000$ |
| 4 | $x_4 = -1/2 \times \sin(252) = 0.47553$ | $y_4 = -1/2 \times \cos(252) = 0.15451$ |
| 5 | $x_5 = -1/2 \times \sin(324) = 0.29389$ | $y_5 = -1/2 \times \cos(324) = -0.40451$ |

In Fig. 2b, the origin of the coordinate system (point 0,0) is located at the top left of the figure at the intersection of the two lines labeled "Ref." The center of the hole circle is offset from the coordinate system origin by distance $X_O$ in the $+x$ direction and by distance $Y_O$ in the $+y$ direction. In practice, the origin of the coordinate system can be chosen at any convenient distance from the hole circle origin. In Fig. 2b, it can be determined by inspection that distance $X_O = Y_O = D/2$. The equations for calculating hole positions of type "B" circles of the Fig. 2b type are the same as in Equation (2a) but with the addition of $X_O$ and $Y_O$ terms, as follows:

$$\theta = \frac{360}{n} = \frac{2\pi}{n} \qquad x_H = -\frac{D}{2}\sin\left((H-1)\theta + \frac{\theta}{2}\right) + X_O \qquad y_H = -\frac{D}{2}\cos\left((H-1)\theta + \frac{\theta}{2}\right) + Y_O \quad (2b)$$

*Example 2(b):* Use the coordinates obtained in Example 2(a) to determine the hole coordinates of a 5-hole circle shown in Fig. 1b with circle diameter = 1. Compare the results to the data in Table 2b.

| Hole | $\theta = 360/5 = 72°$ $\theta/2 = 36°$ $D = 1$ $X_O = D/2 = 0.50000$ | $Y_O = D/2 = 0.50000$ |
|---|---|---|
| 1 | $x_1 = -1/2 \times \sin(36) + 0.50000 = 0.20611$ | $y_1 = -1/2 \times \cos(36) + 0.50000 = 0.09549$ |
| 2 | $x_2 = -1/2 \times \sin(108) + 0.50000 = 0.02447$ | $y_2 = -1/2 \times \cos(108) + 0.50000 = 0.65451$ |
| 3 | $x_3 = -1/2 \times \sin(180) + 0.50000 = 0.50000$ | $y_3 = -1/2 \times \cos(180) + 0.50000 = 1.00000$ |
| 4 | $x_4 = -1/2 \times \sin(252) + 0.50000 = 0.97553$ | $y_4 = -1/2 \times \cos(252) + 0.50000 = 0.65451$ |
| 5 | $x_5 = -1/2 \times \sin(324) + 0.50000 = 0.79389$ | $y_5 = -1/2 \times \cos(324) + 0.50000 = 0.09549$ |

**Adapting Hole Coordinate Equations for Different Geometry.—** Hole coordinate values in Table 1a through Table 2b are obtained using the equations given previously, along with the geometry of the corresponding figures. If the geometry does not match that given in one of the previous figures, hole coordinate values from the tables or equations will be incorrect. Fig. 3 illustrates such a case. Fig. 3 resembles a type "A" hole circle (Fig. 1b) with hole number 2 at the top, and it also resembles a type "B" hole circle (Fig. 2b) in which all holes have been rotated 90° clockwise. A closer look also reveals that the positive $y$ direction in Fig. 3 is opposite that used in Fig. 1b and Fig. 2b. Therefore, to determine the hole coordinates of Fig. 3 it is necessary to create new equations that match the given geometry or to modify the previous equations to match the Fig. 3 geometry.

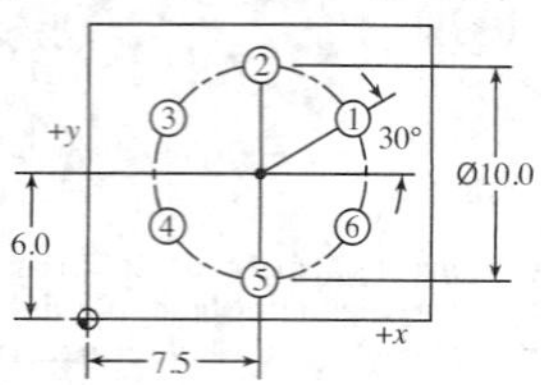


Fig. 3.

## Parallel Bars

*Parallel bars* are used for workpiece support during layout, machining, and inspection operations. Parallel bars are made from either steel, cast iron, or granite, come in a wide range of sizes and are either used alone or in matched pairs. In general, there are four types of parallel bars: Type I, Solid; Type II, Ribbed; Type III, Box; Type IV, Adjustable. Although the parallel bar is a relatively simple tool, it is still considered a precision instrument and must be handled with the same attention to care and handling as other precision ground and finished supporting gages. Most precision parallels are made from heat-treated steel and hardened to 55 to 60 on the C Rockwell scale. Generally, they have a finish of 8 μin Ra with a fine finish free from all grind marks, chatter, or cracks. Granite parallel bars are also very desirable in an environment where thermal expansion is a consideration or where steel or cast iron would not be acceptable. Granite parallel bars are made to a very high degree of accuracy and can be used in a precision inspection setup.

**Type I, Solid.—**Type I parallel bars are designed to be used independently as an individual bar to aid in inspection setups. They can be used as a reference surface or to establish a vertical plane perpendicular to a base plane. They can also be used as an extended support plane, when used in matched pairs, to establish an elevated plane that is parallel to the base plane. Type I parallel bars have four finished sides that shall not vary from a true plane by more than 0.0002 inch per foot (0.005mm/300mm). The adjacent sides will be square to each other within 0.0005 inch (0.012mm). Type I parallel bars will appear similar to Fig. 1 and will meet the tolerance specifications stated in Table 1.

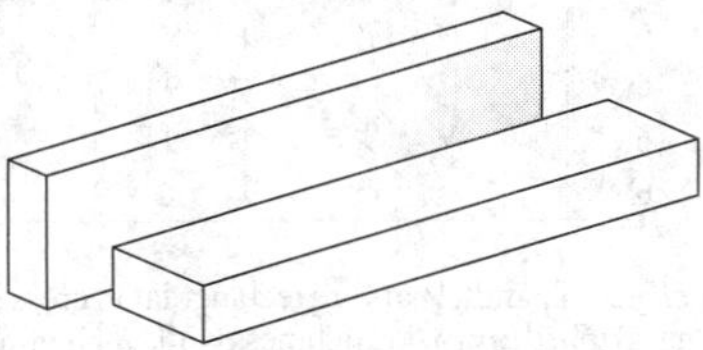

Fig. 1. Parallel Bars, Type 1, Solid

**Table 1. Steel and Cast-Iron Parallel Bar Tolerances, Inch Sizes**

| Size[a] | Tolerance Limits: Straightness (in.) | Parallelism (in.) | Width/Height (in.) | Width/Height Variance Matched Pairs (in.) |
|---|---|---|---|---|
| Inch Sizes | | | | |
| (in.) | (in.) | (in.) | (in.) | (in.) |
| $\frac{1}{8}$ to $\frac{3}{16}$ | 0.002 | 0.0001 | 0.0002 | 0.0002 |
| $\frac{1}{4}$ to $\frac{3}{4}$ | 0.0005 | 0.0001 | 0.0002 | 0.0002 |
| $\frac{3}{4}$ to $1\frac{1}{8}$ | 0.0002 | 0.0001 | 0.0002 | 0.0002 |
| $1\frac{1}{8}$ to $1\frac{1}{2}$ | 0.0002 | 0.00015 | 0.0003 | 0.0003 |
| $1\frac{1}{2}$ to 3 | 0.0002 | 0.0002 | 0.0004 | 0.0004 |
| Millimeter Sizes | | | | |
| (mm) | (mm) | (mm) | (mm) | (mm) |
| 3 to 5 | 0.0500 | 0.0025 | 0.0050 | 0.0050 |
| 6 to 20 | 0.0130 | 0.0025 | 0.0050 | 0.0050 |
| 20 to 25 | 0.0050 | 0.0025 | 0.0050 | 0.0050 |
| 25 to 35 | 0.0050 | 0.0040 | 0.0080 | 0.0080 |
| 35 to 75 | 0.0050 | 0.0050 | 0.0100 | 0.0100 |

[a] Size as applicable to either width or height dimension specified by the manufacturer.

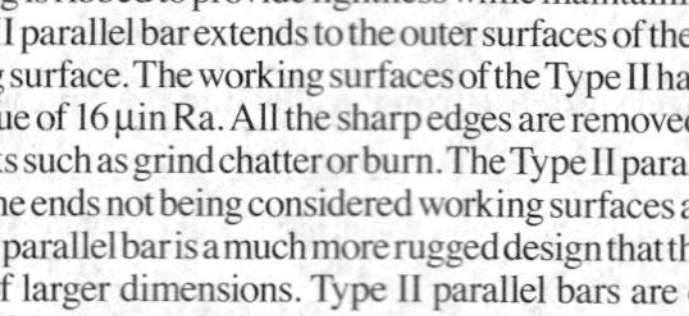

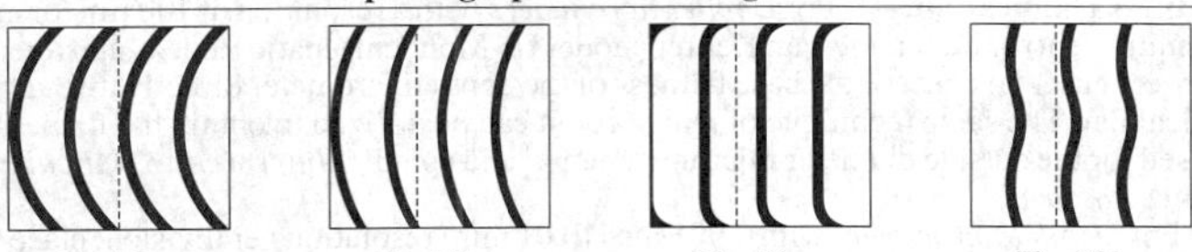

**Table 3b.** *(Continued)* **Indexable Insert Holder Application Guide**

| Tool | Tool Holder Style | Insert Shape | Rake N-Negative P-Positive | Application: Turn | Face | Turn and Face | Turn and Backface | Trace | Groove | Chamfer | Bore | Plane |
|---|---|---|---|---|---|---|---|---|---|---|---|---|
| 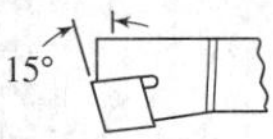  | K | C | N | ● | ● | | | | | | ● | |
| | | | | | | | | | | | | |
| 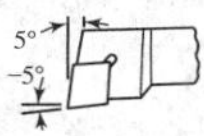  | L | C | N | | | ● | ● | | | | | |
| | | | | | | | | | | | | |
| 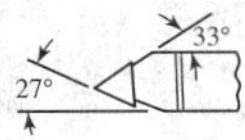  | N | T | N | ● | ● | | | ● | | | | |
| | | | P | ● | ● | | | ● | | | | |
| 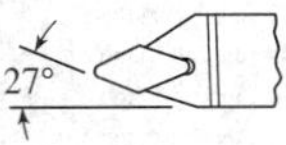  | N | D | N | ● | ● | | | ● | | | | |
| | | | | | | | | | | | | |
| 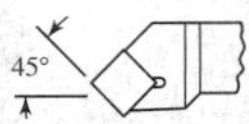  | S | S | N | ● | ● | ● | | ● | | ● | ● | ● |
| | | | P | ● | ● | ● | | ● | | ● | ● | ● |
| 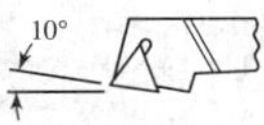  | W | T | N | ● | ● | | | | | | | |
| | | | | | | | | | | | | |

## Sintered Carbide Blanks and Cutting Tools

**Sintered Carbide Blanks.—**As shown in Table 4, American National Standard ANSI B212.1-2002 (Withdrawn) provides standard sizes and designations for eight styles of sintered carbide blanks. These blanks are the unground solid carbide from which either solid or tipped cutting tools are made. Tipped cutting tools are made by brazing a blank onto a shank to produce the cutting tool; these tools differ from carbide *insert* cutting tools which consist of a carbide insert held mechanically in a tool holder. A typical single-point carbide-tipped cutting tool is shown in Fig. 1 on page 837.

**Single-Point, Sintered Carbide-Tipped Tools.—**American National Standard ANSI B212.1-2002 (Withdrawn) covers eight different styles of single-point, carbide-tipped general purpose tools. These styles are designated by the letters A to G inclusive. Styles A, B, F, G, and E with offset point are either right- or left-hand cutting as indicated by the letters R or L. Dimensions of tips and shanks are given in Table 5 to Table 12. For dimensions and tolerances not shown, and for the identification system, dimensions, and tolerances of sintered carbide boring tools, see the Standard.

A number follows the letters of the tool style and hand designation, and, for square shank tools, represents the number of sixteenths of an inch of width, *W*, and height, *H*. With rectangular shanks, the first digit of the number indicates the number of eighths of an inch in the shank width, *W*, and the second digit the number of quarters of an inch in the shank height, *H*. One exception is the $1\frac{1}{2} \times 2$-inch size, which has been arbitrarily assigned the number 90.

**Table 6. American National Standard Style B Carbide-Tipped Tools with 15-Degree Side Cutting Edge Angle** *ANSI B212.1-2002 (Withdrawn)*

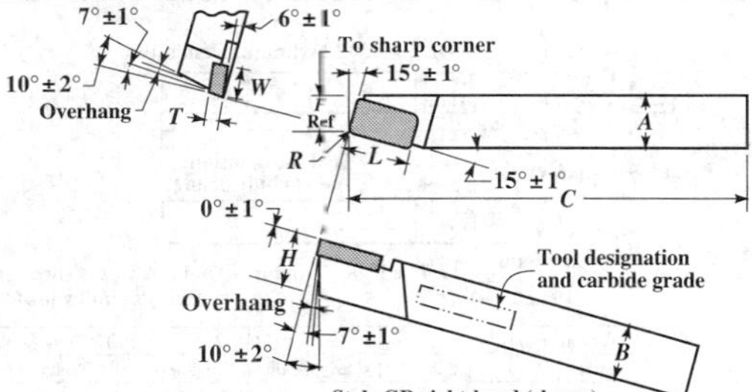


Style GR right-hand (shown)
Style GE left-hand (not shown)

| Designation | | Shank Dimensions | | | Tip Designation[a] | Tip Dimensions | | |
|---|---|---|---|---|---|---|---|---|
| Style BR | Style BL | Width *A* | Height *B* | Length *C* | | Thickness *T* | Width *W* | Length *L* |
| Square Shank | | | | | | | | |
| BR 4 | BL 4 | 1/4 | 1/4 | 2 | 2015 | 1/16 | 5/32 | 1/4 |
| BR 5 | BL 5 | 5/16 | 5/16 | 2 1/4 | 2040 | 3/32 | 3/16 | 5/16 |
| BR 6 | BL 6 | 3/8 | 3/8 | 2 1/2 | 2070 | 3/32 | 1/4 | 1/2 |
| BR 7 | BL 7 | 7/16 | 7/16 | 3 | 2070 | 3/32 | 1/4 | 1/2 |
| BR 8 | BL 8 | 1/2 | 1/2 | 3 1/2 | 2170 | 1/8 | 5/16 | 5/8 |
| BR 10 | BL 10 | 5/8 | 5/8 | 4 | 2230 | 5/32 | 3/8 | 3/4 |
| BR 12 | BL 12 | 3/4 | 3/4 | 4 1/2 | 2310 | 3/16 | 7/16 | 13/16 |
| BR 16 | BL 16 | 1 | 1 | 6 | {3390, 4390 | 1/4 | 9/16 | 1 |
| BR 20 | BL 20 | 1 1/4 | 1 1/4 | 7 | {3460, 4460 | 5/16 | 5/8 | 1 |
| BR 24 | BL 24 | 1 1/2 | 1 1/2 | 8 | {3510, 4510 | 3/8 | 5/8 | 1 |
| Rectangular Shank | | | | | | | | |
| BR 44 | BL 44 | 1/2 | 1 | 6 | 2260 | 3/16 | 5/16 | 5/8 |
| BR 54 | BL 54 | 5/8 | 1 | 6 | {3360, 4360 | 1/4 | 3/8 | 3/4 |
| BR 55 | BL 55 | 5/8 | 1 1/4 | 7 | {3360, 4360 | 1/4 | 3/8 | 3/4 |
| BR 64 | BL 64 | 3/4 | 1 | 6 | {3380, 4380 | 1/4 | 1/2 | 3/4 |
| BR 66 | BL 66 | 3/4 | 1 1/2 | 8 | {3430, 4430 | 5/16 | 7/16 | 15/16 |
| BR 85 | BL 85 | 1 | 1 1/4 | 7 | {3460, 4460 | 5/16 | 5/8 | 1 |
| BR 86 | BL 86 | 1 | 1 1/2 | 8 | {3510, 4510 | 3/8 | 5/8 | 1 |
| BR 88 | BL 88 | 1 | 2 | 10 | {3510, 4510 | 3/8 | 5/8 | 1 |
| BR 90 | BL 90 | 1 1/2 | 2 | 10 | {3540, 4540 | 1/2 | 3/4 | 1 1/4 |

[a] Where a pair of tip numbers is shown, the upper number applies to BR tools, the lower to BL tools. All dimensions are in inches.

**Brazing Carbide Tips to Steel Shanks.**—Sintered carbide tips or blanks are attached to steel shanks by brazing. Shanks usually are made of low-alloy steels having carbon contents ranging from 0.40 to 0.60 percent. Shank Preparation: The carbide tip usually is inserted into a milled recess or seat. When a recess is used, the bottom should be flat to provide a firm even support for the tip. The corner radius of the seat should be somewhat smaller than the radius on the tip to avoid contact and ensure support along each side of the recess. Cleaning: All surfaces to be brazed must be absolutely clean. Surfaces of the tip may be cleaned by grinding lightly or by sand-blasting. Brazing Materials and Equipment: The brazing metal may be copper, naval brass such as Tobin bronze, or silver solder. A flux such as borax is used to protect the clean surfaces and prevent oxidation. Heating may be done in a furnace or by oxy-acetylene torch or an oxy-hydrogen torch. Copper brazing usually is done in a furnace, although an oxy-hydrogen torch with excess hydrogen is sometimes used. Brazing Procedure: One method using a torch is to place a thin sheet material, such as copper foil, around and beneath the carbide tip, the top of which is covered with flux. The flame is applied to the underside of the tool shank, and, when the materials melt, the tip is pressed firmly into its seat with tongs or with the end of a rod. Brazing material in the form of wire or rod may be used to coat or tin the surfaces of the recess after the flux

**Table 12. American National Standard Style G, Offset, Side-Cutting, Carbide-Tipped Tools** *ANSI B212.1-2002 (Withdrawn)*

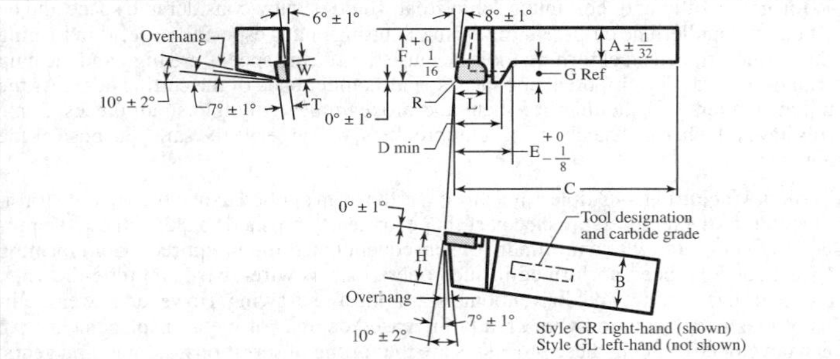


| Designation | | Shank Dimensions | | | | | Tip Designation | Tip Dimensions | | |
|---|---|---|---|---|---|---|---|---|---|---|
| Style GR | Style GL | Width *A* | Height *B* | Length *C* | Offset *G* | Length of Offset *E* | | Thickness *T* | Width *W* | Length *L* |
| | | | | | Square Shank | | | | | |
| GR 8 | GL 8 | 1/2 | 1/2 | 3 1/2 | 1/4 | 1 1/16 | P3170, P4170 | 1/8 | 5/16 | 5/8 |
| GR 10 | GL 10 | 5/8 | 5/8 | 4 | 3/8 | 1 3/8 | P3230, P4230 | 5/32 | 3/8 | 3/4 |
| GR 12 | GL 12 | 3/4 | 3/4 | 4 1/2 | 3/8 | 1 1/2 | P3310, P2310 | 3/16 | 7/16 | 13/16 |
| GR 16 | GL 16 | 1 | 1 | 6 | 1/2 | 1 11/16 | P3390, P4390 | 1/4 | 9/16 | 1 |
| GR 20 | GL 20 | 1 1/4 | 1 1/4 | 7 | 3/4 | 1 13/16 | P3460, P4460 | 5/16 | 5/8 | 1 |
| GR 24 | GL 24 | 1 1/2 | 1 1/2 | 8 | 3/4 | 1 13/16 | P3510, P4510 | 3/8 | 5/8 | 1 |
| | | | | | Rectangular Shank | | | | | |
| GR 44 | GL 44 | 1/2 | 1 | 6 | 1/4 | 1 1/16 | P3260, P4260 | 3/16 | 5/16 | 5/8 |
| GR 55 | GL 55 | 5/8 | 1 1/4 | 7 | 3/8 | 1 3/8 | P3360, P4360 | 1/4 | 3/8 | 3/4 |
| GR 64 | GL 64 | 3/4 | 1 | 6 | 1/2 | 1 7/16 | P3380, P4380 | 1/4 | 1/2 | 3/4 |
| GR 66 | GL 66 | 3/4 | 1 1/2 | 8 | 1/2 | 1 5/8 | P3430, P4430 | 5/16 | 7/16 | 15/16 |
| GR 85 | GL 85 | 1 | 1 1/4 | 7 | 1/2 | 1 11/16 | P3460, P4460 | 5/16 | 5/8 | 1 |
| GR 86 | GL 86 | 1 | 1 1/2 | 8 | 1/2 | 1 11/16 | P3510, P4510 | 3/8 | 5/8 | 1 |
| GR 90 | GL 90 | 1 1/2 | 2 | 10 | 3/4 | 2 1/16 | P3540, P4540 | 1/2 | 3/4 | 1 1/4 |

All dimensions are in inches. Where a pair of tip numbers is shown, the upper number applies to GR tools, the lower to GL tools.

a hardness of 87.5 RA on the Rockwell A scale might be used where a strong grade is required, as for roughing cuts, whereas for light high-speed finishing or other cuts, a hardness of about 92 RA might be preferable. When tungsten carbide is applied to steel, craters or chip cavities are formed back of the cutting edge; hence other carbides have been developed that offer greater resistance to abrasion.

*Tungsten-titanium carbide* (often called "titanium carbide") is adapted to cutting either heat-treated or non-heat-treated steels, cast steel, or any tough material that might form chip cavities. It is also applicable to bronzes, monel metal, aluminum alloys, etc.

*Tungsten-tantalum carbide* or "tantalum carbide" cutting tools are also applicable to steels, bronzes or other tough materials. A hardness of 86.8 RA is recommended by one manufacturer for roughing steel, whereas a grade for finishing might have a hardness ranging from 88.8 to 91.5 RA.

**Table 1. Insert Radius Compensation** *ANSI B212.3-2002 (Withdrawn)*

| | | | | | | |
|---|---|---|---|---|---|---|
| Square Profile | | | | | | |
| B Style[a] Also applies to R Style | 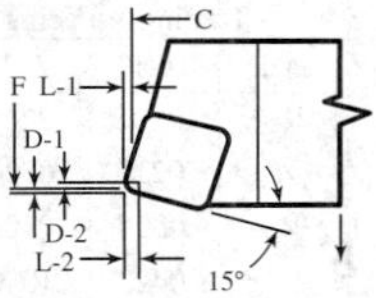 | Turning 15° Lead Angle | | | | |
| | | Rad. | *L*-1 | *L*-2 | *D*-1 | *D*-2 |
| | | 1⁄64 | .0035 | .0191 | .0009 | .0110 |
| | | 1⁄32 | .0070 | .0383 | .0019 | .0221 |
| | | 3⁄64 | .0105 | .0574 | .0028 | .0331 |
| | | 1⁄16 | .0140 | .0765 | .0038 | .0442 |
| D Style[a] Also applies to S Style | 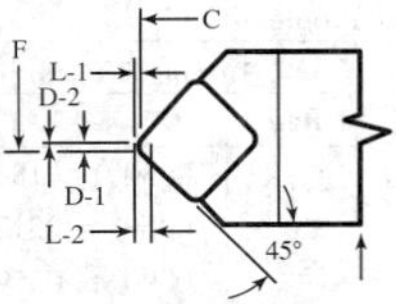 | Turning 45° Lead Angle | | | | |
| | | Rad. | *L*-1 | *L*-2 | *D*-1 | *D*-2 |
| | | 1⁄64 | .0065 | .0221 | .0065 | 0 |
| | | 1⁄32 | .0129 | .0442 | .0129 | 0 |
| | | 3⁄64 | .0194 | .0663 | .0194 | 0 |
| | | 1⁄16 | .0259 | .0884 | .0259 | 0 |
| K Style[a] | 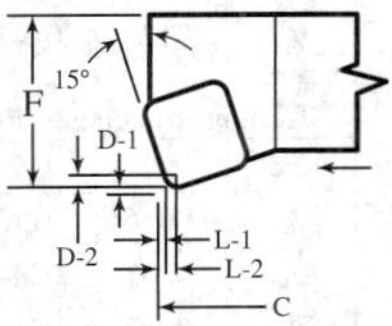 | Facing 15° Lead Angle | | | | |
| | | Rad. | *L*-1 | *L*-2 | *D*-1 | *D*-2 |
| | | 1⁄64 | .0009 | .0110 | .0035 | .0191 |
| | | 1⁄32 | .0019 | .0221 | .0070 | .0383 |
| | | 3⁄64 | .0028 | .0331 | .0105 | .0574 |
| | | 1⁄16 | .0038 | .0442 | .0140 | .0765 |
| Triangle Profile | | | | | | |
| G Style[a] | 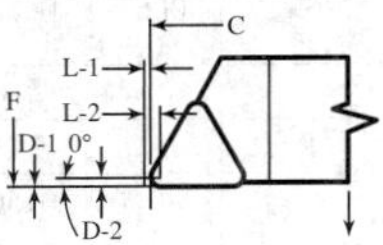 | Turning 0° Lead Angle | | | | |
| | | Rad. | *L*-1 | *L*-2 | *D*-1 | *D*-2 |
| | | 1⁄64 | .0114 | .0271 | 0 | .0156 |
| | | 1⁄32 | .0229 | .0541 | 0 | .0312 |
| | | 3⁄64 | .0343 | .0812 | 0 | .0469 |
| | | 1⁄16 | .0458 | .1082 | 0 | .0625 |
| B Style[a] Also applies to R Style | 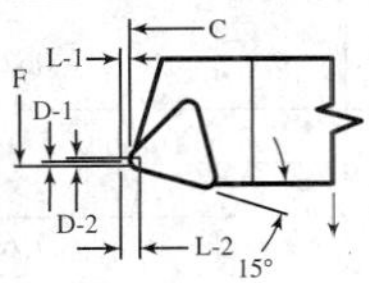 | Turning and Facing 15° Lead Angle | | | | |
| | | Rad. | *L*-1 | *L*-2 | *D*-1 | *D*-2 |
| | | 1⁄64 | .0146 | .0302 | .0039 | .0081 |
| | | 1⁄32 | .0291 | .0604 | .0078 | .0162 |
| | | 3⁄64 | .0437 | .0906 | .0117 | .0243 |
| | | 1⁄16 | .0582 | .1207 | .0156 | .0324 |
| F Style[a] | 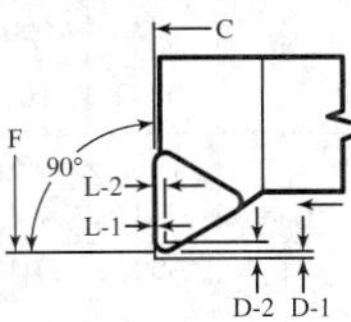 | Facing 90° Lead Angle | | | | |
| | | Rad. | *L*-1 | *L*-2 | *D*-1 | *D*-2 |
| | | 1⁄64 | 0 | .0156 | .0114 | .0271 |
| | | 1⁄32 | 0 | .0312 | .0229 | .0541 |
| | | 3⁄64 | 0 | .0469 | .0343 | .0812 |
| | | 1⁄16 | 0 | .0625 | .0458 | .1082 |

**American National Standard Side Milling Cutters**
*ANSI/ASME B94.19-1997 (R2019)*

| Cutter Diameter | | | Range of Face Widths Nom.[a] | Hole Diameter | | |
|---|---|---|---|---|---|---|
| Nom. | Max. | Min. | | Nom. | Max. | Min. |
| | | | Side Cutters [b] | | | |
| 2 | 2.015 | 1.985 | 3⁄16, 1⁄4, 3⁄8 | 5⁄8 | 0.62575 | 0.6250 |
| 2 1⁄2 | 2.515 | 2.485 | 1⁄4, 3⁄8, 1⁄2 | 7⁄8 | 0.87575 | 0.8750 |
| 3 | 3.015 | 2.985 | 1⁄4, 5⁄16, 3⁄8, 7⁄16, 1⁄2 | 1 | 1.00075 | 1.0000 |
| 4 | 4.015 | 3.985 | 1⁄4, 3⁄8, 1⁄2, 5⁄8, 3⁄4, 7⁄8 | 1 | 1.00075 | 1.0000 |
| 4 | 4.015 | 3.985 | 1⁄2, 5⁄8, 3⁄4 | 1 1⁄4 | 1.2510 | 1.2500 |
| 5 | 5.015 | 4.985 | 1⁄2, 5⁄8, 3⁄4 | 1 | 1.00075 | 1.0000 |
| 5 | 5.015 | 4.985 | 1⁄2, 5⁄8, 3⁄4, 1 | 1 1⁄4 | 1.2510 | 1.2500 |
| 6 | 6.015 | 5.985 | 1⁄2 | 1 | 1.00075 | 1.0000 |
| 6 | 6.015 | 5.985 | 1⁄2, 5⁄8, 3⁄4, 1 | 1 1⁄4 | 1.2510 | 1.2500 |
| 7 | 7.015 | 6.985 | 3⁄4 | 1 1⁄4 | 1.2510 | 1.2500 |
| 7 | 7.015 | 6.985 | 3⁄4 | 1 1⁄2 | 1.5010 | 1.5000 |
| 8 | 8.015 | 7.985 | 3⁄4, 1 | 1 1⁄4 | 1.2510 | 1.2500 |
| 8 | 8.015 | 7.985 | 3⁄4, 1 | 1 1⁄2 | 1.5010 | 1.5000 |
| | | | Staggered-Tooth Side Cutters [c] | | | |
| 2 1⁄2 | 2.515 | 2.485 | 1⁄4, 5⁄16, 3⁄8, 1⁄2 | 7⁄8 | 0.87575 | 0.8750 |
| 3 | 3.015 | 2.985 | 3⁄16, 1⁄4, 5⁄16, 3⁄8 | 1 | 1.00075 | 1.0000 |
| 3 | 3.015 | 2.985 | 1⁄2, 5⁄8, 3⁄4 | 1 1⁄4 | 1.2510 | 1.2500 |
| 4 | 4.015 | 3.985 | 1⁄4, 5⁄16, 3⁄8, 7⁄16, 1⁄2, 5⁄8, 3⁄4 and 7⁄8 | 1 1⁄4 | 1.2510 | 1.2500 |
| 5 | 5.015 | 4.985 | 1⁄2, 5⁄8, 3⁄4 | 1 1⁄4 | 1.2510 | 1.2500 |
| 6 | 6.015 | 5.985 | 3⁄8, 1⁄2, 5⁄8, 3⁄4, 7⁄8, 1 | 1 1⁄4 | 1.2510 | 1.2500 |
| 8 | 8.015 | 7.985 | 3⁄8, 1⁄2, 5⁄8, 3⁄4, 1 | 1 1⁄2 | 1.5010 | 1.5000 |
| | | | Half Side Cutters [d] | | | |
| 4 | 4.015 | 3.985 | 3⁄4 | 1 1⁄4 | 1.2510 | 1.2500 |
| 5 | 5.015 | 4.985 | 3⁄4 | 1 1⁄4 | 1.2510 | 1.2500 |
| 6 | 6.015 | 5.985 | 3⁄4 | 1 1⁄4 | 1.2510 | 1.2500 |

[a] *Tolerances on Face Widths:* For side cutters, +0.002, −0.001 inch; for staggered-tooth side cutters up to 3⁄4-inch face width, inclusive, +0.000 −0.0005 inch, and over 3⁄4 to 1 inch, inclusive, +0.000 −0.0010 inch; and for half side cutters, +0.015, −0.000 inch.

[b] Side milling cutters have straight peripheral teeth and side teeth on both sides.

[c] Staggered-tooth side milling cutters have peripheral teeth of alternate right- and left-hand helix and alternate side teeth.

[d] Half side milling cutters have side teeth on one side only. The peripheral teeth are helical of the same hand as the cut. Made either with right-hand or left-hand cut.

All dimensions are in inches. All cutters are high-speed steel. Side milling cutters are of cylindrical shape, having teeth on the periphery and on one or both sides.

*Hand of Flute Helix:* Milling cutters may have *straight flutes,* which means that their cutting edges are in planes parallel to the cutter axis. Milling cutters with flute helix in one *direction only are* described as having a right-hand helix if the flutes twist away from the observer in a clockwise direction when viewed from either end of the cutter or as having a left-hand helix if the flutes twist away from the observer in a counterclockwise direction when viewed from either end of the cutter. *Staggered tooth cutters* are milling cutters with every other flute of opposite (right- and left-hand) helix.

An illustration describing the various milling cutter elements of both a profile cutter and a form-relieved cutter is given on page 864.

## ANSI Two-Flute, High Helix, Regular-, Long-, and Extra-Long-Length, Single-End End Mills with Weldon Shanks *ANSI/ASME B94.19-1997 (R2019)*

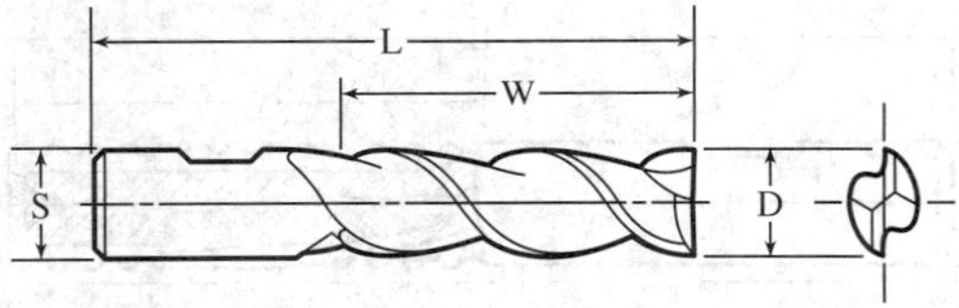


| Cutter Dia., *D* | Regular Mill | | | Long Mill | | | Extra-Long Mill | | |
|---|---|---|---|---|---|---|---|---|---|
| | *S* | *W* | *L* | *S* | *W* | *L* | *S* | *W* | *L* |
| 1/4 | 3/8 | 5/8 | 2 7/16 | 3/8 | 1 1/4 | 3 1/16 | 3/8 | 1 3/4 | 3 9/16 |
| 5/16 | 3/8 | 3/4 | 2 1/2 | 3/8 | 1 3/8 | 3 1/8 | 3/8 | 2 | 3 3/4 |
| 3/8 | 3/8 | 3/4 | 2 1/2 | 3/8 | 1 1/2 | 3 1/4 | 3/8 | 2 1/2 | 4 1/4 |
| 7/16 | 3/8 | 1 | 2 11/16 | 1/2 | 1 3/4 | 3 3/4 | … | … | … |
| 1/2 | 1/2 | 1 1/4 | 3 1/4 | 1/2 | 2 | 4 | 1/2 | 3 | 5 |
| 5/8 | 5/8 | 1 5/8 | 3 3/4 | 5/8 | 2 1/2 | 4 5/8 | 5/8 | 4 | 6 1/8 |
| 3/4 | 3/4 | 1 5/8 | 3 7/8 | 3/4 | 3 | 5 1/4 | 3/4 | 4 | 6 1/4 |
| 7/8 | 7/8 | 1 7/8 | 4 1/8 | … | … | … | … | … | … |
| 1 | 1 | 2 | 4 1/2 | 1 | 4 | 6 1/2 | 1 | 6 | 8 1/2 |
| 1 1/4 | 1 1/4 | 2 | 4 1/2 | 1 1/4 | 4 | 6 1/2 | 1 1/4 | 6 | 8 1/2 |
| 1 1/2 | 1 1/4 | 2 | 4 1/2 | 1 1/4 | 4 | 6 1/2 | 1 1/4 | 8 | 10 1/2 |
| 2 | 1 1/4 | 2 | 4 1/2 | 1 1/4 | 4 | 6 1/2 | … | … | … |

All dimensions are in inches. All cutters are high-speed steel. Right-hand cutters with right-hand helix are standard. Helix angle is greater than 39 degrees.

*Tolerances:* On *D*, +0.003 inch; on *S*, −0.0001 to −0.0005 inch; on *W*, ±1/32 inch; and on *L*, ±1/16 inch.

## Combination Shanks for End Mills *ANSI/ASME B94.19-1997 (R2019)*

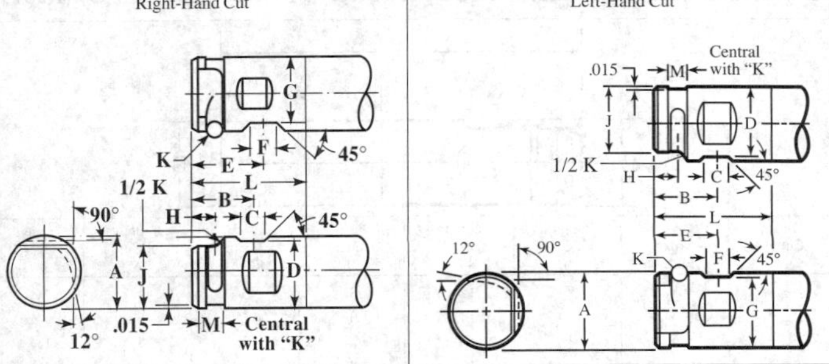


| Dia. *A* | *L*[a] | *B* | *C* | *D* | *E* | *F* | *G* | *H* | *J* | *K* | *M* |
|---|---|---|---|---|---|---|---|---|---|---|---|
| 1 1/2 | 2 11/16 | 1 3/16 | .515 | 1.406 | 1 1/2 | .515 | 1.371 | 9/16 | 1.302 | .377 | 7/16 |
| 2 | 3 1/4 | 1 23/32 | .700 | 1.900 | 1 3/4 | .700 | 1.809 | 5/8 | 1.772 | .440 | 1/2 |
| 2 1/2 | 3 1/2 | 1 15/16 | .700 | 2.400 | 2 | .700 | 2.312 | 3/4 | 2.245 | .503 | 9/16 |

[a] Length of shank.

All dimensions are in inches.

Modified for use as Weldon or Pin Drive shank.

## ANSI Stub-, Regular-, and Long-Length, Four-Flute, Medium Helix, Plain-End, Double-End Miniature End Mills with 3/16-Inch Diameter Straight Shanks

*ANSI/ASME B94.19-1997 (R2019)*

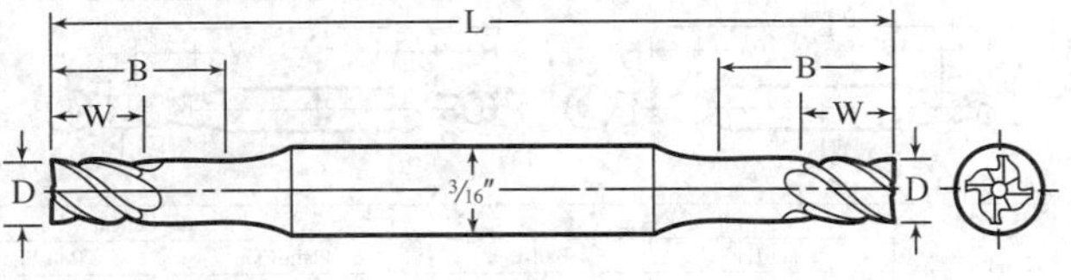


| Dia. D | Stub Length | | Regular Length | |
|---|---|---|---|---|
| | W | L | W | L |
| 1/16 | 3/32 | 2 | 3/16 | 2 1/4 |
| 3/32 | 9/64 | 2 | 9/32 | 2 1/4 |
| 1/8 | 3/16 | 2 | 3/8 | 2 1/4 |
| 5/32 | 15/64 | 2 | 7/16 | 2 1/4 |
| 3/16 | 9/32 | 2 | 1/2 | 2 1/4 |

| Dia. D | Long Length | | |
|---|---|---|---|
| | B | W | L |
| 1/16 | 3/8 | 7/32 | 2 1/2 |
| 3/32 | 1/2 | 9/32 | 2 5/8 |
| 1/8 | 3/4 | 3/4 | 3 1/8 |
| 5/32 | 7/8 | 7/8 | 3 1/4 |
| 3/16 | 1 | 1 | 3 3/8 |

All dimensions are in inches. All cutters are high-speed steel. Right-hand cutters with right-hand helix are standard. Helix angle is greater than 19 degrees but not more than 39 degrees.

*Tolerances:* On *D*, + 0.003 inch (if the shank is the same diameter as the cutting portion, however, then the tolerance on the cutting diameter is − 0.0025 inch.); on *W*, + 1/32, − 1/64 inch; and on *L*, ±1/16 inch.

## American National Standard 60-Degree Single-Angle Milling Cutters with Weldon Shanks *ANSI/ASME B94.19-1997 (R2019)*

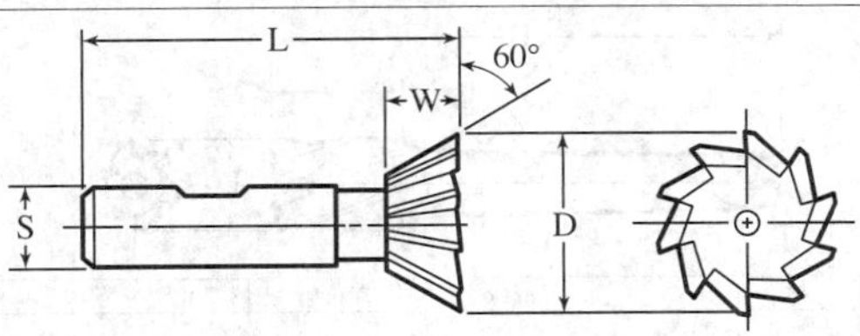


| Dia., D | S | W | L | Dia., D | S | W | L |
|---|---|---|---|---|---|---|---|
| 3/4 | 3/8 | 5/16 | 2 1/8 | 1 7/8 | 7/8 | 13/16 | 3 1/4 |
| 1 3/8 | 5/8 | 9/16 | 2 7/8 | 2 1/4 | 1 | 1 1/16 | 3 3/4 |

All dimensions are in inches. All cutters are high-speed steel. Right-hand cutters are standard.

*Tolerances:* On *D*, ± 0.015 inch; on *S*, − 0.0001 to − 0.0005 inch; on *W*, ± 0.015 inch; and on *L*, ±1/16 inch.

## American National Standard Stub-, Regular-, and Long-Length, Two-Flute, Medium Helix, Plain- and Ball-End, Double-End Miniature End Mills with $^3/_{16}$-Inch Diameter Straight Shanks *ANSI/ASME B94.19-1997 (R2019)*

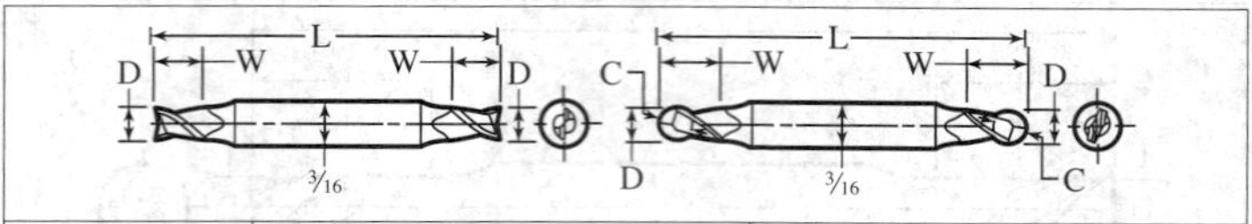


| Dia., *C* and *D* | Stub Length | | | | Regular Length | | | |
|---|---|---|---|---|---|---|---|---|
| | Plain-End | | Ball-End | | Plain-End | | Ball-End | |
| | *W* | *L* | *W* | *L* | *W* | *L* | *W* | *L* |
| 1⁄32 | 3⁄64 | 2 | … | … | 3⁄32 | 2¼ | … | … |
| 3⁄64 | 1⁄16 | 2 | … | … | 9⁄64 | 2¼ | … | … |
| 1⁄16 | 3⁄32 | 2 | 3⁄32 | 2 | 3⁄16 | 2¼ | 3⁄16 | 2¼ |
| 5⁄64 | 1⁄8 | 2 | … | … | 15⁄64 | 2¼ | … | … |
| 3⁄32 | 9⁄64 | 2 | 9⁄64 | 2 | 9⁄32 | 2¼ | 9⁄32 | 2¼ |
| 7⁄64 | 5⁄32 | 2 | … | … | 21⁄64 | 2¼ | … | … |
| 1⁄8 | 3⁄16 | 2 | 3⁄16 | 2 | 3⁄8 | 2¼ | 3⁄8 | 2¼ |
| 9⁄64 | 7⁄32 | 2 | … | … | 13⁄32 | 2¼ | … | … |
| 5⁄32 | 15⁄64 | 2 | 15⁄64 | 2 | 7⁄16 | 2¼ | 7⁄16 | 2¼ |
| 11⁄64 | 1⁄4 | 2 | … | … | 1⁄2 | 2¼ | … | … |
| 3⁄16 | 9⁄32 | 2 | 9⁄32 | 2 | 1⁄2 | 2¼ | 1⁄2 | 2¼ |

| Dia., *D* | Long-Length, Plain-End | | | Dia., *D* | Long-Length, Plain-End | | |
|---|---|---|---|---|---|---|---|
| | *B*[a] | *W* | *L* | | *B*[a] | *W* | *L* |
| 1⁄16 | 3⁄8 | 7⁄32 | 2½ | 5⁄32 | 7⁄8 | 7⁄8 | 3¼ |
| 3⁄32 | 1⁄2 | 9⁄32 | 2⅝ | 3⁄16 | 1 | 1 | 3⅜ |
| 1⁄8 | 3⁄4 | 3⁄4 | 3⅛ | | | | |

[a] *B* is the length below the shank.

All dimensions are in inches. All cutters are high-speed steel. Right-hand cutters with right-hand helix are standard. Helix angle is greater than 19 degrees but not more than 39 degrees.

*Tolerances:* On *C* and *D*, − 0.0015 inch for stub and regular length; + 0.003 inch for long length (if the shank is the same diameter as the cutting portion, however, then the tolerance on the cutting diameter is − 0.0025 inch.); on *W*, + $^1/_{32}$, − $^1/_{64}$ inch; and on *L*, ± $^1/_{16}$ inch.

## American National Standard Multiple-Flute, Helical Series End Mills with Brown & Sharpe Taper Shanks

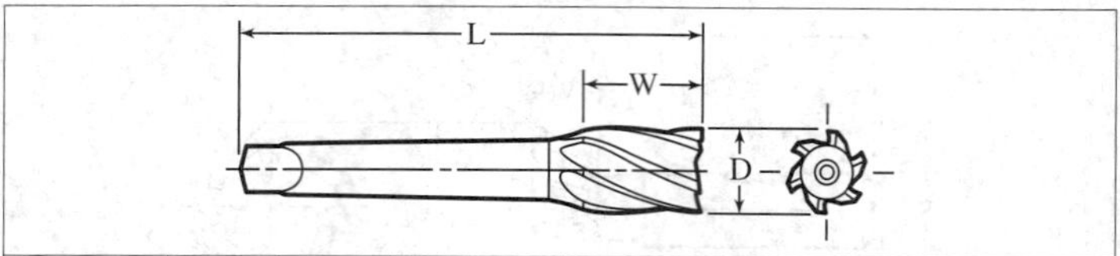


| Dia., *D* | *W* | *L* | Taper No. | Dia., *D* | *W* | *L* | Taper No. |
|---|---|---|---|---|---|---|---|
| ½ | 1 5⁄16 | 4 15⁄16 | 7 | 1¼ | 2 | 7¼ | 9 |
| ¾ | 1¼ | 5¼ | 7 | 1½ | 2¼ | 7½ | 9 |
| 1 | 1⅝ | 5⅝ | 7 | 2 | 2¾ | 8 | 9 |

All dimensions are in inches. All cutters are high-speed steel. Right-hand cutters with right-hand helix are standard. Helix angle is not less than 10 degrees.

No. 5 taper is standard without tang; Nos. 7 and 9 are standard with tang only.

*Tolerances:* On *D*, +0.005 inch; on *W*, ±$^1/_{32}$ inch; and on *L* ±$^1/_{16}$ inch.

For dimensions of B & S taper shanks, see information given on page 1011.

**American National Standard Plain- and Ball-End, Heavy-Duty, Medium Helix, Single-End End Mills with 2-Inch Diameter Shanks** *ANSI/ASME B94.19-1997 (R2019)*

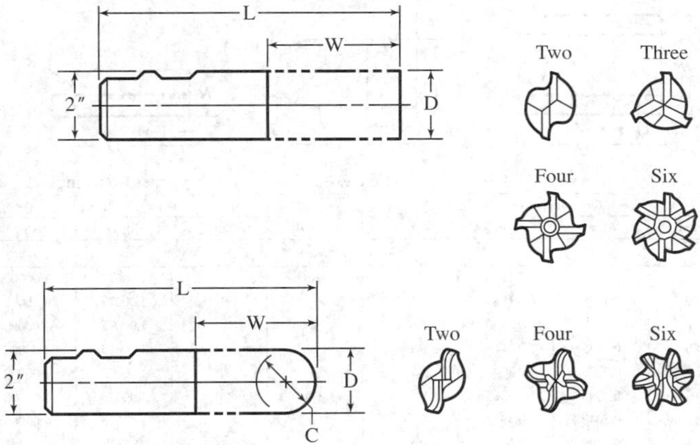


| Dia., *C* and *D* | Plain End | | | Ball End | | |
|---|---|---|---|---|---|---|
| | *W* | *L* | No. of Flutes | *W* | *L* | No. of Flutes |
| 2 | 2 | 5¾ | 2, 4, 6 | … | … | … |
| 2 | 3 | 6¾ | 2, 3 | … | … | … |
| 2 | 4 | 7¾ | 2, 3, 4, 6 | 4 | 7¾ | 6 |
| 2 | … | … | … | 5 | 8¾ | 2, 4 |
| 2 | 6 | 9¾ | 2, 3, 4, 6 | 6 | 9¾ | 6 |
| 2 | 8 | 11¾ | 6 | 8 | 11¾ | 6 |
| 2½ | 4 | 7¾ | 2, 3, 4, 6 | … | … | … |
| 2½ | … | … | … | 5 | 8¾ | 4 |
| 2½ | 6 | 9¾ | 2, 4, 6 | … | … | … |
| 2½ | 8 | 11¾ | 6 | … | … | … |

All dimensions are in inches. All cutters are high-speed steel. Right-hand cutters with right-hand helix are standard. Helix angle is greater than 19 degrees but not more than 39 degrees.

*Tolerances:* On C and D, +0.005 inch for 2, 3, 4 and 6 flutes: on $W$, $\pm\frac{1}{16}$ inch; and on $L$, $\pm\frac{1}{16}$ inch.

**Dimensions of American National Standard Weldon Shanks**
*ANSI/ASME B94.19-1997 (R2019)*

| Shank | | Flat | | Shank | | Flat | |
|---|---|---|---|---|---|---|---|
| Dia. | Length | *X*[a] | Length[b] | Dia. | Length | *X*[a] | Length[b] |
| ⅜ | $1\frac{9}{16}$ | 0.325 | 0.280 | 1 | $2\frac{9}{32}$ | 0.925 | 0.515 |
| ½ | $1\frac{25}{32}$ | 0.440 | 0.330 | 1¼ | $2\frac{9}{32}$ | 1.156 | 0.515 |
| ⅝ | $1\frac{29}{32}$ | 0.560 | 0.400 | 1½ | $2\frac{11}{16}$ | 1.406 | 0.515 |
| ¾ | $2\frac{1}{32}$ | 0.675 | 0.455 | 2 | 3¼ | 1.900 | 0.700 |
| ⅞ | $2\frac{1}{32}$ | 0.810 | 0.455 | 2½ | 3½ | 2.400 | 0.700 |

[a] *X* is distance from bottom of flat to opposite side of shank.

[b] Minimum.

All dimensions are in inches.

Centerline of flat is at half-length of shank except for $1\frac{1}{2}$-, 2- and $2\frac{1}{2}$-inch shanks where it is $1\frac{3}{16}$, $1\frac{27}{32}$ and $1\frac{15}{16}$ from shank end, respectively.

Tolerance on shank diameter, −0.0001 to −0.0005 inch.

**American National Standard Roller Chain Sprocket Milling Cutters** *ANSI/ASME B94.19-1997 (R2019) (Continued)*

| Chain Pitch | Dia. of Roll | No. of Teeth in Sprocket | Dia. of Cutter, *D* | Width of Cutter, *W* | Dia. of Hole, *H* |
|---|---|---|---|---|---|
| 3⁄4 | 0.469 | 6 | 3 1⁄4 | 29⁄32 | 1 |
| 3⁄4 | 0.469 | 7-8 | 3 1⁄4 | 29⁄32 | 1 |
| 3⁄4 | 0.469 | 9-11 | 3 3⁄8 | 29⁄32 | 1 |
| 3⁄4 | 0.469 | 12-17 | 3 3⁄8 | 7⁄8 | 1 |
| 3⁄4 | 0.469 | 18-34 | 3 3⁄8 | 27⁄32 | 1 |
| 3⁄4 | 0.469 | 35 and over | 3 3⁄8 | 13⁄16 | 1 |
| 1 | 0.625 | 6 | 3 7⁄8 | 1 1⁄2 | 1 1⁄4 |
| 1 | 0.625 | 7-8 | 4 | 1 1⁄2 | 1 1⁄4 |
| 1 | 0.625 | 9-11 | 4 1⁄8 | 1 15⁄32 | 1 1⁄4 |
| 1 | 0.625 | 18-34 | 4 1⁄4 | 1 13⁄32 | 1 1⁄4 |
| 1 | 0.625 | 35 and over | 4 1⁄4 | 1 11⁄32 | 1 1⁄4 |
| 1 1⁄4 | 0.750 | 6 | 4 1⁄4 | 1 13⁄16 | 1 1⁄4 |
| 1 1⁄4 | 0.750 | 7-8 | 4 3⁄8 | 1 13⁄16 | 1 1⁄4 |
| 1 1⁄4 | 0.750 | 9-11 | 4 1⁄2 | 1 25⁄32 | 1 1⁄4 |
| 1 1⁄4 | 0.750 | 18-34 | 4 5⁄8 | 1 11⁄16 | 1 1⁄4 |
| 1 1⁄4 | 0.750 | 35 and over | 4 5⁄8 | 1 5⁄8 | 1 1⁄4 |
| 1 1⁄2 | 0.875 | 6 | 4 3⁄8 | 1 13⁄16 | 1 1⁄4 |
| 1 1⁄2 | 0.875 | 7-8 | 4 1⁄2 | 1 13⁄16 | 1 1⁄4 |
| 1 1⁄2 | 0.875 | 9-11 | 4 5⁄8 | 1 25⁄32 | 1 1⁄4 |
| 1 1⁄2 | 0.875 | 12-17 | 4 5⁄8 | 1 3⁄4 | 1 1⁄4 |
| 1 1⁄2 | 0.875 | 18-34 | 4 3⁄4 | 1 11⁄16 | 1 1⁄4 |
| 1 1⁄2 | 0.875 | 35 and over | 4 3⁄4 | 1 5⁄8 | 1 1⁄4 |
| 1 3⁄4 | 1.000 | 6 | 5 | 2 3⁄32 | 1 1⁄2 |
| 1 3⁄4 | 1.000 | 7-8 | 5 1⁄8 | 2 3⁄32 | 1 1⁄2 |
| 1 3⁄4 | 1.000 | 9-11 | 5 1⁄4 | 2 1⁄16 | 1 1⁄2 |
| 1 3⁄4 | 1.000 | 12-17 | 5 3⁄8 | 2 1⁄32 | 1 1⁄2 |
| 1 3⁄4 | 1.000 | 18-34 | 5 1⁄2 | 1 31⁄32 | 1 1⁄2 |
| 1 3⁄4 | 1.000 | 35 and over | 5 1⁄2 | 1 7⁄8 | 1 1⁄2 |
| 2 | 1.125 | 6 | 5 3⁄8 | 2 13⁄32 | 1 1⁄2 |
| 2 | 1.125 | 7-8 | 5 1⁄2 | 2 13⁄32 | 1 1⁄2 |
| 2 | 1.125 | 9-11 | 5 5⁄8 | 2 3⁄8 | 1 1⁄2 |
| 2 | 1.125 | 12-17 | 5 3⁄4 | 2 5⁄16 | 1 1⁄2 |
| 2 | 1.125 | 18-34 | 5 7⁄8 | 2 1⁄4 | 1 1⁄2 |
| 2 | 1.125 | 35 and over | 5 7⁄8 | 2 5⁄32 | 1 1⁄2 |
| 2 1⁄4 | 1.406 | 6 | 5 7⁄8 | 2 11⁄16 | 1 1⁄2 |
| 2 1⁄4 | 1.406 | 7-8 | 6 | 2 11⁄16 | 1 1⁄2 |
| 2 1⁄4 | 1.406 | 9-11 | 6 1⁄4 | 2 21⁄32 | 1 1⁄2 |
| 2 1⁄4 | 1.406 | 12-17 | 6 3⁄8 | 2 19⁄32 | 1 1⁄2 |
| 2 1⁄4 | 1.406 | 18-34 | 6 1⁄2 | 2 15⁄32 | 1 1⁄2 |
| 2 1⁄4 | 1.406 | 35 and over | 6 1⁄2 | 2 13⁄32 | 1 1⁄2 |
| 2 1⁄2 | 1.563 | 6 | 6 3⁄8 | 3 | 1 3⁄4 |
| 2 1⁄2 | 1.563 | 7-8 | 6 5⁄8 | 3 | 1 3⁄4 |
| 2 1⁄2 | 1.563 | 9-11 | 6 3⁄4 | 2 15⁄16 | 1 3⁄4 |
| 2 1⁄2 | 1.563 | 12-17 | 6 7⁄8 | 2 29⁄32 | 1 3⁄4 |
| 2 1⁄2 | 1.563 | 18-34 | 7 | 2 3⁄4 | 1 3⁄4 |
| 2 1⁄2 | 1.563 | 35 and over | 7 1⁄8 | 2 11⁄16 | 1 3⁄4 |
| 3 | 1.875 | 6 | 7 1⁄2 | 3 19⁄32 | 2 |
| 3 | 1.875 | 7-8 | 7 3⁄4 | 3 19⁄32 | 2 |
| 3 | 1.875 | 9-11 | 7 7⁄8 | 3 17⁄32 | 2 |
| 3 | 1.875 | 12-17 | 8 | 3 15⁄32 | 2 |
| 3 | 1.875 | 18-34 | 8 | 3 11⁄32 | 2 |
| 3 | 1.875 | 35 and over | 8 1⁄4 | 3 7⁄32 | 2 |

All dimensions are in inches.

All cutters are high-speed steel and are form relieved.

For keyway dimensions see page 882.

*Tolerances:* Outside diameter, $+{}^1\!/_{16}$, $-{}^1\!/_{16}$ inch; hole diameter, through 1-inch diameter, + 0.00075 inch, above 1-inch diameter and through 2-inch diameter, + 0.0010 inch.

For tooth form, see ANSI/ASME sprocket tooth form table on page 2639.

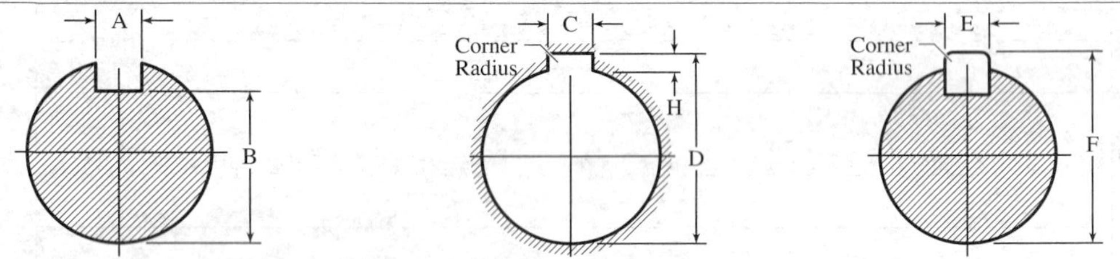

**American National Standard Woodruff Keyseat Cutters—Shank-Type Straight-Teeth and Arbor-Type Staggered-Teeth** *ANSI/ASME B94.19-1997 (R2019)*

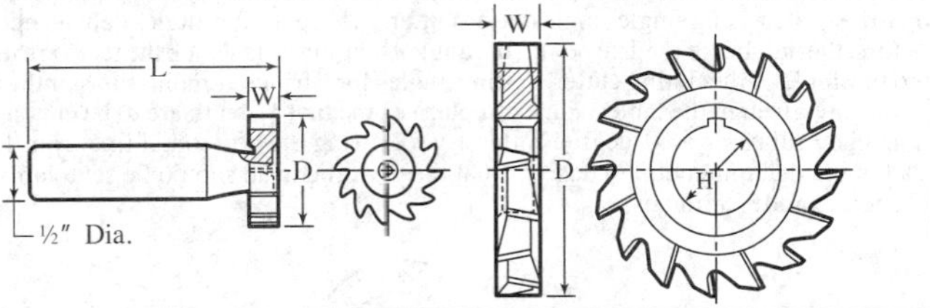


| Shank-type Cutters | | | | | | | | | | | |
|---|---|---|---|---|---|---|---|---|---|---|---|
| Cutter Number | Nom. Dia. of Cutter, *D* | Width of Face, *W* | Length Over-all, *L* | Cutter Number | Nom. Dia. of Cutter, *D* | Width of Face, *W* | Length Over-all, *L* | Cutter Number | Nom. Dia. of Cutter, *D* | Width of Face, *W* | Length Over-all, *L* |
| 202 | ¼ | 1⁄16 | 2 1⁄16 | 506 | ¾ | 5⁄32 | 2 5⁄32 | 809 | 1⅛ | ¼ | 2¼ |
| 202½ | 5⁄16 | 1⁄16 | 2 1⁄16 | 606 | ¾ | 3⁄16 | 2 3⁄16 | 1009 | 1⅛ | 5⁄16 | 2 5⁄16 |
| 302½ | 5⁄16 | 3⁄32 | 2 3⁄32 | 806 | ¾ | ¼ | 2¼ | 610 | 1¼ | 3⁄16 | 2 3⁄16 |
| 203 | ⅜ | 1⁄16 | 2 1⁄16 | 507 | ⅞ | 5⁄32 | 2 5⁄32 | 710 | 1¼ | 7⁄32 | 2 7⁄32 |
| 303 | ⅜ | 3⁄32 | 2 3⁄32 | 607 | ⅞ | 3⁄16 | 2 3⁄16 | 810 | 1¼ | ¼ | 2¼ |
| 403 | ⅜ | ⅛ | 2⅛ | 707 | ⅞ | 7⁄32 | 2 7⁄32 | 1010 | 1¼ | 5⁄16 | 2 5⁄16 |
| 204 | ½ | 1⁄16 | 2 1⁄16 | 807 | ⅞ | ¼ | 2¼ | 1210 | 1¼ | ⅜ | 2⅜ |
| 304 | ½ | 3⁄32 | 2 3⁄32 | 608 | 1 | 3⁄16 | 2 3⁄16 | 811 | 1⅜ | ¼ | 2¼ |
| 404 | ½ | ⅛ | 2⅛ | 708 | 1 | 7⁄32 | 2 7⁄32 | 1011 | 1⅜ | 5⁄16 | 2 5⁄16 |
| 305 | ⅝ | 3⁄32 | 2 3⁄32 | 808 | 1 | ¼ | 2¼ | 1211 | 1⅜ | ⅜ | 2⅜ |
| 405 | ⅝ | ⅛ | 2⅛ | 1008 | 1 | 5⁄16 | 2 5⁄16 | 812 | 1½ | ¼ | 2¼ |
| 505 | ⅝ | 5⁄32 | 2 5⁄32 | 1208 | 1 | ⅜ | 2⅜ | 1012 | 1½ | 5⁄16 | 2 5⁄16 |
| 605 | ⅝ | 3⁄16 | 2 3⁄16 | 609 | 1⅛ | 3⁄16 | 2 3⁄16 | 1212 | 1½ | ⅜ | 2⅜ |
| 406 | ¾ | ⅛ | 2⅛ | 709 | 1⅛ | 7⁄32 | 2 7⁄32 | … | … | … | … |

| Arbor-type Cutters | | | | | | | | | | | |
|---|---|---|---|---|---|---|---|---|---|---|---|
| Cutter Number | Nom. Dia. of Cutter, *D* | Width of Face, *W* | Dia. of Hole, *H* | Cutter Number | Nom. Dia of Cutter, *D* | Width of Face, *W* | Dia. of Hole, *H* | Cutter Number | Nom. Dia. of Cutter, *D* | Width of Face, *W* | Dia. of Hole, *H* |
| 617 | 2⅛ | 3⁄16 | ¾ | 1022 | 2¾ | 5⁄16 | 1 | 1628 | 3½ | ½ | 1 |
| 817 | 2⅛ | ¼ | ¾ | 1222 | 2¾ | ⅜ | 1 | 1828 | 3½ | 9⁄16 | 1 |
| 1017 | 2⅛ | 5⁄16 | ¾ | 1422 | 2¾ | 7⁄16 | 1 | 2028 | 3½ | ⅝ | 1 |
| 1217 | 2⅛ | ⅜ | ¾ | 1622 | 2¾ | ½ | 1 | 2428 | 3½ | ¾ | 1 |
| 822 | 2¾ | ¼ | 1 | 1228 | 3½ | ⅜ | 1 | … | … | … | … |

All dimensions are given in inches. All cutters are high-speed steel.

Shank type cutters are standard with right-hand cut and straight teeth. All sizes have ½-inch diameter straight shank.

Arbor type cutters have staggered teeth.

For Woodruff key and key-slot dimensions, see pages 2563 through 2565.

*Tolerances:* Face with *W* for shank-type cutters: 1⁄16- to 5⁄32-inch face, + 0.0000, −0.0005; 3⁄16 to 7⁄32, − 0.0002, − 0.0007; ¼, −0.0003, −0.0008; 5⁄16, −0.0004, −0.0009; ⅜, − 0.0005, −0.0010 inch. Face width *W* for arbor-type cutters; 3⁄16 inch face, −0.0002, −0.0007; ¼, −0.0003, −0.0008; 5⁄16, −0.0004, −0.0009; ⅜ and over, −0.0005, − 0.0010 inch. Hole size *H:* +0.00075, −0.0000 inch. Diameter *D* for shank type cutters: ¼- through ¾-inch diameter, +0.010, +0.015, ⅞ through 1⅛, +0.012, +0.017; 1¼ through 1½, +0.015, +0.020 inch. These tolerances include an allowance for sharpening. For arbor-type cutters, diameter *D* is furnished 1⁄32 inch larger than listed, and a tolerance of ±0.002 inch applies to the oversize diameter.

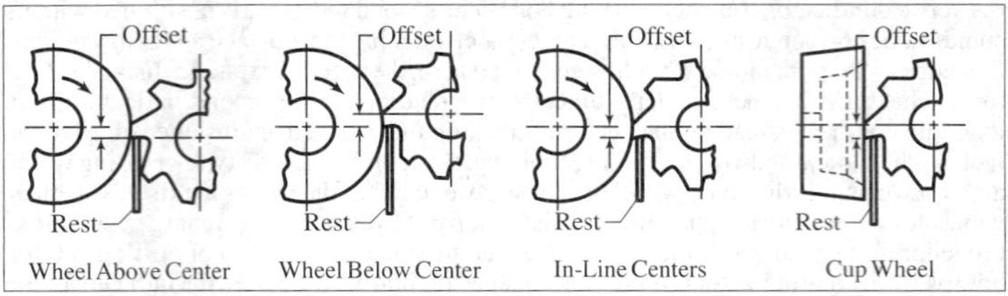

*Radial Relief:* Relief in a radial direction measured in the plane of rotation. It can be measured by the amount of indicator drop at a given radius in a given amount of angular rotation.

*Primary Relief:* The relief immediately behind the cutting edge or margin. Properly called relief.

*Secondary Relief:* An additional relief behind the primary relief.

*Relief Angle:* The angle, measured in a transverse plane, between the relieved surface and a plane tangent to the periphery at the cutting edge.

*Secondary Chamfer:* A slight relieved chamfer adjacent to and following the initial chamfer on a reamer.

*Shank:* The portion of the reamer by which it is held and driven.

*Squared Shank:* A cylindrical shank having a driving square on the back end.

*Starting Radius:* A relieved radius at the entering end of a reamer in place of a chamfer.

*Starting Taper:* A slight relieved taper on the front end of a reamer.

*Straight Shank:* A cylindrical shank.

*Tang:* The flatted end of a taper shank which fits a slot in the socket.

*Taper per Foot:* The difference in diameter between two points 12 inches apart measured along the axis.

*Taper Shank:* A shank made to fit a specific (conical) taper socket.

**Direction of Rotation and Helix.**—The terms "right-hand" and "left-hand" are used to describe both direction of rotation and direction of flute helix or reamers.

*Hand of Rotation (or Hand of Cut): Right-Hand Rotation (or Right-Hand Cut):* When viewed from the cutting end, the reamer must revolve counterclockwise to cut

*Left-hand Rotation (or Left-Hand Cut):* When viewed from the cutting end, the reamer must revolve clockwise to cut

*Hand of Flute Helix: Right-Hand Helix:* When the flutes twist away from the observer in a clockwise direction when viewed from either end of the reamer.

*Left-Hand Helix:* When the flutes twist away from the observer in a counterclockwise direction when viewed from either end of the reamer. The standard reamers on the tables that follow are all right-hand rotation.

**Dimensions of Centers for Reamers and Arbors**

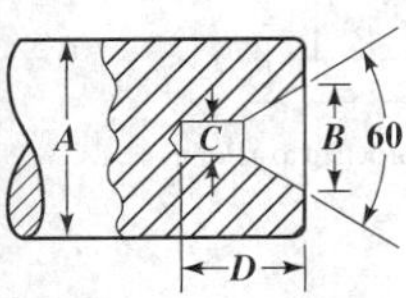


| Arbor. Dia. *A* | Large Center Dia. *B* | Drill No. *C* | Hole Depth *D* | Arbor Dia. *A* | Large Center Dia. *B* | Drill No. *C* | Hole Depth *D* | Arbor Dia. *A* | Large Center Dia. *B* | Drill No. *C* | Hole Depth *D* |
|---|---|---|---|---|---|---|---|---|---|---|---|
| | | | | 3/4 | 3/8 | 25 | 7/16 | 2 1/2 | 11/16 | **J** | 27/32 |
| | | | | 13/16 | 13/32 | 20 | 1/2 | 2 5/8 | 45/64 | **K** | 7/8 |
| | | | | 7/8 | 7/16 | 17 | 17/32 | 2 3/4 | 23/32 | **L** | 29/32 |
| | | | | 15/16 | 15/32 | 12 | 9/16 | 2 7/8 | 47/64 | **M** | 29/32 |
| | | | | 1 | 1/2 | 8 | 19/32 | 3 | 3/4 | **N** | 15/16 |
| | | | | 1 1/8 | 33/64 | 5 | 5/8 | 3 1/8 | 49/64 | **N** | 31/32 |
| | | | | 1 1/4 | 17/32 | 3 | 21/32 | 3 1/4 | 25/32 | **O** | 31/32 |
| | | | | 1 3/8 | 35/64 | 2 | 21/32 | 3 3/8 | 51/64 | **O** | 1 |
| | | | | 1 1/2 | 9/16 | 1 | 11/16 | 3 1/2 | 13/16 | **P** | 1 |
| 1/4 | 1/8 | 55 | 5/32 | … | … | Letter | … | 3 5/8 | 53/64 | **Q** | 1 1/16 |
| 5/16 | 5/32 | 52 | 3/16 | 1 5/8 | 37/64 | **A** | 23/32 | 3 3/4 | 27/32 | **R** | 1 1/16 |
| 3/8 | 3/16 | 48 | 7/32 | 1 3/4 | 19/32 | **B** | 23/32 | 3 7/8 | 55/64 | **R** | 1 1/16 |
| 7/16 | 7/32 | 43 | 1/4 | 1 7/8 | 39/64 | **C** | 3/4 | 4 | 7/8 | **S** | 1 1/8 |
| 1/2 | 1/4 | 39 | 5/16 | 2 | 5/8 | **E** | 3/4 | 4 1/4 | 29/32 | **T** | 1 1/8 |
| 9/16 | 9/32 | 33 | 11/32 | 2 1/8 | 41/64 | **F** | 25/32 | 4 1/2 | 15/16 | **V** | 1 3/16 |
| 5/8 | 5/16 | 30 | 3/8 | 2 1/4 | 21/32 | **G** | 13/16 | 4 3/4 | 31/32 | **W** | 1 1/4 |
| 11/16 | 11/32 | 29 | 13/32 | 2 3/8 | 43/64 | **H** | 27/32 | 5 | 1 | **X** | 1 1/4 |

## Straight-Shank Center Reamers and Machine Countersinks

*ANSI/ASME B94.2-1995 (R2020)*

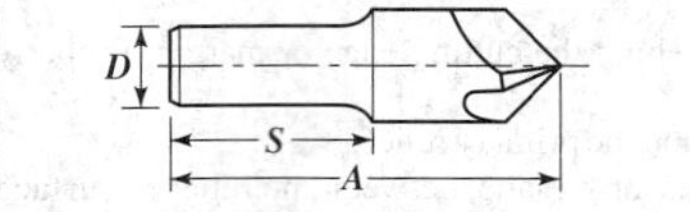


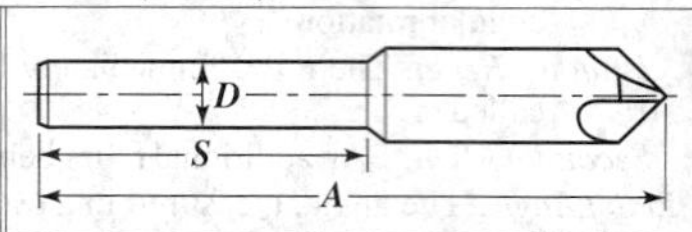


| Center Reamers (Short Countersinks) | | | | Machine Countersinks | | | |
|---|---|---|---|---|---|---|---|
| Dia. of Cut | Approx. Length Overall, *A* | Length of Shank, *S* | Dia. of Shank, *D* | Dia. of Cut | Approx. Length Overall, *A* | Length of Shank, *S* | Dia. of Shank, *D* |
| $\frac{1}{4}$ | $1\frac{1}{2}$ | $\frac{3}{4}$ | $\frac{3}{16}$ | $\frac{1}{2}$ | $3\frac{7}{8}$ | $2\frac{1}{4}$ | $\frac{1}{2}$ |
| $\frac{3}{8}$ | $1\frac{3}{4}$ | $\frac{7}{8}$ | $\frac{1}{4}$ | $\frac{5}{8}$ | 4 | $2\frac{1}{4}$ | $\frac{1}{2}$ |
| $\frac{1}{2}$ | 2 | 1 | $\frac{3}{8}$ | $\frac{3}{4}$ | $4\frac{1}{8}$ | $2\frac{1}{4}$ | $\frac{1}{2}$ |
| $\frac{5}{8}$ | $2\frac{1}{4}$ | 1 | $\frac{3}{8}$ | $\frac{7}{8}$ | $4\frac{1}{4}$ | $2\frac{1}{4}$ | $\frac{1}{2}$ |
| $\frac{3}{4}$ | $2\frac{5}{8}$ | $1\frac{1}{4}$ | $\frac{1}{2}$ | 1 | $4\frac{3}{8}$ | $2\frac{1}{4}$ | $\frac{1}{2}$ |

All dimensions are given in inches. Material is high-speed steel. Reamers and countersinks have 3 or 4 flutes. Center reamers are standard with 60, 82, 90, or 100 degrees included angle. Machine countersinks are standard with either 60 or 82 degrees included angle.

*Tolerances:* On overall length *A*, the tolerance is $\pm\frac{1}{8}$ inch for center reamers in a size range of from $\frac{1}{4}$ to $\frac{3}{8}$ inch, incl., and machine countersinks in a size range of from $\frac{1}{2}$ to $\frac{5}{8}$ inch. incl.; $\pm\frac{3}{16}$ inch for center reamers, $\frac{1}{2}$ to $\frac{3}{4}$ inch, incl.; and machine countersinks, $\frac{3}{4}$ to 1 inch, incl. On shank diameter *D*, the tolerance is −0.0005 to −0.002 inch. On shank length *S*, the tolerance is $\pm\frac{1}{16}$ inch.

**Calculating Countersink and Spot Drill Depths.—**The following formulas are for countersinks and spot drills that have been sharpened to a point, allowances on tool depth will have to be made if the countersink or spot drill is not sharpened to a point.

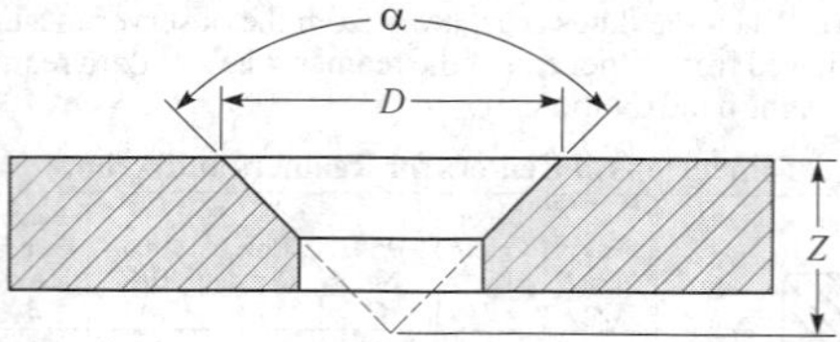


Fig. 1. For Countersinks and Spot Drills that have been Sharpened to a Point

$\alpha$ = countersink angle

$D$ = countersink diameter

$Z$ = depth of tool travel required from surface of part

To find the tool travel "Z" required for any sharp-point countersink or spot drill angle, use either of the following formulas:

$$Z = \frac{D}{2} \cdot \frac{1}{\tan\frac{\alpha}{2}} \quad \text{or} \quad Z = \frac{D}{2} \cdot \cot\frac{\alpha}{2}$$

To find the tool travel "*Z-a*" required for any truncated countersink or spot drill angle, use either of the following formulas:

$$Z - a = \frac{D - T}{2} \cdot \frac{1}{\tan\frac{\alpha}{2}} \quad \text{or} \quad Z - a = \frac{D - T}{2} \cdot \cot\frac{\alpha}{2}$$

## Expansion Chucking Reamers—Straight and Taper Shanks

*ANSI/ASME B94.2-1995 (R2020)*

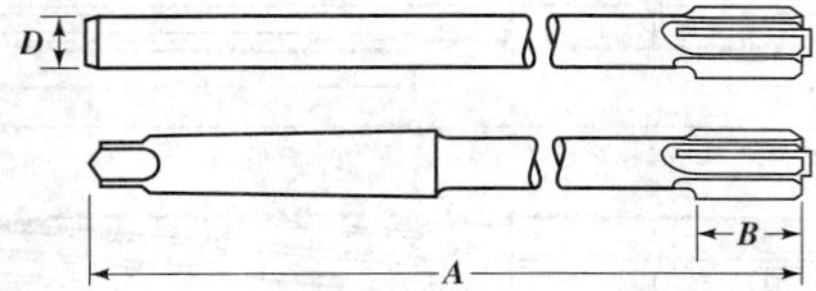


| Dia of Reamer | Length, *A* | Flute Length, *B* | Shank Dia., *D* | | Dia.of Reamer | Length, *A* | Flute Length, *B* | Shank Dia., *D* | |
|---|---|---|---|---|---|---|---|---|---|
| | | | Max. | Min. | | | | Max. | Min. |
| 3⁄8 | 7 | 3⁄4 | 0.3105 | 0.3095 | 1 3⁄32 | 10 1⁄2 | 1 5⁄8 | 0.8745 | 0.8730 |
| 13⁄32 | 7 | 3⁄4 | 0.3105 | 0.3095 | 1 1⁄8 | 11 | 1 3⁄4 | 0.8745 | 0.8730 |
| 7⁄16 | 7 | 7⁄8 | 0.3730 | 0.3720 | 1 5⁄32 | 11 | 1 3⁄4 | 0.8745 | 0.8730 |
| 15⁄32 | 7 | 7⁄8 | 0.3730 | 0.3720 | 1 3⁄16 | 11 | 1 3⁄4 | 0.9995 | 0.9980 |
| 1⁄2 | 8 | 1 | 0.4355 | 0.4345 | 1 7⁄32 | 11 | 1 3⁄4 | 0.9995 | 0.9980 |
| 17⁄32 | 8 | 1 | 0.4355 | 0.4345 | 1 1⁄4 | 11 1⁄2 | 1 7⁄8 | 0.9995 | 0.9980 |
| 9⁄16 | 8 | 1 1⁄8 | 0.4355 | 0.4345 | 1 5⁄16 | 11 1⁄2 | 1 7⁄8 | 0.9995 | 0.9980 |
| 19⁄32 | 8 | 1 1⁄8 | 0.4355 | 0.4345 | 1 3⁄8 | 12 | 2 | 0.9995 | 0.9980 |
| 5⁄8 | 9 | 1 1⁄4 | 0.5620 | 0.5605 | 1 7⁄16 | 12 | 2 | 1.2495 | 1.2480 |
| 21⁄32 | 9 | 1 1⁄4 | 0.5620 | 0.5605 | 1 1⁄2 | 12 1⁄2 | 2 1⁄8 | 1.2495 | 1.2480 |
| 11⁄16 | 9 | 1 1⁄4 | 0.5620 | 0.5605 | 1 9⁄16[a] | 12 1⁄2 | 2 1⁄8 | 1.2495 | 1.2480 |
| 23⁄32 | 9 | 1 1⁄4 | 0.5620 | 0.5605 | 1 5⁄8 | 13 | 2 1⁄4 | 1.2495 | 1.2480 |
| 3⁄4 | 9 1⁄2 | 1 3⁄8 | 0.6245 | 0.6230 | 1 11⁄16[a] | 13 | 2 1⁄4 | 1.2495 | 1.2480 |
| 25⁄32 | 9 1⁄2 | 1 3⁄8 | 0.6245 | 0.6230 | 1 3⁄4 | 13 1⁄2 | 2 3⁄8 | 1.2495 | 1.2480 |
| 13⁄16 | 9 1⁄2 | 1 3⁄8 | 0.6245 | 0.6230 | 1 13⁄16[a] | 13 1⁄2 | 2 3⁄8 | 1.4995 | 1.4980 |
| 27⁄32 | 9 1⁄2 | 1 3⁄8 | 0.6245 | 0.6230 | 1 7⁄8 | 14 | 2 1⁄2 | 1.4995 | 1.4980 |
| 7⁄8 | 10 | 1 1⁄2 | 0.7495 | 0.7480 | 1 15⁄16[a] | 14 | 2 1⁄2 | 1.4995 | 1.4980 |
| 29⁄32 | 10 | 1 1⁄2 | 0.7495 | 0.7480 | 2 | 14 | 2 1⁄2 | 1.4995 | 1.4980 |
| 15⁄16 | 10 | 1 1⁄2 | 0.7495 | 0.7480 | 2 1⁄8[b] | 14 1⁄2 | 2 3⁄4 | … | … |
| 31⁄32 | 10 | 1 1⁄2 | 0.7495 | 0.7480 | 2 1⁄4[b] | 14 1⁄2 | 2 3⁄4 | … | … |
| 1 | 10 1⁄2 | 1 5⁄8 | 0.8745 | 0.8730 | 2 3⁄8[b] | 15 | 3 | … | … |
| 1 1⁄32 | 10 1⁄2 | 1 5⁄8 | 0.8745 | 0.8730 | 2 1⁄2[b] | 15 | 3 | … | … |
| 1 1⁄16 | 10 1⁄2 | 1 5⁄8 | 0.8745 | 0.8730 | … | … | … | … | … |

[a] Straight shank only.

[b] Taper shank only.

All dimensions in inches. Material is high-speed steel. The number of flutes is as follows: 3⁄8- to 15⁄32-inch sizes, 4 to 6; 1⁄2- to 31⁄32-inch sizes, 6 to 8; 1- to 1 11⁄16-inch sizes, 8 to 10; 1 3⁄4- to 1 15⁄16-inch sizes, 8 to 12; 2- to 2 1⁄4-inch sizes, 10 to 12; 2 3⁄8- and 2 1⁄2-inch sizes, 10 to 14. The expansion feature of these reamers provides a means of adjustment that is important in reaming holes to close tolerances. When worn undersize, they may be expanded and reground to the original size.

*Tolerances:* On reamer diameter, 3⁄8- to 1-inch sizes, incl., +0.0001 to +0.0005 inch; over 1-inch size, + 0.0002 to + 0.0006 inch. On length *A* and flute length *B*, 3⁄8- to 1-inch sizes, incl., ±1⁄16 inch; 1 1⁄32- to 2-inch sizes, incl., ±3⁄32 inch; over 2-inch sizes, ±1⁄8 inch.

Taper is Morse taper: No. 1 for sizes 3⁄8 to 19⁄32 inch, incl.; No. 2 for sizes 5⁄8 to 29⁄32 incl.; No. 3 for sizes 15⁄16 to 1 7⁄32, incl.; No. 4 for sizes 1 1⁄4 to 1 5⁄8, incl.; and No. 5 for sizes 1 3⁄4 to 2 1⁄2, incl. For amount of taper, see Table on page 1001.

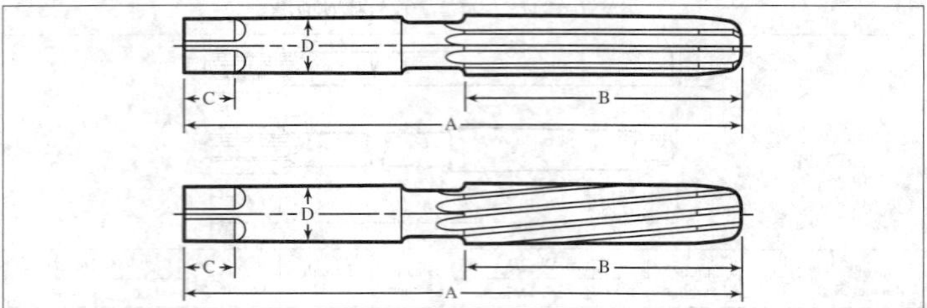

## American National Standard Expansion Hand Reamers—Straight and Helical Flutes, Squared Shank *ANSI/ASME B94.2-1995 (R2020)*

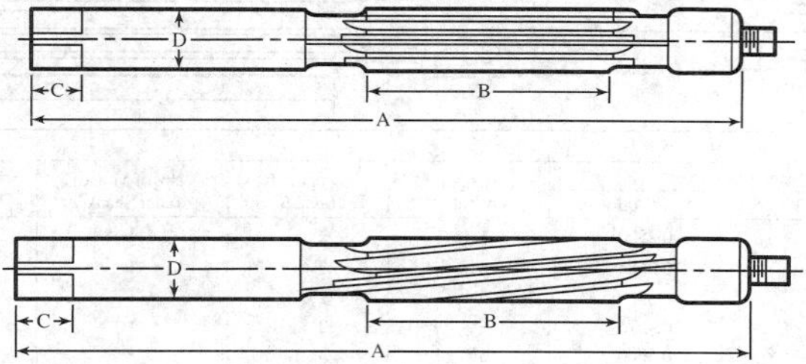


| Reamer Dia. | Length Overall A | | Flute Length B | | Length of Square C | Shank Dia. D | Size of Square | Number of Flutes |
|---|---|---|---|---|---|---|---|---|
| | Max. | Min. | Max. | Min. | | | | |
| **Straight Flutes** | | | | | | | | |
| 1/4 | 4 3/8 | 3 3/4 | 1 3/4 | 1 1/2 | 1/4 | 1/4 | 0.185 | 6 to 8 |
| 5/16 | 4 3/8 | 4 | 1 7/8 | 1 1/2 | 5/16 | 5/16 | 0.235 | 6 to 8 |
| 3/8 | 5 3/8 | 4 1/4 | 2 | 1 3/4 | 3/8 | 3/8 | 0.280 | 6 to 9 |
| 7/16 | 5 3/8 | 4 1/2 | 2 | 1 3/4 | 7/16 | 7/16 | 0.330 | 6 to 9 |
| 1/2 | 6 1/2 | 5 | 2 1/2 | 1 3/4 | 1/2 | 1/2 | 0.375 | 6 to 9 |
| 9/16 | 6 1/2 | 5 3/8 | 2 1/2 | 1 7/8 | 9/16 | 9/16 | 0.420 | 6 to 9 |
| 5/8 | 7 | 5 3/4 | 3 | 2 1/4 | 5/8 | 5/8 | 0.470 | 6 to 9 |
| 11/16 | 7 5/8 | 6 1/4 | 3 | 2 1/2 | 11/16 | 11/16 | 0.515 | 6 to 10 |
| 3/4 | 8 | 6 1/2 | 3 1/2 | 2 5/8 | 3/4 | 3/4 | 0.560 | 6 to 10 |
| 7/8 | 9 | 7 1/2 | 4 | 3 1/8 | 7/8 | 7/8 | 0.655 | 8 to 10 |
| 1 | 10 | 8 3/8 | 4 1/2 | 3 1/8 | 1 | 1 | 0.750 | 8 to 10 |
| 1 1/8 | 10 1/2 | 9 | 4 3/4 | 3 1/2 | 1 | 1 1/8 | 0.845 | 8 to 12 |
| 1 1/4 | 11 | 9 3/4 | 5 | 4 1/4 | 1 | 1 1/4 | 0.935 | 8 to 12 |
| **Helical Flutes** | | | | | | | | |
| 1/4 | 4 3/8 | 3 7/8 | 1 3/4 | 1 1/2 | 1/4 | 1/4 | 0.185 | 6 to 8 |
| 5/16 | 4 3/8 | 4 | 1 3/4 | 1 1/2 | 5/16 | 5/16 | 0.235 | 6 to 8 |
| 3/8 | 6 1/8 | 4 1/4 | 2 | 1 3/4 | 3/8 | 3/8 | 0.280 | 6 to 9 |
| 7/16 | 6 1/4 | 4 1/2 | 2 | 1 3/4 | 7/16 | 7/16 | 0.330 | 6 to 9 |
| 1/2 | 6 1/2 | 5 | 2 1/2 | 1 3/4 | 1/2 | 1/2 | 0.375 | 6 to 9 |
| 5/8 | 8 | 6 | 3 | 2 1/4 | 5/8 | 5/8 | 0.470 | 6 to 9 |
| 3/4 | 8 5/8 | 6 1/2 | 3 1/2 | 2 5/8 | 3/4 | 3/4 | 0.560 | 6 to 10 |
| 7/8 | 9 3/8 | 7 1/2 | 4 | 3 1/8 | 7/8 | 7/8 | 0.655 | 6 to 10 |
| 1 | 10 1/4 | 8 3/8 | 4 1/2 | 3 1/8 | 1 | 1 | 0.750 | 6 to 10 |
| 1 1/4 | 11 3/8 | 9 3/4 | 5 | 4 1/4 | 1 | 1 1/4 | 0.935 | 8 to 12 |

All dimensions are given in inches. Material is carbon steel. Reamers with helical flutes that are left-hand are standard. Expansion hand reamers are primarily designed for work where it is necessary to enlarge reamed holes by a few thousandths. The pilots and guides on these reamers are ground undersize for clearance. The maximum expansion on these reamers is as follows: .006 inch for the 1/4- to 7/16-inch sizes. .010 inch for the 1/2- to 7/8-inch sizes and .012 inch for the 1- to 1 1/4-inch sizes.

*Tolerances:* On length overall *A* and flute length *B*, ±1/16 inch for 1/4- to 1-inch sizes, ± 3/32 inch for 1 1/8-to 1 1/4-inch sizes; on length of square *C*, ±1/32 inch for 1/4- to 1-inch sizes, ± 1/16 inch for 1 1/8-to 1 1/4-inch sizes; on shank diameter *D* −.001 to −.005 inch for 1/4- to 1-inch sizes, −.0015 to −.006 inch for 1 1/8- to 1 1/4-inch sizes; on size of square, −.004 inch for 1/4- to 1/2-inch sizes. −.006 inch for 9/16- to 1-inch sizes, and −.008 inch for 1 1/8- to 1 1/4-inch sizes.

## American National Standard Straight-Shank Rose Chucking and Chucking Reamers— Straight and Helical Flutes, Fractional Sizes *ANSI/ASME B94.2-1995 (R2020)*

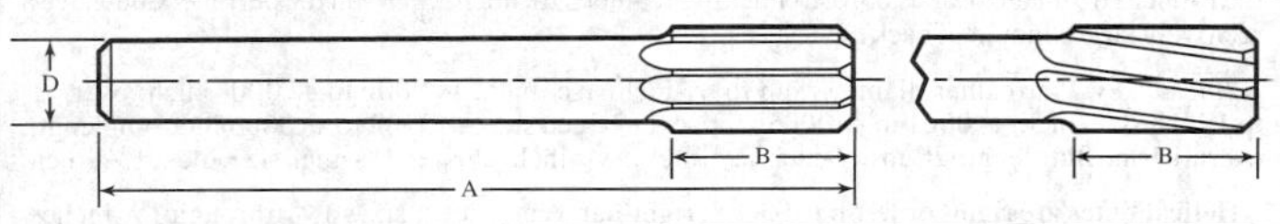


| Reamer Diameter | | Length | Flute | Shank Dia. *D* | | No. of |
|---|---|---|---|---|---|---|
| Chucking | Rose Chucking | Overall *A* | Length *B* | Max. | Min. | Flutes |
| 3⁄64[a] | … | 2½ | ½ | 0.0455 | 0.0445 | 4 |
| 1⁄16 | … | 2½ | ½ | 0.0585 | 0.0575 | 4 |
| 5⁄64 | … | 3 | ¾ | 0.0720 | 0.0710 | 4 |
| 3⁄32 | … | 3 | ¾ | 0.0880 | 0.0870 | 4 |
| 7⁄64 | … | 3½ | ⅞ | 0.1030 | 0.1020 | 4 to 6 |
| ⅛ | ⅛[a] | 3½ | ⅞ | 0.1190 | 0.1180 | 4 to 6 |
| 9⁄64 | … | 4 | 1 | 0.1350 | 0.1340 | 4 to 6 |
| 5⁄32 | … | 4 | 1 | 0.1510 | 0.1500 | 4 to 6 |
| 11⁄64 | … | 4½ | 1⅛ | 0.1645 | 0.1635 | 4 to 6 |
| 3⁄16 | 3⁄16[a] | 4½ | 1⅛ | 0.1805 | 0.1795 | 4 to 6 |
| 13⁄64 | … | 5 | 1¼ | 0.1945 | 0.1935 | 4 to 6 |
| 7⁄32 | … | 5 | 1¼ | 0.2075 | 0.2065 | 4 to 6 |
| 15⁄64 | … | 6 | 1½ | 0.2265 | 0.2255 | 4 to 6 |
| ¼ | ¼[a] | 6 | 1½ | 0.2405 | 0.2395 | 4 to 6 |
| 17⁄64 | … | 6 | 1½ | 0.2485 | 0.2475 | 4 to 6 |
| 9⁄32 | … | 6 | 1½ | 0.2485 | 0.2475 | 4 to 6 |
| 19⁄64 | … | 6 | 1½ | 0.2792 | 0.2782 | 4 to 6 |
| 5⁄16 | 5⁄16[a] | 6 | 1½ | 0.2792 | 0.2782 | 4 to 6 |
| 21⁄64 | … | 6 | 1½ | 0.2792 | 0.2782 | 4 to 6 |
| 11⁄32 | … | 6 | 1½ | 0.2792 | 0.2782 | 4 to 6 |
| 23⁄64 | … | 7 | 1¾ | 0.3105 | 0.3095 | 4 to 6 |
| ⅜ | ⅜[a] | 7 | 1¾ | 0.3105 | 0.3095 | 4 to 6 |
| 25⁄64 | … | 7 | 1¾ | 0.3105 | 0.3095 | 4 to 6 |
| 13⁄32 | … | 7 | 1¾ | 0.3105 | 0.3095 | 4 to 6 |
| 27⁄64 | … | 7 | 1¾ | 0.3730 | 0.3720 | 6 to 8 |
| 7⁄16 | 7⁄16[a] | 7 | 1¾ | 0.3730 | 0.3720 | 6 to 8 |
| 29⁄64 | … | 7 | 1¾ | 0.3730 | 0.3720 | 6 to 8 |
| 15⁄32 | … | 7 | 1¾ | 0.3730 | 0.3720 | 6 to 8 |
| 31⁄64 | … | 8 | 2 | 0.4355 | 0.4345 | 6 to 8 |
| ½ | ½[a] | 8 | 2 | 0.4355 | 0.4345 | 6 to 8 |
| 17⁄32 | … | 8 | 2 | 0.4355 | 0.4345 | 6 to 8 |
| 9⁄16 | … | 8 | 2 | 0.4355 | 0.4345 | 6 to 8 |
| 19⁄32 | … | 8 | 2 | 0.4355 | 0.4345 | 6 to 8 |
| ⅝ | … | 9 | 2¼ | 0.5620 | 0.5605 | 6 to 8 |
| 21⁄32 | … | 9 | 2¼ | 0.5620 | 0.5605 | 6 to 8 |
| 11⁄16 | … | 9 | 2¼ | 0.5620 | 0.5605 | 6 to 8 |
| 23⁄32 | … | 9 | 2¼ | 0.5620 | 0.5605 | 6 to 8 |
| ¾ | … | 9½ | 2½ | 0.6245 | 0.6230 | 6 to 8 |
| 25⁄32 | … | 9½ | 2½ | 0.6245 | 0.6230 | 8 to 10 |
| 13⁄16 | … | 9½ | 2½ | 0.6245 | 0.6230 | 8 to 10 |
| 27⁄32 | … | 9½ | 2½ | 0.6245 | 0.6230 | 8 to 10 |
| ⅞ | … | 10 | 2⅝ | 0.7495 | 0.7480 | 8 to 10 |
| 29⁄32 | … | 10 | 2⅝ | 0.7495 | 0.7480 | 8 to 10 |
| 15⁄16 | … | 10 | 2⅝ | 0.7495 | 0.7480 | 8 to 10 |
| 31⁄32 | … | 10 | 2⅝ | 0.7495 | 0.7480 | 8 to 10 |
| 1 | … | 10½ | 2¾ | 0.8745 | 0.8730 | 8 to 12 |
| 1 1⁄16 | … | 10½ | 2¾ | 0.8745 | 0.8730 | 8 to 12 |
| 1⅛ | … | 11 | 2⅞ | 0.8745 | 0.8730 | 8 to 12 |
| 1 3⁄16 | … | 11 | 2⅞ | 0.9995 | 0.9980 | 8 to 12 |
| 1¼ | … | 11½ | 3 | 0.9995 | 0.9980 | 8 to 12 |
| 1 5⁄16[b] | … | 11½ | 3 | 0.9995 | 0.9980 | 10 to 12 |
| 1⅜ | … | 12 | 3¼ | 0.9995 | 0.9980 | 10 to 12 |
| 1 7⁄16[b] | … | 12 | 3¼ | 1.2495 | 1.2480 | 10 to 12 |
| 1½ | … | 12½ | 3½ | 1.2495 | 1.2480 | 10 to 12 |

[a] Reamer with straight flutes is standard only.

[b] Reamer with helical flutes is standard only.

## American National Standard Arbors for Shell Reamers— Straight and Taper Shanks *ANSI/ASME B94.2-1995 (R2020)*

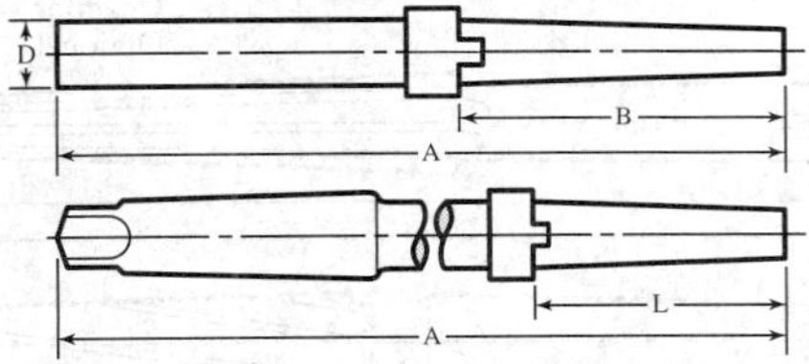


| Arbor Size No. | Overall Length *A* | Approx. Length of Taper *L* | Reamer Size | Taper Shank No.[a] | Straight Shank Dia. *D* | Arbor Size No. | Overall Length *A* | Approx. Length of Taper *L* | Reamer Size | Taper Shank No.[a] | Straight Shank Dia. *D* |
|---|---|---|---|---|---|---|---|---|---|---|---|
| 4 | 9 | 2¼ | ¾ | 2 | ½ | 7 | 11 | 3 | 1⁵⁄₁₆ to 1⅝ | 3 | ⅞ |
| 5 | 9½ | 2½ | ¹³⁄₁₆ to 1 | 2 | ⅝ | 8 | 12 | 3½ | 1¹¹⁄₁₆ to 2 | 4 | 1⅛ |
| 6 | 10 | 2¾ | 1¹⁄₁₆ to 1¼ | 3 | ¾ | 9 | 13 | 3¾ | 2¹⁄₁₆ to 2½ | 4 | 1⅜ |

[a] American National Standard self-holding tapers (see Table 7a on page 1008).

All dimensions are given in inches. These arbors are designed to fit standard shell reamers (see table). End that fits reamer has taper of ⅛ inch per foot.

## Stub Screw Machine Reamers—Helical Flutes *ANSI/ASME B94.2-1995 (R2020)*

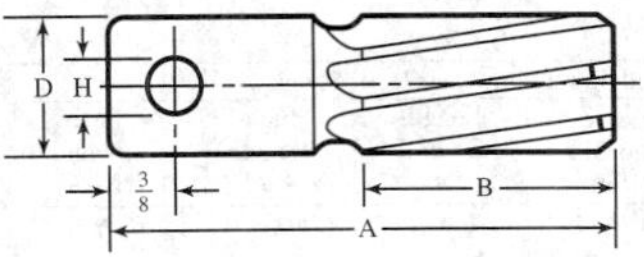


| Series No. | Diameter Range | Length Overall *A* | Length of Flute *B* | Dia. of Shank *D* | Size of Hole *H* | Flute No. | Series No. | Diameter Range | Length Overall *A* | Length of Flute *B* | Dia. of Shank *D* | Size of Hole *H* | Flute No. |
|---|---|---|---|---|---|---|---|---|---|---|---|---|---|
| 00 | .0600-.066 | 1¾ | ½ | ⅛ | ¹⁄₁₆ | 4 | 12 | .3761-.407 | 2½ | 1¼ | ½ | ³⁄₁₆ | 6 |
| 0 | .0661-.074 | 1¾ | ½ | ⅛ | ¹⁄₁₆ | 4 | 13 | .4071-.439 | 2½ | 1¼ | ½ | ³⁄₁₆ | 6 |
| 1 | .0741-.084 | 1¾ | ½ | ⅛ | ¹⁄₁₆ | 4 | 14 | .4391-.470 | 2½ | 1¼ | ½ | ³⁄₁₆ | 6 |
| 2 | .0841-.096 | 1¾ | ½ | ⅛ | ¹⁄₁₆ | 4 | 15 | .4701-.505 | 2½ | 1¼ | ½ | ³⁄₁₆ | 6 |
| 3 | .0961-.126 | 2 | ¾ | ⅛ | ¹⁄₁₆ | 4 | 16 | .5051-.567 | 3 | 1½ | ⅝ | ¼ | 6 |
| 4 | .1261-.158 | 2¼ | 1 | ¼ | ³⁄₃₂ | 4 | 17 | .5671-.630 | 3 | 1½ | ⅝ | ¼ | 6 |
| 5 | .1581-.188 | 2¼ | 1 | ¼ | ³⁄₃₂ | 4 | 18 | .6301-.692 | 3 | 1½ | ⅝ | ¼ | 6 |
| 6 | .1881-.219 | 2¼ | 1 | ¼ | ³⁄₃₂ | 6 | 19 | .6921-.755 | 3 | 1½ | ¾ | ⁵⁄₁₆ | 8 |
| 7 | .2191-.251 | 2¼ | 1 | ¼ | ³⁄₃₂ | 6 | 20 | .7551-.817 | 3 | 1½ | ¾ | ⁵⁄₁₆ | 8 |
| 8 | .2511-.282 | 2¼ | 1 | ⅜ | ⅛ | 6 | 21 | .8171-.880 | 3 | 1½ | ¾ | ⁵⁄₁₆ | 8 |
| 9 | .2821-.313 | 2¼ | 1 | ⅜ | ⅛ | 6 | 22 | .8801-.942 | 3 | 1½ | ¾ | ⁵⁄₁₆ | 8 |
| 10 | .3131-.344 | 2½ | 1¼ | ⅜ | ⅛ | 6 | 23 | .9421-1.010 | 3 | 1½ | ¾ | ⁵⁄₁₆ | 8 |
| 11 | .3441-.376 | 2½ | 1¼ | ⅜ | ⅛ | 6 | … | … | … | … | … | … | … |

All dimensions in inches. Material is high-speed steel.

These reamers are standard with right-hand cut and left-hand helical flutes within the size ranges shown.

*Tolerances:* On diameter of reamer, for sizes 00 to 7, incl., plus .0001 to plus .0004 inch and for sizes 8 to 23, incl., plus .0001 to plus .0005 inch. On overall length *A*, plus or minus ¹⁄₁₆ inch. On length of flute *B*, plus or minus ¹⁄₁₆ inch. On diameter of shank *D*, minus .0005 to minus .002 inch.

## American National Standard Die-Makers' Reamers *ANSI/ASME B94.2-1995 (R2020)*

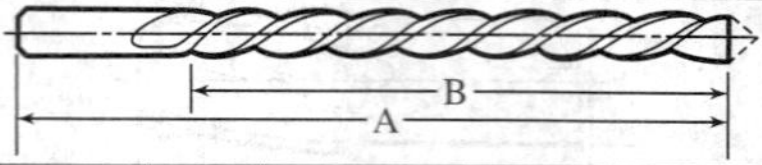


| Letter Size | Diameter Small End | Diameter Large End | Length A | Length B | Letter Size | Diameter Small End | Diameter Large End | Length A | Length B | Letter Size | Diameter Small End | Diameter Large End | Length A | Length B |
|---|---|---|---|---|---|---|---|---|---|---|---|---|---|---|
| AAA | 0.055 | 0.070 | 2¼ | 1⅛ | G | 0.135 | 0.158 | 3 | 1¾ | O | 0.250 | 0.296 | 5 | 3½ |
| AA | 0.065 | 0.080 | 2¼ | 1⅛ | H | 0.145 | 0.169 | 3¼ | 1⅞ | P | 0.275 | 0.327 | 5½ | 4 |
| A | 0.075 | 0.090 | 2¼ | 1⅛ | I | 0.160 | 0.184 | 3¼ | 1⅞ | Q | 0.300 | 0.358 | 6 | 4½ |
| B | 0.085 | 0.103 | 2⅜ | 1⅜ | J | 0.175 | 0.199 | 3¼ | 1⅞ | R | 0.335 | 0.397 | 6½ | 4¾ |
| C | 0.095 | 0.113 | 2½ | 1⅜ | K | 0.190 | 0.219 | 3½ | 2¼ | S | 0.370 | 0.435 | 6¾ | 5 |
| D | 0.105 | 0.126 | 2⅝ | 1⅝ | L | 0.205 | 0.234 | 3½ | 2¼ | T | 0.405 | 0.473 | 7 | 5¼ |
| E | 0.115 | 0.136 | 2¾ | 1⅝ | M | 0.220 | 0.252 | 4 | 2½ | U | 0.440 | 0.511 | 7¼ | 5½ |
| F | 0.125 | 0.148 | 3 | 1¾ | N | 0.235 | 0.274 | 4½ | 3 | … | … | … | … | … |

All dimensions in inches. Material is high-speed steel. These reamers are designed for use in diemaking, have a taper of ¾ degree included angle or 0.013 inch per inch, and have 2 or 3 flutes. Reamers are standard with left-hand spiral flutes.

Tip of reamer may have conical end.

*Tolerances:* On length overall *A* and flute length *B*, ±1⁄16 inch.

## Taper Pin Reamers — Straight and Left-Hand Spiral Flutes, Squared Shank; Left-Hand High-Spiral Flutes, Round Shank *ANSI/ASME B94.2-1995 (R2020)*

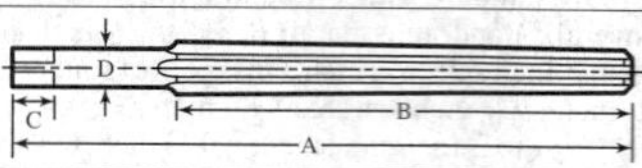


| No. of Taper Pin Reamer | Diameter at Large End of Reamer (Ref.) | Diameter at Small End of Reamer (Ref.) | Overall Length of Reamer *A* | Length of Flute *B* | Length of Square *C*[a] | Diameter of Shank *D* | Size of Square[a] |
|---|---|---|---|---|---|---|---|
| 8/0[b] | 0.0514 | 0.0351 | 1⅝ | 25⁄32 | … | 1⁄16 | … |
| 7/0 | 0.0666 | 0.0497 | 1 13⁄16 | 13⁄16 | 5⁄32 | 5⁄64 | 0.060 |
| 6/0 | 0.0806 | 0.0611 | 1 15⁄16 | 15⁄16 | 5⁄32 | 3⁄32 | 0.070 |
| 5/0 | 0.0966 | 0.0719 | 2 3⁄16 | 1 3⁄16 | 5⁄32 | 7⁄64 | 0.080 |
| 4/0 | 0.1142 | 0.0869 | 2 5⁄16 | 1 5⁄16 | 5⁄32 | 1⁄8 | 0.095 |
| 3/0 | 0.1302 | 0.1029 | 2 5⁄16 | 1 5⁄16 | 5⁄32 | 9⁄64 | 0.105 |
| 2/0 | 0.1462 | 0.1137 | 2 9⁄16 | 1 9⁄16 | 7⁄32 | 5⁄32 | 0.115 |
| 0 | 0.1638 | 0.1287 | 2 15⁄16 | 1 11⁄16 | 7⁄32 | 11⁄64 | 0.130 |
| 1 | 0.1798 | 0.1447 | 2 15⁄16 | 1 11⁄16 | 7⁄32 | 3⁄16 | 0.140 |
| 2 | 0.2008 | 0.1605 | 3 3⁄16 | 1 15⁄16 | 1⁄4 | 13⁄64 | 0.150 |
| 3 | 0.2294 | 0.1813 | 3 11⁄16 | 2 5⁄16 | 1⁄4 | 15⁄64 | 0.175 |
| 4 | 0.2604 | 0.2071 | 4 1⁄16 | 2 9⁄16 | 1⁄4 | 17⁄64 | 0.200 |
| 5 | 0.2994 | 0.2409 | 4 5⁄16 | 2 13⁄16 | 5⁄16 | 5⁄16 | 0.235 |
| 6 | 0.3540 | 0.2773 | 5 7⁄16 | 3 11⁄16 | 3⁄8 | 23⁄64 | 0.270 |
| 7 | 0.4220 | 0.3297 | 6 5⁄16 | 4 7⁄16 | 3⁄8 | 13⁄32 | 0.305 |
| 8 | 0.5050 | 0.3971 | 7 3⁄16 | 5 3⁄16 | 7⁄16 | 7⁄16 | 0.330 |
| 9 | 0.6066 | 0.4805 | 8 5⁄16 | 6 1⁄16 | 9⁄16 | 9⁄16 | 0.420 |
| 10 | 0.7216 | 0.5799 | 9 5⁄16 | 6 13⁄16 | 5⁄8 | 5⁄8 | 0.470 |

[a] Not applicable to high-spiral flute reamers.

[b] Not applicable to straight and left-hand spiral-fluted, squared-shank reamers.

All dimensions in inches. Reamers have a taper of ¼ inch per foot and are made of high-speed steel. Straight flute reamers of carbon steel are also standard. The number of flutes is as follows: 3 or 4, for 7/0 to 4/0 sizes; 4 to 6, for 3/0 to 0 sizes; 5 or 6, for 1 to 5 sizes; 6 to 8, for 6 to 9 sizes; 7 or 8, for the 10 size in the case of straight- and spiral-flute reamers; and 2 or 3, for 8/0 to 8 sizes; 2 to 4, for the 9 and 10 sizes in the case of high-spiral flute reamers.

*Tolerances:* On length overall *A* and flute length *B*, ±1⁄16 inch. On length of square *C*, ±1⁄32 inch. On shank diameter *D*, −.001 to −.005 inch for straight- and spiral-flute reamers and −.0005 to −.002 inch for high-spiral flute reamers. On size of square, −.004 inch for 7/0 to 7 sizes and −.006 inch for 8 to 10 sizes.

**Table 1.** *(Continued)* **ANSI Straight-Shank Twist Drills — Jobbers Length through 17.5 mm, Taper Length through 12.7 mm, and Screw Machine Length through 25.4 mm Diameter** *ANSI/ASME B94.11M-1993*

| Drill Diameter, $D^a$ | | | | Jobbers Length | | | | Taper Length | | | | Screw Machine Length | | | |
|---|---|---|---|---|---|---|---|---|---|---|---|---|---|---|---|
| Fraction No. or Ltr. | | Equivalent | | Flute | | Overall | | Flute | | Overall | | Flute | | Overall | |
| | | | | F | | L | | F | | L | | F | | L | |
| | mm | Decimal Inch | mm | Inch | mm | Inch | mm | Inch | mm | Inch | mm | Inch | mm | Inch | mm |
| | 3.10 | 0.1220 | 3.100 | 1⅝ | 41 | 2¾ | 70 | 2¾ | 70 | 5⅛ | 130 | ⅞ | 22 | 1⅞ | 48 |
| ⅛ | | 0.1250 | 3.175 | 1⅝ | 41 | 2¾ | 70 | 2¾ | 70 | 5⅛ | 130 | ⅞ | 22 | 1⅞ | 48 |
| | 3.20 | 0.1260 | 3.200 | 1⅝ | 41 | 2¾ | 70 | 3 | 76 | 5⅜ | 137 | 15/16 | 24 | 1 15/16 | 49 |
| 30 | | 0.1285 | 3.264 | 1⅝ | 41 | 2¾ | 70 | 3 | 76 | 5⅜ | 137 | 15/16 | 24 | 1 15/16 | 49 |
| | 3.30 | 0.1299 | 3.300 | 1¾ | 44 | 2⅞ | 73 | 3 | 76 | 5⅜ | 137 | 15/16 | 24 | 1 15/16 | 49 |
| | 3.40 | 0.1339 | 3.400 | 1¾ | 44 | 2⅞ | 73 | 3 | 76 | 5⅜ | 137 | 15/16 | 24 | 1 15/16 | 49 |
| 29 | | 0.1360 | 3.454 | 1¾ | 44 | 2⅞ | 73 | 3 | 76 | 5⅜ | 137 | 15/16 | 24 | 1 15/16 | 49 |
| | 3.50 | 0.1378 | 3.500 | 1¾ | 44 | 2⅞ | 73 | 3 | 76 | 5⅜ | 137 | 15/16 | 24 | 1 15/16 | 49 |
| 28 | | 0.1405 | 3.569 | 1¾ | 44 | 2⅞ | 73 | 3 | 76 | 5⅜ | 137 | 15/16 | 24 | 1 15/16 | 49 |
| 9/64 | | 0.1406 | 3.571 | 1¾ | 44 | 2⅞ | 73 | 3 | 76 | 5⅜ | 137 | 15/16 | 24 | 1 15/16 | 49 |
| | 3.60 | 0.1417 | 3.600 | 1⅞ | 48 | 3 | 76 | 3 | 76 | 5⅜ | 137 | 1 | 25 | 2 1/16 | 52 |
| 27 | | 0.1440 | 3.658 | 1⅞ | 48 | 3 | 76 | 3 | 76 | 5⅜ | 137 | 1 | 25 | 2 1/16 | 52 |
| | 3.70 | 0.1457 | 3.700 | 1⅞ | 48 | 3 | 76 | 3 | 76 | 5⅜ | 137 | 1 | 25 | 2 1/16 | 52 |
| 26 | | 0.1470 | 3.734 | 1⅞ | 48 | 3 | 76 | 3 | 76 | 5⅜ | 137 | 1 | 25 | 2 1/16 | 52 |
| 25 | | 0.1495 | 3.797 | 1⅞ | 48 | 3 | 76 | 3 | 76 | 5⅜ | 137 | 1 | 25 | 2 1/16 | 52 |
| | 3.80 | 0.1496 | 3.800 | 1⅞ | 48 | 3 | 76 | 3 | 76 | 5⅜ | 137 | 1 | 25 | 2 1/16 | 52 |
| 24 | | 0.1520 | 3.861 | 2 | 51 | 3⅛ | 79 | 3 | 76 | 5⅜ | 137 | 1 | 25 | 2 1/16 | 52 |
| | 3.90 | 0.1535 | 3.900 | 2 | 51 | 3⅛ | 79 | 3 | 76 | 5⅜ | 137 | 1 | 25 | 2 1/16 | 52 |
| 23 | | 0.1540 | 3.912 | 2 | 51 | 3⅛ | 79 | 3 | 76 | 5⅜ | 137 | 1 | 25 | 2 1/16 | 52 |
| 5/32 | | 0.1562 | 3.967 | 2 | 51 | 3⅛ | 79 | 3 | 76 | 5⅜ | 137 | 1 | 25 | 2 1/16 | 52 |
| 22 | | 0.1570 | 3.988 | 2 | 51 | 3⅛ | 79 | 3⅜ | 86 | 5¾ | 146 | 1 1/16 | 27 | 2⅛ | 54 |
| | 4.00 | 0.1575 | 4.000 | 2⅛ | 54 | 3¼ | 83 | 3⅜ | 86 | 5¾ | 146 | 1 1/16 | 27 | 2⅛ | 54 |
| 21 | | 0.1590 | 4.039 | 2⅛ | 54 | 3¼ | 83 | 3⅜ | 86 | 5¾ | 146 | 1 1/16 | 27 | 2⅛ | 54 |
| 20 | | 0.1610 | 4.089 | 2⅛ | 54 | 3¼ | 83 | 3⅜ | 86 | 5¾ | 146 | 1 1/16 | 27 | 2⅛ | 54 |
| | 4.10 | 0.1614 | 4.100 | 2⅛ | 54 | 3¼ | 83 | 3⅜ | 86 | 5¾ | 146 | 1 1/16 | 27 | 2⅛ | 54 |
| | 4.20 | 0.1654 | 4.200 | 2⅛ | 54 | 3¼ | 83 | 3⅜ | 86 | 5¾ | 146 | 1 1/16 | 27 | 2⅛ | 54 |
| 19 | | 0.1660 | 4.216 | 2⅛ | 54 | 3¼ | 83 | 3⅜ | 86 | 5¾ | 146 | 1 1/16 | 27 | 2⅛ | 54 |
| | 4.30 | 0.1693 | 4.300 | 2⅛ | 54 | 3¼ | 83 | 3⅜ | 86 | 5¾ | 146 | 1 1/16 | 27 | 2⅛ | 54 |
| 18 | | 0.1695 | 4.305 | 2⅛ | 54 | 3¼ | 83 | 3⅜ | 86 | 5¾ | 146 | 1 1/16 | 27 | 2⅛ | 54 |
| 11/64 | | 0.1719 | 4.366 | 2⅛ | 54 | 3¼ | 83 | 3⅜ | 86 | 5¾ | 146 | 1 1/16 | 27 | 2⅛ | 54 |
| 17 | | 0.1730 | 4.394 | 2 3/16 | 56 | 3⅜ | 86 | 3⅜ | 86 | 5¾ | 146 | 1⅛ | 29 | 2 3/16 | 56 |
| | 4.40 | 0.1732 | 4.400 | 2 3/16 | 56 | 3⅜ | 86 | 3⅜ | 86 | 5¾ | 146 | 1⅛ | 29 | 2 3/16 | 56 |
| 16 | | 0.1770 | 4.496 | 2 3/16 | 56 | 3⅜ | 86 | 3⅜ | 86 | 5¾ | 146 | 1⅛ | 29 | 2 3/16 | 56 |
| | 4.50 | 0.1772 | 4.500 | 2 3/16 | 56 | 3⅜ | 86 | 3⅜ | 86 | 5¾ | 146 | 1⅛ | 29 | 2 3/16 | 56 |
| 15 | | 0.1800 | 4.572 | 2 3/16 | 56 | 3⅜ | 86 | 3⅜ | 86 | 5¾ | 146 | 1⅛ | 29 | 2 3/16 | 56 |
| | 4.60 | 0.1811 | 4.600 | 2 3/16 | 56 | 3⅜ | 86 | 3⅜ | 86 | 5¾ | 146 | 1⅛ | 29 | 2 3/16 | 56 |
| 14 | | 0.1820 | 4.623 | 2 3/16 | 56 | 3⅜ | 86 | 3⅜ | 86 | 5¾ | 146 | 1⅛ | 29 | 2 3/16 | 56 |
| 13 | 4.70 | 0.1850 | 4.700 | 2 5/16 | 59 | 3½ | 89 | 3⅜ | 86 | 5¾ | 146 | 1⅛ | 29 | 2 3/16 | 56 |
| 3/16 | | 0.1875 | 4.762 | 2 5/16 | 59 | 3½ | 89 | 3⅜ | 86 | 5¾ | 146 | 1⅛ | 29 | 2 3/16 | 56 |
| 12 | 4.80 | 0.1890 | 4.800 | 2 5/16 | 59 | 3½ | 89 | 3⅝ | 92 | 6 | 152 | 1 3/16 | 30 | 2¼ | 57 |
| 11 | | 0.1910 | 4.851 | 2 5/16 | 59 | 3½ | 89 | 3⅝ | 92 | 6 | 152 | 1 3/16 | 30 | 2¼ | 57 |
| | 4.90 | 0.1929 | 4.900 | 2 7/16 | 62 | 3⅝ | 92 | 3⅝ | 92 | 6 | 152 | 1 3/16 | 30 | 2¼ | 57 |
| 10 | | 0.1935 | 4.915 | 2 7/16 | 62 | 3⅝ | 92 | 3⅝ | 92 | 6 | 152 | 1 3/16 | 30 | 2¼ | 57 |
| 9 | | 0.1960 | 4.978 | 2 7/16 | 62 | 3⅝ | 92 | 3⅝ | 92 | 6 | 152 | 1 3/16 | 30 | 2¼ | 57 |
| | 5.00 | 0.1969 | 5.000 | 2 7/16 | 62 | 3⅝ | 92 | 3⅝ | 92 | 6 | 152 | 1 3/16 | 30 | 2¼ | 57 |
| 8 | | 0.1990 | 5.054 | 2 7/16 | 62 | 3⅝ | 92 | 3⅝ | 92 | 6 | 152 | 1 3/16 | 30 | 2¼ | 57 |

**Table 1.** *(Continued)* **ANSI Straight-Shank Twist Drills — Jobbers Length through 17.5 mm, Taper Length through 12.7 mm, and Screw Machine Length through 25.4 mm Diameter** *ANSI/ASME B94.11M-1993*

| Drill Diameter, D[a] | | | | Jobbers Length | | | | Taper Length | | | | Screw Machine Length | | | |
|---|---|---|---|---|---|---|---|---|---|---|---|---|---|---|---|
| Frac-tion No. or Ltr. | mm | Equivalent | | Flute | | Overall | | Flute | | Overall | | Flute | | Overall | |
| | | Decimal Inch | mm | F | | L | | F | | L | | F | | L | |
| | | | | Inch | mm | Inch | mm | Inch | mm | Inch | mm | Inch | mm | Inch | mm |
| X | | 0.3970 | 10.084 | 3¾ | 95 | 5½ | 130 | … | … | … | … | 1¹⁵⁄₁₆ | 49 | 3⁵⁄₁₆ | 84 |
| | 10.20 | 0.4016 | 10.200 | 3⅞ | 98 | 5¼ | 133 | 4⅜ | 111 | 7 | 178 | 1¹⁵⁄₁₆ | 49 | 3⁵⁄₁₆ | 84 |
| Y | | 0.4040 | 10.262 | 3⅞ | 98 | 5¼ | 133 | … | … | … | … | 1¹⁵⁄₁₆ | 49 | 3⁵⁄₁₆ | 84 |
| ¹³⁄₃₂ | | 0.4062 | 10.317 | 3⅞ | 98 | 5¼ | 133 | 4⅜ | 111 | 7 | 178 | 1¹⁵⁄₁₆ | 49 | 3⁵⁄₁₆ | 84 |
| Z | | 0.4130 | 10.490 | 3⅞ | 98 | 5¼ | 133 | … | … | … | … | 2 | 51 | 3⅜ | 86 |
| | 10.50 | 0.4134 | 10.500 | 3⅞ | 98 | 5¼ | 133 | 4⅝ | 117 | 7¼ | 184 | 2 | 51 | 3⅜ | 86 |
| ²⁷⁄₆₄ | | 0.4219 | 10.716 | 3¹⁵⁄₁₆ | 100 | 5⅜ | 137 | 4⅝ | 117 | 7¼ | 184 | 2 | 51 | 3⅜ | 86 |
| | 10.80 | 0.4252 | 10.800 | 4¹⁄₁₆ | 103 | 5½ | 140 | 4⅝ | 117 | 7¼ | 184 | 2¹⁄₁₆ | 52 | 3⁷⁄₁₆ | 87 |
| | 11.00 | 0.4331 | 11.000 | 4¹⁄₁₆ | 103 | 5½ | 140 | 4⅝ | 117 | 7¼ | 184 | 2¹⁄₁₆ | 52 | 3⁷⁄₁₆ | 87 |
| ⁷⁄₁₆ | | 0.4375 | 11.112 | 4¹⁄₁₆ | 103 | 5½ | 140 | 4⅝ | 117 | 7¼ | 184 | 2¹⁄₁₆ | 52 | 3⁷⁄₁₆ | 87 |
| | 11.20 | 0.4409 | 11.200 | 4³⁄₁₆ | 106 | 5⅝ | 143 | 4¾ | 121 | 7½ | 190 | 2⅛ | 54 | 3⁹⁄₁₆ | 90 |
| | 11.50 | 0.4528 | 11.500 | 4³⁄₁₆ | 106 | 5⅝ | 143 | 4¾ | 121 | 7½ | 190 | 2⅛ | 54 | 3⁹⁄₁₆ | 90 |
| ²⁹⁄₆₄ | | 0.4531 | 11.509 | 4³⁄₁₆ | 106 | 5⅝ | 143 | 4¾ | 121 | 7½ | 190 | 2⅛ | 54 | 3⁹⁄₁₆ | 90 |
| | 11.80 | 0.4646 | 11.800 | 4⁵⁄₁₆ | 110 | 5¾ | 146 | 4¾ | 121 | 7½ | 190 | 2⅛ | 54 | 3⅝ | 92 |
| ¹⁵⁄₃₂ | | 0.4688 | 11.908 | 4⁵⁄₁₆ | 110 | 5¾ | 146 | 4¾ | 121 | 7½ | 190 | 2⅛ | 54 | 3⅝ | 92 |
| | 12.00 | 0.4724 | 12.000 | 4⅜ | 111 | 5⅞ | 149 | 4¾ | 121 | 7¾ | 197 | 2³⁄₁₆ | 56 | 3¹¹⁄₁₆ | 94 |
| | 12.20 | 0.4803 | 12.200 | 4⅜ | 111 | 5⅞ | 149 | 4¾ | 121 | 7¾ | 197 | 2³⁄₁₆ | 56 | 3¹¹⁄₁₆ | 94 |
| ³¹⁄₆₄ | | 0.4844 | 12.304 | 4⅜ | 111 | 5⅞ | 149 | 4¾ | 121 | 7¾ | 197 | 2³⁄₁₆ | 56 | 3¹¹⁄₁₆ | 94 |
| | 12.50 | 0.4921 | 12.500 | 4½ | 114 | 6 | 152 | 4¾ | 121 | 7¾ | 197 | 2¼ | 57 | 3¾ | 95 |
| ½ | | 0.5000 | 12.700 | 4½ | 114 | 6 | 152 | 4¾ | 121 | 7¾ | 197 | 2¼ | 57 | 3¾ | 95 |
| | 12.80 | 0.5039 | 12.800 | 4½ | 114 | 6 | 152 | … | … | … | … | 2⅜ | 60 | 3⅞ | 98 |
| | 13.00 | 0.5118 | 13.000 | 4½ | 114 | 6 | 152 | … | … | … | … | 2⅜ | 60 | 3⅞ | 98 |
| ³³⁄₆₄ | | 0.5156 | 13.096 | 4¹³⁄₁₆ | 122 | 6⅝ | 168 | … | … | … | … | 2⅜ | 60 | 3⅞ | 98 |
| | 13.20 | 0.5197 | 13.200 | 4¹³⁄₁₆ | 122 | 6⅝ | 168 | … | … | … | … | 2⅜ | 60 | 3⅞ | 98 |
| ¹⁷⁄₃₂ | | 0.5312 | 13.492 | 4¹³⁄₁₆ | 122 | 6⅝ | 168 | … | … | … | … | 2⅜ | 60 | 3⅞ | 98 |
| | 13.50 | 0.5315 | 13.500 | 4¹³⁄₁₆ | 122 | 6⅝ | 168 | … | … | … | … | 2⅜ | 60 | 3⅞ | 98 |
| | 13.80 | 0.5433 | 13.800 | 4¹³⁄₁₆ | 122 | 6⅝ | 168 | … | … | … | … | 2½ | 64 | 4 | 102 |
| ³⁵⁄₆₄ | | 0.5469 | 13.891 | 4¹³⁄₁₆ | 122 | 6⅝ | 168 | … | … | … | … | 2½ | 64 | 4 | 102 |
| | 14.00 | 0.5512 | 14.000 | 4¹³⁄₁₆ | 122 | 6⅝ | 168 | … | … | … | … | 2½ | 64 | 4 | 102 |
| | 14.25 | 0.5610 | 14.250 | 4¹³⁄₁₆ | 122 | 6⅝ | 168 | … | … | … | … | 2½ | 64 | 4 | 102 |
| ⁹⁄₁₆ | | 0.5625 | 14.288 | 4¹³⁄₁₆ | 122 | 6⅝ | 168 | … | … | … | … | 2½ | 64 | 4 | 102 |
| | 14.50 | 0.5709 | 14.500 | 4¹³⁄₁₆ | 122 | 6⅝ | 168 | … | … | … | … | 2⅝ | 67 | 4⅛ | 105 |
| ³⁷⁄₆₄ | | 0.5781 | 14.684 | 4¹³⁄₁₆ | 122 | 6⅝ | 168 | … | … | … | … | 2⅝ | 67 | 4⅛ | 105 |
| | 14.75 | 0.5807 | 14.750 | 5³⁄₁₆ | 132 | 7⅛ | 181 | … | … | … | … | 2⅝ | 67 | 4⅛ | 105 |
| | 15.00 | 0.5906 | 15.000 | 5³⁄₁₆ | 132 | 7⅛ | 181 | … | … | … | … | 2⅝ | 67 | 4⅛ | 105 |
| ¹⁹⁄₃₂ | | 0.5938 | 15.083 | 5³⁄₁₆ | 132 | 7⅛ | 181 | … | … | … | … | 2⅝ | 67 | 4⅛ | 105 |
| | 15.25 | 0.6004 | 15.250 | 5³⁄₁₆ | 132 | 7⅛ | 181 | … | … | … | … | 2¾ | 70 | 4¼ | 108 |
| ³⁹⁄₆₄ | | 0.6094 | 15.479 | 5³⁄₁₆ | 132 | 7⅛ | 181 | … | … | … | … | 2¾ | 70 | 4¼ | 108 |
| | 15.50 | 0.6102 | 15.500 | 5³⁄₁₆ | 132 | 7⅛ | 181 | … | … | … | … | 2¾ | 70 | 4¼ | 108 |
| | 15.75 | 0.6201 | 15.750 | 5³⁄₁₆ | 132 | 7⅛ | 181 | … | … | … | … | 2¾ | 70 | 4¼ | 108 |
| ⅝ | | 0.6250 | 15.875 | 5³⁄₁₆ | 132 | 7⅛ | 181 | … | … | … | … | 2¾ | 70 | 4¼ | 108 |
| | 16.00 | 0.6299 | 16.000 | 5³⁄₁₆ | 132 | 7⅛ | 181 | … | … | … | … | 2⅞ | 73 | 4½ | 114 |
| | 16.25 | 0.6398 | 16.250 | 5³⁄₁₆ | 132 | 7⅛ | 181 | … | … | … | … | 2⅞ | 73 | 4½ | 114 |
| ⁴¹⁄₆₄ | | 0.6406 | 16.271 | 5³⁄₁₆ | 132 | 7⅛ | 181 | … | … | … | … | 2⅞ | 73 | 4½ | 114 |
| | 16.50 | 0.6496 | 16.500 | 5³⁄₁₆ | 132 | 7⅛ | 181 | … | … | … | … | 2⅞ | 73 | 4½ | 114 |
| ²¹⁄₃₂ | | 0.6562 | 16.669 | 5³⁄₁₆ | 132 | 7⅛ | 181 | … | … | … | … | 2⅞ | 73 | 4½ | 114 |

**Table 2.** *(Continued)* **ANSI Straight-Shank Twist Drills — Taper Length — Over ½ Inch (12.7 mm) Diameter, Fractional and Metric Sizes** *ANSI/ASME B94.11M-1993*

*B94.11M-1993*

| Diameter of Drill | | | | Flute Length | | Overall Length | | Length of Body | | Minimum Length of Shk. | | Maximum Length of Neck | |
|---|---|---|---|---|---|---|---|---|---|---|---|---|---|
| *D* | | Decimal Inch Equiv. | Millimeter Equiv. | *F* | | *L* | | *B* | | *S* | | *N* | |
| Frac. | mm | | | Inch | mm | Inch | mm | Inch | mm | Inch | mm | Inch | mm |
| | 20.00 | 0.7874 | 20.000 | 6⅛ | 156 | 10 | 254 | 6¼ | 159 | 3⅛ | 79 | ⅝ | 16 |
| 51⁄64 | | 0.7969 | 20.241 | 6⅛ | 156 | 10 | 254 | 6¼ | 159 | 3⅛ | 79 | ⅝ | 16 |
| | 20.50 | 0.8071 | 20.500 | 6⅛ | 156 | 10 | 254 | 6¼ | 159 | 3⅛ | 79 | ⅝ | 16 |
| 13⁄16 | | 0.8125 | 20.638 | 6⅛ | 156 | 10 | 254 | 6¼ | 159 | 3⅛ | 79 | ⅝ | 16 |
| | 21.00 | 0.8268 | 21.000 | 6⅛ | 156 | 10 | 254 | 6¼ | 159 | 3⅛ | 79 | ⅝ | 16 |
| 53⁄64 | | 0.8281 | 21.034 | 6⅛ | 156 | 10 | 254 | 6¼ | 159 | 3⅛ | 79 | ⅝ | 16 |
| 27⁄32 | | 0.8438 | 21.433 | 6⅛ | 156 | 10 | 254 | 6¼ | 159 | 3⅛ | 79 | ⅝ | 16 |
| | 21.50 | 0.8465 | 21.500 | 6⅛ | 156 | 10 | 254 | 6¼ | 159 | 3⅛ | 79 | ⅝ | 16 |
| 55⁄64 | | 0.8594 | 21.829 | 6⅛ | 156 | 10 | 254 | 6¼ | 159 | 3⅛ | 79 | ⅝ | 16 |
| | 22.00 | 0.8661 | 22.000 | 6⅛ | 156 | 10 | 254 | 6¼ | 159 | 3⅛ | 79 | ⅝ | 16 |
| ⅞ | | 0.8750 | 22.225 | 6⅛ | 156 | 10 | 254 | 6¼ | 159 | 3⅛ | 79 | ⅝ | 16 |
| | 22.50 | 0.8858 | 22.500 | 6⅛ | 156 | 10 | 254 | 6¼ | 159 | 3⅛ | 79 | ⅝ | 16 |
| 57⁄64 | | 0.8906 | 22.621 | 6⅛ | 156 | 10 | 254 | 6¼ | 159 | 3⅛ | 79 | ⅝ | 16 |
| | 23.00 | 0.9055 | 23.000 | 6⅛ | 156 | 10 | 254 | 6¼ | 159 | 3⅛ | 79 | ⅝ | 16 |
| 29⁄32 | | 0.9062 | 23.017 | 6⅛ | 156 | 10 | 254 | 6¼ | 159 | 3⅛ | 79 | ⅝ | 16 |
| 59⁄64 | | 0.9219 | 23.416 | 6⅛ | 156 | 10¾ | 273 | 6¼ | 159 | 3⅞ | 98 | ⅝ | 16 |
| | 23.50 | 0.9252 | 23.500 | 6⅛ | 156 | 10¾ | 273 | 6¼ | 159 | 3⅞ | 98 | ⅝ | 16 |
| 15⁄16 | | 0.9375 | 23.812 | 6⅛ | 156 | 10¾ | 273 | 6¼ | 159 | 3⅞ | 98 | ⅝ | 16 |
| | 24.00 | 0.9449 | 24.000 | 6⅜ | 162 | 11 | 279 | 6½ | 165 | 3⅞ | 98 | ⅝ | 16 |
| 61⁄64 | | 0.9531 | 24.209 | 6⅜ | 162 | 11 | 279 | 6½ | 165 | 3⅞ | 98 | ⅝ | 16 |
| | 24.50 | 0.9646 | 24.500 | 6⅜ | 162 | 11 | 279 | 6½ | 165 | 3⅞ | 98 | ⅝ | 16 |
| 31⁄32 | | 0.9688 | 24.608 | 6⅜ | 162 | 11 | 279 | 6½ | 165 | 3⅞ | 98 | ⅝ | 16 |
| | 25.00 | 0.9843 | 25.000 | 6⅜ | 162 | 11 | 279 | 6½ | 165 | 3⅞ | 98 | ⅝ | 16 |
| 63⁄64 | | 0.9844 | 25.004 | 6⅜ | 162 | 11 | 279 | 6½ | 165 | 3⅞ | 98 | ⅝ | 16 |
| 1 | | 1.0000 | 25.400 | 6⅜ | 162 | 11 | 279 | 6½ | 165 | 3⅞ | 98 | ⅝ | 16 |
| | 25.50 | 1.0039 | 25.500 | 6½ | 165 | 11⅛ | 282 | 6⅝ | 168 | 3⅞ | 98 | ⅝ | 16 |
| 1 1⁄64 | | 1.0156 | 25.796 | 6½ | 165 | 11⅛ | 282 | 6⅝ | 168 | 3⅞ | 98 | ⅝ | 16 |
| | 26.00 | 1.0236 | 26.000 | 6½ | 165 | 11⅛ | 282 | 6⅝ | 168 | 3⅞ | 98 | ⅝ | 16 |
| 1 1⁄32 | | 1.0312 | 26.192 | 6½ | 165 | 11⅛ | 282 | 6⅝ | 168 | 3⅞ | 98 | ⅝ | 16 |
| | 26.50 | 1.0433 | 26.560 | 6⅝ | 168 | 11¼ | 286 | 6¾ | 172 | 3⅞ | 98 | ⅝ | 16 |
| 1 3⁄64 | | 1.0469 | 26.591 | 6⅝ | 168 | 11¼ | 286 | 6¾ | 172 | 3⅞ | 98 | ⅝ | 16 |
| 1 1⁄16 | | 1.0625 | 26.988 | 6⅝ | 168 | 11¼ | 286 | 6¾ | 172 | 3⅞ | 98 | ⅝ | 16 |
| | 27.00 | 1.0630 | 27.000 | 6⅝ | 168 | 11¼ | 286 | 6¾ | 172 | 3⅞ | 98 | ⅝ | 16 |
| 1 5⁄64 | | 1.0781 | 27.384 | 6⅞ | 175 | 11½ | 292 | 7 | 178 | 3⅞ | 98 | ⅝ | 16 |
| | 27.50 | 1.0827 | 27.500 | 6⅞ | 175 | 11½ | 292 | 7 | 178 | 3⅞ | 98 | ⅝ | 16 |
| 1 3⁄32 | | 1.0938 | 27.783 | 6⅞ | 175 | 11½ | 292 | 7 | 178 | 3⅞ | 98 | ⅝ | 16 |
| | 28.00 | 1.1024 | 28.000 | 7⅛ | 181 | 11¾ | 298 | 7¼ | 184 | 3⅞ | 98 | ⅝ | 16 |
| 1 7⁄64 | | 1.1094 | 28.179 | 7⅛ | 181 | 11¾ | 298 | 7¼ | 184 | 3⅞ | 98 | ⅝ | 16 |
| | 28.50 | 1.1220 | 28.500 | 7⅛ | 181 | 11¾ | 298 | 7¼ | 184 | 3⅞ | 98 | ⅝ | 16 |
| 1⅛ | | 1.1250 | 28.575 | 7⅛ | 181 | 11¾ | 298 | 7¼ | 184 | 3⅞ | 98 | ⅝ | 16 |
| 1 9⁄64 | | 1.1406 | 28.971 | 7¼ | 184 | 11⅞ | 301 | 7⅜ | 187 | 3⅞ | 98 | ⅝ | 16 |
| | 29.00 | 1.1417 | 29.000 | 7¼ | 184 | 11⅞ | 301 | 7⅜ | 187 | 3⅞ | 98 | ⅝ | 16 |
| 1 5⁄32 | | 1.1562 | 29.367 | 7¼ | 184 | 11⅞ | 301 | 7⅜ | 187 | 3⅞ | 98 | ⅝ | 16 |
| | 29.50 | 1.1614 | 29.500 | 7⅜ | 187 | 12 | 305 | 7½ | 191 | 3⅞ | 98 | ⅝ | 16 |
| 1 11⁄64 | | 1.1719 | 29.766 | 7⅜ | 187 | 12 | 305 | 7½ | 191 | 3⅞ | 98 | ⅝ | 16 |
| | 30.00 | 1.1811 | 30.000 | 7⅜ | 187 | 12 | 305 | 7½ | 191 | 3⅞ | 98 | ⅝ | 16 |
| 1 3⁄16 | | 1.1875 | 30.162 | 7⅜ | 187 | 12 | 305 | 7½ | 191 | 3⅞ | 98 | ⅝ | 16 |
| | 30.50 | 1.2008 | 30.500 | 7½ | 190 | 12⅛ | 308 | 7⅝ | 194 | 3⅞ | 98 | ⅝ | 16 |
| 1 13⁄64 | | 1.2031 | 30.559 | 7½ | 190 | 12⅛ | 308 | 7⅝ | 194 | 3⅞ | 98 | ⅝ | 16 |
| 1 7⁄32 | | 1.2188 | 30.958 | 7½ | 190 | 12⅛ | 308 | 7⅝ | 194 | 3⅞ | 98 | ⅝ | 16 |
| | 31.00 | 1.2205 | 31.000 | 7⅞ | 200 | 12½ | 317 | 8 | 203 | 3⅞ | 98 | ⅝ | 16 |
| 1 15⁄64 | | 1.2344 | 31.354 | 7⅞ | 200 | 12½ | 317 | 8 | 203 | 3⅞ | 98 | ⅝ | 16 |
| | 31.50 | 1.2402 | 31.500 | 7⅞ | 200 | 12½ | 317 | 8 | 203 | 3⅞ | 98 | ⅝ | 16 |

**Table 2.** *(Continued)* **ANSI Straight-Shank Twist Drills — Taper Length — Over 1/2 Inch (12.7 mm) Diameter, Fractional and Metric Sizes** *ANSI/ASME B94.11M-1993*

| Diameter of Drill | | | | Flute Length | | Overall Length | | Length of Body | | Minimum Length of Shk. | | Maximum Length of Neck | |
|---|---|---|---|---|---|---|---|---|---|---|---|---|---|
| *D* | | Decimal Inch Equiv. | Millimeter Equiv. | *F* | | *L* | | *B* | | *S* | | *N* | |
| Frac. | mm | | | Inch | mm | Inch | mm | Inch | mm | Inch | mm | Inch | mm |
| 1 1/4 | | 1.2500 | 31.750 | 7 7/8 | 200 | 12 1/2 | 317 | 8 | 203 | 3 7/8 | 98 | 5/8 | 16 |
| | 32.00 | 1.2598 | 32.000 | 8 1/2 | 216 | 14 1/8 | 359 | 8 5/8 | 219 | 4 7/8 | 124 | 5/8 | 16 |
| | 32.50 | 1.2795 | 32.500 | 8 1/2 | 216 | 14 1/8 | 359 | 8 5/8 | 219 | 4 7/8 | 124 | 5/8 | 16 |
| 1 9/32 | | 1.2812 | 32.542 | 8 1/2 | 216 | 14 1/8 | 359 | 8 5/8 | 219 | 4 7/8 | 124 | 5/8 | 16 |
| | 33.00 | 1.2992 | 33.000 | 8 5/8 | 219 | 14 1/4 | 362 | 8 3/4 | 222 | 4 7/8 | 124 | 5/8 | 16 |
| 1 5/16 | | 1.3125 | 33.338 | 8 5/8 | 219 | 14 1/4 | 362 | 8 3/4 | 222 | 4 7/8 | 124 | 5/8 | 16 |
| | 33.50 | 1.3189 | 33.500 | 8 3/4 | 222 | 14 3/8 | 365 | 8 7/8 | 225 | 4 7/8 | 124 | 5/8 | 16 |
| | 34.00 | 1.3386 | 34.000 | 8 3/4 | 222 | 14 3/8 | 365 | 8 7/8 | 225 | 4 7/8 | 124 | 5/8 | 16 |
| 1 11/32 | | 1.3438 | 34.133 | 8 3/4 | 222 | 14 3/8 | 365 | 8 7/8 | 225 | 4 7/8 | 124 | 5/8 | 16 |
| | 34.50 | 1.3583 | 34.500 | 8 7/8 | 225 | 14 1/2 | 368 | 9 | 229 | 4 7/8 | 124 | 5/8 | 16 |
| 1 3/8 | | 1.3750 | 34.925 | 8 7/8 | 225 | 14 1/2 | 368 | 9 | 229 | 4 7/8 | 124 | 5/8 | 16 |
| | 35.00 | 1.3780 | 35.000 | 9 | 229 | 14 5/8 | 372 | 9 1/8 | 232 | 4 7/8 | 124 | 5/8 | 16 |
| | 35.50 | 1.3976 | 35.500 | 9 | 229 | 14 5/8 | 372 | 9 1/8 | 232 | 4 7/8 | 124 | 5/8 | 16 |
| 1 13/32 | | 1.4062 | 35.717 | 9 | 229 | 14 5/8 | 372 | 9 1/8 | 232 | 4 7/8 | 124 | 5/8 | 16 |
| | 36.00 | 1.4173 | 36.000 | 9 1/8 | 232 | 14 3/4 | 375 | 9 1/4 | 235 | 4 7/8 | 124 | 5/8 | 16 |
| | 36.50 | 1.4370 | 36.500 | 9 1/8 | 232 | 14 3/4 | 375 | 9 1/4 | 235 | 4 7/8 | 124 | 5/8 | 16 |
| 1 7/16 | | 1.4375 | 36.512 | 9 1/8 | 232 | 14 3/4 | 375 | 9 1/4 | 235 | 4 7/8 | 124 | 5/8 | 16 |
| | 37.00 | 1.4567 | 37.000 | 9 1/4 | 235 | 14 7/8 | 378 | 9 3/8 | 238 | 4 7/8 | 124 | 5/8 | 16 |
| 1 15/32 | | 1.4688 | 37.308 | 9 1/4 | 235 | 14 7/8 | 378 | 9 3/8 | 238 | 4 7/8 | 124 | 5/8 | 16 |
| | 37.50 | 1.4764 | 37.500 | 9 3/8 | 238 | 15 | 381 | 9 1/2 | 241 | 4 7/8 | 124 | 5/8 | 16 |
| | 38.00 | 1.4961 | 38.000 | 9 3/8 | 238 | 15 | 381 | 9 1/2 | 241 | 4 7/8 | 124 | 5/8 | 16 |
| 1 1/2 | | 1.5000 | 38.100 | 9 3/8 | 238 | 15 | 381 | 9 1/2 | 241 | 4 7/8 | 124 | 5/8 | 16 |
| 1 9/16 | | 1.5625 | 39.688 | 9 5/8 | 244 | 15 1/4 | 387 | 9 3/4 | 247 | 4 7/8 | 124 | 5/8 | 16 |
| 1 5/8 | | 1.6250 | 41.275 | 9 7/8 | 251 | 15 5/8 | 397 | 10 | 254 | 4 7/8 | 124 | 3/4 | 19 |
| 1 3/4 | | 1.7500 | 44.450 | 10 1/2 | 267 | 16 1/4 | 413 | 10 5/8 | 270 | 4 7/8 | 124 | 3/4 | 19 |

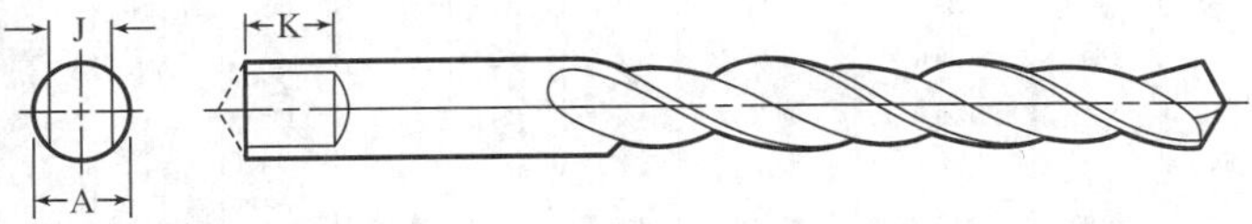


**Table 3. American National Standard Tangs for Straight-Shank Drills** *ANSI/ASME B94.11M-1993*

| Nominal Diameter of Drill Shank, *A* | | Thickness of Tang, *J* | | | | Length of Tang, *K* | |
|---|---|---|---|---|---|---|---|
| | | Inches | | Millimeters | | | |
| Inches | Millimeters | Max. | Min. | Max. | Min. | Inches | Millimeters |
| 1/8 thru 3/16 | 3.18 thru 4.76 | 0.094 | 0.090 | 2.39 | 2.29 | 9/32 | 7.0 |
| over 3/16 thru 1/4 | over 4.76 thru 6.35 | 0.122 | 0.118 | 3.10 | 3.00 | 5/16 | 8.[illegible] |
| over 1/4 thru 5/16 | over 6.35 thru 7.94 | 0.162 | 0.158 | 4.11 | 4.01 | 11/32 | 8.[illegible] |
| over 5/16 thru 3/8 | over 7.94 thru 9.53 | 0.203 | 0.199 | 5.16 | 5.06 | 3/8 | 9.[illegible] |
| over 3/8 thru 15/32 | over 9.53 thru 11.91 | 0.243 | 0.239 | 6.17 | 6.07 | 7/16 | 11.0 |
| over 15/32 thru 9/16 | over 11.91 thru 14.29 | 0.303 | 0.297 | 7.70 | 7.55 | 1/2 | 12.5 |
| over 9/16 thru 21/32 | over 14.29 thru 16.67 | 0.373 | 0.367 | 9.47 | 9.32 | 9/16 | 14.5 |
| over 21/32 thru 3/4 | over 16.67 thru 19.05 | 0.443 | 0.437 | 11.25 | 11.10 | 5/8 | 16.0 |
| over 3/4 thru 7/8 | over 19.05 thru 22.23 | 0.514 | 0.508 | 13.05 | 12.90 | 11/16 | 17.5 |
| over 7/8 thru 1 | over 22.23 thru 25.40 | 0.609 | 0.601 | 15.47 | 15.27 | 3/4 | 19.0 |
| over 1 thru 1 3/16 | over 25.40 thru 30.16 | 0.700 | 0.692 | 17.78 | 17.58 | 13/16 | 20.5 |
| over 1 3/16 thru 1 3/8 | over 30.16 thru 34.93 | 0.817 | 0.809 | 20.75 | 20.55 | 7/8 | 22.0 |

To fit split-sleeve collet-type drill drivers. See page 943.

**Table 9. ANSI Three- and Four-Flute Straight-Shank Core Drills — Fractional Sizes Only** *ANSI/ASME B94.11M-1993*

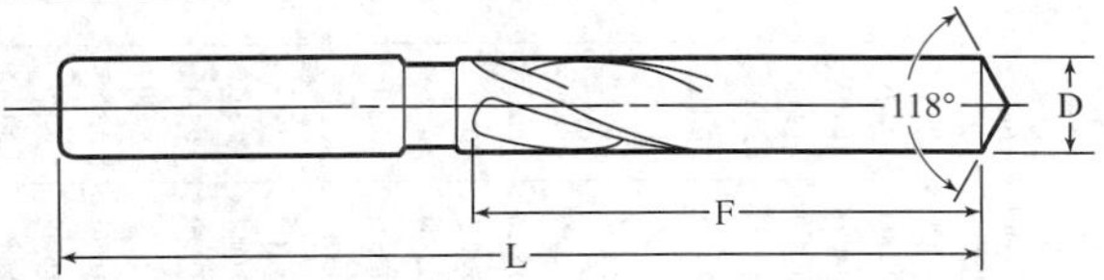


Drill Diameter 11/32″ (8.733 mm) and Smaller

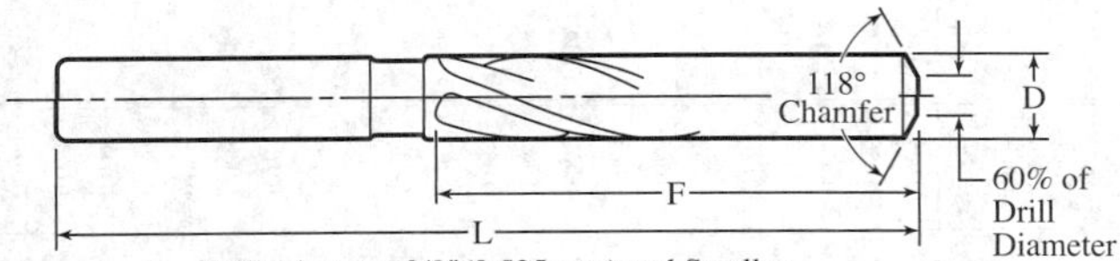


Drill Diameter 3/8″ (9.525 mm) and Smaller

Nominal Shank Size is Same Size as Nominal Drill Size

| Drill Diameter, *D* | | | Three-Flute Drills | | | | Four-Flute Drills | | | |
|---|---|---|---|---|---|---|---|---|---|---|
| | Equivalent | | Flute Length | | Overall Length | | Flute Length | | Overall Length | |
| | | | *F* | | *L* | | *F* | | *L* | |
| Inch | Decimal Inch | mm | Inch | mm | Inch | mm | Inch | mm | Inch | mm |
| 1/4 | 0.2500 | 6.350 | 3 3/4 | 95 | 6 1/8 | 156 | … | … | … | … |
| 9/32 | 0.2812 | 7.142 | 3 7/8 | 98 | 6 1/4 | 159 | … | … | … | … |
| 5/16 | 0.3125 | 7.938 | 4 | 102 | 6 3/8 | 162 | … | … | … | … |
| 11/32 | 0.3438 | 8.733 | 4 1/8 | 105 | 6 1/2 | 165 | … | … | … | … |
| 3/8 | 0.3750 | 9.525 | 4 1/8 | 105 | 6 3/4 | 171 | … | … | … | … |
| 13/32 | 0.4062 | 10.317 | 4 3/8 | 111 | 7 | 178 | … | … | … | … |
| 7/16 | 0.4375 | 11.112 | 4 5/8 | 117 | 7 1/4 | 184 | … | … | … | … |
| 15/32 | 0.4688 | 11.908 | 4 3/4 | 121 | 7 1/2 | 190 | … | … | … | … |
| 1/2 | 0.5000 | 12.700 | 4 3/4 | 121 | 7 3/4 | 197 | 4 3/4 | 121 | 7 3/4 | 197 |
| 17/32 | 0.5312 | 13.492 | 4 3/4 | 121 | 8 | 203 | 4 3/4 | 121 | 8 | 203 |
| 9/16 | 0.5625 | 14.288 | 4 7/8 | 124 | 8 1/4 | 210 | 4 7/8 | 124 | 8 1/4 | 210 |
| 19/32 | 0.5938 | 15.083 | 4 7/8 | 124 | 8 3/4 | 222 | 4 7/8 | 124 | 8 3/4 | 222 |
| 5/8 | 0.6250 | 15.875 | 4 7/8 | 124 | 8 3/4 | 222 | 4 7/8 | 124 | 8 3/4 | 222 |
| 21/32 | 0.6562 | 16.667 | 5 1/8 | 130 | 9 | 229 | 5 1/8 | 130 | 9 | 229 |
| 11/16 | 0.6875 | 17.462 | 5 3/8 | 137 | 9 1/4 | 235 | 5 3/8 | 137 | 9 1/4 | 235 |
| 23/32 | 0.7188 | 18.258 | … | … | … | … | 5 5/8 | 143 | 9 1/2 | 241 |
| 3/4 | 0.7500 | 19.050 | 5 7/8 | 149 | 9 3/4 | 248 | 5 7/8 | 149 | 9 3/4 | 248 |
| 25/32 | 0.7812 | 19.842 | … | … | … | … | 6 | 152 | 9 7/8 | 251 |
| 13/16 | 0.8125 | 20.638 | … | … | … | … | 6 1/8 | 156 | 10 | 25[illegible] |
| 27/32 | 0.8438 | 21.433 | … | … | … | … | 6 1/8 | 156 | 10 | 25[illegible] |
| 7/8 | 0.8750 | 22.225 | … | … | … | … | 6 1/8 | 156 | 10 | 25[illegible] |
| 29/32 | 0.9062 | 23.017 | … | … | … | … | 6 1/8 | 156 | 10 | 25[illegible] |
| 15/16 | 0.9375 | 23.812 | … | … | … | … | 6 1/8 | 156 | 10 3/4 | 27[illegible] |
| 31/32 | 0.9688 | 24.608 | … | … | … | … | 6 3/8 | 162 | 11 | 27[illegible] |
| 1 | 1.0000 | 25.400 | … | … | … | … | 6 3/8 | 162 | 11 | 27[illegible] |
| 1 1/32 | 1.0312 | 26.192 | … | … | … | … | 6 1/2 | 165 | 11 1/8 | 283 |
| 1 1/16 | 1.0625 | 26.988 | … | … | … | … | 6 5/8 | 168 | 11 1/4 | 285 |
| 1 3/32 | 1.0938 | 27.783 | … | … | … | … | 6 7/8 | 175 | 11 1/2 | 292 |
| 1 1/8 | 1.1250 | 28.575 | … | … | … | … | 7 1/8 | 181 | 11 3/4 | 29[illegible] |
| 1 1/4 | 1.2500 | 31.750 | … | … | … | … | 7 7/8 | 200 | 12 1/2 | 3[illegible] |

**Table 15. British Standard Parallel Shank Jobber Series Twist Drills — Standard Metric Sizes** *BS 328: Part 1:1993*

| Diameter | Flute Length | Overall Length | Diameter | Flute Length | Overall Length | Diameter | Flute Length | Overall Length | Diameter | Flute Length | Overall Length |
|---|---|---|---|---|---|---|---|---|---|---|---|
| 0.20 | 2.5 | 19 | 1.75 | 22 | 46 | 5.40 | 57 | 93 | 10.20 | 87 | 133 |
| 0.22 | | | 1.80 | | | 5.50 | | | 10.30 | | |
| 0.25 | 3.0 | 19 | 1.85 | | | 5.60 | | | 10.40 | | |
| 0.28 | | | 1.90 | | | 5.70 | | | 10.50 | | |
| 0.30 | 4 | 19 | 1.95 | 24 | 49 | 5.80 | | | 10.60 | | |
| 0.32 | | | 2.00 | | | 5.90 | | | 10.70 | 94 | 142 |
| 0.35 | | | 2.05 | | | 6.00 | | | 10.80 | | |
| 0.38 | | | 2.10 | | | 6.10 | 63 | 101 | 10.90 | | |
| 0.40 | 5 | 20 | 2.15 | 27 | 53 | 6.20 | | | 11.00 | | |
| 0.42 | | | 2.20 | | | 6.30 | | | 11.10 | | |
| 0.45 | | | 2.25 | | | 6.40 | | | 11.20 | | |
| 0.48 | | | 2.30 | | | 6.50 | | | 11.30 | | |
| 0.50 | 6 | 22 | 2.35 | | | 6.60 | | | 11.40 | | |
| 0.52 | | | 2.40 | 30 | 57 | 6.70 | | | 11.50 | | |
| 0.55 | 7 | 24 | 2.45 | | | 6.80 | 69 | 109 | 11.60 | | |
| 0.58 | | | 2.50 | | | 6.90 | | | 11.70 | | |
| 0.60 | | | 2.55 | | | 7.00 | | | 11.80 | | |
| 0.62 | 8 | 26 | 2.60 | | | 7.10 | | | 11.90 | 101 | 151 |
| 0.65 | | | 2.65 | | | 7.20 | | | 12.00 | | |
| 0.68 | 9 | 28 | 2.70 | 33 | 61 | 7.30 | | | 12.10 | | |
| 0.70 | | | 2.75 | | | 7.40 | | | 12.20 | | |
| 0.72 | | | 2.80 | | | 7.50 | | | 12.30 | | |
| 0.75 | | | 2.85 | | | 7.60 | 75 | 117 | 12.40 | | |
| 0.78 | 10 | 30 | 2.90 | | | 7.70 | | | 12.50 | | |
| 0.80 | | | 2.95 | | | 7.80 | | | 12.60 | | |
| 0.82 | | | 3.00 | | | 7.90 | | | 12.70 | | |
| 0.85 | | | 3.10 | 36 | 65 | 8.00 | | | 12.80 | | |
| 0.88 | 11 | 32 | 3.20 | | | 8.10 | | | 12.90 | | |
| 0.90 | | | 3.30 | | | 8.20 | | | 13.00 | | |
| 0.92 | | | 3.40 | 39 | 70 | 8.30 | | | 13.10 | | |
| 0.95 | | | 3.50 | | | 8.40 | | | 13.20 | | |
| 0.98 | 12 | 34 | 3.60 | | | 8.50 | | | 13.30 | 108 | 160 |
| 1.00 | | | 3.70 | | | 8.60 | 81 | 125 | 13.40 | | |
| 1.05 | | | 3.80 | 43 | 75 | 8.70 | | | 13.50 | | |
| 1.10 | 14 | 36 | 3.90 | | | 8.80 | | | 13.60 | | |
| 1.15 | | | 4.00 | | | 8.90 | | | 13.70 | | |
| 1.20 | 16 | 38 | 4.10 | | | 9.00 | | | 13.80 | | |
| 1.25 | | | 4.20 | | | 9.10 | | | 13.90 | | |
| 1.30 | | | 4.30 | 47 | 80 | 9.20 | | | 14.00 | | |
| 1.35 | 18 | 40 | 4.40 | | | 9.30 | | | 14.25 | 114 | 169 |
| 1.40 | | | 4.50 | | | 9.40 | | | 14.50 | | |
| 1.45 | | | 4.60 | | | 9.50 | | | 14.75 | | |
| 1.50 | | | 4.70 | | | 9.60 | 87 | 133 | 15.00 | | |
| 1.55 | 20 | 43 | 4.80 | 52 | 86 | 9.70 | | | 15.25 | 120 | 178 |
| 1.60 | | | 4.90 | | | 9.80 | | | 15.50 | | |
| 1.65 | | | 5.00 | | | 9.90 | | | 15.75 | | |
| 1.70 | | | 5.10 | | | 10.00 | | | 16.00 | | |
| | | | 5.20 | | | 10.10 | | | | | |
| | | | 5.30 | | | | | | | | |

All dimensions are in millimeters. Tolerances on diameters are given in Table 14.

### Table 16. British Standard Parallel Shank Long Series Twist Drills — Standard Metric Sizes *BS 328: Part 1:1993*

| Diameter | Flute Length | Overall Length |
|---|---|---|
| 2.00 | 56 | 85 |
| 2.05 | | |
| 2.10 | | |
| 2.15 | 59 | 90 |
| 2.20 | | |
| 2.25 | | |
| 2.30 | | |
| 2.35 | | |
| 2.40 | 62 | 95 |
| 2.45 | | |
| 2.50 | | |
| 2.55 | | |
| 2.60 | | |
| 2.65 | | |
| 2.70 | 66 | 100 |
| 2.75 | | |
| 2.80 | | |
| 2.85 | | |
| 2.90 | | |
| 2.95 | | |
| 3.00 | | |
| 3.10 | 69 | 106 |
| 3.20 | | |
| 3.30 | | |
| 3.40 | 73 | 112 |
| 3.50 | | |
| 3.60 | | |
| 3.70 | | |
| 3.80 | 78 | 119 |
| 3.90 | | |
| 4.00 | | |
| 4.10 | | |
| 4.20 | | |
| 4.30 | 82 | 126 |
| 4.40 | | |
| 4.50 | | |
| 4.60 | | |
| 4.70 | | |
| 4.80 | 87 | 132 |
| 4.90 | | |
| 5.00 | | |
| 5.10 | | |
| 5.20 | | |
| 5.30 | | |
| 5.40 | 91 | 139 |
| 5.50 | | |
| 5.60 | | |
| 5.70 | | |
| 5.80 | | |
| 5.90 | | |
| 6.00 | | |
| 6.10 | 97 | 148 |
| 6.20 | | |
| 6.30 | | |
| 6.40 | | |
| 6.50 | | |
| 6.60 | | |
| 6.70 | | |
| 6.80 | 102 | 156 |
| 6.90 | | |
| 7.00 | | |
| 7.10 | | |
| 7.20 | | |
| 7.30 | | |
| 7.40 | | |
| 7.50 | | |
| 7.60 | 109 | 165 |
| 7.70 | | |
| 7.80 | | |
| 7.90 | | |
| 8.00 | | |
| 8.10 | | |
| 8.20 | | |
| 8.30 | | |
| 8.40 | | |
| 8.50 | | |
| 8.60 | 115 | 175 |
| 8.70 | | |
| 8.80 | | |
| 8.90 | | |
| 9.00 | | |
| 9.10 | | |
| 9.20 | | |
| 9.30 | | |
| 9.40 | | |
| 9.50 | | |
| 9.60 | 121 | 184 |
| 9.70 | | |
| 9.80 | | |
| 9.90 | | |
| 10.00 | | |
| 10.10 | | |
| 10.20 | | |
| 10.30 | | |
| 10.40 | | |
| 10.50 | | |
| 10.60 | | |
| 10.70 | 128 | 195 |
| 10.80 | | |
| 10.90 | | |
| 11.00 | | |
| 11.10 | | |
| 11.20 | | |
| 11.30 | | |
| 11.40 | | |
| 11.50 | | |
| 11.60 | | |
| 11.70 | | |
| 11.80 | | |
| 11.90 | 134 | 205 |
| 12.00 | | |
| 12.10 | | |
| 12.20 | | |
| 12.30 | | |
| 12.40 | | |
| 12.50 | | |
| 12.60 | | |
| 12.70 | 134 | 205 |
| 12.80 | | |
| 12.90 | | |
| 13.00 | | |
| 13.10 | | |
| 13.20 | | |
| 13.30 | 140 | 214 |
| 13.40 | | |
| 13.50 | | |
| 13.60 | | |
| 13.70 | | |
| 13.80 | | |
| 13.90 | | |
| 14.00 | | |
| 14.25 | 144 | 220 |
| 14.50 | | |
| 14.75 | | |
| 15.00 | | |
| 15.25 | 149 | 227 |
| 15.50 | | |
| 15.75 | | |
| 16.00 | | |
| 16.25 | 154 | 235 |
| 16.50 | | |
| 16.75 | | |
| 17.00 | | |
| 17.25 | 158 | 241 |
| 17.50 | | |
| 17.75 | | |
| 18.00 | | |
| 18.25 | 162 | 247 |
| 18.50 | | |
| 18.75 | | |
| 19.00 | | |
| 19.25 | 166 | 254 |
| 19.50 | | |
| 19.75 | | |
| 20.00 | | |
| 20.25 | 171 | 261 |
| 20.50 | | |
| 20.75 | | |
| 21.00 | | |
| 21.25 | 176 | 268 |
| 21.50 | | |
| 21.75 | | |
| 22.00 | | |
| 22.25 | | |
| 22.50 | 180 | 275 |
| 22.75 | | |
| 23.00 | | |
| 23.25 | | |
| 23.50 | | |
| 23.75 | 185 | 282 |
| 24.00 | | |
| 24.25 | | |
| 24.50 | | |
| 24.75 | | |
| 25.00 | | |

All dimensions are in millimeters. Tolerances on diameters are given in Table 14.

### Oversize Diameters in Drilling

| Drill Dia., Inch | Amount Oversize, Inch | | | Drill Dia., Inch | Amount Oversize, Inch | | |
|---|---|---|---|---|---|---|---|
| | Average Max. | Mean | Average Min. | | Average Max. | Mean | Average Min. |
| 1⁄16 | 0.002 | 0.0015 | 0.001 | 1⁄2 | 0.008 | 0.005 | 0.003 |
| 1⁄8 | 0.0045 | 0.003 | 0.001 | 3⁄4 | 0.008 | 0.005 | 0.003 |
| 1⁄4 | 0.0065 | 0.004 | 0.0025 | 1 | 0.009 | 0.007 | 0.004 |

Courtesy of The United States Cutting Tool Institute

Some conditions will cause the drilled hole to be undersize. For example, holes drilled in light metals and in other materials having a high coefficient of thermal expansion such as plastics may contract to a size that is smaller than the diameter of the drill as the material surrounding the hole is cooled after having been heated by the drilling. The elastic action of the material surrounding the hole may also cause the drilled hole to be undersize when drilling high-strength materials with a drill that is dull at its outer corner.

The accuracy of the drill point has a great effect on the accuracy of the drilled hole. An inaccurately ground twist drill will produce holes that are excessively oversize. The drill point must be symmetrical; i.e., the point angles must be equal, as well as the lip lengths and the axial height of the lips. Any alterations to the lips or to the chisel edge, such as thinning the web, must be done carefully to preserve the symmetry of the drill point. Adequate relief should be provided behind the chisel edge to prevent heel drag. On conventionally ground drill points, this relief can be estimated by the chisel edge angle.

When drilling a hole, as the drill point starts to enter the workpiece, the drill will be unstable and will tend to wander. Then, as the body of the drill enters the hole, the drill will tend to stabilize. The result of this action is a tendency to drill a bellmouth shape in the hole at the entrance and perhaps beyond. Factors contributing to bellmouthing are: an unsymmetrically ground drill point; a large chisel edge length; inadequate relief behind the chisel edge; runout of the spindle and the chuck; using a slender drill that will bend easily; and lack of rigidity of the machine tool, workpiece, or the setup. Correcting these conditions as required will reduce the tendency for bellmouthing to occur and improve the accuracy of the hole diameter and its straightness. Starting the hole with a short stiff drill, such as a center drill, will quickly stabilize the drill that follows and reduce or eliminate bellmouthing; this procedure should always be used when drilling in a lathe, where the work is rotating. Bellmouthing can also be eliminated almost entirely and the accuracy of the hole improved by using a close-fitting drill jig bushing placed close to the workpiece. Although specific recommendations cannot be made, many cutting fluids will help to increase the accuracy of the diameters of drilled holes. Double-margin twist drills, available in the smaller sizes, will drill a more accurate hole than conventional twist drills having only a single margin at the leading edge of the land. The second land, located on the trailing edge of each land, provides greater stability in the drill bushing and in the hole. These drills are especially useful in drilling intersecting off-center holes. Single and double-margin step drills, also available in the smaller sizes, will produce very accurate drilled holes, which are usually less than 0.002 inch (0.051 mm) larger than the drill size.

**Counterboring.—**Counterboring (called spot-facing if the depth is shallow) is the enlargement of a previously formed hole. Counterbores for screw holes are generally made in sets. Each set contains three counterbores: one with the body of the size of the screw head and the pilot the size of the hole to admit the body of the screw; one with the body the size of the head of the screw and the pilot the size of the tap drill; and the third with the body the size of the body of the screw and the pilot the size of the tap drill. Counterbores are usually provided with helical flutes to provide positive effective rake on the cutting edges. The four flutes are so positioned that the end teeth cut ahead of center to provide a shearing action and eliminate chatter in the cut. Three designs are most common: solid, two-piece, and three-piece. Solid designs have the body, cutter, and pilot all in one piece. Two-piece designs have an integral shank and counterbore cutter, with an interchangeable pilot, and provide true concentricity of the cutter diameter with the shank while

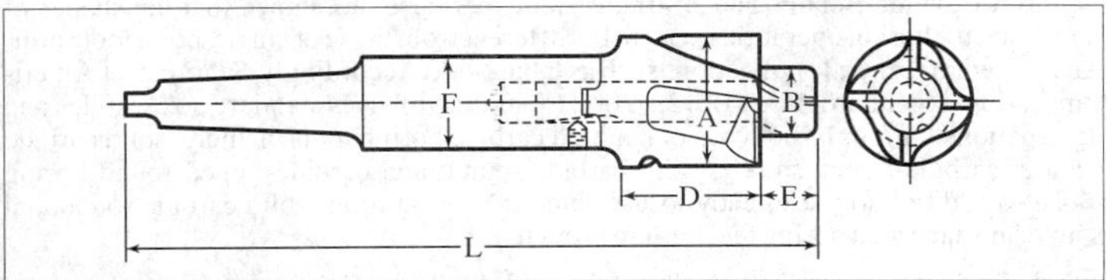

**Counterbore Sizes for Hex-Head Bolts and Nuts.—**Table 3a, page 1689, shows the maximum socket wrench dimensions for standard $\frac{1}{4}$-, $\frac{1}{2}$- and $\frac{3}{4}$-inch drive socket sets. For a given socket size (nominal size equals the maximum width across the flats of nut or bolt head), the dimension $K$ given in the table is the minimum counterbore diameter required to provide socket wrench clearance for access to the bolt or nut.

**Sintered Carbide Boring Tools.—**Industrial experience has shown that the shapes of tools used for boring operations need to be different from those of single-point tools ordinarily used for general applications such as lathe work. Accordingly, Section 5 of American National Standard ANSI B212.1-2002 (Withdrawn) gives standard sizes, styles and designations for four basic types of sintered carbide boring tools, namely: solid carbide square; carbide-tipped square; solid carbide round; and carbide-tipped round boring tools. In addition to these ready-to-use standard boring tools, solid carbide round and square unsharpened boring tool bits are provided.

*Style Designations for Carbide Boring Tools:* Table 1 shows designations used to specify the styles of American National Standard sintered carbide boring tools. The first letter denotes solid (S) or tipped (T). The second letter denotes square (S) or round (R). The side cutting edge angle is denoted by a third letter (A through H) to complete the style designation. Solid square and round bits with the mounting surfaces ground but the cutting edges unsharpened (Table 3) are designated using the same system except that the third letter indicating the side cutting edge angle is omitted.

**Table 1. American National Standard Sintered Carbide Boring Tools — Style Designations** *ANSI B212.1-2002 (Withdrawn)*

| Side Cutting Edge Angle $E$ | | Boring Tool Styles | | | |
|---|---|---|---|---|---|
| Degrees | Designation | Solid Square (SS) | Tipped Square (TS) | Solid Round (SR) | Tipped Round (TR) |
| 0 | A | | TSA | | |
| 10 | B | | TSB | | |
| 30 | C | SSC | TSC | SRC | TRC |
| 40 | D | | TSD | | |
| 45 | E | SSE | TSE | SRE | TRE |
| 55 | F | | TSF | | |
| 90 (0° Rake) | G | | | | TRG |
| 90 (10° Rake) | H | | | | TRH |

*Size Designation of Carbide Boring Tools:* Specific sizes of boring tools are identified by the addition of numbers after the style designation. The first number denotes the diameter or square size in number of $\frac{1}{32}$nds for types SS and SR and in number of $\frac{1}{16}$ths for types TS and TR. The second number denotes length in number of 1/8ths for types SS and SR. For styles TRG and TRH, a letter "U" after the number denotes a semi-finished tool (cutting edges unsharpened). Complete designations for the various standard sizes of carbide boring tools are given in Table 2 through Table 7. In the diagrams in the tables, angles shown without tolerance are ± 1°.

*Examples of Tool Designation:* The designation TSC-8 indicates a carbide-tipped tool (T); square cross section (S); 30-degree side cutting edge angle (C); and $\frac{8}{16}$ or $\frac{1}{2}$ inch square size (8).

The designation SRE-66 indicates a solid carbide tool (S); round cross section (R); 45-degree side cutting edge angle (E); $\frac{6}{32}$ or $\frac{3}{16}$ inch diameter (6); and $\frac{6}{8}$ or $\frac{3}{4}$ inch long (6).

The designation SS-610 indicates a solid carbide tool (S); square cross section (S); $\frac{6}{32}$ or $\frac{3}{16}$ inch square size (6); $\frac{10}{8}$ or $1\frac{1}{4}$ inches long (10).

It should be noted in this last example that the absence of a third letter (from A to H) indicates that the tool has its mounting surfaces ground but that the cutting edges are unsharpened.

**Table 2. ANSI Carbide-Tipped Round General-Purpose Square-End Boring Tools Style TRG with 0° Rake and Style TRH with 10° Rake** *ANSI B212.1-2002 (Withdrawn)*

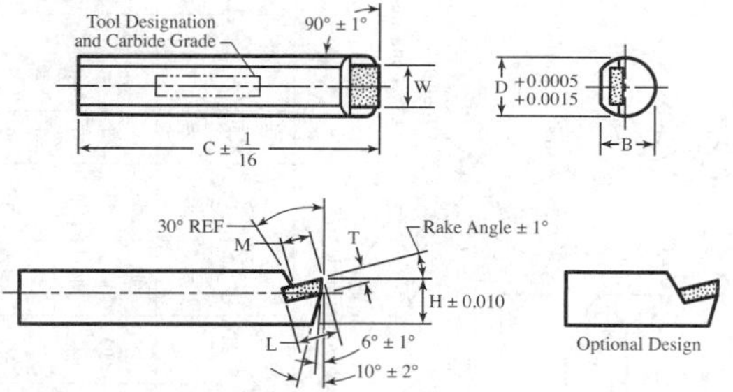


| Tool Designation | | Shank Dimensions, Inches | | | | | Rake Angle Deg. | Tip No. | Tip Dimensions, Inches | | |
|---|---|---|---|---|---|---|---|---|---|---|---|
| Finished | Semi-finished[a] | Dia. *D* | Length *C* | Dim.Over Flat *B* | Nose Height *H* | Setback *M* (Min) | | | *T* | *W* | *L* |
| TRG-5 | TRG-5U | 5/16 | 1½ | 19/64 | 3/16 | 3/16 | 0 | 1025 | 1/16 | ¼ | ¼ |
| TRH-5 | TRH-5U | | | ±.005 | 7/32 | 3/16 | 10 | | | | |
| TRG-6 | TRG-6U | ⅜ | 1¾ | 11/32 | 7/32 | 3/16 | 0 | 1030 | 1/16 | 5/16 | ¼ |
| TRH-6 | TRH-6U | | | ±.010 | ¼ | | 10 | | | | |
| TRG-7 | TRG-7U | 7/16 | 2½ | 13/32 | ¼ | 3/16 | 0 | 1080 | 3/32 | 5/16 | ⅜ |
| TRH-7 | TRH-7U | | | ±.010 | 5/16 | | 10 | | | | |
| TRG-8 | TRG-8U | ½ | 2½ | 15/32 | 9/32 | ¼ | 0 | 1090 | 3/32 | ⅜ | ⅜ |
| TRH-8 | TRH-8U | | | ±.010 | 11/32 | | 10 | | | | |

[a] Semifinished tool will be without Flat (B) and carbide unground on the end.

**Table 3. Solid Carbide Square and Round Boring Tool Bits**

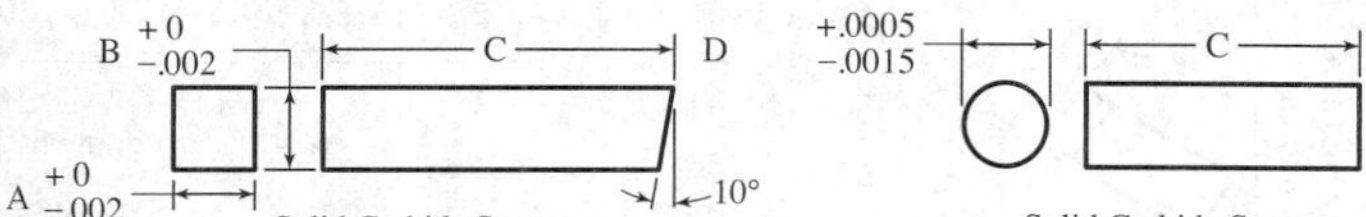


Solid Carbide Square Boring Tool Bits Style SS

Solid Carbide Square Boring Tool Bits Style SR

| Square Bits | | | | Round Bits | | | | | | | | |
|---|---|---|---|---|---|---|---|---|---|---|---|---|
| Tool Designation | *A* | *B* | *C* | Tool Designation | *D* | *C* | Tool Designation | *D* | *C* | Tool Designation | *D* | *C* |
| SS-58 | 5/32 | 5/32 | 1 | SR-33 | 3/32 | ⅜ | SR-55 | 5/32 | ⅝ | SR-88 | ¼ | 1 |
| SS-610 | 3/16 | 3/16 | 1¼ | SR-34 | 3/32 | ½ | SR-64 | 3/16 | ½ | SR-810 | ¼ | 1½ |
| SS-810 | ¼ | ¼ | 1¼ | SR-44 | ⅛ | ½ | SR-66 | 3/16 | ¾ | SR-1010 | 5/16 | 1½ |
| SS-1012 | 5/16 | 5/16 | 1½ | SR-46 | ⅛ | ¾ | SR-69 | 3/16 | 1⅛ | … | … | … |
| SS-1214 | ⅜ | ⅜ | 1¾ | SR-48 | ⅛ | 1 | SR-77 | 7/32 | ⅞ | … | … | … |

All dimensions are in inches.

*Tolerance on Length:* Through 1 inch, + 1/32, −0; over 1 inch, +1/16, −0.

**Table 4. ANSI Solid Carbide Square Boring Tools**
**Style SSC for 60° Boring Bar and Style SSE for 45° Boring Bar**
*ANSI B212.1-2002 (Withdrawn)*

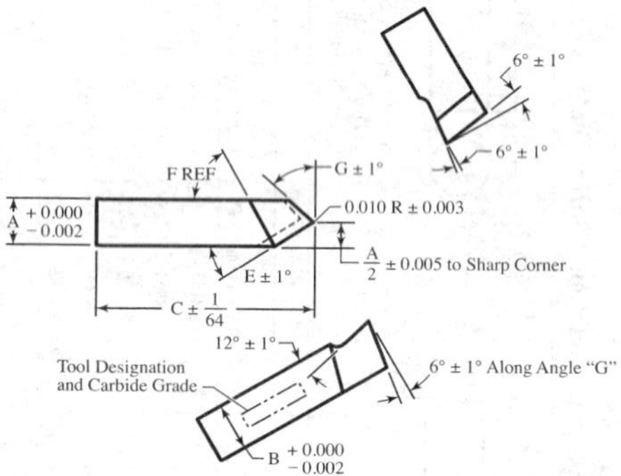


| Tool Designation | Boring Bar Angle, Deg. from Axis | Shank Dimensions, Inches Width A | Height B | Length C | Side Cutting Edge Angle E, Deg. | End Cutting Edge Angle G, Deg. | Shoulder Angle F, Deg. |
|---|---|---|---|---|---|---|---|
| SSC-58 | 60 | 5/32 | 5/32 | 1 | 30 | 38 | 60 |
| SSE-58 | 45 | | | | 45 | 53 | 45 |
| SSC-610 | 60 | 3/16 | 3/16 | 1 1/4 | 30 | 38 | 60 |
| SSE-610 | 45 | | | | 45 | 53 | 45 |
| SSC-810 | 60 | 1/4 | 1/4 | 1 1/4 | 30 | 38 | 60 |
| SSE-810 | 45 | | | | 45 | 53 | 45 |
| SSC-1012 | 60 | 5/16 | 5/16 | 1 1/2 | 30 | 38 | 60 |
| SSE-1012 | 45 | | | | 45 | 53 | 45 |

**Table 5. ANSI Carbide-Tipped Round Boring Tools**
**Style TRC for 60° Boring Bar and Style TRE for 45° Boring Bar**
*ANSI B212.1-2002 (Withdrawn)*

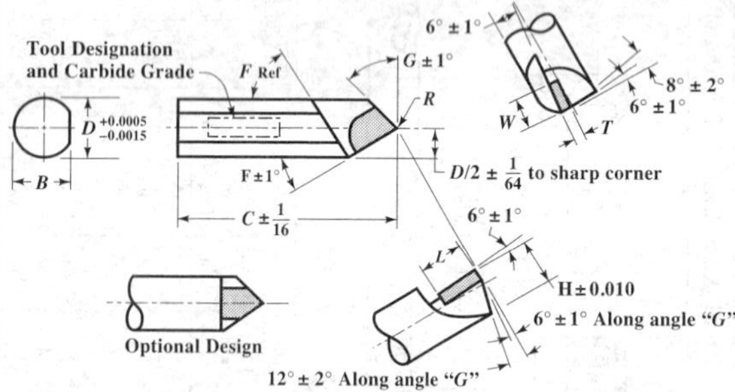


| Tool Designation | Bor. Bar Angle from Axis, Deg. | Shank Dimensions, Inches D | C | B | H | R | Side Cut. Edge Angle E, Deg. | End Cut. Edge Angle G, Deg. | Shoulder Angle F, Deg. | Tip No. | Tip Dimensions, Inches T | W | L |
|---|---|---|---|---|---|---|---|---|---|---|---|---|---|
| TRC-5 | 60 | 5/16 | 1 1/2 | 19/64 ±.005 | 7/32 | 1/64 ±.005 | 30 | 38 | 60 | 2020 | 1/16 | 3/16 | 1/4 |
| TRE-5 | 45 | | | | | | 45 | 53 | 45 | | | | |
| TRC-6 | 60 | 3/8 | 1 3/4 | 11/32 ±.010 | 9/32 | 1/64 | 30 | 38 | 60 | 2040 | 3/32 | 3/16 | 5/16 |
| TRE-6 | 45 | | | | | ±.005 | 45 | 53 | 45 | 2020 | 1/16 | 3/16 | 1/4 |
| TRC-7 | 60 | 7/16 | 2 1/2 | 13/32 ±.010 | 5/16 | 1/32 ±.010 | 30 | 38 | 60 | 2060 | 3/32 | 1/4 | 3/8 |
| TRE-7 | 45 | | | | | | 45 | 53 | 45 | | | | |
| TRC-8 | 60 | 1/2 | 2 1/2 | 15/32 ±.010 | 3/8 | 1/32 ±.010 | 30 | 38 | 60 | 2060 | 3/32 | 1/4 | 3/8 |
| TRE-8 | 45 | | | | | | 45 | 53 | 45 | 2080 | 3/32 | 5/16 | 3/8 |

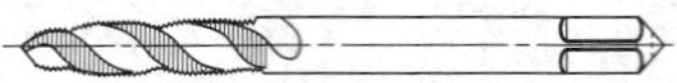


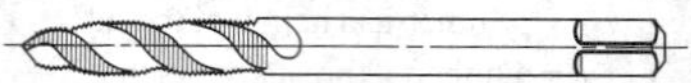

with the letters "D" or "DU," respectively, followed by a limit number. The limit number is determined as follows:

$$\text{D Limit number} = \frac{\text{Tap PD} - \text{Basic PD}}{0.0013}$$

$$\text{DU Limit number} = \frac{\text{Basic PD} - \text{Tap PD}}{0.0013}$$

*Example:* M1.6×0.35 HS G D3

$$\begin{aligned}\text{Maximum tap PD} &= \text{Basic PD} + 0.0039\\ &= 1.6 - (0.64951904 \times 0.35) + 0.0039\\ &= 1.3727 + 0.039\\ &= 1.412\\ \text{Minimum tap PD} &= \text{Maximum tap PD} - 0.015\\ &= 1.412 - 0.015\\ &= 1.397\end{aligned}$$

M6×1 HS G DU4

$$\begin{aligned}\text{Minimum tap PD} &= \text{Basic PD} - 0.052\\ &= 6 - (0.64951904 \times 1.0) - 0.052\\ &= 5.350 - 0.052\\ &= 5.298\\ \text{Maximum tap PD} &= \text{Minimum tap PD} + 0.025\\ &= 5.298 + 0.025\\ &= 5.323\end{aligned}$$

**Definitions of Tap Terms.**—The definitions that follow are taken from ANSI/ASME B94.9 but include only the more important terms. Some tap terms are the same as screw thread terms; therefore, see *Definitions of Screw Threads* starting on page 1945.

*Actual size:* The measured size of an element on an individual part.

*Allowance:* A prescribed difference between the maximum material limits of mating parts. It is the minimum clearance or maximum interference between such parts.

*Basic Size:* The size from which the limits are derived by application of allowance and tolerance.

*Bottom Top:* A tap having a chamfer length of 1 to 2 pitches.

*Chamfer:* Tapering of the threads at the front end of each land or chaser of a tap by cutting away and relieving the crest of the first few teeth to distribute the cutting action over several teeth.

*Chamfer Angle:* Angle formed between the chamfer and the axis of the tap measured in an axial plane at the cutting edge.

*Chamfer Relief:* The gradual degrees in land height from cutting edge to heel on the chamfered portion of the land to provide radial clearance for the cutting edge.

*Chamfer Relief Angle:* Complement of the angle formed between a tangent to the relieved surface at the cutting edge and a radial line to the same point on the cutting edge.

*Classes of Thread:* Designation of the class that determines the specification of the size, allowance, and tolerance to which a given threaded product is to be manufactured. It is not applicable to the tools used for threading.

*Concentric:* Having a common center.

*Crest:* The surface of the thread that joins the flanks of the thread and is farthest from the cylinder or cone from which the thread projects.

*Interrupted Thread Tap:* A tap having an odd number of lands with alternate teeth in the thread helix removed. In some designs, alternate teeth are removed only for a portion of the thread length.

*Land:* One of the threaded sections between the flutes of a tap.

*Lead:* Distance a screw thread advances axially in one complete turn.

*Lead Error:* Deviation from prescribed limits.

*Lead Deviation:* Deviation from the basic nominal lead.

*Progressive Lead Deviation:* (1) On a straight thread, the deviation from a true helix where the thread helix advances uniformly. (2) On a taper thread, the deviation from a true spiral where the thread spiral advances uniformly.

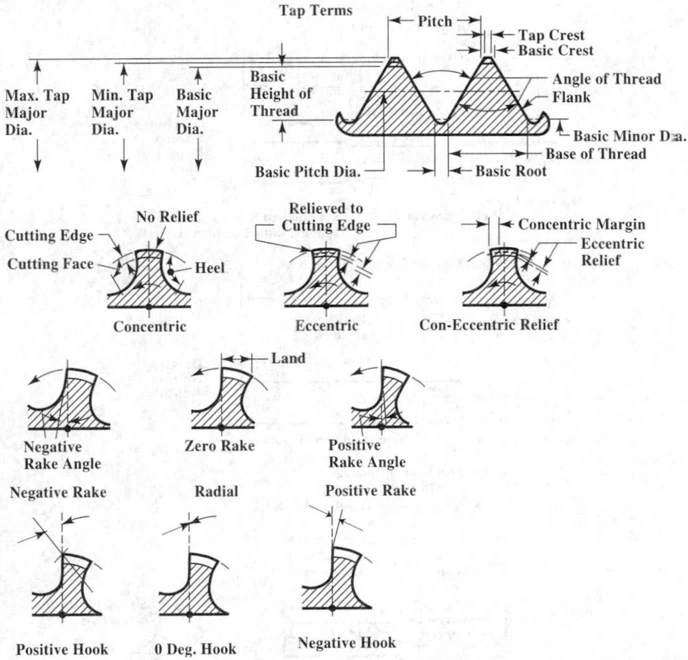


Fig. 3. Tap Terms

*Left-Hand Cut:* Rotation in a clockwise direction from cutting when viewed from the chamfered end of a tap.

*Length of Engagement:* The length of engagement of two mating threads is the axial distance over which two mating threads are designed to contact.

*Length of Thread:* The length of the thread of the tap includes the chamfered threads and the full threads, but does not include an external center. It is indicated by the letter "B" in the illustrations at the heads of the tables.

*Limits:* The limits of size are the applicable maximum and minimum sizes.

*Major Diameter:* On a straight thread, the major diameter is that of the major cylinder. On a taper thread, the major diameter at a given position on the thread axis is that of the major cone at that position.

**Table 5b. Standard Tap Dimensions Tolerances (Ground and Cut Thread)** *ANSI/ASME B94.9-2008 (R2018)*

| Element | Nominal Diameter Range, inch | | Direction | Tolerance, inch | |
|---|---|---|---|---|---|
| | Over | To (inclusive) | | Ground Thread | Cut Thread |
| Length overall, $L$ | 0.5200 | 1.0100 | ± | 0.0300 | 0.0300 |
| | 1.0100 | 2.0000 | ± | 0.0600 | 0.0600 |
| Length of thread, $I$ | 0.0520 | 0.2230 | ± | 0.0500 | 0.0500 |
| | 0.2230 | 0.5100 | ± | 0.0600 | 0.0600 |
| | 0.5100 | 1.5100 | ± | 0.0900 | 0.0900 |
| | 1.5100 | 2.0000 | ± | 0.1300 | 0.1300 |
| Length of thread, $I_2$ | 0.0520 | 1.0100 | ± | 0.0300 | 0.0300 |
| | 1.0100 | 2.0000 | ± | 0.0600 | 0.0600 |
| Diameter of shank, $d_1$ | 0.0520 | 0.2230 | – | 0.0015 | 0.0040 |
| | 0.2230 | 0.6350 | – | 0.0015 | 0.0050 |
| | 0.6350 | 1.0100 | – | 0.0020 | 0.0050 |
| | 1.0100 | 1.5100 | – | 0.0020 | 0.0070 |
| | 1.5100 | 2.0000 | – | 0.0030 | 0.0070 |
| Size of square, $a$ | 0.0520 | 0.5100 | – | 0.0040 | 0.0040 |
| | 0.5100 | 1.0100 | – | 0.0060 | 0.0060 |
| | 1.0100 | 2.0000 | – | 0.0080 | 0.0080 |

**Entry Taper Length.—**Entry taper length is measured on the full diameter of the thread-forming lobes and is the axial distance from the entry diameter position to the theoretical intersection of tap major diameter and entry taper angle. Beveled end threads provided on taps having internal center or incomplete threads retained when external center is removed. Whenever entry taper length is specified in terms of number of threads, this length is measured in number of pitches, $P$.

$$\text{Bottom length} = 1 \sim 2\frac{1}{2}\ \text{pitches}$$
$$\text{Plug length} = 3 \sim 5\ \text{pitches}$$

Entry diameter measured at the thread crest nearest the front of the tap is an appropriate amount smaller than the diameter of the hole drilled for tapping.

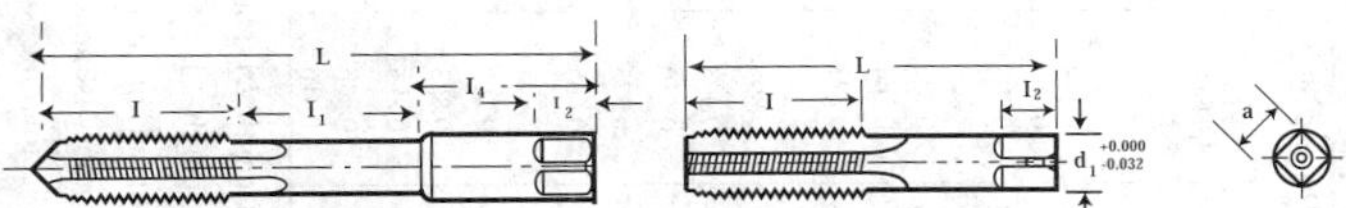


BLANK Design 2 with Optional Neck

BLANK Design 3

Optional Neck and Optional Shortened Thread Length, Ground and Cut Thread (Table 6)

**Table 7. Machine Screw and Fractional Size Ground Thread Dimensions for Screw Thread Insert (STI) Taps** *ANSI/ASME B94.9-2008 (R2018)*

| Nominal Size (STI) | Threads per inch NC | Threads per inch NF | Blank Design No. | Tap Dimensions, inch: Overall length, $L$ | Thread Length, $I$ | Square Length, $l_2$ | Shank Diameter, $d_1$ | Size of Square, $a$ | Table 5a Blank Equivalent (Reference) |
|---|---|---|---|---|---|---|---|---|---|
| 1 | 64 | … | 1 | 1.81 | 0.50 | 0.19 | 0.141 | 0.110 | No. 3 |
| 2 | 56 | 64 | 1 | 1.88 | 0.56 | 0.19 | 0.141 | 0.110 | No. 4 |
| 3 | 48 | 56 | 1 | 1.94 | 0.63 | 0.19 | 0.141 | 0.110 | No. 5 |
| 4 | 40 | 48 | 1 | 2.00 | 0.69 | 0.19 | 0.141 | 0.110 | No. 6 |
| 5 | 40 | … | 1 | 2.13 | 0.75 | 0.25 | 0.168 | 0.131 | No. 8 |
| 6 | 32 | … | 1 | 2.38 | 0.88 | 0.25 | 0.194 | 0.152 | No. 10 |
| | … | 40 | 1 | 2.13 | 0.75 | 0.25 | 0.168 | 0.131 | No. 8 |
| 8 | 32 | 36 | 1 | 2.38 | 0.94 | 0.28 | 0.220 | 0.165 | No. 12 |
| 10 | 24 | 32 | 2 | 2.50 | 1.00 | 0.31 | 0.255 | 0.191 | 1/4 |
| 12 | 24 | … | 2 | 2.72 | 1.13 | 0.38 | 0.318 | 0.238 | 5/16 |
| 1/4 | 20 | 28 | 2 | 2.72 | 1.13 | 0.38 | 0.318 | 0.238 | 5/16 |
| 5/16 | 18 | 24 | 2 | 2.94 | 1.25 | 0.44 | 0.381 | 0.286 | 3/8 |
| 3/8 | 16 | … | 3 | 3.38 | 1.66 | 0.44 | 0.367 | 0.275 | 1/2 |
| | … | 24 | 3 | 3.16 | 1.44 | 0.41 | 0.323 | 0.242 | 7/16 |
| 7/16 | 14 | … | 3 | 3.59 | 1.66 | 0.50 | 0.429 | 0.322 | 9/16 |
| | … | 20 | 3 | 3.38 | 1.66 | 0.44 | 0.367 | 0.275 | 1/2 |
| 1/2 | 13 | … | 3 | 3.81 | 1.81 | 0.56 | 0.480 | 0.360 | 5/8 |
| | … | 20 | 3 | 3.59 | 1.66 | 0.50 | 0.429 | 0.322 | 9/16 |
| 9/16 | 12 | … | 3 | 4.03 | 1.81 | 0.63 | 0.542 | 0.406 | 11/16 |
| | … | 18 | 3 | 3.81 | 1.81 | 0.56 | 0.480 | 0.360 | 5/8 |
| 5/8 | 11 | … | 3 | 4.25 | 2.00 | 0.69 | 0.590 | 0.442 | 3/4 |
| | … | 18 | 3 | 4.03 | 1.81 | 0.63 | 0.542 | 0.406 | 11/16 |
| 3/4 | 10 | … | 3 | 4.69 | 2.22 | 0.75 | 0.697 | 0.523 | 7/8 |
| | … | 16 | 3 | 4.47 | 2.00 | 0.69 | 0.652 | 0.489 | 13/16 |
| 7/8 | 9 | 14 | 3 | 5.13 | 2.50 | 0.81 | 0.800 | 0.600 | 1 |
| 1 | 8 | … | 3 | 5.75 | 2.56 | 1.00 | 1.021 | 0.766 | 1 1/4 |
| | … | 12, 14 NS | 3 | 5.44 | 2.56 | 0.88 | 0.896 | 0.672 | 1 1/8 |
| 1 1/8 | 7 | … | 3 | 6.06 | 3.00 | 1.06 | 1.108 | 0.831 | 1 3/8 |
| | … | 12 | 3 | 5.75 | 2.56 | 1.00 | 1.021 | 0.766 | 1 1/4 |
| 1 1/4 | 7 | … | 3 | 6.38 | 3.00 | 1.13 | 1.233 | 0.925 | 1 1/2 |
| | … | 12 | 3 | 6.06 | 3.00 | 1.06 | 1.108 | 0.831 | 1 3/8 |
| 1 3/8 | 6 | … | 3 | 6.69 | 3.19 | 1.13 | 1.305 | 0.979 | 1 5/8 |
| | … | 12 | 3 | 6.38 | 3.00 | 1.13 | 1.233 | 0.925 | 1 1/2 |
| 1 1/2 | 6 | … | 3 | 7.00 | 3.19 | 1.25 | 1.430 | 1.072 | 1 3/4 |
| | … | 12 | 3 | 6.69 | 3.19 | 1.13 | 1.305 | 0.979 | 1 5/8 |

These threads are larger than nominal size to the extent that the internal thread they produce will accommodate a helical coil screw insert, which at final assembly will accept a screw thread of the nominal size and pitch.

For optional necks, refer to Table 6 using dimensions for equivalent blank sizes.

*Ground Thread Taps:* STI sizes 5/16 inch and smaller have external center on thread end (may be removed on bottom taps); sizes 10 through 5/16 inch will have an external partial cone center on shank end, with the length of the cone center approximately 1/4 of the diameter of shank; sizes larger than 5/16 inch may have internal centers on both the thread and shank ends.

For runout tolerances of tap elements, refer to Table 14 using dimensions for equivalent blank sizes.

For number of flutes, refer to Table 11 using dimensions for equivalent blank sizes.

For general dimension tolerances, refer to Table 5b using Table 5a equivalent blank size.

## Table 11. Standard Number of Flutes (Ground and Cut Thread) *ANSI/ASME B94.9-2008 (R2018)*

| Machine Screw Size, Nom. Fractional Dia. inch | | Nominal Metric Dia. | | TPI/Pitch | | | Straight Flutes | | Spiral Point | | Spiral Point Only | Reg. Spiral Flute | Fast Spiral Flute |
|---|---|---|---|---|---|---|---|---|---|---|---|---|---|
| | | mm | inch | UNC NC | UNF NF | mm | Standard | Optional | Standard | Optional | | | |
| 0 | (0.0600) | M1.6 | 0.0630 | … | 80 | 0.35 | 2 | … | 2 | … | … | … | … |
| 1 | (0.0730) | … | … | 64 | 72 | … | 2 | … | 2 | … | … | … | … |
| 2 | (0.0860) | M2.0 | 0.0787 | 56 | 64 | 0.40 | 3 | 2 | 2 | … | … | … | … |
| 3 | (0.0990) | M 2.5 | 0.0984 | 48 | 56 | 0.45 | 3 | 2 | 2 | … | … | … | 2 |
| 4 | (0.1120) | … | … | 40 | 48 | | 3 | 2 | 2 | … | 2 | 2 | 2 |
| 5 | (0.1250) | M3.0 | 0.1181 | 40 | 44 | 0.50 | 3 | 2 | 2 | … | 2 | 2 | 2 |
| 6 | (0.1380) | M3.5 | 0.1378 | 32 | 40 | 0.60 | 3 | 2 | 2 | … | 2 | 2 | 2 |
| 8 | (0.1640) | M4.0 | 0.1575 | 32 | 36 | 0.70 | 4 | 2/3 | 2 | … | 2 | 2 | 3 |
| 10 | (0.1900) | M4.5 | 0.1772 | 24 | 32 | 0.75 | 4 | 2/3 | 2 | … | 2 | 2 | 3 |
| … | | M5 | 0.1969 | … | … | 0.80 | 4 | 2/3 | 2 | … | 2 | 2 | 3 |
| 12 | (0.2160) | … | … | 24 | 28 | … | 4 | 2/3 | 2 | … | 2 | 2 | 3 |
| 1/4 | (0.2500) | M6 | 0.2362 | 20 | 28 | 1.00 | 4 | 2/3 | 2 | 3 | 2 | 3 (optional) | 3 |
| … | | M7 | 0.2756 | 18 | 24 | 1.00 | 4 | 2/2 | 2 | 3 | 2 | 3 | 3 |
| 5/16 | (0.3125) | M8 | 0.3150 | 18 | 24 | 1.25 | 4 | 2/3 | 2 | 3 | 2 | 3 | 3 |
| 3/8 | (0.3750) | M10 | 0.3937 | 16 | 24 | 1.50 | 4 | 3 | 3 | … | 3 | 3 | 3 |
| 7/16 | (0.4375) | … | … | 14 | 20 | … | 4 | 3 | 3 | … | 3 | 3 | 3 |
| 1/2 | (0.5000) | M12 | 0.4724 | 13 | 20 | 1.75 | 4 | 3 | 3 | … | … | … | 3 |
| 9/16 | (0.5625) | M14 | 0.5512 | 12 | 18 | 2.00 | 4 | … | 3 | … | … | … | … |
| 5/8 | (0.6250) | M16 | 0.6299 | 11 | 18 | 2.00 | 4 | … | 3 | … | … | … | … |
| 3/4 | (0.7500) | … | … | 10 | 16 | … | 4 | … | 3 | … | … | … | … |
| … | | M20 | 0.7874 | … | … | 2.5 | 4 | … | … | … | … | … | … |
| 7/8 | (0.8750) | … | … | 9 | 14 | … | 4 | … | … | … | … | … | … |
| … | | M24 | 0.9449 | … | … | 3.00 | 4 | … | … | … | … | … | … |
| 1 | (1.0000) | … | … | 8 | 12 | … | 4 | … | … | … | … | … | … |
| 1 1/8 | (1.1250) | … | … | 7 | 12 | 4.00 | … | … | … | … | … | … | … |
| … | | M30 | 1.1811 | … | … | 3.50 | 4 | … | … | … | … | … | … |
| 1 1/4 | (1.2500) | … | … | 7 | … | … | 4 | … | … | … | … | … | … |
| … | | … | … | … | 12 | … | 6 | … | … | … | … | … | … |
| 1 3/8 | (1.3750) | … | … | 6 | … | … | 6 | … | … | … | … | … | … |
| … | | … | … | … | 12 | … | 6 | … | … | … | … | … | … |
| … | | M36 | 1.4173 | … | … | 4.00 | 4 | … | … | … | … | … | … |
| 1 1/2 | (1.5000) | … | … | 6 | … | … | 4 | … | … | … | … | … | … |
| | | … | … | … | 12 | … | 6 | … | … | … | … | … | … |
| 1 3/4 | (1.7500) | … | … | 5 | … | … | 6 | … | … | … | … | … | … |
| 2 | (2.0000) | … | … | 4 1/2 | … | … | 6 | … | … | … | … | … | … |

For pulley taps, see Table 12. For taper pipe, see Table 13a. For straight pipe taps, see Table 13a. For STI taps, use number of flutes for blank size equivalent on Table 5a. For optional flutes, see Table 6.

## Table 21. Tap Thread Limits: Fractional Sizes, Ground Thread *ANSI/ASME B94.9-2008 (R2018)* (Unified and American National Thread Forms, Standard Thread Limits)

| Size inch | NC UNC | NF UNF | NS | Major Diameter Basic | Major Diameter Min. | Major Diameter Max. | Pitch Diameter Basic | H1 limit Min. | H1 limit Max. | H2 limit Min. | H2 limit Max. | H3 limit Min. | H3 limit Max. | H4 limit Min. | H4 limit Max. | H5 limit Min. | H5 limit Max. | H6 limit[a] Min. | H6 limit[a] Max. | H7 limit Min. | H7 limit Max. | H8 limit[b] Min. | H8 limit[b] Max. |
|---|---|---|---|---|---|---|---|---|---|---|---|---|---|---|---|---|---|---|---|---|---|---|---|
| ¼ | 20 | … | … | 0.2500 | 0.2532 | 0.2565 | 0.2175 | 0.2175 | 0.2180 | 0.2180 | 0.2185 | 0.2185 | 0.2190 | … | … | 0.2195[a] | 0.2200[a] | … | … | … | … | … | … |
| ¼ | … | 28 | … | 0.2500 | 0.2523 | 0.2546 | 0.2268 | 0.2268 | 0.2273 | 0.2273 | 0.2278 | 0.2278 | 0.2283 | 0.2283 | 0.2288 | … | … | … | … | … | … | … | … |
| 5⁄16 | 18 | … | … | 0.3125 | 0.3161 | 0.3197 | 0.2764 | 0.2764 | 0.2769 | 0.2769 | 02774 | 0.2774 | 0.2779 | … | … | 0.2784[a] | 0.2789[a] | … | … | 0.2794[c] | 0.2799[c] | … | … |
| 5⁄16 | … | 24 | … | 0.3125 | 0.3152 | 0.3179 | 0.2854 | 0.2854 | 0.2859 | 0.2859 | 0.2864 | 0.2864 | 0.2869 | 0.2869 | 0.2874 | … | … | … | … | 0.2884[c] | 0.2889[c] | … | … |
| ⅜ | 16 | … | … | 0.3750 | 0.3790 | 0.3831 | 0.3344 | 0.3344 | 0.3349 | 0.3349 | 0.3354 | 0.3354 | 0.3359 | … | … | 0.3364[a] | 0.3369[a] | … | … | 0.3374[c] | 0.3379[c] | … | … |
| ⅜ | … | 24 | … | 0.3750 | 0.3777 | 0.3804 | 0.3479 | 0.3479 | 0.3484 | 0.3484 | 0.3489 | 0.3489 | 0.3494 | 0.3494 | 0.3499 | … | … | … | … | 0.3509[c] | 0.3514[c] | … | … |
| 7⁄16 | 14 | … | … | 0.4375 | 0.4422 | 0.4468 | 0.3911 | … | … | 0.3916 | 0.3921 | 0.3921 | 0.3926 | … | … | 0.3931[a] | 0.3936[a] | … | … | … | … | 0.3946 | 0.3951 |
| 7⁄16 | … | 20 | … | 0.4375 | 0.4407 | 0.4440 | 0.4050 | … | … | … | … | 0.4060 | 0.4065 | … | … | 0.4070[a] | 0.4075[a] | … | … | … | … | 0.4085 | 0.4090 |
| ½ | 13 | … | … | 0.5000 | 0.5050 | 0.5100 | 0.4500 | 0.4500 | 0.4505 | 0.4505 | 0.4510 | 0.4510 | 0.4515 | … | … | 0.4520[a] | 0.4525[a] | … | … | … | … | 0.4535 | 0.4540 |
| ½ | … | 20 | … | 0.5000 | 0.5032 | 0.5065 | 0.4675 | 0.4675 | 0.4680 | 0.4680 | 0.4685 | 0.4685 | 0.4690 | … | … | 0.4695[a] | 0.4700[a] | … | … | … | … | 0.4710 | 0.4715 |
| 9⁄16 | 12 | … | … | 0.5625 | 0.5679 | 0.5733 | 0.5084 | … | … | … | … | 0.5094 | 0.5099 | … | … | 0.5104[a] | 0.5109[a] | … | … | 0.5114[c] | 0.5119[c] | … | … |
| 9⁄16 | … | 18 | … | 0.5625 | 0.5661 | 0.5697 | 0.5264 | … | … | 0.5269 | 0.5274 | 0.5274 | 0.5279 | … | … | 0.5284[a] | 0.5289[a] | … | … | 0.5294[c] | 0.5299[c] | … | … |
| ⅝ | 11 | … | … | 0.6250 | 0.6309 | 0.6368 | 0.566 | … | … | 0.5665 | 0.567 | 0.567 | 0.5675 | … | … | 0.5680[a] | 0.5685[a] | … | … | 0.5690[c] | 0.5695[c] | … | … |
| ⅝ | … | 18 | … | 0.6250 | 0.6286 | 0.6322 | 0.5889 | … | … | 0.5894 | 0.5899 | 0.5899 | 0.5904 | … | … | 0.5909[a] | 0.5914[a] | … | … | 0.5919[c] | 0.5924[c] | … | … |
| 11⁄16 | … | … | 11 | 0.6875 | 0.6934 | 0.6993 | 0.6285 | … | … | … | … | 0.6295 | 0.6300 | … | … | … | … | … | … | … | … | … | … |
| 11⁄16 | … | … | 16 | 0.6875 | 0.6915 | 0.6956 | 0.6469 | … | … | … | … | 0.6479 | 0.6484 | … | … | … | … | … | … | … | … | … | … |
| ¾ | 10 | … | … | 0.7500 | 0.7565 | 0.7630 | 0.6850 | … | … | 0.6855 | 0.6860 | 0.6860 | 0.6865 | … | … | 0.6870 | 0.6875 | … | … | 0.6880[d] | 0.6885[d] | … | … |
| ¾ | … | 16 | … | 0.7500 | 0.7540 | 0.7581 | 0.7094 | 0.7094 | 0.7099 | 0.7099 | 0.7104 | 0.7104 | 0.7109 | … | … | 0.7114[a] | 0.1119[a] | … | … | 0.7124[d] | 0.4129[d] | … | … |
| ⅞ | 9 | … | … | 0.8750 | 0.8822 | 0.8894 | 0.8028 | … | … | … | … | … | … | 0.8043 | 0.8048 | … | … | 0.8053 | 0.8058 | … | … | … | … |

**Table 21.** *(Continued)* **Tap Thread Limits: Fractional Sizes, Ground Thread** *ANSI/ASME B94.9-2008 (R2018)* **(Unified and American National Thread Forms, Standard Thread Limits)**

| Size inch | NC UNC | NF UNF | NS | Major Diameter | | | Pitch Diameter | | | | | | | | | | | | | | | | |
|---|---|---|---|---|---|---|---|---|---|---|---|---|---|---|---|---|---|---|---|---|---|---|---|
| | | | | | | | | H1 limit | | H2 limit | | H3 limit | | H4 limit | | H5 limit | | H6 limit[a] | | H7 limit | | H8 limit[b] | |
| | | | | Basic | Min. | Max. | Basic | Min. | Max. | Min. | Max. | Min. | Max. | Min. | Max. | Min. | Max. | Min. | Max. | Min. | Max. | Min. | Max. |
| 7/8 | … | 14 | … | 0.8750 | 0.8797 | 0.8843 | 0.8286 | … | … | 0.8291 | 0.8296 | … | … | 0.8301 | 0.8306 | … | … | … | … | … | … | … | … |
| 1 | 8 | … | … | 1.0000 | 1.0082 | 1.0163 | 0.9188 | … | … | … | … | … | … | 0.9203 | 0.9208 | … | … | 0.9213 | 0.9218 | … | … | … | … |
| 1 | … | 12 | … | 1.0000 | 1.0054 | 1.0108 | 0.9459 | … | … | … | … | … | … | 0.9474 | 0.9479 | … | … | … | … | … | … | … | … |
| 1 | … | … | … | 1.0000 | 1.0047 | 1.0093 | 0.9536 | … | … | … | … | … | … | 0.9551 | 0.9556 | … | … | … | … | … | … | … | … |
| 1 1/8 | 7 | | … | 1.1250 | 1.1343 | 1.1436 | 1.0322 | … | … | … | … | … | … | 1.0337 | 1.0342 | … | … | … | … | … | … | … | … |
| 1 1/8 | | 12 | … | 1.1250 | 1.1304 | 1.1358 | 1.0709 | … | … | … | … | … | … | 1.0724 | 1.0729 | … | … | … | … | … | … | … | … |
| 1 1/4 | 7 | | … | 1.2500 | 1.2593 | 1.2686 | 1.1572 | … | … | … | … | … | … | 1.1587 | 1.1592 | … | … | … | … | … | … | … | … |
| 1 1/4 | | 12 | … | 1.2500 | 1.2554 | 1.2608 | 1.1959 | … | … | … | … | … | … | 1.1974 | 1.1979 | … | … | … | … | … | … | … | … |
| 1 3/8 | 6 | | … | 1.3750 | 1.3859 | 1.3967 | 1.2667 | … | … | … | … | … | … | 1.2682 | 1.2687 | … | … | … | … | … | … | … | … |
| 1 3/8 | | 12 | … | 1.3750 | 1.3804 | 1.3858 | 1.3209 | … | … | … | … | … | … | 1.3224 | 1.3229 | … | … | … | … | … | … | … | … |
| 1 1/2 | 6 | | … | 1.5000 | 1.5109 | 1.5217 | 1.3917 | … | … | … | … | … | … | 1.3932 | 1.3937 | … | … | … | … | … | … | … | … |
| 1 1/2 | | 12 | … | 1.5000 | 1.5054 | 1.5108 | 1.4459 | … | … | … | … | … | … | 1.4474 | 1.4479 | … | … | … | … | … | … | … | … |
| 1 3/4 | | 5 | … | 1.7500 | 1.7630 | 1.7760 | 1.6201 | … | … | … | … | … | … | 1.6216 | 1.6221 | … | … | … | … | … | … | … | … |
| 2 | | 4.5 | … | 2.0000 | 2.0145 | 2.0289 | 1.8557 | … | … | … | … | … | … | 1.8572 | 1.8577 | … | … | … | … | … | … | … | … |

[a] Minimum and maximum major diameters are 0.0010 larger than shown.

[b] Minimum and maximum major diameters are 0.0035 larger than shown.

[c] Minimum and maximum major diameters are 0.0020 larger than shown.

[d] Minimum and maximum major diameters are 0.0015 larger than shown.

General notes:

Limits listed in Table 21 are the most commonly used in the industry.

Not all styles of taps are available with all limits listed.

For calculation of limits other than those listed, see formulas and Table 2.

## Table 23a. Straight Pipe Thread Limits: NPS, Ground Thread ANSI Straight Pipe Thread Form (NPSC, NPSM) *ANSI/ASME B94.9-2008 (R2018)*

| Nominal Size, Inches | Threads per Inch, NPS, NPSC, NPSM | Major Diameter | | | Pitch Diameter | | |
|---|---|---|---|---|---|---|---|
| | | Plug at Gaging Notch | Min. *G* | Max. *H* | Plug at Gaging Notch *E* | Min. *K* | Max. *L* |
| ⅛ | 27 | 0.3983 | 0.4022 | 0.4032 | 0.3736 | 0.3746 | 0.3751 |
| ¼ | 18 | 0.5286 | 0.5347 | 0.5357 | 0.4916 | 0.4933 | 0.4938 |
| ⅜ | 18 | 0.6640 | 0.6701 | 0.6711 | 0.6270 | 0.6287 | 0.6292 |
| ½ | 14 | 0.8260 | 0.8347 | 0.8357 | 0.7784 | 0.7806 | 0.7811 |
| ¾ | 14 | 1.0364 | 1.0447 | 1.0457 | 0.9889 | 0.9906 | 0.9916 |
| 1 | 11½ | 1.2966 | 1.3062 | 1.3077 | 1.2386 | 1.2402 | 1.2412 |

Formulas for NPS Ground Thread Taps[a]

| Nominal Size | Major Diameter | | Minor Dia. | Threads per Inch | *A* | *B* |
|---|---|---|---|---|---|---|
| | Min. *G* | Max. *H* | Max. | 27 | 0.0296 | 0.0257 |
| ⅛ | $H-0.0010$ | $(K+A)-0.0010$ | $M-B$ | 18 | 0.0444 | 0.0401 |
| ¼ to ¾ | $H-0.0010$ | $(K+A)-0.0020$ | $M-B$ | 14 | 0.0571 | 0.0525 |
| 1 | $H-0.0015$ | $(K+A)-0.0021$ | $M-B$ | 11½ | 0.0696 | 0.0647 |

[a] In the formulas, *M* equals the actual measured pitch diameter.

All dimensions are given in inches.

*Maximum pitch diameter* of tap is based upon an allowance deducted from the maximum product pitch diameter of NPSC or NPSM, whichever is smaller.

*Minimum pitch diameter* of tap is derived by subtracting the ground thread pitch diameter tolerance for actual equivalent size.

*Lead tolerance:* A maximum lead deviation pf ± 0.0005 inch within any two threads not farther apart than 1 inch.

*Angle Tolerance:* 11½ to 27 threads per inch, plus or minus 30 min. in half angle.

Taps made to the specifications in Table 23a are to be marked NPS and used for NPSC and NPSM.

## Table 23b. Straight Pipe Thread Limits: NPSF Ground Thread ANSI Standard Straight Pipe Thread Form (NPSF) *ANSI/ASME B94.9-2008 (R2018)*

| Nominal Size, Inches | Threads per Inch | Major Diameter | | Pitch Diameter | | | |
|---|---|---|---|---|---|---|---|
| | | Min. *G* | Max. *H* | Plug at Gaging Notch *E* | Min. *K* | Max. *L* | Minor[a] Dia. Flat, Max. |
| $\frac{1}{16}$ | 27 | 0.3008 | 0.3018 | 0.2812 | 0.2772 | 0.2777 | 0.004 |
| ⅛ | 27 | 0.3932 | 0.3942 | 0.3736 | 0.3696 | 0.3701 | 0.004 |
| ¼ | 18 | 0.5239 | 0.5249 | 0.4916 | 0.4859 | 0.4864 | 0.005 |
| ⅜ | 18 | 0.6593 | 0.6603 | 0.6270 | 0.6213 | 0.6218 | 0.005 |
| ½ | 14 | 0.8230 | 0.8240 | 0.7784 | 0.7712 | 0.7717 | 0.005 |
| ¾ | 14 | 1.0335 | 1.0345 | 0.9889 | 0.9817 | 0.9822 | 0.005 |

[a] As specified or sharper.

All dimensions are given in inches.

## Table 23c. Standard Straight Pipe Thread Limits: NPSF Ground Thread Dryseal ANSI Standard Straight Pipe Thread Form (NPSF)

*ANSI/ASME B94.9-2008 (R2018)*

| Formulas for American Dryseal (NPSF) Ground Thread Taps | | | | | |
|---|---|---|---|---|---|
| Nominal Size, Inches | Major Diameter | | Pitch Diameter | | Max. Minor Dia. |
| | Min. $G$ | Max. $H$ | Min. $K$ | Max. $L$ | |
| 1/16 | $H - 0.0010$ | $K + Q - 0.0005$ | $L - 0.0005$ | $E - F$ | $M - Q$ |
| 1/8 | $H - 0.0010$ | $K + Q - 0.0005$ | $L - 0.0005$ | $E - F$ | $M - Q$ |
| 1/4 | $H - 0.0010$ | $K + Q - 0.0005$ | $L - 0.0005$ | $E - F$ | $M - Q$ |
| 3/8 | $H - 0.0010$ | $K + Q - 0.0005$ | $L - 0.0005$ | $E - F$ | $M - Q$ |
| 1/2 | $H - 0.0010$ | $K + Q - 0.0005$ | $L - 0.0005$ | $E - F$ | $M - Q$ |
| 3/4 | $H - 0.0010$ | $K + Q - 0.0005$ | $L - 0.0005$ | $E - F$ | $M - Q$ |

| Values to Use in Formulas | | | | |
|---|---|---|---|---|
| Threads per Inch | $E$ | $F$ | $M$ | $Q$ |
| 27 | Pitch diameter of plug at gaging notch | 0.0035 | Actual measured pitch diameter | 0.0251 |
| 18 | | 0.0052 | | 0.0395 |
| 14 | | 0.0067 | | 0.0533 |

All dimensions are given in inches.

*Lead Tolerance:* A maximum lead deviation of ±0.0005 inch within any two threads not farther apart than 1 inch.

*Angle Tolerance:* Plus or minus 30 min. in half angle for 14 to 27 threads per inch, inclusive.

## Table 23d. Standard Straight Pipe Tap Limits: (NPS) Cut Thread ANSI Straight Pipe Thread Form (NPSC) *ANSI/ASME B94.9-2008 (R2018)*

| Nominal Size | Threads per Inch, NPS, NPSC | Size at Gaging Notch | Pitch Diameter | | Values to Use in Formulas | | |
|---|---|---|---|---|---|---|---|
| | | | Min. | Max. | $A$ | $B$ | $C$ |
| 1/8 | 27 | 0.3736 | 0.3721 | 0.3751 | 0.0267 | 0.0296 | 0.0257 |
| 1/4 | 18 | 0.4916 | 0.4908 | 0.4938 | 0.0408 | 0.0444 | 0.0401 |
| 3/8 | 18 | 0.6270 | 0.6257 | 0.6292 | | | |
| 1/2 | 14 | 0.7784 | 0.7776 | 0.7811 | 0.0535 | 0.0571 | 0.0525 |
| 3/4 | 14 | 0.9889 | 0.9876 | 0.9916 | | | |
| 1 | 11½ | 1.2386 | 1.2372 | 1.2412 | 0.0658 | 0.0696 | 0.0647 |

The following are approximate formulas, in which $M$ = measured pitch diameter in inches:

Major dia., min. = $M + A$

Major dia., max. = $M + B$

Minor dia., max. = $M - C$

*Maximum pitch diameter* of tap is based on an allowance deducted from the maximum product pitch diameter of NPSC.

*Minimum pitch diameter* of tap equals maximum pitch diameter minus the tolerance.

All dimensions are given in inches.

*Lead Tolerance:* ±0.003 inch per inch of thread.

*Angle Tolerance:* For all pitches, tolerance will be ± 45″ for half angle and ± 68″ for full angle. Taps made to these specifications are to be marked NPS and used for NPSC thread form.

Taps made to the specifications in Table 23a are to be marked NPS and used for NPSC.

As the American National Standard straight pipe thread form is to be maintained, the major and minor diameters vary with the pitch diameter. Either a flat or rounded form is allowable at both the crest and the root.

**Table 25. Tap Thread Limits for Screw Thread Inserts (STI), Ground Thread, Machine Screw, and Fractional Size** *ANSI/ASME B94.9-2008 (R2018)*

| Nominal Screw Size STI | Fractional Size STI | Threads Per Inch | | Tap Major Diameter | | Pitch Diameter Limits | | | | | |
|---|---|---|---|---|---|---|---|---|---|---|---|
| | | | | | | 2B | | | 3B | | |
| | | NC | NF | Min. | Max. | H limit | Min. | Max. | H limit | Min. | Max. |
| 1 | … | 64 | … | 0.0948 | 0.0958 | H2 | 0.0837 | 0.0842 | H1 | 0.0832 | 0.0837 |
| 2 | … | 56 | … | 0.1107 | 0.1117 | H2 | 0.0981 | 0.0986 | H1 | 0.0976 | 0.0981 |
| | … | … | 64 | 0.1088 | 0.1088 | H2 | 0.0967 | 0.0972 | H1 | 0.0962 | 0.0967 |
| 3 | … | 48 | … | 0.1289 | 0.1289 | H2 | 0.1131 | 0.1136 | H1 | 0.1126 | 0.1131 |
| | … | … | 56 | 0.1237 | 0.1247 | H2 | 0.1111 | 0.1116 | H1 | 0.1106 | 0.1111 |
| 4 | … | 40 | … | 0.1463 | 0.1473 | H2 | 0.1288 | 0.1293 | H1 | 0.1283 | 0.1288 |
| | … | … | 48 | 0.1409 | 0.1419 | H2 | 0.1261 | 0.1266 | H1 | 0.1256 | 0.1261 |
| 5 | … | 40 | … | 0.1593 | 0.1603 | H2 | 0.1418 | 0.1423 | H1 | 0.1413 | 0.1418 |
| 6 | … | 32 | … | 0.1807 | 0.1817 | H3 | 0.1593 | 0.1598 | H2 | 0.1588 | 0.1593 |
| | … | … | 40 | 0.1723 | 0.1733 | H2 | 0.1548 | 0.1553 | H1 | 0.1543 | 0.1548 |
| 8 | … | 32 | … | 0.2067 | 0.2077 | H3 | 0.1853 | 0.1858 | H2 | 0.1848 | 0.1853 |
| | … | … | 36 | 0.2022 | 0.2032 | H2 | 0.1826 | 0.1831 | H1 | 0.1821 | 0.1826 |
| 10 | … | 24 | … | 0.2465 | 0.2475 | H3 | 0.2180 | 0.2185 | H2 | 0.2175 | 0.2180 |
| | … | … | 32 | 0.2327 | 0.2337 | H3 | 0.2113 | 0.2118 | H2 | 0.2108 | 0.2113 |
| 12 | … | 24 | … | 0.2725 | 0.2735 | H3 | 0.2440 | 0.2445 | H2 | 0.2435 | 0.2440 |
| … | 1/4 | 20 | … | 0.3177 | 0.3187 | H3 | 0.2835 | 0.2840 | H2 | 0.2830 | 0.2835 |
| … | … | … | 28 | 0.2985 | 0.2995 | H3 | 0.2742 | 0.2747 | H2 | 0.2737 | 0.2742 |
| … | 5/16 | 18 | … | 0.3874 | 0.3884 | H4 | 0.3501 | 0.3506 | H3 | 0.3496 | 0.3501 |
| … | … | … | 24 | 0.3690 | 0.3700 | H3 | 0.3405 | 0.3410 | H2 | 0.3400 | 0.3405 |
| … | 3/8 | 16 | … | 0.4592 | 0.4602 | H4 | 0.4171 | 0.4176 | H3 | 0.4166 | 0.4171 |
| … | … | … | 24 | 0.4315 | 0.4325 | H3 | 0.4030 | 0.4035 | H2 | 0.4025 | 0.4030 |
| … | 7/16 | 14 | … | 0.5333 | 0.5343 | H4 | 0.4854 | 0.4859 | H3 | 0.4849 | 0.4854 |
| … | … | … | 20 | 0.5052 | 0.5062 | H4 | 0.4715 | 0.4720 | H3 | 0.4710 | 0.4715 |
| … | 1/2 | 13 | … | 0.6032 | 0.6042 | H4 | 0.5514 | 0.5519 | H3 | 0.5509 | 0.5514 |
| … | … | … | 20 | 0.5677 | 0.5687 | H4 | 0.5340 | 0.5345 | H3 | 0.5335 | 0.5340 |
| … | 9/16 | 12 | … | 0.6741 | 0.6751 | H4 | 0.6182 | 0.6187 | H3 | 0.6117 | 0.6182 |
| … | … | … | 18 | 06374 | 0.6384 | H4 | 0.6001 | 0.6006 | H3 | 0.5996 | 0.6001 |
| … | 5/8 | 11 | … | 0.7467 | 0.7477 | H4 | 0.6856 | 0.6861 | H3 | 0.6851 | 0.6856 |
| … | … | … | 18 | 0.6999 | 0.7009 | H4 | 0.6626 | 0.6631 | H3 | 0.6621 | 0.6626 |
| … | 3/4 | 10 | … | 0.8835 | 0.8850 | H5 | 0.8169 | 0.8174 | H3 | 0.8159 | 0.8164 |
| … | … | … | 18 | 0.8342 | 0.8352 | H4 | 0.7921 | 0.7926 | H3 | 0.7916 | 0.7921 |
| … | 7/8 | 9 | … | 1.0232 | 1.0247 | H5 | 0.9491 | 0.9496 | H3 | 0.9481 | 0.9486 |
| … | … | … | 14 | 0.9708 | 0.9718 | H4 | 0.9234 | 0.9239 | H3 | 0.9224 | 0.9229 |
| … | 1 | 8 | … | 1.1666 | 1.1681 | H6 | 1.0832 | 1.0842 | H4 | 1.0822 | 1.0832 |
| … | … | … | 12 | 1.1116 | 1.1126 | H6 | 1.0562 | 1.0572 | H4 | 1.0552 | 1.0562 |
| … | … | … | 14 NS | 1.0958 | 1.0968 | H6 | 1.0484 | 1.0494 | H4 | 1.0474 | 1.0484 |
| … | 1 1/8 | 7 | … | 1.3151 | 1.3171 | H6 | 1.2198 | 1.2208 | H4 | 1.2188 | 1.2198 |
| … | | | 12 | 1.2366 | 1.2376 | H6 | 1.1812 | 1.1822 | H4 | 1.1802 | 1.1812 |
| … | 1 1/4 | 7 | … | 1.4401 | 1.4421 | H6 | 1.3448 | 1.3458 | H4 | 1.3438 | 1.3448 |
| … | … | … | 12 | 1.3616 | 1.3626 | H6 | 1.3062 | 1.3072 | H4 | 1.3052 | 1.3062 |
| … | 1 3/8 | 6 | … | 1.5962 | 1.5982 | H8 | 1.4862 | 1.4872 | H6 | 1.4852 | 1.4862 |
| … | … | … | 12 | 1.4866 | 1.4876 | H6 | 1.4312 | 1.4322 | H4 | 1.4302 | 1.4312 |
| … | 1 1/2 | 6 | … | 1.7212 | 1.7232 | H8 | 1.6112 | 1.6122 | H6 | 1.6102 | 1.6112 |
| … | … | … | 12 | 1.6116 | 1.6126 | H6 | 1.5562 | 1.5572 | H4 | 1.5552 | 1.5562 |

These taps are over the nominal size to the extent that the internal thread they produce will accommodate a helical coil screw insert, which at final assembly will accept a screw thread of the normal size and pitch.

**Table 27. Dimensions of Acme Threads Taps in Sets of Three Taps**

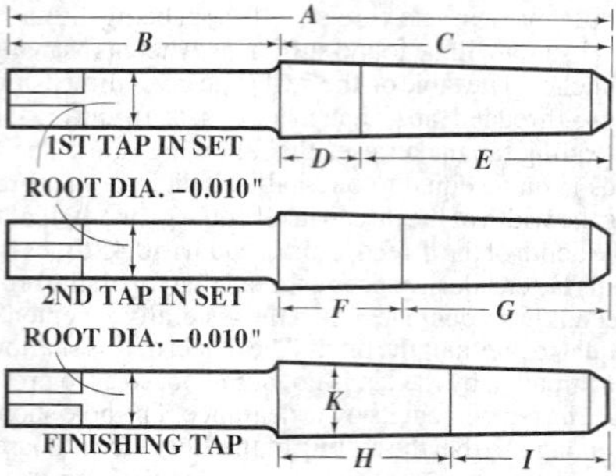


| Nominal Dia. | *A* | *B* | *C* | *D* | *E* | *F* | *G* | *H* | *I* | *K* |
|---|---|---|---|---|---|---|---|---|---|---|
| 1/2 | 4 1/4 | 1 7/8 | 2 3/8 | 1/2 | 1 7/8 | 5/8 | 1 3/4 | 7/8 | 1 1/2 | 0.520 |
| 9/16 | 4 7/8 | 2 1/8 | 2 3/4 | 9/16 | 2 3/16 | 3/4 | 2 | 1 | 1 3/4 | 0.582 |
| 5/8 | 5 1/2 | 2 3/8 | 3 1/8 | 5/8 | 2 1/2 | 7/8 | 2 1/4 | 1 1/8 | 2 | 0.645 |
| 11/16 | 6 | 2 1/2 | 3 1/2 | 3 13/16 | 2 13/16 | 15/16 | 2 9/16 | 1 1/4 | 2 1/4 | 0.707 |
| 3/4 | 6 1/2 | 2 11/16 | 3 13/16 | 11/16 | 3 1/8 | 1 | 2 13/16 | 1 3/8 | 2 7/16 | 0.770 |
| 13/16 | 6 7/8 | 2 13/16 | 4 1/16 | 3/4 | 3 5/16 | 1 1/16 | 3 | 1 7/16 | 2 5/8 | 0.832 |
| 7/8 | 7 1/4 | 3 | 4 1/4 | 3/4 | 3 1/2 | 1 1/8 | 3 1/8 | 1 1/2 | 2 3/4 | 0.895 |
| 15/16 | 7 9/16 | 3 1/8 | 4 7/16 | 13/16 | 3 5/8 | 1 3/16 | 3 1/4 | 1 9/16 | 2 7/8 | 0.957 |
| 1 | 7 7/8 | 3 1/4 | 4 5/8 | 13/16 | 3 13/16 | 1 1/4 | 3 3/8 | 1 5/8 | 3 | 1.020 |
| 1 1/8 | 8 1/2 | 3 9/16 | 4 15/16 | 7/8 | 4 1/16 | 1 5/16 | 3 5/8 | 1 3/4 | 3 3/16 | 1.145 |
| 1 1/4 | 9 | 3 3/4 | 5 1/4 | 15/16 | 4 5/16 | 1 3/8 | 3 7/8 | 1 7/8 | 3 3/8 | 1.270 |
| 1 3/8 | 9 1/2 | 4 | 5 1/2 | 1 | 4 1/2 | 1 7/16 | 4 1/16 | 2 | 3 1/2 | 1.395 |
| 1 1/2 | 10 | 4 1/4 | 5 3/4 | 1 | 4 3/4 | 1 1/2 | 4 1/4 | 2 1/8 | 3 5/8 | 1.520 |
| 1 5/8 | 10 1/2 | 4 1/2 | 6 | 1 | 5 | 1 1/2 | 4 1/2 | 2 1/8 | 3 7/8 | 1.645 |
| 1 3/4 | 11 | 4 3/4 | 6 1/4 | 1 1/16 | 5 3/16 | 1 9/16 | 4 11/16 | 2 1/4 | 4 | 1.770 |
| 1 7/8 | 11 3/8 | 4 7/8 | 6 1/2 | 1 1/16 | 5 7/16 | 1 9/16 | 4 15/16 | 2 1/4 | 4 1/4 | 1.895 |
| 2 | 11 3/4 | 5 | 6 3/4 | 1 1/8 | 5 5/8 | 1 5/8 | 5 1/8 | 2 3/8 | 4 3/8 | 2.020 |
| 2 1/4 | 12 1/2 | 5 1/4 | 7 1/4 | 1 1/8 | 6 1/8 | 1 13/16 | 5 1/2 | 2 1/2 | 4 3/4 | 2.270 |
| 2 1/2 | 13 1/4 | 5 1/2 | 7 3/4 | 1 3/4 | 6 9/16 | 1 7/8 | 5 7/8 | 2 5/8 | 5 1/8 | 2.520 |
| 2 3/4 | 14 | 5 3/4 | 8 1/4 | 1 1/4 | 7 | 2 | 6 1/4 | 2 3/4 | 5 1/2 | 2.770 |
| 3 | 15 | 6 1/4 | 8 3/4 | 1 1/4 | 7 1/2 | 2 | 6 3/4 | 3 | 5 3/4 | 3.020 |

*Diameters of General-Purpose Acme Screw Threads of Classes 2G, 3G, and 4G* may be calculated from pitch = 1/number of threads per inch, and:

minimum diameter = basic major diameter − pitch

maximum diameter = minimum minor diameter + 0.05 × pitch

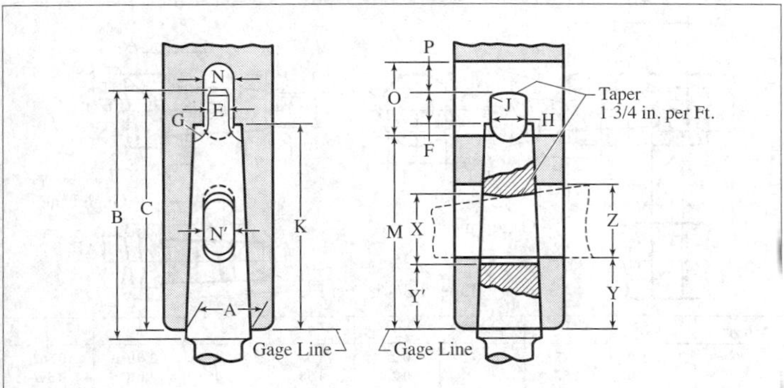

**Table 6. American National Standard Steep Machine Tapers**
*ANSI/ASME B5.10-1994 (R2013)*

| No. of Taper | Taper per Foot[a] | Dia. at Gage Line[b] | Length Along Axis | No. of Taper | Taper per Foot[a] | Dia. at Gage Line[b] | Length Along Axis |
|---|---|---|---|---|---|---|---|
| 5 | 3.500 | 0.500 | 0.6875 | 35 | 3.500 | 1.500 | 2.2500 |
| **10** | **3.500** | **0.625** | **0.8750** | **40** | **3.500** | **1.750** | **2.5625** |
| 15 | 3.500 | 0.750 | 1.0625 | 45 | 3.500 | 2.250 | 3.3125 |
| **20** | **3.500** | **0.875** | **1.3125** | **50** | **3.500** | **2.750** | **4.0000** |
| 25 | 3.500 | 1.000 | 1.5625 | 55 | 3.500 | 3.500 | 5.1875 |
| **30** | **3.500** | **1.250** | **1.8750** | **60** | **3.500** | **4.250** | **6.3750** |

[a] This taper corresponds to an included angle of 16°, 35′, 39.4″.
[b] The basic diameter at gage line is at large end of taper.
All dimensions given in inches.

The tapers numbered 10, 20, 30, 40, 50, and 60 that are printed in heavy-faced type are designated as the "Preferred Series." The tapers numbered 5, 15, 25, 35, 45, and 55 that are printed in light-faced type are designated as the "Intermediate Series."

**Table 7a. American National Standard Self-Holding Tapers — Basic Dimensions** *ANSI/ASME B5.10-1994 (R2013)*

| No. of Taper | Taper per Foot | Dia. at Gage Line[a] *A* | Means of Driving and Holding[a] | Origin of Series |
|---|---|---|---|---|
| .239 | 0.50200 | 0.23922 | Tang Drive with Shank Held in by Friction (See Table 2) (.239 through 7) | Brown & Sharpe Taper Series |
| .299 | 0.50200 | 0.29968 | | |
| .375 | 0.50200 | 0.37525 | | |
| 1 | 0.59858 | 0.47500 | | Morse Taper Series |
| 2 | 0.59941 | 0.70000 | | |
| 3 | 0.60235 | 0.93800 | Tang Drive with Shank Held in by Key (See Table 3) (3 through 6) | |
| 4 | 0.62326 | 1.23100 | | |
| 4½ | 0.62400 | 1.50000 | | |
| 5 | 0.63151 | 1.74800 | | |
| 6 | 0.62565 | 2.49400 | | |
| 7 | 0.62400 | 3.27000 | | |
| 200 | 0.750 | 2.000 | Key Drive with Shank Held in by Key (See Table 4); Key Drive with Shank Held in by Draw-bolt (See Table 5) (200 through 1200) | ¾ Inch per Foot Taper Series |
| 250 | 0.750 | 2.500 | | |
| 300 | 0.750 | 3.000 | | |
| 350 | 0.750 | 3.500 | | |
| 400 | 0.750 | 4.000 | | |
| 450 | 0.750 | 4.500 | | |
| 500 | 0.750 | 5.000 | | |
| 600 | 0.750 | 6.000 | | |
| 800 | 0.750 | 8.000 | | |
| 1000 | 0.750 | 10.000 | | |
| 1200 | 0.750 | 12.000 | | |

[a] See illustrations above Table 2 through Table 5.
All dimensions given in inches.

**Table 12. American National Standard Plug and Ring Gages for Steep Machine Tapers** *ANSI/ASME B5.10-1994 (R2013)*

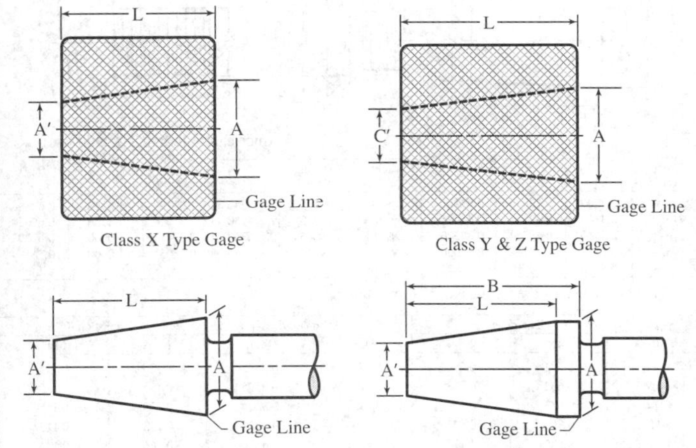


Class X Type Gage — Class Y & Z Type Gage

Class X Type Gage — Class Y & Z Type Gage

| No. of Taper | Taper per Foot[a] (Basic) | Diameter at Gage Line[a] *A* | Tolerances for Diameter A[b] | | | Diameter at Small End[a] *A′* | Length Gage Line to Small End *L* | Overall Length of Gage Body *B* | Dia. of Opening *C* |
|---|---|---|---|---|---|---|---|---|---|
| | | | Class X Gage | Class Y Gage | Class Z Gage | | | | |
| 5 | 3.500 | 0.500 | 0.00004 | 0.00007 | 0.00010 | 0.2995 | 0.6875 | 0.81 | 0.30 |
| 10 | 3.500 | 0.625 | 0.00004 | 0.00007 | 0.00010 | 0.3698 | 0.8750 | 1.00 | 0.36 |
| 15 | 3.500 | 0.750 | 0.00004 | 0.00007 | 0.00010 | 0.4401 | 1.0625 | 1.25 | 0.44 |
| 20 | 3.500 | 0.875 | 0.00006 | 0.00009 | 0.00012 | 0.4922 | 1.3125 | 1.50 | 0.48 |
| 25 | 3.500 | 1.000 | 0.00006 | 0.00009 | 0.00012 | 0.5443 | 1.5625 | 1.75 | 0.53 |
| 30 | 3.500 | 1.250 | 0.00006 | 0.00009 | 0.00012 | 0.7031 | 1.8750 | 2.06 | 0.70 |
| 35 | 3.500 | 1.500 | 0.00006 | 0.00009 | 0.00012 | 0.8438 | 2.2500 | 2.44 | 0.84 |
| 40 | 3.500 | 1.750 | 0.00008 | 0.00012 | 0.00016 | 1.0026 | 2.5625 | 2.75 | 1.00 |
| 45 | 3.500 | 2.250 | 0.00008 | 0.00012 | 0.00016 | 1.2839 | 3.3125 | 3.50 | 1.00 |
| 50 | 3.500 | 2.750 | 0.00010 | 0.00015 | 0.00020 | 1.5833 | 4.0000 | 4.25 | 1.00 |
| 55 | 3.500 | 3.500 | 0.00010 | 0.00015 | 0.00020 | 1.9870 | 5.1875 | 5.50 | 1.00 |
| 60 | 3.500 | 4.250 | 0.00010 | 0.00015 | 0.00020 | 2.3906 | 6.3750 | 6.75 | 2.00 |

[a] The taper per foot and diameter *A* at gage line are basic dimensions. Dimensions in Column *A′* are calculated for reference only.

[b] Tolerances for diameter *A* are plus for plug gages and minus for ring gages.

All dimensions are in inches.

The amounts of taper deviation for Class X, Class Y, and Class Z gages are the same, respectively, as the amounts shown for tolerances on diameter *A*. Taper deviation is the permissible allowance from true taper at any point of diameter in the length of the gage. On taper *plug* gages, this deviation may be applied only in the direction that *decreases* the rate of taper. On taper *ring* gages, this deviation may be applied only in the direction that *increases* the rate of taper. Tolerances on two-decimal dimensions are ±0.010.

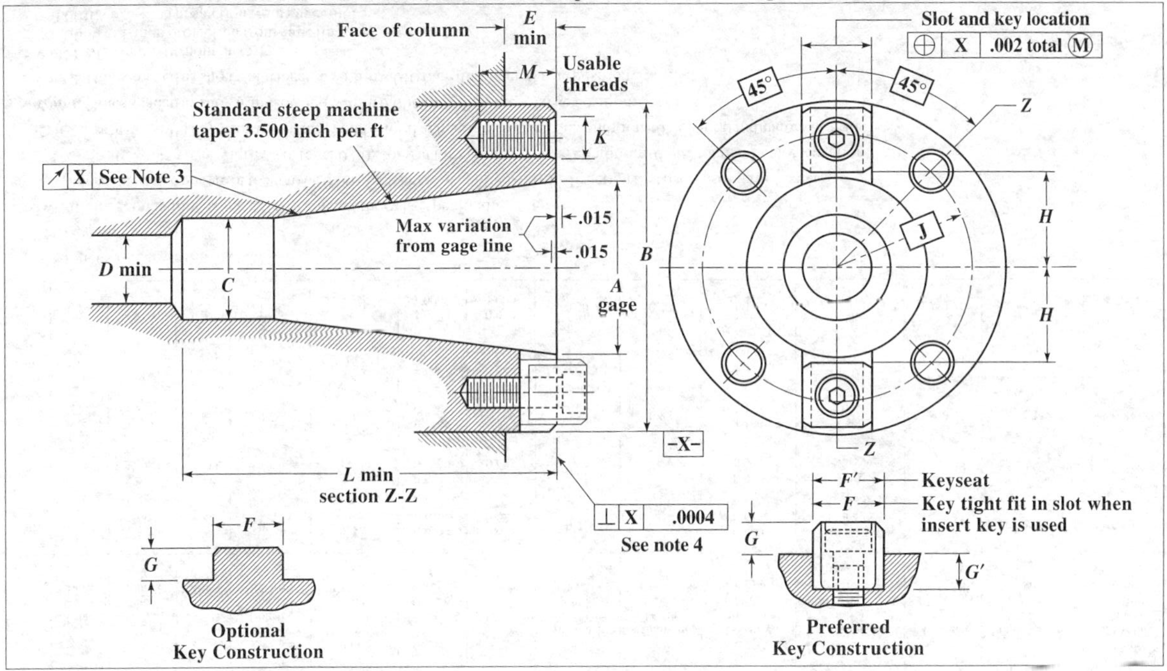

## ARBORS, CHUCKS, AND SPINDLES

### Portable Tool Spindles

**Circular Saw Arbors.**—ASME B107.4-2005 (R2011) "Driving and Spindle Ends for Portable Hand, Air, and Air Electric Tools" calls for a round arbor of ⅝-inch diameter for nominal saw blade diameters of 6 to 8.5 inches, inclusive, and a ¾-inch diameter round arbor for saw blade diameters of 9 to 12 inches, inclusive.

**Spindles for Geared Chucks.**—Recommended threaded and tapered spindles for portable tool geared chucks of various sizes are as given in the following table:

**Recommended Spindle Sizes**

| Chuck Sizes, Inch | Recommended Spindles | |
|---|---|---|
| | Threaded | Taper[a] |
| 3/16 and ¼ Light | ⅜-24 | 1 |
| ¼ and 5/16 Medium | ⅜-24 or ½-20 | 2 Short |
| ⅜ Light | ⅜-24 or ½-20 | 2 |
| ⅜ Medium | ½-20 or ⅝-16 | 2 |
| ½ Light | ½-20 or ⅝-16 | 33 |
| ½ Medium | ⅝-16 or ¾-16 | 6 |
| ⅝ and ¾ Medium | ⅝-16 or ¾-16 | 3 |

[a] Jacobs number.

**Vertical and Angle Portable Tool Grinder Spindles.**—The ⅝-11 spindle with a length of 1⅛ inches shown on page 1028 is designed to permit the use of a jam nut with threaded cup wheels. When a revolving guard is used, the length of the spindle is measured from the wheel bearing surface of the guard. For unthreaded wheels with a ⅞-inch hole, a safety sleeve nut is recommended. The unthreaded wheel with ⅝-inch hole is not recommended because a jam nut alone may not resist the inertia effect when motor power is cut off.

**Straight Grinding Wheel Spindles for Portable Tools.**—Portable grinders with pneumatic or induction electric motors should be designed for the use of organic bond wheels rated 9500 ft per min (48.25 m/s). Light-duty electric grinders may be designed for vitrified wheels rated 6500 ft per min (33.0 m/s). Recommended maximum sizes of wheels of both types are as given in the following table:

**Recommended Maximum Grinding Wheel Sizes for Portable Tools**

| Spindle Size | Maximum Wheel Dimensions | | | |
|---|---|---|---|---|
| | 9500 fpm | | 6500 fpm | |
| | Diameter *D* | Thickness *T* | Diameter *D* | Thickness *T* |
| ⅜-24 × 1⅛ | 2½ | ½ | 4 | ½ |
| ½-13 × 1¾ | 4 | ¾ | 5 | ¾ |
| ⅝-11 × 2⅛ | 8 | 1 | 8 | 1 |
| ⅝-11 × 3⅛ | 6 | 2 | … | … |
| ⅝-11 × 3⅛ | 8 | 1½ | … | … |
| ¾-10 × 3¼ | 8 | 2 | … | … |

Minimum *T* with the first three spindles is about ⅛ inch to accommodate cutting off wheels. Flanges are assumed to be according to ANSI B7.1 and threads to ANSI/ASME B1.1.

## American Standard Square Drives for Portable Air and Electric Tools *ASA B5.38-1958*

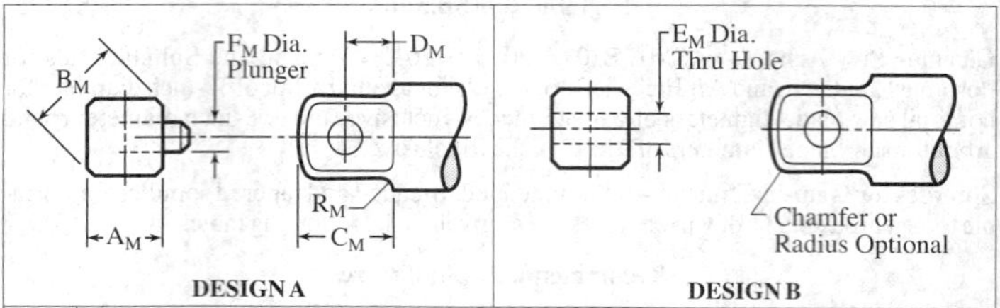


| Drive Size | Design. | $A_M$ Max. | $A_M$ Min. | $B_M$ Max. | $C_M$ Max. | $C_M$ Min. | $D_M$ Max. | $D_M$ Min. | $E_M$ Min. | $F_M$ Max. | $R_M$ Max. |
|---|---|---|---|---|---|---|---|---|---|---|---|
| | | Male End | | | | | | | | | |
| ¼ | A | 0.252 | 0.247 | 0.330 | 0.312 | 0.265 | 0.165 | 0.153 | … | 0.078 | 0.015 |
| ⅜ | A | 0.377 | 0.372 | 0.500 | 0.438 | 0.406 | 0.227 | 0.215 | … | 0.156 | 0.031 |
| ½ | A | 0.502 | 0.497 | 0.665 | 0.625 | 0.531 | 0.321 | 0.309 | … | 0.187 | 0.031 |
| ⅝ | A | 0.627 | 0.622 | 0.834 | 0.656 | 0.594 | 0.321 | 0.309 | … | 0.187 | 0.047 |
| ¾ | B | 0.752 | 0.747 | 1.000 | 0.938 | 0.750 | 0.415 | 0.403 | 0.216 | … | 0.047 |
| 1 | B | 1.002 | 0.997 | 1.340 | 1.125 | 1.000 | 0.602 | 0.590 | 0.234 | … | 0.063 |
| 1½ | B | 1.503 | 1.498 | 1.968 | 1.625 | 1.562 | 0.653 | 0.641 | 0.310 | … | 0.094 |

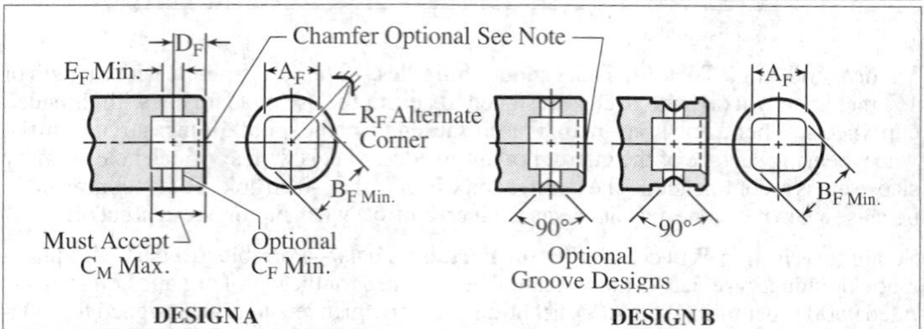


| Drive Size | Design | $A_F$ Max. | $A_F$ Min. | $B_F$ Min. | $D_F$ Max. | $D_F$ Min. | $E_F$ Min. | $R_F$ Max. |
|---|---|---|---|---|---|---|---|---|
| | | Female End | | | | | | |
| ¼ | A | 0.258 | 0.253 | 0.335 | 0.159 | 0.147 | 0.090 | … |
| ⅜ | A | 0.383 | 0.378 | 0.505 | 0.221 | 0.209 | 0.170 | … |
| ½ | A | 0.508 | 0.503 | 0.670 | 0.315 | 0.303 | 0.201 | … |
| ⅝ | A | 0.633 | 0.628 | 0.839 | 0.315 | 0.303 | 0.201 | … |
| ¾ | B | 0.758 | 0.753 | 1.005 | 0.409 | 0.397 | 0.216 | 0.047 |
| 1 | B | 1.009 | 1.004 | 1.350 | 0.596 | 0.584 | 0.234 | 0.062 |
| 1½ | B | 1.510 | 1.505 | 1.983 | 0.647 | 0.635 | 0.310 | 0.125 |

All dimensions in inches.

Incorporating fillet radius ($R_M$) at shoulder of male tang precludes use of minimum diameter cross-hole in socket ($E_F$), unless female drive end is chamfered (shown as optional).

If female drive end is not chamfered, socket cross-hole diameter ($E_F$) is increased to compensate for fillet radius $R_M$, max.

Minimum clearance across flats male to female is 0.001 inch through ¾-inch size; 0.002 inch in 1- and 1½-inch sizes. For impact wrenches, $A_M$ should be held as close to maximum as practical.

$C_F$, min., for both designs A and B should be equal to $C_M$, max.

**Table 1. Designing Data for Surface Broaches**

| Material to be Broached | Depth of Cut per Tooth | | | | Face Angle or Rake, Degrees | Clearance Angle, Degrees | |
|---|---|---|---|---|---|---|---|
| | Roughing[a] | | Finishing | | | | |
| | inch | mm | inch | mm | | Rough | Finish |
| Steel, High Tensile Strength | 0.0015–0.002 | 0.04–0.05 | 0.0005 | 0.013 | 10–12 | 1.5–3 | 0.5–1 |
| Steel, Med.Tensile Strength | 0.0025–0.005 | 0.06–0.13 | 0.0005 | 0.013 | 14–18 | 1.5–3 | 0.5–1 |
| Cast Steel | 0.0025–0.005 | 0.06–0.13 | 0.0005 | 0.013 | 10 | 1.53 | 0.5 |
| Malleable Iron | 0.0025–0.005 | 0.06–0.13 | 0.0005 | 0.013 | 7 | 1.5–3 | 0.5 |
| Cast Iron, Soft | 0.006–0.010 | 0.15–0.25 | 0.0005 | 0.013 | 10–15 | 1.5–3 | 0.5 |
| Cast Iron, Hard | 0.003–0.005 | 0.08–0.13 | 0.0005 | 0.013 | 5 | 1.5–3 | 0.5 |
| Zinc Die Castings | 0.005–0.010 | 0.13–0.25 | 0.0010 | 0.025 | 12[b] | 5 | 2 |
| Cast Bronze | 0.010–0.025 | 0.25–0.64 | 0.0005 | 0.013 | 8 | 0 | 0 |
| Wrought Aluminum Alloys | 0.005–0.010 | 0.13–0.25 | 0.0010 | 0.025 | 15[b] | 3 | 1 |
| Cast Aluminum Alloys | 0.005–0.010 | 0.13–0.25 | 0.0010 | 0.025 | 12[b] | 3 | 1 |
| Magnesium Die Castings | 0.010–0.015 | 0.25–0.38 | 0.0010 | 0.025 | 20[b] | 3 | 1 |

[a] The lower depth-of-cut values for roughing are recommended when work is not very rigid, the tolerance is small, a good finish is required, or length of cut is comparatively short.

[b] In broaching these materials, smooth surfaces for tooth and chip spaces are especially recommended.

**Table 2. Broaching Pressure *P* for Use in Pitch Formulas (2a) and (2b)**

| Material to be Broached | Depth *d* of cut per tooth, inch (mm) | | | | | | | | | | Side-cutting Broaches | | | |
|---|---|---|---|---|---|---|---|---|---|---|---|---|---|---|
| | 0.024 (0.60) | | 0.01 (0.25) | | 0.004 (0.10) | | 0.002 (0.05) | | 0.001 (0.025) | | | | | |
| | Pressure, *P* | | | | | | | | | | Pressure, *P* | | Cut, *d* | |
| | Ton/in² | MPa | Ton/in² | MPa | Ton/in² | MPa | Ton/in² | MPa | Ton/in² | MPa | Ton/in² | MPa | inch | mm |
| Steel, High Tensile Strength | … | … | … | … | | | 250 | 3447 | 312 | 4302 | 200 | 2758 | 0.004 | 0.10 |
| Steel, Med. Tensile Strength | … | … | … | … | 158 | 2179 | 185 | 2551 | 243 | 3351 | 143 | 1972 | 0.006 | 0.15 |
| Cast Steel | … | … | … | … | 128 | 1765 | 158 | 2179 | … | … | 115 | 1586 | 0.006 | 0.15 |
| Malleable Iron | … | … | … | … | 108 | 1489 | 128 | 1765 | … | … | 100 | 1379 | 0.006 | 0.15 |
| Cast Iron | … | … | 115 | 1586 | 115 | 1586 | 143 | 1972 | … | … | 115 | 1586 | 0.020 | 0.51 |
| Cast Brass | … | … | 50 | 689 | 50 | 689 | … | … | … | … | … | … | … | … |
| Brass, Hot-pressed | … | … | 85 | 1172 | 85 | 1172 | … | … | … | … | … | … | … | … |
| Zinc Die Castings | … | … | 70 | 965 | 70 | 965 | … | … | … | … | … | … | … | … |
| Cast Bronze | 35 | 483 | 35 | 483 | … | … | … | … | … | … | … | … | … | … |
| Wrought Aluminum | … | … | 70 | 965 | 70 | 965 | … | … | … | … | … | … | … | … |
| Cast Aluminum | … | … | 85 | 1172 | 85 | 1172 | … | … | … | … | … | … | … | … |
| Magnesium Alloy | 35 | 483 | 35 | 483 | … | … | … | … | … | … | … | … | … | … |

The minimum pitch shown by Formula (1) is based upon the receiving capacity of the chip space. The minimum pitch should not be less than 0.2 inch (5.0 mm) unless a smaller pitch is required for extremely short cuts to provide at least two teeth in contact simultaneously with the part being broached. A reduction below 0.2 inch (5.0 mm) is seldom required in surface broaching but may be necessary in connection with internal broaching.

$$\text{Minimum pitch} = 3\sqrt{LdF} \tag{1}$$

Whether the minimum pitch may be used or not depends upon the power of the available machine. The factor $F$ in the formula provides for the increase in volume as the material is broached into chips. If a broach has adjustable inserts for the finishing teeth, the pitch of the finishing teeth may be smaller than the pitch of the roughing teeth because of the smaller depth $d$ of the cut. The higher value of $F$ for finishing teeth prevents the pitch from becoming too small, so that the spirally curled chips will not be crowded into too small a

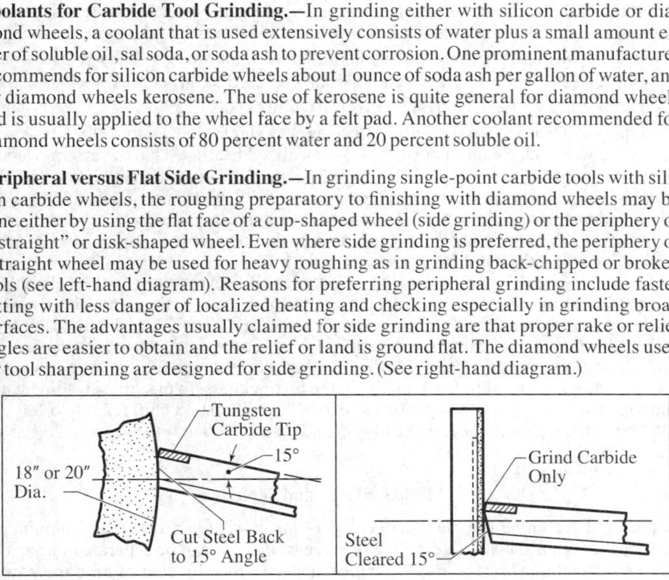

## Revolutions per Minute for Various Cutting Speeds and Diameters

| Dia., Inches | Cutting Speed, Feet per Minute | | | | | | | | | | | |
|---|---|---|---|---|---|---|---|---|---|---|---|---|
| | 225 | 250 | 275 | 300 | 325 | 350 | 375 | 400 | 425 | 450 | 500 | 550 |
| | Revolutions per Minute | | | | | | | | | | | |
| ¼ | 3438 | 3820 | 4202 | 4584 | 4966 | 5348 | 5730 | 6112 | 6493 | 6875 | 7639 | 8403 |
| 5⁄16 | 2750 | 3056 | 3362 | 3667 | 3973 | 4278 | 4584 | 4889 | 5195 | 5501 | 6112 | 6723 |
| 3⁄8 | 2292 | 2546 | 2801 | 3056 | 3310 | 3565 | 3820 | 4074 | 4329 | 4584 | 5093 | 5602 |
| 7⁄16 | 1964 | 2182 | 2401 | 2619 | 2837 | 3056 | 3274 | 3492 | 3710 | 3929 | 4365 | 4802 |
| ½ | 1719 | 1910 | 2101 | 2292 | 2483 | 2675 | 2866 | 3057 | 3248 | 3439 | 3821 | 4203 |
| 9⁄16 | 1528 | 1698 | 1868 | 2037 | 2207 | 2377 | 2547 | 2717 | 2887 | 3056 | 3396 | 3736 |
| 5⁄8 | 1375 | 1528 | 1681 | 1834 | 1987 | 2139 | 2292 | 2445 | 2598 | 2751 | 3057 | 3362 |
| 11⁄16 | 1250 | 1389 | 1528 | 1667 | 1806 | 1941 | 2084 | 2223 | 2362 | 2501 | 2779 | 3056 |
| ¾ | 1146 | 1273 | 1401 | 1528 | 1655 | 1783 | 1910 | 2038 | 2165 | 2292 | 2547 | 2802 |
| 13⁄16 | 1058 | 1175 | 1293 | 1410 | 1528 | 1646 | 1763 | 1881 | 1998 | 2116 | 2351 | 2586 |
| 7⁄8 | 982 | 1091 | 1200 | 1310 | 1419 | 1528 | 1637 | 1746 | 1855 | 1965 | 2183 | 2401 |
| 15⁄16 | 917 | 1019 | 1120 | 1222 | 1324 | 1426 | 1528 | 1630 | 1732 | 1834 | 2038 | 2241 |
| 1 | 859 | 955 | 1050 | 1146 | 1241 | 1337 | 1432 | 1528 | 1623 | 1719 | 1910 | 2101 |
| 1 1⁄16 | 809 | 899 | 988 | 1078 | 1168 | 1258 | 1348 | 1438 | 1528 | 1618 | 1798 | 1977 |
| 1 1⁄8 | 764 | 849 | 933 | 1018 | 1103 | 1188 | 1273 | 1358 | 1443 | 1528 | 1698 | 1867 |
| 1 3⁄16 | 724 | 804 | 884 | 965 | 1045 | 1126 | 1206 | 1287 | 1367 | 1448 | 1609 | 1769 |
| 1¼ | 687 | 764 | 840 | 917 | 993 | 1069 | 1146 | 1222 | 1299 | 1375 | 1528 | 1681 |
| 1 5⁄16 | 654 | 727 | 800 | 873 | 946 | 1018 | 1091 | 1164 | 1237 | 1309 | 1455 | 1601 |
| 1 3⁄8 | 625 | 694 | 764 | 833 | 903 | 972 | 1042 | 1111 | 1181 | 1250 | 1389 | 1528 |
| 1 7⁄16 | 598 | 664 | 730 | 797 | 863 | 930 | 996 | 1063 | 1129 | 1196 | 1329 | 1461 |
| 1½ | 573 | 636 | 700 | 764 | 827 | 891 | 955 | 1018 | 1082 | 1146 | 1273 | 1400 |
| 1 9⁄16 | 550 | 611 | 672 | 733 | 794 | 855 | 916 | 978 | 1039 | 1100 | 1222 | 1344 |
| 1 5⁄8 | 528 | 587 | 646 | 705 | 764 | 822 | 881 | 940 | 999 | 1057 | 1175 | 1293 |
| 1 11⁄16 | 509 | 566 | 622 | 679 | 735 | 792 | 849 | 905 | 962 | 1018 | 1132 | 1245 |
| 1¾ | 491 | 545 | 600 | 654 | 709 | 764 | 818 | 873 | 927 | 982 | 1091 | 1200 |
| 1 13⁄16 | 474 | 527 | 579 | 632 | 685 | 737 | 790 | 843 | 895 | 948 | 1054 | 1159 |
| 1 7⁄8 | 458 | 509 | 560 | 611 | 662 | 713 | 764 | 815 | 866 | 917 | 1019 | 1120 |
| 1 15⁄16 | 443 | 493 | 542 | 591 | 640 | 690 | 739 | 788 | 838 | 887 | 986 | 1084 |
| 2 | 429 | 477 | 525 | 573 | 620 | 668 | 716 | 764 | 811 | 859 | 955 | 1050 |
| 2 1⁄8 | 404 | 449 | 494 | 539 | 584 | 629 | 674 | 719 | 764 | 809 | 899 | 988 |
| 2¼ | 382 | 424 | 468 | 509 | 551 | 594 | 636 | 679 | 721 | 764 | 849 | 933 |
| 2 3⁄8 | 362 | 402 | 442 | 482 | 522 | 563 | 603 | 643 | 683 | 724 | 804 | 884 |
| 2½ | 343 | 382 | 420 | 458 | 496 | 534 | 573 | 611 | 649 | 687 | 764 | 840 |
| 2 5⁄8 | 327 | 363 | 400 | 436 | 472 | 509 | 545 | 582 | 618 | 654 | 727 | 800 |
| 2¾ | 312 | 347 | 381 | 416 | 451 | 486 | 520 | 555 | 590 | 625 | 694 | 763 |
| 2 7⁄8 | 299 | 332 | 365 | 398 | 431 | 465 | 498 | 531 | 564 | 598 | 664 | 730 |
| 3 | 286 | 318 | 350 | 381 | 413 | 445 | 477 | 509 | 541 | 572 | 636 | 700 |
| 3 1⁄8 | 274 | 305 | 336 | 366 | 397 | 427 | 458 | 488 | 519 | 549 | 611 | 672 |
| 3¼ | 264 | 293 | 323 | 352 | 381 | 411 | 440 | 470 | 499 | 528 | 587 | 646 |
| 3 3⁄8 | 254 | 283 | 311 | 339 | 367 | 396 | 424 | 452 | 481 | 509 | 566 | 622 |
| 3½ | 245 | 272 | 300 | 327 | 354 | 381 | 409 | 436 | 463 | 490 | 545 | 600 |
| 3 5⁄8 | 237 | 263 | 289 | 316 | 342 | 368 | 395 | 421 | 447 | 474 | 527 | 579 |
| 3¾ | 229 | 254 | 280 | 305 | 331 | 356 | 382 | 407 | 433 | 458 | 509 | 560 |
| 3 7⁄8 | 221 | 246 | 271 | 295 | 320 | 345 | 369 | 394 | 419 | 443 | 493 | 542 |
| 4 | 214 | 238 | 262 | 286 | 310 | 334 | 358 | 382 | 405 | 429 | 477 | 525 |
| 4¼ | 202 | 224 | 247 | 269 | 292 | 314 | 337 | 359 | 383 | 404 | 449 | 494 |
| 4½ | 191 | 212 | 233 | 254 | 275 | 297 | 318 | 339 | 360 | 382 | 424 | 466 |
| 4¾ | 180 | 201 | 221 | 241 | 261 | 281 | 301 | 321 | 341 | 361 | 402 | 442 |
| 5 | 171 | 191 | 210 | 229 | 248 | 267 | 286 | 305 | 324 | 343 | 382 | 420 |
| 5¼ | 163 | 181 | 199 | 218 | 236 | 254 | 272 | 290 | 308 | 327 | 363 | 399 |
| 5½ | 156 | 173 | 190 | 208 | 225 | 242 | 260 | 277 | 294 | 312 | 347 | 381 |
| 5¾ | 149 | 166 | 182 | 199 | 215 | 232 | 249 | 265 | 282 | 298 | 332 | 365 |
| 6 | 143 | 159 | 174 | 190 | 206 | 222 | 238 | 254 | 270 | 286 | 318 | 349 |
| 6¼ | 137 | 152 | 168 | 183 | 198 | 213 | 229 | 244 | 259 | 274 | 305 | 336 |
| 6½ | 132 | 146 | 161 | 176 | 190 | 205 | 220 | 234 | 249 | 264 | 293 | 322 |
| 6¾ | 127 | 141 | 155 | 169 | 183 | 198 | 212 | 226 | 240 | 254 | 283 | 311 |
| 7 | 122 | 136 | 149 | 163 | 177 | 190 | 204 | 218 | 231 | 245 | 272 | 299 |
| 7¼ | 118 | 131 | 144 | 158 | 171 | 184 | 197 | 210 | 223 | 237 | 263 | 289 |
| 7½ | 114 | 127 | 139 | 152 | 165 | 178 | 190 | 203 | 216 | 229 | 254 | 279 |
| 7¾ | 111 | 123 | 135 | 148 | 160 | 172 | 185 | 197 | 209 | 222 | 246 | 271 |
| 8 | 107 | 119 | 131 | 143 | 155 | 167 | 179 | 191 | 203 | 215 | 238 | 262 |

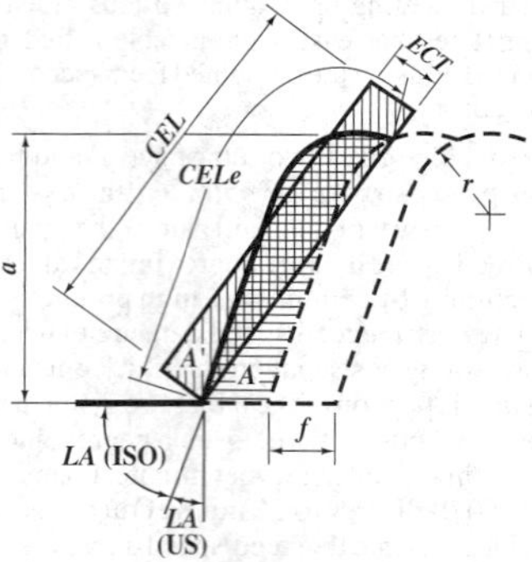


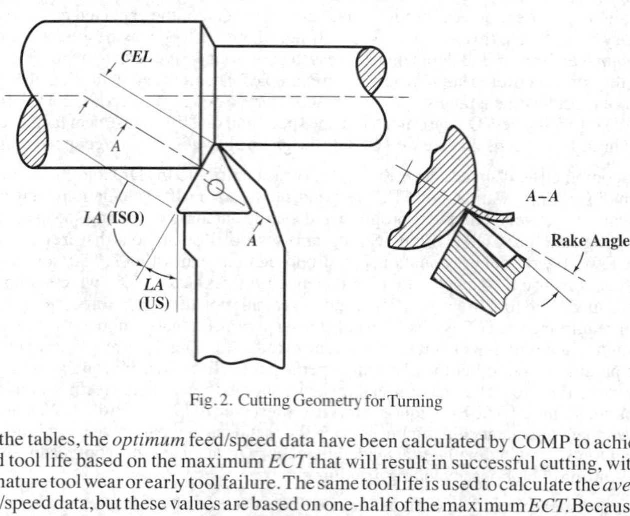

## Table 3. Cutting Speeds and Feeds for Turning Stainless Steels

| Material | Brinell Hardness Number | Tool Material: Uncoated HSS Speed (fpm) | | Uncoated Carbide: Hard Opt. | Uncoated Carbide: Hard Avg. | Uncoated Carbide: Tough Opt. | Uncoated Carbide: Tough Avg. | Coated Carbide: Hard Opt. | Coated Carbide: Hard Avg. | Coated Carbide: Tough Opt. | Coated Carbide: Tough Avg. | Cermet Opt. | Cermet Avg. |
|---|---|---|---|---|---|---|---|---|---|---|---|---|---|
| | | | | f = feed (0.001 in/rev), s = speed (ft/min) | | | | *Metric Units:* **f** × 25.4 = mm/rev, **s** × 0.3048 = m/min | | | | | |
| Free-machining stainless steel (Ferritic): 430F, 430FSe | 135-185 | 110 | **f**<br>**s** | 20<br>480 | 10<br>660 | 36<br>370 | 17<br>395 | 17<br>755 | 8<br>945 | 28<br>640 | 13<br>810 | 7<br>790 | 3<br>995 |
| (Austenitic): 203EZ, 303, 303Se, 303MA, 303Pb, 303Cu, 303 Plus X | 135-185 | 100 | **f**<br>**s** | 13<br>520 | 7<br>640 | 36<br>310 | 17<br>345 | | | 28<br>625 | 13<br>815 | 7<br>695 | 3<br>875 |
| | 225-275 | 80 | | | | | | | | | | | |
| (Martensitic): 416, 416Se, 416 Plus X, 420F, 420FSe, 440F, 440FSe | 135-185 | 110 | **f**<br>**s** | 13<br>520 | 7<br>640 | 36<br>310 | | | | 28<br>625 | 13<br>815 | 7<br>695 | 3<br>875 |
| | 185-240 | 100 | | | | | | | | | | | |
| | 275-325 | 60 | **f**<br>**s** | 13<br>210 | 7<br>260 | 36<br>85 | 17<br>135 | | | 28<br>130 | 13<br>165 | | |
| | 375-425 | 30 | | | | | | | | | | | |
| Stainless steels (Ferritic): 405, 409 429, 430, 434, 436, 442, 446, 502 | 135-185 | 90 | **f**<br>**s** | 20<br>480 | 10<br>660 | 36<br>370 | 17<br>395 | 17<br>755 | 8<br>945 | 28<br>640 | 13<br>810 | 7<br>790 | 3<br>995 |
| (Austenitic): 201, 202, 301, 302, 304, 304L, 305, 308, 321, 347, 348 | 135-185 | 75 | **f**<br>**s** | 13<br>520 | 7<br>640 | 36<br>310 | 17<br>345 | | | 28<br>625 | 13<br>815 | 7<br>695 | 3<br>875 |
| | 225-275 | 65 | | | | | | | | | | | |
| (Austenitic): 302B, 309, 309S, 310, 310S, 314, 316, 316L, 317, 330 | 135-185 | 70 | | | | | | | | | | | |
| (Martensitic): 403, 410, 420, 501 | 135-175 | 95 | | | | | | | | | | | |
| | 175-225 | 85 | | | | | | | | | | | |
| | 275-325 | 55 | **f**<br>**s** | 13<br>210 | 7<br>260 | 36<br>85 | 17<br>135 | | | 28<br>130 | 13<br>165 | 13<br>200† | 7<br>230 |
| | 375-425 | 35 | | | | | | | | | | | |
| (Martensitic): 414, 431, Greek Ascoloy, 440A, 440B, 440C | 225-275 | 55-60 | | | | | | | | | | | |
| | 275-325 | 45-50 | | | | | | | | | | | |
| | 375-425 | 30 | | | | | | | | | | | |
| (Precipitation hardening): 15-5PH, 17-4PH, 17-7PH, AF-71, 17-14CuMo, AFC-77, AM-350, AM-355, AM-362, Custom 455, HNM, PH13-8, PH14-8Mo, PH15-7Mo, Stainless W | 150-200 | 60 | **f**<br>**s** | 13<br>520 | 7<br>640 | 36<br>310 | 17<br>345 | | | 28<br>625 | 13<br>815 | 13<br>695 | 7<br>875 |
| | 275-325 | 50 | | | | | | | | | | | |
| | 325-375 | 40 | **f**<br>**s** | 13<br>195 | 7<br>240 | 36<br>85 | 17<br>155 | | | | | | |
| | 375-450 | 25 | | | | | | | | | | | |

See footnote to Table 1 for more information. The combined feed/speed data in this table are based on tool grades (identified in Table 16) as follows: uncoated carbides, hard = 17, tough = 19; coated carbides, hard = 11, tough = 14; cermet = 7, † = 18.

**Table 4a. Cutting Speeds and Feeds for Turning Ferrous Cast Metals**

| Material | Brinell Hardness Number | HSS Speed (fpm) | | Uncoated Carbide Tough Opt. | Uncoated Carbide Tough Avg. | Coated Carbide Hard Opt. | Coated Carbide Hard Avg. | Coated Carbide Tough Opt. | Coated Carbide Tough Avg. | Ceramic Hard Opt. | Ceramic Hard Avg. | Ceramic Tough Opt. | Ceramic Tough Avg. | Cermet Opt. | Cermet Avg. | CBN Opt. | CBN Avg. |
|---|---|---|---|---|---|---|---|---|---|---|---|---|---|---|---|---|---|
| | | | | f = feed (0.001 in/rev), s = speed (ft/min) *Metric Units:* **f** × 25.4 = mm/rev, **s** × 0.3048 = m/min | | | | | | | | | | | | | |
| Gray Cast Iron | | | | | | | | | | | | | | | | | |
| ASTM Class 20 | 120-150 | 120 | **f** | 28 | 13 | 28 | 13 | 28 | 13 | 15 | 8 | 15 | 8 | 8 | 4 | 24 | 11 |
| ASTM Class 25 | 160-200 | 90 | **s** | 240 | 365 | 665 | 1040 | 585 | 945 | 1490 | 2220 | 1180 | 1880 | 395 | 510 | 8490 | 36380 |
| ASTM Class 30, 35, and 40 | 190-220 | 80 | | | | | | | | | | | | | | | |
| ASTM Class 45 and 50 | 220-260 | 60 | **f** | 28 | 13 | 28 | 13 | 28 | 13 | 11 | 6 | 11 | 6 | 8 | 4 | 24 | 11 |
| ASTM Class 55 and 60 | 250-320 | 35 | **s** | 160 | 245 | 400 | 630 | 360 | 580 | 1440 | 1880 | 1200 | 1570 | 335 | 420 | 1590 | 2200 |
| ASTM Type 1, 1b, 5 (Ni resist) | 100-215 | 70 | **f** | 28 | 13 | | | 28 | 13 | 15 | 8 | 15 | 8 | 8 | 4 | | |
| ASTM Type 2, 3, 6 (Ni resist) | 120-175 | 65 | **s** | 110 | 175 | | | 410 | 575 | 1060 | 1590 | 885 | 1320 | 260 | 325 | | |
| ASTM Type 2b, 4 (Ni resist) | 150-250 | 50 | | | | | | | | | | | | | | | |
| Malleable Iron | | | | | | | | | | | | | | | | | |
| (Ferritic): 32510, 35018 | 110-160 | 130 | **f** | 28 | 13 | 28 | 13 | 28 | 13 | 15 | 8 | 15 | 8 | | | | |
| | | | **s** | 180 | 280 | 730 | 940 | 660 | 885 | 1640 | 2450 | 1410 | 2110 | | | | |
| (Pearlitic): 40010, 43010, 45006, 45008, 48005, 50005 | 160-200 | 95 | **f** | 28 | 13 | 28 | 13 | 28 | 13 | 13 | 7 | 13 | 7 | | | | |
| | 200-240 | 75 | **s** | 125 | 200 | 335 | 505 | 340 | 510 | 1640 | 2310 | 1400 | 1970 | | | | |
| (Martensitic): 53004, 60003, 60004 | 200-255 | 70 | | | | | | | | | | | | | | | |
| (Martensitic): 70002, 70003 | 220-260 | 60 | **f** | 28 | 13 | | | 28 | 13 | 11 | 6 | 11 | 6 | | | | |
| (Martensitic): 80002 | 240-280 | 50 | **s** | 100 | 120 | | | 205 | 250 | 1720 | 2240 | 1460 | 1910 | | | | |
| (Martensitic): 90001 | 250-320 | 30 | | | | | | | | | | | | | | | |

Speeds for HSS (high-speed steel) tools are based on a feed of 0.012 inch/rev and a depth of cut of 0.125 inch; use Table 5c to adjust the given speeds for other feeds and depths of cut. The combined feed/speed data in the remaining columns are based on a depth of cut of 0.1 inch, lead angle of 15 degrees, and nose radius of $^3/_{64}$ inch. Use Table 5a to adjust the given speeds for other feeds, depths of cut, and lead angles; use Table 5b to adjust given speeds for increased tool life up to 180 minutes. Examples are given in the text.

The combined feed/speed data in this table are based on tool grades (identified in Table 16) as follows: uncoated carbides, tough = 15; Coated carbides, hard = 11, tough = 14; ceramics, hard = 2, tough = 3; cermet = 7; CBN = 1.

## Table 7. Cutting Speeds and Feeds for Turning Titanium and Titanium Alloys

| Material | Brinell Hardness Number | Tool Material: HSS Speed (fpm) | | Uncoated Carbide (Tough): f = feed (0.001 in/rev), s = speed (ft/min); *Metric Units:* f × 25.4 = mm/rev, s × 0.3048 = m/min — *Opt.* | *Avg.* |
|---|---|---|---|---|---|
| Commercially Pure and Low Alloyed | | | | | |
| 99.5Ti, 99.5Ti-0.15Pd | 110-150 | 100-105 | f | 28 | 13 |
| | | | s | 55 | 190 |
| 99.1Ti, 99.2Ti, 99.2Ti-0.15Pd, 98.9Ti-0.8Ni-0.3Mo | 180-240 | 85-90 | f | 28 | 13 |
| | | | s | 50 | 170 |
| 99.0 Ti | 250-275 | 70 | f | 20 | 10 |
| | | | s | 75 | 210 |
| Alpha Alloys and Alpha-Beta Alloys | | | | | |
| 5Al-2.5Sn, 8Mn, 2Al-11Sn-5Zr-1Mo, 4Al-3Mo-1V, 5Al-6Sn-2Zr-1Mo, 6Al-2Sn-4Zr-2Mo, 6Al-2Sn-4Zr-6Mo, 6Al-2Sn-4Zr-2Mo-0.25Si | 300-350 | 50 | | | |
| 6Al-4V | 310-350 | 40 | | | |
| 6Al-6V-2Sn, Al-4Mo, | 320-370 | 30 | f | 17 | 8 |
| 8V-5Fe-IAl | 320-380 | 20 | s | 95 | 250 |
| 6Al-4V, 6Al-2Sn-4Zr-2Mo, 6Al-2Sn-4Zr-6Mo, 6Al-2Sn-4Zr-2Mo-0.25Si | 320-380 | 40 | | | |
| 4Al-3Mo-1V, 6Al-6V-2Sn, 7Al-4Mo | 375-420 | 20 | | | |
| I Al-8V-5Fe | 375-440 | 20 | | | |
| Beta Alloys | | | | | |
| 13V-11Cr-3Al, 8Mo-8V-2Fe-3Al, 3Al-8V-6Cr-4Mo-4Zr, 11.5Mo-6Zr-4.5Sn { | 275-350 | 25 | f | 17 | 8 |
| | 375-440 | 20 | s | 55 | 150 |

The speed recommendations for turning with HSS (high-speed steel) tools may be used as starting speeds for milling titanium alloys, using Table 15a to estimate the feed required. Speeds for HSS (high-speed steel) tools are based on a feed of 0.012 inch/rev and a depth of cut of 0.125 inch; use Table 5c to adjust the given speeds for other feeds and depths of cut. The combined feed/speed data in the remaining columns are based on a depth of cut of 0.1 inch, lead angle of 15 degrees, and nose radius of 3⁄64 inch. Use Table 5a to adjust given speeds for other feeds, depths of cut, and lead angles; use Table 5b to adjust given speeds for increased tool life up to 180 minutes. Examples are given in the text. The combined feed/speed data in this table are based on tool grades (identified in Table 16) as follows: uncoated carbide, 15.

## Table 8. Cutting Speeds and Feeds for Turning Light Metals

| Material Description | Material Condition | Tool Material: HSS Speed (fpm) | | Uncoated Carbide (Tough) *Opt.* | Uncoated Carbide (Tough) *Avg.* | Polycrystalline Diamond *Opt.* | Polycrystalline Diamond *Avg.* |
|---|---|---|---|---|---|---|---|
| | | | | f = feed (0.001 in/rev), s = speed (ft/min); *Metric:* f × 25.4 = mm/rev, s × 0.3048 = m/min | | | |
| All wrought and cast magnesium alloys | A, CD, ST, and A | 800 | | | | | |
| All wrought aluminum alloys, including 6061-T651, 5000, 6000, and 7000 series | CD | 600 | | | | | |
| | ST and A | 500 | f | 36 | 17 | | |
| All aluminum sand and permanent mold casting alloys | AC | 750 | s | 2820 | 4570 | | |
| | ST and A | 600 | | | | | |
| Aluminum Die-Casting Alloys | | | | | | | |
| Alloys 308.0 and 319.0 | — | — | f | 36 | 17 | 11 | 8 |
| | | | s | 865 | 1280 | 5890[a] | 8270 |
| Alloys 390.0 and 392.0 | AC | 80 | f | 24 | 11 | 8 | 4 |
| | ST and A | 60 | s | 2010 | 2760 | 4765 | 5755 |
| Alloy 413 | — | — | f | 32 | 15 | 10 | 5 |
| All other aluminum die-casting alloys including alloys 360.0 and 380.0 | ST and A | 100 | s | 430 | 720 | 5085 | 6570 |
| | AC | 125 | f | 36 | 17 | 11 | 6 |
| | | | s | 630 | 1060 | 7560 | 9930 |

**Table 11. Cutting Speeds and Feeds for Milling Plain Carbon and Alloy Steels**

| Material | Brinell Hardness Number [a] | HSS Speed (fpm) | | End Milling HSS Opt. | End Milling HSS Avg. | End Milling Uncoated Carbide Opt. | End Milling Uncoated Carbide Avg. | End Milling Coated Carbide Opt. | End Milling Coated Carbide Avg. | Face Milling Uncoated Carbide Opt. | Face Milling Uncoated Carbide Avg. | Face Milling Coated Carbide Opt. | Face Milling Coated Carbide Avg. | Slit Milling Uncoated Carbide Opt. | Slit Milling Uncoated Carbide Avg. | Slit Milling Coated Carbide Opt. | Slit Milling Coated Carbide Avg. |
|---|---|---|---|---|---|---|---|---|---|---|---|---|---|---|---|---|---|
| | | | f = feed (0.001 in/tooth), s = speed (ft/min) *Metric Units:* f × 25.4 = mm/rev, s × 0.3048 = m/min | | | | | | | | | | | | | | |
| Free-machining plain carbon steels (resulfurized): 1212, 1213, 1215 | 100-150 | 140 | f<br>s | 7<br>45 | 4<br>125 | 7<br>465 | 4<br>735 | 7<br>800 | 4<br>1050 | 39<br>225 | 20<br>335 | 39<br>415 | 20<br>685 | 39<br>265 | 20<br>495 | 39<br>525 | 20<br>830 |
| | 150-200 | 130 | f<br>s | 7<br>35 | 4<br>100 | | | | | | | 39<br>215 | 20<br>405 | | | | |
| (Resulfurized): 1108, 1109, 1115, 1117, 1118, 1120, 1126, 1211 | 100-150 | 130 | f | 730 | 4 | 7 | 4 | 7 | 4 | 39 | 20 | 39 | 20 | 39 | 20 | 39 | 20 |
| | 150-200 | 115 | s | | 85 | 325 | 565 | 465 | 720 | 140 | 220 | 195 | 365 | 170 | 350 | 245 | 495 |
| | 175-225 | 115 | f<br>s | 7<br>30 | 4<br>85 | | | | | | | | | 39<br>185 | 20<br>350 | | |
| (Resulfurized): 1132, 1137, 1139, 1140, 1144, 1146, 1151 | 275-325 | 70 | | | | | | | | | | | | | | | |
| | 325-375 | 45 | f<br>s | 7<br>25 | 4<br>70 | 7<br>210 | 4<br>435 | 7<br>300 | 4<br>560 | 39<br>90 | 20<br>170 | 39<br>175 | 20<br>330 | 39<br>90 | 20<br>235 | 39<br>135 | 20<br>325 |
| | 375-425 | 35 | | | | | | | | | | | | | | | |
| (Leaded): 11L17, 11L18, 12L13, 12L14 | 100-150 | 140 | f | 7 | 4 | | | | | | | 39 | 20 | | | | |
| | 150-200 | 130 | s | 35 | 100 | | | | | | | 215 | 405 | | | | |
| | 200-250 | 110 | f<br>s | 7<br>30 | 4<br>85 | | | | | | | 39<br>185 | 20<br>350 | | | | |
| Plain carbon steels: 1006, 1008, 1009, 1010, 1012, 1015, 1016, 1017, 1018, 1019, 1020, 1021, 1022, 1023, 1024, 1025, 1026, 1513, 1514 | 100-125 | 110 | f<br>s | 7<br>45 | 4<br>125 | 7<br>465 | 4<br>735 | 7<br>800 | 4<br>1050 | 39<br>225 | 20<br>335 | 39<br>415 | 20<br>685 | 39<br>265 | 20<br>495 | 39<br>525 | 20<br>830 |
| | 125-175 | 110 | f<br>s | 7<br>35 | 4<br>100 | | | | | | | 39<br>215 | 20<br>405 | | | | |
| | 175-225 | 90 | f | 7 | 4 | | | | | | | 39 | 20 | | | | |
| | 225-275 | 65 | s | 30 | 85 | | | | | | | 185 | 350 | | | | |

**Table 11.** *(Continued)* **Cutting Speeds and Feeds for Milling Plain Carbon and Alloy Steels**

| Material | Brinell Hardness Number[a] | HSS Speed (fpm) | | End Milling HSS Opt. | End Milling HSS Avg. | End Milling Uncoated Carbide Opt. | End Milling Uncoated Carbide Avg. | End Milling Coated Carbide Opt. | End Milling Coated Carbide Avg. | Face Milling Uncoated Carbide Opt. | Face Milling Uncoated Carbide Avg. | Face Milling Coated Carbide Opt. | Face Milling Coated Carbide Avg. | Slit Milling Uncoated Carbide Opt. | Slit Milling Uncoated Carbide Avg. | Slit Milling Coated Carbide Opt. | Slit Milling Coated Carbide Avg. |
|---|---|---|---|---|---|---|---|---|---|---|---|---|---|---|---|---|---|
| | | | | f = feed (0.001 in/tooth), s = speed (ft/min) *Metric Units:* **f** × 25.4 = mm/rev, **s** × 0.3048 = m/min | | | | | | | | | | | | | |
| Ultra-high-strength steels (not AISI): AMS 6421 (98B37 Mod.), 6422 (98BV40), 6424, 6427, 6428, 6430, 6432, 6433, 6434, 6436, and 6442; 300M, D6ac | 220-300 | 60 | **f** | | | 8 | 4 | 8 | 4 | | | | | | | | |
| | 300-350 | 45 | **s** | | | 165 | 355 | 300 | 480 | | | | | | | | |
| | 350-400 | 20 | **f** | 8 | 4 | 8 | 4 | | | | | 39 | 20 | 39 | 20 | | |
| | | | **s** | 15 | 45 | 150 | 320 | | | | | 130 | 235 | 75 | 175 | | |
| | 43-52 RC | — | **f** | | | 5 | 3 | | | | | | | 39 | 20 | | |
| | | | **s** | | | 20† | 55 | | | | | | | 5 | 15 | | |
| Maraging steels (not AISI): 18% Ni Grades 200, 250, 300, and 350 | 250-325 | 50 | **f** | | | 8 | 4 | 8 | 4 | | | | | | | | |
| | | | **s** | | | 165 | 355 | 300 | 480 | | | | | | | | |
| | 50-52 RC | — | **f** | | | 5 | 3 | | | | | | | 39 | 20 | | |
| | | | **s** | | | 20† | 55 | | | | | | | 5 | 15 | | |
| Nitriding steels (not AISI): Nitralloy 125, 135, 135 Mod., 225, and 230, Nitralloy N, Nitralloy EZ, Nitrex 1 | 200-250 | 60 | **f** | 15 | 8 | 15 | 8 | 15 | 8 | | | 39 | 20 | 39 | 20 | 39 | 20 |
| | | | **s** | 7 | 30 | 105 | 270 | 220 | 450 | | | 295 | 475 | 135 | 305 | 265 | 495 |
| | 300-350 | 25 | **f** | 15 | 8 | 15 | 8 | 15 | 8 | | | 39 | 20 | | | | |
| | | | **s** | 5 | 20 | 40 | 155 | 75 | 225 | | | 175 | 280 | | | | |

[a] Brinell Hardness Number given unless otherwise indicated by RC for hardness on the Rockwell C scale.

For HSS (high-speed steel) tools in the first speed column only, use Table 15a for recommended feed in inches per tooth and depth of cut.

*End Milling:* Table data for end milling are based on a 3-tooth, 20-degree helix angle tool with a diameter of 1.0 inch, an axial depth of cut of 0.2 inch, and a radial depth of cut of 1 inch (full slot). Use Table 15b to adjust speeds for other feeds and axial depths of cut, and Table 15c to adjust speeds if the radial depth of cut is less than the tool diameter. Speeds are valid for all tool diameters.

*Face Milling:* Table data for face milling are based on a 10-tooth, 8-inch diameter face mill, operating with a 15-degree lead angle, $3/64$-inch nose radius, axial depth of cut = 0.1 inch, and radial depth (width) of cut = 6 inches (i.e., width of cut to cutter diameter ratio = $3/4$). These speeds are valid if the cutter axis is above or close to the center line of the workpiece (eccentricity is small). Under these conditions, use Table 15d to adjust speeds for other feeds and axial and radial depths of cut. For larger eccentricity (i.e., when the cutter axis to workpiece center line offset is one half the cutter diameter or more), use the end and side milling adjustment factors (Table 15b and Table 15c) instead of the face milling factors.

*Slit and Slot Milling:* Table data for slit milling are based on an 8-tooth, 10-degree helix angle tool with a cutter width of 0.4 inch, diameter *D* of 4.0 inches, and a depth of cut of 0.6 inch. Speeds are valid for all tool diameters and widths. See the examples in the text for adjustments to the given speeds for other feeds and depths of cut.

Tool life for all tabulated values is approximately 45 minutes; use Table 15e to adjust tool life from 15 to 180 minutes. The combined feed/speed data in this table are based on tool grades (identified in Table 16) as follows: end and slit milling uncoated carbide = 20 except † = 15; face milling uncoated carbide = 19; end, face, and slit milling coated carbide = 10.

## Table 12. Cutting Speeds and Feeds for Milling Tool Steels

| Material | Brinell Hardness Number [a] | HSS Speed (fpm) | | End Milling HSS Opt. | End Milling HSS Avg. | End Milling Uncoated Carbide Opt. | End Milling Uncoated Carbide Avg. | End Milling Coated Carbide Opt. | End Milling Coated Carbide Avg. | Face Milling Uncoated Carbide Opt. | Face Milling Uncoated Carbide Avg. | Face Milling CBN Opt. | Face Milling CBN Avg. | Slit Milling Uncoated Carbide Opt. | Slit Milling Uncoated Carbide Avg. | Slit Milling Coated Carbide Opt. | Slit Milling Coated Carbide Avg. |
|---|---|---|---|---|---|---|---|---|---|---|---|---|---|---|---|---|---|
| | | | | f = feed (0.001 in/tooth), s = speed (ft/min) *Metric Units:* f × 25.4 = mm/rev, s × 0.3048 = m/min | | | | | | | | | | | | | |
| Water hardening: W1, W2, W5 | 150-200 | 85 | | | | | | | | | | | | | | | |
| Shock resisting: S1, S2, S5, S6, S7 | 175-225 | 55 | | | | | | | | | | | | | | | |
| Cold-work, oil hardening: O1, O2, O6, O7 | 175-225 | 50 | f<br>s | 8<br>25 | 4<br>70 | 8<br>235 | 4<br>455 | 8<br>405 | 4<br>635 | 39<br>235 | 20<br>385 | | | 39<br>115 | 20<br>265 | 39<br>245 | 20<br>445 |
| Cold-work, high carbon, high chromium: D2, D3, D4, D5, D7 | 200-250 | 40 | | | | | | | | | | | | | | | |
| Cold-work, air hardening: A2, A3, A8, A9, A10 | 200-250 | 50 | | | | | | | | | | | | | | | |
| A4, A6 | 200-250 | 45 | f<br>s | | | | | | | 39<br>255 | 20<br>385 | | | | | | |
| A7 | 225-275 | 40 | | | | | | | | | | | | | | | |
| Hot-work, chromium type: H10, H11, H12, H13, H14, H19 | 150-200 | 60 | | | | | | | | | | | | | | | |
| | 200-250 | 50 | | | | | | | | | | | | | | | |
| | 325-375 | 30 | f<br>s | 8<br>15 | 4<br>45 | 8<br>150 | 4<br>320 | | | 39<br>130 | 20<br>235 | | | 39<br>75 | 20<br>175 | | |
| | 48-50 RC | — | | | | | | | | | | | | | | | |
| | 50-52 RC | — | f<br>s | | | 5<br>20† | 3<br>55 | | | | | 39<br>50 | 20<br>135 | 39<br>5† | 20<br>15 | | |
| | 52-56 RC | — | | | | | | | | | | | | | | | |
| Hot-work, tungsten and molybdenum types: H21, H22, H23, H24, H25, H26, H41, H42, H43 | 150-200 | 55 | f<br>s | | | | | | | 39<br>255 | 20<br>385 | | | | | | |
| | 200-250 | 45 | | | | | | | | | | | | | | | |
| Special-purpose, low alloy: L2, L3, L6 | 150-200 | 65 | | | | | | | | | | | | | | | |
| Mold: P2, P3, P4, P5, P6 P20, P21 | 100-150 | 75 | f<br>s | 8<br>25 | 4<br>70 | 8<br>235 | 4<br>455 | 8<br>405 | 4<br>635 | 39<br>235 | 20<br>385 | | | 39<br>115 | 20<br>265 | 39<br>245 | 20<br>445 |
| | 150-200 | 60 | | | | | | | | | | | | | | | |
| High-speed steel: M1, M2, M6, M10, T1, T2, T6 | 200-250 | 50 | | | | | | | | | | | | | | | |
| M3-1, M4, M7, M30, M33, M34, M36, M41, M42, M43, M44, M46, M47, T5, T8 | 225-275 | 40 | f<br>s | | | | | | | 39<br>255 | 20<br>385 | | | | | | |
| T15, M3-2 | 225-275 | 30 | | | | | | | | | | | | | | | |

[a] Brinell Hardness Number given unless otherwise indicated by RC for hardness on the Rockwell C scale.

For HSS (high-speed steel) tools in the first speed column only, use Table 15a for recommended feed in inches per tooth and depth of cut.

*End Milling:* Table data for end milling are based on a 3-tooth, 20-degree helix angle tool with a diameter of 1.0 inch, an axial depth of cut of 0.2 inch, and a radial depth of cut of 1 inch (full slot). Use Table 15b to adjust speeds for other feeds and axial depths of cut, and Table 15c to adjust speeds if the radial depth of cut is less than the tool diameter. Speeds are valid for all tool diameters.

*Face Milling:* Table data for face milling are based on a 10-tooth, 8-inch diameter face mill, operating with a 15-degree lead angle, 3⁄64-inch nose radius, axial depth of cut = 0.1 inch, and radial depth (width) of cut = 6 inches (i.e., width of cut to cutter diameter ratio = 3⁄4). These speeds are valid if the cutter axis is above or close to the center line of the workpiece (eccentricity is small). Under these conditions, use Table 15d to adjust speeds for other feeds and axial and radial depths of cut. For larger eccentricity (i.e., when the cutter axis to workpiece center line offset is one half the cutter diameter or more), use the end and side milling adjustment factors (Table 15b and Table 15c) instead of the face milling factors.

*Slit and Slot Milling:* Table data for slit milling are based on an 8-tooth, 10-degree helix angle tool with a cutter width of 0.4 inch, diameter $D$ of 4.0 inches, and a depth of cut of 0.6 inch. Speeds are valid for all tool diameters and widths. See the examples in the text for adjustments to the given speeds for other feeds and depths of cut.

Tool life for all tabulated values is approximately 45 minutes; use Table 15e to adjust tool life from 15 to 180 minutes. The combined feed/speed data in this table are based on tool grades (identified in Table 16) as follows: uncoated carbide = 20, † = 15; coated carbide = 10; CBN = 1.

## Table 13. Cutting Speeds and Feeds for Milling Stainless Steels

| Material | Brinell Hardness Number | HSS Speed (fpm) | | End Milling HSS | | End Milling Uncoated Carbide | | End Milling Coated Carbide | | Face Milling Coated Carbide | | Slit Milling Uncoated Carbide | | Slit Milling Coated Carbide | |
|---|---|---|---|---|---|---|---|---|---|---|---|---|---|---|---|
| | | | | **f** = feed (0.001 in/tooth), **s** = speed (ft/min) *Metric Units:* **f** × 25.4 = mm/rev, **s** × 0.3048 = m/min | | | | | | | | | | | |
| | | | | *Opt.* | *Avg.* | *Opt.* | *Avg.* | *Opt.* | *Avg.* | *Opt.* | *Avg.* | *Opt.* | *Avg.* | *Opt.* | *Avg.* |
| Free-machining stainless steels (Ferritic): 430F, 430FSe | 135-185 | 110 | f | 7 | 4 | 7 | 4 | 7 | 4 | 39 | 20 | 39 | 20 | 39 | 20 |
| | | | s | 30 | 80 | 305 | 780 | 420 | 1240 | 210 | 385 | 120 | 345 | 155 | 475 |
| (Austenitic): 203EZ, 303, 303Se, 303MA, 303Pb, 303Cu, 303 Plus X | 135-185 | 100 | | | | | | | | | | | | | |
| | 225-275 | 80 | f | 7 | 4 | 7 | 4 | | | | | 39 | 20 | | |
| (Martensitic): 416, 416Se, 416 Plus X, 420F, 420FSe, 440F, 440FSe | 135-185 | 110 | s | 20 | 55 | 210 | 585 | | | | | 75 | 240 | | |
| | 185-240 | 100 | | | | | | | | | | | | | |
| | 275-325 | 60 | | | | | | | | | | | | | |
| | 375-425 | 30 | | | | | | | | | | | | | |
| Stainless steels (Ferritic): 405, 409, 429, 430, 434, 436, 442, 446, 502 | 135-185 | 90 | f | 7 | 4 | 7 | 4 | 7 | 4 | 39 | 20 | 39 | 20 | 39 | 20 |
| | | | s | 30 | 80 | 305 | 780 | 420 | 1240 | 210 | 385 | 120 | 345 | 155 | 475 |
| (Austenitic): 201, 202, 301, 302, 304, 304L, 305, 308, 321, 347, 348 | 135-185 | 75 | | | | | | | | | | | | | |
| | 225-275 | 65 | | | | | | | | | | | | | |
| (Austenitic): 302B, 309, 309S, 310, 310S, 314, 316, 316L, 317, 330 | 135-185 | 70 | f | 7 | 4 | 7 | 4 | | | | | 39 | 20 | | |
| | | | s | 20 | 55 | 210 | 585 | | | | | 75 | 240 | | |
| (Martensitic): 403, 410, 420, 501 | 135-175 | 95 | | | | | | | | | | | | | |
| | 175-225 | 85 | | | | | | | | | | | | | |
| | 275-325 | 55 | | | | | | | | | | | | | |
| | 375-425 | 35 | | | | | | | | | | | | | |

**Table 13.** *(Continued)* **Cutting Speeds and Feeds for Milling Stainless Steels**

| Material | Brinell Hardness Number | HSS Speed (fpm) | | End Milling: HSS Opt. | End Milling: HSS Avg. | End Milling: Uncoated Carbide Opt. | End Milling: Uncoated Carbide Avg. | End Milling: Coated Carbide Opt. | End Milling: Coated Carbide Avg. | Face Milling: Coated Carbide Opt. | Face Milling: Coated Carbide Avg. | Slit Milling: Uncoated Carbide Opt. | Slit Milling: Uncoated Carbide Avg. | Slit Milling: Coated Carbide Opt. | Slit Milling: Coated Carbide Avg. |
|---|---|---|---|---|---|---|---|---|---|---|---|---|---|---|---|
| | | | f = feed (0.001 in/tooth), s = speed (ft/min) *Metric Units:* f × 25.4 = mm/rev, s × 0.3048 = m/min | | | | | | | | | | | | |
| Stainless Steels (Martensitic): 414, 431, Greek Ascoloy, 440A, 440B, 440C | 225-275 | 55-60 | | | | | | | | | | | | | |
| | 275-325 | 45-50 | | | | | | | | | | | | | |
| | 375-425 | 30 | | | | | | | | | | | | | |
| (Precipitation hardening): 15-5PH, 17-4PH, 17-7PH, AF-71, 17-14CuMo, AFC-77, AM-350, AM-355, AM-362, Custom 455, HNM, PH13-8, PH14-8Mo, PH15-7Mo, Stainless W | 150-200 | 60 | **f** | 7 | 4 | 7 | 4 | | | | | 39 | 20 | | |
| | | | **s** | 20 | 55 | 210 | 585 | | | | | 75 | 240 | | |
| | 275-325 | 50 | | | | | | | | | | | | | |
| | 325-375 | 40 | | | | | | | | | | | | | |
| | 375-450 | 25 | | | | | | | | | | | | | |

For HSS (high-speed steel) tools in the first speed column only, use Table 15a for recommended feed in inches per tooth and depth of cut.

*End Milling:* Table data for end milling are based on a 3-tooth, 20-degree helix angle tool with a diameter of 1.0 inch, an axial depth of cut of 0.2 inch, and a radial depth of cut of 1 inch (full slot). Use Table 15b to adjust speeds for other feeds and axial depths of cut, and Table 15c to adjust speeds if the radial depth of cut is less than the tool diameter. Speeds are valid for all tool diameters.

*Face Milling:* Table data for face milling are based on a 10-tooth, 8-inch diameter face mill, operating with a 15-degree lead angle, $3/64$-inch nose radius, axial depth of cut = 0.1 inch, and radial depth (width) of cut = 6 inches (i.e., width of cut to cutter diameter ratio = $3/4$). These speeds are valid if the cutter axis is above or close to the center line of the workpiece (eccentricity is small). Under these conditions, use Table 15dto adjust speeds for other feeds and axial and radial depths of cut. For larger eccentricity (i.e., when the cutter axis to workpiece center line offset is one half the cutter diameter or more), use the end and side milling adjustment factors (Table 15b and Table 15c) instead of the face milling factors.

*Slit and Slot Milling:* Table data for slit milling are based on an 8-tooth, 10-degree helix angle tool with a cutter width of 0.4 inch, diameter *D* of 4.0 inch, and a depth of cut of 0.6 inch. Speeds are valid for all tool diameters and widths. See the examples in the text for adjustments to the given speeds for other feeds and depths of cut.

Tool life for all tabulated values is approximately 45 minutes; use Table 15e to adjust tool life from 15 to 180 minutes. The combined feed/speed data in this table are based on tool grades (identified in Table 16) as follows: uncoated carbide = 20; coated carbide = 10.

## Table 14. Cutting Speeds and Feeds for Milling Ferrous Cast Metals

f = feed (0.001 in/tooth), s = speed (ft/min). *Metric Units:* f × 25.4 = mm/rev, s × 0.3048 = m/min

| Material | Brinell Hardness Number | HSS Speed (fpm) | | End Milling HSS Opt. | End Milling HSS Avg. | End Milling Uncoated Carbide Opt. | End Milling Uncoated Carbide Avg. | End Milling Coated Carbide Opt. | End Milling Coated Carbide Avg. | Face Milling Uncoated Carbide Opt. | Face Milling Uncoated Carbide Avg. | Face Milling Coated Carbide Opt. | Face Milling Coated Carbide Avg. | Face Milling Ceramic Opt. | Face Milling Ceramic Avg. | Face Milling CBN Opt. | Face Milling CBN Avg. | Slit Milling Uncoated Carbide Opt. | Slit Milling Uncoated Carbide Avg. | Slit Milling Coated Carbide Opt. | Slit Milling Coated Carbide Avg. |
|---|---|---|---|---|---|---|---|---|---|---|---|---|---|---|---|---|---|---|---|---|---|
| **Gray Cast Iron** | | | | | | | | | | | | | | | | | | | | | |
| ASTM Class 20 | 120-150 | 100 | | | | | | | | | | | | | | | | | | | |
| ASTM Class 25 | 160-200 | 80 | f | 5 | 3 | 5 | 3 | | | 39 | 20 | 39 | 20 | 39 | 20 | 39 | 20 | 39 | 20 | | |
| ASTM Class 30, 35, and 40 | 190-220 | 70 | s | 35 | 90 | 520 | 855 | | | 140 | 225 | 285 | 535 | 1130 | 1630 | 200 | 530 | 205 | 420 | | |
| ASTM Class 45 and 50 | 220-260 | 50 | f | 5 | 3 | 5 | 3 | | | 39 | 20 | 39 | 20 | 39 | 20 | 39 | 20 | 39 | 20 | | |
| ASTM Class 55 and 60 | 250-320 | 30 | s | 30 | 70 | 515 | 1100 | | | 95 | 160 | 185 | 395 | 845 | 1220 | 150 | 400 | 145 | 380 | | |
| ASTM Type 1, 1b, 5 (Ni resist) | 100-215 | 50 | | | | | | | | | | | | | | | | | | | |
| ASTM Type 2, 3, 6 (Ni resist) | 120-175 | 40 | | | | | | | | | | | | | | | | | | | |
| ASTM Type 2b, 4 (Ni resist) | 150-250 | 30 | | | | | | | | | | | | | | | | | | | |
| **Malleable Iron** | | | | | | | | | | | | | | | | | | | | | |
| (Ferritic): 32510, 35018 | 110-160 | 110 | f | 5 | 3 | 5 | 3 | | | 39 | 20 | 39 | 20 | 39 | 20 | | | 39 | 20 | | |
| | | | s | 30 | 70 | 180 | 250 | | | 120 | 195 | 225 | 520 | 490 | 925 | | | 85 | 150 | | |
| (Pearlitic): 40010, 43010, 45006, 45008, 48005, 50005 | 160-200 | 80 | f | 5 | 3 | 5 | 3 | | | 39 | 20 | 39 | 20 | 39 | 20 | | | 39 | 20 | | |
| | 200-240 | 65 | s | 25 | 65 | 150 | 215 | | | 90 | 150 | 210 | 400 | 295 | 645 | | | 70 | 125 | | |
| (Martensitic): 53004, 60003, 60004 | 200-255 | 55 | | | | | | | | | | | | | | | | | | | |
| (Martensitic): 70002, 70003 | 220-260 | 50 | | | | | | | | | | | | | | | | | | | |
| (Martensitic): 80002 | 240-280 | 45 | | | | | | | | | | | | | | | | | | | |
| (Martensitic): 90001 | 250-320 | 25 | | | | | | | | | | | | | | | | | | | |
| **Nodular (Ductile) Iron** | | | | | | | | | | | | | | | | | | | | | |
| (Ferritic): 60-40-18, 65-45-12 | 140-190 | 75 | f | 7 | 4 | 7 | 4 | | | 39 | 20 | 39 | 20 | 39 | 20 | | | 39 | 20 | | |
| | | | s | 15 | 35 | 125 | 240 | | | 100 | 155 | 120 | 255 | 580 | 920 | | | 60 | 135 | | |
| (Ferritic-Pearlitic): 80-55-06 | 190-225 | 60 | | | | | | | | | | | | | | | | | | | |
| | 225-260 | 50 | f | 7 | 4 | 7 | 4 | | | 39 | 20 | 39 | 20 | 39 | 20 | | | 39 | 20 | | |
| (Pearlitic-Martensitic): 100-70-03 | 240-300 | 40 | s | 10 | 30 | 90 | 210 | | | 95 | 145 | 150 | 275 | 170 | 415 | | | 40 | 100 | | |
| (Martensitic): 120-90-02 | 270-330 | 25 | | | | | | | | | | | | | | | | | | | |

**Table 15e. Tool Life Adjustment Factors for Face Milling, End Milling, Drilling, and Reaming**

| Tool Life, $T$ (minutes) | Face Milling with Carbides and Mixed Ceramics | | | End Milling with Carbides and HSS | | | Twist Drilling and Reaming with HSS | | |
|---|---|---|---|---|---|---|---|---|---|
| | $f_s$ | $f_m$ | $f_l$ | $f_s$ | $f_m$ | $f_l$ | $f_s$ | $f_m$ | $f_l$ |
| 15 | 1.69 | 1.78 | 1.87 | 1.10 | 1.23 | 1.35 | 1.11 | 1.21 | 1.30 |
| 45 | 1.00 | 1.00 | 1.00 | 1.00 | 1.00 | 1.00 | 1.00 | 1.00 | 1.00 |
| 90 | 0.72 | 0.70 | 0.67 | 0.94 | 0.89 | 0.83 | 0.93 | 0.89 | 0.85 |
| 180 | 0.51 | 0.48 | 0.45 | 0.69 | 0.69 | 0.69 | 0.87 | 0.80 | 0.72 |

The speeds and feeds given in Table 11 through Table 14 and Table 17 through Table 23 (except for HSS speeds in the first speed column) are based on a 45-minute tool life. To adjust the given speeds to obtain another tool life, multiply the adjusted cutting speed for the 45-minute tool life $V_{45}$ by the tool life factor from this table according to the following rules: for small feeds, where feed $\leq \frac{1}{2} f_{opt}$, the cutting speed for the desired tool life $T$ is $V_T = f_s \times V_{15}$; for medium feeds, where $\frac{1}{2} f_{opt} < \text{feed} < \frac{3}{4} f_{opt}$, $V_T = f_m \times V_{15}$; and for larger feeds, where $\frac{3}{4} f_{opt} \leq \text{feed} \leq f_{opt}$, $V_T = f_l \times V_{15}$. Here, $f_{opt}$ is the largest (*optimum*) feed of the two feed/speed values given in the speed tables or the maximum feed $f_{max}$ obtained from Table 15c, if that table was used in calculating speed adjustment factors.

**Table 16. Cutting Tool Grade Descriptions and Common Vendor Equivalents**

| Grade Description | Tool Identification Code | Approximate Vendor Equivalents: Sandvik Coromant | Kennametal | Seco | Valenite |
|---|---|---|---|---|---|
| Cubic boron nitride | 1 | CB50 | KD050 | CBN20 | VC721 |
| Ceramics | 2 | CC620 | K060 | 480 | — |
| | 3 | CC650 | K090 | 480 | Q32 |
| | 4 (Whiskers) | CC670 | KYON2500 | — | — |
| | 5 (Sialon) | CC680 | KYON2000 | 480 | — |
| | 6 | CC690 | KYON3000 | — | Q6 |
| Cermets | 7 | CT515 | KT125 | CM | VC605 |
| | 8 | CT525 | KT150 | CR | VC610 |
| Polycrystalline | 9 | CD10 | KD100 | PAX20 | VC727 |
| Coated carbides | 10 | GC-A | — | — | — |
| | 11 | GC3015 | KC910 | TP100 | SV310 |
| | 12 | GC235 | KC9045 | TP300 | SV235 |
| | 13 | GC4025 | KC9025 | TP200 | SV325 |
| | 14 | GC415 | KC950 | TP100 | SV315 |
| Uncoated carbides | 15 | H13A | K8, K4H | 883 | VC2 |
| | 16 | S10T | K420, K28 | CP20 | VC7 |
| | 17 | S1P | K45 | CP20 | VC7 |
| | 18 | S30T | — | CP25 | VC5 |
| | 19 | S6 | K21, K25 | CP50 | VC56 |
| | 20 | SM30 | KC710 | CP25 | VC35M |

See Table 2 on page 854 and the section *Cemented Carbides and Other Hard Materials* for more detailed information on cutting tool grades.

The identification codes in column two correspond to the grade numbers given in the footnotes to Table 1 to Table 4b, Table 6 to Table 14, and Table 17 to Table 23.

## Table 17. Speeds and Feeds for Drilling, Reaming, and Threading Plain Carbon and Alloy Steels

| Material | Brinell Hardness Number | Drilling HSS Speed (fpm) | Reaming HSS Speed (fpm) | | Drilling HSS Opt. | Drilling HSS Avg. | Drilling Indexable Insert Coated Carbide Opt. | Drilling Indexable Insert Coated Carbide Avg. | Reaming HSS Opt. | Reaming HSS Avg. | Threading HSS Opt. | Threading HSS Avg. |
|---|---|---|---|---|---|---|---|---|---|---|---|---|
| | | | | f = feed (0.001 in/rev), s = speed (ft/min) *Metric Units:* f × 25.4 = mm/rev, s × 0.3048 = m/min | | | | | | | | |
| Free-machining plain carbon steels (Resulfurized): 1212, 1213, 1215 | 100-150 | 120 | 80 | f<br>s | 21<br>55 | 11<br>125 | 8<br>310 | 4<br>620 | 36<br>140 | 18<br>185 | 83<br>140 | 20<br>185 |
| | 150-200 | 125 | 80 | | | | | | | | | |
| (Resulfurized): 1108, 1109, 1115, 1117, 1118, 1120, 1126, 1211 | 100-150 | 110 | 75 | f<br>s | 16<br>50 | 8<br>95 | 8<br>370 | 4<br>740 | 27<br>105 | 14<br>115 | 83<br>90 | 20<br>115 |
| | 150-200 | 120 | 80 | | | | | | | | | |
| | 175-225 | 100 | 65 | f<br>s | | | 8<br>365 | 4<br>735 | | | | |
| (Resulfurized): 1132, 1137, 1139, 1140, 1144, 1146, 1151 | 275-325 | 70 | 45 | | | | | | | | | |
| | 325-375 | 45 | 30 | | | | | | | | | |
| | 375-425 | 35 | 20 | | | | | | | | | |
| (Leaded): 11L17, 11L18, 12L13, 12L14 | 100-150 | 130 | 83 | | | | | | | | | |
| | 150-200 | 120 | 80 | | | | | | | | | |
| | 200-250 | 90 | 60 | f<br>s | | | 8<br>365 | 4<br>735 | | | | |
| Plain carbon steels: 1006, 1008, 1009, 1010, 1012, 1015, 1016, 1017, 1018, 1019, 1020, 1021, 1022, 1023, 1024, 1025, 1026, 1513, 1514 | 100-125 | 100 | 65 | f<br>s | 21<br>55 | 11<br>125 | 8<br>310 | 4<br>620 | 36<br>140 | 18<br>185 | 83<br>140 | 20<br>185 |
| | 125-175 | 90 | 60 | | | | | | | | | |
| | 175-225 | 70 | 45 | f<br>s | | | 8<br>365 | 4<br>735 | | | | |
| | 225-275 | 60 | 40 | | | | | | | | | |
| Plain carbon steels: 1027, 1030, 1033, 1035, 1036, 1037, 1038, 1039, 1040, 1041, 1042, 1043, 1045, 1046, 1048, 1049, 1050, 1052, 1524, 1526, 1527, 1541 | 125-175 | 90 | 60 | | | | | | | | | |
| | 175-225 | 75 | 50 | | | | | | | | | |
| | 225-275 | 60 | 40 | f<br>s | | | 8<br>365 | 4<br>735 | | | | |
| | 275-325 | 50 | 30 | | | | | | | | | |
| | 325-375 | 35 | 20 | | | | | | | | | |
| | 375-425 | 25 | 15 | | | | | | | | | |

**Table 17.** *(Continued)* **Speeds and Feeds for Drilling, Reaming, and Threading Plain Carbon and Alloy Steels**

| Material | Brinell Hardness Number | Drilling HSS Speed (fpm) | Reaming HSS Speed (fpm) | | Drilling HSS Opt. | Drilling HSS Avg. | Drilling Indexable Insert Coated Carbide Opt. | Drilling Indexable Insert Coated Carbide Avg. | Reaming HSS Opt. | Reaming HSS Avg. | Threading HSS Opt. | Threading HSS Avg. |
|---|---|---|---|---|---|---|---|---|---|---|---|---|
| | | | | | f = feed (0.001 in/rev), s = speed (ft/min) *Metric Units:* f × 25.4 = mm/rev, s × 0.3048 = m/min | | | | | | | |
| Alloy steels: 1330, 1335, 1340, 1345, 4032, 4037, 4042, 4047, 4130, 4135, 4137, 4140, 4142, 4145, 4147, 4150, 4161, 4337, 4340, 50B44, 50B46, 50B50, 50B60, 5130, 5132, 5140, 5145, 5147, 5150, 5160, 51B60, 6150, 81B45, 8630, 8635, 8637, 8640, 8642, 8645, 8650, 8655, 8660, 8740, 9254, 9255, 9260, 9262, 94B30 | 175-225 | 75 (60) | 50 (40) | f | 16 | 8 | 8 | 4 | 26 | 13 | 83 | 20 |
| | | | | s | 75 | 140 | 410 | 685 | 150 | 160 | 125 | 160 |
| | 225-275 | 60 (50) | 40 (30) | f | | | 8 | 4 | | | | |
| | | | | s | | | 355 | 600 | | | | |
| | 275-325 | 45 (35) | 30 (25) | f | 11 | 6 | 8 | 4 | 19 | 10 | 83 | 20 |
| | | | | s | 50 | 85 | 335 | 570 | 95 | 135 | 60 | 95 |
| E51100, E52100: use (HSS speeds) | 325-375 | 30 (30) | 15 (20) | f | | | 8 | 4 | | | | |
| | 375-425 | 20 (20) | 15 (10) | s | | | 310 | 525 | | | | |
| Ultra-high-strength steels (not AISI): AMS 6421 (98B37 Mod.), 6422 (98BV40), 6424, 6427, 6428, 6430, 6432, 6433, 6434, 6436, and 6442; 300M, D6ac | 220-300 | 50 | 30 | f | | | 8 | 4 | | | | |
| | 300-350 | 35 | 20 | s | | | 325 | 545 | | | | |
| | 350-400 | 20 | 10 | f | | | 8 | 4 | | | | |
| | | | | s | | | 270 | 450 | | | | |
| Maraging steels (not AISI): 18% Ni Grade 200, 250, 300, and 350 | 250-325 | 50 | 30 | f | | | 8 | 4 | | | | |
| | | | | s | | | 325 | 545 | | | | |
| Nitriding steels (not AISI): Nitralloy 125, 135, 135 Mod., 225, and 230, Nitralloy N, Nitralloy EZ, Nitrex I | 200-250 | 60 | 40 | f | 16 | 8 | 8 | 4 | 26 | 13 | 83 | 20 |
| | | | | s | 75 | 140 | 410 | 685 | 150 | 160 | 125 | 160 |
| | 300-350 | 35 | 20 | f | | | 8 | 4 | | | | |
| | | | | s | | | 310 | 525 | | | | |

The two leftmost speed columns in this table contain traditional *Handbook* speeds for drilling and reaming with HSS steel tools. The section Feed Rates for Drilling and Reaming contains useful information concerning feeds to use in conjunction with these speeds.

*HSS Drilling and Reaming:* The combined feed/speed data for drilling are based on a 0.60-inch diameter HSS drill with standard drill point geometry (2-flute with 118° tip angle). Speed adjustment factors in Table 22 are used to adjust drilling speeds for other feeds and drill diameters. Examples of using this data are given in the text. The given speeds and feeds for reaming are based on an 8-tooth, $^{25}/_{32}$-inch diameter, 30° lead angle reamer, and a 0.008-inch radial depth of cut. For other feeds, the correct speed can be obtained by interpolation using the given speeds if the desired feed lies in the recommended range (between the given values of *optimum* and *average* feed). If a feed lower than the given *average* value is chosen, the speed should be maintained at the corresponding *average* speed (i.e., the highest of the two speed values given). The cutting speeds for reaming do not require adjustment for tool diameters for standard ratios of radial depth of cut to reamer diameter (i.e., $f_d$ = 1.00). Speed adjustment factors to modify tool life are found in Table 15e.

## Table 20. Speeds and Feeds for Drilling, Reaming, and Threading Ferrous Cast Metals

| Material | Brinell Hardness Number | Drilling | Reaming | | Drilling | | | | | | Reaming | | Threading | |
|---|---|---|---|---|---|---|---|---|---|---|---|---|---|---|
| | | HSS | | | HSS | | Indexable Carbide Insert | | | | HSS | | HSS | |
| | | | | | | | Uncoated | | Coated | | | | | |
| | | Speed (fpm) | | | f = feed (0.001 in/rev), s = speed (ft/min) *Metric Units:* f × 25.4 = mm/rev, s × 0.3048 = m/min | | | | | | | | | |
| | | | | | *Opt.* | *Avg.* | *Opt.* | *Avg.* | *Opt.* | *Avg.* | *Opt.* | *Avg.* | *Opt.* | *Avg.* |
| ASTM Class 20 | 120-150 | 100 | 65 | | | | | | | | | | | |
| ASTM Class 25 | 160-200 | 90 | 60 | f<br>s | 16<br>80 | 8<br>90 | 11<br>85 | 6<br>180 | 11<br>235 | 6<br>485 | 26<br>85 | 13<br>65 | 83<br>90 | 20<br>80 |
| ASTM Class 30, 35, and 40 | 190-220 | 80 | 55 | | | | | | | | | | | |
| ASTM Class 45 and 50 | 220-260 | 60 | 40 | f<br>s | 13<br>50 | 6<br>50 | 11<br>70 | 6<br>150 | 11<br>195 | 6<br>405 | 21<br>50 | 10<br>30 | 83<br>55 | 20<br>45 |
| ASTM Class 55 and 60 | 250-320 | 30 | 20 | | | | | | | | | | | |
| ASTM Type 1, 1b, 5 (Ni resist) | 100-215 | 50 | 30 | | | | | | | | | | | |
| ASTM Type 2, 3, 6 (Ni resist) | 120-175 | 40 | 25 | | | | | | | | | | | |
| ASTM Type 2b, 4 (Ni resist) | 150-250 | 30 | 20 | | | | | | | | | | | |
| Malleable Iron | | | | | | | | | | | | | | |
| (Ferritic): 32510, 35018 | 110-160 | 110 | 75 | f<br>s | 19<br>80 | 10<br>100 | 11<br>85 | 6<br>180 | 11<br>270 | 6<br>555 | 30<br>95 | 16<br>80 | 83<br>100 | 20<br>85 |
| (Pearlitic): 40010, 43010, 45006, 45008, 48005, 50005 | 160-200 | 80 | 55 | f<br>s | 14<br>65 | 7<br>65 | | | 11<br>235 | 6<br>485 | 22<br>65 | 11<br>45 | 83<br>70 | 20<br>60 |
| | 200-240 | 70 | 45 | | | | | | | | | | | |
| (Martensitic): 53004, 60003, 60004 | 200-255 | 55 | 35 | | | | | | | | | | | |
| (Martensitic): 70002, 70003 | 220-260 | 50 | 30 | | | | | | | | | | | |
| (Martensitic): 80002 | 240-280 | 45 | 30 | | | | | | | | | | | |
| (Martensitic): 90001 | 250-320 | 25 | 15 | | | | | | | | | | | |
| Nodular (Ductile) Iron | | | | | | | | | | | | | | |
| (Ferritic): 60-40-18, 65-45-12 | 140-190 | 100 | 65 | f<br>s | 17<br>70 | 9<br>80 | 11<br>85 | 6<br>180 | 11<br>235 | 6<br>485 | 28<br>80 | 14<br>60 | 83<br>80 | 20<br>70 |

## Table 21. Speeds and Feeds for Drilling, Reaming, and Threading Light Metals

| Material | Brinell Hardness Number | Drilling HSS Speed (fpm) | Reaming HSS Speed (fpm) | | Drilling HSS Opt. | Drilling HSS Avg. | Drilling Indexable Insert Uncoated Carbide Opt. | Drilling Indexable Insert Uncoated Carbide Avg. | Reaming HSS Opt. | Reaming HSS Avg. | Threading HSS Opt. | Threading HSS Avg. |
|---|---|---|---|---|---|---|---|---|---|---|---|---|
| | | | | f = feed (0.001 in/rev), s = speed (ft/min) *Metric Units:* f × 25.4 = mm/rev, s × 0.3048 = m/min | | | | | | | | |
| All wrought aluminum alloys, 6061-T651, 5000, 6000, 7000 series | CD | 400 | 400 | | | | | | | | | |
| | ST and A | 350 | 350 | f | 31 | 16 | 11 | 6 | 52 | 26 | 83 | 20 |
| All aluminum sand and permanent mold casting alloys | AC | 500 | 500 | s | 390 | 580 | 3235 | 11370 | 610 | 615 | 635 | 565 |
| | ST and A | 350 | 350 | | | | | | | | | |
| **Aluminum Die-Casting Alloys** | | | | | | | | | | | | |
| Alloys 308.0 and 319.0 | — | — | — | f | 23 | 11 | 11 | 6 | 38 | 19 | 83 | 20 |
| | | | | s | 110 | 145 | 945 | 3325 | 145 | 130 | 145 | 130 |
| Alloys 360.0 and 380.0 | — | — | — | f | 27 | 14 | 11 | 6 | 45 | 23 | 83 | 20 |
| | | | | s | 90 | 125 | 855 | 3000 | 130 | 125 | 130 | 115 |
| Alloys 390.0 and 392.0 | AC | 300 | 300 | | | | | | | | | |
| | ST and A | 70 | 70 | | | | | | | | | |
| Alloys 413 | — | — | — | f | 24 | 12 | 11 | 6 | 40 | 20 | 83 | 20 |
| | ST and A | 45 | 40 | s | 65 | 85 | 555 | 1955 | 85 | 80 | 85 | 80 |
| All other aluminum die-casting alloys | AC | 125 | 100 | f | 27 | 14 | 11 | 6 | 45 | 23 | 83 | 20 |
| | | | | s | 90 | 125 | 855 | 3000 | 130 | 125 | 130 | 115 |
| **Magnesium Alloys** | | | | | | | | | | | | |
| All wrought magnesium alloys | A, CD, ST and A | 500 | 500 | | | | | | | | | |
| All cast magnesium alloys | A, AC, ST and A | 450 | 450 | | | | | | | | | |

Abbreviations designate: A, annealed; AC, as cast; CD, cold drawn; and ST and A, solution treated and aged, respectively. See the footnote to Table 17 for instructions concerning the use of this table. The combined feed/speed data in this table are based on tool grades (identified in Table 16) as follows; uncoated carbide = 15.

$$\boldsymbol{P_c = K_p CQW = 1.69 \times 0.94 \times 1.81 \times 1.30 = 3.74\ \text{kW}}$$

$$\boldsymbol{P_m = \frac{P_c}{E} = \frac{3.74}{0.80} = 4.677\ \text{kW}}$$

Whenever possible the maximum power available on a machine tool should be used when heavy cuts must be taken.

The cutting conditions for utilizing the maximum power should be selected in the following order: 1) select the maximum depth of cut that can be used; 2) select the maximum feed rate that can be used; and 3) estimate the cutting speed that will utilize the maximum power available on the machine. This sequence is based on obtaining the longest tool life of the cutting tool and at the same time obtaining as much production as possible from the machine.

*The life of a cutting tool is most affected by cutting speed, then by feed rate, and least of all by depth of cut.* The maximum metal removal rate that a given machine is capable of machining from a given material is used as the basis for estimating the cutting speed that will utilize all the power available on the machine.

*Example:* A 160–180 BHN gray iron casting that is 6 inches wide is to have ⅛ inch stock removed on a 10 hp milling machine, using an 8 inch diameter, 10 tooth, indexable insert cemented carbide face milling cutter. The feed rate selected for this cutter is 0.012 in/tooth, and all the stock (0.125 inch) will be removed in one cut. Estimate the cutting speed that will utilize the maximum power available on the machine.

$K_p = 0.52$ (From Table 1a)
$C = 1.00$ (From Table 2)
$W = 1.20$ (From Table 3)
$E = 0.80$ (From Table 4)

$$Q_{max} = \frac{P_m E}{K_p CW} = \frac{10 \times 0.80}{0.52 \times 1.00 \times 1.20} = 12.82\ \text{in}^3/\text{min} \qquad \left(P_m = \frac{K_p CQW}{E}\right)$$

$$f_m = \frac{Q_{max}}{wd} = \frac{12.82}{6 \times 0.125} = 17.1\ \text{in/min} \qquad (Q = f_m wd)$$

$$N = \frac{f_{max}}{f_t n_t} = \frac{17}{0.012 \times 10} = 142.4\ \text{rpm} \qquad (f_m = f_t n_t N)$$

$$V = \frac{\pi DN}{12} = \frac{\pi \times 8 \times 142}{12} = 298.3\ \text{fpm} \qquad \left(N = \frac{12V}{\pi D}\right)$$

Calculated cutting speed of 298.3 fpm is correct, but very low for available cutting tools. Today, milling of gray irons is performed, mostly, using cubic boron nitride indexable inserts. Typical range of the cutting speed for milling with large engagement (0.8) and 0.004–0.008 inch per tooth chip load is 2700–2000 sfm. Estimating maximum productivity of cutting should be based on the recommended cutting speeds versus work materials.

***Example:*** **A 3.2 mm deep cut is to be taken on a 200–210 BHN AISI 1050 steel part using a 7.5 kW geared head lathe. The feed rate selected for this job is 0.45 mm/rev. Estimate the cutting speed that will utilize the maximum power available on the lathe.**

$\boldsymbol{K_p = 2.32}$ **(From Table 1b)**
$\boldsymbol{C = 0.92}$ **(From Table 2)**
$\boldsymbol{W = 1.30}$ **(From Table 3)**
$\boldsymbol{E = 0.80}$ **(From Table 4)**

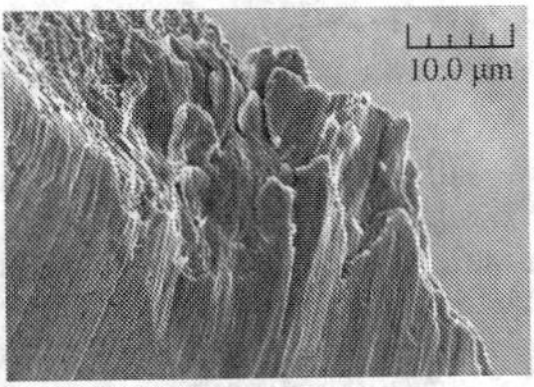




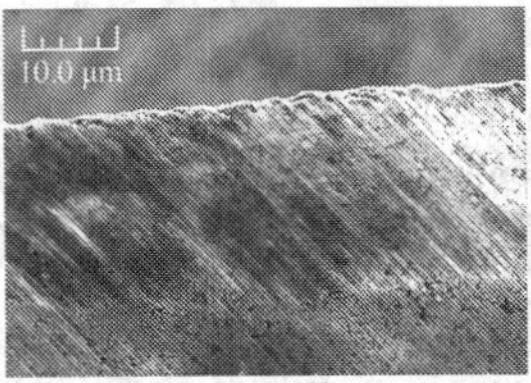

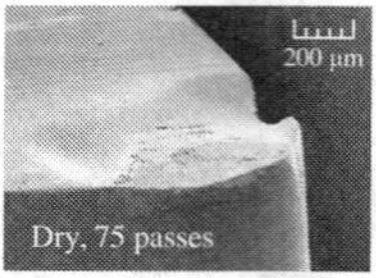


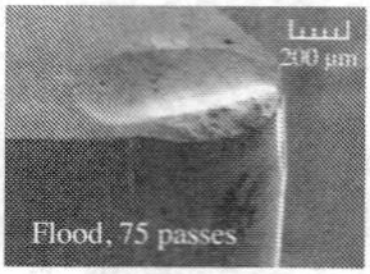


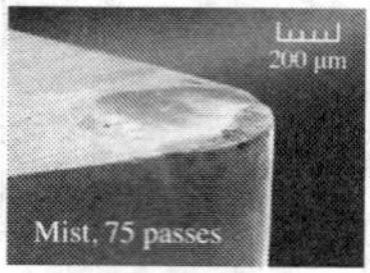

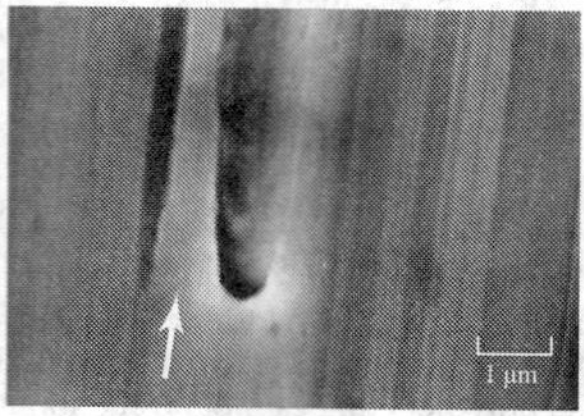

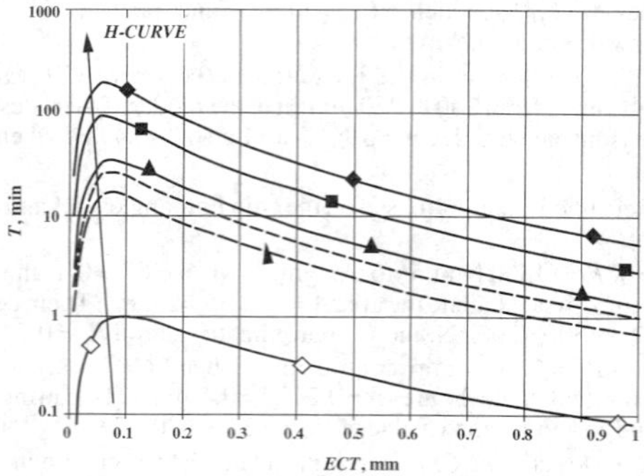


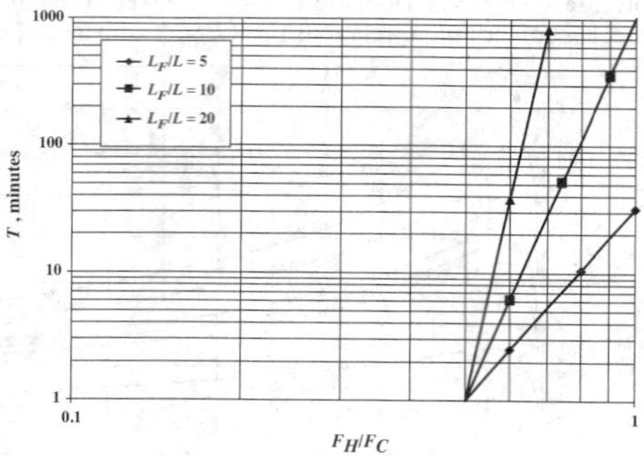

**Table 1. Approximate Cutting Speeds and Feeds for Standard Automatic Screw Machine Tools—Brown and Sharpe**

| Tool | Cut: Width or Depth, Inches | Cut: Dia. of Hole, Inches | Brass[a]: Feed, Inches per Rev. | Mild or Soft Steel: Feed, Inches per Rev. | Mild or Soft Steel: Surface Speed, Feet per Min., Carbon Tools | Mild or Soft Steel: Surface Speed, Feet per Min., H.S.S. Tools | Tool Steel, 0.80–1.00% C: Feed, Inches per Rev. | Tool Steel, 0.80–1.00% C: Surface Speed, Feet per Min., Carbon Tools | Tool Steel, 0.80–1.00% C: Surface Speed, Feet per Min., H.S.S. Tools |
|---|---|---|---|---|---|---|---|---|---|
| Boring tools | 0.005 | … | … | 0.008 | 50 | 110 | 0.004 | 30 | 60 |
| Box tools, roller rest, Single chip finishing | 1/32 | … | 0.012 | 0.010 | 70 | 150 | 0.005 | 40 | 75 |
| | 1/16 | … | 0.010 | 0.008 | 70 | 150 | 0.004 | 40 | 75 |
| | 1/8 | … | 0.008 | 0.007 | 70 | 150 | 0.003 | 40 | 75 |
| | 3/16 | … | 0.008 | 0.006 | 70 | 150 | 0.002 | 40 | 75 |
| | 1/4 | … | 0.006 | 0.005 | 70 | 150 | 0.0015 | 40 | 75 |
| Finishing | 0.005 | … | 0.010 | 0.010 | 70 | 150 | 0.006 | 40 | 75 |
| Center drills | … | Under 1/8 | 0.003 | 0.0015 | 50 | 110 | 0.001 | 30 | 75 |
| | … | Over 1/8 | 0.006 | 0.0035 | 50 | 110 | 0.002 | 30 | 75 |
| Cutoff tools { Angular | … | … | 0.0015 | 0.0006 | 80 | 150 | 0.0004 | 50 | 85 |
| Cutoff tools { Circular | 3/64-1/8 | … | 0.0035 | 0.0015 | 80 | 150 | 0.001 | 50 | 85 |
| Cutoff tools { Straight | 1/16-1/8 | … | 0.0035 | 0.0015 | 80 | 150 | 0.001 | 50 | 85 |
| Stock diameter under 1/8 in. | … | … | 0.002 | 0.0008 | 80 | 150 | 0.0005 | 50 | 85 |
| Dies { Button | … | … | … | … | 30 | … | … | 14 | … |
| Dies { Chaser | … | … | … | … | 30 | 40 | … | 16 | 20 |
| Drills, twist cut | … | 0.02 | 0.0014 | 0.001 | 40 | 60 | 0.0006 | 30 | 45 |
| | … | 0.04 | 0.002 | 0.0014 | 40 | 60 | 0.0008 | 30 | 45 |
| | … | 1/16 | 0.004 | 0.002 | 40 | 60 | 0.0012 | 30 | 45 |
| | … | 3/32 | 0.006 | 0.0025 | 40 | 60 | 0.0016 | 30 | 45 |
| | … | 1/8 | 0.009 | 0.0035 | 40 | 75 | 0.002 | 30 | 60 |
| | … | 3/16 | 0.012 | 0.004 | 40 | 75 | 0.003 | 30 | 60 |
| | … | 1/4 | 0.014 | 0.005 | 40 | 75 | 0.003 | 30 | 60 |
| | … | 5/16 | 0.016 | 0.005 | 40 | 75 | 0.0035 | 30 | 60 |
| | … | 3/8-5/8 | 0.016 | 0.006 | 40 | 85 | 0.004 | 30 | 60 |
| Form tools, circular | 1/8 | … | 0.002 | 0.0009 | 80 | 150 | 0.0006 | 50 | 85 |
| | 1/4 | … | 0.002 | 0.0008 | 80 | 150 | 0.0005 | 50 | 85 |
| | 3/8 | … | 0.0015 | 0.0007 | 80 | 150 | 0.0004 | 50 | 85 |
| | 1/2 | … | 0.0012 | 0.0006 | 80 | 150 | 0.0004 | 50 | 85 |
| | 5/8 | … | 0.001 | 0.0005 | 80 | 150 | 0.0003 | 50 | 85 |
| | 3/4 | … | 0.001 | 0.0005 | 80 | 150 | 0.0003 | 50 | 85 |
| | 1 | … | 0.001 | 0.0004 | 80 | 150 | … | … | … |

*Example 3:* Fig. 2b shows that maximum values of wheel-life occur along the *H*-curve. For the 3 speeds 1800, 2700, and 3600 m/min, maximum wheel lives are approximately 70, 14 and 4 minutes, respectively, at *ECT* around $17 \times 10^{-5}$ through $20 \times 10^{-5}$ mm along the *H*-curve. Left and right of the *H*-curve wheel-life is shorter.

Fig. 2c depicts wheel speed *V* versus *ECT* with wheel-life *T* parameter shown as curves in log-log coordinates, similar to those for the other cutting operations, with the characteristic *H*- and *G*-curves.

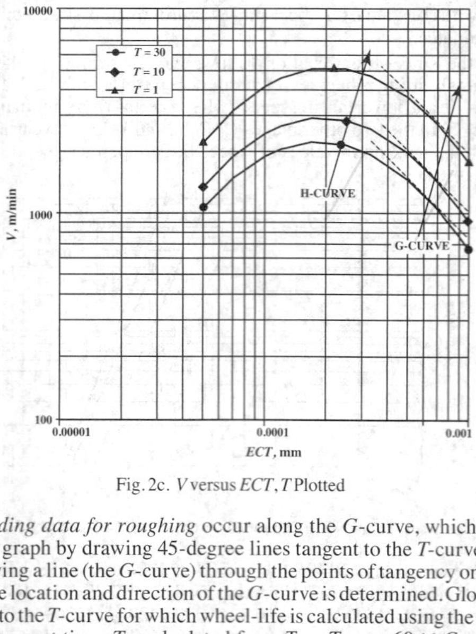


Fig. 2c. *V* versus *ECT*, *T* Plotted

*Optimum grinding data for roughing* occur along the *G*-curve, which is determined from the *V-ECT* graph by drawing 45-degree lines tangent to the *T*-curves, as shown in Fig. 2c, and drawing a line (the *G*-curve) through the points of tangency on the respective *T*-curves, thus the location and direction of the *G*-curve is determined. Globally optimum data correspond to the *T*-curve for which wheel-life is calculated using the corresponding equivalent tooling-cost time, $T_V$, calculated from $T_V = T_{RPL} + 60 \times C_E \div H_R$, minutes, where $T_{RPL}$ is the time required to replace wheel, $C_E$ = cost per wheel dressing = wheel cost + cost per dressing, and $H_R$ is the hourly rate.

*Minimum cost conditions occur* along the *G*-curve; if optimum life $T_O$ was determined at either 10 or 30 minutes then $V_O$ = 1500 or 1100 m/min, respectively, and *ECT* is around $65 - 70 \times 10^{-5}$ mm in both cases. The corresponding optimum values of *SMRR* are $1000 \times 1500 \times 67 \times 10^{-5} = 1000$ and $1000 \times 1100 \times 67 \times 10^{-5} = 740$ mm³/min/mm wheel contact width (1.5 to 1.1 in³/in/min).

Using Fig. 1 we find optimum work speeds for depths of cut $a_r = 0.01$ and 0.005 mm to be $V_w$ = 100 and 75 m/min, and 200 and 150 m/min (330 and 250 fpm, and 660 and 500 fpm) respectively for 10- and 30-minute wheel-life.

These high work speeds are possible using proper dressing conditions, high system stiffness, good grinding fluid quality and wheel composition.

Fig. 3 shows the variation of specific metal removal rate with wheel speed for several materials and a range of *ECT*s at 10- and 30-minutes wheel-life. *ECT* decreases when moving to the left and down along each curve. The two curves for unhardened 1020 steel have the largest values of *SMRR*, and represent the most productive grinding conditions, while the heat resistant alloy Inconel yields the least productive grinding conditions. Each

**Table 1a.** *(Continued)* **Standard Shapes and Inch Size Ranges of Grinding Wheels**
*ANSI B74.2-2003*

| Applications | Size Ranges of Principal Dimensions, Inches | | |
|---|---|---|---|
| | *D* = Dia. | *T* = Thickness | *H* = Hole |

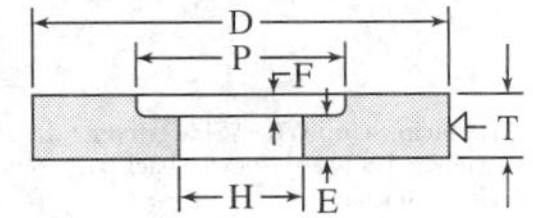


**Type 5. Wheel,** recessed one side For peripheral grinding. Allows wider faced wheels than the available mounting thickness, also grinding clearance for the nut and flange.

| Applications | *D* = Dia. | *T* = Thickness | *H* = Hole |
|---|---|---|---|
| CYLINDRICAL GRINDING<br>Between centers | 12 to 36 | 1½ to 4 | 5 or 12 |
| CYLINDRICAL GRINDING<br>Centerless regulating wheel | 8 to 14 | 3 to 6 | 3 or 5 |
| INTERNAL GRINDING | ⅜ to 4 | ⅜ to 2 | ⅛ to ⅞ |
| SURFACE GRINDING<br>Horizontal spindle machines | 7 to 24 | ¾ to 6 | 1¼ to 12 |

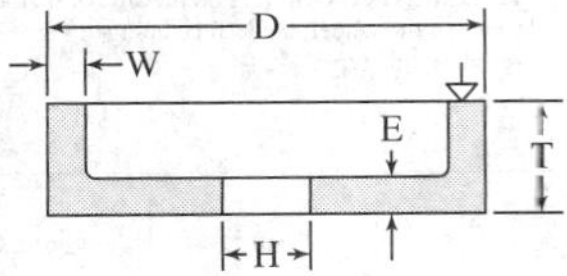


**Type 6. Straight-Cup Wheel** Side grinding wheel, in whose dimensioning the wall thickness (*W*) takes precedence over the diameter of the recess. Hole is ⅝-11UNC-2B threaded for the snagging wheels and ½ or 1¼″ for the tool grinding wheels.

| Applications | *D* = Dia. | *T* = Thickness | *W* = Wall |
|---|---|---|---|
| SNAGGING<br>Portable machines, organic bond only. | 4 to 6 | 2 | ¾ to 1½ |
| TOOL GRINDING<br>Broaches, cutters, mills, reamers, taps, etc. | 2 to 6 | 1 ¼ to 2 | 5/16 or ⅜ |

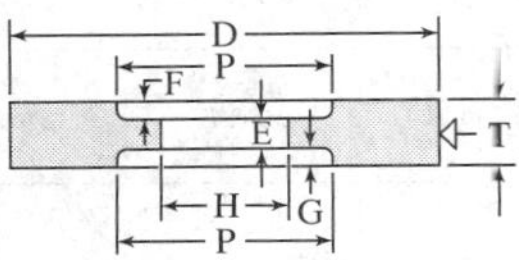


**Type 7. Wheel,** recessed two sides Peripheral grinding. Recesses allow grinding clearance for both flanges and also narrower mounting thickness than overall thickness.

| Applications | *D* = Dia. | *T* = Thickness | *H* = Hole |
|---|---|---|---|
| CYLINDRICAL GRINDING<br>Between centers | 12 to 36 | 1½ to 4 | 5 or 12 |
| CYLINDRICAL GRINDING<br>Centerless regulating wheel | 8 to 14 | 4 to 20 | 3 to 6 |
| SURFACE GRINDING<br>Horizontal spindle machines | 12 to 24 | 2 to 6 | 5 to 12 |

**Table 1a.** *(Continued)* **Standard Shapes and Inch Size Ranges of Grinding Wheels**
*ANSI B74.2-2003*

| Applications | Size Ranges of Principal Dimensions, Inches | | |
|---|---|---|---|
| | *D* = Dia. | *T* = Thickness | *H* = Hole |
| 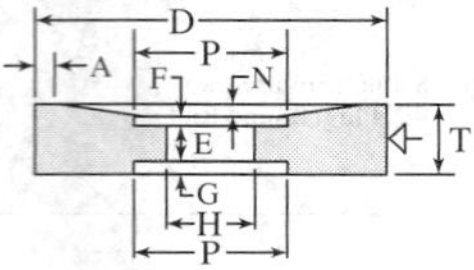 | **Type 24. Wheel, Relieved and Recessed One Side, Recessed Other Side** One side recessed, the other side is relieved to a recess. | | |
| 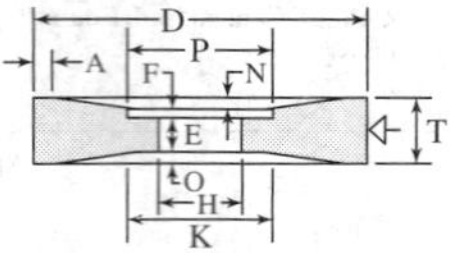 | **Type 25. Wheel, Relieved and Recessed One Side, Relieved Other Side** One side relieved to a flat, the other side relieved to a recess. | | |
| 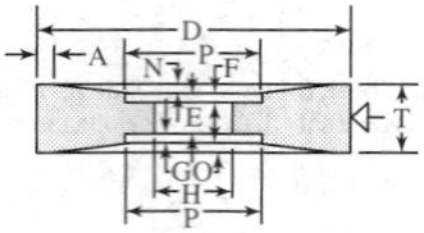 | **Type 26. Wheel, Relieved and Recessed Both Sides** | | |
| CYLINDRICAL GRINDING<br>Between centers, with the periphery of the wheel | 20 to 36 | 2 to 4 | 12 or 20 |
| 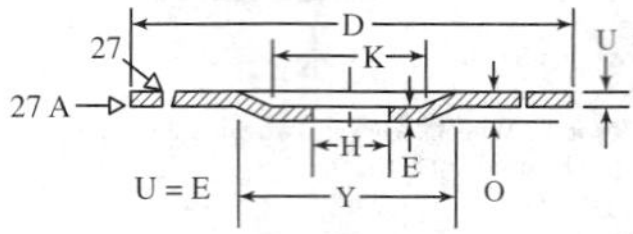 | **TYPES 27 & 27A. Wheel, Depressed Center**<br>27. *Portable Grinding:* Grinding normally done by contact with work at approx. a 15° angle with face of the wheel.<br>27A. *Cutting-off:* Using the periphery as grinding face. | | |
| CUTTING OFF<br>Reinforced organic bonds only | 16 to 30 | *U* = *E* = 5/32 to 1/4 | 1 or 1 1/2 |
| SNAGGING<br>Portable machine | 3 to 9 | *U* = Uniform thickness 1/8 to 3/8 | 3/8 or 7/8 |
| 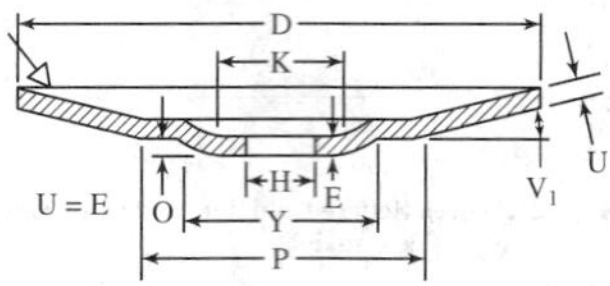 | **Type 28. Wheel, Depressed Center (Saucer-Shaped Grinding Face)** Grinding at approx. 15° angle with wheel face. | | |
| SNAGGING<br>Portable machine | 7 or 9 | *U* = Uniform thickness 1/4 | 7/8 |

Throughout table large open-head arrows indicate grinding surfaces.

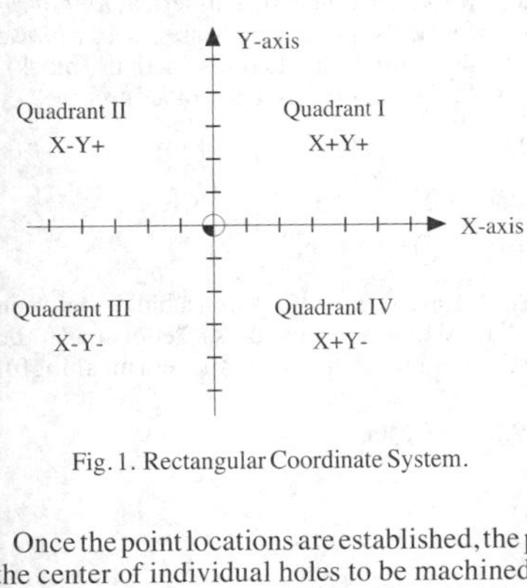


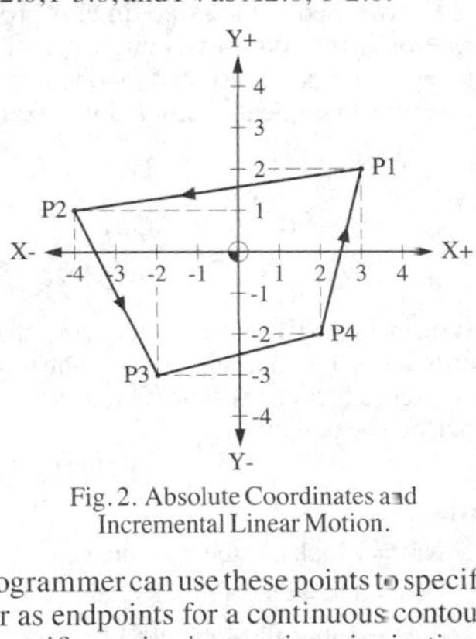

## TABLE OF CONTENTS

## MANUFACTURING PROCESSES

### WELDING

*(Continued)*



### FINISHING OPERATIONS

**Table 2. Minimum Distance between Punched Opening and Edges and Rounded Radius**

| Part is | Form of opening | Minimum Distance | Sketch |
|---|---|---|---|
| Blanked | Circle | $c \geq T$<br>$r \geq 0.5T$ | 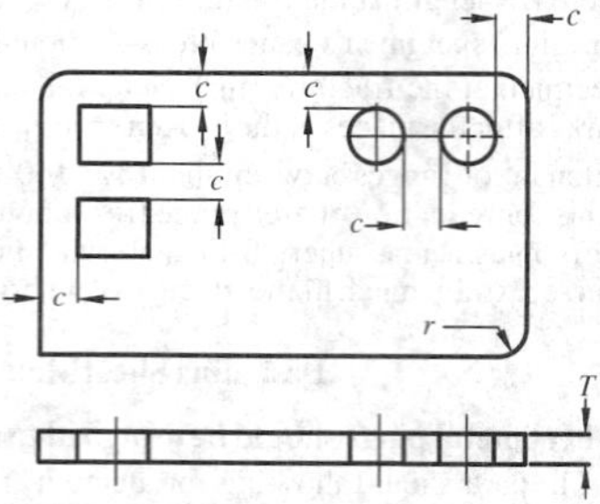 |
| | Rectangle | $c \geq 1.2T$<br>$r \geq 0.5T$ | |
| Bent | Circle | $c \geq 2T$<br>$r \geq (0.5 - 1.0)T$ [a]<br>$r \geq (1.0 - 2.0)T$ [b]<br>$r \geq (2.0 - 3.0)T$ [c] | 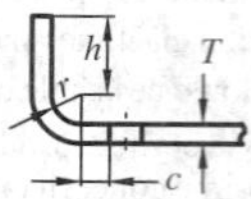 |
| Drawn | Circle | $d \leq (d_1 - 2r)$<br>$D_1 \geq (d_1 - 2r)$<br>$D \geq (D_1 + 3T + d_2)$<br>$c \geq r + 0.5T$ | 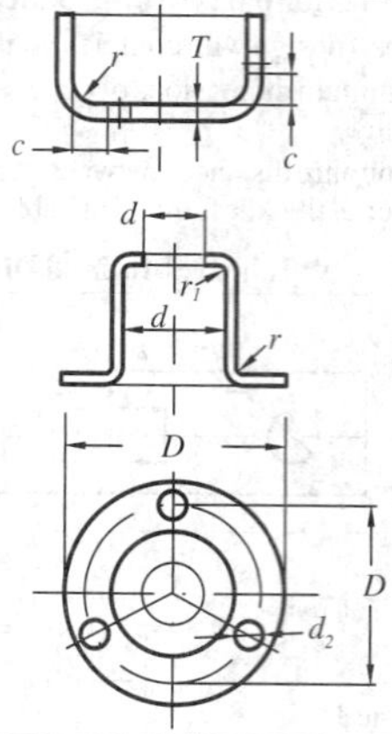 |

[a] Al and brass
[b] Steel
[c] Al-alloy 6000 series

***Design rules for parts produced by bending:***

Minimum bend radius should be used only if it is necessary for correct function of part.

The bend radius should be larger than the thickness of the material.

Use minimum distances between punched opening and bend radius given in Table 2.

Flange length as shown in Table 2 needs to be $h \geq 2T$.

If a part has more than one bend, it is necessary to define technological data.

weld design. Seam types must be selected based on material and pressure. A duct section between adjacent hangers must be able to carry its own weight and to resist the external loads for which it is constructed. Reinforcing members must be able to resist the external deflection of the sheet and their own deflection.

There is a relationship between duct width, reinforcement spacing, reinforcement size, pressure, and sheet thickness. For constant pressure and constant duct size, the thicker sheet allows more distance between reinforcements. The higher the pressure the shorter the spacing between reinforcements. Joints and intermediate reinforcements are labor-intensive and may be more costly than the savings gained by a reduction in wall thickness Thicker duct wall and stronger joints are more cost-effective than using more reinforcement. The following material illustrates various joint designs, used both in duct work and other sheet metal assemblies.

## Sheet Metal Joints

*Plain Lap and Flush Lap:*

Fig. 35. Plain Lap

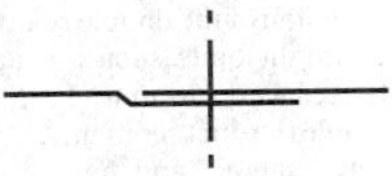

Fig. 36. Flush Lap

The *plain lap* (Fig. 35) and *flush lap* (Fig. 36) are both used for various materials such as galvanized or black iron, copper, stainless steel, aluminum, or other metals, and may be soldered and/or riveted, as well as spot, tack, or solid-welded. Lap dimensions vary with the particular application, and since it is the duty of the drafter to specify straight joints in lengths that use full-sheet sizes, transverse lap dimensions must be known.

*Raw and Flange Corner:*

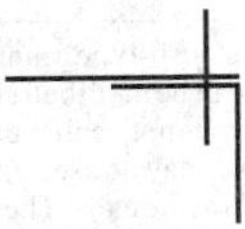

Fig. 37. Raw and Flange Corner

The *raw and flange corner* (Fig. 37) is generally spot-welded, but may be riveted or soldered. For heavy gages it is tack-welded or solid-welded.

*Flange and Flange Corner:*

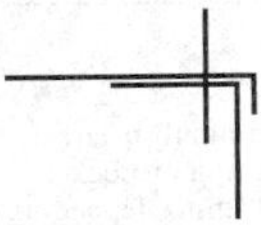

Fig. 38. Flange and Flange Corner

The *flange and flange corner* (Fig. 38) is a refinement of the raw and flange corner. It is particularly useful for heavy-gage duct sections that require flush outside corners and must be field-erected.

*Standing Seam:*

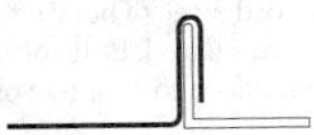

Fig. 39. Standing Seam

The *standing seam* (Fig. 39) is often used for large plenums, or casings. Before the drafter is able to lay out a casing drawing the seam allowance measurements are needed, so that panel sizes can be detailed for economical use of standard sheets. Considering velocity levels, standing seams are considered for duct interiors: 1-inch (25.4-mm) seam is normally applied for duct widths up to 42 inches (1066.8 mm) and $1\frac{1}{2}$-inch (38.1-mm) seam is applied for bigger ducts.

*Groove Seam:*

Fig. 40. Groove Seam

The *groove seam* (Fig. 40) is often used for rectangular or round duct straight joints, or to join some sheets for fittings that are too large to be cut out from standard sheets. It is also known as the pipelock, or flat lock seam.

*Corner Standing Seam:*

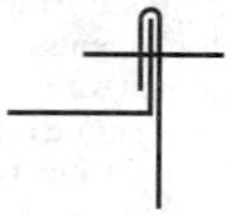

Fig. 41. Corner Standing Seam

The *corner standing seam* (Fig. 41) has applications similar to the standing seam, and also can be used for straight-duct sections. This type of seam is mostly applied at the ends at 8 inches (203 mm) intervals.

*Double Seam:*

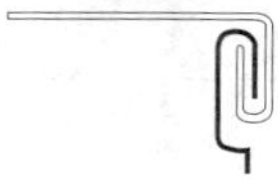

Fig. 42. Double Corner Seam

The *double corner seam* (Fig. 42) at one time was the most commonly used method for duct fabrication. However, although it is seldom used because of the hand operations required for assembly, the double seam can be used advantageously for duct fittings with compound curves. It is called the slide lock seam. Machines are available to automatically close this seam.

*Slide-Corner:*

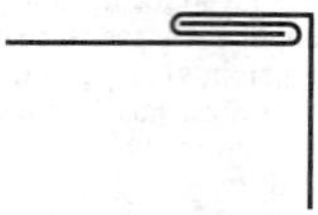

Fig. 43. Slide Corner

The *slide-corner* (Fig. 43) is a large version of the double seam. It is often used for field assembly of straight joints, such as in an existing ceiling space, or other restricted working area where ducts must be built in place. To assemble the duct segments, opposite ends of each seam are merely "entered" and then pushed into position. Ducts are sent to job sites "knocked-down" for more efficient use of shipping space.

*Button Punch Snap Lock:*

Fig. 44. Button Punch Snap Lock

The *button punch snap lock* (Fig. 44) is a flush-type seam that may be soldered or caulked. This seam can be modified slightly for use as a "snap lock." This type of seam is not applicable for aluminum or other soft metals. This seam may be used up to 4 inches (101.6 mm) w.g. by using screws at the ends. The pocket depth should not be smaller than $\frac{5}{8}$ inch (15.88 mm) for 20-, 22- and 26-gage material (0.91-0.45 mm).

*Pittsburgh:*

Fig. 45. Pittsburgh

The *Pittsburgh* (Fig. 45) is the most commonly used seam for standard-gage duct construction. The common pocket depths are $\frac{5}{16}$ inch (7.94 mm) and $\frac{5}{8}$ inch (15.88 mm), depending on the thickness of the sheet.

*Flange:*

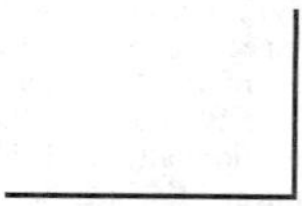

Fig. 46. Flange

The *flange* (Fig. 46) is an end edge stiffener. The drafter must indicate size of flange, direction of bend, degree of bend (if other than 90°) and when full corners are desired. Full corners are generally advisable for collar connections to concrete or masonry wall openings at louvers.

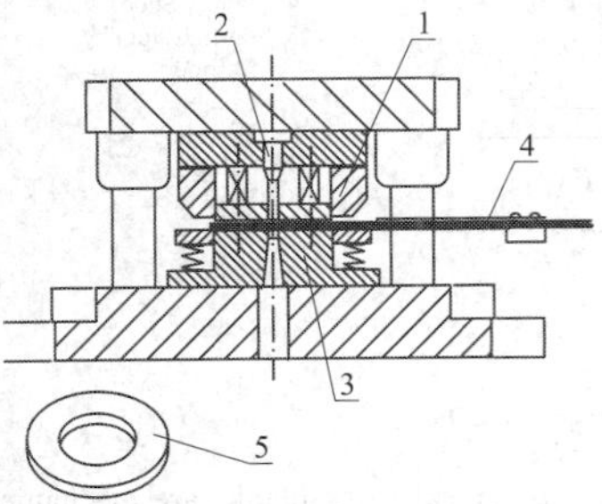


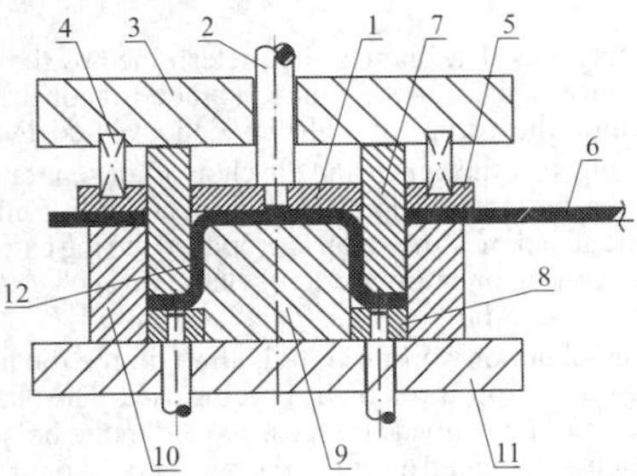

**Multiple-Station Dies.**—Multiple-station dies are arranged so a series of sequential operations is performed with each press stroke. The two types are progressive and transfer dies.

*Progressive Die:* A progressive die (Fig. 63) is used to transform coil stock or strips into a finished part. This transformation is performed progressively by a series of stations aligned in a row; the workpiece is fed from station to station with each stroke, by being attached to the scrap skeleton. Force for the movement is applied to the incoming sheet metal strip or coil and outgoing scrap skeleton by means of rollers. The cost of a progressive die is high, so it is best used for high-volume production.

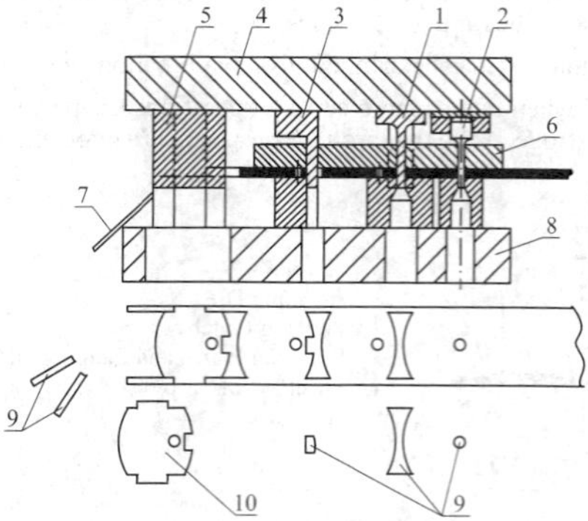


1. Semi-Notching Punch
2. Hole Punch
3. Notching Punch
4. Punch Shoe
5. Parting Punch
6. Stripper
7. Chute
8. Die Shoe
9. Scrap
10. Workpiece

Fig. 63. Progressive Die

*Transfer Die:* In transfer die operations, individual stock blanks are mechanically moved from die station to die station in a single die set. Large workpieces are managed with tandem press lines, in which the stock is moved from press to press where specific operations are performed.

## Types of Dies

There are many types of dies for metalworking. In this section, descriptions and schematic illustrations are given of die types, with a focus on those most used for sheet metal working.

**Blanking Dies.**—Blanking dies are cutting dies. The blank size is determined by the dimension of the die, and the cutting clearance is given on the punch. By mode of construction, blanking dies can be classified as cutoff dies, drop-through dies, and invented dies.

*Cutoff Blanking Die:* In the cutoff die (Fig. 64), the punch and die chop off or separate a flat and/or square profile workpiece from a continuous supply of metal. The trailing edge of a part also forms the leading edge of the next part, so no scrap is produced during cutoff. This type of die also may be used for separating or cutting scrap skeleton. In such cases, the cutoff die is usually a section of a progressive die.

*Drop-Through Blanking Die:* When the blank or workpiece falls through the die and press bed into a pan or conveyer for removal, a drop-through die is being used. The blank drops of its own weight through the die opening, with clearance provided in the bolster plate and press bed. The die block assembly is mounted on either the bolster plate or press bed, and the punch assembly is mounted on the press slide.

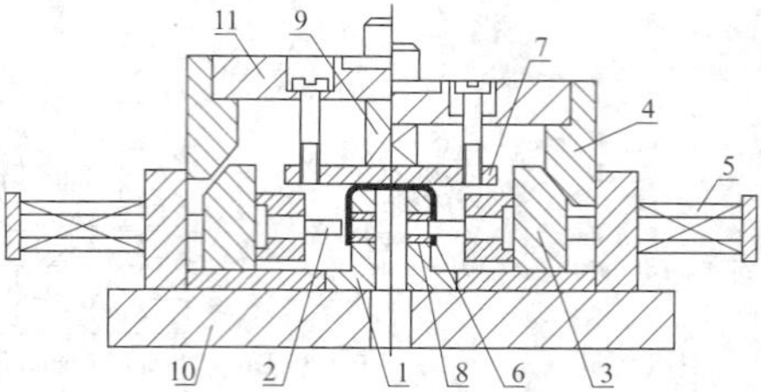


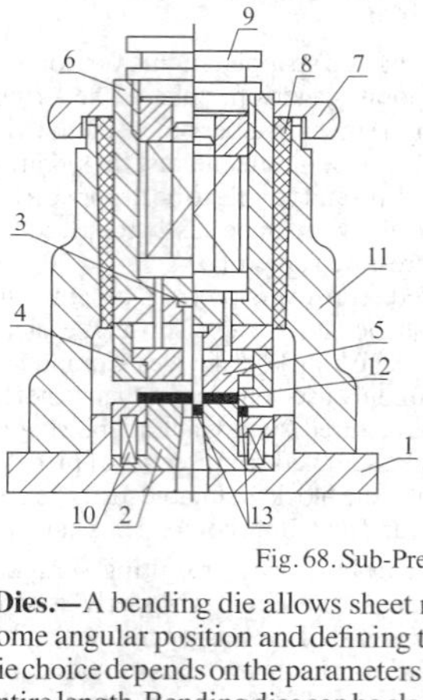

**Table 1.** *(Continued)* **Additional Die Types Commonly Used for Sheet Metalworking**

| | |
|---|---|
| 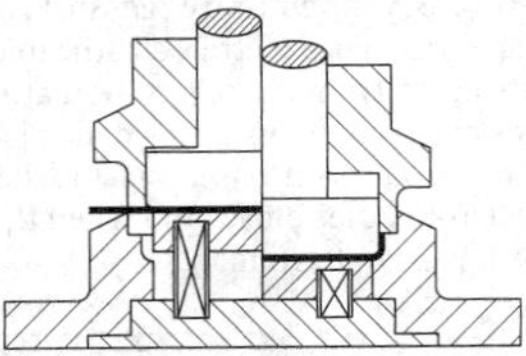 | **Drawing Die:** Used in a double-action press, cuts the blank as the cutting punch enters the die, and then holds the blank in place as the drawing punch pushes the blank into the die. |
| 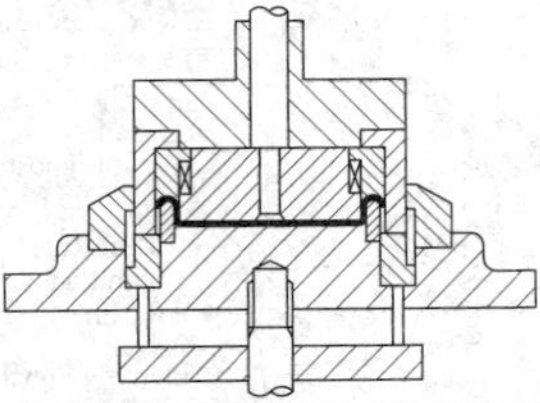 | **Inverted Drawing Die:** Cuts the blank and then performs two drawing operations: the workpiece is drawn into a cylindrical cup, and then it is curled. |

**Steel Rule Dies.**—Steel rule dies (or knife dies) were patented by Robert Gair in 1879 and, as the name implies, have cutting edges made from steel strips of about the same proportions as those used in making graduated rules for measuring purposes. According to J. A. Richards, Sr., of the J. A. Richards Co., Kalamazoo, MI, a pioneer in the field, these dies were first used in the printing and shoemaking industries for cutting out shapes in paper, cardboard, leather, rubber, cork, felt, and similar soft materials. Steel rule dies were later adopted for cutting upholstery material for the automotive and other industries, and for cutting out simple to intricate shapes in sheet metal, including copper, brass, and aluminum. A typical steel rule die, designed to cut a simple circular shape, is shown in Fig. 82, partially cut away to illustrate the construction. This die also cuts a rectangular opening in the workpiece, and pierces four holes, all in one press stroke. Such dies generally cost 25 to 35 percent of the cost of conventional blanking dies, and can be produced in much less time.

The die blocks that hold the steel strips on edge on the press platen or in the die set may be made from plaster, hot lead or type metal, or epoxy resin, all of which can be poured to shape. For light work, $\frac{3}{4}$-in. (19.05 mm) thick, five- or seven-ply maple or birch wood may be used. Narrow slots are cut in this wood with a jig saw to hold the strips vertically. Where greater forces are involved, as with operations on metal sheets, the blocks usually are made from Lignostone densified wood or from metal. In the $\frac{3}{4}$-in. thickness mostly used, metal or medium- and high-density Lignostone-laminated wood, which is available in the $\frac{3}{4}$-in. thickness, is made from about 35 plies of highly compressed lignite wood, bonded with phenolformaldehyde resin, which imparts great density and strength. The material is made in thicknesses up to 6 in. (15.24 cm) and in various widths and lengths.

Steel rule die blocks can carry punches of various shapes to pierce holes in the stock, also projections designed to form strengthening ribs and other shapes in material such as aluminum, at the same time as the die cuts the component to shape. Several dies can be combined or nested, and operated together in a large press, to produce various shapes simultaneously from one sheet of material.

As shown in Fig. 82, the die steel is held in the die block slot on its edge, usually against the flat platen of a die set attached to the moving slide of the press. The sharp, free end of the rule faces toward the workpiece, which is supported by the face of the other die half. This other die half may be flat or may have a punch attached to it, as shown, and it withstands the pressure exerted in the cutting or forming action when the press is operated.

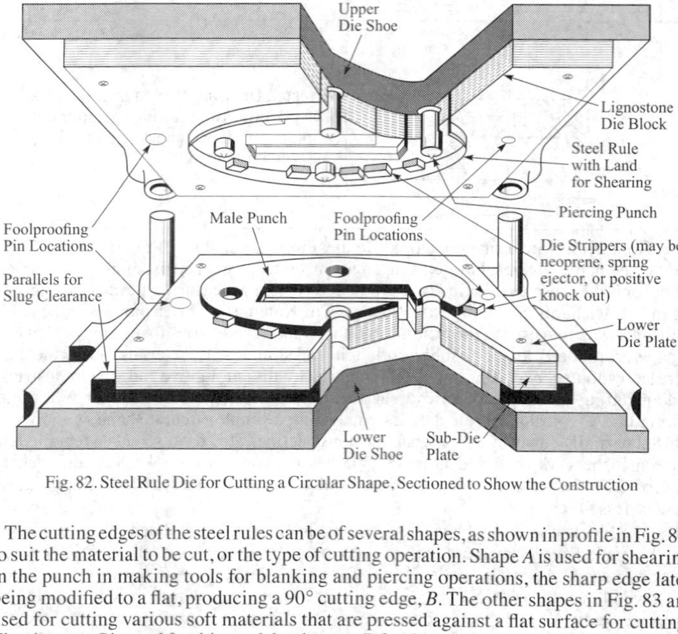


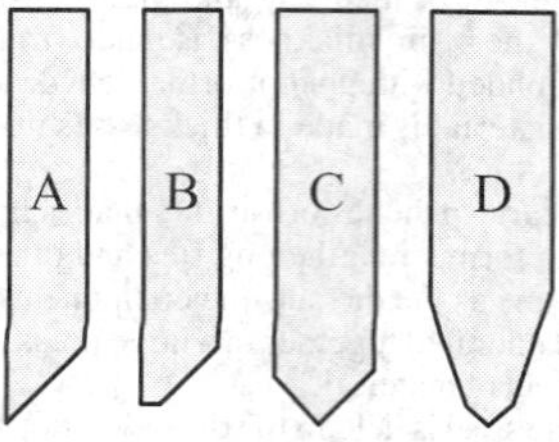

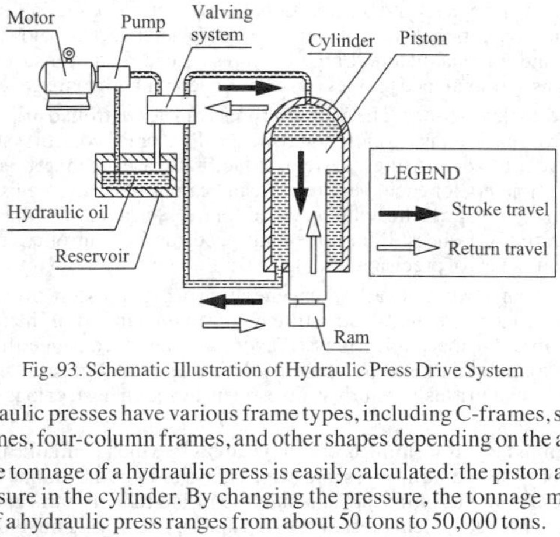

removes metal at a slower rate than electrode negative, which is mostly used for high-speed metal removal with graphite electrodes. Negative polarity is also used for machining carbides, titanium, and refractory alloys using metallic electrodes. Metal removal with graphite electrodes can be as much as 50 percent faster with electrode negative polarity than with electrode positive, but negative polarity results in much faster electrode wear, so it is generally restricted to electrode shapes that can be redressed easily.

Newer generators can provide less than 1 percent wear with either copper or graphite electrodes during roughing operations. Roughing is typically done with a positive-polarity electrode using elevated on times. Some electrodes, particularly micrograin graphites, have a high resistance to wear. Fine-grain, high-density graphites provide better wear characteristics than coarser, less dense grades, and copper-tungsten resists wear better than pure copper electrodes.

*Machine Settings:* For vertical machines, a rule of thumb for power selection on graphite and copper electrodes is 50 to 65 amps per square inch of electrode engagement. For example, an electrode that is $\frac{1}{2}$ in$^2$ (3.2 cm$^2$) might use $0.5 \times 0.5 \times 50 = 12.5$ amps. Although each square inch of electrode surface may be able to withstand higher currents, lower settings should be used with very large jobs or the workpiece may become overheated and it may be difficult to clean up the recast layer. Lower amperage settings are required for electrodes that are thin or have sharp details. The voltage applied across the arc gap between the electrode and the workpiece is ideally about 35 volts, but should be as small as possible to maintain stability of the process.

*Spark Frequency:* Spark frequency is the number of times per second that the current is switched on and off. Higher frequencies are used for finishing operations and for work on cemented carbide, titanium, and copper alloys. The frequency of sparking affects the surface finish produced, low frequencies being used with large spark gaps for rapid metal removal with a rough finish, and higher frequencies with small gaps for finer finishes. High frequency usually increases, and low frequency reduces electrode wear.

*The Duty Cycle:* Electronic units on modern EDM machines provide extremely close control of each stage in the sparking cycle, down to millionths of a second (μs). A typical EDM cycle might last 100 μs. Of this time, the current might be on for 40 μs and off for 60 μs. The relationship between the lengths of the on and off times is called the duty cycle, and it indicates the degree of efficiency of the operation. The duty cycle states the on time as a percentage of the total cycle time, and in the previous example it is 40 percent. Although reducing the off time will increase the duty cycle, factors such as flushing efficiency, electrode and workpiece material, and dielectric condition control the minimum off time. Some EDM units incorporate sensors and fuzzy logic circuits that provide for adaptive control of cutting conditions for unattended operation. Efficiency is also reported as the amount of metal removed, expressed as in$^3$/hr.

In the EDM process, work is done only during the on time, and the longer the on time, the more material is removed in each sparking cycle. Roughing operations use extended on time for high metal-removal rates, resulting in fewer cycles per second, or lower frequency. The resulting craters are broader and deeper so that the surface is rougher and the heat-affected zone (HAZ) on the workpiece is deeper. With positively charged electrodes, the spark moves from the electrode toward the workpiece and the maximum material is removed from the workpiece. However, every spark takes a minute particle from the electrode so that the electrode also is worn away. Finishing electrodes tend to wear much faster than roughing electrodes because more sparks are generated in unit time.

The part of the cycle needed for reionizing the dielectric (the off time) greatly affects the operating speed. Although increasing the off time slows the process, longer off times can increase stability by providing more time for the ejected material to be swept away by the flow of the dielectric fluid, and for deionization of the fluid, so that erratic cycling of the servo-mechanisms that advance and retract the electrode is avoided. In any vertical

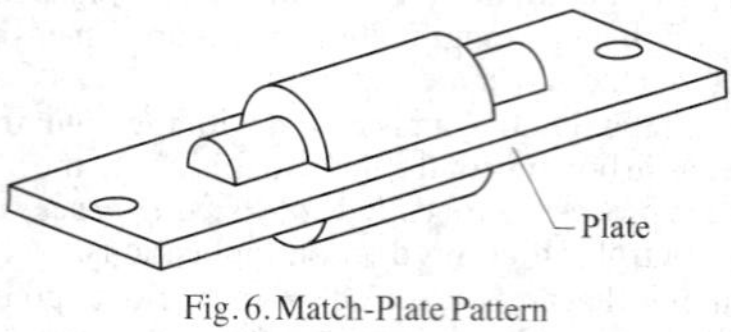

pressure. The procedure is governed by the kind of alloy, the size of the investment cavity, and its contours or shape.

**Investment Removal.**—When the casting has solidified, the investment material is removed by destroying it. Some investments are soluble in water, but those used for ferrous castings are broken by using pneumatic tools, hammers, or by shot or abrasive blasting and tumbling to remove all particles. Gates, sprues, and runners may be removed from the castings by an abrasive cutting wheel or a band saw according to the shape of the cluster and machinability of the material.

**Accuracy of Investment Castings.**—The accuracy of precision investment castings may, in general, compare favorably with that of many machined parts. The overall tolerance varies with the size of the work, the kind of metal and the skill and experience of the operators. Under normal conditions, tolerances may vary from ±0.005 or ±0.006 inch per inch (or mm/mm), down to ±0.0015 to ±0.002 inch per inch (or mm/mm), and even smaller tolerances are possible on very small dimensions. Where tolerances applying to a lengthwise dimension must be smaller than would be normal for the casting process, the casting gate may be placed at one end to permit controlling the length by a grinding operation when the gate is removed.

**Casting Weights and Sizes.**—Investment castings may vary in weight from a fractional part of an ounce up to 75 lb (34 kg) or more. Although the range of weights representing the practice of different firms specializing in investment casting may vary from about $\frac{1}{2}$ pound up to 10 or 20 lb (4.5 or 9.1 kg), a practical limit of 10 or 15 lb (4.5 or 6.8 kg) is common. The length of investment castings ordinarily does not exceed 12 or 15 in. (30 or 38 cm), but much longer parts may be cast. It is possible to cast sections having a thickness of only a few thousandths of an inch, but the preferred minimum thickness, as a general rule, is about 0.020 in. (0.50 mm) for alloys of high castability and 0.040 in. (1 mm) for alloys of low castability.

**Design for Investment Casting.**—As with most casting processes, best results from investment casting are achieved when uniform wall thicknesses between 0.040 and 0.375 in. (1 and 10 mm) are used for both cast components and channels forming runners in the mold. Gradual transition from thick to thin sections is also desirable. It is important that molten metal should not have to pass through a thin section to fill a thick part of the casting. Thin edges should be avoided because of the difficulty of producing them in the wax pattern. Fillets should be used in all internal corners to avoid stress concentrations that usually accompany sharp angles. Thermal contraction usually causes distortion of the casting, and should be allowed for if machining is to be minimized. Machining allowances vary from 0.010 in. (0.25 mm) on small, to 0.04 in. (1 mm) on large parts. With proper arrangement of castings in the mold, grain size and orientation can be controlled and directional solidification can often be used to advantage to ensure desired physical properties in the finished components.

**Casting Milling Cutters by Investment Method.**—Possible applications of precision investment casting in tool manufacture and in other industrial applications are indicated by its use in producing high-speed steel milling cutters of various forms and sizes. Removal of the risers, sand blasting to improve the appearance, and grinding the cutting edges are the only machining operations required. The bore is used as cast. Numerous tests have shown that the life of these cutters compares favorably with high-speed steel cutters made in the usual way.

## Hard Mold Casting

Hard mold casting is permanent mold casting. Two halves of a mold are made from materials such as cast iron, steel, bronze, graphite, or refractory metal alloys. The mold cavity and gating system are machined into the mold and become an integral part of it.

To produce castings with internal cavities, cores made of metal or sand aggregate are placed in the mold prior to casting. Typical core materials are oil-bonded or resin-bonded sand, plaster, graphite, gray iron, low-carbon steel, and hot-work die steel.

the bottom of the overhang. This roughness is particularly high if the angle of the overhang, relative to horizontal, is less than 45 degrees. As a result, the as-printed finish of up-facing surfaces will be better than down-facing overhang surfaces. (Surface finish in SLM material may be improved by subsequent shot peening.) Typical surface finishes of up-facing and side-facing surfaces are equivalent to those of investment castings, so only bearing or mounting surfaces may need to be machined.

Since the rate of heat transfer into loose powder is much less than the rate into solid metal, heat transfer from overhangs is reduced and the cooling rate is lower than in layers in the body of a part. Structural supports printed beneath overhangs (Fig. 26) and connected to the base plate provide a path for increased heat transfer from overhangs. Supports also anchor the printed part to the base and reduce distortion due to differences in temperature throughout the part as it is printed. Without such supports, overhanging layers of the printed parts would deflect upward due to thermal stress. A downside of such supports is that they can interfere with printing more than one layer of parts in one build.

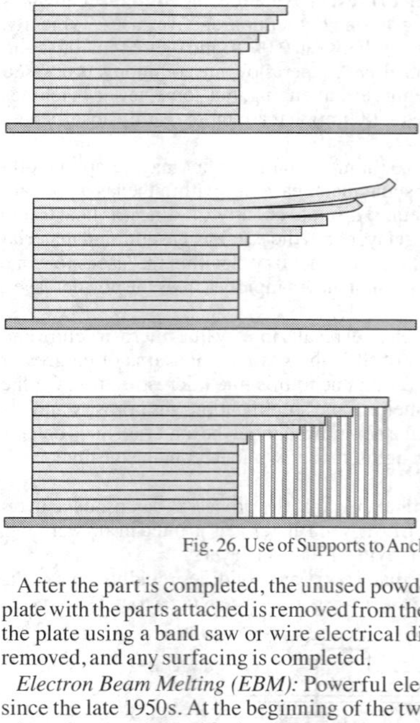

Where 3D printed layers overhang loose powder instead of solid layers, more heat is retained, changing the cooling rate.

When these overhanging layers are longer, they can bend upward during cooling, as a result of thermal stress.

This distortion can be prevented with added structures (to be removed later) to reduce thermal stress, support the overhangs, and anchor the part.

Fig. 26. Use of Supports to Anchor Parts in SLM Process

After the part is completed, the unused powder is collected for future use, and the base plate with the parts attached is removed from the SLM machine. Parts usually are cut from the plate using a band saw or wire electrical discharge machining (EDM), supports are removed, and any surfacing is completed.

*Electron Beam Melting (EBM):* Powerful electron beams have been used for welding since the late 1950s. At the beginning of the twenty-first century, this energy source was adapted to powder bed fusion (PBF) as the electron beam melting (EBM) process. EBM is commonly used to process titanium and superalloy powders into net-shaped parts—providing significant material savings over traditional manufacturing processes.

In EBM, a tungsten filament emits electrons at a high voltage of 60,000 V (60 kV). A focusing coil or electromagnetic lens produces a converging Gaussian beam with a spot size of 0.008–0.04 in. (0.203–1.02 mm) and a power of up to 6,000 W; a deflection coil directs the scanning beam over the surface of the powder layer at a speed of 26,000 ft/s (8,000 m/s). These focus and deflection coils are the electronic counterparts of the f-theta and scanning optics of the laser beam systems in SLM. Schematically, the EBM system is similar to that of SLM; however, while the laser beam is moved optically by deflection

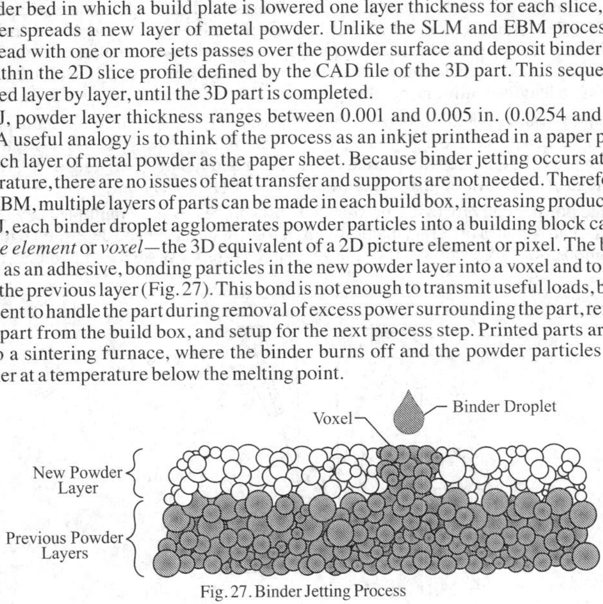

**Contact Tip Recess.—**The dimension labeled contact tip recess in Fig. 2, and indicated as $\frac{1}{8}$ in. (3 mm), should be about $\frac{1}{2}$ in. (13 mm) for a minimum electrode extension of $\frac{3}{4}$ in. (19 mm), for FCAW welding. This dimension is critical for obtaining high-quality welds with all-position electrodes because they have a fast-freezing slag and operate with low to medium current and voltage. If the recess dimension is less than the optimum, the voltage may be lower than the minimum recommended, and if the settings are less than the minimum, the fast-freezing slag may solidify too rapidly, causing excess porosity or "worm tracks" on the weld surface.

The recommended length of electrode extension for all-position FCAW, E71T-1 electrodes is $\frac{3}{4}$ to 1 in. (19 to 25 mm). The size of this extension not only affects the minimum required parameters, but a long electrode extension also ensures preheating of the electrode and allows lower current to be used. Preheating the electrode is further beneficial as it reduces moisture on the electrode surface, and in the electrode flux. When a change is made from the GMAW to the FCAW process, welders must be aware of the influence on weld quality of the electrode extension in the FCAW process.

**Porosity and Worm Tracks.—**As mentioned above, porosity and worm tracks typically result from a combination of incorrect electrode extension, incorrect welding settings, humidity, electrode moisture, refill scale, rust, paint, oils, or poor welding technique. Where humidity levels are high, potential for porosity and worm tracks increases. The FCAW process is less sensitive to mill scale than the GMAW spray transfer mode but mill scale will often cause excess weld porosity. The best way to avoid the effects of mill scale, rust, oil, and surface contaminants is to grind the area to be welded.

Another way to reduce porosity is to keep weaving to a minimum. If the correct size flux-cored electrode is used, weaving can be kept to a minimum for most flux-cored applications. The forehand technique produces the best weld bead surface on fillet weld beads up to $\frac{3}{4}$ in. (19 mm) steel thickness in the flat and horizontal weld positions. On larger single-pass fillet welds, the backhand technique is beneficial because the voltage directed at the weld provides additional weld puddle control to the fluid welds. The backhand technique used for flat and horizontal welds produces a more convex weld bead, reduces potential for porosity, and increases penetration.

If porosity or worm tracks occur, the prime solution is in weld practices that increase heat at the weld, but the following remedies can also be tried. Grind clean the surface to be welded; use recommended electrode extensions; increase current (wire feed rate) decrease voltage; use the backhand welding technique; slow down travel speed, consider use of a different electrode formulation containing increased deoxidizers, avoid weaving; change from argon + $CO_2$ mixture to straight $CO_2$; and provide a protective cover to keep the electrode spool clean and dry.

**Welding with 0.045-Inch (1.2-mm) Diameter All-Position Electrodes.—**Fig. 5 shows wire feed settings for welding of steel with 0.045-in. (1.2-mm) diameter, E71T-1 all-position electrodes using a mixture of argon + 15 to 25 percent $CO_2$ as the shielding gas, and an electrode extension of $\frac{3}{4}$ in. (18 mm). Parameters for vertical up welding, shown at the left in Fig. 5, include setting the wire feed rate at the twelve o'clock position, or about 350 in/min, using 200 to 190 amps, and setting the voltage between 24 and 25 volts. Optimum parameters for flat welding, shown at the right in Fig. 5 include setting the wire feed rate at three o'clock position, or 560 in/min (240 mm/s), and 270 amps at 25 to 27 volts.

**Welding with 0.052-Inch (1.3-mm) Diameter All-Position Electrodes.—**Settings for vertical up and flat welding with all-position E71T-1 electrodes of 0.052-in. (1.3-mm) diameter are seen at the left in Fig. 6. These electrodes are suited to welding steel having thicknesses of $\frac{1}{4}$ in. (6 mm) and thicker. For vertical up welding, the wire feed rate is set between the ten and eleven o'clock positions, or 250 in/min (106 mm/s), with about 200 amps at 25 volts. Flat welding with these electrodes is best done with the wire feed rate set between the two and three o'clock positions, or 490 to 560 in/min (207 to 237 mm/s), giving approximately 300 amps at 28 volts.

Plasma is a high-temperature, ionized, gaseous column that is formed when electrons in the arc collide with the shielding gas molecules. The gas atoms lose one or more electrons, leaving them positively charged. The electrons and the resulting plasma are concentrated at the electrode tip, where they cause the plasma pressure to be at its greatest. The electron density thins out as the electrons travel from the straight-polarity, negatively charged, tungsten electrode across the open arc. As the electrons traverse the arc to the work, the resulting arc column width increases slightly, controlled in part by the electromagnetic forces generated by the current. With the increase in the arc column width, the density and pressure of the plasma decrease. The electrons collide with the work, liberating much heat. The downward pressure of the plasma is exerted against the surface of the weld pool. The gas ions in the plasma are positively charged and greater in mass than the electrons.

In DC−, straight-polarity welding, the positive gas ions are drawn to the negative electrode. The electron flow to the weld ensures that most of the arc heat is generated at the positive work side of the arc. This current setup provides maximum penetration potential, as indicated in Fig. 8. With DC−, straight polarity, the tungsten electrode can carry a higher current and operate at lower temperatures than with the other current arrangements.

*Direct-Current Reverse Polarity (DC+):* With direct-current positive polarity (DC+), the tungsten electrode is connected to the power-source positive terminal so that the electrons flow from the negative work to the positive electrode. As illustrated in Fig. 9, the electrons impinging on the electrode tip reverse the direction of the heat concentration that occurs with straight polarity, as described above. Approximately two-thirds of the heat generated with DC+ reverse polarity is at the electrode tip, and the electrode becomes very hot, even with low current levels. DC+ reverse polarity requires large-diameter electrodes.

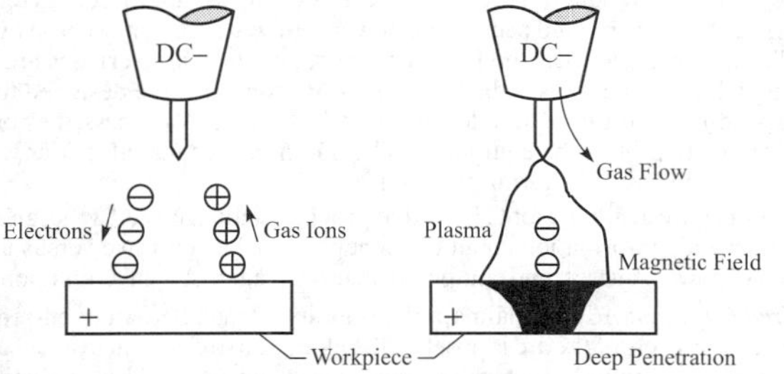


Fig. 8. Straight Polarity (DC−) Provides Highest Electrode Current Capacity and Deepest Penetration Potential

In the current range of 100 to 150 amps, DC+ reverse polarity requires a $\frac{1}{4}$-in. (6.4-mm) diameter electrode. This larger electrode produces a weld puddle almost twice as wide as that produced by a 120-amp, $\frac{1}{16}$-in. (1.6-mm) diameter, DC− straight polarity electrode. Most of the heat is generated at the electrode tip with DC+ reverse polarity, so penetration is much less than with DC− straight polarity. With DC+ reverse polarity, the positive gas ions in the arc plasma are drawn to the negative workpiece where they bombard and break up the surface oxides that form on metals such as aluminum and magnesium. However, the best welding method for aluminum and magnesium is to use alternating current (AC), which combines the benefits of DC− straight and DC+ reverse polarity.

*Alternating Current (AC):* The surface oxides formed on metals such as aluminum and magnesium disturb the arc and reduce the weld quality. Welding of these metals requires DC+ reverse or AC polarity to break up the surface oxides. An alternating current (AC) cycle consists of one-half cycle of straight polarity and one-half cycle of reverse polarity. With alternating current, the cleaning action benefits of the reverse-polarity arc can be combined with the electrode current-carrying capacity of the straight-polarity

electrode and consistent arc stability. Varying polarity has advantages in both gas metal arc and plasma arc welding, but with the keyhole process it allows single-pass, square-groove, full-penetration welds in materials up to $\frac{1}{2}$ in. (12.7 mm) thick.

The VPPA process ensures extremely low levels of porosity in weld areas in aluminum. VPPA welding is often used in the vertical up position for aluminum because it provides superior control of root reinforcement, which tends to be excessive when welding is done in the flat position. Pulsing in the VPPA process when welding aluminum of $\frac{1}{8}$ to $\frac{1}{4}$ in. thickness in the flat position gives satisfactory root profiles. Pulsing gives improved arc control in keyhole welds in both ferrous and nonferrous metals and is beneficial with melt-in fusion welding of thin materials as it provides better control of heat input to the workpiece.

### Plasma Arc Surface Coating

*Plasma Arc Surfacing* uses an arc struck between the electrode and the workpiece, or transferred arc, to apply coatings of other metals or alloys to the workpiece surface. This high-temperature process produces homogeneous welds in which the ionized plasma gas stream melts both the work surface and a stream of powdered alloy or filler wire fed into the arc. Dilution of the base metal can be held below 5 percent if required. With arc temperatures between 25,000 and 50,000° F (14,000 and 28,000° C), deposition occurs rapidly, and a rate of 15 lb/h (6.8 kg/h) of powdered alloy is not unusual. Deposition from wire can be performed at rates up to 28 lb/h (12.7 kg/h), much higher than with oxygen/fuel or gas metal/arc methods.

In the nontransferred arc process used for coating of surfaces, the arc is struck between the electrode and the torch nozzle so that it does not attach to the work surface. This process is sometimes called metal spraying, and is used for building up surfaces for hard facing and for application of anticorrosion and barrier layers. Argon is frequently used as the plasma gas. As the coating material in the form of powder or wire enters the plasma, it is melted thoroughly by the plasma column and is propelled toward the work at high velocity to form a mechanical bond with the work surface. Some 500 different powder combinations are available for this process so that a variety of requirements can be fulfilled and deposition rates up to 100 lb/h (45 kg/h) can be achieved.

The plasma arc process allows parts to be modified or recovered if worn, and surfaces with unique properties can be provided on new or existing components. Low levels of porosity in the deposited metal can be achieved. Metal spraying can be performed manually or automatically, and its use depends primarily on whether a mechanical bond is acceptable. Other factors include the volume of parts to be treated, the time needed for the process and for subsequent finishing, the quality requirements for the finished parts, rejection rates, and costs of consumable materials and energy.

Some systems are available that can use either metal powder or wire as the spray material, and can be operated at higher voltage settings that result in longer plasma arc lengths at temperatures over 10,000° F (5,537° C). With these systems, the plasma velocity is increased to about 12,000 ft/s (3,658 m/s), giving an extremely dense coating with less than 1 percent porosity. Current ranges of 30 to 500 amps are available, and nitrogen is frequently used as the plasma gas, coupled with $CO_2$, nitrogen, or compressed air as the shielding gas. Gas flow rates are between 50 and 350 cubic feet per hour (24 and 165 l/min). Large or small surface areas can be coated at low cost, with minimum heat input, if other aspects of the process are compatible with the product being made.

### Plasma Arc Cutting of Metals

Higher current and gas flow rates than for plasma arc welding are used for the plasma arc cutting (PAC) process, which operates on DC straight polarity, and uses a transferred arc to melt through the material to be cut. The nozzle is positioned close to the work surface

Portable flux feeding units are designed to inject a predetermined amount of the flux powder. The rate of flux flow is accurately regulated by a vibrator type of dispenser with rheostat control. The flux-injection method is applicable either to machine cutting or to a hand-controlled torch. The operating procedure and speed of cutting are practically the same as in cutting mild steel.

**Cutting Cast Iron.—**The cutting of cast iron with the oxyacetylene torch is practicable, although it cannot be cut as readily as steel. The ease of cutting seems to depend largely on the physical character of the cast iron, very soft cast iron being more difficult to cut than harder varieties. The cost is much higher than that for cutting the same thickness of steel, because of the larger preheating flame necessary and the larger oxygen consumption. In spite of this extra cost, however, this method is often economical. The slag from a cast-iron cut contains considerable melted cast iron, whereas in steel, the slag is practically free from particles of the metal, indicating that cast-iron cutting is partly a melting operation. Increased speed and decreased cost often can be obtained by feeding a steel rod, about $\frac{1}{4}$ inch (6.35 mm) in diameter, into the top of the cut, beneath the torch tip. This rod furnishes a large amount of slag that flows over the cut and increases the temperature of the cast iron. Special tips are used because of the larger amounts of heat and oxygen required.

**Mechanically Guided Torches.—**Cutting torches used for cutting openings in plates or blocks or for cutting parts to some definite outline are often guided mechanically or by numerical control. Torches guided by pantograph mechanisms are especially adapted for tracing the outline to be cut from a pattern or drawing. Other designs are preferable for straight-line cutting and one type is designed for circular cutting.

**Cutting Steel Castings.—**When cutting steel castings, care should be taken to prevent burning pockets in the metal when the flame strikes a blowhole. If a blowhole is penetrated, the molten oxide will splash into the cavity and the flame will be diverted. The presence of the blowhole is generally indicated by excessive sparks. The operator should immediately move the torch back along the cut and direct it at an angle so as to strike the metal beneath the blowhole and burn it away if possible beyond the cavity. Cutting in the normal position then may be resumed.

**Thickness of Metal That Can Be Cut.—**The maximum thickness of metal that can be cut by these high-temperature flames depends largely upon the gases used and the pressure of the oxygen, which may be as high as 150 lb/in$^2$ (1034 kPa). The thicker the metal, the higher the pressure required. When using an oxyacetylene flame, it might be practicable to cut iron or steel up to 12 or 14 inches (30.5 or 35.6 cm) in thickness, whereas an oxyhydrogen flame has been used to cut steel plates 24 inches (61 cm) thick. The oxyhydrogen flame will cut thicker material principally because it is longer than the oxyacetylene flame and can penetrate to the full depth of the cut, thus keeping all the oxide in a molten condition so that it can be easily blown out by the oxygen cutting jet. A mechanically guided torch will cut thick material more satisfactorily than a hand-guided torch, because the flame is directed straight into the cut and does not wobble, as it tends to do when the torch is held by hand. With any flame, the cut is less accurate and the kerf wider, as the thickness of the metal increases. When cutting light material, the kerf might be $\frac{1}{16}$ inch (1.59 mm) wide, whereas for heavy stock, it might be $\frac{1}{4}$ or $\frac{3}{8}$ inch (6.35 or 9.5 mm) wide.

## Hard Facing

Hard facing is a method of adding a coating, edge, or point, of a metal or alloy capable of resisting abrasion, corrosion, heat, or impact, to a metal component. The process can be applied equally well to new parts or old worn parts. The most common welding methods used to apply hard-facing materials include the oxyacetylene gas, shielded-metal arc, submerged arc, plasma arc, and inert-gas-shielded arc (consuming and nonconsuming electrode). Such coatings can also be applied by a spraying process, using equipment designed to handle the coating material in the form of a wire or a powder.

**Hard-Facing Materials.—**The first thing to be considered in the selection of a hard-facing material is the type of service the part in question is to undergo. Other considerations include

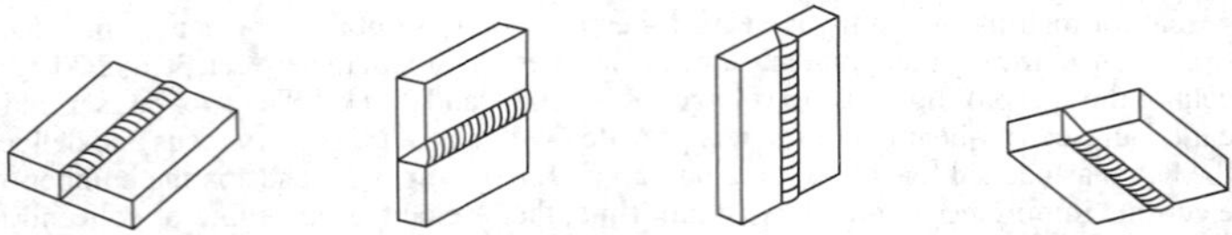

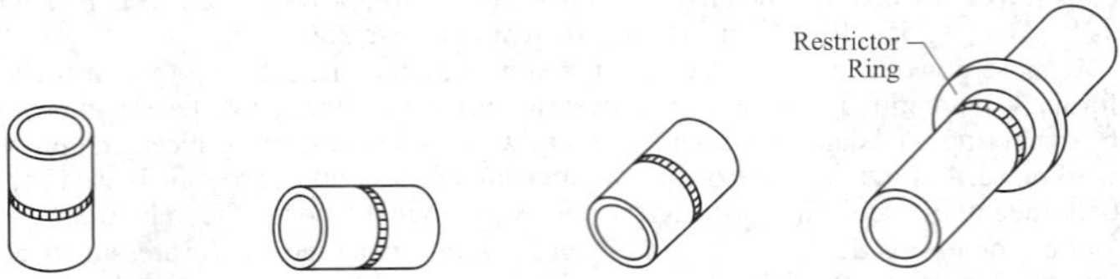

**Complete Weld Fusion.—**It is essential that new weld metal deposits be completely fused with the pipe components, and with metal laid down in successive passes. Factors that can prevent complete fusing are too numerous to list here. Some basic rules that, if followed, will improve weld fusion and quality in MIG welding in the 1G and 5G positions are:

1) The maximum gap at the root should be $\frac{3}{16}$ inch (5 mm).
2) The root land should be $\frac{1}{16}$ to $\frac{3}{32}$ inch (1.6 to 2.4 mm) wide.
3) A bevel angle of 80° inclusive should be used for MIG and flux-cored welding of pipe to provide width for weaving and improve fusion.
4) A 0.035-in MIG electrode should have a minimum short circuit current of 135 amps for fill passes.
5) Tack and root welds should be made in the vertical-down position.
6) Tack welds should be about 1.5 in (38 mm) long by $\frac{1}{16}$ to $\frac{3}{32}$ inch (1.6 to 2.4 mm) thick.
7) Short circuit fill passes should be made in the vertical-up position.
8) With flux-cored electrodes, a minimum of 0.7 inch (18 mm) wire stick-out from the contact tip must be maintained.
9) Current and voltage must be related to the pipe wall thickness.
10) Argon + 25 percent $CO_2$ is recommended for short circuit welding of pipe roots.
11) Use of undiluted $CO_2$ gas will improve MIG weld fusion in fill passes because of the "digging" action of the arc and the increased weld energy.
12) With pipe wall thicknesses of 0.4 inch (10 mm) or greater, preheating to between 400 and 500° F (205 and 260° C) will help make fusion complete.

**Other Methods.—**Pulsed MIG is a viable alternative to flux-cored for all-position welds on 5G pipe, but requires more costly equipment. The pulsed MIG process, however, has few advantages over conventional MIG and flux-cored when the latter are used correctly. Pulsed MIG may have some advantage on mechanized 5G welds and on welding of stainless steel pipe in the 5G position.

Metal-cored electrode wire also has few advantages for pipe welds because they work best with low-energy gas welds, which cancels out the increased current density claimed for them.

On most manual pipe welds, the welder needs time to control and direct the weave to ensure even heating and avoid lack of fusion. Satisfactory welds are often performed at travel speeds of 4 to 12 in/min (0.10–0.31 m/min), giving deposit rates of 3–5 lb/hr (1.36–2.27 kg/hr).

## Pipe Welding Procedure

Because of the variety of parameter combinations that can be used in pipe welding, it is suggested that charts be prepared and displayed in welding booths to remind welders of the basic settings to be used. Examples of such charts for tack, root, fill, and cover passes are included in what follows:

**FCAW 5G (Non-Rotated) MIG Welding of Thick-Walled, Carbon Steel Pipes, Procedure for Root Welding.—**This procedure can be applied to most pipe sizes, and should be given special consideration for 5G (non-rotated) welds on carbon steel pipe with $\frac{3}{8}$-inch (10-mm) wall thickness and thicker.

### Pipe and Weld Data

*Pipe bevel included angle* = 80
*Root face land* = $\frac{3}{32} \pm \frac{1}{32}$ inch (2.4 ± 0.8 mm)
*Root gap between faces* = $\frac{5}{32} \pm \frac{1}{32}$ inch (4 ± 0.8 mm)
*Electrode for root weld* = 0.035-inch (0.9-mm) diameter, E70S-3 flux-cored
*Gas* = argon with 15–25% $CO_2$
*Gas flow rates* = 30 to 40 cubic ft/hr (0.85–1.13 $m^3$/hr)

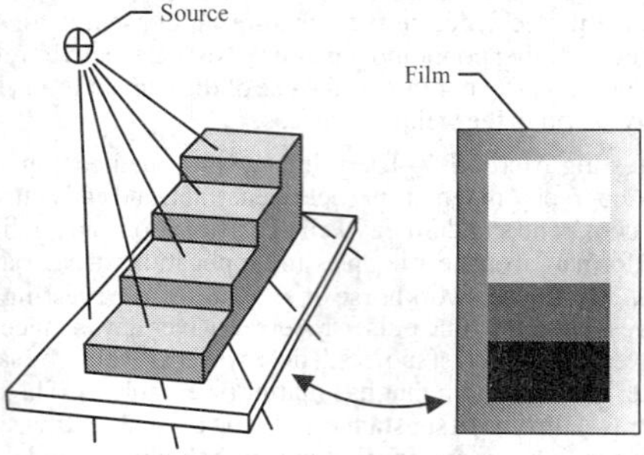


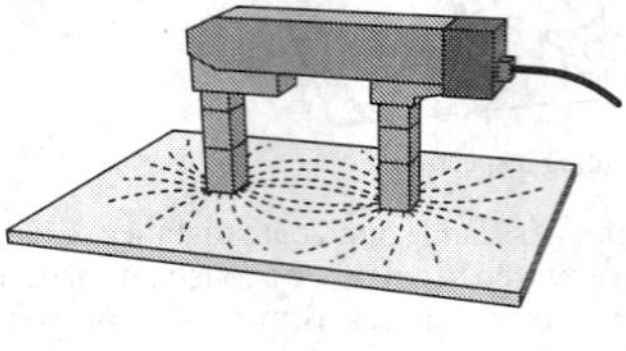

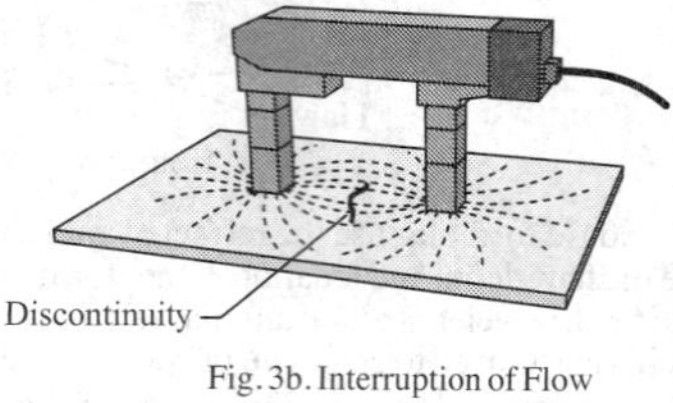

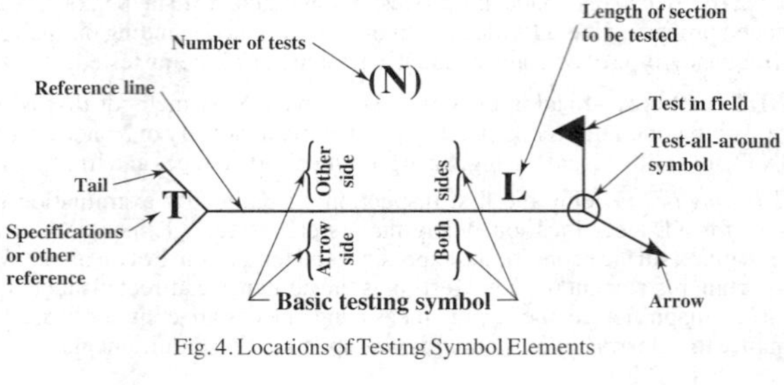

## TABLE OF CONTENTS

## FASTENERS

### METRIC THREADED FASTENERS

*(Continued)*



### HELICAL COIL SCREW THREAD INSERTS



### HELICAL COIL SCREW THREAD INSERTS

*(Continued)*



### BRITISH FASTENERS



### MACHINE SCREWS AND NUTS

$$\text{UNJ}_{\text{torque}} = \left(\frac{d \times n - 0.6495}{d \times n - 0.9743}\right)^2 \times \text{Unified}_{\text{torque}} \tag{10}$$

where $d$ is the basic thread major diameter, and $n$ is the number of threads per inch.

The tensile stress area for metric threads is based on a diameter equivalent to the mean of the pitch diameter and a diameter obtained by subtracting $\frac{1}{6}$ the height of the fundamental thread triangle from the external-thread minor diameter. The Japanese Industrial Standard JIS B 1082 (see also ISO 898/1) defines the stress area of metric screw threads as follows:

$$A_s = \frac{\pi}{4}\left(\frac{d_2 + d_3}{2}\right)^2 \tag{11}$$

In Equation (11), $A_s$ is the stress area of the metric screw thread in mm$^2$; $d_2$ is the pitch diameter of the external thread in mm, given by $d_2 = d - 0.649515 \times P$; and $d_3$ is defined by $d_3 = d_1 - H/6$. Here, $d$ is the nominal bolt diameter; $P$ is the thread pitch; $d_1 = d - 1.082532 \times P$ is the minor diameter of the external thread in mm; and $H = 0.866025 \times P$ is the height of the fundamental thread triangle. Substituting the formulas for $d_2$ and $d_3$ into Equation (11) results in $A_s = 0.7854(d - 0.9382P)^2$.

The stress area, $A_s$, of Unified threads in mm$^2$ is given in JIS B 1082 as:

$$A_s = 0.7854\left(d - \frac{0.9743}{n} \times 25.4\right)^2 \tag{12}$$

**Relation between Torque and Clamping Force.**—The Japanese Industrial Standard JIS B 1803 defines fastener tightening torque $T_f$ as the sum of the bearing surface torque $T_w$ and the shank (threaded) portion torque $T_s$. The relationship between the applied tightening torque and bolt preload $F_{ft}$ is as follows: $T_f = T_s + T_w = K \times F_f \times d$. In the preceding, $d$ is the nominal diameter of the screw thread, and $K$ is the torque coefficient defined as follows:

$$K = \frac{1}{2d}\left(\frac{P}{\pi} + \mu_s d_2 \sec\alpha' + \mu_w D_w\right) \tag{13}$$

where $P$ is the screw thread pitch; $\mu_s$ is the coefficient of friction between threads; $d_2$ is the pitch diameter of the thread; $\mu_w$ is the coefficient of friction between bearing surfaces; $D_w$ is the equivalent diameter of the friction torque bearing surfaces; and $\alpha'$ is the flank angle at the ridge perpendicular section of the thread ridge, defined by $\tan\alpha' = \tan\alpha\cos\beta$, where $\alpha$ is the thread half angle (30°, for example), and $\beta$ is the thread helix, or lead, angle. $\beta$ can be found from $\tan\beta = l \div 2\pi r$, where $l$ is the thread lead, and $r$ is the thread radius (i.e., one-half the nominal diameter $d$). When the bearing surface contact area is circular, $D_w$ can be obtained as follows:

$$D_w = \frac{2}{3} \times \frac{D_o^3 - D_i^3}{D_o^2 - D_i^2} \tag{14}$$

where $D_o$ and $D_i$ are the outside and inside diameters, respectively, of the bearing surface contact area.

The torques attributable to the threaded portion of a fastener, $T_s$, and bearing surfaces of a joint, $T_w$, are as follows:

$$T_s = \frac{F_f}{2}\left(\frac{P}{\pi} + \mu_s d_2 \sec\alpha'\right) \tag{15}$$

$$T_w = \frac{F_f}{2}\mu_w D_w \tag{16}$$

where $F_f, P, \mu, d_2, \alpha', \mu_w$, and $D_w$ are as previously defined.

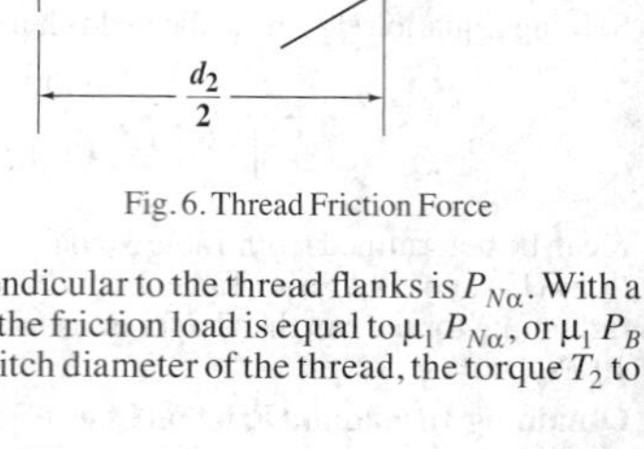

**Table 5. American National Standard Square Lag Screws** *ANSI/ASME B18.2.1-2012 (R2021)*

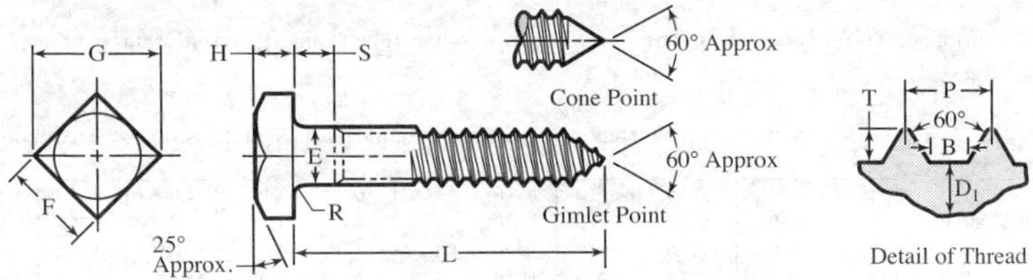


| Nominal Size[a] or Basic Product Dia. | | Body or Shoulder Dia. *E* | | Width Across Flats *F* | | | Width Across Corners *G* | | Height *H* | | | Shoulder Length *S* | Radius of Fillet *R* | Thds. per Inch | Thread Dimensions | | | |
|---|---|---|---|---|---|---|---|---|---|---|---|---|---|---|---|---|---|---|
| | | Max. | Min. | Basic | Max. | Min. | Max. | Min. | Basic | Max. | Min. | Min. | Max. | | Pitch *P* | Flat at Root *B* | Depth of Thd. *T* | Root Dia. $D_1$ |
| No. 10 | 0.1900 | 0.199 | 0.178 | 9⁄32 | 0.281 | 0.271 | 0.398 | 0.372 | 1⁄8 | 0.140 | 0.110 | 0.094 | 0.03 | 11 | 0.091 | 0.039 | 0.035 | 0.120 |
| 1⁄4 | 0.2500 | 0.260 | 0.237 | 3⁄8 | 0.375 | 0.362 | 0.530 | 0.498 | 11⁄64 | 0.188 | 0.156 | 0.094 | 0.03 | 10 | 0.100 | 0.043 | 0.039 | 0.173 |
| 5⁄16 | 0.3125 | 0.324 | 0.298 | 1⁄2 | 0.500 | 0.484 | 0.707 | 0.665 | 13⁄64 | 0.220 | 0.186 | 0.125 | 0.03 | 9 | 0.111 | 0.048 | 0.043 | 0.227 |
| 3⁄8 | 0.3750 | 0.388 | 0.360 | 9⁄16 | 0.562 | 0.544 | 0.795 | 0.747 | 1⁄4 | 0.268 | 0.232 | 0.125 | 0.03 | 7 | 0.143 | 0.062 | 0.055 | 0.265 |
| 7⁄16 | 0.4375 | 0.452 | 0.421 | 5⁄8 | 0.625 | 0.603 | 0.884 | 0.828 | 19⁄64 | 0.316 | 0.278 | 0.156 | 0.03 | 7 | 0.143 | 0.062 | 0.055 | 0.328 |
| 1⁄2 | 0.5000 | 0.515 | 0.482 | 3⁄4 | 0.750 | 0.725 | 1.061 | 0.995 | 21⁄64 | 0.348 | 0.308 | 0.156 | 0.03 | 6 | 0.167 | 0.072 | 0.064 | 0.371 |
| 5⁄8 | 0.6250 | 0.642 | 0.605 | 15⁄16 | 0.938 | 0.906 | 1.326 | 1.244 | 27⁄64 | 0.444 | 0.400 | 0.312 | 0.06 | 5 | 0.200 | 0.086 | 0.077 | 0.471 |
| 3⁄4 | 0.7500 | 0.768 | 0.729 | 1 1⁄8 | 1.125 | 1.088 | 1.591 | 1.494 | 1⁄2 | 0.524 | 0.476 | 0.375 | 0.06 | 4 1⁄2 | 0.222 | 0.096 | 0.085 | 0.579 |
| 7⁄8 | 0.8750 | 0.895 | 0.852 | 1 5⁄16 | 1.312 | 1.269 | 1.856 | 1.742 | 19⁄32 | 0.620 | 0.568 | 0.375 | 0.06 | 4 | 0.250 | 0.108 | 0.096 | 0.683 |
| 1 | 1.0000 | 1.022 | 0.976 | 1 1⁄2 | 1.500 | 1.450 | 2.121 | 1.991 | 21⁄32 | 0.684 | 0.628 | 0.625 | 0.09 | 3 1⁄2 | 0.286 | 0.123 | 0.110 | 0.780 |
| 1 1⁄8 | 1.1250 | 1.149 | 1.098 | 1 11⁄16 | 1.688 | 1.631 | 2.386 | 2.239 | 3⁄4 | 0.780 | 0.720 | 0.625 | 0.09 | 3 1⁄4 | 0.308 | 0.133 | 0.119 | 0.887 |
| 1 1⁄4 | 1.2500 | 1.277 | 1.223 | 1 7⁄8 | 1.875 | 1.812 | 2.652 | 2.489 | 27⁄32 | 0.876 | 0.812 | 0.625 | 0.09 | 3 1⁄4 | 0.308 | 0.133 | 0.119 | 1.012 |

[a] When specifying decimal nominal size, zeros before decimal point and in fourth decimal place are omitted.

All dimensions in inches.

Minimum thread length is $^1/_2$ length of screw plus 0.50 inch, or 6.00 inches, whichever is shorter. Screws too short for the formula thread length shall be threaded as close to the head as practicable.

Thread formulas: Pitch = 1 ÷ threads per inch. Flat at root = 0.4305 × pitch. Depth of single thread = 0.385 × pitch.

## Table 8. American National Standard and Unified Standard Hex Flat Nuts and Flat Jam Nuts and Heavy Hex Flat Nuts and Flat Jam Nuts
*ANSI/ASME B18.2.2-2022*

| Nominal Size or Basic Major Dia. of Thread | | Width Across Flats F | | | Width Across Corners G | | Thickness, Flat Nuts H | | | Thickness, Flat Jam Nuts $H_1$ | | |
|---|---|---|---|---|---|---|---|---|---|---|---|---|
| | | Basic | Max. | Min. | Max. | Min. | Basic | Max. | Min. | Basic | Max. | Min. |
| Hex Flat Nuts and Hex Flat Jam Nuts (Fig. 7) | | | | | | | | | | | | |
| **1 1/8** | **1.1250** | **1 11/16** | **1.688** | **1.631** | **1.949** | **1.859** | **1** | **1.030** | **0.970** | **5/8** | **0.655** | **0.595** |
| **1 1/4** | **1.2500** | **1 7/8** | **1.875** | **1.812** | **2.165** | **2.066** | **1 3/32** | **1.126** | **1.062** | **3/4** | **0.782** | **0.718** |
| **1 3/8** | **1.3750** | **2 1/16** | **2.062** | **1.994** | **2.382** | **2.273** | **1 13/64** | **1.237** | **1.169** | **13/16** | **0.846** | **0.778** |
| **1 1/2** | **1.5000** | **2 1/4** | **2.250** | **2.175** | **2.598** | **2.480** | **1 5/16** | **1.348** | **1.276** | **7/8** | **0.911** | **0.839** |
| Heavy Hex Flat Nuts and Heavy Hex Flat Jam Nuts (Fig. 7) | | | | | | | | | | | | |
| **1 1/8** | **1.1250** | **1 13/16** | **1.812** | **1.756** | **2.093** | **2.002** | **1 1/8** | **1.155** | **1.079** | **5/8** | **0.655** | **0.579** |
| **1 1/4** | **1.2500** | **2** | **2.000** | **1.938** | **2.309** | **2.209** | **1 1/4** | **1.282** | **1.187** | **3/4** | **0.782** | **0.687** |
| **1 3/8** | **1.3750** | **2 3/16** | **2.188** | **2.119** | **2.526** | **2.416** | **1 3/8** | **1.409** | **1.310** | **13/16** | **0.846** | **0.747** |
| **1 1/2** | **1.5000** | **2 3/8** | **2.375** | **2.300** | **2.742** | **2.622** | **1 1/2** | **1.536** | **1.433** | **7/8** | **0.911** | **0.808** |
| **1 3/4** | **1.7500** | **2 3/4** | **2.750** | **2.662** | **3.175** | **3.035** | **1 3/4** | **1.790** | **1.679** | **1** | **1.040** | **0.929** |
| **2** | **2.0000** | **3 1/8** | **3.125** | **3.025** | **3.608** | **3.449** | **2** | **2.044** | **1.925** | **1 1/8** | **1.169** | **1.050** |
| 2 1/4 | 2.2500 | 3 1/2 | 3.500 | 3.388 | 4.041 | 3.862 | 2 1/4 | 2.298 | 2.155 | 1 1/4 | 1.298 | 1.155 |
| 2 1/2 | 2.5000 | 3 7/8 | 3.875 | 3.750 | 4.474 | 4.275 | 2 1/2 | 2.552 | 2.401 | 1 1/2 | 1.552 | 1.401 |
| 2 3/4 | 2.7500 | 4 1/4 | 4.250 | 4.112 | 4.907 | 4.688 | 2 3/4 | 2.806 | 2.647 | 1 5/8 | 1.681 | 1.522 |
| 3 | 3.0000 | 4 5/8 | 4.625 | 4.475 | 5.340 | 5.102 | 3 | 3.060 | 2.893 | 1 3/4 | 1.810 | 1.643 |
| 3 1/4 | 3.2500 | 5 | 5.000 | 4.838 | 5.774 | 5.515 | 3 1/4 | 3.314 | 3.124 | 1 7/8 | 1.939 | 1.748 |
| 3 1/2 | 3.5000 | 5 3/8 | 5.375 | 5.200 | 6.207 | 5.928 | 3 1/2 | 3.568 | 3.370 | 2 | 2.068 | 1.870 |
| 3 3/4 | 3.7500 | 5 3/4 | 5.750 | 5.562 | 6.640 | 6.341 | 3 3/4 | 3.822 | 3.616 | 2 1/8 | 2.197 | 1.990 |
| 4 | 4.0000 | 6 1/8 | 6.125 | 5.925 | 7.073 | 6.755 | 4 | 4.076 | 3.862 | 2 1/4 | 2.326 | 2.112 |

All dimensions are in inches.
**Bold type indicates nuts unified dimensionally with British and Canadian Standards.**
Threads are Unified Coarse-thread series (UNC), Class 2B.

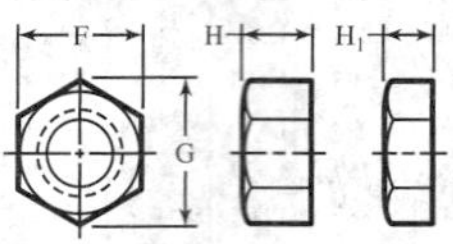


Fig. 7. Hex Flat Nuts, Heavy Hex Flat Nuts, Hex Flat Jam Nuts, and Heavy Hex Flat Jam Nuts (Table 8)

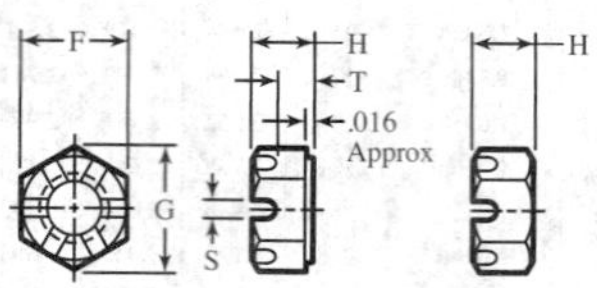


Fig. 8. Hex Slotted Nuts, Heavy Hex Slotted Nuts, and Hex Thick Slotted Nuts (Table 9)

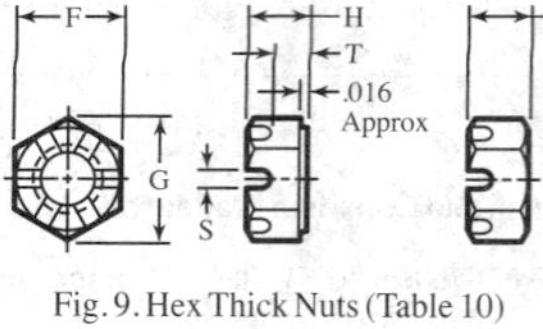


Fig. 9. Hex Thick Nuts (Table 10)

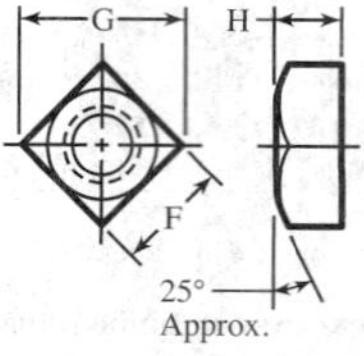


Fig. 10. Square Nuts, Heavy Square Nuts (Table 10)

**Table 1. Wrench Openings for Nuts** *ANSI/ASME B18.2.2-2022, Appendix*

| Max.[a] Width Across Flats of Nut | Wrench Opening[b] Min. | Wrench Opening[b] Max. | Max.[a] Width Across Flats of Nut | Wrench Opening[b] Min. | Wrench Opening[b] Max. | Max.[a] Width Across Flats of Nut | Wrench Opening[b] Min. | Wrench Opening[b] Max. |
|---|---|---|---|---|---|---|---|---|
| 5/32 | 0.158 | 0.163 | 1 1/4 | 1.257 | 1.267 | 2 15/16 | 2.954 | 2.973 |
| 3/16 | 0.190 | 0.195 | 1 5/16 | 1.320 | 1.331 | *3* | *3.016* | *3.035* |
| 7/32 | 0.220 | 0.225 | *1 3/8* | *1.383* | *1.394* | 3 1/8 | 3.142 | 3.162 |
| 1/4 | 0.252 | 0.257 | 1 7/16 | 1.446 | 1.457 | *3 3/8* | *3.393* | *3.414* |
| 9/32 | 0.283 | 0.288 | 1 1/2 | 1.508 | 1.520 | 3 1/2 | 3.518 | 3.540 |
| 5/16 | 0.316 | 0.322 | 1 5/8 | 1.634 | 1.646 | *3 3/4* | *3.770* | *3.793* |
| 11/32 | 0.347 | 0.353 | 1 11/16 | 1.696 | 1.708 | 3 7/8 | 3.895 | 3.918 |
| 3/8 | 0.378 | 0.384 | 1 13/16 | 1.822 | 1.835 | *4 1/8* | *4.147* | *4.172* |
| 7/16 | 0.440 | 0.446 | 1 7/8 | 1.885 | 1.898 | 4 1/4 | 4.272 | 4.297 |
| 1/2 | 0.504 | 0.510 | 2 | 2.011 | 2.025 | 4 1/2 | 4.524 | 4.550 |
| 9/16 | 0.566 | 0.573 | 2 1/16 | 2.074 | 2.088 | 4 5/8 | 4.649 | 4.676 |
| 5/8 | 0.629 | 0.636 | 2 3/16 | 2.200 | 2.215 | *4 7/8* | *4.900* | *4.928* |
| 11/16 | 0.692 | 0.699 | 2 1/4 | 2.262 | 2.277 | 5 | 5.026 | 5.055 |
| 3/4 | 0.755 | 0.763 | 2 3/8 | 2.388 | 2.404 | *5 1/4* | *5.277* | *5.307* |
| 13/16 | 0.818 | 0.826 | *2 7/16* | *2.450* | *2.466* | 5 3/8 | 5.403 | 5.434 |
| 7/8 | 0.880 | 0.888 | 2 9/16 | 2.576 | 2.593 | *5 5/8* | *5.654* | *5.686* |
| 15/16 | 0.944 | 0.953 | *2 5/8* | *2.639* | *2.656* | 5 3/4 | 5.780 | 5.813 |
| 1 | 1.006 | 1.015 | 2 3/4 | 2.766 | 2.783 | *6* | *6.031* | 6.157 |
| 1 1/16 | 1.068 | 1.077 | *2 13/16* | *2.827* | *2.845* | 6 1/8 | 6.065 | 6.192 |
| 1 1/8 | 1.132 | 1.142 | | | | | | |

[a] Wrenches are marked with the "Nominal Size of Wrench," which is equal to the basic or maximum width across flats of the corresponding nut. Minimum wrench opening is (1.005W + 0.001). Tolerance on wrench opening is (0.005W + 0.004) from minimum, where W equals nominal size of wrench.

[b] Openings for 5/32 to 3/8 widths from old ASA B18.2-1960 and italic values are from former ANSI B18.2.2-1972.

All dimensions given in inches.

**Table 2. Clearances for Open End Engineers Wrench (15°)**

| Nominal Wrench Size | | *A* Min. (in.) | *B*[a] Max. (in.) | *C* Min. (in.) | *D* Min. (in.) | *E* Min. (in.) | *F*[b] Max. (in.) | *G* Ref. (in.) | *H*[c] Max. (in.) | *J* Min.[d] in.-lbf |
|---|---|---|---|---|---|---|---|---|---|---|
| 5/32 | 0.156 | 0.220 | 0.250 | 0.390 | 0.160 | 0.250 | 0.200 | 0.030 | 0.094 | 35 |
| 3/16 | 0.188 | 0.250 | 0.280 | 0.430 | 0.190 | 0.270 | 0.230 | 0.030 | 0.172 | 45 |
| 1/4 | 0.250 | 0.280 | 0.340 | 0.530 | 0.270 | 0.310 | 0.310 | 0.030 | 0.172 | 67 |
| 5/16 | 0.313 | 0.380 | 0.470 | 0.660 | 0.280 | 0.390 | 0.390 | 0.050 | 0.203 | 138 |
| 11/32 | 0.344 | 0.420 | 0.500 | 0.750 | 0.340 | 0.450 | 0.450 | 0.050 | 0.203 | 193 |
| 3/8 | 0.375 | 0.420 | 0.500 | 0.780 | 0.360 | 0.450 | 0.520 | 0.050 | 0.219 | 275 |
| 7/16 | 0.438 | 0.470 | 0.590 | 0.890 | 0.420 | 0.520 | 0.640 | 0.050 | 0.250 | 413 |
| 1/2 | 0.500 | 0.520 | 0.640 | 1.000 | 0.470 | 0.580 | 0.660 | 0.050 | 0.266 | 550 |
| 9/16 | 0.563 | 0.590 | 0.770 | 1.130 | 0.520 | 0.660 | 0.700 | 0.050 | 0.297 | 770 |
| 5/8 | 0.625 | 0.640 | 0.830 | 1.230 | 0.550 | 0.700 | 0.700 | 0.050 | 0.344 | 1100 |
| 11/16 | 0.688 | 0.770 | 0.920 | 1.470 | 0.660 | 0.880 | 0.800 | 0.060 | 0.375 | 1375 |
| 3/4 | 0.750 | 0.770 | 0.920 | 1.510 | 0.670 | 0.880 | 0.800 | 0.060 | 0.375 | 1650 |
| 13/16 | 0.813 | 0.910 | 1.120 | 1.660 | 0.720 | 0.970 | 0.860 | 0.060 | 0.406 | 2200 |
| 7/8 | 0.875 | 0.970 | 1.150 | 1.810 | 0.800 | 1.060 | 0.910 | 0.060 | 0.438 | 2475 |
| 15/16 | 0.938 | 0.970 | 1.150 | 1.850 | 0.810 | 1.060 | 0.950 | 0.060 | 0.438 | 3025 |
| 1 | 1.000 | 1.050 | 1.230 | 2.000 | 0.880 | 1.160 | 1.060 | 0.060 | 0.500 | 3575 |
| 1 1/16 | 1.063 | 1.090 | 1.250 | 2.100 | 0.970 | 1.200 | 1.200 | 0.080 | 0.500 | 3850 |
| 1 1/8 | 1.125 | 1.140 | 1.370 | 2.210 | 1.000 | 1.270 | 1.230 | 0.080 | 0.500 | 4400 |
| 1 1/4 | 1.250 | 1.270 | 1.420 | 2.440 | 1.080 | 1.390 | 1.310 | 0.080 | 0.562 | 5775 |
| 1 5/16 | 1.313 | 1.390 | 1.690 | 2.630 | 1.170 | 1.520 | 1.340 | 0.080 | 0.562 | 6600 |
| 1 7/16 | 1.438 | 1.470 | 1.720 | 2.800 | 1.250 | 1.590 | 1.340 | 0.090 | 0.641 | 8250 |
| *1 1/2* | *1.500* | *1.470* | *1.720* | 2.840 | 1.270 | 1.590 | 1.450 | 0.090 | 0.641 | 8500 |
| 1 5/8 | 1.625 | 1.560 | 1.880 | 3.100 | 1.380 | 1.750 | 1.560 | 0.090 | 0.641 | 9000 |

[a] *B* = arc radius created by the swing of the wrench.

[b] *F* = inside arc radius of part.

[c] *H* = thickness of wrench head. (Dimension line not shown.)

[d] *J* = torque that wrench will withstand in inch-pounds. Values updated from ANSI/ASME B107, 100-2020, Flat Wrenches.

**Table 4. Clearances for Box Wrenches — 12-Point Inch and Metric Series**

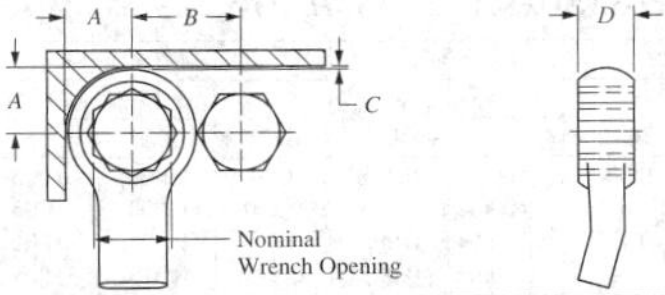


| US Customary (inch) | | | | | | | Metric (mm) | | | | | |
|---|---|---|---|---|---|---|---|---|---|---|---|---|
| Nominal Wrench Opening (in.) | | A Min. (in.) | B Min. (in.) | C Ref.[a] (in.) | Head Thickness D Max., (in.) | Proof Torque (lbf-in.) | Nominal Wrench Opening (mm) | A Min. (mm) | B Min. (mm) | C Ref.[b] (mm) | Head Thickness D Max., (mm) | Proof Torque (N-m) |
| 1⁄8 | (0.125) | 0.179 | 0.219 | 0.030 | 0.172 | 60 | 4 | 4.56 | 6.03 | 0.762 | 4.0 | 12 |
| 5⁄32 | (0.156) | 0.187 | 0.244 | 0.030 | 0.172 | 90 | 5 | 5.26 | 7.29 | 0.762 | 4.6 | 17 |
| 3⁄16 | (0.188) | 0.218 | 0.301 | 0.030 | 0.203 | 150 | 5.5 | 6.66 | 8.97 | 0.762 | 6.0 | 18 |
| 7⁄32 | (0.219) | 0.233 | 0.325 | 0.030 | 0.234 | 165 | 6 | 7.11 | 9.69 | 0.762 | 7.4 | 20 |
| 1⁄4 | (0.250) | 0.269 | 0.378 | 0.030 | 0.295 | 220 | 7 | 7.91 | 11.05 | 0.762 | 7.7 | 27 |
| 9⁄32 | (0.281) | 0.280 | 0.407 | 0.030 | 0.280 | 248 | 8 | 8.26 | 11.98 | 0.762 | 8.2 | 30 |
| 5⁄16 | (0.313) | 0.316 | 0.461 | 0.030 | 0.330 | 275 | 9 | 9.46 | 13.76 | 0.762 | 9.0 | 40 |
| 11⁄32 | (0.344) | 0.336 | 0.499 | 0.030 | 0.335 | 275 | 10 | 10.16 | 15.04 | 0.762 | 9.0 | 71 |
| 3⁄8 | (0.375) | 0.362 | 0.543 | 0.030 | 0.344 | 605 | 11 | 10.71 | 16.15 | 0.762 | 10.0 | 80 |
| 7⁄16 | (0.438) | 0.395 | 0.612 | 0.030 | 0.391 | 715 | 12 | 11.46 | 17.47 | 0.762 | 10.0 | 91 |
| 1⁄2 | (0.500) | 0.442 | 0.694 | 0.030 | 0.394 | 1020 | 13 | 12.31 | 18.89 | 0.762 | 10.5 | 115 |
| 9⁄16 | (0.563) | 0.492 | 0.779 | 0.030 | 0.425 | 1500 | 14 | 12.96 | 20.10 | 0.762 | 11.5 | 158[illegible] |
| 5⁄8 | (0.625) | 0.530 | 0.853 | 0.030 | 0.500 | 2200 | 15 | 13.76 | 21.46 | 0.762 | 11.5 | 200[illegible] |
| 11⁄16 | (0.688) | 0.577 | 0.935 | 0.030 | 0.535 | 2640 | 16 | 14.26 | 22.53 | 0.762 | 12.1 | 24[illegible] |
| 3⁄4 | (0.750) | 0.618 | 1.012 | 0.030 | 0.594 | 2860 | 17 | 15.41 | 24.25 | 0.762 | 12.7 | 26[illegible] |
| 13⁄16 | (0.813) | 0.702 | 1.132 | 0.030 | 0.609 | 3300 | 18 | 15.41 | 24.83 | 0.762 | 12.7 | 30[illegible] |
| 7⁄8 | (0.875) | 0.718 | 1.183 | 0.030 | 0.688 | 3630 | 19 | 16.36 | 26.35 | 0.762 | 14.8 | 32[illegible] |
| 15⁄16 | (0.938) | 0.765 | 1.266 | 0.030 | 0.701 | 4510 | 20 | 17.21 | 27.77 | 0.762 | 14.8 | 34[illegible] |
| 1 | (1.000) | 0.796 | 1.330 | 0.030 | 0.719 | 5390 | 21 | 17.66 | 28.79 | 0.762 | 16.3 | 37[illegible] |
| 1 1⁄16 | (1.063) | 0.874 | 1.445 | 0.030 | 0.790 | 5940 | 22 | 18.56 | 30.27 | 0.762 | 16.3 | 40[illegible] |
| 1 1⁄8 | (1.125) | 0.892 | 1.498 | 0.030 | 0.860 | 6430 | 23 | 19.41 | 31.69 | 0.762 | 16.5 | 45[illegible] |
| 1 3⁄16 | (1.188) | 0.937 | 1.579 | 0.030 | 0.890 | 7200 | 24 | 19.81 | 32.65 | 0.762 | 17.8 | 50[illegible] |
| 1 1⁄4 | (1.250) | 0.983 | 1.661 | 0.030 | 0.940 | 7920 | 25 | 20.86 | 34.24 | 0.762 | 17.9 | 55[illegible] |
| 1 5⁄16 | (1.313) | 1.062 | 1.775 | 0.030 | 0.940 | 8400 | 26 | 12.86 | 26.79 | 0.762 | 18.0 | 60[illegible] |
| 1 3⁄8 | (1.375) | 1.087 | 1.836 | 0.030 | 0.940 | 8970 | 27 | 22.86 | 37.37 | 0.762 | 19.8 | 67[illegible] |
| 1 7⁄16 | (1.438) | 1.144 | 1.929 | 0.030 | 0.953 | 9240 | 28 | 23.41 | 38.49 | 0.762 | 19.8 | 71[illegible] |
| 1 1⁄2 | (1.500) | 1.228 | 2.049 | 0.030 | 1.008 | 10,365 | 29 | 23.41 | 39.06 | 0.762 | 19.8 | 75[illegible] |
| 1 9⁄16 | (1.563) | 1.249 | 2.104 | 0.030 | 1.031 | 11,495 | 30 | 24.51 | 40.73 | 0.762 | 20.0 | 79[illegible] |
| 1 5⁄8 | (1.625) | 1.351 | 2.241 | 0.030 | 1.063 | 12,800 | 31 | 25.06 | 41.85 | 0.762 | 20.5 | 85[illegible] |
| 1 11⁄16 | (1.688) | 1.425 | 2.351 | 0.030 | 1.063 | 13,570 | 32 | 25.66 | 43.03 | 0.762 | 22.0 | 9[illegible]5 |
| 1 3⁄4 | (1.750) | 1.499 | 2.461 | 0.030 | 1.125 | 14,300 | 33 | 25.91 | 43.84 | 0.762 | 22.3 | 9[illegible]0 |
| 1 13⁄16 | (1.813) | 1.499 | 2.496 | 0.030 | 1.125 | 15,100 | 34 | 26.76 | 45.26 | 0.762 | 23.2 | 9[illegible]4 |
| 1 7⁄8 | (1.875) | 1.593 | 2.625 | 0.030 | 1.125 | 15,900 | 36 | 28.81 | 48.47 | 0.762 | 25.1 | 11[illegible]5 |
| 2 | (2.000) | 1.593 | 2.696 | 0.030 | 1.125 | 17,400 | 41 | 32.21 | 54.68 | 0.762 | 25.3 | 15[illegible]9 |
| 2 1⁄16 | (2.063) | 1.687 | 2.825 | 0.030 | 1.234 | 18,200 | 46 | 34.76 | 60.06 | 0.762 | 25.8 | 20[illegible]7 |
| 2 1⁄8 | (2.125) | 1.687 | 2.861 | 0.030 | 1.234 | 19,000 | 50 | 38.76 | 66.33 | 0.762 | 27.6 | 25[illegible]2 |
| 2 3⁄16 | (2.188) | 1.687 | 2.896 | 0.030 | 1.234 | 19,700 | … | … | … | … | … | … |
| 2 1⁄4 | (2.250) | 1.687 | 2.931 | 0.030 | 1.234 | 20,500 | … | … | … | … | … | … |

[a] From SAE Aeronautical Drafting Manual

[b] Converted from SAE Aeronautical Drafting Manual. For details not shown, including material, see ANSI/ASME B107, 100-2020 Flat Wrenches

**Table 1a. American National Standard Type A Plain Washers— Preferred Sizes** *ANSI/ASME B18.21.1-2009 (R2016)*

| Nominal Washer Size[a] | | Series | Inside Diameter | | | Outside Diameter | | | Thickness | | |
|---|---|---|---|---|---|---|---|---|---|---|---|
| | | | | Tolerance | | | Tolerance | | | | |
| | | | Basic | Plus | Minus | Basic | Plus | Minus | Basic | Max. | Min. |
| No. 0 | — | | 0.078 | 0.000 | 0.005 | 0.188 | 0.000 | 0.005 | 0.020 | 0.025 | 0.016 |
| No. 2 | — | | 0.094 | 0.000 | 0.005 | 0.250 | 0.000 | 0.005 | 0.020 | 0.025 | 0.016 |
| No. 4 | — | | 0.125 | 0.008 | 0.005 | 0.312 | 0.008 | 0.005 | 0.032 | 0.040 | 0.025 |
| No. 6 | 0.138 | | 0.156 | 0.008 | 0.005 | 0.375 | 0.015 | 0.005 | 0.049 | 0.065 | 0.036 |
| No. 8 | 0.164 | | 0.188 | 0.008 | 0.005 | 0.438 | 0.015 | 0.005 | 0.049 | 0.065 | 0.036 |
| No. 10 | 0.190 | | 0.219 | 0.008 | 0.005 | 0.500 | 0.015 | 0.005 | 0.049 | 0.065 | 0.036 |
| 3⁄16 | 0.188 | | 0.250 | 0.015 | 0.005 | 0.562 | 0.015 | 0.005 | 0.049 | 0.065 | 0.036 |
| No. 12 | 0.216 | | 0.250 | 0.015 | 0.005 | 0.562 | 0.015 | 0.005 | 0.065 | 0.080 | 0.051 |
| 1⁄4 | 0.250 | N | 0.281 | 0.015 | 0.005 | 0.625 | 0.015 | 0.005 | 0.065 | 0.080 | 0.051 |
| 1⁄4 | 0.250 | W | 0.312 | 0.015 | 0.005 | 0.734[b] | 0.015 | 0.007 | 0.065 | 0.080 | 0.051 |
| 5⁄16 | 0.312 | N | 0.344 | 0.015 | 0.005 | 0.688 | 0.015 | 0.007 | 0.065 | 0.080 | 0.051 |
| 5⁄16 | 0.312 | W | 0.375 | 0.015 | 0.005 | 0.875 | 0.030 | 0.007 | 0.083 | 0.104 | 0.064 |
| 3⁄8 | 0.375 | N | 0.406 | 0.015 | 0.005 | 0.812 | 0.015 | 0.007 | 0.065 | 0.080 | 0.051 |
| 3⁄8 | 0.375 | W | 0.438 | 0.015 | 0.005 | 1.000 | 0.030 | 0.007 | 0.083 | 0.104 | 0.064 |
| 7⁄16 | 0.438 | N | 0.469 | 0.015 | 0.005 | 0.922 | 0.015 | 0.007 | 0.065 | 0.080 | 0.051 |
| 7⁄16 | 0.438 | W | 0.500 | 0.015 | 0.005 | 1.250 | 0.030 | 0.007 | 0.083 | 0.104 | 0.064 |
| 1⁄2 | 0.500 | N | 0.531 | 0.015 | 0.005 | 1.062 | 0.030 | 0.007 | 0.095 | 0.121 | 0.074 |
| 1⁄2 | 0.500 | W | 0.562 | 0.015 | 0.005 | 1.375 | 0.030 | 0.007 | 0.109 | 0.132 | 0.086 |
| 9⁄16 | 0.562 | N | 0.594 | 0.015 | 0.005 | 1.156[b] | 0.030 | 0.007 | 0.095 | 0.121 | 0.074 |
| 9⁄16 | 0.562 | W | 0.625 | 0.015 | 0.005 | 1.469[b] | 0.030 | 0.007 | 0.109 | 0.132 | 0.086 |
| 5⁄8 | 0.625 | N | 0.656 | 0.030 | 0.007 | 1.312 | 0.030 | 0.007 | 0.095 | 0.121 | 0.074 |
| 5⁄8 | 0.625 | W | 0.688 | 0.030 | 0.007 | 1.750 | 0.030 | 0.007 | 0.134 | 0.160 | 0.108 |
| 3⁄4 | 0.750 | N | 0.812 | 0.030 | 0.007 | 1.469 | 0.030 | 0.007 | 0.134 | 0.160 | 0.108 |
| 3⁄4 | 0.750 | W | 0.812 | 0.030 | 0.007 | 2.000 | 0.030 | 0.007 | 0.148 | 0.177 | 0.122 |
| 7⁄8 | 0.875 | N | 0.938 | 0.030 | 0.007 | 1.750 | 0.030 | 0.007 | 0.134 | 0.160 | 0.108 |
| 7⁄8 | 0.875 | W | 0.938 | 0.030 | 0.007 | 2.250 | 0.030 | 0.007 | 0.165 | 0.192 | 0.136 |
| 1 | 1.000 | N | 1.062 | 0.030 | 0.007 | 2.000 | 0.030 | 0.007 | 0.134 | 0.160 | 0.108 |
| 1 | 1.000 | W | 1.062 | 0.030 | 0.007 | 2.500 | 0.030 | 0.007 | 0.165 | 0.192 | 0.136 |
| 1 1⁄8 | 1.125 | N | 1.250 | 0.030 | 0.007 | 2.250 | 0.030 | 0.007 | 0.134 | 0.160 | 0.108 |
| 1 1⁄8 | 1.125 | W | 1.250 | 0.030 | 0.007 | 2.750 | 0.030 | 0.007 | 0.165 | 0.192 | 0.136 |
| 1 1⁄4 | 1.250 | N | 1.375 | 0.030 | 0.007 | 2.500 | 0.030 | 0.007 | 0.165 | 0.192 | 0.136 |
| 1 1⁄4 | 1.250 | W | 1.375 | 0.030 | 0.007 | 3.000 | 0.030 | 0.007 | 0.165 | 0.192 | 0.136 |
| 1 3⁄8 | 1.375 | N | 1.500 | 0.030 | 0.007 | 2.750 | 0.030 | 0.007 | 0.165 | 0.192 | 0.136 |
| 1 3⁄8 | 1.375 | W | 1.500 | 0.045 | 0.010 | 3.250 | 0.045 | 0.010 | 0.180 | 0.213 | 0.153 |
| 1 1⁄2 | 1.500 | N | 1.625 | 0.030 | 0.007 | 3.000 | 0.030 | 0.007 | 0.165 | 0.192 | 0.136 |
| 1 1⁄2 | 1.500 | W | 1.625 | 0.045 | 0.010 | 3.500 | 0.045 | 0.010 | 0.180 | 0.213 | 0.153 |
| 1 5⁄8 | 1.625 | | 1.750 | 0.045 | 0.010 | 3.750 | 0.045 | 0.010 | 0.180 | 0.213 | 0.153 |
| 1 3⁄4 | 1.750 | | 1.875 | 0.045 | 0.010 | 4.000 | 0.045 | 0.010 | 0.180 | 0.213 | 0.153 |
| 1 7⁄8 | 1.875 | | 2.000 | 0.045 | 0.010 | 4.250 | 0.045 | 0.010 | 0.180 | 0.213 | 0.153 |
| 2 | 2.000 | | 2.125 | 0.045 | 0.010 | 4.500 | 0.045 | 0.010 | 0.180 | 0.213 | 0.153 |
| 2 1⁄4 | 2.250 | | 2.375 | 0.045 | 0.010 | 4.750 | 0.045 | 0.010 | 0.220 | 0.248 | 0.193 |
| 2 1⁄2 | 2.500 | | 2.625 | 0.045 | 0.010 | 5.000 | 0.045 | 0.010 | 0.238 | 0.280 | 0.210 |
| 2 3⁄4 | 2.750 | | 2.875 | 0.065 | 0.010 | 5.250 | 0.065 | 0.010 | 0.259 | 0.310 | 0.228 |
| 3 | 3.000 | | 3.125 | 0.065 | 0.010 | 5.500 | 0.065 | 0.010 | 0.284 | 0.327 | 0.249 |

[a] Nominal washer sizes are intended for use with comparable nominal screw or bolt sizes.

[b] The 0.734-inch, 1.156-inch, and 1.469-inch outside diameters avoid washers which could be used in coin operated devices.

All dimensions are in inches.

Preferred sizes are for the most part from series previously designated "Standard Plate" and "SAE." Where common sizes existed in the two series, the SAE size is designated "N" (narrow) and the Standard Plate "W" (wide). These sizes as well as all other sizes of Type A Plain Washers are to be ordered by inside diameter, outside diameter, and thickness dimensions.

Additional selected sizes of Type A Plain Washers are shown in Table 1b.

**Table 2. American National Standard Helical Spring Lock Washers** *ANSI/ASME B18.21.1-2009 (R2016)*

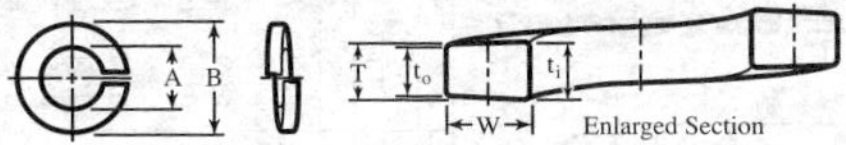


| Nominal Washer Size | | Inside Diameter, A Max. | Inside Diameter, A Min. | Regular O.D., B Max. | Regular Section Width, W | Regular Section Thickness, T[a] | Heavy O.D., B Max. | Heavy Section Width, W | Heavy Section Thickness, T[a] | Extra Duty O.D., B Max. | Extra Duty Section Width, W | Extra Duty Section Thickness, T[a] |
|---|---|---|---|---|---|---|---|---|---|---|---|---|
| No. 2 | 0.086 | 0.094 | 0.088 | 0.172 | 0.035 | 0.020 | 0.182 | 0.040 | 0.025 | 0.208 | 0.053 | 0.027 |
| No. 3 | 0.099 | 0.107 | 0.101 | 0.195 | 0.040 | 0.025 | 0.209 | 0.047 | 0.031 | 0.239 | 0.062 | 0.034 |
| No. 4 | 0.112 | 0.120 | 0.114 | 0.209 | 0.040 | 0.025 | 0.223 | 0.047 | 0.031 | 0.253 | 0.062 | 0.034 |
| No. 5 | 0.125 | 0.133 | 0.127 | 0.236 | 0.047 | 0.031 | 0.252 | 0.055 | 0.040 | 0.300 | 0.079 | 0.045 |
| No. 6 | 0.138 | 0.148 | 0.141 | 0.250 | 0.047 | 0.031 | 0.266 | 0.055 | 0.040 | 0.314 | 0.079 | 0.045 |
| No. 8 | 0.164 | 0.174 | 0.167 | 0.293 | 0.055 | 0.040 | 0.307 | 0.062 | 0.047 | 0.375 | 0.096 | 0.057 |
| No. 10 | 0.190 | 0.200 | 0.193 | 0.334 | 0.062 | 0.047 | 0.350 | 0.070 | 0.056 | 0.434 | 0.112 | 0.068 |
| No. 12 | 0.216 | 0.227 | 0.220 | 0.377 | 0.070 | 0.056 | 0.391 | 0.077 | 0.063 | 0.497 | 0.130 | 0.080 |
| 1⁄4 | 0.250 | 0.260 | 0.252 | 0.487 | 0.109 | 0.062 | 0.489 | 0.110 | 0.077 | 0.533 | 0.132 | 0.084 |
| 5⁄16 | 0.3125 | 0.322 | 0.314 | 0.583 | 0.125 | 0.078 | 0.593 | 0.130 | 0.097 | 0.619 | 0.143 | 0.108 |
| 3⁄8 | 0.375 | 0.385 | 0.377 | 0.680 | 0.141 | 0.094 | 0.688 | 0.145 | 0.115 | 0.738 | 0.170 | 0.123 |
| 7⁄16 | 0.4375 | 0.450 | 0.440 | 0.776 | 0.156 | 0.109 | 0.784 | 0.160 | 0.133 | 0.836 | 0.186 | 0.143 |
| 1⁄2 | 0.500 | 0.512 | 0.502 | 0.869 | 0.171 | 0.125 | 0.879 | 0.176 | 0.151 | 0.935 | 0.204 | 0.162 |
| 9⁄16 | 0.5625 | 0.574 | 0.564 | 0.965 | 0.188 | 0.141 | 0.975 | 0.193 | 0.170 | 1.035 | 0.223 | 0.182 |
| 5⁄8 | 0.625 | 0.641 | 0.628 | 1.073 | 0.203 | 0.156 | 1.087 | 0.210 | 0.189 | 1.151 | 0.242 | 0.202 |
| 11⁄16 | 0.6875 | 0.704 | 0.691 | 1.170 | 0.219 | 0.172 | 1.186 | 0.227 | 0.207 | 1.252 | 0.260 | 0.221 |
| 3⁄4 | 0.750 | 0.766 | 0.753 | 1.265 | 0.234 | 0.188 | 1.285 | 0.244 | 0.226 | 1.355 | 0.279 | 0.241 |
| 13⁄16 | 0.8125 | 0.832 | 0.816 | 1.363 | 0.250 | 0.203 | 1.387 | 0.262 | 0.246 | 1.458 | 0.298 | 0.261 |
| 7⁄8 | 0.875 | 0.894 | 0.878 | 1.459 | 0.266 | 0.219 | 1.489 | 0.281 | 0.266 | 1.571 | 0.322 | 0.285 |
| 15⁄16 | 0.9375 | 0.958 | 0.941 | 1.556 | 0.281 | 0.234 | 1.590 | 0.298 | 0.284 | 1.684 | 0.345 | 0.308 |
| 1 | 1.000 | 1.024 | 1.003 | 1.656 | 0.297 | 0.250 | 1.700 | 0.319 | 0.306 | 1.794 | 0.366 | 0.330 |
| 1 1⁄16 | 1.0625 | 1.087 | 1.066 | 1.751 | 0.312 | 0.266 | 1.803 | 0.338 | 0.326 | 1.905 | 0.389 | 0.352 |
| 1 1⁄8 | 1.125 | 1.153 | 1.129 | 1.847 | 0.328 | 0.281 | 1.903 | 0.356 | 0.345 | 2.013 | 0.411 | 0.375 |
| 1 3⁄16 | 1.1875 | 1.217 | 1.192 | 1.943 | 0.344 | 0.297 | 2.001 | 0.373 | 0.364 | 2.107 | 0.431 | 0.396 |
| 1 1⁄4 | 1.250 | 1.280 | 1.254 | 2.036 | 0.359 | 0.312 | 2.104 | 0.393 | 0.384 | 2.222 | 0.452 | 0.417 |
| 1 5⁄16 | 1.3125 | 1.344 | 1.317 | 2.133 | 0.375 | 0.328 | 2.203 | 0.410 | 0.403 | 2.327 | 0.472 | 0.438 |
| 1 3⁄8 | 1.375 | 1.408 | 1.379 | 2.219 | 0.391 | 0.344 | 2.301 | 0.427 | 0.422 | 2.429 | 0.491 | 0.458 |
| 1 7⁄16 | 1.4375 | 1.472 | 1.442 | 2.324 | 0.406 | 0.359 | 2.396 | 0.442 | 0.440 | 2.530 | 0.509 | 0.478 |
| 1 1⁄2 | 1.500 | 1.534 | 1.504 | 2.419 | 0.422 | 0.375 | 2.491 | 0.458 | 0.458 | 2.627 | 0.526 | 0.496 |

[a] $T$ = mean section thickness = $(t_i + t_o) \div 2$.

## Table 4. American National Standard Internal and External Tooth Lock Washers *ANSI/ASME B18.21.1-2009 (R2016)*

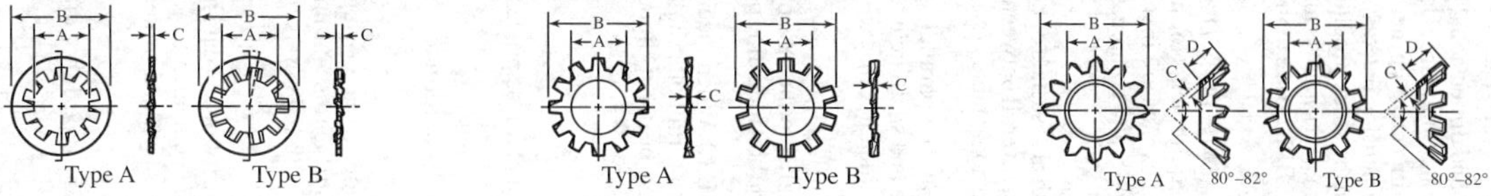


**Internal Tooth Lock Washers**

| | Size | #2 | #3 | #4 | #5 | #6 | #8 | #10 | #12 | 1⁄4 | 5⁄16 | 3⁄8 | 7⁄16 | 1⁄2 | 9⁄16 | 5⁄8 | 11⁄16 | 3⁄4 | 13⁄16 | 7⁄8 | 1 | 1 1⁄8 | 1 1⁄4 |
|---|---|---|---|---|---|---|---|---|---|---|---|---|---|---|---|---|---|---|---|---|---|---|---|
| *A* | Max. | 0.095 | 0.109 | 0.123 | 0.136 | 0.150 | 0.176 | 0.204 | 0.231 | 0.267 | 0.332 | 0.398 | 0.464 | 0.530 | 0.596 | 0.663 | 0.728 | 0.795 | 0.861 | 0.927 | 1.060 | 1.192 | 1.325 |
| | Min. | 0.089 | 0.102 | 0.115 | 0.129 | 0.141 | 0.168 | 0.195 | 0.221 | 0.256 | 0.320 | 0.384 | 0.448 | 0.512 | 0.576 | 0.640 | 0.704 | 0.769 | 0.832 | 0.894 | 1.019 | 1.144 | 1.275 |
| *B* | Max. | 0.200 | 0.232 | 0.270 | 0.280 | 0.295 | 0.340 | 0.381 | 0.410 | 0.478 | 0.610 | 0.692 | 0.789 | 0.900 | 0.985 | 1.071 | 1.166 | 1.245 | 1.315 | 1.410 | 1.637 | 1.830 | 1.975 |
| | Min. | 0.175 | 0.215 | 0.245 | 0.255 | 0.275 | 0.325 | 0.365 | 0.394 | 0.460 | 0.594 | 0.670 | 0.740 | 0.867 | 0.957 | 1.045 | 1.130 | 1.220 | 1.290 | 1.364 | 1.590 | 1.799 | 1.921 |
| *C* | Max. | 0.016 | 0.016 | 0.018 | 0.020 | 0.022 | 0.023 | 0.024 | 0.027 | 0.028 | 0.034 | 0.040 | 0.040 | 0.045 | 0.045 | 0.050 | 0.050 | 0.055 | 0.055 | 0.060 | 0.067 | 0.067 | 0.067 |
| | Min. | 0.010 | 0.010 | 0.012 | 0.014 | 0.016 | 0.018 | 0.018 | 0.020 | 0.023 | 0.028 | 0.032 | 0.032 | 0.037 | 0.037 | 0.042 | 0.042 | 0.047 | 0.047 | 0.052 | 0.059 | 0.059 | 0.059 |

**External Tooth Lock Washers**

| | Size | #2 | #3 | #4 | #5 | #6 | #8 | #10 | #12 | 1⁄4 | 5⁄16 | 3⁄8 | 7⁄16 | 1⁄2 | 9⁄16 | 5⁄8 | 11⁄16 | 3⁄4 | 13⁄16 | 7⁄8 | 1 | 1 1⁄8 | 1 1⁄4 |
|---|---|---|---|---|---|---|---|---|---|---|---|---|---|---|---|---|---|---|---|---|---|---|---|
| *A* | Max. | … | 0.109 | 0.123 | 0.136 | 0.150 | 0.176 | 0.204 | 0.231 | 0.267 | 0.332 | 0.398 | 0.464 | 0.530 | 0.596 | 0.663 | 0.728 | 0.795 | 0.861 | 0.927 | 1.060 | … | … |
| | Min. | … | 0.102 | 0.115 | 0.129 | 0.141 | 0.168 | 0.195 | 0.221 | 0.256 | 0.320 | 0.384 | 0.448 | 0.513 | 0.576 | 0.641 | 0.704 | 0.768 | 0.833 | 0.897 | 1.025 | … | … |
| *B* | Max. | … | 0.235 | 0.260 | 0.285 | 0.320 | 0.381 | 0.410 | 0.475 | 0.510 | 0.610 | 0.694 | 0.760 | 0.900 | 0.985 | 1.070 | 1.155 | 1.260 | 1.315 | 1.410 | 1.620 | … | … |
| | Min. | … | 0.220 | 0.245 | 0.270 | 0.305 | 0.365 | 0.395 | 0.460 | 0.494 | 0.588 | 0.670 | 0.740 | 0.880 | 0.960 | 1.045 | 1.130 | 1.220 | 1.290 | 1.380 | 1.590 | … | … |
| *C* | Max. | … | 0.016 | 0.018 | 0.020 | 0.022 | 0.023 | 0.024 | 0.027 | 0.028 | 0.034 | 0.040 | 0.040 | 0.045 | 0.045 | 0.050 | 0.050 | 0.055 | 0.055 | 0.060 | 0.067 | … | … |
| | Min. | … | 0.010 | 0.012 | 0.014 | 0.016 | 0.018 | 0.018 | 0.020 | 0.023 | 0.028 | 0.032 | 0.032 | 0.037 | 0.037 | 0.042 | 0.042 | 0.047 | 0.047 | 0.052 | 0.059 | … | … |

**Heavy Internal Tooth Lock Washers**

| | Size | 1⁄4 | 5⁄16 | 3⁄8 | 7⁄16 | 1⁄2 | 9⁄16 | 5⁄8 | 3⁄4 | 7⁄8 |
|---|---|---|---|---|---|---|---|---|---|---|
| *A* | Max. | 0.267 | 0.332 | 0.398 | 0.464 | 0.530 | 0.596 | 0.663 | 0.795 | 0.927 |
| | Min. | 0.256 | 0.320 | 0.384 | 0.448 | 0.512 | 0.576 | 0.640 | 0.768 | 0.894 |
| *B* | Max. | 0.536 | 0.607 | 0.748 | 0.858 | 0.924 | 1.034 | 1.135 | 1.265 | 1.447 |
| | Min. | 0.500 | 0.590 | 0.700 | 0.800 | 0.880 | 0.990 | 1.100 | 1.240 | 1.400 |
| *C* | Max. | 0.045 | 0.050 | 0.050 | 0.067 | 0.067 | 0.067 | 0.067 | 0.084 | 0.084 |
| | Min. | 0.035 | 0.040 | 0.042 | 0.050 | 0.055 | 0.055 | 0.059 | 0.070 | 0.075 |

**Countersunk External Tooth Lock Washers**[a]

| | Size | #4 | #6 | #8 | #10 | #12 | 1⁄4 | #16 | 5⁄16 | 3⁄8 | 7⁄16 | 1⁄2 |
|---|---|---|---|---|---|---|---|---|---|---|---|---|
| *A* | Max. | 0.123 | 0.150 | 0.177 | 0.205 | 0.231 | 0.267 | 0.287 | 0.333 | 0.398 | 0.463 | 0.529 |
| | Min. | 0.113 | 0.140 | 0.167 | 0.195 | 0.220 | 0.255 | 0.273 | 0.318 | 0.383 | 0.448 | 0.512 |
| *C* | Max. | 0.019 | 0.021 | 0.021 | 0.025 | 0.025 | 0.025 | 0.028 | 0.028 | 0.034 | 0.045 | 0.045 |
| | Min. | 0.015 | 0.017 | 0.017 | 0.020 | 0.020 | 0.020 | 0.023 | 0.023 | 0.028 | 0.037 | 0.037 |
| *D* | Max. | 0.065 | 0.092 | 0.099 | 0.105 | 0.128 | 0.128 | 0.147 | 0.192 | 0.255 | 0.270 | 0.304 |
| | Min. | 0.050 | 0.082 | 0.083 | 0.088 | 0.118 | 0.113 | 0.137 | 0.165 | 0.242 | 0.260 | 0.294 |

[a] Starting with #4, approx. O.D.'s are: 0.213, 0.289, 0.322, 0.354, 0.421 0.454, 0.505, 0.599, 0.765, 0.867, and 0.976.

All dimensions are given in inches.

verifiable in-process inspection is used in accordance with ANSI/ASME B18.18-2017 (R2021), the final inspection level sample sizes of these respective standards shall apply.

**Compressible Washer-type Direct Tension Indicators.**—*Dimensions:* The dimensions for the two washer types of direct tension indicators, Type A 325 and A 490, shall be in accordance with Table 9.

*Finish:* Unless otherwise specified, direct tension indicators shall be supplied with a plain (as-processed) finish, tinplated, or uncoated. If zinc coatings are required, they shall be in accordance with ASTM F 959.

*Materials and Performance:* Direct tension indicators shall conform to the requirements of ASTM F 959.

*Workmanship:* The workmanship shall be smooth and free of burrs, laps, seams, excess mill scale, and foreign material on bearing surfaces or in protrusion welds, or other defects which would make them unsuitable for intended application.

*Designation:* Compressible washer-type direct tension indicators shall be designated by the following data in the sequence shown: product name, nominal size (fractional or decimal equivalent), type (325 or 490), finish (plain, zinc, or epoxy). See example below:

*Example:* DTI, 1/2 Type 325, Plain Finish

*Identification Symbols:* Grade and source marking and symbols shall conform to the requirements of ASTM F 959.

*Lot Number:* Each direct tension indicator shall be marked with a lot number in accordance with ASTM F 959.

*Quality Assurance:* Unless otherwise specified, products shall be furnished in accordance with ANSI/ASME B18.18 as noted in *Dimensional Characteristics*.

*Dimensional Characteristics:* Direct tension indicators shall conform to the dimensions indicated in Table 9. The designated characteristics defined in Table 9, Note (1) shall be inspected in accordance with ANSI/ASME B18.18. For nondesignated characteristics, the provisions of ANSI/ASME B18.18 shall apply. Should a nondesignated dimension be determined to have a variance, it shall be deemed conforming to this Standard if the user, who is the installer, accepts the variance based on fit, form, and function considerations. Where verifiable in-process inspection is used in accordance with ANSI/ASME B18.18-2017 (R2021), the final inspection level sample sizes of these respective standards shall apply.

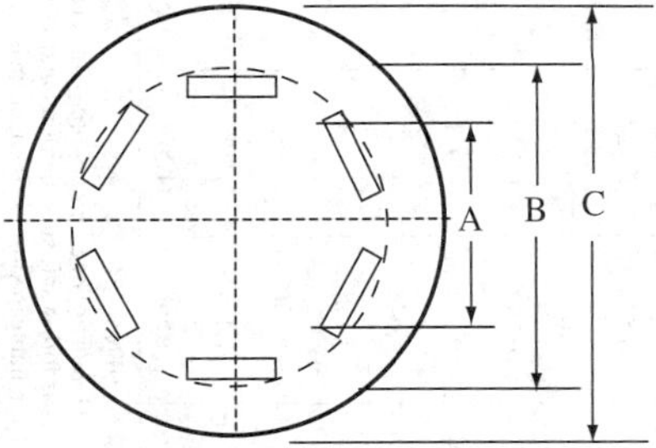


Fig. 3. Compressible Washer Dimensions

**Torx Screws.—**The hexalobular internal driving feature, commonly known as the Torx® screw or sometimes 6-lobe, was developed in the 1960s by Camcar Textron. This screw type has become widely used worldwide in many industries, including vehicles and electronics. The standard internal driving feature has six lobular points of contact, enabling significant turning torque. Many head configurations are available requiring various drivers; refer to type standards such as ISO 10664-2014 and manufacturer documentation.

**Table 24. Hexalobular Internal Driving Feature** *ISO 10664-2014*

Counterbore Depth $C$ = ≤ 0.13 mm (0.005") up to socket number 5. ≤ 0.25 mm (0.010") over socket number 5.

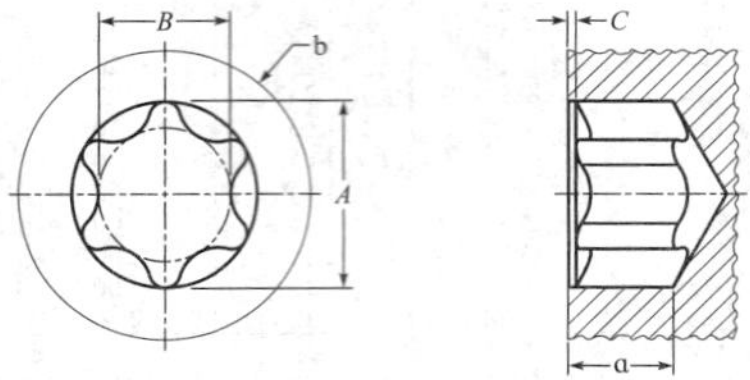


[a] Refer to the relevant product standard for the penetration depth.
[b] Refer to the relevant product standard for the head configuration.

| Socket Number | External Diameter *A* | | Internal Diameter *B* | | Torx Driver Number |
|---|---|---|---|---|---|
| | mm | inch | mm | inch | |
| 1 | 0.9 | 0.035 | 0.6 | 0.024 | T1 |
| 2 | 1.0 | 0.039 | 0.7 | 0.028 | T2 |
| 3 | 1.2 | 0.047 | 0.85 | 0.033 | T3 |
| 4 | 1.35 | 0.053 | 1.0 | 0.039 | T4 |
| 5 | 1.5 | 0.059 | 1.1 | 0.043 | T5 |
| 6 | 1.75 | 0.069 | 1.27 | 0.050 | T6 |
| 7 | 2.1 | 0.083 | 1.5 | 0.059 | T7 |
| 8 | 2.4 | 0.094 | 1.75 | 0.069 | T8 |
| 9 | 2.6 | 0.102 | 1.9 | 0.075 | T9 |
| 10 | 2.8 | 0.110 | 2.05 | 0.081 | T10 |
| 15 | 3.35 | 0.132 | 2.4 | 0.094 | T15 |
| 20 | 3.95 | 0.156 | 2.85 | 0.112 | T20 |
| 25 | 4.5 | 0.177 | 3.25 | 0.128 | T25 |
| 27 | 5.1 | 0.201 | 3.68 | 0.145 | T27 |
| 30 | 5.6 | 0.220 | 4.05 | 0.159 | T30 |
| 40 | 6.75 | 0.266 | 4.85 | 0.191 | T40 |
| 45 | 7.93 | 0.312 | 5.64 | 0.222 | T45 |
| 50 | 8.95 | 0.352 | 6.45 | 0.254 | T50 |
| 55 | 11.35 | 0.447 | 8.05 | 0.317 | T55 |
| 60 | 13.45 | 0.530 | 9.6 | 0.378 | T60 |
| 70 | 15.7 | 0.618 | 11.2 | 0.441 | T70 |
| 80 | 17.75 | 0.699 | 12.8 | 0.504 | T80 |
| 90 | 20.2 | 0.795 | 14.4 | 0.567 | T90 |
| 100 | 22.4 | 0.882 | 16 | 0.630 | T100 |

## Metric Nuts

The American National Standards covering metric were established in cooperation with the Department of Defense so that they could be used by the government for procurement purposes. Extensive information concerning these nuts is given in the following text and tables, but for more complete manufacturing and acceptance specifications, reference should be made to the respective Standards, which may be obtained by non-governmental agencies from the American National Standards Institute, 25 West 43rd Street, 4th Floor, New York, NY 10036, www.ansi.org. Manufacturers should be consulted about items and sizes in stock production.

**Comparison with ISO Standards.—**American National Standards for metric nuts have been coordinated to the extent possible with comparable ISO Standards or proposed Standards, thus: ANSI B18.2.4.1M (Withdrawn) Metric Hex Nuts, Style 1 with ISO 4032; B18.2.4.2M (Withdrawn) Metric Hex Nuts, Style 2 with ISO 4033; B18.2.4.4M

## British Standard General Purpose Studs *BS 2693:Part 1:1956*

| Limits for End Screwed into Component (All threads except BA) | | | | | | | | | | | | | |
|---|---|---|---|---|---|---|---|---|---|---|---|---|---|
| Nom. Dia. *D* | Major Dia. | Thds. per In. | Major Dia. | Effective Diameter | | Minor Diameter | | Thds. per In. | Major Dia. | Effective Diameter | | Minor Dia. | |
| | Max. | | Min. | Max. | Min. | Max. | Min. | | Min. | Max. | Min. | Max. | Min. |
| UN THREADS | | UNF THREADS | | | | | | UNC THREADS | | | | | |
| 1/4 | 0.2500 | 28 | 0.2435 | 0.2294 | 0.2265 | 0.2088 | 0.2037 | 20 | 0.2419 | 0.2201 | 0.2172 | 0.1913 | 0.1849 |
| 5/16 | 0.3125 | 24 | 0.3053 | 0.2883 | 0.2852 | 0.2643 | 0.2586 | 18 | 0.3038 | 0.2793 | 0.2762 | 0.2472 | 0.2402 |
| 3/8 | 0.3750 | 24 | 0.3678 | 0.3510 | 0.3478 | 0.3270 | 0.3211 | 16 | 0.3656 | 0.3375 | 0.3343 | 0.3014 | 0.2936 |
| 7/16 | 0.4375 | 20 | 0.4294 | 0.4084 | 0.4050 | 0.3796 | 0.3729 | 14 | 0.4272 | 0.3945 | 0.3911 | 0.3533 | 0.3447 |
| 1/2 | 0.5000 | 20 | 0.4919 | 0.4712 | 0.4675 | 0.4424 | 0.4356 | 13 | 0.4891 | 0.4537 | 0.4500 | 0.4093 | 0.4000 |
| 9/16 | 0.5625 | 18 | 0.5538 | 0.5302 | 0.5264 | 0.4981 | 0.4907 | 12 | 0.5511 | 0.5122 | 0.5084 | 0.4641 | 0.4542 |
| 5/8 | 0.6250 | 18 | 0.6163 | 0.5929 | 0.5889 | 0.5608 | 0.5533 | 11 | 0.6129 | 0.5700 | 0.5660 | 0.5175 | 0.5069 |
| 3/4 | 0.7500 | 16 | 0.7406 | 0.7137 | 0.7094 | 0.6776 | 0.6693 | 10 | 0.7371 | 0.6893 | 0.6850 | 0.6316 | 0.6200 |
| 7/8 | 0.8750 | 14 | 0.8647 | 0.8332 | 0.8286 | 0.7920 | 0.7828 | 9 | 0.8611 | 0.8074 | 0.8028 | 0.7433 | 0.7306 |
| 1 | 1.0000 | 12 | 0.9886 | 0.9510 | 0.9459 | 0.9029 | 0.8925 | 8 | 0.9850 | 0.9239 | 0.9188 | 0.8517 | 0.8376 |
| 1 1/8 | 1.1250 | 12 | 1.1136 | 1.0762 | 1.0709 | 1.0281 | 1.0176 | 7 | 1.1086 | 1.0375 | 1.0322 | 0.9550 | 0.9393 |
| 1 1/4 | 1.2500 | 12 | 1.2386 | 1.2014 | 1.1959 | 1.1533 | 1.1427 | 7 | 1.2336 | 1.1627 | 1.1572 | 1.0802 | 1.0644 |
| 1 3/8 | 1.3750 | 12 | 1.3636 | 1.3265 | 1.3209 | 1.2784 | 1.2677 | 6 | 1.3568 | 1.2723 | 1.2667 | 1.1761 | 1.1581 |
| 1 1/2 | 1.5000 | 12 | 1.4886 | 1.4517 | 1.4459 | 1.4036 | 1.3928 | 6 | 1.4818 | 1.3975 | 1.3917 | 1.3013 | 1.2832 |
| BS THREADS | | BSF THREADS | | | | | | BSW THREADS | | | | | |
| 1/4 | 0.2500 | 26 | 0.2455 | 0.2280 | 0.2251 | 0.2034 | 0.1984 | 20 | 0.2452 | 0.2206 | 0.2177 | 0.1886 | 0.1831 |
| 5/16 | 0.3125 | 22 | 0.3077 | 0.2863 | 0.2832 | 0.2572 | 0.2517 | 18 | 0.3073 | 0.2798 | 0.2767 | 0.2442 | 0.2383 |
| 3/8 | 0.3750 | 20 | 0.3699 | 0.3461 | 0.3429 | 0.3141 | 0.3083 | 16 | 0.3695 | 0.3381 | 0.3349 | 0.0981 | 0.2919 |
| 7/16 | 0.4375 | 18 | 0.4320 | 0.4053 | 0.4019 | 0.3697 | 0.3635 | 14 | 0.4316 | 0.3952 | 0.3918 | 0.3495 | 0.3428 |
| 1/2 | 0.5000 | 16 | 0.4942 | 0.4637 | 0.4600 | 0.4237 | 0.4172 | 12 | 0.4937 | 0.4503 | 0.4466 | 0.3969 | 0.3897 |
| 9/16 | 0.5625 | 16 | 0.5566 | 0.5263 | 0.5225 | 0.4863 | 0.4797 | 12 | 0.5560 | 0.5129 | 0.5091 | 0.4595 | 0.4521 |
| 5/8 | 0.6250 | 14 | 0.6187 | 0.5833 | 0.5793 | 0.5376 | 0.5305 | 11 | 0.6183 | 0.5708 | 0.5668 | 0.5126 | 0.5050 |
| 3/4 | 0.7500 | 12 | 0.7432 | 0.7009 | 0.6966 | 0.6475 | 0.6398 | 10 | 0.7428 | 0.6903 | 0.6860 | 0.6263 | 0.6182 |
| 7/8 | 0.8750 | 11 | 0.8678 | 0.8214 | 0.8168 | 0.7632 | 0.7551 | 9 | 0.8674 | 0.8085 | 0.8039 | 0.7374 | 0.7288 |
| 1 | 1.0000 | 10 | 0.9924 | 0.9411 | 0.9360 | 0.8771 | 0.8686 | 8 | 0.9920 | 0.9251 | 0.9200 | 0.8451 | 0.8360 |
| 1 1/8 | 1.1250 | 9 | 1.1171 | 1.0592 | 1.0539 | 0.9881 | 0.9792 | 7 | 1.1164 | 1.0388 | 1.0335 | 0.9473 | 0.9376 |
| 1 1/4 | 1.2500 | 9 | 1.2419 | 1.1844 | 1.1789 | 1.1133 | 1.1042 | 7 | 1.2413 | 1.1640 | 1.1585 | 1.0725 | 1.0627 |
| 1 3/8 | 1.3750 | 8 | 1.3665 | 1.3006 | 1.2950 | 1.2206 | 1.2110 | 6 | 1.4906 | 1.3991 | 1.3933 | 1.2924 | 1.2818 |
| 1 1/2 | 1.5000 | 8 | 1.4913 | 1.4258 | 1.4200 | 1.3458 | 1.3360 | … | … | … | … | … | … |

| Limits for End Screwed into Component (BA Threads)[a] | | | | | | | |
|---|---|---|---|---|---|---|---|
| Designation No. | Pitch | Major Diameter | | Effective Diameter | | Minor Diameter | |
| | | Max. | Min. | Max. | Min. | Max. | Min. |
| 2 | 0.8100 mm<br>0.03189 in. | 4.700 mm<br>0.1850 in. | 4.580 mm<br>0.1803 in. | 4.275 mm<br>0.1683 in. | 4.200 mm<br>0.1654 in. | 3.790 mm<br>0.1492 in. | 3.620 mm<br>0.1425 in. |
| 4 | 0.6600 mm<br>0.2598 in. | 3.600 mm<br>0.1417 in. | 3.500 mm<br>0.1378 in. | 3.260 mm<br>0.1283 in. | 3.190 mm<br>0.1256 in. | 2.865 mm<br>0.1128 in. | 2.720 mm<br>0.1071 in. |

[a] Approximate inch equivalents are shown below the dimensions given in mm.

| Minimum Nominal Lengths of Studs[a] | | | | | | | | |
|---|---|---|---|---|---|---|---|---|
| Nom. Stud. Dia. | For Thread Length (Component End) of | | Nom. Stud. Dia. | For Thread Length (Component End) of | | Nom. Stud Dia. | For Thread Length (Component End) of | |
| | 1*D* | 1.5*D* | | 1*D* | 1.5*D* | | 1*D* | 1.5*D* |
| 1/4 | 7/8 | 1 | 9/16 | 2 | 2 3/8 | 1 1/8 | 4 | 4 5/8 |
| 5/16 | 1 1/8 | 1 3/8 | 5/8 | 2 1/4 | 2 5/8 | 1 1/4 | 4 3/4 | 5 1/2 |
| 3/8 | 1 3/8 | 1 5/8 | 3/4 | 2 5/8 | 3 | 1 3/8 | 5 | 5 3/4 |
| 7/16 | 1 5/8 | 1 7/8 | 7/8 | 3 1/8 | 3 5/8 | 1 1/2 | 5 1/4 | 6 |
| 1/2 | 1 3/4 | 2 | 1 | 3 1/2 | 4 | … | … | … |

[a] The Standard also gives preferred and standard lengths of studs: *Preferred* lengths of studs: 7/8, 1, 1 1/8, 1 1/4, 1 3/8, 1 1/2, 1 3/4, 2, 2 1/4, 2 1/2, 2 3/4, 3, 3 1/4, 3 1/2 and for lengths above 3 1/2 the preferred increment is 1/2.

*Standard* lengths of studs: 7/8, 1, 1 1/8, 1 1/4, 1 3/8, 1 1/2, 1 5/8, 1 3/4, 1 7/8, 2, 2 1/8, 2 1/4, 2 3/8, 2 1/2, 2 5/8, 2 3/4, 2 7/8, 3, 3 1/8, 3 1/4, 3 3/8, 3 1/2 and for lengths above 3 1/2 the standard increment is 1/4.

All dimensions are in inches except where otherwise noted.

See page 2109 for interference-fit threads.

## British Standard Black Metal Washers—Metric Series *BS 4320:1968 (Withdrawn)*

| Nom Bolt or Screw Size | Inside Diameter Nom. | Inside Diameter Max. | Inside Diameter Min. | Outside Diameter Nom. | Outside Diameter Max. | Outside Diameter Min. | Thickness Nom. | Thickness Max. | Thickness Min. |
|---|---|---|---|---|---|---|---|---|---|
| NORMAL DIAMETER SIZES (Form E) | | | | | | | | | |
| M 5 | 5.5 | 5.8 | 5.5 | 10.0 | 10.0 | 9.2 | 1.0 | 1.2 | 0.8 |
| M 6 | 6.6 | 7.0 | 6.6 | 12.5 | 12.5 | 11.7 | 1.6 | 1.9 | 1.3 |
| (M 7) | 7.6 | 8.0 | 7.6 | 14.0 | 14.0 | 13.2 | 1.6 | 1.9 | 1.3 |
| M 8 | 9.0 | 9.4 | 9.0 | 17 | 17 | 16.2 | 1.6 | 1.9 | 1.3 |
| M 10 | 11.0 | 11.5 | 11.0 | 21 | 21 | 20.2 | 2.0 | 2.3 | 1.7 |
| M 12 | 14 | 14.5 | 14 | 24 | 24 | 23.2 | 2.5 | 2.8 | 2.2 |
| (M 14) | 16 | 16.5 | 16 | 28 | 28 | 27.2 | 2.5 | 2.8 | 2.2 |
| M 16 | 18 | 18.5 | 18 | 30 | 30 | 29.2 | 3.0 | 3.6 | 2.4 |
| (M 18) | 20 | 20.6 | 20 | 34 | 34 | 32.8 | 3.0 | 3.6 | 2.4 |
| M 20 | 22 | 22.6 | 22 | 37 | 37 | 35.8 | 3.0 | 3.6 | 2.4 |
| (M 22) | 24 | 24.6 | 24 | 39 | 39 | 37.8 | 3.0 | 3.6 | 2.4 |
| M 24 | 26 | 26.6 | 26 | 44 | 44 | 42.8 | 4 | 4.6 | 3.4 |
| (M 27) | 30 | 30.6 | 30 | 50 | 50 | 48.8 | 4 | 4.6 | 3.4 |
| M 30 | 33 | 33.8 | 33 | 56 | 56 | 54.5 | 4 | 4.6 | 3.4 |
| (M 33) | 36 | 36.8 | 36 | 60 | 60 | 58.5 | 5 | 6.0 | 4.0 |
| M 36 | 39 | 39.8 | 39 | 66 | 66 | 64.5 | 5 | 6.0 | 4.0 |
| (M 39) | 42 | 42.8 | 42 | 72 | 72 | 70.5 | 6 | 7.0 | 5.0 |
| M 42 | 45 | 45.8 | 45 | 78 | 78 | 76.5 | 7 | 8.2 | 5.8 |
| (M 45) | 48 | 48.8 | 48 | 85 | 85 | 83 | 7 | 8.2 | 5.8 |
| M 48 | 52 | 53 | 52 | 92 | 92 | 90 | 8 | 9.2 | 6.8 |
| (M 52) | 56 | 57 | 56 | 98 | 98 | 96 | 8 | 9.2 | 6.8 |
| M 56 | 62 | 63 | 62 | 105 | 105 | 103 | 9 | 10.2 | 7.8 |
| (M 60) | 66 | 67 | 66 | 110 | 110 | 108 | 9 | 10.2 | 7.8 |
| M 64 | 70 | 71 | 70 | 115 | 115 | 113 | 9 | 10.2 | 7.8 |
| (M 68) | 74 | 75 | 74 | 120 | 120 | 118 | 10 | 11.2 | 8.8 |
| LARGE DIAMETER SIZES (Form F) | | | | | | | | | |
| M 8 | 9 | 9.4 | 9.0 | 21 | 21 | 20.2 | 1.6 | 1.9 | 1.3 |
| M 10 | 11 | 11.5 | 11 | 24 | 24 | 23.2 | 2 | 2.3 | 1.7 |
| M 12 | 14 | 14.5 | 14 | 28 | 28 | 27.2 | 2.5 | 2.8 | 2.2 |
| (M 14) | 16 | 16.5 | 16 | 30 | 30 | 29.2 | 2.5 | 2.8 | 2.2 |
| M 16 | 18 | 18.5 | 18 | 34 | 34 | 32.8 | 3 | 3.6 | 2.4 |
| (M 18) | 20 | 20.6 | 20 | 37 | 37 | 35.8 | 3 | 3.6 | 2.4 |
| M 20 | 22 | 22.6 | 22 | 39 | 39 | 37.8 | 3 | 3.6 | 2.4 |
| (M 22) | 24 | 24.6 | 24 | 44 | 44 | 42.8 | 3 | 3.6 | 2.4 |
| M 24 | 26 | 26.6 | 26 | 50 | 50 | 48.8 | 4 | 4.6 | 3.4 |
| (M 27) | 30 | 30.6 | 30 | 56 | 56 | 54.5 | 4 | 4.6 | 3.4 |
| M 30 | 33 | 33.8 | 33 | 60 | 60 | 58.5 | 4 | 4.6 | 3.4 |
| (M 33) | 36 | 36.8 | 36 | 66 | 66 | 64.5 | 5 | 6.0 | 4 |
| M 36 | 39 | 39.8 | 39 | 72 | 72 | 70.5 | 5 | 6.0 | 4 |
| (M 39) | 42 | 42.8 | 42 | 77 | 77 | 75.5 | 6 | 7 | 5 |
| EXTRA LARGE DIAMETER SIZES (Form G) | | | | | | | | | |
| M 5 | 5.5 | 5.8 | 5.5 | 15 | 15 | 14.2 | 1.6 | 1.9 | 1.3 |
| M 6 | 6.6 | 7.0 | 6.6 | 18 | 18 | 17.2 | 2 | 2.3 | 1.7 |
| (M 7) | 7.6 | 8.0 | 7.6 | 21 | 21 | 20.2 | 2 | 2.3 | 1.7 |
| M 8 | 9 | 9.4 | 9.0 | 24 | 24 | 23.2 | 2 | 2.3 | 1.7 |
| M 10 | 11 | 11.5 | 11.0 | 30 | 30 | 29.2 | 2.5 | 2.8 | 2.2 |
| M 12 | 14 | 14.5 | 14.0 | 36 | 36 | 34.8 | 3 | 3.6 | 2.4 |
| (M 14) | 16 | 16.5 | 16.0 | 42 | 42 | 40.8 | 3 | 3.6 | 2.4 |
| M 16 | 18 | 18.5 | 18 | 48 | 48 | 46.8 | 4 | 4.6 | 3.4 |
| (M 18) | 20 | 20.6 | 20 | 54 | 54 | 52.5 | 4 | 4.6 | 3.4 |
| M 20 | 22 | 22.6 | 22 | 60 | 60 | 58.5 | 5 | 6.0 | 4 |
| (M 22) | 24 | 24.6 | 24 | 66 | 66 | 64.5 | 5 | 6.0 | 4 |
| M 24 | 26 | 26.6 | 26 | 72 | 72 | 70.5 | 6 | 7 | 5 |
| (M 27) | 30 | 30.6 | 30 | 81 | 81 | 79 | 6 | 7 | 5 |
| M 30 | 33 | 33.8 | 33 | 90 | 90 | 88 | 8 | 9.2 | 6.3 |
| (M 33) | 36 | 36.8 | 36 | 99 | 99 | 97 | 8 | 9.2 | 6.3 |
| M 36 | 39 | 39.8 | 39 | 108 | 108 | 106 | 10 | 11.2 | 8.3 |
| (M39) | 42 | 42.8 | 42 | 117 | 117 | 115 | 10 | 11.2 | 8.3 |

All dimensions are in millimeters.

Nominal bolt or screw sizes shown in parentheses are nonpreferred.

**Diameter of Body.—**The diameter of machine screw bodies is not less than Class 2A thread minimum pitch diameter nor greater than the basic major diameter of the thread. Cross-recessed trim head machine screws not threaded to the head have an 0.062 in. minimum length shoulder under the head with diameter limits as specified in the dimensional tables in the Standard.

**Designation.—**Machine screws are designated by the following data in the sequence shown: Nominal size (number, fraction, or decimal equivalent); threads per inch; nominal length (fraction or decimal equivalent); product name, including head type and driving provision; header point, if desired; material; and protective finish, if required. For example:

$^1/_4 - 20 \times 1^1/_4$ Slotted Pan Head Machine Screw, Steel, Zinc Plated

$6 - 32 \times ^3/_4$ Type IA Cross Recessed Fillister Head Machine Screw, Brass

Machine screw nuts are designated by the following data in the sequence shown: Nominal size (number, fraction, or decimal equivalent); threads per inch; product name; material; and protective finish, if required. For example:

10 – 24 Hexagon Machine Screw Nut, Steel, Zinc Plated

0.138 – 32 Square Machine Screw Nut, Brass

**Table 2. American National Standard Slotted 100-Degree Flat Countersunk Head Machine Screws** *ANSI/ASME B18.6.3-2013*

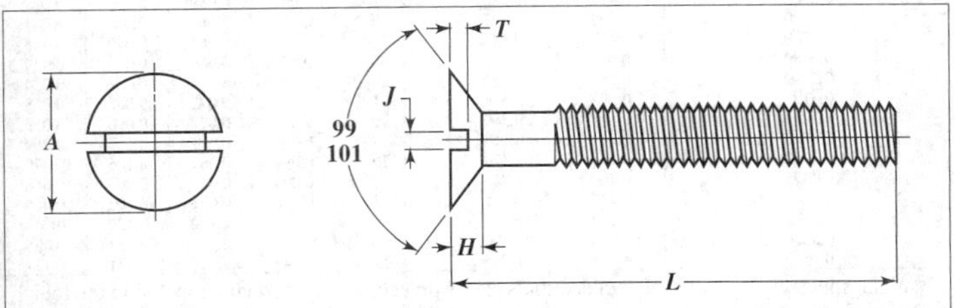


| Nominal Size[a] or Basic Screw Dia. | | Head Dia., *A* | | Head Height, *H* | Slot Width, *J* | | Slot Depth, *T* | |
|---|---|---|---|---|---|---|---|---|
| | | Max. | Min. | Ref. | Max. | Min. | Max. | Min. |
| 0000 | 0.0210 | 0.040 | 0.035 | 0.009 | 0.008 | 0.005 | 0.008 | 0.004 |
| 000 | 0.0340 | 0.060 | 0.055 | 0.014 | 0.012 | 0.008 | 0.011 | 0.007 |
| 00 | 0.0470 | 0.087 | 0.080 | 0.020 | 0.017 | 0.010 | 0.013 | 0.008 |
| 0 | 0.0600 | 0.112 | 0.095 | 0.026 | 0.023 | 0.016 | 0.013 | 0.008 |
| 1 | 0.0730 | 0.137 | 0.118 | 0.031 | 0.026 | 0.019 | 0.016 | 0.010 |
| 2 | 0.0860 | 0.162 | 0.142 | 0.037 | 0.031 | 0.023 | 0.019 | 0.012 |
| 3 | 0.0990 | 0.187 | 0.165 | 0.043 | 0.035 | 0.027 | 0.022 | 0.014 |
| 4 | 0.1120 | 0.212 | 0.188 | 0.049 | 0.039 | 0.031 | 0.024 | 0.017 |
| 6 | 0.1380 | 0.262 | 0.235 | 0.060 | 0.048 | 0.039 | 0.030 | 0.022 |
| 8 | 0.1640 | 0.312 | 0.282 | 0.072 | 0.054 | 0.045 | 0.036 | 0.027 |
| 10 | 0.1900 | 0.362 | 0.329 | 0.083 | 0.060 | 0.050 | 0.042 | 0.031 |
| $^1/_4$ | 0.2500 | 0.477 | 0.437 | 0.110 | 0.075 | 0.064 | 0.055 | 0.042 |
| $^5/_{16}$ | 0.3125 | 0.597 | 0.550 | 0.138 | 0.084 | 0.072 | 0.069 | 0.053 |
| $^3/_8$ | 0.3750 | 0.717 | 0.662 | 0.165 | 0.094 | 0.081 | 0.083 | 0.065 |

[a] When specifying nominal size in decimals, zeros preceding the decimal point and in the fourth decimal place are omitted.

All dimensions are in inches.

## Table 6. American National Standard Slotted Pan Head Machine Screws *ANSI/ASME B18.6.3-2013*

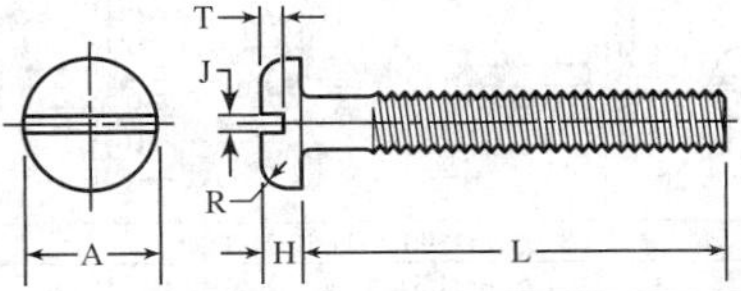


| Nominal Size[a] or Basic Screw Dia. | | Head Dia., *A* Max. | Head Dia., *A* Min. | Head Height, *H* Max. | Head Height, *H* Min. | Head Radius, *R* Max. | Slot Width, *J* Max. | Slot Width, *J* Min. | Slot Depth, *T* Max. | Slot Depth, *T* Min. |
|---|---|---|---|---|---|---|---|---|---|---|
| 0000 | 0.0210 | .042 | .036 | .016 | .010 | .007 | .008 | .004 | .008 | .004 |
| 000 | 0.0340 | .066 | .060 | .023 | .017 | .010 | .012 | .008 | .012 | .008 |
| 00 | 0.0470 | .090 | .082 | .032 | .025 | .015 | .017 | .010 | .016 | .010 |
| 0 | 0.0600 | .116 | .104 | .039 | .031 | .020 | .023 | .016 | .022 | .014 |
| 1 | 0.0730 | .142 | .130 | .046 | .038 | .025 | .026 | .019 | .027 | .018 |
| 2 | 0.0860 | .167 | .155 | .053 | .045 | .035 | .031 | .023 | .031 | .022 |
| 3 | 0.0990 | .193 | .180 | .060 | .051 | .037 | .035 | .027 | .036 | .026 |
| 4 | 0.1120 | .219 | .205 | .068 | .058 | .042 | .039 | .031 | .040 | .030 |
| 5 | 0.1250 | .245 | .231 | .075 | .065 | .044 | .043 | .035 | .045 | .034 |
| 6 | 0.1380 | .270 | .256 | .082 | .072 | .046 | .048 | .039 | .050 | .037 |
| 7 | 0.1510 | .296 | .281 | .089 | .079 | .049 | .048 | .039 | .054 | .041 |
| 8 | 0.1640 | .322 | .306 | .096 | .085 | .052 | .054 | .045 | .058 | .045 |
| 10 | 0.1900 | .373 | .357 | .110 | .099 | .061 | .060 | .050 | .068 | .053 |
| 12 | 0.2160 | .425 | .407 | .125 | .112 | .078 | .067 | .056 | .077 | .061 |
| 14 | 0.2420 | .476 | .457 | .139 | .126 | .087 | .075 | .064 | .085 | .068 |
| 1/4 | 0.2500 | .492 | .473 | .144 | .130 | .087 | .075 | .064 | .087 | .070 |
| 16 | 0.2680 | .528 | .508 | .153 | .139 | .094 | .075 | .064 | .093 | .074 |
| 18 | 0.2940 | .579 | .558 | .168 | .153 | .099 | .084 | .072 | .100 | .080 |
| 5/16 | 0.3125 | .615 | .594 | .178 | .162 | .099 | .084 | .072 | .106 | .085 |
| 20 | 0.3200 | .631 | .608 | .182 | .166 | .121 | .084 | .072 | .108 | .087 |
| 24 | 0.3720 | .734 | .709 | .211 | .193 | .143 | .094 | .081 | .123 | .100 |
| 3/8 | 0.3750 | .740 | .716 | .212 | .195 | .143 | .094 | .081 | .124 | .100 |
| 7/16 | 0.4375 | .863 | .837 | .247 | .228 | .153 | .094 | .081 | .142 | .116 |
| 1/2 | 0.5000 | .987 | .958 | .281 | .260 | .175 | .106 | .091 | .161 | .131 |
| 9/16 | 0.5625 | 1.041 | 1.000 | .315 | .293 | .197 | .118 | .102 | .179 | .146 |
| 5/8 | 0.6250 | 1.172 | 1.125 | .350 | .325 | .219 | .133 | .116 | .197 | .162 |
| 3/4 | 0.7500 | 1.435 | 1.375 | .419 | .390 | .263 | .149 | .131 | .234 | .192 |

[a] Where specifying nominal size in decimals, zeros preceding decimal and in the fourth decimal place are omitted.

All dimensions are in inches.

## Table 7. Nos. 0000, 000 and 00 Threads *ANSI/ASME B18.6.3-2013 Appendix*

| Nominal Size[a] and Threads Per Inch | Series Designat. | External[b] Class | External[b] Major Diameter Max. | External[b] Major Diameter Min. | External[b] Pitch Diameter Max. | External[b] Pitch Diameter Min. | External[b] Pitch Diameter Tol. | External[b] Minor Dia. | Internal[c] Class | Internal[c] Pitch Diameter Min. | Internal[c] Pitch Diameter Max. | Internal[c] Pitch Diameter Tol. | Internal[c] Major Dia. Min. |
|---|---|---|---|---|---|---|---|---|---|---|---|---|---|
| 0000-160 or 0.0210-160 | NS | 2 | .0210 | .0195 | .0169 | .0158 | .0011 | .0128 | 2 | .0169 | .0181 | .0012 | .0210 |
| 000-120 or 0.0340-120 | NS | 2 | .0340 | .0325 | .0286 | 0.272 | .0014 | .0232 | 2 | .0286 | .0300 | .0014 | .034 |
| 00-90 or 0.0470-90 | NS | 2 | .0470 | .0450 | .0398 | .0382 | .0016 | .0326 | 2 | .0398 | .0414 | .0016 | .047 |
| 00-96 or 0.0470-96 | NS | 2 | .0470 | .0450 | .0402 | .0386 | .0016 | .0334 | 2 | .0402 | .0418 | .0016 | .047 |

[a] Where specifying nominal size in decimals, zeros preceding decimal and in the fourth decimal place are omitted.

[b] There is no allowance provided on the external threads.

[c] The minor diameter limits for internal threads are not specified, they being determined by the amount of thread engagement necessary to satisfy the strength requirements and tapping performance in the intended application.

All dimensions are in inches.

## Table 9. American National Standard Slotted Oval Countersunk Head Machine Screws *ANSI/ASME B18.6.3-2013*

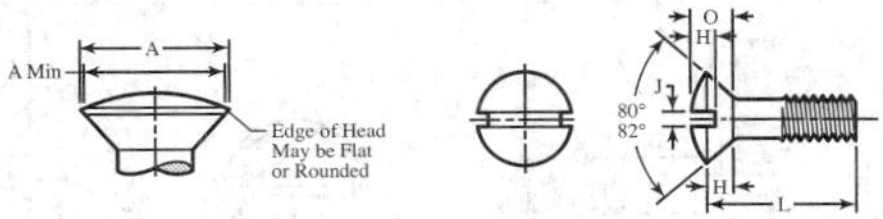


| Nominal Size[a] or Basic Screw Dia. | | Max $L$[b] | Head Dia., $A$ Max., Edge Sharp | Head Dia., $A$ Min., Edge Rounded or Flat | Head Side Height, $H$ Ref. | Total Head Height, $O$ Max. | Total Head Height, $O$ Min. | Slot Width, $J$ Max. | Slot Width, $J$ Min. | Slot Depth, $T$ Max. | Slot Depth, $T$ Min. |
|---|---|---|---|---|---|---|---|---|---|---|---|
| 00 | 0.0470 | … | .087 | .080 | .028 | .042 | .034 | .017 | .010 | .023 | .016 |
| 0 | 0.0600 | 1/8 | .112 | .096 | .035 | .056 | .041 | .023 | .016 | .030 | .025 |
| 1 | 0.0730 | 1/8 | .137 | .120 | .043 | .068 | .052 | .026 | .019 | .038 | .031 |
| 2 | 0.0860 | 1/8 | .162 | .144 | .051 | .080 | .063 | .031 | .023 | .045 | .037 |
| 3 | 0.0990 | 1/8 | .187 | .167 | .059 | .092 | .073 | .035 | .027 | .052 | .043 |
| 4 | 0.1120 | 3/16 | .212 | .191 | .067 | .104 | .084 | .039 | .031 | .059 | .049 |
| 5 | 0.1250 | 3/16 | .237 | .215 | .075 | .116 | .095 | .043 | .035 | .067 | .055 |
| 6 | 0.1380 | 3/16 | .262 | .238 | .083 | .128 | .105 | .048 | .039 | .074 | .060 |
| 7 | 0.1510 | … | .287 | .262 | .091 | .140 | .115 | .048 | .039 | .081 | .066 |
| 8 | 0.1640 | 1/4 | .312 | .285 | .100 | .152 | .126 | .054 | .045 | .088 | .072 |
| 10 | 0.1900 | 5/16 | .362 | .333 | .116 | .176 | .148 | .060 | .050 | .103 | .084 |
| 12 | 0.2160 | 3/8 | .412 | .380 | .132 | .200 | .169 | .067 | .056 | .117 | .096 |
| 14 | 0.2420 | … | .462 | .427 | .148 | .224 | .190 | .075 | .064 | .132 | .108 |
| 1/4 | 0.2500 | 7/16 | .477 | .442 | .153 | .232 | .197 | .075 | .064 | .136 | .112 |
| 16 | 0.2680 | … | .512 | .475 | .164 | .248 | .212 | .075 | .064 | .146 | .120 |
| 18 | 0.2940 | … | .561 | .522 | .180 | .272 | .233 | .084 | .072 | .160 | .132 |
| 5/16 | 0.3125 | 1/2 | .597 | .556 | .191 | .290 | .248 | .084 | .072 | .171 | .141 |
| 20 | 0.3200 | … | .611 | .569 | .196 | .296 | .254 | .094 | .072 | .175 | .144 |
| 24 | 0.3720 | … | .711 | .664 | .228 | .344 | .297 | .094 | .081 | .204 | .168 |
| 3/8 | 0.3750 | 9/16 | .717 | .670 | .230 | .347 | .300 | .094 | .081 | .206 | .170 |
| 7/16 | 0.4375 | 5/8 | .760 | .715 | .223 | .345 | .295 | .094 | .081 | .210 | .174 |
| 1/2 | 0.5000 | 3/4 | .815 | .765 | .223 | .354 | .299 | .106 | .091 | .216 | .176 |
| 9/16 | 0.5625 | … | .932 | .878 | .260 | .410 | .350 | .118 | .102 | .250 | .207 |
| 5/8 | 0.6250 | … | 1.050 | .990 | .298 | .467 | .399 | .133 | .116 | .285 | .235 |
| 3/4 | 0.7500 | … | 1.285 | 1.215 | .372 | .578 | .497 | .149 | .131 | .353 | .293 |

[a] When specifying nominal size in decimals, zeros preceding decimal points and in the fourth decimal place are omitted.

[b] These lengths or shorter are undercut.

All dimensions are in inches.

## Table 10. American National Standard Header Points for Machine Screws before Threading *ANSI/ASME B18.6.3-2013*

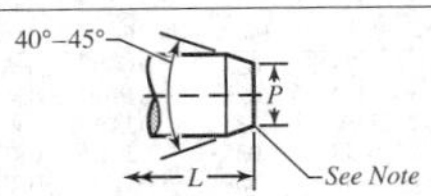


| Nom. Size | Threads per Inch | Max. $P$ | Min. $P$ | Max. $L$ |
|---|---|---|---|---|
| 2 | 56 | 0.057 | 0.050 | 1/2 |
| | 64 | 0.060 | 0.053 | |
| 4 | 40 | 0.074 | 0.065 | 1/2 |
| | 48 | 0.079 | 0.070 | |
| 5 | 40 | 0.086 | 0.076 | 1/2 |
| | 44 | 0.088 | 0.079 | |
| 6 | 32 | 0.090 | 0.080 | 3/4 |
| | 40 | 0.098 | 0.087 | |
| 8 | 32 | 0.114 | 0.102 | 1 |
| | 36 | 0.118 | 0.106 | |
| 10 | 24 | 0.125 | 0.112 | 1 1/4 |
| | 32 | 0.138 | 0.124 | |
| 12 | 24 | 0.149 | 0.134 | 1 3/8 |
| | 28 | 0.156 | 0.141 | |
| 1/4 | 20 | 0.170 | 0.153 | 1 1/2 |
| | 28 | 0.187 | 0.169 | |
| 5/16 | 18 | 0.221 | 0.200 | 1 1/2 |
| | 24 | 0.237 | 0.215 | |
| 3/8 | 16 | 0.270 | 0.244 | 1 1/2 |
| | 24 | 0.295 | 0.267 | |
| 7/16 | 14 | 0.316 | 0.287 | 1 1/2 |
| | 20 | 0.342 | 0.310 | |
| 1/2 | 13 | 0.367 | 0.333 | 1 1/2 |
| | 20 | 0.399 | 0.362 | |

All dimensions in inches. Edges of point may be rounded and end of point need not be flat nor perpendicular to shank. Machine screws normally have plain sheared ends but when specified may have header points, as shown above.

**Table 4. American National Standard Slotted and Cross and Square Recessed Pan Head Metric Machine Screws**
*ANSI/ASME B18.6.7M-1999 (Withdrawn)*

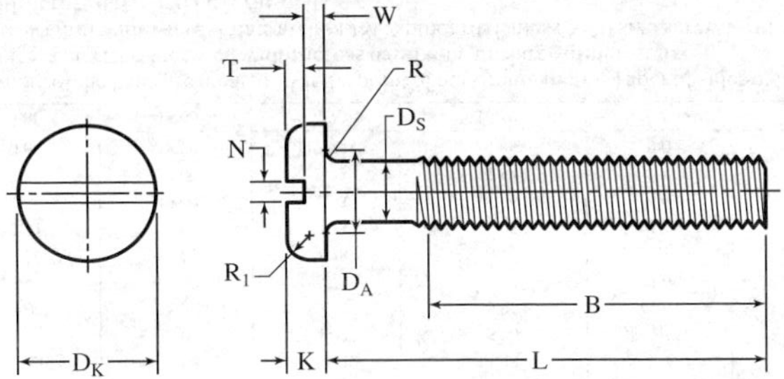


| Nominal Screw Size and Thread Pitch | $D_s$ Body Diameter | | $D_K$ Head Diameter | | Slotted K Head Height | | Slotted $R_1$ Head Radius | Cross and Square Recess K Head Height | | Cross and Square Recess $R_1$ Head Radius | $D_A$ Underhead Fillet Transition Dia. | R Underhead Fillet Radius | N Slot Width | | T Slot Depth | W Unslotted Head Thickness |
|---|---|---|---|---|---|---|---|---|---|---|---|---|---|---|---|---|
| | Max. | Min. | Max. | Min. | Max. | Min. | Max. | Max. | Min. | Ref. | Max. | Min. | Max. | Min. | Min. | Min. |
| M2 × 0.4[a] | 2.00 | 1.65 | 4.0 | 3.7 | 1.3 | 1.1 | 0.8 | 1.6 | 1.4 | 3.2 | 2.6 | 0.1 | 0.7 | 0.5 | 0.5 | 0.4 |
| M2.5 × 0.45 | 2.50 | 2.12 | 5.0 | 4.7 | 1.5 | 1.3 | 1.0 | 2.1 | 1.9 | 4.0 | 3.1 | 0.1 | 0.8 | 0.6 | 0.6 | 0.5 |
| M3 × 0.5 | 3.00 | 2.58 | 5.6 | 5.3 | 1.8 | 1.6 | 1.2 | 2.4 | 2.2 | 5.0 | 3.6 | 0.1 | 1.0 | 0.8 | 0.7 | 0.7 |
| M3.5 × 0.6 | 3.50 | 3.00 | 7.0 | 6.6 | 2.1 | 1.9 | 1.4 | 2.6 | 2.3 | 6.0 | 4.1 | 0.1 | 1.2 | 1.0 | 0.8 | 0.8 |
| M4 × 0.7 | 4.00 | 3.43 | 8.0 | 7.6 | 2.4 | 2.2 | 1.6 | 3.1 | 2.8 | 6.5 | 4.7 | 0.2 | 1.5 | 1.2 | 1.0 | 0.9 |
| M5 × 0.8 | 5.00 | 4.36 | 9.5 | 9.1 | 3.0 | 2.7 | 2.0 | 3.7 | 3.4 | 8.0 | 5.7 | 0.2 | 1.5 | 1.2 | 1.2 | 1.2 |
| M6 × 1 | 6.00 | 5.21 | 12.0 | 11.5 | 3.6 | 3.3 | 2.5 | 4.6 | 4.3 | 10.0 | 6.8 | 0.3 | 1.9 | 1.6 | 1.4 | 1.4 |
| M8 × 1.25 | 8.00 | 7.04 | 16.0 | 15.5 | 4.8 | 4.5 | 3.2 | 6.0 | 5.6 | 13.0 | 9.2 | 0.4 | 2.3 | 2.0 | 1.9 | 1.9 |
| M10 × 1.5 | 10.00 | 8.86 | 20.0 | 19.4 | 6.0 | 5.7 | 4.0 | 7.5 | 7.1 | 16.0 | 11.2 | 0.4 | 2.8 | 2.5 | 2.4 | 2.4 |

[a] This size not specified for Type III square recessed pan heads; Type II cross recess is not specified for any size.

All dimensions in millimeters.

For dimension *B*, see Table 1.

For dimension *L*, see Table 6.

**Table 5. American National Standard Hex and Hex Flange Head Metric Machine Screws** *ANSI/ASME B18.6.7M-1999 (Withdrawn)*

Hex Head

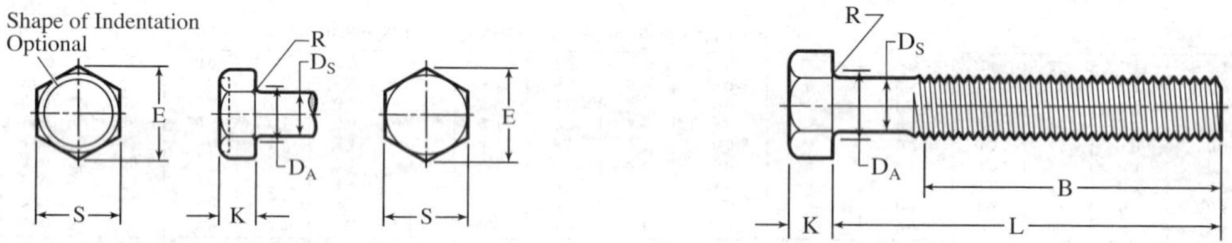


| Nominal Screw Size and Thread Pitch | $D_S$ Body Diameter Max. | $D_S$ Body Diameter Min. | $S$[a] Hex Width Across Flats Max. | $S$[a] Hex Width Across Flats Min. | $E$[a] Hex Width Across Corners Min. | K Head Height Max. | K Head Height Min. | $D_A$ Underhead Fillet Transition Dia. Max. | R Underhead Fillet Radius Min. |
|---|---|---|---|---|---|---|---|---|---|
| M2 × 0.4 | 2.00 | 1.65 | 3.20 | 3.02 | 3.38 | 1.6 | 1.3 | 2.6 | 0.1 |
| M2.5 × 0.45 | 2.50 | 2.12 | 4.00 | 3.82 | 4.28 | 2.1 | 1.8 | 3.1 | 0.1 |
| M3 × 0.5 | 3.00 | 2.58 | 5.00 | 4.82 | 5.40 | 2.3 | 2.0 | 3.6 | 0.1 |
| M3.5 × 0.6 | 3.50 | 3.00 | 5.50 | 5.32 | 5.96 | 2.6 | 2.3 | 4.1 | 0.1 |
| M4 × 0.7 | 4.00 | 3.43 | 7.00 | 6.78 | 7.59 | 3.0 | 2.6 | 4.7 | 0.2 |
| M5 × 0.8 | 5.00 | 4.36 | 8.00 | 7.78 | 8.71 | 3.8 | 3.3 | 5.7 | 0.2 |
| M6 × 1 | 6.00 | 5.21 | 10.00 | 9.78 | 10.95 | 4.7 | 4.1 | 6.8 | 0.3 |
| M8 × 1.25 | 8.00 | 7.04 | 13.00 | 12.73 | 14.26 | 6.0 | 5.2 | 9.2 | 0.4 |
| M10 × 1.5 | 10.00 | 8.86 | 16.00 | 15.73 | 17.62 | 7.5 | 6.5 | 11.2 | 0.4 |
| M12 × 1.75 | 12.00 | 10.68 | 18.00 | 17.73 | 19.86 | 9.0 | 7.8 | 13.2 | 0.4 |
| M10 × 1.5[b] | 10.00 | 8.86 | 15.00 | 14.73 | 16.50 | 7.5 | 6.5 | 11.2 | 0.4 |

[a] Dimensions across flats and across corners of the head are measured at the point of maximum metal. Taper of sides of head (angle between one side and the axis) shall not exceed 2° or 0.10 mm, whichever is greater, the specified width across flats being the large dimension.

[b] The M10 size screws having heads with 15 mm width across flats are not ISO standard. Unless M10 size screws with 15 mm width across flats are specifically ordered, M10 size screws with 16 mm width across flats shall be furnished.

*General Dimensions:* The general dimensions and tolerances for screws and nuts are given in the accompanying tables. Although slotted screw dimensions are given, recessed head screws are also standard and available. Dimensions of recessed head screws are given in BS 4183:1967.

**British Unified Machine Screws and Nuts.**—British Standard BS 1981:1991 covers certain types of machine screws and machine screw nuts for which agreement has been reached with the United States and Canada as to general dimensions for interchangeability. These types are: countersunk, raised-countersunk, pan, and raised-cheese head screws with slotted or recessed heads; small hexagon head screws; and precision and pressed nuts. All have Unified threads. Head shapes are shown on page 1812 and dimensions are given on page 1813.

*Identification:* The above-mentioned screws and nuts that conform to this Standard should have a distinguishing feature applied to identify them as Unified. All recessed head screws are to be identified as Unified by a groove in the form of four arcs of a circle in the upper surface of the head. All hexagon head screws are to be identified as Unified by: 1) a circular recess in the upper surface of the head; 2) a continuous line of circles indented on one or more of the flats of the hexagon and parallel to the screw axis; and 3) at least two contiguous circles indented on the upper surface of the head. All *machine screw* nuts of the pressed type shall be identified as Unified by means of the application of a groove indented in one face of the nut approximately midway between the major diameter of the thread and flats of the square or hexagon. *Slotted head screws* shall be identified as Unified either by a circular recess or by a circular platform or raised portion on the upper surface of the head. *Machine screw nuts* of the *precision type* shall be identified as Unified by either a groove indented on one face of the front approximately midway between the major diameter of the thread and the flats of the hexagon or a continuous line of circles indented on one or more of the flats of the hexagon and parallel to the nut axis.

**Identification Markings for British Standard Unified Machine Screws**

**Recessed and Hexagon Head Screws**

Precision Type        Pressed Type

**Hexagon Machine Screw Nuts**

**Slotted Head Screws**

## British Standard Whitworth (BSW) and Fine (BSF) Machine Screws

*BS 450:1958*

| | Nom. Size of Screw | Basic Dia. *D* | Threads per Inch BSW | Threads per Inch BSF | Dia. of Head *A* Max. | Dia. of Head *A* Min. | Depth of Head *B* Max. | Depth of Head *B* Min. | Width of Slot *H* Max. | Width of Slot *H* Min. | Depth of Slot *J* |
|---|---|---|---|---|---|---|---|---|---|---|---|
| 90° Countersunk Head Screws[a,b] | 1/8 | 0.1250 | 40 | … | 0.219 | 0.201 | 0.056 | … | 0.039 | 0.032 | 0.027 |
| | 3/16 | 0.1875 | 24 | 32[c] | 0.328 | 0.307 | 0.084 | … | 0.050 | 0.042 | 0.041 |
| | 7/32 | 0.2188 | … | 28[c] | 0.383 | 0.360 | 0.098 | … | 0.055 | 0.046 | 0.048 |
| | 1/4 | 0.2500 | 20 | 26 | 0.438 | 0.412 | 0.113 | … | 0.061 | 0.051 | 0.055 |
| | 5/16 | 0.3125 | 18 | 22 | 0.547 | 0.518 | 0.141 | … | 0.071 | 0.061 | 0.069 |
| | 3/8 | 0.3750 | 16 | 20 | 0.656 | 0.624 | 0.169 | … | 0.082 | 0.072 | 0.083 |
| | 7/16 | 0.4375 | 14 | 18 | 0.766 | 0.729 | 0.197 | … | 0.093 | 0.082 | 0.097 |
| | 1/2 | 0.5000 | 12 | 16 | 0.875 | 0.835 | 0.225 | … | 0.104 | 0.092 | 0.111 |
| | 9/16 | 0.5625 | 12[c] | 16[c] | 0.984 | 0.941 | 0.253 | … | 0.115 | 0.103 | 0.125 |
| | 5/8 | 0.6250 | 11 | 14 | 1.094 | 1.046 | 0.281 | … | 0.126 | 0.113 | 0.138 |
| | 3/4 | 0.7500 | 10 | 12 | 1.312 | 1.257 | 0.338 | … | 0.148 | 0.134 | 0.166 |
| Round Head Screws[b] | 1/8 | 0.1250 | 40 | … | 0.219 | 0.206 | 0.087 | 0.082 | 0.039 | 0.032 | 0.048 |
| | 3/16 | 0.1875 | 24 | 32[c] | 0.328 | 0.312[d] | 0.131 | 0.124 | 0.050 | 0.042 | 0.072 |
| | 7/32 | 0.2188 | … | 28[c] | 0.383 | 0.365 | 0.153 | 0.145 | 0.055 | 0.046 | 0.084 |
| | 1/4 | 0.2500 | 20 | 26 | 0.438 | 0.417 | 0.175 | 0.165 | 0.061 | 0.051 | 0.096 |
| | 5/16 | 0.3125 | 18 | 22 | 0.547 | 0.524 | 0.219 | 0.207 | 0.071 | 0.061 | 0.120 |
| | 3/8 | 0.3750 | 16 | 20 | 0.656 | 0.629 | 0.262 | 0.249 | 0.082 | 0.072 | 0.144 |
| | 7/16 | 0.4375 | 14 | 18 | 0.766 | 0.735 | 0.306 | 0.291 | 0.093 | 0.082 | 0.168 |
| | 1/2 | 0.5000 | 12 | 16 | 0.875 | 0.840 | 0.350 | 0.333 | 0.104 | 0.092 | 0.192 |
| | 9/16 | 0.5625 | 12[c] | 16[c] | 0.984 | 0.946 | 0.394 | 0.375 | 0.115 | 0.103 | 0.217 |
| | 5/8 | 0.6250 | 11 | 14 | 1.094 | 1.051 | 0.437 | 0.417 | 0.126 | 0.113 | 0.240 |
| | 3/4 | 0.7500 | 10 | 12 | 1.312 | 1.262 | 0.525 | 0.500 | 0.148 | 0.134 | 0.288 |
| Pan Head Screws[b] | 1/8 | 0.1250 | 40 | … | 0.245 | 0.231 | 0.075 | 0.065 | 0.039 | 0.032 | 0.040 |
| | 3/16 | 0.1875 | 24 | 32[c] | 0.373 | 0.375 | 0.110 | 0.099 | 0.050 | 0.042 | 0.061 |
| | 7/32 | 0.2188 | … | 28[c] | 0.425 | 0.407 | 0.125 | 0.112 | 0.055 | 0.046 | 0.069 |
| | 1/4 | 0.2500 | 20 | 26 | 0.492 | 0.473[e] | 0.144 | 0.130 | 0.061 | 0.051 | 0.078 |
| | 5/16 | 0.3125 | 18 | 22 | 0.615 | 0.594 | 0.178 | 0.162 | 0.071 | 0.061 | 0.095 |
| | 3/8 | 0.3750 | 16 | 20 | 0.740 | 0.716 | 0.212 | 0.195 | 0.082 | 0.072 | 0.112 |
| | 7/16 | 0.4375 | 14 | 18 | 0.863 | 0.838 | 0.247 | 0.227 | 0.093 | 0.082 | 0.129 |
| | 1/2 | 0.5000 | 12 | 16 | 0.987 | 0.958 | 0.281 | 0.260 | 0.104 | 0.092 | 0.145 |
| | 9/16 | 0.5625 | 12[c] | 16[c] | 1.031 | 0.999 | 0.315 | 0.293 | 0.115 | 0.103 | 0.162 |
| | 5/8 | 0.6250 | 11 | 14 | 1.125 | 1.090 | 0.350 | 0.325 | 0.126 | 0.113 | 0.179 |
| | 3/4 | 0.7500 | 10 | 12 | 1.250 | 1.209 | 0.419 | 0.390 | 0.148 | 0.134 | 0.213 |
| Cheese Head Screws[b] | 1/8 | 0.1250 | 40 | … | 0.188 | 0.180 | 0.087 | 0.082 | 0.039 | 0.032 | 0.039 |
| | 3/16 | 0.1875 | 24 | 32[c] | 0.281 | 0.270 | 0.131 | 0.124 | 0.050 | 0.042 | 0.059 |
| | 7/32 | 0.2188 | … | 28[c] | 0.328 | 0.315 | 0.153 | 0.145 | 0.055 | 0.046 | 0.069 |
| | 1/4 | 0.2500 | 20 | 26 | 0.375 | 0.360 | 0.175 | 0.165 | 0.061 | 0.051 | 0.079 |
| | 5/16 | 0.3125 | 18 | 22 | 0.469 | 0.450 | 0.219 | 0.207 | 0.071 | 0.061 | 0.098 |
| | 3/8 | 0.3750 | 16 | 20 | 0.562 | 0.540 | 0.262 | 0.249 | 0.082 | 0.072 | 0.118 |
| | 7/16 | 0.4375 | 14 | 18 | 0.656 | 0.630 | 0.306 | 0.291 | 0.093 | 0.082 | 0.138 |
| | 1/2 | 0.5000 | 12 | 16 | 0.750 | 0.720 | 0.350 | 0.333 | 0.104 | 0.092 | 0.157 |
| | 9/16 | 0.5625 | 12[c] | 16[c] | 0.844 | 0.810 | 0.394 | 0.375 | 0.115 | 0.103 | 0.177 |
| | 5/8 | 0.6250 | 11 | 14 | 0.938 | 0.900 | 0.437 | 0.417 | 0.126 | 0.113 | 0.197 |
| | 3/4 | 0.7500 | 10 | 12 | 1.125 | 1.080 | 0.525 | 0.500 | 0.148 | 0.134 | 0.236 |
| Mushroom Head Screws[b] | 1/8 | 0.1250 | 40 | … | 0.289 | 0.272 | 0.078 | 0.066 | 0.043 | 0.035 | 0.040 |
| | 3/16 | 0.1875 | 24 | 32[c] | 0.448 | 0.425 | 0.118 | 0.103 | 0.060 | 0.050 | 0.061 |
| | 1/4 | 0.2500 | 20 | 26 | 0.573 | 0.546 | 0.150 | 0.133 | 0.075 | 0.064 | 0.079 |
| | 5/16 | 0.3125 | 18 | 22 | 0.698 | 0.666 | 0.183 | 0.162 | 0.084 | 0.072 | 0.096 |
| | 3/8 | 0.3750 | 16 | 20 | 0.823 | 0.787 | 0.215 | 0.191 | 0.094 | 0.081 | 0.112 |

[a] All dimensions, except *J*, given for the 1/8- through 3/8-inch sizes also apply to all the 90° Raised Countersunk Head Screw dimensions given in the Standard.

[b] These screws are also available with recessed heads; dimensions of recess are not given here but may be found in the Standard.

[c] Non-preferred size; avoid use whenever possible.

[d] By arrangement may also be 0.309.

[e] By arrangement may also be 0.468.

All dimensions in inches.

See diagram on page 1812 for a pictorial representation of screws and letter dimensions.

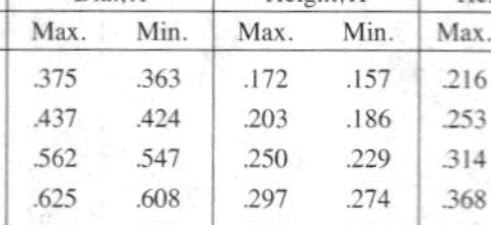

## Table 4. American National Standard Hexagon and Spline Socket Head Cap Screws *ANSI/ASME B18.3-2012*

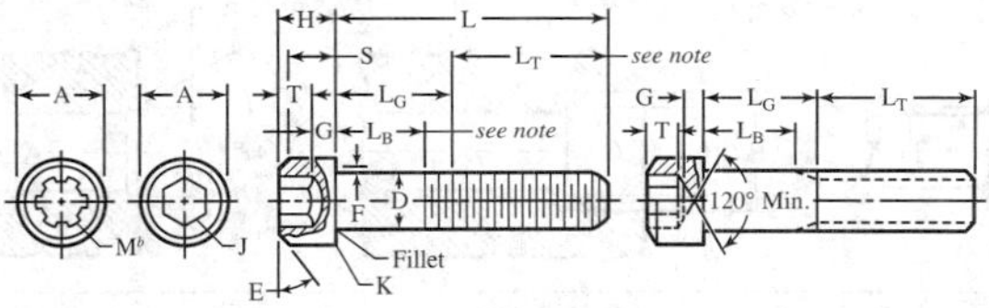


| Nominal Size | Body Diameter, D | | Head Diameter, A | | Head Height, H | | Spline Socket[a] Size, M[b] | Nom. Hex. Socket Size, J | | Fillet Ext., F | Key Engagement,[a] T |
|---|---|---|---|---|---|---|---|---|---|---|---|
| | Max. | Min. | Max. | Min. | Max. | Min. | | | | Max. | |
| 0 | 0.0600 | 0.0568 | 0.096 | 0.091 | 0.060 | 0.057 | 0.060 | | 0.050 | 0.007 | 0.025 |
| 1 | 0.0730 | 0.0695 | 0.118 | 0.112 | 0.073 | 0.070 | 0.072 | 1⁄16 | 0.062 | 0.007 | 0.031 |
| 2 | 0.0860 | 0.0822 | 0.140 | 0.134 | 0.086 | 0.083 | 0.096 | 5⁄64 | 0.078 | 0.008 | 0.038 |
| 3 | 0.0990 | 0.0949 | 0.161 | 0.154 | 0.099 | 0.095 | 0.096 | 5⁄64 | 0.078 | 0.008 | 0.044 |
| 4 | 0.1120 | 0.1075 | 0.183 | 0.176 | 0.112 | 0.108 | 0.111 | 3⁄32 | 0.094 | 0.009 | 0.051 |
| 5 | 0.1250 | 0.1202 | 0.205 | 0.198 | 0.125 | 0.121 | 0.111 | 3⁄32 | 0.094 | 0.010 | 0.057 |
| 6 | 0.1380 | 0.1329 | 0.226 | 0.218 | 0.138 | 0.134 | 0.133 | 7⁄64 | 0.109 | 0.010 | 0.064 |
| 8 | 0.1640 | 0.1585 | 0.270 | 0.262 | 0.164 | 0.159 | 0.168 | 9⁄64 | 0.141 | 0.012 | 0.077 |
| 10 | 0.1900 | 0.1840 | 0.312 | 0.303 | 0.190 | 0.185 | 0.183 | 5⁄32 | 0.156 | 0.014 | 0.090 |
| 1⁄4 | 0.2500 | 0.2435 | 0.375 | 0.365 | 0.250 | 0.244 | 0.216 | 3⁄16 | 0.188 | 0.014 | 0.120 |
| 5⁄16 | 0.3125 | 0.3053 | 0.469 | 0.457 | 0.312 | 0.306 | 0.291 | 1⁄4 | 0.250 | 0.017 | 0.151 |
| 3⁄8 | 0.3750 | 0.3678 | 0.562 | 0.550 | 0.375 | 0.368 | 0.372 | 5⁄16 | 0.312 | 0.020 | 0.182 |
| 7⁄16 | 0.4375 | 0.4294 | 0.656 | 0.642 | 0.438 | 0.430 | 0.454 | 3⁄8 | 0.375 | 0.023 | 0.213 |
| 1⁄2 | 0.5000 | 0.4919 | 0.750 | 0.735 | 0.500 | 0.492 | 0.454 | 3⁄8 | 0.375 | 0.026 | 0.245 |
| 5⁄8 | 0.6250 | 0.6163 | 0.938 | 0.921 | 0.625 | 0.616 | 0.595 | 1⁄2 | 0.500 | 0.032 | 0.307 |
| 3⁄4 | 0.7500 | 0.7406 | 1.125 | 1.107 | 0.750 | 0.740 | 0.620 | 5⁄8 | 0.625 | 0.039 | 0.370 |
| 7⁄8 | 0.8750 | 0.8647 | 1.312 | 1.293 | 0.875 | 0.864 | 0.698 | 3⁄4 | 0.750 | 0.044 | 0.432 |
| 1 | 1.0000 | 0.9886 | 1.500 | 1.479 | 1.000 | 0.988 | 0.790 | 3⁄4 | 0.750 | 0.050 | 0.495 |
| 1 1⁄8 | 1.1250 | 1.1086 | 1.688 | 1.665 | 1.125 | 1.111 | … | 7⁄8 | 0.875 | 0.055 | 0.557 |
| 1 1⁄4 | 1.2500 | 1.2336 | 1.875 | 1.852 | 1.250 | 1.236 | … | 7⁄8 | 0.875 | 0.060 | 0.620 |
| 1 3⁄8 | 1.3750 | 1.3568 | 2.062 | 2.038 | 1.375 | 1.360 | … | 1 | 1.000 | 0.065 | 0.682 |
| 1 1⁄2 | 1.5000 | 1.4818 | 2.250 | 2.224 | 1.500 | 1.485 | … | 1 | 1.000 | 0.070 | 0.745 |
| 1 3⁄4 | 1.7500 | 1.7295 | 2.625 | 2.597 | 1.750 | 1.734 | … | 1 1⁄4 | 1.250 | 0.080 | 0.870 |
| 2 | 2.0000 | 1.9780 | 3.000 | 2.970 | 2.000 | 1.983 | … | 1 1⁄2 | 1.500 | 0.090 | 0.995 |
| 2 1⁄4 | 2.2500 | 2.2280 | 3.375 | 3.344 | 2.250 | 2.232 | … | 1 3⁄4 | 1.750 | 0.100 | 1.120 |
| 2 1⁄2 | 2.5000 | 2.4762 | 3.750 | 3.717 | 2.500 | 2.481 | … | 1 3⁄4 | 1.750 | 0.110 | 1.245 |
| 2 3⁄4 | 2.7500 | 2.7262 | 4.125 | 4.090 | 2.750 | 2.730 | … | 2 | 2.000 | 0.120 | 1.370 |
| 3 | 3.0000 | 2.9762 | 4.500 | 4.464 | 3.000 | 2.979 | … | 2 1⁄4 | 2.250 | 0.130 | 1.495 |
| 3 1⁄4 | 3.2500 | 3.2262 | 4.875 | 4.837 | 3.250 | 3.228 | … | 2 1⁄4 | 2.250 | 0.140 | 1.620 |
| 3 1⁄2 | 3.5000 | 3.4762 | 5.250 | 5.211 | 3.500 | 3.478 | … | 2 3⁄4 | 2.750 | 0.150 | 1.745 |
| 3 3⁄4 | 3.7500 | 3.7262 | 5.625 | 5.584 | 3.750 | 3.727 | … | 2 3⁄4 | 2.750 | 0.160 | 1.870 |
| 4 | 4.0000 | 3.9762 | 6.000 | 5.958 | 4.000 | 3.976 | … | 3 | 3.000 | 0.170 | 1.995 |

[a] Key engagement depths are minimum. Spline socket sizes are nominal.

[b] Spline driven fasteners are removed from the latest revision. For additional information on these fasteners, see ANSI/ASME B18.3-1998.

All dimensions in inches. The body length $L_B$ of the screw is the length of the unthreaded cylindrical portion of the shank. The length of thread, $L_T$, is the distance from the extreme point to the last complete (full form) thread. Standard length increments for screw diameters up to 1 inch are 1⁄16 inch for lengths 1⁄8 through 1⁄4 inch, 1⁄8 inch for lengths 1⁄4 through 1 inch, 1⁄4 inch for lengths 1 through 3 1⁄2 inches, 1⁄2 inch for lengths 3 1⁄2 through 7 inches, 1 inch for lengths 7 through 10 inches and for diameters over 1 inch are 1⁄2 inch for lengths 1 through 7 inches, 1 inch for lengths 7 through 10 inches, and 2 inches for lengths over 10 inches.

Heads may be plain or knurled, and chamfered to an angle *E* of 30 to 45 degrees with the surface of the flat. The thread conforms to the Unified Standard with radius root, Class 3A UNRC and UNRF for screw sizes No. 0 through 1 inch inclusive, Class 2A UNRC and UNRF for over 1 inch through 1 1⁄2 inches inclusive, and Class 2A UNRC for larger sizes. Socket dimensions are given in Table 11. For details not shown, including materials, see ANSI/ASME B18.3-2012 and ANSI/ASME B18.3-1998 (for Spline driven fasteners).

**Table 6. American National Standard Hexagon and Spline Socket Flat Countersunk Head Cap Screws** *ANSI/ASME B18.3-2012*

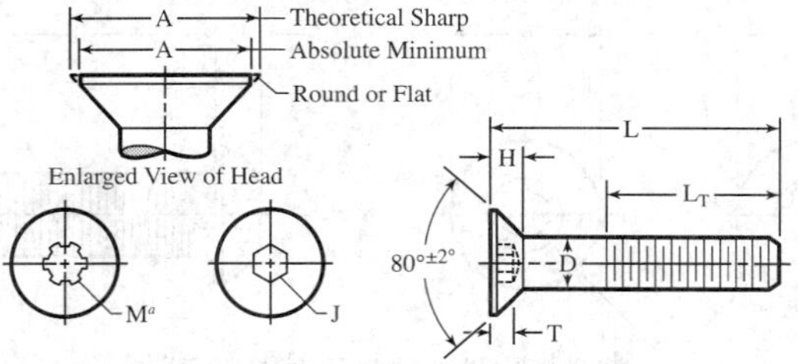


| Nominal Size | Body Diameter | | Head Diameter | | Head Height | Spline Socket Size | Hexagon Socket Size | Key Engagement |
|---|---|---|---|---|---|---|---|---|
| | Max. | Min. | Theoretical Sharp Max. | Abs. Min. | Reference | | Nom. | Min. |
| | *D* | | *A* | | *H* | *M*[a] | *J* | *T* |
| 0 | 0.0600 | 0.0568 | 0.138 | 0.117 | 0.044 | 0.048 | 0.035 | 0.025 |
| 1 | 0.0730 | 0.0695 | 0.168 | 0.143 | 0.054 | 0.060 | 0.050 | 0.031 |
| 2 | 0.0860 | 0.0822 | 0.197 | 0.168 | 0.064 | 0.060 | 0.050 | 0.038 |
| 3 | 0.0990 | 0.0949 | 0.226 | 0.193 | 0.073 | 0.072 | 1/16 | 0.044 |
| 4 | 0.1120 | 0.1075 | 0.255 | 0.218 | 0.083 | 0.072 | 1/16 | 0.055 |
| 5 | 0.1250 | 0.1202 | 0.281 | 0.240 | 0.090 | 0.096 | 5/64 | 0.061 |
| 6 | 0.1380 | 0.1329 | 0.307 | 0.263 | 0.097 | 0.096 | 5/64 | 0.066 |
| 8 | 0.1640 | 0.1585 | 0.359 | 0.311 | 0.112 | 0.111 | 3/32 | 0.076 |
| 10 | 0.1900 | 0.1840 | 0.411 | 0.359 | 0.127 | 0.145 | 1/8 | 0.087 |
| 1/4 | 0.2500 | 0.2435 | 0.531 | 0.480 | 0.161 | 0.183 | 5/32 | 0.111 |
| 5/16 | 0.3125 | 0.3053 | 0.656 | 0.600 | 0.198 | 0.216 | 3/16 | 0.135 |
| 3/8 | 0.3750 | 0.3678 | 0.781 | 0.720 | 0.234 | 0.251 | 7/32 | 0.159 |
| 7/16 | 0.4375 | 0.4294 | 0.844 | 0.781 | 0.234 | 0.291 | 1/4 | 0.159 |
| 1/2 | 0.5000 | 0.4919 | 0.938 | 0.872 | 0.251 | 0.372 | 5/16 | 0.172 |
| 5/8 | 0.6250 | 0.6163 | 1.188 | 1.112 | 0.324 | 0.454 | 3/8 | 0.220 |
| 3/4 | 0.7500 | 0.7406 | 1.438 | 1.355 | 0.396 | 0.454 | 1/2 | 0.220 |
| 7/8 | 0.8750 | 0.8647 | 1.688 | 1.604 | 0.468 | … | 9/16 | 0.248 |
| 1 | 1.0000 | 0.9886 | 1.938 | 1.841 | 0.540 | … | 5/8 | 0.297 |
| 1 1/8 | 1.1250 | 1.1086 | 2.188 | 2.079 | 0.611 | … | 3/4 | 0.325 |
| 1 1/4 | 1.2500 | 1.2336 | 2.438 | 2.316 | 0.683 | … | 7/8 | 0.358 |
| 1 3/8 | 1.3750 | 1.3568 | 2.688 | 2.553 | 0.755 | … | 7/8 | 0.402 |
| 1 1/2 | 1.5000 | 1.4818 | 2.938 | 2.791 | 0.827 | … | 1 | 0.435 |

[a] Spline driven fasteners are removed from the latest revision. For additional information on these fasteners, see ANSI/ASME B18.3-1998.

All dimensions in inches.

The body of the screw is the unthreaded cylindrical portion of the shank where not threaded to the head; the shank being the portion of the screw from the point of juncture of the conical bearing surface and the body to the flat of the point. The length of thread $L_T$ is the distance measured from the extreme point to the last complete (full form) thread.

Standard length increments of No. 0 through 1-inch sizes are as follows: $\frac{1}{16}$ inch for nominal screw lengths of $\frac{1}{8}$ through $\frac{1}{4}$ inch; $\frac{1}{8}$ inch for lengths of $\frac{1}{4}$ through 1 inch; $\frac{1}{4}$ inch for lengths of 1 inch through 3 $\frac{1}{2}$ inches; $\frac{1}{2}$ inch for lengths of 3 $\frac{1}{2}$ through 7 inches; and 1 inch for lengths of 7 through 10 inches, incl. For screw sizes over 1 inch, length increments are: $\frac{1}{2}$ inch for nominal screw lengths of 1 inch through 7 inches; 1 inch for lengths of 7 through 10 inches; and 2 inches for lengths over 10 inches.

Threads shall be Unified external threads with radius root; Class 3A UNRC and UNRF series for sizes No. 0 through 1 inch and Class 2A UNRC and UNRF series for sizes over 1 inch to 1 $\frac{1}{2}$ inches, incl.

For manufacturing details not shown, including materials, see American National Standard ANSI/ASME B18.3-2012 and ANSI/ASME B18.3-1998 (for Spline driven fasteners). Socket dimensions are given in Table 11.

**Table 7. American National Standard Hexagon Socket and Spline Socket Button Head Cap Screws** *ANSI/ASME B18.3-2012*

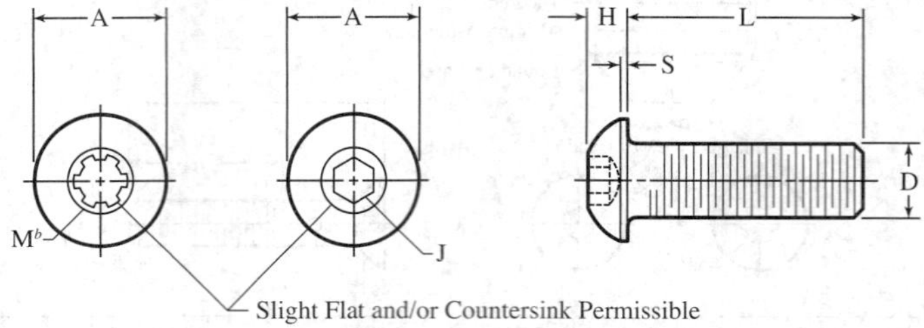


| Nominal Size | Screw Diameter | Head Diameter | | Head Height | | Head Side Height | Spline Socket Size[a] | Hexagon Socket Size[a] | Standard Length |
|---|---|---|---|---|---|---|---|---|---|
| | Basic | Max. | Min. | Max. | Min. | Ref. | Nom. | Nom. | Max. |
| | *D* | *A* | | *H* | | *S* | *M*[b] | *J* | *L* |
| 0 | 0.0600 | 0.114 | 0.104 | 0.032 | 0.026 | 0.010 | 0.048 | 0.035 | ½ |
| 1 | 0.0730 | 0.139 | 0.129 | 0.039 | 0.033 | 0.010 | 0.060 | 0.050 | ½ |
| 2 | 0.0860 | 0.164 | 0.154 | 0.046 | 0.038 | 0.010 | 0.060 | 0.050 | ½ |
| 3 | 0.0990 | 0.188 | 0.176 | 0.052 | 0.044 | 0.010 | 0.072 | 1⁄16 | ½ |
| 4 | 0.1120 | 0.213 | 0.201 | 0.059 | 0.051 | 0.015 | 0.072 | 1⁄16 | ½ |
| 5 | 0.1250 | 0.238 | 0.226 | 0.066 | 0.058 | 0.015 | 0.096 | 5⁄64 | ½ |
| 6 | 0.1380 | 0.262 | 0.250 | 0.073 | 0.063 | 0.015 | 0.096 | 5⁄64 | 5⁄8 |
| 8 | 0.1640 | 0.312 | 0.298 | 0.087 | 0.077 | 0.015 | 0.111 | 3⁄32 | ¾ |
| 10 | 0.1900 | 0.361 | 0.347 | 0.101 | 0.091 | 0.020 | 0.145 | 1⁄8 | 1 |
| ¼ | 0.2500 | 0.437 | 0.419 | 0.132 | 0.122 | 0.031 | 0.183 | 5⁄32 | 1 |
| 5⁄16 | 0.3125 | 0.547 | 0.527 | 0.166 | 0.152 | 0.031 | 0.216 | 3⁄16 | 1 |
| 3⁄8 | 0.3750 | 0.656 | 0.636 | 0.199 | 0.185 | 0.031 | 0.251 | 7⁄32 | 1¼ |
| ½ | 0.5000 | 0.875 | 0.851 | 0.265 | 0.245 | 0.046 | 0.372 | 5⁄16 | 2 |
| 5⁄8 | 0.6250 | 1.000 | 0.970 | 0.331 | 0.311 | 0.062 | 0.454 | 3⁄8 | 2 |

[a] Socket dimensions are given in Table 11.

[b] Spline driven fasteners are removed from the latest revision. For additional information on these fasteners, see ANSI/ASME B18.3-1998.

All dimensions in inches.

These cap screws have been designed and recommended for light fastening applications. They are not suggested for use in critical high-strength applications where socket head cap screws should normally be used.

Standard length increments for socket button head cap screws are as follows: 1⁄16 inch for nominal screw lengths of 1⁄8 through ¼ inch, 1⁄8 inch for nominal screw lengths of ¼ through 1 inch, and ¼ inch for nominal screw lengths of 1 inch through 2 inches. Tolerances on lengths are −0.03 inch for lengths up to 1 inch inclusive. For lengths from 1 through 2 inches, inclusive, length tolerances are −0.04 inch.

The thread conforms to the Unified Standard, Class 3A, with radius root, UNRC and UNRF.

To prevent interference, American National Standard ANSI/ASME B18.3.4M-1986 gives metric dimensional and general requirements for a lower head profile hexagon socket button head cap screw. Because of its design, wrenchability and other design factors are reduced; therefore, B18.3.4M should be reviewed carefully. Available only in metric sizes and with metric threads.

For manufacturing details, including materials, not shown, see American National Standard ANSI/ASME B18.3-2012 and ANSI/ASME B18.3-1998 (for Spline driven fasteners).

**Table 10. American National Standard Hexagon and Spline Socket Set Screw Optional Cup Points** *ANSI/ASME B18.3-2012* and *ANSI/ASME B18.3-1998 (for Spline driven fasteners).*

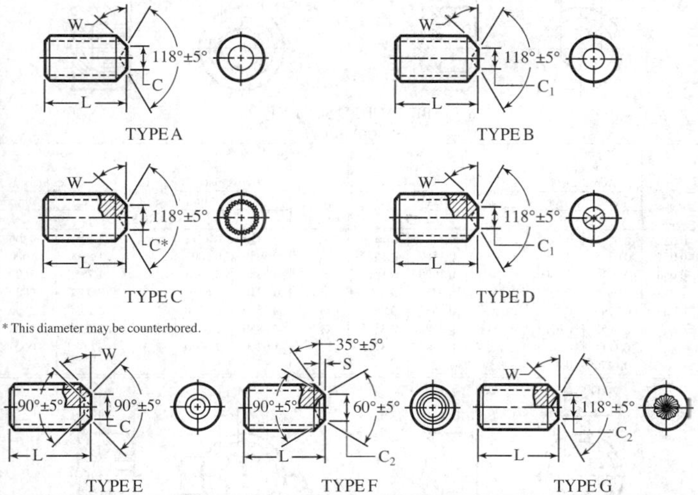


| Nom. Size | Point Dia. | | Point Dia. | | Point Dia. | | Point Length | |
|---|---|---|---|---|---|---|---|---|
| | Max. | Min. | Max. | Min. | Max. | Min. | Max. | Min. |
| | *C* | | *C1* | | *C2* | | *S* | |
| 0 | 0.033 | 0.027 | 0.032 | 0.027 | 0.027 | 0.022 | 0.007 | 0.004 |
| 1 | 0.040 | 0.033 | 0.038 | 0.033 | 0.035 | 0.030 | 0.008 | 0.005 |
| 2 | 0.047 | 0.039 | 0.043 | 0.038 | 0.043 | 0.038 | 0.010 | 0.007 |
| 3 | 0.054 | 0.045 | 0.050 | 0.045 | 0.051 | 0.046 | 0.011 | 0.007 |
| 4 | 0.061 | 0.051 | 0.056 | 0.051 | 0.059 | 0.054 | 0.013 | 0.008 |
| 5 | 0.067 | 0.057 | 0.062 | 0.056 | 0.068 | 0.063 | 0.014 | 0.009 |
| 6 | 0.074 | 0.064 | 0.069 | 0.062 | 0.074 | 0.068 | 0.017 | 0.012 |
| 8 | 0.087 | 0.076 | 0.082 | 0.074 | 0.090 | 0.084 | 0.021 | 0.016 |
| 10 | 0.102 | 0.088 | 0.095 | 0.086 | 0.101 | 0.095 | 0.024 | 0.019 |
| ¼ | 0.132 | 0.118 | 0.125 | 0.114 | 0.156 | 0.150 | 0.027 | 0.022 |
| 5/16 | 0.172 | 0.156 | 0.156 | 0.144 | 0.190 | 0.185 | 0.038 | 0.033 |
| 3/8 | 0.212 | 0.194 | 0.187 | 0.174 | 0.241 | 0.236 | 0.041 | 0.036 |
| 7/16 | 0.252 | 0.232 | 0.218 | 0.204 | 0.286 | 0.281 | 0.047 | 0.042 |
| ½ | 0.291 | 0.270 | 0.250 | 0.235 | 0.333 | 0.328 | 0.054 | 0.049 |
| 5/8 | 0.371 | 0.347 | 0.312 | 0.295 | 0.425 | 0.420 | 0.067 | 0.062 |
| ¾ | 0.450 | 0.425 | 0.375 | 0.357 | 0.523 | 0.518 | 0.081 | 0.076 |
| 7/8 | 0.530 | 0.502 | 0.437 | 0.418 | … | … | … | … |
| 1 | 0.609 | 0.579 | 0.500 | 0.480 | … | … | … | … |
| 1 1/8 | 0.689 | 0.655 | 0.562 | 0.542 | … | … | … | … |
| 1¼ | 0.767 | 0.733 | 0.625 | 0.605 | … | … | … | … |
| 1 3/8 | 0.848 | 0.808 | 0.687 | 0.667 | … | … | … | … |
| 1½ | 0.926 | 0.886 | 0.750 | 0.730 | … | … | … | … |
| 1¾ | 1.086 | 1.039 | 0.875 | 0.855 | … | … | … | … |
| 2 | 1.244 | 1.193 | 1.000 | 0.980 | … | … | … | … |

All dimensions are in inches.
The cup point types shown are those available from various manufacturers.

**Table 13. American National Standard Square Head Set Screws**
*ANSI/ASME B18.6.2-2020*

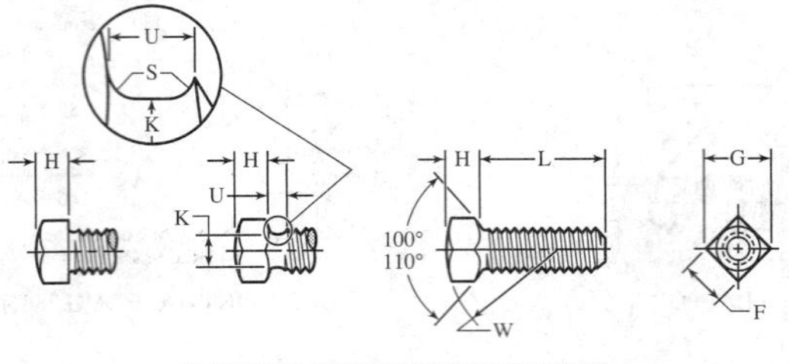


| Nominal Size[a] or Basic Screw Diameter | | Width Across Flats, *F* | | Width Across Corners, *G* | | Head Height, *H* | | Neck Relief Diameter, *K* | | Neck Relief Fillet Rad., *S* | Neck Relief Width, *U* | Head Rad., *W* |
|---|---|---|---|---|---|---|---|---|---|---|---|---|
| | | Max. | Min. | Max. | Min. | Max. | Min. | Max. | Min. | Max. | Min. | Min. |
| 10 | 0.1900 | .188 | .180 | .265 | .247 | .148 | .134 | .145 | .140 | .027 | .083 | 0.48 |
| 1/4 | 0.2500 | .250 | .241 | .354 | .331 | .196 | .178 | .185 | .170 | .032 | .100 | 0.62 |
| 5/16 | 0.3125 | .312 | .302 | .442 | .415 | .245 | .224 | .240 | .225 | .036 | .111 | 0.78 |
| 3/8 | 0.3750 | .375 | .362 | .530 | .497 | .293 | .270 | .294 | .279 | .041 | .125 | 0.94 |
| 7/16 | 0.4375 | .438 | .423 | .619 | .581 | .341 | .315 | .345 | .330 | .046 | .143 | 1.09 |
| 1/2 | 0.5000 | .500 | .484 | .707 | .665 | .389 | .361 | .400 | .385 | .050 | .154 | 1.25 |
| 9/16 | 0.5625 | .562 | .545 | .795 | .748 | .437 | .407 | .454 | .439 | .054 | .167 | 1.41 |
| 5/8 | 0.6250 | .625 | .606 | .884 | .833 | .485 | .452 | .507 | .492 | .059 | .182 | 1.56 |
| 3/4 | 0.7500 | .750 | .729 | 1.060 | 1.001 | .582 | .544 | .620 | .605 | .065 | .200 | 1.88 |
| 7/8 | 0.8750 | .875 | .852 | 1.237 | 1.170 | .678 | .635 | .731 | .716 | .072 | .222 | 2.19 |
| 1 | 1.0000 | 1.000 | .974 | 1.414 | 1.337 | .774 | .726 | .838 | .823 | .081 | .250 | 2.50 |
| 1 1/8 | 1.1250 | 1.125 | 1.096 | 1.591 | 1.505 | .870 | .817 | .939 | .914 | .092 | .283 | 2.81 |
| 1 1/4 | 1.2500 | 1.250 | 1.219 | 1.768 | 1.674 | .966 | .908 | 1.064 | 1.039 | .092 | .283 | 3.12 |
| 1 3/8 | 1.3750 | 1.375 | 1.342 | 1.945 | 1.843 | 1.063 | 1.000 | 1.159 | 1.134 | .109 | .333 | 3.44 |
| 1 1/2 | 1.5000 | 1.500 | 1.464 | 2.121 | 2.010 | 1.159 | 1.091 | 1.284 | 1.259 | .109 | .333 | 3.75 |

*Designation:* Square head set screws are designated by the following data in the sequence shown: Nominal size (number, fraction or decimal equivalent); threads per inch; screw length (fraction or decimal equivalent); product name; point style; material; and protective finish, if required. Examples: 1/4 - 20 × 3/4 Square Head Set Screw, Flat Point, Steel, Cadmium Plated; .500 – 13 × 1.25 Square Head Set Screw, Cone Point, Corrosion Resistant Steel.

*Cone Point Angle, Y:* For the following nominal lengths, or longer, shown according to nominal size, the cone point angle is 90° ± 2°: For size No. 10, 1/4; 1/4, 5/16; 5/16, 3/8; 3/8, 7/16; 7/16, 1/2; 1/2, 9/16; 9/16, 5/8; 5/8, 3/4; 3/4, 7/8; 7/8, 1; 1, 1 1/8; 1 1/8, 1 1/4; 1 1/4, 1 1/2; 1 3/8, 1 5/8; and for 1 1/2, 1 3/4. For shorter screws the *cone point angle is 118° ± 2°.*

*Point Types:* Unless otherwise specified, square head set screws are supplied with cup points. Cup points as furnished by some manufacturers may be externally or internally knurled. Where so specified by the purchaser, screws have cone, dog, half-dog, flat or oval points as given on the following page.

*Point Angle, X:* The point angle is 45°, + 5°, – 0° for screws of the nominal lengths, or longer, given just above for cone point angle, and 30° min. for shorter lengths.

**Table 15. ANSI Hexagon and Spline Socket Set Screws** *ANSI/ASME B18.3-2012* and *ANSI/ASME B18.3-1998 (for Spline driven fasteners).*

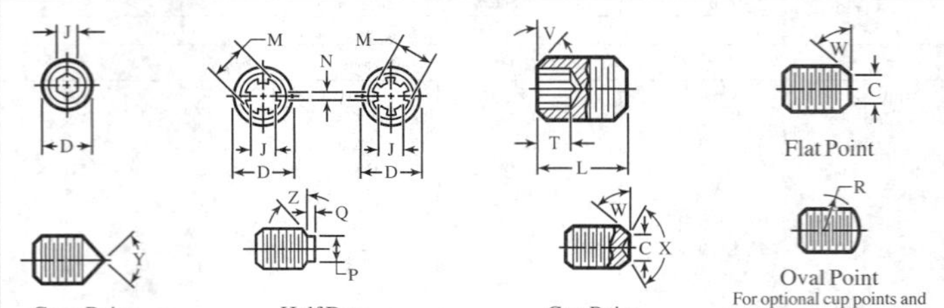


For optional cup points and their dimensions see Table 10.

| Nominal Size or Basic Screw Diameter | | Socket Size Hex. Nom. J | Socket Size Spl. Nom. M | Cup and Flat Point Diameters Max. C | Cup and Flat Point Diameters Min. C | Half Dog Point Dia. Max. P | Half Dog Point Lgth. Max. Q | Oval Point Radius Basic R | Min. Key Engagement Depth Hex. $T_H$[a] | Min. Key Engagement Depth Spl. $T_S$[a] | Lgth. Limit for Angle Y[b] |
|---|---|---|---|---|---|---|---|---|---|---|---|
| 0 | 0.0600 | 0.028 | 0.033 | 0.033 | 0.027 | 0.040 | 0.017 | 0.045 | 0.050 | 0.026 | 0.09 |
| 1 | 0.0730 | 0.035 | 0.033 | 0.040 | 0.033 | 0.049 | 0.021 | 0.055 | 0.060 | 0.035 | 0.09 |
| 2 | 0.0860 | 0.035 | 0.048 | 0.047 | 0.039 | 0.057 | 0.024 | 0.064 | 0.060 | 0.040 | 0.13 |
| 3 | 0.0990 | 0.050 | 0.048 | 0.054 | 0.045 | 0.066 | 0.027 | 0.074 | 0.070 | 0.040 | 0.13 |
| 4 | 0.1120 | 0.050 | 0.060 | 0.061 | 0.051 | 0.075 | 0.030 | 0.084 | 0.070 | 0.045 | 0.19 |
| 5 | 0.1250 | 1⁄16 | 0.072 | 0.067 | 0.057 | 0.083 | 0.033 | 0.094 | 0.080 | 0.055 | 0.19 |
| 6 | 0.1380 | 1⁄16 | 0.072 | 0.074 | 0.064 | 0.092 | 0.038 | 0.104 | 0.080 | 0.055 | 0.19 |
| 8 | 0.1640 | 5⁄64 | 0.096 | 0.087 | 0.076 | 0.109 | 0.043 | 0.123 | 0.090 | 0.080 | 0.25 |
| 10 | 0.1900 | 3⁄32 | 0.111 | 0.102 | 0.088 | 0.127 | 0.049 | 0.142 | 0.100 | 0.080 | 0.25 |
| 1⁄4 | 0.2500 | 1⁄8 | 0.145 | 0.132 | 0.118 | 0.156 | 0.067 | 0.188 | 0.125 | 0.125 | 0.31 |
| 5⁄16 | 0.3125 | 5⁄32 | 0.183 | 0.172 | 0.156 | 0.203 | 0.082 | 0.234 | 0.156 | 0.156 | 0.38 |
| 3⁄8 | 0.3750 | 3⁄16 | 0.216 | 0.212 | 0.194 | 0.250 | 0.099 | 0.281 | 0.188 | 0.188 | 0.44 |
| 7⁄16 | 0.4375 | 7⁄32 | 0.251 | 0.252 | 0.232 | 0.297 | 0.114 | 0.328 | 0.219 | 0.219 | 0.50 |
| 1⁄2 | 0.5000 | 1⁄4 | 0.291 | 0.291 | 0.270 | 0.344 | 0.130 | 0.375 | 0.250 | 0.250 | 0.57 |
| 5⁄8 | 0.6250 | 5⁄16 | 0.372 | 0.371 | 0.347 | 0.469 | 0.164 | 0.469 | 0.312 | 0.312 | 0.75 |
| 3⁄4 | 0.7500 | 3⁄8 | 0.454 | 0.450 | 0.425 | 0.562 | 0.196 | 0.562 | 0.375 | 0.375 | 0.88 |
| 7⁄8 | 0.8750 | 1⁄2 | 0.595 | 0.530 | 0.502 | 0.656 | 0.227 | 0.656 | 0.500 | 0.500 | 1.00 |
| 1 | 1.0000 | 9⁄16 | … | 0.609 | 0.579 | 0.750 | 0.260 | 0.750 | 0.562 | … | 1.13 |
| 1 1⁄8 | 1.1250 | 9⁄16 | … | 0.689 | 0.655 | 0.844 | 0.291 | 0.844 | 0.562 | … | 1.25 |
| 1 1⁄4 | 1.2500 | 5⁄8 | … | 0.767 | 0.733 | 0.938 | 0.323 | 0.938 | 0.625 | … | 1.50 |
| 1 3⁄8 | 1.3750 | 5⁄8 | … | 0.848 | 0.808 | 1.031 | 0.354 | 1.031 | 0.625 | … | 1.63 |
| 1 1⁄2 | 1.5000 | 3⁄4 | … | 0.926 | 0.886 | 1.125 | 0.385 | 1.125 | 0.750 | … | 1.75 |
| 1 3⁄4 | 1.7500 | 1 | … | 1.086 | 1.039 | 1.312 | 0.448 | 1.321 | 1.000 | … | 2.00 |
| 2 | 2.0000 | 1 | … | 1.244 | 1.193 | 1.500 | 0.510 | 1.500 | 1.000 | … | 2.25 |

[a] Reference should be made to the Standard for shortest optimum nominal lengths to which the minimum key engagement depths $T_H$ and $T_S$ apply.

[b] Cone point angle *Y* is 90 degrees plus or minus 2 degrees for these nominal lengths or longer and 118 degrees plus or minus 2 degrees for shorter nominal lengths.

All dimensions are in inches. The thread conforms to the Unified Standard, Class 3A, UNC and UNF series. The socket depth *T* is included in the Standard and some are shown here. The nominal length *L* of all socket type set screws is the total or overall length. For nominal screw lengths of 1⁄16 through 3⁄16 inch (0 through 3 sizes incl.) the standard length increment is 0.06 inch; for lengths 1⁄8 through 1 inch the increment is 1⁄8 inch; for lengths 1 through 2 inches the increment is 1⁄4 inch; for lengths 2 through 6 inches the increment is 1⁄2 inch; for lengths 6 inches and longer the increment is 1 inch. Socket dimensions are given in Table 11.

*Length Tolerance:* The allowable tolerance on length *L* for all set screws of the socket type is ± 0.01 inch for set screws up to 5⁄8 inch long; ± 0.02 inch for screws over 5⁄8 to 2 inches long; ± 0.03 inch for screws over 2 to 6 inches long and ± 0.06 inch for screws over 6 inches long. Socket dimensions are given in Table 11.

For manufacturing details, including materials, not shown, see American National Standard ANSI/ASME B18.3-2012 and ANSI/ASME B18.3-1998 (for Spline driven fasteners).

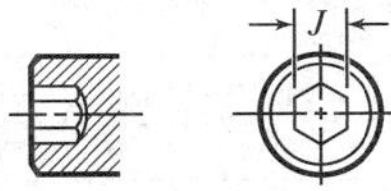


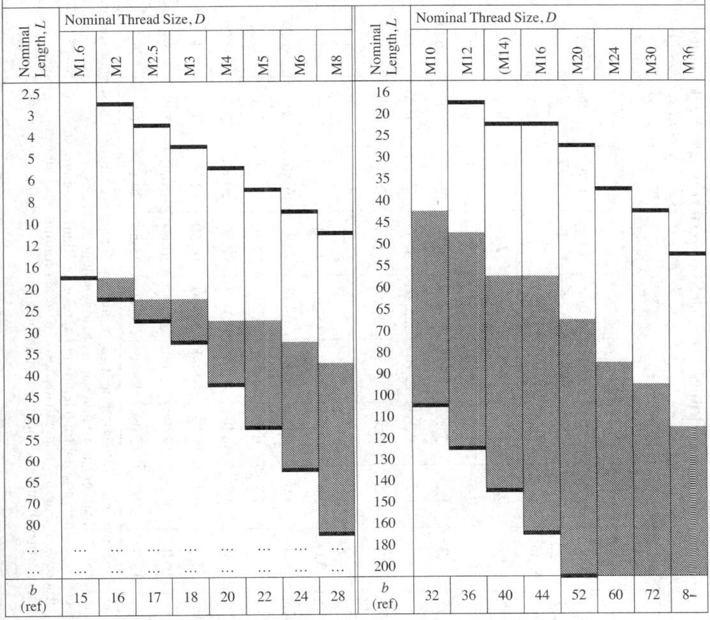

## Table 4. British Standard Hexagon Socket Countersunk and Button Head Screws—Metric Series *BS 4168: Parts 6 and 8:1982 (Withdrawn)*

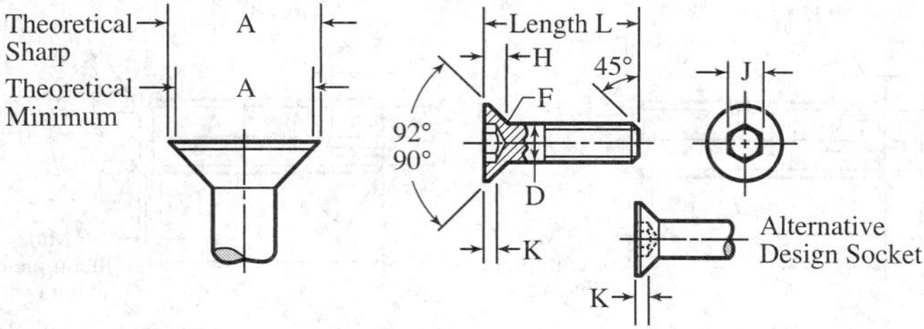


COUNTERSUNK HEAD SCREWS

| Nom. Size | Body Diameter, *D* Max. | Body Diameter, *D* Min. | Head Diameter, *A* Theor. Sharp Max. | Head Diameter, *A* Absolute Min. | Head Height, *H* Ref. | Head Height, *H* Flushness Tolerance | Hexagon Socket Size, *J* Nom. | Key Engagement, *K* Min. | Fillet Radius, *F* Max. |
|---|---|---|---|---|---|---|---|---|---|
| M3 | 3.00 | 2.86 | 6.72 | 5.82 | 1.86 | 0.20 | 2.00 | 1.05 | 0.40 |
| M4 | 4.00 | 3.82 | 8.96 | 7.78 | 2.48 | 0.20 | 2.50 | 1.49 | 0.40 |
| M5 | 5.00 | 4.82 | 11.20 | 9.78 | 3.10 | 0.20 | 3.00 | 1.86 | 0.40 |
| M6 | 6.00 | 5.82 | 13.44 | 11.73 | 3.72 | 0.20 | 4.00 | 2.16 | 0.60 |
| M8 | 8.00 | 7.78 | 17.92 | 15.73 | 4.96 | 0.24 | 5.00 | 2.85 | 0.70 |
| M10 | 10.00 | 9.78 | 22.40 | 19.67 | 6.20 | 0.30 | 6.00 | 3.60 | 0.80 |
| M12 | 12.00 | 11.73 | 26.88 | 23.67 | 7.44 | 0.36 | 8.00 | 4.35 | 1.10 |
| (M14) | 14.00 | 13.73 | 30.24 | 26.67 | 8.12 | 0.40 | 10.00 | 4.65 | 1.10 |
| M16 | 16.00 | 15.73 | 33.60 | 29.67 | 8.80 | 0.45 | 10.00 | 4.89 | 1.10 |
| (M18) | 18.00 | 17.73 | 36.96 | 32.61 | 9.48 | 0.50 | 12.00 | 5.25 | 1.10 |
| M20 | 20.00 | 19.67 | 40.32 | 35.61 | 10.16 | 0.54 | 12.00 | 5.45 | 1.10 |

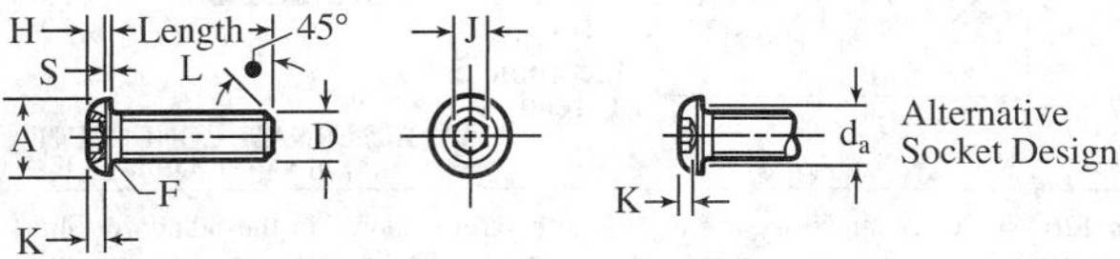


BUTTON HEAD SCREWS

| Nom. Size, *D* | Head Diameter, *A* Max. | Head Diameter, *A* Min. | Head Height, *H* Max. | Head Height, *H* Min. | Head Side Height, *S* Ref. | Hexagon Socket Size, *J* Nom. | Key Engagement, *K* Min. | Fillet Radius *F* Min. | Fillet Radius *da* Max. |
|---|---|---|---|---|---|---|---|---|---|
| M3 | 5.50 | 5.32 | 1.60 | 1.40 | 0.38 | 2.00 | 1.04 | 0.10 | 3.60 |
| M4 | 7.50 | 7.28 | 2.10 | 1.85 | 0.38 | 2.50 | 1.30 | 0.20 | 4.70 |
| M5 | 9.50 | 9.28 | 2.70 | 2.45 | 0.50 | 3.00 | 1.56 | 0.20 | 5.70 |
| M6 | 10.50 | 10.23 | 3.20 | 2.95 | 0.80 | 4.00 | 2.08 | 0.25 | 6.80 |
| M8 | 14.00 | 13.73 | 4.30 | 3.95 | 0.80 | 5.00 | 2.60 | 0.40 | 9.20 |
| M10 | 18.00 | 17.73 | 5.30 | 4.95 | 0.80 | 6.00 | 3.12 | 0.40 | 11.20 |
| M12 | 21.00 | 20.67 | 6.40 | 5.90 | 0.80 | 8.00 | 4.16 | 0.60 | 14.20 |

All dimensions are given in millimeters.

**Table 6. Thread and Point Dimensions for Type C Thread Forming Tapping Screws** *ANSI/ASME B18.6.3-2013, Appendix*

| | | D | | P | | S | | | | L | | | |
|---|---|---|---|---|---|---|---|---|---|---|---|---|---|
| | | | | | | Point Taper Length[c] | | | | Determinant Lengths for Point Taper[c] | | Minimum Practical Nominal Screw Lengths | |
| Nominal Size[a] or Basic Screw Diameter | Threads per inch | Major Diameter | | Point Diameter[b] | | For Short Screws | | For Long Screws | | | | | |
| | | Max. | Min. | Max. | Min. | Max. | Min. | Max. | Min. | 90° Heads | Csk Heads | 90° Heads | Csk Heads |
| 2 0.0860 | 56 | 0.0860 | 0.0813 | 0.068 | 0.061 | 0.062 | 0.045 | 0.080 | 0.062 | 5⁄32 | 3⁄16 | 5⁄32 | 3⁄16 |
| 2 0.0860 | 64 | 0.0860 | 0.0816 | 0.070 | 0.064 | 0.055 | 0.039 | 0.070 | 0.055 | 1⁄8 | 3⁄16 | 1⁄8 | 5⁄32 |
| 3 0.0990 | 48 | 0.0990 | 0.0938 | 0.078 | 0.070 | 0.073 | 0.052 | 0.094 | 0.073 | 3⁄16 | 7⁄32 | 5⁄32 | 7⁄32 |
| 3 0.0990 | 56 | 0.0990 | 0.0942 | 0.081 | 0.074 | 0.062 | 0.045 | 0.080 | 0.062 | 5⁄32 | 3⁄16 | 5⁄32 | 3⁄16 |
| 4 0.1120 | 40 | 0.1120 | 0.1061 | 0.087 | 0.078 | 0.088 | 0.062 | 0.112 | 0.088 | 7⁄32 | 1⁄4 | 3⁄16 | 1⁄4 |
| 4 0.1120 | 48 | 0.1120 | 0.1068 | 0.091 | 0.083 | 0.073 | 0.052 | 0.094 | 0.073 | 3⁄16 | 7⁄32 | 5⁄32 | 7⁄32 |
| 5 0.1250 | 40 | 0.1250 | 0.1191 | 0.100 | 0.091 | 0.088 | 0.062 | 0.112 | 0.088 | 7⁄32 | 9⁄32 | 3⁄16 | 1⁄4 |
| 5 0.1250 | 44 | 0.1250 | 0.1195 | 0.102 | 0.094 | 0.080 | 0.057 | 0.102 | 0.080 | 3⁄16 | 1⁄4 | 3⁄16 | 1⁄4 |
| 6 0.1380 | 32 | 0.1380 | 0.1312 | 0.107 | 0.096 | 0.109 | 0.078 | 0.141 | 0.109 | 1⁄4 | 5⁄16 | 1⁄4 | 5⁄16 |
| 6 0.1380 | 40 | 0.1380 | 0.1321 | 0.113 | 0.104 | 0.088 | 0.062 | 0.112 | 0.088 | 7⁄32 | 9⁄32 | 3⁄16 | 1⁄4 |
| 8 0.1640 | 32 | 0.1640 | 0.1571 | 0.132 | 0.122 | 0.109 | 0.078 | 0.141 | 0.109 | 1⁄4 | 11⁄32 | 1⁄4 | 5⁄16 |
| 8 0.1640 | 36 | 0.1640 | 0.1577 | 0.136 | 0.126 | 0.097 | 0.069 | 0.125 | 0.097 | 7⁄32 | 5⁄16 | 7⁄32 | 9⁄32 |
| 10 0.1900 | 24 | 0.1900 | 0.1818 | 0.148 | 0.135 | 0.146 | 0.104 | 0.188 | 0.146 | 11⁄32 | 7⁄16 | 5⁄16 | 13⁄32 |
| 10 0.1900 | 32 | 0.1900 | 0.1831 | 0.158 | 0.148 | 0.109 | 0.078 | 0.141 | 0.109 | 1⁄4 | 11⁄32 | 1⁄4 | 5⁄16 |
| 12 0.2160 | 24 | 0.2160 | 0.2078 | 0.174 | 0.161 | 0.146 | 0.104 | 0.188 | 0.146 | 11⁄32 | 7⁄16 | 5⁄16 | 13⁄32 |
| 12 0.2160 | 28 | 0.2160 | 0.2085 | 0.180 | 0.168 | 0.125 | 0.089 | 0.161 | 0.125 | 5⁄16 | 13⁄32 | 9⁄32 | 3⁄8 |
| 1⁄4 0.2500 | 20 | 0.2500 | 0.2408 | 0.200 | 0.184 | 0.175 | 0.125 | 0.225 | 0.175 | 13⁄32 | 17⁄32 | 3⁄8 | 1⁄2 |
| 1⁄4 0.2500 | 28 | 0.2500 | 0.2425 | 0.214 | 0.202 | 0.125 | 0.089 | 0.161 | 0.125 | 5⁄16 | 13⁄32 | 9⁄32 | 3⁄8 |
| 5⁄16 0.3125 | 18 | 0.3125 | 0.3026 | 0.257 | 0.239 | 0.194 | 0.139 | 0.250 | 0.194 | 15⁄32 | 19⁄32 | 7⁄16 | 9⁄16 |
| 5⁄16 0.3125 | 24 | 0.3125 | 0.3042 | 0.271 | 0.257 | 0.146 | 0.104 | 0.188 | 0.146 | 11⁄32 | 15⁄32 | 5⁄16 | 15⁄32 |
| 3⁄8 0.3750 | 16 | 0.3750 | 0.3643 | 0.312 | 0.293 | 0.219 | 0.156 | 0.281 | 0.219 | 1⁄2 | 11⁄16 | 15⁄32 | 5⁄8 |
| 3⁄8 0.3750 | 24 | 0.3750 | 0.3667 | 0.333 | 0.319 | 0.146 | 0.104 | 0.188 | 0.146 | 11⁄32 | 1⁄2 | 5⁄16 | 1⁄2 |
| 7⁄16 0.4375 | 14 | 0.4375 | 0.4258 | 0.366 | 0.344 | 0.250 | 0.179 | 0.321 | 0.250 | 19⁄32 | 3⁄4 | 9⁄16 | 23⁄32 |
| 7⁄16 0.4375 | 20 | 0.4375 | 0.4281 | 0.387 | 0.371 | 0.175 | 0.125 | 0.225 | 0.175 | 13⁄32 | 9⁄16 | 3⁄8 | 17⁄32 |
| 1⁄2 0.5000 | 13 | 0.5000 | 0.4876 | 0.423 | 0.399 | 0.269 | 0.192 | 0.346 | 0.269 | 5⁄8 | 25⁄32 | 19⁄32 | 3⁄4 |
| 1⁄2 0.5000 | 20 | 0.5000 | 0.4906 | 0.450 | 0.433 | 0.175 | 0.125 | 0.225 | 0.175 | 13⁄32 | 9⁄16 | 3⁄8 | 17⁄32 |

[a] Where specifying nominal size in decimals, zeros preceding decimal and in the fourth decimal place shall be omitted.

[b] The tabulated values apply to screw blanks before roll threading.

[c] Screws of these nominal lengths and shorter shall have point taper length specified above for short screws. Longer lengths shall have point taper length specified for long screws.

All dimensions are in inches. See Table 1 for thread diagrams. Type C is not recommended for new designs.

Tapered threads shall have unfinished crests.

## Table 7. ANSI Standard Thread and Point Dimensions for Types D, F, G, and T Thread Cutting Tapping Screws *ANSI/ASME B18.6.3-2013*

| Nominal Size[a] or Basic Screw Diameter | Threads per inch | D: Major Diameter Max. | D: Major Diameter Min. | P: Point Diameter[b] Max. | P: Point Diameter[b] Min. | S: Point Taper Length[c] For Short Screws Max. | S: For Short Screws Min. | S: For Long Screws Max. | S: For Long Screws Min. | L: Determinant Lengths for Point Taper[c] 90° Heads | L: Determinant Lengths Csk Heads | L: Minimum Practical Nominal Screw Lengths 90° Heads | L: Minimum Practical Csk Heads |
|---|---|---|---|---|---|---|---|---|---|---|---|---|---|
| 2 0.0860 | 56 | 0.0860 | 0.0813 | 0.068 | 0.061 | 0.062 | 0.045 | 0.080 | 0.062 | 5/32 | 3/16 | 5/32 | 3/16 |
| 2 0.0860 | 64 | 0.0860 | 0.0816 | 0.070 | 0.064 | 0.055 | 0.039 | 0.070 | 0.055 | 1/8 | 3/16 | 1/8 | 5/32 |
| 3 0.0990 | 48 | 0.0990 | 0.0938 | 0.078 | 0.070 | 0.073 | 0.052 | 0.094 | 0.073 | 3/16 | 7/32 | 5/32 | 7/32 |
| 3 0.0990 | 56 | 0.0990 | 0.0942 | 0.081 | 0.074 | 0.062 | 0.045 | 0.080 | 0.062 | 5/32 | 3/16 | 5/32 | 3/16 |
| 4 0.1120 | 40 | 0.1120 | 0.1061 | 0.087 | 0.078 | 0.088 | 0.062 | 0.112 | 0.088 | 7/32 | 1/4 | 3/16 | 1/4 |
| 4 0.1120 | 48 | 0.1120 | 0.1068 | 0.091 | 0.083 | 0.073 | 0.052 | 0.094 | 0.073 | 3/16 | 7/32 | 5/32 | 7/32 |
| 5 0.1250 | 40 | 0.1250 | 0.1191 | 0.100 | 0.091 | 0.088 | 0.062 | 0.112 | 0.088 | 7/32 | 9/32 | 3/16 | 1/4 |
| 5 0.1250 | 44 | 0.1250 | 0.1195 | 0.102 | 0.094 | 0.080 | 0.057 | 0.102 | 0.080 | 3/16 | 1/4 | 3/16 | 1/4 |
| 6 0.1380 | 32 | 0.1380 | 0.1312 | 0.107 | 0.096 | 0.109 | 0.078 | 0.141 | 0.109 | 1/4 | 5/16 | 1/4 | 5/16 |
| 6 0.1380 | 40 | 0.1380 | 0.1321 | 0.113 | 0.104 | 0.088 | 0.062 | 0.112 | 0.088 | 7/32 | 9/32 | 3/16 | 1/4 |
| 8 0.1640 | 32 | 0.1640 | 0.1571 | 0.132 | 0.122 | 0.109 | 0.078 | 0.141 | 0.109 | 1/4 | 11/32 | 1/4 | 5/16 |
| 8 0.1640 | 36 | 0.1640 | 0.1577 | 0.136 | 0.126 | 0.097 | 0.069 | 0.125 | 0.097 | 7/32 | 5/16 | 7/32 | 9/32 |
| 10 0.1900 | 24 | 0.1900 | 0.1818 | 0.148 | 0.135 | 0.146 | 0.104 | 0.188 | 0.146 | 11/32 | 7/16 | 5/16 | 13/32 |
| 10 0.1900 | 32 | 0.1900 | 0.1831 | 0.158 | 0.148 | 0.109 | 0.078 | 0.141 | 0.109 | 1/4 | 11/32 | 1/4 | 5/16 |
| 12 0.2160 | 24 | 0.2160 | 0.2078 | 0.174 | 0.161 | 0.146 | 0.104 | 0.188 | 0.146 | 11/32 | 7/16 | 5/16 | 13/32 |
| 12 0.2160 | 28 | 0.2160 | 0.2085 | 0.180 | 0.168 | 0.125 | 0.089 | 0.161 | 0.125 | 5/16 | 13/32 | 9/32 | 3/8 |
| 1/4 0.2500 | 20 | 0.2500 | 0.2408 | 0.200 | 0.184 | 0.175 | 0.125 | 0.225 | 0.175 | 13/32 | 17/32 | 3/8 | 1/2 |
| 1/4 0.2500 | 28 | 0.2500 | 0.2425 | 0.214 | 0.202 | 0.125 | 0.089 | 0.161 | 0.125 | 5/16 | 13/32 | 9/32 | 3/8 |
| 5/16 0.3125 | 18 | 0.3125 | 0.3026 | 0.257 | 0.239 | 0.194 | 0.139 | 0.250 | 0.194 | 15/32 | 19/32 | 7/16 | 9/16 |
| 5/16 0.3125 | 24 | 0.3125 | 0.3042 | 0.271 | 0.257 | 0.146 | 0.104 | 0.188 | 0.146 | 11/32 | 15/32 | 5/16 | 15/32 |
| 3/8 0.3750 | 16 | 0.3750 | 0.3643 | 0.312 | 0.293 | 0.219 | 0.156 | 0.281 | 0.219 | 1/2 | 11/16 | 15/32 | 5/8 |
| 3/8 0.3750 | 24 | 0.3750 | 0.3667 | 0.333 | 0.319 | 0.146 | 0.104 | 0.188 | 0.146 | 11/32 | 1/2 | 5/16 | 1/2 |
| 7/16 0.4375 | 14 | 0.4375 | 0.4258 | 0.366 | 0.344 | 0.250 | 0.179 | 0.321 | 0.250 | 19/32 | 3/4 | 9/16 | 23/32 |
| 7/16 0.4375 | 20 | 0.4375 | 0.4281 | 0.387 | 0.371 | 0.175 | 0.125 | 0.225 | 0.175 | 13/32 | 9/16 | 3/8 | 17/32 |
| 1/2 0.5000 | 13 | 0.5000 | 0.4876 | 0.423 | 0.399 | 0.269 | 0.192 | 0.346 | 0.269 | 5/8 | 25/32 | 19/32 | 3/4 |
| 1/2 0.5000 | 20 | 0.5000 | 0.4906 | 0.450 | 0.433 | 0.175 | 0.125 | 0.225 | 0.175 | 13/32 | 9/16 | 3/8 | 17/32 |

[a] Where specifying nominal size in decimals, zeros preceding decimal and in the fourth decimal place shall be omitted.

[b] The tabulated values apply to screw blanks before roll threading.

[c] Screws of these nominal lengths and shorter shall have point taper length specified above for short screws. Longer lengths shall have point taper length specified for long screws.

All dimensions are in inches. See Table 2 for thread diagrams.

“Type D” otherwise designated “Type 1.”

“Type T” otherwise designated “Type 23.”

## Table 10. Approximate Pierced or Extruded Hole Sizes for Types AB, B, and BP Steel Thread Forming Screws

| Screw Size | Metal Thickness | Pierced or Extruded Hole Size | Screw Size | Metal Thickness | Pierced or Extruded Hole Size | Screw Size | Metal Thickness | Pierced or Extruded Hole Size |
|---|---|---|---|---|---|---|---|---|
| In Steel, Stainless Steel, Monel Metal, and Brass Sheet Metal | | | | | | | | |
| 4 | 0.015 | 0.086 | 7 | 0.024 | 0.120 | 10 | 0.030 | 0.157 |
| | 0.018 | 0.086 | | 0.030 | 0.120 | | 0.036 | 0.157 |
| | 0.024 | 0.098 | | 0.036 | 0.120 | | 0.048 | 0.157 |
| | 0.030 | 0.098 | | 0.048 | 0.120 | 12 | 0.024 | 0.185 |
| | 0.036 | 0.098 | 8 | 0.018 | 0.136 | | 0.030 | 0.185 |
| 6 | 0.015 | 0.111 | | 0.024 | 0.136 | | 0.036 | 0.185 |
| | 0.018 | 0.111 | | 0.030 | 0.136 | | 0.048 | 0.185 |
| | 0.024 | 0.111 | | 0.036 | 0.136 | 1/4 | 0.030 | 0.209 |
| | 0.030 | 0.111 | | 0.048 | 0.136 | | 0.036 | 0.209 |
| | 0.036 | 0.111 | 10 | 0.018 | 0.157 | | 0.048 | 0.209 |
| 7 | 0.018 | 0.120 | | 0.024 | 0.157 | … | … | … |
| In Aluminum Alloy Sheet Metal | | | | | | | | |
| 4 | 0.024 | 0.086 | 6 | 0.048 | 0.111 | 8 | 0.036 | 0.136 |
| | 0.030 | 0.086 | 7 | 0.024 | 0.120 | | 0.048 | 0.136 |
| | 0.036 | 0.086 | | 0.030 | 0.120 | 10 | 0.024 | 0.157 |
| | 0.048 | 0.086 | | 0.036 | 0.120 | | 0.030 | 0.157 |
| 6 | 0.024 | 0.111 | | 0.048 | 0.120 | | 0.036 | 0.157 |
| | 0.030 | 0.111 | 8 | 0.024 | 0.136 | | 0.048 | 0.157 |
| | 0.036 | 0.111 | | 0.030 | 0.136 | … | … | … |

All dimensions are in inches except whole number screw and drill sizes.

Since conditions differ widely, it may be necessary to vary the hole size to suit a particular application.

## Table 11. Drilled Hole Sizes for Types AB, B, and BP Steel Thread Forming Screws

| Screw Size | Hole Size | Drill Size | Min. Mat'l Thickness | Penetration in Blind Holes | | Screw Size | Hole Size | Drill Size | Min. Mat'l Thickness | Penetration in Blind Holes | |
|---|---|---|---|---|---|---|---|---|---|---|---|
| | | | | Min. | Max. | | | | | Min. | Max. |
| In Plywood (Resin Impregnated) | | | | | | In Asbestos Compositions | | | | | |
| 2 | 0.073 | 49 | 0.125 | 0.188 | 0.500 | 2 | 0.076 | 48 | 0.125 | 0.188 | 0.500 |
| 4 | 0.100 | 39 | 0.188 | 0.250 | 0.625 | 4 | 0.101 | 38 | 0.188 | 0.250 | 0.625 |
| 6 | 0.125 | 1/8 | 0.188 | 0.250 | 0.625 | 6 | 0.120 | 31 | 0.188 | 0.250 | 0.625 |
| 7 | 0.136 | 29 | 0.188 | 0.250 | 0.750 | 7 | 0.136 | 29 | 0.250 | 0.312 | 0.750 |
| 8 | 0.144 | 27 | 0.188 | 0.250 | 0.750 | 8 | 0.147 | 26 | 0.312 | 0.375 | 0.750 |
| 10 | 0.173 | 17 | 0.250 | 0.312 | 1.000 | 10 | 0.166 | 19 | 0.312 | 0.375 | 1.000 |
| 12 | 0.194 | 10 | 0.312 | 0.375 | 1.000 | 12 | 0.196 | 9 | 0.312 | 0.375 | 1.000 |
| 1/4 | 0.228 | 1 | 0.312 | 0.375 | 1.000 | 1/4 | 0.228 | 1 | 0.438 | 0.500 | 1.000 |
| In Aluminum, Magnesium, Zinc, Brass, and Bronze Castings[a] | | | | | | In Phenol Formaldehyde Plastics[a] | | | | | |
| 2 | 0.078 | 47 | … | 0.125 | … | 2 | 0.078 | 47 | … | 0.188 | … |
| 4 | 0.104 | 37 | … | 0.188 | … | 4 | 0.100 | 39 | … | 0.250 | … |
| 6 | 0.128 | 30 | … | 0.250 | … | 6 | 0.128 | 30 | … | 0.250 | … |
| 7 | 0.144 | 27 | … | 0.250 | … | 7 | 0.136 | 29 | … | 0.250 | … |
| 8 | 0.152 | 24 | … | 0.250 | … | 8 | 0.150 | 25 | … | 0.312 | … |
| 10 | 0.177 | 16 | … | 0.250 | … | 10 | 0.177 | 16 | … | 0.312 | … |
| 12 | 0.199 | 8 | … | 0.281 | … | 12 | 0.199 | 8 | … | 0.375 | … |
| 1/4 | 0.234 | 15/64 | … | 0.312 | … | 1/4 | 0.234 | 15/64 | … | 0.375 | … |
| In Cellulose Acetate and Nitrate, and Acrylic and Styrene Resins[a] | | | | | | | | | | | |
| 2 | 0.078 | 47 | … | 0.188 | … | 8 | 0.144 | 27 | … | 0.312 | … |
| 4 | 0.094 | 42 | … | 0.250 | … | 10 | 0.170 | 18 | … | 0.312 | … |
| 6 | 0.120 | 31 | … | 0.250 | … | 12 | 0.191 | 11 | … | 0.375 | … |
| 7 | 0.128 | 30 | … | 0.250 | … | 1/4 | 0.221 | 2 | … | 0.375 | … |

[a] Data below apply to Types B and BP only.

All dimensions are in inches except whole number screw and drill sizes.

Since conditions differ widely, it may be necessary to vary the hole size to suit a particular application.

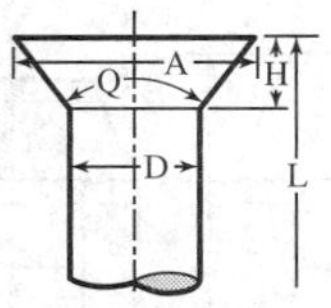

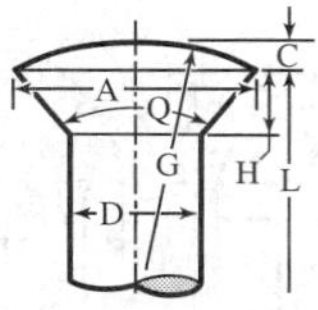

## Table 1b. American National Standard Large Rivet

*ANSI/ASME B18.1.2-1972 (R2021)*

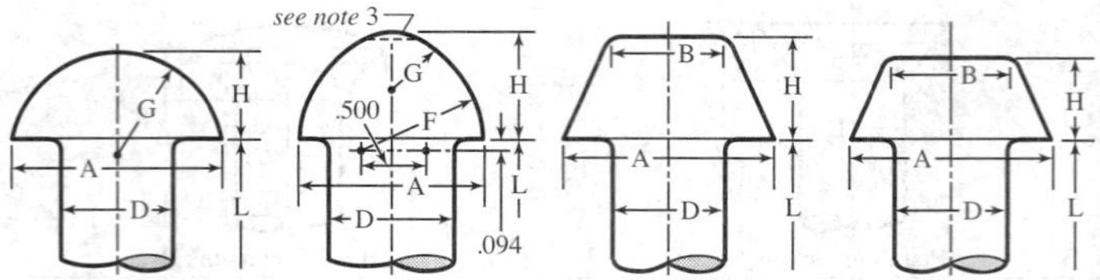


Button Head    High Head    Cone Head    Pan Head

| Nom. Body Dia. D[a] | Head Dia. A Mfd.[b] | Head Dia. A Driven[c] | Height H Mfd.[b] | Height H Driven[c] | Head Dia. A Mfd.[b] | Head Dia. A Driven[c] | Height H Mfd.[b,d] | Height H Driven[c,d] |
|---|---|---|---|---|---|---|---|---|
| | **Button Head** | | | | **High Button Head (Acorn)** | | | |
| ½ | 0.875 | 0.922 | 0.375 | 0.344 | 0.781 | 0.875 | 0.500 | 0.375 |
| ⅝ | 1.094 | 1.141 | 0.469 | 0.438 | 0.969 | 1.062 | 0.594 | 0.453 |
| ¾ | 1.312 | 1.375 | 0.562 | 0.516 | 1.156 | 1.250 | 0.688 | 0.531 |
| ⅞ | 1.531 | 1.594 | 0.656 | 0.609 | 1.344 | 1.438 | 0.781 | 0.609 |
| 1 | 1.750 | 1.828 | 0.750 | 0.688 | 1.531 | 1.625 | 0.875 | 0.688 |
| 1⅛ | 1.969 | 2.062 | 0.844 | 0.781 | 1.719 | 1.812 | 0.969 | 0.766 |
| 1¼ | 2.188 | 2.281 | 0.938 | 0.859 | 1.906 | 2.000 | 1.062 | 0.844 |
| 1⅜ | 2.406 | 2.516 | 1.031 | 0.953 | 2.094 | 2.188 | 1.156 | 0.938 |
| 1½ | 2.625 | 2.734 | 1.125 | 1.031 | 2.281 | 2.375 | 1.250 | 1.000 |
| 1⅝ | 2.844 | 2.969 | 1.219 | 1.125 | 2.469 | 2.562 | 1.344 | 1.094 |
| 1¾ | 3.062 | 3.203 | 1.312 | 1.203 | 2.656 | 2.750 | 1.438 | 1.172 |
| | **Cone Head** | | | | **Pan Head** | | | |
| ½ | 0.875 | 0.922 | 0.438 | 0.406 | 0.800 | 0.844 | 0.350 | 0.328 |
| ⅝ | 1.094 | 1.141 | 0.547 | 0.516 | 1.000 | 1.047 | 0.438 | 0.406 |
| ¾ | 1.312 | 1.375 | 0.656 | 0.625 | 1.200 | 1.266 | 0.525 | 0.484 |
| ⅞ | 1.531 | 1.594 | 0.766 | 0.719 | 1.400 | 1.469 | 0.612 | 0.578 |
| 1 | 1.750 | 1.828 | 0.875 | 0.828 | 1.600 | 1.687 | 0.700 | 0.656 |
| 1⅛ | 1.969 | 2.063 | 0.984 | 0.938 | 1.800 | 1.891 | 0.788 | 0.734 |
| 1¼ | 2.188 | 2.281 | 1.094 | 1.031 | 2.000 | 2.094 | 0.875 | 0.812 |
| 1⅜ | 2.406 | 2.516 | 1.203 | 1.141 | 2.200 | 2.312 | 0.962 | 0.906 |
| 1½ | 2.625 | 2.734 | 1.312 | 1.250 | 2.400 | 2.516 | 1.050 | 0.984 |
| 1⅝ | 2.844 | 2.969 | 1.422 | 1.344 | 2.600 | 2.734 | 1.138 | 1.062 |
| 1¾ | 3.062 | 3.203 | 1.531 | 1.453 | 2.800 | 2.938 | 1.225 | 1.141 |

[a] Tolerance for diameter of body is plus and minus from nominal and for ½-in. size equals +0.020, −0.022; for sizes ⅝ to 1-in., incl., equals +0.030, −.025; for sizes 1⅛ and 1¼-in. equals +0.035; −0.027; for sizes 1⅜ and 1½-in. equals +0.040, −0.030; for sizes 1⅝ and 1¾-in. equals +0.040, −0.037.

[b] Note 1. Basic dimensions of head as manufactured. All dimensions are given in inches. The following formulas give the basic dimensions for manufactured shapes: *Button Head,* $A = 1.750D$; $H = 0.750D$; $G = 0.885D$. *High Button Head,* $A = 1.500D + 0.031$; $H = 0.750D + 0.125$; $F = 0.750D + 0.281$; $G = 0.750D - 0.281$. *Cone Head,* $A = 1.750D$; $B = 0.938D$; $H = 0.875D$. *Pan Head,* $A = 1.600D$; $B = 1.000D$; $H = 0.700D$. Length L is measured parallel to the rivet axis, from the extreme end to the bearing surface plane for flat bearing surface head type rivets, or to the intersection of the head top surface with the head diameter for countersunk head-type rivets.

[c] Note 2. Dimensions of manufactured head after driving and also of driven head.

[d] Note 3. Slight flat permissible within the specified head-height tolerance.

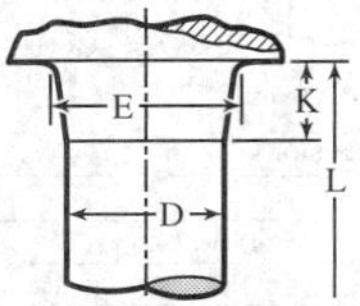

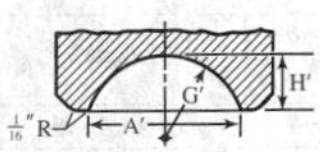

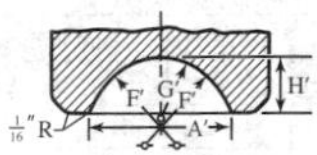

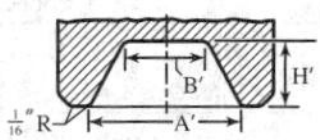

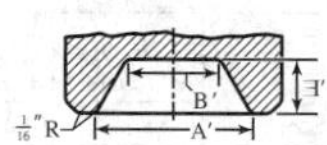

## Table 2a. American National Standard Small Solid Rivets

*ANSI/ASME B18.1.1-1972 (R2021) and Appendix*

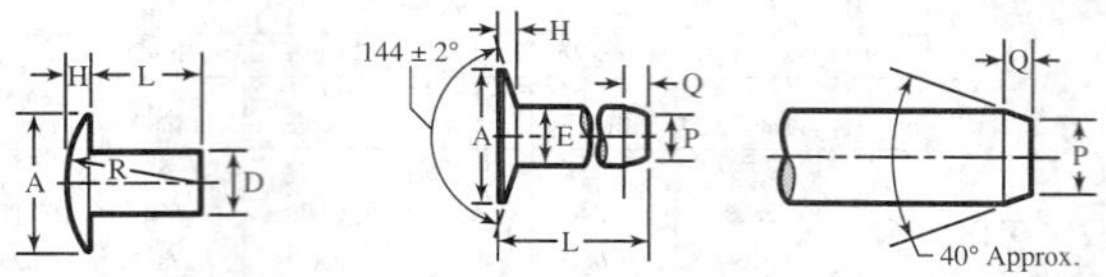


$P = D \times 0.818$

$Q = D \times 0.25$

Truss Head Rivets — Coopers Rivets — Point Dimensions

**Truss Head Rivets**[a]

| Shank Dia.,[b] $D$ Nominal | | Head Dia., $A$ Max. | Head Dia., $A$ Min. | Head Height, $H$ Max. | Head Height, $H$ Min. | Rad. $R$ Approx. | Shank Dia.,[b] $D$ Nominal | | Head Dia., $A$ Max. | Head Dia., $A$ Min. | Head Height, $H$ Max. | Head Height, $H$ Min. | Rad. $R$ Approx. |
|---|---|---|---|---|---|---|---|---|---|---|---|---|---|
| 3⁄32 | 0.094 | 0.226 | 0.206 | 0.038 | 0.026 | 0.239 | 9⁄32 | 0.281 | 0.661 | 0.631 | 0.103 | 0.085 | 0.706 |
| 1⁄8 | 0.125 | 0.297 | 0.277 | 0.048 | 0.036 | 0.314 | 5⁄16 | 0.312 | 0.732 | 0.702 | 0.113 | 0.095 | 0.784 |
| 5⁄32 | 0.156 | 0.368 | 0.348 | 0.059 | 0.045 | 0.392 | 11⁄32 | 0.344 | 0.806 | 0.776 | 0.124 | 0.104 | 0.862 |
| 3⁄16 | 0.188 | 0.442 | 0.422 | 0.069 | 0.055 | 0.470 | 3⁄8 | 0.375 | 0.878 | 0.848 | 0.135 | 0.115 | 0.942 |
| 7⁄32 | 0.219 | 0.515 | 0.495 | 0.080 | 0.066 | 0.555 | 13⁄32 | 0.406 | 0.949 | 0.919 | 0.145 | 0.123 | 1.028 |
| 1⁄4 | 0.250 | 0.590 | 0.560 | 0.091 | 0.075 | 0.628 | 7⁄16 | 0.438 | 1.020 | 0.990 | 0.157 | 0.135 | 1.098 |

[a] All dimensions in inches except where otherwise noted. Length tolerance of rivets is + or – .016 inch. Approximate proportions of rivets: $A = 2.300 \times D$, $H = 0.330 \times D$, $R = 2.512 \times D$.

[b] Tolerances on the nominal shank diameter in inches are given for the following body diameter ranges: 3⁄32. to 5⁄32, plus 0.002, minus 0.004; 3⁄16 to 1⁄4, plus 0.003, minus 0.006; 9⁄32 to 11⁄32, plus 0.004, minus 0.008; and 3⁄8 to 7⁄16, plus 0.005, minus 0.010.

**Coopers Rivets**

| Size No.[a] | Shank Diameter, $E$ Max. | Shank Diameter, $E$ Min. | Head Diameter, $A$ Max. | Head Diameter, $A$ Min. | Head Height, $H$ Max. | Head Height, $H$ Min. | Point Dimensions[b] Dia., $P$ Nom. | Point Dimensions[b] Length, $Q$ Nom. | Length, $L$ Max. | Length, $L$ Min. |
|---|---|---|---|---|---|---|---|---|---|---|
| 1 lb | 0.111 | 0.105 | 0.291 | 0.271 | 0.045 | 0.031 | Not Pointed | | 0.249 | 0.219 |
| 1¼ lb | 0.122 | 0.116 | 0.324 | 0.302 | 0.050 | 0.036 | Not Pointed | | 0.285 | 0.255 |
| 1½ lb | 0.132 | 0.126 | 0.324 | 0.302 | 0.050 | 0.036 | Not Pointed | | 0.285 | 0.255 |
| 1¾ lb | 0.136 | 0.130 | 0.324 | 0.302 | 0.052 | 0.034 | Not Pointed | | 0.318 | 0.284 |
| 2 lb | 0.142 | 0.136 | 0.355 | 0.333 | 0.056 | 0.038 | Not Pointed | | 0.322 | 0.288 |
| 3 lb | 0.158 | 0.152 | 0.386 | 0.364 | 0.058 | 0.040 | 0.123 | 0.062 | 0.387 | 0.353 |
| 4 lb | 0.168 | 0.159 | 0.388 | 0.362 | 0.058 | 0.040 | 0.130 | 0.062 | 0.418 | 0.388 |
| 5 lb | 0.183 | 0.174 | 0.419 | 0.393 | 0.063 | 0.045 | 0.144 | 0.062 | 0.454 | 0.420 |
| 6 lb | 0.206 | 0.197 | 0.482 | 0.456 | 0.073 | 0.051 | 0.160 | 0.094 | 0.498 | 0.457 |
| 7 lb | 0.223 | 0.214 | 0.513 | 0.487 | 0.076 | 0.054 | 0.175 | 0.094 | 0.561 | 0.523 |
| 8 lb | 0.241 | 0.232 | 0.546 | 0.516 | 0.081 | 0.059 | 0.182 | 0.094 | 0.597 | 0.559 |
| 9 lb | 0.248 | 0.239 | 0.578 | 0.548 | 0.085 | 0.063 | 0.197 | 0.094 | 0.601 | 0.563 |
| 10 lb | 0.253 | 0.244 | 0.578 | 0.548 | 0.085 | 0.063 | 0.197 | 0.094 | 0.632 | 0.594 |
| 12 lb | 0.263 | 0.251 | 0.580 | 0.546 | 0.086 | 0.060 | 0.214 | 0.094 | 0.633 | 0.575 |
| 14 lb | 0.275 | 0.263 | 0.611 | 0.577 | 0.091 | 0.065 | 0.223 | 0.094 | 0.670 | 0.612 |
| 16 lb | 0.285 | 0.273 | 0.611 | 0.577 | 0.089 | 0.063 | 0.223 | 0.094 | 0.699 | 0.641 |
| 18 lb | 0.285 | 0.273 | 0.642 | 0.608 | 0.108 | 0.082 | 0.230 | 0.125 | 0.749 | 0.691 |
| 20 lb | 0.316 | 0.304 | 0.705 | 0.671 | 0.128 | 0.102 | 0.250 | 0.125 | 0.769 | 0.711 |
| 3⁄8 in. | 0.380 | 0.365 | 0.800 | 0.762 | 0.136 | 0.106 | 0.312 | 0.125 | 0.840 | 0.778 |

[a] All dimensions in inches except where otherwise noted. Size numbers in pounds refer to the approximate weight of 1000 rivets.

[b] When specified American National Standard Small Solid Rivets may be obtained with points. Point dimensions for belt and coopers rivets are given in the accompanying tables. Formulas for calculating point dimensions of other rivets are given alongside the right diagram in Table 2a.

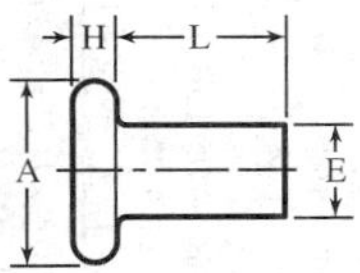


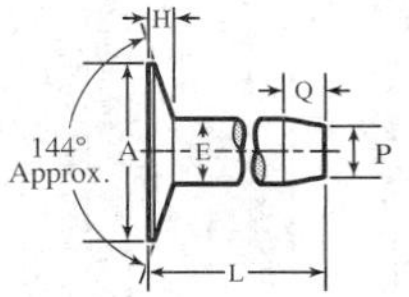

## Table 3b. British Standard Rivets for General Engineering Purposes
*BS 4620:1970 (Withdrawn)*

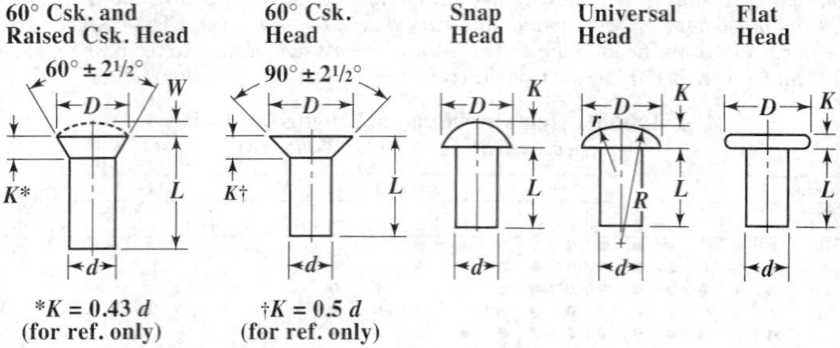


Hot Forged Rivets

| Nom. Shank Dia.[a] *d* | Tol. on Dia. *d* | 60° Csk. and Raised Csk. Head: Nom. Dia. *D* | 60° Csk. and Raised Csk. Head: Height of Raise *W* | Snap Head: Nom. Dia. *D* | Snap Head: Nom. Depth *K* | Universal Head: Nom. Dia. *D* | Universal Head: Nom. Depth *K* | Universal Head: Rad. *R* | Universal Head: Rad. *r* |
|---|---|---|---|---|---|---|---|---|---|
| (14) | | 21 | 2.8 | 22 | 9 | 28 | 5.6 | 42 | 8.4 |
| 16 | ±0.43 | 24 | 3.2 | 25 | 10 | 32 | 6.4 | 48 | 9.6 |
| (18) | | 27 | 3.6 | 28 | 11.5 | 36 | 7.2 | 54 | 11 |
| 20 | | 30 | 4.0 | 32 | 13 | 40 | 8.0 | 60 | 12 |
| (22) | ±0.52 | 33 | 4.4 | 36 | 14 | 44 | 8.8 | 66 | 13 |
| 24 | | 36 | 4.8 | 40 | 16 | 48 | 9.6 | 72 | 14 |
| (27) | | 40 | 5.4 | 43 | 17 | 54 | 10.8 | 81 | 16 |
| 30 | | 45 | 6.0 | 48 | 19 | 60 | 12.0 | 90 | 18 |
| (33) | ±0.62 | 50 | 6.6 | 53 | 21 | 66 | 13.2 | 99 | 20 |
| 36 | | 55 | 7.2 | 58 | 23 | 72 | 14.4 | 108 | 22 |
| (39) | | 59 | 7.8 | 62 | 25 | 78 | 15.6 | 117 | 23 |

Cold Forged Rivets

| Nom. Shank Dia.[a] *d* | Tol. on Dia. *d* | 90° Csk. Head: Nom. Dia. *D* | Snap Head: Nom. Dia. *D* | Snap Head: Nom. Dia. *K* | Universal Head: Nom. Dia. *D* | Universal Head: Nom. Depth *K* | Universal Head: Rad. *R* | Universal Head: Rad. *r* | Flat Head: Nom. Dia. *D* | Flat Head: Nom. Depth *K* |
|---|---|---|---|---|---|---|---|---|---|---|
| 1 | | 2 | 1.8 | 0.6 | 2 | 0.4 | 3.0 | 0.6 | 2 | 0.25 |
| 1.2 | | 2.4 | 2.1 | 0.7 | 2.4 | 0.5 | 3.6 | 0.7 | 2.4 | 0.3 |
| 1.6 | ±0.07 | 3.2 | 2.8 | 1.0 | 3.2 | 0.6 | 4.8 | 1.0 | 3.2 | 0.4 |
| 2 | | 4 | 3.5 | 1.2 | 4 | 0.8 | 6.0 | 1.2 | 4 | 0.5 |
| 2.5 | | 5 | 4.4 | 1.5 | 5 | 1.0 | 7.5 | 1.5 | 5 | 0.6 |
| 3 | | 6 | 5.3 | 1.8 | 6 | 1.2 | 9.0 | 1.8 | 6 | 0.8 |
| (3.5) | | 7 | 6.1 | 2.1 | 7 | 1.4 | 10.5 | 2.1 | 7 | 0.9 |
| 4 | ±0.09 | 8 | 7 | 2.4 | 8 | 1.6 | 12 | 2.4 | 8 | 1.0 |
| 5 | | 10 | 8.8 | 3.0 | 10 | 2.0 | 15 | 3.0 | 10 | 1.3 |
| 6 | | 12 | 10.5 | 3.6 | 12 | 2.4 | 18 | 3.6 | 12 | 1.5 |
| (7) | | 14 | 12.3 | 4.2 | 14 | 2.8 | 21 | 4.2 | 14 | 1.8 |
| 8 | ±0.11 | 16 | 14 | 4.8 | 16 | 3.2 | 24 | 4.8 | 16 | 2 |
| 10 | | 20 | 18 | 6.0 | 20 | 4.0 | 30 | 6 | 20 | 2.5 |
| 12 | | 24 | 21 | 7.2 | 24 | 4.8 | 36 | 7.2 | … | … |
| (14) | ±0.14 | … | 25 | 8.4 | 28 | 5.6 | 42 | 8.4 | … | … |
| 16 | | … | 28 | 9.6 | 32 | 6.4 | 48 | 9.6 | … | … |

[a] All dimensions are in millimeters. Sizes shown in parentheses are nonpreferred.

## Table 4. British Standard Small Rivets for General Purposes

*BS 641:1951 (Withdrawn)*

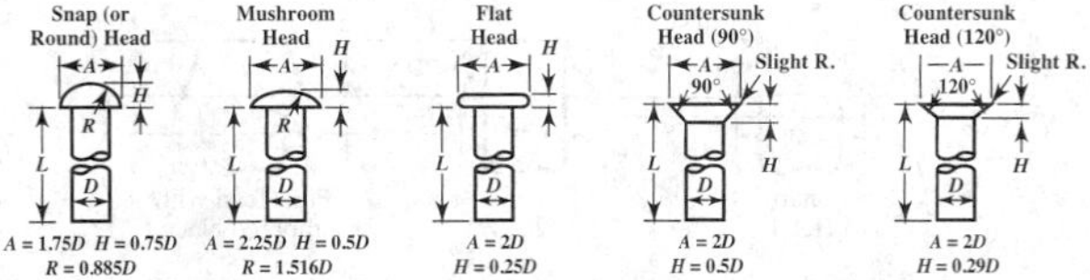


| Nom. Dia.[a] D | Snap (or Round Head) | | | Mushroom Head | | | Flat Head | | Countersunk Head (90°) | | Countersunk Head (120°) | |
|---|---|---|---|---|---|---|---|---|---|---|---|---|
| | Head Dimensions | | | | | | | | | | | |
| | Dia. A | Ht. H | Rad. R | Dia. A | Ht. H | Rad. R | Dia. A | Ht. H | Dia. A | Ht. H | Dia. A | Ht. H |
| 1⁄16 | 0.109 | 0.047 | 0.055 | 0.141 | 0.031 | 0.095 | 0.125 | 0.016 | 0.125 | 0.031 | … | … |
| 3⁄32 | 0.164 | 0.070 | 0.083 | 0.211 | 0.047 | 0.142 | 0.188 | 0.023 | 0.188 | 0.047 | … | … |
| 1⁄8 | 0.219 | 0.094 | 0.111 | 0.281 | 0.063 | 0.189 | 0.250 | 0.031 | 0.250 | 0.063 | 0.250 | 0.0[illegible]6 |
| 5⁄32 | 0.273 | 0.117 | 0.138 | 0.352 | 0.078 | 0.237 | 0.313 | 0.039 | 0.313 | 0.078 | … | … |
| 3⁄16 | 0.328 | 0.141 | 0.166 | 0.422 | 0.094 | 0.284 | 0.375 | 0.047 | 0.375 | 0.094 | 0.375 | 0.0[illegible]4 |
| 1⁄4 | 0.438 | 0.188 | 0.221 | 0.563 | 0.125 | 0.379 | 0.500 | 0.063 | 0.500 | 0.125 | 0.500 | 0.0[illegible]3 |
| 5⁄16 | 0.547 | 0.234 | 0.277 | 0.703 | 0.156 | 0.474 | 0.625 | 0.078 | 0.625 | 0.156 | 0.625 | 0.0[illegible]1 |
| 3⁄8 | 0.656 | 0.281 | 0.332 | 0.844 | 0.188 | 0.568 | 0.750 | 0.094 | 0.750 | 0.188 | 0.750 | 0.1[illegible]9 |
| 7⁄16 | 0.766 | 0.328 | 0.387 | 0.984 | 0.219 | 0.663 | 0.875 | 0.109 | 0.875 | 0.219 | … | … |

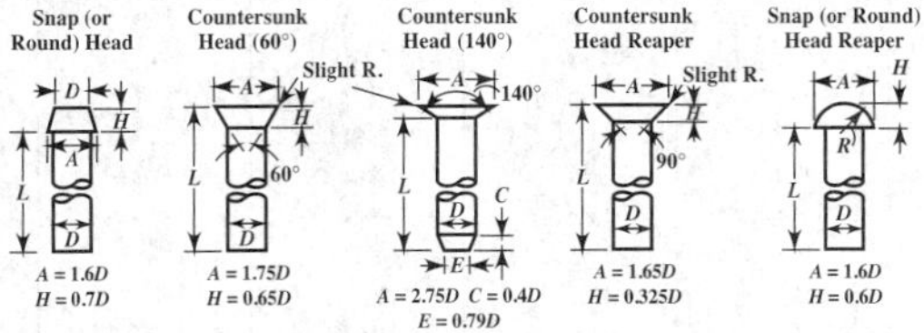


| Nominal Diameter D | | Pan Head | | Countersunk Head (60°) | | Countersunk Head (140°) | | | Countersunk Head Reaper | | Snap (or Round) Head Reaper | |
|---|---|---|---|---|---|---|---|---|---|---|---|---|
| | | Head Dimensions | | | | | | | | | | |
| Inch | Gage No.[b] | Dia. A | Ht. H | Dia. A | Ht. H | Dia. A | Ht. C | Dia. E | Dia. A | Ht. H | Dia. A | Ht. H |
| 0.104 | 12 | … | … | … | … | 0.286 | 0.042 | 0.082 | … | … | … | … |
| 0.116 | 11 | … | … | … | … | 0.319 | 0.046 | 0.092 | … | … | … | … |
| 0.128 | 10 | … | … | … | … | 0.352 | 0.051 | 0.101 | … | … | … | … |
| 0.144 | 9 | … | … | … | … | 0.396 | 0.058 | 0.114 | … | … | … | … |
| 0.160 | 8 | … | … | … | … | 0.440 | 0.064 | 0.126 | … | … | … | … |
| 0.176 | 7 | … | … | … | … | 0.484 | 0.070 | 0.139 | … | … | … | … |
| 3⁄16 | … | 0.300 | 0.131 | 0.328 | 0.122 | … | … | … | … | … | … | … |
| 0.192 | 6 | … | … | … | … | 0.528 | 0.077 | 0.152 | 0.317 | 0.062 | 0.307 | 0.115 |
| 0.202 | … | … | … | … | … | … | … | … | 0.333 | 0.066 | 0.323 | 0.121 |
| 0.212 | 5 | … | … | … | … | 0.583 | 0.085 | 0.167 | 0.350 | 0.069 | 0.339 | 0.127 |
| 0.232 | 4 | … | … | … | … | 0.638 | 0.093 | 0.183 | 0.383 | 0.075 | 0.371 | 0.139 |
| 1⁄4 | … | 0.400 | 0.175 | 0.438 | 0.162 | 0.688 | 0.100 | 0.198 | … | … | … | … |
| 0.252 | 3 | … | … | … | … | … | … | … | 0.416 | 0.082 | 0.403 | 0.151 |
| 5⁄16 | … | 0.500 | 0.219 | 0.547 | 0.203 | 0.859 | 0.125 | 0.247 | … | … | … | … |
| 3⁄8 | … | 0.600 | 0.263 | 0.656 | 0.244 | 10.031 | 0.150 | 0.296 | … | … | … | … |
| 7⁄16 | … | 0.700 | 0.306 | 0.766 | 0.284 | … | … | … | … | … | … | … |

[a] All dimensions in inches unless specified otherwise.

[b] Gage numbers are British Standard Wire Gage (S.W.G.) numbers.

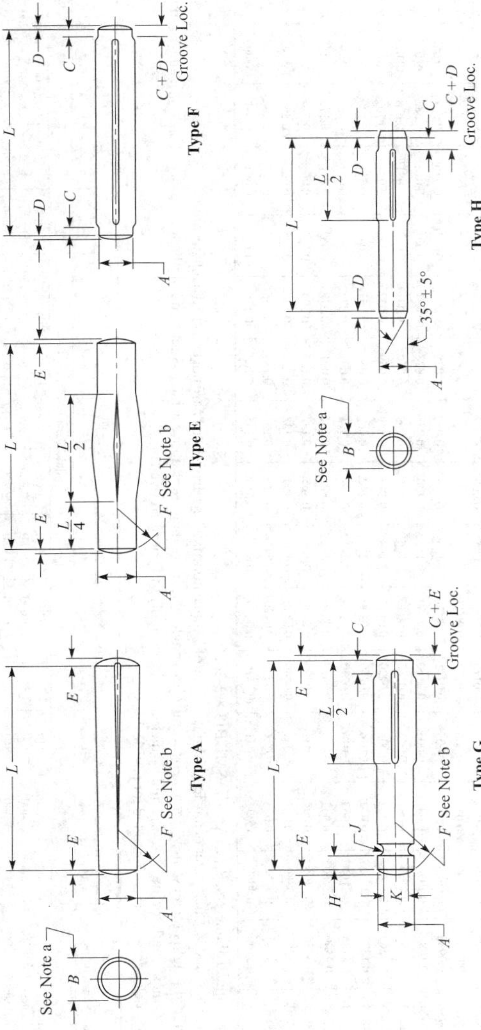

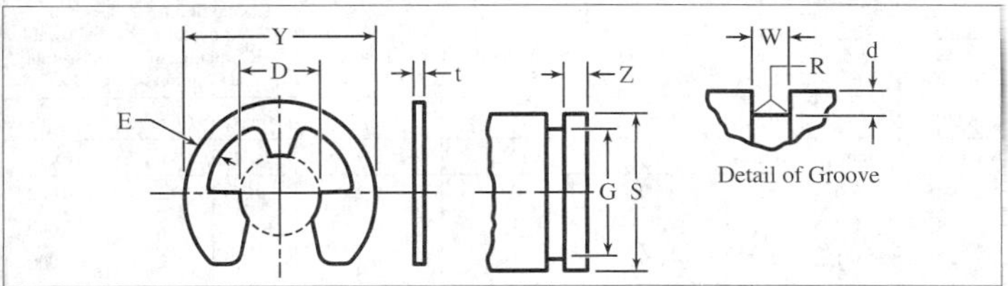

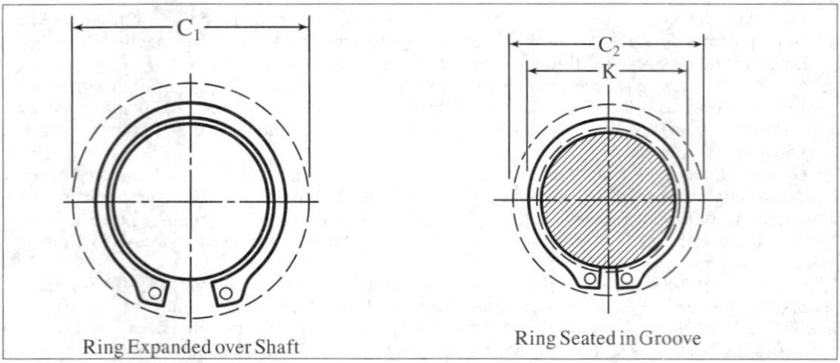

**Table 5. American National Standard Metric Basic Internal Series—3BMI Retaining Rings—Checking and Performance Data** *ANSI/ASME B27.7-1977 (R2017)*

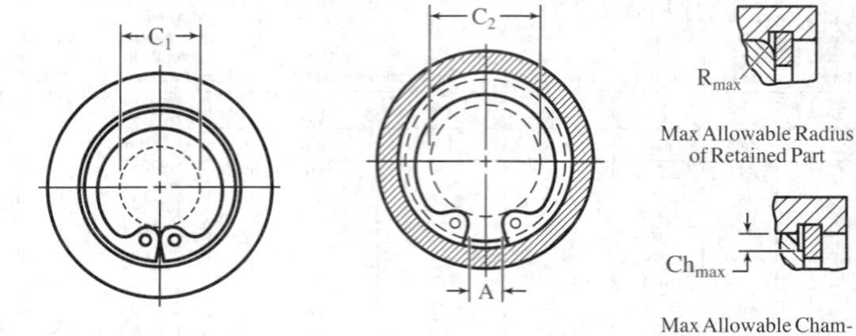


Ring Compressed in Bore — Ring Seated in Groove

| Ring Series and Size No. | Clearance Dia. Ring in Bore | Clearance Dia. Ring in Groove | Gaging Diameter[a] | Allowable Thrust Loads Sharp Corner Abutment | | Maximum Allowable Corner Radii and Chamfers | |
|---|---|---|---|---|---|---|---|
| 3BMI | $C_1$ | $C_2$ | *A* min. | $P_r$[b] | $P_g$[c] | *R* max. | *Ch* max. |
| No. | mm | mm | mm | kN | kN | mm | mm |
| −8 | 4.4 | 4.8 | 1.40 | 2.4 | 1.0 | 0.4 | 0.3 |
| −9 | 4.6 | 5.0 | 1.50 | 4.4 | 1.2 | 0.5 | 0.35 |
| −10 | 5.5 | 6.0 | 1.85 | 4.9 | 1.5 | 0.5 | 0.35 |
| −11 | 5.7 | 6.3 | 1.95 | 5.4 | 2.0 | 0.6 | 0.4 |
| −12 | 6.7 | 7.3 | 2.25 | 5.8 | 2.4 | 0.6 | 0.4 |
| −13 | 6.8 | 7.5 | 2.35 | 8.9 | 2.6 | 0.7 | 0.5 |
| −14 | 6.9 | 7.7 | 2.65 | 9.7 | 3.2 | 0.7 | 0.5 |
| −15 | 7.9 | 8.7 | 2.80 | 10.4 | 3.7 | 0.7 | 0.5 |
| −16 | 8.8 | 9.7 | 2.80 | 11.0 | 4.2 | 0.7 | 0.5 |
| −17 | 9.8 | 10.8 | 3.35 | 11.7 | 4.9 | 0.75 | 0.6 |
| −18 | 10.3 | 11.3 | 3.40 | 12.3 | 5.5 | 0.75 | 0.6 |
| −19 | 11.4 | 12.5 | 3.40 | 13.1 | 6.0 | 0.8 | 0.65 |
| −20 | 11.6 | 12.7 | 3.8 | 13.7 | 6.6 | 0.9 | 0.7 |
| −21 | 12.6 | 13.8 | 4.2 | 14.5 | 7.3 | 0.9 | 0.7 |
| −22 | 13.5 | 14.8 | 4.3 | 22.5 | 8.3 | 0.9 | 0.7 |
| −23 | 14.5 | 15.9 | 4.9 | 23.5 | 8.9 | 1.0 | 0.8 |
| −24 | 15.5 | 16.9 | 5.2 | 24.8 | 9.7 | 1.0 | 0.8 |
| −25 | 16.5 | 18.1 | 6.0 | 25.7 | 11.6 | 1.0 | 0.8 |
| −26 | 17.5 | 19.2 | 5.7 | 26.8 | 12.7 | 1.2 | 1.0 |
| −27 | 17.4 | 19.2 | 5.9 | 33 | 14.0 | 1.2 | 1.0 |
| −28 | 18.2 | 20.0 | 6.0 | 34 | 14.6 | 1.2 | 1.0 |
| −30 | 20.0 | 21.9 | 6.0 | 37 | 16.5 | 1.2 | 1.0 |
| −32 | 22.0 | 23.9 | 7.3 | 39 | 17.6 | 1.2 | 1.0 |
| −34 | 24.0 | 26.1 | 7.6 | 42 | 20.6 | 1.2 | 1.0 |
| −35 | 25.0 | 27.2 | 8.0 | 43 | 22.3 | 1.2 | 1.0 |
| −36 | 26.0 | 28.3 | 8.3 | 44 | 23.9 | 1.2 | 1.0 |
| −37 | 27.0 | 29.3 | 8.4 | 45 | 24.6 | 1.2 | 1.0 |
| −38 | 28.0 | 30.4 | 8.6 | 46 | 26.4 | 1.2 | 1.0 |
| −40 | 29.2 | 31.6 | 9.7 | 62 | 27.7 | 1.7 | 1.3 |
| −42 | 29.7 | 32.2 | 9.0 | 65 | 30.2 | 1.7 | 1.3 |
| −45 | 32.3 | 34.9 | 9.6 | 69 | 33.8 | 1.7 | 1.3 |
| −46 | 33.3 | 36.0 | 9.7 | 71 | 36 | 1.7 | 1.3 |
| −47 | 34.3 | 37.1 | 10.0 | 72 | 38 | 1.7 | 1.3 |
| −48 | 35.0 | 37.9 | 10.5 | 74 | 40 | 1.7 | 1.3 |
| −50 | 36.9 | 40.0 | 12.1 | 77 | 45 | 1.7 | 1.3 |
| −52 | 38.6 | 41.9 | 11.7 | 99 | 50 | 2.0 | 1.6 |
| −55 | 40.8 | 44.2 | 11.9 | 105 | 54 | 2.0 | 1.6 |
| −57 | 42.2 | 45.7 | 12.5 | 109 | 58 | 2.0 | 1.6 |

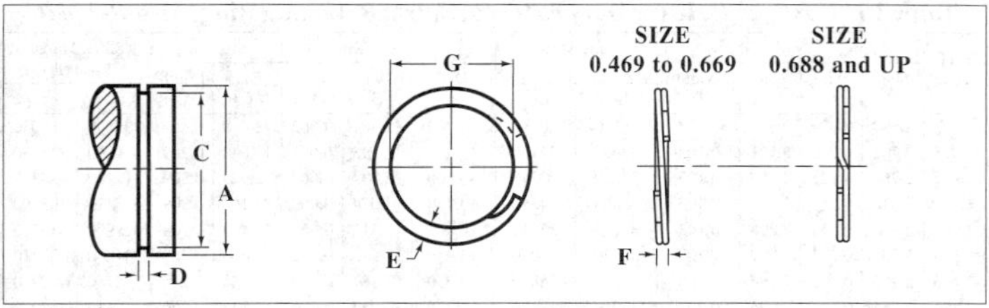


## Table 10. Heavy Duty External Spiral Retaining Rings *MIL-R-27426*

| Shaft Dia. A | Ring Dia. G | Ring Wall E | Groove Dia. C | Groove Width D | Static Thrust Load (lb) Ring | Static Thrust Load (lb) Groove | Shaft Dia. A | Ring Dia. G | Ring Wall E | Groove Dia. C | Groove Width D | Static Thrust Load (lb) Ring | Static Thrust Load (lb) Groove |
|---|---|---|---|---|---|---|---|---|---|---|---|---|---|
| 0.469 | 0.439 | 0.045 | 0.443 | 0.029 | 1880 | 510 | 3.500 | 3.293 | 0.270 | 3.316 | 0.120 | 48820 | 32250 |
| 0.500 | 0.464 | 0.050 | 0.468 | 0.039 | 2530 | 440 | 3.543 | 3.333 | 0.270 | 3.357 | 0.120 | 49420 | 33000 |
| 0.551 | 0.514 | 0.050 | 0.519 | 0.039 | 2790 | 540 | 3.625 | 3.411 | 0.270 | 3.435 | 0.120 | 50560 | 34490 |
| 0.562 | 0.525 | 0.050 | 0.530 | 0.039 | 2840 | 560 | 3.687 | 3.469 | 0.270 | 3.493 | 0.120 | 51430 | 35820 |
| 0.594 | 0.554 | 0.050 | 0.559 | 0.039 | 3000 | 700 | 3.750 | 3.527 | 0.270 | 3.552 | 0.120 | 52310 | 37180 |
| 0.625 | 0.583 | 0.055 | 0.588 | 0.039 | 3160 | 820 | 3.875 | 3.647 | 0.270 | 3.673 | 0.120 | 54050 | 39190 |
| 0.669 | 0.623 | 0.055 | 0.629 | 0.039 | 3380 | 1070 | 3.938 | 3.708 | 0.270 | 3.734 | 0.120 | 54930 | 40230 |
| 0.688 | 0.641 | 0.065 | 0.646 | 0.046 | 4170 | 960 | 4.000 | 3.765 | 0.270 | 3.792 | 0.120 | 55790 | 41660 |
| 0.750 | 0.698 | 0.065 | 0.704 | 0.046 | 4550 | 1250 | 4.250 | 4.037 | 0.270 | 4.065 | 0.120 | 59280 | 39370 |
| 0.781 | 0.727 | 0.065 | 0.733 | 0.046 | 4740 | 1430 | 4.375 | 4.161 | 0.270 | 4.190 | 0.120 | 61020 | 40530 |
| 0.812 | 0.756 | 0.065 | 0.762 | 0.046 | 4930 | 1620 | 4.500 | 4.280 | 0.270 | 4.310 | 0.120 | 62770 | 42810 |
| 0.875 | 0.814 | 0.075 | 0.821 | 0.046 | 5310 | 2000 | 4.750 | 4.518 | 0.270 | 4.550 | 0.120 | 66260 | 47570 |
| 0.938 | 0.875 | 0.075 | 0.882 | 0.046 | 5690 | 2440 | 5.000 | 4.756 | 0.270 | 4.790 | 0.120 | 69740 | 52580 |
| 0.984 | 0.919 | 0.085 | 0.926 | 0.046 | 5970 | 2790 | 5.250 | 4.995 | 0.350 | 5.030 | 0.139 | 83790 | 57830 |
| 1.000 | 0.932 | 0.085 | 0.940 | 0.046 | 6070 | 2950 | 5.500 | 5.228 | 0.350 | 5.265 | 0.139 | 87780 | 64720 |
| 1.023 | 0.953 | 0.085 | 0.961 | 0.046 | 6210 | 3170 | 5.750 | 5.466 | 0.350 | 5.505 | 0.139 | 91770 | 70540 |
| 1.062 | 0.986 | 0.103 | 0.998 | 0.056 | 7010 | 2810 | 6.000 | 5.705 | 0.350 | 5.745 | 0.139 | 95760 | 76610 |
| 1.125 | 1.047 | 0.103 | 1.059 | 0.056 | 7420 | 2890 | 6.250 | 5.938 | 0.418 | 5.985 | 0.174 | 122520 | 82930 |
| 1.188 | 1.105 | 0.103 | 1.118 | 0.056 | 7840 | 3450 | 6.500 | 6.181 | 0.418 | 6.225 | 0.174 | 127420 | 89510 |
| 1.250 | 1.163 | 0.103 | 1.176 | 0.056 | 8250 | 4110 | 6.750 | 6.410 | 0.418 | 6.465 | 0.174 | 132320 | 96330 |
| 1.312 | 1.218 | 0.118 | 1.232 | 0.056 | 8650 | 4810 | 7.000 | 6.648 | 0.418 | 6.705 | 0.174 | 137220 | 103400 |
| 1.375 | 1.277 | 0.118 | 1.291 | 0.056 | 9070 | 5650 | 7.250 | 6.891 | 0.418 | 6.942 | 0.174 | 142130 | 111810 |
| 1.438 | 1.336 | 0.118 | 1.350 | 0.056 | 9490 | 6340 | 7.500 | 7.130 | 0.437 | 7.180 | 0.209 | 176240 | 120170 |
| 1.500 | 1.385 | 0.118 | 1.406 | 0.056 | 9900 | 7060 | 7.750 | 7.368 | 0.437 | 7.420 | 0.209 | 182120 | 128060 |
| 1.562 | 1.453 | 0.128 | 1.468 | 0.068 | 12780 | 6600 | 8.000 | 7.606 | 0.437 | 7.660 | 0.209 | 187990 | 136200 |
| 1.625 | 1.513 | 0.128 | 1.529 | 0.068 | 13290 | 7330 | 8.250 | 7.845 | 0.437 | 7.900 | 0.209 | 193870 | 144590 |
| 1.687 | 1.573 | 0.128 | 1.589 | 0.068 | 13800 | 8190 | 8.500 | 8.083 | 0.437 | 8.140 | 0.209 | 199740 | 153220 |
| 1.750 | 1.633 | 0.128 | 1.650 | 0.068 | 14320 | 8760 | 8.750 | 8.324 | 0.437 | 8.383 | 0.209 | 205620 | 160800 |
| 1.771 | 1.651 | 0.128 | 1.669 | 0.068 | 14490 | 9040 | 9.000 | 8.560 | 0.500 | 8.620 | 0.209 | 211490 | 171250 |
| 1.812 | 1.690 | 0.128 | 1.708 | 0.068 | 14820 | 9440 | 9.250 | 8.798 | 0.500 | 8.860 | 0.209 | 217370 | 180640 |
| 1.875 | 1.751 | 0.158 | 1.769 | 0.068 | 15340 | 9950 | 9.500 | 9.036 | 0.500 | 9.100 | 0.209 | 223240 | 190280 |
| 1.969 | 1.838 | 0.158 | 1.857 | 0.068 | 16110 | 11040 | 9.750 | 9.275 | 0.500 | 9.338 | 0.209 | 229120 | 201140 |
| 2.000 | 1.867 | 0.158 | 1.886 | 0.068 | 16360 | 11420 | 10.000 | 9.508 | 0.500 | 9.575 | 0.209 | 234990 | 212810 |
| 2.062 | 1.932 | 0.168 | 1.946 | 0.086 | 21220 | 11820 | 10.250 | 9.745 | 0.500 | 9.814 | 0.209 | 240870 | 223780 |
| 2.125 | 1.989 | 0.168 | 2.003 | 0.086 | 21870 | 12980 | 10.500 | 9.984 | 0.500 | 10.054 | 0.209 | 246740 | 234490 |
| 2.156 | 2.018 | 0.168 | 2.032 | 0.086 | 22190 | 13390 | 10.750 | 10.221 | 0.500 | 10.293 | 0.209 | 252620 | 246000 |
| 2.250 | 2.105 | 0.168 | 2.120 | 0.086 | 23160 | 14650 | 11.000 | 10.459 | 0.500 | 10.533 | 0.209 | 258490 | 257230 |
| 2.312 | 2.163 | 0.168 | 2.178 | 0.086 | 23790 | 15510 | 11.250 | 10.692 | 0.500 | 10.772 | 0.209 | 264360 | 269270 |
| 2.375 | 2.223 | 0.200 | 2.239 | 0.086 | 24440 | 16170 | 11.500 | 10.934 | 0.562 | 11.011 | 0.209 | 270240 | 281590 |
| 2.437 | 2.283 | 0.200 | 2.299 | 0.086 | 25080 | 16840 | 11.750 | 11.171 | 0.562 | 11.250 | 0.209 | 276120 | 294180 |
| 2.500 | 2.343 | 0.200 | 2.360 | 0.086 | 25730 | 17530 | 12.000 | 11.410 | 0.562 | 11.490 | 0.209 | 281990 | 306450 |
| 2.559 | 2.402 | 0.200 | 2.419 | 0.086 | 26340 | 17940 | 12.250 | 11.647 | 0.562 | 11.729 | 0.209 | 287860 | 319580 |
| 2.625 | 2.464 | 0.200 | 2.481 | 0.086 | 27020 | 18930 | 12.500 | 11.885 | 0.562 | 11.969 | 0.209 | 293740 | 332360 |
| 2.687 | 2.523 | 0.200 | 2.541 | 0.086 | 27650 | 19640 | 12.750 | 12.124 | 0.562 | 12.208 | 0.209 | 299610 | 346030 |

**Table 10.** *(Continued)* **Heavy Duty External Spiral Retaining Rings** *MIL-R-27426*

| Shaft Dia. A | Ring Dia. G | Ring Wall E | Groove Dia. C | Groove Width D | Static Thrust Load (lb) Ring | Static Thrust Load (lb) Groove | Shaft Dia. A | Ring Dia. G | Ring Wall E | Groove Dia. C | Groove Width D | Static Thrust Load (lb) Ring | Static Thrust Load (lb) Groove |
|---|---|---|---|---|---|---|---|---|---|---|---|---|---|
| 2.750 | 2.584 | 0.225 | 2.602 | 0.103 | 32140 | 20380 | 13.000 | 12.361 | 0.662 | 12.448 | 0.209 | 305490 | 359330 |
| 2.875 | 2.702 | 0.225 | 2.721 | 0.103 | 33600 | 22170 | 13.250 | 12.598 | 0.662 | 12.687 | 0.209 | 311360 | 373530 |
| 2.937 | 2.760 | 0.225 | 2.779 | 0.103 | 34320 | 23240 | 13.500 | 12.837 | 0.662 | 12.927 | 0.209 | 317240 | 387340 |
| 3.000 | 2.818 | 0.225 | 2.838 | 0.103 | 35060 | 24340 | 13.750 | 13.074 | 0.662 | 13.166 | 0.209 | 323110 | 402090 |
| 3.062 | 2.878 | 0.225 | 2.898 | 0.103 | 35780 | 25140 | 14.000 | 13.311 | 0.662 | 13.405 | 0.209 | 328990 | 417110 |
| 3.125 | 2.936 | 0.225 | 2.957 | 0.103 | 36520 | 26290 | 14.250 | 13.548 | 0.662 | 13.644 | 0.209 | 334860 | 432410 |
| 3.156 | 2.965 | 0.225 | 2.986 | 0.103 | 36880 | 26860 | 14.500 | 13.787 | 0.750 | 13.884 | 0.209 | 340740 | 447250 |
| 3.250 | 3.054 | 0.225 | 3.076 | 0.103 | 37980 | 28320 | 14.750 | 14.024 | 0.750 | 14.123 | 0.209 | 346610 | 463090 |
| 3.344 | 3.144 | 0.225 | 3.166 | 0.103 | 39080 | 29800 | 15.000 | 14.262 | 0.750 | 14.363 | 0.209 | 352490 | 478450 |
| 3.437 | 3.234 | 0.225 | 3.257 | 0.103 | 40170 | 30980 | | | | | | | |

*Source:* Spirolox Retaining Rings, RSN Series. All dimensions are in inches. Depth of groove $d = (A - C)/2$. Thickness indicated is for unplated rings; add 0.002 to upper tolerance for plated rings. Standard material: carbon spring steel (SAE 1070-1090).

*Ring Thickness, F*: For shaft size 0.469, 0.025; for sizes 0.500 through 0.669, 0.035; for sizes 0.688 through 1.023, 0.042; for sizes 1.062 through 1.500, 0.050; for sizes 1.562 through 2.000, 0.062; for sizes 2.062 through 2.687, 0.078; for sizes 2.750 through 3.437, 0.093; for sizes 3.500 through 5.000, 0.111; for sizes 5.250 through 6.000, 0.127; for sizes 6.250 through 7.250, 0.156; for sizes 7.500 through 15.000, 0.187.

*Ring Free Diameter Tolerances:* For shaft sizes 0.469 through 1.500, +0.000, −0.013; for sizes 1.562 through 2.000, +0.000, −0.020; for sizes 2.062 through 2.687, + 0.000, −0.025; for sizes 2.750 through 3.437, +0.000, −0.030; for sizes 3.500 through 5.000, +0.000, −0.035; for sizes 5.250 through 6.000, +0.000, −0.050; for sizes 6.250 through 7.000, +0.000, −0.060; for sizes 7.250 through 10.000, +0.000, −0.070; for sizes 10.250 through 12.500, +0.000, −0.090; for sizes 12.750 through 15.000, +0.000, −0.110.

*Ring Thickness Tolerances:* For shaft sizes 0.469 through 1.500, ±0.002; for sizes 1.562 through 5.000, ±0.003; for sizes 5.250 through 6.000, ±0.004; for sizes 6.250 through 15.000, ±0.005.

*Groove Diameter Tolerances:* For shaft sizes 0.469 through 0.562, ±0.002; for sizes 0.594 through 1.023, ±0.003; for sizes 1.062 through 1.500, ±0.004; for sizes 1.562 through 2.000, ±0.005; for sizes 2.062 through 5.000, ±0.006; for sizes 5.250 through 6.000, ±0.007; for sizes 6.250 through 10.000, ±0.008; for sizes 10.250 through 12.500, ±0.010; for sizes 12.750 through 15.000, ±0.012.

*Groove Width Tolerances:* For shaft sizes 0.469 through 1.023, +0.003, −0.000; for sizes 1.062 through 2.000, +0.004, −0.000; for sizes 2.062 through 5.000, +0.005, −0.000; for sizes 5.250 through 6.000, +0.006; −0.000; for sizes 6.250 through 7.250, + 0.008, −0.000; for sizes 7.500 through 15.000, +0.008, −0.000.

*Thrust Load Capacity:* The most important criterion in determining which ring is best suited for a specific application is thrust load capacity. The strength of the retaining ring and groove must both be considered when analyzing the thrust load capacity of an application to determine whether the groove or the retaining ring is likely to fail first. When a retaining ring application fails, the fault will usually be with the groove, unless the groove material is of very high strength.

*Ring Material:* The standard materials for spiral-wound retaining rings are SAE 1070 to 1090 carbon spring steels and 18-8 type 302 stainless steels. The 1070 to 1090 carbon spring steels provide high-strength retaining rings at low cost. Type 302 stainless steel withstands ordinary rusting. Other materials are used for specialized applications, such as the type 316 stainless frequently used in the food industry. For high-temperature use, superalloy A286 rings can be used at up to 900° F (482° C) and Inconel X-750 at up to 1200° F (649° C). Other materials, such as 316 stainless steel, 17-7PH and Inconel stainless steels are sometimes used for special-purpose and custom-made rings. Standard ring are typically supplied uncoated, however, special finishes such as cadmium, phosphate, zinc, or black oxide coatings for carbon spring steel rings and passivation of stainless steel rings are available.

## Table 5. American National Standard Types A and B Wing Screws
*ANSI/ASME B18.6.9-2010 (R2017)*

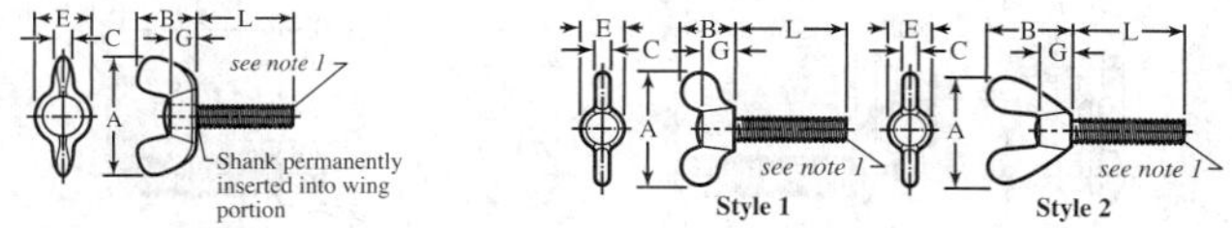


TYPE A TYPE B

| Nominal Size or Basic Major Diameter[a] | Thds. per Inch | Series[b] | Head Blank size (Ref) | A Wing Spread Max. | A Wing Spread Min. | B Wing Height Max. | B Wing Height Min. | C Wing Thickness Max. | C Wing Thickness Min. | E Boss Dia. Max. | E Boss Dia. Min. | G Boss Height Max. | G Boss Height Min. | L Practical Screw Lengths Max. | L Practical Screw Lengths Min. |
|---|---|---|---|---|---|---|---|---|---|---|---|---|---|---|---|
| **Type A** | | | | | | | | | | | | | | | |
| 4 (0.1120) | 40 | Hvy. | AA | 0.72 | 0.59 | 0.41 | 0.28 | 0.11 | 0.07 | 0.33 | 0.29 | 0.14 | 0.10 | 0.75 | 0.25 |
| 6 (0.1380) | 32 | Lgt. | AA | 0.72 | 0.59 | 0.41 | 0.28 | 0.11 | 0.07 | 0.33 | 0.29 | 0.14 | 0.10 | } 0.75 | 0.25 |
| | | **Hvy.** | **A** | **0.91** | **0.78** | **0.47** | **0.34** | **0.14** | **0.10** | **0.43** | **0.39** | **0.18** | **0.14** | | |
| 8 (0.1640) | 32 | **Lgt.** | **A** | **0.91** | **0.78** | **0.47** | **0.34** | **0.14** | **0.10** | **0.43** | **0.39** | **0.18** | **0.14** | } 0.75 | 0.38 |
| | | Hvy. | B | 1.10 | 0.97 | 0.57 | 0.43 | 0.18 | 0.14 | 0.50 | 0.45 | 0.22 | 0.17 | | |
| 10 (0.1900) | 24, 32 | **Lgt.** | **A** | **0.91** | **0.78** | **0.47** | **0.34** | **0.14** | **0.10** | **0.43** | **0.39** | **0.18** | **0.14** | } 1.00 | 0.38 |
| | | Hvy. | B | 1.10 | 0.97 | 0.57 | 0.43 | 0.18 | 0.14 | 0.50 | 0.45 | 0.22 | 0.17 | | |
| 12 (0.2160) | 24 | **Lgt.** | **B** | **1.10** | **0.97** | **0.57** | **0.43** | **0.18** | **0.14** | **0.50** | **0.45** | **0.22** | **0.17** | } 1.00 | 0.38 |
| | | Hvy. | C | 1.25 | 1.12 | 0.66 | 0.53 | 0.21 | 0.17 | 0.58 | 0.51 | 0.25 | 0.20 | | |
| 1/4 (0.2500) | 20 | **Lgt.** | **B** | **1.10** | **0.97** | **0.57** | **0.43** | **0.18** | **0.14** | **0.50** | **0.45** | **0.22** | **0.17** | } 1.50 | 0.50 |
| | | Reg. | C | 1.25 | 1.12 | 0.66 | 0.53 | 0.21 | 0.17 | 0.58 | 0.51 | 0.25 | 0.20 | | |
| | | Hvy. | D | 1.44 | 1.31 | 0.79 | 0.65 | 0.24 | 0.20 | 0.70 | 0.64 | 0.30 | 0.26 | | |
| 5/16 (0.3125) | 18 | **Lgt.** | **C** | **1.25** | **1.12** | **0.66** | **0.53** | **0.21** | **0.17** | **0.58** | **0.51** | **0.25** | **0.20** | } 1.50 | 0.50 |
| | | Reg. | D | 1.44 | 1.31 | 0.79 | 0.65 | 0.24 | 0.20 | 0.70 | 0.64 | 0.30 | 0.26 | | |
| | | Hvy. | E | 1.94 | 1.81 | 1.00 | 0.87 | 0.33 | 0.26 | 0.93 | 0.86 | 0.39 | 0.35 | | |
| 3/8 (0.3750) | 16 | **Lgt.** | **D** | **1.44** | **1.31** | **0.79** | **0.65** | **0.24** | **0.20** | **0.70** | **0.64** | **0.30** | **0.26** | } 2.00 | 0.75 |
| | | Reg. | E | 1.94 | 1.81 | 1.00 | 0.87 | 0.33 | 0.26 | 0.93 | 0.86 | 0.39 | 0.35 | | |
| | | Hvy. | F | 2.76 | 2.62 | 1.44 | 1.31 | 0.40 | 0.34 | 1.19 | 1.13 | 0.55 | 0.51 | | |
| 7/16 (0.4375) | 14 | **Lgt.** | **E** | **1.94** | **1.81** | **1.00** | 0.87 | **0.33** | **0.26** | **0.93** | **0.86** | **0.39** | **0.35** | } 4.00 | 1.00 |
| | | Hvy. | F | 2.76 | 2.62 | 1.44 | 1.31 | 0.40 | 0.34 | 1.19 | 1.13 | 0.55 | 0.51 | | |
| 1/2 (0.5000) | 13 | **Lgt.** | **E** | **1.94** | **1.81** | **1.00** | **0.87** | **0.33** | **0.26** | **0.93** | **0.86** | **0.39** | **0.35** | } 4.00 | 1.00 |
| | | Hvy. | F | 2.76 | 2.62 | 1.44 | 1.31 | 0.40 | 0.34 | 1.19 | 1.13 | 0.55 | 0.51 | | |
| 5/8 (0.6250) | 11 | Hvy. | F | 2.76 | 2.62 | 1.44 | 1.31 | 0.40 | 0.34 | 1.19 | 1.13 | 0.55 | 0.51 | 4.00 | 1.25 |
| **Type B, Style 1** | | | | | | | | | | | | | | | |
| 10 (0.1900) | 24 | … | … | 0.97 | 0.91 | 0.45 | 0.39 | 0.15 | 0.12 | 0.39 | 0.36 | 0.28 | 0.22 | 2.00 | 0.50 |
| 1/4 (0.2500) | 20 | … | … | 1.16 | 1.09 | 0.56 | 0.50 | 0.17 | 0.14 | 0.47 | 0.44 | 0.34 | 0.28 | 3.00 | 0.50 |
| 5/16 (0.3125) | 18 | … | … | 1.44 | 1.38 | 0.67 | 0.61 | 0.18 | 0.15 | 0.55 | 0.52 | 0.41 | 0.34 | 3.00 | 0.50 |
| 3/8 (0.3750) | 16 | … | … | 1.72 | 1.66 | 0.80 | 0.73 | 0.20 | 0.17 | 0.63 | 0.60 | 0.47 | 0.41 | 4.00 | 0.50 |
| 7/16 (0.4375) | 14 | … | … | 2.00 | 1.94 | 0.91 | 0.84 | 0.21 | 0.18 | 0.71 | 0.68 | 0.53 | 0.47 | 3.00 | 1.00 |
| 1/2 (0.5000) | 13 | … | … | 2.31 | 2.22 | 1.06 | 0.94 | 0.23 | 0.20 | 0.79 | 0.76 | 0.62 | 0.50 | 3.00 | 1.00 |
| 5/8 (0.6250) | 11 | … | … | 2.84 | 2.72 | 1.31 | 1.19 | 0.27 | 0.23 | 0.96 | 0.92 | 0.75 | 0.62 | 2.50 | 1.00 |
| **Type B, Style 2** | | | | | | | | | | | | | | | |
| 10 (0.1900) | 24 | … | … | 1.01 | 0.95 | 0.78 | 0.72 | 0.14 | 0.11 | 0.39 | 0.36 | 0.28 | 0.22 | 1.25 | 0.50 |
| 1/4 (0.2500) | 20 | … | … | 1.22 | 1.16 | 0.94 | 0.88 | 0.16 | 0.13 | 0.47 | 0.44 | 0.34 | 0.28 | 2.00 | 0.50 |
| 5/16 (0.3125) | 18 | … | … | 1.43 | 1.37 | 1.09 | 1.03 | 0.17 | 0.14 | 0.55 | 0.52 | 0.41 | 0.34 | 2.00 | 0.50 |
| 3/8 (0.3750) | 16 | … | … | 1.63 | 1.57 | 1.25 | 1.19 | 0.18 | 0.15 | 0.63 | 0.60 | 0.47 | 0.41 | 2.00 | 0.50 |

All dimensions in inches. Sizes shown in **bold face** are preferred.

[1] Plain point, unless alternate point from styles shown in Table 8 is specified by user.

[a] Where specifying nominal size in decimals, zeros in the fourth decimal place are omitted.

[b] Hvy. = Heavy; Lgt. = Light; Reg. = Regular.

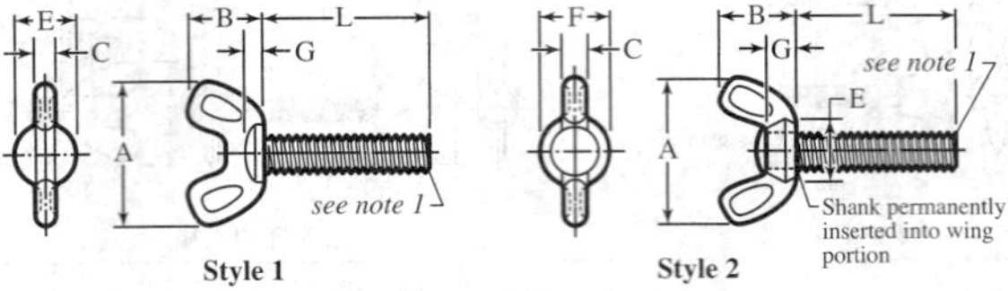


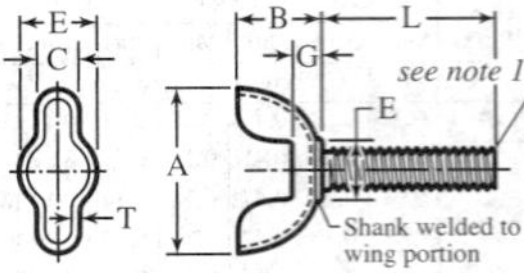

## NAILS, SPIKES, AND WOOD SCREWS

### Standard Wire Nails and Spikes (Size, Length, and Approximate Number to Pound)

| Size of Nail | Length, Inches | Common Wire Nails and Brads Gage | Common Wire Nails and Brads Num/lb | Flooring Brads Gage | Flooring Brads Num/lb | Fence Nails Gage | Fence Nails Num/lb | Casing, Smooth and Barbed Box Gage | Casing, Smooth and Barbed Box Num/lb | Finishing Nails Gage | Finishing Nails Num/lb |
|---|---|---|---|---|---|---|---|---|---|---|---|
| 2 *d* | 1 | 15 | 876 | … | … | … | … | 15 ½ | 1010 | 16 ½ | 1351 |
| 3 *d* | 1 ¼ | 14 | 568 | … | … | … | … | 14 ½ | 635 | 15 ½ | 807 |
| 4 *d* | 1 ½ | 12 ½ | 316 | … | … | … | … | 14 | 473 | 15 | 584 |
| 5 *d* | 1 ¾ | 12 ½ | 271 | … | … | 10 | 142 | 14 | 406 | 15 | 500 |
| 6 *d* | 2 | 11 ½ | 181 | 11 | 157 | 10 | 124 | 12 ½ | 236 | 13 ½ | 309 |
| 7 *d* | 2 ¼ | 11 ½ | 161 | 11 | 139 | 9 | 92 | 12 ½ | 210 | 13 | 238 |
| 8 *d* | 2 ½ | 10 ¼ | 106 | 10 | 99 | 9 | 82 | 11 ½ | 145 | 12 ½ | 189 |
| 9 *d* | 2 ¾ | 10 ¼ | 96 | 10 | 90 | 8 | 62 | 11 ½ | 132 | 12 ½ | 172 |
| 10 *d* | 3 | 9 | 69 | 9 | 69 | 7 | 50 | 10 ½ | 94 | 11 ½ | 121 |
| 12 *d* | 3 ¼ | 9 | 64 | 8 | 54 | 6 | 40 | 10 ½ | 87 | 11 ½ | 113 |
| 16 *d* | 3 ½ | 8 | 49 | 7 | 43 | 5 | 30 | 10 | 71 | 11 | 90 |
| 20 *d* | 4 | 6 | 31 | 6 | 31 | 4 | 23 | 9 | 52 | 10 | 62 |
| 30 *d* | 4 ½ | 5 | 24 | … | … | … | … | 9 | 46 | … | … |
| 40 *d* | 5 | 4 | 18 | … | … | … | … | 8 | 35 | … | … |
| 50 *d* | 5 ½ | 3 | 16 | … | … | … | … | … | … | … | … |
| 60 *d* | 6 | 2 | 11 | … | … | … | … | … | … | … | … |

| Size and Length | | Hinge Nails, Heavy Gage | Hinge Nails, Heavy Num/lb | Hinge Nails, Light Gage | Hinge Nails, Light Num/lb | Clinch Nails Gage | Clinch Nails Num/lb | Barbed Car Nails, Heavy Gage | Barbed Car Nails, Heavy Num/lb | Barbed Car Nails Light Gage | Barbed Car Nails Light Num/lb |
|---|---|---|---|---|---|---|---|---|---|---|---|
| 2 *d* | 1 | … | … | … | … | 14 | 710 | … | … | … | … |
| 3 *d* | 1 ¼ | … | … | … | … | 13 | 429 | … | … | … | … |
| 4 *d* | 1 ½ | 3 | 50 | 6 | 82 | 12 | 274 | 10 | 165 | 12 | 274 |
| 5 *d* | 1 ¾ | 3 | 38 | 6 | 62 | 12 | 235 | 9 | 118 | 10 | 142 |
| 6 *d* | 2 | 3 | 30 | 6 | 50 | 11 | 157 | 9 | 103 | 10 | 124 |
| 7 *d* | 2 ¼ | 00 | 12 | 3 | 25 | 11 | 139 | 8 | 76 | 9 | 92 |
| 8 *d* | 2 ½ | 00 | 11 | 3 | 23 | 10 | 99 | 8 | 69 | 9 | 82 |
| 9 *d* | 2 ¾ | 00 | 10 | 3 | 22 | 10 | 90 | 7 | 54 | 8 | 62 |
| 10 *d* | 3 | 00 | 9 | 3 | 19 | 9 | 69 | 7 | 50 | 8 | 57 |
| 12 *d* | 3 ¼ | … | … | … | … | 9 | 62 | 6 | 42 | 7 | 50 |
| 16 *d* | 3 ½ | … | … | … | … | 8 | 49 | 6 | 35 | 7 | 43 |
| 20 *d* | 4 | … | … | … | … | 7 | 37 | 5 | 26 | 6 | 31 |
| 30 *d* | 4 ½ | … | … | … | … | … | … | 5 | 24 | 6 | 28 |
| 40 *d* | 5 | … | … | … | … | … | … | 4 | 18 | 5 | 21 |
| 50 *d* | 5 ½ | … | … | … | … | … | … | 3 | 15 | 4 | 17 |
| 60 *d* | 6 | … | … | … | … | … | … | 3 | 13 | 4 | 15 |

| Size and Length | | Boat Nails, Heavy Gage | Boat Nails, Heavy Num/lb | Boat Nails, Light Gage | Boat Nails, Light Num/lb | Slating Nails Gage | Slating Nails Num/lb | Spikes Size and Length | | Spikes Gage | Spikes No. to Lb |
|---|---|---|---|---|---|---|---|---|---|---|---|
| 2 *d* | 1 | … | … | … | … | 12 | 411 | | | | |
| 3 *d* | 1 ¼ | … | … | … | … | 10 ½ | 225 | | | | |
| 4 *d* | 1 ½ | ¼ | 44 | 3⁄16 | 82 | 10 ½ | 187 | | | | |
| 5 *d* | 1 ¾ | … | … | … | … | 10 | 142 | 10 *d* | 3 | 6 | 41 |
| 6*d* | 2 | ¼ | 32 | 3⁄16 | 62 | 9 | 103 | 12 *d* | 3 ¼ | 6 | 38 |
| 7 *d* | 2 ¼ | … | … | … | … | … | … | 16 *d* | 3 ½ | 5 | 30 |
| 8 *d* | 2 ½ | ¼ | 26 | 3⁄16 | 50 | … | … | 20 *d* | 4 | 4 | 23 |
| 9 *d* | 2 ¾ | … | … | … | … | … | … | 30 *d* | 4 ½ | 3 | 17 |
| 10 *d* | 3 | ⅜ | 14 | ¼ | 22 | … | … | 40 *d* | 5 | 2 | 13 |
| 12 *d* | 3 ¼ | ⅜ | 13 | ¼ | 20 | … | … | 50 *d* | 5 ½ | 1 | 10 |
| 16 *d* | 3 ½ | ⅜ | 12 | ¼ | 18 | … | … | 60 *d* | 6 | 1 | 8 |
| 20 *d* | 4 | ⅜ | 10 | ¼ | 16 | … | … | … | 7 | 0 | 7 |
| 30 *d* | 4 ½ | … | … | … | … | … | … | … | 8 | 00 | 6 |
| 40 *d* | 5 | … | … | … | … | … | … | … | 9 | 00 | 5 |
| 50 *d* | 5 ½ | … | … | … | … | … | … | … | 10 | ⅜ | 4 |
| 60 *d* | 6 | … | … | … | … | … | … | … | 12 | ⅜ | 3 |

## TABLE OF CONTENTS

## THREADS AND THREADING

### SCREW THREAD SYSTEMS



### UNIFIED SCREW THREADS



### UNIFIED SCREW THREADS

*(Continued)*



### CALCULATING THREAD DIMENSIONS



### METRIC SCREW THREADS



THREADING

## Table 1. American Standard Unified Inch Screw Thread Form Data

| Threads per Inch | Pitch | Depth of Sharp V-Thread | Depth of Int. Thd. and UN Ext. Thd.[a] | Depth of UNR Ext. Thd. | Truncation of Ext. Thd. Root | Truncation of UNR Ext. Thd. Root[b] | Truncation of Ext. Thd. Crest | Truncation of Int. Thd. Root | Truncation of Int. Thd. Crest | Flat at Ext. Thd. Crest and Int. Thd. Root | Basic Flat at Int. Thd. Crest[c] | Maximum Ext. Thd. Root Radius | Addendum of Ext. Thd. |
|---|---|---|---|---|---|---|---|---|---|---|---|---|---|
| $n$ | $P$ | $0.86603P$ | $0.54127P$ | $0.59539P$ | $0.21651P$ | $0.16238P$ | $0.10825P$ | $0.10825P$ | $0.2165P$ | $0.125P$ | $0.25P$ | $0.14434P$ | $0.32476P$ |
| 80 | 0.01250 | 0.01083 | 0.00677 | 0.00744 | 0.00271 | 0.00203 | 0.00135 | 0.00135 | 0.00271 | 0.00156 | 0.00312 | 0.00180 | 0.00406 |
| 72 | 0.01389 | 0.01203 | 0.00752 | 0.00827 | 0.00301 | 0.00226 | 0.00150 | 0.00150 | 0.00301 | 0.00174 | 0.00347 | 0.00200 | 0.00451 |
| 64 | 0.01563 | 0.01353 | 0.00846 | 0.00930 | 0.00338 | 0.00254 | 0.00169 | 0.00169 | 0.00338 | 0.00195 | 0.00391 | 0.00226 | 0.00507 |
| 56 | 0.01786 | 0.01546 | 0.00967 | 0.01063 | 0.00387 | 0.00290 | 0.00193 | 0.00193 | 0.00387 | 0.00223 | 0.00446 | 0.00258 | 0.00580 |
| 48 | 0.02083 | 0.01804 | 0.01128 | 0.01240 | 0.00451 | 0.00338 | 0.00226 | 0.00226 | 0.00451 | 0.00260 | 0.00521 | 0.00301 | 0.00677 |
| 44 | 0.02273 | 0.01968 | 0.01230 | 0.01353 | 0.00492 | 0.00369 | 0.00246 | 0.00246 | 0.00492 | 0.00284 | 0.00568 | 0.00328 | 0.00738 |
| 40 | 0.02500 | 0.02165 | 0.01353 | 0.01488 | 0.00541 | 0.00406 | 0.00271 | 0.00271 | 0.00541 | 0.00312 | 0.00625 | 0.00361 | 0.00812 |
| 36 | 0.02778 | 0.02406 | 0.01504 | 0.01654 | 0.00601 | 0.00451 | 0.00301 | 0.00301 | 0.00601 | 0.00347 | 0.00694 | 0.00401 | 0.00902 |
| 32 | 0.03125 | 0.02706 | 0.01691 | 0.01861 | 0.00677 | 0.00507 | 0.00338 | 0.00338 | 0.00677 | 0.00391 | 0.00781 | 0.00451 | 0.01015 |
| 28 | 0.03571 | 0.03093 | 0.01933 | 0.02126 | 0.00773 | 0.00580 | 0.00387 | 0.00387 | 0.00773 | 0.00446 | 0.00893 | 0.00515 | 0.01160 |
| 27 | 0.03704 | 0.03208 | 0.02005 | 0.02205 | 0.00802 | 0.00601 | 0.00401 | 0.00401 | 0.00802 | 0.00463 | 0.00926 | 0.00535 | 0.01203 |
| 24 | 0.04167 | 0.03608 | 0.02255 | 0.02481 | 0.00902 | 0.00677 | 0.00451 | 0.00451 | 0.00902 | 0.00521 | 0.01042 | 0.00601 | 0.01353 |
| 20 | 0.05000 | 0.04330 | 0.02706 | 0.02977 | 0.01083 | 0.00812 | 0.00541 | 0.00541 | 0.01083 | 0.00625 | 0.01250 | 0.00722 | 0.01624 |
| 18 | 0.05556 | 0.04811 | 0.03007 | 0.03308 | 0.01203 | 0.00902 | 0.00601 | 0.00601 | 0.01203 | 0.00694 | 0.01389 | 0.00802 | 0.01804 |
| 16 | 0.06250 | 0.05413 | 0.03383 | 0.03721 | 0.01353 | 0.01015 | 0.00677 | 0.00677 | 0.01353 | 0.00781 | 0.01562 | 0.00902 | 0.02030 |
| 14 | 0.07143 | 0.06186 | 0.03866 | 0.04253 | 0.01546 | 0.01160 | 0.00773 | 0.00773 | 0.01546 | 0.00893 | 0.01786 | 0.01031 | 0.02320 |
| 13 | 0.07692 | 0.06662 | 0.04164 | 0.04580 | 0.01655 | 0.01249 | 0.00833 | 0.00833 | 0.01665 | 0.00962 | 0.01923 | 0.01110 | 0.02498 |
| 12 | 0.08333 | 0.07217 | 0.04511 | 0.04962 | 0.01804 | 0.01353 | 0.00902 | 0.00902 | 0.01804 | 0.01042 | 0.02083 | 0.01203 | 0.02706 |
| 11½ | 0.08696 | 0.07531 | 0.04707 | 0.05177 | 0.01883 | 0.01412 | 0.00941 | 0.00941 | 0.01883 | 0.01087 | 0.02174 | 0.01255 | 0.02824 |
| 11 | 0.09091 | 0.07873 | 0.04921 | 0.05413 | 0.01968 | 0.01476 | 0.00984 | 0.00984 | 0.01968 | 0.01136 | 0.02273 | 0.01312 | 0.02952 |
| 10 | 0.10000 | 0.08660 | 0.05413 | 0.05954 | 0.02165 | 0.01624 | 0.01083 | 0.01083 | 0.02165 | 0.01250 | 0.02500 | 0.01443 | 0.03248 |
| 9 | 0.11111 | 0.09623 | 0.06014 | 0.06615 | 0.02406 | 0.01804 | 0.01203 | 0.01203 | 0.02406 | 0.01389 | 0.02778 | 0.01604 | 0.03608 |
| 8 | 0.12500 | 0.10825 | 0.06766 | 0.07442 | 0.02706 | 0.02030 | 0.01353 | 0.01353 | 0.02706 | 0.01562 | 0.03125 | 0.01804 | 0.04059 |
| 7 | 0.14286 | 0.12372 | 0.07732 | 0.08506 | 0.03093 | 0.02320 | 0.01546 | 0.01546 | 0.03093 | 0.01786 | 0.03571 | 0.02062 | 0.04639 |
| 6 | 0.16667 | 0.14434 | 0.09021 | 0.09923 | 0.03608 | 0.02706 | 0.01804 | 0.01804 | 0.03608 | 0.02083 | 0.04167 | 0.02406 | 0.05413 |
| 5 | 0.20000 | 0.17321 | 0.10825 | 0.11908 | 0.04330 | 0.03248 | 0.02165 | 0.02165 | 0.04330 | 0.02500 | 0.05000 | 0.02887 | 0.06495 |
| 4½ | 0.22222 | 0.19245 | 0.12028 | 0.13231 | 0.04811 | 0.03608 | 0.02406 | 0.02406 | 0.04811 | 0.02778 | 0.05556 | 0.03208 | 0.07217 |
| 4 | 0.25000 | 0.21651 | 0.13532 | 0.14885 | 0.05413 | 0.04059 | 0.02706 | 0.02706 | 0.05413 | 0.03125 | 0.06250 | 0.03608 | 0.08119 |

[a] Also depth of thread engagement.
[b] Design profile.
[c] Also basic flat at external UN thread root.
All dimensions are in inches.

**Table 3.** *(Continued)* **Standard Series and Selected Combinations — Unified Screw Threads**

| Nominal Size, Threads per Inch, and Series Designation[a] | External[b] | | | | | | | | Internal[b] | | | | | |
|---|---|---|---|---|---|---|---|---|---|---|---|---|---|---|
| | Class | Allowance | Major Diameter | | | Pitch Diameter | | UNR Minor Dia.,[c] Max (Ref.) | Class | Minor Diameter | | Pitch Diameter | | Major Diameter |
| | | | Max[d] | Min | Min[e] | Max[d] | Min | | | Min | Max | Min | Max | Min |
| 1 1/8-14 UNS | 2A | 0.0016 | 1.1234 | 1.1131 | — | 1.0770 | 1.0717 | 1.0383 | 2B | 1.048 | 1.063 | 1.0786 | 1.0855 | 1.1250 |
| 1 1/8-16 UN | 2A | 0.0015 | 1.1235 | 1.1141 | — | 1.0829 | 1.0779 | 1.0491 | 2B | 1.057 | 1.071 | 1.0844 | 1.0910 | 1.1250 |
| | 3A | 0.0000 | 1.1250 | 1.1156 | — | 1.0844 | 1.0806 | 1.0506 | 3B | 1.0570 | 1.0659 | 1.0844 | 1.0893 | 1.1250 |
| 1 1/8-18 UNEF | 2A | 0.0014 | 1.1236 | 1.1149 | — | 1.0875 | 1.0827 | 1.0574 | 2B | 1.065 | 1.078 | 1.0889 | 1.0951 | 1.1250 |
| | 3A | 0.0000 | 1.1250 | 1.1163 | — | 1.0889 | 1.0853 | 1.0588 | 3B | 1.0650 | 1.0730 | 1.0889 | 1.0936 | 1.1250 |
| 1 1/8-20 UN | 2A | 0.0014 | 1.1236 | 1.1155 | — | 1.0911 | 1.0865 | 1.0641 | 2B | 1.071 | 1.082 | 1.0925 | 1.0985 | 1.1250 |
| | 3A | 0.0000 | 1.1250 | 1.1169 | — | 1.0925 | 1.0890 | 1.0655 | 3B | 1.0710 | 1.0786 | 1.0925 | 1.0970 | 1.1250 |
| 1 1/8-24 UNS | 2A | 0.0013 | 1.1237 | 1.1165 | — | 1.0966 | 1.0923 | 1.0741 | 2B | 1.080 | 1.090 | 1.0979 | 1.1035 | 1.1250 |
| 1 1/8-28 UN | 2A | 0.0012 | 1.1238 | 1.1173 | — | 1.1006 | 1.0966 | 1.0813 | 2B | 1.086 | 1.095 | 1.1018 | 1.1070 | 1.1250 |
| | 3A | 0.0000 | 1.1250 | 1.1185 | — | 1.1018 | 1.0988 | 1.0825 | 3B | 1.0860 | 1.0926 | 1.1018 | 1.1057 | 1.1250 |
| 1 3/16-8 UN | 2A | 0.0021 | 1.1854 | 1.1704 | — | 1.1042 | 1.0972 | 1.0366 | 2B | 1.052 | 1.077 | 1.1063 | 1.1154 | 1.1875 |
| | 3A | 0.0000 | 1.1875 | 1.1725 | — | 1.1063 | 1.1011 | 1.0387 | 3B | 1.0520 | 1.0672 | 1.1063 | 1.1131 | 1.1875 |
| 1 3/16-12 UN | 2A | 0.0017 | 1.1858 | 1.1744 | — | 1.1317 | 1.1260 | 1.0866 | 2B | 1.097 | 1.115 | 1.1334 | 1.1409 | 1.1875 |
| | 3A | 0.0000 | 1.1875 | 1.1761 | — | 1.1334 | 1.1291 | 1.0883 | 3B | 1.0970 | 1.1073 | 1.1334 | 1.1390 | 1.1875 |
| 1 3/16-16 UN | 2A | 0.0015 | 1.1860 | 1.1766 | — | 1.1454 | 1.1403 | 1.1116 | 2B | 1.120 | 1.134 | 1.1469 | 1.1535 | 1.1875 |
| | 3A | 0.0000 | 1.1875 | 1.1781 | — | 1.1469 | 1.1431 | 1.1131 | 3B | 1.1200 | 1.1284 | 1.1469 | 1.1518 | 1.1875 |
| 1 3/16-18 UNEF | 2A | 0.0014 | 1.1861 | 1.1774 | — | 1.1500 | 1.1452 | 1.1199 | 2B | 1.127 | 1.140 | 1.1514 | 1.1577 | 1.1875 |
| | 3A | 0.0000 | 1.1875 | 1.1788 | — | 1.1514 | 1.1478 | 1.1213 | 3B | 1.1270 | 1.1355 | 1.1514 | 1.1561 | 1.1875 |
| 1 3/16-20 UN | 2A | 0.0014 | 1.1861 | 1.1780 | — | 1.1536 | 1.1490 | 1.1266 | 2B | 1.133 | 1.145 | 1.1550 | 1.1610 | 1.1875 |
| | 3A | 0.0000 | 1.1875 | 1.1794 | — | 1.1550 | 1.1515 | 1.1280 | 3B | 1.1330 | 1.1411 | 1.1550 | 1.1595 | 1.1875 |
| 1 3/16-28 UN | 2A | 0.0012 | 1.1863 | 1.1798 | — | 1.1631 | 1.1590 | 1.1438 | 2B | 1.149 | 1.157 | 1.1643 | 1.1696 | 1.1875 |
| | 3A | 0.0000 | 1.1875 | 1.1810 | — | 1.1643 | 1.1613 | 1.1450 | 3B | 1.1490 | 1.1551 | 1.1643 | 1.1683 | 1.1875 |
| 1 1/4-7 UNC | 1A | 0.0022 | 1.2478 | 1.2232 | — | 1.1550 | 1.1439 | 1.0777 | 1B | 1.095 | 1.123 | 1.1572 | 1.1716 | 1.2500 |
| | 2A | 0.0022 | 1.2478 | 1.2314 | 1.2232 | 1.1550 | 1.1476 | 1.0777 | 2B | 1.095 | 1.123 | 1.1572 | 1.1668 | 1.2500 |
| | 3A | 0.0000 | 1.2500 | 1.2336 | — | 1.1572 | 1.1517 | 1.0799 | 3B | 1.0950 | 1.1125 | 1.1572 | 1.1644 | 1.2500 |
| 1 1/4-8 UN | 2A | 0.0021 | 1.2479 | 1.2329 | 1.2254 | 1.1667 | 1.1597 | 1.0991 | 2B | 1.115 | 1.140 | 1.1688 | 1.1780 | 1.2500 |
| | 3A | 0.0000 | 1.2500 | 1.2350 | — | 1.1688 | 1.1635 | 1.1012 | 3B | 1.1150 | 1.1297 | 1.1688 | 1.1757 | 1.2500 |
| 1 1/4-10 UNS | 2A | 0.0019 | 1.2481 | 1.2352 | — | 1.1831 | 1.1768 | 1.1290 | 2B | 1.142 | 1.163 | 1.1850 | 1.1932 | 1.2500 |

*Coarse-Thread Series:* This series, UNC/UNRC, is the one most commonly used in the bulk production of bolts, screws, nuts and other general engineering applications. It is also used for threading into lower tensile strength materials such as cast iron, mild steel and softer materials (bronze, brass, aluminum, magnesium and plastics) to obtain the optimum resistance to stripping of the internal thread. It is applicable for rapid assembly or disassembly, or if corrosion or slight damage is possible.

**Table 4a. Coarse-Thread Series, UNC and UNRC—Basic Dimensions**

| Sizes No. or Inches | Basic Major Dia., $D$ | Thds. per Inch, $n$ | Basic Pitch Dia.,[a] $D_2$ | Minor Diameter Ext. Thds.,[c] $d_3$ (Ref.) | Minor Diameter Int. Thds.,[d] $D_1$ | Lead Angle λ at Basic P.D. | | Area of Minor Dia. at $D$-$2h_b$ | Tensile Stress Area[b] |
|---|---|---|---|---|---|---|---|---|---|
| | Inches | | Inches | Inches | Inches | Deg. | Min | Sq. In. | Sq. In. |
| 1 (0.073)[e] | 0.0730 | 64 | 0.0629 | 0.0544 | 0.0561 | 4 | 31 | 0.00218 | 0.00263 |
| 2 (0.086) | 0.0860 | 56 | 0.0744 | 0.0648 | 0.0667 | 4 | 22 | 0.00310 | 0.00370 |
| 3 (0.099)[e] | 0.0990 | 48 | 0.0855 | 0.0741 | 0.0764 | 4 | 26 | 0.00406 | 0.00487 |
| 4 (0.112) | 0.1120 | 40 | 0.0958 | 0.0822 | 0.0849 | 4 | 45 | 0.00496 | 0.00604 |
| 5 (0.125) | 0.1250 | 40 | 0.1088 | 0.0952 | 0.0979 | 4 | 11 | 0.00672 | 0.00796 |
| 6 (0.138) | 0.1380 | 32 | 0.1177 | 0.1008 | 0.1042 | 4 | 50 | 0.00745 | 0.00909 |
| 8 (0.164) | 0.1640 | 32 | 0.1437 | 0.1268 | 0.1302 | 3 | 58 | 0.01196 | 0.0140 |
| 10 (0.190) | 0.1900 | 24 | 0.1629 | 0.1404 | 0.1449 | 4 | 39 | 0.01450 | 0.0175 |
| 12 (0.216)[e] | 0.2160 | 24 | 0.1889 | 0.1664 | 0.1709 | 4 | 1 | 0.0206 | 0.0242 |
| 1⁄4 | 0.2500 | 20 | 0.2175 | 0.1905 | 0.1959 | 4 | 11 | 0.0269 | 0.0318 |
| 5⁄16 | 0.3125 | 18 | 0.2764 | 0.2464 | 0.2524 | 3 | 40 | 0.0454 | 0.0524 |
| 3⁄8 | 0.3750 | 16 | 0.3344 | 0.3005 | 0.3073 | 3 | 24 | 0.0678 | 0.0775 |
| 7⁄16 | 0.4375 | 14 | 0.3911 | 0.3525 | 0.3602 | 3 | 20 | 0.0933 | 0.1063 |
| 1⁄2 | 0.5000 | 13 | 0.4500 | 0.4084 | 0.4167 | 3 | 7 | 0.1257 | 0.1419 |
| 9⁄16 | 0.5625 | 12 | 0.5084 | 0.4633 | 0.4723 | 2 | 59 | 0.162 | 0.182 |
| 5⁄8 | 0.6250 | 11 | 0.5660 | 0.5168 | 0.5266 | 2 | 56 | 0.202 | 0.226 |
| 3⁄4 | 0.7500 | 10 | 0.6850 | 0.6309 | 0.6417 | 2 | 40 | 0.302 | 0.334 |
| 7⁄8 | 0.8750 | 9 | 0.8028 | 0.7427 | 0.7547 | 2 | 31 | 0.419 | 0.462 |
| 1 | 1.0000 | 8 | 0.9188 | 0.8512 | 0.8647 | 2 | 29 | 0.551 | 0.606 |
| 1 1⁄8 | 1.1250 | 7 | 1.0322 | 0.9549 | 0.9704 | 2 | 31 | 0.693 | 0.763 |
| 1 1⁄4 | 1.2500 | 7 | 1.1572 | 1.0799 | 1.0954 | 2 | 15 | 0.890 | 0.969 |
| 1 3⁄8 | 1.3750 | 6 | 1.2667 | 1.1766 | 1.1946 | 2 | 24 | 1.054 | 1.155 |
| 1 1⁄2 | 1.5000 | 6 | 1.3917 | 1.3016 | 1.3196 | 2 | 11 | 1.294 | 1.405 |
| 1 3⁄4 | 1.7500 | 5 | 1.6201 | 1.5119 | 1.5335 | 2 | 15 | 1.74 | 1.90 |
| 2 | 2.0000 | 4 1⁄2 | 1.8557 | 1.7353 | 1.7594 | 2 | 11 | 2.30 | 2.50 |
| 2 1⁄4 | 2.2500 | 4 1⁄2 | 2.1057 | 1.9853 | 2.0094 | 1 | 55 | 3.02 | 3.25 |
| 2 1⁄2 | 2.5000 | 4 | 2.3376 | 2.2023 | 2.2294 | 1 | 57 | 3.72 | 4.00 |
| 2 3⁄4 | 2.7500 | 4 | 2.5876 | 2.4523 | 2.4794 | 1 | 46 | 4.62 | 4.93 |
| 3 | 3.0000 | 4 | 2.8376 | 2.7023 | 2.7294 | 1 | 36 | 5.62 | 5.97 |
| 3 1⁄4 | 3.2500 | 4 | 3.0876 | 2.9523 | 2.9794 | 1 | 29 | 6.72 | 7.10 |
| 3 1⁄2 | 3.500 | 4 | 3.3376 | 3.2023 | 3.2294 | 1 | 22 | 7.92 | 8.33 |
| 3 3⁄4 | 3.7500 | 4 | 3.5876 | 3.4523 | 3.4794 | 1 | 16 | 9.21 | 9.66 |
| 4 | 4.0000 | 4 | 3.8376 | 3.7023 | 3.7294 | 1 | 11 | 10.61 | 11.08 |

[a] British: Effective Diameter.

[b] See formula, pages 1660 and 1668.

[c] Design form for UNR threads. (See figure on page 1951.)

[d] Basic minor diameter.

[e] Secondary sizes.

*Fine-Thread Series:* This series, UNF/UNRF, is suitable for the production of bolts, screws, and nuts and for other applications where the Coarse series is not applicable. External threads of this series have greater tensile stress area than comparable sizes of the Coarse series. The Fine series is suitable when the resistance to stripping of both external

## Table 4c. Extra-Fine-Thread Series, UNEF and UNREF—Basic Dimensions

| Sizes No. or Inches | Basic Major Dia., $D$ | Thds. per Inch, $n$ | Basic Pitch Dia.,[a] $D_2$ | Minor Diameter Ext. Thds.,[c] $d_3$ (Ref.) | Minor Diameter Int. Thds.,[d] $D_1$ | Lead Angle λ at Basic P.D. | | Area of Minor Dia. at $D - 2h_b$ | Tensile Stress Area[b] |
|---|---|---|---|---|---|---|---|---|---|
| | Inches | | Inches | Inches | Inches | Deg. | Min | Sq. In. | Sq. In. |
| **12 (0.216)[e]** | **0.2160** | **32** | **0.1957** | **0.1788** | **0.1822** | **2** | **55** | 0.0242 | 0.0270 |
| **1/4** | **0.2500** | **32** | **0.2297** | **0.2128** | **0.2162** | **2** | **29** | 0.0344 | 0.0379 |
| **5/16** | **0.3125** | **32** | **0.2922** | **0.2753** | **0.2787** | **1** | **57** | 0.0581 | 0.0625 |
| **3/8** | **0.3750** | **32** | **0.3547** | **0.3378** | **0.3412** | **1** | **36** | 0.0878 | 0.0932 |
| **7/16** | **0.4375** | **28** | **0.4143** | **0.3949** | **0.3988** | **1** | **34** | 0.1201 | 0.1274 |
| **1/2** | **0.5000** | **28** | **0.4768** | **0.4574** | **0.4613** | **1** | **22** | 0.162 | 0.170 |
| **9/16** | **0.5625** | **24** | **0.5354** | **0.5129** | **0.5174** | **1** | **25** | 0.203 | 0.214 |
| **5/8** | **0.6250** | **24** | **0.5979** | **0.5754** | **0.5799** | **1** | **16** | 0.256 | 0.268 |
| **11/16[e]** | **0.6875** | **24** | **0.6604** | **0.6379** | **0.6424** | **1** | **9** | 0.315 | 0.329 |
| **3/4** | **0.7500** | **20** | **0.7175** | **0.6905** | **0.6959** | **1** | **16** | 0.369 | 0.386 |
| **13/16[e]** | **0.8125** | **20** | **0.7800** | **0.7530** | **0.7584** | **1** | **10** | 0.439 | 0.458 |
| **7/8** | **0.8750** | **20** | **0.8425** | **0.8155** | **0.8209** | **1** | **5** | 0.515 | 0.536 |
| **15/16[e]** | **0.9375** | **20** | **0.9050** | **0.8780** | **0.8834** | **1** | **0** | 0.598 | 0.620 |
| **1** | **1.0000** | **20** | **0.9675** | **0.9405** | **0.9459** | **0** | **57** | 0.687 | 0.711 |
| **1 1/16[e]** | **1.0625** | **18** | **1.0264** | **0.9964** | **1.0024** | **0** | **59** | 0.770 | 0.799 |
| **1 1/8** | **1.1250** | **18** | **1.0889** | **1.0589** | **1.0649** | **0** | **56** | 0.871 | 0.901 |
| **1 3/16[e]** | **1.1875** | **18** | **1.1514** | **1.1214** | **1.1274** | **0** | **53** | 0.977 | 1.009 |
| **1 1/4** | **1.2500** | **18** | **1.2139** | **1.1839** | **1.1899** | **0** | **50** | 1.090 | 1.123 |
| **1 5/16[e]** | **1.3125** | **18** | **1.2764** | **1.2464** | **1.2524** | **0** | **48** | 1.208 | 1.244 |
| **1 3/8** | **1.3750** | **18** | **1.3389** | **1.3089** | **1.3149** | **0** | **45** | 1.333 | 1.370 |
| **1 7/16[e]** | **1.4375** | **18** | **1.4014** | **1.3714** | **1.3774** | **0** | **43** | 1.464 | 1.503 |
| **1 1/2** | **1.5000** | **18** | **1.4639** | **1.4339** | **1.4399** | **0** | **42** | 1.60 | 1.64 |
| **1 9/16[e]** | **1.5625** | **18** | **1.5264** | **1.4964** | **1.5024** | **0** | **40** | 1.74 | 1.79 |
| **1 5/8** | **1.6250** | **18** | **1.5889** | **1.5589** | **1.5649** | **0** | **38** | 1.89 | 1.94 |
| **1 11/16[e]** | **1.6875** | **18** | **1.6514** | **1.6214** | **1.6274** | **0** | **37** | 2.05 | 2.10 |

[a] British: Effective Diameter.

[b] See formula, pages 1660 and 1668.

[c] Design form for UNR threads. (See figure on page 1951.)

[d] Basic minor diameter.

[e] Secondary sizes.

*Other Threads of Special Diameters, Pitches, and Lengths of Engagement:* Thread data for special combinations of diameter, pitch, and length of engagement not included in selected combinations are also given in the Standard but are not given here. Also, when design considerations require nonstandard pitches or extreme conditions of engagement not covered by the tables, the allowance and tolerances should be derived from the formulas in the Standard. The thread symbol for such special threads is UNS.

**Constant Pitch Series.—**The various constant-pitch series, UN, with 4, 6, 8, 12, 16, 20, 28 and 32 threads per inch, given in Table 3, offer a comprehensive range of diameter-pitch combinations for those purposes where the threads in the Coarse, Fine, and Extra-Fine series do not meet the particular requirements of the design.

When selecting threads from these constant-pitch series, preference should be given wherever possible to those tabulated in the 8-, 12-, or 16-thread series.

*8-Thread Series:* The 8-thread series (8-UN) is a uniform-pitch series for large diameters. Although originally intended for high-pressure-joint bolts and nuts, it is now widely used as a substitute for the Coarse-Thread Series for diameters larger than 1 inch.

2-12 UN-2A(21) or 2.000-12 UN-2A(21)

$^1/_4$-20 UNC-3A-LH(21) or 0.250-20 UNC-3A-LH(21)

For uncoated standard series threads these designations may optionally be supplemented by the addition of the pitch diameter limits of size.

*Examples:*

$^1/_4$-20 UNC-2A(21)

PD 0.2164-0.2127 (Optional for uncoated threads)

**Designating Coated Threads.—**For coated (or plated) Class 2A external threads, the basic (max) major and basic (max) pitch diameters are given followed by the words AFTER COATING. The major and pitch diameter limits of size before coating are also given followed by the words BEFORE COATING.

| | | |
|---|---|---|
| *Example:* | $^3/_4$-10 UNC-2A (21) | |
| | [a]Major dia 0.7500 max<br>PD 0.6850 max | } AFTER COATING |
| | [b]Major dia 0.7482-0.7353<br>PD 0.6832-0.6773 | } BEFORE COATING |

[a] Major and PD values are equal to basic and correspond to those in Table 3 for Class 3A.

[b] Major and PD limits are those in Table 3 for Class 2A.

Certain applications require an allowance for rapid assembly, to permit application of a proper lubricant, or for residual growth due to high-temperature expansion. In such applications where the thread is to be coated and the 2A allowance is not permitted to be consumed by such coating, the thread class symbol is qualified by the addition of the letter G (symbol for allowance) following the class symbol, and the maximum major and maximum pitch diameters are reduced below basic size by the amount of the 2A allowance and followed by the words AFTER COATING. This arrangement ensures that the allowance is maintained. The major and pitch diameter limits of size before coating are also given followed by SPL and BEFORE COATING. For information concerning the designating of this and other special coating conditions reference should be made to ANSI/ASME B1.1-2019.

**Designating UNS Threads.—**UNS screw threads that have special combinations of diameter and pitch with tolerance to Unified formulation have the basic form designation set out first followed always by the limits of size.

**Designating Multiple Start Threads.—**If a screw thread is of multiple start, it is designated by specifying in sequence the nominal size, pitch (in decimals or threads per inch) and lead (in decimals or fractions).

**Other Special Designations.—**For other special designations including threads with modified limits of size or with special lengths of engagement, reference should be made to ANSI/ASME B1.1-2019.

**Hole Sizes for Tapping.—**Hole size limits for tapping Classes 1B, 2B, and 3B threads of various lengths of engagement are given in Table 3 on page 2162.

**Internal Thread Minor Diameter Tolerances.—**Internal thread minor diameter tolerances in Table 3 are based on a length of engagement equal to the nominal diameter. For general applications these tolerances are suitable for lengths of engagement up to $1^1/_2$ diameters. However, some thread applications have lengths of engagement which are greater than $1^1/_2$ diameters or less than the nominal diameter. For such applications it may be advantageous to increase or decrease the tolerance, respectively, as explained in the Tapping Section.

## Table 5. Unified Miniature Screw Threads—Limits of Size and Tolerances

| Size Designation[a] | Pitch | External Threads: Major Dia. Max[b] | External Threads: Major Dia. Min | External Threads: Pitch Dia. Max[b] | External Threads: Pitch Dia. Min | External Threads: Minor Dia. Max[c] | External Threads: Minor Dia. Min[d] | Internal Threads: Minor Dia. Min[b] | Internal Threads: Minor Dia. Max | Internal Threads: Pitch Dia. Min[b] | Internal Threads: Pitch Dia. Max | Internal Threads: Major Dia. Min[e] | Internal Threads: Major Dia. Max[d] | Lead Angle at Basic Pitch Dia. | | Sectional Area at Minor Dia. at D—1.28$p$ |
|---|---|---|---|---|---|---|---|---|---|---|---|---|---|---|---|---|
| | mm | mm | mm | mm | mm | mm | mm | mm | mm | mm | mm | mm | mm | deg | min | sq mm |
| **0.30 UNM** | **0.080** | **0.300** | **0.284** | **0.248** | **0.234** | **0.204** | **0.183** | **0.217** | **0.254** | **0.248** | **0.262** | **0.306** | **0.327** | **5** | **52** | **0.0307** |
| 0.35 UNM | 0.090 | 0.350 | 0.333 | 0.292 | 0.277 | 0.242 | 0.220 | 0.256 | 0.297 | 0.292 | 0.307 | 0.356 | 0.380 | 5 | 37 | 0.0433 |
| **0.40 UNM** | **0.100** | **0.400** | **0.382** | **0.335** | **0.319** | **0.280** | **0.256** | **0.296** | **0.340** | **0.335** | **0.351** | **0.407** | **0.432** | **5** | **26** | **0.0581** |
| 0.45 UNM | 0.100 | 0.450 | 0.432 | 0.385 | 0.369 | 0.330 | 0.306 | 0.346 | 0.390 | 0.385 | 0.401 | 0.457 | 0.482 | 4 | 44 | 0.0814 |
| **0.50 UNM** | **0.125** | **0.500** | **0.479** | **0.419** | **0.401** | **0.350** | **0.322** | **0.370** | **0.422** | **0.419** | **0.437** | **0.509** | **0.538** | **5** | **26** | **0.0908** |
| 0.55 UNM | 0.125 | 0.550 | 0.529 | 0.469 | 0.451 | 0.400 | 0.372 | 0.420 | 0.472 | 0.469 | 0.487 | 0.559 | 0.588 | 4 | 51 | 0.1195 |
| **0.60 UNM** | **0.150** | **0.600** | **0.576** | **0.503** | **0.483** | **0.420** | **0.388** | **0.444** | **0.504** | **0.503** | **0.523** | **0.611** | **0.644** | **5** | **26** | **0.1307** |
| 0.70 UNM | 0.175 | 0.700 | 0.673 | 0.586 | 0.564 | 0.490 | 0.454 | 0.518 | 0.586 | 0.586 | 0.608 | 0.713 | 0.750 | 5 | 26 | 0.1780 |
| **0.80 UNM** | **0.200** | **0.800** | **0.770** | **0.670** | **0.646** | **0.560** | **0.520** | **0.592** | **0.668** | **0.670** | **0.694** | **0.814** | **0.856** | **5** | **26** | **0.232** |
| 0.90 UNM | 0.225 | 0.900 | 0.867 | 0.754 | 0.728 | 0.630 | 0.586 | 0.666 | 0.750 | 0.754 | 0.780 | 0.916 | 0.962 | 5 | 26 | 0.294 |
| **1.00 UNM** | **0.250** | **1.000** | **0.964** | **0.838** | **0.810** | **0.700** | **0.652** | **0.740** | **0.832** | **0.838** | **0.866** | **1.018** | **1.068** | **5** | **26** | **0.363** |
| 1.10 UNM | 0.250 | 1.100 | 1.064 | 0.938 | 0.910 | 0.800 | 0.752 | 0.840 | 0.932 | 0.938 | 0.966 | 1.118 | 1.168 | 4 | 51 | 0.478 |
| **1.20 UNM** | **0.250** | **1.200** | **1.164** | **1.038** | **1.010** | **0.900** | **0.852** | **0.940** | **1.032** | **1.038** | **1.066** | **1.218** | **1.268** | **4** | **23** | **0.608** |
| 1.40 UNM | 0.300 | 1.400 | 1.358 | 1.205 | 1.173 | 1.040 | 0.984 | 1.088 | 1.196 | 1.205 | 1.237 | 1.422 | 1.480 | 4 | 32 | 0.811 |
| | Thds. per in. | inch | inch | inch | inch | inch | inch | inch | inch | inch | inch | inch | inch | deg | min | sq in |
| **0.30 UNM** | **318** | **0.0118** | **0.0112** | **0.0098** | **0.0092** | **0.0080** | **0.0072** | **0.0085** | **0.0100** | **0.0098** | **0.0104** | **0.0120** | **0.0129** | **5** | **52** | **0.0000475** |
| 0.35 UNM | 282 | 0.0138 | 0.0131 | 0.0115 | 0.0109 | 0.0095 | 0.0086 | 0.0101 | 0.0117 | 0.0115 | 0.0121 | 0.0140 | 0.0149 | 5 | 37 | 0.0000671 |
| **0.40 UNM** | **254** | **0.0157** | **0.0150** | **0.0132** | **0.0126** | **0.0110** | **0.0101** | **0.0117** | **0.0134** | **0.0132** | **0.0138** | **0.0160** | **0.0170** | **5** | **26** | **0.0000901** |
| 0.45 UNM | 254 | 0.0177 | 0.0170 | 0.0152 | 0.0145 | 0.0130 | 0.0120 | 0.0136 | 0.0154 | 0.0152 | 0.0158 | 0.0180 | 0.0190 | 4 | 44 | 0.0001262 |
| **0.50 UNM** | **203** | **0.0197** | **0.0189** | **0.0165** | **0.0158** | **0.0138** | **0.0127** | **0.0146** | **0.0166** | **0.0165** | **0.0172** | **0.0200** | **0.0212** | **5** | **26** | **0.0001407** |
| 0.55 UNM | 203 | 0.0217 | 0.0208 | 0.0185 | 0.0177 | 0.0157 | 0.0146 | 0.0165 | 0.0186 | 0.0185 | 0.0192 | 0.0220 | 0.0231 | 4 | 51 | 0.0001852 |
| **0.60 UNM** | **169** | **0.0236** | **0.0227** | **0.0198** | **0.0190** | **0.0165** | **0.0153** | **0.0175** | **0.0198** | **0.0198** | **0.0206** | **0.0240** | **0.0254** | **5** | **26** | **0.000203** |
| 0.70 UNM | 145 | 0.0276 | 0.0265 | 0.0231 | 0.0222 | 0.0193 | 0.0179 | 0.0204 | 0.0231 | 0.0231 | 0.0240 | 0.0281 | 0.0295 | 5 | 26 | 0.000276 |
| **0.80 UNM** | **127** | **0.0315** | **0.0303** | **0.0264** | **0.0254** | **0.0220** | **0.0205** | **0.0233** | **0.0263** | **0.0264** | **0.0273** | **0.0321** | **0.0337** | **5** | **26** | **0.000360** |
| 0.90 UNM | 113 | 0.0354 | 0.0341 | 0.0297 | 0.0287 | 0.0248 | 0.0231 | 0.0262 | 0.0295 | 0.0297 | 0.0307 | 0.0361 | 0.0379 | 5 | 26 | 0.000456 |
| **1.00 UNM** | **102** | **0.0394** | **0.0380** | **0.0330** | **0.0319** | **0.0276** | **0.0257** | **0.0291** | **0.0327** | **0.0330** | **0.0341** | **0.0401** | **0.0420** | **5** | **26** | **0.000563** |
| 1.10 UNM | 102 | 0.0433 | 0.0419 | 0.0369 | 0.0358 | 0.0315 | 0.0296 | 0.0331 | 0.0367 | 0.0369 | 0.0380 | 0.0440 | 0.0460 | 4 | 51 | 0.000741 |
| **1.20 UNM** | **102** | **0.0472** | **0.0458** | **0.0409** | **0.0397** | **0.0354** | **0.0335** | **0.0370** | **0.0406** | **0.0409** | **0.0420** | **0.0480** | **0.0499** | **4** | **23** | **0.000943** |
| 1.40 UNM | 85 | 0.0551 | 0.0535 | 0.0474 | 0.0462 | 0.0409 | 0.0387 | 0.0428 | 0.0471 | 0.0474 | 0.0487 | 0.0560 | 0.0583 | 4 | 32 | 0.001257 |

[a] Sizes shown in bold type are preferred.

[b] This is also the basic dimension.

[c] This limit, in conjunction with root form shown in Table 2, is advocated for use when optical projection methods of gaging are employed. For mechanical gaging the minimum minor diameter of the internal thread is applied.

[d] This limit is provided for reference only. In practice, the form of the threading tool is relied upon for this limit.

[e] This limit is provided for reference only, and is not gaged. For gaging, the maximum major diameter of the external thread is applied.

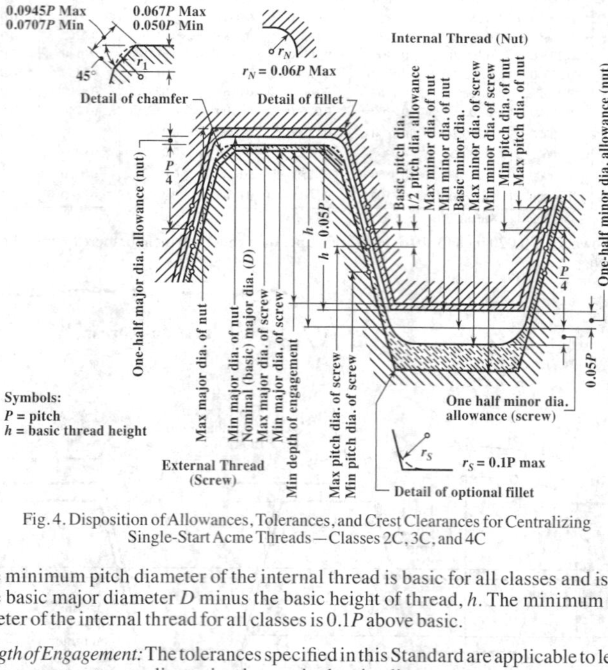

For more complete specifications for production and acceptance of Dryseal pipe threads, see ANSI/ASME B1.20.3 (Inch) and ANSI/ASME B1.20.4 (Withdrawn) (Metric Translation), and for gaging and inspection, see ANSI/ASME B1.20.5 (Inch) and ANSI/ASME B1.20.6M (Withdrawn) (Metric Translation).

*Designation of Dryseal Pipe Threads:* The standard Dryseal pipe threads are designated by specifying in sequence nominal size, thread series symbol, and class:

*Examples:* 1/8-27 NPTF-1; 1/8-27 PTF-SAE SHORT; and 3/8-18 NPTF-1 AFTER PLATING.

**Table 8. Suggested Tap Drill Sizes for Internal Dryseal Pipe Threads**

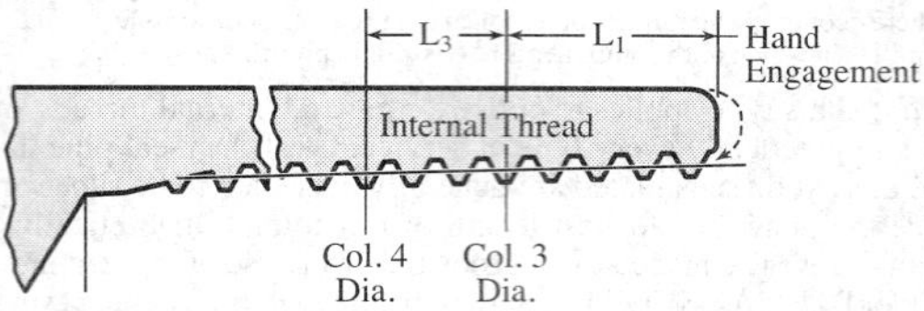


| | | Taper Pipe Thread | | | | Straight Pipe Thread | | |
|---|---|---|---|---|---|---|---|---|
| | | Minor Diameter At Distance | | Drill Size[a] | | Minor Diameter | | |
| Size | Probable Drill Oversize Cut (Mean) | $L_1$ From Large End | $L_1 + L_3$ From Large End | Without Reamer | With Reamer | NPSF | NPSI | Drill Size[a] |
| 1/16-27 | 0.0038 | 0.2443 | 0.2374 | "C" (0.242) | "A" (0.234) | 0.2482 | 0.2505 | "D" (0.246) |
| 1/8-27 | 0.0044 | 0.3367 | 0.3298 | "Q" (0.332) | 21/64 (0.328) | 0.3406 | 0.3429 | "R" (0.339) |
| 1/4-18 | 0.0047 | 0.4362 | 0.4258 | 7/16 (0.438) | 27/64 (0.422) | 0.4422 | 0.4457 | 7/16 (0.438) |
| 3/8-18 | 0.0049 | 0.5708 | 0.5604 | 9/16 (0.562) | 9/16 (0.563) | 0.5776 | 0.5811 | 37/64 (0.578) |
| 1/2-14 | 0.0051 | 0.7034 | 0.6901 | 45/64 (0.703) | 11/16 (0.688) | 0.7133 | 0.7180 | 45/64 (0.703) |
| 3/4-14 | 0.0060 | 0.9127 | 0.8993 | 29/32 (0.906) | 57/64 (0.891) | 0.9238 | 0.9283 | 59/64 (0.922) |
| 1-11 1/2 | 0.0080 | 1.1470 | 1.1307 | 1 9/64 (1.141) | 1 1/8 (1.125) | 1.1600 | 1.1655 | 1 5/32 (1.156) |
| 1 1/4-11 1/2 | 0.0100 | 1.4905 | 1.4742 | 1 31/64 (1.484) | 1 15/32 (1.469) | … | … | … |
| 1 1/2-11 1/2 | 0.0120 | 1.7295 | 1.7132 | 1 23/32 (1.719) | 1 45/64 (1.703) | … | … | … |
| 2-11 1/2 | 0.0160 | 2.2024 | 2.1861 | 2 3/16 (2.188) | 2 11/64 (2.172) | … | … | … |
| 2 1/2-8 | 0.0180 | 2.6234 | 2.6000 | 2 39/64 (2.609) | 2 37/64 (2.578) | … | … | … |
| 3-8 | 0.0200 | 3.2445 | 3.2211 | 3 15/64 (3.234) | 3 13/64 (3.203) | … | … | … |

[a] Some drill sizes listed may not be standard drills.

All dimensions are given in inches.

**Special Dryseal Threads.—**Where design limitations, economy of material, permanent installation, or other limiting conditions prevail, consideration may be given to using a special Dryseal thread series.

*Dryseal Special Short Taper Pipe Thread, PTF-SPL SHORT:* Threads of this series conform in all respects to PTF-SAE SHORT threads except that the full thread length has been further shortened by eliminating one thread at the small end of internal threads or one thread at the large end of external threads.

*Base Screw Shell Gage Tolerances:* Threaded ring gages—"Go," Max. thread size to minus 0.0003 inch; "Not Go," Min. thread size to plus 0.0003 inch. Plain ring gages—"Go," Max. thread OD to minus 0.0002 inch; "Not Go," Min. thread OD to plus 0.0002 inch.

*Socket Screw Shell Gages:* Threaded plug gages—"Go," Min. thread size to plus 0.0003 inch; "Not Go," Max. thread size to minus 0.0003 inch. Plain plug gages—"Go," Min. minor dia. to plus 0.0002 inch; "Not Go," Max. minor dia. to minus 0.0002 inch.

*Check Gages for Base Screw Shell Gages:* Threaded plugs for checking threaded ring gages—"Go," Max. thread size to minus 0.0003 inch; "Not Go," Min. thread size to plus 0.0003 inch.

**Electric Fixture Thread.—**The special straight electric fixture thread consists of a straight thread of the same pitches as the American standard pipe thread, and having the regular American or US standard form; it is used for caps, etc. The male thread is smaller, and the female thread larger than those of the special straight-fixture pipe threads. The male thread assembles with a standard taper female thread, while the female thread assembles with a standard taper male thread. This thread is used when it is desired to have the joint "make up" on a shoulder. The gages used are straight-threaded limit gages.

## Instrument and Microscope Threads

**British Association Standard Thread (BA).—**This form of thread is similar to the Whitworth thread in that the root and crest are rounded (see illustration). The angle, however, is only 47 degrees 30 minutes and the radius of the root and crest are proportionately larger. This thread is used in Great Britain and, to some extent, in other European countries for very small screws. Its use in the United States is practically confined to the manufacture of tools for export. This thread system was originated in Switzerland as a standard for watch and clock screws, and it is sometimes referred to as the "Swiss small screw thread standard." See also *Swiss Screw Thread.*

This screw thread system is recommended by the British Standards Institution for use in preference to the BSW and BSF systems for all screws smaller than $\frac{1}{4}$ inch except that the use of the "0" BA thread be discontinued in favor of the $\frac{1}{4}$-in. BSF. It is further recommended that in the selection of sizes, preference be given to even numbered BA sizes. The thread form is shown by the diagram.

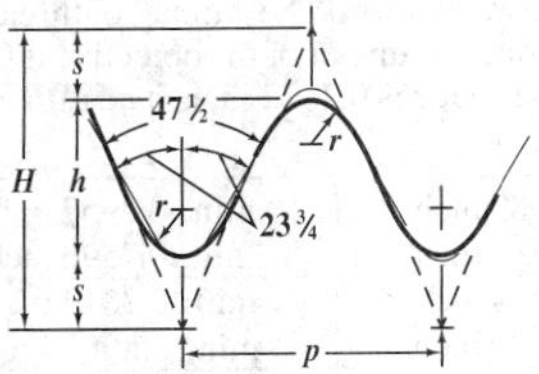


British Association Thread

$$H = 1.13634 \times p$$

$$h = 0.60000 \times p$$

$$r = 0.18083 \times p$$

$$s = 0.26817 \times p$$

It is a symmetrical V-thread, of $47\frac{1}{2}$ degree included angle, having its crests and roots rounded with equal radii, such that the basic depth of the thread is 0.6000 of the pitch Where $p$ = pitch of thread, $H$ = depth of V-thread, $h$ = depth of BA thread, $r$ = radius at root and crest of thread, and $s$ = root and crest truncation.

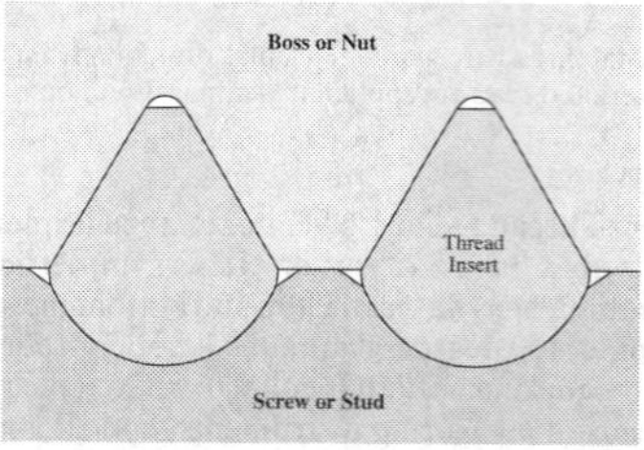

**Harvey Grip Thread.—**The characteristic feature of this thread is that one side inclines 44 degrees from a line at right angles to the axis, whereas the other side has an inclination of only 1 degree. This form of thread is sometimes used when there is considerable resistance or pressure in an axial direction and when it is desirable to reduce the radial or bursting pressure on the nut as much as possible. See *BUTTRESS THREADS* on page 2081.

**Lloyd & Lloyd Thread.—**The Lloyd & Lloyd screw thread is the same as the regular Whitworth screw thread in which the sides of the thread form an angle of 55 degrees with one another. The top and bottom of the thread are rounded.

**Lock-Nut Pipe Thread.—**The lock-nut pipe thread is a straight thread of the largest diameter which can be cut on a pipe. Its form is the same as the American or Briggs standard taper pipe thread. In general, "Go" gages only are required. These consist of a straight-threaded plug representing the minimum female lock-nut thread, and a straight-threaded ring representing the maximum male lock-nut thread. This thread is used only to hold parts together, or to retain a collar on the pipe. It is never used where a tight threaded joint is required.

**Philadelphia Carriage Bolt Thread.—**This is a screw thread for carriage bolts which is somewhat similar to a square thread, but having rounded corners at the top and bottom. The sides of the thread are inclined to an inclusive angle of $3\frac{1}{2}$ degrees. The width of the thread at the top is 0.53 times the pitch.

**SAE Standard Screw Thread.—**The screw thread standard of the Society of Automotive Engineers (SAE) is intended for use in the automotive industries of the United States. The SAE Standard includes a Coarse series, a Fine series, an 8-thread series, a 12-thread series, a 16-thread series, an Extra-fine series, and a Special-pitch series. The Coarse and Fine series, and also the 8-, 12- and 16-thread series, are identical to the corresponding American Standard series. The Extra-fine and Special-pitch series are SAE Standards only.

The American Standard thread *form* (or the form previously known as the US Standard) is applied to all SAE Standard screw threads. The Extra-fine series has a total of six pitches ranging from 32 down to 16 threads per inch. The 16 threads per inch in the Extra-fine series applies to all diameters from $1\frac{3}{4}$ up to 6 inches. This Extra-fine series is intended for use on relatively light sections; on parts requiring fine adjustment; where jar and vibration are important factors; when the thickness of a threaded section is relatively small as in tubing, and where assembly is made without the use of wrenches.

The SAE Special pitches include some which are finer than any in the Extra-fine series. The special pitches apply to a range of diameters extending from No. 10 (0.1900 inch) up to 6 inches. Each diameter has a range of pitches varying from five to eight. For example, a $\frac{1}{4}$-inch diameter has six pitches ranging from 24 to 56 threads per inch, whereas a 6-inch diameter has eight pitches ranging from 4 to 16 threads per inch. These various SAE Standard series are intended to provide adequate screw thread specifications for all uses in the automotive industries.

**Sellers Screw Thread.—**The Sellers screw thread, later known as the 'United States standard thread," and now as the "American Standard," is the most commonly used screw thread in the United States. It was originated by William Sellers, of Philadelphia, and first proposed by him in a paper read before the Franklin Institute, in April, 1864. In 1868, it was adopted by the United States Navy and has since become the generally accepted standard screw thread in the United States.

**Worm Threads.—**The included angle of worm threads range from 29° to 60°; for single-threaded worms 29° is common; multiple-threaded type must have larger helix and thread angles to avoid excessive under-cutting in hobbing the worm-wheel teeth. AGMA recommends 40° included thread angle for triple- and quadruple-thread worms, but many speed reducers and transmissions have 60° thread angles. The 29° angle is the same as the Acme thread, but worm thread depth is greater and widths of the flats at the top and bottom are less. If lead angle is larger than 20°, an increase in included thread angle is desirable. Worm gearing reaches maximum efficiency when lead angle is 45°, thus explaining the 60° thread angle. Thread parts of a 29° worm thread are: $p$ = pitch; $d$ = depth of thread = $0.6866p$; $t$ = width, top of thread = $0.335p$; $b$ = width, bottom of thread = $0.310p$.

**Table for Measuring Whitworth Standard Threads by the Three-wire Method**

| Dia. of Thread | No. of Threads per Inch | Dia. of Wire Used | Dia. Measured over Wires | Dia. of Thread | No. of Threads per Inch | Dia. of Wire Used | Dia. Measured over Wires |
|---|---|---|---|---|---|---|---|
| 1⁄8 | 40 | 0.018 | 0.1420 | 2 1⁄4 | 4 | 0.150 | 2.3247 |
| 3⁄16 | 24 | 0.030 | 0.2158 | 2 3⁄8 | 4 | 0.150 | 2.4497 |
| 1⁄4 | 20 | 0.035 | 0.2808 | 2 1⁄2 | 4 | 0.150 | 2.5747 |
| 5⁄16 | 18 | 0.040 | 0.3502 | 2 5⁄8 | 4 | 0.150 | 2.6997 |
| 3⁄8 | 16 | 0.040 | 0.4015 | 2 3⁄4 | 3 1⁄2 | 0.200 | 2.9257 |
| 7⁄16 | 14 | 0.050 | 0.4815 | 2 7⁄8 | 3 1⁄2 | 0.200 | 3.0507 |
| 1⁄2 | 12 | 0.050 | 0.5249 | 3 | 3 1⁄2 | 0.200 | 3.1757 |
| 9⁄16 | 12 | 0.050 | 0.5874 | 3 1⁄8 | 3 1⁄2 | 0.200 | 3.3007 |
| 5⁄8 | 11 | 0.070 | 0.7011 | 3 1⁄4 | 3 1⁄4 | 0.200 | 3.3905 |
| 11⁄16 | 11 | 0.070 | 0.7636 | 3 3⁄8 | 3 1⁄4 | 0.200 | 3.5155 |
| 3⁄4 | 10 | 0.070 | 0.8115 | 3 1⁄2 | 3 1⁄4 | 0.200 | 3.6405 |
| 13⁄16 | 10 | 0.070 | 0.8740 | 3 5⁄8 | 3 1⁄4 | 0.200 | 3.7655 |
| 7⁄8 | 9 | 0.070 | 0.9187 | 3 3⁄4 | 3 | 0.200 | 3.8495 |
| 15⁄16 | 9 | 0.070 | 0.9812 | 3 7⁄8 | 3 | 0.200 | 3.9745 |
| 1 | 8 | 0.090 | 1.0848 | 4 | 3 | 0.200 | 4.0995 |
| 1 1⁄16 | 8 | 0.090 | 1.1473 | 4 1⁄8 | 3 | 0.200 | 4.2245 |
| 1 1⁄8 | 7 | 0.090 | 1.1812 | 4 1⁄4 | 2 7⁄8 | 0.250 | 4.4846 |
| 1 3⁄16 | 7 | 0.090 | 1.2437 | 4 3⁄8 | 2 7⁄8 | 0.250 | 4.6096 |
| 1 1⁄4 | 7 | 0.090 | 1.3062 | 4 1⁄2 | 2 7⁄8 | 0.250 | 4.7346 |
| 1 5⁄16 | 7 | 0.090 | 1.3687 | 4 5⁄8 | 2 7⁄8 | 0.250 | 4.8596 |
| 1 3⁄8 | 6 | 0.120 | 1.4881 | 4 3⁄4 | 2 3⁄4 | 0.250 | 4.9593 |
| 1 7⁄16 | 6 | 0.120 | 1.5506 | 4 7⁄8 | 2 3⁄4 | 0.250 | 5.0843 |
| 1 1⁄2 | 6 | 0.120 | 1.6131 | 5 | 2 3⁄4 | 0.250 | 5.2093 |
| 1 9⁄16 | 6 | 0.120 | 1.6756 | 5 1⁄8 | 2 3⁄4 | 0.250 | 5.3343 |
| 1 5⁄8 | 5 | 0.120 | 1.6847 | 5 1⁄4 | 2 5⁄8 | 0.250 | 5.4316 |
| 1 11⁄16 | 5 | 0.120 | 1.7472 | 5 3⁄8 | 2 5⁄8 | 0.250 | 5.5566 |
| 1 3⁄4 | 5 | 0.120 | 1.8097 | 5 1⁄2 | 2 5⁄8 | 0.250 | 5.6816 |
| 1 13⁄16 | 5 | 0.120 | 1.8722 | 5 5⁄8 | 2 5⁄8 | 0.250 | 5.8066 |
| 1 7⁄8 | 4 1⁄2 | 0.150 | 1.9942 | 5 3⁄4 | 2 1⁄2 | 0.250 | 5.9011 |
| 1 15⁄16 | 4 1⁄2 | 0.150 | 2.0567 | 5 7⁄8 | 2 1⁄2 | 0.250 | 6.0261 |
| 2 | 4 1⁄2 | 0.150 | 2.1192 | 6 | 2 1⁄2 | 0.250 | 6.1511 |
| 2 1⁄8 | 4 1⁄2 | 0.150 | 2.2442 | … | … | … | … |

All dimensions are given in inches.

*Case 1:* The screw thread or worm is to be milled with a cutter having an included angle equal to the nominal or standard thread angle that is assumed to be the angle in the axial plane. For example, a 60-degree cutter is to be used for milling a thread. In this case, the thread angle in the plane of the axis will exceed 60 degrees by an amount increasing with the lead angle. This variation from the standard angle may be of little or no practical importance if the lead angle is small or if the mating nut (or teeth in worm gearing) is formed to suit the thread as milled.

*Case 2:* The screw thread or worm is to be milled with a cutter reduced to whatever normal angle is equivalent to the standard thread angle in the axial plane. For example, a 29-degree Acme thread is to be milled with a cutter having some angle smaller than 29 degrees (the reduction increasing with the lead angle) to make the thread angle standard in the plane of the axis. Theoretically, the milling cutter angle should always be corrected to suit the normal angle; but if the lead angle is small, such correction may be unnecessary.

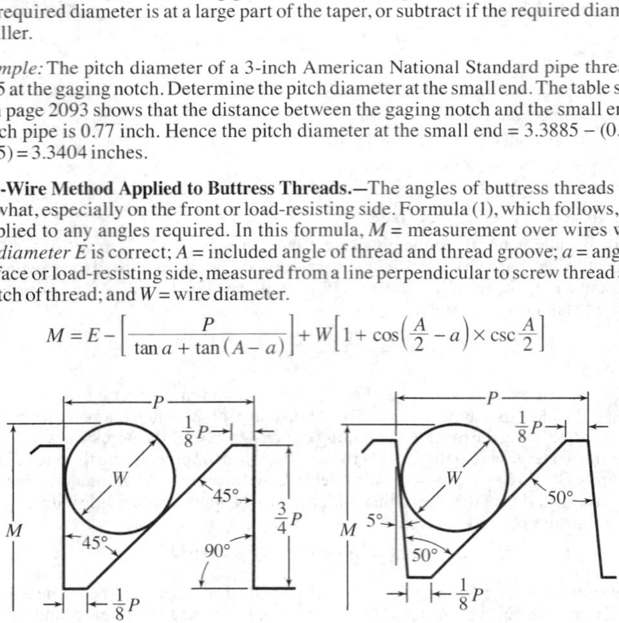

**Table 2.** *(Continued)* **Tap Drill Sizes and Percentage of Thread (Unified Threads)**

| Taps Size Threads per Inch | Drill Size | | | | Percentage of Full Thread | Nut Minor Diameter, min[a] max |
|---|---|---|---|---|---|---|
| | No. or Letter | Fractional Inch | mm | Decimal Inch | | |
| | Y | … | … | 0.4040 | 70 | 0.4043[b] |
| 1/2–12 | … | 27/64 | … | 0.4219 | 72 | 0.4098 |
| | | | | | | 0.4223 |
| 1/2–13 | … | 7/16 | … | 0.4375 | 62 | 0.4167 |
| | … | … | 11 | 0.4331 | 68 | 0.4284 |
| | … | 27/64 | … | 0.4219 | 78 | |
| 1/2–20 | … | 29/64 | … | 0.4531 | 72 | 0.4459 |
| | … | c | … | 0.4492 | 78 | 0.4537 |
| 1/2–27 | … | 15/32 | … | 0.4687 | 65 | 0.4618 |
| | c | … | … | 0.4640 | 75 | 0.4672[b] |
| 9/16–12 | … | 1/2 | … | 0.5000 | 58 | 0.4723 |
| | … | … | 12.5 | 0.4921 | 65 | 0.4843 |
| | … | 31/64 | … | 0.4844 | 72 | |
| 9/16–18 | … | 33/64 | … | 0.5156 | 65 | 0.5024 |
| | … | … | 13 | 0.5118 | 70 | 0.5106 |
| 9/16–27 | … | 17/32 | … | 0.5312 | 65 | 0.5234 |
| | … | c | … | 0.5265 | 75 | 0.5273[b] |
| 5/8–11 | … | … | 14 | 0.5512 | 62 | 0.5266 |
| | … | 35/64 | … | 0.5469 | 66 | 0.5391 |
| | … | 17/32 | … | 0.5312 | 79 | |
| 5/8–18 | … | 37/64 | … | 0.5781 | 65 | 0.5649 |
| | … | … | 14.5 | 0.5709 | 75 | 0.5730 |
| 5/8–27 | … | 19/32 | … | 0.5937 | 65 | 0.5860 |
| | … | c | … | 0.5890 | 75 | 0.5912 |
| 3/4–10 | … | … | 17 | 0.6693 | 62 | 0.6417 |
| | c | 21/32 | … | 0.6563 | 72 | 0.6545[b] |
| | … | … | 16.5 | 0.6496 | 77 | |
| 3/4–16 | … | 45/64 | … | 0.7031 | 58 | 0.6823 |
| | … | … | 17.5 | 0.6890 | 75 | 0.6908 |
| | … | 11/64 | … | 0.6875 | 77 | |
| 3/4–27 | … | 23/32 | … | 0.7187 | 65 | 0.7102 |
| | c | … | … | 0.7140 | 75 | 0.7164[c] |
| 7/8–9 | … | 25/32 | … | 0.7813 | 65 | 0.7547 |
| | … | 49/64 | … | 0.7656 | 76 | 0.7679 |
| 7/8–14 | … | 13/16 | … | 0.8125 | 67 | 0.7977 |
| | … | … | 20.5 | 0.8071 | 73 | 0.8068 |
| 7/8–18 | … | 53/64 | … | 0.8281 | 65 | 0.8149 |
| | c | … | … | 0.8210 | 75 | 0.8223[b] |
| 7/8–27 | … | 27/32 | … | 0.8437 | 65 | 0.8340 |
| | c | … | … | 0.8390 | 75 | 0.8406[c] |
| 1–8 | … | … | 23.0 | 0.9055 | 58 | 0.8647 |
| | … | 57/64 | … | 0.8906 | 67 | 0.8797 |
| | … | 7/8 | … | 0.8750 | 77 | |
| 1–12 | … | 59/64 | … | 0.9219 | 72 | 0.9098 |
| | | | | | | 0.9198 |

[a] Unified or American Standard Threads.

[b] Not based on Unified or American Thread Standards.

[c] Special drill required for this size as the next size smaller gives too great a percentage of thread.

## Feeds for Thread Rolling

| Root Dia. of Blank | Number of Threads per Inch | | | | | | | | | | | | | |
|---|---|---|---|---|---|---|---|---|---|---|---|---|---|---|
| | 72 | 64 | 56 | 48 | 44 | 40 | 36 | 32 | 28 | 24 | 22 | 20 | 18 | 14 |
| | Cross-slide Holders—Feed per Revolution in Inches | | | | | | | | | | | | | |
| 1⁄8 | 0.0045 | 0.0040 | 0.0035 | 0.0030 | 0.0025 | 0.0020 | 0.0015 | 0.0010 | ..... | ..... | ..... | ..... | ..... | ..... |
| 3⁄16 | 0.0050 | 0.0045 | 0.0040 | 0.0035 | 0.0030 | 0.0025 | 0.0020 | 0.0015 | 0.0005 | ..... | ..... | ..... | ..... | ..... |
| 1⁄4 | 0.0055 | 0.0050 | 0.0045 | 0.0040 | 0.0035 | 0.0030 | 0.0025 | 0.0020 | 0.0010 | 0.0005 | 0.0005 | ..... | ..... | ..... |
| 5⁄16 | 0.0060 | 0.0055 | 0.0050 | 0.0045 | 0.0040 | 0.0035 | 0.0030 | 0.0025 | 0.0015 | 0.0010 | 0.0010 | 0.0005 | 0.0005 | ..... |
| 3⁄8 | 0.0065 | 0.0060 | 0.0055 | 0.0050 | 0.0045 | 0.0040 | 0.0035 | 0.0030 | 0.0020 | 0.0015 | 0.0015 | 0.0010 | 0.0010 | 0.0005 |
| 7⁄16 | 0.0070 | 0.0065 | 0.0060 | 0.0055 | 0.0050 | 0.0045 | 0.0040 | 0.0035 | 0.0025 | 0.0020 | 0.0020 | 0.0015 | 0.0015 | 0.0010 |
| 1⁄2 | 0.0075 | 0.0070 | 0.0065 | 0.0060 | 0.0055 | 0.0050 | 0.0045 | 0.0040 | 0.0030 | 0.0025 | 0.0025 | 0.0020 | 0.0020 | 0.0015 |
| 5⁄8 | 0.0080 | 0.0075 | 0.0070 | 0.0065 | 0.0060 | 0.0055 | 0.0050 | 0.0045 | 0.0035 | 0.0030 | 0.0030 | 0.0025 | 0.0025 | 0.0020 |
| 3⁄4 | 0.0085 | 0.0080 | 0.0075 | 0.0070 | 0.0065 | 0.0060 | 0.0055 | 0.0050 | 0.0040 | 0.0035 | 0.0035 | 0.0030 | 0.0030 | 0.0025 |
| 7⁄8 | 0.0090 | 0.0085 | 0.0080 | 0.0075 | 0.0070 | 0.0065 | 0.0060 | 0.0055 | 0.0045 | 0.0040 | 0.0040 | 0.0035 | 0.0035 | 0.0030 |
| 1 | 0.0095 | 0.0090 | 0.0085 | 0.0080 | 0.0075 | 0.0070 | 0.0065 | 0.0060 | 0.0050 | 0.0045 | 0.0045 | 0.0040 | 0.0040 | 0.0035 |
| Root Dia. | Swing Holders—Feed per Revolution in Inches | | | | | | | | | | | | | |
| 1⁄8 | 0.0025 | 0.0020 | 0.0015 | 0.0010 | 0.0005 | ..... | ..... | ..... | ..... | ..... | ..... | ..... | ..... | ..... |
| 3⁄16 | 0.0028 | 0.0025 | 0.0020 | 0.0015 | 0.0008 | 0.0005 | ..... | ..... | ..... | ..... | ..... | ..... | ..... | ..... |
| 1⁄4 | 0.0030 | 0.0030 | 0.0025 | 0.0020 | 0.0010 | 0.0010 | 0.0005 | 0.0005 | 0.0005 | ..... | ..... | ..... | ..... | ..... |
| 5⁄16 | 0.0035 | 0.0035 | 0.0030 | 0.0025 | 0.0015 | 0.0015 | 0.0010 | 0.0010 | 0.0010 | 0.0005 | ..... | ..... | ..... | ..... |
| 3⁄8 | 0.0040 | 0.0040 | 0.0035 | 0.0030 | 0.0020 | 0.0020 | 0.0015 | 0.0015 | 0.0015 | 0.0010 | 0.0005 | 0.0005 | 0.0005 | ..... |
| 7⁄16 | 0.0045 | 0.0045 | 0.0040 | 0.0035 | 0.0030 | 0.0025 | 0.0020 | 0.0020 | 0.0020 | 0.0015 | 0.0010 | 0.0010 | 0.0010 | ..... |
| 1⁄2 | 0.0048 | 0.0048 | 0.0045 | 0.0040 | 0.0035 | 0.0030 | 0.0025 | 0.0025 | 0.0025 | 0.0020 | 0.0015 | 0.0015 | 0.0015 | 0.0005 |
| 5⁄8 | 0.0050 | 0.0050 | 0.0048 | 0.0043 | 0.0040 | 0.0035 | 0.0030 | 0.0030 | 0.0028 | 0.0025 | 0.0020 | 0.0020 | 0.0018 | 0.0010 |
| 3⁄4 | 0.0055 | 0.0052 | 0.0050 | 0.0045 | 0.0043 | 0.0040 | 0.0035 | 0.0035 | 0.0030 | 0.0028 | 0.0025 | 0.0022 | 0.0020 | 0.0013 |
| 7⁄8 | 0.0058 | 0.0055 | 0.0052 | 0.0048 | 0.0045 | 0.0043 | 0.0040 | 0.0038 | 0.0032 | 0.0030 | 0.0028 | 0.0025 | 0.0022 | 0.0015 |
| 1 | 0.0060 | 0.0058 | 0.0054 | 0.0050 | 0.0048 | 0.0047 | 0.0043 | 0.0040 | 0.0035 | 0.0032 | 0.0030 | 0.0028 | 0.0025 | 0.0018 |

teeth on each gear, the increase in pitch diameter per tooth varies according to the pitch. For example, the pitch diameter of a gear having, say, 20 teeth of 4 diametral pitch will be 5 inches; 21 teeth, $5\frac{1}{4}$ inches; and so on, the increase in diameter for each additional tooth being equal to $\frac{1}{4}$ inch for 4 diametral pitch. Similarly, for 2 diametral pitch the variations for successive numbers of teeth would equal $\frac{1}{2}$ inch, and for 10 diametral pitch the variations would equal $\frac{1}{10}$ inch, etc. Where a given center distance must be maintained and no standard diametral pitch can be used, gears should be designed with reference to the gear set center distance procedure discussed in *Gears for Given Center Distance and Ratio* starting on page 2220.

**Table 1. Formulas for Dimensions of Standard Spur Gears**

| To Find | Formula | To Find | Formula |
|---|---|---|---|
| Base Circle Diameter | $D_B = D\cos\phi$ (1) | Number of Teeth | $N = P \times D$ (2a) |
| Circular Pitch | $p = \frac{3.1416D}{N}$ (3a) | | $N = \frac{3.1416D}{p}$ (3b) |
| | $p = \frac{3.1416}{P}$ (3c) | Outside Diameter (Full-depth Teeth) | $D_O = \frac{N+2}{P}$ (4a) |
| Center Distance | $C = \frac{N_P(m_G + 1)}{2P}$ (5a) | | $D_O = \frac{(N+2)p}{3.1416}$ (5b) |
| | $C = \frac{D_P + D_G}{2}$ (5c) | Outside Diameter (American Standard Stub Teeth) | $D_O = \frac{N+1.6}{P}$ (6a) |
| | $C = \frac{N_G + N_P}{2P}$ (6b) | | $D_O = \frac{(N+1.6)p}{3.1416}$ (6c) |
| | $C = \frac{(N_G + N_P)p}{6.2832}$ (6d) | Outside Diameter | $D_O = D + 2a$ (7) |
| Diametral Pitch | $P = \frac{3.1416}{p}$ (8a) | Pitch Diameter | $D = \frac{N}{P}$ (9a) |
| | $P = \frac{N}{D}$ (9b) | | $D = \frac{Np}{3.1416}$ (9c) |
| | $P = \frac{N_P(m_G + 1)}{2C}$ (9d) | Root Diameter[a] | $D_R = D - 2b$ (10) |
| Gear Ratio | $m_G = \frac{N_G}{N_P}$ (11) | Whole Depth | $a + b$ (12) |
| | | Working Depth | $a_G + a_P$ (13) |

[a] See also formulas in Table 2 and Table 4 on pages 2212 and 2216.

**Table 2. Formulas for Tooth Parts, 20-and 25-Degree Involute Full-Depth Teeth ANSI Coarse Pitch Spur Gear Tooth Forms** *ANSI B6.1-1968 (R1974)*

| To Find | Diametral Pitch, $P$, Known | Circular Pitch, $p$, Known |
|---|---|---|
| Addendum | $a = 1.000 \div P$ | $a = 0.3183 \times p$ |
| Dedendum (Preferred) | $b = 1.250 \div P$ | $b = 0.3979 \times p$ |
| (Shaved or Ground Teeth)[a] | $b = 1.350 \div P$ | $b = 0.4297 \times p$ |
| Working Depth | $h_k = 2.000 \div P$ | $h_k = 0.6366 \times p$ |
| Whole Depth (Preferred) | $h_t = 2.250 \div P$ | $h_t = 0.7162 \times p$ |
| (Shaved or Ground Teeth) | $h_t = 2.350 \div P$ | $h_t = 0.7480 \times p$ |
| Clearance (Preferred)[b] | $c = 0.250 \div P$ | $c = 0.0796 \times p$ |
| (Shaved or Ground Teeth) | $c = 0.350 \div P$ | $c = 0.1114 \times p$ |

$$t_{G2} = \frac{N_G}{P}\left(\frac{1.5708}{N_G} + \text{inv}_{\phi_1} - \text{inv}_{\phi}\right) \qquad (8b)$$

The outside diameter of the pinion $D_{PO}$ and the gear $D_{GO}$ are

$$D_{PO} = 2 \times C_1 - D_{GR1} - 2\left(b_c - \frac{1}{P}\right) \qquad (9a)$$

and $$D_{GO} = 2 \times C_1 - D_{PR1} - 2\left(b_c - \frac{1}{P}\right) \qquad (9b)$$

*Example:* Design gears of 8 diametral pitch, 20-degree pressure angle, and 28 and 88 teeth to operate at 7.50-inch center distance. The gears are to be cut with a hob of 0.169-inch addendum.

$$P_1 = \frac{28 + 88}{2 \times 7.50} = 7.7333 \qquad (1)$$

$$\phi_1 = \arccos\left(\frac{7.7333}{8} \times 0.93969\right) = 24.719° \qquad (2)$$

$$D_{P1} = \frac{28}{7.7333} = 3.6207 \text{ in.} \qquad (3a)$$

and $$D_{G1} = \frac{88}{7.7333} = 11.3794 \text{ in.} \qquad (3b)$$

$$D_{PB1} = 3.6207 \times 0.90837 = 3.2889 \text{ in.} \qquad (4a)$$

and $$D_{GB1} = 11.3794 \times 0.90837 = 10.3367 \text{ in.} \qquad (4b)$$

$$t_1 = \frac{1.5708}{7.7333} = 0.20312 \text{ in.} \qquad (5)$$

$$D_{PR1} = \frac{28}{8} - 2 \times 0.1016 = 3.2968 \text{ in.} \qquad (6a)$$

and $$D_{GR1} = \frac{88}{8} - 2 \times (-0.0428) = 11.0856 \text{ in.} \qquad (6b)$$

$$b_{P1} = 0.169 - \left(\frac{0.2454 - \frac{1.5708}{8}}{2 \times 0.36397}\right) = 0.1016 \text{ in.} \qquad (7a)$$

$$b_{G1} = 0.169 - \left(\frac{0.3505 - \frac{1.5708}{8}}{2 \times 0.36397}\right) = -0.0428 \text{ in.} \qquad (7b)$$

$$t_{P2} = \frac{28}{8}\left(\frac{1.5708}{28} + 0.028922 - 0.014904\right) = 0.2454 \text{ in.} \qquad (8a)$$

$$t_{G2} = \frac{88}{8}\left(\frac{1.5708}{88} + 0.028922 - 0.014904\right) - 0.3505 \text{ in.} \qquad (8b)$$

$$D_{PO1} = 2 \times 7.50 - 11.0856 - 2\left(0.169 - \frac{1}{8}\right) = 3.8264 \text{ in.} \qquad (9a)$$

$$D_{GO1} = 2 \times 7.50 - 3.2968 - 2\left(0.169 - \frac{1}{8}\right) = 11.6152 \text{ in.} \qquad (9b)$$

The tooth thickness on the pitch circle can be determined very accurately by means of measurement over wires which are located in tooth spaces that are diametrically opposite or as nearly diametrically opposite as possible. Where measurement over wires is not feasible, the circular or arc tooth thickness can be used in determining the chordal thickness which is the dimension measured with a gear tooth caliper.

**Circular Thickness of Tooth when Outside Diameter has been Enlarged.—**When the outside diameter of a small pinion is not standard but is enlarged to avoid undercut and to improve tooth action, the teeth are located farther out radially relative to the standard pitch diameter and consequently the circular tooth thickness at the standard pitch diameter is increased. To find this increased arc thickness the following formula is used, where $t$ = tooth thickness; $e$ = amount outside diameter is increased over standard; $\phi$ = pressure angle; and $p$ = circular pitch at the standard pitch diameter.

$$t = \frac{p}{2} + e\tan\phi$$

*Example:* The outside diameter of a pinion having 10 teeth of 5 diametral pitch and a pressure angle of $14\frac{1}{2}$ degrees is to be increased by 0.2746 inch. The circular pitch equivalent to 5 diametral pitch is 0.6283 inch. Find the arc tooth thickness at the standard pitch diameter.

$$t = \frac{0.6283}{2} + \left(0.2746 \times \tan 14\frac{1}{2}^\circ\right)$$
$$t = 0.3142 + (0.2746 \times 0.25862) = 0.3852 \text{ inch}$$

**Circular Thickness of Tooth when Outside Diameter has been Reduced.—**If the outside diameter of a gear is reduced, as is frequently done to maintain the standard center distance when the outside diameter of the mating pinion is increased, the circular thickness of the gear teeth at the standard pitch diameter will be reduced.This decreased circular thickness can be found by the following formula where $t$ = circular thickness at the standard pitch diameter; $e$ = amount outside diameter is reduced under standard; $\phi$ = pressure angle; and $p$ = circular pitch.

$$t = \frac{p}{2} - e\tan\phi$$

*Example:* The outside diameter of a gear having a pressure angle of $14\frac{1}{2}$ degrees is to be reduced by 0.2746 inch or an amount equal to the increase in diameter of its mating pinion. The circular pitch is 0.6283 inch. Determine the circular tooth thickness at the standard pitch diameter.

$$t = \frac{0.6283}{2} - \left(0.2746 \times \tan 14\frac{1}{2}^\circ\right)$$
$$t = 0.3142 - (0.2746 \times 0.25862) = 0.2432 \text{ inch}$$

**Chordal Thickness of Tooth when Outside Diameter is Standard.—**To find the chordal or straight line thickness of a gear tooth the following formula can be used where $t_c$ = chordal thickness; $D$ = pitch diameter; and $N$ = number of teeth.

$$t_c = D\sin\left(\frac{90^\circ}{N}\right)$$

*Example:* A pinion has 15 teeth of 3 diametral pitch; the pitch diameter is equal to 15 ÷ 3 or 5 inches. Find the chordal thickness at the standard pitch diameter.

$$t_c = 5\sin\left(\frac{90^\circ}{15}\right) = 5\sin 6^\circ = 5 \times 0.10453 = 0.5226 \text{ inch}$$

## Chordal Thicknesses and Chordal Addenda of Milled, Full-Depth Gear Teeth and of Gear Milling Cutters *(Continued)*

| Circular Pitch | Dimension | Number of Gear Cutter, and Corresponding Number of Teeth | | | | | | | |
|---|---|---|---|---|---|---|---|---|---|
| | | No. 1 135 Teeth | No. 2 55 Teeth | No. 3 35 Teeth | No. 4 26 Teeth | No. 5 21 Teeth | No. 6 17 Teeth | No. 7 14 Teeth | No. 8 12 Teeth |
| 1/4 | T | 0.1250 | 0.1250 | 0.1249 | 0.1249 | 0.1249 | 0.1248 | 0.1247 | 0.1246 |
| | H | 0.0799 | 0.0804 | 0.0809 | 0.0814 | 0.0819 | 0.0824 | 0.0830 | 0.0836 |
| 5/16 | T | 0.1562 | 0.1562 | 0.1562 | 0.1561 | 0.1561 | 0.1560 | 0.1559 | 0.1558 |
| | H | 0.0999 | 0.1006 | 0.1012 | 0.1018 | 0.1023 | 0.1030 | 0.1038 | 0.1045 |
| 3/8 | T | 0.1875 | 0.1875 | 0.1874 | 0.1873 | 0.1873 | 0.1872 | 0.1871 | 0.1870 |
| | H | 0.1199 | 0.1207 | 0.1214 | 0.1221 | 0.1228 | 0.1236 | 0.1245 | 0.1254 |
| 7/16 | T | 0.2187 | 0.2187 | 0.2186 | 0.2186 | 0.2185 | 0.2184 | 0.2183 | 0.2181 |
| | H | 0.1399 | 0.1408 | 0.1416 | 0.1425 | 0.1433 | 0.1443 | 0.1453 | 0.1464 |
| 1/2 | T | 0.2500 | 0.2500 | 0.2499 | 0.2498 | 0.2498 | 0.2496 | 0.2495 | 0.2493 |
| | H | 0.1599 | 0.1609 | 0.1619 | 0.1629 | 0.1638 | 0.1649 | 0.1661 | 0.1673 |
| 9/16 | T | 0.2812 | 0.2812 | 0.2811 | 0.2810 | 0.2810 | 0.2808 | 0.2806 | 0.2804 |
| | H | 0.1799 | 0.1810 | 0.1821 | 0.1832 | 0.1842 | 0.1855 | 0.1868 | 0.1882 |
| 5/8 | T | 0.3125 | 0.3125 | 0.3123 | 0.3123 | 0.3122 | 0.3120 | 0.3118 | 0.3116 |
| | H | 0.1998 | 0.2012 | 0.2023 | 0.2036 | 0.2047 | 0.2061 | 0.2076 | 0.2091 |
| 11/16 | T | 0.3437 | 0.3437 | 0.3436 | 0.3435 | 0.3434 | 0.3432 | 0.3430 | 0.3427 |
| | H | 0.2198 | 0.2213 | 0.2226 | 0.2239 | 0.2252 | 0.2267 | 0.2283 | 0.2300 |
| 3/4 | T | 0.3750 | 0.3750 | 0.3748 | 0.3747 | 0.3747 | 0.3744 | 0.3742 | 0.3740 |
| | H | 0.2398 | 0.2414 | 0.2428 | 0.2443 | 0.2457 | 0.2473 | 0.2491 | 0.2509 |
| 13/16 | T | 0.4062 | 0.4062 | 0.4060 | 0.4059 | 0.4059 | 0.4056 | 0.4054 | 0.4050 |
| | H | 0.2598 | 0.2615 | 0.2631 | 0.2647 | 0.2661 | 0.2679 | 0.2699 | 0.2718 |
| 7/8 | T | 0.4375 | 0.4375 | 0.4373 | 0.4372 | 0.4371 | 0.4368 | 0.4366 | 0.4362 |
| | H | 0.2798 | 0.2816 | 0.2833 | 0.2850 | 0.2866 | 0.2885 | 0.2906 | 0.2927 |
| 15/16 | T | 0.4687 | 0.4687 | 0.4685 | 0.4684 | 0.4683 | 0.4680 | 0.4678 | 0.4674 |
| | H | 0.2998 | 0.3018 | 0.3035 | 0.3054 | 0.3071 | 0.3092 | 0.3114 | 0.3137 |
| 1 | T | 0.5000 | 0.5000 | 0.4998 | 0.4997 | 0.4996 | 0.4993 | 0.4990 | 0.4986 |
| | H | 0.3198 | 0.3219 | 0.3238 | 0.3258 | 0.3276 | 0.3298 | 0.3322 | 0.3346 |
| 1 1/8 | T | 0.5625 | 0.5625 | 0.5623 | 0.5621 | 0.5620 | 0.5617 | 0.5613 | 0.5610 |
| | H | 0.3597 | 0.3621 | 0.3642 | 0.3665 | 0.3685 | 0.3710 | 0.3737 | 0.3764 |
| 1 1/4 | T | 0.6250 | 0.6250 | 0.6247 | 0.6246 | 0.6245 | 0.6241 | 0.6237 | 0.6232 |
| | H | 0.3997 | 0.4023 | 0.4047 | 0.4072 | 0.4095 | 0.4122 | 0.4152 | 0.4182 |
| 1 3/8 | T | 0.6875 | 0.6875 | 0.6872 | 0.6870 | 0.6869 | 0.6865 | 0.6861 | 0.6856 |
| | H | 0.4397 | 0.4426 | 0.4452 | 0.4479 | 0.4504 | 0.4534 | 0.4567 | 0.4600 |
| 1 1/2 | T | 0.7500 | 0.7500 | 0.7497 | 0.7495 | 0.7494 | 0.7489 | 0.7485 | 0.7480 |
| | H | 0.4797 | 0.4828 | 0.4857 | 0.4887 | 0.4914 | 0.4947 | 0.4983 | 0.5019 |
| 1 3/4 | T | 0.8750 | 0.8750 | 0.8746 | 0.8744 | 0.8743 | 0.8737 | 0.8732 | 0.8726 |
| | H | 0.5596 | 0.5633 | 0.5666 | 0.5701 | 0.5733 | 0.5771 | 0.5813 | 0.5855 |
| 2 | T | 1.0000 | 1.0000 | 0.9996 | 0.9994 | 0.9992 | 0.9986 | 0.9980 | 0.9972 |
| | H | 0.6396 | 0.6438 | 0.6476 | 0.6516 | 0.6552 | 0.6596 | 0.6644 | 0.6692 |
| 2 1/4 | T | 1.1250 | 1.1250 | 1.1246 | 1.1242 | 1.1240 | 1.1234 | 1.1226 | 1.1220 |
| | H | 0.7195 | 0.7242 | 0.7285 | 0.7330 | 0.7371 | 0.7420 | 0.7474 | 0.7528 |
| 2 1/2 | T | 1.2500 | 1.2500 | 1.2494 | 1.2492 | 1.2490 | 1.2482 | 1.2474 | 1.2464 |
| | H | 0.7995 | 0.8047 | 0.8095 | 0.8145 | 0.8190 | 0.8245 | 0.8305 | 0.8365 |
| 3 | T | 1.5000 | 1.5000 | 1.4994 | 1.4990 | 1.4990 | 1.4978 | 1.4970 | 1.4960 |
| | H | 0.9594 | 0.9657 | 0.9714 | 0.9774 | 0.9828 | 0.9894 | 0.9966 | 1.0038 |

1) Pinion contact diameter, $d_c$

$$\cos A = \frac{31 \times 0.93969}{3.3 \times 10} = 0.88274 \qquad A = 28°1'30'' \tag{1}$$

$$\tan b = 0.36397 - \frac{31}{23}(0.53227 - 0.36397) = 0.13713 \qquad b = 7°48'26'' \tag{2}$$

$$d_c = \frac{23 \times 0.93969}{10 \times 0.99073} = 2.1815 \text{ inches} \tag{3}$$

2) Gear contact diameter, $D_c$

$$\cos a = \frac{23 \times 0.93963}{2.5 \times 10} = 0.86452 \qquad a = 30°10'20'' \tag{4}$$

$$\tan B = 0.36397 - \frac{23}{31}(0.58136 - 0.36937) = 0.20267 \qquad B = 11°27'26'' \tag{5}$$

$$D_c = \frac{31 \times 0.93969}{10 \times 0.98000} = 2.9725 \text{ inches} \tag{6}$$

3) Contact ratio, $m_f$

$$m_f = \frac{31}{6.28318}(0.53227 - 0.20267) = 1.626 \tag{7a}$$

$$m_f = \frac{23}{6.28318}(0.58136 - 0.13713) = 1.626 \tag{7b}$$

4) Pinion LPSTC, $d_L$

$$\tan c = 0.58136 - \frac{6.28318}{23} = 0.30818 \qquad c = 17°7'41'' \tag{8}$$

$$d_L = \frac{23 \times 0.93969}{10 \times 0.95565} = 2.2616 \text{ inches} \tag{9}$$

5) Gear LPSTC, $D_L$

$$\tan C = 0.53227 - \frac{6.28318}{31} = 0.32959 \qquad C = 18°14'30'' \tag{10}$$

$$D_L = \frac{31 \times 0.93969}{10 \times 0.94974} = 3.0672 \text{ inches} \tag{11}$$

*Application Factor* allows for load fluctuations from the mean load or loads in the load histogram caused by sources external to the gearing.

*Dynamic Factor* allows for load fluctuations arising from contact conditions at the gear mesh.

*Load Distribution* accounts for the increase in local load due to maldistribution of load across the face of the gear tooth caused by deflections, alignment tolerances and helix modifications.

*Minimum Demanded and Actual Safety Factor* The minimum demanded safety factor is agreed between the supplier and the purchaser. The actual safety factor is calculated

*Geometry Factors* allow for the influence of the tooth form, the effect of the fillet and the helix angle on the nominal bending stress for the application of load at the highest point of single pair tooth contact.

*Sensitivity Factor* allows for the sensitivity of the gear material to the presence of notches such as the root fillet.

*Surface Condition Factor* accounts for reduction of the endurance limit due to flaws in the material and the surface roughness of the tooth root fillets.

**ISO TC/600.—**The ISO TC/600 Standard is similar to BS 436: Part 3:1986 (Withdrawn), but is far more comprehensive. For general gear design, the ISO Standard provides a complicated method of arriving at a conclusion similar to that reached by the less complex British Standard. Factors additional to the above that are included in the ISO Standard include the following:

*Application Factor* accounts for dynamic overloads from sources external to the gearing.

*Dynamic Factor* allows for internally generated dynamic loads caused by vibrations of the pinion and wheel against each other.

*Load Distribution* makes allowance for the effects of non-uniform distribution of load across the face width, depending on the mesh alignment error of the loaded gear pair and the mesh stiffness.

*Transverse Load Distribution Factor* takes into account the effect of the load distribution on gear tooth contact stresses.

*Gear Tooth Stiffness Constants* are defined as the load needed to deform one or several meshing gear teeth having 1 mm face width, by an amount of 1 μm (0.00004 in).

*Allowable Contact Stress* is the permissible Hertzian pressure on the gear tooth face.

*Minimum Demanded and Calculated Safety Factors* The minimum demanded safety factor is agreed between the supplier and the customer. The calculated safety factor is the actual safety factor of the gear pair.

*Zone Factor* accounts for the influence on the Hertzian pressure of the tooth flank curvature at the pitch point.

*Elasticity Factor* takes account of the influence of the material properties such as the modulus of elasticity and Poisson's ratio.

*Contact Ratio Factor* accounts for the influence of the transverse contact ratio and the overlap ratio on the specific surface load of the gears.

*Helix Angle Factor* makes allowance for influence of helix angle on surface durability.

*Endurance Limit* is the limit of repeated Hertzian stresses that can be permanently endured by a given material.

*Life Factor* takes account of a higher permissible Hertzian stress if only limited durability is demanded.

*Lubrication Film Factor* refers to when the film of lubricant between the tooth flanks influences the surface load capacity. Factors include the oil viscosity, pitch line velocity and roughness of the tooth flanks.

*Work Hardening Factor* takes account of the increase in surface durability due to meshing a steel wheel with a hardened pinion having smooth tooth surfaces.

*Coefficient of Friction's* mean value depends on the lubricant, surface roughness, lay of surface irregularities, material properties of the tooth flanks, and the force and size of tangential velocities.

*Bulk Temperature Thermal Flash Factor* is dependent on moduli of elasticity and thermal contact coefficients of pinion and wheel materials and geometry of the line of action.

*Welding Factor* accounts for different tooth materials and heat treatments.

*Geometrical Factor* is defined as a function of the gear ratio and the dimensionless parameter on the line of action.

*Integral Temperature Criterion* references the integral temperature of the gears, which depends on the lubricant viscosity and tendency toward scuffing and scoring of the gear materials.

Examination of the above factors shows the similarity in the approach of the British and the ISO standards to that of the ANSI/AGMA standards. Slight variations in the methods used to calculate the factors will result in different allowable stress figures. Experimental work using some of the stressing formulas has shown wide variations and designers must continue to rely on experience to arrive at satisfactory results.

## Standards Nomenclature

All standards are referenced and identified throughout this book by an alphanumeric prefix which designates the organization that administered the development work on the standard, and followed by the specific standard's number.

All standards are reviewed by the relevant committees at regular time intervals, as specified by the overseeing standards organization, to determine whether a standard should be confirmed (reissued without changes other than correction of typographical errors), updated, or removed from service.

The following is for example use only. ANSI B18.8.2-1984, R1994 is a Standard for Taper, Dowel, Straight, Grooved, and Spring Pins. ANSI refers to the American National Standards Institute that is responsible for overseeing the development or approval of the Standard, and B18.8.2 is the number of the Standard. The first date, 1984, indicates the year in which the Standard was issued, and the sequence R1994 (usually set in parentheses) indicates that this Standard was reviewed and reaffirmed in 1994. The current designation of the Standard, ANSI/ASME B18.8.2-2020, indicates that it was revised in 2020; it is ANSI approved; and ASME (American Society of Mechanical Engineers) was the standards body responsible for development of the Standard. This Standard is sometimes also designated simply as ASME B18.8.2-2020.

ISO (International Organization for Standardization) standards use a slightly different format, for example, ISO 5127:2017. The entire ISO reference number consists of a prefix ISO, a serial number, and the year of publication.

Aside from content, ISO standards differ from American National standards in that they are often smaller focused documents, which in turn reference other standards or other parts of the same standard. Unlike the numbering scheme used by ANSI, ISO standards related to a particular topic often do not carry sequential numbers nor are they in consecutive series.

British Standards Institute standards use the following format: BS 1361: 1971 (1986). The first part is the organization prefix BS, followed by the reference number and the date of issue. The number in parenthesis is the date that the Standard was most recently reconfirmed. British Standards may also be designated *withdrawn* (no longer to be used) and *obsolescent* (going out of use, but may be used for servicing older equipment).

| Organization | Web Address | Organization | Web Address |
|---|---|---|---|
| ANSI (American National Standards Institute) | www.ansi.org | ISO (International Organization for Standardization) | www.iso.org |
| ASME (American Society of Mechanical Engineers) | www.asme.org | JSA (Japanese Standards Association Group) | www.jsa.or.jp/en/ |
| BSI (British Standards Institute) | www.bsigroup.com | SAE International (formerly Society of Automotive Engineers) | www.sae.org |
| IEC (International Electrotechnical Commission) | www.iec.ch | SME (Society of Manufacturing Engineers) | www.sme.org |

opposite direction the same amount from a position central with the cutter, and is rotated to line up the cutter with a tooth space at the small end. A trial cut is then taken, which will leave the tooth being milled a little too thick, provided the cutter is thin enough—as it should be—to pass through the small end of the tooth space of the finished gear. This trial tooth is made the proper thickness by rotating the blank toward the cutter. To test the amount of offset, measure the tooth thickness (with a vernier caliper) at the large and small ends. The caliper should be set so that the addendum at the small end is in proper proportion to the addendum at the large end; that is, in the ratio $(C-F)/C$ (see Fig. 2).

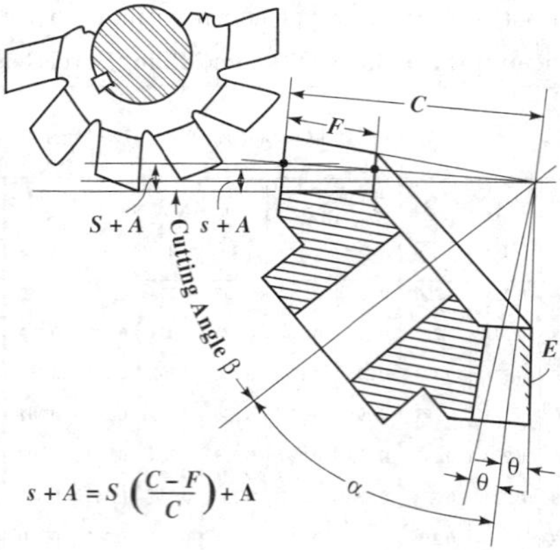


Fig. 2.

In taking these measurements, if the thicknesses at both ends (which should be in this same ratio) are too great, rotate the tooth toward the cutter and take trial cuts until the proper thickness at either the large or small end is obtained. If the large end of the tooth is the right thickness and the small end too thick, the blank was offset too much; inversely, if the small end is correct and the large end too thick, the blank was not set enough off center, and, either way, its position should be changed accordingly. The formula and table previously referred to will enable a properly turned blank to be set accurately enough for general work. The dividing head should be set to the cutting angle β (see Fig. 2), which is found by subtracting the addendum angle θ from the pitch cone angle α. After a bevel gear is cut by the method described, the sides of the teeth at the small end should be filed as indicated by the shade lines at *E*; that is, by filing off a triangular area from the point of the tooth at the large end to the point at the small end, thence down to the pitch line and back diagonally to a point at the large end.

**Circular Thickness, Chordal Thickness, and Chordal Addendum of Milled Bevel Gear Teeth.**—In the formulas that follow, $T$ = circular tooth thickness on pitch circle at large end of tooth; $t$ = circular thickness at small end; $T_c$ and $t_c$ = chordal thickness at large and small ends, respectively; $S_c$ and $s_c$ = chordal addendum at large and small ends, respectively; $D$ = pitch diameter at large end; and $C, F, P, S, s$, and $\alpha$ are as defined on page 2262.

$$T = \frac{1.5708}{P} \qquad T_c = T - \frac{T^3}{6D^2} \qquad S_c = S + \frac{T^2 \cos\alpha}{4D}$$

$$t = \frac{T(C-F)}{C} \qquad t_c = t - \frac{t^3}{6(D-2F\sin\alpha)^2} \qquad s_c = s + \frac{t^2 \cos\alpha}{4(D-2F\sin\alpha)}$$

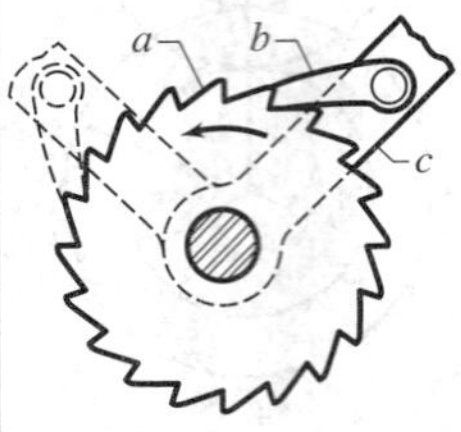

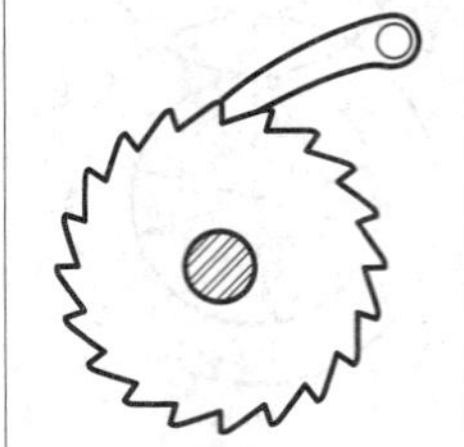
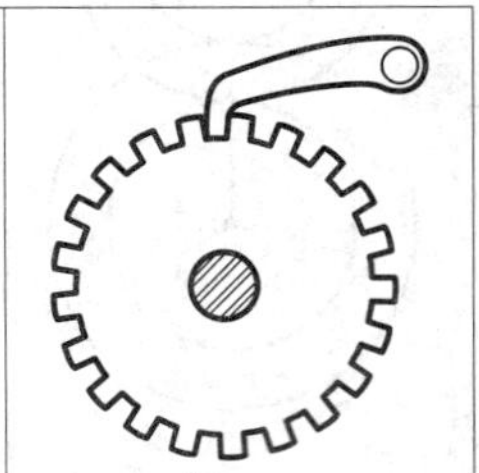
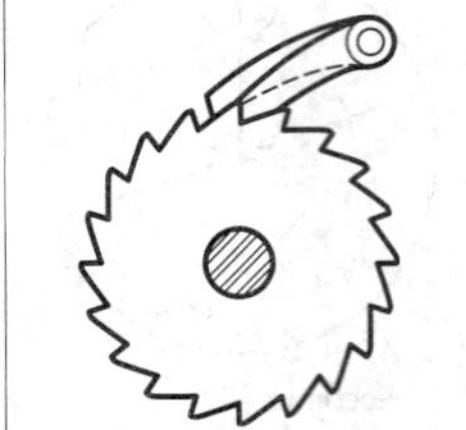
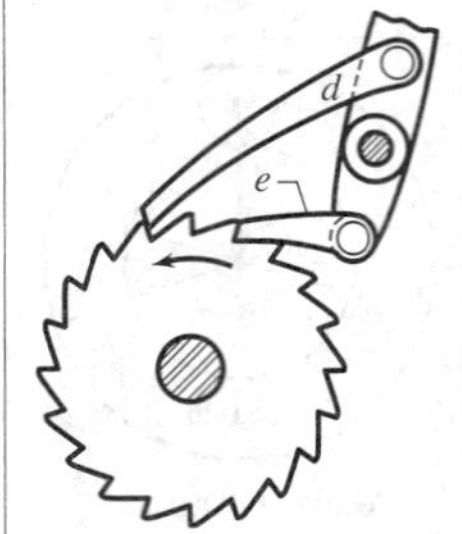

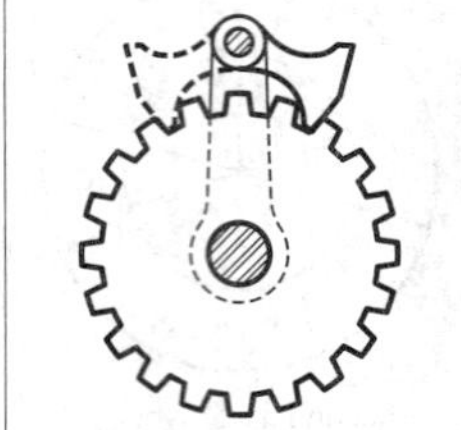
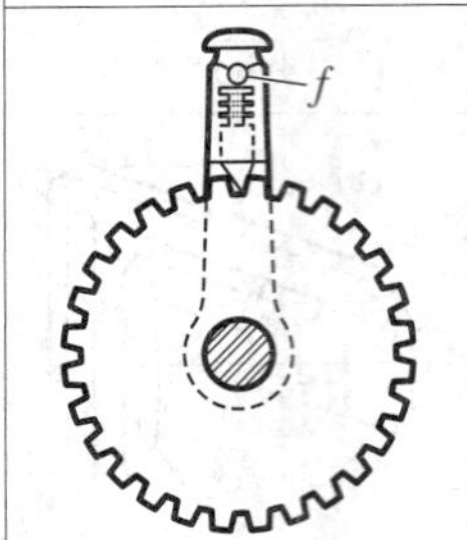

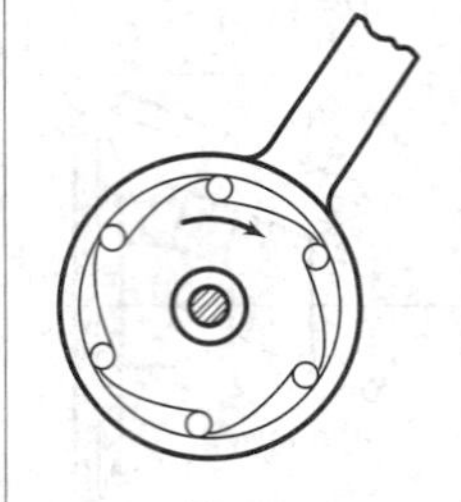
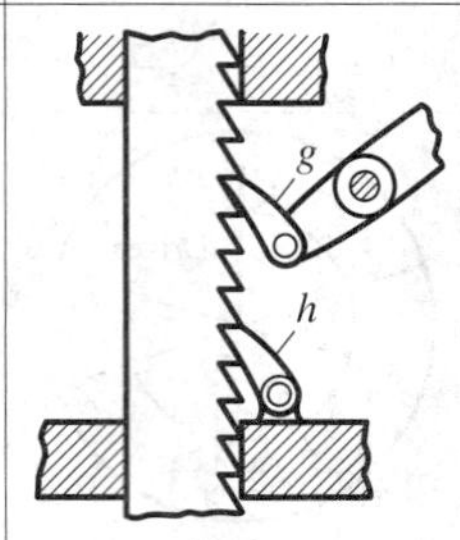

**Table 13. Formulas for Dimensions and Tolerances for All Fit Classes— Metric Module Involute Splines**

| Term | Symbol | Formula | | | |
|---|---|---|---|---|---|
| | | 30-Degree Flat Root 0.5 to 10 module | 30-Degree Fillet Root 0.5 to 10 module | 37.5-Degree Fillet Root 0.5 to 10 module | 45-Degree Fillet Root 0.25 to 2.5 module |
| Pitch Diameter | $D$ | $mZ$ | | | |
| Base Diameter | $DB$ | $mZ \cos \alpha_D$ | | | |
| Circular Pitch | $p$ | $\pi m$ | | | |
| Base Pitch | $p_b$ | $\pi m \cos \alpha_D$ | | | |
| Tooth Thick Mod | $es$ | According to selected fit class, H/h, H/f, H/e, or H/d (see Table 14) | | | |
| Min. Maj. Dia. Int. | $DEI$ min | $m(Z + 1.5)$ | $m(Z + 1.8)$ | $m(Z + 1.4)$ | $m(Z + 1.2)$ |
| Max. Maj. Dia. Int. | $DEI$ max | $DEI$ min + $(T + \lambda)/\tan \alpha_D$ (see Footnote [a]) | | | |
| Form Dia., Int. | $DFI$ | $m(Z + 1) + 2c_F$ | $m(Z + 1) + 2c_F$ | $m(Z + 0.9) + 2c_F$ | $m(Z + 0.8) + 2c_F$ |
| Min. Minor Dia., Int. | $DII$ min | $DFE + 2c_F$ (see Footnote [b]) | | | |
| Max. Minor Dia., Int. | $DII$ max | $DII$ min + $(0.2m^{0.667} - 0.01m^{-0.5})$ (see Footnote [c]) | | | |
| Cir Space Width, | | | | | |
| Basic | $E_{bsc}$ | $0.5\pi m$ | | | |
| Min. Effective | $EV$ min | $0.5\pi m$ | | | |
| Max. Actual | $E$ max | $EV$ min + $(T + \lambda)$ for classes 4, 5, 6, and 7 (see Table 15 for $T + \lambda$) | | | |
| Min. Actual | $E$ min | $EV$ min + $\lambda$ (see text on page 2357 for $\lambda$) | | | |
| Max. Effective | $EV$ max | $E$ max − $\lambda$ (see text on page 2357 for $\lambda$) | | | |
| Max. Major Dia., Ext[d] | $DEE$ max | $m(Z + 1) - es/\tan\alpha_D$ | $m(Z + 1) - es/\tan\alpha_D$ | $m(Z + 0.9) - es/\tan\alpha_D$ | $m(Z + 0.8) - es/\tan\alpha_D$ |
| Min. Major Dia., Ext | $DEE$ min | $DEE$ max − $(0.2m^{0.667} - 0.01m^{-0.5})$ (see Footnote [c]) | | | |
| Form Dia., External | $DFE$ | $2 \times \sqrt{(0.5DB)^2 + \left[0.5D \sin\alpha_D - \dfrac{h_s + (0.5es / \tan\alpha_D)}{\sin\alpha_D}\right]^2}$ | | | |
| Max. Minor Dia., Ext[d] | $DIE$ max | $m(Z - 1.5) - es/\tan\alpha_D$ | $m(Z - 1.8) - es/\tan\alpha_D$ | $m(Z - 1.4) - es/\tan\alpha_D$ | $m(Z - 1.2) - es/\tan\alpha_D$ |
| Min. Minor Dia., Ext | $DIE$ min | $DIE$ max − $(T + \lambda)/\tan \alpha_D$ (see Footnote [a]) | | | |
| Cir Tooth Thick, | | | | | |
| Basic | $S_{bsc}$ | $0.5\pi m$ | | | |
| Max. Effective | $SV$ max | $S_{bsc} - es$ | | | |
| Min. Actual | $S$ min | $SV$ max − $(T + \lambda)$ for classes 4, 5, 6, and 7 (see Table 15 for $T + \lambda$) | | | |
| Max. Actual | $S$ max | $SV$ max − $\lambda$ (see text on page 2357 for $\lambda$) | | | |
| Min. Effective | $SV$ min | $S$ min + $\lambda$ (see text on page 2357 for $\lambda$) | | | |
| Total Tolerance on Circular Space Width or Tooth Thickness | $(T + \lambda)$ | See formulas in Table 15 | | | |
| Machining Tolerance on Circular Space Width or Tooth Thickness | $T$ | $T = (T + \lambda)$ from Table 15; $\lambda$ from text on page 2357. | | | |
| Effective Variation Allowed on Circular Space Width or Tooth Thickness | $\lambda$ | See text on page 2357. | | | |
| Form Clearance | $c_F$ | $0.1m$ | | | |
| Rack Dimension | $h_s$ | $0.6m$ (see Fig. 6a) | $0.6m$ (see Fig. 6b) | $0.55m$ (see Fig. 6c) | $0.5m$ (see Fig. 6d) |

[a] Use $(T + \lambda)$ for class 7 from Table 15

[b] For all types of fit, always use the $DFE$ value corresponding to the H/h fit.

[c] Values of $(0.2m^{0.667} - 0.01m^{-0.5})$ are as follows: for 10 module, 0.93; for 8 module, 0.80; for 6 module, 0.66; for 5 module, 0.58; for 4 module, 0.50; for 3 module, 0.41; for 2.5 module, 0.36; for 2 module, 0.31; for 1.75 module, 0.28; for 1.5 module, 0.25; for 1.25 module, 0.22; for 1 module, 0.19; for 0.75 module, 0.15; for 0.5 module, 0.11; and for 0.25 module, 0.06.

[d] See Table 17 for values of $es/\tan \alpha_D$.

**Table 20. SAE Standard 10-Spline Fittings**

| Nom. Dia. | For All Fits | | | | 10A—Permanent Fit | | | 10B—To Slide, No Load | | | 10C—To Slide Under Load | | |
|---|---|---|---|---|---|---|---|---|---|---|---|---|---|
| | D | | W | | d | | | d | | | d | | |
| | Min. | Max. | Min. | Max. | Min. | Max. | T[a] | Min. | Max. | T[a] | Min. | Max. | T[a] |
| 3/4 | 0.749 | 0.750 | 0.115 | 0.117 | 0.682 | 0.683 | 120 | 0.644 | 0.645 | 183 | 0.607 | 0.608 | 241 |
| 7/8 | 0.874 | 0.875 | 0.135 | 0.137 | 0.795 | 0.796 | 165 | 0.752 | 0.753 | 248 | 0.708 | 0.709 | 329 |
| 1 | 0.999 | 1.000 | 0.154 | 0.156 | 0.909 | 0.910 | 215 | 0.859 | 0.860 | 326 | 0.809 | 0.810 | 430 |
| 1 1/8 | 1.124 | 1.125 | 0.174 | 0.176 | 1.023 | 1.024 | 271 | 0.967 | 0.968 | 412 | 0.910 | 0.911 | 545 |
| 1 1/4 | 1.249 | 1.250 | 0.193 | 0.195 | 1.137 | 1.138 | 336 | 1.074 | 1.075 | 508 | 1.012 | 1.013 | 672 |
| 1 3/8 | 1.374 | 1.375 | 0.213 | 0.215 | 1.250 | 1.251 | 406 | 1.182 | 1.183 | 614 | 1.113 | 1.114 | 813 |
| 1 1/2 | 1.499 | 1.500 | 0.232 | 0.234 | 1.364 | 1.365 | 483 | 1.289 | 1.290 | 732 | 1.214 | 1.215 | 967 |
| 1 5/8 | 1.624 | 1.625 | 0.252 | 0.254 | 1.478 | 1.479 | 566 | 1.397 | 1.398 | 860 | 1.315 | 1.316 | 1135 |
| 1 3/4 | 1.749 | 1.750 | 0.271 | 0.273 | 1.592 | 1.593 | 658 | 1.504 | 1.505 | 997 | 1.417 | 1.418 | 1316 |
| 2 | 1.998 | 2.000 | 0.309 | 0.312 | 1.818 | 1.820 | 860 | 1.718 | 1.720 | 1302 | 1.618 | 1.620 | 1720 |
| 2 1/4 | 2.248 | 2.250 | 0.348 | 0.351 | 2.046 | 2.048 | 1088 | 1.933 | 1.935 | 1647 | 1.821 | 1.823 | 2176 |
| 2 1/2 | 2.498 | 2.500 | 0.387 | 0.390 | 2.273 | 2.275 | 1343 | 2.148 | 2.150 | 2034 | 2.023 | 2.025 | 2688 |
| 3 | 2.998 | 3.000 | 0.465 | 0.468 | 2.728 | 2.730 | 1934 | 2.578 | 2.580 | 2929 | 2.428 | 2.430 | 3869 |
| 3 1/2 | 3.497 | 3.500 | 0.543 | 0.546 | 3.182 | 3.185 | 2632 | 3.007 | 3.010 | 3987 | 2.832 | 2.835 | 5266 |
| 4 | 3.997 | 4.000 | 0.621 | 0.624 | 3.637 | 3.640 | 3438 | 3.437 | 3.440 | 5208 | 3.237 | 3.240 | 6878 |
| 4 1/2 | 4.497 | 4.500 | 0.699 | 0.702 | 4.092 | 4.095 | 4351 | 3.867 | 3.870 | 6591 | 3.642 | 3.645 | 8705 |
| 5 | 4.997 | 5.000 | 0.777 | 0.780 | 4.547 | 4.550 | 5371 | 4.297 | 4.300 | 8137 | 4.047 | 4.050 | 10746 |
| 5 1/2 | 5.497 | 5.500 | 0.855 | 0.858 | 5.002 | 5.005 | 6500 | 4.727 | 4.730 | 9846 | 4.452 | 4.455 | 13003 |
| 6 | 5.997 | 6.000 | 0.933 | 0.936 | 5.457 | 5.460 | 7735 | 5.157 | 5.160 | 11718 | 4.857 | 4.860 | 15475 |

[a] See note at end of Table 21.

**Table 21. SAE Standard 16-Spline Fittings**

| Nom. Dia. | For All Fits | | | | 16A—Permanent Fit | | | 16B—To Slide, No Load | | | 16C—To Slide Under Load | | |
|---|---|---|---|---|---|---|---|---|---|---|---|---|---|
| | D | | W | | d | | | d | | | d | | |
| | Min. | Max. | Min. | Max. | Min. | Max. | T[a] | Min. | Max. | T[a] | Min. | Max. | T[a] |
| 2 | 1.997 | 2.000 | 0.193 | 0.196 | 1.817 | 1.820 | 1375 | 1.717 | 1.720 | 2083 | 1.617 | 1.620 | 2751 |
| 2 1/2 | 2.497 | 2.500 | 0.242 | 0.245 | 2.273 | 2.275 | 2149 | 2.147 | 2.150 | 3255 | 2.022 | 2.025 | 4299 |
| 3 | 2.997 | 3.000 | 0.291 | 0.294 | 2.727 | 2.730 | 3094 | 2.577 | 2.580 | 4687 | 2.427 | 2.430 | 6190 |
| 3 1/2 | 3.497 | 3.500 | 0.340 | 0.343 | 3.182 | 3.185 | 4212 | 3.007 | 3.010 | 6378 | 2.832 | 2.835 | 8426 |
| 4 | 3.997 | 4.000 | 0.389 | 0.392 | 3.637 | 3.640 | 5501 | 3.437 | 3.440 | 8333 | 3.237 | 3.240 | 11005 |
| 4 1/2 | 4.497 | 4.500 | 0.438 | 0.441 | 4.092 | 4.095 | 6962 | 3.867 | 3.870 | 10546 | 3.642 | 3.645 | 13928 |
| 5 | 4.997 | 5.000 | 0.487 | 0.490 | 4.547 | 4.550 | 8595 | 4.297 | 4.300 | 13020 | 4.047 | 4.050 | 17195 |
| 5 1/2 | 5.497 | 5.500 | 0.536 | 0.539 | 5.002 | 5.005 | 10395 | 4.727 | 4.730 | 15754 | 4.452 | 4.455 | 20806 |
| 6 | 5.997 | 6.000 | 0.585 | 0.588 | 5.457 | 5.460 | 12377 | 5.157 | 5.160 | 18749 | 4.857 | 4.860 | 24760 |

[a] *Torque Capacity of Spline Fittings:* The torque capacities of the different spline fittings are given in the columns headed "*T*." The torque capacity, per inch of bearing length at 1000 pounds pressure per square inch on the sides of the spline, may be determined by the following formula, in which $T$ = torque capacity in inch-pounds per inch of length, $N$ = number of splines, $R$ = mean radius or radial distance from center of hole to center of spline, $h$ = depth of spline: $T = 1000\,NRh$

**Table 22. Formulas for Determining Dimensions of SAE Standard Splines**

| No. of Splines | W For All Fits | A Permanent Fit | | B To Slide Without Load | | C To Slide Under Load | |
|---|---|---|---|---|---|---|---|
| | | h | d | h | d | h | d |
| Four | 0.241D[a] | 0.075D | 0.850D | 0.125D | 0.750D | … | … |
| Six | 0.250D | 0.050D | 0.900D | 0.075D | 0.850D | 0.100D | 0.800D |
| Ten | 0.156D | 0.045D | 0.910D | 0.070D | 0.860D | 0.095D | 0.810D |
| Sixteen | 0.098D | 0.045D | 0.910D | 0.070D | 0.860D | 0.095D | 0.810D |

[a] Four splines for fits *A* and *B* only.

The formulas in the table above give the maximum dimensions for *W*, *h*, and *d*, as listed in Table 18 through Table 21 inclusive.

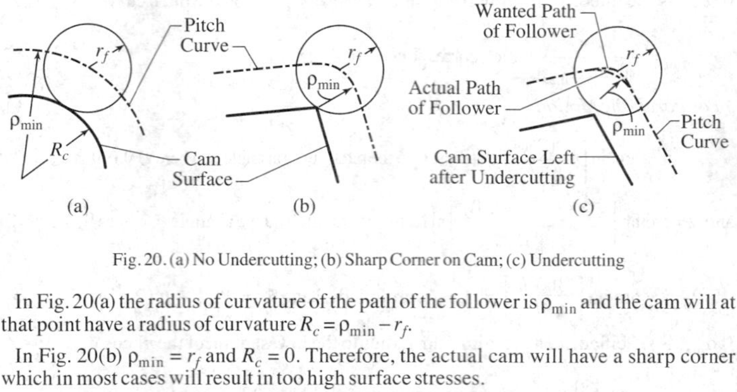

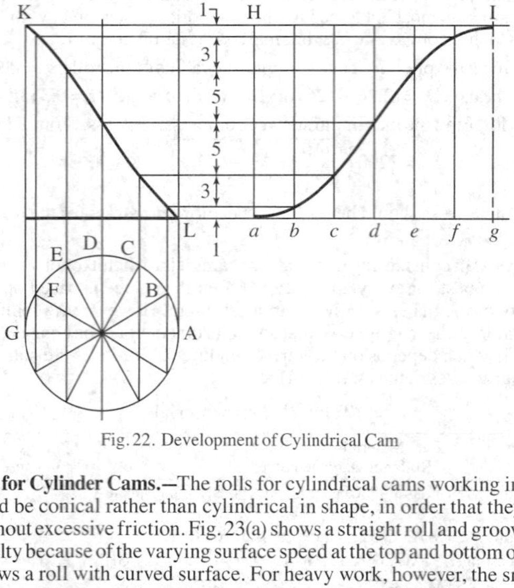

TABLE OF CONTENTS

MACHINE ELEMENTS

## O-RINGS

*(Continued)*



## ROLLED STEEL, WIRE, SHEET METAL, WIRE ROPE



## SHAFT ALIGNMENT



## SHAFT ALIGNMENT

*(Continued)*



## FLUID POWER

# PLAIN BEARINGS

## Introduction

On the following pages are given data and procedures for designing full-film or hydrodynamically lubricated bearings of the journal and thrust types. However, before proceeding to these design methods, it is useful to first review those bearing aspects concerning the types of bearings available; lubricants and lubrication methods; hardness and surface finish; machining methods; seals; retainers; and typical length-to-diameter ratios for various applications.

The following paragraphs preceding the design sections provide guidance in these matters and suggest modifications in allowable loads when other than full-film operating conditions exist in a bearing.

**Classes of Plain Bearings.—**Bearings that provide sliding contact between mating surfaces fall into three general classes: *radial bearings* that support rotating shafts or journals; *thrust bearings* that support axial loads on rotating members; and *guide* or *slipper bearings* that guide moving parts in a straight line. Radial sliding bearings, more commonly called sleeve bearings, may be of several types, the most usual being the plain full journal bearing, which has 360-degree contact with its mating journal, and the partial journal bearing, which has less than 180-degree contact. This latter type is used when the load direction is constant and has the advantages of simplicity, ease of lubrication, and reduced frictional loss.

The relative motions between the parts of plain bearings may take place: 1) as pure sliding without the benefit of a liquid or gaseous lubricating medium between the moving surfaces such as with the dry operation of nylon or Teflon; 2) with hydrodynamic lubrication in which a wedge or film buildup of lubricating medium is produced, with either whole or partial separation of the bearing surfaces; 3) with hydrostatic lubrication in which a lubricating medium is introduced under pressure between the mating surfaces causing a force opposite to the applied load and a lifting or separation of these surfaces; and 4) with a hybrid form or combination of hydrodynamic and hydrostatic lubrication.

Listed below are some of the advantages and disadvantages of sliding contact (plain) bearings as compared with rolling contact (antifriction) bearings.

*Advantages:* 1) require less space; 2) are quieter in operation; 3) are lower in cost, particularly in high-volume production; 4) have greater rigidity; and 5) their life is generally not limited by fatigue.

*Disadvantages:* 1) have higher frictional properties resulting in higher power consumption; 2) are more susceptible to damage from foreign material in lubrication system; 3) have more stringent lubrication requirements; and 4) are more susceptible to damage from interrupted lubrication supply.

**Types of Journal Bearings.—**Many types of journal bearing configurations have been developed; some of these are shown in Fig. 1.

*Circumferential-groove bearings,* Fig. 1(a), have an oil groove extending circumferentially around the bearing. The oil is maintained under pressure in the groove. The groove divides the bearing into two shorter bearings that tend to run at a slightly greater eccentricity. However, the advantage in terms of stability is slight, and this design is most commonly used in reciprocating-load main and connecting-rod bearings because of the uniformity of oil distribution.

Short cylindrical bearings are a better solution than the circumferential-groove bearing for high-speed, low-load service. Often the bearing can be shortened enough to increase the unit loading to a substantial value, causing the shaft to ride at a position of substantial eccentricity in the bearing. Experience has shown that instability rarely results when the shaft eccentricity is greater that 0.6. Very short bearings are not often used for this type of application, because they do not provide a high temporary rotating-load capacity in the event some unbalance should be created in the rotor during service.

*Cylindrical-overshot bearings,* Fig. 1(b), are used where surface speeds of 10,000 fpm or more exist, and where additional oil flow is desired to maintain a reasonable bearing temperature. This bearing has a wide circumferential groove extending from one axial oil groove to the other over the upper half of the bearing. Oil is usually admitted to the trailing-edge oil groove. An inlet orifice is used to control the oil flow. Cooler operation results from the elimination of shearing action over a large section of the upper half of the bearing and, to a great extent, from the additional flow of cool oil over the top half of the bearing.

*Pressure bearings,* Fig. 1(c), employ a groove over the top half of the bearing. The groove terminates at a sharp dam about 45 degrees beyond the vertical in the direction of shaft rotation. Oil is pumped into this groove by shear action from the rotation of the shaft and is then stopped by the dam. In high-speed operation, this situation creates a high oil pressure over the upper half of the bearing. The pressure created in the oil groove and surrounding upper half of the bearing increases the load on the lower half of the bearing. This self-generated load increases the shaft eccentricity. If the eccentricity is increased to 0.6 or greater, stable operation under high-speed, low-load conditions can result. The central oil groove can be extended around the lower half of the bearing, further increasing the effective loading. This design has one primary disadvantage: Dirt in the oil will tend to abrade the sharp edge of the dam and impair ability to create high pressures.

*Multiple-groove bearings,* Fig. 1(d), are sometimes used to provide increased oil flow. The interruptions in the oil film also appear to give this bearing some merit as a stable design.

*Elliptical bearings,* Fig. 1(e), are not truly elliptical, but are formed from two sections of a cylinder. This two-piece bearing has a large clearance in the direction of the split and a smaller clearance in the load direction at right angles to the split. At light loads, the shaft runs eccentric to both halves of the bearing, and hence, the elliptical bearing has a higher oil flow than the corresponding cylindrical bearing. Thus, the elliptical bearing will run cooler and will be more stable than a cylindrical bearing.

*Elliptical-overshot bearings* (not shown) are elliptical bearings in which the upper half is relieved by a wide oil groove connecting the axial oil grooves. They are analogous to cylindrical-overshot bearings.

*Displaced elliptical bearings,* Fig. 1(f), shift the centers of the two bearing arcs in both a horizontal and a vertical direction. This design has greater stiffness than a cylindrical bearing, in both horizontal and vertical directions, with substantially higher oil flow. It has not been extensively used, but offers the prospect of high stability and cool operation.

*Three-lobe bearings,* Fig. 1(g), are made up in cross section of three circular arcs. They are most effective as antioil whip bearings when the centers of curvature of each of the three lobes lie well outside the clearance circle that the shaft center can describe within the bearing. Three axial oil-feed grooves are used. It is a more difficult design to manufacture, because it is almost necessary to make it in three parts instead of two. The bore is machined with shims between each of the three parts. The shims are removed after machining is completed.

*Pivoted-shoe bearings,* Fig. 1(h), are one of the most stable bearings. The bearing surface is divided into three or more segments, each of which is pivoted at the center. In operation, each shoe tilts to form a wedge-shaped oil film, thus creating a force tending to push the shaft toward the center of the bearing. For single direction rotation, the shoes are sometimes pivoted near one end and forced toward the shaft by springs.

*Nutcracker bearings,* Fig. 1(i), consist of two cylindrical half-bearings. The upper half-bearing is free to move in a vertical direction and is forced toward the shaft by a hydraulic cylinder. External oil pressure may be used to create load on the upper half of the bearing through the hydraulic cylinder. Or the high-pressure oil may be obtained from the lower half of the bearing by tapping a hole into the high-pressure oil film, thus creating a self-loading bearing. Either type can increase bearing eccentricity to the point where stable operation can be achieved.

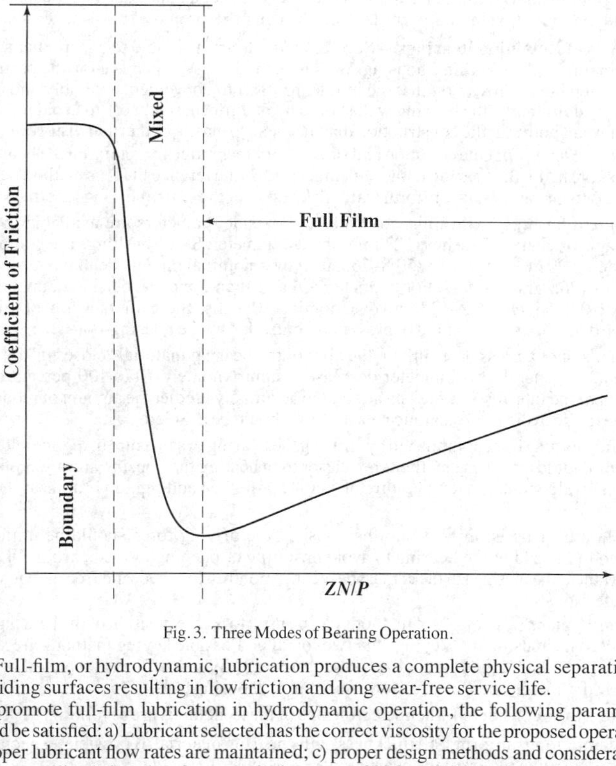

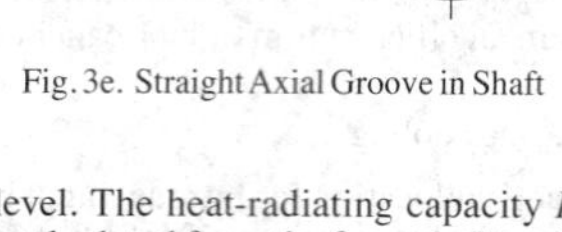

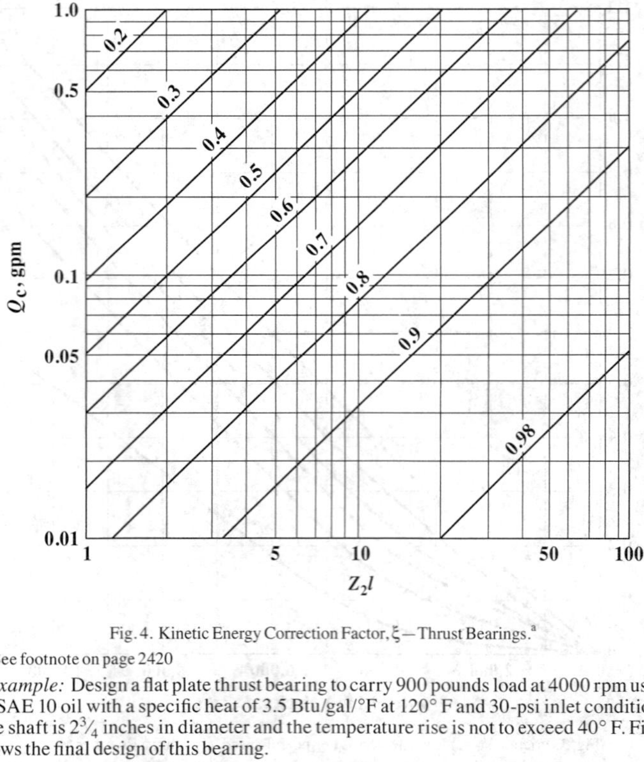

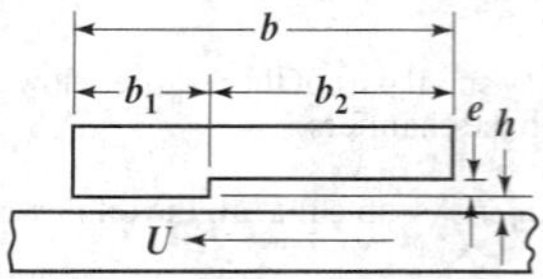

## Table 8. ABEC-5 and RBEC-5 Tolerance Limits for Metric Ball and Roller Bearings *ANSI/ABMA 20-2011 (R2020)*

| INNER RING | | | | | | | | | | | | | |
|---|---|---|---|---|---|---|---|---|---|---|---|---|---|
| Inner Ring Bore Basic Dia., $d$ mm | | $V_{dp}$,[a] max Diameter Series | | $\Delta_{dmp}$[b] | | Radial Runout $K_{ia}$ | Ref. Face Runout with Bore $S_d$ | Axial Runout $S_{ia}$[c] | Width | | | | |
| | | | | | | | | | $\Delta_{Bs}$[d] | | | | $V_{Bs}$[e] |
| Over | Incl. | 7, 8, 9 | 0, 1, 2, 3, 4 | High | Low | max | max | max | All | Normal | Modified[f] | | max |
| 2.5 | 10 | 5 | 4 | 0 | −5 | 4 | 7 | 7 | 0 | −40 | −250 | | 5 |
| 10 | 18 | 5 | 4 | 0 | −5 | 4 | 7 | 7 | 0 | −80 | −250 | | 5 |
| 18 | 30 | 6 | 5 | 0 | −6 | 4 | 8 | 8 | 0 | −120 | −250 | | 5 |
| 30 | 50 | 8 | 6 | 0 | −8 | 5 | 8 | 8 | 0 | −120 | −250 | | 5 |
| 50 | 80 | 9 | 7 | 0 | −9 | 5 | 8 | 8 | 0 | −150 | −250 | | 6 |
| 80 | 120 | 10 | 8 | 0 | −10 | 6 | 9 | 9 | 0 | −200 | −380 | | 7 |
| 120 | 180 | 13 | 10 | 0 | −13 | 8 | 10 | 10 | 0 | −250 | −380 | | 8 |
| 180 | 250 | 15 | 12 | 0 | −15 | 10 | 11 | 13 | 0 | −300 | −500 | | 10 |

[a] Bore ($V_{dp}$) and outside diameter ($V_{Dp}$) variation in a single radial plane.

[b] Single plane mean bore ($\Delta_{dmp}$) and outside diameter ($\Delta_{Dmp}$) deviation from basic. (For a basically tapered bore, $\Delta_{dmp}$ refers only to the theoretical small end of the bore.)

[c] Axial runout of assembled bearing with inner ring $S_{ia}$. Applies to groove-type ball bearings only.

[d] Single bore ($\Delta_{Bs}$) and outer ring ($\Delta_{Cs}$) width variation.

[e] Inner ($V_{Bs}$) and outer ($V_{Cs}$) ring width deviation from basic.

[f] Applies to the rings of single bearings made for paired or stack mounting.

| OUTER RING | | | | | | | | | | | |
|---|---|---|---|---|---|---|---|---|---|---|---|
| Basic Outer Ring Outside Dia., $D$ mm | | $V_{Dp}$,[a, b] max Diameter Series | | $\Delta_{Dmp}$[c] | | Radial Runout $K_{ea}$ | Outside Cylindrical Surface Runout $S_D$[d] | Axial Runout $S_{ea}$[e] | Width | | |
| | | | | | | | | | $\Delta_{Cs}$[f] | | $V_{Cs}$[g] |
| Over | Incl. | 7, 8, 9 | 0, 1, 2, 3, 4 | High | Low | max | max | max | High | Low | max |
| 6 | 18 | 5 | 4 | 0 | −5 | 5 | 8 | 8 | Identical to $\Delta_{Bs}$ of inner ring of same bearing | | 5 |
| 18 | 30 | 6 | 5 | 0 | −6 | 6 | 8 | 8 | | | 5 |
| 30 | 50 | 7 | 5 | 0 | −7 | 7 | 8 | 8 | | | 5 |
| 50 | 80 | 9 | 7 | 0 | −9 | 8 | 8 | 10 | | | 6 |
| 80 | 120 | 10 | 8 | 0 | −10 | 10 | 9 | 11 | | | 8 |
| 120 | 150 | 11 | 8 | 0 | −11 | 11 | 10 | 13 | | | 8 |
| 150 | 180 | 13 | 10 | 0 | −13 | 13 | 10 | 14 | | | 8 |
| 180 | 250 | 15 | 11 | 0 | −15 | 15 | 11 | 15 | | | 10 |

[a] Bore ($V_{dp}$) and outside diameter ($V_{Dp}$) variation in a single radial plane.

[b] No values have been established for capped bearings.

[c] Single plane mean bore ($\Delta_{dmp}$) and outside diameter ($\Delta_{Dmp}$) deviation from basic. (For a basically tapered bore, $\Delta_{dmp}$ refers only to the theoretical small end of the bore.)

[d] Outside cylindrical surface runout with outer ring reference face $S_D$

[e] Axial runout of assembled bearing with outer ring $S_{ea}$.

[f] Single bore ($\Delta_{Bs}$) and outer ring ($\Delta_{Cs}$) width variation.

[g] Inner ($V_{Bs}$) and outer ($V_{Cs}$) ring width deviation from basic.

All units are micrometers, unless otherwise indicated. For sizes beyond range of this table, see *ANSI/ABMA 20-2011 (R2020)*. This table does not cover instrument bearings and tapered roller bearings.

## Table 10. ABEC-9 Tolerance Limits for Metric Ball and Roller Bearing *ANSI/ABMA 20-2011 (R2020)*

INNER RING

| Inner Ring Bore Basic Dia., $d$ mm | | $V_{dp}$,[a] max | $\Delta_{dmp}$[b] | | $\Delta_{ds}$[c] | | Radial Runout $K_{ia}$ | Ref. Face Runout with Bore $S_d$ | Axial Runout of Assembled Bearing with Inner Ring $S_{ia}$[d] | Width $\Delta_{Bs}$[e] | | Width $V_{Bs}$[f] |
|---|---|---|---|---|---|---|---|---|---|---|---|---|
| Over | Incl. | max | High | Low | High | Low | max | max | max | High | Low | max |
| 2.5 | 10 | 2.5 | 0 | −2.5 | 0 | −2.5 | 1.5 | 1.5 | 1.5 | 0 | −40 | 1.5 |
| 10 | 18 | 2.5 | 0 | −2.5 | 0 | −2.5 | 1.5 | 1.5 | 1.5 | 0 | −80 | 1.5 |
| 18 | 30 | 2.5 | 0 | −2.5 | 0 | −2.5 | 2.5 | 1.5 | 2.5 | 0 | −120 | 1.5 |
| 30 | 50 | 2.5 | 0 | −2.5 | 0 | −2.5 | 2.5 | 1.5 | 2.5 | 0 | −120 | 1.5 |
| 50 | 80 | 4 | 0 | −4 | 0 | −4 | 2.5 | 1.5 | 2.5 | 0 | −150 | 1.5 |
| 50 | 80 | 4 | 0 | −4 | 0 | −4 | 2.5 | 1.5 | 2.5 | 0 | −150 | 1.5 |
| 80 | 120 | 5 | 0 | −5 | 0 | −5 | 2.5 | 2.5 | 2.5 | 0 | −200 | 2.5 |
| 120 | 150 | 7 | 0 | −7 | 0 | −7 | 2.5 | 2.5 | 2.5 | 0 | −250 | 2.5 |
| 150 | 180 | 7 | 0 | −7 | 0 | −7 | 5 | 4 | 5 | 0 | −300 | 4 |
| 180 | 250 | 8 | 0 | −8 | 0 | −8 | 5 | 5 | 5 | 0 | −350 | 5 |

[a] Bore ($V_{dp}$) and outside diameter ($V_{Dp}$) variation in a single radial plane.
[b] Single plane mean bore ($\Delta_{dmp}$) and outside diameter ($\Delta_{Dmp}$) deviation from basic. (For a basically tapered bore, $\Delta_{dmp}$ refers to the theoretical small end of the bore.)
[c] Single bore diameter ($\Delta_{ds}$) and outside diameter ($\Delta_{Ds}$) deviation from basic.
[d] Applies to groove-type ball bearings only.
[e] Single bore ($\Delta_{Bs}$) and outer ring ($\Delta_{Cs}$) width variation from basic.
[f] Inner ($V_{Bs}$) and outer ($V_{Cs}$) ring width variation.

OUTER RING

| Basic Outside Diameter, $D$ mm | | $V_{Dp}$[a] | $\Delta_{Dmp}$[b] | | $\Delta_{Ds}$[c] | | Radial Runout $K_{ea}$ | Outside Cylindrical Surface Runout with Outer Ring $S_D$ | Axial Runout of Assembled Bearing with Outer Ring $S_{ea}$ | Width $\Delta_{Cs}$[e] | | Width $V_{Cs}$[f] |
|---|---|---|---|---|---|---|---|---|---|---|---|---|
| Over | Incl. | max | High | Low | High | Low | max | max | max | High | Low | max |
| 6 | 18 | 2.5 | 0 | −2.5 | 0 | −2.5 | 1.5 | 1.5 | 1.5 | Identical to $\Delta_{Bs}$ of inner ring of same bearing | | 1.5 |
| 18 | 30 | 4 | 0 | −4 | 0 | −4 | 2.5 | 1.5 | 2.5 | | | 1.5 |
| 30 | 50 | 4 | 0 | −4 | 0 | −4 | 2.5 | 1.5 | 2.5 | | | 1.5 |
| 50 | 80 | 4 | 0 | −4 | 0 | −4 | 4 | 1.5 | 4 | | | 1.5 |
| 80 | 120 | 5 | 0 | −5 | 0 | −5 | 5 | 2.5 | 5 | | | 1.5 |
| 120 | 150 | 5 | 0 | −5 | 0 | −5 | 5 | 2.5 | 5 | | | 1.5 |
| 150 | 180 | 7 | 0 | −7 | 0 | −7 | 5 | 2.5 | 5 | | | 2.5 |
| 180 | 250 | 8 | 0 | −8 | 0 | −8 | 7 | 4 | 7 | | | 4 |
| 250 | 315 | 8 | 0 | −8 | 0 | −8 | 7 | 5 | 7 | | | 5 |

[a] No values have been established for capped bearings.
All units are micrometers, unless otherwise indicated. For sizes beyond range of this table, see *ANSI/ABMA 20-2011 (R2020)*. This table does not cover instrument bearings.

## Table 11. Tolerance Limits for Single Direction Ball Thrust Bearings — Inch Design *ANSI/ABMA 24.2-1989 (S2020)*[b]

| Bore Diameter[a] $d$, Inches | | Single Plane Mean Bore Dia. Variation, $d$, Inch | | Outside Diameter $D$, Inches | | Single Plane Mean Outside Dia. Variation, $D$, Inch | |
|---|---|---|---|---|---|---|---|
| Over | Incl. | High | Low | Over | Incl. | High | Low |
| 0 | 6.7500 | +0.005 | 0 | 0 | 5.3125 | +0 | −0.002 |
| 6.7500 | 20.0000 | +0.010 | 0 | 5.3125 | 17.3750 | +0 | −0.003 |
| … | … | … | … | 17.3750 | 39.3701 | +0 | −0.004 |

[a] Bore tolerance limits: For bore diameters 0 to 1.8125 inches, inclusive, +0.005, −0.005; over 1.8125 to 12.000 inches, inclusive, +0.010,−0.010; over 12.000 to 20.000, inclusive, +0.0150, −0.0150.
[b] An "S" indicates the standard is maintained under the stabilized maintenance option and satisfies the following criteria: (a) not likely to require revision; (b) not safety or health related; (c) holds status of ANSI/ABMA and has been reaffirmed; (d) at least 10 years have passed since approval or latest revision; and (e) required for use with existing implementations or for reference purposes.

## Table 14. Tolerance Limits for Needle Roller Bearings, Drawn Cup, Without Inner Ring — Inch Types NIB, NIBM, NIY, NIYM, NIH, and NIHM

*ANSI/ABMA 18.2-1982 (S2013)*

| Ring Gage Bore Diameter[a] | | | Basic Bore Diameter under Needle Rollers, $F_w$ | | Allowable Deviation from $F_w$[a] | | Allowable Deviation from Width, $B$ | |
|---|---|---|---|---|---|---|---|---|
| Basic Outside Diameter, $D$ Inch | | Deviation from $D$ Inch | Inch | | Inch | | Inch | |
| Over | Incl. | | Over | Incl. | Low | High | High | Low |
| 0.1875 | 0.9375 | +0.0005 | | | | | | |
| 0.9375 | 4.0000 | −0.0005 | 0.1875 | 0.6875 | +0.0015 | +0.0024 | +0 | −0.0100 |
| | | | 0.6875 | 1.2500 | +0.0005 | +0.0014 | +0 | −0.0100 |
| | | | 1.2500 | 1.3750 | +0.0005 | +0.0015 | +0 | −0.0100 |
| | | | 1.3750 | 1.6250 | +0.0005 | +0.0016 | +0 | −0.0100 |
| For fitting and mounting practice see Table 23. | | | 1.6250 | 1.8750 | +0.0005 | +0.0017 | +0 | −0.0100 |
| | | | 1.8750 | 2.0000 | +0.0006 | +0.0018 | +0 | −0.0100 |
| | | | 2.0000 | 2.5000 | +0.0006 | +0.0020 | +0 | −0.0100 |
| | | | 2.5000 | 3.5000 | +0.0010 | +0.0024 | +0 | −0.0100 |

[a] The bore diameter under needle rollers can be measured only when bearing is pressed into a ring gage, which rounds and sizes the bearing.

## Table 15. Tolerance Limits for Needle Roller Bearings, Drawn Cup, Without Inner Ring — Metric Types NB, NBM, NY, NYM, NH, and NHM

*ANSI/ABMA 18.1-1982 (S2013)*

| Ring Gage Bore Diameter[a] | | | Basic Bore Diameter under Needle Rollers, $F_w$ | | Allowable Deviation from $F_w$[a] | | Allowable Deviation from Width, $B$ | |
|---|---|---|---|---|---|---|---|---|
| Basic Outside Diameter, $D$ | | Deviation from $D$ | | | | | | |
| mm | | | mm | | Micrometers | | Micrometers | |
| Over | Incl. | Micrometers | Over | Incl. | Low | High | High | Low |
| 6 | 10 | −16 | 3 | 6 | +10 | +28 | +0 | −250 |
| 10 | 18 | −20 | 6 | 10 | +13 | +31 | +0 | −250 |
| 30 | 50 | −28 | 18 | 30 | +20 | +41 | +0 | −250 |
| 50 | 80 | −33 | 30 | 50 | +25 | +50 | +0 | −250 |
| … | … | … | 50 | 70 | +30 | +60 | +0 | −250 |

[a] The bore diameter under needle rollers can be measured only when bearing is pressed into a ring gage, which rounds and sizes the bearing.

For fitting and mounting practice, see Table 23.

## Table 16. Tolerance Limits for Needle Roller Bearings, With Cage, Machined Ring, Without Inner Ring—Inch Type NIA *ANSI/ABMA 18.2-1982 (S2013)*

| Basic Outside Diameter, $D$ | | Allowable Deviation From $D$ of Single Mean Diameter, $D_{mp}$ | | Basic Bore Diameter under Needle Rollers, $F_w$ | | Allowable Deviation from $F_w$ | | Allowable Deviation from Width, $B$ | |
|---|---|---|---|---|---|---|---|---|---|
| Inch | | Inch | | Inch | | Inch | | Inch | |
| Over | Incl. | High | Low | Over | Incl. | Low | High | High | Low |
| 0.7500 | 2.0000 | +0 | −0.0005 | 0.3150 | 0.7087 | +0.0008 | +0.0017 | +0 | −0.0050 |
| 2.0000 | 3.2500 | +0 | −0.0006 | 0.7087 | 1.1811 | +0.0009 | +0.0018 | +0 | −0.0050 |
| 3.2500 | 4.7500 | +0 | −0.0008 | 1.1811 | 1.6535 | +0.0010 | +0.0019 | +0 | −0.0050 |
| 4.7500 | 7.2500 | +0 | −0.0010 | 1.6535 | 1.9685 | +0.0010 | +0.0020 | +0 | −0.0050 |
| | | | | 1.9685 | 2.7559 | +0.0011 | +0.0021 | +0 | −0.0050 |
| 7.2500 | 10.2500 | +0 | −0.0012 | 2.7559 | 3.1496 | +0.0011 | +0.0023 | +0 | −0.0050 |
| 10.2500 | 11.1250 | +0 | −0.0014 | 3.1496 | 4.0157 | +0.0012 | +0.0024 | +0 | −0.0050 |
| … | … | … | … | 4.0157 | 4.7244 | +0.0012 | +0.0026 | +0 | −0.0050 |
| … | … | … | … | 4.7244 | 6.2992 | +0.0013 | +0.0027 | +0 | −0.0050 |
| … | … | … | … | 6.2992 | 7.0866 | +0.0013 | +0.0029 | +0 | −0.0050 |
| … | … | … | … | 7.0866 | 7.8740 | +0.0014 | +0.0030 | +0 | −0.0050 |
| … | … | … | … | 7.8740 | 9.2520 | +0.0014 | +0.0032 | +0 | −0.0050 |

For fitting and mounting practice, see Table 24.

popular type is the "cartridge" type of sealed grease ball bearing, many applications have requirements which dictate against them. Often, operating environments may subject bearings to temperatures too high for seals utilized in the more popular designs. If minute leakage or the accumulation of traces of dirt at seal lips cannot be tolerated by the application (as in baking industry machinery), then the selections of bearings must be made with other sealing and lubrication systems in mind.

High shaft speeds generally dictate bearing selection based on the need for cooling, the suppression of churning or aeration of conventional lubricants, and most important of all, the inherent speed limitations of certain bearing types. An example of the latter is the effect of cage design and of the roller-end thrust-flange contact on the lubrication requirements in commercial taper roller bearings, which limit the speed they can endure and the thrust load they can carry. Reference to the manufacturers' catalog and application-design manuals is recommended before making bearing selections.

See *Selecting a Suitable Lubricant* on page 2516 for more information on this topic.

**Type of Mounting.—**Many bearing installations are complicated because the best adapted mounting type was not selected. Similarly, performance, reliability, and maintenance operations are restricted because the mounting was not thoroughly considered. There is no universally adaptable bearing for all needs. Careful reviews of the machine requirements should be made before designs are implemented. In many cases complicated machining, redundant shaft and housings, and use of an oversize bearing can be eliminated if the proper bearing in a well-thought-out mounting is chosen.

Advantage should be taken of the many race variations available in "standard" series of bearings. Puller grooves, tapered sleeves, ranged outer races, split races, fully demountable rolling-element and cage assemblies, flexible mountings, hydraulic removal features, relubrication holes and grooves, and many other innovations are available beyond the obvious advantages which are inherent in the basic bearing types.

**Radial and Axial Clearance.—**In designing the bearing mounting, a major consideration is to provide running clearances consistent with the requirements of the application. Race fits must be expected to absorb some of the original bearing clearance so that allowance should be made for approximately 80 percent of the actual interference showing up in the diameter of the race. This will increase for heavy, stiff housings or for extra light series races shrunk onto solid shafts, while light metal housings (aluminum, magnesium, or sheet metal) and tubular shafts with wall sections less than the race wall thickness will cause a lesser change in the race diameter.

Where the application will impose heat losses through housing or shaft, or where a temperature differential may be expected, allowances must be made in the proper direction to ensure proper operating clearance. Some compromises are required in applications where the indicated modification cannot be fully accommodated without endangering the bearing performance at lower speeds, during starting, or under lower temperature conditions than anticipated. Some leeway can be relied on with ball bearings since they can run with moderate preloads (0.0005 inch or 12.7 μm, max.) without affecting bearing life or temperature rise. Roller bearings, however, have a lesser tolerance for preloading and must be carefully controlled to avoid overheating and resulting self-destruction.

In all critical applications axial and radial clearances should be checked with feeler gages or dial indicators to ensure mounted clearances within tolerances established by the design engineer. Since chips, scores, race misalignment, shaft or housing denting, housing distortion, end cover (closure) off-squareness, and mismatch of rotor and housing and wrong axial and radial clearances in the races, rotor, and housing, as well as variations in temperature can all affect bearing clearances. Careful checks of running clearance is recommended.

For precision applications, taper-sleeve mountings, opposed ball or tapered-roller bearings with adjustable or shimmed closures are employed to provide careful control of radial and/or axial clearances. This practice requires skill and experience as well as the initial assistance of the bearing manufacturer's field engineer.

8) *Groove radii.* If groove radii are smaller than those specified in the bearing types covered, the ability of a bearing to resist fatigue is not improved: however, it is diminished by the use of larger radii.

**Ball Bearing Rating Life.—**According to the American Bearing Manufacturers Association standards the Rating Life $L_{10}$ of a group of apparently identical ball bearings is the life in millions of revolutions that 90 percent of the group will complete or exceed. For a single bearing, $L_{10}$ also refers to the life associated with 90 percent reliability.

*Radial and Angular Contact Ball Bearings:* The magnitude of the Rating Life $L_{10}$ in millions of revolutions, for a radial or angular contact ball bearing application is given by the formula:

$$L_{10} = \left(\frac{C}{P}\right)^3 \qquad (1)$$

where $C$ = basic load rating, newtons (pounds). See Formulas (2), (3a) and (3b)

$P$ = equivalent radial load, newtons (pounds). See Formula (4)

**Table 32. Values of *fc* for Radial and Angular Contact Ball Bearings**

| $\frac{D \cos \alpha}{d_m}$ | Single Row Radial Contact; Single and Double Row Angular Contact, Groove Type[a] | | Double Row Radial Contact Groove Type | | Self-Aligning | |
|---|---|---|---|---|---|---|
| | Values of $f_c$ | | | | | |
| | Metric[b] | Inch[c] | Metric[b] | Inch[c] | Metric[b] | Inch[c] |
| 0.05 | 46.7 | 3550 | 44.2 | 3360 | 17.3 | 1310 |
| 0.06 | 49.1 | 3730 | 46.5 | 3530 | 18.6 | 1420 |
| 0.07 | 51.1 | 3880 | 48.4 | 3680 | 19.9 | 1510 |
| 0.08 | 52.8 | 4020 | 50.0 | 3810 | 21.1 | 1600 |
| 0.09 | 54.3 | 4130 | 51.4 | 3900 | 22.3 | 1690 |
| 0.10 | 55.5 | 4220 | 52.6 | 4000 | 23.4 | 1770 |
| 0.12 | 57.5 | 4370 | 54.5 | 4140 | 25.6 | 1940 |
| 0.14 | 58.8 | 4470 | 55.7 | 4230 | 27.7 | 2100 |
| 0.16 | 59.6 | 4530 | 56.5 | 4290 | 29.7 | 2260 |
| 0.18 | 59.9 | 4550 | 56.8 | 4310 | 31.7 | 2410 |
| 0.20 | 59.9 | 4550 | 56.8 | 4310 | 33.5 | 2550 |
| 0.22 | 59.6 | 4530 | 56.5 | 4290 | 35.2 | 2680 |
| 0.24 | 59.0 | 4480 | 55.9 | 4250 | 36.8 | 2790 |
| 0.26 | 58.2 | 4420 | 55.1 | 4190 | 38.2 | 2910 |
| 0.28 | 57.1 | 4340 | 54.1 | 4110 | 39.4 | 3000 |
| 0.30 | 56.0 | 4250 | 53.0 | 4030 | 40.3 | 3060 |
| 0.32 | 54.6 | 4160 | 51.8 | 3950 | 40.9 | 3110 |
| 0.34 | 53.2 | 4050 | 50.4 | 3840 | 41.2 | 3130 |
| 0.36 | 51.7 | 3930 | 48.9 | 3730 | 41.3 | 3140 |
| 0.38 | 50.0 | 3800 | 47.4 | 3610 | 41.0 | 3110 |
| 0.40 | 48.4 | 3670 | 45.8 | 3480 | 40.4 | 3070 |

[a] A. When calculating the basic load rating for a unit consisting of two similar, single row, radial contact ball bearings, in a duplex mounting, the pair is considered as one, double row, radial contact ball bearing.

B. When calculating the basic load rating for a unit consisting of two, similar, single row, angular contact ball bearings in a duplex mounting, "face-to-face" or "back-to-back," the pair is considered as one, double row, angular contact ball bearing.

C. When calculating the basic load rating for a unit consisting of two or more similar, single angular contact ball bearings mounted "in tandem," properly manufactured and mounted for equal load distribution, the rating of the combination is the number of bearings to the 0.7 power times the rating of a single row ball bearing. If the unit may be treated as a number of individually interchangeable single row bearings, this footnote "C" does not apply.

[b] **Use to obtain *C* in newtons when *D* is given in mm.**

[c] Use to obtain *C* in pounds when *D* is given in inches.

the bearing under consideration and is found by Formula (20); $i$ is the number of rows of balls in the bearing; $Z$ is the number of balls per row in a radial or angular contact bearing or the number of balls in a single row, single direction thrust bearing; $D$ is the ball diameter in millimeters (inches); and $F_r$ is the applied radial load in newtons (pounds).

For radial and angular contact ball bearings with balls not larger than 25.4 mm (1 inch) in diameter, $C$ is found by the formula:

$$C = f_c(i\cos\alpha)^{0.7}Z^{2/3}D^{1.8} \tag{2}$$

and with balls larger than 25.4 mm (1 inch) in diameter $C$ is found by the formula:

$$\mathbf{C = 3.647f_c(i\cos\alpha)^{0.7}Z^{2/3}D^{1.4}\quad (metric)} \tag{3a}$$

$$C = f_c(i\cos\alpha)^{0.7}Z^{2/3}D^{1.4}\text{(inch)} \tag{3b}$$

where $f_c$ = a factor which depends on the geometry of the bearing components, the accuracy to which the various bearing parts are made and the material. Values of $f_c$ are given in Table 32

$i$ = number of rows of balls in the bearing

$\alpha$ = nominal contact angle, degrees

$Z$ = number of balls per row in a radial or angular contact bearing

$D$ = ball diameter, mm (inches)

The magnitude of the equivalent radial load, $P$, in newtons (pounds) for radial and angular contact ball bearings, under combined constant radial and constant thrust loads is given by the formula:

$$P = XF_r + YF_a \tag{4}$$

where $F_r$ = the applied radial load in newtons (pounds)

$F_a$ = the applied axial load in newtons (pounds)

$X$ = radial load factor as given in Table 35

$Y$ = axial load factor as given in Table 35

*Thrust Ball Bearings:* The magnitude of the Rating Life $L_{10}$ in millions of revolutions for a thrust ball bearing application is given by the formula:

$$L_{10} = \left(\frac{C_a}{P_a}\right)^3 \tag{5}$$

where $C_a$ = the basic load rating, newtons (pounds). See Formulas (6) to (10)

$P_a$ = equivalent thrust load, newtons (pounds). See Formula (11)

For single row, single and double direction, thrust ball bearing with balls not larger than 25.4 mm (1 inch) in diameter, $C_a$ is found by the formulas:

$$\text{for } \alpha = 90 \text{ degrees,}\quad C_a = f_c Z^{2/3}D^{1.8} \tag{6}$$

$$\text{for } \alpha \neq 90 \text{ degrees,}\quad C_a = f_c(\cos\alpha)^{0.7}Z^{2/3}D^{1.8}\tan\alpha \tag{7}$$

and with balls larger than 25.4 mm (1 inch) in diameter, $C_a$ is found by the formulas:

$$\mathbf{for\ \alpha = 90\ degrees,\quad C_a = 3.647f_c Z^{2/3}D^{1.4}\ (metric)} \tag{8a}$$

$$C_a = f_c Z^{2/3}D^{1.4}\text{(inch)} \tag{8b}$$

$$\mathbf{for\ \alpha \neq 90\ degrees,\quad C_a = 3.647f_c(\cos\alpha)^{0.7}Z^{2/3}D^{1.4}\tan\alpha\quad (metric)} \tag{9a}$$

$$C_a = f_c(\cos\alpha)^{0.7}Z^{2/3}D^{1.4}\tan\alpha\text{ (inch)} \tag{9b}$$

## COUPLINGS, CLUTCHES, BRAKES

**Drive Couplings.**—An electric motor, steam turbine, and gas turbine are all examples of primary drivers that power pumps, compressors, or fans by transferring torque. To transmit power, these drivers or prime movers are connected to the driven equipment by drive couplings. Typically, shafts of the systems are collinear during operating condition. Coupling type is chosen based on its ability to function as needed, while absorbing the forces involved in the power transmission. As the weakest link in the system, the coupling should preferably fail before the more expensive drive or driven equipment.

**Connecting Shafts.**—Where misalignment is neither present nor desired, shafts may be connected with a simple rigid coupling, made of two halves, securely bolted together. For certain applications, where couplings transmit up to 150 horsepower, rigid, flange-type couplings of appropriate size (as shown in Table 1) can be used. These are known as safety or protected flange couplings, because the bolt heads and nuts are shrouded by the flange. Typically, such couplings are shielded by sheet metal or other covers.

For fractional horsepower and low-power applications, a setscrew may provide the locked connection between the hub and the shaft. A flat surface on the shaft and locking mechanism using setscrew(s) is recommended. High-power applications require a key and usually two setscrews, with one positioned above the key on each hub (drive and driven).

**Table 1. Safety or Protected Flange Couplings**

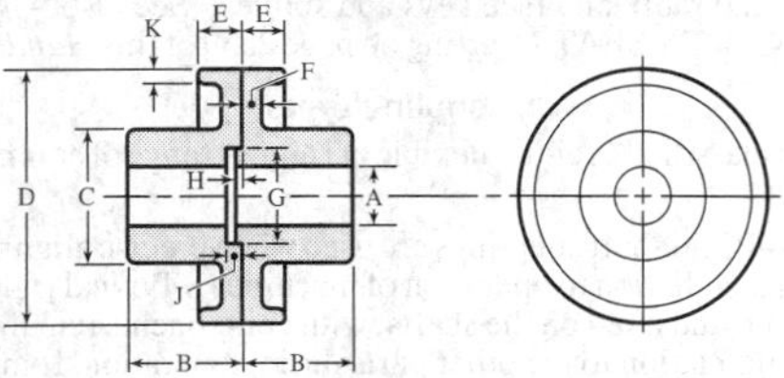


| *A* | *B* | *C* | *D* | *E* | *F* | *G* | *H* | *J* | *K* | Bolts | |
|---|---|---|---|---|---|---|---|---|---|---|---|
| | | | | | | | | | | No. | Dia. |
| 1 | 1¾ | 2¼ | 4 | 11/16 | 5/16 | 1½ | ¼ | 9/32 | ¼ | 5 | ⅜ |
| 1¼ | 2 3/16 | 2¾ | 5 | 13/16 | ⅜ | 1⅞ | ¼ | 9/32 | ¼ | 5 | 7/16 |
| 1½ | 2⅝ | 3⅜ | 6 | 15/16 | 7/16 | 2¼ | ¼ | 9/32 | ¼ | 5 | ½ |
| 1¾ | 3 1/16 | 4 | 7 | 1 1/16 | ½ | 2⅝ | ¼ | 9/32 | ¼ | 5 | 9/16 |
| 2 | 3½ | 4½ | 8 | 1 3/16 | 9/16 | 3 | ¼ | 9/32 | 5/16 | 5 | ⅝ |
| 2¼ | 3 15/16 | 5⅛ | 9 | 1 5/16 | ⅝ | 3⅜ | ¼ | 9/32 | 5/16 | 5 | 11/16 |
| 2½ | 4⅜ | 5⅝ | 10 | 1 7/16 | 11/16 | 3¾ | ¼ | 9/32 | 5/16 | 5 | ¾ |
| 2¾ | 4 13/16 | 6¼ | 11 | 1 9/16 | ¾ | 4⅛ | ¼ | 9/32 | 5/16 | 5 | 13/16 |
| 3 | 5¼ | 6¾ | 12 | 1 11/16 | 13/16 | 4½ | ¼ | 9/32 | ⅜ | 5 | ⅞ |
| 3¼ | 5 11/16 | 7⅜ | 13 | 1 13/16 | ⅞ | 4⅞ | ¼ | 9/32 | ⅜ | 5 | 15/16 |
| 3½ | 6⅛ | 8 | 14 | 1 15/16 | 15/16 | 5¼ | ¼ | 9/32 | ⅜ | 5 | 1 |
| 3¾ | 6 9/16 | 8½ | 15 | 2 1/16 | 1 | 5⅝ | ¼ | 9/32 | ⅜ | 5 | 1 1/16 |
| 4 | 7 | 9 | 16 | 2¼ | 1⅛ | 6 | ¼ | 9/32 | 7/16 | 5 | 1⅛ |
| 4½ | 7⅞ | 10¼ | 18 | 2½ | 1¼ | 6¾ | ¼ | 9/32 | 7/16 | 5 | 1¼ |
| 5 | 8¾ | 11¼ | 20 | 2¾ | 1⅜ | 7½ | ¼ | 9/32 | 7/16 | 5 | 1⅜ |
| 5½ | 8¾ | 11¼ | 20 | 2¾ | 1⅜ | 7½ | ¼ | 9/32 | 7/16 | 5 | 1⅜ |
| 6 | 10½ | 12⅜ | 22 | 2 15/16 | 1½ | 8¼ | 5/16 | 11/32 | ½ | 5 | 1 7/16 |
| 6½ | 11⅜ | 13½ | 24 | 3⅛ | 1⅝ | 9 | 5/16 | 11/32 | ½ | 5 | 1½ |
| 7 | 12¼ | 14⅝ | 26 | 3¼ | 1¾ | 9¾ | 5/16 | 11/32 | 9/16 | 6 | 1½ |
| 7½ | 13⅛ | 15¾ | 28 | 3 7/16 | 1⅞ | 10½ | 5/16 | 11/32 | 9/16 | 6 | 1 9/16 |
| 8 | 14 | 16⅞ | 28 | 3½ | 2 | 10⅞ | 5/16 | 11/32 | ⅝ | 7 | 1½ |
| 8½ | 14⅞ | 18 | 30 | 3 11/16 | 2⅛ | 11¼ | 5/16 | 11/32 | ⅝ | 7 | 1 9/16 |
| 9 | 15¾ | 19⅛ | 31 | 3¾ | 2¼ | 11⅝ | 5/16 | 11/32 | 11/16 | 8 | 1½ |
| 9½ | 16⅝ | 20¼ | 32 | 3 15/16 | 2⅜ | 12 | 5/16 | 11/32 | 11/16 | 8 | 1 9/16 |
| 10 | 17½ | 21⅜ | 34 | 4⅛ | 2½ | 12¾ | 5/16 | 11/32 | ¾ | 8 | 1⅝ |
| 10½ | 18⅜ | 22½ | 35 | 4¼ | 2⅝ | 13⅛ | 5/16 | 11/32 | ¾ | 10 | 1⅝ |
| 11 | 19¼ | 23⅝ | 36 | 4 7/16 | 2¾ | 13½ | 5/16 | 11/32 | ⅞ | 10 | 1 11/16 |
| 11½ | 20⅛ | 24¾ | 37 | 4⅝ | 2⅞ | 13⅞ | 5/16 | 11/32 | ⅞ | 10 | 1¾ |
| 12 | 21 | 25⅞ | 38 | 4 13/16 | 3 | 14¼ | 5/16 | 11/32 | 1 | 10 | 1 13/16 |

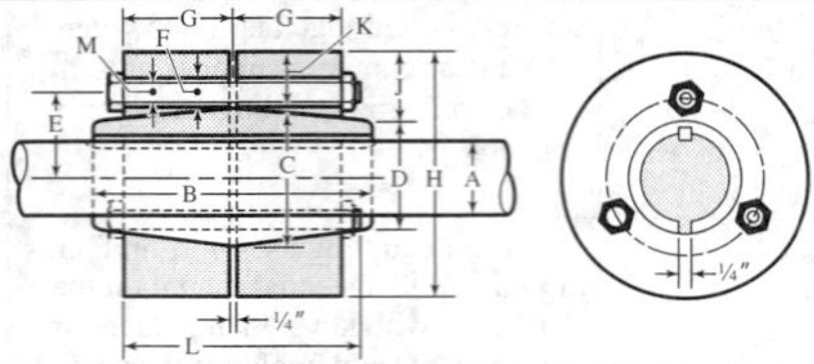


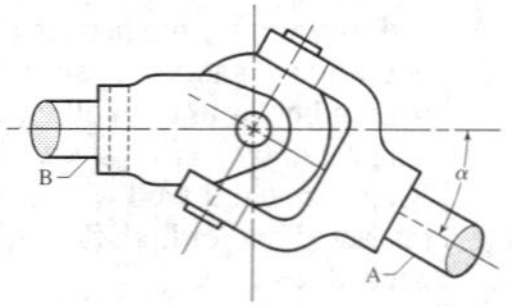

## Proportions of Knuckle Joints

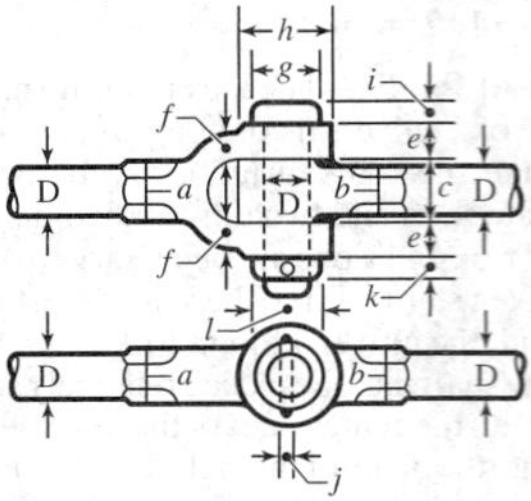


For sizes not given below:

$a = 1.2D$   $h = 2D$
$b = 1.1D$   $i = 0.5D$
$c = 1.2D$   $j = 0.25D$
$e = 0.75D$   $k = 0.5D$
$f = 0.6D$   $l = 1.5D$
$g = 1.5D$

| *D* | *a* | *b* | *c* | *e* | *f* | *g* | *h* | *i* | *j* | *k* | *l* |
|---|---|---|---|---|---|---|---|---|---|---|---|
| ½ | ⅝ | 9/16 | ⅝ | ⅜ | 5/16 | ¾ | 1 | ¼ | ⅛ | ¼ | ¾ |
| ¾ | ⅞ | ¾ | ⅞ | 9/16 | 7/16 | 1⅛ | 1½ | ⅜ | 3/16 | ⅜ | 1⅛ |
| 1 | 1¼ | 1⅛ | 1¼ | ¾ | ⅝ | 1½ | 2 | ½ | ¼ | ½ | 1½ |
| 1¼ | 1½ | 1⅜ | 1½ | 15/16 | ¾ | 1⅞ | 2½ | ⅝ | 5/16 | ⅝ | 1⅞ |
| 1½ | 1¾ | 1⅝ | 1¾ | 1⅛ | ⅞ | 2¼ | 3 | ¾ | ⅜ | ¾ | 2¼ |
| 1¾ | 2⅛ | 2 | 2⅛ | 1 5/16 | 1 1/16 | 2⅝ | 3½ | ⅞ | 7/16 | ⅞ | 2⅝ |
| 2 | 2⅜ | 2¼ | 2⅜ | 1½ | 1 3/16 | 3 | 4 | 1 | ½ | 1 | 3 |
| 2¼ | 2¾ | 2½ | 2¾ | 1 11/16 | 1⅜ | 3⅜ | 4½ | 1⅛ | 9/16 | 1⅛ | 3⅜ |
| 2½ | 3 | 2¾ | 3 | 1⅞ | 1½ | 3¾ | 5 | 1¼ | ⅝ | 1¼ | 3¾ |
| 2¾ | 3¼ | 3 | 3¼ | 2 1/16 | 1⅝ | 4⅛ | 5½ | 1⅜ | 11/16 | 1⅜ | 4⅛ |
| 3 | 3⅝ | 3¼ | 3⅝ | 2¼ | 1 13/16 | 4½ | 6 | 1½ | ¾ | 1½ | 4½ |
| 3¼ | 4 | 3⅝ | 4 | 2 7/16 | 2 | 4⅞ | 6½ | 1⅝ | 13/16 | 1⅝ | 4⅞ |
| 3½ | 4¼ | 3⅞ | 4¼ | 2⅝ | 2⅛ | 5¼ | 7 | 1¾ | ⅞ | 1¾ | 5¼ |
| 3¾ | 4½ | 4⅛ | 4½ | 2 13/16 | 2¼ | 5⅝ | 7½ | 1⅞ | 15/16 | 1⅞ | 5⅝ |
| 4 | 4¾ | 4⅜ | 4¾ | 3 | 2⅜ | 6 | 8 | 2 | 1 | 2 | 6 |
| 4¼ | 5⅛ | 4¾ | 5⅛ | 3 3/16 | 2 9/16 | 6⅜ | 8½ | 2⅛ | 1 1/16 | 2⅛ | 6⅜ |
| 4½ | 5½ | 5 | 5½ | 3⅜ | 2¾ | 6¾ | 9 | 2¼ | 1⅛ | 2¼ | 6¾ |
| 4¾ | 5¾ | 5¼ | 5¾ | 3 9/16 | 2⅞ | 7⅛ | 9½ | 2⅜ | 1 3/16 | 2⅜ | 7⅛ |
| 5 | 6 | 5½ | 6 | 3¾ | 3 | 7½ | 10 | 2½ | 1¼ | 2½ | 7½ |

**Power Transmitting Capacity of Friction Clutches.**—When selecting a clutch for a given class of service, it is advisable to consider any overloads that may be encountered and base the power transmitting capacity of the clutch upon such overloads. When the load varies or is subject to frequent release or engagement, the clutch capacity should be greater than the actual amount of power transmitted. If the power is derived from a gas or gasoline engine, the horsepower rating of the clutch should be 75 or 100 percent greater than that of the engine.

***Power Transmitted by Disk Clutches.***—The approximate amount of power that a disk clutch will transmit may be determined from the following formula, in which $HP$ = horsepower transmitted by the clutch; $\mu$ = coefficient of friction; $r$ = mean radius of engaging surfaces; $F$ = axial force in pounds (spring pressure) holding disks in contact; $N$ = number of frictional surfaces; $S$ = speed of shaft in revolutions per minute:

$$HP = \frac{\mu rFNS}{63,000}$$

## Cast-Iron Friction Clutches

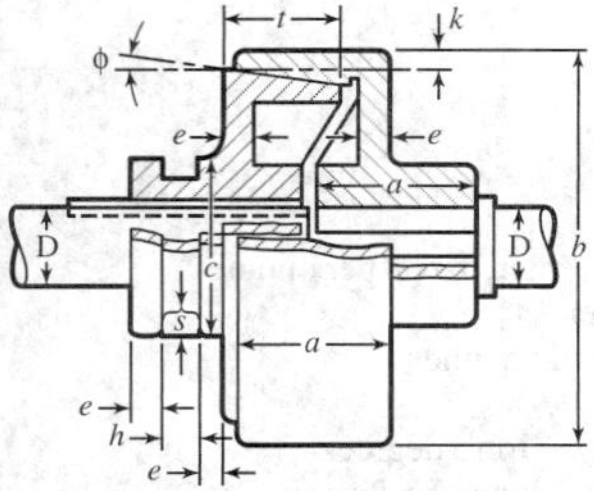


For sizes not given below:

$a = 2D$
$b = 4 \text{ to } 8D$
$c = 2\frac{1}{4}D$
$t = 1\frac{1}{2}D$
$e = \frac{3}{8}D$
$h = \frac{1}{2}D$
$s = \frac{5}{16}D$, nearly
$k = \frac{1}{4}D$

*Note:* The angle φ of the cone may be from 4 to 10 degrees

| *D* | *a* | *b* | *c* | *t* | *e* | *h* | *s* | *k* |
|---|---|---|---|---|---|---|---|---|
| 1 | 2 | 4-8 | 2¼ | 1½ | ⅜ | ½ | 5⁄16 | ¼ |
| 1¼ | 2½ | 5-10 | 2⅞ | 1⅞ | ½ | ⅝ | ⅜ | 5⁄16 |
| 1½ | 3 | 6-12 | 3⅜ | 2¼ | ⅝ | ¾ | ½ | ⅜ |
| 1¾ | 3½ | 7-14 | 4 | 2⅝ | ⅝ | ⅞ | ⅝ | 7⁄16 |
| 2 | 4 | 8-16 | 4½ | 3 | ¾ | 1 | ⅝ | ½ |
| 2¼ | 4½ | 9-18 | 5 | 3⅜ | ⅞ | 1⅛ | ⅝ | 9⁄16 |
| 2½ | 5 | 10-20 | 5⅝ | 3¾ | 1 | 1¼ | ¾ | ⅝ |
| 2¾ | 5½ | 11-22 | 6¼ | 4⅛ | 1 | 1⅜ | ⅞ | 11⁄16 |
| 3 | 6 | 12-24 | 6¾ | 4½ | 1⅛ | 1½ | ⅞ | ¾ |
| 3¼ | 6½ | 13-26 | 7⅜ | 4⅞ | 1¼ | 1⅝ | 1 | 13⁄16 |
| 3½ | 7 | 14-28 | 7⅞ | 5¼ | 1⅜ | 1¾ | 1 | ⅞ |
| 3¾ | 7½ | 15-30 | 8½ | 5⅝ | 1⅜ | 1⅞ | 1¼ | 15⁄16 |
| 4 | 8 | 16-32 | 9 | 6 | 1½ | 2 | 1¼ | 1 |
| 4¼ | 8½ | 17-34 | 9½ | 6⅜ | 1⅝ | 2⅛ | 1⅜ | 1 1⁄16 |
| 4½ | 9 | 18-36 | 10¼ | 6¾ | 1¾ | 2¼ | 1⅜ | 1⅛ |
| 4¾ | 9½ | 19-38 | 10¾ | 7⅛ | 1¾ | 2⅜ | 1½ | 1 3⁄16 |
| 5 | 10 | 20-40 | 11¼ | 7½ | 1⅞ | 2½ | 1½ | 1¼ |
| 5¼ | 10½ | 21-42 | 11¾ | 7⅞ | 2 | 2⅝ | 1⅝ | 1 5⁄16 |
| 5½ | 11 | 22-44 | 12⅜ | 8¼ | 2 | 2¾ | 1¾ | 1⅜ |
| 5¾ | 11½ | 23-46 | 13 | 8⅝ | 2¼ | 2⅞ | 1¾ | 1 7⁄16 |
| 6 | 12 | 24-48 | 13½ | 9 | 2¼ | 3 | 1⅞ | 1½ |

**Frictional Coefficients for Clutch Calculations.—**While the frictional coefficients used by designers of clutches differ somewhat and depend upon variable factors, the following values may be used in clutch calculations: For greasy leather on cast iron about 0.20 or 0.25, leather on metal that is quite oily 0.15; metal and cork on oily metal 0.32; the same on dry metal 0.35; metal on dry metal 0.15; disk clutches having lubricated surfaces 0.10.

**Formulas for Cone Clutches.—**In cone clutch design, different formulas have been developed for determining the horsepower transmitted. These formulas, at first sight, do not seem to agree, there being a variation due to the fact that in some of the formulas the friction clutch surfaces are assumed to engage without slip, whereas, in others, some allowance is made for slip. The following formulas include both of these conditions:

**Table 2. Keyway Dimensions and Tolerances for Metric Square and Rectangular Parallel Keys** *ANSI/ASME B18.25.1M-1996 (Withdrawn)*

| Key size $b \times h$ (mm) | Keyway | | | | | | | | | | | | | | | | |
|---|---|---|---|---|---|---|---|---|---|---|---|---|---|---|---|---|---|
| | Width. | | | | | | | | | | | Depth | | | | Radius, $r$ | |
| | Basic Size[c] | Tolerance[a] and Resulting Fits[b] | | | | | | | | | | Shaft, $t_1$ | | Hub, $t_2$ | | | |
| | | Normal Fit | | | | Close Fit | | Free Fit | | | | | | | | | |
| | | Shaft | | Hub | | Shaft and Hub | | Shaft | | Hub | | | | | | | |
| | | N9 | Fit | JS9 | Fit | P9 | Fit | H9 | Fit | D10 | Fit | Basic Size | Toler-ance | Basic Size | Toler-ance | Min. | Max. |
| 2 × 2 | 2 | −0.004 | 0.010L | +0.0125 | 0.0265L | −0.006 | 0.008L | +0.025 | 0.039L | +0.060 | 0.074L | 1.2 | | 1 | | | |
| 3 × 3 | 3 | −0.029 | 0.029T | −0.0125 | 0.0125T | −0.031 | 0.031T | 0 | 0T | +0.020 | 0.020L | 1.8 | | 1.4 | | 0.08 | 0.16 |
| 4 × 4 | 4 | | | | | | | | | | | 2.5 | | 1.8 | | | |
| 5 × 3 | 5 | | | | | | | | | | | 1.8 | +0.1<br>0 | 1.4 | +0.1<br>0 | | |
| 5 × 5 | 5 | 0<br>−0.030 | 0.018L<br>0.030T | +0.0150<br>−0.0150 | 0.033L<br>0.015T | −0.012<br>−0.042 | 0.006L<br>0.042T | +0.030<br>0 | 0.048L<br>0T | +0.078<br>+0.030 | 0.096L<br>0.030L | 3 | | 2.8 | | | |
| 6 × 4 | 6 | | | | | | | | | | | 2.5 | | 1.8 | | 0.16 | 0.25 |
| 6 × 6 | 6 | | | | | | | | | | | 3.5 | | 2.8 | | | |
| 8 × 5 | 8 | | | | | | | | | | | 3 | | 2.8 | | | |
| 8 × 7 | 8 | | | | | | | | | | | 4 | +0.2<br>0 | 3.3 | +0.1<br>0 | | |
| 10 × 6 | 10 | 0<br>−0.036 | 0.022L<br>0.036T | +0.0180<br>−0.0180 | 0.040L<br>0.018T | −0.015<br>−0.051 | 0.007L<br>0.051T | +0.036<br>0 | 0.058L<br>0T | +0.098<br>+0.040 | 0.120L<br>0.040L | 3.5 | +0.1<br>0 | 2.8 | +0.1<br>0 | | |
| 10 × 8 | 10 | | | | | | | | | | | 5 | +0.2<br>0 | 3.3 | +0.2<br>0 | | |
| 12 × 6 | 12 | | | | | | | | | | | 3.5 | +0.1<br>0 | 2.8 | +0.1<br>0 | | |
| 12 × 8 | 12 | | | | | | | | | | | 5 | +0.2<br>0 | 3.3 | +0.2 | 0.25 | 0.4 |
| 14 × 6 | 14 | 0<br>−0.043 | 0.027L<br>0.043T | +0.0215<br>−0.0215 | 0.0485L<br>0.0215T | −0.018<br>−0.061 | 0.009L<br>0.061T | +0.043<br>0 | 0.070L<br>0T | +0.120<br>+0.050 | 0.147L<br>0.050L | 3.5 | +0.1<br>0 | 2.8 | +0.1<br>0 | | |
| 14 × 9 | 14 | | | | | | | | | | | 5.5 | | 3.8 | | | |
| 16 × 7 | 16 | | | | | | | | | | | 4 | | 3.3 | | | |
| 16 × 10 | 16 | | | | | | | | | | | 6 | +0.2<br>0 | 4.3 | +0.2<br>0 | | |
| 18 × 7 | 18 | | | | | | | | | | | 4 | | 3.3 | | | |
| 18 × 11 | 18 | | | | | | | | | | | 7 | | 4.4 | | | |

**Table 2.** *(Continued)* **Keyway Dimensions and Tolerances for Metric Square and Rectangular Parallel Keys** *ANSI/ASME B18.25.1M-1996 (Withdrawn)*

| Key size $b \times h$ (mm) | Keyway: Width. Basic Size[c] | Tolerance[a] and Resulting Fits[b]: Normal Fit, Shaft, N9 | Normal Fit, Shaft, Fit | Normal Fit, Hub, JS9 | Normal Fit, Hub, Fit | Close Fit, Shaft and Hub, P9 | Close Fit, Shaft and Hub, Fit | Free Fit, Shaft, H9 | Free Fit, Shaft, Fit | Free Fit, Hub, D10 | Free Fit, Hub, Fit | Depth: Shaft, $t_1$, Basic Size | Shaft, $t_1$, Tolerance | Hub, $t_2$, Basic Size | Hub, $t_2$, Tolerance | Radius, $r$, Min. | Radius, $r$, Max. |
|---|---|---|---|---|---|---|---|---|---|---|---|---|---|---|---|---|---|
| 20 × 8 | 20 | | | | | | | | | | | 5 | | 3.3 | | | |
| 20 × 12 | 20 | | | | | | | | | | | 7.5 | | 4.9 | | | |
| 22 × 9 | 22 | | | | | | | | | | | 5.5 | | 3.8 | | | |
| 22 × 14 | 22 | 0 | 0.033L | +0.026 | 0.059L | −0.022 | 0.011L | +0.052 | 0.085L | +0.149 | 0.182L | 9 | | 5.4 | | | |
| 25 × 9 | 25 | −0.052 | 0.052T | −0.026 | 0.026T | −0.074 | 0.074T | 0 | 0T | +0.065 | 0.065L | 5.5 | | 3.8 | | | |
| 25 × 14 | 25 | | | | | | | | | | | 9 | +0.2 | 5.4 | +0.2 | 0.4 | 0.6[c] |
| 28 × 10 | 28 | | | | | | | | | | | 6 | 0 | 4.3 | 0 | | |
| 28 × 16 | 28 | | | | | | | | | | | 10 | | 6.4 | | | |
| 32 × 11 | 32 | | | | | | | | | | | 7[c] | | 4.4 | | | |
| 32 × 18 | 32 | | | | | | | | | | | 11[c] | | 7.4 | | | |
| 36 × 12 | 36 | 0 | 0.039L | +0.031 | 0.070L | −0.026 | 0.013L | +0.062 | 0.101L | +0.180 | 0.219L | 7.5[c] | | 4.9 | | | |
| 36 × 20 | 36 | −0.062 | 0.062T | −0.031 | 0.031T | −0.088 | 0.088T | 0 | 0T | +0.080 | 0.080L | 12 | | 8.4 | | | |
| 40 × 22 | 40 | | | | | | | | | | | 13 | | 9.4 | | 0.7 | 1.0 |
| 45 × 25 | 45 | | | | | | | | | | | 15 | | 10.4 | | | |
| 50 × 28 | 50 | | | | | | | | | | | 17 | | 11.4 | | | |
| 56 × 32 | 56 | | | | | | | | | | | 20 | +0.3 | 12.4 | +0.3 | | |
| 63 × 32 | 63 | 0 | 0.046L | +0.037 | 0.083L | −0.032 | 0.014L | +0.074 | 0.120L | +0.220 | 0.266L | 20 | 0 | 12.4 | 0 | 1.2 | 1.6 |
| 70 × 36 | 70 | −0.074 | 0.074T | −0.037 | 0.037T | −0.106 | 0.106T | 0 | 0T | +0.100 | 0.100L | 22 | | 14.4 | | | |
| 80 × 40 | 80 | | | | | | | | | | | 25 | | 15.4 | | | |
| 90 × 45 | 90 | 0 | 0.054L | +0.0435 | 0.0975L | −0.037 | 0.017L | +0.087 | 0.139L | +0 260 | 0.314L | 28 | | 17.4 | | 2.0 | 2.5 |
| 100 × 50 | 100 | −0.087 | 0.87T | −0.0435 | 0.0435T | −0.1254 | 0.1254T | 0 | 0T | +0.120 | 0.120L | 31 | | 19.5 | | | |

[a] Some of the tolerances are expressed as plus-plus. See *Tolerances* on page 2545 for more information.

[b] Resulting fits: L indicates a clearance between the key and keyway; T indicates an interference between the key and keyway.

[c] Values changed for accuracy from that given in ANSI/ASME 18.25.1M-1996 (Withdrawn).

**Table 4. Keyway Dimensions and Tolerances for Metric Square and Rectangular Parallel Keys**
*ANSI/ASME B18.25.3M-2003 (Withdrawn)* Width Tolerances and Deviations Greater than Basic Size

| | Keyway Width | | | | | | | | | | | Keyway Depth | | | | |
|---|---|---|---|---|---|---|---|---|---|---|---|---|---|---|---|---|
| | | Tolerance and Resulting Fit[a] | | | | | | | | | | Shaft, $t_1$ | | Hub, $t_2$ | | Radius, $r$ |
| | | Normal Fit | | | | Close Fit | | Free Fit | | | | | | | | |
| | | Shaft | | Hub | | Shaft and Hub | | Shaft | | Hub | | | | | | |
| Key Size, $b \times h$ | Nominal | Tolerance | Fit | Tolerance | Fit | Tolerance | Fit | Tolerance | Fit | Tolerance | Fit | Nominal | Tolerance | Nominal | Tolerance | Max. |
| 2×2 | 2 | +0.040 | 0.040L | +0.050 | 0.050L | +0.034 | 0.034L | +0.066 | 0.066L | +0.086 | 0.086L | 1.2 | | 1 | | |
| 3×3 | 3 | +0.010 | 0.030T | +0.025 | 0.015T | −0.008 | 0.032T | +0.040 | 0T | +0.060 | 0.020L | 1.8 | | 1.4 | | 0.16 |
| 4×4 | 4 | | | | | | | | | | | 2.5 | | 1.8 | | |
| 5×3 | 5 | | | | | | | | | | | 1.8 | +0.1 | 1.4 | +0.1 | |
| 5×5 | 5 | +0.045 | 0.045L | +0.060 | 0.060L | +0.035 | 0.035L | +0.075 | 0.075L<br>0T | +0.105 | 0.105L | 3 | 0 | 2.8 | 0 | |
| 6×4 | 6 | +0.015 | 0.030T | +0.015 | 0.015T | −0.005 | 0.040T | +0.045 | | +0.075 | 0.030L | 2.5 | | 1.8 | | |
| 6×6 | 6 | | | | | | | | | | | 3.5 | | 2.8 | | 0.25 |
| 8×5 | 8 | | | | | | | | | | | 3 | | 2.8 | | |
| 8×7 | 8 | | | | | | | | | | | 4 | +0.2<br>0 | 3.3 | +0.2<br>0 | |
| 10×6 | 10 | +0.055<br>+0.015 | 0.055L<br>0.035T | +0.075<br>+0.035 | 0.075L<br>0.015T | +0.040<br>0.000 | 0.040L<br>0.050T | +0.090<br>+0.050 | 0.090L<br>0T | +0.130<br>+0.090 | 0.130L<br>0.040L | 3.5 | +0.1<br>0 | 2.8 | +0.1<br>0 | |
| 10×8 | 10 | | | | | | | | | | | 5 | +0.2<br>0 | 3.3 | +0.2<br>0 | |
| 12×6 | 12 | | | | | | | | | | | 3.5 | +0.1<br>0 | 2.8 | +0.1<br>0 | |
| 12×8 | 12 | +0.080<br>−0.030 | 0.080L<br>0.045T | +0.095<br>+0.055 | 0.095L<br>0.020T | +0.055<br>−0.015 | 0.055L<br>0.060T | +0.135<br>+0.075 | 0.135L<br>0T | +0.185<br>+0.125 | 0.185L<br>0.050L | 5 | +0.2<br>0 | 3.3 | +0.2<br>0 | 0.6 |
| 14×6 | 14 | | | | | | | | | | | 3.5 | +0.1<br>0 | 2.8 | +0.1<br>0 | |
| 14×9 | 14 | | | | | | | | | | | 5.5 | +0.2<br>0 | 3.8 | +0.2<br>0 | |

**Table 3. ANSI Standard Plain and Gib Head Keys** *ANSI/ASME B17.1-1967 (R2013)*

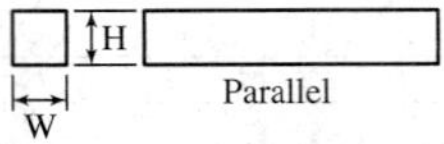


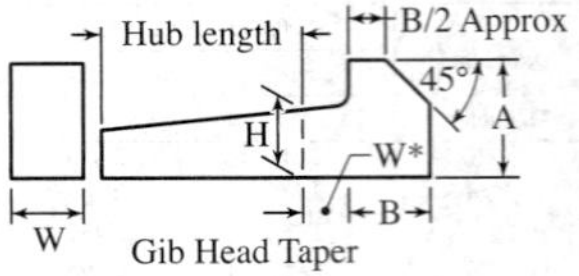


Plain and gib head taper keys have a 1/8″ taper in 12″

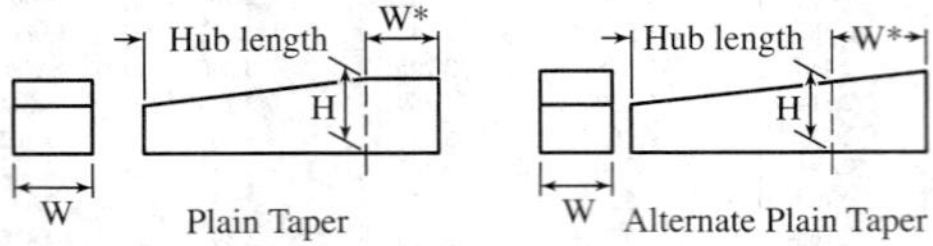


| Key | | | Nominal Key Size Width *W* | | Tolerance | | | |
|---|---|---|---|---|---|---|---|---|
| | | | Over | To (Incl.) | Width, *W* | | Height, *H* | |
| Parallel | Square | Keystock | … | 1¼ | +0.001 | −0.000 | +0.001 | −0.000 |
| | | | 1¼ | 3 | +0.002 | −0.000 | +0.002 | −0.000 |
| | | | 3 | 3½ | +0.003 | −0.000 | +0.003 | −0.000 |
| | | Bar Stock | … | ¾ | +0.000 | −0.002 | +0.000 | −0.002 |
| | | | ¾ | 1½ | +0.000 | −0.003 | +0.000 | −0.003 |
| | | | 1½ | 2½ | +0.000 | −0.004 | +0.000 | −0.004 |
| | | | 2½ | 3½ | +0.000 | −0.006 | +0.000 | −0.006 |
| | Rectangular | Keystock | … | 1¼ | +0.001 | −0.000 | +0.005 | −0.005 |
| | | | 1¼ | 3 | +0.002 | −0.000 | +0.005 | −0.005 |
| | | | 3 | 7 | +0.003 | −0.000 | +0.005 | −0.005 |
| | | Bar Stock | … | ¾ | +0.000 | −0.003 | +0.000 | −0.003 |
| | | | ¾ | 1½ | +0.000 | −0.004 | +0.000 | −0.004 |
| | | | 1½ | 3 | +0.000 | −0.005 | +0.000 | −0.005 |
| | | | 3 | 4 | +0.000 | −0.006 | +0.000 | −0.006 |
| | | | 4 | 6 | +0.000 | −0.008 | +0.000 | −0.008 |
| | | | 6 | 7 | +0.000 | −0.013 | +0.000 | −0.013 |
| Taper | Plain or Gib Head Square or Rectangular | | … | 1¼ | +0.001 | −0.000 | +0.005 | −0.000 |
| | | | 1¼ | 3 | +0.002 | −0.000 | +0.005 | −0.000 |
| | | | 3 | 7 | +0.003 | −0.000 | +0.005 | −0.000 |

| Gib Head Nominal Dimensions | | | | | | | | | | | | | |
|---|---|---|---|---|---|---|---|---|---|---|---|---|---|
| Nominal Key Size Width, *W* | Square | | | Rectangular | | | Nominal Key Size Width, *W* | Square | | | Rectangular | | |
| | *H* | *A* | *B* | *H* | *A* | *B* | | *H* | *A* | *B* | *H* | *A* | *B* |
| ⅛ | ⅛ | ¼ | ¼ | 3⁄32 | 3⁄16 | ⅛ | 1 | 1 | 1⅝ | 1⅛ | ¾ | 1¼ | ⅞ |
| 3⁄16 | 3⁄16 | 5⁄16 | 5⁄16 | ⅛ | ¼ | ¼ | 1¼ | 1¼ | 2 | 17⁄16 | ⅞ | 1⅜ | 1 |
| ¼ | ¼ | 7⁄16 | ⅜ | 3⁄16 | 5⁄16 | 5⁄16 | 1½ | 1½ | 2⅜ | 1¾ | 1 | 1⅝ | 1⅛ |
| 5⁄16 | 5⁄16 | ½ | 7⁄16 | ¼ | 7⁄16 | ⅜ | 1¾ | 1¾ | 2¾ | 2 | 1½ | 2⅜ | 1¾ |
| ⅜ | ⅜ | ⅝ | ½ | ¼ | 7⁄16 | ⅜ | 2 | 2 | 3½ | 2¼ | 1½ | 2⅜ | 1¾ |
| ½ | ½ | ⅞ | ⅝ | ⅜ | ⅝ | ½ | 2½ | 2½ | 4 | 3 | 1¾ | 2¾ | 2 |
| ⅝ | ⅝ | 1 | ¾ | 7⁄16 | ¾ | 9⁄16 | 3 | 3 | 5 | 3½ | 2 | 3½ | 2¼ |
| ¾ | ¾ | 1¼ | ⅞ | ½ | ⅞ | ⅝ | 3½ | 3½ | 6 | 4 | 2½ | 4 | 3 |
| ⅞ | ⅞ | 1⅜ | 1 | ⅝ | 1 | ¾ | … | … | … | … | … | … | … |

All dimensions are given in inches.

*For locating position of dimension *H*. Tolerance does not apply.

For larger sizes the following relationships are suggested as guides for establishing *A* and *B*: $A = 1.8H$ and $B = 1.2H$.

**Table 8. ANSI Standard Woodruff Keys** *ANSI/ASME B17.2-1967 (R2013)*

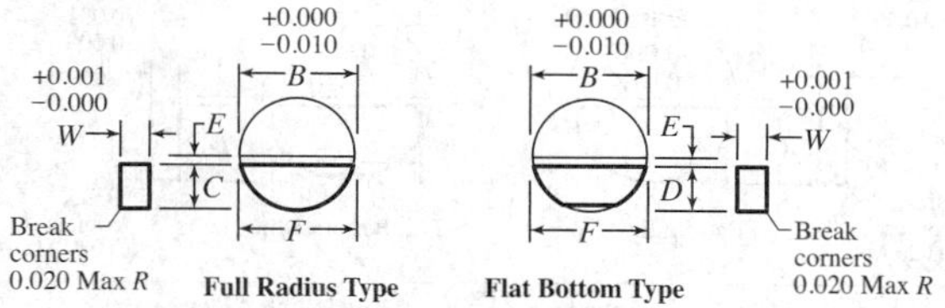


**Full Radius Type** **Flat Bottom Type**

| Key No. | Nominal Key Size W × B | Actual Length F +0.000 −0.010 | Height of Key C Max. | C Min. | D Max. | D Min. | Distance Below Center E |
|---|---|---|---|---|---|---|---|
| 202 | 1⁄16 × 1⁄4 | 0.248 | 0.109 | 0.104 | 0.109 | 0.104 | 1⁄64 |
| 202.5 | 1⁄16 × 5⁄16 | 0.311 | 0.140 | 0.135 | 0.140 | 0.135 | 1⁄64 |
| 302.5 | 3⁄32 × 5⁄16 | 0.311 | 0.140 | 0.135 | 0.140 | 0.135 | 1⁄64 |
| 203 | 1⁄16 × 3⁄8 | 0.374 | 0.172 | 0.167 | 0.172 | 0.167 | 1⁄64 |
| 303 | 3⁄32 × 3⁄8 | 0.374 | 0.172 | 0.167 | 0.172 | 0.167 | 1⁄64 |
| 403 | 1⁄8 × 3⁄8 | 0.374 | 0.172 | 0.167 | 0.172 | 0.167 | 1⁄64 |
| 204 | 1⁄16 × 1⁄2 | 0.491 | 0.203 | 0.198 | 0.194 | 0.188 | 3⁄64 |
| 304 | 3⁄32 × 1⁄2 | 0.491 | 0.203 | 0.198 | 0.194 | 0.188 | 3⁄64 |
| 404 | 1⁄8 × 1⁄2 | 0.491 | 0.203 | 0.198 | 0.194 | 0.188 | 3⁄64 |
| 305 | 3⁄32 × 5⁄8 | 0.612 | 0.250 | 0.245 | 0.240 | 0.234 | 1⁄16 |
| 405 | 1⁄8 × 5⁄8 | 0.612 | 0.250 | 0.245 | 0.240 | 0.234 | 1⁄16 |
| 505 | 5⁄32 × 5⁄8 | 0.612 | 0.250 | 0.245 | 0.240 | 0.234 | 1⁄16 |
| 605 | 3⁄16 × 5⁄8 | 0.612 | 0.250 | 0.245 | 0.240 | 0.234 | 1⁄16 |
| 406 | 1⁄8 × 3⁄4 | 0.740 | 0.313 | 0.308 | 0.303 | 0.297 | 1⁄16 |
| 506 | 5⁄32 × 3⁄4 | 0.740 | 0.313 | 0.308 | 0.303 | 0.297 | 1⁄16 |
| 606 | 3⁄16 × 3⁄4 | 0.740 | 0.313 | 0.308 | 0.303 | 0.297 | 1⁄16 |
| 806 | 1⁄4 × 3⁄4 | 0.740 | 0.313 | 0.308 | 0.303 | 0.297 | 1⁄16 |
| 507 | 5⁄32 × 7⁄8 | 0.866 | 0.375 | 0.370 | 0.365 | 0.359 | 1⁄16 |
| 607 | 3⁄16 × 7⁄8 | 0.866 | 0.375 | 0.370 | 0.365 | 0.359 | 1⁄16 |
| 707 | 7⁄32 × 7⁄8 | 0.866 | 0.375 | 0.370 | 0.365 | 0.359 | 1⁄16 |
| 807 | 1⁄4 × 7⁄8 | 0.866 | 0.375 | 0.370 | 0.365 | 0.359 | 1⁄16 |
| 608 | 3⁄16 × 1 | 0.992 | 0.438 | 0.433 | 0.428 | 0.422 | 1⁄16 |
| 708 | 7⁄32 × 1 | 0.992 | 0.438 | 0.433 | 0.428 | 0.422 | 1⁄16 |
| 808 | 1⁄4 × 1 | 0.992 | 0.438 | 0.433 | 0.428 | 0.422 | 1⁄16 |
| 1008 | 5⁄16 × 1 | 0.992 | 0.438 | 0.433 | 0.428 | 0.422 | 1⁄16 |
| 1208 | 3⁄8 × 1 | 0.992 | 0.438 | 0.433 | 0.428 | 0.422 | 1⁄16 |
| 609 | 3⁄16 × 1 1⁄8 | 1.114 | 0.484 | 0.479 | 0.475 | 0.469 | 5⁄64 |
| 709 | 7⁄32 × 1 1⁄8 | 1.114 | 0.484 | 0.479 | 0.475 | 0.469 | 5⁄64 |
| 809 | 1⁄4 × 1 1⁄8 | 1.114 | 0.484 | 0.479 | 0.475 | 0.469 | 5⁄64 |
| 1009 | 5⁄16 × 1 1⁄8 | 1.114 | 0.484 | 0.479 | 0.475 | 0.469 | 5⁄64 |
| 610 | 3⁄16 × 1 1⁄4 | 1.240 | 0.547 | 0.542 | 0.537 | 0.531 | 5⁄64 |
| 710 | 7⁄32 × 1 1⁄4 | 1.240 | 0.547 | 0.542 | 0.537 | 0.531 | 5⁄64 |
| 810 | 1⁄4 × 1 1⁄4 | 1.240 | 0.547 | 0.542 | 0.537 | 0.531 | 5⁄64 |
| 1010 | 5⁄16 × 1 1⁄4 | 1.240 | 0.547 | 0.542 | 0.537 | 0.531 | 5⁄64 |
| 1210 | 3⁄8 × 1 1⁄4 | 1.240 | 0.547 | 0.542 | 0.537 | 0.531 | 5⁄64 |
| 811 | 1⁄4 × 1 3⁄8 | 1.362 | 0.594 | 0.589 | 0.584 | 0.578 | 3⁄32 |
| 1011 | 5⁄16 × 1 3⁄8 | 1.362 | 0.594 | 0.589 | 0.584 | 0.578 | 3⁄32 |
| 1211 | 3⁄8 × 1 3⁄8 | 1.362 | 0.594 | 0.589 | 0.584 | 0.578 | 3⁄32 |
| 812 | 1⁄4 × 1 1⁄2 | 1.484 | 0.641 | 0.636 | 0.631 | 0.625 | 7⁄64 |
| 1012 | 5⁄16 × 1 1⁄2 | 1.484 | 0.641 | 0.636 | 0.631 | 0.625 | 7⁄64 |
| 1212 | 3⁄8 × 1 1⁄2 | 1.484 | 0.641 | 0.636 | 0.631 | 0.625 | 7⁄64 |

All dimensions are given in inches.

The key numbers indicate normal key dimensions. The last two digits give the nominal diameter *B* in eighths of an inch and the digits preceding the last two give the nominal width *W* in thirty-seconds of an inch.

**Table 9. ANSI Standard Woodruff Keys** *ANSI/ASME B17.2-1967 (R2013)*

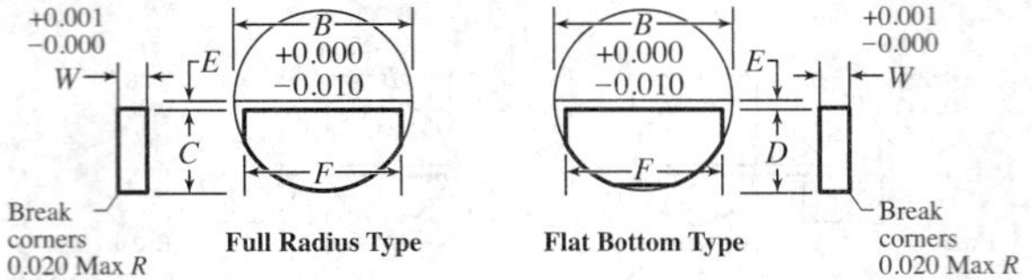


| Key No. | Nominal Key Size $W \times B$ | Actual Length $F$ +0.000 −0.010 | Height of Key $C$ Max. | $C$ Min. | $D$ Max. | $D$ Min. | Distance Below Center $E$ |
|---|---|---|---|---|---|---|---|
| 617-1 | $\frac{3}{16} \times 2\frac{1}{8}$ | 1.380 | 0.406 | 0.401 | 0.396 | 0.390 | $\frac{21}{32}$ |
| 817-1 | $\frac{1}{4} \times 2\frac{1}{8}$ | 1.380 | 0.406 | 0.401 | 0.396 | 0.390 | $\frac{21}{32}$ |
| 1017-1 | $\frac{5}{16} \times 2\frac{1}{8}$ | 1.380 | 0.406 | 0.401 | 0.396 | 0.390 | $\frac{21}{32}$ |
| 1217-1 | $\frac{3}{8} \times 2\frac{1}{8}$ | 1.380 | 0.406 | 0.401 | 0.396 | 0.390 | $\frac{21}{32}$ |
| 617 | $\frac{3}{16} \times 2\frac{1}{8}$ | 1.723 | 0.531 | 0.526 | 0.521 | 0.515 | $\frac{17}{32}$ |
| 817 | $\frac{1}{4} \times 2\frac{1}{8}$ | 1.723 | 0.531 | 0.526 | 0.521 | 0.515 | $\frac{17}{32}$ |
| 1017 | $\frac{5}{16} \times 2\frac{1}{8}$ | 1.723 | 0.531 | 0.526 | 0.521 | 0.515 | $\frac{17}{32}$ |
| 1217 | $\frac{3}{8} \times 2\frac{1}{8}$ | 1.723 | 0.531 | 0.526 | 0.521 | 0.515 | $\frac{17}{32}$ |
| 822-1 | $\frac{1}{4} \times 2\frac{3}{4}$ | 2.000 | 0.594 | 0.589 | 0.584 | 0.578 | $\frac{25}{32}$ |
| 1022-1 | $\frac{5}{16} \times 2\frac{3}{4}$ | 2.000 | 0.594 | 0.589 | 0.584 | 0.578 | $\frac{25}{32}$ |
| 1222-1 | $\frac{3}{8} \times 2\frac{3}{4}$ | 2.000 | 0.594 | 0.589 | 0.584 | 0.578 | $\frac{25}{32}$ |
| 1422-1 | $\frac{7}{16} \times 2\frac{3}{4}$ | 2.000 | 0.594 | 0.589 | 0.584 | 0.578 | $\frac{25}{32}$ |
| 1622-1 | $\frac{1}{2} \times 2\frac{3}{4}$ | 2.000 | 0.594 | 0.589 | 0.584 | 0.578 | $\frac{25}{32}$ |
| 822 | $\frac{1}{4} \times 2\frac{3}{4}$ | 2.317 | 0.750 | 0.745 | 0.740 | 0.734 | $\frac{5}{8}$ |
| 1022 | $\frac{5}{16} \times 2\frac{3}{4}$ | 2.317 | 0.750 | 0.745 | 0.740 | 0.734 | $\frac{5}{8}$ |
| 1222 | $\frac{3}{8} \times 2\frac{3}{4}$ | 2.317 | 0.750 | 0.745 | 0.740 | 0.734 | $\frac{5}{8}$ |
| 1422 | $\frac{7}{16} \times 2\frac{3}{4}$ | 2.317 | 0.750 | 0.745 | 0.740 | 0.734 | $\frac{5}{8}$ |
| 1622 | $\frac{1}{2} \times 2\frac{3}{4}$ | 2.317 | 0.750 | 0.745 | 0.740 | 0.734 | $\frac{5}{8}$ |
| 1228 | $\frac{3}{8} \times 3\frac{1}{2}$ | 2.880 | 0.938 | 0.933 | 0.928 | 0.922 | $\frac{13}{16}$ |
| 1428 | $\frac{7}{16} \times 3\frac{1}{2}$ | 2.880 | 0.938 | 0.933 | 0.928 | 0.922 | $\frac{13}{16}$ |
| 1628 | $\frac{1}{2} \times 3\frac{1}{2}$ | 2.880 | 0.938 | 0.933 | 0.928 | 0.922 | $\frac{13}{16}$ |
| 1828 | $\frac{9}{16} \times 3\frac{1}{2}$ | 2.880 | 0.938 | 0.933 | 0.928 | 0.922 | $\frac{13}{16}$ |
| 2028 | $\frac{5}{8} \times 3\frac{1}{2}$ | 2.880 | 0.938 | 0.933 | 0.928 | 0.922 | $\frac{13}{16}$ |
| 2228 | $\frac{11}{16} \times 3\frac{1}{2}$ | 2.880 | 0.938 | 0.933 | 0.928 | 0.922 | $\frac{13}{16}$ |
| 2428 | $\frac{3}{4} \times 3\frac{1}{2}$ | 2.880 | 0.938 | 0.933 | 0.928 | 0.922 | $\frac{13}{16}$ |

All dimensions are given in inches.

The key numbers indicate nominal key dimensions. The last two digits give the nominal diameter $B$ in eighths of an inch and the digits preceding the last two give the nominal width $W$ in thirty-seconds of an inch.

The key numbers with the –1 designation, while representing the nominal key size have a shorter length $F$ and due to a greater distance below center $E$ are less in height than the keys of the same number without the –1 designation.

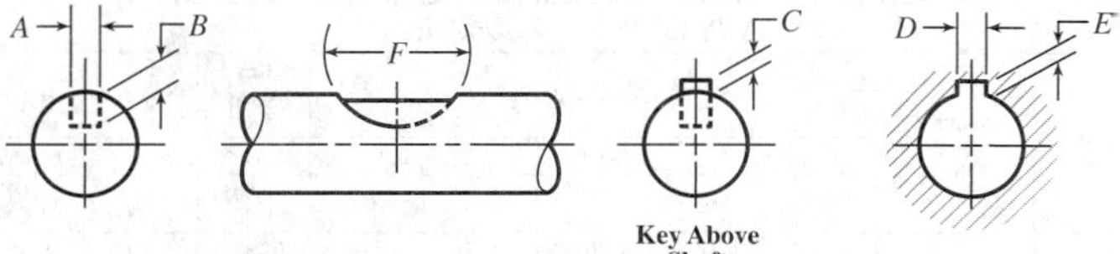


Keyseat—Shaft

Key Above Shaft

Keyseat—Hub

## Table 10. ANSI Keyseat Dimensions for Woodruff Keys

*ANSI/ASME B17.2-1967 (R2013)*

| Key No. | Nominal Size Key | Keyseat—Shaft | | | | | Key Above Shaft | Keyseat—Hub | |
|---|---|---|---|---|---|---|---|---|---|
| | | Width A[a] | | Depth B | Diameter F | | Height C | Width D | Depth E |
| | | Min. | Max. | +0.005 −0.000 | Min. | Max. | +0.005 −0.005 | +0.002 −0.000 | +0.00 −0.00 |
| 202 | 1⁄16 × 1⁄4 | 0.0615 | 0.0630 | 0.0728 | 0.250 | 0.268 | 0.0312 | 0.0635 | 0.037 |
| 202.5 | 1⁄16 × 5⁄16 | 0.0615 | 0.0630 | 0.1038 | 0.312 | 0.330 | 0.0312 | 0.0635 | 0.037 |
| 302.5 | 3⁄32 × 5⁄16 | 0.0928 | 0.0943 | 0.0882 | 0.312 | 0.330 | 0.0469 | 0.0948 | 0.052 |
| 203 | 1⁄16 × 3⁄8 | 0.0615 | 0.0630 | 0.1358 | 0.375 | 0.393 | 0.0312 | 0.0635 | 0.037 |
| 303 | 3⁄32 × 3⁄8 | 0.0928 | 0.0943 | 0.1202 | 0.375 | 0.393 | 0.0469 | 0.0948 | 0.052 |
| 403 | 1⁄8 × 3⁄8 | 0.1240 | 0.1255 | 0.1045 | 0.375 | 0.393 | 0.0625 | 0.1260 | 0.068 |
| 204 | 1⁄16 × 1⁄2 | 0.0615 | 0.0630 | 0.1668 | 0.500 | 0.518 | 0.0312 | 0.0635 | 0.037 |
| 304 | 3⁄32 × 1⁄2 | 0.0928 | 0.0943 | 0.1511 | 0.500 | 0.518 | 0.0469 | 0.0948 | 0.052 |
| 404 | 1⁄8 × 1⁄2 | 0.1240 | 0.1255 | 0.1355 | 0.500 | 0.518 | 0.0625 | 0.1260 | 0.068 |
| 305 | 3⁄32 × 5⁄8 | 0.0928 | 0.0943 | 0.1981 | 0.625 | 0.643 | 0.0469 | 0.0948 | 0.052 |
| 405 | 1⁄8 × 5⁄8 | 0.1240 | 0.1255 | 0.1825 | 0.625 | 0.643 | 0.0625 | 0.1260 | 0.068 |
| 505 | 5⁄32 × 5⁄8 | 0.1553 | 0.1568 | 0.1669 | 0.625 | 0.643 | 0.0781 | 0.1573 | 0.084 |
| 605 | 3⁄16 × 5⁄8 | 0.1863 | 0.1880 | 0.1513 | 0.625 | 0.643 | 0.0937 | 0.1885 | 0.099 |
| 406 | 1⁄8 × 3⁄4 | 0.1240 | 0.1255 | 0.2455 | 0.750 | 0.768 | 0.0625 | 0.1260 | 0.068 |
| 506 | 5⁄32 × 3⁄4 | 0.1553 | 0.1568 | 0.2299 | 0.750 | 0.768 | 0.0781 | 0.1573 | 0.084 |
| 606 | 3⁄16 × 3⁄4 | 0.1863 | 0.1880 | 0.2143 | 0.750 | 0.768 | 0.0937 | 0.1885 | 0.099 |
| 806 | 1⁄4 × 3⁄4 | 0.2487 | 0.2505 | 0.1830 | 0.750 | 0.768 | 0.1250 | 0.2510 | 0.131 |
| 507 | 5⁄32 × 7⁄8 | 0.1553 | 0.1568 | 0.2919 | 0.875 | 0.895 | 0.0781 | 0.1573 | 0.084 |
| 607 | 3⁄16 × 7⁄8 | 0.1863 | 0.1880 | 0.2763 | 0.875 | 0.895 | 0.0937 | 0.1885 | 0.099 |
| 707 | 7⁄32 × 7⁄8 | 0.2175 | 0.2193 | 0.2607 | 0.875 | 0.895 | 0.1093 | 0.2198 | 0.115 |
| 807 | 1⁄4 × 7⁄8 | 0.2487 | 0.2505 | 0.2450 | 0.875 | 0.895 | 0.1250 | 0.2510 | 0.131 |
| 608 | 3⁄16 × 1 | 0.1863 | 0.1880 | 0.3393 | 1.000 | 1.020 | 0.0937 | 0.1885 | 0.099 |
| 708 | 7⁄32 × 1 | 0.2175 | 0.2193 | 0.3237 | 1.000 | 1.020 | 0.1093 | 0.2198 | 0.115 |
| 808 | 1⁄4 × 1 | 0.2487 | 0.2505 | 0.3080 | 1.000 | 1.020 | 0.1250 | 0.2510 | 0.131 |
| 1008 | 5⁄16 × 1 | 0.3111 | 0.3130 | 0.2768 | 1.000 | 1.020 | 0.1562 | 0.3135 | 0.162 |
| 1208 | 3⁄8 × 1 | 0.3735 | 0.3755 | 0.2455 | 1.000 | 1.020 | 0.1875 | 0.3760 | 0.193 |
| 609 | 3⁄16 × 1 1⁄8 | 0.1863 | 0.1880 | 0.3853 | 1.125 | 1.145 | 0.0937 | 0.1885 | 0.099 |
| 709 | 7⁄32 × 1 1⁄8 | 0.2175 | 0.2193 | 0.3697 | 1.125 | 1.145 | 0.1093 | 0.2198 | 0.115 |
| 809 | 1⁄4 × 1 1⁄8 | 0.2487 | 0.2505 | 0.3540 | 1.125 | 1.145 | 0.1250 | 0.2510 | 0.131 |
| 1009 | 5⁄16 × 1 1⁄8 | 0.3111 | 0.3130 | 0.3228 | 1.125 | 1.145 | 0.1562 | 0.3135 | 0.162 |
| 610 | 3⁄16 × 1 1⁄4 | 0.1863 | 0.1880 | 0.4483 | 1.250 | 1.273 | 0.0937 | 0.1885 | 0.099 |
| 710 | 7⁄32 × 1 1⁄4 | 0.2175 | 0.2193 | 0.4327 | 1.250 | 1.273 | 0.1093 | 0.2198 | 0.115 |
| 810 | 1⁄4 × 1 1⁄4 | 0.2487 | 0.2505 | 0.4170 | 1.250 | 1.273 | 0.1250 | 0.2510 | 0.131 |
| 1010 | 5⁄16 × 1 1⁄4 | 0.3111 | 0.3130 | 0.3858 | 1.250 | 1.273 | 0.1562 | 0.3135 | 0.162 |
| 1210 | 3⁄8 × 1 1⁄4 | 0.3735 | 0.3755 | 0.3545 | 1.250 | 1.273 | 0.1875 | 0.3760 | 0.193 |
| 811 | 1⁄4 × 1 3⁄8 | 0.2487 | 0.2505 | 0.4640 | 1.375 | 1.398 | 0.1250 | 0.2510 | 0.131 |
| 1011 | 5⁄16 × 1 3⁄8 | 0.3111 | 0.3130 | 0.4328 | 1.375 | 1.398 | 0.1562 | 0.3135 | 0.162 |

## FLEXIBLE BELTS AND SHEAVES

Flexible belt drives are used in industrial power transmission applications, especially when the speeds of the driver and driven shafts must be different or when shafts must be widely separated. The trend toward higher speed prime movers and the need to achieve a slower, useful driven speed are additional factors favoring the use of belts. Belts have numerous advantages over other means of power transmission; these advantages include overall economy, cleanliness, no need for lubrication, lower maintenance costs, easy installation, dampening of shock loads, and the abilities to be used for clutching and variable-speed power transmission between widely spaced shafts.

### Calculations for Belts and Pulleys

Belt speed plays an important role in the amount of load a friction drive system can transmit. Higher speeds will require higher preloads (increased belt tension) to compensate for the higher centrifugal force. In positive drive (toothed belt) systems, higher speeds generate dynamic forces caused by unavoidable tolerance errors that may result in increased tooth or pin stresses and shorter belt life.

**Pulley Diameters and Drive Ratios.—**Minimum pulley diameters determined by belt manufacturers are based on the minimum radius that a belt can wrap around a pulley without stressing the load-carrying members. For positive drive systems, minimum pulley diameters are also determined by the minimum number of teeth that must be engaged with the sprocket to guarantee the operating load.

Diameters of driving and driven pulleys determine the velocity ratio of the input relative to the output shaft and are derived from the following formulas: for all belt systems, velocity ratio $V = D_{pi}/D_{po}$, and for positive (toothed) drive systems, velocity ratio $V = N_i/N_o$, where $D_{pi}$ is the pitch diameter of the driving pulley, $D_{po}$ is the pitch diameter of the driven pulley, $N_i$ is the number of teeth on the driving pulley, and $N_o$ is the number of teeth on the driven pulley. For most drive systems, a velocity ratio of 8:1 is the largest that should be attempted with a single reduction drive, and 6:1 is a reasonable maximum.

**Wrap Angles and Center-to-Center Distances.—**The radial distance for which the belt is in contact with the pulley surface, or the number of teeth in engagement for positive drive belts, is called the wrap angle. Belt and sprocket combinations should be chosen to ensure a wrap angle of about 120° around the smaller pulley. The wrap angle should not be less than 90°, especially with positive drive belts, because if too few teeth are in engagement, the belt may jump a tooth or pin and timing or synchronization may be lost.

For flat belts, the minimum allowable center-to-center distance (CD) for any belt-and-sprocket combination should be chosen to ensure a minimum wrap angle around the smaller pulley. For high-velocity systems, a good rule of thumb is a minimum CD equal to the sum of the pitch diameter of the larger sprocket and one-half the pitch diameter of the smaller sprocket. This formula ensures a minimum wrap angle of approximately 120°, which is generally sufficient for friction drives and will ensure that positive drive belts do not jump teeth.

**Pulley Center Distances and Belt Lengths.—**Maximum center distances of pulleys should be about 15 to 20 times the pitch diameter of the smaller pulley. Greater spacing requires tight control of the belt tension because a small amount of stretch will cause a large drop in tension. Constant belt tension can be obtained by application of an adjustable tensioning pulley applied to the slack side of the belt. Friction drive systems using flat belts require much more tension than positive drive belt systems.

Belt length can be calculated from: $L = 2C + \pi(D_2 + D_1)/2 + (D_2 - D_1)^2/4C$ for friction drives, and length $L = 2C + \pi(D_2 + D_1)/2 + (D_2 + D_1)^2/4C$ for crossed belt friction belt drives, where $C$ is the center distance, $D_1$ is the pitch diameter of the small pulley, and $D_2$

*Minimum Sheave Size:* The recommended minimum sheave size depends on the rpm of the faster shaft. Minimum sheave diameters for each belt cross section are listed in Table 9.

*Cross Section Selection:* Use the chart (Fig. 4) as a guide to the Classical V-belt cross section for any combination of design horsepower and speed of the faster shaft. When the intersection of the design horsepower and speed of the faster shaft falls near a line between two areas on the chart, the possibilities in both areas should be investigated. Special circumstances (such as space limitations) may lead to a choice of belt cross section different from that indicated in the chart.

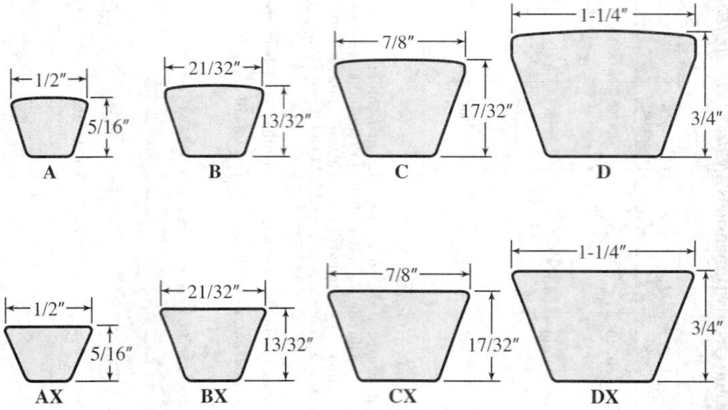


Fig. 3. Classical V-Belt Cross Sections

*Horsepower Ratings:* The horsepower rating formulas for classical V-belts are:

$$\textbf{A}: HP = d_p r\left[1.004 - \frac{1.652}{d_p} - 1.547\times10^{-4}(d_p r)^2 - .02126\log(d_p r)\right] + 1.652r\left(1 - \frac{1}{K_{SR}}\right)$$

$$\textbf{AX}: HP = d_p r\left[1.462 - \frac{2.239}{d_p} - 2.198\times10^{-4}(d_p r)^2 - .04238\log(d_p r)\right] + 2.239r\left(1 - \frac{1}{K_{SR}}\right)$$

$$\textbf{B}: HP = d_p r\left[1.769 - \frac{4.372}{d_p} - 3.081\times10^{-4}(d_p r)^2 - .03658\log(d_p r)\right] + 4.372r\left(1 - \frac{1}{K_{SR}}\right)$$

$$\textbf{BX}: \; HP = d_p r\left[2.051 - \frac{3.532}{d_p} - 3.097\times10^{-4}(d_p r)^2 - .0\,5735\log(d_p r)\right] + 3.532r\left(1 - \frac{1}{K_{SR}}\right)$$

$$\textbf{C}: \; HP = d_p r\left[3.325 - \frac{12.07}{d_p} - 5.828\times10^{-4}(d_p r)^2 - .0\,6886\log(d_p r)\right] + 12.07r\left(1 - \frac{1}{K_{SR}}\right)$$

$$\textbf{CX}: \; HP = d_p r\left[3.272 - \frac{6.655}{d_p} - 5.298\times10^{-4}(d_p r)^2 - .0\,8637\log(d_p r)\right] + 6.655r\left(1 - \frac{1}{K_{SR}}\right)$$

$$\textbf{D}: \; HP = d_p r\left[7.160 - \frac{43.21}{d_p} - 1.384\times10^{-3}(d_p r)^2 - .1\,454\log(d_p r)\right] + 43.21r\left(1 - \frac{1}{K_{SR}}\right)$$

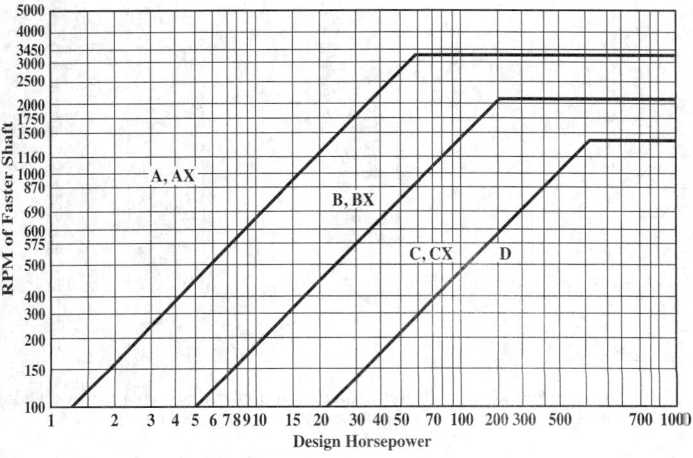


Fig. 4. Selection of Classical V-Belt Cross Sections

In these equations, $d_p$ = pitch diameter of small sheave, in.; $r$ = rpm of the faster shaft divided by 1000; $K_{SR}$ = speed ratio factor given in the accompanying Table 10. These formulas give the basic horsepower rating, corrected for the speed ratio. To obtain the horsepower per belt for an arc of contact other than 180 degrees and for belts shorter or longer than average length, multiply the horsepower obtained from these formulas by the length correction factor (Table 11) and the arc of contact correction factor (Table 12).

**Table 12. Arc of Contact Correction Factors**

| $\frac{D_d - d_d}{C}$ | Arc of Contact, θ, Small Sheave (deg) | Correction Factor V-V | Correction Factor V-Flat[a] | $\frac{D_d - d_d}{C}$ | Arc of Contact, θ Small Sheave (deg) | Correction Factor V-V | Correction Factor V-Flat |
|---|---|---|---|---|---|---|---|
| 0.00 | 180 | 1.00 | 0.75 | 0.80 | 133 | 0.87 | 0.85 |
| 0.10 | 174 | 0.99 | 0.76 | 0.90 | 127 | 0.85 | 0.85 |
| 0.20 | 169 | 0.97 | 0.78 | 1.00 | 120 | 0.82 | 0.82 |
| 0.30 | 163 | 0.96 | 0.79 | 1.10 | 113 | 0.80 | 0.80 |
| 0.40 | 157 | 0.94 | 0.80 | 1.20 | 106 | 0.77 | 0.77 |
| 0.50 | 151 | 0.93 | 0.81 | 1.30 | 99 | 0.73 | 0.73 |
| 0.60 | 145 | 0.91 | 0.83 | 1.40 | 91 | 0.70 | 0.70 |
| 0.70 | 139 | 0.89 | 0.84 | 1.50 | 83 | 0.65 | 0.65 |

[a] A V-flat drive is one using a small sheave and a large diameter flat pulley.

**Double V-Belts ANSI/RMA IP-21.—**Double V-belts or hexagonal belts are used when power input or takeoff is required on both sides of the belt. Designed for use on "serpentine" drives, which consist of sheaves rotating in opposite directions, the belts are available in AA, BB, CC, and DD cross sections and operate in standard classical sheaves. They are specified by cross section and nominal length.

*Belt Cross Sections:* Nominal dimensions of the four cross sections are given in Fig. 5.

*Belt Size Designation:* Double V-belt sizes are identified by a standard belt number, consisting of a letter-numeral combination. The letters identify the cross section; the numbers identify length as shown in Column 1 of Table 13. For example, AA51 indicates an AA cross section and a standard length designation of 51.

**Table 13. Double V-Belt Standard Effective Lengths** *ANSI/RMA IP-21, 1984*

| Standard Length Designation[a] | Standard Effective Length, Cross Section AA | BB | CC | DD | Permissible Deviation from Standard Effective Length | Matching Limits for One Set |
|---|---|---|---|---|---|---|
| 51 | 53.1 | 53.9 | … | … | ±0.7 | 0.15 |
| 55 | … | 57.9 | … | … | ±0.7 | 0.15 |
| 60 | 62.1 | 62.9 | … | … | ±0.7 | 0.15 |
| 68 | 70.1 | 70.9 | … | … | ±0.7 | 0.30 |
| 75 | 77.1 | 77.9 | … | … | ±0.7 | 0.30 |
| 80 | 82.1 | … | … | … | ±0.7 | 0.30 |
| 81 | … | 83.9 | 85.2 | … | ±0.7 | 0.30 |
| 85 | 87.1 | 87.9 | 89.2 | … | ±0.7 | 0.30 |
| 90 | 92.1 | 92.9 | 94.2 | … | ±0.8 | 0.30 |
| 96 | 98.1 | … | 100.2 | … | ±0.8 | 0.30 |
| 97 | … | 99.9 | … | … | ±0.8 | 0.30 |
| 105 | 107.1 | 107.9 | 109.2 | … | ±0.8 | 0.30 |
| 112 | 114.1 | 114.9 | 116.2 | … | ±0.8 | 0.30 |
| 120 | 122.1 | 122.9 | 124.2 | 125.2 | ±0.8 | 0.30 |
| 128 | 130.1 | 130.9 | 132.2 | 133.2 | ±0.8 | 0.30 |
| 144 | … | 146.9 | 148.2 | 149.2 | ±0.8 | 0.30 |
| 158 | … | 160.9 | 162.2 | 163.2 | ±1.0 | 0.45 |
| 173 | … | 175.9 | 177.2 | 178.2 | ±1.0 | 0.45 |
| 180 | … | 182.9 | 184.2 | 185.2 | ±1.0 | 0.45 |
| 195 | … | 197.9 | 199.2 | 200.2 | ±1.1 | 0.45 |
| 210 | … | 212.9 | 214.2 | 215.2 | ±1.1 | 0.45 |
| 240 | … | 241.4 | 242.2 | 242.7 | ±1.3 | 0.45 |
| 270 | … | 271.4 | 272.2 | 272.7 | ±1.6 | 0.60 |
| 300 | … | 301.4 | 302.2 | 302.7 | ±1.6 | 0.60 |
| 330 | … | … | 332.2 | 332.7 | ±2.0 | 0.60 |
| 360 | … | … | 362.2 | 362.7 | ±2.0 | 0.60 |

[a] To specify belt size use the Standard Length Designation prefixed by the letters indicating cross section; for example, BB90.

All dimensions in inches.

*Sheave Dimensions:* Groove angles and dimensions for sheaves and face widths of sheaves for multiple belt drives are given in Table 14, along with various tolerance values.

**Table 14.** *(Continued)* **Double V-Belt Sheave and Groove Dimensions** *ANSI/RMA IP-21, 1984*

| Deep Groove Dimensions[c] | | | | | | | | | | | Drive Design Factors | |
|---|---|---|---|---|---|---|---|---|---|---|---|---|
| Cross Section | Outside Diameter Range | Groove Angle, α ±0.33° | $b_g$ | $h_g$ (Min.) | $2h_d$ | $R_B$ (Min.) | $d_B$ ±0.0005 | $S_g$[a] ±0.025 | $S_e$ | | Minimum Recommended Outside Diameter | $2a_p$ |
| AA | Up through 5.96 | 34 | 0.589 ±0.005 | 0.615 | 0.560 | −0.009 | 0.4375 (7/16) | 0.750 | 0.438 | +0.090 −0.062 | 3.56 | 0.310 |
| | Over 5.96 | 38 | 0.611 ±0.005 | | | −0.008 | | | | | | |
| BB | Up through 7.71 | 34 | 0.747 ±0.006 | 0.730 | 0.710 | +0.007 | 0.5625 (9/16) | 0.875 | 0.562 | +0.120 −0.065 | 6.11 | 0.360 |
| | Over 7.71 | 38 | 0.774 ±0.006 | | | +0.008 | | | | | | |
| CC | Up through 9.00 | 34 | 1.066 ±0.007 | 1.055 | 1.010 | −0.035 | 0.7812 (25/32) | 1.250 | 0.812 | +0.160 −0.070 | 10.01 | 0.610 |
| | Over 9.00 up to and including 13.01 | 36 | 1.085 ±0.007 | | | −0.032 | | | | | | |
| | Over 13.01 | 38 | 1.105 ±0.007 | | | −0.031 | | | | | | |
| DD | Up through 14.42 | 34 | 1.513 ±0.008 | 1.435 | 1.430 | −0.010 | 1.1250 (1⅛) | 1.750 | 1.062 | +0.220 −0.080 | 14.43 | 0.830 |
| | Over 14.42 up to and including 18.43 | 36 | 1.541 ±0.008 | | | −0.009 | | | | | | |
| | Over 18.43 | 38 | 1.569 ±0.008 | | | −0.008 | | | | | | |

[a] Summation of the deviations from $S_g$ for all grooves in any one sheave shall not exceed ±0.050 in. The variation in pitch diameter between the grooves in any one sheave must be within the following limits: Up through 19.9 in. outside diameter and up through 6 grooves: 0.010 in. (add 0.005 in. for each additional groove). 20.0 in. and over on outside diameter and up through 10 grooves: 0.015 in. (add 0.0005 in. for each additional groove). This variation can be obtained easily by measuring the distance across two measuring balls or rods placed diametrically opposite each other in a groove. Comparing this "diameter over balls or rods" measurement between grooves will give the variation in pitch diameter.

[b] The $a_p$ values shown for the A/B combination sheaves are the geometrically derived values. These values may be different from those shown in manufacturer's catalogs.

[c] Deep groove sheaves are intended for drives with belt offset such as quarter-turn or vertical shaft drives.

| Other Sheave Tolerances | | |
|---|---|---|
| Outside Diameter | Radial Runout[a] | Axial Runout[a] |
| Up through 4.0 in. outside diameter ±0.020 in. For each additional inch of outside diameter add ±0.005 in. | Up through 10.0 in. outside diameter ±0.010 in. For each additional inch of outside diameter add 0.0005 in. | Up through 5.0 in. outside diameter 0.005 in. For each additional inch of outside diameter add 0.001 in. |

[a] Total indicator reading.

All dimensions in inches.

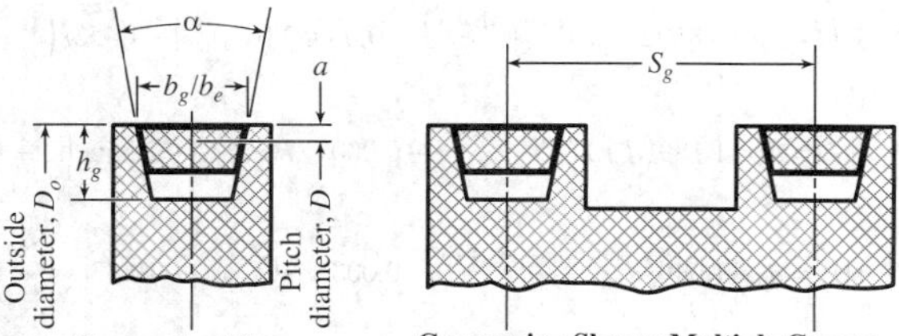


**Companion Sheave** **Companion Sheave Multiple Groove**

Fig. 11. Companion Sheaves

The factors that determine the amount of pitch diameter change on variable-speed sheaves are belt top width, belt thickness, and sheave angle. This pitch diameter change, combined with the selected operating pitch diameters for a sheave, determines the possible speed variation.

The range of output speeds from a variable-speed sheave drive is established by the companion sheave and is a function of the ratio of the pitch diameter of the companion sheave to the maximum and minimum pitch diameters of the variable sheave. Speed variation is usually obtained by varying the center distance between the two sheaves. This type of drive seldom exceeds a speed variation of 3:1.

For a single variable-speed sheave drive, the speed variation

$$\text{Speed variation} = \frac{\text{PD Max}}{\text{PD Min}}\text{(of variable sheave)}$$

For a dual variable-speed sheave drive, which is frequently referred to as a compound drive because both sheaves are variable, the speed variation is

$$\text{Speed variation} = \frac{DR(DN)}{dr(dn)}$$

where $DR$ = max driver PD

$DN$ = max driven PD

$dr$ = min driver PD

$dn$ = min driven PD

With this design, the center distance is generally fixed and speed variation is usually accomplished by mechanically altering the pitch diameter of one sheave. In this type of drive, the other sheave is spring loaded to make an opposite change in the pitch diameter and to provide the correct belt tension. Speed variations of up to 10:1 are common on this type of drive.

*Speed Ratio Adjustment:* All speed ratio changes must be made while the drives are running. Attempting to make adjustments while the unit is stopped creates unnecessary and possibly destructive forces on both the belt and sheaves. In stationary control drives, the belt tension should be released to allow the flanges to adjust without belt force interference.

*Cross Section Selection:* Selection of a variable-speed belt cross section is based on the drive design horsepower and speed variation. Table 33 shows the maximum pitch diameter variation ($2av$) that each cross section can attain.

*Horsepower Ratings:* The general horsepower formulas for variable-speed belts are:

$$\text{1422V}\quad \text{HP} = d_p r\left[0.4907(d_p r)^{-0.09} - \frac{0.8378}{d_p} - 0.000337(d_p r)^2\right] + 0.8378r\left(1 - \frac{1}{K_{SR}}\right)$$

$$1922\text{V}\quad \text{HP} = d_p r\left[0.8502(d_p r)^{-0.09} - \frac{1.453}{d_p} - 0.000538(d_p r)^2\right] + 1.453r\left(1 - \frac{1}{K_{SR}}\right)$$

$$2322\text{V}\quad \text{HP} = d_p r\left[1.189(d_p r)^{-0.09} - \frac{2.356}{d_p} - 0.000777(d_p r)^2\right] + 2.356r\left(1 - \frac{1}{K_{SR}}\right)$$

$$1926\text{V}\quad \text{HP} = d_p r\left[1.046(d_p r)^{-0.09} - \frac{1.833}{d_p} - 0.000589(d_p r)^2\right] + 1.833r\left(1 - \frac{1}{K_{SR}}\right)$$

$$2926\text{V}\quad \text{HP} = d_p r\left[1.769(d_p r)^{-0.09} - \frac{4.189}{d_p} - 0.001059(d_p r)^2\right] + 4.189r\left(1 - \frac{1}{K_{SR}}\right)$$

$$3226\text{V}\quad \text{HP} = d_p r\left[2.073(d_p r)^{-0.09} - \frac{5.236}{d_p} - 0.001217(d_p r)^2\right] + 5.236r\left(1 - \frac{1}{K_{SR}}\right)$$

$$2530\text{V}\quad \text{HP} = d_p r\left[2.395(d_p r)^{-0.09} - \frac{6.912}{d_p} - 0.001148(d_p r)^2\right] + 6.912r\left(1 - \frac{1}{K_{SR}}\right)$$

$$3230\text{V}\quad \text{HP} = d_p r\left[2.806(d_p r)^{-0.09} - \frac{7.854}{d_p} - 0.001520(d_p r)^2\right] + 7.854r\left(1 - \frac{1}{K_{SR}}\right)$$

$$4430\text{V}\quad \text{HP} = d_p r\left[3.454(d_p r)^{-0.09} - \frac{7.854}{d_p} - 0.002196(d_p r)^2\right] + 9.818r\left(1 - \frac{1}{K_{SR}}\right)$$

$$4036\text{V}\quad \text{HP} = d_p r\left[3.566(d_p r)^{-0.09} - \frac{9.687}{d_p} - 0.002060(d_p r)^2\right] + 9.687r\left(1 - \frac{1}{K_{SR}}\right)$$

$$4436\text{V}\quad \text{HP} = d_p r\left[4.041(d_p r)^{-0.09} - \frac{11.519}{d_p} - 0.002297(d_p r)^2\right] + 11.519r\left(1 - \frac{1}{K_{SR}}\right)$$

$$4836\text{V}\quad \text{HP} = d_p r\left[4.564(d_p r)^{-0.09} - \frac{13.614}{d_p} - 0.002634(d_p r)^2\right] + 13.614r\left(1 - \frac{1}{K_{SR}}\right)$$

In these equations, $d_p$ = pitch diameter of small sheave, in.; $r$ = rpm of faster shaft divided by 1000; $K_{SR}$ = speed ratio factor given in the accompanying Table 34. These formulas give the basic horsepower rating, corrected for the speed ratio. To obtain the horsepower for arcs of contact other than 180 degrees and for belts longer or shorter than average length, multiply the horsepower obtained from these formulas by the arc of contact correction factor (Table 36) and the length correction factor (Table 35).

**Table 34. Speed Ratio Correction Factors**

| Speed Ratio[a] | $K_{SR}$ | Speed Ratio[a] | $K_{SR}$ |
|---|---|---|---|
| 1.00-1.01 | 1.0000 | 1.19-1.24 | 1.0719 |
| 1.02-1.04 | 1.0136 | 1.25-1.34 | 1.0875 |
| 1.05-1.08 | 1.0276 | 1.35-1.51 | 1.1036 |
| 1.09-1.12 | 1.0419 | 1.52-1.99 | 1.1202 |
| 1.13-1.18 | 1.0567 | 2.0 and over | 1.1373 |

[a] $D_p/d_p$, where $D_p$ ($d_p$) is the pitch diameter of the large (small) sheave.

**Table 37. SAE V-Belt and Pulley Dimensions**

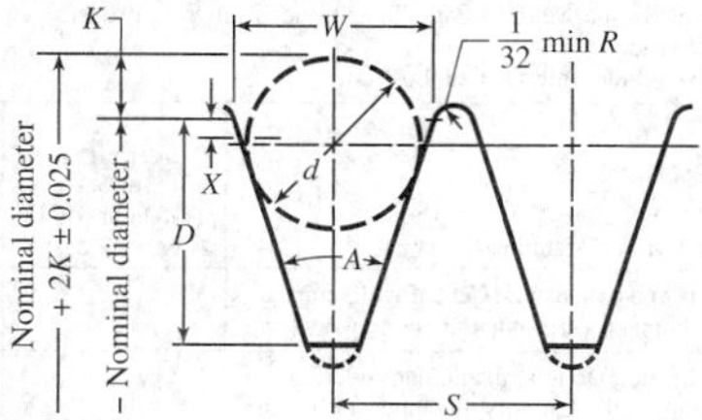


| SAE Size | Recommended Min. Eff Dia[a] | *A* Groove Angle (deg) ±0.5 | *W* Eff. Groove Width | *D* Groove Depth Min | *d* Ball or Rod Dia (±0.0005) | 2*K* 2 × Ball Extension | 2*X*[b] | *S* Groove[c] Spacing (±0.015) |
|---|---|---|---|---|---|---|---|---|
| 0.250 | 2.25 | 36 | 0.248 | 0.276 | 0.2188 | 0.164 | 0.04 | 0.315 |
| 0.315 | 2.25 | 36 | 0.315 | 0.354 | 0.2812 | 0.222 | 0.05 | 0.413 |
| 0.380 | 2.40 | 36 | 0.380 | 0.433 | 0.3125 | 0.154 | 0.06 | 0.541 |
| 0.440 | 2.75 | 36 | 0.441 | 0.512 | 0.3750 | 0.231 | 0.07 | 0.591 |
| 0.500 | 3.00 | 36 | 0.500 | 0.551 | 0.4375 | 0.314 | 0.08 | 0.661 |
| 11⁄16 | 3.00 | 34 | 0.597 | 0.551 | 0.500 | 0.258 | 0.00 | 0.778 |
| | Over 4.00 | 36 | | | | 0.280 | | |
| | Over 6.00 | 38 | | | | 0.302 | | |
| ¾ | 3.00 | 34 | 0.660 | 0.630 | 0.5625 | 0.328 | 0.02 | 0.841 |
| | Over 4.00 | 36 | | | | 0.352 | | |
| | Over 6.00 | 38 | | | | 0.374 | | |
| ⅞ | 3.50 | 34 | 0.785 | 0.709 | 0.6875 | 0.472 | 0.04 | 0.966 |
| | Over 4.50 | 36 | | | | 0.496 | | |
| | Over 6.00 | 38 | | | | 0.520 | | |
| 1 | 4.00 | 34 | 0.910 | 0.827 | 0.8125 | 0.616 | 0.06 | 1.091 |
| | Over 6.00 | 36 | | | | 0.642 | | |
| | Over 8.00 | 38 | | | | 0.666 | | |

All dimensions in inches.

[a] Pulley effective diameters below those recommended should be used with caution, because power transmission and belt life may be reduced.

[b] The *X* dimension is radial; 2*X* is to be subtracted from the effective diameter to obtain "pitch diameter" for speed ratio calculations.

[c] These values are intended for adjacent grooves of the same effective width (*W*). Choice of pulley manufacture or belt design parameter may justify variance from these values. The *S* dimension should be the same on all multiple groove pulleys in a drive using matched belts. 

3) Do not store belts near electrical devices that may generate ozone (transformers, electric motors, etc.).

4) Do not store belts in areas where solvents or chemicals are present in the atmosphere.

Belts should be stored in a cool, dry environment. When stacked on shelves, the stacks should be short enough to avoid excess weight on the bottom belts, which may cause distortion. When stored in containers, the container size and contents should be sufficiently limited to avoid distortion.

*V-Belts:* A common method is to hang the belts on pegs or pin racks. Very long belts stored this way should use sufficiently large pins or crescent-shaped "saddles" to prevent their weight from causing distortion.

*Joined V-belts, Synchronous Belts, V-Ribbed Belts:* Like V-belts, these belts may be stored on pins or saddles with precautions taken to avoid distortion. However, belts of this

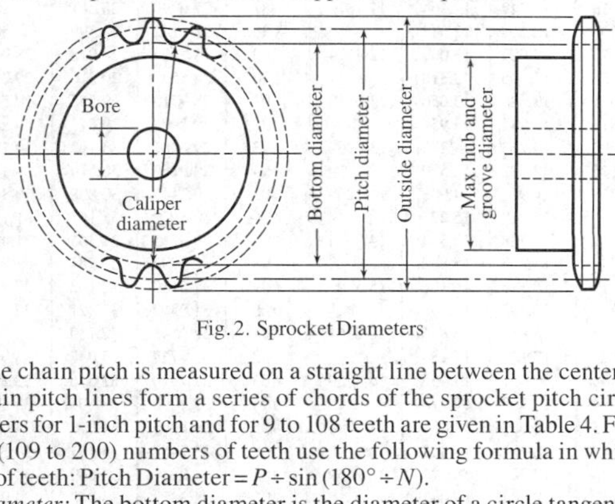

**Proportions of Sprockets.**—Typical proportions of single-strand and multiple-strand cast roller chain sprockets, as provided by the American Chain Association, are shown in Table 7. Typical proportions of roller chain bar-steel sprockets, also provided by this association, are shown in Table 8.

**Table 7. Typical Proportions of Single-Strand and Multiple-Strand Cast Roller Chain Sprockets**

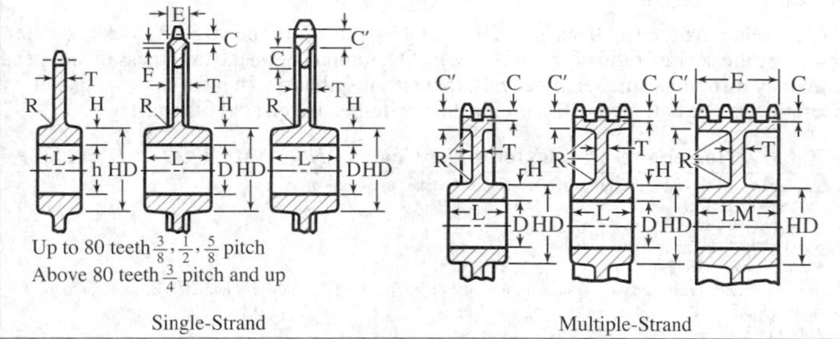


| Sprocket Web Thickness, $T$, for Various Pitches $P$ | | | | | | | | | | | | | | | |
|---|---|---|---|---|---|---|---|---|---|---|---|---|---|---|---|
| Single-Strand | | | | | | | | Multiple-Strand | | | | | | | |
| $P$ | $T$ | $P$ | $T$ | $P$ | $T$ | $P$ | $T$ | $P$ | $T$ | $P$ | $T$ | $P$ | $T$ | $P$ | $T$ |
| 3⁄8 | .312 | 3⁄4 | .437 | 1½ | .625 | 2¼ | 1.000 | 3⁄8 | .375 | 3⁄4 | .500 | 1½ | .750 | 2¼ | 1.125 |
| ½ | .375 | 1 | .500 | 1¾ | .750 | 2½ | 1.125 | ½ | .406 | 1 | .562 | 1¾ | .875 | 2½ | 1.250 |
| 5⁄8 | .406 | 1¼ | .562 | 2 | .875 | 3 | 1.250 | 5⁄8 | .437 | 1¼ | .625 | 2 | 1.000 | 3 | 1.500 |

Formulas for Dimensions of Single and Multiple Sprockets

$H = 0.375 + \frac{D}{6} + 0.01\,PD$

$L = 4H$ for semi-steel castings

$C = 0.5P$

$C' = 0.9P$

$E = 0.625P + 0.93W$

$F = 0.150 + 0.25P$

$G = 2T$

$R = 0.4P$ for single-strand sprockets

$R = 0.5T$ for multiple-strand sprockets

All dimensions in inches. Where $P$ = chain pitch and $W$ = nominal chain width.

**Table 8. Typical Proportions of Roller Chain Bar-Steel Sprockets**

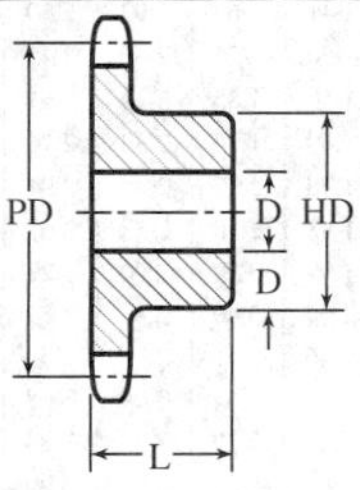


$H = Z + D/6 + 0.01PD$

For $PD$ up to 2 inches, $Z = 0.125$ inch; for 2-4 inches, $Z = 0.187$ inch; for 4-6 inches, 0.25 inch; and for over 6 inches, 0.375 inch.

Hub length $L = 3.3H$, normally, with a minimum of $2.6H$.

Hub diameter $HD = D + 2H$, but not more than the maximum hub diameter $MHD$ given by the formula:

$$MHD = P\left(\cot\frac{180°}{N} - 1\right) - 0.030$$

where

$P$ = Chain pitch, in inches; $N$ = Number of sprocket teeth

When sprocket wheels are designed with spokes, the usual assumptions made in order to determine suitable proportions are as follows: 1) that the maximum torque load acting on a sprocket is the chain tensile strength times the sprocket pitch radius; 2) that the torque load is equally divided between the arms by the rim; and 3) that each arm acts as a cantilever beam.

The arms are generally elliptical in cross section, the major axis twice the minor axis.

**Table 9.** *(Continued)* **Horsepower Ratings for Roller Chains,** *ANSI/ASME B29.1-2011 (R2022)*

| Chain | No. of Teeth Small Spkt. | Revolutions per Minute — Small Sprocket[a] | | | | | | | | | | | | |
|---|---|---|---|---|---|---|---|---|---|---|---|---|---|---|
| | | 25 | 50 | 100 | 120 | 200 | 300 | 400 | 500 | 600 | 800 | 1000 | 1200 | 1400 |
| | | Horsepower Rating | | | | | | | | | | | | |
| 3/4-inch Pitch Standard Single-Strand Roller Chain — No. 60 | 11 | 0.46 | 0.89 | 1.72 | 2.05 | 3.35 | 4.95 | 6.52 | 8.08 | 9.63 | 12.69 | 15.58 | 11.85 | 9.[illegible]1 |
| | 12 | 0.50 | 0.97 | 1.88 | 2.24 | 3.66 | 5.40 | 7.12 | 8.82 | 10.51 | 13.85 | 17.15 | 13.53 | 10.72 |
| | 13 | 0.54 | 1.05 | 2.04 | 2.43 | 3.96 | 5.85 | 7.71 | 9.55 | 11.38 | 15.00 | 18.58 | 15.23 | 12.[illegible]8 |
| | 14 | 0.58 | 1.13 | 2.19 | 2.61 | 4.27 | 6.30 | 8.30 | 10.29 | 12.26 | 16.15 | 20.01 | 17.02 | 13.[illegible]1 |
| | 15 | 0.62 | 1.21 | 2.35 | 2.80 | 4.57 | 6.75 | 8.90 | 11.02 | 13.13 | 17.31 | 21.44 | 18.87 | 14.[illegible]8 |
| | 16 | 0.66 | 1.29 | 2.51 | 2.99 | 4.88 | 7.20 | 9.49 | 11.76 | 14.01 | 18.46 | 22.87 | 20.79 | 16.[illegible]0 |
| | 17 | 0.70 | 1.37 | 2.66 | 3.17 | 5.18 | 7.65 | 10.08 | 12.49 | 14.88 | 19.62 | 24.30 | 22.77 | 18.[illegible]7 |
| | 18 | 0.75 | 1.45 | 2.82 | 3.36 | 5.49 | 8.10 | 10.68 | 13.23 | 15.76 | 20.77 | 25.73 | 24.81 | 19.[illegible]9 |
| | 19 | 0.79 | 1.53 | 2.98 | 3.55 | 5.79 | 8.55 | 11.27 | 13.96 | 16.63 | 21.92 | 27.16 | 26.91 | 21.[illegible]5 |
| | 20 | 0.83 | 1.61 | 3.13 | 3.73 | 6.10 | 9.00 | 11.86 | 14.70 | 17.51 | 23.08 | 28.59 | 29.06 | 23.[illegible]6 |
| | 21 | 0.87 | 1.69 | 3.29 | 3.92 | 6.40 | 9.45 | 12.46 | 15.43 | 18.38 | 24.23 | 30.02 | 31.26 | 24.[illegible]1 |
| | 22 | 0.91 | 1.77 | 3.45 | 4.11 | 6.71 | 9.90 | 13.05 | 16.17 | 19.26 | 25.39 | 31.45 | 33.52 | 26.[illegible]0 |
| | 23 | 0.95 | 1.85 | 3.61 | 4.29 | 7.01 | 10.35 | 13.64 | 16.90 | 20.13 | 26.54 | 32.88 | 35.84 | 28.[illegible]4 |
| | 24 | 0.99 | 1.93 | 3.76 | 4.48 | 7.32 | 10.80 | 14.24 | 17.64 | 21.01 | 27.69 | 34.31 | 38.20 | 30.[illegible]1 |
| | 25 | 1.04 | 2.01 | 3.92 | 4.67 | 7.62 | 11.25 | 14.83 | 18.37 | 21.89 | 28.85 | 35.74 | 40.61 | 32.[illegible]3 |
| | 26 | 1.08 | 2.09 | 4.08 | 4.85 | 7.93 | 11.70 | 15.42 | 19.11 | 22.76 | 30.00 | 37.17 | 43.07 | 34.[illegible]8 |
| | 28 | 1.16 | 2.26 | 4.39 | 5.23 | 8.54 | 12.60 | 16.61 | 20.58 | 24.51 | 32.31 | 40.03 | 47.68 | 38.[illegible] |
| | 30 | 1.24 | 2.42 | 4.70 | 5.60 | 9.15 | 13.50 | 17.79 | 22.05 | 26.26 | 34.62 | 42.89 | 51.09 | 42.[illegible]5 |
| | 32 | 1.33 | 2.58 | 5.02 | 5.98 | 9.76 | 14.40 | 18.98 | 23.52 | 28.01 | 36.92 | 45.75 | 54.50 | 46.[illegible]7 |
| | 35 | 1.45 | 2.82 | 5.49 | 6.54 | 10.67 | 15.75 | 20.76 | 25.72 | 30.64 | 40.39 | 50.03 | 59.60 | 53.[illegible]8 |
| | 40 | 1.66 | 3.22 | 6.27 | 7.47 | 12.20 | 18.00 | 23.73 | 29.39 | 35.02 | 46.16 | 57.18 | 68.12 | 65.2[illegible] |
| | 45 | 1.86 | 3.63 | 7.05 | 8.40 | 13.72 | 20.25 | 26.69 | 33.07 | 39.39 | 51.92 | 64.33 | 76.63 | 77.8[illegible] |
| | | | Type A | | | | | Type B | | | | | Type C | |

| Chain | No. of Teeth Small Spkt. | Revolutions per Minute — Small Sprocket[a] | | | | | | | | | | | | |
|---|---|---|---|---|---|---|---|---|---|---|---|---|---|---|
| | | 25 | 50 | 75 | 100 | 200 | 300 | 400 | 500 | 600 | 700 | 800 | 900 | 1000 |
| | | Horsepower Ratings | | | | | | | | | | | | |
| 1-inch Pitch Standard Single-Strand Roller Chain — No. 80 | 11 | 1.06 | 2.07 | 3.05 | 4.03 | 7.83 | 11.56 | 15.23 | 18.87 | 22.48 | 26.07 | 27.41 | 22.97 | 19.6 |
| | 12 | 1.16 | 2.26 | 3.33 | 4.39 | 8.54 | 12.61 | 16.62 | 20.59 | 24.53 | 28.44 | 31.23 | 26.17 | 22.3[illegible] |
| | 13 | 1.26 | 2.45 | 3.61 | 4.76 | 9.26 | 13.66 | 18.00 | 22.31 | 26.57 | 30.81 | 35.02 | 29.51 | 25.2[illegible] |
| | 14 | 1.35 | 2.63 | 3.89 | 5.12 | 9.97 | 14.71 | 19.39 | 24.02 | 28.62 | 33.18 | 37.72 | 32.98 | 28.1[illegible] |
| | 15 | 1.45 | 2.82 | 4.16 | 5.49 | 10.68 | 15.76 | 20.77 | 25.74 | 30.66 | 35.55 | 40.41 | 36.58 | 31.2[illegible] |
| | 16 | 1.55 | 3.01 | 4.44 | 5.86 | 11.39 | 16.81 | 22.16 | 27.45 | 32.70 | 37.92 | 43.11 | 40.30 | 34.4[illegible] |
| | 17 | 1.64 | 3.20 | 4.72 | 6.22 | 12.10 | 17.86 | 23.54 | 29.17 | 34.75 | 40.29 | 45.80 | 44.13 | 37.6[illegible] |
| | 18 | 1.74 | 3.39 | 5.00 | 6.59 | 12.81 | 18.91 | 24.93 | 30.88 | 36.79 | 42.66 | 48.49 | 48.08 | 41.0[illegible] |
| | 19 | 1.84 | 3.57 | 5.28 | 6.95 | 13.53 | 19.96 | 26.31 | 32.60 | 38.84 | 45.03 | 51.19 | 52.15 | 44.5[illegible] |
| | 20 | 1.93 | 3.76 | 5.55 | 7.32 | 14.24 | 21.01 | 27.70 | 34.32 | 40.88 | 47.40 | 53.88 | 56.32 | 48.0[illegible] |
| | 21 | 2.03 | 3.95 | 5.83 | 7.69 | 14.95 | 22.07 | 29.08 | 36.03 | 42.92 | 49.77 | 56.58 | 60.59 | 51.7[illegible] |
| | 22 | 2.13 | 4.14 | 6.11 | 8.05 | 15.66 | 23.12 | 30.47 | 37.75 | 44.97 | 52.14 | 59.27 | 64.97 | 55.47 |
| | 23 | 2.22 | 4.33 | 6.39 | 8.42 | 16.37 | 24.17 | 31.85 | 39.46 | 47.01 | 54.51 | 61.97 | 69.38 | 59.30 |
| | 24 | 2.32 | 4.52 | 6.66 | 8.78 | 17.09 | 25.22 | 33.24 | 41.18 | 49.06 | 56.88 | 64.66 | 72.40 | 63.21 |
| | 25 | 2.42 | 4.70 | 6.94 | 9.15 | 17.80 | 26.27 | 34.62 | 42.89 | 51.10 | 59.25 | 67.35 | 75.42 | 67.20 |
| | 26 | 2.51 | 4.89 | 7.22 | 9.52 | 18.51 | 27.32 | 36.01 | 44.61 | 53.14 | 61.62 | 70.05 | 78.43 | 71.27 |
| | 28 | 2.71 | 5.27 | 7.77 | 10.25 | 19.93 | 29.42 | 38.78 | 48.04 | 57.23 | 66.36 | 75.44 | 84.47 | 79.65 |
| | 30 | 2.90 | 5.64 | 8.33 | 10.98 | 21.36 | 31.52 | 41.55 | 51.47 | 61.32 | 71.10 | 80.82 | 90.50 | 88.33 |
| | 32 | 3.09 | 6.02 | 8.89 | 11.71 | 22.78 | 33.62 | 44.32 | 54.91 | 65.41 | 75.84 | 86.21 | 96.53 | 97.31 |
| | 35 | 3.38 | 6.58 | 9.72 | 12.81 | 24.92 | 36.78 | 48.47 | 60.05 | 71.54 | 82.95 | 94.29 | 105.58 | 111.31 |
| | 40 | 3.87 | 7.53 | 11.11 | 14.64 | 28.48 | 42.03 | 55.40 | 68.63 | 81.76 | 94.80 | 107.77 | 120.67 | 133.51 |
| | 45 | 4.35 | 8.47 | 12.49 | 16.47 | 32.04 | 47.28 | 62.32 | 77.21 | 91.98 | 106.65 | 121.24 | 135.75 | 150.2[illegible] |
| | | | Type A | | | | | Type B | | | | | Type C | |

Shafting, bearings, and foundations should be suitable to maintain the initial alignment. Periodic maintenance should include an inspection of alignment.

**Example of Roller Chain Drive Design Procedure.—**The selection of a roller chain and sprockets for a specific design requirement is best accomplished by a systematic step-by-step procedure such as is used in the following example.

*Example:* Select a roller chain drive to transmit 10 horsepower from a countershaft to the main shaft of a wire drawing machine. The countershaft is $1^{15}/_{16}$-inches diameter and operates at 1000 rpm. The main shaft is also $1^{15}/_{16}$-inches diameter and must operate between 378 and 382 rpm. Shaft centers, once established, are fixed and by initial calculations must be approximately $22^1/_2$ inches. The load on the main shaft is uneven and presents "peaks," which place it in the heavy shock load category. The input power is supplied by an electric motor. The driving head is fully enclosed and all parts are lubricated from a central system.

*Step 1. Service Factor:* From Table 15 the service factor for heavy shock load and an electric motor drive is 1.5.

*Step 2. Design Horsepower:* The horsepower upon which the chain selection is based (design horsepower) is equal to the specified horsepower multiplied by the service factor, $10 \times 1.5 = 15$ hp.

*Step 3. Chain Pitch and Small Sprocket Size for Single-Strand Drive:* In Table 9 under 1000 rpm, a $^5/_8$-inch pitch chain with a 24-tooth sprocket or a $^3/_4$-inch pitch chain with a 15-tooth sprocket are possible choices.

*Step 4. Check of Chain Pitch and Sprocket Selection:* From Table 10 it is seen that only the 24-tooth sprocket in Step 3 can be bored to fit the $1^{15}/_{16}$-inch diameter main shaft. In Table 9 a $^5/_8$-pitch chain at a small sprocket speed of 1000 rpm is rated at 15.5 hp for a 24-tooth sprocket.

*Step 5. Selection of Large Sprocket:* Since the driver is to operate at 1000 rpm and the driven at a minimum of 378 rpm, the speed ratio 1000/378 = 2.646. Therefore, the large sprocket should have $24 \times 2.646 = 63.5$ (use 63) teeth.

This combination of 24 and 63 teeth will produce a main drive shaft speed of 381 rpm which is within the limitation of 378 to 382 rpm established in the original specification.

*Step 6. Computation of Chain Length:* Since the 24- and 63-tooth sprockets are to be placed on $22^1/_2$-inch centers, the chain length is determined from the formula:

$$L = 2C + \frac{N}{2} + \frac{n}{2} + \left(\frac{N-n}{2\pi}\right)^2 \times \frac{1}{C}$$

where $L$ = chain length in pitches; $C$ = shaft center distance in pitches; $N$ = number of teeth in large sprocket; and $n$ = number of teeth in small sprocket.

$$L = 2 \times 36 + \frac{63+24}{2} + \left(\frac{63-24}{6.28}\right)^2 \times \frac{1}{36} = 116.57 \text{ pitches}$$

*Step 7. Correction of Center Distance:* Since the chain is to couple at a whole number of pitches, 116 pitches will be used and the center distance recomputed based on this figure using the formula on page 2638 where $c$ is the center distance in inches and $P$ is the pitch.

$$c = \frac{P}{8}\left(2L - N - n + \sqrt{(2L - N - n)^2 - 0.810(N-n)^2}\right)$$

$$= \frac{5}{64}\left(2 \times 116 - 63 - 24 + \sqrt{(2 \times 116 - 63 - 24)^2 - 0.810(63-24)^2}\right)$$

$$= \frac{5}{64}(145 + 140.69) = 22.32 \text{ inches, say } 22\frac{3}{8} \text{ inches}$$

**General Leadscrew Design Considerations.—**The following design criteria should be considered when selecting leadscrews.

*Load:* The load is the weight "as seen by the screw." For vertical applications such as lifting or jacking the load equals the weight of the load being moved. For horizontal applications the "load seen" equals the weight of the load times the coefficient of friction of the supporting media. Another way to determine the load in an actual application is to attach a scale to the load and pull it. Base the load on the moving force required, not on the higher starting (breakaway) force.

*Tension and Compression Load:* For reduced screw thread sizes, design for tension loads whenever possible. Compression load designs must factor in column strength to prevent buckling.

*Back-drive:* When a ball screw or acme nut is forced to move linearly along the shaft by an applied force, or gravity in vertical orientation, it forces or back-drives the shaft making it rotate. With their higher efficiency and lower initial start up (breakway) friction value, ball screw assemblies in vertical orientations may back-drive their shafts.

*Driving Torque:* The driving torque ($T_d$) is the amount of force required to rotate the screw and move a load.

$$T_d = \frac{P \times L}{2\pi e}$$

where $T_d$ = driving torque (lb-in)

$P$ = load (lbs)

$L$ = lead of screw (in/revolution)

$e$ = ball bearing screw efficiency (approximately 0.90 for ball screws, 0.40 for acme screws with plastic nuts, and 0.25 for acme screws with bronze nuts.)

*Accuracy:* Several factors affect accuracy. One factor is the lead deviation in inches per foot. Typical values can range from 0.005 in/ft to 0.001 in/ft. Backlash may also be a consideration in applications that require precise positioning in both directions and in CNC applications. A final consideration is accuracy versus time degradation due to screw wear and ball bearing fatigue. This is especially important in systems that require precise positioning but do not have closed-loop feedback controls. Ball screws have an advantage in long term accuracy since the useful life of ball screws is determined by metal fatigue rather than the ordinary wear characteristics of a conventional screw. Due to rolling contact, very little dimensional change occurs over the life of a ball bearing screw, thus eliminating the need for frequent compensating adjustments.

*Lead:* The distance the nut travels with one rotation of the screw.

*Speed:* When rapid travel is required, design for screws with a high lead to reduce shaft speed to avoid rotational and vibrational side effects.

*Life:* The recommended operating loads for ball screw assemblies are based on a predicted life of one million inches of travel, providing the assembly is lubricated. However, it is possible to operate at greater loads if a reduced life is acceptable. The load should not exceed the maximum static load shown in manufacturer design tables.

*Backlash:* Backlash, end play, positional inaccuracy, or lost motion between the nut and the screw can be eliminated by utilizing a preloaded nut assembly. Backlash can also be reduced, at increased cost, in ball screws by changing the conformity ratio (ball track radius/ball diameter) as well as using ogival (gothic) ball tracks. Fig. 3 illustrates the contact points between a preloaded ball screw assembly and the shaft.

*Design Overview:* To design a ball screw assembly many interrelated factors such as load, bearing length, life, and speed must be considered. Changing one parameter will influence others and generally an iterative approach to balancing these requirements is required. An example is the choice of a fine lead versus coarse lead. A fine line lead provides better positioning sensitivity and lower drive torque, but also requires a higher rotary speed to achieve the same linear rate of speed. A coarse lead results in lower rotary speed, but requires a higher drive torque that may require a larger motor and related drive components. Table 1 identifies some of these design interrelationships.

**Table 1. Ball Screw Design Factor Interrelationships**

| Increase In | Affects | How | Increase In | Affects | How |
|---|---|---|---|---|---|
| Screw Length | Critical speed | Decreases | End Mounting Rigidity | Critical speed | Increases |
| | Compression load | Decreases | | Compression load | Increases |
| Screw Diameter | Critical speed | Increases | | System stiffness | Increases |
| | Inertia | Increases | Load | Life | Decreases |
| | Compression load | Increases | Preload | Positioning accuracy | Increases |
| | Stiffness | Increases | | System stiffness | Increases |
| | Spring Rate | Increases | | Drag torque | Increases |
| | Load capacity | Increases | Angular Velocity | Critical speed | Decreases |
| Lead | Drive torque | Increases | Nut length | Load capacity | Increases |
| | Angular velocity | Decreases | | Stiffness | Increases |
| | Load capacity | Increases | Ball diameter | Life | Increases |
| | Positioning accuracy | Decreases | | Stiffness | Increases |
| | Ball diameter | Increases | | Load capacity | Increases |

**Screws, Power Transmission.**—The square form of thread has a somewhat higher efficiency than threads with sloping sides, although when the angle of the thread form is comparatively small, as in the case of an Acme thread, there is little increase in frictional losses. The Acme thread has superseded the square form on many classes of equipment requiring lead screws or other power transmitting screws, because the former has practical advantages in regard to cutting and also in compensating for wear between the screw and nut. Multiple thread screws are much more efficient than single-thread screws, as the efficiency is affected by the helix angle of the thread.

*Force Required to Turn Screw:* In determining the force which must be applied at the end of a given lever-arm in order to turn a screw (or nut surrounding it), there are two conditions to be considered:

1) When rotation is such that the load *resists* the movement of the screw, as in raising a load with a screw jack, $F = L \times \frac{l + 2r\pi\mu}{2r\pi - \mu l} \times \frac{r}{R}$.

2) When rotation is such that the load *assists* the movement of the screw, as in lowering a load, $F = L \times \frac{2r\pi\mu - l}{2r\pi + \mu l} \times \frac{r}{R}$.

In these formulas, $F$ = force applied at end of lever-arm; $L$ = load moved by screw; $R$ = length of lever-arm; $l$ = lead of screw thread; $r$ = mean or pitch radius of screw; $\mu$ = coefficient of friction. If lead $l$ is large in proportion to the diameter so that the helix angle is large, $F$ will have a negative value, indicating that the screw will turn due to the load alone. The efficiency between a screw and nut increases quite rapidly for helix angles up to 10 or 15 degrees (measured from a plane perpendicular to the screw axis). The efficiency remains nearly constant for angles between about 25 and 65 degrees, and the angle of maximum efficiency is between 40 and 50 degrees. A screw will not be self-locking if the efficiency exceeds 50 percent.

**Motor Types According to Variability of Speed.—**Five types of motors classified according to variability of speed are:

*Constant-Speed Motors:* In this type of motor the normal operating speed is constant or practically constant; for example, a synchronous motor, an induction motor with small slip, or a direct-current shunt-wound motor.

*Varying-Speed Motor:* In this type of motor, the speed varies with the load, ordinarily decreasing when the load increases; such as a series-wound or repulsion motor.

*Adjustable-Speed Motor:* In this type of motor, the speed can be varied gradually over a considerable range, but when once adjusted remains practically unaffected by the load; such as a direct-current shunt-wound motor with field resistance control designed for a considerable range of speed adjustment.

The base speed of an adjustable-speed motor is the lowest rated speed obtained at rated load and rated voltage at the temperature rise specified in the rating.

*Adjustable Varying-Speed Motor:* This type of motor is one in which the speed can be adjusted gradually, but when once adjusted for a given load will vary in considerable degree with the change in load; such as a direct-current compound-wound motor adjusted by field control or a wound-rotor induction motor with rheostatic speed control.

*Multispeed Motor:* This type of motor is one which can be operated at any one of two or more definite speeds, each being practically independent of the load; such as a direct-current motor with two armature windings or an induction motor with windings capable of various pole groupings. In the case of multispeed permanent-split capacitor and shaded pole motors, the speeds are dependent upon the load.

**Pull-Up Torque.—**NEMA Standard pull up torques for single-speed, polyphase, squirrel-cage integral-horsepower motors, Designs A and B, with continuous ratings and with rated voltage and frequency applied are as follows: When the locked-rotor torque given in Table 2 is 110 percent or less, the pull-up torque is 90 percent of the locked-rotor torque; when the locked-rotor torque is greater than 110 percent but less than 145 percent, the pull-up torque is 100 percent of full-load torque; and when the locked-rotor torque is 145 percent or more, the pull-up torque is 70 percent of the locked-rotor torque.

For Design C motors, with rated voltage and frequency applied, the pull-up torque is not less than 70 percent of the locked-rotor torque as given in Table 2.

## Types and Characteristics of Electric Motors

**Types of Direct-Current Motors.—**Direct-current motors may be grouped into three general classes: series-wound; shunt-wound; and compound-wound.

In the *series-wound motor* the field windings, which are fixed in the stator frame, and the armature windings, which are placed around the rotor, are connected in series so that all current passing through the armature also passes through the field. In the *shunt-wound motor*, both armature and field are connected across the main power supply so that the armature and field currents are separate. In the *compound-wound motor*, both series and shunt field windings are provided and these may be connected so that the currents in both are flowing in the same direction, called *cumulative compounding*, or so that the currents in each are flowing in opposite directions, called *differential compounding*.

**Characteristics of Series-Wound Direct-Current Motors.—**In the series-wound motor, any increase in load results in more current passing through the armature and the field windings. As the field is strengthened by this increased current, the motor speed decreases. Conversely, as the load is decreased the field is weakened and the speed increases and at very light loads may become excessive. For this reason, series-wound direct-current motors are usually directly connected or geared to the load to prevent "runaway." (A series-wound motor, designated as series-shunt wound, is sometimes provided with a light shunt field winding to prevent dangerously high speeds at light loads.) The increase in armature current with increasing load produces increased torque, so that the series-wound motor is

speeds with as little variation as $1/2$ to 1 percent of full load speed from full load to no load over a line voltage variation of ± 10 percent and over any normal variation in motor temperature and ambient temperature.

*Applications:* These direct-current, adjustable-voltage drives, as they are sometimes called, have been applied successfully to such machine tools as planers, milling machines, boring mills and lathes, as well as to other industrial machines where wide, stepless speed control, uniform speed under all operating conditions, constant torque acceleration and adaptability to automatic operation are required.

**Compound-Wound Motors.**—In the compound-wound motor, the speed variation due to load changes is much less than in the series-wound motor, but greater than in the shunt-wound motor (ranging up to 25 percent from full load to no load). It has a greater starting torque than the shunt-wound motor, is able to withstand heavier overloads, but has a narrower adjustable speed range. Standard motors of this type have a cumulative-compound winding, the differential-compound winding being limited to special applications. They are used where the starting load is very heavy or where the load changes suddenly and violently as with reciprocating pumps, printing presses and punch presses.

**Types of Polyphase Alternating-Current Motors.**—The most widely used polyphase motors are of the induction type. The *"squirrel cage" induction motor* consists of a wound stator which is connected to an external source of alternating-current power and a laminated steel core rotor with a number of heavy aluminum or copper conductors set into the core around its periphery and parallel to its axis. These conductors are connected together at each end of the rotor by a heavy ring, which provides closed paths for the currents induced in the rotor to circulate. The rotor bars form, in effect, a "squirrel cage" from which the motor takes its name.

*Wound-Rotor Type of Induction Motor:* This type has, in addition to a squirrel cage a series of coils set into the rotor which are connected through slip-rings to external variable resistors. By varying the resistance of the wound-rotor circuits, the amount of current flowing in these circuits and hence the speed of the motor can be controlled. Since the rotor of an induction motor is not connected to the power supply, the motor is said to operate by transfer action and is analogous to a transformer with a short-circuited secondary that is free to rotate. Induction motors are built with a wide range of speed and torque characteristics which are discussed under "Operating Characteristics of Squirrel-cage Induction Motors."

*Synchronous Motor:* The other type of polyphase alternating-current motor used industrially is the *synchronous motor.* In contrast to the induction motor, the rotor of the synchronous motor is connected to a direct-current supply which provides a field that rotates in step with the alternating-current field in the stator. After having been brought up to synchronous speed, which is governed by the frequency of the power supply and the number of poles in the rotor, the synchronous motor operates at this constant speed throughout its entire load range.

**Operating Characteristics of Squirrel-Cage Induction Motors.**—In general, squirrel-cage induction motors are simple in design and construction and offer rugged service. They are essentially constant-speed motors, their speed changing very little with load and not being subject to adjustment. They are used for a wide range of industrial applications calling for integral horsepower ratings. According to the NEMA (National Electrical Manufacturers Association) Standards, there are four classes of squirrel-cage induction motors designated respectively as *A*, *B*, *C*, and *D*.

*Design A* motors are not commonly used since Design *B* has similar characteristics with the advantage of lower starting current.

*Design B* motors may be designated as a general purpose type suitable for the majority of polyphase alternating-current applications such as blowers, compressors, drill presses, grinders, hammer mills, lathes, planers, polishers, saws, screw machines, shakers,

## ADHESIVES AND SEALANTS

By strict definition, an adhesive is any substance that fastens or bonds materials to be joined (adherends) by means of surface attachment. The bond durability depends on the strength of the adhesive to the substrate (adhesion) and the strength within the adhesive (cohesion). Besides bonding a joint, an adhesive may serve as a seal against foreign matter. When an adhesive performs both bonding and sealing functions, it is usually referred to as an *adhesive sealant*. Joining materials with adhesives offers significant benefits compared with mechanical methods of uniting two materials.

Among these benefits are that an adhesive distributes a load over an area rather than concentrating it at a point, resulting in a more even distribution of stresses. The adhesive bonded joint is therefore more resistant to flexural and vibrational stresses than, for example, a bolted, riveted, or welded joint. Another benefit is that an adhesive forms a seal as well as a bond. This seal prevents the corrosion that may occur with dissimilar metals, such as aluminum and magnesium, or mechanically fastened joints, by providing a dielectric insulation between the substrates. An adhesive also joins irregularly shaped surfaces more easily than does a mechanical fastener. Other benefits include negligible weight addition and virtually no change to part dimensions or geometry.

Most adhesives are available in liquids, gels, pastes, and tape forms. The growing variety of adhesives available can make the selection of the proper adhesive or sealant a challenging experience. In addition to the technical requirements of the adhesive, time and costs are also important considerations. Proper choice of an adhesive is based on knowledge of the suitability of the adhesive or sealant for the particular substrates. Appropriate surface preparation, curing parameters, and matching the strength and durability characteristics of the adhesive to its intended use are essential. The performance of an adhesive-bonded joint depends on a wide range of these factors, many of them quite complex. Adhesive suppliers can usually offer essential expertise in the area of appropriate selection.

Adhesives can be classified as structural or nonstructural. In general, an adhesive can be considered structural when it is capable of supporting heavy loads; nonstructural when it cannot support such loads. Many adhesives and sealants, under various brand names, may be available for a particular bonding application. It is always advisable to check the adhesive manufacturers' information before making an adhesive sealant selection. Also, testing under end-use conditions is always suggested to help ensure bonded or sealed joints meet or exceed expected performance requirements.

Though not meant to be all-inclusive, the following information correlates the features of some successful adhesive compositions available in the marketplace.

### Bonding Adhesives

Reactive-type bonding adhesives are applied as liquids and react (cure) to solids under appropriate conditions. The cured adhesive is either a thermosetting or thermoplastic polymer. These adhesives are supplied as two-component no-mix, two-component mix, and one-component no-mix types, which are discussed in the following paragraphs.

### Two-Component No-Mix Adhesives

**Types of Adhesives.—***Anaerobic (Urethane Methacrylate Ester) Structural Adhesives:* Anaerobic structural adhesives are mixtures of acrylic esters that remain liquid when exposed to air but harden when confined between metal substrates. These adhesives can be used for large numbers of industrial purposes where high reliability of bond joints is required. Benefits include: no mixing is required (no pot-life or waste problems), flexible/durable bonds are made that withstand thermal cycling, have excellent resistance to solvents and severe environments, and rapid cure at room temperatures (eliminating

## One-Component No-Mix Adhesives

**Types of Adhesives.—***Light-Curable Adhesives:* Light-curing systems use a unique curing mechanism. The adhesives contain photoinitiators that absorb light energy and dissociate to form radicals. These radicals then initiate the polymerization of the polymers, oligomers, and monomers in the adhesive. The photoinitiator acts as a chemical solar cell, converting the light energy into chemical energy for the curing process. Typically, these systems are formulated for use with ultraviolet light sources. However, newer products have been formulated for use with visible light sources.

One of the biggest benefits that light-curing adhesives offer to the manufacturer is the elimination of the work time to work-in-progress trade-off, which is embodied in most adhesive systems. With light-curing systems, the user can take as much time as needed to position the part without fear of the adhesive curing. Upon exposure to the appropriate light source, the adhesive then can be fully cured in less than 1 minute, minimizing the costs associated with work in progress. Adhesives that utilize light as the curing mechanism are often one-part systems with good shelf life, which makes them even more attractive for manufacturing use.

Applications for light-curable adhesives include bonding of glass, and glass to metal, tacking of wires, surface coating, thin-film encapsulation, clear substrate bonding, and potting of components,

*Cyanoacrylate Adhesives (Instant Adhesives):* Cyanoacrylates or instant adhesives are often called Superglue™. Cyanoacrylates are one-part adhesives that cure rapidly, as a result of the presence of surface moisture, to form high-strength bonds, when confined between two substrates. Cyanoacrylates have excellent adhesion to many substrates, including most plastics and they achieve fixture strength in seconds and full strength within 24 hours. These qualities make cyanoacrylates suitable for use in automated production environments. They are available in viscosities ranging from water-thin liquids to thixotropic gels.

Because cyanoacrylates are a relatively mature adhesive family, a wide variety of specialty formulations is now available to help the user address difficult assembly problems. One of the best examples is the availability of polyolefin primers, which allow users to obtain high bond strengths on difficult-to-bond plastics such as polyethylene and polypropylene. One common drawback of cyanoacrylates is that they form a very rigid polymer matrix, resulting in very low peel strengths. To address this problem, formulations have been developed that are rubber-toughened. Although the rubber toughening improves the peel strength of the system to some extent, peel strength remains a weak point for this system, and, therefore, cyanoacrylates are poor candidates for joint designs that require high peel resistance. In manufacturing environments with low relative humidity, the cure of the cyanoacrylate can be significantly retarded.

This problem can be addressed in one of two ways. One approach is to use accelerators that deposit active species on the surface to initiate the cure of the product. The other approach is to use specialty cyanoacrylate formulations that have been engineered to be surface-insensitive. These formulations can cure rapidly even on dry or slightly acidic surfaces.

Applications for cyanoacrylate adhesives include bonding of thermoplastic and thermosetting plastics, rubber, metals, wood, and leather, also strain relief of wires.

*Hot-Melt Adhesives:* Hot-melt adhesives are widely used in assembly applications. In general, hot-melt adhesives permit fixturing speeds that are much faster than can be achieved with water- or solvent-based adhesives. Usually supplied in solid form, hot-melt adhesives liquify when exposed to elevated temperatures. After application, they cool quickly, solidifying and forming a bond between two mating substrates. Hot-melt adhesives have been used successfully for a wide variety of adherends and can greatly reduce both the need for clamping and the length of time for curing. Some drawbacks

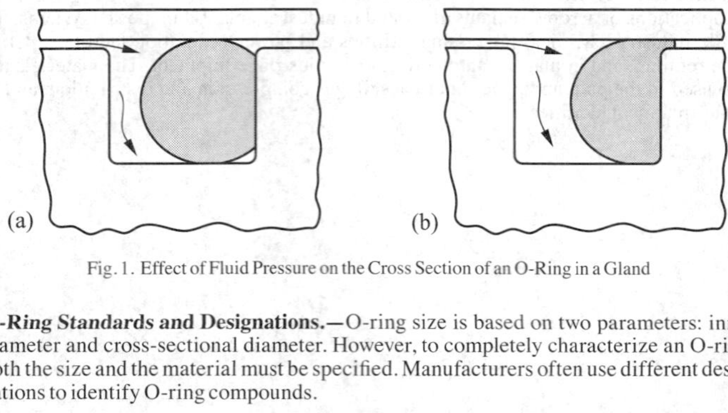

**Table 4. Inch Axial Face Seal Data, Recommended Sizes in Each Range**
*(Based on SAE AS6235A-2015 and 70 Shore A O-Rings)*

| O-Ring | | | | Axial Face Seal Groove | | | | | |
|---|---|---|---|---|---|---|---|---|---|
| | | | | Internal Pressure | | | External Pressure | | |
| Size Code | Cross-Section Diameter | Inner Diameter | | Groove Outer Diameter | Cramping [a] | Gland Fill (No Swell, Straight Sides, No Radii) | Groove Inner Diameter | Stretch | Gland Fill (No Swell, Straight Sides, No Radii) |
| | Nominal | Nominal | ± Tol. | Max/Min | | Min/Max | Min/Max | | Min/Max |
| -006 | 0.070 | 0.114 | 0.005 | | | | 0.107 / 0.112 | 4.1% | 46% / 72% |
| -008 | 0.070 | 0.176 | 0.005 | 0.328 / 0.323 | 0.2% | 60% / 86% | 0.173 / 0.178 | 5.0% | 48% / 71% |
| -028 | 0.070 | 1.364 | 0.013 | 1.504 / 1.499 | −0.7% | 56% / 79% | 1.364 / 1.369 | 0.8% | 54% / 76% |
| -106 | 0.103 | 0.174 | 0.005 | | | | 0.169 / 0.174 | 4.9% | 47% / 67% |
| -108 | 0.103 | 0.237 | 0.005 | 0.452 / 0.447 | −0.9% | 64% / 83% | 0.235 / 0.240 | 4.9% | 49% / 67% |
| -149 | 0.103 | 2.800 | 0.022 | 3.006 / 3.001 | −0.4% | 57% / 75% | 2.800 / 2.805 | 0.5% | 55% / 72% |
| -210 | 0.139 | 0.734 | 0.010 | 1.019 / 1.014 | −0.9% | 62% / 78% | 0.734 / 0.739 | 2.3% | 54% / 70% |
| -247 | 0.139 | 4.609 | 0.030 | 4.887 / 4.877 | −0.4% | 59% / 74% | 4.609 / 4.619 | 0.4% | 57% / 73% |
| -325 | 0.210 | 1.475 | 0.015 | 1.899 / 1.894 | −1.0% | 63% / 77% | 1.475 / 1.480 | 1.6% | 58% / 70% |
| -349 | 0.210 | 4.475 | 0.030 | 4.895 / 4.885 | −0.5% | 61% / 74% | 4.475 / 4.485 | 0.6% | 59% / 72% |
| -425 | 0.275 | 4.475 | 0.033 | 5.025 / 5.015 | −0.6% | 61% / 73% | 4.475 / 4.485 | 0.7% | 58% / 70% |
| -460 | 0.275 | 15.475 | 0.070 | 16.025 / 16.015 | −0.2% | 60% / 72% | 15.475 / 15.485 | 0.2% | 59% / 71% |

Note: Blank table cells indicate O-ring sizes that are not recommended for the given application.
[a] Positive values indicate no cramping.

## Table 5. Metric (mm) Axial Face Seal Data, Recommended Sizes in Each Range
*(Based on ISO 3601-2-2016 and 70 Shore A O-Rings)*

| O-Ring | | | | Axial Face Seal Groove | | | | | | | |
|---|---|---|---|---|---|---|---|---|---|---|---|
| | | | | Internal Pressure | | | | External Pressure | | | |
| Size Code | Cross-Section Diameter | Inner Diameter | | Groove Outer Diameter | Cramping | Gland Fill Min/Max (No Swell, Straight Sides, No Radii) | | Groove Inner Diameter | Stretch | Gland Fill Min/Max (No Swell, Straight Sides, No Radii) | |
| | | | | | | Liquid | Gas & Vac | | | Liquid | Gas & Vac |
| | Nominal | Nominal | ± Tol. | Min/Max | | Min/Max | Min/Max | Min/Max | | Min/Max | Min/Max |
| -004A | 1.78 | 1.78 | 0.13 | | | | | 1.78<br>1.80 | 1.1% | 31%<br>49% | 36%<br>58% |
| -011A | 1.78 | 7.65 | 0.13 | 11.20<br>11.24 | −3.1% | 56%<br>79% | 59%<br>84% | 7.70<br>7.72 | 0.9% | 40%<br>58% | 45%<br>66% |
| -050A | 1.78 | 133.07 | 0.94 | 136.62<br>136.66 | −0.3% | 48%<br>67% | 52%<br>73% | 134.37<br>134.39 | 1.0% | 46%<br>65% | 51%<br>71% |
| -102A | 2.62 | 1.24 | 0.13 | | | | | 1.25<br>1.27 | 2.4% | 38%<br>56% | 45%<br>67% |
| -112A | 2.62 | 12.37 | 0.13 | 17.60<br>17.64 | −2.5% | 63%<br>81% | 67%<br>87% | 12.48<br>12.50 | 1.1% | 50%<br>67% | 57%<br>76% |
| -178A | 2.62 | 247.32 | 1.40 | 252.55<br>252.59 | −0.2% | 57%<br>73% | 63%<br>81% | 248.82<br>248.84 | 0.6% | 56%<br>72% | 62%<br>80% |
| -201A | 3.53 | 4.34 | 0.13 | | | | | 4.37<br>4.39 | 1.2% | 46%<br>61% | 54%<br>71% |
| -212A | 3.53 | 21.82 | 0.25 | 28.88<br>28.92 | −2.1% | 64%<br>79% | 69%<br>85% | 22.03<br>22.05 | 1.1% | 54%<br>68% | 61%<br>77% |
| -284A | 3.53 | 456.06 | 2.16 | 463.12<br>463.16 | −0.1% | 60%<br>73% | 66%<br>80% | 457.56<br>457.58 | 0.3% | 59%<br>72% | 65%<br>80% |
| -318A | 5.33 | 24.77 | 0.25 | 35.43<br>35.47 | −2.3% | 68%<br>80% | 72%<br>85% | 25.00<br>25.02 | 1.0% | 57%<br>69% | 63%<br>76% |
| -395A | 5.33 | 658.88 | 3.05 | 669.54<br>669.58 | −0.1% | 63%<br>74% | 68%<br>80% | 660.38<br>660.40 | 0.2% | 63%<br>73% | 68%<br>80% |
| -425A | 6.99 | 113.67 | 0.84 | 127.63<br>127.67 | −0.7% | 70%<br>80% | 73%<br>84% | 114.78<br>114.80 | 1.0% | 66%<br>77% | 71%<br>82% |
| -475A | 6.99 | 658.88 | 3.05 | 672.84<br>672.88 | −0.1% | 69%<br>79% | 73%<br>83% | 660.38<br>660.40 | 0.2% | 68%<br>78% | 72%<br>83% |

Note: Blank table cells indicate O-ring sizes that are not recommended for the given application.

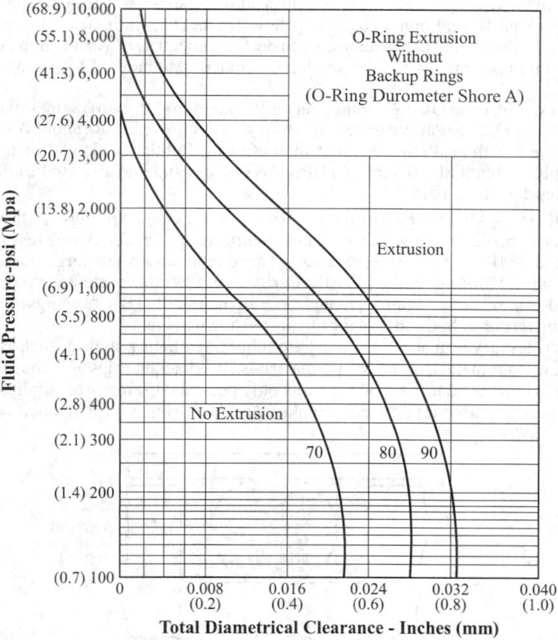


Fig. 4. Extrusion Potential of O-Rings as a Function of Hardness and Clearance Gap

**Clearance Gap.**—When there is relative motion between two parts enclosing an O-ring, there must be a clearance gap between them (see Fig. 2 and Fig. 3). This broadly applies to piston and rod seals. Axial face seal applications also may have a clearance gap, but generally they do not.

For a piston gland (see Fig. 3), the radial clearance gap is the distance between the piston surface and the bore wall. For rod glands (again, see Fig. 3), the gap is the distance between the rod surface and the bore wall. The radial clearance gap on one side is at its maximum value when the piston or rod is side-loaded and touches the bore (zero clearance gap) on the other side. This gap often is expressed in terms of diametral clearance, which is twice the nominal radial clearance; diametral clearance is the same, regardless of piston or rod position.

If the radial clearance gap between the sealing surface and the gland edge is too large, or if the fluid pressure exceeds the deformation limits of the O-ring, it will extrude into the gap, reducing the effective life of the seal. This is especially problematic for soft silicone O-rings, where a 50 percent reduction in the standard clearance gap may be recommended. To prevent O-ring extrusion, reduce the clearance gap by modifying the gland dimensions, lower the system operating pressure, or install backup rings. A harder O-ring material may be used but could result in higher drag, increase the tendency of the seal to leak at low pressures, and necessitate additional dimensional changes.

Fig. 4 indicates effective diametral clearance values that may be used, depending on fluid pressure and O-ring hardness. If conditions of use fall to the right of the curve,

*Example:* Calculate stretch/cramping, squeeze, and gland fill for a -116 size O-ring (-116A in metric, shown in parentheses) in a standard axial face gland with external liquid pressure.

Given: Use gland Standard SAE AS6235A-2015 (ISO 3601-2-2016):

| | | | |
|---|---|---|---|
| $ID_{max} = 0.746$ | $ID_{min} = 0.728$ | $(ID_{max} = 18.95)$ | $(ID_{min} = 18.49)$ |
| $W_{max} = 0.106$ | $W_{min} = 0.1$ | $(W_{max} = 2.71)$ | $(W_{min} = 2.53)$ |
| $WR_{min} = W_{min} = 0.1$ | $WR_{max} = W_{max} = 0.106$ | $(WR_{min} = W_{min} = 2.53)$ | $(WR_{max} = W_{max} = 2.71)$ |
| $IDG_{max} = 0.742$ | $IDG_{min} = 0.737$ | $(IDG_{max} = 18.9)$ | $(IDG_{min} = 18.84)$ |
| $AF_{max} = 0.09$ | $AF_{min} = 0.086$ | $(AF_{max} = 2.1)$ | $(AF_{min} = 2)$ |
| $AG_{min} = 0.141$ | $AG_{max} = 0.151$ | $(AG_{min} = 4)$ | $(AG_{max} = 4.2)$ |

Calculate:

| | | | |
|---|---|---|---|
| $L_{max} = 0.115$ | $L_{min} = 0.112$ | $(L_{max} = 3.125)$ | $(L_{min} = 3.018)$ |
| $SID_{max} = 0.728$ | $SID_{min} = 0.725$ | $(SID_{max} = 18.322)$ | $(SID_{min} = 18.215)$ |
| $Stretch\%_{max} = 2.4\%$ | $Stretch\%_{min} = 1.2\%$ | $(Stretch\%_{max} = 3.8\%)$ | $(Stretch\%_{min} = 2.9\%)$ |
| $ODG_{max} = 1.044$ | $ODG_{min} = 1.019$ | $(ODG_{max} = 27.3)$ | $(ODG_{min} = 26.85)$ |
| $Sq\%_{max} = 19\%$ | $Sq\%_{min} = 10\%$ | $(Sq\%_{max} = 26\%)$ | $(Sq\%_{min} = 17\%)$ |
| $OVol_{max} = 0.024$ | $OVol_{min} = 0.02$ | $(OVol_{max} = 392.5)$ | $(OVol_{min} = 331.98)$ |
| $GVol_{max} = 0.039$ | $GVol_{min} = 0.033$ | $(GVol_{max} = 643.31)$ | $(GVol_{min} = 571.15)$ |
| $Fill\%_{max} = 72\%$ | $Fill\%_{min} = 53\%$ | $(Fill\%_{max} = 69\%)$ | $(Fill\%_{min} = 52\%)$ |

Validate for the -116 SAE O-Ring:

| | | |
|---|---|---|
| $Stretch\%_{max} = 2.4\%$ | $Stretch\%_{min} = 1.2\%$ | Well within SAE limit of 5%. |
| $Sq\%_{min} = 10\%$ | $Sq\%_{max} = 19\%$ | Meets SAE recommendation of 10–20% for this cross section. |
| $Fill\%_{max} = 72\%$ | $Fill\%_{min} = 53\%$ | Within SAE target maximum of 85%. |

Validate for the -116A ISO O-Ring:

| | | |
|---|---|---|
| $(Stretch\%_{max} = 3.8\%)$ | $(Stretch\%_{min} = 2.9\%)$ | Well within ISO limit of 8%. |
| $(Sq\%_{min} = 17\%)$ | $(Sq\%_{max} = 26\%)$ | ISO recommends 19–30% for this cross section. This slight lower limit violation invites caution in critical applications but is not uncommon. |
| $(Fill\%_{max} = 69\%)$ | $(Fill\%_{min} = 52\%)$ | Within ISO target maximum of 85%. |

*PTFE O-Rings in Axial Face Seals:* Axial face seals can work with very stiff O-ring materials. Solid PTFE seals are often chosen because they are highly chemically inert and less costly than perfluorinated elastomer seals with similar chemical resistance. Glands for solid PTFE O-rings often target a squeeze of 10 to 20 percent for O-rings with cross-sectional diameters up to 0.210 in. (5.33 mm), and 10 to 15 percent for cross sections 0.210 to 0.275 in. (5.33 to 6.99 mm).

It is common practice to dimension the groove so that the groove's mean diameter (inner diameter plus one cross-sectional diameter) is equal to the O-ring's free-state mean diameter. Deliberate contact with the inner or outer groove walls also can be targeted as with elastomeric O-rings, but stretch and cramping should be zero or nearly zero. A groove with a full-radius floor is best to support the O-ring and control deformation. The groove width is normally 10 percent greater than the O-ring cross-sectional diameter. If a clearance gap is present, it must not exceed 0.005 in. (0.13 mm).

PTFE O-rings and their glands are extremely vulnerable to nicks, scratches, and particle contamination. These seals must be handled with care and installed in a clean environment.

**Table 12. Dovetail and Half-Dovetail Groove Dimensions for Static Face Seals**
*(Zero Clearance Gap in Gland, 70 Shore A O-Rings)*

| O-Ring | | | Groove Details | | | Dovetail Groove | | | Half-Dovetail Groove | | |
|---|---|---|---|---|---|---|---|---|---|---|---|
| Size Code | Units | Cross-Section Nominal | Wall Angle $\theta \pm 1°$ | Internal Radii | Opening Radii | Squeeze | Depth $L$ | Width $G$ | Squeeze | Depth $L$ | Width $G$ |
| -0XX | inch | 0.070 | 24% | 0.016 | 0.005 | 27% | 0.050–0.052 | 0.055–0.059 | 23% | 0.053–0.055 | 0.064–0.[illegible]6 |
| -0XXA | mm | 1.78 | | 0.40 | 0.13 | | 1.27–1.32 | 1.40–1.50 | | 1.35–1.40 | 1.63–1.68 |
| -1XX | inch | 0.103 | 24% | 0.016 | 0.010 | 21% | 0.081–0.083 | 0.083–0.087 | 19% | 0.083–0.085 | 0.095–0.0[illegible]7 |
| -1XXA | mm | 2.62 | | 0.40 | 0.25 | | 2.06–2.11 | 2.11–2.21 | | 2.11–2.16 | 2.41–2.45 |
| -2XX | inch | 0.139 | 24% | 0.031 | 0.010 | 20% | 0.111–0.113 | 0.113–0.117 | 18% | 0.113–0.115 | 0.124–0.1[illegible]8 |
| -2XXA | mm | 3.53 | | 0.80 | 0.25 | | 2.87–2.92 | 2.87–2.97 | | 2.87–2.92 | 3.15–3.2[illegible] |
| -3XX | inch | 0.210 | 24% | 0.031 | 0.015 | 18% | 0.171–0.173 | 0.171–0.175 | 17% | 0.173–0.176 | 0.190–0.1[illegible]3 |
| -3XXA | mm | 5.33 | | 0.80 | 0.40 | | 4.52–4.57 | 4.01–4.11 | | 4.40–4.47 | 4.83–4.9[illegible] |
| -4XX | inch | 0.275 | 24% | 0.063 | 0.015 | 16% | 0.231–0.234 | 0.231–0.235 | 15% | 0.234–0.238 | 0.255–0.2[illegible]7 |
| -4XXA | mm | 6.99 | | 1.60 | 0.40 | | 5.89–5.94 | 5.54–5.64 | | 5.95–6.05 | 6.48–6.5[illegible] |

Note: Sizes -0XX to -1XXA are not recommended for these grooves.

Note: Sizes -2XX to -2XXA should be used with caution in these grooves.

maximum housing diameter, target a maximum cramping of 1 percent of nominal O-ring dimensions at nominal squeeze in the SAE system or 5 percent in the ISO system for O-rings with an inner diameter less than 250 mm. Larger diameter metric O-rings are allowed only 3 percent cramping. Stretch on the rod is normally limited to 5 percent. It is sometimes acceptable to violate these limits, and standards commonly include such cases for smaller sizes.

*Piston Glands:* To design a piston gland, it is best to start with a known housing bore diameter. Piston glands have a groove cut into the outer diameter of a piston, where the O-ring is stretched slightly around the groove diameter and squeezed radially between the groove floor and housing bore. One exception to this is a pneumatic "floating" piston gland, discussed later. In most piston seal applications, limit maximum installed stretch of most elastomeric O-rings to 5 percent for long seal life. This can be exceeded for some small sizes, and standard gland dimensions include such exceptions.

*Radial Gland Calculations:* Designers usually begin a radial seal design with a preferred cross section and inner diameter selected to work with a given rod or bore. Standards and catalogs can then be used to determine appropriate gland dimensions. In the standards, target squeeze ranges are as follows:

| | | |
|---|---|---|
| SAE Sizes -0XX: 5–30% | SAE Sizes -1XX: 5–21% | SAE Sizes -2XX: 4–20% |
| SAE Sizes -3XX: 4–17% | SAE Sizes -4XX: 7–17% | |
| ISO Sizes -0XXA: 14–35% (static) | 10–24% (pneumatic reciprocating) | 13–27% (hydraulic reciprocating) |
| ISO Sizes -1XXA: 13–30% (static) | 8–22% (pneumatic reciprocating) | 12–24% (hydraulic reciprocating) |
| ISO Sizes -2XXA: 13–30% (static) | 7–20% (pneumatic reciprocating) | 11–22% (hydraulic reciprocating) |
| ISO Sizes -3XXA: 12–28% (static) | 7–18% (pneumatic reciprocating) | 11–20% (hydraulic reciprocating) |
| ISO Sizes -4XXA: 10–25% (static) | 6–17% (pneumatic reciprocating) | 9–19% (hydraulic reciprocating) |

Radial gland depth used to calculate squeeze should take into account any clearance gap and groove eccentricity $E$. The radial clearance gap maximum and minimum are seen when the piston or rod is side-loaded and touching the wall of the bore. While SAE standards include these factors in squeeze range calculations and recommendations, ISO standards do not. Calculation by either method is valid, as long as it is compared with recommended ranges calculated the same way. (Calculated squeeze values given in tables here take clearance gap and eccentricity into account; as a result, some squeeze values shown may be outside ISO recommended ranges.)

Gland data for appropriate size ranges are given in Table 2, Table 3, and Table 6 to Table 11. These dimensions can be customized if desired, but the resulting O-ring stretch, cramping, squeeze, and gland fill must be evaluated to check for potential problems. Precise calculation methods for a radially stretched O-ring cross section are included in the

# ROLLED STEEL, WIRE, SHEET METAL, WIRE ROPE

## Rolled Steel Sections

**Lengths of Angles Bent to Circular Shape.—**To calculate the length of an angle-iron used either inside or outside of a tank or smokestack, the following table of constants may be used: Assume, for example, that a stand-pipe, 20 feet inside diameter, is provided with a 3 by 3 by ⅜ inch angle-iron on the inside at the top. The circumference of a circle 20 feet in diameter is 754 inches. From the table of constants, find the constant for a 3 by 3 by ⅜ inch angle-iron, which is 4.319. The length of the angle then is 754 − 4.319 = 749.681 inches. Should the angle be on the outside, add the constant instead of subtracting it; thus, 754 + 4.319 = 758.319 inches.

| Size of Angle | Const. | Size of Angle | Const. | Size of Angle | Const. |
|---|---|---|---|---|---|
| ¼×2×2 | 2.879 | 5⁄16×3×3 | 4.123 | ½×5×5 | 6.804 |
| 5⁄16×2×2 | 3.076 | ⅜×3×3 | 4.319 | ⅜×6×6 | 7.461 |
| ⅜×2×2 | 3.272 | ½×3×3 | 4.711 | ½×6×6 | 7.854 |
| ¼×2½×2½ | 3.403 | ⅜×3½×3½ | 4.843 | ¾×6×6 | 8.639 |
| 5⁄16×2½×2½ | 3.600 | ½×3½×3½ | 5.235 | ½×8×8 | 9.949 |
| ⅜×2½×2½ | 3.796 | ⅜×4×4 | 5.366 | ¾×8×8 | 10.734 |
| ½×2½×2½ | 4.188 | ½×4×4 | 5.758 | 1×8×8 | 11.520 |
| ¼×3×3 | 3.926 | ⅜×5×5 | 6.414 | … | … |

**Standard Designations of Rolled Steel Shapes.—**Through a joint effort, the American Iron and Steel Institute (AISI) and the American Institute of Steel Construction (AISC) have changed most of the designations for their hot-rolled structural steel shapes. The present designations, standard for steel producing and fabricating industries, should be used when designing, detailing, and ordering steel. The accompanying Table 1 compares the present designations with the previous descriptions.

**Table 1. Hot-Rolled Structural Steel Shape Designations (AISI and AISC)**

| Present Designation | Type of Shape | Previous Designation |
|---|---|---|
| W 24 × 76 | W shape | 24 WF 76 |
| W 14 × 26 | W shape | 14 B 26 |
| S 24 × 100 | S shape | 24 I 100 |
| M 8 × 18.5 | M shape | 8 M 18.5 |
| M 10 × 9 | M shape | 10 JR 9.0 |
| M 8 × 34.3 | M shape | 8 × 8 M 34.3 |
| C 12 × 20.7 | American Standard Channel | 12 [20.7 |
| MC 12 × 45 | Miscellaneous Channel | 12 × 4 [45.0 |
| MC 12 × 10.6 | Miscellaneous Channel | 12 JR [10.6 |
| HP 14 × 73 | HP shape | 14 BP 73 |
| L 6 × 6 × ¾ | Equal Leg Angle | ∠6 × 6 × ¾ |
| L 6 × 4 × ⅝ | Unequal Leg Angle | ∠6 × 4 × ⅝ |
| WT 12 × 38 | Structural Tee cut from W shape | ST 12 WF 38 |
| WT 7 × 13 | Structural Tee cut from W shape | ST 7 B 13 |
| St 12 × 50 | Structural Tee cut from S shape | ST 12 I 50 |
| MT 4 × 9.25 | Structural Tee cut from M shape | ST 4 M 9.25 |
| MT 5 × 4.5 | Structural Tee cut from M shape | ST 5 JR 4.5 |
| MT 4 × 17.15 | Structural Tee cut from M shape | ST 4 M 17.15 |
| PL ½ × 18 | Plate | PL 18 × ½ |
| Bar 1 | Square Bar | Bar 1 |
| Bar 1 ¼ ∅ | Round Bar | Bar 1¼∅ |
| Bar 2½ × ½ | Flat Bar | Bar 2½ × ½ |
| Pipe 4 Std. | Pipe | Pipe 4 Std. |
| Pipe 4 X-Strong | Pipe | Pipe 4 X-Strong |
| Pipe 4 XX-Strong | Pipe | Pipe 4 XX-Strong |
| TS 4 × 4 × .375 | Structural Tubing: Square | Tube 4 × 4 × .375 |
| TS 5 × 3 × .375 | Structural Tubing: Rectangular | Tube 5 × 3 × .375 |
| TS 3 OD × .250 | Structural Tubing: Circular | Tube 3 OD × .250 |

Data taken from the "Manual of Steel Construction," 8th Edition, 1980, with permission of the American Institute of Steel Construction.

## Table 2d. Steel Wide-Flange Sections

Wide-flange sections are designated, in order, by a section letter, nominal depth of the member in inches, and the nominal weight in pounds per foot; thus:

**W 8 × 67**

indicates a wide-flange section having a nominal depth of 8 inches, and a nominal weight per foot of 67 pounds. Actual geometry for each section can be obtained from the values below.

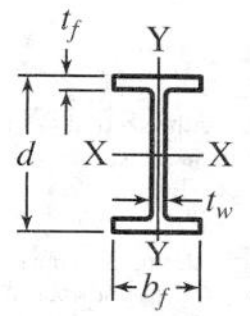


| Designation | Area, $A$ inch² | Depth, $d$ inch | Flange Width, $b_f$ inch | Flange Thickness, $t_f$ inch | Web Thickness, $t_w$ inch | Axis XX $I$ inch⁴ | Axis XX $S$ inch³ | Axis XX $r$ inch | Axis YY $I$ inch⁴ | Axis YY $S$ inch³ | Axis YY $r$ inch |
|---|---|---|---|---|---|---|---|---|---|---|---|
| W 10 × 112 | 32.9 | 11.36 | 10.415 | 1.250 | 0.755 | 716 | 126 | 4.66 | 236 | 45.3 | 2.6[illegible] |
| × 100 | 29.4 | 11.10 | 10.340 | 1.120 | 0.680 | 623 | 112 | 4.60 | 207 | 40.0 | 2.6[illegible] |
| × 88 | 25.9 | 10.84 | 10.265 | 0.990 | 0.605 | 534 | 98.5 | 4.54 | 179 | 34.8 | 2.6[illegible] |
| × 77 | 22.6 | 10.60 | 10.190 | 0.870 | 0.530 | 455 | 85.9 | 4.49 | 154 | 30.1 | 2.6[illegible] |
| × 68 | 20.0 | 10.40 | 10.130 | 0.770 | 0.470 | 394 | 75.7 | 4.44 | 134 | 26.4 | 2.5[illegible] |
| × 60 | 17.6 | 10.22 | 10.080 | 0.680 | 0.420 | 341 | 66.7 | 4.39 | 116 | 23.0 | 2.5[illegible] |
| × 54 | 15.8 | 10.09 | 10.030 | 0.615 | 0.370 | 303 | 60.0 | 4.37 | 103 | 20.6 | 2.5[illegible] |
| × 49 | 14.4 | 9.98 | 10.000 | 0.560 | 0.340 | 272 | 54.6 | 4.35 | 93.4 | 18.7 | 2.5[illegible] |
| × 45 | 13.3 | 10.10 | 8.020 | 0.620 | 0.350 | 248 | 49.1 | 4.32 | 53.4 | 13.3 | 2.0 |
| × 39 | 11.5 | 9.92 | 7.985 | 0.530 | 0.315 | 209 | 42.1 | 4.27 | 45.0 | 11.3 | 1.9[illegible] |
| × 33 | 9.71 | 9.73 | 7.960 | 0.435 | 0.290 | 170 | 35.0 | 4.19 | 36.6 | 9.20 | 1.9[illegible] |
| × 30 | 8.84 | 10.47 | 5.810 | 0.510 | 0.300 | 170 | 32.4 | 4.38 | 16.7 | 5.75 | 1.3[illegible] |
| × 26 | 7.61 | 10.33 | 5.770 | 0.440 | 0.260 | 144 | 27.9 | 4.35 | 14.1 | 4.89 | 1.3[illegible] |
| × 22 | 6.49 | 10.17 | 5.750 | 0.360 | 0.240 | 118 | 23.2 | 4.27 | 11.4 | 3.97 | 1.3[illegible] |
| × 19 | 5.62 | 10.24 | 4.020 | 0.395 | 0.250 | 96.3 | 18.8 | 4.14 | 4.29 | 2.14 | 0.8[illegible]4 |
| × 17 | 4.99 | 10.11 | 4.010 | 0.330 | 0.240 | 81.9 | 16.2 | 4.05 | 3.56 | 1.78 | 0.8[illegible]4 |
| × 15 | 4.41 | 9.99 | 4.000 | 0.270 | 0.230 | 68.9 | 13.8 | 3.95 | 2.89 | 1.45 | 0.8[illegible]0 |
| × 12 | 3.54 | 9.87 | 3.960 | 0.210 | 0.190 | 53.8 | 10.9 | 3.90 | 2.18 | 1.10 | 0.7[illegible]5 |
| W 8 × 67 | 19.7 | 9.00 | 8.280 | 0.935 | 0.570 | 272 | 60.4 | 3.72 | 88.6 | 21.4 | 2.1[illegible] |
| × 58 | 17.1 | 8.75 | 8.220 | 0.810 | 0.510 | 228 | 52.0 | 3.65 | 75.1 | 18.3 | 2.1[illegible] |
| × 48 | 14.1 | 8.50 | 8.110 | 0.685 | 0.400 | 184 | 43.3 | 3.61 | 60.9 | 15.0 | 2.0[illegible] |
| × 40 | 11.7 | 8.25 | 8.070 | 0.560 | 0.360 | 146 | 35.5 | 3.53 | 49.1 | 12.2 | 2.0[illegible] |
| × 35 | 10.3 | 8.12 | 8.020 | 0.495 | 0.310 | 127 | 31.2 | 3.51 | 42.6 | 10.6 | 2.0[illegible] |
| × 31 | 9.13 | 8.00 | 7.995 | 0.435 | 0.285 | 110 | 27.5 | 3.47 | 37.1 | 9.27 | 2.0[illegible] |
| × 28 | 8.25 | 8.06 | 6.535 | 0.465 | 0.285 | 98.0 | 24.3 | 3.45 | 21.7 | 6.63 | 1.6[illegible] |
| × 24 | 7.08 | 7.93 | 6.495 | 0.400 | 0.245 | 82.8 | 20.9 | 3.42 | 18.3 | 5.63 | 1.6[illegible] |
| × 21 | 6.16 | 8.28 | 5.270 | 0.400 | 0.250 | 75.3 | 18.2 | 3.49 | 9.77 | 3.71 | 1.2[illegible] |
| × 18 | 5.26 | 8.14 | 5.250 | 0.330 | 0.230 | 61.9 | 15.2 | 3.43 | 7.97 | 3.04 | 1.2[illegible] |
| × 15 | 4.44 | 8.11 | 4.015 | 0.315 | 0.245 | 48.0 | 11.8 | 3.29 | 3.41 | 1.70 | 0.8[illegible]6 |
| × 13 | 3.84 | 7.99 | 4.000 | 0.255 | 0.230 | 39.6 | 9.91 | 3.21 | 2.73 | 1.37 | 0.8[illegible]3 |
| × 10 | 2.96 | 7.89 | 3.940 | 0.205 | 0.170 | 30.8 | 7.81 | 3.22 | 2.09 | 1.06 | 0.8[illegible]1 |
| W 6 × 25 | 7.34 | 6.38 | 6.080 | 0.455 | 0.320 | 53.4 | 16.7 | 2.70 | 17.1 | 5.61 | 1.5[illegible] |
| × 20 | 5.87 | 6.20 | 6.020 | 0.365 | 0.260 | 41.4 | 13.4 | 2.66 | 13.3 | 4.41 | 1.5[illegible] |
| × 16 | 4.74 | 6.28 | 4.030 | 0.405 | 0.260 | 32.1 | 10.2 | 2.60 | 4.43 | 2.20 | 0.9[illegible]6 |
| × 15 | 4.43 | 5.99 | 5.990 | 0.260 | 0.230 | 29.1 | 9.72 | 2.56 | 9.32 | 3.11 | 1.4[illegible] |
| × 12 | 3.55 | 6.03 | 4.000 | 0.280 | 0.230 | 22.1 | 7.31 | 2.49 | 2.99 | 1.50 | 0.9[illegible]8 |
| × 9 | 2.68 | 5.90 | 3.940 | 0.215 | 0.170 | 16.4 | 5.56 | 2.47 | 2.19 | 1.11 | 0.9[illegible]5 |
| W 5 × 19 | 5.54 | 5.15 | 5.030 | 0.430 | 0.270 | 26.2 | 10.2 | 2.17 | 9.13 | 3.63 | 1.2[illegible] |
| × 16 | 4.68 | 5.01 | 5.000 | 0.360 | 0.240 | 21.3 | 8.51 | 2.13 | 7.51 | 3.00 | 1.2[illegible] |
| W 4 × 13 | 3.83 | 4.16 | 4.060 | 0.345 | 0.280 | 11.3 | 5.46 | 1.72 | 3.86 | 1.90 | 1.0[illegible] |

*Symbols:* I = moment of inertia; S = section modulus; r = radius of gyration.

Data taken from the "Manual of Steel Construction," 8th Edition, 1980, with permission of the American Institute of Steel Construction.

## Table 4. American Standard Steel Channels

American Standard Channels are designated, in order, by a section letter, actual depth in inches, and nominal weight per foot in pounds. Thus:

**C 7 × 14.75**

indicates an American Standard Channel with a depth of 7 inches and a nominal weight of 14.75 pounds per foot.

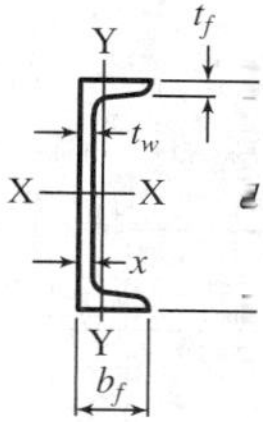


| Designation | Area, $A$ inch² | Depth, $d$ inch | Flange Width, $b_f$ inch | Flange Thickness, $t_f$ inch | Web Thickness, $t_w$ inch | Axis XX $I$ inch⁴ | Axis XX $S$ inch³ | Axis XX $r$ inch | Axis YY $I$ inch⁴ | Axis YY $S$ inch³ | Axis YY $r$ inch | $x$ inch |
|---|---|---|---|---|---|---|---|---|---|---|---|---|
| C 15 × 50 | 14.7 | 15.00 | 3.716 | 0.650 | 0.716 | 404 | 53.8 | 5.24 | 11.0 | 3.78 | 0.867 | .798 |
| × 40 | 11.8 | 15.00 | 3.520 | 0.650 | 0.520 | 349 | 46.5 | 5.44 | 9.23 | 3.37 | 0.886 | .777 |
| × 33.9 | 9.96 | 15.00 | 3.400 | 0.650 | 0.400 | 315 | 42.0 | 5.62 | 8.13 | 3.11 | 0904 | .787 |
| C 12 × 30 | 8.82 | 12.00 | 3.170 | 0.501 | 0.510 | 162 | 27.0 | 4.29 | 5.14 | 2.06 | 0.763 | .674 |
| × 25 | 7.35 | 12.00 | 3.047 | 0.501 | 0.387 | 144 | 24.1 | 4.43 | 4.47 | 1.88 | 0.780 | .674 |
| × 20.7 | 6.09 | 12.00 | 2.942 | 0.501 | 0.282 | 129 | 21.5 | 4.61 | 3.88 | 1.73 | 0.799 | .698 |
| C 10 × 30 | 8.82 | 10.00 | 3.033 | 0.436 | 0.673 | 103 | 20.7 | 3.42 | 3.94 | 1.65 | 0.669 | .649 |
| × 25 | 7.35 | 10.00 | 2.886 | 0.436 | 0.526 | 91.2 | 18.2 | 3.52 | 3.36 | 1.48 | 0.676 | .617 |
| × 20 | 5.88 | 10.00 | 2.739 | 0.436 | 0.379 | 78.9 | 15.8 | 3.66 | 2.81 | 1.32 | 0.692 | .606 |
| × 15.3 | 4.49 | 10.00 | 2.600 | 0.436 | 0.240 | 67.4 | 13.5 | 3.87 | 2.28 | 1.16 | 0.713 | .634 |
| C 9 × 20 | 5.88 | 9.00 | 2.648 | 0.413 | 0.448 | 60.9 | 13.5 | 3.22 | 2.42 | 1.17 | 0.642 | .583 |
| × 15 | 4.41 | 9.00 | 2.485 | 0.413 | 0.285 | 51.0 | 11.3 | 3.40 | 1.93 | 1.01 | 0.661 | .586 |
| × 13.4 | 3.94 | 9.00 | 2.433 | 0.413 | 0.233 | 47.9 | 10.6 | 3.48 | 1.76 | 0.962 | 0.669 | .601 |
| C 8 × 18.75 | 5.51 | 8.00 | 2.527 | 0.390 | 0.487 | 44.0 | 11.0 | 2.82 | 1.98 | 1.01 | 0.599 | .565 |
| × 13.75 | 4.04 | 8.00 | 2.343 | 0.390 | 0.303 | 36.1 | 9.03 | 2.99 | 1.53 | 0.854 | 0.615 | .553 |
| × 11.5 | 3.38 | 8.00 | 2.260 | 0.390 | 0.220 | 32.6 | 8.14 | 3.11 | 1.32 | 0.781 | 0.625 | .571 |
| C 7 × 14.75 | 4.33 | 7.00 | 2.299 | 0.366 | 0.419 | 27.2 | 7.78 | 2.51 | 1.38 | 0.779 | 0.564 | .532 |
| × 12.25 | 3.60 | 7.00 | 2.194 | 0.366 | 0.314 | 24.2 | 6.93 | 2.60 | 1.17 | 0.703 | 0.571 | .525 |
| × 9.8 | 2.87 | 7.00 | 2.090 | 0.366 | 0.210 | 21.3 | 6.08 | 2.72 | 0.968 | 0.625 | 0.581 | .540 |
| C 6 × 13 | 3.83 | 6.00 | 2.157 | 0.343 | 0.437 | 17.4 | 5.80 | 2.13 | 1.05 | 0.642 | 0.525 | .514 |
| × 10.5 | 3.09 | 6.00 | 2.034 | 0.343 | 0.314 | 15.2 | 5.06 | 2.22 | 0.866 | 0.564 | 0.529 | .499 |
| × 8.2 | 2.40 | 6.00 | 1.920 | 0.343 | 0.200 | 13.1 | 4.38 | 2.34 | 0.693 | 0.492 | 0.537 | .511 |
| C 5 × 9 | 2.64 | 5.00 | 1.885 | 0.320 | 0.325 | 8.90 | 3.56 | 1.83 | 0.632 | 0.450 | 0.489 | .478 |
| × 6.7 | 1.97 | 5.00 | 1.750 | 0.320 | 0.190 | 7.49 | 3.00 | 1.95 | 0.479 | 0.378 | 0.493 | .484 |
| C 4 × 7.25 | 2.13 | 4.00 | 1.721 | 0.296 | 0.321 | 4.59 | 2.29 | 1.47 | 0.433 | 0.343 | 0.450 | .459 |
| × 5.4 | 1.59 | 4.00 | 1.584 | 0.296 | 0.184 | 3.85 | 1.93 | 1.56 | 0.319 | 0.283 | 0.449 | .457 |
| C 3 × 6 | 1.76 | 3.00 | 1.596 | 0.273 | 0.356 | 2.07 | 1.38 | 1.08 | 0.305 | 0.268 | 0.416 | .455 |
| × 5 | 1.47 | 3.00 | 1.498 | 0.273 | 0.258 | 1.85 | 1.24 | 1.12 | 0.247 | 0.233 | 0.410 | .438 |
| × 4.1 | 1.21 | 3.00 | 1.410 | 0.273 | 0.170 | 1.66 | 1.10 | 1.17 | 0.197 | 0.202 | 0.404 | .436 |

*Symbols:* $I$ = moment of inertia; $S$ = section modulus; $r$ = radius of gyration; $x$ = distance from center of gravity of section to outer face of structural shape.

Data taken from the "Manual of Steel Construction," 8th Edition, 1980, with permission of the American Institute of Steel Construction.

## Table 6. *(Continued)* Steel Angles with Unequal Legs

| Size inch | Thick-ness inch | Weight per Ft. lb. | Area inch² | Axis XX and Axis YY | | | | Axis YY | | | Axis ZZ | | |
|---|---|---|---|---|---|---|---|---|---|---|---|---|---|
| | | | | *I* inch⁴ | *S* inch³ | *r* inch | *y* inch | *I* inch⁴ | *S* inch³ | *r* inch | *x* inch | *r* inch | Tan *A* |
| 5×3 | 7⁄16 | 11.3 | 3.31 | 8.43 | 2.58 | 1.60 | 1.73 | 2.32 | 1.02 | .837 | .727 | .651 | .361 |
| | 3⁄8 | 9.8 | 2.86 | 7.37 | 2.24 | 1.61 | 1.70 | 2.04 | .888 | .845 | .704 | .654 | .364 |
| | 5⁄16 | 8.2 | 2.40 | 6.26 | 1.89 | 1.61 | 1.68 | 1.75 | .753 | .853 | .681 | .658 | .368 |
| | 1⁄4 | 6.6 | 1.94 | 5.11 | 1.53 | 1.62 | 1.66 | 1.44 | .614 | .861 | .657 | .663 | .371 |
| 4×3½ | 5⁄8 | 14.7 | 4.30 | 6.37 | 2.35 | 1.22 | 1.29 | 4.52 | 1.84 | 1.03 | 1.04 | .719 | .745 |
| | 1⁄2 | 11.9 | 3.50 | 5.32 | 1.94 | 1.23 | 1.25 | 3.79 | 1.52 | 1.04 | 1.00 | .722 | .750 |
| | 7⁄16 | 10.6 | 3.09 | 4.76 | 1.72 | 1.24 | 1.23 | 3.40 | 1.35 | 1.05 | .978 | .724 | .753 |
| | 3⁄8 | 9.1 | 2.67 | 4.18 | 1.49 | 1.25 | 1.21 | 2.95 | 1.17 | 1.06 | .955 | .727 | .755 |
| | 5⁄16 | 7.7 | 2.25 | 3.56 | 1.26 | 1.26 | 1.18 | 2.55 | .994 | 1.07 | .932 | .730 | .757 |
| | 1⁄4 | 6.2 | 1.81 | 2.91 | 1.03 | 1.27 | 1.16 | 2.09 | .808 | 1.07 | .909 | .734 | .759 |
| 4×3 | 5⁄8 | 13.6 | 3.98 | 6.03 | 2.30 | 1.23 | 1.37 | 2.87 | 1.35 | .849 | .871 | .637 | .534 |
| | 1⁄2 | 11.1 | 3.25 | 5.05 | 1.89 | 1.25 | 1.33 | 2.42 | 1.12 | .864 | .827 | .639 | .543 |
| | 7⁄16 | 9.8 | 2.87 | 4.52 | 1.68 | 1.25 | 1.30 | 2.18 | .992 | .871 | .804 | .641 | .547 |
| | 3⁄8 | 8.5 | 2.48 | 3.96 | 1.46 | 1.26 | 1.28 | 1.92 | .866 | .879 | .782 | .644 | .551 |
| | 5⁄16 | 7.2 | 2.09 | 3.38 | 1.23 | 1.27 | 1.26 | 1.65 | .734 | .887 | .759 | .647 | .554 |
| | 1⁄4 | 5.8 | 1.69 | 2.77 | 1.00 | 1.28 | 1.24 | 1.36 | .599 | .896 | .736 | .651 | .558 |
| 3½×3 | 1⁄2 | 10.2 | 3.00 | 3.45 | 1.45 | 1.07 | 1.13 | 2.33 | 1.10 | .881 | .875 | .621 | .714 |
| | 7⁄16 | 9.1 | 2.65 | 3.10 | 1.29 | 1.08 | 1.10 | 2.09 | .975 | .889 | .853 | .622 | .718 |
| | 3⁄8 | 7.9 | 2.30 | 2.72 | 1.13 | 1.09 | 1.08 | 1.85 | .851 | .897 | .830 | .625 | .721 |
| | 5⁄16 | 6.6 | 1.93 | 2.33 | .954 | 1.10 | 1.06 | 1.58 | .722 | .905 | .808 | .627 | .724 |
| | 1⁄4 | 5.4 | 1.56 | 1.91 | .776 | 1.11 | 1.04 | 1.30 | .589 | .914 | .785 | .631 | .727 |
| 3½×2½ | 1⁄2 | 9.4 | 2.75 | 3.24 | 1.41 | 1.09 | 1.20 | 1.36 | .760 | .704 | .705 | .534 | .486 |
| | 7⁄16 | 8.3 | 2.43 | 2.91 | 1.26 | 1.09 | 1.18 | 1.23 | .677 | .711 | .682 | .535 | .491 |
| | 3⁄8 | 7.2 | 2.11 | 2.56 | 1.09 | 1.10 | 1.16 | 1.09 | .592 | .719 | .660 | .537 | .496 |
| | 5⁄16 | 6.1 | 1.78 | 2.19 | .927 | 1.11 | 1.14 | .939 | .504 | .727 | .637 | .540 | .501 |
| | 1⁄4 | 4.9 | 1.44 | 1.80 | .755 | 1.12 | 1.11 | .777 | .412 | .735 | .614 | .544 | .506 |
| 3×2½ | 1⁄2 | 8.5 | 2.50 | 2.08 | 1.04 | .913 | 1.00 | 1.30 | .744 | .722 | .750 | .520 | .667 |
| | 7⁄16 | 7.6 | 2.21 | 1.88 | .928 | .920 | .978 | 1.18 | .664 | .729 | .728 | .521 | .672 |
| | 3⁄8 | 6.6 | 1.92 | 1.66 | .810 | .928 | .956 | 1.04 | .581 | .736 | .706 | .522 | .676 |
| | 5⁄16 | 5.6 | 1.62 | 1.42 | .688 | .937 | .933 | .898 | .494 | .744 | .683 | .525 | .680 |
| | 1⁄4 | 4.5 | 1.31 | 1.17 | .561 | .945 | .911 | .743 | .404 | .753 | .661 | .528 | .684 |
| | 3⁄16 | 3.39 | .996 | .907 | .430 | .954 | .888 | .577 | .310 | .761 | .638 | .533 | .688 |
| 3×2 | 1⁄2 | 7.7 | 2.25 | 1.92 | 1.00 | .924 | 1.08 | .672 | .474 | .546 | .583 | .428 | .414 |
| | 7⁄16 | 6.8 | 2.00 | 1.73 | .894 | .932 | 1.06 | .609 | .424 | .553 | .561 | .429 | .421 |
| | 3⁄8 | 5.9 | 1.73 | 1.53 | .781 | .940 | 1.04 | .543 | .371 | .559 | .539 | .430 | .428 |
| | 5⁄16 | 5.0 | 1.46 | 1.32 | .664 | .948 | 1.02 | .740 | .317 | .567 | .516 | .432 | .435 |
| | 1⁄4 | 4.1 | 1.19 | 1.09 | .542 | .957 | .993 | .392 | .260 | .574 | .493 | .435 | .440 |
| | 3⁄16 | 3.07 | .902 | .842 | .415 | .966 | .970 | .307 | .200 | .583 | .470 | .439 | .446 |
| 2½×2 | 3⁄8 | 5.3 | 1.55 | .912 | .547 | .768 | .831 | .514 | .363 | .577 | .581 | .420 | .614 |
| | 5⁄16 | 4.5 | 1.31 | .788 | .466 | .776 | .809 | .446 | .310 | .584 | .559 | .422 | .620 |
| | 1⁄4 | 3.62 | 1.06 | .654 | .381 | .784 | .787 | .372 | .254 | .592 | .537 | .424 | .626 |
| | 3⁄16 | 2.75 | .809 | .509 | .293 | .793 | .764 | .291 | .196 | .600 | .514 | .427 | .631 |

*Symbols:* $I$ = moment of inertia; $S$ = section modulus; $r$ = radius of gyration; $x$ = distance from center of gravity of section to outer face of structural shape.

Data taken from the "Manual of Steel Construction," 8th Edition, 1980, with permission of the American Institute of Steel Construction.

## Table 7. Aluminum Association Standard Structural Shapes

**I-Beams**

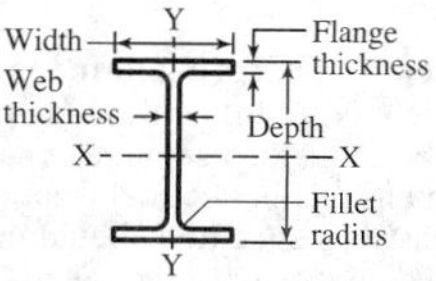


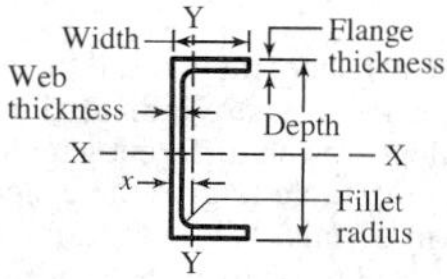


**Channels**

| Depth | Width | Weight per Foot | Area | Flange Thickness | Web Thickness | Fillet Radius | Axis XX *I* | Axis XX *S* | Axis XX *r* | Axis YY *I* | Axis YY *S* | Axis YY *r* | *x* |
|---|---|---|---|---|---|---|---|---|---|---|---|---|---|
| inch | inch | lb. | inch² | inch | inch | inch | inch⁴ | inch³ | inch | inch⁴ | inch³ | inch | inch |
| I-BEAMS | | | | | | | | | | | | | |
| 3.00 | 2.50 | 1.637 | 1.392 | 0.20 | 0.13 | 0.25 | 2.24 | 1.49 | 1.27 | 0.52 | 0.42 | 0.61 | – |
| 3.00 | 2.50 | 2.030 | 1.726 | 0.26 | 0.15 | 0.25 | 2.71 | 1.81 | 1.25 | 0.68 | 0.54 | 0.63 | – |
| 4.00 | 3.00 | 2.311 | 1.965 | 0.23 | 0.15 | 0.25 | 5.62 | 2.81 | 1.69 | 1.04 | 0.69 | 0.73 | – |
| 4.00 | 3.00 | 2.793 | 2.375 | 0.29 | 0.17 | 0.25 | 6.71 | 3.36 | 1.68 | 1.31 | 0.87 | 0.74 | – |
| 5.00 | 3.50 | 3.700 | 3.146 | 0.32 | 0.19 | 0.30 | 13.94 | 5.58 | 2.11 | 2.29 | 1.31 | 0.85 | – |
| 6.00 | 4.00 | 4.030 | 3.427 | 0.29 | 0.19 | 0.30 | 21.99 | 7.33 | 2.53 | 3.10 | 1.55 | 0.95 | – |
| 6.00 | 4.00 | 4.692 | 3.990 | 0.35 | 0.21 | 0.30 | 25.50 | 8.50 | 2.53 | 3.74 | 1.87 | 0.97 | – |
| 7.00 | 4.50 | 5.800 | 4.932 | 0.38 | 0.23 | 0.30 | 42.89 | 12.25 | 2.95 | 5.78 | 2.57 | 1.08 | – |
| 8.00 | 5.00 | 6.181 | 5.256 | 0.35 | 0.23 | 0.30 | 59.69 | 14.92 | 3.37 | 7.30 | 2.92 | 1.18 | – |
| 8.00 | 5.00 | 7.023 | 5.972 | 0.41 | 0.25 | 0.30 | 67.78 | 16.94 | 3.37 | 8.55 | 3.42 | 1.20 | – |
| 9.00 | 5.50 | 8.361 | 7.110 | 0.44 | 0.27 | 0.30 | 102.02 | 22.67 | 3.79 | 12.22 | 4.44 | 1.31 | – |
| 10.00 | 6.00 | 8.646 | 7.352 | 0.41 | 0.25 | 0.40 | 132.09 | 26.42 | 4.24 | 14.78 | 4.93 | 1.42 | – |
| 10.00 | 6.00 | 10.286 | 8.747 | 0.50 | 0.29 | 0.40 | 155.79 | 31.16 | 4.22 | 18.03 | 6.01 | 1.44 | – |
| 12.00 | 7.00 | 11.672 | 9.925 | 0.47 | 0.29 | 0.40 | 255.57 | 42.60 | 5.07 | 26.90 | 7.69 | 1.65 | – |
| 12.00 | 7.00 | 14.292 | 12.153 | 0.62 | 0.31 | 0.40 | 317.33 | 52.89 | 5.11 | 35.48 | 10.14 | 1.71 | – |
| CHANNELS | | | | | | | | | | | | | |
| 2.00 | 1.00 | 0.577 | 0.491 | 0.13 | 0.13 | 0.10 | 0.288 | 0.288 | 0.766 | 0.045 | 0.064 | 0.303 | 0.2[illegible]8 |
| 2.00 | 1.25 | 1.071 | 0.911 | 0.26 | 0.17 | 0.15 | 0.546 | 0.546 | 0.774 | 0.139 | 0.178 | 0.391 | 0.4[illegible]1 |
| 3.00 | 1.50 | 1.135 | 0.965 | 0.20 | 0.13 | 0.25 | 1.41 | 0.94 | 1.21 | 0.22 | 0.22 | 0.47 | 0.[illegible]9 |
| 3.00 | 1.75 | 1.597 | 1.358 | 0.26 | 0.17 | 0.25 | 1.97 | 1.31 | 1.20 | 0.42 | 0.37 | 0.55 | 0.[illegible]2 |
| 4.00 | 2.00 | 1.738 | 1.478 | 0.23 | 0.15 | 0.25 | 3.91 | 1.95 | 1.63 | 0.60 | 0.45 | 0.64 | 0.[illegible]5 |
| 4.00 | 2.25 | 2.331 | 1.982 | 0.29 | 0.19 | 0.25 | 5.21 | 2.60 | 1.62 | 1.02 | 0.69 | 0.72 | 0.[illegible]8 |
| 5.00 | 2.25 | 2.212 | 1.881 | 0.26 | 0.15 | 0.30 | 7.88 | 3.15 | 2.05 | 0.98 | 0.64 | 0.72 | 0.[illegible]3 |
| 5.00 | 2.75 | 3.089 | 2.627 | 0.32 | 0.19 | 0.30 | 11.14 | 4.45 | 2.06 | 2.05 | 1.14 | 0.88 | 0.[illegible]5 |
| 6.00 | 2.50 | 2.834 | 2.410 | 0.29 | 0.17 | 0.30 | 14.35 | 4.78 | 2.44 | 1.53 | 0.90 | 0.80 | 0.[illegible]9 |
| 6.00 | 3.25 | 4.030 | 3.427 | 0.35 | 0.21 | 0.30 | 21.04 | 7.01 | 2.48 | 3.76 | 1.76 | 1.05 | 1.[illegible]2 |
| 7.00 | 2.75 | 3.205 | 2.725 | 0.29 | 0.17 | 0.30 | 22.09 | 6.31 | 2.85 | 2.10 | 1.10 | 0.88 | 0.[illegible]4 |
| 7.00 | 3.50 | 4.715 | 4.009 | 0.38 | 0.21 | 0.30 | 33.79 | 9.65 | 2.90 | 5.13 | 2.23 | 1.13 | 1.[illegible]0 |
| 8.00 | 3.00 | 4.147 | 3.526 | 0.35 | 0.19 | 0.30 | 37.40 | 9.35 | 3.26 | 3.25 | 1.57 | 0.96 | 0.[illegible]3 |
| 8.00 | 3.75 | 5.789 | 4.923 | 0.41 | 0.25 | 0.35 | 52.69 | 13.17 | 3.27 | 7.13 | 2.82 | 1.20 | 1.[illegible]2 |
| 9.00 | 3.25 | 4.983 | 4.237 | 0.35 | 0.23 | 0.35 | 54.41 | 12.09 | 3.58 | 4.40 | 1.89 | 1.02 | 0.[illegible]3 |
| 9.00 | 4.00 | 6.970 | 5.927 | 0.44 | 0.29 | 0.35 | 78.31 | 17.40 | 3.63 | 9.61 | 3.49 | 1.27 | 1.[illegible]5 |
| 10.00 | 3.50 | 6.136 | 5.218 | 0.41 | 0.25 | 0.35 | 83.22 | 16.64 | 3.99 | 6.33 | 2.56 | 1.10 | 1.[illegible]2 |
| 10.00 | 4.25 | 8.360 | 7.109 | 0.50 | 0.31 | 0.40 | 116.15 | 23.23 | 4.04 | 13.02 | 4.47 | 1.35 | 1.[illegible]4 |
| 12.00 | 4.00 | 8.274 | 7.036 | 0.47 | 0.29 | 0.40 | 159.76 | 26.63 | 4.77 | 11.03 | 3.86 | 1.25 | 1.[illegible]4 |
| 12.00 | 5.00 | 11.822 | 10.053 | 0.62 | 0.35 | 0.45 | 239.69 | 39.95 | 4.88 | 25.74 | 7.60 | 1.60 | 1.[illegible]1 |

Structural sections are available in 6061-T6 aluminum alloy. Data supplied by The Aluminum Association.

**Table 5a. Decimal Equivalent of Numbered Drill Sizes**

| Drill Number | Decimal Inch | Decimal mm | Drill Number | Decimal Inch | Decimal mm | Drill Number | Decimal Inch | Decimal mm | Drill Number | Decimal Inch | Decimal mm |
|---|---|---|---|---|---|---|---|---|---|---|---|
| 1 | 0.2280 | 5.791 | 26 | 0.1470 | 3.734 | 51 | 0.0670 | 1.702 | 76 | 0.0200 | 0.508 |
| 2 | 0.2210 | 5.613 | 27 | 0.1440 | 3.658 | 52 | 0.0635 | 1.613 | 77 | 0.0180 | 0.457 |
| 3 | 0.2130 | 5.410 | 28 | 0.1405 | 3.569 | 53 | 0.0595 | 1.511 | 78 | 0.0160 | 0.406 |
| 4 | 0.2090 | 5.309 | 29 | 0.1360 | 3.454 | 54 | 0.0550 | 1.397 | 79 | 0.0145 | 0.368 |
| 5 | 0.2055 | 5.220 | 30 | 0.1285 | 3.264 | 55 | 0.0520 | 1.321 | 80 | 0.0135 | 0.343 |
| 6 | 0.2040 | 5.182 | 31 | 0.1200 | 3.048 | 56 | 0.0465 | 1.181 | 81 | 0.0130 | 0.330 |
| 7 | 0.2010 | 5.105 | 32 | 0.1160 | 2.946 | 57 | 0.0430 | 1.092 | 82 | 0.0125 | 0.318 |
| 8 | 0.1990 | 5.054 | 33 | 0.1130 | 2.870 | 58 | 0.0420 | 1.067 | 83 | 0.0120 | 0.305 |
| 9 | 0.1960 | 4.978 | 34 | 0.1110 | 2.819 | 59 | 0.0410 | 1.041 | 84 | 0.0115 | 0.292 |
| 10 | 0.1935 | 4.915 | 35 | 0.1100 | 2.794 | 60 | 0.0400 | 1.016 | 85 | 0.0110 | 0.280 |
| 11 | 0.1910 | 4.851 | 36 | 0.1065 | 2.705 | 61 | 0.0390 | 0.991 | 86 | 0.0105 | 0.267 |
| 12 | 0.1890 | 4.800 | 37 | 0.1040 | 2.642 | 62 | 0.0380 | 0.965 | 87 | 0.0100 | 0.254 |
| 13 | 0.1850 | 4.700 | 38 | 0.1015 | 2.578 | 63 | 0.0370 | 0.940 | 88 | 0.0095 | 0.241 |
| 14 | 0.1820 | 4.623 | 39 | 0.0995 | 2.527 | 64 | 0.0360 | 0.914 | 89 | 0.0091 | 0.231 |
| 15 | 0.1800 | 4.572 | 40 | 0.0980 | 2.489 | 65 | 0.0350 | 0.889 | 90 | 0.0087 | 0.221 |
| 16 | 0.1770 | 4.496 | 41 | 0.0960 | 2.438 | 66 | 0.0330 | 0.838 | 91 | 0.0083 | 0.211 |
| 17 | 0.1730 | 4.394 | 42 | 0.0935 | 2.375 | 67 | 0.0320 | 0.813 | 92 | 0.0079 | 0.200 |
| 18 | 0.1695 | 4.305 | 43 | 0.0890 | 2.261 | 68 | 0.0310 | 0.787 | 93 | 0.0075 | 0.190 |
| 19 | 0.1660 | 4.216 | 44 | 0.0860 | 2.184 | 69 | 0.0292 | 0.742 | 94 | 0.0071 | 0.180 |
| 20 | 0.1610 | 4.089 | 45 | 0.0820 | 2.083 | 70 | 0.0280 | 0.711 | 95 | 0.0067 | 0.170 |
| 21 | 0.1590 | 4.039 | 46 | 0.0810 | 2.057 | 71 | 0.0260 | 0.660 | 96 | 0.0063 | 0.160 |
| 22 | 0.1570 | 3.988 | 47 | 0.0785 | 1.994 | 72 | 0.0250 | 0.635 | 97 | 0.0059 | 0.150 |
| 23 | 0.1540 | 3.912 | 48 | 0.0760 | 1.930 | 73 | 0.0240 | 0.610 | … | … | … |
| 24 | 0.1520 | 3.861 | 49 | 0.0730 | 1.854 | 74 | 0.0225 | 0.572 | … | … | … |
| 25 | 0.1495 | 3.797 | 50 | 0.0700 | 1.778 | 75 | 0.0210 | 0.533 | … | … | … |

**Table 5b. Decimal Equivalent of Letter Drill Sizes**

| Drill Size | Decimal Inch | Decimal mm | Drill Size | Decimal Inch | Decimal mm | Drill Size | Decimal Inch | Decimal mm | Drill Size | Decimal Inch | Decimal mm | Drill Size | Decimal Inch | Decimal mm |
|---|---|---|---|---|---|---|---|---|---|---|---|---|---|---|
| A | 0.234 | 5.944 | G | 0.261 | 6.629 | M | 0.295 | 7.493 | S | 0.348 | 8.839 | Y | 0.404 | 10.262 |
| B | 0.238 | 6.045 | H | 0.266 | 6.756 | N | 0.302 | 7.671 | T | 0.358 | 9.093 | Z | 0.413 | 10.490 |
| C | 0.242 | 6.147 | I | 0.272 | 6.909 | O | 0.316 | 8.026 | U | 0.368 | 9.347 | … | … | … |
| D | 0.246 | 6.248 | J | 0.277 | 7.036 | P | 0.323 | 8.204 | V | 0.377 | 9.576 | … | … | … |
| E | 0.250 | 6.350 | K | 0.281 | 7.137 | Q | 0.332 | 8.433 | W | 0.386 | 9.804 | … | … | … |
| F | 0.257 | 6.528 | L | 0.290 | 7.366 | R | 0.339 | 8.611 | X | 0.397 | 10.084 | … | … | … |

**Wire Drawing History.—**Wire drawing is the process used for producing wire by drawing a wire or rod of larger diameter through a plate or die provided with a hole which reduces the size to the desired dimension. Previous to wire drawing, wire was made by hammering or beating metal into thin sheets or plates, which were cut into continuous strips. These strips were then afterwards rounded by hammering. There is evidence that wire was produced by a simple hammering process as early as 2000 B.C., and perhaps earlier. The history of the art does not clearly indicate when simple hammering was superseded by the method of drawing through a die, but records are available to show that wire drawing on a commercial scale was practiced in France in 1270, in Germany in 1350, and in England in 1466. The first wire-drawing mill in America was built in 1775, in Norwich, Connecticut, by Nathaniel Niles, who was granted a loan of $1500 by the court for this purpose.

**Shaped Wire.—**Shaped wire is any wire not round in section; it may be square, rectangular, hexagonal, triangular, half-round, oval, etc. Some sections of shaped wire may be either rolled or drawn. Shaped wire is a specialty, subject to agreement between purchaser and manufacturer as to size, finish, temper, and weight of coils and packages.

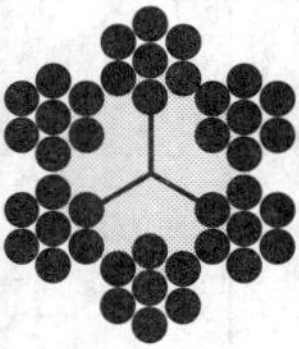

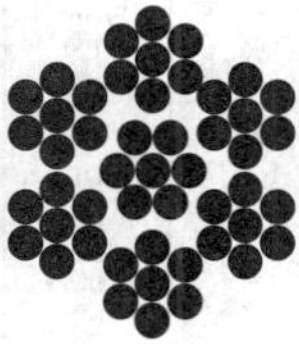

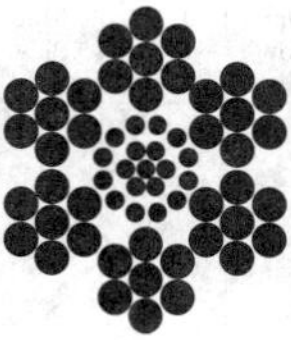

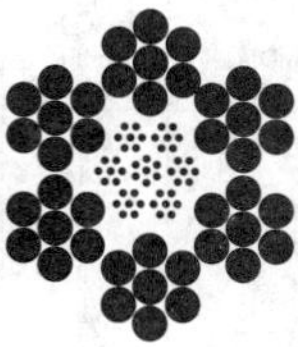

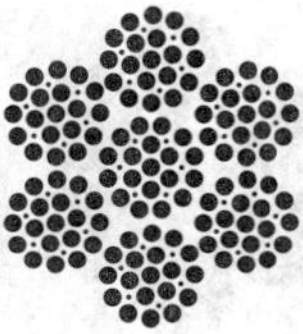

Fig. 2a.
6×25 Filler Wire
with WSC (1×25)

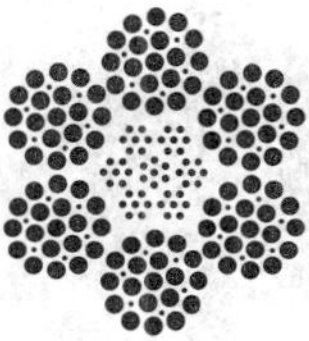

Fig. 2b.
6×25 Filler Wire
with IWRC

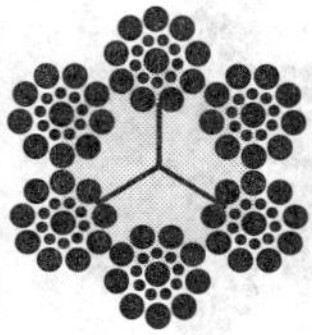

Fig. 2c.
6×19 Seale
with Fiber Core

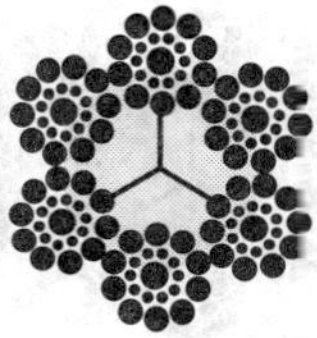

Fig. 2d.
6×21 Seale
with Fiber Core

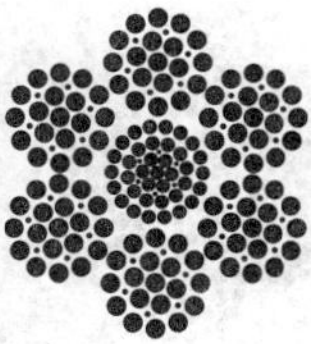

Fig. 2e.
6×25 Filler Wire
with WSC (1×43)

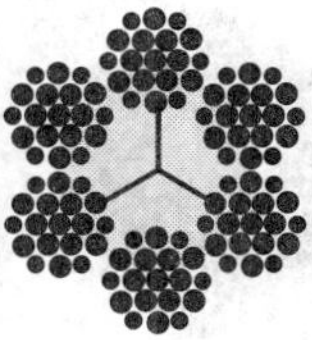

Fig. 2f.
6×19 Warrington
with Fiber Core

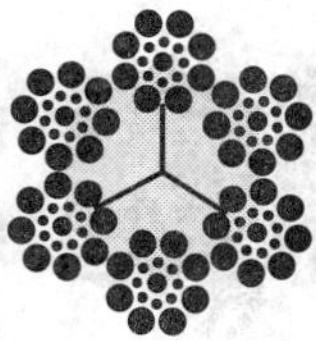

Fig. 2g.
6×17 Seale
with Fiber Core

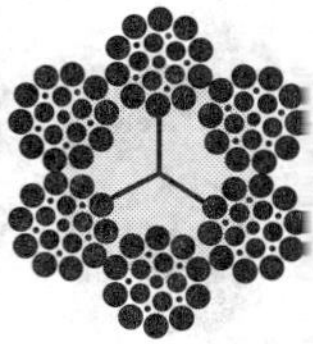

Fig. 2h.
6×21 Filler Wire
with Fiber Core

**Table 3. Weights and Strengths of 6×37 (Extra Flexible Hoisting) Wire Ropes, Preformed and Not Preformed**

| Dia., Inches | Approx. Weight per Ft., Pounds | Breaking Strength, Tons of 2000 Lbs. | | Dia., Inches | Approx. Weight per Ft., Pounds | Breaking Strength, Tons of 2000 Lbs. | |
|---|---|---|---|---|---|---|---|
| | | Impr. Plow Steel | Plow Steel | | | Impr. Plow Steel | Plow Steel |
| 1⁄4 | 0.10 | 2.59 | 2.25 | 1 1⁄2 | 3.49 | 87.9 | 76.4 |
| 5⁄16 | 0.16 | 4.03 | 3.50 | 1 5⁄8 | 4.09 | 103 | 89.3 |
| 3⁄8 | 0.22 | 5.77 | 5.02 | 1 3⁄4 | 4.75 | 119 | 103 |
| 7⁄16 | 0.30 | 7.82 | 6.80 | 1 7⁄8 | 5.45 | 136 | 118 |
| 1⁄2 | 0.39 | 10.2 | 8.85 | 2 | 6.20 | 154 | 134 |
| 9⁄16 | 0.49 | 12.9 | 11.2 | 2 1⁄8 | 7.00 | 173 | 150 |
| 5⁄8 | 0.61 | 15.8 | 13.7 | 2 1⁄4 | 7.85 | 193 | 168 |
| 3⁄4 | 0.87 | 22.6 | 19.6 | 2 1⁄2 | 9.69 | 236 | 205 |
| 7⁄8 | 1.19 | 30.6 | 26.6 | 2 3⁄4 | 11.72 | 284 | 247 |
| 1 | 1.55 | 39.8 | 34.6 | 3 | 14.0 | 335 | 291 |
| 1 1⁄8 | 1.96 | 50.1 | 43.5 | 3 1⁄4 | 16.4 | 390 | 339 |
| 1 1⁄4 | 2.42 | 61.5 | 53.5 | 3 1⁄2 | 19.0 | 449 | 390 |
| 1 3⁄8 | 2.93 | 74.1 | 64.5 | … | … | … | … |

For ropes with steel cores, add 7$^1/_2$ percent to above strengths.

For galvanized ropes, deduct 10 percent from above strengths.

*Source:* Rope diagrams, Bethlehem Steel Co. All data, US Simplified Practice Recommendation 198-50.

As shown in Table 3 and Fig. 3a through Fig. 3h, there are four common types: 6×29 filler wire construction with fiber core and 6×36 filler wire construction with independent wire rope core, a special rope for construction equipment; 6×35 (two operations) construction with fiber core and 6×41 Warrington Seale construction with fiber core, a standard crane rope in this class of rope construction; 6 × 41 filler wire construction with fiber core or

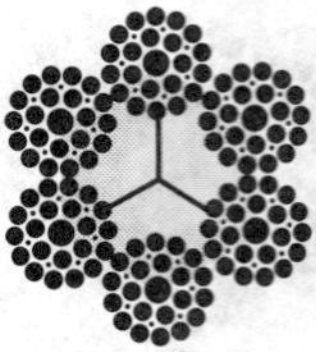

Fig. 3a.
6 × 29 Filler Wire
with Fiber Core

Fig. 3b.
6 × 36 Filler Wire
with IWRC

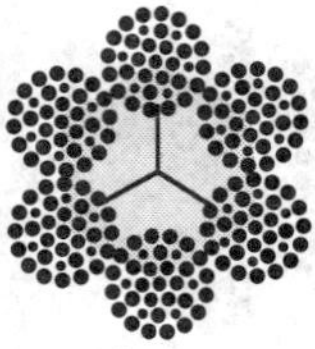

Fig. 3c.
6 × 35 with
Fiber Core

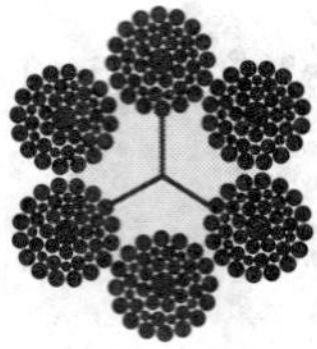

Fig. 3d.
6 × 41 Warrington-Seale
with Fiber Core

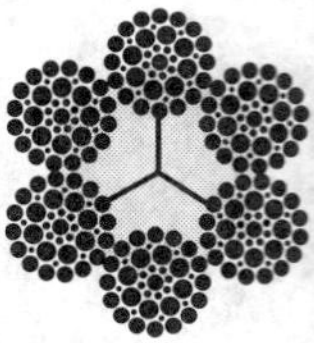

Fig. 3e.
6 × 41 Filler Wire
with Fiber Core

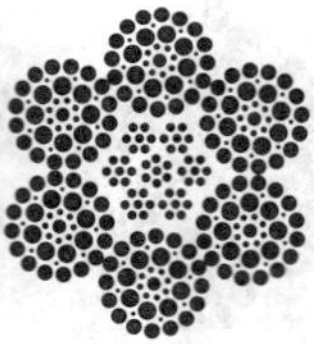

Fig. 3f.
6 × 41 Filler Wire
with IWRC

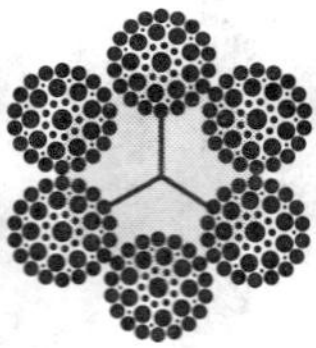

Fig. 3g.
6 × 46 Filler Wire
with Fiber Core

Fig. 3h.
6 × 46 Filler Wire
with IWRC

independent wire core, a special large shovel rope usually furnished in Lang lay; and 6 × 46 filler wire construction with fiber core or independent wire rope core, a special large shovel and dredge rope.

*8 × 19 Class (Special Flexible Hoisting Rope):* This rope is stable and smooth-running, and is especially suitable, because of its flexibility, for high-speed operation with reverse bends. Ropes in this class are available in regular lay with fiber core.

As shown in Table 4 and Fig. 4a through Fig. 4d, there are four common types: 8 × 25 filler wire construction, the most flexible but the least wear resistant rope of the four types; Warrington type in 8 × 19 construction, less flexible than the 8 × 25; 8 × 21 filler wire construction, less flexible than the Warrington; and Seale type in 8 × 19 construction, which has the greatest wear resistance of the four types but is also the least flexible.

**Table 4. Weights and Strengths of 8 × 19 (Special Flexible Hoisting) Wire Ropes, Preformed and Not Preformed**

| Dia., Inches | Approx. Weight per Ft., Pounds | Breaking Strength, Tons of 2000 Lbs. | | Dia., Inches | Approx. Weight per Ft., Pounds | Breaking Strength, Tons of 2000 Lbs. | |
|---|---|---|---|---|---|---|---|
| | | Impr. Plow Steel | Plow Steel | | | Impr. Plow Steel | Plow Steel |
| 1/4 | 0.09 | 2.35 | 2.04 | 3/4 | 0.82 | 20.5 | 17.8 |
| 5/16 | 0.14 | 3.65 | 3.18 | 7/8 | 1.11 | 27.7 | 24.1 |
| 3/8 | 0.20 | 5.24 | 4.55 | 1 | 1.45 | 36.0 | 31.3 |
| 7/16 | 0.28 | 7.09 | 6.17 | 1 1/8 | 1.84 | 45.3 | 39.4 |
| 1/2 | 0.36 | 9.23 | 8.02 | 1 1/4 | 2.27 | 55.7 | 48.4 |
| 9/16 | 0.46 | 11.6 | 10.1 | 1 3/8 | 2.74 | 67.1 | 58.3 |
| 5/8 | 0.57 | 14.3 | 12.4 | 1 1/2 | 3.26 | 79.4 | 69.1 |

For ropes with steel cores, add 7 1/2 percent to above strengths.
For galvanized ropes, deduct 10 percent from above strengths.
*Source:* Rope diagrams, Bethlehem Steel Co. All data, US Simplified Practice Recommendation 198-50.

rope is usually furnished in Lang lay and may be obtained with fiber core or independent wire rope core. The three types shown in Table 6 and Fig. 6a through Fig. 6c are flexible and are designed for hoisting work.

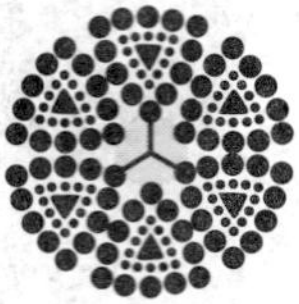

Fig. 6a.
6 × 25 with Fiber Core

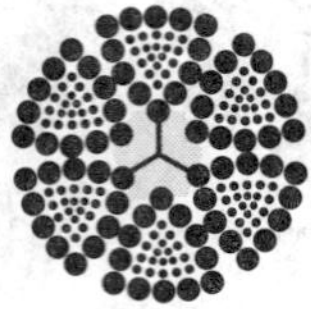

Fig. 6b.
6 × 30 with Fiber Core

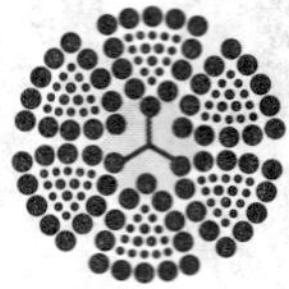

Fig. 6c.
6 × 27 with Fiber Core

**Table 6. Weights and Strengths of Flattened Strand Wire Rope, Preformed and Not Preformed**

| Dia., Inches | Approx. Weight per Ft., Pounds | Breaking Strength, Tons of 2000 Lbs. | | Dia., Inches | Approx. Weight per Ft., Pounds | Breaking Strength, Tons of 2000 Lbs. | |
|---|---|---|---|---|---|---|---|
| | | Impr. Plow Steel | Mild Plow Steel | | | Impr. Plow Steel | Mild Plow Steel |
| 3⁄8[a] | 0.25 | 6.71 | … | 1 3⁄8 | 3.40 | 85.5 | … |
| 1⁄2[a] | 0.45 | 11.8 | 8.94 | 1 1⁄2 | 4.05 | 101 | … |
| 9⁄16[a] | 0.57 | 14.9 | 11.2 | 1 5⁄8 | 4.75 | 118 | … |
| 5⁄8 | 0.70 | 18.3 | 13.9 | 1 3⁄4 | 5.51 | 136 | … |
| 3⁄4 | 1.01 | 26.2 | 19.8 | 2 | 7.20 | 176 | … |
| 7⁄8 | 1.39 | 35.4 | 26.8 | 2 1⁄4 | 9.10 | 220 | … |
| 1 | 1.80 | 46.0 | 34.8 | 2 1⁄2 | 11.2 | 269 | … |
| 1 1⁄8 | 2.28 | 57.9 | 43.8 | 2 3⁄4 | 13.6 | 321 | … |
| 1 1⁄4 | 2.81 | 71.0 | 53.7 | … | … | … | … |

[a] These sizes in Type B only.

Type H is not in US Simplified Practice Recommendation.

*Source:* Rope diagrams, Bethlehem Steel Co. All other data, US Simplified Practice Recommendation 198-50.

*Flat Wire Rope:* This type of wire rope is made up of a number of four-strand rope units placed side by side and stitched together with soft steel sewing wire. These four-strand units are alternately right and left lay to resist warping, curling, or rotating in service. Weights and strengths are shown in Table 7.

**Simplified Practice Recommendations.—**Because the total number of wire rope types is large, manufacturers and users have agreed upon and adopted a US Simplified Practice Recommendation to provide a simplified listing of those kinds and sizes of wire rope which are most commonly used and stocked. These, then, are the types and sizes which are most generally available. Other types and sizes for special or limited uses also may be found in individual manufacturer's catalogs.

**Sizes and Strengths of Wire Rope.—**The data shown in Table 1 through Table 7 have been taken from US Simplified Practice Recommendation 198-50 but do not include those wire ropes shown in that Simplified Practice Recommendation which are intended primarily for marine use.

*Wire Rope Diameter:* The diameter of a wire rope is the diameter of the circle that will just enclose it, hence when measuring the diameter with calipers, care must be taken to obtain the largest outside dimension, taken across the opposite strands, rather than the smallest dimension across opposite "valleys" or "flats." It is standard practice for the nominal diameter to be the minimum with all tolerances taken on the plus side. Limits

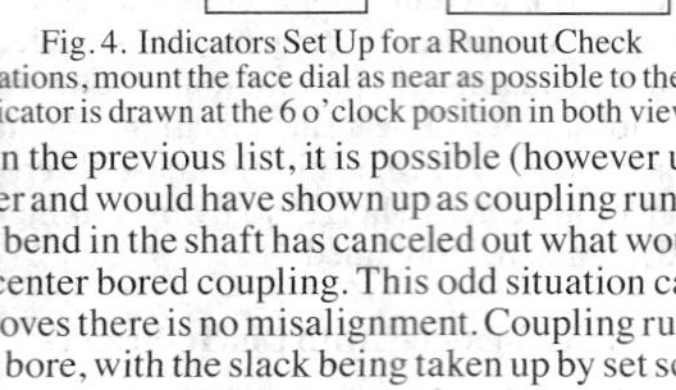

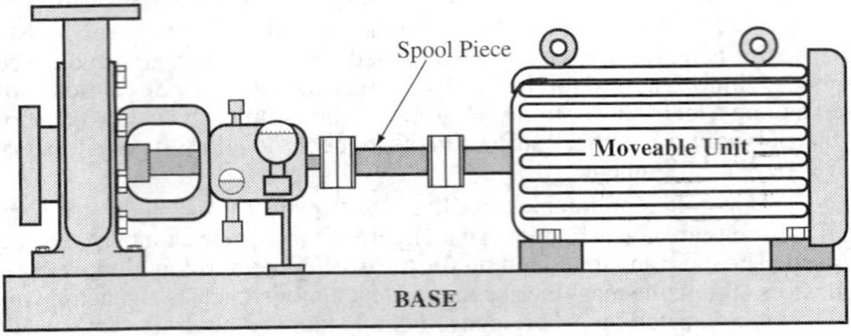


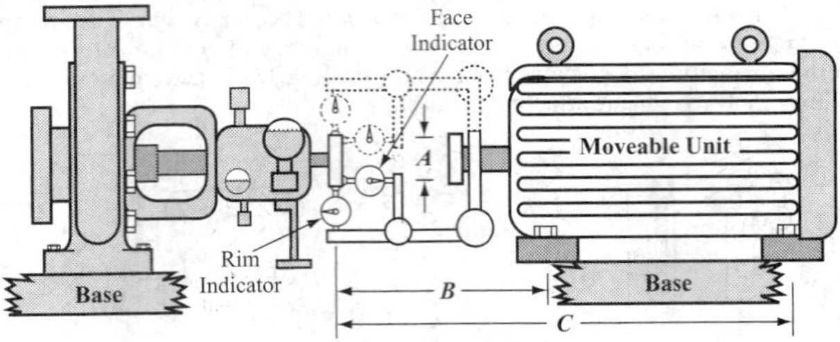

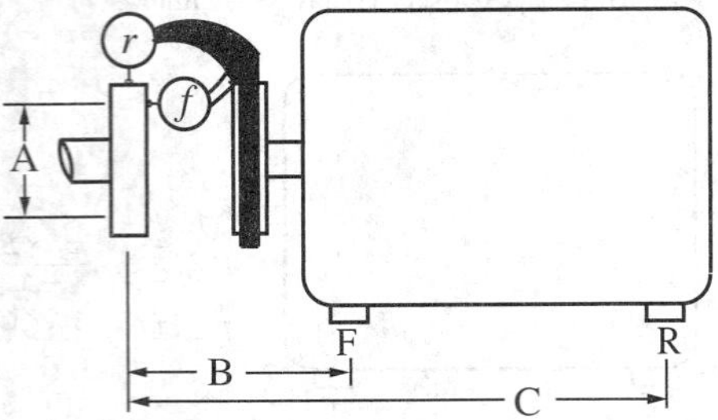


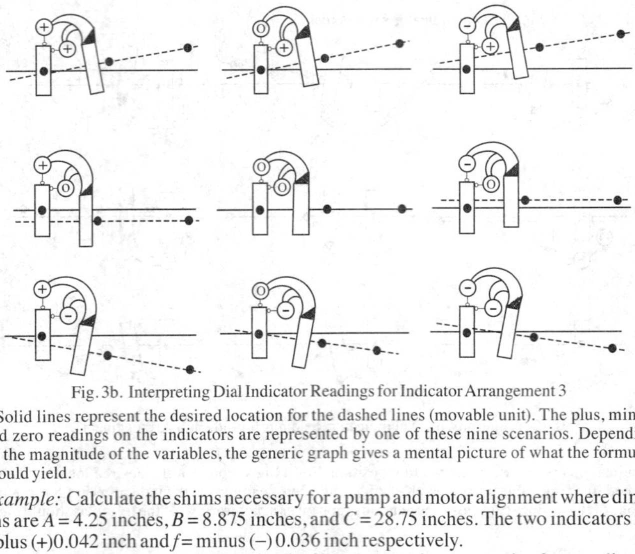

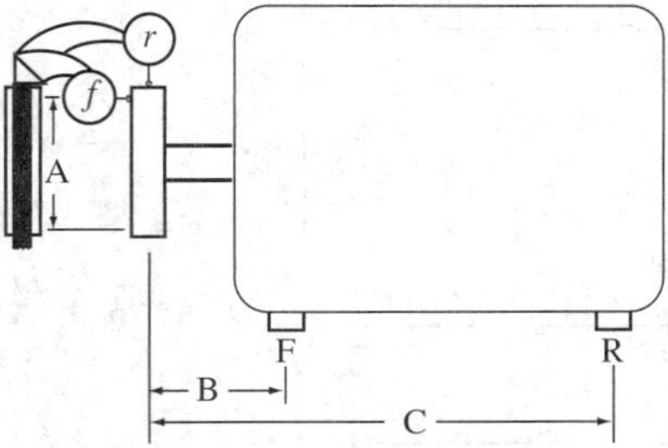


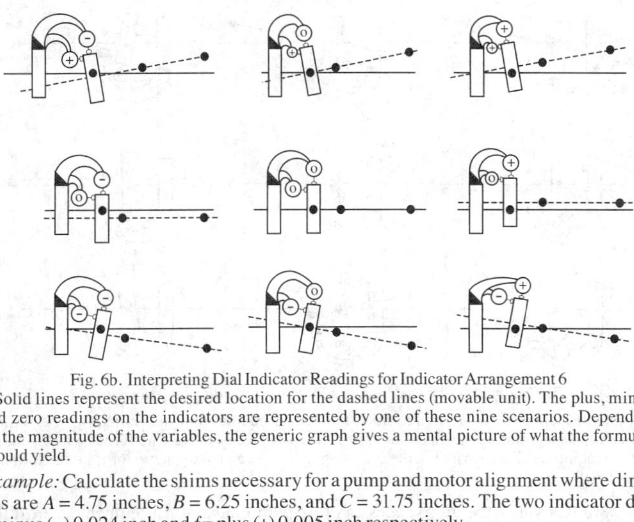

**10th Arrangement.—**Fig. 10a illustrates an arrangement used for taking two separate readings with one dial indicator.

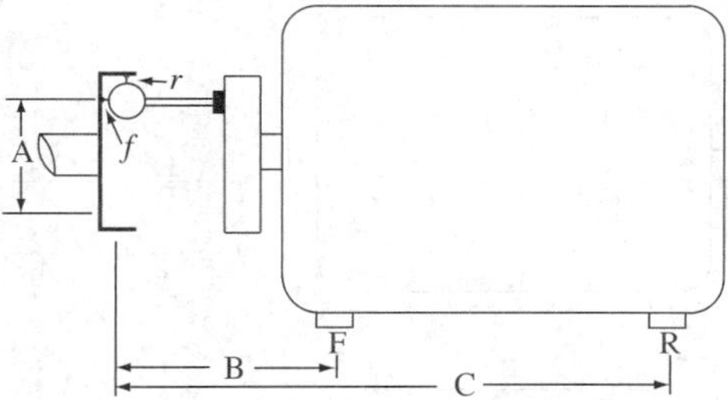


$$F = \frac{\pm r}{-2} + \frac{\pm f \times B}{-A}$$

$$R = \frac{\pm r}{-2} + \frac{\pm f \times C}{-A}$$

Fig. 10a. Dial Indicator Arrangement 10

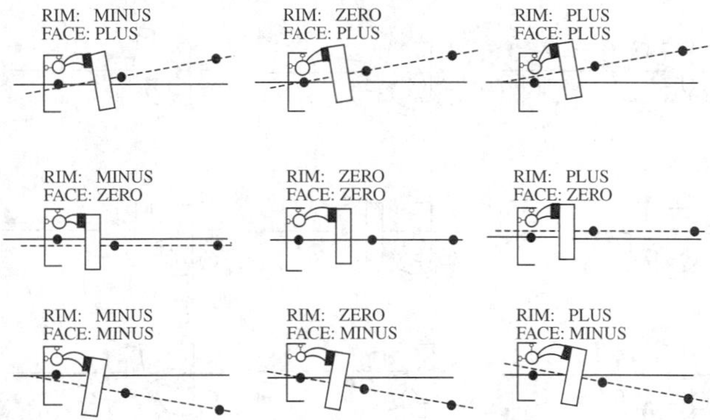


Fig. 10b. Interpreting Dial Indicator Readings for Indicator Arrangement 10

Solid lines represent the desired location for the dashed lines (movable unit). The plus, minus, and zero readings on the indicators are represented by one of these nine scenarios. Depending on the magnitude of the variables, the generic graph gives a mental picture of what the formulas should yield.

*Example:* Calculate the shims necessary for a pump and motor alignment where dimensions are $A = 5.50$ inches, $B = 14.0$ inches, and $C = 35.0$ inches. The indicator readings are $r =$ plus (+) 0.030 inch and $f =$ plus (+) 0.022 inch, respectively.

*Solution:* In Fig. 10b locate the block illustrating the positive rim and positive face condition. The graph suggests both the front and rear motor feet be lowered. The formula solution dictates exactly how much the movable unit is to be lowered at both front and rear shim points:

$$\begin{aligned} F &= \frac{\pm r}{-2} + \frac{\pm f \times B}{-A} \\ &= \frac{0.030}{-2} + \frac{0.022 \times 14}{-5.5} \\ &= -0.015 + (-0.056) \\ &= -0.071 \text{ inch} \end{aligned}$$

$$\begin{aligned} R &= \frac{\pm r}{-2} + \frac{\pm f \times C}{-A} \\ &= \frac{0.030}{-2} + \frac{0.022 \times 35}{-5.5} \\ &= -0.015 + (-0.140) \\ &= -0.155 \text{ inch} \end{aligned}$$

For ideal gases, the relationship between pressure, temperature, and specific volume can be found in the following equations. Specific volume is the volume occupied by a unit mass of a fluid.

$$PV = 1545.3nT \qquad \text{For fps units}$$

$$PV = 8314nT \qquad \text{For SI units}$$

in which $P$ = absolute pressure in pounds per square foot or Pa (N/m$^2$)

$V$ = volume in cubic feet or cubic meter

$T$ = absolute temperature in degrees R or degrees K

$n$ = number of pound moles or kg moles. A mole is the mass of substance, in appropriate units, divided by its molecular weight. The equations above are general forms that apply to any gas that behaves the ideal gas law.

**Viscosity.**—Viscosity is a liquid property that indicates its resistance to flow. It is a measure of a fluid's resistance to a shearing force. High viscosity fluids require a greater force to shear at a given rate than low viscosity fluids. This is the most important property of fluids when designing and analyzing fluid power systems. An ideal fluid has zero viscosity, and gases have nearly zero viscosity. For most fluids, viscosity varies significantly with temperature. It is relatively unaffected by changes in pressure, unless they are extreme. Newtonian fluids are those whose viscosity is independent of velocity gradient.

Viscosity index is the change in viscosity relative to change in temperature. Absolute, or dynamic, viscosity relates shearing stress in a fluid to the velocity gradient. Kinematic viscosity is the ratio of a fluid's absolute viscosity to its mass density. Absolute viscosity is often measured in centipoise (cP), and kinematic viscosity in centistokes (cSt).

Equations and conversion factors for absolute and kinematic viscosity are given below. The absolute and kinematic viscosity of various fluids can be found in Table 4.

*Absolute viscosity:* $\mu = \dfrac{F\,dy}{A\,dv}$ lbf-s/ft$^2$ or N-s/m$^2$ (Pa-s)

Units may be expressed in centipoise, cP

$\text{cP} = 2.09 \times 10^{-5}$ lbf-s/ft$^2$

$\text{cP} = 1 \times 10^{-3}$ Pa-s

*Kinematic viscosity:* $\upsilon = \dfrac{\mu}{\rho}$ ft$^2$/s or m$^2$/s

Units may be expressed in centistokes, cSt

$\text{cSt} = 1.07 \times 10^{-5}$ ft$^2$/s

$\text{cSt} = 1 \times 10^{-6}$ m$^2$/s

**Table 4. Viscosity of Common Fluids at Atmospheric Pressure**

| | Absolute Viscosity | | | Kinematic Viscosity | | |
|---|---|---|---|---|---|---|
| Fluid | lbf-s/ft$^2$ | Pa-s | cP | ft$^2$/s | m$^2$/s | cSt |
| Air | $3.75 \times 10^{-7}$ | $1.80 \times 10^{-5}$ | 0.018 | $1.64 \times 10^{-4}$ | $1.52 \times 10^{-5}$ | 15.24 |
| Nitrogen | $3.75 \times 10^{-7}$ | $1.80 \times 10^{-5}$ | 0.018 | $1.64 \times 10^{-4}$ | $1.52 \times 10^{-5}$ | 15.24 |
| Water | $2.09 \times 10^{-5}$ | $1.00 \times 10^{-3}$ | 1 | $1.07 \times 10^{-5}$ | $1.00 \times 10^{-6}$ | 1 |
| Seawater | $2.26 \times 10^{-5}$ | $1.08 \times 10^{-3}$ | 1.08 | $1.13 \times 10^{-5}$ | $1.05 \times 10^{-6}$ | 1.15 |
| Mercury | $3.34 \times 10^{-5}$ | $1.60 \times 10^{-3}$ | 1.6 | $1.27 \times 10^{-6}$ | $1.18 \times 10^{-7}$ | 0.118 |

Properties based on temperature 68° F or 20° C.

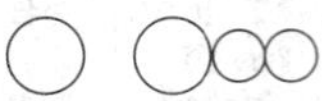 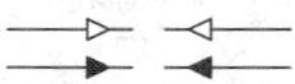 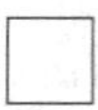 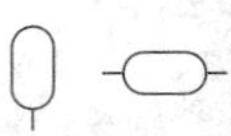 

(NPS) is given, along with the schedule or IPS designation to indicate the full dimensional characteristics of the pipe.

*Sizes:* Plastic pipe size designations are often tied to the outside diameter, as they are for steel pipe according to the NPS system. PVC has its own ASTM pipe sizing standard, while CPVC is available in either ASTM standard sizing or CPVC-CTS (Copper Tube Standard). Dimensions for schedule 40, 80, and 120 PVC pipe are given in Table 12. Alternative sizing methods are the Standard Inside Dimension Ratio (SIDR) method and the Standard Dimension Ratio (SDR) method. The SDR method uses OD and wall thickness, and SIDR uses ID and wall thickness. The SIDR method is useful when designing flow-critical applications. The SDR method is useful when pressure rating is of primary concern. PEX is most commonly sized as tube, which is covered in the section *Plastic Tube* starting on page 2782.

**Table 12. PVC Pipe Dimensions, Schedule 40, 80, and 120, Sizes Up To 12″**

| PVC Pipe Size, inch | Outside Diameter, inch | Inside Diameter, inch | | |
|---|---|---|---|---|
| | | SCH 40 | SCH 80 | SCH 120 |
| 1⁄8 | 0.405 | 0.249 | 0.195 | … |
| 1⁄4 | 0.54 | 0.344 | 0.282 | … |
| 3⁄8 | 0.675 | 0.473 | 0.403 | … |
| 1⁄2 | 0.84 | 0.602 | 0.526 | 0.48 |
| 3⁄4 | 1.05 | 0.804 | 0.722 | 0.69 |
| 1 | 1.315 | 1.029 | 0.936 | 0.891 |
| 1 1⁄4 | 1.66 | 1.36 | 1.255 | 1.204 |
| 1 1⁄2 | 1.9 | 1.59 | 1.476 | 1.423 |
| 2 | 2.375 | 2.047 | 1.913 | 1.845 |
| 2 1⁄2 | 2.875 | 2.445 | 2.29 | 2.239 |
| 3 | 3.5 | 3.042 | 2.864 | 2.758 |
| 3 1⁄2 | 4 | 3.521 | 3.326 | … |
| 4 | 4.5 | 3.998 | 3.786 | 3.574 |
| 5 | 5.563 | 5.016 | 4.768 | … |
| 6 | 6.625 | 6.031 | 5.709 | 5.434 |
| 8 | 8.625 | 7.942 | 7.565 | 7.189 |
| 10 | 10.75 | 9.976 | 9.493 | … |
| 12 | 12.75 | 11.889 | 11.294 | … |

*Pressure Ratings:* Pressure rating of plastic pipe is highly dependent on material, as well as wall thickness. Most commercially available plastic pipe will be rated by the manufacturer. PEX pressure ratings are 160 psi at 74° F and 100 psi at 180° F.

Pressure ratings for polyethylene pipes depend on the material formulation. Pressure ratings of standard sizes of PVC pipe are given in Table 13. Pressure ratings must be derated with elevated temperature. Derating factors for PVC pipe are given in Table 14. Further derating must be done to account for water hammer, and this is often 20 percent.

*Connections:* Some plastic pipe is threaded for use with threaded fittings. Threading normally follows the NPT standard. If the pipe is not threaded, a variety of other connection methods may be used. PVC pipe is usually bonded into socket fittings using solvent cement. PE pipe can be joined using heat fusion and electrofusion. In some cases, tube fittings may be used to connect plastic pipe.

*Expansion with Temperature:* Plastic pipe expands and contracts with temperature by an amount dependent on both temperature change and material. This must be considered

## Table 38. Horsepower Required to Compress Air, Three-Stage Compression

Horsepower required for compressing one cubic foot of free air per minute (isothermally and adiabatically) from atmospheric pressure (14.7 pounds per square inch) to various gage pressures.

Three-stage Compression, initial temperature of air, 60° F, jacket-cooling not considered.

| Gage Pressure, Pounds | Absolute Pressure, Pounds | Number of Atmospheres | Correct Ratio of Cylinder Volumes | Intercooler Gage Pressure, First and Second Stages | Isothermal Compression: Mean Effective Pressure[a] | Isothermal Compression: Horsepower | Adiabatic Compression: Mean Eff. Pressure,[a] Theoretical | Adiabatic Compression: Mean Eff. Pressure plus 15 per cent Friction | Adiabatic Compression: Horsepower Theoretical | Adiabatic Compression: HP plus 15 per cent Friction | Percentage of Saving over Two stage Compression |
|---|---|---|---|---|---|---|---|---|---|---|---|
| 100 | 114.7 | 7.8 | 1.98 | 14.4-42.9 | 30.30 | 0.132 | 33.30 | 38.30 | 0.145 | 0.167 | 5.13 |
| 150 | 164.7 | 11.2 | 2.24 | 18.2-59.0 | 35.59 | 0.155 | 40.30 | 46.50 | 0.175 | 0.202 | 5.12 |
| 200 | 214.7 | 14.6 | 2.44 | 21.2-73.0 | 39.50 | 0.172 | 45.20 | 52.00 | 0.196 | 0.226 | 6.07 |
| 250 | 264.7 | 18.0 | 2.62 | 23.8-86.1 | 42.70 | 0.186 | 49.20 | 56.60 | 0.214 | 0.246 | 6.95 |
| 300 | 314.7 | 21.4 | 2.78 | 26.1-98.7 | 45.30 | 0.197 | 52.70 | 60.70 | 0.229 | 0.264 | 7.23 |
| 350 | 364.7 | 24.8 | 2.92 | 28.2-110.5 | 47.30 | 0.206 | 55.45 | 63.80 | 0.242 | 0.277 | 7.64 |
| 400 | 414.7 | 28.2 | 3.04 | 30.0-121.0 | 49.20 | 0.214 | 58.25 | 66.90 | 0.253 | 0.292 | 8.13 |
| 450 | 464.7 | 31.6 | 3.16 | 31.8-132.3 | 51.20 | 0.223 | 60.40 | 69.40 | 0.263 | 0.302 | 8.35 |
| 500 | 514.7 | 35.0 | 3.27 | 33.4-142.4 | 52.70 | 0.229 | 62.30 | 71.70 | 0.273 | 0.314 | 8.38 |
| 550 | 564.7 | 38.4 | 3.38 | 35.0-153.1 | 53.75 | 0.234 | 65.00 | 74.75 | 0.283 | 0.326 | 8.80 |
| 600 | 614.7 | 41.8 | 3.47 | 36.3-162.3 | 54.85 | 0.239 | 66.85 | 76.90 | 0.291 | 0.334 | 8.85 |
| 650 | 664.7 | 45.2 | 3.56 | 37.6-171.5 | 56.00 | 0.244 | 67.90 | 78.15 | 0.296 | 0.340 | 9.02 |
| 700 | 714.7 | 48.6 | 3.65 | 38.9-180.8 | 57.15 | 0.249 | 69.40 | 79.85 | 0.303 | 0.348 | 9.13 |
| 750 | 764.7 | 52.0 | 3.73 | 40.1-189.8 | 58.10 | 0.253 | 70.75 | 81.40 | 0.309 | 0.355 | ... |
| 800 | 814.7 | 55.4 | 3.82 | 41.4-199.5 | 59.00 | 0.257 | 72.45 | 83.25 | 0.315 | 0.362 | ... |
| 850 | 864.7 | 58.8 | 3.89 | 42.5-207.8 | 60.20 | 0.262 | 73.75 | 84.90 | 0.321 | 0.369 | ... |
| 900 | 914.7 | 62.2 | 3.95 | 43.4-214.6 | 60.80 | 0.265 | 74.80 | 86.00 | 0.326 | 0.375 | ... |
| 950 | 964.7 | 65.6 | 4.03 | 44.6-224.5 | 61.72 | 0.269 | 76.10 | 87.50 | 0.331 | 0.381 | ... |
| 1000 | 1014.7 | 69.0 | 4.11 | 45.7-233.3 | 62.40 | 0.272 | 77.20 | 88.80 | 0.336 | 0.383 | ... |
| 1050 | 1064.7 | 72.4 | 4.15 | 46.3-238.3 | 63.10 | 0.275 | 78.10 | 90.10 | 0.340 | 0.391 | ... |
| 1100 | 1114.7 | 75.8 | 4.23 | 47.5-248.3 | 63.80 | 0.278 | 79.10 | 91.10 | 0.344 | 0.396 | ... |
| 1150 | 1164.7 | 79.2 | 4.30 | 48.5-256.8 | 64.40 | 0.281 | 80.15 | 92.20 | 0.349 | 0.401 | ... |
| 1200 | 1214.7 | 82.6 | 4.33 | 49.0-261.3 | 65.00 | 0.283 | 81.00 | 93.15 | 0.353 | 0.405 | ... |
| 1250 | 1264.7 | 86.0 | 4.42 | 50.3-272.3 | 65.60 | 0.286 | 82.00 | 94.30 | 0.357 | 0.411 | ... |
| 1300 | 1314.7 | 89.4 | 4.48 | 51.3-280.8 | 66.30 | 0.289 | 82.90 | 95.30 | 0.362 | 0.416 | ... |
| 1350 | 1364.7 | 92.8 | 4.53 | 52.0-287.3 | 66.70 | 0.291 | 84.00 | 96.60 | 0.366 | 0.421 | ... |
| 1400 | 1414.7 | 96.2 | 4.58 | 52.6-293.5 | 67.00 | 0.292 | 84.60 | 97.30 | 0.368 | 0.423 | ... |
| 1450 | 1464.7 | 99.6 | 4.64 | 53.5-301.5 | 67.70 | 0.295 | 85.30 | 98.20 | 0.371 | 0.426 | ... |
| 1500 | 1514.7 | 103.0 | 4.69 | 54.3-309.3 | 68.30 | 0.298 | 85.80 | 98.80 | 0.374 | 0.430 | ... |
| 1550 | 1564.7 | 106.4 | 4.74 | 55.0-317.3 | 68.80 | 0.300 | 86.80 | 99.85 | 0.378 | 0.434 | ... |
| 1600 | 1614.7 | 109.8 | 4.79 | 55.8-323.3 | 69.10 | 0.302 | 87.60 | 100.80 | 0.382 | 0.438 | ... |

[a] Mean Effective Pressure (MEP) is defined as that single pressure rise, above atmospheric, which would require the same horsepower as the actual varying pressures during compression.

## SYMBOLS AND ABBREVIATIONS

### Greek Letters and Standard Abbreviations

The Greek letters are frequently used in mathematical expressions and formulas. The Greek alphabet is given below.

| | | | | | | | | | | | |
|---|---|---|---|---|---|---|---|---|---|---|---|
| Α | α | Alpha | Η | η | Eta | Ν | ν | Nu | Τ | τ | Tau |
| Β | β | Beta | Θ | θ ϑ | Theta | Ξ | ξ | Xi | Υ | υ | Upsilon |
| Γ | γ | Gamma | Ι | ι | Iota | Ο | ο | Omicron | Φ | φ | Phi |
| Δ | δ | Delta | Κ | κ | Kappa | Π | π | Pi | Χ | χ | Chi |
| Ε | ε | Epsilon | Λ | λ | Lambda | Ρ | ρ | Rho | Ψ | ψ | Psi |
| Ζ | ζ | Zeta | Μ | μ | Mu | Σ | σ ς | Sigma | Ω | ω | Omega |

### Abbreviations for Scientific and Engineering Terms

*ANSI/ASME Y14.38-2019*

| | | | |
|---|---|---|---|
| Absolute | abs | Decibel | dB |
| Alternating current | ac | Degree | deg or ° |
| Ampere | amp | Degree Centigrade | °C |
| Ampere-hour | amp hr | Degree Fahrenheit | °F |
| **Angstrom unit** | **Å** | Degree Kelvin | K |
| Antilogarithm | antilog | Diameter | dia |
| Arithmetic average | aa | Direct current | dc |
| Atmosphere | atm | Dozen | doz |
| Atomic weight | at wt | Dram | dr |
| Avoirdupois | avdp | Efficiency | eff |
| Barometer | baro | Electric | elec |
| Board-foot (foot board measure) | fbm | Electromotive force | emf |
| Boiler pressure | bopress | Elevation | el |
| Boiling point | bp | Engine | eng |
| Brinell Hardness Number | HBW or BH or BHN | Engineer | engr |
| British thermal unit | Btu | Engineering | engrg |
| Bushel | bu | Equation | eq |
| **Calorie** | **cal** | External | ext |
| Candle | cd | Fluid | fl |
| Center to center | c to c | Foot | ft |
| Centimeter | cm | Foot-candle | fc |
| **Centimeter-gram-second** (system) | **cgs** | Foot-lambert | fL or fl |
| **Chemical** | **chem** | Foot per minute | fpm |
| **Chemically pure** | **cp** | Foot per second | fps |
| Circular | circ | Foot-pound | ft-lb |
| **Circular mil** | **cmil** | Foot-pound-second (system) | fps |
| **Coefficient** | **coef** | Free on board | fob |
| **Cologarithm** | **colog** | Freezing point | fp |
| **Concentrate** | **conc** | Frequency | freq |
| **Conductivity** | **cndct** | Fusion point | fnpt |
| **Constant** | **const** | Gallon | gal |
| **Cord** | **cd** | Gallon per minute | gpm |
| Cosecant | csc | Gallon per second | gps |
| Cosine | cos | Grain | gr |
| **Cost, insurance, and freight** | **cif** | Gram | g |
| Cotangent | cot | Greatest common divisor | gcd |
| **Counter electromotive force** | **cemf** | High pressure | hp |
| Cubic | cu | Horsepower | hp |
| Cubic centimeter | cm$^3$ or cc | Horsepower-hour | hp-hr |
| Cubic foot | ft$^3$ or cu ft | Hour | h or hr |
| Cubic feet per second | ft$^3$ or cfs | Hyperbolic cosine | cosh |
| Cubic inch | in$^3$ or cu in | Hyperbolic sine | sinh |
| Cubic meter | m$^3$ or cu m | Hyperbolic tangent | tanh |
| Cubic millimeter | mm$^3$ or cumm | Inch | in |
| Cubic yard | yd$^3$ or cu yd | Inch per second | in/s or ips |
| **Current density** | **cd** | Inch-pound | in-lb |
| Cylinder | cyl | | |

## Abbreviations for Scientific and Engineering Terms
*ANSI/ASME Y14.38-2019 (Continued)*

| | | | |
|---|---|---|---|
| Indicated horsepower-hour | ihp | Pound-force foot | $\text{lb}_f \cdot \text{ft}$ or lb-ft |
| Intermediate pressure | ip | Pound-force inch | $\text{lb}_f \cdot \text{in}$ or lb-in |
| Internal | intl | Pound-force per square foot | $\text{lb}_f/\text{ft}^2$ or psf |
| Kilovolt-ampere hour | KVA-h or kVah | Pound-force per square inch | $\text{lb}_f/\text{in}^2$ or psi |
| Kilowatt-hour meter | kwhm | Pound per horsepower | lb/hp or php |
| Latitude | lat | Power factor | pf |
| Least common multiple | lcm | Quart | qt |
| Liquid | liq | Reactive volt-ampere meter | rva |
| Logarithm (common) | log | Revolution per minute | r/min or rpm |
| Logarithm (natural) | ln | Revolution per second | r/s or rps |
| Low pressure | lp | Root mean square | rms |
| Lumen per watt | lm/W or lpw | Round | rnd |
| Magnetomotive force | mmf | Secant | sec |
| Mathematics (ical) | math | Second | s or sec |
| Maximum | max | Sine | sin |
| Mean effective pressure | mep | Specific gravity | sp gr |
| Melting point | mp | Specific heat | sp ht |
| Meter | m | Square | sq |
| Meter-kilogram-second (system) | mks | Square centimeter | $\text{cm}^2$ or sq cm |
| **Microfarad** | **µF** | Square foot | $\text{ft}^2$ or sq ft |
| Mile | mi | Square inch | $\text{in}^2$ or sq in |
| Mile per hour | mi/h or mph | Square kilometer | $\text{km}^2$ or sq km |
| **Milliampere** | **mA** | Standard | std |
| Minimum | min | Tangent | tan |
| Molecular weight | mol wt | Temperature | temp |
| Molecule | mol | Tensile strength | ts |
| National Electrical Code | NEC | Versed sine | vers |
| Ounce | oz | Volt | V |
| Ounce-inch | oz-in | Watt | W |
| Pennyweight | dwt | Watt-hour | Wh |
| Pint | pt | Week | wk |
| Potential | pot | Weight | wt |
| Potential difference | pd | Yard | yd |
| Pound | lb | | |

| Alternative abbreviations conforming to the practice of the International Electrotechnical Commission. | | | | | | | |
|---|---|---|---|---|---|---|---|
| Ampere | A | Kilovolt-ampere | kVA | Microfarad | µF | Ohm | Ω |
| Ampere-hour | Ah | Kilowatt | kW | Microwatt | µW | Volt | V |
| Coulomb | C | Kilowatt-hour | kWh | Milliampere | mA | Volt-ampere | VA |
| Farad | F | Megawatt | MW | Millifarad | mF | Volt-coulomb | VC |
| Henry | H | Megaohm | MΩ | Millihenry | mH | Watt | W |
| Joule | J | Microampere | µA | Millivolt | mV | Watthour | Wh |
| Kilovolt | kV | | | | | | |

Note: The bold terms and abbreviations are from previous versions of ANSI/ASME Y14.38-2019 or are otherwise widely used industry acronyms not included in the current standard.

In the two tables above, only the most commonly used terms have been included. These forms are recommended for those whose familiarity with the terms used makes possible a maximum of abbreviations. For others, less contracted combinations made up from this list may be used. For example, the list gives the abbreviation of the term "feet per second" as "fps." To some, however, ft per sec will be more easily understood.

Abbreviations should be used sparingly and only where their meaning will be clear. If there is any doubt, then spell out the term or unit of measurement.

The following points are good practice when preparing engineering documentation. Terms denoting units of measurement should be abbreviated in text only when preceded by the amounts indicated in numerals: "several inches," "one inch," "12 in." A sentence should not begin with a numeral followed by an abbreviation. The use of conventional signs for abbreviations in text should be avoided: use "lb," not "#" or "in," not ″.

Symbols for the chemical elements are listed in the table on page 362.

## Letter Symbols for Mechanics and Time-Related Phenomena

*ANSI/ASME Y10.3M-1984 (Withdrawn)*

| | |
|---|---|
| Acceleration, angular | $\alpha$ (alpha) |
| Acceleration, due to gravity | $g$ |
| Acceleration, linear | $a$ |
| Amplitude[a] | $A$ |
| Angle | $\alpha$ (alpha)<br>$\beta$ (beta)<br>$\gamma$ (gamma)<br>$\theta$ (theta)<br>$\phi$ (phi)<br>$\psi$ (psi) |
| Angle, solid | $\Omega$ (omega) |
| Angular frequency | $\omega$ (omega) |
| Angular momentum | $L$ |
| Angular velocity | $\omega$ (omega) |
| Arc length | $s$ |
| Area | $A$ |
| Axes, through any point[a] | $X$-$X$, $Y$-$Y$, or $Z$-$Z$ |
| Bulk modulus | $K$ |
| Breadth (width) | $b$ |
| Coefficient of expansion, linear[a] | $\alpha$ (alpha) |
| Coefficient of friction | $\mu$ (mu) |
| Concentrated load (same as force) | $F$ |
| Deflection of beam, max[a] | $\delta$ (delta) |
| Density | $\rho$ (rho) |
| Depth | $d$, $\delta$ (delta), or $t$ |
| Diameter | $D$ or $d$ |
| Displacement[a] | $u$, $v$, $w$ |
| Distance, linear[a] | $s$ |
| Eccentricity of application of load[a] | $e$ |
| Efficiency[a] | $\eta$ (eta) |
| Elasticity, modulus of | $E$ |
| Elasticity, modulus of, in shear | $G$ |
| Elongation, total[a] | $\delta$ (delta) |
| Energy, kinetic | $E_k$, $K$, $T$ |
| Energy, potential | $E_P$, $V$, or $\Phi$ (phi) |
| Factor of safety[a] | $N$ or $n$ |
| Force or load, concentrated | $F$ |
| Frequency | $f$ |
| Gyration, radius of[a] | $k$ |

| | |
|---|---|
| Height | $h$ |
| Inertia, moment of | $I$ or $J$ |
| Inertia, polar (area) moment of[a] | $J$ |
| Inertia, product (area) moment of[a] | $I_{xy}$ |
| Length | $L$ or $l$ |
| Load per unit distance[a] | $q$ or $w$ |
| Load, total[a] | $P$ or W |
| Mass | $m$ |
| Moment of force, including bending moment | $M$ |
| Neutral axis, distance to extreme fiber from[a] | $c$ |
| Period | $T$ |
| Poisson's ratio | $\mu$ (mu) or $\nu$ (nu) |
| Power | $P$ |
| Pressure, normal force per unit area | $p$ |
| Radius | $r$ |
| Revolutions per unit of time | $n$ |
| Second moment of area (second axial moment of area) | $I_a$ |
| Second polar moment of area | $I_P$ or $J$ |
| Section modulus | $Z$ |
| Shear force in beam section[a] | $V$ |
| Spring constant (load per unit deflection)[a] | $k$ |
| Statical moment of any area about a given axis[a] | $Q$ |
| Strain, normal | $\varepsilon$ (epsilon) |
| Strain, shear | $\gamma$ (gamma) |
| Stress, concentration factor[a] | $K$ |
| Stress, normal | $\sigma$ (sigma) |
| Stress, shear | $\tau$ (tau) |
| Temperature, absolute[b] | $T$, or $\theta$ (theta) |
| Temperature[b] | $t$, or $\theta$ (theta) |
| Thickness | $d$, $\delta$ (delta), or $t$ |
| Time | $t$ |
| Torque | $T$ |
| Velocity, linear | $v$ |
| Volume | $V$ |
| Wavelength | $\lambda$ (lambda) |
| Weight | $W$ |
| Weight per unit volume | $\gamma$ (gamma) |
| Work | $W$ |

[a] Not specified in Standard

[b] Specified in ANSI Y10.4-1982 (R1988)

### Table 5. Units Outside SI, Accepted for Use with SI

| Name | Symbol | Value in SI Units |
|---|---|---|
| minute | min | 1 min = 60 s |
| hour | h | 1 h = 60 min = 3600 s |
| day | d | 1 d = 24 h = 1440 min = 86400 s |
| liter | L | 1 L = 1 dm$^3$ = $10^{-3}$ m |
| metric ton | t | 1 t = $10^3$ kg = 2205 lb |
| bel | B | 1 B = 10 dB |
| degree (angle) | ° | 1° = π/180 rad |
| minute (angle) | ′ | 1′ = (1/60)° = (π/10800) rad |
| second (angle) | ″ | 1″ = (1/60)′ = (π/6480000) rad |
| electron volt | eV | 1 eV = 1.60218 × $10^{-19}$ J |
| unified atomic mass unit | Da or u | 1 u = 1.66054 × $10^{-27}$ kg |
| astronomical unit | au | 1 au = 1.49598 × $10^{11}$ m |
| nautical mile | nmi | 1 nmi = 1852 m |
| knot | kn | 1 kn = 1 nmi·h$^{-1}$ = 0.514444 m·s$^{-1}$ |
| are | a | 1 a = 100 m$^2$ |
| hectare | ha | 1 ha = 100 a = $10^4$ m$^2$ |
| bar | bar | 1 bar = $10^2$ kPa = $10^5$ Pa |
| ångström | Å | 1 Å = 0.1 nm = $10^{-10}$ m |
| curie | Ci | 1 Ci = 3.7 × $10^{10}$ Bq |
| roentgen | R | 1 R = 2.58 × $10^{-4}$ C·kg$^{-1}$ |
| rad | rad | 1 rad = $10^{-2}$ Gy |
| rem | rem | 1 rem = $10^2$ Sv |

### Table 6. SI Prefixes

| Factor | Name | Symbol | Factor | Name | Symbol |
|---|---|---|---|---|---|
| $10^1$ | deca | da | $10^{-1}$ | deci | d |
| $10^2$ | hecto | h | $10^{-2}$ | centi | c |
| $10^3$ | kilo | k | $10^{-3}$ | milli | m |
| $10^6$ | mega | M | $10^{-6}$ | micro | μ |
| $10^9$ | giga | G | $10^{-9}$ | nano | n |
| $10^{12}$ | tera | T | $10^{-12}$ | pico | p |
| $10^{15}$ | peta | P | $10^{-15}$ | femto | f |
| $10^{18}$ | exa | E | $10^{-18}$ | atto | a |

## Standard of Length and the US Customary Unit System

Among all units of measure, the history of standard of length traces a clear path from the less scientific approach of physical object standards used in past centuries to the today's precise standards, based on physical constants on an atomic level.

The primary Imperial yard was set by the British Weights and Measures Act of 1824. But it was partially destroyed in a fire in 1834, and replaced by a new standard, made of an alloy of copper, tin, and zinc. Between 1845 and 1855, forty copies of the Imperial yard were cast. Bronze yard No. 11 went to the United States, an exact copy of the British Imperial yard, in both form and material.

By an Act of Congress, in 1866, the US legally recognized the meter as a standard of length equal to 39/39.37 = 0.9144 yard; for commercial purposes, 1 meter = 39.37 inches.

## US SYSTEM AND METRIC SYSTEM CONVERSIONS

### Units of Length

**Table 1. Linear Measure Conversion Factors**

**Metric**

1 *kilometer (km)* =
- **1000** meters
- **100,000** centimeters
- **1,000,000** millimeters
- 0.539956 nautical mile
- 0.621371 mile
- 1093.61 yards
- 3280.83 feet
- 39,370.08 inches

1 *meter (m)* =
- **10** decimeters
- **100** centimeters
- **1000** millimeters
- 1.09361 yards
- 3.28084 feet
- 39.37008 inches

1 *decimeter (dm)* = **10** centimeters

1 *centimeter (cm)* =
- **0.01** meter
- **10** millimeters
- 0.0328 foot
- 0.3937 inch

1 *millimeter (mm)* =
- **0.001** meter
- **0.1** centimeter
- **1000** microns
- 0.03937 inch

1 *micrometer* or *micron (μm)* =
- **0.000001** meter = one millionth meter
- **0.0001** centimeter
- **0.001** millimeter
- 0.00003937 inch
- 39.37 micro-inches

**US Customary**

1 *mile (mi)* =
- 0.868976 nautical mile
- **1760** yards
- **5280** feet
- **63,360** inches
- **1.609344** kilometers
- **1609.344** meters
- **160,934.4** centimeters
- **1,609,344** millimeters

1 *yard (yd)* =
- **3** feet
- **36** inches
- **0.9144** meter
- **91.44** centimeter
- **914.4** millimeter

1 *foot (international) (ft)* =
- **12** inches = $\frac{1}{3}$ yard
- **0.3048** meter
- **30.48** centimeter
- **304.8** millimeters

1 *survey foot* =
- 1.000002 international feet
- $^{12}/_{39.37}$ = 0.3048006096012 meter

1 *inch (in)* =
- **1000** mils
- **1,000,000** micro-inches
- **2.54** centimeters
- **25.4** millimeters
- **25,400** microns

1 *mil* =
- **0.001** inch
- **1000** micro-inches
- **0.0254** millimeters

1 *micro-inch (μin)* =
- **0.000001** inch = one millionth inch
- **0.0254** micrometer (micron)

*Note*: Figures in **Bold** indicate exact conversion values

**Surveyors Measure**

1 *mile* = **8** furlongs = **80** chains

1 *furlong* = **10** chains = **220** yards

1 *chain* =
- **4** rods = **22** yards = **66** feet = **100** links

1 *rod* =
- **5.5** yards = **16.5** feet = **25** links
- 5.0292 meters

1 *link* = 7.92 inches

1 *span* = 9 inches

1 *hand* = 4 inches

**Nautical Measure**

1 *league* = 3 nautical miles

1 *nautical mile* =
- 1.1508 statute miles
- **6076.11549** feet
- 1.8516 kilometers

1 *fathom* = **2** yards = **6** feet

1 *knot* = nautical unit of speed =
- 1 nautical mile per hour
- 1.1508 statute miles per hour
- 1.8516 kilometers per hour

## Table 39. Pressure and Stress Conversion Factors

| To Convert ↓ / Multiply By This Factor To Obtain → | atmosphere | pascal ($N/m^2$) | $dyne/cm^2$ | bar | $kg/cm^2$ | $kg/m^2$ | psi ($lb/inch^2$) | $pound/ft^2$ | inch of water | inch of mercury | millimeter of mercury | $ton/ft^2$ (short) |
|---|---|---|---|---|---|---|---|---|---|---|---|---|
| atmosphere | 1 | 101325 | $1.0133 \times 10^{6}$ | 1.01325 | 1.03319076 | 10,331.9076 | 14.6959488 | 2,116.216 | 407.1893 | 29.9212 | 760 | 0.9597354 |
| pascal ($N/m^2$) | $9.8692 \times 10^{-6}$ | 1 | **10** | **$1 \times 10^{-5}$** | $1.01968 \times 10^{-5}$ | 0.101968 | 0.00014504 | 0.02088 | 0.004019 | 0.0002953 | 0.0075 | $9.472 \times 10^{-6}$ |
| $dyne/cm^2$ | $9.8692 \times 10^{-7}$ | **0.1** | 1 | **$1 \times 10^{-6}$** | $1.01968 \times 10^{-6}$ | 0.0101968 | $1.4504 \times 10^{-5}$ | 0.002088 | 0.000402 | $2.95 \times 10^{-5}$ | 0.00075 | $9.472 \times 10^{-7}$ |
| bar | 0.98692327 | **$1 \times 10^{5}$** | **$1 \times 10^{6}$** | 1 | 1.01968 | 10194.8 | 14.5037256 | 2088.5434 | 401.8646 | 29.5299 | 750.06168 | 0.9471852 |
| $kilogram/cm^2$ | 0.96784111 | 98,069.982 | 980,699.83 | 0.9807 | 1 | **10000** | 14.2232691 | 2048.6123 | 394.0945 | 28.9653 | 735.58536 | 0.9289043 |
| $kilogram/meter^2$ | $9.6787 \times 10^{-5}$ | 9.80699 | 98.06998 | $9.807 \times 10^{-5}$ | **0.0001** | 1 | 0.001422 | 0.204823 | 0.039409 | 0.002896 | 0.0735585 | $9.289 \times 10^{-5}$ |
| psi ($lb/inch^2$) | 0.06804596 | 6,894.7572 | 68,947.573 | 0.068947 | 0.07029148 | 703.0446 | 1 | **144** | 27.70768 | 2.03602 | 51.71493 | 0.0653061 |
| $pound/ft^2$ | $4.7254 \times 10^{-4}$ | 47.88025 | 478.80258 | 0.000478 | 0.00048813 | 4.88225 | 0.006944 | 1 | 0.19241 | 0.014139 | 0.3591314 | 0.0004535 |
| inch of water | 0.00245586 | 248.8400 | 2488.4003 | 0.002488 | 0.00253690 | 25.3737 | 0.036091 | 5.19713 | 1 | 0.073482 | 1.866453 | 0.002356 |
| inch of mercury | 0.03342112 | 3386.3949 | 33,863.949 | 0.033863 | 0.03452401 | 345.3039 | 0.491153 | 70.72632 | 13.6087 | 1 | **25.4** | 0.0320754 |
| mm of mercury | 0.00131579 | 133.32236 | 1333.22368 | 0.001333 | 0.00135921 | 13.594615 | 0.019336 | 2.784495 | 0.53577 | 0.03937 | 1 | 0.0012628 |
| $ton/ft^2$ (short) | 0.94508279 | 95760.514 | 957605.14 | 0.957605 | 0.9764854 | 9764.854 | 13.88888 | 2000 | 384.8277 | 28.27801 | 718.2616 | 1 |

## Units of Force

## Table 40. Force Conversion Factors

| To Convert ↓ / Multiply By This Factor To Obtain → | dyne | gram force | joule/cm | newton | $kg_f$ | $lb_f$ | kip | poundal | ounce force |
|---|---|---|---|---|---|---|---|---|---|
| dyne | 1 | 0.00101968 | **0.001** | **0.00001** | $1.01968 \times 10^{-6}$ | $2.24809 \times 10^{-6}$ | $2.2481 \times 10^{-9}$ | $7.233013 \times 10^{-5}$ | $3.59694 \times 10^{-5}$ |
| gram force | 980.7 | 1 | 0.9807 | 0.009807 | **0.001** | 0.0022047 | $2.2047 \times 10^{-6}$ | 0.0709341 | 0.03527521 |
| joule/cm | **1000** | 1.0196798 | 1 | **0.01** | 0.00101968 | 0.002248 | $2.2481 \times 10^{-6}$ | 0.0723301 | 0.03596942 |
| newton | **$1 \times 10^{5}$** | 101.96798 | **100** | 1 | 0.101967982 | 0.2248089 | $2.2481 \times 10^{-4}$ | 7.23301 | 3.596942 |
| kg force | $9.807 \times 10^{5}$ | **1000** | 980.7 | 9.807 | 1 | 2.2047 | 0.0022047 | 70.934129 | 35.2752102 |
| $lb_f$ | $4.4482 \times 10^{5}$ | 453.57627 | 444.822 | 4.44822 | 0.45357626 | 1 | **0.001** | 32.174038 | 16 |
| kip | $4.4482 \times 10^{8}$ | $4.5357 \times 10^{5}$ | $4.4482 \times 10^{5}$ | 4448.2224 | 453.5762688 | **1000** | 1 | 32174.038 | 16000 |
| poundal | 13825.50 | 14.097586 | 13.8255 | 0.1382555 | 0.014097586 | 0.0310809 | $3.1081 \times 10^{-5}$ | 1 | 0.497296 |
| ounce force | 27801.39 | 28.348519 | 27.8013 | 0.278013 | 0.02834852 | 0.06250 | $6.25 \times 10^{-5}$ | 2.010877 | 1 |

Figures in **bold face** indicate the conversion is exact

## Table 41a. Pounds Force to Newtons Conversion

| lbf → ↓ | 0 | 1 | 2 | 3 | 4 | 5 | 6 | 7 | 8 | 9 |
|---|---|---|---|---|---|---|---|---|---|---|
| | newtons | | | | | | | | | |
| 0 | … | 4.448 | 8.896 | 13.345 | 17.793 | 22.241 | 26.689 | 31.138 | 35.586 | 40.0[illegible]4 |
| 10 | 44.482 | 48.930 | 53.379 | 57.827 | 62.275 | 66.723 | 71.172 | 75.620 | 80.068 | 84.5[illegible]6 |
| 20 | 88.964 | 93.413 | 97.861 | 102.309 | 106.757 | 111.206 | 115.654 | 120.102 | 124.550 | 128.9[illegible]8 |
| 30 | 133.447 | 137.895 | 142.343 | 146.791 | 151.240 | 155.688 | 160.136 | 164.584 | 169.032 | 173.4[illegible]1 |
| 40 | 177.929 | 182.377 | 186.825 | 191.274 | 195.722 | 200.170 | 204.618 | 209.066 | 213.515 | 217.9[illegible]3 |
| 50 | 222.411 | 226.859 | 231.308 | 235.756 | 240.204 | 244.652 | 249.100 | 253.549 | 257.997 | 262.4[illegible]5 |
| 60 | 266.893 | 271.342 | 275.790 | 280.238 | 284.686 | 289.134 | 293.583 | 298.031 | 302.479 | 306.9[illegible]7 |
| 70 | 311.376 | 315.824 | 320.272 | 324.720 | 329.168 | 333.617 | 338.065 | 342.513 | 346.961 | 351.4[illegible]0 |
| 80 | 355.858 | 360.306 | 364.754 | 369.202 | 373.651 | 378.099 | 382.547 | 386.995 | 391.444 | 395.8[illegible]2 |
| 90 | 400.340 | 404.788 | 409.236 | 413.685 | 418.133 | 422.581 | 427.029 | 431.478 | 435.926 | 440.3[illegible]4 |
| 100 | 444.822 | 449.270 | 453.719 | 458.167 | 462.615 | 467.063 | 471.512 | 475.960 | 480.408 | 484.8[illegible]6 |

1 pound force = 4.448222 newtons

## Table 41b. Newtons to Pounds Force Conversion

| N → ↓ | 0 | 1 | 2 | 3 | 4 | 5 | 6 | 7 | 8 | 9 |
|---|---|---|---|---|---|---|---|---|---|---|
| | pounds force | | | | | | | | | |
| 0 | … | 0.22481 | 0.44962 | 0.67443 | 0.89924 | 1.12404 | 1.34885 | 1.57366 | 1.79847 | 2.023[illegible]8 |
| 10 | 2.24809 | 2.47290 | 2.69771 | 2.92252 | 3.14732 | 3.37213 | 3.59694 | 3.82175 | 4.04656 | 4.271[illegible]7 |
| 20 | 4.49618 | 4.72099 | 4.94580 | 5.17060 | 5.39541 | 5.62022 | 5.84503 | 6.06984 | 6.29465 | 6.519[illegible]6 |
| 30 | 6.74427 | 6.96908 | 7.19388 | 7.41869 | 7.64350 | 7.86831 | 8.09312 | 8.31793 | 8.54274 | 8.767[illegible]5 |
| 40 | 8.99236 | 9.21716 | 9.44197 | 9.66678 | 9.89159 | 10.1164 | 10.3412 | 10.5660 | 10.7908 | 11.01[illegible]6 |
| 50 | 11.2404 | 11.4653 | 11.6901 | 11.9149 | 12.1397 | 12.3645 | 12.5893 | 12.8141 | 13.0389 | 13.26[illegible]7 |
| 60 | 13.4885 | 13.7133 | 13.9382 | 14.1630 | 14.3878 | 14.6126 | 14.8374 | 15.0622 | 15.2870 | 15.51[illegible]8 |
| 70 | 15.7366 | 15.9614 | 16.1862 | 16.4110 | 16.6359 | 16.8607 | 17.0855 | 17.3103 | 17.5351 | 17.75[illegible]9 |
| 80 | 17.9847 | 18.2095 | 18.4343 | 18.6591 | 18.8839 | 19.1088 | 19.3336 | 19.5584 | 19.7832 | 20.00[illegible]0 |
| 90 | 20.2328 | 20.4576 | 20.6824 | 20.9072 | 21.1320 | 21.3568 | 21.5817 | 21.8065 | 22.0313 | 22.25[illegible]1 |
| 100 | 22.4809 | 22.7057 | 22.9305 | 23.1553 | 23.3801 | 23.6049 | 23.8297 | 24.0546 | 24.2794 | 24.50[illegible]2 |

1 newton = 0.2248089 pound force

# Units of Moment and Torque

## Table 42. Bending Moment or Torque Conversion Factors

| To Convert ↓ | → | dyne-centimeter | kilogram-meter | newton-millimeter | newton-meter | ounce-inch | pound-foot |
|---|---|---|---|---|---|---|---|
| dyne-centimeter | Multiply By Factor | 1 | **1 × 10⁻⁷** | **0.0001** | **1 × 10⁻⁷** | 1.416 × 10⁻⁵ | 7.375 × 10⁻⁸ |
| kilogram-meter | | 9.80665 × 10⁷ | 1 | 9806.65 | 9.80665 | 1388.78818707 | 7.2332717[illegible]2 |
| newton-millimeter | | **10,000** | 0.000101968 | 1 | **0.001** | 0.14161193 | 0.0007375[illegible]2 |
| newton-meter | | **1 × 10⁷** | 0.101967982 | **1000** | 1 | 141.61192894 | 0.7375621[illegible]1 |
| ounce-inch | | 70615.52 | 0.000720052 | 7.061552 | 0.007061552 | 1 | 0.0052083[illegible]3 |
| pound-feet | | 13,558,180 | 0.138250025 | 1355.818 | 1.355818 | 192 | 1 |

Figures in **bold face** indicate the conversion is exact

## Table 43a. Pound-Inches to Newton-Meters Conversion

| lbf-in | N•m | lbf-in | N•m | lbf-in | N•m | lbf-in | N•m | lbf-in | N•m |
|---|---|---|---|---|---|---|---|---|---|
| 100 | 11.298 | 10 | 1.130 | 1 | 0.113 | 0.1 | 0.011 | 0.01 | 0.0[illegible]1 |
| 200 | 22.597 | 20 | 2.260 | 2 | 0.226 | 0.2 | 0.023 | 0.02 | 0.0[illegible]2 |
| 300 | 33.895 | 30 | 3.390 | 3 | 0.339 | 0.3 | 0.034 | 0.03 | 0.0[illegible]3 |
| 400 | 45.194 | 40 | 4.519 | 4 | 0.452 | 0.4 | 0.045 | 0.04 | 0.0[illegible]5 |
| 500 | 56.492 | 50 | 5.649 | 5 | 0.565 | 0.5 | 0.056 | 0.05 | 0.0[illegible]6 |
| 600 | 67.791 | 60 | 6.779 | 6 | 0.678 | 0.6 | 0.068 | 0.06 | 0.0[illegible]7 |
| 700 | 79.089 | 70 | 7.909 | 7 | 0.791 | 0.7 | 0.079 | 0.07 | 0.0[illegible]8 |
| 800 | 90.388 | 80 | 9.039 | 8 | 0.904 | 0.8 | 0.090 | 0.08 | 0.0[illegible]9 |
| 900 | 101.686 | 90 | 10.169 | 9 | 1.017 | 0.9 | 0.102 | 0.09 | 0.0[illegible]0 |
| 1000 | 112.985 | 100 | 11.298 | 10 | 1.130 | 1.0 | 0.113 | 0.10 | 0.0[illegible]1 |

1 pound-inch = 0.1129848 newton-meter

**Table 53.** *(Continued)* **°C → °F and °R Temperature Conversion °F → °C and K**

| K | °C | | °F | °R | K | °C | | °F | °R | K | °C | | °F | °R |
|---|---|---|---|---|---|---|---|---|---|---|---|---|---|---|
| 325.9 | 52.8 | **127** | 260.6 | 720.3 | 357.6 | 84.4 | **184** | 363.2 | 822.9 | 741.5 | 468.3 | **875** | 1607.0 | 2066.7 |
| 326.5 | 53.3 | **128** | 262.4 | 722.1 | 358.2 | 85.0 | **185** | 365.0 | 824.7 | 755.4 | 482.2 | **900** | 1652.0 | 2111.7 |
| 327.0 | 53.9 | **129** | 264.2 | 723.9 | 358.7 | 85.6 | **186** | 366.8 | 826.5 | 769.3 | 496.1 | **925** | 1697.0 | 2156.7 |
| 327.6 | 54.4 | **130** | 266.0 | 725.7 | 359.3 | 86.1 | **187** | 368.6 | 828.3 | 783.2 | 510.0 | **950** | 1742.0 | 2201.7 |
| 328.2 | 55.0 | **131** | 267.8 | 727.5 | 359.8 | 86.7 | **188** | 370.4 | 830.1 | 797.0 | 523.9 | **975** | 1787.0 | 2246.7 |
| 328.7 | 55.6 | **132** | 269.6 | 729.3 | 360.4 | 87.2 | **189** | 372.2 | 831.9 | 810.9 | 537.8 | **1000** | 1832.0 | 2291.7 |
| 329.3 | 56.1 | **133** | 271.4 | 731.1 | 360.9 | 87.8 | **190** | 374.0 | 833.7 | 838.7 | 565.6 | **1050** | 1922.0 | 2381.7 |
| 329.8 | 56.7 | **134** | 273.2 | 732.9 | 361.5 | 88.3 | **191** | 375.8 | 835.5 | 866.5 | 593.3 | **1100** | 2012.0 | 2471.7 |
| 330.4 | 57.2 | **135** | 275.0 | 734.7 | 362.0 | 88.9 | **192** | 377.6 | 837.3 | 894.3 | 621.1 | **1150** | 2102.0 | 2561.7 |
| 330.9 | 57.8 | **136** | 276.8 | 736.5 | 362.6 | 89.4 | **193** | 379.4 | 839.1 | 922.0 | 648.9 | **1200** | 2192.0 | 2651.7 |
| 331.5 | 58.3 | **137** | 278.6 | 738.3 | 363.2 | 90.0 | **194** | 381.2 | 840.9 | 949.8 | 676.7 | **1250** | 2282.0 | 2741.7 |
| 332.0 | 58.9 | **138** | 280.4 | 740.1 | 363.7 | 90.6 | **195** | 383.0 | 842.7 | 977.6 | 704.4 | **1300** | 2372.0 | 2831.7 |
| 332.6 | 59.4 | **139** | 282.2 | 741.9 | 364.3 | 91.1 | **196** | 384.8 | 844.5 | 1005.4 | 732.2 | **1350** | 2462.0 | 2921.7 |
| 333.2 | 60.0 | **140** | 284.0 | 743.7 | 364.8 | 91.7 | **197** | 386.6 | 846.3 | 1033.2 | 760.0 | **1400** | 2552.0 | 3011.7 |
| 333.7 | 60.6 | **141** | 285.8 | 745.5 | 365.4 | 92.2 | **198** | 388.4 | 848.1 | 1060.9 | 787.8 | **1450** | 2642.0 | 3101.7 |
| 334.3 | 61.1 | **142** | 287.6 | 747.3 | 365.9 | 92.8 | **199** | 390.2 | 849.9 | 1088.7 | 815.6 | **1500** | 2732.0 | 3191.7 |
| 334.8 | 61.7 | **143** | 289.4 | 749.1 | 366.5 | 93.3 | **200** | 392.0 | 851.7 | 1116.5 | 843.3 | **1550** | 2822.0 | 3281.7 |
| 335.4 | 62.2 | **144** | 291.2 | 750.9 | 367.0 | 93.9 | **201** | 393.8 | 853.5 | 1144.3 | 871.1 | **1600** | 2912.0 | 3371.7 |
| 335.9 | 62.8 | **145** | 293.0 | 752.7 | 367.6 | 94.4 | **202** | 395.6 | 855.3 | 1172.0 | 898.9 | **1650** | 3002.0 | 3461.7 |
| 336.5 | 63.3 | **146** | 294.8 | 754.5 | 368.2 | 95.0 | **203** | 397.4 | 857.1 | 1199.8 | 926.7 | **1700** | 3092.0 | 3551.7 |
| 337.0 | 63.9 | **147** | 296.6 | 756.3 | 368.7 | 95.6 | **204** | 399.2 | 858.9 | 1227.6 | 954.4 | **1750** | 3182.0 | 3641.7 |
| 337.6 | 64.4 | **148** | 298.4 | 758.1 | 369.3 | 96.1 | **205** | 401.0 | 860.7 | 1255.4 | 982.2 | **1800** | 3272.0 | 3731.7 |
| 338.2 | 65.0 | **149** | 300.2 | 759.9 | 369.8 | 96.7 | **206** | 402.8 | 862.5 | 1283.2 | 1010.0 | **1850** | 3362.0 | 3821.7 |
| 338.7 | 65.6 | **150** | 302.0 | 761.7 | 370.4 | 97.2 | **207** | 404.6 | 864.3 | 1310.9 | 1037.8 | **1900** | 3452.0 | 3911.7 |
| 339.3 | 66.1 | **151** | 303.8 | 763.5 | 370.9 | 97.8 | **208** | 406.4 | 866.1 | 1338.7 | 1065.6 | **1950** | 3542.0 | 400[illegible].7 |
| 339.8 | 66.7 | **152** | 305.6 | 765.3 | 371.5 | 98.3 | **209** | 408.2 | 867.9 | 1366.5 | 1093.3 | **2000** | 3632.0 | 409[illegible].7 |
| 340.4 | 67.2 | **153** | 307.4 | 767.1 | 372.0 | 98.9 | **210** | 410.0 | 869.7 | 1394.3 | 1121.1 | **2050** | 3722.0 | 418[illegible].7 |
| 340.9 | 67.8 | **154** | 309.2 | 768.9 | 372.6 | 99.4 | **211** | 411.8 | 871.5 | 1422.0 | 1148.9 | **2100** | 3812.0 | 427[illegible].7 |
| 341.5 | 68.3 | **155** | 311.0 | 770.7 | 373.2 | 100.0 | **212** | 413.6 | 873.3 | 1449.8 | 1176.7 | **2150** | 3902.0 | 436[illegible].7 |
| 342.0 | 68.9 | **156** | 312.8 | 772.5 | 377.6 | 104.4 | **220** | 428.0 | 887.7 | 1477.6 | 1204.4 | **2200** | 3992.0 | 445[illegible].7 |
| 342.6 | 69.4 | **157** | 314.6 | 774.3 | 383.2 | 110.0 | **230** | 446.0 | 905.7 | 1505.4 | 1232.2 | **2250** | 4082.0 | 454[illegible].7 |
| 343.2 | 70.0 | **158** | 316.4 | 776.1 | 388.7 | 115.6 | **240** | 464.0 | 923.7 | 1533.2 | 1260.0 | **2300** | 4172.0 | 463[illegible].7 |
| 343.7 | 70.6 | **159** | 318.2 | 777.9 | 394.3 | 121.1 | **250** | 482.0 | 941.7 | 1560.9 | 1287.8 | **2350** | 4262.0 | 472[illegible].7 |
| 344.3 | 71.1 | **160** | 320.0 | 779.7 | 408.2 | 135.0 | **275** | 527.0 | 986.7 | 1588.7 | 1315.6 | **2400** | 4352.0 | 481[illegible].7 |
| 344.8 | 71.7 | **161** | 321.8 | 781.5 | 422.0 | 148.9 | **300** | 572.0 | 1031.7 | 1616.5 | 1343.3 | **2450** | 4442.0 | 490[illegible].7 |
| 345.4 | 72.2 | **162** | 323.6 | 783.3 | 435.9 | 162.8 | **325** | 617.0 | 1076.7 | 1644.3 | 1371.1 | **2500** | 4532.0 | 499[illegible].7 |
| 345.9 | 72.8 | **163** | 325.4 | 785.1 | 449.8 | 176.7 | **350** | 662.0 | 1121.7 | 1672.0 | 1398.9 | **2550** | 4622.0 | 508[illegible].7 |
| 346.5 | 73.3 | **164** | 327.2 | 786.9 | 463.7 | 190.6 | **375** | 707.0 | 1166.7 | 1699.8 | 1426.7 | **2600** | 4712.0 | 517[illegible].7 |
| 347.0 | 73.9 | **165** | 329.0 | 788.7 | 477.6 | 204.4 | **400** | 752.0 | 1211.7 | 1727.6 | 1454.4 | **2650** | 4802.0 | 526[illegible].7 |
| 347.6 | 74.4 | **166** | 330.8 | 790.5 | 491.5 | 218.3 | **425** | 797.0 | 1256.7 | 1755.4 | 1482.2 | **2700** | 4892.0 | 535[illegible].7 |
| 348.2 | 75.0 | **167** | 332.6 | 792.3 | 505.4 | 232.2 | **450** | 842.0 | 1301.7 | 1783.2 | 1510.0 | **2750** | 4982.0 | 544[illegible].7 |
| 348.7 | 75.6 | **168** | 334.4 | 794.1 | 519.3 | 246.1 | **475** | 887.0 | 1346.7 | 1810.9 | 1537.8 | **2800** | 5072.0 | 553[illegible].7 |
| 349.3 | 76.1 | **169** | 336.2 | 795.9 | 533.2 | 260.0 | **500** | 932.0 | 1391.7 | 1838.7 | 1565.6 | **2850** | 5162.0 | 562[illegible].7 |
| 349.8 | 76.7 | **170** | 338.0 | 797.7 | 547.0 | 273.9 | **525** | 977.0 | 1436.7 | 1866.5 | 1593.3 | **2900** | 5252.0 | 571[illegible].7 |
| 350.4 | 77.2 | **171** | 339.8 | 799.5 | 560.9 | 287.8 | **550** | 1022.0 | 1481.7 | 1894.3 | 1621.1 | **2950** | 5342.0 | 580[illegible].7 |
| 350.9 | 77.8 | **172** | 341.6 | 801.3 | 574.8 | 301.7 | **575** | 1067.0 | 1526.7 | 1922.0 | 1648.9 | **3000** | 5432.0 | 589[illegible].7 |
| 351.5 | 78.3 | **173** | 343.4 | 803.1 | 588.7 | 315.6 | **600** | 1112.0 | 1571.7 | 2033.2 | 1760.0 | **3200** | 5792.0 | 62[illegible]1.7 |
| 352.0 | 78.9 | **174** | 345.2 | 804.9 | 602.6 | 329.4 | **625** | 1157.0 | 1616.7 | 2144.3 | 1871.1 | **3400** | 6152.0 | 66[illegible]1.7 |
| 352.6 | 79.4 | **175** | 347.0 | 806.7 | 616.5 | 343.3 | **650** | 1202.0 | 1661.7 | 2255.4 | 1982.2 | **3600** | 6512.0 | 69[illegible]1.7 |
| 353.2 | 80.0 | **176** | 348.8 | 808.5 | 630.4 | 357.2 | **675** | 1247.0 | 1706.7 | 2366.5 | 2093.3 | **3800** | 6872.0 | 73[illegible]1.7 |
| 353.7 | 80.6 | **177** | 350.6 | 810.3 | 644.3 | 371.1 | **700** | 1292.0 | 1751.7 | 2477.6 | 2204.4 | **4000** | 7232.0 | 76[illegible]1.7 |
| 354.3 | 81.1 | **178** | 352.4 | 812.1 | 658.2 | 385.0 | **725** | 1337.0 | 1796.7 | 2588.7 | 2315.6 | **4200** | 7592.0 | 80[illegible]1.7 |
| 354.8 | 81.7 | **179** | 354.2 | 813.9 | 672.0 | 398.9 | **750** | 1382.0 | 1841.7 | 2699.8 | 2426.7 | **4400** | 7952.0 | 84[illegible]1.7 |
| 355.4 | 82.2 | **180** | 356.0 | 815.7 | 685.9 | 412.8 | **775** | 1427.0 | 1886.7 | 2810.9 | 2537.8 | **4600** | 8312.0 | 87[illegible]1.7 |
| 355.9 | 82.8 | **181** | 357.8 | 817.5 | 699.8 | 426.7 | **800** | 1472.0 | 1931.7 | 2922.0 | 2648.9 | **4800** | 8672.0 | 91[illegible]1.7 |
| 356.5 | 83.3 | **182** | 359.6 | 819.3 | 713.7 | 440.6 | **825** | 1517.0 | 1976.7 | 3033.2 | 2760.0 | **5000** | 9032.0 | 94[illegible]1.7 |
| 357.0 | 83.9 | **183** | 361.4 | 821.1 | 727.6 | 454.4 | **850** | 1562.0 | 2021.7 | … | … | … | … | … |

Table converts °C → °F and °R, or °F → °C and K. Find "convert from" temperature in **bold** column and read result from °F and °R or °C and K columns. *Example 1:* 183° C = 361.4° F and 821.1° R. *Example 2:* 183° F = 83.9° C and 357.0 K.

## Units of Velocity and Acceleration

### Table 54. Velocity Conversion Factors

| To Convert ↓ / Multiply By Factor To Obtain → | cm/sec | m/sec | km/hr | ft/sec | ft/min | ft/hr | knot[a] | mile/hr |
|---|---|---|---|---|---|---|---|---|
| cm/sec | 1 | **0.01** | **0.036** | 0.032808 | 1.9685 | 118.110236 | 0.01944 | 0.02237 |
| m/sec | **100** | 1 | **3.6** | 3.2808 | 196.8504 | 11811.0236 | 1.94384 | 2.236936 |
| km/hr | 27.77778 | 0.27778 | 1 | 0.911344 | 54.6806 | 3280.8399 | 0.53995 | 0.621371 |
| ft/sec | 30.48 | 0.3048 | 1.09728 | 1 | **60** | **3600** | 0.59248 | 0.681818 |
| ft/min | 0.5080 | 0.00508 | 0.018288 | 0.016667 | 1 | **60** | $9.8 \times 10^{-3}$ | 0.011364 |
| ft/hr | 0.008467 | $8.47 \times 10^{-5}$ | $3.05 \times 10^{-4}$ | $2.78 \times 10^{-4}$ | 0.01666 | 1 | $1.6 \times 10^{-4}$ | $1.89 \times 10^{-4}$ |
| knot | 51.444 | 0.51444 | 1.852 | 1.687808 | 101.2686 | 6076.11549 | 1 | 1.15167 |
| mile/hr | 44.704 | 0.447040 | 1.609344 | 1.466667 | **88** | **5280** | 0.8689 | 1 |

[a] Knot means nautical miles per hour

Figures in **bold face** indicate the conversion is exact

### Table 55. Acceleration Conversion Factors

| To Convert ↓ / Multiply By Factor To Obtain → | $cm/sec^2$ | $m/sec^2$ | $km/hr^2$ | $feet/sec^2$ | $ft/hr^2$ | knot/sec | $miles/hr^2$ |
|---|---|---|---|---|---|---|---|
| $cm/sec^2$ | 1 | **0.01** | 129.6 | 0.0328 | $4.252 \times 10^{5}$ | 0.0194384 | 80.529 |
| $m/sec^2$ | **100** | 1 | 12960 | 3.280 | $4.252 \times 10^{7}$ | 1.943844 | 8052.970 |
| $km/hr^2$ | 0.007716 | $7.72 \times 10^{-5}$ | 1 | $2.532 \times 10^{-4}$ | 3280.84 | 0.0001499 | 0.6213 |
| $ft/sec^2$ | 30.48 | 0.3048 | 3950.20 | 1 | $1.296 \times 10^{7}$ | 0.592483 | 2454.545 |
| $ft/hr^2$ | $2.35 \times 10^{-6}$ | $2.35 \times 10^{-5}$ | $3.048 \times 10^{-4}$ | $7.716 \times 10^{-8}$ | 1 | $4.571 \times 10^{-8}$ | $1.893 \times 10^{-4}$ |
| knot/sec | 51.44444 | 0.514444 | 6667.2 | 1.687809 | $2.187 \times 10^{7}$ | 1 | 4142.8060 |
| $mile/hr^2$ | 0.0124 | 0.000124 | 1.609 | $4.074 \times 10^{-4}$ | **5280** | 0.00024138 | 1 |

Figures in **bold face** indicate the conversion is exact

## Units of Viscosity

### Table 56a. Oil Viscosity Conversion Factors

| | To Convert ↓ / Multiply By This Factor To Obtain → | Poise (*P*) | Centi-poise (*Z*) | Reyn (μ) | Stoke (*S*) | Centistoke (*ν*) |
|---|---|---|---|---|---|---|
| Poise (*P*) | $\frac{\text{dyne-s}}{\text{cm}^2} = \frac{\text{gram mass}}{\text{cm-s}}$ | 1 | 100 | $1.45 \times 10^{-5}$ | $\frac{1}{\rho}$ | $\frac{100}{\rho}$ |
| Centipoise (*Z*) | $\frac{\text{dyne-s}}{100\ \text{cm}^2} = \frac{\text{gram mass}}{100\ \text{cm-s}}$ | 0.01 | 1 | $1.45 \times 10^{-7}$ | $\frac{0.01}{\rho}$ | $\frac{1}{\rho}$ |
| Reyn (μ) | $\frac{\text{lb force-s}}{\text{in}^2}$ | $6.9 \times 10^{4}$ | $6.9 \times 10^{6}$ | 1 | $\frac{6.9 \times 10^{4}}{\rho}$ | $\frac{6.9 \times 10^{6}}{\rho}$ |
| Stoke (*S*) | $\frac{\text{cm}^2}{\text{s}}$ | ρ | 100ρ | $1.45 \times 10^{-5}\ \rho$ | 1 | 100 |
| Centistoke (*ν*) | $\frac{\text{cm}^2}{100\text{s}}$ | 0.01ρ | ρ | $1.45 \times 10^{-7}\ \rho$ | 0.01 | 1 |

### Table 56b. Additional Viscosity Conversion Factors

| Multiply | By | To Obtain | Multiply | By | To Obtain |
|---|---|---|---|---|---|
| centipoise | **0.001** | pascal-second (Pa-s) | pascal-second | **1000** | centipoise |
| centistoke | **0.000001** | $meter^2/second$ ($m^2/s$) | pascal-second | **10** | poise |
| stoke | **0.0001** | $meter^2/second$ ($m^2/s$) | poise | **0.1** | pascal-second (Pa-s) |

ρ = Specific gravity of the oil.

Figures in **bold face** indicate the conversion is exact

## Units of Moment of Inertia and Momentum

### Table 57. Moment of Inertia Conversion Factors

| Multiply | By | To Obtain |
|---|---|---|
| Moment of Inertia and Section Modulus | | |
| moment of inertia [kg•m$^2$] | 23.73036 | pound-foot$^2$ |
| moment of inertia [kg•m$^2$] | 3417.171 | pound-inch$^2$ |
| moment of inertia [lb•ft$^2$] | 0.04214011 | kilogram-meter$^2$ (kg•m$^2$) |
| moment of inertia [lb•inch$^2$] | 0.0002926397 | kilogram-meter$^2$ (kg•m$^2$) |
| moment of section [foot$^4$] | 0.008630975 | meter$^4$ (m$^4$) |
| moment of section [inch$^4$] | 41.62314 | centimeter$^4$ |
| moment of section [meter$^4$] | 115.8618 | foot$^4$ |
| moment of section [centimeter$^4$] | 0.02402510 | inch$^4$ |
| section modulus [foot$^3$] | 0.02831685 | meter$^3$ (m$^3$) |
| section modulus [inch$^3$] | 0.00001638706 | meter$^3$ (m$^3$) |
| section modulus [meter$^3$] | 35.31466 | foot$^3$ |
| section modulus [meter$^3$] | 61,023.76 | inch$^3$ |

### Table 58. Momentum Conversion Factors

| Multiply | By | To Obtain |
|---|---|---|
| Momentum | | |
| kilogram-meter/second | 7.233011 | pound-foot/second |
| kilogram-meter/second | 86.79614 | pound-inch/second |
| pound-foot/second | 0.1382550 | kilogram-meter/second (kg•m/s) |
| pound-inch/second | 0.01152125 | kilogram-meter/second (kg•m/s) |

## Miscellaneous Measuring Units

1 *great gross* = 12 gross = 144 dozen
1 *gross* = 12 dozen = 144 units
1 *dozen* = 12 units

1 *quire* = 24 sheets
1 *ream* = 20 quires = 480 sheets
1 *ream printing paper* = 500 sheets
1 *score* = 20 units

**Ohm's Law.—**The following figure represents basic electrical relationships. This chart has been formatted in such a way that each variable has been related to the other three variables. This figure is simply for reference.

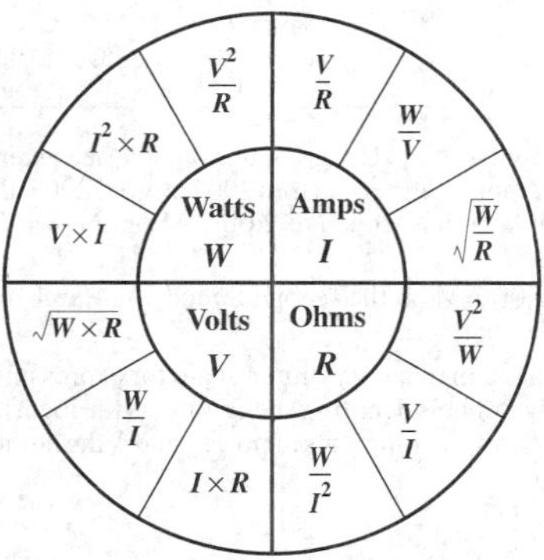


Circular Model of Electrical Relations

*Key to variables:*
$V$ = Voltage (Volts)
$R$ = Resistance (Ohms)
$I$ = Current (Amps)
$W$ = Power (Watts)

# B

## C

# D

# E

# F

# G

# H

# I

## J



## K

# L

# M

# N

# O



# P

# Q



# R

# S

# T

# U

# V

# W

## Y



## Z

## Force

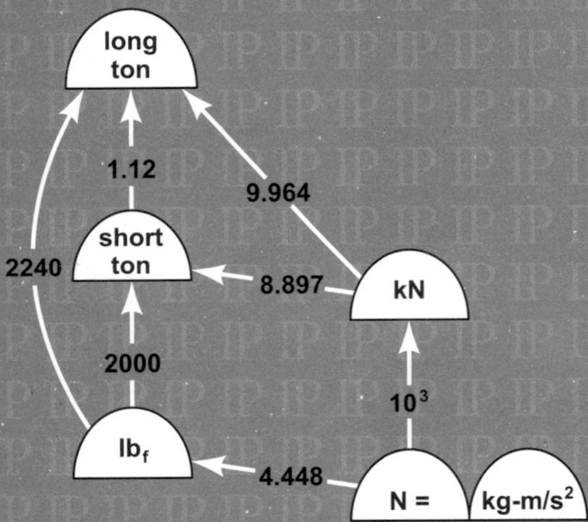


## Pressure

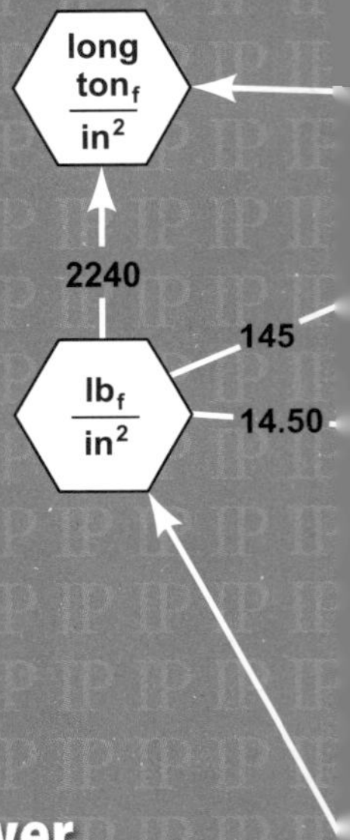


## Energy, Work — Power

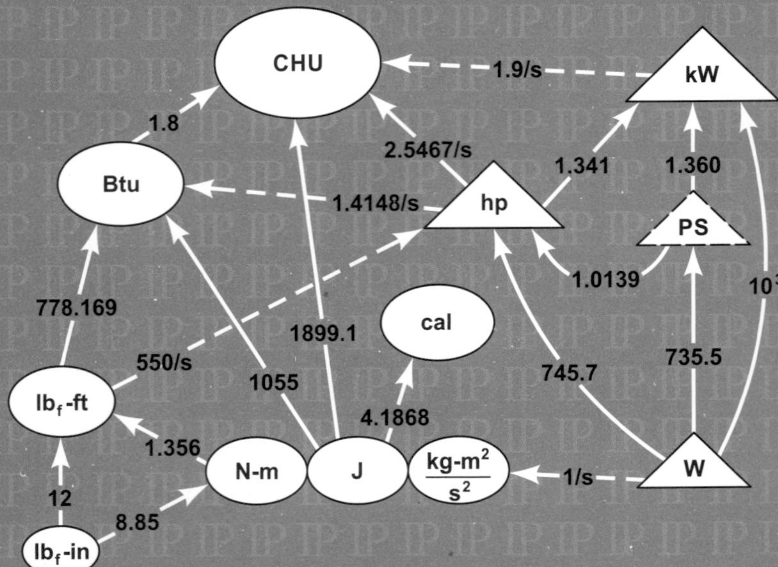